SPIE
Volume 1730

SPIE 1991 Publications Index

Subject/author indexes of all SPIE Proceedings and
SPIE Press books published in 1991

Indexing

250 proceedings and books
8,500 papers
19,000 authors and editors

D1401403

This index is part of **InCite**—SPIE's indexing/abstracting services,
providing access to SPIE's technical information databases.

Published by SPIE—The International Society for Optical Engineering
Bellingham, Washington USA

SPIE Vol. 1730

Library of Congress Catalog Card No. 91-61269
ISBN 0-8194-0903-0
ISSN 1055-6885

SPIE—The International Society for Optical Engineering
P.O. Box 10, Bellingham, Washington 98227-0010 USA
Telephone 206/676-3290 • FAX 206/647-1445

SPIE (Society of Photo-Optical Instrumentation Engineers) is a nonprofit society dedicated to the
advancement of optical and optoelectronic applied science and technology.

SPIE 1991 Publications Index

Contents

Volumes published in 1991 v

List of subject index terms ix

Subject index 1

Author/editor index 433

Publications order information 755

The SPIE 1991 Publications Index references all SPIE Proceedings and SPIE Optical Engineering Press books published between January 1, 1991, and December 31, 1991. The index is divided into two sections, a subject index and author/editor index.

SUBJECT INDEX KEY

Individual paper entry: Each paper title is indexed under appropriate boldface subject terms. Each entry gives the paper title, author(s), volume number, page numbers, and year of publication.

Fiber optic sensors
Fiber optic sensor for nitrates in water (MacCraith, Brian D; Maxwell, J) V1510, 195–203 (1991)

SPIE Proceedings or Press book entry: Each book title is indexed under appropriate boldface subject terms. Each entry gives the book title in italic boldface type, the editor(s), volume number, and year of publication.

Fiber optic sensors
Fiber Optic Smart Structures and Skins IV (Claus, Richard O; Udd, Eric, eds) V1588 (1991)

AUTHOR/EDITOR INDEX KEY

Principal author entry: Each principal author's entry includes coauthors, paper title, volume, page numbers, and year of publication.

MacCraith, Brian D
; Maxwell, J: Fiber optic sensor for nitrates in water, V1510, 195-203 (1991)

SPIE Proceedings or Press book editor or author entry: Each editor or author entry includes coeditors or coauthors, book title in italic boldface type, volume, and year of publication.

Claus, Richard O.
; Udd, Eric; eds.: *Fiber Optic Smart Structures and Skins IV*, V1588 (1991)

Coauthors and coeditors are cross-referenced to the main entry.

Maxwell, J
See MacCraith, Brian D: V1510, 195-203 (1991)

VOLUMES PUBLISHED IN 1991

Listed below by volume number are all of the SPIE Proceedings and SPIE Press book titles included in this index. For prices and order information, see pages 755-762.

Vol.	Title (Editors)
1238	Three-Dimensional Holography: Science, Culture, Education (Markov, Jeong)
1330	Optical Surfaces Resistant to Severe Environments (Musikant)
1332	Optical Testing and Metrology III: Recent Advances in Industrial Optical Inspection (Grover)
1343	X-Ray/EUV Optics for Astronomy, Microscopy, Polarimetry, and Projection Lithography (Hoover, Walker)
1345	Advanced X-Ray/EUV Radiation Sources and Applications (Shenoy, Knauer)
1346	Ultrahigh- and High-Speed Photography, Videography, Photonics, and Velocimetry '90 (Neyer, Shaw, Jaanimagi)
1354	1990 International Lens Design Conference (Lawrence)
1358	19th International Congress on High-Speed Photography and Photonics (Fuller)
1361	Physical Concepts of Materials for Novel Optoelectronic Device Applications I: Materials Growth and Characterization (Razeghi)
1362	Physical Concepts of Materials for Novel Optoelectronic Device Applications II: Device Physics and Applications (Razeghi)
1363	Fiber Optics in the Subscriber Loop (Kahn, Hutcheson)
1364	FDDI, Campus-Wide, and Metropolitan Area Networks (Cudworth, Annamalai, Kasiewicz)
1365	Components for Fiber Optic Applications V (Kopera)
1366	Fiber Optics Reliability: Benign and Adverse Environments IV (Paul, Greenwell)
1367	Fiber Optic and Laser Sensors VIII (Udd, DePaula)
1368	Chemical, Biochemical, and Environmental Fiber Sensors II (Lieberman, Wlodarczyk)
1369	Fiber Optic Systems for Mobile Platforms IV (Lewis, Moore)

Vol.	Title (Editors)
1371	High-Frequency Analog Fiber Optic Systems (Sierak)
1372	Coherent Lightwave Communications: Fifth in a Series (Steele, Sunak)
1373	Fiber Laser Sources and Amplifiers II (Digonnet)
1374	Integrated Optics and Optoelectronics II (Wong)
1376	Laser Noise (Roy)
1377	Excimer Laser Materials Processing and Beam Delivery Systems (Piwczyk)
1378	Optically Activated Switching (Zutavern)
1379	Optics in Agriculture (DeShazer, Meyer)
1380	Biostereometric Technology and Applications (Herron)
1381	Intelligent Robots and Computer Vision IX: Algorithms and Techniques (Casasent)
1382	Intelligent Robots and Computer Vision IX: Neural, Biological, and 3-D Methods (Casasent)
1383	Sensor Fusion III: 3-D Perception and Recognition (Schenker)
1384	High-Speed Inspection Architectures, Barcoding, and Character Recognition (Chen)
1385	Optics, Illumination, and Image Sensing for Machine Vision V (Harding, Uber, Svetkoff, Wittels)
1386	Machine Vision Systems Integration in Industry (Waltz, Batchelor)
1387	Cooperative Intelligent Robotics in Space (Stoney, deFigueiredo)
1388	Mobile Robots V (Chun, Wolfe)
1389	Microelectronic Interconnects and Packages: Optical and Electrical Technologies (Pazaris, Arjavalingam)
1390	Microelectronic Interconnects and Packages: System and Process Integration (Carruthers, Tewksbury)
1391	Laser Technology III (Gajda, Gajda, Wolczak, Romaniuk, Wolinski)
1392	Advanced Techniques for Integrated Circuit Processing (Bondur, Turner)
1393	Rapid Thermal and Related Processing Techniques (Moslehi, Singh)

Vol.	Title (Editors)
1394	Progress In High-Temperature Superconducting Transistors and Other Devices (Narayan, Shaw, Singh)
1396	Applications of Optical Engineering: Proceedings of OE/Midwest '90 (Dubiel, Eppinger, Guzik, Gillespie, Pearson)
1397	8th International Symposium on Gas Flow and Chemical Lasers (Domingo, Orza)
1398	CAN-AM Eastern '90 (Antos, Krisiloff)
1399	Optical Systems in Adverse Environments (Tam, Kuok, Silva)
1400	Optical Fabrication and Testing (Campbell, Lorenzen, Johnson)
1401	Optical Data Storage Technologies (Chua, McCallum)
1402	USSR-CSFR Joint Seminar on Nonlinear Optics in Control, Diagnostics, and Modeling of Biophysical Processes (Akhmanov, Zadkov)
1403	Laser Applications in Life Sciences (Akhmanov, Poroshina)
1406	Image Understanding in the '90s: Building Systems that Work (Mitchell)
1407	Intense Microwave and Particle Beams II (Brandt)
1408	Propagation of High-Energy Laser Beams Through the Earth's Atmosphere II (Wilson, Ulrich)
1409	Nonlinear Optics II (Reintjes, Fisher)
1410	Solid State Lasers II (Dube)
1411	Pulse Power for Lasers III (McDuff)
1412	Gas and Metal Vapor Lasers and Applications (Tittel, Kim)
1413	Short-Pulse High-Intensity Lasers and Applications (Baldis)
1414	Laser Beam Diagnostics (Hindy, Kohanzadeh)
1415	Modeling and Simulation of Laser Systems II (Schnurr)
1416	Laser Radar VI (Becherer)
1417	Free-Space Laser Communication Technologies III (Begley, Seery)
1418	Laser Diode Technology and Applications III (Renner)
1419	Eyesafe Lasers: Components, Systems, and Applications (Johnson)

Vol.	Title (Editors)
1420	Optical Fibers in Medicine VI (Katzir)
1421	Lasers in Urology, Laparoscopy, and General Surgery (Pietrafitta, Watson, Steiner)
1422	Lasers in Dermatology and Tissue Welding (Tan, White, White)
1423	Ophthalmic Technologies (Puliafito)
1424	Lasers in Orthopedic, Dental, and Veterinary Medicine (O'Brien, Wigdor, Trent, Dederich)
1425	Diagnostic and Therapeutic Cardiovascular Interventions (Abela)
1426	Optical Methods for Tumor Treatment and Early Diagnosis: Mechanisms and Techniques (Dougherty)
1427	Laser-Tissue Interaction II (Jacques)
1428	Three-Dimensional Bioimaging Systems and Lasers in the Neurosciences (Smith, Cerullo, Boggan)
1429	Holography, Interferometry, and Optical Pattern Recognition in Biomedicine (Podbielska)
1430	Photon Correlation Spectroscopy: Multicomponent Systems (Schmitz)
1431	Time-Resolved Spectroscopy and Imaging of Tissues (Katzir, Chance)
1432	Biomolecular Spectroscopy II (Birge, Nafie)
1433	Measurement of Atmospheric Gases (Schiff)
1434	Environmental Sensing and Combustion Diagnostics (Santoleri)
1435	Optical Methods for Ultrasensitive Detection and Analysis: Techniques and Applications (Fearey)
1436	Photochemistry and Photoelectrochemistry of Organic and Inorganic Molecular Thin Films (Lawrence, Ramasesha, Wamser, Frank)
1437	Applied Spectroscopy in Material Science (Saperstein)
1439	International Conference on Scientific Optical Imaging (Denton)
1440	Optical Radiation Interaction with Matter (Konov, Libenson, Bonch-Bruevich)
1441	Laser-Induced Damage in Optical Materials: 1990 (Chase, Newman, Guenther, Bennett, Soileau)
1442	7th Mtg in Israel on Optical Engineering (Shladov, Oron)
1443	Medical Imaging V: Image Physics (Schneider)
1444	Medical Imaging V: Image Capture, Formatting, and Display (Kim)
1445	Medical Imaging V: Image Processing (Loew)
1446	Medical Imaging V: PACS Design and Evaluation (Jost)
1447	Charge-Coupled Devices and Solid State Optical Sensors II (Blouke)
1448	Camera and Input Scanner Systems (Chang, Milch)
1449	Electron Image Tubes and Image Intensifiers II (Csorba)
1450	Biomedical Image Processing II (Howard, Bovik)
1451	Nonlinear Image Processing II (Dougherty, Arce, Boncelet)
1452	Image Processing Algorithms and Techniques II (Civanlar, Moorhead, Mitra)
1453	Human Vision, Visual Processing, and Digital Display II (Rogowitz, Allebach, Brill)
1454	Beam Deflection and Scanning Technologies (Beiser, Marshall)
1455	Liquid-Crystal Devices and Materials (Efron, Drzaic)
1456	Large Screen Projection, Avionic, and Helmet-Mounted Displays (Lippert, Bentz, Assenheim, Flasck)
1457	Stereoscopic Displays and Applications II (Merritt, Fisher)
1458	Printing Technologies for Images, Gray Scale, and Color (Dove, Abe, Heinzl)
1459	Extracting Meaning from Complex Data: Processing, Display, Interaction II (Farrell)
1460	Image Handling and Reproduction Systems Integration (Plouffe, Bender)
1461	Practical Holography V (Benton)
1463	Optical/Laser Microlithography IV (Pol)
1464	Integrated Circuit Metrology, Inspection, and Process Control V (Arnold)
1465	Electron-Beam, X-Ray, and Ion-Beam Submicrometer Lithographies for Manufacturing (Peckerar)
1466	Advances in Resist Technology and Processing VIII (Ito)
1467	Thermosense XIII (Baird)
1468	Applications of Artificial Intelligence IX (Trivedi)
1469	Applications of Artificial Neural Networks II (Rogers)
1470	Data Structures and Target Classification (Libby)
1471	Automatic Object Recognition (Sadjadi)
1472	Image Understanding and the Man-Machine Interface III (Pearson, Barrett)
1473	Visual Information Processing: From Neurons to Chips (Mathur, Koch)
1474	Optical Technology for Signal Processing Systems (Bendett)
1475	Monolithic Microwave Integrated Circuits for Sensors, Radar, and Communications Systems (Bhasin, Leonard)
1476	Optical Technology for Microwave Applications V (Yao)
1477	Superconductivity Applications for Infrared and Microwave Devices II (Bhasin, Heinen)
1478	Sensors and Sensor Systems for Guidance and Navigation (Wade, Tuchman)
1479	Surveillance Technologies (Gowrinathan, Mataloni, Schwartz)
1480	Sensors and Sensor Integration (Dean)
1481	Signal and Data Processing of Small Targets 1991 (Drummond)
1482	Acquisition, Tracking, and Pointing V (Stockum, Masten)
1483	Signal and Image Processing Systems Performance Evaluation, Simulation, and Modeling (Bazakos, Nasr)
1484	Growth and Characterization of Materials for Infrared Detectors and Nonlinear Optical Switches (Baars, Longshore)
1485	Reflective and Refractive Optical Materials for Earth and Space Applications (Hale, T B Parsonage, M J Riedl)
1486	Characterization, Propagation, and Simulation of Sources and Backgrounds (Watkins, Clement)
1487	Propagation Engineering: Fourth in a Series (Bissonnette, Miller)
1488	Infrared Imaging Systems: Design, Analysis, Modeling and Testing II (Holst)
1489	Structures Sensing and Control (Breakwell, Varadan)
1490	Future European and Japanese Remote-Sensing Sensors and Programs (Slater)
1491	Remote Sensing of Atmospheric Chemistry (McNeal, McElroy)
1492	Earth and Atmospheric Remote Sensing (Smith, Curran, Watson)
1493	Calibration of Passive Remote Observing Optical and Microwave Instrumentation (Guenther)
1494	Space Astronomical Telescopes and Instruments (Breckinridge, Bely)
1495	Small-Satellite Technology and Applications (Horais)

Vol.	Title (Editors)

1496   10th Annual Symposium on Microlithography (Wiley)

1497   Nonlinear Optics and Materials (Bowden, Cantrell)

1498   Tactical Infrared Systems (Tuttle)

1499   Optical Data Storage '91 (Shull, Imamura, Burke)

1500   Innovative Optics and Phase Conjugate Optics (Ahlers, Tschudi)

1501   Advanced Laser Concepts and Applications (Singer)

1502   Industrial and Scientific Uses of High-Power Lasers (Billon, Fabre)

1503   Excimer Lasers and Applications III (Letardi, Laude)

1504   Fiber-Optic Metrology and Standards (Soares)

1505   Optics for Computers: Architectures and Technologies (Lebreton)

1506   Micro-Optics II (Scheggi)

1507   Holographic Optics III: Principles and Applications (Morris)

1508   Industrial Applications of Holographic and Speckle Measuring Techniques (Jueptner)

1509   Holographic Optical Security Systems (Fagan)

1510   Chemical and Medical Sensors (Wolfbeis)

1511   Fiber Optic Sensors: Engineering and Applications (Bruinsma, Culshaw)

1512   Infrared and Optoelectronic Materials and Devices (Corsi, Naumaan, Baixeras, A J Kreisler)

1513   Glasses for Optoelectronics II (Righini)

1516   International Workshop on Photoinduced Self-Organization Effects in Optical Fiber (Ouellette)

1517   The Marketplace for Industrial Lasers (Belforte, Levitt)

1519   International Conference on Thin Film Physics and Applications (Zhou)

1520   The Laser Marketplace 1991 (Forrest, Levitt)

1521   Image Understanding for Aerospace Applications (Nasr)

1522   Optical Space Communication II (Franz)

1525   Future Trends in Biomedical Applications of Lasers (Svaasand)

1526   Industrial Vision Metrology (El-Hakim)

1527   Current Developments in Optical Design and Optical Engineering (Smith, Fischer)

1528   Nonimaging Optics: Maximum Efficiency Light Transfer (Winston, Holman)

1529   Opthalmic Lens Design and Fabrication (Perrott)

1530   Optical Scatter: Applications, Measurement, and Theory (Stover)

1532   Analysis of Optical Structures (O'Shea)

1533   Optomechanics and Dimensional Stability (Paquin, Vukobratovich)

1534   Diamond Optics IV (Feldman, Holly)

1535   Passive Materials for Optical Elements (Wilkerson)

1536   Optical Materials Technology for Energy Efficiency and Solar Energy Conversion X (Granqvist, Lampert)

1537   Underwater Imaging, Photography, and Visibility (Spinrad)

1538   Airborne Reconnaissance XV (Henkel, Augustyn)

1540   Infrared Technology XVII (Andresen, Scholl, Spiro)

1541   Infrared Sensors: Detectors, Electronics, and Signal Processing (Jayadev)

1542   Active and Adaptive Optical Systems (Ealey)

1544   Miniature and Micro-Optics: Fabrication and System Applications (Veldkamp, Roychoudhuri)

1545   International Conference on the Application and Theory of Periodic Structures (Lerner, McKinney)

1548   Production and Analysis of Polarized X Rays (Siddons)

1549   EUV, X-Ray, and Gamma-Ray Instrumentation for Astronomy II (Siegmund, Rothschild)

1550   X Rays in Materials Analysis II: Novel Applications and Recent Developments (Mills)

1552   Short-Wavelength Radiation Sources (Sprangle)

1554A   Second International Conference on Photomechanics and Speckle Metrology: Speckle Techniques, Birefringence Methods, and Applications to Solid Mechanics (Chiang)

1554B   Second International Conference on Photomechanics and Speckle Metrology: Moire Techniques, Holographic Interferometry, Optical NDT, and Applications to Fluid Mechanics (Chiang)

1555   Computer and Optically Generated Holographic Optics, 4th in a Series (Cindrich, Lee)

1557   Crystal Growth in Space and Related Optical Diagnostics (Lal, Trolinger)

1558   Wave Propagation and Scattering in Varied Media II (Varadan)

1559   Photopolymer Device Physics, Chemistry, and Applications II (Lessard)

1560   Nonlinear Optical Properties of Organic Materials IV (Singer)

1561   Inorganic Crystals for Optics, Electro-Optics, and Frequency Conversion (Bordui)

1562   Devices for Optical Processing (Gookin)

1563   Optical Enhancements to Computing Technology (Neff)

1564   Optical Information Processing Systems and Architectures III (Javidi)

1565   Adaptive Signal Processing (Haykin)

1566   Advanced Signal Processing Algorithms, Architectures, and Implementations II (Luk)

1567   Applications of Digital Image Processing XIV (Tescher)

1568   Image Algebra and Morphological Image Processing II (Gader, Dougherty)

1569   Stochastic and Neural Methods in Signal Processing, Image Processing, and Computer Vision (Chen)

1570   Geometric Methods in Computer Vision (Vemuri)

1572   International Conference on Optical Fibre Sensors in China (Liao, Culshaw)

1574   International Colloquium on Diffractive Optical Elements (Nowak, Zajac)

1579   Advanced Fiber Communications Technologies (Kazovsky)

1583   Integrated Optical Circuits (Wong)

1584   Fiber Optic and Laser Sensors IX (Udd, DePaula)

1588   Fiber Optic Smart Structures and Skins IV (Udd, Claus)

1589   Specialty Fiber Optic Systems for Mobile Platforms (Lewis, Moore)

1590   Submolecular Glass Chemistry and Physics (Bray, Kreidl)

1592   Plastic Optical Fibers (Steele, Kitazawa, Kreidl)

1596   Metallization: Performance and Reliability Issues for VLSI and ULSI (Schwartz, Gildenblat)

1598   Lasers in Microelectronic Manufacturing (Braren)

1605   Visual Communications and Image Processing '91: Visual Communication (Tzou, Koga)

1606   Visual Communications and Image Processing '91: Image Processing (Tzou, Koga)

1621   Optical Memory and Neural Networks (Mikaelian)

SPIE PRESS PUBLICATIONS

Critical Reviews
Invited review papers by leading researchers providing in-depth analysis and review in key optical technologies.

Vol.   Title (Editors)

CR36 Machine Vision Systems Integration (Waltz, Batchelor)

CR37 Standards for Electronic Imaging Systems (Courtot, Nier)

CR38 Infrared Optical Design and Fabrication (Smith, Hartmann)

Institute Series
Peer-reviewed papers by leading experts analyzing the critical issues, future trends, and potential applications of emerging optical technologies.

Vol.   Title (Editors)

IS07   Automatic Object Recognition (Nasr)

IS08   Holography (Greguss, Jeong)

Milestone Series
Selected reprints of milestone papers from the world literature covering important fields in optics.

Vol.   Title (Editors)

MS27 Selected Papers on Ellipsometry (Azzam)

MS28 Selected Papers on Interferometry (Hariharan)

MS29 Selected Papers on Laser Design (Weichel)

MS30 Selected Papers on Free-Space Laser Communications (Begley)

MS31 Selected Papers on Solid State Lasers (Powell)

MS32 Selected Papers on Nonlinear Optics (Brandt)

MS34 Selected Papers on Holographic and Diffractive Lenses and Mirrors (Stone, Thompson)

MS35 Selected Papers on Speckle Metrology (Sirohi)

MS36 Selected Papers on Optical Tolerancing (Wiese)

MS39 Photon Statistics and Coherence in Nonlinear Optics (Perina)

MS40 Computer-Controlled Optical Surfacing (Jones)

MS41 Selected Papers on Automatic Object Recognition (Nasr)

MS42 Selected Papers on Multiple Scattering in Plane Parallel Atmospheres and Oceans: Methods (Kattawar)

MS43 Selected Papers on High-Power Lasers (Soures)

MS44 Selected Papers on Ultrafast Laser Technology (Gosnell, Taylor)

SPIE Press Monographs
Authoritative technical books covering the theory, state of the art, applications, and outlook on topics of special interest to optical scientists and engineers.

Vol.   Title (Editors)

PM05 Optoelectronic Materials and Device Concepts (Razeghi)

Tutorial Texts Series
Tutorial books covering new fields as well as fundamental topics in optical science and technology.

Vol.   Title (Editors)

TT04 An Introduction to Biological and Artificial Neural Networks for Pattern Recognition (Rogers, Kabrisky)

TT05 Single Frequency Semiconductor Lasers (Buus)

TT06 Aberration Theory Made Simple (Mahajan)

TT07 Digital Image Compression Techniques (Rabbani, Jones)

INDEX TERMS

Aberrations—see also lenses; optical design

Absorption—see also optical properties

Acoustics—see also medical imaging; photoacoustics

Acousto-optics—see also Bragg cells; optical processing; signal processing

Actuators—see also optomechanical design

Adaptive optics—see also astronomy; atmospheric optics

Aerosols—see also atmospheric optics; particles

Agriculture

Alignment—see also optomechanical design

Amplifiers—see also optical communications

Annealing—see also microelectronic processing

Antennas—see also arrays; receivers

Architectures—see also fiber optic networks; optical processing; signal processing

Arrays—see also antennas; detectors; lasers, semiconductor; micro-optics; telescopes

Artificial intelligence—see also expert systems; robotics

Aspherics—see also lenses; optical design; optical fabrication

Associative processing—see also optical computing; optical processing

Astronomy—see also lenses; mirrors; optical design; optomechanical design; satellites; space optics; sun; telescopes

Atmospheric optics—see also adaptive optics; aerosols; gases; propagation; thermal blooming; turbulence

Atoms—see also spectroscopy, atomic

Beams—see also lasers; microlithography; propagation

Beamsplitters—see also holography; polarization

Binary optics—see also diffractive optics; micro-optics; holographic optical elements; holography, computer-generated

Biochemistry—see also biology; chemistry; medicine

Biology—see also biochemistry; medicine

Birefringence—see also fiber optics

Bistability—see also integrated optics; integrated optoelectronics; switches

Bragg cells—see also acousto-optics; signal processing

Calibration

Cameras—see also charge-coupled devices; machine vision; imaging systems; image tubes; photography; photography, high-speed; streak cameras; video

Cathode ray tubes—see also displays

Charge-coupled devices—see also astronomy; cameras; detectors

Chemistry—see also biochemistry; laser chemistry

Chromogenics—see also coatings; solar energy

Coatings—see also chromogenics; deposition; optical materials; thin films

Coding—see also image compression; video

Coherence

Collimators

Color—see also colorimetry; printing; visualization

Colorimetry—see also color

Computers

Computer vision—see also image processing; machine vision

Correlation—see also filters; optical processing; pattern recognition

Crystals—see also optical properties

Crystals, liquid—see also displays; spatial light modulators

Deposition—see also coatings; microelectronic processing; thin films

Detectors—see also arrays; charge-coupled devices; infrared; photodetectors; sensors

Diamonds—see also coatings; optical properties; thin films

Diffraction—see also binary optics; gratings; holographic optical elements; holography

Diffractive optics—see binary optics; diffraction; holography, computer-generated; holographic optical elements

Digital processing—see also image processing

Diodes—see also lasers, semiconductor; photodiodes

Displays—see also cathode ray tubes; holography; imaging systems; projection systems; television; three dimensions; video

Doppler effect—see also lidar; radar; velocimetry

Earth—see also remote sensing

Electromagnetic waves

Electron beams—see beams

Electronics—see also optoelectronics

Electro-optics

Ellipsometry

Emission—see also spectroscopy

Endoscopes—see also fiber optics in medicine

Energy

Environmental effects—see also pollution

Epitaxy—see also microelectronic processing

Etalons—see also Fabry-Perot

Etching—see also microelectronic processing; microlithography

Expert systems—see also artificial intelligence

Fabry-Perot

Faraday effect—see also magneto-optics

Ferroelectrics—see also crystals, liquid

Fiber fabrication—see also glass

Fiber optic components

Fiber optic networks—see also optical communications

Fiber optics—see also integrated optics

Fiber optic sensors—see also fiber optics in medicine; gyroscopes; metrology; temperature

Fiber optics in medicine—see also endoscopes; fiber optic sensors; medicine

Films—see also coatings; optical materials; thin films

Filters—see also correlation; image processing; optical processing; pattern recognition

Flows

Fluorescence—see also medical imaging; spectroscopy

Focus—see also alignment; lenses; optical design

Fourier optics

Fourier transforms—see also transforms

Four-wave mixing—see also nonlinear optics; phase conjugation

Fractals

Frequency conversion

Fusion

Gallium arsenide materials—see also integrated optoelectronics; lasers, semiconductor; quantum wells; semiconductors

Gamma rays—see also astronomy

Gases—see also lasers, gas

Geometrical optics—see also optical design

Glass—see also optical properties

Gradient index optics—see also fiber optics; refractive index

Gratings—see also diffraction

Guided waves—see also fiber optics; integrated optics; waveguides

Gyroscopes—see also fiber optic sensors; navigation

Heterodyning

High-speed photography—see photography, high-speed

History

Holographic optical elements—see also binary optics; holography, computer-generated; lenses; micro-optics

Holography

Holography, computer-generated—see also binary optics; holographic optical elements; lenses; optical processing

Image algebra—see also image processing; mathematical morphology

Image analysis—see image processing; image understanding

Image compression—see also coding

Image enhancement

Image formation

Image processing—see also computer vision; filters; image algebra; image compression; image reconstruction/restoration; medical imaging; neural networks; object recognition; optical processing; parallel processing; pattern recognition; visual communications

Image quality—see also image reconstruction/restoration

Image reconstruction/restoration—see also image processing; image quality

Image recording—see also optical data storage; optical recording

Image tubes—see also intensifiers; photography, high-speed; streak cameras

Image understanding

Imaging

Imaging systems—see also cameras; displays; machine vision; projection systems; television; video

Industrial optics—see also lasers, applications; metrology; optical inspection

Information processing—see also optical processing

Information storage—see also optical data storage

Infrared—see also detectors; electro-optics; photodetectors; thermal imaging

Integrated circuits—see also microlithography; semiconductors; silicon

Integrated optics—see also fiber optics; integrated optoelectronics; waveguides

Integrated optoelectronics—see also gallium arsenide materials; integrated optics; semiconductors

Intensifiers—see also image tubes

Interfaces—see also multilayers

Interference

Interferometry—see also fiber optic sensors; holography; moire; nondestructive testing; photomechanics

Ion beams—see beams

Kinoforms—see also holographic optical elements

Laser applications—see also laser chemistry; lidar; microelectronic processing; microlithography; scanning

Laser chemistry—see also chemistry

Laser damage

Laser diagnostics

Laser materials processing—see industrial optics; laser applications; laser chemistry; microelectronic processing; microlithography

Laser-matter interaction—see also plasmas

Laser recording—see also scanning

Lasers—see also resonators; sources

Laser safety—see also laser damage; laser surgery

Lasers, carbon dioxide—see also lasers, gas

Lasers, chemical

Lasers, dye

Lasers, excimer—see also lasers, gas; microlithography

Lasers, fiber

Lasers, free-electron

Lasers, gas

Lasers, helium-neon

Lasers, high-energy

Lasers, high-power—see also lasers, carbon dioxide; lasers, excimer; lasers, gas; lasers, neodymium; lasers, semiconductor; lasers, x-ray

Lasers, neodymium—see also lasers, solid state

Lasers, semiconductor—see also diodes; lasers, high-power; optical communications

Lasers, solid state—see also lasers, neodymium

Lasers, tunable

Laser surgery—see also laser-tissue interaction; medicine

Laser-tissue interaction—see also laser surgery; medicine

Lasers, x-ray

Lenses—see also aspherics; binary optics; holographic optical elements; micro-optics; optical design

Lidar—see also Doppler effect; radar
Lithography—see microlithography
Luminescence—see also fluorescence

Machine vision—see also computer vision
Magneto-optics—see also optical data storage; optical disks; optical recording
Materials—see also coatings; crystals; diamonds; gallium arsenide materials; glass; optical materials; optical properties; photoconductors; photodetectors; photoresists; polymers; semiconductors; silicon; sol-gels; superconductors; thin films
Mathematical morphology—see also image algebra; image processing
Medical imaging—see also acoustics; radiology; tomography; x rays
Medicine—see also biology; fiber optics in medicine; laser surgery; laser-tissue interaction; medical imaging; ophthalmology; radiology; tomography
Meteorology—see also atmospheric optics
Metrology—see also microlithography; nondestructive testing; optical testing
Microelectronic processing—see also annealing; epitaxy; etching; microlithography
Microlithography—see also beams; lasers, excimer; microelectronic processing; photomasks; photoresists; projection systems; silicon; x rays
Micro-optics—see also lenses
Microscopy
Microwaves
Millimeter waves
Mirrors—see also optical design; reflectance; surfaces; telescopes
Modulation
Modulation transfer function—see also optical design
Moire—see also interferometry
Molecules—see also spectroscopy, molecular
Monochromators
Multilayers—see also coatings; interfaces; thin films; x rays
Multiphoton processes—see also spectroscopy
Multiplexing—see also optical communications

Navigation—see also gyroscopes
Neural networks—see also optical computing; pattern recognition
Noise
Nondestructive testing—see also metrology; photomechanics
Nonlinear optics—see also four-wave mixing; phase conjugation; second-harmonic generation

Object recognition—see also neural networks; pattern recognition
Oceanography
Ophthalmology—see also medicine
Optical chaos
Optical communications—see also amplifiers; antennas; fiber optic networks; lasers, semiconductor; modulation; multiplexing; receivers; satellites; space optics; transmission

Optical computing—see also associative processing; neural networks; optical interconnects; optical processing; parallel processing; pattern recognition; spatial light modulators; switches
Optical data storage—see also magneto-optics; optical disks; optical recording
Optical design—see also aberrations; aspherics; coatings; geometrical optics; lenses; modulation transfer function; optical engineering; optical systems; optical transfer function; optomechanical design; standards; stray light; surfaces; telescopes; thin films
Optical disks—see also magneto-optics; optical data storage; optical recording
Optical engineering—see also optical design; optical fabrication; optical testing; optomechanical design
Optical fabrication—see also polishing; optical testing; standards
Optical inspection—see also industrial optics; machine vision; robotics
Optical interconnects—see also fiber optic networks; integrated optics; integrated optoelectronics; optical computing; switches; waveguides
Optical materials—see also coatings; crystals; diamonds; gallium arsenide materials; glass; optical properties; photoconductors; photodetectors; photoresists; polymers; semiconductors; silicon; sol-gels; thin films
Optical processing—see also associative processing; information processing; neural networks; optical computing; parallel processing; pattern recognition
Optical properties—see also optical materials
Optical recording—see also magneto-optics; optical data storage; optical disks
Optical systems—see also imaging systems; optical design; optomechanical design
Optical testing—see also lenses; metrology; mirrors; optical fabrication; surfaces
Optical transfer function—see also optical design
Optoelectronics—see also integrated optoelectronics
Optomechanical design—see also alignment; optical design

Parallel processing—see also optical computing; optical processing
Particles—see also aerosols; pollution
Pattern recognition—see also correlation; filters; image processing; neural networks; object recognition; optical computing; optical processing
Phase
Phase conjugation—see also nonlinear optics; phase
Photoacoustics—see also acoustics; acousto-optics
Photochemistry—see also chemistry; films; holography; photography
Photoconductors
Photodetectors—see also detectors
Photodiodes—see also diodes
Photogrammetry—see also remote sensing
Photography—see also cameras; films; imaging systems; lenses; mirrors; photography, high-speed
Photography, high-speed—see also image tubes; intensifiers; photography; streak cameras

Photomasks—see also microlithography

Photomechanics—see also metrology; nondestructive testing

Photometry—see also radiometry

Photorefraction—see also crystals; refraction; spatial light modulators

Photoresists—see also microlithography

Plasmas—see also laser-matter interaction

Polarimetry—see also polarization

Polarization—see also polarimetry

Pollution—see also environmental efects; particles

Polymers

Printing

Prisms—see also optical design; refraction

Projection systems—see also imaging systems; microlithography

Propagation—see also atmospheric optics; beams; turbulence

Pulses

Q-switching

Quantum wells—see also excitons; gallium arsenide materials; integrated optoelectronics; lasers, semiconductor

Radar—see also lidar

Radiance

Radiation—see also gamma rays; infrared; lasers; microwaves; millimeter waves; radiology; synchrotron radiation; ultraviolet; x rays

Radiology—see also medical imaging; x rays

Radiometry—see also photometry

Raman effect—see also scattering; spectroscopy

Receivers—see also optical communications

Reflectance—see also mirrors; surfaces

Reflectors—see mirrors; telescopes

Refraction—see also photorefraction; refractive index

Refractive index—see also gradient-index optics; lenses; optical design; optical properties; refraction

Remote sensing—see also satellites; sensors; telescopes

Resonators—see also lasers

Robotics—see also artificial intelligence; computer vision; expert systems; industrial optics; machine vision; optical inspection; object recognition; pattern recognition

Satellites—see also astronomy; remote sensing; space optics

Scanning—see also lasers, applications

Scattering—see also Raman effect; surfaces

Second-harmonic generation—see also nonlinear optics

Semiconductors—see also gallium arsenide materials; microelectronic processing; silicon

Sensor fusion—see sensors

Sensors—see also detectors; fiber optic sensors; photodetectors; remote sensing

Signal processing—see also architectures; optical processing

Silicon—see also microelectronic processing; semiconductors

Solar energy—see also chromogenics; coatings

Sol-gels—see also coatings

Sources—see also radiation; sun; synchrotron radiation

Space optics—see also astronomy; robotics; satellites; telescopes

Spatial light modulators—see also acousto-optics; crystals, liquid; optical computing; photorefraction

Speckle

Spectroscopy—see also Raman effect; spectroscopy, atomic; spectroscopy, molecular; spectrum analysis

Spectroscopy, atomic—see also spectroscopy

Spectroscopy, molecular—see also spectroscopy

Spectrum analysis—see also spectroscopy

Standards

Statistical optics

Stray light—see also optical design

Streak cameras—see also image tubes; intensifiers; photography, high-speed

Sun—see also astronomy; sources

Superconductors—see also thin films

Surfaces—see also mirrors; optical testing; scattering

Switches—see also optical interconnects

Synchrotron radiation—see also radiation; sources; x rays

Synthetic apertures—see also radar

Telescopes—see also astronomy; mirrors; satellites; space optics

Television—see also cameras; imaging systems; video

Temperature—see also thermal effects

Thermal blooming—see also atmospheric optics

Thermal effects—see also temperature

Thermal imaging—see also imaging systems; infrared

Thin films—see also coatings; optical materials

Three dimensions—see also displays; object recognition; vision

Tomography—see also medical imaging

Tracking/ranging

Transforms—see also Fourier transforms

Transmission—see also optical communications

Turbulence—see also atmospheric optics

Ultrafast phenomena—see also lasers; photography, high-speed; streak cameras

Ultraviolet—see also astronomy; lasers, excimer; microlithography

Velocimetry—see also Doppler effect; flows

Video—see also cameras; television

Visibility—see also atmospheric optics

Vision—see also biology; color; three dimensions; visualization

Visual communications—see also image processing; television; video

Visualization—see also vision

Water

Wavefronts

Waveguides—see also fiber optics; guided waves; integrated optics; integrated optoelectronics

X rays—see also astronomy; medical imaging; radiology; synchrotron radiation

Subject Index

Aberrations—see also lenses; optical design

Aberration Theory Made Simple (Mahajan, Virendra N.)VTT06(1991)

Aberrations of holographic lens recorded on surface of revolution with shifted pupil (Masajada, Jan)V1574,188-196(1991)

Achromatization of optical waveguide components (Spaulding, Kevin E.; Morris, G. M.)V1507,45-54(1991)

Analysis of the chromatic aberrations of imaging holographic optical elements (Tholl, Hans D.; Luebbers, Hubertus; Stojanoff, Christo G.)V1456,262-273(1991)

Apo-tele lenses with kinoform elements (Gan, Michael A.; Potyemin, Igor S.; Poszinskaja, Irina I.)V1507,116-125(1991)

Astigmatism and field curvature from pin-bars (Kirk, Joseph P.)V1463,282-291(1991)

Calculation of wave aberration in optical systems with holographic optical elements (Yuan, Yanrong)V1354,43-52(1991)

Characteristics of pulsed nuclear-reactor-pumped flowing gas lasers (Neuman, William A.; Fincke, James R.)V1411,28-40(1991)

Comments on the Seidel aberration theory (Kang, Songgao; Lu, Kaichang; Zhu, Yafei)V1527,376-379(1991)

Comparison of the angular resolution limit and SNR of the Hubble Space Telescope and the large ground-based telescopes (Souilhac, Dominique J.; Billerey, Dominique)V1494,503-526(1991)

Correction of image-phase aberrations in MRI with applications (Riek, Jonathan K.; Tekalp, Ahmet M.; Smith, Warren E.; Parker, Kevin J.)V1445,190-197(1991)

Corrections of aberrations using HOEs in UV and visible imaging systems (Yang, Zishao; Rosenbruch, Klaus J.)V1354,323-327(1991)

Design of an anamorphic gradient-index lens to correct astigmatism in diode lasers (Acosta, E.; Gomez-Reino, Carlos; Gonzalez, R. M.)V1401,82-85(1991)

Design of an optimal single-reflective holographic helmet display element (Twardowski, Patrice J.; Meyrueis, Patrick)V1456,164-174(1991)

Disparity between combiners in a double-combiner head-up display (Cohen, Jonathan; Reichert, Abraham)V1456,250-261(1991)

Effect of coma correction on the imaging quality of holographic lenses (Zajac, Marek; Nowak, Jerzy)V1507,73-80(1991)

Effects of higher order aberrations on the process window (Gortych, Joseph E.; Williamson, David M.)V1463,368-381(1991)

Enhancement of Conrady's "D-d" method (Gintner, Henry)V1354,97-102(1991)

Estimation of focused laser beam SBS (stimulated Brillouin scatter) threshold dependence on beam shape and phase aberrations (Clendening, Charles W.)V1415,72-78(1991)

Exact sine condition in the presence of spherical aberration (Shibuya, Masato)V1354,240-245(1991)

Examples of the topographies of the wavefront-variance merit function at different aberration orders (Johnston, Steve C.)V1354,77-82(1991)

Faint object spectrograph early performance (Harms, Richard J.; Fitch, John E.)V1494,49-65(1991)

High-energy laser wavefront sensors (Geary, Joseph M.)V1414,66-79(1991)

High-speed apo-lens with kinoform element (Gan, Michael A.; Potyemin, Igor S.; Perveev, Anatoly F.)V1574,243-249(1991)

Holographic testing canal of adaptive optical systems (Gan, Michael A.; Potyemin, Igor S.)V1507,549-560(1991)

Hubble Space Telescope optics: problems and solutions (Burrows, Christopher J.)V1494,528-533(1991)

Hubble Space Telescope optics status (Burrows, Christopher J.)V1567,284-293(1991)

Lissajous analysis of focus crosstalk in optical disk systems (Grove, Steven L.; Getreuer, Kurt W.; Schell, David L.)V1499,354-359(1991)

Measurement of laser spot quality (Milster, Tom D.; Treptau, Jeffrey P.)V1414,91-96(1991)

Measurement of wave aberrations of intraocular lenses through holographic interferometry (Carretero, L.; Fuentes, Rosa; Fimia-Gil, Antonio)V1507,458-462(1991); V1508,96-100(1991)

Modeling and simulation of systems imaging through atmospheric turbulence (Caponi, Maria Z.)V1415,138-149(1991)

Monochromatic aberrations of an off-axis hologram with carry out pupil (Mikhailov, I. A.)V1238,123-128(1991)

New approach for optimization of optical systems (Bezdidko, Sergey N.)V1574,250-253(1991)

New formulations between spherical aberration and spherical aberration coefficient using the Abbe sine condition (Kang, Songgao; Lu, Kaichang; Zhu, Yafei)V1527,409-412(1991)

New method for calculating third-, fifth-, and seventh-order spherical aberration coefficients and aberration offenses against sine condition (Kang, Songgao; Lu, Kaichang; Zhu, Yafei)V1527,400-405(1991)

Nonmechanical steering of laser beams by multiple aperture antennas: tolerance analysis (Neubert, Wolfgang M.; Leeb, Walter R.; Scholtz, Arpad L.)V1417,122-130(1991)

Off-axis spherical element telescope with binary optic corrector (Brown, Daniel M.; Kathman, Alan D.)V1555,114-127(1991)

Optical aberrations in underwater holography and their compensation (Watson, John; Kilpatrick, J. M.)V1461,245-253(1991)

Optical design with Wood lenses (Caldwell, J. B.)V1354,593-598(1991)

Problems of photogrammetry of moving target in water (Han, Xin Z.)V1537,215-220(1991)

Reducing and magnifying holograms (Fargion, Daniele)VIS08,354-359(1991)

Revision of the Seidel aberration theory for application in range of small viewing field and large aperture (Kang, Songgao; Lu, Kaichang; Zhu, Yafei)V1527,406-408(1991)

Selected Papers on Optical Tolerancing (Wiese, Gary E., ed.)VMS36(1991)

Self-aberration-eliminating interferometer for wavefront measurements (Gorelik, Vladimir S.; Kovalenko, Sergey N.; Turukhano, Boris G.)V1507,488-490(1991)

Simple multidimensional quadratic extrapolation method for the correction of specific aberrations in lens systems (Maxwell, Jonathan; Hull, Chris S.)V1354,277-285(1991)

Sphero-chromatic aberration correction of single holo-lens used as a spectral device (Dubik, Boguslawa; Zajac, Marek)V1574,227-234(1991)

Surface contributions of the wave aberrations up to the eighth degree (Aurin, Friedrich A.)V1527,61-72(1991)

Tale of two underwater lenses (Evans, Clinton E.; Doshi, Rekha)V1537,203-214(1991)

Tertiary spectrum manipulation in apochromats (Maxwell, Jonathan)V1354,408-416(1991)

Thermal aberration analysis of a laser-diode-pumped Nd:YAG laser (Kubota, Shigeo; Oka, Michio; Kaneda, Yushi; Masuda, Hisashi)V1354,572-580(1991)

Time aberrations of combined focusing system of high-speed image converter (Korzhenevich, Irina M.; Kolesov, G. V.)V1358,1090-1095(1991)

Variable phase-shift mask for deep-submicron optical lithography (Terasawa, Tsuneo; Hasegawa, Norio; Imai, Akira; Tanaka, Toshihiko; Katagiri, Souichi)V1463,197-206(1991)

Absorption—see also optical properties

3-D camera based on differential optical absorbance (Houde, Regis; Laurendeau, Denis; Poussart, Denis)V1332,343-354(1991)

About investigation of absolute surface relief by holographic method (Rachkovsky, Leonid I.; Tanin, Leonid V.; Rubanov, Alexander S.)V1461,232-240(1991)

Absorbing materials in multilayer mirrors (Grebenshikov, Sergey V.)V1500,189-193(1991)

Absorption, fluorescence, and stimulated emission in Ti-diffused Er:LiNbO3 waveguides (Brinkmann, R.; Sohler, Wolfgang; Suche, Hubertus)V1362,377-382(1991)

Air quality monitoring with the differential optical absorption spectrometer (Stevens, Robert K.; Conner, Teri L.)V1491,56-67(1991)

Analysis of absorption, scattering, and hemoglobin saturation using phase-modulation spectroscopy (Sevick, Eva M.; Weng, Jian; Maris, Michael B.; Chance, Britton)V1431,264-275(1991)

Analysis of atmospheric trace constituents from high-resolution infrared balloon-borne and ground-based solar absorption spectra (Goldman, Aaron; Murcray, Frank J.; Rinsland, C. P.; Blatherwick, Ronald D.; Murcray, F. H.; Murcray, David G.)V1491,194-202(1991)

Angular confining cavities for photovoltaics (Minano, Juan C.; Luque, Antonio)V1528,58-69(1991)

Application of FM spectroscopy in atmospheric trace gas monitoring: a study of some factors influencing the instrument design (Werle, Peter W.; Josek, K.; Slemr, Franz)V1433,128-135(1991)

Application of tunable diode lasers in control of high-pure-material technologies (Nadezhdinskii, Alexander I.; Stepanov, Eugene V.; Kuznetzov, Andrian I.; Devyatykh, Grigory G.; Maximov, G. A.; Khorshev, V. A.; Shapin, S. M.)V1418,487-495(1991)

Artificial photosynthesis at octane/water interface in the presence of hydrated chlorophyll a oligomer thin film (Volkov, Alexander G.; Gugeshashvili, M. I.; Kandelaki, M. D.; Markin, V. S.; Zelent, B.; Munger, G.; Leblanc, Roger M.)V1436,68-79(1991)

Atmospheric continuum absorption models (Delaye, Corinne T.; Thomas, Michael E.)V1487,291-298(1991)

Bragg grating formation and germanosilicate fiber photosensitivity (Meltz, Gerald; Morey, William W.)V1516,185-199(1991)

Co-immobilization of several dyes on optodes for pH measurements (Boisde, Gilbert; Sebille, Bernard)V1510,80-94(1991)

Data analysis methods for near-infrared spectroscopy of tissue: problems in determining the relative cytochrome aa3 concentration (Cope, Mark; van der Zee, Pieter; Essenpreis, Matthias; Arridge, Simon R.; Delpy, David T.)V1431,251-262(1991)

Deep-UV photolithography linewidth variation from reflective substrates (Dunn, Diana D.; Bruce, James A.; Hibbs, Michael S.)V1463,8-15(1991)

Determination of inhomogeneous trace absorption by using exponential expansion of the absorption Pade-approximant (Dobrego, Kirill V.)V1433,365-374(1991)

Differential time-resolved detection of absorbance changes in composite structures (Nossal, Ralph J.; Bonner, Robert F.)V1431,21-28(1991)

Diode laser spectroscopy of atmospheric pollutants (Nadezhdinskii, Alexander I.; Stepanov, Eugene V.)V1433,202-210(1991)

Double-beam laser absorption spectroscopy: shot noise-limited performance at baseband with a novel electronic noise canceler (Haller, Kurt L.; Hobbs, Philip C.)V1435,298-309(1991)

Effects of atmospheric conditions on the performance of free-space infrared communications (Grotzinger, Timothy L.)V1417,484-495(1991)

Energy bands of graphite and CsC8-GIC, and CO physisorption on graphite basal surface (Yang, Yong; Guo, Jian Q.; Lu, Dong)V1519,444-448(1991)

Excitonic photoabsorption study of AlGaAs/GaAs multiple-quantum-well grown by low-pressure MOCVD (Kwon, O'Dae; Lee, Seung-Won; Choi, Woong-Lim; Kim, Kwang-Il; Jeong, Yoon-Ha)V1361,802-808(1991)

Experiment for testing the closure property in ocean optics (Maffione, Robert A.; Honey, Richard C.; Brown, Robert A.)V1537,115-126(1991)

Fiber defects in Ge-doped fibers: towards a coherent picture (LaRochelle, Sophie)V1516,55-57(1991)

Frequency-modulation absorption spectroscopy for trace species detection: theoretical and experimental comparison among methods (Silver, Joel A.; Bomse, David S.; Stanton, Alan C.)V1435,64-71(1991)

Gradient microbore liquid chromatography with dual-wavelength absorbance detection: tunable analyzers for remote chemical monitoring (Sulya, Andrew W.; Moore, Leslie K.; Synovec, Robert E.)V1434,147-158(1991)

Highly sensitive absorption measurements in organic thin films and optical media (Skumanich, Andrew)V1559,267-277(1991)

Indirect illumination to reduce veiling luminance in seawater (Wells, Willard H.)V1537,2-9(1991)

Influence of atmospheric conditions on two-color temperature discrimination (Mallory, William R.)V1540,365-369(1991)

In-situ surface x-ray scattering of metal monolayers adsorbed at solid-liquid interfaces (Toney, Michael F.; Gordon, Joseph G.; Melroy, Owen R.)V1550,140-150(1991)

Interface properties of a-C:H/a-SiOx:H multilayer (Cui, Jing B.; Zhang, Wei P.; Fang, Rong C.; Wang, Chang S.; Zhou, Guien; Wu, Jan X.)V1519,419-422(1991)

Interferometric analysis of absorbing objects (Korchazhkin, S. V.; Krasnova, L. O.)V1512,195-197(1991)

Investigation of interlevel proximity effects case of the gate level over LOCOS (Festes, Gilles; Chollet, Jean-Paul E.)V1463,245-255(1991)

Light absorption characteristics of the human meniscus: applications for laser ablation (Vangsness, C. T.; Huang, Jay; Smith, Chadwick F.)V1424,16-19(1991)

Light energy conversion with pheophytin a and chlorophyll a monolayers at the optical transparent electrode (Leblanc, Roger M.; Blanchet, P.-F.; Cote, D.; Gugeshashvili, M. I.; Munger, G.; Volkov, Alexander G.)V1436,92-102(1991)

Long-pathlength DOAS (differential optical absorption spectrometer) system for the in situ measurement of xylene in indoor air (Biermann, Heinz W.; Green, Martina; Seiber, James N.)V1433,2-7(1991)

Measurement of color and scattering phenomena of translucent materials (Sjollema, J. I.; den Exter, Ir. T.; Zijp, Jaap R.; Ten Bosch, Jaap J.)V1500,177-188(1991)

Measurement of the tropospheric hydroxyl radical by long-path absorption (Mount, George H.)V1491,26-32(1991)

Measurement of triplet optical densities of organic compounds by means of CW laser excitation (Pavlopoulos, Theodore G.)V1437,168-183(1991)

Mechanism of excimer-laser-induced absorption in fused silica fibers (Artjushenko, Vjacheslav G.; Konov, Vitaly I.; Pashinin, Vladimir P.; Silenok, Alexander S.; Mueller, Gerhard J.; Schaldach, Brita J.; Ulrich, R.)V1420,176-176(1991)

Mechanisms of UV-laser-induced absorption in fused silica fibers (Artjushenko, Vjacheslav G.; Konov, Vitaly I.; Konstantinov, N. Y.; Pashinin, Vladimir P.; Silenok, Alexander S.; Mueller, Gerhard J.; Schaldach, Brita J.; Ulrich, R.; Neuberger, Wolfgang; Castro, Jose L.)V1590,131-136(1991)

Modeling the absorption of intense, short laser pulses in steep density gradients (Alley, W. E.)V1413,89-94(1991)

Model systems for optoelectronic devices based on nonlinear molecular absorption (Speiser, Shammai)V1560,434-442(1991)

Modulation of absorption in DCG (Zhao, Feng; Geng, Wanzhen; Jiang, Lingzhen; Hong, Jing)V1555,297-299(1991)

Multiphoton-absorption-induced structural changes in fused silica (Silin, Andrejs R.)V1513,270-273(1991)

Nonlinear absorbance effects in bacteriorhodopsin (Rayfield, George W.)V1436,150-159(1991)

Open-path tunable diode laser absorption for eddy correlation flux measurements of atmospheric trace gases (Anderson, Stuart M.; Zahniser, Mark S.)V1433,167-178(1991)

Optically induced creation, transformation, and organization of defects and color centers in optical fibers (Russell, Philip S.; Hand, Duncan P.; Chow, Y. T.; Poyntz-Wright, L. J.)V1516,47-54(1991)

Optical nonlinearities of ZnSe thin films (Chen, Lian C.; Zhang, Ji Y.; Fan, Xi W.; Yang, Ai H.; Zheng, Zhu H.)V1519,450-453(1991)

Optical resonances of a semiconductor superlattice in parallel magnetic and electric fields (Pacheco, Monica; Barticevic, Zdenka; Claro, Francisco)V1361,819-826(1991)

Optimal surface characteristics for instruments for use in laser neurosurgery (Heiferman, Kenneth S.; Cramer, K. E.; Walsh, Joseph T.)V1428,128-134(1991)

Photoacoustic absorption spectrum of some rat and bovine tissues in the ultraviolet-visible range (Bernini, Umberto; Russo, Paolo)V1427,398-404(1991)

Photoacoustic characterization of surface absorption (Reicher, David W.; Wilson, Scott R.; Kranenberg, C. F.; Raja, M. Y.; McNeil, John R.; Brueck, Steven R.)V1441,106-112(1991)

Photoinductional change of silicate glasses optical parameters at two-photon laser radiation absorption (Glebov, Leonid B.; Efimov, Oleg M.; Mekryukov, A. M.)V1513,274-282(1991)

Photothermal spectroscopy as a sensitive spectroscopic tool (Tam, Andrew C.)V1435,114-127(1991)

Physical models of second-harmonic generation in optical fibers (Lesche, Bernhard)V1516,125-136(1991)

Plasma heating by ultrashort laser pulse in the regime of anomalous skin effect (Gamaly, Eugene G.; Kiselev, A. Y.; Tikhonchuk, V. T.)V1413,95-106(1991)

Portable and very inexpensive optical fiber sensor for entero-gastric reflux detection (Baldini, Francesco; Falciai, Riccardo; Bechi, Paolo; Cosi, Franco; Bini, Andrea; Milanesi, Francesco)V1510,58-62(1991)

Progress in fiber-remote gas correlation spectrometry (Dakin, John P.; Edwards, Henry O.)V1510,160-169(1991)

Pulsed photothermal spectroscopy applied to lanthanide and actinide speciation (Berg, John M.; Morris, David E.; Clark, David L.; Tait, C. D.; Woodruff, William H.; Van Der Sluys, William G.)V1435,331-337(1991)

Quantum noise reduction in lasers with nonlinear absorbers (Ritsch, Helmut; Zoller, Peter)V1376,206-215(1991)

Removal of adsorbed gases with CO2 snow (Zito, Richard R.)V1494,427-433(1991)

Reversible phase transition and third-order nonlinearity of phthalocyanine derivatives (Suda, Yasumasa; Shigehara, Kiyotaka; Yamada, Akira; Matsuda, Hiro; Okada, Shuji; Masaki, Atsushi; Nakanishi, Hachiro)V1560,75-83(1991)

Self-reflection and self-transmission of pulsed radiation by laser-evaporated media (Furzikov, Nickolay P.)V1415,228-239(1991)

Slow-motion acquisition of laser beam profiles after propagation through gun blast (Kay, Armin V.)V1486,8-16(1991)

Sound generation by laser radiation in air: application to restoration of energy distribution in laser beams (Gurvich, Alexander S.; Myakinin, Vladimir A.; Vorob'ev, Valerii V.; D'yakov, Alexander; Pokasov, Vladimir; Pryanichnikov, Victor)V1408,10-18(1991)

Stratospheric ozone concentration profiles from Spacelab-1 solar occultation infrared absorption spectra (De Maziere, Martine M.; Camy-Peyret, C.; Lippens, C.; Papineau, N.)V1491,288-297(1991)

Study of nighttime NO3 chemistry by differential optical absorption spectroscopy (Plane, John M.; Nien, Chia-Fu)V1433,8-20(1991)

Superheating phenomena in absorbing microdroplets irradiated by pulsed lasers (Armstrong, Robert L.)V1497,132-140(1991)

System for evaluation of trace gas concentration in the atmosphere based on the differential optical absorption spectroscopy technique (Hallstadius, Hans; Uneus, Leif; Wallin, Suante)V1433,36-43(1991)

TiNxOy-Cu coatings for low-emissive solar-selective absorbers (Lazarov, M.; Roehle, B.; Eisenhammer, T.; Sizmann, R.)V1536,183-191(1991)

Tissue ablation by holmium:YSGG laser pulses through saline and blood (van Leeuwen, Ton G.; van der Veen, Maurits J.; Verdaasdonk, Rudolf M.; Borst, Cornelius)V1427,214-219(1991)

Understanding metal-dielectric-metal absorption interference filters using lumped circuit theory and transmission line theory (Pastor, Rickey G.)V1396,501-507(1991)

Use of absorption spectroscopy for refined petroleum product discrimination (Short, Michael)V1480,72-79(1991)

X-ray absorption fine structure of systems in the anharmonic limit (Mustre de Leon, Jose; Conradson, Steven D.; Batistic, I.; Bishop, A. R.; Raistrick, Ian D.; Jackson, W. E.; Brown, George S.)V1550,85-96(1991)

X-ray absorption spectroscopy: how is it done? what can it tell us? (Hayes, Tim)V1550,56-66(1991)

Acoustics—see also medical imaging; photoacoustics

Acoustical and some other nonconventional holography (Greguss, Pal)VIS08,387-401(1991)

Acoustic characterization of optical fiber glasses (Jen, Cheng K.; Neron, C.; Shang, Alain; Abe, Koichi; Bonnell, Lee J.; Kushibiki, J.; Saravanos, C.)V1590,107-119(1991)

Basic use of acoustic speckle pattern for metrology and sea waves study (He, Duo-Min; He, Ming-Shia)V1332,808-819(1991)

Composite cure monitoring with embedded optical fiber sensors (Davis, Andrew; Ohn, Myo M.; Liu, Kexing; Measures, Raymond M.)V1489,33-43(1991)

Continuous recognition of sonar targets using neural networks (Venugopal, Kootala P.; Pandya, Abhijit S.; Sudhakar, Raghavan)V1471,44-53(1991)

Critical frequencies for large-scale resonance signatures from elastic bodies (Werby, Michael F.; Gaunaurd, Guillermo C.)V1471,2-17(1991)

Demonstration of birefringent optical fiber frequency shifter employing torsional acoustic waves (Berwick, Michael; Pannell, Christopher N.; Russell, Philip S.; Jackson, David A.)V1584,364-373(1991)

Demonstration of birefringent optical fibre frequency shifter employing torsional acoustic waves (Berwick, Michael; Pannell, Christopher N.; Russell, Philip S.; Jackson, David A.)V1572,157-162(1991)

Detection of high-frequency elastic waves with embedded ordinary single-mode fibers (Liu, Kexing; Measures, Raymond M.)V1584,226-234(1991)

Effect of acoustic dampers on the excimer laser flow (Zeitoun, David; Tarabelli, D.; Forestier, Bernard M.; Truong, J. P.; Sentis, Marc L.)V1397,585-588(1991)

Fiber optic interferometric ellipsoidal shell hydrophone (Brown, David A.; Garrett, Steven L.; Conte, D. V.; Smith, R. C.; Rothenberg, E.; Young, M.; Rissberger, Ed)V1369,2-8(1991)

Focal-plane image processing using acoustic charge transport technology (Brooks, Jeff W.)V1541,68-72(1991)

General purpose fiber optic hydrophone made of castable epoxy (Garrett, Steven L.; Brown, David A.; Beaton, Brian L.; Wetterskog, Kevin; Serocki, John)V1367,13-29(1991)

Global geometric, sound, and color controls for iconographic displays of scientific data (Smith, Stuart; Grinstein, Georges G.; Pickett, Ronald M.)V1459,192-206(1991)

Holographic soundfield visualization for nondestructive testing of hot surfaces (Crostack, Horst-Artur; Meyer, E. H.; Pohl, Klaus-Juergen)V1508,101-109(1991)

Holophonics: a spread-out of the basic ideas on holography into audio-acoustics (Illenyi, Andras; Jessel, M.)VIS08,39-52(1991)

I-line lithography for highly reproducible fabrication of surface acoustic wave devices (Berek, Stefan; Knauer, Ulrich; Zottl, Helmut)V1463,515-520(1991)

Improved formalism for rough-surface scattering of acoustic and electromagnetic waves (Milder, D. M.)V1558,213-221(1991)

Information retrieval from ultrasonic (acoustical) holograms by moire principle (Greguss, Pal)V1238,421-427(1991)

In-situ film thickness measurements using acoustic techniques (Bhardwaj, Sanjay; Khuri-Yakub, B. T.)V1392,555-562(1991)

Integrating acoustical and optical sensory data for mobile robots (Wang, Gang)V1468,479-482(1991)

Inverse scattering problems in the acoustic resonance region of an underwater target (Gaunaurd, Guillermo C.)V1471,30-41(1991)

Laser ultrasonics: generation and detection considerations for improved signal-to-noise ratio (Wagner, James W.; Deaton, John B.; McKie, Andrew D.; Spicer, James B.)V1332,491-501(1991)

Low-power optical correlator for sonar range finding (Turk, Harris; Leepa, Douglas C.; Snyder, Robert F.; Soos, Jolanta I.; Bronstein, Sam)V1454,344-352(1991)

Numerical wavefront propagation through inhomogeneous media (Zakeri, Gholam-Ali)V1558,103-112(1991)

Phonetic-to-acoustic and acoustic-to-phonetic mapping using recurrent neural networks (Kumar, Vinod V.; Krishnamurthy, Ashok K.; Ahalt, Stanley C.)V1469,484-494(1991)

Photoacoustic microscopy by photodeformation applied to the determination of thermal diffusivity (Balageas, Daniel L.; Boscher, Daniel M.; Deom, Alain A.; Enguehard, Francis; Noirot, Laurence)V1467,278-289(1991)

Q-switched Nd:glass-laser-induced acoustic pulses in lithotripsy (D'yakonov, G. I.; Mikhailov, V. A.; Pak, S. K.; Shcherbakov, I. A.; Andreev, V. G.; Rudenko, O. V.; Sapozhnikov, A. V.)V1421,153-155(1991)

Real-time structural integrity monitoring using a passive quadrature demodulated, localized Michelson optical fiber interferometer capable of simultaneous strain and acoustic emission sensing (Tapanes, Edward)V1588,356-367(1991)

Recent progress in interferometric fiber sensor technology (Kersey, Alan D.)V1367,2-12(1991)

Reduction of acoustic transients in tissue with a 2 um thulium laser (Pinto, Joseph F.; Esterowitz, Leon; Bonner, Robert F.)V1420,242-243(1991)

Relationships between autofocus methods for SAR and self-survey techniques for SONAR (Wahl, Daniel E.; Jakowatz, Charles V.; Ghiglia, Dennis C.; Eichel, Paul H.)V1567,32-40(1991)

Rough-interface scattering without plane waves (Berman, David H.)V1558,191-201(1991)

Some computational results for rough-surface scattering (DeSanto, John A.; Wombell, Richard J.)V1558,202-212(1991)

Sound generation by thermocavitation-induced cw laser in solutions (Rastopov, S. F.; Sukhodolsky, A. T.)V1440,127-134(1991)

Source location of acoustic emissions from atmospheric leakage using neural networks (Barga, Roger S.; Friesel, Mark A.; Meador, Jack L.)V1469,602-611(1991)

Stabilized extrinsic fiber optic Fabry-Perot sensor for surface acoustic wave detection (Tran, Tuan A.; Miller, William V.; Murphy, Kent A.; Vengsarkar, Ashish M.; Claus, Richard O.)V1584,178-186(1991)

Strategies for tunable frequency-selective surfaces (Lakhtakia, Akhlesh)V1489,108-111(1991)

Study and characterization of three-dimensional angular distributions of acoustic scattering from spheroidal targets (George, Jacob; Werby, Michael F.)V1471,66-77(1991)

Study of an opto-ultrasonic technique for cure monitoring (Davis, Andrew; Ohn, Myo M.; Liu, Kexing; Measures, Raymond M.)V1588,264-274(1991)

Target shape and material composition from resonance echoes of submerged elongated elastic targets (Dean, Cleon E.; Werby, Michael F.)V1471,54-65(1991)

Timbre discrimination of signals with identical pitch using neural networks (Sayegh, Samir I.; Pomalaza-Raez, Carlos A.; Tepper, E.; Beer, B. A.)V1569,100-110(1991)

Time development of acoustic bullets, wave-zone form of focus wave modes, and other solutions of the acoustic equations (Moses, Harry E.; Prosser, Reese T.)V1407,354-374(1991)

Ultrasonic NDE (nondestructive evaluation) for composite materials using embedded fiber optic interferometric sensors (Liu, Kexing; Ferguson, Suzanne M.; Davis, Andrew; McEwen, Keith; Measures, Raymond M.)V1398,206-212(1991)

Using sound to extract meaning from complex data (Scaletti, Carla; Craig, Alan B.)V1459,207-219(1991)

Visualizing underwater acoustic matched-field processing (Rosenblum, Lawrence; Kamgar-Parsi, Behzad; Karahalios, Margarida; Heitmeyer, Richard)V1459,274-282(1991)

Wavefront reconstruction of acoustic waves in a variable ocean (Porter, Robert P.; Mourad, Pierre D.; Al-Kurd, Azmi)V1558,91-102(1991)

Acousto-optics—see also Bragg cells; optical processing; signal processing

0.5-GHz cw mode-locked Nd:glass laser (Ling, Junda D.; Yan, Li I.; Liu, YuanQun; Lee, Chi H.; Soos, Jolanta I.)V1454,353-362(1991)

Acoustic mode analysis of multilayered structures for the design of acousto-optic devices (Armenise, Mario N.; Matteo, Annamaria; Passaro, Vittorio M.)V1583,256-267(1991)

Acousto-optic architectures for multidimensional phased-array antenna processing (Riza, Nabeel A.)V1476,144-156(1991)

Acousto-optic color projection system (Hubin, Thomas)V1454,313-322(1991)

Acousto-optic estimation of autocorrelation and spectra using triple correlations and bispectra (Sadler, Brian M.; Giannakis, Georgios B.; Smith, Dale J.)V1476,246-256(1991)

Acousto-optics in integrated-optic devices for optical recording (Petrov, Dmitry V.; Belostotsky, A. L.; Dolgopolov, V. G.; Leonov, A. S.; Fedjukhin, L. A.)V1374,152-159(1991)

AEDC direct-write scene generation test capabilities (Lowry, Heard S.; Elrod, Parker D.; Johnson, R. J.)V1454,453-464(1991)

Automatic target recognition using acousto-optic image correlator (Molley, Perry A.; Kast, Brian A.)V1471,224-232(1991)

Bispectral magnitude and phase recovery using a wide bandwidth acousto-optical processor (Kniffen, Stacy K.; Becker, Michael F.; Powers, Edward J.)V1564,617-627(1991)

Compact low-power acousto-optic range-Doppler-angle processor for a pulsed-Doppler radar (Pape, Dennis R.; Vlannes, Nickolas P.; Patel, Dharmesh P.; Phuvan, Sonlinh)V1476,201-213(1991)

Design and testing of space-domain minimum-average correlation energy filters for 2-D acousto-optic correlators (Connelly, James M.; Vijaya Kumar, B. V. K.; Molley, Perry A.; Stalker, K. T.; Kast, Brian A.)V1564,572-592(1991)

Design of a GaAs acousto-optic correlator for real-time processing (Armenise, Mario N.; Impagnatiello, Fabrizio; Passaro, Vittorio M.; Pansini, Evangelista)V1562,160-171(1991)

Development of wideband 16-channel acousto-optical modulators on the LiNbO3 and TeO2 crystals (Bokov, Lev; Demidov, Anatoly J.; Zadorin, Anatoly; Kushnarev, Igor; Serebrennikov, Leonid J.; Sharangovich, Sergey)V1505,186-198(1991)

DOC II 32-bit digital optical computer: optoelectronic hardware and software (Stone, Richard V.; Zeise, Frederick F.; Guilfoyle, Peter S.)V1563,267-278(1991)

Frequency-dependent optical beam distortion generated by acousto-optic Bragg cells (Pieper, Ronald J.; Poon, Ting-Chung)V1454,324-335(1991)

Generalized phased-array Bragg interaction in anisotropic crystals (Young, Eddie H.; Ho, Huey-Chin C.; Yao, Shi-Kay; Xu, Jieping P.)V1476,178-189(1991)

Highly accurate pattern generation using acousto-optical deflection (Sandstrom, Torbjorn; Tison, James K.)V1463,629-637(1991)

Infrared tellurium two-dimensional acousto-optic processor for synthetic aperture radar (Souilhac, Dominique J.; Billerey, Dominique)V1521,158-174(1991)

Instrumentation to measure the near-IR spectrum of small fruits (Jaenisch, Holger M.; Niedzwiecki, Abraham J.; Cernosek, John D.; Johnson, R. B.; Seeley, John S.; Dull, Gerald G.; Leffler, Richard G.)V1379,162-167(1991)

Integrated optics in optical engineering (Popov, Yury V.)V1399,207-213(1991)

Interferometric acousto-optic receiver results (Gill, E. T.; Tsui, J. B.)V1476,190-200(1991)

Intravascular ultrasound imaging and intracardiac echocardiography: recent developments and future directions (Pandian, Natesa G.)V1425,198-202(1991)

Investigation of higher-order diffraction in a one-crystal 2-D scanner (Melamed, Nathan T.; Gottlieb, Milton S.)V1454,306-312(1991)

Laser-based display technology development at the Naval Ocean Systems Center (Phillips, Thomas E.; Trias, John A.; Lasher, Mark E.; Poirier, Peter M.; Dahlke, Weldon J.; Robinson, Waldo R.)V1454,290-298(1991)

Latest developments in crystal growth and characterization of efficient acousto-optic materials (Paradies, C. J.; Glicksman, M. E.; Jones, O. C.; Kim, G. T.; Lin, Jen T.; Gottlieb, Milton S.; Singh, N. B.)V1561,2-5(1991)

Mercurous halides for long time-delay Bragg cells (Brandt, Gerald B.; Singh, N. B.; Gottlieb, Milton S.)V1454,336-343(1991)

Miniature acousto-optic image correlator (Molley, Perry A.; Sweatt, William C.; Strong, David S.)V1564,610-616(1991)

Modeling of acousto-optic interaction in multilayer guiding structures (Armenise, Mario N.; Matteo, Annamaria; Passaro, Vittorio M.)V1583,289-297(1991)

MTF characteristics of a Scophony scene projector (Schildwachter, Eric F.; Boreman, Glenn D.)V1488,48-57(1991)

Multiwavelength optical switch based on the acousto-optic tunable filter (Liew, Soung C.)V1363,57-61(1991)

Novel acousto-optic photonic switch (Wu, Kuang-Yi; Weverka, Robert T.; Wagner, Kelvin H.; Garvin, Charles G.; Roth, Richard S.)V1563,168-175(1991)

Novel high-speed communication system (Dawber, William N.; Hirst, Peter F.; Condon, Brian P.; Maitland, Arthur; Sutton, Phillip)V1476,81-90(1991)

Novel integrated acousto-optical LiNbO3 device for application in single-laser self-heterodyne systems (Gieschen, Nikolaus; Rocks, Manfred; Olivier, Lutz)V1579,237-248(1991)

Optical A/D conversion based on acousto-optic theta modulation (Li, Yao; Zhang, Yan)V1474,167-173(1991)

Optical Technology for Microwave Applications V (Yao, Shi-Kay, ed.)V1476(1991)

Progress in fiber-remote gas correlation spectrometry (Dakin, John P.; Edwards, Henry O.)V1510,160-169(1991)

Realization of infinite-impulse response filters using acousto-optic cells (Ghosh, Anjan K.)V1564,593-601(1991)

Shape reconstruction from far-field patterns in a stratified ocean environment (Xu, Yongzhi)V1471,78-86(1991)

Sixty-four channel acousto-optical Bragg cells for optical computing applications (Graves, David W.)V1563,229-235(1991)

Sound generation by laser radiation in air: application to restoration of energy distribution in laser beams (Gurvich, Alexander S.; Myakinin, Vladimir A.; Vorob'ev, Valerii V.; D'yakov, Alexander; Pokasov, Vladimir; Pryanichnikov, Victor)V1408,10-18(1991)

Surface-acoustic-wave acousto-optic devices for wide-bandwidth signal processing and switching applications (Garvin, Charles G.; Sadler, Brian M.)V1562,303-318(1991)

TeO2 slow surface acoustic wave Bragg cell (Yao, Shi-Kay)V1476,214-221(1991)

Tracking method of optical tape recorder using acousto-optic scanning (Narahara, Tatsuya; Kumai, Satoshi; Nakao, Takashi; Ozue, Tadashi)V1499,120-128(1991)

Wide-angular aperture acousto-optic Bragg cell (Weverka, Robert T.; Wagner, Kelvin H.)V1562,66-72(1991)

Wideband acousto-optic spectrometer (Chang, I. C.)V1476,257-268(1991)

Actuators—see also optomechanical design

10-mm-thick head mechanism for a stacked optical disk system (Seya, Eiichi; Matsumoto, Kiyoshi; Nihei, Hideki; Ichikawa, Atsushi; Moriyama, Shigeo; Nakamura, Shigeru; Mita, Seiichi)V1499,269-273(1991)

Comparison of analog and digital strategies for automatic vibration control of lightweight space structures (Hong, Suk-Yoon; Varadan, Vasundara V.; Varadan, Vijay K.)V1489,75-83(1991)

Comparisons of deformable-mirror models and influence functions (Hiddleston, H. R.; Lyman, Dwight D.; Schafer, Eric L.)V1542,20-33(1991)

Design of optically activated conventional sensors and actuators (Liu, Kexing)V1398,269-275(1991)

Digital imaging of aircraft dynamic seals: a fiber optics solution (Nwagboso, Christopher O.)V1500,234-245(1991)

Discrete piezoelectric sensors and actuators for active control of two-dimensional spacecraft components (Bayer, Janice I.; Varadan, Vasundara V.; Varadan, Vijay K.)V1480,102-114(1991)

Fitting capability of deformable mirror (Jiang, Wen-Han; Ling, Ning; Rao, Xuejun; Shi, Fan)V1542,130-137(1991)

High-bandwidth control for low-area-density deformable mirrors (How, Jonathan P.; Anderson, Eric H.; Miller, David W.; Hall, Steven R.)V1489,148-162(1991)

High-speed swing arm three-beam optical head (Yak, A. S.; Low, Toh-Siew; Lim, Siak-Piang)V1401,74-81(1991)

New tracking method for two-beam optical heads using continuously grooved disks (Irie, Mitsuru; Takeshita, Nobuo; Fujita, Teruo; Kime, Kenjiro)V1499,360-365(1991)

Photofluidics for integrating fiber sensors with high-authority mechanical actuators (Claus, Richard O.; Murphy, Kent A.; Fogg, Brian R.; Sun, David; Vengsarkar, Ashish M.)V1588,159-168(1991)

Shape control of piezoelectric bimorph mirrors (Burke, Shawn E.; Hubbard, James E.)V1532,207-214(1991)

Super-compact dual-axis optical scanning unit applying a tortional spring resonator driven by a piezoelectric actuator (Goto, Hiroshi; Imanaka, Koichi)V1544,272-281(1991)

Tilt corrector based on spring-type magnetostrictive actuators (Aksinin, V. I.; Apollonov, V. V.; Chetkin, Sergue A.; Kijko, Vadim V.; Muraviev, S. V.; Vdovin, Gleb V.)V1500,93-104(1991)

Use of FEM modes in time-domain servo simulations (Ernst, Charles H.)V1499,129-135(1991)

Adaptive optics—see also astronomy; atmospheric optics

Active and Adaptive Optical Systems (Ealey, Mark A., ed.)V1542(1991)

Active optics system for a 3.5-meter structured mirror (Stepp, Larry M.; Roddier, Nicolas; Dryden, David M.; Cho, Myung K.)V1542,175-185(1991)

Active optics technology: an overview (Ray, Frank B.)V1532,188-206(1991)

Adaptive 4-D IR clutter suppression filtering technique (Aridgides, Athanasios; Fernandez, Manuel F.; Randolph, D.; Ferris, D.)V1481,110-116(1991)

Adaptive optical transfer function modeling (Gaffard, Jean-Paul; Boyer, Corinne; Ledanois, Guy)V1483,92-103(1991); V1542,34-45(1991)

Adaptive optics: a progress review (Hardy, John W.)V1542,2-17(1991)

Adaptive optics for the European very large telescope (Merkle, Fritz; Hubin, Norbert)V1542,283-292(1991)

Adaptive optics system tests at the ESO 3.6-m telescope (Merkle, Fritz; Gehring, G.; Rigaut, Francois; Lena, Pierre J.; Rousset, Gerard; Fontanella, Jean-Claude; Gaffard, Jean-Paul)V1542,308-318(1991)

Adaptive optics, transfer loops modeling (Boyer, Corinne; Gaffard, Jean-Paul; Barrat, Jean-Pierre; Lecluse, Yves)V1483,77-91(1991); V1542,46-61(1991)

Adaptive optics using curvature sensing (Forbes, Fred F.; Roddier, Nicolas)V1542,140-147(1991)

Adaptive structures technology programs for space-based optical systems (Betros, Robert S.; Bronowicki, Allen J.; Manning, Raymond A.)V1542,406-419(1991)

Advanced image intensifier systems for low-light high-speed imaging (Bruno, Theresa L.; Wirth, Allan)V1358,109-116(1991)

Algorithms for wavefront reconstruction out of curvature sensing data (Roddier, Nicolas)V1542,120-129(1991)

Alignment and focus control of a telescope using image sharpening (Jones, Peter A.)V1542,194-204(1991)

Analysis and testing of a soft actuation system for segmented-reflector articulation and isolation (Jandura, Louise; Agronin, Michael L.)V1542,213-224(1991)

Anisoplanatism and the use of laser guide stars (Goad, Larry E.)V1542,100-109(1991)

Astigmatic unstable resonator with an intracavity deformable mirror (Neal, Daniel R.; McMillin, Pat; Michie, Robert B.)V1542,449-458(1991)

Atmospheric turbulence sensing for a multiconjugate adaptive optics system (Johnston, Dustin C.; Welsh, Byron M.)V1542,76-87(1991)

Come-on-plus project: an upgrade of the come-on adaptive optics prototype system (Gendron, Eric; Cuby, Jean-Gabriel; Rigaut, Francois; Lena, Pierre J.; Fontanella, Jean-Claude; Rousset, Gerard; Gaffard, Jean-Paul; Boyer, Corinne; Richard, Jean-Claude; Vittot, M.; Merkle, Fritz; Hubin, Norbert)V1542,297-307(1991)

Comparisons of deformable-mirror models and influence functions (Hiddleston, H. R.; Lyman, Dwight D.; Schafer, Eric L.)V1542,20-33(1991)

Compensation efficiency of an optical adaptive transmitter (Feng, Yue-Zhong; Song, Zhengfang; Gong, Zhi-Ben)V1417,370-372(1991)

Deformable-mirror concept for adaptive optics in space (Kuo, Chin-Po)V1542,420-431(1991)

Dynamic model of deformable adaptive mirror (Glebova, Svetlana N.; Lavrov, Nikolaj A.)V1500,275-280(1991)

Effects of deformable mirror/wavefront sensor separation in laser beam trains (Schafer, Eric L.; Lyman, Dwight D.)V1415,310-316(1991)

Electron trapping materials for adaptive learning in photonic neural networks (Farhat, Nabil H.)V1621,310-319(1991)

Estimation of optimal Kalman filter gain from nonoptimal filter residuals (Chen, Chung-Wen; Huang, Jen-Kuang)V1489,254-265(1991)

Fitting capability of deformable mirror (Jiang, Wen-Han; Ling, Ning; Rao, Xuejun; Shi, Fan)V1542,130-137(1991)

Hartmann-Shack sensor as a component in active optical system to improve the depth resolution of the laser tomographic scanner (Liang, Junzhong; Grimm, B.; Goelz, Stefan; Bille, Josef F.)V1542,543-554(1991)

High-bandwidth control for low-area-density deformable mirrors (How, Jonathan P.; Anderson, Eric H.; Miller, David W.; Hall, Steven R.)V1489,148-162(1991)

Holographic testing canal of adaptive optical systems (Gan, Michael A.; Potyemin, Igor S.)V1507,549-560(1991)

Imaging performance analysis of adaptive optical telescopes using laser guide stars (Welsh, Byron M.)V1542,88-99(1991)

Implementation issues in the control of a flexible mirror testbed (Anderson, Eric H.; How, Jonathan P.)V1542,392-405(1991)

Increasing the isoplanatic patch size with phase-derivative adaptive optics (Feng, Yue-Zhong; Gong, Zhi-Ben; Song, Zhengfang)V1487,356-360(1991)

Johns Hopkins adaptive optics coronagraph (Clampin, Mark; Durrance, Samuel T.; Golimowski, D. A.; Barkhouser, Robert H.)V1542,165-174(1991)

Laser guide stars for adaptive optics systems: Rayleigh scattering experiments (Thompson, Laird A.; Castle, Richard; Carroll, David L.)V1542,110-119(1991)

Latest developments of active optics of the ESO NTT and the implications for the ESO VLT (Noethe, L.; Andreoni, G.; Franza, F.; Giordano, P.; Merkle, Fritz; Wilson, Raymond N.)V1542,293-296(1991)

Liquid-crystal phase modulators for active micro-optic devices (Purvis, Alan; Williams, Geoffrey; Powell, Norman J.; Clark, Michael G.; Wiltshire, Michael C.)V1455,145-149(1991)

Liquid-crystal television optical neural network: architecture, design, and models (Yu, Francis T.)V1455,150-166(1991)

MARTINI: system operation and astronomical performance (Doel, A. P.; Dunlop, C. N.; Major, J. V.; Myers, Richard M.; Sharples, R. M.)V1542,319-326(1991)

Measuring phase errors of an array or segmented mirror with a single far-field intensity distribution (Tyson, Robert K.)V1542,62-75(1991)

Metallic alternative to glass mirrors (active mirrors in aluminum): a review (Rozelot, Jean-Pierre; Leblanc, Jean-M.)V1494,481-490(1991)

MIT multipoint alignment testbed: technology development for optical interferometry (Blackwood, Gary H.; Jacques, Robert; Miller, David W.)V1542,371-391(1991)

Need for active structures in future large IR and sub-mm telescopes (Rapp, Donald)V1542,328-358(1991)

Neural network adaptive optics for the multiple-mirror telescope (Wizinowich, Peter L.; Lloyd-Hart, Michael; McLeod, Brian A.; Colucci, D'nardo; Dekany, Richard G.; Wittman, David; Angel, J. R.; McCarthy, Donald W.; Hulburd, William G.; Sandler, David G.)V1542,148-158(1991)

Optical neurocomputer architectures using spatial light modulators (Robinson, Michael G.; Zhang, Lin; Johnson, Kristina M.)V1469,240-249(1991)

Optical pathlength control in the nanometer regime on the JPL phase-B interferometer testbed (O'Neal, Michael C.; Spanos, John T.)V1542,359-370(1991)

Partially compensated speckle imaging: Fourier phase spectrum estimation (Roggemann, Michael C.; Matson, Charles L.)V1542,477-487(1991)

PCI (phase compensation instability) and minishear (Fried, David L.; Szeto, Roque K.)V1408,150-166(1991)

Performance tests of a 1500 degree-of-freedom adaptive optics system for atmospheric compensation (Cuellar, Louis; Johnson, Paul A.; Sandler, David G.)V1542,468-476(1991)

Primary mirror control system for the Galileo telescope (Bortoletto, Favio; Baruffolo, A.; Bonoli, C.; D'Alessandro, Maurizio; Fantinel, D.; Giudici, G.; Ragazzoni, Roberto; Salvadori, L.; Vanini, P.)V1542,225-235(1991)

Progress report on a five-axis fast guiding secondary for the University of Hawaii 2.2-meter telescope (Cavedoni, Charles P.; Graves, J. E.; Pickles, A. J.)V1542,273-282(1991)

Propagation of High-Energy Laser Beams Through the Earth's Atmosphere II (Ulrich, Peter B.; Wilson, LeRoy E., eds.)V1408(1991)

Prototype high-speed optical delay line for stellar interferometry (Colavita, Mark M.; Hines, Braden E.; Shao, Michael; Klose, George J.; Gibson, B. V.)V1542,205-212(1991)

Quantitative evaluation of optical surfaces using an improved Foucault test approach (Vandenberg, Donald E.; Humbel, William D.; Wertheimer, Alan)V1542,534-542(1991)

Realistic wind effects on turbulence and thermal blooming compensation (Long, Jerry E.; Hills, Louis S.; Gebhardt, Frederick G.)V1408,58-71(1991)

Real-time wavefront correction system using a zonal deformable mirror and a Hartmann sensor (Salmon, J. T.; Bliss, Erlan S.; Long, Theresa W.; Orham, Edward L.; Presta, Robert W.; Swift, Charles D.; Ward, Richard S.)V1542,459-467(1991)

Remote alignment of adaptive optical systems with far-field optimization (Mehta, Naresh C.)V1408,96-111(1991)

Satellite-borne laser for adaptive optics reference (Greenaway, Alan H.)V1494,386-393(1991)

Simplex optimization method for adaptive optics system alignment (Digumarthi, Ramji V.; Mehta, Naresh C.)V1408,136-147(1991)

Solar astronomy with a 19-segment adaptive mirror (Acton, D. S.; Smithson, Robert C.)V1542,159-164(1991)

Solar feature correlation tracker (Rimmele, Thomas; von der Luhe, Oskar; Wiborg, P. H.; Widener, A. L.; Dunn, Richard B.; Spence, G.)V1542,186-193(1991)

Surface control techniques for the segmented primary mirror in the large lunar telescope (Gleckler, Anthony D.; Pflibsen, Kent P.; Ulich, Bobby L.; Smith, Duane D.)V1494,454-471(1991)

Thermal blooming critical power and adaptive optics correction for the ground-based laser (Smith, David C.; Townsend, Sallie S.)V1408,112-118(1991)

Tilt corrector based on spring-type magnetostrictive actuators (Aksinin, V. I.; Apollonov, V. V.; Chetkin, Sergue A.; Kijko, Vadim V.; Muraviev, S. V.; Vdovin, Gleb V.)V1500,93-104(1991)

University of Hawaii adaptive optics system: I. General approach (Roddier, Francois J.; Graves, J. E.; McKenna, Daniel; Northcott, Malcolm J.)V1542,248-253(1991)

University of Hawaii adaptive optics system: II. Computer simulation (Northcott, Malcolm J.)V1542,254-261(1991)

University of Hawaii adaptive optics system: III. Wavefront curvature sensor (Graves, J. E.; McKenna, Daniel)V1542,262-272(1991)

Use of adaptive optics for minimizing atmospheric distortion of optical waves (Lukin, Vladimir P.)V1408,86-95(1991)

Variable wind direction effects on thermal blooming correction (Hills, Louis S.; Long, Jerry E.; Gebhardt, Frederick G.)V1408,41-57(1991)

Wavefront control model of a beam control experiment (Cielinski, Amy J.)V1542,434-448(1991)

Wavefront curvature sensing and compensation methods in adaptive optics (Roddier, Francois J.)V1487,123-128(1991)

Aerosols—see also atmospheric optics; particles

Accuracy of aerosol size measurements and the spectrum of applied laser light (Drobnik, Antoni; Pieszynski, Krzysztof)V1391,378-381(1991)

Aerosol models for optical and IR propagation in the marine atmospheric boundary layer (de Leeuw, Gerrit)V1487,130-159(1991)

Airborne lidar measurements of ozone and aerosols in the summertime Arctic troposphere (Browell, Edward V.)V1491,7-14(1991)

Airborne lidar observations of ozone and aerosols in the wintertime Arctic stratosphere (Browell, Edward V.)V1491,273-281(1991)

Alexandrite laser characterization and airborne lidar developments for water vapor DIAL measurements (Ponsardin, Patrick; Higdon, Noah S.; Grossmann, Benoist E.; Browell, Edward V.)V1492,47-51(1991)

Atmospheric code sensitivity to uncertainties in aerosol optical depth characteristics (Teillet, Philippe M.; Fedosejevs, Gunar; Ahern, Francis J.; Gauthier, Robert P.; Sirois, J.)V1492,213-223(1991)

Balloon-born investigations of total and aerosol attenuation continuous spectra in the stratosphere (Mirzoeva, Larisa A.; Kiseleva, Margaret S.; Reshetnikova, Irina N.)V1540,444-449(1991)

Blurring effect of aerosols on imaging systems (Bissonnette, Luc R.)V1487,333-344(1991)

Convective evaporation of water aerosol droplet irradiated by CO2 laser (Butkovsky, A. V.)V1440,146-152(1991)

Correlation between the aerosol profiles measurements, the meteorological conditions, and the atmospheric IR transmission in a mediterranean marine atmosphere (Tanguy, Mireille; Bonhommet, Herve; Autric, Michel L.; Vigliano, Patrick)V1487,172-183(1991)

Development of 1- and 2-um coherent Doppler lidars for atmospheric sensing (Chan, Kin-Pui; Killinger, Dennis K.)V1492,111-114(1991)

Dynamics of laser light transmission losses in aerosols (Malicka, Marianna; Parma, Ludvik)V1391,181-189(1991)

Effects of thin and subvisible cirrus on HEL far-field intensity calculations at various wavelengths (Harada, Larrene K.)V1408,28-40(1991)

Evaluation of the Navy Oceanic Vertical Aerosol Model using lidar and PMS particle-size spectrometers (Jensen, Douglas R.)V1487,160-171(1991)

Forecasting optical turbulence strength: effects of macroscale meteorology and aerosols (Sadot, Danny; Kopeika, Norman S.)V1442,325-334(1991)

Ground-based CW atmospheric Doppler performance modeling (Becherer, Richard J.; Kahan, Lloyd R.)V1416,306-313(1991)

High-speed microcinematography of aerosols (Lavelle, Stephen P.; Jackman, Louise A.; Nolan, P. F.)V1358,821-830(1991)

High radiometric performance CCD for the third-generation stratospheric aerosol and gas experiment (Delamere, W. A.; Baer, James W.; Ebben, Thomas H.; Flores, James S.; Kleiman, Gary; Blouke, Morley M.; McCormick, M. P.)V1447,204-213(1991)

Investigations of the phenomenon of light scattering on particles of atmospheric aerosols (Drobnik, Antoni)V1391,204-210(1991)

Laser-induced volatilization and ionization of aerosol particles for their mass spectral analysis in real time (Sinha, Mahadeva P.)V1437,150-156(1991)

Lidar backscatter calculations for solid-sphere and layered-sphere aerosols (Youmans, Douglas G.)V1416,151-162(1991)

Measurement of modulation transfer function of desert atmospheres (McDonald, Carlos)V1487,203-219(1991)

Measurement of wind velocity spread: signal-to-noise ratio for heterodyne detection of laser backscatter from aerosol (Fried, David L.; Szeto, Roque K.)V1416,163-176(1991)

Modeling time-dependent obscuration for simulated imaging of dust and smoke clouds (Hoock, Donald W.)V1486,164-175(1991)

Multiple scattering of laser light in dense aerosol (Parma, Ludvik; Malicka, Marianna)V1391,190-198(1991)

Multiscattered lidar returns from atmospheric aerosols (Hutt, Daniel L.; Bissonnette, Luc R.; Durand, Louis-Gilles)V1487,250-261(1991)

NASA's program in lidar remote sensing (Theon, John S.; Vaughan, William W.; Browell, Edward V.; Jones, William D.; McCormick, M. P.; Melfi, Samuel H.; Menzies, Robert T.; Schwemmer, Geary K.; Spinhirne, James D.)V1492,2-23(1991)

Nonlinear laser interactions with saltwater aerosols (Alexander, Dennis R.; Poulain, D. E.; Schaub, Scott A.; Barton, John P.)V1497,90-97(1991)

O3, NO2, NO3, SO2, and aerosol measurements in Beijing (Xue, Qing-yu; Guo, Song; Zhao, Xue-peng; Nieu, Jian-guo; Zhang, Yi-ping)V1491,75-82(1991)

Overall atmospheric MTF and aerosol MTF cutoff (Dror, Itai; Kopeika, Norman S.)V1487,192-202(1991)

Prediction of Cn2 on the basis of macroscale meteorology including aerosols (Sadot, Danny; Kopeika, Norman S.)V1487,40-50(1991)

Prediction of coarse-aerosol statistics according to weather forecast (Gottlieb, J.; Kopeika, Norman S.)V1487,184-191(1991)

Propagation Engineering: Fourth in a Series (Bissonnette, Luc R.; Miller, Walter B., eds.)V1487(1991)

Retrieval and molecule sensitivity studies for the global ozone monitoring experiment and the scanning imaging absorption spectrometer for atmospheric chartography (Chance, Kelly V.; Burrows, John P.; Schneider, Wolfgang)V1491,151-165(1991)

Spherical and nonspherical aerosol and particulate characterization using optical pattern recognition techniques (Marshall, Martin S.; Benner, Robert E.)V1564,121-134(1991)

Stratospheric aerosol and gas experiment III: aerosol and trace gas measurements for the Earth Observing System (McCormick, M. P.; Chu, William P.; Zawodny, J. M.; Mauldin, Lemuel E.; McMaster, Leonard R.)V1491,125-141(1991)

Surface and aerosol models for use in radiative transfer codes (Hart, Quinn J.)V1493,163-174(1991)

Upper-atmosphere research satellite: an overview (McNeal, Robert J.)V1491,84-90(1991)

Agriculture

Advances in R&D in near-infrared spectroscopy in Japan (Kawano, Sumio; Iwamoto, Mutsuo)V1379,2-9(1991)

Analysis of silage composition by near-infrared reflectance spectroscopy (Reeves, James B.; Blosser, Timothy H.; Colenbrander, V. F.)V1379,28-38(1991)

Analysis of vegetation stress and damage from images of the high-resolution airborne pushbroom image spectrograph compact airborne spectrographic imager (Mueksch, Michaela C.)V1399,157-161(1991)

Automated optical grading of timber (Sobey, Peter J.)V1379,168-179(1991)

Comparison of methods to treat nonuniform illumination in images (McDonald, Timothy P.; Chen, Yud-Ren)V1379,89-98(1991)

Corn plant locating by image processing (Jia, Jiancheng; Krutz, Gary W.; Gibson, Harry G.)V1379,246-253(1991)

Damage detection in peanut grade samples using chromaticity and luminance (Dowell, Floyd E.; Powell, J. H.)V1379,136-140(1991)

Detection of citrus freeze damage with natural color and color IR video systems (Blazquez, Carlos H.)V1467,394-401(1991)

Digital restoration of distorted geometric features of pigs (Van der Stuyft, Emmanuel; Goedseels, Vic; Geers, Rony)V1379,189-200(1991)

Effect of image size and contrast on the recognition of insects in radiograms (Schatzki, Thomas F.; Keagy, Pamela M.)V1379,182-188(1991)

Fiber optic lighting system for plant production (St. George, Dennis R.; Feddes, John J.)V1379,69-80(1991)

Fourier analysis of near-infrared spectra (McClure, W. F.)V1379,45-51(1991)

Fresh market carrot inspection by machine vision (Howarth, M. S.; Searcy, Stephen W.)V1379,141-150(1991)

Image analysis applications for grain science (Zayas, Inna Y.; Steele, James L.)V1379,151-161(1991)

Image analysis for estimating the weight of live animals (Schofield, C. P.; Marchant, John A.)V1379,209-219(1991)

Image analysis for vision-based agricultural vehicle guidance (Brown, Neil H.; Wood, Hugh C.; Wilson, James N.)V1379,54-68(1991)

Image processing to locate corn plants (Jia, Jiancheng; Krutz, Gary W.; Gibson, Harry G.)V1396,656-663(1991)

Instrumentation to measure the near-IR spectrum of small fruits (Jaenisch, Holger M.; Niedzwiecki, Abraham J.; Cernosek, John D.; Johnson, R. B.; Seeley, John S.; Dull, Gerald G.; Leffler, Richard G.)V1379,162-167(1991)

Lasers and electro-optic technology in natural resource management (Greer, Jerry D.)V1396,342-352(1991)

Midinfrared backscatter spectra of selected agricultural crops (Narayanan, Ram M.; Green, Steven E.; Alexander, Dennis R.)V1379,116-122(1991)

Model-based image processing for characterizing pigs in scenes (Tillett, Robin D.; Marchant, John A.)V1379,201-208(1991)

Modeling the distribution of optical radiation in diffusely reflecting materials (Birth, Gerald S.)V1379,81-88(1991)

Moisture influence on near-infrared prediction of wheat hardness (Windham, William R.; Gaines, Charles S.; Leffler, Richard G.)V1379,39-44(1991)

Nondestructive determination of the solids content of horticultural products (Birth, Gerald S.; Dull, Gerald G.; Leffler, Richard G.)V1379,10-15(1991)

Optical reflectance sensor for detecting plants (Shropshire, Geoffrey J.; Von Bargen, Kenneth; Mortensen, David A.)V1379,222-235(1991)

Optics in Agriculture (DeShazer, James A.; Meyer, George E., eds.)V1379(1991)

Three-dimensional vision system for peanut pod maturity (Williams, E. J.; Adams, Stephen D.)V1379,236-245(1991)

Use of color, color infrared, black and white films, and video systems in detecting health, stress, and disease in vegetation (Blazquez, Carlos H.)V1379,106-115(1991)

Using computer vision for detecting watercore in apples (Throop, James A.; Rehkugler, Gerald E.; Upchurch, Bruce L.)V1379,124-135(1991)

Using digital images to measure and discriminate small particles in cotton (Taylor, Robert A.; Godbey, Luther C.)V1379,16-27(1991)

Vision methods for inspection of greenhouse poinsettia plants (Meyer, George E.; Troyer, W. W.; Fitzgerald, Jay B.)V1379,99-105(1991)

Alignment—see also optomechanical design

64-Mbit DRAM production with i-line stepper (Shirai, Hisatsugu; Kobayashi, Katsuyoshi; Nakagawa, Kenji)V1463,256-274(1991)

Absolute phasing of segmented mirrors using the polarization phase sensor (Klumpe, Herbert W.; Lajza-Rooks, Barbara A.; Blum, James D.)V1398,95-106(1991); V1532,230-240(1991)

Aligning diamond-turned optics using visible-light interferometry (Figoski, John W.)V1354,540-546(1991)

Alignment and focus control of a telescope using image sharpening (Jones, Peter A.)V1542,194-204(1991)

Alignment of an aspheric mirror subsystem for an advanced infrared catadioptric system (Tingstad, James S.)V1527,194-198(1991)

Alignment of principal axes between birefringent fiber by the spatial technique and its distribution-sensing effect (Zhang, Jinghua; Wang, Chunhua; Huang, Zhaoming)V1572,69-73(1991)

Analysis of alignment in optical interconnection systems (Ghosh, Anjan K.; Beech, Russell S.)V1389,630-641(1991)

Angle measurement by moire interference technique (Singh, Brahm P.; Chitnis, Vijay T.)V1554B,335-338(1991)

Automatic-adjusting optical axis for linear CCD scanner (Chang, Rong-Seng; Chen, Der-Chin)V1527,357-360(1991)

Automatic mask-to-wafer alignment and gap control using moire interferometry (Chitnis, Vijay T.; Kowsalya, S.; Rashmi, Dr.; Kanjilal, A. K.; Narain, Ram)V1332,613-622(1991)

Characterization of automatic overlay measurement technique for sub-half-micron devices (Kawai, Akira; Fujiwara, Keiji; Tsujita, Kouichirou; Nagata, Hitoshi)V1464,267-277(1991)

Computer-aided alignment of a grazing-incidence ring resonator for a visible wavelength free-electron laser (Hudyma, Russell M.; Eigler, Lynne C.)V1354,523-532(1991)

Experimental and simulation studies of alignment marks (Wong, Alfred K.; Doi, Takeshi; Dunn, Diana D.; Neureuther, Andrew R.)V1463,315-323(1991)

Generic model for line-of-sight analysis and calibration (Afik, Zvi; Shammas, A.; Schwartz, Roni; Gal, Eli)V1442,392-398(1991)

"Golden standard" wafer design for optical stepper characterization (Kemp, Kevin G.; King, Charles F.; Wu, Wei; Stager, Charles)V1464,260-266(1991)

High-bandwidth alignment sensing in active optical systems (Kishner, Stanley J.)V1532,215-229(1991)

In-process laser beam position sensing (Chen, Shang-Liang; Li, L.; Modern, P. J.; Steen, William M.)V1502,123-134(1991)

Integrated optics bus access module for intramachine communication (Karioja, Pentti; Tammela, Simo; Hannula, Tapio)V1533,129-137(1991)

Laser alignment modeling using rigorous numerical simulations (Wojcik, Gregory L.; Vaughan, David K.; Mould, John; Leon, Francisco A.; Qian, Qi-de; Lutz, Michael A.)V1463,292-303(1991)

Laser boresighting by second-harmonic generation (Adel, Michael E.; Buckwald, Bob A.; Cabib, Dario)V1442,68-80(1991)

Line-of-sight alignment of a multisensor system (Wilk, Shalom; Goldmunz, Menachem; Shahaf, Nachum; Klein, Yitschak; Goldman, Shmuel; Oren, Ehud)V1442,140-148(1991)

Microdisplacement positioning system for a diffraction grating ruling engine (Yang, Hou-Min; Wang, Xiaolin; Zhang, Yinxian)V1533,185-192(1991)

Micromachined structure for coupling optical fibers to integrated optical waveguides (Goel, Sanjay; Naylor, David L.)V1396,404-410(1991)

Misalignments of airborne laser beams due to mechanical vibrations (Freitas, Jose C.; Abreu, M. A.; Rodrigues, F. C.; Carvalho, Fernando D.)V1399,42-48(1991)

MIT multipoint alignment testbed: technology development for optical interferometry (Blackwood, Gary H.; Jacques, Robert; Miller, David W.)V1542,371-391(1991)

Mix-and-match lithography for half-micrometer technology (Flack, Warren W.; Dameron, David H.)V1465,164-172(1991)

New alignment sensors for wafer stepper (Ota, Kazuya; Magome, Nobutaka; Nishi, Kenji)V1463,304-314(1991)

Optical method of detection for the magnetic alignment of an electron accelerator (Villate, Denis)V1533,193-196(1991)

Optimal subpixel-level IR frame-to-frame registration (Fernandez, Manuel F.; Aridgides, Athanasios; Randolph, D.; Ferris, D.)V1481,172-179(1991)

Origins of asymmetry in spin-cast films over topography (Manske, Loni M.; Graves, David B.)V1463,414-422(1991)

Pointing stability of copper vapor laser with novel off-axis unstable resonator (Lando, Mordechai; Belker, D.; Lerrer, A.; Lotem, Haim; Dikman, A.; Bialolanker, Gabriel; Lavi, S.; Gabay, Shimon)V1442,172-180(1991)

Remote alignment of adaptive optical systems with far-field optimization (Mehta, Naresh C.)V1408,96-111(1991)

Selected performance parameters and functional principles of a new stepper generation (Kliem, Karl-Heinz; Sczepanski, Volker; Michl, Uwe; Hesse, Reiner)V1463,743-751(1991)

Simplex optimization method for adaptive optics system alignment (Digumarthi, Ramji V.; Mehta, Naresh C.)V1408,136-147(1991)

Simultaneous alignment and multiple surface figure testing of optical system components via wavefront aberration measurement and reverse optimization (Lundgren, Mark A.; Wolfe, William L.)V1354,533-539(1991)

Spatial frequency selective error sensing for space-based, wide field-of-view, multiple-aperture imaging systems (Erteza, Ahmed; Schneeberger, Timothy J.)V1527,182-187(1991)

Submicrometer lithographic alignment and overlay strategies (Zaidi, Saleem H.; Naqvi, H. S.; Brueck, Steven R.)V1343,245-255(1991)

Super-accurate positioning technique using diffracted moire signals (Takada, Yutaka; Uchida, Yoshiyuki; Akao, Yasuo; Yamada, Jun; Hattori, Shuzo)V1332,571-576(1991)

System calibration and part alignment for inspection of 2-D electronic circuit patterns (Rodriguez, Arturo A.; Mandeville, Jon R.; Wu, Frederick Y.)V1332,25-35(1991)

Video-based alignment system for x-ray lithography (Hughlett, R. E.; Cooper, Keith A.)V1465,100-110(1991)

Wafer alignment based on existing microstructures (Wyntjes, Geert J.; Hercher, Michael)V1464,539-545(1991)

Wavelength and mode-adjustable source for modal characterization of optical fiber components: application to a new alignment method (Pagnoux, Dominique; Blondy, Jean M.; Facq, Paul)V1504,98-106(1991)

Zero axial irradiance with Rademacher's zone plates (Ojeda-Castaneda, Jorge; Ramirez, G.; Ibarra, J.)V1500,252-255(1991)

Amplifiers—see also optical communications

Absorptive nonlinear semiconductor amplifiers for fast optical switching (Barnsley, Peter E.; Marshall, Ian W.; Fiddyment, Phillip J.; Robertson, Michael J.)V1378,116-126(1991)

Accurate design of multiport low-noise MMICs up to 20 GHz (Willems, David; Bahl, I.; Griffin, Edward)V1475,55-61(1991)

Advances in Erbium-doped fiber amplifiers for optical communications (Zyskind, John L.)V1373,80-92(1991)

Advances in laser pump sources for erbium-doped fiber amplifiers (Henshall, Gordon D.; Hadjifotiou, A.; Baker, R. A.; Warbrick, K. J.)V1418,286-291(1991)

Advances in power MMIC amplifier technology in space communications (Tserng, Hua Q.; Saunier, Paul)V1475,74-85(1991)

Air Force program in coherent semiconductor lasers (Kennett, Ruth D.; Frazier, John C.)V1501,57-68(1991)

Amplification of amplitude modulated signals in a self-pumped photorefractive phase conjugator (Petersen, Paul M.; Buchhave, Preben)V1362,582-585(1991)

Application of Kaband MMIC technology for an Orbiter/ACTS communications experiment (Arndt, George D.; Fink, Patrick W.; Leopold, Louis; Bondyopadhyay, Probir; Shaw, Roland)V1475,231-242(1991)

Applications of fiber amplifiers to high-data-rate nonlinear transmission (Spirit, David M.; Brown, Graeme N.; Marshall, Ian W.; Blank, Lutz C.)V1373,197-208(1991)

Beam quality in laser amplifiers (Martinez-Herrero, R.; Mejias, P. M.)V1397,623-626(1991)

Characteristics of amplification of ultrashort laser pulses in excimer media (Kannari, Fumihiko; Obara, Minoru)V1397,85-89(1991)

Characterization of GRIN-SCH-SQW amplifiers (Haake, John M.; Zediker, Mark S.; Balestra, Chet L.; Krebs, Danny J.; Levy, Joseph L.)V1418,298-308(1991)

Chemically driven pulsed and continuous visible laser amplifiers and oscillators (Gole, James L.; Woodward, J. R.; Cobb, S. H.; Shen, KangKang; Doughty, J. R.)V1397,125-135(1991)

Coherent communication systems research and development at AT&T Bell Laboratories, Solid State Technology Center (Park, Yong K.; Delavaux, Jean-Marc P.; Tench, Robert E.; Cline, Terry W.; Tzeng, Liang D.; Kuo, Chien-yu C.; Wagner, Earl J.; Flores, Carlos F.; Van Eijk, Peter; Pleiss, T. C.; Barski, S.; Owen, B.; Twu, Yih-Jye; Dutta, Niloy K.; Riggs, R. S.; Ogawa, Kinichiro K.)V1372,219-227(1991)

Coherent optical transmission systems with optical amplifiers (Steele, Roger C.; Walker, Nigel G.)V1372,173-187(1991)

Compact PIN-amplifier module for giga bit rates optical interconnection (Suzuki, Tomihiro; Mikamura, Yasuki; Murata, Kazuo; Sekiguchi, Takeshi; Shiga, Nobuo; Murakami, Yasunori)V1389,455-461(1991)

Comparison of gain dependence of different Er-doped fiber structures (Tammela, Simo; Zhan, Xiaowei; Kiiveri, Pauli)V1373,103-110(1991)

Considerations in the design of servo amplifiers for high-performance scanning systems (Bukys, Albert)V1454,186-195(1991)

Design and energy characteristics of a multisegment glass-disk amplifier (Kelly, John H.; Shoup, Milton J.; Skeldon, Mark D.; Bui, Snow T.)V1410,40-46(1991)

Design of a high-power cross-field amplifier at X-band with an internally coupled waveguide (Eppley, Kenneth; Ko, Kwok)V1407,249-259(1991)

Design optimization of three-stage GaAs monolithic optical amplifier using SPICE (Yadav, M. S.; Dumka, D. C.; Ramola, Ramesh C.; Johri, Subodh; Kothari, Harshad S.; Singh, Babu R.)V1362,811-819(1991)

DOC II 32-bit digital optical computer: optoelectronic hardware and software (Stone, Richard V.; Zeise, Frederick F.; Guilfoyle, Peter S.)V1563,267-278(1991)

Efficiency increase in a traveling wave tube by tapering the phase velocity of the wave (Schachter, Levi; Nation, John A.)V1407,44-56(1991)

Efficient, high-power, high-gain Er3+-doped silica fiber amplifier (Massicott, Jennifer F.; Wyatt, Richard; Ainslie, B. J.; Craig-Ryan, S. P.)V1373,93-102(1991)

Erbium-doped fiber amplifiers for future undersea transmission systems (Aspell, Jennifer; Bergano, Neal S.)V1373,188-196(1991)

Excited state cross sections for Er-doped glasses (Zemon, Stanley A.; Lambert, Gary M.; Miniscalco, William J.; Davies, Richard W.; Hall, Bruce T.; Folweiler, Robert C.; Wei, Ta-Sheng; Andrews, Leonard J.; Singh, Mahendra P.)V1373,21-32(1991)

Fabrication of rare-earth-doped optical fiber (DiGiovanni, David J.)V1373,2-8(1991)

Fast-iterative technique for the calculation of frequency-dependent gain in excimer laser amplifiers (Sze, Robert C.)V1412,164-172(1991)

Fiber Laser Sources and Amplifiers II (Digonnet, Michel J., ed.)V1373(1991)

Fidelity of Brillouin amplification with Gaussian input beams (Jones, David C.; Ridley, Kevin D.; Cook, Gary; Scott, Andrew M.)V1500,46-52(1991)

General theoretical approach describing the complete behavior of the erbium-doped fiber amplifier (Marcerou, Jean-Francois; Fevrier, Herve A.; Ramos, Josiane; Auge, Jacques C.; Bousselet, Philippe)V1373,168-186(1991)

High-efficiency Kaband monolithic pseudomorphic HEMT amplifier (Saunier, Paul; Tserng, Hua Q.; Kao, Yung C.)V1475,86-90(1991)

High-efficiency dual-band power amplifier for radar applications (Masliah, Denis A.; Cole, Brad; Platzker, Aryeh; Schindler, Manfred)V1475,113-120(1991)

High-gain high-efficiency TWT (traveling wave tube) amplifiers (Nation, John A.; Ivers, J. D.; Kerslick, G.; Shiffler, Donald; Schachter, Levi)V1407,32-43(1991)

High-power coherent operation of 2-D monolithically integrated master-oscillator power amplifiers (Mehuys, David G.; Welch, David F.; Parke, Ross; Waarts, Robert G.; Hardy, Amos; Scifres, Donald R.)V1418,57-63(1991)

High-power single-element pseudomorphic InGaAs/GaAs/AlGaAs single-quantum-well lasers for pumping Er-doped fiber amplifiers (Larsson, Anders G.; Forouhar, Siamak; Cody, Jeffrey; Lang, Robert J.; Andrekson, Peter A.)V1418,292-297(1991)

Imaging through a low-light-level Raman amplifier (Duncan, Michael D.; Mahon, R.; Tankersley, Lawrence L.; Reintjes, John F.)V1409,127-134(1991)

InGaAs HEMT MMIC low-noise amplifier and doublers for EHF SATCOM ground terminals (Chow, P. D.; Lester, J.; Huang, P.; Jones, William L.)V1475,42-47(1991)

Induction linac-driven relativistic klystron and cyclotron autoresonance maser experiments (Goodman, Daniel L.; Birx, Daniel L.; Danly, Bruce G.)V1407,217-225(1991)

Insertion of emerging GaAs HBT technology in military and communication system applications (McAdam, Bridget A.; Sharma, Arvind K.; Allen, B.; Kintis, M.)V1475,267-274(1991)

Integrated optical preamplifier technology for optical signal processing and optical communication systems (Eichen, Elliot G.; Powazinik, William; Meland, Edmund; Bryant, R.; Rideout, William C.; Schlafer, John; Lauer, Robert B.)V1474,260-267(1991)

Intense Microwave and Particle Beams II (Brandt, Howard E., ed.)V1407(1991)

Large-signal model and signal/noise ratio analysis for Nd3+-doped fiber amplifiers at 1.3 um (Dakss, Mark L.; Miniscalco, William J.)V1373,111-124(1991)

LiNbO3 with rare earths: lasers and amplifiers (Lallier, Eric; Pocholle, Jean-Paul; Papuchon, Michel R.; De Micheli, Marc; Li, M. J.; He, Q.; Ostrowsky, Daniel B.; Grezes-Besset, C.; Pelletier, Emile P.)V1506,71-79(1991)

Liquid-nitrogen-cooled low-noise radiation pulse detector amplifier (Trojnar, Eugeniusz; Trojanowski, Stanislaw; Czechowicz, Roman; Derwiszynski, Mariusz; Kocyba, Krzysztof)V1391,230-237(1991)

Long-pulse modulator-driven cyclotron autoresonance maser and free-electron laser experiments at MIT (Danly, Bruce G.; Hartemann, Frederic V.; Chu, T. S.; Menninger, W. L.; Papavaritis, P.; Pendergast, K. D.; Temkin, Richard J.)V1407,192-201(1991)

Low-noise high-yield octave-band feedback amplifiers to 20 GHz (Minot, Katcha; Cochrane, Mike; Nelson, Bradford; Jones, William L.; Streit, Dwight C.; Liu, Po-Hsin P.)V1475,309-313(1991)

Low-noise MMIC performance in Kaband using ion implantation technology (Mondal, J. P.; Contolatis, T.; Geddes, John J.; Swirhun, S.; Sokolov, Vladimir)V1475,314-318(1991)

Magnetically tapered CARM (cyclotron autoresonance maser) for high power (Wang, Qinsong; McDermott, David B.; Luhmann, Neville C.)V1407,209-216(1991)

Master oscillator-amplifier Nd:YAG laser with a SBS phase-conjugate mirror (Ayral, Jean-Luc; Montel, J.; Huignard, Jean-Pierre)V1500,81-92(1991)

Millimeter wave pseudomorphic HEMT MMIC phased-array components for space communications (Lan, Guey-Liu; Pao, Cheng K.; Wu, Chan-Shin; Mandolia, G.; Hu, M.; Yuan, Steve; Leonard, Regis F.)V1475,184-192(1991)

Modeling direct detection and coherent-detection lightwave communication systems that utilize cascaded erbium-doped fiber amplifiers (Joss, Brian T.; Sunak, Harish R.)V1372,94-117(1991)

Monolithic integrated-circuit charge amplifier and comparator for MAMA readout (Cole, Edward H.; Smeins, Larry G.)V1549,46-51(1991)

Multiple-channel correlated double sampling amplifier hybrid to support a 64 parallel output CCD array (Raanes, Chris A.; McNeill, John A.; Cunningham, Andrew P.)V1358,637-643(1991)

Nd- and Er-doped glass integrated optical amplifiers and lasers (Najafi, S. I.; Wang, Wei-Jian; Orcel, Gerard F.; Albert, Jacques; Honkanen, Seppo; Poyhonen, Pekka; Li, Ming-Jun)V1583,32-36(1991)

Negative energy cyclotron resonance maser (Lednum, Eugene E.; McDermott, David B.; Lin, Anthony T.; Luhmann, Neville C.)V1407,202-208(1991)

New configuration of a generator and regenerative amplifier built on three mirrors (Piotrowski, Jan)V1391,272-278(1991)

Ninety-four GHz InAlAs/InGaAs/InP HEMT low-noise down-converter (Chow, P. D.; Tan, K.; Streit, Dwight C.; Garske, D.; Liu, Po-Hsin P.; Yen, Huan-chun)V1475,48-54(1991)

Noise and gain performance for an Er3+-doped fiber amplifier pumped at 980 nm or 1480 nm (Vendeltorp-Pommer, Helle; Pedersen, Frands B.; Bjarklev, Anders; Hedegaard Povlsen, Joern)V1373,254-265(1991)

Noise characteristics of rare-earth-doped fiber sources and amplifiers (Morkel, Paul R.; Laming, Richard I.; Cowle, Gregory J.; Payne, David N.)V1373,224-233(1991)

Novel selective-plated heatsink, key to compact 2-watt MMIC amplifier (Taylor, Gordon C.; Bechtle, Daniel W.; Jozwiak, Phillip C.; Liu, Shing G.; Camisa, Raymond L.)V1475,103-112(1991)

Optical ISL transmitter design that uses a high-power LD amplifier (Nohara, Mitsuo; Harada, Takashi; Fujise, Masayuki)V1417,338-345(1991)

Optical fiber amplifiers (Ikegami, Tetsuhiko; Nakahara, Motohiro)V1362,350-360(1991)

Optical switches based on semiconductor optical amplifiers (Kalman, Robert F.; Dias, Antonio R.; Chau, Kelvin K.; Goodman, Joseph W.)V1563,34-44(1991)

Optical time-domain reflectometry performance enhancement using erbium-doped fiber amplifiers (Keeble, Peter J.)V1366,39-44(1991)

Optimization of externally modulated analog optical links (Betts, Gary E.; Johnson, Leonard M.; Cox, Charles H.)V1562,281-302(1991)

Performance of a phased array semiconductor laser source for coherent laser communications (Probst, David K.; Rice, Robert R.)V1417,346-357(1991)

Phased-array receiver development using high-performance HEMT MMICs (Liu, Louis; Jones, William L.; Carandang, R.; Lam, Wayne W.; Yonaki, J.; Streit, Dwight C.; Kasody, R.)V1475,193-198(1991)

Power gain characteristics of discharge-excited KrF laser amplifier system (Lee, Choo-Hie; Choi, Boo-Yeon)V1397,91-95(1991)

Preamplifiers of the high-ohmic high-speed photodetector signals (Kovrigin, Yevgeny; Potylitsyn, Yevgeny)V1362,967-975(1991)

Progress in fluoride fiber lasers and amplifiers (France, Paul W.; Brierley, Michael C.)V1373,33-39(1991)

Quantum-well excitonic devices for optical computing (Singh, Jasprit; Bhattacharya, Pallab K.)V1362,586-597(1991)

Rare-earth-doped LiNbO3 waveguide amplifiers and lasers (Sohler, Wolfgang)V1583,110-121(1991)

Relativistic klystron amplifier I: high-power operation (Friedman, Moshe; Serlin, Victor; Lau, Yue Y.; Krall, Jonathan)V1407,2-7(1991)

Relativistic klystron amplifier II: high-frequency operation (Serlin, Victor; Friedman, Moshe; Lau, Yue Y.; Krall, Jonathan)V1407,8-12(1991)

Relativistic klystron amplifier III: dynamical limiting currents, nonlinear beam loading, and conversion efficiency (Colombant, Denis G.; Lau, Yue Y.; Friedman, Moshe; Krall, Jonathan; Serlin, Victor)V1407,13-22(1991)

Relativistic klystron amplifier IV: simulation studies of a coaxial-geometry RKA (Krall, Jonathan; Friedman, Moshe; Lau, Yue Y.; Serlin, Victor)V1407,23-31(1991)

Semiconductor laser amplifiers as all-optical frequency converters (Schunk, Nikolaus; Grosskopf, Gerd; Ludwig, Reinhold; Schnabel, Ronald; Weber, Hans-Georg)V1362,391-397(1991)

Semiconductor laser transmitters for millimeter-wave fiber-optic links (Pan, J. J.)V1476,63-73(1991)

Sensitivity of direct-detection lightwave receivers using optical preamplifiers (Tonguz, Ozan K.; Kazovsky, Leonid G.)V1579,179-183(1991)

Short-pulsed H2-F2 amplifier initiated by optical discharge (Igoshin, Valery I.; Pichugin, Sergei)V1501,150-152(1991)

Single-frequency solid state lasers and amplifiers (Mak, Arthur A.; Orlov, Oleg A.; Ustyugov, Vladimir I.; Vitrishchak, Il'ya B.)V1410,233-243(1991)

Superconducting IR focal plane arrays (Quelle, Fred W.)V1449,157-166(1991)

Theory and simulation of the HARmonic amPlifier Free-Electron Laser (HARP/FEL) (Gregoire, Daniel J.; Harvey, Robin J.; Levush, Baruch)V1552,118-126(1991)

Theory of multimode interactions in cyclotron autoresonance maser amplifiers (Chen, Chiping; Wurtele, Jonathan S.)V1407,183-191(1991)

Transient gain phenomena and gain enhancement in a fast-axial flow CO2 laser amplifier (Sato, Heihachi; Tsuchida, Eiichi; Kasuya, Koichi)V1397,331-338(1991)

Twin traveling wave tube amplifiers driven by a single backward-wave oscillator (Butler, Jennifer M.; Wharton, Charles B.)V1407,57-66(1991)

Ultra-high-frequency InP-based HEMTs for millimeter wave applications (Greiling, Paul T.; Nguyen, Loi D.)V1475,34-41(1991)

Ultralinear low-noise amplifier technology for space communications (Watkins, E. T.; Yu, K. K.; Yau, W.; Wu, Chan-Shin; Yuan, Steve)V1475,62-72(1991)

Ultrashort pulse propagation in visible semiconductor diode laser amplifiers (Tatum, Jim A.; MacFarlane, Duncan L.)V1497,320-325(1991)

Variable-gain MMIC module for space application (Palena, Patricia)V1475,91-102(1991)

Wavelength division and subcarrier system based on Brillouin amplification (Lee, Yang-Hang; Wu, Jingshown; Kao, Ming-Seng; Tsao, Hen-Wai)V1579,155-166(1991)

Wavelength division multiplexers for optical fiber amplifiers (Nagy, Peter A.; Meyer, Tim J.; Tekippe, Vincent J.)V1365,33-37(1991)

Annealing—see also microelectronic processing

Boron depth profiles in a-Si1-xCx:H(B) films after thermal annealing (Liao, Chang G.; Zheng, Zhi H.; Wang, Yong Q.; Yang, Sheng S.)V1519,152-155(1991)

Computer modeling of the dynamics of nanosecond laser annealing of amorphous thin silicon layers (Zhvavyi, S.; Sadovskaya, O.)V1440,8-15(1991)

Dynamics of the optical parameters of molten silicon during nanosecond laser annealing (Boneberg, J.; Yavas, O.; Mierswa, B.; Leiderer, Paul)V1598,84-90(1991)

Growth and characterization of semiconducting Fe-Si2 thin layers on Si(111) (Rizzi, Angela; Moritz, Heiko; Lueth, Hans)V1361,827-833(1991)

High-dose boron implantation and RTP anneal of polysilicon films for shallow junction diffusion sources and interconnects (Raicu, Bruha; Keenan, W. A.; Current, Michael I.; Mordo, David; Brennan, Roger)V1393,161-171(1991)

High-temperature degradation-free rapid thermal annealing of GaAs and InP (Pearton, Stephen J.; Katz, Avishay; Geva, M.)V1393,150-160(1991)

Influence of laser pulse annealing on the depth distribution of Sb recoil atoms in Si (Brylowska, Irena; Paprocki, K.)V1391,164-169(1991)

Integrated rapid isothermal processing (Singh, Rajendra; Sinha, Sanjai; Thakur, Randhir P.; Hsu, N. J.)V1393,78-89(1991)

Low-temperature in-situ dry cleaning process for epitaxial layer multiprocessing (Moslehi, Mehrdad M.)V1393,90-108(1991)

Optimizing the structural and electrical properties of Ba2YCu3O7-delta (Phillips, Julia M.; Siegal, Michael P.)V1394,186-190(1991)

Performance and reliability of ultrathin reoxidized nitrided oxides fabricated by rapid thermal processing (Joshi, A. B.; Lo, G. Q.; Shih, Dennis K.; Kwong, Dim-Lee)V1393,122-149(1991)

Rapid thermal annealing of the through-Ta5Si3 film implantation on GaAs (Huang, Fon-Shan; Chen, W. S.; Hsu, Tzu-min)V1393,172-179(1991)

Rapid thermal processing induced defects and gettering effects in silicon (Hartiti, Bouchaib; Muller, Jean-Claude; Siffert, Paul; Vu, Thuong-Quat)V1393,200-206(1991)

Rapid thermal processing in the manufacturing technology of contacts to InP-based photonic devices (Katz, Avishay)V1393,67-77(1991)

Antennas—see also arrays; receivers

Acousto-optic architectures for multidimensional phased-array antenna processing (Riza, Nabeel A.)V1476,144-156(1991)

Algorithmic sensor failure detection on passive antenna arrays (Chun, Joohwan; Luk, Franklin T.)V1566,483-492(1991)

Ceramic phase-shifters for electronically steerable antenna systems (Selmi, Fathi; Ghodgaonkar, Deepak K.; Hughes, Raymond; Varadan, Vasundara V.; Varadan, Vijay K.)V1489,97-107(1991)

Cost-effective optical switch matrix for microwave phased array (Pan, J. J.; Chau, Seung L.; Li, Wei Z.; Grove, Charles H.)V1476,133-142(1991)

Direct optical phase shifter for phased-array systems (Vawter, Gregory A.; Hietala, Vincent M.; Kravitz, Stanley H.; Meyer, W. J.)V1476,102-106(1991)

Experimental implementation of an optical multiple-aperture antenna for space communications (Neubert, Wolfgang M.; Leeb, Walter R.; Scholtz, Arpad L.)V1522,93-102(1991)

High on-off ratio, ultrafast optical switch for optically controlled phased array (Pan, J. J.; Li, Wei Z.)V1476,122-132(1991)

Integrated circuit active antenna elements for monolithic implementation (Chang, Kai)V1475,164-174(1991)

Integrated optic flat antennae: early applications and design tools (Parriaux, Olivier M.; Sychugov, V. A.)V1583,376-382(1991)

Ka-band MMIC array feed development for deep space applications (Cooley, Thomas W.; Riley, A. L.; Crist, Richard A.; Sukamto, Lin; Jamnejad, V.; Rascoe, Daniel L.)V1475,243-247(1991)

Millimeter wave monolithic antenna and receiver arrays for space-based applications (Rebeiz, Gabriel M.; Ulaby, Fawwaz T.)V1475,199-203(1991)

MMIC: a key technology for future communications satellite antennas (Sorbello, Robert M.; Zaghloul, A. I.; Gupta, R. K.; Geller, B. D.; Assal, F. T.; Potukuchi, J. R.)V1475,175-183(1991)

MMICs for airborne phased arrays (Scalzi, Gary J.; Turtle, John P.; Carr, Paul H.)V1475,2-9(1991)

New concepts in remote sensing and geolocation (Seastone, A. J.)V1495,228-239(1991)

Nonmechanical steering of laser beams by multiple aperture antennas: tolerance analysis (Neubert, Wolfgang M.; Leeb, Walter R.; Scholtz, Arpad L.)V1417,122-130(1991)

Phased-array antenna control by a monolithic photonic integrated circuit (Hietala, Vincent M.; Vawter, Gregory A.; Meyer, W. J.; Kravitz, Stanley H.)V1476,170-175(1991)

Quasioptical MESFET VCOs (Bundy, Scott; Mader, Tom; Popovic, Zoya; Ellinson, Reinold; Hjelme, Dag R.; Surette, Mark R.; Yadlowski, Michael; Mickelson, Alan R.)V1475,319-329(1991)

Status of high-temperature superconducting analog devices (Talisa, Salvador H.)V1477,78-83(1991)

System-level integrated circuit development for phased-array antenna applications (Shalkhauser, Kurt A.; Raquet, Charles A.)V1475,204-209(1991)

True-time-delay steering of dual-band phased-array antenna using laser-switched optical beam forming networks (Ng, Willie W.; Tangonan, Gregory L.; Walston, Andrew; Newberg, Irwin L.; Lee, Jar J.; Bernstein, Norman P.)V1371,205-211(1991)

Variable time delay for RF/microwave signal processing (Toughlian, Edward N.; Zmuda, Henry)V1476,107-121(1991)

Wideband embedded/conformal antenna subsystem concept (Smalanskas, Joseph P.; Valentine, Gary W.; Wolfson, Ronald I.)V1489,2-8(1991)

Architectures—see also fiber optic networks; optical processing; signal processing

Advanced Signal Processing Algorithms, Architectures, and Implementations II (Luk, Franklin T., ed.)V1566(1991)

Architecture and performance of a hardware collision-checking accelerator (Bardin, R. K.; Libby, Vibeke)V1566,394-404(1991)

Architectures and access protocols for multichannel networks (Karol, Mark J.)V1579,2-13(1991)

Arithmetic processor design for the T9000 transputer (Knowles, Simon C.)V1566,230-243(1991)

Bridging issues in DQDB subnetworks (Tantawy, Ahmed N.)V1364,268-276(1991)

Building and campus networks for fiber distributed data interface (McIntosh, Thomas F.)V1364,84-93(1991)

CORDIC processor architectures (Boehme, Johann F.; Timmermann, D.; Hahn, H.; Hosticka, Bedrich J.)V1566,208-219(1991)

CWRUnet: case history of a campus-wide fiber-to-the-desktop network (Neff, Raymond K.; Klingensmith, H. W.; Gumpf, Jeffrey A.; Haigh, Peter J.)V1364,245-256(1991)

Design of a Gaussian elimination architecture for the DOC II processor (Murdocca, Miles J.; Levy, Saul Y.)V1563,255-266(1991)

Digital optical computers at Boulder (Jordan, Harry F.)V1505,87-98(1991)

Distribution fiber FTTH/FTTC trial results and deployment strategies (Coleman, John D.)V1363,2-12(1991)

DS1 mapping considerations for the synchronous optical network (Cubbage, Robert W.; Littlewood, Paul A.)V1363,163-171(1991)

DSP array for real-time adaptive sidelobe cancellation (Rorabaugh, Terry L.; Vaccaro, John J.; Grace, Kevin H.; Pauer, Eric K.)V1566,312-322(1991)

Dynamically reconfigurable optical interconnect architecture for parallel multiprocessor systems (Girard, Mary M.; Husbands, Charles R.; Antoszewska, Reza)V1563,156-167(1991)

Evolution of the DQDB hybrid multiplexing for an integrated service packetized traffic (Gagnaire, A.; Ponsard, Benoit)V1364,277-288(1991)

Fast algorithm and architecture for constrained adaptive sidelobe cancellation (Games, Richard A.; Eastman, Willard L.; Sousa, Michael J.)V1566,323-328(1991)

FDDI, Campus-Wide, and Metropolitan Area Networks (Annamalai, Kadiresan; Cudworth, Stewart K.; Kasiewicz, Allen B., eds.)V1364(1991)

Fiber in the loop: an evolution in services and systems (Engineer, Carl P.)V1363,19-29(1991)

Fiber optics in CATV networks (Wolfe, Ronald; Laor, Herzel)V1363,125-132(1991)

Fiber Optics in the Subscriber Loop (Hutcheson, Lynn D.; Kahn, David A., eds.)V1363(1991)

Impact of fiber-in-the-loop architecture on predicted system reliability (Lee, T. S.)V1366,28-38(1991)

Management of an adaptable-bit-rate video service in a MAN environment (Marini, Michele; Albanese, Andres)V1364,289-294(1991)

Metropolitan area networks: a corner stone in the broadband era (Ghanem, Adel)V1364,312-319(1991)

Modular packaging for FTTC and B-ISDN (Koht, Lowell)V1363,158-162(1991)

Multi-Gb/s optical computer interconnect (Sauer, Jon R.)V1579,49-61(1991)

NAVSEA gigabit optical MAN prototype history and status (Albanese, Andres; Devetzis, Tasco N.; Ippoliti, A, G.; Karr, Michael A.; Maszczak, M. W.; Dorris, H. N.; Davis, James H.)V1364,320-326(1991)

Network powering architecture for fiber-to-the-subscriber systems (Pellerin, Sharon J.)V1363,186-190(1991)

New CATV fiber-to-the-subscriber architectures (Kim, Gary)V1363,133-140(1991)

Optical approaches to overcome present limitations for interconnection and control in parallel electronic architectures (Maurin, T.; Devos, F.)V1505,158-165(1991)

Passive optic solution for an urban rehabilitation topology (Petruziello, David)V1363,30-37(1991)

Potential digital optical computer III architectures: the next generation (Guilfoyle, Peter S.; Morozov, Valentin N.)V1563,279-283(1991)

Proposed one- and two-fiber-to-the-pedestal architectural evolution (Schiffler, Richard A.)V1363,13-18(1991)

Real-time quality inspection system for textile industries (Karkanis, S.; Tsoutsou, K.; Vergados, J.; Dimitriadis, Basile D.)V1500,164-170(1991)

Research and development of a NYNEX switched multi-megabit data service prototype system (Maman, K. H.; Haines, Robert; Chatterjee, Samir)V1364,304-311(1991)

RETINA (RETinally INspired Architecture project) (Caulfield, H. J.; Wilkins, Nathan A.)V1564,496-503(1991)

Soliton ring network (Sauer, Jon R.; Islam, Mohammed N.; Dijaili, Sol P.)V1579,84-97(1991)

Solving clock distribution problems in FDDI concentrators (Li, Gabriel)V1364,72-83(1991)

Systolic array architecture of a new VLSI vision chip (Means, Robert W.)V1566,388-393(1991)

TeraNet: a multigigabit-per-second hybrid circuit/packet-switched lightwave network (Gidron, Rafael; Elby, Stuart D.; Acampora, Anthony S.; Georges, John B.; Lau, Kam-Yin)V1579,40-48(1991)

Topologies for linear lightwave networks (Bala, Krishna; Stern, Thomas E.)V1579,62-73(1991)

Toward low-cost real-time EPLD-based machine vision workstations and target systems (Floeder, Steven P.; Waltz, Frederick M.)V1386,90-101(1991)

VLSI processor for high-performance arithmetic computations (McQuillan, S. E.; McCanny, J. V.)V1566,220-229(1991)

Arrays—see also antennas; detectors; lasers, semiconductor; micro-optics; telescopes

0.36-W cw diffraction-limited-beam operation from phase-locked arrays of antiguides (Mawst, Luke J.; Botez, Dan; Anderson, Eric R.; Jansen, Michael; Ou, Szutsun S.; Sergant, Moshe; Peterson, Gary L.; Roth, Thomas J.; Rozenbergs, John)V1418,353-357(1991)

100-mW four-beam individually addressable monolithic AlGaAs laser diode arrays (Yamaguchi, Takao; Yodoshi, Keiichi; Minakuchi, Kimihide; Tabuchi, Norio; Bessho, Yasuyuki; Inoue, Yasuaki; Komeda, Koji; Mori, Kazushi; Tajiri, Atsushi; Tominaga, Koji)V1418,363-371(1991)

128 x 128 MWIR InSb focal plane and camera system (Parrish, William J.; Blackwell, John D.; Paulson, Robert C.; Arnold, Harold)V1512,68-77(1991)

488 x 640-element hybrid platinum silicide Schottky focal-plane array (Gates, James L.; Connelly, William G.; Franklin, T. D.; Mills, Robert E.; Price, Frederick W.; Wittwer, Timothy Y.)V1540,262-273(1991)

Acousto-optic architectures for multidimensional phased-array antenna processing (Riza, Nabeel A.)V1476,144-156(1991)

Acquisition and tracking performance measurements for a high-speed area array detector system (Short, Ralph C.; Cosgrove, Michael A.; Clark, David L.; Martino, Anthony J.; Park, Hong-Woo; Seery, Bernard D.)V1417,131-141(1991)

Adaptive nonuniformity correction for IR focal-plane arrays using neural networks (Scribner, Dean A.; Sarkady, Kenneth A.; Kruer, Melvin R.; Caulfield, John T.; Hunt, J. D.; Herman, Charles)V1541,100-109(1991)

Adaptive semiconductor laser phased arrays for real-time multiple-access communications (Pan, J. J.; Cordeiro, D.)V1476,157-169(1991)

Advanced infrared focal-plane arrays (Amingual, Daniel)V1512,40-51(1991)

Air Force program in coherent semiconductor lasers (Kennett, Ruth D.; Frazier, John C.)V1501,57-68(1991)

Algorithmic sensor failure detection on passive antenna arrays (Chun, Joohwan; Luk, Franklin T.)V1566,483-492(1991)

All-reflective phased array imaging telescopes (Stuhlinger, Tilman W.)V1354,438-446(1991)

Analysis of upper and lower bounds of the frame noise in linear detector arrays (Jaggi, Sandeep)V1541,152-162(1991)

Analytical studies of large closed-loop arrays (Fikioris, George; Freeman, D. K.; King, Ronold W.; Shen, Hao-Ming; Wu, Tai T.)V1407,295-305(1991)

Application of IR staring arrays to space surveillance (Cantella, Michael J.; Ide, M. H.; O'Donnell, P. J.; Tsaur, Bor-Yeu)V1540,634-652(1991)

Application of adaptive filters to the problem of reducing microphony in arrays of pyroelectric infrared detectors (Carmichael, I. C.; White, Paul R.)V1541,167-177(1991)

Application of low-noise CID imagers in scientific instrumentation cameras (Carbone, Joseph; Hutton, J.; Arnold, Frank S.; Zarnowski, Jeffrey J.; VanGorden, Steve; Pilon, Michael J.; Wadsworth, Mark V.)V1447,229-242(1991)

Applications of charge-coupled and charge-injection devices in analytical spectroscopy (Denton, M. B.)V1447,2-11(1991)

Array detectors in astronomy (Lesser, Michael P.)V1439,144-151(1991)

Automated characterization of Z-technology sensor modules (Gilcrest, Andrew S.)V1541,240-249(1991)

Compact time-delay shifters that are process insensitive (Lesko, Camille; Hill, William A.; Dietrich, Fred; Nelson, William)V1475,330-339(1991)

Computer analysis of signal-to-noise ratio and detection probability for scanning IRCCD arrays (Uda, Gianni; Tofani, Alessandro)V1488,257-266(1991)

Computer simulation of staring-array thermal imagers (Bradley, D. J.; Dennis, Peter N.; Baddiley, C. J.; Murphy, Kevin S.; Carpenter, Stephen R.; Wilson, W. G.)V1488,186-195(1991)

Cost-effective optical switch matrix for microwave phased array (Pan, J. J.; Chau, Seung L.; Li, Wei Z.; Grove, Charles H.)V1476,133-142(1991)

Current status of InGaAs detector arrays for 1-3 um (Olsen, Gregory H.; Joshi, Abhay M.; Ban, Vladimir S.)V1540,596-605(1991)

Decoupled coil detector array in magnetic resonance imaging (Kwiat, Doron; Einav, Shmuel)V1443,2-28(1991)

Design, fabrication, and performance of an integrated optoelectronic cellular array (Hibbs-Brenner, Mary K.; Mukherjee, Sayan D.; Skogen, J.; Grung, B.; Kalweit, Edith; Bendett, Mark P.)V1563,10-20(1991)

Design and performance of a PtSi spectroscopic infrared array and detector head (Cizdziel, Philip J.)V1488,6-27(1991)

Design of array systems using shared symmetry (Miao, Cheng-Hsi)V1354,447-456(1991)

Design optimization of a 10-amplifier coherent array (Zediker, Mark S.; Foresi, James S.; Haake, John M.; Heidel, Jeffrey R.; Williams, Richard A.; Driemeyer, D.; Blackwell, Richard J.; Thomas, G.; Priest, J. A.; Herrmann, Sandy)V1418,309-315(1991)

Development of a large pixel, spectrally optimized, pinned photodiode/ interline CCD detector for the Earth Observing System/Moderate-Resolution Imaging Spectrometer-Tilt Instrument (Ewin, Audrey J.; Jhabvala, Murzy; Shu, Peter K.)V1479,12-20(1991)

Direct optical phase shifter for phased-array systems (Vawter, Gregory A.; Hietala, Vincent M.; Kravitz, Stanley H.; Meyer, W. J.)V1476,102-106(1991)

Dynamic infrared scene projection technology (Mobley, Scottie B.)V1486,325-332(1991)

Edge effects in silicon photodiode arrays (Kenney, Steven B.; Hirleman, Edwin D.)V1480,82-93(1991)

Experimental implementation of an optical multiple-aperture antenna for space communications (Neubert, Wolfgang M.; Leeb, Walter R.; Scholtz, Arpad L.)V1522,93-102(1991)

Experimental study of the resonance of a circular array (Shen, Hao-Ming)V1407,306-315(1991)

Fabrication and optical performance of fractal fiber optics (Cook, Lee M.; Burger, Robert J.)V1449,186-192(1991)

Fiber array optics for electronic imaging (Burger, Robert J.; Greenberg, David A.)V1449,174-185(1991)

Fixed-pattern-noise cancellation in linear pyro arrays (Jain, Subhash C.; Malhotra, H. S.; Sarebahi, K. N.; Bist, K. S.)V1488,410-413(1991)

Frequency-domain imaging using array detectors: present status and prospects for picosecond resolution (Morgan, Chris G.; Murray, J. G.; Mitchell, A. C.)V1525,83-90(1991)

Fundamental array mode operation of semiconductor laser arrays using external spatial filtering (Cherng, Chung-Pin; Osinski, Marek A.)V1418,372-385(1991)

Further performance characteristics of a high-sensitivity 64 x 64 element InSb hybrid focal-plane array (Fischer, Robert C.; Martin, Charles J.; Niblack, Curtis A.; Timlin, Harold A.; Wimmers, James T.)V1494,414-418(1991)

Generalized phased-array Bragg interaction in anisotropic crystals (Young, Eddie H.; Ho, Huey-Chin C.; Yao, Shi-Kay; Xu, Jieping P.)V1476,178-189(1991)

Hartmann-Shack wavefront sensor using a binary optic lenslet array (Kwo, Deborah P.; Damas, George; Zmek, William P.; Haller, Mitch)V1544,66-74(1991)

High-density packaging and interconnect of massively parallel image processors (Carson, John C.; Indin, Ronald)V1541,232-239(1991)

High on-off ratio, ultrafast optical switch for optically controlled phased array (Pan, J. J.; Li, Wei Z.)V1476,122-132(1991)

High-performance 256 x 256 InSb FPA for astronomy (Hoffman, Alan; Randall, David)V1540,297-302(1991)

High-power coherent operation of 2-D monolithically integrated master-oscillator power amplifiers (Mehuys, David G.; Welch, David F.; Parke, Ross; Waarts, Robert G.; Hardy, Amos; Scifres, Donald R.)V1418,57-63(1991)

High-power diffraction-limited phase-locked GaAs/GaAlAs laser diode array (Zhang, Yue-qing; Zhang, Xitian; Piao, Yue-zhi; Li, Dian-en; Wu, Sheng-li; Du, Shu-qin)V1418,444-447(1991)

High-radiometric-performance CCD for the third-generation stratospheric aerosol and gas experiment (Delamere, W. A.; Baer, James W.; Ebben, Thomas H.; Flores, James S.; Kleiman, Gary; Blouke, Morley M.; McCormick, M. P.)V1479,31-40(1991)

High-sensitive thermal video camera with self-scanned 128 InSb linear array (Fujisada, Hiroyuki)V1540,665-676(1991)

Hybrid infrared focal-plane signal and noise modeling (Johnson, Jerris F.; Lomheim, Terrence S.)V1541,110-126(1991)

Image enhancement of infrared uncooled focal plane array imagery (McCauley, Howard M.; Auborn, John E.)V1479,416-422(1991)

Imaging pulse-counting detector systems for space ultraviolet astrophysics missions (Timothy, J. G.)V1494,394-402(1991)

InSb linear multiplexed FPAs for the CRAF/Cassini visible and infrared mapping spectrometer (Niblack, Curtiss A.; Blessinger, Michael A.; Forsthoefel, John J.; Staller, Craig O.; Sobel, Harold R.)V1494,403-413(1991)

Infrared techniques applied to large solar arrays: a ten-year update (Hodor, James R.; Decker, Herman J.; Barney, Jesus J.)V1540,331-337(1991)

ISTS array detector test facility (Thomas, Paul J.; Hollinger, Allan B.; Chu, Kan M.; Harron, John W.)V1488,36-47(1991)

Ka-band MMIC array feed development for deep space applications (Cooley, Thomas W.; Riley, A. L.; Crist, Richard A.; Sukamto, Lin; Jamnejad, V.; Rascoe, Daniel L.)V1475,243-247(1991)

Large-format CCDs for astronomical applications (Geary, John C.)V1439,159-168(1991)

Large staring IRFPAs of HgCdTe on alternative substrates (Kozlowski, Lester J.; Bailey, Robert B.; Cooper, Donald E.; Vural, Kadri; Gertner, E. R.; Tennant, William E.)V1540,250-261(1991)

Linear array camera interface techniques (DeLuca, Dan)V1396,558-565(1991)

Localized wave transmission physics and engineering (Ziolkowski, Richard W.)V1407,375-386(1991)

Low-cost high-performance InSb 256 x 256 infrared camera (Parrish, William J.; Blackwell, John D.; Kincaid, Glen T.; Paulson, Robert C.)V1540,274-284(1991)

Low-insertion-loss, high-precision liquid crystal optical phased array (Cassarly, William J.; Ehlert, John C.; Henry, D.)V1417,110-121(1991)

Low-threshold grating surface-emitting laser arrays (Evans, Gary A.; Carlson, Nils W.; Bour, David P.; Liew, So K.; Amantea, Robert; Wang, Christine A.; Choi, Hong K.; Walpole, James N.; Butler, Jerome K.; Ferguson, W. E.)V1418,406-413(1991)

Millimeter wave pseudomorphic HEMT MMIC phased-array components for space communications (Lan, Guey-Liu; Pao, Cheng K.; Wu, Chan-Shin; Mandolia, G.; Hu, M.; Yuan, Steve; Leonard, Regis F.)V1475,184-192(1991)

MMIC: a key technology for future communications satellite antennas (Sorbello, Robert M.; Zaghloul, A. I.; Gupta, R. K.; Geller, B. D.; Assal, F. T.; Potukuchi, J. R.)V1475,175-183(1991)

MMICs for airborne phased arrays (Scalzi, Gary J.; Turtle, John P.; Carr, Paul H.)V1475,2-9(1991)

Modulation characteristics of high-power phase-locked arrays of antiguides (Anderson, Eric R.; Jansen, Michael; Botez, Dan; Mawst, Luke J.; Roth, Thomas J.; Yang, Jane)V1417,543-549(1991)

Monolithic phased arrays: recent advances (Kinzel, Joseph A.)V1475,158-163(1991)

Monolithic two-dimensional surface-emitting laser diode arrays with 45 degree micromirrors (Jansen, Michael; Yang, Jane; Ou, Szutsun S.; Sergant, Moshe; Mawst, Luke J.; Rozenbergs, John; Wilcox, Jarka Z.; Botez, Dan)V1418,32-39(1991)

Multiwindow method for spectrum estimation and sinusoid detection in an array environment (Onn, Ruth; Steinhardt, Allan O.)V1566,427-438(1991)

New way of making a lunar telescope (Chen, Peter C.)V1494,228-233(1991)

Noise reduction strategy for hybrid IR focal-plane arrays (Fowler, Albert M.; Gatley, Ian)V1541,127-133(1991)

Nonmechanical steering of laser beams by multiple aperture antennas: tolerance analysis (Neubert, Wolfgang M.; Leeb, Walter R.; Scholtz, Arpad L.)V1417,122-130(1991)

OLAS: optical logic array structures (Jones, Robert H.; Hadjinicolaou, M. G.; Musgrave, G.)V1401,138-145(1991)

On-focal-plane-array feature extraction using 3-D artificial neural network (3DANN): Part II (Carson, John C.)V1541,227-231(1991)

On-focal-plane-array feature extraction using a 3-D artificial neural network (3DANN): Part I (Carson, John C.)V1541,141-144(1991)

Operational characteristics of a phase-locked module of relativistic magnetrons (Levine, Jerrold S.; Benford, James N.; Courtney, R.; Harteneck, Bruce D.)V1407,74-82(1991)

Optimum design of optical array used as pseudoconjugator (Xiao, Guohua; Song, Ruhua H.; Hu, Zhiping; Le, Shixiao)V1409,106-113(1991)

Performance of a demonstration system for simultaneous laser beacon tracking and low-data-rate optical communications with multiple platforms (Short, Ralph C.; Cosgrove, Michael A.; Clark, David L.; Oleski, Paul J.)V1417,464-475(1991)

Performance of a phased array semiconductor laser source for coherent laser communications (Probst, David K.; Rice, Robert R.)V1417,346-357(1991)

Phase-shifter technology assessment: prospects and applications (Sokolov, Vladimir)V1475,288-302(1991)

Phased-array antenna control by a monolithic photonic integrated circuit (Hietala, Vincent M.; Vawter, Gregory A.; Meyer, W. J.; Kravitz, Stanley H.)V1476,170-175(1991)

Phased-array receiver development using high-performance HEMT MMICs (Liu, Louis; Jones, William L.; Carandang, R.; Lam, Wayne W.; Yonaki, J.; Streit, Dwight C.; Kasody, R.)V1475,193-198(1991)

Pyroelectric linear array IR detectors with CCD multiplexer (Norkus, Volkmar; Neumann, Norbert; Walther, Ludwig; Hofmann, Guenter; Schieferdecker, Jorg; Krauss, Matthias G.; Budzier, Helmut; Hess, Norbert)V1484,98-105(1991)

Quasioptical MESFET VCOs (Bundy, Scott; Mader, Tom; Popovic, Zoya; Ellinson, Reinold; Hjelme, Dag R.; Surette, Mark R.; Yadlowski, Michael; Mickelson, Alan R.)V1475,319-329(1991)

Recent results in adaptive array detection (Kalson, S. Z.)V1566,406-418(1991)

Relative performance studies for focal-plane arrays (Murphy, Kevin S.; Bradley, D. J.; Dennis, Peter N.)V1488,178-185(1991)

Reporting data for arrays with many elements (Coles, Christopher L.; Phillips, Wayne S.; Vincent, John D.)V1488,327-333(1991)

Revolutionary impact of today's array detector technology on chemical analysis (Radspinner, David A.; Fields, Robert E.; Earle, Colin W.; Denton, M. B.)V1439,2-14(1991)

Si:Ga focal-plane arrays for satellite and ground-based telescopes (Mottier, Patrick; Agnese, Patrick; Lagage, Pierre O.)V1494,419-426(1991); V1512,60-67(1991)

Signal, noise, and readout considerations in the development of amorphous silicon photodiode arrays for radiotherapy and diagnostic x-ray imaging (Antonuk, Larry E.; Boudry, J.; Kim, Chung-Won; Longo, M.; Morton, E. J.; Yorkston, J.; Street, Robert A.)V1443,108-119(1991)

Simulation of sampling effects in FPAs (Cook, Thomas H.; Hall, Charles S.; Smith, Frederick G.; Rogne, Timothy J.)V1488,214-225(1991)

Solutions to modeling of imaging IR systems for missile applications: MICOM imaging IR system performance model-90 (Owen, Philip R.; Dawson, James A.; Borg, Eric J.)V1488,122-132(1991)

Staring sensor MRT measurement and modeling (Mooney, Jonathan M.)V1540,550-564(1991)

Study on the mode and far-field pattern of diode laser-phased arrays (Zhang, Yue-qing; Wu, Sheng-li; Zhu, Lian; Zhang, Xitian; Piao, Yue-zhi; Li, Dian-en)V1400,137-143(1991)

Superconducting IR focal plane arrays (Quelle, Fred W.)V1449,157-166(1991)

Time-derivative adaptive silicon photoreceptor array (Delbruck, Tobi; Mead, Carver A.)V1541,92-99(1991)

Transition from homoclinic to heteroclinic chaos in coupled laser arrays (Otsuka, Kenju)V1497,300-312(1991)

Two-dimensional periodic structures in solid state laser resonator (Okulov, Alexey Y.)V1410,221-232(1991)

Update: high-speed/high-volume radiometric testing of Z-technology focal planes (Johnson, Jerome L.)V1541,210-219(1991)

Use of magneto-optic spatial light modulators and linear detector arrays in inner-product associative memories (Goff, John R.)V1558,466-475(1991)

Variable time delay for RF/microwave signal processing (Toughlian, Edward N.; Zmuda, Henry)V1476,107-121(1991)

Artificial intelligence—see also expert systems; robotics

Acquiring rules of selecting cells by using neural network (Yu, He; Zheng, Xiangjun; Ye, Yizheng; Wang, LiHong)V1469,412-417(1991)

Active pattern recognition based on attributes and reasoning (Chang, Jin W.)V1468,137-146(1991)

AI application in shoe industry CAD/CAM (Wang, He-Chen; Lou, Da-li; Xian, Wu; Song, Xiang; Yong, Jiang)V1386,273-276(1991)

Algorithm development and evaluation on the Multifunction Target Acquisition Processor (Haskett, Michael C.; Lidke, Steve L.)VIS07,14-23(1991)

Analyzing and interpreting pulmonary tomoscintigraphy sequences: realization and perspectives (Forte, Anne-Marie; Bizais, Yves)V1445,409-420(1991)

Application of Dempster-Shafer theory to a novel control scheme for sensor fusion (Murphy, Robin R.)V1569,55-68(1991)

Applications of Artificial Intelligence IX (Trivedi, Mohan M., ed.)V1468(1991)

Applications of learning strategies to pattern recognition (Rizki, Mateen M.; Tamburino, Louis A.; Zmuda, Michael A.)V1469,384-391(1991)

Assumption truth maintenance in model-based ATR algorithm design (Bennett, Laura F.; Johnson, Rubin; Hudson, C. I.)V1470,263-274(1991)

Atomic temporal interval relations in branching time: calculation and application (Anger, Frank D.; Ladkin, Peter B.; Rodriguez, Rita V.)V1468,122-136(1991)

Automated calculation of nonadditive measures for object recognition (Tahani, Hossein; Keller, James M.)V1381,379-389(1991)

Automated instrumentation, evaluation, and diagnostics of automatic target recognition systems (Nasr, Hatem N.)VIS07,202-213(1991)

Automatic analysis of heliotest strips (Langinmaa, Anu)V1468,573-580(1991)

Automatic digitization of contour lines for digital map production (Yla-Jaaski, Juha; Yu, Xiaohan)V1472,201-207(1991)

Automatic method for inspecting plywood shear samples (Avent, R. R.; Conners, Richard W.)V1468,281-295(1991)

Automatic Object Recognition (Nasr, Hatem N., ed.)VIS07(1991)

Automatic reconstruction of buildings from aerial imagery (Sinha, Saravajit S.)V1468,698-709(1991)

Basic manufacturability interval (Billings, Daniel A.)V1468,434-445(1991)

Blackboard architecture for medical image interpretation (Davis, Darryl N.; Taylor, Christopher J.)V1445,421-432(1991)

BUOSHI: a tool for developing active expert systems (Wang, Gang; Dubant, Olivier)V1468,11-15(1991)

Case-based reasoning approach for heuristic search (Krovvidy, Srinivas; Wee, William G.; Han, Chia Y.)V1468,216-226(1991)

CLIPS implementation of a knowledge-based distributed control of an autonomous mobile robot (Bou-Ghannam, Akram A.; Doty, Keith L.)V1468,504-515(1991)

Computer-aided acquisition of design knowledge (Dilger, Werner)V1468,584-595(1991)

Computer interpretation of thallium SPECT studies based on neural network analysis (Wang, David C.; Karvelis, K. C.)V1445,574-575(1991)

Conflict resolution in multi-ES cooperation systems* (Liu, Dayou; Zheng, Fangqing; Ma, Zhifang; Shi, Qiaotin)V1468,37-49(1991)

Connectionist learning systems for control (Baker, Walter L.; Farrell, Jay A.)V1382,181-198(1991)

Considering multiple-surface hypotheses in a Bayesian hierarchy (LaValle, Steven M.; Hutchinson, Seth A.)V1569,2-15(1991)

Constructing attribute classes by example learning: the research of attribute-based knowledge-style pattern recognition (Zhou, Lijia; Song, Hongjun; Zhao, S.)V1469,404-411(1991)

Controller implemented by recording the fuzzy rules by backpropagation neural networks (Ying, Xingren; Zeng, Nan)V1469,846-851(1991)

Control scheme for sensor fusion for navigation of autonomous mobile robots (Murphy, Robin R.)V1383,436-447(1991)

Cooperative Intelligent Robotics in Space (Stoney, William E.; deFigueiredo, Rui J., eds.)V1387(1991)

Coping with complexity in the navigation of an autonomous mobile robot (Dodds, David R.)V1388,448-452(1991)

Deciding not to decide using resource-bounded sensing (Hager, Gregory D.)V1383,379-390(1991)

Design of a cart-pole balancing fuzzy logic controller using a genetic algorithm (Karr, Charles L.)V1468,26-36(1991)

Detecting difficult roads and intersections without map knowledge for robot vehicle navigation (Crisman, Jill D.; Thorpe, Charles E.)V1388,152-164(1991)

Detection and classification of undersea objects using multilayer perceptrons (Shazeer, Dov J.; Bello, M.)V1469,622-636(1991)

Distributed architecture for intelligent robotics (Gouveia, Feliz A.; Barthes, Jean-Paul A.; Oliveira, Eugenio C.)V1468,516-523(1991)

Environment model for mobile robots indoor navigation (Roth-Tabak, Yuval; Weymouth, Terry E.)V1388,453-463(1991)

Example of a Bayes network of relations among visual features (Agosta, John M.)V1569,16-27(1991)

Experiments in real-time visual control (Griswold, Norman C.; Kehtarnavaz, Nasser)V1388,342-349(1991)

Expert system for diagnosis/optimization of microlithography process* (Nicolau, Dan V.; Fulga, Florin; Dusa, Mircea V.)V1468,345-351(1991)

Expert system for fusing weather and doctrinal information used in the intelligence preparation of the battlefield (McWilliams, Gary; Kirby, Steve; Eskridge, Thomas; Newberry, Jeff)V1468,417-428(1991)

Exploiting geometric relationships for object modeling and recognition (Walker, Ellen L.)V1382,353-363(1991)

Exploiting known topologies to navigate with low-computation sensing (Miller, David P.; Gat, Erann)V1383,425-435(1991)

Flat plate project (Wijbrans, Klaas C.; Korsten, Maarten J.)V1386,197-205(1991)

Formalization and implementation of topological visual navigation in two dimensions (Kender, John R.; Park, Il-Pyung; Yang, David)V1388,476-489(1991)

From object structure to object function (Zlateva, Stoyanka D.; Vaina, Lucia M.)V1468,379-393(1991)

Fuzzy ellipsoidal shell clustering algorithm and detection of elliptical shapes (Dave, Rajesh N.; Patel, Kalpesh J.)V1381,320-333(1991)

Fuzzy logic: principles, applications, and perspectives (Zadeh, Lotfi A.)V1468,582-582(1991)

Fuzzy logic and neural networks in artificial intelligence and pattern recognition (Sanchez, Elie)V1569,474-483(1991)

Fuzzy logic approach to multitarget tracking in clutter (Priebe, Russell; Jones, Richard A.)V1482,265-274(1991)

Fuzzy logic controller structures (Yager, Ronald R.)V1381,368-378(1991)

Fuzzy logic for fault diagnosis (Comly, James B.; Bonissone, Piero P.; Dausch, Mark E.)V1381,390-400(1991)

Generation of exploratory schedules in closed loop for enhanced machine learning (Guez, Allon; Ahmad, Ziauddin)V1469,750-755(1991)

Geometric modeling of noisy image objects (Lipari, Charles A.)V1468,905-917(1991)

Geometric property measurement of convex objects using fuzzy sets (Poelzleitner, Wolfgang)V1381,411-422(1991)

Global Hierarchical Opportunistic Scheduling Tool: a system for scheduling based on constraint analysis (Ziebelin, Danielle)V1468,100-109(1991)

Heuristic search approach for mobile robot trap recovery (Zhao, Yilin; BeMent, Spencer L.)V1383,122-130(1991)

Hierarchical Dempster-Shafer evidential reasoning for image interpretation (Andress, Keith M.)V1569,43-54(1991)

High-level parallel architecture for a rule-based system (Karne, Ramesh K.; Sood, Arun K.)V1468,938-949(1991)

High-performance CAM-based Prolog execution scheme (Ali-Yahia, Tahar; Dana, Michel)V1468,950-959(1991)

Image analysis applied to black ice detection (Chen, Yi)V1468,551-562(1991)

Image segmentation using domain constraints (Ward, Matthew O.; Rajasekaran, Suresh)V1381,490-500(1991)

Implementation of a 3-D laser imager-based robot navigation system with location identification (Boltinghouse, Susan T.; Burke, James; Ho, Daniel)V1388,14-29(1991)

Integrating vision and AI in an image processing workstation (Chan, John P.; Batchelor, Bruce G.)V1386,163-170(1991)

Intelligent information system: for automation of airborne early warning crew decision processes (Chin, Hubert H.)V1468,235-244(1991)

Intelligent piloting tools for control of an autonomous mobile robot (Malotaux, Eric; Alimenti, Rodolphe; Bogaert, Marc; Gaspart, Pierre)V1388,372-383(1991)

Intelligent Robots and Computer Vision IX: Algorithms and Techniques (Casasent, David P., ed.)V1381(1991)

Intelligent visual inspection of food products (Chan, John P.; Batchelor, Bruce G.; Harris, I. P.; Perry Beng, S. J.)V1386,171-179(1991)

Intelligent word-based text recognition (Hoenes, Frank; Bleisinger, Rainer; Dengel, Andreas R.)V1384,305-316(1991)

Investigation of methods of combining functional evidence for 3-D object recognition (Stark, Louise; Hall, Lawrence O.; Bowyer, Kevin W.)V1381,334-345(1991)

KEShell: a "rule skeleton + rule body" -based knowledge engineering shell* (Wu, Xindong)V1468,632-639(1991)

Knowledge- and model-based ATR (automatic target recognition) algorithms adaptation (Nasr, Hatem N.; Bazakos, Michael E.; Sadjadi, Firooz A.; Amehdi, Hossien)VIS07,122-129(1991)

Knowledge-based direct 3-D texture segmentation system for confocal microscopic images (Lang, Zhengping; Zhang, Zhen; Scarberry, Randell E.; Shao, Weimin; Sun, Xu-Mei)V1468,826-833(1991)

Knowledge-based nursing diagnosis (Hay, Claudette; Hay, D. R.)V1468,314-322(1991)

Knowledge-based system for analysis of aerial images (Shariat, Hormoz)V1381,306-317(1991)

Knowledge-based system for configuring mixing-machines (Brinkop, Axel; Laudwein, Norbert)V1468,227-234(1991)

LaneLok: an improved Hough transform algorithm for lane sensing using strategic search methods (Kenue, Surender K.; Wybo, David R.)V1468,538-550(1991)

Large-scale networks via self-organizing hierarchical networks (Smotroff, Ira G.; Friedman, David H.; Connolly, Dennis)V1469,544-550(1991)

LCS: a natural language comprehension system (Trigano, Philippe; Talon, Benedicte; Baltazart, Didier; Demko, Christophe; Newstead, Emma)V1468,866-874(1991)

Learning by comparison: improving the task planning capability (del Castillo, Maria D.; Kumpel, Daniel M.)V1468,596-607(1991)

Learning procedure for the recognition of 3-D objects from 2-D images (Bart, Mischa; Buurman, Johannes; Duin, Robert P.)V1381,66-77(1991)

Learning spatially coherent properties of the visual world in connectionist networks (Becker, Suzanna; Hinton, Geoffrey E.)V1569,218-226(1991)

Linear programming solutions to problems in logical inference and space-variant image restoration (Digumarthi, Ramji V.; Payton, Paul M.; Barrett, Eamon B.)V1472,128-138(1991)

Logical account of a terminological tool (Bresciani, Paolo)V1468,245-255(1991)

Machine verification of traced signatures (Krishnan, Ganapathy; Jones, David E.)V1468,563-572(1991)

Medical image understanding system based on Dempster-Shafer reasoning (Chen, Shiuh-Yung; Lin, Wei-Chung; Chen, Chin-Tu)V1445,386-397(1991)

Mobile Robots V (Chun, Wendell H.; Wolfe, William J., eds.)V1388(1991)

Mobile robot system for the handicapped (Palakal, Mathew J.; Chien, Yung-Ping; Chittajallu, Siva K.; Xue, Qing L.)V1468,456-466(1991)

Model-based task planning system for a space laboratory environment (Chi, Sung-Do; Zeigler, Bernard P.; Cellier, Francois E.)V1387,182-193(1991)

Motion analysis for visually-guided navigation (Hildreth, Ellen C.)V1382,167-180(1991)

Multiagent collaboration for experimental calibration of an autonomous mobile robot (Vachon, Bertrand; Berge-Cherfaoui, Veronique)V1468,483-492(1991)

Multilevel qualitative reasoning in CMOS circuit analysis (Kaul, Neeraj; Biswas, Gautam)V1468,204-215(1991)

Multiple-hypothesis-based multiple-sensor spatial data fusion algorithm (Leung, Dominic S.; Williams, D. S.)V1471,314-325(1991)

Multiple-sensor cueing using a heuristic search (David, Philip)V1468,1000-1009(1991)

Multiple-target tracking in a cluttered environment and intelligent track record (Tomasini, Bernard; Cassassolles, Emmanuel; Poyet, Patrice; Maynard de Lavalette, Guy M.; Siffredi, Brigitte)V1468,60-71(1991)

Neural networks for robot navigation (Pandya, Abhijit S.; Luebbers, Paul G.)V1468,802-811(1991)

Neural networks for the recognition of skilled arm and hand movements (Vaina, Lucia M.; Tuncer, Temel E.)V1468,990-999(1991)

New search method based on hash table and heuristic search method (Wang, He-Chen; Xian, Wu; Luo, Da L.; Song, Xiang)V1384,133-136(1991)

New techniques for repertory grid analysis (Liseth, Ole J.; Bezdek, James C.; Ford, Kenneth M.; Adams-Webber, Jack R.)V1468,256-267(1991)

Object segmentation algorithm for use in recognizing 3-D partially occluded objects* (Fan, Kuo-Chin; Chang, Chia-Yuan)V1468,674-684(1991)

On-line visual prosthesis for a decision maker (Ligomenides, Panos A.)V1382,145-156(1991)

Parallel path planning in unknown terrains (Prassler, Erwin E.; Milios, Evangelos E.)V1388,2-13(1991)

Parallel reduced-instruction-set-computer architecture for real-time symbolic pattern matching (Parson, Dale E.)V1468,960-971(1991)

Path planning algorithm for a mobile robot* (Fan, Kuo-Chin; Lui, Po-Chang)V1468,1010-1021(1991)

Pattern recognition, neural networks, and artificial intelligence (Bezdek, James C.)V1468,924-935(1991)

Plan-behavior interaction in autonomous navigation (Lim, William Y.; Eilbert, James L.)V1388,464-475(1991)

Planning, acting, and sensor fusion (Firby, R. J.)V1383,483-489(1991)

Practical approach to knowledge base verification (Preece, Alun D.; Shinghal, Rajjan)V1468,608-619(1991)

Production environment implementation of the stereo extraction of cartographic features using computer vision and knowledge base systems in DMA's digital production system (Gruenewald, Maria M.; Hinchman, John H.)V1468,843-852(1991)

Propagation of variances in belief networks (Neapolitan, Richard E.)V1468,333-344(1991)

Prototype expert system for preventive control in power plants (Jiang, Dareng; Han, Chia Y.; Wee, William G.)V1468,16-25(1991)

Recursive computation of a wire-frame representation of a scene from dynamic stereo using belief functions (Tirumalai, Arun P.; Schunck, Brian G.; Jain, Ramesh C.)V1569,28-42(1991)

Relating binary and continuous problem entropy to backpropagation network architecture (Smith, A.; Dagli, Cihan H.)V1469,551-562(1991)

Representing sentence information (Perkins, Walton A.)V1468,854-865(1991)

Research and improvement of Vogl's acceleration algorithm (Song, Hongjun; Zhou, Lijia)V1469,581-584(1991)

Robot vision system for obstacle avoidance planning (Attolico, Giovanni; Caponetti, Laura; Chiaradia, Maria T.; Distante, Arcangelo)V1388,50-61(1991)

Robust self-calibration and evidential reasoning for building environment maps (Tirumalai, Arun P.; Schunck, Brian G.; Jain, Ramesh C.)V1383,345-358(1991)

Routing in distributed information systems (Ras, Zbigniew W.)V1470,76-87(1991)

Scheduler's assistant: a tool for intelligent scheduling (Griffin, Neal L.)V1468,110-121(1991)

Segmentation using models of expected structure (Shemlon, Stephen; Liang, Tajen; Cho, Kyugon; Dunn, Stanley M.)V1381,470-481(1991)

Selected Papers on Automatic Object Recognition (Nasr, Hatem N., ed.)VMS41(1991)

Sensor-based identification of control parameters for intelligent gripping (Wood, Hugh C.; Vaidyanathan, C. S.)V1387,245-254(1991)

Sensor-knowledge-command fusion paradigm for man-machine systems (Lee, Sukhan; Schenker, Paul S.; Park, Jun S.)V1383,391-402(1991)

Space-time system architecture for the neural optical computing (Lo, Yee-Man V.)V1382,199-208(1991)

Step towards optimal topology of communication networks (Michalewicz, Zbigniew)V1470,112-122(1991)

Stochastic neural nets and vision (Fall, Thomas C.)V1468,778-785(1991)

Task decomposition, distribution, and localization for intelligent robot coordination (Kountouris, Vasilios G.; Stephanou, Harry E.)V1387,169-180(1991)

Ten years of failure in automatic time tables scheduling at the UTC (Trigano, Philippe; Boufflet, Jean-Paul; Newstead, Emma)V1468,408-416(1991)

Terrain acquisition algorithm for an autonomous mobile robot with finite-range sensors (Smith, John M.; Choo, Chang Y.; Nasrabadi, Nasser M.)V1468,493-501(1991)

Terrain classification in navigation of an autonomous mobile robot (Dodds, David R.)V1388,82-89(1991)

Three-dimensional laser radar simulation for autonomous spacecraft landing (Reiley, Michael F.; Carmer, Dwayne C.; Pont, W. F.)V1416,295-303(1991)

Three-dimensional reconstruction of pulmonary blood vessels by using anatomical knowledge base (Inaoka, Noriko; Suzuki, Hideo; Mori, Masaki; Takabatake, Hirotsugu; Suzuki, Akira)V1450,2-12(1991)

Towards a general formula for analogical learning leading to more autonomous systems (Cooke, Daniel E.; Patterson, Dan W.; Starks, Scott A.)V1381,299-305(1991)

Towards a versatile control system for mobile robots (Noreils, Fabrice R.)V1388,384-396(1991)

Towards integrated autonomous systems (Jain, Ramesh C.; Roth-Tabak, Yuval)V1468,188-201(1991)

Transformation from tristimulus RGB to Munsell notation HVC in a colored computer vision system (Jin, Guofan; Zhu, Zimin; Yu, Xinglong)V1569,507-512(1991)

Two-view vision system for 3-D texture recovery (Yu, Xiaohan; Yla-Jaaski, Juha)V1468,834-842(1991)

Use of an expert system to predict thunderstorms and severe weather (Passner, Jeffrey E.; Lee, Robert R.)V1468,2-10(1991)

Using real-time stereopsis for mobile robot control (Bonasso, R. P.; Nishihara, H. K.)V1387,237-244(1991)

Validity criterion for compact and separate fuzzy partitions and its justification (Xie, Xuanli; Beni, Gerardo)V1381,401-410(1991)

Vehicle path planning via dual-world representations (Peck, Alex N.; Breul, Harry T.)V1388,30-38(1991)

Aspherics—see also lenses; optical design; optical fabrication

Alignment of an aspheric mirror subsystem for an advanced infrared catadioptric system (Tingstad, James S.)V1527,194-198(1991)

Applications of diamond-turned null reflectors for generalized aspheric metrology (McCann, James T.)V1332,843-849(1991)

Aspheric surface testing techniques (Stahl, H. P.)V1332,66-76(1991)

Aspheric testing using null mirrors (Murty, Mantravady V.; Kumar, Vas; von Handorf, Robert J.)V1332,107-114(1991)

Creation of aspheric beryllium optical surfaces directly in the hot isostatic pressing consolidation process (Gildner, Donald; Marder, James M.)V1485,46-53(1991)

Design and analysis of aspherical multilayer imaging x-ray microscope (Shealy, David L.; Jiang, Wu; Hoover, Richard B.)V1343,122-132(1991)

Design and optical performance of beryllium assessment mirrors (Thomas, Brigham B.; Maxey, L. C.; Miller, Arthur C.)V1485,20-30(1991)

Fabrication of a fast, aspheric beryllium mirror (Paquin, Roger A.; Gardopee, George J.)V1485,39-45(1991)

Interferometer for testing of general aspherics using computer-generated holograms (Arnold, Steven M.; Jain, Anil K.)V1396,473-480(1991)

Manufacture of fast, aspheric, bare beryllium optics for radiation hard, spaceborne sensor systems (Sweeney, Michael N.)V1485,116-127(1991)

Optical aspheric surface profiler using phase shift interferometry (Sasaki, Kenji; Ono, Akira)V1332,97-106(1991)

Optical design of high-aperture aspherical projection lens (Osawa, Atsuo; Fukuda, Kyohei; Hirata, Kouji)V1354,337-343(1991)

Robotic-based fabrication system for aspheric reflectors (Zimmerman, Jerrold; Jones, Robert A.; Rupp, Wiktor J.)VCR38,184-192(1991)

Role of aspherics in zoom lens design (Betensky, Ellis I.)V1354,656-662(1991)Sola ASL in Spectralite and polycarbonate aspheric lens designs (Machol, Steven)V1529,45-56(1991)

Sola ASL in Spectralite strikes the perfect balance between cosmetics and optics (Machol, Steven; Modglin, Luan)V1529,38-44(1991)

Thick, fine-grained beryllium optical coatings (Murray, Brian W.; Ulph, Eric; Richard, Peter N.)V1485,106-115(1991)

Zoom lens with aspherical lens for camcorder (Yatsu, Masahiko; Deguchi, Masaharu; Maruyama, Takesuke)V1354,663-668(1991)

Associative processing—see also optical computing; optical processing

Binary phase-only filter associative memory (Kane, Jonathan S.; Hemmer, Philip R.; Woods, Charles L.; Khoury, Jehad)V1564,511-520(1991)

Design of a motionless head for parallel readout optical disk (Marchand, Philippe J.; Krishnamoorthy, Ashok V.; Ambs, Pierre; Gresser, Julien; Esener, Sadik C.; Lee, Sing H.)V1505,38-49(1991)

Hetero-association for pattern translation (Yu, Francis T.; Lu, Taiwei; Yang, Xiangyang)V1507,210-221(1991)

Ho-Kashyap CAAP 1:1 associative processors (Casasent, David P.; Telfer, Brian A.)V1382,158-166(1991)

Holographic associative memory with bipolar features (Wang, Xu-Ming; Mu, Guoguang)V1558,518-528(1991)

MSE and hierarchical optical associative processor system (Casasent, David P.; Chien, Sung-II)V1382,304-310(1991)

Multifunctional hybrid neural network: real-time laboratory results (Natarajan, Sanjay S.; Casasent, David P.; Smokelin, John-Scott)V1564,474-488(1991)

New optoelectronic implementation of 2-D associative memory (He, Anzhi; Zhang, Li; Zhang, Jiajun)V1563,208-212(1991)

Optical higher order double-layer associative memory (Lam, David T.; Carroll, John E.)V1505,104-114(1991)

Optical implementation of associative memory based on parallel rank-one interconnections (Jeon, Ho-In; Abushagur, Mustafa A.; Caulfield, H. J.)V1564,522-535(1991)

Projection methods for evaluation of Hopfield-type CAM models (Berus, Tomasz; Macukow, Bohdan)V1564,562-570(1991)

Pseudodeep hologram and its properties (Denisyuk, Yuri N.; Ganzherli, N. M.)V1238,2-12(1991)

Astronomy—see also lenses; mirrors; optical design; optomechanical design; satellites; space optics; sun; telescopes

4096 x 4096 pixel CCD mosaic imager for astronomical applications (Geary, John C.; Luppino, Gerard A.; Bredthauer, Richard A.; Hlivak, Robert J.; Robinson, Lloyd B.)V1447,264-273(1991)

Accretion dynamics and polarized x-ray emission of magnetized neutron stars (Arons, Jonathan)V1548,2-12(1991)

Acquisition and processing of digital images and spectra in astronomy (Lee, Terence J.)V1439,152-158(1991)

Active and Adaptive Optical Systems (Ealey, Mark A., ed.)V1542(1991)

Active optics system for a 3.5-meter structured mirror (Stepp, Larry M.; Roddier, Nicolas; Dryden, David M.; Cho, Myung K.)V1542,175-185(1991)

Adaptive optics: a progress review (Hardy, John W.)V1542,2-17(1991)

Adaptive optics for the European very large telescope (Merkle, Fritz; Hubin, Norbert)V1542,283-292(1991)

Adaptive optics system tests at the ESO 3.6-m telescope (Merkle, Fritz; Gehring, G.; Rigaut, Francois; Lena, Pierre J.; Rousset, Gerard; Fontanella, Jean-Claude; Gaffard, Jean-Paul)V1542,308-318(1991)

Algorithms for wavefront reconstruction out of curvature sensing data (Roddier, Nicolas)V1542,120-129(1991)

Anisoplanatism and the use of laser guide stars (Goad, Larry E.)V1542,100-109(1991)

Aperture synthesis in astronomical radio-interferometry using maximum entropy on the mean (Le Besnerais, Guy; Navaza, Jorge; Demoment, Guy)V1569,386-395(1991)

Application-specific integrated-circuit-based multianode microchannel array readout system (Smeins, Larry G.; Stechman, John M.; Cole, Edward H.)V1549,59-65(1991)

Array detectors in astronomy (Lesser, Michael P.)V1439,144-151(1991)

Back-illuminated large-format Loral CCDs (Lesser, Michael P.)V1447,177-182(1991)

Blazing of transmission gratings for astronomical use (Neviere, Michel)V1545,11-18(1991)

CCD camera for an autoguider (Schempp, William V.)V1448,129-133(1991)

CCD focal-plane imaging detector for the JET-X instrument on spectrum R-G (Wells, Alan A.; Castelli, C. M.; Holland, Andrew D.; McCarthy, Kieran J.; Spragg, J. E.; Whitford, C. H.)V1549,357-373(1991)

Characteristics of a high-pressure gas proportional counter filled with xenon (Sakurai, Hirohisa; Ramsey, Brian D.)V1549,20-27(1991)

Come-on-plus project: an upgrade of the come-on adaptive optics prototype system (Gendron, Eric; Cuby, Jean-Gabriel; Rigaut, Francois; Lena, Pierre J.; Fontanella, Jean-Claude; Rousset, Gerard; Gaffard, Jean-Paul; Boyer, Corinne; Richard, Jean-Claude; Vittot, M.; Merkle, Fritz; Hubin, Norbert)V1542,297-307(1991)

Dead-time effects in microchannel-plate imaging detectors (Zombeck, Martin V.; Fraser, George W.)V1549,90-100(1991)

Development of a low-pass far-infrared filter for lunar observer horizon sensor application (Mobasser, Sohrab; Horwitz, Larry S.; Griffith, O'Dale)V1540,764-774(1991)

EUV, X-Ray, and Gamma-Ray Instrumentation for Astronomy II (Siegmund, Oswald H.; Rothschild, Richard E., eds.)V1549(1991)

Evaluation of a CCD star camera at Table Mountain Observatory (Strikwerda, Thomas E.; Fisher, H. L.; Frank, L. J.; Kilgus, Charles C.; Gray, Connie B.; Barnes, Donald L.)V1478,13-23(1991)

FIR lasers as local oscillators in submillimeter astronomy (Roeser, Hans-Peter; van der Wal, Peter)V1501,194-197(1991)

Faint object spectrograph early performance (Harms, Richard J.; Fitch, John E.)V1494,49-65(1991)

Fixing the Hubble Space Telescope (Crocker, James H.)V1494,2-8(1991)

Foundation, excavation, and radiation-shielding concepts for a 16-m large lunar telescope (Chua, Koon M.; Johnson, Stewart W.)V1494,119-134(1991)

Four-meter lunar engineering telescope (Peacock, Keith; Giannini, Judith A.; Kilgus, Charles C.; Bely, Pierre Y.; May, B. S.; Cooper, Shannon A.; Schlimm, Gerard H.; Sounder, Charles; Ormond, Karen A.; Cheek, Eric A.)V1494,147-159(1991)

Further performance characteristics of a high-sensitivity 64 x 64 element InSb hybrid focal-plane array (Fischer, Robert C.; Martin, Charles J.; Niblack, Curtiss A.; Timlin, Harold A.; Wimmers, James T.)V1494,414-418(1991)

Future directions in focal-plane signal processing for spaceborne scientific imagers (Fossum, Eric R.)V1541,62-67(1991)

Goldhelox: a project to view the x-ray sun (Fair, Melody)V1549,182-192(1991)

Ground systems and operations concepts for the Space Infrared Telescope Facility (Miller, Richard B.)V1540,38-46(1991)

Hard x-ray imaging via crystal diffraction: first results of reflectivity measurements (Frontera, Filippo; De Chiara, P.; Gambaccini, M.; Landini, G.; Pasqualini, G.)V1549,113-119(1991)

High-dynamic-range MCP structures (Slater, David C.; Timothy, J. G.)V1549,68-80(1991)

High-performance 256 x 256 InSb FPA for astronomy (Hoffman, Alan; Randall, David)V1540,297-302(1991)

High-resolution astronomical observations using deconvolution from wavefront sensing (Michau, Vincent; Marais, T.; Laurent, Jean; Primot, Jerome; Fontanella, Jean-Claude; Tallon, M.; Fuensalida, Jesus J.)V1487,64-71(1991)

High-resolution decoding of multianode microchannel array detectors (Kasle, David B.; Morgan, Jeffrey S.)V1549,52-58(1991)

High-resolution imaging with multilayer soft x-ray, EUV, and FUV telescopes of modest aperture and cost (Walker, Arthur B.; Lindblom, Joakim F.; Timothy, J. G.; Hoover, Richard B.; Barbee, Troy W.; Baker, Phillip C.; Powell, Forbes R.)V1494,320-333(1991)

Hubble extra-solar planet interferometer (Shao, Michael)V1494,347-356(1991)

Imager for gamma-ray astronomy: balloon prototype (Di Cocco, Guido; Labanti, Claudio; Malaguti, Giuseppe; Rossi, Elio; Schiavone, Filomena; Spizzichino, A.; Traci, A.; Bird, A. J.; Carter, T.; Dean, Anthony J.; Gomm, A. J.; Grant, K. J.; Corba, M.; Quadrini, E.; Rossi, M.; Villa, G. E.; Swinyard, Bruce M.)V1549,102-112(1991)

Imaging performance analysis of adaptive optical telescopes using laser guide stars (Welsh, Byron M.)V1542,88-99(1991)

Imaging the sun in hard x-rays: spatial and rotating modulation collimators (Campbell, Jonathan W.; Davis, John M.; Emslie, A. G.)V1549,155-179(1991)

Infrared and the search for extrasolar planets (Meinel, Aden B.; Meinel, Marjorie P.)V1540,196-201(1991)

Infrared Technology XVII (Andresen, Bjorn F.; Scholl, Marija S.; Spiro, Irving J., eds.)V1540(1991)

Initial performance of the high-speed photometer (Richards, Evan; Percival, Jeff; Nelson, Matthew; Hatter, Edward; Fitch, John E.; White, Richard L.)V1494,40-48(1991)

Instrument design and test results of the new all-reflection spatial heterodyne spectrometer (Bush, Brett C.; Cotton, Daniel M.; Vickers, James S.; Chakrabarti, Supriya)V1549,376-384(1991)

Large-format CCDs for astronomical applications (Geary, John C.)V1439,159-168(1991)

Laser guide stars for adaptive optics systems: Rayleigh scattering experiments (Thompson, Laird A.; Castle, Richard; Carroll, David L.)V1542,110-119(1991)

Lunar dust: implications for astronomical observatories (Johnson, Stewart W.; Chua, Koon M.; Burns, Jack O.; Slane, Frederic A.)V1494,194-207(1991)

MARTINI: system operation and astronomical performance (Doel, A. P.; Dunlop, C. N.; Major, J. V.; Myers, Richard M.; Sharples, R. M.)V1542,319-326(1991)

MIC photon counting detector (Fordham, John L.; Bellis, J. G.; Bone, David A.; Norton, Timothy J.)V1449,87-98(1991)

Microchannel-plate detectors for space-based astronomy (Crocker, James H.; Cox, Colin R.; Ray, Knute A.; Sen, Amit)V1494,434-439(1991)

Mission design for the Space Infrared Telescope Facility (Kwok, Johnny H.; Osmolovsky, Michael G.)V1540,27-37(1991)

MIT multipoint alignment testbed: technology development for optical interferometry (Blackwood, Gary H.; Jacques, Robert; Miller, David W.)V1542,371-391(1991)

Monolithic integrated-circuit charge amplifier and comparator for MAMA readout (Cole, Edward H.; Smeins, Larry G.)V1549,46-51(1991)

Multilayer mirrors and filters for imaging the earth's inner magnetosphere (Schulze, Dean W.; Sandel, Bill R.; Broadfoot, A. L.)V1549,319-328(1991)

Need for active structures in future large IR and sub-mm telescopes (Rapp, Donald)V1542,328-358(1991)

Neural network adaptive optics for the multiple-mirror telescope (Wizinowich, Peter L.; Lloyd-Hart, Michael; McLeod, Brian A.; Colucci, D'nardo; Dekany, Richard G.; Wittman, David; Angel, J. R.; McCarthy, Donald W.; Hulburd, William G.; Sandler, David G.)V1542,148-158(1991)

Next-generation space telescope: a large UV-IR successor to HST (Illingworth, Garth)V1494,86-97(1991)

Noise reduction strategy for hybrid IR focal-plane arrays (Fowler, Albert M.; Gatley, Ian)V1541,127-133(1991)

Objectives for the Space Infrared Telescope Facility (Spehalski, Richard J.; Werner, Michael J.)V1540,2-14(1991)

One-dimensional proportional counters for x-ray all-sky monitors (Matsuoka, Masaru; Yamauchi, Masamitsu; Nakamura, Haruo; Kondo, M.; Kawai, N.; Yoshida, A.; Imai, Tohru)V1549,2-8(1991)

Optical design of an off-axis low-distortion UV telescope (Richardson, E. H.)V1494,314-319(1991)

Optical system design for a lunar optical interferometer (Colavita, Mark M.; Shao, Michael; Hines, Braden E.; Levine, Bruce M.; Gershman, Robert)V1494,168-181(1991)

Performance characteristics of the imaging MAMA detector systems for SOHO, STIS, and FUSE/Lyman (Timothy, J. G.)V1549,221-233(1991)

Performance of low-resistance microchannel-plate stacks (Siegmund, Oswald H.; Stock, Joseph)V1549,81-89(1991)

Performance of microstrip proportional counters for x-ray astronomy on spectrum-roentgen-gamma (Budtz-Jorgensen, Carl; Bahnsen, Axel; Christensen, Finn E.; Madsen, M. M.; Olesen, C.; Schnopper, Herbert W.)V1549,429-437(1991)

Performance of milliKelvin Si bolometers as x-ray and exotic particle detectors (Zammit, C. C.; Sumner, Timothy J.; Lea, M. J.; Fozooni, P.; Hepburn, I. D.)V1549,274-282(1991)

Polarization of emission lines from relativistic accretion disk (Chen, Kaiyou)V1548,23-33(1991)

Progress report on a five-axis fast guiding secondary for the University of Hawaii 2.2-meter telescope (Cavedoni, Charles P.; Graves, J. E.; Pickles, A. J.)V1542,273-282(1991)

SALSA: a synthesis array for lunar submillimeter astronomy (Mahoney, Michael J.; Marsh, Kenneth A.)V1494,182-193(1991)

Scientific results from the Hubble Space Telescope fine-guidance sensors (Taff, Laurence G.)V1494,66-77(1991)

Semiactive telescope for the French PRONAOS submillimetric mission (Duran, Michel; Luquet, Philippe; Buisson, F.; Cousin, B.)V1494,357-376(1991)

Simple 180o field-of-view F-theta all-sky camera (Andreic, Zeljko)V1500,293-304(1991)

SIRTF focal-plane technologies (Capps, Richard W.; Bothwell, Mary)V1540,47-50(1991)

SIRTF stray light analysis (Elliott, David G.; St. Clair Dinger, Ann)V1540,63-67(1991)

Soft x-ray windows for position-sensitive proportional counters (Viitanen, Veli-Pekka; Nenonen, Seppo A.; Sipila, Heikki; Mutikainen, Risto)V1549,28-34(1991)

SOHO space satellite: UV instrumentation (Poland, Arthur I.; Domingo, Vicente)V1343,310-318(1991)

Solar astronomy with a 19-segment adaptive mirror (Acton, D. S.; Smithson, Robert C.)V1542,159-164(1991)

Solar feature correlation tracker (Rimmele, Thomas; von der Luhe, Oskar; Wiborg, P. H.; Widener, A. L.; Dunn, Richard B.; Spence, G.)V1542,186-193(1991)

Space Astronomical Telescopes and Instruments (Bely, Pierre Y.; Breckinridge, James B., eds.)V1494(1991)

Space Infrared Telescope Facility cryogenic and optical technology (Mason, Peter V.; Kiceniuk, T.; Plamondon, Joseph A.; Petrick, Walt)V1540,88-96(1991)

Space Infrared Telescope Facility science instruments overview (Bothwell, Mary)V1540,15-26(1991)

Space Infrared Telescope Facility structural design requirements (MacNeal, Paul D.; Lou, Michael C.; Chen, Gun-Shing)V1540,68-85(1991)

Space Infrared Telescope Facility telescope overview (Schember, Helene R.; Manhart, Paul K.; Guiar, Cecilia N.; Stevens, James H.)V1540,51-62(1991)

Status of the stellar x-ray polarimeter for the Spectrum-X-Gamma mission (Kaaret, Philip E.; Novick, Robert; Shaw, Ping-Shine; Hanany, Shaul; Liu, Yee; Fleischman, Judith R.; Sunyaev, Rashid; Lapshov, I.; Weisskopf, Martin C.; Elsner, Ronald F.; Ramsey, Brian D.; Silver, Eric H.; Ziock, Klaus P.; Costa, Enrico; Piro, Luigi; Soffitta, Paolo; Manzo, Giuseppe; Giarrusso, Salvatore; Santangelo, Andrea E.; Scarsi, Livio; Fraser, George W.; Pearson, James F.; Lees, John E.; Perola, G. C.; Massaro, Enrico; Matt, Giorgio)V1548,106-117(1991)

Stray light issues for background-limited infrared telescope operation (Scholl, Marija S.; Scholl, James W.)V1540,109-118(1991)

Study of gas scintillation proportional counter physics using synchrotron radiation (Bavdaz, Marcos; Favata, Fabio; Smith, Alan; Parmar, A. N.)V1549,35-44(1991)

Submicron structures—promising filters in EUV: a review (Gruntman, Michael A.)V1549,385-394(1991)

Survey of hard x-ray imaging concepts currently proposed for viewing solar flares (Campbell, Jonathan W.; Davis, John M.; Emslie, A. G.)V1343,359-375(1991)

System concepts for a large UV/optical/IR telescope on the moon (Nein, Max E.; Davis, Billy)V1494,98-110(1991)

System design for lunar-based optical and submillimeter interferometers (Gershman, Robert; Mahoney, Michael J.; Rayman, M. D.; Shao, Michael; Snyder, Gerald C.)V1494,160-167(1991)

Theoretical models for stellar x-ray polarization in compact objects (Meszaros, Peter)V1548,13-22(1991)

Thermal systems analysis for the Space Infrared Telescope Facility dewar (Bhandari, Pradeep; Petrick, Stanley W.; Schember, Helene R.)V1540,97-108(1991)

Transmittances of thin polymer films and their suitability as a supportive substrate for a soft x-ray solar filter (Williams, Memorie; Hansen, Evan; Reyes-Mena, Arturo; Allred, David D.)V1549,147-154(1991)

UCSD high-energy x-ray timing experiment cosmic ray particle anticoincidence detector (Hink, Paul L.; Rothschild, Richard E.; Pelling, Michael R.; MacDonald, Daniel R.; Gruber, Duane E.)V1549,193-202(1991)

UCSD high-energy x-ray timing experiment magnetic shield design and test results (Rothschild, Richard E.; Pelling, Michael R.; Hink, Paul L.)V1549,120-133(1991)

University of Hawaii adaptive optics system: I. General approach (Roddier, Francois J.; Graves, J. E.; McKenna, Daniel; Northcott, Malcolm J.)V1542,248-253(1991)

University of Hawaii adaptive optics system: II. Computer simulation (Northcott, Malcolm J.)V1542,254-261(1991)

Virtual-phase charge-coupled device imaging system for astronomy (Khvilivitzky, A. T.; Zuev, A. G.; Rybakov, M. I.; Kiryan, G. V.; Berezin, V. Y.)V1447,64-68(1991)

Wide-field-of-view star tracker camera (Lewis, Isabella T.; Ledebuhr, Arno G.; Axelrod, Timothy S.; Kordas, Joseph F.; Hills, Robert F.)V1478,2-12(1991)

X-ray interferometric observatory (Martin, Christopher)V1549,203-220(1991)

Atmospheric optics—see also adaptive optics; aerosols; gases; propagation; thermal blooming; turbulence

2.9 micron laser source for use in the two-photon/laser-induced fluorescence detection of atmospheric OH (Bradshaw, John D.; van Dijk, Cornelius A.)V1433,81-91(1991)

AcidMODES: a major field study to evaluate regional-scale air pollution models (Ching, Jason K.; Bowne, Norman E.)V1491,360-370(1991)

Adaptive optical transfer function modeling (Gaffard, Jean-Paul; Boyer, Corinne)V1483,92-103(1991)

Adaptive optics, transfer loops modeling (Boyer, Corinne; Gaffard, Jean-Paul; Barrat, Jean-Pierre; Lecluse, Yves)V1483,77-91(1991); V1542,46-61(1991)

Advances in tunable diode laser technology for atmospheric monitoring applications (Wall, David L.)V1433,94-103(1991)

Aerodyne research mobile infrared methane monitor (McManus, J. B.; Kebabian, Paul L.; Kolb, Charles E.)V1433,330-339(1991)

Aerosol models for optical and IR propagation in the marine atmospheric boundary layer (de Leeuw, Gerrit)V1487,130-159(1991)

Airborne lidar elastic scattering, fluorescent scattering, and differential absorption observations (Uthe, Edward E.)V1479,393-402(1991)

Airborne lidar measurements of ozone and aerosols in the summertime Arctic troposphere (Browell, Edward V.)V1491,7-14(1991)

Airborne lidar observations of ozone and aerosols in the wintertime Arctic stratosphere (Browell, Edward V.)V1491,273-281(1991)

Airborne tunable diode laser sensor for high-precision concentration and flux measurements of carbon monoxide and methane (Sachse, Glen W.; Collins, Jim E.; Hill, G. F.; Wade, L. O.; Burney, Lewis G.; Ritter, J. A.)V1433,157-166(1991)

Aircraft laser infrared absorption spectrometer (ALIAS) for polar ozone studies (Webster, Chris R.; May, R. D.)V1540,187-194(1991)

Air quality monitoring with the differential optical absorption spectrometer (Stevens, Robert K.; Conner, Teri L.)V1491,56-67(1991)

Alexandrite laser characterization and airborne lidar developments for water vapor DIAL measurements (Ponsardin, Patrick; Higdon, Noah S.; Grossmann, Benoist E.; Browell, Edward V.)V1492,47-51(1991)

Analysis of atmospheric trace constituents from high-resolution infrared balloon-borne and ground-based solar absorption spectra (Goldman, Aaron; Murcray, Frank J.; Rinsland, C. P.; Blatherwick, Ronald D.; Murcray, F. H.; Murcray, David G.)V1491,194-202(1991)

Applicability of open-path monitors at Superfund sites (Padgett, Joseph; Pritchett, Thomas H.)V1433,352-364(1991)

Application of FM spectroscopy in atmospheric trace gas monitoring: a study of some factors influencing the instrument design (Werle, Peter W.; Josek, K.; Slemr, Franz)V1433,128-135(1991)

Application of backscatter absorption gas imaging to the detection of chemicals related to drug production (Kulp, Thomas J.; Garvis, Darrel G.; Kennedy, Randall B.; McRae, Thomas J.)V1479,352-363(1991)

Application of the NO/O3 chemiluminescence technique to measurements of reactive nitrogen species in the stratosphere (Fahey, David W.)V1433,212-223(1991)

ATLID: the first preoperational ATmospheric LIDar for the European polar platform (Lange, Robert; Endemann, Martin J.; Reiland, Werner; Krawczyk, Rodolphe; Hofer, Bruno)V1492,24-37(1991)

Atmospheric code sensitivity to uncertainties in aerosol optical depth characteristics (Teillet, Philippe M.; Fedosejevs, Gunar; Ahern, Francis J.; Gauthier, Robert P.; Sirois, J.)V1492,213-223(1991)

Atmospheric continuum absorption models (Delaye, Corinne T.; Thomas, Michael E.)V1487,291-298(1991)

Atmospheric effects on laser systems (Au, Robert H.)V1399,8-15(1991)

Atmospheric infrared sounder on the Earth Observing System: in-orbit spectral calibration (Aumann, Hartmut H.)V1540,176-186(1991)

Atmospheric laser-transmission tables simply generated (Mallory, William R.)V1540,359-364(1991)

Atmospheric propagation effects on pattern recognition by neural networks (Giever, John C.; Hoock, Donald W.)V1486,302-313(1991)

Atmospheric turbulence sensing for a multiconjugate adaptive optics system (Johnston, Dustin C.; Welsh, Byron M.)V1542,76-87(1991)

Atmospheric visibility monitoring for planetary optical communications (Cowles, Kelly A.)V1487,272-279(1991)

Autonomous guidance, navigation, and control bridging program plan (McSwain, G. G.; Fernandes, Stan T.; Doane, Kent B.)V1478,228-238(1991)

Balloon-born investigations of total and aerosol attenuation continuous spectra in the stratosphere (Mirzoeva, Larisa A.; Kiseleva, Margaret S.; Reshetnikova, Irina N.)V1540,444-449(1991)

BEST: a new satellite for a climatology study in the tropics (Orgeret, Marc)V1490,14-22(1991)

Blurring effect of aerosols on imaging systems (Bissonnette, Luc R.)V1487,333-344(1991)

Calibration for the SAGE III/EOS instruments (Chu, William P.; McCormick, M. P.; Zawodny, J. M.; McMaster, Leonard R.)V1491,243-250(1991)

Characterization of simulated and open-air atmospheric turbulence (Razdan, Anil K.; Singh, Brahm P.; Chopra, S.; Modi, M. B.)V1558,384-388(1991)

Characterization, Propagation, and Simulation of Sources and Backgrounds (Watkins, Wendell R.; Clement, Dieter, eds.)V1486(1991)

Characterization of the atmospheric modulation transfer function using the target contrast characterizer (Watkins, Wendell R.; Billingsley, Daniel R.; Palacios, Fernando R.; Crow, Samuel B.; Jordan, Jay B.)V1486,17-24(1991)

Characterization of tropospheric methane through space-based remote sensing (Ashcroft, Peter)V1491,48-55(1991)

Chemical amplifier for peroxy radical measurements based on luminol chemiluminescence (Cantrell, Chris A.; Shetter, Richard E.; Lind, John A.; Gilliland, Curt A.; Calvert, Jack G.)V1433,263-268(1991)

Clouds as calibration targets for AVHRR reflected-solar channels: results from a two-year study at NOAA/NESDIS (Abel, Peter)V1493,195-206(1991)

Coherent launch-site atmospheric wind sounder (Targ, Russell; Hawley, James G.; Otto, Robert G.; Kavaya, Michael J.)V1478,211-227(1991)

Collisional effects in laser detection of tropospheric OH (Crosley, David R.)V1433,58-68(1991)

Compact measurement system for the simultaneous determination of NO, NO2, NOy, and O3 using a small aircraft (Walega, James G.; Dye, James E.; Grahek, Frank E.; Ridley, Brian K.)V1433,232-241(1991)

Comparison of laser radar transmittance for the five atmospheric models (Au, Robert H.)V1487,280-290(1991)

Comparison of the angular resolution limit and SNR of the Hubble Space Telescope and the large ground-based telescopes (Souilhac, Dominique J.; Billerey, Dominique)V1494,503-526(1991)

Comparison of time and frequency multiplexing techniques in multicomponent FM spectroscopy (Muecke, Robert J.; Werle, Peter W.; Slemr, Franz; Prettl, William)V1433,136-144(1991)

Compensation efficiency of an optical adaptive transmitter (Feng, Yue-Zhong; Song, Zhengfang; Gong, Zhi-Ben)V1417,370-372(1991)

Complementary experiments for tether dynamics analysis (Wingo, Dennis R.; Bankston, Cheryl D.)V1495,123-133(1991)

Convective evaporation of water aerosol droplet irradiated by CO2 laser (Butkovsky, A. V.)V1440,146-152(1991)

Correlation between the aerosol profiles measurements, the meteorological conditions, and the atmospheric IR transmission in a mediterranean marine atmosphere (Tanguy, Mireille; Bonhommet, Herve; Autric, Michel L.; Vigliano, Patrick)V1487,172-183(1991)

Cryogenic limb array etalon spectrometer: experiment description (Roche, Aidan E.; Kumer, John B.)V1491,91-103(1991)

Detection and parameter estimation of atmospheric turbulence by ground-based and airborne CO2 Doppler lidars (Pogosov, Gregory A.; Akhmanov, Sergei A.; Gordienko, Vyacheslav M.; Kosovsky, L. A.; Kurochkin, Nikolai N.; Priezzhev, Alexander V.)V1416,115-124(1991)

Detection of stratospheric ozone trends by ground-based microwave observations (Connor, Brian J.; Parrish, Alan; Tsou, Jung-Jung)V1491,218-230(1991)

Determination of inhomogeneous trace absorption by using exponential expansion of the absorption Pade-approximant (Dobrego, Kirill V.)V1433,365-374(1991)

Determination of the altitude of the nitric acid layer from very high resolution, ground-based IR solar spectra (Blatherwick, Ronald D.; Murcray, Frank J.; Murcray, David G.; Locker, M. H.)V1491,203-210(1991)

Determination of the temperature of a single particle heated by a highly concentrated laser beam (Herve, Philippe; Bednarczyk, Sophie; Masclet, Philippe)V1487,387-395(1991)

Developing a long-path diode array spectrometer for tropospheric chemistry studies (Lanni, Thomas R.; Demerjian, Kenneth L.)V1433,21-24(1991)

Development of 1- and 2-um coherent Doppler lidars for atmospheric sensing (Chan, Kin-Pui; Killinger, Dennis K.)V1492,111-114(1991)

Development of an airborne excimer-based UV-DIAL for monitoring ozone and sulfur dioxide in the lower troposphere (Bristow, Michael P.; Diebel, D. E.; Bundy, Donald H.; Edmonds, Curtis M.; Turner, Ruldopha M.; McElroy, James L.)V1491,68-74(1991)

Development of high resolution statistically nonstationary infrared earthlimb radiance scenes (Strugala, Lisa A.; Newt, J. E.; Futterman, Walter I.; Schweitzer, Eric L.; Herman, Bruce J.; Sears, Robert D.)V1486,176-187(1991)

DOAS (differential optical absorption spectroscopy) urban pollution measurements (Stevens, Robert K.; Vossler, T. L.)V1433,25-35(1991)

Earth and Atmospheric Remote Sensing (Curran, Robert J.; Smith, James A.; Watson, Ken, eds.)V1492(1991)

Earth Observing System (Wilson, Stan; Dozier, Jeff)V1491,117-124(1991)

Edge technique: a new method for atmospheric wind measurements with lidar (Korb, C. L.; Gentry, Bruce M.)V1416,177-182(1991)

Effects of thin and subvisible cirrus on HEL far-field intensity calculations at various wavelengths (Harada, Larrene K.)V1408,28-40(1991)

Equipment development for an atmospheric-transmission measurement campaign (Ruiz, Domingo; Kremer, Paul J.)V1417,212-222(1991)

ESA Earth observation polar platform program (Rast, Michael; Readings, C. J.)V1490,51-58(1991)

Evaluation of the Navy Oceanic Vertical Aerosol Model using lidar and PMS particle-size spectrometers (Jensen, Douglas R.)V1487,160-171(1991)

Experimental investigation of image degradation created by a high-velocity flow field (Couch, Lori L.; Kalin, David A.; McNeal, Terry)V1486,417-423(1991)

Far-field pattern of laser diodes as function of the relative atmospheric humidity (Freitas, Jose C.; Carvalho, Fernando D.; Rodrigues, F. C.; Abreu, M. A.; Marcal, Joao P.)V1399,16-23(1991)

Far-IR Fabry-Perot spectrometer for OH measurements (Pickett, Herbert M.; Peterson, Dean B.)V1491,308-313(1991)

Fast-response water vapor and carbon dioxide sensor (Kohsiek, W.)V1511,114-119(1991)

Fluctuation variation of a CO2 laser pulse intensity during its interaction with a cloud (Almaev, R. K.; Semenov, L. P.; Slesarev, A. G.; Volkovitsky, O. A.)V1397,831-834(1991)

Focusing infrared laser beams on targets in space without using adaptive optics (McKechnie, Thomas S.)V1408,119-135(1991)

Forecasting optical turbulence strength: effects of macroscale meteorology and aerosols (Sadot, Danny; Kopeika, Norman S.)V1442,325-334(1991)

Forward-looking IR and lidar atmospheric propagation in the infrared field program (Koenig, George G.; Bissonnette, Luc R.)V1487,240-249(1991)

Frequency modulation spectroscopy for chemical sensing of the environment (Cooper, David E.; Riris, Haris; van der Laan, Jan E.)V1433,120-127(1991)

Functional reconstruction predictions of uplink whole beam Strehl ratios in the presence of thermal blooming (Enguehard, S.; Hatfield, Brian)V1408,186-191(1991)

Future European and Japanese Remote-Sensing Sensors and Programs (Slater, Philip N., ed.)V1490(1991)

Generation of IR sky background images (Levesque, Martin P.)V1486,200-209(1991)

Global ozone monitoring by occultation of stars (Bertaux, J. L.)V1490,133-145(1991)

GLOBE backscatter: climatologies and mission results (Menzies, Robert T.; Post, Madison J.)V1416,139-146(1991)

GLOB(MET)SAT: French proposals for monitoring global change and weather from the polar orbit (Durpaire, Jean-Pierre; Ratier, A.; Dagras, C.)V1490,23-38(1991)

Graphical interface for multispectral simulation, scene generation, and analysis (Sikes, Terry L.; Kreiss, William T.)V1479,199-211(1991)

Ground-based CW atmospheric Doppler performance modeling (Becherer, Richard J.; Kahan, Lloyd R.)V1416,306-313(1991)

Ground-based lidar for long-term and network measurements of ozone (McDermid, I. S.; Schmoe, Martha S.; Walsh, T. D.)V1491,175-181(1991)

Ground-based microwave radiometry of ozone (Kaempfer, Niklaus A.; Bodenmann, P.; Peter, Reto)V1491,314-322(1991)

Ground-based microwave remote sensing of water vapor in the mesosphere and stratosphere (Croskey, Charles L.; Olivero, John J.; Martone, Joseph P.)V1491,323-334(1991)

Ground-based monitoring of water vapor in the middle atmosphere: the NRL water-vapor millimeter-wave spectrometer (Bevilacqua, Richard M.; Schwartz, Philip R.; Pauls, Thomas A.; Waltman, William B.; Thacker, Dorsey L.)V1491,231-242(1991)

Ground-to-space multiline propagation at 1.3 um (Crawford, Douglas P.; Harada, Larrene K.)V1408,167-177(1991)

High-radiometric-performance CCD for the third-generation stratospheric aerosol and gas experiment (Delamere, W. A.; Baer, James W.; Ebben, Thomas H.; Flores, James S.; Kleiman, Gary; Blouke, Morley M.; McCormick, M. P.)V1479,31-40(1991)

High-resolution multichannel mm-wave radiometer for the detection of stratospheric ClO (Gerber, Louis; Kaempfer, Niklaus A.)V1491,211-217(1991)

High-resolution studies of atmospheric IR emission spectra (Murcray, Frank J.; Murcray, F. H.; Goldman, Aaron; Blatherwick, Ronald D.; Murcray, David G.)V1491,282-287(1991)

Images of turbulent, absorbing-emitting atmospheres and their application to windshear detection (Watt, David W.; Philbrick, Daniel A.)V1467,357-368(1991)

Imaging performance analysis of adaptive optical telescopes using laser guide stars (Welsh, Byron M.)V1542,88-99(1991)

Improved limb atmospheric spectrometer and retroreflector in-space for ADEOS (Sasano, Yasuhiro; Asada, Kazuya; Sugimoto, Nobuo; Yokota, Tatsuya; Suzuki, Makoto; Minato, Atsushi; Matsuzaki, Akiyoshi; Akimoto, Hajime)V1490,233-242(1991)

Influence of atmospheric conditions on two-color temperature discrimination (Mallory, William R.)V1540,365-369(1991)

Infrared coherent lidar systems for wind velocity measurements (Gordienko, Vyacheslav M.; Akhmanov, Sergei A.; Bersenev, V. I.; Kosovsky, L. A.; Kurochkin, Nikolai N.; Priezzhev, Alexander V.; Pogosov, Gregory A.; Putivskii, Yu. Y.)V1416,102-114(1991)

Infrared lidars for atmospheric remote sensing (Menzies, Robert T.)V1540,160-163(1991)

Infrared lidar windshear detection for commercial aircraft and the edge technique: a new method for atmospheric wind measurement (Targ, Russell; Bowles, Roland L.; Korb, C. L.; Gentry, Bruce M.; Souilhac, Dominique J.)V1521,144-157(1991)

Infrared spectrometer for ground-based profiling of atmospheric temperature and humidity (Shaw, Joseph A.; Churnside, James H.; Westwater, Edward R.)V1540,681-686(1991)

In-situ measurement of methyl bromide in indoor air using long-path FTIR spectroscopy (Green, Martina; Seiber, James N.; Biermann, Heinz W.)V1433,270-274(1991)

Interferometric measurements of a high-velocity mixing/shear layer (Peters, Bruce R.; Kalin, David A.)V1486,410-416(1991)

Interferometric monitor for greenhouse gasses for ADEOS (Tsuno, Katsuhiko; Kameda, Yoshihiko; Kondoh, Kayoko; Hirai, Shoichi)V1490,222-232(1991)

Interframe registration and preprocessing of image sequences (Tajbakhsh, Shahram; Boyce, James F.)V1521,14-22(1991)

Investigations of the phenomenon of light scattering on particles of atmospheric aerosols (Drobnik, Antoni)V1391,204-210(1991)

Laser Radar VI (Becherer, Richard J., ed.)V1416(1991)

Laser set for the investigations of the NO2 contents in atmosphere (Makuchowski, Jozef; Pokora, Ludwik J.; Ujda, Zbigniew; Wawer, Janusz)V1391,348-350(1991)

Lidar evaluation of the propagation environment (Uthe, Edward E.; Livingston, John M.)V1487,228-239(1991)

Lidar for expendable launch vehicles (Lee, Michael S.)V1480,23-34(1991)

Lidar profiles of atmospheric structure properties (Philbrick, Charles R.)V1492,76-84(1991)

Lidar wind shear detection for commercial aircraft (Targ, Russell; Bowles, Roland L.)V1416,130-138(1991)

Liquid droplet supercritical explosion in the field of CO2 laser radiation and influence of plasma chemical reactions on initiation of optical breakdown in air (Budnik, A. P.; Popov, A. G.)V1440,135-145(1991)

Long-path differential absorption measurements of tropospheric molecules (Harder, Jerald W.; Mount, George H.)V1491,33-42(1991)

Long-path intracavity laser for the measurement of atmospheric trace gases (McManus, J. B.; Kolb, Charles E.)V1433,340-351(1991)

Measurement of atmospheric composition by the ATMOS instrument from Table Mountain Observatory (Gunson, Michael R.; Irion, Fredrick W.)V1491,335-346(1991)

Measurement of Atmospheric Gases (Schiff, Harold I., ed.)V1433(1991)

Measurement of modulation transfer function of desert atmospheres (McDonald, Carlos)V1487,203-219(1991)

Measurement of peroxyacetyl nitrate, NO2, and NOx by using a gas chromatograph with a luminol-based detector (Drummond, John W.; Mackay, Gervase I.; Schiff, Harold I.)V1433,242-252(1991)

Measurement of the tropospheric hydroxyl radical by long-path absorption (Mount, George H.)V1491,26-32(1991)

Measurement of tropospheric carbon monoxide using gas filter radiometers (Reichle, Henry G.)V1491,15-25(1991)

Measurement of wind velocity spread: signal-to-noise ratio for heterodyne detection of laser backscatter from aerosol (Fried, David L.; Szeto, Roque K.)V1416,163-176(1991)

Measurements of atmospheric transmittance of CO2 laser radiation (Aref'ev, Vladimir N.)V1397,827-830(1991)

Measurements of nitrous acid, nitrate radicals, formaldehyde, and nitrogen dioxide for the Southern California Air Quality Study by differential optical absorption spectroscopy (Winer, Arthur M.; Biermann, Heinz W.)V1433,44-55(1991)

Measuring tropospheric ozone using differential absorption lidar technique (Proffitt, Michael H.; Langford, A. O.)V1491,2-6(1991)

Michelson interferometer for passive atmosphere sounding (Posselt, Winfried)V1490,114-125(1991)

Microwave limb sounder experiments for UARS and EOS (Waters, Joe W.)V1491,104-109(1991)

Mission study overview of Japanese polar orbiting platform program (Moriyama, Takashi; Nakayama, Kimihiko; Homma, M.; Haruyama, Yukio)V1490,310-316(1991)

Modeling and Simulation of Laser Systems II (Schnurr, Alvin D., ed.)V1415(1991)

Modeling turbulent transport in laser beam propagation (Wallace, James)V1408,19-27(1991)

Multiple-scattering effects on pulse propagation through burning particles (Ma, Yushieh; Varadan, Vijay K.; Varadan, Vasundara V.)V1487,220-225(1991)

Multiplex Fabry-Perot interferometer (Snell, Hilary E.; Hays, Paul B.)V1492,403-407(1991)

Multiscattered lidar returns from atmospheric aerosols (Hutt, Daniel L.; Bissonnette, Luc R.; Durand, Louis-Gilles)V1487,250-261(1991)

NASA's program in lidar remote sensing (Theon, John S.; Vaughan, William W.; Browell, Edward V.; Jones, William D.; McCormick, M. P.; Melfi, Samuel H.; Menzies, Robert T.; Schwemmer, Geary K.; Spinhirne, James D.)V1492,2-23(1991)

Narrowband alexandrite laser injection seeded with frequency-dithered diode laser (Schwemmer, Geary K.; Lee, Hyo S.; Prasad, Coorg R.)V1492,52-62(1991)

Natural terrain infrared radiance statistics in a wind field (Ruizhong, Rao; Song, Zhengfang)V1486,390-395(1991)

Network for the detection of stratospheric change (Kurylo, Michael J.)V1491,168-174(1991)

New spectroscopic instrumentation for measurement of stratospheric trace species by remote sensing of scattered skylight (Mount, George H.; Jakoubek, Roger O.; Sanders, Ryan W.; Harder, Jerald W.; Solomon, Susan; Winkler, Richard; Thompson, Thomas; Harrop, Walter)V1491,188-193(1991)

Numerical experiments in propagation with wind velocity fluctuation (Carlson, Lawrence W.)V1408,203-211(1991)

O3, NO2, NO3, SO2, and aerosol measurements in Beijing (Xue, Qing-yu; Guo, Song; Zhao, Xue-peng; Nieu, Jian-guo; Zhang, Yi-ping)V1491,75-82(1991)

Observations of uplink and retroreflected scintillation in the Relay Mirror Experiment (Lightsey, Paul A.; Anspach, Joel E.; Sydney, Paul F.)V1482,209-222(1991)

Open-path tunable diode laser absorption for eddy correlation flux measurements of atmospheric trace gases (Anderson, Stuart M.; Zahniser, Mark S.)V1433,167-178(1991)

Optical fiber sensing of corona discharges (Woolsey, G. A.; Lamb, D. W.; Woerner, M. C.)V1584,243-253(1991)

Optical filters for the wind imaging interferometer (Sellar, R. G.; Gault, William A.; Karp, Christopher K.)V1479,140-155(1991)

Optical velocity sensor for air data applications (Smart, Anthony E.)V1480,62-71(1991)

Optics for tunable diode laser spectrometers (Riedel, Wolfgang J.)V1433,179-189(1991)

Overall atmospheric MTF and aerosol MTF cutoff (Dror, Itai; Kopeika, Norman S.)V1487,192-202(1991)

Overview of the spectroscopy of the atmosphere using far-infrared emission experiment (Russell, James M.)V1491,142-150(1991)

Partially compensated speckle imaging: Fourier phase spectrum estimation (Roggemann, Michael C.; Matson, Charles L.)V1542,477-487(1991)

Path integral approach to thermal blooming (Enguehard, S.; Hatfield, Brian)V1408,178-185(1991)

PCI (phase compensation instability) and minishear (Fried, David L.; Szeto, Roque K.)V1408,150-166(1991)

Performance tests of a 1500 degree-of-freedom adaptive optics system for atmospheric compensation (Cuellar, Louis; Johnson, Paul A.; Sandler, David G.)V1542,468-476(1991)

Plasma parameter determination formed under the influence of CO_2 laser radiation on the obstacle in the air using optical methods (Vas'kovsky, Yu. M.; Gordeeva, I. A.; Korenev, A. S.; Rovinsky, R. E.; Cenina, I. S.; Shirokova, I. P.)V1440,229-240(1991)

Polarization and directionality of the Earth's reflectances: the POLDER instrument (Lorsignol, Jean; Hollier, Pierre A.; Deshayes, Jean-Pierre)V1490,155-163(1991)

Potential of tunable lasers for optimized dual-color laser ranging (Lund, Glenn I.; Gaignebet, Jean)V1492,166-175(1991)

Prediction of coarse-aerosol statistics according to weather forecast (Gottlieb, J.; Kopeika, Norman S.)V1487,184-191(1991)

Prediction of thermal-image quality as a function of weather forecast (Shushan, A.; Meninberg, Y.; Levy, I.; Kopeika, Norman S.)V1487,300-311(1991)

Propagation Engineering: Fourth in a Series (Bissonnette, Luc R.; Miller, Walter B., eds.)V1487(1991)

Propagation of High-Energy Laser Beams Through the Earth's Atmosphere II (Ulrich, Peter B.; Wilson, LeRoy E., eds.)V1408(1991)

Pulse propagation in random media (Ishimaru, Akira)V1558,127-129(1991)

Radiative transfer in the cloudy atmosphere: modeling radiative transport (Gabriel, Philip; Stephens, Graeme L.; Tsay, Si-Chee)V1558,76-90(1991)

Recent advances in CO_2 laser catalysts (Upchurch, Billy T.; Schryer, David R.; Brown, K. G.; Kielin, E. J.; Hoflund, Gar B.; Gardner, Steven D.)V1416,21-29(1991)

Recent lidar measurements of stratospheric ozone and temperature within the Network for the Detection of Stratospheric Change (McGee, Thomas J.; Ferrare, Richard; Butler, James J.; Frost, Robert L.; Gross, Michael; Margitan, James)V1491,182-187(1991)

Relay Mirror Experiment scoring analysis and the effects of atmospheric turbulence (Sydney, Paul F.; Dillow, Michael A.; Anspach, Joel E.; Kervin, Paul W.; Lee, Terence B.)V1482,196-208(1991)

Remote Sensing of Atmospheric Chemistry (McElroy, James L.; McNeal, Robert J., eds.)V1491(1991)

Results of calibrations of the NOAA-11 AVHRR made by reference to calibrated SPOT imagery at White Sands, N.M. (Nianzeng, Che; Grant, Barbara G.; Flittner, David E.; Slater, Philip N.; Biggar, Stuart F.; Jackson, Ray D.; Moran, M. S.)V1493,182-194(1991)

Retrieval and molecule sensitivity studies for the global ozone monitoring experiment and the scanning imaging absorption spectrometer for atmospheric chartography (Chance, Kelly V.; Burrows, John P.; Schneider, Wolfgang)V1491,151-165(1991)

Review of the physics of small-scale thermal blooming in uplink propagation (Enguehard, S.; Hatfield, Brian)V1415,128-137(1991)

Scanning imaging absorption spectrometer for atmospheric chartography (Burrows, John P.; Chance, Kelly V.)V1490,146-154(1991)

Scattering and thermal emission from spatially inhomogeneous atmospheric rain and cloud (Jin, Ya-Qiu)V1487,324-332(1991)

Selected Papers on Multiple Scattering in Plane Parallel Atmospheres and Oceans: Methods (Kattawar, George W., ed.)VMS42(1991)

Ship signature measurements for tactical decision-aid input (Cooper, Alfred W.; Milne, Edmund A.; Crittenden, Eugene C.; Walker, Philip L.; Moore, E.; Lentz, William J.)V1486,37-46(1991)

Simulation of partially obscured scenes using the radiosity method (Borel, Christoph C.; Gerstl, Siegfried A.)V1486,271-277(1991)

Simulation of vertical profiles of extinction and backscatter coefficients in very low stratus clouds and subcloud regions (Kilmer, Neal H.; Rachele, Henry)V1487,109-122(1991)

Sound generation by laser radiation in air: application to restoration of energy distribution in laser beams (Gurvich, Alexander S.; Myakinin, Vladimir A.; Vorob'ev, Valerii V.; D'yakov, Alexander; Pokasov, Vladimir; Pryanichnikov, Victor)V1408,10-18(1991)

Space-based sensing of atmospheric conditions over data-void regions (Behunek, Jan L.; Vonder Haar, Thomas H.)V1479,93-100(1991)

Spectroscopic observations of CO_2, CH_4, N_2O, and CO from Kitt Peak, 1979-1990 (Livingston, William C.; Wallace, Lloyd V.)V1491,43-47(1991)

Stratospheric aerosol and gas experiment III: aerosol and trace gas measurements for the Earth Observing System (McCormick, M. P.; Chu, William P.; Zawodny, J. M.; Mauldin, Lemuel E.; McMaster, Leonard R.)V1491,125-141(1991)

Stratospheric ozone concentration profiles from Spacelab-1 solar occultation infrared absorption spectra (De Maziere, Martine M.; Camy-Peyret, C.; Lippens, C.; Papineau, N.)V1491,288-297(1991)

Stratospheric spectroscopy with the far-infrared spectrometer: overview and recent results (Traub, Wesley A.; Chance, Kelly V.; Johnson, David G.; Jucks, Kenneth W.)V1491,298-307(1991)

Stratospheric wind infrared limb sounder (Rider, David M.; McCleese, Daniel J.)V1540,142-147(1991)

Study of nighttime NO_3 chemistry by differential optical absorption spectroscopy (Plane, John M.; Nien, Chia-Fu)V1433,8-20(1991)

Summary of atmospheric chemistry observations from the Antarctic and Arctic aircraft campaigns (Tuck, Adrian F.)V1491,252-272(1991)

System for evaluation of trace gas concentration in the atmosphere based on the differential optical absorption spectroscopy technique (Hallstadius, Hans; Uneus, Leif; Wallin, Suante)V1433,36-43(1991)

Technique for measuring atmospheric effects on image metrics (Crow, Samuel B.; Watkins, Wendell R.; Palacios, Fernando R.; Billingsley, Daniel R.)V1486,333-344(1991)

Thermal blooming critical power and adaptive optics correction for the ground-based laser (Smith, David C.; Townsend, Sallie S.)V1408,112-118(1991)

Thresholds of plasma arising under the pulse CO_2 laser radiation interaction with an obstacle in air and energetic balance of the process (Babaeva, N. A.; Vas'kovsky, Yu. M.; Zhavoronkov, M. I.; Rovinsky, R. E.; Rjabinkina, V. A.)V1440,260-269(1991)

Tropospheric emission spectrometer for the Earth Observing System (Glavich, Thomas A.; Beer, Reinhard)V1540,148-159(1991)

Tunable diode laser spectrometer for high-precision concentration and ratio measurements of long-lived atmospheric gases (Fried, Alan; Drummond, James R.; Henry, Bruce; Fox, Jack)V1433,145-156(1991)

Tunable diode laser systems for trace gas monitoring (Mackay, Gervase I.; Karecki, David R.; Schiff, Harold I.)V1433,104-119(1991)

Turbulence at the inner scale (Tatarskii, V. I.)V1408,2-9(1991)

Universal multifractal theory and observations of land and ocean surfaces, and of clouds (Lavallee, Daniel; Lovejoy, Shaun; Schertzer, Daniel)V1558,60-75(1991)

Upper-atmosphere research satellite: an overview (McNeal, Robert J.)V1491,84-90(1991)

Use of a multibeam transmitter for significant improvement in signal-dynamic-range reduction and near-range coverage for incoherent lidar systems (Zhao, Yanzeng; Hardesty, R. M.; Post, Madison J.)V1492,85-90(1991)

Use of an expert system to predict thunderstorms and severe weather (Passner, Jeffrey E.; Lee, Robert R.)V1468,2-10(1991)

Use of chemiluminescence techniques in portable, lightweight, highly sensitive instruments for measuring NO_2, NO_x, and O_3 (Drummond, John W.; Topham, L. A.; Mackay, Gervase I.; Schiff, Harold I.)V1433,224-231(1991)

Use of satellite data to determine the distribution of ozone in the troposphere (Fishman, Jack; Watson, Catherine E.; Brackett, Vincent G.; Fakhruzzaman, Khan; Veiga, Robert)V1491,348-359(1991)

Variable emissivity plates under a three-dimensional sky background (Meitzler, Thomas J.; Gonda, Teresa G.; Jones, Jack C.; Reynolds, William R.)V1486,380-389(1991)

Variable wind direction effects on thermal blooming correction (Hills, Louis S.; Long, Jerry E.; Gebhardt, Frederick G.)V1408,41-57(1991)

Visible extinction measurements in rain and snow using a forward-scatter meter (Hutt, Daniel L.; Oman, James)V1487,312-323(1991)

VUV/photofragmentation laser-induced fluorescence sensor for the measurement of atmospheric ammonia (Sandholm, Scott T.; Bradshaw, John D.)V1433,69-80(1991)

Water continuum in the 15- to 25-um spectral region: evidence for H_2O_2 in the atmosphere (Devir, Adam D.; Neumann, M.; Lipson, Steven G.; Oppenheim, Uri P.)V1442,347-359(1991)

Wavemeter for tuning solid state lasers (Goad, Joseph H.; Rinsland, Pamela L.; Kist, Edward H.; Irick, Steven C.)V1410,107-115(1991)

Wave-optic model to determine image quality through supersonic boundary and mixing layers (Lawson, Shelah M.; Clark, Rodney L.; Banish, Michele R.; Crouse, Randy F.)V1488,268-278(1991)

Wedge imaging spectrometer: application to drug and pollution law enforcement (Elerding, George T.; Thunen, John G.; Woody, Loren M.)V1479,380-392(1991)

Wide-spectral-range transmissometer used for fog measurements (Turner, Vernon; Trowbridge, Christian A.)V1487,262-271(1991)

Wide bandwidth spectral measurements of atmospheric tilt turbulence (Tiszauer, Detlev H.; Smith, Richard C.)V1408,72-83(1991)

Atoms—see also spectroscopy, atomic

Calculations on the Hanle effect with phase and amplitude fluctuating laser fields (Bergeman, Thomas H.; Ryan, Robert E.)V1376,54-67(1991)

Corrections to the Golden Rule (Fearn, Heidi; Lamb, Willis E.)V1497,245-254(1991)

Gain and threshold in noninversion lasers (Scully, Marlan O.; Zhu, Shi-Yao; Narducci, Lorenzo M.; Fearn, Heidi)V1497,264-276(1991)

Nonlinear theory of a three-level laser with microwave coupling: numerical calculation (Fearn, Heidi; Lamb, Willis E.; Scully, Marlan O.)V1497,283-290(1991)

Optical coherent transients induced by time-delayed fluctuating pulses (Finkelstein, Vladimir; Berman, Paul R.)V1376,68-79(1991)

Parametric pulse breakup due to population pulsations in three-level systems (DiMarco, Steven F.; Cantrell, Cyrus D.)V1497,178-187(1991)

QED theory of excess spontaneous emission noise (Milonni, Peter W.)V1376,164-169(1991)

Beams—see also lasers; microlithography; propagation

5X reticle fabrication using MEBES multiphase virtual address and AZ5206 resist (Milner, Kathy S.; Chipman, Paul S.)V1496,180-196(1991)

Adaptive beamforming using recursive eigenstructure updating with subspace constraint (Yu, Kai-Bor)V1565,288-295(1991)

Advanced concepts of electron-beam-pumped excimer lasers (Tittel, Frank K.; Canarelli, P.; Dane, C. B.; Hofmann, Thomas; Sauerbrey, Roland A.; Sharp, Tracy E.; Szabo, Gabor; Wilson, William L.; Wisoff, P. J.; Yamaguchi, Shigeru)V1397,21-29(1991)

Application and integration of a focused ion beam circuit repair system (Lange, John A.; Allen, Chris)V1465,50-56(1991)

Application of an electron-beam scattering parameter extraction method for proximity correction in direct-write electron-beam lithography (Weiss, Rudolf M.; Sills, Robert M.)V1465,192-200(1991)

Application of rf superconductivity to high-brightness and high-gradient ion beam accelerators (Delayen, Jean R.; Bohn, Courtlandt L.; Roche, C. T.)V1407,524-534(1991)

Applications of synchroscan and dual-sweep streak camera techniques to free-electron laser experiments (Lumpkin, Alex H.)V1552,42-49(1991)

Backside-thinned CCDs for keV electron detection (Ravel, Mihir K.; Reinheimer, Alice L.)V1447,109-122(1991)

Beam-coupling by stimulated Brillouin scattering (Falk, Joel; Chu, Raijun; Kanefsky, Morton; Hua, Xuelei)V1409,83-86(1991)

Beam divergence from sharp emitters in a general longitudinal magnetic field (Lau, Yue Y.; Colombant, Denis G.; Pilloff, Mark D.)V1407,635-646(1991); V1552,182-184(1991)

Beam quality in laser amplifiers (Martinez-Herrero, R.; Mejias, P. M.)V1397,623-626(1991)

Beam shaping system (MacAndrew, J. A.; Humphries, Mark R.; Welford, W. T.; Golby, John A.; Dickinson, P. H.; Wheeler, J. R.)V1500,172-176(1991)

Bessel beam generation: theory and application proposal (Jabczynski, Jan K.)V1391,254-258(1991)

Charge buildup in polypropylene thin films (Ding, Hai)V1519,847-856(1991)

Chemically amplified resists for x-ray and e-beam lithography (Berry, Amanda K.; Graziano, Karen A.; Thompson, Stephen D.; Taylor, James W.; Suh, Doowon; Plumb, Dean)V1465,210-220(1991)

Color- and intensity-balanced multichannel optical beamsplitter (Gilo, Mordechai; Rabinovitch, Kopel)V1442,90-104(1991)

Compact, low-power precision beam-steering mirror (DeWeerd, Herman)V1454,207-214(1991)

Continuous inertial confinement fusion for the generation of very intense plasma jets (Winterberg, F.)V1407,322-325(1991)

Dependence of output beam profile on launching conditions in fiber-optic beam delivery systems for Nd:YAG lasers (Su, Daoning; Boechat, Alvaro A.; Jones, Julian D.)V1502,41-51(1991)

Design and fabrication of soft x-ray photolithography experimental beam line at Beijing National Synchrotron Radiation Laboratory (Zhou, Changxin)V1465,26-33(1991)

Design and performance of a small two-axis high-bandwidth steering mirror (Loney, Gregory C.)V1454,198-206(1991)

DESIRE technology with electron-beam resists: fundamentals, experiments, and simulation (Nicolau, Dan V.; Fulga, Florin; Dusa, Mircea V.)V1465,282-288(1991)

Development of the Aurora high-power microwave source (Huttlin, George A.; Conrad, D. B.; Gavnoudias, S.; Judy, Daniel C.; Lazard, Carl J.; Litz, Marc S.; Pereira, Nino R.; Weidenheimer, Douglas M.)V1407,147-158(1991)

Diffraction effects in directed radiation beams (Hafizi, Bahman; Sprangle, Phillip)V1407,316-321(1991)

Duke storage ring FEL program (Litvinenko, Vladimir N.; Madey, John M.; Benson, Stephen V.; Burnham, B.; Wu, Y.)V1552,2-6(1991)

E-beam data compaction method for large-capacity mask ROM production (Kanemaru, Toyomi; Nakajima, Takashi; Igarashi, Tadanao; Masuda, Rika; Orita, Nobuyuki)V1496,118-123(1991)

Effects of deformable mirror/wavefront sensor separation in laser beam trains (Schafer, Eric L.; Lyman, Dwight D.)V1415,310-316(1991)

Efficient e-beam sustained Ar:Xe laser (Botma, H.; Peters, Peter J.; Witteman, Wilhelmus J.)V1397,573-576(1991)

Electron-beam-addressed lithium niobate spatial light modulator (Hillman, Robert L.; Melnik, George A.; Tsakiris, Todd N.; Leard, Francis L.; Jurgilewicz, Robert P.; Warde, Cardinal)V1562,136-142(1991)

Electron-beam lithography for the microfabrication of OEICs (Engel, Herbert; Doeldissen, Walter)V1506,60-64(1991)

Electron-beam metrology: the European initiative (Jackman, James J.)V1464,71-80(1991)

Electron beam lithographic fabrication of computer-generated holograms (West, Andrew A.; Smith, Robin W.)V1507,158-167(1991)

Electron-Beam, X-Ray, and Ion-Beam Submicrometer Lithographies for Manufacturing (Peckerar, Martin, ed.)V1465(1991)

Electron trapping and acceleration in a modified elongated betatron (Song, Yuanxu Y.; Fisher, Amnon; Prohaska, Robert M.; Rostoker, Norman)V1407,430-441(1991)

Electrostatic focusing and RFQ (radio frequency quadrupole) matching system for a low-energy H- beam (Guharay, Samar K.; Allen, C. K.; Reiser, Martin P.)V1407,610-619(1991)

Estimation of focused laser beam SBS (stimulated Brillouin scatter) threshold dependence on beam shape and phase aberrations (Clendening, Charles W.)V1415,72-78(1991)

Evaluation of a high-resolution negative-acting electron-beam resist GMC for photomask manufacturing (Chen, Wen-Chih; Novembre, Anthony E.)V1496,266-283(1991)

Experiments on the beam breakup instability in long-pulse electron beam transport through cavity systems (Menge, Peter R.; Bosch, Robert A.; Gilgenbach, Ronald M.; Choi, J. J.; Ching, Hong; Spencer, Thomas A.)V1407,578-588(1991)

Fidelity of Brillouin amplification with Gaussian input beams (Jones, David C.; Ridley, Kevin D.; Cook, Gary; Scott, Andrew M.)V1500,46-52(1991)

Fine undercut control in bilayer PMMA-P(MMA-MAA) resist system for e-beam lithography with submicrometer resolution (Bogdanov, Alexei L.; Andersson, Eva K.)V1465,324-329(1991)

FIR optical cavity oscillation is observed with the AT&T Bell Laboratories free-electron laser (Shaw, Earl D.; Chichester, Robert J.; La Porta, A.)V1552,14-23(1991)

Focused ion beam induced deposition: a review (Melngailis, John)V1465,36-49(1991)

Free-electron lasers as light sources for basic research (van Amersfoort, P. W.)V1504,25-36(1991)

Free propagation of high-order moments of laser beam intensity distribution (Sanchez, Miguel; Hernandez Neira, Jose Luis; Delgado, J.; Calvo, G.)V1397,635-638(1991)

Hierarchical proximity effect correction for e-beam direct writing of 64-Mbit DRAM (Misaka, Akio; Hashimoto, Kazuhiko; Kawamoto, M.; Yamashita, H.; Matsuo, Takahiro; Sakashita, Toshihiko; Harafuji, Kenji; Nomura, Noboru)V1465,174-184(1991)

High-gain high-efficiency TWT (traveling wave tube) amplifiers (Nation, John A.; Ivers, J. D.; Kerslick, G.; Shiffler, Donald; Schachter, Levi)V1407,32-43(1991)

High-power particle beams for gas lasers (Mesyats, Gennady A.)V1411,2-14(1991)

High-power pulsed and repetitively pulsed electron-beam-controlled discharge CO laser systems (Ionin, A. A.)V1502,95-102(1991)

High-power switching with electron-beam-controlled semiconductors (Brinkmann, Ralf P.; Schoenbach, Karl H.; Roush, Randy A.; Stoudt, David C.; Lakdawala, Vishnu K.; Gerdin, Glenn A.)V1378,203-208(1991)

High-power uhf rectenna for energy recovery in the HCRF (high-current radio frequency) system (Genuario, Ralph; Koert, Peter)V1407,553-565(1991)

High-sensitivity and high-dry-etching durability positive-type electron-beam resist (Tamura, Akira; Yonezawa, Masaji; Sato, Mitsuyoshi; Fujimoto, Yoshiaki)V1465,271-281(1991)

Hitachi e-beam lithography tools for advanced applications (Colbran, William V.)V1496,90-96(1991)

Holographic precompensation for one-way transmission of diffraction-limited beams through diffusing media (Pan, Anpei; Marhic, Michel E.)V1461,8-16(1991)

Induction linac-driven relativistic klystron and cyclotron autoresonance maser experiments (Goodman, Daniel L.; Birx, Daniel L.; Danly, Bruce G.)V1407,217-225(1991)

Influence of voltage pulse parameters on the beam current density distribution of large-aperture electron gun (Badziak, Jan; Dzwigalski, Zygmunt)V1391,250-253(1991)

Intense Microwave and Particle Beams II (Brandt, Howard E., ed.)V1407(1991)

Inverse bremsstrahlung acceleration in an electrostatic wave (Kim, Shang H.)V1407,620-634(1991)

Investigation into the characteristics of a-C:H films irradiated by electron beam (Gu, Shu L.; He, Yu L.; Wang, Zhi C.)V1519,175-178(1991)

Investigation of higher-order diffraction in a one-crystal 2-D scanner (Melamed, Nathan T.; Gottlieb, Milton S.)V1454,306-312(1991)

Investigation on the DLC films prepared by dual-ion-beam-sputtering deposition (Wang, Tianmin; Wang, Weijie; Liu, Guidng; Huang, Liangpu; Luo, Chuntai; Liu, Dingquan; Xu, Ming; Yang, Yimin)V1519,890-900(1991)

Investigation on the effect of electron-beam acceleration voltage and electron-beam sharpness on 0.2-um patterns (Moniwa, Akemi; Okazaki, Shinji)V1465,154-163(1991)

Investigations of cumulative beam breakup in radio-frequency linacs (Bohn, Courtlandt L.; Delayen, Jean R.)V1407,566-577(1991)

Ion beam modification of glasses (Mazzoldi, Paolo; Carnera, Alberto; Caccavale, F.; Granozzi, G.; Bertoncello, R.; Battaglin, Giancarlo; Boscolo-Boscoletto, A.; Polato, P.)V1513,182-197(1991)

Ion-ripple laser as an advanced coherent radiation source (Chen, Kuan-Ren; Dawson, John M.)V1552,185-196(1991)

Ku-band radiation in the UNM backward-wave oscillator experiment (Schamiloglu, Edl; Gahl, John M.; McCarthy, G.)V1407,242-248(1991)

Large-aperture (80-cm diameter) phase plates for beam smoothing on Nova (Woods, Bruce W.; Thomas, Ian M.; Henesian, Mark A.; Dixit, Sham N.; Powell, Howard T.)V1410,47-54(1991)

Large electron accelerators powered by intense relativistic electron beams (Friedman, Moshe; Serlin, Victor; Lau, Yue Y.; Krall, Jonathan)V1407,474-478(1991)

Large-scale production of a broadband antireflection coating on ophthalmic lenses with ion-assisted deposition (Andreani, F.; Luridiana, S.; Mao, Shu Z.)V1519,63-69(1991)

Laser action of Xe and Ne pumped by electron beam (Tarasenko, Victor F.)V1412,185-196(1991)

Laser beam diagnostics: a conference overview (Sadowski, Thomas J.)V1414,136-140(1991)

Laser beam propagation through inhomogeneous amplifying media (Martinez-Herrero, R.; Mejias, P. M.)V1397,619-622(1991)

Laser linac: nondiffractive beam and gas-loading effects (Bochove, Erik J.; Moore, Gerald T.; Scully, Marlan O.; Wodkiewicz, K.)V1497,338-347(1991)

Laser linac in vacuum: assessment of a high-energy particle accelerator (Moore, Gerald T.; Bochove, Erik J.; Scully, Marlan O.)V1497,328-337(1991)

Laser photocathode development for high-current electron source (Moustaizis, Stavros D.; Fotakis, Costas; Girardeau-Montaut, Jean-Pierre)V1552,50-56(1991)

Long-pulse electron gun for laser applications (Bayless, John R.; Burkhart, Craig P.)V1411,42-46(1991)

Los Alamos photoinjector-driven free-electron laser (O'Shea, Patrick G.)V1552,36-41(1991)

Magnetically tapered CARM (cyclotron autoresonance maser) for high power (Wang, Qinsong; McDermott, David B.; Luhmann, Neville C.)V1407,209-216(1991)

Magnetic field effects on plasma-filled backward-wave oscillators (Lin, Anthony T.)V1407,234-241(1991)

Matrix representation of multimode beam transformation (Alda, Javier; Porras, Miguel A.; Bernabeu, Eusebio)V1527,240-251(1991)

Measurement of synchrotron beam polarization (Singman, Leif V.; Davis, Brent A.; Holmberg, D. L.; Blake, Richard L.; Hockaday, Robert G.)V1548,80-92(1991)

Measurement of the gain in a XeF (C-A) laser pumped by a coaxial e-beam (Bastiaens, H. M.; Peters, Peter J.; Witteman, Wilhelmus J.)V1397,77-80(1991)

Misalignments of airborne laser beams due to mechanical vibrations (Freitas, Jose C.; Abreu, M. A.; Rodrigues, F. C.; Carvalho, Fernando D.)V1399,42-48(1991)

Mix-and-match lithography for half-micrometer technology (Flack, Warren W.; Dameron, David H.)V1465,164-172(1991)

Modular implementations of linearly constrained beamformers (Liu, Tsung-Ching; Van Veen, Barry D.)V1566,419-426(1991)

Multifrequency beamspace Root-MUSIC: an experimental evaluation (Zoltowski, Michael D.; Kautz, Gregory M.; Silverstein, Seth D.)V1566,452-463(1991)

Multimode laser beams behaviour through variable reflectivity mirrors (Porras, Miguel A.; Alda, Javier; Bernabeu, Eusebio)V1397,645-648(1991)

Multiphoton resonance ionization of molecules desorbed from surfaces by ion beams (Winograd, Nicholas; Hrubowchak, D. M.; Ervin, M. H.; Wood, M. C.)V1435,2-11(1991)

Multiple scattered electron-beam effect in electron-beam lithography (Saitou, Norio; Iwasaki, Teruo; Murai, Fumio)V1465,185-191(1991)

Mushroom-shaped gates defined by e-beam lithography down to 80-nm gate lengths and fabrication of pseudomorphic HEMTs with a dry-etched gate recess (Huelsmann, Axel; Kaufel, G.; Raynor, Brian; Koehler, Klaus; Schweizer, T.; Braunstein, Juergen; Schlechtweg, M.; Tasker, Paul J.; Jakobus, Theo F.)V1465,201-208(1991)

Nanometer scale focused ion beam vacuum lithography using an ultrathin oxide resist (Harriott, Lloyd R.; Temkin, Henryk; Chu, C. H.; Wang, Yuh-Lin; Hsieh, Y. F.; Hamm, Robert A.; Panish, Morton B.; Wade, H. H.)V1465,57-63(1991)

Negative deuterium ion thermal energy measurements in a volume ion source (Bacal, Marthe; Courteille, C.; Devynck, Pascal; Jones-King, Yolanda D.; Leroy, Renan; Stern, Raul A.)V1407,605-609(1991)

Negative energy cyclotron resonance maser (Lednum, Eugene E.; McDermott, David B.; Lin, Anthony T.; Luhmann, Neville C.)V1407,202-208(1991)

Negative ions from magnetically insulated diodes (Prohaska, Robert M.; Fisher, Amnon; Rostoker, Norman)V1407,598-604(1991)

Nondegenerate two-wave mixing in shaped microparticle suspensions (Pizzoferrato, R.; De Spirito, M.; Zammit, Ugo; Marinelli, M.; Rogovin, Dan N.; Scholl, James F.)V1409,192-201(1991)

Nonlinear bremsstrahlung in nonequilibrium relativistic beam-plasma systems (Brandt, Howard E.)V1407,326-353(1991)

Nonlinear wakefield generation and relativistic optical guiding of intense laser pulses in plasmas (Esarey, Eric; Sprangle, Phillip; Ting, Antonio C.)V1407,407-417(1991)

Novel modes in a-power GRIN (Ojeda-Castaneda, Jorge; Szwaykowski, P.)V1500,246-251(1991)

Operation of an L-band relativistic magnetron at 100 hz (Smith, Richard R.; Benford, James N.; Cooksey, N. J.; Aiello, Norm; Levine, Jerrold S.; Harteneck, Bruce D.)V1407,83-91(1991)

Optical methods for laser beams control (Mak, Arthur A.; Soms, Leonid N.)V1415,110-119(1991)

Optical quality of a combined aerodynamic window (Du, Keming; Franek, Joachim; Loosen, Peter; Zefferer, H.; Shen, Junquan)V1397,639-643(1991)

Optimization of discharge parameters of an e-beam sustained repetitively pulsed CO2 laser (Beth, Mark-Udo; Hall, Thomas; Mayerhofer, Wilhelm)V1397,577-580(1991)

Parametric simulation studies and injection phase locking of relativistic magnetrons (Chen, Chiping; Chan, Hei-Wai; Davidson, Ronald C.)V1407,105-112(1991)

Paraxial electron imaging system (Kegelman, Thomas D.)V1454,2-10(1991)

Passive optical fibre sensor based on Cerenkov effect (Wang, Yao-Cai; Shi, Yi-Wei; Jiang, Hong-Tao)V1572,32-37(1991)

Phase conjugation in LiF and NaF color center crystals (Basiev, Tasoltan T.; Zverev, Peter G.; Mirov, Sergey B.; Pal, Suranjan)V1500,65-71(1991)

Phase stability of a standing-wave free-electron laser (Sharp, William M.; Rangarajan, G.; Sessler, Andrew M.; Wurtele, Jonathan S.)V1407,535-545(1991)

Photomask fabrication utilizing a Philips/Cambridge vector scan e-beam system (McCutchen, William C.)V1496,97-106(1991)

Planar-grating klystron experiment (Xu, Yian-sun; Jackson, Jonathan A.; Price, Edwin P.; Walsh, John E.)V1407,648-652(1991)

Plasma betatron without gas breakdown (Ishizuka, Hiroshi; Yee, K.; Fisher, Amnon; Rostoker, Norman)V1407,442-455(1991)

Preparation of thin superconducting YBCO films by ion-beam mixing (Fan, Xiang J.; Pen, You G.; Guo, Huai X.; Li, Hong T.; Liu, Chang; Jiang, Chang Z.; Pen, Zhi L.)V1519,805-807(1991)

Production of x-rays by the interaction of charged particle beams with periodic structures and crystalline materials (Rule, Donald W.; Fiorito, Ralph B.; Piestrup, Melvin A.; Gary, Charles K.; Maruyama, Xavier K.)V1552,240-251(1991)

Progress toward steady-state high-efficiency vircators (Poulsen, Peter; Pincosy, Phillip A.; Morrison, Jasper J.)V1407,172-182(1991)

Propagation invariance of laser beam parameters through optical systems (Martinez-Herrero, R.; Mejias, P. M.; Hernandez Neira, Jose Luis; Sanchez, Miguel)V1397,627-630(1991)

Propagation of plasma beams across the magnetic field (Rahman, Hafiz-ur; Yur, Gung; White, R. S.; Wessel, Frank J.; Song, Joshua J.; Rostoker, Norman)V1407,589-597(1991)

Proximity effect correction on MEBES for 1x mask fabrication: lithography issues and tradeoffs at 0.25 micron (Muray, Andrew; Dean, Robert L.)V1496,171-179(1991)

Pulsed electron-beam testing of optical surfaces (Murray, Brian W.; Johnson, Edward A.)V1330,2-24(1991)

Pulsed-power considerations for electron-beam-pumped krypton-fluoride lasers for inertial confinement fusion applications (Rose, Evan A.; McDonald, Thomas E.; Rosocha, Louis A.; Harris, David B.; Sullivan, J. A.; Smith, I. D.)V1411,15-27(1991)

Quantum extension of Child-Langmuir law (Lau, Yue Y.; Chernin, David P.; Colombant, Denis G.; Ho, Ping-Tong)V1407,546-552(1991)

Radio frequency pulse compression experiments at SLAC (Farkas, Zoltan D.; Lavine, T. L.; Menegat, A.; Miller, Roger H.; Nantista, C.; Spalek, G.; Wilson, P. B.)V1407,502-511(1991)

Recent developments on the NRL Modified Betatron Accelerator (Golden, Jeffry; Len, Lek K.; Smith, Tab J.; Dialetis, Demos; Marsh, S. J.; Smith, Kevin; Mathew, Joseph; Loschialpo, Peter; Seto, Lloyd; Chang, Jeng-Hsien; Kapetanakos, Christos A.)V1407,418-429(1991)

Reduction of beam breakup growth by cavity cross-couplings in recirculating accelerators (Colombant, Denis G.; Lau, Yue Y.; Chernin, David P.)V1407,484-495(1991)

Relativistic klystron amplifier I: high-power operation (Friedman, Moshe; Serlin, Victor; Lau, Yue Y.; Krall, Jonathan)V1407,2-7(1991)

Relativistic klystron amplifier II: high-frequency operation (Serlin, Victor; Friedman, Moshe; Lau, Yue Y.; Krall, Jonathan)V1407,8-12(1991)

Relativistic klystron amplifier III: dynamical limiting currents, nonlinear beam loading, and conversion efficiency (Colombant, Denis G.; Lau, Yue Y.; Friedman, Moshe; Krall, Jonathan; Serlin, Victor)V1407,13-22(1991)

Relativistic klystron amplifier IV: simulation studies of a coaxial-geometry RKA (Krall, Jonathan; Friedman, Moshe; Lau, Yue Y.; Serlin, Victor)V1407,23-31(1991)

Relativistic klystron research for future linear colliders (Westenskow, Glen A.; Houck, Timothy L.; Ryne, Robert D.)V1407,496-501(1991)

Research on influence of substrate to crystal growth by ion beams (Yin, Dong; Guan, Wen X.; Sun, Shu Z.; Zhang, Zhao A.)V1519,164-166(1991)

Research on the temperature of thin film under ion beam bombarding (Zhao, Yun F.; Sun, Zhu Z.; Pang, Shi J.)V1519,411-414(1991)

Resist patterning for sub-quarter-micrometer device fabrications (Chiong, Kaolin G.; Hohn, Fritz J.)V1465,221-236(1991)

Resonant and nonresonant ionization in sputtered initiated laser ionization spectrometry (Havrilla, George J.; Nicholas, Mark; Bryan, Scott R.; Pruett, J. G.)V1435,12-18(1991)

Review of nondiffracting Bessel beams (LaPointe, Michael R.)V1527,258-276(1991)

Saturnus: the UCLA compact infrared free-electron laser project (Dodd, James W.; Aghamir, F.; Barletta, W. A.; Cline, David B.; Hartman, Steven C.; Katsouleas, Thomas C.; Kolonko, J.; Park, Sanghyun; Pellegrini, Claudio; Terrien, J. C.; Davis, J. G.; Joshi, Chan J.; Luhmann, Neville C.; McDermott, David B.; Ivanchenkov, S. N.; Lachin, Yu Y.; Varfolomeev, A. A.)V1407,467-473(1991)

Scanning beam switch experiment for intense rf power generation (Humphries, Stanley; Babcock, Steven; Wilson, J. M.; Adler, Richard A.)V1407,512-523(1991)

Shared aperture for two beams of different wavelength using the Talbot effect (Hector, Scott D.; Swanson, Gary J.)V1555,200-213(1991)

Short-Wavelength Radiation Sources (Sprangle, Phillip, ed.)V1552(1991)

Simple beam-propagation measurements on ion lasers (Guggenhiemer, Steven; Wright, David L.)V1414,12-20(1991)

Slow-motion acquisition of laser beam profiles after propagation through gun blast (Kay, Armin V.)V1486,8-16(1991)

Smooth diamond films by reactive ion-beam polishing (Bovard, Bertrand G.; Zhao, Tianji; Macleod, H. A.)V1534,216-222(1991)

Some issues on beam breakup in linear accelerators (Lau, Yue Y.; Colombant, Denis G.)V1407,479-483(1991)

Sophisticated masks (Pease, R. F.; Owen, Geraint; Browning, Raymond; Hsieh, Robert L.; Lee, Julienne Y.; Maluf, Nadim I.; Berglund, C. N.)V1496,234-238(1991)

Sound generation by laser radiation in air: application to restoration of energy distribution in laser beams (Gurvich, Alexander S.; Myakinin, Vladimir A.; Vorob'ev, Valerii V.; D'yakov, Alexander; Pokasov, Vladimir; Pryanichnikov, Victor)V1408,10-18(1991)

Spatial characterization of high-power multimode laser beams (Serna, Julio; Martinez-Herrero, R.; Mejias, P. M.)V1397,631-634(1991)

Status of the proof-of-concept experiment for the spiral line induction accelerator (Bailey, Vernon L.; Corcoran, Patrick; Edighoffer, J. A.; Fockler, J.; Lidestri, Joseph P.; Putnam, Sidney D.; Tiefenback, Michael G.)V1407,400-406(1991)

Superradiant Raman free-electron lasers (Tsui, King H.)V1407,281-284(1991)

System design considerations for a production-grade, ESR-based x-ray lithography beamline (Kovacs, Stephen; Melore, Dan; Cerrina, Franco; Cole, Richard K.)V1465,88-99(1991)

Tandem betatron accelerator (Keinigs, Rhon K.)V1407,456-466(1991)

Theory and simulation of the HARmonic amPlifier Free-Electron Laser (HARP/FEL) (Gregoire, Daniel J.; Harvey, Robin J.; Levush, Baruch)V1552,118-126(1991)

Theory of multimode interactions in cyclotron autoresonance maser amplifiers (Chen, Chiping; Wurtele, Jonathan S.)V1407,183-191(1991)

Twin traveling wave tube amplifiers driven by a single backward-wave oscillator (Butler, Jennifer M.; Wharton, Charles B.)V1407,57-66(1991)

Wet-developed, high-aspect-ratio resist patterns by 20-keV e-beam lithography (Weill, Andre P.; Amblard, Gilles R.; Lalanne, Frederic P.; Panabiere, Jean-Pierre)V1465,264-270(1991)

Whenever two beams interfere, one fringe equals one wave in the plane of interference, always (Williamson, Ray)V1527,252-257(1991)

Beamsplitters—see also holography; polarization

Fabrication of multiphase optical elements for weighted array spot generation (McKee, Paul; Wood, David; Dames, Mark P.; Dix, C.)V1461,17-23(1991)

Polarizing optics for the soft x-ray regime: whispering-gallery mirrors and multilayer beamsplitters (Braud, John P.)V1548,69-72(1991)

V-groove diffraction grating for use in an FUV spatial heterodyne interferometer (Cotton, Daniel M.; Bach, Bernhard W.; Bush, Brett C.; Chakrabarti, Supriya)V1549,313-318(1991)

Binary optics—see also diffractive optics; micro-optics; holographic optical elements; holography, computer-generated

Analysis and performance limits of diamond-turned diffractive lenses for the 3-5 and 8-12 micrometer regions (Riedl, Max J.; McCann, James T.)VCR38,153-163(1991)

Binary optic interconnects: design, fabrication and limits on implementation (Wong, Vincent V.; Swanson, Gary J.)V1544,123-133(1991)

Hartmann-Shack wavefront sensor using a binary optic lenslet array (Kwo, Deborah P.; Damas, George; Zmek, William P.; Haller, Mitch)V1544,66-74(1991)

High-speed binary optic microlens array in GaAs (Motamedi, M. E.; Southwell, William H.; Anderson, Robert J.; Hale, Leonard G.; Gunning, William J.; Holz, Michael)V1544,33-44(1991)

Light scattering from binary optics (Ricks, Douglas W.; Ajmera, Ramesh)V1555,89-100(1991)

Microreflective elements for integrated planar optics interconnects (Sheng, Yunlong; Delisle, Claude; Moreau, Louis; Song, Li; Lessard, Roger A.; Arsenault, Henri H.)V1559,222-228(1991)

Miniature and Micro-Optics: Fabrication and System Applications (Roychoudhuri, Chandrasekhar; Veldkamp, Wilfrid B., eds.)V1544(1991)

Off-axis spherical element telescope with binary optic corrector (Brown, Daniel M.; Kathman, Alan D.)V1555,114-127(1991)

Optimal phase-only correlation filters in colored scene noise (Vijaya Kumar, B. V. K.; Liang, Victor; Juday, Richard D.)V1555,138-145(1991)

Orthogonal cylindrical diffractive lens for parallel readout optical disk system (Urquhart, Kristopher S.; Marchand, Philippe J.; Lee, Sing H.; Esener, Sadik C.)V1555,214-223(1991)

Process error limitations on binary optics performance (Cox, James A.; Fritz, Bernard S.; Werner, Thomas R.)V1555,80-88(1991)

Silicon microlenses for enhanced optical coupling to silicon focal planes (Motamedi, M. E.; Griswold, Marsden P.; Knowlden, Robert E.)V1544,22-32(1991)

Simple design considerations for binary optical holographic elements (Franck, Jerome B.; Hodgkin, Van A.)V1555,63-70(1991)

Techniques for designing hybrid diffractive optical systems (Brown, Daniel M.; Pitalo, Stephen K.)V1527,73-84(1991)

Testing binary optics: accurate high-precision efficiency measurements of microlens arrays in the visible (Holz, Michael; Stern, Margaret B.; Medeiros, Shirley; Knowlden, Robert E.)V1544,75-89(1991)

Upper bound on the diffraction efficiency of phase-only array generators (Mait, Joseph N.)V1555,53-62(1991)

Biochemistry—see also biology; chemistry; medicine

Adaptation of coherent Raman methods for the investigation of biological samples in vivo (Lau, A.; Pfeiffer, M.; Werncke, W.)V1403,212-220(1991)

Advances in polarized fluorescence depletion measurement of cell membrane protein rotation (Barisas, B. G.; Rahman, N. A.; Londo, T. R.; Herman, J. R.; Roess, Debrah A.)V1432,52-63(1991)

Advances in structural studies of viruses by Raman spectroscopy (Towse, Stacy A.; Benevides, James M.; Thomas, George J.)V1403,6-14(1991)

Amorphous microcellular polytetrafluoroethylene foam film (Tang, Chong Z.)V1519,842-846(1991)

Analysis of neuropeptides using capillary zone electrophoresis with multichannel fluorescence detection (Sweedler, Jonathan V.; Shear, Jason B.; Fishman, Harvey A.; Zare, Richard N.; Scheller, Richard H.)V1439,37-46(1991)

Anisotropic polarizability and diffusion of protein in water solutions studied by laser light scattering (Petrova, G. P.; Petrusevich, Yu. M.; Borisov, B. A.)V1403,387-389(1991)

Antitumor drugs as photochemotherapeutic agents (Andreoni, Alessandra; Colasanti, Alberto; Kisslinger, Annamaria; Malatesta, Vincenzo; Mastrocinque, Michele; Roberti, Giuseppe)V1525,351-366(1991)

Background correction in multiharmonic Fourier transform fluorescence lifetime spectroscopy (Swift, Kerry M.; Mitchell, George W.)V1431,171-178(1991)

Biochemical measurement of bilirubin with an evanescent wave optical sensor (Poscio, Patrick; Depeursinge, Ch.; Emery, Y.; Parriaux, Olivier M.; Voirin, G.)V1510,112-117(1991)

Biomedical applications of laser photoionization (Xiong, Xiaoxiong; Moore, Larry J.; Fassett, John D.; O'Haver, Thomas C.)V1435,188-196(1991)

Biomolecular Spectroscopy II (Birge, Robert R.; Nafie, Laurence A., eds.)V1432(1991)

Cellular and extracellular effects of soft laser irradiation on an erythrocytes suspension (Skopinov, S. A.; Zakharov, S. D.; Volf, E. B.; Perov, S. N.; Panasenko, N. A.)V1403,676-679(1991)

Characterization of a vesicle distribution in equilibrium with larger aggregates by accurate static and dynamic laser light scattering measurements (Cantu, Laura; Corti, Mario; Lago, Paolo; Musolino, Mario)V1430,144-159(1991)

Characterization of fluorescence-labeled DNA by time-resolved fluorescence spectroscopy (Seidel, Claus; Rittinger, K.; Cortes, J.; Goody, R. S.; Koellner, Malte; Wolfrum, Juergen M.; Greulich, Karl O.)V1432,105-116(1991)

Characterization of motion of probes and networks in gels by laser light scattering under sample rotation (Kobayasi, Syoyu)V1403,296-305(1991)

Characterization of photobiophysical properties of sensitizers used in photodynamic therapy (Roeder, Beate; Naether, Dirk)V1525,377-384(1991)

Characterization of the fluorescence lifetimes of the ionic species found in aqueous solutions of hematoporphyrin IX as a function of pH (Nadeau, Pierre; Pottier, R.; Szabo, Arthur G.; Brault, Daniel; Vever-Bizet, C.)V1398,151-161(1991)

Chemical analysis of human urinary and renal calculi by Raman laser fiber-optics method (Hong, Nguyen T.; Phat, Darith; Plaza, Pascal; Daudon, Michel; Dao, Nguyen Q.)V1525,132-142(1991)

Chemical and Medical Sensors (Wolfbeis, Otto S., ed.)V1510(1991)

Chemical, Biochemical, and Environmental Fiber Sensors II (Lieberman, Robert A.; Wlodarczyk, Marek T., eds.)V1368(1991)

Clinical fluorescence diagnosis of human bladder carcinoma following low-dose photofrin injection (Baert, Luc; Berg, Roger; van Damme, B.; D'Hallewin, Mich A.; Johansson, Jonas; Svanberg, Katarina; Svanberg, Sune)V1525,385-390(1991)

Cluster model of protein molecule (Chikishev, A. Y.; Khurgin, Yu. I.; Romanovsky, Yuri M.; Shidlovskaya, E. G.)V1403,517-521(1991)

Coherent holographic Raman spectroscopy of molecules (Ivanov, Anatoliy A.; Koroteev, Nikolai I.; Fishman, A. I.)V1403,174-184(1991)

Conformational analysis and circular dichroism of bilirubin, the yellow pigment of jaundice (Lightner, David A.; Person, Richard; Peterson, Blake; Puzicha, Gisbert; Pu, Yu-Ming; Bojadziev, Stefan)V1432,2-13(1991)

Conformational analysis of organic molecules in liquids with polarization-sensitive coherent anti-Stokes Raman scattering spectroscopy (Ivanov, Anatoliy A.; Koroteev, Nikolai I.; Mironov, S. F.; Fishman, A. I.)V1403,243-245(1991)

Crystallo-optic diagnostics method of the soft laser-induced effects in biological fluids (Skopinov, S. A.; Yakovleva, S. V.)V1403,680-681(1991)

Cytochrome c at charged interfaces studied by resonance Raman and surface-enhanced resonance Raman spectroscopy (Hildebrandt, Peter)V1403,102-111(1991)

Damping in the models for molecular dynamics (Netrebko, N. V.; Romanovsky, Yuri M.; Shidlovskaya, E. G.; Tereshko, V. M.)V1403,512-516(1991)

Dehydrogenase enzyme/coenzyme/substrate interactions (Hester, R. E.; Austin, J. C.)V1403,15-21(1991)

Detection of DNA sequence symmetries using parallel micro-optical devices (Christens-Barry, William A.; Terry, David H.; Boone, Bradley G.)V1564,177-188(1991)

Determination of association parameters for the adsorption of mono- and divalent cations at the lipid membrane surface (Ermakov, Yu. A.; Cherny, V. V.)V1403,278-279(1991)

Development of laser-induced fluorescence detection to assay DNA damage (Sharma, Minoti; Freund, Harold G.)V1435,280-291(1991)

Diagnostics of functional state of blood by registration of light scattering intensity variations due to the reversible aggregation of red blood cells (Tukhvatulin, R. T.; Vaulin, P. P.)V1403,390-391(1991)

Differential Raman spectroscopic study of the interaction of nickel (II) cation with adenine nucleotides (Mojzes, Peter)V1403,167-171(1991)

Direct high-spatial-resolution SIMS (secondary ion mass spectrometry) imaging of labeled nucleosides in human chromosomes (Hallegot, Philippe; Girod, C.; LeBeau, M. M.; Levi-Setti, Riccardo)V1396,311-315(1991)

Direct measurement of vibrational energy relaxation in photoexcited deoxyhemoglobin using picosecond Raman spectroscopy (Hopkins, John B.; Xu, Xiaobing; Lingle, Robert; Zhu, Huiping; Yu, Soo-chang)V1432,221-226(1991)

Distribution of fluorescent probe molecules throughout the phospholipid membrane depth (Nemkovich, N. A.; Rubinov, A. N.; Savvidi, M. G.; Tomin, V. I.; Shcherbatska, Nina V.)V1403,578-581(1991)

DNA photoproducts formed using high-intensity 532 nm laser radiation (Kochevar, Irene E.; Hefetz, Yaron; Dunn, D. A.; Buckley, L. M.; Hillenkamp, Franz)V1403,756-763(1991)

Dynamics of components of biomembranes (Chorvat, Dusan; Shvec, Peter)V1403,659-666(1991)

Dynamics of nonlinear excitations in soft-quasi-one-dimensional molecular chains (Makeev, V. Y.; Poponin, Vladimir P.; Schzeglov, V. A.)V1403,522-527(1991)

Dynamics of ultrafast photoprocesses in Zn-octaethylporphyrin and Zn-octaethylchlorin pi-monoanions (Chirvony, V. S.; Sinyakov, G. N.; Gadonas, R.; Krasauskas, V.; Pelakauskas, A.)V1403,504-506(1991)

Dynamic vs static bending rigidities for DNA and M13 virus (Schurr, J. M.; Song, Lu; Kim, Ug-Sung)V1403,248-257(1991)

Effect of counterion distribution on the electrostatic component of the persistence length of flexible linear polyions (Klearman, Debbie; Schmitz, Kenneth S.)V1430,236-255(1991)

Effects of incorporation and functioning of integral proteins on the physical characteristics of lipid bilayers (Hianik, T.; Kavechansky, J.; Piknova, Barbora)V1402,93-96(1991)

Effects of photodynamic treatment on DNA (Oleinick, Nancy L.; Agarwal, Munna L.; Antunez, Antonio R.; Clay, Marian E.; Evans, Helen H.; Harvey, Ella Jo; Rerko, Ronald M.; Xue, Liang-yan)V1427,90-100(1991)

Efficacy of photodynamic killing with membrane associated and internalized photosensitizer molecules (Allison, Beth; Jiang, Frank N.; Levy, Julia G.)V1426,200-207(1991)

Elasticity of biomembranes studied by dynamic light scattering (Fujime, Satoru; Miyamoto, Shigeaki)V1403,306-315(1991)

Electrophoretic mobility and conformational changes of DNA in agarose gels (Chu, Benjamin; Wang, Zhulun)V1403,258-267(1991)

Electrophoretic mobility patterns of collagen following laser welding (Bass, Lawrence S.; Moazami, Nader; Pocsidio, Joanne O.; Oz, Mehmet C.; LoGerfo, Paul; Treat, Michael R.)V1422,123-127(1991)

Enantio-selective optrode for optical isomers of biologically active amines using a new lipophilic aromatic carrier (He, Huarui; Uray, Georg; Wolfbeis, Otto S.)V1510,95-103(1991)

Energy circulation effect between excited singlet- and triplet-state systems in biopigment molecules (Ketsle, G. A.; Letuta, S. N.)V1403,622-630(1991)

Evaluation of Nile Blue E chalcogen analogs as PDT agents (Foley, James W.; Cincotta, Louis; Cincotta, Anthony H.)V1426,208-215(1991)

Excimer formation and singlet-singlet energy transfer of organoluminophores in the premicellar and micellar-polyelectrolytes solutions (Bisenbaev, A. K.; Levshin, L. V.; Saletsky, A. M.)V1403,606-610(1991)

Excitation energy relaxation in model aggregates of photosynthetic pigments upon picosecond laser excitation (Chirvony, V. S.; Zenkevich, E. I.; Gadonas, R.; Krasauskas, V.; Pelakauskas, A.)V1403,638-640(1991)

Experimental basis of the nonlinear models of the internal DNA dynamics (Yakushevich, L. V.)V1403,507-508(1991)

Femtosecond processes in allophycocyanin trimers (Khoroshilov, E. V.; Kryukov, I. V.; Kryukov, P. G.; Sharkov, A. V.; Gillbro, T.)V1403,431-433(1991)

Femtosecond spectroscopy of acidified and neutral bacteriorhodopsin (Kobayashi, Takayoshi T.; Terauchi, Mamoru; Kouyama, Tsutomu; Yoshizawa, Masayuki; Taiji, Makoto)V1403,407-416(1991)

Fiber optic biosensors: the situation of the European market (Scheggi, Annamaria V.; Mignani, Anna G.)V1510,40-45(1991)

Fiber optic evanescent wave biosensor (Duveneck, G.; Ehrat, M.; Widmer, H. M.)V1510,138-145(1991)

Fiber optic measurement of intracellular pH in intact rat liver using pH-sensitive dyes (Felberbauer, Franz; Graf, Juerg)V1510,63-71(1991)

Flow injection analysis with bioluminescence-based fiber-optic biosensors (Blum, Loic J.; Gautier, Sabine; Coulet, Pierre R.)V1510,46-52(1991)

Fluorescence detected energy transport in selectively excited porphyrin dimers (Rempel, U.; von Maltzan, B.; von Borczykowski, C.)V1403,631-634(1991)

Fluorescence imaging in photodynamic therapy (MacRobert, Alexander J.; Phillips, David)V1439,79-87(1991)

Fluorescence line-narrowing spectroscopy in the study of chemical carcinogenesis (Jankowiak, Ryszard; Jeong, H.; Small, Gerald J.)V1435,203-213(1991)

Fluorescence spectrochronography of protein intramolecular dynamics (Chikishev, A. Y.; Ladokhin, Alexey S.; Shkurinov, A. P.)V1403,448-456(1991)

Four-photon spectroscopy of excited molecules: relaxation pathways and rates from high-excited states of 1,4-diphenylbutadiene molecules (Bogdanov, V. L.; Kulya, S. V.; Spiro, A. G.)V1403,470-474(1991)

Fractal dynamics of fluorescence energy transfer in biomembranes (Dewey, T. G.)V1432,64-75(1991)

Frequency-domain fluorescence spectroscopy: instrumentation and applications to the biosciences (Lakowicz, Joseph R.; Gryczynski, Ignacy; Malak, Henryk; Johnson, Michael L.; Laczko, Gabor; Wiczk, Wieslaw M.; Szmacinski, Henryk; Kusba, Jozef)V1435,142-160(1991)

Frequency spectra of erythrocyte membrane flickering measured by laser light scattering (Bek, A. M.; Kononenko, Vadim L.)V1403,384-386(1991)

Generation of free radicals in high-intensity laser photolysis of organic microcyclic compounds: time-resolved spectroscopy and EPR study (Angelov, D.; Gantchev, Ts.; Grabner, G.; Getoff, N.; Keskinova, E.; Shopova, Maria)V1403,572-574(1991)

High-power UV laser photolysis of nucleosides: final product analysis (Angelov, D.; Berger, M.; Cadet, J.; Ballini, Jean-Pierre; Keskinova, E.; Vigny, Paul)V1403,575-577(1991)

Holographic nonlinear Raman spectroscopy of large molecules of biological importance (Ivanov, Anatoliy A.; Koroteev, Nikolai I.; Fishman, A. I.)V1429,132-144(1991); V1432,141-153(1991)

Imaging spectroscope with an optical recombination system (Li, Wenchong; Ma, Chunhua)V1428,242-248(1991)

Importance of pulsed laser intensity in porphyrin-sensitized NADH photo-oxidation (Kirveliene, V.; Rotomskis, Richardas; Juodka, B.; Piskarskas, Algis S.)V1403,582-584(1991)

Influence of hematoporphyrin on the mechanical properties of the lipid bilayer membranes (Hianik, T.; Masarikova, D.; Zhorina, L. V.; Poroshina, Marina Y.; Chernyaeva, E. B.)V1402,85-88(1991)

Integrated optic device for biochemical sensing (Boiarski, Anthony A.; Ridgway, Richard W.; Miller, Larry S.; Bhullar, B. S.)V1368,264-272(1991)

Interaction of phthalocyanine photodynamic treatment with ionophores and lysosomotrophic agents (Oleinick, Nancy L.; Varnes, Marie E.; Clay, Marian E.; Menegay, Harry J.)V1426,235-243(1991)

Inter- and intramolecular processes in metalloporphyrins: study by transient absorption and resonance Raman and coherent anti-Stokes Raman scattering (Apanasevich, P. A.; Chirvony, V. S.; Kruglik, S. G.; Kvach, V. V.; Orlovich, V. A.)V1403,195-211(1991)

Intercalation between antitumor anthracyclines and DNA as probed by resonance and surface-enhanced Raman spectroscopy (Smulevich, G.; Mantini, A. R.; Casu, M.; Marzocchi, M. P.)V1403,125-127(1991)

Intermolecular contacts within sickle hemoglobin fibers (Watowich, Stanley J.; Gross, Leon J.; Josephs, Robert)V1396,316-323(1991)

Interpretation of the resonance Raman spectra of hemoproteins and model compounds (Solovyov, K. N.; Gladkov, L. L.)V1403,132-133(1991)

Intracellular location, picosecond kinetics, and light-induced reactions of photosensitizing porphyrins (Schneckenburger, Herbert; Seidlitz, Harold K.; Wessels, Jurina; Rueck, Angelika C.)V1403,646-652(1991)

Intracellular uptake and ultrastructural phototoxic effects of sulfonated chlor-aluminum phthalocyanine on bladder tumor cells in vitro (Miller, Kurt; Reich, Ella; Grau, T.)V1426,378-383(1991)

Intramolecular processes of excitation energy redistribution in metalloporphyrins: examination by transient absorption and resonance Raman and coherent anti-Stokes Raman scattering spectroscopies (Apanasevich, P. A.; Chyrvony, V. S.; Kruglik, S. G.; Kvach, V. V.; Orlovich, V. A.)V1403,240-242(1991)

Investigation of single biological cells and chromosomes by confocal Raman microspectroscopy (Puppels, G. J.; Otto, C.; de Mul, F. F.; Greve, Jan)V1403,146-146(1991)

Investigation of the photodynamic properties of some chlorophyll a derivatives: the effect of doxorubicin on the chlorine e6 photosensitized death of Ehrlich carcinoma cells (Chekulayev, V.; Shevchuk, Igor; Kahru, A.; Mihkelsoo, V. T.; Kallikorm, A. P.)V1426,367-377(1991)

Investigation of vesicle-capsular plague antigen complex formation by elastic laser radiation scattering (Guseva, N. P.; Maximova, I. L.; Romanov, S. V.; Shubochkin, L. P.; Tatarintsev, S. V.)V1403,332-334(1991)

In vivo energy transfer studies in photosynthetic systems by subpicosecond timing (Shreve, A. P.; Trautman, Jay K.; Owens, T. G.; Frank, Harry A.; Albricht, Andreas C.)V1403,394-399(1991)

Laser Applications in Life Sciences (Akhmanov, Sergei A.; Poroshina, Marina Y., eds.)V1403(1991)

Laser-based flow cytometric analysis of genotoxicity of humans exposed to ionizing radiation during the Chernobyl accident (Jensen, Ronald H.; Bigbee, William L.; Langlois, Richard G.; Grant, Stephen G.; Pleshanov, Pavel G.; Chirkov, Andre A.; Pilinskaya, Maria A.)V1403,372-380(1991)

Laser Doppler diagnostics of vorticity and phenomenological description of the flows of dilute polymer solutions in model tubes (Fedoseeva, E. V.; Polyakova, M. S.)V1403,355-358(1991)

Laser Doppler measurements of ameboid cytoplasmic streaming and problems of mathematical modeling of intracellular hydrodynamics (Priezzhev, Alexander V.; Proskurin, S. G.; Romanovsky, Yuri M.)V1402,107-113(1991)

Laser-induced luminescence of singlet molecular oxygen: generation by drugs and pigments of biological importance (Egorov, S. Y.; Krasnovsky, A. A.)V1403,611-621(1991)

Laser investigation of molecular dynamic processes of the relaxational spectra formation (Neporent, B. S.)V1403,600-605(1991)

Laser light scattering studies of biological gels (Burne, P. M.; Sellen, D. B.)V1403,288-295(1991)

Laser microphotolysis of biological objects with the application of UV solid state lasers (Arutunian, A. H.; Hovanessian, V. A.; Sarkissian, K. A.)V1402,102-106(1991)

Laser nephelometry of erythrocytes in shear flows (Firsov, N. N.; Priezzhev, Alexander V.; Stepanian, A. S.)V1403,350-354(1991)

Laser neural tissue interactions using bilayer membrane models (VanderMeulen, David L.; Khoka, Mustafa; Spears, Kenneth G.)V1428,84-90(1991)

Laser spectroscopy of carotenoids in plant bio-objects (Kozlova, T. G.; Lobacheva, M. I.; Pravdin, A. B.; Romakina, M. Y.; Sinichkin, Yury P.; Tuchin, Valery V.)V1403,159-160(1991)

Laser spectroscopy of proton dynamics in hydrogen bonds (Hochstrasser, Robin M.; Oppenlaender, A.; Pierre, M.; Rambaud, C.; Silbey, R.; Skinner, J. L.; Trommsdorff, H. P.; Vial, J.-C.)V1403,221-229(1991)

Lateral mobility of biological membrane components (Chorvat, Dusan; Shvec, Peter; Kvasnichka, P.; Shipocz, Tibor; Jarkovska, B.)V1402,89-92(1991)

LEAF: a fiber-optic fluorometer for field measurement of chlorophyll fluorescence (Mazzinghi, Piero)V1510,187-194(1991)

Localization of the active site of an enzyme, bacterial luciferase, using two-quantum affinity modification (Benimetskaya, L. Z.; Gitelzon, I. I.; Kozionov, Andrew L.; Novozhilov, S. Y.; Petushkov, V. N.; Rodionova, N. S.; Stockman, M. I.)V1525,242-245(1991)

Manipulation of single-DNA molecules and measurements of their elastic properties under an optical microscope (Bustamante, Carlos; Finzi, Laura; Sebring, Page E.; Smith, Steven B.)V1435,179-187(1991)

Measurement of biological tissue metabolism using phase modulation spectroscopic technology (Weng, Jian; Zhang, M. Z.; Simons, K.; Chance, Britton)V1431,161-170(1991)

Measurement of bleeding sap flow velocity in xylem bundle of herbs by laser probing (Romanovsky, Yuri M.; Stepanian, A. S.; Shogenov, Yu. H.)V1403,359-362(1991)

Method of laser fluorescence microphotolysis (Shvec, Peter; Kvasnichka, P.; Shipocz, Tibor; Chorvat, Dusan)V1402,78-81(1991)

Microwave-optical study of an As(III) derivative of Eco RI methylase (Maki, August H.; Tsao, Desiree H.)V1432,119-128(1991)

Molecular aggregates of quinuclidine and chlorophyll a (Korppi-Tommola, Jouko E.; Hakkarainen, Aulis; Helenius, Vesa F.)V1403,457-465(1991)

Molecular dynamics of stilbene molecule under laser excitation (Vachev, V. D.; Zadkov, Victor N.)V1403,487-496(1991)

Molecular spectroscopy of biological molecules: Raman, NMR, and CD study of monophosphate dinucleosides at different degrees of protonation (Bertoluzza, Alessandro; Fagnano, C.; Morelli, M. A.; Tosi, M. R.; Tugnoli, V.)V1403,150-152(1991)

Multiple quasi-elastic circular-polarized light scattering by brownian moving aspherical particles (Korolevich, A. N.; Khairullina, A. Y.)V1403,364-371(1991)

Naphthalocyanines relevant to the search for second-generation PDT sensitizers (Sounik, James R.; Rihter, Boris D.; Ford, William E.; Rodgers, Michael A.; Kenney, Malcolm E.)V1426,340-349(1991)

New sensitizers for PDT (Morgan, Alan R.; Garbo, Greta M.; Krivak, T.; Mastroianni, Marta; Petousis, Nikolaos H.; St Clair, T.; Weisenberger, M.; van Lier, Johan E.)V1426,350-355(1991)

Nile Blue derivatives as lysosomotropic photosensitizers (Lin, Chi-Wei; Shulok, Janine R.; Kirley, S. D.; Cincotta, Louis; Foley, James W.)V1426,216-227(1991)

Nonlinear optical properties of bacteriorhodopsin: assignment of the third-order polarizability based on two-photon absorption spectroscopy (Birge, Robert R.; Masthay, M. B.; Stuart, Jeffrey A.; Tallent, Jack R.; Zhang, Chian-Fan)V1432,129-140(1991)

Nonlinear properties of oriented purple membrane films derived from second-harmonic generation under picosecond excitation: prospect of electro-optical measurements of ultrafast photoelectric respon (Sharkov, A. V.; Gillbro, T.)V1403,434-438(1991)

Nonresonant background suppression in coherent anti-Stokes Raman scattering spectra of dissolved molecules (Lucassen, G. W.; Scholten, T. A.; de Boey, W. P.; de Mul, F. F.; Greve, Jan)V1403,185-194(1991)

Nucleic acid base specific quenching of coumarin-120-derivative in nucleotid-conjugates-photoinduced electron transfer (Seidel, Claus)V1432,91-104(1991)

Optical Kerr effect in chain carbonyl compounds (Blaszczak, Zdzislaw)V1403,509-511(1991)

Optical correlation studies of biological objects (Angelsky, Oleg V.; Maksimyak, Peter P.)V1403,667-673(1991)

Optical imaging of cortical activity in the living brain (Ratzlaff, Eugene H.; Grinvald, Amiran)V1439,88-94(1991)

Optical Methods for Tumor Treatment and Early Diagnosis: Mechanisms and Techniques (Dougherty, Thomas J., ed.)V1426(1991)

Optical Methods for Ultrasensitive Detection and Analysis: Techniques and Applications (Fearey, Bryan L., ed.)V1435(1991)

Optical quantification of sodium, potassium, and calcium ions in diluted human plasma based on ion-selective liquid membranes (Spichiger, Ursula E.; Seiler, Kurt; Wang, Kemin; Suter, Gaby; Morf, Werner E.; Simon, Wilhelm)V1510,118-130(1991)

Optical sensors for process monitoring in biotechnology (Ploetz, F.; Schelp, C.; Anders, K.; Eberhardt, F.; Scheper, Thomas-Helmut; Bueckmann, F.)V1510,224-230(1991)

Pericyclic photochemical ring-opening reactions are complete in picoseconds: a time-resolved UV resonance Raman study (Reid, Philip J.; Doig, Stephen J.; Mathies, Richard A.)V1432,172-183(1991)

Pharmacokinetics of Photofrin II distribution in man (Kessel, David; Nseyo, Unyime O.; Schulz, Veronique; Sykes, Elizabeth)V1426,180-187(1991)

Photon correlation spectroscopic studies of filamentous actin networks (Newman, Jay E.; San Biagio, Pier L.; Schick, Kenneth L.)V1430,89-108(1991)

Photon correlation spectroscopy of chromaffin granules lysis (Ermakov, Yu. A.; Engel, J.; Donath, E.)V1403,338-339(1991)

Photoproduct formation of endogenous protoporphyrin and its photodynamic activity (Koenig, Karsten; Schneckenburger, Herbert; Rueck, Angelika C.; Auchter, S.)V1525,412-419(1991)

Photosensitive liposomes as potential drug delivery vehicles for photodynamic therapy (Morgan, Chris G.; Mitchell, A. C.; Chowdhary, R. K.)V1525,391-396(1991)

Photosensitization is required for antiretroviral activity of hypericin (Carpenter, Susan; Tossberg, John; Kraus, George A.)V1426,228-234(1991)

Phthalocyanines as phototherapeutic agents for tumors (Jori, Guilio; Reddi, Elena; Biolo, Roberta; Polo, Laura; Valduga, Giuliana)V1525,367-376(1991)

Picosecond absorption and circular dichroism studies of proteins (Simon, John D.; Xie, Xiaoliang; Dunn, Robert C.)V1432,211-220(1991)

Picosecond kinetics and Sn S1 absorption spectra of retinoids and carotenoids (Bondarev, S. L.; Tikhomirov, S. A.; Bachilo, S. M.)V1403,497-499(1991)

Picosecond laser cross: linking histones to DNA in chromatin: implication in studying histone/DNA interactions (Angelov, D.; Dimitrov, S.; Keskinova, E.; Pashev, I.; Russanova, V.; Stefanovsky, Yu.)V1403,230-239(1991)

Picosecond orientational dynamics of complex molecules studied by incoherent light three-wave mixing (Apanasevich, P. A.; Kozich, V. P.; Vodchitz, A. I.; Kontsevoy, B. L.)V1403,475-477(1991)

Picosecond reaction dynamics in photosynthetic and proton pumping systems: picosecond time-resolved Raman spectroscopy of electronic and vibrationally excited states (Atkinson, George H.)V1403,50-58(1991)

Picosecond time-resolved fluorescence spectroscopy of phytochrome and stentorin (Song, Pill-Soon)V1403,590-599(1991)

Picosecond time-resolved resonance Raman spectroscopy of bacteriorhodopsin: structure and kinetics of the J, K, and KL intermediates (Doig, Stephen J.; Reid, Philip J.; Mathies, Richard A.)V1432,184-196(1991)

Planar waveguide optical immunosensors (Choquette, Steven J.; Locascio-Brown, Laurie; Durst, Richard A.)V1368,258-263(1991)

Polarimetric optical fiber sensor for biochemical measurements (Heideman, Rene; Blikman, Albert; Koster, Rients; Kooyman, Rob P.; Greve, Jan)V1510,131-137(1991)

Polynuclear membranes as a substrate for obtaining surface-enhanced Raman scattering films (Oleynikov, V. A.; Sokolov, K. V.; Hodorchenko, P. V.; Nabiev, I. R.)V1403,164-166(1991)

Possibility of resolution of internal macromolecular relaxation by dynamic light scattering (Timchenko, A. A.; Griko, N. B.; Serdyuk, I. N.)V1403,340-343(1991)

Provitamin D photoisomerization kinetics upon picosecond laser irradiation: role of previtamin conformational nonequilibrium (Terenetskaya, I. P.; Repeyev, Yu A.)V1403,500-503(1991)

Quasi-elastic light scattering spectroscopy of single biological cells under a microscope (Tanaka, Toyoichi; Nishio, Izumi; Peetermans, Joyce; Gorti, Sridhar)V1403,280-287(1991)

Radiolabeled red blood cells for the direct measurement of the blood flow kinetics in experimental tumors after photodynamic therapy (Paquette, Benoit; Rousseau, Jacques; Ouellet, Rene; van Lier, Johan E.)V1426,362-366(1991)

Raman and FT-IR characterization of biologically relevant Langmuir-Blodgett films (Katayama, Norihisa; Fukui, Masahiko; Ozaki, Yukihiro; Araki, Toshinari; Yokoi, Seiichi; Iriyama, Keiji)V1403,147-149(1991)

Raman and multichannel Raman spectroscopy of biological systems (Bertoluzza, Alessandro; Caramazza, R.; Fagnano, C.)V1403,40-49(1991)

Raman spectroscopy of biological molecules: uncharged phospholipid/polyamine interactions in the presence of bivalent cations (Bertoluzza, Alessandro; Bonora, S.; Fini, G.; Morelli, M. A.)V1403,153-155(1991)

Rapid sequencing of DNA based on single-molecule detection (Soper, Steven A.; Davis, Lloyd M.; Fairfield, Frederick R.; Hammond, Mark L.; Harger, Carol A.; Jett, James H.; Keller, Richard A.; Morrone, Barbara L.; Martin, John C.; Nutter, Harvey L.; Shera, E. B.; Simpson, Daniel J.)V1435,168-178(1991)

Recent developments by femtosecond spectroscopy in biological ultrafast free radical reactions (Gauduel, Yann; Pommeret, Stanislas; Yamada, Noelle; Antonetti, Andre)V1403,417-426(1991)

Resonance Raman scattering from the primary electron donor in photosynthetic reaction centers from Rhodobacter sphaeroides (Bocian, David F.)V1432,166-171(1991)

Resonance Raman spectra of hematoporphyrin derivative (Golubeva, N. G.; Wang, Litszin)V1403,134-138(1991)

Resonance Raman spectra of transient species of a respiration enzyme detected with an artificial cardiovascular system and Raman/absorption simultaneous measurement system (Kitagawa, Teizo; Ogura, Takashi)V1403,563-571(1991)

Resonance Raman studies of photosynthetic membrane proteins (Lutz, Marc; Mattioli, Tony; Moenne-Loccoz, Pierre; Zhou, Qing; Robert, Bruno)V1403,59-65(1991)

Resonance Raman studies of the peptide bond: implications for the geometry of the electronic-excited state and the nature of the vibronic linewidth (Hudson, Bruce S.)V1403,27-36(1991)

Resonance Raman study of Zn-porphyrin pi-anions (Gurinovich, G. P.; Kruglik, S. G.; Kvach, V. V.; Terekhov, S. N.)V1403,139-141(1991)

Role of counterion size and distribution on the electrostatic component to the persistence length of wormlike polyions (Schmitz, Kenneth S.)V1430,216-235(1991)

Rotational diffusion of receptors for epidermal growth factor measured by time-resolved phosphorescence depolarization (Zidovetzki, Raphael; Johnson, David A.; Arndt-Jovin, Donna J.; Jovin, Thomas M.)V1432,76-81(1991)

Sensitive fiber-optic immunoassay (Walczak, Irene M.; Love, Walter F.; Slovacek, Rudolf E.)V1420,2-12(1991)

Sequence dependence of the length of the B to Z junctions in DNA (Peticolas, Warner L.; Dai, Z.; Thomas, G. A.)V1403,22-26(1991)

Simulation of spin label behavior on a model surface (Balabaev, N. K.; Lemak, A. S.; Fushman, D. A.; Mironova, Yu. V.)V1402,53-69(1991)

Single-color laser-induced fluorescence detection and capillary gel electrophoresis for DNA sequencing (Chen, Da Yong; Swerdlow, H.; Harke, H.; Zhang, Jian Z.; Dovichi, Norman J.)V1435,161-167(1991)

Solution conformation of biomolecules from infrared vibrational circular dichroism spectroscopy (Diem, Max)V1432,28-36(1991)

Spectral properties of the polaron model of a protein (Balabaev, N. K.; Lakhno, V. D.)V1403,478-486(1991)

Spectroscopic properties of the potentiometric probe merocyanine-540 in solutions and liposomes (Ehrenberg, Benjamin; Pevzner, Eliyahu)V1432,154-163(1991)

Spectroscopic studies of second-generation sensitizers and their photochemical reactions in liposomes and cells (Ehrenberg, Benjamin; Gross, Eitan; Lavi, Adina; Johnson, Fred M.; Malik, Zvi)V1426,244-251(1991)

Spectroscopy of proteins on surfaces: implications for protein orientation and protein/protein interactions (Cotton, Therese M.; Rospendowski, Bernard; Schlegel, Vicki; Uphaus, Robert A.; Wang, Danli L.; Eng, Lars H.; Stankovich, Marion T.)V1403,93-101(1991)

Statistical signal processing in Raman spectroscopy of biological samples (Praus, Petr; Stepanek, Josef)V1403,76-84(1991)

Structure and dynamics of the active site of peroxidases as revealed by resonance Raman spectroscopy (Smulevich, G.; English, A. M.; Spiro, T. G.)V1403,440-447(1991)

Studies of the excited states of biological systems using UV-excited resonance Raman and picosecond transient Raman spectroscopy (Gustafson, Terry L.; Iwata, Koichi; Weaver, William L.; Huston, Lisa A.; Benson, Ronda L.)V1403,545-554(1991)

Studies of yeast cell oxygenation and energetics by laser fluorometry of reduced nicotinamide adenine dinucleotide (Pan, Fu-shih; Chen, Stephen; Mintzer, Robert; Chen, Chin-Tu; Schumacker, Paul)V1396,5-8(1991)

Study of aggregation phenomenon of hematoporphyrin derivative by laser microphotolysis (Arutunian, A. H.; Hovanessian, V. A.; Sarkissian, K. A.)V1403,585-587(1991)

Study of biological objects in the reflected light with the help of an analogous fiber optic biophotometer (Gaiduk, Mark I.; Grigoryants, V. V.; Chernousova, I. V.; Menenkov, V. D.)V1403,674-675(1991)

Subpicosecond electron transfer in reaction centers of photosynthetic bacteria (Shuvalov, V. A.; Ganago, A. O.; Shkuropatov, A. Y.; Klevanik, A. V.)V1403,400-406(1991)

Surface-enhanced Raman spectroscopy in the structural studies of biomolecules: the state of the art (Nabiev, I. R.; Sokolov, K. V.; Efremov, R. G.; Chumanov, G. D.)V1403,85-92(1991)

Surface-enhanced Raman spectroscopy of adenosine and 5'AMP: evolution in time (Sanches-Cortes, S.; Garcia-Ramos, J. V.)V1403,142-145(1991)

Surface-enhanced hyper-Raman and near-IR FT-Raman studies of biomolecules (Yu, Nai-Teng; Nie, Shuming)V1403,112-124(1991)

Surface-enhanced resonance hyper-Raman spectra of bacteriorhodopsin adsorbed on silver colloids (Baranov, A. V.; Nabiev, I. R.)V1403,128-131(1991)

Syntheses of porphyrin and chlorin dimers for photodynamic therapy (Pandey, Ravindra K.; Vicente, M. G.; Shiau, Fuu-Yau; Dougherty, Thomas J.; Smith, Kevin M.)V1426,356-361(1991)

Synthetic approaches to long-wavelength photosensitizers for photodynamic therapy (Shiau, Fuu-Yau; Pandey, Ravindra K.; Dougherty, Thomas J.; Smith, Kevin M.)V1426,330-339(1991)

Theory of non-Condon femtosecond quantum beats in electronic transitions of dye molecules (Mazurenko, Yuri T.; Smirnov, V. V.)V1403,466-469(1991)

Three-Dimensional Bioimaging Systems and Lasers in the Neurosciences (Boggan, James E.; Cerullo, Leonard J.; Smith, Louis C., eds.)V1428(1991)

Time-resolved infrared studies of the dynamics of ligand binding to cytochrome c oxidase (Dyer, R. B.; Peterson, Kristen A.; Stoutland, Page O.; Einarsdottir, Oloef; Woodruff, William H.)V1432,197-204(1991)

Time-resolved polarization luminescence spectroscopy of hematoporphyrin in liposomes (Chernyaeva, E. B.; Golubeva, N. A.; Koroteev, Nikolai I.; Lobanov, O. V.; Vardanyan, A. G.)V1402,7-10(1991)

Time resolution of events in an enzyme's active site at 4 K and 300 K using resonance Raman spectroscopy (Carey, Paul R.; Kim, Munsok; Tonge, Peter J.)V1403,37-39(1991)

Transient and persistent hole-burning of photosystem II preparations (Hayes, John M.; Tang, D.; Jankowiak, Ryszard; Small, Gerald J.)V1435,258-266(1991)

Transitions in model membranes (Earnshaw, J. C.; Winch, P. J.)V1403,316-325(1991)

Translational dynamics of immune system components by fluorescence recovery after photobleaching (Shvec, Peter; Miklovicova, J.; Hunakova, L.; Chorvath, B.)V1403,635-637(1991)

Transverse concentrational profiles in dilute erythrocytes suspension flows in narrow channels measured by integral Doppler anemometer (Kononenko, Vadim L.; Shimkus, J. K.)V1403,381-383(1991)

Tripyrroledimethine-derived ("texaphyrin"-type) macrocycles: potential photosensitizers which absorb in the far-red spectral region (Sessler, Jonathan L.; Hemmi, Gregory; Maiya, Bhaskar G.; Harriman, Anthony; Judy, Millard M.; Boriak, Richard; Matthews, James L.; Ehrenberg, Benjamin; Malik, Zvi; Nitzan, Yeshayahu; Rueck, Angelika C.)V1426,318-329(1991)

Two-photon excitation in laser scanning fluorescence microscopy (Strickler, James H.; Webb, Watt W.)V1398,107-118(1991)

Ultrafast and not-so-fast dynamics of cytochrome oxidase: the ligand shuttle and its possible functional significance (Woodruff, William H.; Dyer, R. B.; Einarsdottir, Oloef; Peterson, Kristen A.; Stoutland, Page O.; Bagley, K. A.; Palmer, Graham; Schoonover, J. R.; Kliger, David S.; Goldbeck, Robert A.; Dawes, T. D.; Martin, Jean-Louis; Lambry, J.-C.; Atherton, Stephen J.; Hubig, Stefan M.)V1432,205-210(1991)

Use of fluorescence spectroscopy to elucidate structural features of the nicotinic acetylcholine receptor (Johnson, David A.; Valenzuela, C. F.)V1432,82-90(1991)

USSR-CSFR Joint Seminar on Nonlinear Optics in Control, Diagnostics, and Modeling of Biophysical Processes (Akhmanov, Sergei A.; Zadkov, Victor N., eds.)V1402(1991)

UV resonance Raman spectroscopic study of Trp residues in a hydrophobic environment (Efremov, R. G.; Feofanov, A. V.; Nabiev, I. R.)V1403,161-163(1991)

Vibrational Raman optical activity of biological molecules (Barron, L. D.; Gargaro, A. R.; Hecht, L.; Wen, Z. Q.; Hug, W.)V1403,66-75(1991)

Vibrational spectroscopy of biological molecules: halocompound/nucleic acid component interactions (Bottura, Giorgio; Filippetti, P.; Tinti, A.)V1403,156-158(1991)

Visualization of electron transfer interactions of membrane proteins (Kawato, Suguru)V1429,127-131(1991)

Zero-angle scattering of light in oriented organic liquids: classical and quantum states for both linear and nonlinear scattering (Arakelian, Sergei M.; Chilingarian, Yu. S.)V1403,326-331(1991)

Biology—see also biochemistry; medicine

Background correction in multiharmonic Fourier transform fluorescence lifetime spectroscopy (Swift, Kerry M.; Mitchell, George W.)V1431,171-178(1991)

Behavioral observations in thermal imaging of the big brown bat: Eptesicus fuscus (Kirkwood, James J.; Cartwright, Anne)V1467,369-371(1991)

Biological basis for space-variant sensor design I: parameters of monkey and human spatial vision (Rojer, Alan S.; Schwartz, Eric L.)V1382,132-144(1991)

Biomechanical research of joints: IV. the biohinge of primates (Zhang, Renxiang; Yu, Jie; Lan, Zu-yun; Qu, Wen-ji; Zhang, Hong-zi; Zhang, Kui; Zhang, Liang)V1380,116-121(1991)

Body shape changes in the elderly and the influence of density assumptions on segment inertia parameters (Jensen, Robert K.; Fletcher, P.; Abraham, C.)V1380,124-136(1991)

Circular intensity differential scattering measurements in the soft x-ray region of the spectrum (~16 EV to 500 EV) (Maestre, Marcos F.; Bustamante, Carlos; Snyder, Patricia A.; Rowe, Ednor M.; Hansen, Roger W.)V1548,179-187(1991)

Confocal microscopy for the biological and material sciences: principle, applications, limitations (Brakenhoff, G. J.; van der Voort, H. T.; Visscher, Koen)V1439,121-127(1991)

Degradation of cholesterol crystals in macrophages: the role of phospholipids (Koren, Eugen; Koscec, Mirna; Fugate, Robert D.)V1428,214-223(1991)

Design of eye movement monitoring system for practical environment (Nakamura, Hiroyuki; Kobayashi, Hitoshi; Taya, Katsuo; Ishigami, Shigenobu)V1456,226-238(1991)

Determining range information from self-motion: the template model (Sobey, Peter J.)V1382,123-131(1991)

Development of a synchrotron CCD-based area detector for structural biology (Kalata, Kenneth; Phillips, Walter C.; Stanton, Martin J.; Li, Youli)V1345,270-280(1991)

Focal-plane architectures and signal processing (Jayadev, T. S.)V1541,163-166(1991)

Fusion of human vision system with mid-range IR image processing displays (Forsyth, William B.; Lewis, H. G.)V1472,18-25(1991)

Hearing preservation using CO2 laser for acoustic nerve tumors (Grutsch, James; Heiferman, Kenneth S.; Cerullo, Leonard J.)V1428,136-145(1991)

Instrumentation for simultaneous kinetic imaging of multiple fluorophores in single living cells (Morris, Stephen J.; Beatty, Diane M.; Welling, Larry W.; Wiegmann, Thomas B.)V1428,148-158(1991)

Investigation on phase biological micro-objects with a holographic interferometric microscope on the basis of the photorefractive Bi12TiO20 crystal (Tontchev, Dimitar A.; Zhivkova, Svetla; Miteva, Margarita G.; Grigoriev, Ivo D.; Ivanov, I.)V1429,76-80(1991)

Laser/light tissue interaction: on the mechanism of optical breakdown (Siomos, Konstadinos)V1525,154-161(1991)

Measurement of light scattering from cells using an inverted infrared optical trap (Wright, William H.; Sonek, Gregory J.; Numajiri, Yasuyuki; Berns, Michael W.)V1427,279-287(1991)

Measurements of teeth using the Reflex Microscope (Teaford, Mark F.)V1380,33-44(1991)

Medical prosthetic applications of growth simulations in four-dimensional facial morphology (Sadler, Lewis L.; Chen, Xiaoming; Fyler, Ann)V1380,137-146(1991)

Observation of living cells by x-ray microscopy with a laser-plasma x-ray source (Tomie, Toshihisa; Shimizu, Hazime; Majima, T.; Yamada, Mitsuo; Kanayama, Toshihiko; Yano, M.; Kondo, H.)V1552,254-263(1991)

Optical property measurements in turbid media using frequency-domain photon migration (Tromberg, Bruce J.; Svaasand, Lars O.; Tsay, Tsong-Tseh; Haskell, Richard C.; Berns, Michael W.)V1525,52-58(1991)

Packing geometry of human cone photoreceptors: variations with eccentricity and evidence for local anisotropy (Sloan, Kenneth R.; Curcio, Christine A.)V1453,124-133(1991)

Pattern recognition, attention, and information bottlenecks in the primate visual system (Van Essen, David; Olshausen, B.; Anderson, Clifford H.; Gallant, J. T. L.)V1473,17-28(1991)

Photoacoustic absorption spectrum of some rat and bovine tissues in the ultraviolet-visible range (Bernini, Umberto; Russo, Paolo)V1427,398-404(1991)

Photosensitized receptor inactivation with He-Ne laser: preliminary results (Arber, Simon; Rymer, William Z.; Crumrine, David)V1428,23-29(1991)

Principles of phase-resolved optical measurements (Jacques, Steven L.)V1525,143-153(1991)

Promising applications of scanning electron-beam-pumped laser devices in medicine and biology (Katsap, Victor N.; Koshevoy, Alexander V.; Meerovich, Gennady A.; Ulasjuk, Vladimir N.)V1420,259-265(1991)

Properties of optical waves in turbid media (Svaasand, Lars O.; Tromberg, Bruce J.)V1525,41-51(1991)

Quantitative analysis of three-dimensional landmark coordinate data (Richtsmeier, Joan T.)V1380,12-23(1991)

Quantitative microscope for image cytometry (Jaggi, Bruno; Poon, Steven S.; Pontifex, Brian; Fengler, John J.; Marquis, Jacques; Palcic, Branko)V1448,89-97(1991)

Quantitative thermal gradient imaging of biological surfaces (Swanson, Curtis J.; Wingard, Christopher J.)V1467,372-383(1991)

Real-time simulation of the retina allowing visualization of each processing stage (Teeters, Jeffrey L.; Werblin, Frank S.)V1472,6-17(1991); V1473,102-113(1991)

Recent applications of biostereometrics in research and education (Herron, Robin E.)V1380,2-5(1991)

Receptive fields and the theory of discriminant operators (Gupta, Madan M.; Hungenahally, Suresh K.)V1382,87-98(1991)

Retina-like image acquisition system with wide-range light adaptation (Chang, Po-Rong; Yeh, Bao-Fuh)V1606,456-469(1991)

Self-organized criticality in neural networks (Makarenko, Vladimir I.; Kirillov, A. B.)V1469,843-845(1991)

Silicon retina with adaptive photoreceptors (Mahowald, Misha A.)V1473,52-58(1991)

Study of measuring skin blood flow using speckle counting (Liu, Ying; Ma, Shining; Du, Fuli; Peng, Xiang; Ye, Shenhua)V1554A,610-612(1991)

Study of microbial growth I: by diffraction (Williams, Gareth T.; Bahuguna, Ramendra D.; Arteaga, Humberto; Le Joie, Elaine N.)V1332,802-804(1991)

Study of microbial growth II: by holographic interferomery (Bahuguna, Ramendra D.; Williams, Gareth T.; Pour, I. K.; Raman, R.)V1332,805-807(1991)

Surface digitizing of anatomical subjects with DIGIBOT-4 (Koch, Stephen; Koch, Eric)V1380,163-170(1991)

Use of a cooled CCD camera for confocal light microscopy (Masters, Barry R.)V1448,98-105(1991)

Use of polarization to separate on-axis scattered and unscattered light in red blood cells (Sardar, Dhiraj K.; Nemati, Babak; Barrera, Frederick J.)V1427,374-380(1991)

Vesalius project: interactive computers in anatomical instruction (McCracken, Thomas O.; Roper, Stephen D.; Spurgeon, Thomas L.)V1380,6-10(1991)

X-ray holography for sequencing DNA (Yorkey, Thomas J.; Brase, James M.; Trebes, James E.; Lane, Stephen M.; Gray, Joe W.)V1345,255-259(1991)

Birefringence—see also fiber optics

Actual light deflections in regions of crack tips and their influence on measurements in photomechanics (Hecker, Friedrich W.; Pindera, Jerzy T.; Wen, Baicheng)V1554A,151-162(1991)

All-fiber pressure sensor up to 100 MPa (Bock, Wojtek J.; Wolinski, Tomasz R.; Domanski, Andrzej W.)V1511,250-254(1991)

Analysis of diffracted stress fields around a noncharged borehole with dynamic photoelasticity and gauges (Zhu, Zhenhai; Qu, Guangjian; Yang, Yongqi; Shang, Jian)V1554A,472-481(1991)

Application of electro-optic modulator in photomechanics (Zhang, Yuan P.)V1554B,669-678(1991)

Application of photoelastic coating method on elastoplastic stress analysis of rotation disk (Dong, Benhan; Gao, Penfei; Wang, Ju)V1554A,400-406(1991)

Automatization of measurement and processing of experimental data in photoelasticity (Zhavoronok, I. V.; Nemchinov, V. V.; Litvin, S. A.; Skanavi, A. M.; Pavlov, V. V.; Evsenev, V. S.)V1554A,371-379(1991)

Coupled-mode analysis of dynamic polarization volume holograms (Huang, Tizhi; Wagner, Kelvin H.)V1559,377-384(1991)

Demonstration of birefringent optical fibre frequency shifter employing torsional acoustic waves (Berwick, Michael; Pannell, Christopher N.; Russell, Philip S.; Jackson, David A.)V1572,157-162(1991)

Dynamic photoelasticity applied to crack-branching investigations (Hammami, Slimane; Cottron, M.; Lagarde, Alexis)V1554A,136-142(1991)

Elasto-plastic contact between rollers (Chu, Kunliang; Li, Penghui)V1554A,192-195(1991)

Experimental research on high-bifringence fiber phase retarder (Yao, Minyu; Zhang, Xinyu)V1572,148-150(1991)

Frequency stabilization of AlxGa1-xAs/GaAs lasers using magnetically induced birefringence in an atomic vapor (Lee, W. D.; Campbell, Joe C.)V1365,96-101(1991)

Ground-based experiments on the growth and characterization of L-arginine phosphate crystals (Rao, S. M.; Cao, C.; Batra, Ashok K.; Lal, Ravindra B.; Mookherji, Tripty K.)V1557,283-292(1991)

Hybride fiber-optic temperature sensors on the base of LiNbO3 and LiNbO3:Ti waveguides (Goering, Rolf)V1511,275-280(1991)

Hybrid method to analyze the stress state in piecewise homogeneous two-dimensional objects (Laermann, Karl-Hans)V1554A,143-150(1991)

Integrated photoelasticity for residual stresses in glass specimens of complicated shape (Aben, Hillar K.; Idnurm, S. J.; Josepson, J. I.; Kell, K.-J. E.; Puro, A. E.)V1554A,298-309(1991)

Investigation of strain birefringence and wavefront distortion in 001 plates of KD2PO4 (De Yoreo, James J.; Woods, Bruce W.)V1561,50-58(1991)

Nanosecond time-resolved natural and magnetic circular dichroism spectroscopy of protein dynamics (Goldbeck, Robert A.; Bjorling, Sophie; Kliger, David S.)V1432,14-27(1991)

Nonlinear absorbance effects in bacteriorhodopsin (Rayfield, George W.)V1436,150-159(1991)

Normally incident plane waves on a chiral slab with linear property variations (Lakhtakia, Akhlesh; Varadan, Vijay K.; Varadan, Vasundara V.)V1558,120-126(1991)

Optical Kerr effect in chain carbonyl compounds (Blaszczak, Zdzislaw)V1403,509-511(1991)

Optical Kerr effect in liquid and gaseous carbon dioxide (Blaszczak, Zdzislaw; Gauden, Pawel)V1391,156-163(1991)

Optical isodyne measurements in fracture mechanics (Pindera, Jerzy T.; Wen, Baicheng)V1554A,196-205(1991)

Photo-induced refractive-index changes and birefringence in optically nonlinear polyester (Shi, Youngqiang; Steier, William H.; Yu, Luping; Chen, Mai; Dalton, Larry R.)V1559,118-126(1991)

Photoanisotropic polymeric media and their application in optical devices (Barachevsky, Valery A.)V1559,184-193(1991)

Photoelastic investigation of statics, kinetics, and dynamics of crack formation in transparent models and natural structural elements (Taratorin, B. I.; Sakharov, V. N.; Komlev, O. U.; Stcherbakov, V. N.; Starchevsky, A. V.)V1554A,449-456(1991)

Photoelastic investigation of stress waves using models of viscoelastic materials (Dmokhovskij, A. V.; Filippov, I. G.; Skropkin, S. A.; Kobakhidze, T. G.)V1554A,323-330(1991)

Photoelastic sensors for automatic control system of dam safety (Konwerska-Hrabowska, Joanna; Kryszczynski, Tadeusz; Tomaszewicz, Tomasz; Lietz, J.; Mazurkiewicz, Wojciech)V1554A,388-399(1991)

Photoelastic study of friction at multipoint contacts (Dally, James W.; Chen, Yung-Mien)V1554A,434-443(1991)

Photoelastic transducer for high-temperature applications (Redner, Alex S.; Adamovsky, Grigory; Wesson, Laurence N.)V1332,775-782(1991)

Polarization-maintaining single-mode fibers: measurement and prediction of fundamental characteristics (Sasek, Ladislav; Vohryzek, Jachym)V1504,147-154(1991)

Polarization-maintaining single-mode fibers with layered core (Sasek, Ladislav; Vohryzek, Jachym; Sochor, Vaclav; Paulicka, Ivan; van Nhac, Nguyen; Franek, Alexandr)V1572,151-156(1991)

Scattered-light optical isodynes: basis for 3-D isodyne stress analysis (Pindera, Jerzy T.)V1554A,458-471(1991)

Second Intl Conf on Photomechanics and Speckle Metrology: Speckle Techniques, Birefringence Methods, and Applications to Solid Mechanics (Chiang, Fu-Pen, ed.)V1554A(1991)

Synthesis and property research of birefringent polymers with predicted optical-mechanical parameters (Askadskij, A. A.; Marshalkovich, A. S.; Latysh, E. G.; Goleneva, L. M.; Pastukhov, A. V.; Sidorova, G. I.)V1554A,426-431(1991)

Technique for measuring stress-induced birefringence (Heidel, Jeffrey R.; Zediker, Mark S.)V1418,240-247(1991)

Temperature dependence of optical Kerr effect in aromatic ethers (Blaszczak, Zdzislaw)V1391,146-155(1991)

Two-dimensional flow quantitative visualization by the hybrid method of flow birefringence and boundary integration (Chen, Yuhai; Jia, Youquan)V1554B,566-572(1991)

Use of dome (meniscus) lenses to eliminate birefringence and tensile stresses in spatial filters for the Nova laser (Pitts, John H.; Kyrazis, Demos T.; Seppala, Lynn G.; Bumpas, Stanley E.)V1441,71-81(1991)

Use of scattered-light photoelasticity at crack tips (Ravi-Chandar, K.)V1554A,228-238(1991)

Whole-field stress fringe compensation using photoelastic carrier shifting and optical information processing (Zhang, Xi; Wang, Baishi; Li, Yao W.)V1554A,444-448(1991)

Bistability—see also integrated optics; integrated optoelectronics; switches

Bistability of the Sn donor in AlxGa1-xAs and GaAs under pressure studied by Mossbauer spectroscopy (Gibart, Pierre; Williamson, Don L.)V1362,938-950(1991)

Bistable optical switching in GaAs multiple-quantum-well epitaxial etalons (Oudar, Jean-Louis; Sfez, B. G.; Kuszelewicz, Robert)V1361,490-498(1991)

Intercomparison of homogeneous laser models with transverse effects (Bandy, Donna K.; Hunter, L. W.; Jones, Darlena J.)V1497,142-152(1991)

Laser-induced phase transitions in liquid crystals and distributed feedback-fluctuations, energy exchange, and instabilities: squeezed polarized states and intensity correlations (Arakelian, Sergei M.; Chilingarian, Yu. S.; Alaverdian, R. B.; Alodjants, A. P.; Drnoian, V. E.; Karaian, A. S.)V1402,175-191(1991)

Magnetocapacitance and photoluminescence spectroscopy studies of charge storage, bistability, and energy relaxation effects in resonant tunneling devices (Eaves, Lawrence; Hayes, David; Leadbeater, M. L.; Simmonds, P. E.; Skolnick, Maurice S.)V1362,520-533(1991)

Modal analysis of grating-induced optical bistability (Vitrant, Guy; Vincent, Patrick; Reinisch, Raymond; Neviere, Michel)V1545,225-231(1991)

Module for optical parallel logic using bistable optically addressed ferroelectric spatial light modulators (Guibert, L.; Killinger, M.; de Bougrenet de la Tocnaye, Jean-Louis M.)V1505,99-103(1991)

Multistability, instability, and chaos for intracavity magneto-optic modulation output (Yang, Darang; Song, Ruhua H.; Hu, Zhiping; Le, Shixiao)V1417,440-450(1991)

Nonlinear optical components with liquid crystals (Dumitru, Mihaela A.; Honciuc, Maria; Sterian, Livia)V1500,339-348(1991)

Optical bistability and signal competition in active cavity with photochromic nonlinearity of bacteriorhodopsin (Taranenko, Victor B.; Vasnetsov, M. V.)V1621,169-179(1991)

Optical phase-conjugate resonators, bistabilities, and applications (Venkateswarlu, Putcha; Dokhanian, Mostafa; Sekhar, Prayaga C.; George, M. C.; Jagannath, H.)V1332,245-266(1991)

Optoelectronic and optical bistabilities of photocurrent and photoluminescence at low-temperature avalanche breakdown in GaAs epitaxial films (Ryabushkin, Oleg A.; Platonov, N. S.; Sablikov, V. A.; Sergeyev, V. I.; Bader, Vladimir A.)V1362,75-79(1991)

Perturbation theory for optical bistability of prism and grating couplers and comparison with rigorous method (Akhouayri, Hassan; Vincent, Patrick; Neviere, Michel)V1545,140-144(1991)

Photocryosar: bistable element for optoelectronic computing (Ryabushkin, Oleg A.; Bader, Vladimir A.)V1505,67-74(1991)

Polarization-based all-optical bistable element (Domanski, Andrzej W.; Karpierz, Miroslaw A.; Strojewski, Dariusz)V1505,59-66(1991)

Semiconductor waveguides for optical switching (Laval, Suzanne)V1362,82-92(1991)

Shallow-deep bistability behavior of the DX-centers in n-AlxGa1-xAs and the EL2-defects in n-GaAs (Kadri, Abderrahmane; Portal, Jean-Claude)V1362,930-937(1991)

Total internal reflection studies of a ferroelectric liquid crystal/anisotropic solid interface (Zhuang, Zhiming; Clark, Noel A.; Meadows, Michael R.)V1455,105-109(1991)

Bragg cells—see also acousto-optics; signal processing

Development of wideband 16-channel acousto-optical modulators on the LiNbO3 and TeO2 crystals (Bokov, Lev; Demidov, Anatoly J.; Zadorin, Anatoly; Kushnarev, Igor; Serebrennikov, Leonid J.; Sharangovich, Sergey)V1505,186-198(1991)

Frequency-dependent optical beam distortion generated by acousto-optic Bragg cells (Pieper, Ronald J.; Poon, Ting-Chung)V1454,324-335(1991)

Generalized phased-array Bragg interaction in anisotropic crystals (Young, Eddie H.; Ho, Huey-Chin C.; Yao, Shi-Kay; Xu, Jieping P.)V1476,178-189(1991)

Guided-wave magneto-optic and acousto-optic Bragg cells for RF signal processing (Tsai, Chen S.)V1562,55-65(1991)

Impact of device characteristics on optical processor design (Turpin, Terry M.)V1562,2-10(1991)

Instrumentation concepts for multiplexed Bragg grating sensors (Giesler, Leslie E.; Dunphy, James R.; Morey, William W.; Meltz, Gerald; Glenn, William H.)V1480,138-142(1991)

Latest developments in crystal growth and characterization of efficient acousto-optic materials (Paradies, C. J.; Glicksman, M. E.; Jones, O. C.; Kim, G. T.; Lin, Jen T.; Gottlieb, Milton S.; Singh, N. B.)V1561,2-5(1991)

Mercurous halides for long time-delay Bragg cells (Brandt, Gerald B.; Singh, N. B.; Gottlieb, Milton S.)V1454,336-343(1991)

Novel high-speed communication system (Dawber, William N.; Hirst, Peter F.; Condon, Brian P.; Maitland, Arthur; Sutton, Phillip)V1476,81-90(1991)

Sixty-four channel acousto-optical Bragg cells for optical computing applications (Graves, David W.)V1563,229-235(1991)

TeO2 slow surface acoustic wave Bragg cell (Yao, Shi-Kay)V1476,214-221(1991)

Wide-angular aperture acousto-optic Bragg cell (Weverka, Robert T.; Wagner, Kelvin H.)V1562,66-72(1991)

Wideband acousto-optic spectrometer (Chang, I. C.)V1476,257-268(1991)

Calibration

Actively controlled multiple-sensor system for feature extraction (Daily, Michael J.; Silberberg, Teresa M.)V1472,85-96(1991)

Adaptive nonuniformity correction for IR focal-plane arrays using neural networks (Scribner, Dean A.; Sarkady, Kenneth A.; Kruer, Melvin R.; Caulfield, John T.; Hunt, J. D.; Herman, Charles)V1541,100-109(1991)

Atmospheric infrared sounder on the Earth Observing System: in-orbit spectral calibration (Aumann, Hartmut H.)V1540,176-186(1991)

Blackbody radiators for field calibration (Cross, Edward F.; Mauritz, F. L.; Bixler, H. A.; Kaegi, E. M.; Wiemokly, Gary D.)V1540,756-763(1991)

Calibrated intercepts for solar radiometers used in remote sensor calibration (Gellman, David I.; Biggar, Stuart F.; Slater, Philip N.; Bruegge, Carol J.)V1493,175-180(1991)

Calibration for the medium-resolution imaging spectrometer (Baudin, Gilles; Chessel, Jean-Pierre; Cutter, Mike A.; Lobb, Daniel R.)V1493,16-27(1991)

Calibration of a CCD camera on a hybrid coordinate measuring machine for industrial metrology (Bruzzone, Elisabetta; Mangili, Fulvia)V1526,96-112(1991)

Calibration of EOS multispectral imaging sensors and solar irradiance variability (Mecherikunnel, Ann)V1493,292-302(1991)

Calibration of Passive Remote Observing Optical and Microwave Instrumentation (Guenther, Bruce W., ed.)V1493(1991)

Calibration procedures for the space vision system experiment (MacLean, Steve G.; Pinkney, H. F.)V1526,113-122(1991)

Clouds as calibration targets for AVHRR reflected-solar channels: results from a two-year study at NOAA/NESDIS (Abel, Peter)V1493,195-206(1991)

Cryogenic radiometers and intensity-stabilized lasers for EOS radiometric calibrations (Foukal, Peter; Hoyt, Clifford C.; Jauniskis, L.)V1493,72-79(1991)

Edge effects in silicon photodiode arrays (Kenney, Steven B.; Hirleman, Edwin D.)V1480,82-93(1991)

Enhancements to the radiometric calibration facility for the Clouds and the Earth's Radiant Energy System instruments (Folkman, Mark A.; Jarecke, Peter J.; Darnton, Lane A.)V1493,255-266(1991)

Evaluation of the NOAA-11 solar backscatter ultraviolet radiometer, Mod 2 (SBUV/2): inflight calibration (Weiss, Howard; Cebula, Richard P.; Laamann, K.; Hudson, Robert D.)V1493,80-90(1991)

Flight solar calibrations using the mirror attenuator mosaic: low-scattering mirror (Lee, Robert B.)V1493,267-280(1991)

In-flight calibration of a helicopter-mounted Daedalus multispectral scanner (Balick, Lee K.; Golanics, Charles J.; Shines, Janet E.; Biggar, Stuart F.; Slater, Philip N.)V1493,215-223(1991)

Interactive tools for extraction of cartographic calibration data from aerial photography (Hunt, Bobby R.; Ryan, Thomas W.; Gifford, E.)V1472,190-200(1991)

Large integrating sphere of prelaunch calibration system for Japanese Earth Resources Satellite optical sensors (Suzuki, Naoshi; Narimatsu, Yoshito; Nagura, Riichi; Sakuma, Fumihiro; Ono, Akira)V1493,48-57(1991)

New device for interactive image-guided surgery (Galloway, Robert L.; Edwards, Charles A.; Thomas, Judith G.; Schreiner, Steven; Maciunas, Robert J.)V1444,9-18(1991)

On-board calibration device for a wide field-of-view instrument (Krawczyk, Rodolphe; Chessel, Jean-Pierre; Durpaire, Jean-Pierre; Durieux, Alain; Churoux, Pascal; Briottet, Xavier)V1493,2-15(1991)

Phase-measuring fiber optic ESPI system: phase-step calibration and error sources (Joenathan, Charles; Khorana, Brij M.)V1554B,56-63(1991)

Prelaunch calibration system for optical sensors of Japanese Earth Resources Satellite (Sakuma, Fumihiro; Ono, Akira)V1493,37-47(1991)

Pressure modulator infrared radiometer (PMIRR) optical system alignment and performance (Chrisp, Michael P.; Macenka, Steve A.)V1540,213-218(1991)

Radiometric calibration of SPOT 2 HRV: a comparison of three methods (Biggar, Stuart F.; Dinguirard, Magdeleine C.; Gellman, David I.; Henry, Patrice; Jackson, Ray D.; Moran, M. S.; Slater, Philip N.)V1493,155-162(1991)

Radiometric calibration of an airborne multispectral scanner (Markham, Brian L.; Ahmad, Suraiya P.; Jackson, Ray D.; Moran, M. S.; Biggar, Stuart F.; Gellman, David I.; Slater, Philip N.)V1493,207-214(1991)

Radiometric calibration of space-borne Fourier transform infrared-emission spectrometers: proposed scenario for European Space Agency's Michelson interferometer for passive atmospheric sounding (Lamarre, Daniel; Giroux, Jean)V1493,28-36(1991)

Radiometric calibration plan for the Clouds and the Earth's Radiant Energy System scanning instruments (Jarecke, Peter J.; Folkman, Mark A.; Darnton, Lane A.)V1493,244-254(1991)

Reflectance stability analysis of Spectralon diffuse calibration panels (Bruegge, Carol J.; Stiegman, Albert E.; Coulter, Daniel R.; Hale, Robert R.; Diner, David J.; Springsteen, Arthur W.)V1493,132-142(1991)

Results of calibrations of the NOAA-11 AVHRR made by reference to calibrated SPOT imagery at White Sands, N.M. (Nianzeng, Che; Grant, Barbara G.; Flittner, David E.; Slater, Philip N.; Biggar, Stuart F.; Jackson, Ray D.; Moran, M. S.)V1493,182-194(1991)

Solar-diffuser panel and ratioing radiometer approach to satellite sensor on-board calibration (Slater, Philip N.; Palmer, James M.)V1493,100-105(1991)

State-of-the-art transfer radiometer for testing and calibration of FLIR test equipment (Kopolovich, Zvi; Naor, Yoram; Cabib, Dario; Johnson, W. Todd; Sapir, Eyal)V1540,565-577(1991)

Stray light effects on calibrations using a solar diffuser (Palmer, James M.)V1493,143-154(1991)

Study of gas scintillation proportional counter physics using synchrotron radiation (Bavdaz, Marcos; Favata, Fabio; Smith, Alan; Parmar, A. N.)V1549,35-44(1991)

Surface and aerosol models for use in radiative transfer codes (Hart, Quinn J.)V1493,163-174(1991)

Technique for improving the calibration of large-area sphere sources (Walker, James H.; Cromer, Chris L.; McLean, James T.)V1493,224-230(1991)

Uniformity and transmission of proportional counter window materials for use with AXAF (Flanagan, Kathryn A.; Austin, G. K.; Cobuzzi, J. C.; Goddard, R.; Hughes, John P.; McLaughlin, Edward R.; Podgorski, William A.; Rose, V.; Roy, Adrian G.; Zombeck, Martin V.; Markert, Thomas H.; Bauer, J.; Isobe, T.; Schattenburg, Mark L.)V1549,395-407(1991)

Cameras—see also charge-coupled devices; machine vision; imaging systems; image tubes; photography; photography, high-speed; streak cameras; video

200,000-frame-per-second drum camera with nanosecond synchronized laser illumination (Briscoe, Dennis)V1346,319-323(1991)

3-D camera based on differential optical absorbance (Houde, Regis; Laurendeau, Denis; Poussart, Denis)V1332,343-354(1991)

A 1.3-megapixel-resolution portable CCD electronic still camera (Jackson, Todd A.; Bell, Cynthia S.)V1448,2-12(1991)

Acquisition of very high resolution images using stereo cameras (Aizawa, Kiyoharu; Komatsu, Takashi; Saito, Takahiro)V1605,318-328(1991)

Advanced imaging system for high-precision, high-resolution CCD imaging (Doherty, Peter E.; Sims, Gary R.)V1448,118-128(1991)

Airborne infrared and visible sensors used for law enforcement and drug interdiction (Aikens, David M.; Young, William R.)V1479,435-444(1991)

Camera and Input Scanner Systems (Chang, Win-Chyi; Milch, James R., eds.)V1448(1991)

Camera calibration using distance invariance principles (Raju, G. V.; Rudraraju, Prasad)V1385,50-56(1991)

CCD camera for an autoguider (Schempp, William V.)V1448,129-133(1991)

Day/night aerial surveillance system for fishery patrol (Uhl, Bernd)V1538,140-147(1991)

Design and performance of a PtSi spectroscopic infrared array and detector head (Cizdziel, Philip J.)V1488,6-27(1991)

Development of picosecond x-ray framing camera (Chang, Zenghu; Hou, Xun; Zhang, Xiaoqiu; Gong, Meixia; Niu, Lihong; Yong, Hongru; Liu, Xiouqin; Lei, Zhiyuan)V1358,614-618(1991)

Development of subnanosecond framing cameras in IOFAN (Ludikov, V. V.; Prokhorov, Alexander M.; Chevokin, Victor K.)V1346,418-436(1991)

Direct-drive film magazines (Lewis, George R.)V1538,167-179(1991)

DSP-based stabilization systems for LOROP cameras (Quinn, James)V1538,150-166(1991)Electronically gated airborne video camera (Sturz, Richard A.)V1538,77-80(1991)

Evaluation of a CCD star camera at Table Mountain Observatory (Strikwerda, Thomas E.; Fisher, H. L.; Frank, L. J.; Kilgus, Charles C.; Gray, Connie B.; Barnes, Donald L.)V1478,13-23(1991)

High-performance InSb 256 x 256 infrared camera (Blackwell, John D.; Parrish, William J.; Kincaid, Glen T.)V1479,324-334(1991)

High-performance digital color video camera (Parulski, Kenneth A.; Benamati, Brian L.; D'Luna, Lionel J.; Shelley, Paul R.)V1448,45-58(1991)

High-resolution, low-light, image-intensified CCD camera (Tanaka, Satoru C.; Silvey, Tom; Long, Greg; Braze, Bill)V1448,21-26(1991)

High-resolution CCD still/video and still-still/video systems (Kee, Richard C.)V1448,13-20(1991)

High-resolution fully 3-D mapping of human surfaces by laser array camera and data representations (Bae, Kyongtae T.; Altschuler, Martin D.)V1380,171-178(1991)

High-resolution image digitizing camera for use in quantitative coronary arteriography (Muser, Markus H.; Leemann, Thomas; Anliker, M.)V1448,106-112(1991)

High-sensitive thermal video camera with self-scanned 128 InSb linear array (Fujisada, Hiroyuki)V1540,665-676(1991)

High-speed, high-resolution image reading technique using multi-area sensors (Uehira, Kazutake)V1448,182-190(1991)

High-throughput narrowband 83.4-nm self-filtering camera (Zukic, Muamer; Torr, Douglas G.; Torr, Marsha R.)V1549,234-244(1991)

HST phase retrieval: a parameter estimation (Lyon, Richard G.; Miller, Peter E.; Grusczak, Anthony)V1567,317-326(1991)

Imaging capabilities of small satellites: Indian experience (Alex, T. K.)V1495,52-58(1991)

Improved precision/resolution by camera movement (Taylor, Geoff L.; Derksen, Grant)V1526,27-34(1991)

In-flight performance of the faint object camera of the Hubble Space Telescope (Greenfield, Perry E.; Paresce, Francesco; Baxter, David; Hodge, P.; Hook, R.; Jakobsen, P.; Jedrzejewski, Robert; Nota, Anatonella; Sparks, W. B.; Towers, Nigel M.; Laurance, R. J.; Macchetto, F.)V1494,16-39(1991)

ISOCAM: a camera for the ISO satellite optical bench development (Auternaud, Danielle)V1488,64-72(1991)

Linear LED arrays for film annotation and for high-speed high-resolution printing on ordinary paper (Wareberg, P. G.; Scholes, R.; Taylor, R.)V1538,112-123(1991)

Linear array camera interface techniques (DeLuca, Dan)V1396,558-565(1991)

Low-cost high-performance InSb 256 x 256 infrared camera (Parrish, William J.; Blackwell, John D.; Kincaid, Glen T.; Paulson, Robert C.)V1540,274-284(1991)

Measurements with a 35-psec gate time microchannel plate camera (Bell, Perry M.; Kilkenny, Joseph D.; Hanks, Roy L.; Landen, Otto L.)V1346,456-464(1991)

Model for image sensing and digitization in machine vision (Subbarao, Murali; Nikzad, Arman)V1385,70-84(1991)

New 2/3-inch MF image pick-up tubes for HDTV camera (Kobayashi, Akira; Ishikawa, Masayoshi; Suzuki, Takayoshi; Ikeya, Morihiro; Shimomoto, Yasuharu)V1449,148-156(1991)

MOMS-02/Spacelab D-2: a high-resolution multispectral stereo scanner for the second German Spacelab mission (Ackermann, F.; Bodechtel, Joh; Lanzl, Franz; Meissner, D.; Seige, Peter; Winkenbach, H.; Zilger, Johannes)V1490,94-101(1991)

New approach to synchroballistic photography (McDowell, Maurice W.; Klee, H. W.; Griffith, Derek J.)V1358,227-236(1991)

New stereo laser triangulation device for specular surface inspection (Samson, Marc; Dufour, Marc L.)V1332,314-322(1991)

Optics, Illumination, and Image Sensing for Machine Vision V (Harding, Kevin G.; Svetkoff, Donald J.; Uber, Gordon T.; Wittels, Norman, eds.)V1385(1991)

Phase-shift moire camera for real-time measurements of three-dimensional shape information (Turney, Jerry L.; Lysogorski, Charles; Gottschalk, Paul G.; Chiu, Arnold H.)V1380,53-63(1991)

Proposals for a computer-controlled orbital scanning camera for remote image aquisition (Nwagboso, Christopher O.)V1454,111-122(1991)

PtSi camera: performance model validation (Meidan, Moshe; Schwartz, Roni; Sher, Assaf; Zhaiek, Sasson; Gal, Eli; Neugarten, Michael L.; Afik, Zvi; Baer, C.)V1540,729-737(1991)

Selectable one-to-four-port, very high speed 512 x 512 charge-injection device (Zarnowski, Jeffrey J.; Williams, Bryn; Pace, M.; Joyner, M.; Carbone, Joseph; Borman, C.; Arnold, Frank S.; Wadsworth, Mark V.)V1447,191-201(1991)

Semiautomatic calibration of the general camera model for stereovision (Sung, Eric; Singh, Harcharan; Tan, Daniel H.)V1385,57-69(1991)

Signal processing LSI system for digital still camera (Watanabe, Mikio; Saito, Osamu; Okamoto, Satoru; Ito, Kenji; Moronaga, Kenji; Hayashi, Kenkichi; Nishi, Seiki)V1452,27-36(1991)

Simple 180o field-of-view F-theta all-sky camera (Andreic, Zeljko)V1500,293-304(1991)

Spatial characterization of YAG power laser beam (Grevey, D. F.; Badawi, K. F.)V1502,32-40(1991)

SPOT 4 HRVIR instrument and future high-resolution stereo instruments (Fratter, C.; Reulet, Jean-Francois; Jouan, Jacky)V1490,59-73(1991)

Stereoscopic video and the quest for virtual reality: an annotated bibliography of selected topics (Starks, Michael R.)V1457,327-342(1991)

Sub-100 psec x-ray gating cameras for ICF imaging applications (Kilkenny, Joseph D.; Bell, Perry M.; Hammel, Bruce A.; Hanks, Roy L.; Landen, Otto L.; McEwan, Thomas E.; Montgomery, David S.; Turner, R. E.; Wiedwald, Douglas J.; Bradley, David K.)V1358,117-133(1991)

Super-high-speed reflex-type moving image camera (Drozhbin, Yu. A.; Trofimenko, Vladimir V.)V1358,454-456(1991)

Tactical reconnaissance mission survivability requirements (Lareau, Andy G.; Collins, Ross)V1538,81-98(1991)

TDI camera for industrial applications (Castro, Peter; Gittings, J.; Choi, Yauho J.)V1448,134-139(1991)

Transient radiometric measurements with a PtSi IR camera (Konopka, Wayne L.; Soel, Michael A.; Celentano, A.; Calia, V.)V1488,355-365(1991)

Ultrafast optical-mechanical camera (Drozhbin, Yu. A.; Trofimenko, Vladimir V.; Chernova, T. I.)V1358,451-453(1991)

Use of high-resolution TV camera in photomechanics (Yatagai, Toyohiko; Ino, Tomomi)V1554B,646-649(1991)

Validated CCD camera model (Johnson-Cole, Helen; Clark, Rodney L.)V1488,203-211(1991)

Velocity profiling in linear and rotational systems (Crabtree, Daniel L.)V1482,458-472(1991)

Wide-field-of-view star tracker camera (Lewis, Isabella T.; Ledebuhr, Arno G.; Axelrod, Timothy S.; Kordas, Joseph F.; Hills, Robert F.)V1478,2-12(1991)

Cathode ray tubes—see also displays

Document viewing: display requirements in image management (van Overbeek, Thomas T.)V1454,406-413(1991)

Image signal modulation and noise analysis of CRT displays (Burns, Peter D.)V1454,392-398(1991)

Importance of phosphor persistence characteristics in reducing visual distress symptoms in VDT users (Hayosh, Thomas D.)V1454,399-405(1991)

Physical and psychophysical evaluation of CRT noise performance (Ji, Tinglan; Roehring, Hans; Blume, Hartwig R.; Seeley, George W.; Browne, Michael P.)V1444,136-150(1991)

Charge-coupled devices—see also astronomy; cameras; detectors

1536 x 1024 CCD image sensor (Wong, Kwok Y.; Torok, Georgia R.; Chang, Win-Chyi; Meisenzahl, Eric J.)V1447,283-287(1991)

4096 x 4096 pixel CCD mosaic imager for astronomical applications (Geary, John C.; Luppino, Gerard A.; Bredthauer, Richard A.; Hlivak, Robert J.; Robinson, Lloyd B.)V1447,264-273(1991)

A 1.3-megapixel-resolution portable CCD electronic still camera (Jackson, Todd A.; Bell, Cynthia S.)V1448,2-12(1991)

Acquisition and processing of digital images and spectra in astronomy (Lee, Terence J.)V1439,152-158(1991)

Advanced imaging system for high-precision, high-resolution CCD imaging (Doherty, Peter E.; Sims, Gary R.)V1448,118-128(1991)

Algorithm for the generation of look-up range table in 3-D sensing (Su, Xianyu; Zhou, Wen-Sheng)V1332,355-357(1991)

Analog CCD processors for image filtering (Yang, Woodward)V1473,114-127(1991)

Analysis of neuropeptides using capillary zone electrophoresis with multichannel fluorescence detection (Sweedler, Jonathan V.; Shear, Jason B.; Fishman, Harvey A.; Zare, Richard N.; Scheller, Richard H.)V1439,37-46(1991)

Analytical Raman and atomic spectroscopies using charge-coupled-device detection (Bilhorn, Robert B.; Ferris, Nancy S.)V1439,15-24(1991)

Application of electron-sensitive CCD for taking off the time-dispersed pictures from image tube phosphor screens (Bryukhnevitch, G. I.; Dalinenko, I. N.; Kuz'min, G. A.; Libenson, B. N.; Malyarov, A. V.; Moskalev, B. B.; Postovalov, V. E.; Prokhorov, Alexander M.; Schelev, Mikhail Y.)V1358,739-749(1991)

Application of low-noise CID imagers in scientific instrumentation cameras (Carbone, Joseph; Hutton, J.; Arnold, Frank S.; Zarnowski, Jeffrey J.; VanGorden, Steve; Pilon, Michael J.; Wadsworth, Mark V.)V1447,229-242(1991)

Applications of charge-coupled and charge-injection devices in analytical spectroscopy (Denton, M. B.)V1447,2-11(1991)

Automatic-adjusting optical axis for linear CCD scanner (Chang, Rong-Seng; Chen, Der-Chin)V1527,357-360(1991)

Automatic data processing of speckle fringe pattern (Dupre, Jean-Christophe; Lagarde, Alexis)V1554A,766-771(1991)

Back-illuminated 1024 x 1024 quadrant readout imager: operation and screening test results (Marsh, Harry H.; Hayes, Raymond; Blouke, Morley M.; Yang, Fanling H.)V1447,298-309(1991)

Back-illuminated large-format Loral CCDs (Lesser, Michael P.)V1447,177-182(1991)

Backside-thinned CCDs for keV electron detection (Ravel, Mihir K.; Reinheimer, Alice L.)V1447,109-122(1991)

Calibration of a CCD camera on a hybrid coordinate measuring machine for industrial metrology (Bruzzone, Elisabetta; Mangili, Fulvia)V1526,96-112(1991)

Camera and Input Scanner Systems (Chang, Win-Chyi; Milch, James R., eds.)V1448(1991)

CCD camera for an autoguider (Schempp, William V.)V1448,129-133(1991)

CCD focal-plane imaging detector for the JET-X instrument on spectrum R-G (Wells, Alan A.; Castelli, C. M.; Holland, Andrew D.; McCarthy, Kieran J.; Spragg, J. E.; Whitford, C. H.)V1549,357-373(1991)

CCD performance model (Dial, O. E.)V1479,2-11(1991)

CCD performance model for airborne reconnaissance (Donn, Matthew; Waeber, Bruce)V1538,189-200(1991)

CCD star sensors for Indian remote sensing satellites (Alex, T. K.; Rao, V. K.)V1478,101-105(1991)

Characteristics of MSM detectors for meander channel CCD imagers on GaAs (Kosel, Peter B.; Bozorgebrahimi, Nercy; Iyer, J.)V1541,48-59(1991)

Characterization and effect of system noise in a differential angle of arrival measurement device (Waldie, Arthur H.; Drexler, James J.; Qualtrough, John A.; Soules, David B.; Eaton, Frank D.; Peterson, William A.; Hines, John R.)V1487,103-108(1991)

Charge-coupled-device-addressed liquid-crystal light valve: an update (Efron, Uzi; Byles, W. R.; Goodwin, Norman W.; Forber, Richard A.; Sayyah, Keyvan; Wu, Chiung S.; Welkowsky, Murray S.)V1455,237-247(1991)

Charge-Coupled Devices and Solid State Optical Sensors II (Blouke, Morley M., ed.)V1447(1991)

Colorimetric characterization of CCD sensors by spectrophotometry (Daligault, Laurence; Glasser, Jean)V1512,124-130(1991)

Computer analysis of signal-to-noise ratio and detection probability for scanning IRCCD arrays (Uda, Gianni; Tofani, Alessandro)V1488,257-266(1991)

Design, analysis, and testing of a CCD array mounting structure (Sultana, John A.; O'Neill, Mark B.)V1532,27-38(1991)

Design and characterization of a space-variant CCD sensor (Kreider, Gregory; Van der Spiegel, Jan; Born, I.; Claeys, Cor L.; Debusschere, Ingrid; Sandini, Giulio; Dario, Paolo)V1381,242-249(1991)

Design and operational characteristics of a PV 001 image tube incorporated with EB CCD readout (Bryukhnevitch, G. I.; Dalinenko, I. N.; Ivanov, K. N.; Kaidalov, S. A.; Kuz'min, G. A.; Moskalev, B. B.; Naumov, S. K.; Pischelin, E. V.; Postovalov, V. E.; Prokhorov, Alexander M.; Schelev, Mikhail Y.)V1449,109-115(1991)

Detection with a charge-coupled device in atomic emission spectroscopy (Bilhorn, Robert B.)V1448,74-80(1991)

Development of a large pixel, spectrally optimized, pinned photodiode/interline CCD detector for the Earth Observing System/Moderate-Resolution Imaging Spectrometer-Tilt Instrument (Ewin, Audrey J.; Jhabvala, Murzy; Shu, Peter K.)V1479,12-20(1991)

Development of a synchrotron CCD-based area detector for structural biology (Kalata, Kenneth; Phillips, Walter C.; Stanton, Martin J.; Li, Youli)V1345,270-280(1991)

Digital charge-coupled device color TV system for endoscopy (Vishnevsky, G. I.; Berezin, V. Y.; Lazovsky, L. Y.; Vydrevich, M. G.; Rivkind, V. L.; Zhemerov, B. N.)V1447,34-43(1991)

Digital imaging microscopy: the marriage of spectroscopy and the solid state CCD camera (Jovin, Thomas M.; Arndt-Jovin, Donna J.)V1439,109-120(1991)

Direct view thermal imager (Reinhold, Ralph R.)V1447,251-262(1991)

Displacement damage in Si imagers for space applications (Dale, Cheryl J.; Marshall, Paul W.)V1447,70-86(1991)

Displacement measurement using grating images detected by CCD image sensor (Hane, Kazuhiro; Grover, Chandra P.)V1332,584-590(1991)

Effects of proton damage on charge-coupled devices (Janesick, James R.; Soli, George; Elliott, Tom; Collins, Stewart A.)V1447,87-108(1991)

Evaluation of a CCD star camera at Table Mountain Observatory (Strikwerda, Thomas E.; Fisher, H. L.; Frank, L. J.; Kilgus, Charles C.; Gray, Connie B.; Barnes, Donald L.)V1478,13-23(1991)

Fluorescence imaging of latent fingerprints with a cooled charge-coupled-device detector (Pomeroy, Robert S.; Baker, Mark E.; Radspinner, David A.; Denton, M. B.)V1439,60-65(1991)

Focal-plane processing algorithm and architecture for laser speckle interferometry (Tomlinson, Harold W.; Weir, Michael P.; Michon, G. J.; Possin, G. E.; Chovan, J.)V1541,178-186(1991)

Frequency-domain imaging using array detectors: present status and prospects for picosecond resolution (Morgan, Chris G.; Murray, J. G.; Mitchell, A. C.)V1525,83-90(1991)

Geometrical and radiometrical signal transfer characteristics of a color CCD camera with 21-million pixels (Lenz, Reimar K.; Lenz, Udo)V1526,123-132(1991)

Hadamard transform Raman imaging (Morris, Michael D.; Govil, Anurag; Liu, Kei-Lee; Sheng, Rong-Sheng)V1439,95-101(1991)

High-density CCD neurocomputer chip for accurate real-time segmentation of noisy images (Roth, Michael W.; Thompson, K. E.; Kulp, R. L.; Alvarez, E. B.)V1469,25-36(1991)

High-performance digital color video camera (Parulski, Kenneth A.; Benamati, Brian L.; D'Luna, Lionel J.; Shelley, Paul R.)V1448,45-58(1991)

High-precision digital charge-coupled device TV system (Vishnevsky, G. I.; Ioffe, S. A.; Berezin, V. Y.; Rybakov, M. I.; Mikhaylov, A. V.; Belyaev, L. V.)V1448,69-72(1991)

High-radiometric-performance CCD for the third-generation stratospheric aerosol and gas experiment (Delamere, W. A.; Baer, James W.; Ebben, Thomas H.; Flores, James S.; Kleiman, Gary; Blouke, Morley M.; McCormick, M. P.)V1447,204-213; V1479,31-40(1991)

High-resolution, low-light, image-intensified CCD camera (Tanaka, Satoru C.; Silvey, Tom; Long, Greg; Braze, Bill)V1448,21-26(1991)

High-resolution CCD still/video and still-still/video systems (Kee, Richard C.)V1448,13-20(1991)

High-resolution image digitizing camera for use in quantitative coronary arteriography (Muser, Markus H.; Leemann, Thomas; Anliker, M.)V1448,106-112(1991)

High-speed, high-resolution image reading technique using multi-area sensors (Uehira, Kazutake)V1448,182-190(1991)

High-speed CCD video camera (Germer, Rudolf K.; Meyer-Ilse, Werner)V1358,346-350(1991)

High-speed readout CCDs (Ball, K.; Burt, D. J.; Smith, Graham W.)V1358,409-420(1991)

High-speed signal processing architectures using charge-coupled devices (Boddu, Jayabharat; Udpa, Satish S.; Udpa, L.; Chan, Shiu Chuen M.)V1562,251-262(1991)

Image analysis of discrete and continuous systems: film and CCD sensors (Kriss, Michael A.)V1398,4-14(1991)

Impact of device characteristics on optical processor design (Turpin, Terry M.)V1562,2-10(1991)

Infrared background measurements at White Sands Missile Range, NM (Troyer, David E.; Fouse, Timothy; Murdaugh, William O.; Zammit, Michael G.; Rogers, Stephen B.; Skrzypczak, J. A.; Colley, Charles B.; Taczak, William J.)V1486,396-409(1991)

Interaction of algorithm and implementation for analog VLSI stereo vision (Hakkarainen, J. M.; Little, James J.; Lee, Hae-Seung; Wyatt, John L.)V1473,173-184(1991)

Intl Conf on Scientific Optical Imaging (Denton, M. B., ed.)V1439(1991)

IR CCD staring imaging system (Zhou, Qibo)V1540,677-680(1991)

ISTS array detector test facility (Thomas, Paul J.; Hollinger, Allan B.; Chu, Kan M.; Harron, John W.)V1488,36-47(1991)

Large-aperture CCD x-ray detector for protein crystallography using a fiber-optic taper (Strauss, Michael G.; Westbrook, Edwin M.; Naday, Istvan; Coleman, T. A.; Westbrook, M. L.; Travis, D. J.; Sweet, Robert M.; Pflugrath, J. W.; Stanton, Martin J.)V1447,12-27(1991)

Large-format 1280 x 1024 full-frame CCD image sensor with a lateral-overflow drain and transparent gate electrode (Stevens, Eric G.; Kosman, Steven L.; Cassidy, John C.; Chang, Win-Chyi; Miller, Wesley A.)V1447,274-282(1991)

Large-format CCDs for astronomical applications (Geary, John C.)V1439,159-168(1991)

Linear array camera interface techniques (DeLuca, Dan)V1396,558-565(1991)

Long-wavelength GexSi1-x/Si heterojunction infrared detectors and focal-plane arrays (Tsaur, Bor-Yeu; Chen, C. K.; Marino, S. A.)V1540,580-595(1991)

Megapixel CCD thinning/backside progress at SAIC (Schaeffer, A. R.; Varian, Richard H.; Cover, John R.; Larsen, Robert G.)V1447,165-176(1991)

MIC photon counting detector (Fordham, John L.; Bellis, J. G.; Bone, David A.; Norton, Timothy J.)V1449,87-98(1991)

Model of a thinned CCD (Blouke, Morley M.)V1439,136-143(1991)

Multiple-channel correlated double sampling amplifier hybrid to support a 64 parallel output CCD array (Raanes, Chris A.; McNeill, John A.; Cunningham, Andrew P.)V1358,637-643(1991)

New developments in CCD imaging devices for low-level confocal light imaging (Masters, Barry R.)V1428,169-176(1991)

New method for doing flat-field intensity calibrations of multiplexed ITT Streak Cameras (Hugenberg, Keith F.)V1346,390-397(1991)

NIR/CCD Raman spectroscopy: second battle of a revolution (McCreery, Richard L.)V1439,25-36(1991)

Noise performance of microchannel plate imaging systems (McCammon, Kent G.; Hagans, Karla G.; Hankla, A.)V1346,398-403(1991)

Notch and large-area CCD imagers (Bredthauer, Richard A.; Pinter, Jeff H.; Janesick, James R.; Robinson, Lloyd B.)V1447,310-315(1991)

One-dimensional CCD linear array readout device (Borodin, A. M.; Ivanov, K. N.; Naumov, S. K.; Philippov, S. A.; Postovalov, V. E.; Prokhorov, Alexander M.; Stepanov, M. S.; Schelev, Mikhail Y.)V1358,756-758(1991)

Performance of a thinned back-illuminated CCD coupled to a confocal microscope for low-light-level fluorescence imaging (Masters, Barry R.)V1447,56-63(1991)

Performance of high-dynamic-range CCD arrays with various epilayer structures (Smith, Dale J.; Harrison, Lorna J.; Pellegrino, John M.; Simon, Deborah R.)V1562,242-250(1991)

Performance tests of large CCDs (Robinson, Lloyd B.; Brown, William E.; Gilmore, Kirk; Stover, Richard J.; Wei, Mingzhi; Geary, John C.)V1447,214-228(1991)

Polarization and directionality of the Earth's reflectances: the POLDER instrument (Lorsignol, Jean; Hollier, Pierre A.; Deshayes, Jean-Pierre)V1490,155-163(1991)

Progress with PN-CCDs for the XMM satellite mission (Braueninger, Heinrich; Hauff, D.; Lechner, P.; Lutz, G.; Kink, W.; Meidinger, Norbert; Metzner, G.; Predehl, Peter; Reppin, C.; Strueder, Lothar; Truemper, Joachim; Kendziorra, E.; Staubert, R.; Radeka, V.; Rehak, P.; Rescia, S.; Bertuccio, G.; Gatti, E.; Longoni, Antonio; Sampietro, Marco; Findeis, N.; Holl, P.; Kemmer, J.; von Zanthier, C.)V1549,330-339(1991)

Quantitative microscope for image cytometry (Jaggi, Bruno; Poon, Steven S.; Pontifex, Brian; Fengler, John J.; Marquis, Jacques; Palcic, Branko)V1448,89-97(1991)

Quantum efficiency model for p+-doped back-illuminated CCD imager (Huang, Chin M.; Kosicki, Bernard B.; Theriault, Joseph R.; Gregory, J. A.; Burke, Barry E.; Johnson, Brett W.; Hurley, Edward T.)V1447,156-164(1991)

Radiation concerns for the Solar-A soft x-ray telescope (Acton, Loren W.; Morrison, Mons D.; Janesick, James R.; Elliott, Tom)V1447,123-139(1991)

Reimaging system for evaluating high-resolution charge-coupled-device arrays (Chambers, Robert J.; Warren, David W.; Lawrie, David J.; Lomheim, Terrence S.; Luu, K. T.; Shima, Ralph M.; Schlegel, J. D.)V1488,312-326(1991)

Relative performance studies for focal-plane arrays (Murphy, Kevin S.; Bradley, D. J.; Dennis, Peter N.)V1488,178-185(1991)

Remote media vision-based computer input device (Arabnia, Hamid R.; Chen, Ching-Yi)V1606,917-925(1991)

Selectable one-to-four-port, very high speed 512 x 512 charge-injection device (Zarnowski, Jeffrey J.; Williams, Bryn; Pace, M.; Joyner, M.; Carbone, Joseph; Borman, C.; Arnold, Frank S.; Wadsworth, Mark V.)V1447,191-201(1991)

Signal processing for low-light-level, high-precision CCD imaging (McCurnin, Thomas W.; Schooley, Larry C.; Sims, Gary R.)V1448,225-236(1991)

Simplified model of the back surface of a charge-coupled device (Blouke, Morley M.; Delamere, W. A.; Womack, G.)V1447,142-155(1991)

Slow-motion acquisition of laser beam profiles after propagation through gun blast (Kay, Armin V.)V1486,8-16(1991)

Small-satellite sensors for multispectral space surveillance (Kostishack, Daniel F.)V1495,214-227(1991)

Space-based visible surveillance experiment (Dyjak, Charles P.; Harrison, David C.)V1479,42-56(1991)

Space telescope imaging spectrograph 2048 CCD and its characteristics (Delamere, W. A.; Ebben, Thomas H.; Murata-Seawalt, Debbie; Blouke, Morley M.; Reed, Richard; Woodgate, Bruce E.)V1447,288-297(1991); V1479,21-30(1991)

Speckle measurement for 3-D surface movement (Hilbig, Jens; Ritter, Reinold)V1554A,588-592(1991)

Tale of two underwater lenses (Evans, Clinton E.; Doshi, Rekha)V1537,203-214(1991)

TDI camera for industrial applications (Castro, Peter; Gittings, J.; Choi, Yauho J.)V1448,134-139(1991)

Thinned backside-illuminated cooled CCDs for UV and VUV applications (Dalinenko, I. N.; Kuz'min, G. A.; Malyarov, A. V.; Prokhorov, Alexander M.; Schelev, Mikhail Y.)V1449,167-172(1991)

Three-dimensional imaging using TDI CCD sensors (Fenster, Aaron; Holdsworth, David W.; Drangova, Maria)V1447,28-33(1991)

Time-resolved x-ray scattering studies using CCD detectors (Clarke, Roy; Dos Passos, Waldemar; Lowe, Walter P.; Rodricks, Brian G.; Brizard, Christine M.)V1345,101-114(1991)

Two-dimensional electron-bombarded CCD readout device for picosecond electron-optical information system (Ivanov, K. N.; Krutikov, N. I.; Naumov, S. K.; Pischelin, E. V.; Semenov, V. A.; Stepanov, M. S.; Postovalov, V. E.; Prokhorov, Alexander M.; Schelev, Mikhail Y.)V1358,732-738(1991)

Two-dimensional electron gas charge-coupled devices (Fossum, Eric R.; Song, Jong I.; Rossi, David V.)V1447,202-203(1991)

Update on focal-plane image processing research (Kemeny, Sabrina E.; Eid, El-Sayed I.; Mendis, Sunetra; Fossum, Eric R.)V1447,243-250(1991)

Use of a cooled CCD camera for confocal light microscopy (Masters, Barry R.)V1448,98-105(1991)

Validated CCD camera model (Johnson-Cole, Helen; Clark, Rodney L.)V1488,203-211(1991)

Virtual-phase charge-coupled device image sensors for industrial and scientific applications (Khvilivitzky, A. T.; Berezin, V. Y.; Lazovsky, L. Y.; Tataurschikov, S. S.; Pisarevsky, A. N.; Vydrevich, M. G.; Kossov, V. G.)V1447,184-190(1991)

Virtual-phase charge-coupled device imaging system for astronomy (Khvilivitzky, A. T.; Zuev, A. G.; Rybakov, M. I.; Kiryan, G. V.; Berezin, V. Y.)V1447,64-68(1991)

X-ray detector for time-resolved studies (Rodricks, Brian G.; Brizard, Christine M.; Clarke, Roy; Lowe, Walter P.)V1550,18-26(1991)

Chemistry—see also biochemistry; laser chemistry

2.9 micron laser source for use in the two-photon/laser-induced fluorescence detection of atmospheric OH (Bradshaw, John D.; van Dijk, Cornelius A.)V1433,81-91(1991)

AcidMODES: a major field study to evaluate regional-scale air pollution models (Ching, Jason K.; Bowne, Norman E.)V1491,360-370(1991)

Advanced portable four-wavelength NIR analyzer for rapid chemical composition analysis (Malinen, Jouko; Hyvarinen, Timo S.)V1510,204-209(1991)

Advances in analytical chemistry (Arendale, William F.; Congo, Richard T.; Nielsen, Bruce J.)V1434,159-170(1991)

Advances in R&D in near-infrared spectroscopy in Japan (Kawano, Sumio; Iwamoto, Mutsuo)V1379,2-9(1991)

Advances in Resist Technology and Processing VIII (Ito, Hiroshi, ed.)V1466(1991)

Airborne chemical contamination of a chemically amplified resist (MacDonald, Scott A.; Clecak, Nicholas J.; Wendt, H. R.; Willson, C. G.; Snyder, C. D.; Knors, C. J.; Deyoe, N. B.; Maltabes, John G.; Morrow, James R.; McGuire, Anne E.; Holmes, Steven J.)V1466,2-12(1991)

Air quality monitoring with the differential optical absorption spectrometer (Stevens, Robert K.; Conner, Teri L.)V1491,56-67(1991)

Aluminum metal combustion in water revealed by high-speed microphotography (Tao, William C.; Frank, Alan M.; Clements, Rochelle E.; Shepherd, Joseph E.)V1346,300-310(1991)

Analysis of acid-base indicators covalently bound on glass supports (Baldini, Francesco; Bacci, Mauro; Bracci, Susanna)V1368,210-217(1991)

Analysis of silage composition by near-infrared reflectance spectroscopy (Reeves, James B.; Blosser, Timothy H.; Colenbrander, V. F.)V1379,28-38(1991)

Antibody-based fiber optic sensors for environmental and process-control applications (Barnard, Steven M.; Walt, David R.)V1368,86-92(1991)

Application aspects of the Si-CARL bilayer process (Sebald, Michael; Berthold, Joerg; Beyer, Michael; Leuschner, Rainer; Noelscher, Christoph; Scheler, Ulrich; Sezi, Recai; Ahne, Hellmut; Birkle, Siegfried)V1466,227-237(1991)

Application of the NO/O3 chemiluminescence technique to measurements of reactive nitrogen species in the stratosphere (Fahey, David W.)V1433,212-223(1991)

Brownian dynamics simulation of polarized light scattering from wormlike chains (Allison, Stuart A.)V1430,50-64(1991)

Calibration for the SAGE III/EOS instruments (Chu, William P.; McCormick, M. P.; Zawodny, J. M.; McMaster, Leonard R.)V1491,243-250(1991)

Chemical amplifier for peroxy radical measurements based on luminol chemiluminescence (Cantrell, Chris A.; Shetter, Richard E.; Lind, John A.; Gilliland, Curt A.; Calvert, Jack G.)V1433,263-268(1991)

Chemical, Biochemical, and Environmental Fiber Sensors II (Lieberman, Robert A.; Wlodarczyk, Marek T., eds.)V1368(1991)

Chemical symmetry: developers that look like bleach agents for holography (Bjelkhagen, Hans I.; Phillips, Nicholas J.; Ce, Wang)V1461,321-328(1991)

Chemistry of the Konica Dry Color System (Suda, Yoshihiko; Ohbayashi, Keiji; Onodera, Kaoru)V1458,76-78(1991)

Chlorine sensing by optical techniques (Momin, S. A.; Narayanaswamy, R.)V1510,180-186(1991)

Co-immobilization of several dyes on optodes for pH measurements (Boisde, Gilbert; Sebille, Bernard)V1510,80-94(1991)

Compact measurement system for the simultaneous determination of NO, NO2, NOy, and O3 using a small aircraft (Walega, James G.; Dye, James E.; Grahek, Frank E.; Ridley, Brian K.)V1433,232-241(1991)

Determination of the altitude of the nitric acid layer from very high resolution, ground-based IR solar spectra (Blatherwick, Ronald D.; Murcray, Frank J.; Murcray, David G.; Locker, M. H.)V1491,203-210(1991)

Developing a long-path diode array spectrometer for tropospheric chemistry studies (Lanni, Thomas R.; Demerjian, Kenneth L.)V1433,21-24(1991)

Diffusing wave spectroscopy studies of gelling systems (Horne, David S.)V1430,166-180(1991)

Diffusion of spherical probes in aqueous systems containing the semiflexible polymer hydroxypropylcellulose (Mustafa, Mazidah B.; Russo, Paul S.)V1430,132-141(1991)

Direct exchange of metal ions onto silica waveguides (Petersen, James V.; Dessy, Raymond E.)V1368,61-72(1991)

Dissolution inhibition mechanism of ANR photoresists: crosslinking vs. -OH site consumption (Thackeray, James W.; Orsula, George W.; Rajaratnam, Martha M.; Sinta, Roger F.; Herr, Daniel J.; Pavelchek, Edward K.)V1466,39-52(1991)

Dissolution of poly(p-hydroxystyrene): molecular weight effects (Long, Treva; Rodriguez, Ferdinand)V1466,188-198(1991)

Drilled optical fiber sensors: a novel single-fiber sensor (Lipson, David; McLeaster, Kevin D.; Cohn, Brian; Fischer, Robert E.)V1368,36-43(1991)

Dynamical structure factor of a solution of charged rod-like polymers in the isotropic phase (Maeda, T.; Doi, Masao)V1403,268-277(1991)

Dynamic light scattering from strongly interacting multicomponent systems: salt-free polyelectrolyte solutions (Sedlak, Marian; Amis, Eric J.; Konak, Cestmir; Stepanek, Petr)V1430,191-202(1991)

Dynamic light scattering studies of resorcinol formaldehyde gels as precursors of organic aerogels (Cotts, Patricia M.; Pekala, Rick)V1430,181-190(1991)

Dynamics of stiff macromolecules in concentrated polymer solutions: model of statistical reorientations (Brazhnik, O. D.; Khokhlov, A. R.)V1402,70-77(1991)

Effect of alternating magnetic fields on the properties of water systems (Berezin, M. V.; Levshin, L. V.; Saletsky, A. M.)V1403,335-337(1991)

Enantio-selective optode for the B-blocker propranolol (He, Huarui; Uray, Georg; Wolfbeis, Otto S.)V1368,175-180(1991)

Environmental Sensing and Combustion Diagnostics (Santoleri, Joseph J., ed.)V1434(1991)

Evaluation of polymeric thin film waveguides as chemical sensors (Bowman, Elizabeth M.; Burgess, Lloyd W.)V1368,239-250(1991)

Exposure dose optimization for a positive resist containing polyfunctional photoactive compound (Trefonas, Peter; Mack, Chris A.)V1466,117-131(1991)

Far-IR Fabry-Perot spectrometer for OH measurements (Pickett, Herbert M.; Peterson, Dean B.)V1491,308-313(1991)

Feasibility of optically sensing two parameters simultaneously using one indicator (Wolfbeis, Otto S.)V1368,218-222(1991)

Fiber optic based chemical sensor system for in-situ process measurements using the photothermal effect (Walker, Karl-Heinz; Sontag, Heinz)V1510,212-217(1991)

Fiber optic fluorescence sensors based on sol-gel entrapped dyes (MacCraith, Brian D.; Ruddy, Vincent; Potter, C.; McGilp, J. F.; O'Kelley, B.)V1510,104-109(1991)

Fiber optic liquid crystalline microsensor for temperature measurement in high magnetic field (Domanski, Andrzej W.; Kostrzewa, Stanislaw)V1510,72-77(1991)

Fiber optic sensor for nitrates in water (MacCraith, Brian D.; Maxwell, J.)V1510,195-203(1991)

Fiber optic sensor probe for in-situ surface-enhanced Raman monitoring (Vo-Dinh, Tuan; Stokes, D. L.; Li, Ying-Sing; Miller, Gordon H.)V1368,203-209(1991)

Flow optrodes for chemical analysis (Berman, Richard J.; Burgess, Lloyd W.)V1368,25-35(1991)

Fluorescence-based optrodes for alkali ions based on the use of ion carriers and lipophilic acid/base indicators (He, Huarui; Wolfbeis, Otto S.)V1368,165-171(1991)

Frequency modulation spectroscopy for chemical sensing of the environment (Cooper, David E.; Riris, Haris; van der Laan, Jan E.)V1433,120-127(1991)

Generalized characteristic model for lithography: application to negative chemically amplified resists (Ziger, David H.; Mack, Chris A.; Distasio, Romelia)V1466,270-282(1991)

Gradient microbore liquid chromatography with dual-wavelength absorbance detection: tunable analyzers for remote chemical monitoring (Sulya, Andrew W.; Moore, Leslie K.; Synovec, Robert E.)V1434,147-158(1991)

High-Tg nonlinear optical polymer: poly(N-MNA acrylamide) (Herman, Warren N.; Rosen, Warren A.; Sperling, L. H.; Murphy, C. J.; Jain, H.)V1560,206-213(1991)

Holographic interferometry in corrosion studies of metals: I. Theoretical aspects (Habib, Khaled J.)V1332,193-204(1991)

Holographic interferometry in corrosion studies of metals: II. Applications (Habib, Khaled J.)V1332,205-215(1991)

Hydrogen peroxide and organic peroxides in the marine environment (Heikes, Brian G.; Miller, William L.; Lee, Meehye)V1433,253-262(1991)

Identification of modes in dynamic scattering on ternary polymer mixtures (Akcasu, A. Z.)V1430,142-143(1991)

Improvement in the performance of evanescent wave chemical sensors by special waveguide structures (Stewart, George; Culshaw, Brian; Clark, Douglas F.; Andonovic, Ivan)V1368,230-238(1991)

Influence of gas composition on XeCl laser performance (Jursich, Gregory M.; Von Drusek, William A.; Mulderink, Ken; Olchowka, V.; Reid, John; Brimacombe, Robert K.)V1412,115-122(1991)

Investigation of hot electron emission in MOS structure under avalanche conditions (Solonko, Alexander G.)V1435,360-365(1991)

In-situ CARS detection of H2 in the CVD of Si3N4 (Hay, Stephen O.; Veltri, R. D.; Lee, W. Y.; Roman, Ward C.)V1435,352-358(1991)

In-situ structural studies of the underpotential deposition of copper onto an iodine-covered platinum surface using x-ray standing waves (Bommarito, G. M.; Acevedo, D.; Rodriguez, J. R.; Abruna, H. D.)V1550,156-170(1991)

In-situ surface x-ray scattering of metal monolayers adsorbed at solid-liquid interfaces (Toney, Michael F.; Gordon, Joseph G.; Melroy, Owen R.)V1550,140-150(1991)

Kinetic studies of phosgene reduction via in situ Fourier transform infrared analysis (Farquharson, Stuart; Chauvel, J. P.)V1434,135-146(1991)

Laser probe mass spectrometry (Campana, Joseph E.)V1437,138-149(1991)

Luminescence probes in aqueous micellar solutions (Vecher, Jaroslav)V1402,97-101(1991)

Measurement of atmospheric composition by the ATMOS instrument from Table Mountain Observatory (Gunson, Michael R.; Irion, Fredrick W.)V1491,335-346(1991)

Measurement of peroxyacetyl nitrate, NO2, and NOx by using a gas chromatograph with a luminol-based detector (Drummond, John W.; Mackay, Gervase I.; Schiff, Harold I.)V1433,242-252(1991)

Measurement of the tropospheric hydroxyl radical by long-path absorption (Mount, George H.)V1491,26-32(1991)

Mechanism of dissolution inhibition of novolak-diazoquinone resist (Furuta, Mitsuhiro; Asaumi, Shingo; Yokota, Akira)V1466,477-484(1991)

Multiphoton resonance ionization of molecules desorbed from surfaces by ion beams (Winograd, Nicholas; Hrubowchak, D. M.; Ervin, M. H.; Wood, M. C.)V1435,2-11(1991)

Multiple mode and multiple source coupling into polymer thin-film waveguides (Potter, B. L.; Walker, D. S.; Greer, L.; Saavedra, Steven S.; Reichert, William M.)V1368,251-257(1991)

Negative chemical amplification resist systems based on polyhydroxystyrenes and N-substituted imides or aldehydes (Ito, Hiroshi; Schildknegt, Klaas; Mash, Eugene A.)V1466,408-418(1991)

Negative resist systems using acid-catalyzed pinacol rearrangement reaction in a phenolic resin matrix (Uchino, Shou-ichi; Iwayanagi, Takao; Ueno, Takumi; Hayashi, Nobuaki)V1466,429-435(1991)

Network for the detection of stratospheric change (Kurylo, Michael J.)V1491,168-174(1991)

New aqueous base-developable negative-tone photoresist based on furans (Fahey, James T.; Frechet, Jean M.)V1466,67-74(1991)

New developments in the field of chemical infrared fiber sensors (Kellner, Robert A.; Taga, Karim)V1510,232-241(1991)

New luminescent metal complex for pH transduction in optical fiber sensing: application to a CO2-sensitive device (Moreno-Bondi, Maria C.; Orellana, Guillermo; Camara, Carmen; Wolfbeis, Otto S.)V1368,157-164(1991)

New problems of femtosecond time-domain CARS (coherent antistokes Raman spectroscopy) of large molecules (Kolomoitsev, D. V.; Nikitin, S. Y.)V1402,31-43(1991)

Nondestructive determination of the solids content of horticultural products (Birth, Gerald S.; Dull, Gerald G.; Leffler, Richard G.)V1379,10-15(1991)

Nonmetallic acid generators for i-line and g-line chemically amplified resists (Brunsvold, William R.; Montgomery, Warren; Hwang, Bao)V1466,368-376(1991)

Novel acid-hardening positive photoresist technology (Graziano, Karen A.; Thompson, Stephen D.; Winkle, Mark R.)V1466,75-88(1991)

Novel novolak resins using substituted phenols for high-performance positive photoresist (Kajita, Toru; Ota, Toshiyuki; Nemoto, Hiroaki; Yumoto, Yoshiji; Miura, Takao)V1466,161-173(1991)

Novel quinonediazide-sensitized photoresist system for i-line and deep-UV lithography (Fukunaga, Seiki; Kitaori, Tomoyuki; Koyanagi, Hiroo; Umeda, Shin'ichi; Nagasawa, Kohtaro)V1466,446-457(1991)

Novolak design for high-resolution positive photoresists (IV): tandem-type novolak resin for high-performance positive photoresists (Hanabata, Makoto; Oi, F.; Furuta, Akihiro)V1466,132-140(1991)

Novolak resin design concept for high-resolution positive resists (Noguchi, Tsutomu; Tomita, Hidemi)V1466,149-160(1991)

Onium salt structure/property relationships in poly(4-tert-butyloxycarbonyloxystyrene) deep-UV resists (Schwartzkopf, George; Niazy, Nagla N.; Das, Siddhartha; Surendran, Geetha; Covington, John B.)V1466,26-38(1991)

On-line optical determination of water in ethanol (Kessler, Manfred A.)V1510,218-223(1991)

Optical fiber interferometric sensors for chemical detection (Butler, Michael A.)V1368,46-54(1991)

Overview of planar waveguide techniques for chemical sensing (Burgess, Lloyd W.)V1368,224-229(1991)

Photon Correlation Spectroscopy: Multicomponent Systems (Schmitz, Kenneth S., ed.)V1430(1991)

Photon correlation spectroscopy: technique and instrumentation (Thomas, John C.)V1430,2-18(1991)

Physical effects in time-domain CARS (coherent antistokes Raman spectroscopy) of molecular gases (Kolomoitsev, D. V.; Nikitin, S. Y.)V1402,11-30(1991)

Polyvinylphenols protected with tetrahydropyranyl group in chemical amplification positive deep-UV resist systems (Hayashi, Nobuaki; Schlegel, Leo; Ueno, Takumi; Shiraishi, Hiroshi; Iwayanagi, Takao)V1466,377-383(1991)

Preliminary field demonstration of a fiber optic trichloroethylene sensor (Angel, S. M.; Langry, Kevin; Colston, B. W.; Roe, Jeffrey N.; Daley, Paul F.; Milanovich, Fred P.)V1368,98-104(1991)

Preliminary lithographic characteristics of an all-organic chemically amplified resist formulation for single-layer deep-UV lithography (Nalamasu, Omkaram; Reichmanis, Elsa; Cheng, May; Pol, Victor; Kometani, Janet M.; Houlihan, Frank M.; Neenan, Thomas X.; Bohrer, M. P.; Mixon, D. A.; Thompson, Larry F.; Takemoto, Cliff H.)V1466,13-25(1991)

Preparations and properties of novel positive photosensitive polyimides (Hayase, Rumiko H.; Kihara, Naoko; Oyasato, Naohiko; Matake, S.; Oba, Masayuki)V1466,438-445(1991)

Process latitude for the chemical amplification resists AZ PF514 and AZ PN114 (Eckes, Charlotte; Pawlowski, Georg; Przybilla, Klaus J.; Meier, Winfried; Madore, Michel; Dammel, Ralph)V1466,394-407(1991)

Progress in DUV resins (Przybilla, Klaus J.; Roeschert, Heinz; Spiess, Walter; Eckes, Charlotte; Chatterjee, Subhankar; Khanna, Dinesh N.; Pawlowski, Georg; Dammel, Ralph)V1466,174-187(1991)

Quasielastic and electrophoretic light scattering studies of polyelectrolyte-micelle complexes (Rigsbee, Daniel R.; Dubin, Paul L.)V1430,203-215(1991)

Recent lidar measurements of stratospheric ozone and temperature within the Network for the Detection of Stratospheric Change (McGee, Thomas J.; Ferrare, Richard; Butler, James J.; Frost, Robert L.; Gross, Michael; Margitan, James)V1491,182-187(1991)

Recent progress in fiber optic pH sensing (Baldini, Francesco)V1368,184-190(1991)

Recent progress in intrinsic fiber optic chemical sensing (Lieberman, Robert A.)V1368,15-24(1991)

Relationship between conjugation length and third-order nonlinearity in bis-donor substituted diphenyl polyenes (Spangler, Charles W.; Havelka, Kathleen O.; Becker, Mark W.; Kelleher, Tracy A.; Cheng, Lap Tak A.)V1560,139-147(1991)

Remote Sensing of Atmospheric Chemistry (McElroy, James L.; McNeal, Robert J., eds.)V1491(1991)

Revolutionary impact of today's array detector technology on chemical analysis (Radspinner, David A.; Fields, Robert E.; Earle, Colin W.; Denton, M. B.)V1439,2-14(1991)

Simulation of spin label behavior on a model surface (Balabaev, N. K.; Lemak, A. S.; Fushman, D. A.; Mironova, Yu. V.)V1402,53-69(1991)

Single-component chemically amplified resist materials for electron-beam and x-ray lithography (Novembre, Anthony E.; Tai, Woon W.; Kometani, Janet M.; Hanson, James E.; Nalamasu, Omkaram; Taylor, Gary N.; Reichmanis, Elsa; Thompson, Larry F.)V1466,89-99(1991)

Smectic liquid crystals modified by tail and core fluorination (Twieg, Robert J.; Betterton, K.; DiPietro, Richard; Gravert, D.; Nguyen, Cattien; Nguyen, H. T.; Babeau, A.; Destrade, C.)V1455,86-96(1991)

Structural effects of DNQ-PAC backbone on resist lithographic properties (Uenishi, Kazuya; Kawabe, Yasumasa; Kokubo, Tadayoshi; Slater, Sydney G.; Blakeney, Andrew J.)V1466,102-116(1991)

Structure/property relationships for molecular second-order nonlinear optics (Marder, Seth R.; Cheng, Lap Tak A.; Tiemann, Bruce G.; Beratan, David N.)V1560,86-97(1991)

Studies of dissolution inhibition mechanism of DNQ-novolak resist (II): effect of extended ortho-ortho bond in novolak (Honda, Kenji; Beauchemin, Bernard T.; Fitzgerald, Edward A.; Jeffries, Alfred T.; Tadros, Sobhy P.; Blakeney, Andrew J.; Hurditch, Rodney J.; Tan, Shiro; Sakaguchi, Shinji)V1466,141-148(1991)

Studies of structures and phase transitions of Langmuir monolayers using synchrotron radiation (Dutta, Pulak)V1550,134-139(1991)

Study of the chemically amplifiable resist materials for electron-beam lithography (Koyanagi, Hiroo; Umeda, Shin'ichi; Fukunaga, Seiki; Kitaori, Tomoyuki; Nagasawa, Kohtaro)V1466,346-361(1991)

Summary of atmospheric chemistry observations from the Antarctic and Arctic aircraft campaigns (Tuck, Adrian F.)V1491,252-272(1991)

Superradiance and exciton dynamics in molecular aggregates (Fidder, Henk; Terpstra, Jacob; Wiersma, Douwe A.)V1403,530-544(1991)

Synthesis and nonlinear optical activity of cumulenes (Ermer, Susan P.; Lovejoy, Steven M.; Leung, Doris; Spitzer, Ronnie; Hansen, Glenn A.; Stone, Richard E.)V1560,120-129(1991)

Synthetic approaches to long-wavelength photosensitizers for photodynamic therapy (Shiau, Fuu-Yau; Pandey, Ravindra K.; Dougherty, Thomas J.; Smith, Kevin M.)V1426,330-339(1991)

Theoretical insight into the quadratic nonlinear optical response of organics: derivatives of pyrene and triaminotrinitrobenzene (Bredas, Jean-Luc; Dehu, C.; Meyers, F.; Zyss, Joseph)V1560,98-110(1991)

Upper-atmosphere research satellite: an overview (McNeal, Robert J.)V1491,84-90(1991)

Use of chemiluminescence techniques in portable, lightweight, highly sensitive instruments for measuring NO2, NOx, and O3 (Drummond, John W.; Topham, L. A.; Mackay, Gervase I.; Schiff, Harold I.)V1433,224-231(1991)

Use of the Boulder model for the design of high-polarization fluorinated ferroelectric liquid crystals (Wand, Michael D.; Vohra, Rohini; Thurmes, William; Walba, David M.; Geelhaar, Thomas; Littwitz, Brigitte)V1455,97-104(1991)

USSR-CSFR Joint Seminar on Nonlinear Optics in Control, Diagnostics, and Modeling of Biophysical Processes (Akhmanov, Sergei A.; Zadkov, Victor N., eds.)V1402(1991)

VUV/photofragmentation laser-induced fluorescence sensor for the measurement of atmospheric ammonia (Sandholm, Scott T.; Bradshaw, John D.)V1433,69-80(1991)

Chromogenics—see also coatings; solar energy

Cation intercalation in electrochromic NiOx films (Scarminio, J.; Gorenstein, Annette; Decker, Franco; Passerini, S.; Pileggi, R.; Scrosati, Bruno)V1536,70-80(1991)

Detection by mirage effect of the counter-ion flux between an electrochrome and a liquid electrolyte: application to WO3, Prussian blue, and lutetium diphthalocyanine films (Plichon, V.; Giron, J. C.; Delboulbe, J. P.; Lerbet, F.)V1536,37-47(1991)

Development of laminated nickel/manganese oxide and nickel/niobium oxide electrochromic devices (Ma, Y. P.; Yu, Phillip C.; Lampert, Carl M.)V1536,93-103(1991)

Electrochromic materials for smart window applications (Ashrit, Pandurang V.; Bader, G.; Girouard, Fernand E.; Truong, Vo-Van)V1401,119-129(1991)

Electrochromic properties and temperature dependence of chemically deposited Ni(OH)x thin films (Fantini, Marcia C.; Bezerra, George H.; Carvalho, C. R.; Gorenstein, Annette)V1536,81-92(1991)

Electrochromism in cobalt oxyhydroxide thin films (Gorenstein, Annette; Polo Da Fonseca, C. N.; Torresi, R. M.)V1536,104-115(1991)

Light scattering properties of new materials for glazing applications (Bergkvist, Mikael; Roos, Arne)V1530,352-362(1991)

Nondiffractive optically variable security devices (van Renesse, Rudolf L.)V1509,113-125(1991)

Optical Materials Technology for Energy Efficiency and Solar Energy Conversion X (Lampert, Carl M.; Granqvist, Claes G., eds.)V1536(1991)

Progress on the variable reflectivity electrochromic window (Goldner, Ronald B.; Arntz, Floyd O.; Berera, G.; Haas, Terry E.; Wei, G.; Wong, Kwok-keung; Yu, Phillip C.)V1536,63-69(1991)

Review on electrochromic devices for automotive glazing (Demiryont, Hulya)V1536,2-28(1991)

Solid state ionics and optical materials technology for energy efficiency, solar energy conversion, and environment control (Lusis, Andrejs R.)V1536,116-124(1991)

Structure and properties of electrochromic WO3 produced by sol-gel methods (Bell, J. M.; Green, David C.; Patterson, A.; Smith, Geoffrey B.; MacDonald, K. A.; Lee, K. D.; Kirkup, L.; Cullen, J. D.; West, B. O.; Spiccia, L.; Kenny, M. J.; Wielunski, L. S.)V1536,29-36(1991)

Study on the special properties of electrochromic film of a-WO3 (Luo, Zhongkuan)V1489,124-134(1991)

Transparent storage layers for H+ and Li+ ions prepared by sol-gel technique (Valla, Bruno; Tonazzi, Juan C.; Macedo, Marcelo A.; Dall'Antonia, L. H.; Aegerter, Michel A.; Gomes, M. A.; Bulhoes, Luis O.)V1536,48-62(1991)

Coatings—see also chromogenics; deposition; optical materials; thin films

Advanced broadband baffle materials (Seals, Roland D.)V1485,78-87(1991)

Analysis and design of binary gratings for broadband, low-reflectivity surfaces (Moharam, M. G.)V1485,254-259(1991)

Antireflection coating standards of ophthalmic resin lens materials (Porden, Mark)V1529,115-123(1991)

Application of photoelastic coating method on elastoplastic stress analysis of rotation disk (Dong, Benhan; Gao, Penfei; Wang, Ju)V1554A,400-406(1991)

Applications of pulsed photothermal deflection technique in the study of laser-induced damage in optical coatings (Wu, Zhouling; Reichling, M.; Fan, Zheng X.; Wang, Zhi-Jiang)V1441,214-227(1991)

Beryllium and titanium cost-adjustment report (Owen, John; Ulph, Eric)V1485,128-137(1991)

Broadband, antireflection coating designs for large-aperture infrared windows (Balasubramanian, Kunjithapa; Le, Tam V.; Guenther, Karl H.; Kumar, Vas)V1485,245-253(1991)

Coating development for high-energy KrF excimer lasers (Boyer, James D.; Mauro, Billie R.; Sanders, Virgil E.)V1377,92-98(1991)

Coating thickness gauge (Honda, Tatsuro; Matsui, Kenichi)V1540,709-716(1991)

Color- and intensity-balanced multichannel optical beamsplitter (Gilo, Mordechai; Rabinovitch, Kopel)V1442,90-104(1991)

Correlation of surface topography and coating damage with changes in the responsivity of silicon PIN photodiodes (Huffaker, Diana L.; Walser, Rodger M.; Becker, Michael F.)V1441,365-380(1991)

Design considerations for multilayer-coated Schwarzchild objectives for the XUV (Kortright, Jeffrey B.; Underwood, James H.)V1343,95-103(1991)

Development of damage resistant optics for KrF excimer lasers (Boyer, James D.; Mauro, Billie R.; Sanders, Virgil E.)V1441,255-261(1991)

Differential coating objective (Goldfain, Ervin)V1527,126-133(1991)

Durable, nonchanging, metal-dielectric and all-dielectric mirror coatings (Guenther, Karl H.; Balasubramanian, Kunjithapa; Hu, X. Q.)V1485,240-244(1991)

Effective refractive indices of three-phase optical coatings (Ma, Yushieh; Varadan, Vijay K.; Varadan, Vasundara V.)V1558,138-142(1991)

Effect of the space environment on thermal control coatings (Harada, Yoshiro; Mell, Richard J.; Wilkes, Donald R.)V1330,90-101(1991)

Effects of polishing materials on the laser damage threshold of optical coatings (Crase, Robert J.)V1441,381-389(1991)

Fabrication and characterization of microwave-plasma-assisted chemical vapor deposited dielectric coatings (Wood, Roger M.; Greenham, A. C.; Nichols, B. A.; Nourshargh, Noorallah; Lewis, Keith L.)V1441,316-326(1991)

Fabrication and performance of CdSe/CdS/ZnS photoconductor for liquid-crystal light valve (Zhuang, Song Lin; Jiang, Yingqui; Qiu, Yinggang; Gu, Lingjuan; Cai, Zhonghua; Chen, Wei)V1558,28-33(1991)

Fast full-erasure laser-addressed smectic liquid-crystal light valve (Zhuang, Song Lin; Qiu, Yinggang; Jiang, Yingqui; Tu, Yijun; Chen, Wei)V1558,149-153(1991)

High-frequency fiber optic phase modulator using piezoelectric polymer coating (Imai, Masaaki M.; Yano, T.; Ohtsuka, Yoshihiro)V1371,13-20(1991)

High laser-damage threshold and low-cost sol-gel-coated epoxy-replicated mirrors (Floch, Herve G.; Berger, Michel; Novaro, Marc; Thomas, Ian M.)V1441,304-315(1991)

High-reflective multilayers as narrowband VUV filters (Zukic, Muamer; Torr, Douglas G.)V1485,216-227(1991)

High-threshold highly reflective coatings at 1064 nm (Rainer, Frank; De Marco, Frank P.; Hunt, John T.; Morgan, A. J.; Mott, Leonard P.; Marcelja, Frane; Greenberg, Michael R.)V1441,247-254(1991)

Holographic mirrors laminated into windshields for automotive head-up display and solar protective glazing applications (Beeck, Manfred-Andreas; Frost, Thorsten; Windeln, Wilbert)V1507,394-406(1991)

Infrared thermal-wave studies of coatings and composites (Favro, Lawrence D.; Ahmed, Tasdiq; Crowther, D. J.; Jin, Huijia J.; Kuo, Pao K.; Thomas, Robert L.; Wang, X.)V1467,290-294(1991)

In-situ CARS detection of H2 in the CVD of Si3N4 (Hay, Stephen O.; Veltri, R. D.; Lee, W. Y.; Roman, Ward C.)V1435,352-358(1991)

Intl Conf on Thin Film Physics and Applications (Zhou, Shixun; Wang, Yonglin, eds.)V1519(1991)

Investigation of Ti-Al-Mo-V alloy nitride coatings by ARC technique (Wang, Ren; Yang, Guang Y.; Wu, Bei X.; Fu, Bao W.; Zhan, Yun C.; Zhang, Yun H.)V1519,146-151(1991)

Investigation of neutral atom and ion emission during laser conditioning of multilayer HfO2-SiO2 coatings (Schildbach, M. A.; Chase, Lloyd L.; Hamza, Alex V.)V1441,287-293(1991)

Investigation of process sensitivity for electron-beam evaporation of beryllium (Egert, Charles M.; Schmoyer, D. D.; Nordin, C. W.; Berry, A.)V1485,64-77(1991)

Ion-assisted deposition of graded index silicon oxynitride coatings (Al-Jumaily, Ghanim A.; Gagliardi, F. J.; McColl, P.; Mizerka, Larry J.)V1441,360-365(1991)

Large-scale production of a broadband antireflection coating on ophthalmic lenses with ion-assisted deposition (Andreani, F.; Luridiana, S.; Mao, Shu Z.)V1519,63-69(1991)

Light-absorbing, lightweight beryllium baffle materials (Murray, Brian W.; Floyd, Dennis R.; Ulph, Eric)V1485,88-95(1991)

Light scattered by coated paper (Marx, Egon; Song, J. F.; Vorburger, Theodore V.; Lettieri, Thomas R.)V1332,826-834(1991)

Low-frequency chiral coatings (Ro, Ru-Yen; Varadan, Vasundara V.; Varadan, Vijay K.)V1489,46-55(1991)

Materials for optimal multilayer coating (Grebenshikov, Sergey V.)V1519,302-307(1991)

Mirror fabrication for full-wafer laser technology (Webb, David J.; Benedict, Melvin K.; Bona, Gian-Luca; Buchmann, Peter L.; Daetwyler, K.; Dietrich, H. P.; Moser, A.; Sasso, G.; Vettiger, Peter; Voegeli, O.)V1418,231-239(1991)

Modifications of optical properties with ceramic coatings (Besmann, Theodore M.; Abdel-Latif, A. I.)V1330,78-89(1991)

Morphology and laser damage studies by atomic force microscopy of e-beam evaporation deposited antireflection and high-reflection coatings (Tesar, Aleta A.; Balooch, M.; Shotts, K. W.; Siekhaus, Wigbert J.)V1441,228-236(1991)

Multiple-pulse laser damage to thin film optical coating (Li, Zhong Y.; Li, Cheng F.; Guo, Ju P.)V1519,374-379(1991)

New coating technology and ion source (Yan, Yi S.)V1519,192-193(1991)

New ion-beam sources and their applications to thin film physics (Wei, David T.; Kaufman, Harold R.)V1519,47-55(1991)

Next generation thermal control coatings (Grieser, James L.; Swisher, Richard L.; Phipps, James A.; Pelleymounter, Douglas R.; Hildreth, Eugene N.)V1330,111-118(1991)

Novel cathodic arc plasma PVD system with column target for the deposition of TiN film and other metallic films (Liu, Wei Y.; Li, Yu M.; Cui, Zhan J.; He, Tian X.)V1519,172-174(1991)

Novel perfluorinated antireflective and protective coating for KDP and other optical materials (Thomas, Ian M.; Campbell, John H.)V1441,294-303(1991)

Ophthalmic antireflection coatings with same residual reflective colors on ophthalmic optics with different refractive indices (Jin, Tianfeng; Yuan, Youxin)V1529,132-137(1991)

Opthalmic Lens Design and Fabrication (Perrott, Colin M., ed.)V1529(1991)

Optical and environmentally protective coatings for potassium dihydrogen phosphate harmonic converter crystals (Thomas, Ian M.)V1561,70-82(1991)

Optical characterization of damage resistant kilolayer rugate filters (Elder, Melanie L.; Jancaitis, Kenneth S.; Milam, David; Campbell, John H.)V1441,237-246(1991)

Optical coatings to reduce temperature sensitivity of polarization-maintaining fibers for smart structures and skins (Zhang, Feng; Lit, John W.)V1588,100-109(1991)

Optical properties of granular Sn films with coating Al (Wu, Guang M.; Qian, Zheng X.)V1519,315-320(1991)

Optical properties of some ion-assisted deposited oxides (Andreani, F.; Luridiana, S.; Mao, Shu Z.)V1519,18-22(1991)

Optical Surfaces Resistant to Severe Environments (Musikant, Solomon, ed.)V1330(1991)

Origins of asymmetry in spin-cast films over topography (Manske, Loni M.; Graves, David B.)V1463,414-422(1991)

Performance of longitudinal-mode KD*P Pockels cells with transparent conductive coatings (Skeldon, Mark D.; Jin, Michael S.; Smith, Douglas H.; Bui, Snow T.)V1410,116-124(1991)

Performance of multilayer-coated figured optics for soft x-rays near the diffraction limit (Raab, Eric L.; Tennant, Donald M.; Waskiewicz, Warren K.; MacDowell, Alastair A.; Freeman, Richard R.)V1343,104-109(1991)

Phase-shifting and other challenges in optical mask technology (Lin, Burn J.)V1496,54-79(1991)

Photothermal displacement spectroscopy of optical coatings (Su, Xing; Fan, Zheng X.)V1519,80-84(1991)

Plastic photochromic eyewear: a status report (Crano, John C.; Elias, Richard C.)V1529,124-131(1991)

Preparation of Pb1-xGexTe crystal with high refractive index for IR coating (Zhang, Su Y.; Xu, Bu Y.; Zhang, Feng S.; Yan, Yixun)V1519,508-513(1991)

Reactive ion-beam-sputtering of fluoride coatings for the UV/VUV range (Schink, Harald; Kolbe, J.; Zimmermann, F.; Ristau, Detleu; Welling, Herbert)V1441,327-338(1991)

Reactive low-voltage ion plating of optical coatings on ophthalmic lenses (Balasubramanian, Kunjithapa; Richmond, Jeff; Hu, X. Q.; Guenther, Karl H.)V1529,106-114(1991)

Reduction of the standing wave effect in positive photoresist using an antireflection coating (Mehrotra, R.; Mathur, B. P.; Sharan, Sunil)V1463,487-491(1991)

Reflective and Refractive Optical Materials for Earth and Space Applications (Riedl, Max J.; Hale, Robert R.; Parsonage, Thomas B., eds.)V1485(1991)

Scattering in paper coatings (Hyvarinen, Timo S.; Sumen, Juha)V1530,325-334(1991)

Scattering measurements of optical coatings in high-power lasers (Chen, Yi-Sheng)V1332,115-120(1991)

Silicon/silicon oxide and silicon/silicon nitride multilayers for XUV optical applications (Boher, Pierre; Houdy, Philippe; Hennet, L.; Delaboudiniere, Jean-Pierre; Kuehne, Mikhael; Mueller, Peter; Li, Zhigang; Smith, David J.)V1343,39-55(1991)

Soft x-ray multilayers fabricated by electron-beam deposition (Sudoh, Masaaki; Yokoyama, Ryouhei; Sumiya, Mitsuo; Yamamoto, Masaki; Yanagihara, Mihiro; Namioka, Takeshi)V1343,14-24(1991)

Solar EUV/FUV line polarimetry: observational parameters and theoretical considerations (Fineschi, Silvano; Hoover, Richard B.; Fontenla, Juan; Walker, Arthur B.)V1343,376-388(1991)

Solution for anomalous scattering of bare HIP Be and CVD SiC mirrors (Vernold, Cynthia L.)V1530,130-143(1991)

Space station atomic-oxygen-resistant coatings (Grieser, James L.; Freeland, Alan W.; Fink, Jeffrey D.; Meinke, Gary E.; Hildreth, Eugene N.)V1330,102-110(1991)

SPAT studies of near-surface defects in silicon induced by BF2+ and F++B+ implantation (Li, Xiao Q.; Lin, Cheng L.; Zou, Shi C.; Weng, Hei M.; Han, Xue D.)V1519,14-17(1991)

Studies on antireflection technology of end surfaces for coherent fibre bundles (Qian, Anping; Dong, Linjun)V1572,144-147(1991)

Study on different proportion W-Ti (C) binary alloy carbide thin film (Zhang, Yun H.; Wu, Bei X.; Yang, Guang Y.; Wang, Ren)V1519,729-734(1991)

Study on the mechanism of ZnS antireflecting coating with high strength (Yu, Ju X.; Tang, Jia T.)V1519,308-314(1991)

Subsurface polishing damage of fused silica: nature and effect on laser damage of coated surfaces (Tesar, Aleta A.; Brown, Norman J.; Taylor, John R.; Stolz, Christopher J.)V1441,154-172(1991)

Suppression of columnar-structure formation in Mo-Si layered synthetic microstructures (Niibe, Masahito; Hayashida, Masami; Iizuka, Takashi; Miyake, Akira; Watanabe, Yutaka; Takahashi, Rie; Fukuda, Yasuaki)V1343,2-13(1991)

Surface nitride synthesis by multipulse excimer laser irradiation (D'Anna, Emilia; Leggieri, Gilberto; Luches, Armando; Martino, M.; Perrone, A.; Majni, G.; Mengucci, P.; Drigo, A. V.; Mihailescu, Ion N.)V1503,256-268(1991)

Thermal plastic metal coatings on optical fiber sensors (Sirkis, James S.; Dasgupta, Abhijit)V1588,88-99(1991)

Thermal stress modeling for diamond-coated optical windows (Klein, Claude A.)V1441,488-509(1991)

Thick, fine-grained beryllium optical coatings (Murray, Brian W.; Ulph, Eric; Richard, Peter N.)V1485,106-115(1991)

Time-resolved infrared radiometry of multilayer organic coatings using surface and subsurface heating (Maclachlan Spicer, J. W.; Kerns, W. D.; Aamodt, Leonard C.; Murphy, John C.)V1467,311-321(1991)

Total internal reflection studies of a ferroelectric liquid crystal/anisotropic solid interface (Zhuang, Zhiming; Clark, Noel A.; Meadows, Michael R.)V1455,105-109(1991)

Understanding of the abnormal wavelength effect of overcoats (Wu, Zhouling; Reichling, M.; Fan, Zheng X.; Wang, Zhi-Jiang)V1441,200-213(1991)

Use of antireflective coatings in deep-UV lithography (Sethi, Satyendra A.; Distasio, Romelia; Ziger, David H.; Lamb, James E.; Flaim, Tony)V1463,30-40(1991)

Very high reflective all-dielectric coatings for high-power CO2 lasers (Berger, R. M.; Chmelir, M.; Reedy, Herman E.; Chambers, Jack P.)V1397,611-618(1991)

Why and how to coat ophthalmic lenses (Guenther, Karl H.)V1529,96-105(1991)

X-ray evaluation on residual stresses in vapor-deposited hard coatings (Xu, Kewei; Chen, Jin; Gao, Runsheng; He, Jia W.; Zhao, Cheng; Li, Shi Z.)V1519,765-770(1991)

Coding—see also image compression; video

3-D TV: joined identification of global motion parameters for stereoscopic sequence coding (Tamtaoui, Ahmed; Labit, Claude)V1605,720-731(1991)

45-Mbps multichannel TV coding system (Matsumoto, Shuichi; Hamada, Takahiro; Saito, Masahiro; Murakami, Hitomi)V1605,37-46(1991)

Adaptive coding method of x-ray mammograms (Baskurt, Atilla; Magnin, Isabelle E.; Bremond, Alain; Charvet, Pierre Y.)V1444,240-249(1991)

Adaptive perceptual quantization for video compression (Puri, Atul; Aravind, R.)V1605,297-300(1991)

Analysis of barcode digitization techniques (Boles, John A.; Hems, Randall K.)V1384,195-204(1991)

Analysis of one-dimensional barcode (Wang, Ynjiun P.; Pavlidis, Theo; Swartz, Jerome)V1384,145-160(1991)

Analysis of optimum-frame-rate in low-bit-rate video coding (Takishima, Yasuhiro; Wada, M.; Murakami, Hitomi)V1605,635-645(1991)

Arithmetic coding model for compression of LANDSAT images (Perez, Arnulfo; Kamata, Seiichiro; Kawaguchi, Eiji)V1605,879-884(1991)

Bayesian approach to segmentation of temporal dynamics in video data (Jones, Coleen T.; Sauer, Ken D.)V1605,522-533(1991)

Block-adaptive quantization of multiple-frame motion field (Lavagetto, Fabio; Leonardi, Riccardo)V1605,534-545(1991)

Block arithmetic coding of contour images (Kim, Kyoil; Kim, Jonglak; Kim, Taejeong)V1605,851-862(1991)

Characteristic analysis of color information based on (R,G,B)-> (H,V,C) color space transformation (Gan, Qing; Miyahara, Makaoto; Kotani, Kazunori)V1605,374-381(1991)

Cheops: a modular processor for scalable video coding (Bove, V. M.; Watlington, John)V1605,886-893(1991)

Classified transform coding of images using two-channel conjugate vector quantization (Nam, J. Y.; Rao, K. R.)V1605,202-213(1991)

Classified vector quantizer based on minimum-distance partitioning (Kim, Dong S.; Lee, Sang U.)V1605,190-201(1991)

Coding of motion vectors for motion-compensated predictive/interpolative video coder (Chen, Cheng-Tie; Jeng, Fure-Ching)V1605,812-821(1991)

Color-coding reproduction of 3-D object with rainbow holography (Fan, Cheng; Jiang, Chaochuan; Guo, Lu Rong)V1461,51-55(1991)

Color/texture analysis and synthesis for model-based human image coding (Ishibashi, Satoshi; Kishino, Fumio)V1605,242-252(1991)

Compaction of color images with arithmetic coding (Iwahashi, Masahiro; Masuda, Shun-ichi)V1605,844-850(1991)

Compact motion representation based on global features for semantic image sequence coding (Labit, Claude; Nicolas, Henri)V1605,697-708(1991)

Comparison of directionally based and nondirectionally based subband image coders (Bamberger, Roberto H.; Smith, Mark J.)V1605,757-768(1991)

Construction of efficient variable-length codes with clear synchronizing codewords for digital video applications (Lei, Shaw-Min)V1605,863-873(1991)

Digital holography as a useful model in diffractive optics (Wyrowski, Frank)V1507,128-135(1991)

Digital Image Compression Techniques (Jones, Paul W.; Rabbani, Majid)VTT07(1991)

Digital video codec for medium bitrate transmission (Ebrahimi, Touradj; Dufaux, Frederic; Moccagatta, Iole; Campbell, T. G.; Kunt, Murat)V1605,2-15(1991)

Edge-based subband image coding technique for encoding the upper-frequency bands (Mohsenian, Nader; Nasrabadi, Nasser M.)V1605,781-792(1991)

Effects of M-transform for bit-error resilement in the adaptive DCT coding (Yamane, Nobumoto; Morikawa, Yoshitaka; Hamada, Hiroshi)V1605,679-686(1991)

Efficient odd max quantizer for use in transform image coding (Hauser, Neal A.; Mitchell, Harvey B.)V1605,428-433(1991)

Enhancement of transform coding by nonlinear interpolation (Wu, Siu W.; Gersho, Allen)V1605,487-498(1991)

Entropy coding for wavelet transform of image and its application for motion picture coding (Ohta, Mutsumi; Yano, Mitsuharu; Nishitani, Takao)V1605,456-466(1991)

Enumerative modulation coding with arbitrary constraints and postmodulation error correction coding for data storage systems (Mansuripur, Masud)V1499,72-86(1991)

Estimation and prediction of object-oriented segmentation for video predictive coding (Brofferio, Sergio C.; Comunale, Domenico; Tubaro, Stefano)V1605,500-510(1991)

Estimation of three-dimensional motion in a 3-DTV image sequence (Dugelay, Jean-Luc)V1605,688-696(1991)

Fast access to reduced-resolution subsamples of high-resolution images (Isaacson, Joel S.)V1460,80-91(1991)

Fast finite-state codebook design algorithm for vector quantization (Chang, Ruey-Feng; Chen, Wen-Tsuen)V1605,172-178(1991)

Full-frame entropy coding for radiological image compression (Lo, Shih-Chung B.; Krasner, Brian H.; Mun, Seong Ki; Horii, Steven C.)V1444,265-271(1991)

HDTV compression with vector quantization of transform coefficients (Wu, Siu W.; Gersho, Allen)V1605,73-84(1991)

Hierarchical block motion estimation for video subband coding (Jeon, Joon-hyeon; Hahm, Cheul-hee; Kim, Jae-Kyoon)V1605,954-962(1991)

Hierarchical image decomposition based on modeling of convex hulls corresponding to a set of order statistic filters (Vepsalainen, Ari M.; Linnainmaa, Seppo; Yli-Harja, Olli P.)V1568,2-13(1991)

High-density two-dimensional barcode (Wang, Ynjiun P.; Pavlidis, Theo; Swartz, Jerome)V1384,169-175(1991)

High-fidelity subband coding for superhigh-resolution images (Saito, Takahiro; Higuchi, Hirofumi; Komatsu, Takashi)V1605,382-393(1991)

Highly efficient entropy coding of multilevel images using a modified arithmetic code (Chen, Yan-Ping; Yasuda, Yasuhiko)V1605,822-831(1991)

High-resolution color image coding scheme for office systems (Koshi, Yutaka; Kunitake, Setsu; Suzuki, Kazuhiro; Kamizawa, Koh; Yamasaki, Toru; Miyake, Hidetaka)V1605,362-373(1991)

High-resolution decoding of multinode microchannel array detectors (Kasle, David B.; Morgan, Jeffrey S.)V1549,52-58(1991)

High-speed programmable digitizer for real-time video compression experiments (Cox, Norman R.)V1605,906-915(1991)

High-speed two-dimensional pyramid image coding method and its implementation (Sahinoglou, Haralambos; Cabrera, Sergio D.)V1605,793-804(1991)

Holotag: a novel holographic label (Soares, Oliverio D.; Bernardo, Luis M.; Pinto, M. I.; Morais, F. V.)V1332,166-184(1991)

Human facial motion modeling, analysis, and synthesis for video compression (Huang, Thomas S.; Reddy, Subhash C.; Aizawa, Kiyoharu)V1605,234-241(1991)

Hybrid coder for image sequences using detailed motion estimates (Nickel, Michael; Husoy, John H.)V1605,963-971(1991)

Image coding using adaptive-blocksize Princen-Bradley transform (Mochizuki, Takashi; Yano, Mitsuharu; Nishitani, Takao)V1605,434-444(1991)

Image segmentation based on ULCS color difference (Horita, Yuukou; Miyahara, Makoto)V1606,607-620(1991)

Image vector quantization with block-adaptive scalar prediction (Gupta, Smita; Gersho, Allen)V1605,179-189(1991)

Influence of input information coding for correlation operations (Maze, Sylvie; Joffre, Pascal; Refregier, Philippe)V1505,20-31(1991)

Invariant phase-only filters for phase-encoded inputs (Kallman, Robert R.; Goldstein, Dennis H.)V1564,330-347(1991)

Iterative motion estimation method using triangular patches for motion compensation (Nakaya, Yuichiro; Harashima, Hiroshi)V1605,546-557(1991)

Laplacian pyramid coding of prediction error images (Stiller, Christoph; Lappe, Dirk)V1605,47-57(1991)

Lapped orthogonal transform for motion-compensated video compression (Lynch, William E.; Reibman, Amy R.)V1605,285-296(1991)

Line coding for high-speed fiber optic transmission systems (Subramanian, K. R.; Dubey, V. K.; Low, J. P.; Tan, L. S.)V1364,190-201(1991)

Method to convert image resolution using M-band-extended QMF banks (Kawashima, Masahisa; Tominaga, Hideyoshi)V1605,107-111(1991)

Model-based coding of facial images based on facial muscle motion through isodensity maps (So, Ikken; Nakamura, Osamu; Minami, Toshi)V1605,263-272(1991)

Model for packet image communication in a centralized distribution system (Torbey, Habib H.; Zhang, Zhensheng)V1605,650-666(1991)

Motion affine models identification and application to television image coding (Sanson, Henri)V1605,570-581(1991)

Motion-compensated priority discrete cosine transform coding of image sequences (Efstratiadis, Serafim N.; Huang, Yunming G.; Xiong, Z.; Galatsanos, Nikolas P.; Katsaggelos, Aggelos K.)V1605,16-25(1991)

Motion-compensated wavelet transform coding for color video compression (Zhang, Ya-Qin; Zafar, Sohail)V1605,301-316(1991)

Motion compensation by block matching and vector postprocessing in subband coding of TV signals at 15 Mbit/s (Lallauret, Fabrice; Barba, Dominique)V1605,26-36(1991)

Motion field estimation for complex scenes (Driessen, Johannes N.; Biemond, Jan)V1605,511-521(1991)

Motion video coding for packet-switching networks: an integrated approach (Gilge, Michael; Gusella, Riccardo)V1605,592-603(1991)

Multiscale morphological region coding (Macq, Benoit M.; Ronse, Christian; Van Dongen, V.)V1606,165-173(1991)

New subband scheme for super-HDTV coding (Tanimoto, Masayuki; Yamada, Akio; Naito, Yoichi)V1605,394-405(1991)

Overlapping block transform for offset-sampled image compression (Morikawa, Yoshitaka; Yamane, Nobumoto; Hamada, Hiroshi)V1605,445-455(1991)

Parallel processing approach to transform-based image coding (Normile, James; Wright, Dan; Chu, Ke-Chiang; Yeh, Chia L.)V1452,480-484(1991)

Performance comparison of various data codes in Z-CAV optical recording (Lee, Tzuo-chang; Chen, Di)V1499,87-103(1991)

Performance evaluation of different neural network training algorithms in error control coding (Hussain, Mukhtar; Bedi, Jatinder S.)V1469,697-707(1991)

Performance evaluation of subband coding and optimization of its filter coefficients (Katto, Jiro; Yasuda, Yasuhiko)V1605,95-106(1991)

Performance of Reed-Solomon codes in mulichannel CPFSK coherent optical communications (Wu, Jyh-Horng; Wu, Jingshown)V1579,195-209(1991)

Performance of pseudo-orthogonal codes in temporal, spatial, and spectral code division multiple access systems (Mendez, Antonio J.; Gagliardi, Robert M.)V1364,163-169(1991)

Postprocessing of video sequence using motion-dependent median filters (Lee, Ching-Long; Jeng, Bor S.; Ju, Rong-Hauh; Huang, Huang-Cheng; Kan, Kou-Sou; Huang, Jei-Shyong; Liu, Tsann-Shyong)V1606,728-734(1991)

Probabilistic model for quadtree representation of binary images (Chou, Chun-Hsien; Chu, Chih-Peng)V1605,832-843(1991)

Reed-Solomon encoder/decoder application using a neural network (Hussain, Mukhtar; Bedi, Jatinder S.)V1469,463-471(1991)

Reversible image data compression based on HINT (hierarchical interpolation) decorrelation and arithmetic coding (Roos, Paul; Viergever, Max A.)V1444,283-290(1991)

Robust image coding with a model of adaptive retinal processing (Narayanswamy, Ramkumar; Alter-Gartenberg, Rachel; Huck, Friedrich O.)V1385,93-103(1991)

Scientific data compression for space: a modified block truncation coding algorithm (Lu, Wei-Wei; Gough, M. P.; Davies, Peter N.)V1470,197-205(1991)

Secret transmission method of character data in motion picture communication (Tanaka, Kiyoshi; Nakamura, Yasuhiro; Matsui, Kineo)V1605,646-649(1991)

Selective edge detection based on harmonic oscillator wave functions (Kawakami, Hajimu)V1468,156-166(1991)

Signal extension and noncausal filtering for subband coding of images (Martucci, Stephen A.)V1605,137-148(1991)

Signal loss recovery in DCT-based image and video codecs (Wang, Yao; Zhu, Qin-Fan)V1605,667-678(1991)

Some fundamental experiments in subband coding of images (Aase, Sven O.; Ramstad, Tor A.)V1605,734-744(1991)

Spatial light modulator on the base of shape memory effect (Antonov, Victor A.; Shelyakov, Alexander V.)V1474,116-123(1991)

Static and dynamic spatial resolution in image coding: an investigation of eye movements (Stelmach, Lew B.; Tam, Wa J.; Hearty, Paul J.)V1453,147-152(1991)

Statistically optimized PR-QMF design (Caglar, Hakan; Liu, Yipeng; Akansu, Ali N.)V1605,86-94(1991)

Study of binary image compression using universal coding (Nakano, Yasuhiko; Chiba, Hirotaka; Okada, Yoshiyuki; Yoshida, Shigeru; Mori, Masahiro)V1605,874-878(1991)

Subband coding of video using energy-adaptive arithmetic coding and statistical feedback-free rate control (Popat, Ashok C.; Nicoulin, Andre; Basso, Andrea; Li, Wei; Kunt, Murat)V1605,940-953(1991)

Subband video-coding algorithm and its feasibility on a transputer video coder (Brofferio, Sergio C.; Marcozzi, Elena; Mori, Luigi; Raveglia, Dalmazio)V1605,894-905(1991)

Subjective evaluation of scale-space image coding (de Ridder, Huib)V1453,31-42(1991)

Subsampled device-independent interchange color spaces (Kasson, James M.; Plouffe, Wil)V1460,11-19(1991)

Subsampled vector quantization with nonlinear estimation using neural network approach (Sun, Huifang; Manikopoulos, Constantine N.; Hsu, Hwei P.)V1605,214-220(1991)

Sunset: a hardware-oriented algorithm for lossless compression of gray-scale images (Langdon, Glen G.)V1444,272-282(1991)

Superhigh-definition image processing on a parallel signal processing system (Fujii, Tetsurou; Sawabe, Tomoko; Ohta, Naohisa; Ono, Sadayasu)V1605,339-350(1991)

Temporal projection for motion estimation and motion compensating interpolation (Robert, Philippe)V1605,558-569(1991)

Thin family: a new barcode concept (Allais, David C.)V1384,161-168(1991)

Three-dimensional subband decompositions for hierarchical video coding (Bosveld, Frank; Lagendijk, Reginald L.; Biemond, Jan)V1605,769-780(1991)

Transmission of the motion of a walker by model-based image coding (Kimoto, Tadahiko; Yasuda, Yasuhiko)V1605,253-262(1991)

Tree-structured vector quantization with input-weighted distortion measures (Cosman, Pamela C.; Oehler, Karen; Heaton, Amanda A.; Gray, Robert M.)V1605,162-171(1991)

Two-layer pyramid image coding scheme for interworking of video services in ATM (Sikora, Thomas; Tan, T. K.; Pang, Khee K.)V1605,624-634(1991)

Two new image compression methods utilizing mathematical morphology (Vepsalainen, Ari M.; Toivanen, Pekka J.)V1606,282-293(1991)

Use of a human visual model in subband coding of color video signal with adaptive chrominance signal vector quantization (Barba, Dominique; Hanen, Jose)V1605,408-419(1991)

VLSI implementation of a buffer, universal quantizer, and frame-rate-control processor (Uwabu, H.; Kakii, Eiji; Lacombe, R.; Maruyama, Masanori; Fujiwara, Hiroshi)V1605,928-937(1991)

Variable-bit-rate HDTV coding algorithm for ATM environments for B-ISDN (Kinoshita, Taizo; Nakahashi, Tomoko; Takizawa, Masaaki)V1605,604-613(1991)

Variable-blocksize transform coding of four-color printed images (Kaup, Andre; Aach, Til)V1605,420-427(1991)

Vector quantization of image pyramids with the ECPNN algorithm (de Garrido, Diego P.; Pearlman, William A.; Finamore, Weiler A.)V1605,221-232(1991)

Video compression algorithm with adaptive bit allocation and quantization (Viscito, Eric; Gonzales, Cesar A.)V1605,58-72(1991)

Visual Communications and Image Processing '91: Visual Communication (Tzou, Kou-Hu; Koga, Toshio, eds.)V1605(1991)

Visual factors and image analysis in the encoding of high-quality still images (Algazi, V. R.; Reed, Todd R.; Ford, Gary E.; Estes, Robert R.)V1605,329-338(1991)

Windowed motion compensation (Watanabe, Hiroshi; Singhal, Sharad)V1605,582-589(1991)

Coherence

64-Mbit DRAM production with i-line stepper (Shirai, Hisatsugu; Kobayashi, Katsuyoshi; Nakagawa, Kenji)V1463,256-274(1991)

Air Force program in coherent semiconductor lasers (Kennett, Ruth D.; Frazier, John C.)V1501,57-68(1991)

Coherence and optical Kerr nonlinearity (Depoortere, Marc)V1504,133-139(1991)

Coherence in single and multiple scattering of light from randomly rough surfaces (Gu, Zu-Han; Maradudin, Alexei A.; Mendez, Eugenio R.)V1530,60-70(1991)

Coherent coupling of lasers using a photorefractive ring oscillator (Luo, Jhy-Ming)V1409,100-105(1991)

Coherent detection in confocal microscopy (Wilson, Tony)V1439,104-108(1991)

Coherent optical modulation for antenna remoting (Fitzmartin, Daniel J.; Gels, Robert G.; Balboni, Edmund J.)V1476,56-62(1991)

Compact optical neuro-processors (Paek, Eung Gi; Wullert, John R.; Von Lehman, A.; Patel, J. S.; Martin, R.)V1621,340-350(1991)

Illumination coherence effects in laser-speckle imaging (Voelz, David G.; Idell, Paul S.; Bush, Keith A.)V1416,260-265(1991)

Imaging inside scattering media: chronocoherent imaging (Spears, Kenneth G.; Kume, Stewart M.; Winakur, Eric)V1429,2-8(1991)

Large dynamic range electronically scanned "white-light" interferometer with optical fiber Young's structure (Chen, Shiping; Rogers, Alan J.; Meggitt, Beverley T.)V1511,67-77(1991)

Nondegenerate two-wave mixing in shaped microparticle suspensions (Pizzoferrato, R.; De Spirito, M.; Zammit, Ugo; Marinelli, M.; Rogovin, Dan N.; Scholl, James F.)V1409,192-201(1991)

Optical phase-locked loop for free-space laser communications with heterodyne detection (Win, Moe Z.; Chen, Chien-Chung; Scholtz, Robert A.)V1417,42-52(1991)

Optimization of partial coherence for half-micron i-line lithography (Canestrari, Paolo; Degiorgis, Giorgio A.; De Natale, Paolo; Gazzaruso, Lucia; Rivera, Giovanni)V1463,446-455(1991)

Photon Statistics and Coherence in Nonlinear Optics (Perina, Jan, ed.)VMS39(1991)

Collimators

Fungal testing of diode laser collimators (de Lourdes Quinta, Maria; Freitas, Jose C.; Rodrigues, F. C.; Silva, Jorge A.)V1399,24-29(1991)

Imaging the sun in hard x-rays: spatial and rotating modulation collimators (Campbell, Jonathan W.; Davis, John M.; Emslie, A. G.)V1549,155-179(1991)

Color—see also colorimetry; printing; visualization

3M's Dry Silver technology: an ideal media for electronic imaging (Morgan, David A.)V1458,62-67(1991)

Acousto-optic color projection system (Hubin, Thomas)V1454,313-322(1991)

Advances in color laser printing (Tompkins, Neal)V1458,154-154(1991)

Alignment and amplification as determinants of expressive color (Jacobson, Nathaniel; Bender, Walter J.; Feldman, Uri)V1453,70-80(1991)

Apparent contrast and surface color in complex scenes (Arend, Lawrence)V1453,412-421(1991)

Arithmetic coding model for color images processed by error diffusion (Matsushiro, Nobuhito; Asada, Osamu; Tsuji, Kenzo)V1452,21-26(1991)

Automatic inspection system for full-color printed matter (Meguro, Shin-Ichi; Nunotani, Masakatu; Tanimizu, Katsuyuki; Sano, Mutsuo; Ishii, Akira)V1384,27-37(1991)

Characteristic analysis of color information based on (R,G,B)-> (H,V,C) color space transformation (Gan, Qing; Miyahara, Makaoto; Kotani, Kazunori)V1605,374-381(1991)

Chemistry of the Konica Dry Color System (Suda, Yoshihiko; Ohbayashi, Keiji; Onodera, Kaoru)V1458,76-78(1991)

Chromaticity and color fidelity of images with multicolor rainbow holograms (Zhang, Jingfang; Yu, Meiwen; Tang, Shunqing; Zhu, Zhengfang)V1238,401-405(1991)

Color analysis of nonlinear-phase-modulation method for density pseudocolor encoding technique in medical application (Liu, Dingyu; Yang, Xiaobo; Liu, Changjun; Zhang, Honguo)V1443,191-196(1991)

Color and Grassmann-Cayley coordinates of shape (Petrov, A. P.)V1453,342-352(1991)

Color- and intensity-balanced multichannel optical beamsplitter (Gilo, Mordechai; Rabinovitch, Kopel)V1442,90-104(1991)

Color character recognition method based on a model of human visual processing (Yamaba, Kazuo; Miyake, Yoichi)V1453,290-299(1991)

Color-coding reproduction of 3-D object with rainbow holography (Fan, Cheng; Jiang, Chaochuan; Guo, Lu Rong)V1461,51-55(1991)

Color correction using principle components (Trussell, Henry J.; Vrhel, Michael J.)V1452,2-9(1991)

Color-encoded depth: an image enhancement tool (Bieman, Leonard H.)V1385,229-238(1991)

Color handling in the image retrieval system Imagine (Dal Degan, Nevaino; Lancini, R.; Migliorati, Pierangelo; Pozzi, S.)V1606,934-940(1991)

Color hard copy: a self-tuning color correction algorithm based on a colorimetric model (Petschik, Benno)V1458,108-114(1991)

Colorimetric characterization of CCD sensors by spectrophotometry (Daligault, Laurence; Glasser, Jean)V1512,124-130(1991)

Colorimetry, normal human vision, and visual display (Thornton, William A.)V1456,219-225(1991)

Color-invariant character recognition and character-background color identification by multichannel matched filter (Campos, Juan; Millan, Maria S.; Yzuel, Maria J.; Ferreira, Carlos)V1564,189-198(1991)

Color printing technologies (Sahni, Omesh)V1458,4-16(1991)

Color quantization aspects in stereopsis (Hebbar, Prashant D.; McAllister, David F.)V1457,233-241(1991)

Color reflection holograms with photopolymer plates (Zhang, Jingfang; Ma, Chunrong; Lang, Hengyuan)V1238,306-310(1991)

Color segmentation using MDL clustering (Wallace, Richard S.; Suenaga, Yasuhito)V1381,436-446(1991)

Color space analysis of road detection algorithms (Crisman, Jill D.)V1569,492-506(1991)

Color standards for electronic imaging (McDowell, David Q.)VCR37,40-53(1991)

Color/texture analysis and synthesis for model-based human image coding (Ishibashi, Satoshi; Kishino, Fumio)V1605,242-252(1991)

Compaction of color images with arithmetic coding (Iwahashi, Masahiro; Masuda, Shun-ichi)V1605,844-850(1991)

Computer model for predicting underwater color images (Palowitch, Andrew W.; Jaffe, Jules S.)V1537,128-139(1991)

Decorrelation of color images using total color difference (Zheng, Joe; Valavanis, Kimon P.; Gauch, John M.)V1606,1037-1047(1991)

Denisyuk hologram recording with simultaneous use of all spectral components of the white light (Kostyljov, Ghennadij D.)V1238,316-319(1991)

Dense stereo correspondence using color (Jordan, John R.; Bovik, Alan C.)V1382,111-122(1991)

Design of an imaging spectrometer for observing ocean color (Weng, Zhicheng; Chen, Zhiyong; Cong, Xiaojie)V1527,338-348(1991)

Digital charge-coupled device color TV system for endoscopy (Vishnevsky, G. I.; Berezin, V. Y.; Lazovsky, L. Y.; Vydrevich, M. G.; Rivkind, V. L.; Zhemerov, B. N.)V1447,34-43(1991)

Display holograms in Du Pont's OmniDex films (Zager, Stephen A.; Weber, Andrew M.)V1461,58-67(1991)

Eurosprint proofing system (Froehlich, Helmut H.)V1458,51-60(1991)

Evolving JPEG color data compression standard (Mitchell, Joan L.; Pennebaker, William B.)VCR37,68-97(1991)

Finding distinctive colored regions in images (Syeda, Tanveer F.)V1381,574-581(1991)

Gamut mapping computer-generated imagery (Wallace, William E.; Stone, Maureen C.)V1460,20-28(1991)

High-resolution color image coding scheme for office systems (Koshi, Yutaka; Kunitake, Setsu; Suzuki, Kazuhiro; Kamizawa, Koh; Yamasaki, Toru; Miyake, Hidetaka)V1605,362-373(1991)

Hologram as means of color reproduction (Vlasov, N. G.; Vorobjov, S. P.; Karpova, S. G.)V1238,332-337(1991)

Holographic characteristics of IAE and PFG-01 photoplates for colored pulsed holography (Vorzobova, N. D.; Rjabova, R. V.; Schvarzvald, A. I.)V1238,476-477(1991)

Holographic pseudocoloring of schlieren images (Rodriguez-Vera, Ramon; Olivares-Perez, A.; Morales-Romero, Arquimedes A.)V1507,416-424(1991)

Image Handling and Reproduction Systems Integration (Bender, Walter J.; Plouffe, Wil, eds.)V1460(1991)

Images: from a printer's perspective (Sarkar, N. R.)V1458,42-50(1991)

Image segmentation based on ULCS color difference (Horita, Yuukou; Miyahara, Makoto)V1606,607-620(1991)

Inspecting colored objects using gray-level vision systems (Plummer, A. P.)VCR36,64-77(1991)

Integrated color management: the key to color control in electronic imaging and graphic systems (Taylor, Joann M.)V1460,2-10(1991)

ISDN audio color-graphics teleconferencing system (Oyaizu, Ikuro; Tanaka, Kiyoto; Yamaguchi, Toshikazu; Miyabo, Katsuaki; Takahashi, Mamoru)V1606,990-1001(1991)

Large and small color differences: predicting them from hue scaling (Chan, Hoover; Abramov, Israel; Gordon, James)V1453,381-389(1991)

Measurement of color and scattering phenomena of translucent materials (Sjollema, J. I.; den Exter, Ir. T.; Zijp, Jaap R.; Ten Bosch, Jaap J.)V1500,177-188(1991)

Method of preprocessing color images using a Peano curve on a Transputer array (Lambert, Robin A.; Batchelor, Bruce G.)V1381,582-588(1991)

Monochromatic and two-color recording of holographic portraits with the use of pulsed lasers (Vorzobova, N. D.; Sizov, V. N.; Rjabova, R. V.)V1238,462-464(1991)

Motion-compensated wavelet transform coding for color video compression (Zhang, Ya-Qin; Zafar, Sohail)V1605,301-316(1991)

Multilayer OPC for one-shot two-color printer (Sakai, Katsuo)V1458,179-191(1991)

Neural networks for halftoning of color images (Ling, Daniel T.; Just, Dieter)V1452,10-20(1991)

New approach to palette selection for color images (Balasubramanian, Raja; Allebach, Jan P.)V1453,58-69(1991)

New design of the illuminating system for transmission film copy (Pesl, Ales A.)V1448,218-224(1991)

New method of adjusting color of pseudocolor encoding image (Cai, Hai-Tao; Chen, Zhen-Pei)V1567,703-708(1991)

Nondiffractive optically variable security devices (van Renesse, Rudolf L.)V1509,113-125(1991)

Novel monolithic chip-integrated color spectrometer: the distributed-wavelength filter component (Holm-Kennedy, James W.; Tsang, Koon Wing; Sze, Wah Wai; Jiang, Fenglai; Yang, Datong)V1527,322-331(1991)

Organization of a system for managing the text and images that describe an art collection (Mintzer, Fred; McFall, John D.)V1460,38-49(1991)

Photometric models in multispectral machine vision (Brill, Michael H.)V1453,369-380(1991)

Physics and psychophysics of color reproduction (Giorgianni, Edward)V1458,2-3(1991)

Possibility of keeping color picture in an image converter camera (Zhao, Zongyao)V1358,1252-1256(1991)

Preliminary review of imaging standards (Ren, Victor; Hatfield, Donald J.; Deacutis, Martin)VCR37,54-67(1991)

Printing Technologies for Images, Gray Scale, and Color (Dove, Derek B.; Abe, Takao; Heinzl, Joachim L., eds.)V1458(1991)

Progress in true-color holography (Jeong, Tung H.; Wesly, Edward J.)V1238,298-305(1991)

Pseudocolor reflection hologram properties recorded using monochrome photographic materials (Vanin, V. A.; Vorobjov, S. P.)V1238,324-331(1991)

Pulsed laser system for recording large-scale color hologram (Bespalov, V. G.; Krylov, Vitaly N.; Sizov, V. N.)V1238,457-461(1991)

Quantization of color image components in the DCT domain (Peterson, Heidi A.; Peng, Hui; Morgan, J. H.; Pennebaker, William B.)V1453,210-222(1991)

Recent advances in color reflection holography (Hubel, Paul M.)V1461,167-174(1991)

Reconstruction of quincunx-coded image sequences using vector median (Oistamo, Kai; Neuvo, Yrjo A.)V1606,735-742(1991)

Relief-phase colored hologram registration (Galpern, A. D.; Smaev, V. P.; Paramonov, A. A.; Kiriencko, Yu. A.)V1238,320-323(1991)

Simulating watercolor by modeling diffusion, pigment, and paper fibers (Small, David)V1460,140-146(1991)

Simulation of parvocellular demultiplexing (Martinez-Uriegas, Eugenio)V1453,300-313(1991)

Single-pixel measurements on LCDs (Jenkins, A. J.)V1506,188-193(1991)

Stacked STN LCDs for true-color projection systems (Gulick, Paul E.; Conner, Arlie R.)V1456,76-82(1991)

Stereovision and color segmentation for autonomous navigation (Sung, Eric)V1388,176-187(1991)

Strategies for the color character recognition by optical multichannel correlation (Millan, Maria S.; Yzuel, Maria J.; Campos, Juan; Ferreira, Carlos)V1507,183-197(1991)

Study of thermal dye diffusion (Koshizuka, Kunihiro; Abe, Takao)V1458,97-104(1991)

Subsampled device-independent interchange color spaces (Kasson, James M.; Plouffe, Wil)V1460,11-19(1991)

Summary of color definition activity in the graphic arts (McDowell, David Q.)V1460,29-35(1991)

Supervised color constancy for machine vision (Novak, Carol L.; Shafer, Steven A.)V1453,353-368(1991)

Systems considerations in color printing (Roetling, Paul G.)V1458,17-24(1991)

Theory of color correction by use of chromatic magnification (Ames, Alan J.)V1354,286-290(1991)

Thermal dye transfer color hard-copy image stability (Newmiller, Chris)V1458,92-96(1991)

Thin-film technology in high-resolution, high-density AC plasma displays (Andreadakis, Nicholas C.)V1456,310-315(1991)

Three-layer material for the registration of colored holograms (Smaev, V. P.; Galpern, A. D.; Kiriencko, Yu. A.)V1238,311-315(1991)

Transformation from tristimulus RGB to Munsell notation HVC in a colored computer vision system (Jin, Guofan; Zhu, Zimin; Yu, Xinglong)V1569,507-512(1991)

Trends in color hard-copy technology in Japan (Abe, Takao)V1458,29-40(1991)

Trends in color hard copy (Testan, Peter)V1458,25-28(1991)

Two-colour ratio pyrometer with optical fiber (Wang, Yutian; Shi, Jinshan; Li, Zhiquan)V1572,192-196(1991)

Use of a human visual model in subband coding of color video signal with adaptive chrominance signal vector quantization (Barba, Dominique; Hanen, Jose)V1605,408-419(1991)

Using color to segment images of 3-D scenes (Healey, Glenn)V1468,814-825(1991)

Variable-blocksize transform coding of four-color printed images (Kaup, Andre; Aach, Til)V1605,420-427(1991)

Visual processing, transformability of primaries, and visual efficiency of display devices (Thornton, William A.)V1453,390-401(1991)

Window-based elaboration language for picture processing and painting (Kamoshida, Minoru; Enomoto, Hajime; Miyamura, Isao)V1606,951-960(1991)

Colorimetry—see also color

Color hard copy: a self-tuning color correction algorithm based on a colorimetric model (Petschik, Benno)V1458,108-114(1991)

Colorimetric calibration for scanners and media (Hung, Po-Chieh)V1448,164-174(1991)

Colorimetry, normal human vision, and visual display (Thornton, William A.)V1456,219-225(1991)

Organization of a system for managing the text and images that describe an art collection (Mintzer, Fred; McFall, John D.)V1460,38-49(1991)

Recent advances in color reflection holography (Hubel, Paul M.)V1461,167-174(1991)

Remote colorimetry and its applications (Sheffer, Dan; Ben-Shalom, Ami; Devir, Adam D.)V1493,232-243(1991)

Research on optical fiber colorimeter (Cao, Zheng-Ping; Huang, Yue-Huai)V1572,38-41(1991)

Visual processing, transformability of primaries, and visual efficiency of display devices (Thornton, William A.)V1453,390-401(1991)

Computers

Accurate image simulation by hemisphere projection (Bian, Buming; Wittels, Norman)V1453,333-340(1991)

Adaptive isosurface generation in a distortion-rate framework (Ning, Paul C.; Hesselink, Lambertus B.)V1459,11-21(1991)

Algorithms and architectures for implementing large-velocity filter banks (Stocker, Alan D.; Jensen, Preben D.)V1481,140-155(1991)

Alignment and amplification as determinants of expressive color (Jacobson, Nathaniel; Bender, Walter J.; Feldman, Uri)V1453,70-80(1991)

All-optical interconnection networks (Ghafoor, Arif)V1390,454-466(1991)

Analysis and representation of complex structures in separated flows (Helman, James L.; Hesselink, Lambertus B.)V1459,88-96(1991)

Applications of aerial photography to law enforcement and disaster assessment: a consideration of the state-of-the-art (Cox, William J.; Biache, Andrew)V1479,364-369(1991)

Applications of a minimum sum path algorithm implemented on the connection machine (Rosenfeld, J. P.; Krecker, Donald K.; Hord, R. M.)V1406,147-147(1991)

Applications of the massively parallel machine, the MasPar MP-1, to Earth sciences (Fischer, James R.; Strong, James P.; Dorband, John E.; Tilton, James C.)V1492,229-238(1991)

Automation in optics manufacturing (Pollicove, Harvey M.; Moore, Duncan T.)V1354,482-486(1991)

Autostereoscopic (3-D without glasses) display for personal computer applications (Eichenlaub, Jesse B.)V1398,48-51(1991)

Blink comparison techniques applied to medical images (Craine, Eric R.; Craine, Brian L.)V1444,389-399(1991)

Brownian dynamics simulation of polarized light scattering from wormlike chains (Allison, Stuart A.)V1430,50-64(1991)

Capturing multimedia design knowledge using TYRO, the constraint-based designer's apprentice (MacNeil, Ronald L.)V1460,94-102(1991)

Challenges of using advanced multichip packaging for next generation spaceborne computers (Moravec, Thomas J.)V1390,195-201(1991)

Comparison of 2-D planar and 3-D perspective display formats in multidimensional data visualization (Merwin, David H.; Wickens, Christopher D.)V1456,211-218(1991)

Complexity of computing reachable workspaces for redundant manipulators (Alameldin, Tarek K.; Palis, Michael A.; Rajasekaran, Sanguthevar; Badler, Norman I.)V1381,217-225(1991)

Computer-aided acquisition of design knowledge (Dilger, Werner)V1468,584-595(1991)

Computer-aided alignment of a grazing-incidence ring resonator for a visible wavelength free-electron laser (Hudyma, Russell M.; Eigler, Lynne C.)V1354,523-532(1991)

Computer-aided forensic facial reconstruction (Evenhouse, Raymond J.; Rasmussen, Mary; Sadler, Lewis L.)V1380,147-156(1991)

Computer-aided photorefractive pattern recognition (Sun, Ching-Cherng; Chang, Ming-Wen; Yeh, Smile; Cheng, Nai-Jen)V1564,199-210(1991)

Computer animation method for simulating polymer flow for injection-molded parts (Perry, Meg W.; Rumbaugh, Richard C.; Frost, David P.)V1459,155-156(1991)

Computer-assisted surgical planning and automation of laser delivery systems (Zamorano, Lucia J.; Dujovny, Manuel; Dong, Ada; Kadi, A. M.)V1428,59-75(1991)

Computer-Controlled Optical Surfacing (Jones, Robert A., ed.)VMS40(1991)

Convergence of video and computing (Carlson, Curtis R.)V1472,2-5(1991)

Converting non-interlaced to interlaced images in YIQ and HSI color spaces (Welch, Eric B.; Moorhead, Robert J.; Owens, John K.)V1453,235-243(1991)

Cramer-Rao bound for multiple-target tracking (Daum, Frederick E.)V1481,341-344(1991)

Current status and future research of the Delft 'supercomputer' project (Frietman, Edward E.; Dekker, L.; van Nifterick, W.; Demeester, Piet M.; van Daele, Peter; Smit, W.)V1390,434-453(1991)

CWRUnet: case history of a campus-wide fiber-to-the-desktop network (Neff, Raymond K.; Klingensmith, H. W.; Gumpf, Jeffrey A.; Haigh, Peter J.)V1364,245-256(1991)

Design of eye movement monitoring system for practical environment (Nakamura, Hiroyuki; Kobayashi, Hitoshi; Taya, Katsuo; Ishigami, Shigenobu)V1456,226-238(1991)

Development and test of the Starlab control system (LaMont, Douglas V.; Mar, Lim O.; Rodden, Jack J.)V1482,2-12(1991)

Development of an image processing system on a second-generation RISC workstation (Ryan, Martin J.; Kapp, Oscar H.)V1396,335-339(1991)

Development of computer-aided functions in clinical neurosurgery with PACS (Mukasa, Minoru; Aoki, Makoto; Satoh, Minoru; Kowada, Masayoshi; Kikuchi, K.)V1446,253-265(1991)

Digital image processing for the early localization of cancer (Kelmar, Cheryl M.)V1426,47-57(1991)

Digital map databases in support of avionic display systems (Trenchard, Michael E.; Lohrenz, Maura C.; Rosche, Henry; Wischow, Perry B.)V1456,318-326(1991)

Distributing the server function in a multiring PAC system (Lynne, Kenton J.)V1446,177-187(1991)

Dynamic allocation of buffer space in the bridge of two interconnected token rings (Das, Alok K.; Muhuri, K.)V1364,61-69(1991)

Dynamics of wormlike chains: theory and computer simulations (Aragon, Sergio R.; Luo, Rolland)V1430,65-84(1991)

Electrical characteristics of lossy interconnections for high-performance computer applications (Deutsch, Alina; Kopcsay, Gerard V.; Ranieri, V. A.; Cataldo, J. K.; Galligan, E. A.; Graham, W. S.; McGouey, R. P.; Nunes, S. L.; Paraszczak, Jurij R.; Ritsko, J. J.; Serino, R. J.; Shih, D. Y.; Wilczynski, Janusz S.)V1389,161-176(1991)

Electrical characterization of the interconnects inside a computer (Rubin, Barry)V1389,314-328(1991)

End-to-end model of a diode-pumped Nd:YAG pulsed laser (Mayer, Richard C.; Dreisewerd, Douglas W.)V1415,248-258(1991)

Evaluation of multiresolution elastic matching using MRI data (Gee, Jim C.; Reivich, Martin; Bilaniuk, L.; Hackney, D.; Zimmerman, R.; Kovacic, Stane; Bajcsy, Ruzena R.)V1445,226-234(1991)

Extracting Meaning from Complex Data: Processing, Display, Interaction II (Farrell, Edward J., ed.)V1459(1991)

Fiber channel: the next standard peripheral interface and more (Cummings, Roger)V1364,170-177(1991)

Generic models for rapid calculation of target signatures (Rushmeier, Holly E.; Rodriguez, Leonard J.)V1486,210-216(1991)

Graphical interface for multispectral simulation, scene generation, and analysis (Sikes, Terry L.; Kreiss, William T.)V1479,199-211(1991)

High-density interconnect technology for VAX 9000 system (Deshpande, Ujwal A.; Howell, Gelston; Shamouilian, Shamouil)V1390,489-501(1991)

High-level PC-based laser system modeling (Taylor, Michael S.)V1415,300-309(1991)

High-level design of digital computers using optical logic arrays (Murdocca, Miles J.)V1474,176-187(1991)

High-level parallel architecture for a rule-based system (Karne, Ramesh K.; Sood, Arun K.)V1468,938-949(1991)

High-performance FDDI NIU for streaming voice, video, and data (Bergman, Larry A.; Hartmayer, Ron; Wu, Wennie H.; Cassell, P.; Edgar, G.; Lambert, James L.; Mancini, Richard; Jeng, J.; Pardo, C.; Halloran, Frank; Martinez, James C.)V1364,14-21(1991)

High-power copper vapor laser development (Aoki, Nobutada; Kimura, Hironobu; Konagai, Chikara; Shirayama, Shimpey; Miyazawa, Tatsuo; Takahashi, Tomoyuki)V1412,2-11(1991)

High-speed vision system based on computer graphics models (Baur, Charles; Beer, Simon)V1385,85-92(1991)

Holographic display of computer simulations (Andrews, John R.; Stinehour, Judith E.; Lean, Meng H.; Potyondy, David O.; Wawrzynek, Paul A.; Ingraffea, Anthony R.; Rainsdon, Michael D.)V1461,110-123(1991)

Human visual performance model for crewstation design (Larimer, James O.; Prevost, Michael P.; Arditi, Aries R.; Azueta, Steven; Bergen, James R.; Lubin, Jeffrey)V1456,196-210(1991)

Image computing requirements for the 1990s: from multimedia to medicine (Gove, Robert J.; Lee, Woobin; Kim, Yongmin; Alexander, Thomas)V1444,318-333(1991)

Imagetool: image processing on the Sun workstation (Zander, Mark E.)V1567,9-14(1991)

Infrared systems design from an operational requirement using a hypercard-based program (Harris, William R.)V1488,156-164(1991)

Integrated image processing and tracker performance prediction workstation (Schneeberger, Timothy J.; McIntire, Harold D.)V1567,2-8(1991)

Integrated processor architecture for multisensor signal processing (Nasburg, Robert E.; Stillman, Steve M.; Nguyen, M. T.)V1481,84-95(1991)

Interactive graphics system for multivariate data display (Becker, Richard A.; Cleveland, William S.; Shyu, William M.; Wilks, Allan R.)V1459,48-56(1991)

Internal protocol assistant for distributed systems (Leu, Fang Y.; Chang, Shi-Kuo)V1468,620-631(1991)

Jitter considerations for FDDI PMD (Fukuoka, Takashi; Tejika, Yasuhiro; Takada, Hisashi; Takahashi, Hidenori; Hamasaki, Yiji)V1364,40-48(1991)

Knowledge-based approach to fault diagnosis and control in distributed process environments (Chung, Kwangsue; Tou, Julius T.)V1468,323-332(1991)

Large Screen Projection, Avionic, and Helmet-Mounted Displays (Assenheim, Harry M.; Flasck, Richard A.; Lippert, Thomas M.; Bentz, Jerry, eds.)V1456(1991)

Latest advances in CAD data interfacing: a standardization project of ISO/TC 172/SC1 task group "optical database" (Wise, Timothy D.; Wieder, Eckart)V1346,79-85(1991)

LCS: a natural language comprehension system (Trigano, Philippe; Talon, Benedicte; Baltazart, Didier; Demko, Christophe; Newstead, Emma)V1468,866-874(1991)

Learning by comparison: improving the task planning capability (del Castillo, Maria D.; Kumpel, Daniel M.)V1468,596-607(1991)

Low-cost real-time hardware in the loop FCS performance evaluation (Cifarelli, Salvatore; Magrini, Sandro)V1482,480-490(1991)

Maximum entropy method applied to deblurring images on a MasPar MP-1 computer (Bonavito, N. L.; Dorband, John E.; Busse, Tim)V1406,138-146(1991)

MDIS (medical diagnostic imaging support) workstation issues: clinical perspective (Smith, Donald V.; Smith, Suzy; Cawthon, Michael A.)V1444,357-362(1991)

Microcomputer-based image processing system for CT/MRI scans: hardware configuration and software capacity (Cheng, Andrew Y.; Ho, Wai-Chin; Kwok, John C.; Yu, Peter K.)V1444,400-406(1991)

Microcomputer-based real-time monitoring and control of single-wafer processing (Hauser, John R.; Gyurcsik, Ronald S.)V1392,340-351(1991)

Microcomputer-based workstation for simulation and analysis of background and target IR signatures (Reeves, Richard C.; Schaibly, John H.)V1486,85-101(1991)

Militarized infrared touch panels (Hough, Stewart E.; Stanley, Pamela S.)V1456,240-249(1991)

Miniature signal processor for surveillance sensor applications (Jacobi, William J.; Jensen, Preben D.; Teneketges, Nicholas J.; Wadsworth, Leo A.)V1479,111-119(1991)

Miniaturized low-power parallel processor for space applications (Jacobi, William J.; Jensen, Preben D.; Teneketges, Nicholas J.; Wadsworth, Leo A.)V1495,205-213(1991)

Modeling and visualization of scattered volumetric data (Nielson, Gregory M.; Dierks, Tim)V1459,22-33(1991)

Modular FDDI bridge and concentrator (Coden, Michael H.; Bulusu, Dutt V.; Ramsey, Brian; Sztuka, Edward; Morrow, Joel)V1364,22-39(1991)

Multisensor analysis tool (Gerlach, Francis W.; Cook, Daniel B.)V1488,134-143(1991)

Networking of an electron microscope laboratory internally and to the internet (Zmola, Carl; Kapp, Oscar H.)V1396,331-334(1991)

Network visualization: user interface issues (Becker, Richard A.; Eick, Stephen G.; Miller, Eileen O.; Wilks, Allan R.)V1459,150-154(1991)

New approach to the simulation of optical manufacturing processes (Oinen, Donald E.; Billow, Nick W.)V1354,487-493(1991)

New method for identifying features of an image on a digital video display (Doyle, Michael D.)V1380,86-95(1991)

NIRATAM-NATO infrared air target model (Noah, Meg A.; Kristl, Joseph; Schroeder, John W.; Sandford, B. P.)V1479,275-282(1991)

Octree optimization (Globus, Al)V1459,2-10(1991)

One-dimensional and two-dimensional computer models of industrial CO laser (Iyoda, Mitsuhiro; Murota, Tomoya; Akiyama, Mamoru; Sato, Shunichi)V1415,342-349(1991)

Optical data communication compel the design of a new class of storage media (Frietman, Edward E.; Dekker, L.; van Nifterick, W.)V1401,19-26(1991)

Optical design with physical optics using GLAD (Lawrence, George N.)V1354,126-135(1991)

Optical glass selection using computerized data base (Fischer, Robert E.; Thomas, Michael J.; Hudyma, Russell M.)V1535,78-88(1991)

Optimization of athermal systems (Benham, Paul; Kidger, Michael J.)V1354,120-125(1991)

Optimization using the OSLO and Super-OSLO programs (Sinclair, Douglas C.)V1354,116-119(1991)

Options for campus fiber networks (Henderson, Byron B.; Green, Emily N.)V1364,235-244(1991)

Organization of a system for managing the text and images that describe an art collection (Mintzer, Fred; McFall, John D.)V1460,38-49(1991)

Overview of CODE V development (Harris, Thomas I.)V1354,104-111(1991)

Packaging technology for the NEC SX-3 supercomputers (Murano, Hiroshi; Watari, Toshihiko)V1390,78-90(1991)

Pattern recognition, neural networks, and artificial intelligence (Bezdek, James C.)V1468,924-935(1991)

Practical use of generalized simulated annealing optimization on microcomputers (Hearn, Gregory K.)V1354,186-191(1991)

Predictive control for 4-D guidance (Ilie, Stiharu-Alexe)V1482,491-501(1991)

Project DaVinci (Winarsky, Norman; Alexander, Joanna R.)V1459,67-68(1991)

Prototype expert system for preventive control in power plants (Jiang, Dareng; Han, Chia Y.; Wee, William G.)V1468,16-25(1991)

Rank-down method for automatic lens design (Ooki, Hiroshi)V1354,171-176(1991)

Reconnaissance mission planning (Fishell, Wallace G.; Fox, Alex J.)V1538,5-13(1991)

SAFENET II: The Navy's FDDI-based computer network standard (Paige, Jeffrey L.; Howard, Edward A.)V1364,7-13(1991)

Separation of function in the ASAP software package (Johnston, Steve C.; Greynolds, Alan W.)V1354,136-141(1991)

Signal processor for space-based visible sensing (Anderson, James C.; Downs, G. S.; Trepagnier, Pierre C.)V1479,78-92(1991)

Simulation-based PACS development (Stut, W. J.; van Steen, M. R.; Groenewegen, L. P.; Ratib, Osman M.; Bakker, Albert R.)V1446,396-404(1991)

Simulation of spin label behavior on a model surface (Balabaev, N. K.; Lemak, A. S.; Fushman, D. A.; Mironova, Yu. V.)V1402,53-69(1991)

Simulator for developing mobile robot control systems (Roning, Juha J.; Riekki, Jukka P.; Kemppainen, Seppo)V1388,350-360(1991)

Small-computer program for optical design and analysis written in "C" (Beckmann, Leo H.)V1354,254-261(1991)

Software concept for the new Zeiss interferometer (Doerband, Bernd; Wiedmann, Wolfgang; Wegmann, Ulrich; Kuebler, C. W.; Freischlad, Klaus R.)V1332,664-672(1991)

SYNOPSYS—a lens design computer program package (Dilworth, Donald C.)V1354,112-115(1991)

System for making scientific videotapes (Appino, Perry A.; Farrell, Edward J.)V1459,157-165(1991)

Techniques and strategies for data integration in mineral resource assessment (Trautwein, Charles M.; Dwyer, John L.)V1492,338-338(1991)

Ten years of failure in automatic time tables scheduling at the UTC (Trigano, Philippe; Boufflet, Jean-Paul; Newstead, Emma)V1468,408-416(1991)

Theory and molecular dynamics simulation of one-photon electronic excitation of multiatomic molecules (Grishanin, B. A.; Vachev, V. D.; Zadkov, Victor N.)V1402,44-52(1991)

Using the ACR/NEMA standard with TCP/IP and Ethernet (Chimiak, William J.; Williams, Rodney C.)V1446,93-99(1991)

Vesalius project: interactive computers in anatomical instruction (McCracken, Thomas O.; Roper, Stephen D.; Spurgeon, Thomas L.)V1380,6-10(1991)

Virtual environment for the exploration of three-dimensional steady flows (Bryson, Steve T.; Levit, Creon C.)V1457,161-168(1991)

Virtual environment technology (Zeltzer, David L.)V1459,86-86(1991)

Visual thinking in organizational analysis (Grantham, Charles E.)V1459,77-84(1991)

Warping of a computerized 3-D atlas to match brain image volumes for quantitative neuroanatomical and functional analysis (Evans, Alan C.; Dai, W.; Collins, L.; Neelin, Peter; Marrett, Sean)V1445,236-246(1991)

WEBERSAT: a low-cost imaging satellite (Twiggs, Robert J.; Reister, K. R.)V1495,12-18(1991)

Computer vision—see also image processing; machine vision

Adaptive imager: a real-time locally adaptive edge enhancement system (Strang, Steven E.)V1384,246-256(1991)

Adaptive neural methods for multiplexing oriented edges (Marshall, Jonathan A.)V1382,282-291(1991)

Adaptive snakes: control of damping and material parameters (Samadani, Ramin)V1570,202-213(1991)

Adaptive surface reconstruction (Terzopoulos, Demetri; Vasilescu, Manuela)V1383,257-264(1991)

Algorithm for quality inspection of characters printed on chip resistors (Numagami, Yasuhiko; Hattori, Yasuyuki; Nakamura, Osamu; Minami, Toshi)V1606,970-979(1991)

Analysis and simulation of an inhibitive directional selective unit for computer vision (Fong, David Y.)V1606,941-950(1991)

Analysis and visualization of heart motion (Chen, Chang W.; Huang, Thomas S.; Arrott, Matthew)V1450,231-242(1991)

Analysis of optical flow estimation using epipolar plane images (Rangachar, Ramesh M.; Hong, Tsai-Hong; Herman, Martin; Luck, Randall L.)V1382,376-385(1991)

ANN-implemented robust vision model (Teng, Chungte; Ligomenides, Panos A.)V1382,74-86(1991)

Application of a multilayer network in image object classification (Tang, Yonghong; Wee, William G.; Han, Chia Y.)V1469,113-120(1991)

Application of generalized radial basis functions to the problem of object recognition (Thau, Robert S.)V1469,37-47(1991)

Application of perceptron to the detecting of particle motion (Li, Jie-gu; Yuan, Qiang)V1469,178-187(1991)

Applications of Artificial Intelligence IX (Trivedi, Mohan M., ed.)V1468(1991)

Architecture for a multiprocessing system based on data flow processing elements in a MAXbus system (Bulsink, Bennie J.; Klok, Frits H.)V1384,215-227(1991)

Artificial neural system approach to IR target identification (Holland, Orgal T.; Tarr, Tomas; Farsaie, Ali; Fuller, James M.)V1469,102-112(1991)

Aspect networks: using multiple views to learn and recognize 3-D objects (Seibert, Michael; Waxman, Allen M.)V1383,10-19(1991)

Assumption truth maintenance in model-based ATR algorithm design (Bennett, Laura F.; Johnson, Rubin; Hudson, C. I.)V1470,263-274(1991)

Automated registration of terrain range images using surface feature level sets (Wheeler, Frederick W.; Vaz, Richard F.; Cyganski, David)V1606,78-85(1991)

Automatic acquisition of movement information by a knowledge-based recognition approach (Bae, Kyongtae T.; Altschuler, Martin D.)V1380,108-115(1991)

Automatic analysis system for three-dimensional angiograms (Higgins, William E.; Spyra, Wolfgang J.; Karwoski, Ronald A.; Ritman, Erik L.)V1445,276-286(1991)

Automatic and operator-assisted solid modeling of objects for automatic recognition (Stenstrom, J. R.; Connolly, C. I.)V1470,275-281(1991)

Automatic digitization of contour lines for digital map production (Yla-Jaaski, Juha; Yu, Xiaohan)V1472,201-207(1991)

Automatic inspection of optical fibers (Silberberg, Teresa M.)V1472,150-156(1991)

Automatic method for inspecting plywood shear samples (Avent, R. R.; Conners, Richard W.)V1468,281-295(1991)

Automatic reconstruction of buildings from aerial imagery (Sinha, Saravajit S.)V1468,698-709(1991)

Automatic segmentation of brain images: selection of region extraction methods (Gong, Leiguang; Kulikowski, Casimir A.; Mezrich, Reuben S.)V1450,144-153(1991)

Automatic target detection for surveillance (Ramesh, Nagarajan; Sethi, Ishwar K.; Cheung, Huey)V1468,72-80(1991)

BRICORK: an automatic machine with image processing for the production of corks (Davies, Roger; Correia, Bento A.; Carvalho, Fernando D.; Rodrigues, F. C.)V1459,283-291(1991)

Characteristic pattern matching based on morphology (Zhao, Dongming)V1606,86-96(1991)

Characteristic views and perspective aspect graphs of quadric-surfaced solids (Chen, Shuang; Freeman, Herbert)V1383,2-9(1991)

Classification of tissue-types by combining relaxation labeling with edge detection (Adiseshan, Prakash; Faber, Tracy L.)V1445,128-132(1991)

Clustering algorithms for a PC-based hardware implementation of the unsupervised classifier for the shuttle ice detection system (Jaggi, Sandeep)V1451,289-297(1991)

Collaborative processing to extract myocardium from a sequence of two-dimensional echocardiograms (Revankar, Shriram; Sher, David B.; Rosenthal, Steven)V1459,268-273(1991)

Color segmentation using MDL clustering (Wallace, Richard S.; Suenaga, Yasuhito)V1381,436-446(1991)

Combined approach for large-scale pattern recognition with translational, rotational, and scaling invariances (Xu, Qing; Inigo, Rafael M.; McVey, Eugene S.)V1471,378-389(1991)

Combined edge- and region-based method for range image segmentation (Koivunen, Visa; Pietikainen, Matti)V1381,501-512(1991)

Comparison of mono- and stereo-camera systems for autonomous vehicle tracking (Kehtarnavaz, Nasser; Griswold, Norman C.; Eem, J. K.)V1468,467-478(1991)

Computational model of an integrated vision system (Uttal, William; Shepherd, Thomas; Lovell, Robb E.; Dayanand, Sriram)V1453,258-269(1991)

Computer vision system for automated inspection of molded plastic print wheels (Hu, Yong-Lin; Wee, William G.; Gruver, William A.; Han, Chia Y.)V1468,653-661(1991)

Computer vision systems: integration of software architectures (Bohling, Edward H.; O'Connor, R. P.)V1406,164-168(1991)

Computing image flow and scene depth: an estimation-theoretic fusion-based framework (Singh, Ajit)V1383,122-140(1991)

Computing motion parameters from sparse multisensor range data for telerobotics (Vemuri, Baba C.; Skofteland, G.)V1383,97-108(1991)

Continuous-time segmentation networks (Harris, John G.)V1473,161-172(1991)

Contour estimation using global shape constraints and local forces (Deng, Keqiang; Wilson, Joseph N.)V1570,227-233(1991)

Convexity-based method for extracting object parts from 3-D surfaces (Vaina, Lucia M.; Zlateva, Stoyanka D.)V1468,710-719(1991)

Coupled depth-slope model based on augmented Lagrangian techniques (Suter, David)V1570,129-139(1991)

Data-driven parallel architecture for syntactic pattern recognition (Tseng, Chien-Chao; Hwang, Shu-Yuen)V1384,257-268(1991)

Decomposing morphological structure element into neighborhood configurations (Gong, Wei; Shi, Qinyun; Cheng, Minde)V1606,153-164(1991)

Decorrelation of color images using total color difference (Zheng, Joe; Valavanis, Kimon P.; Gauch, John M.)V1606,1037-1047(1991)

Deformable surfaces: a free-form shape representation (Delingette, Herve; Hebert, Martial; Ikeuchi, Katsushi)V1570,21-30(1991)

Deformable templates, robust statistics, and Hough transforms (Yuille, Alan L.; Peterson, Carsten; Honda, Ko)V1570,166-174(1991)

Dense-depth map from multiple views (Attolico, Giovanni; Caponetti, Laura; Chiaradia, Maria T.; Distante, Arcangelo; Stella, Ettore)V1383,34-46(1991)

Determination of flint wheel orientation for the automated assembly of lighters (Safabakhsh, Reza)V1472,185-189(1991)

Development of a smart workstation for use in mammography (Giger, Maryellen L.; Nishikawa, Robert M.; Doi, Kunio; Yin, Fang-Fang; Vyborny, Carl J.; Schmidt, Robert A.; Metz, Charles E.; Wu, Yuzheng; MacMahon, Heber; Yoshimura, Hitoshi)V1445,101-103(1991)

Development of criteria to compare model-based texture analysis methods (Soh, Young-Sung; Murthy, S. N.; Huntsberger, Terrance L.)V1381,561-573(1991)

Differential properties from adaptive thin-plate splines (Sinha, Saravajit S.)V1570,64-74(1991)

Direct computation of geometric features from motion disparities and shading (Weinshall, Daphna)V1570,274-285(1991)

Direct method for reconstructing shape from shading (Oliensis, John; Dupuis, Paul)V1570,116-128(1991)

Dynamically reconfigurable multiprocessor system for high-order-bidirectional-associative-memory-based image recognition (Wu, Chwan-Hwa; Roland, David A.)V1471,210-221(1991)

Dynamic end-to-end model testbed for IR detection algorithms (Iannarilli, Frank J.; Wohlers, Martin R.)V1483,66-76(1991)

Efficient computation of various types of skeletons (Vincent, Luc M.)V1445,297-311(1991)

Efficient extraction of local myocardial motion with optical flow and a resolution hierarchy (Srikantan, Geetha; Sher, David B.; Newberger, Ed)V1459,258-267(1991)

Efficient object contour tracing in a quadtree encoded image (Kumar, G. N.; Nandhakumar, N.)V1468,884-895(1991)

Efficient system for 3-D object recognition (Sobh, Tarek M.; Alameldin, Tarek K.)V1383,359-366(1991)

Efficient visual representation and reconstruction from generalized curvature measures (Barth, Erhardt; Caelli, Terry M.; Zetzsche, Christoph)V1570,86-95(1991)

End-to-end model for detection performance evaluation against scenario-specific targets (Iannarilli, Frank J.; Wohlers, Martin R.)V1488,226-236(1991)

End-to-end scenario-generating model for IRST performance analysis (Iannarilli, Frank J.; Wohlers, Martin R.)V1481,187-197(1991)

Energy-based segmentation of very sparse range surfaces (Lerner, Mark; Boult, Terrance E.)V1383,277-284(1991)

Energy functions for regularization algorithms (Delingette, Herve; Hebert, Martial; Ikeuchi, Katsushi)V1570,104-115(1991)

Error analysis on target localization from two projection images (Lee, Byung-Uk; Adler, John R.; Binford, Thomas O.)V1380,96-107(1991)

Evaluation of a pose estimation algorithm using single perspective view (Chandra, T.; Abidi, Mongi A.)V1382,409-426(1991)

Experiences with transputer systems for high-speed image processing (Kille, Knut; Ahlers, Rolf-Juergen; Schneider, B.)V1386,76-83(1991)

Experimental methodology for performance characterization of a line detection algorithm (Kanungo, Tapas; Jaisimha, Mysore Y.; Haralick, Robert M.; Palmer, John)V1385,104-112(1991)

Experiments with perceptual grouping (Shiu, Yiu C.)V1381,130-141(1991)

Extraction of hierarchical structures from complicated 2-D shapes (Han, Joon H.; Kim, Myung J.; Cho, Kwang J.)V1381,122-129(1991)

Extraction of human stomach using computational geometry (Aisaka, Kazuo; Arai, Kiyoshi; Tsutsui, Kumiko; Hashizume, Akihide)V1445,312-317(1991)

Face recognition based on depth maps and surface curvature (Gordon, Gaile G.)V1570,234-247(1991)

Fast algorithm for size analysis of irregular pore areas (Yuan, Li-Ping)V1451,125-136(1991)

Fast and precise method to extract vanishing points (Coelho, Christopher; Straforini, Marco; Campani, Marco)V1388,398-408(1991)

Fast piecewise-constant approximation of images (Radha, Hayder; Vetterli, Martin; Leonardi, Riccardo)V1605,475-486(1991)

Feature correspondence in multiple sensor data fusion (Broida, Ted J.)V1383,635-651(1991)

Finding distinctive colored regions in images (Syeda, Tanveer F.)V1381,574-581(1991)

First-order differential technique for optical flow (Campani, Marco; Straforini, Marco; Verri, Alessandro)V1388,409-414(1991)

FLIPS: Friendly Lisp Image Processing System (Gee, Shirley J.)V1472,38-45(1991)

Fourier cross-correlation and invariance transformation for affine groups (Segman, Joseph)V1606,788-802(1991)

From points to surfaces (Fua, Pascal; Sander, Peter T.)V1570,286-296(1991)

From voxel to curvature (Monga, Olivier; Ayache, Nicholas; Sander, Peter T.)V1570,382-390(1991)

Fusing human and machine skills for remote robotic operations (Schenker, Paul S.; Kim, Won S.; Venema, Steven; Bejczy, Antal K.)V1383,202-223(1991)

Fusion of multiple views of multiple reference points using a parallel distributed processing approach (Wolfe, William J.; Magee, Michael)V1383,20-25(1991)

Fusion of stereo views: estimating structure and motion using a robust method (Weng, Juyang; Cohen, Paul)V1383,321-332(1991)

Generalization of Lloyd's algorithm for image segmentation (Morii, Fujiki)V1381,545-552(1991)

Genetic algorithm approach to visual model-based halftone pattern design (Chu, Chee-Hung H.; Kottapalli, M. S.)V1606,470-481(1991)

Geometric Methods in Computer Vision (Vemuri, Baba C., ed.)V1570(1991)

Geometric modeling of noisy image objects (Lipari, Charles A.)V1468,905-917(1991)

Global minima via dynamic programming: energy minimizing active contours (Chandran, Sharat; Maejima, Tsukasa; Miyazaki, Sanae)V1570,391-402(1991)

Grasp-oriented sensing and control (Grupen, Roderic A.; Weiss, Richard S.; Oskard, David N.)V1383,189-201(1991)

Hand-eye coordination for grasping moving objects (Allen, Peter K.; Yoshimi, Billibon; Timcenko, Alexander; Michelman, Paul)V1383,176-188(1991)

Harmonic oscillator model of early visual image processing (Yang, Jian; Reeves, Adam J.)V1606,520-530(1991)

Heterogeneous parallel processor for a model-based vision system (Segal, Andrew C.)V1396,601-614(1991)

Hierarchical Dempster-Shafer evidential reasoning for image interpretation (Andress, Keith M.)V1569,43-54(1991)

Hierarchical decomposition and axial representation of shape (Rom, Hillel; Medioni, Gerard)V1570,262-273(1991)

Hierarchical fusion of geometric constraints for image segmentation (Seetharaman, Guna S.; Chu, Chee-Hung H.)V1383,582-588(1991)

Hierarchical target representation for autonomous recognition using distributed sensors (Luo, Ren C.; Kay, Michael G.)V1383,537-544(1991)

Human face recognition by P-type Fourier descriptor (Aibara, Tsunehiro; Ohue, Kenji; Matsuoka, Yasushi)V1606,198-203(1991)

Human movement analysis with image processing in real time (Fauvet, E.; Paindavoine, M.; Cannard, F.)V1358,620-630(1991)

Human Vision, Visual Processing, and Digital Display II (Rogowitz, Bernice E.; Brill, Michael H.; Allebach, Jan P., eds.)V1453(1991)

Image analysis applied to black ice detection (Chen, Yi)V1468,551-562(1991)

Image-processing system based on algorithmically dedicated functional units (Tozzi, Clesio L.; Castanho, Jose Eduardo C.; Gutierrez da Costa, Henrique S.)V1384,124-132(1991)

Image recognition, learning, and control in a cellular automata network (Raghavan, Raghu)V1469,89-101(1991)

Image representation by group theoretic approach (Segman, Joseph; Zeevi, Yehoshua Y.)V1606,97-109(1991)

Image representation by integrating curvatures and Delaunay triangulations (Wu, Chengke; Mohr, Roger)V1570,362-370(1991)

Image Understanding and the Man-Machine Interface III (Barrett, Eamon B.; Pearson, James J., eds.)V1472(1991)

Implementing neural-morphological operations using programmable logic (Shih, Frank Y.; Moh, Jenlong)V1382,99-110(1991)

Improved adaptive resonance theory (Shih, Frank Y.; Moh, Jenlong)V1382,26-36(1991)

Integrated vision system for object identification and localization using 3-D geometrical models (Bidlack, Clint R.; Trivedi, Mohan M.)V1468,270-280(1991)

Integrating acoustical and optical sensory data for mobile robots (Wang, Gang)V1468,479-482(1991)

Integration of a computer vision system with an IBM 7535 robot (Gonzalez, Orlando; Johnson, Carroll; Starks, Scott A.)V1381,284-291(1991)

Integration of edge- and region-based techniques for range image segmentation (Al-Hujazi, Ezzet H.; Sood, Arun K.)V1381,589-599(1991)

Intelligent Robots and Computer Vision IX: Algorithms and Techniques (Casasent, David P., ed.)V1381(1991)

Intelligent Robots and Computer Vision IX: Neural, Biological, and 3-D Methods (Casasent, David P., ed.)V1382(1991)

Interaction of algorithm and implementation for analog VLSI stereo vision (Hakkarainen, J. M.; Little, James J.; Lee, Hae-Seung; Wyatt, John L.)V1473,173-184(1991)

Interpolation of stereo data using Lagrangian polynomials (Bachnak, Rafic A.; Yamout, Jihad S.)V1457,27-36(1991)

Invariant feature matching in parameter space with application to line features (Hecker, Y. C.; Bolle, Ruud M.)V1570,298-314(1991)

Invariant reconstruction of 3-D curves and surfaces (Stevenson, Robert L.; Delp, Edward J.)V1382,364-375(1991)

Investigation of methods of combining functional evidence for 3-D object recognition (Stark, Louise; Hall, Lawrence O.; Bowyer, Kevin W.)V1381,334-345(1991)

Iso-precision scaling of digitized mammograms to facilitate image analysis (Karssemeijer, Nico; van Erning, Leon J.)V1445,166-177(1991)

Issues in parallelism in object recognition (Bhandarkar, Suchendra M.; Suk, Minsoo)V1384,234-245(1991)

Iterative neural networks for skeletonization and thinning (Krishnapuram, Raghu J.; Chen, Ling-Fan)V1382,271-281(1991)

Knowledge representation by dynamic competitive learning techniques (Racz, Janos; Klotz, Tamas)V1469,778-783(1991)

Labeled object identification for the mobile servicing system on the space station (Zakaria, Marwan F.; Ng, Terence K.)V1386,121-127(1991)

Landmark recognition using motion-derived scene structures (Sadjadi, Firooz A.)V1521,98-105(1991)

Learning procedure for the recognition of 3-D objects from 2-D images (Bart, Mischa; Buurman, Johannes; Duin, Robert P.)V1381,66-77(1991)

Less interclass disturbance learning for unsupervised neural computing (Liu, Lurng-Kuo; Ligomenides, Panos A.)V1606,496-507(1991)

Linear lattice architectures that utilize the central limit for image analysis, Gaussian operators, sine, cosine, Fourier, and Gabor transforms (Ben-Arie, Jezekiel)V1606,823-838(1991)

Matching 3-D smooth surfaces with their 2-D projections using 3-D distance maps (Lavallee, Stephane; Szeliski, Richard; Brunie, Lionel)V1570,322-336(1991)

Matching in image/object dual spaces (Zhang, Yaonan)V1526,195-202(1991)

Mathematical theories of shape: do they model perception? (Mumford, David)V1570,2-10(1991)

Medical image understanding system based on Dempster-Shafer reasoning (Chen, Shiuh-Yung; Lin, Wei-Chung; Chen, Chin-Tu)V1445,386-397(1991)

Medical Imaging V: Image Processing (Loew, Murray H., ed.)V1445(1991)

Method of preprocessing color images using a Peano curve on a Transputer array (Lambert, Robin A.; Batchelor, Bruce G.)V1381,582-588(1991)

Microscopic feature extraction from optical sections of contracting cardiac muscle cells recorded at high speed (Roos, Kenneth P.; Lake, David S.; Lubell, Bradford A.)V1428,159-168(1991)

Miscibility matrices explain the behavior of gray-scale textures generated by Gibbs random fields (Elfadel, Ibrahim M.; Picard, Rosalind W.)V1381,524-535(1991)

Mobile Robots V (Chun, Wendell H.; Wolfe, William J., eds.)V1388(1991)

Model-based boundary detection in echocardiography using dynamic programming technique (Dong, LiXin; Pelle, Gabriel; Brun, Philip; Unser, Michael A.)V1445,178-187(1991)

Model-based surface classification (Newman, Timothy S.; Flynn, Patrick J.; Jain, Anil K.)V1570,250-261(1991)

Model-based vision: an operational reality? (Mundy, Joseph L.)V1567,124-141(1991)

Model-based vision using geometric hashing (Akerman, Alexander; Patton, Ronald)V1406,30-39(1991)

Model generation and partial matching of left ventricular boundaries (Tehrani, Saeid; Weymouth, Terry E.; Mancini, G. B.)V1445,434-445(1991)

Model group indexing for recognition (Clemens, David T.; Jacobs, David W.)V1381,30-42(1991)

Modeling inner and outer plexiform retinal processing using nonlinear coupled resistive networks (Andreou, Andreas G.; Boahen, Kwabena A.)V1453,270-281(1991)

Modeling nonhomogeneous 3-D objects for thermal and visual image synthesis (Karthik, Sankaran; Nandhakumar, N.; Aggarwal, Jake K.)V1468,686-697(1991)

Modeling of the texture structural components using 2-D deterministic random fields (Francos, Joseph M.; Meiri, A. Z.; Porat, Boaz)V1606,553-565(1991)

Motion analysis for visually-guided navigation (Hildreth, Ellen C.)V1382,167-180(1991)

Motion estimation without correspondences and object tracking over long time sequences (Goldgof, Dmitry B.; Lee, Hua; Huang, Thomas S.)V1383,109-121(1991)

Multilevel evidence fusion for the recognition of 3-D objects: an overview of computer vision research at IBM/T.J. Watson (Bolle, Ruud M.; Califano, Andrea; Kender, John R.; Kjeldsen, Rick; Mohan, Rakesh)V1383,305-318(1991)

Multiple resonant boundary contour system (Lehar, Steve M.; Worth, Andrew J.)V1469,50-62(1991)

Multiresolution range-guided stereo matching (Tate, Kevin J.; Li, Ze-Nian)V1383,491-502(1991)

Multitask neurovision processor with extensive feedback and feedforward connections (Gupta, Madan M.; Knopf, George K.)V1606,482-495(1991)

Negative fuse network (Liu, Shih-Chii; Harris, John G.)V1473,185-193(1991)

Neural-network-aided design for image processing (Vitsnudel, Ilia; Ginosar, Ran; Zeevi, Yehoshua Y.)V1606,1086-1091(1991)

Neural-network-based vision processing for autonomous robot guidance (Pomerleau, Dean A.)V1469,121-128(1991)

Neural edge detector (Enab, Yehia M.)V1382,292-303(1991)

Neural model for feature matching in stereo vision (Wang, Shengrui; Poussart, Denis; Gagne, Simon)V1382,37-48(1991)

Neural net selection of features for defect inspection (Sasaki, Kenji; Casasent, David P.; Natarajan, Sanjay S.)V1384,228-233(1991)

Neural network model of dynamic form perception: implications of retinal persistence and extraretinal sharpening for the perception of moving boundaries (Ogmen, Haluk)V1606,350-359(1991)

Neural networks for medical image segmentation (Lin, Wei-Chung; Tsao, Chen-Kuo; Chen, Chin-Tu; Feng, Yu-Jen)V1445,376-385(1991)

Neural networks for robot navigation (Pandya, Abhijit S.; Luebbers, Paul G.)V1468,802-811(1991)

Neural network vision integration with learning (Toborg, Scott T.)V1469,77-88(1991)

New cooperative edge linking (Bonnin, Patrick; Zavidovique, Bertrand)V1381,142-152(1991)

New insights into correlation-based template matching (Ooi, James; Rao, Kashi)V1468,740-751(1991)

New method for designing face image classifiers using 3-D CG model (Akamatsu, Shigeru; Sasaki, Tsutomu; Masui, Nobuhiko; Fukamachi, Hideo; Suenaga, Yasuhito)V1606,204-216(1991)

Noise-tolerant texture classification and image segmentation (Kjell, Bradley P.; Wang, Pearl Y.)V1381,553-560(1991)

Nonlinear optical flow estimation and segmentation (Geurtz, Alexander M.)V1567,110-121(1991)

Object-oriented strategies for a vision dimensional metrology system (Pizzi, Nicolino J.; El-Hakim, Sabry F.)V1468,296-304(1991)

Object segmentation algorithm for use in recognizing 3-D partially occluded objects (Fan, Kuo-Chin; Chang, Chia-Yuan)V1468,674-684(1991)

Oh say, can you see? The physiology of vision (Young, Richard A.)V1453,92-123(1991)

On seeing spaghetti: a novel self-adjusting seven-parameter Hough space for analyzing flexible extruded objects (Kender, John R.; Kjeldsen, Rick)V1570,315-321(1991)

Optical flow techniques for moving target detection (Russo, Paul; Markandey, Vishal; Bui, Trung H.; Shrode, David)V1383,60-71(1991)

Orientation-based differential geometric representations for computer vision applications (Taubes, C. H.; Liang, Ping)V1570,96-102(1991)

Packet-switching algorithm for SIMD computers and its application to parallel computer vision (Maresca, Massimo)V1384,206-214(1991)

Parallel algorithm for volumetric segmentation (Liou, Shih-Ping; Jain, Ramesh C.)V1381,447-458(1991)

Parameter studies for Markov random field models of early vision (Daily, Michael J.)V1473,138-152(1991)

Parametric optical flow without correspondence for moving sensors (Whitten, Gary E.)V1468,167-175(1991)

Part-task training with a helmet-integrated display simulator system (Casey, Curtis J.; Melzer, James E.)V1456,175-178(1991)

Pattern spectrum morphology for texture discrimination and object recognition (Lee, Bonita G.; Tom, Victor T.)V1381,80-91(1991)

Photometric models in multispectral machine vision (Brill, Michael H.)V1453,369-380(1991)

Physically based and probabilistic models for computer vision (Szeliski, Richard; Terzopoulos, Demetri)V1570,140-152(1991)

Pixel level data fusion: from algorithm to chip (Mathur, Bimal P.; Wang, H. T.; Liu, Shih-Chii; Koch, Christof; Luo, Jin)V1473,153-160(1991)

Polynomial approach for morphological operations on 2-D and 3-D images (Bhattacharya, Prabir; Qian, Kai)V1383,530-536(1991)

POPS: parallel opportunistic photointerpretation system (Howard, Michael D.)V1471,422-427(1991)

Practical VLSI realization of morphological operations (Chang, Yiher; Ansari, Nirwan)V1606,839-850(1991)

Primary set of characteristic views for 3-D objects (Chen, Shuang; Freeman, Herbert)V1570,352-361(1991)

Probabilistic modeling of surfaces (Szeliski, Richard)V1570,154-165(1991)

Production environment implementation of the stereo extraction of cartographic features using computer vision and knowledge base systems in DMA's digital production system (Gruenewald, Maria M.; Hinchman, John H.)V1468,843-852(1991)

Pyramid nets for computer vision (Stinson, Michael C.)V1386,53-61(1991)

Radar image understanding for complex space objects (Hemler, Paul F.)V1381,55-65(1991)

Range data from stereo images of edge points (Lim, Hong-Seh)V1382,434-442(1991)

Range image-based object detection and localization for HERMIES III mobile robot (Sluder, John C.; Bidlack, Clint R.; Abidi, Mongi A.; Trivedi, Mohan M.; Jones, Judson P.; Sweeney, Frank J.)V1468,642-652(1991)

Real-time architecture based on the image processing module family (Kimura, Shigeru; Murakami, Yoshiyuki; Matsuda, Hikaru)V1483,10-17(1991)

Real-time region hierarchy and identification algorithm (Reihani, Kamran; Thompson, Wiley E.)V1449,99-108(1991)

Realization of the Zak-Gabor representation of images (Assaleh, Khaled T.; Zeevi, Yehoshua Y.; Gertner, Izidor)V1606,532-540(1991)

Recognition, tracking, and pose estimation of arbitrarily shaped 3-D objects in cluttered intensity and range imagery (Gottschalk, Paul G.; Turney, Jerry L.; Chiu, Arnold H.; Mudge, Trevor N.)V1383,84-96(1991)

Recognition and positioning of rigid objects using algebraic moment invariants (Taubin, Gabriel; Cooper, David B.)V1570,175-186(1991)

Recognition in face space (Turk, Matthew A.; Pentland, Alexander P.)V1381,43-54(1991)

Recognition of a moving planar shape in space from two perspective images (Li, Zhi-yong; Sun, Zhong-kang; Shen, Zhen-kang)V1472,97-105(1991)

Recognition of contacts between objects in the presence of uncertainties (Xiao, Jing)V1470,134-145(1991)

Recognizing human eyes (Hallinan, Peter W.)V1570,214-226(1991)

Recursive computation of a wire-frame representation of a scene from dynamic stereo using belief functions (Tirumalai, Arun P.; Schunck, Brian G.; Jain, Ramesh C.)V1569,28-42(1991)

Remote media vision-based computer input device (Arabnia, Hamid R.; Chen, Ching-Yi)V1606,917-925(1991)

Representing the dynamics of the occluding contour (Seales, W. B.; Dyer, Charles R.)V1383,47-58(1991)

Robust regression in computer vision (Meer, Peter; Mintz, Doron)V1381,424-435(1991)

Robust self-calibration and evidential reasoning for building environment maps (Tirumalai, Arun P.; Schunck, Brian G.; Jain, Ramesh C.)V1383,345-358(1991)

Role of computer graphics in space telerobotics: preview and predictive displays (Bejczy, Antal K.; Venema, Steven; Kim, Won S.)V1387,365-377(1991)

Rotation and scale invariant pattern recognition using a multistaged neural network (Minnix, Jay I.; McVey, Eugene S.; Inigo, Rafael M.)V1606,241-251(1991)

Salient contour extraction for target recognition (Rao, Kashi; Liou, James J.)V1482,293-306(1991)

Segmentation of orientation maps by an integration of edge- and region-based methods (Distante, Arcangelo; D'Orazio, Tiziana; Stella, Ettore)V1381,513-523(1991)

Segmentation using range data and structured light (Hu, Gongzhu)V1381,482-489(1991)

Segmentation via fusion of edge and needle map (Ahn, Hong-Young; Tou, Julius T.)V1468,896-904(1991)

Semiautomatic medical image segmentation using knowledge of anatomic shape (Brinkley, James F.)V1445,78-87(1991)

Sensor Fusion III: 3-D Perception and Recognition (Schenker, Paul S., ed.)V1383(1991)

Sensor fusion using K-nearest neighbor concepts (Scott, David R.; Flachs, Gerald M.; Gaughan, Patrick T.)V1383,367-378(1991)

Shape metrics from curvature-scale space and curvature-tuned smoothing (Dudek, Gregory)V1570,75-85(1991)

Shape reconstruction and object recognition using angles in an image (Fukada, Youji)V1381,111-121(1991)

Shape representation and nonrigid motion tracking using deformable superquadrics (Metaxas, Dimitri; Terzopoulos, Demetri)V1570,12-20(1991)

Software development tools for implementing vision systems on multiprocessors (Choudhary, Alok; Ranka, Sanjay)V1406,148-161(1991)

Spatial and temporal surface interpolation using wavelet bases (Pentland, Alexander P.)V1570,43-62(1991)

Spatio-temporal curvature measures for flow-field analysis (Zetzsche, Christoph; Barth, Erhardt; Berkmann, J.)V1570,337-350(1991)

Statistical and neural network classifiers in model-based 3-D object recognition (Newton, Scott C.; Nutter, Brian S.; Mitra, Sunanda)V1382,209-218(1991)

Statistical approach to model matching (Wells, William M.)V1381,22-29(1991)

Stereo matching, error detection, and surface reconstruction (Stewart, Charles V.)V1383,285-296(1991)

Stereo vision for planetary rovers: stochastic modeling to near-real-time implementation (Matthies, Larry H.)V1570,187-200(1991)

Stochastic field-based object recognition in computer vision (Zhu, Dongping; Beex, A. A.; Conners, Richard W.)V1569,174-181(1991)

Stochastic neural nets and vision (Fall, Thomas C.)V1468,778-785(1991)

Structural identity in visual-perceptual recognition (Ligomenides, Panos A.)V1382,14-25(1991)

Structure and motion of entire polyhedra (Sobh, Tarek M.; Alameldin, Tarek K.)V1388,425-431(1991)

Structured light: theory and practice and practice and practice... (Keizer, Richard L.; Jun, Heesung; Dunn, Stanley M.)V1406,88-97(1991)

Structure of a scene from two and three projections (Tommasi, Tullio)V1383,26-33(1991)

Studies in robust approaches to object detection in high-clutter background (Shirvaikar, Mukul V.; Trivedi, Mohan M.)V1468,52-59(1991)

Supervised color constancy for machine vision (Novak, Carol L.; Shafer, Steven A.)V1453,353-368(1991)

Supervised pixel classification using a feature space derived from an artificial visual system (Baxter, Lisa C.; Coggins, James M.)V1381,459-469(1991)

Surface reconstruction method using deformable templates (Wang, Yuan-Fang; Wang, Jih-Fang)V1383,265-276(1991)

Surface reconstruction with discontinuities (Lee, David T.; Shiau, Jyh-Jen H.)V1383,297-304(1991)

Technique for ground/image truthing using a digital map to reduce the number of required measurements (Der, Sandor Z.; Dome, G. J.; Rusche, Gerald A.)V1483,167-176(1991)

Technology transfer in image understanding (Kohl, Charles A.)V1406,18-29(1991)

Texture boundary classification using Gabor elementary functions (Dunn, Dennis F.; Higgins, William E.; Maida, Anthony; Wakeley, Joseph)V1606,541-552(1991)

Three-dimensional gauging with stereo computer vision (Wong, Kam W.; Ke, Ying; Lew, Michael; Obaidat, Mohammed T.)V1526,17-26(1991)

Three-dimensional object recognition using multiple sensors (Hackett, Jay K.; Lavoie, Matt J.; Shah, Mubarak A.)V1383,611-622(1991)

Three-dimensional object representation by array grammars (Wang, Patrick S.)V1381,210-216(1991)

Three-dimensional orientation from texture using Gabor wavelets (Super, Boaz J.; Bovik, Alan C.)V1606,574-586(1991)

Three-dimensional position determination from motion (Nashman, Marilyn; Chaconas, Karen)V1383,166-175(1991)

Three-dimensional reconstruction using virtual planes and horopters (Grosso, Enrico; Sandini, Giulio; Tistarelli, Massimo)V1570,371-381(1991)

Three-dimensional scene interpretation through information fusion (Shen, Sylvia S.)V1382,427-433(1991)

Three-dimensional scene reconstruction using optimal information fusion (Hong, Lang)V1383,333-344(1991)

Three-dimensional tolerance verification for computer vision systems (Griffin, Paul M.; Taboada, John)V1381,292-298(1991)

Toros: an image processing system for measuring consignments of wood (Correia, Bento A.; Davies, Roger; Carvalho, Fernando D.; Rodrigues, F. C.)V1567,15-24(1991)

Toward computing the aspect graph of deformable generalized cylinders (Wilkins, Belinda; Goldgof, Dmitry B.; Bowyer, Kevin W.)V1468,662-673(1991)

Towards integrated autonomous systems (Jain, Ramesh C.; Roth-Tabak, Yuval)V1468,188-201(1991)

Transformation from tristimulus RGB to Munsell notation HVC in a colored computer vision system (Jin, Guofan; Zhu, Zimin; Yu, Xinglong)V1569,507-512(1991)

Using color to segment images of 3-D scenes (Healey, Glenn)V1468,814-825(1991)

Vision-based model of artificial texture perception (Landraud, Anne M.)V1453,314-320(1991)

Vision for automated imagery exploitation (Ahlquist, Gregory C.)V1381,2-8(1991)

Visualization of image from 2-D strings using visual reasoning (Li, Xiao-Rong; Chang, Shi-Kuo)V1468,720-731(1991)

Visual motion detection: emulation of retinal peripheral visual field (Gupta, Madan M.; Digney, Bruce L.)V1381,346-356(1991)

Visual surveillance system based on spatio-temporal model of moving objects in industrial workroom environments (Motamed, Cina; Schmitt, Alain)V1606,961-969(1991)

Workstation recognition using a constrained edge-based Hough transform for mobile robot navigation (Vaughn, David L.; Arkin, Ronald C.)V1383,503-514(1991)

X-ray inspection utilizing knowledge-based feature isolation with a neural network classifier (Nolan, Adam R.; Hu, Yong-Lin; Wee, William G.)V1472,157-164(1991)

Correlation—see also filters; optical processing; pattern recognition

Acousto-optic estimation of autocorrelation and spectra using triple correlations and bispectra (Sadler, Brian M.; Giannakis, Georgios B.; Smith, Dale J.)V1476,246-256(1991)

Adaptive 4-D IR clutter suppression filtering technique (Aridgides, Athanasios; Fernandez, Manuel F.; Randolph, D.; Ferris, D.)V1481,110-116(1991)

Advanced in-plane rotation-invariant filter results (Ravichandran, Gopalan; Casasent, David P.)V1567,466-479(1991)

Advances in the optical design of miniaturized optical correlators (Gebelein, Rolin J.; Connely, Shawn W.; Foo, Leslie D.)V1564,452-463(1991)

Amplitude-encoded phase-only filters for pattern recognition: influence of the bleaching procedure (Campos, Juan; Janowska-Dmoch, Bozena; Styczynski, K.; Turon, F.; Yzuel, Maria J.; Chalasinska-Macukow, Katarzyna)V1574,141-147(1991)

Analysis and experimental performance of reduced-resolution binary phase-only filters (Kozaitis, Samuel P.; Tepedelenlioglu, N.; Foor, Wesley E.)V1564,373-383(1991)

Analysis of Bragg diffraction in optical memories and optical correlators (Gheen, Gregory)V1564,135-142(1991)

Application of canonical correlation analysis in detection in presence of spatially correlated noise (Chen, Wei G.; Reilly, James P.; Wong, Kon M.)V1566,464-475(1991)

Application of neural networks for the synthesis of binary correlation filters for optical pattern recognition (Mahalanobis, Abhijit; Nadar, Mariappan S.)V1469,292-302(1991)

Application of optical signal processing: fingerprint identification (Fielding, Kenneth H.; Horner, Joseph L.; Makekau, Charles K.)V1564,224-230(1991)

Automated approach to the correlation of defect locations to electrical test results to determine yield reducing defects (Slama, M. M.; Patterson, Angela C.)V1464,602-609(1991)

Automatic target recognition using acousto-optic image correlator (Molley, Perry A.; Kast, Brian A.)V1471,224-232(1991)

Bipolar correlations in composite circular harmonic filters (Leclerc, Luc; Sheng, Yunlong; Arsenault, Henri H.)V1564,78-85(1991)

Brownian dynamics simulation of polarized light scattering from wormlike chains (Allison, Stuart A.)V1430,50-64(1991)

Case study of design trade-offs for ternary phase-amplitude filters (Flannery, David L.; Phillips, William E.; Reel, Richard L.)V1564,65-77(1991)

Characterization of simulated and open-air atmospheric turbulence (Razdan, Anil K.; Singh, Brahm P.; Chopra, S.; Modi, M. B.)V1558,384-388(1991)

Compact, one-lens JTC using a transmissive amorphous silicon FLC-SLM (LAPS-SLM) (Haemmerli, Jean-Francois; Iwaki, Tadao; Yamamoto, Shuhei)V1564,275-284(1991)

Compact joint transform correlators in planar-integrated packages (Ghosh, Anjan K.)V1564,231-235(1991)

Comparison of optically addressed spatial light modulators (Hudson, Tracy D.; Kirsch, James C.; Gregory, Don A.)V1474,101-111(1991)

Composite image joint transform correlator (Mendlovic, David; Konforti, Naim; Deutsch, Meir; Marom, Emanuel)V1442,182-192(1991)

Computer-generated holograms for optical data processing (Casasent, David P.)V1544,101-107(1991)

Concept for the subresolution measurement of earthquake strain fields using SPOT panchromatic imagery (Crippen, Robert E.; Blom, Ronald G.)V1492,370-377(1991)

Constructing an optimal binary phase-only filter using a genetic algorithm (Calloway, David L.)V1564,395-402(1991)

Correction of magneto-optic device phase errors in optical correlators through filter design modifications (Downie, John D.; Reid, Max B.; Hine, Butler P.)V1564,308-319(1991)

Correlation-based optical numeric processors (Casasent, David P.; Woodford, Paul)V1563,112-119(1991)

Correlation between the detection and interpretation of image features (Fuhrman, Carl R.; King, Jill L.; Obuchowski, Nancy A.; Rockette, Howard E.; Sashin, Donald; Harris, Kathleen M.; Gur, David)V1446,422-429(1991)

Correlation model for a class of medical images (Zhang, Ya-Qin; Loew, Murray H.; Pickholtz, Raymond L.)V1445,367-373(1991)

Correlations between time required for radiological diagnoses, readers' performance, display environments, and difficulty of cases (Gur, David; Rockette, Howard E.; Sumkin, Jules H.; Hoy, Ronald J.; Feist, John H.; Thaete, F. L.; King, Jill L.; Slasky, B. S.; Miketic, Linda M.; Straub, William H.)V1446,284-288(1991)

Decorrelation of color images using total color difference (Zheng, Joe; Valavanis, Kimon P.; Gauch, John M.)V1606,1037-1047(1991)

Design and testing of space-domain minimum-average correlation energy filters for 2-D acousto-optic correlators (Connelly, James M.; Vijaya Kumar, B. V. K.; Molley, Perry A.; Stalker, K. T.; Kast, Brian A.)V1564,572-592(1991)

Design and testing of three-level optimal correlation filters (Hendrix, Charles D.; Vijaya Kumar, B. V. K.; Stalker, K. T.; Kast, Brian A.; Shori, Raj K.)V1564,2-13(1991)

Design of a GaAs acousto-optic correlator for real-time processing (Armenise, Mario N.; Impagnatiello, Fabrizio; Passaro, Vittorio M.; Pansini, Evangelista)V1562,160-171(1991)

Digital-signal-processor-based inspection of populated surface-mount technology printed circuit boards (Hartley, David A.; Hobson, Clifford A.; Lilley, F.)V1567,277-282(1991)

Digital speckle correlation search method and its application (Zhou, Xingeng; Gao, Jianxing)V1554A,886-895(1991)

Distortion- and intensity-invariant optical correlation filter system (Rahmati, Mohammad; Hassebrook, Laurence G.; Bhushan, M.)V1567,480-489(1991)

Dynamic light scattering from strongly interacting multicomponent systems: salt-free polyelectrolyte solutions (Sedlak, Marian; Amis, Eric J.; Konak, Cestmir; Stepanek, Petr)V1430,191-202(1991)

Effect of experimental design on sample size (Rockette, Howard E.; Obuchowski, Nancy A.; Gur, David; Good, Walter F.)V1446,276-283(1991)

Effects of thresholding in multiobject binary joint transform correlation (Javidi, Bahram; Wang, Jianping; Tang, Qing)V1564,212-223(1991)

Electro-optical image processing architecture for implementing image algebra operations (Coffield, Patrick C.)V1568,137-148(1991)

Elliptical coordinate transformed phase-only filter for shift and scale invariant pattern recognition (Garcia, Javier; Ferreira, Carlos; Szoplik, Tomasz)V1574,133-140(1991)

Empirical performance of binary phase-only synthetic discriminant functions (Carhart, Gary W.; Draayer, Bret F.; Billings, Paul A.; Giles, Michael K.)V1564,348-362(1991)

Experimental comparison of optical binary phase-only filter and high-pass matched filter correlation (Leib, Kenneth G.; Brandstetter, Robert W.; Drake, Marvin D.; Franks, Glen B.; Siewert, Ronald O.)V1483,140-154(1991)

Experimental investigations of the autocorrelation function for inhomogeneous scatterers (Singh, Brahm P.; Chopra, S.)V1558,317-321(1991)

Experimental measurements and electron correlation theory of third-order nonlinear optical processes in linear chains (Heflin, James R.; Cai, Yongming; Zhou, Qihou L.; Garito, Anthony F.)V1560,2-12(1991)

Fast holographic correlator for machine vision systems (Nekrasov, Victor V.)V1507,170-174(1991)

Full-complex spatial filtering with a phase mostly DMD (Florence, James M.; Juday, Richard D.)V1558,487-498(1991)

High-resolution submicron particle sizing by dynamic light scattering (Nicoli, David F.)V1430,19-36(1991)

Hybrid digital/optical ATR system (Goodwin, David B.; Cappiello, Gregory G.; Coppeta, David A.; Govignon, Jacques P.)V1564,536-549(1991)

Influence of input information coding for correlation operations (Maze, Sylvie; Joffre, Pascal; Refregier, Philippe)V1505,20-31(1991)

Infrared/microwave correlation measurements (Norgard, John D.; Metzger, Don W.; Cleary, John C.; Seifert, Michael)V1540,699-708(1991)

Initial key word OCR filter results (Casasent, David P.; Iyer, Anand K.; Ravichandran, Gopalan)V1384,324-337(1991)

Interference effects and the occurrence of blind spots in coherent optical processors (Christie, Simon; Cai, Xian-Yang; Kvasnik, Frank)V1507,202-209(1991)

Intermodulation effects in pure phase-only correlation method (Chalasinska-Macukow, Katarzyna; Turon, F.; Yzuel, Maria J.; Campos, Juan)V1564,285-293(1991)

Invariant phase-only filters for phase-encoded inputs (Kallman, Robert R.; Goldstein, Dennis H.)V1564,330-347(1991)

Inverse filtering technique for the synthesis of distortion-invariant optical correlation filters (Shen, Weisheng; Zhang, Shen; Tao, Chunkan)V1567,691-697(1991)

Joint transform correlator using nonlinear ferroelectric liquid-crystal spatial light modulator (Kohler, A.; Fracasso, B.; Ambs, Pierre; de Bougrenet de la Tocnaye, Jean-Louis M.)V1564,236-243(1991)

Localized feature selection to maximize discrimination (Duell, Kenneth A.; Freeman, Mark O.)V1564,22-33(1991)

Low-power optical correlator for sonar range finding (Turk, Harris; Leepa, Douglas C.; Snyder, Robert F.; Soos, Jolanta I.; Bronstein, Sam)V1454,344-352(1991)

Matched spatial filtering by feature-extracted reference patterns using cross-correlated signals (Kamemaru, Shun-ichi; Yano, Jun-ichi; Itoh, Haruyasu)V1564,143-154(1991)

Material testing by the laser speckle strain gauge (Yamaguchi, Ichirou; Kobayashi, Koichi)V1554A,240-249(1991)

Matrix optical system for plane-point correlation (Curatu, Eugen O.)V1527,368-375(1991)

Measurement of fluid velocity fields using digital correlation techniques (Matthys, Donald R.; Gilbert, John A.; Puliparambil, Joseph T.)V1332,850-861(1991)

Miniature acousto-optic image correlator (Molley, Perry A.; Sweatt, William C.; Strong, David S.)V1564,610-616(1991)

Multiplexed binary phase-only circular harmonic filters (Sheng, Yunlong; Leclerc, Luc; Arsenault, Henri H.)V1564,320-329(1991)

Neural optoelectronic correlator for pattern recognition (Figue, J.; Refregier, Philippe; Rajbenbach, Henri J.; Huignard, Jean-Pierre)V1564,550-561(1991)

New insights into correlation-based template matching (Ooi, James; Rao, Kashi)V1468,740-751(1991)

New track-to-track association logic for almost identical multiple sensors (Malakian, Kourken; Vidmar, Anthony)V1481,315-328(1991)

Noise and discrimination performance of the MINACE optical correlation filter (Ravichandran, Gopalan; Casasent, David P.)V1471,233-248(1991)

Noise on multiple-tau photon correlation data (Schaetzel, Klaus; Peters, Rainer)V1430,109-115(1991)

Normalization of correlations (Kast, Brian A.; Dickey, Fred M.)V1564,34-42(1991)

Object-enhanced optical correlation (Scholl, Marija S.; Shumate, Michael S.; Udomkesmalee, Suraphol)V1564,165-176(1991)

On the determination of optimal window for registration of nonlinear distributed images (Lure, Yuan-Ming F.)V1452,292-302(1991)

Optical correlation filters for large-class OCR applications (Casasent, David P.; Iyer, Anand K.; Gopalaswamy, Srinivasan)V1470,208-219(1991)

Optical correlation filters to locate destination address blocks in OCR (Casasent, David P.; Ravichandran, Gopalan)V1384,344-354(1991)

Optical correlation using a phase-only liquid-crystal-over-silicon spatial light modulator (Potter, Duncan J.; Ranshaw, M. J.; Al-Chalabi, Adil O.; Fancey, Norman E.; Sillitto, Richard M.; Vass, David G.)V1564,363-372(1991)

Optical correlator field demonstration (Kirsch, James C.; Gregory, Don A.; Hudson, Tracy D.; Loudin, Jeffrey A.; Crowe, William M.)V1482,69-78(1991)

Optical correlator field test results (Hudson, Tracy D.; Gregory, Don A.; Kirsch, James C.; Loudin, Jeffrey A.; Crowe, William M.)V1564,54-64(1991)

Optical correlators in texture analysis (Honkonen, Veijo; Jaaskelainen, Timo; Parkkinen, Jussi P.)V1564,43-51(1991)

Optical correlator techniques applied to robotic vision (Hine, Butler P.; Reid, Max B.; Downie, John D.)V1564,416-426(1991)

Optical correlator vision system for a manufacturing robot assembly cell (Brandstetter, Robert W.; Fonneland, Nils J.; Zanella, R.; Yearwood, M.)V1385,173-189(1991)

Optical implementation of neocognitron and its applications to radar signature discrimination (Chao, Tien-Hsin; Stoner, William W.)V1558,505-517(1991)

Optical inference processing techniques for scene analysis (Casasent, David P.)V1564,504-510(1991)

Optical Information Processing Systems and Architectures III (Javidi, Bahram, ed.)V1564(1991)

Optically implementable algorithm for convolution/correlation of long data streams (Zhang, Yan; Li, Yao)V1474,188-198(1991)

Optical processing of wire-frame models for object recognition (Kozaitis, Samuel P.; Cofer, Rufus H.)V1471,249-254(1991)

Optimal correlation filters for implementation on deformable mirror devices (Vijaya Kumar, B. V. K.; Carlson, Daniel W.)V1558,476-486(1991)

Optimal phase-only correlation filters in colored scene noise (Vijaya Kumar, B. V. K.; Liang, Victor; Juday, Richard D.)V1555,138-145(1991)

Optimal subpixel-level IR frame-to-frame registration (Fernandez, Manuel F.; Aridgides, Athanasios; Randolph, D.; Ferris, D.)V1481,172-179(1991)

Pattern detection using a modified composite filter with nonlinear joint transform correlator (Vallmitjana, Santiago; Juvells, Ignacio; Carnicer, Arturo; Campos, Juan)V1564,266-274(1991)

Performance evaluation methods for multiple-target-tracking algorithms (Fridling, Barry E.; Drummond, Oliver E.)V1481,371-383(1991)

Performance limitations of miniature optical correlators (Crandall, Charles M.; Giles, Michael K.; Clark, Natalie)V1564,98-109(1991)

Performance of pyramidal phase-only filtering of infrared imagery (Kozaitis, Samuel P.; Petrilak, Robert)V1564,403-413(1991)

Photon correlation spectroscopic studies of filamentous actin networks (Newman, Jay E.; San Biagio, Pier L.; Schick, Kenneth L.)V1430,89-108(1991)

Photon correlation spectroscopy: technique and instrumentation (Thomas, John C.)V1430,2-18(1991)

Photopolymer elements for an optical correlator system (Brandstetter, Robert W.; Fonneland, Nils J.)V1559,308-320(1991)

Planetary lander guidance using binary phase-only filters (Reid, Max B.; Hine, Butler P.)V1564,384-394(1991)

Point target detection, location, and track initiation: initial optical lab results (Carender, Neil H.; Casasent, David P.)V1481,35-48(1991)

Position-, scale-, and rotation-invariant photorefractive correlator (Ryan, Vincent; Fielding, Kenneth H.)V1564,86-97(1991)

Purely real correlation filters (Mahalanobis, Abhijit; Song, Sewoong)V1564,14-21(1991)

Quantization analysis of the binary joint transform correlator in the presence of nonlinear compression (Javidi, Bahram; Wang, Jianping)V1564,254-265(1991)

Radiologists' confidence in detecting abnormalities on chest images and their subjective judgments of image quality (King, Jill L.; Gur, David; Rockette, Howard E.; Curtin, Hugh D.; Obuchowski, Nancy A.; Thaete, F. L.; Britton, Cynthia A.; Metz, Charles E.)V1446,268-275(1991)

Real-time nonlinear optical correlator in speckle metrology (Ogiwara, Akifumi; Ohtsubo, Junji)V1564,294-305(1991)

Relaxation properties and learning paradigms in complex systems (Basti, Gianfranco; Perrone, Antonio; Morgavi, Giovanna)V1469,719-736(1991)

Rotation- and scale-invariant joint transform correlator using FLC-SLMs (Mitsuoka, Yasuyuki; Iwaki, Tadao; Yamamoto, Shuhei)V1564,244-252(1991)

Semiautomatic x-ray inspection system (Amladi, Nandan G.; Finegan, Michael K.; Wee, William G.)V1472,165-176(1991)

Solar feature correlation tracker (Rimmele, Thomas; von der Luhe, Oskar; Wiborg, P. H.; Widener, A. L.; Dunn, Richard B.; Spence, G.)V1542,186-193(1991)

Space-time correlation properties and their applications of dynamic speckles after propagation through an imaging system and double-random modulation (Ma, Shining; Liu, Ying; Du, Fuli)V1554A,645-648(1991)

Speckle measurement for 3-D surface movement (Hilbig, Jens; Ritter, Reinold)V1554A,588-592(1991)

Statistical initial orbit determination (Taff, Laurence G.; Belkin, Barry; Schweiter, G. A.; Sommar, K.)V1481,440-448(1991)

Strategies for the color character recognition by optical multichannel correlation (Millan, Maria S.; Yzuel, Maria J.; Campos, Juan; Ferreira, Carlos)V1507,183-197(1991)

Studies of correlation of molecular structure under preparation conditions for noncrystalline selenium thin films with aid of computer simulation (Popov, A.; Vasiljeva, Natalja V.)V1519,37-42(1991)

Surveillance test bed for SDIO (Wesley, Michael; Osterheld, Robert; Kyser, Jeff; Farr, Michele; Vandergriff, Linda J.)V1481,209-220(1991)

Theory of electronic projection correlation and its application in measurement of rigid body displacement and rotation (Li, Xide; Tan, Yushan)V1554B,661-668(1991)

Throughput comparison of optical and digital correlators for automatic target recognition (Huang, Chao H.; Gheen, Gregory; Washwell, Edward R.)V1564,427-438(1991)

Using liquid-crystal TVs in Vander Lugt optical correlators (Clark, Natalie; Crandall, Charles M.; Giles, Michael K.)V1564,439-451(1991)

Using photon correlation spectroscopy to study polymer coil internal dynamic behavior (Selser, James C.; Ellis, Albert R.; Schaller, J. K.; McKiernan, M. L.; Devanand, Krisha)V1430,85-88(1991)

Van der Lugt optical correlation for the measurement of leak rates of hermetically sealed packages (Fitzpatrick, Colleen M.; Mueller, Edward P.)V1332,185-192(1991)

Visualization and comparison of simulation results in computational fluid dynamics (Felger, Wolfgang; Astheimer, Peter)V1459,222-231(1991)

Crystals—see also optical properties

Advantages of drawing crystal-core fibers in microgravity (Shlichta, Paul J.; Nerad, Bruce A.)V1557,10-23(1991)

Application of high-rate crystal growth technique to single crystals of nucleic acid bases (Zachova, J.; Shtepanek, J.; Zaitseva, N. P.)V1402,216-222(1991)

Bulk darkening of flux-grown KTiOPO4 (Rockafellow, David R.; Teppo, Edward A.; Jacco, John C.)V1561,112-118(1991)

Characteristics of domain formation and poling in potassium niobate (Jarman, Richard H.; Johnson, Barry C.)V1561,33-42(1991)

Characterization of the Bridgman crystal growth process by radiographic imaging (Fripp, Archibald L.; Debnam, W. J.; Woodell, G. W.; Berry, R. F.; Simchick, Richard T.; Sorokach, S. K.; Barber, Patrick G.)V1557,236-244(1991)

Commercial crystal growth in space (Wilcox, William R.)V1557,31-41(1991)

Compact spaceflight solution crystal-growth system (Trolinger, James D.; Lal, Ravindra B.; Vikram, Chandra S.; Witherow, William K.)V1557,250-258(1991)

Comparison of laser-induced damage of optical crystals from the USA and USSR (Soileau, M. J.; Wei, Tai-Huei; Said, Ali A.; Chapliev, N. I.; Garnov, Sergei V.; Epifanov, Alexandre S.)V1441,10-15(1991)

Computerized detection and identification of the types of defects on crystal blanks (Bow, Sing T.; Chen, Pei)V1396,646-655(1991)

Continuous automatic scanning picosecond optical parametric source using MgO LiNbO3 in the 700-2200 nm (He, Huijuan; Lu, Yutian; Dong, Jingyuan; Zhao, Quingchun)V1409,18-23(1991)

Crystal growth and characterization of rare-earth-doped gallates of alkaline earth and lanthanum (Ryba-Romanowski, Witold; Golab, Stanislaw; Berkowski, Marek)V1391,2-5(1991)

Crystal growth by solute diffusion in Earth orbit (Lind, M. D.; Nielsen, K. F.)V1557,259-270(1991)

Crystal Growth in Space and Related Optical Diagnostics (Trolinger, James D.; Lal, Ravindra B., eds.)V1557(1991)

Crystallization and optothermal characteristics of germanate glasses (Montenero, Angelo; Gnappi, G.; Bertolotti, Mario; Sibilia, C.; Fazio, E.; Liakhou, G.)V1513,234-242(1991)

Crystallization of hydrogenated amorphous silicon film and its fractal structure (Lin, Hong Y.; Yang, Dao M.; Li, Ying X.)V1519,210-213(1991)

Crystal separation from mother solution and conservation under microgravity conditions using inert liquid (Regel, L. L.; Vedernikov, A. A.; Queeckers, P.; Legros, J. C.)V1557,182-191(1991)

Crystal surface analysis using matrix textural features classified by a probabilistic neural network (Sawyer, Curry R.; Quach, Viet; Nason, Donald; van den Berg, Lodewijk)V1567,254-263(1991)

Degradation of cholesterol crystals in macrophages: the role of phospholipids (Koren, Eugen; Koscec, Mirna; Fugate, Robert D.)V1428,214-223(1991)

Design and development of a transparent Bridgman furnace (Wells, Mark E.; Groff, Mary B.)V1557,71-77(1991)

Design and development of the Zeolite Crystal Growth Facility (Fiske, Michael R.)V1557,78-85(1991)

Development of a laser Doppler system for measurement of velocity fields in PVT crystal growth systems (Jones, O. C.; Glicksman, M. E.; Lin, Jen T.; Kim, G. T.; Singh, N. B.)V1557,202-208(1991)

Development of wideband 16-channel acousto-optical modulators on the LiNbO3 and TeO2 crystals (Bokov, Lev; Demidov, Anatoly J.; Zadorin, Anatoly; Kushnarev, Igor; Serebrennikov, Leonid J.; Sharangovich, Sergey)V1505,186-198(1991)

Display holograms of crystals recording (Smirnova, S. N.; Marchenko, S. N.)V1238,370-370(1991)

Electro-optic illuminating module (Pesl, Ales A.)V1454,299-305(1991)

Fabrication of laser materials by laser-heated pedestal growth (Chang, Robert S.; Sengupta, Sonnath; Shaw, Leslie B.; Djeu, Nick)V1410,125-132(1991)

Flight experiment to investigate microgravity effects on solidification phenomena of selected materials (Maag, Carl R.; Hansen, Patricia A.)V1557,24-30(1991)

GaN single-crystal films on silicon substrates grown by MOVPE (Nagatomo, Takao; Ochiai, Ichiro; Ookoshi, Shigeo; Omoto, Osamu)V1519,90-95(1991)

Generalized phased-array Bragg interaction in anisotropic crystals (Young, Eddie H.; Ho, Huey-Chin C.; Yao, Shi-Kay; Xu, Jieping P.)V1476,178-189(1991)

Generation of ultrashort ultraviolet optical pulses using sum-frequency in LBO crystals (Guo, Ting; Qiu, Peixia; Lin, Fucheng)V1409,24-27(1991)

Ground-based experiments on the growth and characterization of L-arginine phosphate crystals (Rao, S. M.; Cao, C.; Batra, Ashok K.; Lal, Ravindra B.; Mookherji, Tripty K.)V1557,283-292(1991)

Growth, surface passivation, and characterization of CdSe microcrystallites in glass with respect to their application in nonlinear optics (Woggon, Ulrike; Rueckmann, I.; Kornack, J.; Mueller, Matthias; Cesnulevicius, J.; Kolenda, Jonas; Petrauskas, Mendogas)V1362,888-898(1991)

Hard x-ray imaging via crystal diffraction: first results of reflectivity measurements (Frontera, Filippo; De Chiara, P.; Gambaccini, M.; Landini, G.; Pasqualini, G.)V1549,113-119(1991)

High-energy x-ray diffraction (Freund, Andreas K.)V1345,234-244(1991)

High-resolution synchrotron x-radiation diffraction imaging of crystals grown in microgravity and closely related terrestrial crystals (Steiner, Bruce W.; Dobbyn, Ronald C.; Black, David; Burdette, Harold; Kuriyama, Masao; Spal, Richard; van den Berg, Lodewijk; Fripp, Archibald L.; Simchick, Richard T.; Lal, Ravindra B.; Batra, Ashok K.; Matthiesen, David; Ditchek, Brian M.)V1557,156-167(1991)

Highly efficient optical parametric oscillators (Marshall, Larry R.; Hays, Alan D.; Kasinski, Jeff; Burnham, Ralph L.)V1419,141-152(1991)

Holographic devices using photo-induced effect in nondestructive testing techniques (Dovgalenko, George Y.; Onischenko, Yuri I.)V1559,479-486(1991)

Holographic instrumentation for monitoring crystal growth in space (Trolinger, James D.; Lal, Ravindra B.; Batra, Ashok K.)V1332,151-165(1991)

Holographic optical switching with photorefractive crystals (Song, Q. W.; Lee, Mowchen C.; Talbot, Peter J.)V1558,143-148(1991)

Influence of diamond turning and surface cleaning processes on the degradation of KDP crystal surfaces (Kozlowski, Mark R.; Thomas, Ian M.; Edwards, Gary; Stanion, Ken; Fuchs, Baruch A.; Latanich, L.)V1561,59-69(1991)

Influence of surface passivation on the optical bleaching of CdSe microcrystallites embedded in glass (Rueckmann, I.; Woggon, Ulrike; Kornack, J.; Mueller, Matthias; Cesnulevicius, J.; Kolenda, Jonas; Petrauskas, Mendogas)V1513,78-85(1991)

Inorganic Crystals for Optics, Electro-Optics, and Frequency Conversion (Bordui, Peter F., ed.)V1561(1991)

In-situ measurement technique for solution growth in compound semiconductors (Inatomi, Yuko; Kuribayashi, Kazuhiko)V1557,132-139(1991)

In-situ observation of crystal growth in microgravity by high-resolution microscopies (Tsukamoto, Katsuo; Onuma, Kazuo)V1557,112-123(1991)

Interface demarcation in Bridgman-Stockbarger crystal growth of II-VI compounds (Gillies, Donald C.; Lehoczky, S. L.; Szofran, Frank R.; Su, Ching-Hua; Larson, David J.)V1484,2-10(1991)

Investigation of strain birefringence and wavefront distortion in 001 plates of KD2PO4 (De Yoreo, James J.; Woods, Bruce W.)V1561,50-58(1991)

Investigation on phase biological micro-objects with a holographic interferometric microscope on the basis of the photorefractive Bi12TiO20 crystal (Tontchev, Dimitar A.; Zhivkova, Svetla; Miteva, Margarita G.; Grigoriev, Ivo D.; Ivanov, I.)V1429,76-80(1991)

Investigation on the anomalous structure of the nanocrystal Ti and Zr films (Shi, W.; Kong, J.; Shen, H.; Du, G.; Yao, W.; Qi, Zhen Z.)V1519,138-141(1991)

Ionic conductivity and damage mechanisms in KTiOPO4 crystals (Morris, Patricia A.; Crawford, Michael K.; Roelofs, Mark G.; Bierlein, John D.; Baer, Thomas M.)V1561,104-111(1991)

Large zeolites: why and how to grow in space (Sacco, Albert)V1557,6-9(1991)

Laser beam deflection: a method to investigate convection in vapor growth experiments (Lenski, Harald; Braun, Michael)V1557,124-131(1991)

Latest developments in crystal growth and characterization of efficient acousto-optic materials (Paradies, C. J.; Glicksman, M. E.; Jones, O. C.; Kim, G. T.; Lin, Jen T.; Gottlieb, Milton S.; Singh, N. B.)V1561,2-5(1991)

Linear and nonlinear optical properties of substituted pyrrolo[1,2-a]quinolines (van Hulst, Niek F.; Heesink, Gerard J.; Bolger, Bouwe; Kelderman, E.; Verboom, W.; Reinhoudt, D. N.)V1361,581-588(1991)

Liquid as a deformed crystal: the model of a liquid structure (Yakovlev, Evgeni B.)V1440,36-49(1991)

Liquid dendrites growth at laser-induced melting in a NaCl volume (Gorbunov, A. V.)V1440,78-82(1991)

Mapping crystal defects with a digital scanning ultramicroscope (Springer, John M.; Silberman, Enrique; Kroes, Roger L.; Reiss, Don)V1557,192-196(1991)

Measuring residual accelerations in the Spacelab environment (Witherow, William K.)V1557,42-52(1991)

Multiphoton excited emission in zinc selenide and other crystals (Prokhorov, Kirill A.; Prokhorov, Alexander M.; Djibladze, Merab I.; Kekelidze, George N.; Gorelik, Vladimir S.)V1501,80-84(1991)

NASA microgravity materials science program (Sokolowski, Robert S.)V1557,2-5(1991)

New optical approaches to the quantitative characterization of crystal growth, segregation, and defect formation (Carlson, D. J.; Wargo, Michael J.; Cao, X. Z.; Witt, August F.)V1557,140-146(1991)

New organic crystal material for SHG, 2-cyano-3-(3,4-methylene dioxy phenyl)-2-propionic acid ethyl ester (Mori, Yasushi; Sano, Kenji; Todori, Kenji; Kawamonzen, Yosiaki)V1560,310-314(1991)

Nonlinear optical properties of N-(4-nitro-2-pyridinyl)-phenylalaninol single crystals (Sutter, Kurt; Hulliger, J.; Knoepfle, G.; Saupper, N.; Guenter, Peter)V1560,296-301(1991)

Nonlinear optical properties of new KTiOPO4 isostructures (Phillips, Mark L.; Harrison, William T.; Stucky, Galen D.)V1561,84-92(1991)

Nonlinear optical properties of xanthone derivatives (Imanishi, Yasuo; Itoh, Yuzo; Kakuta, Atsushi; Mukoh, Akio)V1361,570-580(1991)

Novel technique for efficient wave mixing in photorefractive materials (Mathey, P.; Launay, Jean C.; Pauliat, G.; Roosen, Gerald)V1500,26-33(1991)

Optical and environmentally protective coatings for potassium dihydrogen phosphate harmonic converter crystals (Thomas, Ian M.)V1561,70-82(1991)

Optical diagnostics of mercuric iodide crystal growth (Burger, Arnold; Morgan, S. H.; Silberman, Enrique; Nason, Donald)V1557,245-249(1991)

Optical investigation of microcrystals in glasses (Ferrara, M.; Lugara, M.; Moro, C.; Cingolani, R.; De Blasi, C.; Manno, D.; Righini, Giancarlo C.)V1513,130-136(1991)

Optical memory in electro-optical crystals (Berezhnoy, Anatoly A.; Popov, Yury V.)V1401,44-49(1991)

Optical properties of KDP crystals grown at high growth rates (Zaitseva, N. P.; Ganikhanov, Ferous S.; Katchalov, O. V.; Efimkov, V. F.; Pastukhov, S. A.; Sobolev, V. B.)V1402,223-230(1991)

Optical research of elastic anisotropy of monocrystals with a cube structure (Polyak, Alexander; Kostin, Ivan K.)V1554A,553-556(1991)

Organic salts with large electro-optic coefficients (Perry, Joseph W.; Marder, Seth R.; Perry, Kelly J.; Sleva, E. T.; Yakymyshyn, Christopher P.; Stewart, Kevin R.; Boden, Eugene P.)V1560,302-309(1991)

Particle image velocimetry experiments for the IML-I spaceflight (Trolinger, James D.; Lal, Ravindra B.; Batra, Ashok K.; McIntosh, D.)V1557,98-109(1991)

Performance of longitudinal-mode KD*P Pockels cells with transparent conductive coatings (Skeldon, Mark D.; Jin, Michael S.; Smith, Douglas H.; Bui, Snow T.)V1410,116-124(1991)

Phase-locking and unstability of light waves in Raman-active crystals (Azarenkov, Aleksey N.; Altshuler, Grigori B.; Belashenkov, Nickolay R.; Inochkin, Mickle V.; Karasev, Viatcheslav B.; Kozlov, Sergey A.)V1409,154-164(1991)

Phase conjugation in LiF and NaF color center crystals (Basiev, Tasoltan T.; Zverev, Peter G.; Mirov, Sergey B.; Pal, Suranjan)V1500,65-71(1991)

Photorefractive two-wave mixing characteristics for image amplification in diffusion-driven media (Gilbreath, G. C.; Clement, Anne E.; Fugera, S. N.; Mizell, Gregory J.)V1409,87-99(1991)

Point defects in KTP and their possible role in laser damage (Scripsick, Michael P.; Edwards, Gary J.; Halliburton, Larry E.; Belt, Roger F.; Kappers, Lawrence A.)V1561,93-103(1991)

Polarization-state mixing in multiple-beam diffraction and its application to solving the phase problem (Shen, Qun)V1550,27-33(1991)

Polarizing x-ray optics for synchrotron radiation (Hart, Michael)V1548,46-55(1991)

Preparation of Pb1-xGexTe crystal with high refractive index for IR coating (Zhang, Su Y.; Xu, Bu Y.; Zhang, Feng S.; Yan, Yixun)V1519,508-513(1991)

Production of sapphire domes by the growth of near-net-shape single crystals (Biderman, Shlomo; Horowitz, Atara; Einav, Yehezkel; Ben-Amar, Gabi; Gazit, Dan; Stern, Adin; Weiss, Matania)V1535,27-34(1991)

Production of x-rays by the interaction of charged particle beams with periodic structures and crystalline materials (Rule, Donald W.; Fiorito, Ralph B.; Piestrup, Melvin A.; Gary, Charles K.; Maruyama, Xavier K.)V1552,240-251(1991)

Real-time quantitative imaging for semiconductor crystal growth, control, and characterization (Wargo, Michael J.)V1557,271-282(1991)

Real-time speckle photography using photorefractive Bi12SiO20 crystal (Nakagawa, Kiyoshi; Minemoto, Takumi)V1508,191-200(1991)

Recent progress in the growth and characterization of large Ge single crystals for IR optics and microelectronics (Azoulay, Moshe; Gafni, Gabriella; Roth, Michael)V1535,35-45(1991)

Reflectance anisotropy spectrometer for real-time crystal growth investigations (Acher, O.; Benferhat, Ramdane; Drevillon, Bernard; Razeghi, Manijeh)V1361,1156-1163(1991)

Registration of 3-D holograms of diamond crystals (Marchenko, S. N.; Smirnova, S. N.)V1238,371-371(1991)

Research on influence of substrate to crystal growth by ion beams (Yin, Dong; Guan, Wen X.; Sun, Shu Z.; Zhang, Zhao A.)V1519,164-166(1991)

Semiconductor-doped glasses: nonlinear and electro-optical properties (Ekimov, A. I.; Kudryavtsev, I. A.; Chepick, D. I.; Shumilov, S. K.)V1513,123-129(1991)

Simulation of optical diagnostics for crystal growth: models and results (Banish, Michele R.; Clark, Rodney L.; Kathman, Alan D.; Lawson, Shelah M.)V1557,209-221(1991)

Single crystallinity and oxygen diffusion in high-quality YBa2Cu3O7-delta films (Wu, Zi L.; Wei, M. Z.; Chen, Y. X.; Ren, Cong X.; Zhang, J. H.)V1519,618-624(1991)

Spectral hole burning of strongly confined CdSe quantum dots (Spiegelberg, Christine; Henneberger, Fritz; Puls, J.)V1362,951-958(1991)

Status of the stellar x-ray polarimeter for the Spectrum-X-Gamma mission (Kaaret, Philip E.; Novick, Robert; Shaw, Ping-Shine; Hanany, Shaul; Liu, Yee; Fleischman, Judith R.; Sunyaev, Rashid; Lapshov, I.; Weisskopf, Martin C.; Elsner, Ronald F.; Ramsey, Brian D.; Silver, Eric H.; Ziock, Klaus P.; Costa, Enrico; Piro, Luigi; Soffitta, Paolo; Manzo, Giuseppe; Giarrusso, Salvatore; Santangelo, Andrea E.; Scarsi, Livio; Fraser, George W.; Pearson, James F.; Lees, John E.; Perola, G. C.; Massaro, Enrico; Matt, Giorgio)V1548,106-117(1991)

Steady-state modeling of large-diameter crystal growth using baffles (Sahai, Viveik; Williamson, John W.; Overfelt, Tony)V1557,60-70(1991)

Structural and optical properties of semiconducting microcrystallite-doped SiO2 glass films prepared by rf-sputtering (Tsunetomo, Keiji; Shimizu, Ryuichiro; Yamamoto, Masaki; Osaka, Yukio)V1513,93-104(1991)

Structural investigations of the (Si1-x,Gex)O2 single-crystal thin films by x-ray photoelectron spectroscopy (Sorokina, Svetlana; Dikov, Juriy)V1519,128-133(1991)

Study of crystal structure of vacuum-evaporated Ag-Cu thin film (Sun, Da M.)V1519,688-691(1991)

Thermal lensing and frequency chirp in a heated CdTe modulator crystal and its effects on laser radar performance (Eng, Richard S.; Kachelmyer, Alan L.; Harris, Neville W.)V1416,70-85(1991)

Two-color interferometry using a detuned frequency-doubling crystal (Koch, Karl W.; Moore, Gerald T.)V1516,67-74(1991)

Vectorial photoelectric effect at 2.69 keV (Shaw, Ping-Shine; Hanany, Shaul; Liu, Yee; Church, Eric D.; Fleischman, Judith R.; Kaaret, Philip E.; Novick, Robert; Santangelo, A.)V1548,118-131(1991)

Void-crack interaction in aluminum single crystal (Li, X. M.; Chiang, Fu-Pen)V1554A,285-296(1991)

Zn3P2: new material for optoelectronic devices (Misiewicz, Jan; Szatkowski, Jan; Mirowska, N.; Gumienny, Zbigniew; Placzek-Popko, E.)V1561,6-18(1991)

Crystals, liquid—see also displays; spatial light modulators

Active matrix LCDs driven by two- and three-terminal switches: a comparison (den Boer, Willem; Yaniv, Zvi)V1455,248-248(1991)

Addressing factors for polymer-dispersed liquid-crystal displays (Margerum, J. D.; Lackner, Anna M.; Erdmann, John H.; Sherman, E.)V1455,27-38(1991)

Application of the phase and amplitude modulating properties of LCTVs (Kirsch, James C.; Loudin, Jeffrey A.; Gregory, Don A.)V1474,90-100(1991)

Characteristics of a ferroelectric liquid crystal spatial light modulator with a dielectric mirror (Kato, Naoki; Sekura, Rieko; Yamanaka, Junko; Ebihara, Teruo; Yamamoto, Shuhei)V1455,190-205(1991)

Charge-coupled-device-addressed liquid-crystal light valve: an update (Efron, Uzi; Byles, W. R.; Goodwin, Norman W.; Forber, Richard A.; Sayyah, Keyvan; Wu, Chiung S.; Welkowsky, Murray S.)V1455,237-247(1991)

Charge separation in functionalized tetrathiafulvalene derivatives (Fox, Marye A.; Pan, Horng-Lon)V1436,2-7(1991)

Comparative study of the dielectric and optical response of PDLC films (Seekola, Desmond; Kelly, Jack R.)V1455,19-26(1991)

Complex researches optically controlled liquid-crystal spatial-time light modulators on the photoconductivity organic polymer basis (Groznov, Michail A.)V1500,281-289(1991)

Computer-generated holograms optimized for illumination with partially coherent light using silicon backplane spatial light modulators as the recording device (O'Brien, Dominic C.; Mears, Robert J.)V1505,32-37(1991)

Droplet-size effects in light scattering from polymer-dispersed liquid-crystal films (Montgomery, G. P.; West, John L.; Tamura-Lis, Winifred)V1455,45-53(1991)

Droplet-size polydispersity in polymer-dispersed liquid-crystal films (Vaz, Nuno A.; Smith, George W.; VanSteenkiste, T. H.; Montgomery, G. P.)V1455,110-122(1991)

Dynamic light scattering from a side-chain liquid crystalline polymer in a nematic solvent (Devanand, Krisha)V1430,160-164(1991)

Effect of polymer mixtures on the performance of PDLC films (Heavin, Scott D.; Fung, Bing M.; Mears, Richard B.; Sluss, James J.; Batchman, Theodore E.)V1455,12-18(1991)

Efficiency of liquid-crystal light valves as polarization rotators (Collings, Neil; Xue, Wei; Pedrini, G.)V1505,12-19(1991)

Fiber optic liquid crystal high-pressure sensor (Wolinski, Tomasz R.; Bock, Wojtek J.)V1511,281-288(1991)

Fiber optic liquid crystalline microsensor for temperature measurement in high magnetic field (Domanski, Andrzej W.; Kostrzewa, Stanislaw)V1510,72-77(1991)

Hologram: liquid-crystal composites (Ingwall, Richard T.; Adams, Timothy)V1555,279-290(1991)

Hybrid modulation properties of the Epson LCTV (Kirsch, James C.; Loudin, Jeffrey A.; Gregory, Don A.)V1558,432-441(1991)

Influence of phototransformed molecules on optical properties of finite cholesteric liquid-crystal cell (Pinkevich, Igor P.; Reshetnyak, Victor Y.; Reznikov, Yuriy)V1455,122-133(1991)

Investigation of fringing fields in liquid-crystal devices (Powell, Norman J.; Kelsall, Robert W.; Love, G. D.; Purvis, Alan)V1545,19-30(1991)

Laser-induced phase transitions in liquid crystals and distributed feedback-fluctuations, energy exchange, and instabilities: squeezed polarized states and intensity correlations (Arakelian, Sergei M.; Chilingarian, Yu. S.; Alaverdian, R. B.; Alodjants, A. P.; Drnoian, V. E.; Karaian, A. S.)V1402,175-191(1991)

Light budget and optimization strategies for display applications of dichroic nematic droplet/polymer films (Drzaic, Paul S.)V1455,255-263(1991)

Liquid circular polarizer in laser system (Cesarz, Tadeusz; Klosowicz, Stanislaw; Zmija, Jozef)V1391,244-249(1991)

Liquid-Crystal Devices and Materials (Drzaic, Paul S.; Efron, Uzi, eds.)V1455(1991)

Liquid-crystal devices in planar optics (Armitage, David; Ticknor, Anthony J.)V1455,206-212(1991).

Liquid-crystal-doped polymers as volume holographic elements (Chen, Geng-Sheng; Brady, David J.)V1562,128-135(1991)

Liquid-crystal light valves for projection displays (Shields, Steven E.; Bleha, William P.)V1455,225-236(1991)

Liquid-crystal phase modulators for active micro-optic devices (Purvis, Alan; Williams, Geoffrey; Powell, Norman J.; Clark, Michael G.; Wiltshire, Michael C.)V1455,145-149(1991)

Liquid-crystal phase modulator used in DSPI (Kadono, Hirofumi; Toyooka, Satoru)V1554A,628-638(1991)

Liquid crystals for lasercom applications (Tan, Chin; Carlson, Robert T.)V1417,391-401(1991)

Liquid-crystal-television-based optical-digital processor for measurement of shortening velocity in single rat heart cells (Yelamarty, Rao V.; Yu, Francis T.; Moore, Russell L.; Cheung, Joseph Y.)V1398,170-179(1991)

Liquid-crystal television optical neural network: architecture, design, and models (Yu, Francis T.)V1455,150-166(1991)

Microwatt all-optical switches, array memories, and flip-flops (Wang, Chang H.; Lloyd, Ashley D.; Wherrett, Brian S.)V1505,130-140(1991)

Molecular anchoring at the droplet wall in PDLC materials (Crawford, Gregory P.; Ondris-Crawford, Renate; Doane, J. W.)V1455,2-11(1991)

New thin-film transistor structure and its processing method for liquid-crystal displays (Kuo, Yue)V1456,288-299(1991)

Nonlinear optical and piezoelectric behavior of liquid-crystalline elastomers (Hirschmann, Harald; Meier, Wolfgang; Finkelmann, Heino)V1559,27-38(1991)

Nonlinear optical components with liquid crystals (Dumitru, Mihaela A.; Honciuc, Maria; Sterian, Livia)V1500,339-348(1991)

Novel supertwisted nematic liquid-crystal-display operating modes and electro-optical performance of generally twisted nematic configurations (Schadt, Martin)V1455,214-224(1991)

Optical enhancements of joint-Fourier-transform correlator by image subtraction (Perez, Osvaldo; Karim, Mohammad A.)V1471,255-264(1991)

Optical fiber units with ferroelectric liquid crystals for optical computing (Domanski, Andrzej W.; Roszko, Marcin; Sierakowski, Marek W.)V1362,907-915(1991)

Optically addressed spatial light modulator with nipin aSi:H layers and bistable ferroelectric liquid crystal (Cambon, P.; Killinger, M.; de Bougrenet de la Tocnaye, Jean-Louis M.)V1562,116-125(1991)

Optical neural network: architecture, design, and models (Yu, Francis T.)V1558,390-405(1991)

Optical neural networks based on liquid-crystal light valves and photorefractive crystals (Owechko, Yuri; Soffer, Bernard H.)V1455,136-144(1991)

Optical second-harmonic generation by polymer-dispersed liquid-crystal films (Yuan, Haiji J.; Li, Le; Palffy-Muhoray, Peter)V1455,73-83(1991)

Optical thresholding with a liquid crystal light valve (Shariv, I.; Friesem, Asher A.)V1442,258-263(1991)

Performance of NCAP projection displays (Jones, Philip J.; Tomita, Akira; Wartenberg, Mark)V1456,6-14(1991)

Phase conjugation by four-wave mixing in nematic liquid crystals (Almeida, Silverio P.; Varamit, Srisuda P.)V1500,34-45(1991)

Polarization dependence and uniformity of FLC layers for phase modulation (Biernacki, Paul D.; Brown, Tyler; Freeman, Mark O.)V1455,167-178(1991)

Polychromatic neural networks (Yu, Francis T.; Yang, Xiangyang; Gregory, Don A.)V1558,450-458(1991)

Polymer-dispersed liquid-crystal shutters for IR imaging (McCargar, James W.; Doane, J. W.; West, John L.; Anderson, Thomas W.)V1455,54-60(1991)

Possibility of liquid crystal display panels for a space-saving PACS workstation (Komori, Masaru; Minato, Kotaro; Takahashi, Takashi; Nakano, Yoshihisa; Sakurai, Tsunetaro)V1444,334-337(1991)

Real-time holography using the high-resolution LCTV-SLM (Hashimoto, Nobuyuki; Morokawa, Shigeru; Kitamura, Kohei)V1461,291-302(1991)

Scattering liquid crystal in optical attenuator applications (Karppinen, Arto; Kopola, Harri K.; Myllyla, Risto A.)V1455,179-189(1991)

Schottky diode silicon liquid-crystal light valve (Sayyah, Keyvan; Efron, Uzi; Forber, Richard A.; Goodwin, Norman W.; Reif, Philip G.)V1455,249-254(1991)

Selection devices for field-sequential stereoscopic displays: a brief history (Lipton, Lenny)V1457,274-282(1991)

Single-pixel measurements on LCDs (Jenkins, A. J.)V1506,188-193(1991)

Smectic liquid crystals modified by tail and core fluorination (Twieg, Robert J.; Betterton, K.; DiPietro, Richard; Gravert, D.; Nguyen, Cattien; Nguyen, H. T.; Babeau, A.; Destrade, C.)V1455,86-96(1991)

Stacked STN LCDs for true-color projection systems (Gulick, Paul E.; Conner, Arlie R.)V1456,76-82(1991)

Surface temperature and shear stress measurement using liquid crystals (Toy, Norman; Savory, Eric; Disimile, Peter J.)V1489,112-123(1991)

Switching speeds in NCAP displays: dependence on collection angle and wavelength (Reamey, Robert H.; Montoya, Wayne; Wartenberg, Mark)V1455,39-44(1991)

Three-dimensional automatic precision measurement system by liquid-crystal plate on moire topography (Arai, Yasuhiko; Yekozeki, Shunsuke; Yamada, Tomoharu)V1554B,266-274(1991)

Threshold of structural transition in nematic drops with normal boundary conditions in AC electric field (Bodnar, Vladimir G.; Koval'chuk, Alexandr V.; Lavrentovich, Oleg D.; Pergamenshchik, V. M.; Sergan, V. V.)V1455,61-72(1991)

Total internal reflection studies of a ferroelectric liquid crystal/anisotropic solid interface (Zhuang, Zhiming; Clark, Noel A.; Meadows, Michael R.)V1455,105-109(1991)

Transmissive analogue SLM using a chiral smectic liquid crystal switched by CdSe TFTs (Crossland, William A.; Davey, A. B.; Sparks, Adrian P.; Lee, Michael J.; Wright, S. W.; Judge, C. P.)V1455,264-273(1991)

Triplet-sensitized reactions of some main chain liquid-crystalline polyaryl cinnamates (Subramanian, P.; Creed, David; Hoyle, Charles E.; Venkataram, Krishnan)V1559,461-469(1991)

Use of the Boulder model for the design of high-polarization fluorinated ferroelectric liquid crystals (Wand, Michael D.; Vohra, Rohini; Thurmes, William; Walba, David M.; Geelhaar, Thomas; Littwitz, Brigitte)V1455,97-104(1991)

Using liquid-crystal TVs in Vander Lugt optical correlators (Clark, Natalie; Crandall, Charles M.; Giles, Michael K.)V1564,439-451(1991)

Deposition—see also coatings; microelectronic processing; thin films

Antireflection coatings of sputter-deposited SnOxFy and SnNxFy (Yin, Zhiqiang; Stjerna, B. A.; Granqvist, Claes G.)V1536,149-157(1991)

Application of electrodeposition processes to advanced package fabrication (Krongelb, Sol; Dukovic, John O.; Komsa, M. L.; Mehdizadeh, S.; Romankiw, Lubomyr T.; Andricacos, P. C.; Pfeiffer, A. T.; Wong, K.)V1389,249-256(1991)

Atomic layer growth of zinc oxide and zinc sulphide (Sanders, Brian W.; Kitai, Adrian H.)V1398,81-87(1991)

Basic mechanisms and application of the laser-induced forward transfer for high-Tc superconducting thin film deposition (Fogarassy, Eric)V1394,169-179(1991)

Bi(Pb)-Sr-Ca-Cu-O superconducting films prepared by chemical spray deposition (Li, Chang J.; Liu, Li M.; Yao, Qi)V1519,779-787(1991)

Broadband multilayer coated blazed grating for x-ray wavelengths below 0.6 nm (den Boggende, Antonius J.; Bruijn, Marcel P.; Verhoeven, Jan; Zeijlemaker, H.; Puik, Eric J.; Padmore, Howard A.)V1345,189-197(1991)

Condensation mechanisms and properties of rf-sputtered a-Si:H (Ligachev, Valery A.; Filikov, V. A.; Gordeev, V. N.)V1519,214-219(1991)

Continuous TEM observation of diamond nucleus growth by side-view method (Goto, Yasuyuki; Kurihara, Kazuaki; Sawamoto, Yumiko; Kitakohji, Toshisuke)V1534,49-58(1991)

Control of oxygen incorporation and lifetime measurement in Si1-xGex epitaxial films grown by rapid thermal chemical vapor deposition (Sturm, James C.; Schwartz, P. V.; Prinz, Erwin J.; Magee, Charles W.)V1393,252-259(1991)

Deposition- controlled uniformity of multilayer mirrors (Jankowski, Alan F.; Makowiecki, Daniel M.; McKernan, M. A.; Foreman, R. J.; Patterson, R. G.)V1343,32-38(1991)

Deposition of a-Si:H using a supersonically expanding argon plasma (Meeusen, G. J.; Qing, Z.; Wilbers, A. T.; Schram, D. C.)V1519,252-257(1991)

Deposition of silica coatings on Incoloy 800H substrates using a high-power laser (Fellowes, Fiona C.; Steen, William M.)V1502,213-222(1991)

Diagnostics of a DC plasma torch (Russell, Derrek; Taborek, Peter)V1534,14-23(1991)

Diamond growth on the (110) surface (Yarbrough, Walter A.)V1534,90-104(1991)

Diamond-like carbon thin films prepared by rf-plasma CVD (Jiang, Jie; Liu, Chen Z.)V1519,717-724(1991)

Diamond Optics IV (Feldman, Albert; Holly, Sandor, eds.)V1534(1991)

Direct measurement of H-atom sticking coefficient during diamond film growth (Martin, L. R.)V1534,175-182(1991)

Dual ion-beam sputtering: a new coating technology for the fabrication of high-power CO2 laser mirrors (Daugy, Eric; Pointu, Bernard; Villela, Gerard; Vincent, Bernard)V1502,203-212(1991)

Effect of laser irradiation on superconducting properties of laser-deposited YBa2Cu3O7 thin films (Singh, Rajiv K.; Bhattacharya, Deepika; Narayan, Jagdish; Jahncke, Catherine; Paesler, Michael A.)V1394,203-213(1991)

Effects of TMSb/TEGa ratios on epilayer properties of gallium antimonide grown by low-pressure MOCVD (Wu, T. S.; Su, Yan K.; Juang, F. S.; Li, N. Y.; Gan, K. J.)V1361,23-33(1991)

Efficient laser cleaning of small particulates using pulsed laser irradiation synchronized with liquid-film deposition (Tam, Andrew C.; Zapka, Werner; Ziemlich, Winfrid)V1598,13-18(1991)

Epitaxial regrowth of silicon on sapphire by rapid isothermal processing (Madarazo, R.; Pedrine, A. G.; Sol, A. A.; Baranauskas, Vitor)V1393,270-277(1991)

Establishment of new criterion aiding the control of antireflection coating semiconductor diodes (Lu, Yu C.; Li, Da Y.; Chen, Jian G.; Luo, Bin)V1519,463-466(1991)

Excimer laser deposition and characterization of tin and tin-oxide films (Borsella, E.; De Padova, P.; Larciprete, Rosanna)V1503,312-320(1991)

Experimental study of the optical properties of LTCVD SiO2 (Aharoni, Herzl; Swart, Pieter L.)V1442,118-125(1991)

Fabrication and characterization of Si-based soft x-ray mirrors (Schmiedeskamp, Bernt; Heidemann, B.; Kleineberg, Ulf; Kloidt, Andreas; Kuehne, Mikhael; Mueller, H.; Mueller, Peter; Nolting, Kerstin; Heinzmann, Ulrich)V1343,64-72(1991)

Fabrication and performance at 1.33 nm of a 0.24-um period multilayer grating (Berrouane, H.; Khan Malek, Chantal; Andre, Jean-Michel; Lesterlin, L.; Ladan, F. R.; Rivoira, R.; Lepetre, Yves; Barchewitz, Robert J.)V1343,428-436(1991)

Fabrication of multilayer Bragg-Fresnel zone plates for the soft x-ray range (Khan Malek, Chantal; Moreno, T.; Guerin, Philippe; Ladan, F. R.; Rivoira, R.; Barchewitz, Robert J.)V1343,56-61(1991)

Focused ion beam induced deposition: a review (Melngailis, John)V1465,36-49(1991)

Formation of SiO2 film on plastic substrate by liquid-phase-deposition method (Kitaoka, Masaki; Honda, Hisao; Yoshida, Harunobu; Takigawa, Akio; Kawahara, Hideo)V1519,109-114(1991)

Formation of boron nitride and silicon nitride bilayer films by ion-beam-enhanced deposition (Feng, Yi P.; Jiang, Bing Y.; Yang, Gen Q.; Huang, Wei; Zheng, Zhi H.; Liu, Xiang H.; Zou, Shi C.)V1519,440-443(1991)

Formation of heterostructure devices in a multichamber processing environment with in-vacuo surface analysis diagnostics and in-situ process monitoring (Lucovsky, Gerald; Kim, Sang S.; Fitch, J. T.; Wang, Cheng)V1392,605-616(1991)

Formation of titanium nitride films by Xe+ ion-beam-enhanced deposition in a N2 gas environment (Wang, Xi; Yang, Gen Q.; Liu, Xiang H.; Zheng, Zhi H.; Huang, Wei; Zhou, Zu Y.; Zou, Shi C.)V1519,740-743(1991)

Fundamental studies of chemical-vapor-deposition diamond growth processes (Shaw, Robert W.; Whitten, W. B.; Ramsey, J. M.; Heatherly, Lee)V1534,170-174(1991)

Growth and properties of GaxIn1-xAs (x<O.47) on InP by MOCVD (Du, MingZe; Yuan, JinShan; Jin, Yixin; Zhou, Tian Ming; Hong, Jiang; Hong, ChunRong; Zhang, BaoLin)V1361,699-705(1991)

Growth and properties of YBCO thin films by metal-organic chemical vapor deposition and plasma-enhanced MOCVD (Zhao, Jing-Fu; Li, Y. Q.; Chern, Chyi S.; Huang, W.; Norris, Peter E.; Gallois, B.; Kear, B. H.; Lu, P.; Kulesha, G. A.; Cosandey, F.)V1362,135-143(1991)

Growth of oxide superconducting thin films by plasma-enhanced MOCVD (Kanehori, Keiichi; Sugii, Nobuyuki)V1394,238-243(1991)

Growth of PbTiO3 films by photo-MOCVD (Shimizu, Masaru; Katayama, Takuma; Fujimoto, Masashi; Shiosaki, Tadashi)V1519,122-127(1991)

Growth of rf-sputtered selenium thin films (Yuan, Xiang L.; Min, Szuk W.; Fang, Zhi Y.; Yu, Da W.; Qi, Lei)V1519,167-171(1991)

Growth of ZnSe-ZnTe strained-layer supperlattices by atmospheric pressure MOCVD on transparent substrate CaF2 (111) (Pan, Chuan K.; Jiang, F. Y.; Fan, Guang H.; Ma, Y. Z.; Fan, Xi W.)V1519,645-651(1991)

High-performance metal/SiO2/InSb capacitor fabricated by photoenhanced chemical vapor deposition (Sun, Tai-Ping; Lee, Si-Chen; Liu, Kou-Chen; Pang, Yen-Ming; Yang, Sheng-Jenn)V1361,1033-1037(1991)

High-resolution electron microscopy of diamond film growth defects and their interactions (Shechtman, Dan; Farabaugh, Edward N.; Robins, Lawrence H.; Feldman, Albert; Hutchison, Jerry L.)V1534,26-43(1991)

Incorporation of As into HgCdTe grown by MOCVD (He, Jin; Yu, Zhen Z.; Ma, Ke J.; Jia, Pei M.; Yang, Jian R.; Shen, Shou Z.; Chen, Wei M.; Yang, Ji M.)V1519,499-507(1991)

Infrared transparent conductive coatings deposited by activated reactive evaporation (Marcovitch, Orna; Zipin, Hedva; Klein, Z.; Lubezky, I.)V1442,58-58(1991)

In-situ investigation of the low-pressure MOCVD growth of III-V compounds using reflectance anisotropy measurements (Drevillon, Bernard; Razeghi, Manijeh)V1361,200-212(1991)

Integrated optical devices with silicon oxynitride prepared by plasma-enhanced chemical vapor deposition (PECVD) on Si and GaAs substrates (Peters, Dethard; Mueller, Joerg)V1362,338-349(1991)

Intl Conf on Thin Film Physics and Applications (Zhou, Shixun; Wang, Yonglin, eds.)V1519(1991)

Investigation of uniform deposition of GaInAsP quantum wells by MOCVD (Puetz, Norbert; Miner, Carla J.; Hingston, G.; Moore, Chris J.; Watt, B.; Hillier, Glen)V1361,692-698(1991)

Investigation on the DLC films prepared by dual-ion-beam-sputtering deposition (Wang, Tianmin; Wang, Weijie; Liu, Guidng; Huang, Liangpu; Luo, Chuntai; Liu, Dingquan; Xu, Ming; Yang, Yimin)V1519,890-900(1991)

Ion-beam-sputtering deposition and etching of high-Tc YBCO superconducting thin films (Zhao, Xing R.; Hao, Jian H.; Zhou, Fang Q.; Sun, Han D.; Wang, Lingjie; Yi, Xin J.)V1519,772-774(1991)

Large-scale production of a broadband antireflection coating on ophthalmic lenses with ion-assisted deposition (Andreani, F.; Luridiana, S.; Mao, Shu Z.)V1519,63-69(1991)

Laser-assisted deposition of thin films onto transparent substrates from liquid-phase organometallic precursor: iron acetylacetonate (Shafeev, George A.; Laude, Lucien D.)V1503,321-329(1991)

Laser-induced gold deposition for thin-film circuit repair (Baum, Thomas H.; Comita, Paul B.; Kodas, Toivo T.)V1598,122-131(1991)

Laser-induced mass transfer simulation and experiment (Shestakov, S. D.; Kotov, Gennady A.; Hekalo, A. V.; Migitko, I. A.; Bajkov, A. V.; Yurkevith, B. M.)V1440,423-435(1991)

Laser-induced metal deposition and laser cutting techniques for fixing IC design errors (Shaver, David C.; Doran, S. P.; Rothschild, Mordechai; Sedlacek, J. H.)V1596,46-50(1991)

Laser chemistry of dimethylcadmium adsorbed on silicon: 308- versus 222-nm laser excitation (Simonov, Alexander P.; Varakin, Vladimir N.; Panesh, Anatoly M.; Lunchev, V. A.)V1503,330-337(1991)

LCVD fabrication of polycrystalline Si pressure sensor (Zhang, Wei; Wang, Xiao-Ru)V1572,15-17(1991)

Low-resistivity contacts to silicon using selective RTCVD of germanium (Grider, Douglas T.; Ozturk, Mehmet C.; Wortman, Jim J.; Littlejohn, Michael A.; Zhong, Y.)V1393,229-239(1991)

Microfabrication techniques for semiconductor lasers (Tamanuki, Takemasa; Tadokoro, T.; Morito, K.; Koyama, Fumio; Iga, Kenichi)V1361,614-617(1991)

Microstructure and superconducting properties of BiSrCaCuO thin films (Wessels, Bruce W.; Zhang, Jiyue; DiMeo, Frank; Richeson, D. S.; Marks, Tobin J.; DeGroot, D. C.; Kannewurf, C. R.)V1394,232-237(1991)

MOCVD of TlBaCaCuO: structure-property relations and progress toward device applications (Hamaguchi, Norihito; Gardiner, R.; Kirlin, Peter S.)V1394,244-254(1991)

Modeling of photochemical vapor deposition of epitaxial silicon using an ArF excimer laser (Fowler, Burt; Lian, S.; Krishnan, S.; Jung, Le-Tien; Li, C.; Banerjee, Sanjay)V1598,108-117(1991)

Multichamber rapid thermal processing (Rosser, Paul J.; Moynagh, P.; Affolter, K. B.)V1393,49-66(1991)

Multiple photo-assisted CVD of thin-film materials for III-V device technology (Nissim, Yves I.; Moison, Jean M.; Houzay, Francoise; Lebland, F.; Licoppe, C.; Bensoussan, M.)V1393,216-228(1991)

Narrow (0.1 um to 0.5 um) copper lines for ultra-large-scale integration technology (Shacham-Diamand, Yosef Y.)V1442,11-19(1991)

New coating technology and ion source (Yan, Yi S.)V1519,192-193(1991)

New deposition system for the preparation of doped a-Si:H (Wu, Zhao P.; Chen, Ru G.; Wang, Yonglin)V1519,194-198(1991)

New ion-beam sources and their applications to thin film physics (Wei, David T.; Kaufman, Harold R.)V1519,47-55(1991)

New mixed sputtering-plasma CVD technique for the deposition of diamondlike films (Demichelis, Francesca; Giachello, G.; Pirri, C. F.; Tagliaferro, Alberto)V1534,140-147(1991)

Novel cathodic arc plasma PVD system with column target for the deposition of TiN film and other metallic films (Liu, Wei Y.; Li, Yu M.; Cui, Zhan J.; He, Tian X.)V1519,172-174(1991)

Optical emission spectroscopy in diamond-like carbon film deposition by glow discharge (Zhang, Wei P.; Chen, Jing; Fang, Rong C.; Hu, Ke L.)V1519,680-682(1991)

Optical properties of amorphous hydrogenated carbon layers (Stenzel, Olaf; Schaarschmidt, Guenther; Roth, Sylvia; Schmidt, Guenther; Scharff, Wolfram)V1534,148-157(1991)

Optical properties of oxide films prepared by ion-beam-sputter deposition (Tang, Xue F.; Fan, Zheng X.; Wang, Zhi-Jiang)V1519,96-98(1991)

Optical properties of some ion-assisted deposited oxides (Andreani, F.; Luridiana, S.; Mao, Shu Z.)V1519,18-22(1991)

Optimum design of phase gratings for diffractive optical elements obtained by thin-film deposition (Beretta, Stefano; Cairoli, Massimo; Viardi, Marzia)V1544,2-9(1991)

Photoluminescence and surface photovoltaic spectra of strained InP on GaAs by MOCVD (Zhuang, Weihua; Chen, Chao; Teng, Da; Yu, Jin-zhong; Li, Yu Z.)V1361,980-986(1991)

Plasma parameters in microwave-plasma-assisted chemical vapor deposition of diamond (Weimer, Wayne A.; Cerio, Frank M.; Johnson, Curtis E.)V1534,9-13(1991)

Polishing of filament-assisted CVD diamond films (Hickey, Carolyn F.; Thorpe, Thomas P.; Morrish, Arthur A.; Butler, James E.; Vold, C.; Snail, Keith A.)V1534,67-76(1991)

Preferential growth in ion-beam-enhanced deposition of Ti(C,N) films (Chen, You S.; Sun, Yi L.; Zhang, Fu M.; Mou, Hai C.)V1519,56-62(1991)

Preparation of SiO2 film utilizing equilibrium reaction in aqueous solution (Kawahara, Hideo; Sakai, Y.; Goda, Takuji; Hishinuma, Akihiro; Takemura, Kazuo)V1513,198-203(1991)

Pressure effects in the microwave plasma growth of polycrystalline diamond (Harker, Alan B.; DeNatale, Jeffrey F.)V1534,2-8(1991)

Progress In High-Temperature Superconducting Transistors and Other Devices (Narayan, Jagdish; Shaw, David T.; Singh, Rajendra, eds.)V1394(1991)

Pulsed-laser deposition of YBa2Cu3O7-x thin films: processing, properties, and performance (Muenchausen, Ross E.; Foltyn, Stephen R.; Wu, Xin D.; Dye, Robert C.; Nogar, Nicholas S.; Carim, A. H.; Heidelbach, F.; Cooke, D. W.; Taber, R. C.; Quinn, Rod K.)V1394,221-229(1991)

Pulsed-laser deposition of oxides over large areas (Greer, James A.; Van Hook, H. J.)V1377,79-90(1991)

Recent progress on research of materials for optoelectronic device applications in China (Chen, Liang-Hui; Kong, Mei-Ying; Wang, Yi-Ming)V1361,60-73(1991)

Reduced thermal budget processing of high-Tc superconducting thin films and related materials by MOCVD (Sinha, Sanjai; Singh, Rajendra; Hsu, N. J.; Ng, J. T.; Chou, P.; Narayan, Jagdish)V1394,266-276(1991)

Reflectance anisotropy spectrometer for real-time crystal growth investigations (Acher, O.; Benferhat, Ramdane; Drevillon, Bernard; Razeghi, Manijeh)V1361,1156-1163(1991)

Research of Cr2O3 thin film deposited by arc discharge plasma deposition as heat-radiation absorbent in electric vacuum devices (Deng, Hong; Wang, Xiang D.; Yuan, Lei)V1519,735-739(1991)

Role of buffer layers in the laser-ablated films on metallic substrates (Shaw, David T.; Narumi, E.; Yang, F.; Patel, Sushil)V1394,214-220(1991)

Selection, growth, and characterization of materials for MBE-produced x-ray optics (Kearney, Patrick A.; Slaughter, Jon M.; Falco, Charles M.)V1343,25-31(1991)

Selective deposition of polycrystalline SixGe1-x by rapid thermal processing (Ozturk, Mehmet C.; Zhong, Y.; Grider, Douglas T.; Sanganeria, M.; Wortman, Jim J.; Littlejohn, Michael A.)V1393,260-269(1991)

Selective low-temperature chemical vapor deposition of copper from new copper(I) compounds (Jain, Ajay; Shin, H. K.; Chi, Kai-Ming; Hampden-Smith, Mark J.; Kodas, Toivo T.; Farkas, Janos; Paffett, M. F.; Farr, J. D.)V1596,23-33(1991)

Selective metal deposition using low-dose focused ion-beam patterning (Kubena, Randall L.; Stratton, F. P.; Mayer, T. M.)V1392,595-597(1991)

Si-based epitaxial growth by rapid thermal processing chemical vapor deposition (Jung, K. H.; Hsieh, T. Y.; Kwong, Dim-Lee; Spratt, D. B.)V1393,240-251(1991)

Silicon/silicon oxide and silicon/silicon nitride multilayers for XUV optical applications (Boher, Pierre; Houdy, Philippe; Hennet, L.; Delaboudiniere, Jean-Pierre; Kuehne, Mikhael; Mueller, Peter; Li, Zhigang; Smith, David J.)V1343,39-55(1991)

Silicon carbide layers produced by rapid thermal chemical vapor deposition (Ruddell, F. H.; McNeill, D.; Armstrong, Brian M.; Gamble, Harold S.)V1361,159-170(1991)

Single-wafer integrated processing as a manufacturing tool using rapid thermal chemical vapor deposition technology (Kermani, Ahmad)V1393,109-119(1991)

Soft and hard x-ray reflectivities of multilayers fabricated by alternating-material sputter deposition (Takenaka, Hisataka; Ishii, Yoshikazu; Kinoshita, Hiroo; Kurihara, Kenji)V1345,213-224(1991)

Soft x-ray multilayers fabricated by electron-beam deposition (Sudoh, Masaaki; Yokoyama, Ryouhei; Sumiya, Mitsuo; Yamamoto, Masaki; Yanagihara, Mihiro; Namioka, Takeshi)V1343,14-24(1991)

SPEEDIE: a profile simulator for etching and deposition (McVittie, James P.; Rey, J. C.; Bariya, A. J.; IslamRaja, M. M.; Cheng, L. Y.; Ravi, S.; Saraswat, Krishna C.)V1392,126-138(1991)

Stoichiometry of laser-deposited Bi-Sr-Ca-Cu-O films on silicon and mass spectrometric investigations of superconductors (Becker, J. S.; Lorenz, M.; Dietze, H.-J.)V1598,227-238(1991)

Structural and electrical properties of epitaxial YBCO films on Si (Fork, David K.; Barrera, A.; Phillips, Julia M.; Newman, N.; Fenner, David B.; Geballe, T. H.; Connell, G.A. N.; Boyce, James B.)V1394,202-202(1991)

Studies of InSb metal oxide semiconductor structure fabricated by photo-CVD using Si2H6 and N2O (Huang, C. J.; Su, Yan K.; Leu, R. L.)V1519,70-73(1991)

Study of HgCdTe/CdTe interface structure grown by metal-organic chemical vapor deposition (Ma, Ke J.; Yu, Zhen Z.; Yanh, Jian R.; Shen, Shou Z.; He, Jin; Chen, Wei M.; Song, Xiang Y.)V1519,489-493(1991)

Study of phase transition VO2 thin film (Gao, Jian C.; Lin, Zhi H.; Han, Li Y.)V1519,159-163(1991)

Study of properties of a-Si1-xGex:H prepared by SAP-CVD method (Wang, Yi-Ming; Jing, Lian-hua; Pang, Da-wen)V1361,325-335(1991)

Suppression of columnar-structure formation in Mo-Si layered synthetic microstructures (Niibe, Masahito; Hayashida, Masami; Iizuka, Takashi; Miyake, Akira; Watanabe, Yutaka; Takahashi, Rie; Fukuda, Yasuaki)V1343,2-13(1991)

Systematic studies on transition layers of carbides between CVD diamond films and substrates of strong carbide-forming elements (Jiang, Xiang-Liu; Zhang, Fang Q.; Li, Jiang-Qi; Yang, Bin; Chen, Guang-Hua)V1534,207-213(1991)

Ta/Al alloy thin film medium-power attenuator (Yang, Bang C.; Jia, Yu M.)V1519,156-158(1991)

Temperature dependence of resistance of diamond film synthesized by microwave plasma CVD (Yang, Bang C.; Gou, Li; Jia, Yu M.; Ran, Jun G.; Zheng, Chang Q.; Tang, Xia)V1519,864-865(1991)

Three materials soft x-ray mirrors: theory and application (Boher, Pierre; Hennet, L.; Houdy, Philippe)V1345,198-212(1991)

Transmission electron microscopy, photoluminescence, and capacitance spectroscopy on GaAs/Si grown by metal organic chemical vapor deposition (Bremond, Georges E.; Said, Hicham; Guillot, Gerard; Meddeb, Jaafar; Pitaval, M.; Draidia, Nasser; Azoulay, Rozette)V1361,732-743(1991)

Vapor phase deposition of transition metal fluoride glasses (Boulard, B.; Jacoboni, Charles)V1513,204-208(1991)

Versatility of metal organic chemical vapor deposition process for fabrication of high-quality YBCO superconducting thin films (Chern, Chyi S.; Zhao, Jing-Fu; Li, Y. Q.; Norris, Peter E.; Kear, B. H.; Gallois, B.)V1394,255-265(1991)

X-ray evaluation on residual stresses in vapor-deposited hard coatings (Xu, Kewei; Chen, Jin; Gao, Runsheng; He, Jia W.; Zhao, Cheng; Li, Shi Z.)V1519,765-770(1991)

Detectors—see also arrays; charge-coupled devices; infrared; photodetectors; sensors

488 x 640-element hybrid platinum silicide Schottky focal-plane array (Gates, James L.; Connelly, William G.; Franklin, T. D.; Mills, Robert E.; Price, Frederick W.; Wittwer, Timothy Y.)V1540,262-273(1991)

Advanced Si IR detectors using molecular beam epitaxy (Lin, TrueLon; Jones, E. W.; George, T.; Ksendzov, Alexander; Huberman, M. L.)V1540,135-139(1991)

Advanced infrared detector materials (Mullin, John B.)V1512,144-154(1991)

Advanced portable four-wavelength NIR analyzer for rapid chemical composition analysis (Malinen, Jouko; Hyvarinen, Timo S.)V1510,204-209(1991)

Analysis of the flying light spot experiment on SPRITE detector (Gu, Bo-qi; Feng, Wen-qing)V1488,443-446(1991)

Analysis of upper and lower bounds of the frame noise in linear detector arrays (Jaggi, Sandeep)V1541,152-162(1991)

Analytical Raman and atomic spectroscopies using charge-coupled-device detection (Bilhorn, Robert B.; Ferris, Nancy S.)V1439,15-24(1991)

Application of canonical correlation analysis in detection in presence of spatially correlated noise (Chen, Wei G.; Reilly, James P.; Wong, Kon M.)V1566,464-475(1991)

Application of nonimaging optical concentrators to infrared energy detection (Ning, Xiaohui)V1528,88-92(1991)

Application of transmission electron detection to x-ray mask calibrations and inspection (Postek, Michael T.; Larrabee, Robert D.; Keery, William J.; Marx, Egon)V1464,35-47(1991)

Application of YBa2Cu3O7-x thin film in high-Tc semiconducting infrared detector (Zhou, Bing; Chen, Ju X.; Shi, Bao A.; Wu, Ru J.; Gong, Shuxing)V1519,454-456(1991)

Applications of diffractive optics to uncooled infrared imagers (Cox, James A.)V1540,606-611(1991)

Application-specific integrated-circuit-based multianode microchannel array readout system (Smeins, Larry G.; Stechman, John M.; Cole, Edward H.)V1549,59-65(1991)

Array detectors in astronomy (Lesser, Michael P.)V1439,144-151(1991)

Assessment of the optimum operating conditions for 2-D focal-plane-array systems (Bourne, Robert W.; Jefferys, E. A.; Murphy, Kevin S.)V1488,73-79(1991)

Automated characterization of Z-technology sensor modules (Gilcrest, Andrew S.)V1541,240-249(1991)

Back-illuminated large-format Loral CCDs (Lesser, Michael P.)V1447,177-182(1991)

Backside-thinned CCDs for keV electron detection (Ravel, Mihir K.; Reinheimer, Alice L.)V1447,109-122(1991)

Battlefield training in impaired visibility (Gammarino, Rudolph R.; Surhigh, James W.)V1419,115-125(1991)

Bidirectional transmittance distribution function measurements on ZnSe and on ZnS Cleartran (Melozzi, Mauro; Mazzoni, Alessandro; Curti, G.)V1512,178-188(1991)

Carbon dioxide eyesafe laser rangefinders (Powell, Richard K.; Berdanier, Barry N.; McKay, James)V1419,126-140(1991)

CCD focal-plane imaging detector for the JET-X instrument on spectrum R-G (Wells, Alan A.; Castelli, C. M.; Holland, Andrew D.; McCarthy, Kieran J.; Spragg, J. E.; Whitford, C. H.)V1549,357-373(1991)

Cerenkov background radiation in imaging detectors (Rosenblatt, Edward I.; Beaver, Edward A.; Cohen, R. D.; Linsky, J. B.; Lyons, R. W.)V1449,72-86(1991)

Characteristics of a high-pressure gas proportional counter filled with xenon (Sakurai, Hirohisa; Ramsey, Brian D.)V1549,20-27(1991)

Charge-Coupled Devices and Solid State Optical Sensors II (Blouke, Morley M., ed.)V1447(1991)

Coherent detection in confocal microscopy (Wilson, Tony)V1439,104-108(1991)

Compensated digital readout family (Ludwig, David E.; Skow, Michael)V1541,73-82(1991)

Computer analysis of signal-to-noise ratio and detection probability for scanning IRCCD arrays (Uda, Gianni; Tofani, Alessandro)V1488,257-266(1991)

Critical look at AlGaAs/GaAs multiple-quantum-well infrared detectors for thermal imaging applications (Adams, Frank W.; Cuff, K. F.; Gal, George; Harwit, Alex; Whitney, Raymond L.)V1541,24-37(1991)

Current instrument status of the airborne visible/infrared imaging spectrometer (AVIRIS) (Eastwood, Michael L.; Sarture, Charles M.; Chrien, Thomas G.; Green, Robert O.; Porter, Wallace M.)V1540,164-175(1991)

Current status of InGaAs detector arrays for 1-3 um (Olsen, Gregory H.; Joshi, Abhay M.; Ban, Vladimir S.)V1540,596-605(1991)

Current transport in charge injection devices (Wu, Chao-Wen; Lin, Hao-Hsiung)V1362,768-777(1991)

Dead-time effects in microchannel-plate imaging detectors (Zombeck, Martin V.; Fraser, George W.)V1549,90-100(1991)

Decoupled coil detector array in magnetic resonance imaging (Kwiat, Doron; Einav, Shmuel)V1443,2-28(1991)

Design and performance of an automatic gain control system for the high-energy x-ray timing experiment (Pelling, Michael R.; Rothschild, Richard E.; MacDonald, Daniel R.; Hertel, Robert H.; Nishiie, Edward S.)V1549,134-146(1991)

Detection of infrared, free-electron laser radiation (Kimmitt, Maurice F.)V1501,86-96(1991)

Detector perturbation of ocean radiance measurements (Macdonald, Burns; Helliwell, William S.; Sanborn, James; Voss, Kenneth J.)V1537,104-114(1991)

Development and analysis of a simple model for an IR sensor (Ballik, Edward A.; Wan, William)V1488,249-256(1991)

Development of a large pixel, spectrally optimized, pinned photodiode/interline CCD detector for the Earth Observing System/Moderate-Resolution Imaging Spectrometer-Tilt Instrument (Ewin, Audrey J.; Jhabvala, Murzy; Shu, Peter K.)V1479,12-20(1991)

Development of a synchrotron CCD-based area detector for structural biology (Kalata, Kenneth; Phillips, Walter C.; Stanton, Martin J.; Li, Youli)V1345,270-280(1991)

Development of laser-induced fluorescence detection to assay DNA damage (Sharma, Minoti; Freund, Harold G.)V1435,280-291(1991)

Digital radiology with solid state linear x-ray detectors (Munier, Bernard; Prieur-Drevon, P.; Chabbal, Jean)V1447,44-55(1991)

Edge effects in silicon photodiode arrays (Kenney, Steven B.; Hirleman, Edwin D.)V1480,82-93(1991)

Electron Image Tubes and Image Intensifiers II (Csorba, Illes P., ed.)V1449(1991)

Eyesafe diode laser rangefinder technology (Perger, Andreas; Metz, Jurgen; Tiedeke, J.; Rille, Eduard P.)V1419,75-83(1991)

Eyesafe high-pulse-rate laser progress at Hughes (Stultz, Robert D.; Nieuwsma, Daniel E.; Gregor, Eduard)V1419,64-74(1991)

Fixed-pattern-noise cancellation in linear pyro arrays (Jain, Subhash C.; Malhotra, H. S.; Sarebahi, K. N.; Bist, K. S.)V1488,410-413(1991)

Fluorescence imaging of latent fingerprints with a cooled charge-coupled-device detector (Pomeroy, Robert S.; Baker, Mark E.; Radspinner, David A.; Denton, M. B.)V1439,60-65(1991)

Focal-plane image processing using acoustic charge transport technology (Brooks, Jeff W.)V1541,68-72(1991)

Future directions in focal-plane signal processing for spaceborne scientific imagers (Fossum, Eric R.)V1541,62-67(1991)

High-dynamic-range MCP structures (Slater, David C.; Timothy, J. G.)V1549,68-80(1991)

High-performance InGaAs PIN and APD (avalanche photdiode) detectors for 1.54 um eyesafe rangefinding (Olsen, Gregory H.; Ackley, Donald A.; Hladky, J.; Spadafora, J.; Woodruff, K. M.; Lange, M. J.; Van Orsdel, Brian T.; Forrest, Stephen R.; Liu, Y.)V1419,24-31(1991)

High-radiometric-performance CCD for the third-generation stratospheric aerosol and gas experiment (Delamere, W. A.; Baer, James W.; Ebben, Thomas H.; Flores, James S.; Kleiman, Gary; Blouke, Morley M.; McCormick, M. P.)V1479,31-40(1991)

High-resolution, two-dimensional imaging, microchannel-plate detector for use on a sounding rocket experiment (Bush, Brett C.; Cotton, Daniel M.; Siegmund, Oswald H.; Chakrabarti, Supriya; Harris, Walter; Clarke, John T.)V1549,290-301(1991)

High-resolution decoding of multianode microchannel array detectors (Kasle, David B.; Morgan, Jeffrey S.)V1549,52-58(1991)

High-resolution imaging with multilayer soft x-ray, EUV, and FUV telescopes of modest aperture and cost (Walker, Arthur B.; Lindblom, Joakim F.; Timothy, J. G.; Hoover, Richard B.; Barbee, Troy W.; Baker, Phillip C.; Powell, Forbes R.)V1494,320-333(1991)

High-Tc bolometer developments for planetary missions (Brasunas, John C.; Lakew, Brook)V1477,166-173(1991)

High-Tc superconducting infrared bolometric detector (Cole, Barry E.)V1394,126-138(1991)

HTC microbolometer for far-infrared detection (Barholm-Hansen, Claus; Levinsen, Mogens T.)V1512,218-225(1991)

Hybrid phototube with Si target (van Geest, Lambertus K.; Stoop, Karel W.)V1449,121-134(1991)

Image enhancement of infrared uncooled focal plane array imagery (McCauley, Howard M.; Auborn, John E.)V1479,416-422(1991)

Imager for gamma-ray astronomy: balloon prototype (Di Cocco, Guido; Labanti, Claudio; Malaguti, Giuseppe; Rossi, Elio; Schiavone, Filomena; Spizzichino, A.; Traci, A.; Bird, A. J.; Carter, T.; Dean, Anthony J.; Gomm, A. J.; Grant, K. J.; Corba, M.; Quadrini, E.; Rossi, M.; Villa, G. E.; Swinyard, Bruce M.)V1549,102-112(1991)

Imaging gas scintillation proportional counters for ASTRO-D (Ohashi, T.; Makishima, K.; Ishida, M.; Tsuru, T.; Tashiro, M.; Mihara, Teruyoshi; Kohmura, Y.; Inoue, Hiroyuki)V1549,9-19(1991)

Imaging pulse-counting detector systems for space ultraviolet astrophysics missions (Timothy, J. G.)V1494,394-402(1991)

Industrial applications of spectroscopic imaging (Miller, Richard M.; Birmingham, John J.; Cummins, Philip G.; Singleton, Scott)V1439,66-78(1991)

In-flight performance of the faint object camera of the Hubble Space Telescope (Greenfield, Perry E.; Paresce, Francesco; Baxter, David; Hodge, P.; Hook, R.; Jakobsen, P.; Jedrzejewski, Robert; Nota, Anatonella; Sparks, W. B.; Towers, Nigel M.; Laurance, R. J.; Macchetto, F.)V1494,16-39(1991)

Infrared and the search for extrasolar planets (Meinel, Aden B.; Meinel, Marjorie P.)V1540,196-201(1991)

Infrared detector arrays with integrating cryogenic read-out electronics (Engemann, Detlef; Faymonville, Rudolf; Felten, Rainer; Frenzl, Otto)V1362,710-720(1991)

Infrared detectors from YBaCuO thin films (Zhou, Fang Q.; Sun, Han D.; Zhao, Xing R.; Wang, Lingjie; Yi, Xin J.)V1477,178-181(1991)

Infrared focal-plane design for the comet rendezvous/asteroid flyby and Cassini visible and infrared mapping spectrometers (Staller, Craig O.; Niblack, Curtiss A.; Evans, Thomas G.; Blessinger, Michael A.; Westrick, Anthony)V1540,219-230(1991)

Infrared Imaging Systems: Design, Analysis, Modeling and Testing II (Holst, Gerald C., ed.)V1488(1991)

Infrared optical response of superconducting YBaCuO thin films (Sun, Han D.; Zhou, Fang Q.; Zhao, Xing R.; Wang, Lingjie; Yi, Xin J.)V1477,174-177(1991)

Infrared Sensors: Detectors, Electronics, and Signal Processing (Jayadev, T. S., ed.)V1541(1991)

Intl Conf on Scientific Optical Imaging (Denton, M. B., ed.)V1439(1991)

ISTS array detector test facility (Thomas, Paul J.; Hollinger, Allan B.; Chu, Kan M.; Harron, John W.)V1488,36-47(1991)

Large-aperture CCD x-ray detector for protein crystallography using a fiber-optic taper (Strauss, Michael G.; Westbrook, Edwin M.; Naday, Istvan; Coleman, T. A.; Westbrook, M. L.; Travis, D. J.; Sweet, Robert M.; Pflugrath, J. W.; Stanton, Martin J.)V1447,12-27(1991)

Large staring IRFPAs of HgCdTe on alternative substrates (Kozlowski, Lester J.; Bailey, Robert B.; Cooper, Donald E.; Vural, Kadri; Gertner, E. R.; Tennant, William E.)V1540,250-261(1991)

Lidar for expendable launch vehicles (Lee, Michael S.)V1480,23-34(1991)

Lightweight surveillance FLIR (Fawcett, James M.)V1498,82-91(1991)

Long-wave infrared detectors based on III-V materials (Maserjian, Joseph L.)V1540,127-134(1991)

Long-wavelength GexSi1-x/Si heterojunction infrared detectors and focal-plane arrays (Tsaur, Bor-Yeu; Chen, C. K.; Marino, S. A.)V1540,580-595(1991)

Long-wavelength lasers and detectors fabricated on InP/GaAs superheteroepitaxial wafer (Aiga, Masao; Omura, Etsuji E.)V1418,217-222(1991)

Low-temperature operation of silicon drift detectors (Sumner, Timothy J.; Roe, S.; Rochester, G. K.; Hall, G.; Evensen, Per; Avset, B. S.)V1549,265-273(1991)

Megapixel CCD thinning/backside progress at SAIC (Schaeffer, A. R.; Varian, Richard H.; Cover, John R.; Larsen, Robert G.)V1447,165-176(1991)

MIC photon counting detector (Fordham, John L.; Bellis, J. G.; Bone, David A.; Norton, Timothy J.)V1449,87-98(1991)

Microchannel-plate detectors for space-based astronomy (Crocker, James H.; Cox, Colin R.; Ray, Knute A.; Sen, Amit)V1494,434-439(1991)

Model-based analysis of 3-D spatial-temporal IR clutter suppression filtering (Chan, David S.)V1481,117-128(1991)

Monolithic epitaxial IV-VI compound IR-sensor arrays on Si substrates for the SWIR, MWIR and LWIR range (Zogg, Hans; Masek, Jiri; Maissen, Clau; Hoshino, Taizo J.; Blunier, Stefan; Tiwari, A. N.)V1361,1079-1086(1991)

Multilayered superconducting tunnel junctions for use as high-energy-resolution x-ray detectors (Rippert, Edward D.; Song, S. N.; Ketterson, John B.; Ulmer, Melville P.)V1549,283-288(1991)

Multimode IRST/FLIR design issues (Armstrong, George R.; Oakley, Philip J.; Ranat, Bhadrayu M.)VCR38,120-141(1991)

Multiplex and multichannel detection of near-infrared Raman scattering (Chase, Bruce D.)V1439,47-57(1991)

Multiplication-based analog motion detection chip (Moore, Andrew; Koch, Christof)V1473,66-75(1991)

Nb tunnel junctions as x-ray spectrometers (Rando, Nicola; Peacock, Anthony J.; Foden, Clare; van Dordrecht, Axel; Engelhardt, Ralph; Lumley, John M.; Pereira, Carl)V1549,340-356(1991)

Noise mechanisms of high-temperature superconducting infrared detectors (Khalil, Ali E.)V1477,148-158(1991)

Nonlinear optic in-situ diagnostics of a crystalline film in molecular-beam-epitaxy devices (Krasnov, Victor F.; Musher, Semion L.; Prots, V. I.; Rubenchik, Aleksandr M.; Ryabchenko, Vladimir E.; Stupak, Mikhail F.)V1506,179-187(1991)

Nonselective thermal detectors of radiation (Pankratov, Nickolai A.)V1540,432-443(1991)

Novel doping superlattice-based PbTe-IR detector device (Oswald, Josef; Pippan, Manfred; Tranta, Beate; Bauer, Guenther E.)V1362,534-543(1991)

Numerical investigation of effect of dynamic range and nonlinearity of detector on phase-stepping holographic interferometry (Fang, Qiang; Luo, Xiangyang; Tan, Yushan)V1332,216-222(1991)

One-dimensional proportional counters for x-ray all-sky monitors (Matsuoka, Masaru; Yamauchi, Masamitsu; Nakamura, Haruo; Kondo, M.; Kawai, N.; Yoshida, A.; Imai, Tohru)V1549,2-8(1991)

Optical detector prepared by high-Tc superconducting thin film (Wang, Lingjie; Zhou, Fang Q.; Zhao, Xing R.; Sun, Han D.; Yi, Xin J.)V1540,738-741(1991)

Optical dividers for quadrant avalanche photodiode detectors (Green, Samuel I.)V1417,496-512(1991)

Optimization of point source detection (Friedenberg, Abraham)V1442,60-65(1991)

Optimum structure learning algorithms for competitive learning neural network (Uchiyama, Toshio; Sakai, Mitsuhiro; Saito, Tomohide; Nakamura, Taichi)V1451,192-203(1991)

Optoelectronic Gabor detector for transient signals (Zhang, Yan; Li, Yao; Tolimieri, Richard; Kanterakis, Emmanuel G.; Katz, Al; Lu, X. J.; Caviris, Nicholas P.)V1481,23-31(1991)

ORION semiconductor optical detectors: research and development (Khryapov, V. T.; Ponomarenko, Vladimir P.; Butkevitch, V. G.; Taubkin, I. I.; Stafeev, V. I.; Popov, S. A.; Osipov, V. V.)V1540,412-423(1991)

Pacing elements of IR system design (Zissis, George J.)VCR38,44-54(1991)

Performance characteristics of Y-Ba-Cu-O microwave superconducting detectors (Shewchun, John; Marsh, P. F.)V1477,115-138(1991)

Performance characteristics of the imaging MAMA detector systems for SOHO, STIS, and FUSE/Lyman (Timothy, J. G.)V1549,221-233(1991)

Performance of a thinned back-illuminated CCD coupled to a confocal microscope for low-light-level fluorescence imaging (Masters, Barry R.)V1447,56-63(1991)

Performance of infrared systems under field conditions (Chrzanowski, Krzysztof)V1512,78-83(1991)

Performance of low-resistance microchannel-plate stacks (Siegmund, Oswald H.; Stock, Joseph)V1549,81-89(1991)

Performance of microstrip proportional counters for x-ray astronomy on spectrum-roentgen-gamma (Budtz-Jorgensen, Carl; Bahnsen, Axel; Christensen, Finn E.; Madsen, M. M.; Olesen, C.; Schnopper, Herbert W.)V1549,429-437(1991)

Performance of milliKelvin Si bolometers as x-ray and exotic particle detectors (Zammit, C. C.; Sumner, Timothy J.; Lea, M. J.; Fozooni, P.; Hepburn, I. D.)V1549,274-282(1991)

Performance tests of large CCDs (Robinson, Lloyd B.; Brown, William E.; Gilmore, Kirk; Stover, Richard J.; Wei, Mingzhi; Geary, John C.)V1447,214-228(1991)

Photovoltaic HgCdTe MWIR-detector arrays on (100)CdZnTe/(100)GaAs grown by hot-wall-beam epitaxy (Gresslehner, Karl-Heinz; Schirz, W.; Humenberger, Josef; Sitter, Helmut; Andorfer, J.; Lischka, Klaus)V1361,1087-1093(1991)

Planar InGaAs APD (avalanche photodiode) for eyesafe laser rangefinding applications (Webb, Paul P.)V1419,17-23(1991)

Possible enhancement in bolometric response using free-standing film of YBa2Cu3Ox (Ng, Hon K.; Kilibarda, S.)V1477,15-19(1991)

Progress with PN-CCDs for the XMM satellite mission (Braeuninger, Heinrich; Hauff, D.; Lechner, P.; Lutz, G.; Kink, W.; Meidinger, Norbert; Metzner, G.; Predehl, Peter; Reppin, C.; Strueder, Lothar; Truemper, Joachim; Kendziorra, E.; Staubert, R.; Radeka, V.; Rehak, P.; Rescia, S.; Bertuccio, G.; Gatti, E.; Longoni, Antonio; Sampietro, Marco; Findeis, N.; Holl, P.; Kemmer, J.; von Zanthier, C.)V1549,330-339(1991)

Quantum efficiency and crosstalk of an improved backside-illuminated indium antimonide focal plane array (Bloom, I.; Nemirovsky, Yael)V1442,286-297(1991)

Quantum efficiency model for p+-doped back-illuminated CCD imager (Huang, Chin M.; Kosicki, Bernard B.; Theriault, Joseph R.; Gregory, J. A.; Burke, Barry E.; Johnson, Brett W.; Hurley, Edward T.)V1447,156-164(1991)

Receivers for eyesafe laser rangefinders: an overview (Crawford, Ian D.)V1419,9-16(1991)

Recent developments using GaAs as an x-ray detector (Sumner, Timothy J.; Grant, S. M.; Bewick, A.; Li, J. P.; Spooner, N. J.; Smith, K.; Beaumont, Steven P.)V1549,256-264(1991)

Recent results in adaptive array detection (Kalson, S. Z.)V1566,406-418(1991)

Regular doping structures: a Si-based, quantum-well infrared detector (Koch, J. F.)V1362,544-552(1991)

Revolutionary impact of today's array detector technology on chemical analysis (Radspinner, David A.; Fields, Robert E.; Earle, Colin W.; Denton, M. B.)V1439,2-14(1991)

Robust CFAR detection using order statistic processors for Weibull-distributed clutter (Nagle, Daniel T.; Saniie, Jafar)V1481,49-63(1991)

Selective edge detection based on harmonic oscillator wave functions (Kawakami, Hajimu)V1468,156-166(1991)

Sensors and Sensor Integration (Dean, Peter D., ed.)V1480(1991)

Signal and background models in nonstandard IR systems (Snyder, John L.)V1498,52-63(1991)

Signal processing for low-light-level, high-precision CCD imaging (McCurnin, Thomas W.; Schooley, Larry C.; Sims, Gary R.)V1448,225-236(1991)

Silicon/PVDF integrated double detector: application to obstacle detection in automotive (Simonne, John J.; Pham, Vui V.; Esteve, Daniel; Alaoui-Amine, Mohammed; Bousbiat, Essaid)V1374,107-115(1991)

Silicon x-ray array detector on spectrum-x-gamma satellite (Sipila, Heikki; Huttunen, Pekka; Kamarainen, Veikko J.; Vilhu, Osmi; Kurki, Jouko; Leppelmeier, Gilbert W.; Taylor, Ivor; Niemela, Arto; Laegsgaard, Erik; Sunyaev, Rashid)V1549,246-255(1991)

Simplified model of the back surface of a charge-coupled device (Blouke, Morley M.; Delamere, W. A.; Womack, G.)V1447,142-155(1991)

Simulation of a neural network for decentralized detection of a signal in noise (Amirmehrabi, Hamid; Viswanathan, R.)V1396,252-265(1991)

Sine wave measurements of SPRITE detector MTF (Barnard, Kenneth J.; Boreman, Glenn D.; Plogstedt, Allen E.; Anderson, Barry K.)V1488,426-431(1991)

SIRE (sight-integrated ranging equipment): an eyesafe laser rangefinder for armored vehicle fire control systems (Keeter, Howard S.; Gudmundson, Glen A.; Woodall, Milton A.)V1419,84-93(1991)

Soft x-ray windows for position-sensitive proportional counters (Viitanen, Veli-Pekka; Nenonen, Seppo A.; Sipila, Heikki; Mutikainen, Risto)V1549,28-34(1991)

Space telescope imaging spectrograph 2048 CCD and its characteristics (Delamere, W. A.; Ebben, Thomas H.; Murata-Seawalt, Debbie; Blouke, Morley M.; Reed, Richard; Woodgate, Bruce E.)V1479,21-30(1991)

Spectroscopy and polarimetry capabilities of the INTEGRAL imager: Monte Carlo simulation results (Swinyard, Bruce M.; Malaguti, Giuseppe; Caroli, Ezio; Dean, Anthony J.; Di Cocco, Guido)V1548,94-105(1991)

SPRITE detector characterization through impulse response testing (Anderson, Barry K.; Boreman, Glenn D.; Barnard, Kenneth J.; Plogstedt, Allen E.)V1488,416-425(1991)

Superconducting IR focal plane arrays (Quelle, Fred W.)V1449,157-166(1991)

Superconducting bolometers: high-Tc and low-Tc (Richards, Paul L.)V1477,2-6(1991)

Surveillance Technologies (Gowrinathan, Sankaran; Mataloni, Raymond J.; Schwartz, Stanley J., eds.)V1479(1991)

Synthesis of a maneuver detector and adaptive gain tracking filter (Gardner, Kenneth R.; Kasky, Thomas J.)V1481,254-260(1991)

Technique for improving the calibration of large-area sphere sources (Walker, James H.; Cromer, Chris L.; McLean, James T.)V1493,224-230(1991)

Temperature-monitored/controlled silicon photodiodes for standardization (Eppeldauer, George)V1479,71-77(1991)

Thinned backside-illuminated cooled CCDs for UV and VUV applications (Dalinenko, I. N.; Kuz'min, G. A.; Malyarov, A. V.; Prokhorov, Alexander M.; Schelev, Mikhail Y.)V1449,167-172(1991)

Two applications for microlens arrays: detector fill-factor improvement and laser diode collimation (D'Amato, Dante P.; Centamore, Robert M.)V1544,166-177(1991)

Two-dimensional encoding of images using discrete reticles (Wellfare, Michael R.)V1478,33-40(1991)

Two-dimensional surface strain measurement based on a variation of Yamaguchi's laser-speckle strain gauge (Barranger, John P.)V1332,757-766(1991)

UCSD high-energy x-ray timing experiment cosmic ray particle anticoincidence detector (Hink, Paul L.; Rothschild, Richard E.; Pelling, Michael R.; MacDonald, Daniel R.; Gruber, Duane E.)V1549,193-202(1991)

UCSD high-energy x-ray timing experiment magnetic shield design and test results (Rothschild, Richard E.; Pelling, Michael R.; Hink, Paul L.)V1549,120-133(1991)

Update: high-speed/high-volume radiometric testing of Z-technology focal planes (Johnson, Jerome L.)V1541,210-219(1991)

WEBERSAT: measuring micrometeorite impacts in a polar orbit (Evans, Phillip R.)V1495,149-156(1991)

X-ray detector for time-resolved studies (Rodricks, Brian G.; Brizard, Christine M.; Clarke, Roy; Lowe, Walter P.)V1550,18-26(1991)

Diamonds—see also coatings; optical properties; thin films

Continuous TEM observation of diamond nucleus growth by side-view method (Goto, Yasuyuki; Kurihara, Kazuaki; Sawamoto, Yumiko; Kitakohji, Toshisuke)V1534,49-58(1991)

Critical-point phonons of diamond (Klein, Claude A.; Hartnett, Thomas M.; Robinson, Clifford J.)V1534,117-138(1991)

CVD diamond as an optical material for adverse environments (Snail, Keith A.)V1330,46-64(1991)

Determination of the optical constants of thin chemical-vapor-deposited diamond windows from 0.5 to 6.5 eV (Robins, Lawrence H.; Farabaugh, Edward N.; Feldman, Albert)V1534,105-116(1991)

Diagnostics of a DC plasma torch (Russell, Derrek; Taborek, Peter)V1534,14-23(1991)

Diamond growth on the (110) surface (Yarbrough, Walter A.)V1534,90-104(1991)

Diamond multichip modules (McSheery, Tracy D.)V1563,21-26(1991)

Diamond Optics IV (Feldman, Albert; Holly, Sandor, eds.)V1534(1991)

Diamond windows for the infrared: fact and fallacy (Klein, Claude A.)VCR38,218-257(1991)

Direct measurement of H-atom sticking coefficient during diamond film growth (Martin, L. R.)V1534,175-182(1991)

Effects of interfacial modifications on diamond film adhesion (DeNatale, Jeffrey F.; Flintoff, John F.; Harker, Alan B.)V1534,44-48(1991)

Excimer laser processing of diamond-like films (Ageev, Vladimir P.; Glushko, T. N.; Dorfman, V. F.; Kuzmichov, A. V.; Pypkin, B. N.)V1503,453-462(1991)

Fundamental studies of chemical-vapor-deposition diamond growth processes (Shaw, Robert W.; Whitten, W. B.; Ramsey, J. M.; Heatherly, Lee)V1534,170-174(1991)

Fusion of diamond phases of graphite in laser shock waves (Bugrov, N. V.; Zakharov, N. S.)V1440,416-422(1991)

High-resolution electron microscopy of diamond film growth defects and their interactions (Shechtman, Dan; Farabaugh, Edward N.; Robins, Lawrence H.; Feldman, Albert; Hutchison, Jerry L.)V1534,26-43(1991)

High-temperature Raman scattering behavior in diamond (Herchen, Harald; Cappelli, Mark A.)V1534,158-168(1991)

Microstructures and domain size effects in diamond films characterized by Raman spectroscopy (Nemanich, Robert J.; Bergman, Larry; LeGrice, Yvonne M.; Turner, K. F.; Humphreys, T. P.)V1437,2-12(1991)

Morphological phenomena of CVD diamond (Part II) (Zhu, Wei; Messier, Russell F.; Badzian, Andrzej R.)V1534,230-242(1991)

New diamond activities at Osaka University (Hiraki, Akio)V1534,198-198(1991)

New mixed sputtering-plasma CVD technique for the deposition of diamondlike films (Demichelis, Francesca; Giachello, G.; Pirri, C. F.; Tagliaferro, Alberto)V1534,140-147(1991)

New preparation method and properties of diamondlike carbon films (Yu, Bing Kun; Chen, Xao Min)V1534,223-229(1991)

Optical emission spectroscopy of diamond-producing plasmas (Plano, Linda S.)V1437,13-23(1991)

Optical properties of DC arc-discharge plasma CVD diamond (Trombetta, John M.; Hoggins, James T.; Klocek, Paul; McKenna, T. A.)V1534,77-88(1991)

Optical properties of GaAs, GaP, and CVD diamond (Klocek, Paul; Hoggins, James T.; McKenna, T. A.; Trombetta, John M.; Boucher, Maurice W.)V1498,147-157(1991)

Optical properties of ZnS/diamond composites (Xue, L. A.; Noh, T. W.; Sievers, A. J.; Raj, Rishi)V1534,183-196(1991)

Photoconductive switching for high-power microwave generation (Pocha, Michael D.; Hofer, Wayne W.)V1378,2-9(1991)

Plasma parameters in microwave-plasma-assisted chemical vapor deposition of diamond (Weimer, Wayne A.; Cerio, Frank M.; Johnson, Curtis E.)V1534,9-13(1991)

Polishing of filament-assisted CVD diamond films (Hickey, Carolyn F.; Thorpe, Thomas P.; Morrish, Arthur A.; Butler, James E.; Vold, C.; Snail, Keith A.)V1534,67-76(1991)

Pressure effects in the microwave plasma growth of polycrystalline diamond (Harker, Alan B.; DeNatale, Jeffrey F.)V1534,2-8(1991)

Properties of CVD diamond for optical applications (Gray, Kevin J.; Lu, Grant)V1534,60-66(1991)

Properties of diamonds with varying isotopic composition (Banholzer, William; Fulghum, Stephen)V1501,163-176(1991)

Smooth diamond films by reactive ion-beam polishing (Bovard, Bertrand G.; Zhao, Tianji; Macleod, H. A.)V1534,216-222(1991)

Soft x-ray, optical, and thermal properties of hard carbon films (Alvey, Mark D.)V1330,39-45(1991)

Study of impurities in CVD diamond using cathodoluminescence (Nishimura, Kazuhito; Ma, Jing S.; Yokota, Yoshihiro; Mori, Yusuke; Kotsuka, Hiroshi; Hirao, Takashi; Kitabatake, Makoto; Deguchi, Masahiro; Ogawa, Kazuo; Ning, Gang; Tomimori, Hiroshi; Hiraki, Akio)V1534,199-206(1991)

Systematic studies on transition layers of carbides between CVD diamond films and substrates of strong carbide-forming elements (Jiang, Xiang-Liu; Zhang, Fang Q.; Li, Jiang-Qi; Yang, Bin; Chen, Guang-Hua)V1534,207-213(1991)

Temperature dependence of resistance of diamond film synthesized by microwave plasma CVD (Yang, Bang C.; Gou, Li; Jia, Yu M.; Ran, Jun G.; Zheng, Chang Q.; Tang, Xia)V1519,864-865(1991)

Vibrational Raman characterization of hard-carbon and diamond films (Ager, Joel W.; Veirs, D. K.; Marchon, Bruno; Cho, Namhee; Rosenblatt, Gern M.)V1437,24-31(1991)

X-ray diffraction from materials under extreme pressures (Brister, Keith)V1550,2-10(1991)

Diffraction—see also binary optics; gratings; holographic optical elements; holography

Analysis of Bragg diffraction in optical memories and optical correlators (Gheen, Gregory)V1564,135-142(1991)

Analysis of polarization properties of shallow metallic gratings by an extended Rayleigh-Fano theory (Koike, Masato; Namioka, Takeshi)V1545,88-94(1991)

Application of error diffusion in diffractive optics (Weissbach, Severin; Wyrowski, Frank)V1507,149-152(1991)

Applications of powder diffraction in materials science using synchrotron radiation (Hart, Michael)V1550,11-17(1991)

Backward diffraction of the light by the phase-transmission holographic grating (Markov, Vladimir B.; Shishkov, Vladimir F.)V1238,41-43(1991)

Binary optics in the '90s (Gallagher, Neal C.)V1396,722-733(1991)

Blazed zone plates for the 10-um spectral region (Hutley, M. C.)V1574,2-7(1991)

Boundary diffraction wave in imaging by small holograms (Mulak, Grazyna)V1574,266-271(1991)

Bragg diffraction with multiple internal reflections (Markov, Vladimir B.; Shishkov, Vladimir F.)V1238,30-40(1991)

Calculation of diffraction efficiencies for spherical and cylindrical holographic lenses (Defosse, Yves; Renotte, Yvon L.; Lion, Yves F.)V1507,277-287(1991)

Characteristic wave theory for volume holographic gratings with arbitrarily slanted fringes (Chernov, Boris C.)V1507,302-309(1991)

Combined x-ray absorption spectroscopy and x-ray powder diffraction (Dent, Andrew J.; Derbyshire, Gareth E.; Greaves, G. N.; Ramsdale, Christine A.; Couves, J. W.; Jones, Richard; Catlow, C. R.; Thomas, John M.)V1550,97-107(1991)

Comparison of rigorous and approximate methods of analyzing holographic gratings diffraction (Chernov, Boris C.)V1238,44-53(1991)

Computational experiment for computer-generated optical elements (Golub, Mikhail A.; Doskolovich, Leonid L.; Kazanskiy, Nikolay L.; Kharitonov, Sergey I.; Orlova, Natalia G.; Sisakian, Iosif N.; Soifer, Victor A.)V1500,194-206(1991)

Computer-driven Fourier diffractometer (Blocki, Narcyz; Daszkiewicz, Marek; Galas, Jacek)V1562,172-183(1991)

Computer-generated diffraction gratings in optical region (Silvennoinen, Raimo; Hamalainen, Rauno M.; Rasanen, Jari)V1574,84-88(1991)

Design of achromatized hybrid diffractive lens systems (Londono, Carmina; Clark, Peter P.)V1354,30-37(1991)

Diffraction analysis of beams for barcode scanning (Eastman, Jay M.; Quinn, Anna M.)V1384,185-194(1991)

Diffraction analysis of optical disk readout signal deterioration caused by mark-size fluctuation (Honguh, Yoshinori)V1527,315-321(1991)

Diffraction by one-dimensional or two-dimensional periodic arrays of conducting plates (Petit, Roger; Bouchitte, G.; Tayeb, Gerard; Zolla, F.)V1545,31-41(1991)

Diffraction effects in directed radiation beams (Hafizi, Bahman; Sprangle, Phillip)V1407,316-321(1991)

Diffraction effects in stimulated Raman scattering (Scalora, Michael; Haus, Joseph W.)V1497,153-164(1991)

Diffraction performance calculations in lens design (Malacara Hernandez, Daniel)V1354,2-14(1991)

Diffractive doublet corrected on-axis at two wavelengths (Farn, Michael W.; Goodman, Joseph W.)V1354,24-29(1991)

Diffractive optical elements for optoelectronic interconnections (Streibl, Norbert)V1574,34-47(1991)

Diffractive optics for x rays: the state of the art (Michette, Alan G.)V1574,8-21(1991)

Diffractive properties of surface-relief microstructures (Qu, Dong-Ning; Burge, Ronald E.; Yuan, X.)V1506,152-159(1991)

Digital holography as a useful model in diffractive optics (Wyrowski, Frank)V1507,128-135(1991)

DOE design and manufacture at CSEM (Buczek, Harthmuth; Mayor, J. M.; Regnault, P.)V1574,48-57(1991)

Double dispersion from dichromated gelatin volume transmission gratings (Sheat, Dennis E.; Chamberlin, Giles R.; McCartney, David J.)V1461,35-38(1991)

Effective holographic grating model to analyze thick holograms (Belendez, A.; Pascual, I.; Fimia-Gil, Antonio)V1507,268-276(1991)

Electron-beam-written reflection diffractive microlenses for oblique incidence (Shiono, Teruhiro; Ogawa, Hisahito)V1545,232-240(1991)

Fabrication and properties of chalcogenide IR diffractive elements (Ewen, Peter J.; Slinger, Christopher W.; Zakery, A.; Zekak, A.; Owen, A. E.)V1512,101-111(1991)

Fabrication of multilayer Bragg-Fresnel zone plates for the soft x-ray range (Khan Malek, Chantal; Moreno, T.; Guerin, Philippe; Ladan, F. R.; Rivoira, R.; Barchewitz, Robert J.)V1343,56-61(1991)

Fabrication of phase structures with continuous and multilevel profile for diffraction optics (Poleshchuk, Alexander G.)V1574,89-100(1991)

Finite thickness effect of a zone plate on focusing hard x-rays (Yun, Wen-Bing; Chrzas, John J.; Viccaro, P. J.)V1345,146-164(1991)

Focal length measurement using diffraction at a grating (Sirohi, Rajpal S.; Kumar, Harish; Jain, Narinder K.)V1332,50-55(1991)

Focusators at letters diffraction design (Golub, Mikhail A.; Doskolovich, Leonid L.; Kazanskiy, Nikolay L.; Kharitonov, Sergey I.; Sisakian, Iosif N.; Soifer, Victor A.)V1500,211-221(1991)

Free-space simulator for laser transmission (Inagaki, Keizo; Nohara, Mitsuo; Araki, Ken'ichi; Fujise, Masayuki; Furuhama, Yoji)V1417,160-169(1991)

Frequency-dependent optical beam distortion generated by acousto-optic Bragg cells (Pieper, Ronald J.; Poon, Ting-Chung)V1454,324-335(1991)

Fresnel zone plate moire patterns and its metrological applications (Jaroszewicz, Zbigniew)V1574,154-158(1991)

Gaussian scaling laws for diffraction: top-hat irradiance and Gaussian beam propagation through a paraxial optical train (Townsend, Sallie S.; Cunningham, Philip R.)V1415,154-194(1991)

Grating efficiency theory versus experimental data in extreme situations (Neviere, Michel; den Boggende, Antonius J.; Padmore, Howard A.; Hollis, K.)V1545,76-87(1991)

High-energy x-ray diffraction (Freund, Andreas K.)V1345,234-244(1991)

High-resolution synchrotron x-radiation diffraction imaging of crystals grown in microgravity and closely related terrestrial crystals (Steiner, Bruce W.; Dobbyn, Ronald C.; Black, David; Burdette, Harold; Kuriyama, Masao; Spal, Richard; van den Berg, Lodewijk; Fripp, Archibald L.; Simchick, Richard T.; Lal, Ravindra B.; Batra, Ashok K.; Matthiesen, David; Ditchek, Brian M.)V1557,156-167(1991)

Holographic diffraction gratings on the base of chalcogenide semiconductors (Indutnyi, I. Z.; Robur, I.; Romanenko, Peter F.; Stronski, Alexander V.)V1555,243-253(1991)

Holographic disk scanner for active infrared sensors (Kawauchi, Yoshikazu; Toyoda, Ryuuichi; Kimura, Minoru; Kawata, Koichi)V1555,224-227(1991)

Integral equation method for biperiodic diffraction structures (Dobson, David C.; Cox, James A.)V1545,106-113(1991)

Intercomparison of homogeneous laser models with transverse effects (Bandy, Donna K.; Hunter, L. W.; Jones, Darlena J.)V1497,142-152(1991)

Intl Colloquium on Diffractive Optical Elements (Nowak, Jerzy; Zajac, Marek, eds.)V1574(1991)

Intl Conf on the Application and Theory of Periodic Structures (Lerner, Jeremy M.; McKinney, Wayne R., eds.)V1545(1991)

Inverse-grating diffraction problems in the coupled-wave analysis (Kuittinen, Markku; Jaaskelainen, Timo)V1507,258-267(1991)

Inverse grating diffraction problems (Jaaskelainen, Timo; Kuittinen, Markku)V1574,272-281(1991)

Investigation of higher-order diffraction in a one-crystal 2-D scanner (Melamed, Nathan T.; Gottlieb, Milton S.)V1454,306-312(1991)

Ion treatment in technology of diffraction optical elements (Shevchenko, N. P.; Megorskaja, K. D.; Reshetnikova, Irina N.)V1574,66-71(1991)

Kinetics of surface ordering: Pb on Ni(001) (Eng, Peter J.; Stephens, Peter; Tse, Teddy)V1550,110-121(1991)

Mathematical simulation of composite optical systems loaded with active medium (Apollonova, O. V.; Elkin, Nickolai N.; Korjov, M. Y.; Korotkov, V. A.; Likhanskii, Vladimir V.; Napartovich, Anatoly P.; Troshchiev, V. E.)V1501,108-119(1991)

Measurement of light scattering from cells using an inverted infrared optical trap (Wright, William H.; Sonek, Gregory J.; Numajiri, Yasuyuki; Berns, Michael W.)V1427,279-287(1991)

New effects in sound generation in organic dye solutions (Altshuler, Grigori B.; Belashenkov, Nickolay R.; Karasev, Viatcheslav B.; Okishev, Andrey V.)V1440,116-126(1991)

New iterative algorithm for the design of phase-only gratings (Farn, Michael W.)V1555,34-42(1991)

Noncontact technique for the measurement of linear displacement using chirped diffraction gratings (Spillman, William B.; Fuhr, Peter L.)V1332,591-601(1991)

Optical analysis of segmented aircraft windows (Jones, Mike I.; Jones, Mark S.)V1498,110-127(1991)

Optical diffractometry with directionally variable incident light wave (Daszkiewicz, Marek)V1562,184-191(1991)

Optimization of reconstruction geometry for maximum diffraction efficiency in HOE: the influence of recording material (Belendez, A.; Pascual, I.; Fimia-Gil, Antonio)V1574,77-83(1991)

Optimum numerical aperture for optical projection microlithography (Lin, Burn J.)V1463,42-53(1991)

Phase synthesis of elongated holographic diffraction gratings (Turukhano, Boris G.; Gorelik, Vladimir S.; Turukhano, Nikulina)V1500,290-292(1991)

Pixelgram: an application of electron-beam lithography for the security printing industry (Lee, Robert A.)V1509,48-54(1991)

Planar diffractive optical elements prepared by electron-beam lithography (Urban, Frantisek; Matejka, Frantisek)V1574,58-65(1991)

Polarization-state mixing in multiple-beam diffraction and its application to solving the phase problem (Shen, Qun)V1550,27-33(1991)

Process-dependent kinoform performance (Cox, James A.; Fritz, Bernard S.; Werner, Thomas R.)V1507,100-109(1991)

Production of x-rays by the interaction of charged particle beams with periodic structures and crystalline materials (Rule, Donald W.; Fiorito, Ralph B.; Piestrup, Melvin A.; Gary, Charles K.; Maruyama, Xavier K.)V1552,240-251(1991)

Real-time x-ray studies of semiconductor device structures (Clarke, Roy; Dos Passos, Waldemar; Chan, Yi-Jen; Pavlidis, Dimitris; Lowe, Walter P.; Rodricks, Brian G.; Brizard, Christine M.)V1361,2-12(1991)

Relative merits of bulk and surface relief diffracting components (Hutley, M. C.)V1574,294-302(1991)

Rigorous diffraction theory of binary optical interconnects (Vasara, Antti H.; Noponen, Eero; Turunen, Jari P.; Miller, J. M.; Taghizadeh, Mohammad R.; Tuovinen, Jussi)V1507,224-238(1991)

Scattering from objects near a rough surface (Rino, Charles L.; Ngo, Hoc D.)V1558,339-350(1991)

Selected Papers on Holographic and Diffractive Lenses and Mirrors (Stone, Thomas W.; Thompson, Brian J., eds.)VMS34(1991)

Simple beam-propagation measurements on ion lasers (Guggenhiemer, Steven; Wright, David L.)V1414,12-20(1991)

Simulation of local layer-thickness deviation on multilayer diffraction (Guo, S. P.; He, X. C.; Redko, S. V.; Wu, Z. Q.)V1519,400-404(1991)

Small-angle scattering measurement (Gu, Zu-Han; Dummer, Richard S.)V1558,368-378(1991)

Small-signal gain for parabolic profile beams in free-electron lasers (Elliott, C. J.)V1552,175-181(1991)

Study of microbial growth I: by diffraction (Williams, Gareth T.; Bahuguna, Ramendra D.; Arteaga, Humberto; Le Joie, Elaine N.)V1332,802-804(1991)

Study of structural imperfections in epitaxial beta-SiC layers by method of x-ray differential diffractometry (Baranov, Igor M.; Kutt, R. N.; Nikitina, Irina P.)V1361,1110-1115(1991)

Synthesis method applied to the problem of diffraction by gratings: the method of fictitious sources (Tayeb, Gerard; Petit, Roger; Cadilhac, M.)V1545,95-105(1991)

Techniques for designing hybrid diffractive optical systems (Brown, Daniel M.; Pitalo, Stephen K.)V1527,73-84(1991)

Two-dimensional periodic structures in solid state laser resonator (Okulov, Alexey Y.)V1410,221-232(1991)

Two-wave coupled-wave theory of the polarizing properties of volume phase gratings (Tholl, Hans D.)V1555,101-111(1991)

Understanding metrology of polysilicon gates through reflectance measurements and simulation (Tadros, Karim H.; Neureuther, Andrew R.; Guerrieri, Roberto)V1464,177-186(1991)

Unsymmetrical spectrum of reflective hologram grating (Dahe, Liu; Liang, Zhujian; Tang, Weiguo)V1507,310-315(1991)

Upper bound on the diffraction efficiency of phase-only array generators (Mait, Joseph N.)V1555,53-62(1991)

Use of diffracted light from latent images to improve lithography control (Hickman, Kirt C.; Gaspar, Susan M.; Bishop, Kenneth P.; Naqvi, H. S.; McNeil, John R.; Tipton, Gary D.; Stallard, Brian R.; Draper, B. L.)V1464,245-257(1991)

Use of rigorous vector coupled-wave theory for designing and tolerancing surface-relief diffractive components for magneto-optical heads (Haggans, Charles W.; Kostuk, Raymond K.)V1499,293-302(1991)

X-ray diffraction from materials under extreme pressures (Brister, Keith)V1550,2-10(1991)

X-ray diffraction study of GaSb/AlSb strained-layer-superlattices grown on miscut (100) substrates (Macrander, Albert T.; Schwartz, Gary P.; Gualteri, Gregory J.; Gilmer, George)V1550,122-133(1991)

Zone plate of anisotropic profile (Wardosanidze, Zurab V.)V1574,109-120(1991)

Diffractive optics—see binary optics; diffraction; holography, computer-generated; holographic optical elements

Digital processing—see also image processing

Acquisition and processing of digital images and spectra in astronomy (Lee, Terence J.)V1439,152-158(1991)

Analysis of barcode digitization techniques (Boles, John A.; Hems, Randall K.)V1384,195-204(1991)

Automated grading of venous beading: an algorithm and parallel implementation (Shen, Zhijiang; Gregson, Peter H.; Cheng, Heng-Da; Kozousek, V.)V1606,632-640(1991)

Capabilities of the optical microscope (Bradbury, Savile)V1439,128-134(1991)

Computer-aided speckle interferometry: Part II—an alternative approach using spectral amplitude and phase information (Chen, Duanjun; Chiang, Fu-Pen; Tan, Yushan; Don, H. S.)V1554A,706-717(1991)

Detection of unresolved target tracks in infrared imagery (Rajala, Sarah A.; Nolte, Loren W.; Aanstoos, James V.)V1606,360-371(1991)

Digital halftoning using a blue-noise mask (Mitsa, Theophano; Parker, Kevin J.)V1452,47-56(1991)

Digital halftoning using a generalized Peano scan (Agui, Takeshi; Nagae, Takanori; Nakajima, Masayuki)V1606,912-916(1991)

Digital replication of chest radiographs without altering diagnostic observer performance (Flynn, Michael J.; Davies, Eric; Spizarny, David; Beute, Gordon; Peterson, Ed; Eyler, William R.; Gross, Barry; Chen, Ji)V1444,172-179(1991)

Digital restoration of distorted geometric features of pigs (Van der Stuyft, Emmanuel; Goedseels, Vic; Geers, Rony)V1379,189-200(1991)

Digital speckle correlation search method and its application (Zhou, Xingeng; Gao, Jianxing)V1554A,886-895(1991)

Digitize your films without losing resolution (Kallhammer, Jan-Erik O.)V1358,631-636(1991)

Direct digital image transfer gateway (Mun, In K.; Kim, Y. S.; Mun, Seong Ki)V1444,232-237(1991)

Discrete analog and digital devices using fiber-optic logic elements (Petrov, Mikhail P.; Miridonov, S. V.)V1621,402-413(1991)

Display nonlinearity in digital image processing for visual communications (Peli, Eli)V1606,508-519(1991)

Evaluation of interference fringe pattern on spatially curved objects (Laermann, Karl-Hans)V1554A,522-528(1991)

Experiences with transputer systems for high-speed image processing (Kille, Knut; Ahlers, Rolf-Juergen; Schneider, B.)V1386,76-83(1991)

Fast-digital multiplication using multizero neural networks (Hu, Chia-Lun J.)V1469,586-591(1991)

Frequency modulation of printed gratings as a protection against copying (Spannenburg, S.)V1509,88-104(1991)

Ground-based PIV and numerical flow visualization results from the surface-tension-driven convection experiment (Pline, Alexander D.; Wernet, Mark P.; Hsieh, Kwang-Chung)V1557,222-234(1991)

High-dynamic-range fiber gyro with all-digital signal processing (Lefevre, Herve C.; Martin, Philippe; Morisse, J.; Simonpietri, Pascal; Vivenot, P.; Arditty, Herve J.)V1367,72-80(1991)

High-resolution computer-aided moire (Sciammarella, Cesar A.; Bhat, Gopalakrishna K.)V1554B,162-173(1991)

High-speed programmable digitizer for real-time video compression experiments (Cox, Norman R.)V1605,906-915(1991)

Images: from a printer's perspective (Sarkar, N. R.)V1458,42-50(1991)

In-line wafer inspection using 100-megapixel-per-second digital image processing technology (Dickerson, Gary; Wallace, Rick P.)V1464,584-595(1991)

Measuring and display system of a marathon runner by real-time digital image processing (Sasaki, Nobuyuki; Namikawa, Iwao)V1606,1002-1013(1991)

Motion estimation in digital angiographic images using skeletons (Kwak, J. Y.; Efstratiadis, Serafim N.; Katsaggelos, Aggelos K.; Sahakian, Alan V.; Sullivan, Barry J.; Swiryn, Steven; Hueter, David C.; Frohlich, Thomas)V1396,32-44(1991)

Multimode fiber optic rotation sensor with low-cost digital signal processing (Johnson, Dean R.; Fredricks, Ronald J.; Vuong, S. C.; Dembinski, David T.; Sabri, Sehbaz H.)V1367,140-154(1991)

New method of thinning photoelastic interference fringes in image processing (Zhang, Yuan P.)V1554A,862-866(1991)

Optical associative memory for nontraditional architecture digital computers and database management systems (Burtsev, Vsevolod S.; Fyodorov, Vyatcheslav B.)V1621,215-226(1991)

Pacifist's guide to optical computers (Caulfield, H. J.)V1564,632-632(1991)

Programmable processor for multidimensional digital signal processing (Abdelrazik, Mohamed B.)V1606,812-822(1991)

Quality assessment of video image capture systems (Rowberg, Alan H.; Lian, Jing)V1444,125-127(1991)

Range of measurement of computer-aided speckle interferometry (Chen, Duanjun; Li, Shen; Hsu, T. Y.; Chiang, Fu-Pen)V1554A,922-931(1991)

Real-time edge extraction by active defocusing (Hung, Yau Y.; Zhu, Qiuming; Shi, Dahuan; Tang, Shou-Hong)V1332,332-342(1991)

Removing vertical lines generated when x-ray images are digitized (Oyama, Yoshiro; Tani, Yuichiro; Shigemura, Naoshi; Abe, Toshio; Matsuda, Koyo; Kubota, Shigeto; Inami, Takashi)V1444,413-423(1991)

Scaling of digital shapes with subpixel boundary estimation (Koplowitz, Jack)V1470,167-174(1991)

Signal processing LSI system for digital still camera (Watanabe, Mikio; Saito, Osamu; Okamoto, Satoru; Ito, Kenji; Moronaga, Kenji; Hayashi, Kenkichi; Nishi, Seiki)V1452,27-36(1991)

Study of TEM micrographs of thin-film cross-section replica using spectral analysis (Mei, Ting; Liu, Xu; Tang, Jinfa; Gu, Peifu)V1554A,570-578(1991)

Superhigh-definition image communication: an application perspective (Kohli, Jagdish C.)V1605,351-361(1991)

Theory of electronic projection correlation and its application in measurement of rigid body displacement and rotation (Li, Xide; Tan, Yushan)V1554B,661-668(1991)

Training image collection at CECOM's Center for Night Vision and Electro-Optics (Harr, Richard W.)V1483,231-239(1991)

Ultrahigh-sensitivity moire interferometry (Han, Bongtae)V1554B,399-411(1991)

Vascular parameters from angiographic images (Close, Robert A.; Duckwiler, Gary R.; Vinuela, Fernando; Dion, Jacques E.)V1444,196-203(1991)

Vibration modal analysis using stroboscopic digital speckle pattern interferometry (Wang, Xizhou; Tan, Yushan)V1554A,907-914(1991)

Diodes—see also lasers, semiconductors; photodiodes

1.25-Gb/s wideband LED driver design using active matching techniques (Gershman, Vladimir)V1474,75-82(1991)

1.55-um superluminescent diode for a fiber optic gyroscope (Kashima, Yasumasa; Matoba, Akio; Kobayashi, Masao; Takano, Hiroshi)V1365,102-107(1991)

10-watt cw diode laser bar efficiently fiber-coupled to a 381 um diameter fiber-optic connector (Willing, Steven L.; Worland, Phil; Harnagel, Gary L.; Endriz, John G.)V1418,358-362(1991)

Coplanar SIMMWIC circuits (Luy, Johann-Freidrich; Strohm, Karl M.; Buechler, J.)V1475,129-139(1991)

Current-voltage characteristics of resonant tunneling diodes (Sen, Susanta; Nag, B. R.; Midday, S.)V1362,750-759(1991)

Diode pumping of tunable Cr-doped lasers (Scheps, Richard)V1410,190-194(1991)

Establishment of new criterion aiding the control of antireflection coating semiconductor diodes (Lu, Yu C.; Li, Da Y.; Chen, Jian G.; Luo, Bin)V1519,463-466(1991)

Fabrication of high-radiance LEDs by epitaxial lift-off (Pollentier, Ivan K.; Ackaert, A.; De Dobbelaere, P.; Buydens, Luc; van Daele, Peter; Demeester, Piet M.)V1361,1056-1062(1991)

Glass requirements for encapsulating metallurgical diodes (Ali, Mir A.; Meldrum, Gerald L.; Krieger, Jeffry M.)V1513,215-223(1991)

Heterostructure photosensitive memory (Manasson, Vladimir A.; Sadovnik, Lev S.; Chen, Ray T.)V1559,194-201(1991)

High-efficiency UV and blue emitting devices prepared by MOVPE and low-energy electron-beam irradiation treatment (Akasaki, Isamu; Amano, H.)V1361,138-149(1991)

High-intensity soft-flash x-ray generator utilizing a low-vacuum diode (Isobe, Hiroshi; Sato, Eiichi; Shikoda, Arimitsu; Takahashi, Kei; Tamakawa, Yoshiharu; Yanagisawa, Toru)V1358,471-478(1991)

High-voltage picosecond pulse generation using avalanche diodes (McEwan, Thomas E.; Hanks, Roy L.)V1346,465-470(1991)

Indirect stimulated emission at room temperature in the visible range (Rinker, Michael; Kalt, Heinz; Lu, Yin-Cheng; Bauser, Elizabeth; Koehler, Klaus)V1362,14-23(1991)

InGaAs/AlGaAs vertical optical modulators and sources on a transparent GaAs substrate (Buydens, Luc; Demeester, Piet M.; De Dobbelaere, P.; van Daele, Peter)V1362,50-58(1991)

KTP waveguides for frequency upconversion of strained-layer InGaAs laser diodes (Risk, William P.; Nadler, Ch. K.)V1561,130-134(1991)

Laser effects on fibrin clot response by human meniscal fibrochondrocytes in organ culture (Forman, Scott K.; Oz, Mehmet C.; Wong, Edison; Treat, Michael R.; Kiernan, Howard)V1424,2-6(1991)

Molecular beam epitaxy/liquid phase epitaxy hybrid growth for GaAs-LED on Si (Minemura, Tetsuroh; Yazawa, Y.; Asano, J.; Unno, T.)V1361,344-353(1991)

Multilayer InSb diodes grown by molecular beam epitaxy for near-ambient temperature operation (Ashley, Timothy; Dean, A. B.; Elliott, Charles T.; Houlton, M. R.; McConville, C. F.; Tarry, H. A.; Whitehouse, Colin R.)V1361,238-244(1991)

Negative ions from magnetically insulated diodes (Prohaska, Robert M.; Fisher, Amnon; Rostoker, Norman)V1407,598-604(1991)

New process for improving reverse characteristics of platinum silicide Schottky barrier power diodes (Zhao, Shu L.; Li, Yuan J.; Yu, Jia F.; Yang, Ya L.; Liu, Shu Q.; Fan, Ya F.; Bao, Xiu Y.; Li, Zheng Q.)V1519,275-280(1991)

Optoelectronic compare-and-exchange switches based on BILED circuits (Mao, Xianjun; Liu, Shutian; Wang, Ruibo)V1563,58-63(1991)

Optoelectronic devices for fiber-optic sensor interface systems (Hong, C. S.; Hager, Harold E.; Capron, Barbara; Mantz, Joseph L.; Beranek, Mark W.; Huggins, Raymond W.; Chan, Eric Y.; Voitek, Mark; Griffith, David M.; Livezey, Darrell L.; Scharf, Bruce R.)V1418,177-187(1991)

Progress toward steady-state high-efficiency vircators (Poulsen, Peter; Pincosy, Phillip A.; Morrison, Jasper J.)V1407,172-182(1991)

Pulsed-power considerations for electron-beam-pumped krypton-fluoride lasers for inertial confinement fusion applications (Rose, Evan A.; McDonald, Thomas E.; Rosocha, Louis A.; Harris, David B.; Sullivan, J. A.; Smith, I. D.)V1411,15-27(1991)

Quantum extension of Child-Langmuir law (Lau, Yue Y.; Chernin, David P.; Colombant, Denis G.; Ho, Ping-Tong)V1407,546-552(1991)

Resonant tunneling in microcrystalline silicon quantum box diode (Tsu, Raphael; Ye, Qui-Yi; Nicollian, Edward H.)V1361,232-235(1991)

Surface plasmon enhanced light emission in GaAs/AlGaAs light emitting diodes (Gornik, Erich; Koeck, A.; Thanner, C.; Korte, Lutz)V1362,1-13(1991)

Thirty-two-channel LED array spectrometer module with compact optomechanical construction (Malinen, Jouko; Keranen, Heimo; Hannula, Tapio; Hyvarinen, Timo S.)V1533,122-128(1991)

Displays—see also cathode ray tubes; holography; imaging systems; projection systems; television; three dimensions; video

3-DTV research and development in Europe (Sand, Ruediger)V1457,76-84(1991)

Active matrix LCDs driven by two- and three-terminal switches: a comparison (den Boer, Willem; Yaniv, Zvi)V1455,248-248(1991)

Addressing factors for polymer-dispersed liquid-crystal displays (Margerum, J. D.; Lackner, Anna M.; Erdmann, John H.; Sherman, E.)V1455,27-38(1991)

Analysis of the chromatic aberrations of imaging holographic optical elements (Tholl, Hans D.; Luebbers, Hubertus; Stojanoff, Christo G.)V1456,262-273(1991)

Application of the edge-lit format to holographic stereograms (Farmer, William J.; Benton, Stephen A.; Klug, Michael A.)V1461,215-226(1991)

Atomic layer growth of zinc oxide and zinc sulphide (Sanders, Brian W.; Kitai, Adrian H.)V1398,81-87(1991)

Automatic adjustment of display window (gray level) for MR images using a neural network (Ohhashi, Akinami; Yamada, Shinichi; Haruki, Kazuhito; Hatano, Hisaaki; Fujii, Yumi; Yamaguchi, Koujiro; Ogata, Hakaru)V1444,63-74(1991)

Autostereoscopic (3-D without glasses) display for personal computer applications (Eichenlaub, Jesse B.)V1398,48-51(1991)

Basic principles of stereographic software development (Hodges, Larry F.)V1457,9-17(1991)

Charge-coupled-device-addressed liquid-crystal light valve: an update (Efron, Uzi; Byles, W. R.; Goodwin, Norman W.; Forber, Richard A.; Sayyah, Keyvan; Wu, Chiung S.; Welkowsky, Murray S.)V1455,237-247(1991)

Clinical experience with a stereoscopic image workstation (Henri, Christopher J.; Collins, D. L.; Pike, G. B.; Olivier, A.; Peters, Terence M.)V1444,306-317(1991)

Colorimetry, normal human vision, and visual display (Thornton, William A.)V1456,219-225(1991)

Color quantization aspects in stereopsis (Hebbar, Prashant D.; McAllister, David F.)V1457,233-241(1991)

Combat vehicle stereo HMD (Rallison, Richard D.; Schicker, Scott R.)V1456,179-190(1991)

Compact zoom lens for stereoscopic television (Scheiwiller, Peter M.; Murphy, S. P.; Dumbreck, Andrew A.)V1457,2-8(1991)

Comparison of 2-D planar and 3-D perspective display formats in multidimensional data visualization (Merwin, David H.; Wickens, Christopher D.)V1456,211-218(1991)

Comparison of 3-D display formats for CAD applications (McWhorter, Shane W.; Hodges, Larry F.; Rodriguez, Walter E.)V1457,85-90(1991)

Comparison of stereoscopic cursors for the interactive manipulation of B-splines (Barham, Paul T.; McAllister, David F.)V1457,18-26(1991)

Comparison of three-dimensional surface rendering techniques (Thomas, Judith G.; Galloway, Robert L.; Edwards, Charles A.; Haden, Gerald L.; Maciunas, Robert J.)V1444,379-388(1991)

Computational model for the stereoscopic optics of a head-mounted display (Robinett, Warren; Rolland, Jannick P.)V1457,140-160(1991)

Contouring using gratings created on a LCD panel (Asundi, Anand K.; Wong, C. M.)V1400,80-85(1991)

Convergence of video and computing (Carlson, Curtis R.)V1472,2-5(1991)

Correlation between the detection and interpretation of image features (Fuhrman, Carl R.; King, Jill L.; Obuchowski, Nancy A.; Rockette, Howard E.; Sashin, Donald; Harris, Kathleen M.; Gur, David)V1446,422-429(1991)

Correlations between time required for radiological diagnoses, readers' performance, display environments, and difficulty of cases (Gur, David; Rockette, Howard E.; Sumkin, Jules H.; Hoy, Ronald J.; Feist, John H.; Thaete, F. L.; King, Jill L.; Slasky, B. S.; Miketic, Linda M.; Straub, William H.)V1446,284-288(1991)

Depth cueing for visual search and cursor positioning (Reinhart, William F.)V1457,221-232(1991)

Description and performance of a 256 x 256 electrically heated pixel IR scene generator (Lake, Stephen P.; Pritchard, Alan P.; Sturland, Ian M.; Murray, Anthony R.; Prescott, Anthony J.; Gough, David W.)V1486,286-293(1991)

Designing the right visor (Gilboa, Pini)V1456,154-163(1991)

Design of an optimal single-reflective holographic helmet display element (Twardowski, Patrice J.; Meyrueis, Patrick)V1456,164-174(1991)

Design rules for pseudocolor transmission holographic stereograms (Andrews, John R.)V1507,407-415(1991)

Detection of contraband brought into the United States by aircraft and other transportation methods: a changing problem (Bruder, Joseph A.; Greneker, E. F.; Nathanson, F. E.; Henneberger, T. C.)V1479,316-321(1991)

Development of a large-screen high-definition laser video projection system (Clynick, Tony)V1456,51-57(1991)

Development of a stereoscopic three-dimensional drawing application (Carver, Donald E.; McAllister, David F.)V1457,54-65(1991)

Digital map databases in support of avionic display systems (Trenchard, Michael E.; Lohrenz, Maura C.; Rosche, Henry; Wischow, Perry B.)V1456,318-326(1991)

Disparity between combiners in a double-combiner head-up display (Cohen, Jonathan; Reichert, Abraham)V1456,250-261(1991)

Display for advanced research and training: an inexpensive answer to tactical simulation (Thomas, Melvin L.; Reining, Gale; Kelly, George)V1456,65-75(1991)

Display systems for medical imaging (Erdekian, Vahram V.; Trombetta, Steven P.)V1444,151-158(1991)

Document viewing: display requirements in image management (van Overbeek, Thomas T.)V1454,406-413(1991)

Droplet-size polydispersity in polymer-dispersed liquid-crystal films (Vaz, Nuno A.; Smith, George W.; VanSteenkiste, T. H.; Montgomery, G. P.)V1455,110-122(1991)

Ecological approach to partial binocular overlap (Melzer, James E.; Moffitt, Kirk W.)V1456,124-131(1991)

Effect of experimental design on sample size (Rockette, Howard E.; Obuchowski, Nancy A.; Gur, David; Good, Walter F.)V1446,276-283(1991)

Effect of polymer mixtures on the performance of PDLC films (Heavin, Scott D.; Fung, Bing M.; Mears, Richard B.; Sluss, James J.; Batchman, Theodore E.)V1455,12-18(1991)

Effect of viewing distance and disparity on perceived depth (Gooding, Linda; Miller, Michael E.; Moore, Jana; Kim, Seong-Han)V1457,259-266(1991)

Effects of alternate pictorial pathway displays and stereo 3-D presentation on simulated transport landing approach performance (Busquets, Anthony M.; Parrish, Russell V.; Williams, Steven P.)V1457,91-102(1991)

Electro-optical autostereoscopic displays using large cylindrical lenses (Hattori, Tomohiko)V1457,283-289(1991)

Extracting Meaning from Complex Data: Processing, Display, Interaction II (Farrell, Edward J., ed.)V1459(1991)

Fast full-erasure laser-addressed smectic liquid-crystal light valve (Zhuang, Song Lin; Qiu, Yinggang; Jiang, Yingqui; Tu, Yijun; Chen, Wei)V1558,149-153(1991)

Fusion of human vision system with mid-range IR image processing displays (Forsyth, William B.; Lewis, H. G.)V1472,18-25(1991)

Global geometric, sound, and color controls for iconographic displays of scientific data (Smith, Stuart; Grinstein, Georges G.; Pickett, Ronald M.)V1459,192-206(1991)

Helmet-mounted sight and display testing (Boehm, Hans-Dieter V.; Schreyer, H.; Schranner, R.)V1456,95-123(1991)

High-resolution display using a laser-addressed ferroelectric liquid-crystal light valve (Nakajima, Hajime; Kisaki, Jyunko; Tahata, Shin; Horikawa, Tsuyoshi; Nishi, Kazuro)V1456,29-39(1991)

High-speed integrated rendering algorithm for interpreting multiple-variable 3-D data (Miyazawa, Tatsuo)V1459,36-47(1991)

Holographic display of computer simulations (Andrews, John R.; Stinehour, Judith E.; Lean, Meng H.; Potyondy, David O.; Wawrzynek, Paul A.; Ingraffea, Anthony R.; Rainsdon, Michael D.)V1461,110-123(1991)

Holographic mirrors laminated into windshields for automotive head-up display and solar protective glazing applications (Beeck, Manfred-Andreas; Frost, Thorsten; Windeln, Wilbert)V1507,394-406(1991)

Human Vision, Visual Processing, and Digital Display II (Rogowitz, Bernice E.; Brill, Michael H.; Allebach, Jan P., eds.)V1453(1991)

Human visual performance model for crewstation design (Larimer, James O.; Prevost, Michael P.; Arditi, Aries R.; Azueta, Steven; Bergen, James R.; Lubin, Jeffrey)V1456,196-210(1991)

Humidity dependence of ceramic substrate electroluminescent devices (Young, Richard; Kitai, Adrian H.)V1398,71-80(1991)

Image annotation under X Windows (Pothier, Steven)V1472,46-53(1991)

Image quality metrics for volumetric laser displays (Williams, Rodney D.; Donohoo, Daniel)V1457,210-220(1991)

Image signal modulation and noise analysis of CRT displays (Burns, Peter D.)V1454,392-398(1991)

Interaction of objects in a virtual environment: a two-point paradigm (Bryson, Steve T.)V1457,180-187(1991)

Interactive analysis of transient field data (Dickinson, Robert R.)V1459,166-176(1991)

Interactive graphics system for multivariate data display (Becker, Richard A.; Cleveland, William S.; Shyu, William M.; Wilks, Allan R.)V1459,48-56(1991)

Large Screen Projection, Avionic, and Helmet-Mounted Displays (Assenheim, Harry M.; Flasck, Richard A.; Lippert, Thomas M.; Bentz, Jerry, eds.)V1456(1991)

Laser-based display technology development at the Naval Ocean Systems Center (Phillips, Thomas E.; Trias, John A.; Lasher, Mark E.; Poirier, Peter M.; Dahlke, Weldon J.; Robinson, Waldo R.)V1454,290-298(1991)

Light budget and optimization strategies for display applications of dichroic nematic droplet/polymer films (Drzaic, Paul S.)V1455,255-263(1991)

Liquid-crystal light valves for projection displays (Shields, Steven E.; Bleha, William P.)V1455,225-236(1991)

Luminance asymmetry in stereo TV images (Beldie, Ion P.; Kost, Bernd)V1457,242-247(1991)

MDIS (medical diagnostic imaging support) workstation issues: clinical perspective (Smith, Donald V.; Smith, Suzy; Cawthon, Michael A.)V1444,357-362(1991)

Medical Imaging V: Image Capture, Formatting, and Display (Kim, Yongmin, ed.)V1444(1991)

Militarized infrared touch panels (Hough, Stewart E.; Stanley, Pamela S.)V1456,240-249(1991)

Modeling and visualization of scattered volumetric data (Nielson, Gregory M.; Dierks, Tim)V1459,22-33(1991)

Molecular anchoring at the droplet wall in PDLC materials (Crawford, Gregory P.; Ondris-Crawford, Renate; Doane, J. W.)V1455,2-11(1991)

Network visualization: user interface issues (Becker, Richard A.; Eick, Stephen G.; Miller, Eileen O.; Wilks, Allan R.)V1459,150-154(1991)

New approach to palette selection for color images (Balasubramanian, Raja; Allebach, Jan P.)V1453,58-69(1991)

New thin-film transistor structure and its processing method for liquid-crystal displays (Kuo, Yue)V1456,288-299(1991)

Novel supertwisted nematic liquid-crystal-display operating modes and electro-optical performance of generally twisted nematic configurations (Schadt, Martin)V1455,214-224(1991)

Observer performance in dynamic displays: effect of frame rate on visual signal detection in noisy images (Whiting, James S.; Honig, David A.; Carterette, Edward; Eigler, Neal)V1453,165-175(1991)

Optical design of dual-combiner head-up displays (Kirkham, Anthony J.)V1354,310-315(1991)

PACS reading time comparision: the workstation versus alternator for ultrasound (Horii, Steven C.; Garra, Brian S.; Mun, Seong Ki; Singer, Jon; Zeman, Robert K.; Levine, Betty A.; Fielding, Robert; Lo, Ben)V1446,475-480(1991)

Part-task training with a helmet-integrated display simulator system (Casey, Curtis J.; Melzer, James E.)V1456,175-178(1991)

Perceptual noise measurement of displays (Chakraborty, Dev P.; Pfeiffer, Douglas E.; Brikman, Inna)V1443,183-190(1991)

Perceptual training with cues for hazard detection in off-road driving (Merritt, John O.; CuQlock-Knopp, V. G.)V1457,133-138(1991)

Positioning accuracy of a virtual stereographic pointer in a real stereoscopic video world (Drascic, David; Milgram, Paul)V1457,302-313(1991)

Possibility of liquid crystal display panels for a space-saving PACS workstation (Komori, Masaru; Minato, Kotaro; Takahashi, Takashi; Nakano, Yoshihisa; Sakurai, Tsunetaro)V1444,334-337(1991)

Practical low-cost stereo head-mounted display (Pausch, Randy; Dwivedi, Pramod; Long, Allan C.)V1457,198-208(1991)

Progress in autostereoscopic display technology at Dimension Technologies Inc. (Eichenlaub, Jesse B.)V1457,290-301(1991)

Projection screens for high-definition television (Kirkpatrick, Michael D.; Mihalakis, George M.)V1456,40-47(1991)

RadGSP: a medical image display and user interface for UWGSP3 (Yee, David K.; Lee, Woobin; Kim, Dong-Lok; Haass, Clark D.; Rowberg, Alan H.; Kim, Yongmin)V1444,292-305(1991)

Radiologists' confidence in detecting abnormalities on chest images and their subjective judgments of image quality (King, Jill L.; Gur, David; Rockette, Howard E.; Curtin, Hugh D.; Obuchowski, Nancy A.; Thaete, F. L.; Britton, Cynthia A.; Metz, Charles E.)V1446,268-275(1991)

Real-time holographic display: improvements using a multichannel acousto-optic modulator and holographic optical elements (St.-Hilaire, Pierre; Benton, Stephen A.; Lucente, Mark; Underkoffler, John S.; Yoshikawa, Hiroshi)V1461,254-261(1991)

Recent advances in color reflection holography (Hubel, Paul M.)V1461,167-174(1991)

Remote driving: one eye or two (Bryant, Keith; Ince, Ilhan)V1457,120-132(1991)

Schottky diode silicon liquid-crystal light valve (Sayyah, Keyvan; Efron, Uzi; Forber, Richard A.; Goodwin, Norman W.; Reif, Philip G.)V1455,249-254(1991)

Selection devices for field-sequential stereoscopic displays: a brief history (Lipton, Lenny)V1457,274-282(1991)

Simultaneous graphics and multislice raster image display for interactive image-guided surgery (Edwards, Charles A.; Galloway, Robert L.; Thomas, Judith G.; Schreiner, Steven; Maciunas, Robert J.)V1444,38-46(1991)

Single-pixel measurements on LCDs (Jenkins, A. J.)V1506,188-193(1991)

Slice plane generation for three-dimensional image viewing using multiprocessing (Ho, Bruce K.; Ma, Marco; Chuang, Keh-Shih)V1445,95-100(1991)

Solid models for CT/MR image display: accuracy and utility in surgical planning (Mankovich, Nicholas J.; Yue, Alvin; Ammirati, Mario; Kioumehr, Farhad; Turner, Scott)V1444,2-8(1991)

Some effects on depth-position and course-prediction judgments in 2-D and 3-D displays (Miller, Robert H.; Beaton, Robert J.)V1457,248-258(1991)

Standards for flat panel display systems (Greeson, James C.)VCR37,146-158(1991)

Stereoscopic Displays and Applications II (Merritt, John O.; Fisher, Scott S., eds.)V1457(1991)

Stereoscopic versus orthogonal view displays for performance of a remote manipulation task (Spain, Edward H.; Holzhausen, Klause-Peter)V1457,103-110(1991)

Survey of helmet tracking technologies (Ferrin, Frank J.)V1456,86-94(1991)

Switching speeds in NCAP displays: dependence on collection angle and wavelength (Reamey, Robert H.; Montoya, Wayne; Wartenberg, Mark)V1455,39-44(1991)

Teleoperator performance with virtual window display (Cole, Robert E.; Merritt, John O.; Coleman, Richard; Ikehara, Curtis)V1457,111-119(1991)

Thin-film technology in high-resolution, high-density AC plasma displays (Andreadakis, Nicholas C.)V1456,310-315(1991)

Three-dimensional display of MRI data in neurosurgery: segmentation and rendering aspects (Barillot, Christian; Lachmann, F.; Gibaud, Bernard; Scarabin, Jean-Marie)V1445,54-65(1991)

Three-dimensional grating images (Takahashi, Susumu; Toda, Toshiki; Iwata, Fujio)V1461,199-205(1991)

Three-dimensional moving-image display by modulated coherent optical fibers: a proposal (Hoshino, Hideshi; Sato, Koki)V1461,227-231(1991)

Use of flicker-free television products for stereoscopic display applications (Woods, Andrew J.; Docherty, Tom; Koch, Rolf)V1457,322-326(1991)

Use of the Boulder model for the design of high-polarization fluorinated ferroelectric liquid crystals (Wand, Michael D.; Vohra, Rohini; Thurmes, William; Walba, David M.; Geelhaar, Thomas; Littwitz, Brigitte)V1455,97-104(1991)

User benefits of visualization with 3-D stereoscopic displays (Wichansky, Anna M.)V1457,267-271(1991)

Virtual environment for the exploration of three-dimensional steady flows (Bryson, Steve T.; Levit, Creon C.)V1457,161-168(1991)

Virtual environment system for simulation of leg surgery (Pieper, Steve; Delp, Scott; Rosen, Joseph; Fisher, Scott S.)V1457,188-197(1991)

Visual characteristics of LED display pushbuttons for avionic applications (Vanni, Paolo; Isoldi, Felice)V1456,300-309(1991)

Visual field information in nap-of-the-earth flight by teleoperated helmet-mounted displays (Grunwald, Arthur J.; Kohn, S.; Merhav, S. J.)V1456,132-153(1991)

Visualization tool for human-machine interface designers (Prevost, Michael P.; Banda, Carolyn P.)V1459,58-66(1991)

Visual processing, transformability of primaries, and visual efficiency of display devices (Thornton, William A.)V1453,390-401(1991)

Waveguide holography and its applications (Huang, Qiang; Caulfield, H. J.)V1461,303-312(1991)

ZnS:Mn thin film electroluminescent display devices using hafnium dioxide as insulating layer (Hsu, C. T.; Li, J. W.; Liu, C. S.; Su, Yan K.; Wu, T. S.; Yokoyama, M.)V1519,391-395(1991)

Doppler effect—see also lidar; radar; velocimetry

Application of low-coherence optical fiber Doppler anemometry to fluid-flow measurement: optical system considerations (Boyle, William J.; Grattan, Kenneth T.; Palmer, Andrew W.; Meggitt, Beverley T.)V1511,51-56(1991)

Application of neural networks to range-Doppler imaging (Wu, Xiaoqing; Zhu, Zhaoda)V1569,484-490(1991)

Compact low-power acousto-optic range-Doppler-angle processor for a pulsed-Doppler radar (Pape, Dennis R.; Vlannes, Nickolas P.; Patel, Dharmesh P.; Phuvan, Sonlinh)V1476,201-213(1991)

Depth dependent laser Doppler perfusion measurements: theory and instrumentation (Koelink, M. H.; de Mul, F. F.; Greve, Jan; Graaff, Reindert; Dassel, A. C.; Aarnouds, J. G.)V1403,347-349(1991)

Development of a laser Doppler system for measurement of velocity fields in PVT crystal growth systems (Jones, O. C.; Glicksman, M. E.; Lin, Jen T.; Kim, G. T.; Singh, N. B.)V1557,202-208(1991)

Doppler-guided retrograde catheterization system (Frazin, Leon J.; Vonesh, Michael J.; Chandran, Krishnan B.; Khasho, Fouad; Lanza, George M.; Talano, James V.; McPherson, David D.)V1425,207-207(1991)

Experimental investigations of the autocorrelation function for inhomogeneous scatterers (Singh, Brahm P.; Chopra, S.)V1558,317-321(1991)

Improvement in spatial resolution of a forward-scatter laser Doppler velocimeter (Mozumdar, Subir; Bond, Robert L.)V1584,254-261(1991)

In-vivo blood flow velocity measurements using the self-mixing effect in a fiber-coupled semiconductor laser (Koelink, M. H.; Slot, M.; de Mul, F. F.; Greve, Jan; Graaff, Reindert; Dassel, A. C.; Aarnouds, J. G.)V1511,120-128(1991)

Laser Doppler diagnostics of vorticity and phenomenological description of the flows of dilute polymer solutions in model tubes (Fedoseeva, E. V.; Polyakova, M. S.)V1403,355-358(1991)

Laser Doppler flowmetry in neurosurgery (Fasano, Victor A.; Urciuoli, Rosa; Bolognese, Paolo; Fontanella, Marco)V1428,2-12(1991)

Laser application in chosen maritime economy divisions (Kirkiewicz, Jozef)V1391,351-360(1991)

Maximum likelihood estimation of differential delay and differential Doppler (Greene, Herbert G.; MacMullan, Jay)V1470,98-102(1991)

Miniature laser Doppler anemometer for sensor concepts (Damp, Stephan)V1418,459-470(1991)

Monte Carlo simulations and measurements of signals in laser Doppler flowmetry on human skin (Koelink, M. H.; de Mul, F. F.; Greve, Jan; Graaff, Reindert; Dassel, A. C.; Aarnouds, J. G.)V1431,63-72(1991)

Pulp blood flow assessment in human teeth by laser Doppler flowmetry (Pettersson, Hans; Oberg, Per A.)V1424,116-119(1991)

Three-dimensional color Doppler imaging of the carotid artery (Picot, Paul A.; Rickey, Daniel W.; Mitchell, J. R.; Rankin, Richard N.; Fenster, Aaron)V1444,206-213(1991)

Transverse concentrational profiles in dilute erythrocytes suspension flows in narrow channels measured by integral Doppler anemometer (Kononenko, Vadim L.; Shimkus, J. K.)V1403,381-383(1991)

Earth—see also remote sensing

Applications of laser ranging to ocean, ice, and land topography (Degnan, John J.)V1492,176-186(1991)

Applications of the massively parallel machine, the MasPar MP-1, to Earth sciences (Fischer, James R.; Strong, James P.; Dorband, John E.; Tilton, James C.)V1492,229-238(1991)

ATHENA: a high-resolution wide-area coverage commercial remote sensing system (Claybaugh, William R.; Megill, L. R.)V1495,81-94(1991)

Coastal survey with a multispectral video system (Niedrauer, Terren M.)V1492,240-251(1991)

Concept for the subresolution measurement of earthquake strain fields using SPOT panchromatic imagery (Crippen, Robert E.; Blom, Ronald G.)V1492,370-377(1991)

Design considerations for EOS direct broadcast (Vermillion, Charles H.; Chan, Paul H.)V1492,224-228(1991)

Eagle-class small satellite for LEO applications (O'Neil, Jason; Goralczyk, Steven M.)V1495,72-80(1991)

Earth and Atmospheric Remote Sensing (Curran, Robert J.; Smith, James A.; Watson, Ken, eds.)V1492(1991)

Earth Observing System (Wilson, Stan; Dozier, Jeff)V1491,117-124(1991)

ESA Earth observation polar platform program (Rast, Michael; Readings, C. J.)V1490,51-58(1991)

Forest decline model development with LANDSAT TM, SPOT, and DEM DATA (Brockhaus, John A.; Campbell, Michael V.; Khorram, Siamak; Bruck, Robert I.; Stallings, Casson)V1492,200-205(1991)

Future European and Japanese Remote-Sensing Sensors and Programs (Slater, Philip N., ed.)V1490(1991)

Imaging capabilities of small satellites: Indian experience (Alex, T. K.)V1495,52-58(1991)

In situ monitoring for hydrocarbons using fiber optic chemical sensors (Klainer, Stanley M.; Thomas, Johnny R.; Dandge, Dileep K.; Frank, Chet A.; Butler, Marcus S.; Arman, Helen; Goswami, Kisholoy)V1434,119-126(1991)

Integration of diverse remote sensing data sets for geologic mapping and resource exploration (Kruse, Fred A.; Dietz, John B.)V1492,326-337(1991)

Japanese mission overview of JERS and ASTER programs (Yamaguchi, Yasushi; Tsu, Hiroji; Sato, Isao)V1490,324-334(1991)

Laser-induced fluorescence in contaminated soils (Lurk, Paul W.; Cooper, Stafford S.; Malone, Philip G.; Olsen, R. S.; Lieberman, Stephen H.)V1434,114-118(1991)

Medium-resolution imaging spectrometer (Baudin, Gilles; Bessudo, Richard; Cutter, Mike A.; Lobb, Daniel R.; Bezy, Jean L.)V1490,102-113(1991)

Mission overview of ADEOS program (Iwasaki, Nobuo; Hara, Norikazu; Kajii, Makoto; Tange, Yoshio; Miyachi, Yuji; Sato, Ryota; Inoue, Kouichi)V1490,192-199(1991)

MOMS-02/Spacelab D-2: a high-resolution multispectral stereo scanner for the second German Spacelab mission (Ackermann, F.; Bodechtel, Joh; Lanzl, Franz; Meissner, D.; Seige, Peter; Winkenbach, H.; Zilger, Johannes)V1490,94-101(1991)

Optical design of the moderate-resolution imaging spectrometer-tilt for the Earth Observing System (Maymon, Peter W.)V1492,286-297(1991)

Optical mapping instrument (Bagot, K. H.)V1490,126-132(1991)

Polarization and directionality of the Earth's reflectances: the POLDER instrument (Lorsignol, Jean; Hollier, Pierre A.; Deshayes, Jean-Pierre)V1490,155-163(1991)

Remote sensing of coastal environmental hazards (Huh, Oscar K.; Roberts, Harry H.; Rouse, Lawrence J.)V1492,378-386(1991)

Remote sensing of volcanic ash hazards to aircraft (Rose, William I.; Schneider, David J.)V1492,387-390(1991)

Remote spectral identification of surface aggregates by thermal imaging techniques: progress report (Scholen, Douglas E.; Clerke, William H.; Burns, Gregory S.)V1492,358-369(1991)

Role of orbital observations in detecting and monitoring geological hazards: prospects for the future (Pieri, David C.)V1492,410-417(1991)

ScaRaB Earth radiation budget scanning radiometer (Monge, J.L.; Kandel, Robert S.; Pakhomov, L. A.; Bauche, B.)V1490,84-93(1991)

Scanning imaging absorption spectrometer for atmospheric chartography (Burrows, John P.; Chance, Kelly V.)V1490,146-154(1991)

Single-mode fiber Mach-Zehnder interferometer as an earth strain sensor (Wang, An; Xie, Haiming)V1572,440-443(1991)

Spectral stratigraphy (Lang, Harold R.)V1492,351-357(1991)

Techniques and strategies for data integration in mineral resource assessment (Trautwein, Charles M.; Dwyer, John L.)V1492,338-338(1991)

Territorial analysis by fusion of LANDSAT and SAR data (Vernazza, Gianni L.; Dambra, Carlo; Parizzi, Francesco; Roli, Fabio; Serpico, Sebastiano B.)V1492,206-212(1991)

TREIS: a concept for a user-affordable, user-friendly radar satellite system for tropical forest monitoring (Raney, R. K.; Specter, Christine N.)V1492,298-306(1991)

Using digital-scanned aerial photography for wetlands delineation (Anderson, John E.; Roos, Maurits)V1492,252-262(1991)

Electromagnetic waves

Analysis of multiple-multipole scattering by time-resolved spectroscopy and spectrometry (Frank, Klaus; Hoeper, J.; Zuendorf, J.; Tauschek, D.; Kessler, Manfred; Wiesner, J.; Wokaun, Alexander J.)V1431,2-11(1991)

Analytical studies of large closed-loop arrays (Fikioris, George; Freeman, D. K.; King, Ronold W.; Shen, Hao-Ming; Wu, Tai T.)V1407,295-305(1991)

Broadband electromagnetic environment monitoring using semiconductor electroabsorption modulators (Pappert, Stephen A.; Lin, S. C.; Orazi, Richard J.; McLandrich, Matthew N.; Yu, Paul K.; Li, S. T.)V1476,282-293(1991)

Chirality and its applications to engineered materials (Varadan, Vasundara V.; Varadan, Vijay K.)V1558,156-181(1991)

Comparison of negative and positive polarity reflex diode microwave source (Litz, Marc S.; Huttlin, George A.; Lazard, Carl J.; Golden, Jeffry; Pereira, Nino R.; Hahn, Terry D.)V1407,159-166(1991)

Diffractive properties of surface-relief microstructures (Qu, Dong-Ning; Burge, Ronald E.; Yuan, X.)V1506,152-159(1991)

Dimension and time effects caused by nonlocal scattering of laser radiation from a rough metal surface (Dolgina, A. N.; Kovalev, A. A.; Kondratenko, P. S.)V1440,342-353(1991)

Double-resonant tunneling via surface plasmons in a metal-semiconductor system (Sreseli, Olga M.; Belyakov, Ludvig V.; Goryachev, D. N.; Rumyantsev, B. L.; Yaroshetskii, Ilya D.)V1440,326-331(1991)

Effective properties of electromagnetic wave propagation in some composite media (Artola, Michel; Cessenat, Michel)V1558,14-21(1991)

Electromagnetic scattering from a finite cylinder with complex permittivity (Murphy, Robert A.; Christodoulou, Christos G.; Phillips, Ronald L.)V1558,295-305(1991)

Experimental study of chiral composites (Ro, Ru-Yen; Varadan, Vasundara V.; Varadan, Vijay K.)V1558,269-287(1991)

Experimental study of electromagnetic missiles from a hyperboloidal lens (Shen, Hao-Ming; Wu, Tai T.; Myers, John M.)V1407,286-294(1991)

Experimental study of the resonance of a circular array (Shen, Hao-Ming)V1407,306-315(1991)

Experimental verification of grating theory for surface-emitting structures (Ayekavadi, Raj; Yeh, C. S.; Butler, Jerome K.; Evans, Gary A.; Stabile, Paul J.; Rosen, Arye)V1418,74-85(1991)

GigaHertz RMS current sensors for electromagnetic compatibility testing (Mitchell, Gordon L.; Saaski, Elric W.; Pace, John W.)V1367,266-272(1991)

Heterophase isotopic SF6 molecule separation in the surface electromagnetic wavefield (Bordo, V. G.; Ershov, I. A.; Kravchenko, V. A.; Petrov, Yu. N.)V1440,364-369(1991)

High-power uhf rectenna for energy recovery in the HCRF (high-current radio frequency) system (Genuario, Ralph; Koert, Peter)V1407,553-565(1991)

Hydrodynamic evolution of picosecond laser plasmas (Landen, Otto L.; Vu, Brian-Tinh; Stearns, Daniel G.; Alley, W. E.)V1413,120-130(1991)

Improved formalism for rough-surface scattering of acoustic and electromagnetic waves (Milder, D. M.)V1558,213-221(1991)

Intense Microwave and Particle Beams II (Brandt, Howard E., ed.)V1407(1991)

Investigations of cumulative beam breakup in radio-frequency linacs (Bohn, Courtlandt L.; Delayen, Jean R.)V1407,566-577(1991)

IR laser-light backscattering by an arbitrarily shaped dielectric object with rough surface (Wu, Zhensen; Cheng, Denghui)V1558,251-257(1991)

Kinetic stability analysis of the extraordinary mode perturbations in a cylindrical magnetron (Uhm, Han S.; Chen, H. C.; Stark, Robert A.)V1407,113-127(1991)

Laser-induced generation of surface periodic structures resulting from the waveguide mode interaction (Bazakutsa, P. V.; Maslennikov, V. L.; Sychugov, V. A.; Yakovlev, V. A.)V1440,370-376(1991)

Light scattering from gold-coated ground glass and chemically etched surfaces (Ruiz-Cortes, Victor; Mendez, Eugenio R.; Gu, Zu-Han; Maradudin, Alexei A.)V1558,222-232(1991)

Long-pulse modulator-driven cyclotron autoresonance maser and free-electron laser experiments at MIT (Danly, Bruce G.; Hartemann, Frederic V.; Chu, T. S.; Menninger, W. L.; Papavaritis, P.; Pendergast, K. D.; Temkin, Richard J.)V1407,192-201(1991)

Magnetically tapered CARM (cyclotron autoresonance maser) for high power (Wang, Qinsong; McDermott, David B.; Luhmann, Neville C.)V1407,209-216(1991)

Measurement of SEW phase velocity by optical heterodyning method (Libenson, Michail N.; Makin, Vladimir S.; Trubaev, Vladimir V.)V1440,354-356(1991)

Negative energy cyclotron resonance maser (Lednum, Eugene E.; McDermott, David B.; Lin, Anthony T.; Luhmann, Neville C.)V1407,202-208(1991)

Nonlinear bremsstrahlung in nonequilibrium relativistic beam-plasma systems (Brandt, Howard E.)V1407,326-353(1991)

Nonoptical noncoherent imaging in industrial testing (Delecki, Z. A.; Barakat, M. A.)V1526,157-167(1991)

Parametric simulation studies and injection phase locking of relativistic magnetrons (Chen, Chiping; Chan, Hei-Wai; Davidson, Ronald C.)V1407,105-112(1991)

Peculiarities of the formation of periodic structures on silicon under millisecond laser radiation (Kokin, A. N.; Libenson, Michail N.; Minaev, Sergei M.)V1440,338-341(1991)

Polarization dependence of light scattered from rough surfaces with steep slopes (O'Donnell, Kevin A.; Knotts, Micheal E.)V1558,362-367(1991)

Radio frequency pulse compression experiments at SLAC (Farkas, Zoltan D.; Lavine, T. L.; Menegat, A.; Miller, Roger H.; Nantista, C.; Spalek, G.; Wilson, P. B.)V1407,502-511(1991)

Scanning beam switch experiment for intense rf power generation (Humphries, Stanley; Babcock, Steven; Wilson, J. M.; Adler, Richard A.)V1407,512-523(1991)

Scattering of waves from dense discrete random media: theory and applications in remote sensing (Tsang, Leung; Ding, Kung-Hau; Kong, Jin A.; Winebrenner, Dale P.)V1558,260-268(1991)

Short-pulse electromagnetics for sensing applications (Felsen, Leopold B.; Vecchi, G.; Carin, L.; Bertoni, H. L.)V1471,154-162(1991)

Silicon calorimeter for high-power microwave measurements (Lazard, Carl J.; Pereira, Nino R.; Huttlin, George A.; Litz, Marc S.)V1407,167-171(1991)

Some computational results for rough-surface scattering (DeSanto, John A.; Wombell, Richard J.)V1558,202-212(1991)

Sparse random distribution of noninteracting small chiral spheres in a chiral host medium (Lakhtakia, Akhlesh; Varadan, Vijay K.; Varadan, Vasundara V.)V1558,22-27(1991)

Spectrum of surface electromagnetic waves in CdxHg1-xTe crystals at 0.3 K < T < 77 K (Vertiy, Alexey A.; Beletskii, N. N.; Gorbatyuk, I. N.; Ivanchenko, I. V.; Popenko, N. A.; Rarenko, I. M.; Tarapov, Sergey I.)V1361,1070-1078(1991)

Surface structures formation by pulse heating of metals in oxidized environment (Bazhenov, V. V.; Libenson, Michail N.; Makin, Vladimir S.; Trubaev, Vladimir V.)V1440,332-337(1991)

Theoretical modeling of chiral composites (Apparao, R. T.; Varadan, Vasundara V.; Varadan, Vijay K.)V1558,2-13(1991)

Theory of multimode interactions in cyclotron autoresonance maser amplifiers (Chen, Chiping; Wurtele, Jonathan S.)V1407,183-191(1991)

Understanding metrology of polysilicon gates through reflectance measurements and simulation (Tadros, Karim H.; Neureuther, Andrew R.; Guerrieri, Roberto)V1464,177-186(1991)

Wave interactions with continuous fractal layers (Kim, Yun J.; Jaggard, Dwight L.)V1558,113-119(1991)

Wave Propagation and Scattering in Varied Media II (Varadan, Vijay K., ed.)V1558(1991)

Electron beams—see beams

Electronics—see also optoelectronics

Advanced multichip module packaging and interconnect issues for GaAs signal processors operating above 1 GHz clock rates (Gilbert, Barry K.; Thompson, R.; Fokken, G.; McNeff, W.; Prentice, Jeffrey A.; Rowlands, David O.; Staniszewski, A.; Walters, W.; Zahn, S.; Pan, George W.)V1390,235-248(1991)

All-fiber closed-loop gyroscope with self-calibration (Ecke, Wolfgang; Schroeter, Siegmund; Schwotzer, Guenter; Willsch, Reinhardt)V1511,57-66(1991)

Analysis and modeling of uniformly- and nonuniformly-coupled lossy lines for interconnections and packaging in hybrid and monolithic circuits (Orhanovic, Neven; Hayden, Leonard A.; Tripathi, Vijai K.)V1389,273-284(1991)

Analysis of skin effect loss in high-frequency interconnects with finite metallization thickness (Kiang, Jean-Fu)V1389,340-351(1991)

Answer to the dynamic (fretting effect) and static (oxide) behavior of electric contact surfaces: based on a five-year infrared thermographic study (Paez-Leon, Cristobal J.; Patino, Antonio R.; Aguillon, Luis)V1467,188-194(1991)

Application-specific integrated-circuit-based multianode microchannel array readout system (Smeins, Larry G.; Stechman, John M.; Cole, Edward H.)V1549,59-65(1991)

Application of electrodeposition processes to advanced package fabrication (Krongelb, Sol; Dukovic, John O.; Komsa, M. L.; Mehdizadeh, S.; Romankiw, Lubomyr T.; Andricacos, P. C.; Pfeiffer, A. T.; Wong, K.)V1389,249-256(1991)

Application of real-time holographic interferometry in the nondestructive inspection of electronic parts and assemblies (Wood, Craig P.; Trolinger, James D.)V1332,122-131(1991)

Bessel beam generation: theory and application proposal (Jabczynski, Jan K.)V1391,254-258(1991)

CAD in new areas of the package and interconnect design space (McBride, Dennis J.)V1390,330-335(1991)

CAE tools for verifying high-performance digital systems (Rubin, Lawrence M.)V1390,336-358(1991)

Challenges of using advanced multichip packaging for next generation spaceborne computers (Moravec, Thomas J.)V1390,195-201(1991)

Chaotic nature of mesh networks with distributed routing (Rucinski, Andrzej; Drexel, Peter G.; Dziurla, Barbara)V1390,388-398(1991)

Characterization, modeling, and design of dielectric resonators based on thin ceramic tape (Morris, Jacqueline H.; Belopolsky, Yakov)V1389,236-248(1991)

Color standards for electronic imaging (McDowell, David Q.)VCR37,40-53(1991)

Comparison of optical and electronic 3-dimensional circuits (Stirk, Charles W.; Psaltis, Demetri)V1389,580-593(1991)

Compensated digital readout family (Ludwig, David E.; Skow, Michael)V1541,73-82(1991)

Compensation of the phase shift in the optical fibre current sensor (Ye, Miaoyuan; Hu, Shichuang)V1572,483-486(1991)

Computer-Aided System Interconnect Design (CASID) in multipackage environment (Chandrasekhar, N. S.; Sundari, V.; Vengal, Jacob V.; Rao, P. N.)V1390,523-536(1991)

Continuous-time segmentation networks (Harris, John G.)V1473,161-172(1991)

Depopulation kinetics of electron traps in thin oxynitride films (Wong, H.; Cheng, Y. C.; Yang, Bing L.; Liu, Bai Y.)V1519,494-498(1991)

Design, simulation model, and measurements for high-density interconnections (Shrivastava, Udy A.)V1389,122-137(1991)

Design and performance of an automatic gain control system for the high-energy x-ray timing experiment (Pelling, Michael R.; Rothschild, Richard E.; MacDonald, Daniel R.; Hertel, Robert H.; Nishiie, Edward S.)V1549,134-146(1991)

Design and process impact on thin-film interconnection performance (Rinne, Glenn A.; Hwang, Lih-Tyng; Adema, G. M.; King, Donald A.; Turlik, Iwona)V1389,110-121(1991)

Determination of electrostatic potentials and charge distributions in bulk and at interfaces by electron microscopy techniques (Hugsted, B.; Gjonnes, K.; Tafto, J.; Gjonnes, Jon; Matsuhata, H.)V1361,751-757(1991)

Development of an optical fiber and photoelectric coupling V/F converter for 5.4-MV impulse generator (Guan, Genzhi)V1572,487-491(1991)

Dispersion of picosecond pulses propagating on microstrip interconnections on semiconductor integrated-circuit substrates (Pasik, Michael F.; Cangellaris, Andreas C.; Prince, John L.)V1389,297-301(1991)

Distortion characteristic of transient signals through bend discontinuity of high-speed integrated curcuits (Huang, Wei-Xu; Wing, Omar)V1389,199-204(1991)

Dual-channel current sensor capable of simultaneously measuring two currents (Bush, Simon P.; Jackson, David A.)V1584,103-109(1991)

Effects of conductor losses on cross-talk in multilevel-coupled VLSI interconnections (van Deventer, T. E.; Katehi, Linda P.; Cangellaris, Andreas C.)V1389,285-296(1991)

Effects of packaging and interconnect technology on testability of printed wiring boards (Hughes, Joseph L.; Pahlajrai, Prem)V1389,87-97(1991)

Efficient and accurate dynamic analysis of microstrip integrated circuits (Rahal Arabi, Tawfik R.; Murphy, Arthur T.; Sarkar, Tapan K.)V1389,302-313(1991)

Electrical characteristics of lossy interconnections for high-performance computer applications (Deutsch, Alina; Kopcsay, Gerard V.; Ranieri, V. A.; Cataldo, J. K.; Galligan, E. A.; Graham, W. S.; McGouey, R. P.; Nunes, S. L.; Paraszczak, Jurij R.; Ritsko, J. J.; Serino, R. J.; Shih, D. Y.; Wilczynski, Janusz S.)V1389,161-176(1991)

Electrical characterization of the interconnects inside a computer (Rubin, Barry)V1389,314-328(1991)

Electronic-digital detection system for an optical fiber current sensor (Guedes Valente, Luiz C.; Kawase, Liliana R.; Afonso, Jose A.; Kalinowski, Hypolito J.)V1584,96-102(1991)

Electronic materials basic research program managed by the Advanced Technology Directorate of the U.S. Army Strategic Defense Command (Martin, William D.)V1559,10-17(1991)

Electronic properties of Si-doped nipi structures in GaAs (Fong, C. Y.; Gallup, R. F.; Nelson, J. S.)V1361,479-488(1991)

Electronic properties of mercury-based type-III superlattices (Guldner, Yves; Manasses, J.)V1361,657-668(1991)

Evaluation of electrolytic capacitors for high-peak current pulse duty (Harris, Kevin; McDuff, G. G.; Burkes, Tom R.)V1411,87-99(1991)

FFT measuring method for magneto-optical ac current measurement (Hu, Shichuang; Ye, Miaoyuan; Qu, Gen)V1572,492-496(1991)

Fiber optic magnetic field and current sensor using magneto-birefringence of dense ferrofluid thin films (Pan, Yingtain; Liu, Xiande; Du, Chongwu; Li, Zai Q.)V1572,477-482(1991)

Fiber optic magnetic field sensors based on Faraday effect in new materials (Nikitin, Petr I.; Grigorenko, A. N.; Konov, Vitaly I.; Savchuk, A. I.)V1584,124-134(1991)

Fractal description of computer interconnections (Christie, Phillip; Styer, Stephen B.)V1390,359-367(1991)

Frequency domain evaluation of the accuracy of lumped element models for RLC transmission lines (Delbare, William; Dhaene, Tom; Vanhauwermeiren, Luc; De Zutter, Daniel)V1389,257-272(1991)

GE high-density interconnect: a solution to the system interconnect problem (Adler, Michael S.)V1390,504-508(1991)

High-density chip-to-chip interconnect system for GaAs semiconductor devices (Wigginton, Stewart C.; Davidson, Scott E.; Harting, William L.)V1390,560-567(1991)

High-density interconnect technology for VAX 9000 system (Deshpande, Ujwal A.; Howell, Gelston; Shamouilian, Shamouil)V1390,489-501(1991)

High-density multichip interconnect: military packaging for the 1990s (Trask, Philip A.)V1390,223-234(1991)

High-power particle beams for gas lasers (Mesyats, Gennady A.)V1411,2-14(1991)

High-speed electronic memory video recording (Thomas, Don L.)V1448,140-147(1991)

High-temperature superconductor junction technology (Simon, Randy W.)V1477,184-191(1991)

Imagery technology database (Courtot, Marilyn E.; Nier, Michael)VCR37,221-246(1991)

Influence of voltage pulse parameters on the beam current density distribution of large-aperture electron gun (Badziak, Jan; Dzwigalski, Zygmunt)V1391,250-253(1991)

Intelligent magnetometer with photoelectric sampler (Wang, Defang; Xu, Yan; Zhu, Minjun)V1572,514-516(1991)

Intensity-type fiber optic electric current sensor (Carome, Edward F.; Kubulins, Vilnis E.; Flanagan, Roger L.; Shamray-Bertaud, Patricia)V1584,110-117(1991)

Interconnect and packaging technology in the '90s (Seraphim, Donald; Barr, Donald E.)V1389,39-54(1991); V1390,39-54(1991)

Interconnection problems in VLSI random access memory chip (Rayapati, Venkatapathi N.; Mukhedkar, Dinkar)V1389,98-109(1991)

Machine Vision Systems Integration (Batchelor, Bruce G.; Waltz, Frederick M., eds.)VCR36(1991)

Magnet-sensitive optical fiber and its application in current sensor system (Yu, Tong; Li, Qin; Chen, Rongsheng; Yan, Jin-Li)V1572,469-471(1991); V1584,135-137(1991)

Magneto-optical apparatus with comparator for the measurement of large direct current (Zhang, Zhipeng; Zhao, Zhi; Chong, Baoxin)V1572,464-468(1991)

Measurements and characterization of multiple-coupled interconnection lines in hybrid and monolithic integrated circuits (Hayden, Leonard A.; Jong, Jyh-Ming; Rettig, John B.; Tripathi, Vijai K.)V1389,205-214(1991)

Metallic-glass-coated optical fibers as magnetic-field sensors (Larson, Donald C.; Bibby, Y. W.; Tyagi, S.)V1572,517-522(1991)

Methods for comparative analysis of waveform degradation in electrical and optical high-performance interconnections (Merkelo, Henri; McCredie, B. D.; Veatch, M. S.; Quinn, D. L.; Dorneich, M.; Doi, Yutaka)V1390,91-163(1991)

Microelectronic Interconnects and Packages: Optical and Electrical Technologies (Arjavalingam, Gnanalingam; Pazaris, James, eds.)V1389(1991)

Microelectronic Interconnects and Packages: System and Process Integration (Carruthers, John R.; Tewksbury, Stuart K., eds.)V1390(1991)

Miniature HiBi current sensor (Chu, W. W.; McStay, Daniel; Rogers, Alan J.)V1572,523-527(1991)

Mixed-type optical fiber current sensor (Wu, Gengsheng; Yang, Fan; Li, Wen; Liao, Yan-Biao)V1572,497-502(1991)

Modeling progress and trends in electrical interconnects (Prince, John L.; Cangellaris, Andreas C.; Palusinski, Olgierd A.)V1390,271-285(1991)

Multidisciplinary analysis and design of printed wiring boards (Fulton, Robert E.; Hughes, Joseph L.; Scott, Waymond R.; Umeagukwu, Charles; Yeh, Chao-Pin)V1389,144-155(1991)

Novel microstructures for low-distortion chip-to-chip interconnects (Blennemann, Heinrich C.; Pease, R. F.)V1389,215-235(1991)

Optical fiber magneto-optic current sensor (Zhao, Huafeng; Zhang, Peng-Gang; Liao, Yan-Biao)V1572,503-507(1991)

Optimization of pulsed laser power supply system (Alci, Mustafa; Yilbas, Bekir S.; Danisman, Kenan; Ciftlikli, Cebrail; Altuner, Mehmet)V1411,100-106(1991)

Optimum design of transducer for fiber-optic magnetometer (Zou, Kun; Cheng, Yuqi)V1572,472-476(1991)

Packaging technology for GaAs MMIC (monolithic microwave integrated circuits) modules (Tomimuro, Hisashi)V1390,214-222(1991)

Packaging technology for the NEC SX-3 supercomputers (Murano, Hiroshi; Watari, Toshihiko)V1390,78-90(1991)

Partitioning: the payoff role in telecommunication system design (Wilson, Donald K.)V1390,537-547(1991)

Projection direct imaging for high-density interconnection and printed circuit manufacture (Bergstrom, Neil G.)V1390,509-512(1991)

Properties of liquid-nitrogen-cooled electronic elements (Trojnar, Eugeniusz; Trojanowski, Stanislaw; Czechowicz, Roman; Derwiszynski, Mariusz; Kocyba, Krzysztof)V1391,238-243(1991)

Quick prototyping center for hybrid-wafer-scale integration (HWSI) multichip modules (Chandra, S.; Lee, Yung-Cheng)V1390,548-559(1991)

Recent progress in artificial vision (Normann, Richard A.)V1423,40-45(1991)

Sensitivity study of printed wiring board plated-through-holes copper barrel voids (Yeh, Chao-Pin; Umeagukwu, Charles; Fulton, Robert E.; Teat, William)V1389,187-198(1991)

Shot noise limited optical measurements at baseband with noisy lasers (Hobbs, Philip C.)V1376,216-221(1991)

Skin effect in high-speed ULSI/VLSI packages (Hwang, Lih-Tyng; Turlik, Iwona)V1390,249-260(1991)

Standardization efforts for the preservation of electronic imagery (Adelstein, Peter Z.; Storm, William D.)VCR37,159-179(1991)

Standards for electronic imaging for facsimile systems (Urban, Stephen J.)VCR37,113-145(1991)

Standards for electronic imaging for graphic arts systems (Dunn, S. T.; Dunn, Patrice M.)VCR37,98-112(1991)

Standards for Electronic Imaging Systems (Nier, Michael; Courtot, Marilyn E., eds.)VCR37(1991)

Standards for flat panel display systems (Greeson, James C.)VCR37,146-158(1991)

State-of-the-art multichip modules for avionics (Hagge, John K.)V1390,175-194(1991)

Switching noise in a medium-film copper/polyimide multichip module (Sandborn, Peter A.; Hashemi, Seyed H.; Weigler, William)V1389,177-186(1991)

System interconnection of high-density multichip modules (Krusius, J. P.)V1390,261-270(1991)

System interconnect issues for subnanosecond signal transmission (Moresco, Larry L.)V1390,202-213(1991)

System issues for multichip packaging (Sage, Maurice G.; Hartley, Neil)V1390,302-310(1991)

Temperature compensation of a highly birefringent optical fiber current sensor (McStay, Daniel; Chu, W. W.; Rogers, Alan J.)V1584,118-123(1991)

Termination for minimal reflection of high-speed pulse propagation along multiple-coupled microstrip lines (Kuo, Jen-Tsai; Tzuang, Ching-Kuang C.)V1389,156-160(1991)

Theoretical analysis and design on optical fiber magneto-optic current sensing head (Zhang, Peng-Gang; Zhao, Huafeng; Liao, Yan-Biao)V1572,528-533(1991)

Three-dimensional capacitance modeling of advanced multilayer interconnection technologies (Edelstein, Daniel C.)V1389,352-360(1991)

Thyratron-switched, L-C inverter, prepulse-sustainer, laser discharge circuit (Pacala, Thomas J.; Tranis, Art; Laudenslager, James B.; Kinley, Fred G.)V1411,69-79(1991)

Transmission of high-speed voltage waves at the junction of three transmission lines (Sakagami, Iwata; Kaji, Akihiro)V1389,329-339(1991)

Transparent phase-shifting mask with multistage phase shifter and comb-shaped shifter (Watanabe, Hisashi; Todokoro, Yoshihiro; Hirai, Yoshihiko; Inoue, Morio)V1463,101-110(1991)

V-line: a new interconnect for packaging and microwave applications (Schutt-Aine, Jose E.; Lee, Jin-Fa)V1389,138-143(1991)

Wafer scale integration modular packaging (Brewer, Joe E.; French, Larry E.)V1390,164-174(1991)

Electro-optics

A/D conversion of microwave signals using a hybrid optical/electronic technique (Bell, John A.; Hamilton, Michael C.; Leep, David A.; Taylor, Henry F.; Lee, Y.-H.)V1476,326-329(1991)

Airborne electro-optical sensor: performance predictions and design considerations (Mishra, R. K.; Pillai, A. M.; Sheshadri, M. R.; Sarma, C. G.)V1482,138-145(1991)

Airborne Reconnaissance XV (Augustyn, Thomas W.; Henkel, Paul A., eds.)V1538(1991)

Application of electro-optic modulator in photomechanics (Zhang, Yuan P.)V1554B,669-678(1991)

CCD performance model for airborne reconnaissance (Donn, Matthew; Waeber, Bruce)V1538,189-200(1991)

Comparison of Kodak Professional Digital Camera System images to conventional film, still video, and freeze-frame images (Kent, Richard A.; McGlone, John T.; Zoltowski, Norbert W.)V1448,27-44(1991)

Developmental test and evaluation plans for the advanced tactical air reconnaissance system (Minor, John L.; Jenquin, Michael J.)V1538,18-39(1991)

Dielectric relaxation studies of x2 dye containing polystyrene films (Schen, Michael A.; Mopsik, Fred)V1560,315-325(1991)

Droplet-size polydispersity in polymer-dispersed liquid-crystal films (Vaz, Nuno A.; Smith, George W.; VanSteenkiste, T. H.; Montgomery, G. P.)V1455,110-122(1991)

Dye orientation in organic guest host systems on ferroelectric polymers (Osterfeld, Martin; Knabke, Gerhard; Franke, Hilmar)V1559,49-55(1991)

Electrical and nonlinear optical properties of zirconium phosphonate multilayer assemblies (Katz, Howard E.; Schilling, M. L.; Ungashe, S.; Putvinski, T. M.; Scheller, G. E.; Chidsey, C. E.; Wilson, William L.)V1560,370-376(1991)

Electro-optical autostereoscopic displays using large cylindrical lenses (Hattori, Tomohiko)V1457,283-289(1991)

Electro-optical effects in semiconductor superlattices (Voos, Michel; Voisin, Paul)V1361,416-423(1991)

Electro-optical light modulation in novel azo-dye-substituted poled polymers (Shuto, Yoshito; Amano, Michiyuki; Kaino, Toshikuni)V1560,184-195(1991)

Electro-optical transducer employing liquid crystal target for processing images in real-time scale (Ignatosyan, S. S.; Simonov, V. P.; Stepanov, Boris M.)V1358,100-108(1991)

Electro-optic coefficients in electric-field poled-polymer waveguides (Smith, Barton A.; Herminghaus, Stephan; Swalen, Jerome D.)V1560,400-405(1991)

Electro-optic illuminating module (Pesl, Ales A.)V1454,299-305(1991)

Electro-optic measurements of dye/polymer systems (Wang, Chin H.; Guan, H. W.; Zhang, J. F.)V1559,39-48(1991)

Electro-optic modulator for high-speed Nd:YAG laser communication (Petsch, Thomas)V1522,72-82(1991)

Electro-optic resonant modulator for coherent optical communication (Robinson, Deborah L.; Chen, Chien-Chung; Hemmati, Hamid)V1417,421-430(1991)

Electrochromic properties of poly(pyrrole)/dodecylbenzenesulfonate (Peres, Rosa C.; De Paoli, Marco-Aurelio; Panero, Stefania; Scrosati, Bruno)V1559,151-158(1991)

Electronic f-theta correction for hologon deflector systems (Whitman, Tony; Araghi, Mehdi N.)V1454,426-433(1991)

Experimental verification of grating theory for surface-emitting structures (Ayekavadi, Raj; Yeh, C. S.; Butler, Jerome K.; Evans, Gary A.; Stabile, Paul J.; Rosen, Arye)V1418,74-85(1991)

Graphical interface for multispectral simulation, scene generation, and analysis (Sikes, Terry L.; Kreiss, William T.)V1479,199-211(1991)

Heterostructure photosensitive memory (Manasson, Vladimir A.; Sadovnik, Lev S.; Chen, Ray T.)V1559,194-201(1991)

Impact of tactical maneuvers on EO sensor imagery (Hanson, David S.)V1538,48-63(1991)

Laptop image transmission equipment (Mocenter, Michael M.)V1538,132-139(1991)

Liquid-Crystal Devices and Materials (Drzaic, Paul S.; Efron, Uzi, eds.)V1455(1991)

LiNbO3-based multichannel electro-optical light modulators (Kiselyov, Boris S.; Mikaelian, Andrei L.; Novoselov, B. A.; Shkitin, Vladimir A.; Arkhontov, L. B.; Evtikhiev, Nickolay N.)V1621,126-137(1991)

Low-cost real-time hardware in the loop FCS performance evaluation (Cifarelli, Salvatore; Magrini, Sandro)V1482,480-490(1991)

Low-intensity conflict aircraft systems (Henkel, Paul A.)V1538,2-4(1991)

Methods and applications for intensity stabilization of pulsed and cw lasers from 257 nm to 10.6 microns (Miller, Peter J.)V1376,180-191(1991)

Microcomputer-based workstation for simulation and analysis of background and target IR signatures (Reeves, Richard C.; Schaibly, John H.)V1486,85-101(1991)

Microwave high-dynamic-range EO modulators (Pan, J. J.; Garafalo, David A.)V1371,21-35(1991)

Millimeter-wave signal generation and control using optical heterodyne techniques and electro-optic devices (Thaniyavarn, Suwat; Abbas, Gregory L.; Dougherty, William A.)V1371,250-251(1991)

Multimode approach to optical fiber components and sensors (Johnstone, Walter; Thursby, G.; Culshaw, Brian; Murray, S.; Gill, M.; McDonach, Alaster; Moodie, D. G.; Fawcett, G. M.; Stewart, George; McCallion, Kevin J.)V1506,145-149(1991)

Multiple-mode reconfigurable electro-optic switching network for optical fiber sensor array (Chen, Ray T.; Wang, Michael R.; Jannson, Tomasz; Baumbick, Robert J.)V1374,223-236(1991)

Multiplexed mid-wavelength IR long linear photoconductive focal-plane arrays (Kreider, James F.; Preis, Mark K.; Roberts, Peter C.; Owen, Larry D.; Scott, Walter M.; Walmsley, Charles F.; Quin, Alan)V1488,376-388(1991)

Multisensor analysis tool (Gerlach, Francis W.; Cook, Daniel B.)V1488,134-143(1991)

Nonlinear properties of poled-polymer films: SHG and electro-optic measurements (Morichere, D.; Dumont, Michel L.; Levy, Yves; Gadret, G.; Kajzar, Francois)V1560,214-225(1991)

Novel supertwisted nematic liquid-crystal-display operating modes and electro-optical performance of generally twisted nematic configurations (Schadt, Martin)V1455,214-224(1991)

Optical delay tester (Wakana, Shin-ichi; Nagai, Toshiaki; Hama, Soichi; Goto, Yoshiro)V1479,283-290(1991)

Optically nonlinear polymeric devices (Moehlmann, Gustaaf R.; Horsthuis, Winfried H.; Mertens, Hans W.; Diemeer, Mart B.; Suyten, F. M.; Hendriksen, B.; Duchet, Christian; Fabre, P.; Brot, Christian; Copeland, J. M.; Mellor, J. R.; van Tomme, E.; van Daele, Peter; Baets, Roel G.)V1560,426-433(1991)

Optical sampling and demultiplexing applied to A/D conversion (Bell, John A.; Hamilton, Michael C.; Leep, David A.)V1562,276-280(1991)

Optical second-harmonic generation by polymer-dispersed liquid-crystal films (Yuan, Haiji J.; Li, Le; Palffy-Muhoray, Peter)V1455,73-83(1991)

Organic electro-optic devices for optical interconnnection (Lipscomb, George F.; Lytel, Richard S.; Ticknor, Anthony J.; Van Eck, Timothy E.; Girton, Dexter G.; Ermer, Susan P.; Valley, John F.; Kenney, John T.; Binkley, E. S.)V1560,388-399(1991)

Organic salts with large electro-optic coefficients (Perry, Joseph W.; Marder, Seth R.; Perry, Kelly J.; Sleva, E. T.; Yakymyshyn, Christopher P.; Stewart, Kevin R.; Boden, Eugene P.)V1560,302-309(1991)

Photon-induced charge separation in molecular systems studied by time-resolved microwave conductivity: molecular optoelectric switches (Warman, John M.; Jonker, Stephan A.; de Haas, Matthijs P.; Verhoeven, Jan W.; Paddon-Row, Michael N.)V1559,159-170(1991)

Photopolymer Device Physics, Chemistry, and Applications II (Lessard, Roger A., ed.)V1559(1991)

Poled polyimides as thermally stable electro-optic polymer (Wu, Jeong W.; Valley, John F.; Stiller, Marc A.; Ermer, Susan P.; Binkley, E. S.; Kenney, John T.; Lipscomb, George F.; Lytel, Richard S.)V1560,196-205(1991)

Polymer waveguide systems for nonlinear and electro-optic applications (Pantelis, Philip; Hill, Julian R.; Kashyap, Raman)V1559,2-9(1991)

Reflection-mode polymeric interference modulators (Yankelevich, Diego; Knoesen, Andre; Eldering, Charles A.; Kowel, Stephen T.)V1560,406-415(1991)

Schottky diode silicon liquid-crystal light valve (Sayyah, Keyvan; Efron, Uzi; Forber, Richard A.; Goodwin, Norman W.; Reif, Philip G.)V1455,249-254(1991)

Selection devices for field-sequential stereoscopic displays: a brief history (Lipton, Lenny)V1457,274-282(1991)

Signal processor for space-based visible sensing (Anderson, James C.; Downs, G. S.; Trepagnier, Pierre C.)V1479,78-92(1991)

Slapper detonator flyer microphotography with a multiframe Kerr cell and Cranz-Schardin camera (McDaniel, Olin K.)V1358,1164-1179(1991)

Sled tracking system (Downey, George A.; Fountain, H. W.; Riding, Thomas J.; Eggleston, James; Hopkins, Michael; Adams, Billy)V1482,40-47(1991)

Static and dynamic measurements using electro-optic holography (Pryputniewicz, Ryszard J.)V1554B,790-798(1991)

Study of electronic shearing speckle technique (Qin, Yuwen; Wang, Jinqi; Ji, Xinhua)V1554A,739-746(1991)

Surface field measurement of photoconductive power switches using the electro-optic Kerr effect (Sardesai, Harshad P.; Nunnally, William C.; Williams, Paul F.)V1378,237-248(1991)

Tactical reconnaissance mission survivability requirements (Lareau, Andy G.; Collins, Ross)V1538,81-98(1991)

Thin films of solid electrolytes and studies of their surface (Pan, Xiao R.; Gu, Zhi X.)V1519,85-89(1991)

What is MRT and how do I get one? (Hoover, Carl W.; Webb, Curtis M.)V1488,280-288(1991)

Ellipsometry

Analysis and control of semiconductor crystal growth with reflectance-difference spectroscopy and spectroellipsometry (Aspnes, David E.)V1361,551-561(1991)

Ellipsometric studies of the optical anisotropy of GdBa2Cu3O7-x epitaxial films (Liu, Ansheng; Keller, Ole)V1512,226-231(1991)

High-speed digital ellipsometer for the study of fiber optic sensor systems (Saxena, Indu)V1367,367-373(1991)

Nanosecond time-resolved natural and magnetic circular dichroism spectroscopy of protein dynamics (Goldbeck, Robert A.; Bjorling, Sophie; Kliger, David S.)V1432,14-27(1991)

New syndioregic main-chain, nonlinear optical polymers, and their ellipsometric characterization (Lindsay, Geoffrey A.; Nee, Soe-Mie F.; Hoover, James M.; Stenger-Smith, John D.; Henry, Ronald A.; Kubin, R. F.; Seltzer, Michael D.)V1560,443-453(1991)

Selected Papers on Ellipsometry (Azzam, Rasheed M., ed.)VMS27(1991)

Study of PEO on LTI carbon surfaces by ellipsometry and tribometry (Wang, Jinyu; Stroup, Eric; Wang, Xing F.; Andrade, Joseph D.)V1519,835-841(1991)

Thickness measurement of combined a-Si and Ti films on c-Si using a monochromatic ellipsometer (Yoo, Chue-San; Jans, Jan C.)V1464,393-403(1991)

Emission—see also spectroscopy

Beam divergence from sharp emitters in a general longitudinal magnetic field (Lau, Yue Y.; Colombant, Denis G.; Pilloff, Mark D.)V1552,182-184(1991)

Cavity-QED-enhanced spontaneous emission and lasing in liquid droplets (Campillo, Anthony J.; Eversole, J. D.; Lin, H.-B.; Merritt, C. D.)V1497,78-89(1991)

Comparative analysis of external factors' influences on the GaP light-emitting p-n-junctions (Rizikov, Igor V.; Svechnikov, Georgy S.; Bulyarsky, Sergey V.; Ambrozevich, Alexander S.)V1362,664-673(1991)

Determination of the temperature of a single particle heated by a highly concentrated laser beam (Herve, Philippe; Bednarczyk, Sophie; Masclet, Philippe)V1487,387-395(1991)

Developing a trial burn plan (Smith, Walter S.; Wong, Tony; Williams, Gary L.; Brintle, David G.)V1434,14-25(1991)

Earthlimb emission analysis of spectral infrared rocket experiment data at 2.7 micrometers: a ten-year update (Sharma, Ramesh D.; Healey, Rebecca J.)V1540,314-320(1991)

Effect of the glass composition on the emission band of erbium-doped active fibers (Cognolato, Livio; Gnazzo, Angelantonio; Sordo, Bruno; Cocito, Guiseppe)V1579,249-256(1991)

Efficient population of low-vibrational-number electronic states of excimer molecules: the argon dimer (Efthimiopoulos, Tom)V1503,430-437(1991)

Electronic structure of Ge(001) 2x1 by different angle-resolved photoemission techniques: EDC, CFS and CIS (Kipp, Lutz; Manzke, Recardo; Skibowski, Michael)V1361,794-801(1991)

Emission of the 1.54um Er-related peaks by impact excitation of Er atoms in InP and its characteristics (Isshiki, Hideo; Kobayashi, Hitoshi; Yugo, Shigemi; Saito, Riichiro; Kimura, Tadamasa; Ikoma, Toshiaki)V1361,223-227(1991)

Emissivity of silicon wafers during rapid thermal processing (Vandenabeele, Peter; Maex, Karen)V1393,316-336(1991)

Environmental Sensing and Combustion Diagnostics (Santoleri, Joseph J., ed.)V1434(1991)

Evolution of a space-charge layer, its instability, and ignition of arc gas discharge under photoemission from a target into a gas (Meshalkin, E. A.)V1440,211-221(1991)

External factors' influences on AIIIBV light-emitting structures (Svechnikov, Georgy S.; Rizikov, Igor V.)V1362,674-683(1991)

Extractive sampling systems for continuous emissions monitors (White, John R.)V1434,104-112(1991)

Far-IR magneto-emission study of the quantum-hall state and breakdown of the quantum-hall effect (Raymond, Andre; Chaubet, C.; Razeghi, Manijeh)V1362,275-281(1991)

Further observations of vectorial effects in the x-ray photoemission from caesium iodide (Fraser, George W.; Lees, John E.; Pearson, James F.)V1343,438-456(1991)

Gaseous incinerator emissions analysis by FTIR (Fourier transform infrared) spectroscopy (Herget, William F.; Demirgian, Jack)V1434,39-45(1991)

High-emittance surfaces for high-temperature space radiator applications (Banks, Bruce A.; Rutledge, Sharon K.; Hotes, Deborah)V1330,66-77(1991)

Hot carrier photoeffects in inhomogeneous semiconductors and their applications to light detectors (Amosova, L. P.; Marmur, I. Y.; Oksman, Ya. A.; Ashmontas, S.; Gradauskas, I.; Shirmulis, E.)V1440,406-413(1991)

Incinerator technology overview (Santoleri, Joseph J.)V1434,2-13(1991)

Indirect stimulated emission at room temperature in the visible range (Rinker, Michael; Kalt, Heinz; Lu, Yin-Cheng; Bauser, Elizabeth; Koehler, Klaus)V1362,14-23(1991)

Influence of photon energy on the photoemission from ultrafast electronic materials (Ghatak, Kamakhya P.; Ghoshal, Ardhendhu; Bhattacharyya, Sankar; Mondal, Manabendra)V1346,471-489(1991)

Infrared monitoring of combustion (Bates, Stephen C.; Morrison, Philip W.; Solomon, Peter R.)V1434,28-38(1991)

Investigation of hot electron emission in MOS structure under avalanche conditions (Solonko, Alexander G.)V1435,360-365(1991)

Kinetic stability analysis of the extraordinary mode perturbations in a cylindrical magnetron (Uhm, Han S.; Chen, H. C.; Stark, Robert A.)V1407,113-127(1991)

Laser-produced continua for studies in the XUV (Carroll, P. K.; O'Sullivan, Gerard D.)V1503,416-427(1991)

Multicomponent analysis using established techniques (Dillehay, David L.)V1434,56-66(1991)

Multiphoton excited emission in zinc selenide and other crystals (Prokhorov, Kirill A.; Prokhorov, Alexander M.; Djibladze, Merab I.; Kekelidze, George N.; Gorelik, Vladimir S.)V1501,80-84(1991)

Nonlinear laser interactions with saltwater aerosols (Alexander, Dennis R.; Poulain, D. E.; Schaub, Scott A.; Barton, John P.)V1497,90-97(1991)

Observation of tunneling emission from a single-quantum-well using deep-level transient spectroscopy (Letartre, Xavier; Stievenard, Didier)V1361,1144-1155(1991)

Optical emission spectroscopy in diamond-like carbon film deposition by glow discharge (Zhang, Wei P.; Chen, Jing; Fang, Rong C.; Hu, Ke L.)V1519,680-682(1991)

Optical emission spectroscopy of diamond-producing plasmas (Plano, Linda S.)V1437,13-23(1991)

Photoelectric effect from CsI by polarized soft x-rays (Shaw, Ping-Shine; Church, Eric D.; Hanany, Shaul; Liu, Yee; Fleischman, Judith R.; Kaaret, Philip E.; Novick, Robert; Manzo, Giuseppe)V1343,485-499(1991)

Photoemission from periodic structure of graded superlattices under magnetic field (Ghatak, Kamakhya P.)V1545,282-293(1991)

Photoemission from quantum-confined structure of nonlinear optical materials (Ghatak, Kamakhya P.; Biswas, Shambhu N.)V1484,136-148(1991)

Photoemission under three-photon excitation in a NEA GaAs photocathode (Wang, Liming; Hou, Xun; Cheng, Zhao)V1415,120-126(1991)

Progress toward steady-state high-efficiency vircators (Poulsen, Peter; Pincosy, Phillip A.; Morrison, Jasper J.)V1407,172-182(1991)

QED theory of excess spontaneous emission noise (Milonni, Peter W.)V1376,164-169(1991)

Realistic model for battlefield fire plume simulation (Bruce, Dorothy)V1486,231-236(1991)

Soft x-ray emission characteristics from laser plasma sources (Chen, Shisheng; Xu, Zhizhan; Li, Yao-lin; Wang, Xiaofang; Qian, Aidi D.)V1552,288-295(1991)

Spontaneous emission noise reduction of a laser output by extracavity destructive interference (Diels, Jean-Claude; Lai, Ming)V1376,198-205(1991)

Surface plasmon enhanced light emission in GaAs/AlGaAs light emitting diodes (Gornik, Erich; Koeck, A.; Thanner, C.; Korte, Lutz)V1362,1-13(1991)

Tunable cyclotron-resonance laser in germanium (Kremser, Christian; Unterrainer, Karl; Gornik, Erich; Strasser, G.; Pidgeon, Carl R.)V1501,69-79(1991)

Endoscopes—see also fiber optics in medicine

Bifunctional irrigation liquid as an ideal energy converter for laser lithotripsy with nanosecond laser pulses (Reichel, Erich; Schmidt-Kloiber, Heinz; Paltauf, Guenther; Groke, Karl)V1421,129-133(1991)

Digital charge-coupled device color TV system for endoscopy (Vishnevsky, G. I.; Berezin, V. Y.; Lazovsky, L. Y.; Vydrevich, M. G.; Rivkind, V. L.; Zhemerov, B. N.)V1447,34-43(1991)

Discrimination between urinary tract tissue and urinary stones by fiber-optic-pulsed photothermal radiometry method in vivo (Daidoh, Yuichiro; Arai, Tsunenori; Suda, Akira; Kikuchi, Makoto; Komine, Yukikuni; Murai, Masaru; Nakamura, Hiroshi)V1421,120-123(1991)

Effect of surface boundary on time-resolved reflectance: measurements with a prototype endoscopic catheter (Jacques, Steven L.; Flock, Stephen T.)V1431,12-20(1991)

Endocular ophthalmoscope: miniaturization and optical imaging quality (Rol, Pascal O.; Beck, Dominik; Niederer, Peter)V1423,84-88(1991)

Endoscopic removal of PMMA in hip revision surgery with a CO2 laser (Sazy, John; Kollmer, Charles; Uppal, Gurvinder S.; Lane, Gregory J.; Sherk, Henry H.)V1424,50-50(1991)

Endoscopic tissue autofluorescence measurements in the upper aerodigestive tract and the bronchi (Braichotte, D.; Wagnieres, G.; Monnier, Philippe; Savary, M.; Bays, Roland; van den Bergh, Hubert; Chatelain, Andre)V1525,212-218(1991)

Fiber optic image guide rods as ultrathin endoscopy (Kociszewski, Longin; Pysz, Dariusz)V1420,212-217(1991)

Holographic high-resolution endoscopic image recording (Bjelkhagen, Hans I.)V1396,93-98(1991)

Holography in endoscopy: illuminating dark holes with Gabor's principle (von Bally, Gert)VIS08,335-346(1991)

Imaging in digestive videoendoscopy (Guadagni, Stefano; Nadeau, Theodore R.; Lombardi, Loreto; Pistoia, Francesco; Pistoia, Maria A.)V1420,178-182(1991)

Improved instrumentation for photodynamic fluorescence detection of cancer (Baumgartner, R.; Heil, P.; Jocham, D.; Kriegmair, M.; Stepp, Herbert; Unsoeld, Eberhard)V1525,246-248(1991)

Light distributor for endoscopic photochemotherapy (Lenz, P.)V1525,192-195(1991)

Lung imaging fluorescence endoscope: development and experimental prototype (Palcic, Branko; Lam, Stephen; MacAulay, Calum; Hung, Jacklyn; Jaggi, Bruno; Radjinia, Massud; Pon, Alfred; Profio, A. E.)V1448,113-117(1991)

Micro-optical elements in holography (von Bally, Gert; Dirksen, D.; Zou, Y.)V1507,66-72(1991)

New approach for endoscopic stereotactic brain surgery using high-power laser (Otsuki, Taisuke; Yoshimoto, Takashi)V1420,220-224(1991)

New image diagnosis system with ultrathin endoscope and clinical results (Tsumanuma, Takashi; Toriya, T.; Tanaka, T.; Shamoto, N.; Seto, K.; Sanada, Kazuo; Okazaki, A.; Okazaki, M.)V1420,193-198(1991)

Optimal fluorescence imaging of atherosclerotic human tissue (Davenport, Carolyn M.; Alexander, Andrew L.; Gmitro, Arthur F.)V1425,16-27(1991)

Photodetection of early cancer in the upper aerodigestive tract and the bronchi using photofrin II and colorectal adenocarcinoma with fluoresceinated monoclonal antibodies (Wagnieres, G.; Braichotte, D.; Chatelain, Andre; Depeursinge, Ch.; Monnier, Philippe; Savary, M.; Fontolliet, Ch.; Calmes, J.-M.; Givel, J.-C.; Chapuis, G.; Folli, S.; Pelegrin, A.; Buchegger, F.; Mach, J.-P.; van den Bergh, Hubert)V1525,219-236(1991)

Potentials for pulsed YAG:Nd laser application to endoscopic surgery (Manenkov, Alexander A.; Denisov, N. N.; Bagdasarov, V. H.; Starkovsky, A. N.; Yurchenko, S. V.; Kornilov, Yu. M.; Mikaberidze, V. M.; Sarkisov, S. E.)V1420,254-258(1991)

Practical considerations for effective microendoscopy (Papaioannou, Thanassis; Papazoglou, Theodore G.; Daykhovsky, Leon; Gershman, Alex; Segalowitz, Jacob; Reznik, G.; Beeder, Clain; Chandra, Mudjianto; Grundfest, Warren S.)V1420,203-211(1991)

Three-dimensional imaging laparoscope (Jones, Edwin R.; McLaurin, A. P.; Mason, J. L.)V1457,318-321(1991)

Trends in holographic endoscopy (Podbielska, Halina)V1429,207-213(1991)

Energy

Energy storage efficiency and population dynamics in flashlamp-pumped sensitized erbium glass laser (Lukac, Matjaz)V1419,55-62(1991)

Experimental investigation of 1.06 um laser interaction with Al target in air (Gang, Yuan)V1415,225-227(1991)

Small-capacity low-cost (Ni-H2) design concept for commercial, military, and higher volume aerospace applications (Wheeler, James R.; Cook, William D.; Smith, Ron)V1495,280-285(1991)

Environmental effects—see also pollution

Aging behavior of low-strength fused silica fibers (Yuce, Hakan H.; Key, P. L.; Chandan, Harish C.)V1366,120-128(1991)

Airborne chemical contamination of a chemically amplified resist (MacDonald, Scott A.; Clecak, Nicholas J.; Wendt, H. R.; Willson, C. G.; Snyder, C. D.; Knors, C. J.; Deyoe, N. B.; Maltabes, John G.; Morrow, James R.; McGuire, Anne E.; Holmes, Steven J.)V1466,2-12(1991)

Analysis of fiber damage and field failures in fiber-grip-type mechanical fiber optic splices (Kiss, Gabor D.; Pellegrino, Anthony; Leopold, Simon)V1366,223-234(1991)

Analysis of vegetation stress and damage from images of the high-resolution airborne pushbroom image spectrograph compact airborne spectrographic imager (Mueksch, Michaela C.)V1399,157-161(1991)

Application of surface effects externally to alter optical and electronic properties of existing optoelectronic devices: a ten-year update (Hava, Shlomo)V1540,350-358(1991)

Applications of resonance ionization spectroscopy for semiconductor, environmental and biomedical analysis, and for DNA sequencing (Arlinghaus, Heinrich F.; Spaar, M. T.; Thonnard, N.; McMahon, A. W.; Jacobson, K. B.)V1435,26-35(1991)

Are optical fiber sensors intrinsically, inherently, or relatively safe? (McGeehin, Peter)V1504,75-79(1991)

Atmospheric effects on laser systems (Au, Robert H.)V1399,8-15(1991)

Attacking dimensional instability problems in graphite/epoxy structures (Krumweide, Gary C.; Brand, Richard A.)V1533,252-261(1991)

Autonomous mobile laser complex (Fakhrutdinov, I. H.; Avdoshin, A. P.; Moshin, J. N.; Poltavsky, V. V.)V1399,98-106(1991)

Characterization of macrobend sensitivity of step index optical fibers used in intensity sensors (York, Jim F.; Nelson, Gary W.; Varshneya, Deepak)V1584,308-319(1991)

Characterization of the dimensional stability of advanced metallic materials using an optical test bench structure (Hsieh, Cheng; O'Donnell, Timothy P.)V1533,240-251(1991)

Characterization, Propagation, and Simulation of Sources and Backgrounds (Watkins, Wendell R.; Clement, Dieter, eds.)V1486(1991)

Chemical-vapor-deposited silicon and silicon carbide optical substrates for severe environments (Goela, Jitendra S.; Pickering, Michael A.; Taylor, Raymond L.)V1330,25-38(1991)

Clutter metrics in infrared target acquisition (Tidhar, Gil; Rotman, Stanley R.)V1442,310-324(1991)

Commercial crystal growth in space (Wilcox, William R.)V1557,31-41(1991)

Comparison of performances on analog fiber optic links equipped with either noncontact or physical-contact connectors (Giannini, Jean-Pierre; Delay, P.; Duval, A.; Wtodkiewiez, A.)V1366,215-222(1991)

Comparison of transient analog data fiber optic links (Clark, Wally T.; Darrow, K. W.; Skipper, M. C.)V1371,258-265(1991)

Corrosion of ZnSe by alternating high voltage in a saline solution (King, Joseph A.)V1535,216-223(1991)

CVD diamond as an optical material for adverse environments (Snail, Keith A.)V1330,46-64(1991)

Cyclic fatigue behavior of silica fiber (Rogers, Harvey N.)V1366,112-117(1991)

Design considerations for air-to-air laser communications (Casey, William L.; Doughty, Glenn R.; Marston, Robert K.; Muhonen, John)V1417,89-98(1991)

Developing a trial burn plan (Smith, Walter S.; Wong, Tony; Williams, Gary L.; Brintle, David G.)V1434,14-25(1991)

Displacement damage in Si imagers for space applications (Dale, Cheryl J.; Marshall, Paul W.)V1447,70-86(1991)

Durable, nonchanging, metal-dielectric and all-dielectric mirror coatings (Guenther, Karl H.; Balasubramanian, Kunjithapa; Hu, X. Q.)V1485,240-244(1991)

Effect of earthquake motion on the mechanical reliability of optical cables (Hart, Patrick W.; Tucker, Russ; Yuce, Hakan H.; Varachi, John P.; Wieczorek, Casey J.; DeVito, Anthony)V1366,334-342(1991)

Effect of the space environment on thermal control coatings (Harada, Yoshiro; Mell, Richard J.; Wilkes, Donald R.)V1330,90-101(1991)

Effects of atmospheric conditions on the performance of free-space infrared communications (Grotzinger, Timothy L.)V1417,484-495(1991)

Effects of environmental and installation-specific factors on process gas delivery via mass-flow controller with an emphasis on real-time behavior (Gray, David E.; Benjamin, Neil M.; Chapman, Brian N.)V1392,402-410(1991)

Effects of ionizing radiation on fiber optic systems and components for use in mobile platforms (Reddy, Mahesh C.; Krinsky, Jeffrey A.)V1369,107-113(1991)

Effects of proton damage on charge-coupled devices (Janesick, James R.; Soli, George; Elliott, Tom; Collins, Stewart A.)V1447,87-108(1991)

Effects of the lunar environment on optical telescopes and instruments (Johnson, Charles L.; Dietz, Kurtis L.)V1494,208-218(1991)

Electromagnetic enviromental effects on shipboard fiber optic installations (Bucholz, Roger C.)V1369,19-23(1991)

Environmentally insensitive commercial pressure sensor (Wlodarczyk, Marek T.)V1368,121-131(1991)

Environmental testing of a Q-switched Nd:YLF laser and a Nd:YAG ring laser (Robinson, Deborah L.)V1417,562-572(1991)

Environments stressful to optical materials in low earth orbit (Musikant, Solomon; Malloy, W. J.)V1330,119-130(1991)

Expert system for fusing weather and doctrinal information used in the intelligence preparation of the battlefield (McWilliams, Gary; Kirby, Steve; Eskridge, Thomas; Newberry, Jeff)V1468,417-428(1991)

Extended environmental performance of attitude and heading reference grade fiber optic rotation sensors (Chin, Gene H.; Cordova, Amado; Goldner, Eric L.)V1367,107-120(1991)

Far-field pattern of laser diodes as function of the relative atmospheric humidity (Freitas, Jose C.; Carvalho, Fernando D.; Rodrigues, F. C.; Abreu, M. A.; Marcal, Joao P.)V1399,16-23(1991)

Fiber construction for improved mechanical reliabiltiy (Roberts, Daniel R.; Cuellar, Enrique; Kennedy, Michael T.; Tomita, Akira)V1366,129-135(1991)

Fiber optic link performance in the presence of reflection-induced intensity noise (Radcliffe, Jerry K.)V1366,361-371(1991)

Fiber optic network for mining seismology (Lach, Zbigniew; Zientkiewicz, Jacek K.)V1364,209-220(1991)

Fiber optics in liquid propellant rocket engine environments (Delcher, Ray C.; Dinnsen, Doug K.; Barkhoudarian, S.)V1369,114-120(1991)

Fiber optics network for the adverse coal mining environment (Zientkiewicz, Jacek K.; Lach, Zbigniew)V1366,45-56(1991)

Fiber Optics Reliability: Benign and Adverse Environments IV (Greenwell, Roger A.; Paul, Dilip K., eds.)V1366(1991)

Fiber Optic Systems for Mobile Platforms IV (Lewis, Norris E.; Moore, Emery L., eds.)V1369(1991)

Finite element analysis of deformation in large optics due to space environment radiation (Merzbacher, Celia I.; Friebele, E. J.; Ruller, Jacqueline A.; Matic, P.)V1533,222-228(1991)

Flight experiment to investigate microgravity effects on solidification phenomena of selected materials (Maag, Carl R.; Hansen, Patricia A.)V1557,24-30(1991)

Free-Space Laser Communication Technologies III (Begley, David L.; Seery, Bernard D., eds.)V1417(1991)

Fungal testing of diode laser collimators (de Lourdes Quinta, Maria; Freitas, Jose C.; Rodrigues, F. C.; Silva, Jorge A.)V1399,24-29(1991)

Generation of realistic IR images of tactical targets in obscured environments (Greenleaf, William G.; Siniard, Sheri M.; Tait, Mary B.)V1486,364-375(1991)

High-impact shock testing of fiber optic components (Brown, Gair D.; Ingold, Joseph P.; Spence, Scott E.; Paxton, Jack G.)V1366,351-360(1991)

Humidity dependence of ceramic substrate electroluminescent devices (Young, Richard; Kitai, Adrian H.)V1398,71-80(1991)

Ignition risks of fiber optic systems (Hills, Peter C.; Samson, Peter J.)V1589,110-119(1991)

Impact of relative humidity on mechanical test results for optical fiber (Dreyer, Donald R.; Saikkonen, Stuart L.; Hanson, Thomas A.; Linchuck, Barry A.)V1366,372-379(1991)

Influence of dose rate on radiation-induced loss in optical fibers (Henschel, Henning; Koehn, Otmar; Schmidt, Hans U.)V1399,49-63(1991)

Integrated optics displacement sensor (d'Alessandro, Antonio; De Sario, Marco; D'Orazio, Antonella; Petruzzelli, Vincenzo)V1366,313-323(1991)

Ionic effects on silica optical fiber strength and models for fatigue (Rondinella, Vincenzo V.; Matthewson, M. J.)V1366,77-84(1991)

Ionizing radiation-induced attenuation in optical fibers at multiple wavelengths and temperature extremes (Krinsky, Jeffrey A.; Reddy, Mahesh C.)V1366,191-203(1991)

Large zeolites: why and how to grow in space (Sacco, Albert)V1557,6-9(1991)

Laser-induced volatilization and ionization of aerosol particles for their mass spectral analysis in real time (Sinha, Mahadeva P.)V1437,150-156(1991)

Laser sensing in the iron-making blast furnace (Scott, Chris J.)V1399,137-144(1991)

Long-term performance of fiber optic cable plant in Navy ships (Brown, Gair D.; Ingold, Joseph P.; Paxton, Jack G.)V1589,58-68(1991)

Long-term reliability tests on fused biconical taper couplers (Moore, Douglas R.)V1366,241-250(1991)

Lunar dust: implications for astronomical observatories (Johnson, Stewart W.; Chua, Koon M.; Burns, Jack O.; Slane, Frederic A.)V1494,194-207(1991)

Measuring residual accelerations in the Spacelab environment (Witherow, William K.)V1557,42-52(1991)

Mechanical performance and reliability of Corning Titan SMF CPC5 fiber after exposure to a variety of environments (Vethanayagam, Thirukumar K.; Linchuck, Barry A.; Dreyer, Donald R.; Toler, J. R.; Amos, Lynn G.; Taylor, Donna L.)V1366,343-350(1991)

Mechanical testing and reliability considerations of fusion splices (Wei, Ta-Sheng; Yuce, Hakan H.; Varachi, John P.; Pellegrino, Anthony)V1366,235-240(1991)

Method and device that prevent target sensors from being radiation overexposed in the presence of a nuclear blast (Holubowicz, Kazimierz S.)V1456,274-285(1991)

Misalignments of airborne laser beams due to mechanical vibrations (Freitas, Jose C.; Abreu, M. A.; Rodrigues, F. C.; Carvalho, Fernando D.)V1399,42-48(1991)

Modeling time-dependent obscuration for simulated imaging of dust and smoke clouds (Hoock, Donald W.)V1486,164-175(1991)

Modifications of optical properties with ceramic coatings (Besmann, Theodore M.; Abdel-Latif, A. I.)V1330,78-89(1991)

Moisture- and water-induced crack growth in optical materials (Cranmer, David C.; Freiman, Stephen W.; White, Grady S.; Raynes, Alan S.)V1330,152-163(1991)

NASA microgravity materials science program (Sokolowski, Robert S.)V1557,2-5(1991)

Neutron irradiation effects on fibers operating at 1.3 um and 1.55 um (Singh, Anjali; Banerjee, Pranab K.; Mitra, Shashanka S.)V1366,184-190(1991)

Next generation thermal control coatings (Grieser, James L.; Swisher, Richard L.; Phipps, James A.; Pelleymounter, Douglas R.; Hildreth, Eugene N.)V1330,111-118(1991)

Novel fiber optic coupler/repeater for military systems (Glista, Andrew S.)V1369,24-34(1991)

Optical cable reliability: lessons learned from post-mortem analyses (Kilmer, Joyce P.; DeVito, Anthony; Yuce, Hakan H.; Wieczorek, Casey J.; Varachi, John P.; Anderson, William T.)V1366,85-91(1991)

Optical characterization of photolithographic metal grids (Osmer, Kurt A.; Jones, Mike I.)V1498,138-146(1991)

Optical computer-assisted tomography realized by coherent detection imaging incorporating laser heterodyne method for biomedical applications (Inaba, Humio; Toida, Masahiro; Ichimura, Tsutomu)V1399,108-115(1991)

Optical fiber radiation damage measurements (Ediriweera, Sanath R.; Kvasnik, Frank)V1399,64-75(1991)

Optically triggered GaAs thyristor switches: integrated structures for environmental hardening (Carson, Richard F.; Weaver, Harry T.; Hughes, Robert C.; Zipperian, Thomas E.; Brennan, Thomas M.; Hammons, B. E.)V1378,84-94(1991)

Optical materials for use under extreme service conditions (Glebov, Leonid B.; Petrovskii, Gurii T.; Tshavelev, Oleg S.)V1399,200-206(1991)

Optical space-to-ground link availability assessment and diversity requirements (Chapman, William W.; Fitzmaurice, Michael W.)V1417,63-74(1991)

Optical stability of diffuse reflectance materials in space (Hale, Robert R.)V1485,173-182(1991)

Optics in adverse environments (Macleod, H. A.)V1399,2-6(1991)

Optical Surfaces Resistant to Severe Environments (Musikant, Solomon, ed.)V1330(1991)

Optical Systems in Adverse Environments (Kuok, M. H.; Silva, Donald E.; Tam, Siu-Chung, eds.)V1399(1991)

Overview of fiber optics in the natural space environment (Barnes, Charles E.; Dorsky, Leonard; Johnston, Alan R.; Bergman, Larry A.; Stassinopoulos, E.)V1366,9-16(1991)

Pod-mounted MIL-STD-2179B recorder (Kessler, William D.; Abeille, Pierre; Sulzer, Jean-Francois)V1538,104-111(1991)

Polarimetric sensor strain sensitivity in different thermal operating conditions (De Maria, Letizia; Escobar Rojo, Priscilla; Martinelli, Mario; Pistoni, Natale C.)V1366,304-312(1991)

Process control capability using a diaphragm photochemical dispense system (Cambria, Terrell D.; Merrow, Scott F.)V1466,670-675(1991)

Production of laser simulation systems for adverse environments (Rodrigues, F. C.; Simao, Jose V.; Oliveira, Joao; Freitas, Jose C.; Carvalho, Fernando D.)V1399,90-97(1991)

Pulsed electron-beam testing of optical surfaces (Murray, Brian W.; Johnson, Edward A.)V1330,2-24(1991)

Radiation concerns for the Solar-A soft x-ray telescope (Acton, Loren W.; Morrison, Mons D.; Janesick, James R.; Elliott, Tom)V1447,123-139(1991)

Radiation effects on GaAs optical system FET devices (Kanofsky, Alvin S.; Spector, Magaly; Remke, Ronald L.; Witmer, Steve B.)V1374,48-58(1991)

Radiation effects on bend-insensitive fibers at 1300 nm and 1500 nm (Karbassiyoon, Kamran; Greenwell, Roger A.; Scott, David M.; Spencer, Robert A.)V1366,178-183(1991)

Radiation effects on dynamical behavior of LiNbO3 switching devices (Kanofsky, Alvin S.; Minford, William J.; Watson, James E.)V1374,59-66(1991)

Radiation effects on various optical components for the Mars Observer Spacecraft (Lowry, Jay H.; Iffrig, C. D.)V1330,132-141(1991)

Realistic model for battlefield fire plume simulation (Bruce, Dorothy)V1486,231-236(1991)

Real-time automatic inspection under adverse conditions (Carvalho, Fernando D.; Correia, Fernando C.; Freitas, Jose C.; Rodrigues, F. C.)V1399,130-136(1991)

Recording and analysis of (high-frequency) sinusoidal vibrations using computerized TV-holography (Ellingsrud, Svein; Lokberg, Ole J.; Pedersen, Hans M.)V1399,30-41(1991)

Reflectance stability analysis of Spectralon diffuse calibration panels (Bruegge, Carol J.; Stiegman, Albert E.; Coulter, Daniel R.; Hale, Robert R.; Diner, David J.; Springsteen, Arthur W.)V1493,132-142(1991)

Reliability assurance of optoelectronic devices in the local loop (Koelbl, Roy S.)V1366,17-25(1991)

Reliability improvement methods for sapphire fiber temperature sensors (Schietinger, Chuck W.; Adams, Bruce)V1366,284-293(1991)

Reliability of fiber optic position sensors (Park, Eric D.; Swafford, William J.; Lamb, Bryan K.)V1366,294-303(1991)

Reliability of planar optical couplers (Matthews, James E.; Cucalon, Antoine)V1366,206-214(1991)

Remote colorimetry and its applications (Sheffer, Dan; Ben-Shalom, Ami; Devir, Adam D.)V1493,232-243(1991)

Removal of adsorbed gases with CO_2 snow (Zito, Richard R.)V1494,427-433(1991)

Scatter and roughness measurements on optical surfaces exposed to space (Schmitt, Dirk-Roger; Swoboda, Helmut; Rosteck, Helmut)V1530,104-110(1991)

Sensitivity analysis of Navy tactical decision-aid FLIR performance codes (Milne, Edmund A.; Cooper, Alfred W.; Reategui, Rodolfo; Walker, Philip L.)V1486,151-161(1991)

Sensitivity of polarization-maintaining fibers to temperature variations (Ruffin, Paul B.; Sung, C. C.)V1478,160-167(1991)

Simulation study to characterize thermal infrared sensor false alarms (Sabol, Bruce M.; Mixon, Harold D.)V1486,258-270(1991)

Soft x-ray, optical, and thermal properties of hard carbon films (Alvey, Mark D.)V1330,39-45(1991)

Solid state lasers for field application (Motenko, Boris; Ermakov, Boris A.; Berezin, Boris)V1399,78-81(1991)

Space experiments using small satellites (Schor, Matthew J.)V1495,146-148(1991)

Space station atomic-oxygen-resistant coatings (Grieser, James L.; Freeland, Alan W.; Fink, Jeffrey D.; Meinke, Gary E.; Hildreth, Eugene N.)V1330,102-110(1991)

Strength and fatigue of optical fibers at different temperatures (Biswas, Dipak R.)V1366,71-76(1991)

Studies on defocus in thermal imaging systems (Venkateswara Rao, B.)V1399,145-156(1991)

Target acquisition model appropriate for dynamically changing scenarios (Rotman, Stanley R.; Gordon, E. S.)V1442,335-346(1991)

Temperature stress testing of laser modules for the uncontrolled environment (Su, Pin)V1366,94-106(1991)

Theoretical limits of dimensional stability for space structures (Dolgin, Benjamin P.; Moacanin, Jovan; O'Donnell, Timothy P.)V1533,229-239(1991)

Theory of stressed fiber lifetime calculations (Kapron, Felix P.)V1366,136-143(1991)

Thermal analysis of a small expendable tether satellite package (Randorf, Jeffrey A.)V1495,259-267(1991)

Thermal and radiometric modeling of terrain backgrounds (Conant, John A.; Hummel, John R.)V1486,217-230(1991)

Use of Fokker-Planck equations for the statistical properties of laser light (Jung, Peter; Risken, H.)V1376,82-93(1991)

Use of fatigue measurements for fiber lifetime prediction (Yuce, Hakan H.; Kapron, Felix P.)V1366,144-156(1991)

Wind tunnel model aircraft attitude and motion analysis (Mostafavi, Hassan)V1483,104-111(1991)

You can't just plug it in: digital image networks/picture archiving and communication systems installation (Gelish, Anthony)V1444,363-372(1991)

Epitaxy—see also microelectronic processing

Advanced InGaAs/InP p-type pseudomorphic MODFET (Malzahn, Eric; Heuken, Michael; Gruetzmacher, Dettev; Stollenwerk, M.; Heime, Klaus)V1362,199-204(1991)

Advanced Si IR detectors using molecular beam epitaxy (Lin, TrueLon; Jones, E. W.; George, T.; Ksendzov, Alexander; Huberman, M. L.)V1540,135-139(1991)

Application of epitaxial lift-off to optoelectronic material studies (Price, Garth L.; Usher, Brian F.)V1361,543-550(1991)

Applications of GaAs grade-period doping superlattice for negative-differential-resistance device (Liu, W. C.; Sun, C. Y.; Lour, W. S.; Guo, D. F.; Lee, Y. S.)V1519,640-644(1991)

Characterization of GaAs thin films grown by molecular beam epitaxy on Si-on-insulator (Zhu, Wen H.; Lin, Cheng L.; Yu, Yue H.; Li, Aizhen; Zou, Shi C.; Hemment, Peter L.)V1519,423-427(1991)

Control of oxygen incorporation and lifetime measurement in $Si_{1-x}Ge_x$ epitaxial films grown by rapid thermal chemical vapor deposition (Sturm, James C.; Schwartz, P. V.; Prinz, Erwin J.; Magee, Charles W.)V1393,252-259(1991)

Current state of gas-source molecular beam epitaxy for growth of optoelectronic materials (Pessa, Markus; Hakkarainen, T.; Keskinen, Jari; Rakennus, K.; Salokatve, A.; Zhang, G.; Asonen, Harry M.)V1361,529-542(1991)

Differentiation of the nonradiative recombination properties of the two interfaces of molecular beam epitaxy grown GaAs-GaAlAs quantum wells (Sermage, Bernard; Gerard, Jean M.; Bergomi, Lorenzo; Marzin, Jean Y.)V1361,131-135(1991)

Epitaxial growth and photoluminescence investigations of InP/InAs quantum well grown by hydride vapor phase epitaxy (Banvillet, Henri; Gil, E.; Vasson, A. M.; Cadoret, R.; Tabata, A.; Benyattou, Taha; Guillot, Gerard)V1361,972-979(1991)

Epitaxial growth of gallium arsenide from elemental arsenic (Chu, Ting L.; Chu, Shirley S.; Green, Richard F.; Cerny, C. L.)V1361,523-528(1991)

Epitaxial growth of the semiconducting silicide FeSi2 on silicon (Chevrier, Joel S.; Thanh, V. L.; Derrien, J.)V1512,278-288(1991)

Epitaxial liftoff technology (Yablonovitch, Eli)V1563,8-9(1991)

Epitaxial regrowth of silicon on sapphire by rapid isothermal processing (Madarazo, R.; Pedrine, A. G.; Sol, A. A.; Baranauskas, Vitor)V1393,270-277(1991)

Excitonic photoabsorption study of AlGaAs/GaAs multiple-quantum-well grown by low-pressure MOCVD (Kwon, O'Dae; Lee, Seung-Won; Choi, Woong-Lim; Kim, Kwang-Il; Jeong, Yoon-Ha)V1361,802-808(1991)

Fabrication of high-radiance LEDs by epitaxial lift-off (Pollentier, Ivan K.; Ackaert, A.; De Dobbelaere, P.; Buydens, Luc; van Daele, Peter; Demeester, Piet M.)V1361,1056-1062(1991)

Far-IR studies of moderately doped molecular beam epitaxy grown GaAs on Si(100) (Morley, Stefan; Eickhoff, Thomas; Zahn, Dietrich R.; Richter, W.; Woolf, D.; Westwood, D. I.; Williams, R. H.)V1361,213-222(1991)

Formation and electronic properties of epitaxial erbium silicide (Nguyen, Tan T.; Veuillen, J. Y.)V1512,289-298(1991)

Growth and characterization of ZnSe and ZnTe grown on GaAs by hot-wall epitaxy (Hingerl, Kurt; Pesek, Andreas; Sitter, Helmut; Krost, Alois; Zahn, Dietrich R.; Richter, W.; Kudlek, Gotthard; Gutowski, Juergen)V1361,383-393(1991)

Growth and characterization of ultrathin SimGen strained-layer superlattices (Presting, Hartmut; Jaros, Milan; Abstreiter, Gerhard)V1512,250-277(1991)

Growth by liquid phase epitaxy and characterization of GaInAsSb and InAsSbP alloys for mid-infrared applications (2-3 um) (Tournie, Eric; Lazzari, J. L.; Mani, Habib; Pitard, F.; Alibert, Claude L.; Joullie, Andre F.)V1361,641-656(1991)

Growth dynamics of lattice-matched and strained layer III-V compounds in molecular beam epitaxy (Joyce, Bruce A.; Zhang, J.; Foxon, C. T.; Vvedensky, D. D.; Shitara, T.; Myers-Beaghton, A. K.)V1361,13-22(1991)

Growth of CdTe-CdMnTe heterostructures by molecular beam epitaxy (Bicknell-Tassius, Robert N.)V1484,11-18(1991)

Heteroepitaxial growth of InP and GaInAs on GaAs substrates using nonhydride sources (Chu, Shirley S.; Chu, Ting L.; Yoo, C. H.; Smith, G. L.)V1361,1020-1025(1991)

Heteroepitaxy of II-VI and IV-VI semiconductors on Si substrates (Zogg, Hans; Tiwari, A. N.; Blunier, Stefan; Maissen, Clau; Masek, Jiri)V1361,406-413(1991)

High-efficiency UV and blue emitting devices prepared by MOVPE and low-energy electron-beam irradiation treatment (Akasaki, Isamu; Amano, H.)V1361,138-149(1991)

High-performance GaAs on silicon technology for VLSI, MMICs, and optical interconnects (Christou, Aristos)V1361,354-361(1991)

High-quality heavily strained II-VI quantum well (Li, Jie; He, Li; Tang, Wen G.; Shan, W.; Yuan, Shi X.)V1519,660-664(1991)

InGaAs/GaAs interdigitated metal-semiconductor-metal (IMSM) photodetectors operational at 1.3 um grown by molecular beam epitaxy (Elman, Boris S.; Chirravuri, Jagannath; Choudhury, A. N.; Silletti, Andrew; Negri, A. J.; Powers, J.)V1362,610-616(1991)

Investigation of defects in HgCdTe epi-films grown from Te solutions (Wang, Yue; Tang, Zhi J.; Zhuang, Wei S.; He, Jing F.)V1519,428-433(1991)

Laser-induced thermal desorption studies of surface reaction kinetics (George, Steven M.; Coon, P. A.; Gupta, P.; Wise, M. L.)V1437,157-165(1991)

Lattice-mismatched elemental and compound semiconductor heterostructures for 2-D and 3-D applications (Lee, El-Hang)V1362,499-509(1991)

Low-substrate temperature molecular beam epitaxy growth and thermal stability of strained InGaAs/GaAs single-quantum-wells (Elman, Boris S.; Koteles, Emil S.; Melman, Paul; Rothman, Mark A.)V1361,362-372(1991)

Magneto-optical studies of n-type Hg0.622Cd0.378Te grown by molecular beam epitaxy (Liu, Wei J.; Liu, Pu L.; Shi, Guo L.; Zhu, Jing-Bing; Yuan, Shi X.; Xie, Qin X.; He, Li)V1519,415-418(1991)

Material for future InP-based optoelectronics: InGaAsP versus InGaAlAs (Quillec, Maurice)V1361,34-46(1991)

Metal-organic molecular beam epitaxy of II-VI materials (Summers, Christopher J.; Wagner, Brent K.; Benz, Rudolph G.; Rajavel, D.)V1512,170-176(1991)

Molecular beam epitaxial growth of ZnSe-ZnS strained-layer superlattices (Shen, Ai D.; Cui, Jie; Wang, Hai L.; Wang, Zhi-Jiang)V1519,656-659(1991)

Molecular beam epitaxy/liquid phase epitaxy hybrid growth for GaAs-LED on Si (Minemura, Tetsuroh; Yazawa, Y.; Asano, J.; Unno, T.)V1361,344-353(1991)

Molecular beam epitaxy GaAs on Si: material and devices for optical interconnects (Panayotatos, Paul; Georgakilas, Alexandros; Mourrain, Jean-Loic; Christou, Aristos)V1361,1100-1109(1991)

Molecular beam epitaxy of CdTe and HgCdTe on large-area Si(100) (Sporken, R.; Lange, M. D.; Faurie, Jean-Pierre)V1512,155-163(1991)

Monolithic epitaxial IV-VI compound IR-sensor arrays on Si substrates for the SWIR, MWIR and LWIR range (Zogg, Hans; Masek, Jiri; Maissen, Clau; Hoshino, Taizo J.; Blunier, Stefan; Tiwari, A. N.)V1361,1079-1086(1991)

MOVPE technology in device applications for telecommunication (Moss, Rodney H.)V1361,1170-1181(1991)

Multialkali photocathodes grown by molecular beam epitaxy technique (Dubovoi, I. A.; Chernikov, A. S.; Prokhorov, Alexander M.; Schelev, Mikhail Y.; Ushakov, V. K.)V1358,134-138(1991)

Multilayer InSb diodes grown by molecular beam epitaxy for near-ambient temperature operation (Ashley, Timothy; Dean, A. B.; Elliott, Charles T.; Houlton, M. R.; McConville, C. F.; Tarry, H. A.; Whitehouse, Colin R.)V1361,238-244(1991)

New concept for multiwafer production of highly uniform III-V layers for optoelectronic applications by MOVPE (Heyen, Meino)V1362,146-153(1991)

New materials for high-performance III-V ICs and OEICs: an industrial approach (Martin, Gerard M.; Frijlink, P. M.)V1362,67-74(1991)

New semiconductor material AlxInAsySb/InAs: LPE synthesis and properties (Charykov, N. A.; Litvak, Alexandr M.; Moiseev, K. D.; Yakovlev, Yurii P.)V1512,198-203(1991)

Novel GaP/InP strained heterostructures: growth, characterization,and technological perspectives (Recio, Miguel; Ruiz, Ana; Melendez, J.; Rodriguez, Jose M.; Armelles, Gaspar; Dotor, M. L.; Briones, Fernando)V1361,469-478(1991)

Novel narrow-gap semiconductor systems (Stradling, R. A.)V1361,630-640(1991)

Novel optoelectronic devices and integrated circuits using epitaxial lift-off (Demeester, Piet M.; Pollentier, Ivan K.; Buydens, Luc; van Daele, Peter)V1361,987-998(1991)

O/I-MBE: formation of highly ordered phthalocyanine/semiconductor junctions by molecular-beam epitaxy: photoelectrochemical characterization (Armstrong, Neal R.; Nebesny, Ken W.; Collins, Greg E.; Lee, Paul A.; Chau, Lai K.; Arbour, Claude; Parkinson, Bruce)V1559,18-26(1991)

Optical characterization of Hg1-xCdxTe/CdTe/GaAs multilayers grown by molecular beam epitaxy (Liu, Wei J.; Liu, Pu L.; Shi, Guo L.; Zhu, Jing-Bing; He, Li; Xie, Qin X.; Yuan, Shi X.)V1519,481-488(1991)

Optical characterization of InP epitaxial layers on different substrates (Jiao, Kaili L.; Zheng, J. P.; Kwok, Hoi-Sing; Anderson, Wayne A.)V1361,776-783(1991)

Optical properties of molecular beam epitaxy grown ZnTe epilayers (Kudlek, Gotthard; Presser, Nazmir; Gutowski, Juergen; Mathine, David L.; Kobayashi, Masakazu; Gunshor, Robert L.)V1361,150-158(1991)

PbEuSeTe/Pb1-xSnxTe buried heterostructure diode lasers grown by molecular beam epitaxy (Feit, Zeev; Kostyk, D.; Woods, R. J.; Mak, Paul S.)V1512,164-169(1991)

Photovoltaic HgCdTe MWIR-detector arrays on (100)CdZnTe/(100)GaAs grown by hot-wall-beam epitaxy (Gresslehner, Karl-Heinz; Schirz, W.; Humenberger, Josef; Sitter, Helmut; Andorfer, J.; Lischka, Klaus)V1361,1087-1093(1991)

Production considerations necessary to produce large quantities of optoelectronic devices by MOCVD epitaxy (Boldish, Steven I.)V1449,51-64(1991)

Properties of ZnSe/ZnS grown by MOVPE on a rotating substrate (Soellner, Joerg; Heuken, Michael; Heime, Klaus)V1361,963-971(1991)

Recent progress in device-oriented II-VI research at the University of Wuerzburg (Landwehr, Gottfried; Waag, Andreas; Hofmann, K.; Kallis, N.; Bicknell-Tassius, Robert N.)V1362,282-290(1991)

Recent progress on research of materials for optoelectronic device applications in China (Chen, Liang-Hui; Kong, Mei-Ying; Wang, Yi-Ming)V1361,60-73(1991)

Selectively grown InxGa1-xAs and InxGa1-xP structures: locally resolved stoichiometry determination by Raman spectroscopy (Geurts, Jean; Finders, J.; Kayser, O.; Opitz, B.; Maassen, M.; Westphalen, R.; Balk, P.)V1361,744-750(1991)

Si-based epitaxial growth by rapid thermal processing chemical vapor deposition (Jung, K. H.; Hsieh, T. Y.; Kwong, Dim-Lee; Spratt, D. B.)V1393,240-251(1991)

Stabilization of CdxHg1-xTe heterointerfaces (Clifton, Paul A.; Brown, Paul D.)V1361,1063-1069(1991)

Strained semiconductors and heterostructures: synthesis and applications (Bhattacharya, Pallab K.; Singh, Jasprit)V1361,394-405(1991)

Study of GaAs/AlGaAs quantum-well structures grown by MOVPE using tertiarybutylarsine (Lee, Hyung G.; Kim, HyungJun; Park, S. H.; Langer, Dietrich W.)V1361,893-900(1991)

Study of structural imperfections in epitaxial beta-SiC layers by method of x-ray differential diffractometry (Baranov, Igor M.; Kutt, R. N.; Nikitina, Irina P.)V1361,1110-1115(1991)

Threshold current and carrier lifetime in MOVPE regrown 1.5 um GaInAsP buried ridge structure lasers (Tischel, M.; Rosenzweig, M.; Hoffmann, Axel; Venghaus, Herbert; Fidorra, F.)V1361,917-926(1991)

Tunneling spectroscopy at nanometer scale in molecular beam epitaxy grown (Al)GaAs multilayers (Albrektsen, O.; Koenraad, Paul; Salemink, Huub W.)V1361,338-342(1991)

Etalons—see also Fabry-Perot

Bistable optical switching in GaAs multiple-quantum-well epitaxial etalons (Oudar, Jean-Louis; Sfez, B. G.; Kuszelewicz, Robert)V1361,490-498(1991)

Fabry-Perot etalon as a CO2 laser Q-switch modulator (Walocha, Jerzy)V1391,286-289(1991)

Fiber optic based miniature high-temperature probe exploiting coherence-tuned signal recovery via multimode laser diode illumination (Gerges, Awad S.; Jackson, David A.)V1504,233-236(1991)

High-average-power narrow-band KrF excimer laser (Wakabayashi, Osamu; Kowaka, Masahiko; Kobayashi, Yukio)V1463,617-628(1991)

Etching—see also microelectronic processing; microlithography

Ablation of ITO and TO films from glass substrates (Meringdal, Frode; Slinde, Harald)V1503,292-298(1991)

Advanced Techniques for Integrated Circuit Processing (Bondur, James; Turner, Terry R., eds.)V1392(1991)

Application of adaptive network theory to dry-etch monitoring and control (Deshmukh, V. G.; Hope, D. A.; Cox, T. I.; Hydes, A. J.)V1392,352-360(1991)

Applications of dry etching to InP-based laser fabrication (Hayes, Todd R.; Kim, Sung J.; Green, Christian A.)V1418,190-202(1991)

Characteristics of gate oxide surface material after exposure to magnetron-enhanced reactive ion etching plasma (Webb, Jennifer M.; Amini, Zahra H.)V1392,47-54(1991)

Chlorine or bromine chemistry in reactive ion etching Si-trench etching (Rangelow, Iwilo W.; Fichelscher, Andreas)V1392,240-245(1991)

Contamination and damage of silicon surfaces during magnetron-enhanced reactive ion etching in a single-wafer system (Tan, Swie-In; Colavito, D. B.)V1392,106-118(1991)

Current trends and issues for low-damage dry etching of optoelectronic devices (Hu, Evelyn L.)V1361,512-522(1991)

Dry development and plasma durability of resists: melt viscosity and self-diffusion effects (Paniez, Patrick J.; Joubert, Olivier P.; Pons, Michel J.; Oberlin, Jean C.; Weill, Andre P.)V1466,583-591(1991)

Dry etching for silylated resist development (Laporte, Philippe; Van den hove, Luc; Melaku, Yosias)V1392,196-207(1991)

Dry etching of high-aspect ratio contact holes (Wiepking, Mark; LeVan, M.; Mayo, Phyllis)V1392,139-150(1991)

Enhanced etching of InP by cycling with sputter etching and reactive ion etching (Demos, Alexandros T.; Fogler, H. S.; Pang, Stella W.; Elta, Michael E.)V1392,291-297(1991)

Enhanced process control of submicron contact definition (Ostrout, Wayne H.; Hunkler, Sean; Ward, Steven D.)V1392,151-164(1991)

Etch conditions of photolithographic holograms (Guo, Yongkang; Guo, Lu Rong; Zhang, Xiao-Chun)V1461,97-100(1991)

Etch tailoring through flexible end-point detection (Angell, David; Oehrlein, Gottleib S.)V1392,543-550(1991)

Evaluation of low-pressure silicon dry-etch processes with regard to low-substrate degradation (Engelhardt, Manfred)V1392,38-46(1991)

Fabrication of microlenses by laser-assisted chemical etching (Gratrix, Edward J.; Zarowin, Charles B.)V1544,238-243(1991)

Focused ion-beam vacuum lithography of InP with an ultrathin native oxide resist (Wang, Yuh-Lin; Temkin, Henryk; Harriott, Lloyd R.; Hamm, Robert A.)V1392,588-594(1991)

Formation of heterostructure devices in a multichamber processing environment with in-vacuo surface analysis diagnostics and in-situ process monitoring (Lucovsky, Gerald; Kim, Sang S.; Fitch, J. T.; Wang, Cheng)V1392,605-616(1991)

High-resolution tri-level process by downstream-microwave rf-biased etching (Rangelow, Iwilo W.)V1392,180-184(1991)

High-sensitivity and high-dry-etching durability positive-type electron-beam resist (Tamura, Akira; Yonezawa, Masaji; Sato, Mitsuyoshi; Fujimoto, Yoshiaki)V1465,271-281(1991)

Honeywell's submicron polysilicon gate process (Chan, Lap S.; Hertog, Craig K.; Youngner, D. W.)V1392,232-239(1991)

Improvement in dry etching of tungsten features (Heitzmann, Michel; Laporte, Philippe; Tabouret, Evelyne)V1392,272-279(1991)

Influence of sheath properties on the profile evolution in reactive ion etching processes (Fichelscher, Andreas; Rangelow, Iwilo W.; Stamm, A.)V1392,77-83(1991)

In-situ measurements of radicals and particles in a selective silicon oxide etching plasma (Singh, Jyothi)V1392,474-486(1991)

Instantaneous etch rate measurement of thin transparent films by interferometry for use in an algorithm to control a plasma etcher (Mishurda, Helen L.; Hershkowitz, Noah)V1392,563-569(1991)

Ion-beam-sputtering deposition and etching of high-Tc YBCO superconducting thin films (Zhao, Xing R.; Hao, Jian H.; Zhou, Fang Q.; Sun, Han D.; Wang, Lingjie; Yi, Xin J.)V1519,772-774(1991)

Laser-induced etched grating on InP for integrated optical circuit elements (Grebel, Haim; Pien, P.)V1583,331-337(1991)

LH electron cyclotron resonance plasma source (Kretschmer, K.-H.; Lorenz, Gerhard; Castrischer, G.; Kessler, I.; Baumann, P.)V1392,246-252(1991)

Magnetically enhanced reactive ion etching of submicron silicon trenches (Cooper, Kent; Nguyen, Bich-Yen; Lin, Jung-Hui; Roman, Bernard J.; Tobin, Phil; Ray, Wayne)V1392,253-264(1991)

Measuring for thickness distribution of recording layer of PLH (Zhang, Xiao-Chun; Guo, Lu Rong; Guo, Yongkang)V1461,93-96(1991)

Microfabrication techniques for semiconductor lasers (Tamanuki, Takemasa; Tadokoro, T.; Morito, K.; Koyama, Fumio; Iga, Kenichi)V1361,614-617(1991)

Monitoring and control of rf electrical parameters near plasma loads (Rummel, Paul)V1392,411-420(1991)

MTF of photolithographic hologram (Guo, Lu Rong; Guo, Yongkang; Zhang, Xiao-Chun)V1392,119-123(1991)

Multichamber reactive ion etching processing for III-V optoelectronic devices (Rothman, Mark A.; Thompson, John A.; Armiento, Craig A.)V1392,598-604(1991)

Nanometer scale focused ion beam vacuum lithography using an ultrathin oxide resist (Harriott, Lloyd R.; Temkin, Henryk; Chu, C. H.; Wang, Yuh-Lin; Hsieh, Y. F.; Hamm, Robert A.; Panish, Morton B.; Wade, H. H.)V1465,57-63(1991)

New gas-phase etching method for preparation of polarization-maintaining fibers (Matejec, Vlastimil; Sasek, Ladislav; Gotz, J.; Ivanov, G. A.; Koreneva, N. A.; Grigor'yants, Vil V.)V1513,174-179(1991)

New process for improving reverse characteristics of platinum silicide Schottky barrier power diodes (Zhao, Shu L.; Li, Yuan J.; Yu, Jia F.; Yang, Ya L.; Liu, Shu Q.; Fan, Ya F.; Bao, Xiu Y.; Li, Zheng Q.)V1519,275-280(1991)

Novel surface imaging masking technique for high-aspect-ratio dry etching applications (Calabrese, Gary C.; Abali, Livingstone N.; Bohland, John F.; Pavelchek, Edward K.; Sricharoenchaikit, Prasit; Vizvary, Gerald; Bobbio, Stephen M.; Smith, Patrick)V1466,528-537(1991)

Oxygen plasma etching of silylated resist in top-imaging lithographic process (Dijkstra, Han J.)V1466,592-603(1991)

Oxygen reactive ion etching of polymers: profile evolution and process mechanisms (Pilz, Wolfgang; Janes, Joachim; Muller, Karl P.; Pelka, Joachim)V1392,84-94(1991)

Pattern etching and selective growth of GaAs by in-situ electron-beam lithography using an oxidized thin layer (Akita, K.; Sugimoto, Yoshimasa; Taneya, M.; Hiratani, Y.; Ohki, Y.; Kawanishi, Hidenori; Katayama, Yoshifumi)V1392,576-587(1991)

Photoelectrochemical etching of n-InP producing antireflecting structures for solar cells (Soltz, David; Cescato, Lucila H.; Decker, Franco)V1536,268-276(1991)

Plasma diagnostics as inputs to the modeling of the oxygen reactive ion etching of multilevel resist structures (Hope, D. A.; Hydes, A. J.; Cox, T. I.; Deshmukh, V. G.)V1392,185-195(1991)

Polysilicon etching for nanometer-scale features (Lajzerowicz, Jean; Tedesco, Serge V.; Pierrat, Christophe; Muyard, D.; Taccussel, M. C.; Laporte, Philippe)V1392,222-231(1991)

Progress in the study of development-free vapor photolithography (Hong, Xiao-Yin; Liu, Dan; Li, Zhong-Zhe; Xiao, Ji-Quang; Dong, Gui-Rong)V1466,546-557(1991)

Progress of an advanced diffusion source plasma reactor (Benjamin, Neil M.; Chapman, Brian N.; Boswell, Rod W.)V1392,95-105(1991)

Radial ion energy measurements in an electron cyclotron resonance reactor (O'Neill, James A.; Holber, William M.; Caughman, John)V1392,516-528(1991)

Reactive ion etching of InP and its optical assessment (MacLeod, Roderick W.; Sotomayor Torres, Clivia M.; Razeghi, Manijeh; Stanley, C. R.; Wilkinson, Chris D.)V1361,562-567(1991)

Reactive ion etching of deep isolation trenches using sulfur hexafluoride, chlorine, helium, and oxygen (Krawiec, Theresa M.; Giammarco, Nicholas J.)V1392,265-271(1991)

Real-time automation of a dry etching system (McCafferty, Robert H.)V1392,331-339(1991)

Real-time monitoring and control of plasma etching (Butler, Stephanie W.; McLaughlin, Kevin J.; Edgar, Thomas F.; Trachtenberg, Isaac)V1392,361-372(1991)

Simulation of ion-enhanced dry-etch processes (Pelka, Joachim)V1392,55-66(1991)

Single-crystal silicon trench etching for fabrication of highly integrated circuits (Engelhardt, Manfred)V1392,210-221(1991)

SPEEDIE: a profile simulator for etching and deposition (McVittie, James P.; Rey, J. C.; Bariya, A. J.; IslamRaja, M. M.; Cheng, L. Y.; Ravi, S.; Saraswat, Krishna C.)V1392,126-138(1991)

Spin-on-glass/phosphosilicate glass etchback planarization process for 1.0 um CMOS technology (Bogle-Rohwer, Elizabeth; Nulty, James E.; Chu, Wileen; Cohen, Andrew)V1392,280-290(1991)

Surface imaging on the basis of phenolic resin: experiments and simulation (Bauch, Lothar; Jagdhold, Ulrich A.; Dreger, Helge H.; Bauer, Joachim J.; Hoeppner, Wolfgang W.; Erzgraeber, Hartmut H.; Mehliss, Georg G.)V1466,510-519(1991)

Thermo-activated photoetching of PMMA in direct writing by laser beam (Maslenitsyn, S. F.; Svetovoy, V. B.)V1440,436-441(1991)

Two-layer 1.2-micron pitch multilevel metal demonstrator using resist patterning by surface imaging and dry development (Martin, Brian; Snowden, Ian M.; Mortimer, Simon H.)V1463,584-594(1991)

Unframed via interconnection of nonplanar device structures (Kim, Manjin J.)V1596,12-22(1991)

Vertical oxide etching without inducing change in critical dimensions (Nagy, Andrew)V1392,165-179(1991)

Expert systems—see also artificial intelligence

Active pattern recognition based on attributes and reasoning (Chang, Jin W.)V1468,137-146(1991)

Analysis of multidimensional confocal images (Samarabandu, J. K.; Acharya, Raj S.; Edirisinghe, Chandima D.; Cheng, Ping-Chin; Kim, Hyo-Gun; Lin, T. H.; Summers, R. G.; Musial, C. E.)V1450,296-322(1991)

Analyzing and interpreting pulmonary tomoscintigraphy sequences: realization and perspectives (Forte, Anne-Marie; Bizais, Yves)V1445,409-420(1991)

Applications of Artificial Intelligence IX (Trivedi, Mohan M., ed.)V1468(1991)

Artificial neural network for supervised learning based on residual analysis (Chan, Keith C.; Vieth, John O.; Wong, Andrew K.)V1469,359-372(1991)

Attempt to develop a zoom-lens-design expert system (Weng, Zhicheng; Chen, Zhiyong; Yang, Yu-Hong; Ren, Tao; Cong, Xiaojie; Yao, Yuchuan; He, Fengling; Li, Yuan-Yuan)V1527,349-356(1991)

Basic manufacturability interval (Billings, Daniel A.)V1468,434-445(1991)

BUOSHI: a tool for developing active expert systems (Wang, Gang; Dubant, Olivier)V1468,11-15(1991)

Capturing multimedia design knowledge using TYRO, the constraint-based designer's apprentice (MacNeil, Ronald L.)V1460,94-102(1991)

Case-based reasoning approach for heuristic search (Krovvidy, Srinivas; Wee, William G.; Han, Chia Y.)V1468,216-226(1991)

CLIPS implementation of a knowledge-based distributed control of an autonomous mobile robot (Bou-Ghannam, Akram A.; Doty, Keith L.)V1468,504-515(1991)

Computer-aided acquisition of design knowledge (Dilger, Werner)V1468,584-595(1991)

Computerized system for clinical diagnosis of melanoma (Ferrario, Mario; Barbieri, Fabio)V1450,108-117(1991)

Constructing attribute classes by example learning: the research of attribute-based knowledge-style pattern recognition (Zhou, Lijia; Song, Hongjun; Zhao, S.)V1469,404-411(1991)

Design of a cart-pole balancing fuzzy logic controller using a genetic algorithm (Karr, Charles L.)V1468,26-36(1991)

Expert system and process optimization techniques for real-time monitoring and control of plasma processes (Cheng, Jie; Qian, Zhaogang; Irani, Keki B.; Etemad, H.; Elta, Michael E.)V1392,373-384(1991)

Expert system for diagnosis/optimization of microlithography process* (Nicolau, Dan V.; Fulga, Florin; Dusa, Mircea V.)V1468,345-351(1991)

Expert system for fusing weather and doctrinal information used in the intelligence preparation of the battlefield (McWilliams, Gary; Kirby, Steve; Eskridge, Thomas; Newberry, Jeff)V1468,417-428(1991)

Expert systems in lens design (Dilworth, Donald C.)V1354,359-370(1991)

Fault management of a fiber optic LAN (Spencer, Paul E.; Zaharakis, Steven C.; Denton, Richard T.)V1364,228-234(1991)

Future of global optimization in optical design (Sturlesi, Doron; O'Shea, Donald C.)V1354,54-68(1991)

Fuzzy logic: principles, applications, and perspectives (Zadeh, Lotfi A.)V1468,582-582(1991)

Generalization of the problem of correspondence in long-range motion and the proposal for a solution (Stratton, Norman A.; Vaina, Lucia M.)V1468,176-185(1991)

High-level parallel architecture for a rule-based system (Karne, Ramesh K.; Sood, Arun K.)V1468,938-949(1991)

Infrared systems design from an operational requirement using a hypercard-based program (Harris, William R.)V1488,156-164(1991)

Integrated oceanographic image understanding system (Lybanon, Matthew; Peckinpaugh, Sarah; Holyer, Ronald J.; Cambridge, Vivian)V1406,180-189(1991)

Intelligent information system: for automation of airborne early warning crew decision processes (Chin, Hubert H.)V1468,235-244(1991)

Intelligent packaging and material handling (Hall, Ernest L.; Shell, Richard; Slutzky, Gale D.)V1381,162-170(1991)

Internal protocol assistant for distributed systems (Leu, Fang Y.; Chang, Shi-Kuo)V1468,620-631(1991)

KEShell: a "rule skeleton + rule body" -based knowledge engineering shell* (Wu, Xindong)V1468,632-639(1991)

Knowledge-based environment for optical system design (Johnson, R. B.)V1354,346-358(1991)

Knowledge-based nursing diagnosis (Hay, Claudette; Hay, D. R.)V1468,314-322(1991)

Knowledge-based process planning and line design in robotized assembly (Delchambre, Alain)V1468,367-378(1991)

Knowledge-based system for configuring mixing-machines (Brinkop, Axel; Laudwein, Norbert)V1468,227-234(1991)

Knowledge-based system using a neural network (Szabo, Raisa R.; Pandya, Abhijit S.; Szabo, Bela)V1468,794-801(1991)

LCS: a natural language comprehension system (Trigano, Philippe; Talon, Benedicte; Baltazart, Didier; Demko, Christophe; Newstead, Emma)V1468,866-874(1991)

Learning by comparison: improving the task planning capability (del Castillo, Maria D.; Kumpel, Daniel M.)V1468,596-607(1991)

Lighting and optics expert system for machine vision (Novini, Amir R.)V1386,2-9(1991)

Logical account of a terminological tool (Bresciani, Paolo)V1468,245-255(1991)

Microcomputer-based image processing system for CT/MRI scans: II. Expert system (Kwok, John C.; Yu, Peter K.; Cheng, Andrew Y.; Ho, Wai-Chin)V1445,446-455(1991)

Model-based labeling, analysis, and three-dimensional visualization from two-dimensional medical images (Arata, Louis K.; Dhawan, Atam P.; Thomas, Stephen R.)V1446,465-474(1991)

Model-based vision system for automatic recognition of structures in dental radiographs (Acharya, Raj S.; Samarabandu, J. K.; Hausmann, E.; Allen, K. A.)V1450,170-177(1991)

Multilevel qualitative reasoning in CMOS circuit analysis (Kaul, Neeraj; Biswas, Gautam)V1468,204-215(1991)

Multiple-target tracking in a cluttered environment and intelligent track record (Tomasini, Bernard; Cassassolles, Emmanuel; Poyet, Patrice; Maynard de Lavalette, Guy M.; Siffredi, Brigitte)V1468,60-71(1991)

Multisensor fusion using the sensor algorithm research expert system (Bullock, Michael E.; Miltonberger, Thomas W.; Reinholdtsen, Paul A.; Wilson, Kathleen)V1471,291-302(1991)

Natural language parsing in a hybrid connectionist-symbolic architecture (Mueller, Adrian; Zell, Andreas)V1468,875-881(1991)

New techniques for repertory grid analysis (Liseth, Ole J.; Bezdek, James C.; Ford, Kenneth M.; Adams-Webber, Jack R.)V1468,256-267(1991)

Optical Distributed Inference Network (ODIN) (Bostel, Ashley J.; McOwan, Peter W.; Hall, Trevor J.)V1385,165-172(1991)

Optical laser intelligent verification expert system (Jones, Robert H.)V1401,86-93(1991)

Parallel optical information, concept, and response evolver: POINCARE (Caulfield, H. J.; Caulfield, Kimberly)V1469,232-239(1991)

Parallel rule inferencing for automatic target recognition (Pacelli, Jean L.; Geyer, Steve L.; Ramsey, Timothy S.)V1472,76-84(1991)

Photolith analysis and control system (Srikanth, Usha; Sundararajan, Srikanth)V1468,429-433(1991)

Practical approach to knowledge base verification (Preece, Alun D.; Shinghal, Rajjan)V1468,608-619(1991)

Problem solving of optical design by R graph with bidirectional search mechanics (Chang, Rong-Seng; Chen, Der-Chin)V1354,379-385(1991)

Production environment implementation of the stereo extraction of cartographic features using computer vision and knowledge base systems in DMA's digital production system (Gruenewald, Maria M.; Hinchman, John H.)V1468,843-852(1991)

Propagation of variances in belief networks (Neapolitan, Richard E.)V1468,333-344(1991)

Prototype expert system for preventive control in power plants (Jiang, Dareng; Han, Chia Y.; Wee, William G.)V1468,16-25(1991)

Quantitative analysis of cardiac imaging using expert systems (Dreyer, Keith J.; Simko, Joseph; Held, A. C.)V1445,398-408(1991)

Research and improvement of Vogl's acceleration algorithm (Song, Hongjun; Zhou, Lijia)V1469,581-584(1991)

Routing in distributed information systems (Ras, Zbigniew W.)V1470,76-87(1991)

Scheduler's assistant: a tool for intelligent scheduling (Griffin, Neal L.)V1468,110-121(1991)

Ten years of failure in automatic time tables scheduling at the UTC (Trigano, Philippe; Boufflet, Jean-Paul; Newstead, Emma)V1468,408-416(1991)

Two-valued neural logic network (Hsu, L. S.; Loe, K. F.; Chan, Sing C.; Teh, H. H.)V1469,197-207(1991)

Use of a knowledge-based system for the valuation of unlisted shares (Long, J. A.; Manousos, S. N.)V1468,446-454(1991)

Use of an expert system to predict thunderstorms and severe weather (Passner, Jeffrey E.; Lee, Robert R.)V1468,2-10(1991)

Using an expert system to interface mainframe computing resources with an interactive video system (Carey, Raymond; Wible, Sheryl F.; Gaynor, Wayne H.; Hendry, Timothy G.)V1464,500-507(1991)

Visualization of image from 2-D strings using visual reasoning (Li, Xiao-Rong; Chang, Shi-Kuo)V1468,720-731(1991)

Visualization tool for human-machine interface designers (Prevost, Michael P.; Banda, Carolyn P.)V1459,58-66(1991)

Fabry-Perot

All-silicon Fabry-Perot modulator based on thermo-optic effect (Rendina, Ivo; Cocorullo, Giuseppe)V1583,338-343(1991)

Amplitude-modulated laser-driven fiber-optic RF interferometric strain sensor (Schoenwald, Jeffrey S.)V1418,450-458(1991)

Analysis of the self-oscillation phenomenon of fiber optically addressed silicon microresonators (Rao, Yun-Jiang; Culshaw, Brian)V1506,126-133(1991)

Application of Fabry-Perot velocimetry to hypervelocity impact experiments (Chau, Henry H.; Osher, John E.)V1346,103-112(1991)

Asymmetric Fabry-Perot modulators for optical signal processing and optical computing applications (Kilcoyne, M. K.; Whitehead, Mark; Coldren, Larry A.)V1389,422-454(1991)

Compact optical associative memory (Burns, Thomas J.; Rogers, Steven K.; Kabrisky, Matthew; Vogel, George A.)V1469,208-218(1991)

Data transmission at 1.3 um using silicon spatial light modulator (Xiao, Xiaodong; Goel, Kamal K.; Sturm, James C.; Schwartz, P. V.)V1476,301-304(1991)

Decay of the nonlinear susceptibility components in main-chain functionalized poled polymers (Meyrueix, Remi; LeCompte, J. P.; Tapolsky, Gilles)V1560,454-466(1991)

Demodulation of a fiber Fabry-Perot strain sensor using white light interferometry (Zuliani, Gary; Hogg, W. D.; Liu, Kexing; Measures, Raymond M.)V1588,308-313(1991)

Development of a fiber Fabry-Perot strain gauge (Hogg, W. D.; Janzen, Doug; Valis, Tomas; Measures, Raymond M.)V1588,300-307(1991)

Electro-optic resonant modulator for coherent optical communication (Robinson, Deborah L.; Chen, Chien-Chung; Hemmati, Hamid)V1417,421-430(1991)

Far-IR Fabry-Perot spectrometer for OH measurements (Pickett, Herbert M.; Peterson, Dean B.)V1491,308-313(1991)

Fiber optic based vortex shedder flow meter (Chu, Beatrice C.; Newson, Trevor P.; Jackson, David A.)V1504,251-257(1991)

High-efficiency vertical-cavity lasers and modulators (Coldren, Larry A.; Corzine, Scott W.; Geels, Randall S.; Gossard, Arthur C.; Law, K. K.; Merz, James L.; Scott, Jeffrey W.; Simes, Robert J.; Yan, Ran H.)V1362,24-37(1991)

Interferometric technique for the concurrent determination of thermo-optic and thermal expansion coefficients (Jewell, John M.; Askins, Charles G.; Aggarwal, Ishwar D.)V1441,38-44(1991)

Line-imaging Fabry-Perot interferometer (Mathews, Allen R.; Warnes, Richard H.; Hemsing, Willard F.; Whittemore, Gerald R.)V1346,122-132(1991)

Multiorder etalon sounder for vertical temperature profiling: technique and performance analysis (Wang, Jin-Xue; Hays, Paul B.; Drayson, S. R.)V1492,391-402(1991)

Multiplex Fabry-Perot interferometer (Snell, Hilary E.; Hays, Paul B.)V1492,403-407(1991)

Novel angular discriminator for spatial tracking in free-space laser communications (Fung, Jackie S.)V1417,224-232(1991)

Optical bistability and signal competition in active cavity with photochromic nonlinearity of bacteriorhodopsin (Taranenko, Victor B.; Vasnetsov, M. V.)V1621,169-179(1991)

Reflection-mode polymeric interference modulators (Yankelevich, Diego; Knoesen, Andre; Eldering, Charles A.; Kowel, Stephen T.)V1560,406-415(1991)

Silicon modulator for integrated optics (Cocorullo, Giuseppe; Della Corte, Francesco G.; Rendina, Ivo; Cutolo, Antonello)V1374,132-137(1991)

Striped Fabry-Perots: improved efficiency for velocimetry (McMillan, Charles F.; Steinmetz, Lloyd L.)V1346,113-121(1991)

Threshold current density of InGaAsP/InP surface-emitting laser diodes with hemispherical resonator (Jing, Xing-Liang; Zhang, Yong-Tao; Chen, Yi-Xin)V1418,434-441(1991)

Uses of Fabry-Perot velocimeters in studies of high explosives detonation (Breithaupt, R. D.; Tarver, Craig M.)V1346,96-102(1991)

Faraday effect—see also magneto-optics

Optical technique for the compensation of the temperature-dependent Verdet constant in Faraday rotation magnetometers (Hamid, Sohail; Tatam, Ralph P.)V1511,78-89(1991)

Waveguide-based fiber optic magnetic field sensor with directional sensitivity (Sohlstrom, Hans B.; Svantesson, Kjell G.)V1511,142-148(1991)

Ways of the high-speed increasing of magneto-optical spatial light modulators (Randoshkin, Vladimir V.)V1469,796-803(1991)

Ferroelectrics—see also crystals, liquid

Characteristics of domain formation and poling in potassium niobate (Jarman, Richard H.; Johnson, Barry C.)V1561,33-42(1991)

Dye orientation in organic guest host systems on ferroelectric polymers (Osterfeld, Martin; Knabke, Gerhard; Franke, Hilmar)V1559,49-55(1991)

Growth mechanism of orientated PLZT thin films sputtered on glass substrate (Zhang, Rui T.; Ge, Ming; Luo, Wei G.)V1519,757-760(1991)

Properties and applications of ferroelectric and piezoelectric thin films (Shiosaki, Tadashi)V1519,694-703(1991)

Specific properties of ferroelectric thin films (Liu, Wei G.)V1519,704-706(1991)

Standing spin wave modes in permalloy-FeCr multilayer films (Chen, H. Y.; Luo, Y. Q.)V1519,761-764(1991)

Fiber fabrication—see also glass

Active glasses prepared by the sol-gel method including islands of CdS or silver (Reisfeld, Renata; Minti, Harry; Eyal, Marek)V1513,360-367(1991)

Advantages of drawing crystal-core fibers in microgravity (Shlichta, Paul J.; Nerad, Bruce A.)V1557,10-23(1991)

Analysis and modeling of low-loss fused fiber couplers (Das, Alok K.; Pandit, Malay K.)V1365,74-85(1991)

Annealing properties of proton-exchanged waveguides in LiNbO3 fabricated using stearic acid (Pun, Edwin Y.; Loi, K. K.; Zhao, S.; Chung, P. S.)V1583,102-108(1991)

Bragg grating formation and germanosilicate fiber photosensitivity (Meltz, Gerald; Morey, William W.)V1516,185-199(1991)

Characterization of planar optical waveguides by K+ ion exchange in glass at 1.152 and 1.523 um (Yip, Gar L.; Kishioka, Kiyoshi; Xiang, Feng; Chen, J. Y.)V1583,14-18(1991)

Comparison of photorefractive effects and photogenerated components in polarization-maintaining fibers (Kanellopoulos, S. E.; Guedes Valente, Luiz C.; Handerek, Vincent A.; Rogers, Alan J.)V1516,200-210(1991)

Defect centers and photoinduced self-organization in Ge-doped silica core fiber (Tsai, T. E.; Griscom, David L.)V1516,14-28(1991)

Development of optical waveguides by sol-gel techniques for laser patterning (Schmidt, Helmut; Krug, Herbert; Kasemann, Reiner; Tiefensee, Frank)V1590,36-43(1991)

Dynamics of phase-grating formation in optical fibers (An, Sunghyuck; Sipe, John E.)V1516,175-184(1991)

Dynamics of self-organized x(2) gratings in optical fibers (Kamal, Avais; Terhune, Robert W.; Weinberger, Doreen A.)V1516,137-153(1991)

Experimental and theoretical investigation of surface- and bulk-induced attenuation in solution-deposited waveguides (Roncone, Ronald L.; Burke, James J.; Weisenbach, Lori; Zelinski, Brian J.)V1590,14-25(1991)

Fabrication of large multimode glass waveguides by dry silver ion exchange in vacuum (Tammela, Simo; Pohjonen, Harri; Honkanen, Seppo; Tervonen, Ari)V1583,37-42(1991)

Fabrication of rare-earth-doped optical fiber (DiGiovanni, David J.)V1373,2-8(1991)

Fiber construction for improved mechanical reliabiltiy (Roberts, Daniel R.; Cuellar, Enrique; Kennedy, Michael T.; Tomita, Akira)V1366,129-135(1991)

Fiber defects in Ge-doped fibers: towards a coherent picture (LaRochelle, Sophie)V1516,55-57(1991)

Fiber optic fluorescence sensors based on sol-gel entrapped dyes (MacCraith, Brian D.; Ruddy, Vincent; Potter, C.; McGilp, J. F.; O'Kelley, B.)V1510,104-109(1991)

Frequency doubling in optical fibers: a complex puzzle (Margulis, Walter; Carvalho, Isabel C.; Lesche, Bernhard)V1516,60-66(1991)

Fully planar proton-exchanged lithium niobate waveguides with grating (Zhang, H.; Li, Ming-Jun; Najafi, S. I.; Schwelb, Otto)V1583,83-89(1991)

Glass-ceramic fiber optic sensors (Romaniuk, Ryszard S.; Stepien, Ryszard)V1368,73-84(1991)

Glasses for Optoelectronics II (Righini, Giancarlo C., ed.)V1513(1991)

Glass waveguides by ion exchange with ionic masking (Wang, Wei-Jian; Li, Ming-Jun; Honkanen, Seppo; Najafi, S. I.; Tervonen, Ari)V1513,434-440(1991)

Graded-index polymer optical fiber by new random copolymerization technique (Koike, Yasuhiro; Hondo, Yukie; Nihei, Eisuke)V1592,62-72(1991)

Gradient-index fiber-optic preforms by a sol-gel method (Banash, Mark A.; Caldwell, J. B.; Che, Tessie M.; Mininni, Robert M.; Soskey, Paul R.; Warden, Victor N.; Pope, Edward J.)V1590,8-13(1991)

H:LiNbO3 optical waveguides made from pyrophosphoric acid (Ziling, C. C.; Pokrovskii, L.; Terpugov, N. V.; Kuneva, M.; Savatinova, Ivanka T.; Armenise, Mario N.)V1583,90-101(1991)

High-silica cascaded three-waveguide couplers for wideband filtering by flame hydrolysis on Si (Barbarossa, Giovanni; Laybourn, Peter J.)V1583,122-128(1991)

High-silica low-loss three-waveguide couplers on Si by flame hydrolysis deposition (Barbarossa, Giovanni; Laybourn, Peter J.)V1513,37-43(1991)

Influence of processing variables on the optical properties of SiO2-TiO2 planar waveguides (Weisenbach, Lori; Zelinski, Brian J.; O'Kelly, John; Morreale, Jeanne; Roncone, Ronald L.; Burke, James J.)V1590,50-58(1991)

Infrared transmitting glasses and fibers for chemical analysis (Hilton, Albert R.)V1437,54-59(1991)

Inhomogeneous line broadening of optical transitions in Nd3+ and Er3+ doped preforms and fibers (Briancon, Anne-Marie; Jacquier, Bernard; Gacon, Jean-Claude; Le Sergent, Christian; Marcerou, Jean-Francois)V1373,9-20(1991)

Integrated Optical Circuits (Wong, Ka-Kha, ed.)V1583(1991)

Intl Workshop on Photoinduced Self-Organization Effects in Optical Fiber (Ouellette, Francois, ed.)V1516(1991)

Ion beam modification of glasses (Mazzoldi, Paolo; Carnera, Alberto; Caccavale, F.; Granozzi, G.; Bertoncello, R.; Battaglin, Giancarlo; Boscolo-Boscoletto, A.; Polato, P.)V1513,182-197(1991)

Laser-fabricated fiber optical taps for interconnects and optical data processing devices (Imen, Kamran; Lee, Changhun H.; Yang, Y. Y.; Allen, Susan D.; Ghosh, Anjan K.)V1365,60-64(1991)

Low-profile fibers for embedded smart structure applications (Vengsarkar, Ashish M.; Murphy, Kent A.; Gunther, Michael F.; Plante, Angela J.; Claus, Richard O.)V1588,2-13(1991)

Low-temperature ion exchange of dried gels for potential waveguide fabrication in glasses (Risen, William M.; Morse, Ted F.; Tsagaropoulos, George)V1590,44-49(1991)

Melt-processed calcium aluminate fibers: structural and optical properties (Wallenberger, Frederick T.; Weston, Norman E.; Brown, Sherman D.)V1484,116-124(1991)

Melt processing of calcium aluminate fibers with sapphirelike infrared transmission (Wallenberger, Frederick T.; Koutsky, J. A.; Brown, Sherman D.)V1590,72-82(1991)

Mixed-convection effects during the drawing of optical fibers (Papamichael, Haris; Miaoulis, Ioannis)V1590,122-130(1991)

Multimode stripe waveguides for optical interconnections (Maile, Michael; Weidel, Edgar)V1563,188-196(1991)

Nd- and Er-doped glass integrated optical amplifiers and lasers (Najafi, S. I.; Wang, Wei-Jian; Orcel, Gerard F.; Albert, Jacques; Honkanen, Seppo; Poyhonen, Pekka; Li, Ming-Jun)V1583,32-36(1991)

New gas-phase etching method for preparation of polarization-maintaining fibers (Matejec, Vlastimil; Sasek, Ladislav; Gotz, J.; Ivanov, G. A.; Koreneva, N. A.; Grigor'yants, Vil V.)V1513,174-179(1991)

Nonlinear optical properties of a fiber with an organic core crystal grown from solution (Ohmi, Toshihiko; Yoshikawa, Nobuo; Sakai, Koji; Koike, Tomoyuki; Umegaki, Shinsuke)V1361,606-612(1991)

Optical fibers for UV applications (Fabian, Heinz; Grzesik, Ulrich; Woerner, K.-H.; Klein, Karl-Friedrich)V1513,168-173(1991)

Optically induced creation, transformation, and organization of defects and color centers in optical fibers (Russell, Philip S.; Hand, Duncan P.; Chow, Y. T.; Poyntz-Wright, L. J.)V1516,47-54(1991)

Optical properties of glass materials obtained by inorganic sol-gel synthesis (Glebov, Leonid B.; Evstropiev, Sergei K.; Petrovskii, Gurii T.; Shashkin, Viktor S.)V1513,224-231(1991)

Phase-matched second-harmonic generation of infrared wavelengths in optical fibers (Kashyap, Raman; Davey, Steven T.; Williams, Doug L.)V1516,164-174(1991)

Photoinduced second-harmonic generation and luminescence of defects in Ge-doped silica fibers (Krol, Denise M.; Atkins, Robert M.; Lemaire, Paul J.)V1516,38-46(1991)

Photolithographic processing of integrated optic devices in glasses (Mendoza, Edgar A.; Gafney, Harry D.; Morse, David L.)V1583,43-51(1991)

Photosensitivity in optical fibers: detection, characterization, and application to the fabrication of in-core fiber index gratings (Malo, Bernard; Bilodeau, Francois; Johnson, Derwyn C.; Skinner, Iain M.; Hill, Kenneth O.; Morse, Ted F.; Kilian, Arnd; Reinhart, Larry J.; Oh, Kyunghwan)V1590,83-93(1991)

Physical models of second-harmonic generation in optical fibers (Lesche, Bernhard)V1516,125-136(1991)

Planar and strip optical waveguides by sol-gel method and laser densification (Guglielmi, Massimo; Colombo, Paolo; Mancinelli Degli Esposti, Luca; Righini, Giancarlo C.; Pelli, Stefano)V1513,44-49(1991)

Polarization-maintaining single-mode fibers with layered core (Sasek, Ladislav; Vohryzek, Jachym; Sochor, Vaclav; Paulicka, Ivan; van Nhac, Nguyen; Franek, Alexandr)V1572,151-156(1991)

Polymeric optical waveguides for device applications (McFarland, Michael J.; Beeson, Karl W.; Horn, Keith A.; Nahata, Ajay; Wu, Chengjiu; Yardley, James T.)V1583,344-354(1991)

Preparation of SiO2 film utilizing equilibrium reaction in aqueous solution (Kawahara, Hideo; Sakai, Y.; Goda, Takuji; Hishinuma, Akihiro; Takemura, Kazuo)V1513,198-203(1991)

Progress in fluoride fiber lasers and amplifiers (France, Paul W.; Brierley, Michael C.)V1373,33-39(1991)

Properties and processing of the TeX glasses (Zhang, Xhang H.; Ma, Hong-Li; Lucas, Jacques)V1513,209-214(1991)

Proton exchange: past, present, and future (Jackel, Janet L.)V1583,54-63(1991)

Rare-earth-doped LiNbO3 waveguide amplifiers and lasers (Sohler, Wolfgang)V1583,110-121(1991)

SHG in fiber: is a high-conversion efficiency possible? (Ouellette, Francois)V1516,113-114(1991)

Sol-gel overview: transparent, microporous silica, its synthesis and characterization (Klein, Lisa C.)V1590,2-7(1991)

Spectroscopic characteristics of Eu-doped aluminosilicate optical fiber preform (Oh, Kyunghwan; Morse, Ted F.; Reinhart, Larry J.; Kilian, Arnd)V1590,94-100(1991)

Spectroscopic properties of Er3+-doped glasses for the realization of active waveguides by ion-exchange technique (Cognolato, Livio; De Bernardi, Carlo; Ferraris, M.; Gnazzo, Angelantonio; Morasca, Salvatore; Scarano, Domenica)V1513,368-377(1991)

Stress analysis in ion-exchanged waveguides by using a polarimetric technique (Gonella, Francesco; Mazzi, Giulio; Quaranta, Alberto)V1513,425-433(1991)

Study using nuclear techniques of waveguides produced by electromigration processes (Battaglin, Giancarlo; De Marchi, Giovanna; Losacco, Aurora M.; Mazzoldi, Paolo; Miotello, Antonio; Quaranta, Alberto; Valentini, Antonio)V1513,441-450(1991)

Submolecular Glass Chemistry and Physics (Bray, Phillip; Kreidl, Norbert J., eds.)V1590(1991)

Supported sol-gel thin-film glasses embodying laser dyes II: three-layered waveguide assemblies (Haruvy, Yair; Heller, Adam; Webber, Stephen E.)V1590,59-70(1991)

UV spectroscopy of optical fibers and preforms (Williams, Doug L.; Davey, Steven T.; Kashyap, Raman; Armitage, J. R.; Ainslie, B. J.)V1516,29-37(1991)

Vapor phase deposition of transition metal fluoride glasses (Boulard, B.; Jacoboni, Charles)V1513,204-208(1991)

Very high temperature fiber processing and testing through the use of ultrahigh solar energy concentration (Jacobson, Benjamin A.; Gleckman, Philip; Holman, Robert L.; Sagie, Daniel; Winston, Roland)V1528,82-85(1991)

Waveguide formation by laser irradiation of sol-gel coatings (Zaugg, Thomas C.; Fabes, Brian D.; Weisenbach, Lori; Zelinski, Brian J.)V1590,26-35(1991)

What we can learn about second-harmonic generation in germanosilicate glass from the analogous effect in semiconductor-doped glasses (Lawandy, Nabil M.)V1516,99-112(1991)

Fiber optic components

1-W cw separate confinement InGaAsP/InP (lamda = 1.3 um) laser diodes and their coupling with optical fibers (Garbuzov, Dmitriy Z.; Goncharov, S. E.; Il'in, Y. V.; Mikhailov, A. V.; Ovchinnikov, Alexander V.; Pikhtin, N. A.; Tarasov, I. S.)V1418,386-393(1991)

10-watt cw diode laser bar efficiently fiber-coupled to a 381 um diameter fiber-optic connector (Willing, Steven L.; Worland, Phil; Harnagel, Gary L.; Endriz, John G.)V1418,358-362(1991)

Aerospace resource document: fiber optic interconnection hardware (Little, William R.)V1589,20-23(1991)

Analysis and modeling of low-loss fused fiber couplers (Das, Alok K.; Pandit, Malay K.)V1365,74-85(1991)

Analysis of fiber damage and field failures in fiber-grip-type mechanical fiber optic splices (Kiss, Gabor D.; Pellegrino, Anthony; Leopold, Simon)V1366,223-234(1991)

Applications and characteristics of polished polarization-splitting couplers (Lefevre, Herve C.; Simonpietri, Pascal; Martin, Philippe)V1365,65-73(1991)

Applications and testing of an opto-mechanical switch (Thevenot, Clarel; Newhouse, Mark A.; Annunziata, Frank A.)V1398,250-260(1991)

Applications of high speed silicon bipolar ICs in fiber optic systems (LaBelle, Gary L.; McDonald, Mark D.)V1365,116-121(1991)

Applications of optical switches in fiber optic communication networks (Hanson, Daniel; Gosset, Nathalie M.)V1363,48-56(1991)

Channel waveguide Mach-Zehnder interferometer for wavelength splitting and combining (Tervonen, Ari; Poyhonen, Pekka; Honkanen, Seppo; Tahkokorpi, Markku T.)V1513,71-75(1991)

Characterization of micro-optical components fabricated by deep-etch x-ray lithography (Goettert, Jost; Mohr, Jurgen)V1506,170-178(1991)

Comparison of alternative modulation techniques for microwave optical links (Kasemset, Dumrong; Ackerman, Edward I.; Wanuga, Stephen; Herczfeld, Peter R.; Daryoush, Afshin S.)V1371,104-114(1991)

Comparison of performances on analog fiber optic links equipped with either noncontact or physical-contact connectors (Giannini, Jean-Pierre; Delay, P.; Duval, A.; Wtodkiewiez, A.)V1366,215-222(1991)

Components and applications for high-speed optical analog links (Johnson, Peter T.; Debney, Brian T.; Carter, Andrew C.)V1371,87-97(1991)

Components for Fiber Optic Applications V (Kopera, Paul M., ed.)V1365(1991)

Computer-generated optical elements for fiber's mode selection and launching (Golub, Mikhail A.; Sisakian, Iosif N.; Soifer, Victor A.)V1365,156-165(1991)

Cost-effective optical switch matrix for microwave phased array (Pan, J. J.; Chau, Seung L.; Li, Wei Z.; Grove, Charles H.)V1476,133-142(1991)

Design, fabrication, and testing of a 7-bit binary fiber optic delay line (Goutzoulis, Anastasios P.; Davies, D. K.)V1371,182-194(1991)

Design and optimization of demultiplexer in ion-exchanged glass waveguides (Mazzola, M.; Montrosset, Ivo; Fincato, Antonello)V1365,2-12(1991)

Design of adaptive optical equalizers for fiber optic communication systems (Ghosh, Anjan K.; Barner, Jim; Paparao, Palacharla; Allen, Susan D.; Imen, Kamran)V1371,170-181(1991)

Dynamic allocation of buffer space in the bridge of two interconnected token rings (Das, Alok K.; Muhuri, K.)V1364,61-69(1991)

Effect of earthquake motion on the mechanical reliability of optical cables (Hart, Patrick W.; Tucker, Russ; Yuce, Hakan H.; Varachi, John P.; Wieczorek, Casey J.; DeVito, Anthony)V1366,334-342(1991)

Examining cable plant bandwidth for FDDI (Hayes, James E.)V1364,115-119(1991)

Excimer laser machining of optical fiber taps (Coyle, Richard J.; Serafino, Anthony J.; Grimes, Gary J.; Bortolini, James R.)V1412,129-137(1991)

Fabrication and optical performance of fractal fiber optics (Cook, Lee M.; Burger, Robert J.)V1449,186-192(1991)

Fast, epoxiless bonding system for fiber optic connectors (Lee, Nicholas A.)V1365,139-143(1991)

FDDI components for workstation interconnection (Anderson, Stephen J.; Bulusu, Dutt V.; Racette, James; Scholl, Frederick W.; Zack, Tim; Abbott, Peter G.)V1364,94-100(1991)

FDDI network cabling (Stevens, R. S.)V1364,101-114(1991)

Fiber array optics for electronic imaging (Burger, Robert J.; Greenberg, David A.)V1449,174-185(1991)

Fiber construction for improved mechanical reliabiltiy (Roberts, Daniel R.; Cuellar, Enrique; Kennedy, Michael T.; Tomita, Akira)V1366,129-135(1991)

Fiber optic controls for aircraft engines: issues and implications (Dasgupta, Samhita; Poppel, Gary L.; Anderson, William P.)V1374,211-222(1991)

Fiber optic link performance in the presence of reflection-induced intensity noise (Radcliffe, Jerry K.)V1366,361-371(1991)

Fiber optics development at McDonnell Douglas (Udd, Eric; Clark, Timothy E.; Joseph, Alan A.; Levy, Ram L.; Schwab, Scott D.; Smith, Herb G.; Balestra, Chet L.; Todd, John R.; Marcin, John)V1418,134-152(1991)

Fiber Optics Reliability: Benign and Adverse Environments IV (Greenwell, Roger A.; Paul, Dilip K., eds.)V1366(1991)

Fused 3 x 3 single-mode fiber-optic couplers with stable characteristics for fiber interferometric sensors (Xie, Tonglin)V1572,132-136(1991)

GaInAs PIN photodetectors on semi-insulating substrates (Crawford, Deborah L.; Wey, Y. G.; Bowers, John E.; Hafich, Michael J.; Robinson, Gary Y.)V1371,138-141(1991)

GRIN fiber lens connectors (Gomez-Reino, Carlos; Linares, Jesus)V1332,468-473(1991)

Hermaphroditic small tactical connector for single-fiber applications (Darden, Bruce V.; LeFevre, B. G.; Kalomiris, Vasilios E.)V1474,300-308(1991)

High-dynamic-range, low-noise analog optical links using external modulators: analysis and demonstration (Betts, Gary E.; Johnson, Leonard M.; Cox, Charles H.)V1371,252-257(1991)

High-dynamic-range fiber optic link using external modulator diode pumped Nd:YAG lasers (Childs, Richard B.; O'Byrne, Vincent A.)V1371,223-232(1991)

High-efficient fiber-to-stripe waveguide coupler (Domanski, Andrzej W.; Roszko, Marcin; Sierakowski, Marek W.)V1362,844-852(1991)

High-frequency fiber optic phase modulator using piezoelectric polymer coating (Imai, Masaaki M.; Yano, T.; Ohtsuka, Yoshihiro)V1371,13-20(1991)

High-impact shock testing of fiber optic components (Brown, Gair D.; Ingold, Joseph P.; Spence, Scott E.; Paxton, Jack G.)V1366,351-360(1991)

High-isolation single-taper filters (Jones, Michael G.; Moore, Douglas R.)V1365,43-52(1991)

High on-off ratio, ultrafast optical switch for optically controlled phased array (Pan, J. J.; Li, Wei Z.)V1476,122-132(1991)

High-power laser/optical-fiber-coupling device (Falciai, Riccardo; Pascucci, Tania)V1506,120-125(1991)

High-quality 2 x 2 and 4 x 4 wavelength-flatted couplers for sensor applications (Xu, Jisen; Zhang, Yi; Zhang, Xiang)V1572,137-139(1991)

High-resolution integrated optic holographic wavelength division multiplexer (Liu, William Y.; Strzelecki, Eva M.; Lin, Freddie S.; Jannson, Tomasz)V1365,20-24(1991)

High-silica cascaded three-waveguide couplers for wideband filtering by flame hydrolysis on Si (Barbarossa, Giovanni; Laybourn, Peter J.)V1583,122-128(1991)

High-silica low-loss three-waveguide couplers on Si by flame hydrolysis deposition (Barbarossa, Giovanni; Laybourn, Peter J.)V1513,37-43(1991)

High-speed polymer optical fiber networks (Bulusu, Dutt V.; Zack, Tim; Scholl, Frederick W.; Coden, Michael H.; Steele, Robert E.; Miller, Gregory D.; Lynn, Mark A.)V1364,49-60(1991)

Holographic microlenses for optical fiber interconnects (Galloway, Peter C.; Dobson, Peter J.)V1365,131-138(1991)

Implementation of FDDI in the intelligent wiring hub (Tarrant, Peter J.; Truman, Alan K.)V1364,2-6(1991)

Instrumentation concepts for multiplexed Bragg grating sensors (Giesler, Leslie E.; Dunphy, James R.; Morey, William W.; Meltz, Gerald; Glenn, William H.)V1480,138-142(1991)

Integrated "Byte-to-light" solution for fiber optic data communication (Kubinec, James J.; Somerville, James A.; Chown, David P.; Birch, Martin J.)V1364,130-143(1991)

Integrated-optical modulators for bandpass analog links (Johnson, Leonard M.; Betts, Gary E.; Roussell, Harold V.)V1371,2-7(1991)

Integrated optics bus access module for intramachine communication (Karioja, Pentti; Tammela, Simo; Hannula, Tapio)V1533,129-137(1991)

Integrated optics for fiber optic sensors (Minford, William J.; DePaula, Ramon P.)V1367,46-52(1991)

K-ion-exchange waveguide directional coupler sensor (Chen, Zheng; Dai, Ji Zhi)V1572,129-131(1991)

Laser-fabricated fiber optical taps for interconnects and optical data processing devices (Imen, Kamran; Lee, Changhun H.; Yang, Y. Y.; Allen, Susan D.; Ghosh, Anjan K.)V1365,60-64(1991)

Light coupling characteristics of planar microlens (Oikawa, Masahiro; Nemoto, Hiroyuki; Hamanaka, Kenjiro; Imanishi, Hideki; Kishimoto, Takashi)V1544,226-237(1991)

Linearized external modulator for analog applications (Trisno, Yudhi S.; Chen, Lian K.; Huber, David R.)V1371,8-12(1991)

Long-term performance of fiber optic cable plant in Navy ships (Brown, Gair D.; Ingold, Joseph P.; Paxton, Jack G.)V1589,58-68(1991)

Long-term reliability tests on fused biconical taper couplers (Moore, Douglas R.)V1366,241-250(1991)

Loss modeling of amodal fiber-to-fiber interconnects (Jennings, Kurt L.; Miller, Gregory D.)V1369,36-42(1991)

Low-back-reflection, low-loss fiber switch (Roberts, Harold A.; Emmons, David J.; Beard, Michael S.; Lu, Liang-ju)V1363,62-69(1991)

Low-cost and compact fiber-to-laser coupling with micro-Fresnel lens (Ogata, Shiro; Yoneda, Masahiro; Maeda, Tetsuo; Imanaka, Koichi)V1544,92-100(1991)

Magneto-optical optical fiber switch (Gajda, Jerzy K.; Niesterowicz, Andrzej)V1391,329-331(1991)

Mechanical testing and reliability considerations of fusion splices (Wei, Ta-Sheng; Yuce, Hakan H.; Varachi, John P.; Pellegrino, Anthony)V1366,235-240(1991)

Method to find the transimpedance gain of optical receivers using measured S-parameters (Saad, Ricardo E.; Souza, Rui F.)V1371,142-148(1991)

Microwave fiber optic RF/IF link (Pan, J. J.)V1371,195-204(1991)

Microwave high-dynamic-range EO modulators (Pan, J. J.; Garafalo, David A.)V1371,21-35(1991)

MMIC compatible photodetector design and characterization (Dallabetta, Kyle A.; de La Chapelle, Michael; Lawrence, Robert C.)V1371,116-127(1991)

Modular FDDI bridge and concentrator (Coden, Michael H.; Bulusu, Dutt V.; Ramsey, Brian; Sztuka, Edward; Morrow, Joel)V1364,22-39(1991)

Multichannel data acquistion system for assessing reliability of fiber optic couplers (Bussard, Anne B.; Biller, M.; Serfas, D. A.)V1366,380-386(1991)

Multimode approach to optical fiber components and sensors (Johnstone, Walter; Thursby, G.; Culshaw, Brian; Murray, S.; Gill, M.; McDonach, Alaster; Moodie, D. G.; Fawcett, G. M.; Stewart, George; McCallion, Kevin J.)V1506,145-149(1991)

Multiwavelength optical switch based on the acousto-optic tunable filter (Liew, Soung C.)V1363,57-61(1991)

Naval fiber optic system development program (Blackwell, Luther)V1589,69-82(1991)

Near-term applications of optical switching in the metropolitan networks (King, F. D.; Tremblay, Yves)V1364,295-303(1991)

Novel configuration of a two-mode-interference polarization splitter with a buffer layer (Antuofermo, Pasquale; Losacco, Aurora M.; De Pascale, Olga)V1583,143-149(1991)

Novel fiber optic coupler/repeater for military systems (Glista, Andrew S.)V1369,24-34(1991)

Novel high-speed dual-wavelength InAlAs/InGaAs graded superlattice

Schottky barrier photodiode for 0.8- and 1.3-um detection (Hwang, Kiu C.; Li, Sheng S.; Kao, Yung C.)V1371,128-137(1991)

Novel optical-fiber-based conical scan tracking device (Johann, Ulrich A.; Pribil, Klaus; Sontag, Heinz)V1522,243-252(1991)

Optical components for future fiber optic systems (Chapelle, Walter E.)V1396,389-395(1991)

Optical design of a high-power fiber-optic coupler (English, R. Edward; Halpin, John M.; House, F. A.; Paris, Robert D.)V1527,174-179(1991)

Optical fiber amplifiers (Ikegami, Tetsuhiko; Nakahara, Motohiro)V1362,350-360(1991)

Optical fiber demultiplexer for telecommunications (Falciai, Riccardo; Scheggi, Annamaria V.; Cosi, Franco; Cao, J. Y.)V1365,38-42(1991)

Optical fiber filter comprising a single-coupler fiber ring (or loop) and a double-coupler fiber mirror (Ja, (Yu) Frank H.)V1372,48-61(1991)

Optical frequency shifter based on stimulated Brillouin scattering in birefringent optical fiber (Duffy, Christopher J.; Tatam, Ralph P.)V1511,155-165(1991)

Optical performance of wavelength division multiplexers made by ion-exchange in glass (Nissim, Carlos; Beguin, Alain; Laborde, Pascale; Lerminiaux, Christian)V1365,13-19(1991)

Optical rotary connector for transfer of data signals from fiber optic sensors plasing on rotary objects (Svechnikov, Georgy S.; Shapar, Vladimir N.)V1589,24-31(1991)

Packaging considerations of fiber-optic laser sources (Heikkinen, Veli; Tukkiniemi, Kari; Vahakangas, Jouko; Hannula, Tapio)V1533,115-121(1991)

Passive components for the subscriber loop (Morrel, William G.)V1363,40-47(1991)

Photonic computer-aided design tools for high-speed optical modulators (Liu, Pao-Lo)V1371,46-55(1991)

Photonics technology for aerospace applications (Figueroa, Luis; Hong, C. S.; Miller, Glen E.; Porter, Charles R.; Smith, David K.)V1418,153-176(1991)

Physical-connection compliance testing for FDDI (Baldwin, Christopher)V1364,120-129(1991)

Plastic fiber couplers using simple polishing techniques (Bougas, V.; Kalymnios, Demetrios)V1504,298-302(1991)

Plastic star coupler (Yuuki, Hayato; Ito, Takeharu; Sugimoto, Tetsuo)V1592,2-11(1991)

Polymer-based electro-optic modulators: fabrication and performance (Haas, David R.; Man, Hong-Tai; Teng, Chia-Chi; Chiang, Kophu P.; Yoon, Hyun N.; Findakly, Talal K.)V1371,56-67(1991)

Protection with heat-shrinkable sleeves for optical fiber arc fusion splicing (Trunk, Jonah; Moreira, Roberto P.; Monteiro, Ricardo M.)V1365,124-130(1991)

Recent advances in single-mode 1 x n splitters using high-silica optical waveguide circuit technology (Schmidt, Kevin M.; Sumida, Shin S.; Miyashita, Tadashi M.)V1396,744-752(1991)

Reliability of planar optical couplers (Matthews, James E.; Cucalon, Antoine)V1366,206-214(1991)

Rugged 20-km fiber optic link for 2-18-GHz communications (Buckley, Robert H.; Lyons, E. R.; Goga, George)V1371,212-222(1991)

Scattering liquid crystal in optical attenuator applications (Karppinen, Arto; Kopola, Harri K.; Myllyla, Risto A.)V1455,179-189(1991)

Single-mode MxN star couplers fabricated using fused biconical taper techniques (Daniel, Hani E.; Moore, Douglas R.)V1365,53-59(1991)

Solving clock distribution problems in FDDI concentrators (Li, Gabriel)V1364,72-83(1991)

Specialty Fiber Optic Systems for Mobile Platforms (Lewis, Norris E.; Moore, Emery L., eds.)V1589(1991)

Splicing plastic optical fibers (Carson, Susan D.; Salazar, Robert A.)V1592,134-138(1991)

Transmission characteristics of multimode optical fiber with multisplices (Das, Alok K.; Mandal, Anup K.)V1365,144-155(1991)

Velocity-matched III-V travelling wave electro-optic modulator (Wang, S. Y.; Tan, Michael T.; Houng, Y. M.)V1371,98-103(1991)

Velocity-matched electro-optic modulator (Bridges, William B.; Sheehy, Finbar T.; Schaffner, James H.)V1371,68-77(1991)

Wavelength and mode-adjustable source for modal characterization of optical fiber components: application to a new alignment method (Pagnoux, Dominique; Blondy, Jean M.; Facq, Paul)V1504,98-106(1991)

Wavelength dependence of proton-exchanged LiNbO3 integrated optic directional couplers from 1.5um - 1.65um (Feuerstein, Robert J.; Januar, Indra; Mickelson, Alan R.; Sauer, Jon R.)V1583,196-201(1991)

Wavelength division multiplexers for optical fiber amplifiers (Nagy, Peter A.; Meyer, Tim J.; Tekippe, Vincent J.)V1365,33-37(1991)

Wavelength division multiplexing based on mode-selective coupling (Ouellette, Francois; Duguay, Michel A.)V1365,25-32(1991)

Wide-band analog frequency modulation of optic signals using indirect techniques (Fitzmartin, Daniel J.; Balboni, Edmund J.; Gels, Robert G.)V1371,78-86(1991)

Fiber optic networks—see also optical communications

34-Mb/s TDM photonic switching system (Yuan, Weitao; Zha, Kaide; Guo, Yili; Zhou, Bing-Kun)V1572,78-83(1991)

4-channel, 662-Mb/s medium-density WDM system with Fabry-Perot laser diodes for subscriber loop applications (Wang, Lon A.; Chapuran, Thomas E.; Menendez, Ronald C.)V1363,85-91(1991)

500 MHz baseband fiber optic transmission system for medical imaging applications (Cheng, Xin; Huang, H. K.)V1364,204-208(1991)

Advanced Fiber Communications Technologies (Kazovsky, Leonid G., ed.)V1579(1991)

Applications of optical switches in fiber optic communication networks (Hanson, Daniel; Gosset, Nathalie M.)V1363,48-56(1991)

Architectures and access protocols for multichannel networks (Karol, Mark J.)V1579,2-13(1991)

Bandwidth, throughput, and information capacity of fiber optic networks for space systems (Choudry, Amar)V1369,121-125(1991)

Behavior of WDM system for intensity modulation (Pierre, Guillaume; Jarret, Bertrand; Brun, Eric)V1511,201-211(1991)

Bridging issues in DQDB subnetworks (Tantawy, Ahmed N.)V1364,268-276(1991)

Building and campus networks for fiber distributed data interface (McIntosh, Thomas F.)V1364,84-93(1991)

Campus fiber optic enterprise networks (Weeks, Richard A.)V1364,222-227(1991)

Cascaded optical modulators: any advantages? (Sierak, Paul)V1398,238-249(1991)

Code Division Multiple Access system candidate for integrated modular avionics (Mendez, Antonio J.; Gagliardi, Robert M.)V1369,67-71(1991)

Connection system designed for plastic optical fiber local area networks (Cirillo, James R.; Jennings, Kurt L.; Lynn, Mark A.; Messuri, Dominic A.; Steele, Robert E.)V1592,53-59(1991)

Cost aspects of narrowband and broadband passive optical networks (Jones, J. R.; Sharpe, Randall B.)V1363,106-118(1991)

Crossconnects in a SONET network (Bootman, Steven R.)V1363,142-148(1991)

Crosstalk in direct-detection optical fiber FDMA networks (Hamdy, Walid M.; Humblet, Pierre A.)V1579,184-194(1991)

CWRUnet: case history of a campus-wide fiber-to-the-desktop network (Neff, Raymond K.; Klingensmith, H. W.; Gumpf, Jeffrey A.; Haigh, Peter J.)V1364,245-256(1991)

Definition and evaluation of the data-link layer of PACnet (Alsafadi, Yasser H.; Martinez, Ralph; Sanders, William H.)V1446,129-140(1991)

Designing enhanced maintainability fiber-optic networks (Schrever, Koen; De Vilder, Jan; Rolain, Yves; Voet, Marc R.; Barel, Alain R.)V1572,107-112(1991)

Distribution fiber FTTH/FTTC trial results and deployment strategies (Coleman, John D.)V1363,2-12(1991)

DS1 mapping considerations for the synchronous optical network (Cubbage, Robert W.; Littlewood, Paul A.)V1363,163-171(1991)

Dynamic allocation of buffer space in the bridge of two interconnected token rings (Das, Alok K.; Muhuri, K.)V1364,61-69(1991)

Efficient multiaccess scheme for linear lightwave networks (Kovacevic, Milan; Gerla, Mario)V1579,74-83(1991)

Evolution of fiber-to-the-curb networks toward broadband capabilities (Menendez, Ronald C.; Lu, Kevin W.; Rizzo, Annmarie; Lemberg, Howard L.)V1363,97-105(1991)

Evolution of the DQDB hybrid multiplexing for an integrated service packetized traffic (Gagnaire, A.; Ponsard, Benoit)V1364,277-288(1991)

Examining cable plant bandwidth for FDDI (Hayes, James E.)V1364,115-119(1991)

Extending HIPPI at 800-mega-bits-per-second over serial links using HOT ROD technology (Annamalai, Kadiresan)V1364,178-189(1991)

Fault management of a fiber optic LAN (Spencer, Paul E.; Zaharakis, Steven C.; Denton, Richard T.)V1364,228-234(1991)

FDDI, Campus-Wide, and Metropolitan Area Networks (Annamalai, Kadiresan; Cudworth, Stewart K.; Kasiewicz, Allen B., eds.)V1364(1991)

FDDI components for workstation interconnection (Anderson, Stephen J.; Bulusu, Dutt V.; Racette, James; Scholl, Frederick W.; Zack, Tim; Abbott, Peter G.)V1364,94-100(1991)

FDDI network cabling (Stevens, R. S.)V1364,101-114(1991)

Fiber access maintenance leverages (Sinnott, Heather; MacLeod, Wade)V1363,196-200(1991)

Fiber channel: the next standard peripheral interface and more (Cummings, Roger)V1364,170-177(1991)

Fiber hub in a second-generation ethernet system at Taylor University (Rowan, Paul)V1364,262-266(1991)

Fiber in the loop: an evolution in services and systems (Engineer, Carl P.)V1363,19-29(1991)

Fiber optic backbone for a submarine combat system (Van Metre, Richard; Curran, Mark E.; Cole, George R.; Williams, Ken)V1369,9-18(1991)

Fiber optic network for mining seismology (Lach, Zbigniew; Zientkiewicz, Jacek K.)V1364,209-220(1991)

Fiber optic sensor networks (Culshaw, Brian)V1511,168-178(1991)

Fiber optics in CATV networks (Wolfe, Ronald; Laor, Herzel)V1363,125-132(1991)

Fiber optics in a broadband bus architecture (Bagg, John C.; Ohlhaber, Jack)V1369,126-137(1991)

Fiber Optics in the Subscriber Loop (Hutcheson, Lynn D.; Kahn, David A., eds.)V1363(1991)

Fiber optics network for the adverse coal mining environment (Zientkiewicz, Jacek K.; Lach, Zbigniew)V1366,45-56(1991)

Fiber Optic Systems for Mobile Platforms IV (Lewis, Norris E.; Moore, Emery L., eds.)V1369(1991)

Flow-control mechanism for distributed systems (Maitan, Jacek)V1470,88-97(1991)

High-frequency 1.3 um InGaAsP semi-insulating buried crescent lasers for analog applications (Cheng, Wood-Hi; Appelbaum, Ami; Huang, Rong-Ting; Renner, Daniel; Cioffi, Ken R.)V1418,279-283(1991)

High-performance FDDI NIU for streaming voice, video, and data (Bergman, Larry A.; Hartmayer, Ron; Wu, Wennie H.; Cassell, P.; Edgar, G.; Lambert, James L.; Mancini, Richard; Jeng, J.; Pardo, C.; Halloran, Frank; Martinez, James C.)V1364,14-21(1991)

High-speed polymer optical fiber networks (Bulusu, Dutt V.; Zack, Tim; Scholl, Frederick W.; Coden, Michael H.; Steele, Robert E.; Miller, Gregory D.; Lynn, Mark A.)V1364,49-60(1991)

Hybrid fiber optic/electrical network for launch vehicles (Clark, Timothy E.; Curran, Mark E.)V1369,98-106(1991)

Impact of fiber-in-the-loop architecture on predicted system reliability (Lee, T. S.)V1366,28-38(1991)

Impact of fiber backscatter on loop video transmission without optical isolator (Das, Santanu K.; Ocenasek, Josef)V1363,172-176(1991)

Implementation of FDDI in the intelligent wiring hub (Tarrant, Peter J.; Truman, Alan K.)V1364,2-6(1991)

In-line testing for fiber subscriber loop applications (Jiang, Jing-Wen; So, Vincent; Lessard, Michel; Vella, Paul J.)V1363,191-195(1991)

Integrated "Byte-to-light" solution for fiber optic data communication (Kubinec, James J.; Somerville, James A.; Chown, David P.; Birch, Martin J.)V1364,130-143(1991)

Integrated-optic interconnects and fiber-optic WDM data links based on volume holography (Jannson, Tomasz; Lin, Freddie S.; Moslehi, Behzad M.; Shirk, Kevin W.)V1555,159-176(1991)

Integration of a coherent optical receiver with adaptive image rejection capability (Lachs, Gerard; Zaidi, Syed M.; Singh, Amit K.; Henning, Rudolf E.; Trascritti, D.; Kim, H.; Bhattacharya, Pallab K.; Pamulapati, J.; McCleer, P. J.; Haddad, George J.; Peng, S.)V1474,248-259(1991)

Jitter considerations for FDDI PMD (Fukuoka, Takashi; Tejika, Yasuhiro; Takada, Hisashi; Takahashi, Hidenori; Hamasaki, Yiji)V1364,40-48(1991)

Line coding for high-speed fiber optic transmission systems (Subramanian, K. R.; Dubey, V. K.; Low, J. P.; Tan, L. S.)V1364,190-201(1991)

Local area network applications of plastic optical fiber (Cirillo, James R.; Jennings, Kurt L.; Lynn, Mark A.; Messuri, Dominic A.; Steele, Robert E.)V1592,42-52(1991)

Low-back-reflection, low-loss fiber switch (Roberts, Harold A.; Emmons, David J.; Beard, Michael S.; Lu, Liang-ju)V1363,62-69(1991)

Management of an adaptable-bit-rate video service in a MAN environment (Marini, Michele; Albanese, Andres)V1364,289-294(1991)

Metropolitan area networks: a corner stone in the broadband era (Ghanem, Adel)V1364,312-319(1991)

Modular FDDI bridge and concentrator (Coden, Michael H.; Bulusu, Dutt V.; Ramsey, Brian; Sztuka, Edward; Morrow, Joel)V1364,22-39(1991)

Modular packaging for FTTC and B-ISDN (Koht, Lowell)V1363,158-162(1991)

Multibit optical sensor networking (Pervez, Anjum)V1511,220-231(1991)

Multichannel fiber optic broadband video communication system for monitoring CT/MR examinations (Huang, H. K.; Kangarloo, Hooshang; Tecotzky, Raymond H.; Cheng, Xin; Vanderweit, Don)V1444,214-220(1991)

Multichannel optical data link (Ota, Yusuke; Swartz, Robert G.)V1364,146-152(1991)

Multiple communication networks for a radiological PACS (Wong, Albert W.; Stewart, Brent K.; Lou, Shyh-Liang; Chan, Kelby K.; Huang, H. K.)V1446,73-80(1991)

Multiwavelength optical switch based on the acousto-optic tunable filter (Liew, Soung C.)V1363,57-61(1991)

Naval fiber optic system development program (Blackwell, Luther)V1589,69-82(1991)

NAVSEA gigabit optical MAN prototype history and status (Albanese, Andres; Devetzis, Tasco N.; Ippoliti, A, G.; Karr, Michael A.; Maszczak, M. W.; Dorris, H. N.; Davis, James H.)V1364,320-326(1991)

Near-term applications of optical switching in the metropolitan networks (King, F. D.; Tremblay, Yves)V1364,295-303(1991)

Network powering architecture for fiber-to-the-subscriber systems (Pellerin, Sharon L.)V1363,186-190(1991)

New CATV fiber-to-the-subscriber architectures (Kim, Gary)V1363,133-140(1991)

Novel approach to optical frequency synthesis in coherent lightwave systems (Fernando, P. N.; Fake, M.; Seeds, A. J.)V1372,152-163(1991)

Novel fiber optic coupler/repeater for military systems (Glista, Andrew S.)V1369,24-34(1991)

Novel integrated acousto-optical LiNbO3 device for application in single-laser self-heterodyne systems (Gieschen, Nikolaus; Rocks, Manfred; Olivier, Lutz)V1579,237-248(1991)

Optical bus protocol for a distributed-shared-memory multiprocessor (Davis, Martin H.; Ramachandran, Umakishore)V1563,176-187(1991)

Optical subscriber line transmission system to support an ISDN primary-rate interface (Wataya, Hideo; Tsuchiya, Toshiyuki)V1363,72-84(1991)

Options for campus fiber networks (Henderson, Byron B.; Green, Emily N.)V1364,235-244(1991)

Optoelectronic devices for fiber-optic sensor interface systems (Hong, C. S.; Hager, Harold E.; Capron, Barbara; Mantz, Joseph L.; Beranek, Mark W.; Huggins, Raymond W.; Chan, Eric Y.; Voitek, Mark; Griffith, David M.; Livezey, Darrell L.; Scharf, Bruce R.)V1418,177-187(1991)

Passive components for the subscriber loop (Morrel, William G.)V1363,40-47(1991)

Passive optic solution for an urban rehabilitation topology (Petruziello, David)V1363,30-37(1991)

Performance analysis of lightwave packet communication networks (Ramaswamy, Raju)V1364,153-162(1991)

Performance of pseudo-orthogonal codes in temporal, spatial, and spectral code division multiple access systems (Mendez, Antonio J.; Gagliardi, Robert M.)V1364,163-169(1991)

Physical-connection compliance testing for FDDI (Baldwin, Christopher)V1364,120-129(1991)

Planning for fiber optic use at the University of Massachusetts (Sailer, Donald R.)V1364,257-261(1991)

Plastic star coupler (Yuuki, Hayato; Ito, Takeharu; Sugimoto, Tetsuo)V1592,2-11(1991)

Polarization-diversity fiber networks (Cullen, Thomas J.)V1372,164-172(1991)

Policy issues affecting telephone company opportunities in CATV and deployment of fiber optic cable (Keane, William K.)V1363,184-185(1991)

Proposed one- and two-fiber-to-the-pedestal architectural evolution (Schiffler, Richard A.)V1363,13-18(1991)

Prospects for the development and application of plastic optical fibers (Groh, Werner; Kuder, James E.; Theis, Juergen)V1592,20-30(1991)

Recent progress of coherent lightwave systems at Bellcore (Sessa, William B.; Welter, Rudy; Wagner, Richard E.; Maeda, Mari W.)V1372,208-218(1991)

Reconfiguration phase in rearrangeable multihop lightwave networks (Acampora, Anthony S.; Labourdette, Jean-Francois P.)V1579,30-39(1991)

Reliability assurance of optoelectronic devices in the local loop (Koelbl, Roy S.)V1366,17-25(1991)

Research and development of a NYNEX switched multi-megabit data service prototype system (Maman, K. H.; Haines, Robert; Chatterjee, Samir)V1364,304-311(1991)

Rings in a SONET network (Clendening, Steven J.)V1363,149-157(1991)

Routing in distributed information systems (Ras, Zbigniew W.)V1470,76-87(1991)

SAFENET II: The Navy's FDDI-based computer network standard (Paige, Jeffrey L.; Howard, Edward A.)V1364,7-13(1991)

SONET inter-vendor compatibility (Bowmaster, Thomas A.; Cockings, Orville R.; Swanson, Robert A.)V1363,119-124(1991)

STARNET: an integrated services broadband optical network with physical star topology (Poggiolini, Pierluigi T.; Kazovsky, Leonid G.)V1579,14-29(1991)

Single-mode MxN star couplers fabricated using fused biconical taper techniques (Daniel, Hani S.; Moore, Douglas R.)V1365,53-59(1991)

Soliton ring network (Sauer, Jon R.; Islam, Mohammed N.; Dijaili, Sol P.)V1579,84-97(1991)

Solving clock distribution problems in FDDI concentrators (Li, Gabriel)V1364,72-83(1991)

Specialty Fiber Optic Systems for Mobile Platforms (Lewis, Norris E.; Moore, Emery L., eds.)V1589(1991)

Spread spectrum technique for passive multiplexing of interferometric optical fiber sensors (Uttamchandani, Deepak G.; Al-Raweshidy, H. S.)V1511,212-219(1991)

TeraNet: a multigigabit-per-second hybrid circuit/packet-switched lightwave network (Gidron, Rafael; Elby, Stuart D.; Acampora, Anthony S.; Georges, John B.; Lau, Kam-Yin)V1579,40-48(1991)

Time-division optical microarea networks (Prucnal, Paul R.; Johns, Steven T.; Krol, Mark F.; Stacy, John L.)V1389,462-476(1991)

Topologies for linear lightwave networks (Bala, Krishna; Stern, Thomas E.)V1579,62-73(1991)

Use of fiber optic communications and control for a tethered undersea vehicle (Caimi, Frank M.; Neely, Jerry; Grossman, Barry G.; Alavie, A. T.)V1589,90-99(1991)

Wavelength dependence of proton-exchanged LiNbO3 integrated optic directional couplers from 1.5um - 1.65um (Feuerstein, Robert J.; Januar, Indra; Mickelson, Alan R.; Sauer, Jon R.)V1583,196-201(1991)

Wavelength division multiplexing of services in a fiber-to-the-home system (Unterleitner, Fred C.)V1363,92-96(1991)

Fiber optics—see also integrated optics

1.25-Gb/s wideband LED driver design using active matching techniques (Gershman, Vladimir)V1474,75-82(1991)

1.55-um broadband fiber sources pumped near 980 nm (Wysocki, Paul F.; Kalman, Robert F.; Digonnet, Michel J.; Kim, Byoung-Yoon)V1373,66-77(1991)

7th Mtg in Israel on Optical Engineering (Oron, Moshe; Shladov, Itzhak, eds.)V1442(1991)

Acoustic characterization of optical fiber glasses (Jen, Cheng K.; Neron, C.; Shang, Alain; Abe, Koichi; Bonnell, Lee J.; Kushibiki, J.; Saravanos, C.)V1590,107-119(1991)

Advanced Fiber Communications Technologies (Kazovsky, Leonid G., ed.)V1579(1991)

Advances in Erbium-doped fiber amplifiers for optical communications (Zyskind, John L.)V1373,80-92(1991)

Advances in laser pump sources for erbium-doped fiber amplifiers (Henshall, Gordon D.; Hadjifotiou, A.; Baker, R. A.; Warbrick, K. J.)V1418,286-291(1991)

Aging behavior of low-strength fused silica fibers (Yuce, Hakan H.; Key, P. L.; Chandan, Harish C.)V1366,120-128(1991)

Analysis of the self-oscillation phenomenon of fiber optically addressed silicon microresonators (Rao, Yun-Jiang; Culshaw, Brian)V1506,126-133(1991)

Angularly-polished optical fiber tips (Bunch, Robert M.; Caughey, Joseph P.)V1396,411-416(1991)

Application of fiber optic delay lines and semiconductor optoelectronics to microwave signal processing (Taylor, Henry F.)V1371,150-160(1991)

Applications of fiber amplifiers to high-data-rate nonlinear transmission (Spirit, David M.; Brown, Graeme N.; Marshall, Ian W.; Blank, Lutz C.)V1373,197-208(1991)

Automatic inspection of optical fibers (Silberberg, Teresa M.)V1472,150-156(1991)

A very unique pre-1970 fiber optic instrumentation system (Miller, Glen E.)V1589,2-10(1991)

Bandwidth measurements of polymer optical fibers (Karim, Douglas P.)V1592,31-41(1991)

CAN-AM Eastern '90 (Antos, Ronald L.; Krisiloff, Allen J., eds.)V1398(1991)

Characteristics of MSM detectors for meander channel CCD imagers on GaAs (Kosel, Peter B.; Bozorgebrahimi, Nercy; Iyer, J.)V1541,48-59(1991)

Characterization of GRIN-SCH-SQW amplifiers (Haake, John M.; Zediker, Mark S.; Balestra, Chet L.; Krebs, Danny J.; Levy, Joseph L.)V1418,298-308(1991)

Chemical analysis of human urinary and renal calculi by Raman laser fiber-optics method (Hong, Nguyen T.; Phat, Darith; Plaza, Pascal; Daudon, Michel; Dao, Nguyen Q.)V1525,132-142(1991)

Coherent detection: n-ary PPM versus PCM (Cryan, Robert A.; Unwin, Rodney T.; Massarella, Alistair J.; Sibley, Martin J.; Garrett, Ian)V1372,64-71(1991)

Coherent Lightwave Communications: Fifth in a Series (Steele, Roger C.; Sunak, Harish R., eds.)V1372(1991)

Coherent optical fiber communications (Meissner, P.)V1522,182-193(1991)

Comparison of Gauss' and Petermann's formulas for real single-mode fibers by far-field pattern technique (Pospisilova, Marie; Schneiderova, Martina)V1504,287-291(1991)

Comparison of gain dependence of different Er-doped fiber structures (Tammela, Simo; Zhan, Xiaowei; Kiiveri, Pauli)V1373,103-110(1991)

Comparison of transient analog data fiber optic links (Clark, Wally T.; Darrow, K. W.; Skipper, M. C.)V1371,258-265(1991)

Criteria for accurate cutback attenuation measurements (Haigh, N. R.; Linton, R. S.; Johnson, R.; Grigsby, R.)V1366,259-264(1991)

Cyclic fatigue behavior of silica fiber (Rogers, Harvey N.)V1366,112-117(1991)

Data transmission at 1.3 um using silicon spatial light modulator (Xiao, Xiaodong; Goel, Kamal K.; Sturm, James C.; Schwartz, P. V.)V1476,301-304(1991)

Dependence of output beam profile on launching conditions in fiber-optic beam delivery systems for Nd:YAG lasers (Su, Daoning; Boechat, Alvaro A.; Jones, Julian D.)V1502,41-51(1991)

Design, fabrication, and testing of a 7-bit binary fiber optic delay line (Goutzoulis, Anastasios P.; Davies, D. K.)V1371,182-194(1991)

Design of adaptive optical equalizers for fiber optic communication systems (Ghosh, Anjan K.; Barner, Jim; Paparao, Palacharla; Allen, Susan D.; Imen, Kamran)V1371,170-181(1991)

Design of an optical fiber power supplying link (Wojcik, Waldemar; Smolarz, Andrzej)V1504,292-297(1991)

Development of fly-by-light systems for commercial aircraft (Todd, John R.; Yount, Larry J.)V1369,72-78(1991)

Digital imaging of aircraft dynamic seals: a fiber optics solution (Nwagboso, Christopher O.)V1500,234-245(1991)

Discrete analog and digital devices using fiber-optic logic elements (Petrov, Mikhail P.; Miridonov, S. V.)V1621,402-413(1991)

Effect of color removal on optical fiber reliability (Kennedy, Michael T.; Cuellar, Enrique; Roberts, Daniel R.)V1366,167-176(1991)

Effect of the glass composition on the emission band of erbium-doped active fibers (Cognolato, Livio; Gnazzo, Angelantonio; Sordo, Bruno; Cocito, Guiseppe)V1579,249-256(1991)

Effects of ionizing radiation on fiber optic systems and components for use in mobile platforms (Reddy, Mahesh C.; Krinsky, Jeffrey A.)V1369,107-113(1991)

Efficiency of a 5V/5-mW power by light power supply for avionics applications (Sherman, Bradley D.; Mendez, Antonio J.; Morookian, John-Michael)V1369,60-66(1991)

Efficient, high-power, high-gain Er3+-doped silica fiber amplifier (Massicott, Jennifer F.; Wyatt, Richard; Ainslie, B. J.; Craig-Ryan, S. P.)V1373,93-102(1991)

Endoscopic tissue autofluorescence measurements in the upper aerodigestive tract and the bronchi (Braichotte, D.; Wagnieres, G.; Monnier, Philippe; Savary, M.; Bays, Roland; van den Bergh, Hubert; Chatelain, Andre)V1525,212-218(1991)

Erbium-doped fiber amplifiers for future undersea transmission systems (Aspell, Jennifer; Bergano, Neal S.)V1373,188-196(1991)

Evaluation of parameters in stimulated backward Brillouin scattering (de Oliveira, C. A.; Jen, Cheng K.)V1590,101-106(1991)

Excess-loss dependence of field shape of cladding modes in single-mode fibers (Chen, Mei-Xia)V1504,274-280(1991)

Excimer laser micromachining for passive fiber coupling to polymeric waveguide devices (Booth, Bruce L.; Hohman, James L.; Keating, Kenneth B.; Marchegiano, Joseph E.; Witman, Sandy L.)V1377,57-63(1991)

Excited state cross sections for Er-doped glasses (Zemon, Stanley A.; Lambert, Gary M.; Miniscalco, William J.; Davies, Richard W.; Hall, Bruce T.; Folweiler, Robert C.; Wei, Ta-Sheng; Andrews, Leonard J.; Singh, Mahendra P.)V1373,21-32(1991)

Fabrication of light-guiding devices and fiber-coupling structures by the LIGA process (Rogner, Arnd; Ehrfeld, Wolfgang)V1506,80-91(1991)

Fabrication of rare-earth-doped optical fiber (DiGiovanni, David J.)V1373,2-8(1991)

Fiber coupled heterodyne interferometric displacement sensor (Nerheim, Noble M.)V1542,523-533(1991)

Fiber Laser Sources and Amplifiers II (Digonnet, Michel J., ed.)V1373(1991)

Fiber optic based multiprobe system for intraoperative monitoring of brain functions (Mayevsky, Avraham; Flamm, E. S.; Pennie, W.; Chance, Britton)V1431,303-313(1991)

Fiber optic delay lines for wideband signal processing (Taylor, Henry F.; Gweon, S.; Fang, S. P.; Lee, Chung E.)V1562,264-275(1991)

Fiber optic lighting system for plant production (St. George, Dennis R.; Feddes, John J.)V1379,69-80(1991)

Fiber optic link for millimeter-wave communication satellites (Polifko, David M.; Malone, Steven A.; Daryoush, Afshin S.; Kunath, Richard R.)V1476,91-99(1991)

Fiber optic measurement of intracellular pH in intact rat liver using pH-sensitive dyes (Felberbauer, Franz; Graf, Juerg)V1510,63-71(1991)

Fiber optic measurement standards (Pollitt, Stuart)V1504,80-87(1991)

Fiber Optic Metrology and Standards (Soares, Oliverio D., ed.)V1504(1991)

Fiber optics development at McDonnell Douglas (Udd, Eric; Clark, Timothy E.; Joseph, Alan A.; Levy, Ram L.; Schwab, Scott D.; Smith, Herb G.; Balestra, Chet L.; Todd, John R.; Marcin, John)V1418,134-152(1991)

Fiber optics in liquid propellant rocket engine environments (Delcher, Ray C.; Dinnsen, Doug K.; Barkhoudarian, S.)V1369,114-120(1991)

Fiber Optics Reliability: Benign and Adverse Environments IV (Greenwell, Roger A.; Paul, Dilip K., eds.)V1366(1991)

Fiber Optic Systems for Mobile Platforms IV (Lewis, Norris E.; Moore, Emery L., eds.)V1369(1991)

Fiber probe for ring pattern (Yang, Xueyu; Spears, Kenneth G.)V1367,382-386(1991)

Fluoride fiber lasers (Smart, Richard G.; Carter, John N.; Tropper, Anne C.; Hanna, David C.)V1373,158-165(1991)

FM-cavity-dumped Nd-doped fiber laser (Zenteno, Luis A.; Po, Hong)V1373,246-253(1991)

Free-surface temperature measurement of shock-loaded tin using ultrafast infrared pyrometry (Mondot, Michel; Remiot, Christian)V1558,351-361(1991)

French proposal for IEC/TC 86/WG 4 OTDR calibration (Gauthier, Francis)V1504,55-65(1991)

Frequency doubling, absorption, and grating formation in glass fibers: effective defects or defective effects? (Russell, Philip S.; Poyntz-Wright, L. J.; Hand, Duncan P.)V1373,126-139(1991)

Frequency up-conversion in Pr3+ doped fibers (Gomes, Anderson S.; de Araujo, Cid B.; Moraes, E. S.; Opalinska, M. M.; Gouveia-Neto, A. S.)V1579,257-263(1991)

General theoretical approach describing the complete behavior of the erbium-doped fiber amplifier (Marcerou, Jean-Francois; Fevrier, Herve A.; Ramos, Josiane; Auge, Jacques C.; Bousselet, Philippe)V1373,168-186(1991)

Geometric measurement of optical fibers with pulse-counting method (Nie, Qiuhua; Nelson, John C.; Fleming, Simon C.)V1332,409-420(1991)

Germanium photodiodes calibration as standards of optical fiber systems power measurements (Campos, Joaquin; Corredera, Pedro; Pons, Alicia A.; Corrons, Antonio)V1504,66-74(1991)

High-fidelity phase conjugation generated by holograms: application to imaging through multimode fibers (Pan, Anpei; Marhic, Michel E.; Epstein, Max)V1396,99-106(1991)

High-Frequency Analog Fiber Optic Systems (Sierak, Paul, ed.)V1371(1991)

High-dynamic-range, low-noise analog optical links using external modulators: analysis and demonstration (Betts, Gary E.; Johnson, Leonard M.; Cox, Charles H.)V1371,252-257(1991)

High-power single-element pseudomorphic InGaAs/GaAs/AlGaAs single-quantum-well lasers for pumping Er-doped fiber amplifiers (Larsson, Anders G.; Forouhar, Siamak; Cody, Jeffrey; Lang, Robert J.; Andrekson, Peter A.)V1418,292-297(1991)

High-power transmission through step-index multimode fibers (Setchell, Robert E.; Meeks, Kent D.; Trott, Wayne M.; Klingsporn, Paul E.; Berry, Dante M.)V1441,61-70(1991)

High-selectivity spectral multiplexers-demultiplexers usable in optical telecommunications obtained from multidielectric coatings at the end of optical fibers (Richier, R.; Amra, Claude)V1504,202-210(1991)

High-speed polarimetric measurements for fiber-optic communications (Calvani, Riccardo A.; Caponi, Renato; Piglia, Roberto)V1504,258-263(1991)

High-speed polymer optical fiber networks (Bulusu, Dutt V.; Zack, Tim; Scholl, Frederick W.; Coden, Michael H.; Steele, Robert E.; Miller, Gregory D.; Lynn, Mark A.)V1364,49-60(1991)

Holographic precompensation for one-way transmission of diffraction-limited beams through diffusing media (Pan, Anpei; Marhic, Michel E.)V1461,8-16(1991)

Ignition risks of fiber optic systems (Hills, Peter C.; Samson, Peter J.)V1589,110-119(1991)

Impact of relative humidity on mechanical test results for optical fiber (Dreyer, Donald R.; Saikkonen, Stuart L.; Hanson, Thomas A.; Linchuck, Barry A.)V1366,372-379(1991)

Indentation experiments on silica optical fibers (Lin, Bochien; Matthewson, M. J.; Nelson, G. J.)V1366,157-166(1991)

Influence of dose rate on radiation-induced loss in optical fibers (Henschel, Henning; Koehn, Otmar; Schmidt, Hans U.)V1399,49-63(1991)

Infrared transmitting glasses and fibers for chemical analysis (Hilton, Albert R.)V1437,54-59(1991)

Inhomogeneous line broadening of optical transitions in Nd3+ and Er3+ doped preforms and fibers (Briancon, Anne-Marie; Jacquier, Bernard; Gacon, Jean-Claude; Le Sergent, Christian; Marcerou, Jean-Francois)V1373,9-20(1991)

Interferometric system for the inspection and measurement of the quality of optical fiber ends (Corredera, Pedro; Pons, Alicia A.; Campos, Joaquin; Corrons, Antonio)V1504,281-286(1991)

Interpretation of measured spectral attenuation curves of optical fibers by deconvolution with source spectrum (Hoefle, Wolfgang)V1504,140-146(1991)

Intl Workshop on Photoinduced Self-Organization Effects in Optical Fiber (Ouellette, Francois, ed.)V1516(1991)

Intrapulse stimulated Raman scattering and ultrashort solitons in optical fibers (Headley, Clifford; Agrawal, Govind P.; Reardon, A. C.)V1497,197-201(1991)

Ionic effects on silica optical fiber strength and models for fatigue (Rondinella, Vincenzo V.; Matthewson, M. J.)V1366,77-84(1991)

Ionizing radiation-induced attenuation in optical fibers at multiple wavelengths and temperature extremes (Krinsky, Jeffrey A.; Reddy, Mahesh C.)V1366,191-203(1991)

Large-signal model and signal/noise ratio analysis for Nd3+-doped fiber amplifiers at 1.3 um (Dakss, Mark L.; Miniscalco, William J.)V1373,111-124(1991)

Laser optical fiber high-speed camera (Xia, Sheng-jie; Yang, Ye-min; Tang, Di-zhu)V1358,43-45(1991)

Light distributor for endoscopic photochemotherapy (Lenz, P.)V1525,192-195(1991)

Linear position sensing by light exchange between two lossy waveguides (Sultan, Michel F.; O'Rourke, Michael J.)V1584,212-219(1991)

Long-term fiber reliability (Saifi, Mansoor A.)V1366,58-70(1991)

Low-loss L-band microwave fiber optic link for control of a T/R module (Wanuga, Stephen; Ackerman, Edward I.; Kasemset, Dumrong; Hogue, David W.; Chinn, Stephen R.)V1374,97-106(1991)

Measurement of mode field diameter and fiber bending loss (Kiang, Ying J.; Stigliani, Daniel J.)V1366,252-258(1991)

Measurement of nonlinear constants in photosensitive glass optical fibers (Oesterberg, Ulf L.)V1504,107-109(1991)

Measurement of physical parameters with special fibers (Shiota, Alan T.; Inada, Koichi)V1504,90-97(1991)

Measurement of radiation-induced attenuation in optical fibers by optical-time-domain reflectometry (Looney, Larry D.; Lyons, Peter B.)V1474,132-137(1991)

Mechanical performance and reliability of Corning Titan SMF CPC5 fiber after exposure to a variety of environments (Vethanayagam, Thirukumar K.; Linchuck, Barry A.; Dreyer, Donald R.; Toler, J. R.; Amos, Lynn G.; Taylor, Donna L.)V1366,343-350(1991)

Mechanism of excimer-laser-induced absorption in fused silica fibers (Artjushenko, Vjacheslav G.; Konov, Vitaly I.; Pashinin, Vladimir P.; Silenok, Alexander S.; Mueller, Gerhard J.; Schaldach, Brita J.; Ulrich, R.)V1420,176-176(1991)

Mechanisms of UV-laser-induced absorption in fused silica fibers (Artjushenko, Vjacheslav G.; Konov, Vitaly I.; Konstantinov, N. Y.; Pashinin, Vladimir P.; Silenok, Alexander S.; Mueller, Gerhard J.; Schaldach, Brita J.; Ulrich, R.; Neuberger, Wolfgang; Castro, Jose L.)V1590,131-136(1991)

Melt-processed calcium aluminate fibers: structural and optical properties (Wallenberger, Frederick T.; Weston, Norman E.; Brown, Sherman D.)V1484,116-124(1991)

Micro-Optics II (Scheggi, Annamaria V., ed.)V1506(1991)

Microwave fiber optic RF/IF link (Pan, J. J.)V1371,195-204(1991)

Microwave fiber optic link with DFB lasers (Huff, David B.; Blauvelt, Henry A.)V1371,244-249(1991)

Modal noise reduction in analog fiber optic links by superposition of high-frequency modulation (Pepeljugoski, Petar K.; Lau, Kam-Yin)V1371,233-243(1991)

Modeling direct detection and coherent-detection lightwave communication systems that utilize cascaded erbium-doped fiber amplifiers (Joss, Brian T.; Sunak, Harish R.)V1372,94-117(1991)

Modeling of three-level laser superfluorescent fiber sources (Kalman, Robert F.; Digonnet, Michel J.; Wysocki, Paul F.)V1373,209-222(1991)

Model of second-harmonic generation in glass fibers based on multiphoton ionization interference effects (Anderson, Dana Z.; Mizrahi, Victor; Sipe, John E.)V1516,154-161(1991)

Multigigabit solitary-wave propagation in both the normal and anomalous dispersion regions of optical fibers (Potasek, M. J.; Tabor, Mark)V1579,232-236(1991)

Narrow linewidth fiber laser sources (Cowle, Gregory J.; Reekie, Laurence; Morkel, Paul R.; Payne, David N.)V1373,54-65(1991)

Nd:YAG-laser-based time-domain reflectometry measurements of the intrinsic reflection signature from PMMA fiber splices (Lawson, Christopher M.; Michael, Robert R.; Dressel, Earl M.; Harmony, David W.)V1592,73-83(1991)

Neutron irradiation effects on fibers operating at 1.3 um and 1.55 um (Singh, Anjali; Banerjee, Pranab K.; Mitra, Shashanka S.)V1366,184-190(1991)

New measurement techniques for modal power distribution in fibers (Golub, Mikhail A.; Sisakian, Iosif N.; Soifer, Victor A.; Uvarov, G. V.)V1366,273-282(1991)

Noise and gain performance for an Er3+-doped fiber amplifier pumped at 980 nm or 1480 nm (Vendeltorp-Pommer, Helle; Pedersen, Frands B.; Bjarklev, Anders; Hedegaard Povlsen, Joern)V1373,254-265(1991)

Noise characteristics of rare-earth-doped fiber sources and amplifiers (Morkel, Paul R.; Laming, Richard I.; Cowle, Gregory J.; Payne, David N.)V1373,224-233(1991)

Novel approach for the refractive index gradient measurement in microliter volumes using fiber-optic technology (Synovec, Robert E.; Renn, Curtiss N.)V1435,128-139(1991)

Novel modes in a-power GRIN (Ojeda-Castaneda, Jorge; Szwaykowski, P.)V1500,246-251(1991)

Novel plastic image-transmitting fiber (Suzuki, Fumio)V1592,150-157(1991)

One-dimensional proportional counters for x-ray all-sky monitors (Matsuoka, Masaru; Yamauchi, Masamitsu; Nakamura, Haruo; Kondo, M.; Kawai, N.; Yoshida, A.; Imai, Tohru)V1549,2-8(1991)

Optical cable reliability: lessons learned from post-mortem analyses (Kilmer, Joyce P.; DeVito, Anthony; Yuce, Hakan H.; Wieczorek, Casey J.; Varachi, John P.; Anderson, William T.)V1366,85-91(1991)

Optical data storage in photosensitive fibers (Campbell, Robert J.; Kashyap, Raman)V1499,160-164(1991)

Optical fiber filter comprising a single-coupler fiber ring (or loop) and a double-coupler fiber mirror (Ja, (Yu) Frank H.)V1372,48-61(1991)

Optical fiber measurements and standardization: status and perspectives (Di Vita, P.)V1504,38-46(1991)

Optical fiber radiation damage measurements (Ediriweera, Sanath R.; Kvasnik, Frank)V1399,64-75(1991)

Optical fibers for UV applications (Fabian, Heinz; Grzesik, Ulrich; Woerner, K.-H.; Klein, Karl-Friedrich)V1513,168-173(1991)

Optical fibers for magneto-optical recording (Opsasnick, Michael N.; Stancil, Daniel D.; White, Sean T.; Tsai, Ming-Horn)V1499,276-280(1991)

Optical fiber units with ferroelectric liquid crystals for optical computing (Domanski, Andrzej W.; Roszko, Marcin; Sierakowski, Marek W.)V1362,907-915(1991)

Optical time-domain reflectometry performance enhancement using erbium-doped fiber amplifiers (Keeble, Peter J.)V1366,39-44(1991)

Optimization of gain-switched diode lasers for high-speed fiber optics (MacFarlane, Duncan L.; Tatum, Jim A.)V1365,88-95(1991)

Optomechanical M x N fiber-optic matrix switch (Pesavento, Gerry A.)V1474,57-61(1991)

OTDR calibration for attenuation measurement (Moeller, Werner; Heitmann, Walter; Reich, M.)V1504,47-54(1991)

Overview of fiber optics in the natural space environment (Barnes, Charles E.; Dorsky, Leonard; Johnston, Alan R.; Bergman, Larry A.; Stassinopoulos, E.)V1366,9-16(1991)

Phase conjugation as a probe for noncentrosymmetry grating formation in organics (Charra, Fabrice; Nunzi, Jean-Michel; Messier, Jean)V1516,211-219(1991)

Phase measuring fiber optic electronic speckle pattern interferometer (Joenathan, Charles; Khorana, Brij M.)V1396,155-163(1991)

Phenomenological description of self-organized x(2) grating formation in centrosymmetric doped optical fibers (Chmela, Pavel)V1516,116-124(1991)

Photoinduced effects in optical waveguides (Dianov, Evgeni M.; Kazansky, Peter G.; Stepanov, D. Y.)V1516,81-98(1991)

Photoinduced self-organization in optical fiber: some answered and unanswered questions (Ouellette, Francois; Gagnon, Daniel; LaRochelle, Sophie; Poirier, Michel)V1516,2-13(1991)

Photosensitive germanosilicate preforms and fibers (Williams, Doug L.; Davey, Steven T.; Kashyap, Raman; Armitage, J. R.; Ainslie, B. J.)V1513,158-167(1991)

Plastic fiber applications for lighting of airports and buildings (Jaquet, Patrick J.)V1592,165-172(1991)

Plastic-optical-fiber-based photonic switch (Grimes, Gary J.; Blyler, Lee L.; Larson, Allen L.; Farleigh, Scott E.)V1592,139-149(1991)

Plastic Optical Fibers (Kitazawa, Mototaka; Kreidl, John F.; Steele, Robert E., eds.)V1592(1991)

Plastic optical fibers for automotive applications (Suganuma, Heiroku; Matsunaga, Tadayo)V1592,12-17(1991)

Polarization effects in fiber lasers: phenomena, theory, and applications (Lin, Jin T.; Gambling, William A.)V1373,42-53(1991)

Polarization-maintaining single-mode fibers: measurement and prediction of fundamental characteristics (Sasek, Ladislav; Vohryzek, Jachym)V1504,147-154(1991)

Power-by-light flight control: an EMI-immune backup (Todd, John R.)V1589,48-53(1991)

Progress in fluoride fiber lasers and amplifiers (France, Paul W.; Brierley, Michael C.)V1373,33-39(1991)

Progress in organic third-order nonlinear optical materials (Kuzyk, Mark G.)V1436,160-168(1991)

Radiation effects on bend-insensitive fibers at 1300 nm and 1500 nm (Karbassiyoon, Kamran; Greenwell, Roger A.; Scott, David M.; Spencer, Robert A.)V1366,178-183(1991)

Random imperfections of multimode fiber and the signal dispersion (Heimrath, Adam E.; Bielak, Leslaw)V1366,265-272(1991)

Rare-earth-doped fluoride glasses for active optical fiber applications (Adam, Jean-Luc; Smektala, Frederic; Denoue, Emmanuel; Lucas, Jacques)V1513,150-157(1991)

Recent developments in plastic optical fibers: 135-C heat-resistant fibers and image transfer fibers (Kitazawa, Mototaka)V1369,44-47(1991)

Recirculating fiber optical RF-memory loop in countermeasure systems (Even-Or, Baruch; Lipsky, S.; Markowitz, Raymond; Herczfeld, Peter R.; Daryoush, Afshin S.; Saedi, Reza)V1371,161-169(1991)

Reliability considerations for fiber optic systems in telecommunications (Shelton, Douglas S.)V1366,2-8(1991)

Remote Raman spectroscopy using diode lasers and fiber-optic probes (Angel, S. M.; Myrick, Michael L.; Vess, Thomas M.)V1435,72-81(1991)

Remote visual monitoring of seal performance in aircraft jacks using fiber optics (Nwagboso, Christopher O.; Whomes, Terence L.; Davies, P. B.)V1386,30-41(1991)

Research on a curved optical fiber (Eisenbach, Shlomo; Bodenheimer, Joseph S.)V1442,242-251(1991)

Rugged 20-km fiber optic link for 2-18-GHz communications (Buckley, Robert H.; Lyons, E. R.; Goga, George)V1371,212-222(1991)

Self-action of supremely short light pulses in fibers (Azarenkov, Aleksey N.; Altshuler, Grigori B.; Kozlov, Sergey A.)V1409,166-177(1991)

Sensitivity of polarization-maintaining fibers to temperature variations (Ruffin, Paul B.; Sung, C. C.)V1478,160-167(1991)

Simulation and design of integrated optical waveguide devices by the BPM (Yip, Gar L.)V1583,240-248(1991)

Single-fiber wideband transmission systems for HDTV production applications (Cheng, Xin; Levin, Paul A.; Nguyen, Hat)V1363,177-181(1991)

Single-mode fiber optic rotary joint for aircraft applications (Lewis, Warren H.; Miller, Michael B.)V1369,79-86(1991)

Single-mode fibers to single-mode waveguides coupling with minimum Fresnel back-reflection (Sneh, Anat; Ruschin, Shlomo; Marom, Emanuel)V1442,252-257(1991)

Specialty Fiber Optic Systems for Mobile Platforms (Lewis, Norris E.; Moore, Emery L., eds.)V1589(1991)

Spectroscopical studies of the ionizing-radiation-induced damage in optical fibers (Ediriweera, Sanath R.; Kvasnik, Frank)V1504,110-117(1991)

Spectrum thermal stability of Nd- and Er-doped fiber sources (Wysocki, Paul F.; Fesler, Kenneth A.; Liu, K.; Digonnet, Michel J.; Kim, Byoung-Yoon)V1373,234-245(1991)

Strength and fatigue of optical fibers at different temperatures (Biswas, Dipak R.)V1366,71-76(1991)

Study of different optical fibers (Boudot, C.; Vastra, I.)V1502,72-82(1991)

Study of stick slip behavior in interface friction using optical fiber pull-out experiment (Tsai, Kun-Hsieh; Kim, Kyung-Suk)V1554A,529-541(1991)

Temperature compensation of a highly birefringent optical fiber current sensor (McStay, Daniel; Chu, W. W.; Rogers, Alan J.)V1584,118-123(1991)

Temperature stress testing of laser modules for the uncontrolled environment (Su, Pin)V1366,94-106(1991)

Test of photovoltaic model of photoinduced second-harmonic generation in optical fibers (Dianov, Evgeni M.; Kazansky, Peter G.; Krautschik, Christof G.; Stepanov, D. Y.)V1516,75-80(1991)

Theory of stressed fiber lifetime calculations (Kapron, Felix P.)V1366,136-143(1991)

Three-dimensional moving-image display by modulated coherent optical fibers: a proposal (Hoshino, Hideshi; Sato, Koki)V1461,227-231(1991)

Thulium-doped silica fiber lasers (Tropper, Anne C.; Smart, Richard G.; Perry, Ian R.; Hanna, David C.; Lincoln, John; Brocklesby, Bill)V1373,152-157(1991)

Time dependence of laser-induced surface breakdown in fused silica at 355 nm in the nanosecond regime (Albagli, Douglas; Izatt, Joseph A.; Hayes, Gary B.; Banish, Bryan; Janes, G. Sargent; Itzkan, Irving; Feld, Michael S.)V1441,146-153(1991)

True-time-delay steering of dual-band phased-array antenna using laser-switched optical beam forming networks (Ng, Willie W.; Tangonan, Gregory L.; Walston, Andrew; Newberg, Irwin L.; Lee, Jar J.; Bernstein, Norman P.)V1371,205-211(1991)

Two-color interferometry using a detuned frequency-doubling crystal (Koch, Karl W.; Moore, Gerald T.)V1516,67-74(1991)

Ultrafast pulse generation in fiber lasers (Kafka, James D.; Baer, Thomas M.)V1373,140-149(1991)

Ultrahigh resolution OTDR using streak camera technology (Sawaki, Akihiro; Miwa, Mitsuharu; Roehrenbeck, Paul W.)V1366,324-331(1991)

Use of fatigue measurements for fiber lifetime prediction (Yuce, Hakan H.; Kapron, Felix P.)V1366,144-156(1991)

Uses of fiber optics to enhance and extend the capabilities of holographic interferometry (Gilbert, John A.; Dudderar, Thomas D.)VIS08,146-159(1991)

Wavelength and temperature dependence of bending loss in monomode optical fibers (Morgan, Russell D.; Jones, Julian D.; Harper, Philip G.; Barton, James S.)V1504,118-124(1991)

Fiber optic sensors—see also fiber optics in medicine; gyroscopes; metrology; temperature

1.55-um superluminescent diode for a fiber optic gyroscope (Kashima, Yasumasa; Matoba, Akio; Kobayashi, Masao; Takano, Hiroshi)V1365,102-107(1991)

Absolute interferometer for manufacturing applications (Tucker, Michael R.; Christenson, Eric)V1367,289-299(1991)

Activities at the Smart Structures Research Institute (Gardiner, Peter T.)V1588,314-324(1991)

Advanced approaches in luminescence and Raman spectroscopy (Vo-Dinh, Tuan)V1435,197-202(1991)

Advances in analytical chemistry (Arendale, William F.; Congo, Richard T.; Nielsen, Bruce J.)V1434,159-170(1991)

Algorithm for a novel fiber optic weigh-in-motion sensor system (Tobin, Kenneth W.; Muhs, Jeffry D.)V1589,102-109(1991)

Alignment of principal axes between birefringent fiber by the spatial technique and its distribution-sensing effect (Zhang, Jinghua; Wang, Chunhua; Huang, Zhaoming)V1572,69-73(1991)

All-fiber closed-loop gyroscope with self-calibration (Ecke, Wolfgang; Schroeter, Siegmund; Schwotzer, Guenter; Willsch, Reinhardt)V1511,57-66(1991)

All-fiber pressure sensor up to 100 MPa (Bock, Wojtek J.; Wolinski, Tomasz R.; Domanski, Andrzej W.)V1511,250-254(1991)

All-fiber temperature sensing system by using polarization-maintaining fiber coupler as beamsplitter (Fang, Yin; Rong, Jian; Sheng, Kemin; Qin, Pingling)V1572,197-200(1991)

All-optical-fiber- and general-halogen-lamp-based remote measuring system for CH4 (Cheng, Yuqi; Zou, Kun; Shan, Xuekang)V1572,392-395(1991)

All-optical data-input device based on fiber optic interferometric strain gauges (Fuerstenau, Norbert; Schmidt, Walter)V1367,357-366(1991)

Amplitude-modulated laser-driven fiber-optic RF interferometric strain sensor (Schoenwald, Jeffrey S.)V1418,450-458(1991)

Analysis of acid-base indicators covalently bound on glass supports (Baldini, Francesco; Bacci, Mauro; Bracci, Susanna)V1368,210-217(1991)

Analysis of macro-model composites with Fabry-Perot fiber-optic sensors (Fogg, Brian R.; Miller, William V.; Lesko, John J.; Carman, Gregory P.; Vengsarkar, Ashish M.; Reifsnider, Kenneth L.; Claus, Richard O.)V1588,14-25(1991)

Analysis on the characteristics of the optical fiber compensation network for the intensity modulation optical fiber sensors (Zhong, Xian-Xin; Li, Jianshu; Fu, Xin; Huang, Shang-Lian)V1572,84-87(1991)

Anamorphosor for scintillating plastic optical fiber applications (Chiron, Bernard)V1592,158-164(1991)

Antibody-based fiber optic sensors for environmental and process-control applications (Barnard, Steven M.; Walt, David R.)V1368,86-92(1991)

Application issues of fiber-optic sensors in aircraft structures (Lu, Zhuo J.; Blaha, Franz A.)V1588,276-281(1991)

Application of analog fiber optic position sensors to flight control systems (Miller, Glen E.)V1367,165-173(1991)

Application of an optical fibre sensor for weighing the truckload (Zhang, L. B.; Shi, T. Y.; Zhang, J.)V1572,240-242(1991)

Application of fiber optic sensors in pavement maintenance (Shadaram, Mehdi; Solehjou, Amin; Nazarian, Soheil)V1332,487-490(1991)

Application of fiber optic thermometry to the monitoring of winding temperatures in medium- and large-power transformers (Wickersheim, Kenneth A.)V1584,3-14(1991)

Application of low-coherence optical fiber Doppler anemometry to fluid-flow measurement: optical system considerations (Boyle, William J.; Grattan, Kenneth T.; Palmer, Andrew W.; Meggitt, Beverley T.)V1511,51-56(1991)

Application of the plastic optical fiber in fiber-optic sensors (Chen, Yaosheng; Xiao, Wen; Xue, Mingqiu)V1572,124-128(1991)

Are optical fiber sensors intrinsically, inherently, or relatively safe? (McGeehin, Peter)V1504,75-79(1991)

Automated fiber optic moisture sensor system (Herskowitz, Gerald J.; Mezhoudi, Mohcene)V1368,55-60(1991)

Bare fiber temperature sensor (Soares, Edmundo A.; Dantas, Tarcisio M.)V1367,261-265(1991)

Basic principle and system research of self-emission fiber optic temperature sensor (Nie, Chao-Jiang; Yu, Jing-ming; Wu, Fang D.; Gao, Yu-ping)V1584,87-93(1991)

Behavior of WDM system for intensity modulation (Pierre, Guillaume; Jarret, Bertrand; Brun, Eric)V1511,201-211(1991)

Bile optical fiber sensor: the method for entero-gastric reflux detection (Falciai, Riccardo; Baldini, Francesco; Bechi, Paolo; Cosi, Franco; Bini, Andrea)V1572,424-427(1991)

Biochemical measurement of bilirubin with an evanescent wave optical sensor (Poscio, Patrick; Depeursinge, Ch.; Emery, Y.; Parriaux, Olivier M.; Voirin, G.)V1510,112-117(1991)

Characteristics of a multimode laser diode source in several types of dual-interferometer configuration (Ning, Yanong N.; Grattan, Kenneth T.; Palmer, Andrew W.; Meggitt, Beverley T.)V1367,347-356(1991)

Characterization of fluorescent plastic optical fibers for x-ray beam detection (Laguesse, Michel F.; Bourdinaud, Michel J.)V1592,96-107(1991)

Characterization of macrobend sensitivity of step index optical fibers used in intensity sensors (York, Jim F.; Nelson, Gary W.; Varshneya, Deepak)V1584,308-319(1991)

Characterization of time-division-multiplexed digital optical position transducer (Varshneya, Deepak; Lapierre, A.)V1584,188-201(1991)

Chemical and Medical Sensors (Wolfbeis, Otto S., ed.)V1510(1991)

Chemo-optical microsensing systems (Lambeck, Paul V.)V1511,100-113(1991)

Chemical, Biochemical, and Environmental Fiber Sensors II (Lieberman, Robert A.; Wlodarczyk, Marek T., eds.)V1368(1991)

Chlorine sensing by optical techniques (Momin, S. A.; Narayanaswamy, R.)V1510,180-186(1991)

Clinical measuring system for the form and position errors of circular workpieces using optical fiber sensors (Tan, Jiubin; Qiang, Xifu; Ding, Xuemei)V1572,552-557(1991)

Closed-loop fiber-optic gyroscope (Fang, Zhen-he; Huang, Shao-ming; Shen, Yu-qing)V1572,342-346(1991)

Coherence and optical Kerr nonlinearity (Depoortere, Marc)V1504,133-139(1991)

Co-immobilization of several dyes on optodes for pH measurements (Boisde, Gilbert; Sebille, Bernard)V1510,80-94(1991)

Common-path optical fiber heterodyne interferometric current sensor (Bartlett, Steven C.; Farahi, Faramarz; Jackson, David A.)V1504,247-250(1991)

Compact probe for all-fiber optically addressed silicon cantilever microresonators (Rao, Yun-Jiang; Uttamchandani, Deepak G.; Culshaw, Brian)V1572,287-292(1991)

Compensation mechanism of an optical fiber turning reflective sensor (Yuan, Libo; Shou, Reilan; Qiu, Anping; Lu, Zhiyi)V1572,258-263(1991)

Compensation of the phase shift in the optical fibre current sensor (Ye, Miaoyuan; Hu, Shichuang)V1572,483-486(1991)

Composite cure monitoring with embedded optical fiber sensors (Davis, Andrew; Ohn, Myo M.; Liu, Kexing; Measures, Raymond M.)V1489,33-43(1991)

Composite-embedded fiber-optic data links and related material/connector issues (Morgan, Robert E.; Ehlers, Sandy L.; Jones, Katharine J.)V1588,189-197(1991)

Composite damage assessment employing an optical neural network processor and an embedded fiber-optic sensor array (Grossman, Barry G.; Gao, Xing; Thursby, Michael H.)V1588,64-75(1991)

Cooperative implementation of a high-temperature acoustic sensor (Baldini, S. E.; Nowakowski, Edward; Smith, Herb G.; Friebele, E. J.; Putnam, Martin A.; Rogowski, Robert S.; Melvin, Leland D.; Claus, Richard O.; Tran, Tuan A.; Holben, Milford S.)V1588,125-131(1991)

Coupled fiber ring interferometer array: theory (Wang, An; Xie, Haiming)V1572,365-369(1991)

Critical dimension control using development end point detection for wafers with multilayer structures (Hagi, Toshio; Okuda, Yoshimitsu; Ohkuma, Tohru)V1464,215-221(1991)

Damage assessment in composites with embedded optical fiber sensors (Measures, Raymond M.; Liu, Kexing; LeBlanc, Michel; McEwen, Keith; Shankar, K.; Tennyson, R. C.; Ferguson, Suzanne M.)V1489,86-96(1991)

Damage detection in woven-composite materials using embedded fiber-optic sensors (Bonniau, Philippe; Chazelas, Jean; Lecuellet, Jerome; Gendre, Francois; Turpin, Marc; Le Pesant, Jean-Pierre; Brevignon, Michele)V1588,52-63(1991)

Delayed fluorescence of eosin-bound protein: a probe for measurement of slow-rotational mobility (Yao, Jialing; McStay, Daniel; Rogers, Alan J.; Quinn, Peter J.)V1572,428-433(1991)

Demodulation of a fiber Fabry-Perot strain sensor using white light interferometry (Zuliani, Gary; Hogg, W. D.; Liu, Kexing; Measures, Raymond M.)V1588,308-313(1991)

Demonstration of birefringent optical fiber frequency shifter employing torsional acoustic waves (Berwick, Michael; Pannell, Christopher N.; Russell, Philip S.; Jackson, David A.)V1572,157-162; V1584,364-373(1991)

Depolarized fiber optic gyro for future tactical applications (Bramson, Michael D.)V1367,155-160(1991)

Designing enhanced maintainability fiber-optic networks (Schrever, Koen; De Vilder, Jan; Rolain, Yves; Voet, Marc R.; Barel, Alain R.)V1572,107-112(1991)

Design of optically activated conventional sensors and actuators (Liu, Kexing)V1398,269-275(1991)

Detection of general anesthetics using a fluorescence-based sensor: incorporation of a single-fiber approach (Abrams, Susan B.; McDonald, Hillary L.; Yager, Paul)V1420,13-21(1991)

Detection of high-frequency elastic waves with embedded ordinary single-mode fibers (Liu, Kexing; Measures, Raymond M.)V1584,226-234(1991)

Development of a fiber Fabry-Perot strain gauge (Hogg, W. D.; Janzen, Doug; Valis, Tomas; Measures, Raymond M.)V1588,300-307(1991)

Development of a fibre optics flowmeter (She, K.; Greated, C. A.; Easson, William J.; Skyner, D.; Xu, M. C.)V1572,581-587(1991)

Development of an interferometric fiber optic sensor using diode laser (Brewer, Donald R.; Joenathan, Charles; Bibby, Yu Wang; Khorana, Brij M.)V1396,430-434(1991)

Development of an optical fiber and photoelectric coupling V/F converter for 5.4-MV impulse generator (Guan, Genzhi)V1572,487-491(1991)

Development of chemical sensors using plastic optical fiber (Zhou, Quan; Tabacco, Mary B.; Rosenblum, Karl W.)V1592,108-113(1991)

Development of fly-by-light systems for commercial aircraft (Todd, John R.; Yount, Larry J.)V1369,72-78(1991)

Diagnostic technique for electrical power equipment using fluorescent fiber (Kurosawa, Kiyoshi; Sawa, Takeshi; Sawada, Hisashi; Tanaka, Akira; Wakatsuki, Noboru)V1368,150-156(1991)

Diaphragm size and sensitivity for fiber optic pressure sensors (He, Gang; Cuomo, Frank W.; Zuckerwar, Allan)V1584,152-156(1991)

Direct exchange of metal ions onto silica waveguides (Petersen, James V.; Dessy, Raymond E.)V1368,61-72(1991)

Direct fluoroimmunoassay in Langmuir-Blodgett films of immunoglobulin G (Turko, Illarion V.; Lepesheva, Galina I.; Chashchin, Vadim L.)V1572,419-423(1991)

Direct readout of dynamic phase changes in a fiber-optic homodyne interferometer (Jin, Wei; Uttamchandani, Deepak G.; Culshaw, Brian)V1504,125-132(1991)

Distributed-effect optical fiber sensors for trusses and plates (Reichard, Karl M.; Lindner, Douglas K.)V1480,115-125(1991)

Distributed-fibre-optic methane gas concentration detection (Shi, Yi-Wei; Wang, Yao-Cai; Jiang, Hong-Tao; Yao, Cheng-Shan; Wu, Zhen-Chun)V1572,308-312(1991)

Distributed fiber optic pressure sensor (Luo, Fei; Yan, Muolin; Huang, Shang-Lian)V1367,221-224(1991)

Distributed optical fiber sensing (Rogers, Alan J.)V1504,2-24(1991); V1506,2-24(1991) ;1507,2-24(1991)

Drilled optical fiber sensors: a novel single-fiber sensor (Lipson, David; McLeaster, Kevin D.; Cohn, Brian; Fischer, Robert E.)V1368,36-43(1991)

Dual-channel current sensor capable of simultaneously measuring two currents (Bush, Simon P.; Jackson, David A.)V1584,103-109(1991)

Dual-eigenstate polarization preserving fiber optic sensor (Yu, Dong X.; Storti, George M.)V1584,236-242(1991)

Dynamical and real-time measurement of the fringe visibility of optical fiber interferometer (Fang, Xian-chen; Guo, Jian)V1572,52-55(1991)

Dynamic holography application in fiber optic interferometry (Kozhevnikov, Nikolai M.; Barmenkov, Yuri O.)V1584,387-395(1991)

Dynamic range limits in field determination of fluorescence using fiber optic sensors (Chudyk, Wayne; Pohlig, Kenneth)V1368,105-114(1991)

Effect of liquid on partially removed cladding SM fiber and its application to sensors (Das, Alok K.; Mandal, Anup K.; Pandit, Malay K.)V1572,572-580(1991)

Electromagnetic enviromental effects on shipboard fiber optic installations (Bucholz, Roger C.)V1369,19-23(1991)

Electronic-digital detection system for an optical fiber current sensor (Guedes Valente, Luiz C.; Kawase, Liliana R.; Afonso, Jose A.; Kalinowski, Hypolito J.)V1584,96-102(1991)

Electronic polarimetric detection system for optical fiber sensor application (Brooking, Nicholas L.; Guedes Valente, Luiz C.; Kawase, Liliana R.; Afonso, Jose A.)V1572,88-93(1991)

Embedded fiber-optic sensors in large structures (Udd, Eric)V1588,178-181(1991)

Emerging technology in fiber optic sensors (Dyott, Richard B.)V1396,709-717(1991)

Enantio-selective optode for the B-blocker propranolol (He, Huarui; Uray, Georg; Wolfbeis, Otto S.)V1368,175-180(1991)

Environmentally insensitive commercial pressure sensor (Wlodarczyk, Marek T.)V1368,121-131(1991)

Evaluation of commercial fiber optic sensors in a marine boiler room (Musselman, Martin L.)V1589,56-57(1991)

Evaluation of polymeric thin film waveguides as chemical sensors (Bowman, Elizabeth M.; Burgess, Lloyd W.)V1368,239-250(1991)

Excess noise in fiber gyroscope sources (Burns, William K.; Moeller, Robert P.; Dandridge, Anthony D.)V1367,87-92(1991)

Excitation efficiency of an optical fiber core source (Egalon, Claudio O.; Rogowski, Robert S.; Tai, Alan C.)V1489,9-16(1991)

Experimental developments in the RFOG (Kaiser, Todd J.; Cardarelli, Donato; Walsh, Joseph)V1367,121-126(1991)

Experimental research on high-bifringence fiber phase retarder (Yao, Minyu; Zhang, Xinyu)V1572,148-150(1991)

Experimental research on optical fibre microbending sensors in on-line measuring of deep-hole drilling bit wear (Yang, Zhiguo; Zhong, Hengyong; Cheng, Jubing; Wang, Youguan)V1572,252-257(1991)

Extended environmental performance of attitude and heading reference grade fiber optic rotation sensors (Chin, Gene H.; Cordova, Amado; Goldner, Eric L.)V1367,107-120(1991)

Fabry-Perot fiber-optic sensors in full-scale fatigue testing on an F-15 aircraft (Murphy, Kent A.; Gunther, Michael F.; Vengsarkar, Ashish M.; Claus, Richard O.)V1588,134-142(1991)

Faraday-effect magnetic field sensors based on substituted iron garnets (Deeter, Merritt N.; Rose, Allen H.; Day, Gordon W.)V1367,243-248(1991)

Fast-response water vapor and carbon dioxide sensor (Kohsiek, W.)V1511,114-119(1991)

Feasibility of optically sensing two parameters simultaneously using one indicator (Wolfbeis, Otto S.)V1368,218-222(1991)

FFT measuring method for magneto-optical ac current measurement (Hu, Shichuang; Ye, Miaoyuan; Qu, Gen)V1572,492-496(1991)

Fiber microbend sensor and instrumentation for fluid-level measurement (Wang, Yutian; Jiang, G.; Yu, J.)V1572,230-234(1991)

Fiber Optic and Laser Sensors VIII (DePaula, Ramon P.; Udd, Eric, eds.)V1367(1991)

Fiber Optic and Laser Sensors IX (DePaula, Ramon P.; Udd, Eric, eds.)V1584(1991)

Fiber optic based chemical sensor system for in-situ process measurements using the photothermal effect (Walker, Karl-Heinz; Sontag, Heinz)V1510,212-217(1991)

Fiber optic based miniature high-temperature probe exploiting coherence-tuned signal recovery via multimode laser diode illumination (Gerges, Awad S.; Jackson, David A.)V1504,233-236(1991)

Fiber optic based vortex shedder flow meter (Chu, Beatrice C.; Newson, Trevor P.; Jackson, David A.)V1504,251-257(1991)

Fiber optic biosensors: the situation of the European market (Scheggi, Annamaria V.; Mignani, Anna G.)V1510,40-45(1991)

Fiber optic controls for aircraft engines: issues and implications (Dasgupta, Samhita; Poppel, Gary L.; Anderson, William P.)V1374,211-222(1991)

Fiber optic damage detection for an aircraft leading edge (Measures, Raymond M.; LeBlanc, Michel; Hogg, W. D.; McEwen, Keith; Park, B. K.)V1332,431-443(1991)

Fiber optic differential pressure sensor for leak-detection system (Sun, Xiaohan; Zhang, Mingde; Wan, Suiren; Wang, Shuhua)V1572,243-247(1991)

Fiber optic displacement sensor for measurement of thin film thickness (Wang, Jianhua)V1572,264-267(1991)

Fiber optic evanescent wave biosensor (Duveneck, G.; Ehrat, M.; Widmer, H. M.)V1510,138-145(1991)

Fiber optic Fabry-Perot sensors for high-speed heat transfer measurements (Kidd, S. R.; Sinha, P. G.; Barton, James S.; Jones, Julian D.)V1504,180-190(1991)

Fiber optic fluorescence sensors based on sol-gel entrapped dyes (MacCraith, Brian D.; Ruddy, Vincent; Potter, C.; McGilp, J. F.; O'Kelley, B.)V1510,104-109(1991)

Fiber optic gyroscopes in Japan (Kurokawa, Akihiro; Hayakawa, Yoshiaki)V1504,156-164(1991)

Fiber optic high-temperature sensor for applications in iron and steel industries (Hao, Tianyou; Zhou, Feng-Shen; Xie, Xiou-Qioun; Hu, Ji-Wu; Wang, Wei-Yen)V1584,32-38(1991)

Fiber optic ice sensors for refrigerators (Paone, Nicola; Rossi, G.)V1511,129-139(1991)

Fiber optic interferometric ellipsoidal shell hydrophone (Brown, David A.; Garrett, Steven L.; Conte, D. V.; Smith, R. C.; Rothenberg, E.; Young, M.; Rissberger, Ed)V1369,2-8(1991)

Fiber optic interferometric sensors using multimode fibers (Ignatyev, Alexander V.; Galkin, S. L.; Nikolaev, V. A.; Strigalev, V. E.)V1584,336-345(1991)

Fiber optic interferometric x-ray dosimeter (Barone, Fabrizio; Bernini, Umberto; Conti, M.; Del Guerra, Alberto; Di Fiore, Luciano; Maddalena, P.; Milano, L.; Russo, G.; Russo, Paolo)V1584,304-307(1991)

Fiber optic liquid crystal high-pressure sensor (Wolinski, Tomasz R.; Bock, Wojtek J.)V1511,281-288(1991)

Fiber optic liquid crystalline microsensor for temperature measurement in high magnetic field (Domanski, Andrzej W.; Kostrzewa, Stanislaw)V1510,72-77(1991)

Fiber optic magnetic field and current sensor using magneto-birefringence of dense ferrofluid thin films (Pan, Yingtain; Liu, Xiande; Du, Chongwu; Li, Zai Q.)V1572,477-482(1991)

Fiber optic magnetic field sensors based on Faraday effect in new materials (Nikitin, Petr I.; Grigorenko, A. N.; Konov, Vitaly I.; Savchuk, A. I.)V1584,124-134(1991)

Fiber optic magnetic field sensor using spectral modulation encoding (Lequime, Michael; Meunier, Carole; Giovannini, Hugues)V1367,236-242(1991)

Fiber optic magnetometer with stable and linear output (Zhang, Wei; Zhang, Zhipeng)V1572,458-463(1991)

Fiber Optic Metrology and Standards (Soares, Oliverio D., ed.)V1504(1991)

Fiber optic multiple sensor for simultaneous measurements of temperature and vibrations (Brenci, Massimo; Mencaglia, Andrea; Mignani, Anna G.; Barbero, V.; Cimbrico, P. L.; Pessino, P.)V1572,318-324(1991)

Fiber optic photoplethysmograph (Bokun, Leszek J.; Domanski, Andrzej W.)V1420,93-99(1991)

Fiber optic position transducers for aircraft controls (Glomb, Walter L.)V1367,162-164(1991)

Fiber optic pressure and temperature sensor for down-hole applications (Lequime, Michael; Lecot, C.; Jouve, Philippe; Pouleau, J.)V1511,244-249(1991)

Fiber optic pressure sensor (Ingold, Joseph P.; Sun, Mei H.; Bigelow, Russell N.)V1589,83-89(1991)

Fiber optic pressure sensor system for gas turbine engine control (Wesson, Laurence N.; Cabato, Nellie L.; Pine, Nicholson L.; Bird, Victor J.)V1367,204-213(1991)

Fiber optic remote Fourier transform infrared spectroscopy (Druy, Mark A.; Glatkowski, Paul J.; Stevenson, William A.)V1584,48-52(1991)

Fiber optic rotary position sensors for vehicle and propulsion controls (Gardiner, Peter T.)V1374,200-210(1991)

Fiber optics development at McDonnell Douglas (Udd, Eric; Clark, Timothy E.; Joseph, Alan A.; Levy, Ram L.; Schwab, Scott D.; Smith, Herb G.; Balestra, Chet L.; Todd, John R.; Marcin, John)V1418,134-152(1991)

Fiber optic sensing technique employing rf-modulated interferometry (Eustace, John G.; Coghlan, Gregory A.; Yorka, Christian M.; Carome, Edward F.; Adamovsky, Grigory)V1584,320-327(1991)

Fiber optic sensor applied to measure high temperature under high-pressure condition (Xiao, Wen; Chen, Yaosheng; Gao, Wei; Xue, Mingqiu)V1572,170-174(1991)

Fiber optic sensor considerations and developments for smart structures (Measures, Raymond M.)V1588,282-299(1991)

Fiber optic sensor for ammonia vapors of variable temperature (Potyrailo, Radislav A.; Golubkov, Sergei P.; Borsuk, Pavel S.)V1572,434-438(1991)

Fiber optic sensor for nitrates in water (MacCraith, Brian D.; Maxwell, J.)V1510,195-203(1991)

Fiber optic sensor for plasma current diagnostics in tokamaks (Kozhevnikov, Nikolai M.; Barmenkov, Yuri O.; Belyakov, V. A.; Medvedev, A. A.; Razdobarin, G. T.)V1584,138-144(1991)

Fiber optic sensor for simultaneous measurement of strain and temperature (Vengsarkar, Ashish M.; Michie, W. C.; Jankovic, Lilja; Culshaw, Brian; Claus, Richard O.)V1367,249-260(1991)

Fiber optic sensor for the study of temperature and structural integrity of PZT: epoxy composite materials (Vishnoi, Gargi; Pillai, P.K. C.; Goel, T. C.)V1572,94-100(1991)

Fiber optic sensor networks (Culshaw, Brian)V1511,168-178(1991)

Fiber optic sensor probe for in-situ surface-enhanced Raman monitoring (Vo-Dinh, Tuan; Stokes, D. L.; Li, Ying-Sing; Miller, Gordon H.)V1368,203-209(1991)

Fiber Optic Sensors: Engineering and Applications (Bruinsma, Anastasius J.; Culshaw, Brian, eds.)V1511(1991)

Fiber optic sensors for heat transfer studies (Farahi, Faramarz; Jones, Julian D.; Jackson, David A.)V1584,53-61(1991)

Fiber optic sensors for process monitoring and control (Marcus, Michael A.)V1398,194-205(1991)

Fiber Optic Smart Structures and Skins IV (Claus, Richard O.; Udd, Eric, eds.)V1588(1991)

Fiber optic smart structures: structures that see the light (Measures, Raymond M.)V1332,377-398(1991)

Fiber optic speed sensor for advanced gas turbine engine control (Varshneya, Deepak; Maida, John L.; Overstreet, Mark A.)V1367,181-191(1991)

Fiber Optic Systems for Mobile Platforms IV (Lewis, Norris E.; Moore, Emery L., eds.)V1369(1991)

Fiber optic technique for simultaneous measurement of strain and temperature variations in composite materials (Michie, W. C.; Culshaw, Brian; Roberts, Scott S.; Davidson, Roger)V1588,342-355(1991)

Fiber optic temperature probe system for inner body (Liu, Bo; Deng, Xingzhong; Cao, Wei; Cheng, Xianping; Xie, Tuqiang; Zhong, Zugen)V1572,211-215(1991)

Fiber optic temperature sensor based on bend losses (Liu, Rui-Fu; Xi, Xiao-chun; Li, Wei-min; Tian, Da-chao)V1572,180-184(1991)

Fiber optic temperature sensor for aerospace applications (Jensen, Stephen C.; Tilstra, Shelle D.; Barnabo, Geoffrey A.; Thomas, David C.; Phillips, Richard W.)V1369,87-95(1991)

Fiber optic thermometer using Fourier transform spectroscopy (Beheim, Glenn; Sotomayor, Jorge L.; Flatico, Joseph M.; Azar, Massood T.)V1584,64-71(1991)

Fiber sensor design for turbine engines (Tobin, Kenneth W.; Beshears, David L.; Turley, W. D.; Lewis, Wilfred; Noel, Bruce W.)V1584,23-31(1991)

Field testing of a fiber optic rotor temperature monitor for power generators (Brown, Stewart K.; Mannik, Len)V1584,15-22(1991)

Field test results on fiber optic pressure transmitter system (Berthold, John W.)V1584,39-47(1991)

Fizeau-type of gradient-index rod lens interferometer by using semiconductor laser (Ming, Hai; Sun, Yuesheng; Ren, Baorui; Xie, Jianping; Nakajima, Toshinori)V1572,27-31(1991)

Flow injection analysis with bioluminescence-based fiber-optic biosensors (Blum, Loic J.; Gautier, Sabine; Coulet, Pierre R.)V1510,46-52(1991)

Flow optrodes for chemical analysis (Berman, Richard J.; Burgess, Lloyd W.)V1368,25-35(1991)

Fluorescence-based fiber optic temperature sensor for aerospace applications (Tilstra, Shelle D.)V1589,32-37(1991)

Fluorescence-based optrodes for alkali ions based on the use of ion carriers and lipophilic acid/base indicators (He, Huarui; Wolfbeis, Otto S.)V1368,165-171(1991)

Frequency-derived distributed optical fiber sensing: backscatter analysis (Rogers, Alan J.; Handerek, Vincent A.; Parvaneh, Farhad)V1511,190-200(1991)

Fused 3 x 3 single-mode fiber-optic couplers with stable characteristics for fiber interferometric sensors (Xie, Tonglin)V1572,132-136(1991)

General purpose fiber optic hydrophone made of castable epoxy (Garrett, Steven L.; Brown, David A.; Beaton, Brian L.; Wetterskog, Kevin; Serocki, John)V1367,13-29(1991)

GigaHertz RMS current sensors for electromagnetic compatibility testing (Mitchell, Gordon L.; Saaski, Elric W.; Pace, John W.)V1367,266-272(1991)

Glass-ceramic fiber optic sensors (Romaniuk, Ryszard S.; Stepien, Ryszard)V1368,73-84(1991)

Gradient microbore liquid chromatography with dual-wavelength absorbance detection: tunable analyzers for remote chemical monitoring (Sulya, Andrew W.; Moore, Leslie K.; Synovec, Robert E.)V1434,147-158(1991)

Heat monitoring by fiber-optic microswitches (Mencaglia, Andrea; Brenci, Massimo; Falciai, Riccardo; Guzzi, D.; Pascucci, Tania)V1506,140-144(1991)

High-accurate optical fiber liquid-level sensor (Sun, Dexing; Chen, Shouliu; Pan, Chao; Jin, Henghuan)V1572,508-513(1991)

High-dynamic-range fiber gyro with all-digital signal processing (Lefevre, Herve C.; Martin, Philippe; Morisse, J.; Simonpietri, Pascal; Vivenot, P.; Arditty, Herve J.)V1367,72-80(1991)

High-hydrostatic-pressure sensor using elliptical-core optical fibers (Bock, Wojtek J.; Wolinski, Tomasz R.; Fontaine, Marie)V1584,157-161(1991)

Highly efficient plastic optical fluorescent fibers and sensors (Chiron, Bernard)V1592,86-95(1991)

High-quality 2 x 2 and 4 x 4 wavelength-flatted couplers for sensor applications (Xu, Jisen; Zhang, Yi; Zhang, Xiang)V1572,137-139(1991)

High-speed digital ellipsometer for the study of fiber optic sensor systems (Saxena, Indu)V1367,367-373(1991)

High stability optical fiber sensor system (Chen, Yu-Feng; Liang, Yue-kun; Bian, Hong)V1572,113-117(1991)

High-temperature optical sensor for displacement measurement (Ebbeni, Jean P.)V1504,268-272(1991)

Hollow waveguides for sensor applications (Saggese, Steven J.; Harrington, James A.; Sigel, George H.)V1368,2-14(1991)

Holographic coupler for fiber optic Sagnac interferometer (Zou, Yunlu; Hsu, Dahsiung; Wang, Ben; Tao, Huiying)V1238,452-456(1991)

Hybride fiber-optic temperature sensors on the base of LiNbO3 and LiNbO3:Ti waveguides (Goering, Rolf)V1511,275-280(1991)

Important position of optical fibre technology in shipboard equipment construction (Zhao, Enyi)V1572,65-68(1991)

Improvement in the performance of evanescent wave chemical sensors by special waveguide structures (Stewart, George; Culshaw, Brian; Clark, Douglas F.; Andonovic, Ivan)V1368,230-238(1991)

Improvement of specular reflection pyrometer (Wen, Lin Ying; Hua, Yun)V1367,300-302(1991)

Infrared fiber optic sensors: new applications in biology and medicine (Swairjo, Manal; Rothschild, Kenneth J.; Nappi, Bruce; Lane, Alan; Gold, Harris)V1437,60-65(1991)

In-line Fabry-Perot interferometric temperature sensor with digital signal processing (Yeh, Yunhae; Lee, J. H.; Lee, Chung E.; Taylor, Henry F.)V1584,72-78(1991)

In-line fiber Fabry-Perot interferometer with high-reflectance internal mirrors (Lee, Chung E.; Gibler, William N.; Atkins, Robert A.; Taylor, Henry F.)V1584,396-399(1991)

In-situ characterization of resin chemistry with infrared transmitting optical fibers and infrared spectroscopy (Druy, Mark A.; Glatkowski, Paul J.; Stevenson, William A.)V1437,66-74(1991)

In situ monitoring for hydrocarbons using fiber optic chemical sensors (Klainer, Stanley M.; Thomas, Johnny R.; Dandge, Dileep K.; Frank, Chet A.; Butler, Marcus S.; Arman, Helen; Goswami, Kisholoy)V1434,119-126(1991)

Integrated optical device in a fiber gyroscope (Rasch, Andreas; Goering, Rolf; Karthe, Wolfgang; Schroeter, Siegmund; Ecke, Wolfgang; Schwotzer, Guenter; Willsch, Reinhardt)V1511,149-154(1991)

Integrated optic components for advanced turbine engine control systems (Emo, Stephen M.; Kinney, Terrance R.; Wong, Ka-Kha)V1374,266-276(1991)

Integrated optic device for biochemical sensing (Boiarski, Anthony A.; Ridgway, Richard W.; Miller, Larry S.; Bhullar, B. S.)V1368,264-272(1991)

Integrated Optics and Optoelectronics II (Wong, Ka-Kha, ed.)V1374(1991)

Integrated optics displacement sensor (d'Alessandro, Antonio; De Sario, Marco; D'Orazio, Antonella; Petruzzelli, Vincenzo)V1366,313-323(1991)

Integrated optics for fiber optic sensors (Minford, William J.; DePaula, Ramon P.)V1367,46-52(1991)

Intelligent composites containing measuring fiber-optic networks for continuous self-diagnosis (Sansonetti, Pierre; Lequime, Michael; Engrand, D.; Guerin, J. J.; Davidson, Roger; Roberts, Scott S.; Fornari, B.; Martinelli, Mario; Escobar Rojo, Priscilla; Gusmeroli, Valeria; Ferdinand, Pierre; Plantey, J.; Crowther, Margaret F.; Culshaw, Brian; Michie, W. C.)V1588,198-209(1991)

Intelligent magnetometer with photoelectric sampler (Wang, Defang; Xu, Yan; Zhu, Minjun)V1572,514-516(1991)

Intelligent optical fiber sensor system for measurement of gas concentration (Pan, Jingming; Yin, Zongming)V1572,403-405(1991)

Intensity-type fiber optic electric current sensor (Carome, Edward F.; Kubulins, Vilnis E.; Flanagan, Roger L.; Shamray-Bertaud, Patricia)V1584,110-117(1991)

Interferometric fiber-optic sensing using a multimode laser diode source (Gerges, Awad S.; Newson, Trevor P.; Jackson, David A.)V1504,176-179(1991)

Interferometric fiber optic accelerometer (Brown, David A.; Garrett, Steven L.)V1367,282-288(1991)

Interferometric fiber optic gyroscopes for today's market (LaViolette, Kerry D.; Bossler, Franklin B.)V1398,213-218(1991)

Interferometric fiber optic sensors for use with composite materials (Measures, Raymond M.; Valis, Tomas; Liu, Kexing; Hogg, W. D.; Ferguson, Suzanne M.; Tapanes, Edward)V1332,421-430(1991)

Interferometric optical fiber sensors for absolute measurement of displacement and strain (Kersey, Alan D.; Berkoff, Timothy A.; Dandridge, Anthony D.)V1511,40-50(1991)

Interferometric signal processing schemes for the measurement of strain (Berkoff, Timothy A.; Kersey, Alan D.)V1588,169-176(1991)

Intl Conf on Optical Fibre Sensors in China (Culshaw, Brian; Liao, Yan-Biao, eds.)V1572(1991)

Investigation of optical fibers as sensors for condition monitoring of composite materials (Nielsen, Peter L.)V1588,229-240(1991)

Investigation of optical fibre sensor for remote measuring optical activity (Li, Yuchuan; Xiong, Guiguang; Yu, Guoping; Jiang, Zhiying; Wang, Fang)V1572,382-385(1991)

In-vivo blood flow velocity measurements using the self-mixing effect in a fiber-coupled semiconductor laser (Koelink, M. H.; Slot, M.; de Mul, F. F.; Greve, Jan; Graaff, Reindert; Dassel, A. C.; Aarnouds, J. G.)V1511,120-128(1991)

Key issues in selecting plastic optical fibers used in novel medical sensors (Kosa, Nadhir B.)V1592,114-121(1991)

K-ion-exchange waveguide directional coupler sensor (Chen, Zheng; Dai, Ji Zhi)V1572,129-131(1991)

Large-dynamic-range elecronically scanned "white-light" interferometer with optical fiber Young's structure (Chen, Shiping; Rogers, Alan J.; Meggitt, Beverley T.)V1504,191-201(1991)

Large dynamic range electronically scanned "white-light" interferometer with optical fiber Young's structure (Chen, Shiping; Rogers, Alan J.; Meggitt, Beverley T.)V1511,67-77(1991)

Laser speckle and optical fiber sensors for micromovements monitoring in biotissues (Tuchin, Valery V.; Ampilogov, Andrey V.; Bogoroditsky, Alexander G.; Rabinovich, Emmanuil M.; Ryabukho, Vladimir P.; Ul'yanov, Sergey S.; V'yushkin, Maksim E.)V1420,81-92(1991); V1429,62-73(1991)

LCVD fabrication of polycrystalline Si pressure sensor (Zhang, Wei; Wang, Xiao-Ru)V1572,15-17(1991)

LEAF: a fiber-optic fluorometer for field measurement of chlorophyll fluorescence (Mazzinghi, Piero)V1510,187-194(1991)

Long-term reliability and performance testing of fiber optic sensors for engineering applications (Weinberger, Alex; Weinberger, Ervin)V1367,30-45(1991)

Loss tolerant, self-monitoring fiber optic discrete position sensor (Pokorski, Joseph D.)V1398,219-229(1991)

Low-cost fiber optic sensing systems using spatial division multiplexing (Paton, Barry E.)V1332,446-455(1991)

Low-cost in-soil organic contaminant sensor (Brossia, Charles E.; Wu, Samuel C.)V1368,115-120(1991)

Low-drift fiber-optic gyro for earth-rate applications (Dyott, Richard B.; Huang, Yung Y.; Jannush, D. A.; Morrison, Steve A.)V1482,439-443(1991)

Lower cost fiber optic vibration sensors (Carome, Edward F.; Kubulins, Vilnis E.; Flanagan, Roger L.)V1589,133-138(1991)

Low-frequency fiber optic magnetic field sensors (Nader-Rezvani, Navid; Claus, Richard O.; Sarrafzadeh, A. K.)V1584,405-414(1991)

Low-loss Y-couplers for fiber optic gyro applications (Page, Jerry L.)V1374,287-293(1991)

Low-profile fibers for embedded smart structure applications (Vengsarkar, Ashish M.; Murphy, Kent A.; Gunther, Michael F.; Plante, Angela J.; Claus, Richard O.)V1588,2-13(1991)

Magnetic field sensitivity of depolarized fiber optic gyros (Blake, James N.)V1367,81-86(1991)

Magneto-optical apparatus with comparator for the measurement of large direct current (Zhang, Zhipeng; Zhao, Zhi; Chong, Baoxin)V1572,464-468(1991)

Magnet-sensitive optical fiber and its application in current sensor system (Yu, Tong; Li, Qin; Chen, Rongsheng; Yan, Jin-Li)V1572,469-471(1991); V1584,135-137(1991)

Measurement of fibre Verdet constant with twist method (Dong, Xiaopeng; Hu, Hao; Qian, Jingren)V1572,56-60(1991)

Measurement of polarization model dispersion and mode-coupling parameter of a polarization-maintaining fiber (Huang, Zhaoming; Wang, Chunhua; Zhang, Jinghua)V1572,140-143(1991)

Measurement of small strain of a solid body by two-frequency laser optical fiber sensor (Fang, Yin; Sheng, Kemin)V1572,453-456(1991)

Measurement of surface roughness using optical fibre sensor and microcomputer (Fan, Dapeng; Zhang, Honghai; Chen, Jihong; Chen, Riyao)V1572,11-14(1991)

Measuring photon pathlengths by quasielastic light scattering in a multiply scattering medium (Nossal, Ralph J.; Schmitt, Joseph M.)V1430,37-47(1991)

Mechanical properties of composite materials containing embedded fiber-optic sensors (Roberts, Scott S.; Davidson, Roger)V1588,326-341(1991)

Metal-embedded optical fiber pressure sensor (Kidwell, J. J.; Berthold, John W.)V1367,192-196(1991)

Metallic-glass-coated optical fibers as magnetic-field sensors (Larson, Donald C.; Bibby, Y. W.; Tyagi, S.)V1572,517-522(1991)

Method for embedding optical fibers and optical fiber sensors in metal parts and structures (Lee, Chung E.; Alcoz, J. J.; Gibler, William N.; Atkins, Robert A.; Taylor, Henry F.)V1588,110-116(1991)

Microbend pressure sensor for high-temperature environments (Majercak, David; Sernas, Valentinas; Polymeropoulos, Constantine E.; Sigel, George H.)V1584,162-169(1991)

Microdisplacement fiber sensor using two-frequency interferometry (Tedjojuwono, Ken K.)V1584,146-151(1991)

Microinteraction of optical fibers embedded in laminated composites (Singh, Hemant; Sirkis, James S.; Dasgupta, Abhijit)V1588,76-85(1991)

Microwave warming of biological tissue and its control by IR fiber thermometry (Drizlikh, S.; Zur, A.; Moser, Frank; Katzir, Abraham)V1420,53-62(1991)

Miniature HiBi current sensor (Chu, W. W.; McStay, Daniel; Rogers, Alan J.)V1572,523-527(1991)

Minimum detectable changes in Rayleigh backscatter from distributed fiber sensors (Garside, Brian K.; Park, R. E.)V1588,150-158(1991)

Mixed-type optical fiber current sensor (Wu, Gengsheng; Yang, Fan; Li, Wen; Liao, Yan-Biao)V1572,497-502(1991)

Mode-selective fiber sensors operating with computer-generated optical elements (Golub, Mikhail A.; Sisakyan, Iosiph N.; Soifer, Victor A.; Uvarov, G. V.)V1572,101-106(1991)

Model of an axially strained weakly guiding optical fiber modal pattern (Egalon, Claudio O.; Rogowski, Robert S.)V1588,241-254(1991)

Model of a thin-film optical fiber fluorosensor (Egalon, Claudio O.; Rogowski, Robert S.)V1368,134-149(1991)

Monitoring of tissue temperature during microwave hyperthermia utilizing a fiber optic liquid crystalline microsensor (Domanski, Andrzej W.; Kostrzewa, Stanislaw; Hliniak, Andrzej)V1420,72-80(1991)

Multi-analog track fiber-coupled position sensor (Huggins, Raymond W.)V1367,174-180(1991)

Multibit optical sensor networking (Pervez, Anjum)V1511,220-231(1991)

Multifunction multichannel remote-reading optical fiber sensor system (Zheng, Gang; Tian, Qian; Liang, Jinwen)V1572,299-303(1991)

Multimode approach to optical fiber components and sensors (Johnstone, Walter; Thursby, G.; Culshaw, Brian; Murray, S.; Gill, M.; McDonach, Alaster; Moodie, D. G.; Fawcett, G. M.; Stewart, George; McCallion, Kevin J.)V1506,145-149(1991)

Multimode fiber-optic Mach-Zehnder interferometric strain sensor (Wang, An; Xie, Haiming)V1572,444-449(1991)

Multimode fiber-optic temperature sensor system based on dual-wavelength difference absorption principle (Zhang, Zaixuan; Lin, Dan; Fang, Xiao; Jing, Shangzhong)V1572,201-204(1991)

Multimode fiber optic rotation sensor with low-cost digital signal processing (Johnson, Dean R.; Fredricks, Ronald J.; Vuong, S. C.; Dembinski, David T.; Sabri, Sehbaz H.)V1367,140-154(1991)

Multiple-channel sensing with fiber specklegrams (Wu, Shudong; Yin, Shizhuo; Rajan, Sumati; Yu, Francis T.)V1584,415-424(1991)

Multiple-mode reconfigurable electro-optic switching network for optical fiber sensor array (Chen, Ray T.; Wang, Michael R.; Jannson, Tomasz; Baumbick, Robert J.)V1374,223-236(1991)

Multiple fiber optic probe for several sensing applications (Dhadwal, Harbans S.; Ansari, Rafat R.)V1584,262-272(1991)

Multiple mode and multiple source coupling into polymer thin-film waveguides (Potter, B. L.; Walker, D. S.; Greer, L.; Saavedra, Steven S.; Reichert, William M.)V1368,251-257(1991)

Multiplexed approach for the fiber optic gyro inertial measurement unit (Page, Jerry L.)V1367,93-102(1991)

Multiplexing of remote all-fiber Michelson interferometers with lead insensitivity (Santos, Jose L.; Farahi, Faramarz; Newson, Trevor P.; Leite, Antonio P.; Jackson, David A.)V1511,179-189(1991)

Multipurpose fiber optic sensor with sloped tip (Melnik, Ivan S.; Krivokhizha, A. M.; Ptashnik, O. V.)V1572,118-122(1991)

Near-infrared fiber optic temperature sensor (Schoen, Christian; Sharma, Shiv K.; Seki, Arthur; Angel, S. M.)V1584,79-86(1991)

Near real-time operation of a centimeter-scale distributed fiber sensing system (Garside, Brian K.)V1332,399-408(1991)

Neural control of smart electromagnetic structures (Thursby, Michael H.; Yoo, Kisuck; Grossman, Barry G.)V1588,218-228(1991)

Neural networks for smart structures with fiber optic sensors (Sayeh, Mohammad R.; Viswanathan, R.; Dhali, Shirshak K.)V1396,417-429(1991)

New developments in the field of chemical infrared fiber sensors (Kellner, Robert A.; Taga, Karim)V1510,232-241(1991)

New luminescent metal complex for pH transduction in optical fiber sensing: application to a CO2-sensitive device (Moreno-Bondi, Maria C.; Orellana, Guillermo; Camara, Carmen; Wolfbeis, Otto S.)V1368,157-164(1991)

New method for detection of blood coagulation using fiber-optic sensor (Fediay, Sergey G.; Kuznetzov, Alexsey V.)V1420,41-43(1991)

New modulation scheme for optical fiber point temperature sensor (Farahi, Faramarz)V1504,237-246(1991)

New robot slip sensor using optical fibre and its application (Chen, Jinjiang)V1572,284-286(1991)

Noncontact measurement of microscopic displacement and vibration by means of fiber optics bundle (Toba, Eiji; Shimosaka, Tetsuya; Shimazu, Hideto)V1584,353-363(1991)

Novel analog phase tracker for interferometric fiber optic sensor applications (Berkoff, Timothy A.; Kersey, Alan D.; Moeller, Robert P.)V1367,53-58(1991)

Novel applications of monomode fiber tapers (Payne, Frank P.; Mackenzie, H. S.)V1504,165-175(1991)

Novel fiber-optic interferometer with high sensitivity and common-mode compensation (Chen, Xiaoguang)V1572,332-336(1991)

Novel noncontact sensor for surface topography measurements using fiber optics (Butler, Clive; Gregoriou, Gregorios)V1584,282-293(1991)

Novel optical processing scheme for interferometric vibration measurement using a low-coherence source with a fiber optic probe (Weir, Kenneth; Boyle, William J.; Palmer, Andrew W.; Grattan, Kenneth T.; Meggitt, Beverley T.)V1584,220-225(1991)

Novel signal processing scheme for ruby-fluorescence-based fiber-optic temperature sensor (Zhang, Zhiyi; Grattan, Kenneth T.; Palmer, Andrew W.)V1511,264-274(1991)

Novel system for measuring extinction ratio on polarization-maintaining fibers and their devices (Xu, Sen-lu; Sheng, Lie-yi; Zhu, Lie-wei)V1367,303-308(1991)

On-line optical determination of water in ethanol (Kessler, Manfred A.)V1510,218-223(1991)

Optical coatings to reduce temperature sensitivity of polarization-maintaining fibers for smart structures and skins (Zhang, Feng; Lit, John W.)V1588,100-109(1991)

Optical encoders using pseudo-random-binary-sequence scales (Johnston, James S.; Romer, A. E.; Beales, M. S.)V1589,126-132(1991)

Optical fiber densimeter for water in oil (Liu, Rui-Fu; Xi, Xiao-chun; Liang, Chen)V1572,399-402(1991)

Optical fiber interferometric sensors for chemical detection (Butler, Michael A.)V1368,46-54(1991)

Optical fiber magneto-optic current sensor (Zhao, Huafeng; Zhang, Peng-Gang; Liao, Yan-Biao)V1572,503-507(1991)

Optical fiber pressure transducer with improved sensitivity and linearity (Chen, Xiaobao; Chen, Qianmei)V1572,226-229(1991)

Optical fiber refractometer and its application in the sugar industry (Ma, Junxian; Yang, Shuwen)V1572,377-381(1991)

Optical fiber sensing of corona discharges (Woolsey, G. A.; Lamb, D. W.; Woerner, M. C.)V1584,243-253(1991)

Optical fiber sensor for ammonia monitoring in the blood (Yan, Hongtao; Li, Hanjie; Li, Yonghong)V1572,396-398(1991)

Optical fiber sensor for temperature measurement from 600 to 1900 C in gas turbine engines (Tregay, George W.; Calabrese, Paul R.; Kaplin, Peter L.; Finney, Mark J.)V1589,38-47(1991)

Optical fibre PH sensor based on immobilized indicator (Cui, DaFu; Cao, Qiang; Han, JingHong; Cai, Jine; Li, YaTing; Zhu, ZeMin; Fan, Jie; Gao, Ning)V1572,386-391(1991)

Optical fibre image sensor (Li, Aizhong)V1572,548-551(1991)

Optical fibre interferometer for monitoring tool wear (Zheng, S. X.; McBride, R.; Hale, K. F.; Jones, Barry E.; Barton, James S.; Jones, Julian D.)V1572,359-364(1991)

Optical fibre system with CCLID for the transmission of two-dimensional images (Li, Shaohui; Wen, Shengping; Zhang, Zhipeng; Lu, Feizhen)V1572,543-547(1991)

Optical frequency shifter based on stimulated Brillouin scattering in birefringent optical fiber (Duffy, Christopher J.; Tatam, Ralph P.)V1511,155-165(1991)

Optical heterodyne fiber-coil deformation sensor operating in a wide dynamic range (Ohtsuka, Yoshihiro; Nishi, Y.; Sawae, S.; Tanaka, Satoshi)V1572,347-352(1991)

Optically powered sensor system using conventional electrical sensors (Nieuwkoop, E.; Kapsenberg, Th.; Steenvoorden, G. K.; Bruinsma, Anastasius J.)V1511,255-263(1991)

Optically powered thermistor with optical fiber link (Shi, Jinshan; Wang, Yutian)V1572,175-179(1991)

Optically self-excited miniature fixed-beam resonator sensor (Gu, Lizhong; Ma, Jiancheng; Wang, Jiazhen)V1572,450-452(1991)

Optical sensors embedded in composite materials (Bocquet, Jean-Claud; Lecoy, Pierre; Baptiste, Didier)V1588,210-217(1991)

Optical sensors for process monitoring in biotechnology (Ploetz, F.; Schelp, C.; Anders, K.; Eberhardt, F.; Scheper, Thomas-Helmut; Bueckmann, F.)V1510,224-230(1991)

Optical technique for the compensation of the temperature-dependent Verdet constant in Faraday rotation magnetometers (Hamid, Sohail; Tatam, Ralph P.)V1511,78-89(1991)

Optimizing the performance of a frequency-division distributed-optical-fiber sensing system (Leung, Chung-yee; Wu, Jiunn-Shyong; Ho, M. Y.; Chen, Kuang-yi)V1572,566-571(1991)

Optimum design of transducer for fiber-optic magnetometer (Zou, Kun; Cheng, Yuqi)V1572,472-476(1991)

Outlook of fiber-optic gyroscope (Hayakawa, Yoshiaki; Kurokawa, Akihiro)V1572,353-358(1991)

Overview of planar waveguide techniques for chemical sensing (Burgess, Lloyd W.)V1368,224-229(1991)

Parallel coherence receiver for quasidistributed optical sensor (Sansonetti, Pierre; Guerin, J. J.; Lequime, Michael; Debrie, J.)V1588,143-149(1991)

Parallel network for optical fiber sensors (Jiang, Desheng; Ye, Qizheng; Zhang, Shengpei; Li, Faxian)V1572,313-317(1991)

Passive laser phase noise suppression technique for fiber interferometers (Kersey, Alan D.; Berkoff, Timothy A.)V1367,310-318(1991)

Passive optical fibre sensor based on Cerenkov effect (Wang, Yao-Cai; Shi, Yi-Wei; Jiang, Hong-Tao)V1572,32-37(1991)

Performance comparison of various low-cost multimode fiber optic rotation rate sensor designs (Fredricks, Ronald J.; Johnson, Dean R.; Sabri, Sehbaz H.; Yu, Ming H.)V1367,127-139(1991)

Performance of an optoelectronic probe used with coordinate measuring machines (Shams, Iden; Butler, Clive)V1589,120-125(1991)

Phase-strain-temperature model for structurally embedded interferometric optical fiber strain sensors with applications (Sirkis, James S.)V1588,26-43(1991)

Phase compensation of PZT in an optical fibre Mach-Zehnder interferometer (Liu, Yanbing; Zhang, Jinru)V1572,61-64(1991)

Photofluidics for integrating fiber sensors with high-authority mechanical actuators (Claus, Richard O.; Murphy, Kent A.; Fogg, Brian R.; Sun, David; Vengsarkar, Ashish M.)V1588,159-168(1991)

Photon correlation spectroscopy and electrophoretic light scattering using optical fibers (Macfadyen, Allan J.; Jennings, B. R.)V1367,319-328(1991)

Photonics technology for aerospace applications (Figueroa, Luis; Hong, C. S.; Miller, Glen E.; Porter, Charles R.; Smith, David K.)V1418,153-176(1991)

Planar waveguide optical immunosensors (Choquette, Steven J.; Locascio-Brown, Laurie; Durst, Richard A.)V1368,258-263(1991)

Plastic Optical Fibers (Kitazawa, Mototaka; Kreidl, John F.; Steele, Robert E., eds.)V1592(1991)

Polarimetric monomode optical fibre sensor for monitoring tool wear (Zheng, S. X.; Hale, K. F.; Jones, Barry E.)V1572,268-272(1991)

Polarimetric optical fiber pressure sensor (Li, Luksun; Kerr, Anthony; Giles, Ian P.)V1584,170-177(1991)

Polarimetric optical fiber sensor for biochemical measurements (Heideman, Rene; Blikman, Albert; Koster, Rients; Kooyman, Rob P.; Greve, Jan)V1510,131-137(1991)

Polarimetric sensor strain sensitivity in different thermal operating conditions (De Maria, Letizia; Escobar Rojo, Priscilla; Martinelli, Mario; Pistoni, Natale C.)V1366,304-312(1991)

Polyimide-coated embedded optical fiber sensors (Nath, Dilip K.; Nelson, Gary W.; Griffin, Stephen E.; Harrington, C. T.; He, Yi-Fei; Reinhart, Larry J.; Paine, D. C.; Morse, Ted F.)V1489,17-32(1991)

Portable and very inexpensive optical fiber sensor for entero-gastric reflux detection (Baldini, Francesco; Falciai, Riccardo; Bechi, Paolo; Cosi, Franco; Bini, Andrea; Milanesi, Francesco)V1510,58-62(1991)

Portable fiber optic current sensor (Shafir, Ehud; Ben-Kish, A.; Konforti, Naim; Tur, Moshe)V1442,236-241(1991)

Potential for integrated optical circuits in advanced aircraft with fiber optic control and monitoring systems (Baumbick, Robert J.)V1374,238-250(1991)

Practicable fiber optic displacement sensor with subnanometer resolution (Xu, Jing)V1367,214-220(1991)

Predetection correlation in a spread-spectrum multiplexing system for fiber optic interferometers (Al-Raweshidy, H. S.; Uttamchandani, Deepak G.)V1367,329-336(1991)

Preliminary field demonstration of a fiber optic trichloroethylene sensor (Angel, S. M.; Langry, Kevin; Colston, B. W.; Roe, Jeffrey N.; Daley, Paul F.; Milanovich, Fred P.)V1368,98-104(1991)

Problems and solutions in fiber-optic amplitude-modulated sensors (Brenci, Massimo; Mencaglia, Andrea; Mignani, Anna G.)V1504,212-220(1991)

Process monitoring and control with fiber optics (Marcus, Michael A.)V1368,191-202(1991)

Proton-exchange X-cut lithium tantalate fiber optic gyro chips (Wong, Ka-Kha; Killian, Kevin M.; Dimitrov-Kuhl, K. P.; Long, Margaret; Fleming, J. T.; van de Vaart, Herman)V1374,278-286(1991)

Pulp blood flow assessment in human teeth by laser Doppler flowmetry (Pettersson, Hans; Oberg, Per A.)V1424,116-119(1991)

Ratiometric fiber optic sensor utilizing a fused biconically tapered coupler (Booysen, Andre; Spammer, Stephanus J.; Swart, Pieter L.)V1584,273-279(1991)

Real-time structural integrity monitoring using a passive quadrature demodulated, localized Michelson optical fiber interferometer capable of simultaneous strain and acoustic emission sensing (Tapanes, Edward)V1588,356-367(1991)

Recent developing status of fiber optic sensors in China (Xu, Sen-lu; Luo, Gei-peng; Xu, Wei-dong)V1367,59-69(1991)

Recent developments in fiber optic and laser sensors for flow, surface vibration, rotation, and velocity measurements (Arik, Engin B.)V1584,202-211(1991)

Recent developments in fiber optic magnetostrictive sensors (Bucholtz, Frank; Dagenais, Dominique M.; Koo, Kee P.; Vohra, Sandeep T.)V1367,226-235(1991)

Recent developments in fiber optic ring laser gyros (Smith, S. P.; Zarinetchi, F.; Ezekiel, Shaoul)V1367,103-106(1991)

Recent progress in fiber optic pH sensing (Baldini, Francesco)V1368,184-190(1991)

Recent progress in interferometric fiber sensor technology (Kersey, Alan D.)V1367,2-12(1991)

Recent progress in intrinsic fiber optic chemical sensing (Lieberman, Robert A.)V1368,15-24(1991)

Reduced cost coil windings for interferometric fiber-optic gyro sensors (Smith, Ronald H.)V1478,145-149(1991)

Reflective fiber temperature sensor using a bimetallic transducer (Liu, Rui-Fu; Jiang, G.; Jiang, H. Y.; Xi, Xiao-chun; Song, W. S.)V1572,189-191(1991)

Reflective optical fibre displacement sensor (Lu, Xiaoming; Ren, Xin; Wang, Peizheng; Chi, Rongsheng)V1572,248-251(1991)

Reflective optical sensor system for measurement of intracranial pressure (Zoghi, Behbood; Rastegar, Sohi)V1420,63-71(1991)

Reliability improvement methods for sapphire fiber temperature sensors (Schietinger, Chuck W.; Adams, Bruce)V1366,284-293(1991)

Reliability of fiber optic position sensors (Park, Eric D.; Swafford, William J.; Lamb, Bryan K.)V1366,294-303(1991)

Research of distributed-fiber-optic pressure sensor (Lu, Xiaoming; Ren, Xin; Chen, Yu-bao; Chi, Rongsheng)V1572,304-307(1991)

Research of optical fiber pyrometric sensor for gas-making furnace (Deng, Xingzhong; Zhong, Zugen; Cheng, Xianping; Liu, Bo; Cao, Wei)V1572,220-223(1991)

Research of signal transmission and processing of fiber-optic pyrometer (Cheng, Xianping; Deng, Xingzhong; Xie, Tuqiang; Cao, Wei; Liu, Bo)V1572,216-219(1991)

Research on optical fiber colorimeter (Cao, Zheng-Ping; Huang, Yue-Huai)V1572,38-41(1991)

Research on the characteristics of temperature drift in fiber-optic gyroscope (Luo, Gei-peng; Xu, Sen-lu)V1572,337-341(1991)

Research on the photoelectricity-based swing measuring system (Xu, Ken; Zhang, Zhipeng)V1572,235-239(1991)

Results of a portable fiber optic weigh-in-motion system (Muhs, Jeffry D.; Jordan, John K.; Scudiere, M. B.; Tobin, Kenneth W.)V1584,374-386(1991)

Review of the Fiber Optic Control System Integration program (Baumbick, Robert J.)V1589,12-19(1991)

Role of adhesion in optical-fiber-based smart composite structures and its implementation in strain analysis for the modeling of an embedded optical fiber (DiFrancia, Celene; Claus, Richard O.; Ward, T. C.)V1588,44-49(1991)

Sapphire fiber interferometer for microdisplacement measurements at high temperatures (Murphy, Kent A.; Fogg, Brian R.; Wang, George Z.; Vengsarkar, Ashish M.; Claus, Richard O.)V1588,117-124(1991)

Scintillating plastic optical fiber radiation detectors in high-energy particle physics (Bross, Alan D.)V1592,122-132(1991)

Selected Papers on Interferometry (Hariharan, P., ed.)VMS28(1991)

Sensitive fiber-optic immunoassay (Walczak, Irene M.; Love, Walter F.; Slovacek, Rudolf E.)V1420,2-12(1991)

Signal processing in fiber-optic interferometer with FM light sources (Chi, Jifu; Chang, Meitung)V1572,74-77(1991)

Single-mode fiber Mach-Zehnder interferometer as an earth strain sensor (Wang, An; Xie, Haiming)V1572,440-443(1991)

Small single-sensor for temperature, flow, and pressure measurement (Sun, Mei H.; Kamal, Arvind)V1420,44-52(1991)

Smart civil structures: an overview (Huston, Dryver R.)V1588,182-188(1991)

Spacially periodical stress method for measuring the beat length of a highly birefringent optical fiber (Zhang, Zhongxian; Gao, Hangjun; Nan, Zhilin)V1572,5-10(1991)

Special optical fibers for sensors (Inada, Koichi)V1572,163-168(1991)

Specialty Fiber Optic Systems for Mobile Platforms (Lewis, Norris E.; Moore, Emery L., eds.)V1589(1991)

Spread spectrum technique for passive multiplexing of interferometric optical fiber sensors (Uttamchandani, Deepak G.; Al-Raweshidy, H. S.)V1511,212-219(1991)

Stability studies of optical fiber pressure sensor (Fang, X. J.; Wang, A. B.; Se, H.; Jin, X. D.; Jang, T.; Lin, J. X.)V1572,279-283(1991)

Stabilized extrinsic fiber optic Fabry-Perot sensor for surface acoustic wave detection (Tran, Tuan A.; Miller, William V.; Murphy, Kent A.; Vengsarkar, Ashish M.; Claus, Richard O.)V1584,178-186(1991)

Strain sensing using a fiber-optic Bragg grating (Melle, Serge M.; Liu, Kexing; Measures, Raymond M.)V1588,255-263(1991)

Structures Sensing and Control (Breakwell, John; Varadan, Vijay K., eds.)V1489(1991)

Studies of blood gas determination and intelligent image (Wang, Chihcheng; Jin, Xi)V1572,406-409(1991)

Studies of displacement sensing based on the deformation loss of an optical fiber ring (Sheng, Lie-yi; Li, Shenghong; Xu, Sen-lu; Zhu, Lie-wei)V1572,273-278(1991)

Studies on antireflection technology of end surfaces for coherent fibre bundles (Qian, Anping; Dong, Linjun)V1572,144-147(1991)

Study of OFS for gas-liquid two-phase flow (Liao, Yan-Biao; Lai, Shurong; Zhao, Huafeng; Wu, Gengsheng)V1584,400-404(1991)

Study of an opto-ultrasonic technique for cure monitoring (Davis, Andrew; Ohn, Myo M.; Liu, Kexing; Measures, Raymond M.)V1588,264-274(1991)

Study on an optical fibre detector in fabric edge control (Huang, Yue-Huai; Mu, Lemin)V1572,539-542(1991)

Study on optical fibre sensor for on-line correlation velocity measurement (Xiang, Tingyuan; Zheng, Yingna; Huang, Nanmin; Fan, Xinrui)V1572,372-376(1991)

Study on pyrometer with double Y-type optical fibers (Li, Zhiquan; Shi, Jinshan; Wang, Yutian)V1572,185-188(1991)

Study on quasi-instantaneous converse piezoelectric effect of piezoelectric ceramics with the sinusoidal phase-modulating interferometer using optical fibers (Li, Naiji; Wang, Huiwen; Yang, Yang; Gu, Shenghua)V1572,47-51(1991)

Surface contouring using TV holography (Atcha, Hashim; Tatam, Ralph P.; Buckberry, Clive H.; Davies, Jeremy C.; Jones, Julian D.)V1504,221-232(1991)

Surface inspection using optical fiber sensor (Abe, Makoto; Ohta, Shigekata; Sawabe, Masaji)V1332,366-376(1991)

Surface reflection coefficient correction technique for a microdisplacement OFS (Wei, Cailin)V1572,42-46(1991)

Symmetric 3 x 3 coupler based demodulator for fiber optic interferometric sensors (Brown, David A.; Cameron, C. B.; Keolian, Robert M.; Gardner, David L.; Garrett, Steven L.)V1584,328-335(1991)

Temperature and fading effects of fiber-optic dosimeters for radiotherapy (Bueker, Harald; Gripp, S.; Haesing, Friedrich W.)V1572,410-418(1991)

Temperature and strain sensing using monomode optical fiber (Farahi, Faramarz; Jackson, David A.)V1511,234-243(1991)

Theoretical analysis and design on optical fiber magneto-optic current sensing head (Zhang, Peng-Gang; Zhao, Huafeng; Liao, Yan-Biao)V1572,528-533(1991)

Theoretical analysis for a class of fiber-optic acceleration sensors (Zou, Zi-Li; Lou, Pei-De; Tang, Minguang)V1572,18-26(1991)

Theoretical analysis of two-mode, elliptical-core optical fiber sensors (Shaw, J. K.; Vengsarkar, Ashish M.; Claus, Richard O.)V1367,337-346(1991)

Thermal plastic metal coatings on optical fiber sensors (Sirkis, James S.; Dasgupta, Abhijit)V1588,88-99(1991)

Three-dimensional fibre-optic position sensor (Yang, Q.; Butler, Clive)V1572,558-563(1991)

Three-dimensional interferometric and fiber-optic displacement measuring probe (Liang, Dawei; Fraser Monteiro, L.; Fraser Monteiro, M. L.; Lu, Boyin)V1511,90-97(1991)

Time-addressing of coherence-tuned optical fibre sensors based on a multimode laser diode (Santos, Jose L.; Jackson, David A.)V1572,325-330(1991)

Time characteristics in HTS rare-earth-doped optical fiber at high temperature (Yang, Yang; Li, Naiji; Wang, Huiwen)V1572,205-210(1991)

Transmissive serial interferometric fiber-optic sensor array (Chen, Xiaoguang; Tang, Weizhong; Zhou, Wen)V1572,294-298(1991)

Two-colour ratio pyrometer with optical fiber (Wang, Yutian; Shi, Jinshan; Li, Zhiquan)V1572,192-196(1991)

Two-mode elliptical-core fiber sensors for measurement of strain and temperature (Wang, Anbo; Wang, Zhiguang; Vengsarkar, Ashish M.; Claus, Richard O.)V1584,294-303(1991)

Ultrasonic NDE (nondestructive evaluation) for composite materials using embedded fiber optic interferometric sensors (Liu, Kexing; Ferguson, Suzanne M.; Davis, Andrew; McEwen, Keith; Measures, Raymond M.)V1398,206-212(1991)

Universal light source for optical fiber sensors (Semenov, Alexandr T.; Elenkrig, Boris B.; Logozinskii, Valerii N.)V1584,348-352(1991)

U-shaped fiber-optic refractive-index sensor and its applications (Takeo, Takashi; Hattori, Hajime)V1544,282-286(1991)

Use of 3 X 3 integrated optic polarizer/splitters for a smart aerospace plane structure (Seshamani, Ramani; Alex, T. K.)V1489,56-64(1991)

Use of laser diodes and monomode optical fiber in electronic speckle pattern interferometry (Atcha, Hashim; Tatam, Ralph P.)V1584,425-434(1991)

Use of optical sensors and signal processing in gas turbine engines (Davinson, Ian)V1374,251-265(1991)

Using a fiber-optic pulse sensor in magnetic resonance imaging (Henning, Michael R.; Gerdt, David W.; Spraggins, Thomas)V1420,34-40(1991)

UV optical fiber distributed temperature sensor (Paton, Andrew T.; Scott, Chris J.)V1367,274-281(1991)

Venous occlusion plethysmography based on fiber-optic sensor using the microbending principle (Stenow, Eric; Rohman, H.; Eriksson, L.-E.; Oberg, Per A.)V1420,29-33(1991)

Vibration sensing in flexible structures using a distributed-effect modal domain optical fiber sensor (Reichard, Karl M.; Lindner, Douglas K.; Claus, Richard O.)V1489,218-229(1991)

Visible/infrared integrated double detector: application to obstacle detection in automotive (Phase 2) (Simonne, John J.; Pham, Vui V.; Esteve, Daniel; Clot, Jean; Mahrane, Achour; Beconne, Jean P.)V1589,139-147(1991)

Waveguide-based fiber optic magnetic field sensor with directional sensitivity (Sohlstrom, Hans B.; Svantesson, Kjell G.)V1511,142-148(1991)

Wavelength-based sensor for the measurement of small angles (Depoortere, Marc; Ebbeni, Jean P.)V1504,264-267(1991)

Wavelength-encoded fiber optic angular displacement sensor (Spillman, William B.; Rudd, Robert E.; Hoff, Frederick G.; Patriquin, Douglas R.; Lord, Jeffrey R.)V1367,197-203(1991)

Wavelength-multiplexed fiber optic position encoder for aircraft control systems (Beheim, Glenn; Krasowski, Michael J.; Sotomayor, Jorge L.; Fritsch, Klaus; Flatico, Joseph M.; Bathurst, Richard L.; Eustace, John G.; Anthan, Donald J.)V1369,50-59(1991)

Wavelength distribution optical fiber sensor (Luo, Nan; Wang, Changgui; Xiao, Wen; Zhao, Yanhan; Chen, Yaosheng)V1572,2-4(1991)

Width measurement of cold-rolling strip (Qi, GuangXue; Deng, Zesheng)V1572,536-538(1991)

Fiber optics in medicine—see also endoscopes; fiber optic sensors; medicine

Bile optical fiber sensor: the method for entero-gastric reflux detection (Falciai, Riccardo; Baldini, Francesco; Bechi, Paolo; Cosi, Franco; Bini, Andrea)V1572,424-427(1991)

Bundle of tapered fibers for the transmission of high-power excimer laser pulses (Hitzler, Hermine; Leclerc, Norbert; Pfleiderer, Christoph; Wolfrum, Juergen M.; Greulich, Karl O.; Klein, Karl-Friedrich)V1503,355-362(1991)

Characterization and uses of plastic hollow fibers for CO_2 laser energy transmission (Gannot, Israel; Dror, Jacob; Dahan, Reuben; Alaluf, M.; Croitoru, Nathan I.)V1442,156-161(1991)

Clinical applications of pulmonary artery oximetry (Barker, Steven J.)V1420,22-28(1991)

Combined guidance technique using angioscope and fluoroscope images for CO laser angioplasty: in-vivo animal experiment (Arai, Tsunenori; Mizuno, Kyoichi; Sakurada, Masami; Miyamoto, Akira; Arakawa, Koh; Kurita, Akira; Suda, Akira; Kikuchi, Makoto; Nakamura, Haruo; Utsumi, Atsushi; Akai, Yoshiro; Takeuchi, Kiyoshi)V1425,191-195(1991)

Comparison of silica-core optical fibers (McCann, Brian P.)V1420,116-125(1991)

Comparison of the ablation of polymethylmethacrylate by two fiber-optic-compatible infrared lasers (Garino, Jonathan P.; Nazarian, David; Froimson, Mark I.; Grelsamer, Ronald P.; Treat, Michael R.)V1424,43-47(1991)

Core-clad silver halide fibers for CO2 laser power transmission (Paiss, Idan; Moser, Frank; Katzir, Abraham)V1420,141-148(1991)

Cryosurgical ablation of the prostate (Cohen, Jeffrey K.)V1421,45-45(1991)

Delayed fluorescence of eosin-bound protein: a probe for measurement of slow-rotational mobility (Yao, Jialing; McStay, Daniel; Rogers, Alan J.; Quinn, Peter J.)V1572,428-433(1991)

Detection of general anesthetics using a fluorescence-based sensor: incorporation of a single-fiber approach (Abrams, Susan B.; McDonald, Hillary L.; Yager, Paul)V1420,13-21(1991)

Diagnosis of atherosclerotic tissue by resonance fluorescence spectroscopy (Neu, Walter; Haase, Karl K.; Tischler, Christian; Nyga, Ralf; Karsch, Karl R.)V1425,28-36(1991)

Direct fluoroimmunoassay in Langmuir-Blodgett films of immunoglobulin G (Turko, Illarion V.; Lepesheva, Galina I.; Chashchin, Vadim L.)V1572,419-423(1991)

Early diagnosis of lung cancer (Saccomanno, Geno; Bechtel, Joel J.)V1426,2-12(1991)

Effect of surface boundary on time-resolved reflectance: measurements with a prototype endoscopic catheter (Jacques, Steven L.; Flock, Stephen T.)V1431,12-20(1991)

Elaboration of excimer lasers dosimetry for bone and meniscus cutting and drilling using optical fibers (Jahn, Renate; Dressel, Martin; Neu, Walter; Jungbluth, Karl-Heinz)V1424,23-32(1991)

Fiber design for interstitial laser treatment (Beuthan, Jurgen; Mueller, Gerhard J.; Schaldach, Brita J.; Zur, Ch.)V1420,234-241(1991)

Fiber fragmentation during laser lithotripsy (Flemming, G.; Brinkmann, Ralf E.; Strunge, Ch.; Engelhardt, R.)V1421,146-152(1991)

Fiber optic image guide rods as ultrathin endoscopy (Kociszewski, Longin; Pysz, Dariusz)V1420,212-217(1991)

Fiber optic photoplethysmograph (Bokun, Leszek J.; Domanski, Andrzej W.)V1420,93-99(1991)

Fused silica fibers for the delivery of high-power UV radiation (Artjushenko, Vjacheslav G.; Konov, Vitaly I.; Pashinin, Vladimir P.; Silenok, Alexander S.; Blinov, Leonid M.; Solomatin, A. M.; Shilov, I. P.; Volodko, V. V.; Mueller, Gerhard J.; Schaldach, Brita J.; Ulrich, R.; Neuberger, Wolfgang)V1420,149-156(1991)

High-strength optical fiber for medical applications (Krohn, David A.; Maklad, Mokhtar S.; Bacon, Fredrick)V1420,126-135(1991)

Hollow curved Al2O3 waveguides for CO2 laser surgery (Gregory, Christopher C.; Harrington, James A.; Altkorn, Robert I.; Haidle, Rudy H.; Helenowski, Tomasz)V1420,169-175(1991)

Imaging in digestive videoendoscopy (Guadagni, Stefano; Nadeau, Theodore R.; Lombardi, Loreto; Pistoia, Francesco; Pistoia, Maria A.)V1420,178-182(1991)

Infrared cables and catheters for medical applications (Artjushenko, Vjacheslav G.; Ivchenko, N.; Konov, Vitaly I.; Kryukov, A. P.; Krupchitsky, Vladimir P.; Kuznetcov, R.; Lerman, A. A.; Litvinenko, E. G.; Nabatov, A. O.; Plotnichenko, V. G.; Prokhorov, Alexander M.; Pylnov, I. L.; Tsibulya, Andrew B.; Vojtsekhovsky, V. V.; Ashraf, N.; Neuberger, Wolfgang; Moran, Kelly B.; Mueller, Gerhard J.; Schaldach, Brita J.)V1420,157-168(1991)

Integral prism-tipped optical fibers (Friedl, Stephan E.; Kunz, Warren F.; Mathews, Eric D.; Abela, George S.)V1425,134-141(1991)

Intl Conf on Optical Fibre Sensors in China (Culshaw, Brian; Liao, Yan-Biao, eds.)V1572(1991)

Key issues in selecting plastic optical fibers used in novel medical sensors (Kosa, Nadhir B.)V1592,114-121(1991)

Laser energy repartition inside metal, sapphire, and quartz surgical laser tips (Seka, Wolf D.; Golding, Douglas J.; Klein, B.; Lanzafame, Raymond J.; Rogers, David W.)V1398,162-169(1991)

Laser-induced fluorescence imaging of coronary arteries for open-heart surgery applications (Taylor, Roderick S.; Gladysz, D.; Brown, D.; Higginson, Lyall A.)V1420,183-192(1991)

Laser speckle and optical fiber sensors for micromovements monitoring in biotissues (Tuchin, Valery V.; Ampilogov, Andrey V.; Bogoroditsky, Alexander G.; Rabinovich, Emmanuil M.; Ryabukho, Vladimir P.; Ul'yanov, Sergey S.; V'yushkin, Maksim E.)V1420,81-92(1991); V1429,62-73(1991)

Lesion-specific laser catheters for angioplasty (Murphy-Chutorian, Douglas)V1420,244-248(1991)

Microwave warming of biological tissue and its control by IR fiber thermometry (Drizlikh, S.; Zur, A.; Moser, Frank; Katzir, Abraham)V1420,53-62(1991)

Monitoring of tissue temperature during microwave hyperthermia utilizing a fiber optic liquid crystalline microsensor (Domanski, Andrzej W.; Kostrzewa, Stanislaw; Hliniak, Andrzej)V1420,72-80(1991)

New approach for endoscopic stereotactic brain surgery using high-power laser (Otsuki, Taisuke; Yoshimoto, Takashi)V1420,220-224(1991)

New image diagnosis system with ultrathin endoscope and clinical results (Tsumanuma, Takashi; Toriya, T.; Tanaka, T.; Shamoto, N.; Seto, K.; Sanada, Kazuo; Okazaki, A.; Okazaki, M.)V1420,193-198(1991)

New method for detection of blood coagulation using fiber-optic sensor (Fediay, Sergey G.; Kuznetzov, Alexsey V.)V1420,41-43(1991)

Optical characteristics of sapphire laser scalpels analysed by ray-tracing (Verdaasdonk, Rudolf M.; Borst, Cornelius)V1420,136-140(1991)

Optical fiber sensor for ammonia monitoring in the blood (Yan, Hongtao; Li, Hanjie; Li, Yonghong)V1572,396-398(1991)

Optical fibers in artificial joint (Yu, Jie; Huang, Jianmin; Zhang, Kui; Zhang, Liang)V1420,266-270(1991)

Optical Fibers in Medicine VI (Katzir, Abraham, ed.)V1420(1991)

Percutaneous coronary angioscopy during coronary angioplasty: clinical findings and implications (Ramee, Stephen R.; White, Christopher J.; Mesa, Juan E.; Jain, Ashit; Collins, Tyrone J.)V1420,199-202(1991)

Photodynamic therapy of malignant brain tumors: supplementary postoperative light delivery by implanted optical fibers: field fractionation (Muller, Paul J.; Wilson, Brian C.)V1426,254-265(1991)

Plastic hollow fibers employed for CO2 laser power transmission in oral surgery (Calderon, S.; Gannot, Israel; Dror, Jacob; Dahan, Reuben; Croitoru, Nathan I.)V1420,108-115(1991)

Potentials for pulsed YAG:Nd laser application to endoscopic surgery (Manenkov, Alexander A.; Denisov, N. N.; Bagdasarov, V. H.; Starkovsky, A. N.; Yurchenko, S. V.; Kornilov, Yu. M.; Mikaberidze, V. M.; Sarkisov, S. E.)V1420,254-258(1991)

Power transmission for silica fiber laser delivery systems (McCann, Brian P.)V1398,230-237(1991)

Practical considerations for effective microendoscopy (Papaioannou, Thanassis; Papazoglou, Theodore G.; Daykhovsky, Leon; Gershman, Alex; Segalowitz, Jacob; Reznik, G.; Beeder, Clain; Chandra, Mudjianto; Grundfest, Warren S.)V1420,203-211(1991)

Q-switching and pulse shaping with IR lasers (Brinkmann, Ralf E.; Bauer, K.)V1421,134-139(1991)

Ray-tracing of optically modified fiber tips for laser angioplasty (Verdaasdonk, Rudolf M.; Borst, Cornelius)V1425,102-109(1991)

Reflective optical sensor system for measurement of intracranial pressure (Zoghi, Behbood; Rastegar, Sohi)V1420,63-71(1991)

Sensitive fiber-optic immunoassay (Walczak, Irene M.; Love, Walter F.; Slovacek, Rudolf E.)V1420,2-12(1991)

Small single-sensor for temperature, flow, and pressure measurement (Sun, Mei H.; Kamal, Arvind)V1420,44-52(1991)

Studies of blood gas determination and intelligent image (Wang, Chihcheng; Jin, Xi)V1572,406-409(1991)

Study of biological objects in the reflected light with the help of an analogous fiber optic biophotometer (Gaiduk, Mark I.; Grigoryants, V. V.; Chernousova, I. V.; Menenkov, V. D.)V1403,674-675(1991)

Temperature and fading effects of fiber-optic dosimeters for radiotherapy (Bueker, Harald; Gripp, S.; Haesing, Friedrich W.)V1572,410-418(1991)

Ultrafast imaging of vascular tissue ablation by an XeCl excimer laser (Neu, Walter; Nyga, Ralf; Tischler, Christian; Haase, Karl K.; Karsch, Karl R.)V1425,37-44(1991)

Venous occlusion plethysmography based on fiber-optic sensor using the microbending principle (Stenow, Eric; Rohman, H.; Eriksson, L.-E.; Oberg, Per A.)V1420,29-33(1991)

Films—see also coatings; optical materials; thin films

Amorphous microcellular polytetrafluoroethylene foam film (Tang, Chong Z.)V1519,842-846(1991)

Antihumidity dichromated gelatin holographic recording material (Guo, Lu Rong; Dai, Chao M.; Guo, Yongkang; Cai, Tiequan)V1555,293-296(1991)

Applications of diamond-like carbon films for write-once optical recording (Armeyev, V. Y.; Arslanbekov, A. H.; Chapliev, N. I.; Konov, Vitaly I.; Ralchenko, V. G.; Strelnitsky, V. E.)V1621,2-10(1991)

Behavior of a thin liquid film under thermal stimulation: theory and applications to infrared interferometry (Ledoyen, Fernand; Lewandowski, Jacques; Cormier, Maurice)V1507,328-338(1991)

Characterization of DC-PVA films for holographic recording materials (Leclere, Philippe; Renotte, Yvon L.; Lion, Yves F.)V1507,339-344(1991)

Collinear asymmetrical polymer waveguide modulator (Chen, Ray T.; Sadovnik, Lev S.)V1559,449-460(1991)

Comparison of dichromated gelatin and Du Pont HRF-700 photopolymer as media for holographic notch filters (Salter, Jeffery L.; Loeffler, Mary F.)V1555,268-278(1991)

Computer and Optically Generated Holographic Optics; 4th in a Series (Cindrich, Ivan; Lee, Sing H., eds.)V1555(1991)

Decay of the nonlinear susceptibility components in main-chain functionalized poled polymers (Meyrueix, Remi; LeCompte, J. P.; Tapolsky, Gilles)V1560,454-466(1991)

Development and experimental investigation of a copying procedure for the reproduction of large-format transmissive holograms (Kubitzek, Ruediger; Froelich, Klaus; Tropartz, Stephan; Stojanoff, Christo G.)V1507,365-372(1991)

Development and investigation of dichromated gelatin film for the fabrication of large-format holograms operating at 400-900 nm (Tropartz, Stephan; Brasseur, Olivier; Kubitzek, Ruediger; Stojanoff, Christo G.)V1507,345-353(1991)

Direct-drive film magazines (Lewis, George R.)V1538,167-179(1991)

Display holograms in Du Pont's OmniDex films (Zager, Stephen A.; Weber, Andrew M.)V1461,58-67(1991)

Dry photopolymer embossing: novel photoreplication technology for surface relief holographic optical elements (Shvartsman, Felix P.)V1507,383-391(1991)

Electro-optic measurements of dye/polymer systems (Wang, Chin H.; Guan, H. W.; Zhang, J. F.)V1559,39-48(1991)

Electrophotographic properties of thiophene derivatives as charge transport material (Kuroda, Masami; Kawate, K.; Nabeta, Osamu; Furusho, N.)V1458,155-161(1991)

Eurosprint proofing system (Froehlich, Helmut H.)V1458,51-60(1991)

High-capacity optical spatial switch based on reversible holograms (Mikaelian, Andrei L.; Salakhutdinov, Viktor K.; Vsevolodov, N. N.; Dyukova, T. V.)V1621,148-157(1991)

High-sensitive photopolymer for large-size holograms (Ishikawa, Toshiharu; Kuwabara, Y.; Koseki, Kenichi; Yamaoka, Tsuguo)V1461,73-78(1991)

Hologram: liquid-crystal composites (Ingwall, Richard T.; Adams, Timothy)V1555,279-290(1991)

Holographic spectral selectors and filters based on phase gratings and planar waveguides (Bazhenov, V. Y.; Burykin, N. M.; Soskin, M. S.; Taranenko, Victor B.; Vasnetsov, M. V.)V1574,148-153(1991)

Holographic transmission elements using improved photopolymer films (Gambogi, William J.; Gerstadt, William A.; Mackara, Steven R.; Weber, Andrew M.)V1555,256-267(1991)

Laser image recording on a metal/polymer medium (Erokhovets, Valerii K.; Larchenko, Yu. V.; Leonov, A. M.; Tkachenko, Vadim V.)V1621,227-236(1991)

Light budget and optimization strategies for display applications of dichroic nematic droplet/polymer films (Drzaic, Paul S.)V1455,255-263(1991)

Linear LED arrays for film annotation and for high-speed high-resolution printing on ordinary paper (Wareberg, P. G.; Scholes, R.; Taylor, R.)V1538,112-123(1991)

Manufacturing and reproduction of holographic optical elements in dichromated gelatin films for operation in the infrared (Stojanoff, Christo G.; Tropartz, Stephan; Brasseur, Olivier; Kubitzek, Ruediger)V1485,274-280(1991)

Modulation of absorption in DCG (Zhao, Feng; Geng, Wanzhen; Jiang, Lingzhen; Hong, Jing)V1555,297-299(1991)

New design of the illuminating system for transmission film copy (Pesl, Ales A.)V1448,218-224(1991)

New developing process for PVCz holograms (Yamagishi, Yasuo; Ishitsuka, Takeshi; Kuramitsu, Yoko; Yoneda, Yasuhiro)V1461,68-72(1991)

New method of making HOE by copying CGH on NGD (Guo, Lu Rong; Cheng, Xiaoxue; Guo, Yongkang; Hsu, Ping)V1555,300-303(1991)

New thermistor material for thermistor bolometer: material preparation and characterization (Umadevi, P.; Nagendra, C. L.; Thutupalli, G. K.; Mahadevan, K.; Yadgiri, G.)V1484,125-135(1991); V1485,195-205(1991)

Non-gelatin-dichromated holographic film (Guo, Lu Rong; Cheng, Qirei; Wang, Kuoping)V1461,91-92(1991)

Nonlinear optical properties of phenosafranin-doped substrates (Speiser, Shammai)V1559,238-244(1991)

Optical information recording on vitreous semiconductors with a thermoplastic method of visualization (Panasyuk, L. M.; Forsh, A. A.)V1621,74-82(1991)

Optical recording material based on bacteriorhodopsin modified with hydroxylamine (Vsevolodov, N. N.; Dyukova, T. V.; Druzhko, A. B.; Shakhbazyan, V. Y.)V1621,11-20(1991)

Optical scanning system for a CCD telecine for HDTV (Kurtz, Andrew F.; Kessler, David)V1448,191-205(1991)

Optimal characteristics of rheology and electric field in deformable polymer films of optoelectronic image formation devices (Tarasov, Victor A.; Kuleshov, Nickolay B.; Novoselets, Mikhail K.; Sarkisov, Sergey S.)V1559,331-342(1991)

Peculiarities of anisotropic photopolymerization in films (Krongauz, Vadim V.; Schmelzer, E. R.)V1559,354-376(1991)

Performance of the Multi-Spectral Solar Telescope Array VI: performance and characteristics of the photographic films (Hoover, Richard B.; Walker, Arthur B.; DeForest, Craig E.; Allen, Maxwell J.; Lindblom, Joakim F.)V1343,175-188(1991)

Phase-conjugate interferometry by using dye-doped polymer films (Nakagawa, Kazuo; Egami, Chikara; Suzuki, Takayoshi; Fujiwara, Hirofumi)V1332,267-273(1991)

Photoanisotropic incoherent-to-coherent conversion using five-wave mixing (Huang, Tizhi; Wagner, Kelvin H.)V1562,44-54(1991)

Photochemical and thermal treatment of dichromated gelatin film for the manufacturing of holographic optical elements for operation in the IR (Stojanoff, Christo G.; Schuette, H.; Brasseur, Olivier; Kubitzek, Ruediger; Tropartz, Stephan)V1559,321-330(1991)

Photolithographic imaging of computer-generated holographic optical elements (Shvartsman, Felix P.; Oren, Moshe)V1555,71-79(1991)

Preliminary lithographic characteristics of an all-organic chemically amplified resist formulation for single-layer deep-UV lithography (Nalamasu, Omkaram; Reichmanis, Elsa; Cheng, May; Pol, Victor; Kometani, Janet M.; Houlihan, Frank M.; Neenan, Thomas X.; Bohrer, M. P.; Mixon, D. A.; Thompson, Larry F.; Takemoto, Cliff H.)V1466,13-25(1991)

Real-time holographic recording material: NGD (Guo, Lu Rong; Wang, Kuoping; Cheng, Qirei)V1555,291-292(1991)

Reliability of contrast and dissolution-rate-derived parameters as predictors of photoresist performance (Spragg, Peggy M.; Hurditch, Rodney J.; Toukhy, Medhat A.; Helbert, John N.; Malhotra, Sandeep)V1466,283-296(1991)

Studies of InSb metal oxide semiconductor structure fabricated by photo-CVD using Si2H6 and N2O (Huang, C. J.; Su, Yan K.; Leu, R. L.)V1519,70-73(1991)

Thermomagnetic recording on amorphous ferrimagnetic films (Aleksandrov, K. S.; Berman, G. P.; Frolov, G. I.; Seredkin, V. A.)V1621,51-61(1991)

Volume holographic optics recording in photopolymerizable layers (Boiko, Yuri B.)V1507,318-327(1991)

Wavelength shifting and bandwidth broadening in DCG (Corlatan, Dorina; Schaefer, Martin; Anders, Gerhard)V1507,354-364(1991)

Filters—see also correlation; image processing; optical processing; pattern recognition

Accuracy of the output peak localization in two-dimensional matched filtering (Chalasinska-Macukow, Katarzyna)V1391,295-302(1991)

Adaptive 4-D IR clutter suppression filtering technique (Aridgides, Athanasios; Fernandez, Manuel F.; Randolph, D.; Ferris, D.)V1481,110-116(1991)

Adaptive beamforming using recursive eigenstructure updating with subspace constraint (Yu, Kai-Bor)V1565,288-295(1991)

Adaptive deconvolution based on spectral decomposition (Ahlen, Anders; Sternad, Mikael)V1565,130-142(1991)

Adaptive filters and blind equalizers for mixed-phase channels (Tugnait, Jitendra K.)V1565,209-220(1991)

Adaptive image sequence noise filtering methods (Katsaggelos, Aggelos K.; Kleihorst, Richard P.; Efstratiadis, Serafim N.; Lagendijk, Reginald L.)V1606,716-727(1991)

Adaptive morphological filter for image processing (Cheng, Fulin; Venetsanopoulos, Anastasios N.)V1483,49-59(1991)

Adaptive morphological multiresolution decomposition (Salembier, Philippe; Jaquenoud, Laurent)V1568,26-37(1991)

Adaptive multistage weighted order-statistic filters for image restoration (Yin, Lin; Astola, Jaakko; Neuvo, Yrjo A.)V1451,216-227(1991)

Adaptive Signal Processing (Haykin, Simon, ed.)V1565(1991)

Advanced in-plane rotation-invariant filter results (Ravichandran, Gopalan; Casasent, David P.)V1567,466-479(1991)

Algorithms and architectures for implementing large-velocity filter banks (Stocker, Alan D.; Jensen, Preben D.)V1481,140-155(1991)

Amplitude-encoded phase-only filters for pattern recognition: influence of the bleaching procedure (Campos, Juan; Janowska-Dmoch, Bozena; Styczynski, K.; Turon, F.; Yzuel, Maria J.; Chalasinska-Macukow, Katarzyna)V1574,141-147(1991)

Analysis and experimental performance of reduced-resolution binary phase-only filters (Kozaitis, Samuel P.; Tepedelenlioglu, N.; Foor, Wesley E.)V1564,373-383(1991)

Analysis of noise attenuation in morphological image processing (Koskinen, Lasse; Astola, Jaakko; Neuvo, Yrjo A.)V1451,102-113(1991)

Angle-only tracking and prediction of boost vehicle position (Tsai, Ming-Jer; Rogal, Fannie A.)V1481,281-291(1991)

Application of a blind-deconvolution restoration technique to space imagery (Lewis, Tom R.; Mitra, Sunanda)V1565,221-226(1991)

Application of adaptive filters to the problem of reducing microphony in arrays of pyroelectric infrared detectors (Carmichael, I. C.; White, Paul R.)V1541,167-177(1991)

Application of median-type filtering to image segmentation in electrophoresis (Wang, Qiaofei; Neuvo, Yrjo A.)V1450,47-58(1991)

Application of neural networks for the synthesis of binary correlation filters for optical pattern recognition (Mahalanobis, Abhijit; Nadar, Mariappan S.)V1469,292-302(1991)

Application of the half-filter method to the flash radiography using a neutral filter in the range of x-rays (Gerstenmayer, Jean-Louis; Vibert, Patrick)V1346,286-292(1991)

Backward consistency concept and a new decomposition of the error propagation dynamics in RLS algorithms (Slock, Dirk T.)V1565,14-24(1991)

Bayesian iterative method for blind deconvolution (Neri, Alessandro; Scarano, Gaetano; Jacovitti, Giovanni)V1565,196-208(1991)

Binary phase-only filter associative memory (Kane, Jonathan S.; Hemmer, Philip R.; Woods, Charles L.; Khoury, Jehad)V1564,511-520(1991)

Bipolar correlations in composite circular harmonic filters (Leclerc, Luc; Sheng, Yunlong; Arsenault, Henri H.)V1564,78-85(1991)

Blind equalization (Proakis, John G.; Nikias, Chrysostomos L.)V1565,76-87(1991)

Blind equalization and deconvolution (Bellini, Sandro)V1565,88-101(1991)

Bounds on the performance of optimal four-dimensional filters for detection of low-contrast IR point targets (Wohlers, Martin R.)V1481,129-139(1991)

Case study of design trade-offs for ternary phase-amplitude filters (Flannery, David L.; Phillips, William E.; Reel, Richard L.)V1564,65-77(1991)

Characterization, modeling, and design of dielectric resonators based on thin ceramic tape (Morris, Jacqueline H.; Belopolsky, Yakov)V1389,236-248(1991)

Class of GOS (generalized order-statistic) filters, similar to the median, that provide edge enhancement (Longbotham, Harold G.; Barsalou, Norman)V1451,36-47(1991)

Coding of digital TV by motion-compensated Gabor decomposition (Dufaux, Frederic; Ebrahimi, Touradj; Geurtz, Alexander M.; Kunt, Murat)V1567,362-379(1991)

Coherent detection in confocal microscopy (Wilson, Tony)V1439,104-108(1991)

Color-invariant character recognition and character-background color identification by multichannel matched filter (Campos, Juan; Millan, Maria S.; Yzuel, Maria J.; Ferreira, Carlos)V1564,189-198(1991)

Comparative performance study of several blind equalization algorithms (Shynk, John J.; Gooch, Richard P.; Krishnamurthy, Giridhar; Chan, Christina K.)V1565,102-117(1991)

Comparison of dichromated gelatin and Du Pont HRF-700 photopolymer as media for holographic notch filters (Salter, Jeffery L.; Loeffler, Mary F.)V1555,268-278(1991)

Conditional-expectation-based implementation of the optimal mean-square binary morphological filter (Dougherty, Edward R.; Mathew, A.; Swarnakar, Vivek)V1451,137-147(1991)

Constructing an optimal binary phase-only filter using a genetic algorithm (Calloway, David L.)V1564,395-402(1991)

Coplanar wavequide microwave filter of YBa2Cu3O7 (Chew, Wilbert; Riley, A. L.; Rascoe, Daniel L.; Hunt, Brian D.; Foote, Marc C.; Cooley, Thomas W.; Bajuk, Louis J.)V1477,95-100(1991)

Correction of magneto-optic device phase errors in optical correlators through filter design modifications (Downie, John D.; Reid, Max B.; Hine, Butler P.)V1564,308-319(1991)

Crosstalk in direct-detection optical fiber FDMA networks (Hamdy, Walid M.; Humblet, Pierre A.)V1579,184-194(1991)

Custom-made filters in digital image analysis system (Ostrowski, Tomasz)V1391,264-266(1991)

Decision-directed entropy-based adaptive filtering (Myler, Harley R.; Weeks, Arthur R.; Van Dyke-Lewis, Michelle)V1565,57-68(1991)

Design and performance analysis of optoelectronic adaptive infinite-impulse response filters (Ghosh, Anjan K.)V1565,69-73(1991)

Design and testing of space-domain minimum-average correlation energy filters for 2-D acousto-optic correlators (Connelly, James M.; Vijaya Kumar, B. V. K.; Molley, Perry A.; Stalker, K. T.; Kast, Brian A.)V1564,572-592(1991)

Design and testing of three-level optimal correlation filters (Hendrix, Charles D.; Vijaya Kumar, B. V. K.; Stalker, K. T.; Kast, Brian A.; Shori, Raj K.)V1564,2-13(1991)

Design of M-band filter banks based on wavelet transform (Yaou, Ming-Haw; Chang, Wen-Thong)V1605,149-159(1991)

Design of minimum MAE generalized stack filters for image processing (Zeng, Bing; Gabbouj, Moncef; Neuvo, Yrjo A.)V1606,443-454(1991)

Design of novel integrated optic devices utilizing depressed index waveguides (Lopez-Amo, Manuel; Menendez-Valdes, Pedro; Sanz, Inmaculada; Muriel, Miguel A.)V1374,74-85(1991)

Design of parallel multiresolution filter banks by simulated annealing (Li, Wei; Basso, Andrea; Popat, Ashok C.; Nicoulin, Andre; Kunt, Murat)V1605,124-136(1991)

Design optimization of optical filters for space applications (Annapurna, M. N.; Nagendra, C. L.; Thutupalli, G. K.)V1485,260-271(1991)

Detection of moving subpixel targets in infrared clutter with space-time filtering (Braunreiter, Dennis C.; Banh, Nam D.)V1481,73-83(1991)

Detection of phase objects in transparent liquids using nonlinear coupling in BaTiO3 crystal (Siahmakoun, Azad; Shen, Xuanguo)V1396,535-538(1991)

Development of a low-pass far-infrared filter for lunar observer horizon sensor application (Mobasser, Sohrab; Horwitz, Larry S.; Griffith, O'Dale)V1540,764-774(1991)

Distortion- and intensity-invariant optical correlation filter system (Rahmati, Mohammad; Hassebrook, Laurence G.; Bhushan, M.)V1567,480-489(1991)

Domain-variant gray-scale morphology (Kraus, Eugene J.)V1451,171-178(1991)

Elliptical coordinate transformed phase-only filter for shift and scale invariant pattern recognition (Garcia, Javier; Ferreira, Carlos; Szoplik, Tomasz)V1574,133-140(1991)

Empirical performance of binary phase-only synthetic discriminant functions (Carhart, Gary W.; Draayer, Bret F.; Billings, Paul A.; Giles, Michael K.)V1564,348-362(1991)

Estimation of optimal Kalman filter gain from nonoptimal filter residuals (Chen, Chung-Wen; Huang, Jen-Kuang)V1489,254-265(1991)

Estimation of prospective locations in mature hydrocarbon producing areas (Isaksen, Tron)V1452,270-291(1991)

Existing gap between theory and application of blind equalization (Ding, Zhi; Johnson, C. R.)V1565,154-165(1991)

Experimental system for detecting lung nodules by chest x-ray image processing (Suzuki, Hideo; Inaoka, Noriko; Takabatake, Hirotsugu; Mori, Masaki; Natori, Hiroshi; Suzuki, Akira)V1450,99-107(1991)

Fast dilation and erosion of time-varying grey-valued images with uncertainty (Laplante, Phillip A.; Giardina, Charles R.)V1568,295-302(1991)

Filter and window assemblies for high-power insertion device synchrotron radiation sources (Khounsary, Ali M.; Viccaro, P. J.; Kuzay, Tuncer M.)V1345,42-54(1991)

Finite-precision error analysis of a QR-decomposition-based lattice predictor (Syed, Mushtaq A.; Mathews, V. J.)V1565,25-34(1991)

Full-complex modulation with two one-parameter SLMs (Juday, Richard D.; Florence, James M.)V1558,499-504(1991)

Fusion of multiple-sensor imagery based on target motion characteristics (Tsao, Tien-Ren J.; Libert, John M.)V1470,37-47(1991)

Global modeling approach for multisensor problems (Chung, Yi-Nung; Emre, Erol; Gustafson, Donald L.)V1481,306-314(1991)

Hexagonal sampling and filtering for target detection with a scanning E-O sensor (Sperling, I.; Drummond, Oliver E.; Reed, Irving S.)V1481,2-11(1991)

Hierarchical image decomposition based on modeling of convex hulls corresponding to a set of order statistic filters (Vepsalainen, Ari M.; Linnainmaa, Seppo; Yli-Harja, Olli P.)V1568,2-13(1991)

High-contrast composite infrared filters (Borisevich, Nikolai A.; Zamkovets, A. D.; Ponyavina, A. N.)V1500,222-231(1991)

High-isolation single-taper filters (Jones, Michael G.; Moore, Douglas R.)V1365,43-52(1991)

High-reflective multilayers as narrowband VUV filters (Zukic, Muamer; Torr, Douglas G.)V1485,216-227(1991)

High-selectivity spectral multiplexers-demultiplexers usable in optical telecommunications obtained from multidielectric coatings at the end of optical fibers (Richier, R.; Amra, Claude)V1504,202-210(1991)

Hole spectrum: model-based optimization of morphological filters (Dougherty, Edward R.; Haralick, Robert M.)V1568,224-232(1991)

Holographic filter for coherent radiation (Shi, Dexiu; Xing, Xiaozheng; Wolbarsht, Myron L.)V1419,40-49(1991)

Holographic notch filter (Owen, Harry)V1555,228-235(1991)

Holographic spectral selectors and filters based on phase gratings and planar waveguides (Bazhenov, V. Y.; Burykin, N. M.; Soskin, M. S.; Taranenko, Victor B.; Vasnetsov, M. V.)V1574,148-153(1991)

Hubble extra-solar planet interferometer (Shao, Michael)V1494,347-356(1991)

Image coding using adaptive-blocksize Princen-Bradley transform (Mochizuki, Takashi; Yano, Mitsuharu; Nishitani, Takao)V1605,434-444(1991)

Image reconstruction of IDS filter response (Vaezi, Matt M.; Bavarian, Behnam; Healey, Glenn)V1606,803-809(1991)

Image restoration based on perception-related cost functions (Palmieri, Francesco; Croteau, R. E.)V1451,24-35(1991)

Implementation of an angle-only tracking filter (Allen, Ross R.; Blackman, Samuel S.)V1481,292-303(1991)

Incremental model for target maneuver estimation (Chang, Wen-Thong; Lin, Shao-An)V1481,242-253(1991)

Initial key word OCR filter results (Casasent, David P.; Iyer, Anand K.; Ravichandran, Gopalan)V1384,324-337(1991)

Instrumentation to measure the near-IR spectrum of small fruits (Jaenisch, Holger M.; Niedzwiecki, Abraham J.; Cernosek, John D.; Johnson, R. B.; Seeley, John S.; Dull, Gerald G.; Leffler, Richard G.)V1379,162-167(1991)

Intensity edge detection with stack filters (Yoo, Jisang; Bouman, Charles A.; Delp, Edward J.; Coyle, Edward J.)V1451,58-69(1991)

Interference effects and the occurrence of blind spots in coherent optical processors (Christie, Simon; Cai, Xian-Yang; Kvasnik, Frank)V1507,202-209(1991)

Intermodulation effects in pure phase-only correlation method (Chalasinska-Macukow, Katarzyna; Turon, F.; Yzuel, Maria J.; Campos, Juan)V1564,285-293(1991)

Invariant phase-only filters for phase-encoded inputs (Kallman, Robert R.; Goldstein, Dennis H.)V1564,330-347(1991)

Inverse filtering technique for the synthesis of distortion-invariant optical correlation filters (Shen, Weisheng; Zhang, Shen; Tao, Chunkan)V1567,691-697(1991)

Investigation of new filtering schemes for computerized detection of lung nodules (Yoshimura, Hitoshi; Giger, Maryellen L.; Matsumoto, Tsuneo; Doi, Kunio; MacMahon, Heber; Montner, Steven M.)V1445,47-51(1991)

Laboratory development of a nonlinear optical tracking filter (Block, Kenneth L.; Whitworth, Ernest E.; Bergin, Joseph E.)V1483,62-65(1991)

Learning filter systems with maximum correlation and maximum separation properties (Lenz, Reiner; Osterberg, Mats)V1469,784-795(1991)

L-filter design using the gradient search algorithm (Roy, Sumit)V1451,254-256(1991)

Lippmann volume holographic filters for Rayleigh line rejection in Raman spectroscopy (Rich, Chris C.; Cook, David M.)V1461,2-7(1991)

Ll-filters in CFAR (constant false-alarm rate) detection (Mahmood Reza, Syed; Willett, Peter K.)V1451,298-308(1991)

LMMSE restoration of blurred and noisy image sequences (Ozkan, Mehmet K.; Sezan, M. I.; Erdem, A. T.; Tekalp, Ahmet M.)V1606,743-754(1991)

Localized feature selection to maximize discrimination (Duell, Kenneth A.; Freeman, Mark O.)V1564,22-33(1991)

LUM filters for smoothing and sharpening (Boncelet, Charles G.; Hardie, Russell C.; Hakami, M. R.; Arce, Gonzalo R.)V1451,70-74(1991)

Marriage between digital holography and optical pattern recognition (Wyrowski, Frank; Bernhardt, Michael)V1555,146-153(1991)

Massively parallel implementation for real-time Gabor decomposition (Dufaux, Frederic; Ebrahimi, Touradj; Kunt, Murat)V1606,851-864(1991)

Matched-filter algorithm for subpixel spectral detection in hyperspectral image data (Borough, Howard C.)V1541,199-208(1991)

Method and device that prevent target sensors from being radiation overexposed in the presence of a nuclear blast (Holubowicz, Kazimierz S.)V1456,274-285(1991)

Model-based analysis of 3-D spatial-temporal IR clutter suppression filtering (Chan, David S.)V1481,117-128(1991)

Morphological pattern-spectra-based Tau-opening optimization (Dougherty, Edward R.; Haralick, Robert M.; Chen, Yidong; Li, Bo; Agerskov, Carsten; Jacobi, Ulrik; Sloth, Poul H.)V1606,141-152(1991)

Morphological pyramid with alternating sequential filters (Morales, Aldo W.; Acharya, Raj S.)V1452,258-269(1991)

Motion-compensated subsampling of HDTV (Belfor, Ricardo A.; Lagendijk, Reginald L.; Biemond, Jan)V1605,274-284(1991)

Multilayer mirrors and filters for imaging the earth's inner magnetosphere (Schulze, Dean W.; Sandel, Bill R.; Broadfoot, A. L.)V1549,319-328(1991)

Multiline holographic notch filters (Ning, Xiaohui; Masso, Jon D.)V1545,125-129(1991)

Multiple-target tracking using the SME filter with polar coordinate measurements (Sastry, C. R.; Kamen, Edward W.)V1481,261-280(1991)

Multiplexed binary phase-only circular harmonic filters (Sheng, Yunlong; Leclerc, Luc; Arsenault, Henri H.)V1564,320-329(1991)

Narrowband optical interference filters (Cotton, John M.; Casey, William L.)V1417,525-536(1991)

New algorithms for adaptive median filters (Hwang, Humor; Haddad, Richard A.)V1606,400-407(1991)

New approach to image coding using 1-D subband filtering (Yu, Tian-Hu; Mitra, Sanjit K.)V1452,420-429(1991)

New approach to obtain uniform thickness ZnS thin film interference filters (Mei, Yuan S.; Shang, Shi X.; Shan, Jin A.; Sun, Jian G.)V1519,370-373(1991)

New estimation architecture for multisensor data fusion (Covino, Joseph M.; Griffiths, Barry E.)V1478,114-125(1991)

New type of modified trimmed mean filter (Wu, Wen-Rong; Kundu, Amlan)V1451,13-23(1991)

Noise reduction in heart movies by motion-compensated filtering (Reinen, Tor A.)V1606,755-763(1991)

Noise reduction in ultrasound images using multiple linear regression in a temporal context (Olstad, Bjorn)V1451,269-281(1991)

Nonlinear Image Processing II (Arce, Gonzalo R.; Boncelet, Charles G.; Dougherty, Edward R., eds.)V1451(1991)

Nonlinear regression for signal processing (Restrepo, Alfredo)V1451,258-268(1991)

Normalization of correlations (Kast, Brian A.; Dickey, Fred M.)V1564,34-42(1991)

Novel monolithic chip-integrated color spectrometer: the distributed-wavelength filter component (Holm-Kennedy, James W.; Tsang, Koon Wing; Sze, Wah Wai; Jiang, Fenglai; Yang, Datong)V1527,322-331(1991)

Novel nonlinear filter for image enhancement (Yu, Tian-Hu; Mitra, Sanjit K.; Kaiser, James F.)V1452,303-309(1991)

Numerical design of parallel multiresolution filter banks for image coding applications (Popat, Ashok C.; Li, Wei; Kunt, Murat)V1567,341-353(1991)

Numerical stability issues in fast least-squares adaptive algorithms (Regalia, Phillip A.)V1565,2-13(1991)

Object-enhanced optical correlation (Scholl, Marija S.; Shumate, Michael S.; Udomkesmalee, Suraphol)V1564,165-176(1991)

On-line arithmetic for recurrence problems (Ercegovac, Milos D.)V1566,263-274(1991)

Optical Hartley-transform-based adaptive filter (Abushagur, Mustafa A.; Berinato, Robert J.)V1564,602-609(1991)

Optical characterization of damage resistant kilolayer rugate filters (Elder, Melanie L.; Jancaitis, Kenneth S.; Milam, David; Campbell, John H.)V1441,237-246(1991)

Optical correlation filters to locate destination address blocks in OCR (Casasent, David P.; Ravichandran, Gopalan)V1384,344-354(1991)

Optical correlation using a phase-only liquid-crystal-over-silicon spatial light modulator (Potter, Duncan J.; Ranshaw, M. J.; Al-Chalabi, Adil O.; Fancey, Norman E.; Sillitto, Richard M.; Vass, David G.)V1564,363-372(1991)

Optical correlator field test results (Hudson, Tracy D.; Gregory, Don A.; Kirsch, James C.; Loudin, Jeffrey A.; Crowe, William M.)V1564,54-64(1991)

Optical correlators in texture analysis (Honkonen, Veijo; Jaaskelainen, Timo; Parkkinen, Jussi P.)V1564,43-51(1991)

Optical filters for the wind imaging interferometer (Sellar, R. G.; Gault, William A.; Karp, Christopher K.)V1479,140-155(1991)

Optical inference processing techniques for scene analysis (Casasent, David P.)V1564,504-510(1991)

Optimal distortion-invariant quadratic filters (Gheen, Gregory)V1564,112-120(1991)

Optimal generalized weighted-order-statistic filters (Yin, Lin; Astola, Jaakko; Neuvo, Yrjo A.)V1606,431-442(1991)

Optimal phase-only correlation filters in colored scene noise (Vijaya Kumar, B. V. K.; Liang, Victor; Juday, Richard D.)V1555,138-145(1991)

Optimal subpixel-level IR frame-to-frame registration (Fernandez, Manuel F.; Aridgides, Athanasios; Randolph, D.; Ferris, D.)V1481,172-179(1991)

Optimum intensity-dependent spread filters in image processing (Vaezi, Matt M.; Bavarian, Behnam; Healey, Glenn)V1452,57-63(1991)

Optimum receiver structure and filter design for MPAM optical space communication systems (Al-Ramli, Intesar F.)V1522,111-123(1991)

Order-statistic filters on matrices of images (Wilson, Stephen S.)V1451,242-253(1991)

Parallel DC notch filter (Kwok, Kam-cheung; Chan, Ming-kam)V1567,709-719(1991)

Parallel uses for serial arithmetic in signal processors (Owens, Robert M.; Irwin, Mary J.)V1566,252-262(1991)

Pattern detection using a modified composite filter with nonlinear joint transform correlator (Vallmitjana, Santiago; Juvells, Ignacio; Carnicer, Arturo; Campos, Juan)V1564,266-274(1991)

Performance of pyramidal phase-only filtering of infrared imagery (Kozaitis, Samuel P.; Petrilak, Robert)V1564,403-413(1991)

Performance of the Multi-Spectral Solar Telescope Array IV: the soft x-ray and extreme ultraviolet filters (Lindblom, Joakim F.; O'Neal, Ray H.; Walker, Arthur B.; Powell, Forbes R.; Barbee, Troy W.; Hoover, Richard B.; Powell, Stephen F.)V1343,544-557(1991)

Phase-conjugate optical preprocessing filter for small-target tracking (Block, Kenneth L.; Whitworth, Ernest E.; Bergin, Joseph E.)V1481,32-34(1991)

Photothermoplastic spatial filters for optical pattern recognition (Isaev, Urkaly T.; Akaev, Askar A.; Kutanov, Askar A.)V1507,198-201(1991)

Physically based and probabilistic models for computer vision (Szeliski, Richard; Terzopoulos, Demetri)V1570,140-152(1991)

Planetary lander guidance using binary phase-only filters (Reid, Max B.; Hine, Butler P.)V1564,384-394(1991)

Postprocessing of video sequence using motion-dependent median filters (Lee, Ching-Long; Jeng, Bor S.; Ju, Rong-Hauh; Huang, Huang-Cheng; Kan, Kou-Sou; Huang, Jei-Shyong; Liu, Tsann-Shyong)V1606,728-734(1991)

Precision tracking of small target in IR systems (Lu, Huanzhang; Sun, Zhong-kang)V1481,398-405(1991)

Purely real correlation filters (Mahalanobis, Abhijit; Song, Sewoong)V1564,14-21(1991)

Pyramid median filtering by block threshold decomposition (Zhou, Hongbing; Zeng, Bing; Neuvo, Yrjo A.)V1451,2-12(1991)

Real-time one-pass distortion correction (Oldekop, Erik; Siahmakoun, Azad)V1396,174-177(1991)

Realization of infinite-impulse response filters using acousto-optic cells (Ghosh, Anjan K.)V1564,593-601(1991)

Reconstruction of quincunx-coded image sequences using vector median (Oistamo, Kai; Neuvo, Yrjo A.)V1606,735-742(1991)

Recurrent neural network application to image filtering: 2-D Kalman filtering approach (Swiniarski, Roman W.; Dzielinski, Andrzej; Skoneczny, Slawomir; Butler, Michael P.)V1451,234-241(1991)

Recursive total-least-squares adaptive filtering (Dowling, Eric M.; DeGroat, Ronald D.)V1565,35-46(1991)

Reduction of blocking artifacts using motion-compensated spatial-temporal filtering (Chu, Frank J.; Yeh, Chia L.)V1452,38-46(1991)

Residue-producing E-filters and their applications in medical image analysis (Preston, Kendall)V1450,59-70(1991)

Restoration of distorted depth maps calculated from stereo sequences (Damour, Kevin; Kaufman, Howard)V1452,78-89(1991)

Schur RLS adaptive filtering using systolic arrays (Strobach, Peter)V1565,307-322(1991)

Signal and Data Processing of Small Targets 1991 (Drummond, Oliver E., ed.)V1481(1991)

Signal enhancement in noise- and clutter-corrupted images using adaptive predictive filtering techniques (Soni, Tarun; Zeidler, James R.; Rao, Bhaskar D.; Ku, Walter H.)V1565,338-344(1991)

Signal extension and noncausal filtering for subband coding of images (Martucci, Stephen A.)V1605,137-148(1991)

Signal processing with neural networks: throwing off the yoke of linearity (Hecht-Nielsen, Robert)V1541,146-151(1991)

Soft morphological filters (Koskinen, Lasse; Astola, Jaakko; Neuvo, Yrjo A.)V1568,262-270(1991)

Solid matrix Christiansen filters (Milanovic, Zoran; Jacobson, Michael R.; Macleod, H. A.)V1535,160-170(1991)

Some fundamental experiments in subband coding of images (Aase, Sven O.; Ramstad, Tor A.)V1605,734-744(1991)

Some properties of the two-dimensional pseudomedian filter (Schulze, Mark A.; Pearce, John A.)V1451,48-57(1991)

Spatial frequency filtering on the basis of the nonlinear optics phenomena (Kudryavtseva, Anna D.; Tcherniega, Nicolaii V.; Brekhovskikh, Galina L.; Sokolovskaya, Albina I.)V1385,190-199(1991)

Spatiotemporal filtering of digital angiographic image sequences corrupted by quantum mottle (Chan, Cheuk L.; Sullivan, Barry J.; Sahakian, Alan V.; Katsaggelos, Aggelos K.; Frohlich, Thomas; Byrom, Ernest)V1450,208-217(1991)

Spectrometer on a chip: InP-based integrated grating spectrograph for wavelength-multiplexed optical processing (Soole, Julian B.; Scherer, Axel; LeBlanc, Herve P.; Andreadakis, Nicholas C.; Bhat, Rajaram; Koza, M. A.)V1474,268-276(1991)

Statistical morphology (Yuille, Alan L.; Vincent, Luc M.; Geiger, Davi)V1568,271-282(1991)

Status of high-temperature superconducting analog devices (Talisa, Salvador H.)V1477,78-83(1991)

Strategies for the color character recognition by optical multichannel correlation (Millan, Maria S.; Yzuel, Maria J.; Campos, Juan; Ferreira, Carlos)V1507,183-197(1991)

Subband decomposition procedure for quincunx sampling grids (Kim, Chai W.; Ansari, Rashid)V1605,112-123(1991)

Submicron structures—promising filters in EUV: a review (Gruntman, Michael A.)V1549,385-394(1991)

Synthesis of a maneuver detector and adaptive gain tracking filter (Gardner, Kenneth R.; Kasky, Thomas J.)V1481,254-260(1991)

Synthesis of large-aperture interference fields (Turukhano, Boris G.)V1500,305-308(1991)

Terrain and target segmentation using coherent laser radar (Renhorn, Ingmar G.; Letalick, Dietmar; Millnert, Mille)V1480,35-45(1991)

Texture boundary classification using Gabor elementary functions (Dunn, Dennis F.; Higgins, William E.; Maida, Anthony; Wakeley, Joseph)V1606,541-552(1991)

Three-dimensional morphology for target detection (Patterson, Tim J.)V1471,358-368(1991)

Trade-offs between pseudohexagonal and pseudocubic filters (Preston, Kendall)V1451,75-90(1991)

Transmittances of thin polymer films and their suitability as a supportive substrate for a soft x-ray solar filter (Williams, Memorie; Hansen, Evan; Reyes-Mena, Arturo; Allred, David D.)V1549,147-154(1991)

Understanding metal-dielectric-metal absorption interference filters using lumped circuit theory and transmission line theory (Pastor, Rickey G.)V1396,501-507(1991)

Use of dome (meniscus) lenses to eliminate birefringence and tensile stresses in spatial filters for the Nova laser (Pitts, John H.; Kyrazis, Demos T.; Seppala, Lynn G.; Bumpas, Stanley E.)V1441,71-81(1991)

Use of threshold decomposition theory to derive basic properties of median filters (Hawley, Robert W.; Gallagher, Neal C.)V1451,91-100(1991)

Using structuring-element libraries to design suboptimal morphological filters (Loce, Robert P.; Dougherty, Edward R.)V1568,233-246(1991)

Wavelength-dispersive and filtering applications of volume holographic optical elements (Jannson, Tomasz; Rich, Chris C.; Sadovnik, Lev S.)V1545,42-63(1991)

White-light transmission holographic interferometry using chromatic corrective filters (Grover, Chandra P.)V1332,132-141(1991)

Flows

Analysis and representation of complex structures in separated flows (Helman, James L.; Hesselink, Lambertus B.)V1459,88-96(1991)

Analysis of dynamic characteristics of nonstationary gas streams using interferometry techniques (Abrukov, Victor S.; Ilyin, Stanislav V.)V1554B,540-543(1991)

Analysis of results from high-speed photogrammetry of flow tracers in blast waves (Dewey, John M.; McMillin, Douglas J.)V1358,246-253(1991)

Application of cylindrical blast waves to impact studies of materials (Parry, David J.; Stewardson, H. R.; Ahmad, S. H.; Al-Maliky, Noori S.)V1358,1057-1064(1991)

Application of high-speed infrared emission spectroscopy in reacting flows (Klingenberg, Guenter; Rockstroh, Helmut)V1358,851-858(1991)

Application of holographic interferometry to a three-dimensional flow field (Doerr, Stephen E.)V1554B,544-555(1991)

Application of low-coherence optical fiber Doppler anemometry to fluid-flow measurement: optical system considerations (Boyle, William J.; Grattan, Kenneth T.; Palmer, Andrew W.; Meggitt, Beverley T.)V1511,51-56(1991)

Applications of laser techniques in fluid mechanics (Chan, W. K.; Liu, C. Y.; Wong, Y. W.)V1399,82-89(1991)

Assessment of a discharge-excited supersonic free jet as a laser medium (Kannari, Fumihiko; Sato, F.; Obara, Minoru)V1397,493-497(1991)

Chemical gas-dynamic mixing CO2 laser pumped by the reactions between N2O and CO (Doroschenko, V. M.; Kudriavtsev, N. N.; Sukhov, A. M.)V1397,503-511(1991)

CO2 coupled-mode, CO, and other lasers with supersonic cooling of gas mixture (Kudriavtsev, E. M.)V1397,475-484(1991)

Collapsing cavities in reactive and nonreactive media (Bourne, Neil K.; Field, John E.)V1358,1046-1056(1991)

Comparison of high-current discharges with axial and transverse gas flow for UV ion lasers (Babin, Sergei A.; Kuklin, A. E.)V1397,589-592(1991)

Computer modeling of unsteady gas-dynamical and optical phenomena in low-temperature laser plasma (Kanevsky, M. F.; Bolshov, L. A.; Chernov, S. Y.; Vorobjev, V. A.)V1440,154-165(1991)

Computing image flow and scene depth: an estimation-theoretic fusion-based framework (Singh, Ajit)V1383,122-140(1991)

Debris plume phenomenology for laser-material interaction in high-speed flowfields (Reilly, James P.)V1397,661-674(1991)

Development of a fibre optics flowmeter (She, K.; Greated, C. A.; Easson, William J.; Skyner, D.; Xu, M. C.)V1572,581-587(1991)

Development of image processing techniques for applications in flow visualization and analysis (Disimile, Peter J.; Shoe, Bridget; Toy, Norman; Savory, Eric; Tahouri, Bahman)V1489,66-74(1991)

Diagnostic of the reaction behaviour of insensitive high explosives under jet attack (Held, Manfred)V1358,1021-1028(1991)

Dynamics of near-surface plasma formation and laser absorption waves under the action of microsecond laser radiation with different wavelengths on absorbing condensed media (Min'ko, L. Y.; Chumakov, A. N.)V1440,166-178(1991)

Effect of acoustic dampers on the excimer laser flow (Zeitoun, David; Tarabelli, D.; Forestier, Bernard M.; Truong, J. P.; Sentis, Marc L.)V1397,585-588(1991)

Effects of environmental and installation-specific factors on process gas delivery via mass-flow controller with an emphasis on real-time behavior (Gray, David E.; Benjamin, Neil M.; Chapman, Brian N.)V1392,402-410(1991)

Evolution of high-speed photography and photonics techniques in detonics experiments (Cavailler, Claude)V1358,210-226(1991)

Experimental investigation on the flow behavior of liquid aluminum inside pressure-die-casting dies using high-speed photography (Jiang, Xuping; Wu, Guobing)V1358,1237-1244(1991)

Experimental research on the casing-shaped charge (Gao, Er-xin)V1358,1115-1119(1991)

Fast-flow gas-dynamic effects in high-pulse repetition-rate excimer lasers (Delaporte, Philippe C.; Fontaine, Bernard L.; Forestier, Bernard M.; Sentis, Marc L.)V1397,485-492(1991)

First-order differential technique for optical flow (Campani, Marco; Straforini, Marco; Verri, Alessandro)V1388,409-414(1991)

Flow behavior of thermoset molding compound in conventional PDIP molds (Gonzalez, Marcelo S.; Mena, Manolo G.)V1390,568-579(1991)

Fluid-dynamic perturbations in gas lasers (Horton, T. E.)V1397,549-554(1991)

Fluid-flow-rate metrology: laboratory uncertainties and traceabilities (Mattingly, G. E.)V1392,386-401(1991)

Focus-error detection from far-field image flow (Ishibashi, Hiromichi; Tanaka, Shin-ichi; Moriya, Mitsuro)V1499,340-347(1991)

Focusing of shock waves in water and its observation by the schlieren method (Isuzugawa, Kohji; Horiuchi, Makoto; Okumura, Yoshiyuki)V1358,1003-1010(1991)

Fusion of diamond phases of graphite in laser shock waves (Bugrov, N. V.; Zakharov, N. S.)V1440,416-422(1991)

Gas-dynamically cooled CO laser with rf-excitation: design and performance (von Buelow, H.; Zeyfang, E.)V1397,499-502(1991)

Ground-based PIV and numerical flow visualization results from the surface-tension-driven convection experiment (Pline, Alexander D.; Wernet, Mark P.; Hsieh, Kwang-Chung)V1557,222-234(1991)

Health monitoring of rocket engines using image processing (Disimile, Peter J.; Shoe, Bridget; Toy, Norman)V1483,39-48(1991)

High-speed photographic study of a cavitation bubble (Soh, W. K.)V1358,1011-1015(1991)

High-speed time-resolved holographic interferometer using solid state shutters (Racca, Roberto G.; Dewey, John M.)V1358,932-939(1991)

Holographic interferometric observation of shock wave focusing to extracorporeal shock wave lithotripsy (Takayama, Kazuyoshi; Obara, Tetsuro; Onodera, Osamu)V1358,1180-1190(1991)

Holographic visualization of hypervelocity explosive events (Cullis, I. C.; Parker, Richard J.; Sewell, Derek)V1358,52-64(1991)

Instantaneous measurement of density from double-simultaneous interferograms (Desse, Jean-Michel; Pegneaux, Jean-Claude)V1358,766-774(1991)

Interactions of laser-induced cavitation bubbles with a rigid boundary (Ward, Barry; Emmony, David C.)V1358,1035-1045(1991)

Investigation of diesel injection jets using high-speed photography and speed holography (Eisfeld, Fritz)V1358,660-671(1991)

Kalman-based computation of optical flow fields (Viswanath, Harsha C.; Jones, Richard A.)V1482,275-284(1991)

Laser-based flow cytometric analysis of genotoxicity of humans exposed to ionizing radiation during the Chernobyl accident (Jensen, Ronald H.; Bigbee, William L.; Langlois, Richard G.; Grant, Stephen G.; Pleshanov, Pavel G.; Chirkov, Andre A.; Pilinskaya, Maria A.)V1403,372-380(1991)

Laser Doppler diagnostics of vorticity and phenomenological description of the flows of dilute polymer solutions in model tubes (Fedoseeva, E. V.; Polyakova, M. S.)V1403,355-358(1991)

Laser generation of Stoneley waves at liquid-solid boundaries (Ward, Barry; Emmony, David C.)V1358,1228-1236(1991)

Laser light sheet investigation into transonic external aerodynamics (Towers, Catherine E.; Towers, David P.; Judge, Thomas R.; Bryanston-Cross, Peter J.)V1358,952-965(1991)

Laser nephelometry of erythrocytes in shear flows (Firsov, N. N.; Priezzhev, Alexander V.; Stepanian, A. S.)V1403,350-354(1991)

Laser surface treatment: numerical simulation of thermocapillary flows (Roux, Agnes; Cipriani, Francois D.)V1397,693-696(1991)

Low-cost high-quality range camera system (Sewell, Derek)V1358,1209-1214(1991)

Measurement of bleeding sap flow velocity in xylem bundle of herbs by laser probing (Romanovsky, Yuri M.; Stepanian, A. S.; Shogenov, Yu. H.)V1403,359-362(1991)

Measurement of fluid velocity fields using digital correlation techniques (Matthys, Donald R.; Gilbert, John A.; Puliparambil, Joseph T.)V1332,850-861(1991)

Measurement of particles and drops in combusting flows (Ereaut, Peter R.)V1554B,556-565(1991)

Measurement of the temperature field in confined jet impingement using phase-stepping video holography (Dobbins, B. N.; He, Shi P.; Jambunathan, K.; Kapasi, S.; Wang, Liu Sheng; Button, B. L.)V1554B,586-592(1991)

Multipass holographic interferometry for low-density gas flow analysis (Surget, Jean; Dunet, G.)V1358,65-72(1991)

Natural pixel decomposition for interferometric tomography (Cha, Dong J.; Cha, Soyoung S.)V1554B,600-609(1991)

New apparatus and method for fluid composition monitoring and control (Urmson, John)V1392,421-427(1991)

New method of increasing the sensitivity of Schlieren interferometer using two Wollaston prisms and its application to flow field (Yan, Da-Peng; He, Anzhi; Yang, Zu Q.; Zhu, Yi Yun)V1554B,636-640(1991)

Nonlinear optical flow estimation and segmentation (Geurtz, Alexander M.)V1567,110-121(1991)

Nonlinear signal processing using integration of fluid dynamics equations (Eidelman, Shmuel; Grossmann, William; Friedman, Aharon)V1567,439-450(1991)

Numerical inversion method for determining aerodynamic effects on particulate exhaust plumes from onboard irradiance data (Cousins, Daniel)V1467,402-409(1991)

Numerical simulation of the combustion chamber of a chemical laser (Saatdjian, E.; Caressa, J. P.; Andre, Jean-Claude)V1397,535-538(1991)

Object-oriented data management for interactive visual analysis of three-dimensional fluid-flow models (Walther, Sandra S.; Peskin, Richard L.)V1459,232-243(1991)

Optical flow techniques for moving target detection (Russo, Paul; Markandey, Vishal; Bui, Trung H.; Shrode, David)V1383,60-71(1991)

Optical quality of a combined aerodynamic window (Du, Keming; Franek, Joachim; Loosen, Peter; Zefferer, H.; Shen, Junquan)V1397,639-643(1991)

Optical techniques for determination of normal shock position in supersonic flows for aerospace applications (Adamovsky, Grigory; Eustace, John G.)V1332,750-756(1991)

Optic flow: multiple instantaneous rigid motions (Zhuang, Xinhua; Wang, Tao; Zhang, Peng)V1569,434-445(1991)

Orthogonal shadowgraphic nanolite stations (Celens, Eduard A.; Chabotier, A.)V1358,1103-1114(1991)

Parametric optical flow without correspondence for moving sensors (Whitten, Gary E.)V1468,167-175(1991)

Particle analysis in liquid flow by the registration of elastic light scattering in the condition of laser beam scanning (Dubrovsky, V.; Grinevich, A. E.; Ossin, A. B.)V1403,344-346(1991)

Performance characteristics of premixing gas-dynamic laser utilizing liquid C6H6 and liquid N2O (Yokozawa, T.; Nakajima, H.; Yamaguchi, S.; Ebina, K.; Ohara, M.; Kanazawa, Hirotaka; Yuasa, M.; Komatsu, Katsuhiko; Hara, Hiroshi)V1397,513-517(1991)

Plasma motion velocity along laser beam and continuous optical discharge in gas flow (Budnik, A. P.; Gus'kov, K. G.; Raizer, Yu. P.; Surjhikov, S. T.)V1397,721-724(1991)

Precise uncoupling theory to study gain of gas-dynamic laser (Bahram-pour, A. R.; Mehdizadeh, E.; Bolorizadeh, M. A.; Shojaey, M.)V1397,539-542(1991)

Propagation of the spherical short-duration shock wave in a straight tunnel (Ahn, Jae W.; Song, So-Young; Lee, Jun Wung; Yang, Joon Mook)V1358,269-277(1991)

Recent developments in fiber optic and laser sensors for flow, surface vibration, rotation, and velocity measurements (Arik, Engin B.)V1584,202-211(1991)

Recovering 3-D translation of a rigid surface by a binocular observer using moments (Al-Hudaithi, Aziz; Udpa, Satish S.)V1567,490-501(1991)

Shocks and other nonlinear filtering applied to image processing (Osher, Stanley; Rudin, Leonid I.)V1567,414-431(1991)

Simultaneous imaging and interferometric turbule visualization in a high-velocity mixing/shear layer (Kalin, David A.; Saylor, Danny A.; Street, Troy A.)V1358,780-787(1991)

Spatial and electrical characteristics of capacitively ballasted rf laser discharges (Baker, Howard J.; Laidler, Ian)V1397,545-548(1991)

Spatio-temporal curvature measures for flow-field analysis (Zetzsche, Christoph; Barth, Erhardt; Berkmann, J.)V1570,337-350(1991)

Structure and motion of entire polyhedra (Sobh, Tarek M.; Alameldin, Tarek K.)V1388,425-431(1991)

Study of OFS for gas-liquid two-phase flow (Liao, Yan-Biao; Lai, Shurong; Zhao, Huafeng; Wu, Gengsheng)V1584,400-404(1991)

Surface temperature and shear stress measurement using liquid crystals (Toy, Norman; Savory, Eric; Disimile, Peter J.)V1489,112-123(1991)

Telescope enclosure flow visualization (Forbes, Fred F.; Wong, Woon-Yin; Baldwin, Jack; Siegmund, Walter A.; Limmongkol, Siriluk; Comfort, Charles H.)V1532,146-160(1991)

Transverse concentratial profiles in dilute erythrocytes suspension flows in narrow channels measured by integral Doppler anemometer (Kononenko, Vadim L.; Shimkus, J. K.)V1403,381-383(1991)

Twenty-five years of aerodynamic research with IR imaging (Gartenberg, Ehud; Roberts, A. S.)V1467,338-356(1991)

Two-dimensional flow quantitative visualization by the hybrid method of flow birefringence and boundary integration (Chen, Yuhai; Jia, Youquan)V1554B,566-572(1991)

Using MRI to calculate cardiac velocity fields (Santiago, Peter; Slade, James N.)V1445,555-563(1991)

Using fixation for direct recovery of motion and shape in the general case (Taalebinezhaad, M. A.)V1388,199-209(1991)

Velocity measurements in molten pools during high-power laser interaction with metals (Caillibotte, Georges; Kechemair, Didier; Sabatier, Lilian)V1412,209-211(1991)

Virtual environment for the exploration of three-dimensional steady flows (Bryson, Steve T.; Levit, Creon C.)V1457,161-168(1991)

Visualization and comparison of simulation results in computational fluid dynamics (Felger, Wolfgang; Astheimer, Peter)V1459,222-231(1991)

Visualization of impingement field of real-rocket-exhausted jets by using moire deflectometry (He, Anzhi; Yan, Da-Peng; Miao, Peng-Cheng; Wang, Hai-Lin)V1554B,429-434(1991)

Visual observation and numerical analysis on the reaction zone structure of a supersonic-flow CO chemical laser (Masuda, Wataru)V1397,531-534(1991)

Fluorescence—see also medical imaging; spectroscopy

2.9 micron laser source for use in the two-photon/laser-induced fluorescence detection of atmospheric OH (Bradshaw, John D.; van Dijk, Cornelius A.)V1433,81-91(1991)

Absorption, fluorescence, and stimulated emission in Ti-diffused Er:LiNbO3 waveguides (Brinkmann, R.; Sohler, Wolfgang; Suche, Hubertus)V1362,377-382(1991)

Advances in polarized fluorescence depletion measurement of cell membrane protein rotation (Barisas, B. G.; Rahman, N. A.; Londo, T. R.; Herman, J. R.; Roess, Debrah A.)V1432,52-63(1991)

Analysis of fluorescence kinetics and the computing of the decay time distribution (Gakamsky, D. M.; Goldin, A. A.; Petrov, E. P.; Rubinov, A. N.)V1403,641-643(1991)

Analysis of neuropeptides using capillary zone electrophoresis with multichannel fluorescence detection (Sweedler, Jonathan V.; Shear, Jason B.; Fishman, Harvey A.; Zare, Richard N.; Scheller, Richard H.)V1439,37-46(1991)

Artificial photosynthesis at octane/water interface in the presence of hydrated chlorophyll a oligomer thin film (Volkov, Alexander G.; Gugeshashvili, M. I.; Kandelaki, M. D.; Markin, V. S.; Zelent, B.; Munger, G.; Leblanc, Roger M.)V1436,68-79(1991)

Automatic counting of chromosome fragments for the determination of radiation dose (Smith, Warren E.; Leung, Billy C.; Leary, James F.)V1398,142-150(1991)

Background correction in multiharmonic Fourier transform fluorescence lifetime spectroscopy (Swift, Kerry M.; Mitchell, George W.)V1431,171-178(1991)

Basis and applicaton of evanescent fluorescence measurement (Yuan, Y. F.; Heavens, Oliver S.)V1519,434-439(1991)

Biomolecular Spectroscopy II (Birge, Robert R.; Nafie, Laurence A., eds.)V1432(1991)

Characterization of fluorescence-labeled DNA by time-resolved fluorescence spectroscopy (Seidel, Claus; Rittinger, K.; Cortes, J.; Goody, R. S.; Koellner, Malte; Wolfrum, Juergen M.; Greulich, Karl O.)V1432,105-116(1991)

Characterization of fluorescent plastic optical fibers for x-ray beam detection (Laguesse, Michel F.; Bourdinaud, Michel J.)V1592,96-107(1991)

Characterization of the fluorescence lifetimes of the ionic species found in aqueous solutions of hematoporphyrin IX as a function of pH (Nadeau, Pierre; Pottier, R.; Szabo, Arthur G.; Brault, Daniel; Vever-Bizet, C.)V1398,151-161(1991)

Chemical and Medical Sensors (Wolfbeis, Otto S., ed.)V1510(1991)

Chlorine sensing by optical techniques (Momin, S. A.; Narayanaswamy, R.)V1510,180-186(1991)

Clinical fluorescence diagnosis of human bladder carcinoma following low-dose photofrin injection (Baert, Luc; Berg, Roger; van Damme, B.; D'Hallewin, Mich A.; Johansson, Jonas; Svanberg, Katarina; Svanberg, Sune)V1525,385-390(1991)

Collisional effects in laser detection of tropospheric OH (Crosley, David R.)V1433,58-68(1991)

Confocal redox fluorescence microscopy for the evaluation of corneal hypoxia (Masters, Barry R.; Kriete, Andres; Kukulies, Joerg)V1431,218-223(1991)

Decrease in total fluorescence from human arteries with increasing beta-carotene content (Ye, Biqing; Abela, George S.)V1425,45-54(1991)

Delayed fluorescence of eosin-bound protein: a probe for measurement of slow-rotational mobility (Yao, Jialing; McStay, Daniel; Rogers, Alan J.; Quinn, Peter J.)V1572,428-433(1991)

Detection of atheroma using Photofrin II and laser-induced fluorescence spectroscopy (Vari, Sandor G.; Papazoglou, Theodore G.; van der Veen, Maurits J.; Papaioannou, Thanassis; Fishbein, Michael C.; Chandra, Mudjianto; Beeder, Clain; Shi, Wei-Qiang; Grundfest, Warren S.)V1426,58-65(1991)

Detection of general anesthetics using a fluorescence-based sensor: incorporation of a single-fiber approach (Abrams, Susan B.; McDonald, Hillary L.; Yager, Paul)V1420,13-21(1991)

Development of laser-induced fluorescence detection to assay DNA damage (Sharma, Minoti; Freund, Harold G.)V1435,280-291(1991)

Diagnostic technique for electrical power equipment using fluorescent fiber (Kurosawa, Kiyoshi; Sawa, Takeshi; Sawada, Hisashi; Tanaka, Akira; Wakatsuki, Noboru)V1368,150-156(1991)

Digital image processing for the early localization of cancer (Kelmar, Cheryl M.)V1426,47-57(1991)

Direct fluoroimmunoassay in Langmuir-Blodgett films of immunoglobulin G (Turko, Illarion V.; Lepesheva, Galina I.; Chashchin, Vadim L.)V1572,419-423(1991)

Distribution of fluorescent probe molecules throughout the phospholipid membrane depth (Nemkovich, N. A.; Rubinov, A. N.; Savvidi, M. G.; Tomin, V. I.; Shcherbatska, Nina V.)V1403,578-581(1991)

Dynamic range limits in field determination of fluorescence using fiber optic sensors (Chudyk, Wayne; Pohlig, Kenneth)V1368,105-114(1991)

Dynamics of components of biomembranes (Chorvat, Dusan; Shvec, Peter)V1403,659-666(1991)

Dynamics of wormlike chains: theory and computer simulations (Aragon, Sergio R.; Luo, Rolland)V1430,65-84(1991)

Early detection of dysplasia in colon and bladder tissue using laser-induced fluorescence (Rava, Richard P.; Richards-Kortum, Rebecca; Fitzmaurice, Maryann; Cothren, Robert M.; Petras, Robert; Sivak, Michael V.; Levine, Howard)V1426,68-78(1991)

Effectiveness of porphyrin-like compounds in photodynamic damage of atherosclerotic plaque (Zalessky, Viacheslav N.; Bobrov, Vladimir; Michalkin, Igor; Trunov, Vitaliy)V1426,162-169(1991)

Enantio-selective optrode for optical isomers of biologically active amines using a new lipophilic aromatic carrier (He, Huarui; Uray, Georg; Wolfbeis, Otto S.)V1510,95-103(1991)

Endoscopic fluorescence detection of early lung cancer (Profio, A. E.; Balchum, Oscar J.; Lam, Stephen)V1426,44-46(1991)

Endoscopic tissue autofluorescence measurements in the upper aerodigestive tract and the bronchi (Braichotte, D.; Wagnieres, G.; Monnier, Philippe; Savary, M.; Bays, Roland; van den Bergh, Hubert; Chatelain, Andre)V1525,212-218(1991)

Excimer-laser-induced fluorescence spectroscopy of human arteries during laser ablation (Abel, B.; Hippler, Horst; Koerber, B.; Morguet, A.; Neu, Walter)V1525,110-118(1991)

Fiber optic based multiprobe system for intraoperative monitoring of brain functions (Mayevsky, Avraham; Flamm, E. S.; Pennie, W.; Chance, Britton)V1431,303-313(1991)

Fiber optic fluorescence sensors based on sol-gel entrapped dyes (MacCraith, Brian D.; Ruddy, Vincent; Potter, C.; McGilp, J. F.; O'Kelley, B.)V1510,104-109(1991)

Fiber optic measurement of intracellular pH in intact rat liver using pH-sensitive dyes (Felberbauer, Franz; Graf, Juerg)V1510,63-71(1991)

Finding a single molecule in a haystack: laser spectroscopy of solids from (square root of)N to N=1 (Moerner, William E.; Ambrose, William P.)V1435,244-251(1991)

Fluctuations in atomic fluorescence induced by laser noise (Vemuri, Gautam)V1376,34-46(1991)

Fluorescence-based fiber optic temperature sensor for aerospace applications (Tilstra, Shelle S.)V1589,32-37(1991)

Fluorescence-based optrodes for alkali ions based on the use of ion carriers and lipophilic acid/base indicators (He, Huarui; Wolfbeis, Otto S.)V1368,165-171(1991)

Fluorescence characteristics of atherosclerotic plaque and malignant tumors (Andersson-Engels, Stefan; Baert, Luc; Berg, Roger; D'Hallewin, Mich A.; Johansson, Jonas; Stenram, U.; Svanberg, Katarina; Svanberg, Sune)V1426,31-43(1991)

Fluorescence detected energy transport in selectively excited porphyrin dimers (Rempel, U.; von Maltzan, B.; von Borczykowski, C.)V1403,631-634(1991)

Fluorescence detection of tumors: studies on the early diagnosis of microscopic lesions in preclinical and clinical studies (Mang, Thomas S.; McGinnis, Carolyn; Crean, David H.; Khan, S.; Liebow, Charles)V1426,97-110(1991)

Fluorescence imaging in photodynamic therapy (MacRobert, Alexander J.; Phillips, David)V1439,79-87(1991)

Fluorescence imaging of latent fingerprints with a cooled charge-coupled-device detector (Pomeroy, Robert S.; Baker, Mark E.; Radspinner, David A.; Denton, M. B.)V1439,60-65(1991)

Fluorescence line-narrowing spectroscopy in the study of chemical carcinogenesis (Jankowiak, Ryszard; Jeong, H.; Small, Gerald J.)V1435,203-213(1991)

Fluorescence properties of Cu+ ion in borate and phosphate glasses (Boutinaud, P.; Parent, C.; Le Flem, Gille; Moine, Bernard; Pedrini, Christian; Duloisy, E.)V1590,168-178(1991)

Fluorescence spectrochronography of protein intramolecular dynamics (Chikishev, A. Y.; Ladokhin, Alexey S.; Shkurinov, A. P.)V1403,448-456(1991)

Fluorescence spectroscopy of normal and atheromatous human aorta: optimum illumination wavelength (Alexander, Andrew L.; Davenport, Carolyn M.; Gmitro, Arthur F.)V1425,6-15(1991)

Fluorescent imaging (Thompson, Jill C.)V1482,253-257(1991)

Fractal dynamics of fluorescence energy transfer in biomembranes (Dewey, T. G.)V1432,64-75(1991)

Frequency-domain fluorescence spectroscopy: instrumentation and applications to the biosciences (Lakowicz, Joseph R.; Gryczynski, Ignacy; Malak, Henryk; Johnson, Michael L.; Laczko, Gabor; Wiczk, Wieslaw M.; Szmacinski, Henryk; Kusba, Jozef)V1435,142-160(1991)

Frequency-doubled alexandrite laser for tissue differentiation in angioplasty (Scheu, M.; Engelhardt, R.)V1425,63-69(1991)

Highly efficient plastic optical fluorescent fibers and sensors (Chiron, Bernard)V1592,86-95(1991)

Hole-burning and picosecond time-resolved spectroscopy of isolated molecular clusters (Wittmeyer, Stacey A.; Kaziska, Andrew J.; Shchuka, Maria I.; Topp, Michael R.)V1435,267-278(1991)

Hydrogen peroxide and organic peroxides in the marine environment (Heikes, Brian G.; Miller, William L.; Lee, Meehye)V1433,253-262(1991)

Image analysis of two impinging jets using laser-induced fluorescence and smoke flow visualization (Shoe, Bridget; Disimile, Peter J.; Savory, Eric; Toy, Norman; Tahouri, Bahman)V1521,34-45(1991)

Imaging of tumors by time-delayed laser-induced fluorescence (Kohl, M.; Neukammer, Jorg; Sukowski, U.; Rinneberg, Herbert H.; Sinn, H.-J.; Friedrich, E. A.; Graschew, Georgi; Schlag, Peter M.; Woehrle, D.)V1525,26-34(1991)

Improved instrumentation for photodynamic fluorescence detection of cancer (Baumgartner, R.; Heil, P.; Jocham, D.; Kriegmair, M.; Stepp, Herbert; Unsoeld, Eberhard)V1525,246-248(1991)

Instrumentation for simultaneous kinetic imaging of multiple fluorophores in single living cells (Morris, Stephen J.; Beatty, Diane M.; Welling, Larry W.; Wiegmann, Thomas B.)V1428,148-158(1991)

Knowledge-driven image analysis of cell structures (Nederlof, Michel A.; Witkin, Andrew; Taylor, D. L.)V1428,233-241(1991)

Langmuir-Blodgett films of immunoglobulin G and direct immunochemical sensing (Turko, Illarion V.; Pikuleva, Irene A.; Yurkevich, Igor S.; Chashchin, Vadim L.)V1510,53-56(1991)

Laser diagnostic techniques in a resonant incinerator (Cadou, Christopher P.; Logan, Pamela; Karagozian, Ann; Marchant, Roy; Smith, Owen I.)V1434,67-77(1991)

Laser-excited fluorescence and fluorescence probes for diagnosing bulk damage in cable insulation (Ordonez, Ishmael D.; Crafton, J.; Murdock, R. H.; Hatfield, Lynn L.; Menzel, E. R.)V1437,184-193(1991)

Laser-induced fluorescence imaging of coronary arteries for open-heart surgery applications (Taylor, Roderick S.; Gladysz, D.; Brown, D.; Higginson, Lyall A.)V1420,183-192(1991)

Laser-induced fluorescence in contaminated soils (Lurk, Paul W.; Cooper, Stafford S.; Malone, Philip G.; Olsen, R. S.; Lieberman, Stephen H.)V1434,114-118(1991)

Laser-induced fluorescence of biological tissue (Dietel, W.; Dorn, P.; Zenk, W.; Zielinski, M.)V1403,653-658(1991)

Laser-induced fluorescence spectroscopy of pathologically enlarged prostate gland in vitro (Chandra, Mudjianto; Gershman, Alex; Papazoglou, Theodore G.; Bender, Leon; Danoff, Dudley; Papaioannou, Thanassis; Vari, Sandor G.; Coons, Gregory; Grundfest, Warren S.)V1421,68-71(1991)

Laser neural tissue interactions using bilayer membrane models (VanderMeulen, David L.; Khoka, Mustafa; Spears, Kenneth G.)V1428,84-90(1991)

Lateral mobility of biological membrane components (Chorvat, Dusan; Shvec, Peter; Kvasnichka, P.; Shipocz, Tibor; Jarkovska, B.)V1402,89-92(1991)

LEAF: a fiber-optic fluorometer for field measurement of chlorophyll fluorescence (Mazzinghi, Piero)V1510,187-194(1991)

Low-dose magnetic-field-immune biplanar fluoroscopy for neurosurgery (Ramos, P. A.; Lawson, Michael A.; Wika, Kevin G.; Allison, Stephen W.; Quate, E. G.; Molloy, J. A.; Ritter, Rogers C.; Gillies, George T.)V1443,160-170(1991)

Lung imaging fluorescence endoscope: development and experimental prototype (Palcic, Branko; Lam, Stephen; MacAulay, Calum; Hung, Jacklyn; Jaggi, Bruno; Radjinia, Massud; Pon, Alfred; Profio, A. E.)V1448,113-117(1991)

Manipulation of single-DNA molecules and measurements of their elastic properties under an optical microscope (Bustamante, Carlos; Finzi, Laura; Sebring, Page E.; Smith, Steven B.)V1435,179-187(1991)

Measurement of fluorescence spectra and quantum yields of 193 nm ArF laser photoablation of the cornea and synthetic lenticules (Milne, Peter J.; Zika, Rod G.; Parel, Jean-Marie; Denham, David B.; Penney, Carl M.)V1423,122-129(1991)

Method of laser fluorescence microphotolysis (Shvec, Peter; Kvasnichka, P.; Shipocz, Tibor; Chorvat, Dusan)V1402,78-81(1991)

Microlocalization of Photofrin in neoplastic lesions (Korbelik, Mladen; Krosl, Gorazd; Lam, Stephen; Chaplin, David J.; Palcic, Branko)V1426,172-179(1991)

Microscopic fluorescence spectroscopy and diagnosis (Schneckenburger, Herbert; Seidlitz, Harold K.; Wessels, Jurina; Strauss, Wolfgang; Rueck, Angelika C.)V1525,91-98(1991)

Model of a thin-film optical fiber fluorosensor (Egalon, Claudio O.; Rogowski, Robert S.)V1368,134-149(1991)

NADH-fluorescence in medical diagnostics: first experimental results (Schramm, Werner; Naundorf, M.)V1525,237-241(1991)

New method for tissue indentification: resonance fluorescence spectroscopy (Neu, Walter)V1525,124-131(1991)

New results in dosimetry of laser radiation in medical treatment (Beuthan, Jurgen; Hagemann, Roland; Mueller, Gerhard J.; Schaldach, Brita J.; Zur, Ch.)V1420,225-233(1991)

Novel signal processing scheme for ruby-fluorescence-based fiber-optic temperature sensor (Zhang, Zhiyi; Grattan, Kenneth T.; Palmer, Andrew W.)V1511,264-274(1991)

Nucleic acid base specific quenching of coumarin-120-derivative in nucleotid-conjugates-photoinduced electron transfer (Seidel, Claus)V1432,91-104(1991)

On-line optical determination of water in ethanol (Kessler, Manfred A.)V1510,218-223(1991)

Optical diagnostics of mercuric iodide crystal growth (Burger, Arnold; Morgan, S. H.; Silberman, Enrique; Nason, Donald)V1557,245-249(1991)

Optical imaging of cortical activity in the living brain (Ratzlaff, Eugene H.; Grinvald, Amiran)V1439,88-94(1991)

Optical sensors for process monitoring in biotechnology (Ploetz, F.; Schelp, C.; Anders, K.; Eberhardt, F.; Scheper, Thomas-Helmut; Bueckmann, F.)V1510,224-230(1991)

Optimal fluorescence imaging of atherosclerotic human tissue (Davenport, Carolyn M.; Alexander, Andrew L.; Gmitro, Arthur F.)V1425,16-27(1991)

PDT of rat mammary adenocarcinoma in vitro and in a rat dorsal-skin-flap window chamber using Photofrin and chloroaluminum-sulfonated phthalocyanine (Flock, Stephen T.; Jacques, Steven L.; Small, Susan M.; Stern, Scott J.)V1427,77-89(1991)

Performance of a thinned back-illuminated CCD coupled to a confocal microscope for low-light-level fluorescence imaging (Masters, Barry R.)V1447,56-63(1991)

Photochemistry and photophysics of stilbene and diphenylpolyene surfactants in supported multilayer films (Spooner, Susan P.; Whitten, David G.)V1436,82-91(1991)

Photodegradation of a laser dye in a silica gel matrix (Glab, Wallace L.; Bistransin, Mark; Borst, Walter L.)V1497,389-395(1991)

Photodetection of early cancer in the upper aerodigestive tract and the bronchi using photofrin II and colorectal adenocarcinoma with fluoresceinated monoclonal antibodies (Wagnieres, G.; Braichotte, D.; Chatelain, Andre; Depeursinge, Ch.; Monnier, Philippe; Savary, M.; Fontolliet, Ch.; Calmes, J.-M.; Givel, J.-C.; Chapuis, G.; Folli, S.; Pelegrin, A.; Buchegger, F.; Mach, J.-P.; van den Bergh, Hubert)V1525,219-236(1991)

Photophysics of 1,3,5-triaryl-2-pyrazolines (Sahyun, Melville R.; Crooks, G. P.; Sharma, D. K.)V1436,125-133(1991)

Photo resonance excitation and ionization characteristics of atoms by pulsed laser irradiation (Tanazawa, Takeshi; Adachi, Hajime A.; Nakahara, Ktsuhiko; Nittoh, Koichi; Yoshida, Toshifumi; Yoshida, Tadashi; Matsuda, Yasuhiko)V1435,310-321(1991)

Picosecond time-resolved fluorescence spectroscopy of phytochrome and stentorin (Song, Pill-Soon)V1403,590-599(1991)

Principles of optical dosimetry: fluorescence diagnostics (van Gemert, Martin J.)V1525,100-109(1991)

Quantitative analysis at the molecular level of laser/neural tissue interactions using a liposome model system (VanderMeulen, David L.; Misra, Prabhakar; Michael, Jason; Spears, Kenneth G.; Khoka, Mustafa)V1428,91-98(1991)

Rapid sequencing of DNA based on single-molecule detection (Soper, Steven A.; Davis, Lloyd M.; Fairfield, Frederick R.; Hammond, Mark L.; Harger, Carol A.; Jett, James H.; Keller, Richard A.; Morrone, Barbara L.; Martin, John C.; Nutter, Harvey L.; Shera, E. B.; Simpson, Daniel J.)V1435,168-178(1991)

Reflectance at the red edge as a sensitive indicator of the damage of trees and its correlation to the state of the photosynthetic system (Ruth, Bernhard)V1521,131-142(1991)

Semiconductor lasers in analytical chemistry (Patonay, Gabor; Antoine, Miquel D.; Boyer, A. E.)V1435,52-63(1991)

Sensitive fiber-optic immunoassay (Walczak, Irene M.; Love, Walter F.; Slovacek, Rudolf E.)V1420,2-12(1991)

Signal-to-noise performance in cesium iodide x-ray fluorescent screens (Hillen, Walter; Eckenbach, W.; Quadflieg, P.; Zaengel, Thomas T.)V1443,120-131(1991)

Single-color laser-induced fluorescence detection and capillary gel electrophoresis for DNA sequencing (Chen, Da Yong; Swerdlow, H.; Harke, H.; Zhang, Jian Z.; Dovichi, Norman J.)V1435,161-167(1991)

Spectral and time-resolved measurements of pollutants on water surface by an XeCl laser fluorosensor (Barbini, Roberto; Fantoni, Roberta; Palucci, Antonio; Ribezzo, Sergio; van der Steen, Hendricus J.)V1503,363-374(1991)

Study of fluorescent glass-ceramics (Qiu, Guanming; DuanMu, Qingduo)V1513,396-407(1991)

Time-gated fluorescence spectroscopy and imaging of porphyrins and phthalocyanines (Cubeddu, Rinaldo; Canti, Gianfranco L.; Taroni, Paola; Valentini, G.)V1525,17-25(1991)

Time-resolved fluorescence of normal and atherosclerotic arteries (Pradhan, Asima; Das, Bidyut B.; Liu, C. H.; Alfano, Robert R.; O'Brien, Kenneth M.; Stetz, Mark L.; Scott, John J.; Deckelbaum, Lawrence I.)V1425,2-5(1991)

Time-resolved photon echo and fluorescence anisotropy study of organically doped sol-gel glasses (Chronister, Eric L.; L'Esperance, Drew M.; Pelo, John; Middleton, John; Crowell, Robert A.)V1559,56-64(1991)

Translational dynamics of immune system components by fluorescence recovery after photobleaching (Shvec, Peter; Miklovicova, J.; Hunakova, L.; Chorvath, B.)V1403,635-637(1991)

Two-photon excitation in laser scanning fluorescence microscopy (Strickler, James H.; Webb, Watt W.)V1398,107-118(1991)

Ultraviolet-laser-induced fluorescence imaging sensor (Thompson, Jill C.)V1479,412-414(1991)

Upconversion intensity and multiphonon relaxation of Er3+-doped glasses (Tanabe, Setsuhisa; Hirao, Kazuyuki; Soga, Naohiro; Hanada, Teiichi)V1513,340-348(1991)

Use of fluorescence spectroscopy to elucidate structural features of the nicotinic acetylcholine receptor (Johnson, David A.; Valenzuela, C. F.)V1432,82-90(1991)

UV-fluorescence spectroscopic technique in the diagnosis of breast, ovarian, uterus, and cervix cancer (Das, Bidyut B.; Glassman, W. L. S.; Alfano, Robert R.; Cleary, Joseph; Prudente, R.; Celmer, E.; Lubicz, Stephanie)V1427,368-373(1991)

Vanderbilt Free-Electron Laser Center for Biomedical and Materials Research (Tolk, Norman H.; Brau, Charles A.; Edwards, Glenn S.; Margaritondo, Giorgio; McKinley, J. T.)V1552,7-13(1991)

Variability in thickness measurements using x-ray fluorescence technique (Baltazar, Inmaculada C.; Mena, Manolo G.)V1392,670-680(1991)

VUV/photofragmentation laser-induced fluorescence sensor for the measurement of atmospheric ammonia (Sandholm, Scott T.; Bradshaw, John D.)V1433,69-80(1991)

Focus—see also alignment; lenses; optical design

64-Mbit DRAM production with i-line stepper (Shirai, Hisatsugu; Kobayashi, Katsuyoshi; Nakagawa, Kenji)V1463,256-274(1991)

Athermalization of IR optical systems (Rogers, Philip J.)VCR38,69-94(1991)

Auto-focus video camera system with bag-type lens (Sugiura, Norio; Morita, Shinzo)V1358,442-446(1991)

Comparison of synthetic-aperture radar autofocus techniques: phase gradient versus subaperture (Calloway, Terry M.; Jakowatz, Charles V.; Thompson, Paul A.; Eichel, Paul H.)V1566,353-364(1991)

Computational experiment for computer-generated optical elements (Golub, Mikhail A.; Doskolovich, Leonid L.; Kazanskiy, Nikolay L.; Kharitonov, Sergey I.; Orlova, Natalia G.; Sisakian, Iosif N.; Soifer, Victor A.)V1500,194-206(1991)

Computer-generated diffractive elements focusing light into arbitrary line segments (Jaroszewicz, Zbigniew; Kolodziejczyk, Andrzej)V1555,236-240(1991)

Computing illumination-bundle focusing by lens systems (Forkner, John F.)V1354,210-215(1991)

Differential spot-size focus servo (Milster, Tom D.; Wang, Mark S.; Froehlich, Fred F.; Kann, J. L.; Treptau, Jeffrey P.; Erwin, K. E.)V1499,348-353(1991)

Direct laser beam diagnostics (van Gilse, Jan; Koczera, Stanley; Greby, Daniel F.)V1414,45-54(1991)

Effects of focus misregistration on optical disk performance (Finkelstein, Blair I.; Childers, Ed R.)V1499,438-449(1991)

Electrostatic focusing and RFQ (radio frequency quadrupole) matching system for a low-energy H- beam (Guharay, Samar K.; Allen, C. K.; Reiser, Martin P.)V1407,610-619(1991)

Finite thickness effect of a zone plate on focusing hard x-rays (Yun, Wen-Bing; Chrzas, John J.; Viccaro, P. J.)V1345,146-164(1991)

Focus-error detection from far-field image flow (Ishibashi, Hiromichi; Tanaka, Shin-ichi; Moriya, Mitsuro)V1499,340-347(1991)

Focusators at letters diffraction design (Golub, Mikhail A.; Doskolovich, Leonid L.; Kazanskiy, Nikolay L.; Kharitonov, Sergey I.; Sisakian, Iosif N.; Soifer, Victor A.)V1500,211-221(1991)

Focus considerations with high-numerical-aperture widefield lenses (Leebrick, David H.)V1463,275-280(1991)

Focus sensing method with improved pattern noise rejection (Marshall, Daniel R.)V1499,332-339(1991)

High-resolution microchannel plate image tube development (Johnson, C. B.; Patton, Stanley B.; Bender, E.)V1449,2-12(1991)

Interferential diagnosis of self-focusing of Q-switched YAG laser in liquid (Lu, Jian-Feng; Wang, Chang Xing; Miao, Peng-Cheng; Ni, Xiao W.; He, Anzhi)V1415,220-224(1991)

Investigations of the focal shift of the high-power cw YAG laser beam (Jabczynski, Jan K.; Mindak, Marek K.)V1391,109-116(1991)

Kinoforms with increased depth of focus (Koronkevich, Voldemar P.; Palchikova, Irena G.)V1507,110-115(1991)

Laser beam scanner for uniform halftones (Ando, Toshinori)V1458,128-132(1991)

Lissajous analysis of focus crosstalk in optical disk systems (Grove, Steven L.; Getreuer, Kurt W.; Schell, David L.)V1499,354-359(1991)

Massively parallel synthetic-aperture radar autofocus (Mastin, Gary A.; Plimpton, Steven J.; Ghiglia, Dennis C.)V1566,341-352(1991)

Monocular passive range sensing using defocus information (Prasad, K. V.; Mammone, Richard J.)V1385,280-291(1991)

New tracking method for two-beam optical heads using continuously grooved disks (Irie, Mitsuru; Takeshita, Nobuo; Fujita, Teruo; Kime, Kenjiro)V1499,360-365(1991)

Novel device for increasing the laser pulse intensity in multiphoton ionization mass spectrometry (Liang, Dawei; Fraser Monteiro, L.; Fraser Monteiro, M. L.)V1501,129-134(1991)

Precision high-power solid state laser diagnostics for target-irradiation studies and target-plane irradiation modeling (Wegner, Paul J.; Henesian, Mark A.)V1414,162-174(1991)

Rectangular focus spots with uniform intensity profile formed by computer-generated holograms (Hossfeld, Jens; Jaeger, Erwin; Tschudi, Theo T.; Churin, Evgeny G.; Koronkevich, Voldemar P.)V1574,159-166(1991)

Relationships between autofocus methods for SAR and self-survey techniques for SONAR (Wahl, Daniel E.; Jakowatz, Charles V.; Ghiglia, Dennis C.; Eichel, Paul H.)V1567,32-40(1991)

Shape-from-focus: surface reconstruction of hybrid surfaces (Chen, Su-Shing; Tang, Wu-bin; Xu, Jian-hua)V1569,446-450(1991)

Short pulse self-focusing (Strickland, Donna; Corkum, Paul B.)V1413,54-58(1991)

Studies on defocus in thermal imaging systems (Venkateswara Rao, B.)V1399,145-156(1991)

Theory of transient self-focusing (Hsia, Kangmin; Cantrell, Cyrus D.)V1497,166-177(1991)

Time aberrations of combined focusing system of high-speed image converter (Korzhenevich, Irina M.; Kolesov, G. V.)V1358,1090-1095(1991)

Transparency and blur as selective cues for complex visual information (Colby, Grace; Scholl, Laura)V1460,114-125(1991)

Vector beam propagation using Hertz vectors (Milsted, Carl S.; Cantrell, Cyrus D.)V1497,202-215(1991)

Verification of tracking servo signal simulation from scanning tunneling microscope surface profiles (Karis, Thomas E.; Best, Margaret E.; Logan, John A.; Lyerla, James R.; Lynch, Robert T.; McCormack, R. P.)V1499,366-376(1991)

Wave-optics analysis of fast-beam focusing (Shih, Chun-Ching)V1415,150-153(1991)

Fourier optics—see also image processing

Advances in the optical design of miniaturized optical correlators (Gebelein, Rolin J.; Connely, Shawn W.; Foo, Leslie D.)V1564,452-463(1991)

Digital imaging of Giemsa-banded human chromosomes: eigenanalysis and the Fourier phase reconstruction (Jericevic, Zeljko; McGavran, Loris; Smith, Louis C.)V1428,200-213(1991)

Fourier processing with binary spatial light modulators (Hossack, William J.; Vass, David G.; Underwood, Ian)V1564,697-702(1991)

Holographic-coordinate-transform-based system for direct Fourier tomographic reconstruction (Huang, Qiang; Freeman, Mark O.)V1564,644-655(1991)

Low-insertion-loss, high-precision liquid crystal optical phased array (Cassarly, William J.; Ehlert, John C.; Henry, D.)V1417,110-121(1991)

Optical design and optimization with physical optics (Lawrence, George N.; Moore, Kenneth E.)V1354,15-22(1991)

Optical design with physical optics using GLAD (Lawrence, George N.)V1354,126-135(1991)

Optical processing of wire-frame models for object recognition (Kozaitis, Samuel P.; Cofer, Rufus H.)V1471,249-254(1991)

Real-time optical processor for increasing resolution beyond the diffraction limit (Dhadwal, Harbans S.; Noel, Eric)V1564,664-673(1991)

Shape recognition in the Fourier domain (Udomkesmalee, Suraphol)V1564,464-472(1991)

Wave-optic model to determine image quality through supersonic boundary and mixing layers (Lawson, Shelah M.; Clark, Rodney L.; Banish, Michele R.; Crouse, Randy F.)V1488,268-278(1991)

Fourier transforms—see also transforms

Analysis of exhaust from clean-fuel vehicles using FTIR spectroscopy (Rieger, Paul L.; Maddox, Christine E.)V1433,290-301(1991)

Analysis of measurement principle of moire interferometer using Fourier method (Wang, Hai-Lin; Miao, Peng-Cheng; He, Anzhi)V1527,419-422(1991)

Analysis of molecular adsorbates by laser-induced thermal desorption (McIver, Robert T.; Hemminger, John C.; Parker, D.; Li, Y.; Land, Donald P.; Pettiette-Hall, C. L.)V1437,124-128(1991)

Analytical design of curved holographic optical elements for Fourier transform (Talatinian, A.; Pluta, Mieczyskaw)V1574,205-217(1991)

Applications of the bispectrum in radar signature analysis and target identification (Jouny, Ismail; Garber, Frederick D.; Moses, Randolph L.; Walton, Eric K.)V1471,142-153(1991)

Auto exhaust gas analysis by FTIR spectroscopy (Herget, William F.; Lowry, Steven R.)V1433,275-289(1991)

Automatic data processing of speckle fringe pattern (Dupre, Jean-Christophe; Lagarde, Alexis)V1554A,766-771(1991)

Bayesian signal reconstruction from Fourier transform magnitude and x-ray crystallography (Doerschuk, Peter C.)V1569,70-79(1991)

Best fit ellipse for cell shape analysis (Wali, Rahman; Colef, Michael; Barba, Joseph)V1606,665-674(1991)

Carrier pattern analysis of moire interferometry using Fourier transform (Morimoto, Yoshiharu; Post, Daniel; Gascoigne, Harold E.)V1554B,493-502(1991)

Cone array of Fourier lenses for contouring applications (Goldfain, Ervin)V1527,210-215(1991)

Displacement measurement of track on magnetic tape by Fourier transform grid method (Sogabe, Yasushi; Morimoto, Yoshiharu; Murata, Shigeki)V1554B,289-297(1991)

Does a Fourier holographic lens exist? (Blank, R.; Friesem, Asher A.)V1442,26-30(1991)

Evaluation of interference patterns: a comparison of methods (Kreis, Thomas M.; Geldmacher, Juergen)V1554B,718-724(1991)

Fourier cross-correlation and invariance transformation for affine groups (Segman, Joseph)V1606,788-802(1991)

Fourier-transform holographic microscope (Haddad, Waleed S.; Cullen, David; Solem, Johndale C.; Longworth, James W.; McPherson, Armon; Boyer, Keith; Rhodes, Charles K.)V1448,81-88(1991)

Fourier transform method to determine human corneal endothelial morphology (Masters, Barry R.; Lee, Yimkul; Rhodes, William T.)V1429,82-90(1991)

Frequency-modulated reticle trackers in the pupil plane (Taylor, James S.; Krapels, Keith A.)V1478,41-49(1991)

Fringe analysis in moire interferometry (Asundi, Anand K.)V1554B,472-480(1991)

FTIR: fundamentals and applications in the analysis of dilute vehicle exhaust (Gierczak, Christine A.; Andino, J. M.; Butler, James W.; Heiser, G. A.; Jesion, G.; Korniski, T. J.)V1433,315-328(1991)

High-resolution computer-aided moire (Sciammarella, Cesar A.; Bhat, Gopalakrishna K.)V1554B,162-173(1991)

Image restoration algorithms based on the bispectrum (Kang, Moon G.; Lay, Kuen-Tsair; Katsaggelos, Aggelos K.)V1606,408-418(1991)

In-plane ESPI measurements in hostile environments (Preater, Richard W.; Swain, Robin C.)V1554A,727-738(1991)

Infrared fiber optic sensors: new applications in biology and medicine (Swairjo, Manal; Rothschild, Kenneth J.; Nappi, Bruce; Lane, Alan; Gold, Harris)V1437,60-65(1991)

In situ measurement of methyl bromide in indoor air using long-path FTIR spectroscopy (Green, Martina; Seiber, James N.; Biermann, Heinz W.)V1433,270-274(1991)

Laser beam modeling in optical storage systems (Treptau, Jeffrey P.; Milster, Tom D.; Flagello, Donis G.)V1415,317-321(1991)

Laser probe mass spectrometry (Campana, Joseph E.)V1437,138-149(1991)

Linear lattice architectures that utilize the central limit for image analysis, Gaussian operators, sine, cosine, Fourier, and Gabor transforms (Ben-Arie, Jezekiel)V1606,823-838(1991)

Matrix-assisted laser desorption by Fourier transform mass spectrometry (Nuwaysir, Lydia M.; Wilkins, Charles L.)V1437,112-123(1991)

Measurement of interfacial tension by automated video techniques (Deason, Vance A.; Miller, R. L.; Watkins, Arthur D.; Ward, Michael B.; Barrett, K. B.)V1332,868-876(1991)

Metal vapor gain media based on multiphoton dissociation of organometallics (Samoriski, Brian; Wiedeger, S.; Villarica, M.; Chaiken, Joseph)V1412,12-18(1991)

New phase measurement for nonmonotonical fringe patterns (Tang, Shou-Hong; Hung, Yau Y.; Zhu, Qiuming)V1332,731-737(1991)

Objective assessment of clinical computerized thermal images (Anbar, Michael)V1445,479-484(1991)

Optical enhancements of joint-Fourier-transform correlator by image subtraction (Perez, Osvaldo; Karim, Mohammad A.)V1471,255-264(1991)

Photoelastic sensors for automatic control system of dam safety (Konwerska-Hrabowska, Joanna; Kryszczynski, Tadeusz; Tomaszewicz, Tomasz; Lietz, J.; Mazurkiewicz, Wojciech)V1554A,388-399(1991)

Radiometric calibration of space-borne Fourier transform infrared-emission spectrometers: proposed scenario for European Space Agency's Michelson interferometer for passive atmospheric sounding (Lamarre, Daniel; Giroux, Jean)V1493,28-36(1991)

Rotation invariant object classification using fast Fourier transform features (Celenk, Mehmet; Datari, Srinivasa R.)V1468,752-763(1991)

Target identification by means of impulse radar (Abrahamsson, S.; Brusmark, B.; Gaunaurd, Guillermo C.; Strifors, Hans C.)V1471,130-141(1991)

Three-dimensional surface inspection using interferometric grating and 2-D FFT-based technique (Hung, Yau Y.; Tang, Shou-Hong; Zhu, Qiuming)V1332,696-703(1991)

Use of a Fourier transform spectrometer as a remote sensor at Superfund sites (Russwurm, George M.; Kagann, Robert H.; Simpson, Orman A.; McClenny, William A.)V1433,302-314(1991)

Four-wave mixing—see also nonlinear optics; phase conjugation

Analysis of Brillouin-enhanced four-wave mixing phase-conjugate systems (Dansereau, Jeffrey P.; Hills, Louis S.; Mani, Siva A.)V1409,67-82(1991)

Beam-coupling by stimulated Brillouin scattering (Falk, Joel; Chu, Raijun; Kanefsky, Morton; Hua, Xuelei)V1409,83-86(1991)

Degenerate four-wave mixing and optical nonlinearities in quantum-confined CdSe microcrystallites (Park, S. H.; Casey, Michael P.; Falk, Joel)V1409,9-13(1991)

Degenerate four-wave mixing using wave pump beams near the critical angle: two distinct behaviors (Malouin, Christian; Song, Li; Thibault, S.; Denariez-Roberge, Marguerite M.; Lessard, Roger A.)V1559,385-392(1991)

Measurements of second hyperpolarisabilities of diphenylpolyenes by means of phase-conjugate interferometry (Persoons, Andre P.; Van Wonterghem, Bruno M.; Tackx, Peter)V1409,220-229(1991)

Nonlinear Optics II (Fisher, Robert A.; Reintjes, John F., eds.)V1409(1991)

Optical associative memories based on time-delayed four-wave mixing (Belov, M. N.; Manykin, Edward A.)V1621,268-279(1991)

Phase conjugation by four-wave mixing in nematic liquid crystals (Almeida, Silverio P.; Varamit, Srisuda P.)V1500,34-45(1991)

Phase conjugation in LiF and NaF color center crystals (Basiev, Tasoltan T.; Zverev, Peter G.; Mirov, Sergey B.; Pal, Suranjan)V1500,65-71(1991)

Position-, scale-, and rotation-invariant photorefractive correlator (Ryan, Vincent; Fielding, Kenneth H.)V1564,86-97(1991)

Spectral characteristics of Brillouin-enhanced four-wave mixing for pulsed and CW inputs (Lebow, Paul S.; Ackerman, John R.)V1409,60-66(1991)

Subattomole detection in the condensed phase by nonlinear laser spectroscopy based on degenerate four-wave mixing (Tong, William G.)V1435,90-94(1991)

TEA CO2 laser mirror by degenerate four-wave mixing (Vigroux, Luc M.; Bourdet, Gilbert L.; Cassard, Philippe; Ouhayoun, Michel O.)V1500,74-79(1991)

Theoretical study of ultrafast dephasing by four-wave mixing (Hoerner, Claudine; Lavoine, J. P.; Villaeys, A. A.)V1362,863-869(1991)

Third-order nonlinear optical characterization of side-chain copolymers (Norwood, Robert A.; Sounik, James R.; Popolo, J.; Holcomb, D. R.)V1560,54-65(1991)

Fractals

Bonding and nonequilibrium crystallization of a-C:H/a-Se and a-C:H/KCl (He, Da-Ren; Ji, Xiuyan; Wang, Ruo Bao; Liu, Qihai; Wang, Wangdi; Liu, Maili; Chen, Weizong; Liu, Zhiyuan; Ji, Wanxi; Zhang, Renji)V1362,696-701(1991)

Computer simulation of magnetic resonance images using fractal-grown brain slices (Cheng, Shirley N.)V1396,9-14(1991)

Crystallization of hydrogenated amorphous silicon film and its fractal structure (Lin, Hong Y.; Yang, Dao M.; Li, Ying X.)V1519,210-213(1991)

Dimension and lacunarity measurement of IR images using Hilbert scanning (Moghaddam, Baback; Hintz, Kenneth J.; Stewart, Clayton V.)V1486,115-126(1991)

Discrete-angle radiative transfer in a multifractal medium (Davis, Anthony B.; Lovejoy, Shaun; Schertzer, Daniel)V1558,37-59(1991)

Estimation of plastic strain by fractal (Dai, YuZhong; Chiang, Fu-Pen)V1332,767-774(1991)

Fabrication and optical performance of fractal fiber optics (Cook, Lee M.; Burger, Robert J.)V1449,186-192(1991)

Fractal-based multifeature texture description (Kasparis, Takis; Tzannes, Nicolaos S.; Bassiouni, Mostafa; Chen, Qing)V1521,46-54(1991)

Fractal analysis as a means for the quantification of intramandibular trabecular bone loss from dental radiographs (Doyle, Michael D.; Rabin, Harold; Suri, Jasjit S.)V1380,227-235(1991)

Fractal description of computer interconnections (Christie, Phillip; Styer, Stephen B.)V1390,359-367(1991)

Fractal dynamics of fluorescence energy transfer in biomembranes (Dewey, T. G.)V1432,64-75(1991)

Fractal image compression and texture analysis (Moghaddam, Baback; Hintz, Kenneth J.; Stewart, Clayton V.)V1406,42-57(1991)

Fractal patterns in the human retina and their physiological correlates (Masters, Barry R.)V1380,218-226(1991)

Fractal simulation of aggregation in magnetic thin film (Wang, Yi P.; Wang, Yu J.)V1519,605-608(1991)

Fractional Brownian motion and its fractal dimension estimation (Zhang, Peng; Martinez, Andrew B.; Barad, Herbert S.)V1569,398-409(1991)

Guest-host polymer fibers and fractal clusters for nonlinear optics (Kuzyk, Mark G.; Andrews, Mark P.; Paek, Un-Chul; Dirk, Carl W.)V1560,44-53(1991)

Measurement of fractal dimension using 3-D technique (Chuang, Keh-Shih; Valentino, Daniel J.; Huang, H. K.)V1445,341-347(1991)

Natural texture analysis in a multiscale context using fractal dimension (Kpalma, Kidiyo; Bruno, Alain; Haese-Coat, Veronique)V1606,55-66(1991)

Pattern recognition using w-orbit finite automata (Liu, Ying; Ma, Hede)V1606,226-240(1991)

Practical approach to fractal-based image compression (Pentland, Alexander P.; Horowitz, Bradley)V1605,467-474(1991)

Signal processing for nonlinear systems (Broomhead, David S.)V1565,228-243(1991)

Terrain classification in navigation of an autonomous mobile robot (Dodds, David R.)V1388,82-89(1991)

Universal multifractal theory and observations of land and ocean surfaces, and of clouds (Lavallee, Daniel; Lovejoy, Shaun; Schertzer, Daniel)V1558,60-75(1991)

Wave interactions with continuous fractal layers (Kim, Yun J.; Jaggard, Dwight L.)V1558,113-119(1991)

Frequency conversion

Bulk darkening of flux-grown KTiOPO4 (Rockafellow, David R.; Teppo, Edward A.; Jacco, John C.)V1561,112-118(1991)

Design and construction of a wideband efficient electro-optic modulator (Harris, Neville W.; Sobolewski, J. M.; Summers, Charles L.; Eng, Richard S.)V1416,59-69(1991)

Emerging laser technologies (Forrest, Gary T.)V1520,37-57(1991)

Frequency doubling and optical parametric oscillation with potassium niobate (Polzik, Eugene S.; Kimble, H. J.)V1561,143-146(1991)

Frequency response study of traps in III-V compound semiconductors (Kachwalla, Zain)V1361,784-793(1991)

High-efficiency tunable mid-infrared generation in KNbO3 (Guyer, Dean R.; Bosenberg, W. R.; Braun, Frank D.)V1409,14-17(1991)

Influence of diamond turning and surface cleaning processes on the degradation of KDP crystal surfaces (Kozlowski, Mark R.; Thomas, Ian M.; Edwards, Gary; Stanion, Ken; Fuchs, Baruch A.; Latanich, L.)V1561,59-69(1991)

Inhomogeneous quarter-wave transformers for a waveguide electro-optic modulator (Harris, Neville W.; Wong, D. M.)V1416,86-99(1991)

KTP waveguides for frequency upconversion of strained-layer InGaAs laser diodes (Risk, William P.; Nadler, Ch. K.)V1561,130-134(1991)

Measurement of nonlinear optical coefficients by phase-matched harmonic generation (Eckardt, Robert C.; Byer, Robert L.)V1561,119-127(1991)

Modeling of large-aperture third-harmonic frequency conversion of high-power Nd:glass laser systems (Henesian, Mark A.; Wegner, Paul J.; Speck, David R.; Bibeau, Camille; Ehrlich, Robert B.; Laumann, Curt W.; Lawson, Janice K.; Weiland, Timothy L.)V1415,90-103(1991)

Nonlinear optical properties of new KTiOPO4 isostructures (Phillips, Mark L.; Harrison, William T.; Stucky, Galen D.)V1561,84-92(1991)

Nonlinear optic frequency converters with lowered sensibility to spectral width of laser radiation (Volosov, Vladimir D.)V1500,105-110(1991)

Progress in diode laser upconversion (Dixon, G. J.; Kean, P. N.)V1561,147-150(1991)

Quasi-phase-matched frequency conversion in lithium niobate and lithium tantalate waveguides (Lim, Eric J.; Matsumoto, S.; Bortz, M. L.; Fejer, Martin M.)V1561,135-142(1991)

Semiconductor laser amplifiers as all-optical frequency converters (Schunk, Nikolaus; Grosskopf, Gerd; Ludwig, Reinhold; Schnabel, Ronald; Weber, Hans-Georg)V1362,391-397(1991)

Fusion

Beam control and power conditioning of GEKKO glass laser system (Nakatsuka, Masahiro; Jitsuno, Takahisa; Kanabe, Tadashi; Urushihara, Shinji; Miyanaga, Noriaki; Nakai, Sadao S.)V1411,108-115(1991)

Continuous inertial confinement fusion for the generation of very intense plasma jets (Winterberg, F.)V1407,322-325(1991)

Development of third-harmonic output beam diagnostics on NOVA (Laumann, Curt W.; Caird, John A.; Smith, James R.; Horton, R. L.; Nielsen, Norman D.)V1414,151-160(1991)

Diode-pumped, high-power, solid-state lasers (Yamanaka, Masanobu; Naito, Kenta; Nakatsuka, Masahiro; Yamanaka, Tatsuhiko; Nakai, Sadao S.; Yamanaka, Chiyoe)V1501,30-39(1991)

Evaluation of dynamic range for LLNL streak cameras using high-contrast pulses and "pulse podiatry" on the NOVA laser system (Richards, James B.; Weiland, Timothy L.; Prior, John A.)V1346,384-389(1991)

High-power lasers and the production of energy by inertial fusion (Velarde, G.; Perlado, J. M.; Aragones, J. M.; Honrubia, J. J.; Martinez-Val, J. M.; Minguez, E.)V1502,242-257(1991)

High-precision measurements of the 24-beam UV-OMEGA laser (Jaanimagi, Paul A.; Hestdalen, C.; Kelly, John H.; Seka, Wolf D.)V1358,337-343(1991)

Industrial and Scientific Uses of High-Power Lasers (Billon, Jean P.; Fabre, Edouard, eds.)V1502(1991)

Laser application for fusion using volume ignition and smoothing by suppression of pulsation (Hora, Heinrich; Aydin, M.; Kasotakis, G.; Stening, R. L.)V1502,258-269(1991)

Output pulse and energy capabilities of the PHEBUS laser facility (Fleurot, Noel A.; Andre, Michel L.; Estraillier, P.; Friart, Daniel; Gouedard, C.; Rouyer, C.; Thebault, J. P.; Thiell, Gaston; Veron, Didier)V1502,230-241(1991)

Precision high-power solid state laser diagnostics for target-irradiation studies and target-plane irradiation modeling (Wegner, Paul J.; Henesian, Mark A.)V1414,162-174(1991)

Pulsed-power considerations for electron-beam-pumped krypton-fluoride lasers for inertial confinement fusion applications (Rose, Evan A.; McDonald, Thomas E.; Rosocha, Louis A.; Harris, David B.; Sullivan, J. A.; Smith, I. D.)V1411,15-27(1991)

Recent advances in excimer laser technology at Los Alamos (Bigio, Irving J.; Czuchlewski, Stephen J.; McCown, Andrew W.; Taylor, Antoinette J.)V1397,47-53(1991)

Sub-100 psec x-ray gating cameras for ICF imaging applications (Kilkenny, Joseph D.; Bell, Perry M.; Hammel, Bruce A.; Hanks, Roy L.; Landen, Otto L.; McEwan, Thomas E.; Montgomery, David S.; Turner, R. E.; Wiedwald, Douglas J.; Bradley, David K.)V1358,117-133(1991)

Temporal fiducial for picosecond streak cameras in laser fusion experiments (Mens, Alain; Gontier, D.; Giraud, P.; Thebault, J. P.)V1358,878-887(1991)

Timing between streak cameras with a precision of 10 ps (Lerche, Richard A.)V1346,376-383(1991)

Upgrade of the LLNL Nova laser for inertial confinement fusion (Murray, John R.; Trenholme, J. B.; Hunt, John T.; Frank, D. N.; Lowdermilk, W. H.; Storm, E.)V1410,28-39(1991)

Gallium arsenide materials—see also integrated optoelectronics; lasers, semiconductor; quantum wells; semiconductors

0.36-W cw diffraction-limited-beam operation from phase-locked arrays of antiguides (Mawst, Luke J.; Botez, Dan; Anderson, Eric R.; Jansen, Michael; Ou, Szutsun S.; Sergant, Moshe; Peterson, Gary L.; Roth, Thomas J.; Rozenbergs, John)V1418,353-357(1991)

1-W cw separate confinement InGaAsP/InP (lamda = 1.3 um) laser diodes and their coupling with optical fibers (Garbuzov, Dmitriy Z.; Goncharov, S. E.; Il'in, Y. V.; Mikhailov, A. V.; Ovchinnikov, Alexander V.; Pikhtin, N. A.; Tarasov, I. S.)V1418,386-393(1991)

100-mW four-beam individually addressable monolithic AlGaAs laser diode arrays (Yamaguchi, Takao; Yodoshi, Keiichi; Minakuchi, Kimihide; Tabuchi, Norio; Bessho, Yasuyuki; Inoue, Yasuaki; Komeda, Koji; Mori, Kazushi; Tajiri, Atsushi; Tominaga, Koji)V1418,363-371(1991)

Advanced InGaAs/InP p-type pseudomorphic MODFET (Malzahn, Eric; Heuken, Michael; Gruetzmacher, Dettev; Stollenwerk, M.; Heime, Klaus)V1362,199-204(1991)

Advanced multichip module packaging and interconnect issues for GaAs signal processors operating above 1 GHz clock rates (Gilbert, Barry K.; Thompson, R.; Fokken, G.; McNeff, W.; Prentice, Jeffrey A.; Rowlands, David O.; Staniszewski, A.; Walters, W.; Zahn, S.; Pan, George W.)V1390,235-248(1991)

Advances in multiple-quantum-well IR detectors (Bloss, Walter L.; O'Loughlin, Michael J.; Rosenbluth, Mary)V1541,2-10(1991)

Application of AlGaAs/GaAs superlattice for negative-differential-resistance transistor (Liu, W. C.; Lour, W. S.; Sun, C. Y.; Lee, Y. S.; Guo, D. F.)V1519,670-674(1991)

Application of epitaxial lift-off to optoelectronic material studies (Price, Garth L.; Usher, Brian F.)V1361,543-550(1991)

Asymmetric superlattices for microwave detection (Syme, Richard T.; Kelly, Michael J.; Condie, Angus; Dale, Ian)V1362,467-476(1991)

Atomic scale simulation of the growth of CdTe layers on GaAs (Rouhani, Mehdi D.; Laroussi, M.; Gue, A. M.; Esteve, Daniel)V1361,954-962(1991)

Attenuation of leaky waves in GaAs/AlGaAs MQW waveguides formed on a GaAs substrate (Kubica, Jacek M.)V1506,134-139(1991)

Balanced optical mixer integrated in InGaAlAs/InP for coherent receivers (Caldera, Claudio; De Bernardi, Carlo; Destefanis, Giovanni; Meliga, Marina; Morasca, Salvatore; Rigo, Cesare F.; Stano, Alessandro)V1372,82-87(1991)

Band-structure dependence of impact ionization: bulk semiconductors, strained Ge/Si alloys, and multiple-quantum-well avalanche photodetectors (Czajkowski, Igor K.; Allam, Jeremy; Adams, Alfred R.)V1362,179-190(1991)

Behaviour of a single quantum-well under deep level transient spectroscopy (DLTS) measurement: a new theoretical model (Letartre, Xavier; Stievenard, Didier; Barbier, E.)V1362,778-789(1991)

Bias dependence of the hole tunneling time in AlAs/GaAs resonant tunneling structures (Van Hoof, Chris A.; Goovaerts, Etienne; Borghs, Gustaaf)V1362,291-300(1991)

Bistability of the Sn donor in AlxGa1-xAs and GaAs under pressure studied by Mossbauer spectroscopy (Gibart, Pierre; Williamson, Don L.)V1362,938-950(1991)

Bistable optical switching in GaAs multiple-quantum-well epitaxial etalons (Oudar, Jean-Louis; Sfez, B. G.; Kuszelewicz, Robert)V1361,490-498(1991)

Characteristics of MSM detectors for meander channel CCD imagers on GaAs (Kosel, Peter B.; Bozorgebrahimi, Nercy; Iyer, J.)V1541,48-59(1991)

Characterization and switching study of an optically controlled GaAs switch (Stoudt, David C.; Mazzola, Michael S.; Griffiths, Scott F.)V1378,280-285(1991)

Characterization of picosecond GaAs metal-semiconductor-metal photodetectors (Rosenzweig, Josef; Moglestue, C.; Axmann, A.; Schneider, Joachim J.; Huelsmann, Axel; Lambsdorff, M.; Kuhl, Juergen; Klingenstein, M.; Leier, H.; Forchel, Alfred W.)V1362,168-178(1991)

Cold to hot electron transition devices (Chang, C. Y.)V1362,978-983(1991)

Compact PIN-amplifier module for giga bit rates optical interconnection (Suzuki, Tomihiro; Mikamura, Yasuki; Murata, Kazuo; Sekiguchi, Takeshi; Shiga, Nobuo; Murakami, Yasunori)V1389,455-461(1991)

Comparison of MESFET and HEMT MMIC technologies using a compact Kaband voltage-controlled oscillator (Swirhun, S.; Geddes, John J.; Sokolov, Vladimir; Bosch, D.; Gawronski, M. J.; Anholt, R.)V1475,303-308(1991)

Corbino-capacitance technique for contactless measurements on conducting layers: application to persistent photoconductivity (Hansen, Ole P.; Kristensen, Anders; Bruus, Henrik; Razeghi, Manijeh)V1362,192-198(1991)

Critical look at AlGaAs/GaAs multiple-quantum-well infrared detectors for thermal imaging applications (Adams, Frank W.; Cuff, K. F.; Gal, George; Harwit, Alex; Whitney, Raymond L.)V1541,24-37(1991)

Current-voltage characteristics of resonant tunneling diodes (Sen, Susanta; Nag, B. R.; Midday, S.)V1362,750-759(1991)

Current technologies for very high performance VLSI ICs (Perea, Ernesto H.)V1362,477-483(1991)

Cyclotron resonance and photoluminescence in GaAs in a microwave field (Ashkinadze, B. M.; Bel'Kov, V. V.; Krasinskaya, A. G.)V1361,866-873(1991)

Deep-level configuration of GaAs:Si:Cu: a material for a new type of optoelectronic switch (Schoenbach, Karl H.; Schulz, Hans-Joachim; Lakdawala, Vishnu K.; Kimpel, B. M.; Brinkmann, Ralf P.; Germer, Rudolf K.; Barevadia, Gordon R.)V1362,428-435(1991)

Deep levels in III-V compounds, heterostructures, and superlattices (Bourgoin, J. C.; Feng, S. L.; von Bardeleben, H. J.)V1361,184-194(1991)

Design of a GaAs acousto-optic correlator for real-time processing (Armenise, Mario N.; Impagnatiello, Fabrizio; Passaro, Vittorio M.; Pansini, Evangelista)V1562,160-171(1991)

Design optimization of a 10-amplifier coherent array (Zediker, Mark S.; Foresi, James S.; Haake, John M.; Heidel, Jeffrey R.; Williams, Richard A.; Driemeyer, D.; Blackwell, Richard J.; Thomas, G.; Priest, J. A.; Herrmann, Sandy)V1418,309-315(1991)

Design optimization of three-stage GaAs monolithic optical amplifier using SPICE (Yadav, M. S.; Dumka, D. C.; Ramola, Ramesh C.; Johri, Subodh; Kothari, Harshad S.; Singh, Babu R.)V1362,811-819(1991)

Differentiation of the nonradiative recombination properties of the two interfaces of molecular beam epitaxy grown GaAs-GaAlAs quantum wells (Sermage, Bernard; Gerard, Jean M.; Bergomi, Lorenzo; Marzin, Jean Y.)V1361,131-135(1991)

Effects of radiation on millimeter wave monolithic integrated circuits (Meulenberg, A.; Hung, Hing-Loi A.; Singer, J. L.; Anderson, Wallace T.)V1475,280-285(1991)

Efficient optical waveguide modulation based on Wannier-Stark localization in a InGaAs-InAlAs superlattice (Bigan, Erwan; Allovon, Michel; Carre, Madeleine; Carenco, Alain; Voisin, Paul)V1362,553-558(1991)

Electronic properties of Si-doped nipi structures in GaAs (Fong, C. Y.; Gallup, R. F.; Nelson, J. S.)V1361,479-488(1991)

Energy-level structure and electron transitions of GaAs:Cr optoelectronic materials (Schulz, Hans-Joachim; Schoenbach, Karl H.; Kimpel, B. M.; Brinkmann, Ralf P.)V1361,834-847(1991)

Energy and momentum relaxation of electrons in GaAs quantum-wells: effect of nondrifting hot phonons and interface roughness (Gupta, Rita P.; Balkan, N.; Ridley, Brian K.; Emeny, M. T.)V1362,798-803(1991)

Energy levels of GaAs/A1xGa1-xAs double-barrier quantum wells (Chen, Yong; Neu, G.; Deparis, C.; Massies, J.)V1361,860-865(1991)

Epitaxial growth of gallium arsenide from elemental arsenic (Chu, Ting L.; Chu, Shirley S.; Green, Richard F.; Cerny, C. L.)V1361,523-528(1991)

Evaluation of diode laser failure mechanisms and factors influencing reliability (Baumann, John A.; Shepard, Allan H.; Waters, Robert G.; Yellen, Steven L.; Harding, Charleton M.; Serreze, Harvey B.)V1418,328-337(1991)

Excitonic photoabsorption study of AlGaAs/GaAs multiple-quantum-well grown by low-pressure MOCVD (Kwon, O'Dae; Lee, Seung-Won; Choi, Woong-Lim; Kim, Kwang-Il; Jeong, Yoon-Ha)V1361,802-808(1991)

Exciton-polariton photoluminescence in ultrapure GaAs (Zhilyaev, Yuri V.; Rossin, Victor V.; Rossina, Tatiana V.; Travnikov, V. V.)V1361,848-859(1991)

Fabrication of high-radiance LEDs by epitaxial lift-off (Pollentier, Ivan K.; Ackaert, A.; De Dobbelaere, P.; Buydens, Luc; van Daele, Peter; Demeester, Piet M.)V1361,1056-1062(1991)

Far-IR emission spectroscopy on electron-beam-irradiated A1GaAs/GaAs heterostructures (Diessel, Edgar; Sigg, Hans; von Klitzing, Klaus)V1361,1094-1099(1991)

Far-IR magneto-emission study of the quantum-hall state and breakdown of the quantum-hall effect (Raymond, Andre; Chaubet, C.; Razeghi, Manijeh)V1362,275-281(1991)

Far-IR studies of moderately doped molecular beam epitaxy grown GaAs on Si(100) (Morley, Stefan; Eickhoff, Thomas; Zahn, Dietrich R.; Richter, W.; Woolf, D.; Westwood, D. I.; Williams, R. H.)V1361,213-222(1991)

GaAs/GaAlAs superlattice avalanche photodiode at wavelength L = 0.8 um (Kibeya, Saidi; Orsal, Bernard; Alabedra, Robert; Lippens, D.)V1501,97-106(1991)

GaAs monolithic RF modules for SARSAT distress beacons (Cauley, Mike A.)V1475,275-279(1991)

GaAs opto-thyristor for pulsed power applications (Hur, Jung H.; Hadizad, Peyman; Zhao, Hanmin; Hummel, Steven R.; Dapkus, P. D.; Fetterman, Harold R.; Gundersen, Martin A.)V1378,95-100(1991)

GaInAs PIN photodetectors on semi-insulating substrates (Crawford, Deborah L.; Wey, Y. G.; Bowers, John E.; Hafich, Michael J.; Robinson, Gary Y.)V1371,138-141(1991)

Generalization of Bragg reflector geometry: application to (Ga,Al)As - (Ca,Sr)F2 reflectors (Fontaine, Chantal; Requena, Philippe; Munoz-Yague, Antonio)V1362,59-66(1991)

Graded AlxGa1-xAs photoconductive devices for high-efficiency picosecond optoelectronic switching (Morse, Jeffrey D.; Mariella, Raymond P.; Dutton, Robert W.)V1378,55-59(1991)

Heteroepitaxial growth of InP and GaInAs on GaAs substrates using nonhydride sources (Chu, Shirley S.; Chu, Ting L.; Yoo, C. H.; Smith, G. L.)V1361,1020-1025(1991)

High-conducting p+-InGaAs toplayers produced by simultaneous diffusion of Zn and Cd (Gruska, Bernd; Ambree, P.; Wandel, K.; Wielsch, U.)V1361,758-764(1991)

High contrast ratio InxGa1-xAs/GaAs multiple-quantum-well spatial light modulators (Harwit, Alex; Fernandez, R.; Eades, Wendell D.)V1541,38-47(1991)

High-density chip-to-chip interconnect system for GaAs semiconductor devices (Wigginton, Stewart C.; Davidson, Scott E.; Harting, William L.)V1390,560-567(1991)

High-efficiency Kaband monolithic pseudomorphic HEMT amplifier (Saunier, Paul; Tserng, Hua Q.; Kao, Yung C.)V1475,86-90(1991)

High-frequency 1.3 um InGaAsP semi-insulating buried crescent lasers for analog applications (Cheng, Wood-Hi; Appelbaum, Ami; Huang, Rong-Ting; Renner, Daniel; Cioffi, Ken R.)V1418,279-283(1991)

High-performance GaAs on silicon technology for VLSI, MMICs, and optical interconnects (Christou, Aristos)V1361,354-361(1991)

High-power diffraction-limited phase-locked GaAs/GaAlAs laser diode array (Zhang, Yue-qing; Zhang, Xitian; Piao, Yue-zhi; Li, Dian-en; Wu, Sheng-li; Du, Shu-qin)V1418,444-447(1991)

High-power visible semiconductor lasers (Ishikawa, Masayuki M.; Itaya, Kazuhiko; Okajima, Masaki; Hatakoshi, Gen-ichi)V1418,344-352(1991)

High-resolution spectral characterization of high-power laser diodes (Dorsch, Friedhelm)V1418,477-486(1991)

High-speed GaAs metal-semiconductor-metal photodetectors with sub-0.1um finger width and finger spacing (Chou, Stephen Y.; Liu, Yue; Fischer, Paul B.)V1474,243-247(1991)

High-speed binary optic microlens array in GaAs (Motamedi, M. E.; Southwell, William H.; Anderson, Robert J.; Hale, Leonard G.; Gunning, William J.; Holz, Michael)V1544,33-44(1991)

Hot carrier relaxation in bulk InGaAs and quantum-wells (Gregory, Andrew; Phillips, Richard T.; Majumder, Fariduddin A.)V1362,268-274(1991)

Hot electron instabilities and light emission in GaAs quantum wells (Balkan, N.; Ridley, Brian K.)V1361,927-934(1991)

III-V monolithic resonant photoreceiver on silicon substrate for long-wavelength operation (Aboulhouda, S.; Razeghi, Manijeh; Vilcot, Jean-Pierre; Decoster, Didier; Francois, M.; Maricot, S.)V1362,494-498(1991)

Indirect stimulated emission at room temperature in the visible range (Rinker, Michael; Kalt, Heinz; Lu, Yin-Cheng; Bauser, Elizabeth; Koehler, Klaus)V1362,14-23(1991)

Influence of the p-type doping of the InP cladding layer on the threshold current density in 1.5 um QW lasers (Sermage, Bernard; Blez, M.; Kazmierski, Christophe; Ougazzaden, A.; Mircea, Andrei; Bouley, J. C.)V1362,617-622(1991)

InGaAs-GaAs strained layer lasers: physics and reliability (Coleman, James J.; Waters, Robert G.; Bour, David P.)V1418,318-327(1991)

InGaAs/AlGaAs vertical optical modulators and sources on a transparent GaAs substrate (Buydens, Luc; Demeester, Piet M.; De Dobbelaere, P.; van Daele, Peter)V1362,50-58(1991)

InGaAs/GaAs interdigitated metal-semiconductor-metal (IMSM) photodetectors operational at 1.3 um grown by molecular beam epitaxy (Elman, Boris S.; Chirravuri, Jagannath; Choudhury, A. N.; Silletti, Andrew; Negri, A. J.; Powers, J.)V1362,610-616(1991)

InGaAs/InP distributed feedback quantum-well lasers (Temkin, Henryk; Logan, Ralph A.; Tanbun-Ek, Tawaee)V1418,88-98(1991)

InGaAs/InP monolithic photoreceivers for 1.3-1.5 um optical fiber transmission (Scavennec, Andre; Billard, M.; Blanconnier, P.; Caquot, E.; Carer, P.; Giraudet, Louis; Nguyen, L.; Lugiez, F.; Praseuth, Jean-Pierre)V1362,331-337(1991)

InGaAs HEMT MMIC low-noise amplifier and doublers for EHF SATCOM ground terminals (Chow, P. D.; Lester, J.; Huang, P.; Jones, William L.)V1475,42-47(1991)

InP-based quantum-well infrared photodetectors (Gunapala, S. D.; Levine, Barry F.; Ritter, D.; Hamm, Robert A.; Panish, Morton B.)V1541,11-23(1991)

Insertion of emerging GaAs HBT technology in military and communication system applications (McAdam, Bridget A.; Sharma, Arvind K.; Allen, B.; Kintis, M.)V1475,267-274(1991)

In-situ measurement technique for solution growth in compound semiconductors (Inatomi, Yuko; Kuribayashi, Kazuhiko)V1557,132-139(1991)

Integrated optics intensity modulators in the GaAs/AlGaAs system (Khan, M. A.; Naumaan, Ahmed; Van Hove, James M.)V1396,753-759(1991)

Investigation of uniform deposition of GaInAsP quantum wells by MOCVD (Puetz, Norbert; Miner, Carla J.; Hingston, G.; Moore, Chris J.; Watt, B.; Hillier, Glen)V1361,692-698(1991)

KTP waveguides for frequency upconversion of strained-layer InGaAs laser diodes (Risk, William P.; Nadler, Ch. K.)V1561,130-134(1991)

Laser diodes with gain-coupled distributed optical feedback (Nakano, Yoshiaki; Luo, Yi; Tada, Kunio)V1418,250-260(1991)

Laser Diode Technology and Applications III (Renner, Daniel, ed.)V1418(1991)

Laser drilling vias in GaAs wafers (Riley, Susan; Schick, Larry A.)V1598,118-120(1991)

Light scatter variations with respect to wafer orientation in GaAs (Brown, Jeff L.)V1530,299-305(1991)

Long-wavelength GaAs quantum-well infrared photodetectors (Levine, Barry F.)V1362,163-167(1991)

Long-wavelength lasers and detectors fabricated on InP/GaAs superheteroepitaxial wafer (Aiga, Masao; Omura, Etsuji E.)V1418,217-222(1991)

Low-noise high-yield octave-band feedback amplifiers to 20 GHz (Minot, Katcha; Cochrane, Mike; Nelson, Bradford; Jones, William L.; Streit, Dwight C.; Liu, Po-Hsin P.)V1475,309-313(1991)

Low-substrate temperature molecular beam epitaxy growth and thermal stability of strained InGaAs/GaAs single-quantum-wells (Elman, Boris S.; Koteles, Emil S.; Melman, Paul; Rothman, Mark A.)V1361,362-372(1991)

Low-threshold room temperature continuous-wave operation of a GaAs single-quantum-well mushroom structure surface-emitting laser (Yang, Y. J.; Dziura, T. G.; Wang, S. C.; Fernandez, R.; Du, G.; Wang, Shyh)V1418,414-421(1991)

Material for future InP-based optoelectronics: InGaAsP versus InGaAlAs (Quillec, Maurice)V1361,34-46(1991)

Measurements of the InxGa1-xAs/GaAs critical layer thickness (Andersson, Thorvald G.; Ekenstedt, M. J.; Kulakovskii, Vladimir D.; Wang, S. M.; Yao, J. Y.)V1361,434-442(1991)

Microfabrication techniques for semiconductor lasers (Tamanuki, Takemasa; Tadokoro, T.; Morito, K.; Koyama, Fumio; Iga, Kenichi)V1361,614-617(1991)

Microscopic origin of the shallow-deep transition of impurity levels in III-V and II-VI semiconductors (Chadi, D. J.)V1361,228-231(1991)

Microwave monolithic integrated circuits for high-data-rate satellite communications (Turner, Elbert L.; Hill, William A.)V1475,248-256(1991)

Modulation doping and delta doping of III-V compound semiconductors (Hendriks, Peter; Zwaal, E. A.; Haverkort, Jos E.; Wolter, Joachim H.)V1362,217-227(1991)

Molecular beam epitaxy/liquid phase epitaxy hybrid growth for GaAs-LED on Si (Minemura, Tetsuroh; Yazawa, Y.; Asano, J.; Unno, T.)V1361,344-353(1991)

Molecular beam epitaxy GaAs on Si: material and devices for optical interconnects (Panayotatos, Paul; Georgakilas, Alexandros; Mourrain, Jean-Loic; Christou, Aristos)V1361,1100-1109(1991)

Monolithic GaAs integrated circuit millimeter wave imaging sensors (Weinreb, Sander)V1475,25-31(1991)

Monolithic microwave integrated circuit activities in ESA-ESTEC (Gatti, Giuliano)V1475,10-24(1991)

Monolithic Microwave Integrated Circuits for Sensors, Radar, and Communications Systems (Leonard, Regis F.; Bhasin, Kul B., eds.)V1475(1991)

Mushroom-shaped gates defined by e-beam lithography down to 80-nm gate lengths and fabrication of pseudomorphic HEMTs with a dry-etched gate recess (Huelsmann, Axel; Kaufel, G.; Raynor, Brian; Koehler, Klaus; Schweizer, T.; Braunstein, Juergen; Schlechtweg, M.; Tasker, Paul J.; Jakobus, Theo F.)V1465,201-208(1991)

New approaches to ultrasensitive magnetic resonance (Bowers, C. R.; Buratto, S. K.; Carson, Paul; Cho, H. M.; Hwang, J. Y.; Mueller, L.; Pizarro, P. J.; Shykind, David; Weitekamp, Daniel P.)V1435,36-50(1991)

New structure and method for fabricating InP/InGaAsP buried heterostructure semiconductor lasers (Holmstrom, Roger P.; Meland, Edmund; Powazinik, William)V1418,223-230(1991)

Ninety-four GHz InAlAs/InGaAs/InP HEMT low-noise down-converter (Chow, P. D.; Tan, K.; Streit, Dwight C.; Garske, D.; Liu, Po-Hsin P.; Yen, Huan-chun)V1475,48-54(1991)

Nonlinear optical gain in InGaAs/InGaAsP quantum-wells (Rosenzweig, M.; Moehrle, M.; Dueser, H.; Tischel, M.; Heitz, R.; Hoffmann, Axel)V1362,876-887(1991)

Novel distributed feedback structure for surface-emitting semiconductor lasers (Mahbobzadeh, Mohammad; Osinski, Marek A.)V1418,25-31(1991)

Novel high-speed dual-wavelength InAlAs/InGaAs graded superlattice Schottky barrier photodiode for 0.8- and 1.3-um detection (Hwang, Kiu C.; Li, Sheng S.; Kao, Yung C.)V1371,128-137(1991)

Novel optoelectronic devices and integrated circuits using epitaxial lift-off (Demeester, Piet M.; Pollentier, Ivan K.; Buydens, Luc; van Daele, Peter)V1361,987-998(1991)

Observation of tunneling emission from a single-quantum-well using deep-level transient spectroscopy (Letartre, Xavier; Stievenard, Didier)V1361,1144-1155(1991)

Ohmic and Schottky contacts to GaSb (Wu, T. S.; Su, Yan K.; Juang, F. S.; Li, N. Y.; Gan, K. J.)V1519,263-268(1991)

Optically activated GaAs MMIC switch for microwave and millimeter wave applications (Paolella, Arthur; Madjar, Asher; Herczfeld, Peter R.; Sturzebecher, Dana)V1378,195-202(1991)

Optically Activated Switching (Zutavern, Fred J., ed.)V1378(1991)

Optically triggered GaAs thyristor switches: integrated structures for environmental hardening (Carson, Richard F.; Weaver, Harry T.; Hughes, Robert C.; Zipperian, Thomas E.; Brennan, Thomas M.; Hammons, B. E.)V1378,84-94(1991)

Optical nonlinearities due to long-lived electron-hole plasmas (Dawson, Philip; Galbraith, Ian; Kucharska, Alicia I.; Foxon, C. T.)V1362,384-390(1991)

Optical properties of GaAs, GaP, and CVD diamond (Klocek, Paul; Hoggins, James T.; McKenna, T. A.; Trombetta, John M.; Boucher, Maurice W.)V1498,147-157(1991)

Optical properties of InGaAs/InP strained quantum wells (Abram, Richard A.; Wood, Andrew C.; Robbins, D. J.)V1361,424-433(1991)

Optical resonances of a semiconductor superlattice in parallel magnetic and electric fields (Pacheco, Monica; Barticevic, Zdenka; Claro, Francisco)V1361,819-826(1991)

Optimization of reflection electro-absorption modulators (Pezeshki, Bardia; Thomas, Dominique; Harris, James S.)V1362,559-565(1991)

Optimization of strained layer InGaAs/GaAs quantum-well lasers (Fu, Richard J.; Hong, C. S.; Chan, Eric Y.; Booher, Dan J.; Figueroa, Luis)V1418,108-115(1991)

Optoelectronic and optical bistabilities of photocurrent and photoluminescence at low-temperature avalanche breakdown in GaAs epitaxial films (Ryabushkin, Oleg A.; Platonov, N. S.; Sablikov, V. A.; Sergeyev, V. I.; Bader, Vladimir A.)V1362,75-79(1991)

Optoelectronic master chip for optical computing (Lang, Robert J.; Kim, Jae H.; Larsson, Anders G.; Nouhi, Akbar; Cody, Jeffrey; Lin, Steven H.; Psaltis, Demetri; Tiberio, Richard C.; Porkolab, Gyorgy A.; Wolf, Edward D.)V1563,2-7(1991)

Oxide removal from GaAs(100) by atomic hydrogen (Schaefer, Juergen A.; Persch, V.; Stock, S.; Allinger, Thomas; Goldmann, A.)V1361,1026-1032(1991)

Packaging technology for GaAs MMIC (monolithic microwave integrated circuits) modules (Tomimuro, Hisashi)V1390,214-222(1991)

Pattern etching and selective growth of GaAs by in-situ electron-beam lithography using an oxidized thin layer (Akita, K.; Sugimoto, Yoshimasa; Taneya, M.; Hiratani, Y.; Ohki, Y.; Kawanishi, Hidenori; Katayama, Yoshifumi)V1392,576-587(1991)

Performances of gallium arsenide on silicon substrate photoconductive detectors (Constant, Monique T.; Boussekey, Luc; Decoster, Didier; Vilcot, Jean-Pierre)V1362,156-162(1991)

Phase-shifter technology assessment: prospects and applications (Sokolov, Vladimir)V1475,288-302(1991)

Photocurrent response to picosecond pulses in semiconductors: application to EL2 in gallium arsenide (Pugnet, Michel; Collet, Jacques; Nardo, Laurent)V1361,96-108(1991)

Photoemission under three-photon excitation in a NEA GaAs photocathode (Wang, Liming; Hou, Xun; Cheng, Zhao)V1415,120-126(1991)

Photoluminescence and deep-level transient spectroscopy of DX-centers in selectively silicon-doped GaAs-AlAs superlattices (Ababou, Soraya; Benyattou, Taha; Marchand, Jean J.; Mayet, Louis; Guillot, Gerard; Mollot, Francis; Planel, Richard)V1361,706-711(1991)

Photoluminescence and surface photovoltaic spectra of strained InP on GaAs by MOCVD (Zhuang, Weihua; Chen, Chao; Teng, Da; Yu, Jin-zhong; Li, Yu Z.)V1361,980-986(1991)

Photoquenching and characterization studies in a bulk optically controlled GaAs semiconductor switch (Lakdawala, Vishnu K.; Schoenbach, Karl H.; Roush, Randy A.; Barevadia, Gordon R.; Mazzola, Michael S.)V1378,259-270(1991)

Physical Concepts of Materials for Novel Optoelectronic Device Applications I: Materials Growth and Characterization (Razeghi, Manijeh, ed.)V1361(1991)

Physical Concepts of Materials for Novel Optoelectronic Device Applications II: Device Physics and Applications (Razeghi, Manijeh, ed.)V1362(1991)

Picosecond photocurrent measurements of negative differential velocity in GaAs/AlAs superlattices (Minot, Christophe; Le Person, H.; Mollot, Francis; Palmier, Jean F.)V1362,301-308(1991)

Predicting diode laser performance (Lim, G.; Park, Youngsoh; Zmudzinski, C. A.; Zory, Peter S.; Miller, L. M.; Cockerill, T. M.; Coleman, James J.; Hong, C. S.; Figueroa, Luis)V1418,123-131(1991)

Quasi-three-dimensional electron systems and superlattices in wide parabolic wells: fabrication and physics (Shayegan, Mansour)V1362,228-239(1991)

Radiation effects on GaAs optical system FET devices (Kanofsky, Alvin S.; Spector, Magaly; Remke, Ronald L.; Witmer, Steve B.)V1374,48-58(1991)

Radiative processes in quantum-confined structures (Merz, James L.; Holtz, Per O.)V1361,76-88(1991)

Raman scattering determination of nonpersistent optical control of electron density in a heterojunction (Richards, David R.; Fasol, Gerhard; Ploog, Klaus H.)V1361,246-254(1991)

Rapid isothermal process technology for optoelectronic applications (Singh, Rajendra)V1418,203-216(1991)

Recent developments using GaAs as an x-ray detector (Sumner, Timothy J.; Grant, S. M.; Bewick, A.; Li, J. P.; Spooner, N. J.; Smith, K.; Beaumont, Steven P.)V1549,256-264(1991)

Rise time and recovery of GaAs photoconductive semiconductor switches (Zutavern, Fred J.; Loubriel, Guillermo M.; O'Malley, Marty W.; McLaughlin, Dan L.; Helgeson, Wes D.)V1378,271-279(1991)

Shallow-deep bistability behavior of the DX-centers in n-AlxGa1-xAs and the EL2-defects in n-GaAs (Kadri, Abderrahmane; Portal, Jean-Claude)V1362,930-937(1991)

Strained quantum-well leaky-mode diode laser arrays (Shiau, T. H.; Sun, Shang-zhu; Schaus, Christian F.; Zheng, Kang; Hadley, G. R.; Hohimer, John P.)V1418,116-122(1991)

Structure optimization of selectively doped heterojunctions: evidences for a magnetically induced Wigner solidification (Etienne, Bernard; Paris, E.; Dorin, C.; Thierry-Mieg, V.; Williams, F. I.; Glattly, D. C.; Deville, G.; Andrei, E. Y.; Probst, O.)V1362,256-267(1991)

Study of GaAs/AlGaAs quantum-well structures grown by MOVPE using tertiarybutylarsine (Lee, Hyung G.; Kim, HyungJun; Park, S. H.; Langer, Dietrich W.)V1361,893-900(1991)

Study of novel oscillations in degenerate GaAs/A1GaAs quantum-wells using electro-optic voltage probing (Tsui, Ernest S.; Vickers, Anthony J.)V1362,804-810(1991)

Subnanosecond, high-voltage photoconductive switching in GaAs (Druce, Robert L.; Pocha, Michael D.; Griffin, Kenneth L.; O'Bannon, Jim)V1378,43-54(1991)

Surface plasmon enhanced light emission in GaAs/AlGaAs light emitting diodes (Gornik, Erich; Koeck, A.; Thanner, C.; Korte, Lutz)V1362,1-13(1991)

Technique for measuring stress-induced birefringence (Heidel, Jeffrey R.; Zediker, Mark S.)V1418,240-247(1991)

Temporal model of optically initiated GaAs avalanche switches (Falk, R. A.; Adams, Jeff C.)V1378,70-81(1991)

Theory and experiments on passive negative feedback pulse control in active/passive mode-locked solid state lasers (Agnesi, Antoniangelo; Fogliani, Manlio F.; Reali, Giancarlo C.; Kubecek, Vaclav)V1415,242-247(1991)

Theory of optical-phonon interactions in a rectangular quantum wire (Stroscio, Michael A.; Kim, K. W.; Littlejohn, Michael A.)V1362,566-579(1991)

Threshold current and carrier lifetime in MOVPE regrown 1.5 um GaInAsP buried ridge structure lasers (Tischel, M.; Rosenzweig, M.; Hoffmann, Axel; Venghaus, Herbert; Fidorra, F.)V1361,917-926(1991)

Threshold current density of InGaAsP/InP surface-emitting laser diodes with hemispherical resonator (Jing, Xing-Liang; Zhang, Yong-Tao; Chen, Yi-Xin)V1418,434-441(1991)

Time development of AlGaAs single-quantum-well laser facet temperature on route to catastrophical breakdown (Tang, Wade C.; Rosen, Hal J.; Vettiger, Peter; Webb, David J.)V1418,338-342(1991)

Time-resolved luminescence experiments on modulation n-doped GaAs quantum wells (Lopez, C.; Meseguer, Francisco; Sanchez-Dehesa, Jose; Ruehle, Wolfgang W.; Ploog, Klaus H.)V1361,89-95(1991)

Transient of electrostatic potential at GaAs/AlAs heterointerfaces characterized by x-ray photoemission spectroscopy (Hirakawa, Kazuhiko; Hashimoto, Y.; Ikoma, Toshiaki)V1361,255-261(1991)

Transmission electron microscopy, photoluminescence, and capacitance spectroscopy on GaAs/Si grown by metal organic chemical vapor deposition (Bremond, Georges E.; Said, Hicham; Guillot, Gerard; Meddeb, Jaafar; Pitaval, M.; Draidia, Nasser; Azoulay, Rozette)V1361,732-743(1991)

Transport time and single-particle relaxation time in two-dimensional semiconductors (Gold, Alfred)V1362,309-313(1991)

Tunneling spectroscopy at nanometer scale in molecular beam epitaxy grown (Al)GaAs multilayers (Albrektsen, O.; Koenraad, Paul; Salemink, Huub W.)V1361,338-342(1991)

Two-dimensional electron gas charge-coupled devices (Fossum, Eric R.; Song, Jong I.; Rossi, David V.)V1447,202-203(1991)

Use of admittance spectroscopy to probe the DX-centers in AlGaAs (Subramanian, S.; Chakravarty, S.; Anand, S.; Arora, B. M.)V1362,205-216(1991)

Vertical-cavity surface-emitting semiconductor lasers: present status and future prospects (Osinski, Marek A.)V1418,2-24(1991)

Gamma rays—see also astronomy

EUV, X-Ray, and Gamma-Ray Instrumentation for Astronomy II (Siegmund, Oswald H.; Rothschild, Richard E., eds.)V1549(1991)

Imager for gamma-ray astronomy: balloon prototype (Di Cocco, Guido; Labanti, Claudio; Malaguti, Giuseppe; Rossi, Elio; Schiavone, Filomena; Spizzichino, A.; Traci, A.; Bird, A. J.; Carter, T.; Dean, Anthony J.; Gomm, A. J.; Grant, K. J.; Corba, M.; Quadrini, E.; Rossi, M.; Villa, G. E.; Swinyard, Bruce M.)V1549,102-110(1991)

Performance and spectroscopic characterization of irradiated Nd:YAG (Rose, Todd S.; Fincher, Curtis L.; Fields, Renny A.)V1561,43-49(1991)

Spectroscopy and polarimetry capabilities of the INTEGRAL imager: Monte Carlo simulation results (Swinyard, Bruce M.; Malaguti, Giuseppe; Caroli, Ezio; Dean, Anthony J.; Di Cocco, Guido)V1548,94-105(1991)

Gases—see also lasers, gas

2.9 micron laser source for use in the two-photon/laser-induced fluorescence detection of atmospheric OH (Bradshaw, John D.; van Dijk, Cornelius A.)V1433,81-91(1991)

Adsorption of bromine in CuBr laser (Wang, Yongjiang; Shi, Shuyi; Qian, Yujun; Pan, Bailiang; Ding, Xiande; Guo, Qiang; Chen, Lideng; Fan, Ruixing)V1412,67-71(1991)

Aerodyne research mobile infrared methane monitor (McManus, J. B.; Kebabian, Paul L.; Kolb, Charles E.)V1433,330-339(1991)

Airborne lidar observations of ozone and aerosols in the wintertime Arctic stratosphere (Browell, Edward V.)V1491,273-281(1991)

Airborne tunable diode laser sensor for high-precision concentration and flux measurements of carbon monoxide and methane (Sachse, Glen W.; Collins, Jim E.; Hill, G. F.; Wade, L. O.; Burney, Lewis G.; Ritter, J. A.)V1433,157-166(1991)

All-optical-fiber- and general-halogen-lamp-based remote measuring system for CH4 (Cheng, Yuqi; Zou, Kun; Shan, Xuekang)V1572,392-395(1991)

Analysis of atmospheric trace constituents from high-resolution infrared balloon-borne and ground-based solar absorption spectra (Goldman, Aaron; Murcray, Frank J.; Rinsland, C. P.; Blatherwick, Ronald D.; Murcray, F. H.; Murcray, David G.)V1491,194-202(1991)

Analysis of dynamic characteristics of nonstationary gas streams using interferometry techniques (Abrukov, Victor S.; Ilyin, Stanislav V.)V1554B,540-543(1991)

Analysis of exhaust from clean-fuel vehicles using FTIR spectroscopy (Rieger, Paul L.; Maddox, Christine E.)V1433,290-301(1991)

Applicability of open-path monitors at Superfund sites (Padgett, Joseph; Pritchett, Thomas H.)V1433,352-364(1991)

Application of FM spectroscopy in atmospheric trace gas monitoring: a study of some factors influencing the instrument design (Werle, Peter W.; Josek, K.; Slemr, Franz)V1433,128-135(1991)

Application of tunable diode laser spectroscopy to the real-time analysis of engine oil economy (Carduner, Keith R.; Colvin, A. D.; Leong, D. Y.; Schuetzle, Dennis; Mackay, Gervase I.)V1433,190-201(1991)

Auto exhaust gas analysis by FTIR spectroscopy (Herget, William F.; Lowry, Steven R.)V1433,275-289(1991)

Characteristics of a high-pressure gas proportional counter filled with xenon (Sakurai, Hirohisa; Ramsey, Brian D.)V1549,20-27(1991)

Characterization of tropospheric methane through space-based remote sensing (Ashcroft, Peter)V1491,48-55(1991)

Compact measurement system for the simultaneous determination of NO, NO2, NOy, and O3 using a small aircraft (Walega, James G.; Dye, James E.; Grahek, Frank E.; Ridley, Brian K.)V1433,232-241(1991)

Comparison of time and frequency multiplexing techniques in multicomponent FM spectroscopy (Muecke, Robert J.; Werle, Peter W.; Slemr, Franz; Prettl, William)V1433,136-144(1991)

Determination of inhomogeneous trace absorption by using exponential expansion of the absorption Pade-approximant (Dobrego, Kirill V.)V1433,365-374(1991)

Determination of the altitude of the nitric acid layer from very high resolution, ground-based IR solar spectra (Blatherwick, Ronald D.; Murcray, Frank J.; Murcray, David G.; Locker, M. H.)V1491,203-210(1991)

Development of an airborne excimer-based UV-DIAL for monitoring ozone and sulfur dioxide in the lower troposphere (Bristow, Michael P.; Diebel, D. E.; Bundy, Donald H.; Edmonds, Curtis M.; Turner, Ruldopha M.; McElroy, James L.)V1491,68-74(1991)

Distributed-fibre-optic methane gas concentration detection (Shi, Yi-Wei; Wang, Yao-Cai; Jiang, Hong-Tao; Yao, Cheng-Shan; Wu, Zhen-Chun)V1572,308-312(1991)

DOAS (differential optical absorption spectroscopy) urban pollution measurements (Stevens, Robert K.; Vossler, T. L.)V1433,25-35(1991)

Emissions monitoring by infrared photoacoustic spectroscopy (Jalenak, Wayne)V1434,46-54(1991)

Evolution of an excimer laser gas mix (Boardman, A. D.; Hodgson, Elizabeth M.; Spence, A. J.; Richardson, A. D.; Richardson, M. B.)V1503,160-166(1991)

Evolution of a space-charge layer, its instability, and ignition of arc gas discharge under photoemission from a target into a gas (Meshalkin, E. A.)V1440,211-221(1991)

Far-IR Fabry-Perot spectrometer for OH measurements (Pickett, Herbert M.; Peterson, Dean B.)V1491,308-313(1991)

Fast-response water vapor and carbon dioxide sensor (Kohsiek, W.)V1511,114-119(1991)

Fiber optic sensor for ammonia vapors of variable temperature (Potyrailo, Radislav A.; Golubkov, Sergei P.; Borsuk, Pavel S.)V1572,434-438(1991)

Field and laboratory studies of Fourier transform infrared spectroscopy in continuous emissions monitoring applications (Plummer, Grant M.)V1434,78-89(1991)

Flash soft radiography: its adaption to the study of breakup mechanisms of liquid jets into a high-density gas (Krehl, Peter; Warken, D.)V1358,162-173(1991)

Formation of titanium nitride films by Xe+ ion-beam-enhanced deposition in a N2 gas environment (Wang, Xi; Yang, Gen Q.; Liu, Xiang H.; Zheng, Zhi H.; Huang, Wei; Zhou, Zu Y.; Zou, Shi C.)V1519,740-743(1991)

Frequency-modulation absorption spectroscopy for trace species detection: theoretical and experimental comparison among methods (Silver, Joel A.; Bomse, David S.; Stanton, Alan C.)V1435,64-71(1991)

FTIR: fundamentals and applications in the analysis of dilute vehicle exhaust (Gierczak, Christine A.; Andino, J. M.; Butler, James W.; Heiser, G. A.; Jesion, G.; Korniski, T. J.)V1433,315-328(1991)

Gaseous incinerator emissions analysis by FTIR (Fourier transform infrared) spectroscopy (Herget, William F.; Demirgian, Jack)V1434,39-45(1991)

Gas flow effects in pulse avalanche discharge XeCl excimer laser (Lou, Qihong)V1397,103-106(1991)

Gas handling technology for excimer lasers (Turner, Robert E.; Remo, John L.; Bradford, Elaine; Dietz, Alvin)V1377,99-106(1991)

Gas sensor based on an integrated optical interferometer (Brandenburg, Albrecht; Edelhaeuser, Rainer; Hutter, Frank)V1510,148-159(1991)

German ATMOS program (Puls, Juergen)V1490,2-13(1991)

Ground-based lidar for long-term and network measurements of ozone (McDermid, I. S.; Schmoe, Martha S.; Walsh, T. D.)V1491,175-181(1991)

Ground-based microwave radiometry of ozone (Kaempfer, Niklaus A.; Bodenmann, P.; Peter, Reto)V1491,314-322(1991)

Heterophase isotopic SF6 molecule separation in the surface electromagnetic wavefield (Bordo, V. G.; Ershov, I. A.; Kravchenko, V. A.; Petrov, Yu. N.)V1440,364-369(1991)

High radiometric performance CCD for the third-generation stratospheric aerosol and gas experiment (Delamere, W. A.; Baer, James W.; Ebben, Thomas H.; Flores, James S.; Kleiman, Gary; Blouke, Morley M.; McCormick, M. P.)V1447,204-213(1991)

High-resolution multichannel mm-wave radiometer for the detection of stratospheric ClO (Gerber, Louis; Kaempfer, Niklaus A.)V1491,211-217(1991)

High-speed photography applied for the investigation of the dynamics of free falling evaporating droplets (Guryashkin, L. P.; Stasenko, A. L.)V1358,974-978(1991)

Hydrogen peroxide and organic peroxides in the marine environment (Heikes, Brian G.; Miller, William L.; Lee, Meehye)V1433,253-262(1991)

Imaging gas scintillation proportional counters for ASTRO-D (Ohashi, T.; Makishima, K.; Ishida, M.; Tsuru, T.; Tashiro, M.; Mihara, Teruyoshi; Kohmura, Y.; Inoue, Hiroyuki)V1549,9-19(1991)

Incinerator technology overview (Santoleri, Joseph J.)V1434,2-13(1991)

Influence of gas composition and pressure on the pulse duration of electron beam controlled discharge XeCl laser (Badziak, Jan; Drazek, Wieslaw; Dubicki, Adam; Perlinski, Leszek)V1397,81-84(1991)

Influence of gas composition on XeCl laser performance (Jursich, Gregory M.; Von Drusek, William A.; Mulderink, Ken; Olchowka, V.; Reid, John; Brimacombe, Robert K.)V1412,115-122(1991)

Intelligent optical fiber sensor system for measurement of gas concentration (Pan, Jingming; Yin, Zongming)V1572,403-405(1991)

Interferometric monitor for greenhouse gasses for ADEOS (Tsuno, Katsuhiko; Kameda, Yoshihiko; Kondoh, Kayoko; Hirai, Shoichi)V1490,222-232(1991)

Laser linac: nondiffractive beam and gas-loading effects (Bochove, Erik J.; Moore, Gerald T.; Scully, Marlan O.; Wodkiewicz, K.)V1497,338-347(1991)

Laser set for the investigations of the NO2 contents in atmosphere (Makuchowski, Jozef; Pokora, Ludwik J.; Ujda, Zbigniew; Wawer, Janusz)V1391,348-350(1991)

Long-path differential absorption measurements of tropospheric molecules (Harder, Jerald W.; Mount, George H.)V1491,33-42(1991)

Long-path intracavity laser for the measurement of atmospheric trace gases (McManus, J. B.; Kolb, Charles E.)V1433,340-351(1991)

Measurement of Atmospheric Gases (Schiff, Harold I., ed.)V1433(1991)

Measurement of peroxyacetyl nitrate, NO2, and NOx by using a gas chromatograph with a luminol-based detector (Drummond, John W.; Mackay, Gervase I.; Schiff, Harold I.)V1433,242-252(1991)

Measurement of tropospheric carbon monoxide using gas filter radiometers (Reichle, Henry G.)V1491,15-25(1991)

Measurements of gas flow and gas constituent in a wind-tunnel-type excimer laser under high-repetition-rate operations (Kasuya, Koichi; Horioka, K.; Hikida, N.; Murazi, M.; Nakata, K.; Kawakita, Y.; Miyai, Y.; Kato, S.; Yoshida, S.; Ideno, S.)V1397,67-70(1991)

Measurements of nitrous acid, nitrate radicals, formaldehyde, and nitrogen dioxide for the Southern California Air Quality Study by differential optical absorption spectroscopy (Winer, Arthur M.; Biermann, Heinz W.)V1433,44-55(1991)

Measuring tropospheric ozone using differential absorption lidar technique (Proffitt, Michael H.; Langford, A. O.)V1491,2-6(1991)

Mechanisms of ionization for gas adjoining to plasma in intense laser beam (Fisher, Vladimir)V1440,179-187(1991)

Michelson interferometer for passive atmosphere sounding (Posselt, Winfried)V1490,114-125(1991)

Multicomponent analysis using established techniques (Dillehay, David L.)V1434,56-66(1991)

Network for the detection of stratospheric change (Kurylo, Michael J.)V1491,168-174(1991)

New spectroscopic instrumentation for measurement of stratospheric trace species by remote sensing of scattered skylight (Mount, George H.; Jakoubek, Roger O.; Sanders, Ryan W.; Harder, Jerald W.; Solomon, Susan; Winkler, Richard; Thompson, Thomas; Harrop, Walter)V1491,188-193(1991)

O3, NO2, NO3, SO2, and aerosol measurements in Beijing (Xue, Qing-yu; Guo, Song; Zhao, Xue-peng; Nieu, Jian-guo; Zhang, Yi-ping)V1491,75-82(1991)

Open-path tunable diode laser absorption for eddy correlation flux measurements of atmospheric trace gases (Anderson, Stuart M.; Zahniser, Mark S.)V1433,167-178(1991)

Optics for tunable diode laser spectrometers (Riedel, Wolfgang J.)V1433,179-189(1991)

Optoelectronic LED-photodiode pairs for moisture and gas sensors in the spectral range 1.8-4.8 um (Yakovlev, Yurii P.; Baranov, Alexej N.; Imenkov, Albert N.; Mikhailova, Maya P.)V1510,170-177(1991)

Overview of the halogen occultation experiment (Russell, James M.)V1491,110-116(1991)

Overview of the spectroscopy of the atmosphere using far-infrared emission experiment (Russell, James M.)V1491,142-150(1991)

Physical effects in time-domain CARS (coherent antistokes Raman spectroscopy) of molecular gases (Kolomoitsev, D. V.; Nikitin, S. Y.)V1402,11-30(1991)

Preparation of tin oxide and insulating oxide thin films for multilayered gas sensors (Feng, Chang D.; Shimizu, Yasuhiro; Egashira, Makoto)V1519,8-13(1991)

Progress in fiber-remote gas correlation spectrometry (Dakin, John P.; Edwards, Henry O.)V1510,160-169(1991)

Progress in gas purifiers for industrial excimer lasers (Naylor, Graham A.; Kearsley, Andrew J.)V1377,107-112(1991)

Remote Sensing of Atmospheric Chemistry (McElroy, James L.; McNeal, Robert J., eds.)V1491(1991)

Retrieval and molecule sensitivity studies for the global ozone monitoring experiment and the scanning imaging absorption spectrometer for atmospheric chartography (Chance, Kelly V.; Burrows, John P.; Schneider, Wolfgang)V1491,151-165(1991)

Sensitivity and selectivity enhancement in semiconductor gas sensors (Egashira, Makoto; Shimizu, Yasuhiro)V1519,467-476(1991)

Spectroscopic observations of CO2, CH4, N2O, and CO from Kitt Peak, 1979-1990 (Livingston, William C.; Wallace, Lloyd V.)V1491,43-47(1991)

Stratospheric aerosol and gas experiment III: aerosol and trace gas measurements for the Earth Observing System (McCormick, M. P.; Chu, William P.; Zawodny, J. M.; Mauldin, Lemuel E.; McMaster, Leonard R.)V1491,125-141(1991)

Stratospheric spectroscopy with the far-infrared spectrometer: overview and recent results (Traub, Wesley A.; Chance, Kelly V.; Johnson, David G.; Jucks, Kenneth W.)V1491,298-307(1991)

Studies of blood gas determination and intelligent image (Wang, Chihcheng; Jin, Xi)V1572,406-409(1991)

Study of gas scintillation proportional counter physics using synchrotron radiation (Bavdaz, Marcos; Favata, Fabio; Smith, Alan; Parmar, A. N.)V1549,35-44(1991)

System for evaluation of trace gas concentration in the atmosphere based on the differential optical absorption spectroscopy technique (Hallstadius, Hans; Uneus, Leif; Wallin, Suante)V1433,36-43(1991)

Time-space resolved optical study of the plasma produced by laser ablation of Ge: the role of oxygen pressure (Vega, Fidel; Solis, Javier; Afonso, Carmen N.)V1397,807-811(1991)

Tunable diode laser spectrometer for high-precision concentration and ratio measurements of long-lived atmospheric gases (Fried, Alan; Drummond, James R.; Henry, Bruce; Fox, Jack)V1433,145-156(1991)

Tunable diode laser systems for trace gas monitoring (Mackay, Gervase I.; Karecki, David R.; Schiff, Harold I.)V1433,104-119(1991)

Use of a Fourier transform spectrometer as a remote sensor at Superfund sites (Russwurm, George M.; Kagann, Robert H.; Simpson, Orman A.; McClenny, William A.)V1433,302-314(1991)

Use of chemiluminescence techniques in portable, lightweight, highly sensitive instruments for measuring NO2, NOx, and O3 (Drummond, John W.; Topham, L. A.; Mackay, Gervase I.; Schiff, Harold I.)V1433,224-231(1991)

Use of satellite data to determine the distribution of ozone in the troposphere (Fishman, Jack; Watson, Catherine E.; Brackett, Vincent G.; Fakhruzzaman, Khan; Veiga, Robert)V1491,348-359(1991)

Visualization study on pool boiling heat transfer (Kamei, Shuya; Hirata, Masaru)V1358,979-983(1991)

Geometrical optics—see also optical design

Aberration Theory Made Simple (Mahajan, Virendra N.)VTT06(1991)

Binary optics from a ray-tracing point of view (Southwell, William H.)V1354,38-42(1991)

Chirality and its applications to engineered materials (Varadan, Vasundara V.; Varadan, Vijay K.)V1558,156-181(1991)

Image plane tilt in optical systems (Sasian, Jose M.)V1527,85-95(1991)

Integration of geometrical and physical optics (Lawrence, George N.; Moore, Kenneth E.)V1415,322-329(1991)

Investigations of the focal shift of the high-power cw YAG laser beam (Jabczynski, Jan K.; Mindak, Marek K.)V1391,109-116(1991)

Moving M2 mirror without pointing offset (Ragazzoni, Roberto; Bortoletto, Favio)V1542,236-246(1991)

Nonimaging Optics: Maximum Efficiency Light Transfer (Winston, Roland; Holman, Robert L., eds.)V1528(1991)

Resonances and internal electric energy density in droplets (Hill, Steven C.; Barber, Peter W.; Chowdhury, Dipakbin Q.; Khaled, Elsayed-Esam M.; Mazumder, Mohiuddin)V1497,16-27(1991)

Thermodynamics of light concentrators (Ries, Harald; Smestad, Greg P.; Winston, Roland)V1528,7-14(1991)

Two-dimensional control of radiant intensity by use of nonimaging optics (Kuppenheimer, John D.; Lawson, Robert I.)V1528,93-103(1991)

Wave Propagation and Scattering in Varied Media II (Varadan, Vijay K., ed.)V1558(1991)

Glass—see also optical properties

Active glasses prepared by the sol-gel method including islands of CdS or silver (Reisfeld, Renata; Minti, Harry; Eyal, Marek)V1513,360-367(1991)

Advanced processes for the shaping of the Zerodur glass ceramic (Marx, Thomas A.)V1398,94-94(1991)

Advanced processing of the Zerodur R glass ceramic (Marx, Thomas A.)V1535,130-135(1991)

Advantages of drawing crystal-core fibers in microgravity (Shlichta, Paul J.; Nerad, Bruce A.)V1557,10-23(1991)

Analysis of the refractive-index profile in ion-exchanged waveguides (Righini, Giancarlo C.; Pelli, Stefano; Saracini, R.; Battaglin, Giancarlo; Scaglione, Antonio)V1513,418-424(1991)

Application and machining of Zerodur for optical purposes (Reisert, Norbert)V1400,171-177(1991)

Application of statistical design in materials development and production (Pouskouleli, G.; Wheat, T. A.; Ahmad, A.; Varma, Sudhanshu; Prasad, S. E.)V1590,179-190(1991)

Buried-glass waveguides by ion exchange through ionic barrier (Li, Ming-Jun; Honkanen, Seppo; Wang, Wei-Jian; Najafi, S. I.; Tervonen, Ari; Poyhonen, Pekka)V1506,52-57(1991)

Characterization, modeling, and design optimization of integrated optical waveguide devices in glass (Yip, Gar L.)V1513,26-36(1991)

Characterization of planar optical waveguides by K+ ion exchange in glass at 1.152 and 1.523 um (Yip, Gar L.; Kishioka, Kiyoshi; Xiang, Feng; Chen, J. Y.)V1583,14-18(1991)

Characterization of subsurface damage in glass and metal optics (Polvani, Robert S.; Evans, Chris J.)V1441,173-173(1991)

Correction of secondary and higher-order spectrum using special materials (Mercado, Romeo I.)V1535,184-198(1991)

Cr3+ tunable laser glass (Izumitani, Tetsuro; Zou, Xuolu; Wang, Y.)V1535,150-159(1991)

Crystallization and optothermal characteristics of germanate glasses (Montenero, Angelo; Gnappi, G.; Bertolotti, Mario; Sibilia, C.; Fazio, E.; Liakhou, G.)V1513,234-242(1991)

Damage to polymer-coated glass surfaces by small-particle impact (Chaudhri, Mohammad M.; Smith, Alan L.)V1358,683-689(1991)

Defect centers and photoinduced self-organization in Ge-doped silica core fiber (Tsai, T. E.; Griscom, David L.)V1516,14-28(1991)

Delayed elasticity in Zerodur at room temperature (Pepi, John W.; Golini, Donald)V1533,212-221(1991)

Diode-pumped Er3+ glass laser at 2.7 um (Yanagita, Hiroaki; Toratani, Hisayoshi; Yamashita, Toshiharu T.; Masuda, Isao)V1513,386-395(1991)

Dual-core ion-exchanged glass waveguides (Li, Ming-Jun; Honkanen, Seppo; Wang, Wei-Jian; Leonelli, Richard; Najafi, S. I.)V1513,410-417(1991)

Effect of the glass composition on the emission band of erbium-doped active fibers (Cognolato, Livio; Gnazzo, Angelantonio; Sordo, Bruno; Cocito, Guiseppe)V1579,249-256(1991)

Electrical and optical properties of porous glass (Rysiakiewicz-Pasek, Ewa; Marczuk, Krystyna)V1513,283-290(1991)

Fabrication of large multimode glass waveguides by dry silver ion exchange in vacuum (Tammela, Simo; Pohjonen, Harri; Honkanen, Seppo; Tervonen, Ari)V1583,37-42(1991)

Fatigue-resistant coating of SiO2 glass (Tomozawa, Minoru; Han, Won-Taek; Davis, Kenneth M.)V1590,160-167(1991)

Fiber defects in Ge-doped fibers: towards a coherent picture (LaRochelle, Sophie)V1516,55-57(1991)

Fluorescence properties of Cu+ ion in borate and phosphate glasses (Boutinaud, P.; Parent, C.; Le Flem, Gille; Moine, Bernard; Pedrini, Christian; Duloisy, E.)V1590,168-178(1991)

Formation of optical elements by photothermo-induced crystallization of glass (Glebov, Leonid B.; Nikonorov, Nikolai V.; Petrovskii, Gurii T.; Kharchenko, Mikhail V.)V1440,24-35(1991)

Frequency doubling in optical fibers: a complex puzzle (Margulis, Walter; Carvalho, Isabel C.; Lesche, Bernhard)V1516,60-66(1991)

Frequency up-conversion in Pr3+ doped fibers (Gomes, Anderson S.; de Araujo, Cid B.; Moraes, E. S.; Opalinska, M. M.; Gouveia-Neto, A. S.)V1579,257-263(1991)

Glasses for Optoelectronics II (Righini, Giancarlo C., ed.)V1513(1991)

Glasses including quantum dots of cadmium sulfide, silver, and laser dyes (Reisfeld, Renata; Eyal, Marek; Chernyak, Valery; Jorgensen, Christian K.)V1590,215-228(1991)

Glass requirements for encapsulating metallurgical diodes (Ali, Mir A.; Meldrum, Gerald L.; Krieger, Jeffry M.)V1513,215-223(1991)

Glass waveguides by ion exchange with ionic masking (Wang, Wei-Jian; Li, Ming-Jun; Honkanen, Seppo; Najafi, S. I.; Tervonen, Ari)V1513,434-440(1991)

Grating splitter for glass waveguide (Jin, Guoliang; Shen, Ronggui; Ying, Zaisheng)V1513,50-55(1991)

Growth, surface passivation, and characterization of CdSe microcrystallites in glass with respect to their application in nonlinear optics (Woggon, Ulrike; Rueckmann, I.; Kornack, J.; Mueller, Matthias; Cesnulevicius, J.; Kolenda, Jonas; Petrauskas, Mendogas)V1362,888-898(1991)

Guided-wave nonlinear spectroscopy of silica glasses (Aleksandrov, I. V.; Nesterova, Z. V.; Petrovskii, Gurii T.)V1513,309-312(1991)

High-resolution microchannel plate image tube development (Johnson, C. B.; Patton, Stanley B.; Bender, E.)V1449,2-12(1991)

High-silica low-loss three-waveguide couplers on Si by flame hydrolysis deposition (Barbarossa, Giovanni; Laybourn, Peter J.)V1513,37-43(1991)

High-temperature Raman spectra of nioboborate glass melts (Yang, Quanzu; Wang, Zhongcai; Wang, Shizhuo)V1513,264-269(1991)

IR reflectance spectroscopy and AES investigation of titanium ion-beam-doped silica (Belostotsky, Vladimir I.; Solinov, Vladimir F.)V1513,313-318(1991)

IR-spectroscopical investigations on the glass structure of porous and sintered compacts of colloidal silica gels (Clasen, Rolf; Hornfeck, M.; Theiss, W.)V1513,243-254(1991)

Influence of surface passivation on the optical bleaching of CdSe microcrystallites embedded in glass (Rueckmann, I.; Woggon, Ulrike; Kornack, J.; Mueller, Matthias; Cesnulevicius, J.; Kolenda, Jonas; Petrauskas, Mendogas)V1513,78-85(1991)

Infrared transmitting glasses and fibers for chemical analysis (Hilton, Albert R.)V1437,54-59(1991)

Integrated optical Mach-Zehnder interferometers in glass (Lefebvre, P.; Vahid-Shahidi, A.; Albert, Jacques; Najafi, S. I.)V1583,221-225(1991)

Investigation of optical parameters of silicate glasses in case of before-threshold effect of laser radiation (Bosyi, O. N.; Efimov, Oleg M.; Mekryukov, A. M.)V1440,57-62(1991)

Ion beam milling of fused silica for window fabrication (Wilson, Scott R.; Reicher, David W.; Kranenberg, C. F.; McNeil, John R.; White, Patricia L.; Martin, Peter M.; McCready, David E.)V1441,82-86(1991)

Ion beam modification of glasses (Mazzoldi, Paolo; Carnera, Alberto; Caccavale, F.; Granozzi, G.; Bertoncello, R.; Battaglin, Giancarlo; Boscolo-Boscoletto, A.; Polato, P.)V1513,182-197(1991)

Ion-exchanged waveguides in glass: simulation and experiment (Wolf, Barbara; Fabricius, Norbert; Foss, Wolfgang; Dorsel, Andreas N.)V1506,40-51(1991)

Ion-exchanged waveguides in semiconductor-doped glasses (Righini, Giancarlo C.; Pelli, Stefano; De Blasi, C.; Fagherazzi, Giuliano; Manno, D.)V1513,105-111(1991)

Ion-exchange strengthening of high-average-power phosphate laser glass (Lee, Huai-Chuan; Meissner, Helmuth E.)V1441,87-103(1991)

Laser glasses (Lunter, Sergei G.; Dymnikov, Alexander A.; Przhevuskii, Alexander K.; Fedorov, Yurii K.)V1513,349-359(1991)

Laser heating and evaporation of glass and glass-borning materials and its application for creating MOC (Veiko, Vadim P.; Yakovlev, Evgeni B.; Frolov, V. V.; Chujko, V. A.; Kromin, A. K.; Abbakumov, M. O.; Shakola, A. T.; Fomichov, P. A.)V1544,152-163(1991)

Laser-induced diffusion and second-harmonic generation in glasses (Aksenov, V. P.)V1440,377-383(1991)

Low-temperature viscosity measurements of infrared transmitting halide glasses (Seddon, Angela B.; Cardoso, A. V.)V1513,255-263(1991)

Measurement of the sizes of the semiconductor crystallites in colored glasses through neutron scattering (Banfi, Giampiero P.; Degiorgio, Vittorio; Rennie, A. R.; Righini, Giancarlo C.)V1361,874-880(1991)

Melt processing of calcium aluminate fibers with sapphirelike infrared transmission (Wallenberger, Frederick T.; Koutsky, J. A.; Brown, Sherman D.)V1590,72-82(1991)

Multiphoton-absorption-induced structural changes in fused silica (Silin, Andrejs R.)V1513,270-273(1991)

Multiple-pulse damage of BK-7 glass (Yoshida, Kunio; Yoshida, Hiroshi; Noda, T.; Nakai, Sadao S.)V1441,9-9(1991)

New glasses for optics and optoelectronics (Morian, Hans F.)V1400,146-157(1991)

New possibilities of photosensitive glasses for three-dimensional phase hologram recording (Glebov, Leonid B.; Nikonorov, Nikolai V.; Panysheva, Elena I.; Petrovskii, Gurii T.; Savvin, Vladimir V.; Tunimanova, Irina V.; Tsekhomskii, Victor A.)V1621,21-32(1991)

Nonlinear optical properties of chalcogenide As-S, As-Se glasses (Andriesh, A.; Bertolotti, Mario; Chumach, V.; Fazio, E.; Ferrari, A.; Liakhou, G.; Sibilia, C.)V1513,137-147(1991)

Nonlinear optical properties of germanium diselenide glasses (Haro-Poniatowski, Emmanuel; Guasti, Manuel F.)V1513,86-92(1991)

Nonlinear optical susceptibility of CuCl quantum dot glass (Justus, Brian L.; Seaver, Mark E.; Ruller, Jacqueline A.; Campillo, Anthony J.)V1409,2-8(1991)

Nonlinear optical transmission of an integrated optical bent coupler in semiconductor-doped glass (Guntau, Matthias; Possner, Torsten; Braeuer, Andreas; Dannberg, Peter)V1513,112-122(1991)

Optical effects induced in oxide glasses by irradiation (Araujo, Roger J.; Borrelli, Nicholas F.)V1590,138-145(1991)

Optical glass selection using computerized data base (Fischer, Robert E.; Thomas, Michael J.; Hudyma, Russell M.)V1535,78-88(1991)

Optical investigation of microcrystals in glasses (Ferrara, M.; Lugara, M.; Moro, C.; Cingolani, R.; De Blasi, C.; Manno, D.; Righini, Giancarlo C.)V1513,130-136(1991)

Optically induced creation, transformation, and organization of defects and color centers in optical fibers (Russell, Philip S.; Hand, Duncan P.; Chow, Y. T.; Poyntz-Wright, L. J.)V1516,47-54(1991)

Optical performance of angle-dependent light-control glass (Maeda, Koichi; Ishizuka, S.; Tsujino, T.; Yamamoto, H.; Takigawa, Akio)V1536,138-148(1991)

Optical properties: a trip through the glass map (Marker, Alexander J.)V1535,60-77(1991)

Optical properties of glass materials obtained by inorganic sol-gel synthesis (Glebov, Leonid B.; Evstropiev, Sergei K.; Petrovskii, Gurii T.; Shashkin, Viktor S.)V1513,224-231(1991)

Passive Materials for Optical Elements (Wilkerson, Gary W., ed.)V1535(1991)

Photochemical generation of gradient indices in glass (Mendoza, Edgar A.; Gafney, Harry D.; Morse, David L.)V1378,139-144(1991)

Photoinduced second-harmonic generation and luminescence of defects in Ge-doped silica fibers (Krol, Denise M.; Atkins, Robert M.; Lemaire, Paul J.)V1516,38-46(1991)

Photoinduced self-organization in optical fiber: some answered and unanswered questions (Ouellette, Francois; Gagnon, Daniel; LaRochelle, Sophie; Poirier, Michel)V1516,2-13(1991)

Photoinductional change of silicate glasses optical parameters at two-photon laser radiation absorption (Glebov, Leonid B.; Efimov, Oleg M.; Mekryukov, A. M.)V1513,274-282(1991)

Photolithographic processing of integrated optic devices in glasses (Mendoza, Edgar A.; Gafney, Harry D.; Morse, David L.)V1583,43-51(1991)

Photosensitive germanosilicate preforms and fibers (Williams, Doug L.; Davey, Steven T.; Kashyap, Raman; Armitage, J. R.; Ainslie, B. J.)V1513,158-167(1991)

Photosensitivity in optical fibers: detection, characterization, and application to the fabrication of in-core fiber index gratings (Malo, Bernard; Bilodeau, Francois; Johnson, Derwyn C.; Skinner, Iain M.; Hill, Kenneth O.; Morse, Ted F.; Kilian, Arnd; Reinhart, Larry J.; Oh, Kyunghwan)V1590,83-93(1991)

Physical and optical properties of organically modified silicates doped with laser and NLO dyes (Capozzi, Carol A.; Pye, L. D.)V1513,320-329(1991)

Planar and strip optical waveguides by sol-gel method and laser densification (Guglielmi, Massimo; Colombo, Paolo; Mancinelli Degli Esposti, Luca; Righini, Giancarlo C.; Pelli, Stefano)V1513,44-49(1991)

Planar optical waveguides on glasses and glass-ceramic materials (Glebov, Leonid B.; Nikonorov, Nikolai V.; Petrovskii, Gurii T.)V1513,56-70(1991)

Porous glass structure as revealed by capacitance and conductance measurements (Marczuk, Krystyna; Prokopovich, Ludvig P.; Roizin, Yacov O.; Rysiakiewicz-Pasek, Ewa; Sviridov, Victor N.)V1513,291-296(1991)

Possibility of a "lithium glass" state appearance in the Li3Sc2-xFex(PO4)3 superionic solid solutions (Michailov, Vladimir I.; Sigaryov, Sergei E.; Terziev, Vladimir G.)V1590,203-214(1991)

Preparation of SiO2 film utilizing equilibrium reaction in aqueous solution (Kawahara, Hideo; Sakai, Y.; Goda, Takuji; Hishinuma, Akihiro; Takemura, Kazuo)V1513,198-203(1991)

Propagation and diffusion characteristics of optical waveguides made by electric field-assisted k+-ion exchange in glass (Noutsios, Peter C.; Yip, Gar L.; Kishioka, Kiyoshi)V1374,14-22(1991)

Properties and processing of the TeX glasses (Zhang, Xhang H.; Ma, Hong-Li; Lucas, Jacques)V1513,209-214(1991)

Radiation-resistant optical glasses (Marker, Alexander J.; Hayden, Joseph S.; Speit, Burkhard)V1485,160-172(1991)

Rare-earth-doped fluoride glasses for active optical fiber applications (Adam, Jean-Luc; Smektala, Frederic; Denoue, Emmanuel; Lucas, Jacques)V1513,150-157(1991)

Reluctant glass formers and their applications in lens design (Johnson, R. B.; Feng, Chen; Ethridge, Edwin C.)V1527,2-18(1991); V1535,231-247(1991)

Review of measurement systems for evaluating thermal expansion homogeneity of Corning Code 7971 ULETM (Hagy, Henry E.)V1533,198-211(1991)

Scattering properties of ZrF4-based glasses prepared by the gas film levitation technique (Lopez, Adolphe; Baniel, P.; Gall, Pascal; Granier, Jean E.)V1590,191-202(1991)

Semiconductor-doped glass as a nonlinear material (Speit, Burkhard; Remitz, K. E.; Neuroth, Norbert N.)V1361,1128-1131(1991)

Semiconductor-doped glasses: nonlinear and electro-optical properties (Ekimov, A. I.; Kudryavtsev, I. A.; Chepick, D. I.; Shumilov, S. K.)V1513,123-129(1991)

Solid matrix Christiansen filters (Milanovic, Zoran; Jacobson, Michael R.; Macleod, H. A.)V1535,160-170(1991)

Spectroscopic properties of Er3+-doped glasses for the realization of active waveguides by ion-exchange technique (Cognolato, Livio; De Bernardi, Carlo; Ferraris, M.; Gnazzo, Angelantonio; Morasca, Salvatore; Scarano, Domenica)V1513,368-377(1991)

Sputtering of silicate glasses (Kai, Teruhiko; Takebe, Hiromichi; Morinaga, Kenji)V1519,99-103(1991)

Stress analysis in ion-exchanged waveguides by using a polarimetric technique (Gonella, Francesco; Mazzi, Giulio; Quaranta, Alberto)V1513,425-433(1991)

Stripe waveguides by Cs+- and K+-exchange in neodymium-doped soda silicate glasses for laser application (Possner, Torsten; Ehrt, Doris; Sargsjan, Geworg; Unger, Clemens)V1513,378-385(1991)

Structural and optical properties of chalcogenide glass-ceramics (Boehm, Leah; Assaf, Haim)V1535,171-181(1991)

Structural and optical properties of semiconducting microcrystallite-doped SiO2 glass films prepared by rf-sputtering (Tsunetomo, Keiji; Shimizu, Ryuichiro; Yamamoto, Masaki; Osaka, Yukio)V1513,93-104(1991)

Study of fluorescent glass-ceramics (Qiu, Guanming; DuanMu, Qingduo)V1513,396-407(1991)

Study of regularities and the mechanism of intrinsic optical breakdown of glasses (Glebov, Leonid B.; Efimov, Oleg M.)V1440,50-56(1991)

Study using nuclear techniques of waveguides produced by electromigration processes (Battaglin, Giancarlo; De Marchi, Giovanna; Losacco, Aurora M.; Mazzoldi, Paolo; Miotello, Antonio; Quaranta, Alberto; Valentini, Antonio)V1513,441-450(1991)

Submolecular Glass Chemistry and Physics (Bray, Phillip; Kreidl, Norbert J., eds.)V1590(1991)

Synthesis and properties of sol-gel-derived AgClxBr1-x colloid containing sodium alumo borosilicate glasses (Mennig, Martin; Schmidt, Helmut; Fink, Claudia)V1590,152-159(1991)

Test of photovoltaic model of photoinduced second-harmonic generation in optical fibers (Dianov, Evgeni M.; Kazansky, Peter G.; Krautschik, Christof G.; Stepanov, D. Y.)V1516,75-80(1991)

Theoretical models in optics of glass-composite materials (Kuchinskii, Sergei A.; Dostenko, Alexander V.)V1513,297-308(1991)

Thermal relaxation of tellurium-halide-based glasses (Ma, Hong-Li; Zhang, Xhang H.; Lucas, Jacques)V1590,146-151(1991)

Tilt-tolerant dual-band interferometer objective lens (Jones, Mike I.)V1498,158-162(1991)

Time-resolved photon echo and fluorescence anisotropy study of organically doped sol-gel glasses (Chronister, Eric L.; L'Esperance, Drew M.; Pelo, John; Middleton, John; Crowell, Robert A.)V1559,56-64(1991)

Upconversion in rare-earth-doped fluoride glasses (van Dongen, A. M.; Oomen, Emmanuel W.; le Gall, Perrine M.)V1513,330-339(1991)

Upconversion intensity and multiphonon relaxation of Er3+-doped glasses (Tanabe, Setsuhisa; Hirao, Kazuyuki; Soga, Naohiro; Hanada, Teiichi)V1513,340-348(1991)

UV spectroscopy of optical fibers and preforms (Williams, Doug L.; Davey, Steven T.; Kashyap, Raman; Armitage, J. R.; Ainslie, B. J.)V1516,29-37(1991)

Vapor phase deposition of transition metal fluoride glasses (Boulard, B.; Jacoboni, Charles)V1513,204-208(1991)

Verdet constant and its dispersion in optical glasses (Westenberger, Gerhard; Hoffmann, Hans J.; Jochs, Werner W.; Przybilla, Gudrun)V1535,113-120(1991)

WDS measurement of thallium-diffused glass waveguides and numerical simulation: a comparison (Bourhis, Jean-Francois; Guerin, Philippe; Teral, Stephane; Haux, Denys; Di Maggio, Michel)V1583,383-388(1991)

XPS, IR, and Mossbauer studies of lithium phosphate glasses containing iron oxides (Wang, Guomei; Lei, Jiaheng; Yun, Huaishun; Guo, Liping; Jin, Baohui)V1590,229-236(1991)

Gradient index optics—see also fiber optics; refractive index

Actual light deflections in regions of crack tips and their influence on measurements in photomechanics (Hecker, Friedrich W.; Pindera, Jerzy T.; Wen, Baicheng)V1554A,151-162(1991)

Analysis of the refractive-index profile in ion-exchanged waveguides (Righini, Giancarlo C.; Pelli, Stefano; Saracini, R.; Battaglin, Giancarlo; Scaglione, Antonio)V1513,418-424(1991)

Crack-tip deformation field measurements using coherent gradient sensing (Tippur, Hareesh V.; Krishnaswamy, Sridhar; Rosakis, Ares J.)V1554A,176-191(1991)

Design of an anamorphic gradient-index lens to correct astigmatism in diode lasers (Acosta, E.; Gomez-Reino, Carlos; Gonzalez, R. M.)V1401,82-85(1991)

Dual-core ion-exchanged glass waveguides (Li, Ming-Jun; Honkanen, Seppo; Wang, Wei-Jian; Leonelli, Richard; Najafi, S. I.)V1513,410-417(1991)

Experimental characterization of microlenses for WDM transmission systems (Zoboli, Maurizio; Bassi, Paolo)V1506,160-169(1991)

Fizeau-type of gradient-index rod lens interferometer by using semiconductor laser (Ming, Hai; Sun, Yuesheng; Ren, Baorui; Xie, Jianping; Nakajima, Toshinori)V1572,27-31(1991)

Graded-index polymer optical fiber by new random copolymerization technique (Koike, Yasuhiro; Hondo, Yukie; Nihei, Eisuke)V1592,62-72(1991)

Gradient-index fiber-optic preforms by a sol-gel method (Banash, Mark A.; Caldwell, J. B.; Che, Tessie M.; Mininni, Robert M.; Soskey, Paul R.; Warden, Victor N.; Pope, Edward J.)V1590,8-13(1991)

GRIN fiber lens connectors (Gomez-Reino, Carlos; Linares, Jesus)V1332,468-473(1991)

Heat monitoring by fiber-optic microswitches (Mencaglia, Andrea; Brenci, Massimo; Falciai, Riccardo; Guzzi, D.; Pascucci, Tania)V1506,140-144(1991)

High-power laser/optical-fiber-coupling device (Falciai, Riccardo; Pascucci, Tania)V1506,120-125(1991)

Ion-assisted deposition of graded index silicon oxynitride coatings (Al-Jumaily, Ghanim A.; Gagliardi, F. J.; McColl, P.; Mizerka, Larry J.)V1441,360-365(1991)

Novel modes in a-power GRIN (Ojeda-Castaneda, Jorge; Szwaykowski, P.)V1500,246-251(1991)

Optical design with Wood lenses (Caldwell, J. B.)V1354,593-598(1991)

Photochemical generation of gradient indices in glass (Mendoza, Edgar A.; Gafney, Harry D.; Morse, David L.)V1378,139-144(1991)

Restrictions on the geometry of the reference wave in holocoupler devices (Calvo, Maria L.; De Pedraza-Velasco, L.)V1507,288-301(1991)

Study of different optical fibers (Boudot, C.; Vastra, I.)V1502,72-82(1991)

Systematic approach to axial gradient lens design (Wang, David Y.; Moore, Duncan T.)V1354,599-605(1991)

Thermal aberration analysis of a laser-diode-pumped Nd:YAG laser (Kubota, Shigeo; Oka, Michio; Kaneda, Yushi; Masuda, Hisashi)V1354,572-580(1991)

Zoom lens design using GRIN materials (Tsuchida, Hirofumi; Aoki, Norihiko; Hyakumura, Kazushi; Yamamoto, Kimiaki)V1354,246-251(1991)

Gratings—see also diffraction

Analysis and design of binary gratings for broadband, infrared, low-reflectivity surfaces (Moharam, M. G.)V1485,254-259(1991)

Analysis of images of periodic structures obtained by Photon Scanning Tunneling Microscopy (Goudonnet, Jean-Pierre; Salomon, L.; de Fornel, F.; Adam, P.; Bourillot, E.; Neviere, Michel; Guerin, Philippe)V1545,130-139(1991)

Analysis of moire deflectometry by wave optics (Wang, Hai-Lin; Miao, Peng-Cheng; Yan, Da-Peng; He, Anzhi)V1545,268-273(1991)

Analysis of polarization properties of shallow metallic gratings by an extended Rayleigh-Fano theory (Koike, Masato; Namioka, Takeshi)V1545,88-94(1991)

Analysis of the displacements of cylindrical shells by moire techniques (Laermann, Karl-Hans)V1554B,248-256(1991)

Anomaly reduction in gratings (Hoose, John F.; Loewen, Erwin G.)V1545,189-199(1991)

Apodized outcouplers for unstable resonators (Budzinski, Christel; Grunwald, Ruediger; Pinz, Ingo; Schaefer, Dieter; Schoennagel, Horst)V1500,264-274(1991)

Backward diffraction of the light by the phase-transmission holographic grating (Markov, Vladimir B.; Shishkov, Vladimir F.)V1238,41-43(1991)

Binary optic interconnects: design, fabrication and limits on implementation (Wong, Vincent V.; Swanson, Gary J.)V1544,123-133(1991)

Binary optics in the '90s (Gallagher, Neal C.)V1396,722-733(1991)

Blazing of transmission gratings for astronomical use (Neviere, Michel)V1545,11-18(1991)

Bragg diffraction with multiple internal reflections (Markov, Vladimir B.; Shishkov, Vladimir F.)V1238,30-40(1991)

Bragg grating formation and germanosilicate fiber photosensitivity (Meltz, Gerald; Morey, William W.)V1516,185-199(1991)

Broadband multilayer coated blazed grating for x-ray wavelengths below 0.6 nm (den Boggende, Antonius J.; Bruijn, Marcel P.; Verhoeven, Jan; Zeijlemaker, H.; Puik, Eric J.; Padmore, Howard A.)V1345,189-197(1991)

Characteristic wave theory for volume holographic gratings with arbitrarily slanted fringes (Chernov, Boris C.)V1507,302-309(1991)

Chromeless phase-shifted masks: a new approach to phase-shifting masks (Toh, Kenny K.; Dao, Giang T.; Singh, Rajeev; Gaw, Henry T.)V1496,27-53(1991)

Comparison of rigorous and approximate methods of analyzing holographic gratings diffraction (Chernov, Boris C.)V1238,44-53(1991)

Computer-generated diffraction gratings in optical region (Silvennoinen, Raimo; Hamalainen, Rauno M.; Rasanen, Jari)V1574,84-88(1991)

Contouring using gratings created on a LCD panel (Asundi, Anand K.; Wong, C. M.)V1400,80-85(1991)

Coordinate measuring system for 2-D scanners (Bukatin, Vladimir V.)V1454,283-288(1991)

Crack-tip deformation field measurements using coherent gradient sensing (Tippur, Hareesh V.; Krishnaswamy, Sridhar; Rosakis, Ares J.)V1554A,176-191(1991)

Design and analysis of the reflection grating arrays for the X-Ray Multi-Mirror Mission (Atkinson, Dennis P.; Bixler, Jay V.; Geraghty, Paul; Hailey, Charles J.; Klingmann, Jeffrey L.; Montesanti, Richard C.; Kahn, Steven M.; Paerels, F. B.)V1343,530-541(1991)

Design of multiple-beam gratings for far-IR applications (Kinrot, O.; Friesem, Asher A.)V1442,106-108(1991)

Design of spherical varied line-space gratings for a high-resolution EUV spectrometer (Harada, Tatsuo; Kita, Toshiaki; Bowyer, C. S.; Hurwitz, Mark)V1545,2-10(1991)

Designs for two-dimensional nonseparable array generators (Mait, Joseph N.)V1555,43-52(1991)

Deviated-plane varied-line-space grating spectrograph (Edelstein, Jerry)V1545,145-148(1991)

Diffraction by one-dimensional or two-dimensional periodic arrays of conducting plates (Petit, Roger; Bouchitte, G.; Tayeb, Gerard; Zolla, F.)V1545,31-41(1991)

Diffractive properties of surface-relief microstructures (Qu, Dong-Ning; Burge, Ronald E.; Yuan, X.)V1506,152-159(1991)

Dislocation mechanism of periodical relief formation in laser-irradiated silicon (Shandybina, Galina)V1440,384-389(1991)

Displacement measurement of track on magnetic tape by Fourier transform grid method (Sogabe, Yasushi; Morimoto, Yoshiharu; Murata, Shigeki)V1554B,289-297(1991)

Displacement measurement using grating images detected by CCD image sensor (Hane, Kazuhiro; Grover, Chandra P.)V1332,584-590(1991)

Double dispersion from dichromated gelatin volume transmission gratings (Sheat, Dennis E.; Chamberlin, Giles R.; McCartney, David J.)V1461,35-38(1991)

Double-resonant tunneling via surface plasmons in layered gratings (Sreseli, Olga M.; Belyakov, Ludvig V.; Goryachev, D. N.; Rumyantsev, B. L.; Yaroshetskii, Ilya D.)V1545,149-158(1991)

Dynamics of phase-grating formation in optical fibers (An, Sunghyuck; Sipe, John E.)V1516,175-184(1991)

Dynamics of self-organized x(2) gratings in optical fibers (Kamal, Avais; Terhune, Robert W.; Weinberger, Doreen A.)V1516,137-153(1991)

Effective holographic grating model to analyze thick holograms (Belendez, A.; Pascual, I.; Fimia-Gil, Antonio)V1507,268-276(1991)

Error transfer function for grating interferometer (Hao, De-Fu)V1545,261-265(1991)

EUV performance of a multilayer-coated high-density toroidal grating (Keski-Kuha, Ritva A.; Thomas, Roger J.; Neupert, Werner M.; Condor, Charles E.; Gum, Jeffrey S.)V1343,566-575(1991)

Evaluation of a new electrical resistance shear-strain gauge using moire interferometry (Ifju, Peter G.)V1554B,420-428(1991)

Fabrication and performance at 1.33 nm of a 0.24-um period multilayer grating (Berrouane, H.; Khan Malek, Chantal; Andre, Jean-Michel; Lesterlin, L.; Ladan, F. R.; Rivoira, R.; Lepetre, Yves; Barchewitz, Robert J.)V1343,428-436(1991)

Fabrication of toroidal and coma-corrected toroidal diffraction gratings from spherical master gratings using elastically deformable substrates: a progress report (Huber, Martin C.; Timothy, J. G.; Morgan, Jeffrey S.; Lemaitre, Gerard R.; Tondello, Giuseppe; Naletto, Giampiero)V1494,472-480(1991)

Focal length measurement using diffraction at a grating (Sirohi, Rajpal S.; Kumar, Harish; Jain, Narinder K.)V1332,50-55(1991)

Focus grating coupler construction optics: theory, design, and tolerancing (Moore, Kenneth E.; Lawrence, George N.)V1354,581-587(1991)

Frequency modulation of printed gratings as a protection against copying (Spannenburg, S.)V1509,88-104(1991)

Fringe-scanning moire system using a servo-controlled grating (Kurokawa, Haruhisa; Ichikawa, Naoki; Yajima, Nobuyuki)V1332,643-654(1991)

Generalized Talbot effect (Wang, Hai-Lin; Miao, Peng-Cheng; Yan, Da-Peng; Ni, Xiao W.; He, Anzhi)V1545,274-277(1991)

Geometrical moire method for displaying directly the stress field of a circular disk (Wang, Ji-Zhong; Wang, Yun-Shan; Yin, Yuan-Cheng; Wu, Rui-Lan)V1554B,188-192(1991)

Grating beam splitting polarizer using multilayer resist method (Aoyama, Shigeru; Yamashita, Tsukasa)V1545,241-250(1991)

Grating efficiency theory versus experimental data in extreme situations (Neviere, Michel; den Boggende, Antonius J.; Padmore, Howard A.; Hollis, K.)V1545,76-87(1991)

Grating line shape characterization using scatterometry (Bishop, Kenneth P.; Gaspar, Susan M.; Milner, Lisa-Michelle; Naqvi, H. S.; McNeil, John R.)V1545,64-73(1991)

High-average-power narrow-band KrF excimer laser (Wakabayashi, Osamu; Kowaka, Masahiko; Kobayashi, Yukio)V1463,617-628(1991)

Holographic crossed gratings: their nature and applications (Miler, Miroslav; Aubrecht, I.)V1574,22-33(1991)

Holographic diffraction gratings on the base of chalcogenide semiconductors (Indutnyi, I. Z.; Robur, I.; Romanenko, Peter F.; Stronski, Alexander V.)V1555,243-253(1991)

Holographic disk scanner for active infrared sensors (Kawauchi, Yoshikazu; Toyoda, Ryuuichi; Kimura, Minoru; Kawata, Koichi)V1555,224-227(1991)

Instrumentation concepts for multiplexed Bragg grating sensors (Giesler, Leslie E.; Dunphy, James R.; Morey, William W.; Meltz, Gerald; Glenn, William H.)V1480,138-142(1991)

Integral equation method for biperiodic diffraction structures (Dobson, David C.; Cox, James A.)V1545,106-113(1991)

Interference grating with an arbitrary transparent-to-opaque groove ratio (Tavassoly, M. T.)V1527,392-399(1991)

Interference phenomenon with correlated masks and its application (Grover, Chandra P.; Hane, Kazuhiro)V1332,624-631(1991)

Interlaminar deformations on the cylindrical surface of a hole in laminated composites by moire interferometry (Boeman, Raymond G.)V1554B,323-330(1991)

Intl Colloquium on Diffractive Optical Elements (Nowak, Jerzy; Zajac, Marek, eds.)V1574(1991)

Intl Conf on the Application and Theory of Periodic Structures (Lerner, Jeremy M.; McKinney, Wayne R., eds.)V1545(1991)

Inverse-grating diffraction problems in the coupled-wave analysis (Kuittinen, Markku; Jaaskelainen, Timo)V1507,258-267(1991)

Inverse grating diffraction problems (Jaaskelainen, Timo; Kuittinen, Markku)V1574,272-281(1991)

Laser-induced diffusion and second-harmonic generation in glasses (Aksenov, V. P.)V1440,377-383(1991)

Laser-induced etched grating on InP for integrated optical circuit elements (Grebel, Haim; Pien, P.)V1583,331-337(1991)

Lau imaging (Cartwright, Steven L.)V1396,481-487(1991)

Light-induced volume-phase holograms for cold neutrons (Ibel, Konrad; Matull, Ralph; Rupp, Romano A.; Eschkoetter, Peter; Hehmann, Joerg)V1559,393-402(1991)

Matching of Bragg condition of holographic phase gratings in 1.3-1.5um region (Silvennoinen, Raimo; Hamalainen, Rauno M.)V1574,261-265(1991)

Microdisplacement positioning system for a diffraction grating ruling engine (Yang, Hou-Min; Wang, Xiaolin; Zhang, Yinxian)V1533,185-192(1991)

Micrograting device using electron-beam lithography (Yamashita, Tsukasa; Aoyama, Shigeru)V1507,81-93(1991)

Modal analysis of grating-induced optical bistability (Vitrant, Guy; Vincent, Patrick; Reinisch, Raymond; Neviere, Michel)V1545,225-231(1991)

Moire sensor as an automatic feeler gauge (Konwerska-Hrabowska, Joanna; Kryszczynski, Tadeusz; Smolka, M.; Olbrysz, P.; Widomski, L.)V1554B,225-232(1991)

Multifunction grating for signal detection of optical disk (Gupta, Mool C.; Peng, Song-Tsuen)V1499,303-306(1991)

New iterative algorithm for the design of phase-only gratings (Farn, Michael W.)V1555,34-42(1991)

New technique for rapid making specimen: gratings for moire interferometry (Jia, Youquan; Kang, Yilan; Du, Ji)V1554B,331-334(1991)

Noncontact technique for the measurement of linear displacement using chirped diffraction gratings (Spillman, William B.; Fuhr, Peter L.)V1332,591-601(1991)

Nonlinear optical processing using phase grating (Sadovnik, Lev S.; Demichovskaya, Olga; Chen, Ray T.)V1545,200-208(1991)

Novel grating methods for optical inspection (Asundi, Anand K.)V1554B,708-715(1991)

Novel method to fabricate corrugation for distributed-feedback lasers using a grating photomask (Okai, Makoto O.; Harada, Tatsuo)V1545,218-224(1991)

Optical data storage in photosensitive fibers (Campbell, Robert J.; Kashyap, Raman)V1499,160-164(1991)

Optimization of grating depth and layer thicknesses for DFB lasers (Jiang, Ching-Long; Agarwal, Rajiv; Kuwamoto, Hide; Huang, Rong-Ting; Appelbaum, Ami; Renner, Daniel; Su, Chin B.)V1418,261-271(1991)

Optimum design of phase gratings for diffractive optical elements obtained by thin-film deposition (Beretta, Stefano; Cairoli, Massimo; Viardi, Marzia)V1544,2-9(1991)

Peculiarities of the formation of periodic structures on silicon under millisecond laser radiation (Kokin, A. N.; Libenson, Michail N.; Minaev, Sergei M.)V1440,338-341(1991)

Performance of a variable-line-spaced master reflection grating for use in the reflection grating spectrometer on the x-ray multimirror mission (Bixler, Jay V.; Hailey, Charles J.; Mauche, C. W.; Teague, Peter F.; Thoe, Robert S.; Kahn, Steven M.; Paerels, F. B.)V1549,420-428(1991)

Phase conjugation as a probe for noncentrosymmetry grating formation in organics (Charra, Fabrice; Nunzi, Jean-Michel; Messier, Jean)V1516,211-219(1991)

Phase synthesis of elongated holographic diffraction gratings (Turukhano, Boris G.; Gorelik, Vladimir S.; Turukhano, Nikulina)V1500,290-292(1991)

Photoemission from periodic structure of graded superlattices under magnetic field (Ghatak, Kamakhya P.)V1545,282-293(1991)

Pixelgram: an application of electron-beam lithography for the security printing industry (Lee, Robert A.)V1509,48-54(1991)

Polarization conversion through the excitation of electromagnetic modes on a grating (Bryan-Brown, G. P.; Elston, S. J.; Sambles, J. R.)V1545,167-178(1991)

Process-dependent kinoform performance (Cox, James A.; Fritz, Bernard S.; Werner, Thomas R.)V1507,100-109(1991)

Production of gratings with desired pitch using the Talbot effect (Li, Wen L.; Zhong, An)V1554B,275-280(1991)

Projection moire using PSALM (Asundi, Anand K.)V1554B,257-265(1991)

Reflection spectrum of multiple chirped gratings (Shellan, Jeffrey B.; Yeh, Pochi A.)V1545,179-188(1991)

Relative merits of bulk and surface relief diffracting components (Hutley, M. C.)V1574,294-302(1991)

Restrictions on the geometry of the reference wave in holocoupler devices (Calvo, Maria L.; De Pedraza-Velasco, L.)V1507,288-301(1991)

Rigorous diffraction theory of binary optical interconnects (Vasara, Antti H.; Noponen, Eero; Turunen, Jari P.; Miller, J. M.; Taghizadeh, Mohammad R.; Tuovinen, Jussi)V1507,224-238(1991)

Rigorous electromagnetic modeling of diffractive optical elements (Johnson, Eric G.; Kathman, Alan D.)V1545,209-216(1991)

Scatter properties of gratings at ultraviolet and visible wavelengths (Hoose, John F.; Olson, Jeffrey)V1545,160-166(1991)

Self-organization of spontaneous structures in photosensitive layers (Miloslavsky, V. K.; Ageev, L. A.; Lymar, V. I.)V1440,90-96(1991)

Shadow method of scale with a-posteriori increase of measurements sensitivity (Spornik, Nikolai M.; Serenko, M. Y.)V1508,162-169(1991)

Shared aperture for two beams of different wavelength using the Talbot effect (Hector, Scott D.; Swanson, Gary J.)V1555,200-213(1991)

Spectrometer on a chip: InP-based integrated grating spectrograph for wavelength-multiplexed optical processing (Soole, Julian B.; Scherer, Axel; LeBlanc, Herve P.; Andreadakis, Nicholas C.; Bhat, Rajaram; Koza, M. A.)V1474,268-276(1991)

Strain sensing using a fiber-optic Bragg grating (Melle, Serge M.; Liu, Kexing; Measures, Raymond M.)V1588,255-263(1991)

Submicrometer lithographic alignment and overlay strategies (Zaidi, Saleem H.; Naqvi, H. S.; Brueck, Steven R.)V1343,245-255(1991)

Synthesis method applied to the problem of diffraction by gratings: the method of fictitious sources (Tayeb, Gerard; Petit, Roger; Cadilhac, M.)V1545,95-105(1991)

Temperature stability of Bragg grating resonant frequency (Bradley, Eric M.; Rybka, Theodore W.; Yu, Paul K.)V1418,272-278(1991)

Three-dimensional contour of crack tips using a grating method (Andresen, Klaus; Kamp, B.; Ritter, Reinold)V1554A,93-100(1991)

Three-dimensional shape restoration using virtual grating phase detection from deformed grating (Zhou, Shaoxiang; Jiang, Jinyou; Wang, Qimin)V1358,788-792(1991)

Tunability of cascaded grating that is used in distributed feedback laser (Rahnavard, Mohammad H.; Moheimany, O. R.; Abiri-Jahromi, H.)V1367,374-381(1991)

Two-wave coupled-wave theory of the polarizing properties of volume phase gratings (Tholl, Hans D.)V1555,101-111(1991)

Unsymmetrical spectrum of reflective hologram grating (Dahe, Liu; Liang, Zhujian; Tang, Weiguo)V1507,310-315(1991)

Upper bound on the diffraction efficiency of phase-only array generators (Mait, Joseph N.)V1555,53-62(1991)

Use of rigorous vector coupled-wave theory for designing and tolerancing surface-relief diffractive components for magneto-optical heads (Haggans, Charles W.; Kostuk, Raymond K.)V1499,293-302(1991)

V-groove diffraction grating for use in an FUV spatial heterodyne interferometer (Cotton, Daniel M.; Bach, Bernhard W.; Bush, Brett C.; Chakrabarti, Supriya)V1549,313-318(1991)

Wavelength shifting and bandwidth broadening in DCG (Corlatan, Dorina; Schaefer, Martin; Anders, Gerhard)V1507,354-364(1991)

Whole field displacement and strain rosettes by grating objective speckle method (Tu, Meirong; Gielisse, Peter J.; Xu, Wei)V1554A,593-601(1991)

XMM space telescope: development plan for the lightweight replicated x-ray gratings (Montesanti, Richard C.; Atkinson, Dennis P.; Edwards, David F.; Klingmann, Jeffrey L.)V1343,558-565(1991)

X-ray multilayer-coated reflection gratings: theory and applications (Neviere, Michel; den Boggende, Antonius J.)V1545,116-124(1991)

Guided waves—see also fiber optics; integrated optics; waveguides

Analysis of thin-film losses from guided wave attenuations with photothermal deflection technique (Liu, Xu; Tang, Jinfa; Pelletier, Emile P.)V1554A,558-569(1991)

Degenerate four-wave mixing using wave pump beams near the critical angle: two distinct behaviors (Malouin, Christian; Song, Li; Thibault, S.; Denariez-Roberge, Marguerite M.; Lessard, Roger A.)V1559,385-392(1991)

Design optimization of a 10-amplifier coherent array (Zediker, Mark S.; Foresi, James S.; Haake, John M.; Heidel, Jeffrey R.; Williams, Richard A.; Driemeyer, D.; Blackwell, Richard J.; Thomas, G.; Priest, J. A.; Herrmann, Sandy)V1418,309-315(1991)

Experimental verification of grating theory for surface-emitting structures (Ayekavadi, Raj; Yeh, C. S.; Butler, Jerome K.; Evans, Gary A.; Stabile, Paul J.; Rosen, Arye)V1418,74-85(1991)

Fiber optic evanescent wave biosensor (Duveneck, G.; Ehrat, M.; Widmer, H. M.)V1510,138-145(1991)

Fundamental array mode operation of semiconductor laser arrays using external spatial filtering (Cherng, Chung-Pin; Osinski, Marek A.)V1418,372-385(1991)

Guided-wave magneto-optic and acousto-optic Bragg cells for RF signal processing (Tsai, Chen S.)V1562,55-65(1991)

Guided-wave nonlinear optics in 2-docosylamino-5-nitropyridine Langmuir-Blodgett films (Bosshard, Christian; Kuepfer, Manfred; Floersheimer, M.; Guenter, Peter)V1560,344-352(1991)

Hybride fiber-optic temperature sensors on the base of LiNbO3 and LiNbO3:Ti waveguides (Goering, Rolf)V1511,275-280(1991)

Linear and nonlinear optical effects in polymer waveguides (Braeuer, Andreas; Bartuch, Ulrike; Zeisberger, M.; Bauer, T.; Dannberg, Peter)V1559,470-478(1991)

LiTaO3 and LiNbO3:Ti responses to ionizing radiation (Padden, Richard J.; Taylor, Edward W.; Sanchez, Anthony D.; Berry, J. N.; Chapman, S. P.; DeWalt, Steve A.; Wong, Ka-Kha)V1474,148-159(1991)

Localized waves in complex environments (Ziolkowski, Richard W.; Besieris, Ioannis M.; Shaarawi, Amr M.)V1407,387-397(1991)

Novel technique for the measurement of coupling length in directional couplers (Cheng, Hsing C.; Ramaswamy, Ramu V.)V1474,291-297(1991)

Radiation-induced crosstalk in guided-wave devices (Taylor, Edward W.; Padden, Richard J.; Sanchez, Anthony D.; Chapman, S. P.; Berry, J. N.; DeWalt, Steve A.)V1474,126-131(1991)

Temperature stability of Bragg grating resonant frequency (Bradley, Eric M.; Rybka, Theodore W.; Yu, Paul K.)V1418,272-278(1991)

Gyroscopes—see also fiber optic sensors; navigation

1.55-um superluminescent diode for a fiber optic gyroscope (Kashima, Yasumasa; Matoba, Akio; Kobayashi, Masao; Takano, Hiroshi)V1365,102-107(1991)

All-fiber closed-loop gyroscope with self-calibration (Ecke, Wolfgang; Schroeter, Siegmund; Schwotzer, Guenter; Willsch, Reinhardt)V1511,57-66(1991)

Closed-loop fiber-optic gyroscope (Fang, Zhen-he; Huang, Shao-ming; Shen, Yu-qing)V1572,342-346(1991)

Depolarized fiber optic gyro for future tactical applications (Bramson, Michael D.)V1367,155-160(1991)

Deterministic and noise-induced phase jumps in the ring laser gyroscope (Chyba, Thomas H.)V1376,132-142(1991)

Excess noise in fiber gyroscope sources (Burns, William K.; Moeller, Robert P.; Dandridge, Anthony D.)V1367,87-92(1991)

Experimental developments in the RFOG (Kaiser, Todd J.; Cardarelli, Donato; Walsh, Joseph)V1367,121-126(1991)

Extended environmental performance of attitude and heading reference grade fiber optic rotation sensors (Chin, Gene H.; Cordova, Amado; Goldner, Eric L.)V1367,107-120(1991)

Fiber Optic and Laser Sensors VIII (DePaula, Ramon P.; Udd, Eric, eds.)V1367(1991)

High-dynamic-range fiber gyro with all-digital signal processing (Lefevre, Herve C.; Martin, Philippe; Morisse, J.; Simonpietri, Pascal; Vivenot, P.; Arditty, Herve J.)V1367,72-80(1991)

Improved ring laser gyro navigator (Hadfield, Michael J.; Stiles, Tom; Seidel, David; Miller, William G.; Hensley, David; Wisotsky, Steve; Foote, Michael; Gregory, Rob)V1478,126-144(1991)

Integrated optical device in a fiber gyroscope (Rasch, Andreas; Goering, Rolf; Karthe, Wolfgang; Schroeter, Siegmund; Ecke, Wolfgang; Schwotzer, Guenter; Willsch, Reinhardt)V1511,149-154(1991)

Integrated optics for fiber optic sensors (Minford, William J.; DePaula, Ramon P.)V1367,46-52(1991)

Integrated optics in optical engineering (Popov, Yury V.)V1399,207-213(1991)

Interferometric fiber optic gyroscopes for today's market (LaViolette, Kerry D.; Bossler, Franklin B.)V1398,213-218(1991)

Low-drift fiber-optic gyro for earth-rate applications (Dyott, Richard B.; Huang, Yung Y.; Jannush, D. A.; Morrison, Steve A.)V1482,439-443(1991)

Low-loss Y-couplers for fiber optic gyro applications (Page, Jerry L.)V1374,287-293(1991)

Magnetic field sensitivity of depolarized fiber optic gyros (Blake, James N.)V1367,81-86(1991)

Multimode fiber optic rotation sensor with low-cost digital signal processing (Johnson, Dean R.; Fredricks, Ronald J.; Vuong, S. C.; Dembinski, David T.; Sabri, Sehbaz H.)V1367,140-154(1991)

Multiplexed approach for the fiber optic gyro inertial measurement unit (Page, Jerry L.)V1367,93-102(1991)

Novel analog phase tracker for interferometric fiber optic sensor applications (Berkoff, Timothy A.; Kersey, Alan D.; Moeller, Robert P.)V1367,53-58(1991)

Outlook of fiber-optic gyroscope (Hayakawa, Yoshiaki; Kurokawa, Akihiro)V1572,353-358(1991)

Performance comparison of various low-cost multimode fiber optic rotation rate sensor designs (Fredricks, Ronald J.; Johnson, Dean R.; Sabri, Sehbaz H.; Yu, Ming H.)V1367,127-139(1991)

Proton-exchange X-cut lithium tantalate fiber optic gyro chips (Wong, Ka-Kha; Killian, Kevin M.; Dimitrov-Kuhl, K. P.; Long, Margaret; Fleming, J. T.; van de Vaart, Herman)V1374,278-286(1991)

Recent developments in fiber optic ring laser gyros (Smith, S. P.; Zarinetchi, F.; Ezekiel, Shaoul)V1367,103-106(1991)

Reduced cost coil windings for interferometric fiber-optic gyro sensors (Smith, Ronald H.)V1478,145-149(1991)

Research on the characteristics of temperature drift in fiber-optic gyroscope (Luo, Gei-peng; Xu, Sen-lu)V1572,337-341(1991)

Heterodyning

10.6 um laser frequency stabilization system with two optical circuits (Hu, Yu; Li, Xian; Zhu, Dayong; Ye, Naiqun; Pen, Shengyang)V1409,230-239(1991)

Cellular vibration measurement with a noninvasive optical system (Khanna, Shyam M.)V1429,9-20(1991)

Coherent laser radar for target classification (Kranz, Wolfgang)V1479,270-274(1991)

Common-path optical fiber heterodyne interferometric current sensor (Bartlett, Steven C.; Farahi, Faramarz; Jackson, David A.)V1504,247-250(1991)

Fiber coupled heterodyne interferometric displacement sensor (Nerheim, Noble M.)V1542,523-533(1991)

High-speed polarimetric measurements for fiber-optic communications (Calvani, Riccardo A.; Caponi, Renato; Piglia, Roberto)V1504,258-263(1991)

Homodyne PSK receivers with laser diode sources (Mecherle, G. S.; Henderson, Robert J.)V1417,99-107(1991)

Instrument design and test results of the new all-reflection spatial heterodyne spectrometer (Bush, Brett C.; Cotton, Daniel M.; Vickers, James S.; Chakrabarti, Supriya)V1549,376-384(1991)

Low-data-rate coherent optical link demonstration using frequency-stabilized solid-state lasers (Chen, Chien-Chung; Win, Moe Z.; Marshall, William K.; Lesh, James R.)V1417,170-181(1991)

Measurement of SEW phase velocity by optical heterodyning method (Libenson, Michail N.; Makin, Vladimir S.; Trubaev, Vladimir V.)V1440,354-356(1991)

Measurement of the axial eye length and retinal thickness by laser Doppler interferometry (Hitzenberger, Christoph K.; Fercher, Adolf F.; Juchem, M.)V1423,46-50(1991); V1429,21-25(1991)

Millimeter-wave signal generation and control using optical heterodyne techniques and electro-optic devices (Thaniyavarn, Suwat; Abbas, Gregory L.; Dougherty, William A.)V1371,250-251(1991)

Novel integrated acousto-optical LiNbO3 device for application in single-laser self-heterodyne systems (Gieschen, Nikolaus; Rocks, Manfred; Olivier, Lutz)V1579,237-248(1991)

Optical computer-assisted tomography realized by coherent detection imaging incorporating laser heterodyne method for biomedical applications (Inaba, Humio; Toida, Masahiro; Ichimura, Tsutomu)V1399,108-115(1991)

Optical profilometry using spatial heterodyne interferometric methods (Frankowski, Gottfried)V1500,114-123(1991)

Optical tomography by heterodyne holographic interferometry (Vukicevic, Dalibor; Neger, Theo; Jaeger, Helmut; Woisetschlaeger, Jakob; Philipp, Harald)VIS08,160-193(1991)

Pulp blood flow assessment in human teeth by laser Doppler flowmetry (Pettersson, Hans; Oberg, Per A.)V1424,116-119(1991)

Quantum-limited quasiparticle mixers at 100 GHz (Mears, Carl A.; Hu, Qing; Richards, Paul L.; Worsham, A.; Prober, Daniel E.; Raisanen, Antti)V1477,221-233(1991)

Realization of heterodyne acquisition and tracking with diode lasers at lambda=1.55 um (Hueber, Martin F.; Leeb, Walter R.; Scholtz, Arpad L.)V1522,259-267(1991)

Solid-state lidar measurements at 1 and 2 um (Killinger, Dennis K.; Chan, Kin-Pui)V1416,125-128(1991)

V-groove diffraction grating for use in an FUV spatial heterodyne interferometer (Cotton, Daniel M.; Bach, Bernhard W.; Bush, Brett C.; Chakrabarti, Supriya)V1549,313-318(1991)

High-speed photography—see photography, high-speed

History

Appreciation of the Society of Motion Picture and Television Engineers: the originators of the Congress (Lunn, George H.)V1358,1161-1163(1991)

PACS: "Back to the Future" (Carey, Bruce; Seshadri, Sridhar B.; Arenson, Ronald L.)V1446,414-419(1991)

Holographic optical elements—see also binary optics; holography, computer-generated; lenses; micro-optics

Aberrations of holographic lens recorded on surface of revolution with shifted pupil (Masajada, Jan)V1574,188-196(1991)

About contradictions among different criteria for evaluation of image (interference) elements width changes (Wolczak, Bohdan K.)V1391,318-328(1991)

Analysis of the chromatic aberrations of imaging holographic optical elements (Tholl, Hans D.; Luebbers, Hubertus; Stojanoff, Christo G.)V1456,262-273(1991)

Analytical design of curved holographic optical elements for Fourier transform (Talatinian, A.; Pluta, Mieczyskaw)V1574,205-217(1991)

Aplanatic holographic systems (Fiala, Pavel; Jerie, Tomas)V1574,179-187(1991)

Apo-tele lenses with kinoform elements (Gan, Michael A.; Potyemin, Igor S.; Poszinskaja, Irina I.)V1507,116-125(1991)

Binary optics from a ray-tracing point of view (Southwell, William H.)V1354,38-42(1991)

Binary optics in lens design (Kathman, Alan D.; Pitalo, Stephen K.)V1354,297-309(1991)

Blazed zone plates for the 10-um spectral region (Hutley, M. C.)V1574,2-7(1991)

Calculation of diffraction efficiencies for spherical and cylindrical holographic lenses (Defosse, Yves; Renotte, Yvon L.; Lion, Yves F.)V1507,277-287(1991)

Calculation of wave aberration in optical systems with holographic optical elements (Yuan, Yanrong)V1354,43-52(1991)

Combat vehicle stereo HMD (Rallison, Richard D.; Schicker, Scott R.)V1456,179-190(1991)

Comparison of dichromated gelatin and Du Pont HRF-700 photopolymer as media for holographic notch filters (Salter, Jeffery L.; Loeffler, Mary F.)V1555,268-278(1991)

Computer and Optically Generated Holographic Optics; 4th in a Series (Cindrich, Ivan; Lee, Sing H., eds.)V1555(1991)

Computer-generated holograms fabricated on photopolymer for optical interconnect applications (Oren, Moshe)V1389,527-534(1991)

Corrections of aberrations using HOEs in UV and visible imaging systems (Yang, Zishao; Rosenbruch, Klaus J.)V1354,323-327(1991)

Design, fabrication, and integration of holographic dispersive solar concentrator for terrestrial applications (Stojanoff, Christo G.; Kubitzek, Ruediger; Tropartz, Stephan; Froehlich, K.; Brasseur, Olivier)V1536,206-214(1991)

Design and fabrication of large-format holographic lenses (Stojanoff, Christo G.; Kubitzek, Ruediger; Tropartz, Stephan; Brasseur, Olivier; Froehlich, K.)V1527,48-60(1991)

Design of an optimal single-reflective holographic helmet display element (Twardowski, Patrice J.; Meyrueis, Patrick)V1456,164-174(1991)

Design of optical systems with HOE by DEMOS program (Gan, Michael A.; Zhdanov, Dmitriy D.; Novoselskiy, Vadim V.; Ustinov, Sergey I.; Fiodorov, Alexander O.; Potyemin, Igor S.; Bezdidko, Sergey N.)V1574,254-260(1991)

Design of some achromatic imaging hybrid diffractive-refractive lenses (Twardowski, Patrice J.; Meyrueis, Patrick)V1507,55-65(1991)

Design of two- and three-element diffractive telescopes (Buralli, Dale A.; Morris, G. M.)V1354,292-296(1991)

Development, manufacturing, and integration of holographic optical elements for laser Doppler velocimetry applications (Stojanoff, Christo G.; Tholl, Hans D.; Luebbers, Hubertus; Windeln, Wilbert)V1507,426-434(1991)

Diffractive optical elements for optoelectronic interconnections (Streibl, Norbert)V1574,34-47(1991)

Diffractive optics for x rays: the state of the art (Michette, Alan G.)V1574,8-21(1991)

DOE design and manufacture at CSEM (Buczek, Harthmuth; Mayor, J. M.; Regnault, P.)V1574,48-57(1991)

Does a Fourier holographic lens exist? (Blank, R.; Friesem, Asher A.)V1442,26-30(1991)

Double dispersion from dichromated gelatin volume transmission gratings (Sheat, Dennis E.; Chamberlin, Giles R.; McCartney, David J.)V1461,35-38(1991)

Dry photopolymer embossing: novel photoreplication technology for surface relief holographic optical elements (Shvartsman, Felix P.)V1507,383-391(1991)

Effective holographic grating model to analyze thick holograms (Belendez, A.; Pascual, I.; Fimia-Gil, Antonio)V1507,268-276(1991)

Effect of coma correction on the imaging quality of holographic lenses (Zajac, Marek; Nowak, Jerzy)V1507,73-80(1991)

Electron beam lithographic fabrication of computer-generated holograms (West, Andrew A.; Smith, Robin W.)V1507,158-167(1991)

Elements with long-focal depth and high-lateral resolution (Davidson, Nir; Friesem, Asher A.; Hasman, Erez)V1442,22-25(1991)

Etch conditions of photolithographic holograms (Guo, Yongkang; Guo, Lu Rong; Zhang, Xiao-Chun)V1461,97-100(1991)

Fabrication and properties of chalcogenide IR diffractive elements (Ewen, Peter J.; Slinger, Christopher W.; Zakery, A.; Zekak, A.; Owen, A. E.)V1512,101-111(1991)

Fabrication of multiphase optical elements for weighted array spot generation (McKee, Paul; Wood, David; Dames, Mark P.; Dix, C.)V1461,17-23(1991)

Fabrication of phase structures with continuous and multilevel profile for diffraction optics (Poleshchuk, Alexander G.)V1574,89-100(1991)

Fan-out elements by multiple-beam recording in volume holograms (Herzig, Hans-Peter; Ehbets, Peter; Prongue, Damien; Daendliker, Rene)V1507,247-255(1991)

Fast holographic correlator for machine vision systems (Nekrasov, Victor V.)V1507,170-174(1991)

Fourier processing with binary spatial light modulators (Hossack, William J.; Vass, David G.; Underwood, Ian)V1564,697-702(1991)

Fresnel zone plate moire patterns and its metrological applications (Jaroszewicz, Zbigniew)V1574,154-158(1991)

High-efficiency reflective holograms: subangstrom spectral selectors (Goncharov, V. F.; Popov, Alexander P.; Veniaminov, Andrei V.)V1238,97-102(1991)

Holographic characterization of DYE-PVA films studied at 442 nm for optical elements fabrication (Couture, Jean J.)V1559,429-435(1991)

Holographic data storage on the photothermoplastic tape carrier (Akaev, Askar A.; Zhumaliev, K. M.; Jamankyzov, N.)V1621,182-193(1991)

Holographic deflectors for graphic arts applications: an overview (Kramer, Charles J.)V1454,68-100(1991)

Holographic filter for coherent radiation (Shi, Dexiu; Xing, Xiaozheng; Wolbarsht, Myron L.)V1419,40-49(1991)

Holographic microlenses for optical fiber interconnects (Galloway, Peter C.; Dobson, Peter J.)V1365,131-138(1991)

Holographic mirrors laminated into windshields for automotive head-up display and solar protective glazing applications (Beeck, Manfred-Andreas; Frost, Thorsten; Windeln, Wilbert)V1507,394-406(1991)

Holographic notch filter (Owen, Harry)V1555,228-235(1991)

Holographic optical backplane for boards interconnection (Sebillotte-Caron, Claudine)V1389,600-611(1991)

Holographic optical backplane hardware implementation for parallel and distributed processors (Kim, Richard C.; Lin, Freddie S.)V1474,27-38(1991)

Holographic optical elements as laser irradiation sensor components (Leib, Kenneth G.; Pernick, Benjamin J.)V1532,261-270(1991)

Holographic optical elements by dry photopolymer embossing (Shvartsman, Felix P.)V1461,313-320(1991)

Holographic optical elements for free-space clock distribution (Zarschizky, Helmut; Karstensen, Holger; Gerndt, Christian; Klement, Ekkehard; Schneider, Hartmut W.)V1389,484-495(1991)

Holographic optical profilometer (Hasman, Erez; Davidson, Nir; Friesem, Asher A.)V1442,372-377(1991)

Holographic optical switching with photorefractive crystals (Song, Q. W.; Lee, Mowchen C.; Talbot, Peter J.)V1558,143-148(1991)

Holographic optical system to copy holographic optical elements (Pascual, I.; Belendez, A.; Fimia-Gil, Antonio)V1507,373-378(1991)

Holographic Optics III: Principles and Applications (Morris, G. M., ed.)V1507(1991)

Holographic recordings on 2-hydroxyethyl methacrylate and applications of water-immersed holograms (Yacoubian, Araz; Savant, Gajendra D.; Aye, Tin M.)V1559,403-409(1991)

Holographic space-variant prism arrays for free-space data permutation in digital optical networks (Kobolla, Harald; Sauer, Frank; Schwider, Johannes; Streibl, Norbert; Voelkel, Reinhard)V1507,175-182(1991)

Holographic spectral selectors and filters based on phase gratings and planar waveguides (Bazhenov, V. Y.; Burykin, N. M.; Soskin, M. S.; Taranenko, Victor B.; Vasnetsov, M. V.)V1574,148-153(1991)

Holographic testing canal of adaptive optical systems (Gan, Michael A.; Potyemin, Igor S.)V1507,549-560(1991)

Holographic transmission elements using improved photopolymer films (Gambogi, William J.; Gerstadt, William A.; Mackara, Steven R.; Weber, Andrew M.)V1555,256-267(1991)

Intl Colloquium on Diffractive Optical Elements (Nowak, Jerzy; Zajac, Marek, eds.)V1574(1991)

Inverse-grating diffraction problems in the coupled-wave analysis (Kuittinen, Markku; Jaaskelainen, Timo)V1507,258-267(1991)

Investigation of imaging quality of Fourier holographic lens (Zajac, Marek; Nowak, Jerzy)V1574,197-204(1991)

Ion treatment in technology of diffraction optical elements (Shevchenko, N. P.; Megorskaja, K. D.; Reshetnikova, Irina N.)V1574,66-71(1991)

Kinetics of formation of holographic structure of a hologram mirror in dichromated gelatin (Kzuzhilin, Yu E.; Mel'nichenko, Yu. B.; Shilov, V. V.)V1238,200-205(1991)

Kinoforms with increased depth of focus (Koronkevich, Voldemar P.; Palchikova, Irena G.)V1507,110-115(1991)

Light scattering holographic optical elements formation in photopolymerizable layers (Boiko, Yuri B.; Granchak, Vasilij M.)V1507,544-548(1991)

Lippmann volume holographic filters for Rayleigh line rejection in Raman spectroscopy (Rich, Chris C.; Cook, David M.)V1461,2-7(1991)

Liquid-crystal-doped polymers as volume holographic elements (Chen, Geng-Sheng; Brady, David J.)V1562,128-135(1991)

Local microholograms recording on the moving photothermoplastic disk (Kutanov, Askar A.; Akaev, Askar A.; Abdrisaev, Baktybek D.; Snimshikov, Igor A.)V1507,94-98(1991)

Manufacturing and reproduction of holographic optical elements in dichromated gelatin films for operation in the infrared (Stojanoff, Christo G.; Tropartz, Stephan; Brasseur, Olivier; Kubitzek, Ruediger)V1485,274-280(1991)

Measuring for thickness distribution of recording layer of PLH (Zhang, Xiao-Chun; Guo, Lu Rong; Guo, Yongkang)V1461,93-96(1991)

Methylene-blue-sensitized gelatin used for the fabrication of holographic optical elements in the near-infrared (Cappolla, Nadia; Lessard, Roger A.)V1389,612-620(1991)

Moire interferometry of sticking film using white light (Luo, Zhishan; Zhang, Souyi; Tuo, Sueshang)V1554B,339-342(1991)

Monochromatic aberrations of an off-axis hologram with carry out pupil (Mikhailov, I. A.)V1238,123-128(1991)

Nd:YAG laser machining with multilevel resist kinoforms (Ekberg, Mats; Larsson, Michael; Bolle, Aldo; Hard, Sverker)V1527,236-239(1991)

New method of making HOE by copying CGH on NGD (Guo, Lu Rong; Cheng, Xiaoxue; Guo, Yongkang; Hsu, Ping)V1555,300-303(1991)

New method of recording holographic optical elements applied to optical interconnection in VLSI (Zhao, Feng; Sun, Junyong; Geng, Wanzhen; Jiang, Lingzhen; Hong, Jing)V1461,262-264(1991)

New transducer for displacement measurement (Han, Zhao-Jin)V1555,304-308(1991)

Optical interconnect and neural network applications based on a simplified holographic N4 recording technique (Lin, Freddie S.; Lu, Taiwei; Kostrzewski, Andrew A.; Chou, Hung)V1558,406-413(1991)

Optical interconnections based on waveguide holograms (Putilin, Andrei N.)V1621,93-101(1991)

Optimization of reconstruction geometry for maximum diffraction efficiency in HOE: the influence of recording material (Belendez, A.; Pascual, I.; Fimia-Gil, Antonio)V1574,77-83(1991)

Optimum design of phase gratings for diffractive optical elements obtained by thin-film deposition (Beretta, Stefano; Cairoli, Massimo; Viardi, Marzia)V1544,2-9(1991)

Phase-circular hologram as a laser beam splitter (Janicijevic, Lj.; Jonoska, M.)V1574,167-178(1991)

Photochemical and thermal treatment of dichromated gelatin film for the manufacturing of holographic optical elements for operation in the IR (Stojanoff, Christo G.; Schuette, H.; Brasseur, Olivier; Kubitzek, Ruediger; Tropartz, Stephan)V1559,321-330(1991)

Photopolymer elements for an optical correlator system (Brandstetter, Robert W.; Fonneland, Nils J.)V1559,308-320(1991)

Photopolymers for optical devices in the USSR (Mikaelian, Andrei L.; Barachevsky, Valery A.)V1559,246-257(1991)

Planar diffractive optical elements prepared by electron-beam lithography (Urban, Frantisek; Matejka, Frantisek)V1574,58-65(1991)

Polarization holographic elements (Kakichashvili, Shermazan D.)V1574,101-108(1991)

Polarizing holographic optical elements for optical data storage (Ono, Yuzo)V1555,177-181(1991)

Possibility of associative reconstruction of reflection hologram in phase-conjugate interferometer (Lazaruk, A. M.; Rubanov, Alexander S.; Serebryakova, L. M.)V1621,114-124(1991)

Practical design considerations and performance characteristics of high-numerical-aperture holographic lenses (Kostuk, Raymond K.)V1461,24-34(1991)

Practical Holography V (Benton, Stephen A., ed.)V1461(1991)

Process-dependent kinoform performance (Cox, James A.; Fritz, Bernard S.; Werner, Thomas R.)V1507,100-109(1991)

Real-time holographic display: improvements using a multichannel acousto-optic modulator and holographic optical elements (St.-Hilaire, Pierre; Benton, Stephen A.; Lucente, Mark; Underkoffler, John S.; Yoshikawa, Hiroshi)V1461,254-261(1991)

Recent advances in optical pickup head with holographic optical elements (Kato, Makoto; Kadowaki, Shin-ichi; Komma, Yoshiaki; Hori, Yoshikazu)V1507,36-44(1991)

Reflection holographic optical elements in silver-halide-sensitized gelatin (Pascual, I.; Belendez, A.; Fimia-Gil, Antonio)V1574,72-76(1991)

Relative merits of bulk and surface relief diffracting components (Hutley, M. C.)V1574,294-302(1991)

Rigorous electromagnetic modeling of diffractive optical elements (Johnson, Eric G.; Kathman, Alan D.)V1545,209-216(1991)

Selected Papers on Holographic and Diffractive Lenses and Mirrors (Stone, Thomas W.; Thompson, Brian J., eds.)VMS34(1991)

Simple design considerations for binary optical holographic elements (Franck, Jerome B.; Hodgkin, Van A.)V1555,63-70(1991)

Some applications of a HOE-based desensitized interferometer in materials research (Boone, Pierre M.; Jacquot, Pierre M.)V1554A,512-521(1991)

Spectrally nonselective holographic objective (Wardosanidze, Zurab V.)V1574,218-226(1991)

Spectrally nonselective reflection holograms (Kakichashvili, Shermazan D.; Wardosanidze, Zurab V.)V1238,134-137(1991)

Sphero-chromatic aberration correction of single holo-lens used as a spectral device (Dubik, Boguslawa; Zajac, Marek)V1574,227-234(1991)

Study of the structure of reflection holgrams on a dichromate gelatin layer (Meshalkina, M. N.; Smirnov, V. V.)V1238,195-199(1991)

Synthesis of large-aperture interference fields (Turukhano, Boris G.)V1500,305-308(1991)

Three-beam CD optical pickup using a holographic optical element (Yoshida, Yoshio; Miyake, Takahiro; Kurata, Yukio; Ishikawa, Toshio)V1401,58-65(1991)

Ultra-high scanning holographic deflector unit (Kramer, Charles J.; Szalowski, Rafal; Watkins, Mark)V1454,434-446(1991)

Using hybrid refractive-diffractive elements in infrared Petzval objectives (Wood, Andrew P.)V1354,316-322(1991)

Volume holographic optics recording in photopolymerizable layers (Boiko, Yuri B.)V1507,318-327(1991)

Waveguide holography and its applications (Huang, Qiang; Caulfield, H. J.)V1461,303-312(1991)

Wavelength-dispersive and filtering applications of volume holographic optical elements (Jannson, Tomasz; Rich, Chris C.; Sadovnik, Lev S.)V1545,42-63(1991)

Wavelength-sensitive holographic optical interconnects (Lin, Freddie S.; Zaleta, David E.; Jannson, Tomasz)V1474,45-56(1991)

Wavelength shift in DCG holograms (Lelievre, Sylviane; Pawluczyk, Romuald)V1559,288-297(1991)

Zone plate of anisotropic profile (Wardosanidze, Zurab V.)V1574,109-120(1991)

Holography

About investigation of absolute surface relief by holographic method (Rachkovsky, Leonid I.; Tanin, Leonid V.; Rubanov, Alexander S.)V1461,232-240(1991)

Acoustical and some other nonconventional holography (Greguss, Pal)VIS08,387-401(1991)

Advanced hot stamping foil-based OVD technology: an overview about security applications (Reinhart, Werner)V1509,67-72(1991)

Advanced materials characterization by means of moire techniques (Aymerich, Francesco; Ginesu, Francesco; Priolo, Pierluigi)V1554B,304-314(1991)

Aesthetic message of holography (Zec, Peter)V1238,355-364(1991)

Analysis of Bragg diffraction in optical memories and optical correlators (Gheen, Gregory)V1564,135-142(1991)

Analysis of powertrain noise and vibration using interferometry (Griffen, Christopher T.; Fryska, Slawomir T.; Bernier, Paul R.)V1554B,754-766(1991)

Analysis of the Focar-type silver-halide heterogeneous media (Andreeva, Olga V.)V1238,231-234(1991)

Angular sensitivity of holograms with a reference speckle wave (Darsky, Alexei M.; Markov, Vladimir B.)V1238,54-61(1991)

Antihumidity dichromated gelatin holographic recording material (Guo, Lu Rong; Dai, Chao M.; Guo, Yongkang; Cai, Tiequan)V1555,293-296(1991)

Application of holographic interferometry to a three-dimensional flow field (Doerr, Stephen E.)V1554B,544-555(1991)

Application of real-time holographic interferometry in the nondestructive inspection of electronic parts and assemblies (Wood, Craig P.; Trolinger, James D.)V1332,122-131(1991)

Application of spatially resolving holographic interferometry to plasma diagnostics (Neger, Theo; Jaeger, Helmut; Philipp, Harald; Pretzler, Georg; Widmann, Klaus; Woisetschlaeger, Jakob)V1507,476-487(1991)

Application of the edge-lit format to holographic stereograms (Farmer, William J.; Benton, Stephen A.; Klug, Michael A.)V1461,215-226(1991)

Applications of holographic gratings in x-ray mask metrology (Hansen, Matthew E.)V1396,78-79(1991)

Applications of Optical Engineering: Proceedings of OE/Midwest '90 (Dubiel, Mary K.; Eppinger, Hans E.; Gillespie, Richard E.; Guzik, Rudolph P.; Pearson, James E., eds.)V1396(1991)

Applied research of optical fiber pulsed-laser holographic interferometry (Wang, Guozhi; Zhu, Guangkuan)V1554B,119-125(1991)

Approach for applying holographic interferometry to large deformations and modifications (Schumann, Walter)V1507,526-537(1991)

Automated approach to the correlation of defect locations to electrical test results to determine yield reducing defects (Slama, M. M.; Patterson, Angela C.)V1464,602-609(1991)

Automated systems for quantitative analysis of holograms (Pryputniewicz, Ryszard J.)VIS08,215-246(1991)

Automatic heterodyning of fiber-optic speckle pattern interferometry (Valera Robles, Jesus D.; Harvey, David; Jones, Julian D.)V1508,170-179(1991)

Backward diffraction of the light by the phase-transmission holographic grating (Markov, Vladimir B.; Shishkov, Vladimir F.)V1238,41-43(1991)

Behavior of a thin liquid film under thermal stimulation: theory and applications to infrared interferometry (Ledoyen, Fernand; Lewandowski, Jacques; Cormier, Maurice)V1507,328-338(1991)

Binary phase-only filter associative memory (Kane, Jonathan S.; Hemmer, Philip R.; Woods, Charles L.; Khoury, Jehad)V1564,511-520(1991)

Biomedical applications of video-speckle techniques (Lokberg, Ole J.)V1525,9-16(1991)

Boundary diffraction wave in imaging by small holograms (Mulak, Grazyna)V1574,266-271(1991)

Bragg diffraction with multiple internal reflections (Markov, Vladimir B.; Shishkov, Vladimir F.)V1238,30-40(1991)

Bridging the gap between Soviet and Western holography (Phillips, Nicholas J.)VIS08,206-214(1991)

Broad-range holographic contouring of diffuse surfaces by dual-beam illumination: study of two related techniques (Rastogi, Pramod K.; Pflug, Leopold)V1554B,48-55(1991)

CCD holographic phase and intensity measurement of laser wavefront (Wickham, Michael; Munch, Jesper)V1414,80-90(1991)

Characteristic curves and phase-exposition characteristics of holographic photomaterials (Churaev, A. L.; Artyomova, V. V.)V1238,158-165(1991)

Characteristics of a dynamic holographic sensor for shape control of a large reflector (Welch, Sharon S.; Cox, David E.)V1480,2-10(1991)

Characteristic wave theory for volume holographic gratings with arbitrarily slanted fringes (Chernov, Boris C.)V1507,302-309(1991)

Characterization of DC-PVA films for holographic recording materials (Leclere, Philippe; Renotte, Yvon L.; Lion, Yves F.)V1507,339-344(1991)

Chemical symmetry: developers that look like bleach agents for holography (Bjelkhagen, Hans I.; Phillips, Nicholas J.; Ce, Wang)V1461,321-328(1991)

Chromaticity and color fidelity of images with multicolor rainbow holograms (Zhang, Jingfang; Yu, Meiwen; Tang, Shunqing; Zhu, Zhengfang)V1238,401-405(1991)

Circular encoding in large-area multiexposure holography (Aliaga, R.; Choi, Peter; Chuaqui, Hernan H.)V1358,1257-1264(1991)

Coherent length and holography (Jeong, Tung H.; Feng, Qiang; Wesly, Edward J.; Qu, Zhi-Min)V1396,60-70(1991)

Color-coding reproduction of 3-D object with rainbow holography (Fan, Cheng; Jiang, Chaochuan; Guo, Lu Rong)V1461,51-55(1991)

Color reflection holograms with photopolymer plates (Zhang, Jingfang; Ma, Chunrong; Lang, Hengyuan)V1238,306-310(1991)

Compact spaceflight solution crystal-growth system (Trolinger, James D.; Lal, Ravindra B.; Vikram, Chandra S.; Witherow, William K.)V1557,250-258(1991)

Comparison of detection systems in time-of-flight transillumination imaging (Hayes, Jeffrey A.; Sullivan, Barry J.)V1396,324-330(1991)

Comparison of rigorous and approximate methods of analyzing holographic gratings diffraction (Chernov, Boris C.)V1238,44-53(1991)

Computer and Optically Generated Holographic Optics; 4th in a Series (Cindrich, Ivan; Lee, Sing H., eds.)V1555(1991)

Computer-controlled pulse laser system for dynamic holography (Chen, Terry Y.; Ju, M. S.; Lee, C. Y.; Lo, M. P.)V1554B,92-98(1991)

Corneal topography: the dark side of the moon (Bores, Leo D.)V1423,28-39(1991); V1429,217-228(1991)

Coupled-mode analysis of dynamic polarization volume holograms (Huang, Tizhi; Wagner, Kelvin H.)V1559,377-384(1991)

Dark stability of holograms and holographic investigation of slow diffusion in polymers (Veniaminov, Andrei V.)V1238,266-270(1991)

Deformation measurement of the bone fixed with external fixator using holographic interferometry (Kojima, Arata; Ogawa, Ryokei; Izuchi, N.; Yamamoto, M.; Nishimoto, T.; Matsumoto, Toshiro)V1429,162-171(1991)

Denisyuk hologram recording with simultaneous use of all spectral components of the white light (Kostyljov, Ghennadij D.)V1238,316-319(1991)

Design characters and radiation parameters of a gas laser as a holographic tool (Lypsky, Volodymyr V.)V1238,390-395(1991)

Design rules for pseudocolor transmission holographic stereograms (Andrews, John R.)V1507,407-415(1991)

Determination of displacements, strain, and rotations from holographic interferometry data using a 2-D fringe order function (Albertazzi, Armando A.)V1554B,64-74(1991)

Determination of the adhesive load by holographic interferometry using the results of FEM calculations (Bischof, Thomas; Jueptner, Werner P.)V1508,90-95(1991)

Determination of the homogeneous degree of deformation of the bearing cap of the motor by reflection holography (Cao, Zhengyuan; Cheng, Fang)V1554B,81-85(1991)

Development and experimental investigation of a copying procedure for the reproduction of large-format transmissive holograms (Kubitzek, Ruediger; Froelich, Klaus; Tropartz, Stephan; Stojanoff, Christo G.)V1507,365-372(1991)

Development and investigation of dichromated gelatin film for the fabrication of large-format holograms operating at 400-900 nm (Tropartz, Stephan; Brasseur, Olivier; Kubitzek, Ruediger; Stojanoff, Christo G.)V1507,345-353(1991)

Development of new beamsplitter system for real-time high-speed holographic interferometry (Yamamoto, Yoshitaka)V1358,940-951(1991)

Didactic problems of using holographic means of education (Naumov, B. L.)V1238,365-367(1991)

Difference holographic interferometry: an overview (Fuzessy, Zoltan; Gyi'mesi, Ferenc)VIS08,194-204(1991)

Diffuse radiation unsteady transformation by thermal dynamic holograms (Berezinskaya, Aleksandra M.; Dukhovniy, Anatoliy M.)V1238,80-84(1991)

Displacement analysis of the interior walls of a pipe using panoramic holointerferometry (Gilbert, John A.; Matthys, Donald R.; Hendren, Christelle M.)V1554B,128-134(1991)

Display and applied holography in culture development (Markov, Vladimir B.)VIS08,268-304(1991)

Display holograms in Du Pont's OmniDex films (Zager, Stephen A.; Weber, Andrew M.)V1461,58-67(1991)

Display holograms of crystals recording (Smirnova, S. N.; Marchenko, S. N.)V1238,370-370(1991)

Display holography commercial prospects (Vanin, V. A.; Trukhanov, V. A.)V1238,372-380(1991)

Display holography with pulsed ruby laser: recording and copying (Kliot-Dashinskaya, I. M.; Mikhailova, V. I.; Paltsev, G. P.; Strigun, V. L.)V1238,465-469(1991)

Double-exposure phase-shifting holography applied to particle velocimetry (Lai, Tianshu; Tan, Yushan)V1554B,580-585(1991)

Dynamic holographic interconnects: experimental study using photothermoplastics with improved cycling properties (Gravey, Philippe; Moisan, Jean-Yves)V1507,239-246(1991)

Dynamic holographic microphasometry (Kozhevnikov, Nikolai M.)V1507,509-516(1991)

Dynamic holography application in fiber optic interferometry (Kozhevnikov, Nikolai M.; Barmenkov, Yuri O.)V1584,387-395(1991)

Dynamic particle holographic instrument (Wang, Guozhi; Feng, San; Wang, Zhengrong; Wang, Shuyan)V1358,73-81(1991)

Edible holography: the application of holographic techniques to food processing (Begleiter, Eric)V1461,102-109(1991)

Effective holographic grating model to analyze thick holograms (Belendez, A.; Pascual, I.; Fimia-Gil, Antonio)V1507,268-276(1991)

Effective transmission holograms produced on CK-type photoresist (Bulygin, A. R.)V1238,248-252(1991)

Electro-optical system for the nondestructive evaluation of bioengineering materials (Sciammarella, Cesar A.; Bhat, Gopalakrishna K.; Albertazzi, Armando A.)V1429,183-194(1991)

Eliminating system errors of a large-aperture and high-sensitivity moire deflector by real-time holography (Yan, Da-Peng; Wang, Hai-Lin; Miao, Peng-Cheng; He, Anzhi)V1527,442-447(1991)

Engineering fabrication technology of multiplexed hologram in single-mode optical waveguide (Lin, Freddie S.; Chen, Jenkins C.; Nguyen, Cong T.; Liu, William Y.)V1461,39-50(1991)

Enhanced holographic recording capabilities for dynamic applications (Hough, Gary R.; Gustafson, D. M.; Thursby, William R.)V1346,194-199(1991)

Etch conditions of photolithographic holograms (Guo, Yongkang; Guo, Lu Rong; Zhang, Xiao-Chun)V1461,97-100(1991)

Evaluation of interference fringe pattern on spatially curved objects (Laermann, Karl-Hans)V1554A,522-528(1991)

Experiments in holographic video imaging (Benton, Stephen A.)VIS08,247-267(1991)

Flat holographic stereograms synthesized from computer-generated images by using LiNbO3 crystal (Qu, Zhi-Min; Liu, Jinsheng; Xu, Liangying)V1238,406-411(1991)

Focal-plane processing algorithm and architecture for laser speckle interferometry (Tomlinson, Harold W.; Weir, Michael P.; Michon, G. J.; Possin, G. E.; Chovan, J.)V1541,178-186(1991)

Four-dimensional photosensitive materials: applications for shaping, time-domain holographic dissection, and scanning of picosecond pulses (Saari, Peeter M.; Kaarli, Rein K.; Sonajalg, Heiki)V1507,500-508(1991)

Four cases of engineering vibration studies using pulsed ESPI (Mendoza Santoyo, Fernando; Shellabear, Michael C.; Tyrer, John R.)V1508,143-152(1991)

Fourier-transform holographic microscope (Haddad, Waleed S.; Cullen, David; Solem, Johndale C.; Longworth, James W.; McPherson, Armon; Boyer, Keith; Rhodes, Charles K.)V1448,81-88(1991)

Fraud involving OVDs and possible security measures (Schuurman, Dirk)V1509,126-130(1991)

Fringe quality in pulsed TV-holography (Spooren, Rudie)V1508,118-127(1991)

Gray level distortions induced by 3-D reflective hologram (Odinokov, S. B.; Poddubnaya, T. E.; Rozhkov, Oleg V.; Yakimovich, A. P.)V1238,109-117(1991)

Herbarium holographicum (Boone, Pierre M.)VIS08,370-380(1991)

Heterogeneous recording media (Sukhanov, Vitaly I.)V1238,226-230(1991)

High-capacity optical spatial switch based on reversible holograms (Mikaelian, Andrei L.; Salakhutdinov, Viktor K.; Vsevolodov, N. N.; Dyukova, T. V.)V1621,148-157(1991)

High-efficiency holograms on the basis of polarization recording by means of beams from opposite directions (Shatalin, Igor D.)V1238,68-73(1991)

High-efficiency reflective holograms: subangstrom spectral selectors (Goncharov, V. F.; Popov, Alexander P.; Veniaminov, Andrei V.)V1238,97-102(1991)

High-fidelity phase conjugation generated by holograms: application to imaging through multimode fibers (Pan, Anpei; Marhic, Michel E.; Epstein, Max)V1396,99-106(1991)

High-resolution photographic material of Institute of Atomic Energy for holography in intersecting beams (Ryabova, R. V.; Barinova, E. S.; Myatezh, O. V.; Zaborov, A. N.; Romash, E. V.; Chernyi, D. I.)V1238,166-170(1991)

High-sensitive layers of dichromated gelatin for hologram recording by continuous wave and pulsed laser radiation (Kalyashora, L. N.; Michailova, A. G.; Pavlov, A. P.; Rykov, V. S.; Guba, B. S.)V1238,189-194(1991)

High-sensitive photopolymer for large-size holograms (Ishikawa, Toshiharu; Kuwabara, Y.; Koseki, Kenichi; Yamaoka, Tsuguo)V1461,73-78(1991)

High-speed holographic interferometric study of the propagation of the electrohydraulic shock wave (Jia, Youquan; Sun, Yongda)V1554B,135-138(1991)

HOLIDDO: an interferometer for space experiments (Mary, Joel; Bernard, Yves; Lefaucheux, Francoise; Gonzalez, Francois)V1557,147-155(1991)

Holo-interferometric patterns recorded through a panoramic annular lens (Gilbert, John A.; Greguss, Pal; Kransteuber, Amy S.)V1238,412-420(1991)

Hologram: liquid-crystal composites (Ingwall, Richard T.; Adams, Timothy)V1555,279-290(1991)

Hologram as means of color reproduction (Vlasov, N. G.; Vorobjov, S. P.; Karpova, S. G.)V1238,332-337(1991)

Holograms in optical computing (Caulfield, H. J.)VIS08,54-61(1991)

Holographically protected printwork (Coblijn, Alexander B.)V1509,73-78(1991)

Holographic art (Bryskin, V. Z.; Prostev, A.)V1238,368-369(1991)

Holographic associative memory of biological systems (Gariaev, Peter P.; Chudin, Viktor I.; Komissarov, Gennady G.; Berezin, Andrey A.; Vasiliev, Antoly A.)V1621,280-291(1991)

Holographic associative memory with bipolar features (Wang, Xu-Ming; Mu, Guoguang)V1558,518-528(1991)

Holographic center high-mounted stoplight (Smith, Ronald T.)V1461,186-198(1991)

Holographic characteristics of IAE and PFG-01 photoplates for colored pulsed holography (Vorzobova, N. D.; Rjabova, R. V.; Schvarzvald, A. I.)V1238,476-477(1991)

Holographic characterization of DYE-PVA films studied at 442 nm for optical elements fabrication (Couture, Jean J.)V1559,429-435(1991)

Holographic coupler for fiber optic Sagnac interferometer (Zou, Yunlu; Hsu, Dahsiung; Wang, Ben; Tao, Huiying)V1238,452-456(1991)

Holographic crossed gratings: their nature and applications (Miler, Miroslav; Aubrecht, I.)V1574,22-33(1991)

Holographic devices using photo-induced effect in nondestructive testing techniques (Dovgalenko, George Y.; Onischenko, Yuri I.)V1559,479-486(1991)

Holographic diffraction gratings on the base of chalcogenide semiconductors (Indutnyi, I. Z.; Robur, I.; Romanenko, Peter F.; Stronski, Alexander V.)V1555,243-253(1991)

Holographic display of computer simulations (Andrews, John R.; Stinehour, Judith E.; Lean, Meng H.; Potyondy, David O.; Wawrzynek, Paul A.; Ingraffea, Anthony R.; Rainsdon, Michael D.)V1461,110-123(1991)

Holographic high-resolution endoscopic image recording (Bjelkhagen, Hans I.)V1396,93-98(1991)

Holographic in-plane displacement measurement in cracked specimens in plane stress state (Sainov, Ventseslav C.; Simova, Eli S.)V1461,175-183(1991)

Holographic instrumentation for monitoring crystal growth in space (Trolinger, James D.; Lal, Ravindra B.; Batra, Ashok K.)V1332,151-165(1991)

Holographic interferometry in corrosion studies of metals: I. Theoretical aspects (Habib, Khaled J.)V1332,193-204(1991)

Holographic interferometry in corrosion studies of metals: II. Applications (Habib, Khaled J.)V1332,205-215(1991)

Holographic measurement of deformation using carrier fringe and FFT techniques (Quan, C.; Judge, Thomas R.; Bryanston-Cross, Peter J.)V1507,463-475(1991)

Holographic measurement of the angular error of a table moving along a slideway (Matsuda, Kiyofumi; Tenjimbayashi, Koji)V1332,230-235(1991)

Holographic nonlinear Raman spectroscopy of large molecules of biological importance (Ivanov, Anatoliy A.; Koroteev, Nikolai I.; Fishman, A. I.)V1429,132-144(1991); V1432,141-153(1991)

Holographic Optical Security Systems (Fagan, William F., ed.)V1509(1991)

Holographic optical security systems (Fagan, William F.)VIS08,381-386(1991)

Holographic optical system to copy holographic optical elements (Pascual, I.; Belendez, A.; Fimia-Gil, Antonio)V1507,373-378(1991)

Holographic Optics III: Principles and Applications (Morris, G. M., ed.)V1507(1991)

Holographic photoelasticity applied to ceramics fracture (Guo, Maolin; Wang, Yigong; Zhao, Yin; Du, Shanyi)V1554A,310-312(1991)

Holographic portraits (Bjelkhagen, Hans I.)VIS08,347-353(1991)

Holographic precompensation for one-way transmission of diffraction-limited beams through diffusing media (Pan, Anpei; Marhic, Michel E.)V1461,8-16(1991)

Holographic pseudocoloring of schlieren images (Rodriguez-Vera, Ramon; Olivares-Perez, A.; Morales-Romero, Arquimedes A.)V1507,416-424(1991)

Holographic recording in photorefractive media containing bacteriorhodopsin (Kozhevnikov, Nikolai M.; Barmenkov, Yuri O.; Lipovskaya, Margarita J.)V1507,517-524(1991)

Holographic recordings on 2-hydroxyethyl methacrylate and applications of water-immersed holograms (Yacoubian, Araz; Savant, Gajendra D.; Aye, Tin M.)V1559,403-409(1991)

Holographic soundfield visualization for nondestructive testing of hot surfaces (Crostack, Horst-Artur; Meyer, E. H.; Pohl, Klaus-Juergen)V1508,101-109(1991)

Holographic technique inculcation economic assessment (Savrukov, N. T.)V1238,382-389(1991)

Holographic visualization of hypervelocity explosive events (Cullis, I. C.; Parker, Richard J.; Sewell, Derek)V1358,52-64(1991)

Holography (Greguss, Pal; Jeong, Tung H., eds.)VIS08(1991)

Holography and education (Jeong, Tung H.)VIS08,360-369(1991)

Holography and education in the United States (Jeong, Tung H.)V1238,351-354(1991)

Holography and relativity (Abramson, Nils H.)VIS08,2-32(1991)

Holography as a tool for widespread industrial applications: analysis and comments (Smigielski, Paul)V1508,38-49(1991)

Holography in endoscopy: illuminating dark holes with Gabor's principle (von Bally, Gert)VIS08,335-346(1991)

Holography in museums of the Ukraine (Markov, Vladimir B.; Mironyuk, G. I.)V1238,340-347(1991)

Holography, Interferometry, and Optical Pattern Recognition in Biomedicine (Podbielska, Halina, ed.)V1429(1991)

Holography in the '90s (Jeong, Tung H.)V1396,718-721(1991)

Holography of human pathologic specimens with continuous-beam lasers through plastination (Myers, M. B.; Bickley, Harmon)V1461,242-244(1991)

Holography with a single picosecond pulse (Abramson, Nils H.)V1332,224-229(1991)

Holophonics: a spread-out of the basic ideas on holography into audio-acoustics (Illenyi, Andras; Jessel, M.)VIS08,39-52(1991)

Image formation through inhomogeneities (Leith, Emmett N.; Chen, Chiaohsiang; Cunha, Andre)V1396,80-84(1991)

Imaging inside scattering media: chronocoherent imaging (Spears, Kenneth G.; Kume, Stewart M.; Winakur, Eric)V1429,2-8(1991)

Improvement of recordable depth of field in far-field holography for analysis of particle size (Lai, Tianshu; Tan, Yushan)V1461,286-290(1991)

Industrial Applications of Holographic and Speckle Measuring Techniques (Jueptner, Werner P., ed.)V1508(1991)

Industrial applications of self-diffraction phenomena in holography on photorefractive crystals (Dovgalenko, George Y.; Yeskov-Soskovetz, Vladimir M.; Cherkashin, G. V.)V1508,110-115(1991)

Influence of fixing stress on the sensitivity of HNDT (Yang, Yu X.; Tan, Yushan)V1554B,768-773(1991)

Information capacity of holograms with reference speckle wave (Darsky, Alexei M.; Markov, Vladimir B.)V1509,36-46(1991)

Information extracting and application for the combining objective speckle and reflection holography (Cao, Zhengyuan; Cheng, Fang)V1332,358-364(1991)

Information retrieval from ultrasonic (acoustical) holograms by moire principle (Greguss, Pal)V1238,421-427(1991)

Instant measurement of phase-characteristic curve (Sadovnik, Lev S.; Rich, Chris C.)V1559,424-428(1991)

Integrated-optic interconnects and fiber-optic WDM data links based on volume holography (Jannson, Tomasz; Lin, Freddie S.; Moslehi, Behzad M.; Shirk, Kevin W.)V1555,159-176(1991)

Interferometric analysis for nondestructive product monitoring (Cohn, Gerald E.; Domanik, Richard A.)V1396,131-142(1991)

Interpretation of holographic and shearographic fringes for estimating the size and depth of debonds in laminated plates (Shang, H. M.; Chau, Fook S.; Tay, C. J.; Toh, Siew-Lok)V1554B,680-691(1991)

Investigation of diesel injection jets using high-speed photography and speed holography (Eisfeld, Fritz)V1358,660-671(1991)

Investigation of fiber-reinforced-plastics-based components by means of holographic interferometry (Jueptner, Werner P.; Bischof, Thomas)V1400,69-79(1991)

Investigation of the arsenic sulphide films for relief-phase holograms (Yusupov, I. Y.; Mikhailov, M. D.; Herke, R. R.; Goray, L. I.; Mamedov, S. B.; Yakovuk, O. A.)V1238,240-247(1991)

Investigation of the possibility of creating a multichannel photodetector based on the avalanche MRS-structure (Sadyigov, Z. Y.; Gasanov, A. G.; Yusipov, N. Y.; Golovin, V. M.; Gulanian, Emin H.; Vinokurov, Y. S.; Simonov, A. V.)V1621,158-168(1991)

Kinetics of formation of holographic structure of a hologram mirror in dichromated gelatin (Kzuzhilin, Yu E.; Mel'nichenko, Yu. B.; Shilov, V. V.)V1238,200-205(1991)

Large-viewing-angle rainbow hologram by holographic phase conjugation (Fan, Chechung; Jiang, Chaochuan; Guo, Lu Rong)V1461,265-269(1991)

Large one-step holographic stereogram (Honda, Toshio; Kang, Der-Kuan; Shimura, Kei; Enomoto, H.; Yamaguchi, Masahiro; Ohyama, Nagaaki)V1461,156-166(1991)

Laser speckle and its temporal variability: the implications for biomedical holography (Briers, J. D.)V1429,48-54(1991)

Le Musee de l'Holographie de Paris and its activities: 1980-1990 (Christakis, Anne-Marie)V1238,348-350(1991)

Light-induced volume-phase holograms for cold neutrons (Ibel, Konrad; Matull, Ralph; Rupp, Romano A.; Eschkoetter, Peter; Hehmann, Joerg)V1559,393-402(1991)

Matching of Bragg condition of holographic phase gratings in 1.3-1.5um region (Silvennoinen, Raimo; Hamalainen, Rauno M.)V1574,261-265(1991)

Mathematical modeling and standardization in holography (Brown, Simon)V1509,79-86(1991)

Measurement of Poisson's ratio of nonmetallic materials by laser holographic interferometry (Zhu, Jian T.)V1554B,148-154(1991)

Measurement of particles and drops in combusting flows (Ereaut, Peter R.)V1554B,556-565(1991)

Measurement of strains by means of electro-optics holography (Sciammarella, Cesar A.; Bhat, Gopalakrishna K.; Albertazzi, Armando A.)V1396,143-154(1991)

Measurement of the piezoelectric effect in bone using quasiheterodyne holographic interferometry (Ovryn, Benjie; Haacke, E. M.)V1429,172-182(1991)

Measurement of thermal deformation of square plate by using holographic interferometry (Kang, Dae I.; Kwon, Young H.; Ko, Yeong-Uk)V1554B,139-147(1991)

Measurement of the temperature field in confined jet impingement using phase-stepping video holography (Dobbins, B. N.; He, Shi P.; Jambunathan, K.; Kapasi, S.; Wang, Liu Sheng; Button, B. L.)V1554B,586-592(1991)

Medical applications of holographic stereograms (Tsujiuchi, Jumpei)V1238,398-400(1991)

Medical applications of holography (von Bally, Gert)V1525,2-8(1991)

Metal-ion-doped polymer systems for real-time holographic recording (Lessard, Roger A.; Changkakoti, Rupak; Manivannan, Gurusamy)V1559,438-448(1991)

Micro-optical elements in holography (von Bally, Gert; Dirksen, D.; Zou, Y.)V1507,66-72(1991)

Modal analysis of musical instruments with holographic interferometry (Rossing, Thomas D.; Hampton, D. S.)V1396,108-121(1991)

Modeling and simulation of systems imaging through atmospheric turbulence (Caponi, Maria Z.)V1415,138-149(1991)

Modern holographic studio (Bryskin, V. Z.; Krylov, Vitaly N.; Staselko, D. I.)V1238,448-451(1991)

Modern optical method for strain separation in photoplasticity (Zhang, Zuxun)V1554A,482-487(1991)

Modulation of absorption in DCG (Zhao, Feng; Geng, Wanzhen; Jiang, Lingzhen; Hong, Jing)V1555,297-299(1991)

Moire-holographic evaluation of the strain fields at the weld toes of welded structures (Pappalettere, Carmine; Trentadue, Bartolo; Monno, Giuseppe)V1554B,99-105(1991)

Monochromatic aberrations of an off-axis hologram with carry out pupil (Mikhailov, I. A.)V1238,123-128(1991)

Monochromatic and two-color recording of holographic portraits with the use of pulsed lasers (Vorzobova, N. D.; Sizov, V. N.; Rjabova, R. V.)V1238,462-464(1991)

MTF of photolithographic hologram (Guo, Lu Rong; Guo, Yongkang; Zhang, Xiao-Chun)V1392,119-123(1991)

Multiline holographic notch filters (Ning, Xiaohui; Masso, Jon D.)V1545,125-129(1991)

Multipass holographic interferometry for low-density gas flow analysis (Surget, Jean; Dunet, G.)V1358,65-72(1991)

Multiplexed binary phase-only circular harmonic filters (Sheng, Yunlong; Leclerc, Luc; Arsenault, Henri H.)V1564,320-329(1991)

Multipulse polarized-holographic set-up for isodromic fringe registration in the field of running stress waves (Khesin, G. L.; Sakharov, V. N.; Zhavoronok, I. V.; Rottenkolber, Hans; Schorner, Jurgen)V1554B,86-90(1991)

New developing process for PVCz holograms (Yamagishi, Yasuo; Ishitsuka, Takeshi; Kuramitsu, Yoko; Yoneda, Yasuhiro)V1461,68-72(1991)

New holographic overlays (Hopwood, Anthony I.)V1509,26-35(1991)

New holographic system for measuring vibration (Cai, Yunliang)V1554B,75-80(1991)

New method of recording holographic optical elements applied to optical interconnection in VLSI (Zhao, Feng; Sun, Junyong; Geng, Wanzhen; Jiang, Lingzhen; Hong, Jing)V1461,262-264(1991)

New multisource contouring method (Rachkovsky, Leonid I.; Drobot, Igor L.; Tanin, Leonid V.)V1507,450-457(1991)

New options of holographic metrology (Steinbichler, Hans)V1507,435-445(1991)

New possibilities of photosensitive glasses for three-dimensional phase hologram recording (Glebov, Leonid B.; Nikonorov, Nikolai V.; Panysheva, Elena I.; Petrovskii, Gurii T.; Savvin, Vladimir V.; Tunimanova, Irina V.; Tsekhomskii, Victor A.)V1621,21-32(1991)

New technique for characterizing holographic recording materials (Leclere, Philippe; Renotte, Yvon L.; Lion, Yves F.)V1559,298-307(1991)

New way of optical interconnection for VLSI (Zhao, Feng; Geng, Wanzhen; Jiang, Lingzhen; Hong, Jing)V1555,241-242(1991)

Non-gelatin-dichromated holographic film (Guo, Lu Rong; Cheng, Qirei; Wang, Kuoping)V1461,91-92(1991)

Nondestructive evaluation of turbine blades vibrating in resonant modes (Sciammarella, Cesar A.; Ahmadshahi, Mansour A.)V1554B,743-753(1991)

Nondestructive testing of printed circuit board by phase-shifting interferometry (Lu, Yue-guang; Jiang, Lingzhen; Zou, Lixun; Zhao, Xia; Sun, Junyong)V1332,287-291(1991)

Numerical investigation of effect of dynamic range and nonlinearity of detector on phase-stepping holographic interferometry (Fang, Qiang; Luo, Xiangyang; Tan, Yushan)V1332,216-222(1991)

Observation of reflection holograms (Vorobjov, S. P.)V1238,62-67(1991)

One-dimensional hologram recording in thick photolayer disk (Mikaelian, Andrei L.; Gulanian, Emin H.; Kretlov, Boris S.; Molchanova, L. V.; Semichev, V. A.)V1621,194-203(1991)

Optical aberrations in underwater holography and their compensation (Watson, John; Kilpatrick, J. M.)V1461,245-253(1991)

Optical/electronical hybrid three-layer neural network for pattern recognition (Zhang, Yanxin; Shen, Jinyuan)V1469,303-307(1991)

Optical Memory and Neural Networks (Mikaelian, Andrei L., ed.)V1621(1991)

Optical neural network with reconfigurable holographic interconnection (Kirk, Andrew G.; Kendall, G. D.; Imam, H.; Hall, Trevor J.)V1621,320-327(1991)

Optical nondestructive examination for honeycomb structure (Zhu, Jian T.)V1554B,774-784(1991)

Optical research of elastic anisotropy of monocrystals with a cube structure (Polyak, Alexander; Kostin, Ivan K.)V1554A,553-556(1991)

Optical testing by dynamic holographic interferometry with photorefractive crystals and computer image processing (Vlad, Ionel V.; Popa, Dragos; Petrov, M. P.; Kamshilin, Alexei A.)V1332,236-244(1991)

Optical tomography by heterodyne holographic interferometry (Vukicevic, Dalibor; Neger, Theo; Jaeger, Helmut; Woisetschlaeger, Jakob; Philipp, Harald)VIS03,160-193(1991)

Parameters of recording media for 3-D holograms (Mikhailov, I. A.)V1238,140-143(1991)

PC-based determination of 3-D deformation using holographic interferometry (Schad, Hanspeter; Schweizer, Edwin; Jodlbauer, Heibert)V1507,446-449(1991)

PFG-04 photographic plates based on the nonhardened dichromated gelatin for recordng color reflection holograms (Artemjev, S. V.; Koval, G. I.; Obyknovennaja, I. E.; Cherkasov, A. S.; Shevtsov, M. K.)V1238,206-210(1991)

Phase-conjugate interferometry by using dye-doped polymer films (Nakagawa, Kazuo; Egami, Chikara; Suzuki, Takayoshi; Fujiwara, Hirofumi)V1332,267-273(1991)

Phase-measurement of interferometry techniques for nondestructive testing (Creath, Katherine)V1554B,701-707(1991)

Phase-stepping technique in holography (Wen, Zheng; Tan, Yushan)V1461,278-285(1991)

Phase synthesis of elongated holographic diffraction gratings (Turukhano, Boris G.; Gorelik, Vladimir S.; Turukhano, Nikulina)V1500,290-292(1991)

Photoanisotropic incoherent-to-coherent conversion using five-wave mixing (Huang, Tizhi; Wagner, Kelvin H.)V1562,44-54(1991)

Photoelasticity, holography, and moire for strain-stress-state studies of NPS, HPS underground and machinery structures: comparative analysis of physical modeling and numerical methods (Khesin, G. L.)V1554A,313-322(1991)

Photothermoplastic media with organic and inorganic photosemiconductors for hologram recording (Zaichenko, O. V.; Komarov, Vyacheslav A.; Kuzmin, I. V.)V1238,271-274(1991)

Physicochemistry of holographic recording and processing in dichromated gelatin (Sherstyuk, Valentin P.; Maloletov, Sergei M.; Kondratenko, Nina A.)V1238,211-217(1991)

Possibility of associative reconstruction of reflection hologram in phase-conjugate interferometer (Lazaruk, A. M.; Rubanov, Alexander S.; Serebryakova, L. M.)V1621,114-124(1991)

Practical design considerations and performance characteristics of high-numerical-aperture holographic lenses (Kostuk, Raymond K.)V1461,24-34(1991)

Practical Holography V (Benton, Stephen A., ed.)V1461(1991)

Practical use of reference mirror rotation in holographic interferometry (Boxler, Lawrence H.; Brown, Glen; Western, Arthur B.)V1396,85-92(1991)

Primary research for mechanism of forming PLH (Guo, Lu Rong; Zhang, Xiao-Chun; Guo, Yongkang)V1463,534-538(1991)

Problems of production of display holograms (Rumyantsev, V. A.)V1238,381-381(1991)

Production of reflection relief holograms with asymmetric shape of slits profile and the examination of their spectral characteristics (Bulygin, A. R.)V1238,129-133(1991)

Programmable holographic scanning (Zhengmin, Li; Li, Min; Tang, Jinfa)V1401,66-73(1991)

Progress in true-color holography (Jeong, Tung H.; Wesly, Edward J.)V1238,298-305(1991)

Properties of volume reflection silver-halide gelatin holograms (Kosobokova, N. L.; Usanov, Yu. Y.; Shevtsov, M. K.)V1238,183-188(1991)

Pseudocolor reflection hologram properties recorded using monochrome photographic materials (Vanin, V. A.; Vorobjov, S. P.)V1238,324-331(1991)

Pseudodeep hologram and its properties (Denisyuk, Yuri N.; Ganzherli, N. M.)V1238,2-12(1991)

Pulsed hologram recording by nanosecond and subnanosecond small-dimensioned laser system (Bespalov, V. G.; Dikasov, A. B.)V1238,470-475(1991)

Pulsed holographic recording of very high speed transient events (Steckenrider, John S.; Ehrlich, Michael J.; Wagner, James W.)V1554B,106-112(1991)

Pulsed laser system for recording large-scale color hologram (Bespalov, V. G.; Krylov, Vitaly N.; Sizov, V. N.)V1238,457-461(1991)

Quantitative analysis of contact deformation using holographic interferometry and speckle photography (Joo, Jin W.; Kwon, Young H.; Park, Seung O.)V1554B,113-118(1991)

Quantum holography (Granik, Alex T.; Caulfield, H. J.)VIS08,33-38(1991)

Quantum holography and neurocomputer architectures (Schempp, Walter)VIS08,62-144(1991)

Rapid procedure for obtaining time-average interferograms of vibrating bodies (Rapoport, Eliezer; Bar, Doron; Shiloh, Klara)V1442,383-391(1991)

Real-time holographic display: improvements using a multichannel acousto-optic modulator and holographic optical elements (St.-Hilaire, Pierre; Benton, Stephen A.; Lucente, Mark; Underkoffler, John S.; Yoshikawa, Hiroshi)V1461,254-261(1991)

Real-time holographic microscope with nonlinear optics (Goldfain, Ervin)V1527,199-209(1991)

Real-time holographic recording material: NGD (Guo, Lu Rong; Wang, Kuoping; Cheng, Qirei)V1555,291-292(1991)

Real-time holography using the high-resolution LCTV-SLM (Hashimoto, Nobuyuki; Morokawa, Shigeru; Kitamura, Kohei)V1461,291-302(1991)

Real-time interferometry by optical phase conjugation in dye-doped film (Fujiwara, Hirofumi)V1507,492-499(1991)

Recent advances in color reflection holography (Hubel, Paul M.)V1461,167-174(1991)

Recent advances in white-light display holography at CSIO, Chandigarh (Aggarwal, Anil K.; Kaura, Sushil K.; Chhachhia, D. P.)V1238,18-27(1991)

Recent developments on holography in China (Hsu, Dahsiung; Jiao, Jiangzhong; Tao, Huiying; Long, Pin)V1238,13-17(1991)

Reconstruction algorithm of displacement field by using holographic image (Chen, Ke-long; Xu, Zhu; Wen, Zhen-chu; Chen, Yu-zhuo)V1385,206-213(1991)

Reconstruction of three-dimensional displacement fields by carrier holography (Xu, Zhu; Chen, Ke-long; Wen, Zhen-chu; Yang, Han-Guo; He, Xiao-yuan; Zhang, Bao-he)V1399,172-177(1991)

Recording and analysis of (high-frequency) sinusoidal vibrations using computerized TV-holography (Ellingsrud, Svein; Lokberg, Ole J.; Pedersen, Hans M.)V1399,30-41(1991)

Recording mechanism and postpolymerizing self-amplification of holograms (Gulnazarov, Eduard S.; Smirnova, Tatiana N.; Tikhonov, Evgenij A.)V1238,235-239(1991)

Reduced size holography (Fargion, Daniele)V1238,428-438(1991)

Reducing and magnifying holograms (Fargion, Daniele)VIS08,354-359(1991)

Reducing errors in bending tests of ceramic materials (Furgiuele, Franco M.; Lamberti, Antonio; Pagnotta, Leonard)V1554A,275-284(1991)

Reflection silver-halide gelatin holograms (Usanov, Yu. Y.; Vavilova, Ye. A.; Kosobokova, N. L.; Shevtsov, M. K.)V1238,178-182(1991)

Refractive properties of TGS aqueous solution for two-color interferometry (Vikram, Chandra S.; Witherow, William K.; Trolinger, James D.)V1557,197-201(1991)

Registration of 3-D holograms of diamond crystals (Marchenko, S. N.; Smirnova, S. N.)V1238,371-371(1991)

Relief-phase colored hologram registration (Galpern, A. D.; Smaev, V. P.; Paramonov, A. A.; Kiriencko, Yu. A.)V1238,320-323(1991)

Relief holograms recording on liquid photopolymerizable layers (Boiko, Yuri B.; Granchak, Vasilij M.; Dilung, Iosiph I.; Solovjev, Vladimir S.; Sisakian, Iosif N.; Soifer, Victor A.)V1238,253-257(1991)

Research of fast stages of latent image formation in holographic photoemulsions influenced by ultrashort radiation pulses (Starobogatov, Igor O.; Nicolaev, S. D.)V1238,153-157(1991)

Resolution enhancement in digital in-line holography (Hua, Lifan; Xie, Gong-Wie; Shaw, David T.; Scott, Peter D.)V1385,142-151(1991)

Resolution problems in holography (Banyasz, Istvan)V1574,282-293(1991)

Restrictions on the geometry of the reference wave in holocoupler devices (Calvo, Maria L.; De Pedraza-Velasco, L.)V1507,288-301(1991)

Scanning method of receiving high-effective reflective holograms on bichromate gelatin (Imedadze, Theodore S.; Kakichashvili, Shermazan D.)V1238,439-441(1991)

Scattering parameters of volume phase hologram: Bragg's approach (Gubanov, V. A.; Kiselyov, Boris S.)V1621,102-113(1991)

Search for short-lived particles using holography (Brucker, E. B.)V1461,206-214(1991)

Second Intl Conf on Photomechanics and Speckle Metrology: Moire Techniques, Holographic Interferometry, Optical NDT, and Applications to Fluid Mechanics (Chiang, Fu-Pen, ed.)V1554B(1991)

Self-diffraction for active stabilization of interference field for reflection hologram recording (Kalyashov, E. V.; Tyutchev, M. V.)V1238,442-446(1991)

Shadow method of scale with a-posteriori increase of measurements sensitivity (Spornik, Nikolai M.; Serenko, M. Y.)V1508,162-169(1991)

Shear-lens photography for holographic stereograms (Molteni, William J.)V1461,132-141(1991)

Signal and noise characteristics of space-time light modulators (Komarov, Vyacheslav A.; Melnik, N. E.)V1238,275-285(1991)

Simple technique for 3-D displacements measurement using synthesis of holographic interferometry and speckle photography (Kwon, Young H.; Park, Seung O.; Park, B. C.)V1554A,639-644(1991)

Simulation of optical diagnostics for crystal growth: models and results (Banish, Michele R.; Clark, Rodney L.; Kathman, Alan D.; Lawson, Shelah M.)V1557,209-221(1991)

Some cases of applying photothermoplastic carriers in holographic interferometry and speckle photography (Anikin, V. I.; Dementiev, I. V.; Zhurminskii, Igor L.; Nasedkhina, N. V.; Panasiuk, L. M.)V1238,286-296(1991)

Some new developments in display holography (Hegedus, Zoltan S.; Hariharan, P.)V1238,480-488(1991)

Some principles for formation of self-developing dichromate media (Sherstyuk, Valentin P.; Malov, Alexander N.; Maloletov, Sergei M.; Kalinkin, Vyacheslav V.)V1238,218-223(1991)

Spectrally nonselective reflection holograms (Kakichashvili, Shermazan D.; Wardosanidze, Zurab V.)V1238,134-137(1991)

Stabilization of ill-posed nonlinear regression model and its application to interferogram reduction (Slepicka, James S.; Cha, Soyoung S.)V1554B,574-579(1991)

Static and dynamic measurements using electro-optic holography (Pryputniewicz, Ryszard J.)V1554B,790-798(1991)

Stereoscopic holographic cinematography (Albe, Felix; Smigielski, Paul)V1358,1098-1102(1991)

Study of LiNbO3 in optical associative memory (Yuan, Xiao L.; Sayeh, Mohammad R.)V1396,178-187(1991)

Study of electronic stage of the Herschel effect in holographic emulsions with different types of chemical sensitization (Mikhailov, V. N.; Grinevitskaya, O. V.; Zagorskaya, Z. A.; Mikhailova, V. I.)V1238,144-152(1991)

Study of human cardiac cycle using holographic interferometry (Brown, Glen; Boxler, Lawrence H.; Chun, Patrick K.; Western, Arthur B.)V1396,164-173(1991)

Study of microbial growth II: by holographic interferomery (Bahuguna, Ramendra D.; Williams, Gareth T.; Pour, I. K.; Raman, R.)V1332,805-807(1991)

Study of plate vibrations by moire holography (Blanco, J.; Fernandez, J. L.; Doval, A. F.; Lopez, C.; Pino, F.; Perez-Amor, Mariano)V1508,180-190(1991)

Study of the fracture process using laser holographic interferometry and image analysis (Castro-Montero, Alberto; Shah, S. P.; Bjelkhagen, Hans I.)V1396,122-130(1991)

Study of the structure of reflection holgrams on a dichromate gelatin layer (Meshalkina, M. N.; Smirnov, V. V.)V1238,195-199(1991)

Study of welding residual stress by means of laser holography interference (Yang, Jiahua)V1554B,155-160(1991)

Substrate-mode holograms for board-to-board two-way communications (Huang, Yang-Tung; Kostuk, Raymond K.)V1474,39-44(1991)

Synthesized holograms in medicine and industry (Tsujiuchi, Jumpei)VIS08,326-334(1991)

Synthetic wavelength interferometry for the extension of the dynamic range (Skudayski, Ulf; Jueptner, Werner P.)V1508,68-72(1991)

Teaching holography in an art school environment: the program at the School of the Art Institute of Chicago (Wesly, Edward J.)V1396,71-77(1991)

Theoretical background and practical processing for art and technical work in dichromated gelatin holography (Coblijn, Alexander B.)VIS08,305-324(1991)

Three-dimensional display of inside of human body by holographic stereogram (Sato, Koki; Akiyama, Iwaki; Shoji, Hideo; Sumiya, Daigaku; Wada, Tuneyo; Katsuma, Hidetoshi; Itoh, Koichi)V1461,124-131(1991)

Three-dimensional grating images (Takahashi, Susumu; Toda, Toshiki; Iwata, Fujio)V1461,199-205(1991)

Three-dimensional holograms in polarization-sensitive media (Kakichashvili, Shermazan D.; Kilosanidze, Barbara N.)V1238,74-79(1991)

Three-dimensional holography of nonstationary waves (Mazurenko, Yuri T.)V1238,85-96(1991)

Three-Dimensional Holography: Science, Culture, Education (Jeong, Tung H.; Markov, Vladimir B., eds.)V1238(1991)

Three-layer material for the registration of colored holograms (Smaev, V. P.; Galpern, A. D.; Kirienco, Yu. A.)V1238,311-315(1991)

Transer function of a 3-D reflective hologram (Odinokov, S. B.; Poddubnaya, T. E.; Rozhkov, Oleg V.)V1238,103-108(1991)

Trends in holographic endoscopy (Podbielska, Halina)V1429,207-213(1991)

Tunable holographic interferometer using photorefractive crystal (Yang, Guanglu; Siahmakoun, Azad)V1396,552-556(1991)

TV holography and image processing in practical use (Lokberg, Ole J.; Ellingsrud, Svein; Vikhagen, Eiolf)V1332,142-150(1991)

Ultra-fine-grain silver-halide photographic materials for holography on the flexible film base (Logak, L. G.; Fassakhova, H. H.; Antonova, N. E.; Minina, L. A.; Gainutdinov, R. K.)V1238,171-176(1991)

Ultragram: a generalized holographic stereogram (Halle, Michael W.; Benton, Stephen A.; Klug, Michael A.; Underkoffler, John S.)V1461,142-155(1991)

Unsymmetrical spectrum of reflective hologram grating (Dahe, Liu; Liang, Zhujian; Tang, Weiguo)V1507,310-315(1991)

Uses of fiber optics to enhance and extend the capabilities of holographic interferometry (Gilbert, John A.; Dudderar, Thomas D.)VIS08,146-159(1991)

Using dynamic holography for iron fibers with submicron diameter and high velocity (Xie, Gong-Wie; Hua, Lifan; Patel, Sushil; Scott, Peter D.; Shaw, David T.)V1385,132-141(1991)

Valuable security features in a competitive banking environment: does security attract criminals instead of customers? (Brongers, J. D.)V1509,105-112(1991)

Van der Lugt optical correlation for the measurement of leak rates of hermetically sealed packages (Fitzpatrick, Colleen M.; Mueller, Edward P.)V1332,185-192(1991)

Vibration analysis of the tympanic membrane with a ventilation tube and a perforation by holography (Maeta, Manabu; Kawakami, Shinichiro; Ogawara, Toshiaki; Masuda, Yu)V1429,152-161(1991)

Visible-laser-light-induced polymerization: an overview of photosensitive formulations (Fouassier, Jean-Pierre)V1559,76-88(1991)

Vision sensor system for the environment of disaster (Kawauchi, Yoshikazu; Kawata, Koichi)V1507,538-543(1991)

Volume holograms in liquid photopolymerizable layers (Boiko, Yuri B.; Granchak, Vasilij M.; Dilung, Iosiph I.; Mironchenko, Vladislav Y.)V1238,258-265(1991)

Wavefront reconstruction of acoustic waves in a variable ocean (Porter, Robert P.; Mourad, Pierre D.; Al-Kurd, Azmi)V1558,91-102(1991)

Waveguide hologram star couplers (Caulfield, H. J.; Johnson, R. B.; Huang, Qiang)V1555,154-158(1991)

Waveguide holography and its applications (Huang, Qiang; Caulfield, H. J.)V1461,303-312(1991)

Wavelength shift in DCG holograms (Lelievre, Sylviane; Pawluczyk, Romuald)V1559,288-297(1991)

Wavelength shifting and bandwidth broadening in DCG (Corlatan, Dorina; Schaefer, Martin; Anders, Gerhard)V1507,354-364(1991)

White-light transmission holographic interferometry using chromatic corrective filters (Grover, Chandra P.)V1332,132-141(1991)

X-ray holography for sequencing DNA (Yorkey, Thomas J.; Brase, James M.; Trebes, James E.; Lane, Stephen M.; Gray, Joe W.)V1345,255-259(1991)

Holography, computer-generated—see also binary optics; holographic optical elements; lenses; optical processing

Accuracy of CGH encoding schemes for optical data processing (Casasent, David P.; Coetzee, Frans M.; Natarajan, Sanjay S.; Xu, Tianning; Yu, Daming; Liu, Hua-Kuang)V1555,23-33(1991)

Application of error diffusion in diffractive optics (Weissbach, Severin; Wyrowski, Frank)V1507,149-152(1991)

Axial apodization using polar curves (Ojeda-Castaneda, Jorge; Rodriguez-Montero, Ponciano)V1500,256-261(1991)

Binary optics in the '90s (Gallagher, Neal C.)V1396,722-733(1991)

CAD system for CGHs and laser beam lithography (Yatagai, Toyohiko; Geiser, Martial; Tian, Ronglong; Tian, Xingkang; Onda, Hajime)V1555,8-12(1991)

Computational experiment for computer-generated optical elements (Golub, Mikhail A.; Doskolovich, Leonid L.; Kazanskiy, Nikolay L.; Kharitonov, Sergey I.; Orlova, Natalia G.; Sisakian, Iosif N.; Soifer, Victor A.)V1500,194-206(1991)

Computer and Optically Generated Holographic Optics; 4th in a Series (Cindrich, Ivan; Lee, Sing H., eds.)V1555(1991)

Computer-generated diffraction gratings in optical region (Silvennoinen, Raimo; Hamalainen, Rauno M.; Rasanen, Jari)V1574,84-88(1991)

Computer-generated diffractive elements focusing light into arbitrary line segments (Jaroszewicz, Zbigniew; Kolodziejczyk, Andrzej)V1555,236-240(1991)

Computer-generated holograms fabricated on photopolymer for optical interconnect applications (Oren, Moshe)V1389,527-534(1991)

Computer-generated holograms for optical data processing (Casasent, David P.)V1544,101-107(1991)

Computer-generated holograms of diffused objects (Fimia-Gil, Antonio; Navarro, Maria T.; Egozcue, Juan J.)V1507,153-157(1991)

Computer-generated holograms of linear segments (Navarro, Maria T.; Egozcue, Juan J.; Fimia-Gil, Antonio)V1507,142-148(1991)

Computer-generated holograms optimized for illumination with partially coherent light using silicon backplane spatial light modulators as the recording device (O'Brien, Dominic C.; Mears, Robert J.)V1505,32-37(1991)

Computer-generated optical elements for fiber's mode selection and launching (Golub, Mikhail A.; Sisakian, Iosif N.; Soifer, Victor A.)V1365,156-165(1991)

Design and fabrication of computer-generated holographic fan-out elements for a matrix/matrix interconnection scheme (Kirk, Andrew G.; Imam, H.; Bird, K.; Hall, Trevor J.)V1574,121-132(1991)

Design of CGH for special image formation (Tang, Yaw-Tzong; Lee, Wai-Hon; Chang, Ming-Wen)V1555,194-199(1991)

Design of a motionless head for parallel readout optical disk (Marchand, Philippe J.; Krishnamoorthy, Ashok V.; Ambs, Pierre; Gresser, Julien; Esener, Sadik C.; Lee, Sing H.)V1505,38-49(1991)

Designs for two-dimensional nonseparable array generators (Mait, Joseph N.)V1555,43-52(1991)

Digital holography as a useful model in diffractive optics (Wyrowski, Frank)V1507,128-135(1991)

DOE design and manufacture at CSEM (Buczek, Harthmuth; Mayor, J. M.; Regnault, P.)V1574,48-57(1991)

Electron beam lithographic fabrication of computer-generated holograms (West, Andrew A.; Smith, Robin W.)V1507,158-167(1991)

Fabrication of computer-generated holograms for the interconnection of single-mode devices (Baettig, Rainer K.; Toms, Dennis J.; Guest, Clark C.)V1563,93-102(1991)

Focusators at letters diffraction design (Golub, Mikhail A.; Doskolovich, Leonid L.; Kazanskiy, Nikolay L.; Kharitonov, Sergey I.; Sisakian, Iosif N.; Soifer, Victor A.)V1500,211-221(1991)

Focusing and collimating laser diodes and laser diode arrays (Coetzee, Frans M.; Casasent, David P.)V1544,108-122(1991)

Holographic-coordinate-transform-based system for direct Fourier tomographic reconstruction (Huang, Qiang; Freeman, Mark O.)V1564,644-655(1991)

Holographic image reconstruction from interferograms of laser-illuminated complex targets (Wlodawski, Mitchell; Nowakowski, Jerzy)V1416,241-249(1991)

Holographic labeling for automated identification (McOwan, Peter W.; Powell, A. K.; Burge, Ronald E.)V1384,75-82(1991)

Holographic optical interconnects for multichip modules (Feldman, Michael R.)V1390,427-433(1991)

Holotag: a novel holographic label (Soares, Oliverio D.; Bernardo, Luis M.; Pinto, M. I.; Morais, F. V.)V1332,166-184(1991)

Hough transform computer-generated holograms: new output format (Carender, Neil H.; Casasent, David P.; Coetzee, Frans M.; Yu, Daming)V1555,182-193(1991)

Interferometer for testing aspheric surfaces with electron-beam computer-generated holograms (Gemma, Takashi; Hideshima, Masayuki; Taya, Makoto; Watanabe, Nobuko)V1332,77-84(1991)

Interferometer for testing of general aspherics using computer-generated holograms (Arnold, Steven M.; Jain, Anil K.)V1396,473-480(1991)

Marriage between digital holography and optical pattern recognition (Wyrowski, Frank; Bernhardt, Michael)V1555,146-153(1991)

Mode-selective fiber sensors operating with computer-generated optical elements (Golub, Mikhail A.; Sisakyan, Iosiph N.; Soifer, Victor A.; Uvarov, G. V.)V1572,101-106(1991)

Optical Distributed Inference Network (ODIN) (Bostel, Ashley J.; McOwan, Peter W.; Hall, Trevor J.)V1385,165-172(1991)

Optimal algorithms for holographic image reconstruction of an object with a glint using nonuniform illumination (Nowakowski, Jerzy; Wlodawski, Mitchell)V1416,229-240(1991)

Phase-only encoding method for complex wavefronts (Urquhart, Kristopher S.; Lee, Sing H.)V1555,13-22(1991)

Photolithographic imaging of computer-generated holographic optical elements (Shvartsman, Felix P.; Oren, Moshe)V1555,71-79(1991)

Position detection using computer-generated holograms (Leseberg, Detlef)V1500,171-171(1991)

Process error limitations on binary optics performance (Cox, James A.; Fritz, Bernard S.; Werner, Thomas R.)V1555,80-88(1991)

Quantization and sampling considerations of computer-generated hologram for optical interconnection (Long, Pin; Hsu, Dahsiung)V1461,270-277(1991)

Rectangular focus spots with uniform intensity profile formed by computer-generated holograms (Hossfeld, Jens; Jaeger, Erwin; Tschudi, Theo T.; Churin, Evgeny G.; Koronkevich, Voldemar P.)V1574,159-166(1991)

Spatial frequency analysis for the computer-generated holography of 3-D objects (Tommasi, Tullio; Bianco, Bruno)V1507,136-141(1991)

Synthetic holographic beamsplitters for integrated optics (Saarinen, Jyrki V.; Huttunen, Juhani; Vasara, Antti H.; Noponen, Eero; Turunen, Jari P.; Salin, Arto U.)V1555,128-137(1991)

Three-dimensional grating images (Takahashi, Susumu; Toda, Toshiki; Iwata, Fujio)V1461,199-205(1991)

Upper bound on the diffraction efficiency of phase-only array generators (Mait, Joseph N.)V1555,53-62(1991)

Image algebra—see also image processing; mathematical morphology

Approximation-based video tracking system (Deng, Keqiang; Wilson, Joseph N.)V1568,304-312(1991)

Decomposition and inversion of von Neumann-like convolution operators (Manseur, Zohra Z.; Wilson, David C.)V1568,164-173(1991)

Electro-optical image processing architecture for implementing image algebra operations (Coffield, Patrick C.)V1568,137-148(1991)

Extension of Rader's algorithm for high-speed multidimensional autocorrelation (Rinaldo, R.; Bernardini, Riccardo; Cortelazzo, Guido M.)V1606,773-787(1991)

Heterogeneous matrix products (Ritter, Gerhard X.)V1568,92-100(1991)

Image Algebra and Morphological Image Processing II (Gader, Paul D.; Dougherty, Edward R., eds.)V1568(1991)

Image algebra preprocessor for the MasPar parallel computer (Meyer, Trevor E.; Davidson, Jennifer L.)V1568,125-136(1991)

Introduction to Image Algebra Ada (Wilson, Joseph N.)V1568,101-112(1991)

Max-polynomials and template decomposition (Li, Dong)V1568,149-156(1991)

Object-oriented language for image and vision execution (Flickner, Myron D.; Lavin, Mark A.; Das, Sujata)V1568,113-124(1991)

Statistical image algebra: a Bayesian approach (Davidson, Jennifer L.; Cressie, N. A.)V1569,288-297(1991)

Template decomposition and inversion over hexagonally sampled images (Lucas, Dean; Gibson, Laurie)V1568,157-163(1991)

Template learning in morphological neural nets (Davidson, Jennifer L.; Sun, K.)V1568,176-187(1991)

Image analysis—see image processing; image understanding

Image compression—see also coding

140-Mbit/s HDTV coding using subband and hybrid techniques (Amor, Hamed; Wietzke, Joachim)V1567,578-588(1991)

3-D TV: joined identification of global motion parameters for stereoscopic sequence coding (Tamtaoui, Ahmed; Labit, Claude)V1605,720-731(1991)

Adaptive coding method of x-ray mammograms (Baskurt, Atilla; Magnin, Isabelle E.; Bremond, Alain; Charvet, Pierre Y.)V1444,240-249(1991)

Adaptive perceptual quantization for video compression (Puri, Atul; Aravind, R.)V1605,297-300(1991)

Arithmetic coding model for color images processed by error diffusion (Matsushiro, Nobuhito; Asada, Osamu; Tsuji, Kenzo)V1452,21-26(1991)

Arithmetic coding model for compression of LANDSAT images (Perez, Arnulfo; Kamata, Seiichiro; Kawaguchi, Eiji)V1605,879-884(1991)

Block arithmetic coding of contour images (Kim, Kyoil; Kim, Jonglak; Kim, Taejeong)V1605,851-862(1991)

Camera zoom/pan estimation and compensation for video compression (Tse, Yi-tong; Baker, Richard L.)V1452,468-479(1991)

Cascade coding with error-constrained relative entropy decoding (Tzannes, Alexis P.; Tzannes, Michael A.; Tzannes, Marcos C.; Tzannes, Nicolaos S.)V1452,497-500(1991)

Classified transform coding of images using two-channel conjugate vector quantization (Nam, J. Y.; Rao, K. R.)V1605,202-213(1991)

Classified vector quantizer based on minimum-distance partitioning (Kim, Dong S.; Lee, Sang U.)V1605,190-201(1991)

Coding of digital TV by motion-compensated Gabor decomposition (Dufaux, Frederic; Ebrahimi, Touradj; Geurtz, Alexander M.; Kunt, Murat)V1567,362-379(1991)

Combined-transform coding scheme for medical images (Zhang, Ya-Qin; Loew, Murray H.; Pickholtz, Raymond L.)V1445,358-366(1991)

Compaction of color images with arithmetic coding (Iwahashi, Masahiro; Masuda, Shun-ichi)V1605,844-850(1991)

Comparison of image compression techniques for high quality based on properties of visual perception (Algazi, V. R.; Reed, Todd R.)V1567,589-598(1991)

Construction of efficient variable-length codes with clear synchronizing codewords for digital video applications (Lei, Shaw-Min)V1605,863-873(1991)

Digital Image Compression Techniques (Jones, Paul W.; Rabbani, Majid)VTT07(1991)

Digital map databases in support of avionic display systems (Trenchard, Michael E.; Lohrenz, Maura C.; Rosche, Henry; Wischow, Perry B.)V1456,318-326(1991)

Digital picture processing for the transmission of HDTV: the progress in Europe (Le Pannerer, Yves-Marie; Tourtier, Philippe)V1567,556-565(1991)

Discrete-cosine-transform-based image compression applied to dermatology (Cookson, John P.; Sneiderman, Charles; Rivera, Christopher)V1444,374-378(1991)

Edge-based block matching technique for video motion estimation (Han, Richard Y.; Zakhor, Avideh)V1452,395-408(1991)

Effects of M-transform for bit-error resilement in the adaptive DCT coding (Yamane, Nobumoto; Morikawa, Yoshitaka; Hamada, Hiroshi)V1605,679-686(1991)

Efficient odd max quantizer for use in transform image coding (Hauser, Neal A.; Mitchell, Harvey B.)V1605,428-433(1991)

Enhancement of transform coding by nonlinear interpolation (Wu, Siu W.; Gersho, Allen)V1605,487-498(1991)

Ericsson digital recce management system (Johansson, Micael; Engman, Per; Johansson, Bo G.)V1538,180-188(1991)

Error free and transparent coding of images using approximations by splines (Algazi, V. R.; Maurincomme, Eric; Ford, Gary E.)V1452,364-370(1991)

Evaluation of medical image compression by Gabor elementary functions (Anderson, Mary P.; Brown, David G.; Loew, Murray H.)V1444,407-412(1991)

Evolving JPEG color data compression standard (Mitchell, Joan L.; Pennebaker, William B.)VCR37,68-97(1991)

Fast finite-state codebook design algorithm for vector quantization (Chang, Ruey-Feng; Chen, Wen-Tsuen)V1605,172-178(1991)

Fractal image compression and texture analysis (Moghaddam, Baback; Hintz, Kenneth J.; Stewart, Clayton V.)V1406,42-57(1991)

Full-frame entropy coding for radiological image compression (Lo, Shih-Chung B.; Krasner, Brian H.; Mun, Seong Ki; Horii, Steven C.)V1444,265-271(1991)

HDTV compression with vector quantization of transform coefficients (Wu, Siu W.; Gersho, Allen)V1605,73-84(1991)

Hierarchical motion-compensated interframe DPCM algorithm for low-bit-rate coding (Xie, Kan; Van Eycken, Luc; Oosterlinck, Andre J.)V1567,380-389(1991)

High-speed programmable digitizer for real-time video compression experiments (Cox, Norman R.)V1605,906-915(1991)

High-speed two-dimensional pyramid image coding method and its implementation (Sahinoglou, Haralambos; Cabrera, Sergio D.)V1605,793-804(1991)

Hilbert scanning arithmetic coding for multispectral image compression (Perez, Arnulfo; Kamata, Seiichiro; Kawaguchi, Eiji)V1567,354-361(1991)

Human facial motion modeling, analysis, and synthesis for video compression (Huang, Thomas S.; Reddy, Subhash C.; Aizawa, Kiyoharu)V1605,234-241(1991)

Image coding based on two-channel conjugate vector quantization (Nam, J. Y.; Rao, K. R.)V1452,485-496(1991)

Image compression for digital video tape recording with high-speed playback capability (Wu, Siu W.; Gersho, Allen)V1452,352-363(1991)

Image enhancement using nonuniform sampling (Bender, Walter J.; Rosenberg, Charles)V1460,59-70(1991)

Image Processing Algorithms and Techniques II (Civanlar, Mehmet R.; Mitra, Sanjit K.; Moorhead, Robert J., eds.)V1452(1991)

Image vector quantization with block-adaptive scalar prediction (Gupta, Smita; Gersho, Allen)V1605,179-189(1991)

Information preserving image data compression (Tavakoli, Nassrin)V1452,371-382(1991)

Iterative procedures for reduction of blocking effects in transform image coding (Rosenholtz, Ruth E.; Zakhor, Avideh)V1452,116-126(1991)

Lapped orthogonal transform for motion-compensated video compression (Lynch, William E.; Reibman, Amy R.)V1605,285-296(1991)

Legibility of compressed document images at various spatial resolutions (Kidd, Robert C.; Zalinski, Charles M.; Nadel, Jerome I.; Klein, Robert D.)V1454,414-424(1991)

Model for packet image communication in a centralized distribution system (Torbey, Habib H.; Zhang, Zhensheng)V1605,650-666(1991)

Motion-compensated subsampling of HDTV (Belfor, Ricardo A.; Lagendijk, Reginald L.; Biemond, Jan)V1605,274-284(1991)

Motion-compensated wavelet transform coding for color video compression (Zhang, Ya-Qin; Zafar, Sohail)V1605,301-316(1991)

MPEG (moving pictures expert group) video compression algorithm: a review (Le Gall, Didier J.)V1452,444-457(1991)

N-dimensional Hilbert scanning and its application to data compression (Perez, Arnulfo; Kamata, Seiichiro; Kawaguchi, Eiji)V1452,430-441(1991)

Neural networks in bandwidth compression (Habibi, Ali)V1567,334-340(1991)

New approach to image coding using 1-D subband filtering (Yu, Tian-Hu; Mitra, Sanjit K.)V1452,420-429(1991)

New method for chain coding based on convolution (Qing, Kent P.; Means, Robert W.)V1567,390-396(1991)

Nonuniform visual pattern image sequence coding (Silsbee, Peter L.; Bovik, Alan C.)V1452,409-419(1991)

Numerical design of parallel multiresolution filter banks for image coding applications (Popat, Ashok C.; Li, Wei; Kunt, Murat)V1567,341-353(1991)

Object-oriented representation of image space by puzzletrees (Dengel, Andreas R.)V1606,20-30(1991)

Observer detection of image degradation caused by irreversible data compression processes (Chen, Ji; Flynn, Michael J.; Gross, Barry; Spizarny, David)V1444,256-264(1991)

Overlapping block transform for offset-sampled image compression (Morikawa, Yoshitaka; Yamane, Nobumoto; Hamada, Hiroshi)V1605,445-455(1991)

Pattern recognition using w-orbit finite automata (Liu, Ying; Ma, Hede)V1606,226-240(1991)

"Perfect" displays and "perfect" image compression in space and time (Klein, Stanley A.; Carney, Thom)V1453,190-205(1991)

Performance of an HDTV codec adopting transform and motion compensation techniques (Barbero, Marzio; Cucchi, Silvio; Muratori, Mario)V1567,566-577(1991)

Practical approach to fractal-based image compression (Pentland, Alexander P.; Horowitz, Bradley)V1605,467-474(1991)

Present state of HDTV coding in Japan and future prospect (Murakami, Hitomi)V1567,544-555(1991)

Prioritized DCT (discrete cosine transform) image coding (Huang, Yunming G.; Dreizen, Howard M.)V1396,624-637(1991)

Quantization of color image components in the DCT domain (Peterson, Heidi A.; Peng, Hui; Morgan, J. H.; Pennebaker, William B.)V1453,210-222(1991)

Rapid enhancement and compression of image data (Balram, Nikhil; Moura, Jose M.)V1606,374-385(1991)

Recursive scaled DCT (Hou, Hsieh S.)V1567,402-412(1991)

Reduction of blocking artifacts using motion-compensated spatial-temporal filtering (Chu, Frank J.; Yeh, Chia L.)V1452,38-46(1991)

Residual VQ (vector quantizaton) with state prediction: a new method for image coding (Kossentini, Faouzi; Smith, Mark J.; Barnes, Christopher)V1452,383-394(1991)

Reversible compression of industrial radiographs using multiresolution decorrelation (Chen, Keshi; Ramabadran, Tenkasi V.)V1567,397-401(1991)

Reversible image data compression based on HINT (hierarchical interpolation) decorrelation and arithmetic coding (Roos, Paul; Viergever, Max A.)V1444,283-290(1991)

Scaled discrete cosine transform algorithms for JPEG and MPEG implementations on fused multiply/add architectures (Feig, Ephraim; Linzer, Elliot)V1452,458-467(1991)

Scientific data compression for space: a modified block truncation coding algorithm (Lu, Wei-Wei; Gough, M. P.; Davies, Peter N.)V1470,197-205(1991)

Signal processing LSI system for digital still camera (Watanabe, Mikio; Saito, Osamu; Okamoto, Satoru; Ito, Kenji; Moronaga, Kenji; Hayashi, Kenkichi; Nishi, Seiki)V1452,27-36(1991)

Space-filling curves for image compression (Moghaddam, Baback; Hintz, Kenneth J.; Stewart, Clayton V.)V1471,414-421(1991)

Standards for electronic imaging for facsimile systems (Urban, Stephen J.)VCR37,113-145(1991)

Storing and managing three-dimensional digital medical image information (Chapman, Michael A.; Denby, N.)V1526,190-194(1991)

Study of binary image compression using universal coding (Nakano, Yasuhiko; Chiba, Hirotaka; Okada, Yoshiyuki; Yoshida, Shigeru; Mori, Masahiro)V1605,874-878(1991)

Study on objective fidelity for progressive image transmission systems (Lin, Eleanor; Lu, Cheng-Chang)V1453,206-209(1991)

Subjective evaluation of scale-space image coding (de Ridder, Huib)V1453,31-42(1991)

Subsampled vector quantization with nonlinear estimation using neural network approach (Sun, Huifang; Manikopoulos, Constantine N.; Hsu, Hwei P.)V1605,214-220(1991)

Sunset: a hardware-oriented algorithm for lossless compression of gray-scale images (Langdon, Glen G.)V1444,272-282(1991)

Tree-structured vector quantization with input-weighted distortion measures (Cosman, Pamela C.; Oehler, Karen; Heaton, Amanda A.; Gray, Robert M.)V1605,162-171(1991)

Two new image compression methods utilizing mathematical morphology (Vepsalainen, Ari M.; Toivanen, Pekka J.)V1606,282-293(1991)

Update on focal-plane image processing research (Kemeny, Sabrina E.; Eid, El-Sayed I.; Mendis, Sunetra; Fossum, Eric R.)V1447,243-250(1991)

Use of a human visual model in subband coding of color video signal with adaptive chrominance signal vector quantization (Barba, Dominique; Hanen, Jose)V1605,408-419(1991)

Use of coordinate mapping as a method for image data reduction (Abbott, Mark A.; Messner, Richard A.)V1381,272-282(1991)

Vector quantization of image pyramids with the ECPNN algorithm (de Garrido, Diego P.; Pearlman, William A.; Finamore, Weiler A.)V1605,221-232(1991)

Video compression algorithm with adaptive bit allocation and quantization (Viscito, Eric; Gonzales, Cesar A.)V1605,58-72(1991)

Visual Communications and Image Processing '91: Image Processing (Tzou, Kou-Hu; Koga, Toshio, eds.)V1606(1991)

Visualization and volumetric compression (Chan, Kelby K.; Lau, Christina C.; Chuang, Keh-Shih; Morioka, Craig A.)V1444,250-255(1991)

Image enhancement

Adaptive filtering and enhancement method of speckle pattern (Yang, Yibing; Wang, Houshu)V1554A,781-788(1991)

Antialiasing-warped imagery using lookup-table-based methods for adaptive resampling (Walterman, Michael T.; Weinhaus, Frederick M.)V1567,204-214(1991)

Application of super-resolution techniques to passive millimeter-wave images (Gleed, David G.; Lettington, Alan H.)V1567,65-72(1991)

Applications of cellular logic image processing (Denber, Michel J.)V1398,29-38(1991)

Approach to real-time contrast enhancement for airborne reconnaissance applications (Henson, Michael A.; Petro, David)V1396,582-589(1991)

Brightness and contrast adjustments for different tissue densities in digital chest radiographs (McNitt-Gray, Michael F.; Taira, Ricky K.; Eldredge, Sandra L.; Razavi, Mahmood)V1445,468-478(1991)

Circular polarization effects in the light scattering from single and suspensions of dinoflagellates (Shapiro, Daniel B.; Hunt, Arlon J.; Quinby-Hunt, Mary S.; Hull, Patricia G.)V1537,30-41(1991)

Color-encoded depth: an image enhancement tool (Bieman, Leonard H.)V1385,229-238(1991)

Edge detection with subpixel accuracy (Koplowitz, Jack; Lee, Xiaobing)V1471,452-463(1991)

Enhancement of dental x-ray images by two-channel image processing (Mitra, Sanjit K.; Yu, Tian-Hu)V1445,156-165(1991)

Enhancement of transform coding by nonlinear interpolation (Wu, Siu W.; Gersho, Allen)V1605,487-498(1991)

Generic modular imaging IR signal processor (Auborn, John E.; Harris, William R.)V1483,2-9(1991)

Gray scale and resolution enhancement capabilities of edge emitter imaging stations (Leksell, David; Kun, Zoltan K.; Asars, Juris A.; Phillips, Norman J.; Brandt, Gerald B.; Stringer, J. T.; Matty, T. C.; Gigante, Joseph R.)V1458,133-144(1991)

High-resolution reconstruction of line drawings (Wong, Ping W.; Koplowitz, Jack)V1398,39-47(1991)

Image and modality control issues in the objective evaluation of manipulation techniques for digital chest images (Rehm, Kelly; Seeley, George W.; Dallas, William J.)V1445,24-35(1991)

Image annotation under X Windows (Pothier, Steven)V1472,46-53(1991)

Image contrast enhancement via blurred weighted adaptive histogram equalization (Gauch, John M.)V1606,386-399(1991)

Image enhancement of infrared uncooled focal plane array imagery (McCauley, Howard M.; Auborn, John E.)V1479,416-422(1991)

Image enhancement using nonuniform sampling (Bender, Walter J.; Rosenberg, Charles)V1460,59-70(1991)

Imagery super-resolution: emerging prospects (Hunt, Bobby R.)V1567,600-608(1991)

Information-theoretic approach to optimal quantization (Lorenzo, Maximo; Der, Sandor Z.; Moulton, Joseph R.)V1483,118-137(1991)

Investigation of self-aligned phase-shifting reticles by simulation techniques (Noelscher, Christoph; Mader, Leonhard)V1463,135-150(1991)

L-filter design using the gradient search algorithm (Roy, Sumit)V1451,254-256(1991)

LUM filters for smoothing and sharpening (Boncelet, Charles G.; Hardie, Russell C.; Hakami, M. R.; Arce, Gonzalo R.)V1451,70-74(1991)

Massively parallel image restoration (Menon, Murali M.)V1471,185-190(1991)

Modified Laplacian enhancement of low-resolution digital images (Naylor, David C.; Daemi, M. F.)V1567,522-532(1991)

New edge-enhancement method based on the deviation of the local image grey center (Zhao, Mingsheng; Shen, Zhen-kang; Chen, Huihuang)V1471,464-473(1991)

New method of adjusting color of pseudocolor encoding image (Cai, Hai-Tao; Chen, Zhen-Pei)V1567,703-708(1991)

New perspectives on image enhancement for the visually impaired (Peli, Eli)V1382,49-59(1991)

New type of modified trimmed mean filter (Wu, Wen-Rong; Kundu, Amlan)V1451,13-23(1991)

Noise reduction in ultrasound images using multiple linear regression in a temporal context (Olstad, Bjorn)V1451,269-281(1991)

Nonlinear signal processing using integration of fluid dynamics equations (Eidelman, Shmuel; Grossmann, William; Friedman, Aharon)V1567,439-450(1991)

Novel nonlinear filter for image enhancement (Yu, Tian-Hu; Mitra, Sanjit K.; Kaiser, James F.)V1452,303-309(1991)

On the determination of optimal window for registration of nonlinear distributed images (Lure, Yuan-Ming F.)V1452,292-302(1991)

Optimization of morphological structuring elements for angiogram enhancement (Andress, Keith M.; Wilson, David L.)V1445,6-10(1991)

Photorefractive two-wave mixing characteristics for image amplification in diffusion-driven media (Gilbreath, G. C.; Clement, Anne E.; Fugera, S. N.; Mizell, Gregory J.)V1409,87-99(1991)

Quantitative evaluation of image enhancement algorithms (Lu, Hong-Qian)V1453,223-234(1991)

Quantitative performance evaluation of the EM algorithm applied to radiographic images (Brailean, James C.; Giger, Maryellen L.; Chen, Chin-Tu; Sullivan, Barry J.)V1450,40-46(1991)

Real-time optical processor for increasing resolution beyond the diffraction limit (Dhadwal, Harbans S.; Noel, Eric)V1564,664-673(1991)

Resolution enhancement in digital in-line holography (Hua, Lifan; Xie, Gong-Wie; Shaw, David T.; Scott, Peter D.)V1385,142-151(1991)

Resolution enhancement of blurred star field images by maximally sparse restoration (Jeffs, Brian D.; Gunsay, Metin; Dougal, John)V1567,511-521(1991)

Scaling of digital shapes with subpixel boundary estimation (Koplowitz, Jack)V1470,167-174(1991)

Shocks and other nonlinear filtering applied to image processing (Osher, Stanley; Rudin, Leonid I.)V1567,414-431(1991)

Subjective evaluation of image enhancements in improving the visibility of pathology in chest radiographs (Plessis, Brigitte; Goldberg, Morris; Belanger, Garry; Hickey, Nancy M.)V1445,539-554(1991)

Using information from multiple low-resolution images to increase resolution (Gagne, Philippe; Arsenault, Henri H.)V1564,656-663(1991)

Wiener-matrix image restoration beyond the sampling passband (Rahman, Zia-ur; Alter-Gartenberg, Rachel; Fales, Carl L.; Huck, Friedrich O.)V1488,237-248(1991)

Image formation

Acquisition of very high resolution images using stereo cameras (Aizawa, Kiyoharu; Komatsu, Takashi; Saito, Takahiro)V1605,318-328(1991)

Adaptive isosurface generation in a distortion-rate framework (Ning, Paul C.; Hesselink, Lambertus B.)V1459,11-21(1991)

Anatomical constraints for neuromagnetic source models (George, John S.; Lewis, Paul S.; Ranken, D. M.; Kaplan, L.; Wood, C. C.)V1443,37-51(1991)

Basic principles of stereographic software development (Hodges, Larry F.)V1457,9-17(1991)

Binary and phase-shifting image design for optical lithography (Liu, Yong; Zakhor, Avideh)V1463,382-399(1991)

Characterization, Propagation, and Simulation of Sources and Backgrounds (Watkins, Wendell R.; Clement, Dieter, eds.)V1486(1991)

Comparison of three efficient-detail-synthesis methods for modeling using under-sampled data (Iannarilli, Frank J.; Wohlers, Martin R.)V1486,314-324(1991)

Digital shearing laser interferometry for heterodyne array phasing (Cederquist, Jack N.; Fienup, James R.; Marron, Joseph C.; Schulz, Timothy J.; Seldin, J. H.)V1416,266-277(1991)

Discussion on spectral sampling of imaging spectrometer (Han, Xin Z.)V1538,99-102(1991)

Dynamic infrared scene projection technology (Mobley, Scottie B.)V1486,325-332(1991)

Dynamic sea image generation (Levesque, Martin P.)V1486,294-300(1991)

Effects of microscan operation on staring infrared sensor imagery (Blommel, Fred P.; Dennis, Peter N.; Bradley, D. J.)V1540,653-664(1991)

Equalization radiography with radiation quality modulation (Geluk, Ronald J.; Vlasbloem, Hugo)V1443,143-152(1991)

Estimation of scene correlation lengths (Futterman, Walter I.; Schweitzer, Eric L.; Newt, J. E.)V1486,127-140(1991)

General principles of constructing the nuclei of nonlinear IR-image transformations (Nesteruk, Vsevolod F.)V1540,468-476(1991)

Generation of synthetic stereo views from digital terrain models and digitized photographs (Bethel, James S.)V1457,49-53(1991)

Geometrical and radiometrical signal transfer characteristics of a color CCD camera with 21-million pixels (Lenz, Reimar K.; Lenz, Udo)V1526,123-132(1991)

Image acquisition: quality control for document image scanners (Gershbock, Richard)VCR37,20-39(1991)

Implementing early visual processing in analog VLSI: light adaptation (Mann, James)V1473,128-136(1991)

Improved precision/resolution by camera movement (Taylor, Geoff L.; Derksen, Grant)V1526,27-34(1991)

Linewidth measurement comparison between a photometric optical microscope and a scanning electron microscope backed with Monte Carlo trajectory computations (Nunn, John W.; Turner, Nicholas P.)V1464,50-61(1991)

Magnetic resonance reconstruction from projections using half the data (Noll, Douglas C.; Pauly, John M.; Nishimura, Dwight G.; Macovski, Albert)V1443,29-36(1991)

Medical Imaging V: Image Physics (Schneider, Roger H., ed.)V1443(1991)

Modeling ultrasound speckle formation and its dependence on imaging system's response (Rao, Navalgund A.; Zhu, Hui)V1443,81-95(1991)

Monte Carlo modeling of secondary electron signals from heterogeneous specimens with nonplanar surfaces (Russ, John C.; Dudley, Bruce W.; Jones, Susan K.)V1464,10-21(1991)

More realistic and efficient algorithm for the drawing of 3-D space-filling molecular models (Wang, Yanqun)V1606,1027-1036(1991)

Nonoptical noncoherent imaging in industrial testing (Delecki, Z. A.; Barakat, M. A.)V1526,157-167(1991)

Optical image transformation including IR region for information conformity of their formation and perception processes based on Fibonacci polynomials and series (Miroshnikov, Mikhail M.; Nesteruk, Vsevolod F.)V1540,477-487(1991)

Optimal characteristics of rheology and electric field in deformable polymer films of optoelectronic image formation devices (Tarasov, Victor A.; Kuleshov, Nickolay B.; Novoselets, Mikhail K.; Sarkisov, Sergey S.)V1559,331-342(1991)

Peculiarities of anisotropic photopolymerization in films (Krongauz, Vadim V.; Schmelzer, E. R.)V1559,354-376(1991)

Real-time simulation of the retina allowing visualization of each processing stage (Teeters, Jeffrey L.; Werblin, Frank S.)V1472,6-17(1991); V1473,102-113(1991)

Realistic model for battlefield fire plume simulation (Bruce, Dorothy)V1486,231-236(1991)

Retina-like image acquisition system with wide-range light adaptation (Chang, Po-Rong; Yeh, Bao-Fuh)V1606,456-469(1991)

Scan-free echo imaging of dynamic objects (Soumekh, Mehrdad)V1443,96-106(1991)

Short-scan fan beam algorithm for noncircular detector orbits (Zeng, Gengsheng L.; Gullberg, Grant T.)V1445,332-340(1991)

Signal, noise, and readout considerations in the development of amorphous silicon photodiode arrays for radiotherapy and diagnostic x-ray imaging (Antonuk, Larry E.; Boudry, J.; Kim, Chung-Won; Longo, M.; Morton, E. J.; Yorkston, J.; Street, Robert A.)V1443,108-119(1991)

Signal-to-noise performance in cesium iodide x-ray fluorescent screens (Hillen, Walter; Eckenbach, W.; Quadflieg, P.; Zaengel, Thomas T.)V1443,120-131(1991)

Silicon retina with adaptive photoreceptors (Mahowald, Misha A.)V1473,52-58(1991)

Simultaneous graphics and multislice raster image display for interactive image-guided surgery (Edwards, Charles A.; Galloway, Robert L.; Thomas, Judith G.; Schreiner, Steven; Maciunas, Robert J.)V1444,38-46(1991)

Stereoscopic ray tracing of implicitly defined surfaces (Devarajan, Ravinder; McAllister, David F.)V1457,37-48(1991)

Thermal model for real-time textured IR background simulation (Bernstein, Uri; Keller, Catherine E.)V1486,345-351(1991)

Three-dimensional real-time ultrasonic imaging using ellipsoidal backprojection (Anderson, Forrest L.)V1443,62-80(1991)

Training image collection at CECOM's Center for Night Vision and Electro-Optics (Harr, Richard W.)V1483,231-239(1991)

Validated CCD camera model (Johnson-Cole, Helen; Clark, Rodney L.)V1488,203-211(1991)

Image processing—see also computer vision; filters; image algebra; image compression; image reconstruction/restoration; medical imaging; neural networks; object recognition; optical processing; parallel processing; pattern recognition; visual communications

Accuracy and high-speed technique for autoprocessing of Young's fringes (Chen, Wenyi; Tan, Yushan)V1554A,879-885(1991)

Acquisition and processing of digital images and spectra in astronomy (Lee, Terence J.)V1439,152-158(1991)

Adaptive detection of subpixel targets using multiband frame sequences (Stocker, Alan D.; Yu, Xiaoli; Winter, Edwin M.; Hoff, Lawrence E.)V1481,156-169(1991)

Adaptive filtering and enhancement method of speckle pattern (Yang, Yibing; Wang, Houshu)V1554A,781-788(1991)

Adaptive imager: a real-time locally adaptive edge enhancement system (Strang, Steven E.)V1384,246-256(1991)

Adaptive isosurface generation in a distortion-rate framework (Ning, Paul C.; Hesselink, Lambertus B.)V1459,11-21(1991)

Adaptive morphological filter for image processing (Cheng, Fulin; Venetsanopoulos, Anastasios N.)V1483,49-59(1991)

Adaptive morphological multiresolution decomposition (Salembier, Philippe; Jaquenoud, Laurent)V1568,26-37(1991)

Adaptive typography for dynamic mapping environments (Bardon, Didier)V1460,126-139(1991)

Aircraft exterior scratch measurement system using machine vision (Sarr, Dennis P.)V1472,177-184(1991)

Analog CCD processors for image filtering (Yang, Woodward)V1473,114-127(1991)

Analog CMOS IC for object position and orientation (Standley, David L.; Horn, Berthold K.)V1473,194-201(1991)

Analysis of noise attenuation in morphological image processing (Koskinen, Lasse; Astola, Jaakko; Neuvo, Yrjo A.)V1451,102-113(1991)

Analyzing and interpreting pulmonary tomoscintigraphy sequences: realization and perspectives (Forte, Anne-Marie; Bizais, Yves)V1445,409-420(1991)

ANN approach for 2-D echocardiographic image processing: application of neocognitron model to LV boundary formation (Wan, Liqun; Li, Dapeng; Wee, William G.; Han, Chia Y.; Porembka, D. T.)V1469,432-440(1991)

Application of a multilayer network in image object classification (Tang, Yonghong; Wee, William G.; Han, Chia Y.)V1469,113-120(1991)

Application of back-propagation to the recognition of handwritten digits using morphologically derived features (Hepp, Daniel J.)V1451,228-233(1991)

Application of median-type filtering to image segmentation in electrophoresis (Wang, Qiaofei; Neuvo, Yrjo A.)V1450,47-58(1991)

Application of morphological pseudoconvolutions to scanning-tunneling and atomic force microscopy (Dougherty, Edward R.; Weisman, Andrew; Mizes, Howard; Miller, Robert J.)V1567,88-99(1991)

Application of the phase and amplitude modulating properties of LCTVs (Kirsch, James C.; Loudin, Jeffrey A.; Gregory, Don A.)V1474,90-100(1991)

Application of wavelet-type functions in image processing (Segall, Ilana; Zeevi, Yehoshua Y.)V1606,1048-1058(1991)

Applications of aerial photography to law enforcement and disaster assessment: a consideration of the state-of-the-art (Cox, William J.; Biache, Andrew)V1479,364-369(1991)

Applications of Artificial Neural Networks II (Rogers, Steven K., ed.)V1469(1991)

Applications of cellular logic image processing (Denber, Michel J.)V1398,29-38(1991)

Applications of Digital Image Processing XIV (Tescher, Andrew G., ed.)V1567(1991)

Applications of text-image editing (Bagley, Steven C.; Kopec, Gary E.)V1460,71-79(1991)

Approach to invariant object recognition on grey-level images by exploiting neural network models (Rybak, Ilya A.; Golovan, Alexander V.; Gusakova, Valentina I.)V1469,472-482(1991)

Architecture for surveillance in real time using nonlinear image processing hardware (Pais, Cassiano P.; Carvalho, Fernando D.; Silvestre, Victor M.)V1451,282-288(1991)

Artificial neural net learns the Sobel operators (and more) (Weller, Scott W.)V1469,219-224(1991)

Artificial neural network and image processing using the Adaptive Solutions' architecture (Baker, Thomas E.)V1452,502-511(1991)

Artificial neural system approach to IR target identification (Holland, Orgal T.; Tarr, Tomas; Farsaie, Ali; Fuller, James M.)V1469,102-112(1991)

ATR performance modeling for building multiscenario adaptive systems (Nasr, Hatem N.)V1483,112-117(1991)

Automated labeling of coronary arterial tree segments in angiographic projection data (Dumay, Adrie C.; Gerbrands, Jan J.; Geest, Rob J.; Verbruggen, Patricia E.; Reiber, J. H.)V1445,38-46(1991)

Automated optical grading of timber (Sobey, Peter J.)V1379,168-179(1991)

Automated registration of terrain range images using surface feature level sets (Wheeler, Frederick W.; Vaz, Richard F.; Cyganski, David)V1606,78-85(1991)

Automated three-dimensional registration of medical images (Neiw, Han-Min; Chen, Chin-Tu; Lin, Wei-Chung; Pelizzari, Charles A.)V1445,259-264(1991)

Automatic 3-D reconstruction of vascular geometry from two orthogonal angiographic image sequences (Close, Robert A.)V1445,513-522(1991)

Automatic analysis of heliotest strips (Langinmaa, Anu)V1468,573-580(1991)

Automatic inspection of optical fibers (Silberberg, Teresa M.)V1472,150-156(1991)

Automatic inspection system for full-color printed matter (Meguro, Shin-Ichi; Nunotani, Masakatu; Tanimizu, Katsuyuki; Sano, Mutsuo; Ishii, Akira)V1384,27-37(1991)

Automatic Object Recognition (Sadjadi, Firooz A., ed.)V1471(1991)

Automatic object recognition: critical issues and current approaches (Sadjadi, Firooz A.)V1471,303-313(1991)

Automatic target detection for surveillance (Ramesh, Nagarajan; Sethi, Ishwar K.; Cheung, Huey)V1468,72-80(1991)

Automatic target recognition using acousto-optic image correlator (Molley, Perry A.; Kast, Brian A.)V1471,224-232(1991)

Automatic vehicle model identification (Varga, Margaret J.; Radford, John C.)V1381,92-100(1991)

Bayesian approach to segmentation of temporal dynamics in video data (Jones, Coleen T.; Sauer, Ken D.)V1605,522-533(1991)

Bayesian estimation of smooth object motion using data from direction-sensitive velocity sensors (Fong, David Y.; Pomalaza-Raez, Carlos A.)V1569,156-161(1991)

Biomedical Image Processing II (Bovik, Alan C.; Howard, Vyvyan, eds.)V1450(1991)

Bispectral magnitude and phase recovery using a wide bandwidth acousto-optical processor (Kniffen, Stacy K.; Becker, Michael F.; Powers, Edward J.)V1564,617-627(1991)

Black and white cinefilm (ASA 5) applied to dynamic laser speckle and its processing (He, Anzhi; Yu, Gui Ying; Ni, Xiao W.; Fan, Hua Ying)V1554A,747-749(1991)

Block-adaptive quantization of multiple-frame motion field (Lavagetto, Fabio; Leonardi, Riccardo)V1605,534-545(1991)

Block arithmetic coding of contour images (Kim, Kyoil; Kim, Jonglak; Kim, Taejeong)V1605,851-862(1991)

Bounds on the performance of optimal four-dimensional filters for detection of low-contrast IR point targets (Wohlers, Martin R.)V1481,129-139(1991)

Brightness and contrast adjustments for different tissue densities in digital chest radiographs (McNitt-Gray, Michael F.; Taira, Ricky K.; Eldredge, Sandra L.; Razavi, Mahmood)V1445,468-478(1991)

BRICORK: an automatic machine with image processing for the production of corks (Davies, Roger; Correia, Bento A.; Carvalho, Fernando D.; Rodrigues, F. C.)V1459,283-291(1991)

Building an infrastructure for system integration: machine vision standards (Bieman, Leonard H.; Peyton, James A.)VCR36,3-19(1991)

Calibration issues in the measurement of ocular movement and position using computer image processing (Weeks, Arthur R.; Myler, Harley R.; Jolson, Alfred S.)V1567,77-87(1991)

Case studies in machine vision integration (Ahlers, Rolf-Juergen)VCR36,56-63(1991)

Centroid tracking of range-Doppler images (Kachelmyer, Alan L.; Nordquist, David P.)V1416,184-198(1991)

Classification of tissue-types by combining relaxation labeling with edge detection (Adiseshan, Prakash; Faber, Tracy L.)V1445,128-132(1991)

Class of GOS (generalized order-statistic) filters, similar to the median, that provide edge enhancement (Longbotham, Harold G.; Barsalou, Norman)V1451,36-47(1991)

Clinical image-intensifier-based volume CT imager for angiography (Ning, Ruola; Barsotti, John B.; Kido, Daniel K.; Kruger, Robert A.)V1443,236-249(1991)

Clustering algorithms for a PC-based hardware implementation of the unsupervised classifier for the shuttle ice detection system (Jaggi, Sandeep)V1451,289-297(1991)

Color analysis of nonlinear-phase-modulation method for density pseudocolor encoding technique in medical application (Liu, Dingyu; Yang, Xiaobo; Liu, Changjun; Zhang, Honguo)V1443,191-196(1991)

Color character recognition method based on a model of human visual processing (Yamaba, Kazuo; Miyake, Yoichi)V1453,290-299(1991)

Colorimetric calibration for scanners and media (Hung, Po-Chieh)V1448,164-174(1991)

Color segmentation using MDL clustering (Wallace, Richard S.; Suenaga, Yasuhito)V1381,436-446(1991)

Combined edge- and region-based method for range image segmentation (Koivunen, Visa; Pietikainen, Matti)V1381,501-512(1991)

Combined-transform coding scheme for medical images (Zhang, Ya-Qin; Loew, Murray H.; Pickholtz, Raymond L.)V1445,358-366(1991)

Comparative evaluation of neural-based versus conventional segmentors (Daniell, Cindy E.; Kemsley, David; Bouyssounouse, Xavier)V1471,436-451(1991)

Comparative study of texture measurements for cellular organelle recognition (Grenier, Marie-Claude; Durand, Louis-Gilles; de Guise, J.)V1450,154-169(1991)

Comparison of backpropagation neural networks and statistical techniques for analysis of geological features in Landsat imagery (Parikh, Jo Ann; DaPonte, John S.; Damodaran, Meledath; Karageorgiou, Angelos; Podaras, Petros)V1469,526-538(1991)

Comparison of methods to treat nonuniform illumination in images (McDonald, Timothy P.; Chen, Yud-Ren)V1379,89-98(1991)

Comparison of morphological and conventional edge detectors in medical imaging applications (Kaabi, Lotfi; Loloyan, Mansur; Huang, H. K.)V1445,11-23(1991)

Comparison of neural network classifiers to quadratic classifiers for sensor fusion (Brown, Joseph R.; Bergondy, Daniel; Archer, Susan J.)V1469,539-543(1991)

Comparison of three efficient-detail-synthesis methods for modeling using under-sampled data (Iannarilli, Frank J.; Wohlers, Martin R.)V1486,314-324(1991)

Composite PET and MRI for accurate localization and metabolic modeling: a very useful tool for research and clinic (Bidaut, Luc M.)V1445,66-77(1991)

Computational model of the imaging process in scanning-x microscopy (Gallarda, Harry S.; Jain, Ramesh C.)V1464,459-473(1991)

Computer vision system for automated inspection of molded plastic print wheels (Hu, Yong-Lin; Wee, William G.; Gruver, William A.; Han, Chia Y.)V1468,653-661(1991)

Concept for the subresolution measurement of earthquake strain fields using SPOT panchromatic imagery (Crippen, Robert E.; Blom, Ronald G.)V1492,370-377(1991)

Converting non-interlaced to interlaced images in YIQ and HSI color spaces (Welch, Eric B.; Moorhead, Robert J.; Owens, John K.)V1453,235-243(1991)

Corn plant locating by image processing (Jia, Jiancheng; Krutz, Gary W.; Gibson, Harry G.)V1379,246-253(1991)

Correction of image-phase aberrations in MRI with applications (Riek, Jonathan K.; Tekalp, Ahmet M.; Smith, Warren E.; Parker, Kevin J.)V1445,190-197(1991)

Correlation model for a class of medical images (Zhang, Ya-Qin; Loew, Murray H.; Pickholtz, Raymond L.)V1445,367-373(1991)

CT image processing for hardwood log inspection (Zhu, Dongping; Conners, Richard W.; Araman, Philip A.)V1567,232-243(1991)

Custom-made filters in digital image analysis system (Ostrowski, Tomasz)V1391,264-266(1991)

Decision-directed entropy-based adaptive filtering (Myler, Harley R.; Weeks, Arthur R.; Van Dyke-Lewis, Michelle)V1565,57-68(1991)

Decomposing morphological structure element into neighborhood configurations (Gong, Wei; Shi, Qinyun; Cheng, Minde)V1606,153-164(1991)

Decomposition and inversion of von Neumann-like convolution operators (Manseur, Zohra Z.; Wilson, David C.)V1568,164-173(1991)

Decorrelation of color images using total color difference (Zheng, Joe; Valavanis, Kimon P.; Gauch, John M.)V1606,1037-1047(1991)

Deformable markers: mathematical morphology for active contour models control (Rougon, Nicolas F.; Preteux, Francoise)V1568,78-89(1991)

Deformation measurements of the human tympanic membrane under static pressure using automated moire topography (Dirckx, Joris J.; Decraemer, Willem F.)V1429,34-38(1991)

Design and analysis of the closed-loop pointing system of a scientific satellite (Heyler, Gene A.; Garlick, Dean S.; Yionoulis, Steve M.)V1481,198-208(1991)

Design considerations for EOS direct broadcast (Vermillion, Charles H.; Chan, Paul H.)V1492,224-228(1991)

Design of minimum MAE generalized stack filters for image processing (Zeng, Bing; Gabbouj, Moncef; Neuvo, Yrjo A.)V1606,443-454(1991)

Detection and classification of undersea objects using multilayer perceptrons (Shazeer, Dov J.; Bello, M.)V1469,622-636(1991)

Detection and visualization of porosity in industrial CT scans of aluminum die castings (Andrews, Lee T.; Klingler, Joseph W.; Schindler, Jeffery A.; Begeman, Michael S.; Farron, Donald; Vaughan, Bobbi; Riggs, Bud; Cestaro, John)V1459,125-135(1991)

Detection of liver metastisis using the backpropagation algorithm and linear discriminant analysis (DaPonte, John S.; Parikh, Jo Ann; Katz, David A.)V1469,441-450(1991)

Detection of surface-laid minefields using a hierarchical image processing algorithm (McFee, John E.; Russell, Kevin L.; Ito, Mabo R.)V1567,42-52(1991)

Detection of targets in terrain clutter by using multispectral infrared image processing (Hoff, Lawrence E.; Evans, John R.; Bunney, Laura)V1481,98-109(1991)

Determination of flint wheel orientation for the automated assembly of lighters (Safabakhsh, Reza)V1472,185-189(1991)

Determination of the homogeneous degree of deformation of the bearing cap of the motor by reflection holography (Cao, Zhengyuan; Cheng, Fang)V1554B,81-85(1991)

Development of a low-cost VME-based Nth order 2-D warper (Messner, Richard A.; Whitney, Erich C.)V1381,261-271(1991)

Development of a low-cost computed tomography image processing system (Hughes, Simon H.; Slocum, Robert E.)V1396,575-581(1991)

Development of an image processing system on a second-generation RISC workstation (Ryan, Martin J.; Kapp, Oscar H.)V1396,335-339(1991)

Development of a smart workstation for use in mammography (Giger, Maryellen L.; Nishikawa, Robert M.; Doi, Kunio; Yin, Fang-Fang; Vyborny, Carl J.; Schmidt, Robert A.; Metz, Charles E.; Wu, Yuzheng; MacMahon, Heber; Yoshimura, Hitoshi)V1445,101-103(1991)

Development of computer-aided functions in clinical neurosurgery with PACS (Mukasa, Minoru; Aoki, Makoto; Satoh, Minoru; Kowada, Masayoshi; Kikuchi, K.)V1446,253-265(1991)

Development of criteria to compare model-based texture analysis methods (Soh, Young-Sung; Murthy, S. N.; Huntsberger, Terrance L.)V1381,561-573(1991)

Development of image processing techniques for applications in flow visualization and analysis (Disimile, Peter J.; Shoe, Bridget; Toy, Norman; Savory, Eric; Tahouri, Bahman)V1489,66-74(1991)

Diagnostic image processing of remote operating seals for aerospace application (Nwagboso, Christopher O.)V1521,55-63(1991)

Digital halftoning using a generalized Peano scan (Agui, Takeshi; Nagae, Takanori; Nakajima, Masayuki)V1606,912-916(1991)

Digital imaging of aircraft dynamic seals: a fiber optics solution (Nwagboso, Christopher O.)V1500,234-245(1991)

Digital restoration of scanning electrochemical microscope images (Bartels, Keith A.; Lee, Chongmok; Bovik, Alan C.; Bard, Allen J.)V1450,30-39(1991)

Display nonlinearity in digital image processing for visual communications (Peli, Eli)V1606,508-519(1991)

Distributed industrial vision systems (Neve, Peter)VCR36,78-92(1991)

Distribution of the pattern spectrum mean for convex base images (Dougherty, Edward R.; Sand, Francis M.)V1451,114-124(1991)

Domain-variant gray-scale morphology (Kraus, Eugene J.)V1451,171-178(1991)

Dual-band optical system for IR multicolor signal processing (Kimball, Paulette R.; Fraser, James C.; Johnson, Jeffrey P.; Siegel, Andrew M.)V1540,687-698(1991)

Dynamic integration of visual cues for position estimation (Das, Subhodev; Ahuja, Narendra)V1382,341-352(1991)

Edge detection with subpixel accuracy (Koplowitz, Jack; Lee, Xiaobing)V1471,452-463(1991)

Effects of microscan operation on staring infrared sensor imagery (Blommel, Fred P.; Dennis, Peter N.; Bradley, D. J.)V1540,653-664(1991)

Efficient computation of various types of skeletons (Vincent, Luc M.)V1445,297-311(1991)

Efficient object contour tracing in a quadtree encoded image (Kumar, G. N.; Nandhakumar, N.)V1468,884-895(1991)

Efficient recognition with high-order neural networks (Carlson, Rolf; Jeffries, Clark)V1469,684-696(1991)

Efficient software techniques for morphological image processing with desktop computers (Zmuda, Michael A.; Tamburino, Louis A.; Rizki, Mateen M.)V1470,183-189(1991)

Efficient transformation algorithm for 3-D images (Vepsalainen, Ari M.; Rantala, Aarne E.)V1452,64-75(1991)

Electro-optical image processing architecture for implementing image algebra operations (Coffield, Patrick C.)V1568,137-148(1991)

Electro-optical transducer employing liquid crystal target for processing images in real-time scale (Ignatosyan, S. S.; Simonov, V. P.; Stepanov, Boris M.)V1358,100-108(1991)

Endoscopic inspection using a panoramic annular lens (Matthys, Donald R.; Gilbert, John A.; Puliparambil, Joseph T.)V1554B,736-742(1991)

Enhancement of dental x-ray images by two-channel image processing (Mitra, Sanjit K.; Yu, Tian-Hu)V1445,156-165(1991)

Estimation of linear parametric models of non-Gaussian discrete random fields (Tugnait, Jitendra K.)V1452,204-215(1991)

Evaluation of image tracker algorithms (Marshall, William C.)V1483,207-218(1991)

Evaluation of multiresolution elastic matching using MRI data (Gee, Jim C.; Reivich, Martin; Bilaniuk, L.; Hackney, D.; Zimmerman, R.; Kovacic, Stane; Bajcsy, Ruzena R.)V1445,226-234(1991)

Experiences with transputer systems for high-speed image processing (Kille, Knut; Ahlers, Rolf-Juergen; Schneider, B.)V1386,76-83(1991)

Experimental methodology for performance characterization of a line detection algorithm (Kanungo, Tapas; Jaisimha, Mysore Y.; Haralick, Robert M.; Palmer, John)V1385,104-112(1991)

Experiments with perceptual grouping (Shiu, Yiu C.)V1381,130-141(1991)

Extracting characters from illustration document by relaxation (Takeda, Haruo)V1386,128-134(1991)

Fast algorithm for size analysis of irregular pore areas (Yuan, Li-Ping)V1451,125-136(1991)

Fast dilation and erosion of time-varying grey-valued images with uncertainty (Laplante, Phillip A.; Giardina, Charles R.)V1568,295-302(1991)

Fast identification of images using neural networks (Min, Kwang-Shik; Min, Hisook L.)V1469,129-136(1991)

Fast piecewise-constant approximation of images (Radha, Hayder; Vetterli, Martin; Leonardi, Riccardo)V1605,475-486(1991)

Faying surface-gap measurement of aircraft structures for shim fabrication and installation (Sarr, Dennis P.; Jurick, Thomas W.)V1469,506-514(1991)

Field-portable laser beam diagnostics (Forrest, Gary T.)V1414,55-64(1991)

First direct measurements of transformation strains in crack-tip zone (Dadkhah, Mahyar S.; Marshall, David B.; Morris, Winfred L.)V1554A,164-175(1991)

First programmable digital optical processor: optical cellular logic image processor (Craig, Robert G.; Wherrett, Brian S.; Walker, Andrew C.; McKnight, D. J.; Redmond, Ian R.; Snowdon, John F.; Buller, G. S.; Restall, Edward J.; Wilson, R. A.; Wakelin, S.; McArdle, N.; Meredith, P.; Miller, J. M.; Taghizadeh, Mohammad R.; Mackinnon, G.; Smith, S. D.)V1505,76-86(1991)

FLIPS: Friendly Lisp Image Processing System (Gee, Shirley J.)V1472,38-45(1991)

Focal-plane image processing using acoustic charge transport technology (Brooks, Jeff W.)V1541,68-72(1991)

Fourier transform method to determine human corneal endothelial morphology (Masters, Barry R.; Lee, Yimkul; Rhodes, William T.)V1429,82-90(1991)

Fractal-based multifeature texture description (Kasparis, Takis; Tzannes, Nicolaos S.; Bassiouni, Mostafa; Chen, Qing)V1521,46-54(1991)

From object structure to object function (Zlateva, Stoyanka D.; Vaina, Lucia M.)V1468,379-393(1991)

Full-complex spatial filtering with a phase mostly DMD (Florence, James M.; Juday, Richard D.)V1558,487-498(1991)

Fusion of human vision system with mid-range IR image processing displays (Forsyth, William B.; Lewis, H. G.)V1472,18-25(1991)

Fuzzy logic approach to multitarget tracking in clutter (Priebe, Russell; Jones, Richard A.)V1482,265-274(1991)

Generalization of Lloyd's algorithm for image segmentation (Morii, Fujiki)V1381,545-552(1991)

Generalized neural networks for tactical target image segmentation (Tarr, Gregory L.; Rogers, Steven K.; Kabrisky, Matthew; Oxley, Mark E.; Priddy, Kevin L.)V1469,2-11(1991)

Generalizing from a small set of training exemplars for handwritten digit recognition (Simon, Wayne E.; Carter, Jeffrey R.)V1469,592-601(1991)

Generic modular imaging IR signal processor (Auborn, John E.; Harris, William R.)V1483,2-9(1991)

Genetic algorithm approach to visual model-based halftone pattern design (Chu, Chee-Hung H.; Kottapalli, M. S.)V1606,470-481(1991)

Graphical interface for multispectral simulation, scene generation, and analysis (Sikes, Terry L.; Kreiss, William T.)V1479,199-211(1991)

Graph theoretic approach to segmentation of MR images (Wu, Zhenyu; Leahy, Richard M.)V1450,120-132(1991)

Grouping and forming quantitative descriptions of image features by a novel parallel algorithm (Ben-Arie, Jezekiel; Huddleston, James)V1606,2-19(1991)

Hadamard transform-based object recognition using an array processor (Celenk, Mehmet; Moiz, Saifuddin)V1468,764-775(1991)

Hardware implementation of a neural network performing multispectral image fusion (Kagel, Joseph H.)V1469,659-664(1991)

Harmonic oscillator model of early visual image processing (Yang, Jian; Reeves, Adam J.)V1606,520-530(1991)

Health monitoring of rocket engines using image processing (Disimile, Peter J.; Shoe, Bridget; Toy, Norman)V1483,39-48(1991)

Hierarchical image decomposition based on modeling of convex hulls corresponding to a set of order statistic filters (Vepsalainen, Ari M.; Linnainmaa, Seppo; Yli-Harja, Olli P.)V1568,2-13(1991)

Hierarchical motion-compensated deinterlacing (Woods, John W.; Han, Soo-Chul)V1605,805-810(1991)

Hierarchical symmetry segmentation (Laird, Alan; Miller, James)V1381,536-544(1991)

High-accuracy target tracking algorithm based on deviation vector of the local window grey center (Shen, Zhen-kang; Zhao, Mingsheng)V1482,325-336(1991)

High-accuracy tracking algorithm based on iterating tracking window center towards the target's center (Shen, Zhen-kang; Han, Zhenduo; Zhao, Mingsheng)V1482,337-347(1991)

High-density packaging and interconnect of massively parallel image processors (Carson, John C.; Indin, Ronald)V1541,232-239(1991)

High-performance digital color video camera (Parulski, Kenneth A.; Benamati, Brian L.; D'Luna, Lionel J.; Shelley, Paul R.)V1448,45-58(1991)

High-precision digital charge-coupled device TV system (Vishnevsky, G. I.; Ioffe, S. A.; Berezin, V. Y.; Rybakov, M. I.; Mikhaylov, A. V.; Belyaev, L. V.)V1448,69-72(1991)

High-resolution image digitizing camera for use in quantitative coronary arteriography (Muser, Markus H.; Leemann, Thomas; Anliker, M.)V1448,106-112(1991)

High-resolution texture analysis for cytology (Hallouche, Farid; Adams, Alan E.; Hinton, Oliver R.)V1445,504-512(1991)

High spatial and temporal resolution in the optical investigation of biological objects (Damm, Tobias; Kempe, M.; Stamm, Uwe; Stolberg, K. P.; Wabnitz, H.)V1403,686-694(1991)

High-speed fine-motion tracking of some parts of a target (Casals, Alicia; Amat, Josep; Quesada, Jose L.; Sanchez, Luis)V1482,317-324(1991)

High-speed laser speckle photography (Huntley, Jonathan M.; Field, John E.)V1554A,756-765(1991)

Holotag: a novel holographic label (Soares, Oliverio D.; Bernardo, Luis M.; Pinto, M. I.; Morais, F. V.)V1332,166-184(1991)

Hough transform implementation using an analog associative network (Arrue, Begona C.; Inigo, Rafael M.)V1469,420-431(1991)

Human face recognition by P-type Fourier descriptor (Aibara, Tsunehiro; Ohue, Kenji; Matsuoka, Yasushi)V1606,198-203(1991)

Human movement analysis with image processing in real time (Fauvet, E.; Paindavoine, M.; Cannard, F.)V1358,620-630(1991)

Human Vision, Visual Processing, and Digital Display II (Rogowitz, Bernice E.; Brill, Michael H.; Allebach, Jan P., eds.)V1453(1991)

Hybrid bipixel structuring element decomposition and Euclidean morphological transforms (Zhou, Ziheng; Venetsanopoulos, Anastasios N.)V1606,309-319(1991)

Hybrid method to analyze the stress state in piecewise homogeneous two-dimensional objects (Laermann, Karl-Hans)V1554A,143-150(1991)

Hybrid pattern recognition system for the robotic vision (Li, Yulin; Zhao, Mingjun; Zhao, Li)V1385,200-205(1991)

Hybrid solution for high-speed target acquisition and identification systems (Udomkesmalee, Suraphol; Scholl, Marija S.; Shumate, Michael S.)V1468,81-91(1991)

Illumination and viewing methods for machine vision (Uber, Gordon T.; Harding, Kevin G.)VCR36,20-33(1991)

Image Algebra and Morphological Image Processing II (Gader, Paul D.; Dougherty, Edward R., eds.)V1568(1991)

Image analysis applied to black ice detection (Chen, Yi)V1468,551-562(1991)

Image analysis for estimating the weight of live animals (Schofield, C. P.; Marchant, John A.)V1379,209-219(1991)

Image analysis for vision-based agricultural vehicle guidance (Brown, Neil H.; Wood, Hugh C.; Wilson, James N.)V1379,54-68(1991)

Image analysis of discrete and continuous systems: film and CCD sensors (Kriss, Michael A.)V1398,4-14(1991)

Image computing requirements for the 1990s: from multimedia to medicine (Gove, Robert J.; Lee, Woobin; Kim, Yongmin; Alexander, Thomas)V1444,318-333(1991)

Image contrast enhancement via blurred weighted adaptive histogram equalization (Gauch, John M.)V1606,386-399(1991)

Image-display optimization using clinical history (Nodine, Calvin F.; Brikman, Inna; Kundel, Harold L.; Douglas, A.; Seshadri, Sridhar B.; Arenson, Ronald L.)V1444,56-62(1991)

Image Handling and Reproduction Systems Integration (Bender, Walter J.; Plouffe, Wil, eds.)V1460(1991)

Image noise smoothing based on nonparametric statistics (Chuang, Keh-Shih; Huang, H. K.)V1445,496-503(1991)

Image Processing Algorithms and Techniques II (Civanlar, Mehmet R.; Mitra, Sanjit K.; Moorhead, Robert J., eds.)V1452(1991)

Image processing methodology for optimizing the quality of corks in the punching process (Davies, Roger; Correia, Bento A.; Carvalho, Fernando D.)V1567,244-253(1991)

Image-processing system based on algorithmically dedicated functional units (Tozzi, Clesio L.; Castanho, Jose Eduardo C.; Gutierrez da Costa, Henrique S.)V1384,124-132(1991)

Image processing system for brain and neural tissue (Sun, Bingrong; Xu, Jiafang)V1606,1022-1026(1991)

Image processing techniques for aquisition of faint point laser targets in space (Garcia-Prieto, Rafael; Spiero, Francois; Popescu, Alexandru F.)V1522,267-276(1991)

Image processing utilizing an APL interface (Zmola, Carl; Kapp, Oscar H.)V1396,51-55(1991)

Image quality measurements with a neural brightness perception model (Grogan, Timothy A.; Wu, Mei)V1453,16-30(1991)

Image restoration based on perception-related cost functions (Palmieri, Francesco; Croteau, R. E.)V1451,24-35(1991)

Image segmentation based on ULCS color difference (Horita, Yuukou; Miyahara, Makaoto)V1606,607-620(1991)

Image segmentation using domain constraints (Ward, Matthew O.; Rajasekaran, Suresh)V1381,490-500(1991)

Imagetool: image processing on the Sun workstation (Zander, Mark E.)V1567,9-14(1991)

Image Understanding and the Man-Machine Interface III (Barrett, Eamon B.; Pearson, James J., eds.)V1472(1991)

Image Understanding for Aerospace Applications (Nasr, Hatem N., ed.)V1521(1991)

Impact of tactical maneuvers on EO sensor imagery (Hanson, David S.)V1538,48-63(1991)

Industrial vision systems based on a digital image correlation chip (Masaki, Ichiro)V1473,90-94(1991)

In-line wafer inspection using 100-megapixel-per-second digital image processing technology (Dickerson, Gary; Wallace, Rick P.)V1464,584-595(1991)

Inspecting colored objects using gray-level vision systems (Plummer, A. P.)VCR36,64-77(1991)

Instantaneous velocity field measurement of objects in coaxial rotation using digital image velocimetry (Cho, Young-Chung; Park, Hong-Woo)V1346,160-171(1991)

Integrated image processing and tracker performance prediction workstation (Schneeberger, Timothy J.; McIntire, Harold D.)V1567,2-8(1991)

Integrating vision and AI in an image processing workstation (Chan, John P.; Batchelor, Bruce G.)V1386,163-170(1991)

Integration of edge- and region-based techniques for range image segmentation (Al-Hujazi, Ezzet H.; Sood, Arun K.)V1381,589-599(1991)

Intelligent Robots and Computer Vision IX: Algorithms and Techniques (Casasent, David P., ed.)V1381(1991)

Intelligent Robots and Computer Vision IX: Neural, Biological, and 3-D Methods (Casasent, David P., ed.)V1382(1991)

Intensity edge detection with stack filters (Yoo, Jisang; Bouman, Charles A.; Delp, Edward J.; Coyle, Edward J.)V1451,58-69(1991)

Intensity interpolation for branching in reconstructing three-dimensional objects from serial cross-sections (Liang, Cheng-Chung; Chen, Chin-Tu; Lin, Wei-Chung)V1445,456-467(1991)

Interaction of algorithm and implementation for analog VLSI stereo vision (Hakkarainen, J. M.; Little, James J.; Lee, Hae-Seung; Wyatt, John L.)V1473,173-184(1991)

Interactive image processing in swallowing research (Dengel, Gail A.; Robbins, JoAnne; Rosenbek, John C.)V1445,88-94(1991)

Interactive tools for extraction of cartographic calibration data from aerial photography (Hunt, Bobby R.; Ryan, Thomas W.; Gifford, E.)V1472,190-200(1991)

Interframe registration and preprocessing of image sequences (Tajbakhsh, Shahram; Boyce, James F.)V1521,14-22(1991)

Intraluminal laser atherectomy with ultrasound and electromagnetic guidance (Gregory, Kenton W.; Aretz, H. T.; Martinelli, Michael A.; Ledet, Earl G.; Hatch, G. F.; Gregg, Richard E.; Sedlacek, Tomas; Haase, W. C.)V1425,217-225(1991)

Invariant pattern recognition via higher order preprocessing and backprop (Davis, Jon P.; Schmidt, William A.)V1469,804-811(1991)

Investigation of new filtering schemes for computerized detection of lung nodules (Yoshimura, Hitoshi; Giger, Maryellen L.; Matsumoto, Tsuneo; Doi, Kunio; MacMahon, Heber; Montner, Steven M.)V1445,47-51(1991)

ISDN audio color-graphics teleconferencing system (Oyaizu, Ikuro; Tanaka, Kiyoto; Yamaguchi, Toshikazu; Miyabo, Katsuaki; Takahashi, Mamoru)V1606,990-1001(1991)

Iso-precision scaling of digitized mammograms to facilitate image analysis (Karssemeijer, Nico; van Erning, Leon J.)V1445,166-177(1991)

Iterative motion estimation method using triangular patches for motion compensation (Nakaya, Yuichiro; Harashima, Hiroshi)V1605,546-557(1991)

Iterative neural networks for skeletonization and thinning (Krishnapuram, Raghu J.; Chen, Ling-Fan)V1382,271-281(1991)

Kalman-based computation of optical flow fields (Viswanath, Harsha C.; Jones, Richard A.)V1482,275-284(1991)

Knowledge-based direct 3-D texture segmentation system for confocal microscopic images (Lang, Zhengping; Zhang, Zhen; Scarberry, Randell E.; Shao, Weimin; Sun, Xu-Mei)V1468,826-833(1991)

Laboratory development of a nonlinear optical tracking filter (Block, Kenneth L.; Whitworth, Ernest E.; Bergin, Joseph E.)V1483,62-65(1991)

Landmark-based partial shape recognition by a BAM neural network (Liu, Xianjun; Ansari, Nirwan)V1606,1069-1079(1991)

Landmark recognition using motion-derived scene structures (Sadjadi, Firooz A.)V1521,98-105(1991)

LaneLok: an improved Hough transform algorithm for lane sensing using strategic search methods (Kenue, Surender K.; Wybo, David R.)V1468,538-550(1991)

Laplacian pyramid coding of prediction error images (Stiller, Christoph; Lappe, Dirk)V1605,47-57(1991)

Laptop image transmission equipment (Mocenter, Michael M.)V1538,132-139(1991)

Large one-step holographic stereogram (Honda, Toshio; Kang, Der-Kuan; Shimura, Kei; Enomoto, H.; Yamaguchi, Masahiro; Ohyama, Nagaaki)V1461,156-166(1991)

Laser disk: a practicum on effective image management (Myers, Richard F.)V1460,50-58(1991)

Laser metrological measurement of transient strain fields in Hopkinson-bar experiments (Vogel, Dietmar; Michel, Bernd; Totzauer, Werner F.; Schreppel, Ulrich; Clos, Rainer)V1554A,262-274(1991)

Learning filter systems with maximum correlation and maximum separation properties (Lenz, Reiner; Osterberg, Mats)V1469,784-795(1991)

Legibility of compressed document images at various spatial resolutions (Kidd, Robert C.; Zalinski, Charles M.; Nadel, Jerome I.; Klein, Robert D.)V1454,414-424(1991)

Lipschitz lattices and numerical morphology (Serra, Jean C.)V1568,54-65(1991)

Local spectrum analysis of medical images (Baskurt, Atilla; Peyrin, Francoise; Min, Zhu-Yue; Goutte, Robert)V1445,485-495(1991)

LUM filters for smoothing and sharpening (Boncelet, Charles G.; Hardie, Russell C.; Hakami, M. R.; Arce, Gonzalo R.)V1451,70-74(1991)

Machine verification of traced signatures (Krishnan, Ganapathy; Jones, David E.)V1468,563-572(1991)

Machine vision feedback for on-line correction of manufacturing processes: a control formulation (Taylor, Andrew T.; Wang, Paul K.)V1567,220-231(1991)

Machine Vision Systems Integration (Batchelor, Bruce G.; Waltz, Frederick M., eds.)VCR36(1991)

Machine Vision Systems Integration in Industry (Batchelor, Bruce G.; Waltz, Frederick K., eds.)V1386(1991)

Marker-controlled picture segmentation applied to electrical logging images (Rivest, Jean-Francois; Beucher, Serge; Delhomme, J.)V1451,179-190(1991)

Markov random fields on a SIMD machine for global region labelling (Budzban, Gregory M.; DeCatrel, John M.)V1470,175-182(1991)

Massively parallel image restoration (Menon, Murali M.)V1471,185-190(1991)

Max-polynomials and template decomposition (Li, Dong)V1568,149-156(1991)

Measurement of fractal dimension using 3-D technique (Chuang, Keh-Shih; Valentino, Daniel J.; Huang, H. K.)V1445,341-347(1991)

Measurement of residual stresses in plastic materials by electronic shearography (Long, Kah W.; Hung, Yau Y.; Der Hovanesian, Joseph)V1554A,116-123(1991)

Measuring and display system of a marathon runner by real-time digital image processing (Sasaki, Nobuyuki; Namikawa, Iwao)V1606,1002-1013(1991)

Medical Imaging V: Image Processing (Loew, Murray H., ed.)V1445(1991)

Method of preprocessing color images using a Peano curve on a Transputer array (Lambert, Robin A.; Batchelor, Bruce G.)V1381,582-588(1991)

Microcomputer-based image processing system for CT/MRI scans: II. Expert system (Kwok, John C.; Yu, Peter K.; Cheng, Andrew Y.; Ho, Wai-Chin)V1445,446-455(1991)

Microcomputer-based image processing system for CT/MRI scans: hardware configuration and software capacity (Cheng, Andrew Y.; Ho, Wai-Chin; Kwok, John C.; Yu, Peter K.)V1444,400-406(1991)

Microscopic feature extraction from optical sections of contracting cardiac muscle cells recorded at high speed (Roos, Kenneth P.; Lake, David S.; Lubell, Bradford A.)V1428,159-168(1991)

MIMD (multiple instruction multiple data) multiprocessor system for real-time image processing (Pirsch, Peter; Jeschke, Hartwig)V1452,544-555(1991)

Minimum resolution for human face detection and identification (Samal, Ashok)V1453,81-89(1991)

Miscibility matrices explain the behavior of gray-scale textures generated by Gibbs random fields (Elfadel, Ibrahim M.; Picard, Rosalind W.)V1381,524-535(1991)

Mission verification systems for FMS applications (Flaherty, Marty)V1538,64-68(1991)

MITAS: multisensor imaging technology for airborne surveillance (Thomas, John)V1470,65-74(1991)

Model-based boundary detection in echocardiography using dynamic programming technique (Dong, LiXin; Pelle, Gabriel; Brun, Philip; Unser, Michael A.)V1445,178-187(1991)

Model-based coding of facial images based on facial muscle motion through isodensity maps (So, Ikken; Nakamura, Osamu; Minami, Toshi)V1605,263-272(1991)

Model-based halftoning (Pappas, Thrasyvoulos N.; Neuhoff, David L.)V1453,244-255(1991)

Model-based image processing for characterizing pigs in scenes (Tillett, Robin D.; Marchant, John A.)V1379,201-208(1991)

Modeling nonhomogeneous 3-D objects for thermal and visual image synthesis (Karthik, Sankaran; Nandhakumar, N.; Aggarwal, Jake K.)V1468,686-697(1991)

Modeling of local neural networks of the visual cortex and applications to image processing (Rybak, Ilya A.; Shevtsova, Natalia A.; Podladchikova, Lubov N.)V1469,737-748(1991)

Model of the left ventricle 3-D global motion: application to MRI data (Friboulet, Denis; Magnin, Isabelle E.; Mathieu, Christophe; Revel, D.; Amiel, Michel)V1445,106-117(1991)

Moire interferometry for subdynamic tests in normal light environment (Luo, Zhishan; Luo, Chanzou; Hu, Qinghua; Mu, Zongxue)V1554B,523-528(1991)

Morphological algorithms for modeling Gaussian image features (Bhagvati, Chakravarthy; Marineau, Peter; Skolnick, Michael M.; Sternberg, Stanley)V1606,112-119(1991)

Morphological pattern-spectra-based Tau-opening optimization (Dougherty, Edward R.; Haralick, Robert M.; Chen, Yidong; Li, Bo; Agerskov, Carsten; Jacobi, Ulrik; Sloth, Poul H.)V1606,141-152(1991)

Morphological processing to reduce shading and illumination effects (Casasent, David P.; Schaefer, Roland H.; Kokaj, Jahja O.)V1385,152-164(1991)

Motion-compensated priority discrete cosine transform coding of image sequences (Efstratiadis, Serafim N.; Huang, Yunming G.; Xiong, Z.; Galatsanos, Nikolas P.; Katsaggelos, Aggelos K.)V1605,16-25(1991)

Motion affine models identification and application to television image coding (Sanson, Henri)V1605,570-581(1991)

Motion compensation by block matching and vector postprocessing in subband coding of TV signals at 15 Mbit/s (Lallauret, Fabrice; Barba, Dominique)V1605,26-36(1991)

Motion estimation in digital angiographic images using skeletons (Kwak, J. Y.; Efstratiadis, Serafim N.; Katsaggelos, Aggelos K.; Sahakian, Alan V.; Sullivan, Barry J.; Swiryn, Steven; Hueter, David C.; Frohlich, Thomas)V1396,32-44(1991)

Multiple resonant boundary contour system (Lehar, Steve M.; Worth, Andrew J.)V1469,50-62(1991)

Multiresolution segmentation of forward-looking IR and SAR imagery using neural networks (Beck, Hal E.; Bergondy, Daniel; Brown, Joseph R.; Sari-Sarraf, Hamed)V1381,600-609(1991)

Multiscale morphological region coding (Macq, Benoit M.; Ronse, Christian; Van Dongen, V.)V1606,165-173(1991)

Multisensor fusion methodologies compared (Swan, John; Shields, Frank J.)V1483,219-230(1991)

Multisensor image processing (de Salabert, Arturo; Pike, T. K.; Sawyer, F. G.; Jones-Parry, I. H.; Rye, A. J.; Oddy, C. J.; Johnson, D. G.; Mason, D.; Wielogorski, A. L.; Plassard, T.; Serpico, Sebastiano B.; Hindley, N.)V1521,74-88(1991)

Multispectral image segmentation of breast pathology (Hornak, Joseph P.; Blaakman, Andre; Rubens, Deborah; Totterman, Saara)V1445,523-533(1991)

Multitarget adaptive gate tracker with linear prediction (Liu, Zhili)V1482,285-292(1991)

Multitarget detection and estimation parallel algorithm (Krikelis, A.)V1482,307-316(1991)

Natural texture analysis in a multiscale context using fractal dimension (Kpalma, Kidiyo; Bruno, Alain; Haese-Coat, Veronique)V1606,55-66(1991)

Near-real-time biplanar fluoroscopic tracking system for the video tumor fighter (Lawson, Michael A.; Wika, Kevin G.; Gillies, George T.; Ritter, Rogers C.)V1445,265-275(1991)

Negative fuse network (Liu, Shih-Chii; Harris, John G.)V1473,185-193(1991)

Neural-network-aided design for image processing (Vitsnudel, Ilia; Ginosar, Ran; Zeevi, Yehoshua Y.)V1606,1086-1091(1991)

Neural-network-based image processing of human corneal endothelial micrograms (Hasegawa, Akira; Zhang, Wei; Itoh, Kazuyoshi; Ichioka, Yoshiki)V1558,414-421(1991)

Neural network edge detector (Spreeuwers, Luuk J.)V1451,204-215(1991)

Neural networks and model-based approaches to object identification (Perlovsky, Leonid I.)V1606,1080-1085(1991)

Neural networks for ATR parameters adaptation (Amehdi, Hossien; Nasr, Hatem N.)V1483,177-184(1991)

Neural networks for medical image segmentation (Lin, Wei-Chung; Tsao, Chen-Kuo; Chen, Chin-Tu; Feng, Yu-Jen)V1445,376-385(1991)

Neural network vision integration with learning (Toborg, Scott T.)V1469,77-88(1991)

New algorithms for adaptive median filters (Hwang, Humor; Haddad, Richard A.)V1606,400-407(1991)

New concepts in mathematical morphology: the topographical and differential distance functions (Preteux, Francoise; Merlet, Nicolas)V1568,66-77(1991)

New cooperative edge linking (Bonnin, Patrick; Zavidovique, Bertrand)V1381,142-152(1991)

New edge-enhancement method based on the deviation of the local image grey center (Zhao, Mingsheng; Shen, Zhen-kang; Chen, Huihuang)V1471,464-473(1991)

New human vision system model for spatio-temporal image signals (Matsui, Toshikazu; Hirahara, Shuzo)V1453,282-289(1991)

New inverse synthetic aperture radar algorithm for translational motion compensation (Bocker, Richard P.; Henderson, Thomas B.; Jones, Scott A.; Frieden, B. R.)V1569,298-310(1991)

New method of target acquisition in the presence of clutter (Tidhar, Gil; Rotman, Stanley R.)V1486,188-199(1991)

New method of thinning photoelastic interference fringes in image processing (Zhang, Yuan P.)V1554A,862-866(1991)

New parallel algorithms for thinning of binary images (Bhattacharya, Prabir; Lu, Xun)V1468,734-739(1991)

New trends in morphological algorithms (Vincent, Luc M.)V1451,158-170(1991)

New type of IR to visible real-time image converter: design and fabrication (Sun, Fang-kui; Yang, Mao-hua; Gao, Shao-hong; Zhao, Shi-jie)V1488,2-5(1991)

Noise-tolerant texture classification and image segmentation (Kjell, Bradley P.; Wang, Pearl Y.)V1381,553-560(1991)

Nondestructive evaluation of turbine blades vibrating in resonant modes (Sciammarella, Cesar A.; Ahmadshahi, Mansour A.)V1554B,743-753(1991)

Nonlinear Image Processing II (Arce, Gonzalo R.; Boncelet, Charles G.; Dougherty, Edward R., eds.)V1451(1991)

Nonlinear regression for signal processing (Restrepo, Alfredo)V1451,258-268(1991)

Nonparametric dominant point detection (Ansari, Nirwan; Huang, Kuo-Wei)V1606,31-42(1991)

Novel block segmentation and processing for Chinese-English document (Chien, Bing-Shan; Jeng, Bor S.; Sun, San-Wei; Chang, Gan-How; Shyu, Keh-Haw; Shih, Chun-Hsi)V1606,588-598(1991)

Novel regular-array ASIC architecture for 2-D ROS sorting (van Swaaij, Michael F.; Catthoor, Francky V.; De Man, Hugo J.)V1606,901-910(1991)

Object extraction method for image synthesis (Inoue, Seiki)V1606,43-54(1991)

Objective assessment of clinical computerized thermal images (Anbar, Michael)V1445,479-484(1991)

Object-oriented representation of image space by puzzletrees (Dengel, Andreas R.)V1606,20-30(1991)

Object segmentation algorithm for use in recognizing 3-D partially occluded objects* (Fan, Kuo-Chin; Chang, Chia-Yuan)V1468,674-684(1991)

On-focal-plane-array feature extraction using 3-D artificial neural network (3DANN): Part II (Carson, John C.)V1541,227-231(1991)

Optical image segmentor using wavelet filtering techniques as the front-end of a neural network classifier (Veronin, Christopher P.; Rogers, Steven K.; Kabrisky, Matthew; Welsh, Byron M.; Priddy, Kevin L.; Ayer, Kevin W.)V1469,281-291(1991)

Optical processing of wire-frame models for object recognition (Kozaitis, Samuel P.; Cofer, Rufus H.)V1471,249-254(1991)

Optics, Illumination, and Image Sensing for Machine Vision V (Harding, Kevin G.; Svetkoff, Donald J.; Uber, Gordon T.; Wittels, Norman, eds.)V1385(1991)

Optimal generalized weighted-order-statistic filters (Yin, Lin; Astola, Jaakko; Neuvo, Yrjo A.)V1606,431-442(1991)

Optimization and evaluation of an image intensifier TV system for digital chest imaging (Angelhed, Jan-Erik; Mansson, Lars G.; Kheddache, Susanne)V1444,159-170(1991)

Optimization of morphological structuring elements for angiogram enhancement (Andress, Keith M.; Wilson, David L.)V1445,6-10(1991)

Optimum intensity-dependent spread filters in image processing (Vaezi, Matt M.; Bavarian, Behnam; Healey, Glenn)V1452,57-63(1991)

Parallel algorithm for volumetric segmentation (Liou, Shih-Ping; Jain, Ramesh C.)V1381,447-458(1991)

Parallel constructs for three-dimensional registration on a SIMD (single-instruction stream/multiple-data stream) processor (Morioka, Craig A.; Chan, Kelby K.; Huang, H. K.)V1445,534-538(1991)

Parallel DC notch filter (Kwok, Kam-cheung; Chan, Ming-kam)V1567,709-719(1991)

Parallel rule inferencing for automatic target recognition (Pacelli, Jean L.; Geyer, Steve L.; Ramsey, Timothy S.)V1472,76-84(1991)

Parameter studies for Markov random field models of early vision (Daily, Michael J.)V1473,138-152(1991)

Partially compensated speckle imaging: Fourier phase spectrum estimation (Roggemann, Michael C.; Matson, Charles L.)V1542,477-487(1991)

Passive and active sensors for autonomous space applications (Tchoryk, Peter; Gleichman, Kurt W.; Carmer, Dwayne C.; Morita, Yuji; Trichel, Milton; Gilbert, R. K.)V1479,164-182(1991)

Perceiving the coherent movements of spatially separated features (Mowafy, Lyn; Lappin, Joseph S.)V1453,177-187(1991)

Performance analysis through memory of a proposed parallel architecture for the efficient use of memory in image processing applications (Faruque, Abdullah; Fong, David Y.)V1606,865-877(1991)

Performance characterization of vision algorithms (Haralick, Robert M.; Ramesh, Visvanathan)V1406,2-16(1991)

Performance evaluation of a texture-based segmentation algorithm (Sadjadi, Firooz A.)V1483,185-195(1991)

Performance of a neural-network-based 3-D object recognition system (Rak, Steven J.; Kolodzy, Paul J.)V1471,177-184(1991)

Personal verification system with high tolerance of poor-quality fingerprints (Sasakawa, Koichi; Isogai, Fumihiko; Ikebata, Sigeki)V1386,265-272(1991)

Phase-conjugate optical preprocessing filter for small-target tracking (Block, Kenneth L.; Whitworth, Ernest E.; Bergin, Joseph E.)V1481,32-34(1991)

Pixel level data fusion: from algorithm to chip (Mathur, Bimal P.; Wang, H. T.; Liu, Shih-Chii; Koch, Christof; Luo, Jin)V1473,153-160(1991)

Possibilities of the fringe pattern learning system VARNA (Nedkova, Rumiana)V1429,214-216(1991)

Practical VLSI realization of morphological operations (Chang, Yiher; Ansari, Nirwan)V1606,839-850(1991)

Practical method for automatic detection of brightest ridge line of photomechanical fringe pattern (Ding, Zu-Quan; Yuan, Xun-Hua)V1554A,898-906(1991)

Precision tracking of small target in IR systems (Lu, Huanzhang; Sun, Zhong-kang)V1481,398-405(1991)

Preliminary tests of maximum likelihood image reconstruction method on 3-D real data and some practical considerations for the data corrections (Liu, Yi-Hwa; Holmes, Timothy J.; Koshy, Matthew)V1428,191-199(1991)

Probabilistic model for quadtree representation of binary images (Chou, Chun-Hsien; Chu, Chih-Peng)V1605,832-843(1991)

Programmable command interpreter to automate image processing of IR thermography (Hughett, Paul)V1467,416-426(1991)

Projectile velocity and spin rate by image processing of synchro-ballistic photography (Hughett, Paul)V1346,237-248(1991)

Projection moire as a tool for the automated determination of surface topography (Cardenas-Garcia, Jaime F.; Zheng, S.; Shen, F. Z.)V1554B,210-224(1991)

Prototype neural network pattern recognition testbed (Worrell, Steven W.; Robertson, James A.; Varner, Thomas L.; Garvin, Charles G.)V1382,219-227(1991)

Pyramid median filtering by block threshold decomposition (Zhou, Hongbing; Zeng, Bing; Neuvo, Yrjo A.)V1451,2-12(1991)

Quadtree decomposition of binary structuring elements (Shoji, Kenji)V1451,148-157(1991)

Quantitative analysis of cardiac imaging using expert systems (Dreyer, Keith J.; Simko, Joseph; Held, A. C.)V1445,398-408(1991)

RadGSP: a medical image display and user interface for UWGSP3 (Yee, David K.; Lee, Woobin; Kim, Dong-Lok; Haass, Clark D.; Rowberg, Alan H.; Kim, Yongmin)V1444,292-305(1991)

Random mapping network for tactical target reacquisition after loss of track (Church, Susan D.; Burman, Jerry A.)V1471,192-199(1991)

Rapid system integration with symbolic programming (Walsh, Peter M.)V1386,84-89(1991)

Realization of the Zak-Gabor representation of images (Assaleh, Khaled T.; Zeevi, Yehoshua Y.; Gertner, Izidor)V1606,532-540(1991)

Real-time architecture based on the image processing module family (Kimura, Shigeru; Murakami, Yoshiyuki; Matsuda, Hikaru)V1483,10-17(1991)

Real-time edge extraction by active defocusing (Hung, Yau Y.; Zhu, Qiuming; Shi, Dahuan; Tang, Shou-Hong)V1332,332-342(1991)

Real-time quantitative imaging for semiconductor crystal growth, control, and characterization (Wargo, Michael J.)V1557,271-282(1991)

Recognition of partially occluded objects using B-spline representation (Salari, Ezzatollah; Balaji, Sridhar)V1384,115-123(1991)

Reconstructing MR images from incomplete Fourier data using the maximum entropy method (Fielden, John; Kwong, Henry Y.; Wilbrink, Jacob)V1445,145-154(1991)

Recovering absolute depth and motion of multiple objects from intensity images (Jiang, Fan; Schunck, Brian G.)V1569,162-173(1991)

Reduction of blocking artifacts using motion-compensated spatial-temporal filtering (Chu, Frank J.; Yeh, Chia L.)V1452,38-46(1991)

Registration of medical images by coincident bit counting (Chiang, John Y.; Sullivan, Barry J.)V1396,15-26(1991)

Residue-producing E-filters and their applications in medical image analysis (Preston, Kendall)V1450,59-70(1991)

Resolution improvement for in-vivo magnetic resonance spectroscopic images (Plevritis, Sylvia; Macovski, Albert)V1445,118-127(1991)

Robust regression in computer vision (Meer, Peter; Mintz, Doron)V1381,424-435(1991)

Robust statistical method for background extraction in image segmentation (Rodriguez, Arturo A.; Mitchell, O. R.)V1569,182-199(1991)

Rotation invariant object classification using fast Fourier transform features (Celenk, Mehmet; Datari, Srinivasa R.)V1468,752-763(1991)

Rule-based system to reconstruct 3-D tree structure from two views (Liu, Iching; Sun, Ying)V1606,67-77(1991)

SADARM status report (DiNardo, Anthony J.)V1479,228-248(1991)

Scale-space features for object detection (Kjell, Bradley P.; Sood, Arun K.; Topkar, V. A.)V1468,148-155(1991)

Scientific data compression for space: a modified block truncation coding algorithm (Lu, Wei-Wei; Gough, M. P.; Davies, Peter N.)V1470,197-205(1991)

Segmentation of orientation maps by an integration of edge- and region-based methods (Distante, Arcangelo; D'Orazio, Tiziana; Stella, Ettore)V1381,513-523(1991)

Segmentation using models of expected structure (Shemlon, Stephen; Liang, Tajen; Cho, Kyugon; Dunn, Stanley M.)V1381,470-481(1991)

Segmentation using range data and structured light (Hu, Gongzhu)V1381,482-489(1991)

Segmentation via fusion of edge and needle map (Ahn, Hong-Young; Tou, Julius T.)V1468,896-904(1991)

Selected Papers on Automatic Object Recognition (Nasr, Hatem N., ed.)VMS41(1991)

Selective edge detection based on harmonic oscillator wave functions (Kawakami, Hajimu)V1468,156-166(1991)

Semiautomatic medical image segmentation using knowledge of anatomic shape (Brinkley, James F.)V1445,78-87(1991)

Semiautomatic x-ray inspection system (Amladi, Nandan G.; Finegan, Michael K.; Wee, William G.)V1472,165-176(1991)

Shape recognition combining mathematical morphology and neural networks (Schmitt, Michel; Mattioli, Juliette)V1469,392-403(1991)

Shape registration by morphological operations (Chang, Long-Wen; Tsai, Mong-Jean)V1606,120-131(1991)

Short-scan fan beam algorithm for noncircular detector orbits (Zeng, Gengsheng L.; Gullberg, Grant T.)V1445,332-340(1991)

Signal and Image Processing Systems Performance Evaluation, Simulation, and Modeling (Nasr, Hatem N.; Bazakos, Michael E., eds.)V1483(1991)

Silicon retina with adaptive photoreceptors (Mahowald, Misha A.)V1473,52-58(1991)

Simple iterative method for finding the foe using depth-is-positive constraint (Michael, David J.)V1388,234-245(1991)

Simulation of infrared backgrounds using two-dimensional models (Cadzow, James A.; Wilkes, D. M.; Peters, Richard A.; Li, Xingkang; Patel, Jamshed N.)V1486,352-363(1991)

Small-target acquisition and typing by AASAP (Huguenin, Robert L.; Tahmoush, Donald J.)V1481,64-72(1991)

Some properties of the two-dimensional pseudomedian filter (Schulze, Mark A.; Pearce, John A.)V1451,48-57(1991)

Spatial frequency filtering on the basis of the nonlinear optics phenomena (Kudryavtseva, Anna D.; Tcherniega, Nicolaii V.; Brekhovskikh, Galina L.; Sokolovskaya, Albina I.)V1385,190-199(1991)

Spatiotemporal filtering of digital angiographic image sequences corrupted by quantum mottle (Chan, Cheuk L.; Sullivan, Barry J.; Sahakian, Alan V.; Katsaggelos, Aggelos K.; Frohlich, Thomas; Byrom, Ernest)V1450,208-217(1991)

SPOT system and defence applications (Bernard, Christian)V1521,66-73(1991)

State-space search as high-level control for machine vision (Hwang, Shu-Yuen)V1386,145-156(1991)

Stochastic and Neural Methods in Signal Processing, Image Processing, and Computer Vision (Chen, Su-Shing, ed.)V1569(1991)

Stochastic neural nets and vision (Fall, Thomas C.)V1468,778-785(1991)

Stress intensity factors of edge-cracked semi-infinite plates under transient thermal loading (Wang, Wei-Chung; Chen, Tsai-Lin; Hwang, Chi-Hung)V1554A,124-135(1991)

Studies in robust approaches to object detection in high-clutter background (Shirvaikar, Mukul V.; Trivedi, Mohan M.)V1468,52-59(1991)

Subband video-coding algorithm and its feasibility on a transputer video coder (Brofferio, Sergio C.; Marcozzi, Elena; Mori, Luigi; Raveglia, Dalmazio)V1605,894-905(1991)

Subjective evaluation of image enhancements in improving the visibility of pathology in chest radiographs (Plessis, Brigitte; Goldberg, Morris; Belanger, Garry; Hickey, Nancy M.)V1445,539-549(1991)

Subpixel measurement of image features based on paraboloid surface fit (Gleason, Shaun S.; Hunt, Martin A.; Jatko, W. B.)V1386,135-144(1991)

Superhigh-definition image processing on a parallel signal processing system (Fujii, Tetsurou; Sawabe, Tomoko; Ohta, Naohisa; Ono, Sadayasu)V1605,339-350(1991)

Supervised pixel classification using a feature space derived from an artificial visual system (Baxter, Lisa C.; Coggins, James M.)V1381,459-469(1991)

Surface definition technique for clinical imaging (Liao, Wen-gen; Simovsky, Ilya; Li, Andrew; Kramer, David M.; Kaufman, Leon; Rhodes, Michael L.)V1444,47-55(1991)

Survey of parallel architectures used for three image processing algorithms (Smith, Ross)V1396,615-623(1991)

Systolic array architecture of a new VLSI vision chip (Means, Robert W.)V1566,388-393(1991)

Target detection using co-occurrence matrix segmentation and its hardware implementation (Auborn, John E.; Fuller, James M.; McCauley, Howard M.)V1482,246-252(1991)

Target detection using multilayer feedforward neural networks (Scherf, Alan V.; Scott, Peter A.)V1469,63-68(1991)

Task performance based on the posterior probability of maximum-entropy reconstructions obtained with MEMSYS 3 (Myers, Kyle J.; Hanson, Kenneth M.)V1443,172-182(1991)

Template decomposition and inversion over hexagonally sampled images (Lucas, Dean; Gibson, Laurie)V1568,157-163(1991)

Temporal projection for motion estimation and motion compensating interpolation (Robert, Philippe)V1605,558-569(1991)

Theoretical approach to hyperacuity tests based on resolution criteria for two-line images (Mondal, Pronab K.; Calvo, Maria L.; Chevalier, Margarita L.; Lakshminarayanan, Vasudevan)V1429,108-116(1991)

Three-dimensional contour of crack tips using a grating method (Andresen, Klaus; Kamp, B.; Ritter, Reinold)V1554A,93-100(1991)

Three-dimensional display of MRI data in neurosurgery: segmentation and rendering aspects (Barillot, Christian; Lachmann, F.; Gibaud, Bernard; Scarabin, Jean-Marie)V1445,54-65(1991)

Three-dimensional display of inside of human body by holographic stereogram (Sato, Koki; Akiyama, Iwaki; Shoji, Hideo; Sumiya, Daigaku; Wada, Tuneyo; Katsuma, Hidetoshi; Itoh, Koichi)V1461,124-131(1991)

Three-dimensional image processing method to compensate for depth-dependent light attenuation in images from a confocal microscope (Aslund, Nils R.; Liljeborg, Anders; Oldmixon, E. H.; Ulfsparre, M.)V1450,329-337(1991)

Three-dimensional laser radar simulation for autonomous spacecraft landing (Reiley, Michael F.; Carmer, Dwayne C.; Pont, W. F.)V1416,295-303(1991)

Three-dimensional motion analysis and structure recovering by multistage Hough transform (Nakajima, Shigeyoshi; Zhou, Mingyong; Hama, Hiromitsu; Yamashita, Kazumi)V1605,709-719(1991)

Three-dimensional reconstruction from cone beam projection by a block iterative technique (Peyrin, Francoise; Goutte, Robert; Amiel, Michel)V1443,268-279(1991)

Three-dimensional reconstruction of pulmonary blood vessels by using anatomical knowledge base (Inaoka, Noriko; Suzuki, Hideo; Mori, Masaki; Takabatake, Hirotsugu; Suzuki, Akira)V1450,2-12(1991)

Time-derivative adaptive silicon photoreceptor array (Delbruck, Tobi; Mead, Carver A.)V1541,92-99(1991)

Tissue volume determinations from brain MRI images: a phantom study (Chandra, Ramesh; Rusinek, Henry)V1445,133-144(1991)

Toros: an image processing system for measuring consignments of wood (Correia, Bento A.; Davies, Roger; Carvalho, Fernando D.; Rodrigues, F. C.)V1567,15-24(1991)

Toward low-cost real-time EPLD-based machine vision workstations and target systems (Floeder, Steven P.; Waltz, Frederick M.)V1386,90-101(1991)

Trade-offs between pseudohexagonal and pseudocubic filters (Preston, Kendall)V1451,75-90(1991)

Transmission of the motion of a walker by model-based image coding (Kimoto, Tadahiko; Yasuda, Yasuhiko)V1605,253-262(1991)

Transparency and blur as selective cues for complex visual information (Colby, Grace; Scholl, Laura)V1460,114-125(1991)

Transputer-based parallel algorithms for automatic object recognition (Bison, Paolo G.; Braggiotti, Alberto; Grinzato, Ermanno G.)V1471,369-377(1991)

TV holography and image processing in practical use (Lokberg, Ole J.; Ellingsrud, Svein; Vikhagen, Eiolf)V1332,142-150(1991)

Two-dimensional boundary inspection using autoregressive model (Wani, M. A.; Batchelor, Bruce G.)V1384,83-89(1991)

Ultrahigh-sensitivity moire interferometry (Han, Bongtae)V1554B,399-411(1991)

Update on focal-plane image processing research (Kemeny, Sabrina E.; Eid, El-Sayed I.; Mendis, Sunetra; Fossum, Eric R.)V1447,243-250(1991)

Use of anatomical knowledge to register 3-D blood vessel data derived from DSA with MR images (Hill, Derek L.; Hawkes, David J.; Hardingham, Charles R.)V1445,348-357(1991)

Use of coordinate mapping as a method for image data reduction (Abbott, Mark A.; Messner, Richard A.)V1381,272-282(1991)

Use of high-resolution TV camera in photomechanics (Yatagai, Toyohiko; Ino, Tomomi)V1554B,646-649(1991)

User interface development for semiautomated imagery exploitation (O'Connor, R. P.; Bohling, Edward H.)V1472,26-37(1991)

Using digital-scanned aerial photography for wetlands delineation (Anderson, John E.; Roos, Maurits)V1492,252-262(1991)

Using local orientation and hierarchical spatial feature matching for the robust recognition of objects (Seitz, Peter; Lang, Graham K.)V1606,252-259(1991)

Using morphology in document image processing (Concepcion, Vicente P.; Grzech, Matthew P.; D'Amato, Donald P.)V1606,132-140(1991)

Using MRI to calculate cardiac velocity fields (Santago, Peter; Slade, James N.)V1445,555-563(1991)

Using structuring-element libraries to design suboptimal morphological filters (Loce, Robert P.; Dougherty, Edward R.)V1568,233-246(1991)

Utilizing the central limit theorem for parallel multiple-scale image processing with neural architectures (Ben-Arie, Jezekiel)V1569,227-238(1991)

Variable-blocksize transform coding of four-color printed images (Kaup, Andre; Aach, Til)V1605,420-427(1991)

Vector and scalar field interpretation (Farrell, Edward J.; Aukrust, Trond; Oberhuber, Josef M.)V1459,136-147(1991)

Video-rate image remapper for the PC/AT bus (Weiman, Carl F.; Weber, Robert G.)V1386,102-110(1991)

Vision sensing and processing system for monitoring and control of welding and other high-luminosity processes (Agapakis, John E.; Bolstad, Jon O.)V1385,32-38(1991)

Visual aspects of picture storage (Brettel, Hans)V1401,50-55(1991)

Visual Communications and Image Processing '91: Image Processing (Tzou, Kou-Hu; Koga, Toshio, eds.)V1606(1991)

Visual Communications and Image Processing '91: Visual Communication (Tzou, Kou-Hu; Koga, Toshio, eds.)V1605(1991)

Visual Information Processing: From Neurons to Chips (Mathur, Bimal P.; Koch, Christof, eds.)V1473(1991)

VLSI implementable neural networks for target tracking (Himes, Glenn S.; Inigo, Rafael M.; Narathong, Chiewcharn)V1469,671-682(1991)

VLSI implementation of a buffer, universal quantizer, and frame-rate-control processor (Uwabu, H.; Kakii, Eiji; Lacombe, R.; Maruyama, Masanori; Fujiwara, Hiroshi)V1605,928-937(1991)

Volumetric-intravascular imaging by high-frequency ultrasound (Sehgal, Chandra M.; Chandrasekaran, K.; Pandian, Natesa G.)V1425,226-233(1991)

Warping of a computerized 3-D atlas to match brain image volumes for quantitative neuroanatomical and functional analysis (Evans, Alan C.; Dai, W.; Collins, L.; Neelin, Peter; Marrett, Sean)V1445,236-246(1991)

Weighted-outer-product associative neural network (Ji, Han-Bing)V1606,1060-1068(1991)

Window-based elaboration language for picture processing and painting (Kamoshida, Minoru; Enomoto, Hajime; Miyamura, Isao)V1606,951-960(1991)

Windowed motion compensation (Watanabe, Hiroshi; Singhal, Sharad)V1605,582-589(1991)

Wind tunnel model aircraft attitude and motion analysis (Mostafavi, Hassan)V1483,104-111(1991)

X-ray inspection utilizing knowledge-based feature isolation with a neural network classifier (Nolan, Adam R.; Hu, Yong-Lin; Wee, William G.)V1472,157-164(1991)

Image quality—see also image reconstruction/restoration

Accurate image simulation by hemisphere projection (Bian, Buming; Wittels, Norman)V1453,333-340(1991)

Acquisition of very high resolution images using stereo cameras (Aizawa, Kiyoharu; Komatsu, Takashi; Saito, Takahiro)V1605,318-328(1991)

Adaptive notch filter for removal of coherent noise from infrared scanner data (Jaggi, Sandeep)V1541,134-140(1991)

Adaptive optics using curvature sensing (Forbes, Fred F.; Roddier, Nicolas)V1542,140-147(1991)

Adaptive typography for dynamic mapping environments (Bardon, Didier)V1460,126-139(1991)

Advanced real-time scanning concept for full dynamics recording, high image quality, and superior measurement accuracy (Lindstrom, Kjell M.; Wallin, Bo)V1488,389-398(1991)

Alignment and focus control of a telescope using image sharpening (Jones, Peter A.)V1542,194-204(1991)

Application of a blind-deconvolution restoration technique to space imagery (Lewis, Tom R.; Mitra, Sunanda)V1565,221-226(1991)

Astigmatism and field curvature from pin-bars (Kirk, Joseph P.)V1463,282-291(1991)

Blurring effect of aerosols on imaging systems (Bissonnette, Luc R.)V1487,333-344(1991)

Capabilities of the optical microscope (Bradbury, Savile)V1439,128-134(1991)

Characteristic analysis of color information based on (R,G,B)-> (H,V,C) color space transformation (Gan, Qing; Miyahara, Makoto; Kotani, Kazunori)V1605,374-381(1991)

Chemistry of the Konica Dry Color System (Suda, Yoshihiko; Ohbayashi, Keiji; Onodera, Kaoru)V1458,76-78(1991)

Clinical aspects of quality assurance in PACS (Marglin, Stephen I.)V1444,83-86(1991)

Clinical evaluation of a 2K x 2K workstation for primary diagnosis in pediatric radiology (Razavi, Mahmood; Sayre, James W.; Simons, Margaret A.; Hamedaninia, Azar; Boechat, Maria I.; Hall, Theodore R.; Kangarloo, Hooshang; Taira, Ricky K.; Chuang, Keh-Shih; Kashifian, Payam)V1446,24-34(1991)

Color quantization aspects in stereopsis (Hebbar, Prashant D.; McAllister, David F.)V1457,233-241(1991)

Comparison of 3-D display formats for CAD applications (McWhorter, Shane W.; Hodges, Larry F.; Rodriguez, Walter E.)V1457,85-90(1991)

Comparison of Kodak Professional Digital Camera System images to conventional film, still video, and freeze-frame images (Kent, Richard A.; McGlone, John T.; Zoltowski, Norbert W.)V1448,27-44(1991)

Comparison of directionally based and nondirectionally based subband image coders (Bamberger, Roberto H.; Smith, Mark J.)V1605,757-768(1991)

Comparison of images on the basis of structural features (Markov, Vladimir B.; Shishkov, Vladimir F.; Voroshnin, A. B.)V1238,118-122(1991)

Complex phase error and motion estimation in synthetic aperture radar imaging (Soumekh, Mehrdad; Yang, H.)V1452,104-113(1991)

Computed estimation of visual acuity after laser refractive keratectomy (Rol, Pascal O.; Parel, Jean-Marie; Hanna, Khalil)V1423,89-93(1991)

Converting non-interlaced to interlaced images in YIQ and HSI color spaces (Welch, Eric B.; Moorhead, Robert J.; Owens, John K.)V1453,235-243(1991)

Correlation between the detection and interpretation of image features (Fuhrman, Carl R.; King, Jill L.; Obuchowski, Nancy A.; Rockette, Howard E.; Sashin, Donald; Harris, Kathleen M.; Gur, David)V1446,422-429(1991)

Correlations between time required for radiological diagnoses, readers' performance, display environments, and difficulty of cases (Gur, David; Rockette, Howard E.; Sumkin, Jules H.; Hoy, Ronald J.; Feist, John H.; Thaete, F. L.; King, Jill L.; Slasky, B. S.; Miketic, Linda M.; Straub, William H.)V1446,284-288(1991)

Design characteristics and point-spread function evaluation of bifocal intraocular lenses (Rol, Pascal O.)V1423,15-19(1991)

Design of M-band filter banks based on wavelet transform (Yaou, Ming-Haw; Chang, Wen-Thong)V1605,149-159(1991)

Design of achromatized hybrid diffractive lens systems (Londono, Carmina; Clark, Peter P.)V1354,30-37(1991)

Design of parallel multiresolution filter banks by simulated annealing (Li, Wei; Basso, Andrea; Popat, Ashok C.; Nicoulin, Andre; Kunt, Murat)V1605,124-136(1991)

Digital replication of chest radiographs without altering diagnostic observer performance (Flynn, Michael J.; Davies, Eric; Spizarny, David; Beute, Gordon; Peterson, Ed; Eyler, William R.; Gross, Barry; Chen, Ji)V1444,172-179(1991)

Display systems for medical imaging (Erdekian, Vahram V.; Trombetta, Steven P.)V1444,151-158(1991)

Document viewing: display requirements in image management (van Overbeek, Thomas T.)V1454,406-413(1991)

Ecological approach to partial binocular overlap (Melzer, James E.; Moffitt, Kirk W.)V1456,124-131(1991)

Edge-profile, materials, and protective coating effects on image quality (Doi, Takeshi; Tadros, Karim H.; Kuyel, Birol; Neureuther, Andrew R.)V1464,336-345(1991)

Effect of experimental design on sample size (Rockette, Howard E.; Obuchowski, Nancy A.; Gur, David; Good, Walter F.)V1446,276-283(1991)

Effect of viewing distance and disparity on perceived depth (Gooding, Linda; Miller, Michael E.; Moore, Jana; Kim, Seong-Han)V1457,259-266(1991)

Effects of higher order aberrations on the process window (Gortych, Joseph E.; Williamson, David M.)V1463,368-381(1991)

Elements of real-space imaging: a proposed taxonomy (Naimark, Michael)V1457,169-179(1991)

Endocular ophthalmoscope: miniaturization and optical imaging quality (Rol, Pascal O.; Beck, Dominik; Niederer, Peter)V1423,84-88(1991)

Entropy coding for wavelet transform of image and its application for motion picture coding (Ohta, Mutsumi; Yano, Mitsuharu; Nishitani, Takao)V1605,456-466(1991)

Evaluation of the 3-D radon transform algorithm for cone beam reconstruction (Grangeat, Pierre; Le Masson, Patrick; Melennec, Pierre; Sire, Pascal)V1445,320-331(1991)

Evaluation of the effect of noise on subjective image quality (Barten, Peter G.)V1453,2-15(1991)

Gray level distortions induced by 3-D reflective hologram (Odinokov, S. B.; Poddubnaya, T. E.; Rozhkov, Oleg V.; Yakimovich, A. P.)V1238,109-117(1991)

High-bandwidth recording in a hostile environment (Darg, David A.; Kikuchi, Akira)V1538,124-131(1991)

High-fidelity subband coding for superhigh-resolution images (Saito, Takahiro; Higuchi, Hirofumi; Komatsu, Takashi)V1605,382-393(1991)

High-quality image recorder and evaluation (Suzuki, Masayuki; Shiraiwa, Yoshinobu; Minoura, Kazuo)V1454,370-381(1991)

High-resolution color image coding scheme for office systems (Koshi, Yutaka; Kunitake, Setsu; Suzuki, Kazuhiro; Kamizawa, Koh; Yamasaki, Toru; Miyake, Hidetaka)V1605,362-373(1991)

High-resolution microchannel plate image tube development (Johnson, C. B.; Patton, Stanley B.; Bender, E.)V1449,2-12(1991)

Human recognition of infrared images II (Sanders, Jeffrey S.; Currin, Michael S.)V1488,144-155(1991)

Human visual performance model for crewstation design (Larimer, James O.; Prevost, Michael P.; Arditi, Aries R.; Azueta, Steven; Bergen, James R.; Lubin, Jeffrey)V1456,196-210(1991)

Identification of a class of space-variant image blurs (Ozkan, Mehmet K.; Tekalp, Ahmet M.; Sezan, M. I.)V1452,146-156(1991)

Image analysis of discrete and continuous systems: film and CCD sensors (Kriss, Michael A.)V1398,4-14(1991)

Image quality, dollars, and very low contrast documents (Weideman, William E.)V1454,382-390(1991)

Image quality evaluation of multifocal intraocular lenses (Silberman, Donn M.)V1423,20-28(1991)

Image quality measurements with a neural brightness perception model (Grogan, Timothy A.; Wu, Mei)V1453,16-30(1991)

Image quality metrics for volumetric laser displays (Williams, Rodney D.; Donohoo, Daniel)V1457,210-220(1991)

Images: from a printer's perspective (Sarkar, N. R.)V1458,42-50(1991)

Image signal modulation and noise analysis of CRT displays (Burns, Peter D.)V1454,392-398(1991)

Importance of phosphor persistence characteristics in reducing visual distress symptoms in VDT users (Hayosh, Thomas D.)V1454,399-405(1991)

Improved image quality from retroreflective screens by spectral smearing (Stevens, Richard F.)V1358,1265-1267(1991)

Improved phantom for quality control of laser scanner digitizers in PACS (Halpern, Ethan J.; Esser, Peter D.)V1444,104-115(1991)

Improved precision/resolution by camera movement (Taylor, Geoff L.; Derksen, Grant)V1526,27-34(1991)

Investigation of imaging quality of Fourier holographic lens (Zajac, Marek; Nowak, Jerzy)V1574,197-204(1991)

Legibility of compressed document images at various spatial resolutions (Kidd, Robert C.; Zalinski, Charles M.; Nadel, Jerome I.; Klein, Robert D.)V1454,414-424(1991)

Light-pulse-induced background in image intensifiers with MCPs (Dashevsky, Boris E.)V1449,25-29(1991)

MCP image intensifier in the 100-KeV to 1-MeV x-ray range (Veaux, Jacqueline; Cavailler, Claude; Gex, Jean-Pierre; Hauducoeur, Alain; Hivernage, M.)V1449,13-24(1991)

Measuring the geometric accuracy of a very accurate, large, drum-type scanner (Wolber, John W.)V1448,175-180(1991)

Medical Imaging V: Image Capture, Formatting, and Display (Kim, Yongmin, ed.)V1444(1991)

Method of estimating light loadings power and repetition frequency effects on the image quality of gated image intensifier tubes (Dashevsky, Boris E.)V1449,65-70(1991)

Method to convert image resolution using M-band-extended QMF banks (Kawashima, Masahisa; Tominaga, Hideyoshi)V1605,107-111(1991)

Model-based halftoning (Pappas, Thrasyvoulos N.; Neuhoff, David L.)V1453,244-255(1991)

Modeling and suppression of amplitude artifacts due to z motion in MR imaging (Riek, Jonathan K.; Mitsa, Theophano; Parker, Kevin J.; Smith, Warren E.; Tekalp, Ahmet M.; Szumowski, J.)V1398,130-141(1991)

Modeling data acquisition and the effects of patient motion in magnetic resonance imaging (Riek, Jonathan K.; Smith, Warren E.; Tekalp, Ahmet M.; Parker, Kevin J.; Mitsa, Theophano; Szumowski, J.)V1445,198-206(1991)

Monogon laser scanner with no line wobble (Beiser, Leo)V1454,33-36(1991)

Motion-compensated subsampling of HDTV (Belfor, Ricardo A.; Lagendijk, Reginald L.; Biemond, Jan)V1605,274-284(1991)

Multiresponse imaging system design for improved resolution (Alter-Gartenberg, Rachel; Fales, Carl L.; Huck, Friedrich O.; Rahman, Zia-ur; Reichenbach, Stephen E.)V1605,745-756(1991)

Narrowband optical interference filters (Cotton, John M.; Casey, William L.)V1417,525-536(1991)

Need for quality assurance related to PACS (Rowberg, Alan H.)V1444,80-82(1991)

New human vision system model for spatio-temporal image signals (Matsui, Toshikazu; Hirahara, Shuzo)V1453,282-289(1991)

New image diagnosis system with ultrathin endoscope and clinical results (Tsumanuma, Takashi; Toriya, T.; Tanaka, T.; Shamoto, N.; Seto, K.; Sanada, Kazuo; Okazaki, A.; Okazaki, M.)V1420,193-198(1991)

New subband scheme for super-HDTV coding (Tanimoto, Masayuki; Yamada, Akio; Naito, Yoichi)V1605,394-405(1991)

Normal-contrast lith (Netz, Yoel; Hoffman, Arnold)V1458,61-61(1991)

Numerical calculation of image motion and vibration modulation transfer function (Hadar, Ofer; Fisher, Moshe; Kopeika, Norman S.)V1482,79-91(1991)

Observer detection of image degradation caused by irreversible data compression processes (Chen, Ji; Flynn, Michael J.; Gross, Barry; Spizarny, David)V1444,256-264(1991)

On-board calibration device for a wide field-of-view instrument (Krawczyk, Rodolphe; Chessel, Jean-Pierre; Durpaire, Jean-Pierre; Durieux, Alain; Churoux, Pascal; Briottet, Xavier)V1493,2-15(1991)

Optical aberrations in underwater holography and their compensation (Watson, John; Kilpatrick, J. M.)V1461,245-253(1991)

"Perfect" displays and "perfect" image compression in space and time (Klein, Stanley A.; Carney, Thom)V1453,190-205(1991)

Performance evaluation of a vision dimension metrology system (El-Hakim, Sabry F.; Westmore, David B.)V1526,56-67(1991)

Performance evaluation of subband coding and optimization of its filter coefficients (Katto, Jiro; Yasuda, Yasuhiko)V1605,95-106(1991)

Phase-shift mask pattern accuracy requirements and inspection technology (Wiley, James N.; Fu, Tao-Yi; Tanaka, Takashi; Takeuchi, Susumu; Aoyama, Satoshi; Miyazaki, Junji; Watakabe, Yaichiro)V1464,346-355(1991)

Precise method for measuring the edge acuity of electronically (digitally) printed hard-copy images (White, William)V1398,24-28(1991)

Prediction of thermal-image quality as a function of weather forecast (Shushan, A.; Meninberg, Y.; Levy, I.; Kopeika, Norman S.)V1487,300-311(1991)

Psychometrics for quantitative print quality studies (Parker, James D.)V1398,15-23(1991)

Quality assessment of video image capture systems (Rowberg, Alan H.; Lian, Jing)V1444,125-127(1991)

Quality assurance from a manufacturer's standpoint (Anderson, William J.)V1444,128-133(1991)

Quality assurance of PACS systems with laser film digitizers (Esser, Peter D.; Halpern, Ethan J.)V1444,100-103(1991)

Quantitative evaluation of image enhancement algorithms (Lu, Hong-Qian)V1453,223-234(1991)

Quantization of color image components in the DCT domain (Peterson, Heidi A.; Peng, Hui; Morgan, J. H.; Pennebaker, William B.)V1453,210-222(1991)

Radiologists' confidence in detecting abnormalities on chest images and their subjective judgments of image quality (King, Jill L.; Gur, David; Rockette, Howard E.; Curtin, Hugh D.; Obuchowski, Nancy A.; Thaete, F. L.; Britton, Cynthia A.; Metz, Charles E.)V1446,268-275(1991)

Real-time video signal processing by generalized DDA and control memories: three-dimensional rotation and mapping (Hama, Hiromitsu; Yamashita, Kazumi)V1606,878-890(1991)

Relationships between autofocus methods for SAR and self-survey techniques for SONAR (Wahl, Daniel E.; Jakowatz, Charles V.; Ghiglia, Dennis C.; Eichel, Paul H.)V1567,32-40(1991)

Resolution limits for high-resolution imaging lidar (Idell, Paul S.)V1416,250-259(1991)

Resolution problems in holography (Banyasz, Istvan)V1574,282-293(1991)

Scanner analyzer target (Simonis, Roland)V1454,364-369(1991)

Secret transmission method of character data in motion picture communication (Tanaka, Kiyoshi; Nakamura, Yasuhiro; Matsui, Kineo)V1605,646-649(1991)

Sensor line-of-sight stabilization (Cooper, C. J.)V1498,39-51(1991)

Signal and background models in nonstandard IR systems (Snyder, John L.)V1498,52-63(1991)

Signal extension and noncausal filtering for subband coding of images (Martucci, Stephen A.)V1605,137-148(1991)

Some fundamental experiments in subband coding of images (Aase, Sven O.; Ramstad, Tor A.)V1605,734-744(1991)

Specifications for image stabilization systems (Hilkert, James M.; Bowen, Max L.; Wang, Joe)V1498,24-38(1991)

Standardization of image quality measurements of medical x-ray image intensifier systems (Sandrik, John M.)VCR37,180-206(1991)

Statistically optimized PR-QMF design (Caglar, Hakan; Liu, Yipeng; Akansu, Ali N.)V1605,86-94(1991)

Study on objective fidelity for progressive image transmission systems (Lin, Eleanor; Lu, Cheng-Chang)V1453,206-209(1991)

Subband decomposition procedure for quincunx sampling grids (Kim, Chai W.; Ansari, Rashid)V1605,112-123(1991)

Subjective evaluation of scale-space image coding (de Ridder, Huib)V1453,31-42(1991)

Subsampled device-independent interchange color spaces (Kasson, James M.; Plouffe, Wil)V1460,11-19(1991)

Task performance based on the posterior probability of maximum-entropy reconstructions obtained with MEMSYS 3 (Myers, Kyle J.; Hanson, Kenneth M.)V1443,172-182(1991)

Technical and clinical evaluations of a 2048 x 2048-matrix digital radiography system for gastrointestinal examinations (Ogura, Toshihiro; Masuda, Yukihisa; Fujita, Hiroshi; Inoue, Nobuo; Yonekura, Fukuo; Miyagi, Yoshihiro; Takatsu, Kazuaki; Akahira, Katsuyoshi; Tsuruta, Shigehiko; Kamiya, Masami; Takahashi, Fumitaka; Oda, Kazuyuki; Ikeda, Shigeyuki; Koike, Kouichi)V1443,153-157(1991)

Thermal dye transfer color hard-copy image stability (Newmiller, Chris)V1458,92-96(1991)

Use of the Society of Motion Picture and Television Engineers test pattern in picture archiving and communication systems (PACS) (Gray, Joel E.)V1444,118-124(1991)

Using information from multiple low-resolution images to increase resolution (Gagne, Philippe; Arsenault, Henri H.)V1564,656-663(1991)

Wave-optic model to determine image quality through supersonic boundary and mixing layers (Lawson, Shelah M.; Clark, Rodney L.; Banish, Michele R.; Crouse, Randy F.)V1488,268-278(1991)

What is MRT and how do I get one? (Hoover, Carl W.; Webb, Curtis M.)V1488,280-288(1991)

Wiener-matrix image restoration beyond the sampling passband (Rahman, Zia-ur; Alter-Gartenberg, Rachel; Fales, Carl L.; Huck, Friedrich O.)V1488,237-248(1991)

Image reconstruction/restoration—see also image processing; image quality

Adaptive image sequence noise filtering methods (Katsaggelos, Aggelos K.; Kleihorst, Richard P.; Efstratiadis, Serafim N.; Lagendijk, Reginald L.)V1606,716-727(1991)

Adaptive projection technique for CT images refinement (Kuo, Shyh-Shiaw; Mammone, Richard J.)V1606,641-652(1991)

Application of wavelet-type functions in image processing (Segall, Ilana; Zeevi, Yehoshua Y.)V1606,1048-1058(1991)

Adaptive multistage weighted order-statistic filters for image restoration (Yin, Lin; Astola, Jaakko; Neuvo, Yrjo A.)V1451,216-227(1991)

Adaptive surface reconstruction (Terzopoulos, Demetri; Vasilescu, Manuela)V1383,257-264(1991)

Aperture synthesis in astronomical radio-interferometry using maximum entropy on the mean (Le Besnerais, Guy; Navaza, Jorge; Demoment, Guy)V1569,386-395(1991)

Application of a blind-deconvolution restoration technique to space imagery (Lewis, Tom R.; Mitra, Sunanda)V1565,221-226(1991)

Applications of Digital Image Processing XIV (Tescher, Andrew G., ed.)V1567(1991)

Artificial scenes and simulated imaging (Reichenbach, Stephen E.; Park, Stephen K.; Alter-Gartenberg, Rachel; Rahman, Zia-ur)V1569,422-433(1991)

Bayesian signal reconstruction from Fourier transform magnitude and x-ray crystallography (Doerschuk, Peter C.)V1569,70-79(1991)

Binary and phase-shifting image design for optical lithography (Liu, Yong; Zakhor, Avideh)V1463,382-399(1991)

Biomedical applications of video-speckle techniques (Lokberg, Ole J.)V1525,9-16(1991)

Blur identification and image restoration with the expectation-maximization algorithm (Durack, Donald L.)V1487,72-83(1991)

Complex phase error and motion estimation in synthetic aperture radar imaging (Soumekh, Mehrdad; Yang, H.)V1452,104-113(1991)

Computer-generated correlated noise images for various statistical distributions (Wenaas, Holly; Weeks, Arthur R.; Myler, Harley R.)V1569,410-421(1991)

Conditional-expectation-based implementation of the optimal mean-square binary morphological filter (Dougherty, Edward R.; Mathew, A.; Swarnakar, Vivek)V1451,137-147(1991)

Continuous-time segmentation networks (Harris, John G.)V1473,161-172(1991)

Coupled depth-slope model based on augmented Lagrangian techniques (Suter, David)V1570,129-139(1991)

Dense-depth map from multiple views (Attolico, Giovanni; Caponetti, Laura; Chiaradia, Maria T.; Distante, Arcangelo; Stella, Ettore)V1383,34-46(1991)

Differential properties from adaptive thin-plate splines (Sinha, Saravajit S.)V1570,64-74(1991)

Digital imaging of Giemsa-banded human chromosomes: eigenanalysis and the Fourier phase reconstruction (Jericevic, Zeljko; McGavran, Loris; Smith, Louis C.)V1428,200-213(1991)

Digital restoration of scanning electrochemical microscope images (Bartels, Keith A.; Lee, Chongmok; Bovik, Alan C.; Bard, Allen J.)V1450,30-39(1991)

Direct method for reconstructing shape from shading (Oliensis, John; Dupuis, Paul)V1570,116-128(1991)

Edge-based subband image coding technique for encoding the upper-frequency bands (Mohsenian, Nader; Nasrabadi, Nasser M.)V1605,781-792(1991)

Edge detection with subpixel accuracy (Koplowitz, Jack; Lee, Xiaobing)V1471,452-463(1991)

Efficient visual representation and reconstruction from generalized curvature measures (Barth, Erhardt; Caelli, Terry M.; Zetzsche, Christoph)V1570,86-95(1991)

Energy-based segmentation of very sparse range surfaces (Lerner, Mark; Boult, Terrance E.)V1383,277-284(1991)

Errors inherent in the restoration of imagery acquired through remotely sensed refractive interfaces and scattering media (Schmalz, Mark S.)V1479,183-198(1991)

Evaluation of a pose estimation algorithm using single perspective view (Chandra, T.; Abidi, Mongi A.)V1382,409-426(1991)

Evaluation of the 3-D radon transform algorithm for cone beam reconstruction (Grangeat, Pierre; Le Masson, Patrick; Melennec, Pierre; Sire, Pascal)V1445,320-331(1991)

Fast full-erasure laser-addressed smectic liquid-crystal light valve (Zhuang, Song Lin; Qiu, Yinggang; Jiang, Yingqui; Tu, Yijun; Chen, Wei)V1558,149-153(1991)

First-order differential technique for optical flow (Campani, Marco; Straforini, Marco; Verri, Alessandro)V1388,409-414(1991)

General method for accelerating simulated annealing algorithms for Bayesian image restoration (Bilbro, Griff L.)V1569,88-98(1991)

High-density CCD neurocomputer chip for accurate real-time segmentation of noisy images (Roth, Michael W.; Thompson, K. E.; Kulp, R. L.; Alvarez, E. B.)V1469,25-36(1991)

High-resolution reconstruction of line drawings (Wong, Ping W.; Koplowitz, Jack)V1398,39-47(1991)

Hole spectrum: model-based optimization of morphological filters (Dougherty, Edward R.; Haralick, Robert M.)V1568,224-232(1991)

Holographic-coordinate-transform-based system for direct Fourier tomographic reconstruction (Huang, Qiang; Freeman, Mark O.)V1564,644-655(1991)

Holographic image reconstruction from interferograms of laser-illuminated complex targets (Wlodawski, Mitchell; Nowakowski, Jerzy)V1416,241-249(1991)

Holographic pseudocoloring of schlieren images (Rodriguez-Vera, Ramon; Olivares-Perez, A.; Morales-Romero, Arquimedes A.)V1507,416-424(1991)

HST image processing: an overview of algorithms for image restoration (Gonsalves, Robert A.; Nisenson, Peter)V1567,294-307(1991)

HST image processing: how does it work and what are the problems? (White, Richard L.; Hanisch, Robert J.)V1567,308-316(1991)

HST phase retrieval: a parameter estimation (Lyon, Richard G.; Miller, Peter E.; Gruszczak, Anthony)V1567,317-326(1991)

Identification and restoration of images with out-of-focus blurs (Liu, Ke; Quan, Jun; Yang, Jing-Yu; Cheng, Yong-Qing)V1567,720-728(1991)

Identification of a class of space-variant image blurs (Ozkan, Mehmet K.; Tekalp, Ahmet M.; Sezan, M. I.)V1452,146-156(1991)

Image coding using adaptive-blocksize Princen-Bradley transform (Mochizuki, Takashi; Yano, Mitsuharu; Nishitani, Takao)V1605,434-444(1991)

Image Processing Algorithms and Techniques II (Civanlar, Mehmet R.; Mitra, Sanjit K.; Moorhead, Robert J., eds.)V1452(1991)

Image reconstruction of IDS filter response (Vaezi, Matt M.; Bavarian, Behnam; Healey, Glenn)V1606,803-809(1991)

Image restoration algorithms based on the bispectrum (Kang, Moon G.; Lay, Kuen-Tsair; Katsaggelos, Aggelos K.)V1606,408-418(1991)

Image restoration and identification using the EM and adaptive RAP algorithms (Kuo, Shyh-Shiaw; Mammone, Richard J.)V1606,419-430(1991)

Image restoration based on perception-related cost functions (Palmieri, Francesco; Croteau, R. E.)V1451,24-35(1991)

Imaging of subsurface regions of random media by remote sensing (Barbour, Randall L.; Graber, Harry L.; Aronson, Raphael; Lubowsky, Jack)V1431,52-62(1991)

Incorporation of structural CT and MR images in PET image reconstruction (Chen, Chin-Tu; Ouyang, Xiaolong; Ordonez, Caesar; Hu, Xiaoping; Wong, Wing H.; Metz, Charles E.)V1445,222-225(1991)

Intensity interpolation for branching in reconstructing three-dimensional objects from serial cross-sections (Liang, Cheng-Chung; Chen, Chin-Tu; Lin, Wei-Chung)V1445,456-467(1991)

Invariant reconstruction of 3-D curves and surfaces (Stevenson, Robert L.; Delp, Edward J.)V1382,364-375(1991)

Iterative algorithms with fast-convergence rates in nonlinear image restoration (Zervakis, Michael E.; Venetsanopoulos, Anastasios N.)V1452,90-103(1991)

Iterative procedures for reduction of blocking effects in transform image coding (Rosenholtz, Ruth E.; Zakhor, Avideh)V1452,116-126(1991)

Linear feature SNR enhancement in radon transform space (Meckley, John R.)V1569,375-385(1991)

Linear programming solutions to problems in logical inference and space-variant image restoration (Digumarthi, Ramji V.; Payton, Paul M.; Barrett, Eamon B.)V1472,128-138(1991)

LMMSE restoration of blurred and noisy image sequences (Ozkan, Mehmet K.; Sezan, M. I.; Erdem, A. T.; Tekalp, Ahmet M.)V1606,743-754(1991)

Magnetic resonance reconstruction from projections using half the data (Noll, Douglas C.; Pauly, John M.; Nishimura, Dwight G.; Macovski, Albert)V1443,29-36(1991)

Massively parallel image restoration (Menon, Murali M.)V1471,185-190(1991)

Maximum entropy method applied to deblurring images on a MasPar MP-1 computer (Bonavito, N. L.; Dorband, John E.; Busse, Tim)V1406,138-146(1991)

MAP image reconstruction using intensity and line processes for emission tomography data (Yan, Xiao-Hong; Leahy, Richard M.)V1452,158-169(1991)

MAP segmentation of magnetic resonance images using mean field annealing (Logenthiran, Ambalavaner; Snyder, Wesley E.; Santago, Peter; Link, Kerry M.)V1452,225-243(1991)

Minimum cross-entropy algorithm for image reconstruction from incomplete projection (Zhuang, Tian-ge; Zhu, Yang-ming; Zhang, Xiao L.)V1606,697-704(1991)

Multiresponse imaging system design for improved resolution (Alter-Gartenberg, Rachel; Fales, Carl L.; Huck, Friedrich O.; Rahman, Zia-ur; Reichenbach, Stephen E.)V1605,745-756(1991)

Near-infrared imaging in vivo: imaging of Hb oxygenation in living tissues (Araki, Ryuichiro; Nashimoto, Ichiro)V1431,321-332(1991)

New algorithm and an efficient parallel implementation of the expectation maximization technique in PET (positron emission tomography) imaging (Buyukkoc, Cagatay; Persiano, G.)V1452,170-179(1991)

Noise reduction in ultrasound images using multiple linear regression in a temporal context (Olstad, Bjorn)V1451,269-281(1991)

Nonlinear approach to the 3-D reconstruction of microscopic objects (Wu, Xiangchen; Schwarzmann, Peter)V1450,278-285(1991)

Numerical calculation of image motion and vibration modulation transfer function (Hadar, Ofer; Fisher, Moshe; Kopeika, Norman S.)V1482,79-91(1991)

Operator/system communication: an optimizing decision tool (Sobh, Tarek M.; Alameldin, Tarek K.)V1388,524-535(1991)

Optic flow: multiple instantaneous rigid motions (Zhuang, Xinhua; Wang, Tao; Zhang, Peng)V1569,434-445(1991)

Optimal algorithms for holographic image reconstruction of an object with a glint using nonuniform illumination (Nowakowski, Jerzy; Wlodawski, Mitchell)V1416,229-240(1991)

Optimal regularization parameter estimation for image restoration (Reeves, Stanley J.; Mersereau, Russell M.)V1452,127-138(1991)

Optoelectronic implementation of filtered-back-projection tomography algorithm (Lu, Tongxin; Udpa, Satish S.; Udpa, L.)V1564,704-713(1991)

PET reconstruction using multisensor fusion techniques (Acharya, Raj S.; Gai, Nevile)V1445,207-221(1991)

Phase retrieval for the Hubble Space Telescope using iterative propagation algorithms (Fienup, James R.)V1567,327-332(1991)

Postprocessing of video sequence using motion-dependent median filters (Lee, Ching-Long; Jeng, Bor S.; Ju, Rong-Hauh; Huang, Huang-Cheng; Kan, Kou-Sou; Huang, Jei-Shyong; Liu, Tsann-Shyong)V1606,728-734(1991)

Practical approach to fractal-based image compression (Pentland, Alexander P.; Horowitz, Bradley)V1605,467-474(1991)

Preliminary tests of maximum likelihood image reconstruction method on 3-D real data and some practical considerations for the data corrections (Liu, Yi-Hwa; Holmes, Timothy J.; Koshy, Matthew)V1428,191-199(1991)

Properties of different estimates of the regularizing parameter for the least-squares image restoration problem (Galatsanos, Nikolas P.; Katsaggelos, Aggelos K.)V1396,590-600(1991)

Pyramid median filtering by block threshold decomposition (Zhou, Hongbing; Zeng, Bing; Neuvo, Yrjo A.)V1451,2-12(1991)

Rapid enhancement and compression of image data (Balram, Nikhil; Moura, Jose M.)V1606,374-385(1991)

Reconstruction algorithm of displacement field by using holographic image (Chen, Ke-long; Xu, Zhu; Wen, Zhen-chu; Chen, Yu-zhuo)V1385,206-213(1991)

Reconstruction during camera fixation (Raviv, Daniel)V1382,312-319(1991)

Reconstruction methods for infrared absorption imaging (Arridge, Simon R.; van der Zee, Pieter; Cope, Mark; Delpy, David T.)V1431,204-215(1991)

Reconstruction of quincunx-coded image sequences using vector median (Oistamo, Kai; Neuvo, Yrjo A.)V1606,735-742(1991)

Reconstruction of three-dimensional displacement fields by carrier holography (Xu, Zhu; Chen, Ke-long; Wen, Zhen-chu; Yang, Han-Guo; He, Xiao-yuan; Zhang, Bao-he)V1399,172-177(1991)

Recovering epipolar geometry in 3-D vision systems (Schenk, Anton F.; Toth, Charles)V1457,66-73(1991)

Recovery for a real image from its Hartley transform modulus only (Dong, Bizhen; Gu, Benyuan; Yang, Guo-Zhen)V1429,117-126(1991)

Recurrent neural network application to image filtering: 2-D Kalman filtering approach (Swiniarski, Roman W.; Dzielinski, Andrzej; Skoneczny, Slawomir; Butler, Michael P.)V1451,234-241(1991)

Refinement of EM (expectation maximization) restored images (Kuo, Shyh-Shiaw; Mammone, Richard J.)V1452,192-202(1991)

Resolution enhancement of blurred star field images by maximally sparse restoration (Jeffs, Brian D.; Gunsay, Metin; Dougal, John)V1567,511-521(1991)

Resolution enhancement of CT images (Kuo, Shyh-Shiaw; Mammone, Richard J.)V1450,18-29(1991)

Resolution improvement for in-vivo magnetic resonance spectroscopic images (Plevritis, Sylvia; Macovski, Albert)V1445,118-127(1991)

Resolution limits for high-resolution imaging lidar (Idell, Paul S.)V1416,250-259(1991)

Restoration of distorted depth maps calculated from stereo sequences (Damour, Kevin; Kaufman, Howard)V1452,78-89(1991)

Restoration of spatially variant motion blurs in sequential imagery (Trussell, Henry J.; Fogel, Sergei)V1452,139-145(1991)

Restoration of subpixel detail using the regularized pseudoinverse of the imaging operator (Abbiss, John B.; Brames, Bryan J.)V1566,365-375(1991)

Robust regularized image restoration (Kwon, Taek M.; Zervakis, Michael E.)V1569,317-328(1991)

Rule-based system to reconstruct 3-D tree structure from two views (Liu, Iching; Sun, Ying)V1606,67-77(1991)

Scaling of digital shapes with subpixel boundary estimation (Koplowitz, Jack)V1470,167-174(1991)

Shape-from-focus: surface reconstruction of hybrid surfaces (Chen, Su-Shing; Tang, Wu-bin; Xu, Jian-hua)V1569,446-450(1991)

Signal enhancement in noise- and clutter-corrupted images using adaptive predictive filtering techniques (Soni, Tarun; Zeidler, James R.; Rao, Bhaskar D.; Ku, Walter H.)V1565,338-344(1991)

Signal loss recovery in DCT-based image and video codecs (Wang, Yao; Zhu, Qin-Fan)V1605,667-678(1991)

Simulated annealing image reconstruction for an x-ray coded source tomograph (El Alaoui, Mohsine; Magnin, Isabelle E.; Amiel, Michel)V1569,80-87(1991)

Simultaneous object estimation and image reconstruction in a Bayesian setting (Hanson, Kenneth M.)V1452,180-191(1991)

Some analytical and statistical properties of Fisher information (Frieden, B. R.)V1569,311-316(1991)

Spatial frequency filtering on the basis of the nonlinear optics phenomena (Kudryavtseva, Anna D.; Tcherniega, Nicolaii V.; Brekhovskikh, Galina L.; Sokolovskaya, Albina I.)V1385,190-199(1991)

Spline-based tomographic reconstruction method (Guedon, Jean-Pierre; Bizais, Yves)V1443,214-225(1991)

Stereo matching, error detection, and surface reconstruction (Stewart, Charles V.)V1383,285-296(1991)

Stochastic and Neural Methods in Signal Processing, Image Processing, and Computer Vision (Chen, Su-Shing, ed.)V1569(1991)

Structure of a scene from two and three projections (Tommasi, Tullio)V1383,26-33(1991)

Surface reconstruction method using deformable templates (Wang, Yuan-Fang; Wang, Jih-Fang)V1383,265-276(1991)

Surface reconstruction with discontinuities (Lee, David T.; Shiau, Jyh-Jen H.)V1383,297-304(1991)

Task performance based on the posterior probability of maximum-entropy reconstructions obtained with MEMSYS 3 (Myers, Kyle J.; Hanson, Kenneth M.)V1443,172-182(1991)

Three-dimensional reconstruction from cone beam projection by a block iterative technique (Peyrin, Francoise; Goutte, Robert; Amiel, Michel)V1443,268-279(1991)

Three-dimensional reconstruction from cone beam projections (Ohishi, Satoru; Yamaguchi, Masahiro; Ohyama, Nagaaki; Honda, Toshio)V1443,280-285(1991)

Three-dimensional reconstruction from optical flow using temporal integration (Rangachar, Ramesh M.; Hong, Tsai-Hong; Herman, Martin; Luck, Randall L.)V1382,331-340(1991)

Three-dimensional reconstruction of liver from 2-D tomographic slices (Chou, Jin-Shin; Chen, Chin-Tu; Giger, Maryellen L.; Kahn, Charles E.; Bae, Kyongtae T.; Lin, Wei-Chung)V1396,45-50(1991)

Three-dimensional reconstruction using virtual planes and horopters (Grosso, Enrico; Sandini, Giulio; Tistarelli, Massimo)V1570,371-381(1991)

Three-dimensional scene reconstruction using optimal information fusion (Hong, Lang)V1383,333-344(1991)

Toward tactile sensor-based exploration in a robotic environment (Gadagkar, Hrishikesh P.; Trivedi, Mohan M.)V1383,142-150(1991)

Transfer function considerations for the CERES (cloud's and earth's radiant energy system) scanning radiometer (Manalo, Natividad D.; Smith, G. L.; Barkstrom, Bruce R.)V1521,106-116(1991)

Use of cross-validation as a stopping rule in emission tomography image reconstruction (Coakley, Kevin J.; Llacer, Jorge)V1443,226-233(1991)

Using correlated CT images in compensation for attenuation in PET (positron emission tomography) image reconstruction (Yu, Xiaolin; Chen, Chin-Tu; Bartlett, R.; Pelizzari, Charles A.; Ordonez, Caesar)V1396,56-58(1991)

Visualization of 3-D phase structure in confocal and conventional microscopy (Cogswell, Carol J.; Sheppard, Colin J.)V1450,323-328(1991)

Wiener-matrix image restoration beyond the sampling passband (Rahman, Zia-ur; Alter-Gartenberg, Rachel; Fales, Carl L.; Huck, Friedrich O.)V1488,237-248(1991)

X-ray projection microscopy and cone-beam microtomography (Wang, Ge; Lin, T. H.; Cheng, Ping-Chin; Shinozaki, D. M.; Newberry, S. P.)V1398,180-190(1991)

Image recording—see also optical data storage; optical recording

Automatic data processing of speckle fringe pattern (Dupre, Jean-Christophe; Lagarde, Alexis)V1554A,766-771(1991)

Bidirectional printing method for a thermal ink transfer printer (Nagato, Hitoshi; Ohno, Tadayoshi)V1458,84-91(1991)

Electronic f-theta correction for hologon deflector systems (Whitman, Tony; Araghi, Mehdi N.)V1454,426-433(1991)

Elements of real-space imaging: a proposed taxonomy (Naimark, Michael)V1457,169-179(1991)

Ericsson digital recce management system (Johansson, Micael; Engman, Per; Johansson, Bo G.)V1538,180-188(1991)

High-bandwidth recording in a hostile environment (Darg, David A.; Kikuchi, Akira)V1538,124-131(1991)

High-quality image recorder and evaluation (Suzuki, Masayuki; Shiraiwa, Yoshinobu; Minoura, Kazuo)V1454,370-381(1991)

High-resolution high-speed laser recording and printing using low-speed polygons (Razzaghi-Masoud, Mahmoud)V1458,145-153(1991)

High-speed electronic memory video recording (Thomas, Don L.)V1448,140-147(1991)

High-speed video instrumentation system (Gorenflo, Ronald L.; Stockum, Larry A.; Barnett, Brett)V1346,42-53(1991)

Holographic deflectors for graphic arts applications: an overview (Kramer, Charles J.)V1454,68-100(1991)

Holographic high-resolution endoscopic image recording (Bjelkhagen, Hans I.)V1396,93-98(1991)

Image acquisition unit for the Mayo/IBM PACS project (Reardon, Frank J.; Salutz, James R.)V1446,481-491(1991)

Laptop image transmission equipment (Mocenter, Michael M.)V1538,132-139(1991)

Large-aperture CCD x-ray detector for protein crystallography using a fiber-optic taper (Strauss, Michael G.; Westbrook, Edwin M.; Naday, Istvan; Coleman, T. A.; Westbrook, M. L.; Travis, D. J.; Sweet, Robert M.; Pflugrath, J. W.; Stanton, Martin J.)V1447,12-27(1991)

Laser disk: a practicum on effective image management (Myers, Richard F.)V1460,50-58(1991)

Linear LED arrays for film annotation and for high-speed high-resolution printing on ordinary paper (Wareberg, P. G.; Scholes, R.; Taylor, R.)V1538,112-123(1991)

Low-cost, low-risk approach to tactical reconnaissance (Beving, James E.; Fishell, Wallace G.)V1538,14-17(1991)

MCP image intensifier in the 100-KeV to 1-MeV x-ray range (Veaux, Jacqueline; Cavailler, Claude; Gex, Jean-Pierre; Hauducoeur, Alain; Hivernage, M.)V1449,13-24(1991)

Method of caustics for anisotropic materials (Rossmanith, H. P.)V1554A,835-849(1991)

New image-capturing techniques for a high-speed motion analyzer (Balch, Kris S.)V1346,2-23(1991)

On-line acquisition of CT and MRI studies from multiple scanners (Weinberg, Wolfram S.; Loloyan, Mansur; Chan, Kelby K.)V1446,430-435(1991)

Pod-mounted MIL-STD-2179B recorder (Kessler, William D.; Abeille, Pierre; Sulzer, Jean-Francois)V1538,104-111(1991)

Poly(bis-alkylthio-acetylen): a dual-mode laser-sensitive material (Baumann, Reinhard; Bargon, Joachim; Roth, Hans-Klaus)V1463,638-643(1991)

Practical videography (Sturz, Richard A.)V1346,64-67(1991)

Stationary platen 2-axis scanner (Schermer, Mack J.)V1454,257-264(1991)

System for making scientific videotapes (Appino, Perry A.; Farrell, Edward J.)V1459,157-165(1991)

Three-dimensional grating images (Takahashi, Susumu; Toda, Toshiki; Iwata, Fujio)V1461,199-205(1991)

Ultra-high scanning holographic deflector unit (Kramer, Charles J.; Szalowski, Rafal; Watkins, Mark)V1454,434-446(1991)

Image tubes—see also intensifiers; photography, high-speed; streak cameras

328 MHz synchroscan streak camera (Schelev, Mikhail Y.; Serdyuchenko, Yuri N.; Vaschenko, G. O.)V1358,569-573(1991)

Advanced image intensifier systems for low-light high-speed imaging (Bruno, Theresa L.; Wirth, Allan)V1358,109-116(1991)

Application of electron-sensitive CCD for taking off the time-dispersed pictures from image tube phosphor screens (Bryukhnevitch, G. I.; Dalinenko, I. N.; Kuz'min, G. A.; Libenson, B. N.; Malyarov, A. V.; Moskalev, B. B.; Postovalov, V. E.; Prokhorov, Alexander M.; Schelev, Mikhail Y.)V1358,739-749(1991)

C 850X picosecond high-resolution streak camera (Mens, Alain; Dalmasso, J. M.; Sauneuf, Richard; Verrecchia, R.; Roth, J. M.; Tomasini, F.; Miehe, Joseph A.; Rebuffie, Jean-Claude)V1358,315-328(1991)

Disk-cathode flash x-ray tube driven by a repetitive type of Blumlein pulser (Sato, Eiichi; Kimura, Shingo; Isobe, Hiroshi; Takahashi, Kei; Tamakawa, Yoshiharu; Yanagisawa, Toru)V1358,146-153(1991)

Femtosecond streak tube (Kinoshita, Katsuyuki; Suyama, Motohiro; Inagaki, Yoshinori; Ishihara, Y.; Ito, Masuo)V1358,490-496(1991)

First results on a developmental deflection tube and its associated electronics for streak camera applications (Froehly, Claude; Laucournet, A.; Miehe, Joseph A.; Rebuffie, Jean-Claude; Roth, J. M.; Tomasini, F.)V1358,532-540(1991)

Fourth-generation motion analyzer (Balch, Kris S.)V1358,373-397(1991)

Fundamental studies for the high-intensity long-duration flash x-ray generator for biomedical radiography (Sato, Eiichi; Isobe, Hiroshi; Takahashi, Kei; Tamakawa, Yoshiharu; Yanagisawa, Toru)V1358,193-200(1991)

Gating techniques for imaging digicon tubes (Swanberg, Norman; Urbach, Michael K.; Ginaven, Robert O.)V1346,249-267(1991)

High-intensity soft-flash x-ray generator utilizing a low-vacuum diode (Isobe, Hiroshi; Sato, Eiichi; Shikoda, Arimitsu; Takahashi, Kei; Tamakawa, Yoshiharu; Yanagisawa, Toru)V1358,471-478(1991)

High-precision measurements of the 24-beam UV-OMEGA laser (Jaanimagi, Paul A.; Hestdalen, C.; Kelly, John H.; Seka, Wolf D.)V1358,337-343(1991)

Improved version of the PIF01 streak image tube (Degtyareva, V. P.; Fedotov, V. I.; Moskalev, B. B.; Postovalov, V. E.; Prokhorov, Alexander M.; Schelev, Mikhail Y.; Soldatov, N. F.)V1358,524-531(1991)

Kilohertz range pulsed x-ray generator having a hot cathode triode (Sato, Eiichi; Shikoda, Arimitsu; Isobe, Hiroshi; Tamakawa, Yoshiharu; Yanagisawa, Toru; Honda, Keiji; Yokota, Yoshiharu)V1358,479-487(1991)

Multialkali photocathodes grown by molecular beam epitaxy technique (Dubovoi, I. A.; Chernikov, A. S.; Prokhorov, Alexander M.; Schelev, Mikhail Y.; Ushakov, V. K.)V1358,134-138(1991)

Multiframing image converter camera (Frontov, H. N.; Serdyuchenko, Yuri N.)V1358,311-314(1991)

New electron optic for high-speed single-frame and streak image intensifier tubes (Dashevsky, Boris E.)V1358,561-568(1991)

New framing image tube with high-spatial-resolution (Chang, Zenghu; Hou, Xun; Zhang, Yongfeng; Zhu, Wenhua; Niu, Lihong; Liu, Xiouqin)V1358,541-545(1991)

New streak tubes of the P500 series: features and experimental results (Rebuffie, Jean-Claude; Mens, Alain)V1358,511-523(1991)

New streak tube with femtosecond time resolution (Feldman, G. G.; Ilyna, T. A.; Korjenevitch, I. N.; Syrtzev, V. N.)V1358,497-502(1991)

Picosecond intensifier gating with a plated webbing cathode underlay (Thomas, Stan W.; Trevino, Jimmy)V1358,84-90(1991)

Profile-related time resolution for a femtosecond streak tube (Liu, Yueping; Sibbett, Wilson)V1358,503-510(1991)

Recent advances in gated x-ray imaging at LLNL (Wiedwald, Douglas J.; Bell, Perry M.; Kilkenny, Joseph D.; Bonner, R.; Montgomery, David S.)V1346,449-455(1991)

Recent research and development in electron image tubes/cameras/systems (Prokhorov, Alexander M.; Schelev, Mikhail Y.)V1358,280-289(1991)

Repetitive flash x-ray generator as an energy transfer source utilizing a compact-glass body diode (Shikoda, Arimitsu; Sato, Eiichi; Kimura, Shingo; Isobe, Hiroshi; Takahashi, Kei; Tamakawa, Yoshiharu; Yanagisawa, Toru)V1358,154-161(1991)

Repetitive flash x-ray generator operated at low-dose rates for a medical x-ray television system (Sato, Eiichi; Isobe, Hiroshi; Takahashi, Kei; Tamakawa, Yoshiharu; Yanagisawa, Toru)V1358,462-470(1991)

Repetitive flash x-ray generator utilizing an enclosed-type diode with a ring-shaped graphite cathode (Isobe, Hiroshi; Sato, Eiichi; Kimura, Shingo; Tamakawa, Yoshiharu; Yanagisawa, Toru; Honda, Keiji; Yokota, Yoshiharu)V1358,201-208(1991)

Review of ITT/EOPD's special purpose photosensitive devices and technologies (Johnson, C. B.)V1396,360-376(1991)

Some comparative results of two approaches in computer simulation of electron lenses for streak image tubes (Degtyareva, V. P.; Ivanov, V. Y.; Ignatov, A. M.; Kolesnikov, Sergey V.; Kulikov, Yu. V.; Monastyrski, M. A.; Niu, Hanben; Schelev, Mikhail Y.)V1358,546-548(1991)

Soviet-American image converter cameras "PROSCHEN" (Briscoe, Dennis; Shrivastava, Chinmaya A.; Nebeker, Sidney J.; Hsu, S.; Lozovoi, V. I.; Postovalov, V. E.; Prokhorov, Alexander M.; Schelev, Mikhail Y.; Serdyuchenko, Yuri N.; Vaschenko, G. O.)V1358,329-336(1991)

Streak camera phosphors: response to ultrashort excitation (Jaanimagi, Paul A.; Hestdalen, C.)V1346,443-448(1991)

Subnanosecond intensifier gating using heavy and mesh cathode underlays (Thomas, Stan W.; Shimkunas, Alex R.; Mauger, Philip E.)V1358,91-99(1991)

Techniques for capturing over 10,000 images/second with intensified imagers (Balch, Kris S.)V1358,358-372(1991)

Theoretical and experimental performance evaluations of Picoframe framing cameras (Liu, Yueping; Sibbett, Wilson; Walker, David R.)V1358,300-310(1991)

Two-dimensional electron-bombarded CCD readout device for picosecond electron-optical information system (Ivanov, K. N.; Krutikov, N. I.; Naumov, S. K.; Pischelin, E. V.; Semenov, V. A.; Stepanov, M. S.; Postovalov, V. E.; Prokhorov, Alexander M.; Schelev, Mikhail Y.)V1358,732-738(1991)

ULTRANAC: a new programmable image converter framing camera (Garfield, Brian R.; Riches, Mark J.)V1358,290-299(1991)

Veiling glare in the F4111 image intensifier (Acharya, Mukund; Bunch, Robert M.)V1396,377-388(1991)

Wide-aperture x-ray image converter tubes (Dashevsky, Boris E.; Podvyaznikov, V. A.; Prokhorov, Alexander M.; Chevokin, Victor K.)V1346,437-442(1991)

Image understanding

Accurate image simulation by hemisphere projection (Bian, Buming; Wittels, Norman)V1453,333-340(1991)

Actively controlled multiple-sensor system for feature extraction (Daily, Michael J.; Silberberg, Teresa M.)V1472,85-96(1991)

Analysis and visualization of heart motion (Chen, Chang W.; Huang, Thomas S.; Arrott, Matthew)V1450,231-242(1991)

Analysis of multidimensional confocal images (Samarabandu, J. K.; Acharya, Raj S.; Edirisinghe, Chandima D.; Cheng, Ping-Chin; Kim, Hyo-Gun; Lin, T. H.; Summers, R. G.; Musial, C. E.)V1450,296-322(1991)

Analysis of optical flow estimation using epipolar plane images (Rangachar, Ramesh M.; Hong, Tsai-Hong; Herman, Martin; Luck, Randall L.)V1382,376-385(1991)

Anthropomorphic classification using three-dimensional Fourier descriptor (Lee, Nahm S.; Park, Kyung S.)V1450,133-143(1991)

Apparent contrast and surface color in complex scenes (Arend, Lawrence)V1453,412-421(1991)

Application of a discrete-space representation to three-dimensional medical imaging (Toennies, Klaus D.; Tronnier, Uwe)V1444,19-25(1991)

Application of generalized radial basis functions to the problem of object recognition (Thau, Robert S.)V1469,37-47(1991)

Application of image processing technology to 2-D signal processing (Meckley, John R.)V1406,72-73(1991)

Application of mathematical morphology to the automated determination of microstructural characteristics of composites (Chackalackal, Mathew S.; Basart, John P.)V1568,347-356(1991)

Application of the nonrigid shape matching algorithm to volumetric cardiac images (Goldof, Dmitry B.; Chandra, Kambhamettu)V1450,264-276(1991)

Applications of Digital Image Processing XIV (Tescher, Andrew G., ed.)V1567(1991)

Artificial neural network models for image understanding (Kulkarni, Arun D.; Byars, P.)V1452,512-522(1991)

Assessment of the information content of patterns: an algorithm (Daemi, M. F.; Beurle, R. L.)V1567,621-631(1991)

Assumption truth maintenance in model-based ATR algorithm design (Bennett, Laura F.; Johnson, Rubin; Hudson, C. I.)V1470,263-274(1991)

Automated labeling of coronary arterial tree segments in angiographic projection data (Dumay, Adrie C.; Gerbrands, Jan J.; Geest, Rob J.; Verbruggen, Patricia E.; Reiber, J. H.)V1445,38-46(1991)

Automatic digitization of contour lines for digital map production (Yla-Jaaski, Juha; Yu, Xiaohan)V1472,201-207(1991)

Automatic method for inspecting plywood shear samples (Avent, R. R.; Conners, Richard W.)V1468,281-295(1991)

Automatic object recognition: critical issues and current approaches (Sadjadi, Firooz A.)V1471,303-313(1991)

Automatic reconstruction of buildings from aerial imagery (Sinha, Saravajit S.)V1468,698-709(1991)

Automatic searching center measurement of profile of a line (Qiang, Xue-li; Yang, Qing)V1567,670-679(1991)

Automatic segmentation of brain images: selection of region extraction methods (Gong, Leiguang; Kulikowski, Casimir A.; Mezrich, Reuben S.)V1450,144-153(1991)

Automatic segmentation of microvessels using textural analysis (Albert, Thomas; O'Connor, Carol; Harris, Patrick D.)V1450,84-89(1991)

Automatic wall motion detection in the left ventricle using ultrasonic images (Olstad, Bjorn)V1450,243-254(1991)

Autonomous navigation of structured city roads (Aubert, Didier; Kluge, Karl; Thorpe, Charles E.)V1388,141-151(1991)

Background characterization using a second-order moment function (Scholl, Marija S.; Udomkesmalee, Suraphol)V1468,92-98(1991)

Biomedical Image Processing II (Bovik, Alan C.; Howard, Vyvyan, eds.)V1450(1991)

Biomedical structure recognition by successive approximations (Venturi, Giovanni; Dellepiane, Silvana G.; Vernazza, Gianni L.)V1606,217-225(1991)

Blackboard architecture for medical image interpretation (Davis, Darryl N.; Taylor, Christopher J.)V1445,421-432(1991)

Brain surface maps from 3-D medical images (Lu, Jiuhuai; Hansen, Eric W.; Gazzaniga, Michael S.)V1459,117-124(1991)

BRICORK: an automatic machine with image processing for the production of corks (Davies, Roger; Correia, Bento A.; Carvalho, Fernando D.; Rodrigues, F. C.)V1459,283-291(1991)

Cellular-automata-based learning network for pattern recognition (Tzionas, Panagiotis; Tsalides, Ph.; Thanailakis, A.)V1606,269-280(1991)

Characteristic pattern matching based on morphology (Zhao, Dongming)V1606,86-96(1991)

Clue derivation and selection activities in a robot vision system (Reihani, Kamran; Thompson, Wiley E.)V1468,305-312(1991)

Collaborative processing to extract myocardium from a sequence of two-dimensional echocardiograms (Revankar, Shriram; Sher, David B.; Rosenthal, Steven)V1459,268-273(1991)

Color and Grassmann-Cayley coordinates of shape (Petrov, A. P.)V1453,342-352(1991)

Combined edge- and region-based method for range image segmentation (Koivunen, Visa; Pietikainen, Matti)V1381,501-512(1991)

Comparison of surface and volume presentation of multislice biomedical images (Browne, Mark A.; Jolleys, Glenn D.; Joyner, David J.)V1450,338-349(1991)

Computational model of an integrated vision system (Uttal, William; Shepherd, Thomas; Lovell, Robb E.; Dayanand, Sriram)V1453,258-269(1991)

Computer animation method for simulating polymer flow for injection-molded parts (Perry, Meg W.; Rumbaugh, Richard C.; Frost, David P.)V1459,155-156(1991)

Computerized system for clinical diagnosis of melanoma (Ferrario, Mario; Barbieri, Fabio)V1450,108-117(1991)

Computer vision systems: integration of software architectures (Bohling, Edward H.; O'Connor, R. P.)V1406,164-168(1991)

Connectivity-preserving morphological image transformations (Bloomberg, Dan S.)V1606,320-334(1991)

Contextual image understanding of airport photographs (Nasr, Hatem N.)V1521,24-33(1991)

Continuous-time segmentation networks (Harris, John G.)V1473,161-172(1991)

Crystal surface analysis using matrix textural features classified by a probabilistic neural network (Sawyer, Curry R.; Quach, Viet; Nason, Donald; van den Berg, Lodewijk)V1567,254-263(1991)

Decision support system for capillaroscopic images (Tascini, Guido; Puliti, Paolo; Zingaretti, Primo)V1450,178-185(1991)

Detecting difficult roads and intersections without map knowledge for robot vehicle navigation (Crisman, Jill D.; Thorpe, Charles E.)V1388,152-164(1991)

Detection and visualization of porosity in industrial CT scans of aluminum die castings (Andrews, Lee T.; Klingler, Joseph W.; Schindler, Jeffery A.; Begeman, Michael S.; Farron, Donald; Vaughan, Bobbi; Riggs, Bud; Cestaro, John)V1459,125-135(1991)

Detection of unresolved target tracks in infrared imagery (Rajala, Sarah A.; Nolte, Loren W.; Aanstoos, James V.)V1606,360-371(1991)

Development of criteria to compare model-based texture analysis methods (Soh, Young-Sung; Murthy, S. N.; Huntsberger, Terrance L.)V1381,561-573(1991)

Different aspects of backgrounds in various spectral bands (Ben-Shalom, Ami; Devir, Adam D.; Ribak, Erez N.; Talmore, Eli T.; Balfour, L. S.; Brandman, N.)V1486,238-257(1991)

Discrete random set models for shape synthesis and analysis (Goutsias, John I.; Wen, Chuanju)V1606,174-185(1991)

Divergence as a measure of visual recognition: bias and errors caused by small samples (Lau, Manhot; Okagaki, Takashi)V1606,705-713(1991)

Dynamically reconfigurable multiprocessor system for high-order-bidirectional-associative-memory-based image recognition (Wu, Chwan-Hwa; Roland, David A.)V1471,210-221(1991)

Dynamic range data acquisition and pose estimation for 3-D regular objects (Marszalec, Janusz A.; Heikkila, Tapio A.; Jarviluoma, Markku)V1382,443-452(1991)

EGOLOGY: psychological spatial breakthrough for social redirection—multidisciplinary spatial focus for individuals/humankind (Thompson, Robert A.; Thompson, Louise A.)V1469,451-462(1991)

Edge detection with subpixel accuracy (Koplowitz, Jack; Lee, Xiaobing)V1471,452-463(1991)

Effect of image size and contrast on the recognition of insects in radiograms (Schatzki, Thomas F.; Keagy, Pamela M.)V1379,182-188(1991)

Efficient extraction of local myocardial motion with optical flow and a resolution hierarchy (Srikantan, Geetha; Sher, David B.; Newberger, Ed)V1459,258-267(1991)

Electro-optical system for the nondestructive evaluation of bioengineering materials (Sciammarella, Cesar A.; Bhat, Gopalakrishna K.; Albertazzi, Armando A.)V1429,183-194(1991)

Evaluation of a pose estimation algorithm using single perspective view (Chandra, T.; Abidi, Mongi A.)V1382,409-426(1991)

Experiences with a parallel architecture for image analysis (Evans, Robert H.; Williams, Elmer F.; Brant, Karl)V1406,201-202(1991)

Experimental system for detecting lung nodules by chest x-ray image processing (Suzuki, Hideo; Inaoka, Noriko; Takabatake, Hirotsugu; Mori, Masaki; Natori, Hiroshi; Suzuki, Akira)V1450,99-107(1991)

Extracting local stretching from left ventricle angiography data (Mishra, Sanjoy K.; Goldof, Dmitry B.)V1450,218-230(1991)

Extracting Meaning from Complex Data: Processing, Display, Interaction II (Farrell, Edward J., ed.)V1459(1991)

Extraction of human stomach using computational geometry (Aisaka, Kazuo; Arai, Kiyoshi; Tsutsui, Kumiko; Hashizume, Akihide)V1445,312-317(1991)

Fast algorithm for obtaining dense depth maps for high-speed navigation (Khalili, Payman; Jain, Ramesh C.)V1388,210-221(1991)

Fast and precise method to extract vanishing points (Coelho, Christopher; Straforini, Marco; Campani, Marco)V1388,398-408(1991)

Feature discrimination using multiband classification techniques (Ivey, Jim; Fairchild, Scott; Peterson, James R.; Stahl, Charles G.)V1567,170-178(1991)

Feature selection technique for classification of hyperspectral AVIRIS data (Shen, Sylvia S.; Trang, Bonnie Y.)V1567,188-193(1991)

Finding distinctive colored regions in images (Syeda, Tanveer F.)V1381,574-581(1991)

Focusing on targets through exclusion (Mueller, Walter; Olson, James; Martin, Andrew; Hinchman, John H.)V1472,66-75(1991)

Fractal-based multifeature texture description (Kasparis, Takis; Tzannes, Nicolaos S.; Bassiouni, Mostafa; Chen, Qing)V1521,46-54(1991)

Fractal image compression and texture analysis (Moghaddam, Baback; Hintz, Kenneth J.; Stewart, Clayton V.)V1406,42-57(1991)

From voxel to curvature (Monga, Olivier; Ayache, Nicholas; Sander, Peter T.)V1570,382-390(1991)

Fusion of human vision system with mid-range IR image processing displays (Forsyth, William B.; Lewis, H. G.)V1472,18-25(1991)

Fuzzy ellipsoidal shell clustering algorithm and detection of elliptical shapes (Dave, Rajesh N.; Patel, Kalpesh J.)V1381,320-333(1991)

Generalization of Lloyd's algorithm for image segmentation (Morii, Fujiki)V1381,545-552(1991)

Generalizing from a small set of training exemplars for handwritten digit recognition (Simon, Wayne E.; Carter, Jeffrey R.)V1469,592-601(1991)

Geometric modeling of noisy image objects (Lipari, Charles A.)V1468,905-917(1991)

Geometric property measurement of binary images by polynomial representation (Bhattacharya, Prabir; Qian, Kai)V1468,918-922(1991)

Grouping and forming quantitative descriptions of image features by a novel parallel algorithm (Ben-Arie, Jezekiel; Huddleston, James)V1606,2-19(1991)

Hierarchical Dempster-Shafer evidential reasoning for image interpretation (Andress, Keith M.)V1569,43-54(1991)

Hierarchical neural net with pyramid data structures for region labeling of images (Rosten, David P.; Yuen, P.; Hunt, Bobby R.)V1472,118-127(1991)

Hierarchical symmetry segmentation (Laird, Alan; Miller, James)V1381,536-544(1991)

High-resolution texture analysis for cytology (Hallouche, Farid; Adams, Alan E.; Hinton, Oliver R.)V1445,504-512(1991)

High-speed sensor-based systems for mobile robotics (Kayaalp, Ali E.; Moezzi, Saied; Davies, Henry C.)V1406,98-109(1991)

Human face recognition by P-type Fourier descriptor (Aibara, Tsunehiro; Ohue, Kenji; Matsuoka, Yasushi)V1606,198-203(1991)

Image analysis applied to black ice detection (Chen, Yi)V1468,551-562(1991)

Image analysis for DNA sequencing (Palaniappan, K.; Huang, Thomas S.)V1450,186-197(1991)

Image analysis for estimating the weight of live animals (Schofield, C. P.; Marchant, John A.)V1379,209-219(1991)

Image analysis of two impinging jets using laser-induced fluorescence and smoke flow visualization (Shoe, Bridget; Disimile, Peter J.; Savory, Eric; Toy, Norman; Tahouri, Bahman)V1521,34-45(1991)

Image analysis using attributed fuzzy tournament matching algorithm (Shaout, Adnan)V1381,357-367(1991)

Image analysis using threshold reduction (Bloomberg, Dan S.)V1568,38-52(1991)

Image-display optimization using clinical history (Nodine, Calvin F.; Brikman, Inna; Kundel, Harold L.; Douglas, A.; Seshadri, Sridhar B.; Arenson, Ronald L.)V1444,56-62(1991)

Image representation by group theoretic approach (Segman, Joseph; Zeevi, Yehoshua Y.)V1606,97-109(1991)

Image resampling in remote sensing and image visualization applications (Trainer, Thomas J.; Sun, Fang-Kuo)V1567,650-658(1991)

Image Understanding and the Man-Machine Interface III (Barrett, Eamon B.; Pearson, James J., eds.)V1472(1991)

Image Understanding for Aerospace Applications (Nasr, Hatem N., ed.)V1521(1991)

Image Understanding in the '90s: Building Systems that Work (Mitchell, Brian T., ed.)V1406(1991)

INS integrated motion analysis for autonomous vehicle navigation (Roberts, Barry; Bazakos, Michael E.)V1521,2-13(1991)

Integrated neural network system for histological image understanding (Refenes, A. N.; Jain, N.; Alsulaiman, Mansour M.)V1386,62-74(1991)

Integrated oceanographic image understanding system (Lybanon, Matthew; Peckinpaugh, Sarah; Holyer, Ronald J.; Cambridge, Vivian)V1406,180-189(1991)

Integration of diverse remote sensing data sets for geologic mapping and resource exploration (Kruse, Fred A.; Dietz, John B.)V1492,326-337(1991)

Integration of edge- and region-based techniques for range image segmentation (Al-Hujazi, Ezzet H.; Sood, Arun K.)V1381,589-599(1991)

Intensity edge detection with stack filters (Yoo, Jisang; Bouman, Charles A.; Delp, Edward J.; Coyle, Edward J.)V1451,58-69(1991)

Interactive tools for assisting the extraction of cartographic features (Hunt, Bobby R.; Ryan, Thomas W.; Sementilli, P.; DeKruger, D.)V1472,208-218(1991)

Interpolation of stereo data using Lagrangian polynomials (Bachnak, Rafic A.; Yamout, Jihad S.)V1457,27-36(1991)

Investigation of methods of combining functional evidence for 3-D object recognition (Stark, Louise; Hall, Lawrence O.; Bowyer, Kevin W.)V1381,334-345(1991)

Knowledge-based system for analysis of aerial images (Shariat, Hormoz)V1381,306-317(1991)

Landmark recognition using motion-derived scene structures (Sadjadi, Firooz A.)V1521,98-105(1991)

LANELOK: an algorithm for extending the lane sensing operating range to 100-feet (Kenue, Surender K.)V1388,222-233(1991)

Learnability of min-max pattern classifiers (Yang, Ping-Fai; Maragos, Petros)V1606,294-308(1991)

Learning object shapes from examples (Shariat, Hormoz)V1567,194-203(1991)

Linear lattice architectures that utilize the central limit for image analysis, Gaussian operators, sine, cosine, Fourier, and Gabor transforms (Ben-Arie, Jezekiel)V1606,823-838(1991)

Linear programming for learning in neural networks (Raghavan, Raghu)V1472,139-148(1991)

Linear programming solutions to problems in logical inference and space-variant image restoration (Digumarthi, Ramji V.; Payton, Paul M.; Barrett, Eamon B.)V1472,128-138(1991)

Linear resection, intersection, and perspective-independent model matching in photogrammetry: theory (Barrett, Eamon B.; Brill, Michael H.; Haag, Nils N.; Payton, Paul M.)V1567,142-169(1991)

Local spectrum analysis of medical images (Baskurt, Atilla; Peyrin, Francoise; Min, Zhu-Yue; Goutte, Robert)V1445,485-495(1991)

Low-level image segmentation via texture recognition (Patel, Devesh; Stonham, T. J.)V1606,621-629(1991)

Matching in image/object dual spaces (Zhang, Yaonan)V1526,195-202(1991)

Maximum entropy method applied to deblurring images on a MasPar MP-1 computer (Bonavito, N. L.; Dorband, John E.; Busse, Tim)V1406,138-146(1991)

Maximum likelihood estimation of affine-modeled image motion (Shaltaf, Samir J.; Namazi, Nader M.)V1567,609-620(1991)

Measurements of lightness: dependence on the position of a white in the field of view (McCann, John J.; Savoy, Robert L.)V1453,402-411(1991)

Medical image understanding system based on Dempster-Shafer reasoning (Chen, Shiuh-Yung; Lin, Wei-Chung; Chen, Chin-Tu)V1445,386-397(1991)

Method for removing background regions from moving images (Fujimoto, Tsuyoshi; Shoman, Mineo; Hase, Masahiko)V1606,599-606(1991)

Microcomputer-based workstation for simulation and analysis of background and target IR signatures (Reeves, Richard C.; Schaibly, John H.)V1486,85-101(1991)

Minimum resolution for human face detection and identification (Samal, Ashok)V1453,81-89(1991)

Model-based automatic target recognition development tools (Nasr, Hatem N.; Amehdi, Hossien)V1471,283-290(1991)

Model-based labeling, analysis, and three-dimensional visualization from two-dimensional medical images (Arata, Louis K.; Dhawan, Atam P.; Thomas, Stephen R.)V1446,465-474(1991)

Model-based morphology (Haralick, Robert M.; Dougherty, Edward R.; Katz, Philip L.)V1472,108-117(1991)

Model-based system for automatic target recognition (Verly, Jacques G.; Delanoy, Richard L.; Dudgeon, Dan E.)V1471,266-282(1991)

Model-based vision system for automatic recognition of structures in dental radiographs (Acharya, Raj S.; Samarabandu, J. K.; Hausmann, E.; Allen, K. A.)V1450,170-177(1991)

Modeling of the texture structural components using 2-D deterministic random fields (Francos, Joseph M.; Meiri, A. Z.; Porat, Boaz)V1606,553-565(1991)

Multiscale analysis based on mathematical morphology (Lu, Yi; Vogt, Robert C.)V1568,14-25(1991)

Multisensor fusion using the sensor algorithm research expert system (Bullock, Michael E.; Miltonberger, Thomas W.; Reinholdtsen, Paul A.; Wilson, Kathleen)V1471,291-302(1991)

Multistage object recognition using dynamical-link graph matching (Flaton, Kenneth A.)V1469,137-148(1991)

Natural texture analysis in a multiscale context using fractal dimension (Kpalma, Kidiyo; Bruno, Alain; Haese-Coat, Veronique)V1606,55-66(1991)

Neural-network-based vision processing for autonomous robot guidance (Pomerleau, Dean A.)V1469,121-128(1991)

Neural network for improving terrain elevation measurement from stereo images (Jordan, Michael)V1567,179-187(1991)

Neural network labeling of the Gulf Stream (Lybanon, Matthew; Molinelli, Eugene J.; Muncill, G.; Pepe, Kevin)V1469,637-647(1991)

Neural network model of dynamic form perception: implications of retinal persistence and extraretinal sharpening for the perception of moving boundaries (Ogmen, Haluk)V1606,350-359(1991)

New edge-enhancement method based on the deviation of the local image grey center (Zhao, Mingsheng; Shen, Zhen-kang; Chen, Huihuang)V1471,464-473(1991)

New formulation of reflection coefficient in propagation theory for imaging inhomogeneous media and for nonuniform waveguides (Berger, Henry; Del Bosque-Izaguirre, Delma)V1521,117-130(1991)

New insights into correlation-based template matching (Ooi, James; Rao, Kashi)V1468,740-751(1991)

New method for designing face image classifiers using 3-D CG model (Akamatsu, Shigeru; Sasaki, Tsutomu; Masui, Nobuhiko; Fukamachi, Hideo; Suenaga, Yasuhito)V1606,204-216(1991)

New parallel algorithms for thinning of binary images (Bhattacharya, Prabir; Lu, Xun)V1468,734-739(1991)

Noise-tolerant texture classification and image segmentation (Kjell, Bradley P.; Wang, Pearl Y.)V1381,553-560(1991)

Nonparametric dominant point detection (Ansari, Nirwan; Huang, Kuo-Wei)V1606,31-42(1991)

Nonuniform image motion estimation in transformed-domain (Namazi, Nader M.; Lipp, John I.)V1567,659-669(1991)

Novel block segmentation and processing for Chinese-English document (Chien, Bing-Shan; Jeng, Bor S.; Sun, San-Wei; Chang, Gan-How; Shyu, Keh-Haw; Shih, Chun-Hsi)V1606,588-598(1991)

Novel method for the computer analysis of cryomicroscopic images (Chen, Wei; Gong, Yiming; Wang, Ruli; Xu, Yayong)V1450,198-205(1991)

Oh say, can you see? The physiology of vision (Young, Richard A.)V1453,92-123(1991)

Parallel rule inferencing for automatic target recognition (Pacelli, Jean L.; Geyer, Steve L.; Ramsey, Timothy S.)V1472,76-84(1991)

Parametric optical flow without correspondence for moving sensors (Whitten, Gary E.)V1468,167-175(1991)

Performance evaluation of a texture-based segmentation algorithm (Sadjadi, Firooz A.)V1483,185-195(1991)

Photometric models in multispectral machine vision (Brill, Michael H.)V1453,369-380(1991)

Pixel level data fusion: from algorithm to chip (Mathur, Bimal P.; Wang, H. T.; Liu, Shih-Chii; Koch, Christof; Luo, Jin)V1473,153-160(1991)

POPS: parallel opportunistic photointerpretation system (Howard, Michael D.)V1471,422-427(1991)

Production environment implementation of the stereo extraction of cartographic features using computer vision and knowledge base systems in DMA's digital production system (Gruenewald, Maria M.; Hinchman, John H.)V1468,843-852(1991)

Qualitative three-dimensional shape from stereo (Wildes, Richard P.)V1382,453-463(1991)

Radar image understanding for complex space objects (Hemler, Paul F.)V1381,55-65(1991)

Range data from stereo images of edge points (Lim, Hong-Seh)V1382,434-442(1991)

Realization of the Zak-Gabor representation of images (Assaleh, Khaled T.; Zeevi, Yehoshua Y.; Gertner, Izidor)V1606,532-540(1991)

Real-time inspection of pavement by moire patterns (Guralnick, Sidney A.; Suen, Eric S.)V1396,664-677(1991)

Real-time region hierarchy and identification algorithm (Reihani, Kamran; Thompson, Wiley E.)V1449,99-108(1991)

Recognition of a moving planar shape in space from two perspective images (Li, Zhi-yong; Sun, Zhong-kang; Shen, Zhen-kang)V1472,97-105(1991)

Recognition of handwritten katakana in a frame using moment invariants based on neural network (Agui, Takeshi; Takahashi, Hiroki; Nakajima, Masayuki; Nagahashi, Hiroshi)V1606,188-197(1991)

Reconstructing visible surfaces (Schenk, Anton F.; Toth, Charles)V1526,78-89(1991)

Recovering 3-D translation of a rigid surface by a binocular observer using moments (Al-Hudaithi, Aziz; Udpa, Satish S.)V1567,490-501(1991)

Recursion and feedback in image algebra (Ritter, Gerhard X.; Davidson, Jennifer L.)V1406,74-86(1991)

Reflectance at the red edge as a sensitive indicator of the damage of trees and its correlation to the state of the photosynthetic system (Ruth, Bernhard)V1521,131-142(1991)

Representing three-dimensional shapes for visual recognition (Hoffman, Donald)V1445,2-4(1991)

Rigid body motion estimation using a sequence of images from a static camera (Raju, G. V.; Rudraraju, Prasad)V1388,415-424(1991)

Robust stereo vision (Sinha, Saravajit S.; Moezzi, Saied; Schunck, Brian G.)V1385,259-267(1991)

Rotation and scale invariant pattern recognition using a multistaged neural network (Minnix, Jay I.; McVey, Eugene S.; Inigo, Rafael M.)V1606,241-251(1991)

Salient contour extraction for target recognition (Rao, Kashi; Liou, James J.)V1482,293-306(1991)

Scene description: an iterative approach (Mulgaonkar, Prasanna G.; Decurtins, Jeff; Cowan, Cregg K.)V1382,320-330(1991)

SCORPIUS: a successful image understanding implementation (Wilkes, Scott C.)V1406,110-121(1991)

SCORPIUS: final report (Onishi, Randall M.)V1472,56-65(1991)

SCORPIUS: lessons learned in developing a successful image understanding system (Gee, Shirley J.)V1406,190-200(1991)

SCORPIUS: lessons learned in managing an image understanding system (Onishi, Randall M.; Bogdanowicz, Julius F.; Watanabe, Miki)V1406,171-178(1991)

Segmentation of orientation maps by an integration of edge- and region-based methods (Distante, Arcangelo; D'Orazio, Tiziana; Stella, Ettore)V1381,513-523(1991)

Segmentation using models of expected structure (Shemlon, Stephen; Liang, Tajen; Cho, Kyugon; Dunn, Stanley M.)V1381,470-481(1991)

Segmentation using range data and structured light (Hu, Gongzhu)V1381,482-489(1991)

Segmentation via fusion of edge and needle map (Ahn, Hong-Young; Tou, Julius T.)V1468,896-904(1991)

Semiautomated detection and measurement of glomerular basement membrane from electron micrographs (Ong, Sim-Heng; Giam, S. T.; Jayasooriah, Mr.; Sinniah, R.)V1445,564-573(1991)

Shape from shading with circular symmetry (Gennert, Michael A.)V1385,256-258(1991)

Shape reconstruction and object recognition using angles in an image (Fukada, Youji)V1381,111-121(1991)

Spatial and temporal surface interpolation using wavelet bases (Pentland, Alexander P.)V1570,43-62(1991)

Speed ranges accommodated by network architectures of elementary velocity estimators (Courellis, Spiridon H.; Marmarelis, Vasilis Z.)V1606,336-349(1991)

Sperm motion analysis (Salari, Valiollah)V1450,255-263(1991)

Stereo matching by energy function minimization (Gennert, Michael A.; Ren, Biao; Yuille, Alan L.)V1385,268-279(1991)

Structure and motion of entire polyhedra (Sobh, Tarek M.; Alameldin, Tarek K.)V1388,425-431(1991)

Survey: omnifont-printed character recognition (Tian, Qi; Zhang, Peng; Alexander, Thomas; Kim, Yongmin)V1606,260-268(1991)

Target cuing: a heterogeneous neural network approach (McCauley, Howard M.)V1469,69-76(1991)

Target detection performance using 3-D laser radar images (Green, Thomas J.; Shapiro, Jeffrey H.; Menon, Murali M.)V1471,328-341(1991)

Technique for ground/image truthing using a digital map to reduce the number of required measurements (Der, Sandor Z.; Dome, G. J.; Rusche, Gerald A.)V1483,167-176(1991)

Technology transfer in image understanding (Kohl, Charles A.)V1406,18-29(1991)

Territorial analysis by fusion of LANDSAT and SAR data (Vernazza, Gianni L.; Dambra, Carlo; Parizzi, Francesco; Roli, Fabio; Serpico, Sebastiano B.)V1492,206-212(1991)

Texture boundary classification using Gabor elementary functions (Dunn, Dennis F.; Higgins, William E.; Maida, Anthony; Wakeley, Joseph)V1606,541-552(1991)

Texture classification and neural network methods (Visa, Ari J.)V1469,820-831(1991)

Texture discrimination using wavelets (Carter, Patricia H.)V1567,432-438(1991)

Theoretical approach to hyperacuity tests based on resolution criteria for two-line images (Mondal, Pronab K.; Calvo, Maria L.; Chevalier, Margarita L.; Lakshminarayanan, Vasudevan)V1429,108-116(1991)

Three-dimensional CT image segmentation by volume growing (Zhu, Dongping; Conners, Richard W.; Araman, Philip A.)V1606,685-696(1991)

Three-dimensional description of symmetric objects from range images (Alvertos, Nicolas; D'Cunha, Ivan)V1382,388-396(1991)

Three-dimensional face model reproduction method using multiview images (Nagashima, Yoshio; Agawa, Hiroshi; Kishino, Fumio)V1606,566-573(1991)

Three-dimensional reconstruction of pulmonary blood vessels by using anatomical knowledge base (Inaoka, Noriko; Suzuki, Hideo; Mori, Masaki; Takabatake, Hirotsugu; Suzuki, Akira)V1450,2-12(1991)

Three-dimensional scene interpretation through information fusion (Shen, Sylvia S.)V1382,427-433(1991)

Three-dimensional target recognition from fusion of dense range and intensity images (Ramirez, Manuel; Mitra, Sunanda)V1567,632-637(1991)

Topographic mapping for stereo and motion processing (Mallot, Hanspeter A.; Zielke, Thomas; Storjohann, Kai; von Seelen, Werner)V1382,397-408(1991)

Two-view vision system for 3-D texture recovery (Yu, Xiaohan; Yla-Jaaski, Juha)V1468,834-842(1991)

Use of heterogeneous distributed memory parallel systems in image processing (Pinfold, Wilfred)V1406,132-137(1991)

Using color to segment images of 3-D scenes (Healey, Glenn)V1468,814-825(1991)

Using digital images to measure and discriminate small particles in cotton (Taylor, Robert A.; Godbey, Luther C.)V1379,16-27(1991)

Validation of vision-based obstacle detection algorithms for low-altitude helicopter flight (Suorsa, Raymond E.; Sridhar, Banavar)V1388,90-103(1991)

Video browsing using brightness data (Otsuji, Kiyotaka; Tonomura, Yoshinobu; Ohba, Yuji)V1606,980-989(1991)

Vision-based model of artificial texture perception (Landraud, Anne M.)V1453,314-320(1991)

Vision for automated imagery exploitation (Ahlquist, Gregory C.)V1381,2-8(1991)

Visualization of liver in 3-D (Chen, Chin-Tu; Chou, Jin-Shin; Giger, Maryellen L.; Kahn, Charles E.; Bae, Kyongtae T.; Lin, Wei-Chung)V1444,75-77(1991)

X-ray inspection utilizing knowledge-based feature isolation with a neural network classifier (Nolan, Adam R.; Hu, Yong-Lin; Wee, William G.)V1472,157-164(1991)

Imaging

1536 x 1024 CCD image sensor (Wong, Kwok Y.; Torok, Georgia R.; Chang, Win-Chyi; Meisenzahl, Eric J.)V1447,283-287(1991)

3M's Dry Silver technology: an ideal media for electronic imaging (Morgan, David A.)V1458,62-67(1991)

Adaptive optical transfer function modeling (Gaffard, Jean-Paul; Boyer, Corinne)V1483,92-103(1991)

Adaptive optics, transfer loops modeling (Boyer, Corinne; Gaffard, Jean-Paul; Barrat, Jean-Pierre; Lecluse, Yves)V1483,77-91(1991); V1542,46-61(1991)

Analytic approach to centroid performance analysis (Schultz, Kenneth I.)V1416,199-208(1991)

Applications of Optical Engineering: Proceedings of OE/Midwest '90 (Dubiel, Mary K.; Eppinger, Hans E.; Gillespie, Richard E.; Guzik, Rudolph P.; Pearson, James E., eds.)V1396(1991)

a-Si:H TFT-driven high-gray-scale contact image sensor with a ground-mesh-type multiplex circuit (Kobayashi, Kenichi; Abe, Tsutomu; Miyake, Hiroyuki; Kashimura, Hirotsugu; Ozawa, Takashi; Hamano, Toshihisa; Fennell, Leonard E.; Turner, William D.; Weisfield, Richard L.)V1448,157-163(1991)

Automatic recognition of multidimensional objects buried in layered elastic background media (Ayme-Bellegarda, Eveline J.; Habashy, Tarek M.; Bellegarda, Jerome R.)V1471,18-29(1991)

B-transformation and Fibonacci-transformations of optical images for the information conformity of their perception (Miroshnikov, Mikhail M.; Nesteruk, Vsevolod F.)V1500,322-333(1991)

Back-illuminated 1024 x 1024 quadrant readout imager: operation and screening test results (Marsh, Harry H.; Hayes, Raymond; Blouke, Morley M.; Yang, Fanling H.)V1447,298-309(1991)

Backside-thinned CCDs for keV electron detection (Ravel, Mihir K.; Reinheimer, Alice L.)V1447,109-122(1991)

Boundary diffraction wave in imaging by small holograms (Mulak, Grazyna)V1574,266-271(1991)

Building an infrastructure for system integration: machine vision standards (Bieman, Leonard H.; Peyton, James A.)VCR36,3-19(1991)

CAN-AM Eastern '90 (Antos, Ronald L.; Krisiloff, Allen J., eds.)V1398(1991)

Centroid tracking of range-Doppler images (Kachelmyer, Alan L.; Nordquist, David P.)V1416,184-198(1991)

Cerenkov background radiation in imaging detectors (Rosenblatt, Edward I.; Beaver, Edward A.; Cohen, R. D.; Linsky, J. B.; Lyons, R. W.)V1449,72-86(1991)

Characterization of the atmospheric modulation transfer function using the target contrast characterizer (Watkins, Wendell R.; Billingsley, Daniel R.; Palacios, Fernando R.; Crow, Samuel B.; Jordan, Jay B.)V1486,17-24(1991)

Characterization of the Bridgman crystal growth process by radiographic imaging (Fripp, Archibald L.; Debnam, W. J.; Woodell, G. W.; Berry, R. F.; Simchick, Richard T.; Sorokach, S. K.; Barber, Patrick G.)V1557,236-244(1991)

Color standards for electronic imaging (McDowell, David Q.)VCR37,40-53(1991)

Confocal microscopy for the biological and material sciences: principle, applications, limitations (Brakenhoff, G. J.; van der Voort, H. T.; Visscher, Koen)V1439,121-127(1991)

Dead-time effects in microchannel-plate imaging detectors (Zombeck, Martin V.; Fraser, George W.)V1549,90-100(1991)

Defocused white light speckle method for object contouring (Asundi, Anand K.)V1385,239-245(1991)

Design and operational characteristics of a PV 001 image tube incorporated with EB CCD readout (Bryukhnevitch, G. I.; Dalinenko, I. N.; Ivanov, K. N.; Kaidalov, S. A.; Kuz'min, G. A.; Moskalev, B. B.; Naumov, S. K.; Pischelin, E. V.; Postovalov, V. E.; Prokhorov, Alexander M.; Schelev, Mikhail Y.)V1449,109-115(1991)

Design considerations for infrared imaging radiometers (Fraedrich, Douglas S.)V1486,2-7(1991)

Development of the water-window imaging x-ray microscope (Hoover, Richard B.; Shealy, David L.; Baker, Phillip C.; Barbee, Troy W.; Walker, Arthur B.)V1435,338-351(1991)

Double-passage imaging through turbulence (Dainty, J. C.; Mavroidis, Theo; Solomon, Christopher J.)V1487,2-9(1991)

Dynamic sea image generation (Levesque, Martin P.)V1486,294-300(1991)

Electronically gated airborne video camera (Sturz, Richard A.)V1538,77-80(1991)

Electron Image Tubes and Image Intensifiers II (Csorba, Illes P., ed.)V1449(1991)

Experiments in holographic video imaging (Benton, Stephen A.)VIS08,247-267(1991)

Fast algorithm for image-based ranging (Menon, P. K.; Chatterji, Gano B.; Sridhar, Banavar)V1478,190-200(1991)

Ferrofluid film bearing for enhancement of rotary scanner performance (Cheever, Charles J.; Li, Zhixin; Raj, K.)V1454,139-151(1991)

Fiber array optics for electronic imaging (Burger, Robert J.; Greenberg, David A.)V1449,174-185(1991)

Fluorescence imaging of latent fingerprints with a cooled charge-coupled-device detector (Pomeroy, Robert S.; Baker, Mark E.; Radspinner, David A.; Denton, M. B.)V1439,60-65(1991)

Fluorescent imaging (Thompson, Jill C.)V1482,253-257(1991)

Gamut mapping computer-generated imagery (Wallace, William E.; Stone, Maureen C.)V1460,20-28(1991)

Generation of IR sky background images (Levesque, Martin P.)V1486,200-209(1991)

Hadamard transform Raman imaging (Morris, Michael D.; Govil, Anurag; Liu, Kei-Lee; Sheng, Rong-Sheng)V1439,95-101(1991)

Hard x-ray imaging via crystal diffraction: first results of reflectivity measurements (Frontera, Filippo; De Chiara, P.; Gambaccini, M.; Landini, G.; Pasqualini, G.)V1549,113-119(1991)

Health monitoring of rocket engines using image processing (Disimile, Peter J.; Shoe, Bridget; Toy, Norman)V1483,39-48(1991)

High-dynamic-range MCP structures (Slater, David C.; Timothy, J. G.)V1549,68-80(1991)

High-fidelity phase conjugation generated by holograms: application to imaging through multimode fibers (Pan, Anpei; Marhic, Michel E.; Epstein, Max)V1396,99-106(1991)

High-performance 400-DPI A4-size contact image sensor module for scanner and G4 fax applications (Yeh, Long-Ching; Wu, Way-Chen; Tang, Ru-Shyah; Tsai, Yong-Song; Chiang, Eugene; Chiao, Pat; Chu, Chu-Lin; Wang, Weng-Lyang)V1527,361-367(1991)

High-resolution astronomical observations using deconvolution from wavefront sensing (Michau, Vincent; Marais, T.; Laurent, Jean; Primot, Jerome; Fontanella, Jean-Claude; Tallon, M.; Fuensalida, Jesus J.)V1487,64-71(1991)

High-resolution imaging interferometer (Ballangrud, Ase; Jaeger, Tycho C.; Wang, Gunnar)V1521,89-96(1991)

High-resolution microchannel plate image tube development (Johnson, C. B.; Patton, Stanley B.; Bender, E.)V1449,2-12(1991)

High-resolution MS-type Saticon pick-up tube with optimized electron optical properties (Barden, Raimund; Mertelmeier, Thomas; Traupe, U.)V1449,136-147(1991)

High-resolution synchrotron x-radiation diffraction imaging of crystals grown in microgravity and closely related terrestrial crystals (Steiner, Bruce W.; Dobbyn, Ronald C.; Black, David; Burdette, Harold; Kuriyama, Masao; Spal, Richard; van den Berg, Lodewijk; Fripp, Archibald L.; Simchick, Richard T.; Lal, Ravindra B.; Batra, Ashok K.; Matthiesen, David; Ditchek, Brian M.)V1557,156-167(1991)

High-resolution, two-dimensional imaging, microchannel-plate detector for use on a sounding rocket experiment (Bush, Brett C.; Cotton, Daniel M.; Siegmund, Oswald H.; Chakrabarti, Supriya; Harris, Walter; Clarke, John T.)V1549,290-301(1991)

High-throughput narrowband 83.4-nm self-filtering camera (Zukic, Muamer; Torr, Douglas G.; Torr, Marsha R.)V1549,234-244(1991)

HOLIDDO: an interferometer for space experiments (Mary, Joel; Bernard, Yves; Lefaucheux, Francoise; Gonzalez, Francois)V1557,147-155(1991)

Hybrid phototube with Si target (van Geest, Lambertus K.; Stoop, Karel W.)V1449,121-134(1991)

Illumination coherence effects in laser-speckle imaging (Voelz, David G.; Idell, Paul S.; Bush, Keith A.)V1416,260-265(1991)

Image annotation under X Windows (Pothier, Steven)V1472,46-53(1991)

Image formation through inhomogeneities (Leith, Emmett N.; Chen, Chiaohsiang; Cunha, Andre)V1396,80-84(1991)

Imager for gamma-ray astronomy: balloon prototype (Di Cocco, Guido; Labanti, Claudio; Malaguti, Giuseppe; Rossi, Elio; Schiavone, Filomena; Spizzichino, A.; Traci, A.; Bird, A. J.; Carter, T.; Dean, Anthony J.; Gomm, A. J.; Grant, K. J.; Corba, M.; Quadrini, E.; Rossi, M.; Villa, G. E.; Swinyard, Bruce M.)V1549,102-112(1991)

Imagery technology database (Courtot, Marilyn E.; Nier, Michael)VCR37,221-246(1991)

Imaging capabilities of small satellites: Indian experience (Alex, T. K.)V1495,52-58(1991)

Imaging science in the 1990s (Beck, Robert N.)V1396,688-695(1991)

Imaging the sun in hard x-rays: spatial and rotating modulation collimators (Campbell, Jonathan W.; Davis, John M.; Emslie, A. G.)V1549,155-179(1991)

Imaging through a low-light-level Raman amplifier (Duncan, Michael D.; Mahon, R.; Tankersley, Lawrence L.; Reintjes, John F.)V1409,127-134(1991)

Impact on the medium MTF by model estimation of b (Estep, Leland; Arnone, Robert A.)V1537,89-96(1991)

Industrial applications of spectroscopic imaging (Miller, Richard M.; Birmingham, John J.; Cummins, Philip G.; Singleton, Scott)V1439,66-78(1991)

Infrared and the search for extrasolar planets (Meinel, Aden B.; Meinel, Marjorie P.)V1540,196-201(1991)

Interaction of exposure time and system noise with angle-of-arrival measurements (Eaton, Frank D.; Peterson, William A.; Hines, John R.; Waldie, Arthur H.; Drexler, James J.; Qualtrough, John A.; Soules, David B.)V1487,84-90(1991)

Intl Conf on Scientific Optical Imaging (Denton, M. B., ed.)V1439(1991)

Large-format 1280 x 1024 full-frame CCD image sensor with a lateral-overflow drain and transparent gate electrode (Stevens, Eric G.; Kosman, Steven L.; Cassidy, John C.; Chang, Win-Chyi; Miller, Wesley A.)V1447,274-282(1991)

Large-format CCDs for astronomical applications (Geary, John C.)V1439,159-168(1991)

Laser diagnostic techniques in a resonant incinerator (Cadou, Christopher P.; Logan, Pamela; Karagozian, Ann; Marchant, Roy; Smith, Owen I.)V1434,67-77(1991)

Laser Image Detection System (LIDS): laser-based imaging (Hilton, Peter J.; Gabric, R.; Walternberg, P. T.)V1385,27-31(1991)

Laser Radar VI (Becherer, Richard J., ed.)V1416(1991)

Lau imaging (Cartwright, Steven L.)V1396,481-487(1991)

Lens evaluation for electronic photography (Bell, Cynthia S.)V1448,59-68(1991)

Less interclass disturbance learning for unsupervised neural computing (Liu, Lurng-Kuo; Ligomenides, Panos A.)V1606,496-507(1991)

Light-pulse-induced background in image intensifiers with MCPs (Dashevsky, Boris E.)V1449,25-29(1991)

Matched-filter algorithm for subpixel spectral detection in hyperspectral image data (Borough, Howard C.)V1541,199-208(1991)

Megapixel CCD thinning/backside progress at SAIC (Schaeffer, A. R.; Varian, Richard H.; Cover, John R.; Larsen, Robert G.)V1447,165-176(1991)

METEOSAT second-generation program (Markland, Chris A.)V1490,39-50(1991)

Modeling time-dependent obscuration for simulated imaging of dust and smoke clouds (Hoock, Donald W.)V1486,164-175(1991)

Mode-locked Nd:YLF laser for precision range-Doppler imaging (Simpson, Thomas B.; Malley, Michael M.; Sutton, George W.; Doft, F.)V1416,2-9(1991)

Multispectral imaging with frequency-modulated reticles (Sanders, Jeffrey S.; Halford, Carl E.)V1478,52-63(1991)

New 2/3-inch MF image pick-up tubes for HDTV camera (Kobayashi, Akira; Ishikawa, Masayoshi; Suzuki, Takayoshi; Ikeya, Morihiro; Shimomoto, Yasuharu)V1449,148-156(1991)

New approach to palette selection for color images (Balasubramanian, Raja; Allebach, Jan P.)V1453,58-69(1991)

Notch and large-area CCD imagers (Bredthauer, Richard A.; Pinter, Jeff H.; Janesick, James R.; Robinson, Lloyd B.)V1447,310-315(1991)

Novel plastic image-transmitting fiber (Suzuki, Fumio)V1592,150-157(1991)

Optical fibre image sensor (Li, Aizhong)V1572,548-551(1991)

Optimum numerical aperture for optical projection microlithography (Lin, Burn J.)V1463,42-53(1991)

Panoramic security (Greguss, Pal)V1509,55-66(1991)

Pattern recognition in autoradiography of optical materials (Ashurov, Mukhsin K.; Gafitullina, Dilyara)V1550,50-54(1991)

Phase object imaging inside the airy disc (Tychinsky, Vladimir P.)V1392,570-572(1991)

Preliminary review of imaging standards (Ren, Victor; Hatfield, Donald J.; Deacutis, Martin)VCR37,54-67(1991)

Printing Technologies for Images, Gray Scale, and Color (Dove, Derek B.; Abe, Takao; Heinzl, Joachim L., eds.)V1458(1991)

Progress in autostereoscopic display technology at Dimension Technologies Inc. (Eichenlaub, Jesse B.)V1457,290-301(1991)

Projection screens for high-definition television (Kirkpatrick, Michael D.; Mihalakis, George M.)V1456,40-47(1991)

Quantum efficiency model for p+-doped back-illuminated CCD imager (Huang, Chin M.; Kosicki, Bernard B.; Theriault, Joseph R.; Gregory, J. A.; Burke, Barry E.; Johnson, Brett W.; Hurley, Edward T.)V1447,156-164(1991)

Radiometric calibration of an airborne multispectral scanner (Markham, Brian L.; Ahmad, Suraiya P.; Jackson, Ray D.; Moran, M. S.; Biggar, Stuart F.; Gellman, David I.; Slater, Philip N.)V1493,207-214(1991)

Real-time region hierarchy and identification algorithm (Reihani, Kamran; Thompson, Wiley E.)V1449,99-108(1991)

Reconnaissance and imaging sensor test facilities at Eglin Air Force Base (Pratt, Stephen R.; Tucker, Robert)V1538,40-45(1991)

Recovery of aliased signals for high-resolution digital imaging (John, Sarah)V1385,39-47(1991)

Remote alignment of adaptive optical systems with far-field optimization (Mehta, Naresh C.)V1408,96-111(1991)

Robust image coding with a model of adaptive retinal processing (Narayanswamy, Ramkumar; Alter-Gartenberg, Rachel; Huck, Friedrich O.)V1385,93-103(1991)

Selectable one-to-four-port, very high speed 512 x 512 charge-injection device (Zarnowski, Jeffrey J.; Williams, Bryn; Pace, M.; Joyner, M.; Carbone, Joseph; Borman, C.; Arnold, Frank S.; Wadsworth, Mark V.)V1447,191-201(1991)

Shape reconstruction from far-field patterns in a stratified ocean environment (Xu, Yongzhi)V1471,78-86(1991)

Simplified pupil enlargement technique (Radl, Bruce M.)V1457,314-317(1991)

Simulation of scenes, sensors, and tracker for space-based acquisition, tracking, and pointing experiments (DeYoung, David B.)V1415,13-21(1991)

Single-chip imager and feature extractor (Tanner, John E.; Luo, Jin)V1473,76-87(1991)

Single-lens moire contouring method (Harding, Kevin G.; Kaltenbacher, Eric; Bieman, Leonard H.)V1385,246-255(1991)

Some effects on depth-position and course-prediction judgments in 2-D and 3-D displays (Miller, Robert H.; Beaton, Robert J.)V1457,248-258(1991)

Standards for Electronic Imaging Systems (Nier, Michael; Courtot, Marilyn E., eds.)VCR37(1991)

Standards for flat panel display systems (Greeson, James C.)VCR37,146-158(1991)

Stereoscopic Displays and Applications II (Merritt, John O.; Fisher, Scott S., eds.)V1457(1991)

Summary of color definition activity in the graphic arts (McDowell, David Q.)V1460,29-35(1991)

Superconducting IR focal plane arrays (Quelle, Fred W.)V1449,157-166(1991)

TDI camera for industrial applications (Castro, Peter; Gittings, J.; Choi, Yauho J.)V1448,134-139(1991)

Thinned backside-illuminated cooled CCDs for UV and VUV applications (Dalinenko, I. N.; Kuz'min, G. A.; Malyarov, A. V.; Prokhorov, Alexander M.; Schelev, Mikhail Y.)V1449,167-172(1991)

Three-dimensional simulation of optical lithography (Toh, Kenny K.; Neureuther, Andrew R.)V1463,356-367(1991)

Toward an ESA strategy for optical interferometry outside the atmosphere: recommendations by the ESA space interferometry study team (Noordam, Jan E.)V1494,344-346(1991)

Transferred electron photocathode with greater than 5% quantum efficiency beyond 1 micron (Costello, Kenneth A.; Davis, Gary A.; Weiss, Robert; Aebi, Verle W.)V1449,40-50(1991)

Transparent phase-shifting mask with multistage phase shifter and comb-shaped shifter (Watanabe, Hisashi; Todokoro, Yoshihiro; Hirai, Yoshihiko; Inoue, Morio)V1463,101-110(1991)

Ultraviolet-laser-induced fluorescence imaging sensor (Thompson, Jill C.)V1479,412-414(1991)

Ultraviolet light imaging technology and applications (Yokoi, Takane; Suzuki, Kenji; Oba, Koichiro)V1449,30-39(1991)

Uniform field laser illuminator for remote sensing (Di Benedetto, John A.; Capelle, Gene; Lutz, Stephen S.)V1492,115-125(1991)

Use of color, color infrared, black and white films, and video systems in detecting health, stress, and disease in vegetation (Blazquez, Carlos H.)V1379,106-115(1991)

Using dynamic holography for iron fibers with submicron diameter and high velocity (Xie, Gong-Wie; Hua, Lifan; Patel, Sushil; Scott, Peter D.; Shaw, David T.)V1385,132-141(1991)

Variable phase-shift mask for deep-submicron optical lithography (Terasawa, Tsuneo; Hasegawa, Norio; Imai, Akira; Tanaka, Toshihiko; Katagiri, Souichi)V1463,197-206(1991)

Virtual-phase charge-coupled device image sensors for industrial and scientific applications (Khvilivitzky, A. T.; Berezin, V. Y.; Lazovsky, L. Y.; Tataurschikov, S. S.; Pisarevsky, A. N.; Vydrevich, M. G.; Kossov, V. G.)V1447,184-190(1991)

Wiener-matrix image restoration beyond the sampling passband (Rahman, Zia-ur; Alter-Gartenberg, Rachel; Fales, Carl L.; Huck, Friedrich O.)V1488,237-248(1991)

X-ray holography for sequencing DNA (Yorkey, Thomas J.; Brase, James M.; Trebes, James E.; Lane, Stephen M.; Gray, Joe W.)V1345,255-259(1991)

X-ray interferometric observatory (Martin, Christopher)V1549,203-220(1991)

Imaging systems—see also cameras; displays; machine vision; projection systems; television; video

3-DTV research and development in Europe (Sand, Ruediger)V1457,76-84(1991)

4096 x 4096 pixel CCD mosaic imager for astronomical applications (Geary, John C.; Luppino, Gerard A.; Bredthauer, Richard A.; Hlivak, Robert J.; Robinson, Lloyd B.)V1447,264-273(1991)

A 1.3-megapixel-resolution portable CCD electronic still camera (Jackson, Todd A.; Bell, Cynthia S.)V1448,2-12(1991)

Adaptive typography for dynamic mapping environments (Bardon, Didier)V1460,126-139(1991)

Advanced imaging system for high-precision, high-resolution CCD imaging (Doherty, Peter E.; Sims, Gary R.)V1448,118-128(1991)

Airborne imaging system performance model (Redus, Wesley D.)V1498,2-23(1991)

Airborne infrared and visible sensors used for law enforcement and drug interdiction (Aikens, David M.; Young, William R.)V1479,435-444(1991)

Allowable delay time of images with motion parallax and high-speed image generation (Satoh, Takanori; Tomono, Akira; Kishino, Fumio)V1606,1014-1021(1991)

Analysis of the chromatic aberrations of imaging holographic optical elements (Tholl, Hans D.; Luebbers, Hubertus; Stojanoff, Christo G.)V1456,262-273(1991)

Application of low-noise CID imagers in scientific instrumentation cameras (Carbone, Joseph; Hutton, J.; Arnold, Frank S.; Zarnowski, Jeffrey J.; VanGorden, Steve; Pilon, Michael J.; Wadsworth, Mark V.)V1447,229-242(1991)

Applications of charge-coupled and charge-injection devices in analytical spectroscopy (Denton, M. B.)V1447,2-11(1991)

Applications of high-resolution still video cameras to ballistic imaging (Snyder, Donald R.; Kosel, Frank M.)V1346,216-225(1991)

Applications of text-image editing (Bagley, Steven C.; Kopec, Gary E.)V1460,71-79(1991)

Assessment of the optimum operating conditions for 2-D focal-plane-array systems (Bourne, Robert W.; Jefferys, E. A.; Murphy, Kevin S.)V1488,73-79(1991)

Beam Deflection and Scanning Technologies (Beiser, Leo; Marshall, Gerald F., eds.)V1454(1991)

Blurring effect of aerosols on imaging systems (Bissonnette, Luc R.)V1487,333-344(1991)

Buried object remote detection technology for law enforcement (Del Grande, Nancy K.; Clark, Greg A.; Durbin, Philip F.; Fields, David J.; Hernandez, Jose E.; Sherwood, Robert J.)V1479,335-351(1991)

Camera and Input Scanner Systems (Chang, Win-Chyi; Milch, James R., eds.)V1448(1991)

Charge-Coupled Devices and Solid State Optical Sensors II (Blouke, Morley M., ed.)V1447(1991)

Cheops: a modular processor for scalable video coding (Bove, V. M.; Watlington, John)V1605,886-893(1991)

Clinical evaluation of a 2K x 2K workstation for primary diagnosis in pediatric radiology (Razavi, Mahmood; Sayre, James W.; Simons, Margaret A.; Hamedaninia, Azar; Boechat, Maria I.; Hall, Theodore R.; Kangarloo, Hooshang; Taira, Ricky K.; Chuang, Keh-Shih; Kashifian, Payam)V1446,24-34(1991)

Coherent digital/optical system for automatic target recognition (Derr, John I.; Ghaffari, Tammy G.)V1406,127-128(1991)

Color handling in the image retrieval system Imagine (Dal Degan, Nevaino; Lancini, R.; Migliorati, Pierangelo; Pozzi, S.)V1606,934-940(1991)

Comparison of electro-optic diagnostic systems (Hagans, Karla G.; Sargis, Paul G.)V1346,404-408(1991)

Comparison of Kodak Professional Digital Camera System images to conventional film, still video, and freeze-frame images (Kent, Richard A.; McGlone, John T.; Zoltowski, Norbert W.)V1448,27-44(1991)

Comparison of mono- and stereo-camera systems for autonomous vehicle tracking (Kehtarnavaz, Nasser; Griswold, Norman C.; Eem, J. K.)V1468,467-478(1991)

Comparison of stereoscopic cursors for the interactive manipulation of B-splines (Barham, Paul T.; McAllister, David F.)V1457,18-26(1991)

Comparison of three efficient-detail-synthesis methods for modeling using under-sampled data (Iannarilli, Frank J.; Wohlers, Martin R.)V1486,314-324(1991)

Comparison of worldwide opinions on the costs and benefits of PACS (van Gennip, Elisabeth M.; van Poppel, Bas M.; Bakker, Albert R.; Ottes, Fenno P.)V1446,442-450(1991)

Complementary experiments for tether dynamics analysis (Wingo, Dennis R.; Bankston, Cheryl D.)V1495,123-133(1991)

Computer-aided performance evaluation system for the on-board data compression system in HIRIS (Qian, Shen-en; Wang, Ruqin; Li, Shuqiu; Dai, Yisong)V1483,196-206(1991)

Computer model for predicting underwater color images (Palowitch, Andrew W.; Jaffe, Jules S.)V1537,128-139(1991)

Contextual image understanding of airport photographs (Nasr, Hatem N.)V1521,24-33(1991)

Control of night vision pilotage systems (Heaton, Mark W.; Ewing, William S.)V1482,444-457(1991)

Definition and evaluation of the data-link layer of PACnet (Alsafadi, Yasser H.; Martinez, Ralph; Sanders, William H.)V1446,129-140(1991)

Design equations for a polygon laser scanner (Beiser, Leo)V1454,60-66(1991)

Design of an automated imaging system for use in a space experiment (Hartz, William G.; Bozzolo, Nora G.; Lewis, Catherine C.; Pestak, Christopher J.)V1398,52-60(1991)

Design of an optimal single-reflective holographic helmet display element (Twardowski, Patrice J.; Meyrueis, Patrick)V1456,164-174(1991)

Developmental test and evaluation plans for the advanced tactical air reconnaissance system (Minor, John L.; Jenquin, Michael J.)V1538,18-39(1991)

Development of high resolution statistically nonstationary infrared earthlimb radiance scenes (Strugala, Lisa A.; Newt, J. E.; Futterman, Walter I.; Schweitzer, Eric L.; Herman, Bruce J.; Sears, Robert D.)V1486,176-187(1991)

Development of the marine-aggregated-particle profiling and enumerating rover (Costello, David K.; Carder, Kendall L.; Steward, Robert G.)V1537,161-172(1991)

Diagnostic value model for the evaluation of PACS: physician ratings of the importance of prompt image access and the utilization of a display station in an intensive care unit (Kundel, Harold L.; Seshadri, Sridhar B.; Arenson, Ronald L.)V1446,297-300(1991)

Digital charge-coupled device color TV system for endoscopy (Vishnevsky, G. I.; Berezin, V. Y.; Lazovsky, L. Y.; Vydrevich, M. G.; Rivkind, V. L.; Zhemerov, B. N.)V1447,34-43(1991)

Direct digital image transfer gateway (Mun, In K.; Kim, Y. S.; Mun, Seong Ki)V1444,232-237(1991)

Document viewing: display requirements in image management (van Overbeek, Thomas T.)V1454,406-413(1991)

Dynamic spatial reconstructor: a high-speed, stop-action, 3-D, digital radiographic imager of moving internal organs and blood (Jorgensen, Steven M.; Whitlock, S. V.; Thomas, Paul J.; Roessler, R. W.; Ritman, Erik L.)V1346,180-191(1991)

European activities towards a hospital-integrated PACS based on open systems (Kouwenberg, Jef M.; Ottes, Fenno P.; Bakker, Albert R.)V1446,357-361(1991)

Excimer laser photolithography with a 1:1 broadband catadioptric optics (Zhang, Yudong; Lu, Dunwu; Zou, Haixing; Wang, Zhi-Jiang)V1463,456-463(1991)

Experiences with a comprehensive hospital information system that incorporates image management capabilities (Dayhoff, Ruth E.; Maloney, Daniel L.; Kuzmak, Peter; Shepard, Barclay M.)V1446,323-329(1991)

Experimental investigation of image degradation created by a high-velocity flow field (Couch, Lori L.; Kalin, David A.; McNeal, Terry)V1486,417-423(1991)

Experimental system using an interactive drawing input method (Nagano, Yasutada; Kanechika, Hideaki; Tanaka, Satoshi; Maehara, Hideaki; Maeda, Akira)V1605,614-623(1991)

Fast access to reduced-resolution subsamples of high-resolution images (Isaacson, Joel S.)V1460,80-91(1991)

Filled-arm Fizeau telescope (Synnott, Stephen P.)V1494,334-343(1991)

Fourier-transform holographic microscope (Haddad, Waleed S.; Cullen, David; Solem, Johndale C.; Longworth, James W.; McPherson, Armon; Boyer, Keith; Rhodes, Charles K.)V1448,81-88(1991)

Global geometric, sound, and color controls for iconographic displays of scientific data (Smith, Stuart; Grinstein, Georges G.; Pickett, Ronald M.)V1459,192-206(1991)

High-accuracy capacitive position sensing for low-inertia actuators (Stokes, Brian P.)V1454,223-229(1991)

High-performance digital color video camera (Parulski, Kenneth A.; Benamati, Brian L.; D'Luna, Lionel J.; Shelley, Paul R.)V1448,45-58(1991)

High-performance InSb 256 x 256 infrared camera (Blackwell, John D.; Parrish, William J.; Kincaid, Glen T.)V1479,324-334(1991)

High-precision video frame grabber for computed tomography (Drawert, Bruce M.; Slocum, Robert E.)V1396,566-567(1991)

High-resolution airborne multisensor system (Prutzer, Steven; Biron, David G.; Quist, Theodore M.)V1480,46-61(1991)

High-resolution CCD still/video and still-still/video systems (Kee, Richard C.)V1448,13-20(1991)

High-resolution display using a laser-addressed ferroelectric liquid-crystal light valve (Nakajima, Hajime; Kisaki, Jyunko; Tahata, Shin; Horikawa, Tsuyoshi; Nishi, Kazuro)V1456,29-39(1991)

High-resolution, low-light, image-intensified CCD camera (Tanaka, Satoru C.; Silvey, Tom; Long, Greg; Braze, Bill)V1448,21-26(1991)

High-resolution teleradiology applications within the hospital (Jost, R. G.; Blaine, G. J.; Kocher, Thomas E.; Muka, Edward; Whiting, Bruce R.)V1446,2-9(1991)

High-spatial-resolution FLIR (Tucker, Christopher J.; Mitchell, Robert J.)V1498,92-98(1991)

High-speed hardware architecture for high-definition videotex system (Maruyama, Mitsuru; Sakamoto, Hideki; Ishibashi, Yutaka; Nishimura, Kazutoshi)V1605,916-927(1991)

High-speed, high-resolution image reading technique using multi-area sensors (Uehira, Kazutake)V1448,182-190(1991)

High-speed nonsilver lithographic system for laser direct imaging (DoMinh, Thap)V1458,68-68(1991)

High-speed video instrumentation system (Gorenflo, Ronald L.; Stockum, Larry A.; Barnett, Brett)V1346,42-53(1991)

Human recognition of infrared images II (Sanders, Jeffrey S.; Currin, Michael S.)V1488,144-155(1991)

Hypoechoic media: a landmark for intravascular ultrasonic imaging (Gussenhoven, Elma J.; Bom, Nicolaas; Li, Wenguang; van Urk, Hero; Pietermann, Herman; van Suylen, Robert J.; Salem, H. K.)V1425,203-206(1991)

Illumination and viewing methods for machine vision (Uber, Gordon T.; Harding, Kevin G.)VCR36,20-33(1991)

Image analysis for diagnostics in photonic switching (Morrison, Richard L.)V1396,568-574(1991)

Image analysis of discrete and continuous systems: film and CCD sensors (Kriss, Michael A.)V1398,4-14(1991)

Image computing requirements for the 1990s: from multimedia to medicine (Gove, Robert J.; Lee, Woobin; Kim, Yongmin; Alexander, Thomas)V1444,318-333(1991)

Image delivery performance of a CT/MR PACS module applied in neuroradiology (Lou, Shyh-Liang; Loloyan, Mansur; Weinberg, Wolfram S.; Valentino, Daniel J.; Lufkin, Robert B.; Hanafee, William; Bentson, John R.; Jabour, Bradly; Huang, H. K.)V1446,302-311(1991)

Image display and background analysis with the Naval Postgraduate School infrared search and target designation system (Cooper, Alfred W.; Lentz, William J.; Baca, Michael J.; Bernier, J. D.)V1486,47-57(1991)

Image Handling and Reproduction Systems Integration (Bender, Walter J.; Plouffe, Wil, eds.)V1460(1991)

Image quality, dollars, and very low contrast documents (Weideman, William E.)V1454,382-390(1991)

Imager of METEOSAT second generation (Hollier, Pierre A.)V1490,74-81(1991)

Image signal modulation and noise analysis of CRT displays (Burns, Peter D.)V1454,392-398(1991)

Image Understanding in the '90s: Building Systems that Work (Mitchell, Brian T., ed.)V1406(1991)

Imaging autotracker technology for guided missile systems (Hammon, Ricky K.; Helton, Monte K.)V1482,258-264(1991)

Imaging model for underwater range-gated imaging systems (Strand, Michael P.)V1537,151-160(1991)

Imaging spectroscope with an optical recombination system (Li, Wenchong; Ma, Chunhua)V1428,242-248(1991)

Implementing early vision algorithms in analog hardware: an overview (Koch, Christof)V1473,2-16(1991)

Improved IR image generator for real-time scene simulation (Keller, Catherine E.; Stenger, Anthony J.; Bernstein, Uri)V1486,278-285(1991)

Improved real-time volumetric ultrasonic imaging system (Pavy, Henry G.; Smith, Stephen W.; von Ramm, Olaf T.)V1443,54-61(1991)

Infrared Imaging Systems: Design, Analysis, Modeling and Testing II (Holst, Gerald C., ed.)V1488(1991)

Initial experiences with PACS in a clinical and research environment (Honeyman, Janice C.; Staab, Edward V.; Frost, Meryll M.)V1446,362-368(1991)

Integrated color management: the key to color control in electronic imaging and graphic systems (Taylor, Joann M.)V1460,2-10(1991)

Interaction of objects in a virtual environment: a two-point paradigm (Bryson, Steve T.)V1457,180-187(1991)

Interactive graphics system for locating plunge electrodes in cardiac MRI images (Laxer, Cary; Johnson, G. A.; Kavanagh, Katherine M.; Simpson, Edward V.; Ideker, Raymond E.; Smith, William M.)V1444,190-195(1991)

IR CCD staring imaging system (Zhou, Qibo)V1540,677-680(1991)

Johns Hopkins adaptive optics coronagraph (Clampin, Mark; Durrance, Samuel T.; Golimowski, D. A.; Barkhouser, Robert H.)V1542,165-174(1991)

Laser disk: a practicum on effective image management (Myers, Richard F.)V1460,50-58(1991)

Laser guide stars for adaptive optics systems: Rayleigh scattering experiments (Thompson, Laird A.; Castle, Richard; Carroll, David L.)V1542,110-119(1991)

Laser range-gated underwater imaging including polarization discrimination (Swartz, Barry A.; Cummings, James D.)V1537,42-56(1991)

Lightweight surveillance FLIR (Fawcett, James M.)V1498,82-91(1991)

Linear resection, intersection, and perspective-independent model matching in photogrammetry: theory (Barrett, Eamon B.; Brill, Michael H.; Haag, Nils N.; Payton, Paul M.)V1567,142-169(1991)

Linear resonant approach to scanning (Confer, Charles L.; Burrer, Gordon J.)V1454,215-222(1991)

Low-cost computed tomography system using an image intensifier (Hoeft, Gregory L.; Hughes, Simon H.; Slocum, Robert E.)V1396,638-645(1991)

Low light level imaging systems application considerations and calculations (Caudle, Dennis E.)V1346,54-63(1991)

Machine vision system for ore sizing (Eichelberger, Christopher L.; Blair, Steven M.; Khorana, Brij M.)V1396,678-686(1991)

Machine Vision Systems Integration (Batchelor, Bruce G.; Waltz, Frederick M., eds.)VCR36(1991)

Management system for a PACS network in a hospital environment (Mattheus, Rudy A.; Temmerman, Yvan; Verhellen, P.; Osteaux, Michel)V1446,341-351(1991)

MDIS (medical diagnostic imaging support) workstation issues: clinical perspective (Smith, Donald V.; Smith, Suzy; Cawthon, Michael A.)V1444,357-362(1991)

Measuring and display system of a marathon runner by real-time digital image processing (Sasaki, Nobuyuki; Namikawa, Iwao)V1606,1002-1013(1991)

Medical diagnostic imaging support systems for military medicine (Goeringer, Fred)V1444,340-350(1991)

Medical Imaging V: PACS Design and Evaluation (Jost, R. G., ed.)V1446(1991)

Microcomputer-based image processing system for CT/MRI scans: hardware configuration and software capacity (Cheng, Andrew Y.; Ho, Wai-Chin; Kwok, John C.; Yu, Peter K.)V1444,400-406(1991)

Modeling and simulation of systems imaging through atmospheric turbulence (Caponi, Maria Z.)V1415,138-149(1991)

Monogon laser scanner with no line wobble (Beiser, Leo)V1454,33-36(1991)

Monte Carlo calculation of light distribution in an integrating cavity illuminator (Kaplan, Martin)V1448,206-217(1991)

Multichannel fiber optic broadband video communication system for monitoring CT/MR examinations (Huang, H. K.; Kangarloo, Hooshang; Tecotzky, Raymond H.; Cheng, Xin; Vanderweit, Don)V1444,214-220(1991)

Multicomputer performance evaluation tool and its application to the Mayo/IBM image archival system (Pavicic, Mark J.; Ding, Yingjai)V1446,370-378(1991)

Multiresponse imaging system design for improved resolution (Alter-Gartenberg, Rachel; Fales, Carl L.; Huck, Friedrich O.; Rahman, Zia-ur; Reichenbach, Stephen E.)V1605,745-756(1991)

Multisensor fusion methodologies compared (Swan, John; Shields, Frank J.)V1483,219-230(1991)

New concept for a highly compact imaging Fourier transform spectrometer (Simeoni, Denis)V1479,127-138(1991)

New design of the illuminating system for transmission film copy (Pesl, Ales A.)V1448,218-224(1991)

New family of 1:1 catadioptric broadband deep-UV high-Na lithography lenses (Zhang, Yudong; Zou, Haixing; Wang, Zhi-Jiang)V1463,688-694(1991)

New image-capturing techniques for a high-speed motion analyzer (Balch, Kris S.)V1346,2-23(1991)

New stereo laser triangulation device for specular surface inspection (Samson, Marc; Dufour, Marc L.)V1332,314-322(1991)

Novel technical advances provide easy solutions to tough motion analysis problems (Brown, Michael J.)V1346,24-32(1991)

Object-oriented model for medical image database (Aubry, Florent; Bizais, Yves; Gibaud, Bernard; Forte, Anne-Marie; Chameroy, Virginie; Di Paola, Robert; Scarabin, Jean-Marie)V1446,168-176(1991)

Operational infrastructure for a clinical PACS (Boehme, Johannes M.; Chimiak, William J.; Choplin, Robert H.; Maynard, C. D.)V1446,312-317(1991)

Optical design of an off-axis low-distortion UV telescope (Richardson, E. H.)V1494,314-319(1991)

Optical mapping instrument (Bagot, K. H.)V1490,126-132(1991)

Optical scanning system for a CCD telecine for HDTV (Kurtz, Andrew F.; Kessler, David)V1448,191-205(1991)

Organization of a system for managing the text and images that describe an art collection (Mintzer, Fred; McFall, John D.)V1460,38-49(1991)

Overlay and matching strategy for large-area lithography (Holbrook, David S.; Donaher, J. C.)V1463,475-486(1991)

Overview of ASTER design concept (Fujisada, Hiroyuki; Ono, Akira)V1490,244-254(1991)

PACS and teleradiology for on-call support of abdominal imaging (Horii, Steven C.; Garra, Brian S.; Mun, Seong Ki; Zeman, Robert K.; Levine, Betty A.; Fielding, Robert)V1446,10-15(1991)

PACS modeling and development: requirements versus reality (Seshadri, Sridhar B.; Kishore, Sheel; Khalsa, Satjeet S.; Feingold, Eric; Arenson, Ronald L.)V1446,388-395(1991)

Paraxial electron imaging system (Kegelman, Thomas D.)V1454,2-10(1991)

Passive and active sensors for autonomous space applications (Tchoryk, Peter; Gleichman, Kurt W.; Carmer, Dwayne C.; Morita, Yuji; Trichel, Milton; Gilbert, R. K.)V1479,164-182(1991)

Percutaneous coronary angioscopy during coronary angioplasty: clinical findings and implications (Ramee, Stephen R.; White, Christopher J.; Mesa, Juan E.; Jain, Ashit; Collins, Tyrone J.)V1420,199-202(1991)

Performance characteristics of an ultrafast network for PACS (Stewart, Brent K.; Lou, Shyh-Liang; Wong, Albert W.; Chan, Kelby K.; Huang, H. K.)V1446,141-153(1991)

Performance characteristics of the imaging MAMA detector systems for SOHO, STIS, and FUSE/Lyman (Timothy, J. G.)V1549,221-233(1991)

Performance characteristics of the Mayo/IBM PACS (Persons, Kenneth R.; Gehring, Dale G.; Pavicic, Mark J.; Ding, Yingjai)V1446,60-72(1991)

Physics and psychophysics of color reproduction (Giorgianni, Edward)V1458,2-3(1991)

Practical videography (Sturz, Richard A.)V1346,64-67(1991)

Precision high-power solid state laser diagnostics for target-irradiation studies and target-plane irradiation modeling (Wegner, Paul J.; Henesian, Mark A.)V1414,162-174(1991)

Qualitative approach to medical image databases (Bizais, Yves; Gibaud, Bernard; Forte, Anne-Marie; Aubry, Florent; Di Paola, Robert; Scarabin, Jean-Marie)V1446,156-167(1991)

Quality assurance from a manufacturer's standpoint (Anderson, William J.)V1444,128-133(1991)

Quantitative microscope for image cytometry (Jaggi, Bruno; Poon, Steven S.; Pontifex, Brian; Fengler, John J.; Marquis, Jacques; Palcic, Branko)V1448,89-97(1991)

Radiology workstation for mammography: preliminary observations, eyetracker studies, and design (Beard, David V.; Johnston, R. E.; Pisano, E.; Hemminger, Bradley M.; Pizer, Stephen M.)V1446,289-296(1991)

Recent advances in gated x-ray imaging at LLNL (Wiedwald, Douglas J.; Bell, Perry M.; Kilkenny, Joseph D.; Bonner, R.; Montgomery, David S.)V1346,449-455(1991)

Reliability issues in PACS (Taira, Ricky K.; Chan, Kelby K.; Stewart, Brent K.; Weinberg, Wolfram S.)V1446,451-458(1991)

Sampling system for in vivo ultrasound images (Jensen, Jorgen A.; Mathorne, Jan)V1444,221-231(1991)

SCORPIUS: a successful image understanding implementation (Wilkes, Scott C.)V1406,110-121(1991)

SCORPIUS: final report (Onishi, Randall M.)V1472,56-65(1991)

SCORPIUS: lessons learned in developing a successful image understanding system (Gee, Shirley J.)V1406,190-200(1991)

SCORPIUS: lessons learned in managing an image understanding system (Onishi, Randall M.; Bogdanowicz, Julius F.; Watanabe, Miki)V1406,171-178(1991)

Selection of image acquisition methods (Donnelly, Joseph J.)V1444,351-356(1991)

Sensor line-of-sight stabilization (Cooper, C. J.)V1498,39-51(1991)

Signal processing for low-light-level, high-precision CCD imaging (McCurnin, Thomas W.; Schooley, Larry C.; Sims, Gary R.)V1448,225-236(1991)

Signa Tutor: results and future directions (Rundle, Debra A.; Watson, Carolyn K.; Seshadri, Sridhar B.; Wehrli, Felix W.)V1446,379-387(1991)

Simulating watercolor by modeling diffusion, pigment, and paper fibers (Small, David)V1460,140-146(1991)

Simulation-based PACS development (Stut, W. J.; van Steen, M. R.; Groenewegen, L. P.; Ratib, Osman M.; Bakker, Albert R.)V1446,396-404(1991)

Sled tracking system (Downey, George A.; Fountain, H. W.; Riding, Thomas J.; Eggleston, James; Hopkins, Michael; Adams, Billy)V1482,40-47(1991)

Space telescope imaging spectrograph 2048 CCD and its characteristics (Delamere, W. A.; Ebben, Thomas H.; Murata-Seawalt, Debbie; Blouke, Morley M.; Reed, Richard; Woodgate, Bruce E.)V1447,288-297(1991)

Specifications for image stabilization systems (Hilkert, James M.; Bowen, Max L.; Wang, Joe)V1498,24-38(1991)

SPOT 4 HRVIR instrument and future high-resolution stereo instruments (Fratter, C.; Reulet, Jean-Francois; Jouan, Jacky)V1490,59-73(1991)

SPRITE detector characterization through impulse response testing (Anderson, Barry K.; Boreman, Glenn D.; Barnard, Kenneth J.; Plogstedt, Allen E.)V1488,416-425(1991)

Standardization efforts for the preservation of electronic imagery (Adelstein, Peter Z.; Storm, William D.)VCR37,159-179(1991)

Standards for electronic imaging for facsimile systems (Urban, Stephen J.)VCR37,113-145(1991)

Standards for electronic imaging for graphic arts systems (Dunn, S. T.; Dunn, Patrice M.)VCR37,98-112(1991)

Standards for Electronic Imaging Systems (Nier, Michael; Courtot, Marilyn E., eds.)VCR37(1991)

Standards for image input devices: review and forecast (Gilblom, David L.)VCR37,3-19(1991)

STARCON: a reconfigurable fieldable signal processing system (Brandt, S. A.; Budenske, John)V1406,122-126(1991)

Structured light: theory and practice and practice and practice... (Keizer, Richard L.; Jun, Heesung; Dunn, Stanley M.)V1406,88-97(1991)

Study on image-stabilizing reflecting prisms in the case of a finite angular perturbation (Yao, Wu; Lian, Tongshu)V1527,448-455(1991)

Study on objective fidelity for progressive image transmission systems (Lin, Eleanor; Lu, Cheng-Chang)V1453,206-209(1991)

Subsampled device-independent interchange color spaces (Kasson, James M.; Plouffe, Wil)V1460,11-19(1991)

Superhigh-definition image communication: an application perspective (Kohli, Jagdish C.)V1605,351-361(1991)

Systems analysis and design for next generation high-speed video systems (Snyder, Donald R.; Rowe, W. J.)V1346,226-236(1991)

Systems considerations in color printing (Roetling, Paul G.)V1458,17-24(1991)

Technique for measuring atmospheric effects on image metrics (Crow, Samuel B.; Watkins, Wendell R.; Palacios, Fernando R.; Billingsley, Daniel R.)V1486,333-344(1991)

Teleradiology in the local environment (Staab, Edward V.; Honeyman, Janice C.; Frost, Meryll M.; Bidgood, W. D.)V1446,16-22(1991)

Temporal adaptation of multimedia scripts (Robin, Laura)V1460,103-113(1991)

Theoretical and practical aspects of real-time Fourier imaging (Grimard, Dennis S.; Terry, Fred L.; Elta, Michael E.)V1392,535-542(1991)

Thermal signature training for military observers (LaFollette, Robert; Horger, John D.)V1488,289-299(1991)

Three-dimensional analysis framework and measurement methodology for imaging system noise (D'Agostino, John A.; Webb, Curtis M.)V1488,110-121(1991)

Three-dimensional color Doppler imaging of the carotid artery (Picot, Paul A.; Rickey, Daniel W.; Mitchell, J. R.; Rankin, Richard N.; Fenster, Aaron)V1444,206-213(1991)

Three-dimensional real-time ultrasonic imaging using ellipsoidal backprojection (Anderson, Forrest L.)V1443,62-80(1991)

Tools for designing industrial vision systems (Batchelor, Bruce G.)VCR36,138-166(1991)

Transient radiometric measurements with a PtSi IR camera (Konopka, Wayne L.; Soel, Michael A.; Celentano, A.; Calia, V.)V1488,355-365(1991)

Ultrahigh- and High-Speed Photography, Videography, Photonics, and Velocimetry '90 (Jaanimagi, Paul A.; Neyer, Barry T.; Shaw, Larry L., eds.)V1346(1991)

Underwater Imaging, Photography, and Visibility (Spinrad, Richard W., ed.)V1537(1991)

Underwater laser scanning system (Austin, Roswell W.; Duntley, Seibert Q.; Ensminger, Richard L.; Petzold, Theodore J.; Smith, Raymond C.)V1537,57-73(1991)

Update on the C2NVEO FLIR90 and ACQUIRE sensor performance models (Scott, Luke B.; Tomkinson, David M.)V1488,99-109(1991)

Vector and scalar field interpretation (Farrell, Edward J.; Aukrust, Trond; Oberhuber, Josef M.)V1459,136-147(1991)

Virtual-phase charge-coupled device imaging system for astronomy (Khvilivitzky, A. T.; Zuev, A. G.; Rybakov, M. I.; Kiryan, G. V.; Berezin, V. Y.)V1447,64-68(1991)

Visual inspection system using multidirectional 3-D imager (Koezuka, Tetsuo; Kakinoki, Yoshikazu; Hashinami, Shinji; Nakashima, Masato)V1332,323-331(1991)

WEBERSAT: a low-cost imaging satellite (Twiggs, Robert J.; Reister, K. R.)V1495,12-18(1991)

Wedge imaging spectrometer: application to drug and pollution law enforcement (Elerding, George T.; Thunen, John G.; Woody, Loren M.)V1479,380-392(1991)

X-window-based 2K display workstation (Weinberg, Wolfram S.; Hayrapetian, Alek; Cho, Paul S.; Valentino, Daniel J.; Taira, Ricky K.; Huang, H. K.)V1446,35-39(1991)

You can't just plug it in: digital image networks/picture archiving and communication systems installation (Gelish, Anthony)V1444,363-372(1991)

Industrial optics—see also lasers, applications; metrology; optical inspection

Absolute interferometer for manufacturing applications (Tucker, Michael R.; Christenson, Eric)V1367,289-299(1991)

Accuracy/repeatability test for a video photogrammetric measurement (Gustafson, Peter C.)V1526,36-41(1991)

Advances in the practical application of schlieren photography in industry (Herbrich, Horst R.)V1358,24-28(1991)

Application of fiber optic thermometry to the monitoring of winding temperatures in medium- and large-power transformers (Wickersheim, Kenneth A.)V1584,3-14(1991)

Application of high-speed photography to chip refining (Stationwala, Mustafa I.; Miller, Charles E.; Atack, Douglas; Karnis, A.)V1358,237-245(1991)

Automated photogrammetric surface reconstruction with structured light (Maas, Hans-Gerd)V1526,70-77(1991)

Automated visual imaging interface for the plant floor (Wutke, John R.)V1386,180-184(1991)

Calibration of a CCD camera on a hybrid coordinate measuring machine for industrial metrology (Bruzzone, Elisabetta; Mangili, Fulvia)V1526,96-112(1991)

Deformation study of the Klang Gates Dam with multispectral analysis method (Hassan, Azmi; Subari, Mustofa; Som, Zainal A.)V1526,142-156(1991)

Design optimization of moire interferometers for rapid 3-D manufacturing inspection (Dubowsky, Steven; Holly, Krisztina J.; Murray, Annie L.; Wander, Joseph M.)V1386,10-20(1991)

Fiber optic high-temperature sensor for applications in iron and steel industries (Hao, Tianyou; Zhou, Feng-Shen; Xie, Xiou-Qioun; Hu, Ji-Wu; Wang, Wei-Yen)V1584,32-38(1991)

Field testing of a fiber optic rotor temperature monitor for power generators (Brown, Stewart K.; Mannik, Len)V1584,15-22(1991)

Field test results on fiber optic pressure transmitter system (Berthold, John W.)V1584,39-47(1991)

Fundamentals of on-line gauging for machine vision (Novini, Amir R.)V1526,2-16(1991)

High-accuracy edge-matching with an extension of the MPGC-matching algorithm (Gruen, Armin; Stallmann, Dirk)V1526,42-55(1991)

Implementation of a 3-D stereovision system for the production of customized orthotic accessories (Daher, Reinhard; McAdam, Wylie; Pizey, Gordon)V1526,90-93(1991)

Improved precision/resolution by camera movement (Taylor, Geoff L.; Derksen, Grant)V1526,27-34(1991)

Industrial Vision Metrology (El-Hakim, Sabry F., ed.)V1526(1991)

Inspection of a class of industrial objects using a dense range map and CAD model (Ailisto, Heikki J.; Paakkari, Jussi; Moring, Ilkka)V1384,50-59(1991)

Knowledge-based approach to fault diagnosis and control in distributed process environments (Chung, Kwangsue; Tou, Julius T.)V1468,323-332(1991)

Knowledge-based process planning and line design in robotized assembly (Delchambre, Alain)V1468,367-378(1991)

Laser sensing in the iron-making blast furnace (Scott, Chris J.)V1399,137-144(1991)

Machine Vision Systems integration in Industry (Batchelor, Bruce G.; Waltz, Frederick M., eds.)V1386(1991)

New scheme for real-time visual inspection of bearing roller (Zhang, Yu; Xian, Wu; Li, Li-Ping; Hall, Ernest L.; Tu, James Z.)V1384,60-65(1991)

Nonoptical noncoherent imaging in industrial testing (Delecki, Z. A.; Barakat, M. A.)V1526,157-167(1991)

One-dimensional and two-dimensional computer models of industrial CO laser (Iyoda, Mitsuhiro; Murota, Tomoya; Akiyama, Mamoru; Sato, Shunichi)V1415,342-349(1991)

Optical correlator vision system for a manufacturing robot assembly cell (Brandstetter, Robert W.; Fonneland, Nils J.; Zanella, R.; Yearwood, M.)V1385,173-189(1991)

Parameter estimation for process control with neural networks (Samad, Tariq; Mathur, Anoop)V1469,766-777(1991)

Process monitoring during CO2 laser cutting (Jorgensen, Henning; Olsen, Flemming O.)V1412,198-208(1991)

Real-time classification of wooden boards (Poelzleitner, Wolfgang; Schwingshakl, Gert)V1384,38-49(1991)

Remote visual monitoring of seal performance in aircraft jacks using fiber optics (Nwagboso, Christopher O.; Whomes, Terence L.; Davies, P. B.)V1386,30-41(1991)

Spectral signature analysis for industrial inspection (Rauchmiller, Robert F.; Vanderbok, Raymond S.)V1384,100-114(1991)

Three-dimensional gauging with stereo computer vision (Wong, Kam W.; Ke, Ying; Lew, Michael; Obaidat, Mohammed T.)V1526,17-26(1991)

Ultraviolet reflector materials for solar detoxification of hazardous waste (Jorgensen, Gary J.; Govindarajan, Rangaprasad)V1536,194-205(1991)

Vision sensing and processing system for monitoring and control of welding and other high-luminosity processes (Agapakis, John E.; Bolstad, Jon O.)V1385,32-38(1991)

Visual inspection machine for solder joints using tiered illumination (Takagi, Yuuji; Hata, Seiji; Hibi, Susumu; Beutel, Wilhelm)V1386,21-29(1991)

What industry needs in a high-power excimer laser (Lankard, John R.)V1377,2-5(1991)

Information processing—see also optical processing

Application of neural networks to group technology (Caudell, Thomas P.; Smith, Scott D.; Johnson, G. C.; Wunsch, Donald C.)V1469,612-621(1991)

Basic manufacturability interval (Billings, Daniel A.)V1468,434-445(1991)

Basic principles of stereographic software development (Hodges, Larry F.)V1457,9-17(1991)

Capturing multimedia design knowledge using TYRO, the constraint-based designer's apprentice (MacNeil, Ronald L.)V1460,94-102(1991)

Color analysis of nonlinear-phase-modulation method for density pseudocolor encoding technique in medical application (Liu, Dingyu; Yang, Xiaobo; Liu, Changjun; Zhang, Hongguo)V1443,191-196(1991)

Convergence of video and computing (Carlson, Curtis R.)V1472,2-5(1991)

Design considerations for a high-resolution film scanner for teleradiology applications (Kocher, Thomas E.; Whiting, Bruce R.)V1446,459-464(1991)

Distributed architecture for intelligent robotics (Gouveia, Feliz A.; Barthes, Jean-Paul A.; Oliveira, Eugenio C.)V1468,516-523(1991)

Ericsson digital recce management system (Johansson, Micael; Engman, Per; Johansson, Bo G.)V1538,180-188(1991)

EXCON: a graphics-based experiment-control manager (Khan, Mumit; Anderson, Paul D.; Cerrina, Franco)V1465,315-323(1991)

Experimental comparison of optical binary phase-only filter and high-pass matched filter correlation (Leib, Kenneth G.; Brandstetter, Robert W.; Drake, Marvin D.; Franks, Glen B.; Siewert, Ronald O.)V1483,140-154(1991)

Folder management on a multimodality PACS display station (Feingold, Eric; Seshadri, Sridhar B.; Arenson, Ronald L.)V1446,211-216(1991)

Imagery technology database (Courtot, Marilyn E.; Nier, Michael)VCR37,221-246(1991)

Information-theoretic approach to optimal quantization (Lorenzo, Maximo; Der, Sandor Z.; Moulton, Joseph R.)V1483,118-137(1991)

Knowledge-based approach to fault diagnosis and control in distributed process environments (Chung, Kwangsue; Tou, Julius T.)V1468,323-332(1991)

Low-cost spacecraft buses for remote sensing applications (Harvey, Edwin L.; Cullen, Robert M.)V1495,134-145(1991)

Multitask neurovision processor with extensive feedback and feedforward connections (Gupta, Madan M.; Knopf, George K.)V1606,482-495(1991)

Network visualization: user interface issues (Becker, Richard A.; Eick, Stephen G.; Miller, Eileen O.; Wilks, Allan R.)V1459,150-154(1991)

New estimation architecture for multisensor data fusion (Covino, Joseph M.; Griffiths, Barry E.)V1478,114-125(1991)

Object-oriented data management for interactive visual analysis of three-dimensional fluid-flow models (Walther, Sandra S.; Peskin, Richard L.)V1459,232-243(1991)

Object-oriented model for medical image database (Aubry, Florent; Bizais, Yves; Gibaud, Bernard; Forte, Anne-Marie; Chameroy, Virginie; Di Paola, Robert; Scarabin, Jean-Marie)V1446,168-176(1991)

Optical principles of information processing in supercomputer architecture (Burtsev, Vsevolod S.)V1621,380-387(1991)

Optical processing and hybrid neural nets (Casasent, David P.)V1469,256-267(1991)

Optimum structure learning algorithms for competitive learning neural network (Uchiyama, Toshio; Sakai, Mitsuhiro; Saito, Tomohide; Nakamura, Taichi)V1451,192-203(1991)

Parallel data fusion on a hypercube multiprocessor (Davis, Paul B.; Cate, D.; Abidi, Mongi A.)V1383,515-529(1991)

Pattern recognition, attention, and information bottlenecks in the primate visual system (Van Essen, David; Olshausen, B.; Anderson, Clifford H.; Gallant, J. T. L.)V1473,17-28(1991)

Pegasus air-launched space booster payload interfaces and processing procedures for small optical payloads (Mosier, Marty R.; Harris, Gary N.; Whitmeyer, Charlie)V1495,177-192(1991)

Perceptual alternation of ambiguous patterns: a model based on an artificial neural network (Riani, Massimo; Masulli, Francesco; Simonotto, Enrico)V1469,166-177(1991)

Picture archiving and communications systems protocol based on ISO-OSI standard (Martinez, Ralph; Nam, Jiseung; Dallas, William J.; Osada, Masakazu; McNeill, Kevin M.; Ozeki, Takeshi; Komatsu, Ken-ichi)V1446,100-107(1991)

Prefetching: PACS image management optimization using HIS/RIS information (Lodder, Herman; van Poppel, Bas M.; Bakker, Albert R.)V1446,227-233(1991)

Progress and perspectives on optical information processing in Japan (Ishihara, Satoshi)V1389,68-74(1991); V1390,68-74(1991)

Project DaVinci (Winarsky, Norman; Alexander, Joanna R.)V1459,67-68(1991)

Real-time simulation of the retina allowing visualization of each processing stage (Teeters, Jeffrey L.; Werblin, Frank S.)V1472,6-17(1991); V1473,102-113(1991)

Receptive fields and the theory of discriminant operators (Gupta, Madan M.; Hungenahally, Suresh K.)V1382,87-98(1991)

Resist parameter extraction with graphical user interface in X (Chiu, Anita S.; Ferguson, Richard A.; Doi, Takeshi; Wong, Alfred K.; Tam, Nelson; Neureuther, Andrew R.)V1466,641-652(1991)

Role of orbital observations in detecting and monitoring geological hazards: prospects for the future (Pieri, David C.)V1492,410-417(1991)

Self-organized criticality in neural networks (Makarenko, Vladimir I.; Kirillov, A. B.)V1469,843-845(1991)

Short-scan fan beam algorithm for noncircular detector orbits (Zeng, Gengsheng L.; Gullberg, Grant T.)V1445,332-340(1991)

Small-satellite constellation for many uses (Seiberling, Walter E.; Traxler-Lee, Laura A.; Collins, Sean K.)V1495,32-41(1991)

State estimation for distributed systems with sensing delay (Alexander, Harold L.)V1470,103-111(1991)

Target lifetimes in natural resource management (Greer, Jerry D.)V1538,69-76(1991)

Techniques and strategies for data integration in mineral resource assessment (Trautwein, Charles M.; Dwyer, John L.)V1492,338-338(1991)

Temporal adaptation of multimedia scripts (Robin, Laura)V1460,103-113(1991)

Theories in distributed decision fusion: comparison and generalization (Thomopoulos, Stelios C.)V1383,623-634(1991)

Three-dimensional visualization and quantification of evolving amorphous objects (Silver, Deborah E.; Zabusky, Norman J.)V1459,97-108(1991)

Two-dimensional flow quantitative visualization by the hybrid method of flow birefringence and boundary integration (Chen, Yuhai; Jia, Youquan)V1554B,566-572(1991)

Use of spline expansions and regularization in the unfolding of data from spaceborne sensors (Fisher, Thornton R.; Perez, Joseph D.)V1479,212-225(1991)

Using sound to extract meaning from complex data (Scaletti, Carla; Craig, Alan B.)V1459,207-219(1991)

Using the ACR/NEMA standard with TCP/IP and Ethernet (Chimiak, William J.; Williams, Rodney C.)V1446,93-99(1991)

Virtual environment technology (Zeltzer, David L.)V1459,86-86(1991)

Visualization of manufacturing process data in N-dimensional spaces: a reanalysis of the data (Fulop, Ann C.; Allen, Donald M.; Deffner, Gerhard)V1459,69-76(1991)

Visualizing underwater acoustic matched-field processing (Rosenblum, Lawrence; Kamgar-Parsi, Behzad; Karahalios, Margarida; Heitmeyer, Richard)V1459,274-282(1991)

Visual thinking in organizational analysis (Grantham, Charles E.)V1459,77-84(1991)

Visual workbench for analyzing the behavior of dynamical systems (Cahoon, Peter)V1459,244-253(1991)

Information storage—see also optical data storage

Advanced thin films for optical storage (Gan, Fuxi)V1519,530-538(1991)

Applications of Z-Plane memory technology to high-frame rate imaging systems (Shanken, Stuart N.; Ludwig, David E.)V1346,210-215(1991)

Clinical aspects of the Mayo/IBM PACS project (Forbes, Glenn S.; Morin, Richard L.; Pavlicek, William)V1446,318-322(1991)

Comparison of worldwide opinions on the costs and benefits of PACS (van Gennip, Elisabeth M.; van Poppel, Bas M.; Bakker, Albert R.; Ottes, Fenno P.)V1446,442-450(1991)

Detailed description of the Mayo/IBM PACS (Gehring, Dale G.; Persons, Kenneth R.; Rothman, Melvyn L.; Salutz, James R.; Morin, Richard L.)V1446,248-252(1991)

Diagnostic report acquisition unit for the Mayo/IBM PACS project (Brooks, Everett G.; Rothman, Melvyn L.)V1446,217-226(1991)

Diagnostic value model for the evaluation of PACS: physician ratings of the importance of prompt image access and the utilization of a display station in an intensive care unit (Kundel, Harold L.; Seshadri, Sridhar B.; Arenson, Ronald L.)V1446,297-300(1991)

Digital film library implementation (Kishore, Sheel; Khalsa, Satjeet S.; Seshadri, Sridhar B.; Arenson, Ronald L.)V1446,188-194(1991)

Distributing the server function in a multiring PAC system (Lynne, Kenton J.)V1446,177-187(1991)

DRILL: a standardized radiology-teaching knowledge base (Rundle, Debra A.; Evers, K.; Seshadri, Sridhar B.; Arenson, Ronald L.)V1446,405-413(1991)

Effects of transmission errors on medical images (Pronios, Nikolaos-John B.; Yovanof, Gregory S.)V1446,108-128(1991)

Ericsson digital recce management system (Johansson, Micael; Engman, Per; Johansson, Bo G.)V1538,180-188(1991)

European activities towards a hospital-integrated PACS based on open systems (Kouwenberg, Jef M.; Ottes, Fenno P.; Bakker, Albert R.)V1446,357-361(1991)

Evaluation of total workstation CT interpretation quality: a single-screen pilot study (Beard, David V.; Perry, John R.; Muller, K.; Misra, Ram B.; Brown, P.; Hemminger, Bradley M.; Johnston, R. E.; Mauro, M.; Jaques, P. F.; Schiebler, M.)V1446,52-58(1991)

Experiences with a comprehensive hospital information system that incorporates image management capabilities (Dayhoff, Ruth E.; Maloney, Daniel L.; Kuzmak, Peter; Shepard, Barclay M.)V1446,323-329(1991)

Expert system for diagnosis/optimization of microlithography process* (Nicolau, Dan V.; Fulga, Florin; Dusa, Mircea V.)V1468,345-351(1991)

Fast access to reduced-resolution subsamples of high-resolution images (Isaacson, Joel S.)V1460,80-91(1991)

Folder management on a multimodality PACS display station (Feingold, Eric; Seshadri, Sridhar B.; Arenson, Ronald L.)V1446,211-216(1991)

High-bandwidth recording in a hostile environment (Darg, David A.; Kikuchi, Akira)V1538,124-131(1991)

High-density memory packaging technology high-speed imaging applications (Frew, Dean L.)V1346,200-209(1991)

High-speed swing arm three-beam optical head (Yak, A. S.; Low, Toh-Siew; Lim, Siak-Piang)V1401,74-81(1991)

Holographic associative memory of biological systems (Gariaev, Peter P.; Chudin, Viktor I.; Komissarov, Gennady G.; Berezin, Andrey A.; Vasiliev, Antoly A.)V1621,280-291(1991)

Hospital-integrated PACS at the University Hospital of Geneva (Ratib, Osman M.; Ligier, Yves; Hochstrasser, Denis; Scherrer, Jean-Raoul)V1446,330-340(1991)

Image acquisition unit for the Mayo/IBM PACS project (Reardon, Frank J.; Salutz, James R.)V1446,481-491(1991)

Image delivery performance of a CT/MR PACS module applied in neuroradiology (Lou, Shyh-Liang; Loloyan, Mansur; Weinberg, Wolfram S.; Valentino, Daniel J.; Lufkin, Robert B.; Hanafee, William; Bentson, John R.; Jabour, Bradly; Huang, H. K.)V1446,302-311(1991)

Initial experiences with PACS in a clinical and research environment (Honeyman, Janice C.; Staab, Edward V.; Frost, Meryll M.)V1446,362-368(1991)

Laser-induced phase transition in crystal InSb films as used in optical storage (Sun, Yang; Li, Cheng F.; Deng, He; Gan, Fuxi)V1519,554-558(1991)

Management system for a PACS network in a hospital environment (Mattheus, Rudy A.; Temmerman, Yvan; Verhellen, P.; Osteaux, Michel)V1446,341-351(1991)

Medical Imaging V: PACS Design and Evaluation (Jost, R. G., ed.)V1446(1991)

Multicomputer performance evaluation tool and its application to the Mayo/IBM image archival system (Pavicic, Mark J.; Ding, Yingjai)V1446,370-378(1991)

Multi-media PACS integrated with HIS/RIS employing magneto-optical disks (Umeda, Tokuo; Inamura, Kiyonari; Inamoto, Kazuo; Kondoh, Hiroshi P.; Kozuka, Takahiro)V1446,199-210(1991)

Multiple communication networks for a radiological PACS (Wong, Albert W.; Stewart, Brent K.; Lou, Shyh-Liang; Chan, Kelby K.; Huang, H. K.)V1446,73-80(1991)

Off-line image exchange between two PACS modules using the "ISAC" magneto-optical disk (Minato, Kotaro; Komori, Masaru; Nakano, Yoshihisa; Yonekura, Yoshiharu; Sasayama, Satoshi; Takahashi, Takashi; Konishi, Junji; Abe, Mituyuki; Sato, Kazuhiro; Hosoba, Minoru)V1446,195-198(1991)

OLAS: optical logic array structures (Jones, Robert H.; Hadjinicolaou, M. G.; Musgrave, G.)V1401,138-145(1991)

On enhancing the performance of the ACR-NEMA protocol (Maydell, Ursula M.; Hassanein, Hossam S.; Deng, Shuang)V1446,81-92(1991)

One-year clinical experience with a fully digitized nuclear medicine department: organizational and economical aspects (Anema, P. C.; de Graaf, C. N.; Wilmink, J. B.; Hall, David; Hoekstra, A.; van Rijk, P. P.; Van Isselt, J. W.; Viergever, Max A.)V1446,352-356(1991)

On-line acquisition of CT and MRI studies from multiple scanners (Weinberg, Wolfram S.; Loloyan, Mansur; Chan, Kelby K.)V1446,430-435(1991)

Operational infrastructure for a clinical PACS (Boehme, Johannes M.; Chimiak, William J.; Choplin, Robert H.; Maynard, C. D.)V1446,312-317(1991)

Optical Data Storage Technologies (Chua, Soo-Jin; McCallum, John C., eds.)V1401(1991)

PACS and teleradiology for on-call support of abdominal imaging (Horii, Steven C.; Garra, Brian S.; Mun, Seong Ki; Zeman, Robert K.; Levine, Betty A.; Fielding, Robert)V1446,10-15(1991)

PACS: "Back to the Future" (Carey, Bruce; Seshadri, Sridhar B.; Arenson, Ronald L.)V1446,414-419(1991)

PACS for GU radiology (Hayrapetian, Alek; Barbaric, Zoran L.; Weinberg, Wolfram S.; Chan, Kelby K.; Loloyan, Mansur; Taira, Ricky K.; Huang, H. K.)V1446,243-247(1991)

PACS modeling and development: requirements versus reality (Seshadri, Sridhar B.; Kishore, Sheel; Khalsa, Satjeet S.; Feingold, Eric; Arenson, Ronald L.)V1446,388-395(1991)

PACS reading time comparision: the workstation versus alternator for ultrasound (Horii, Steven C.; Garra, Brian S.; Mun, Seong Ki; Singer, Jon; Zeman, Robert K.; Levine, Betty A.; Fielding, Robert; Lo, Ben)V1446,475-480(1991)

Performance characteristics of the Mayo/IBM PACS (Persons, Kenneth R.; Gehring, Dale G.; Pavicic, Mark J.; Ding, Yingjai)V1446,60-72(1991)

Photochemical hole burning in rigidly coupled polyacenes (Iannone, Mark A.; Salt, Kimberly L.; Scott, Gary W.; Yamashita, Tomihiro)V1559,172-183(1991)

Picture archiving and communications systems protocol based on ISO-OSI standard (Martinez, Ralph; Nam, Jiseung; Dallas, William J.; Osada, Masakazu; McNeill, Kevin M.; Ozeki, Takeshi; Komatsu, Ken-ichi)V1446,100-107(1991)

Pod-mounted MIL-STD-2179B recorder (Kessler, William D.; Abeille, Pierre; Sulzer, Jean-Francois)V1538,104-111(1991)

Prefetching: PACS image management optimization using HIS/RIS information (Lodder, Herman; van Poppel, Bas M.; Bakker, Albert R.)V1446,227-233(1991)

Preliminary results of a PACS implementation (Kishore, Sheel; Khalsa, Satjeet S.; Seshadri, Sridhar B.; Arenson, Ronald L.)V1446,236-242(1991)

Present status and future directions of the Mayo/IBM PACS project (Morin, Richard L.; Forbes, Glenn S.; Gehring, Dale G.; Salutz, James R.; Pavlicek, William)V1446,436-441(1991)

Project DaVinci (Winarsky, Norman; Alexander, Joanna R.)V1459,67-68(1991)

Qualitative approach to medical image databases (Bizais, Yves; Gibaud, Bernard; Forte, Anne-Marie; Aubry, Florent; Di Paola, Robert; Scarabin, Jean-Marie)V1446,156-167(1991)

Radiology workstation for mammography: preliminary observations, eyetracker studies, and design (Beard, David V.; Johnston, R. E.; Pisano, E.; Hemminger, Bradley M.; Pizer, Stephen M.)V1446,289-296(1991)

Rapid display of radiographic images (Cox, Jerome R.; Moore, Stephen M.; Whitman, Robert A.; Blaine, G. J.; Jost, R. G.; Karlsson, L. M.; Monsees, Thomas L.; Hassen, Gregory L.; David, Timothy C.)V1446,40-51(1991)

Reconnaissance mission planning (Fishell, Wallace G.; Fox, Alex J.)V1538,5-13(1991)

Recording and erasing characteristics of GeSbTe-based phase change thin films (Hou, Li S.; Zhu, Chang X.; Gu, Dong H.)V1519,548-553(1991)

Reliability issues in PACS (Taira, Ricky K.; Chan, Kelby K.; Stewart, Brent K.; Weinberg, Wolfram S.)V1446,451-458(1991)

Representing sentence information (Perkins, Walton A.)V1468,854-865(1991)

Routing in distributed information systems (Ras, Zbigniew W.)V1470,76-87(1991)

Simulation-based PACS development (Stut, W. J.; van Steen, M. R.; Groenewegen, L. P.; Ratib, Osman M.; Bakker, Albert R.)V1446,396-404(1991)

Storing and managing three-dimensional digital medical image information (Chapman, Michael A.; Denby, N.)V1526,190-194(1991)

Thin film magnetic recording (Chen, Yi-Xin)V1519,539-547(1991)

Using the ACR/NEMA standard with TCP/IP and Ethernet (Chimiak, William J.; Williams, Rodney C.)V1446,93-99(1991)

X-window-based 2K display workstation (Weinberg, Wolfram S.; Hayrapetian, Alek; Cho, Paul S.; Valentino, Daniel J.; Taira, Ricky K.; Huang, H. K.)V1446,35-39(1991)

Infrared—see also detectors; electro-optics; photodetectors; thermal imaging

1-W cw separate confinement InGaAsP/InP (lamda = 1.3 um) laser diodes and their coupling with optical fibers (Garbuzov, Dmitriy Z.; Goncharov, S. E.; Il'in, Y. V.; Mikhailov, A. V.; Ovchinnikov, Alexander V.; Pikhtin, N. A.; Tarasov, I. S.)V1418,386-393(1991)

128 x 128 MWIR InSb focal plane and camera system (Parrish, William J.; Blackwell, John D.; Paulson, Robert C.; Arnold, Harold)V1512,68-77(1991)

488 x 640-element hybrid platinum silicide Schottky focal-plane array (Gates, James L.; Connelly, William G.; Franklin, T. D.; Mills, Robert E.; Price, Frederick W.; Wittwer, Timothy Y.)V1540,262-273(1991)

640 x 480 MOS PtSi IR sensor (Sauer, Donald J.; Shallcross, Frank V.; Hsueh, Fu-Lung; Meray, G. M.; Levine, Peter A.; Gilmartin, Harvey R.; Villani, Thomas S.; Esposito, Benjamin J.; Tower, John R.)V1540,285-296(1991)

7th Mtg in Israel on Optical Engineering (Oron, Moshe; Shladov, Itzhak, eds.)V1442(1991)

Accuracies in FLIR test equipment (Fourier, Ron)V1442,109-117(1991)

Accurate temperature measurement in thermography: an overview of relevant features, parameters, and definitions (Hamrelius, Torbjorn)V1467,448-457(1991)

Adaptive 4-D IR clutter suppression filtering technique (Aridgides, Athanasios; Fernandez, Manuel F.; Randolph, D.; Ferris, D.)V1481,110-116(1991)

Adaptive nonuniformity correction for IR focal-plane arrays using neural networks (Scribner, Dean A.; Sarkady, Kenneth A.; Kruer, Melvin R.; Caulfield, John T.; Hunt, J. D.; Herman, Charles)V1541,100-109(1991)

Adaptive notch filter for removal of coherent noise from infrared scanner data (Jaggi, Sandeep)V1541,134-140(1991)

Advanced infrared detector materials (Mullin, John B.)V1512,144-154(1991)

Advanced infrared focal-plane arrays (Amingual, Daniel)V1512,40-51(1991)

Advanced infrared optically black baffle materials (Seals, Roland D.; Egert, Charles M.; Allred, David D.)V1330,164-177(1991)

Advanced portable four-wavelength NIR analyzer for rapid chemical composition analysis (Malinen, Jouko; Hyvarinen, Timo S.)V1510,204-209(1991)

Advanced Si IR detectors using molecular beam epitaxy (Lin, TrueLon; Jones, E. W.; George, T.; Ksendzov, Alexander; Huberman, M. L.)V1540,135-139(1991)

Advanced visible and near-IR radiometer for ADEOS (Iwasaki, Nobuo; Tange, Yoshio; Miyachi, Yuji; Inoue, Kouichi; Kadowaki, Tomoko; Tanaka, Hirokazu; Michioka, Hidekazu)V1490,216-221(1991)

Advances in multiple-quantum-well IR detectors (Bloss, Walter L.; O'Loughlin, Michael J.; Rosenbluth, Mary)V1541,2-10(1991)

Advances in R&D in near-infrared spectroscopy in Japan (Kawano, Sumio; Iwamoto, Mutsuo)V1379,2-9(1991)

AEDC direct-write scene generation test capabilities (Lowry, Heard S.; Elrod, Parker D.; Johnson, R. J.)V1454,453-464(1991)

Airborne and spaceborne thermal multispectral remote sensing (Watanabe, Hiroshi; Sano, Masaharu; Mills, F.; Chang, Sheng-Huei; Masuda, Shoichi)V1490,317-323(1991)

Airborne imaging system performance model (Redus, Wesley D.)V1498,2-23(1991)

Airborne seeker evaluation and test system (Jollie, William B.)V1482,92-103(1991)

Aircraft laser infrared absorption spectrometer (ALIAS) for polar ozone studies (Webster, Chris R.; May, R. D.)V1540,187-194(1991)

Analysis and performance limits of diamond-turned diffractive lenses for the 3-5 and 8-12 micrometer regions (Riedl, Max J.; McCann, James T.)VCR38,153-163(1991)

Analysis of silage composition by near-infrared reflectance spectroscopy (Reeves, James B.; Blosser, Timothy H.; Colenbrander, V. F.)V1379,28-38(1991)

Analysis of the flying light spot experiment on SPRITE detector (Gu, Bo-qi; Feng, Wen-qing)V1488,443-446(1991)

Analysis of upper and lower bounds of the frame noise in linear detector arrays (Jaggi, Sandeep)V1541,152-162(1991)

Anodic oxides on HgZnTe (Eger, David; Zigelman, Alex)V1484,48-54(1991)

Anomalous optical response of YBa2Cu3O7-x thin films during superconducting transitions (Xi, Xiaoxing; Venkatesan, T.; Etemad, Shahab; Hemmick, D.; Li, Q.)V1477,20-25(1991)

Answer to the dynamic (fretting effect) and static (oxide) behavior of electric contact surfaces: based on a five-year infrared thermographic study (Paez-Leon, Cristobal J.; Patino, Antonio R.; Aguillon, Luis)V1467,188-194(1991)

Application of adaptive filters to the problem of reducing microphony in arrays of pyroelectric infrared detectors (Carmichael, I. C.; White, Paul R.)V1541,167-177(1991)

Application of high-speed infrared emission spectroscopy in reacting flows (Klingenberg, Guenter; Rockstroh, Helmut)V1358,851-858(1991)

Application of high-speed mirror chronograph 3CX-1 to plasma investigations in visible and medium infrared spectrum ranges (Drozhbin, Yu. A.; Zvorykin, V. D.; Polyansky, S. V.; Sychugov, G. V.; Trofimenko, Vladimir V.; Yarova, A. G.)V1358,1029-1034(1991)

Application of IR photorecorders based on ionization chambers for fast processes investigation (Egorov, V. V.; Lazarchuk, V. P.; Murugov, V. M.; Sheremetyev, Yu. N.)V1358,984-991(1991)

Application of IR staring arrays to space surveillance (Cantella, Michael J.; Ide, M. H.; O'Donnell, P. J.; Tsaur, Bor-Yeu)V1540,634-652(1991)

Application of nonimaging optical concentrators to infrared energy detection (Ning, Xiaohui)V1528,88-92(1991)

Application of YBa2Cu3O7-x thin film in high-Tc semiconducting infrared detector (Zhou, Bing; Chen, Ju X.; Shi, Bao A.; Wu, Ru J.; Gong, Shuxing)V1519,454-456(1991)

Applications and development of IR techniques for building research in Finland (Kaasinen, Harri I.; Kauppi, Ari; Nykanen, Esa)V1467,90-98(1991)

Applications of diffractive optics to uncooled infrared imagers (Cox, James A.)V1540,606-611(1991)

Applications of tridimensional heat calibration to a thermographic nondestructive evaluation station (Maldague, Xavier; Fortin, Louis; Picard, J.)V1467,239-251(1991)

Assessment of the optimum operating conditions for 2-D focal-plane-array systems (Bourne, Robert W.; Jefferys, E. A.; Murphy, Kevin S.)V1488,73-79(1991)

ASTER calibration concept (Ono, Akira; Sakuma, Fumihiro)V1490,285-298(1991)

Athermalization of IR optical systems (Rogers, Philip J.)VCR38,69-94(1991)

Athermalized FLIR optics (Rogers, Philip J.)V1354,742-751(1991)

Atmospheric infrared sounder on the Earth Observing System: in-orbit spectral calibration (Aumann, Hartmut H.)V1540,176-186(1991)

Automated characterization of Z-technology sensor modules (Gilcrest, Andrew S.)V1541,240-249(1991)

Automatic active athermalization of infrared optical systems (Kuerbitz, Gunther)V1540,612-621(1991)

AutoSPEC image evaluation laboratory (Brown, James C.; Webb, Curtis M.; Bell, Paul A.; Washington, Randolph T.; Riordan, Richard J.)V1488,300-311(1991)

Balloon-born investigations of total and aerosol attenuation continuous spectra in the stratosphere (Mirzoeva, Larisa A.; Kiseleva, Margaret S.; Reshetnikova, Irina N.)V1540,444-449(1991)

Battlefield training in impaired visibility (Gammarino, Rudolph R.; Surhigh, James W.)V1419,115-125(1991)

Behavioral observations in thermal imaging of the big brown bat: Eptesicus fuscus (Kirkwood, James J.; Cartwright, Anne)V1467,369-371(1991)

Bidirectional transmittance distribution function measurements on ZnSe and on ZnS Cleartran (Melozzi, Mauro; Mazzoni, Alessandro; Curti, G.)V1512,178-188(1991)

Bounds on the performance of optimal four-dimensional filters for detection of low-contrast IR point targets (Wohlers, Martin R.)V1481,129-139(1991)

Broadband, antireflection coating designs for large-aperture infrared windows (Balasubramanian, Kunjithapa; Le, Tam V.; Guenther, Karl H.; Kumar, Vas)V1485,245-253(1991)

Buried-steam-line temperature and heat loss calculation (MacDavid, Jacob H.)V1467,11-17(1991)

Catalog of infrared and cryo-optical properties for selected materials (Heaney, James B.; Alley, Phillip W.; Bradley, Scott E.)V1485,140-159(1991)

CCD holographic phase and intensity measurement of laser wavefront (Wickham, Michael; Munch, Jesper)V1414,80-90(1991)

Characterization of surface contaminants using infrared microspectroscopy (Blair, Dianna S.; Ward, Kenneth J.)V1437,76-79(1991)

Characterization of the atmospheric modulation transfer function using the target contrast characterizer (Watkins, Wendell R.; Billingsley, Daniel R.; Palacios, Fernando R.; Crow, Samuel B.; Jordan, Jay B.)V1486,17-24(1991)

Characterization, Propagation, and Simulation of Sources and Backgrounds (Watkins, Wendell R.; Clement, Dieter, eds.)V1486(1991)

Choice of means for the adaptation of infrared systems figures of merit (Yakushenkov, Yuri G.)V1540,455-459(1991)

Clutter metrics in infrared target acquisition (Tidhar, Gil; Rotman, Stanley R.)V1442,310-324(1991)

Coating thickness gauge (Honda, Tatsuro; Matsui, Kenichi)V1540,709-716(1991)

Combination of mechanical athermalization with manual in IR zoom telescope (Chen, Ruiyi; Zheng, Dayue; Zhou, Xiuli; Zhang, Xingde)V1540,724-728(1991)

Compact far-infrared free-electron laser (Ride, Sally K.; Golightly, W.)V1552,128-137(1991)

Comparison of flash-pumped Cr;Tm 2-um laser action in garnet hosts (Quarles, Gregory J.; Pinto, Joseph F.; Esterowitz, Leon; Kokta, Milan R.)V1410,165-174(1991)

Comparison of some algorithms commonly used in IR pyrometry: a computer simulation (Barani, Gianni; Tofani, Alessandro)V1467,458-468(1991)

Computer analysis of signal-to-noise ratio and detection probability for scanning IRCCD arrays (Uda, Gianni; Tofani, Alessandro)V1488,257-266(1991)

Contrast, size, and orientation-invariant target detection in infrared imagery (Zhou, Yi-Tong; Crawshaw, Richard D.)V1471,404-411(1991)

Cooling system for short-wave infrared radiometer of JERS-1 optical sensor (Ohmori, Yasuhiro; Arakawa, Isao; Nakai, Akira; Tsubosaka, Kazuyoshi)V1490,177-183(1991)

Correction of inherent scan nonplanarity in the Boeing infrared sensor calibration facility (Chase, Robert P.)V1533,138-149(1991)

Correlation between images in the long-wave infrared and short-wave infrared of natural ground terrain (Agassi, Eyal; Wilner, Kalman; Ben-Yosef, Nissim)V1442,126-132(1991)

Correlation between the aerosol profiles measurements, the meteorological conditions, and the atmospheric IR transmission in a mediterranean marine atmosphere (Tanguy, Mireille; Bonhommet, Herve; Autric, Michel L.; Vigliano, Patrick)V1487,172-183(1991)

Corrosion evaluation of coated sheet metal by means of thermography and image analysis (Jernberg, Per)V1467,295-302(1991)

Critical look at AlGaAs/GaAs multiple-quantum-well infrared detectors for thermal imaging applications (Adams, Frank W.; Cuff, K. F.; Gal, George; Harwit, Alex; Whitney, Raymond L.)V1541,24-37(1991)

Current instrument status of the airborne visible/infrared imaging spectrometer (AVIRIS) (Eastwood, Michael L.; Sarture, Charles M.; Chrien, Thomas G.; Green, Robert O.; Porter, Wallace M.)V1540,164-175(1991)

Current status of InGaAs detector arrays for 1-3 um (Olsen, Gregory H.; Joshi, Abhay M.; Ban, Vladimir S.)V1540,596-605(1991)

Description and calibration of a fully automated infrared scatterometer (Mainguy, Stephan; Olivier, Michel; Josse, Michel A.; Guidon, Michel)V1530,269-282(1991)

Description and performance of a 256 x 256 electrically heated pixel IR scene generator (Lake, Stephen P.; Pritchard, Alan P.; Sturland, Ian M.; Murray, Anthony R.; Prescott, Anthony J.; Gough, David W.)V1486,286-293(1991)

Design and performance of a 486 x 640 pixel platinum silicide IR imaging system (Clark, David L.; Berry, Joseph R.; Compagna, Gary L.; Cosgrove, Michael A.; Furman, Geoffrey G.; Heydweiller, James R.; Honickman, Harris; Rehberg, Raymond A.; Sorlie, Paul H.; Nelson, Edward T.)V1540,303-311(1991)

Design and performance of a PtSi spectroscopic infrared array and detector head (Cizdziel, Philip J.)V1488,6-27(1991)

Design considerations for infrared imaging radiometers (Fraedrich, Douglas S.)V1486,2-7(1991)

Design of an athermalized three-fields-of-view infrared sensor (Wickholm, David R.)V1488,58-63(1991)

Design of an IR non-lens, or how I buried 100 mm of germanium (Aikens, David M.)V1485,183-194(1991)

Design of compact IR zoom telescope (Chen, Ruiyi; Zhou, Xiuli; Zhang, Xingde)V1540,717-723(1991)

Design of multiple-beam gratings for far-IR applications (Kinrot, O.; Friesem, Asher A.)V1442,106-108(1991)

Detection of citrus freeze damage with natural color and color IR video systems (Blazquez, Carlos H.)V1467,394-401(1991)

Detection of infrared, free-electron laser radiation (Kimmitt, Maurice F.)V1501,86-96(1991)

Detection of moving subpixel targets in infrared clutter with space-time filtering (Braunreiter, Dennis C.; Banh, Nam D.)V1481,73-83(1991)

Detection of targets in terrain clutter by using multispectral infrared image processing (Hoff, Lawrence E.; Evans, John R.; Bunney, Laura)V1481,98-109(1991)

Detection of unresolved target tracks in infrared imagery (Rajala, Sarah A.; Nolte, Loren W.; Aanstoos, James V.)V1606,360-371(1991)

Determination of FLIR LOS stabilization errors (Pinsky, Howard J.)V1488,334-342(1991)

Development and analysis of a simple model for an IR sensor (Ballik, Edward A.; Wan, William)V1488,249-256(1991)

Development of a fire and forget imaging infrared seeker missile simulation (Hall, Charles S.; Alongi, Robert E.; Fortner, Russ L.; Fraser, Laurie K.)V1483,29-38(1991)

Development of a low-pass far-infrared filter for lunar observer horizon sensor application (Mobasser, Sohrab; Horwitz, Larry S.; Griffith, O'Dale)V1540,764-774(1991)

Development of an IR projector using a deformable mirror device (Lanteigne, David J.)V1486,376-379(1991)

Development of high resolution statistically nonstationary infrared earthlimb radiance scenes (Strugala, Lisa A.; Newt, J. E.; Futterman, Walter I.; Schweitzer, Eric L.; Herman, Bruce J.; Sears, Robert D.)V1486,176-187(1991)

Devices for generation and detection of subnanosecond IR and FIR radiation pulses (Beregulin, Eugene V.; Ganichev, Sergey D.; Yaroshetskii, Ilya D.; Lang, Peter T.; Schatz, Wolfgan; Renk, Karl F.)V1362,853-862(1991)

Diamond windows for the infrared: fact and fallacy (Klein, Claude A.)VCR38,218-257(1991)

Diffuse tomography (Gruenbaum, F. A.; Kohn, Philip D.; Latham, Geoff A.; Singer, Jay R.; Zubelli, Jorge P.)V1431,232-238(1991)

Dimension and lacunarity measurement of IR images using Hilbert scanning (Moghaddam, Baback; Hintz, Kenneth J.; Stewart, Clayton V.)V1486,115-126(1991)

Direct view thermal imager (Reinhold, Ralph R.)V1447,251-262(1991)

Discussion of the standard practice for the location of wet insulation in roofing systems using infrared imaging (ASTM C1153-90) (Sopko, Victor)V1467,83-89(1991)

Dual-band optical system for IR multicolor signal processing (Kimball, Paulette R.; Fraser, James C.; Johnson, Jeffrey P.; Siegel, Andrew M.)V1540,687-698(1991)

Dynamic end-to-end model testbed for IR detection algorithms (Iannarilli, Frank J.; Wohlers, Martin R.)V1483,66-76(1991)

Dynamic infrared scene projection technology (Mobley, Scottie B.)V1486,325-332(1991)

Earthlimb emission analysis of spectral infrared rocket experiment data at 2.7 micrometers: a ten-year update (Sharma, Ramesh D.; Healey, Rebecca J.)V1540,314-320(1991)

Effective electron mass in narrow-band-gap IR materials under different physical conditions (Ghatak, Kamakhya P.; Biswas, Shambhu N.)V1484,149-166(1991)

Effects of microscan operation on staring infrared sensor imagery (Blommel, Fred P.; Dennis, Peter N.; Bradley, D. J.)V1540,653-664(1991)

Electromagnetically carrier depleted IR photodetector (Djuric, Zoran G.; Piotrowski, Jozef)V1540,622-632(1991)

Electron-beam-addressed membrane light modulator for IR scene projection (Horsky, Thomas N.; Schiller, Craig M.; Genetti, George J.; O'Mara, Daniel M.; Hamnett, Whitney S.; Warde, Cardinal)V1540,527-532(1991)

Electronic and optical properties of silicide/silicon IR detectors (Cabanski, Wolfgang A.; Schulz, Max J.)V1484,81-97(1991)

Ellipsometric studies of the optical anisotropy of GdBa2Cu3O7-x epitaxial films (Liu, Ansheng; Keller, Ole)V1512,226-231(1991)

Emissions monitoring by infrared photoacoustic spectroscopy (Jalenak, Wayne)V1434,46-54(1991)

End-to-end model for detection performance evaluation against scenario-specific targets (Iannarilli, Frank J.; Wohlers, Martin R.)V1488,226-236(1991)

End-to-end scenario-generating model for IRST performance analysis (Iannarilli, Frank J.; Wohlers, Martin R.)V1481,187-197(1991)

Enhanced thematic mapper cold focal plane: design and testing (Yang, Bing T.)V1488,399-409(1991)

Evaluation of the IR signature of dynamic air targets (Porta, Paola M.)V1540,508-518(1991)

Experimental and theoretical studies of second-phase scattering in IR transmitting ZnS-based windows (Chen, William W.; Dunn, Bruce S.; Zhang, Jimin)V1535,199-208(1991)

Experimental simulation of IR signatures (Mulero, Manuel A.; Barreiros, Manuel A.)V1540,519-526(1991)

Experimental study of the optical properties of LTCVD SiO2 (Aharoni, Herzl; Swart, Pieter L.)V1442,118-125(1991)

Fabrication and properties of chalcogenide IR diffractive elements (Ewen, Peter J.; Slinger, Christopher W.; Zakery, A.; Zekak, A.; Owen, A. E.)V1512,101-111(1991)

Fiber optic remote Fourier transform infrared spectroscopy (Druy, Mark A.; Glatkowski, Paul J.; Stevenson, William A.)V1584,48-52(1991)

Field and laboratory studies of Fourier transform infrared spectroscopy in continuous emissions monitoring applications (Plummer, Grant M.)V1434,78-89(1991)

Field documentation and client presentation of IR inspections on new masonry structures (McMullan, Phillip C.)V1467,66-74(1991)

Firefly system concept (Nichols, Joseph D.)V1540,202-206(1991)

FIR lasers as local oscillators in submillimeter astronomy (Roeser, Hans-Peter; van der Wal, Peter)V1501,194-197(1991)

Flaw dynamics and vibrothermographic-thermoelastic nondestructive evaluation of advanced composite materials (Tenek, Lazarus H.; Henneke, Edmund G.)V1467,252-263(1991)

Flight test integration and evaluation of the LANTIRN system on the F-15E (Presuhn, Gary G.; Zeis, Joseph E.)V1479,249-258(1991)

Focal-plane architectures and signal processing (Jayadev, T. S.)V1541,163-166(1991)

Focusing infrared laser beams on targets in space without using adaptive optics (McKechnie, Thomas S.)V1408,119-135(1991)

Forward-looking IR and lidar atmospheric propagation in the infrared field program (Koenig, George G.; Bissonnette, Luc R.)V1487,240-249(1991)

Fourier analysis of near-infrared spectra (McClure, W. F.)V1379,45-51(1991)

Free-surface temperature measurement of shock-loaded tin using ultrafast infrared pyrometry (Mondot, Michel; Remiot, Christian)V1558,351-361(1991)

Functional row of pyroelectric sensors for infrared devices used in ecological monitoring (Savinykh, Viktor P.; Glushko, A. A.)V1540,450-454(1991)

Further performance characteristics of a high-sensitivity 64 x 64 element InSb hybrid focal-plane array (Fischer, Robert C.; Martin, Charles J.; Niblack, Curtiss A.; Timlin, Harold A.; Wimmers, James T.)V1494,414-418(1991)

Gaseous incinerator emissions analysis by FTIR (Fourier transform infrared) spectroscopy (Herget, William F.; Demirgian, Jack)V1434,39-45(1991)

General principles of constructing the nuclei of nonlinear IR-image transformations (Nesteruk, Vsevolod F.)V1540,468-476(1991)

Generation of IR sky background images (Levesque, Martin P.)V1486,200-209(1991)

Generation of realistic IR images of tactical targets in obscured environments (Greenleaf, William G.; Siniard, Sheri M.; Tait, Mary B.).V1486,364-375(1991)

Generic models for rapid calculation of target signatures (Rushmeier, Holly E.; Rodriguez, Leonard J.)V1486,210-216(1991)

Global approach toward the evaluation of thermal infrared countermeasures (Verlinde, Patrick F.; Proesmans, Marc)V1486,58-65(1991)

Ground systems and operations concepts for the Space Infrared Telescope Facility (Miller, Richard B.)V1540,38-46(1991)

Growth and Characterization of Materials for Infrared Detectors and Nonlinear Optical Switches (Longshore, Randolph E.; Baars, Jan W., eds.)V1484(1991)

Growth by liquid phase epitaxy and characterization of GaInAsSb and InAsSbP alloys for mid-infrared applications (2-3 um) (Tournie, Eric; Lazzari, J. L.; Mani, Habib; Pitard, F.; Alibert, Claude L.; Joullie, Andre F.)V1361,641-656(1991)

Growth of CdTe-CdMnTe heterostructures by molecular beam epitaxy (Bicknell-Tassius, Robert N.)V1484,11-18(1991)

High-contrast composite infrared filters (Borisevich, Nikolai A.; Zamkovets, A. D.; Ponyavina, A. N.)V1500,222-231(1991)

High-efficiency tunable mid-infrared generation in KNbO3 (Guyer, Dean R.; Bosenberg, W. R.; Braun, Frank D.)V1409,14-17(1991)

High-fill-factor monolithic infrared image sensor (Kimata, Masafumi; Yutani, Naoki; Yagi, Hirofumi; Nakanishi, Junji; Tsubouchi, Natsuro; Seto, Toshiki)V1540,238-249(1991)

High-performance 256 x 256 InSb FPA for astronomy (Hoffman, Alan; Randall, David)V1540,297-302(1991)

High-performance FLIR testing using reflective-target technology (McHugh, Stephen W.)V1540,775-780(1991)

High-performance InSb 256 x 256 infrared camera (Blackwell, John D.; Parrish, William J.; Kincaid, Glen T.)V1479,324-334(1991)

High-performance IR thermography system based on Class II Thermal Imaging Common Modules (Bell, Ian G.)V1467,438-447(1991)

High-performance, wide-magnification-range IR zoom telescope with automatic compensation for temperature effects (Shechterman, Mark)V1442,276-285(1991)

High-resolution airborne multisensor system (Prutzer, Steven; Biron, David G.; Quist, Theodore M.)V1480,46-61(1991)

High-resolution spectral analysis and modeling of infrared ocean surface radiometric clutter (McGlynn, John D.; Ellis, Kenneth K.; Kryskowski, David)V1486,141-150(1991)

High-resolution studies of atmospheric IR emission spectra (Murcray, Frank J.; Murcray, F. H.; Goldman, Aaron; Blatherwick, Ronald D.; Murcray, David G.)V1491,282-287(1991)

High-resolution thermal imager with a field-of-view of 112 degrees (Matsushita, Tadashi; Suzuki, Hiroshi; Wakabayashi, Satoshi; Tajime, Toru)V1488,368-375(1991)

High-sensitive thermal video camera with self-scanned 128 InSb linear array (Fujisada, Hiroyuki)V1540,665-676(1991)

High-spatial-resolution FLIR (Tucker, Christopher J.; Mitchell, Robert J.)V1498,92-98(1991)

High-Tc bolometer developments for planetary missions (Brasunas, John C.; Lakew, Brook)V1477,166-173(1991)

High-Tc superconducting infrared bolometric detector (Cole, Barry E.)V1394,126-138(1991)

History of infrared optics (Johnson, R. B.; Feng, Chen)VCR38,3-18(1991)

Hollow and dielectric waveguides for infrared spectroscopic applications (Saggese, Steven J.; Harrington, James A.; Sigel, George H.)V1437,44-53(1991)

HTC microbolometer for far-infrared detection (Barholm-Hansen, Claus; Levinsen, Mogens T.)V1512,218-225(1991)

Human recognition of infrared images II (Sanders, Jeffrey S.; Currin, Michael S.)V1488,144-155(1991)

Hybrid infrared focal-plane signal and noise modeling (Johnson, Jerris F.; Lomheim, Terrence S.)V1541,110-126(1991)

Image display and background analysis with the Naval Postgraduate School infrared search and target designation system (Cooper, Alfred W.; Lentz, William J.; Baca, Michael J.; Bernier, J. D.)V1486,47-57(1991)

Image enhancement of infrared uncooled focal plane array imagery (McCauley, Howard M.; Auborn, John E.)V1479,416-422(1991)

Images of turbulent, absorbing-emitting atmospheres and their application to windshear detection (Watt, David W.; Philbrick, Daniel A.)V1467,357-368(1991)

Importance of dispersion tolerances in infrared lens design (Korniski, Ronald J.; Thompson, Kevin P.)V1354,402-407(1991)

Improved IR image generator for real-time scene simulation (Keller, Catherine E.; Stenger, Anthony J.; Bernstein, Uri)V1486,278-285(1991)

Improvement in detection of small wildfires (Sleigh, William J.)V1540,207-212(1991)

Incorporation of time-dependent thermodynamic models and radiation propagation models into IR 3-D synthetic image generation models (Schott, John R.; Raqueno, Rolando; Salvaggio, Carl; Kraus, Eugene J.)V1540,533-549(1991)

Infrared and Optoelectronic Materials and Devices (Naumaan, Ahmed; Corsi, Carlo; Baixeras, Joseph M.; Kreisler, Alain J., eds.)V1512(1991)

Infrared and the search for extrasolar planets (Meinel, Aden B.; Meinel, Marjorie P.)V1540,196-201(1991)

Infrared background measurements at White Sands Missile Range, NM (Troyer, David E.; Fouse, Timothy; Murdaugh, William O.; Zammit, Michael G.; Rogers, Stephen B.; Skrzypczak, J. A.; Colley, Charles B.; Taczak, William J.)V1486,396-409(1991)

Infrared cables and catheters for medical applications (Artjushenko, Vjacheslav G.; Ivchenko, N.; Konov, Vitaly I.; Kryukov, A. P.; Krupchitsky, Vladimir P.; Kuznetcov, R.; Lerman, A. A.; Litvinenko, E. G.; Nabatov, A. O.; Plotnichenko, V. G.; Prokhorov, Alexander M.; Pylnov, I. L.; Tsibulya, Andrew B.; Vojtsekhovsky, V. V.; Ashraf, N.; Neuberger, Wolfgang; Moran, Kelly B.; Mueller, Gerhard J.; Schaldach, Brita J.)V1420,157-168(1991)

Infrared clutter measurements of marine backgrounds (Schwering, Piet B.)V1486,25-36(1991)

Infrared detection of moist areas in monumental buildings based on thermal inertia analysis (Grinzato, Ermanno G.; Mazzoldi, Andrea)V1467,75-82(1991)

Infrared detector arrays with integrating cryogenic read-out electronics (Engemann, Detlef; Faymonville, Rudolf; Felten, Rainer; Frenzl, Otto)V1362,710-720(1991)

Infrared detectors from YBaCuO thin films (Zhou, Fang Q.; Sun, Han D.; Zhao, Xing R.; Wang, Lingjie; Yi, Xin J.)V1477,178-181(1991)

Infrared focal-plane design for the comet rendezvous/asteroid flyby and Cassini visible and infrared mapping spectrometers (Staller, Craig O.; Niblack, Curtiss A.; Evans, Thomas G.; Blessinger, Michael A.; Westrick, Anthony)V1540,219-230(1991)

Infrared Imaging Systems: Design, Analysis, Modeling and Testing II (Holst, Gerald C., ed.)V1488(1991)

Infrared instrumental complex for remote measurement of ocean surface temperature distribution (Miroshnikov, Mikhail M.; Minyeev, V. N.; Povarkov, V. I.; Samkov, V. M.; Solovyev, V. I.)V1540,496-505(1991)

Infrared in the USSR: brief historical survey of infrared development in the Soviet Union (Miroshnikov, Mikhail M.)V1540,372-400(1991)

Infrared lidars for atmospheric remote sensing (Menzies, Robert T.)V1540,160-163(1991)

Infrared microanalysis of contaminants at grazing incidence (Reffner, John A.)V1437,89-94(1991)

Infrared/microwave correlation measurements (Norgard, John D.; Metzger, Don W.; Cleary, John C.; Seifert, Michael)V1540,699-708(1991)

Infrared monitoring of combustion (Bates, Stephen C.; Morrison, Philip W.; Solomon, Peter R.)V1434,28-38(1991)

Infrared Optical Design and Fabrication (Hartmann, Rudolf; Smith, Warren J., eds.)VCR38(1991)

Infrared optical response of superconducting YBaCuO thin films (Sun, Han D.; Zhou, Fang Q.; Zhao, Xing R.; Wang, Lingjie; Yi, Xin J.)V1477,174-177(1991)

Infrared photodetector based on the photofluxonic effect in superconducting thin films (Kadin, Alan M.; Leung, Michael; Smith, Andrew D.; Murduck, J. M.)V1477,156-165(1991)

Infrared reflectivity: a tool for bond investigation in II-VI ternaries (Granger, Rene')V1484,39-46(1991)

Infrared refractive-index measurement results for single-crystal and polycrystal germanium (Hilton, Albert R.)V1498,128-137(1991)

Infrared Sensors: Detectors, Electronics, and Signal Processing (Jayadev, T. S., ed.)V1541(1991)

Infrared simulation of missile dome heating (Rich, Brian W.)V1540,781-786(1991)

Infrared Space Observatory optical subsystem (Singer, Christian; Massoni, Jean A.; Mossbacher, Bernard; Cinotti, Ciro)V1494,255-264(1991)

Infrared spectrometer for ground-based profiling of atmospheric temperature and humidity (Shaw, Joseph A.; Churnside, James H.; Westwater, Edward R.)V1540,681-686(1991)

Infrared systems design from an operational requirement using a hypercard-based program (Harris, William R.)V1488,156-164(1991)

Infrared techniques applied to large solar arrays: a ten-year update (Hodor, James R.; Decker, Herman J.; Barney, Jesus J.)V1540,331-337(1991)

Infrared Technology XVII (Andresen, Bjorn F.; Scholl, Marija S.; Spiro, Irving J., eds.)V1540(1991)

Infrared thermal-wave studies of coatings and composites (Favro, Lawrence D.; Ahmed, Tasdiq; Crowther, D. J.; Jin, Huijia J.; Kuo, Pao K.; Thomas, Robert L.; Wang, X.)V1467,290-294(1991)

Infrared thermographic analysis of snow ski tracks (Roberts, Charles C.)V1467,207-218(1991)

Infrared-thermography-based pipeline leak detection systems (Weil, Gary J.; Graf, Richard J.)V1467,18-33(1991)

Infrared transmitting glasses and fibers for chemical analysis (Hilton, Albert R.)V1437,54-59(1991)

Infrared window damage measured by reflective scatter (Bernt, Marvin L.; Stover, John C.)V1530,42-49(1991)

Infrared zoom lenses in the eighties and beyond (Mann, Allen)V1540,338-349(1991)

InP-based quantum-well infrared photodetectors (Gunapala, S. D.; Levine, Barry F.; Ritter, D.; Hamm, Robert A.; Panish, Morton B.)V1541,11-23(1991)

InSb linear multiplexed FPAs for the CRAF/Cassini visible and infrared mapping spectrometer (Niblack, Curtiss A.; Blessinger, Michael A.; Forsthoefel, John J.; Staller, Craig O.; Sobel, Harold R.)V1494,403-413(1991)

In-situ characterization of resin chemistry with infrared transmitting optical fibers and infrared spectroscopy (Druy, Mark A.; Glatkowski, Paul J.; Stevenson, William A.)V1437,66-74(1991)

In-situ measurement technique for solution growth in compound semiconductors (Inatomi, Yuko; Kuribayashi, Kazuhiko)V1557,132-139(1991)

Integrating thermography into the Palisades Nuclear Plant's electrical predictive maintenance program (Ridley, W. C.)V1467,51-58(1991)

Interface demarcation in Bridgman-Stockbarger crystal growth of II-VI compounds (Gillies, Donald C.; Lehoczky, S. L.; Szofran, Frank R.; Su, Ching-Hua; Larson, David J.)V1484,2-10(1991)

Interference visualization of infrared images (Vlasov, Nikolai G.; Korchazhkin, S. V.; Manikalo, V. V.)V1358,1018-1020(1991)

Intrinsic carrier concentration and effective masses in Hg1-xMnxTe (Rogalski, Antoni)V1512,189-194(1991)

Introducing multiple-dynamic-windows in thermal imaging (Bales, Maurice; Boulton, Herbert)V1467,195-206(1991)

Ion implantation and diffusion for electrical junction formation in HgCdTe (Bubulac, Lucia O.)V1484,67-71(1991)

IR CCD staring imaging system (Zhou, Qibo)V1540,677-680(1991)

IR/MMW fusion ATR (automatic target recognition) (Thiede, Edwin C.)VIS07,24-35(1991)

IR objective with internal scan mirror (Eisenberg, Shai; Menache, Ram)V1442,133-138(1991)

IR-spectroscopical investigations on the glass structure of porous and sintered compacts of colloidal silica gels (Clasen, Rolf; Hornfeck, M.; Theiss, W.)V1513,243-254(1991)

Is it worth it?— statistics of corporate-based IR program results (Johnson, Peter F.)V1467,47-50(1991)

ISOCAM: a camera for the ISO satellite optical bench development (Auternaud, Danielle)V1488,64-72(1991)

Key technologies for IR zoom lenses: aspherics and athermalization (Nory, Pierre)VCR38,142-152(1991)

Kinetic studies of phosgene reduction via in situ Fourier transform infrared analysis (Farquharson, Stuart; Chauvel, J. P.)V1434,135-146(1991)

Large staring IRFPAs of HgCdTe on alternative substrates (Kozlowski, Lester J.; Bailey, Robert B.; Cooper, Donald E.; Vural, Kadri; Gertner, E. R.; Tennant, William E.)V1540,250-261(1991)

Lens design for the infrared (Fischer, Robert E.)VCR38,19-43(1991)

Lightweight surveillance FLIR (Fawcett, James M.)V1498,82-91(1991)

Linear resonant approach to scanning (Confer, Charles L.; Burrer, Gordon J.)V1454,215-222(1991)

Long-wave infrared detectors based on III-V materials (Maserjian, Joseph L.)V1540,127-134(1991)

Long-wavelength GaAs/AlxGa1-xAs quantum-well infrared photodetectors (Levine, Barry F.; Bethea, Clyde G.; Stayt, J. W.; Glogovsky, K. G.; Leibenguth, R. E.; Gunapala, S. D.; Pei, S. S.; Kuo, Jenn-Ming)V1540,232-237(1991)

Long-wavelength GaAs quantum-well infrared photodetectors (Levine, Barry F.)V1362,163-167(1991)

Long-wavelength GexSi1-x/Si heterojunction infrared detectors and focal-plane arrays (Tsaur, Bor-Yeu; Chen, C. K.; Marino, S. A.)V1540,580-595(1991)

Low-cost high-performance InSb 256 x 256 infrared camera (Parrish, William J.; Blackwell, John D.; Kincaid, Glen T.; Paulson, Robert C.)V1540,274-284(1991)

Low-temperature viscosity measurements of infrared transmitting halide glasses (Seddon, Angela B.; Cardoso, A. V.)V1513,255-263(1991)

Magnetoconcentration nonequilibrium IR photodetectors (Piotrowski, Jozef; Djuric, Zoran G.)V1512,84-90(1991)

Manufacturing and reproduction of holographic optical elements in dichromated gelatin films for operation in the infrared (Stojanoff, Christo G.; Tropartz, Stephan; Brasseur, Olivier; Kubitzek, Ruediger)V1485,274-280(1991)

Materials technology for SIRTF (Coulter, Daniel R.; Dolgin, Benjamin P.; Rainen, R.; O'Donnell, Timothy P.)V1540,119-126(1991)

Matrix-assisted laser desorption by Fourier transform mass spectrometry (Nuwaysir, Lydia M.; Wilkins, Charles L.)V1437,112-123(1991)

Measurement of point spread function of thermal imager (Ryu, Zee Man)V1467,469-474(1991)

Mechanical cooler development program for ASTER (Kawada, Masakuni; Fujisada, Hiroyuki)V1490,299-308(1991)

Microcomputer-based workstation for simulation and analysis of background and target IR signatures (Reeves, Richard C.; Schaibly, John H.)V1486,85-101(1991)

Microlens array for staring infrared imager (Werner, Thomas R.; Cox, James A.; Swanson, S.; Holz, Michael)V1544,46-57(1991)

Midinfrared backscatter spectra of selected agricultural crops (Narayanan, Ram M.; Green, Steven E.; Alexander, Dennis R.)V1379,116-122(1991)

Militarized infrared touch panels (Hough, Stewart E.; Stanley, Pamela S.)V1456,240-249(1991)

Miscellaneous modulation transfer function effects relating to sample summing (Kennedy, Howard V.)V1488,165-176(1991)

Mission design for the Space Infrared Telescope Facility (Kwok, Johnny H.; Osmolovsky, Michael G.)V1540,27-37(1991)

Model-based analysis of 3-D spatial-temporal IR clutter suppression filtering (Chan, David S.)V1481,117-128(1991)

Modeling of pumping kinetics of an iodine photodissociation laser with long pumping pulse (Rohlena, Karel; Beranek, J.; Masek, Karel)V1415,259-268(1991)

Moisture influence on near-infrared prediction of wheat hardness (Windham, William R.; Gaines, Charles S.; Leffler, Richard G.)V1379,39-44(1991)

Monolithic epitaxial IV-VI compound IR-sensor arrays on Si substrates for the SWIR, MWIR and LWIR range (Zogg, Hans; Masek, Jiri; Maissen, Clau; Hoshino, Taizo J.; Blunier, Stefan; Tiwari, A. N.)V1361,1079-1086(1991)

MTF characteristics of a Scophony scene projector (Schildwachter, Eric F.; Boreman, Glenn D.)V1488,48-57(1991)

Multicomponent analysis using established techniques (Dillehay, David L.)V1434,56-66(1991)

Multimode IRST/FLIR design issues (Armstrong, George R.; Oakley, Philip J.; Ranat, Bhadrayu M.)VCR38,120-141(1991)

Multiplex and multichannel detection of near-infrared Raman scattering (Chase, Bruce D.)V1439,47-57(1991)

Multiplexed mid-wavelength IR long linear photoconductive focal-plane arrays (Kreider, James F.; Preis, Mark K.; Roberts, Peter C.; Owen, Larry D.; Scott, Walter M.; Walmsley, Charles F.; Quin, Alan)V1488,376-388(1991)

Narcissus in current generation FLIR systems (Ford, Eric H.; Hasenauer, David M.)VCR38,95-119(1991)

Natural terrain infrared radiance statistics in a wind field (Ruizhong, Rao; Song, Zhengfang)V1486,390-395(1991)

Near-infrared optical monitoring of cardiac oxygen sufficiency through thoracic wall without open-chest surgery (Kakihana, Yasuyuki; Tamura, Mamoru)V1431,314-321(1991)

New developments in the field of chemical infrared fiber sensors (Kellner, Robert A.; Taga, Karim)V1510,232-241(1991)

New method of target acquisition in the presence of clutter (Tidhar, Gil; Rotman, Stanley R.)V1486,188-199(1991)

New thermistor material for thermistor bolometer: material preparation and characterization (Umadevi, P.; Nagendra, C. L.; Thutupalli, G. K.; Mahadevan, K.; Yadgiri, G.)V1484,125-135(1991)

New type of IR to visible real-time image converter: design and fabrication (Sun, Fang-kui; Yang, Mao-hua; Gao, Shao-hong; Zhao, Shi-jie)V1488,2-5(1991)

NIRATAM-NATO infrared air target model (Noah, Meg A.; Kristl, Joseph; Schroeder, John W.; Sandford, B. P.)V1479,275-282(1991)

NIR/CCD Raman spectroscopy: second battle of a revolution (McCreery, Richard L.)V1439,25-36(1991)

Noise mechanisms of high-temperature superconducting infrared detectors (Khalil, Ali E.)V1477,148-158(1991)

Noise reduction strategy for hybrid IR focal-plane arrays (Fowler, Albert M.; Gatley, Ian)V1541,127-133(1991)

Noncontact lifetime characterization technique for LWIR HgCdTe using transient millimeter-wave reflectance (Schimert, Thomas R.; Tyan, John; Barnes, Scott L.; Kenner, Vern E.; Brouns, Austin J.; Wilson, H. L.)V1484,19-30(1991)

Nondestructive determination of the solids content of horticultural products (Birth, Gerald S.; Dull, Gerald G.; Leffler, Richard G.)V1379,10-15(1991)

Nonselective thermal detectors of radiation (Pankratov, Nickolai A.)V1540,432-443(1991)

Novel doping superlattice-based PbTe-IR detector device (Oswald, Josef; Pippan, Manfred; Tranta, Beate; Bauer, Guenther E.)V1362,534-543(1991)

Numerical evaluation of the efficiency of camouflage systems in the thermal infrared (Proesmans, Marc; Verlinde, Patrick S.)V1486,102-114(1991)

Numerical inversion method for determining aerodynamic effects on particulate exhaust plumes from onboard irradiance data (Cousins, Daniel)V1467,402-409(1991)

Objective assessment of clinical computerized thermal images (Anbar, Michael)V1445,479-484(1991)

Objectives for the Space Infrared Telescope Facility (Spehalski, Richard J.; Werner, Michael J.)V1540,2-14(1991)

On-focal-plane-array feature extraction using a 3-D artificial neural network (3DANN): Part I (Carson, John C.)V1541,141-144(1991)

Optical characterization of photolithographic metal grids (Osmer, Kurt A.; Jones, Mike I.)V1498,138-146(1991)

Optical detector prepared by high-Tc superconducting thin film (Wang, Lingjie; Zhou, Fang Q.; Zhao, Xing R.; Sun, Han D.; Yi, Xin J.)V1540,738-741(1991)

Optical image transformation including IR region for information conformity of their formation and perception processes based on Fibonacci polynomials and series (Miroshnikov, Mikhail M.; Nesteruk, Vsevolod F.)V1540,477-487(1991)

Optical materials for infrared range of spectrum (Petrovskii, Gurii T.)V1540,401-411(1991)

Optical materials for the infrared (Wolfe, William L.)VCR38,55-68(1991)

Optical response in high-temperature superconducting thin films (Thiede, David A.)V1484,72-80(1991)

Optimal subpixel-level IR frame-to-frame registration (Fernandez, Manuel F.; Aridgides, Athanasios; Randolph, D.; Ferris, D.)V1481,172-179(1991)

Optimization of a gimbal-scanned infrared seeker (Williams, Elmer F.; Evans, Robert H.; Brant, Karl; Stockum, Larry A.)V1482,104-111(1991)

Optimization of point source detection (Friedenberg, Abraham)V1442,60-65(1991)

Optimum choice of anamorphic ratio and boost filter parameters for a SPRITE-based infrared sensor (Fredin, Per)V1488,432-442(1991)

ORION semiconductor optical detectors: research and development (Khryapov, V. T.; Ponomarenko, Vladimir P.; Butkevitch, V. G.; Taubkin, I. I.; Stafeev, V. I.; Popov, S. A.; Osipov, V. V.)V1540,412-423(1991)

Overpressure proof testing of large infrared windows for aircraft applications (Pruszynski, Charles J.)V1498,163-170(1991)

Pacing elements of IR system design (Zissis, George J.)VCR38,44-54(1991)

Parametric analysis of target/decoy performance (Kerekes, John P.)V1483,155-166(1991)

Performance of infrared systems under field conditions (Chrzanowski, Krzysztof)V1512,78-83(1991)

Performance of pyramidal phase-only filtering of infrared imagery (Kozaitis, Samuel P.; Petrilak, Robert)V1564,403-413(1991)

Phase-matched second-harmonic generation of infrared wavelengths in optical fibers (Kashyap, Raman; Davey, Steven T.; Williams, Doug L.)V1516,164-174(1991)

Photoacoustic microscopy by photodeformation applied to the determination of thermal diffusivity (Balageas, Daniel L.; Boscher, Daniel M.; Deom, Alain A.; Enguehard, Francis; Noirot, Laurence)V1467,278-289(1991)

Photochemical and thermal treatment of dichromated gelatin film for the manufacturing of holographic optical elements for operation in the IR (Stojanoff, Christo G.; Schuette, H.; Brasseur, Olivier; Kubitzek, Ruediger; Tropartz, Stephan)V1559,321-330(1991)

Photoconductivity decay method for determining minority carrier lifetime of p-type HgCdTe (Reichman, Joseph)V1484,31-38(1991)

Photovoltaic HgCdTe MWIR-detector arrays on (100)CdZnTe/(100)GaAs grown by hot-wall-beam epitaxy (Gresslehner, Karl-Heinz; Schirz, W.; Humenberger, Josef; Sitter, Helmut; Andorfer, J.; Lischka, Klaus)V1361,1087-1093(1991)

Physical foundations of high-speed thermomagnetic tuning of spatial-temporal structure of far-IR and submillimeter beams (Shvartsburg, Alexandre B.)V1488,28-35(1991)

Polymer-dispersed liquid-crystal shutters for IR imaging (McCargar, James W.; Doane, J. W.; West, John L.; Anderson, Thomas W.)V1455,54-60(1991)

Possible enhancement in bolometric response using free-standing film of YBa2Cu3Ox (Ng, Hon K.; Kilibarda, S.)V1477,15-19(1991)

Postprocessing of thermograms in infrared nondestructive testing (Vavilov, Vladimir P.; Maldague, Xavier; Saptzin, V. M.)V1540,488-495(1991)

Precision tracking of small target in IR systems (Lu, Huanzhang; Sun, Zhong-kang)V1481,398-405(1991)

Predicting electronic component lifetime using thermography (Moy, Richard Q.; Vargas, Raymund; Eubanks, Charles)V1467,154-160(1991)

Pressure modulator infrared radiometer (PMIRR) optical system alignment and performance (Chrisp, Michael P.; Macenka, Steve A.)V1540,213-218(1991)

Production of sapphire domes by the growth of near-net-shape single crystals (Biderman, Shlomo; Horowitz, Atara; Einav, Yehezkel; Ben-Amar, Gabi; Gazit, Dan; Stern, Adin; Weiss, Matania)V1535,27-34(1991)

Programmable command interpreter to automate image processing of IR thermography (Hughett, Paul)V1467,416-426(1991)

Properties of the new UCSB free-electron lasers (Ramian, Gerald)V1552,57-68(1991)

Proposed conversion of the McMath Telescope to 4.0-meter aperture for solar observations in the IR (Livingston, William C.)V1494,498-502(1991)

PtSi camera: performance model validation (Meidan, Moshe; Schwartz, Roni; Sher, Assaf; Zhaiek, Sasson; Gal, Eli; Neugarten, Michael L.; Afik, Zvi; Baer, C.)V1540,729-737(1991)

Pyroelectric linear array IR detectors with CCD multiplexer (Norkus, Volkmar; Neumann, Norbert; Walther, Ludwig; Hofmann, Guenter; Schieferdecker, Jorg; Krauss, Matthias G.; Budzier, Helmut; Hess, Norbert)V1484,98-105(1991)

Qualitative and quantitative evaluation of moisture in thermal insulation by using thermography (Vavilov, Vladimir P.; Ivanov, A. I.; Sengulye, A. A.)V1467,230-233(1991)

Quantitative evaluation of cavities and inclusions in solids using IR thermography (Madrid, Angel)V1467,322-336(1991)

Quantitative measurement of thermal parameters over large areas using pulse-video thermography (Hobbs, Chris P.; Kenway-Jackson, Damian; Milne, James M.)V1467,264-277(1991)

Quantitative thermal gradient imaging of biological surfaces (Swanson, Curtis J.; Wingard, Christopher J.)V1467,372-383(1991)

Quantum efficiency and crosstalk of an improved backside-illuminated indium antimonide focal plane array (Bloom, I.; Nemirovsky, Yael)V1442,286-297(1991)

Radiometric versus thermometric calibration of IR test systems: which is best? (Richardson, Philip I.)V1488,80-88(1991)

Real-time temperature measurement on PCB:s, hybrids, and microchips (Wallin, Bo)V1467,180-187(1991)

Recent progress in the growth and characterization of large Ge single crystals for IR optics and microelectronics (Azoulay, Moshe; Gafni, Gabriella; Roth, Michael)V1535,35-45(1991)

Recognition criterion for two-dimensional minimum resolvable temperature difference (Kennedy, Howard V.)V1488,196-202(1991)

Reconstruction methods for infrared absorption imaging (Arridge, Simon R.; van der Zee, Pieter; Cope, Mark; Delpy, David T.)V1431,204-215(1991)

Reflectors for efficient and uniform distribution of radiation for lighting and infrared based on nonimaging optics (Cai, Wen; Gordon, Jeff M.; Kashin, Peter; Rabl, Ari)V1528,118-128(1991)

Refractive-index interpolation fit criterion for materials used in optical design (Korniski, Ronald J.)VCR38,193-217(1991)

Regular doping structures: a Si-based, quantum-well infrared detector (Koch, J. F.)V1362,544-552(1991)

Relative performance studies for focal-plane arrays (Murphy, Kevin S.; Bradley, D. J.; Dennis, Peter N.)V1488,178-185(1991)

Remote temperature sensing of a pulsed thermionic cathode (Del Grande, J. M.)V1467,427-437(1991)

Reporting data for arrays with many elements (Coles, Christopher L.; Phillips, Wayne S.; Vincent, John D.)V1488,327-333(1991)

Research on enhancing signal and SNR in laser/IR inspection of solder joints quality (Xiong, Zhengjun; Cheng, Xuezhong; Liu, Xiande)V1467,410-415(1991)

Review of some non-LTE high-altitude CO2 4.3-micrometer background effects: a ten-year update (Kumer, John B.)V1540,321-330(1991)

SADARM status report (DiNardo, Anthony J.)V1479,228-248(1991)

Scanning infrared earth sensor for INSAT-II (Alex, T. K.; Kamalakar, J. A.)V1478,106-111(1991)

Scattering contribution to the error budget of an emissive IR calibration sphere (Chalupa, John; Cobb, W. K.; Murdock, Tom L.)V1530,343-351(1991)

Selective photodetectors: a view from the USSR (Ovsyuk, Victor N.; Svitashev, Konstantin K.)V1540,424-431(1991)

Sensitivity analysis of Navy tactical decision-aid FLIR performance codes (Milne, Edmund A.; Cooper, Alfred W.; Reategui, Rodolfo; Walker, Philip L.)V1486,151-161(1991)

Sensor fusion approach to optimization for human perception: an observer-optimized tricolor IR target locating sensor (Miller, Walter E.)V1482,224-233(1991)

Sensor line-of-sight stabilization (Cooper, C. J.)V1498,39-51(1991)

Ship signature measurements for tactical decision-aid input (Cooper, Alfred W.; Milne, Edmund A.; Crittenden, Eugene C.; Walker, Philip L.; Moore, E.; Lentz, William J.)V1486,37-46(1991)

Short-wavelength infrared subsystem design status of ASTER (Akasaka, Akira; Ono, Makoto; Sakurai, Yasushi; Hayashida, Bun)V1490,269-277(1991)

Si:Ga focal-plane arrays for satellite and ground-based telescopes (Mottier, Patrick; Agnese, Patrick; Lagage, Pierre O.)V1494,419-426(1991); V1512,60-67(1991)

Signal and background models in nonstandard IR systems (Snyder, John L.)V1498,52-63(1991)

Signature prediction models for flir target recognition (Velten, Vincent J.)VIS07,98-107(1991)

Simulation of infrared backgrounds using two-dimensional models (Cadzow, James A.; Wilkes, D. M.; Peters, Richard A.; Li, Xingkang; Patel, Jamshed N.)V1486,352-363(1991)

Simulation of sampling effects in FPAs (Cook, Thomas H.; Hall, Charles S.; Smith, Frederick G.; Rogne, Timothy J.)V1488,214-225(1991)

Simulation study to characterize thermal infrared sensor false alarms (Sabol, Bruce M.; Mixon, Harold D.)V1486,258-270(1991)

Simultaneous active/passive IR vehicle detection (Baum, Jerrold E.; Rak, Steven J.)V1416,209-220(1991)

Sine wave measurements of SPRITE detector MTF (Barnard, Kenneth J.; Boreman, Glenn D.; Plogstedt, Allen E.; Anderson, Barry K.)V1488,426-431(1991)

SIRTF focal-plane technologies (Capps, Richard W.; Bothwell, Mary)V1540,47-50(1991)

SIRTF stray light analysis (Elliott, David G.; St. Clair Dinger, Ann)V1540,63-67(1991)

Solution conformation of biomolecules from infrared vibrational circular dichroism spectroscopy (Diem, Max)V1432,28-36(1991)

Solutions to modeling of imaging IR systems for missile applications: MICOM imaging IR system performance model-90 (Owen, Philip R.; Dawson, James A.; Borg, Eric J.)V1488,122-132(1991)

So now what?— things to do if your IR program stops producing results (Lucier, Ronald D.)V1467,59-62(1991)

Soviet IR imagers and their applications: short state of the art (Vavilov, Vladimir P.)V1540,460-465(1991)

Space Infrared Telescope Facility cryogenic and optical technology (Mason, Peter V.; Kiceniuk, T.; Plamondon, Joseph A.; Petrick, Walt)V1540,88-96(1991)

Space Infrared Telescope Facility science instruments overview (Bothwell, Mary)V1540,15-26(1991)

Space Infrared Telescope Facility structural design requirements (MacNeal, Paul D.; Lou, Michael C.; Chen, Gun-Shing)V1540,68-85(1991)

Space Infrared Telescope Facility telescope overview (Schember, Helene R.; Manhart, Paul K.; Guiar, Cecilia N.; Stevens, James H.)V1540,51-62(1991)

SPRITE detector characterization through impulse response testing (Anderson, Barry K.; Boreman, Glenn D.; Barnard, Kenneth J.; Plogstedt, Allen E.)V1488,416-425(1991)

Staring sensor MRT measurement and modeling (Mooney, Jonathan M.)V1540,550-564(1991)

State-of-the-art transfer radiometer for testing and calibration of FLIR test equipment (Kopolovich, Zvi; Naor, Yoram; Cabib, Dario; Johnson, W. Todd; Sapir, Eyal)V1540,565-577(1991)

Status and needs of infrared optical property information for optical designers (Wolfe, William L.)V1354,696-741(1991)

Status report on thermographer certification (Baird, George S.; Mack, Russell T.)V1467,63-63(1991)

Stratospheric ozone concentration profiles from Spacelab-1 solar occultation infrared absorption spectra (De Maziere, Martine M.; Camy-Peyret, C.; Lippens, C.; Papineau, N.)V1491,288-297(1991)

Stratospheric spectroscopy with the far-infrared spectrometer: overview and recent results (Traub, Wesley A.; Chance, Kelly V.; Johnson, David G.; Jucks, Kenneth W.)V1491,298-307(1991)

Stratospheric wind infrared limb sounder (Rider, David M.; McCleese, Daniel J.)V1540,142-147(1991)

Stray light issues for background-limited infrared telescope operation (Scholl, Marija S.; Scholl, James W.)V1540,109-118(1991)

Structural design considerations for the Space Infrared Telescope Facility (MacNeal, Paul D.; Lou, Michael C.)V1494,236-254(1991)

Structural design of the large deployable reflector (Satter, Celeste M.; Lou, Michael C.)V1494,279-300(1991)

Submicron thin-film metal-oxide-metal infrared detectors (Wilke, Ingrid; Moix, Dominique; Herrmann, W.; Kneubuhl, F. K.)V1442,2-10(1991)

Superconducting bolometers: high-Tc and low-Tc (Richards, Paul L.)V1477,2-6(1991)

Superconductivity Applications for Infrared and Microwave Devices II (Heinen, Vernon O.; Bhasin, Kul B., eds.)V1477(1991)

Supervision of self-heating in peat stockpiles by aerial thermography (Tervo, Matti; Kauppinen, Timo)V1467,161-168(1991)

Surveying and damping heat loss from machines with high surface temperatures: thermography as a tool (Perch-Nielsen, Thomas; Paulsen, Otto; Drivsholm, Christian)V1467,169-179(1991)

Surveying the elements of successful infrared predictive maintenance programs (Snell, John R.; Spring, Robert W.)V1467,2-10(1991)

Tactical Infrared Systems (Tuttle, Jerry W., ed.)V1498(1991)

Target acquisition model appropriate for dynamically changing scenarios (Rotman, Stanley R.; Gordon, E. S.)V1442,335-346(1991)

Target detection using co-occurrence matrix segmentation and its hardware implementation (Auborn, John E.; Fuller, James M.; McCauley, Howard M.)V1482,246-252(1991)

Technique for measuring atmospheric effects on image metrics (Crow, Samuel B.; Watkins, Wendell R.; Palacios, Fernando R.; Billingsley, Daniel R.)V1486,333-344(1991)

Technology trends for high-performance windows (Askinazi,.Joel)V1498,100-109(1991)

Temperature chamber FLIR and missile test system (Johnson, W. Todd; Lavi, Moshe; Sapir, Eyal)V1488,343-354(1991)

Temperature rise due to dynamic crack growth in Beta-C titanium (Zehnder, Alan T.; Kallivayalil, Jacob A.)V1554A,48-59(1991)

Thermal analysis of masonry block buildings during construction (Allen, Lee R.; Semanovich, Sharon A.)V1467,99-103(1991)

Thermal analysis of the bottle forming process (Wilson, Jeannie S.)V1467,219-228(1991)

Thermal and radiometric modeling of terrain backgrounds (Conant, John A.; Hummel, John R.)V1486,217-230(1991)

Thermal compensation of infrared achromatic objectives with three optical materials (Rayces, Juan L.; Lebich, Lan)V1354,752-759(1991)

Thermal diagnostics for monitoring welding parameters in real time (Fuchs, Elizabeth A.; Mahin, K. W.; Ortega, A. R.; Bertram, L. A.; Williams, Dean R.; Pomplun, Alan R.)V1467,136-149(1991)

Thermal infrared imagery from the Geoscan Mk II scanner and its calibration: two case histories from Nevada—Ludwig Skarn (Yerington District) & Virginia City (Lyon, Ronald J.; Honey, Frank R.)V1492,339-350(1991)

Thermal infrared subsystem design status of ASTER (Aoki, Yutaka; Ohmae, Hirokazu; Kitamura, Shin-ichi)V1490,278-284(1991)

Thermal model for real-time textured IR background simulation (Bernstein, Uri; Keller, Catherine E.)V1486,345-351(1991)

Thermal sensing of fireball plumes (Toossi, Reza)V1467,384-393(1991)

Thermal signature training for military observers (LaFollette, Robert; Horger, John D.)V1488,289-299(1991)

Thermal systems analysis for the Space Infrared Telescope Facility dewar (Bhandari, Pradeep; Petrick, Stanley W.; Schember, Helene R.)V1540,97-108(1991)

Thermographic analysis of the anisotropy in the thermal conductivity of composite materials (Burleigh, Douglas D.; De La Torre, William)V1467,303-310(1991)

Thermographic monitoring of lubricated couplings (Wurzbach, Richard N.)V1467,41-46(1991)

Thermography and complementary method: a tool for cost-effective measures in retrofitting buildings (Lyberg, Mats D.; Ljungberg, Sven-Ake)V1467,104-115(1991)

Thermosense XIII (Baird, George S., ed.)V1467(1991)

Time-resolved infrared radiometry of multilayer organic coatings using surface and subsurface heating (Maclachlan Spicer, J. W.; Kerns, W. D.; Aamodt, Leonard C.; Murphy, John C.)V1467,311-321(1991)

Time-resolved infrared studies of the dynamics of ligand binding to cytochrome c oxidase (Dyer, R. B.; Peterson, Kristen A.; Stoutland, Page O.; Einarsdottir, Oloef; Woodruff, William H.)V1432,197-204(1991)

Time-resolved videothermography at above-frame-rate frequencies (Shepard, Steven M.; Sass, David T.; Imirowicz, Thomas P.; Meng, A.)V1467,234-238(1991)

Transient radiometric measurements with a PtSi IR camera (Konopka, Wayne L.; Soel, Michael A.; Celentano, A.; Calia, V.)V1488,355-365(1991)

Twenty-five years of aerodynamic research with IR imaging (Gartenberg, Ehud; Roberts, A. S.)V1467,338-356(1991)

Two-dimensional encoding of images using discrete reticles (Wellfare, Michael R.)V1478,33-40(1991)

Universal equation for IR thermometer and its applications (Liu, Jian; Bao, Xue-Cheng; Zhang, Cai-Gen; Zhang, You-Wen)V1540,744-755(1991)

Unsupervised target detection in a single IR image frame (Zhou, Yi-Tong)V1567,502-510(1991)

Update on the C2NVEO FLIR90 and ACQUIRE sensor performance models (Scott, Luke B.; Tomkinson, David M.)V1488,99-109(1991)

Use of Fourier transform spectroscopy in combustion effluent monitoring (Howe, Gordon S.; McIntosh, Bruce C.)V1434,90-103(1991)

Using IR thermography as a manufacturing tool to analyze and repair defects in printed circuit boards (Fike, Daniel K.)V1467,150-153(1991)

Utility gains through infrared predictive maintenance (Black, James E.)V1467,34-40(1991)

Validated CCD camera model (Johnson-Cole, Helen; Clark, Rodney L.)V1488,203-211(1991)

Variable emissivity plates under a three-dimensional sky background (Meitzler, Thomas J.; Gonda, Teresa G.; Jones, Jack C.; Reynolds, William R.)V1486,380-389(1991)

What is MRT and how do I get one? (Hoover, Carl W.; Webb, Curtis M.)V1488,280-288(1991)

Integrated circuits—see also microlithography; semiconductors; silicon

Accurate and efficient simulation of MMIC layouts (Wu, Doris I.; Chang, David C.)V1475,140-150(1991)

Accurate design of multiport low-noise MMICs up to 20 GHz (Willems, David; Bahl, I.; Griffin, Edward)V1475,55-61(1991)

Advanced lithographic methods for contact patterning on severe topography (Chu, Ron; Greeneich, James S.; Katz, Barton A.; Lin, Hwang-Kuen; Huang, Dong-Tsair)V1465,238-243(1991)

Advanced multichip module packaging and interconnect issues for GaAs signal processors operating above 1 GHz clock rates (Gilbert, Barry K.; Thompson, R.; Fokken, G.; McNeff, W.; Prentice, Jeffrey A.; Rowlands, David O.; Staniszewski, A.; Walters, W.; Zahn, S.; Pan, George W.)V1390,235-248(1991)

Advances in power MMIC amplifier technology in space communications (Tserng, Hua Q.; Saunier, Paul)V1475,74-85(1991)

Analysis and modeling of uniformly- and nonuniformly-coupled lossy lines for interconnections and packaging in hybrid and monolithic circuits (Orhanovic, Neven; Hayden, Leonard A.; Tripathi, Vijai K.)V1389,273-284(1991)

Application and integration of a focused ion beam circuit repair system (Lange, John A.; Allen, Chris)V1465,50-56(1991)

Application of electrodeposition processes to advanced package fabrication (Krongelb, Sol; Dukovic, John O.; Komsa, M. L.; Mehdizadeh, S.; Romankiw, Lubomyr T.; Andricacos, P. C.; Pfeiffer, A. T.; Wong, K.)V1389,249-256(1991)

Application of Kaband MMIC technology for an Orbiter/ACTS communications experiment (Arndt, George D.; Fink, Patrick W.; Leopold, Louis; Bondyopadhyay, Probir; Shaw, Roland)V1475,231-242(1991)

Application of thermal wave technology to thickness and grain size monitoring of aluminum films (Opsal, Jon L.)V1596,120-131(1991)

Applications of an automated particle detection and identification system in VLSI wafer processing (Hattori, Takeshi; Koyata, Sakuo)V1464,367-376(1991)

Applications of high speed silicon bipolar ICs in fiber optic systems (LaBelle, Gary L.; McDonald, Mark D.)V1365,116-121(1991)

Applications of Z-Plane memory technology to high-frame rate imaging systems (Shanken, Stuart N.; Ludwig, David E.)V1346,210-215(1991)

Application-specific integrated-circuit-based multianode microchannel array readout system (Smeins, Larry G.; Stechman, John M.; Cole, Edward H.)V1549,59-65(1991)

Automated approach to the correlation of defect locations to electrical test results to determine yield reducing defects (Slama, M. M.; Patterson, Angela C.)V1464,602-609(1991)

Automated visual inspection system for IC bonding wires using morphological processing (Tsukahara, Hiroyuki; Nakashima, Masato; Sugawara, Takehisa)V1384,15-26(1991)

Automated wafer inspection in the manufacturing line (Harrigan, Jeanne E.; Stoller, Meryl D.)V1464,596-601(1991)

Bridge-type optoelectronic sample and hold circuit (Sun, C. K.; Wu, Chao-Chia C.; Chang, Ching T.; Yu, Paul K.; McKnight, William H.)V1476,294-300(1991)

Buried-ridge-stripe lasers monolithically integrated with butt-coupled passive waveguides for OEIC (Remiens, D.; Hornung, V.; Rose, B.; Robein, D.)V1362,323-330(1991)

CAD in new areas of the package and interconnect design space (McBride, Dennis J.)V1390,330-335(1991)

Challenges of using advanced multichip packaging for next generation spaceborne computers (Moravec, Thomas J.)V1390,195-201(1991)

Compact time-delay shifters that are process insensitive (Lesko, Camille; Hill, William A.; Dietrich, Fred; Nelson, William)V1475,330-339(1991)

Comparison of MESFET and HEMT MMIC technologies using a compact Kaband voltage-controlled oscillator (Swirhun, S.; Geddes, John J.; Sokolov, Vladimir; Bosch, D.; Gawronski, M. J.; Anholt, R.)V1475,303-308(1991)

Computational model of the imaging process in scanning-x microscopy (Gallarda, Harry S.; Jain, Ramesh C.)V1464,459-473(1991)

Coplanar SIMMWIC circuits (Luy, Johann-Freidrich; Strohm, Karl M.; Buechler, J.)V1475,129-139(1991)

Coplanar waveguide InP-based HEMT MMICs for microwave and millimeter wave applications (Chou, Chia-Shing; Litvin, K.; Larson, Larry E.; Rosenbaum, Steven E.; Nguyen, Loi D.; Mishra, Umesh K.; Lui, M.; Thompson, M.; Ngo, Catherine M.; Melendes, M.)V1475,151-156(1991)

Current technologies for very high performance VLSI ICs (Perea, Ernesto H.)V1362,477-483(1991)

Design and process impact on thin-film interconnection performance (Rinne, Glenn A.; Hwang, Lih-Tyng; Adema, G. M.; King, Donald A.; Turlik, Iwona)V1389,110-121(1991)

Design methodology for dark-field phase-shifted masks (Toh, Kenny K.; Dao, Giang T.; Gaw, Henry T.; Neureuther, Andrew R.; Fredrickson, Larry R.)V1463,402-413(1991)

Design of an analog VLSI chip for a neural network target tracker (Narathong, Chiewcharn; Inigo, Rafael M.)V1452,523-531(1991)

Design, simulation model, and measurements for high-density interconnections (Shrivastava, Udy A.)V1389,122-137(1991)

Development of a high-speed high-precision laser plotter (Tachihara, Satoru; Miyoshi, Tamihiro)V1527,305-314(1991)

Digital-signal-processor-based inspection of populated surface-mount technology printed circuit boards (Hartley, David A.; Hobson, Clifford A.; Lilley, F.)V1567,277-282(1991)

Dispersion of picosecond pulses propagating on microstrip interconnections on semiconductor integrated-circuit substrates (Pasik, Michael F.; Cangellaris, Andreas C.; Prince, John L.)V1389,297-301(1991)

Distortion characteristic of transient signals through bend discontinuity of high-speed integrated curcuits (Huang, Wei-Xu; Wing, Omar)V1389,199-204(1991)

Dynamic behavior of internal elements of high-frequency integrated circuits studied by time-resolved optical-beam-induced current (OBIC) method (Bergner, Harald; Hempel, Klaus; Stamm, Uwe)V1362,484-493(1991)

Effect of Cu at Al grain boundaries on electromigration behavior in Al thin films (Frear, Darrel R.; Michael, J. R.; Kim, C.; Romig, A. D.; Morris, J. W.)V1596,72-82(1991)

Effects of conductor losses on cross-talk in multilevel-coupled VLSI interconnections (van Deventer, T. E.; Katehi, Linda P.; Cangellaris, Andreas C.)V1389,285-296(1991)

Effects of packaging and interconnect technology on testability of printed wiring boards (Hughes, Joseph L.; Pahlajrai, Prem)V1389,87-97(1991)

Effects of radiation on millimeter wave monolithic integrated circuits (Meulenberg, A.; Hung, Hing-Loi A.; Singer, J. L.; Anderson, Wallace T.)V1475,280-285(1991)

Efficient and accurate dynamic analysis of microstrip integrated circuits (Rahal Arabi, Tawfik R.; Murphy, Arthur T.; Sarkar, Tapan K.)V1389,302-313(1991)

Efficient use of data structures for digital monopulse feature extraction (McEachern, Robert; Eckhardt, Andrew J.; Nauda, Alexander)V1470,226-232(1991)

Electromigration in VLSI metallization (Kwok, Thomas Y.)V1596,60-71(1991)

Electromigration physical modeling of failure in thin film structures (Lloyd, James R.)V1596,106-117(1991)

Electron-beam lithography for the microfabrication of OEICs (Engel, Herbert; Doeldissen, Walter)V1506,60-64(1991)

Electronic/photonic inversion channel technology for optoelectronic ICs and photonic switching (Taylor, Geoff W.; Cooke, Paul W.; Kiely, Philip A.; Claisse, Paul R.; Sargood, Stephen K.; Doctor, D. P.; Vang, T.; Evaldsson, P.; Daryanani, Sonu L.)V1476,2-13(1991)

EXCON: a graphics-based experiment-control manager (Khan, Mumit; Anderson, Paul D.; Cerrina, Franco)V1465,315-323(1991)

GE high-density interconnect: a solution to the system interconnect problem (Adler, Michael S.)V1390,504-508(1991)

Generation and sampling of high-repetition-rate/high-frequency electrical waveforms in microstrip circuits by picosecond optoelectronic technique (Lee, Chi H.)V1390,377-387(1991)

Growth and characterization of ultrathin SimGen strained-layer superlattices (Presting, Hartmut; Jaros, Milan; Abstreiter, Gerhard)V1512,250-277(1991)

High-density chip-to-chip interconnect system for GaAs semiconductor devices (Wigginton, Stewart C.; Davidson, Scott E.; Harting, William L.)V1390,560-567(1991)

High-density circuit approach for low-cost MMIC circuits (Bauhahn, Paul E.; Geddes, John J.)V1475,122-128(1991)

High-density memory packaging technology high-speed imaging applications (Frew, Dean L.)V1346,200-209(1991)

High-density multichip interconnect: military packaging for the 1990s (Trask, Philip A.)V1390,223-234(1991)

High-dynamic-range mixer using novel balun structure (Bharj, Sarjit; Taylor, Gordon C.; Denlinger, E. J.; Milgazo, H.)V1475,340-349(1991)

High-efficiency dual-band power amplifier for radar applications (Masliah, Denis A.; Cole, Brad; Platzker, Aryeh; Schindler, Manfred)V1475,113-120(1991)

High-efficiency Kaband monolithic pseudomorphic HEMT amplifier (Saunier, Paul; Tserng, Hua Q.; Kao, Yung C.)V1475,86-90(1991)

High-performance GaAs on silicon technology for VLSI, MMICs, and optical interconnects (Christou, Aristos)V1361,354-361(1991)

High-resolution decoding of multianode microchannel array detectors (Kasle, David B.; Morgan, Jeffrey S.)V1549,52-58(1991)

High-temperature superconductor junction technology (Simon, Randy W.)V1477,184-191(1991)

Hybrid wafer scale optoelectronic integration (Lockwood, Harry F.)V1389,55-67(1991); V1390,55-67(1991)

III-V semiconductor integrated optoelectronics for optical computing (Wada, Osamu)V1362,598-607(1991)

Image registration for automated inspection of 2-D electronic circuit patterns (Rodriguez, Arturo A.; Mandeville, Jon R.)V1384,2-14(1991)

Improved planarization techniques applied to a low dielectric constant polyimide used in multilevel metal ICs (Chang, Li-Hsin; Goodner, Ray)V1596,34-45(1991)

InGaAs HEMT MMIC low-noise amplifier and doublers for EHF SATCOM ground terminals (Chow, P. D.; Lester, J.; Huang, P.; Jones, William L.)V1475,42-47(1991)

In-line wafer inspection using 100-megapixel-per-second digital image processing technology (Dickerson, Gary; Wallace, Rick P.)V1464,584-595(1991)

Integrated circuit active antenna elements for monolithic implementation (Chang, Kai)V1475,164-174(1991)

Integrated Circuit Metrology, Inspection, and Process Control V (Arnold, William H., ed.)V1464(1991)

Integrated circuits with three-dimensional optical interconnections: an element base of neural networks (Gulyaev, Yuri V.; Elinson, Matvey I.; Kopylov, Yuri L.; Perov, Polievkt I.)V1621,84-92(1991)

Interconnect and packaging technology in the '90s (Seraphim, Donald; Barr, Donald E.)V1389,39-54(1991); V1390,39-54(1991)

Interconnection problems in VLSI random access memory chip (Rayapati, Venkatapathi N.; Mukhedkar, Dinkar)V1389,98-109(1991)

Interfacial adhesive strength measurement in a multilayered two-level metal device structure (Siddiqui, Humayun R.; Ryan, Vivian; Shimer, Julie A.)V1596,139-157(1991)

Ka-band MMIC array feed development for deep space applications (Cooley, Thomas W.; Riley, A. L.; Crist, Richard A.; Sukamto, Lin; Jamnejad, V.; Rascoe, Daniel L.)V1475,243-247(1991)

Laser-induced metal deposition and laser cutting techniques for fixing IC design errors (Shaver, David C.; Doran, S. P.; Rothschild, Mordechai; Sedlacek, J. H.)V1596,46-50(1991)

Localization of hot spots in silicon devices with a laser scanning microscope (Bergner, Harald; Krause, A.; Stamm, Uwe)V1361,723-731(1991)

Low-noise high-yield octave-band feedback amplifiers to 20 GHz (Minot, Katcha; Cochrane, Mike; Nelson, Bradford; Jones, William L.; Streit, Dwight C.; Liu, Po-Hsin)V1475,309-313(1991)

Measurements and characterization of multiple-coupled interconnection lines in hybrid and monolithic integrated circuits (Hayden, Leonard A.; Jong, Jyh-Ming; Rettig, John B.; Tripathi, Vijai K.)V1389,205-214(1991)

Metallization: Performance and Reliability Issues for VLSI and ULSI (Gildenblat, Gennady S.; Schwartz, Gary P., eds.)V1596(1991)

Methods for comparative analysis of waveform degradation in electrical and optical high-performance interconnections (Merkelo, Henri; McCredie, B. D.; Veatch, M. S.; Quinn, D. L.; Dorneich, M.; Doi, Yutaka)V1390,91-163(1991)

Microelectronic Interconnects and Packages: Optical and Electrical Technologies (Arjavalingam, Gnanalingam; Pazaris, James, eds.)V1389(1991)

Microelectronic Interconnects and Packages: System and Process Integration (Carruthers, John R.; Tewksbury, Stuart K., eds.)V1390(1991)

Millimeter-wave and optoelectronic applications of heterostructure integrated circuits (Pavlidis, Dimitris)V1362,450-466(1991)

Millimeter wave pseudomorphic HEMT MMIC phased-array components for space communications (Lan, Guey-Liu; Pao, Cheng K.; Wu, Chan-Shin; Mandolia, G.; Hu, M.; Yuan, Steve; Leonard, Regis F.)V1475,184-192(1991)

Mixed application MMIC technologies: progress in combining RF, digital, and photonic circuits (Swirhun, S.; Bendett, Mark P.; Sokolov, Vladimir; Bauhahn, Paul E.; Sullivan, Charles T.; Mactaggart, R.; Mukherjee, Sayan D.; Hibbs-Brenner, Mary K.; Mondal, J. P.)V1475,223-230(1991)

MMIC compatible photodetector design and characterization (Dallabetta, Kyle A.; de La Chapelle, Michael; Lawrence, Robert C.)V1371,116-127(1991)

MMICs for airborne phased arrays (Scalzi, Gary J.; Turtle, John P.; Carr, Paul H.)V1475,2-9(1991)

Modeling inner and outer plexiform retinal processing using nonlinear coupled resistive networks (Andreou, Andreas G.; Boahen, Kwabena A.)V1453,270-281(1991)

Modeling progress and trends in electrical interconnects (Prince, John L.; Cangellaris, Andreas C.; Palusinski, Olgierd A.)V1390,271-285(1991)

Moire image overlapping method for PCB inspection designator (Chang, Rong-Seng; Hu, Yeu-Jent)V1567,216-219(1991)

Monolithic GaAs integrated circuit millimeter wave imaging sensors (Weinreb, Sander)V1475,25-31(1991)

Monolithic integrated-circuit charge amplifier and comparator for MAMA readout (Cole, Edward H.; Smeins, Larry G.)V1549,46-51(1991)

Monolithic integration of a semiconductor ring laser and a monitoring photodetector (Krauss, Thomas; Laybourn, Peter J.)V1583,150-152(1991)

Monolithic microwave integrated circuit activities in ESA-ESTEC (Gatti, Giuliano)V1475,10-24(1991)

Monolithic Microwave Integrated Circuits for Sensors, Radar, and Communications Systems (Leonard, Regis F.; Bhasin, Kul B., eds.)V1475(1991)

Monolithic phased arrays: recent advances (Kinzel, Joseph A.)V1475,158-163(1991)

Narrow (0.1 um to 0.5 um) copper lines for ultra-large-scale integration technology (Shacham-Diamand, Yosef Y.)V1442,11-19(1991)

Neural net selection of features for defect inspection (Sasaki, Kenji; Casasent, David P.; Natarajan, Sanjay S.)V1384,228-233(1991)

Neural network design for channel routing (Zargham, Mehdi R.; Sayeh, Mohammad R.)V1396,202-208(1991)

New 0.54 aperture i-line wafer stepper with field-by-field leveling combined with global alignment (van den Brink, Martin A.; Katz, Barton A.; Wittekoek, Stefan)V1463,709-724(1991)

New materials for high-performance III-V ICs and OEICs: an industrial approach (Martin, Gerard M.; Frijlink, P. M.)V1362,67-74(1991)

New way of optical interconnection for VLSI (Zhao, Feng; Geng, Wanzhen; Jiang, Lingzhen; Hong, Jing)V1555,241-242(1991)

Ninety-four GHz InAlAs/InGaAs/InP HEMT low-noise down-converter (Chow, P. D.; Tan, K.; Streit, Dwight C.; Garske, D.; Liu, Po-Hsin P.; Yen, Huan-chun)V1475,48-54(1991)

Novel microstructures for low-distortion chip-to-chip interconnects (Blennemann, Heinrich C.; Pease, R. F.)V1389,215-235(1991)

Novel monolithic chip-integrated color spectrometer: the distributed-wavelength filter component (Holm-Kennedy, James W.; Tsang, Koon Wing; Sze, Wah Wai; Jiang, Fenglai; Yang, Datong)V1527,322-331(1991)

Novel selective-plated heatsink, key to compact 2-watt MMIC amplifier (Taylor, Gordon C.; Bechtle, Daniel W.; Jozwiak, Phillip C.; Liu, Shing G.; Camisa, Raymond L.)V1475,103-112(1991)

Numerical simulation of thick-linewidth measurements by reflected light (Wojcik, Gregory L.; Mould, John; Monteverde, Robert J.; Prochazka, Jaroslav J.; Frank, John R.)V1464,187-203(1991)

Observation of stress voids and grain structure in laser-annealed aluminum using focused ion-beam microscopy (Pramanik, Dipankar; Jain, Vivek)V1596,132-138(1991)

Optical laser intelligent verification expert system (Jones, Robert H.)V1401,86-93(1991)

Optically coupled 3-D common memory with GaAs on Si structure (Hirose, Masataka; Takata, H.; Koyanagi, Mitsumasa)V1362,316-322(1991)

Optically powered optoelectronic integrated circuits for optical interconnects (Brown, Julia J.; Gardner, J. T.; Forrest, Stephen R.)V1474,236-242(1991)

Optical techniques for microwave monolithic circuit characterization (Hung, Hing-Loi A.; Li, Ming-Guang; Lee, Chi H.)V1476,276-281(1991)

Optoelectronic Materials and Device Concepts (Razeghi, Manijeh)VPM05(1991)

Packaging technology for GaAs MMIC (monolithic microwave integrated circuits) modules (Tomimuro, Hisashi)V1390,214-222(1991)

Packaging technology for the NEC SX-3 supercomputers (Murano, Hiroshi; Watari, Toshihiko)V1390,78-90(1991)

Performance enhancement in future communications satellites with MMIC technology insertion (Hung, Hing-Loi A.; Mahle, Christoph E.)V1475,212-222(1991)

Phased-array antenna control by a monolithic photonic integrated circuit (Hietala, Vincent M.; Vawter, Gregory A.; Meyer, W. J.; Kravitz, Stanley H.)V1476,170-175(1991)

Phased-array receiver development using high-performance HEMT MMICs (Liu, Louis; Jones, William L.; Carandang, R.; Lam, Wayne W.; Yonaki, J.; Streit, Dwight C.; Kasody, R.)V1475,193-198(1991)

Phase-shifting photolithography applicable to real IC patterns (Yanagishita, Yuichiro; Ishiwata, Naoyuki; Tabata, Yasuko; Nakagawa, Kenji; Shigematsu, Kazumasa)V1463,207-217(1991)

Photodetectors: how to integrate them with microelectronic and optical devices (Decoster, Didier; Vilcot, Jean-Pierre)V1362,959-966(1991)

Photolith analysis and control system (Srikanth, Usha; Sundararajan, Srikanth)V1468,429-433(1991)

Projection direct imaging for high-density interconnection and printed circuit manufacture (Bergstrom, Neil G.)V1390,509-512(1991)

Quasioptical MESFET VCOs (Bundy, Scott; Mader, Tom; Popovic, Zoya; Ellinson, Reinold; Hjelme, Dag R.; Surette, Mark R.; Yadlowski, Michael; Mickelson, Alan R.)V1475,319-329(1991)

Quick prototyping center for hybrid-wafer-scale integration (HWSI) multichip modules (Chandra, S.; Lee, Yung-Cheng)V1390,548-559(1991)

Selective low-temperature chemical vapor deposition of copper from new copper(I) compounds (Jain, Ajay; Shin, H. K.; Chi, Kai-Ming; Hampden-Smith, Mark J.; Kodas, Toivo T.; Farkas, Janos; Paffett, M. F.; Farr, J. D.)V1596,23-33(1991)

Shadow masked growth for the fabrication of photonic integrated circuits (Demeester, Piet M.; Moerman, Ingrid; Zhu, Youcai; van Daele, Peter; Thomson, J.)V1361,1132-1143(1991)

Silicon-on-silicon microsystem in plastic packages (Novak, Agneta; Glaes, Anders; Blom, Claes; Hentzell, Hans; Hodges, Charles; Kalhur, Farzeen)V1389,80-86(1991)

Single-chip imager and feature extractor (Tanner, John E.; Luo, Jin)V1473,76-87(1991)

Skin effect in high-speed ULSI/VLSI packages (Hwang, Lih-Tyng; Turlik, Iwona)V1390,249-260(1991)

Some fundamental issues on metallization in VLSI (Ferry, David K.; Kozicki, M. N.; Raupp, Gregory B.)V1596,2-11(1991)

Spin-on glasses in the silicon IC: plague or panacea? (Lifshitz, N.; Pinto, Mark R.)V1596,96-105(1991)

State-of-the-art multichip modules for avionics (Hagge, John K.)V1390,175-194(1991)

Statistical approach to optimizing advanced low-voltage SEM operation (Apostolakis, Peter J.)V1464,406-412(1991)

Stress-induced voiding in aluminum alloy metallizations (Sullivan, Timothy D.; Ryan, James G.; Riendeau, J. R.; Bouldin, Dennis P.)V1596,83-95(1991)

Switching noise in a medium-film copper/polyimide multichip module (Sandborn, Peter A.; Hashemi, Seyed H.; Weigler, William)V1389,177-186(1991)

System interconnection of high-density multichip modules (Krusius, J. P.)V1390,261-270(1991)

System interconnect issues for subnanosecond signal transmission (Moresco, Larry L.)V1390,202-213(1991)

System issues for multichip packaging (Sage, Maurice G.; Hartley, Neil)V1390,302-310(1991)

System-level integrated circuit development for phased-array antenna applications (Shalkhauser, Kurt A.; Raquet, Charles A.)V1475,204-209(1991)

Technique of assessing contact ohmicity and their relevance to heterostructure devices (Harrison, H. B.; Reeves, Geoffrey K.)V1596,52-59(1991)

Termination for minimal reflection of high-speed pulse propagation along multiple-coupled microstrip lines (Kuo, Jen-Tsai; Tzuang, Ching-Kuang C.)V1389,156-160(1991)

Three-dimensional capacitance modeling of advanced multilayer interconnection technologies (Edelstein, Daniel C.)V1389,352-360(1991)

Two-dimensional electron gas charge-coupled devices (Fossum, Eric R.; Song, Jong I.; Rossi, David V.)V1447,202-203(1991)

Ultra-high-frequency InP-based HEMTs for millimeter wave applications (Greiling, Paul T.; Nguyen, Loi D.)V1475,34-41(1991)

Ultralinear low-noise amplifier technology for space communications (Watkins, E. T.; Yu, K. K.; Yau, W.; Wu, Chan-Shin; Yuan, Steve)V1475,62-72(1991)

Unframed via interconnection of nonplanar device structures (Kim, Manjin J.)V1596,12-22(1991)

Update on focal-plane image processing research (Kemeny, Sabrina E.; Eid, El-Sayed I.; Mendis, Sunetra; Fossum, Eric R.)V1447,243-250(1991)

Use of diffracted light from latent images to improve lithography control (Hickman, Kirt C.; Gaspar, Susan M.; Bishop, Kenneth P.; Naqvi, H. S.; McNeil, John R.; Tipton, Gary D.; Stallard, Brian R.; Draper, B. L.)V1464,245-257(1991)

Using IR thermography as a manufacturing tool to analyze and repair defects in printed circuit boards (Fike, Daniel K.)V1467,150-153(1991)

Variable-gain MMIC module for space application (Palena, Patricia)V1475,91-102(1991)

V-line: a new interconnect for packaging and microwave applications (Schutt-Aine, Jose E.; Lee, Jin-Fa)V1389,138-143(1991)

Wafer scale integration modular packaging (Brewer, Joe E.; French, Larry E.)V1390,164-174(1991)

X-ray laminography analysis of ultra-fine-pitch solder connections on ultrathin boards (Adams, John A.)V1464,484-497(1991)

Integrated optics—see also fiber optics; integrated optoelectronics; waveguides

Achromatization of optical waveguide components (Spaulding, Kevin E.; Morris, G. M.)V1507,45-54(1991)

Acousto-optics in integrated-optic devices for optical recording (Petrov, Dmitry V.; Belostotsky, A. L.; Dolgopolov, V. G.; Leonov, A. S.; Fedjukhin, L. A.)V1374,152-159(1991)

All-silicon Fabry-Perot modulator based on thermo-optic effect (Rendina, Ivo; Cocorullo, Giuseppe)V1583,338-343(1991)

Attenuation of leaky waves in GaAs/AlGaAs MQW waveguides formed on a GaAs substrate (Kubica, Jacek M.)V1506,134-139(1991)

Balanced optical mixer integrated in InGaAlAs/InP for coherent receivers (Caldera, Claudio; De Bernardi, Carlo; Destefanis, Giovanni; Meliga, Marina; Morasca, Salvatore; Rigo, Cesare F.; Stano, Alessandro)V1372,82-87(1991)

Boundary diffraction wave in imaging by small holograms (Mulak, Grazyna)V1574,266-271(1991)

Broadband electromagnetic environment monitoring using semiconductor electroabsorption modulators (Pappert, Stephen A.; Lin, S. C.; Orazi, Richard J.; McLandrich, Matthew N.; Yu, Paul K.; Li, S. T.)V1476,282-293(1991)

Broadband microwave and millimeter-wave EOMs with ultraflat frequency response (Pan, J. J.; Li, Yi Q.)V1476,22-31(1991)

Buried-glass waveguides by ion exchange through ionic barrier (Li, Ming-Jun; Honkanen, Seppo; Wang, Wei-Jian; Najafi, S. I.; Tervonen, Ari; Poyhonen, Pekka)V1506,52-57(1991)

Characterization, modeling, and design optimization of integrated optical waveguide devices in glass (Yip, Gar L.)V1513,26-36(1991)

Characterization of proton-exchanged and annealed proton-exchanged optical waveguides in z-cut LiNbO3 (Nikolopoulos, John; Yip, Gar L.)V1374,30-36(1991)

Characterization of the photorefractive effect in Ti:LiNbO3 stripe waveguides (Volk, Raimund; Sohler, Wolfgang)V1362,820-826(1991)

Chemo-optical microsensing systems (Lambeck, Paul V.)V1511,100-113(1991)

Degenerate four-wave mixing using wave pump beams near the critical angle: two distinct behaviors (Malouin, Christian; Song, Li; Thibault, S.; Denariez-Roberge, Marguerite M.; Lessard, Roger A.)V1559,385-392(1991)

Design and fabrication of amplitude modulators for CATV (Mahapatra, Amaresh; Cooper, Ronald F.)V1374,296-299(1991)

Design criteria of an integrated optics microdisplacement sensor (d'Alessandro, Antonio; De Sario, Marco; D'Orazio, Antonella; Petruzzelli, Vincenzo)V1332,554-562(1991)

Design of a GaAs Fresnel lens array for optical signal-processing devices (Armenise, Mario N.; Impagnatiello, Fabrizio; Passaro, Vittorio M.)V1374,86-96(1991)

Design of novel integrated optic devices utilizing depressed index waveguides (Lopez-Amo, Manuel; Menendez-Valdes, Pedro; Sanz, Inmaculada; Muriel, Miguel A.)V1374,74-85(1991)

Design of photonic switches for optimizing performances of interconnection networks (Armenise, Mario N.; Castagnolo, Beniamino)V1374,186-197(1991)

Diffusion and solubility of ion-implanted Nd and Er in LiNbO3 (Buchal, Christoph; Mohr, S.)V1361,881-892(1991)

Engineering fabrication technology of multiplexed hologram in single-mode optical waveguide (Lin, Freddie S.; Chen, Jenkins C.; Nguyen, Cong T.; Liu, William Y.)V1461,39-50(1991)

Fabrication of light-guiding devices and fiber-coupling structures by the LIGA process (Rogner, Arnd; Ehrfeld, Wolfgang)V1506,80-91(1991)

Free-space optical TDM switch (Goel, Kamal K.; Prucnal, Paul R.; Stacy, John L.; Krol, Mark F.; Johns, Steven T.)V1476,314-319(1991)

Grating splitter for glass waveguide (Jin, Guoliang; Shen, Ronggui; Ying, Zaisheng)V1513,50-55(1991)

High-efficient fiber-to-stripe waveguide coupler (Domanski, Andrzej W.; Roszko, Marcin; Sierakowski, Marek W.)V1362,844-852(1991)

High-resolution integrated optic holographic wavelength division multiplexer (Liu, William Y.; Strzelecki, Eva M.; Lin, Freddie S.; Jannson, Tomasz)V1365,20-24(1991)

High-spatial-resolution and high-sensitivity interferometric optical-time-domain reflectometer (Kobayashi, Masaru; Noda, Juichi; Takada, Kazumasa; Taylor, Henry F.)V1474,278-284(1991)

Hybrid wafer scale optoelectronic integration (Lockwood, Harry F.)V1389,55-67(1991); V1390,55-67(1991)

Integrated free-space optics (Jahns, Juergen)V1354,588-592(1991)

Integrated Optical Circuits (Wong, Ka-Kha, ed.)V1583(1991)

Integrated optical device in a fiber gyroscope (Rasch, Andreas; Goering, Rolf; Karthe, Wolfgang; Schroeter, Siegmund; Ecke, Wolfgang; Schwotzer, Guenter; Willsch, Reinhardt)V1511,149-154(1991)

Integrated optical devices for high-data-rate serial-to-parallel conversion (Verber, Carl M.; Kenan, Richard P.; Tan, Ronson K.; Bao, Y.)V1374,68-73(1991)

Integrated optical Mach-Zehnder interferometers in glass (Lefebvre, P.; Vahid-Shahidi, A.; Albert, Jacques; Najafi, S. I.)V1583,221-225(1991)

Integrated-optical modulators for bandpass analog links (Johnson, Leonard M.; Betts, Gary E.; Roussell, Harold V.)V1371,2-7(1991)

Integrated optical preamplifier technology for optical signal processing and optical communication systems (Eichen, Elliot G.; Powazinik, William; Meland, Edmund; Bryant, R.; Rideout, William C.; Schlafer, John; Lauer, Robert B.)V1474,260-267(1991)

Integrated optic components for advanced turbine engine control systems (Emo, Stephen M.; Kinney, Terrance R.; Wong, Ka-Kha)V1374,266-276(1991)

Integrated optic device for biochemical sensing (Boiarski, Anthony A.; Ridgway, Richard W.; Miller, Larry S.; Bhullar, B. S.)V1368,264-272(1991)

Integrated optic flat antennae: early applications and design tools (Parriaux, Olivier M.; Sychugov, V. A.)V1583,376-382(1991)

Integrated Optics and Optoelectronics II (Wong, Ka-Kha, ed.)V1374(1991)

Integrated optics bus access module for intramachine communication (Karioja, Pentti; Tammela, Simo; Hannula, Tapio)V1533,129-137(1991)

Integrated optics displacement sensor (d'Alessandro, Antonio; De Sario, Marco; D'Orazio, Antonella; Petruzzelli, Vincenzo)V1366,313-323(1991)

Integrated optics for fiber optic sensors (Minford, William J.; DePaula, Ramon P.)V1367,46-52(1991)

Integrated optics in optical engineering (Popov, Yury V.)V1399,207-213(1991)

Integrated optics intensity modulators in the GaAs/AlGaAs system (Khan, M. A.; Naumaan, Ahmed; Van Hove, James M.)V1396,753-759(1991)

Integrated optics sensor on silicon for the measurement of displacement, force, and refractive index (Ulbers, Gerd)V1506,99-110(1991)

Integrated optics temperature sensor (d'Alessandro, Antonio; De Sario, Marco; D'Orazio, Antonella; Petruzzelli, Vincenzo)V1399,184-191(1991)

Ion-exchanged waveguides: current status (Srivastava, Ramakant; Ramaswamy, Ramu V.)V1583,2-13(1991)

Ion-exchanged waveguides in glass: simulation and experiment (Wolf, Barbara; Fabricius, Norbert; Foss, Wolfgang; Dorsel, Andreas N.)V1506,40-51(1991)

Issues affecting the characterization of integrated optical devices subjected to ionizing radiation (Hickernell, Robert K.; Sanford, N. A.; Christensen, David H.)V1474,138-147(1991)

Laser-induced etched grating on InP for integrated optical circuit elements (Grebel, Haim; Pien, P.)V1583,331-337(1991)

Length-minimization design considerations in photonic integrated circuits incorporating directional couplers (Boyd, Joseph T.; Radens, Carl J.; Kauffman, Michael T.)V1374,138-143(1991)

LiNbO3 with rare earths: lasers and amplifiers (Lallier, Eric; Pocholle, Jean-Paul; Papuchon, Michel R.; De Micheli, Marc; Li, M. J.; He, Q.; Ostrowsky, Daniel B.; Grezes-Besset, C.; Pelletier, Emile P.)V1506,71-79(1991)

LiTaO3 and LiNbO3:Ti responses to ionizing radiation (Padden, Richard J.; Taylor, Edward W.; Sanchez, Anthony D.; Berry, J. N.; Chapman, S. P.; DeWalt, Steve A.; Wong, Ka-Kha)V1474,148-159(1991)

Lithium niobate proton-exchange technology for phase-amplitude modulators (Varasi, Mauro; Vannucci, Antonello; Signorazzi, Mario)V1583,165-169(1991)

Lithium niobate waveguide devices: present performance and future applications (Bulmer, Catherine H.)V1583,176-183(1991)

Low-coherence optical reflectometry of laser diode waveguides (Boisrobert, Christian Y.; Franzen, Douglas L.; Danielson, Bruce L.; Christensen, David H.)V1474,285-290(1991)

Low-loss L-band microwave fiber optic link for control of a T/R module (Wanuga, Stephen; Ackerman, Edward I.; Kasemset, Dumrong; Hogue, David W.; Chinn, Stephen R.)V1374,97-106(1991)

Low-loss waveguides on silicon substrates for photonic circuits (Davis, Richard L.; Lee, Sae H.)V1474,20-26(1991)

Low-loss Y-couplers for fiber optic gyro applications (Page, Jerry L.)V1374,287-293(1991)

Magneto-optical reading and writing integrated heads: a way to a multigigabyte multi-rigid-disk drive (Renard, Stephane; Valette, Serge)V1499,238-247(1991)

Mass-producible optical guided-wave devices fabricated by photopolymerization (Hosokawa, Hayami; Horie, Noriyoshi; Yamashita, Tsukasa)V1559,229-237(1991)

Micromachined structure for coupling optical fibers to integrated optical waveguides (Goel, Sanjay; Naylor, David L.)V1396,404-410(1991)

Micro-Optics II (Scheggi, Annamaria V., ed.)V1506(1991)

Modeling of traveling wave Ti:LiNbO3 waveguide modulators (Rocca, Corrado; Montrosset, Ivo; Gollinucci, Stefano; Ghione, Giovanni)V1372,39-47(1991)

Monolithic integration of a semiconductor ring laser and a monitoring photodetector (Krauss, Thomas; Laybourn, Peter J.)V1583,150-152(1991)

Nd- and Er-doped glass integrated optical amplifiers and lasers (Najafi, S. I.; Wang, Wei-Jian; Orcel, Gerard F.; Albert, Jacques; Honkanen, Seppo; Poyhonen, Pekka; Li, Ming-Jun)V1583,32-36(1991)

New integrated optic TE/TM splitter made on LiNbO3 isotropic cut (Duchet, Christian; Flaaronning, Nils; Brot, Christian; Sarrabay, Laurence)V1372,72-81(1991)

Novel integrated acousto-optical LiNbO3 device for application in single-laser self-heterodyne systems (Gieschen, Nikolaus; Rocks, Manfred; Olivier, Lutz)V1579,237-248(1991)

Novel technique for analysis and design of diffused Ti:LiNbO3 optical planar waveguides (Lopez-Higuera, Jose M.; Lopez-Amo, Manuel; Muriel, Miguel A.)V1374,144-151(1991)

Novel technique for the measurement of coupling length in directional couplers (Cheng, Hsing C.; Ramaswamy, Ramu V.)V1474,291-297(1991)

Optical carrier modulation by integrated optical devices in lithium niobate (Rasch, Andreas; Buss, Wolfgang; Goering, Rolf; Steinberg, Steffen; Karthe, Wolfgang)V1522,83-92(1991)

Optical interconnects: a solution to very high speed integrated circuits and systems (Chen, Ray T.)V1374,162-175(1991)

Optical Technology for Microwave Applications V (Yao, Shi-Kay, ed.)V1476(1991)

Optical Technology for Signal Processing Systems (Bendett, Mark P., ed.)V1474(1991)

Optimization of externally modulated analog optical links (Betts, Gary E.; Johnson, Leonard M.; Cox, Charles H.)V1562,281-302(1991)

Organic electro-optic devices for optical interconnnection (Lipscomb, George F.; Lytel, Richard S.; Ticknor, Anthony J.; Van Eck, Timothy E.; Girton, Dexter G.; Ermer, Susan P.; Valley, John F.; Kenney, John T.; Binkley, E. S.)V1560,388-399(1991)

Passive integrated optics in optical sensor systems (Parriaux, Olivier M.)V1506,111-119(1991)

Passive optical silica-on-silicon waveguide components (McGoldrick, Elizabeth; Hubbard, Steven D.; Maxwell, Graeme D.; Thomas, N.)V1374,118-125(1991)

Phase-bias tuning and extinction ratio improvement of Mach-Zehnder interferometer (Baranov, D. V.; Zolotov, Evgeny M.; Pelekhaty, V. M.; Tavlykaev, R. F.)V1583,389-394(1991)

Photochemical formation of polymeric optical waveguides and devices for optical interconnection applications (Beeson, Karl W.; Horn, Keith A.; McFarland, Michael J.; Wu, Chengjiu; Yardley, James T.)V1374,176-185(1991)

Photoneuron: dynamically reconfigurable information processing control element utilizing embedded-fiber waveguide interconnects (Glista, Andrew S.)V1563,139-155(1991)

Potential for integrated optical circuits in advanced aircraft with fiber optic control and monitoring systems (Baumbick, Robert J.)V1374,238-250(1991)

Propagation and diffusion characteristics of optical waveguides made by electric field-assisted k+-ion exchange in glass (Noutsios, Peter C.; Yip, Gar L.; Kishioka, Kiyoshi)V1374,14-22(1991)

Proton-diffused channel waveguides on Y-cut LiNbO3 using a self-aligned SiO2-cap diffusion method (Son, Yung-Sung; Lee, Hyung-Jae; Yi, Sang-Yoon; Shin, Sang-Yung)V1374,23-29(1991)

Proton exchange in LiTaO3 with different stoichiometric composition (Savatinova, Ivanka T.; Kuneva, M.; Levi, Zelma; Atuchin, V.; Ziling, K.; Armenise, Mario N.)V1374,37-46(1991)

Proton-exchange X-cut lithium tantalate fiber optic gyro chips (Wong, Ka-Kha; Killian, Kevin M.; Dimitrov-Kuhl, K. P.; Long, Margaret; Fleming, J. T.; van de Vaart, Herman)V1374,278-286(1991)

Radiation effects on dynamical behavior of LiNbO3 switching devices (Kanofsky, Alvin S.; Minford, William J.; Watson, James E.)V1374,59-66(1991)

Radiation effects on GaAs optical system FET devices (Kanofsky, Alvin S.; Spector, Magaly; Remke, Ronald L.; Witmer, Steve B.)V1374,48-58(1991)

Recent development of proton-exchanged waveguides and devices in lithium niobate using phosphoric acid (Pun, Edwin Y.)V1374,2-13(1991)

Silica optical integrated devices (Kobayashi, Soichi; Sumida, Shin S.; Miyashita, Tadashi M.)V1374,300-303(1991)

Silicon modulator for integrated optics (Cocorullo, Giuseppe; Della Corte, Francesco G.; Rendina, Ivo; Cutolo, Antonello)V1374,132-137(1991)

Silicon/PVDF integrated double detector: application to obstacle detection in automotive (Simonne, John J.; Pham, Vui V.; Esteve, Daniel; Alaoui-Amine, Mohammed; Bousbiat, Essaid)V1374,107-115(1991)

Simple and accurate technique to determine substrate indices by the multilayer Brewster angle measurement (Xiang, Feng; Yip, Gar L.)V1583,271-277(1991)

Simulation and design of integrated optical waveguide devices by the BPM (Yip, Gar L.)V1583,240-248(1991)

Spectrometer on a chip: InP-based integrated grating spectrograph for wavelength-multiplexed optical processing (Soole, Julian B.; Scherer, Axel; LeBlanc, Herve P.; Andreadakis, Nicholas C.; Bhat, Rajaram; Koza, M. A.)V1474,268-276(1991)

State-of-the-art multichip modules for avionics (Hagge, John K.)V1390,175-194(1991)

Symmetry properties of reverse-delta-beta directional couplers (Smith, Terrance L.; Misemer, David K.; Attanasio, Daniel V.; Crow, Gretchen L.; Smith, Wiley K.; Swierczek, Mary J.; Watson, James E.)V1512,92-100(1991)

Synthetic holographic beamsplitters for integrated optics (Saarinen, Jyrki V.; Huttunen, Juhani; Vasara, Antti H.; Noponen, Eero; Turunen, Jari P.; Salin, Arto U.)V1555,128-137(1991)

Techniques for integrating 3-D optical systems (Brenner, Karl-Heinz)V1544,263-270(1991)

Temperature and modulation dependence of spectral linewidth in distributed Bragg reflector laser diodes (Suzaki, Shinzoh; Ishii, Masanori; Watanabe, Tsutomu; Shiota, Alan T.)V1374,126-131(1991)

Three-dimensional integration of optical systems (Brenner, Karl-Heinz)V1506,94-98(1991)

Ultralinear electro-optic modulators for microwave fiber-optic communications (Pan, J. J.; Li, Wei Z.; Li, Yi Q.)V1476,32-43(1991)

Use of 3 X 3 integrated optic polarizer/splitters for a smart aerospace plane structure (Seshamani, Ramani; Alex, T. K.)V1489,56-64(1991)

Use of optical sensors and signal processing in gas turbine engines (Davinson, Ian)V1374,251-265(1991)

Wavelength dependence of proton-exchanged LiNbO3 integrated optic directional couplers from 1.5um - 1.65um (Feuerstein, Robert J.; Januar, Indra; Mickelson, Alan R.; Sauer, Jon R.)V1583,196-201(1991)

Integrated optoelectronics—see also gallium arsenide materials; integrated optics; semiconductors

Architecture of an integrated computer-aided design system for optoelectronics (Kiamilev, Fouad E.; Fan, J.; Catanzaro, Brian E.; Esener, Sadik C.; Lee, Sing H.)V1390,311-329(1991)

Asymmetric superlattices for microwave detection (Syme, Richard T.; Kelly, Michael J.; Condie, Angus; Dale, Ian)V1362,467-476(1991)

Bridge-type optoelectronic sample and hold circuit (Sun, C. K.; Wu, Chao-Chia C.; Chang, Ching T.; Yu, Paul K.; McKnight, William H.)V1476,294-300(1991)

Buried-ridge-stripe lasers monolithically integrated with butt-coupled passive waveguides for OEIC (Remiens, D.; Hornung, V.; Rose, B.; Robein, D.)V1362,323-330(1991)

Compact joint transform correlators in planar-integrated packages (Ghosh, Anjan K.)V1564,231-235(1991)

Compact magneto-optical disk head integrated with chip elements (Yamanaka, Yutaka; Katayama, Ryuichi; Komatsu, Y.; Ishikawa, S.; Itoh, M.; Ono, Yuzo)V1499,263-268(1991)

Compact PIN-amplifier module for giga bit rates optical interconnection (Suzuki, Tomihiro; Mikamura, Yasuki; Murata, Kazuo; Sekiguchi, Takeshi; Shiga, Nobuo; Murakami, Yasunori)V1389,455-461(1991)

Current status and future research of the Delft 'supercomputer' project (Frietman, Edward E.; Dekker, L.; van Nifterick, W.; Demeester, Piet M.; van Daele, Peter; Smit, W.)V1390,434-453(1991)

Current technologies for very high performance VLSI ICs (Perea, Ernesto H.)V1362,477-483(1991)

Design, fabrication, and performance of an integrated optoelectronic cellular array (Hibbs-Brenner, Mary K.; Mukherjee, Sayan D.; Skogen, J.; Grung, B.; Kalweit, Edith; Bendett, Mark P.)V1563,10-20(1991)

Design optimization of a 10-amplifier coherent array (Zediker, Mark S.; Foresi, James S.; Haake, John M.; Heidel, Jeffrey R.; Williams, Richard A.; Driemeyer, D.; Blackwell, Richard J.; Thomas, G.; Priest, J. A.; Herrmann, Sandy)V1418,309-315(1991)

Device concepts for SiGe optoelectronics (Kasper, Erich; Presting, Hartmut)V1361,302-312(1991)

Devices for Optical Processing (Gookin, Debra M., ed.)V1562(1991)

Devices for optoelectronic integrated systems (Drabik, Timothy J.)V1562,194-203(1991)

Diamond multichip modules (McSheery, Tracy D.)V1563,21-26(1991)

Direct optical phase shifter for phased-array systems (Vawter, Gregory A.; Hietala, Vincent M.; Kravitz, Stanley H.; Meyer, W. J.)V1476,102-106(1991)

Effects of optoelectronic device characteristics on the performance and design of POEM systems (Esener, Sadik C.; Kiamilev, Fouad E.; Krishnamoorthy, Ashok V.; Marchand, Philippe J.)V1562,11-20(1991)

Electronic/photonic inversion channel technology for optoelectronic ICs and photonic switching (Taylor, Geoff W.; Cooke, Paul W.; Kiely, Philip A.; Claisse, Paul R.; Sargood, Stephen K.; Doctor, D. P.; Vang, T.; Evaldsson, P.; Daryanani, Sonu L.)V1476,2-13(1991)

Electronics/photonics technology: vision and reality (Ross, Ian M.)V1389,2-26(1991); V1390,2-26(1991)

Electro-optic polymer devices for optical interconnects (Lytel, Richard S.; Binkley, E. S.; Girton, Dexter G.; Kenney, John T.; Lipscomb, George F.; Ticknor, Anthony J.; Van Eck, Timothy E.)V1389,547-558(1991)

Epitaxial liftoff technology (Yablonovitch, Eli)V1563,8-9(1991)

Flying head read/write characteristics using a monolithically integrated laser diode/photodiode at a wavelength of 1.3 um (Ukita, Hiroo; Katagiri, Yoshitada; Nakada, Hiroshi)V1499,248-262(1991)

Free-space board-to-board optical interconnections (Tsang, Dean Z.)V1563,66-71(1991)

Free-space optical interconnect using microlasers and modulator arrays (Dickinson, Alex G.; Downs, Maralene M.; LaMarche, R. E.; Prise, Michael E.)V1389,503-514(1991)

GaAs/GaAlAs integrated optoelectronic transmitter for microwave applications (Yap, Daniel; Narayanan, Authi A.; Rosenbaum, Steven E.; Chou, Chia-Shing; Hooper, W. W.; Quen, R. W.; Walden, Robert H.)V1418,471-476(1991)

Gas sensor based on an integrated optical interferometer (Brandenburg, Albrecht; Edelhaeuser, Rainer; Hutter, Frank)V1510,148-159(1991)

Generation and sampling of high-repetition-rate/high-frequency electrical waveforms in microstrip circuits by picosecond optoelectronic technique (Lee, Chi H.)V1390,377-387(1991)

High-density waveguide modulator arrays for parallel interconnection (Bristow, Julian P.; Mukherjee, Sayan D.; Khan, M. N.; Hibbs-Brenner, Mary K.; Sullivan, Charles T.; Kalweit, Edith)V1389,535-546(1991)

High-speed optical interconnects for parallel processing and neural networks (Barnes, Nigel; Healey, Peter; McKee, Paul; O'Neill, Alan; Rejman-Greene, Marek A.; Scott, Geoff; Smith, David W.; Webb, Roderick P.; Wood, David)V1389,477-483(1991)

Holographic optical backplane for boards interconnection (Sebillotte-Caron, Claudine)V1389,600-611(1991)

Holographic optical elements for free-space clock distribution (Zarschizky, Helmut; Karstensen, Holger; Gerndt, Christian; Klement, Ekkehard; Schneider, Hartmut W.)V1389,484-495(1991)

Holographic optical interconnects for multichip modules (Feldman, Michael R.)V1390,427-433(1991)

Hybrid wafer scale optoelectronic integration (Lockwood, Harry F.)V1389,55-67(1991); V1390,55-67(1991)

III-V monolithic resonant photoreceiver on silicon substrate for long-wavelength operation (Aboulhouda, S.; Razeghi, Manijeh; Vilcot, Jean-Pierre; Decoster, Didier; Francois, M.; Maricot, S.)V1362,494-498(1991)

III-V semiconductor integrated optoelectronics for optical computing (Wada, Osamu)V1362,598-607(1991)

Improvements in self electro-optic effect devices: toward system implementation (Morgan, Robert A.)V1562,213-227(1991)

InGaAs/InP monolithic photoreceivers for 1.3-1.5 um optical fiber transmission (Scavennec, Andre; Billard, M.; Blanconnier, P.; Caquot, E.; Carer, P.; Giraudet, Louis; Nguyen, L.; Lugiez, F.; Praseuth, Jean-Pierre)V1362,331-337(1991)

Integrated micro-optics for computing and switching applications (Jahns, Juergen)V1544,246-262(1991)

Integrated optical channel waveguides in silicon using SiGe alloys (Splett, Armin O.; Schmidtchen, Joachim; Schueppert, B.; Petermann, Klaus)V1362,827-833(1991)

Integrated optical devices with silicon oxynitride prepared by plasma-enhanced chemical vapor deposition (PECVD) on Si and GaAs substrates (Peters, Dethard; Mueller, Joerg)V1362,338-349(1991)

Integrated packaging of optical backplanes (Jahns, Juergen)V1389,523-526(1991)

Integration of a coherent optical receiver with adaptive image rejection capability (Lachs, Gerard; Zaidi, Syed M.; Singh, Amit K.; Henning, Rudolf E.; Trascritti, D.; Kim, H.; Bhattacharya, Pallab K.; Pamulapati, J.; McCleer, P. J.; Haddad, George I.; Peng, S.)V1474,248-259(1991)

Interactive Design and Electro-optic Analysis Liaise (IDEAL) (Alukaidey, Talib A.; Pettiford, Alvin A.)V1390,513-522(1991)

Intercircuit optical interconnects using quantum-well modulators (Goodfellow, Robert C.; Goodwin, Martin J.; Moseley, Andrew J.)V1389,594-599(1991)

Interconnection requirements in avionic systems (Vergnolle, Claude; Houssay, Bruno)V1389,648-658(1991)

Laser Diode Technology and Applications III (Renner, Daniel, ed.)V1418(1991)

Laser diodes with gain-coupled distributed optical feedback (Nakano, Yoshiaki; Luo, Yi; Tada, Kunio)V1418,250-260(1991)

Liquid-crystal devices in planar optics (Armitage, David; Ticknor, Anthony J.)V1455,206-212(1991)

Long-wavelength lasers and detectors fabricated on InP/GaAs superheteroepitaxial wafer (Aiga, Masao; Omura, Etsuji E.)V1418,217-222(1991)

Low-threshold room temperature continuous-wave operation of a GaAs single-quantum-well mushroom structure surface-emitting laser (Yang, Y. J.; Dziura, T. G.; Wang, S. C.; Fernandez, R.; Du, G.; Wang, Shyh)V1418,414-421(1991)

Massive connectivity and SEEDs (Chirovsky, Leo M.)V1562,228-241(1991)

Massive optical interconnections for computer applications (Neff, John A.)V1389,27-38(1991); V1390,27-38(1991)

Methods for comparative analysis of waveform degradation in electrical and optical high-performance interconnections (Merkelo, Henri; McCredie, B. D.; Veatch, M. S.; Quinn, D. L.; Dorneich, M.; Doi, Yutaka)V1390,91-163(1991)

Microelectronic Interconnects and Packages: Optical and Electrical Technologies (Arjavalingam, Gnanalingam; Pazaris, James, eds.)V1389(1991)

Microelectronic Interconnects and Packages: System and Process Integration (Carruthers, John R.; Tewksbury, Stuart K., eds.)V1390(1991)

Microreflective elements for integrated planar optics interconnects (Sheng, Yunlong; Delisle, Claude; Moreau, Louis; Song, Li; Lessard, Roger A.; Arsenault, Henri H.)V1559,222-228(1991)

Microwave control using a high-gain bias-free optoelectronic switch (Freeman, J. L.; Ray, Sankar; West, David L.; Thompson, Alan G.; LaGasse, M. J.)V1476,320-325(1991)

Millimeter-wave and optoelectronic applications of heterostructure integrated circuits (Pavlidis, Dimitris)V1362,450-466(1991)

Mirror fabrication for full-wafer laser technology (Webb, David J.; Benedict, Melvin K.; Bona, Gian-Luca; Buchmann, Peter L.; Daetwyler, K.; Dietrich, H. P.; Moser, A.; Sasso, G.; Vettiger, Peter; Voegeli, O.)V1418,231-239(1991)

Mixed application MMIC technologies: progress in combining RF, digital, and photonic circuits (Swirhun, S.; Bendett, Mark P.; Sokolov, Vladimir; Bauhahn, Paul E.; Sullivan, Charles T.; Mactaggart, R.; Mukherjee, Sayan D.; Hibbs-Brenner, Mary K.; Mondal, J. P.)V1475,223-230(1991)

Molecular beam epitaxy GaAs on Si: material and devices for optical interconnects (Panayotatos, Paul; Georgakilas, Alexandros; Mourrain, Jean-Loic; Christou, Aristos)V1361,1100-1109(1991)

Multiprocessor optical bus (Kostuk, Raymond K.; Huang, Yang-Tung; Kato, Masayuki)V1389,515-522(1991)

Novel distributed feedback structure for surface-emitting semiconductor lasers (Mahbobzadeh, Mohammad; Osinski, Marek A.)V1418,25-31(1991)

Optical approaches to overcome present limitations for interconnection and control in parallel electronic architectures (Maurin, T.; Devos, F.)V1505,158-165(1991)

Optical Enhancements to Computing Technology (Neff, John A., ed.)V1563(1991)

Optically coupled 3-D common memory with GaAs on Si structure (Hirose, Masataka; Takata, H.; Koyanagi, Mitsumasa)V1362,316-322(1991)

Optically coupled three-dimensional memory system (Koyanagi, Mitsumasa)V1390,467-476(1991)

Optical receivers in ECL for 1 GHz parallel links (Wieland, Joerg B.; Melchior, Hans M.)V1389,659-664(1991)

Optical switches based on semiconductor optical amplifiers (Kalman, Robert F.; Dias, Antonio R.; Chau, Kelvin K.; Goodman, Joseph W.)V1563,34-44(1991)

Optical techniques for microwave monolithic circuit characterization (Hung, Hing-Loi A.; Li, Ming-Guang; Lee, Chi H.)V1476,276-281(1991)

Optical Technology for Microwave Applications V (Yao, Shi-Kay, ed.)V1476(1991)

Optoelectronically implemented neural network for early visual processing (Mehanian, Courosh; Aull, Brian F.; Nichols, Kirby B.)V1469,275-280(1991)

Optoelectronic approach to optical parallel processing based on the photonic parallel memory (Matsuda, Kenichi; Shibata, Jun)V1562,21-29(1991)

Optoelectronic compare-and-exchange switches based on BILED circuits (Mao, Xianjun; Liu, Shutian; Wang, Ruibo)V1563,58-63(1991)

Optoelectronic master chip for optical computing (Lang, Robert J.; Kim, Jae H.; Larsson, Anders G.; Nouhi, Akbar; Cody, Jeffrey; Lin, Steven H.; Psaltis, Demetri; Tiberio, Richard C.; Porkolab, Gyorgy A.; Wolf, Edward D.)V1563,2-7(1991)

Optoelectronic Materials and Device Concepts (Razeghi, Manijeh)VPM05(1991)

Optoelectronic neuron (Pankove, Jacques I.; Radehaus, C.; Borghs, Gustaaf)V1361,620-627(1991)

Optoelectronic neuron arrays (Psaltis, Demetri; Lin, Steven H.)V1562,204-212(1991)

Optoelectronic sampling receiver for time-division multiplexed signal processing applications (Yaseen, Mohammed; Walker, Stuart D.)V1562,319-326(1991)

Packaging issues for free-space optically interconnected multiprocessors (Ozguz, Volkan H.; Esener, Sadik C.; Lee, Sing H.)V1390,477-488(1991)

Photodetectors: how to integrate them with microelectronic and optical devices (Decoster, Didier; Vilcot, Jean-Pierre)V1362,959-966(1991)

Photonic Multichip Packaging (PMP) using electro-optic organic materials and devices (McDonald, John F.; Vlannes, Nickolas P.; Lu, Toh-Ming; Wnek, Gary E.; Nason, Theodore C.; You, Lu)V1390,286-301(1991)

Physical Concepts of Materials for Novel Optoelectronic Device Applications II: Device Physics and Applications (Razeghi, Manijeh, ed.)V1362(1991)

Pipelined communications on optical busses (Guo, Zicheng; Melhem, Rami G.; Hall, Richard W.; Chiarulli, Donald M.; Levitan, Steven P.)V1390,415-426(1991)

Proposed electro-optic package with bidirectional lensed coupling (Rajasekharan, K.; Michalka, Timothy)V1389,568-579(1991)

Quantum-well devices for optics in digital systems (Miller, David A.)V1389,496-502(1991)

Refractive index of multiple-quantum-well waveguides subject to impurity induced disordering using boron and fluorine (Hansen, Stein I.; Marsh, John H.; Roberts, John S.; Jeynes, C.)V1362,361-369(1991)

Ring resonators for microwave optoelectronics (Gopalakrishnan, G. K.; Fairchild, B. W.; Yeh, C. L.; Park, C. S.; Chang, Kai; Weichold, Mark H.; Taylor, Henry F.)V1476,270-275(1991)

Self-routing interconnection structures using coincident pulse techniques (Chiarulli, Donald M.; Levitan, Steven P.; Melhem, Rami G.)V1390,403-414(1991)

Surface-acoustic-wave acousto-optic devices for wide-bandwidth signal processing and switching applications (Garvin, Charles G.; Sadler, Brian M.)V1562,303-318(1991)

Temperature stability of Bragg grating resonant frequency (Bradley, Eric M.; Rybka, Theodore W.; Yu, Paul K.)V1418,272-278(1991)

Time-division optical microarea networks (Prucnal, Paul R.; Johns, Steven T.; Krol, Mark F.; Stacy, John L.)V1389,462-476(1991)

Two new developments for optoelectronic bus systems (Jiang, Jie; Kraemer, Udo)V1505,166-174(1991)

Vertical 3-D integration of silicon waveguides in a Si-SiO2-Si-SiO2-Si structure (Soref, Richard A.; Namavar, Fereydoon; Cortesi, Elisabetta; Friedman, Lionel R.; Lareau, Richard)V1389,408-421(1991)

Vertical-cavity surface-emitters for optoelectronic integration (Geels, Randall S.; Corzine, Scott W.; Coldren, Larry A.)V1418,46-56(1991)

Vertical-cavity surface-emitting lasers and arrays for optical interconnect and optical computing applications (Wang, S. C.; Dziura, T. G.; Yang, Y. J.)V1563,27-33(1991)

Vertical-cavity surface-emitting semiconductor lasers: present status and future prospects (Osinski, Marek A.)V1418,2-24(1991)

Waveguiding characteristics in polyimide films with different chemistry of formation (Chakravorty, Kishore K.; Chien, Chung-Ping)V1389,559-567(1991)

Intensifiers—see also image tubes

Advanced image intensifier systems for low-light high-speed imaging (Bruno, Theresa L.; Wirth, Allan)V1358,109-116(1991)

Clinical image-intensifier-based volume CT imager for angiography (Ning, Ruola; Barsotti, John B.; Kido, Daniel K.; Kruger, Robert A.)V1443,236-249(1991)

CT imaging with an image intensifier: using a radiation therapy simulator as a CT scanner (Silver, Michael D.; Nishiki, Masayuki; Tochimura, Katsumi; Arita, Masataka; Drawert, Bruce M.; Judd, Thomas C.)V1443,250-260(1991)

Electron Image Tubes and Image Intensifiers II (Csorba, Illes P., ed.)V1449(1991)

Fourth-generation motion analyzer (Balch, Kris S.)V1358,373-397(1991)

Gatable photonic detector and its image processing (Kume, Hidehiro; Nakamura, Haruhito; Suzuki, Makoto)V1358,1144-1155(1991)

Hybrid phototube with Si target (van Geest, Lambertus K.; Stoop, Karel W.)V1449,121-134(1991)

Intensified multispectral imaging measuring in the spatial, frequency, and time domains with a single instrument (Kennedy, Benjamin J.)V1346,68-74(1991)

Light-pulse-induced background in image intensifiers with MCPs (Dashevsky, Boris E.)V1449,25-29(1991)

Low-cost computed tomography system using an image intensifier (Hoeft, Gregory L.; Hughes, Simon H.; Slocum, Robert E.)V1396,638-645(1991)

Low light level imaging systems application considerations and calculations (Caudle, Dennis E.)V1346,54-63(1991)

MCP image intensifier in the 100-KeV to 1-MeV x-ray range (Veaux, Jacqueline; Cavailler, Claude; Gex, Jean-Pierre; Hauducoeur, Alain; Hivernage, M.)V1449,13-24(1991)

MIC photon counting detector (Fordham, John L.; Bellis, J. G.; Bone, David A.; Norton, Timothy J.)V1449,87-98(1991)

Method of estimating light loadings power and repetition frequency effects on the image quality of gated image intensifier tubes (Dashevsky, Boris E.)V1449,65-70(1991)

New method for doing flat-field intensity calibrations of multiplexed ITT Streak Cameras (Hugenberg, Keith F.)V1346,390-397(1991)

Noise performance of microchannel plate imaging systems (McCammon, Kent G.; Hagans, Karla G.; Hankla, A.)V1346,398-403(1991)

Optimization and evaluation of an image intensifier TV system for digital chest imaging (Angelhed, Jan-Erik; Mansson, Lars G.; Kheddache, Susanne)V1444,159-170(1991)

Picosecond intensifier gating with a plated webbing cathode underlay (Thomas, Stan W.; Trevino, Jimmy)V1358,84-90(1991)

Practical videography (Sturz, Richard A.)V1346,64-67(1991)

Simulation of scenes, sensors, and tracker for space-based acquisition, tracking, and pointing experiments (DeYoung, David B.)V1415,13-21(1991)

Streak camera phosphors: response to ultrashort excitation (Jaanimagi, Paul A.; Hestdalen, C.)V1346,443-448(1991)

Subnanosecond intensifier gating using heavy and mesh cathode underlays (Thomas, Stan W.; Shimkunas, Alex R.; Mauger, Philip E.)V1358,91-99(1991)

Target acquisition modeling based on human visual system performance (Valeton, J. M.; van Meeteren, Aart)V1486,68-84(1991)

Techniques for capturing over 10,000 images/second with intensified imagers (Balch, Kris S.)V1358,358-372(1991)

Ultraviolet light imaging technology and applications (Yokoi, Takane; Suzuki, Kenji; Oba, Koichiro)V1449,30-39(1991)

Veiling glare in the F4111 image intensifier (Acharya, Mukund; Bunch, Robert M.)V1396,377-388(1991)

Interfaces—see also multilayers

Enhanced backscattering of s- and p-polarized light from particles above a substrate (Greffet, Jean-Jacques; Sentenac, Anne)V1558,288-294(1991)

Interface cracks close to free surfaces: a caustic study (Rossmanith, H. P.; Beer, Rudolf J.; Knasmillner, R. E.)V1554A,850-860(1991)

Interface properties of a-C:H/a-SiOx:H multilayer (Cui, Jing B.; Zhang, Wei P.; Fang, Rong C.; Wang, Chang S.; Zhou, Guien; Wu, Jan X.)V1519,419-422(1991)

Interface stress at thin film semiconductor heterostructures (Nishino, Taneo)V1519,382-390(1991)

Multimedia courseware in an open-systems environment: a DoD strategy (Welsch, Lawrence A.)VCR37,207-220(1991)

Rough-interface scattering without plane waves (Berman, David H.)V1558,191-201(1991)

Study of HgCdTe/CdTe interface structure grown by metal-organic chemical vapor deposition (Ma, Ke J.; Yu, Zhen Z.; Yanh, Jian R.; Shen, Shou Z.; He, Jin; Chen, Wei M.; Song, Xiang Y.)V1519,489-493(1991)

Study of photoelectric transformation process at p-PAn/n-Si interface (Yuan, Ren K.; Liu, Yu X.; Yuan, Hong; Wang, Yong B.; Zheng, Xiang Q.; Xu, Jian)V1519,396-399(1991)

Study of stick slip behavior in interface friction using optical fiber pull-out experiment (Tsai, Kun-Hsieh; Kim, Kyung-Suk)V1554A,529-541(1991)

User interfaces for automated visual inspection systems (Waltz, Frederick M.)VCR36,105-137(1991)

Interference

90 degree optical hybrid for coherent receivers (Garreis, Reiner)V1522,210-219(1991)

Chromeless phase-shifted masks: a new approach to phase-shifting masks (Toh, Kenny K.; Dao, Giang T.; Singh, Rajeev; Gaw, Henry T.)V1496,27-53(1991)

Coordinate measuring system for 2-D scanners (Bukatin, Vladimir V.)V1454,283-288(1991)

Evaluation of interference patterns: a comparison of methods (Kreis, Thomas M.; Geldmacher, Juergen)V1554B,718-724(1991)

Interference effects and the occurrence of blind spots in coherent optical processors (Christie, Simon; Cai, Xian-Yang; Kvasnik, Frank)V1507,202-209(1991)

Interference grating with an arbitrary transparent-to-opaque groove ratio (Tavassoly, M. T.)V1527,392-399(1991)

Interference visualization of infrared images (Vlasov, Nikolai G.; Korchazhkin, S. V.; Manikalo, V. V.)V1358,1018-1020(1991)

Interferential diagnosis of self-focusing of Q-switched YAG laser in liquid (Lu, Jian-Feng; Wang, Chang Xing; Miao, Peng-Cheng; Ni, Xiao W.; He, Anzhi)V1415,220-224(1991)

Novel configuration of a two-mode-interference polarization splitter with a buffer layer (Antuofermo, Pasquale; Losacco, Aurora M.; De Pascale, Olga)V1583,143-149(1991)

Optical nondestructive examination for honeycomb structure (Zhu, Jian T.)V1554B,774-784(1991)

Spontaneous emission noise reduction of a laser output by extracavity destructive interference (Diels, Jean-Claude; Lai, Ming)V1376,198-205(1991)

Synthesis of large-aperture interference fields (Turukhano, Boris G.)V1500,305-308(1991)

Whenever two beams interfere, one fringe equals one wave in the plane of interference, always (Williamson, Ray)V1527,252-257(1991)

Interferometry—see also fiber optic sensors; holography; moire; nondestructive testing; photomechanics

About investigation of absolute surface relief by holographic method (Rachkovsky, Leonid I.; Tanin, Leonid V.; Rubanov, Alexander S.)V1461,232-240(1991)

Absolute interferometer for manufacturing applications (Tucker, Michael R.; Christenson, Eric)V1367,289-299(1991)

Absolute interferometric testing of spherical surfaces (Truax, Bruce E.)V1400,61-68(1991)

Absolute measurement of spherical surfaces (Creath, Katherine; Wyant, James C.)V1332,2-7(1991)

Absolute phasing of segmented mirrors using the polarization phase sensor (Klumpe, Herbert W.; Lajza-Rooks, Barbara A.; Blum, James D.)V1398,95-106(1991)

Accumulating displacement fields from different steps in laser or white-light speckle methods (Shao, C. A.; King, H. J.; Wang, Yeong-Kang; Chiang, Fu-Pen)V1554A,613-618(1991)

Accuracy and high-speed technique for autoprocessing of Young's fringes (Chen, Wenyi; Tan, Yushan)V1554A,879-885(1991)

Active stabilization of interferometers by two-frequency phase modulation (Gorelik, Vladimir S.; Kovalenko, Sergey N.; Turukhano, Boris G.)V1507,379-382(1991)

Active vibration filtering: application to an optical delay line for stellar interferometer (Koehler, Bertrand; Bourlon, Philippe M.; Dugue, Michel)V1489,177-188(1991)

Aerodynamic interferograms of explosive field (Miao, Peng-Cheng; Wang, Hai-Lin; Yan, Da-Peng; He, Anzhi)V1554B,641-644(1991)

Aligning diamond-turned optics using visible-light interferometry (Figoski, John W.)V1354,540-546(1991)

All-optical data-input device based on fiber optic interferometric strain gauges (Fuerstenau, Norbert; Schmidt, Walter)V1367,357-366(1991)

Analysis of dynamic characteristics of nonstationary gas streams using interferometry techniques (Abrukov, Victor S.; Ilyin, Stanislav V.)V1554B,540-543(1991)

Analysis of macro-model composites with Fabry-Perot fiber-optic sensors (Fogg, Brian R.; Miller, William V.; Lesko, John J.; Carman, Gregory P.; Vengsarkar, Ashish M.; Reifsnider, Kenneth L.; Claus, Richard O.)V1588,14-25(1991)

Analysis of measurement principle of moire interferometer using Fourier method (Wang, Hai-Lin; Miao, Peng-Cheng; He, Anzhi)V1527,419-422(1991)

Analysis of powertrain noise and vibration using interferometry (Griffen, Christopher T.; Fryska, Slawomir T.; Bernier, Paul R.)V1554B,754-766(1991)

Angle encoding with the folded Brewster interferometer (Young, Niels O.)V1454,235-244(1991)

Angle measurement by moire interference technique (Singh, Brahm P.; Chitnis, Vijay T.)V1554B,335-338(1991)

Application of Fabry-Perot velocimetry to hypervelocity impact experiments (Chau, Henry H.; Osher, John E.)V1346,103-112(1991)

Application of holographic interferometry to a three-dimensional flow field (Doerr, Stephen E.)V1554B,544-555(1991)

Application of low-coherence optical fiber Doppler anemometry to fluid-flow measurement: optical system considerations (Boyle, William J.; Grattan, Kenneth T.; Palmer, Andrew W.; Meggitt, Beverley T.)V1511,51-56(1991)

Application of phase-stepping speckle interferometry to shape and deformation measurement of a 3-D surface (Dobbins, B. N.; He, Shi P.; Kapasi, S.; Wang, Liu Sheng; Button, B. L.; Wu, Xiao-Ping)V1554A,772-780(1991)

Application of real-time holographic interferometry in the nondestructive inspection of electronic parts and assemblies (Wood, Craig P.; Trolinger, James D.)V1332,122-131(1991)

Application of spatially resolving holographic interferometry to plasma diagnostics (Neger, Theo; Jaeger, Helmut; Philipp, Harald; Pretzler, Georg; Widmann, Klaus; Woisetschlaeger, Jakob)V1507,476-487(1991)

Applied research of optical fiber pulsed-laser holographic interferometry (Wang, Guozhi; Zhu, Guangkuan)V1554B,119-125(1991)

Approach for applying holographic interferometry to large deformations and modifications (Schumann, Walter)V1507,526-537(1991)

ASK transmitter for high-bit-rate systems (Schiellerup, Gert; Pedersen, Rune J.)V1372,27-38(1991)

Automated fringe analysis for moire interferometry (Brownell, John B.; Parker, Richard J.)V1554B,481-492(1991)

Automated measurement of 3-D shapes by a dual-beam digital speckle interferometric technique (Shi, Dahuan; Qin, Jing; Hung, Yau Y.)V1554A,680-689(1991)

Automated systems for quantitative analysis of holograms (Pryputniewicz, Ryszard J.)VIS08,215-246(1991)

Automatic heterodyning of fiber-optic speckle pattern interferometry (Valera Robles, Jesus D.; Harvey, David; Jones, Julian D.)V1508,170-179(1991)

Automatic, high-resolution analysis of low-noise fringes (Lassahn, Gordon D.)V1332,690-695(1991)

Automatic mask-to-wafer alignment and gap control using moire interferometry (Chitnis, Vijay T.; Kowsalya, S.; Rashmi, Dr.; Kanjilal, A. K.; Narain, Ram)V1332,613-622(1991)

Automatization of measurement and processing of experimental data in photoelasticity (Zhavoronok, I. V.; Nemchinov, V. V.; Litvin, S. A.; Skanavi, A. M.; Pavlov, V. V.; Evsenev, V. S.)V1554A,371-379(1991)

Behavior of a thin liquid film under thermal stimulation: theory and applications to infrared interferometry (Ledoyen, Fernand; Lewandowski, Jacques; Cormier, Maurice)V1507,328-338(1991)

Broad-range holographic contouring of diffuse surfaces by dual-beam illumination: study of two related techniques (Rastogi, Pramod K.; Pflug, Leopold)V1554B,48-55(1991)

Carrier pattern analysis of moire interferometry using Fourier transform (Morimoto, Yoshiharu; Post, Daniel; Gascoigne, Harold E.)V1554B,493-502(1991)

CCD holographic phase and intensity measurement of laser wavefront (Wickham, Michael; Munch, Jesper)V1414,80-90(1991)

Cellular vibration measurement with a noninvasive optical system (Khanna, Shyam M.)V1429,9-20(1991)

Channel waveguide Mach-Zehnder interferometer for wavelength splitting and combining (Tervonen, Ari; Poyhonen, Pekka; Honkanen, Seppo; Tahkokorpi, Markku T.)V1513,71-75(1991)

Characteristics of a multimode laser diode source in several types of dual-interferometer configuration (Ning, Yanong N.; Grattan, Kenneth T.; Palmer, Andrew W.; Meggitt, Beverley T.)V1367,347-356(1991)

Common-path optical fiber heterodyne interferometric current sensor (Bartlett, Steven C.; Farahi, Faramarz; Jackson, David A.)V1504,247-250(1991)

Computer-aided speckle interferometry: Part II—an alternative approach using spectral amplitude and phase information (Chen, Duanjun; Chiang, Fu-Pen; Tan, Yushan; Don, H. S.)V1554A,706-717(1991)

Computer-controlled pulse laser system for dynamic holography (Chen, Terry Y.; Ju, M. S.; Lee, C. Y.; Lo, M. P.)V1554B,92-98(1991)

Computerized vibration analysis of hot objects (Lokberg, Ole J.; Rosvold, Geir O.; Malmo, Jan T.; Ellingsrud, Svein)V1508,153-160(1991)

Contouring by DSPI for surface inspection (Paoletti, Domenica; Schirripa Spagnolo, Giuseppe)V1554A,660-667(1991)

Conversion of schlieren systems to high-speed interferometers (Anderson, Roland C.; Milton, James E.)V1358,992-1002(1991)

Corneal topography: the dark side of the moon (Bores, Leo D.)V1423,28-39(1991); V1429,217-228(1991)

Correspondence in damage phenomena and R-curve behavior in ceramics and geomaterials using moire interferometry (Perry, Kenneth E.; Epstein, Jonathan S.; May, G. B.; Shull, J. E.)V1554A,209-227(1991)

Coupled fiber ring interferometer array: theory (Wang, An; Xie, Haiming)V1572,365-369(1991)

Crack-tip deformation field measurements using coherent gradient sensing (Tippur, Hareesh V.; Krishnaswamy, Sridhar; Rosakis, Ares J.)V1554A,176-191(1991)

Curvature of bending shells by moire interferometry (Wang, Lingli; Yun, Dazhen)V1554B,436-440(1991)

Deformation measurement of the bone fixed with external fixator using holographic interferometry (Kojima, Arata; Ogawa, Ryokei; Izuchi, N.; Yamamoto, M.; Nishimoto, T.; Matsumoto, Toshiro)V1429,162-171(1991)

Demodulation of a fiber Fabry-Perot strain sensor using white light interferometry (Zuliani, Gary; Hogg, W. D.; Liu, Kexing; Measures, Raymond M.)V1588,308-313(1991)

Design and application of a moire interferometer (He, Anzhi; Wang, Hai-Lin; Yan, Da-Peng; Miao, Peng-Cheng)V1527,334-337(1991)

Design criteria of an integrated optics microdisplacement sensor (d'Alessandro, Antonio; De Sario, Marco; D'Orazio, Antonella; Petruzzelli, Vincenzo)V1332,554-562(1991)

Design optimization of moire interferometers for rapid 3-D manufacturing inspection (Dubowsky, Steven; Holly, Krisztina J.; Murray, Annie L.; Wander, Joseph M.)V1386,10-20(1991)

Determination of displacements, strain, and rotations from holographic interferometry data using a 2-D fringe order function (Albertazzi, Armando A.)V1554B,64-74(1991)

Determination of the adhesive load by holographic interferometry using the results of FEM calculations (Bischof, Thomas; Jueptner, Werner P.)V1508,90-95(1991)

Determination of thermal stresses in a bimaterial specimen by moire interferometry (Kang, Yilan; Jia, Youquan; Du, Ji)V1554B,514-522(1991)

Development of an interferometric fiber optic sensor using diode laser (Brewer, Donald R.; Joenathan, Charles; Bibby, Yu Wang; Khorana, Brij M.)V1396,430-434(1991)

Development of new beamsplitter system for real-time high-speed holographic interferometry (Yamamoto, Yoshitaka)V1358,940-951(1991)

Difference holographic interferometry: an overview (Fuzessy, Zoltan; Gyi'mesi, Ferenc)VIS08,194-204(1991)

Digital shearing laser interferometry for heterodyne array phasing (Cederquist, Jack N.; Fienup, James R.; Marron, Joseph C.; Schulz, Timothy J.; Seldin, J. H.)V1416,266-277(1991)

Digital Talbot interferometer (Tam, Siu-Chung; Silva, Donald E.; Wong, H. L.)V1400,38-48(1991)

Direct readout of dynamic phase changes in a fiber-optic homodyne interferometer (Jin, Wei; Uttamchandani, Deepak G.; Culshaw, Brian)V1504,125-132(1991)

Displacement analysis of the interior walls of a pipe using panoramic holointerferometry (Gilbert, John A.; Matthys, Donald R.; Hendren, Christelle M.)V1554B,128-134(1991)

Displacements and strains in thick-walled composite rings subjected to external pressure using moire interferometry (Gascoigne, Harold E.; Abdallah, Mohamed G.)V1554B,315-322(1991)

DSPI: a tool for analyzing thermal strain on ceramic and composite materials (Hoefling, Roland; Aswendt, Petra; Totzauer, Werner F.; Jueptner, Werner P.)V1508,135-142(1991)

Dynamical and real-time measurement of the fringe visibility of optical fiber interferometer (Fang, Xian-chen; Guo, Jian)V1572,52-55(1991)

Dynamic holography application in fiber optic interferometry (Kozhevnikov, Nikolai M.; Barmenkov, Yuri O.)V1584,387-395(1991)

Dynamic two-beam speckle interferometry (Ahmadshahi, Mansour A.; Krishnaswamy, Sridhar; Nemat-Nasser, Siavouche)V1554A,620-627(1991)

Electronic shearography versus ESPI in nondestructive evaluation (Hung, Yau Y.)V1554B,692-700(1991)

Electronic speckle pattern interferometry for 3-D dynamic deformation analysis in industrial environments (Guelker, Gerd; Haack, Olaf; Hinsch, Klaus D.; Hoelscher, Claudia; Kuls, Juergen; Platen, Winfried)V1500,124-134(1991)

Electro-optical system for the nondestructive evaluation of bioengineering materials (Sciammarella, Cesar A.; Bhat, Gopalakrishna K.; Albertazzi, Armando A.)V1429,183-194(1991)

Error transfer function for grating interferometer (Hao, De-Fu)V1545,261-265(1991)

Evaluation of a new electrical resistance shear-strain gauge using moire interferometry (Ifju, Peter G.)V1554B,420-428(1991)

Experimental characterization of microlenses for WDM transmission systems (Zoboli, Maurizio; Bassi, Paolo)V1506,160-169(1991)

Experimental investigation of dynamic mixed-mode fracture initiation (Lambros, John M.; Mason, James J.; Rosakis, Ares J.)V1554A,70-83(1991)

Experimental investigation of speckle-size distribution (Markhvida, Igor V.; Tanin, Leonid V.; Drobot, Igor L.)V1508,128-134(1991)

Fiber Optic and Laser Sensors VIII (DePaula, Ramon P.; Udd, Eric, eds.)V1367(1991)

Fiber Optic and Laser Sensors IX (DePaula, Ramon P.; Udd, Eric, eds.)V1584(1991)

Fiber optic based vortex shedder flow meter (Chu, Beatrice C.; Newson, Trevor P.; Jackson, David A.)V1504,251-257(1991)

Fiber optic Fabry-Perot sensors for high-speed heat transfer measurements (Kidd, S. R.; Sinha, P. G.; Barton, James S.; Jones, Julian D.)V1504,180-190(1991)

Fiber optic interferometric ellipsoidal shell hydrophone (Brown, David A.; Garrett, Steven L.; Conte, D. V.; Smith, R. C.; Rothenberg, E.; Young, M.; Rissberger, Ed)V1369,2-8(1991)

Fiber optic interferometric sensors using multimode fibers (Ignatyev, Alexander V.; Galkin, S. L.; Nikolaev, V. A.; Strigalev, V. E.)V1584,336-345(1991)

Fiber optic interferometric x-ray dosimeter (Barone, Fabrizio; Bernini, Umberto; Conti, M.; Del Guerra, Alberto; Di Fiore, Luciano; Maddalena, P.; Milano, L.; Russo, G.; Russo, Paolo)V1584,304-307(1991)

Fiber optic sensing technique employing rf-modulated interferometry (Eustace, John G.; Coghlan, Gregory A.; Yorka, Christian M.; Carome, Edward F.; Adamovsky, Grigory)V1584,320-327(1991)

Fiber Optic Smart Structures and Skins IV (Claus, Richard O.; Udd, Eric, eds.)V1588(1991)

Fiber optic thermometer using Fourier transform spectroscopy (Beheim, Glenn; Sotomayor, Jorge L.; Flatico, Joseph M.; Azar, Massood T.)V1584,64-71(1991)

Filled-arm Fizeau telescope (Synnott, Stephen P.)V1494,334-343(1991)

Fizeau-type of gradient-index rod lens interferometer by using semiconductor laser (Ming, Hai; Sun, Yuesheng; Ren, Baorui; Xie, Jianping; Nakajima, Toshinori)V1572,27-31(1991)

Focal-plane processing algorithm and architecture for laser speckle interferometry (Tomlinson, Harold W.; Weir, Michael P.; Michon, G. J.; Possin, G. E.; Chovan, J.)V1541,178-186(1991)

Four cases of engineering vibration studies using pulsed ESPI (Mendoza Santoyo, Fernando; Shellabear, Michael C.; Tyrer, John R.)V1508,143-152(1991)

Fringe analysis in moire interferometry (Asundi, Anand K.)V1554B,472-480(1991)

Fringe formation in speckle shearing interferometry (Xu, Boqin; Wu, Xiao-Ping)V1554A,789-799(1991)

Fringe quality in pulsed TV-holography (Spooren, Rudie)V1508,118-127(1991)

Gas sensor based on an integrated optical interferometer (Brandenburg, Albrecht; Edelhaeuser, Rainer; Hutter, Frank)V1510,148-159(1991)

General purpose fiber optic hydrophone made of castable epoxy (Garrett, Steven L.; Brown, David A.; Beaton, Brian L.; Wetterskog, Kevin; Serocki, John)V1367,13-29(1991)

Geometrical and radiometrical signal transfer characteristics of a color CCD camera with 21-million pixels (Lenz, Reimar K.; Lenz, Udo)V1526,123-132(1991)

Geometrical moire method for displaying directly the stress field of a circular disk (Wang, Ji-Zhong; Wang, Yun-Shan; Yin, Yuan-Cheng; Wu, Rui-Lan)V1554B,188-192(1991)

Graduating rules checking up by laser interferometry (Miron, Nicolae; Sporea, Dan G.)V1500,334-338(1991)

High-bandwidth alignment sensing in active optical systems (Kishner, Stanley J.)V1532,215-229(1991)

High-precision interferometric testing of spherical mirrors with long radius of curvature (Freischlad, Klaus R.; Kuechel, Michael F.; Wiedmann, Wolfgang; Kaiser, Winfried; Mayer, Max)V1332,8-17(1991)

High-resolution imaging interferometer (Ballangrud, Ase; Jaeger, Tycho C.; Wang, Gunnar)V1521,89-96(1991)

High-sensitivity interferometric technique for strain measurements (Voloshin, Arkady S.; Bastawros, Adel F.)V1400,50-60(1991)

High-spatial-resolution and high-sensitivity interferometric optical-time-domain reflectometer (Kobayashi, Masaru; Noda, Juichi; Takada, Kazumasa; Taylor, Henry F.)V1474,278-284(1991)

High-speed holographic interferometric study of the propagation of the electrohydraulic shock wave (Jia, Youquan; Sun, Yongda)V1554B,135-138(1991)

High-speed time-resolved holographic interferometer using solid state shutters (Racca, Roberto G.; Dewey, John M.)V1358,932-939(1991)

HOLIDDO: an interferometer for space experiments (Mary, Joel; Bernard, Yves; Lefaucheux, Francoise; Gonzalez, Francois)V1557,147-155(1991)

Holographic coupler for fiber optic Sagnac interferometer (Zou, Yunlu; Hsu, Dahsiung; Wang, Ben; Tao, Huiying)V1238,452-456(1991)

Holographic in-plane displacement measurement in cracked specimens in plane stress state (Sainov, Ventseslav C.; Simova, Eli S.)V1461,175-183(1991)

Holographic instrumentation for monitoring crystal growth in space (Trolinger, James D.; Lal, Ravindra B.; Batra, Ashok K.)V1332,151-165(1991)

Holographic interferometric analysis of the bovine cornea expansion (Foerster, Werner; Kasprzak, Henryk; von Bally, Gert; Busse, H.)V1429,146-151(1991)

Holographic interferometric observation of shock wave focusing to extracorporeal shock wave lithotripsy (Takayama, Kazuyoshi; Obara, Tetsuro; Onodera, Osamu)V1358,1180-1190(1991)

Holographic interferometry analysis of sealed, disposable containers for internal defects (Cohn, Gerald E.; Domanik, Richard A.)V1429,195-206(1991)

Holographic interferometry in corrosion studies of metals: I. Theoretical aspects (Habib, Khaled J.)V1332,193-204(1991)

Holographic interferometry in corrosion studies of metals: II. Applications (Habib, Khaled J.)V1332,205-215(1991)

Holographic interferometry of the corneal surface (Friedlander, Miles H.; Mulet, Miguel; Buzard, Kurt A.; Granet, Nicole; Baker, Phillip C.)V1423,62-69(1991); V1429,229-236(1991)

Holographic measurement of deformation using carrier fringe and FFT techniques (Quan, C.; Judge, Thomas R.; Bryanston-Cross, Peter J.)V1507,463-475(1991)

Holographic measurement of the angular error of a table moving along a slideway (Matsuda, Kiyofumi; Tenjimbayashi, Koji)V1332,230-235(1991)

Holographic Optics III: Principles and Applications (Morris, G. M., ed.)V1507(1991)

Holographic photoelasticity applied to ceramics fracture (Guo, Maolin; Wang, Yigong; Zhao, Yin; Du, Shanyi)V1554A,310-312(1991)

Holography as a tool for widespread industrial applications: analysis and comments (Smigielski, Paul)V1508,38-49(1991)

Holography, Interferometry, and Optical Pattern Recognition in Biomedicine (Podbielska, Halina, ed.)V1429(1991)

Holo-interferometric patterns recorded through a panoramic annular lens (Gilbert, John A.; Greguss, Pal; Kransteuber, Amy S.)V1238,412-420(1991)

Hubble extra-solar planet interferometer (Shao, Michael)V1494,347-356(1991)

Industrial Applications of Holographic and Speckle Measuring Techniques (Jueptner, Werner P., ed.)V1508(1991)

Industrial applications of self-diffraction phenomena in holography on photorefractive crystals (Dovgalenko, George Y.; Yeskov-Soskovetz, Vladimir M.; Cherkashin, G. V.)V1508,110-115(1991)

Influence of fixing stress on the sensitivity of HNDT (Yang, Yu X.; Tan, Yushan)V1554B,768-773(1991)

In-line Fabry-Perot interferometric temperature sensor with digital signal processing (Yeh, Yunhae; Lee, J. H.; Lee, Chung E.; Taylor, Henry F.)V1584,72-78(1991)

In-line fiber Fabry-Perot interferometer with high-reflectance internal mirrors (Lee, Chung E.; Gibler, William N.; Atkins, Robert A.; Taylor, Henry F.)V1584,396-399(1991)

Innovative Optics and Phase Conjugate Optics (Ahlers, Rolf-Juergen; Tschudi, Theo T., eds.)V1500(1991)

In-plane displacement measurement on rotating components using pulsed laser ESPI (electronic speckle pattern interferometry) (Preater, Richard W.)V1399,164-171(1991)

In-plane ESPI measurements in hostile environments (Preater, Richard W.; Swain, Robin C.)V1554A,727-738(1991)

In-situ interferometric measurements in a rapid thermal processor (Dilhac, Jean-Marie R.; Ganibal, Christian; Nolhier, N.; Amat, L.)V1393,349-353(1991)

In-situ measurement of piston jitter in a ring resonator (Cunningham, Philip R.; Hay, Stephen O.; Francis, Denise M.; Trott, G.E.)V1414,97-129(1991)

In-situ monitoring of silylation mechanisms by laser interferometry (Pierrat, Christophe; Paniez, Patrick J.; Martin, P.)V1466,248-256(1991)

In-situ observation of crystal growth in microgravity by high-resolution microscopies (Tsukamoto, Katsuo; Onuma, Kazuo)V1557,112-123(1991)

Instantaneous etch rate measurement of thin transparent films by interferometry for use in an algorithm to control a plasma etcher (Mishurda, Helen L.; Hershkowitz, Noah)V1392,563-569(1991)

Instantaneous measurement of density from double-simultaneous interferograms (Desse, Jean-Michel; Pegneaux, Jean-Claude)V1358,766-774(1991)

Instrument design and test results of the new all-reflection spatial heterodyne spectrometer (Bush, Brett C.; Cotton, Daniel M.; Vickers, James S.; Chakrabarti, Supriya)V1549,376-384(1991)

Integrated optical Mach-Zehnder interferometers in glass (Lefebvre, P.; Vahid-Shahidi, A.; Albert, Jacques; Najafi, S. I.)V1583,221-225(1991)

Integrated optics sensor on silicon for the measurement of displacement, force, and refractive index (Ulbers, Gerd)V1506,99-110(1991)

Interactions of laser-induced cavitation bubbles with a rigid boundary (Ward, Barry; Emmony, David C.)V1358,1035-1045(1991)

Interference detection of plasma of laser field interaction with optical thin films (Ni, Xiao W.; Lu, Jian-Feng; He, Anzhi; Ma, Zi; Zhou, Jiu L.)V1554B,632-635(1991)

Interference measurement of thermal effect induced by optical pumping on YAG crystal rod (Luo, Bikai; Ni, Xiao W.; Zhang, Qi; He, Anzhi)V1554A,542-546(1991)

Interference phenomenon with correlated masks and its application (Grover, Chandra P.; Hane, Kazuhiro)V1332,624-631(1991)

Interferometer accuracy and precision (Selberg, Lars A.)V1400,24-32(1991)

Interferometer for testing aspheric surfaces with electron-beam computer-generated holograms (Gemma, Takashi; Hideshima, Masayuki; Taya, Makoto; Watanabe, Nobuko)V1332,77-84(1991)

Interferometer for testing of general aspherics using computer-generated holograms (Arnold, Steven M.; Jain, Anil K.)V1396,473-480(1991)

Interferometric analysis for nondestructive product monitoring (Cohn, Gerald E.; Domanik, Richard A.)V1396,131-142(1991)

Interferometric analysis of absorbing objects (Korchazhkin, S. V.; Krasnova, L. O.)V1512,195-197(1991)

Interferometric fiber optic accelerometer (Brown, David A.; Garrett, Steven L.)V1367,282-288(1991)

Interferometric fiber optic gyroscopes for today's market (LaViolette, Kerry D.; Bossler, Franklin B.)V1398,213-218(1991)

Interferometric fiber-optic sensing using a multimode laser diode source (Gerges, Awad S.; Newson, Trevor P.; Jackson, David A.)V1504,176-179(1991)

Interferometric fiber optic sensors for use with composite materials (Measures, Raymond M.; Valis, Tomas; Liu, Kexing; Hogg, W. D.; Ferguson, Suzanne M.; Tapanes, Edward)V1332,421-430(1991)

Interferometric measurement for the plasma produced by Q-switched laser in air near the surface of an Al Target (Lu, Jian-Feng; Ni, Xiao W.; Wang, Hong Y.; He, Anzhi)V1554B,593-597(1991)

Interferometric measurement of in-plane motion (Hercher, Michael; Wyntjes, Geert J.)V1332,602-612(1991)

Interferometric measurements of a high-velocity mixing/shear layer (Peters, Bruce R.; Kalin, David A.)V1486,410-416(1991)

Interferometric moire analysis of wood and paper structures (Hyzer, Jim B.; Shih, J.; Rowlands, Robert E.)V1554B,371-382(1991)

Interferometric monitor for greenhouse gasses for ADEOS (Tsuno, Katsuhiko; Kameda, Yoshihiko; Kondoh, Kayoko; Hirai, Shoichi)V1490,222-232(1991)

Interferometric optical fiber sensors for absolute measurement of displacement and strain (Kersey, Alan D.; Berkoff, Timothy A.; Dandridge, Anthony D.)V1511,40-50(1991)

Interferometric system for the inspection and measurement of the quality of optical fiber ends (Corredera, Pedro; Pons, Alicia A.; Campos, Joaquin; Corrons, Antonio)V1504,281-286(1991)

Interferometry, streak photography, and stereo photography of laser-driven miniature flying plates (Paisley, Dennis L.; Montoya, Nelson I.; Stahl, David B.; Garcia, Ismel A.)V1358,760-765(1991)

Interferometry with laser diodes (Hariharan, P.)V1400,2-10(1991)

Interlaminar deformations on the cylindrical surface of a hole in laminated composites by moire interferometry (Boeman, Raymond G.)V1554B,323-330(1991)

Intl Conf on Optical Fibre Sensors in China (Culshaw, Brian; Liao, Yan-Biao, eds.)V1572(1991)

Investigation of fiber-reinforced-plastics-based components by means of holographic interferometry (Jueptner, Werner P.; Bischof, Thomas)V1400,69-79(1991)

Investigation on phase biological micro-objects with a holographic interferometric microscope on the basis of the photorefractive Bi12TiO20 crystal (Tontchev, Dimitar A.; Zhivkova, Svetla; Miteva, Margarita G.; Grigoriev, Ivo D.; Ivanov, I.)V1429,76-80(1991)

Large-aperture high-accuracy lateral shearing interferometer utilizing a Twyman-Green interferometer (He, Anzhi; Wang, Hai-Lin; Miao, Peng-Cheng; Yan, Da-Peng)V1527,423-426(1991)

Large-dynamic-range electronically scanned "white-light" interferometer with optical fiber Young's structure (Chen, Shiping; Rogers, Alan J.; Meggitt, Beverley T.)V1504,191-201(1991); V1511,67-77(1991)

Laser speckle and optical fiber sensors for micromovements monitoring in biotissues (Tuchin, Valery V.; Ampilogov, Andrey V.; Bogoroditsky, Alexander G.; Rabinovich, Emmanuil M.; Ryabukho, Vladimir P.; Ul'yanov, Sergey S.; V'yushkin, Maksim E.)V1420,81-92(1991); V1429,62-73(1991)

Line-imaging Fabry-Perot interferometer (Mathews, Allen R.; Warnes, Richard H.; Hemsing, Willard F.; Whittemore, Gerald R.)V1346,122-132(1991)

Linearity of coherence probe metrology: simulation and experiment (Davidson, Mark P.; Monahan, Kevin M.; Monteverde, Robert J.)V1464,155-176(1991)

Liquid-crystal phase modulator used in DSPI (Kadono, Hirofumi; Toyooka, Satoru)V1554A,628-638(1991)

Low-frequency fiber optic magnetic field sensors (Nader-Rezvani, Navid; Claus, Richard O.; Sarrafzadeh, A. K.)V1584,405-414(1991)

Measurement of Poisson's ratio of nonmetallic materials by laser holographic interferometry (Zhu, Jian T.)V1554B,148-154(1991)

Measurement of residual stress during implant resistance welding of plastics (Park, Joon B.; Benatar, Avraham)V1554B,357-370(1991)

Measurement of the axial eye length and retinal thickness by laser Doppler interferometry (Hitzenberger, Christoph K.; Fercher, Adolf F.; Juchem, M.)V1423,46-50(1991); V1429,21-25(1991)

Measurement of the piezoelectric effect in bone using quasiheterodyne holographic interferometry (Ovryn, Benjie; Haacke, E. M.)V1429,172-182(1991)

Measurement of thermal deformation of square plate by using holographic interferometry (Kang, Dae I.; Kwon, Young H.; Ko, Yeong-Uk)V1554B,139-147(1991)

Measurement of the temperature field in confined jet impingement using phase-stepping video holography (Dobbins, B. N.; He, Shi P.; Jambunathan, K.; Kapasi, S.; Wang, Liu Sheng; Button, B. L.)V1554B,586-592(1991)

Measurement of wave aberrations of intraocular lenses through holographic interferometry (Carretero, L.; Fuentes, Rosa; Fimia-Gil, Antonio)V1507,458-462(1991); V1508,96-100(1991)

Measurements of second hyperpolarisabilities of diphenylpolyenes by means of phase-conjugate interferometry (Persoons, Andre P.; Van Wonterghem, Bruno M.; Tackx, Peter)V1409,220-229(1991)

Michelson interferometer for passive atmosphere sounding (Posselt, Winfried)V1490,114-125(1991)

Microdisplacement fiber sensor using two-frequency interferometry (Tedjojuwono, Ken K.)V1584,146-151(1991)

Microdisplacement positioning system for a diffraction grating ruling engine (Yang, Hou-Min; Wang, Xiaolin; Zhang, Yinxian)V1533,185-192(1991)

Microwave interferometric measurements of process plasma density (Cheah, Chun-Wah; Cecchi, Joseph L.; Stevens, J. L.)V1392,487-497(1991)

MIT multipoint alignment testbed: technology development for optical interferometry (Blackwood, Gary H.; Jacques, Robert; Miller, David W.)V1542,371-391(1991)

Modal analysis of musical instruments with holographic interferometry (Rossing, Thomas D.; Hampton, D. S.)V1396,108-121(1991)

Modular removable precision mechanism for alignment of an FUV spatial heterodyne interferometer (Tom, James L.; Cotton, Daniel M.; Bush, Brett C.; Chung, Ray; Chakrabarti, Supriya)V1549,308-312(1991)

Moire interferometry for subdynamic tests in normal light environment (Luo, Zhishan; Luo, Chanzou; Hu, Qinghua; Mu, Zongxue)V1554B,523-528(1991)

Moire interferometry of sticking film using white light (Luo, Zhishan; Zhang, Souyi; Tuo, Sueshang)V1554B,339-342(1991)

Moire pattern and interferogram transform (Lu, Zhen-Wu)V1554B,383-388(1991)

Moire-shift interferometer measurements of the shape of human and cat tympanic membrane (Decraemer, Willem F.; Dirckx, Joris J.)V1429,26-33(1991)

Multichannel chromatic interferometry: metrology applications (Tribillon, Gilbert M.; Calatroni, Jose; Sandoz, Patrick)V1332,632-642(1991)

Multimode fiber-optic Mach-Zehnder interferometric strain sensor (Wang, An; Xie, Haiming)V1572,444-449(1991)

Multipass holographic interferometry for low-density gas flow analysis (Surget, Jean; Dunet, G.)V1358,65-72(1991)

Multiplexing of remote all-fiber Michelson interferometers with lead insensitivity (Santos, Jose L.; Farahi, Faramarz; Newson, Trevor P.; Leite, Antonio P.; Jackson, David A.)V1511,179-189(1991)

Multipulse polarized-holographic set-up for isodromic fringe registration in the field of running stress waves (Khesin, G. L.; Sakharov, V. N.; Zhavoronok, I. V.; Rottenkolber, Hans; Schorner, Jurgen)V1554B,86-90(1991)

Natural pixel decomposition for interferometric tomography (Cha, Dong J.; Cha, Soyoung S.)V1554B,600-609(1991)

New concept for a highly compact imaging Fourier transform spectrometer (Simeoni, Denis)V1479,127-138(1991)

New holographic system for measuring vibration (Cai, Yunliang)V1554B,75-80(1991)

New method of increasing the sensitivity of Schlieren interferometer using two Wollaston prisms and its application to flow field (Yan, Da-Peng; He, Anzhi; Yang, Zu Q.; Zhu, Yi Yun)V1554B,636-640(1991)

New method of thinning photoelastic interference fringes in image processing (Zhang, Yuan P.)V1554A,862-866(1991)

New modulation scheme for optical fiber point temperature sensor (Farahi, Faramarz)V1504,237-246(1991)

New options of holographic metrology (Steinbichler, Hans)V1507,435-445(1991)

New phase measurement for nonmonotonical fringe patterns (Tang, Shou-Hong; Hung, Yau Y.; Zhu, Qiuming)V1332,731-737(1991)

New technique for rapid making specimen: gratings for moire interferometry (Jia, Youquan; Kang, Yilan; Du, Ji)V1554B,331-334(1991)

New way of making a lunar telescope (Chen, Peter C.)V1494,228-233(1991)

New Zeiss interferometer (Kuechel, Michael F.)V1332,655-663(1991)

No-blind-area one-photograph HNDT tire analyzer (Ge, Fang X.; Xiong, Xian M.)V1554B,785-789(1991)

Nondestructive testing of printed circuit board by phase-shifting interferometry (Lu, Yue-guang; Jiang, Lingzhen; Zou, Lixun; Zhao, Xia; Sun, Junyong)V1332,287-291(1991)

Novel analog phase tracker for interferometric fiber optic sensor applications (Berkoff, Timothy A.; Kersey, Alan D.; Moeller, Robert P.)V1367,53-58(1991)

Novel fiber-optic interferometer with high sensitivity and common-mode compensation (Chen, Xiaoguang)V1572,332-336(1991)

Novel interferometer setup for evaluating the sum of surface contributions to transmitted wavefront distortion (Williamson, Ray)V1527,188-193(1991)

Novel optical processing scheme for interferometric vibration measurement using a low-coherence source with a fiber optic probe (Weir, Kenneth; Boyle, William J.; Palmer, Andrew W.; Grattan, Kenneth T.; Meggitt, Beverley T.)V1584,220-225(1991)

Numerical investigation of effect of dynamic range and nonlinearity of detector on phase-stepping holographic interferometry (Fang, Qiang; Luo, Xiangyang; Tan, Yushan)V1332,216-222(1991)

Optical aspheric surface profiler using phase shift interferometry (Sasaki, Kenji; Ono, Akira)V1332,97-106(1991)

Optical Fabrication and Testing (Campbell, Duncan R.; Johnson, Craig W.; Lorenzen, Manfred, eds.)V1400(1991)

Optical fiber interferometric sensors for chemical detection (Butler, Michael A.)V1368,46-54(1991)

Optical fibre interferometer for monitoring tool wear (Zheng, S. X.; McBride, R.; Hale, K. F.; Jones, Barry E.; Barton, James S.; Jones, Julian D.)V1572,359-364(1991)

Optical figure testing of prototype mirrors for JPL's precision segmented-reflector program (Hochberg, Eric B.)V1542,511-522(1991)

Optical filters for the wind imaging interferometer (Sellar, R. G.; Gault, William A.; Karp, Christopher K.)V1479,140-155(1991)

Optical implementation of winner-take-all models of neural networks (Kobzev, E. F.; Vorontsov, Michael A.)V1402,165-174(1991)

Optical pathlength control in the nanometer regime on the JPL phase-B interferometer testbed (O'Neal, Michael C.; Spanos, John T.)V1542,359-370(1991)

Optical profilometry using spatial heterodyne interferometric methods (Frankowski, Gottfried)V1500,114-123(1991)

Optical research of elastic anisotropy of monocrystals with a cube structure (Polyak, Alexander; Kostin, Ivan K.)V1554A,553-556(1991)

Optical surface microtopography using phase-shifting Nomarski microscope (Shimada, Wataru; Sato, Tadamitu; Yatagai, Toyohiko)V1332,525-529(1991)

Optical system design for a lunar optical interferometer (Colavita, Mark M.; Shao, Michael; Hines, Braden E.; Levine, Bruce M.; Gershman, Robert)V1494,168-181(1991)

Optical technique for the compensation of the temperature-dependent Verdet constant in Faraday rotation magnetometers (Hamid, Sohail; Tatam, Ralph P.)V1511,78-89(1991)

Optical Testing and Metrology III: Recent Advances in Industrial Optical Inspection (Grover, Chandra P., ed.)V1332(1991)

Optical testing by dynamic holographic interferometry with photorefractive crystals and computer image processing (Vlad, Ionel V.; Popa, Dragos; Petrov, M. P.; Kamshilin, Alexei A.)V1332,236-244(1991)

Optical testing with wavelength scanning interferometer (Okada, Katsuyuki; Tsujiuchi, Jumpei)V1400,33-37(1991)

Optical tomography by heterodyne holographic interferometry (Vukicevic, Dalibor; Neger, Theo; Jaeger, Helmut; Woisetschlaeger, Jakob; Philipp, Harald)VIS08,160-193(1991)

Optimal control of positive optical photoresist development (Carroll, Thomas A.; Ramirez, W. F.)V1464,222-231(1991)

Origins of asymmetry in spin-cast films over topography (Manske, Loni M.; Graves, David B.)V1463,414-422(1991)

Overview of the spectroscopy of the atmosphere using far-infrared emission experiment (Russell, James M.)V1491,142-150(1991)

Passive laser phase noise suppression technique for fiber interferometers (Kersey, Alan D.; Berkoff, Timothy A.)V1367,310-318(1991)

PC-based determination of 3-D deformation using holographic interferometry (Schad, Hanspeter; Schweizer, Edwin; Jodlbauer, Heibert)V1507,446-449(1991)

Phase-bias tuning and extinction ratio improvement of Mach-Zehnder interferometer (Baranov, D. V.; Zolotov, Evgeny M.; Pelekhaty, V. M.; Tavlykaev, R. F.)V1583,389-394(1991)

Phase compensation of PZT in an optical fibre Mach-Zehnder interferometer (Liu, Yanbing; Zhang, Jinru)V1572,61-64(1991)

Phase-conjugate interferometry by using dye-doped polymer films (Nakagawa, Kazuo; Egami, Chikara; Suzuki, Takayoshi; Fujiwara, Hirofumi)V1332,267-273(1991)

Phase-conjugate Twyman-Green interferometer for testing conicoidal surfaces (Shukla, Ram P.; Dokhanian, Mostafa; Venkateswarlu, Putcha; George, M. C.)V1332,274-286(1991)

Phase-map unwrapping: a comparison of some traditional methods and a presentation of a new approach (Owner-Petersen, Mette)V1508,73-82(1991)

Phase-measurement of interferometry techniques for nondestructive testing (Creath, Katherine)V1554B,701-707(1991)

Phase measuring fiber optic electronic speckle pattern interferometer (Joenathan, Charles; Khorana, Brij M.)V1396,155-163(1991)

Phase-measuring fiber optic ESPI system: phase-step calibration and error sources (Joenathan, Charles; Khorana, Brij M.)V1554B,56-63(1991)

Phase-shifting hand-held diffraction moire interferometer (Deason, Vance A.; Ward, Michael B.)V1554B,390-398(1991)

Phase-strain-temperature model for structurally embedded interferometric optical fiber strain sensors with applications (Sirkis, James S.)V1588,26-43(1991)

Phenomenon of fringe patterns from speckle photogram and interferometry within a film (Bo, Weiyun; Wang, Rupeng)V1554A,750-754(1991)

Photoelastic modeling of linear and nonlinear creep using two- and three-dimensional models (Vardanjan, G. S.; Musatov, L. G.)V1554A,496-502(1991)

Physical mechanism of high-power pulse laser ophthalmology: experimental research (He, Anzhi; Lu, Jian-Feng; Ni, Xiao W.; Li, Yong N.)V1423,117-120(1991)

Polarization approach to high-sensitivity moire interferometry (Salbut, Leszek A.; Patorski, Krzysztof; Kujawinska, Malgorzata)V1554B,451-460(1991)

Possibilities of the fringe pattern learning system VARNA (Nedkova, Rumiana)V1429,214-216(1991)

Possibility of associative reconstruction of reflection hologram in phase-conjugate interferometer (Lazaruk, A. M.; Rubanov, Alexander S.; Serebryakova, L. M.)V1621,114-124(1991)

Practical use of reference mirror rotation in holographic interferometry (Boxler, Lawrence H.; Brown, Glen; Western, Arthur B.)V1396,85-92(1991)

Predetection correlation in a spread-spectrum multiplexing system for fiber optic interferometers (Al-Raweshidy, H. S.; Uttamchandani, Deepak G.)V1367,329-336(1991)

Prototype high-speed optical delay line for stellar interferometry (Colavita, Mark M.; Hines, Braden E.; Shao, Michael; Klose, George J.; Gibson, B. V.)V1542,205-212(1991)

Pulsed holographic and speckle interferometry using Hopkinson loading technique to investigate the dynamical deformation on plates (Han, Lei; Wu, Xiao-Ping; Hu, Shisheng)V1358,793-803(1991)

Quantitative analysis of contact deformation using holographic interferometry and speckle photography (Joo, Jin W.; Kwon, Young H.; Park, Seung O.)V1554B,113-118(1991)

Range of measurement of computer-aided speckle interferometry (Chen, Duanjun; Li, Shen; Hsu, T. Y.; Chiang, Fu-Pen)V1554A,922-931(1991)

Rapid procedure for obtaining time-average interferograms of vibrating bodies (Rapoport, Eliezer; Bar, Doron; Shiloh, Klara)V1442,383-391(1991)

Real-time inspection of pavement by moire patterns (Guralnick, Sidney A.; Suen, Eric S.)V1396,664-677(1991)

Real-time interferometry by optical phase conjugation in dye-doped film (Fujiwara, Hirofumi)V1507,492-499(1991)

Real-time structural integrity monitoring using a passive quadrature demodulated, localized Michelson optical fiber interferometer capable of simultaneous strain and acoustic emission sensing (Tapanes, Edward)V1588,356-367(1991)

Real-time wavefront measurement with lambda/10 fringe spacing for the optical shop (Freischlad, Klaus R.; Kuechel, Michael F.; Schuster, Karl-Heinz; Wegmann, Ulrich; Kaiser, Winfried)V1332,18-24(1991)

Recent progress in interferometric fiber sensor technology (Kersey, Alan D.)V1367,2-12(1991)

Recording of laser radiation power distribution with the use of nonlinear optical effects (Korchazhkin, S. V.; Krasnova, L. O.)V1358,966-972(1991)

Reduced cost coil windings for interferometric fiber-optic gyro sensors (Smith, Ronald H.)V1478,145-149(1991)

Reducing errors in bending tests of ceramic materials (Furgiuele, Franco M.; Lamberti, Antonio; Pagnotta, Leonard)V1554A,275-284(1991)

Refractive properties of TGS aqueous solution for two-color interferometry (Vikram, Chandra S.; Witherow, William K.; Trolinger, James D.)V1557,197-201(1991)

Residual stress evaluation using shearography with large shear displacements (Hathaway, Richard B.; Der Hovanesian, Joseph; Hung, Yau Y.)V1554B,725-735(1991)

Review of interferogram analysis methods (Malacara Hernandez, Daniel)V1332,678-689(1991)

Review of phase-measuring interferometry (Stahl, H. P.)V1332,704-719(1991)

Rigid lightweight optical bench for a spaceborne FUV spatial heterodyne interferometer (Tom, James L.; Cotton, Daniel M.; Bush, Brett C.; Chung, Ray; Chakrabarti, Supriya)V1549,302-307(1991)

SALSA: a synthesis array for lunar submillimeter astronomy (Mahoney, Michael J.; Marsh, Kenneth A.)V1494,182-193(1991)

Sapphire fiber interferometer for microdisplacement measurements at high temperatures (Murphy, Kent A.; Fogg, Brian R.; Wang, George Z.; Vengsarkar, Ashish M.; Claus, Richard O.)V1588,117-124(1991)

Second Intl Conf on Photomechanics and Speckle Metrology: Moire Techniques, Holographic Interferometry, Optical NDT, and Applications to Fluid Mechanics (Chiang, Fu-Pen, ed.)V1554B(1991)

Selected Papers on Interferometry (Hariharan, P., ed.)VMS28(1991)

Selected Papers on Speckle Metrology (Sirohi, Rajpal S., ed.)VMS35(1991)

Self-aberration-eliminating interferometer for wavefront measurements (Gorelik, Vladimir S.; Kovalenko, Sergey N.; Turukhano, Boris G.)V1507,488-490(1991)

Self-referencing Mach-Zehnder interferometer as a laser system diagnostic (Feldman, Mark; Mockler, Daniel J.; English, R. Edward; Byrd, Jerry L.; Salmon, J. T.)V1542,490-501(1991)

Shadow method of scale with a-posteriori increase of measurements sensitivity (Spornik, Nikolai M.; Serenko, M. Y.)V1508,162-169(1991)

Signal processing in fiber-optic interferometer with FM light sources (Chi, Jifu; Chang, Meitung)V1572,74-77(1991)

Simple technique for 3-D displacements measurement using synthesis of holographic interferometry and speckle photography (Kwon, Young H.; Park, Seung O.; Park, B. C.)V1554A,639-644(1991)

Simplified VISAR system (Sweatt, William C.; Stanton, Philip L.; Crump, O. B.)V1346,151-159(1991)

Simulation of optical diagnostics for crystal growth: models and results (Banish, Michele R.; Clark, Rodney L.; Kathman, Alan D.; Lawson, Shelah M.)V1557,209-221(1991)

Simultaneous imaging and interferometric turbule visualization in a high-velocity mixing/shear layer (Kalin, David A.; Saylor, Danny A.; Street, Troy A.)V1358,780-787(1991)

Single-mode fiber Mach-Zehnder interferometer as an earth strain sensor (Wang, An; Xie, Haiming)V1572,440-443(1991)

Software concept for the new Zeiss interferometer (Doerband, Bernd; Wiedmann, Wolfgang; Wegmann, Ulrich; Kuebler, C. W.; Freischlad, Klaus R.)V1332,664-672(1991)

Some applications of a HOE-based desensitized interferometer in materials research (Boone, Pierre M.; Jacquot, Pierre M.)V1554A,512-521(1991)

Some cases of applying photothermoplastic carriers in holographic interferometry and speckle photography (Anikin, V. I.; Dementiev, I. V.; Zhurminskii, Igor L.; Nasedkhina, N. V.; Panasiuk, L. M.)V1238,286-296(1991)

Spatial-carrier phase-shifting technique of fringe pattern analysis (Kujawinska, Malgorzata; Schmidt, Joanna)V1508,61-67(1991)

Spatial phase-shifting techniques of fringe pattern analysis in photomechanics (Kujawinska, Malgorzata; Wojiak, Joanna)V1554B,503-513(1991)

Spatial techniques of fringe pattern analysis in interferometry (Kujawinska, Malgorzata; Spik, Andrzej)V1391,303-312(1991)

Stabilization of ill-posed nonlinear regression model and its application to interferogram reduction (Slepicka, James S.; Cha, Soyoung S.)V1554B,574-579(1991)

Static and dynamic measurements using electro-optic holography (Pryputniewicz, Ryszard J.)V1554B,790-798(1991)

Station-keeping strategy for multiple-spacecraft interferometry (DeCou, Anthony B.)V1494,440-451(1991)

Statistical interferometry based on the statistics of speckle phase (Kadono, Hirofumi; Toyooka, Satoru)V1554A,718-726(1991)

Strategies for unwrapping noisy interferograms in phase-sampling interferometry (Andrae, P.; Mieth, Ulrike; Osten, Wolfgang)V1508,50-60(1991)

Striped Fabry-Perots: improved efficiency for velocimetry (McMillan, Charles F.; Steinmetz, Lloyd L.)V1346,113-121(1991)

Study of displacement and residual displacement field of an interface crack by moire interferometry (Wang, Y. Y.; Chiang, Fu-Pen; Barsoum, Roshdy S.; Chou, S. T.)V1554B,344-356(1991)

Study of electronic shearing speckle technique (Qin, Yuwen; Wang, Jinqi; Ji, Xinhua)V1554A,739-746(1991)

Study of human cardiac cycle using holographic interferometry (Brown, Glen; Boxler, Lawrence H.; Chun, Patrick K.; Western, Arthur B.)V1396,164-173(1991)

Study of microbial growth II: by holographic interferomery (Bahuguna, Ramendra D.; Williams, Gareth T.; Pour, I. K.; Raman, R.)V1332,805-807(1991)

Study of the fracture process using laser holographic interferometry and image analysis (Castro-Montero, Alberto; Shah, S. P.; Bjelkhagen, Hans I.)V1396,122-130(1991)

Study of welding residual stress by means of laser holography interference (Yang, Jiahua)V1554B,155-160(1991)

Study on quasi-instantaneous converse piezoelectric effect of piezoelectric ceramics with the sinusoidal phase-modulating interferometer using optical fibers (Li, Naiji; Wang, Huiwen; Yang, Yang; Gu, Shenghua)V1572,47-51(1991)

Submicron linewidth measurement using an interferometric optical profiler (Creath, Katherine)V1464,474-483(1991)

Surface contouring using electronic speckle pattern interferometry (Kerr, David; Rodriguez-Vera, Ramon; Mendoza Santoyo, Fernando)V1554A,668-679(1991)

Surface contouring using TV holography (Atcha, Hashim; Tatam, Ralph P.; Buckberry, Clive H.; Davies, Jeremy C.; Jones, Julian D.)V1504,221-232(1991)

Symmetric 3 x 3 coupler based demodulator for fiber optic interferometric sensors (Brown, David A.; Cameron, C. B.; Keolian, Robert M.; Gardner, David L.; Garrett, Steven L.)V1584,328-335(1991)

Synchronous phase-extraction technique and its applications (Hung, Yau Y.; Tang, Shou-Hong; Jin, Guofan; Zhu, Qiuming)V1332,738-747(1991)

Synthetic wavelength interferometry for the extension of the dynamic range (Skudayski, Ulf; Jueptner, Werner P.)V1508,68-72(1991)

System design for lunar-based optical and submillimeter interferometers (Gershman, Robert; Mahoney, Michael J.; Rayman, M. D.; Shao, Michael; Snyder, Gerald C.)V1494,160-167(1991)

Testing of PZT shifters for interferometric measurements (Schmit, Joanna; Piatkowski, Tadeusz)V1391,313-317(1991)

Thermal strain measurements of solder joints in electronic packaging using moire interferometry (Woychik, Charls G.; Guo, Yi F.)V1554B,461-470(1991)

Three-dimensional interferometric and fiber-optic displacement measuring probe (Liang, Dawei; Fraser Monteiro, L.; Fraser Monteiro, M. L.; Lu, Boyin)V1511,90-97(1991)

Three-dimensional surface inspection using interferometric grating and 2-D FFT-based technique (Hung, Yau Y.; Tang, Shou-Hong; Zhu, Qiuming)V1332,696-703(1991)

Tilt-tolerant dual-band interferometer objective lens (Jones, Mike I.)V1498,158-162(1991)

Toward an ESA strategy for optical interferometry outside the atmosphere: recommendations by the ESA space interferometry study team (Noordam, Jan E.)V1494,344-346(1991)

Tunable holographic interferometer using photorefractive crystal (Yang, Guanglu; Siahmakoun, Azad)V1396,552-556(1991)

Two-dimensional dynamic neural network optical system with simplest types of large-scale interactions (Vorontsov, Michael A.; Zheleznykh, N. I.; Larichev, A. V.)V1402,154-164(1991)

Ultrahigh-sensitivity moire interferometry (Han, Bongtae)V1554B,399-411(1991)

Ultrasonic NDE (nondestructive evaluation) for composite materials using embedded fiber optic interferometric sensors (Liu, Kexing; Ferguson, Suzanne M.; Davis, Andrew; McEwen, Keith; Measures, Raymond M.)V1398,206-212(1991)

Universal interferometer with synthesized reference wave (Spornik, Nikolai M.; Tujev, A. F.)V1508,83-88(1991)

Use of Fourier transform spectroscopy in combustion effluent monitoring (Howe, Gordon S.; McIntosh, Bruce C.)V1434,90-103(1991)

Use of laser diodes and monomode optical fiber in electronic speckle pattern interferometry (Atcha, Hashim; Tatam, Ralph P.)V1584,425-434(1991)

Uses of Fabry-Perot velocimeters in studies of high explosives detonation (Breithaupt, R. D.; Tarver, Craig M.)V1346,96-102(1991)

Uses of fiber optics to enhance and extend the capabilities of holographic interferometry (Gilbert, John A.; Dudderar, Thomas D.)VIS08,146-159(1991)

Velocity interferometry of miniature flyer plates with subnanosecond time resolution (Paisley, Dennis L.; Montoya, Nelson I.; Stahl, David B.; Garcia, Ismel A.; Hemsing, Willard F.)V1346,172-178(1991)

V-groove diffraction grating for use in an FUV spatial heterodyne interferometer (Cotton, Daniel M.; Bach, Bernhard W.; Bush, Brett C.; Chakrabarti, Supriya)V1549,313-318(1991)

Vibration analysis of the tympanic membrane with a ventilation tube and a perforation by holography (Maeta, Manabu; Kawakami, Shinichiro; Ogawara, Toshiaki; Masuda, Yu)V1429,152-161(1991)

Vibration-insensitive moire interferometry system for off-table applications (Guo, Yi F.)V1554B,412-419(1991)

Vibration modal analysis using stroboscopic digital speckle pattern interferometry (Wang, Xizhou; Tan, Yushan)V1554A,907-914(1991)

VISAR: displacement-mode data reduction (Hemsing, Willard F.)V1346,141-150(1991)

VISAR: line-imaging interferometer (Hemsing, Willard F.; Mathews, Allen R.; Warnes, Richard H.; Whittemore, Gerald R.)V1346,133-140(1991)

Wavemeter for tuning solid state lasers (Goad, Joseph H.; Rinsland, Pamela L.; Kist, Edward J.; Irick, Steven C.)V1410,107-115(1991)

White-light moire phase-measuring interferometry (Stahl, H. P.)V1332,720-730(1991)

White-light speckle applied to composites (Guo, Maolin; Zhao, Yin; Wu, Zhongming; Du, Shanyi)V1554A,657-658(1991)

White-light transmission holographic interferometry using chromatic corrective filters (Grover, Chandra P.)V1332,132-141(1991)

X-ray interferometric observatory (Martin, Christopher)V1549,203-220(1991)

Ion beams—see beams

Kinoforms—see also holographic optical elements

High-resolution micro-objective lens with kinoform corrector (Lenkova, Galina A.; Churin, Evgeny G.)V1574,235-242(1991)

High-speed apo-lens with kinoform element (Gan, Michael A.; Potyemin, Igor S.; Perveev, Anatoly F.)V1574,243-249(1991)

Laser applications—see also laser chemistry; lidar; microelectronic processing; microlithography; scanning

Ablation of ITO and TO films from glass substrates (Meringdal, Frode; Slinde, Harald)V1503,292-298(1991)

Absorption behavior of ceramic materials irradiated with excimer lasers (Toenshoff, Hans K.; Gedrat, Olaf)V1377,38-44(1991)

Accuracy of aerosol size measurements and the spectrum of applied laser light (Drobnik, Antoni; Pieszynski, Krzysztof)V1391,378-381(1991)

Adaptive optics: a progress review (Hardy, John W.)V1542,2-17(1991)

Advanced Laser Concepts and Applications (Singer, Sidney, ed.)V1501(1991)

Advanced technology development for high-fluence laser applications (Novaro, Marc; Andre, Michel L.)V1501,183-193(1991)

Advances in color laser printing (Tompkins, Neal)V1458,154-154(1991)

Analysis of laser welding process with the mathematical model GMDH (De Iorio, I.; Sergi, Vincenzo; Tagliaferri, V.)V1397,787-790(1991)

Antitumor drugs as photochemotherapeutic agents (Andreoni, Alessandra; Colasanti, Alberto; Kisslinger, Annamaria; Malatesta, Vincenzo; Mastrocinque, Michele; Roberti, Giuseppe)V1525,351-366(1991)

Applications of excimer lasers in microelectronics (Yu, Chang; Sandhu, Gurtej S.; Mathews, V. K.; Doan, Trung T.)V1598,186-197(1991)

Applications of excimer lasers to combustion research (Davis, Steven J.; Allen, Mark G.)V1377,113-118(1991)

Applications of laser ranging to ocean, ice, and land topography (Degnan, John J.)V1492,176-186(1991)

Applications of laser techniques in fluid mechanics (Chan, W. K.; Liu, C. Y.; Wong, Y. W.)V1399,82-89(1991)

Aspects of keyhole/melt interaction in high-speed laser welding (Beck, M.; Berger, Peter; Dausinger, Friedrich; Huegel, Helmut)V1397,769-774(1991)

Carbon dioxide laser effects on caries-like lesions of dental enamel (Featherstone, John D.; Zhang, S. H.; Shariati, M.; McCormack, S. M.)V1424,145-149(1991)

Characteristics of sprinklers and water spray mists for fire safety (Jackman, Louise A.; Lavelle, Stephen P.; Nolan, P. F.)V1358,831-842(1991)

Chick chorioallantoic membrane for the study of synergistic effects of hyperthermia and photodynamic therapy (Kimel, Sol; Svaasand, Lars O.; Hammer-Wilson, Marie J.; Gottfried, Varda; Berns, Michael W.)V1525,341-350(1991)

Clinical fluorescence diagnosis of human bladder carcinoma following low-dose photofrin injection (Baert, Luc; Berg, Roger; van Damme, B.; D'Hallewin, Mich A.; Johansson, Jonas; Svanberg, Katarina; Svanberg, Sune)V1525,385-390(1991)

Compact free-electron laser at the Los Alamos National Laboratory (Chan, Kwok-Chi D.; Meier, Karl L.; Nguyen, Dinh C.; Sheffield, Richard L.; Wang, Tai-San; Warren, Roger W.; Wilson, William; Young, Lloyd M.)V1552,69-78(1991)

Comparative studies on hyperthermia induced by laser light, microwaves, and ultrasonics (Greguss, Pal)V1525,313-324(1991)

Comparison of welding results with stable and unstable resonators (Franek, Joachim; Du, Keming; Pflueger, Silke; Imhoff, Ralf; Loosen, Peter)V1397,791-795(1991)

Computational and experimental progress on laser-activated gas avalanche switches for broadband, high-power electromagnetic pulse generation (Mayhall, David J.; Yee, Jick H.; Villa, Francesco)V1378,101-114(1991)

Computer-controlled pulse laser system for dynamic holography (Chen, Terry Y.; Ju, M. S.; Lee, C. Y.; Lo, M. P.)V1554B,92-98(1991)

Controlled structuring of polymer surfaces by UV laser irradiation (Bahners, Thomas; Kesting, Wolfgang; Schollmeyer, Eckhard)V1503,206-214(1991)

Cryogenic radiometers and intensity-stabilized lasers for EOS radiometric calibrations (Foukal, Peter; Hoyt, Clifford C.; Jauniskis, L.)V1493,72-79(1991)

Cumulative effect and cutting quality improvement of XeCl laser ablation of PMMA (Lou, Qihong; Guo, Hongping)V1598,221-226(1991)

Current state of research and technical application of lasers in Slovakia: their special use in biophysical research at Comenius University in Bratislava (Chorvat, Dusan; Hianik, T.)V1402,194-197(1991)

Deformation of a gamma/gamma WASPALOY after laser shock (Bourda, C.; Puig, Thierry T.; Decamps, B.; Condat, M.)V1502,148-159(1991)

Deposition of silica coatings on Incoloy 800H substrates using a high-power laser (Fellowes, Fiona C.; Steen, William M.)V1502,213-222(1991)

Development of a high-speed high-precision laser plotter (Tachihara, Satoru; Miyoshi, Tamihiro)V1527,305-314(1991)

Development of a large-screen high-definition laser video projection system (Clynick, Tony)V1456,51-57(1991)

Development of a laser Doppler system for measurement of velocity fields in PVT crystal growth systems (Jones, O. C.; Glicksman, M. E.; Lin, Jen T.; Kim, G. T.; Singh, N. B.)V1557,202-208(1991)

Development of a precision high-speed flying spot position detector (Jodoin, Ronald E.; Loce, Robert P.; Nowak, William J.; Costanza, Daniel W.)V1398,61-70(1991)

Development of laser bonding as a manufacturing process for inner lead bonding (Hayward, James D.)V1598,164-174(1991)

Dynamics of the optical parameters of molten silicon during nanosecond laser annealing (Boneberg, J.; Yavas, O.; Mierswa, B.; Leiderer, Paul)V1598,84-90(1991)

Economic factors for a free-space laser communication system (Begley, David L.; Marshalek, Robert G.)V1522,234-242(1991)

Effective temperatures of polymer laser ablation (Furzikov, Nickolay P.)V1503,231-235(1991)

Effect of laser irradiation on superconducting properties of laser-deposited YBa2Cu3O7 thin films (Singh, Rajiv K.; Bhattacharya, Deepika; Narayan, Jagdish; Jahncke, Catherine; Paesler, Michael A.)V1394,203-213(1991)

Efficient laser cleaning of small particulates using pulsed laser irradiation synchronized with liquid-film deposition (Tam, Andrew C.; Zapka, Werner; Ziemlich, Winfrid)V1598,13-18(1991)

Emerging laser technologies (Forrest, Gary T.)V1520,37-57(1991)

Energy balance in high-power CO2 laser welding (Del Bello, Umberto; Rivela, Cristina; Cantello, Maichi; Penasa, Mauro)V1502,104-116(1991)

European industrial laser market (Mayer, Arnold)V1517,15-27(1991)

Excess carrier profile in a moving Gaussian spot illumination semiconductor panel and its use as a laser sensor (Rahnavard, Mohammad H.; Bakhtazad, A.)V1385,123-130(1991)

Excimer laser ceramic and metal surface alloying applications (Hontzopoulos, Elias I.; Zervaki, A.; Zergioti, Y.; Hourdakis, G.; Raptakis, E.; Giannacopoulos, A.; Fotakis, Costas)V1397,761-768(1991)

Excimer laser machining of optical fiber taps (Coyle, Richard J.; Serafino, Anthony J.; Grimes, Gary J.; Bortolini, James R.)V1412,129-137(1991)

Excimer Laser Materials Processing and Beam Delivery Systems (Piwczyk, Bernhard P., ed.)V1377(1991)

Excimer laser micromachining for passive fiber coupling to polymeric waveguide devices (Booth, Bruce L.; Hohman, James L.; Keating, Kenneth B.; Marchegiano, Joseph E.; Witman, Sandy L.)V1377,57-63(1991)

Excimer laser patterning of flexible materials (Kollia, Z.; Hontzopoulos, Elias I.)V1503,215-222(1991)

Excimer laser processing of ceramics and fiber-reinforced polymers assisted by a diagnostic system (Geiger, Manfred; Lutz, Norbert; Biermann, Stephan)V1503,238-248(1991)

Excimer laser processing of diamond-like films (Ageev, Vladimir P.; Glushko, T. N.; Dorfman, V. F.; Kuzmichov, A. V.; Pypkin, B. N.)V1503,453-462(1991)

Excimer Lasers and Applications III (Letardi, Tommaso; Laude, Lucien D., eds.)V1503(1991)

Excimer laser surface treatment of ceramics (Hourdakis, G.; Hontzopoulos, Elias I.; Tsetsekou, A.; Zampetakis, Th.; Stournaras, C.)V1503,249-255(1991)

Experimental investigations in laser microsoldering (Hartmann, Martin; Bergmann, Hans W.; Kupfer, Roland)V1598,175-185(1991)

Experimental study of the parameters of the laser-induced plasma observed in welding of iron targets with continuous high-power CO2 lasers (Poueyo, Anne; Sabatier, Lilian; Deshors, G.; Fabbro, Remy; de Frutos, Angel M.; Bermejo, Dionisio; Orza, Jose M.)V1502,140-147(1991)

Experiments on convection in laser-melted pools (Caillibotte, Georges; Kechemair, Didier; Sabatier, Lilian)V1502,117-122(1991)

Eyesafe laser application in military and law enforcement training (Mosbrooker, Michael L.)V1419,107-114(1991)

Fiber Optic and Laser Sensors VIII (DePaula, Ramon P.; Udd, Eric, eds.)V1367(1991)

Fiber Optic and Laser Sensors IX (DePaula, Ramon P.; Udd, Eric, eds.)V1584(1991)

Free-electron lasers as light sources for basic research (van Amersfoort, P. W.)V1504,25-36(1991)

Future trends in the medical laser industry (Harris, David M.)V1396,696-698(1991)

Highly conductive amorphous-ferrite formed by excimer laser material processing (Kashiwabara, S.; Watanabe, Kazuhiro; Fujimoto, R.)V1397,803-806(1991)

High-peak-power Nd:glass laser facilities for end users (Hunt, John T.)V1410,2-14(1991)

High-resolution display using a laser-addressed ferroelectric liquid-crystal light valve (Nakajima, Hajime; Kisaki, Jyunko; Tahata, Shin; Horikawa, Tsuyoshi; Nishi, Kazuro)V1456,29-39(1991)

High-resolution high-speed laser recording and printing using low-speed polygons (Razzaghi-Masoud, Mahmoud)V1458,145-153(1991)

High-speed nonsilver lithographic system for laser direct imaging (DoMinh, Thap)V1458,68-68(1991)

Hydrodynamic efficiency as determined from implosion experiments at wavelength = 0.26 um (Koenig, Michel; Fabre, Edouard; Malka, Victor; Hammerling, Peter; Michard, Alain; Boudenne, Jean-Michel; Fews, Peter)V1502,338-342(1991)

IC rewiring by laser microchemistry (Pelous, G.; Guern, Yves; Gobleid, D.; David, J.; Chion, A.; Tonneau, Didier)V1598,149-158(1991)

Image analysis of two impinging jets using laser-induced fluorescence and smoke flow visualization (Shoe, Bridget; Disimile, Peter J.; Savory, Eric; Toy, Norman; Tahouri, Bahman)V1521,34-45(1991)

Image quality metrics for volumetric laser displays (Williams, Rodney D.; Donohoo, Daniel)V1457,210-220(1991)

Implementation of a 3-D laser imager-based robot navigation system with location identification (Boltinghouse, Susan T.; Burke, James; Ho, Daniel)V1388,14-29(1991)

Industrial and Scientific Uses of High-Power Lasers (Billon, Jean P.; Fabre, Edouard, eds.)V1502(1991)

Industrial applications of metal vapor lasers (Gokay, Cem)V1412,28-31(1991)

Industrial laser market (Belforte, David A.)V1520,93-117(1991)

Industrial lasers in Japan (Karube, Norio)V1517,1-14(1991)

Industrial lasers in the United States (Rutt, James)V1517,28-43(1991)

Influence of laser melting on microstructure and properties of M2 high-speed tool steel (Kusinski, Jan P.)V1391,387-392(1991)

Influence of the distance between the welding head and the surface material of the weld shape (Boudot, C.; Vastra, I.)V1502,177-189(1991)

Influence of the shielding gas internal bore welding (Vastra, I.; Boudot, C.)V1502,190-202(1991)

Influencing adherence properties of polymers by excimer laser radiation (Breuer, J.; Metev, S.; Sepold, Gerd; Krueger, G.; Hennemann, O. D.)V1503,223-230(1991)

In-process laser beam position sensing (Chen, Shang-Liang; Li, L.; Modern, P. J.; Steen, William M.)V1502,123-134(1991)

In-situ growth of Y1Ba2Cu3O7-x thin films using XeCl excimer and Nd:YAG lasers (Gerri, Mireille; Marine, W.; Mathey, Yves; Sentis, Marc L.; Delaporte, Philippe C.; Fontaine, Bernard L.; Forestier, Bernard M.)V1503,280-291(1991)

Inspection system using still vision for a rotating laser-textured dull roll (Torao, Akira; Uchida, Hiroyuki; Moriya, Susumu; Ichikawa, Fumihiko; Kataoka, Kenji; Wakui, Tsuneyoshi)V1358,843-850(1991)

Integration of diagnostics in high-power laser systems for optimization of laser material processing (Biermann, Stephan; Geiger, Manfred)V1415,330-341(1991)

Interferometry, streak photography, and stereo photography of laser-driven miniature flying plates (Paisley, Dennis L.; Montoya, Nelson I.; Stahl, David B.; Garcia, Ismel A.)V1358,760-765(1991)

Investigations on excimer-laser-treated Cu/Cr contact materials (Schubert, Emil; Rosiwal, S.; Bergmann, Hans W.)V1503,299-309(1991)

Job shop market (Belforte, David A.)V1517,176-196(1991)

KrF laser ablation of polyurethane (Kueper, Stephan; Brannon, James H.)V1598,27-35(1991)

Large-scale industrial application for excimer lasers: via-hole-drilling by photoablation (Bachmann, Friedrich G.)V1361,500-511(1991); V1377,18-29(1991)

Laser application in chosen maritime economy divisions (Kirkiewicz, Jozef)V1391,351-360(1991)

Laser applications for multisensor systems (Smolka, Greg L.; Strother, George T.; Ott, Carl)V1498,70-80(1991)

Laser-based triangulation techniques in optical inspection of industrial structures (Clarke, Timothy A.; Grattan, Kenneth T.; Lindsey, N. E.)V1332,474-486(1991)

Laser beam scanner for uniform halftones (Ando, Toshinori)V1458,128-132(1991)

Laser drilling vias in GaAs wafers (Riley, Susan; Schick, Larry A.)V1598,118-120(1991)

Laser-formed structures to facilitate TAB bonding (Ledermann, Peter G.; Johnson, Glen W.; Ritter, Mark B.)V1598,160-163(1991)

Laser-generated 3-D prototypes (DeAngelis, Franco E.)V1598,61-70(1991)

Laser generation of Stoneley waves at liquid-solid boundaries (Ward, Barry; Emmony, David C.)V1358,1228-1236(1991)

Laser heating and evaporation of glass and glass-borning materials and its application for creating MOC (Veiko, Vadim P.; Yakovlev, Evgeni B.; Frolov, V. V.; Chujko, V. A.; Kromin, A. K.; Abbakumov, M. O.; Shakola, A. T.; Fomichov, P. A.)V1544,152-163(1991)

Laser high-speed photography systems used to ammunition measures and tests (Wang, Yuren)V1346,331-337(1991)

Laser Image Detection System (LIDS): laser-based imaging (Hilton, Peter J.; Gabric, R.; Walternberg, P. T.)V1385,27-31(1991)

Laser-induced fluorescence imaging of coronary arteries for open-heart surgery applications (Taylor, Roderick S.; Gladysz, D.; Brown, D.; Higginson, Lyall A.)V1420,183-192(1991)

Laser-induced gold deposition for thin-film circuit repair (Baum, Thomas H.; Comita, Paul B.; Kodas, Toivo T.)V1598,122-131(1991)

Laser-induced metal deposition and laser cutting techniques for fixing IC design errors (Shaver, David C.; Doran, S. P.; Rothschild, Mordechai; Sedlacek, J. H.)V1596,46-50(1991)

Laser light sheet investigation into transonic external aerodynamics (Towers, Catherine E.; Towers, David P.; Judge, Thomas R.; Bryanston-Cross, Peter J.)V1358,952-965(1991)

Laser linac in vacuum: assessment of a high-energy particle accelerator (Moore, Gerald T.; Bochove, Erik J.; Scully, Marlan O.)V1497,328-337(1991)

Laser linac: nondiffractive beam and gas-loading effects (Bochove, Erik J.; Moore, Gerald T.; Scully, Marlan O.; Wodkiewicz, K.)V1497,338-347(1991)

Laser market in the 1990s (Levitt, Morris R.)V1520,1-36(1991)

LAser Microprobe Mass Spectrometry (LAMMS) in dental science: basic principles, instrumentation, and applications (Rechmann, Peter; Tourmann, J. L.; Kaufmann, Raimund)V1424,106-115(1991)

Laser optical fiber high-speed camera (Xia, Sheng-jie; Yang, Ye-min; Tang, Di-zhu)V1358,43-45(1991)

Laser processes for repair of thin-film wiring (Wassick, Thomas A.)V1598,141-148(1991)

Laser process for personalization and repair of multichip modules (Mueller, Heinrich G.; Galanakis, Claire T.; Sommerfeldt, Scott C.; Hirsch, Tom J.; Miracky, Robert F.)V1598,132-140(1991)

Lasers and electro-optic technology in natural resource management (Greer, Jerry D.)V1396,342-352(1991)

Laser sensing in the iron-making blast furnace (Scott, Chris J.)V1399,137-144(1991)

Laser set for the investigations of the NO2 contents in atmosphere (Makuchowski, Jozef; Pokora, Ludwik J.; Ujda, Zbigniew; Wawer, Janusz)V1391,348-350(1991)

Lasers in dentistry (Keller, Ulrich)V1525,282-288(1991)

Lasers in Microelectronic Manufacturing (Braren, Bodil, ed.)V1598(1991)

Lasers in Orthopedic, Dental, and Veterinary Medicine (Dederich, Douglas N.; O'Brien, Stephen J.; Trent, Ava M.; Wigdor, Harvey A., eds.)V1424(1991)

Laser systems for precision micromachining (Nowicki, Marian; Niechoda, Zygmunt J.)V1391,370-377(1991)

Laser Technology III (Wolinski, Wieslaw; Wolczak, Bohdan K.; Gajda, Jerzy K.; Gajda, Danuta; Romaniuk, Ryszard S., eds.)V1391(1991)

Laser ultrasonics: generation and detection considerations for improved signal-to-noise ratio (Wagner, James W.; Deaton, John B.; McKie, Andrew D.; Spicer, James B.)V1332,491-501(1991)

Laser welding of INCONEL 600 (Daurelio, Giuseppe; Dionoro, G.; Memola Capece Minutolo, F.)V1397,783-786(1991)

LEAF: a fiber-optic fluorometer for field measurement of chlorophyll fluorescence (Mazzinghi, Piero)V1510,187-194(1991)

Light-induced polishing of evaporating surface (Tokarev, Vladimir N.; Konov, Vitaly I.)V1503,269-278(1991)

Low-power laser effects in equine traumatology and postsurgically (Antikas, Theo G.)V1424,186-197(1991)

Market for high-power CO2 lasers (Fortino, Dennis J.)V1517,61-84(1991)

Market for industrial excimer lasers (Zaal, Gerard)V1517,100-116(1991)

Market for low-power CO2 lasers (Wheeler, John P.)V1517,44-60(1991)

Market for multiaxis laser machine tools (Ream, Stanley L.)V1517,136-149(1991)

Markets for marking systems (Austin, Patrick D.)V1517,150-175(1991)

Material influence on cutting and drilling of metals using copper vapor lasers (Kupfer, Roland; Bergmann, Hans W.; Lingenauer, Marion)V1598,46-60(1991)

Measurement of interfacial tension by automated video techniques (Deason, Vance A.; Miller, R. L.; Watkins, Arthur D.; Ward, Michael B.; Barrett, K. B.)V1332,868-876(1991)

Measurement of particles and drops in combusting flows (Ereaut, Peter R.)V1554B,556-565(1991)

Measurements of the atmospheric turbulence spectrum and intermittency using laser scintillation (Frehlich, Rod G.)V1487,10-18(1991)

Medical marketplace (Moretti, Michael)V1520,118-131(1991)

Melt dynamics and surface deformation in processing with laser radiation (Kreutz, Ernst W.; Pirch, Norbert)V1502,160-176(1991)

Microprocessor-based laser range finder (Rao, M. K.; Tam, Siu-Chung)V1399,116-121(1991)

Mid-infrared laser applications (Daly, John G.)V1419,94-99(1991)

Modeling of high-power laser welding (Schuoecker, Dieter)V1397,745-751(1991)

Modular streak camera for laser ranging (Prochazka, Ivan; Hamal, Karel; Schelev, Mikhail Y.; Lozovoi, V. I.; Postovalov, V. E.)V1358,574-577(1991)

NADH-fluorescence in medical diagnostics: first experimental results (Schramm, Werner; Naundorf, M.)V1525,237-241(1991)

Nd:YAG laser machining with multilevel resist kinoforms (Ekberg, Mats; Larsson, Michael; Bolle, Aldo; Hard, Sverker)V1527,236-239(1991)

Nd:YAG laser market (Llewellyn, Steven A.)V1517,85-99(1991)

Near-UV laser ablation of doped polymers (Ihlemann, Juergen; Bolle, Matthias; Luther, Klaus; Troe, Juergen)V1361,1011-1019(1991)

New approaches to signal-to-noise ratio optimization in background-limited photothermal measurements (Rice, Patrick D.; Thorne, John B.; Bobbitt, Donald R.)V1435,104-113(1991)

New CO2 laser equipped with high-peak pulse power and high-speed drilling process (Hongu, Hitoshi; Karasaki, Hidehiko)V1501,198-204(1991)

No-blind-area one-photograph HNDT tire analyzer (Ge, Fang X.; Xiong, Xian M.)V1554B,785-789(1991)

Numerical modeling of laser-matter interaction in high-intensity laser applications (Ocana, Jose L.)V1397,813-820(1991)

Optical delay tester (Wakana, Shin-ichi; Nagai, Toshiaki; Hama, Soichi; Goto, Yoshiro)V1479,283-290(1991)

Percutaneous laser balloon coagulation of accessory pathways (McMath, Linda P.; Schuger, Claudio D.; Crilly, Richard J.; Spears, J. R.)V1425,165-171(1991)

Photoablative etching of materials for optoelectronic integrated devices (Lemoine, Patrick; Magan, John D.; Blau, Werner)V1377,45-56(1991)

Poly(bis-alkylthio-acetylen): a dual-mode laser-sensitive material (Baumann, Reinhard; Bargon, Joachim; Roth, Hans-Klaus)V1463,638-643(1991)

Present and future trends of laser materials processing in Japan (Matsunawa, Akira)V1502,60-71(1991)

Producing a uniform excimer laser beam for materials processing applications (Bunis, Jenifer L.; Abele, C. C.; Campbell, James D.; Caudle, George F.)V1377,30-36(1991)

Production of laser simulation systems for adverse environments (Rodrigues, F. C.; Simao, Jose V.; Oliveira, Joao; Freitas, Jose C.; Carvalho, Fernando D.)V1399,90-97(1991)

Promising applications of scanning electron-beam-pumped laser devices in medicine and biology (Katsap, Victor N.; Koshevoy, Alexander V.; Meerovich, Gennady A.; Ulasjuk, Vladimir N.)V1420,259-265(1991)

Properties of the new UCSB free-electron lasers (Ramian, Gerald)V1552,57-68(1991)

Pulsed-laser deposition of YBa2Cu3O7-x thin films: processing, properties, and performance (Muenchausen, Ross E.; Foltyn, Stephen R.; Wu, Xin D.; Dye, Robert C.; Nogar, Nicholas S.; Carim, A. H.; Heidelbach, F.; Cooke, D. W.; Taber, R. C.; Quinn, Rod K.)V1394,221-229(1991)

Quality control of laser cutting process by surface morphology (Sergi, Vincenzo)V1397,775-781(1991)

Real-time actuating of laser power and scanning velocity for thermal regulation during laser hardening (Bataille, F.; Kechemair, Didier; Houdjal, J.)V1502,135-139(1991)

Recent developments of x-ray lithography in Canada (Chaker, Mohamed; Boily, S.; Ginovker, A.; Jean, A.; Kieffer, J. C.; Mercier, P. P.; Pepin, Henri; Leung, Pak K.; Currie, John F.; Lafontaine, H.)V1465,16-25(1991)

Reflow soldering of fine-pitch devices using a Nd:YAG laser (Glynn, Thomas J.; Flanagan, Aiden J.; Redfern, R. M.)V1598,200-205(1991)

Remote 3-D laser topographic mapping with dental application (Altschuler, Bruce R.)V1380,238-247(1991)

Role of buffer layers in the laser-ablated films on metallic substrates (Shaw, David T.; Narumi, E.; Yang, F.; Patel, Sushil)V1394,214-220(1991)

Root resection of endodontically treated teeth by Erbium:YAG laser radiation (Paghdiwala, A. F.)V1424,150-159(1991)

Satellite-borne laser for adaptive optics reference (Greenaway, Alan H.)V1494,386-393(1991)

Scientific laser trends (Messenger, Heather)V1520,58-92(1991)

Selected Papers on Free-Space Laser Communications (Begley, David L., ed.)VMS30(1991)

Sheet-metal cutting market (Berkhahn, Glenn)V1517,117-135(1991)

Simulation of the process of laser beam machining based on the model of surface vaporization (Stankiewicz, Maria; Zachorowski, Jan)V1391,174-180(1991)

Site-specific laser modification (cleavage) of oligodeoxynucleotides (Benimetskaya, L. Z.; Bulychev, N. V.; Kozionov, Andrew L.; Koshkin, A. A.; Lebedev, A. V.; Novozhilov, S. Y.; Stockman, M. I.)V1525,210-211(1991)

Some aspects of the application of the laser flash method for investigations on the thermal diffusivity of porous materials (Drobnik, Antoni; Rozniakowski, Kazimierz; Wojtatowicz, Tomasz W.)V1391,361-369(1991)

Space-qualified streak camera for the Geodynamic Laser Ranging System (Johnson, C. B.; Abshire, James B.; Zagwodzki, Thomas W.; Hunkler, L. T.; Letzring, Samuel A.; Jaanimagi, Paul A.)V1346,340-370(1991)

Structures of laser hardening of high-speed steel SW7M (Adamiak, S.; Bylica, A.; Kuzma, Marian)V1391,382-386(1991)

Studies on laser dynamic precision measurement of fine-wire diameters (Bao, Liangbi; Chen, Fuyao; Wu, Shixiong; Xu, Jiangtong; Guan, Zhilian)V1332,862-867(1991)

Surface absorptance in CO2 laser steel processing (Covelli, L.; De Iorio, I.; Tagliaferri, V.)V1397,797-802(1991)

Surface layers of metals alloyed with a pulsed laser (Pawlak, Ryszard; Raczynski, Tomasz)V1391,170-173(1991)

Surface nitride synthesis by multipulse excimer laser irradiation (D'Anna, Emilia; Leggieri, Gilberto; Luches, Armando; Martino, M.; Perrone, A.; Majni, G.; Mengucci, P.; Drigo, A. V.; Mihailescu, Ion N.)V1503,256-268(1991)

The EU 194 project: industrial applications of high-power CO2 cw lasers (Quenzer, A.; Beyer, Eckhard; Orza, Jose M.; Ricciardi, G.; Russell, J. D.; Sanz Justes, Pedro; Serafetinides, A.; Schuoecker, Dieter; Thorstensen, B.)V1397,753-759(1991)

The Laser Marketplace 1991 (Forrest, Gary T.; Levitt, Morris R., eds.)V1520(1991)

The Marketplace for Industrial Lasers (Belforte, David A.; Levitt, Morris R., eds.)V1517(1991)

Thermal regulation applied to CO2 laser self-quenching of complex geometry workpieces (Bataille, F.; Kechemair, Didier; Pawlovski, C.; Houdjal, R.)V1397,839-842(1991)

Thin-film selective multishot ablation at 248 nm (Hunger, Hans E.; Pietsch, H.; Petzoldt, Stefan; Matthias, Eckart)V1598,19-26(1991)

Three-dimensional lithography: laser modeling using photopolymers (Heller, Timothy B.)V1454,272-282(1991)

Threshold measurements in laser-assisted particle removal (Lee, Shyan J.; Imen, Kamran; Allen, Susan D.)V1598,2-12(1991)

Time-space resolved optical study of the plasma produced by laser ablation of Ge: the role of oxygen pressure (Vega, Fidel; Solis, Javier; Afonso, Carmen N.)V1397,807-811(1991)

Tomographic imaging using picosecond pulses of light (Hebden, Jeremy C.; Kruger, Robert A.; Wong, K. S.)V1443,294-300(1991)

Total internal reflection mirrors fabricated in polymeric optical waveguides via excimer laser ablation (Trewhella, Jeannine M.; Oprysko, Modest M.)V1377,64-72(1991)

Toward the optimization of passive ground targets in spaceborne laser ranging (Lund, Glenn I.; Renault, Herve)V1492,153-165(1991)

Track-while-image in the presence of background (Mentle, Robert E.; Shapiro, Jeffrey H.)V1471,342-353(1991)

U.S. government market (Speser, Philip L.)V1520,132-150(1991)

USSR-CSFR Joint Seminar on Nonlinear Optics in Control, Diagnostics, and Modeling of Biophysical Processes (Akhmanov, Sergei A.; Zadkov, Victor N., eds.)V1402(1991)

Variation of magnetic properties of iron-contained quartzite irradiated by CO2 laser (Mukhamedgalieva, Anel F.)V1502,223-225(1991)

Laser chemistry—see also chemistry

Analysis of molecular adsorbates by laser-induced thermal desorption (McIver, Robert T.; Hemminger, John C.; Parker, D.; Li, Y.; Land, Donald P.; Pettiette-Hall, C. L.)V1437,124-128(1991)

Applied Spectroscopy in Material Science (Saperstein, David D., ed.)V1437(1991)

Biomedical applications of laser photoionization (Xiong, Xiaoxiong; Moore, Larry J.; Fassett, John D.; O'Haver, Thomas C.)V1435,188-196(1991)

Determination of gasoline fuel properties by Raman spectroscopy (Clarke, Richard H.; Chung, W. M.; Wang, Q.; De Jesus, Stephen T.)V1437,198-204(1991)

DNA photoproducts formed using high-intensity 532 nm laser radiation (Kochevar, Irene E.; Hefetz, Yaron; Dunn, D. A.; Buckley, L. M.; Hillenkamp, Franz)V1403,756-763(1991)

Excimer laser deposition and characterization of tin and tin-oxide films (Borsella, E.; De Padova, P.; Larciprete, Rosanna)V1503,312-320(1991)

Excimer laser fabrication of waveguide devices (Stiller, Marc A.)V1377,73-78(1991)

Excimer Laser Materials Processing and Beam Delivery Systems (Piwczyk, Bernhard P., ed.)V1377(1991)

High-efficiency resonance ionization mass spectrometric analysis by external laser cavity enhancement techniques (Johnson, Stephen G.; Rios, E. L.; Miller, Charles M.; Fearey, Bryan L.)V1435,292-297(1991)

Hole-burning spectroscopy of phthalocyanine Langmuir-Blodget films (Adamec, F.; Ambroz, M.; Dian, J.; Vacha, M.; Hala, J.; Balog, P.; Brynda, E.)V1402,82-84(1991)

Importance of pulsed laser intensity in porphyrin-sensitized ÑADH photo-oxidation (Kirveliene, V.; Rotomskis, Richardas; Juodka, B.; Piskarskas, Algis S.)V1403,582-584(1991)

Laser Applications in Life Sciences (Akhmanov, Sergei A.; Poroshina, Marina Y., eds.)V1403(1991)

Laser-assisted deposition of thin films onto transparent substrates from liquid-phase organometallic precursor: iron acetylacetonate (Shafeev, George A.; Laude, Lucien D.)V1503,321-329(1991)

Laser chemistry of dimethylcadmium adsorbed on silicon: 308- versus 222-nm laser excitation (Simonov, Alexander P.; Varakin, Vladimir N.; Panesh, Anatoly M.; Lunchev, V. A.)V1503,330-337(1991)

Laser desorption jet cooling spectroscopy (de Vries, Mattanjah S.; Hunziker, Heinrich E.; Meijer, Gerard; Wendt, H. R.)V1437,129-137(1991)

Laser Doppler measurements of ameboid cytoplasmic streaming and problems of mathematical modeling of intracellular hydrodynamics (Priezzhev, Alexander V.; Proskurin, S. G.; Romanovsky, Yuri M.)V1402,107-113(1991)

Laser-induced thermal desorption studies of surface reaction kinetics (George, Steven M.; Coon, P. A.; Gupta, P.; Wise, M. L.)V1437,157-165(1991)

Laser microphotolysis of biological objects with the application of UV solid state lasers (Arutunian, A. H.; Hovanessian, V. A.; Sarkissian, K. A.)V1402,102-106(1991)

Measurement of trace isotopes by photon burst mass spectrometry (Fairbank, William M.; Hansen, C. S.; LaBelle, R. D.; Pan, X. J.; Chamberlin, E. P.; Fearey, Bryan L.; Gritzo, R. E.; Keller, Richard A.; Miller, Charles M.; Oona, H.)V1435,86-89(1991)

Measurement of triplet optical densities of organic compounds by means of CW laser excitation (Pavlopoulos, Theodore G.)V1437,168-183(1991)

Metal vapor gain media based on multiphoton dissociation of organometallics (Samoriski, Brian; Wiedeger, S.; Villarica, M.; Chaiken, Joseph)V1412,12-18(1991)

Method of laser fluorescence microphotolysis (Shvec, Peter; Kvasnichka, P.; Shipocz, Tibor; Chorvat, Dusan)V1402,78-81(1991)

Molecular dynamics of stilbene molecule under laser excitation (Vachev, V. D.; Zadkov, Victor N.)V1403,487-496(1991)

Nonlinear absorption of organic compounds in the picosecond laser radiation field (Arutunian, A. H.; Hovanessian, V. A.; Sarkissian, K. A.)V1402,2-6(1991)

Nonlinear properties of oriented purple membrane films derived from second-harmonic generation under picosecond excitation: prospect of electro-optical measurements of ultrafast photoelectric respon (Sharkov, A. V.; Gillbro, T.)V1403,434-438(1991)

Novel approach for the refractive index gradient measurement in microliter volumes using fiber-optic technology (Synovec, Robert E.; Renn, Curtiss N.)V1435,128-139(1991)

Optical Methods for Ultrasensitive Detection and Analysis: Techniques and Applications (Fearey, Bryan L., ed.)V1435(1991)

Photoablative etching of materials for optoelectronic integrated devices (Lemoine, Patrick; Magan, John D.; Blau, Werner)V1377,45-56(1991)

Photochemical changes of rare-earth valent state in gamma-irradiated CaF2:Pr crystals by the excimer laser radiation: investigation and application (Lukishova, Svetlana G.; Obidin, Alexey Z.; Vartapetov, Serge K.; Veselovsky, Igor A.; Osiko, Anatoly V.; Tulajkova, Tamara V.; Ter-Mikirtychev, Valery V.; Mendez, Nestor R.)V1503,338-345(1991)

Photo resonance excitation and ionization characteristics of atoms by pulsed laser irradiation (Tanazawa, Takeshi; Adachi, Hajime A.; Nakahara, Ktsuhiko; Nittoh, Koichi; Yoshida, Toshifumi; Yoshida, Tadashi; Matsuda, Yasuhiko)V1435,310-321(1991)

Picosecond laser cross: linking histones to DNA in chromatin: implication in studying histone/DNA interactions (Angelov, D.; Dimitrov, S.; Keskinova, E.; Pashev, I.; Russanova, V.; Stefanovsky, Yu.)V1403,230-239(1991)

Poly(bis-alkylthio-acetylen): a dual-mode laser-sensitive material (Baumann, Reinhard; Bargon, Joachim; Roth, Hans-Klaus)V1463,638-643(1991)

Provitamin D photoisomerization kinetics upon picosecond laser irradiation: role of previtamin conformational nonequilibrium (Terenetskaya, I. P.; Repeyev, Yu A.)V1403,500-503(1991)

Pulsed-laser deposition of oxides over large areas (Greer, James A.; Van Hook, H. J.)V1377,79-90(1991)

Pulsed photothermal spectroscopy applied to lanthanide and actinide speciation (Berg, John M.; Morris, David E.; Clark, David L.; Tait, C. D.; Woodruff, William H.; Van Der Sluys, William G.)V1435,331-337(1991)

Second-harmonic generation as an in situ diagnostic for corrosion (Klenerman, David; Spowage, K.; Walpole, B.)V1437,95-102(1991)

Semiconductor lasers in analytical chemistry (Patonay, Gabor; Antoine, Miquel D.; Boyer, A. E.)V1435,52-63(1991)

Study of aggregation phenomenon of hematoporphyrin derivative by laser microphotolysis (Arutunian, A. H.; Hovanessian, V. A.; Sarkissian, K. A.)V1403,585-587(1991)

Subattomole detection in the condensed phase by nonlinear laser spectroscopy based on degenerate four-wave mixing (Tong, William G.)V1435,90-94(1991)

Symmetrization analysis of lattice-vibrational modes and study of Raman-IR spectra for B-BaB2O4 (Hong, Shuili)V1437,194-197(1991)

Theory and molecular dynamics simulation of one-photon electronic excitation of multiatomic molecules (Grishanin, B. A.; Vachev, V. D.; Zadkov, Victor N.)V1402,44-52(1991)

UV laser-induced photofragmentation and photoionization of dimethylcadmium chemisorbed on silicon (Simonov, Alexander P.; Varakin, Vladimir N.; Panesh, Anatoly M.)V1436,20-30(1991)

Laser damage

Absorption calorimetry and laser-induced damage threshold measurements of antireflective-coated ZnSe and metal mirrors at 10.6 um (Rahe, Manfred; Oertel, E.; Reinhardt, L.; Ristau, Detleu; Welling, Herbert)V1441,113-126(1991)

Advances in clinical percutaneous excimer laser angioplasty (Viligiardi, Riccardo; Pini, Roberto; Salimbeni, Renzo; Galiberti, Sandra)V1425,72-74(1991)

Analysis of kinetic rate modeling of thermal damage in laser-irradiated tissue (Rastegar, Sohi; Glenn, T.)V1427,300-306(1991)

Applications of pulsed photothermal deflection technique in the study of laser-induced damage in optical coatings (Wu, Zhouling; Reichling, M.; Fan, Zheng X.; Wang, Zhi-Jiang)V1441,214-227(1991)

AR layer properties for high-power laser prepared by neutral-solution processing (Wu, Fang F.; Su, Kai L.)V1519,347-349(1991)

Boiling process in PMMA irradiated by CO2, DF and HF laser radiations (Joeckle, Rene C.; Rapp, Gerard; Sontag, Andre)V1397,683-687(1991)

Bulk darkening of flux-grown KTiOPO4 (Rockafellow, David R.; Teppo, Edward J.; Jacco, John C.)V1561,112-118(1991)

Changes in collagen birefringence: a quantitative histologic marker of thermal damage in skin (Thomsen, Sharon L.; Cheong, Wai-Fung; Pearce, John A.)V1422,34-42(1991)

Characterization of subsurface damage in glass and metal optics (Polvani, Robert S.; Evans, Chris J.)V1441,173-173(1991)

Closed-form onset threshold analysis of defect-driven surface and bulk laser damage (O'Connell, Robert M.)V1441,406-419(1991)

Coating development for high-energy KrF excimer lasers (Boyer, James D.; Mauro, Billie R.; Sanders, Virgil E.)V1377,92-98(1991)

Comparison of laser-induced damage of optical crystals from the USA and USSR (Soileau, M. J.; Wei, Tai-Huei; Said, Ali A.; Chapliev, N. I.; Garnov, Sergei V.; Epifanov, Alexandre S.)V1441,10-15(1991)

Correlation between the laser-induced breakdown threshold in solids, liquids, and gases (Bettis, Jerry R.)V1441,521-534(1991)

Correlation of surface topography and coating damage with changes in the responsivity of silicon PIN photodiodes (Huffaker, Diana L.; Walser, Rodger M.; Becker, Michael F.)V1441,365-380(1991)

Damage induced by pulsed IR laser radiation at transitions between different tissues (Frenz, Martin; Greber, Charlotte M.; Romano, Valerio; Forrer, Martin; Weber, Heinz P.)V1427,9-15(1991)

Damage resistant optics for a megajoule solid state laser (Campbell, John H.; Rainer, Frank; Kozlowski, Mark R.; Wolfe, C. R.; Thomas, Ian M.; Milanovich, Fred P.)V1441,444-456(1991)

Damage testing of optical components for high-power excimer lasers (Mann, Klaus R.; Gerhardt, Harald)V1503,176-184(1991)

Damage to InAs surface from long-pulse 10.6um laser radiation (Kovalev, Valeri I.)V1441,536-540(1991)

Determination of SBS-induced damage limits in large fused silica optics for intense, time-varying laser pulses (Kyrazis, Demos T.; Weiland, Timothy L.)V1441,469-477(1991)

Detonation mechanism of breakdowns in dielectrics (Bobilkov, G. P.; Genkin, Vladimir N.)V1440,98-104(1991)

Development and implementation of MIS-36477 laser damage certification of designator optical components (Mordaunt, David W.; Nieuwsma, Daniel E.)V1441,27-30(1991)

Development of damage resistant optics for KrF excimer lasers (Boyer, James D.; Mauro, Billie R.; Sanders, Virgil E.)V1441,255-261(1991)

Dispersion of n2 in solids (Sheik-Bahae, Mansoor; Hutchings, David C.; Hagan, David J.; Soileau, M. J.; Van Stryland, Eric W.)V1441,430-443(1991)

Dosimetry for lasers and light in dermatology: Monte Carlo simulations of 577nm-pulsed laser penetration into cutaneous vessels (Jacques, Steven L.; Keijzer, Marleen)V1422,2-13(1991)

Effect of ionizing radiations and thermal treatment on the infrared transmittance of polycrystal CsI (Miley, George H.; Barnouin, O.; Procoli, Alfredo)V1441,16-26(1991)

Effect of polarization on two-color laser-induced damage (Becker, Wilhelm; McIver, John K.; Guenther, Arthur H.)V1441,541-552(1991)

Effect of spatial distribution of irradiated sites on injury selectivity in vascular tissue (Cheong, Wai-Fung; Morrison, Paul R.; Trainor, S. W.; Kurban, Amal K.; Tan, Oon T.)V1422,19-26(1991)

Effects of polishing materials on the laser damage threshold of optical coatings (Crase, Robert J.)V1441,381-389(1991)

Excimer laser optics (Morozov, N. V.; Sagitov, S. I.; Sergeyev, P. B.)V1441,557-564(1991)

Finite element analysis of the transient behavior of optical components under irradiation (Borik, Stefan; Giesen, Adolf)V1441,420-429(1991)

Heating and damage of thin metal films under the conditions of disturbed equilibrium (Minaeva, E. M.; Libenson, Michail N.)V1440,63-70(1991)

High laser-damage threshold and low-cost sol-gel-coated epoxy-replicated mirrors (Floch, Herve G.; Berger, Michel; Novaro, Marc; Thomas, Ian M.)V1441,304-315(1991)

Highly damage-resistant reflectors for 248 nm formed by fluorides multilayers (Izawa, Takao; Yamamura, N.; Uchimura, R.; Hashimoto, I.; Yakuoh, T.; Owadano, Yoshirou; Matsumoto, Y.; Yano, M.)V1441,339-344(1991)

High-power transmission through step-index multimode fibers (Setchell, Robert E.; Meeks, Kent D.; Trott, Wayne M.; Klingsporn, Paul E.; Berry, Dante M.)V1441,61-70(1991)

High-threshold highly reflective coatings at 1064 nm (Rainer, Frank; De Marco, Frank P.; Hunt, John T.; Morgan, A. J.; Mott, Leonard P.; Marcelja, Frane; Greenberg, Michael R.)V1441,247-254(1991)

Histological distinction of mechanical and thermal defects produced by nanosecond laser pulses in striated muscle at 1064 nm (Gratzl, Thomas; Dohr, Gottfried; Schmidt-Kloiber, Heinz; Reichel, Erich)V1427,55-62(1991)

Histopathologic assessment of water-dominated photothermal effects produced with laser irradiation (Thomsen, Sharon L.; Cheong, Wai-Fung; Pearce, John A.)V1422,14-18(1991)

Importance of pulse duration in laser-tissue interactions: a histological study (Kurban, Amal K.; Morrison, Paul R.; Trainor, S. W.; Cheong, Wai-Fung; Yasuda, Yukio; Tan, Oon T.)V1422,43-49(1991)

Inorganic Crystals for Optics, Electro-Optics, and Frequency Conversion (Bordui, Peter F., ed.)V1561(1991)

Interaction of 1064 nm photons with the Al2O3(1120) surface (Schildbach, M. A.; Hamza, Alex V.)V1441,139-145(1991)

Investigation of neutral atom and ion emission during laser conditioning of multilayer HfO2-SiO2 coatings (Schildbach, M. A.; Chase, Lloyd L.; Hamza, Alex V.)V1441,287-293(1991)

Ion beam milling of fused silica for window fabrication (Wilson, Scott R.; Reicher, David W.; Kranenberg, C. F.; McNeil, John R.; White, Patricia L.; Martin, Peter M.; McCready, David E.)V1441,82-86(1991)

Ion-exchange strengthening of high-average-power phosphate laser glass (Lee, Huai-Chuan; Meissner, Helmuth E.)V1441,87-103(1991)

Ionic conductivity and damage mechanisms in KTiOPO4 crystals (Morris, Patricia A.; Crawford, Michael K.; Roelofs, Mark G.; Bierlein, John D.; Baer, Thomas M.)V1561,104-111(1991)

Kinetic models for coagulation processes: determination of rate coefficients in vivo (Pearce, John A.; Cheong, Wai-Fung; Pandit, Kirit; McMurray, Tom J.; Thomsen, Sharon L.)V1422,27-33(1991)

Laser conditioning and electronic defects of HfO2 and SiO2 thin films (Kozlowski, Mark R.; Staggs, Michael C.; Rainer, Frank; Stathis, J. H.)V1441,269-282(1991)

Laser-Induced Damage in Optical Materials: 1990 (Bennett, Harold E.; Chase, Lloyd L.; Guenther, Arthur H.; Newman, Brian; Soileau, M. J., eds.)V1441(1991)

Laser-induced damage of diamond films (Read, Harold E.; Merker, M.; Gurtman, G. A.; Wilson, Russell S.)V1441,345-359(1991)

Laser-induced destruction of solids due to photo-excited carriers recombination (Grigor'ev, N. N.; Kudykina, T. A.; Tomchuk, P. M.)V1440,105-111(1991)

Laser interaction with solids (Emmony, David C.; Clark, Stuart E.; Kerr, Noel C.; Omar, Basil A.)V1397,651-659(1991)

Laser Raman measurements of dielectric coatings as a function of temperature (Exarhos, Gregory J.; Hess, Nancy J.; Ryan, Samantha)V1441,190-199(1991)

Lasers in Dermatology and Tissue Welding (Tan, Oon T.; White, Rodney A.; White, John V., eds.)V1422(1991)

Laser-Tissue Interaction II (Jacques, Steven L., ed.)V1427(1991)

Low-energy hydrodynamic mechanism of laser destruction of thin films (Sarnakov, S. M.)V1440,112-114(1991)

Mathematical modeling for laser PUVA treatment of psoriasis (Medvedev, Boris A.; Tuchin, Valery V.; Yaroslavsky, Ilya V.)V1422,73-84(1991)

Measurement of nonlinear optical coefficients by phase-matched harmonic generation (Eckardt, Robert C.; Byer, Robert L.)V1561,119-127(1991)

Measurement of temperature distributions after pulsed IR radiation impact in biological tissue models with fluorescent thin films (Romano, Valerio; Greber, Charlotte M.; Frenz, Martin; Forrer, Martin; Weber, Heinz P.)V1427,16-26(1991)

Mechanical effects induced by high-power HF laser pulses on different materials under normal atmospheric conditions (Paolacci, Sylvie; Hugenschmidt, Manfred; Bournot, Philippe)V1397,705-708(1991)

Mechanism of injurious effect of excimer (308 nm) laser on the cell (Nevorotin, Alexey J.; Kallikorm, A. P.; Zeltzer, G. L.; Kull, Mart M.; Mihkelsoo, V. T.)V1427,381-397(1991)

Monte Carlo calculations of laser-induced free-electron heating in SiO2 (Arnold, Douglas; Cartier, E.; Fischetti, Massimo V.)V1441,478-487(1991)

Morphology and laser damage studies by atomic force microscopy of e-beam evaporation deposited antireflection and high-reflection coatings (Tesar, Aleta A.; Balooch, M.; Shotts, K. W.; Siekhaus, Wigbert J.)V1441,228-236(1991)

Multiple-pulse damage of BK-7 glass (Yoshida, Kunio; Yoshida, Hiroshi; Noda, T.; Nakai, Sadao S.)V1441,9-9(1991)

Multiple-pulse laser damage to thin film optical coating (Li, Zhong Y.; Li, Cheng F.; Guo, Ju P.)V1519,374-379(1991)

Multipulse laser-induced failure prediction for Mo metal mirrors (Becker, Michael F.; Ma, Chun-Chi; Walser, Rodger M.)V1441,174-187(1991)

Multishot ablation of thin films: sensitive detection of film/substrate transition by shockwave monitoring (Hunger, Hans E.; Petzoldt, Stefan; Pietsch, H.; Reif, J.; Matthias, Eckart)V1441,283-286(1991)

New definition of laser damage threshold of thin film (Ni, Xiao W.; Lu, Jian-Feng; He, Anzhi; Ma, Zi; Zhou, Jiu L.)V1527,437-441(1991)

New definition of laser damage threshold of thin film: optical breakdown threshold (Ni, Xiao W.; Lu, Jian-Feng; He, Anzhi; Ma, Zi; Zhou, Jiu L.)V1519,365-369(1991)

Novel perfluorinated antireflective and protective coating for KDP and other optical materials (Thomas, Ian M.; Campbell, John H.)V1441,294-303(1991)

Optical characterization of damage resistant kilolayer rugate filters (Elder, Melanie L.; Jancaitis, Kenneth S.; Milam, David; Campbell, John H.)V1441,237-246(1991)

Optical Radiation Interaction with Matter (Bonch-Bruevich, Aleksei M.; Konov, Vitaly I.; Libenson, Michail N., eds.)V1440(1991)

Phase-locking and unstability of light waves in Raman-active crystals (Azarenkov, Aleksey N.; Altshuler, Grigori B.; Belashenkov, Nickolay R.; Inochkin, Mickle V.; Karasev, Viatcheslav B.; Kozlov, Sergey A.)V1409,154-164(1991)

Photodestruction of organic compounds exposed to pulsed VUV irradiation (Belov, Sergei N.; Vangonen, Albert I.; Levina, Olga V.; Puhov, Anatoly M.)V1503,503-509(1991)

Physics of multishot laser damage to optical materials (Manenkov, Alexander A.; Nechitailo, V. S.)V1441,392-405(1991)

Point defects in KTP and their possible role in laser damage (Scripsick, Michael P.; Edwards, Gary J.; Halliburton, Larry E.; Belt, Roger F.; Kappers, Lawrence A.)V1561,93-103(1991)

Practical and theoretical considerations on the use of quantitative histology to measure thermal damage of collagen in cardiovascular tissues (Thomsen, Sharon L.)V1425,110-113(1991)

Prepulse suppression using a self-induced ultrashort pulse plasma mirror (Gold, David; Nathel, Howard; Bolton, Paul R.; White, William E.; Van Woerkom, Linn D.)V1413,41-52(1991)

Pulsed CO2 laser-material interaction: mechanical coupling and reflected and scattered radiation (Prat, Ch.; Autric, Michel L.; Inglesakis, Georges; Astic, Dominique)V1397,701-704(1991)

Quantitative analysis of a CO2 laser beam by PMMA burn patterns (Joeckle, Rene C.; Koeneke, Axel; Sontag, Andre)V1397,679-682(1991)

Quartz gauge and ballistic pendulum measurements of the mechanical impulse imparted to a target by a laser pulse (David, Jean; Wettling, J. C.; Combis, P.; Nierat, G.; Rostaing, M.)V1397,697-700(1991)

Reactive ion-beam-sputtering of fluoride coatings for the UV/VUV range (Schink, Harald; Kolbe, J.; Zimmermann, F.; Ristau, Detleu; Welling, Herbert)V1441,327-338(1991)

Role of valence-band excitation in laser ablation of KCl (Haglund, Richard F.; Tang, Kai; Bunton, P. H.; Wang, Ling-jun)V1441,127-138(1991)

Selected Papers on High Power Lasers (Soures, John M., ed.)VMS43(1991)

Simulations of intracavity laser heating of particles (Linford, Gary J.)V1415,196-210(1991)

Slow thermodeformation of metals with fast laser heating (Baloshin, Yu. A.; Yurevich, V. I.; Sud'enkov, Yu. V.)V1440,71-77(1991)

Subsurface polishing damage of fused silica: nature and effect on laser damage of coated surfaces (Tesar, Aleta A.; Brown, Norman J.; Taylor, John R.; Stolz, Christopher J.)V1441,154-172(1991)

Thermal analysis of multifacet-mirror ring resonator for XUV free-electron lasers (McVey, Brian D.; Goldstein, John C.; McFarland, Robert D.; Newnam, Brian E.)V1441,457-468(1991)

Thermal process of laser-induced damage in optical thin films (Fan, Zheng X.; Wu, Zou L.; Shi, Zeng R.)V1519,359-364(1991)

Thermal stress modeling for diamond-coated optical windows (Klein, Claude A.)V1441,488-509(1991)

Time dependence of laser-induced surface breakdown in fused silica at 355 nm in the nanosecond regime (Albagli, Douglas; Izatt, Joseph A.; Hayes, Gary B.; Banish, Bryan; Janes, G. Sargent; Itzkan, Irving; Feld, Michael S.)V1441,146-153(1991)

Toward an explanation of laser-induced stimulation and damage of cell cultures (Friedmann, Harry; Lubart, Rachel; Laulicht, Israel; Rochkind, Simeone)V1427,357-362(1991)

Two-photon absorption calculations in HgCdTe (Nathan, Vaidya)V1441,553-556(1991)

Understanding of the abnormal wavelength effect of overcoats (Wu, Zhouling; Reichling, M.; Fan, Zheng X.; Wang, Zhi-Jiang)V1441,200-213(1991)

Use of dome (meniscus) lenses to eliminate birefringence and tensile stresses in spatial filters for the Nova laser (Pitts, John H.; Kyrazis, Demos T.; Seppala, Lynn G.; Bumpas, Stanley E.)V1441,71-81(1991)

Laser diagnostics

Analysis of thin-film losses from guided wave attenuations with photothermal deflection technique (Liu, Xu; Tang, Jinfa; Pelletier, Emile P.)V1554A,558-569(1991)

Applications of excimer lasers to combustion research (Davis, Steven J.; Allen, Mark G.)V1377,113-118(1991)

Beam characterization and measurement of propagation attributes (Sasnett, Michael W.; Johnston, Timothy J.)V1414,21-32(1991)

CCD holographic phase and intensity measurement of laser wavefront (Wickham, Michael; Munch, Jesper)V1414,80-90(1991)

Compact spaceflight solution crystal-growth system (Trolinger, James D.; Lal, Ravindra B.; Vikram, Chandra S.; Witherow, William K.)V1557,250-258(1991)

Computer-assisted surgical planning and automation of laser delivery systems (Zamorano, Lucia J.; Dujovny, Manuel; Dong, Ada; Kadi, A. M.)V1428,59-75(1991)

Confocal redox fluorescence microscopy for the evaluation of corneal hypoxia (Masters, Barry R.; Kriete, Andres; Kukulies, Joerg)V1431,218-223(1991)

Decrease in total fluorescence from human arteries with increasing beta-carotene content (Ye, Biqing; Abela, George S.)V1425,45-54(1991)

Development of third-harmonic output beam diagnostics on NOVA (Laumann, Curt W.; Caird, John A.; Smith, James R.; Horton, R. L.; Nielsen, Norman D.)V1414,151-160(1991)

Diagnosis of atherosclerotic tissue by resonance fluorescence spectroscopy (Neu, Walter; Haase, Karl K.; Tischler, Christian; Nyga, Ralf; Karsch, Karl R.)V1425,28-36(1991)

Diagnostic instrumentation suite for the characterization of two coupled lasers (Suter, Kevin J.; Oliver, Jeffrey W.; Cunningham, Philip R.)V1414,33-65(1991)

Diagnostics of a compact UV-preionized XeCl laser with BCl3 halogen donor (Peet, Viktor E.; Treshchalov, Alexei B.; Slivinskij, E. V.)V1412,138-148(1991)

Direct laser beam diagnostics (van Gilse, Jan; Koczera, Stanley; Greby, Daniel F.)V1414,45-54(1991)

Discrimination between urinary tract tissue and urinary stones by fiber-optic-pulsed photothermal radiometry method in vivo (Daidoh, Yuichiro; Arai, Tsunenori; Suda, Akira; Kikuchi, Makoto; Komine, Yukikuni; Murai, Masaru; Nakamura, Hiroshi)V1421,120-123(1991)

Early detection of dysplasia in colon and bladder tissue using laser-induced fluorescence (Rava, Richard P.; Richards-Kortum, Rebecca; Fitzmaurice, Maryann; Cothren, Robert M.; Petras, Robert; Sivak, Michael V.; Levine, Howard)V1426,68-78(1991)

Endoscopic fluorescence detection of early lung cancer (Profio, A. E.; Balchum, Oscar J.; Lam, Stephen)V1426,44-46(1991)

Experimental method for gas kinetic temperature measurements in a thermal plasma (Reynolds, Larry D.; Shaw, C. B.)V1554B,622-631(1991)

Far-field and wavefront characterization of a high-power semiconductor laser for free-space optical communications (Cornwell, Donald M.; Saif, Babak N.)V1417,431-439(1991)

Field-portable laser beam diagnostics (Forrest, Gary T.)V1414,55-64(1991)

Fluorescence characteristics of atherosclerotic plaque and malignant tumors (Andersson-Engels, Stefan; Baert, Luc; Berg, Roger; D'Hallewin, Mich A.; Johansson, Jonas; Stenram, U.; Svanberg, Katarina; Svanberg, Sune)V1426,31-43(1991)

Fluorescence detection of tumors: studies on the early diagnosis of microscopic lesions in preclinical and clinical studies (Mang, Thomas S.; McGinnis, Carolyn; Crean, David H.; Khan, S.; Liebow, Charles)V1426,97-110(1991)

High-energy laser wavefront sensors (Geary, Joseph M.)V1414,66-79(1991)

High-fidelity diagnostic beam sampling of a tunable high-energy laser (Mitchell, John H.; Cohen, Harold L.; Hanley, Stephen T.)V1414,141-150(1991)

High-sensitivity sensor of gases based on IR tunable diode lasers for human exhalation monitoring (Moskalenko, Konstantin L.; Nadezhdinskii, Alexander I.; Stepanov, Eugene V.)V1426,121-132(1991)

Improvement of recordable depth of field in far-field holography for analysis of particle size (Lai, Tianshu; Tan, Yushan)V1461,286-290(1991)

In-plane ESPI measurements in hostile environments (Preater, Richard W.; Swain, Robin C.)V1554A,727-738(1991)

In-situ CARS detection of H2 in the CVD of Si3N4 (Hay, Stephen O.; Veltri, R. D.; Lee, W. Y.; Roman, Ward C.)V1435,352-358(1991)

In-situ measurement of piston jitter in a ring resonator (Cunningham, Philip R.; Hay, Stephen O.; Francis, Denise M.; Trott, G.E.)V1414,97-129(1991)

Integration of diagnostics in high-power laser systems for optimization of laser material processing (Biermann, Stephan; Geiger, Manfred)V1415,330-341(1991)

Intraoperative metastases detection by laser-induced fluorescence spectroscopy (Vari, Sandor G.; Papazoglou, Theodore G.; van der Veen, Maurits J.; Fishbein, Michael C.; Young, J. D.; Chandra, Mudjianto; Papaioannou, Thanassis; Beeder, Clain; Shi, Wei-Qiang; Grundfest, Warren S.)V1426,111-120(1991)

Laser beam deflection: a method to investigate convection in vapor growth experiments (Lenski, Harald; Braun, Michael)V1557,124-131(1991)

Laser Beam Diagnostics (Hindy, Robert N.; Kohanzadeh, Youssef, eds.)V1414(1991)

Laser beam diagnostics: a conference overview (Sadowski, Thomas J.)V1414,136-140(1991)

Laser beam width, divergence, and propagation factor: status and experience with the draft standard (Fleischer, John M.)V1414,2-11(1991)

Laser diagnostic techniques in a resonant incinerator (Cadou, Christopher P.; Logan, Pamela; Karagozian, Ann; Marchant, Roy; Smith, Owen I.)V1434,67-77(1991)

Laser Doppler flowmetry in neurosurgery (Fasano, Victor A.; Urciuoli, Rosa; Bolognese, Paolo; Fontanella, Marco)V1428,2-12(1991)

Laser-excited fluorescence and fluorescence probes for diagnosing bulk damage in cable insulation (Ordonez, Ishmael D.; Crafton, J.; Murdock, R. H.; Hatfield, Lynn L.; Menzel, E. R.)V1437,184-193(1991)

Laser-induced fluorescence in contaminated soils (Lurk, Paul W.; Cooper, Stafford S.; Malone, Philip G.; Olsen, R. S.; Lieberman, Stephen H.)V1434,114-118(1991)

Laser-target diagnostics instrumentation system (Bousek, Ronald R.)V1414,175-184(1991)

Light scattering matrices of a densely packed binary system of hard spheres (Maksimova, I. L.; Shubochkin, L. P.)V1403,749-751(1991)

Measurement of laser spot quality (Milster, Tom D.; Treptau, Jeffrey P.)V1414,91-96(1991)

Measuring device for space-temporal characteristics of technological lasers radiation (Mnatsakanyan, Eduard A.; Andreev, V. G.; Bestalanny, S. I.; Velikoselsky, V.V.; Grinyakin, A.P.; Grinyakin, V.P.; Kolesnikov, S.A.; Rozhnov, U.V.)V1414,130-133(1991)

Monte Carlo simulations and measurements of signals in laser Doppler flowmetry on human skin (Koelink, M. H.; de Mul, F. F.; Greve, Jan; Graaff, Reindert; Dassel, A. C.; Aarnouds, J. G.)V1431,63-72(1991)

Near-infrared time-resolved spectroscopy and fast scanning spectrophotometry in ischemic human forearm (Ferrari, Marco; De Blasi, Roberto A.; Bruscaglioni, Piero; Barilli, Marco; Carraresi, Luca; Gurioli, G. M.; Quaglia, Enrico; Zaccanti, Giovanni)V1431,276-283(1991)

Need for laser beam diagnostics (Farnworth, Chuck)V1441,2-8(1991)

Negative deuterium ion thermal energy measurements in a volume ion source (Bacal, Marthe; Courteille, C.; Devynck, Pascal; Jones-King, Yolanda D.; Leroy, Renan; Stern, Raul A.)V1407,605-609(1991)

New method for detection of blood coagulation using fiber-optic sensor (Fediay, Sergey G.; Kuznetzov, Alexsey V.)V1420,41-43(1991)

New results in human eye laser diagnostics (Shubochkin, L. P.; Tuchin, Valery V.)V1403,720-731(1991)

Optical Methods for Tumor Treatment and Early Diagnosis: Mechanisms and Techniques (Dougherty, Thomas J., ed.)V1426(1991)

Precision high-power solid state laser diagnostics for target-irradiation studies and target-plane irradiation modeling (Wegner, Paul J.; Henesian, Mark A.)V1414,162-174(1991)

Quantitative analysis of hemoglobin oxygenation state of rat head by time-resolved photometry using picosecond laser pulse at 1064 nm (Nomura, Yasutomo; Tamura, Mamoru)V1431,102-109(1991)

Research and development of optical measurement techniques for aerospace propulsion research: a NASA/Lewis Research Center perspective (Lesco, Daniel J.)V1554B,530-539(1991)

Research for mechanical properties of rock and clay by laser speckle photography (Wu, Ruiqi)V1554A,690-695(1991)

Self-referencing Mach-Zehnder interferometer as a laser system diagnostic (Feldman, Mark; Mockler, Daniel J.; English, R. Edward; Byrd, Jerry L.; Salmon, J. T.)V1542,490-501(1991)

Simple beam-propagation measurements on ion lasers (Guggenhiemer, Steven; Wright, David L.)V1414,12-20(1991)

Study of measuring skin blood flow using speckle counting (Liu, Ying; Ma, Shining; Du, Fuli; Peng, Xiang; Ye, Shenhua)V1554A,610-612(1991)

Time-of-flight breast imaging system: spatial resolution performance (Hebden, Jeremy C.; Kruger, Robert A.)V1431,225-231(1991)

Time-Resolved Spectroscopy and Imaging of Tissues (Chance, Britton; Katzir, Abraham, eds.)V1431(1991)

Time-resolved transillumination for medical diagnostics (Berg, Roger; Andersson-Engels, Stefan; Jarlman, Olof; Svanberg, Sune)V1431,110-119(1991)

Toward phase noise reduction in a Nd:YLF laser using electro-optic feedback control (Brown, David L.; Seka, Wolf D.; Letzring, Samuel A.)V1410,209-214(1991)

UV-fluorescence spectroscopic technique in the diagnosis of breast, ovarian, uterus, and cervix cancer (Das, Bidyut B.; Glassman, W. L. S.; Alfano, Robert R.; Cleary, Joseph; Prudente, R.; Celmer, E.; Lubicz, Stephanie)V1427,368-373(1991)

UV waveguide gas laser for biological and medical diagnostic methods (Kushlevsky, S. V.; Patrin, V. V.; Provorov, Alexander S.; Salmin, V. V.)V1403,799-800(1991)

Laser materials processing—see industrial optics; laser applications; laser chemistry; microelectronic processing; microlithography

Laser-matter interaction—see also plasmas

Anisotropic local melting of semiconductors under light pulse irradiation (Fattakhov, Yakh'ya V.; Khaibullin, Ildus B.; Bayazitov, Rustem M.)V1440,16-23(1991)

Boiling process in PMMA irradiated by CO2, DF and HF laser radiations (Joeckle, Rene C.; Rapp, Gerard; Sontag, Andre)V1397,683-687(1991)

Certain results of the interaction of laser light with various media (Drobnik, Antoni)V1391,211-214(1991)

Characteristic features of melting surface of tin produced with laser pulse (Rozniakowski, Kazimierz)V1391,215-223(1991)

Computer modeling of unsteady gas-dynamical and optical phenomena in low-temperature laser plasma (Kanevsky, M. F.; Bolshov, L. A.; Chernov, S. Y.; Vorobjev, V. A.)V1440,154-165(1991)

Convective evaporation of water aerosol droplet irradiated by CO_2 laser (Butkovsky, A. V.)V1440,146-152(1991)

Debris plume phenomenology for laser-material interaction in high-speed flowfields (Reilly, James P.)V1397,661-674(1991)

Detonation mechanism of breakdowns in dielectrics (Bobilkov, G. P.; Genkin, Vladimir N.)V1440,98-104(1991)

Dimension and time effects caused by nonlocal scattering of laser radiation from a rough metal surface (Dolgina, A. N.; Kovalev, A. A.; Kondratenko, P. S.)V1440,342-353(1991)

Dislocation mechanism of periodical relief formation in laser-irradiated silicon (Shandybina, Galina)V1440,384-389(1991)

Dynamics of near-surface plasma formation and laser absorption waves under the action of microsecond laser radiation with different wavelengths on absorbing condensed media (Min'ko, L. Y.; Chumakov, A. N.)V1440,166-178(1991)

Electron density measurements of laser-induced surface plasma by means of a beam deflection technique (Michaelis, Alexander; Uhlenbusch, Juergen; Vioel, Wolfgang)V1397,709-712(1991)

Energy release in interactions of laser pulse with solid fuels and metals (Vorobyov, A. Y.; Libenson, Michail N.)V1440,197-205(1991)

Evolution of a space-charge layer, its instability, and ignition of arc gas discharge under photoemission from a target into a gas (Meshalkin, E. A.)V1440,211-221(1991)

Formation of large-scale relief on a target surface under multiple-pulsed action of laser radiation (Brailovsky, A. B.; Gaponov, Sergey V.; Dorofeev, I. A.; Lutschin, V. I.; Semenov, V. E.)V1440,84-89(1991)

Fusion of diamond phases of graphite in laser shock waves (Bugrov, N. V.; Zakharov, N. S.)V1440,416-422(1991)

Heating and damage of thin metal films under the conditions of disturbed equilibrium (Minaeva, E. M.; Libenson, Michail N.)V1440,63-70(1991)

Heterophase isotopic SF_6 molecule separation in the surface electromagnetic wavefield (Bordo, V. G.; Ershov, I. A.; Kravchenko, V. A.; Petrov, Yu. N.)V1440,364-369(1991)

Hydrodynamic efficiency as determined from implosion experiments at wavelength = 0.26 um (Koenig, Michel; Fabre, Edouard; Malka, Victor; Hammerling, Peter; Michard, Alain; Boudenne, Jean-Michel; Fews, Peter)V1502,338-342(1991)

Industrial and Scientific Uses of High-Power Lasers (Billon, Jean P.; Fabre, Edouard, eds.)V1502(1991)

Influence of a high-power laser beam on electrophysical and photoelectrical properties of epitaxial films of CdxHg1-xTe (x is approximately equal to 0.2) (Szeregij, E. M.; Ugrin, J. O.; Virt, I. S.; Abeynayake, C.; Kuzma, Marian)V1391,199-204(1991)

Influence of defects on dynamics of semiconductors (Ge, Si, GaAs) heating by laser radiation (Moin, M.)V1440,2-7(1991)

Influence of laser irradiation on the kinetics of oxide layer growth and their optical characteristics (Uglov, A.; Krivonogov, Yu.)V1440,310-320(1991)

Influence of laser pulse annealing on the depth distribution of Sb recoil atoms in Si (Brylowska, Irena; Paprocki, K.)V1391,164-169(1991)

Interaction of ultra-bright lasers with matter: program of the French Commissariat a l'Energie Atomique (Andre, Michel L.; Coutant, Jacques; Dautray, Robert; Decroisette, Michel; Lompre, Louis A.; Naudy, Michel; Manus, Claude; Mainfray, Gerard L.; Migus, Arnold; Normand, Didier; Sauteret, Christian; Watteau, Jean P.)V1502,286-298(1991)

Interference effect on laser trimming and layer thickness optimization (Sun, Yunlong)V1598,91-97(1991)

Intracavity spectroscopy measurements of atom and ion densities in near-surface laser plasma (Burakov, V. S.; Lopasov, V. P.; Naumenkov, P. A.; Raikov, S. N.)V1440,270-276(1991)

Investigation of high-temperature superconductors under the effect of picosecond laser pulses (Agranat, M. B.; Ashitkov, S. I.; Granovsky, A. B.; Kuznetsov, V. I.)V1440,397-400(1991)

Investigation of optical parameters of silicate glasses in case of before-threshold effect of laser radiation (Bosyi, O. N.; Efimov, Oleg M.; Mekryukov, A. M.)V1440,57-62(1991)

Kinetic properties of adsorbed particles photostimulated migration upon the surface of ionic-covalent-type semiconductors (Kluyev, V. G.; Kushnir, M. A.; Latyshev, A. N.; Voloshina, T. V.)V1440,303-308(1991)

Laser-induced destruction of solids due to photo-excited carriers recombination (Grigor'ev, N. N.; Kudykina, T. A.; Tomchuk, P. M.)V1440,105-111(1991)

Laser-induced diffusion and second-harmonic generation in glasses (Aksenov, V. P.)V1440,377-383(1991)

Laser-induced generation of surface periodic structures resulting from the waveguide mode interaction (Bazakutsa, P. V.; Maslennikov, V. L.; Sychugov, V. A.; Yakovlev, V. A.)V1440,370-376(1991)

Laser-induced mass transfer simulation and experiment (Shestakov, S. D.; Kotov, Gennady A.; Hekalo, A. V.; Migitko, I. A.; Bajkov, A. V.; Yurkevith, B. M.)V1440,423-435(1991)

Laser-induced optical effects in light-sensitive complexes (Kotov, Gennady A.; Filippov, N.; Shandybina, Galina)V1440,321-324(1991)

Laser interaction with solids (Emmony, David C.; Clark, Stuart E.; Kerr, Noel C.; Omar, Basil A.)V1397,651-659(1991)

Laser surface treatment: numerical simulation of thermocapillary flows (Roux, Agnes; Cipriani, Francois D.)V1397,693-696(1991)

Laser Technology III (Wolinski, Wieslaw; Wolczak, Bohdan K.; Gajda, Jerzy K.; Gajda, Danuta; Romaniuk, Ryszard S., eds.)V1391(1991)

Liquid as a deformed crystal: the model of a liquid structure (Yakovlev, Evgeni B.)V1440,36-49(1991)

Liquid dendrites growth at laser-induced melting in a NaCl volume (Gorbunov, A. V.)V1440,78-82(1991)

Liquid droplet supercritical explosion in the field of CO_2 laser radiation and influence of plasma chemical reactions on initiation of optical breakdown in air (Budnik, A. P.; Popov, A. G.)V1440,135-145(1991)

Low-energy hydrodynamic mechanism of laser destruction of thin films (Sarnakov, S. M.)V1440,112-114(1991)

Luminescence and ionization of krypton by multiphotonic excitation near the 3P1 resonant state (Saissac, M.; Berejny, P.; Millet, P.; Yousfi, M.; Salamero, Y.)V1397,739-742(1991)

Magnetic fields generated in laser-produced plasmas (Shainoga, I. S.; Shentsev, N. I.; Jakovlev, N. S.)V1440,277-290(1991)

Measurement of SEW phase velocity by optical heterodyning method (Libenson, Michail N.; Makin, Vladimir S.; Trubaev, Vladimir V.)V1440,354-356(1991)

Mechanical effects induced by high-power HF laser pulses on different materials under normal atmospheric conditions (Paolacci, Sylvie; Hugenschmidt, Manfred; Bournot, Philippe)V1397,705-708(1991)

Mechanisms of ionization for gas adjoining to plasma in intense laser beam (Fisher, Vladimir)V1440,179-187(1991)

Modeling of electron density produced by femtosecond laser on metallic photocathodes (Girardeau-Montaut, Jean-Pierre; Girardeau-Montaut, Claire)V1502,331-335(1991)

Near-surface plasma initiation model for short laser pulses (Vas'kovsky, Yu. M.; Rovinsky, R. E.; Fedjushin, B. T.; Filatova, S. A.)V1440,241-249(1991)

New effects in sound generation in organic dye solutions (Altshuler, Grigori B.; Belashenkov, Nickolay R.; Karasev, Viatcheslav B.; Okishev, Andrey V.)V1440,116-126(1991)

New photovoltaic effect in semiconductor junctions n+/p (Dominguez Ferrari, E.; Encinas Sanz, F.; Guerra Perez, J. M.)V1397,725-727(1991)

Nonlinear effects in hidden picture amplification and contrast improvement in polymer electron and Roentgenoresist PMMA (Aleksandrov, A. P.; Genkin, Vladimir N.; Myl'nikov, M. Y.; Rukhman, N. V.)V1440,442-453(1991)

Numerical modeling of laser-matter interaction in high-intensity laser applications (Ocana, Jose L.)V1397,813-820(1991)

Numerical simulation of energy transport mechanisms in high-intensity laser-matter interaction experiments (Ocana, Jose L.)V1502,299-310(1991)

Opacity studies with laser-produced plasma as an x-ray source (Eidmann, K.; Lanig, E. M.; Schwanda, W.; Sigel, Richard; Tsakiris, George D.)V1502,320-330(1991)

Optical Kerr effect in liquid and gaseous carbon dioxide (Blaszczak, Zdzislaw; Gauden, Pawel)V1391,156-163(1991)

Optical Radiation Interaction with Matter (Bonch-Bruevich, Aleksei M.; Konov, Vitaly I.; Libenson, Michail N., eds.)V1440(1991)

Peculiarities of metal surface heating by repetitively pulsed CO_2 laser radiation with the blowing (Anisimov, N. R.; Buzykin, O. G.; Zayakin, A. A.; Makarov, V. V.; Makashev, N. K.; Nosachev, L. V.; Frolov, I. P.)V1440,206-210(1991)

Peculiarities of the formation of periodic structures on silicon under millisecond laser radiation (Kokin, A. N.; Libenson, Michail N.; Minaev, Sergei M.)V1440,338-341(1991)

Photodestruction of organic compounds exposed to pulsed VUV irradiation (Belov, Sergei N.; Vangonen, Albert I.; Levina, Olga V.; Puhov, Anatoly M.)V1503,503-509(1991)

Photodissociation of single-adsorbed molecules of cesium halogenides (Zandberg, E. Y.; Knat'ko, M. V.; Paleev, V. I.; Sushchikh, M. M.)V1440,292-302(1991)

Plasma motion velocity along laser beam and continuous optical discharge in gas flow (Budnik, A. P.; Gus'kov, K. G.; Raizer, Yu. P.; Surjhikov, S. T.)V1397,721-724(1991)

Plasma parameter determination formed under the influence of CO2 laser radiation on the obstacle in the air using optical methods (Vas'kovsky, Yu. M.; Gordeeva, I. A.; Korenev, A. S.; Rovinsky, R. E.; Cenina, I. S.; Shirokova, I. P.)V1440,229-240(1991)

Pulsed CO2 laser-material interaction: mechanical coupling and reflected and scattered radiation (Prat, Ch.; Autric, Michel L.; Inglesakis, Georges; Astic, Dominique)V1397,701-704(1991)

Quantitative analysis of a CO2 laser beam by PMMA burn patterns (Joeckle, Rene C.; Koeneke, Axel; Sontag, Andre)V1397,679-682(1991)

Quartz gauge and ballistic pendulum measurements of the mechanical impulse imparted to a target by a laser pulse (David, Jean; Wettling, J. C.; Combis, P.; Nierat, G.; Rostaing, M.)V1397,697-700(1991)

Resonant IR laser-induced diffusion of oxygen in silicon (Artsimovich, M. V.; Baranov, A. N.; Krivov, V. V.; Kudriavtsev, E. M.; Lotkova, E. N.; Makeev, B. H.; Mogilnik, I. F.; Pavlovich, V. N.; Romanuk, B. N.; Soroka, V. I.; Tokarevski, V. V.; Zotov, S. D.)V1397,729-733(1991)

Role of defects in the ablation of wide-bandgap materials (Dickinson, J. T.; Langford, S. C.; Jensen, L. C.)V1598,72-83(1991)

Screening properties of the erosion torch and pressure oscillations at a laser-irradiated target (Kuznetsov, L. I.)V1440,222-228(1991)

Self-organization of spontaneous structures in photosensitive layers (Miloslavsky, V. K.; Ageev, L. A.; Lymar, V. I.)V1440,90-96(1991)

Shortwave radiation induced by CO2 laser pulse interaction with aluminum target (Golovin, A. F.; Golub, A. P.; Zemtsov, S. S.; Fedyushin, B. T.)V1440,250-259(1991)

Simulation of the process of laser beam machining based on the model of surface vaporization (Stankiewicz, Maria; Zachorowski, Jan)V1391,174-180(1991)

Slow thermodeformation of metals with fast laser heating (Baloshin, Yu. A.; Yurevich, V. I.; Sud'enkov, Yu. V.)V1440,71-77(1991)

Sound generation by thermocavitation-induced cw laser in solutions (Rastopov, S. F.; Sukhodolsky, A. T.)V1440,127-134(1991)

Spectroscopic determination of the parameters of an iron plasma produced by a CO2 laser (de Frutos, Angel M.; Sabatier, Lilian; Poueyo, Anne; Fabbro, Remy; Bermejo, Dionisio; Orza, Jose M.)V1397,717-720(1991)

Spectroscopic measurements of plasmas temperature and density during high-energy pulsed laser-materials interaction processes (Astic, Dominique; Vigliano, Patrick; Autric, Michel L.; Inglesakis, Georges; Prat, Ch.)V1397,713-716(1991)

Spontaneous magnetic field diffusion from laser plasma (Bodulinsky, V. K.; Kondratenko, P. S.)V1440,392-396(1991)

Study of regularities and the mechanism of intrinsic optical breakdown of glasses (Glebov, Leonid B.; Efimov, Oleg M.)V1440,50-56(1991)

Study of the interaction of a high-power laser radiation and a transparent liquid (Alloncle, Anne P.; Viernes, Jacques; Dufresne, Daniel; Clement, X.; Guerin, Jean M.; Testud, P.)V1397,675-678(1991)

Surface layers of metals alloyed with a pulsed laser (Pawlak, Ryszard; Raczynski, Tomasz)V1391,170-173(1991)

Surface structures formation by pulse heating of metals in oxidized environment (Bazhenov, V. V.; Libenson, Michail N.; Makin, Vladimir S.; Trubaev, Vladimir V.)V1440,332-337(1991)

Temperature dependence of optical Kerr effect in aromatic ethers (Blaszczak, Zdzislaw)V1391,146-155(1991)

Test of Geltman theory of multiple ionization of xenon by intense laser pulses (Bruzzese, Riccardo; Berardi, V.; de Lisio, C.; Solimeno, Salvatore; Spinelli, N.)V1397,735-738(1991)

Thermocapillary effects in a melted pool during laser surface treatment (Morvan, D.; Cipriani, Francois D.; Dufresne, Daniel; Garino, A.)V1397,689-692(1991)

Thermocapillary mechanism of laser pulse alloying of a metal surface layer (Kostrubiec, Franciszek)V1391,224-227(1991)

Thermoelectric voltage in slant-angle-deposited metallic films (Verechshagin, I. I.; Oksman, Ya. A.)V1440,401-405(1991)

Thresholds of plasma arising under the pulse CO2 laser radiation interaction with an obstacle in air and energetic balance of the process (Babaeva, N. A.; Vas'kovsky, Yu. M.; Zhavoronkov, M. I.; Rovinsky, R. E.; Rjabinkina, V. A.)V1440,260-269(1991)

X-ray and optical diagnostics of a 100-fs laser-produced plasma (Geindre, Jean-Paul; Audebert, Patrick; Chenais-Popovics, Claude; Gauthier, Jean-Claude J.; Benattar, Rene; Chambaret, J. P.; Mysyrowicz, Andre; Antonetti, Andre)V1502,311-318(1991)

Laser recording—see also scanning

Beam Deflection and Scanning Technologies (Beiser, Leo; Marshall, Gerald F., eds.)V1454(1991)

Butterfly line scanner: rotary twin reflective deflector that desensitizes scan-line jitter to wobble of the rotational axis (Marshall, Gerald F.; Vettese, Thomas; Carosella, John H.)V1454,37-45(1991)

Coordinate measuring system for 2-D scanners (Bukatin, Vladimir V.)V1454,283-288(1991)

Design equations for a polygon laser scanner (Beiser, Leo)V1454,60-66(1991)

Fabrication of micro-optical components by laser beam writing in photoresist (Gale, Michael T.; Lang, Graham K.; Raynor, Jeffrey M.; Schuetz, Helmut)V1506,65-70(1991)

Highly accurate pattern generation using acousto-optical deflection (Sandstrom, Torbjorn; Tison, James K.)V1463,629-637(1991)

Holographic photoelasticity applied to ceramics fracture (Guo, Maolin; Wang, Yigong; Zhao, Yin; Du, Shanyi)V1554A,310-312(1991)

Laser image recording on a metal/polymer medium (Erokhovets, Valerii K.; Larchenko, Yu. V.; Leonov, A. M.; Tkachenko, Vadim V.)V1621,227-236(1991)

Monogon laser scanner with no line wobble (Beiser, Leo)V1454,33-36(1991)

Organo-metallic thin films for erasable photochromatic laser discs (Hua, Zhong Y.; Chen, G. R.; Wang, Z. H.)V1519,2-7(1991)

Overview of a high-performance polygonal scanner subsystem (Rynkowski, Gerald)V1454,102-110(1991)

Lasers—see also resonators; sources

10.6 um laser frequency stabilization system with two optical circuits (Hu, Yu; Li, Xian; Zhu, Dayong; Ye, Naiqun; Pen, Shengyang)V1409,230-239(1991)

Advanced Laser Concepts and Applications (Singer, Sidney, ed.)V1501(1991)

Astigmatic unstable resonator with an intracavity deformable mirror (Neal, Daniel R.; McMillin, Pat; Michie, Robert B.)V1542,449-458(1991)

Calculations on the Hanle effect with phase and amplitude fluctuating laser fields (Bergeman, Thomas H.; Ryan, Robert E.)V1376,54-67(1991)

Cavity-QED-enhanced spontaneous emission and lasing in liquid droplets (Campillo, Anthony J.; Eversole, J. D.; Lin, H.-B.; Merritt, C. D.)V1497,78-89(1991)

Chaotic behavior of a Raman ring laser (Englund, John C.)V1497,292-299(1991)

Coherent coupling of lasers using a photorefractive ring oscillator (Luo, Jhy-Ming)V1409,100-105(1991)

Correlation of speckle pattern generated by TEM10 laser mode (Grzegorzewski, Bronislaw; Kowalczyk, M.; Mallek, Janusz)V1391,290-294(1991)

Effect of magnetic quantization on the pulse power output of spontaneous emission for II-VI and IV-VI lasers operated at low temperature (Ghatak, Kamakhya P.; Mitra, Bhaswati; Biswas, Shambhu N.)V1411,137-148(1991)

Electromagnetic field calculations for a tightly focused laser beam incident upon a microdroplet: applications to nonlinear optics (Barton, John P.; Alexander, Dennis R.)V1497,64-77(1991)

Emerging laser technologies (Forrest, Gary T.)V1520,37-57(1991)

Enhanced Schawlow-Townes linewidth in lasers with nonorthogonal transverse eigenmodes (Mussche, Paul L.; Siegman, Anthony E.)V1376,153-163(1991)

European industrial laser market (Mayer, Arnold)V1517,15-27(1991)

Eyesafe Lasers: Components, Systems, and Applications (Johnson, Anthony M., ed.)V1419(1991)

Fluctuations in atomic fluorescence induced by laser noise (Vemuri, Gautam)V1376,34-46(1991)

Gain and threshold in noninversion lasers (Scully, Marlan O.; Zhu, Shi-Yao; Narducci, Lorenzo M.; Fearn, Heidi)V1497,264-276(1991)

General jump model for laser noise: non-Markovian phase and frequency jumps (Levine, Alfred M.; Ozizmir, Ercument; Zaibel, Reuben; Prior, Yehiam)V1376,47-53(1991)

Higher order effect of regular pumping in lasers and masers (Zhu, Shi-Yao; Scully, Marlan O.; Su, Chang)V1497,255-262(1991)

Holographic interferometric analysis of the bovine cornea expansion (Foerster, Werner; Kasprzak, Henryk; von Bally, Gert; Busse, H.)V1429,146-151(1991)

Industrial laser market (Belforte, David A.)V1520,93-117(1991)

Industrial lasers in the United States (Rutt, James)V1517,28-43(1991)

Influence of atomic decay on micromaser operation (Zhu, Shi-Yao)V1497,240-244(1991)

Integration of geometrical and physical optics (Lawrence, George N.; Moore, Kenneth E.)V1415,322-329(1991)

Intelligent laser systems: adaptive compensation of phase distortions in nonlinear system with two-dimensional feedback (Vorontsov, Michael A.; Larichev, A. V.)V1409,260-266(1991)

Intercomparison of homogeneous laser models with transverse effects (Bandy, Donna K.; Hunter, L. W.; Jones, Darlena J.)V1497,142-152(1991)

Job shop market (Belforte, David A.)V1517,176-196(1991)

Laser glasses (Lunter, Sergei G.; Dymnikov, Alexander A.; Przhevuskii, Alexander K.; Fedorov, Yurii K.)V1513,349-359(1991)

Laser market in the 1990s (Levitt, Morris R.)V1520,1-36(1991)

Laser Noise (Roy, Rajarshi, ed.)V1376(1991)

Laser Technology III (Wolinski, Wieslaw; Wolczak, Bohdan K.; Gajda, Jerzy K.; Gajda, Danuta; Romaniuk, Ryszard S., eds.)V1391(1991)

Market for high-power CO2 lasers (Fortino, Dennis J.)V1517,61-84(1991)

Market for industrial excimer lasers (Zaal, Gerard)V1517,100-116(1991)

Market for low-power CO2 lasers (Wheeler, John P.)V1517,44-60(1991)

Market for multiaxis laser machine tools (Ream, Stanley L.)V1517,136-149(1991)

Mathematical simulation of composite optical systems loaded with active medium (Apollonova, O. V.; Elkin, Nickolai N.; Korjov, M. Y.; Korotkov, V. A.; Likhanskii, Vladimir V.; Napartovich, Anatoly P.; Troshchiev, V. E.)V1501,108-119(1991)

Measuring device for space-temporal characteristics of technological lasers radiation (Mnatsakanyan, Eduard A.; Andreev, V. G.; Bestalanny, S. I.; Velikoselsky, V.V.; Grinyakin, A.P.; Grinyakin, V.P.; Kolesnikov, S.A.; Rozhnov, U.V.)V1414,130-133(1991)

Methods and applications for intensity stabilization of pulsed and cw lasers from 257 nm to 10.6 microns (Miller, Peter J.)V1376,180-191(1991)

Modeling and Simulation of Laser Systems II (Schnurr, Alvin D., ed.)V1415(1991)

Modeling the absorption of intense, short laser pulses in steep density gradients (Alley, W. E.)V1413,89-94(1991)

Nd:YAG laser market (Llewellyn, Steven A.)V1517,85-99(1991)

Nonlinear Optics and Materials (Cantreil, Cyrus D.; Bowden, Charles M., eds.)V1497(1991)

Nonlinear theory of a three-level laser with microwave coupling: numerical calculation (Fearn, Heidi; Lamb, Willis E.; Scully, Marlan O.)V1497,283-290(1991)

Numerical study of the onset of chaos in coupled resonators (Rogers, Mark E.; Rought, Nathan W.)V1415,24-37(1991)

Observations of uplink and retroreflected scintillation in the Relay Mirror Experiment (Lightsey, Paul A.; Anspach, Joel E.; Sydney, Paul F.)V1482,209-222(1991)

Optical coherent transients induced by time-delayed fluctuating pulses (Finkelstein, Vladimir; Berman, Paul R.)V1376,68-79(1991)

Optical methods for laser beams control (Mak, Arthur A.; Soms, Leonid N.)V1415,110-119(1991)

Overview of experimental investigations of laser bandwidth effects in nonlinear optics (Elliott, Daniel S.)V1376,22-33(1991)

Pockels cell driver with high repetition rate (Brzezinski, Ryszard)V1391,279-285(1991)

QED theory of excess spontaneous emission noise (Milonni, Peter W.)V1376,164-169(1991)

Quantum noise reduction in lasers with nonlinear absorbers (Ritsch, Helmut; Zoller, Peter)V1376,206-215(1991)

Recording of laser radiation power distribution with the use of nonlinear optical effects (Korchazhkin, S. V.; Krasnova, L. O.)V1358,966-972(1991)

Scientific laser trends (Messenger, Heather)V1520,58-92(1991)

Selected Papers on Laser Design (Weichel, Hugo, ed.)VMS29(1991)

Selected Papers on Ultrafast Laser Technology (Gosnell, Timothy R.; Taylor, Antoinette J., eds.)VMS44(1991)

Sheet-metal cutting market (Berkhahn, Glenn)V1517,117-135(1991)

Shot noise limited optical measurements at baseband with noisy lasers (Hobbs, Philip C.)V1376,216-221(1991)

Similarity and difference between degenerate parametric oscillators and two-photon correlated-spontaneous-emission lasers (Zhu, Shi-Yao; Scully, Marlan O.)V1497,277-282(1991)

Single-ion spectroscopy (Bergquist, James C.; Wineland, D. J.; Itano, W. M.; Diedrich, F.; Raizen, M. G.; Elsner, Frank)V1435,82-85(1991)

Six years of transendoscopic Nd:YAG application in large animals (Tate, Lloyd P.; Glasser, Mardi)V1424,209-217(1991)

Solar-powered blackbody-pumped lasers (Christiansen, Walter H.; Sirota, J. M.)V1397,821-825(1991)

Spatial characterization of YAG power laser beam (Grevey, D. F.; Badawi, K. F.)V1502,32-40(1991)

Spontaneous emission noise reduction of a laser output by extracavity destructive interference (Diels, Jean-Claude; Lai, Ming)V1376,198-205(1991)

The Laser Marketplace 1991 (Forrest, Gary T.; Levitt, Morris R., eds.)V1520(1991)

The Marketplace for Industrial Lasers (Belforte, David A.; Levitt, Morris R., eds.)V1517(1991)

Theory of laser noise (Lax, Melvin)V1376,2-20(1991)

Time-addressing of coherence-tuned optical fibre sensors based on a multimode laser diode (Santos, Jose L.; Jackson, David A.)V1572,325-330(1991)

Transition from homoclinic to heteroclinic chaos in coupled laser arrays (Otsuka, Kenju)V1497,300-312(1991)

U.S. government market (Speser, Philip L.)V1520,132-150(1991)

Use of Fokker-Planck equations for the statistical properties of laser light (Jung, Peter; Risken, H.)V1376,82-93(1991)

Visible-laser-light-induced polymerization: an overview of photosensitive formulations (Fouassier, Jean-Pierre)V1559,76-88(1991)

Laser safety—see also laser damage; laser surgery

Battlefield training in impaired visibility (Gammarino, Rudolph R.; Surhigh, James W.)V1419,115-125(1991)

Carbon dioxide eyesafe laser rangefinders (Powell, Richard K.; Berdanier, Barry N.; McKay, James)V1419,126-140(1991)

Energy storage efficiency and population dynamics in flashlamp-pumped sensitized erbium glass laser (Lukac, Matjaz)V1419,55-62(1991)

Eyesafe diode laser rangefinder technology (Perger, Andreas; Metz, Jurgen; Tiedeke, J.; Rille, Eduard P.)V1419,75-83(1991)

Eyesafe high-pulse-rate laser progress at Hughes (Stultz, Robert D.; Nieuwsma, Daniel E.; Gregor, Eduard)V1419,64-74(1991)

Eyesafe laser application in military and law enforcement training (Mosbrooker, Michael L.)V1419,107-114(1991)

Eyesafe Lasers: Components, Systems, and Applications (Johnson, Anthony M., ed.)V1419(1991)

Fiber fragmentation during laser lithotripsy (Flemming, G.; Brinkmann, Ralf E.; Strunge, Ch.; Engelhardt, R.)V1421,146-152(1991)

Highly efficient optical parametric oscillators (Marshall, Larry R.; Hays, Alan D.; Kasinski, Jeff; Burnham, Ralph L.)V1419,141-152(1991)

High-performance InGaAs PIN and APD (avalanche photdiode) detectors for 1.54 um eyesafe rangefinding (Olsen, Gregory H.; Ackley, Donald A.; Hladky, J.; Spadafora, J.; Woodruff, K. M.; Lange, M. J.; Van Orsdel, Brian T.; Forrest, Stephen R.; Liu, Y.)V1419,24-31(1991)

High-repetition-rate eyesafe rangefinders (Corcoran, Vincent J.)V1419,160-169(1991)

High repetition rate Q-switched erbium glass lasers (Hamlin, Scott J.; Myers, John D.; Myers, Michael J.)V1419,100-106(1991)

High-speed short-range laser rangefinder (Gielen, Robert M.; Slegtenhorst, Ronald P.)V1419,153-159(1991)

Holographic filter for coherent radiation (Shi, Dexiu; Xing, Xiaozheng; Wolbarsht, Myron L.)V1419,40-49(1991)

Mid-infrared laser applications (Daly, John G.)V1419,94-99(1991)

New results in dosimetry of laser radiation in medical treatment (Beuthan, Jurgen; Hagemann, Roland; Mueller, Gerhard J.; Schaldach, Brita J.; Zur, Ch.)V1420,225-233(1991)

Optimal surface characteristics for instruments for use in laser neurosurgery (Heiferman, Kenneth S.; Cramer, K. E.; Walsh, Joseph T.)V1428,128-134(1991)

Passive Q-switching of eyesafe Er:glass lasers (Denker, Boris I.; Maksimova, G. V.; Osiko, Vyacheslav V.; Prokhorov, Alexander M.; Sverchkov, Sergey E.; Sverchkov, Yuri E.; Horvath, Zoltan G.)V1419,50-54(1991)

Planar InGaAs APD (avalanche photodiode) for eyesafe laser rangefinding applications (Webb, Paul P.)V1419,17-23(1991)

Receivers for eyesafe laser rangefinders: an overview (Crawford, Ian D.)V1419,9-16(1991)

Safety of medical excimer lasers with an emphasis on compressed gases (Sliney, David H.; Clapham, Terrance N.)V1423,157-162(1991)

SIRE (sight-integrated ranging equipment): an eyesafe laser rangefinder for armored vehicle fire control systems (Keeter, Howard S.; Gudmundson, Glen A.; Woodall, Milton A.)V1419,84-93(1991)

Testing laser eye protection (Labo, Jack A.; Mayo, Michael W.)V1419,32-39(1991)

What is eye safe? (Franks, James K.)V1419,2-8(1991)

Lasers, carbon dioxide—see also lasers, gas

8th Intl Symp on Gas Flow and Chemical Lasers (Domingo, Concepcion; Orza, Jose M., eds.)V1397(1991)

Absorption calorimetry and laser-induced damage threshold measurements of antireflective-coated ZnSe and metal mirrors at 10.6 um (Rahe, Manfred; Oertel, E.; Reinhardt, L.; Ristau, Detleu; Welling, Herbert)V1441,113-126(1991)

Apodized outcouplers for unstable resonators (Budzinski, Christel; Grunwald, Ruediger; Pinz, Ingo; Schaefer, Dieter; Schoennagel, Horst)V1500,264-274(1991)

Blackbody-pumped CO2 lasers using Gaussian and waveguide cavities (Chang, Jim J.; Christiansen, Walter H.)V1412,150-163(1991)

Carbon dioxide eyesafe laser rangefinders (Powell, Richard K.; Berdanier, Barry N.; McKay, James)V1419,126-140(1991)

Carbon dioxide laser effects on caries-like lesions of dental enamel (Featherstone, John D.; Zhang, S. H.; Shariati, M.; McCormack, S. M.)V1424,145-149(1991)

Characteristics of a compact 12 kW transverse-flow CO2 laser with rf-excitation (Wildermuth, Eberhard; Walz, B.; Wessel, K.; Schock, Wolfram)V1397,367-371(1991)

Characteristics of a downstream-mixing CO2 gas-dynamic laser (Hashimoto, Takashi; Nakano, Susumu; Hachijin, Michio; Komatsu, Katsuhiko; Hara, Hiroshi)V1397,519-522(1991)

Characterization and uses of plastic hollow fibers for CO2 laser energy transmission (Gannot, Israel; Dror, Jacob; Dahan, Reuben; Alaluf, M.; Croitoru, Nathan I.)V1442,156-161(1991)

Chemical gas-dynamic mixing CO2 laser pumped by the reactions between N2O and CO (Doroschenko, V. M.; Kudriavtsev, N. N.; Sukhov, A. M.)V1397,503-511(1991)

Clinical evaluation of tumor promotion by CO2 laser (Braun, Robert E.; Liebow, Charles)V1424,138-144(1991)

CO2 coupled-mode, CO, and other lasers with supersonic cooling of gas mixture (Kudriavtsev, E. M.)V1397,475-484(1991)

CO2 lasers and phase conjugation (Sherstobitov, Vladimir E.; Kalinin, Victor P.; Goryachkin, Dmitriy A.; Romanov, Nikolay A.; Dimakov, Sergey A.; Kuprenyuk, Victor I.)V1415,79-89(1991)

CO2 partial matricectomy in the treatment of ingrown toenails (Uppal, Gurvinder S.; Sherk, Henry H.; Black, Johnathan D.; Rhodes, Anthony; Sazy, John; Lane, Gregory J.)V1424,51-52(1991)

Comparison of carbon monoxide and carbon dioxide laser-tissue interaction (Waters, Ruth A.; Thomas, J. M.; Clement, R. M.; Ledger, N. R.)V1427,336-343(1991)

Coupled CO2 lasers (Antyuhov, V.; Bondarenko, Alexander V.; Glova, Alexander F.; Golubentzev, A. A.; Danshikov, E.; Kachurin, O. R.; Lebedev, Fedor V.; Likhanskii, Vladimir V.; Napartovich, Anatoly P.; Pis'menny, Vladislav D.; Yarzev, V.; Yaztsev, Vladimir P.)V1397,355-365(1991); V1415,48-59(1991)

Design and performance characteristics of a compact CO2 Doppler lidar transmitter (Pearson, Guy N.)V1416,147-150(1991)

Design, construction, and operation of 65 kilowatt carbon dioxide electric discharge coaxial laser device (Reilly, James P.; Lander, Mike L.; Maxwell, K.; Hull, R. J.)V1397,339-354(1991)

Detection and parameter estimation of atmospheric turbulence by ground-based and airborne CO2 Doppler lidars (Pogosov, Gregory A.; Akhmanov, Sergei A.; Gordienko, Vyacheslav M.; Kosovsky, L. A.; Kurochkin, Nikolai N.; Priezzhev, Alexander V.)V1416,115-124(1991)

Direct laser beam diagnostics (van Gilse, Jan; Koczera, Stanley; Greby, Daniel F.)V1414,45-54(1991)

Double-mode CO2 laser with complex cavity for ultrasensitive sub-Doppler spectroscopy (Kurochkin, Vadim Y.; Petrovsky, Victor N.; Protsenko, Evgeniy D.; Rururkin, Alexander N.; Golovchenko, Anotoly M.)V1435,322-330(1991)

Double-pulsed TEA CO2 laser (Li, Xiang Ying; Shi, Zhenbang; Sun, Ning)V1412,246-251(1991)

Dry/wet wire reflectivities at 10.6 um (Young, Donald S.; Osche, Gregory R.; Fisher, Kirk L.; Lok, Y. F.)V1416,221-228(1991)

Dual ion-beam sputtering: a new coating technology for the fabrication of high-power CO2 laser mirrors (Daugy, Eric; Pointu, Bernard; Villela, Gerard; Vincent, Bernard)V1502,203-212(1991)

Electron density measurements of laser-induced surface plasma by means of a beam deflection technique (Michaelis, Alexander; Uhlenbusch, Juergen; Vioel, Wolfgang)V1397,709-712(1991)

Endoscopic removal of PMMA in hip revision surgery with a CO2 laser (Sazy, John; Kollmer, Charles; Uppal, Gurvinder S.; Lane, Gregory J.; Sherk, Henry H.)V1424,50-50(1991)

Energy balance in high-power CO2 laser welding (Del Bello, Umberto; Rivela, Cristina; Cantello, Maichi; Penasa, Mauro)V1502,104-116(1991)

Estimation of CO2/N2 mixing region by the small-signal gain measurement (Ohue, Hiroshi; Kasahara, Eiji; Uemura, Kamon; Mito, Keiichi)V1397,527-530(1991)

European industrial laser market (Mayer, Arnold)V1517,15-27(1991)

Experimental and theoretical description of DC transverse glow-discharge excitation of high-power convective CO2 laser and stabilization of its output power (Konefal, Janusz)V1391,135-143(1991)

Experimental study of the parameters of the laser-induced plasma observed in welding of iron targets with continuous high-power CO2 lasers (Poueyo, Anne; Sabatier, Lilian; Deshors, G.; Fabbro, Remy; de Frutos, Angel M.; Bermejo, Dionisio; Orza, Jose M.)V1502,140-147(1991)

Fabry-Perot etalon as a CO2 laser Q-switch modulator (Walocha, Jerzy)V1391,286-289(1991)

FIR lasers as local oscillators in submillimeter astronomy (Roeser, Hans-Peter; van der Wal, Peter)V1501,194-197(1991)

Flowfield in a CO2-N2 gas-dynamics laser with staggered nozzles (Tarabelli, D.; Zeitoun, David; Imbert, Michel P.)V1397,523-526(1991)

Fluctuation variation of a CO2 laser pulse intensity during its interaction with a cloud (Almaev, R. K.; Semenov, L. P.; Slesarev, A. G.; Volkovitsky, O. A.)V1397,831-834(1991)

Gas and Metal Vapor Lasers and Applications (Kim, Jin J.; Tittel, Frank K., eds.)V1412(1991)

Hearing preservation using CO2 laser for acoustic nerve tumors (Grutsch, James; Heiferman, Kenneth S.; Cerullo, Leonard J.)V1428,136-145(1991)

High-power CO2 laser excited by 2.45 GHz microwave discharges (Freisinger, Bernhard; Pauls, Markus; Schaefer, Johannes H.; Uhlenbusch, Juergen)V1397,311-318(1991)

High-power unstable resonator CO laser (Sato, Shunichi; Takahashi, Kunimitsu; Tanaka, Ikuzo; Noda, Osama; Kuribayashi, Shizuma; Imatake, Shigenori; Kondo, Motoe)V1397,421-425(1991)

High-repetition-rate, multijoule transversely excited atmospheric CO2 laser (Hatanaka, Hidekazu; Kawahara, Nobuo; Midorikawa, Katsumi; Tashiro, Hideo; Obara, Minoru)V1397,379-382(1991)

Industrial lasers in Japan (Karube, Norio)V1517,1-14(1991)

Industrial lasers in the United States (Rutt, James)V1517,28-43(1991)

Inhomogeneous quarter-wave transformers for a waveguide electro-optic modulator (Harris, Neville W.; Wong, D. M.)V1416,86-99(1991)

Job shop market (Belforte, David A.)V1517,176-196(1991)

Laser-induced medium perturbation in pulsed CO2 lasers (Dente, Gregory C.; Walter, Robert F.; Gardner, Daniel C.)V1397,403-408(1991)

Lidar wind shear detection for commercial aircraft (Targ, Russell; Bowles, Roland L.)V1416,130-138(1991)

Long-time operation of a 5kW cw CO laser (Noda, Osama; Kuribayashi, Shizuma; Imatake, Shigenori; Kondo, Motoe; Sato, Shunichi; Takahashi, Kunimitsu)V1397,427-432(1991)

Market for high-power CO2 lasers (Fortino, Dennis J.)V1517,61-84(1991)

Market for low-power CO2 lasers (Wheeler, John P.)V1517,44-60(1991)

Markets for marking systems (Austin, Patrick D.)V1517,150-175(1991)

Measurements of atmospheric transmittance of CO2 laser radiation (Aref'ev, Vladimir N.)V1397,827-830(1991)

Miniature pulsed TEA CO2 laser with a 20-Hz repetition frequency (Kocyba, Krzysztof)V1391,61-64(1991)

Modifications to the LP-140 pulsed CO2 laser for lidar use (Jaenisch, Holger M.; Spiers, Gary D.)V1411,127-136(1991)

Multikilowatt transverse-flow CO2 laser with variable reflectivity mirrors (Serri, Laura; Maggi, C.; Garifo, Luciano; De Silvestri, S.; Magni, Vittorio C.; Svelto, Orazio)V1397,469-472(1991)

Multipass unstable negative branch resonator with a spatial filter for a transverse-flow CO2 laser (Rabczuk, G.)V1397,387-389(1991)

New CO2 laser equipped with high-peak pulse power and high-speed drilling process (Hongu, Hitoshi; Karasaki, Hidehiko)V1501,198-204(1991)

New construction of sealed-off CO2 laser (Wolinski, Wieslaw; Wolski, Radoslaw; Janulewicz, K. A.)V1397,391-393(1991)

Newly developed excitation circuit for kHz pulsed lasers (Yasuoka, Koichi; Ishii, Akira; Tamagawa, Tohru; Ohshima, Iwao)V1412,32-37(1991)

Optimization of affecting parameters in relation to pulsed CO2 laser design (Danisman, Kenan; Yilbas, Bekir S.; Altuner, Mehmet; Ciftlikli, Cebrail)V1412,218-226(1991)

Optimization of discharge parameters of an e-beam sustained repetitively pulsed CO2 laser (Beth, Mark-Udo; Hall, Thomas; Mayerhofer, Wilhelm)V1397,577-580(1991)

Output characteristics of a transverse-flow cw CO2 modular laser (Kukiello, P.; Rabczuk, G.)V1397,417-419(1991)

Performance of CO2 and CO diffusively cooled rf-excited strip-line lasers with different electrode materials and gas composition (Yatsiv, Shaul; Gabay, Amnon; Sterman, Baruch; Sintov, Yoav)V1397,319-329(1991)

Plastic hollow fibers employed for CO2 laser power transmission in oral surgery (Calderon, Israel; Dror, Jacob; Dahan, Reuben; Croitoru, Nathan I.)V1420,108-115(1991)

Probe measurements in a CO2 laser plasma (Leys, C.; Sona, P.; Muys, Peter F.)V1397,399-402(1991)

Process monitoring during CO2 laser cutting (Jorgensen, Henning; Olsen, Flemming O.)V1412,198-208(1991)

Promotional effects of CO2 laser on neoplastic lesions in hamsters (Kingsbury, Jeffrey S.; Margarone, Joseph E.; Satchidanand, S.; Liebow, Charles)V1427,363-367(1991)

Pulsed CO2 laser for intra-articular cartilage vaporization and subchondral bone perforation in horses (Nixon, Alan J.; Roth, Jerry E.; Krook, Lennart)V1424,198-208(1991)

Pulsed CO2 laser-material interaction: mechanical coupling and reflected and scattered radiation (Prat, Ch.; Autric, Michel L.; Inglesakis, Georges; Astic, Dominique)V1397,701-704(1991)

Real-time actuating of laser power and scanning velocity for thermal regulation during laser hardening (Bataille, F.; Kechemair, Didier; Houdjal, R.)V1502,135-139(1991)

Recent advances in CO2 laser catalysts (Upchurch, Billy T.; Schryer, David R.; Brown, K. G.; Kielin, E. J.; Hoflund, Gar B.; Gardner, Steven D.)V1416,21-29(1991)

Regular and chaotic pulsations of radiation intensity in a CO2 laser with modulated parameters (Bondarenko, Alexander V.; Glova, Alexander F.; Kozlov, Sergei N.; Lebedev, Fedor V.; Likhanskii, Vladimir V.; Napartovich, Anatoly P.; Pis'menny, Vladislav D.; Yaztsev, Vladimir P.)V1376,117-127(1991)

Resonators for coaxial slow-flow CO2 lasers (Habich, Uwe; Bauer, Axel; Loosen, Peter; Plum, Heinz-Dieter)V1397,383-386(1991)

Review of the performances of industrial 1.5-kW and 3-kW CO2 lasers, mobile on gantry or 2-D machines (Aubert, Philippe; Beck, Rasmus)V1502,52-59(1991)

Sheet-metal cutting market (Berkhahn, Glenn)V1517,117-135(1991)

Spatial and electrical characteristics of capacitively ballasted rf laser discharges (Baker, Howard J.; Laidler, Ian)V1397,545-548(1991)

Spatial dynamics of picosecond CO2 laser pulses produced by optical switching in Ge (Pogorelsky, Igor V.; Fisher, A. S.; Veligdan, James T.; Russell, J.)V1413,21-31(1991)

Spectroscopic determination of the parameters of an iron plasma produced by a CO2 laser (de Frutos, Angel M.; Sabatier, Lilian; Poueyo, Anne; Fabbro, Remy; Bermejo, Dionisio; Orza, Jose M.)V1397,717-720(1991)

Spectroscopic measurements of plasmas temperature and density during high-energy pulsed laser-materials interaction processes (Astic, Dominique; Vigliano, Patrick; Autric, Michel L.; Inglesakis, Georges; Prat, Ch.)V1397,713-716(1991)

Study of characteristics of a multikilowatt CO2 laser operating in flow-closed cycle: a semiempirical model (Konefal, Janusz)V1397,409-415(1991)

Surface absorptance in CO2 laser steel processing (Covelli, L.; De Iorio, I.; Tagliaferri, V.)V1397,797-802(1991)

TEA CO2 laser mirror by degenerate four-wave mixing (Vigroux, Luc M.; Bourdet, Gilbert L.; Cassard, Philippe; Ouhayoun, Michel O.)V1500,74-79(1991)

Technique of CO2 laser arthroscopic surgery (Meller, Menachem M.; Sherk, Henry H.; Rhodes, Anthony; Sazy, John; Uppal, Gurvinder S.; Lane, Gregory J.)V1424,60-61(1991)

Temperature response of biological tissues to nonablative pulsed CO2 laser irradiation (van Gemert, Martin J.; Brugmans, Marco J.; Gijsbers, Geert H.; Kemper, J.; van der Meulen, F. W.; Nijdam, D. C.)V1427,316-319(1991)

TE two-frequency pulsed laser at 0.337 um and 10.6 um with a plasma cathode (Branzalov, Peter P.)V1412,236-245(1991)

The Marketplace for Industrial Lasers (Belforte, David A.; Levitt, Morris R., eds.)V1517(1991)

Thermal lensing effect in fast-axial flow CO2 lasers (Moissl, M.; Paul, R.; Breining, K.; Giesen, Adolf; Huegel, Helmut)V1397,395-398(1991)

Thermal regulation applied to CO2 laser self-quenching of complex geometry workpieces (Bataille, F.; Kechemair, Didier; Pawlovski, C.; Houdjal, R.)V1397,839-842(1991)

Transendoscopic and freehand use of flexible hollow fibers for carbon dioxide laser surgery in the upper airway of the horse: a preliminary report (Palmer, Scott E.)V1424,218-220(1991)

Transient gain phenomena and gain enhancement in a fast-axial flow CO2 laser amplifier (Sato, Heihachi; Tsuchida, Eiichi; Kasuya, Koichi)V1397,331-338(1991)

Transverse-flow cw CO2 modular laser: preliminary investigation (Kukiello, P.; Rabczuk, G.)V1391,93-97(1991)

Two-photon absorption calculations in HgCdTe (Nathan, Vaidya)V1441,553-556(1991)

Variable reflectivity output coupler for improvement of the beam quality of a fast-axial flow CO2 laser (Sona, P.; Muys, Peter F.; Leys, C.; Sherman, Glenn H.)V1397,373-377(1991)

Vascular tissue welding of the CO2 laser: limitations (Dalsing, Michael C.; Kruepper, Peter; Cikrit, Dolores F.)V1422,98-102(1991)

Very high reflective all-dielectric coatings for high-power CO2 lasers (Berger, R. M.; Chmelir, M.; Reedy, Herman E.; Chambers, Jack P.)V1397,611-618(1991)

Lasers, chemical

30-Torr pulsed singlet oxygen generator (Endo, Masamori; Arai, S.; Yamashita, T.; Uchiyama, Taro)V1397,267-270(1991)

8th Intl Symp on Gas Flow and Chemical Lasers (Domingo, Concepcion; Orza, Jose M., eds.)V1397(1991)

Arc heater for thermal driven HF/DF chemical laser (Sontag, Andre; Baronnet, Jean M.)V1397,287-290(1991)

Chemical generation of electronically excited nitrogen N2(A3 sigma +u) and lasers on electronic transitions (Dvoryankin, A. N.; Makarov, V. N.)V1397,177-180(1991)

Chemically driven pulsed and continuous visible laser amplifiers and oscillators (Gole, James L.; Woodward, J. R.; Cobb, S. H.; Shen, KangKang; Doughty, J. R.)V1397,125-135(1991)

Chemical oxygen-iodine laser: flow diagnostics and overall qualification (Georges, Eric; Barraud, Roger; Mouthon, Alain)V1397,243-246(1991)

Continuous chemical lasers of visible region (Dvoryankin, A. N.)V1397,145-152(1991)

Design and performance of an atmospheric pressure HF chemical laser (Chuchem, D.; Kalisky, Yehoshua Y.; Amit, M.; Smilanski, Israel)V1397,277-281(1991)

Development of chemical oxygen-iodine laser for industrial application (Fujii, H.; Iizuka, Masahiro; Muro, Mikio; Kuchiki, Hirotsuna; Atsuta, Toshio)V1397,213-220(1991)

Experiment and modeling of O2(1 delta) generation in a bubble-column-type reactor for chemically pumped iodine lasers (Aharon, O.; Elior, A.; Herskowitz, M.; Lebiush, E.; Rosenwaks, Salman)V1397,251-255(1991)

Frequency up-conversion of a discharge-pumped molecular fluorine laser by stimulated Raman scattering in H2 (Kakehata, Masayuki; Hashimoto, Etsu; Kannari, Fumihiko; Obara, Minoru)V1397,185-189(1991)

Ground-to-space multiline propagation at 1.3 um (Crawford, Douglas P.; Harada, Larrene K.)V1408,167-177(1991)

High-power chemical oxygen-iodine laser (Barnault, B.; Barraud, Roger; Forestier, L.; Georges, Eric; Louvet, Y.; Mouthon, Alain; Ory, M.; Pigache, Daniel R.)V1397,231-234(1991)

High-power chemical oxygen-iodine lasers and applications (Yoshida, S.; Shimizu, K.)V1397,205-212(1991)

Influence of mixing on the characteristics of the oxygen-iodine laser (Dvoryankin, A. N.; Kulagin, Yu. A.; Kudryavtcev, N. Y.)V1397,247-250(1991)

Kinetics and yield of Xe excitation as third body in the process of N(4S) atom recombination (Ivanov, U. V.; Pravilov, A. M.; Smirnova, L. G.; Vilesov, A. F.)V1397,181-184(1991)

Mixing diagnostic in a cw DF chemical laser operating at high-cavity pressure (Voignier, Francois; Merat, Frederic; Brunet, Henri)V1397,297-301(1991)

New emission spectra from chemically excited oxygen and potentiality as a visible chemical laser (Bacis, Roger; Bonnet, Jean C.; Bouvier, A. J.; Churassy, S.; Grozet, P.; Erba, B.; Georges, Eric; Jouvet, C.; Lamarre, J.; Louvet, Y.; Nota, M.; Pigache, Daniel R.; Ross, A. J.; Setra, M.)V1397,173-176(1991)

Numerical simulation of the combustion chamber of a chemical laser (Saatdjian, E.; Caressa, J. P.; Andre, Jean-Claude)V1397,535-538(1991)

Photoinitiation of chemical lasers using radiation from a cylindrical surface discharge (Beverly, R. E.)V1397,581-584(1991)

Population of high-vibrational levels of the iodine ground state in its dissociation process by singlet oxygen (Barnault, B.; Joly, V.; Pigache, Daniel R.)V1397,257-260(1991)

Possibility of long population inversion in active media for IR chemical lasers (Barmashenko, B. D.; Kochelap, Viatcheslav A.)V1397,303-307(1991)

Potential IF chemical laser (Leporcq, B.; Verdier, C.; Arbus, R.; Lemoine, F.)V1397,153-156(1991)

Progress toward the demonstration of a visible (blue) chemical laser (Herbelin, John M.)V1397,161-167(1991)

Pulsed chemical oxygen-iodine laser (Yuryshev, N. N.)V1397,221-230(1991)

Pulsed HF chemical laser using a VUV phototriggered discharge (Brunet, Henri; Mabru, Michel; Rocca Serra, J.; Vannier, C.)V1397,273-276(1991)

Pulse shape effects in a twin cell DF laser (Gorton, E. K.; Parcell, E. W.; Cross, P. H.)V1397,291-295(1991)

Reactive-flow modeling of the H/NF2/BiF reaction system (Acebal, Robert; Dansereau, Jeffrey P.; Jones, R.; Malins, Robert J.; Schreiber, H.; Smith, W.; Taylor, S.; Duncan, William A.; Patterson, S.)V1397,191-196(1991)

Red emitter resulting from O2 (a1 delta g): a new lasing species? (Zhuang, Qi; Cui, T. J.; Xie, X. B.; Sang, Fengting; Yuan, Q. N.; Zhang, Rongyao; Yang, H. P.; Li, Li; Zhu, Q. S.; Zhang, Cunhao)V1397,157-160(1991)

Second harmonic generation of chemical oxygen-iodine laser (Matsuzaka, Fumio; Nigawara, Kazushige; Terasawa, Ken; Uchiyama, Taro)V1397,239-242(1991)

Short-pulsed H2-F2 amplifier initiated by optical discharge (Igoshin, Valery I.; Pichugin, Sergei)V1501,150-152(1991)

Small-signal gain in the oxygen-iodine laser (Bohn, Willy L.; Truesdell, Keith; Latham, William P.; Avizonis, Petras V.)V1397,235-238(1991)

Studies of short-wavelength chemical lasers: enhanced emission of Pb atoms following detonation of lead azide via a supersonic nozzle (Bar, I.; Ben-Porat, T.; Cohen, A.; Heflinger, Dov; Miron, G.; Tzuk, Y.; Rosenwaks, Salman)V1397,169-172(1991)

Study of the production of S2(B) from the recombination of sulfur atoms (Prigent, Pascale; Brunet, Henri)V1397,197-201(1991)

Study of visible chemical lasers: modeling of an IF chemical laser within the F-NH3-IF system (Sang, Fengting; Zhuang, Qi; Wang, Chengdong; Feng, Hao; Zhang, Cunhao)V1412,252-257(1991)

Surface quenching of singlet delta oxygen (Crannage, R. P.; Johnson, Daniel E.; Dorko, Ernest A.)V1397,261-265(1991)

Theoretical study of the high-power pulsed vibrational overtone HF chemical laser (Ashidate, Shu-ichi; Takashima, Toshiaki; Kannari, Fumihiko; Obara, Minoru)V1397,283-286(1991)

Visual observation and numerical analysis on the reaction zone structure of a supersonic-flow CO chemical laser (Masuda, Wataru)V1397,531-534(1991)

Lasers, dye

9H-Indolo(1,2-f) phenanthridinium hemicyanines: a new group of high-efficiency broadband generating laser dyes (Kotowski, Tomasz; Skubiszak, Wojciech; Soroka, Jacek A.; Soroka, Krystyna B.; Stacewicz, Tadeusz)V1391,6-11(1991)

Clinical applications of PDT in urology: present and near future (Shumaker, Bryan P.)V1426,293-300(1991)

Deterministic and quantum noise in dye lasers (Raymer, M. G.; Gadomski, Wojciech; Hodges, Steven E.; Adkison, D.)V1376,128-131(1991)

Electron microscopic study on black pig skin irradiated with pulsed dye laser (504 nm) (Yasuda, Yukio; Tan, Oon T.; Kurban, Amal K.; Tsukada, Sadao)V1422,50-55(1991)

Endobronchial occlusive disease: Nd:YAG or PDT? (Regal, Anne-Marie; Takita, Hiroshi)V1426,271-278(1991)

Energy of the lowest triplet states of rhodamines in ethanolic solutions (Targowski, Piotr; Zietek, Bernard)V1391,12-17(1991)

Excimer-dye laser system for diagnosis and therapy of cancer (Pokora, Ludwik J.; Puzewicz, Zbigniew)V1503,467-478(1991)

Histochemical identification of malignant and premalignant lesions (Liebow, Charles; Maloney, M. J.)V1426,22-30(1991)

Imaging of tumors by time-delayed laser-induced fluorescence (Kohl, M.; Neukammer, Jorg; Sukowski, U.; Rinneberg, Herbert H.; Sinn, H.-J.; Friedrich, E. A.; Graschew, Georgi; Schlag, Peter M.; Woehrle, D.)V1525,26-34(1991)

In-line power meter for use during laser angioplasty (Smith, Roy E.; Milnes, Peter; Edwards, David H.; Mitchell, David C.; Wood, Richard F.)V1425,116-121(1991)

Ion yields from strong optical field ionization experiments using 100-femtosecond laser pulses (Fittinghoff, David N.; Bolton, Paul R.; Chang, Britton; Van Woerkom, Linn D.; White, William E.)V1413,81-88(1991)

Long-path intracavity laser for the measurement of atmospheric trace gases (McManus, J. B.; Kolb, Charles E.)V1433,340-351(1991)

New method for tissue indentification: resonance fluorescence spectroscopy (Neu, Walter)V1525,124-131(1991)

Novel applications of monomode fiber tapers (Payne, Frank P.; Mackenzie, H. S.)V1504,165-175(1991)

N-substituted 1,8-naphthalimide derivatives as high-efficiency laser-dye: dependence of dye laser emission of protonated solvent (Martin, E.; Pardo, A.; Poyato, J. M.; Weigand, R.; Guerra, J. M.)V1397,835-838(1991)

Photodegradation of a laser dye in a silica gel matrix (Glab, Wallace L.; Bistransin, Mark; Borst, Walter L.)V1497,389-395(1991)

Photoionization of rhodamines in etanolic solution at 77 K (Zietek, Bernard; Targowski, Piotr)V1391,18-23(1991)

Preliminary results of laser-assisted sealing of hand-sewn canine esophageal anastomoses (Auteri, Joseph S.; Oz, Mehmet C.; Sanchez, Juan A.; Bass, Lawrence S.; Jeevanandam, Valluvan; Williams, Mathew R.; Smith, Craig R.; Treat, Michael R.)V1421,182-184(1991)

Statistics of laser switch-on (San Miguel, Maximino)V1376,272-283(1991)

Theoretical analysis of stone fragmentation rates (Sterenborg, H. J.)V1525,170-176(1991)

Tissue interactions of ball-tipped and multifiber catheters (Mitchell, David C.; Smith, Roy E.; Walters, Tena K.; Murray, Alan; Wood, Richard F.)V1427,181-188(1991)

Ultrafast streak camera evaluations of phase noise from an actively stabilized colliding-pulse-mode-locked ring dye laser (Walker, David R.; Sleat, William E.; Evans, J.; Sibbett, Wilson)V1358,860-867(1991)

Uniform field laser illuminator for remote sensing (Di Benedetto, John A.; Capelle, Gene; Lutz, Stephen S.)V1492,115-125(1991)

Lasers, excimer—see also lasers, gas; microlithography

100-Hz KrF laser plasma x-ray source (Turcu, I. C.; Gower, Malcolm C.; Reason, C. J.; Huntington, P.; Schulz, M.; Michette, Alan G.; Bijkerk, Fred; Louis, Eric; Tallents, Gregory J.; Al-Hadithi, Yas; Batani, D.)V1503,391-405(1991)

2.5 kHz high-repetition-rate XeCl excimer laser (Ishikawa, Ken; Takagi, S.; Okamoto, N.; Kakizaki, K.; Sato, S.; Goto, T.)V1397,55-58(1991)

8th Intl Symp on Gas Flow and Chemical Lasers (Domingo, Concepcion; Orza, Jose M., eds.)V1397(1991)

Ablation of hard dental tissues with an ArF-pulsed excimer laser (Neev, Joseph; Raney, Daniel; Whalen, William E.; Fujishige, Jack T.; Ho, Peter D.; McGrann, John V.; Berns, Michael W.)V1427,162-172(1991)

Ablation of ITO and TO films from glass substrates (Meringdal, Frode; Slinde, Harald)V1503,292-298(1991)

Absorption behavior of ceramic materials irradiated with excimer lasers (Toenshoff, Hans K.; Gedrat, Olaf)V1377,38-44(1991)

Advanced concepts of electron-beam-pumped excimer lasers (Tittel, Frank K.; Canarelli, P.; Dane, C. B.; Hofmann, Thomas; Sauerbrey, Roland A.; Sharp, Tracy E.; Szabo, Gabor; Wilson, William L.; Wisoff, P. J.; Yamaguchi, Shigeru)V1397,21-29(1991)

Advances in clinical percutaneous excimer laser angioplasty (Viligiardi, Riccardo; Pini, Roberto; Salimbeni, Renzo; Galiberti, Sandra)V1425,72-74(1991)

Analysis of an adjustable slit design for correcting astigmatism (Clapham, Terrance N.; D'Arcy, John; Bechtel, Lorne; Glockler, Hermann; Munnerlyn, Charles R.; McDonnell, Peter J.; Garbus, Jenny)V1423,2-7(1991)

Applications of excimer lasers in microelectronics (Yu, Chang; Sandhu, Gurtej S.; Mathews, V. K.; Doan, Trung T.)V1598,186-197(1991)

Applications of excimer lasers to combustion research (Davis, Steven J.; Allen, Mark G.)V1377,113-118(1991)

Applications of laser plasmas in XUV photoabsorption spectroscopy (Kennedy, Eugene T.; Costello, John T.; Mosnier, Jean-Paul)V1503,406-415(1991)

Argon fluoride excimer laser source for sub-0.25 mm optical lithography (Sandstrom, Richard L.)V1463,610-616(1991)

Atomic xenon recombination laser excited by thermal ionizing radiation from a magnetoplasma compressor and discharge (Kamrukov, A. S.; Kozlov, N. P.; Opekan, A. G.; Protasov, Yu. S.; Rudoi, I. G.; Soroka, A. M.)V1503,438-452(1991)

Axial and transverse displacement tolerances during excimer laser surgery for myopia (Shimmick, John K.; Munnerlyn, Charles R.; Clapham, Terrance N.; McDonald, Marguerite B.)V1423,140-153(1991)

Bundle of tapered fibers for the transmission of high-power excimer laser pulses (Hitzler, Hermine; Leclerc, Norbert; Pfleiderer, Christoph; Wolfrum, Juergen M.; Greulich, Karl O.; Klein, Karl-Friedrich)V1503,355-362(1991)

Characteristics of amplification of ultrashort laser pulses in excimer media (Kannari, Fumihiko; Obara, Minoru)V1397,85-89(1991)

Characterization of a subpicosecond XeF (C->A) excimer laser (Hofmann, Thomas; Sharp, Tracy E.; Dane, C. B.; Wisoff, P. J.; Wilson, William L.; Tittel, Frank K.; Szabo, Gabor)V1412,84-90(1991)

Clinical experience with an excimer laser angioplasty system (Golobic, Robert A.; Bohley, Thomas K.; Wells, Lisa D.; Sanborn, Timothy A.)V1425,84-92(1991)

Coating development for high-energy KrF excimer lasers (Boyer, James D.; Mauro, Billie R.; Sanders, Virgil E.)V1377,92-98(1991)

Cold-laser microsurgery of the retina with a syringe-guided 193 nm excimer laser (Lewis, Aaron; Palanker, Daniel; Hemo, Itzhak; Pe'er, Jacob; Zauberman, Hanan)V1423,98-102(1991)

Comparative study of excimer and erbium:YAG lasers for ablation of structural components of the knee (Vari, Sandor G.; Shi, Wei-Qiang; van der Veen, Maurits J.; Fishbein, Michael C.; Miller, J. M.; Papaioannou, Thanassis; Grundfest, Warren S.)V1424,33-42(1991)

Comparison of 248-nm line narrowing resonator optics for deep-UV lithography lasers (Kahlert, Hans-Juergen; Rebhan, Ulrich; Lokai, Peter; Basting, Dirk)V1463,604-609(1991)

Controlled structuring of polymer surfaces by UV laser irradiation (Bahners, Thomas; Kesting, Wolfgang; Schollmeyer, Eckhard)V1503,206-214(1991)

Corneal refractive surgery using an ultraviolet (213 nm) solid state laser (Ren, Qiushi; Gailitis, Raymond P.; Thompson, Keith P.; Penney, Carl M.; Lin, J. T.; Waring, George O.)V1423,129-139(1991)

Cumulative effect and cutting quality improvement of XeCl laser ablation of PMMA (Lou, Qihong; Guo, Hongping)V1598,221-226(1991)

Damage testing of optical components for high-power excimer lasers (Mann, Klaus R.; Gerhardt, Harald)V1503,176-184(1991)

Design considerations for high-power industrial excimer lasers (Jetter, Heinz-Leonard)V1503,48-52(1991)

Development of damage resistant optics for KrF excimer lasers (Boyer, James D.; Mauro, Billie R.; Sanders, Virgil E.)V1441,255-261(1991)

Diagnostics of a compact UV-preionized XeCl laser with BCl3 halogen donor (Peet, Viktor E.; Treshchalov, Alexei B.; Slivinskij, E. V.)V1412,138-148(1991)

Discharge studies with a high-efficiency XeCl excimer laser (Trentelman, M.; Ekelmans, G. B.; van Goor, Frederik A.; Witteman, Wilhelmus J.)V1397,115-118(1991)

Discharge technology for excimer lasers of high-average power (Witteman, Wilhelmus J.; Ekelmans, G. B.; Trentelman, M.; van Goor, Frederik A.)V1397,37-45(1991)

Dose control for short exposures in excimer laser lithography (Hollman, Richard F.)V1377,119-125(1991)

Effect of acoustic dampers on the excimer laser flow (Zeitoun, David; Tarabelli, D.; Forestier, Bernard M.; Truong, J. P.; Sentis, Marc L.)V1397,585-588(1991)

Efficient e-beam sustained Ar:Xe laser (Botma, H.; Peters, Peter J.; Witteman, Wilhelmus J.)V1397,573-576(1991)

Efficient population of low-vibrational-number electronic states of excimer molecules: the argon dimer (Efthimiopoulos, Tom)V1503,430-437(1991)

Elaboration of excimer lasers dosimetry for bone and meniscus cutting and drilling using optical fibers (Jahn, Renate; Dressel, Martin; Neu, Walter; Jungbluth, Karl-Heinz)V1424,23-32(1991)

Evolution of an excimer laser gas mix (Boardman, A. D.; Hodgson, Elizabeth M.; Spence, A. J.; Richardson, A. D.; Richardson, M. B.)V1503,160-166(1991)

Excimer and nitrogen lasers with low-average power for technology applications (Pokora, Ludwik J.)V1397,111-114(1991)

Excimer-dye laser system for diagnosis and therapy of cancer (Pokora, Ludwik J.; Puzewicz, Zbigniew)V1503,467-478(1991)

Excimer laser ceramic and metal surface alloying applications (Hontzopoulos, Elias I.; Zervaki, A.; Zergioti, Y.; Hourdakis, G.; Raptakis, E.; Giannacopoulos, A.; Fotakis, Costas)V1397,761-768(1991)

Excimer laser cutting of corneal transplants (Tamkivi, Raivo; Schotter, Leo L.; Pakhomova, Tat'yana A.)V1503,375-378(1991)

Excimer laser deposition and characterization of tin and tin-oxide films (Borsella, E.; De Padova, P.; Larciprete, Rosanna)V1503,312-320(1991)

Excimer laser development at the ENEA Frascati Centre: discharge instabilities study (Bollanti, Sarah; Di Lazzaro, Paolo; Flora, Francesco; Fu, Shufen; Giordano, Gualtiero; Letardi, Tommaso; Lisi, Nicola; Schina, Giovanni; Zheng, Cheng-En)V1503,80-87(1991)

Excimer laser fabrication of waveguide devices (Stiller, Marc A.)V1377,73-78(1991)

Excimer laser in arthroscopic surgery (Koort, Hans J.)V1424,53-59(1991)

Excimer laser machining of optical fiber taps (Coyle, Richard J.; Serafino, Anthony J.; Grimes, Gary J.; Bortolini, James R.)V1412,129-137(1991)

Excimer Laser Materials Processing and Beam Delivery Systems (Piwczyk, Bernhard P., ed.)V1377(1991)

Excimer laser micromachining for passive fiber coupling to polymeric waveguide devices (Booth, Bruce L.; Hohman, James L.; Keating, Kenneth B.; Marchegiano, Joseph E.; Witman, Sandy L.)V1377,57-63(1991)

Excimer laser optics (Morozov, N. V.; Sagitov, S. I.; Sergeyev, P. B.)V1441,557-564(1991)

Excimer laser patterning of flexible materials (Kollia, Z.; Hontzopoulos, Elias I.)V1503,215-222(1991)

Excimer laser performance under various microwave excitation conditions (Klingenberg, Hans H.; Gekat, Frank)V1503,140-145(1991)

Excimer laser photolithography with a 1:1 broadband catadioptric optics (Zhang, Yudong; Lu, Dunwu; Zou, Haixing; Wang, Zhi-Jiang)V1463,456-463(1991)

Excimer laser processing of ceramics and fiber-reinforced polymers assisted by a diagnostic system (Geiger, Manfred; Lutz, Norbert; Biermann, Stephan)V1503,238-248(1991)

Excimer laser processing of diamond-like films (Ageev, Vladimir P.; Glushko, T. N.; Dorfman, V. F.; Kuzmichov, A. V.; Pypkin, B. N.)V1503,453-462(1991)

Excimer laser projector for microelectronics applications (Rumsby, Phil T.; Gower, Malcolm C.)V1598,36-45(1991)

Excimer Lasers and Applications III (Letardi, Tommaso; Laude, Lucien D., eds.)V1503(1991)

Excimer lasers for deep-UV lithography (Elliott, David J.; Sengupta, Uday K.)V1377,6-17(1991)

Excimer laser surface treatment of ceramics (Hourdakis, G.; Hontzopoulos, Elias I.; Tsetsekou, A.; Zampetakis, Th.; Stournaras, C.)V1503,249-255(1991)

Excimer laser with sealed x-ray preionizer (Atjezhev, Vladimir V.; Belov, Sergey R.; Bukreev, Viacheslav S.; Vartapetov, Serge K.; Zhukov, Alexander N.; Ziganshin, Ilnur T.; Prokhorov, Alexander M.; Soldatkin, Alexey E.; Stepanov, Yuri D.)V1503,197-199(1991)

Excitation of an excimer laser with microwave resonator (Huenermann, Lucia; Meyer, Rudolph; Richter, Franz; Schnase, Alexander)V1503,134-139(1991)

Experimental studies of an XeCl laser with UV preionization perpendicular and parallel to the electrode surfaces (Nassisi, Vincenzo)V1503,115-125(1991)

Fast-flow gas-dynamic effects in high-pulse repetition-rate excimer lasers (Delaporte, Philippe C.; Fontaine, Bernard L.; Forestier, Bernard M.; Sentis, Marc L.)V1397,485-492(1991)

Fast-iterative technique for the calculation of frequency-dependent gain in excimer laser amplifiers (Sze, Robert C.)V1412,164-172(1991)

Fused silica fibers for the delivery of high-power UV radiation (Artjushenko, Vjacheslav G.; Konov, Vitaly I.; Pashinin, Vladimir P.; Silenok, Alexander S.; Blinov, Leonid M.; Solomatin, A. M.; Shilov, I. P.; Volodko, V. V.; Mueller, Gerhard J.; Schaldach, Brita J.; Ulrich, R.; Neuberger, Wolfgang)V1420,149-156(1991)

Gas and Metal Vapor Lasers and Applications (Kim, Jin J.; Tittel, Frank K., eds.)V1412(1991)

Gas flow effects in pulse avalanche discharge XeCl excimer laser (Lou, Qihong)V1397,103-106(1991)

Gas handling technology for excimer lasers (Turner, Robert E.; Remo, John L.; Bradford, Elaine; Dietz, Alvin)V1377,99-106(1991)

High-average-power narrow-band KrF excimer laser (Wakabayashi, Osamu; Kowaka, Masahiko; Kobayashi, Yukio)V1463,617-628(1991)

High-average power XeCl laser with surface corona-discharge preionization (Nagai, Haruhiko; Haruta, K.; Sato, Y.; Inoue, M.; Suzuki, A.)V1397,31-36(1991)

High-gain tunable laser medium: XeF-doped Ar crystals (Zerza, Gerald; Kometer, R.; Sliwinski, G.; Schwentner, N.)V1397,107-110(1991)

Highly conductive amorphous-ferrite formed by excimer laser material processing (Kashiwabara, S.; Watanabe, Kazuhiro; Fujimoto, R.)V1397,803-806(1991)

High-power KrF lasers (Key, Michael H.; Bailly-Salins, Rene; Edwards, B.; Harvey, Erol C.; Hirst, Graeme J.; Hooker, Chris J.; Kidd, A. K.; Madraszek, E. M.; Rodgers, P. A.; Ross, Ian N.; Shaw, M. J.; Steyer, M.)V1397,9-17(1991)

Hollow-tube-guide for UV-power laser beams (Kubo, Uichi; Okada, Kasuyuki; Hashishin, Yuichi)V1420,102-107(1991)

Imaging of excimer laser vascular tissue ablation by ultrafast photography (Nyga, Ralf; Neu, Walter; Preisack, M.; Haase, Karl K.; Karsch, Karl R.)V1525,119-123(1991)

Impedance of a UV preionized excimer laser (Nassisi, Vincenzo)V1527,291-304(1991)

Importance of nonequilibrium vibrational kinetics of HCl in XeCl laser modeling (Capitelli, Mario; Gorse, Claudine; Longo, Savino; Bretagne, J.; Estocq, Emmanuel)V1503,126-131(1991)

Improvement of focus and exposure latitude by the use of phase-shifting masks for DUV applications (Op de Beeck, Maaike; Tokui, Akira; Fujinaga, Masato; Yoshioka, Nobuyuki; Kamon, Kazuya; Hanawa, Tetsuro; Tsukamoto, Katsuhiro)V1463,180-196(1991)

Industrial excimer and CO2 TEA lasers with kilowatt average output power (von Bergmann, Hubertus M.; Swart, Pieter L.)V1397,63-66(1991)

Influence of gas composition and pressure on the pulse duration of electron beam controlled discharge XeCl laser (Badziak, Jan; Drazek, Wieslaw; Dubicki, Adam; Perlinski, Leszek)V1397,81-84(1991)

Influence of gas composition on XeCl laser performance (Jursich, Gregory M.; Von Drusek, William A.; Mulderink, Ken; Olchowka, V.; Reid, John; Brimacombe, Robert K.)V1412,115-122(1991)

Influencing adherence properties of polymers by excimer laser radiation (Breuer, J.; Metev, S.; Sepold, Gerd; Krueger, G.; Hennemann, O. D.)V1503,223-230(1991)

Investigation of microwave-pumped excimer and rare-gas laser transitions (Klingenberg, Hans H.; Gekat, Frank)V1412,103-114(1991)

Investigations of an excimer laser working with a four-component gaseous mixture He-Kr:Xe-HCl (Iwanejko, Leszek; Pokora, Ludwik J.)V1391,105-108(1991)

Investigations on excimer-laser-treated Cu/Cr contact materials (Schubert, Emil; Rosiwal, S.; Bergmann, Hans W.)V1503,299-309(1991)

In-vitro ablation of fibrocartilage by XeCl excimer laser (Buchelt, Martin; Papaioannou, Thanassis; Fishbein, Michael C.; Peters, Werner; Beeder, Clain; Grundfest, Warren S.)V1420,249-253(1991)

Keratorefractive procedures (Seiler, Theo)V1525,280-281(1991)

Kilowatt-range high-repetition-rate excimer lasers (Borisov, V. M.; Khristoforov, O. B.; Kirykhin, Yu. B.; Kuznetsov, S. G.; Stepanov, Yu. Y.; Vinokhodov, A. Y.)V1503,40-47(1991)

Kinetics and electrical modeling of a long-pulse high-efficiency XeCl laser with double discharge and fast magnetic switch (Kobhio, M. N.; Fontaine, Bernard L.; Hueber, Jean-Marc; Delaporte, Philippe C.; Forestier, Bernard M.; Sentis, Marc L.)V1397,555-558(1991)

KrF laser ablation of polyurethane (Kueper, Stephan; Brannon, James H.)V1598,27-35(1991)

Large-scale industrial application for excimer lasers: via-hole-drilling by photoablation (Bachmann, Friedrich G.)V1361,500-511(1991); V1377,18-29(1991)

Large volume XeCl laser with longitudinal gas flow: experimental results and theoretical analysis (Bollanti, Sarah; Di Lazzaro, Paolo; Flora, Francesco; Giordano, Gualtiero; Letardi, Tommaso; Lisi, Nicola; Schina, Giovanni; Zheng, Cheng-En)V1397,97-102(1991)

Laser chemistry of dimethylcadmium adsorbed on silicon: 308- versus 222-nm laser excitation (Simonov, Alexander P.; Varakin, Vladimir N.; Panesh, Anatoly M.; Lunchev, V. A.)V1503,330-337(1991)

Laser guide stars for adaptive optics systems: Rayleigh scattering experiments (Thompson, Laird A.; Castle, Richard; Carroll, David L.)V1542,110-119(1991)

Laser-induced shock wave effects on red blood cells (Flotte, Thomas J.; Frisoli, Joan K.; Goetschkes, Margaret; Doukas, Apostolos G.)V1427,36-44(1991)

Laser plasma XUV sources: a role for excimer lasers? (Bijkerk, Fred; Shevelko, A. P.)V1503,380-390(1991)

Laser-produced continua for studies in the XUV (Carroll, P. K.; O'Sullivan, Gerard D.)V1503,416-427(1991)

Light-induced polishing of evaporating surface (Tokarev, Vladimir N.; Konov, Vitaly I.)V1503,269-278(1991)

Low-loss line-narrowed excimer oscillator for projection photolithography: experiments and simulation (Volkov, Gennady S.; Zaroslov, D. Y.)V1503,146-153(1991)

L-shell x-ray spectroscopy of laser-produced plasmas in the 1-keV region (Batani, D.; Giulietti, Antonio; Palladino, Libero; Tallents, Gregory J.; Turcu, I. C.)V1503,479-491(1991)

Magnetic pulse compression in the prepulse circuit for a 1 kW, 1kHz XeCl excimer laser (Ekelmans, G. B.; van Goor, Frederik A.; Trentelman, M.; Witteman, Wilhelmus J.)V1397,569-572(1991)

Magnetic pulse compressor for high-power excimer discharge laser pumping (Vizir, V. A.; Kiryukhin, Yu. B.; Manylov, V. I.; Nosenko, S. P.; Sychev, S. P.; Chervyakov, V. V.; Shubkin, N. G.)V1411,63-68(1991)

Market for industrial excimer lasers (Zaal, Gerard)V1517,100-116(1991)

Measurement of the gain in a XeF (C-A) laser pumped by a coaxial e-beam (Bastiaens, H. M.; Peters, Peter J.; Witteman, Wilhelmus J.)V1397,77-80(1991)

Measurements of gas flow and gas constituent in a wind-tunnel-type excimer laser under high-repetition-rate operations (Kasuya, Koichi; Horioka, K.; Hikida, N.; Murazi, M.; Nakata, K.; Kawakita, Y.; Miyai, Y.; Kato, S.; Yoshida, S.; Ideno, S.)V1397,67-70(1991)

Mechanism of excimer-laser-induced absorption in fused silica fibers (Artjushenko, Vjacheslav G.; Konov, Vitaly I.; Pashinin, Vladimir P.; Silenok, Alexander S.; Mueller, Gerhard J.; Schaldach, Brita J.; Ulrich, R.)V1420,176-176(1991)

Mechanism of injurious effect of excimer (308 nm) laser on the cell (Nevorotin, Alexey J.; Kallikorm, A. P.; Zeltzer, G. L.; Kull, Mart M.; Mihkelsoo, V. T.)V1427,381-397(1991)

Medical marketplace (Moretti, Michael)V1520,118-131(1991)

Modeling of a long-pulse high-efficiency XeCl laser with double-discharge and fast magnetic switch (Kobhio, M. N.; Fontaine, Bernard L.; Hueber, Jean-Marc; Delaporte, Philippe C.; Forestier, Bernard M.; Sentis, Marc L.)V1503,88-97(1991)

Modeling of photochemical vapor deposition of epitaxial silicon using an ArF excimer laser (Fowler, Burt; Lian, S.; Krishnan, S.; Jung, Le-Tien; Li, C.; Banerjee, Sanjay)V1598,108-117(1991)

Modifications to achieve subnanosecond jitter from standard commercial excimer lasers (Rust, Kenneth R.)V1411,80-86(1991)

Multishot ablation of thin films: sensitive detection of film/substrate transition by shockwave monitoring (Hunger, Hans E.; Petzoldt, Stefan; Pietsch, H.; Reif, J.; Matthias, Eckart)V1441,283-286(1991)

Multispot scanning exposure system for excimer laser stepper (Yoshitake, Yasuhiro; Oshida, Yoshitada; Tanimoto, Tetsuzou; Tanaka, Minoru; Yoshida, Minoru)V1463,678-687(1991)

Multistage XeCl excimer system "Cactus" and some investigations of stimulated scattering in liquids (Karpov, V. I.; Korobkin, V. V.; Naboichenko, A. K.; Dolgolenko, D. A.)V1503,492-502(1991)

Novel excimer laser beam delivery technique using binary masks (Liu, Yung S.; Levinson, R. M.; Rose, J. W.)V1377,126-133(1991)

Novel high-resolution large-field scan-and-repeat projection lithography system (Jain, Kanti)V1463,666-677(1991)

Optical engineering of an excimer laser ophthalmic surgery system (Yoder, Paul R.)V1442,162-171(1991)

Optical/Laser Microlithography IV (Pol, Victor, ed.)V1463(1991)

Output characteristics of a multikilowatt repetitively pulsed XeF laser (Walter, Robert F.; Palumbo, L. J.; Townsend, Steven W.; Tannen, Peter D.)V1397,71-76(1991)

Output power stabilization of a XeCl excimer laser by HCl gas injection (Ogura, Satoshi; Kawakubo, Yukio; Sasaki, Kouji; Kubota, Yoshiyuki; Miki, Atsushi)V1412,123-128(1991)

Parametric study of a high-average-power XeCl laser (Godard, Bruno; Estocq, Emmanuel; Joulain, Franck; Murer, Pierre; Stehle, Marc X.; Bonnet, Jean C.; Pigache, Daniel R.)V1503,71-77(1991)

Parametric study of small-volume long-pulse x-ray preionized XeCl laser with double-discharge and fast magnetic switch (Hueber, Jean-Marc; Fontaine, Bernard L.; Kobhio, M. N.; Delaporte, Philippe C.; Forestier, Bernard M.; Sentis, Marc L.)V1503,62-70(1991)

Percutaneous peripheral excimer laser angioplasty: immediate success rate and short-term outcome (Visona, Adriana; Liessi, Guido; Bonanome, Andrea; Lusiani, Luigi; Miserocchi, Luigi; Pagnan, Antonio)V1425,75-83(1991)

Performance characteristics of a discharge-pumped XeCl laser driven by a pulse forming network containing nonlinear ferroelectric capacitors (Fairlie, S. A.; Smith, Paul W.)V1411,56-62(1991)

Photoablative etching of materials for optoelectronic integrated devices (Lemoine, Patrick; Magan, John D.; Blau, Werner)V1377,45-56(1991)

Photochemical changes of rare-earth valent state in gamma-irradiated CaF2:Pr crystals by the excimer laser radiation: investigation and application (Lukishova, Svetlana G.; Obidin, Alexey Z.; Vartapetov, Serge K.; Veselovsky, Igor A.; Osiko, Anatoly V.; Tulajkova, Tamara V.; Ter-Mikirtychev, Valery V.; Mendez, Nestor R.)V1503,338-345(1991)

Photodestruction of organic compounds exposed to pulsed VUV irradiation (Belov, Sergei N.; Vangonen, Albert I.; Levina, Olga V.; Puhov, Anatoly M.)V1503,503-509(1991)

Pico- and femtosecond pulses in the UV and XUV (Dinev, S.; Dreischuh, A.)V1403,427-430(1991)

Polymerization kinetics of mono- and multifunctional monomers initiated by high-intensity laser pulses: dependence of rate on peak-pulse intensity and chemical structure (Hoyle, Charles E.; Sundell, Per-Erik; Trapp, Martin A.; Kang, Deokman; Sheng, D.; Nagarajan, Rajamani)V1559,202-213(1991)

Polysilyne resists for 193 nm excimer laser lithography (Kunz, Roderick R.; Bianconi, Patricia A.; Horn, Mark W.; Paladugu, R. R.; Shaver, David C.; Smith, David A.; Freed, Charles A.)V1466,218-226(1991)

Possibility of short-pulses generation in excimer lasers by self-injection (Badziak, Jan; Dubicki, Adam; Piotrowski, Jan)V1391,117-126(1991)

Power gain characteristics of discharge-excited KrF laser amplifier system (Lee, Choo-Hie; Choi, Boo-Yeon)V1397,91-95(1991)

Processing of bioceramic implants with excimer laser (Koort, Hans J.)V1427,173-180(1991)

Processing of polytetrafluoroethylene with high-power VUV laser radiation (Basting, Dirk; Sowada, Ulrich; Voss, F.; Oesterlin, Peter)V1412,80-83(1991)

Producing a uniform excimer laser beam for materials processing applications (Bunis, Jenifer L.; Abele, C. C.; Campbell, James D.; Caudle, George F.)V1377,30-36(1991)

Progress in gas purifiers for industrial excimer lasers (Naylor, Graham A.; Kearsley, Andrew J.)V1377,107-112(1991)

Prototype of an excimer laser for microprocessing (Iwanejko, Leszek; Pokora, Ludwik J.; Wolinski, Wieslaw)V1391,98-100(1991)

Pulse compression of KrF laser radiation by stimulated scattering (Alimpiev, Sergei S.; Bukreev, Viacheslav S.; Kusakin, Vladimir I.; Likhansky, Sergey V.; Obidin, Alexey Z.; Vartapetov, Serge K.; Veselovsky, Igor A.)V1503,154-158(1991)

Pulsed-laser deposition of oxides over large areas (Greer, James A.; Van Hook, H. J.)V1377,79-90(1991)

Pulsed microwave excitation of rare-gas halide mixtures (Gekat, Frank; Klingenberg, Hans H.)V1411,47-54(1991)

Pulsed-power considerations for electron-beam-pumped krypton-fluoride lasers for inertial confinement fusion applications (Rose, Evan A.; McDonald, Thomas E.; Rosocha, Louis A.; Harris, David B.; Sullivan, J. A.; Smith, I. D.)V1411,15-27(1991)

Recent advances in excimer laser technology at Los Alamos (Bigio, Irving J.; Czuchlewski, Stephen J.; McCown, Andrew W.; Taylor, Antoinette J.)V1397,47-53(1991)

Recent progress towards multikilowatt output (Mueller-Horsche, Elmar; Oesterlin, Peter; Basting, Dirk)V1503,28-39(1991)

Review of the multikilohertz performance of the CHIRP laser and components (Fieret, Jim; Green, J. M.; Heath, R.; O'Key, Michael A.; Osborne, Michael R.; Osbourn, S. J.; Taylor, Arthur F.; Winfield, R. J.)V1503,53-61(1991)

Safety of medical excimer lasers with an emphasis on compressed gases (Sliney, David H.; Clapham, Terrance N.)V1423,157-162(1991)

Self-aligned synthesis of titanium silicide by multipulse excimer laser irradiation (Craciun, Valentin; Craciun, Doina; Mihailescu, Ion N.; Kuzmichov, A. V.; Konov, Vitaly I.; Uglov, S. A.)V1392,625-628(1991)

Spectral and time-resolved measurements of pollutants on water surface by an XeCl laser fluorosensor (Barbini, Roberto; Fantoni, Roberta; Palucci, Antonio; Ribezzo, Sergio; van der Steen, Hendricus J.)V1503,363-374(1991)

Study of high-average-power excimer laser with circulation loop (Godard, Bruno; Estocq, Emmanuel; Stehle, Marc X.; Bonnet, Jean C.; Pigache, Daniel R.)V1397,59-62(1991)

Super small excimer laser (Vill, Arnold A.; Salk, Ants A.; Berik, Irina K.)V1503,110-114(1991)

Surface nitride synthesis by multipulse excimer laser irradiation (D'Anna, Emilia; Leggieri, Gilberto; Luches, Armando; Martino, M.; Perrone, A.; Majni, G.; Mengucci, P.; Drigo, A. V.; Mihailescu, Ion N.)V1503,256-268(1991)

Temperature increase during in vitro 308 nm excimer laser ablation of porcine aortic tissue (Gijsbers, Geert H.; Sprangers, Rene L.; van den Broecke, Duco G.; van Wieringen, Niek; Brugmans, Marco J.; van Gemert, Martin J.)V1425,94-101(1991)

Theoretical and experimental investigations on pressure-wave reflections and attenuation in high-power excimer lasers (Holzwarth, Achim; Griebsch, Juergen; Berger, Peter)V1503,98-109(1991)

Thomson scattering diagnostics of discharge plasmas in an excimer laser (Yamakoshi, H.; Kato, M.; Kubo, Y.; Enokizono, H.; Uchino, K.; Muraoka, K.; Takahashi, A.; Maeda, Mitsuo)V1397,119-122(1991)

Thyratron-switched, L-C inverter, prepulse-sustainer, laser discharge circuit (Pacala, Thomas J.; Tranis, Art; Laudenslager, James B.; Kinley, Fred G.)V1411,69-79(1991)

Total internal reflection mirrors fabricated in polymeric optical waveguides via excimer laser ablation (Trewhella, Jeannine M.; Oprysko, Modest M.)V1377,64-72(1991)

Ultrafast imaging of vascular tissue ablation by an XeCl excimer laser (Neu, Walter; Nyga, Ralf; Tischler, Christian; Haase, Karl K.; Karsch, Karl R.)V1425,37-44(1991)

Uniform field laser illuminator for remote sensing (Di Benedetto, John A.; Capelle, Gene; Lutz, Stephen S.)V1492,115-125(1991)

Use of excimer lasers in medicine: applications, problems, and dangers (Mommsen, Jens; Stuermer, Martin)V1503,348-354(1991)

UV-VIS solid state excimer laser: XeF in crystalline argon (Zerza, Gerald; Knopp, F.; Kometer, R.; Sliwinski, G.; Schwentner, N.)V1410,202-208(1991)

Visible and UV gas lasers with high-current radiative discharges excitation (Kamrukov, A. S.; Kozlov, N. P.; Protasov, Yu. S.)V1397,137-144(1991)

What industry needs in a high-power excimer laser (Lankard, John R.)V1377,2-5(1991)

XeCl laser with LC-circuit excitation research (Anufrik, S. S.; Znosko, K. F.; Kurgansky, A. D.)V1391,87-92(1991)

X-ray preionization for high-repetition-rate discharge excimer lasers (van Goor, Frederik A.; Witteman, Wilhelmus J.)V1412,91-102(1991)

Lasers, fiber

1.55-um broadband fiber sources pumped near 980 nm (Wysocki, Paul F.; Kalman, Robert F.; Digonnet, Michel J.; Kim, Byoung-Yoon)V1373,66-77(1991)

Absorption, fluorescence, and stimulated emission in Ti-diffused Er:LiNbO3 waveguides (Brinkmann, R.; Sohler, Wolfgang; Suche, Hubertus)V1362,377-382(1991)

Application of a linear picosecond streak camera to the investigation of a 1.55 um mode-locked Er3+ fiber laser (de Souza, Eunezio A.; Cruz, C. H.; Scarparo, Marco A.; Prokhorov, Alexander M.; Postovalov, V. E.; Vorobiev, N. S.; Schelev, Mikhail Y.)V1358,556-560(1991)

Applied research of optical fiber pulsed-laser holographic interferometry (Wang, Guozhi; Zhu, Guangkuan)V1554B,119-125(1991)

Diode-pumped Er3+ glass laser at 2.7 um (Yanagita, Hiroaki; Toratani, Hisayoshi; Yamashita, Toshiharu T.; Masuda, Isao)V1513,386-395(1991)

Excited state cross sections for Er-doped glasses (Zemon, Stanley A.; Lambert, Gary M.; Miniscalco, William J.; Davies, Richard W.; Hall, Bruce T.; Folweiler, Robert C.; Wei, Ta-Sheng; Andrews, Leonard J.; Singh, Mahendra P.)V1373,21-32(1991)

Fiber Laser Sources and Amplifiers II (Digonnet, Michel J., ed.)V1373(1991)

Fluoride fiber lasers (Smart, Richard G.; Carter, John N.; Tropper, Anne C.; Hanna, David C.)V1373,158-165(1991)

FM-cavity-dumped Nd-doped fiber laser (Zenteno, Luis A.; Po, Hong)V1373,246-253(1991)

Frequency up-conversion in Pr3+ doped fibers (Gomes, Anderson S.; de Araujo, Cid B.; Moraes, E. S.; Opalinska, M. M.; Gouveia-Neto, A. S.)V1579,257-263(1991)

Mode frequency behaviour in distributed-feedback lasers (Szczepanski, Pawel; Sklodowska, Malgorzata; Wolinski, Wieslaw)V1391,72-78(1991)

Modeling of three-level laser superfluorescent fiber sources (Kalman, Robert F.; Digonnet, Michel J.; Wysocki, Paul F.)V1373,209-222(1991)

Narrow linewidth fiber laser sources (Cowle, Gregory J.; Reekie, Laurence; Morkel, Paul R.; Payne, David N.)V1373,54-65(1991)

Nd- and Er-doped glass integrated optical amplifiers and lasers (Najafi, S. I.; Wang, Wei-Jian; Orcel, Gerard F.; Albert, Jacques; Honkanen, Seppo; Poyhonen, Pekka; Li, Ming-Jun)V1583,32-36(1991)

Noise characteristics of rare-earth-doped fiber sources and amplifiers (Morkel, Paul R.; Laming, Richard I.; Cowle, Gregory J.; Payne, David N.)V1373,224-233(1991)

Nonlinear operation of fiber distributed-feedback lasers (Szczepanski, Pawel; Sklodowska, Malgorzata; Wolinski, Wieslaw)V1391,65-71(1991)

Packaging considerations of fiber-optic laser sources (Heikkinen, Veli; Tukkiniemi, Kari; Vahakangas, Jouko; Hannula, Tapio)V1533,115-121(1991)

Polarization effects in fiber lasers: phenomena, theory, and applications (Lin, Jin T.; Gambling, William A.)V1373,42-53(1991)

Progress in fluoride fiber lasers and amplifiers (France, Paul W.; Brierley, Michael C.)V1373,33-39(1991)

Rare-earth-doped LiNbO3 waveguide amplifiers and lasers (Sohler, Wolfgang)V1583,110-121(1991)

Spectrum thermal stability of Nd- and Er-doped fiber sources (Wysocki, Paul F.; Fesler, Kenneth A.; Liu, K.; Digonnet, Michel J.; Kim, Byoung-Yoon)V1373,234-245(1991)

Thulium-doped silica fiber lasers (Tropper, Anne C.; Smart, Richard G.; Perry, Ian R.; Hanna, David C.; Lincoln, John; Brocklesby, Bill)V1373,152-157(1991)

Ultrafast pulse generation in fiber lasers (Kafka, James D.; Baer, Thomas M.)V1373,140-149(1991)

Upconversion in rare-earth-doped fluoride glasses (van Dongen, A. M.; Oomen, Emmanuel W.; le Gall, Perrine M.)V1513,330-339(1991)

Lasers, free-electron

Applications of synchroscan and dual-sweep streak camera techniques to free-electron laser experiments (Lumpkin, Alex H.)V1552,42-49(1991)

Applications of the FEL to NLO spectroscopy of semiconductors (Pidgeon, Carl R.)V1501,178-182(1991)

Beam divergence from sharp emitters in a general longitudinal magnetic field (Lau, Yue Y.; Colombant, Denis G.; Pilloff, Mark D.)V1552,182-184(1991)

Compact far-infrared free-electron laser (Ride, Sally K.; Golightly, W.)V1552,128-137(1991)

Compact free-electron laser at the Los Alamos National Laboratory (Chan, Kwok-Chi D.; Meier, Karl L.; Nguyen, Dinh C.; Sheffield, Richard L.; Wang, Tai-San; Warren, Roger W.; Wilson, William; Young, Lloyd M.)V1552,69-78(1991)

Compact, free-electron laser devices (Ciocci, Franco; Bartolini, R.; Dattoli, Giuseppe; Dipace, A.; Doria, Andrea; Gallerano, Gian P.; Kimmitt, Maurice F.; Messina, G.; Renieri, Alberto; Sabia, E.; Walsh, John E.)V1501,154-162(1991)

Comparison of integrated numerical experiments with accelerator and FEL experiments (Thode, Lester E.; Carlsten, Bruce E.; Chan, Kwok-Chi D.; Cooper, Richard K.; Elliott, C. J.; Gitomer, Steven J.; Goldstein, John C.; Jones, M. E.; McVey, Brian D.; Schmitt, Mark J.; Takeda, H.; Tokar, Robert L.; Wang, Tai-San; Young, Lloyd M.)V1552,87-106(1991)

Detection of infrared, free-electron laser radiation (Kimmitt, Maurice F.)V1501,86-96(1991)

Development of an XUV-IR free-electron laser user facility for scientific research and industrial applications (Newnam, Brian E.; Warren, Roger W.; Conradson, Steven D.; Goldstein, John C.; McVey, Brian D.; Schmitt, Mark J.; Elliott, C. J.; Burns, M. J.; Carlsten, Bruce E.; Chan, Kwok-Chi D.; Johnson, W. J.; Wang, Tai-San; Sheffield, Richard L.; Meier, Karl L.; Olsher, R. H.; Scott, Marion L.; Griggs, J. E.)V1552,154-174(1991)

Duke storage ring FEL program (Litvinenko, Vladimir N.; Madey, John M.; Benson, Stephen V.; Burnham, B.; Wu, Y.)V1552,2-6(1991)

Efficiency increase in a traveling wave tube by tapering the phase velocity of the wave (Schachter, Levi; Nation, John A.)V1407,44-56(1991)

FIR optical cavity oscillation is observed with the AT&T Bell Laboratories free-electron laser (Shaw, Earl D.; Chichester, Robert J.; La Porta, A.)V1552,14-23(1991)

Free-electron lasers as light sources for basic research (van Amersfoort, P. W.)V1500,2-13(1991); V1501,2-13(1991); 1502,2-13(1991)

High-power uhf rectenna for energy recovery in the HCRF (high-current radio frequency) system (Genuario, Ralph; Koert, Peter)V1407,553-565(1991)

Inverse bremsstrahlung acceleration in an electrostatic wave (Kim, Shang H.)V1407,620-634(1991)

Ion-ripple laser as an advanced coherent radiation source (Chen, Kuan-Ren; Dawson, John M.)V1552,185-196(1991)

Long-pulse modulator-driven cyclotron autoresonance maser and free-electron laser experiments at MIT (Danly, Bruce G.; Hartemann, Frederic V.; Chu, T. S.; Menninger, W. L.; Papavaritis, P.; Pendergast, K. D.; Temkin, Richard J.)V1407,192-201(1991)

Los Alamos photoinjector-driven free-electron laser (O'Shea, Patrick G.)V1552,36-41(1991)

Numerical studies of resonators with on-axis holes in mirrors for FEL applications (Keselbrener, Michel; Ruschin, Shlomo; Lissak, Boaz; Gover, Avraham)V1415,38-47(1991)

Optical method of detection for the magnetic alignment of an electron accelerator (Villate, Denis)V1533,193-196(1991)

Phase stability of a standing-wave free-electron laser (Sharp, William M.; Rangarajan, G.; Sessler, Andrew M.; Wurtele, Jonathan S.)V1407,535-545(1991)

Properties of the new UCSB free-electron lasers (Ramian, Gerald)V1552,57-68(1991)

Saturnus: the UCLA compact infrared free-electron laser project (Dodd, James W.; Aghamir, F.; Barletta, W. A.; Cline, David B.; Hartman, Steven C.; Katsouleas, Thomas C.; Kolonko, J.; Park, Sanghyun; Pellegrini, Claudio; Terrien, J. C.; Davis, J. G.; Joshi, Chan J.; Luhmann, Neville C.; McDermott, David B.; Ivanchenkov, S. N.; Lachin, Yu Y.; Varfolomeev, A. A.)V1407,467-473(1991)

Short-Wavelength Radiation Sources (Sprangle, Phillip, ed.)V1552(1991)

Simulation of free-electron lasers in the presence of measured magnetic field errors of the undulator (Marable, William P.; Tang, Cha-Mei; Esarey, Eric)V1552,80-86(1991)

Small-signal gain for parabolic profile beams in free-electron lasers (Elliott, C. J.)V1552,175-181(1991)

Spontaneous emission from free-electron lasers (Schmitt, Mark J.)V1552,107-117(1991)

Stanford picosecond FEL center (Swent, Richard L.; Schwettman, H. A.; Smith, T. I.)V1552,24-35(1991)

Stimulated relativistic harmonic generation from intense laser interactions with beams and plasmas (Sprangle, Phillip; Esarey, Eric)V1552,147-153(1991)

Superradiant Raman free-electron lasers (Tsui, King H.)V1407,281-284(1991)

Synchrotron radiation lasers (Hirshfield, Jay L.; Park, Gun-Sik)V1552,138-146(1991)

Theory and simulation of the HARmonic amPlifier Free-Electron Laser (HARP/FEL) (Gregoire, Daniel J.; Harvey, Robin J.; Levush, Baruch)V1552,118-126(1991)

Theory of multimode interactions in cyclotron autoresonance maser amplifiers (Chen, Chiping; Wurtele, Jonathan S.)V1407,183-191(1991)

Thermal analysis of multifacet-mirror ring resonator for XUV free-electron lasers (McVey, Brian D.; Goldstein, John C.; McFarland, Robert D.; Newnam, Brian E.)V1441,457-468(1991)

Vanderbilt Free-Electron Laser Center for Biomedical and Materials Research (Tolk, Norman H.; Brau, Charles A.; Edwards, Glenn S.; Margaritondo, Giorgio; McKinley, J. T.)V1552,7-13(1991)

XUV free-electron laser-based projection lithography systems (Newnam, Brian E.)V1343,214-228(1991)

Lasers, gas

200,000-frame-per-second drum camera with nanosecond synchronized laser illumination (Briscoe, Dennis)V1346,319-323(1991)

8th Intl Symp on Gas Flow and Chemical Lasers (Domingo, Concepcion; Orza, Jose M., eds.)V1397(1991)

Adsorption of bromine in CuBr laser (Wang, Yongjiang; Shi, Shuyi; Qian, Yujun; Pan, Bailiang; Ding, Xiande; Guo, Qiang; Chen, Lideng; Fan, Ruixing)V1412,67-71(1991)

Alternated or simultaneous sealed-off room temperature CO/CO2 laser tuning by chemical reactions (Masychev, Victor I.; Alejnikov, Vladislav S.)V1412,227-235(1991)

Analysis of pyrotechnic devices by laser-illuminated high-speed photography (Dosser, Larry R.; Stark, Margaret A.)V1346,293-299(1991)

Applications of copper vapor laser lighting in high-speed motion analysis (Hogan, Daniel C.)V1346,324-330(1991)

Argon laser vascular tissue fusion: current status and future perspectives (White, Rodney A.; Kopchok, George E.)V1422,103-110(1991)

Assessment of a discharge-excited supersonic free jet as a laser medium (Kannari, Fumihiko; Sato, F.; Obara, Minoru)V1397,493-497(1991)

Autonomous mobile laser complex (Fakhrutdinov, I. H.; Avdoshin, A. P.; Moshin, J. N.; Poltavsky, V. V.)V1399,98-106(1991)

Cavityless dielectric-waveguide-mode generation in a weakly amplifying gaseous medium (Mel'nikov, Lev Y.; Kochelap, Viatcheslav A.; Izmailov, I. A.)V1397,603-610(1991)

Characteristics of a downstream-mixing CO2 gas-dynamic laser (Hashimoto, Takashi; Nakano, Susumu; Hachijin, Michio; Komatsu, Katsuhiko; Hara, Hiroshi)V1397,519-522(1991)

Characteristics of pulsed nuclear-reactor-pumped flowing gas lasers (Neuman, William A.; Fincke, James R.)V1411,28-40(1991)

Chemical gas-dynamic mixing CO2 laser pumped by the reactions between N2O and CO (Doroschenko, V. M.; Kudriavtsev, N. N.; Sukhov, A. M.)V1397,503-511(1991)

Chemical problems of high-power sealed-off carbon monoxide lasers (Alejnikov, Vladislav S.; Masychev, Victor I.; Karpetscki, V. V.)V1412,173-184(1991)

CO2 coupled-mode, CO, and other lasers with supersonic cooling of gas mixture (Kudriavtsev, E. M.)V1397,475-484(1991)

Coagulation and precise ablation of biotissues by pulsed sealed-off carbon monoxide laser (Masychev, Victor I.; Alejnikov, Vladislav S.; Klimenko, Vladimir I.)V1427,344-356(1991)

Combined guidance technique using angioscope and fluoroscope images for CO laser angioplasty: in-vivo animal experiment (Arai, Tsunenori; Mizuno, Kyoichi; Sakurada, Masami; Miyamoto, Akira; Arakawa, Koh; Kurita, Akira; Suda, Akira; Kikuchi, Makoto; Nakamura, Haruo; Utsumi, Atsushi; Akai, Yoshiro; Takeuchi, Kiyoshi)V1425,191-195(1991)

Comparison of high-current discharges with axial and transverse gas flow for UV ion lasers (Babin, Sergei A.; Kuklin, A. E.)V1397,589-592(1991)

Copper vapor laser precision processing (Nikonchuk, Michael O.)V1412,38-49(1991)

Design characters and radiation parameters of a gas laser as a holographic tool (Lypsky, Volodymyr V.)V1238,390-395(1991)

Development of solid state pulse power supply for copper vapor laser (Fujii, Takashi; Nemoto, Koshichi; Ishikawa, Rikio; Hayashi, Kazuo; Noda, Etsuo)V1412,50-57(1991)

Development of the aerodynamic window for high-power CO laser (Kuribayashi, Shizuna; Noda, Osama; Ogino, M.; Imatake, Shigenori; Kondo, Motoe; Sato, Shunichi; Takahashi, Kunimitsu)V1397,439-443(1991)

Devices for generation and detection of subnanosecond IR and FIR radiation pulses (Beregulin, Eugene V.; Ganichev, Sergey D.; Yaroshetskii, Ilya D.; Lang, Peter T.; Schatz, Wolfgan; Renk, Karl F.)V1362,853-862(1991)

Efficacy of argon-laser-mediated hot-balloon angioplasty (Sakurada, Masami; Miyamoto, Akira; Mizuno, Kyoichi; Nozaki, Youichi; Tabata, Hirotsugu; Etsuda, Hirokuni; Kurita, Akira; Nakamura, Haruo; Arai, Tsunenori; Suda, Akira; Kikuchi, Makoto; Watanabe, Tamishige; Utsumi, Atsushi; Akai, Yoshiro; Takeuchi, Kiyoshi)V1425,158-164(1991)

Electrical discharges investigation in gas-pulsed laser (Persephonis, Peter; Giannetas, B.; Parthenios, J.; Georgiades, C.)V1503,185-196(1991)

Estimation of CO2/N2 mixing region by the small-signal gain measurement (Ohue, Hiroshi; Kasahara, Eiji; Uemura, Kamon; Mito, Keiichi)V1397,527-530(1991)

Excimer and nitrogen lasers with low-average power for technology applications (Pokora, Ludwik J.)V1397,111-114(1991)

Experimental and computer-modeled results of titanium sapphire lasers pumped by copper vapor lasers (Knowles, Martyn R.; Webb, Colin E.; Naylor, Graham A.)V1410,195-201(1991)

Flowfield in a CO2-N2 gas-dynamics laser with staggered nozzles (Tarabelli, D.; Zeitoun, David; Imbert, Michel P.)V1397,523-526(1991)

Fluid-dynamic perturbations in gas lasers (Horton, T. E.)V1397,549-554(1991)

Focal plane intensity distribution of copper vapor laser with different unstable resonators (Nikonchuk, Michael O.; Polyakov, Igor V.)V1412,72-78(1991)

Gas and Metal Vapor Lasers and Applications (Kim, Jin J.; Tittel, Frank K., eds.)V1412(1991)

Gas-dynamically cooled CO laser with rf-excitation: design and performance (von Buelow, H.; Zeyfang, E.)V1397,499-502(1991)

Gas-dynamically cooled CO laser with rf-excitation: optical performance (Zeyfang, E.; von Buelow, H.; Stoehr, M.)V1397,449-452(1991)

High-power copper vapor laser development (Aoki, Nobutada; Kimura, Hironobu; Konagai, Chikara; Shirayama, Shimpey; Miyazawa, Tatsuo; Takahashi, Tomoyuki)V1412,2-11(1991)

High-power electron beam controlled discharge N2O laser (Frolov, K.; Ionin, A. A.; Kelner, M.; Sinitsin, D. V.; Suchkov, A. F.; Zhivukcin, I.)V1397,461-468(1991)

High-power nanosecond pulse iodine laser provided with SBS mirror (Dolgopolov, Y. V.; Kirillov, Gennadi A.; Kochemasov, G. G.; Kulikov, Stanislav M.; Murugov, V. M.; Pevny, S. N.; Sukharev, S. A.)V1412,267-275(1991)

High-power particle beams for gas lasers (Mesyats, Gennady A.)V1411,2-14(1991)

High-power pulsed and repetitively pulsed electron-beam-controlled discharge CO laser systems (Ionin, A. A.)V1502,95-102(1991)

High-repetition-rate x-ray preionization source (van Goor, Frederik A.)V1397,563-568(1991)

Industrial applications of metal vapor lasers (Gokay, Cem)V1412,28-31(1991)

Industrial excimer and CO2 TEA lasers with kilowatt average output power (von Bergmann, Hubertus M.; Swart, Pieter L.)V1397,63-66(1991)

Ion argon laser with metal-ceramic discharge tube: construction, technology, and gas-pumping effect (Kesik, Jerzy; Siejca, Antoni; Sokolowski, Maciej)V1391,34-41(1991)

Laser action of Xe and Ne pumped by electron beam (Tarasenko, Victor F.)V1412,185-196(1991)

Long-pulse electron gun for laser applications (Bayless, John R.; Burkhart, Craig P.)V1411,42-46(1991)

Low-pressure nitrogen laser pulse repetition frequency to 100 Hz (Makuchowski, Jozef; Pokora, Ludwik J.; Wawer, Janusz)V1391,79-86(1991)

Material influence on cutting and drilling of metals using copper vapor lasers (Kupfer, Roland; Bergmann, Hans W.; Lingenauer, Marion)V1598,46-60(1991)

Mechanical effects induced by high-power HF laser pulses on different materials under normal atmospheric conditions (Paolacci, Sylvie; Hugenschmidt, Manfred; Bournot, Philippe)V1397,705-708(1991)

Metal vapor gain media based on multiphoton dissociation of organometallics (Samoriski, Brian; Wiedeger, S.; Villarica, M.; Chaiken, Joseph)V1412,12-18(1991)

Modeling of pumping kinetics of an iodine photodissociation laser with long pumping pulse (Rohlena, Karel; Beranek, J.; Masek, Karel)V1415,259-268(1991)

Modifications to the LP-140 pulsed CO2 laser for lidar use (Jaenisch, Holger M.; Spiers, Gary D.)V1411,127-136(1991)

Modified off-axis unstable resonator for copper vapor laser (Lando, Mordechai; Belker, D.; Lerrer, A.; Lotem, Haim; Dikman, A.; Bialolanker, Gabriel; Lavi, S.; Gabay, Shimon)V1412,19-26(1991)

Multiple-spectral structure of the 578.2 nm line for copper vapor laser (Wang, Yongjiang; Pan, Bailiang; Ding, Xiande; Qian, Yujun; Shi, Shuyi)V1412,60-66(1991)

New generation of copper vapor lasers for high-speed photography (Walder, Brian T.)V1358,811-820(1991)

One-dimensional and two-dimensional computer models of industrial CO laser (Iyoda, Mitsuhiro; Murota, Tomoya; Akiyama, Mamoru; Sato, Shunichi)V1415,342-349(1991)

Optomechanical considerations for stable lasers (Horton, T. E.; Cason, Charles; Dezenberg, George J.)V1416,10-20(1991)

Performance characteristics of premixing gas-dynamic laser utilizing liquid C6H6 and liquid N2O (Yokozawa, T.; Nakajima, H.; Yamaguchi, S.; Ebina, A.; Ohara, M.; Kanazawa, Hirotaka; Yuasa, M.; Komatsu, Katsuhiko; Hara, Hiroshi)V1397,513-517(1991)

Performance of CO2 and CO diffusively cooled rf-excited strip-line lasers with different electrode materials and gas composition (Yatsiv, Shaul; Gabay, Amnon; Sterman, Baruch; Sintov, Yoav)V1397,319-329(1991)

Plasma motion velocity along laser beam and continuous optical discharge in gas flow (Budnik, A. P.; Gus'kov, K. G.; Raizer, Yu. P.; Surjhikov, S. T.)V1397,721-724(1991)

Pointing stability of copper vapor laser with novel off-axis unstable resonator (Lando, Mordechai; Belker, D.; Lerrer, A.; Lotem, Haim; Dikman, A.; Bialolanker, Gabriel; Lavi, S.; Gabay, Shimon)V1442,172-180(1991)

Power stabilization of high-power industrial CO laser using gas exchange (Kanazawa, Hirotaka; Yamaguchi, Naohito; Nakajima, Takuro; Hamano, Yasunori; Satani, Ryoichi; Taira, Tatsuji)V1397,445-448(1991)

Precise uncoupling theory to study gain of gas-dynamic laser (Bahram-pour, A. R.; Mehdizadeh, E.; Bolorizadeh, M. A.; Shojaey, M.)V1397,539-542(1991)

Pulse Power for Lasers III (McDuff, G. G., ed.)V1411(1991)

Scaling of self-sustained discharge-excited cw CO laser (Sato, Shunichi; Takahashi, Kunimitsu; Noda, Osama; Kuribayashi, Shizuma; Imatake, Shigenori; Kondo, Motoe)V1397,433-437(1991)

Selected Papers on High Power Lasers (Soures, John M., ed.)VMS43(1991)

Superhigh-gain gas laser (Zykov, L. I.; Kormer, S. B.; Kulikov, Stanislav M.; Sukharev, S. A.; Sckapa, A. F.)V1412,258-266(1991)

Supersonic electron beam controlled discharge CO laser (Ionin, A. A.; Kotkov, A. A.; Minkovsky, M. G.; Sinitsin, D. V.)V1397,453-456(1991)

TE two-frequency pulsed laser at 0.337 um and 10.6 um with a plasma cathode (Branzalov, Peter P.)V1412,236-245(1991)

The EU 194 project: industrial applications of high-power CO2 cw lasers (Quenzer, A.; Beyer, Eckhard; Orza, Jose M.; Ricciardi, G.; Russell, J. D.; Sanz Justes, Pedro; Serafetinides, A.; Schuoecker, Dieter; Thorstensen, B.)V1397,753-759(1991)

Theory of solar-powered, cavityless waveguide laser (Mel'nikov, Lev Y.; Kochelap, Viatcheslav A.; Izmailov, I. A.)V1501,144-149(1991)

Thyristor driven pulser for multikilowatt lasers (Swart, Pieter L.; von Bergmann, Hubertus M.)V1397,559-562(1991)

Two-dimensional computer modeling of discharge-excited CO gas flow (Iyoda, Mitsuhiro; Murota, Tomoya; Akiyama, Mamoru; Sato, Shunichi)V1397,457-460(1991)

UV waveguide gas laser for biological and medical diagnostic methods (Kushlevsky, S. V.; Patrin, V. V.; Provorov, Alexander S.; Salmin, V. V.)V1403,799-800(1991)

Lasers, helium-neon

Aerodyne research mobile infrared methane monitor (McManus, J. B.; Kebabian, Paul L.; Kolb, Charles E.)V1433,330-339(1991)

Deterministic and noise-induced phase jumps in the ring laser gyroscope (Chyba, Thomas H.)V1376,132-142(1991)

Effectiveness of porphyrin-like compounds in photodynamic damage of atherosclerotic plaque (Zalessky, Viacheslav N.; Bobrov, Vladimir; Michalkin, Igor; Trunov, Vitaliy)V1426,162-169(1991)

Effects of He-Ne regional irradiation on 53 cases in the field of pediatric surgery (Guo, Jing-Zhen)V1422,136-139(1991)

Helium-neon effects of laser radiation in rats infected with thromboxane B2 (Juri, Hugo; Palma, J. A.; Campana, Vilma; Gavotto, A.; Lapin, R.; Yung, S.; Lillo, J.)V1422,128-135(1991)

Helium neon laser optics: scattered light measurements and process control (Perilloux, Bruce E.)V1530,255-262(1991)

Noise in He:Ne lasers near threshold (Singh, Surendra; Mortazavi, Mansour; Phillips, K. J.; Young, M. R.)V1376,143-152(1991)

Photosensitized receptor inactivation with He-Ne laser: preliminary results (Arber, Simon; Rymer, William Z.; Crumrine, David)V1428,23-29(1991)

Underwater laser scanning system (Austin, Roswell W.; Duntley, Seibert Q.; Ensminger, Richard L.; Petzold, Theodore J.; Smith, Raymond C.)V1537,57-73(1991)

Vibration analysis of the tympanic membrane with a ventilation tube and a perforation by holography (Maeta, Manabu; Kawakami, Shinichiro; Ogawara, Toshiaki; Masuda, Yu)V1429,152-161(1991)

Lasers, high-energy

Effects of deformable mirror/wavefront sensor separation in laser beam trains (Schafer, Eric L.; Lyman, Dwight D.)V1415,310-316(1991)

High-peak-power Nd:glass laser facilities for end users (Hunt, John T.)V1410,2-14(1991)

Laser-target diagnostics instrumentation system (Bousek, Ronald R.)V1414,175-184(1991)

Long-pulse electron gun for laser applications (Bayless, John R.; Burkhart, Craig P.)V1411,42-46(1991)

Selected Papers on Ultrafast Laser Technology (Gosnell, Timothy R.; Taylor, Antoinette J., eds.)VMS44(1991)

Lasers, high-power—see also lasers, carbon dioxide; lasers, excimer; lasers, gas; lasers, neodymium; lasers, semiconductor; lasers, x-ray

100-mW four-beam individually addressable monolithic AlGaAs laser diode arrays (Yamaguchi, Takao; Yodoshi, Keiichi; Minakuchi, Kimihide; Tabuchi, Norio; Bessho, Yasuyuki; Inoue, Yasuaki; Komeda, Koji; Mori, Kazushi; Tajiri, Atsushi; Tominaga, Koji)V1418,363-371(1991)

Active-passive colliding pulse mode-locked Nd:YAG laser (Li, Shiying; Chen, Shisheng; Lin, Li-Huang; Wang, Shijie; Bin, Ouyang; Xu, Zhizhan)V1410,215-220(1991)

AR layer properties for high-power laser prepared by neutral-solution processing (Wu, Fang F.; Su, Kai J.)V1519,347-349(1991)

Characterization of a subpicosecond XeF (C->A) excimer laser (Hofmann, Thomas; Sharp, Tracy E.; Dane, C. B.; Wisoff, P. J.; Wilson, William L.; Tittel, Frank K.; Szabo, Gabor)V1412,84-90(1991)

Chemical problems of high-power sealed-off carbon monoxide lasers (Alejnikov, Vladislav S.; Masychev, Victor I.; Karpetscki, V. V.)V1412,173-184(1991)

Coupled CO2 lasers (Antyuhov, V.; Bondarenko, Alexander V.; Glova, Alexander F.; Golubenzev, A. A.; Danshikov, E.; Kachurin, O. R.; Lebedev, Fedor V.; Likhanskii, Vladimir V.; Napartovich, Anatoly P.; Pis'menny, Vladislav D.; Yarzev, V.)V1415,48-59(1991)

Design considerations for high-power industrial excimer lasers (Jetter, Heinz-Leonard)V1503,48-52(1991)

Development of solid state pulse power supply for copper vapor laser (Fujii, Takashi; Nemoto, Koshichi; Ishikawa, Rikio; Hayashi, Kazuo; Noda, Etsuo)V1412,50-57(1991)

Development of third-harmonic output beam diagnostics on NOVA (Laumann, Curt W.; Caird, John A.; Smith, James R.; Horton, R. L.; Nielsen, Norman D.)V1414,151-160(1991)

Diode-pumped, high-power, solid-state lasers (Yamanaka, Masanobu; Naito, Kenta; Nakatsuka, Masahiro; Yamanaka, Tatsuhiko; Nakai, Sadao S.; Yamanaka, Chiyoe)V1501,30-39(1991)

Direct laser beam diagnostics (van Gilse, Jan; Koczera, Stanley; Greby, Daniel F.)V1414,45-54(1991)

Evaluation of electrolytic capacitors for high-peak current pulse duty (Harris, Kevin; McDuff, G. G.; Burkes, Tom R.)V1411,87-99(1991)

Excimer Lasers and Applications III (Letardi, Tommaso; Laude, Lucien D., eds.)V1503(1991)

Fast-iterative technique for the calculation of frequency-dependent gain in excimer laser amplifiers (Sze, Robert C.)V1412,164-172(1991)

Frequency stability of a solid state mode-locked laser system (Simpson, Thomas B.; Doft, F.; Malley, Michael M.; Sutton, George W.; Day, Timothy)V1410,133-140(1991)

Fundamental array mode operation of semiconductor laser arrays using external spatial filtering (Cherng, Chung-Pin; Osinski, Marek A.)V1418,372-385(1991)

Gas and Metal Vapor Lasers and Applications (Kim, Jin J.; Tittel, Frank K., eds.)V1412(1991)

Generation of ultrashort high-average-power passively mode-locked pulses from a Nd:YLF laser with a nonlinear external-coupled cavity at high-repetition rates (Chee, Joseph K.; Kong, Mo-Nga; Liu, Jia-ming)V1413,14-20(1991)

High-energy laser wavefront sensors (Geary, Joseph M.)V1414,66-79(1991)

High-fidelity diagnostic beam sampling of a tunable high-energy laser (Mitchell, John H.; Cohen, Harold L.; Hanley, Stephen T.)V1414,141-150(1991)

High-level PC-based laser system modeling (Taylor, Michael S.)V1415,300-309(1991)

High-peak-power Nd:glass laser facilities for end users (Hunt, John T.)V1410,2-14(1991)

High-power coherent diode lasers (Botez, Dan; Mawst, Luke J.; Jansen, Michael; Anderson, Eric R.; Ou, Szutsun S.; Sergant, Moshe; Peterson, Gary L.; Roth, Thomas J.; Rozenbergs, John)V1474,64-74(1991)

High-power copper vapor laser development (Aoki, Nobutada; Kimura, Hironobu; Konagai, Chikara; Shirayama, Shimpey; Miyazawa, Tatsuo; Takahashi, Tomoyuki)V1412,2-11(1991)

High-power diffraction-limited phase-locked GaAs/GaAlAs laser diode array (Zhang, Yue-qing; Zhang, Xitian; Piao, Yue-zhi; Li, Dian-en; Wu, Sheng-li; Du, Shu-qin)V1418,444-447(1991)

High-power lasers and the production of energy by inertial fusion (Velarde, G.; Perlado, J. M.; Aragones, J. M.; Honrubia, J. J.; Martinez-Val, J. M.; Minguez, E.)V1502,242-257(1991)

High-power nanosecond pulse iodine laser provided with SBS mirror (Dolgopolov, Y. V.; Kirillov, Gennadi A.; Kochemasov, G. G.; Kulikov, Stanislav M.; Murugov, V. M.; Pevny, S. N.; Sukharev, S. A.)V1412,267-275(1991)

High-repetition-rate pseudospark switches for pulsed high-power lasers (Bickel, P.; Christiansen, Jens; Frank, Klaus; Goertler, Andreas; Hartmann, Werner; Kozlik, Claudius; Wiesneth, Peter)V1503,167-175(1991)

Industrial and Scientific Uses of High-Power Lasers (Billon, Jean P.; Fabre, Edouard, eds.)V1502(1991)

Integration of diagnostics in high-power laser systems for optimization of laser material processing (Biermann, Stephan; Geiger, Manfred)V1415,330-341(1991)

Interaction of ultra-bright lasers with matter: program of the French Commissariat a l'Energie Atomique (Andre, Michel L.; Coutant, Jacques; Dautray, Robert; Decroisette, Michel; Lompre, Louis A.; Naudy, Michel; Manus, Claude; Mainfray, Gerard L.; Migus, Arnold; Normand, Didier; Sauteret, Christian; Watteau, Jean P.)V1502,286-298(1991)

Interferential diagnosis of self-focusing of Q-switched YAG laser in liquid (Lu, Jian-Feng; Wang, Chang Xing; Miao, Peng-Cheng; Ni, Xiao W.; He, Anzhi)V1415,220-224(1991)

Interferometric measurement for the plasma produced by Q-switched laser in air near the surface of an Al Target (Lu, Jian-Feng; Ni, Xiao W.; Wang, Hong Y.; He, Anzhi)V1554B,593-597(1991)

Ion-ripple laser as an advanced coherent radiation source (Chen, Kuan-Ren; Dawson, John M.)V1552,185-196(1991)

Kilowatt-range high-repetition-rate excimer lasers (Borisov, V. M.; Khristoforov, O. B.; Kirykhin, Yu. B.; Kuznetsov, S. G.; Stepanov, Yu. Y.; Vinokhodov, A. Y.)V1503,40-47(1991)

Laser application for fusion using volume ignition and smoothing by suppression of pulsation (Hora, Heinrich; Aydin, M.; Kasotakis, G.; Stening, R. L.)V1502,258-269(1991)

Laser Diode Technology and Applications III (Renner, Daniel, ed.)V1418(1991)

Life test of high-power Matsushita BTRS laser diodes (Holcomb, Terry L.)V1417,328-337(1991)

Magnetic pulse compressor for high-power excimer discharge laser pumping (Vizir, V. A.; Kiryukhin, Yu. B.; Manylov, V. I.; Nosenko, S. P.; Sychev, S. P.; Chervyakov, V. V.; Shubkin, N. G.)V1411,63-68(1991)

Market for high-power CO2 lasers (Fortino, Dennis J.)V1517,61-84(1991)

Modeling of high-power laser welding (Schuoecker, Dieter)V1397,745-751(1991)

Modeling of large-aperture third-harmonic frequency conversion of high-power Nd:glass laser systems (Henesian, Mark A.; Wegner, Paul J.; Speck, David R.; Bibeau, Camille; Ehrlich, Robert B.; Laumann, Curt W.; Lawson, Janice K.; Weiland, Timothy L.)V1415,90-103(1991)

Modeling of pumping kinetics of an iodine photodissociation laser with long pumping pulse (Rohlena, Karel; Beranek, J.; Masek, Karel)V1415,259-268(1991)

Modeling the pedestal in a chirped-pulse-amplification laser (Chuang, Yung-Ho; Peatross, J.; Meyerhofer, David D.)V1413,32-40(1991)

Modifications to achieve subnanosecond jitter from standard commercial excimer lasers (Rust, Kenneth R.)V1411,80-86(1991)

New opportunities with intense ultra-short-pulse lasers (Richardson, Martin C.)V1410,15-25(1991)

Novel distributed feedback structure for surface-emitting semiconductor lasers (Mahbobzadeh, Mohammad; Osinski, Marek A.)V1418,25-31(1991)

Numerical simulation of energy transport mechanisms in high-intensity laser-matter interaction experiments (Ocana, Jose L.)V1502,299-310(1991)

One-dimensional and two-dimensional computer models of industrial CO laser (Iyoda, Mitsuhiro; Murota, Tomoya; Akiyama, Mamoru; Sato, Shunichi)V1415,342-349(1991)

Optimization of pulsed laser power supply system (Alci, Mustafa; Yilbas, Bekir S.; Danisman, Kenan; Ciftlikli, Cebrail; Altuner, Mehmet)V1411,100-106(1991)

Output power stabilization of a XeCl excimer laser by HCl gas injection (Ogura, Satoshi; Kawakubo, Yukio; Sasaki, Kouji; Kubota, Yoshiyuki; Miki, Atsushi)V1412,123-128(1991)

Output pulse and energy capabilities of the PHEBUS laser facility (Fleurot, Noel A.; Andre, Michel L.; Estraillier, P.; Friart, Daniel; Gouedard, C.; Rouyer, C.; Thebault, J. P.; Thiell, Gaston; Veron, Didier)V1502,230-241(1991)

Parametric study of a high-average-power XeCl laser (Godard, Bruno; Estocq, Emmanuel; Joulain, Franck; Murer, Pierre; Stehle, Marc X.; Bonnet, Jean C.; Pigache, Daniel R.)V1503,71-77(1991)

Parametric study of small-volume long-pulse x-ray preionized XeCl laser with double-discharge and fast magnetic switch (Hueber, Jean-Marc; Fontaine, Bernard L.; Kobhio, M. N.; Delaporte, Philippe C.; Forestier, Bernard M.; Sentis, Marc L.)V1503,62-70(1991)

Performance characteristics of a discharge-pumped XeCl laser driven by a pulse forming network containing nonlinear ferroelectric capacitors (Fairlie, S. A.; Smith, Paul W.)V1411,56-62(1991)

Plasma heating by ultrashort laser pulse in the regime of anomalous skin effect (Gamaly, Eugene G.; Kiselev, A. Y.; Tikhonchuk, V. T.)V1413,95-106(1991)

Plasma interaction with powerful laser beams (Giulietti, Antonio; Giulietti, Danilo; Willi, Oswald)V1502,270-283(1991)

Properties of the new UCSB free-electron lasers (Ramian, Gerald)V1552,57-68(1991)

Pulse Power for Lasers III (McDuff, G. G., ed.)V1411(1991)

Pump-probe investigation of picosecond laser-gas target interactions (Durfee, C.; Milchberg, Howard M.)V1413,78-80(1991)

Recent progress towards multikilowatt output (Mueller-Horsche, Elmar; Oesterlin, Peter; Basting, Dirk)V1503,28-39(1991)

Review of the multikilohertz performance of the CHIRP laser and components (Fieret, Jim; Green, J. M.; Heath, R.; O'Key, Michael A.; Osborne, Michael R.; Osbourn, S. J.; Taylor, Arthur F.; Winfield, R. J.)V1503,53-61(1991)

Saturnus: the UCLA compact infrared free-electron laser project (Dodd, James W.; Aghamir, F.; Barletta, W. A.; Cline, David B.; Hartman, Steven C.; Katsouleas, Thomas C.; Kolonko, J.; Park, Sanghyun; Pellegrini, Claudio; Terrien, J. C.; Davis, J. G.; Joshi, Chan J.; Luhmann, Neville C.; McDermott, David B.; Ivanchenkov, S. N.; Lachin, Yu Y.; Varfolomeev, A. A.)V1407,467-473(1991)

Scattering measurements of optical coatings in high-power lasers (Chen, Yi-Sheng)V1332,115-120(1991)

Second-harmonic generation of mode-locked Nd:YAG and Nd:YLF lasers using LiB3O5 (Reed, Murray K.; Tyminski, Jacek K.; Bischel, William K.)V1410,179-184(1991)

Selected Papers on High Power Lasers (Soures, John M., ed.)VMS43(1991)

Self-reflection and self-transmission of pulsed radiation by laser-evaporated media (Furzikov, Nickolay P.)V1415,228-239(1991)

Short-Pulse High-Intensity Lasers and Applications (Baldis, Hector A., ed.)V1413(1991)

Stimulated relativistic harmonic generation from intense laser interactions with beams and plasmas (Sprangle, Phillip; Esarey, Eric)V1552,147-153(1991)

Upgrade of the LLNL Nova laser for inertial confinement fusion (Murray, John R.; Trenholme, J. B.; Hunt, John T.; Frank, D. N.; Lowdermilk, W. H.; Storm, E.)V1410,28-39(1991)

Velocity measurements in molten pools during high-power laser interaction with metals (Caillibotte, Georges; Kechemair, Didier; Sabatier, Lilian)V1412,209-211(1991)

Vertical-cavity surface-emitting semiconductor lasers: present status and future prospects (Osinski, Marek A.)V1418,2-24(1991)

What industry needs in a high-power excimer laser (Lankard, John R.)V1377,2-5(1991)

X-ray and optical diagnostics of a 100-fs laser-produced plasma (Geindre, Jean-Paul; Audebert, Patrick; Chenais-Popovics, Claude; Gauthier, Jean-Claude J.; Benattar, Rene; Chambaret, J. P.; Mysyrowicz, Andre; Antonetti, Andre)V1502,311-318(1991)

Lasers, neodymium—see also lasers, solid state

0.5-GHz cw mode-locked Nd:glass laser (Ling, Junda D.; Yan, Li I.; Liu, YuanQun; Lee, Chi H.; Soos, Jolanta I.)V1454,353-362(1991)

1W cw diode-pumped Nd:YAG laser for coherent space communication systems (Johann, Ulrich A.; Seelert, Wolf)V1522,158-168(1991)

Acoustic effects of Q-switched Nd:YAG on crystalline lens (Porindla, Sridhar N.; Rylander, Henry G.; Welch, Ashley J.)V1427,267-272(1991)

Active-passive colliding pulse mode-locked Nd:YAG laser (Li, Shiying; Chen, Shisheng; Lin, Li-Huang; Wang, Shijie; Bin, Ouyang; Xu, Zhizhan)V1410,215-220(1991)

Arthroscopic contact Nd:YAG laser meniscectomy: surgical technique and clinical follow-up (O'Brien, Stephen J.; Miller, Drew V.; Fealy, Stephen V.; Gibney, Mary A.; Kelly, Anne M.)V1424,62-75(1991)

Beam control and power conditioning of GEKKO glass laser system (Nakatsuka, Masahiro; Jitsuno, Takahisa; Kanabe, Tadashi; Urushihara, Shinji; Miyanaga, Noriaki; Nakai, Sadao S.)V1411,108-115(1991)

Bifunctional irrigation liquid as an ideal energy converter for laser lithotripsy with nanosecond laser pulses (Reichel, Erich; Schmidt-Kloiber, Heinz; Paltauf, Guenther; Groke, Karl)V1421,129-133(1991)

Blue laser devices for optical data storage (Lenth, Wilfried; Kozlovsky, William J.; Risk, William P.)V1499,308-313(1991)

BPSK homodyne and DPSK heterodyne receivers for free-space communication with ND:host lasers (Bopp, Matthias; Huether, Gerhard; Spatscheck, Thomas; Specker, Harald; Wiesmann, Theo J.)V1522,199-209(1991)

Changes in optical density of normal vessel wall and lipid atheromatous plaque after Nd:YAG laser irradiation (Schwarzmaier, Hans-Joachim; Heintzen, Matthias P.; Zumdick, Mathias; Kaufmann, Raimund; Wolbarsht, Myron L.)V1427,128-133(1991)

Clinical experience in applying endoscopic Nd:YAG laser to treat 451 esophagostenotic cases (Wang, Rui-zhong; Wang, Zhen-he; Lu, Kuang-sheng; Yang, Xiao-zhi; Lu, Bo-kao)V1421,203-207(1991)

Definition of interorbit link optical terminals with diode-pumped Nd:host laser technology (Marini, Andrea E.; Della Torre, Antonio)V1522,222-233(1991)

Dependence of output beam profile on launching conditions in fiber-optic beam delivery systems for Nd:YAG lasers (Su, Daoning; Boechat, Alvaro A.; Jones, Julian D.)V1502,41-51(1991)

Design of a diode-pumped Nd:YAG laser communication system (Sontag, Heinz; Johann, Ulrich A.; Pribil, Klaus)V1417,573-587(1991)

Deterministic fluctuations in an intracavity-coupled solid state laser (Bracikowski, Christopher; Roy, Rajarshi)V1376,103-116(1991)

Diffraction-limited Nd:glass and alexandrite lasers using graded reflectivity mirror unstable resonators (Snell, Kevin J.; Duplain, Gaetan; Parent, Andre; Labranche, Bruno; Galarneau, Pierre)V1410,99-106(1991)

Diode-pumped Nd:YAG laser transmitter for free-space optical communications (Nava, Enzo; Re Garbagnati, Giuseppe; Garbi, Maurizio; Marchiori, Livio D.; Marini, Andrea E.)V1417,307-315(1991)

Effect of Nd:YAG laser on dentinal bond strength (Dederich, Douglas N.; Tulip, John)V1424,134-137(1991)

Electronic system for generation of laser pulses by self-injection method (Brzezinski, Ryszard; Piotrowski, Jan)V1391,127-134(1991)

Electro-optic modulator for high-speed Nd:YAG laser communication (Petsch, Thomas)V1522,72-82(1991)

Endoscopic YAG laser coagulation for early prostate cancer (McNicholas, Thomas A.; O'Donoghue, Neil)V1421,56-67(1991)

End-to-end model of a diode-pumped Nd:YAG pulsed laser (Mayer, Richard C.; Dreisewerd, Douglas W.)V1415,248-258(1991)

Energy conversion efficiency during optical breakdown (Grad, Ladislav; Diaci, J.; Mozina, Janez)V1525,206-209(1991)

Experimental investigation of 1.06 um laser interaction with Al target in air (Gang, Yuan)V1415,225-227(1991)

Frequency stability of a solid state mode-locked laser system (Simpson, Thomas B.; Doft, F.; Malley, Michael M.; Sutton, George W.; Day, Timothy)V1410,133-140(1991)

Generation of ultrashort high-average-power passively mode-locked pulses from a Nd:YLF laser with a nonlinear external-coupled cavity at high-repetition rates (Chee, Joseph K.; Kong, Mo-Nga; Liu, Jia-ming)V1413,14-20(1991)

Geoscience laser ranging system design and performance predictions (Anderson, Kent L.)V1492,142-152(1991)

High-dynamic-range fiber optic link using external modulator diode pumped Nd:YAG lasers (Childs, Richard B.; O'Byrne, Vincent A.)V1371,223-232(1991)

High-energy Nd:glass laser for oncology (Bouchenkov, Vyatcheslav A.; Utenkov, Boris I.; Zajtsev, V. K.; Bayanov, Valentin I.; Serebryakov, Victor A.)V1410,244-247; V1427,405-408(1991)

High-peak-power Nd:glass laser facilities for end users (Hunt, John T.)V1410,2-14(1991)

High-power diode-pumped solid state lasers for optical space communications (Koechner, Walter; Burnham, Ralph L.; Kasinski, Jeff; Bournes, Patrick A.; DiBiase, Don; Le, Khoa; Marshall, Larry R.; Hays, Alan D.)V1522,169-179(1991)

Hyperthermia treatment of spontaneously occurring oral cavity tumors using a computer-controlled Nd:YAG laser system (Panjehpour, Masoud; Overholt, Bergein F.; Frazier, Donita L.; Klebanow, Edward R.)V1424,179-185(1991)

Industrial lasers in Japan (Karube, Norio)V1517,1-14(1991)

Influence of helium, oxygen, nitrogen, and room air environment in determining Nd-YAG laser/brain tissue interaction (Chavantes, Maria C.; Vinas, Federico; Zamorano, Lucia J.; Dujovny, Manuel; Dragovic, Ljubisa)V1428,13-22(1991)

Injection chaining of diode-pumped single-frequency ring lasers for free-space communication (Cheng, Emily A.; Kane, Thomas J.; Wallace, Richard W.; Cornwell, Donald M.)V1417,300-306(1991)

Interstitial laser coagulation of the prostate: experimental studies (McNicholas, Thomas A.; Steger, Adrian C.; Bown, Stephen G.; O'Donoghue, Neil)V1421,30-35(1991)

Job shop market (Belforte, David A.)V1517,176-196(1991)

Laser-assisted skin closure at 1.32 microns: the use of a software-driven medical laser system (Dew, Douglas K.; Hsu, Tung M.; Hsu, Long S.; Halpern, Steven J.; Michaels, Charles E.)V1422,111-115(1991)

Laser photocathode development for high-current electron source (Moustaizis, Stavros D.; Fotakis, Costas; Girardeau-Montaut, Jean-Pierre)V1552,50-56(1991)

Laser range-gated underwater imaging including polarization discrimination (Swartz, Barry A.; Cummings, James D.)V1537,42-56(1991)

LiNbO3 with rare earths: lasers and amplifiers (Lallier, Eric; Pocholle, Jean-Paul; Papuchon, Michel R.; De Micheli, Marc; Li, M. J.; He, Q.; Ostrowsky, Daniel B.; Grezes-Besset, C.; Pelletier, Emile P.)V1506,71-79(1991)

Liquid circular polarizer in laser system (Cesarz, Tadeusz; Klosowicz, Stanislaw; Zmija, Jozef)V1391,244-249(1991)

Local interstitial hyperthermia in malignant brain tumors using a low-power Nd:YAG laser (Bettag, Martin; Ulrich, Frank; Kahn, Thomas; Seitz, R.)V1525,409-411(1991)

Markets for marking systems (Austin, Patrick D.)V1517,150-175(1991)

Master oscillator-amplifier Nd:YAG laser with a SBS phase-conjugate mirror (Ayral, Jean-Luc; Montel, J.; Huignard, Jean-Pierre)V1500,81-92(1991)

Modeling of large-aperture third-harmonic frequency conversion of high-power Nd:glass laser systems (Henesian, Mark A.; Wegner, Paul J.; Speck, David R.; Bibeau, Camille; Ehrlich, Robert B.; Laumann, Curt W.; Lawson, Janice K.; Weiland, Timothy L.)V1415,90-103(1991)

Mode-locked Nd:YLF laser for precision range-Doppler imaging (Simpson, Thomas B.; Malley, Michael M.; Sutton, George W.; Doft, F.)V1416,2-9(1991)

Nd-doped lasers with widely variable pulsewidths (Lin, Li-Huang; Ge, Wen; Kang, Yilan; Chen, Shisheng; Bin, Ouyang; Li, Shiying; Wang, Shijie)V1410,65-71(1991)

Nd-glass microspherical cavity laser induced by cavity QED effects (Wang, Yuzhu; Li, Yongqing; Liu, Yashu; Lu, Baolong)V1501,40-48(1991)

Nd:host laser-based optical terminal development study for intersatellite links (Marini, Andrea E.; Della Torre, Antonio; Popescu, Alexandru F.)V1417,200-211(1991)

Nd:YAG laser market (Llewellyn, Steven A.)V1517,85-99(1991)

Nd:YAG laser treatment in patients with prostatic adenocarcinoma stage A (Gaboardi, Franco; Dotti, Ernesto; Bozzola, Andrea; Galli, Luigi)V1421,50-55(1991)

New opportunities with intense ultra-short-pulse lasers (Richardson, Martin C.)V1410,15-25(1991)

Optical Space Communication II (Franz, Juergen, ed.)V1522(1991)

Performance and spectroscopic characterization of irradiated Nd:YAG (Rose, Todd S.; Fincher, Curtis L.; Fields, Renny A.)V1561,43-49(1991)

Performance of longitudinal-mode KD*P Pockels cells with transparent conductive coatings (Skeldon, Mark D.; Jin, Michael S.; Smith, Douglas H.; Bui, Snow T.)V1410,116-124(1991)

Photon-noise reduction experiments with a Q-switched Nd:YAG laser (Kumar, Prem; Huang, Jianming; Aytur, Orhan)V1376,192-197(1991)

Positive-branch unstable resonator for Nd:YAG laser with Q-switching (Marczak, Jan; Rycyk, Antoni; Sarzynski, Antoni)V1391,48-51(1991)

Potentials for pulsed YAG:Nd laser application to endoscopic surgery (Manenkov, Alexander A.; Denisov, N. N.; Bagdasarov, V. H.; Starkovsky, A. N.; Yurchenko, S. V.; Kornilov, Yu. M.; Mikaberidze, V. M.; Sarkisov, S. E.)V1420,254-258(1991)

Progress in diode laser upconversion (Dixon, G. J.; Kean, P. N.)V1561,147-150(1991)

Q-switched Nd:glass-laser-induced acoustic pulses in lithotripsy (D'yakonov, G. I.; Mikhailov, V. A.; Pak, S. K.; Shcherbakov, I. A.; Andreev, V. G.; Rudenko, O. V.; Sapozhnikov, A. V.)V1421,153-155(1991)

Recanalization of azoospermia due to a Mullerian duct cyst by Nd:YAG laser (Gaboardi, Franco; Bozzola, Andrea; Zago, Tiziano; Gulfi, Gildo M.; Galli, Luigi)V1421,73-77(1991)

Reflow soldering of fine-pitch devices using a Nd:YAG laser (Glynn, Thomas J.; Flanagan, Aiden J.; Redfern, R. M.)V1598,200-205(1991)

Removal of small particles from surfaces by pulsed laser irradiation: observations and a mechanism (Kelley, J. D.; Stuff, Michael I.; Hovis, Floyd E.; Linford, Gary J.)V1415,211-219(1991)

Second-generation high-data-rate interorbit link based on diode-pumped Nd:YAG laser technology (Sontag, Heinz; Johann, Ulrich A.; Pribil, Klaus)V1522,61-69(1991)

Second-harmonic generation of mode-locked Nd:YAG and Nd:YLF lasers using LiB3O5 (Reed, Murray K.; Tyminski, Jacek K.; Bischel, William K.)V1410,179-184(1991)

Selected Papers on High Power Lasers (Soures, John M., ed.)VMS43(1991)

Selected Papers on Solid State Lasers (Powell, Richard C., ed.)VMS31(1991)

Shot-noise-limited high-power green laser for higher density optical disk systems (Kubota, Shigeo; Oka, Michio)V1499,314-323(1991)

Single-frequency Nd:YAG laser development for space communication (Letterer, Rudolf; Wallmeroth, Klaus)V1522,154-157(1991)

Six years of transendoscopic Nd:YAG application in large animals (Tate, Lloyd P.; Glasser, Mardi)V1424,209-217(1991)

SOLACOS: a diode-pumped Nd:YAG laser breadboard for coherent space communication system verification (Pribil, Klaus; Johann, Ulrich A.; Sontag, Heinz)V1522,36-47(1991)

Solid State Lasers II (Dube, George, ed.)V1410(1991)

Space-qualified laser transmitter for lidar applications (Chang, John H.; Reithmaier, Karl D.)V1492,38-42(1991)

Stage III endobronchial squamous cell cancer: survival after Nd:YAG laser combined with photodynamic therapy versus Nd:YAG laser or photodynamic therapy alone (McCaughan, James S.; Barabash, Rostislav D.; Hawley, Philip)V1426,279-287(1991)

Stimulatory effects of Nd:YAG lasers on canine articular cartilage (Lane, Gregory J.; Sherk, Henry H.; Kollmer, Charles; Uppal, Gurvinder S.; Rhodes, Anthony; Sazy, John; Black, Johnathan D.; Lee, Steven)V1424,7-11(1991)

Test results on pulsed cesium amalgam flashlamps for solid state laser pumping (Witting, Harald L.)V1410,90-98(1991)

The Marketplace for Industrial Lasers (Belforte, David A.; Levitt, Morris R., eds.)V1517(1991)

Theory and experiments on passive negative feedback pulse control in active/passive mode-locked solid state lasers (Agnesi, Antoniangelo; Fogliani, Manlio F.; Reali, Giancarlo C.; Kubecek, Vaclav)V1415,242-247(1991)

Thermal effects in diode-laser-pumped monolithic Nd:glass lasers (Schmitt, Randal L.; Spence, Paul A.; Scerbak, David G.)V1410,55-64(1991)

Time-resolved fluorescence of normal and atherosclerotic arteries (Pradhan, Asima; Das, Bidyut B.; Liu, C. H.; Alfano, Robert R.; O'Brien, Kenneth M.; Stetz, Mark L.; Scott, John J.; Deckelbaum, Lawrence I.)V1425,2-5(1991)

Time-resolved x-ray absorption spectroscopy apparatus using laser plasma as an x-ray source (Yoda, Osamu; Miyashita, Atsumi; Murakami, Kouichi; Aoki, Sadao; Yamaguchi, Naohiro)V1503,463-466(1991)

Toward phase noise reduction in a Nd:YLF laser using electro-optic feedback control (Brown, David L.; Seka, Wolf D.; Letzring, Samuel A.)V1410,209-214(1991)

Two-frequency picosecond laser with electro-optical feedback (Vorobiev, N. S.; Konoplev, O. A.)V1358,895-901(1991)

Unstable resonator with a super-Gaussian dielectric mirror for Nd:YAG Q-switched laser (Firak, Jozef; Marczak, Jan; Sarzynski, Antoni)V1391,42-47(1991)

Upgrade of the LLNL Nova laser for inertial confinement fusion (Murray, John R.; Trenholme, J. B.; Hunt, John T.; Frank, D. N.; Lowdermilk, W. H.; Storm, E.)V1410,28-39(1991)

Lasers, semiconductor—see also diodes; lasers, high-power; optical communications

0.36-W cw diffraction-limited-beam operation from phase-locked arrays of antiguides (Mawst, Luke J.; Botez, Dan; Anderson, Eric R.; Jansen, Michael; Ou, Szutsun S.; Sergant, Moshe; Peterson, Gary L.; Roth, Thomas J.; Rozenbergs, John)V1418,353-357(1991)

1-W cw separate confinement InGaAsP/InP (lamda = 1.3 um) laser diodes and their coupling with optical fibers (Garbuzov, Dmitriy Z.; Goncharov, S. E.; Il'in, Y. V.; Mikhailov, A. V.; Ovchinnikov, Alexander V.; Pikhtin, N. A.; Tarasov, I. S.)V1418,386-393(1991)

10-watt cw diode laser bar efficiently fiber-coupled to a 381 um diameter fiber-optic connector (Willing, Steven L.; Worland, Phil; Harnagel, Gary L.; Endriz, John G.)V1418,358-362(1991)

100-mW four-beam individually addressable monolithic AlGaAs laser diode arrays (Yamaguchi, Takao; Yodoshi, Keiichi; Minakuchi, Kimihide; Tabuchi, Norio; Bessho, Yasuyuki; Inoue, Yasuaki; Komeda, Koji; Mori, Kazushi; Tajiri, Atsushi; Tominaga, Koji)V1418,363-371(1991)

4-channel, 662-Mb/s medium-density WDM system with Fabry-Perot laser diodes for subscriber loop applications (Wang, Lon A.; Chapuran, Thomas E.; Menendez, Ronald C.)V1363,85-91(1991)

A 120-Mbit/s QPPM high-power semiconductor transmitter performance and reliability (Greulich, Peter; Hespeler, Bernd; Spatscheck, Thomas)V1417,358-369(1991)

Active mode-locking of external cavity semiconductor laser with 1 GHz repetition rate (Wang, Xianhua; Chen, Guofu; Liu, D.; Xu, L.; Ruan, S.)V1358,775-779(1991)

Adaptive semiconductor laser phased arrays for real-time multiple-access communications (Pan, J. J.; Cordeiro, D.)V1476,157-169(1991)

Advances in laser pump sources for erbium-doped fiber amplifiers (Henshall, Gordon D.; Hadjifotiou, A.; Baker, R. A.; Warbrick, K. J.)V1418,286-291(1991)

Air Force program in coherent semiconductor lasers (Kennett, Ruth D.; Frazier, John C.)V1501,57-68(1991)

Amplified quantum fluctuation as a mechanism for generating ultrashort pulses in semiconductor lasers (Yuan, Ruixi; Taylor, Henry F.)V1497,313-319(1991)

Analysis and measurement of the external modulation of modelocked laser diodes (relative noise performance) (Lam, Benson C.; Kellner, Albert L.; Campion, David C.; Costa, Joannes M.; Yu, Paul K.)V1371,36-45(1991)

Application of tunable diode lasers in control of high-pure-material technologies (Nadezhdinskii, Alexander I.; Stepanov, Eugene V.; Kuznetzov, Andrian I.; Devyatykh, Grigory G.; Maximov, G. A.; Khorshev, V. A.; Shapin, S. M.)V1418,487-495(1991)

Applications of dry etching to InP-based laser fabrication (Hayes, Todd R.; Kim, Sung J.; Green, Christian A.)V1418,190-202(1991)

Beam position noise and other fundamental noise processes that affect optical storage (Levenson, Marc D.)V1376,259-271(1991)

Blue laser devices for optical data storage (Lenth, Wilfried; Kozlovsky, William J.; Risk, William P.)V1499,308-313(1991)

Buried-ridge-stripe lasers monolithically integrated with butt-coupled passive waveguides for OEIC (Remiens, D.; Hornung, V.; Rose, B.; Robein, D.)V1362,323-330(1991)

Carrier recovery and filtering in optical BPSK systems using external cavity semiconductor lasers (Pires, Joao J.; Rocha, Jose R.)V1372,118-127(1991)

Characteristics of a multimode laser diode source in several types of dual-interferometer configuration (Ning, Yanong N.; Grattan, Kenneth T.; Palmer, Andrew W.; Meggitt, Beverley T.)V1367,347-356(1991)

Characterization of GRIN-SCH-SQW amplifiers (Haake, John M.; Zediker, Mark S.; Balestra, Chet L.; Krebs, Danny J.; Levy, Joseph L.)V1418,298-308(1991)

Characterizing frequency chirps and phase fluctuations during noisy laser transients (Abraham, Neal B.)V1376,284-293(1991)

Comparative life test of 0.8-um laser diodes for SILEX under NRZ and QPPM modulation (Menke, Bodo; Loeffler, Roland)V1417,316-327(1991)

Comparison of simulation and experimental measurements of avalanche photodiode receiver performance (Mecherle, G. S.; Henderson, Robert J.)V1417,537-542(1991)

Components for Fiber Optic Applications V (Kopera, Paul M., ed.)V1365(1991)

Computer-controlled two-segment DFB local oscillator laser tuning for multichannel coherent systems (Johnson, Peter T.; Hankey, Judith; Debney, Brian T.)V1372,188-199(1991)

Design of an anamorphic gradient-index lens to correct astigmatism in diode lasers (Acosta, E.; Gomez-Reino, Carlos; Gonzalez, R. M.)V1401,82-85(1991)

Development of an interferometric fiber optic sensor using diode laser (Brewer, Donald R.; Joenathan, Charles; Bibby, Yu Wang; Khorana, Brij M.)V1396,430-434(1991)

Diode laser spectroscopy of atmospheric pollutants (Nadezhdinskii, Alexander I.; Stepanov, Eugene V.)V1433,202-210(1991)

Diode pumping of tunable Cr-doped lasers (Scheps, Richard)V1410,190-194(1991)

Effect of a longitudinal magnetic field on the diffusion coefficient of the minority carriers in solid state junction lasers (Ghatak, Kamakhya P.; Mitra, Bhaswati; Biswas, Shambhu N.)V1415,281-297(1991)

Effect of input pulse shape on FSK optical coherent communication system (Shadaram, Mehdi; John, Eugine)V1365,108-115(1991)

Efficient, uniform transverse coupling of diode arrays to laser rods using nonimaging optics (Minnigh, Stephen W.; Knights, Mark G.; Avidor, Joel M.; Chicklis, Evan P.)V1528,129-134(1991)

Evaluation of diode laser failure mechanisms and factors influencing reliability (Baumann, John A.; Shepard, Allan H.; Waters, Robert G.; Yellen, Steven L.; Harding, Charleton M.; Serreze, Harvey B.)V1418,328-337(1991)

Experimental determination of recombination mechanisms in strained and unstrained quantum-well lasers (Rideout, William C.)V1418,99-107(1991)

Eyesafe diode laser rangefinder technology (Perger, Andreas; Metz, Jurgen; Tiedeke, J.; Rille, Eduard P.)V1419,75-83(1991)

Fabrication of unstable resonator diode lasers (Largent, Craig C.; Gallant, David J.; Yang, Jane; Allen, Michael S.; Jansen, Michael)V1418,40-45(1991)

Far-field and wavefront characterization of a high-power semiconductor laser for free-space optical communications (Cornwell, Donald M.; Saif, Babak N.)V1417,431-439(1991)

Far-field pattern of laser diodes as function of the relative atmospheric humidity (Freitas, Jose C.; Carvalho, Fernando D.; Rodrigues, F. C.; Abreu, M. A.; Marcal, Joao P.)V1399,16-23(1991)

Fast-laser-power control for high-density optical disk systems (Satoh, Hiroharu; Kinoshita, Yoshio; Okao, Keiichi; Tanaka, Masahiko; Nagatani, Hiroyuki; Honguh, Yoshinori)V1499,324-329(1991)

Feedback noise in single-mode semiconductor lasers (Lenstra, Daan; Cohen, Julius S.)V1376,245-258(1991)

Fine structure characteristics of semiconductor laser diode coupled to an external cavity (Ghiasi, Ali; Gopinath, Anand)V1583,170-174(1991)

Flying head read/write characteristics using a monolithically integrated laser diode/photodiode at a wavelength of 1.3 um (Ukita, Hiroo; Katagiri, Yoshitada; Nakada, Hiroshi)V1499,248-262(1991)

Focusing and collimating laser diodes and laser diode arrays (Coetzee, Frans M.; Casasent, David P.)V1544,108-122(1991)

Free-space board-to-board optical interconnections (Tsang, Dean Z.)V1563,66-71(1991)

Free-space optical interconnect using microlasers and modulator arrays (Dickinson, Alex G.; Downs, Maralene M.; LaMarche, R. E.; Prise, Michael E.)V1389,503-514(1991)

Frequency-modulation absorption spectroscopy for trace species detection: theoretical and experimental comparison among methods (Silver, Joel A.; Bomse, David S.; Stanton, Alan C.)V1435,64-71(1991)

Frequency stabilization of AlxGa1-xAs/GaAs lasers using magnetically induced birefringence in an atomic vapor (Lee, W. D.; Campbell, Joe C.)V1365,96-101(1991)

Fundamental array mode operation of semiconductor laser arrays using external spatial filtering (Cherng, Chung-Pin; Osinski, Marek A.)V1418,372-385(1991)

Fungal testing of diode laser collimators (de Lourdes Quinta, Maria; Freitas, Jose C.; Rodrigues, F. C.; Silva, Jorge A.)V1399,24-29(1991)

GaAs/GaAlAs integrated optoelectronic transmitter for microwave applications (Yap, Daniel; Narayanan, Authi A.; Rosenbaum, Steven E.; Chou, Chia-Shing; Hooper, W. W.; Quen, R. W.; Walden, Robert H.)V1418,471-476(1991)

Generation of 1-GHz ps optical pulses by direct modulation of a 1.532-um DFB-LD (Nie, Chao-Jiang; Li, Ying; Wu, Fang D.; Chen, Ying-Li; Li, Qi Hua, Yi-Min)V1579,264-267(1991)

Generic applications for Si and GaAs optical switching devices utilizing semiconductor lasers as an optical source (Rosen, Arye; Stabile, Paul J.; Zutavern, Fred J.; Loubriel, Guillermo M.)V1378,187-194(1991)

GSFC conceptual design study for an intersatellite optical multiple access communication system (Fox, Neil D.; Maynard, William L.; Clarke, Ernest S.; Bruno, Ronald C.)V1417,452-463(1991)

Heterodyne acquisition and tracking in a free-space diode laser link (Hueber, Martin F.; Scholtz, Arpad L.; Leeb, Walter R.)V1417,233-239(1991)

High-current, high-bandwidth laser diode current driver (Copeland, David J.; Zimmerman, Robert K.)V1417,412-420(1991)

High-efficiency vertical-cavity lasers and modulators (Coldren, Larry A.; Corzine, Scott W.; Geels, Randall S.; Gossard, Arthur C.; Law, K. K.; Merz, James L.; Scott, Jeffrey W.; Simes, Robert J.; Yan, Ran H.)V1362,24-37(1991)

High-frequency 1.3 um InGaAsP semi-insulating buried crescent lasers for analog applications (Cheng, Wood-Hi; Appelbaum, Ami; Huang, Rong-Ting; Renner, Daniel; Cioffi, Ken R.)V1418,279-283(1991)

High-power coherent diode lasers (Botez, Dan; Mawst, Luke J.; Jansen, Michael; Anderson, Eric R.; Ou, Szutsun S.; Sergant, Moshe; Peterson, Gary L.; Roth, Thomas J.; Rozenbergs, John)V1474,64-74(1991)

High-power coherent operation of 2-D monolithically integrated master-oscillator power amplifiers (Mehuys, David G.; Welch, David F.; Parke, Ross; Waarts, Robert G.; Hardy, Amos; Scifres, Donald R.)V1418,57-63(1991)

High-power diffraction-limited phase-locked GaAs/GaAlAs laser diode array (Zhang, Yue-qing; Zhang, Xitian; Piao, Yue-zhi; Li, Dian-en; Wu, Sheng-li; Du, Shu-qin)V1418,444-447(1991)

High-power/high-pulse repetition frequency (PRF) pulse generation using a laser-diode-activated photoconductive GaAs switch (Kim, A. H.; Zeto, Robert J.; Youmans, Robert J.; Kondek, Christine D.; Weiner, Maurice; Lalevic, Bogoliub)V1378,173-178(1991)

High-power laser arrays for optical computing (Zucker, Erik P.; Craig, Richard R.; Mehuys, David G.; Nam, Derek W.; Welch, David F.; Scifres, Donald R.)V1563,223-228(1991)

High-power single-element pseudomorphic InGaAs/GaAs/AlGaAs single-quantum-well lasers for pumping Er-doped fiber amplifiers (Larsson, Anders G.; Forouhar, Siamak; Cody, Jeffrey; Lang, Robert J.; Andrekson, Peter A.)V1418,292-297(1991)

High-power visible semiconductor lasers (Ishikawa, Masayuki M.; Itaya, Kazuhiko; Okajima, Masaki; Hatakoshi, Gen-ichi)V1418,344-352(1991)

High-repetition-rate eyesafe rangefinders (Corcoran, Vincent J.)V1419,160-169(1991)

High-resolution spectral characterization of high-power laser diodes (Dorsch, Friedhelm)V1418,477-486(1991)

High-sensitivity sensor of gases based on IR tunable diode lasers for human exhalation monitoring (Moskalenko, Konstantin L.; Nadezhdinskii, Alexander I.; Stepanov, Eugene V.)V1426,121-132(1991)

High-speed short-range laser rangefinder (Gielen, Robert M.; Slegtenhorst, Ronald P.)V1419,153-159(1991)

Homodyne PSK receivers with laser diode sources (Mecherle, G. S.; Henderson, Robert J.)V1417,99-107(1991)

Importance of nonlinear gain in semiconductor lasers (Agrawal, Govind P.; Gray, George R.)V1497,444-455(1991)

Influence of the p-type doping of the InP cladding layer on the threshold current density in 1.5 um QW lasers (Sermage, Bernard; Blez, M.; Kazmierski, Christophe; Ougazzaden, A.; Mircea, Andrei; Bouley, J. C.)V1362,617-622(1991)

InGaAs-GaAs strained layer lasers: physics and reliability (Coleman, James J.; Waters, Robert G.; Bour, David P.)V1418,318-327(1991)

InGaAs/InP distributed feedback quantum-well lasers (Temkin, Henryk; Logan, Ralph A.; Tanbun-Ek, Tawaee)V1418,88-98(1991)

InGaAsP/InP distributed-feedback lasers for long-wavelength optical communication systems, lambda=1.55 um: electrical and optical noises study (Orsal, Bernard; Alabedra, Robert; Signoret, P.; Letellier, H.)V1512,112-123(1991)

Interferometric fiber-optic sensing using a multimode laser diode source (Gerges, Awad S.; Newson, Trevor P.; Jackson, David A.)V1504,176-179(1991)

Interferometry with laser diodes (Hariharan, P.)V1400,2-10(1991)

In-vivo blood flow velocity measurements using the self-mixing effect in a fiber-coupled semiconductor laser (Koelink, M. H.; Slot, M.; de Mul, F. F.; Greve, Jan; Graaff, Reindert; Dassel, A. C.; Aarnouds, J. G.)V1511,120-128(1991)

Laser beam modeling in optical storage systems (Treptau, Jeffrey P.; Milster, Tom D.; Flagello, Donis G.)V1415,317-321(1991)

Laser diodes with gain-coupled distributed optical feedback (Nakano, Yoshiaki; Luo, Yi; Tada, Kunio)V1418,250-260(1991)

Laser Diode Technology and Applications III (Renner, Daniel, ed.)V1418(1991)

Laser galvo-angle-encoder with zero added inertia (Hercher, Michael; Wyntjes, Geert J.)V1454,230-234(1991)

Laser link performance improvements with wideband microwave impedance matching (Baldwin, David L.; Sokolov, Vladimir; Bauhahn, Paul E.)V1476,46-55(1991)

Laser Noise (Roy, Rajarshi, ed.)V1376(1991)

Laser radar based on diode lasers (Levenstein, Harold)V1416,30-43(1991)

Lasers applied to photoelastic stress measurements (Lukasiewicz, Stan)V1554A,349-358(1991)

Lateral-mode discrimination in surface-emitting DBR lasers with cylindrical symmetry (Gong, Xue-Mei; Chan, Andrew K.; Taylor, Henry F.)V1418,422-433(1991)

Life test of high-power Matsushita BTRS laser diodes (Holcomb, Terry L.)V1417,328-337(1991)

Long-wavelength lasers and detectors fabricated on InP/GaAs superheteroepitaxial wafer (Aiga, Masao; Omura, Etsuji E.)V1418,217-222(1991)

Low-coherence optical reflectometry of laser diode waveguides (Boisrobert, Christian Y.; Franzen, Douglas L.; Danielson, Bruce L.; Christensen, David H.)V1474,285-290(1991)

Low-frequency intensity fluctuations in external cavity semiconductor lasers (McInerney, John G.)V1376,236-244(1991)

Low-threshold grating surface-emitting laser arrays (Evans, Gary A.; Carlson, Nils W.; Bour, David P.; Liew, So K.; Amantea, Robert; Wang, Christine A.; Choi, Hong K.; Walpole, James N.; Butler, Jerome K.; Ferguson, W. E.)V1418,406-413(1991)

Low-threshold room temperature continuous-wave operation of a GaAs single-quantum-well mushroom structure surface-emitting laser (Yang, Y. J.; Dziura, T. G.; Wang, S. C.; Fernandez, R.; Du, G.; Wang, Shyh)V1418,414-421(1991)

Measurement of the axial eye length and retinal thickness by laser Doppler interferometry (Hitzenberger, Christoph K.; Fercher, Adolf F.; Juchem, M.)V1423,46-50(1991); V1429,21-25(1991)

Mechanically stable external cavity for laser diodes (Yaeli, Joseph; Streifer, William S.; Cross, P. S.; Rozhyki, Alicia; Scifres, Donald R.; Welch, David F.)V1442,378-382(1991)

Microfabrication techniques for semiconductor lasers (Tamanuki, Takemasa; Tadokoro, T.; Morito, K.; Koyama, Fumio; Iga, Kenichi)V1361,614-617(1991)

Microprocessor-based laser range finder (Rao, M. K.; Tam, Siu-Chung)V1399,116-121(1991)

Microwave fiber optic link with DFB lasers (Huff, David B.; Blauvelt, Henry A.)V1371,244-249(1991)

Miniature laser Doppler anemometer for sensor concepts (Damp, Stephan)V1418,459-470(1991)

Mirror fabrication for full-wafer laser technology (Webb, David J.; Benedict, Melvin K.; Bona, Gian-Luca; Buchmann, Peter L.; Daetwyler, K.; Dietrich, H. P.; Moser, A.; Sasso, G.; Vettiger, Peter; Voegeli, O.)V1418,231-239(1991)

Modeling of coupled grating surface-emitting diodes (Salvi, Theodore C.; Shakir, Sami A.)V1418,64-73(1991)

Modulation characteristics of high-power phase-locked arrays of antiguides (Anderson, Eric R.; Jansen, Michael; Botez, Dan; Mawst, Luke J.; Roth, Thomas J.; Yang, Jane)V1417,543-549(1991)

Monolithic integration of a semiconductor ring laser and a monitoring photodetector (Krauss, Thomas; Laybourn, Peter J.)V1583,150-152(1991)

Monolithic two-dimensional surface-emitting laser diode arrays with 45 degree micromirrors (Jansen, Michael; Yang, Jane; Ou, Szutsun S.; Sergant, Moshe; Mawst, Luke J.; Rozenbergs, John; Wilcox, Jarka Z.; Botez, Dan)V1418,32-39(1991)

Multiple-diode laser optomechanical issues (Jackson, John E.; Armentrout, Ben A.; Buck, J. P.; Chenoweth, Amos J.; Elliott, G. A.; Fox, Allen M.; Ganley, J. T.; Gray, W. C.; Jett, L. L.; Johnson, Kevin M.; Kelsey, J. F.; Minelli, R. J.; Rose, G. E.; Shepherd, W. J.; Zino, Joseph D.)V1533,75-86(1991)

New structure and method for fabricating InP/InGaAsP buried heterostructure semiconductor lasers (Holmstrom, Roger P.; Meland, Edmund; Powazinik, William)V1418,223-230(1991)

Noise in semiconductor lasers and its impact on optical communication systems (Agrawal, Govind P.)V1376,224-235(1991)

Nonlinear phenomena in semiconductor lasers (Otsuka, Kenju)V1497,432-443(1991)

Novel device for short-pulse generation using optoelectronic feedback (Phelan, Paul J.; Hegarty, John; Elsasser, Wolfgang E.)V1362,623-630(1991)

Novel distributed feedback structure for surface-emitting semiconductor lasers (Mahbobzadeh, Mohammad; Osinski, Marek A.)V1418,25-31(1991)

Novel method to fabricate corrugation for distributed-feedback lasers using a grating photomask (Okai, Makoto O.; Harada, Tatsuo)V1545,218-224(1991)

Optical fiber amplifiers (Ikegami, Tetsuhiko; Nakahara, Motohiro)V1362,350-360(1991)

Optical leak detection of oxygen using IR laser diodes (Disimile, Peter J.; Fox, Curtis F.; Toy, Norman)V1492,64-75(1991)

Optical properties of InGaAs/InP strained quantum wells (Abram, Richard A.; Wood, Andrew C.; Robbins, D. J.)V1361,424-433(1991)

Optical pumping of a solid state laser with a high-frequency train of pumping pulsed (Czechowicz, Roman; Kopczynski, Krzysztof)V1391,52-60(1991)

Optimization of gain-switched diode lasers for high-speed fiber optics (MacFarlane, Duncan L.; Tatum, Jim A.)V1365,88-95(1991)

Optimization of grating depth and layer thicknesses for DFB lasers (Jiang, Ching-Long; Agarwal, Rajiv; Kuwamoto, Hide; Huang, Rong-Ting; Appelbaum, Ami; Renner, Daniel; Su, Chin B.)V1418,261-271(1991)

Optimization of strained layer InGaAs/GaAs quantum-well lasers (Fu, Richard J.; Hong, C. S.; Chan, Eric Y.; Booher, Dan J.; Figueroa, Luis)V1418,108-115(1991)

Optoelectronic master chip for optical computing (Lang, Robert J.; Kim, Jae H.; Larsson, Anders G.; Nouhi, Akbar; Cody, Jeffrey; Lin, Steven H.; Psaltis, Demetri; Tiberio, Richard C.; Porkolab, Gyorgy A.; Wolf, Edward D.)V1563,2-7(1991)

PbEuSeTe/Pb1-xSnxTe buried heterostructure diode lasers grown by molecular beam epitaxy (Feit, Zeev; Kostyk, D.; Woods, R. J.; Mak, Paul S.)V1512,164-169(1991)

Performance of a phased array semiconductor laser source for coherent laser communications (Probst, David K.; Rice, Robert R.)V1417,346-357(1991)

Predicting diode laser performance (Lim, G.; Park, Youngsoh; Zmudzinski, C. A.; Zory, Peter S.; Miller, L. M.; Cockerill, T. M.; Coleman, James J.; Hong, C. S.; Figueroa, Luis)V1418,123-131(1991)

Preliminary results with sutured colonic anastomoses reinforced with dye-enhanced fibrinogen and a diode laser (Libutti, Steven K.; Williams, Mathew R.; Oz, Mehmet C.; Forde, Kenneth A.; Bass, Lawrence S.; Weinstein, Samuel; Auteri, Joseph S.; Treat, Michael R.; Nowygrod, Roman)V1421,169-172(1991)

Progress in planarized vertical-cavity surface-emitting laser devices and arrays (Morgan, Robert A.; Chirovsky, Leo M.; Focht, Marlin W.; Guth, Gregory O.; Asom, Moses T.; Leibenguth, R. E.; Robinson, K. C.; Lee, Yong H.; Jewell, Jack L.)V1562,149-159(1991)

Progress of surface-emitting lasers in Japan (Ogura, Mutsuo)V1418,396-405(1991)

Reduction of spectral linewidth and FM noise in semiconductor lasers by application of optical feedback (Li, Hua; Park, J. D.; Seo, Dongsun; Marin, L. D.; McInerney, John G.; Telle, H. R.)V1376,172-179(1991)

Reliability analysis of a multiple-laser-diode beacon for intersatellite links (Mauroschat, Andreas)V1417,513-524(1991)

Remote Raman spectroscopy using diode lasers and fiber-optic probes (Angel, S. M.; Myrick, Michael L.; Vess, Thomas M.)V1435,72-81(1991)

Second harmonic generation of diode laser radiation in KNbO3 (van Hulst, Niek F.; Heesink, Gerard J.; de Leeuw, H.; Bolger, Bouwe)V1362,631-646(1991)

Semiconductor laser amplifiers as all-optical frequency converters (Schunk, Nikolaus; Grosskopf, Gerd; Ludwig, Reinhold; Schnabel, Ronald; Weber, Hans-Georg)V1362,391-397(1991)

Semiconductor laser diode with weak optical feedback: self-coupling effects on P-I characteristics (Milani, Marziale; Mazzoleni, S.; Brivio, Franca)V1474,83-97(1991)

Semiconductor lasers for coherent lightwave communcation (Dutta, Niloy K.)V1372,4-12(1991)

Semiconductor lasers in analytical chemistry (Patonay, Gabor; Antoine, Miquel D.; Boyer, A. E.)V1435,52-63(1991)

Semiconductor laser transmitters for millimeter-wave fiber-optic links (Pan, J. J.)V1476,63-73(1991)

Single Frequency Semiconductor Lasers (Buus, Jens)VTT05(1991)

Some characteristics of 3.2 um injection lasers based on InAsSb/InAsSbP system (Mani, Habib; Joullie, Andre F.)V1362,38-48(1991)

Spectral and modulation characteristics of tunable multielectrode DBR lasers (Ferreira, Mario F.; Rocha, Jose R.; Pinto, Joao L.)V1372,14-26(1991)

Stable broad near-field (single-lateral mode) semiconductor laser (Chua, Soo-Jin; Leow, S. K.; Ng, T. B.; Chong, Tow C.; Kanhere, R.)V1401,96-102(1991)

Statistics of laser switch-on (San Miguel, Maximino)V1376,272-283(1991)

Strained quantum-well leaky-mode diode laser arrays (Shiau, T. H.; Sun, Shang-zhu; Schaus, Christian F.; Zheng, Kang; Hadley, G. R.; Hohimer, John P.)V1418,116-122(1991)

Study on the mode and far-field pattern of diode laser-phased arrays (Zhang, Yue-qing; Wu, Sheng-li; Zhu, Lian; Zhang, Xitian; Piao, Yue-zhi; Li, Dian-en)V1400,137-143(1991)

Subminiature package external cavity laser (Fatah, Rebwar M.; Cox, Maurice K.; Bird, David M.; Cameron, Keith H.)V1501,120-128(1991)

Surface-emitting, distributed-feedback laser as a source for laser radar (Akkapeddi, Prasad R.; Macomber, Steven H.)V1416,44-49(1991)

Tapered DBR for improving characteristics of DFB and DBR lasers (Rahnavard, Mohammad H.; Bakhtazad, A.; Abiri-Jahromi, H.)V1562,327-337(1991)

Temperature and modulation dependence of spectral linewidth in distributed Bragg reflector laser diodes (Suzaki, Shinzoh; Ishii, Masanori; Watanabe, Tsutomu; Shiota, Alan T.)V1374,126-131(1991)

Temperature stress testing of laser modules for the uncontrolled environment (Su, Pin)V1366,94-106(1991)

Threshold current and carrier lifetime in MOVPE regrown 1.5 um GaInAsP buried ridge structure lasers (Tischel, M.; Rosenzweig, M.; Hoffmann, Axel; Venghaus, Herbert; Fidorra, F.)V1361,917-926(1991)

Threshold current density of InGaAsP/InP surface-emitting laser diodes with hemispherical resonator (Jing, Xing-Liang; Zhang, Yong-Tao; Chen, Yi-Xin)V1418,434-441(1991)

Time development of AlGaAs single-quantum-well laser facet temperature on route to catastrophical breakdown (Tang, Wade C.; Rosen, Hal J.; Vettiger, Peter; Webb, David J.)V1418,338-342(1991)

Time-integrating optical raster spectrum analyzer using semiconductor laser as input modulator (Larkin, Alexander I.; Matveev, Alexander; Mironov, Yury)V1621,414-423(1991)

Top-side electroluminescence: a failure analysis technique to view electroluminescence along a laser channel (Blow, Victor O.; Giewont, Kenneth J.)V1366,107-111(1991)

Triggering GaAs lock-on switches with laser diode arrays (Loubriel, Guillermo M.; Buttram, Malcolm T.; Helgeson, Wes D.; McLaughlin, Dan L.; O'Malley, Marty W.; Zutavern, Fred J.; Rosen, Arye; Stabile, Paul J.)V1378,179-186(1991)

Tunability of cascaded grating that is used in distributed feedback laser (Rahnavard, Mohammad H.; Moheimany, O. R.; Abiri-Jahromi, H.)V1367,374-381(1991)

Tunable cyclotron-resonance laser in germanium (Kremser, Christian; Unterrainer, Karl; Gornik, Erich; Strasser, G.; Pidgeon, Carl R.)V1501,69-79(1991)

Two applications for microlens arrays: detector fill-factor improvement and laser diode collimation (D'Amato, Dante P.; Centamore, Robert M.)V1544,166-177(1991)

Two-dimensional edge- and surface-emitting semiconductor laser arrays for optically activated switching (Evans, Gary A.; Rosen, Arye; Stabile, Paul J.; Bour, David P.; Carlson, Nils W.; Connolly, John C.)V1378,146-161(1991)

Ultrashort pulse propagation in visible semiconductor diode laser amplifiers (Tatum, Jim A.; MacFarlane, Duncan L.)V1497,320-325(1991)

Universal light source for optical fiber sensors (Semenov, Alexandr T.; Elenkrig, Boris B.; Logozinskii, Valerii N.)V1584,348-352(1991)

Vertical cavity lasers for optical interconnects (Jewell, Jack L.; Lee, Yong H.; McCall, Samuel L.; Scherer, Axel; Harbison, James P.; Florez, Leigh T.; Olsson, N. A.; Tucker, Rodney S.; Burrus, Charles A.; Sandroff, Claude J.)V1389,401-407(1991)

Vertical-cavity surface-emitters for optoelectronic integration (Geels, Randall S.; Corzine, Scott W.; Coldren, Larry A.)V1418,46-56(1991)

Vertical-cavity surface-emitting lasers and arrays for optical interconnect and optical computing applications (Wang, S. C.; Dziura, T. G.; Yang, Y. J.)V1563,27-33(1991)

Vertical-cavity surface-emitting semiconductor lasers: present status and future prospects (Osinski, Marek A.)V1418,2-24(1991)

Lasers, solid-state—see also lasers, neodymium

10-W Ho laser for surgery (Bouchenkov, Vyatcheslav A.; Utenkov, Boris I.; Antipenko, Boris M.; Berezin, Juri D.; Berezin, U. D.; Malinin, Boris G.; Serebryakov, Victor A.)V1410,185-188(1991); V1427,409-412(1991)

2.014-micron Cr;Tm:YAG: optimization of doping concentration for flash lamp operation (Bar-Joseph, Dan)V1142-147(1991)

2.8-um Er3+:YLiF4 laser resonantly pumped at 970 nm (Stoneman, Robert C.; Esterowitz, Leon; Lynn, J. G.)V1410,148-155(1991)

Alexandrite laser and blind lithotripsy: initial experience—first clinical results (Mattioli, Stefano; Cremona, M.; Benaim, George; Ferrario, Angelo)V1421,114-119(1991)

Beam quality measurements and improvement in solar-pumped laser systems (Bernstein, Hana; Thompson, George A.; Yogev, Amnon; Oron, Moshe)V1442,81-88(1991)

Brightness enhancement of solid state laser oscillators in single-mode lasing using novel inside-resonator optical elements with radially variable transmission (Lukishova, Svetlana G.; Mendez, Nestor R.; Ter-Mikirtychev, Valery V.; Tulajkova, Tamara V.)V1527,380-391(1991)

Coherent launch-site atmospheric wind sounder (Targ, Russell; Hawley, James G.; Otto, Robert G.; Kavaya, Michael J.)V1478,211-227(1991)

Compact Raman instrumentation for process and environmental monitoring (Carrabba, Michael M.; Spencer, Kevin M.; Rauh, R. D.)V1434,127-134(1991)

Comparative performance of infrared solid state lasers in laser lithotripsy (D'yakonov, G. I.; Konov, Vitaly I.; Mikhailov, V. A.; Nikolaev, D. A.; Pak, S. K.; Shcherbakov, I. A.)V1421,156-162(1991)

Comparative study of excimer and erbium:YAG lasers for ablation of structural components of the knee (Vari, Sandor G.; Shi, Wei-Qiang; van der Veen, Maurits J.; Fishbein, Michael C.; Miller, J. M.; Papaioannou, Thanassis; Grundfest, Warren S.)V1424,33-42(1991)

Comparative study of gelatin ablation by free-running and Q-switch modes of Er:YAG laser (Konov, Vitaly I.; Kulevsky, Lev A.; Lukashev, Alexei V.; Pashinin, Vladimir P.; Silenok, Alexander S.)V1427,232-242(1991)

Comparison of flash-pumped Cr;Tm 2-um laser action in garnet hosts (Quarles, Gregory J.; Pinto, Joseph F.; Esterowitz, Leon; Kokta, Milan R.)V1410,165-174(1991)

Comparison of the ablation of polymethylmethacrylate by two fiber-optic-compatible infrared lasers (Garino, Jonathan P.; Nazarian, David; Froimson, Mark I.; Grelsamer, Ronald P.; Treat, Michael R.)V1424,43-47(1991)

Cr3+ tunable laser glass (Izumitani, Tetsuro; Zou, Xuolu; Wang, Y.)V1535,150-159(1991)

Cr,Er:YSGG laser as an instrument for dental surgery (D'yakonov, G. I.; Konov, Vitaly I.; Mikhailov, V. A.; Pak, S. K.; Shcherbakov, J. A.; Ershova, N. I.; Maksimovskiy, Y. V.)V1424,81-86(1991)

Damage resistant optics for a megajoule solid state laser (Campbell, John H.; Rainer, Frank; Kozlowski, Mark R.; Wolfe, C. R.; Thomas, Ian M.; Milanovich, Fred P.)V1441,444-456(1991)

Design and energy characteristics of a multisegment glass-disk amplifier (Kelly, John H.; Shoup, Milton J.; Skeldon, Mark D.; Bui, Snow T.)V1410,40-46(1991)

Designing and application of solid state lasers for streak cameras calibration (Babushkin, A. V.; Vorobiev, N. S.; Prokhorov, Alexander M.; Schelev, Mikhail Y.)V1346,410-417(1991)

Diffraction-limited Nd:glass and alexandrite lasers using graded reflectivity mirror unstable resonators (Snell, Kevin J.; Duplain, Gaetan; Parent, Andre; Labranche, Bruno; Galarneau, Pierre)V1410,99-106(1991)

Diode-pumped, high-power, solid-state lasers (Yamanaka, Masanobu; Naito, Kenta; Nakatsuka, Masahiro; Yamanaka, Tatsuhiko; Nakai, Sadao S.; Yamanaka, Chiyoe)V1501,30-39(1991)

Diode pumping of tunable Cr-doped lasers (Scheps, Richard)V1410,190-194(1991)

Dual wavelengths (750/375 nm) laser lithotripsy (Steiner, Rudolf W.; Meier, Thomas H.)V1421,124-128(1991)

Effective laser with active element rectangular geometry (Danilov, Alexander A.; Nikirui, Ernest Y.; Osiko, Vyacheslav V.; Polushkin, Valery G.; Sorokin, Svjatoslav N.; Timoshechkin, M. I.)V1362,916-920(1991)

Effect of a longitudinal magnetic field on the diffusion coefficient of the minority carriers in solid state junction lasers (Ghatak, Kamakhya P.; Mitra, Bhaswati; Biswas, Shambhu N.)V1415,281-297(1991)

Emerging laser technologies (Forrest, Gary T.)V1520,37-57(1991)

Energy storage efficiency and population dynamics in flashlamp-pumped sensitized erbium glass laser (Lukac, Matjaz)V1419,55-62(1991)

Environmental testing of a Q-switched Nd:YLF laser and a Nd:YAG ring laser (Robinson, Deborah L.)V1417,562-572(1991)

Erbium laser ablation of bone: effect of water content (Walsh, Joseph T.; Hill, D. A.)V1427,27-33(1991)

Experimental and computer-modeled results of titanium sapphire lasers pumped by copper vapor lasers (Knowles, Martyn R.; Webb, Colin E.; Naylor, Graham A.)V1410,195-201(1991)

Experimental and first clinical results with the alexandrite laser lithotripter (Miller, Kurt; Weber, Hans M.; Rueschoff, Josef; Hautmann, Richard E.)V1421,108-113(1991)

Experimental simulation of holmium laser action on biological tissues (Konov, Vitaly I.; Prokhorov, Alexander M.; Silenok, Alexander S.; Tsarkova, O. G.; Tsvetkov, V. B.; Shcherbakov, I. A.)V1427,220-231(1991)

Fabrication of laser materials by laser-heated pedestal growth (Chang, Robert S.; Sengupta, Sonnath; Shaw, Leslie B.; Djeu, Nick)V1410,125-132(1991)

Frequency-doubled alexandrite laser for tissue differentiation in angioplasty (Scheu, M.; Engelhardt, R.)V1425,63-69(1991)

Future technologies for lidar/DIAL remote sensing (Allario, Frank; Barnes, Norman P.; Storm, Mark E.)V1492,92-110(1991)

Highly efficient optical parametric oscillators (Marshall, Larry R.; Hays, Alan D.; Kasinski, Jeff; Burnham, Ralph L.)V1419,141-152(1991)

High-performance picosecond and femtosecond solid-state lasers with feedback-controlled passive mode-locking (Komarov, Konstantin P.; Kuch'yanov, Aleksandr S.; Ugozhayev, Vladimir D.)V1501,135-143(1991)

High repetition rate Q-switched erbium glass lasers (Hamlin, Scott J.; Myers, John D.; Myers, Michael J.)V1419,100-106(1991)

Holmium:YAG and erbium:YAG laser interaction with hard and soft tissue (Charlton, Andrew; Dickinson, Mark R.; King, Terence A.; Freemont, Anthony J.)V1427,189-197(1991)

Hyperthermia treatment using a computer-controlled Nd:YAG laser system in combination with surface cooling (Panjehpour, Masoud; Wilke, August; Frazier, Donita L.; Overholt, Bergein F.)V1427,307-315(1991)

Interaction of erbium laser radiation with corneal tissue (Wannop, Neil M.; Charlton, Andrew; Dickinson, Mark R.; King, Terence A.)V1423,163-166(1991)

Interference measurement of thermal effect induced by optical pumping on YAG crystal rod (Luo, Bikai; Ni, Xiao W.; Zhang, Qi; He, Anzhi)V1554A,542-546(1991)

Investigations of the focal shift of the high-power cw YAG laser beam (Jabczynski, Jan K.; Mindak, Marek K.)V1391,109-116(1991)

Laser angioplasty with lensed fibers and a holmium:YAG laser in iliac artery occlusions (White, Christopher J.; Ramee, Stephen R.; Mesa, Juan E.; Collins, Tyrone J.; Kotmel, Robert F.; Godfrey, Maureen A.)V1425,130-133(1991)

Laser-flash photographic studies of Er:YAG laser ablation of water (Jacques, Steven L.; Gofstein, Gary)V1427,63-67(1991)

Laser transmitter for lidar in-space technology experiment (Chang, John H.; Cimolino, Marc C.; Petros, Mulugeta)V1492,43-46(1991)

Low-data-rate coherent optical link demonstration using frequency-stabilized solid-state lasers (Chen, Chien-Chung; Win, Moe Z.; Marshall, William K.; Lesh, James R.)V1417,170-181(1991)

Mechanisms for the cathodoluminescence of cerium-doped yttrium aluminum garnet phosphors (Rotman, Stanley R.; Aizenberg, G. E.; Gordon, E. S.; Tuller, H. L.; Warde, Cardinal)V1442,205-215(1991)

Medical marketplace (Moretti, Michael)V1520,118-131(1991)

Mid-infrared laser applications (Daly, John G.)V1419,94-99(1991)

New approach to creation of stable and unstable active resonators for high-power solid state lasers (Apollonov, V. V.; Chetkin, Sergue A.; Kislov, V. I.; Vdovin, Gleb V.)V1502,83-94(1991)

New concept of a compact multiwavelength solid state laser for laser-induced shock wave lithotripsy (Steiger, Erwin)V1421,140-146(1991)

New opportunities with intense ultra-short-pulse lasers (Richardson, Martin C.)V1410,15-25(1991)

Nonimaging concentrators for diode-pumped slab lasers (Lacovara, Phil; Gleckman, Philip; Holman, Robert L.; Winston, Roland)V1528,135-141(1991)

Numerical model of Q-switched solid state laser (Altshuler, Grigori B.; Kargin, Igor U.; Khloponin, Leonid V.; Khramov, Valery)V1415,269-280(1991)

Operation of a YLF-Er laser at near infrared (0.85 to 2.9 um) (Tkachuk, T. M.; Shumilin, V. V.)V1403,805-808(1991)

Optical pumping of a solid state laser with a high-frequency train of pumping pulsed (Czechowicz, Roman; Kopczynski, Krzysztof)V1391,52-60(1991)

Optical requirements for light-activated switches (McIntyre, Iain A.; Giorgi, David M.; Hargis, David E.; Zucker, Oved S.)V1378,162-172(1991)

Passive Q-switching of eyesafe Er:glass lasers (Denker, Boris I.; Maksimova, G. V.; Osiko, Vyacheslav V.; Prokhorov, Alexander M.; Sverchkov, Sergey E.; Sverchkov, Yuri E.; Horvath, Zoltan G.)V1419,50-54(1991)

Percutaneous lumbar discectomy using Ho:YAG laser (Black, Johnathan D.; Sherk, Henry H.; Uppal, Gurvinder S.; Sazy, John; Meller, Menachem M.; Rhodes, Anthony; Lane, Gregory J.)V1424,20-22(1991)

Photoablation using the Holmium:YAG laser: a laboratory and clinical study (Waidhauser, Erich; Markmiller, U.; Enders, S.; Hessel, Stefan F.; Ulrich, Frank; Beck, Oskar J.; Feld, Michael S.)V1428,75-83(1991)

Pulsed YLF-Ho laser emission at 0.75 to 2.9 um (Tkachuk, A. M.; Petrova, Maria V.; Shumilin, V. V.)V1403,801-804(1991)

Pulse formation and characteristics of the cw mode-locked titanium-doped sapphire laser (Zschocke, Wolfgang; Stamm, Uwe; Heumann, Ernst; Ledig, Mario; Guenzel, Uwe; Kvapil, Jiri; Koselja, Michal P.; Kubelka, Jiri)V1501,49-56(1991)

Q-switching and pulse shaping with IR lasers (Brinkmann, Ralf E.; Bauer, K.)V1421,134-139(1991)

Reduction of acoustic transients in tissue with a 2 um thulium laser (Pinto, Joseph F.; Esterowitz, Leon; Bonner, Robert F.)V1420,242-243(1991)

Removal of dental filling materials by Er:YAG laser radiation (Hibst, Raimund; Keller, Ulrich)V1424,120-126(1991)

Root resection of endodontically treated teeth by Erbium:YAG laser radiation (Paghdiwala, A. F.)V1424,150-159(1991)

Selected Papers on Solid State Lasers (Powell, Richard C., ed.)VMS31(1991)

Single-frequency solid state lasers and amplifiers (Mak, Arthur A.; Orlov, Oleg A.; Ustyugov, Vladimir I.; Vitrishchak, Il'ya B.)V1410,233-243(1991)

Solid State Lasers II (Dube, George, ed.)V1410(1991)

Solid state lasers for field application (Motenko, Boris; Ermakov, Boris A.; Berezin, Boris)V1399,78-81(1991)

Solid state lasers for planetary exploration (Greene, Ben; Taubman, Matthew; Watts, Jeffrey W.; Gaither, Gary L.)V1492,126-139(1991)

Solid state lasers with passive mode-locking and negative feedback for picosecond spectroscopy (Danelius, R.; Grigonis, R.; Piskarskas, Algis S.; Podenas, D.; Sirutkaitis, V.)V1402,198-208(1991)

Solid-state lidar measurements at 1 and 2 um (Killinger, Dennis K.; Chan, Kin-Pui)V1416,125-128(1991)

Soviet developments in solid state lasers (Prokhorov, Alexander M.; Shcherbakov, I. A.)V1410,70-88(1991)

Spontaneous emission amplification and the utmost energy characteristics of solid state lasers on the base of optically dense media (Danilov, Alexander A.; Sorokin, Svjatoslav N.)V1362,647-654(1991)

Stable picosecond solid state YA103:Nd3+ laser for streak cameras dynamic evaluation (Babushkin, A. V.)V1358,888-894(1991)

Stone/ureter identification during alexandrite laser lithotripsy (Scheu, M.; Flemming, G.; Engelhardt, R.)V1421,100-107(1991)

Suppression of relaxation oscillations in flash-pumped 2-um lasers (Pinto, Joseph F.; Esterowitz, Leon)V1410,175-178(1991)

Test results on pulsed cesium amalgam flashlamps for solid state laser pumping (Witting, Harald L.)V1410,90-98(1991)

Ti:Al2O3 laser with an intracavity phase-conjugate mirror (Jankiewicz, Zdzislaw; Szydlak, J.; Skorczakowski, M.; Zendzian, W.; Dvornikov, S. S.; Kondratiuk, N. V.; Skripko, G. A.)V1391,101-104(1991)

Tissue ablation by holmium:YSGG laser pulses through saline and blood (van Leeuwen, Ton G.; van der Veen, Maurits J.; Verdaasdonk, Rudolf M.; Borst, Cornelius)V1427,214-219(1991)

Tooth pulp reaction following Er:YAG laser application (Keller, Ulrich; Hibst, Raimund)V1424,127-133(1991)

Transient stimulated Raman scattering: theory and experiments of pulse shortening and phase conjugation properties (Agnesi, Antoniangelo; Reali, Giancarlo C.; Kubecek, Vaclav)V1415,104-109(1991)

Two-dimensional periodic structures in solid state laser resonator (Okulov, Alexey Y.)V1410,221-232(1991)

Ultrashort pulse generation in solid state lasers (Fujimoto, James G.; Schulz, Peter A.; Fan, Tso Y.)V1413,2-13(1991)

Uniform energy discharge for pulsed lasers (Koschmann, Eric C.)V1411,118-126(1991)

Unusually fast energy transfer in solid state crystals and glasses (Rotman, Stanley R.; Maoz, O.; Arnon, S.; Kaczelnik, F.; Felus, Y.; Weiss, Aryeh M.; Reisfeld, Renata; Eyal, Marek; Hartmann, F. X.)V1442,194-204(1991)

Upconversion-pumped IR (2.8-2.9 microns) lasing of Er3+ in garnets (Pollack, S. A.; Chang, David B.; Birnbaum, Milton; Kokta, Milan R.)V1410,156-164(1991)

UV-VIS solid state excimer laser: XeF in crystalline argon (Zerza, Gerald; Knopp, F.; Kometer, R.; Sliwinski, G.; Schwentner, N.)V1410,202-208(1991)

Wavemeter for tuning solid state lasers (Goad, Joseph H.; Rinsland, Pamela L.; Kist, Edward H.; Irick, Steven C.)V1410,107-115(1991)

Lasers, tunable

Advances in tunable diode laser technology for atmospheric monitoring applications (Wall, David L.)V1433,94-103(1991)

Airborne tunable diode laser sensor for high-precision concentration and flux measurements of carbon monoxide and methane (Sachse, Glen W.; Collins, Jim E.; Hill, G. F.; Wade, L. O.; Burney, Lewis G.; Ritter, J. A.)V1433,157-166(1991)

Alexandrite laser characterization and airborne lidar developments for water vapor DIAL measurements (Ponsardin, Patrick; Higdon, Noah S.; Grossmann, Benoist E.; Browell, Edward V.)V1492,47-51(1991)

Alternated or simultaneous sealed-off room temperature CO/CO2 laser tuning by chemical reactions (Masychev, Victor I.; Alejnikov, Vladislav S.)V1412,227-235(1991)

Application of tunable diode laser spectroscopy to the real-time analysis of engine oil economy (Carduner, Keith R.; Colvin, A. D.; Leong, D. Y.; Schuetzle, Dennis; Mackay, Gervase I.)V1433,190-201(1991)

Characterization of a subpicosecond XeF (C->A) excimer laser (Hofmann, Thomas; Sharp, Tracy E.; Dane, C. B.; Wisoff, P. J.; Wilson, William L.; Tittel, Frank K.; Szabo, Gabor)V1412,84-90(1991)

Continuous automatic scanning picosecond optical parametic source using MgO LiNbO3 in the 700-2200 nm (He, Huijuan; Lu, Yutian; Dong, Jingyuan; Zhao, Quingchun)V1409,18-23(1991)

Development of an XUV-IR free-electron laser user facility for scientific research and industrial applications (Newnam, Brian E.; Warren, Roger W.; Conradson, Steven D.; Goldstein, John C.; McVey, Brian D.; Schmitt, Mark J.; Elliott, C. J.; Burns, M. J.; Carlsten, Bruce E.; Chan, Kwok-Chi D.; Johnson, W. J.; Wang, Tai-San; Sheffield, Richard L.; Meier, Karl L.; Olsher, R. H.; Scott, Marion L.; Griggs, J. E.)V1552,154-174(1991)

Diode pumping of tunable Cr-doped lasers (Scheps, Richard)V1410,190-194(1991)

Experimental and computer-modeled results of titanium sapphire lasers pumped by copper vapor lasers (Knowles, Martyn R.; Webb, Colin E.; Naylor, Graham A.)V1410,195-201(1991)

Frequency modulation spectroscopy for chemical sensing of the environment (Cooper, David E.; Riris, Haris; van der Laan, Jan E.)V1433,120-127(1991)

Imaging inside scattering media: chronocoherent imaging (Spears, Kenneth G.; Kume, Stewart M.; Winakur, Eric)V1429,2-8(1991)

Ion-ripple laser as an advanced coherent radiation source (Chen, Kuan-Ren; Dawson, John M.)V1552,185-196(1991)

Measurement of Atmospheric Gases (Schiff, Harold I., ed.)V1433(1991)

Narrowband alexandrite laser injection seeded with frequency-dithered diode laser (Schwemmer, Geary K.; Lee, Hyo S.; Prasad, Coorg R.)V1492,52-62(1991)

Open-path tunable diode laser absorption for eddy correlation flux measurements of atmospheric trace gases (Anderson, Stuart M.; Zahniser, Mark S.)V1433,167-178(1991)

Optics for tunable diode laser spectrometers (Riedel, Wolfgang J.)V1433,179-189(1991)

Potential of tunable lasers for optimized dual-color laser ranging (Lund, Glenn I.; Gaignebet, Jean)V1492,166-175(1991)

Solid-state lidar measurements at 1 and 2 um (Killinger, Dennis K.; Chan, Kin-Pui)V1416,125-128(1991)

Stanford picosecond FEL center (Swent, Richard L.; Schwettman, H. A.; Smith, T. I.)V1552,24-35(1991)

Tunable diode laser spectrometer for high-precision concentration and ratio measurements of long-lived atmospheric gases (Fried, Alan; Drummond, James R.; Henry, Bruce; Fox, Jack)V1433,145-156(1991)

Tunable diode laser systems for trace gas monitoring (Mackay, Gervase I.; Karecki, David R.; Schiff, Harold I.)V1433,104-119(1991)

Ultrashort pulse generation in solid state lasers (Fujimoto, James G.; Schulz, Peter A.; Fan, Tso Y.)V1413,2-13(1991)

Laser surgery—see also laser-tissue interaction; medicine

10-W Ho laser for surgery (Bouchenkov, Vyatcheslav A.; Utenkov, Boris I.; Antipenko, Boris M.; Berezin, Juri D.; Berezin, U. D.; Malinin, Boris G.; Serebryakov, Victor A.)V1410,185-188(1991); V1427,409-412(1991)

Advances in clinical percutaneous excimer laser angioplasty (Viligiardi, Riccardo; Pini, Roberto; Salimbeni, Renzo; Galiberti, Sandra)V1425,72-74(1991)

Alexandrite laser and blind lithotripsy: initial experience—first clinical results (Mattioli, Stefano; Cremona, M.; Benaim, George; Ferrario, Angelo)V1421,114-119(1991)

Analysis of an adjustable slit design for correcting astigmatism (Clapham, Terrance N.; D'Arcy, John; Bechtel, Lorne; Glockler, Hermann; Munnerlyn, Charles R.; McDonnell, Peter J.; Garbus, Jenny)V1423,2-7(1991)

Analysis of kinetic rate modeling of thermal damage in laser-irradiated tissue (Rastegar, Sohi; Glenn, T.)V1427,300-306(1991)

Application of conjugated heparin-albumin microparticles with laser-balloon angioplasty: a potential method for reducing adverse biologic reactivity after angioplasty (Kundu, Sourav K.; McMath, Linda P.; Zaidan, Jonathan T.; Spears, J. R.)V1425,142-148(1991)

Applications of lasers in laparoscopic cholecystectomy: technical considerations and future directions (Lanzafame, Raymond J.)V1421,189-196(1991)

Argon dye photocoagulator for microsurgery of the interior structure of the eye (Wolinski, Wieslaw; Kazmirowski, Antoni; Kesik, Jerzy; Korobowicz, Witold; Spytkowski, Wojciech)V1391,334-340(1991)

Argon laser vascular tissue fusion: current status and future perspectives (White, Rodney A.; Kopchok, George E.)V1422,103-110(1991)

Arthroscopic contact Nd:YAG laser meniscectomy: surgical technique and clinical follow-up (O'Brien, Stephen J.; Miller, Drew V.; Fealy, Stephen V.; Gibney, Mary A.; Kelly, Anne M.)V1424,62-75(1991)

Axial and transverse displacement tolerances during excimer laser surgery for myopia (Shimmick, John K.; Munnerlyn, Charles R.; Clapham, Terrance N.; McDonald, Marguerite B.)V1423,140-153(1991)

Bifunctional irrigation liquid as an ideal energy converter for laser lithotripsy with nanosecond laser pulses (Reichel, Erich; Schmidt-Kloiber, Heinz; Paltauf, Guenther; Groke, Karl)V1421,129-133(1991)

Biomechanical and structural studies of rabbits carotid arteries after endovascular laser exposition (Purinsh, Juris; Elksninsh, N.; Dzenis, J.; Tomass, V.; Teivans, A.; Freimanis, R.; Garsha, I.; Ozolinsh, H.)V1525,289-308(1991)

Bladder outlet obstruction treated with transurethral ultrasonic aspiration (Malloy, Terrence P.)V1421,46-46(1991)

Caries selective ablation by pulsed lasers (Hennig, Thomas; Rechmann, Peter; Pilgrim, C.; Schwarzmaier, Hans-Joachim; Kaufmann, Raimund)V1424,99-105(1991)

Clinical evaluation of tumor promotion by CO2 laser (Braun, Robert E.; Liebow, Charles)V1424,138-144(1991)

Clinical experience in applying endoscopic Nd:YAG laser to treat 451 esophagostenotic cases (Wang, Rui-zhong; Wang, Zhen-he; Lu, Kuang-sheng; Yang, Xiao-zhi; Lu, Bo-kao)V1421,203-207(1991)

Clinical experience with an excimer laser angioplasty system (Golobic, Robert A.; Bohley, Thomas K.; Wells, Lisa D.; Sanborn, Timothy A.)V1425,84-92(1991)

CO2 partial matricectomy in the treatment of ingrown toenails (Uppal, Gurvinder S.; Sherk, Henry H.; Black, Johnathan D.; Rhodes, Anthony; Sazy, John; Lane, Gregory J.)V1424,51-52(1991)

Coagulation and precise ablation of biotissues by pulsed sealed-off carbon monoxide laser (Masychev, Victor I.; Alejnikov, Vladislav S.; Klimenko, Vladimir I.)V1427,344-356(1991)

Cold-laser microsurgery of the retina with a syringe-guided 193 nm excimer laser (Lewis, Aaron; Palanker, Daniel; Hemo, Itzhak; Pe'er, Jacob; Zauberman, Hanan)V1423,98-102(1991)

Combined guidance technique using angioscope and fluoroscope images for CO laser angioplasty: in-vivo animal experiment (Arai, Tsunenori; Mizuno, Kyoichi; Sakurada, Masami; Miyamoto, Akira; Arakawa, Koh; Kurita, Akira; Suda, Akira; Kikuchi, Makoto; Nakamura, Haruo; Utsumi, Atsushi; Akai, Yoshiro; Takeuchi, Kiyoshi)V1425,191-195(1991)

Comparative performance of infrared solid state lasers in laser lithotripsy (D'yakonov, G. I.; Konov, Vitaly I.; Mikhailov, V. A.; Nikolaev, D. A.; Pak, S. K.; Shcherbakov, I. A.)V1421,156-162(1991)

Comparative study of excimer and erbium:YAG lasers for ablation of structural components of the knee (Vari, Sandor G.; Shi, Wei-Qiang; van der Veen, Maurits J.; Fishbein, Michael C.; Miller, J. M.; Papaioannou, Thanassis; Grundfest, Warren S.)V1424,33-42(1991)

Comparison of silica-core optical fibers (McCann, Brian P.)V1420,116-125(1991)

Comparison of the ablation of polymethylmethacrylate by two fiber-optic-compatible infrared lasers (Garino, Jonathan P.; Nazarian, David; Froimson, Mark I.; Grelsamer, Ronald P.; Treat, Michael R.)V1424,43-47(1991)

Comparison of the excimer laser with the erbium yttrium aluminum garnet laser for applications in osteotomy (Li, Zhao-zhang; Van De Merwe, Willem P.; Reinisch, Lou)V1427,152-161(1991)

Comparison of thermal and optical techniques for describing light interaction with vascular grafts, sutures, and thrombus (Obremski, Susan M.; LaMuraglia, Glenn M.; Bruggemann, Ulrich H.; Anderson, R. R.)V1427,327-334(1991)

Computed estimation of visual acuity after laser refractive keratectomy (Rol, Pascal O.; Parel, Jean-Marie; Hanna, Khalil)V1423,89-93(1991)

Core-clad silver halide fibers for CO2 laser power transmission (Paiss, Idan; Moser, Frank; Katzir, Abraham)V1420,141-148(1991)

Corneal and retinal energy density with various laser beam delivery systems and contact lenses (Dewey, David)V1423,105-116(1991)

Corneal epithelium, visual acuity, and laser refractive keratectomy (Simon, Gabriel; Parel, Jean-Marie; Kervick, Gerard N.; Rol, Pascal O.; Hanna, Khalil; Thompson, Keith P.)V1423,154-156(1991)

Corneal refractive surgery using an ultraviolet (213 nm) solid state laser (Ren, Qiushi; Gailitis, Raymond P.; Thompson, Keith P.; Penney, Carl M.; Lin, J. T.; Waring, George O.)V1423,129-139(1991)

Correlation tracking: a new technology applied to laser photocoagulation (Forster, Albert A.)V1423,103-104(1991)

Cr,Er:YSGG laser as an instrument for dental surgery (D'yakonov, G. I.; Konov, Vitaly I.; Mikhailov, V. A.; Pak, S. K.; Shcherbakov, I. A.; Ershova, N. I.; Maksimovskiy, Y. V.)V1424,81-86(1991)

Current and future use of lasers in vascular neurosurgery (Chavantes, Maria C.; Zamorano, Lucia J.)V1428,99-127(1991)

Diagnostic and Therapeutic Cardiovascular Interventions (Abela, George S., ed.)V1425(1991)

Discrimination between urinary tract tissue and urinary stones by fiber-optic-pulsed photothermal radiometry method in vivo (Daidoh, Yuichiro; Arai, Tsunenori; Suda, Akira; Kikuchi, Makoto; Komine, Yukikuni; Murai, Masaru; Nakamura, Hiroshi)V1421,120-123(1991)

Dual wavelengths (750/375 nm) laser lithotripsy (Steiner, Rudolf W.; Meier, Thomas H.)V1421,124-128(1991)

Effect of coagulation on laser light distribution in myocardial tissue (Agah, Ramtin; Sheth, Devang; Motamedi, Massoud E.)V1425,172-179(1991)

Effects of He-Ne regional irradiation on 53 cases in the field of pediatric surgery (Guo, Jing-Zhen)V1422,136-139(1991)

Effects of pressure rise on cw laser ablation of tissue (LeCarpentier, Gerald L.; Motamedi, Massoud E.; Welch, Ashley J.)V1427,273-278(1991)

Efficacy of argon-laser-mediated hot-balloon angioplasty (Sakurada, Masami; Miyamoto, Akira; Mizuno, Kyoichi; Nozaki, Youichi; Tabata, Hirotsugu; Etsuda, Hirokuni; Kurita, Akira; Nakamura, Haruo; Arai, Tsunenori; Suda, Akira; Kikuchi, Makoto; Watanabe, Tamishige; Utsumi, Atsushi; Akai, Yoshiro; Takeuchi, Kiyoshi)V1425,158-164(1991)

Elaboration of excimer lasers dosimetry for bone and meniscus cutting and drilling using optical fibers (Jahn, Renate; Dressel, Martin; Neu, Walter; Jungbluth, Karl-Heinz)V1424,23-32(1991)

Electrophoretic mobility patterns of collagen following laser welding (Bass, Lawrence S.; Moazami, Nader; Pocsidio, Joanne O.; Oz, Mehmet C.; LoGerfo, Paul; Treat, Michael R.)V1422,123-127(1991)

Endoscopic removal of PMMA in hip revision surgery with a CO2 laser (Sazy, John; Kollmer, Charles; Uppal, Gurvinder S.; Lane, Gregory J.; Sherk, Henry H.)V1424,50-50(1991)

Endoscopic YAG laser coagulation for early prostate cancer (McNicholas, Thomas A.; O'Donoghue, Neil)V1421,56-67(1991)

Erbium laser ablation of bone: effect of water content (Walsh, Joseph T.; Hill, D. A.)V1421,27-33(1991)

Excimer laser cutting of corneal transplants (Tamkivi, Raivo; Schotter, Leo L.; Pakhomova, Tat'yana A.)V1503,375-378(1991)

Excimer laser in arthroscopic surgery (Koort, Hans J.)V1424,53-59(1991)

Experimental and first clinical results with the alexandrite laser lithotripter (Miller, Kurt; Weber, Hans M.; Rueschoff, Josef; Hautmann, Richard E.)V1421,108-113(1991)

Experimental simulation of holmium laser action on biological tissues (Konov, Vitaly I.; Prokhorov, Alexander M.; Silenok, Alexander S.; Tsarkova, O. G.; Tsvetkov, V. B.; Shcherbakov, I. A.)V1427,220-231(1991)

Fiber design for interstitial laser treatment (Beuthan, Jurgen; Mueller, Gerhard J.; Schaldach, Brita J.; Zur, Ch.)V1420,234-241(1991)

Fiber fragmentation during laser lithotripsy (Flemming, G.; Brinkmann, Ralf E.; Strunge, Ch.; Engelhardt, R.)V1421,146-152(1991)

Fragmentation methods in laser lithotripsy (Jiang, Zhi X.; Whitehurst, Colin; King, Terence A.)V1421,88-99(1991)

Frequency-doubled alexandrite laser for tissue differentiation in angioplasty (Scheu, M.; Engelhardt, R.)V1425,63-69(1991)

Future Trends in Biomedical Applications of Lasers (Svaasand, Lars O., ed.)V1525(1991)

Grueneisen-stress-induced ablation of biological tissue (Dingus, Ronald S.; Scammon, R. J.)V1427,45-54(1991)

Hearing preservation using CO2 laser for acoustic nerve tumors (Grutsch, James; Heiferman, Kenneth S.; Cerullo, Leonard J.)V1428,136-145(1991)

High-energy Nd:glass laser for oncology (Bouchenkov, Vyatcheslav A.; Utenkov, Boris I.; Zaitsev, V. K.; Bayanov, Valentin I.; Serebryakov, Victor A.)V1410,244-247; V1427,405-408(1991)

Histochemical identification of malignant and premalignant lesions (Liebow, Charles; Maloney, M. J.)V1426,22-30(1991)

Hollow curved Al2O3 waveguides for CO2 laser surgery (Gregory, Christopher C.; Harrington, James A.; Altkorn, Robert I.; Haidle, Rudy H.; Helenowski, Tomasz)V1420,169-175(1991)

Hollow-tube-guide for UV-power laser beams (Kubo, Uichi; Okada, Kasuyuki; Hashishin, Yuichi)V1420,102-107(1991)

Holmium:YAG and erbium:YAG laser interaction with hard and soft tissue (Charlton, Andrew; Dickinson, Mark R.; King, Terence A.; Freemont, Anthony J.)V1427,189-197(1991)

Influence of helium, oxygen, nitrogen, and room air environment in determining Nd-YAG laser/brain tissue interaction (Chavantes, Maria C.; Vinas, Federico; Zamorano, Lucia J.; Dujovny, Manuel; Dragovic, Ljubisa)V1428,13-22(1991)

Infrared cables and catheters for medical applications (Artjushenko, Vjacheslav G.; Ivchenko, N.; Konov, Vitaly I.; Kryukov, A. P.; Krupchitsky, Vladimir P.; Kuznetcov, R.; Lerman, A. A.; Litvinenko, E. G.; Nabatov, A. O.; Plotnichenko, V. G.; Prokhorov, Alexander M.; Pylnov, I. L.; Tsibulya, Andrew B.; Vojtsekhovsky, V. V.; Ashraf, N.; Neuberger, Wolfgang; Moran, Kelly B.; Mueller, Gerhard J.; Schaldach, Brita J.)V1420,157-168(1991)

Infrared tissue ablation: consequences of liquefaction (Zweig, A. D.)V1427,2-8(1991)

Integral prism-tipped optical fibers (Friedl, Stephan E.; Kunz, Warren F.; Mathews, Eric D.; Abela, George S.)V1425,134-141(1991)

Interaction of erbium laser radiation with corneal tissue (Wannop, Neil M.; Charlton, Andrew; Dickinson, Mark R.; King, Terence A.)V1423,163-166(1991)

Interstitial laser coagulation of the prostate: experimental studies (McNicholas, Thomas A.; Steger, Adrian C.; Bown, Stephen G.; O'Donoghue, Neil)V1421,30-35(1991)

Intraluminal laser atherectomy with ultrasound and electromagnetic guidance (Gregory, Kenton W.; Aretz, H. T.; Martinelli, Michael A.; Ledet, Earl G.; Hatch, G. F.; Gregg, Richard E.; Sedlacek, Tomas; Haase, W. C.)V1425,217-225(1991)

Intraoperative clinical use of low-power laser irradiation following surgical treatment of the tethered spinal cord (Rochkind, Simeone; Alon, M.; Ouaknine, G. E.; Weiss, S.; Avram, J.; Razon, N.; Lubart, Rachel; Friedmann, Harry)V1428,52-58(1991)

Keratorefractive procedures (Seiler, Theo)V1525,280-281(1991)

KTP-532 laser ablation of urethral strictures (Malloy, Terrence P.)V1421,72-72(1991)

KTP-532 laser utilization in endoscopic pelvic lymphadenectomy (Malloy, Terrence P.)V1421,78-78(1991)

Laboratory and clinical data on wound healing by low-power laser from the Medical Institute of Vilafortuny, Spain (Trelles, Mario A.; Mayayo, E.; Resa, A. M.; Rigau, J.; Calvo, G.)V1403,781-798(1991)

Laparoscopically guided bilateral pelvic lymphadenectomy (Gershman, Alex; Danoff, Dudley; Chandra, Mudjianto; Grundfest, Warren S.)V1421,186-188(1991)

Laparoscopic appendectomy (Richards, Kent F.; Christensen, Brent J.)V1421,198-202(1991)

Laparoscopic applications of laser-activated tissue glues (Bass, Lawrence S.; Oz, Mehmet C.; Auteri, Joseph S.; Williams, Mathew R.; Rosen, Jeffrey; Libutti, Steven K.; Eaton, Alexander M.; Lontz, John F.; Nowygrod, Roman; Treat, Michael R.)V1421,164-168(1991)

Laparoscopic use of laser and monopolar electrocautery (Hunter, John G.)V1421,173-183(1991)

Laser angioplasty with lensed fibers and a holmium:YAG laser in iliac artery occlusions (White, Christopher J.; Ramee, Stephen R.; Mesa, Juan E.; Collins, Tyrone J.; Kotmel, Robert F.; Godfrey, Maureen A.)V1425,130-133(1991)

Laser-assisted skin closure at 1.32 microns: the use of a software-driven medical laser system (Dew, Douglas K.; Hsu, Tung M.; Hsu, Long S.; Halpern, Steven J.; Michaels, Charles E.)V1422,111-115(1991)

Laser Doppler flowmetry in neurosurgery (Fasano, Victor A.; Urciuoli, Rosa; Bolognese, Paolo; Fontanella, Marco)V1428,2-12(1991)

Laser effects on fibrin clot response by human meniscal fibrochondrocytes in organ culture (Forman, Scott K.; Oz, Mehmet C.; Wong, Edison; Treat, Michael R.; Kiernan, Howard)V1424,2-6(1991)

Laser energy repartition inside metal, sapphire, and quartz surgical laser tips (Seka, Wolf D.; Golding, Douglas J.; Klein, B.; Lanzafame, Raymond J.; Rogers, David W.)V1398,162-169(1991)

Laser lithotripsy with a pulsed dye laser: correlation between threshold energy and optical properties (Sterenborg, H. J.; van Swol, Christiaan F.; Biganic, Dane D.; van Gemert, Martin J.)V1427,256-266(1991)

Lasers in dentistry (Keller, Ulrich)V1525,282-288(1991)

Lasers in Dermatology and Tissue Welding (Tan, Oon T.; White, Rodney A.; White, John V., eds.)V1422(1991)

Lasers in Orthopedic, Dental, and Veterinary Medicine (Dederich, Douglas N.; O'Brien, Stephen J.; Trent, Ava M.; Wigdor, Harvey A., eds.)V1424(1991)

Lasers in Urology, Laparoscopy, and General Surgery (Watson, Graham M.; Steiner, Rudolf W.; Pietrafitta, Joseph J., eds.)V1421(1991)

Laser surgery for selected small animal soft-tissue conditions (Bartels, Kenneth E.)V1424,164-170(1991)

Laser surgical unit for photoablative and photothermal keratoplasty (Cartlidge, Andy G.; Parel, Jean-Marie; Yokokura, Takashi; Lowery, Joseph A.; Kobayashi, K.; Nose, I.; Lee, William; Simon, Gabriel; Denham, David B.)V1423,167-174(1991)

Laser-Tissue Interaction II (Jacques, Steven L., ed.)V1427(1991)

Lesion-specific laser catheters for angioplasty (Murphy-Chutorian, Douglas)V1420,244-248(1991)

Light absorption characteristics of the human meniscus: applications for laser ablation (Vangsness, C. T.; Huang, Jay; Smith, Chadwick F.)V1424,16-19(1991)

Measurement of fluorescence spectra and quantum yields of 193 nm ArF laser photoablation of the cornea and synthetic lenticules (Milne, Peter J.; Zika, Rod G.; Parel, Jean-Marie; Denham, David B.; Penney, Carl M.)V1423,122-129(1991)

Measurement of temperature distributions after pulsed IR radiation impact in biological tissue models with fluorescent thin films (Romano, Valerio; Greber, Charlotte M.; Frenz, Martin; Forrer, Martin; Weber, Heinz P.)V1427,16-26(1991)

Mechanical and acoustic analysis in ultrasonic angioplasty (Detwiler, Paul W.; Watkins, James F.; Rose, Eric A.; Ratner, A.; Vu, Louis P.; Severinsky, J. Y.; Rosenschein, Uri)V1425,149-155(1991)

Mechanisms of ultraviolet and mid-infrared tissue ablation (Hibst, Raimund)V1525,162-169(1991)

Nd:YAG laser irradiation in the treatment of upper tract urothelial tumors (Gaboardi, Franco; Melodia, Tommaso; Bozzola, Andrea; Galli, Luigi)V1421,79-85(1991)

Nd:YAG laser treatment in patients with prostatic adenocarcinoma stage A (Gaboardi, Franco; Dotti, Ernesto; Bozzola, Andrea; Galli, Luigi)V1421,50-55(1991)

New approaches to local destruction of tumors: interstitial laser hyperthermia and photodynamic therapy (Bown, Stephen G.)V1525,325-330(1991)

New approach for endoscopic stereotactic brain surgery using high-power laser (Otsuki, Taisuke; Yoshimoto, Takashi)V1420,220-224(1991)

New concept of a compact multiwavelength solid state laser for laser-induced shock wave lithotripsy (Steiger, Erwin)V1421,140-146(1991)

Noninvasive ultrasound-produced volume lesion in prostate (Foster, Richard S.)V1421,47-47(1991)

Ophthalmic Technologies (Puliafito, Carmen A., ed.)V1423(1991)

Optical and mechanical parameter detection of calcified plaque for laser angioplasty (Stetz, Mark L.; O'Brien, Kenneth M.; Scott, John J.; Baker, Glenn S.; Deckelbaum, Lawrence I.)V1425,55-62(1991)

Optical characteristics of sapphire laser scalpels analysed by ray-tracing (Verdaasdonk, Rudolf M.; Borst, Cornelius)V1420,136-140(1991)

Optical Fibers in Medicine VI (Katzir, Abraham, ed.)V1420(1991)

Optimal surface characteristics for instruments for use in laser neurosurgery (Heiferman, Kenneth S.; Cramer, K. E.; Walsh, Joseph T.)V1428,128-134(1991)

Peak pressures and temperatures within laser-ablated tissues (Furzikov, Nickolay P.; Dmitriev, A. C.; Lekhtsier, Eugeny N.; Orlov, M. Y.; Semyenov, Alexander D.; Tyurin, Vladimir S.)V1427,288-297(1991)

Percutaneous lumbar discectomy using Ho:YAG laser (Black, Johnathan D.; Sherk, Henry H.; Uppal, Gurvinder S.; Sazy, John; Meller, Menachem M.; Rhodes, Anthony; Lane, Gregory J.)V1424,20-22(1991)

Percutaneous peripheral excimer laser angioplasty: immediate success rate and short-term outcome (Visona, Adriana; Liessi, Guido; Bonanome, Andrea; Lusiani, Luigi; Miserocchi, Luigi; Pagnan, Antonio)V1425,75-83(1991)

Perspectives of powerful laser technique for medicine (Konov, Vitaly I.; Prokhorov, Alexander M.; Shcherbakov, I. A.)V1525,250-252(1991)

Photoablation (Doerschel, Klaus; Mueller, Gerhard J.)V1525,253-279(1991)

Photoablation using the Holmium:YAG laser: a laboratory and clinical study (Waidhauser, Erich; Markmiller, U.; Enders, S.; Hessel, Stefan F.; Ulrich, Frank; Beck, Oskar J.; Feld, Michael S.)V1428,75-83(1991)

Photodynamic treatment of lens epithelial cells for cataract surgery (Lingua, Robert W.; Parel, Jean-Marie; Simon, Gabriel; Li, Kam)V1423,58-61(1991)

Photosensitized receptor inactivation with He-Ne laser: preliminary results (Arber, Simon; Rymer, William Z.; Crumrine, David)V1428,23-29(1991)

Physical mechanism of high-power pulse laser ophthalmology: experimental research (He, Anzhi; Lu, Jian-Feng; Ni, Xiao W.; Li, Yong N.)V1423,117-120(1991)

Plastic hollow fibers employed for CO_2 laser power transmission in oral surgery (Calderon, S.; Gannot, Israel; Dror, Jacob; Dahan, Reuben; Croitoru, Nathan I.)V1420,108-115(1991)

Positron emission tomography of laser-induced interstitial hyperthermia in cerebral gliomas (Ulrich, Frank; Bettag, Martin; Langen, K. J.)V1428,135-135(1991)

Potentials for pulsed YAG:Nd laser application to endoscopic surgery (Manenkov, Alexander A.; Denisov, N. N.; Bagdasarov, V. H.; Starkovsky, A. N.; Yurchenko, S. V.; Kornilov, Yu. M.; Mikaberidze, V. M.; Sarkisov, S. E.)V1420,254-258(1991)

Power transmission for silica fiber laser delivery systems (McCann, Brian P.)V1398,230-237(1991)

Preliminary evaluation of collagen as a component in the thermally induced 'weld' (Lemole, G. M.; Anderson, R. R.; DeCoste, Sue)V1422,116-122(1991)

Preliminary experience with laser reinforcement of vascular anastomoses (Oz, Mehmet C.; Bass, Lawrence S.; Williams, Mathew R.; Benvenisty, Alan I.; Hardy, Mark A.; Libutti, Steven K.; Eaton, Alexander M.; Treat, Michael R.; Nowygrod, Roman)V1422,147-150(1991)

Preliminary results of laser-assisted sealing of hand-sewn canine esophageal anastomoses (Auteri, Joseph S.; Oz, Mehmet C.; Sanchez, Juan A.; Bass, Lawrence S.; Jeevanandam, Valluvan; Williams, Mathew R.; Smith, Craig R.; Treat, Michael R.)V1421,182-184(1991)

Preliminary results with sutured colonic anastomoses reinforced with dye-enhanced fibrinogen and a diode laser (Libutti, Steven K.; Williams, Mathew R.; Oz, Mehmet C.; Forde, Kenneth A.; Bass, Lawrence S.; Weinstein, Samuel; Auteri, Joseph S.; Treat, Michael R.; Nowygrod, Roman)V1421,169-172(1991)

Preliminary stress/strain analysis of laser-soldered and -sutured vascular tissue (Ashton, Robert C.; Oz, Mehmet C.; Lontz, John F.; Lemole, Gerald M.)V1422,151-155(1991)

Promotional effects of CO_2 laser on neoplastic lesions in hamsters (Kingsbury, Jeffrey S.; Margarone, Joseph E.; Satchidanand, S.; Liebow, Charles)V1427,363-367(1991)

Pulsed CO_2 laser for intra-articular cartilage vaporization and subchondral bone perforation in horses (Nixon, Alan J.; Roth, Jerry E.; Krook, Lennart)V1424,198-208(1991)

Pulsed lasers in dentistry: sense or nonsense? (Koort, Hans J.; Frentzen, Matthias)V1424,87-98(1991)

Q-switched Nd:glass-laser-induced acoustic pulses in lithotripsy (D'yakonov, G. I.; Mikhailov, V. A.; Pak, S. K.; Shcherbakov, I. A.; Andreev, V. G.; Rudenko, O. V.; Sapozhnikov, A. V.)V1421,153-155(1991)

Q-switching and pulse shaping with IR lasers (Brinkmann, Ralf E.; Bauer, K.)V1421,134-139(1991)

Range of modalities for prostate therapy (Watson, Graham M.)V1421,2-5(1991)

Ray-tracing of optically modified fiber tips for laser angioplasty (Verdaasdonk, Rudolf M.; Borst, Cornelius)V1425,102-109(1991)

Recanalization of azoospermia due to a Mullerian duct cyst by Nd:YAG laser (Gaboardi, Franco; Bozzola, Andrea; Zago, Tiziano; Gulfi, Gildo M.; Galli, Luigi)V1421,73-77(1991)

Reduction of acoustic transients in tissue with a 2 um thulium laser (Pinto, Joseph F.; Esterowitz, Leon; Bonner, Robert F.)V1420,242-243(1991)

Remote Raman spectroscopic imaging of human artery wall (Phat, Darith; Vuong, Phat N.; Plaza, Pascal; Cheilan, Francis; Dao, Nguyen Q.)V1525,196-205(1991)

Removal of dental filling materials by Er:YAG laser radiation (Hibst, Raimund; Keller, Ulrich)V1424,120-126(1991)

Restenosis after hot-tip laser-balloon angioplasty: histologic evaluation of the samples removed by Simpson atherectomy (Barbieri, Enrico; Tanganelli, Pietro; Taddei, Giuseppe; Perbellini, Antonio; Attino, Vito; Destro, Gianni; Zardini, Piero)V1425,122-127(1991)

Safety of medical excimer lasers with an emphasis on compressed gases (Sliney, David H.; Clapham, Terrance N.)V1423,157-162(1991)

Selective tumor destruction with photodynamic therapy: exploitation of photodynamic thresholds (Barr, Hugh)V1525,331-340(1991)

Simple analytical model for laser-induced tissue ablation (Kukreja, Lalit M.; Braun, R.; Hess, P.)V1427,243-254(1991)

Skin closure with dye-enhanced laser welding and fibrinogen (Wider, Todd M.; Libutti, Steven K.; Greenwald, Daniel P.; Oz, Mehmet C.; Yager, Jeffrey S.; Treat, Michael R.; Hugo, Norman E.)V1422,56-61(1991)

Stage III endobronchial squamous cell cancer: survival after Nd:YAG laser combined with photodynamic therapy versus Nd:YAG laser or photodynamic therapy alone (McCaughan, James S.; Barabash, Rostislav D.; Hawley, Philip)V1426,279-287(1991)

Stimulatory effects of Nd:YAG lasers on canine articular cartilage (Lane, Gregory J.; Sherk, Henry H.; Kollmer, Charles; Uppal, Gurvinder S.; Rhodes, Anthony; Sazy, John; Black, Johnathan D.; Lee, Steven)V1424,7-11(1991)

Stone/ureter identification during alexandrite laser lithotripsy (Scheu, M.; Flemming, G.; Engelhardt, R.)V1421,100-107(1991)

Study of bone ablation dynamics with sequenced pulses (Izatt, Joseph A.; Albagli, Douglas; Itzkan, Irving; Feld, Michael S.)V1427,110-116(1991)

Sutureless cataract incision closure using laser-activated tissue glues (Eaton, Alexander M.; Bass, Lawrence S.; Libutti, Steven K.; Schubert, Herman D.; Treat, Michael R.)V1423,52-57(1991)

Technique of CO2 laser arthroscopic surgery (Meller, Menachem M.; Sherk, Henry H.; Rhodes, Anthony; Sazy, John; Uppal, Gurvinder S.; Lane, Gregory J.)V1424,60-61(1991)

Temperature increase during in vitro 308 nm excimer laser ablation of porcine aortic tissue (Gijsbers, Geert H.; Sprangers, Rene L.; van den Broecke, Duco G.; van Wieringen, Niek; Brugmans, Marco J.; van Gemert, Martin J.)V1425,94-101(1991)

Temperature response of biological tissues to nonablative pulsed CO2 laser irradiation (van Gemert, Martin J.; Brugmans, Marco J.; Gijsbers, Geert H.; Kemper, J.; van der Meulen, F. W.; Nijdam, D. C.)V1427,316-319(1991)

Theoretical analysis of stone fragmentation rates (Sterenborg, H. J.)V1525,170-176(1991)

Three-Dimensional Bioimaging Systems and Lasers in the Neurosciences (Boggan, James E.; Cerullo, Leonard J.; Smith, Louis C., eds.)V1428(1991)

Tissue interactions of ball-tipped and multifiber catheters (Mitchell, David C.; Smith, Roy E.; Walters, Tena K.; Murray, Alan; Wood, Richard F.)V1427,181-188(1991)

Toward an explanation of laser-induced stimulation and damage of cell cultures (Friedmann, Harry; Lubart, Rachel; Laulicht, Israel; Rochkind, Simeone)V1427,357-362(1991)

Transendoscopic and freehand use of flexible hollow fibers for carbon dioxide laser surgery in the upper airway of the horse: a preliminary report (Palmer, Scott E.)V1424,218-220(1991)

Transurethral laser prostatectomy using a right-angle laser delivery system (Johnson, Douglas E.; Levinson, A. K.; Greskovich, Frank J.; Cromeens, Douglas M.; Ro, Jae Y.; Costello, Anthony J.; Wishnow, Kenneth I.)V1421,36-41(1991)

Transurethral ultrasound-guided laser-induced prostatectomy (Babayan, Richard K.; Roth, Robert A.)V1421,42-44(1991)

Unresolved issues in excimer laser corneal surgery (Trokel, Stephen L.)V1423,94-97(1991)

Use of excimer lasers in medicine: applications, problems, and dangers (Mommsen, Jens; Stuermer, Martin)V1503,348-354(1991)

Use of krypton laser stimulation in the treatment of dry eye syndrome (Kecik, Tadeusz; Switka-Wieclawska, Iwona; Ciszewska, Joanna; Portacha, Lidia)V1391,341-345(1991)

Vascular tissue welding of the CO2 laser: limitations (Dalsing, Michael C.; Kruepper, Peter; Cikrit, Dolores F.)V1422,98-102(1991)

Wavelength selection in laser arthroscopy (Black, Johnathan D.; Sherk, Henry H.; Meller, Menachem M.; Uppal, Gurvinder S.; Divan, James; Sazy, John; Rhodes, Anthony; Lane, Gregory J.)V1424,12-15(1991)

Laser-tissue interaction—see also laser surgery; medicine

10-W Ho laser for surgery (Bouchenkov, Vyatcheslav A.; Utenkov, Boris I.; Antipenko, Boris M.; Berezin, U. D.; Malinin, Boris G.; Serebryakov, Victor A.)V1427,409-412(1991)

2.014-micron Cr;Tm:YAG: optimization of doping concentration for flash lamp operation (Bar-Joseph, Dan)V1410,142-147(1991)

Ablation of hard dental tissues with an ArF-pulsed excimer laser (Neev, Joseph; Raney, Daniel; Whalen, William E.; Fujishige, Jack T.; Ho, Peter D.; McGrann, John V.; Berns, Michael W.)V1427,162-172(1991)

Acoustic effects of Q-switched Nd:YAG on crystalline lens (Porindla, Sridhar N.; Rylander, Henry G.; Welch, Ashley J.)V1427,267-272(1991)

Analysis of kinetic rate modeling of thermal damage in laser-irradiated tissue (Rastegar, Sohi; Glenn, T.)V1427,300-306(1991)

Argon laser vascular tissue fusion: current status and future perspectives (White, Rodney A.; Kopchok, George E.)V1422,103-110(1991)

Biomechanical and structural studies of rabbits carotid arteries after endovascular laser exposition (Purinsh, Juris; Elksninsh, N.; Dzenis, J.; Tomass, V.; Teivans, A.; Freimanis, R.; Garsha, I.; Ozolinsh, H.)V1525,289-308(1991)

Biomedical applications of laser technology at Los Alamos (Bigio, Irving J.; Loree, Thomas R.)V1403,776-780(1991)

Cellular and extracellular effects of soft laser irradiation on an erythrocytes suspension (Skopinov, S. A.; Zakharov, S. D.; Volf, E. B.; Perov, S. N.; Panasenko, N. A.)V1403,676-679(1991)

Changes in optical density of normal vessel wall and lipid atheromatous plaque after Nd:YAG laser irradiation (Schwarzmaier, Hans-Joachim; Heintzen, Matthias P.; Zumdick, Mathias; Kaufmann, Raimund; Wolbarsht, Myron L.)V1427,128-133(1991)

Characterization of photobiophysical properties of sensitizers used in photodynamic therapy (Roeder, Beate; Naether, Dirk)V1525,377-384(1991)

Characterization of the fluorescence lifetimes of the ionic species found in aqueous solutions of hematoporphyrin IX as a function of pH (Nadeau, Pierre; Pottier, R.; Szabo, Arthur G.; Brault, Daniel; Vever-Bizet, C.)V1398,151-161(1991)

Chick chorioallantoic membrane for the study of synergistic effects of hyperthermia and photodynamic therapy (Kimel, Sol; Svaasand, Lars O.; Hammer-Wilson, Marie J.; Gottfried, Varda; Berns, Michael W.)V1525,341-350(1991)

Clinical applications of PDT in urology: present and near future (Shumaker, Bryan P.)V1426,293-300(1991)

Clinical evaluation of tumor promotion by CO2 laser (Braun, Robert E.; Liebow, Charles)V1424,138-144(1991)

Clinical optical dose measurement for PDT: invasive and noninvasive techniques (Bays, Roland; Winterhalter, L.; Funakubo, H.; Monnier, Philippe; Savary, M.; Wagnieres, G.; Braichotte, D.; Chatelain, Andre; van den Bergh, Hubert; Svaasand, Lars O.; Burckhardt, C. W.)V1525,397-408(1991)

Coagulation and precise ablation of biotissues by pulsed sealed-off carbon monoxide laser (Masychev, Victor I.; Alejnikov, Vladislav S.; Klimenko, Vladimir I.)V1427,344-356(1991)

Comparative studies on hyperthermia induced by laser light, microwaves, and ultrasonics (Greguss, Pal)V1525,313-324(1991)

Comparative study of excimer and erbium:YAG lasers for ablation of structural components of the knee (Vari, Sandor G.; Shi, Wei-Qiang; van der Veen, Maurits J.; Fishbein, Michael C.; Miller, J. M.; Papaioannou, Thanassis; Grundfest, Warren S.)V1424,33-42(1991)

Comparative study of gelatin ablation by free-running and Q-switch modes of Er:YAG laser (Konov, Vitaly I.; Kulevsky, Lev A.; Lukashev, Alexei V.; Pashinin, Vladimir P.; Silenok, Alexander S.)V1427,232-242(1991)

Comparison of carbon monoxide and carbon dioxide laser-tissue interaction (Waters, Ruth A.; Thomas, J. M.; Clement, R. M.; Ledger, N. R.)V1427,336-343(1991)

Comparison of the excimer laser with the erbium yttrium aluminum garnet laser for applications in osteotomy (Li, Zhao-zhang; Van De Merwe, Willem P.; Reinisch, Lou)V1427,152-161(1991)

Comparison of thermal and optical techniques for describing light interaction with vascular grafts, sutures, and thrombus (Obremski, Susan M.; LaMuraglia, Glenn M.; Bruggemann, Ulrich H.; Anderson, R. R.)V1427,327-334(1991)

Crystallo-optic diagnostics method of the soft laser-induced effects in biological fluids (Skopinov, S. A.; Yakovleva, S. V.)V1403,680-681(1991)

Current status of photodynamic therapy for human cancer (Marcus, Stuart L.)V1426,301-310(1991)

Damage induced by pulsed IR laser radiation at transitions between different tissues (Frenz, Martin; Greber, Charlotte M.; Romano, Valerio; Forrer, Martin; Weber, Heinz P.)V1427,9-15(1991)

Dependence of photodynamic threshold dose on treatment parameters in normal rat liver in vivo (Farrell, Thomas J.; Wilson, Brian C.; Patterson, Michael S.; Chow, Rowena)V1426,146-155(1991)

Depth dependent laser Doppler perfusion measurements: theory and instrumentation (Koelink, M. H.; de Mul, F. F.; Greve, Jan; Graaff, Reindert; Dassel, A. C.; Aarnouds, J. G.)V1403,347-349(1991)

Detection of atheroma using Photofrin II and laser-induced fluorescence spectroscopy (Vari, Sandor G.; Papazoglou, Theodore G.; van der Veen, Maurits J.; Papaioannou, Thanassis; Fishbein, Michael C.; Chandra, Mudjianto; Beeder, Clain; Shi, Wei-Qiang; Grundfest, Warren S.)V1426,58-65(1991)

Determination of the optical penetration depth in tumors from biopsy samples (Lenz, P.)V1525,183-191(1991)

Development of a novel in-vivo drug/in-vitro light system to investigate mechanisms of cell killing with photodynamic therapy (Hampton, James A.; Selman, Steven H.)V1426,134-145(1991)

Diagnostic and Therapeutic Cardiovascular Interventions (Abela, George S., ed.)V1425(1991)

Digital image processing for the early localization of cancer (Kelmar, Cheryl M.)V1426,47-57(1991)

Dosimetry for lasers and light in dermatology: Monte Carlo simulations of 577nm-pulsed laser penetration into cutaneous vessels (Jacques, Steven L.; Keijzer, Marleen)V1422,2-13(1991)

Drug-target interactions on a single living cell: an approach by optical microspectroscopy (Manfait, Michel; Morjani, Hamid; Millot, Jean-Marc; Debal, Vincent; Angiboust, Jean-Francois; Nabiev, I. R.)V1403,695-707(1991)

Effectiveness of porphyrin-like compounds in photodynamic damage of atherosclerotic plaque (Zalessky, Viacheslav N.; Bobrov, Vladimir; Michalkin, Igor; Trunov, Vitaliy)V1426,162-169(1991)

Effect of coagulation on laser light distribution in myocardial tissue (Agah, Ramtin; Sheth, Devang; Motamedi, Massoud E.)V1425,172-179(1991)

Effect of Nd:YAG laser on dentinal bond strength (Dederich, Douglas N.; Tulip, John)V1424,134-137(1991)

Effect of spatial distribution of irradiated sites on injury selectivity in vascular tissue (Cheong, Wai-Fung; Morrison, Paul R.; Trainor, S. W.; Kurban, Amal K.; Tan, Oon T.)V1422,19-26(1991)

Effects of He-Ne regional irradiation on 53 cases in the field of pediatric surgery (Guo, Jing-Zhen)V1422,136-139(1991)

Effects of photodynamic treatment on DNA (Oleinick, Nancy L.; Agarwal, Munna L.; Antunez, Antonio R.; Clay, Marian E.; Evans, Helen H.; Harvey, Ella Jo; Rerko, Ronald M.; Xue, Liang-yan)V1427,90-100(1991)

Effects of pressure rise on cw laser ablation of tissue (LeCarpentier, Gerald L.; Motamedi, Massoud E.; Welch, Ashley J.)V1427,273-278(1991)

Efficacy of photodynamic killing with membrane associated and internalized photosensitizer molecules (Allison, Beth; Jiang, Frank N.; Levy, Julia G.)V1426,200-207(1991)

Electron microscopic study on black pig skin irradiated with pulsed dye laser (504 nm) (Yasuda, Yukio; Tan, Oon T.; Kurban, Amal K.; Tsukada, Sadao)V1422,50-55(1991)

Electrophoretic mobility patterns of collagen following laser welding (Bass, Lawrence S.; Moazami, Nader; Pocsidio, Joanne O.; Oz, Mehmet C.; LoGerfo, Paul; Treat, Michael R.)V1422,123-127(1991)

Endobronchial occlusive disease: Nd:YAG or PDT? (Regal, Anne-Marie; Takita, Hiroshi)V1426,271-278(1991)

Endoscopic tissue autofluorescence measurements in the upper aerodigestive tract and the bronchi (Braichotte, D.; Wagnieres, G.; Monnier, Philippe; Savary, M.; Bays, Roland; van den Bergh, Hubert; Chatelain, Andre)V1525,212-218(1991)

Erbium laser ablation of bone: effect of water content (Walsh, Joseph T.; Hill, D. A.)V1427,27-33(1991)

Evaluation of Nile Blue E chalcogen analogs as PDT agents (Foley, James W.; Cincotta, Louis; Cincotta, Anthony H.)V1426,208-215(1991)

Excimer-dye laser system for diagnosis and therapy of cancer (Pokora, Ludwik J.; Puzewicz, Zbigniew)V1503,467-478(1991)

Excimer-laser-induced fluorescence spectroscopy of human arteries during laser ablation (Abel, B.; Hippler, Horst; Koerber, B.; Morguet, A.; Neu, Walter)V1525,110-118(1991)

Experimental simulation of holmium laser action on biological tissues (Konov, Vitaly I.; Prokhorov, Alexander M.; Silenok, Alexander S.; Tsarkova, O. G.; Tsvetkov, V. B.; Shcherbakov, I. A.)V1427,220-231(1991)

Future Trends in Biomedical Applications of Lasers (Svaasand, Lars O., ed.)V1525(1991)

Grueneisen-stress-induced ablation of biological tissue (Dingus, Ronald S.; Scammon, R. J.)V1427,45-54(1991)

Helium-neon effects of laser radiation in rats infected with thromboxane B2 (Juri, Hugo; Palma, J. A.; Campana, Vilma; Gavotto, A.; Lapin, R.; Yung, S.; Lillo, J.)V1422,128-135(1991)

High-energy Nd:glass laser for oncology (Bouchenkov, Vyatcheslav A.; Utenkov, Boris I.; Zaitsev, V. K.; Bayanov, Valentin I.; Serebryakov, Victor A.)V1410, 244-247; V1427,405-408(1991)

Histological distinction of mechanical and thermal defects produced by nanosecond laser pulses in striated muscle at 1064 nm (Gratzl, Thomas; Dohr, Gottfried; Schmidt-Kloiber, Heinz; Reichel, Erich)V1427,55-62(1991)

Histopathologic assessment of water-dominated photothermal effects produced with laser irradiation (Thomsen, Sharon L.; Cheong, Wai-Fung; Pearce, John A.)V1422,14-18(1991)

Holmium:YAG and erbium:YAG laser interaction with hard and soft tissue (Charlton, Andrew; Dickinson, Mark R.; King, Terence A.; Freemont, Anthony J.)V1427,189-197(1991)

Hyperthermia treatment of spontaneously occurring oral cavity tumors using a computer-controlled Nd:YAG laser system (Panjehpour, Masoud; Overholt, Bergein F.; Frazier, Donita L.; Klebanow, Edward R.)V1424,179-185(1991)

Hyperthermia treatment using a computer-controlled Nd:YAG laser system in combination with surface cooling (Panjehpour, Masoud; Wilke, August; Frazier, Donita L.; Overholt, Bergein F.)V1427,307-315(1991)

Imaging of excimer laser vascular tissue ablation by ultrafast photography (Nyga, Ralf; Neu, Walter; Preisack, M.; Haase, Karl K.; Karsch, Karl R.)V1525,119-123(1991)

Importance of pulse duration in laser-tissue interactions: a histological study (Kurban, Amal K.; Morrison, Paul R.; Trainor, S. W.; Cheong, Wai-Fung; Yasuda, Yukio; Tan, Oon T.)V1422,43-49(1991)

Indirect spectroscopic detection of singlet oxygen during photodynamic therapy (Tromberg, Bruce J.; Dvornikov, Tatiana; Berns, Michael W.)V1427,101-108(1991)

Influence of helium, oxygen, nitrogen, and room air environment in determining Nd-YAG laser/brain tissue interaction (Chavantes, Maria C.; Vinas, Federico; Zamorano, Lucia J.; Dujovny, Manuel; Dragovic, Ljubisa)V1428,13-22(1991)

Influence of hematoporphyrin on the mechanical properties of the lipid bilayer membranes (Hianik, T.; Masarikova, D.; Zhorina, L. V.; Poroshina, Marina Y.; Chernyaeva, E. B.)V1402,85-88(1991)

Influence of the laser-induced temperature rise in photodynamic therapy (Gottschalk, Wolfgang; Hengst, Joachim; Sroka, Ronald; Unsoeld, Eberhard)V1427,320-326(1991)

Infrared tissue ablation: consequences of liquefaction (Zweig, A. D.)V1427,2-8(1991)

Interaction of erbium laser radiation with corneal tissue (Wannop, Neil M.; Charlton, Andrew; Dickinson, Mark R.; King, Terence A.)V1423,163-166(1991)

Interaction of phthalocyanine photodynamic treatment with ionophores and lysosomotrophic agents (Oleinick, Nancy L.; Varnes, Marie E.; Clay, Marian E.; Menegay, Harry J.)V1426,235-243(1991)

Intracellular location, picosecond kinetics, and light-induced reactions of photosensitizing porphyrins (Schneckenburger, Herbert; Seidlitz, Harold K.; Wessels, Jurina; Rueck, Angelika C.)V1403,646-652(1991)

Intracellular uptake and ultrastructural phototoxic effects of sulfonated chlor-aluminum phthalocyanine on bladder tumor cells in vitro (Miller, Kurt; Reich, Ella; Grau, N.)V1426,378-383(1991)

Intraoperative clinical use of low-power laser irradiation following surgical treatment of the tethered spinal cord (Rochkind, Simeone; Alon, M.; Ouaknine, G. E.; Weiss, S.; Avram, J.; Razon, N.; Lubart, Rachel; Friedmann, Harry)V1428,52-58(1991)

Investigation of the photodynamic properties of some chlorophyll a derivatives: the effect of doxorubicin on the chlorine e6 photosensitized death of Ehrlich carcinoma cells (Chekulayev, V.; Shevchuk, Igor; Kahru, A.; Mihkelsoo, V. T.; Kallikorm, A. P.)V1426,367-377(1991)

In-vitro study of the effects of Congo Red on the ablation of atherosclerotic plaque (Beyerbacht, Hugo P.; Aggarwal, Shanti J.; Jansen, E. D.; Welch, Ashley J.)V1427,117-127(1991)

In-vivo optical attenuation in normal rat brain and its implication in PDT (Chen, Qun; Wilson, Brian C.; Patterson, Michael S.; Chopp, Michael; Hetzel, Fred W.)V1426,156-161(1991)

Kinetic models for coagulation processes: determination of rate coefficients in vivo (Pearce, John A.; Cheong, Wai-Fung; Pandit, Kirit; McMurray, Tom J.; Thomsen, Sharon L.)V1422,27-33(1991)

Laboratory and clinical data on wound healing by low-power laser from the Medical Institute of Vilafortuny, Spain (Trelles, Mario A.; Mayayo, E.; Resa, A. M.; Rigau, J.; Calvo, G.)V1403,781-798(1991)

Laparoscopic applications of laser-activated tissue glues (Bass, Lawrence S.; Oz, Mehmet C.; Auteri, Joseph S.; Williams, Mathew R.; Rosen, Jeffrey; Libutti, Steven K.; Eaton, Alexander M.; Lontz, John F.; Nowygrod, Roman; Treat, Michael R.)V1421,164-168(1991)

Laser Applications in Life Sciences (Akhmanov, Sergei A.; Poroshina, Marina Y., eds.)V1403(1991)

Laser-assisted skin closure at 1.32 microns: the use of a software-driven medical laser system (Dew, Douglas K.; Hsu, Tung M.; Hsu, Long S.; Halpern, Steven J.; Michaels, Charles E.)V1422,111-115(1991)

Laser effects on fibrin clot response by human meniscal fibrochondrocytes in organ culture (Forman, Scott K.; Oz, Mehmet C.; Wong, Edison; Treat, Michael R.; Kiernan, Howard)V1424,2-6(1991)

Laser energy repartition inside metal, sapphire, and quartz surgical laser tips (Seka, Wolf D.; Golding, Douglas J.; Klein, B.; Lanzafame, Raymond J.; Rogers, David W.)V1398,162-169(1991)

Laser-flash photographic studies of Er:YAG laser ablation of water (Jacques, Steven L.; Gofstein, Gary)V1427,63-67(1991); V1525,309-312(1991)

Laser-induced fluorescence of biological tissue (Dietel, W.; Dorn, P.; Zenk, W.; Zielinski, M.)V1403,653-658(1991)

Laser-induced shock wave effects on red blood cells (Flotte, Thomas J.; Frisoli, Joan K.; Goetschkes, Margaret; Doukas, Apostolos G.)V1427,36-44(1991)

Laser/light tissue interaction: on the mechanism of optical breakdown (Siomos, Konstadinos)V1525,154-161(1991)

Laser neural tissue interactions using bilayer membrane models (VanderMeulen, David L.; Khoka, Mustafa; Spears, Kenneth G.)V1428,84-90(1991)

Laser photochemotherapy of psoriasis (Tuchin, Valery V.; Utz, Sergey R.; Barabanov, Alexander J.; Dovzansky, S. I.; Ulyanov, A. N.; Aravin, Vladislav A.; Khomutova, T. G.)V1422,85-96(1991)

Laser photodynamic therapy of cancer: the chorioallantoic membrane model for measuring damage to blood vessels in-vivo (Gottfried, Varda; Lindenbaum, Ella S.; Kimel, Sol; Hammer-Wilson, Marie J.; Berns, Michael W.)V1442,218-229(1991)

Lasers in dentistry (Keller, Ulrich)V1525,282-288(1991)

Lasers in Dermatology and Tissue Welding (Tan, Oon T.; White, Rodney A.; White, John V., eds.)V1422(1991)

Lasers in Orthopedic, Dental, and Veterinary Medicine (Dederich, Douglas N.; O'Brien, Stephen J.; Trent, Ava M.; Wigdor, Harvey A., eds.)V1424(1991)

Lasers in Urology, Laparoscopy, and General Surgery (Watson, Graham M.; Steiner, Rudolf W.; Pietrafitta, Joseph J., eds.)V1421(1991)

Laser surgery for selected small animal soft-tissue conditions (Bartels, Kenneth E.)V1424,164-170(1991)

Laser-Tissue Interaction II (Jacques, Steven L., ed.)V1427(1991)

Light absorption characteristics of the human meniscus: applications for laser ablation (Vangsness, C. T.; Huang, Jay; Smith, Chadwick F.)V1424,16-19(1991)

Light distributor for endoscopic photochemotherapy (Lenz, P.)V1525,192-195(1991)

Light dosimetry in vivo in interstitial photodynamic therapy of human tumors (Reynes, Anne M.; Diebold, Simon; Lignon, Dominique; Granjon, Yves; Guillemin, Francois)V1525,177-182(1991)

Local interstitial hyperthermia in malignant brain tumors using a low-power Nd:YAG laser (Bettag, Martin; Ulrich, Frank; Kahn, Thomas; Seitz, R.)V1525,409-411(1991)

Mathematical modeling for laser PUVA treatment of psoriasis (Medvedev, Boris A.; Tuchin, Valery V.; Yaroslavsky, Ilya V.)V1422,73-84(1991)

Mathematical model of laser PUVA psoriasis treatment (Medvedev, Boris A.; Tuchin, Valery V.; Yaroslavsky, Ilya V.)V1403,682-685(1991)

Mathematical models of laser/tissue interactions for treatment and diagnosis in ophthalmology (Zheltov, Georgi I.; Glazkov, V. N.; Kirkovsky, A. N.; Podol'tsev, A. S.)V1403,752-753(1991)

Measurement of temperature distributions after pulsed IR radiation impact in biological tissue models with fluorescent thin films (Romano, Valerio; Greber, Charlotte M.; Frenz, Martin; Forrer, Martin; Weber, Heinz P.)V1427,16-26(1991)

Mechanism of injurious effect of excimer (308 nm) laser on the cell (Nevorotin, Alexey J.; Kallikorm, A. P.; Zeltzer, G. L.; Kull, Mart M.; Mihkelsoo, V. T.)V1427,381-397(1991)

Mechanisms of ultraviolet and mid-infrared tissue ablation (Hibst, Raimund)V1525,162-169(1991)

Microlocalization of Photofrin in neoplastic lesions (Korbelik, Mladen; Krosl, Gorazd; Lam, Stephen; Chaplin, David J.; Palcic, Branko)V1426,172-179(1991)

Multiphoton photobiologic effects (Hasan, Tayyaba)V1427,70-76(1991)

NADH-fluorescence in medical diagnostics: first experimental results (Schramm, Werner; Naundorf, M.)V1525,237-241(1991)

Naphthalocyanines relevant to the search for second-generation PDT sensitizers (Sounik, James R.; Rihter, Boris D.; Ford, William E.; Rodgers, Michael A.; Kenney, Malcolm E.)V1426,340-349(1991)

Nd:YAG laser irradiation in the treatment of upper tract urothelial tumors (Gaboardi, Franco; Melodia, Tommaso; Bozzola, Andrea; Galli, Luigi)V1421,79-85(1991)

New approaches to local destruction of tumors: interstitial laser hyperthermia and photodynamic therapy (Bown, Stephen G.)V1525,325-330(1991)

New sensitizers for PDT (Morgan, Alan R.; Garbo, Greta M.; Krivak, T.; Mastroianni, Marta; Petousis, Nikolaos H.; St Clair, T.; Weisenberger, M.; van Lier, Johan E.)V1426,350-355(1991)

Nile Blue derivatives as lysosomotropic photosensitizers (Lin, Chi-Wei; Shulok, Janine R.; Kirley, S. D.; Cincotta, Louis; Foley, James W.)V1426,216-227(1991)

Nonlinear optical effects under laser pulse interaction with tissues (Altshuler, Grigori B.; Belikov, Andrey V.; Erofeev, Andrey V.)V1427,141-150(1991)

Ophthalmic Technologies (Puliafito, Carmen A., ed.)V1423(1991)

Optical Methods for Tumor Treatment and Early Diagnosis: Mechanisms and Techniques (Dougherty, Thomas J., ed.)V1426(1991)

PDT of rat mammary adenocarcinoma in vitro and in a rat dorsal-skin-flap window chamber using Photofrin and chloroaluminum-sulfonated phthalocyanine (Flock, Stephen T.; Jacques, Steven L.; Small, Susan M.; Stern, Scott J.)V1427,77-89(1991)

Peak pressures and temperatures within laser-ablated tissues (Furzikov, Nickolay P.; Dmitriev, A. C.; Lekhtsier, Eugeny N.; Orlov, M. Y.; Semyenov, Alexander D.; Tyurin, Vladimir S.)V1427,288-297(1991)

Perspectives of powerful laser technique for medicine (Konov, Vitaly I.; Prokhorov, Alexander M.; Shcherbakov, I. A.)V1525,250-252(1991)

Pharmacokinetics of Photofrin II distribution in man (Kessel, David; Nseyo, Unyime O.; Schulz, Veronique; Sykes, Elizabeth)V1426,180-187(1991)

Photoablation (Doerschel, Klaus; Mueller, Gerhard J.)V1525,253-279(1991)

Photodynamic therapy in the prophylactic management of bladder cancer (Nseyo, Unyime O.; Lundahl, Scott L.; Merrill, Daniel C.)V1426,287-292(1991)

Photodynamic therapy of malignant brain tumors: supplementary postoperative light delivery by implanted optical fibers: field fractionation (Muller, Paul J.; Wilson, Brian C.)V1426,254-265(1991)

Photodynamic therapy of pet animals with spontaneously occurring head and neck carcinomas (Beck, Elsa R.; Hetzel, Fred W.)V1426,311-315(1991)

Photodynamic treatment of lens epithelial cells for cataract surgery (Lingua, Robert W.; Parel, Jean-Marie; Simon, Gabriel; Li, Kam)V1423,58-61(1991)

Photoproduct formation of endogeneous protoporphyrin and its photodynamic activity (Koenig, Karsten; Schneckenburger, Herbert; Rueck, Angelika C.; Auchter, S.)V1525,412-419(1991)

Phthalocyanines as phototherapeutic agents for tumors (Jori, Guilio; Reddi, Elena; Biolo, Roberta; Polo, Laura; Valduga, Giuliana)V1525,367-376(1991)

Physical processes of laser tissue ablation (Furzikov, Nickolay P.)V1403,764-775(1991)

Practical and theoretical considerations on the use of quantitative histology to measure thermal damage of collagen in cardiovascular tissues (Thomsen, Sharon L.)V1425,110-113(1991)

Preliminary evaluation of collagen as a component in the thermally induced 'weld' (Lemole, G. M.; Anderson, R. R.; DeCoste, Sue)V1422,116-122(1991)

Preliminary experience with laser reinforcement of vascular anastomoses (Oz, Mehmet C.; Bass, Lawrence S.; Williams, Mathew R.; Benvenisty, Alan I.; Hardy, Mark A.; Libutti, Steven K.; Eaton, Alexander M.; Treat, Michael R.; Nowygrod, Roman)V1422,147-150(1991)

Preliminary results of laser-assisted sealing of hand-sewn canine esophageal anastomoses (Auteri, Joseph S.; Oz, Mehmet C.; Sanchez, Juan A.; Bass, Lawrence S.; Jeevanandam, Valluvan; Williams, Mathew R.; Smith, Craig R.; Treat, Michael R.)V1421,144-146(1991)

Preliminary results with sutured colonic anastomoses reinforced with dye-enhanced fibrinogen and a diode laser (Libutti, Steven K.; Williams, Mathew R.; Oz, Mehmet C.; Forde, Kenneth A.; Bass, Lawrence S.; Weinstein, Samuel; Auteri, Joseph S.; Treat, Michael R.; Nowygrod, Roman)V1421,169-172(1991)

Preliminary stress/strain analysis of laser-soldered and -sutured vascular tissue (Ashton, Robert C.; Oz, Mehmet C.; Lontz, John F.; Lemole, Gerald M.)V1422,151-155(1991)

Processing of bioceramic implants with excimer laser (Koort, Hans J.)V1427,173-180(1991)

Promotional effects of CO2 laser on neoplastic lesions in hamsters (Kingsbury, Jeffrey S.; Margarone, Joseph E.; Satchidanand, S.; Liebow, Charles)V1427,363-367(1991)

Pulsed CO2 laser for intra-articular cartilage vaporization and subchondral bone perforation in horses (Nixon, Alan J.; Roth, Jerry E.; Krook, Lennart)V1424,198-208(1991)

Quantitative analysis at the molecular level of laser/neural tissue interactions using a liposome model system (VanderMeulen, David L.; Misra, Prabhakar; Michael, Jason; Spears, Kenneth G.; Khoka, Mustafa)V1428,91-98(1991)

Radiolabeled red blood cells for the direct measurement of the blood flow kinetics in experimental tumors after photodynamic therapy (Paquette, Benoit; Rousseau, Jacques; Ouellet, Rene; van Lier, Johan E.)V1426,362-366(1991)

Raman and FT-IR studies of ocular tissues (Ozaki, Yukihiro; Mizuno, Aritake)V1403,710-719(1991)

Rational anatomical treatment of basal cell carcinoma with photodynamic therapy (Keller, Gregory S.)V1426,266-270(1991)

Reduction of acoustic transients in tissue with a 2 um thulium laser (Pinto, Joseph F.; Esterowitz, Leon; Bonner, Robert F.)V1420,242-243(1991)

Remote Raman spectroscopic imaging of human artery wall (Phat, Darith; Vuong, Phat N.; Plaza, Pascal; Cheilan, Francis; Dao, Nguyen Q.)V1525,196-205(1991)

Role of various wavelengths in phototherapy (Lubart, Rachel; Wollman, Yoram; Friedmann, Harry; Rochkind, Simeone; Laulicht, Israel)V1422,140-146(1991)

Selective tumor destruction with photodynamic therapy: exploitation of photodynamic thresholds (Barr, Hugh)V1525,331-340(1991)

Simple analytical model for laser-induced tissue ablation (Kukreja, Lalit M.; Braun, R.; Hess, P.)V1427,243-254(1991)

Six years of transendoscopic Nd:YAG application in large animals (Tate, Lloyd P.; Glasser, Mardi)V1424,209-217(1991)

Skin closure with dye-enhanced laser welding and fibrinogen (Wider, Todd M.; Libutti, Steven K.; Greenwald, Daniel P.; Oz, Mehmet C.; Yager, Jeffrey S.; Treat, Michael R.; Hugo, Norman E.)V1422,56-61(1991)

Spectral-luminescence analysis as method of tissue states evaluation and laser influence on tissues before and after transplantation (Loshchenov, V. B.; Baryshev, M. V.; Svystushkin, V. M.; Ovchinnikov, U. M.; Babyn, A. V.; Schaldach, Brita J.; Mueller, Gerhard J.)V1420,271-281(1991)

Spectroscopic studies of second-generation sensitizers and their photochemical reactions in liposomes and cells (Ehrenberg, Benjamin; Gross, Eitan; Lavi, Adina; Johnson, Fred M.; Malik, Zvi)V1426,244-251(1991)

Stage III endobronchial squamous cell cancer: survival after Nd:YAG laser combined with photodynamic therapy versus Nd:YAG laser or photodynamic therapy alone (McCaughan, James S.; Barabash, Rostislav D.; Hawley, Philip)V1426,279-287(1991)

Stimulatory effects of Nd:YAG lasers on canine articular cartilage (Lane, Gregory J.; Sherk, Henry H.; Kollmer, Charles; Uppal, Gurvinder S.; Rhodes, Anthony; Sazy, John; Black, Johnathan D.; Lee, Steven)V1424,7-11(1991)

Studies on the absence of photodynamic mechanism in the normal pancreas (Mang, Thomas S.; Wieman, T. J.; Crean, David H.)V1426,188-199(1991)

Study of bone ablation dynamics with sequenced pulses (Izatt, Joseph A.; Albagli, Douglas; Itzkan, Irving; Feld, Michael S.)V1427,110-116(1991)

Sutureless cataract incision closure using laser-activated tissue glues (Eaton, Alexander M.; Bass, Lawrence S.; Libutti, Steven K.; Schubert, Herman D.; Treat, Michael R.)V1423,52-57(1991)

Syntheses of porphyrin and chlorin dimers for photodynamic therapy (Pandey, Ravindra K.; Vicente, M. G.; Shiau, Fuu-Yau; Dougherty, Thomas J.; Smith, Kevin M.)V1426,356-361(1991)

Synthetic approaches to long-wavelength photosensitizers for photodynamic therapy (Shiau, Fuu-Yau; Pandey, Ravindra K.; Dougherty, Thomas J.; Smith, Kevin M.)V1426,330-339(1991)

Temperature distributions in laser-irradiated tissues (Valderrama, Giuseppe L.; Fredin, Leif G.; Berry, Michael J.; Dempsey, B. P.; Harpole, George M.)V1427,200-213(1991)

Temperature response of biological tissues to nonablative pulsed CO2 laser irradiation (van Gemert, Martin J.; Brugmans, Marco J.; Gijsbers, Geert H.; Kemper, J.; van der Meulen, F. W.; Nijdam, D. C.)V1427,316-319(1991)

Temperature variations of reflection, transmission, and fluorescence of the arterial wall (Chambettaz, Francois; Clivaz, Xavier; Marquis-Weible, Fabienne D.; Salathe, R. P.)V1427,134-140(1991)

Time-resolved polarization luminescence spectroscopy of hematoporphyrin in liposomes (Chernyaeva, E. B.; Golubeva, N. A.; Koroteev, Nikolai I.; Lobanov, O. V.; Vardanyan, A. G.)V1402,7-10(1991)

Time-resolved transillumination for medical diagnostics (Berg, Roger; Andersson-Engels, Stefan; Jarlman, Olof; Svanberg, Sune)V1431,110-119(1991)

Tissue ablation by holmium:YSGG laser pulses through saline and blood (van Leeuwen, Ton G.; van der Veen, Maurits J.; Verdaasdonk, Rudolf M.; Borst, Cornelius)V1427,214-219(1991)

Tissue interactions of ball-tipped and multifiber catheters (Mitchell, David C.; Smith, Roy E.; Walters, Tena K.; Murray, Alan; Wood, Richard F.)V1427,181-188(1991)

Tooth pulp reaction following Er:YAG laser application (Keller, Ulrich; Hibst, Raimund)V1424,127-133(1991)

Toward an explanation of laser-induced stimulation and damage of cell cultures (Friedmann, Harry; Lubart, Rachel; Laulicht, Israel; Rochkind, Simeone)V1427,357-362(1991)

Tripyrroledimethine-derived ("texaphyrin"-type) macrocycles: potential photosensitizers which absorb in the far-red spectral region (Sessler, Jonathan L.; Hemmi, Gregory; Maiya, Bhaskar G.; Harriman, Anthony; Judy, Millard M.; Boriak, Richard; Matthews, James L.; Ehrenberg, Benjamin; Malik, Zvi; Nitzan, Yeshayahu; Rueck, Angelika C.)V1426,318-329(1991)

Use of chloro-aluminum sulfonated phthalocyanine as a photosensitizer in the treatment of malignant tumors in dogs and cats (Peavy, George M.; Klein, Mary K.; Newman, H. C.; Roberts, Walter G.; Berns, Michael W.)V1424,171-178(1991)

Use of excimer lasers in medicine: applications, problems, and dangers (Mommsen, Jens; Stuermer, Martin)V1503,348-354(1991)

Use of polarization to separate on-axis scattered and unscattered light in red blood cells (Sardar, Dhiraj K.; Nemati, Babak; Barrera, Frederick J.)V1427,374-380(1991)

Vascular tissue welding of the CO2 laser: limitations (Dalsing, Michael C.; Kruepper, Peter; Cikrit, Dolores F.)V1422,98-102(1991)

Wavelength selection in laser arthroscopy (Black, Johnathan D.; Sherk, Henry H.; Meller, Menachem M.; Uppal, Gurvinder S.; Divan, James; Sazy, John; Rhodes, Anthony; Lane, Gregory J.)V1424,12-15(1991)

Lasers, x-ray

Development of an XUV-IR free-electron laser user facility for scientific research and industrial applications (Newnam, Brian E.; Warren, Roger W.; Conradson, Steven D.; Goldstein, John C.; McVey, Brian D.; Schmitt, Mark J.; Elliott, C. J.; Burns, M. J.; Carlsten, Bruce E.; Chan, Kwok-Chi D.; Johnson, W. J.; Wang, Tai-San; Sheffield, Richard L.; Meier, Karl L.; Olsher, R. H.; Scott, Marion L.; Griggs, J. E.)V1552,154-174(1991)

Experiments with SPRITE 12 ps facility (Tallents, Gregory J.; Key, Michael H.; Norreys, P.; Jacoby, J.; Kodama, R.; Tragin, N.; Baldis, Hector A.; Dunn, James; Brown, D.)V1413,70-76(1991)

New opportunities with intense ultra-short-pulse lasers (Richardson, Martin C.)V1410,15-25(1991)

Optical diagnostics of line-focused laser-produced plasmas (Lin, Li-Huang; Chen, Shisheng; Jiang, Z. M.; Ge, Wen; Qian, Aidi D.; Bin, Ouyang; Li, Yongchun L.; Kang, Yilan; Xu, Zhizhan)V1346,490-501(1991)

Optically ionized plasma recombination x-ray lasers (Amendt, Peter; Eder, David C.; Wilks, S. C.; Dunning, M. J.; Keane, Christopher J.)V1413,59-69(1991)

Theory and simulation of Raman scattering in intense short-pulse laser-plasma interactions (Wilks, S. C.; Kruer, William L.; Langdon, A. B.; Amendt, Peter; Eder, David C.; Keane, Christopher J.)V1413,131-137(1991)

Transition radiation in foil stack and x-ray laser (Yan, Zu Q.; Ruan, Ke F.)V1519,183-191(1991)

Tutorial on x-ray lasers (Silfvast, William T.)V1397,3-8(1991)

Lenses— see also aspherics; binary optics; holographic optical elements; micro-optics; optical design

1990 International Lens Design Conference lens design problems: the design of a NonLens (Clark, Peter P.; Londono, Carmina)V1354,555-569(1991)

1990 Intl Lens Design Conf (Lawrence, George N., ed.)V1354(1991)

Absolute measurement of spherical surfaces (Creath, Katherine; Wyant, James C.)V1332,2-7(1991)

Accurate method for measuring oblique astigmatism and oblique power of ophthalmic lenses (Wihardjo, Erning; Silva, Donald E.)V1529,57-62(1991)

Achromatization of optical waveguide components (Spaulding, Kevin E.; Morris, G. M.)V1507,45-54(1991)

Analysis on image performance of a moisture-absorbed plastic singlet for an optical disk (Tanaka, Yuki; Miyamae, Hiroshi)V1354,395-401(1991)

Apo-tele lenses with kinoform elements (Gan, Michael A.; Potyemin, Igor S.; Poszinskaja, Irina I.)V1507,116-125(1991)

Athermalized FLIR optics (Rogers, Philip J.)V1354,742-751(1991)

Attempt to develop a zoom-lens-design expert system (Weng, Zhicheng; Chen, Zhiyong; Yang, Yu-Hong; Ren, Tao; Cong, Xiaojie; Yao, Yuchuan; He, Fengling; Li, Yuan-Yuan)V1527,349-356(1991)

Automatic inspection technique for optical surface flaws (Yang, Guoguang; Gao, Wenliang; Cheng, Shangyi)V1332,56-63(1991)

Binary optics from a ray-tracing point of view (Southwell, William H.)V1354,38-42(1991)

Binary optics in lens design (Kathman, Alan D.; Pitalo, Stephen K.)V1354,297-309(1991)

Binary optics in the '90s (Gallagher, Neal C.)V1396,722-733(1991)

Blazed zone plates for the 10-um spectral region (Hutley, M. C.)V1574,2-7(1991)

Calculation method of the Strehl Definition for decentral optical systems (Zhuang, Song Lin; Chen, Huai'an)V1354,252-253(1991)

Calculation of diffraction efficiencies for spherical and cylindrical holographic lenses (Defosse, Yves; Renotte, Yvon L.; Lion, Yves F.)V1507,277-287(1991)

Calculation of wave aberration in optical systems with holographic optical elements (Yuan, Yanrong)V1354,43-52(1991)

Capabilities of simple lenses in a free-space perfect shuffle (Miller, Andrew S.; Sawchuk, Alexander A.)V1563,81-92(1991)

Color vision enhancement with spectacles (Perrott, Colin M.)V1529,31-36(1991)

Compact nonimaging lens with totally internally reflecting facets (Parkyn, William A.; Pelka, David G.)V1528,70-81(1991)

Compact zoom lens for stereoscopic television (Scheiwiller, Peter M.; Murphy, S. P.; Dumbreck, Andrew A.)V1457,2-8(1991)

Computing illumination-bundle focusing by lens systems (Forkner, John F.)V1354,210-215(1991)

Cone array of Fourier lenses for contouring applications (Goldfain, Ervin)V1527,210-215(1991)

Corneal and retinal energy density with various laser beam delivery systems and contact lenses (Dewey, David)V1423,105-116(1991)

Correction of secondary and higher-order spectrum using special materials (Mercado, Romeo I.)V1535,184-198(1991)

Corrections of aberrations using HOEs in UV and visible imaging systems (Yang, Zishao; Rosenbruch, Klaus J.)V1354,323-327(1991)

Current Developments in Optical Design and Optical Engineering (Fischer, Robert E.; Smith, Warren J., eds.)V1527(1991)

Design characteristics and point-spread function evaluation of bifocal intraocular lenses (Rol, Pascal O.)V1423,15-19(1991)

Designing apochromatic telescope objectives with liquid lenses (Sigler, Robert D.)V1535,89-112(1991)

Design of a GaAs Fresnel lens array for optical signal-processing devices (Armenise, Mario N.; Impagnatiello, Fabrizio; Passaro, Vittorio M.)V1374,86-96(1991)

Design of achromatized hybrid diffractive lens systems (Londono, Carmina; Clark, Peter P.)V1354,30-37(1991)

Design of an IR non-lens, or how I buried 100 mm of germanium (Aikens, David M.)V1485,183-194(1991)

Design of double-Gauss objective by using concentric lenses (Juang, Jeng-Dang; Chang, Ming-Wen)V1354,273-276(1991)

Design of high-performance aplanatic achromats for the near-ultraviolet waveband (Al-Baho, Tareq I.; Learner, R. C.; Maxwell, Jonathan)V1354,417-428(1991)

Design of nonimaging lenses and lens-mirror combinations (Minano, Juan C.; Gonzalez, Juan C.)V1528,104-115(1991)

Design of some achromatic imaging hybrid diffractive-refractive lenses (Twardowski, Patrice J.; Meyrueis, Patrick)V1507,55-65(1991)

Design of two- and three-element diffractive telescopes (Buralli, Dale A.; Morris, G. M.)V1354,292-296(1991)

Designs of two-glass apochromats and superachromats (Mercado, Romeo I.)V1354,262-272(1991)

Developments in projection lenses for HDTV (Rudolph, John D.)V1456,15-28(1991)

Differential coating objective (Goldfain, Ervin)V1527,126-133(1991)

Diffraction performance calculations in lens design (Malacara Hernandez, Daniel)V1354,2-14(1991)

Diffractive doublet corrected on-axis at two wavelengths (Farn, Michael W.; Goodman, Joseph W.)V1354,24-29(1991)

Diffractive optics for x rays: the state of the art (Michette, Alan G.)V1574,8-21(1991)

Effect of coma correction on the imaging quality of holographic lenses (Zajac, Marek; Nowak, Jerzy)V1507,73-80(1991)

Electron-beam-written reflection diffractive microlenses for oblique incidence (Shiono, Teruhiro; Ogawa, Hisahito)V1545,232-240(1991)

Electro-optical autostereoscopic displays using large cylindrical lenses (Hattori, Tomohiko)V1457,283-289(1991)

Electrostatic focusing and RFQ (radio frequency quadrupole) matching system for a low-energy H- beam (Guharay, Samar K.; Allen, C. K.; Reiser, Martin P.)V1407,610-619(1991)

Endoscopic inspection using a panoramic annular lens (Matthys, Donald R.; Gilbert, John A.; Puliparambil, Joseph T.)V1554B,736-742(1991)

Enhancement of Conrady's "D-d" method (Gintner, Henry)V1354,97-102(1991)

Exact sine condition in the presence of spherical aberration (Shibuya, Masato)V1354,240-245(1991)

Examples of the topographies of the wavefront-variance merit function at different aberration orders (Johnston, Steve C.)V1354,77-82(1991)

Existence of local minima in lens design (Kidger, Michael J.; Leamy, Paul T.)V1354,69-76(1991)

Experimental study of electromagnetic missiles from a hyperboloidal lens (Shen, Hao-Ming; Wu, Tai T.; Myers, John M.)V1407,286-294(1991)

Expert systems in lens design (Dilworth, Donald C.)V1354,359-370(1991)

Fabrication and characterization of semiconductor microlens arrays (Diadiuk, Vicky; Liau, Zong-Long; Walpole, James N.)V1354,496-500(1991)

Fabrication and performance of a one-to-one erect imaging microlens array for fax (Bellman, Robert H.; Borrelli, Nicholas F.; Mann, L. G.; Quintal, J. M.)V1544,209-217(1991)

Fabrication of microlenses by laser-assisted chemical etching (Gratrix, Edward J.; Zarowin, Charles B.)V1544,238-243(1991)

Fast, inexpensive, diffraction-limited cylindrical microlenses (Snyder, James J.; Reichert, Patrick)V1544,146-151(1991)

Fiber probe for ring pattern (Yang, Xueyu; Spears, Kenneth G.)V1367,382-386(1991)

Finite element analysis of large lenses for the Keck telescope high-resolution echelle spectrograph (Bigelow, Bruce C.)V1532,15-26(1991)

Focal length measurement using diffraction at a grating (Sirohi, Rajpal S.; Kumar, Harish; Jain, Narinder K.)V1332,50-55(1991)

Focus considerations with high-numerical-aperture widefield lenses (Leebrick, David H.)V1463,275-280(1991)

Formation of optical elements by photothermo-induced crystallization of glass (Glebov, Leonid B.; Nikonorov, Nikolai V.; Petrovskii, Gurii T.; Kharchenko, Mikhail V.)V1440,24-35(1991)

Future of global optimization in optical design (Sturlesi, Doron; O'Shea, Donald C.)V1354,54-68(1991)

Global optimization using the y-ybar diagram (Brown, Daniel M.)V1527,19-25(1991)

Global status of diffraction optics as the basis for an intraocular lens (Isaacson, William B.)V1529,71-83(1991)

GRIN fiber lens connectors (Gomez-Reino, Carlos; Linares, Jesus)V1332,468-473(1991)

High-speed binary optic microlens array in GaAs (Motamedi, M. E.; Southwell, William H.; Anderson, Robert J.; Hale, Leonard G.; Gunning, William J.; Holz, Michael)V1544,33-44(1991)

Holo-interferometric patterns recorded through a panoramic annular lens (Gilbert, John A.; Greguss, Pal; Kransteuber, Amy S.)V1238,412-420(1991)

Homogeneous thin film lens on LiNbO3 (Jiang, Pisu; Laybourn, Peter J.; Righini, Giancarlo C.)V1362,899-906(1991)

Image quality evaluation of multifocal intraocular lenses (Silberman, Donn M.)V1423,20-28(1991)

Improved plastic molding technology for ophthalmic lens and contact lens (Galic, George J.; Maus, Steven)V1529,13-21(1991)

Infrared zoom lenses in the eighties and beyond (Mann, Allen)V1540,338-349(1991)

Interdependence of design, optical evaluation, and visual performance of ophthalmic lenses (Freeman, Michael H.)V1529,2-12(1991)

Investigation of fringing fields in liquid-crystal devices (Powell, Norman J.; Kelsall, Robert W.; Love, G. D.; Purvis, Alan)V1545,19-30(1991)

IR objective with internal scan mirror (Eisenberg, Shai; Menache, Ram)V1442,133-138(1991)

Key technologies for IR zoom lenses: aspherics and athermalization (Nory, Pierre)VCR38,142-152(1991)

Kinoforms with increased depth of focus (Koronkevich, Voldemar P.; Palchikova, Irena G.)V1507,110-115(1991)

Large one-step holographic stereogram (Honda, Toshio; Kang, Der-Kuan; Shimura, Kei; Enomoto, H.; Yamaguchi, Masahiro; Ohyama, Nagaaki)V1461,156-166(1991)

Laser ablation of refractive micro-optic lenslet arrays (Bartley, James A.; Goltsos, William)V1544,140-145(1991)

Lens design for the infrared (Fischer, Robert E.)VCR38,19-43(1991)

Lens evaluation for electronic photography (Bell, Cynthia S.)V1448,59-68(1991)

Lens for microlithography (Hsieh, Hung-yu; Wagner, Jerome F.)V1396,467-472(1991)

Lens with maximum power occupying a given cylindrical volume (Reardon, Patrick J.; Chipman, Russell A.)V1354,234-239(1991)

Light coupling characteristics of planar microlens (Oikawa, Masahiro; Nemoto, Hiroyuki; Hamanaka, Kenjiro; Imanishi, Hideki; Kishimoto, Takashi)V1544,226-237(1991)

Linearized ray-trace analysis (Redding, David C.; Breckenridge, William G.)V1354,216-221(1991)

Low-cost and compact fiber-to-laser coupling with micro-Fresnel lens (Ogata, Shiro; Yoneda, Masahiro; Maeda, Tetsuo; Imanaka, Koichi)V1544,92-100(1991)

Measurement of wave aberrations of intraocular lenses through holographic interferometry (Carretero, L.; Fuentes, Rosa; Fimia-Gil, Antonio)V1507,458-462(1991); V1508,96-100(1991)

Micrograting device using electron-beam lithography (Yamashita, Tsukasa; Aoyama, Shigeru)V1507,81-93(1991)

Microlens array fabricated in surface relief with high numerical aperture (Lau, Hon W.; Davies, Neil A.; McCormick, Malcolm)V1544,178-188(1991)

Microlens array for modification of SLM devices (Kathman, Alan D.; Temmen, Mark G.; Scott, Miles L.)V1544,58-65(1991)

Microlens array for staring infrared imager (Werner, Thomas R.; Cox, James A.; Swanson, S.; Holz, Michael)V1544,46-57(1991)

Microlens arrays in integral photography and optical metrology (Davies, Neil A.; McCormick, Malcolm; Lau, Hon W.)V1544,189-198(1991)

Micro-optic lens for data storage (Milster, Tom D.; Trusty, Robert M.; Wang, Mark S.; Froehlich, Fred F.; Erwin, J. K.)V1499,286-292(1991)

Micro-optic studies using photopolymers (Phillips, Nicholas J.; Barnett, Christopher A.)V1544,10-21(1991)

Miniature and Micro-Optics: Fabrication and System Applications (Roychoudhuri, Chandrasekhar; Veldkamp, Wilfrid B., eds.)V1544(1991)

Moire measurements using a panoramic annular lens (Gilbert, John A.; Matthys, Donald R.; Lehner, David L.)V1554B,202-209(1991)

Monochromatic quartet: a search for the global optimum (O'Shea, Donald C.)V1354,548-554(1991)

Moulding process for contact lens (Skipper, Richard S.; Shepherd, David W.)V1529,22-30(1991)

MTF optimization in lens design (Rimmer, Matthew P.; Bruegge, Thomas J.; Kuper, Thomas G.)V1354,83-91(1991)

New family of 1:1 catadioptric broadband deep-UV high-Na lithography lenses (Zhang, Yudong; Zou, Haixing; Wang, Zhi-Jiang)V1463,688-694(1991)

New i-line lens for half-micron lithography (Takahashi, Kazuhiro; Ohta, Masakatsu; Kojima, Toshiyuki; Noguchi, Miyoko)V1463,696-708(1991)

New method for detection of blood coagulation using fiber-optic sensor (Fediay, Sergey G.; Kuznetzov, Alexsey V.)V1420,41-43(1991)

New methods for economic production of prisms and lenses (Richter, G.)V1400,11-23(1991)

Nonlinearity and lens design (Tatian, Berge)V1354,154-164(1991)

Nonlinear model of the optimal tolerance design for a lens system (Zhuang, Song Lin; Qu, Zhijin)V1354,177-179(1991)

Opthalmic Lens Design and Fabrication (Perrott, Colin M., ed.)V1529(1991)

Ophthalmic lenses testing by moire deflectometry (Yonte, T.; Quiroga, J.; Alda, J.; Bernabeu, Eusebio)V1554B,233-241(1991)

Optical design and optimization with physical optics (Lawrence, George N.; Moore, Kenneth E.)V1354,15-22(1991)

Optical design of dual-combiner head-up displays (Kirkham, Anthony J.)V1354,310-315(1991)

Optical design of high-aperture aspherical projection lens (Osawa, Atsuo; Fukuda, Kyohei; Hirata, Kouji)V1354,337-343(1991)

Optical properties of diffractive, bifocal intraocular lenses (Larsson, Michael; Beckman, Claes; Nystrom, Alf; Hard, Sverker; Sjostrand, Johan)V1529,63-70(1991)

Optics for vector scanning (Ehrmann, Jonathan S.)V1454,245-256(1991)

Optimization of athermal systems (Benham, Paul; Kidger, Michael J.)V1354,120-125(1991)

Optimization of original lens structure type from optical lens data base (Yang, Daren; Jiang, Huilin; Li, Gongde; Yang, Huamin)V1527,456-461(1991)

Optimization of the optical transfer function (Kidger, Michael J.; Benham, Paul)V1354,92-96(1991)

Optimization of the Seidel image errors by bending of lenses using a 4th-degree merit function (Aurin, Friedrich A.)V1354,180-185(1991)

Overview of CODE V development (Harris, Thomas I.)V1354,104-111(1991)

Passive Materials for Optical Elements (Wilkerson, Gary W., ed.)V1535(1991)

Performance and measurements of refractive microlens arrays (Feldblum, Avi Y.; Nijander, Casimir R.; Townsend, Wesley P.; Mayer-Costa, Carlos M.)V1544,200-208(1991)

Photodynamic treatment of lens epithelial cells for cataract surgery (Lingua, Robert W.; Parel, Jean-Marie; Simon, Gabriel; Li, Kam)V1423,58-61(1991)

Plastic lens array with the function of forming unit magnification erect image using roof prisms (Takahashi, Eietsu; Tanji, Shigeo; Tanaka, Akira; Hayashi, Yuji)V1527,145-154(1991)

Practical design considerations and performance characteristics of high-numerical-aperture holographic lenses (Kostuk, Raymond K.)V1461,24-34(1991)

Practical use of generalized simulated annealing optimization on microcomputers (Hearn, Gregory K.)V1354,186-191(1991)

Principles of optimization in lens design developed at the Institute of Optics in Berlin (West) (Kross, Juergen)V1354,165-170(1991)

Process-dependent kinoform performance (Cox, James A.; Fritz, Bernard S.; Werner, Thomas R.)V1507,100-109(1991)

Quadric surfaces: some derivations and applications (Larkin, Eric W.)V1354,222-231(1991)

Rank-down method for automatic lens design (Ooki, Hiroshi)V1354,171-176(1991)

Ray tracing through progressive ophthalmic lenses (Bourdoncle, Bernard; Chauveau, J. P.; Mercier, Jean-Louis M.)V1354,194-199(1991)

Reactive low-voltage ion plating of optical coatings on ophthalmic lenses (Balasubramanian, Kunjithapa; Richmond, Jeff; Hu, X. Q.; Guenther, Karl H.)V1529,106-114(1991)

Real-time wavefront measurement with lambda/10 fringe spacing for the optical shop (Freischlad, Klaus R.; Kuechel, Michael F.; Schuster, Karl-Heinz; Wegmann, Ulrich; Kaiser, Winfried)V1332,18-24(1991)

Reluctant glass formers and their applications in lens design (Johnson, R. B.; Feng, Chen; Ethridge, Edwin C.)V1527,2-18(1991); V1535,231-247(1991)

Rigorous electromagnetic modeling of diffractive optical elements (Johnson, Eric G.; Kathman, Alan D.)V1545,209-216(1991)

Role of aspherics in zoom lens design (Betensky, Ellis I.)V1354,656-662(1991)**Selected Papers on Holographic and Diffractive Lenses and Mirrors** (Stone, Thomas W.; Thompson, Brian J., eds.)VMS34(1991)

Shear-lens photography for holographic stereograms (Molteni, William J.)V1461,132-141(1991)

Silicon microlenses for enhanced optical coupling to silicon focal planes (Motamedi, M. E.; Griswold, Marsden P.; Knowlden, Robert E.)V1544,22-32(1991)

Simple high-precision extinction method for measuring refractive index of transparent materials (Nee, Soe-Mie F.; Bennett, Harold E.)V1441,31-37(1991)

Simple method for finding the valid ray-surface intersection (Freeman, David E.)V1354,200-209(1991)

Simple multidimensional quadratic extrapolation method for the correction of specific aberrations in lens systems (Maxwell, Jonathan; Hull, Chris S.)V1354,277-285(1991)

Small-computer program for optical design and analysis written in "C" (Beckmann, Leo H.)V1354,254-261(1991)

Small-signal gain for parabolic profile beams in free-electron lasers (Elliott, C. J.)V1552,175-181(1991)

Sola ASL in Spectralite and polycarbonate aspheric lens designs (Machol, Steven)V1529,45-56(1991)

Sola ASL in Spectralite strikes the perfect balance between cosmetics and optics (Machol, Steven; Modglin, Luan)V1529,38-44(1991)

Stability considerations in relay lens design for optical communications (Gardam, Allan; Jonas, Reginald P.)V1417,381-390(1991)

Stray-light reduction in a WFOV star tracker lens (Lewis, Isabella T.; Ledebuhr, Arno G.; Axelrod, Timothy S.; Ruddell, Scott A.)V1530,306-324(1991)

SYNOPSYS—a lens design computer program package (Dilworth, Donald C.)V1354,112-115(1991)

Tale of two underwater lenses (Evans, Clinton E.; Doshi, Rekha)V1537,203-214(1991)

Techniques for designing hybrid diffractive optical systems (Brown, Daniel M.; Pitalo, Stephen K.)V1527,73-84(1991)

Tertiary spectrum manipulation in apochromats (Maxwell, Jonathan)V1354,408-416(1991)

Testing binary optics: accurate high-precision efficiency measurements of microlens arrays in the visible (Holz, Michael; Stern, Margaret B.; Medeiros, Shirley; Knowlden, Robert E.)V1544,75-89(1991)

Theory of color correction by use of chromatic magnification (Ames, Alan J.)V1354,286-290(1991)

Theory of two-component zoom systems (Oskotsky, Mark L.)V1527,37-47(1991)

Thermal compensation of infrared achromatic objectives with three optical materials (Rayces, Juan L.; Lebich, Lan)V1354,752-759(1991)

Tilt-tolerant dual-band interferometer objective lens (Jones, Mike I.)V1498,158-162(1991)

Towards global optimization with adaptive simulated annealing (Forbes, Greg W.; Jones, Andrew E.)V1354,144-153(1991)

Two applications for microlens arrays: detector fill-factor improvement and laser diode collimation (D'Amato, Dante P.; Centamore, Robert M.)V1544,166-177(1991)

Ultra-high-performance zoom lens for the visible waveband (Neil, Iain A.)V1354,684-694(1991)

Use of dome (meniscus) lenses to eliminate birefringence and tensile stresses in spatial filters for the Nova laser (Pitts, John H.; Kyrazis, Demos T.; Seppala, Lynn G.; Bumpas, Stanley E.)V1441,71-81(1991)

Using hybrid refractive-diffractive elements in infrared Petzval objectives (Wood, Andrew P.)V1354,316-322(1991)

Video Hartmann wavefront diagnostic that incorporates a monolithic microlens array (Toeppen, John S.; Bliss, Erlan S.; Long, Theresa W.; Salmon, J. T.)V1544,218-225(1991)

Wave-optics analysis of fast-beam focusing (Shih, Chun-Ching)V1415,150-153(1991)

Why and how to coat ophthalmic lenses (Guenther, Karl H.)V1529,96-105(1991)

Zoom lens design using GRIN materials (Tsuchida, Hirofumi; Aoki, Norihiko; Hyakumura, Kazushi; Yamamoto, Kimiaki)V1354,246-251(1991)

Zoom lenses with a single moving element (Johnson, R. B.; Feng, Chen)V1354,676-683(1991)

Zoom lens with aspherical lens for camcorder (Yatsu, Masahiko; Deguchi, Masaharu; Maruyama, Takesuke)V1354,663-668(1991)

Lidar—see also Doppler effect; radar

Airborne lidar elastic scattering, fluorescent scattering, and differential absorption observations (Uthe, Edward E.)V1479,393-402(1991)

Airborne lidar measurements of ozone and aerosols in the summertime Arctic troposphere (Browell, Edward V.)V1491,7-14(1991)

Airborne lidar observations of ozone and aerosols in the wintertime Arctic stratosphere (Browell, Edward V.)V1491,273-281(1991)

Alexandrite laser characterization and airborne lidar developments for water vapor DIAL measurements (Ponsardin, Patrick; Higdon, Noah S.; Grossmann, Benoist E.; Browell, Edward V.)V1492,47-51(1991)

Analytic approach to centroid performance analysis (Schultz, Kenneth I.)V1416,199-208(1991)

Application of backscatter absorption gas imaging to the detection of chemicals related to drug production (Kulp, Thomas J.; Garvis, Darrel G.; Kennedy, Randall B.; McRae, Thomas G.)V1479,352-363(1991)

Application of laser radar to autonomous spacecraft landing (Gleichman, Kurt W.; Tchoryk, Peter; Sampson, Robert E.)V1416,286-294(1991)

ATLID: the first preoperational ATmospheric LIDar for the European polar platform (Lange, Robert; Endemann, Martin J.; Reiland, Werner; Krawczyk, Rodolphe; Hofer, Bruno)V1492,24-37(1991)

Atmospheric effects on laser systems (Au, Robert H.)V1399,8-15(1991)

Atmospheric laser-transmission tables simply generated (Mallory, William R.)V1540,359-364(1991)

BEST: a new satellite for a climatology study in the tropics (Orgeret, Marc)V1490,14-22(1991)

Centroid tracking of range-Doppler images (Kachelmyer, Alan L.; Nordquist, David P.)V1416,184-198(1991)

Coherent laser radar for target classification (Kranz, Wolfgang)V1479,270-274(1991)

Coherent launch-site atmospheric wind sounder (Targ, Russell; Hawley, James G.; Otto, Robert G.; Kavaya, Michael J.)V1478,211-227(1991)

Comparison of laser radar transmittance for the five atmospheric models (Au, Robert H.)V1487,280-290(1991)

Design and construction of a wideband efficient electro-optic modulator (Harris, Neville W.; Sobolewski, J. M.; Summers, Charles L.; Eng, Richard S.)V1416,59-69(1991)

Design and performance characteristics of a compact CO2 Doppler lidar transmitter (Pearson, Guy N.)V1416,147-150(1991)

Detection and parameter estimation of atmospheric turbulence by ground-based and airborne CO2 Doppler lidars (Pogosov, Gregory A.; Akhmanov, Sergei A.; Gordienko, Vyacheslav M.; Kosovsky, L. A.; Kurochkin, Nikolai N.; Priezzhev, Alexander V.)V1416,115-124(1991)

Development of 1- and 2-um coherent Doppler lidars for atmospheric sensing (Chan, Kin-Pui; Killinger, Dennis K.)V1492,111-114(1991)

Development of an airborne excimer-based UV-DIAL for monitoring ozone and sulfur dioxide in the lower troposphere (Bristow, Michael P.; Diebel, D. E.; Bundy, Donald H.; Edmonds, Curtis M.; Turner, Ruldopha M.; McElroy, James L.)V1491,68-74(1991)

Digital shearing laser interferometry for heterodyne array phasing (Cederquist, Jack N.; Fienup, James R.; Marron, Joseph C.; Schulz, Timothy J.; Seldin, J. H.)V1416,266-277(1991)

Dry/wet wire reflectivities at 10.6 um (Young, Donald S.; Osche, Gregory R.; Fisher, Kirk L.; Lok, Y. F.)V1416,221-228(1991)

Earth and Atmospheric Remote Sensing (Curran, Robert J.; Smith, James A.; Watson, Ken, eds.)V1492(1991)

Edge technique: a new method for atmospheric wind measurements with lidar (Korb, C. L.; Gentry, Bruce M.)V1416,177-182(1991)

Forward-looking IR and lidar atmospheric propagation in the infrared field program (Koenig, George G.; Bissonnette, Luc R.)V1487,240-249(1991)

Future technologies for lidar/DIAL remote sensing (Allario, Frank; Barnes, Norman P.; Storm, Mark E.)V1492,92-110(1991)

GLOBE backscatter: climatologies and mission results (Menzies, Robert T.; Post, Madison J.)V1416,139-146(1991)

Ground-based CW atmospheric Doppler performance modeling (Becherer, Richard J.; Kahan, Lloyd R.)V1416,306-313(1991)

Ground-based lidar for long-term and network measurements of ozone (McDermid, I. S.; Schmoe, Martha S.; Walsh, T. D.)V1491,175-181(1991)

Infrared coherent lidar systems for wind velocity measurements (Gordienko, Vyacheslav M.; Akhmanov, Sergei A.; Bersenev, V. I.; Kosovsky, L. A.; Kurochkin, Nikolai N.; Priezzhev, Alexander V.; Pogosov, Gregory A.; Putivskii, Yu. Y.)V1416,102-114(1991)

Infrared lidars for atmospheric remote sensing (Menzies, Robert T.)V1540,160-163(1991)

Infrared lidar windshear detection for commercial aircraft and the edge technique: a new method for atmospheric wind measurement (Targ, Russell; Bowles, Roland L.; Korb, C. L.; Gentry, Bruce M.; Souilhac, Dominique J.)V1521,144-157(1991)

Laser applications for multisensor systems (Smolka, Greg L.; Strother, George T.; Ott, Carl)V1498,70-80(1991)

Laser Radar VI (Becherer, Richard J., ed.)V1416(1991)

Laser radar based on diode lasers (Levenstein, Harold)V1416,30-43(1991)

Laser remote sensing of natural water organics (Babichenko, Sergey M.; Poryvkina, Larisa)V1492,319-323(1991)

Laser transmitter for lidar in-space technology experiment (Chang, John H.; Cimolino, Marc C.; Petros, Mulugeta)V1492,43-46(1991)

Laser velocimetry applications (Soreide, David C.; McGarvey, John A.)V1416,280-285(1991)

Lidar backscatter calculations for solid-sphere and layered-sphere aerosols (Youmans, Douglas G.)V1416,151-162(1991)

Lidar evaluation of the propagation environment (Uthe, Edward E.; Livingston, John M.)V1487,228-239(1991)

Lidar for expendable launch vehicles (Lee, Michael S.)V1480,23-34(1991)

Lidar profiles of atmospheric structure properties (Philbrick, Charles R.)V1492,76-84(1991)

Lidar wind shear detection for commercial aircraft (Targ, Russell; Bowles, Roland L.)V1416,130-138(1991)

Measurement of wind velocity spread: signal-to-noise ratio for heterodyne detection of laser backscatter from aerosol (Fried, David L.; Szeto, Roque K.)V1416,163-176(1991)

Measuring tropospheric ozone using differential absorption lidar technique (Proffitt, Michael H.; Langford, A. O.)V1491,2-6(1991)

Model-based system for automatic target recognition (Verly, Jacques G.; Delanoy, Richard L.; Dudgeon, Dan E.)V1471,266-282(1991)

Modeling and experiments with a subsea laser radar system (Bjarnar, Morten L.; Klepsvik, John O.; Nilsen, Jan E.)V1537,74-88(1991)

Mode-locked Nd:YLF laser for precision range-Doppler imaging (Simpson, Thomas B.; Malley, Michael M.; Sutton, George W.; Doft, F.)V1416,2-9(1991)

Modifications to the LP-140 pulsed CO2 laser for lidar use (Jaenisch, Holger M.; Spiers, Gary D.)V1411,127-136(1991)

Multiscattered lidar returns from atmospheric aerosols (Hutt, Daniel L.; Bissonnette, Luc R.; Durand, Louis-Gilles)V1487,250-261(1991)

Narrowband alexandrite laser injection seeded with frequency-dithered diode laser (Schwemmer, Geary K.; Lee, Hyo S.; Prasad, Coorg R.)V1492,52-62(1991)

NASA's program in lidar remote sensing (Theon, John S.; Vaughan, William W.; Browell, Edward V.; Jones, William D.; McCormick, M. P.; Melfi, Samuel H.; Menzies, Robert T.; Schwemmer, Geary K.; Spinhirne, James D.)V1492,2-23(1991)

Optical power budget and device time-constant considerations in undersea laser-based sensor design (Leatham, James G.)V1537,194-202(1991)

Optomechanical considerations for stable lasers (Horton, T. E.; Cason, Charles; Dezenberg, George J.)V1416,10-20(1991)

Raman lidar for measuring backscattering in the China Sea (Liu, Zhi-Shen; Zhang, Jin-Long; Chen, Wen-Zhong; Huang, Xiao-Sheng; Ma, Jun)V1558,379-383(1991)

Recent lidar measurements of stratospheric ozone and temperature within the Network for the Detection of Stratospheric Change (McGee, Thomas J.; Ferrare, Richard; Butler, James J.; Frost, Robert L.; Gross, Michael; Margitan, James)V1491,182-187(1991)

Resolution limits for high-resolution imaging lidar (Idell, Paul S.)V1416,250-259(1991)

SAW real-time Doppler analysis (Martin, Tom A.)V1416,52-58(1991)

Simultaneous active/passive IR vehicle detection (Baum, Jerrold E.; Rak, Steven J.)V1416,209-220(1991)

Solid-state lidar measurements at 1 and 2 um (Killinger, Dennis K.; Chan, Kin-Pui)V1416,125-128(1991)

Space-qualified laser transmitter for lidar applications (Chang, John H.; Reithmaier, Karl D.)V1492,38-42(1991)

Surface-emitting, distributed-feedback laser as a source for laser radar (Akkapeddi, Prasad R.; Macomber, Steven H.)V1416,44-49(1991)

Target detection performance using 3-D laser radar images (Green, Thomas J.; Shapiro, Jeffrey H.; Menon, Murali M.)V1471,328-341(1991)

Thermal lensing and frequency chirp in a heated CdTe modulator crystal and its effects on laser radar performance (Eng, Richard S.; Kachelmyer, Alan L.; Harris, Neville W.)V1416,70-85(1991)

Three-dimensional laser radar simulation for autonomous spacecraft landing (Reiley, Michael F.; Carmer, Dwayne C.; Pont, W. F.)V1416,295-303(1991)

Track-while-image in the presence of background (Mentle, Robert E.; Shapiro, Jeffrey H.)V1471,342-353(1991)

Use of a multibeam transmitter for significant improvement in signal-dynamic-range reduction and near-range coverage for incoherent lidar systems (Zhao, Yanzeng; Hardesty, R. M.; Post, Madison J.)V1492,85-90(1991)

Wavemeter for tuning solid state lasers (Goad, Joseph H.; Rinsland, Pamela L.; Kist, Edward H.; Irick, Steven C.)V1410,107-115(1991)

Lithography—see microlithography

Luminescence

Advanced approaches in luminescence and Raman spectroscopy (Vo-Dinh, Tuan)V1435,197-202(1991)

Application of the NO/O3 chemiluminescence technique to measurements of reactive nitrogen species in the stratosphere (Fahey, David W.)V1433,212-223(1991)

Atomic layer growth of zinc oxide and zinc sulphide (Sanders, Brian W.; Kitai, Adrian H.)V1398,81-87(1991)

Compact measurement system for the simultaneous determination of NO, NO2, NOy, and O3 using a small aircraft (Walega, James G.; Dye, James E.; Grahek, Frank E.; Ridley, Brian K.)V1433,232-241(1991)

Cyclotron resonance and photoluminescence in GaAs in a microwave field (Ashkinadze, B. M.; Bel'Kov, V. V.; Krasinskaya, A. G.)V1361,866-873(1991)

Electrical and optical properties of As- and Li-doped ZnSe films (Hingerl, Kurt; Lilja, J.; Toivonen, M.; Pessa, Markus; Jantsch, Wolfgang; As, D. J.; Rothemund, W.; Juza, P.; Sitter, Helmut)V1361,943-953(1991)

Emission of the 1.54um Er-related peaks by impact excitation of Er atoms in InP and its characteristics (Isshiki, Hideo; Kobayashi, Hitoshi; Yugo, Shigemi; Saito, Riichiro; Kimura, Tadamasa; Ikoma, Toshiaki)V1361,223-227(1991)

Energy-level structure and electron transitions of GaAs:Cr optoelectronic materials (Schulz, Hans-Joachim; Schoenbach, Karl H.; Kimpel, B. M.; Brinkmann, Ralf P.)V1361,834-847(1991)

Epitaxial growth and photoluminescence investigations of InP/InAs quantum well grown by hydride vapor phase epitaxy (Banvillet, Henri; Gil, E.; Vasson, A. M.; Cadoret, R.; Tabata, A.; Benyattou, Taha; Guillot, Gerard)V1361,972-979(1991)

Exciton-polariton photoluminescence in ultrapure GaAs (Zhilyaev, Yuri V.; Rossin, Victor V.; Rossina, Tatiana V.; Travnikov, V. V.)V1361,848-859(1991)

Exciton spectroscopy of semiconductor materials used in laser elements (Nasibov, Alexander S.; Markov, L. S.; Fedorov, D. L.; Shapkin, P. V.; Korostelin, Y. V.; Machintsev, G. A.)V1361,901-908(1991)

Feasibility of luminescence-eliminated anti-Stokes Raman spectroscopy (Ishibashi, Taka-aki; Hamaguchi, Hiro-o)V1403,555-562(1991)

Fiber optic evanescent wave biosensor (Duveneck, G.; Ehrat, M.; Widmer, H. M.)V1510,138-145(1991)

Flow injection analysis with bioluminescence-based fiber-optic biosensors (Blum, Loic J.; Gautier, Sabine; Coulet, Pierre R.)V1510,46-52(1991)

Frequency-domain fluorescence spectroscopy: instrumentation and applications to the biosciences (Lakowicz, Joseph R.; Gryczynski, Ignacy; Malak, Henryk; Johnson, Michael L.; Laczko, Gabor; Wiczk, Wieslaw M.; Szmacinski, Henryk; Kusba, Jozef)V1435,142-160(1991)

Growth and characterization of ZnSe and ZnTe grown on GaAs by hot-wall epitaxy (Hingerl, Kurt; Pesek, Andreas; Sitter, Helmut; Krost, Alois; Zahn, Dietrich R.; Richter, W.; Kudlek, Gotthard; Gutowski, Juergen)V1361,383-393(1991)

Humidity dependence of ceramic substrate electroluminescent devices (Young, Richard; Kitai, Adrian H.)V1398,71-80(1991)

Hydrogen peroxide and organic peroxides in the marine environment (Heikes, Brian G.; Miller, William L.; Lee, Meehye)V1433,253-262(1991)

Laser-induced luminescence of singlet molecular oxygen: generation by drugs and pigments of biological importance (Egorov, S. Y.; Krasnovsky, A. A.)V1403,611-621(1991)

Luminescence and chemical potential of solar cells (Smestad, Greg P.; Ries, Harald)V1536,234-245(1991)

Luminescence and ionization of krypton by multiphotonic excitation near the 3P1 resonant state (Saissac, M.; Berejny, P.; Millet, P.; Yousfi, M.; Salamero, Y.)V1397,739-742(1991)

Luminescence molulation for the characterization of radiation damage within scintillator material (Bayer, Eberhard G.)V1361,195-199(1991)

Luminescence probes in aqueous micellar solutions (Vecher, Jaroslav)V1402,97-101(1991)

Luminescence studies in the process of preparation high-Tc superconducting films with excimer laser ablation (Fan, Yong C.; An, Cheng W.; Lu, Dong S.; Li, Zai Q.)V1519,813-817(1991)

Measurement of peroxyacetyl nitrate, NO2, and NOx by using a gas chromatograph with a luminol-based detector (Drummond, John W.; Mackay, Gervase I.; Schiff, Harold I.)V1433,242-252(1991)

Mechanisms for the cathodoluminescence of cerium-doped yttrium aluminum garnet phosphors (Rotman, Stanley R.; Aizenberg, G. E.; Gordon, E. S.; Tuller, H. L.; Warde, Cardinal)V1442,205-215(1991)

Microwave-optical study of an As(III) derivative of Eco RI methylase (Maki, August H.; Tsao, Desiree H.)V1432,119-128(1991)

Multiphoton excited emission in zinc selenide and other crystals (Prokhorov, Kirill A.; Prokhorov, Alexander M.; Djibladze, Merab I.; Kekelidze, George N.; Gorelik, Vladimir S.)V1501,80-84(1991)

New approach to colored thin film electroluminescence (Xu, Xu R.; Lei, Gang; Xu, Zheng; Shen, Meng Y.)V1519,525-528(1991)

New luminescent metal complex for pH transduction in optical fiber sensing: application to a CO2-sensitive device (Moreno-Bondi, Maria C.; Orellana, Guillermo; Camara, Carmen; Wolfbeis, Otto S.)V1368,157-164(1991)

Nondiffractive optically variable security devices (van Renesse, Rudolf L.)V1509,113-125(1991)

Nonstoichiometry effect on mercury thiogallate luminescence (Yelisseyev, Alexander P.; Sinyakova, Elena F.)V1512,204-212(1991)

Optical characterization of InP epitaxial layers on different substrates (Jiao, Kaili L.; Zheng, J. P.; Kwok, Hoi-Sing; Anderson, Wayne A.)V1361,776-783(1991)

Optoelectronic and optical bistabilities of photocurrent and photoluminescence at low-temperature avalanche breakdown in GaAs epitaxial films (Ryabushkin, Oleg A.; Platonov, N. S.; Sablikov, V. A.; Sergeyev, V. I.; Bader, Vladimir A.)V1362,75-79(1991)

Photoluminescence and deep-level transient spectroscopy of DX-centers in selectively silicon-doped GaAs-AlAs superlattices (Ababou, Soraya; Benyattou, Taha; Marchand, Jean J.; Mayet, Louis; Guillot, Gerard; Mollot, Francis; Planel, Richard)V1361,706-711(1991)

Photoluminescence and surface photovoltaic spectra of strained InP on GaAs by MOCVD (Zhuang, Weihua; Chen, Chao; Teng, Da; Yu, Jin-zhong; Li, Yu Z.)V1361,980-986(1991)

Scanning exciton microscopy and single-molecule resolution and detection (Kopelman, Raoul; Tan, Weihong; Lewis, Aaron; Lieberman, Klony)V1435,96-101(1991)

Spectral-luminescence analysis as method of tissue states evaluation and laser influence on tissues before and after transplantation (Loshchenov, V. B.; Baryshev, M. V.; Svystushkin, V. M.; Ovchinnikov, U. M.; Babyn, A. V.; Schaldach, Brita J.; Mueller, Gerhard J.)V1420,271-281(1991)

Study of impurities in CVD diamond using cathodoluminescence (Nishimura, Kazuhito; Ma, Jing S.; Yokota, Yoshihiro; Mori, Yusuke; Kotsuka, Hiroshi; Hirao, Takashi; Kitabatake, Makoto; Deguchi, Masahiro; Ogawa, Kazuo; Ning, Gang; Tomimori, Hiroshi; Hiraki, Akio)V1534,199-206(1991)

Study of the production of S2(B) from the recombination of sulfur atoms (Prigent, Pascale; Brunet, Henri)V1397,197-201(1991)

Time-resolved luminescence experiments on modulation n-doped GaAs quantum wells (Lopez, C.; Meseguer, Francisco; Sanchez-Dehesa, Jose; Ruehle, Wolfgang W.; Ploog, Klaus H.)V1361,89-95(1991)

Top-side electroluminescence: a failure analysis technique to view electroluminescence along a laser channel (Blow, Victor O.; Giewont, Kenneth J.)V1366,107-111(1991)

Tunneling recombination of carriers at type-II interface in GaInAsSb-GaSb heterostructures (Titkov, A. N.; Yakovlev, Yurii P.; Baranov, Alexej N.; Cheban, V. N.)V1361,669-673(1991)

Use of chemiluminescence techniques in portable, lightweight, highly sensitive instruments for measuring NO2, NOx, and O3 (Drummond, John W.; Topham, L. A.; Mackay, Gervase I.; Schiff, Harold I.)V1433,224-231(1991)

Machine vision—see also computer vision

10 MHz multichannel image detection and processing system (Damstra, Geert C.; Eenink, A. H.)V1358,644-651(1991)

3-D camera based on differential optical absorbance (Houde, Regis; Laurendeau, Denis; Poussart, Denis)V1332,343-354(1991)

Absolute range measurement system for real-time 3-D vision (Wood, Christopher M.; Shaw, Michael M.; Harvey, David M.; Hobson, Clifford A.; Lalor, Michael J.; Atkinson, John T.)V1332,301-313(1991)

Accurate image simulation by hemisphere projection (Bian, Buming; Wittels, Norman)V1453,333-340(1991)

Active stereo inspection using computer solids models (Nurre, Joseph H.)V1381,171-176(1991)

Aircraft exterior scratch measurement system using machine vision (Sarr, Dennis P.)V1472,177-184(1991)

Analog CMOS IC for object position and orientation (Standley, David L.; Horn, Berthold K.)V1473,194-201(1991)

Analog retina model for detecting moving objects against a moving background (Searfus, Robert M.; Eeckman, Frank H.; Colvin, Michael E.; Axelrod, Timothy S.)V1473,95-101(1991)

Application of Dempster-Shafer theory to a novel control scheme for sensor fusion (Murphy, Robin R.)V1569,55-68(1991)

Application of neural networks to group technology (Caudell, Thomas P.; Smith, Scott D.; Johnson, G. C.; Wunsch, Donald C.)V1469,612-621(1991)

Applications of Artificial Neural Networks II (Rogers, Steven K., ed.)V1469(1991)

Architecture and performance of a hardware collision-checking accelerator (Bardin, R. K.; Libby, Vibeke)V1566,394-404(1991)

Architecture for surveillance in real time using nonlinear image processing hardware (Pais, Cassiano P.; Carvalho, Fernando D.; Silvestre, Victor M.)V1451,282-288(1991)

Automated dimensional inspection of cars in crash tests with digital photogrammetry (Beyer, Horst A.)V1526,134-141(1991)

Automated visual imaging interface for the plant floor (Wutke, John R.)V1386,180-184(1991)

Automated visual inspection of printed wiring boards (Chin, Roland T.; Iverson, Rolf D.)VCR36,93-104(1991)

Automated visual inspection system for IC bonding wires using morphological processing (Tsukahara, Hiroyuki; Nakashima, Masato; Sugawara, Takehisa)V1384,15-26(1991)

Automatically inspecting gross features of machined objects using three-dimensional depth data (Marshall, Andrew D.)V1386,243-254(1991)

Automatic analysis of heliotest strips (Langinmaa, Anu)V1468,573-580(1991)

Automatic inspection system for full-color printed matter (Meguro, Shin-Ichi; Nunotani, Masakatu; Tanimizu, Katsuyuki; Sano, Mutsuo; Ishii, Akira)V1384,27-37(1991)

Automatic method for inspecting plywood shear samples (Avent, R. R.; Conners, Richard W.)V1468,281-295(1991)

Autonomous navigation in a dynamic environment (Davies, Henry C.; Kayaalp, Ali E.; Moezzi, Saied)V1388,165-175(1991)

Autonomous navigation of structured city roads (Aubert, Didier; Kluge, Karl; Thorpe, Charles E.)V1388,141-151(1991)

Binary object analysis hardware area parameter acceleration (Seitzler, Thomas M.)V1567,25-30(1991)

Biological basis for space-variant sensor design I: parameters of monkey and human spatial vision (Rojer, Alan S.; Schwartz, Eric L.)V1382,132-144(1991)

Biological basis for space-variant sensor design II: implications for VLSI sensor design (Rojer, Alan S.; Schwartz, Eric L.)V1386,44-52(1991)

BRICORK: an automatic machine with image processing for the production of corks (Davies, Roger; Correia, Bento A.; Carvalho, Fernando D.; Rodrigues, F. C.)V1459,283-291(1991)

Building an infrastructure for system integration: machine vision standards (Bieman, Leonard H.; Peyton, James A.)VCR36,3-19(1991)

Calibration procedures for the space vision system experiment (MacLean, Steve G.; Pinkney, H. F.)V1526,113-122(1991)

Camera calibration using distance invariance principles (Raju, G. V.; Rudraraju, Prasad)V1385,50-56(1991)

Case studies in machine vision integration (Ahlers, Rolf-Juergen)VCR36,56-63(1991)

Color character recognition method based on a model of human visual processing (Yamaba, Kazuo; Miyake, Yoichi)V1453,290-299(1991)

Color-encoded depth: an image enhancement tool (Bieman, Leonard H.)V1385,229-238(1991)

Combined approach for large-scale pattern recognition with translational, rotational, and scaling invariances (Xu, Qing; Inigo, Rafael M.; McVey, Eugene S.)V1471,378-389(1991)

Computer vision system for automated inspection of molded plastic print wheels (Hu, Yong-Lin; Wee, William G.; Gruver, William A.; Han, Chia Y.)V1468,653-661(1991)

Cooperative Intelligent Robotics in Space (Stoney, William E.; deFigueiredo, Rui J., eds.)V1387(1991)

Corn plant locating by image processing (Jia, Jiancheng; Krutz, Gary W.; Gibson, Harry G.)V1379,246-253(1991)

Damage detection in peanut grade samples using chromaticity and luminance (Dowell, Floyd E.; Powell, J. H.)V1379,136-140(1991)

Data Structures and Target Classification (Libby, Vibeke, ed.)V1470(1991)

Deformation study of the Klang Gates Dam with multispectral analysis method (Hassan, Azmi; Subari, Mustofa; Som, Zainal A.)V1526,142-156(1991)

Dense stereo correspondence using color (Jordan, John R.; Bovik, Alan C.)V1382,111-122(1991)

Depth determination using complex logarithmic mapping (Bartlett, Sandra L.; Jain, Ramesh C.)V1382,3-13(1991)

Design and characterization of a space-variant CCD sensor (Kreider, Gregory; Van der Spiegel, Jan; Born, I.; Claeys, Cor L.; Debusschere, Ingrid; Sandini, Giulio; Dario, Paolo)V1381,242-249(1991)

Detecting difficult roads and intersections without map knowledge for robot vehicle navigation (Crisman, Jill D.; Thorpe, Charles E.)V1388,152-164(1991)

Determination of flint wheel orientation for the automated assembly of lighters (Safabakhsh, Reza)V1472,185-189(1991)

Determining range information from self-motion: the template model (Sobey, Peter J.)V1382,123-131(1991)

Development of a low-cost VME-based Nth order 2-D warper (Messner, Richard A.; Whitney, Erich C.)V1381,261-271(1991)

Digital imaging of aircraft dynamic seals: a fiber optics solution (Nwagboso, Christopher O.)V1500,234-245(1991)

Digital restoration of distorted geometric features of pigs (Van der Stuyft, Emmanuel; Goedseels, Vic; Geers, Rony)V1379,189-200(1991)

Distributed industrial vision systems (Neve, Peter)VCR36,78-92(1991)

Dynamic integration of visual cues for position estimation (Das, Subhodev; Ahuja, Narendra)V1382,341-352(1991)

Dynamic range data acquisition and pose estimation for 3-D regular objects (Marszalec, Janusz A.; Heikkila, Tapio A.; Jarviluoma, Markku)V1382,443-452(1991)

Efficient object contour tracing in a quadtree encoded image (Kumar, G. N.; Nandhakumar, N.)V1468,884-895(1991)

Error analysis of combined stereo/optical-flow passive ranging (Barniv, Yair)V1479,259-267(1991)

Essential for success: people as part of the system (Fosdick, Jerilyn J.)VCR36,34-45(1991)

Experiments in real-time visual control (Griswold, Norman C.; Kehtarnavaz, Nasser)V1388,342-349(1991)

Exploiting geometric relationships for object modeling and recognition (Walker, Ellen L.)V1382,353-363(1991)

Extracting characters from illustration document by relaxation (Takeda, Haruo)V1386,128-134(1991)

Fast algorithm for obtaining dense depth maps for high-speed navigation (Khalili, Payman; Jain, Ramesh C.)V1388,210-221(1991)

Fast holographic correlator for machine vision systems (Nekrasov, Victor V.)V1507,170-174(1991)

Faying surface-gap measurement of aircraft structures for shim fabrication and installation (Sarr, Dennis P.; Jurick, Thomas W.)V1469,506-514(1991)

Flat plate project (Wijbrans, Klaas C.; Korsten, Maarten J.)V1386,197-205(1991)

Flexible gray-level vision system based on multiple cell-feature description and generalized Hough transform (Sano, Mutsuo; Ishii, Akira; Meguro, Shin-Ichi)V1381,101-110(1991)

Flexible object-centered illuminator (Uber, Gordon T.)V1385,2-7(1991)

FLIPS: Friendly Lisp Image Processing System (Gee, Shirley J.)V1472,38-45(1991)

Fresh market carrot inspection by machine vision (Howarth, M. S.; Searcy, Stephen W.)V1379,141-150(1991)

Fundamentals of on-line gauging for machine vision (Novini, Amir R.)V1526,2-16(1991)

Fusion of multiple fixations with a space-variant sensor: conditional optimality of maximum-resolution blending (Rojer, Alan S.; Schwartz, Eric L.)V1381,250-260(1991)

Fusion of multiple-sensor imagery based on target motion characteristics (Tsao, Tien-Ren J.; Libert, John M.)V1470,37-47(1991)

HelpMate autonomous mobile robot navigation system (King, Steven J.; Weiman, Carl F.)V1388,190-198(1991)

Hierarchical symmetry segmentation (Laird, Alan; Miller, James)V1381,536-544(1991)

Hierarchical terrain representations for off-road navigation (Gowdy, Jay W.; Stentz, Anthony; Hebert, Martial)V1388,131-140(1991)

High-accuracy edge-matching with an extension of the MPGC-matching algorithm (Gruen, Armin; Stallmann, Dirk)V1526,42-55(1991)

High-density CCD neurocomputer chip for accurate real-time segmentation of noisy images (Roth, Michael W.; Thompson, K. E.; Kulp, R. L.; Alvarez, E. B.)V1469,25-36(1991)

High-resolution fully 3-D mapping of human surfaces by laser array camera and data representations (Bae, Kyongtae T.; Altschuler, Martin D.)V1380,171-178(1991)

High-speed fine-motion tracking of some parts of a target (Casals, Alicia; Amat, Josep; Quesada, Jose L.; Sanchez, Luis)V1482,317-324(1991)

High-Speed Inspection Architectures, Barcoding, and Character Recognition (Chen, Michael J., ed.)V1384(1991)

High-speed sensor-based systems for mobile robotics (Kayaalp, Ali E.; Moezzi, Saied; Davies, Henry C.)V1406,98-109(1991)

High-speed vision system based on computer graphics models (Baur, Charles; Beer, Simon)V1385,85-92(1991)

Holographic interferometry analysis of sealed, disposable containers for internal defects (Cohn, Gerald E.; Domanik, Richard A.)V1429,195-206(1991)

Hough transform implementation using an analog associative network (Arrue, Begona C.; Inigo, Rafael M.)V1469,420-431(1991)

Hybrid pattern recognition system for the robotic vision (Li, Yulin; Zhao, Mingjun; Zhao, Li)V1385,200-205(1991)

Hybrid system for computing reachable workspaces for redundant manipulators (Alameldin, Tarek K.; Badler, Norman I.; Sobh, Tarek M.)V1386,112-120(1991)

Illumination and viewing methods for machine vision (Uber, Gordon T.; Harding, Kevin G.)VCR36,20-33(1991)

Image analysis applications for grain science (Zayas, Inna Y.; Steele, James L.)V1379,151-161(1991)

Image analysis for DNA sequencing (Palaniappan, K.; Huang, Thomas S.)V1450,186-197(1991)

Image processing to locate corn plants (Jia, Jiancheng; Krutz, Gary W.; Gibson, Harry G.)V1396,656-663(1991)

Image registration for automated inspection of 2-D electronic circuit patterns (Rodriguez, Arturo A.; Mandeville, Jon R.)V1384,2-14(1991)

Image Understanding and the Man-Machine Interface III (Barrett, Eamon B.; Pearson, James J., eds.)V1472(1991)

Image Understanding in the '90s: Building Systems that Work (Mitchell, Brian T., ed.)V1406(1991)

Implementation of a 3-D laser imager-based robot navigation system with location identification (Boltinghouse, Susan T.; Burke, James; Ho, Daniel)V1388,14-29(1991)

Implementation of a 3-D stereovision system for the production of customized orthotic accessories (Daher, Reinhard; McAdam, Wylie; Pizey, Gordon)V1526,90-93(1991)

Implementing early vision algorithms in analog hardware: an overview (Koch, Christof)V1473,2-16(1991)

Implementing early visual processing in analog VLSI: light adaptation (Mann, James)V1473,128-136(1991)

Industrial Vision Metrology (El-Hakim, Sabry F., ed.)V1526(1991)

Industrial vision systems based on a digital image correlation chip (Masaki, Ichiro)V1473,90-94(1991)

Innovative architectural and theoretical considerations yield efficient fuzzy logic controller VLSI design (Basehore, Paul; Yestrebsky, Joseph T.)V1470,190-196(1991)

INS integrated motion analysis for autonomous vehicle navigation (Roberts, Barry; Bazakos, Michael E.)V1521,2-13(1991)

Inspecting colored objects using gray-level vision systems (Plummer, A. P.)VCR36,64-77(1991)

Inspection of a class of industrial objects using a dense range map and CAD model (Ailisto, Heikki J.; Paakkari, Jussi; Moring, Ilkka)V1384,50-59(1991)

Integrated approach to machine vision application development (Rosenthal, Steven; Stahlberg, Larry)V1386,158-162(1991)

Integrating vision and AI in an image processing workstation (Chan, John P.; Batchelor, Bruce G.)V1386,163-170(1991)

Integration of an application accelerator for high-speed inspection (Kille, Knut; Ahlers, Rolf-Juergen; Hager, K.)V1386,222-227(1991)

Intelligent material handling: use of vision (Dickerson, Stephen L.; Lee, Kok-Meng; Lee, Eun Ho; Single, Thomas; Li, Da-ren)V1381,201-209(1991)

Intelligent packaging and material handling (Hall, Ernest L.; Shell, Richard; Slutzky, Gale D.)V1381,162-170(1991)

Intelligent Robots and Computer Vision IX: Algorithms and Techniques (Casasent, David P., ed.)V1381(1991)

Intelligent vision process for robot manipulation (Chen, Alexander Y.; Chen, Eugene Y.)V1381,226-239(1991)

Intelligent visual inspection of food products (Chan, John P.; Batchelor, Bruce G.; Harris, I. P.; Perry Beng, S. J.)V1386,171-179(1991)

Interaction of algorithm and implementation for analog VLSI stereo vision (Hakkarainen, J. M.; Little, James J.; Lee, Hae-Seung; Wyatt, John L.)V1473,173-184(1991)

Knowledge-driven image analysis of cell structures (Nederlof, Michel A.; Witkin, Andrew; Taylor, D. L.)V1428,233-241(1991)

LANELOK: an algorithm for extending the lane sensing operating range to 100-feet (Kenue, Surender K.)V1388,222-233(1991)

Laser Image Detection System (LIDS): laser-based imaging (Hilton, Peter J.; Gabric, R.; Walternberg, P. T.)V1385,27-31(1991)

Learnability of min-max pattern classifiers (Yang, Ping-Fai; Maragos, Petros)V1606,294-308(1991)

Lighting and optics expert system for machine vision (Novini, Amir R.)V1386,2-9(1991)

Machine verification of traced signatures (Krishnan, Ganapathy; Jones, David E.)V1468,563-572(1991)

Machine vision applications of image invariants: real-time processing experiments (Payton, Paul M.; Haines, Barry K.; Smedley, Kirk G.; Barrett, Eamon B.)V1406,58-71(1991)

Machine vision feedback for on-line correction of manufacturing processes: a control formulation (Taylor, Andrew T.; Wang, Paul K.)V1567,220-231(1991)

Machine vision inspection of fluorescent lamps (Bains, Narinder; David, Frank)V1386,232-242(1991)

Machine vision platform requirements for successful implementation and support in the semiconductor assembly manufacturing environment (LeBeau, Christopher J.)V1386,228-231(1991)

Machine vision system for ore sizing (Eichelberger, Christopher L.; Blair, Steven M.; Khorana, Brij M.)V1396,678-686(1991)

Machine Vision Systems Integration (Batchelor, Bruce G.; Waltz, Frederick M., eds.)VCR36(1991)

Machine Vision Systems Integration in Industry (Batchelor, Bruce G.; Waltz, Frederick M., eds.)V1386(1991)

Markov random fields on a SIMD machine for global region labelling (Budzban, Gregory M.; DeCatrel, John M.)V1470,175-182(1991)

Minimum resolution for human face detection and identification (Samal, Ashok)V1453,81-89(1991)

Mobile Robots V (Chun, Wendell H.; Wolfe, William J., eds.)V1388(1991)

Mobile system for measuring retroreflectance of traffic signs (Lumia, John J.)V1385,15-26(1991)

Model-based system for automatic target recognition (Verly, Jacques G.; Delanoy, Richard L.; Dudgeon, Dan E.)V1471,266-282(1991)

Model for image sensing and digitization in machine vision (Subbarao, Murali; Nikzad, Arman)V1385,70-84(1991)

Modular algorithms for depth characterization of object surfaces with ultrasonic sensors (Brule, Michel J.; Soucy, L.)V1388,432-441(1991)

Monocular passive range sensing using defocus information (Prasad, K. V.; Mammone, Richard J.)V1385,280-291(1991)

Morphological processing to reduce shading and illumination effects (Casasent, David P.; Schaefer, Roland H.; Kokaj, Jahja O.)V1385,152-164(1991)

Multiplication-based analog motion detection chip (Moore, Andrew; Koch, Christof)V1473,66-75(1991)

Multisensor image processing (de Salabert, Arturo; Pike, T. K.; Sawyer, F. G.; Jones-Parry, I. H.; Rye, A. J.; Oddy, C. J.; Johnson, D. G.; Mason, D.; Wielogorski, A. L.; Plassard, T.; Serpico, Sebastiano B.; Hindley, N.)V1521,74-88(1991)

Multisensor object segmentation using a neural network (Gaughan, Patrick T.; Flachs, Gerald M.; Jordan, Jay B.)V1469,812-819(1991)

Multistage object recognition using dynamical-link graph matching (Flaton, Kenneth A.)V1469,137-148(1991)

Multitask neural network for vision machine systems (Gupta, Madan M.; Knopf, George K.)V1382,60-73(1991)

Network compensation for missing sensors (Ahumada, Albert J.; Mulligan, Jeffrey B.)V1453,134-146(1991)

Neural-network-based vision processing for autonomous robot guidance (Pomerleau, Dean A.)V1469,121-128(1991)

Neural networks and model-based approaches to object identification (Perlovsky, Leonid I.)V1606,1080-1085(1991)

New human vision system model for spatio-temporal image signals (Matsui, Toshikazu; Hirahara, Shuzo)V1453,282-289(1991)

New insights into correlation-based template matching (Ooi, James; Rao, Kashi)V1468,740-751(1991)

New method for sensor data fusion in machine vision (Wang, Yuan-Fang)V1570,31-42(1991)

New scheme for real-time visual inspection of bearing roller (Zhang, Yu; Xian, Wu; Li, Li-Ping; Hall, Ernest L.; Tu, James Z.)V1384,60-65(1991)

Noise statistics of ratiometric signal processing systems (Svetkoff, Donald J.; Xydis, Thomas G.; Kilgus, Donald B.)V1385,113-122(1991)

Object classification for obstacle avoidance (Regensburger, Uwe; Graefe, Volker)V1388,112-119(1991)

Object detection in real-time (Solder, Ulrich; Graefe, Volker)V1388,104-111(1991)

Object recognition using coding schemes (Sadjadi, Firooz A.)V1471,428-434(1991)

Off-line machine recognition of forgeries (Randolph, David; Krishnan, Ganapathy)V1386,255-264(1991)

On-line visual prosthesis for a decision maker (Ligomenides, Panos A.)V1382,145-156(1991)

Operator-coached machine vision for space telerobotics (Bon, Bruce; Wilcox, Brian H.; Litwin, Todd; Gennery, Donald B.)V1387,337-342(1991)

Optical correlator techniques applied to robotic vision (Hine, Butler P.; Reid, Max B.; Downie, John D.)V1564,416-426(1991)

Optical correlator vision system for a manufacturing robot assembly cell (Brandstetter, Robert W.; Fonneland, Nils J.; Zanella, R.; Yearwood, M.)V1385,173-189(1991)

Optical image segmentor using wavelet filtering techniques as the front-end of a neural network classifier (Veronin, Christopher P.; Rogers, Steven K.; Kabrisky, Matthew; Welsh, Byron M.; Priddy, Kevin L.; Ayer, Kevin W.)V1469,281-291(1991)

Optical target location using machine vision in space robotics tasks (Sklair, Cheryl W.; Gatrell, Lance B.; Hoff, William A.; Magee, Michael)V1387,380-391(1991)

Optics, Illumination, and Image Sensing for Machine Vision V (Harding, Kevin G.; Svetkoff, Donald J.; Uber, Gordon T.; Wittels, Norman, eds.)V1385(1991)

Optics in Agriculture (DeShazer, James A.; Meyer, George E., eds.)V1379(1991)

Optoelectronically implemented neural network for early visual processing (Mehanian, Courosh; Aull, Brian F.; Nichols, Kirby B.)V1469,275-280(1991)

Parameter studies for Markov random field models of early vision (Daily, Michael J.)V1473,138-152(1991)

Performance characterization of vision algorithms (Haralick, Robert M.; Ramesh, Visvanathan)V1406,2-16(1991)

Performance evaluation of a vision dimension metrology system (El-Hakim, Sabry F.; Westmore, David B.)V1526,56-67(1991)

Personal verification system with high tolerance of poor-quality fingerprints (Sasakawa, Koichi; Isogai, Fumihiko; Ikebata, Sigeki)V1386,265-272(1991)

Polynomial regression analysis for estimating motion from image sequences (Frau, Juan; Llario, Vicenc; Oliver, Gabriel)V1388,329-340(1991)

POPS: parallel opportunistic photointerpretation system (Howard, Michael D.)V1471,422-427(1991)

Production quality control problems (Doney, Thomas A.)V1381,9-20(1991)

Proposals for a computer-controlled orbital scanning camera for remote image aquisition (Nwagboso, Christopher O.)V1454,111-122(1991)

Qualitative three-dimensional shape from stereo (Wildes, Richard P.)V1382,453-463(1991)

Range data analysis using cross-stripe structured-light system (Kim, Whoi-Yul)V1385,216-228(1991)

Rapid system integration with symbolic programming (Walsh, Peter M.)V1386,84-89(1991)

Real-time automatic inspection under adverse conditions (Carvalho, Fernando D.; Correia, Fernando C.; Freitas, Jose C.; Rodrigues, F. C.)V1399,130-136(1991)

Real-time classification of wooden boards (Poelzleitner, Wolfgang; Schwingshakl, Gert)V1384,38-49(1991)

Real-time edge extraction by active defocusing (Hung, Yau Y.; Zhu, Qiuming; Shi, Dahuan; Tang, Shou-Hong)V1332,332-342(1991)

Real-time motion detection using an analog VLSI zero-crossing chip (Bair, Wyeth; Koch, Christof)V1473,59-65(1991)

Real-time quality inspection system for textile industries (Karkanis, S.; Tsoutsou, K.; Vergados, J.; Dimitriadis, Basile D.)V1500,164-170(1991)

Recent advances in the development and transfer of machine vision technologies for space (deFigueiredo, Rui J.; Pendleton, Thomas W.)V1387,330-336(1991)

Receptive fields and the theory of discriminant operators (Gupta, Madan M.; Hungenahally, Suresh K.)V1382,87-98(1991)

Recognition and tracking of moving objects (El-Konyaly, Sayed H.; Enab, Yehia M.; Soltan, Hesham)V1388,317-328(1991)

Recognition of a moving planar shape in space from two perspective images (Li, Zhi-yong; Sun, Zhong-kang; Shen, Zhen-kang)V1472,97-105(1991)

Recognition of movement object collision (Chang, Hsiao T.; Sun, Geng-tian; Zhang, Yan)V1388,442-446(1991)

Recognition of partially occluded objects using B-spline representation (Salari, Ezzatollah; Balaji, Sridhar)V1384,115-123(1991)

Reconstructing visible surfaces (Schenk, Anton F.; Toth, Charles)V1526,78-89(1991)

Reconstruction during camera fixation (Raviv, Daniel)V1382,312-319(1991)

Recovering epipolar geometry in 3-D vision systems (Schenk, Anton F.; Toth, Charles)V1457,66-73(1991)

Recovery of aliased signals for high-resolution digital imaging (John, Sarah)V1385,39-47(1991)

Remote visual monitoring of seal performance in aircraft jacks using fiber optics (Nwagboso, Christopher O.; Whomes, Terence L.; Davies, P. B.)V1386,30-41(1991)

Retina-like image acquisition system with wide-range light adaptation (Chang, Po-Rong; Yeh, Bao-Fuh)V1606,456-469(1991)

Rigid body motion estimation using a sequence of images from a static camera (Raju, G. V.; Rudraraju, Prasad)V1388,415-424(1991)

Robot vision system for obstacle avoidance planning (Attolico, Giovanni; Caponetti, Laura; Chiaradia, Maria T.; Distante, Arcangelo)V1388,50-61(1991)

Robust image coding with a model of adaptive retinal processing (Narayanswamy, Ramkumar; Alter-Gartenberg, Rachel; Huck, Friedrich O.)V1385,93-103(1991)

Robust stereo vision (Sinha, Saravajit S.; Moezzi, Saied; Schunck, Brian G.)V1385,259-267(1991)

Salient contour extraction for target recognition (Rao, Kashi; Liou, James J.)V1482,293-306(1991)

Scene description: an iterative approach (Mulgaonkar, Prasanna G.; Decurtins, Jeff; Cowan, Cregg K.)V1382,320-330(1991)

SCORPIUS: a successful image understanding implementation (Wilkes, Scott C.)V1406,110-121(1991)

Segmentation via fusion of edge and needle map (Ahn, Hong-Young; Tou, Julius T.)V1468,896-904(1991)

Semiautomatic calibration of the general camera model for stereovision (Sung, Eric; Singh, Harcharan; Tan, Daniel H.)V1385,57-69(1991)

Shape from shading with circular symmetry (Gennert, Michael A.)V1385,256-258(1991)

Shortage of system integrators (Braggins, Donald W.)VCR56,46-55(1991)

Simple iterative method for finding the foe using depth-is-positive constraint (Michael, David J.)V1388,234-245(1991)

State-space search as high-level control for machine vision (Hwang, Shu-Yuen)V1386,145-156(1991)

Stereo matching by energy function minimization (Gennert, Michael A.; Ren, Biao; Yuille, Alan L.)V1385,268-279(1991)

Stereovision and color segmentation for autonomous navigation (Sung, Eric)V1388,176-187(1991)

Stereo vision: a neural network application to constraint satisfaction problem (Mousavi, Madjid S.; Schalkoff, Robert J.)V1382,228-239(1991)

SUB-3D high-resolution pose measurement system (Hudgens, Jeffrey C.; Tesar, Delbert; Sklar, Michael E.)V1387,271-282(1991)

Subpixel measurement of image features based on paraboloid surface fit (Gleason, Shaun S.; Hunt, Martin A.; Jatko, W. B.)V1386,135-144(1991)

Supervised color constancy for machine vision (Novak, Carol L.; Shafer, Steven A.)V1453,353-368(1991)

System calibration and part alignment for inspection of 2-D electronic circuit patterns (Rodriguez, Arturo A.; Mandeville, Jon R.; Wu, Frederick Y.)V1332,25-35(1991)

Test of a vision-based autonomous space station robotic task (Castellano, Anthony R.; Hwang, Vincent S.; Stoney, William E.)V1387,343-350(1991)

Three-dimensional machine vision using line-scan sensors (Godber, Simon X.; Robinson, Max; Evans, Paul)V1526,170-189(1991)

Three-dimensional reconstruction from optical flow using temporal integration (Rangachar, Ramesh M.; Hong, Tsai-Hong; Herman, Martin; Luck, Randall L.)V1382,331-340(1991)

Three-dimensional vision: requirements and applications in a space environment (Noseworthy, J. R.; Gerhardt, Lester A.)V1387,26-37(1991)

Three-dimensional vision system for peanut pod maturity (Williams, E. J.; Adams, Stephen D.)V1379,236-245(1991)

Tissue identification in MR images by adaptive cluster analysis (Gutfinger, Dan; Hertzberg, Efrat M.; Tolxdorff, T.; Greensite, F.; Sklansky, Jack)V1445,288-296(1991)

Tools for designing industrial vision systems (Batchelor, Bruce G.)VCR36,138-166(1991)

Topographic mapping for stereo and motion processing (Mallot, Hanspeter A.; Zielke, Thomas; Storjohann, Kai; von Seelen, Werner)V1382,397-408(1991)

Toward computing the aspect graph of deformable generalized cylinders (Wilkins, Belinda; Goldgof, Dmitry B.; Bowyer, Kevin W.)V1468,662-673(1991)

Toward low-cost real-time EPLD-based machine vision workstations and target systems (Floeder, Steven P.; Waltz, Frederick M.)V1386,90-101(1991)

Towards integrated autonomous systems (Jain, Ramesh C.; Roth-Tabak, Yuval)V1468,188-201(1991)

Transputer-based parallel algorithms for automatic object recognition (Bison, Paolo G.; Braggiotti, Alberto; Grinzato, Ermanno G.)V1471,369-377(1991)

Two-dimensional boundary inspection using autoregressive model (Wani, M. A.; Batchelor, Bruce G.)V1384,83-89(1991)

Two stage object identification system in the Delft intelligent assembly cell (Buurman, Johannes; Bierhuizen, David J.)V1386,185-196(1991)

Two-view vision system for 3-D texture recovery (Yu, Xiaohan; Yla-Jaaski, Juha)V1468,834-842(1991)

User interfaces for automated visual inspection systems (Waltz, Frederick M.)VCR36,105-137(1991)

Using color to segment images of 3-D scenes (Healey, Glenn)V1468,814-825(1991)

Using computer vision for detecting watercore in apples (Throop, James A.; Rehkugler, Gerald E.; Upchurch, Bruce L.)V1379,124-135(1991)

Using fixation for direct recovery of motion and shape in the general case (Taalebinezhaad, M. A.)V1388,199-209(1991)

Using local orientation and hierarchical spatial feature matching for the robust recognition of objects (Seitz, Peter; Lang, Graham K.)V1606,252-259(1991)

Validation of vision-based obstacle detection algorithms for low-altitude helicopter flight (Suorsa, Raymond E.; Sridhar, Banavar)V1388,90-103(1991)

Video-image-based neural network guidance system with adaptive view-angles for autonomous vehicles (Luebbers, Paul G.; Pandya, Abhijit S.)V1469,756-765(1991)

Video-rate image remapper for the PC/AT bus (Weiman, Carl F.; Weber, Robert G.)V1386,102-110(1991)

Vision-based strip inspection hardware for metal production (Seitzler, Thomas M.)V1567,73-76(1991)

Vision sensing and processing system for monitoring and control of welding and other high-luminosity processes (Agapakis, John E.; Bolstad, Jon O.)V1385,32-38(1991)

Visual Information Processing: From Neurons to Chips (Mathur, Bimal P.; Koch, Christof, eds.)V1473(1991)

Visual inspection machine for solder joints using tiered illumination (Takagi, Yuuji; Hata, Seiji; Hibi, Susumu; Beutel, Wilhelm)V1386,21-29(1991)

Weighted least-squared error method for object localization (Wu, Hsiang-Lung)V1384,90-99(1991)

Magneto-optics—see also optical data storage; optical disks; optical recording

Advanced magneto-optic spatial light modulator device development (Ross, William E.; Lambeth, David N.)V1562,93-102(1991)

Advanced thin films for optical storage (Gan, Fuxi)V1519,530-538(1991)

Compact magneto-optical disk head integrated with chip elements (Yamanaka, Yutaka; Katayama, Ryuichi; Komatsu, Y.; Ishikawa, S.; Itoh, M.; Ono, Yuzo)V1499,263-268(1991)

Correction of magneto-optic device phase errors in optical correlators through filter design modifications (Downie, John D.; Reid, Max B.; Hine, Butler P.)V1564,308-319(1991)

Development of a high-performance 86-mm MO disk by using polycarbonate substrate (Ito, Katsunori; Shibata, Yasumasa; Honda, Katsunori; Tsukahara, Makoto; Kojima, Kotaro; Kimura, Masakatsu; Fujisima, Toshihiko; Yoshinaga, Kazuomi; Kanai, Toshitaka; Shimizu, Keijiro)V1499,382-385(1991)

Directly overwritable magneto-optical disk with light-power-modulation method using no initializing magnet (Tsutsumi, K.; Nakaki, Y.; Fukami, T.; Tokunaga, T.)V1499,55-61(1991)

Direct overwritable magneto-optical exchange-coupled multilayered disk by laser power modulation recording (Saito, Jun; Akasaka, Hideki)V1499,44-54(1991)

Direct overwriting system by light-intensity modulation using triple-layer disks (Maeda, Fumisada; Arai, Masayuki; Owa, Hideo; Takahashi, Hiroo; Kaneko, Masahiko)V1499,62-69(1991)

Dispersion of nonlinear magnetostatic surface waves on thin films (Boardman, A. D.; Nikitov, S. A.; Wang, Qi; Bao, Jia S.; Cai, Ying S.; Shen, Janice)V1519,609-615(1991)

DyFeCo magneto-optical disks with a Ce-SiO2 protective film (Naitou, Kazunori; Numata, Takehiko; Nakashima, Kazuo; Maeda, Miyozo; Koshino, Nagaaki)V1499,386-392(1991)

Electronic interface for high-frame-rate electrically addressed spatial light modulators (Kozaitis, Samuel P.; Kirschner, K.; Kelly, E.; Been, D.; Delgado, J.; Velez, E.; Alkindy, A.; Al-Houra, H.; Ali, F.)V1474,112-115(1991)

Faraday-effect magnetic field sensors based on substituted iron garnets (Deeter, Merritt N.; Rose, Allen H.; Day, Gordon W.)V1367,243-248(1991)

Far-IR magneto-emission study of the quantum-hall state and breakdown of the quantum-hall effect (Raymond, Andre; Chaubet, C.; Razeghi, Manijeh)V1362,275-281(1991)

FFT measuring method for magneto-optical ac current measurement (Hu, Shichuang; Ye, Miaoyuan; Qu, Gen)V1572,492-496(1991)

Fiber optic magnetic field and current sensor using magneto-birefringence of dense ferrofluid thin films (Pan, Yingtain; Liu, Xiande; Du, Chongwu; Li, Zai Q.)V1572,477-482(1991)

Fiber optic magnetic field sensor using spectral modulation encoding (Lequime, Michael; Meunier, Carole; Giovannini, Hugues)V1367,236-242(1991)

Fractal simulation of aggregation in magnetic thin film (Wang, Yi P.; Wang, Yu J.)V1519,605-608(1991)

Growth mechanism of orientated PLZT thin films sputtered on glass substrate (Zhang, Rui T.; Ge, Ming; Luo, Wei G.)V1519,757-760(1991)

Guided-wave magneto-optic and acousto-optic Bragg cells for RF signal processing (Tsai, Chen S.)V1562,55-65(1991)

High-speed large-capacity optical disk using pit-edge recording and MCAV method (Maeda, Takeshi; Saito, Atsushi; Sugiyama, Hisataka; Ojima, Masahiro; Arai, Shinichi; Shigematu, Kazuo)V1499,414-418(1991)

High-speed recording technologies for a magneto-optical disk system (Sukeda, Hirofumi; Tsuchinaga, Hiroyuki; Tanaka, Satoshi; Niihara, Toshio; Nakamura, Shigeru; Mita, Seiichi; Yamada, Yukinori; Ohta, Norio; Fukushima, Mitsugi)V1499,419-425(1991)

Improved plastic molding technology for magneto-optical disk substrate (Galic, George J.)V1499,539-546(1991)

Magnetically induced super resolution in a novel magneto-optical disk (Aratani, Katsuhisa; Fukumoto, Atsushi; Ohta, Masumi; Kaneko, Masahiko; Watanabe, Kenjirou)V1499,209-215(1991)

Magnetic and electronic properties of Co/Pd superlattices (Victora, Randall H.; MacLaren, J. M.)V1499,378-381(1991)

Magnetic circular dichroism measurements of benzene in the 9-eV region with synchrotron radiation (Snyder, Patricia A.; Munger, Robert; Hansen, Roger W.; Rowe, Ednor M.)V1548,188-196(1991)

Magnetic circular dichroism studies with soft x-rays (Tjeng, L. H.; Rudolf, P.; Meigs, G.; Sette, Francesco; Chen, Chien-Te; Idzerda, Y. U.)V1548,160-167(1991)

Magnetic steering of energy flow of linear and nonlinear magnetostatic waves in ferrimagnetic films (Wang, Qi; Bao, Jia S.; Boardman, A. D.)V1519,589-596(1991)

Magnetoconcentration nonequilibrium IR photodetectors (Piotrowski, Jozef; Djuric, Zoran G.)V1512,84-90(1991)

Magneto-optical apparatus with comparator for the measurement of large direct current (Zhang, Zhipeng; Zhao, Zhi; Chong, Baoxin)V1572,464-468(1991)

Magneto-optical Kerr rotation of thin ferromagnetic films (Zhai, H. R.; Xu, Y. B.; Lu, M.; Miao, Y. Z.; Hogue, K. L.; Naik, H. M.; Ahamd, M.; Dunifer, G. L.)V1519,575-579(1991)

Magneto-optical linear multichannel light modulator for recording of one-dimensional holograms (Abakumov, D. M.)V1621,138-147(1991)

Magneto-optical optical fiber switch (Gajda, Jerzy K.; Niesterowicz, Andrzej)V1391,329-331(1991)

Magneto-optical reading and writing integrated heads: a way to a multigigabyte multi-rigid-disk drive (Renard, Stephane; Valette, Serge)V1499,238-247(1991)

Magneto-optical signal detection with elliptically polarized light (Nakao, Takeshi; Arimoto, Akira A.; Takahashi, Masahiko)V1499,433-437(1991)

Magneto-optical studies of n-type Hg0.622Cd0.378Te grown by molecular beam epitaxy (Liu, Wei J.; Liu, Pu L.; Shi, Guo L.; Zhu, Jing-Bing; Yuan, Shi X.; Xie, Qin X.; He, Li)V1519,415-418(1991)

Magneto-optic data storage in the '90s (Funkenbusch, Arnold W.)V1396,699-708(1991)

Measurement of fibre Verdet constant with twist method (Dong, Xiaopeng; Hu, Hao; Qian, Jingren)V1572,56-60(1991)

Multilayer mirrors and filters for imaging the earth's inner magnetosphere (Schulze, Dean W.; Sandel, Bill R.; Broadfoot, A. L.)V1549,319-328(1991)

Multistability, instability, and chaos for intracavity magneto-optic modulation output (Yang, Darang; Song, Ruhua H.; Hu, Zhiping; Le, Shixiao)V1417,440-450(1991)

Multistable magnetostatic waves in thin film (Boardman, A. D.; Nikitov, S. A.; Wang, Qi; Bao, Jia S.; Cai, Ying S.)V1519,597-604(1991)

New approach to high-density recording on a magneto-optical disk (Fujita, Goro; Urakawa, Yoshinori; Yamagami, Tamotsu; Watanabe, Tetsu)V1499,426-432(1991)

New results on proton nuclear magnetic resonance of a-SiN:H films (Wang, Ji S.; Xu, Chun F.)V1519,857-859(1991)

Optical Data Storage '91 (Burke, James J.; Shull, Thomas A.; Imamura, Nobutake, eds.)V1499(1991)

Optical fiber magneto-optic current sensor (Zhao, Huafeng; Zhang, Peng-Gang; Liao, Yan-Biao)V1572,503-507(1991)

Optimization of quadrilayer structures for various magneto-optical recording materials (He, Ping; McGahan, William A.; Woollam, John A.)V1499,401-411(1991)

Photoemission from periodic structure of graded superlattices under magnetic field (Ghatak, Kamakhya P.)V1545,282-293(1991)

Polarized x-ray absorption spectroscopy for the study of superconductors and magnetic materials (Ramanathan, Mohan; Alp, Esen E.; Mini, Susan M.; Salem-Sugui, S.; Bommannavar, A.)V1548,168-178(1991)

Preparation and magneto-optical properties of NdDyFeCoTi amorphous films (Zhang, Si J.; Yang, Xiao Y.; Li, Xiao L.; Zhang, Feng P.)V1519,744-751(1991)

Preparation of PbTiO3 thin film by dc single-target magnetron sputtering (Yang, Bang C.; Wang, Ju Y.; Jia, Yu M.; Huang, Yong L.)V1519,725-728(1991)

Recent developments in fiber optic magnetostrictive sensors (Bucholtz, Frank; Dagenais, Dominique M.; Koo, Kee P.; Vohra, Sandeep T.)V1367,226-235(1991)

Simulation of free-electron lasers in the presence of measured magnetic field errors of the undulator (Marable, William P.; Tang, Cha-Mei; Esarey, Eric)V1552,80-86(1991)

Small polaron conduction in amorphous CoMnNiO thin film (Tan, Hui; Tao, Ming D.; Gin, Dong; Han, Ying; Lin, Cheng L.)V1519,752-756(1991)

Soft x-ray resonant magnetic scattering study of thin films and multilayers (Kao, Chi-Chang; Johnson, Erik D.; Hastings, Jerome B.; Siddons, D. P.; Vettier, C.)V1548,149-157(1991)

Study on different proportion W-Ti (C) binary alloy carbide thin film (Zhang, Yun H.; Wu, Bei X.; Yang, Guang Y.; Wang, Ren)V1519,729-734(1991)

Super resolution in a magneto-optical disk with an active mask (Fukumoto, Atsushi; Aratani, Katsuhisa; Yoshimura, Shunji; Udagawa, Toshiki; Ohta, Masumi; Kaneko, Masahiko)V1499,216-225(1991)

Theoretical analysis and design on optical fiber magneto-optic current sensing head (Zhang, Peng-Gang; Zhao, Huafeng; Liao, Yan-Biao)V1572,528-533(1991)

Theoretical investigation of near-edge phenomena in magnetic systems (Carra, Paolo)V1548,35-44(1991)

Thin film magnetic recording (Chen, Yi-Xin)V1519,539-547(1991)

Three-beam overwritable magneto-optic disk drive (Watabe, Akinori; Yamada, Ichiro; Yamamoto, Manabu; Katoh, Kikuji)V1499,226-235(1991)

UCSD high-energy x-ray timing experiment magnetic shield design and test results (Rothschild, Richard E.; Pelling, Michael R.; Hink, Paul L.)V1549,120-133(1991)

Use of magneto-optic spatial light modulators and linear detector arrays in inner-product associative memories (Goff, John R.)V1558,466-475(1991)

Waveguide-based fiber optic magnetic field sensor with directional sensitivity (Sohlstrom, Hans B.; Svantesson, Kjell G.)V1511,142-148(1991)

Wave-vector-dependent magneto-optics in semiconductors (De Salvo, Edmondo; Girlanda, Raffaello)V1362,870-875(1991)

Ways of the high-speed increasing of magneto-optical spatial light modulators (Randoshkin, Vladimir V.)V1469,796-803(1991)

Materials—see also coatings; crystals; diamonds; gallium arsenide materials; glass; optical materials; optical properties; photoconductors; photodetectors; photoresists; polymers; semiconductors; silicon; sol-gels; superconductors; thin films

9H-Indolo(1,2-f) phenanthridinium hemicyanines: a new group of high-efficiency broadband generating laser dyes (Kotowski, Tomasz; Skubiszak, Wojciech; Soroka, Jacek A.; Soroka, Krystyna B.; Stacewicz, Tadeusz)V1391,6-11(1991)

Absorption behavior of ceramic materials irradiated with excimer lasers (Toenshoff, Hans K.; Gedrat, Olaf)V1377,38-44(1991)

Advanced materials characterization by means of moire techniques (Aymerich, Francesco; Ginesu, Francesco; Priolo, Pierluigi)V1554B,304-314(1991)

Aging and surface instability in high-Tc superconductors (Larkins, Grover L.; Jones, W. K.; Lu, Q.; Levay, C.; Albaijes, D.)V1477,26-33(1991)

Analysis of macro-model composites with Fabry-Perot fiber-optic sensors (Fogg, Brian R.; Miller, William V.; Lesko, John J.; Carman, Gregory P.; Vengsarkar, Ashish M.; Reifsnider, Kenneth L.; Claus, Richard O.)V1588,14-25(1991)

Anomalous optical response of YBa2Cu3O7-x thin films during superconducting transitions (Xi, Xiaoxing; Venkatesan, T.; Etemad, Shahab; Hemmick, D.; Li, Q.)V1477,20-25(1991)

Application issues of fiber-optic sensors in aircraft structures (Lu, Zhuo J.; Blaha, Franz A.)V1588,276-281(1991)

Applications of powder diffraction in materials science using synchrotron radiation (Hart, Michael)V1550,11-17(1991)

Applied Spectroscopy in Material Science (Saperstein, David D., ed.)V1437(1991)

Attacking dimensional instability problems in graphite/epoxy structures (Krumweide, Gary C.; Brand, Richard A.)V1533,252-261(1991)

Ceramic phase-shifters for electronically steerable antenna systems (Selmi, Fathi; Ghodgaonkar, Deepak K.; Hughes, Raymond; Varadan, Vasundara V.; Varadan, Vijay K.)V1489,97-107(1991)

Characterization and enhancement of the damping within composite beams (FitzSimons, Philip M.; Trahan, Daniel J.)V1489,230-242(1991)

Characterization of the dimensional stability of advanced metallic materials using an optical test bench structure (Hsieh, Cheng; O'Donnell, Timothy P.)V1533,240-251(1991)

Chirality and its applications to engineered materials (Varadan, Vasundara V.; Varadan, Vijay K.)V1558,156-181(1991)

Combined x-ray absorption spectroscopy and x-ray powder diffraction (Dent, Andrew J.; Derbyshire, Gareth E.; Greaves, G. N.; Ramsdale, Christine A.; Couves, J. W.; Jones, Richard; Catlow, C. R.; Thomas, John M.)V1550,97-107(1991)

Comparison of analog and digital strategies for automatic vibration control of lightweight space structures (Hong, Suk-Yoon; Varadan, Vasundara V.; Varadan, Vijay K.)V1489,75-83(1991)

Comparison of flash-pumped Cr;Tm 2-um laser action in garnet hosts (Quarles, Gregory J.; Pinto, Joseph F.; Esterowitz, Leon; Kokta, Milan R.)V1410,165-174(1991)

Composite cure monitoring with embedded optical fiber sensors (Davis, Andrew; Ohn, Myo M.; Liu, Kexing; Measures, Raymond M.)V1489,33-43(1991)

Composite damage assessment employing an optical neural network processor and an embedded fiber-optic sensor array (Grossman, Barry G.; Gao, Xing; Thursby, Michael H.)V1588,64-75(1991)

Composite-embedded fiber-optic data links and related material/connector issues (Morgan, Robert E.; Ehlers, Sandy L.; Jon   Katharine J.)V1588,189-197(1991)

Computer-aided moire strain analysis on thermoplastic models at critical temperature (Barillot, Marc; Jacquot, Pierre M.; Di Chirico, Giuseppe)V1554A,867-878(1991)

Construction of the 16-meter large lunar telescope (Omar, Husam A.)V1494,135-146(1991)

Cooperative implementation of a high-temperature acoustic sensor (Baldini, S. E.; Nowakowski, Edward; Smith, Herb G.; Friebele, E. J.; Putnam, Martin A.; Rogowski, Robert S.; Melvin, Leland D.; Claus, Richard O.; Tran, Tuan A.; Holben, Milford S.)V1588,125-131(1991)

Crystal Growth in Space and Related Optical Diagnostics (Trolinger, James D.; Lal, Ravindra R., eds.)V1557(1991)

Crystal separation from mother solution and conservation under microgravity conditions using inert liquid (Regel, L. L.; Vedernikov, A. A.; Queeckers, P.; Legros, J. C.)V1557,182-191(1991)

Damage assessment in composites with embedded optical fiber sensors (Measures, Raymond M.; Liu, Kexing; LeBlanc, Michel; McEwen, Keith; Shankar, K.; Tennyson, R. C.; Ferguson, Suzanne M.)V1489,86-96(1991)

Damage detection in woven-composite materials using embedded fiber-optic sensors (Bonniau, Philippe; Chazelas, Jean; Lecuellet, Jerome; Gendre, Francois; Turpin, Marc; Le Pesant, Jean-Pierre; Brevignon, Michele)V1588,52-63(1991)

Deformation of a gamma/gamma WASPALOY after laser shock (Bourda, C.; Puig, Thierry T.; Decamps, B.; Condat, M.)V1502,148-159(1991)

Depth profiling resonance ionization mass spectrometry of electronic materials (Downey, Stephen W.; Emerson, A. B.)V1435,19-25(1991)

Development and testing of lightweight composite reflector panels (Helms, Richard G.; Porter, Christopher C.; Kuo, Chin-Po; Tsuyuki, Glenn T.)V1532,64-80(1991)

Development of a high-performance 86-mm MO disk by using polycarbonate substrate (Ito, Katsunori; Shibata, Yasumasa; Honda, Katsunori; Tsukahara, Makoto; Kojima, Kotaro; Kimura, Masakatsu; Fujisima, Toshihiko; Yoshinaga, Kazuomi; Kanai, Toshitaka; Shimizu, Keijiro)V1499,382-385(1991)

Distorted local environment about Zn and transition metals on the copper sites in YBa2Cu3O7 (Bridges, Frank; Li, Guoguang; Boyce, James B.; Claeson, Tord)V1550,76-84(1991)

DSPI: a tool for analyzing thermal strain on ceramic and composite materials (Hoefling, Roland; Aswendt, Petra; Totzauer, Werner F.; Jueptner, Werner P.)V1508,135-142(1991)

Effects of material combinations on the bimaterial fracture behavior (Wang, Wei-Chung; Chen, Jin-Tzaih)V1554A,60-69(1991)

Electronic materials basic research program managed by the Advanced Technology Directorate of the U.S. Army Strategic Defense Command (Martin, William D.)V1559,10-17(1991)

Energy of the lowest triplet states of rhodamines in ethanolic solutions (Targowski, Piotr; Zietek, Bernard)V1391,12-17(1991)

Excimer laser patterning of flexible materials (Kollia, Z.; Hontzopoulos, Elias I.)V1503,215-222(1991)

Excimer laser processing of ceramics and fiber-reinforced polymers assisted by a diagnostic system (Geiger, Manfred; Lutz, Norbert; Biermann, Stephan)V1503,238-248(1991)

Excimer laser surface treatment of ceramics (Hourdakis, G.; Hontzopoulos, Elias I.; Tsetsekou, A.; Zampetakis, Th.; Stournaras, C.)V1503,249-255(1991)

Fabrication and nonlinear optical properties of mixed and layered colloidal particles (McDonald, Joseph K.; LaiHing, Kenneth)V1497,367-370(1991)

Fiber optic damage detection for an aircraft leading edge (Measures, Raymond M.; LeBlanc, Michel; Hogg, W. D.; McEwen, Keith; Park, B. K.)V1332,431-443(1991)

Fiber optic sensor for the study of temperature and structural integrity of PZT: epoxy composite materials (Vishnoi, Gargi; Pillai, P.K. C.; Goel, T. C.)V1572,94-100(1991)

Fiber Optic Smart Structures and Skins IV (Claus, Richard O.; Udd, Eric, eds.)V1588(1991)

Fiber optic smart structures: structures that see the light (Measures, Raymond M.)V1332,377-398(1991)

Fiber optic technique for simultaneous measurement of strain and temperature variations in composite materials (Michie, W. C.; Culshaw, Brian; Roberts, Scott S.; Davidson, Roger)V1588,342-355(1991)

Flaw dynamics and vibrothermographic-thermoelastic nondestructive evaluation of advanced composite material (Tenek, Lazarus H.; Henneke, Edmund G.)V1467,252-263(1991)

Flight experiment to investigate microgravity effects on solidification phenomena of selected materials (Maag, Carl R.; Hansen, Patricia A.)V1557,24-30(1991)

Flow-field velocity measurements for nonisothermal systems (Johnson, Edward J.; Hyer,   V.; Culotta, Paul W.; Clark, Ivan O.)V1557,168-179(1991)

High-order nonlinear susceptibilities (Hammond, Richard T.)V1409,148-153(1991)

History of and potential for optical bonding agents in the visible (Magyar, James T.)V1535,55-58(1991)

Influence of the orthotropy of the ductile materials and the stress-assisted diffusion on the caustics (Papadopoulos, George A.)V1554A,826-834(1991)

Infrared thermal-wave studies of coatings and composites (Favro, Lawrence D.; Ahmed, Tasdiq; Crowther, D. J.; Jin, Huijia J.; Kuo, Pao K.; Thomas, Robert L.; Wang, X.)V1467,290-294(1991)

In-situ characterization of resin chemistry with infrared transmitting optical fibers and infrared spectroscopy (Druy, Mark A.; Glatkowski, Paul J.; Stevenson, William A.)V1437,66-74(1991)

Intelligent composites containing measuring fiber-optic networks for continuous self-diagnosis (Sansonetti, Pierre; Lequime, Michael; Engrand, D.; Guerin, J. J.; Davidson, Roger; Roberts, Scott S.; Fornari, B.; Martinelli, Mario; Escobar Rojo, Priscilla; Gusmeroli, Valeria; Ferdinand, Pierre; Plantey, J.; Crowther, Margaret F.; Culshaw, Brian; Michie, W. C.)V1588,198-209(1991)

Interaction of cracks, waves, and contacts: a photomechanics study (Rossmanith, H. P.)V1554A,2-28(1991); V1554B,2-28(1991)

Interface cracks close to free surfaces: a caustic study (Rossmanith, H. P.; Beer, Rudolf J.; Knasmillner, R. E.)V1554A,850-860(1991)

Interferometric fiber optic sensors for use with composite materials (Measures, Raymond M.; Valis, Tomas; Liu, Kexing; Hogg, W. D.; Ferguson, Suzanne M.; Tapanes, Edward)V1332,421-430(1991)

Interferometric moire analysis of wood and paper structures (Hyzer, Jim B.; Shih, J.; Rowlands, Robert E.)V1554B,371-382(1991)

Investigation of optical fibers as sensors for condition monitoring of composite materials (Nielsen, Peter L.)V1588,229-240(1991)

Investigations on excimer-laser-treated Cu/Cr contact materials (Schubert, Emil; Rosiwal, S.; Bergmann, Hans W.)V1503,299-309(1991)

Large zeolites: why and how to grow in space (Sacco, Albert)V1557,6-9(1991)

Laser beam deflection: a method to investigate convection in vapor growth experiments (Lenski, Harald; Braun, Michael)V1557,124-131(1991)

Lightweight composite mirrors: present and future challenges (Brand, Richard A.; Spinar, Karen K.)V1532,57-63(1991)

Low-frequency chiral coatings (Ro, Ru-Yen; Varadan, Vasundara V.; Varadan, Vijay K.)V1489,46-55(1991)

Low-temperature deformation measurement of materials by the use of laser speckle photography method (Nakahara, Sumio; Hisada, Shigeyoshi; Fujita, Takeyoshi; Sugihara, Kiyoshi)V1554A,602-609(1991)

Magnetic and electronic properties of Co/Pd superlattices (Victora, Randall H.; MacLaren, J. M.)V1499,378-381(1991)

Mapping crystal defects with a digital scanning ultramicroscope (Springer, John M.; Silberman, Enrique; Kroes, Roger L.; Reiss, Don)V1557,192-196(1991)

Materials for optimal multilayer coating (Grebenshikov, Sergey V.)V1519,302-307(1991)

Measurement of Poisson's ratio of nonmetallic materials by laser holographic interferometry (Zhu, Jian T.)V1554B,148-154(1991)

Measuring residual accelerations in the Spacelab environment (Witherow, William K.)V1557,42-52(1991)

Mechanical properties of composite materials containing embedded fiber-optic sensors (Roberts, Scott S.; Davidson, Roger)V1588,326-341(1991)

Melt dynamics and surface deformation in processing with laser radiation (Kreutz, Ernst W.; Pirch, Norbert)V1502,160-176(1991)

Mercurous halides for long time-delay Bragg cells (Brandt, Gerald B.; Singh, N. B.; Gottlieb, Milton S.)V1454,336-343(1991)

Method for embedding optical fibers and optical fiber sensors in metal parts and structures (Lee, Chung E.; Alcoz, J. J.; Gibler, William N,; Atkins, Robert A.; Taylor, Henry F.)V1588,110-116(1991)

Method of caustics for anisotropic materials (Rossmanith, H. P.)V1554A,835-849(1991)

Microinteraction of optical fibers embedded in laminated composites (Singh, Hemant; Sirkis, James S.; Dasgupta, Abhijit)V1588,76-85(1991)

Modeling of InP metal organic chemical vapor deposition (Black, Linda R.; Clark, Ivan O.; Kui, Jianming; Jesser, William A.)V1557,54-59(1991)

NASA microgravity materials science program (Sokolowski, Robert S.)V1557,2-5(1991)

New low-density, high-porosity lithium hydride-beryllium hydride foam: properties and applications to x-ray astronomy (Maienschein, Jon L.; Barry, Patrick E.; McMurphy, Frederick E.; Bowers, John S.)V1343,477-484(1991)

New optical approaches to the quantitative characterization of crystal growth, segregation, and defect formation (Carlson, D. J.; Wargo, Michael J.; Cao, X. Z.; Witt, August F.)V1557,140-146(1991)

Nonlinear electromagnetic field response of high-Tc superconducting microparticle composites (Haus, Joseph W.; Chung-Yau, F.; Bowden, Charles M.)V1497,382-388(1991)

Nonlinear optical properties of suspensions of microparticles: electrostrictive effect and enhanced backscatter (Kang, Chih-Chieh; Fiddy, Michael A.)V1497,372-381(1991)

Optical bonding agents for IR and UV refracting elements (Pellicori, Samuel F.)V1535,48-54(1991)

Optical neural networks using electron trapping materials (Jutamulia, Suganda; Storti, George M.; Seiderman, William M.)V1558,442-447(1991)

Optical recording materials (Savant, Gajendra D.; Jannson, Joanna)V1461,79-90(1991)

Optical sensors embedded in composite materials (Bocquet, Jean-Claud; Lecoy, Pierre; Baptiste, Didier)V1588,210-217(1991)

Photoelastic investigation of statics, kinetics, and dynamics of crack formation in transparent models and natural structural elements (Taratorin, B. I.; Sakharov, V. N.; Komlev, O. U.; Stcherbakov, V. N.; Starchevsky, A. V.)V1554A,449-456(1991)

Photoelastic stress analysis of bridge bearings (Allison, Ian M.)V1554A,332-340(1991)

Photoionization of rhodamines in etanolic solution at 77 K (Zietek, Bernard; Targowski, Piotr)V1391,18-23(1991)

Polyimide-coated embedded optical fiber sensors (Nath, Dilip K.; Nelson, Gary W.; Griffin, Stephen E.; Harrington, C. T.; He, Yi-Fei; Reinhart, Larry J.; Paine, D. C.; Morse, Ted F.)V1489,17-32(1991)

Present and future trends of laser materials processing in Japan (Matsunawa, Akira)V1502,60-71(1991)

Quantitative evaluation of cavities and inclusions in solids using IR thermography (Madrid, Angel)V1467,322-336(1991)

Real-time quantitative imaging for semiconductor crystal growth, control, and characterization (Wargo, Michael J.)V1557,271-282(1991)

Requirements of a solar diffuser and measurements of some candidate materials (Guzman, Carmen T.; Palmer, James M.; Slater, Philip N.; Bruegge, Carol J.; Miller, Edward A.)V1493,120-131(1991)

Role of adhesion in optical-fiber-based smart composite structures and its implementation in strain analysis for the modeling of an embedded optical fiber (DiFrancia, Celene; Claus, Richard O.; Ward, T. C.)V1588,44-49(1991)

Selection, growth, and characterization of materials for MBE-produced x-ray optics (Kearney, Patrick A.; Slaughter, Jon M.; Falco, Charles M.)V1343,25-31(1991)

Slow thermodeformation of metals with fast laser heating (Baloshin, Yu. A.; Yurevich, V. I.; Sud'enkov, Yu. V.)V1440,71-77(1991)

Solution of optimization problems of perforated and box-shaped structures by photoelasticity and numerical methods (Shvej, E. M.; Latysh, E. G.; Morgunov, A. N.; Mikhalchenko, O. E.; Stchetinin, A. L.; Volokh, K. Y.)V1554A,488-495(1991)

Space experiments using small satellites (Schor, Matthew J.)V1495,146-148(1991)

Strain measurements on soft materials application to cloth and papers (Bremand, Fabrice J.; Lagarde, Alexis)V1554B,650-660(1991)

Strategies for tunable frequency-selective surfaces (Lakhtakia, Akhlesh)V1489,108-111(1991)

Structures Sensing and Control (Breakwell, John; Varadan, Vijay K., eds.)V1489(1991)

Study of an opto-ultrasonic technique for cure monitoring (Davis, Andrew; Ohn, Myo M.; Liu, Kexing; Measures, Raymond M.)V1588,264-274(1991)

Study of fluorescent glass-ceramics (Qiu, Guanming; DuanMu, Qingduo)V1513,396-407(1991)

Study of oxidization process in real time using speckle correlation (Muramatsu, Mikiya; Guedes, G. H.; Matsuda, Kiyofumi; Barnes, Thomas H.)V1332,792-797(1991)

Theoretical models in optics of glass-composite materials (Kuchinskii, Sergei A.; Dostenko, Alexander V.)V1513,297-308(1991)

Thermographic analysis of the anisotropy in the thermal conductivity of composite materials (Burleigh, Douglas D.; De La Torre, William)V1467,303-310(1991)

Use of 3 X 3 integrated optic polarizer/splitters for a smart aerospace plane structure (Seshamani, Ramani; Alex, T. K.)V1489,56-64(1991)

Variation of magnetic properties of iron-contained quartzite irradiated by CO2 laser (Mukhamedgalieva, Anel F.)V1502,223-225(1991)

Wideband embedded/conformal antenna subsystem concept (Smalanskas, Joseph P.; Valentine, Gary W.; Wolfson, Ronald I.)V1489,2-8(1991)

X-ray absorption fine structure of systems in the anharmonic limit (Mustre de Leon, Jose; Conradson, Steven D.; Batistic, I.; Bishop, A. R.; Raistrick, Ian D.; Jackson, W. E.; Brown, George S.)V1550,85-96(1991)

X-ray absorption spectroscopy: how is it done? what can it tell us? (Hayes, Tim)V1550,56-66(1991)

X Rays in Materials Analysis II: Novel Applications and Recent Developments (Mills, Dennis M., ed.)V1550(1991)

Mathematical morphology—see also image algebra; image processing

Adaptive morphological filter for image processing (Cheng, Fulin; Venetsanopoulos, Anastasios N.)V1483,49-59(1991)

Adaptive morphological multiresolution decomposition (Salembier, Philippe; Jaquenoud, Laurent)V1568,26-37(1991)

Analysis of noise attenuation in morphological image processing (Koskinen, Lasse; Astola, Jaakko; Neuvo, Yrjo A.)V1451,102-113(1991)

Application of back-propagation to the recognition of handwritten digits using morphologically derived features (Hepp, Daniel J.)V1451,228-233(1991)

Application of mathematical morphology to the automated determination of microstructural characteristics of composites (Chackalackal, Mathew S.; Basart, John P.)V1568,347-356(1991)

Application of morphological pseudoconvolutions to scanning-tunneling and atomic force microscopy (Dougherty, Edward R.; Weisman, Andrew; Mizes, Howard; Miller, Robert J.)V1567,88-99(1991)

Automated visual inspection system for IC bonding wires using morphological processing (Tsukahara, Hiroyuki; Nakashima, Masato; Sugawara, Takehisa)V1384,15-26(1991)

Characteristic pattern matching based on morphology (Zhao, Dongming)V1606,86-96(1991)

Comparison of morphological and conventional edge detectors in medical imaging applications (Kaabi, Lotfi; Loloyan, Mansur; Huang, H. K.)V1445,11-23(1991)

Conditional-expectation-based implementation of the optimal mean-square binary morphological filter (Dougherty, Edward R.; Mathew, A.; Swarnakar, Vivek)V1451,137-147(1991)

Connectivity-preserving morphological image transformations (Bloomberg, Dan S.)V1606,320-334(1991)

CT image processing for hardwood log inspection (Zhu, Dongping; Conners, Richard W.; Araman, Philip A.)V1567,232-243(1991)

Custom-made filters in digital image analysis system (Ostrowski, Tomasz)V1391,264-266(1991)

Decomposing morphological structure element into neighborhood configurations (Gong, Wei; Shi, Qinyun; Cheng, Minde)V1606,153-164(1991)

Deformable markers: mathematical morphology for active contour models control (Rougon, Nicolas F.; Preteux, Francoise)V1568,78-89(1991)

Discrete random set models for shape synthesis and analysis (Goutsias, John I.; Wen, Chuanju)V1606,174-185(1991)

Distribution of the pattern spectrum mean for convex base images (Dougherty, Edward R.; Sand, Francis M.)V1451,114-124(1991)

Domain-variant gray-scale morphology (Kraus, Eugene J.)V1451,171-178(1991)

Efficient software techniques for morphological image processing with desktop computers (Zmuda, Michael A.; Tamburino, Louis A.; Rizki, Mateen M.)V1470,183-189(1991)

Fast algorithm for size analysis of irregular pore areas (Yuan, Li-Ping)V1451,125-136(1991)

Fast dilation and erosion of time-varying grey-valued images with uncertainty (Laplante, Phillip A.; Giardina, Charles R.)V1568,295-302(1991)

Geometrical and morphological image processing algorithm (Reihani, Kamran; Thompson, Wiley E.)V1452,319-329(1991)

Geometric property measurement of binary images by polynomial representation (Bhattacharya, Prabir; Qian, Kai)V1468,918-922(1991)

Gradient descent techniques for feature detection template generation (Pont, W. F.; Gader, Paul D.)V1568,247-260(1991)

Hierarchical image decomposition based on modeling of convex hulls corresponding to a set of order statistic filters (Vepsalainen, Ari M.; Linnainmaa, Seppo; Yli-Harja, Olli P.)V1568,2-13(1991)

Hole spectrum: model-based optimization of morphological filters (Dougherty, Edward R.; Haralick, Robert M.)V1568,224-232(1991)

Hybrid bipixel structuring element decomposition and Euclidean morphological transforms (Zhou, Ziheng; Venetsanopoulos, Anastasios N.)V1606,309-319(1991)

Image Algebra and Morphological Image Processing II (Gader, Paul D.; Dougherty, Edward R., eds.)V1568(1991)

Image analysis applications for grain science (Zayas, Inna Y.; Steele, James L.)V1379,151-161(1991)

Image analysis using threshold reduction (Bloomberg, Dan S.)V1568,38-52(1991)

Implementing neural-morphological operations using programmable logic (Shih, Frank Y.; Moh, Jenlong)V1382,99-110(1991)

Interpolation of stereo data using Lagrangian polynomials (Bachnak, Rafic A.; Yamout, Jihad S.)V1457,27-36(1991)

Learnability of min-max pattern classifiers (Yang, Ping-Fai; Maragos, Petros)V1606,294-308(1991)

Lipschitz lattices and numerical morphology (Serra, Jean C.)V1568,54-65(1991)

Mathematical theories of shape: do they model perception? (Mumford, David)V1570,2-10(1991)

Model-based morphology (Haralick, Robert M.; Dougherty, Edward R.; Katz, Philip L.)V1472,108-117(1991)

Morphological algorithms for modeling Gaussian image features (Bhagvati, Chakravarthy; Marineau, Peter; Skolnick, Michael M.; Sternberg, Stanley)V1606,112-119(1991)

Morphological feature-set optimization using the genetic algorithm (Trenkle, John M.; Schlosser, Steve; Vogt, Robert C.)V1568,212-223(1991)

Morphological pattern-spectra-based Tau-opening optimization (Dougherty, Edward R.; Haralick, Robert M.; Chen, Yidong; Li, Bo; Agerskov, Carsten; Jacobi, Ulrik; Sloth, Poul H.)V1606,141-152(1991)

Morphological processing for the analysis of disordered structures (Casasent, David P.; Sturgill, Robert; Schaefer, Roland H.)V1567,683-690(1991)

Morphological processing to reduce shading and illumination effects (Casasent, David P.; Schaefer, Roland H.; Kokaj, Jahja O.)V1385,152-164(1991)

Morphological pyramid with alternating sequential filters (Morales, Aldo W.; Acharya, Raj S.)V1452,258-269(1991)

Morphological segmentation and 3-D rendering of the brain in magnetic resonance imaging (Connor, William H.; Diaz, Pedro J.)V1568,327-334(1991)

Morphologic edge detection in range images (Gupta, Sundeep; Krishnapuram, Raghu J.)V1568,335-346(1991)

Multiscale analysis based on mathematical morphology (Lu, Yi; Vogt, Robert C.)V1568,14-25(1991)

Multiscale morphological region coding (Macq, Benoit M.; Ronse, Christian; Van Dongen, V.)V1606,165-173(1991)

New concepts in mathematical morphology: the topographical and differential distance functions (Preteux, Francoise; Merlet, Nicolas)V1568,66-77(1991)

New trends in morphological algorithms (Vincent, Luc M.)V1451,158-170(1991)

Nonlinear Image Processing II (Arce, Gonzalo R.; Boncelet, Charles G.; Dougherty, Edward R., eds.)V1451(1991)

Optical gray-scale morphology for target detection (Casasent, David P.; Schaefer, Roland H.)V1568,313-326(1991)

Optical inference processing techniques for scene analysis (Casasent, David P.)V1564,504-510(1991)

Optimization of morphological structuring elements for angiogram enhancement (Andress, Keith M.; Wilson, David L.)V1445,6-10(1991)

Order-statistic filters on matrices of images (Wilson, Stephen S.)V1451,242-253(1991)

Pattern spectrum morphology for texture discrimination and object recognition (Lee, Bonita G.; Tom, Victor T.)V1381,80-91(1991)

Phenomenological description of self-organized x(2) grating formation in centrosymmetric doped optical fibers (Chmela, Pavel)V1516,116-124(1991)

Polynomial approach for morphological operations on 2-D and 3-D images (Bhattacharya, Prabir; Qian, Kai)V1383,530-536(1991)

Practical VLSI realization of morphological operations (Chang, Yiher; Ansari, Nirwan)V1606,839-850(1991)

Quadtree decomposition of binary structuring elements (Shoji, Kenji)V1451,148-157(1991)

Recursion and feedback in image algebra (Ritter, Gerhard X.; Davidson, Jennifer L.)V1406,74-86(1991)

Set discrimination analysis tools for grey-level morphological operators (Vogt, Robert C.)V1568,200-211(1991)

Shape recognition combining mathematical morphology and neural networks (Schmitt, Michel; Mattioli, Juliette)V1469,392-403(1991)

Shape registration by morphological operations (Chang, Long-Wen; Tsai, Mong-Jean)V1606,120-131(1991)

Soft morphological filters (Koskinen, Lasse; Astola, Jaakko; Neuvo, Yrjo A.)V1568,262-270(1991)

Statistical morphology (Yuille, Alan L.; Vincent, Luc M.; Geiger, Davi)V1568,271-282(1991)

Template learning in morphological neural nets (Davidson, Jennifer L.; Sun, K.)V1568,176-187(1991)

Three-dimensional morphology for target detection (Patterson, Tim J.)V1471,358-368(1991)

Trade-offs between pseudohexagonal and pseudocubic filters (Preston, Kendall)V1451,75-90(1991)

Two inverse problems in mathematical morphology (Schmitt, Michel)V1568,283-294(1991)

Two new image compression methods utilizing mathematical morphology (Vepsalainen, Ari M.; Toivanen, Pekka J.)V1606,282-293(1991)

Unsupervised training of structuring elements (Wilson, Stephen S.)V1568,188-199(1991)

Using morphology in document image processing (Concepcion, Vicente P.; Grzech, Matthew P.; D'Amato, Donald P.)V1606,132-140(1991)

Using structuring-element libraries to design suboptimal morphological filters (Loce, Robert P.; Dougherty, Edward R.)V1568,233-246(1991)

Medical imaging—see also acoustics; radiology; tomography; x rays

500 MHz baseband fiber optic transmission system for medical imaging applications (Cheng, Xin; Huang, H. K.)V1364,204-208(1991)

Adaptive coding method of x-ray mammograms (Baskurt, Atilla; Magnin, Isabelle E.; Bremond, Alain; Charvet, Pierre Y.)V1444,240-249(1991)

Adaptive projection technique for CT images refinement (Kuo, Shyh-Shiaw; Mammone, Richard J.)V1606,641-652(1991)

Analysis and visualization of heart motion (Chen, Chang W.; Huang, Thomas S.; Arrott, Matthew)V1450,231-242(1991)

Analysis of absorption, scattering, and hemoglobin saturation using phase-modulation spectroscopy (Sevick, Eva M.; Weng, Jian; Maris, Michael B.; Chance, Britton)V1431,264-275(1991)

Analysis of multidimensional confocal images (Samarabandu, J. K.; Acharya, Raj S.; Edirisinghe, Chandima D.; Cheng, Ping-Chin; Kim, Hyo-Gun; Lin, T. H.; Summers, R. G.; Musial, C. E.)V1450,296-322(1991)

Analyzing and interpreting pulmonary tomoscintigraphy sequences: realization and perspectives (Forte, Anne-Marie; Bizais, Yves)V1445,409-420(1991)

Anatomical constraints for neuromagnetic source models (George, John S.; Lewis, Paul S.; Ranken, D. M.; Kaplan, L.; Wood, C. C.)V1443,37-51(1991)

ANN approach for 2-D echocardiographic image processing: application of neocognitron model to LV boundary formation (Wan, Liqun; Li, Dapeng; Wee, William G.; Han, Chia Y.; Porembka, D. T.)V1469,432-440(1991)

Anthropomorphic classification using three-dimensional Fourier descriptor (Lee, Nahm S.; Park, Kyung S.)V1450,133-143(1991)

Application of a discrete-space representation to three-dimensional medical imaging (Toennies, Klaus D.; Tronnier, Uwe)V1444,19-25(1991)

Application of median-type filtering to image segmentation in electrophoresis (Wang, Qiaofei; Neuvo, Yrjo A.)V1450,47-58(1991)

Application of the nonrigid shape matching algorithm to volumetric cardiac images (Goldgof, Dmitry B.; Chandra, Kambhamettu)V1450,264-276(1991)

Automated cyst recognition from x-ray photographs (Nedkova, Rumiana; Delchev, Georgy)V1429,105-107(1991)

Automated extraction of vascular information from angiographic images using a vessel-tracking algorithm (Alperin, Noam; Hoffmann, Kenneth R.; Doi, Kunio)V1396,27-31(1991)

Automated grading of venous beading: an algorithm and parallel implementation (Shen, Zhijiang; Gregson, Peter H.; Cheng, Heng-Da; Kozousek, V.)V1606,632-640(1991)

Automated labeling of coronary arterial tree segments in angiographic projection data (Dumay, Adrie C.; Gerbrands, Jan J.; Geest, Rob J.; Verbruggen, Patricia E.; Reiber, J. H.)V1445,38-46(1991)

Automated techniques for quality assurance of radiological image modalities (Goodenough, David J.; Atkins, Frank B.; Dyer, Stephen M.)V1444,87-99(1991)

Automated three-dimensional registration of medical images (Neiw, Han-Min; Chen, Chin-Tu; Lin, Wei-Chung; Pelizzari, Charles A.)V1445,259-264(1991)

Automatic 3-D reconstruction of vascular geometry from two orthogonal angiographic image sequences (Close, Robert A.)V1445,513-522(1991)

Automatic acquisition of movement information by a knowledge-based recognition approach (Bae, Kyongtae T.; Altschuler, Martin D.)V1380,108-115(1991)

Automatic adjustment of display window (gray level) for MR images using a neural network (Ohhashi, Akinami; Yamada, Shinichi; Haruki, Kazuhito; Hatano, Hisaaki; Fujii, Yumi; Yamaguchi, Koujiro; Ogata, Hakaru)V1444,63-74(1991)

Automatic analysis system for three-dimensional angiograms (Higgins, William E.; Spyra, Wolfgang J.; Karwoski, Ronald A.; Ritman, Erik L.)V1445,276-286(1991)

Automatic counting of chromosome fragments for the determination of radiation dose (Smith, Warren E.; Leung, Billy C.; Leary, James F.)V1398,142-150(1991)

Automatic recognition of bone for x-ray bone densitometry (Shepp, Larry A.; Vardi, Y.; Lazewatsky, J.; Libeau, James; Stein, Jay A.)V1452,216-224(1991)

Automatic segmentation of brain images: selection of region extraction methods (Gong, Leiguang; Kulikowski, Casimir A.; Mezrich, Reuben S.)V1450,144-153(1991)

Automatic segmentation of microvessels using textural analysis (Albert, Thomas; O'Connor, Carol; Harris, Patrick D.)V1450,84-89(1991)

Automatic wall motion detection in the left ventricle using ultrasonic images (Olstad, Bjorn)V1450,243-254(1991)

Ballistic imaging of biomedical samples using picosecond optical Kerr gate (Wang, LeMing; Liu, Y.; Ho, Ping-Pei; Alfano, Robert R.)V1431,97-101(1991)

Best fit ellipse for cell shape analysis (Wali, Rahman; Colef, Michael; Barba, Joseph)V1606,665-674(1991)

Biomechanical research of joints: IV. the biohinge of primates (Zhang, Renxiang; Yu, Jie; Lan, Zu-yun; Qu, Wen-ji; Zhang, Hong-zi; Zhang, Kui; Zhang, Liang)V1380,116-121(1991)

Biomedical applications of video-speckle techniques (Lokberg, Ole J.)V1525,9-16(1991)

Biomedical Image Processing II (Bovik, Alan C.; Howard, Vyvyan, eds.)V1450(1991)

Biomedical structure recognition by successive approximations (Venturi, Giovanni; Dellepiane, Silvana G.; Vernazza, Gianni L.)V1606,217-225(1991)

Biospeckle phenomena and their applications to blood-flow measurements (Aizu, Yoshihisa; Asakura, Toshimitsu)V1431,239-250(1991)

Biostereometric Technology and Applications (Herron, Robin E., ed.)V1380(1991)

Blackboard architecture for medical image interpretation (Davis, Darryl N.; Taylor, Christopher J.)V1445,421-432(1991)

Blink comparison techniques applied to medical images (Craine, Eric R.; Craine, Brian L.)V1444,389-399(1991)

Body shape changes in the elderly and the influence of density assumptions on segment inertia parameters (Jensen, Robert K.; Fletcher, P.; Abraham, C.)V1380,124-136(1991)

Brain surface maps from 3-D medical images (Lu, Jiuhuai; Hansen, Eric W.; Gazzaniga, Michael S.)V1459,117-124(1991)

Brightness and contrast adjustments for different tissue densities in digital chest radiographs (McNitt-Gray, Michael F.; Taira, Ricky K.; Eldredge, Sandra L.; Razavi, Mahmood)V1445,468-478(1991)

Cellular vibration measurement with a noninvasive optical system (Khanna, Shyam M.)V1429,9-20(1991)

Channeling radiation as an x-ray source for angiography, x-ray lithography, molecular structure determination, and elemental analysis (Uberall, Herbert; Faraday, Bruce J.; Maruyama, Xavier K.; Berman, Barry L.)V1552,198-213(1991)

Classification of tissue-types by combining relaxation labeling with edge detection (Adiseshan, Prakash; Faber, Tracy L.)V1445,128-132(1991)

Clinical aspects of quality assurance in PACS (Marglin, Stephen I.)V1444,83-86(1991)

Clinical aspects of the Mayo/IBM PACS project (Forbes, Glenn S.; Morin, Richard L.; Pavlicek, William)V1446,318-322(1991)

Clinical evaluation of a 2K x 2K workstation for primary diagnosis in pediatric radiology (Razavi, Mahmood; Sayre, James W.; Simons, Margaret A.; Hamedaninia, Azar; Boechat, Maria I.; Hall, Theodore R.; Kangarloo, Hooshang; Taira, Ricky K.; Chuang, Keh-Shih; Kashifian, Payam)V1446,24-34(1991)

Clinical experience with a stereoscopic image workstation (Henri, Christopher J.; Collins, D. L.; Pike, G. B.; Olivier, A.; Peters, Terence M.)V1444,306-317(1991)

Clinical fluorescence diagnosis of human bladder carcinoma following low-dose photofrin injection (Baert, Luc; Berg, Roger; van Damme, B.; D'Hallewin, Mich A.; Johansson, Jonas; Svanberg, Katarina; Svanberg, Sune)V1525,385-390(1991)

Clinical image-intensifier-based volume CT imager for angiography (Ning, Ruola; Barsotti, John B.; Kido, Daniel K.; Kruger, Robert A.)V1443,236-249(1991)

Collaborative processing to extract myocardium from a sequence of two-dimensional echocardiograms (Revankar, Shriram; Sher, David B.; Rosenthal, Steven)V1459,268-273(1991)

Color analysis of nonlinear-phase-modulation method for density pseudocolor encoding technique in medical application (Liu, Dingyu; Yang, Xiaobo; Liu, Changjun; Zhang, Honguo)V1443,191-196(1991)

Combined-transform coding scheme for medical images (Zhang, Ya-Qin; Loew, Murray H.; Pickholtz, Raymond L.)V1445,358-366(1991)

Compact open-architecture computed radiography system (Huang, H. K.; Lim, Art J.; Kangarloo, Hooshang; Eldredge, Sandra L.; Loloyan, Mansur; Chuang, Keh-Shih)V1443,198-202(1991)

Comparative study of texture measurements for cellular organelle recognition (Grenier, Marie-Claude; Durand, Louis-Gilles; de Guise, J.)V1450,154-169(1991)

Comparison of detection systems in time-of-flight transillumination imaging (Hayes, Jeffrey A.; Sullivan, Barry J.)V1396,324-330(1991)

Comparison of morphological and conventional edge detectors in medical imaging applications (Kaabi, Lotfi; Loloyan, Mansur; Huang, H. K.)V1445,11-23(1991)

Comparison of surface and volume presentation of multislice biomedical images (Browne, Mark A.; Jolleys, Glenn D.; Joyner, David J.)V1450,338-349(1991)

Comparison of three-dimensional surface rendering techniques (Thomas, Judith G.; Galloway, Robert L.; Edwards, Charles A.; Haden, Gerald L.; Maciunas, Robert J.)V1444,379-388(1991)

Comparison of worldwide opinions on the costs and benefits of PACS (van Gennip, Elisabeth M.; van Poppel, Bas M.; Bakker, Albert R.; Ottes, Fenno P.)V1446,442-450(1991)

Composite PET and MRI for accurate localization and metabolic modeling: a very useful tool for research and clinic (Bidaut, Luc M.)V1445,66-77(1991)

Computer-assisted surgical planning and automation of laser delivery systems (Zamorano, Lucia J.; Dujovny, Manuel; Dong, Ada; Kadi, A. M.)V1428,59-75(1991)

Computer interpretation of thallium SPECT studies based on neural network analysis (Wang, David C.; Karvelis, K. C.)V1445,574-575(1991)

Computerized system for clinical diagnosis of melanoma (Ferrario, Mario; Barbieri, Fabio)V1450,108-117(1991)

Computer simulation of magnetic resonance images using fractal-grown brain slices (Cheng, Shirley N.)V1396,9-14(1991)

Computer vision system for the detection and characterization of masses for use in mammographic screening programs (Yin, Fang-Fang; Giger, Maryellen L.; Doi, Kunio; Vyborny, Carl J.; Schmidt, Robert A.; Metz, Charles E.)V1396,2-4(1991)

Cone beam for medical imaging and NDE (Smith, Bruce D.)V1450,13-17(1991)

Confocal light microscopy of the living in-situ ocular lens: two- and three-dimensional imaging (Masters, Barry R.)V1443,288-293(1991)

Confocal redox fluorescence microscopy for the evaluation of corneal hypoxia (Masters, Barry R.; Kriete, Andres; Kukulies, Joerg)V1431,218-223(1991)

Constructing topologically connected surfaces for the comprehensive analysis of 3-D medical structures (Kalvin, Alan D.; Cutting, Court B.; Haddad, Betsy; Noz, Marilyn E.)V1445,247-258(1991)

Contrast-agent-enhanced magnetic resonance imaging: early detection of neoplastic lesions of the CNS (Carvlin, Mark J.; Rosa, Louis; Rajan, Sunder; Francisco, John)V1426,13-21(1991)

Correction of image-phase aberrations in MRI with applications (Riek, Jonathan K.; Tekalp, Ahmet M.; Smith, Warren E.; Parker, Kevin J.)V1445,190-197(1991)

Correlation model for a class of medical images (Zhang, Ya-Qin; Loew, Murray H.; Pickholtz, Raymond L.)V1445,367-373(1991)

CT imaging with an image intensifier: using a radiation therapy simulator as a CT scanner (Silver, Michael D.; Nishiki, Masayuki; Tochimura, Katsumi; Arita, Masataka; Drawert, Bruce M.; Judd, Thomas C.)V1443,250-260(1991)

Data analysis methods for near-infrared spectroscopy of tissue: problems in determining the relative cytochrome aa3 concentration (Cope, Mark; van der Zee, Pieter; Essenpreis, Matthias; Arridge, Simon R.; Delpy, David T.)V1431,251-262(1991)

Decision support system for capillaroscopic images (Tascini, Guido; Puliti, Paolo; Zingaretti, Primo)V1450,178-185(1991)

Decoupled coil detector array in magnetic resonance imaging (Kwiat, Doron; Einav, Shmuel)V1443,2-28(1991)

Definition and evaluation of the data-link layer of PACnet (Alsafadi, Yasser H.; Martinez, Ralph; Sanders, William H.)V1446,129-140(1991)

Deformation measurement of the bone fixed with external fixator using holographic interferometry (Kojima, Arata; Ogawa, Ryokei; Izuchi, N.; Yamamoto, M.; Nishimoto, T.; Matsumoto, Toshiro)V1429,162-171(1991)

Deformation measurements of the human tympanic membrane under static pressure using automated moire topography (Dirckx, Joris J.; Decraemer, Willem F.)V1429,34-38(1991)

Design and analysis of a water window imaging x-ray microscope (Hoover, Richard B.; Baker, Phillip C.; Shealy, David L.; Brinkley, B. R.; Walker, Arthur B.; Barbee, Troy W.)V1426,84-96(1991)

Design considerations for a high-resolution film scanner for teleradiology applications (Kocher, Thomas E.; Whiting, Bruce R.)V1446,459-464(1991)

Detailed description of the Mayo/IBM PACS (Gehring, Dale G.; Persons, Kenneth R.; Rothman, Melvyn L.; Salutz, James R.; Morin, Richard L.)V1446,248-252(1991)

Detection and localization of absorbers in scattering media using frequency-domain principles (Berndt, Klaus W.; Lakowicz, Joseph R.)V1431,149-160(1991)

Detection of liver metastisis using the backpropagation algorithm and linear discriminant analysis (DaPonte, John S.; Parikh, Jo Ann; Katz, David A.)V1469,441-450(1991)

Development of a neural network for early detection of renal osteodystrophy (Cheng, Shirley N.; Chan, Heang-Ping; Adler, Ronald; Niklason, Loren T.; Chang, Chair-Li)V1450,90-98(1991)

Development of a smart workstation for use in mammography (Giger, Maryellen L.; Nishikawa, Robert M.; Doi, Kunio; Yin, Fang-Fang; Vyborny, Carl J.; Schmidt, Robert A.; Metz, Charles E.; Wu, Yuzheng; MacMahon, Heber; Yoshimura, Hitoshi)V1445,101-103(1991)

Development of computer-aided functions in clinical neurosurgery with PACS (Mukasa, Minoru; Aoki, Makoto; Satoh, Minoru; Kowada, Masayoshi; Kikuchi, K.)V1446,253-265(1991)

Diagnostic and Therapeutic Cardiovascular Interventions (Abela, George S., ed.)V1425(1991)

Diagnostic report acquisition unit for the Mayo/IBM PACS project (Brooks, Everett G.; Rothman, Melvyn L.)V1446,217-226(1991)

Diagnostic value model for the evaluation of PACS: physician ratings of the importance of prompt image access and the utilization of a display station in an intensive care unit (Kundel, Harold L.; Seshadri, Sridhar B.; Arenson, Ronald L.)V1446,297-300(1991)

Diffuse tomography (Gruenbaum, F. A.; Kohn, Philip D.; Latham, Geoff A.; Singer, Jay R.; Zubelli, Jorge P.)V1431,232-238(1991)

Diffusion of intensity modulated near-infrared light in turbid media (Fishkin, Joshua B.; Gratton, Enrico; vandeVen, Martin J.; Mantulin, William W.)V1431,122-135(1991)

Digital film library implementation (Kishore, Sheel; Khalsa, Satjeet S.; Seshadri, Sridhar B.; Arenson, Ronald L.)V1446,188-194(1991)

Digital image processing for the early localization of cancer (Kelmar, Cheryl M.)V1426,47-57(1991)

Digital imaging microscopy: the marriage of spectroscopy and the solid state CCD camera (Jovin, Thomas M.; Arndt-Jovin, Donna J.)V1439,109-120(1991)

Digital imaging of Giemsa-banded human chromosomes: eigenanalysis and the Fourier phase reconstruction (Jericevic, Zeljko; McGavran, Loris; Smith, Louis C.)V1428,200-213(1991)

Digital radiology with solid state linear x-ray detectors (Munier, Bernard; Prieur-Drevon, P.; Chabbal, Jean)V1447,44-55(1991)

Digital replication of chest radiographs without altering diagnostic observer performance (Flynn, Michael J.; Davies, Eric; Spizarny, David; Beute, Gordon; Peterson, Ed; Eyler, William R.; Gross, Barry; Chen, Ji)V1444,172-179(1991)

Direct digital image transfer gateway (Mun, In K.; Kim, Y. S.; Mun, Seong Ki)V1444,232-237(1991)

Discrete-cosine-transform-based image compression applied to dermatology (Cookson, John P.; Sneiderman, Charles; Rivera, Christopher)V1444,374-378(1991)

Display systems for medical imaging (Erdekian, Vahram V.; Trombetta, Steven P.)V1444,151-158(1991)

Distributing the server function in a multiring PAC system (Lynne, Kenton J.)V1446,177-187(1991)

DRILL: a standardized radiology-teaching knowledge base (Rundle, Debra A.; Evers, K.; Seshadri, Sridhar B.; Arenson, Ronald L.)V1446,405-413(1991)

Effects of transmission errors on medical images (Pronios, Nikolaos-John B.; Yovanof, Gregory S.)V1446,108-128(1991)

Efficient computation of various types of skeletons (Vincent, Luc M.)V1445,297-311(1991)

Efficient extraction of local myocardial motion with optical flow and a resolution hierarchy (Srikantan, Geetha; Sher, David B.; Newberger, Ed)V1459,258-267(1991)

Electro-optical system for the nondestructive evaluation of bioengineering materials (Sciammarella, Cesar A.; Bhat, Gopalakrishna K.; Albertazzi, Armando A.)V1429,183-194(1991)

Endoscopic fluorescence detection of early lung cancer (Profio, A. E.; Balchum, Oscar J.; Lam, Stephen)V1426,44-46(1991)

Enhancement of dental x-ray images by two-channel image processing (Mitra, Sanjit K.; Yu, Tian-Hu)V1445,156-165(1991)

Equalization radiography with radiation quality modulation (Geluk, Ronald J.; Vlasbloem, Hugo)V1443,143-152(1991)

European activities towards a hospital-integrated PACS based on open systems (Kouwenberg, Jef M.; Ottes, Fenno P.; Bakker, Albert R.)V1446,357-361(1991)

Evaluation of a moving slit technique for mammography (Rosenthal, Marc S.; Sashin, Donald; Herron, John M.; Maitz, Glenn S.; Boyer, Joseph W.; Gur, David)V1443,132-142(1991)

Evaluation of facial palsy by moire topography (Inokuchi, Ikuo; Kawakami, Shinichiro; Maeta, Manabu; Masuda, Yu)V1429,39-45(1991)

Evaluation of medical image compression by Gabor elementary functions (Anderson, Mary P.; Brown, David G.; Loew, Murray H.)V1444,407-412(1991)

Evaluation of multiresolution elastic matching using MRI data (Gee, Jim C.; Reivich, Martin; Bilaniuk, L.; Hackney, D.; Zimmerman, R.; Kovacic, Stane; Bajcsy, Ruzena R.)V1445,226-234(1991)

Evaluation of the 3-D radon transform algorithm for cone beam reconstruction (Grangeat, Pierre; Le Masson, Patrick; Melennec, Pierre; Sire, Pascal)V1445,320-331(1991)

Evaluation of total workstation CT interpretation quality: a single-screen pilot study (Beard, David V.; Perry, John R.; Muller, K.; Misra, Ram B.; Brown, P.; Hemminger, Bradley M.; Johnston, R. E.; Mauro, M.; Jaques, P. F.; Schiebler, M.)V1446,52-58(1991)

Excimer-laser-induced fluorescence spectroscopy of human arteries during laser ablation (Abel, B.; Hippler, Horst; Koerber, B.; Morguet, A.; Neu, Walter)V1525,110-118(1991)

Experiences with a comprehensive hospital information system that incorporates image management capabilities (Dayhoff, Ruth E.; Maloney, Daniel L.; Kuzmak, Peter; Shepard, Barclay M.)V1446,323-329(1991)

Experimental study of migration depth for the photons measured at sample surface (Cui, Weijia; Kumar, Chellappa; Chance, Britton)V1431,180-191(1991)

Experimental system for detecting lung nodules by chest x-ray image processing (Suzuki, Hideo; Inaoka, Noriko; Takabatake, Hirotsugu; Mori, Masaki; Natori, Hiroshi; Suzuki, Akira)V1450,99-107(1991)

Extracting local stretching from left ventricle angiography data (Mishra, Sanjoy K.; Goldgof, Dmitry B.)V1450,218-230(1991)

Extraction of human stomach using computational geometry (Aisaka, Kazuo; Arai, Kiyoshi; Tsutsui, Kumiko; Hashizume, Akihide)V1445,312-317(1991)

Extraction of the foveal center and lesion boundary from fundus images (Ishaq, Naseem; Taylor, Kenneth; Steliou, Kypros; Delaney, William)V1381,153-159(1991)

Fiber optic based multiprobe system for intraoperative monitoring of brain functions (Mayevsky, Avraham; Flamm, E. S.; Pennie, W.; Chance, Britton)V1431,303-313(1991)

Fiber optic image guide rods as ultrathin endoscopy (Kociszewski, Longin; Pysz, Dariusz)V1420,212-217(1991)

Fluorescence imaging in photodynamic therapy (MacRobert, Alexander J.; Phillips, David)V1439,79-87(1991)

Folder management on a multimodality PACS display station (Feingold, Eric; Seshadri, Sridhar B.; Arenson, Ronald L.)V1446,211-216(1991)

Fourier transform method to determine human corneal endothelial morphology (Masters, Barry R.; Lee, Yimkul; Rhodes, William T.)V1429,82-90(1991)

Fractal analysis as a means for the quantification of intramandibular trabecular bone loss from dental radiographs (Doyle, Michael D.; Rabin, Harold; Suri, Jasjit S.)V1380,227-235(1991)

Fractal patterns in the human retina and their physiological correlates (Masters, Barry R.)V1380,218-226(1991)

Frequency-domain imaging using array detectors: present status and prospects for picosecond resolution (Morgan, Chris G.; Murray, J. G.; Mitchell, A. C.)V1525,83-90(1991)

Frequency-domain measurements of changes of optical pathlength during spreading depression in a rodent brain model (Maris, Michael B.; Mayevsky, Avraham; Sevick, Eva M.; Chance, Britton)V1431,136-148(1991)

From voxel to curvature (Monga, Olivier; Ayache, Nicholas; Sander, Peter T.)V1570,382-390(1991)

Full-frame entropy coding for radiological image compression (Lo, Shih-Chung B.; Krasner, Brian H.; Mun, Seong Ki; Horii, Steven C.)V1444,265-271(1991)

Future Trends in Biomedical Applications of Lasers (Svaasand, Lars O., ed.)V1525(1991)

Graph theoretic approach to segmentation of MR images (Wu, Zhenyu; Leahy, Richard M.)V1450,120-132(1991)

High-power x-ray generation using transition radiation (Piestrup, Melvin A.; Boyers, D. G.; Pincus, Cary I.; Li, Qiang; Harris, J. L.; Bergstrom, J. C.; Caplan, H. S.; Silzer, R. M.; Skopik, D. M.; Moran, M. J.; Maruyama, Xavier K.)V1552,214-239(1991)

High-precision digital charge-coupled device TV system (Vishnevsky, G. I.; Ioffe, S. A.; Berezin, V. Y.; Rybakov, M. I.; Mikhaylov, A. V.; Belyaev, L. V.)V1448,69-72(1991)

High-resolution image digitizing camera for use in quantitative coronary arteriography (Muser, Markus H.; Leemann, Thomas; Anliker, M.)V1448,106-112(1991)

High-resolution teleradiology applications within the hospital (Jost, R. G.; Blaine, G. J.; Kocher, Thomas E.; Muka, Edward; Whiting, Bruce R.)V1446,2-9(1991)

High-resolution texture analysis for cytology (Hallouche, Farid; Adams, Alan E.; Hinton, Oliver R.)V1445,504-512(1991)

High spatial and temporal resolution in the optical investigation of biological objects (Damm, Tobias; Kempe, M.; Stamm, Uwe; Stolberg, K. P.; Wabnitz, H.)V1403,686-694(1991)

Histochemical identification of malignant and premalignant lesions (Liebow, Charles; Maloney, M. J.)V1426,22-30(1991)

HIV detection by in-situ hybridization based on confocal reflected light microscopy (Smith, Louis C.; Jericevic, Zeljko; Cuellar, Roland; Paddock, Stephen W.; Lewis, Dorothy E.)V1428,224-232(1991)

Holographic interferometric analysis of the bovine cornea expansion (Foerster, Werner; Kasprzak, Henryk; von Bally, Gert; Busse, H.)V1429,146-151(1991)

Holography, Interferometry, and Optical Pattern Recognition in Biomedicine (Podbielska, Halina, ed.)V1429(1991)

Holography of human pathologic specimens with continuous-beam lasers through plastination (Myers, M. B.; Bickley, Harmon)V1461,242-244(1991)

Hospital-integrated PACS at the University Hospital of Geneva (Ratib, Osman M.; Ligier, Yves; Hochstrasser, Denis; Scherrer, Jean-Raoul)V1446,330-340(1991)

Hypoechoic media: a landmark for intravascular ultrasonic imaging (Gussenhoven, Elma J.; Bom, Nicolaas; Li, Wenguang; van Urk, Hero; Pietermann, Herman; van Suylen, Robert J.; Salem, H. K.)V1425,203-206(1991)

Image acquisition unit for the Mayo/IBM PACS project (Reardon, Frank J.; Salutz, James R.)V1446,481-491(1991)

Image analysis for DNA sequencing (Palaniappan, K.; Huang, Thomas S.)V1450,186-197(1991)

Image and modality control issues in the objective evaluation of manipulation techniques for digital chest images (Rehm, Kelly; Seeley, George W.; Dallas, William J.)V1445,24-35(1991)

Image computing requirements for the 1990s: from multimedia to medicine (Gove, Robert J.; Lee, Woobin; Kim, Yongmin; Alexander, Thomas)V1444,318-333(1991)

Image delivery performance of a CT/MR PACS module applied in neuroradiology (Lou, Shyh-Liang; Loloyan, Mansur; Weinberg, Wolfram S.; Valentino, Daniel J.; Lufkin, Robert B.; Hanafee, William; Bentson, John R.; Jabour, Bradly; Huang, H. K.)V1446,302-311(1991)

Image-display optimization using clinical history (Nodine, Calvin F.; Brikman, Inna; Kundel, Harold L.; Douglas, A.; Seshadri, Sridhar B.; Arenson, Ronald L.)V1444,56-62(1991)

Image noise smoothing based on nonparametric statistics (Chuang, Keh-Shih; Huang, H. K.)V1445,496-503(1991)

Image processing system for brain and neural tissue (Sun, Bingrong; Xu, Jiafang)V1606,1022-1026(1991)

Image understanding, visualization, registration, and data fusion of biomedical brain images (Gerson, Nahum D.; Cappelletti, John D.; Hinds, Stuart C.; Glenn, Marcus E.)V1406,129-129(1991)

Imaging in digestive videoendoscopy (Guadagni, Stefano; Nadeau, Theodore R.; Lombardi, Loreto; Pistoia, Francesco; Pistoia, Maria A.)V1420,178-182(1991)

Imaging inside scattering media: chronocoherent imaging (Spears, Kenneth G.; Kume, Stewart M.; Winakur, Eric)V1429,2-8(1991)

Imaging of excimer laser vascular tissue ablation by ultrafast photography (Nyga, Ralf; Neu, Walter; Preisack, M.; Haase, Karl K.; Karsch, Karl R.)V1525,119-123(1991)

Imaging of subsurface regions of random media by remote sensing (Barbour, Randall L.; Graber, Harry L.; Aronson, Raphael; Lubowsky, Jack)V1431,52-62(1991); V1431,192-203(1991)

Imaging of tumors by time-delayed laser-induced fluorescence (Kohl, M.; Neukammer, Jorg; Sukowski, U.; Rinneberg, Herbert H.; Sinn, H.-J.; Friedrich, E. A.; Graschew, Georgi; Schlag, Peter M.; Woehrle, D.)V1525,26-34(1991)

Improved instrumentation for photodynamic fluorescence detection of cancer (Baumgartner, R.; Heil, P.; Jocham, D.; Kriegmair, M.; Stepp, Herbert; Unsoeld, Eberhard)V1525,246-248(1991)

Improved phantom for quality control of laser scanner digitizers in PACS (Halpern, Ethan J.; Esser, Peter D.)V1444,104-115(1991)

Improved real-time volumetric ultrasonic imaging system (Pavy, Henry G.; Smith, Stephen W.; von Ramm, Olaf T.)V1443,54-61(1991)

Incorporation of structural CT and MR images in PET image reconstruction (Chen, Chin-Tu; Ouyang, Xiaolong; Ordonez, Caesar; Hu, Xiaoping; Wong, Wing H.; Metz, Charles E.)V1445,222-225(1991)

Initial experiences with PACS in a clinical and research environment (Honeyman, Janice C.; Staab, Edward V.; Frost, Meryll M.)V1446,362-368(1991)

Instrumentation for simultaneous kinetic imaging of multiple fluorophores in single living cells (Morris, Stephen J.; Beatty, Diane M.; Welling, Larry W.; Wiegmann, Thomas B.)V1428,148-158(1991)

Intensity interpolation for branching in reconstructing three-dimensional objects from serial cross-sections (Liang, Cheng-Chung; Chen, Chin-Tu; Lin, Wei-Chung)V1445,456-467(1991)

Interactive graphics system for locating plunge electrodes in cardiac MRI images (Laxer, Cary; Johnson, G. A.; Kavanagh, Katherine M.; Simpson, Edward V.; Ideker, Raymond E.; Smith, William M.)V1444,190-195(1991)

Interactive image processing in swallowing research (Dengel, Gail A.; Robbins, JoAnne; Rosenbek, John C.)V1445,88-94(1991)

Intraluminal laser atherectomy with ultrasound and electromagnetic guidance (Gregory, Kenton W.; Aretz, H. T.; Martinelli, Michael A.; Ledet, Earl G.; Hatch, G. F.; Gregg, Richard E.; Sedlacek, Tomas; Haase, W. C.)V1425,217-225(1991)

Intraoperative endovascular ultrasonography (Eton, Darwin; Ahn, Samuel S.; Baker, J. D.; Pensabene, Joseph; Yeatman, Lawrence S.; Moore, Wesley S.)V1425,182-187(1991)

Intravascular ultrasound imaging and intracardiac echocardiography: recent developments and future directions (Pandian, Natesa G.)V1425,198-202(1991)

Investigation of new filtering schemes for computerized detection of lung nodules (Yoshimura, Hitoshi; Giger, Maryellen L.; Matsumoto, Tsuneo; Doi, Kunio; MacMahon, Heber; Montner, Steven M.)V1445,47-51(1991)

In-vivo intravascular ultrasound in human ileo-femoral vessels (Tabbara, Marwan R.; Cavaye, Douglas; Kopchok, George E.; White, Rodney A.)V1425,208-216(1991)

Iso-precision scaling of digitized mammograms to facilitate image analysis (Karssemeijer, Nico; van Erning, Leon J.)V1445,166-177(1991)

Knowledge-based direct 3-D texture segmentation system for confocal microscopic images (Lang, Zhengping; Zhang, Zhen; Scarberry, Randell E.; Shao, Weimin; Sun, Xu-Mei)V1468,826-833(1991)

Knowledge-driven image analysis of cell structures (Nederlof, Michel A.; Witkin, Andrew; Taylor, D. L.)V1428,233-241(1991)

Laser-induced fluorescence imaging of coronary arteries for open-heart surgery applications (Taylor, Roderick S.; Gladysz, D.; Brown, D.; Higginson, Lyall A.)V1420,183-192(1991)

Laser speckle and its temporal variability: the implications for biomedical holography (Briers, J. D.)V1429,48-54(1991)

Liquid-crystal-television-based optical-digital processor for measurement of shortening velocity in single rat heart cells (Yelamarty, Rao V.; Yu, Francis T.; Moore, Russell L.; Cheung, Joseph Y.)V1398,170-179(1991)

Local spectrum analysis of medical images (Baskurt, Atilla; Peyrin, Francoise; Min, Zhu-Yue; Goutte, Robert)V1445,485-495(1991)

Low-dose magnetic-field-immune biplanar fluoroscopy for neurosurgery (Ramos, P. A.; Lawson, Michael A.; Wika, Kevin G.; Allison, Stephen W.; Quate, E. G.; Molloy, J. A.; Ritter, Rogers C.; Gillies, George T.)V1443,160-170(1991)

Lung imaging fluorescence endoscope: development and experimental prototype (Palcic, Branko; Lam, Stephen; MacAulay, Calum; Hung, Jacklyn; Jaggi, Bruno; Radjinia, Massud; Pon, Alfred; Profio, A. E.)V1448,113-117(1991)

Magnetic resonance reconstruction from projections using half the data (Noll, Douglas C.; Pauly, John M.; Nishimura, Dwight G.; Macovski, Albert)V1443,29-36(1991)

Management system for a PACS network in a hospital environment (Mattheus, Rudy A.; Temmerman, Yvan; Verhellen, P.; Osteaux, Michel)V1446,341-351(1991)

MAP segmentation of magnetic resonance images using mean field annealing (Logenthiran, Ambalavaner; Snyder, Wesley E.; Santago, Peter; Link, Kerry M.)V1452,225-243(1991)

MDIS (medical diagnostic imaging support) workstation issues: clinical perspective (Smith, Donald V.; Smith, Suzy; Cawthon, Michael A.)V1444,357-362(1991)

Measurement of biological tissue metabolism using phase modulation spectroscopic technology (Weng, Jian; Zhang, M. Z.; Simons, K.; Chance, Britton)V1431,161-170(1991)

Measurement of fractal dimension using 3-D technique (Chuang, Keh-Shih; Valentino, Daniel J.; Huang, H. K.)V1445,341-347(1991)

Measurement of the axial eye length and retinal thickness by laser Doppler interferometry (Hitzenberger, Christoph K.; Fercher, Adolf F.; Juchem, M.)V1429,21-25(1991)

Measurement of the piezoelectric effect in bone using quasiheterodyne holographic interferometry (Ovryn, Benjie; Haacke, E. M.)V1429,172-182(1991)

Mechanical and acoustic analysis in ultrasonic angioplasty (Detwiler, Paul W.; Watkins, James F.; Rose, Eric A.; Ratner, A.; Vu, Louis P.; Severinsky, J. Y.; Rosenschein, Uri)V1425,149-155(1991)

Medical applications of holographic stereograms (Tsujiuchi, Jumpei)V1238,398-400(1991)

Medical applications of holography (von Bally, Gert)V1525,2-8(1991)

Medical diagnostic imaging support systems for military medicine (Goeringer, Fred)V1444,340-350(1991)

Medical image understanding system based on Dempster-Shafer reasoning (Chen, Shiuh-Yung; Lin, Wei-Chung; Chen, Chin-Tu)V1445,386-397(1991)

Medical Imaging V: Image Capture, Formatting, and Display (Kim, Yongmin, ed.)V1444(1991)

Medical Imaging V: Image Physics (Schneider, Roger H., ed.)V1443(1991)

Medical Imaging V: Image Processing (Loew, Murray H., ed.)V1445(1991)

Medical Imaging V: PACS Design and Evaluation (Jost, R. G., ed.)V1446(1991)

Microcomputer-based image processing system for CT/MRI scans: II. Expert system (Kwok, John C.; Yu, Peter K.; Cheng, Andrew Y.; Ho, Wai-Chin)V1445,446-455(1991)

Microcomputer-based image processing system for CT/MRI scans: hardware configuration and software capacity (Cheng, Andrew Y.; Ho, Wai-Chin; Kwok, John C.; Yu, Peter K.)V1444,400-406(1991)

Microscopic feature extraction from optical sections of contracting cardiac muscle cells recorded at high speed (Roos, Kenneth P.; Lake, David S.; Lubell, Bradford A.)V1428,159-168(1991)

Microscopic fluorescence spectroscopy and diagnosis (Schneckenburger, Herbert; Seidlitz, Harold K.; Wessels, Jurina; Strauss, Wolfgang; Rueck, Angelika C.)V1525,91-98(1991)

MIMS: a medical image management system (Badaoui, Said; Aubry, Florent)V1567,31-31(1991)

Model-based boundary detection in echocardiography using dynamic programming technique (Dong, LiXin; Pelle, Gabriel; Brun, Philip; Unser, Michael A.)V1445,178-187(1991)

Model-based labeling, analysis, and three-dimensional visualization from two-dimensional medical images (Arata, Louis K.; Dhawan, Atam P.; Thomas, Stephen R.)V1446,465-474(1991)

Model-based vision system for automatic recognition of structures in dental radiographs (Acharya, Raj S.; Samarabandu, J. K.; Hausmann, E.; Allen, K. A.)V1450,170-177(1991)

Model generation and partial matching of left ventricular boundaries (Tehrani, Saeid; Weymouth, Terry E.; Mancini, G. B.)V1445,434-445(1991)

Modeling and suppression of amplitude artifacts due to z motion in MR imaging (Riek, Jonathan K.; Mitsa, Theophano; Parker, Kevin J.; Smith, Warren E.; Tekalp, Ahmet M.; Szumowski, J.)V1398,130-141(1991)

Modeling data acquisition and the effects of patient motion in magnetic resonance imaging (Riek, Jonathan K.; Smith, Warren E.; Tekalp, Ahmet M.; Parker, Kevin J.; Mitsa, Theophano; Szumowski, J.)V1445,198-206(1991)

Modeling ultrasound speckle formation and its dependence on imaging system's response (Rao, Navalgund A.; Zhu, Hui)V1443,81-95(1991)

Model of the left ventricle 3-D global motion: application to MRI data (Friboulet, Denis; Magnin, Isabelle E.; Mathieu, Christophe; Revel, D.; Amiel, Michel)V1445,106-117(1991)

Moire-shift interferometer measurements of the shape of human and cat tympanic membrane (Decraemer, Willem F.; Dirckx, Joris J.)V1429,26-33(1991)

Monte Carlo and diffusion calculations of photon migration in noninfinite highly scattering media (Haselgrove, John C.; Leigh, John S.; Yee, Conway; Wang, Nai-Guang; Maris, Michael B.; Chance, Britton)V1431,30-41(1991)

Morphological segmentation and 3-D rendering of the brain in magnetic resonance imaging (Connor, William H.; Diaz, Pedro J.)V1568,327-334(1991)

Motion estimation in digital angiographic images using skeletons (Kwak, J. Y.; Efstratiadis, Serafim N.; Katsaggelos, Aggelos K.; Sahakian, Alan V.; Sullivan, Barry J.; Swiryn, Steven; Hueter, David C.; Frohlich, Thomas)V1396,32-44(1991)

Multichannel fiber optic broadband video communication system for monitoring CT/MR examinations (Huang, H. K.; Kangarloo, Hooshang; Tecotzky, Raymond H.; Cheng, Xin; Vanderweit, Don)V1444,214-220(1991)

Multicomputer performance evaluation tool and its application to the Mayo/IBM image archival system (Pavicic, Mark J.; Ding, Yingjai)V1446,370-378(1991)

Multilayer monochromator for synchrotron radiation angiography (Baron, Alfred Q.; Barbee, Troy W.; Brown, George S.)V1343,84-94(1991)

Multi-media PACS integrated with HIS/RIS employing magneto-optical disks (Umeda, Tokuo; Inamura, Kiyonari; Inamoto, Kazuo; Kondoh, Hiroshi P.; Kozuka, Takahiro)V1446,199-210(1991)

Multiple communication networks for a radiological PACS (Wong, Albert W.; Stewart, Brent K.; Lou, Shyh-Liang; Chan, Kelby K.; Huang, H. K.)V1446,73-80(1991)

Multispectral image segmentation of breast pathology (Hornak, joseph P.; Blaakman, Andre; Rubens, Deborah; Totterman, Saara)V1445,523-533(1991)

NADH-fluorescence in medical diagnostics: first experimental results (Schramm, Werner; Naundorf, M.)V1525,237-241(1991)

Near-infrared imaging in vivo: imaging of Hb oxygenation in living tissues (Araki, Ryuichiro; Nashimoto, Ichiro)V1431,321-332(1991)

Near-infrared optical monitoring of cardiac oxygen sufficiency through thoracic wall without open-chest surgery (Kakihana, Yasuyuki; Tamura, Mamoru)V1431,314-321(1991)

Near-infrared time-resolved spectroscopy and fast scanning spectrophotometry in ischemic human forearm (Ferrari, Marco; De Blasi, Roberto A.; Bruscaglioni, Piero; Barilli, Marco; Carraresi, Luca; Gurioli, G. M.; Quaglia, Enrico; Zaccanti, Giovanni)V1431,276-283(1991)

Near-real-time biplanar fluoroscopic tracking system for the video tumor fighter (Lawson, Michael A.; Wika, Kevin G.; Gillies, George T.; Ritter, Rogers C.)V1445,265-275(1991)

Need for quality assurance related to PACS (Rowberg, Alan H.)V1444,80-82(1991)

Neural networks for medical image segmentation (Lin, Wei-Chung; Tsao, Chen-Kuo; Chen, Chin-Tu; Feng, Yu-Jen)V1445,376-385(1991)

New 2/3-inch MF image pick-up tubes for HDTV camera (Kobayashi, Akira; Ishikawa, Masayoshi; Suzuki, Takayoshi; Ikeya, Morihiro; Shimomoto, Yasuharu)V1449,148-156(1991)

New approach for endoscopic stereotactic brain surgery using high-power laser (Otsuki, Taisuke; Yoshimoto, Takashi)V1420,220-224(1991)

New developments in CCD imaging devices for low-level confocal light imaging (Masters, Barry R.)V1428,169-176(1991)

New device for interactive image-guided surgery (Galloway, Robert L.; Edwards, Charles A.; Thomas, Judith G.; Schreiner, Steven; Maciunas, Robert J.)V1444,9-18(1991)

New image diagnosis system with ultrathin endoscope and clinical results (Tsumanuma, Takashi; Toriya, T.; Tanaka, T.; Shamoto, N.; Seto, K.; Sanada, Kazuo; Okazaki, A.; Okazaki, M.)V1420,193-198(1991)

New method for constructing 3-D liver from CT images (Sun, Yung-Nien; Chen, Jiann-Jone; Lin, Xi-Zhang; Mao, Chi-Wu)V1606,653-664(1991)

New method for identifying features of an image on a digital video display (Doyle, Michael D.)V1380,86-95(1991)

New method for tissue indentification: resonance fluorescence spectroscopy (Neu, Walter)V1525,124-131(1991)

Noise reduction in heart movies by motion-compensated filtering (Reinen, Tor A.)V1606,755-763(1991)

Noise reduction in ultrasound images using multiple linear regression in a temporal context (Olstad, Bjorn)V1451,269-281(1991)

Noninvasive hemoglobin oxygenation monitor and computed tomography by NIR spectrophotometry (Oda, Ichiro; Ito, Yasunobu; Eda, Hideo; Tamura, Tomomi; Takada, Michinosuke; Abumi, Rentaro; Nagai, Katumi; Nakagawa, Hachiro; Tamura, Masahide)V1431,284-293(1991)

Noninvasive measurement of regional cerebrovascular oxygen saturation in humans using optical spectroscopy (McCormick, Patrick W.; Stewart, Melville; Lewis, Gary)V1431,294-302(1991)

Nonlinear approach to the 3-D reconstruction of microscopic objects (Wu, Xiangchen; Schwarzmann, Peter)V1450,278-285(1991)

Novel method for the computer analysis of cryomicroscopic images (Chen, Wei; Gong, Yiming; Wang, Ruli; Xu, Yayong)V1450,198-205(1991)

Objective assessment of clinical computerized thermal images (Anbar, Michael)V1445,479-484(1991)

Object-oriented model for medical image database (Aubry, Florent; Bizais, Yves; Gibaud, Bernard; Forte, Anne-Marie; Chameroy, Virginie; Di Paola, Robert; Scarabin, Jean-Marie)V1446,168-176(1991)

Observer detection of image degradation caused by irreversible data compression processes (Chen, Ji; Flynn, Michael J.; Gross, Barry; Spizarny, David)V1444,256-264(1991)

Observer performance in dynamic displays: effect of frame rate on visual signal detection in noisy images (Whiting, James S.; Honig, David A.; Carterette, Edward; Eigler, Neal)V1453,165-175(1991)

Off-line image exchange between two PACS modules using the "ISAC" magneto-optical disk (Minato, Kotaro; Komori, Masaru; Nakano, Yoshihisa; Yonekura, Yoshiharu; Sasayama, Satoshi; Takahashi, Takashi; Konishi, Junji; Abe, Mituyuki; Sato, Kazuhiro; Hosoba, Minoru)V1446,195-198(1991)

On enhancing the performance of the ACR-NEMA protocol (Maydell, Ursula M.; Hassanein, Hossam S.; Deng, Shuang)V1446,81-92(1991)

On-line acquisition of CT and MRI studies from multiple scanners (Weinberg, Wolfram S.; Loloyan, Mansur; Chan, Kelby K.)V1446,430-435(1991)

One-year clinical experience with a fully digitized nuclear medicine department: organizational and economical aspects (Anema, P. C.; de Graaf, C. N.; Wilmink, J. B.; Hall, David; Hoekstra, A.; van Rijk, P. P.; Van Isselt, J. W.; Viergever, Max A.)V1446,352-356(1991)

Operational infrastructure for a clinical PACS (Boehme, Johannes M.; Chimiak, William J.; Choplin, Robert H.; Maynard, C. D.)V1446,312-317(1991)

Optical computer-assisted tomography realized by coherent detection imaging incorporating laser heterodyne method for biomedical applications (Inaba, Humio; Toida, Masahiro; Ichimura, Tsutomu)V1399,108-115(1991)

Optical Fibers in Medicine VI (Katzir, Abraham, ed.)V1420(1991)

Optical imaging of cortical activity in the living brain (Ratzlaff, Eugene H.; Grinvald, Amiran)V1439,88-94(1991)

Optical property measurements in turbid media using frequency-domain photon migration (Tromberg, Bruce J.; Svaasand, Lars O.; Tsay, Tsong-Tseh; Haskell, Richard C.; Berns, Michael W.)V1525,52-58(1991)

Optimal fluorescence imaging of atherosclerotic human tissue (Davenport, Carolyn M.; Alexander, Andrew L.; Gmitro, Arthur F.)V1425,16-27(1991)

Optimization and evaluation of an image intensifier TV system for digital chest imaging (Angelhed, Jan-Erik; Mansson, Lars G.; Kheddache, Susanne)V1444,159-170(1991)

Optimization of morphological structuring elements for angiogram enhancement (Andress, Keith M.; Wilson, David L.)V1445,6-10(1991)

PACS and teleradiology for on-call support of abdominal imaging (Horii, Steven C.; Garra, Brian S.; Mun, Seong Ki; Zeman, Robert K.; Levine, Betty A.; Fielding, Robert)V1446,10-15(1991)

PACS: "Back to the Future" (Carey, Bruce; Seshadri, Sridhar B.; Arenson, Ronald L.)V1446,414-419(1991)

PACS for GU radiology (Hayrapetian, Alek; Barbaric, Zoran L.; Weinberg, Wolfram S.; Chan, Kelby K.; Loloyan, Mansur; Taira, Ricky K.; Huang, H. K.)V1446,243-247(1991)

PACS reading time comparision: the workstation versus alternator for ultrasound (Horii, Steven C.; Garra, Brian S.; Mun, Seong Ki; Singer, Jon; Zeman, Robert K.; Levine, Betty A.; Fielding, Robert; Lo, Ben)V1446,475-480(1991)

Parallel constructs for three-dimensional registration on a SIMD (single-instruction stream/multiple-data stream) processor (Morioka, Craig A.; Chan, Kelby K.; Huang, H. K.)V1445,534-538(1991)

Pattern recognition in pulmonary computerized tomography images using Markovian modeling (Preteux, Francoise; Moubarak, Michel; Grenier, Philippe)V1450,72-83(1991)

Percutaneous coronary angioscopy during coronary angioplasty: clinical findings and implications (Ramee, Stephen R.; White, Christopher J.; Mesa, Juan E.; Jain, Ashit; Collins, Tyrone J.)V1420,199-202(1991)

Performance characteristics of an ultrafast network for PACS (Stewart, Brent K.; Lou, Shyh-Liang; Wong, Albert W.; Chan, Kelby K.; Huang, H. K.)V1446,141-153(1991)

Performance characteristics of the Mayo/IBM PACS (Persons, Kenneth R.; Gehring, Dale G.; Pavicic, Mark J.; Ding, Yingjai)V1446,60-72(1991)

Performance of a thinned back-illuminated CCD coupled to a confocal microscope for low-light-level fluorescence imaging (Masters, Barry R.)V1447,56-63(1991)

PET reconstruction using multisensor fusion techniques (Acharya, Raj S.; Gai, Nevile)V1445,207-221(1991)

Photodetection of early cancer in the upper aerodigestive tract and the bronchi using photofrin II and colorectal adenocarcinoma with fluoresceinated monoclonal antibodies (Wagnieres, G.; Braichotte, D.; Chatelain, Andre; Depeursinge, Ch.; Monnier, Philippe; Savary, M.; Fontolliet, Ch.; Calmes, J.-M.; Givel, J.-C.; Chapuis, G.; Folli, S.; Pelegrin, A.; Buchegger, F.; Mach, J.-P.; van den Bergh, Hubert)V1525,219-236(1991)

Photon dynamics in tissue imaging (Chance, Britton; Haselgrove, John C.; Wang, Nai-Guang; Maris, Michael B.; Sevick, Eva M.)V1525,68-82(1991)

Photon migration in a model of the head measured using time- and frequency-domain techniques: potentials of spectroscopy and imaging (Sevick, Eva M.; Chance, Britton)V1431,84-96(1991)

Physical and psychophysical evaluation of CRT noise performance (Ji, Tinglan; Roehring, Hans; Blume, Hartwig R.; Seeley, George W.; Browne, Michael P.)V1444,136-150(1991)

Picture archiving and communications systems protocol based on ISO-OSI standard (Martinez, Ralph; Nam, Jiseung; Dallas, William J.; Osada, Masakazu; McNeill, Kevin M.; Ozeki, Takeshi; Komatsu, Ken-ichi)V1446,100-107(1991)

Positron emission tomography of laser-induced interstitial hyperthermia in cerebral gliomas (Ulrich, Frank; Bettag, Martin; Langen, K. J.)V1428,135-135(1991)

Possibility of liquid crystal display panels for a space-saving PACS workstation (Komori, Masaru; Minato, Kotaro; Takahashi, Takashi; Nakano, Yoshihisa; Sakurai, Tsunetaro)V1444,334-337(1991)

Potential usefulness of a video printer for producing secondary images from digitized chest radiographs (Nishikawa, Robert M.; MacMahon, Heber; Doi, Kunio; Bosworth, Eric)V1444,180-189(1991)

Practical considerations for effective microendoscopy (Papaioannou, Thanassis; Papazoglou, Theodore G.; Daykhovsky, Leon; Gershman, Alex; Segalowitz, Jacob; Reznik, G.; Beeder, Clain; Chandra, Mudjianto; Grundfest, Warren S.)V1420,203-211(1991)

Precise individualized armature for ear reconstruction (Evenhouse, Raymond J.; Chen, Xiaoming)V1380,248-253(1991)

Prefetching: PACS image management optimization using HIS/RIS information (Lodder, Herman; van Poppel, Bas M.; Bakker, Albert R.)V1446,227-233(1991)

Preliminary results of a PACS implementation (Kishore, Sheel; Khalsa, Satjeet S.; Seshadri, Sridhar B.; Arenson, Ronald L.)V1446,236-242(1991)

Preliminary tests of maximum likelihood image reconstruction method on 3-D real data and some practical considerations for the data corrections (Liu, Yi-Hwa; Holmes, Timothy J.; Koshy, Matthew)V1428,191-199(1991)

Present status and future directions of the Mayo/IBM PACS project (Morin, Richard L.; Forbes, Glenn S.; Gehring, Dale G.; Salutz, James R.; Pavlicek, William)V1446,436-441(1991)

Principles of optical dosimetry: fluorescence diagnostics (van Gemert, Martin J.)V1525,100-109(1991)

Properties of optical waves in turbid media (Svaasand, Lars O.; Tromberg, Bruce J.)V1525,41-51(1991)

Qualitative approach to medical image databases (Bizais, Yves; Gibaud, Bernard; Forte, Anne-Marie; Aubry, Florent; Di Paola, Robert; Scarabin, Jean-Marie)V1446,156-167(1991)

Quality assessment of video image capture systems (Rowberg, Alan H.; Lian, Jing)V1444,125-127(1991)

Quality assurance from a manufacturer's standpoint (Anderson, William J.)V1444,128-133(1991)

Quality assurance of PACS systems with laser film digitizers (Esser, Peter D.; Halpern, Ethan J.)V1444,100-103(1991)

Quantitative analysis of cardiac imaging using expert systems (Dreyer, Keith J.; Simko, Joseph; Held, A. C.)V1445,398-408(1991)

Quantitative performance evaluation of the EM algorithm applied to radiographic images (Brailean, James C.; Giger, Maryellen L.; Chen, Chin-Tu; Sullivan, Barry J.)V1450,40-46(1991)

RadGSP: a medical image display and user interface for UWGSP3 (Yee, David K.; Lee, Woobin; Kim, Dong-Lok; Haass, Clark D.; Rowberg, Alan H.; Kim, Yongmin)V1444,292-305(1991)

Radiative tetrahedral lattices (Driver, Jesse W.; Buckalew, Chris)V1459,109-116(1991)

Radiology workstation for mammography: preliminary observations, eyetracker studies, and design (Beard, David V.; Johnston, R. E.; Pisano, E.; Hemminger, Bradley M.; Pizer, Stephen M.)V1446,289-296(1991)

Rapid display of radiographic images (Cox, Jerome R.; Moore, Stephen M.; Whitman, Robert A.; Blaine, G. J.; Jost, R. G.; Karlsson, L. M.; Monsees, Thomas L.; Hassen, Gregory L.; David, Timothy C.)V1446,40-51(1991)

Real-time optical scanning system for measurement of chest volume changes during anesthesia (Duffy, Neil D.; Drummond, Gordon D.; McGowan, Steve; Dessesard, Pascal)V1380,46-52(1991)

Reconstructing MR images from incomplete Fourier data using the maximum entropy method (Fielden, John; Kwong, Henry Y.; Wilbrink, Jacob)V1445,145-154(1991)

Reconstruction methods for infrared absorption imaging (Arridge, Simon R.; van der Zee, Pieter; Cope, Mark; Delpy, David T.)V1431,204-215(1991)

Reduced defocus degradation in a system for high-speed three-dimensional digital microscopy (Quesenberry, Laura A.; Morris, V. A.; Neering, Ian R.; Taylor, Stuart R.)V1428,177-190(1991)

Registration of medical images by coincident bit counting (Chiang, John Y.; Sullivan, Barry J.)V1396,15-26(1991)

Reliability issues in PACS (Taira, Ricky K.; Chan, Kelby K.; Stewart, Brent K.; Weinberg, Wolfram S.)V1446,451-458(1991)

Remote Raman spectroscopic imaging of human artery wall (Phat, Darith; Vuong, Phat N.; Plaza, Pascal; Cheilan, Francis; Dao, Nguyen Q.)V1525,196-205(1991)

Removing vertical lines generated when x-ray images are digitized (Oyama, Yoshiro; Tani, Yuichiro; Shigemura, Naoshi; Abe, Toshio; Matsuda, Koyo; Kubota, Shigeto; Inami, Takashi)V1444,413-423(1991)

Representing three-dimensional shapes for visual recognition (Hoffman, Donald)V1445,2-4(1991)

Residue-producing E-filters and their applications in medical image analysis (Preston, Kendall)V1450,59-70(1991)

Resolution enhancement of CT images (Kuo, Shyh-Shiaw; Mammone, Richard J.)V1450,18-29(1991)

Resolution improvement for in-vivo magnetic resonance spectroscopic images (Plevritis, Sylvia; Macovski, Albert)V1445,118-127(1991)

Reversible image data compression based on HINT (hierarchical interpolation) decorrelation and arithmetic coding (Roos, Paul; Viergever, Max A.)V1444,283-290(1991)

Sampling system for in vivo ultrasound images (Jensen, Jorgen A.; Mathorne, Jan)V1444,221-231(1991)

Scan-free echo imaging of dynamic objects (Soumekh, Mehrdad)V1443,96-106(1991)

Selection of image acquisition methods (Donnelly, Joseph J.)V1444,351-356(1991)

Semiautomated detection and measurement of glomerular basement membrane from electron micrographs (Ong, Sim-Heng; Giam, S. T.; Jayasooriah, Mr.; Sinniah, R.)V1445,564-573(1991)

Semiautomatic medical image segmentation using knowledge of anatomic shape (Brinkley, James F.)V1445,78-87(1991)

Short-scan fan beam algorithm for noncircular detector orbits (Zeng, Gengsheng L.; Gullberg, Grant T.)V1445,332-340(1991)

Signal, noise, and readout considerations in the development of amorphous silicon photodiode arrays for radiotherapy and diagnostic x-ray imaging (Antonuk, Larry E.; Boudry, J.; Kim, Chung-Won; Longo, M.; Morton, E. J.; Yorkston, J.; Street, Robert A.)V1443,108-119(1991)

Signal-to-noise performance in cesium iodide x-ray fluorescent screens (Hillen, Walter; Eckenbach, W.; Quadflieg, P.; Zaengel, Thomas T.)V1443,120-131(1991)

Signa Tutor: results and future directions (Rundle, Debra A.; Watson, Carolyn K.; Seshadri, Sridhar B.; Wehrli, Felix W.)V1446,379-387(1991)

Simulation-based PACS development (Stut, W. J.; van Steen, M. R.; Groenewegen, L. P.; Ratib, Osman M.; Bakker, Albert R.)V1446,396-404(1991)

Simulation of time-resolved optical-CT imaging (Yamada, Yukio; Hasegawa, Yasuo)V1431,73-82(1991)

Simultaneous graphics and multislice raster image display for interactive image-guided surgery (Edwards, Charles A.; Galloway, Robert L.; Thomas, Judith G.; Schreiner, Steven; Maciunas, Robert J.)V1444,38-46(1991)

Simultaneous patient translation during CT scanning (Crawford, Carl R.; King, Kevin F.)V1443,203-213(1991)

Slice plane generation for three-dimensional image viewing using multiprocessing (Ho, Bruce K.; Ma, Marco; Chuang, Keh-Shih)V1445,95-100(1991)

Solid models for CT/MR image display: accuracy and utility in surgical planning (Mankovich, Nicholas J.; Yue, Alvin; Ammirati, Mario; Kioumehr, Farhad; Turner, Scott)V1444,2-8(1991)

Spatiotemporal filtering of digital angiographic image sequences corrupted by quantum mottle (Chan, Cheuk L.; Sullivan, Barry J.; Sahakian, Alan V.; Katsaggelos, Aggelos K.; Frohlich, Thomas; Byrom, Ernest)V1450,208-217(1991)

Speckle photography for investigation of bones supported by different fixing devices (Kasprzak, Henryk; Podbielska, Halina; Pennig, Dietmar)V1429,55-61(1991)

Sperm motion analysis (Salari, Valiollah)V1450,255-263(1991)

Spline-based tomographic reconstruction method (Guedon, Jean-Pierre; Bizais, Yves)V1443,214-225(1991)

Standardization of image quality measurements of medical x-ray image intensifier systems (Sandrik, John M.)VCR37,180-206(1991)

Stationary platen 2-axis scanner (Schermer, Mack J.)V1454,257-264(1991)

Stereotactic multibeam radiation therapy system in a PACS environment (Fresne, Francoise; Le Gall, G.; Barillot, Christian; Gibaud, Bernard; Manens, J. P.; Toumoulin, Christine; Lemoine, D.; Chenal, C.; Scarabin, Jean-Marie)V1444,26-36(1991)

Storing and managing three-dimensional digital medical image information (Chapman, Michael A.; Denby, N.)V1526,190-194(1991)

Structured light: theory and practice and practice and practice... (Keizer, Richard L.; Jun, Heesung; Dunn, Stanley M.)V1406,88-97(1991)

Study of human cardiac cycle using holographic interferometry (Brown, Glen; Boxler, Lawrence H.; Chun, Patrick K.; Western, Arthur B.)V1396,164-173(1991)

Subjective evaluation of image enhancements in improving the visibility of pathology in chest radiographs (Plessis, Brigitte; Goldberg, Morris; Belanger, Garry; Hickey, Nancy M.)V1445,539-554(1991)

Sunset: a hardware-oriented algorithm for lossless compression of gray-scale images (Langdon, Glen G.)V1444,272-282(1991)

Surface definition technique for clinical imaging (Liao, Wen-gen; Simovsky, Ilya; Li, Andrew; Kramer, David M.; Kaufman, Leon; Rhodes, Michael L.)V1444,47-55(1991)

Synthesized holograms in medicine and industry (Tsujiuchi, Jumpei)VIS08,326-334(1991)

Task performance based on the posterior probability of maximum-entropy reconstructions obtained with MEMSYS 3 (Myers, Kyle J.; Hanson, Kenneth M.)V1443,172-182(1991)

Technical and clinical evaluations of a 2048 x 2048-matrix digital radiography system for gastrointestinal examinations (Ogura, Toshihiro; Masuda, Yukihisa; Fujita, Hiroshi; Inoue, Nobuo; Yonekura, Fukuo; Miyagi, Yoshihiro; Takatsu, Kazuaki; Akahira, Katsuyoshi; Tsuruta, Shigehiko; Kamiya, Masami; Takahashi, Fumitaka; Oda, Kazuyuki; Ikeda, Shigeyuki; Koike, Kouichi)V1443,153-157(1991)

Teleradiology in the local environment (Staab, Edward V.; Honeyman, Janice C.; Frost, Meryll M.; Bidgood, W. D.)V1446,16-22(1991)

Three-Dimensional Bioimaging Systems and Lasers in the Neurosciences (Boggan, James E.; Cerullo, Leonard J.; Smith, Louis C., eds.)V1428(1991)

Three-dimensional color Doppler imaging of the carotid artery (Picot, Paul A.; Rickey, Daniel W.; Mitchell, J. R.; Rankin, Richard N.; Fenster, Aaron)V1444,206-213(1991)

Three-dimensional confocal microscopy of the living cornea and ocular lens (Masters, Barry R.)V1450,286-294(1991)

Three-dimensional display of inside of human body by holographic stereogram (Sato, Koki; Akiyama, Iwaki; Shoji, Hideo; Sumiya, Daigaku; Wada, Tuneyo; Katsuma, Hidetoshi; Itoh, Koichi)V1461,124-131(1991)

Three-dimensional display of MRI data in neurosurgery: segmentation and rendering aspects (Barillot, Christian; Lachmann, F.; Gibaud, Bernard; Scarabin, Jean-Marie)V1445,54-65(1991)

Three-dimensional image processing method to compensate for depth-dependent light attenuation in images from a confocal microscope (Aslund, Nils R.; Liljeborg, Anders; Oldmixon, E. H.; Ulfsparre, M.)V1450,329-337(1991)

Three-dimensional imaging laparoscope (Jones, Edwin R.; McLaurin, A. P.; Mason, J. L.)V1457,318-321(1991)

Three-dimensional imaging using TDI CCD sensors (Fenster, Aaron; Holdsworth, David W.; Drangova, Maria)V1447,28-33(1991)

Three-dimensional magnetic resonance imaging of the head (Keeler, Elaine K.; Oyen, Ordean J.)V1380,24-32(1991)

Three-dimensional real-time ultrasonic imaging using ellipsoidal backprojection (Anderson, Forrest L.)V1443,62-80(1991)

Three-dimensional reconstruction from cone beam projections (Ohishi, Satoru; Yamaguchi, Masahiro; Ohyama, Nagaaki; Honda, Toshio)V1443,280-285(1991)

Three-dimensional reconstruction from cone beam projection by a block iterative technique (Peyrin, Francoise; Goutte, Robert; Amiel, Michel)V1443,268-279(1991)

Three-dimensional reconstruction of liver from 2-D tomographic slices (Chou, Jin-Shin; Chen, Chin-Tu; Giger, Maryellen L.; Kahn, Charles E.; Bae, Kyongtae T.; Lin, Wei-Chung)V1396,45-50(1991)

Three-dimensional reconstruction of pulmonary blood vessels by using anatomical knowledge base (Inaoka, Noriko; Suzuki, Hideo; Mori, Masaki; Takabatake, Hirotsugu; Suzuki, Akira)V1450,2-12(1991)

Time-gated fluorescence spectroscopy and imaging of porphyrins and phthalocyanines (Cubeddu, Rinaldo; Canti, Gianfranco L.; Taroni, Paola; Valentini, G.)V1525,17-25(1991)

Time-of-flight breast imaging system: spatial resolution performance (Hebden, Jeremy C.; Kruger, Robert A.)V1431,225-231(1991)

Time-resolved reflectance spectroscopy (Jacques, Steven L.; Flock, Stephen T.)V1525,35-40(1991)

Time-Resolved Spectroscopy and Imaging of Tissues (Chance, Britton; Katzir, Abraham, eds.)V1431(1991)

Time-resolved transillumination for medical diagnostics (Berg, Roger; Andersson-Engels, Stefan; Jarlman, Olof; Svanberg, Sune)V1431,110-119(1991)

Tissue identification in MR images by adaptive cluster analysis (Gutfinger, Dan; Hertzberg, Efrat M.; Tolxdorff, T.; Greensite, F.; Sklansky, Jack)V1445,288-296(1991)

Tissue volume determinations from brain MRI images: a phantom study (Chandra, Ramesh; Rusinek, Henry)V1445,133-144(1991)

Tomographic imaging using picosecond pulses of light (Hebden, Jeremy C.; Kruger, Robert A.; Wong, K. S.)V1443,294-300(1991)

Trends in holographic endoscopy (Podbielska, Halina)V1429,207-213(1991)

Triangulating between parallel splitting contours using a simplicial algorithm (Miranda, Rick; McCracken, Thomas O.; Fedde, Chris)V1380,210-217(1991)

Tumor detection using time-resolved light transillumination (Berg, Roger; Andersson-Engels, Stefan; Jarlman, Olof; Svanberg, Sune)V1525,59-67(1991)

Ultrasonic b-scan image compounding technique for prosthetic socket design (Xue, Kefu; He, Ping; Fu, Huimin; Bismar, Hisham)V1606,675-684(1991)

Use of a cooled CCD camera for confocal light microscopy (Masters, Barry R.)V1448,98-105(1991)

Use of anatomical knowledge to register 3-D blood vessel data derived from DSA with MR images (Hill, Derek L.; Hawkes, David J.; Hardingham, Charles R.)V1445,348-357(1991)

Use of cross-validation as a stopping rule in emission tomography image reconstruction (Coakley, Kevin J.; Llacer, Jorge)V1443,226-233(1991)

Use of the Society of Motion Picture and Television Engineers test pattern in picture archiving and communication systems (PACS) (Gray, Joel E.)V1444,118-124(1991)

Using a fiber-optic pulse sensor in magnetic resonance imaging (Henning, Michael R.; Gerdt, David W.; Spraggins, Thomas)V1420,34-40(1991)

Using correlated CT images in compensation for attenuation in PET (positron emission tomography) image reconstruction (Yu, Xiaolin; Chen, Chin-Tu; Bartlett, R.; Pelizzari, Charles A.; Ordonez, Caesar)V1396,56-58(1991)

Using MRI to calculate cardiac velocity fields (Santiago, Peter; Slade, James N.)V1445,555-563(1991)

Using the ACR/NEMA standard with TCP/IP and Ethernet (Chimiak, William J.; Williams, Rodney C.)V1446,93-99(1991)

Vascular parameters from angiographic images (Close, Robert A.; Duckwiler, Gary R.; Vinuela, Fernando; Dion, Jacques E.)V1444,196-203(1991)

Vibration analysis of the tympanic membrane with a ventilation tube and a perforation by holography (Maeta, Manabu; Kawakami, Shinichiro; Ogawara, Toshiaki; Masuda, Yu)V1429,152-161(1991)

Virtual environment system for simulation of leg surgery (Pieper, Steve; Delp, Scott; Rosen, Joseph; Fisher, Scott S.)V1457,188-197(1991)

Visualization and volumetric compression (Chan, Kelby K.; Lau, Christina C.; Chuang, Keh-Shih; Morioka, Craig A.)V1444,250-255(1991)

Visualization of 3-D phase structure in confocal and conventional microscopy (Cogswell, Carol J.; Sheppard, Colin J.)V1450,323-328(1991)

Visualization of electron transfer interactions of membrane proteins (Kawato, Suguru)V1429,127-131(1991)

Visualization of liver in 3-D (Chen, Chin-Tu; Chou, Jin-Shin; Giger, Maryellen L.; Kahn, Charles E.; Bae, Kyongtae T.; Lin, Wei-Chung)V1444,75-77(1991)

Volumetric-intravascular imaging by high-frequency ultrasound (Sehgal, Chandra M.; Chandrasekaran, K.; Pandian, Natesa G.)V1425,226-233(1991)

Warping of a computerized 3-D atlas to match brain image volumes for quantitative neuroanatomical and functional analysis (Evans, Alan C.; Dai, W.; Collins, L.; Neelin, Peter; Marrett, Sean)V1445,236-246(1991)

X-ray photogrammetry of the hip revisited (Turner-Smith, Alan R.; White, Steven P.; Bulstrode, Christopher)V1380,75-84(1991)

X-ray projection microscopy and cone-beam microtomography (Wang, Ge; Lin, T. H.; Cheng, Ping-Chin; Shinozaki, D. M.; Newberry, S. P.)V1398,180-190(1991)

X-window-based 2K display workstation (Weinberg, Wolfram S.; Hayrapetian, Alek; Cho, Paul S.; Valentino, Daniel J.; Taira, Ricky K.; Huang, H. K.)V1446,35-39(1991)

You can't just plug it in: digital image networks/picture archiving and communication systems installation (Gelish, Anthony)V1444,363-372(1991)

ZD multipurpose neurosurgical image-guided localizing unit: experience in 103 consecutive cases of open stereotaxis (Zamorano, Lucia J.; Dujovny, Manuel)V1428,30-51(1991)

Medicine—see also biology; fiber optics in medicine; laser surgery; laser-tissue interaction; medical imaging; ophthalmology; radiology; tomography

10-W Ho laser for surgery (Bouchenkov, Vyatcheslav A.; Utenkov, Boris I.; Antipenko, Boris M.; Berezin, U. D.; Malinin, Boris G.; Serebryakov, Victor A.)V1427,409-412(1991)

Advances in clinical percutaneous excimer laser angioplasty (Viligiardi, Riccardo; Pini, Roberto; Salimbeni, Renzo; Galiberti, Sandra)V1425,72-74(1991)

Alexandrite laser and blind lithotripsy: initial experience—first clinical results (Mattioli, Stefano; Cremona, M.; Benaim, George; Ferrario, Angelo)V1421,114-119(1991)

Analysis of multiple-multipole scattering by time-resolved spectroscopy and spectrometry (Frank, Klaus; Hoeper, J.; Zuendorf, J.; Tauschek, D.; Kessler, Manfred; Wiesner, J.; Wokaun, Alexander J.)V1431,2-11(1991)

Antitumor drugs as photochemotherapeutic agents (Andreoni, Alessandra; Colasanti, Alberto; Kisslinger, Annamaria; Malatesta, Vincenzo; Mastrocinque, Michele; Roberti, Giuseppe)V1525,351-366(1991)

Application of a discrete-space representation to three-dimensional medical imaging (Toennies, Klaus D.; Tronnier, Uwe)V1444,19-25(1991)

Application of conjugated heparin-albumin microparticles with laser-balloon angioplasty: a potential method for reducing adverse biologic reactivity after angioplasty (Kundu, Sourav K.; McMath, Linda P.; Zaidan, Jonathan T.; Spears, J. R.)V1425,142-148(1991)

Applications of lasers in laparoscopic cholecystectomy: technical considerations and future directions (Lanzafame, Raymond J.)V1421,189-196(1991)

Argon laser vascular tissue fusion: current status and future perspectives (White, Rodney A.; Kopchok, George E.)V1422,103-110(1991)

Arthroscopic contact Nd:YAG laser meniscectomy: surgical technique and clinical follow-up (O'Brien, Stephen J.; Miller, Drew V.; Fealy, Stephen V.; Gibney, Mary A.; Kelly, Anne M.)V1424,62-75(1991)

Bifunctional irrigation liquid as an ideal energy converter for laser lithotripsy with nanosecond laser pulses (Reichel, Erich; Schmidt-Kloiber, Heinz; Paltauf, Guenther; Groke, Karl)V1421,129-133(1991)

Biochemical measurement of bilirubin with an evanescent wave optical sensor (Poscio, Patrick; Depeursinge, Ch.; Emery, Y.; Parriaux, Olivier M.; Voirin, G.)V1510,112-117(1991)

Biomechanical and structural studies of rabbits carotid arteries after endovascular laser exposition (Purinsh, Juris; Elksninsh, N.; Dzenis, J.; Tomass, V.; Teivans, A.; Freimanis, R.; Garsha, I.; Ozolinsh, H.)V1525,289-308(1991)

Biomechanics of the cornea (Buzard, Kurt A.; Hoeltzel, David A.)V1423,70-81(1991)

Biostereometric Technology and Applications (Herron, Robin E., ed.)V1380(1991)

Bladder outlet obstruction treated with transurethral ultrasonic aspiration (Malloy, Terrence P.)V1421,46-46(1991)

CAN-AM Eastern '90 (Antos, Ronald L.; Krisiloff, Allen J., eds.)V1398(1991)

Carbon dioxide laser effects on caries-like lesions of dental enamel (Featherstone, John D.; Zhang, S. H.; Shariati, M.; McCormack, S. M.)V1424,145-149(1991)

Caries selective ablation by pulsed lasers (Hennig, Thomas; Rechmann, Peter; Pilgrim, C.; Schwarzmaier, Hans-Joachim; Kaufmann, Raimund)V1424,99-105(1991)

Changes in collagen birefringence: a quantitative histologic marker of thermal damage in skin (Thomsen, Sharon L.; Cheong, Wai-Fung; Pearce, John A.)V1422,34-42(1991)

Changes in optical density of normal vessel wall and lipid atheromatous plaque after Nd:YAG laser irradiation (Schwarzmaier, Hans-Joachim; Heintzen, Matthias P.; Zumdick, Mathias; Kaufmann, Raimund; Wolbarsht, Myron L.)V1427,128-133(1991)

Characterization of photobiophysical properties of sensitizers used in photodynamic therapy (Roeder, Beate; Naether, Dirk)V1525,377-384(1991)

Chemical and Medical Sensors (Wolfbeis, Otto S., ed.)V1510(1991)

Chick chorioallantoic membrane for the study of synergistic effects of hyperthermia and photodynamic therapy (Kimel, Sol; Svaasand, Lars O.; Hammer-Wilson, Marie J.; Gottfried, Varda; Berns, Michael W.)V1525,341-350(1991)

Clinical applications of PDT in urology: present and near future (Shumaker, Bryan P.)V1426,293-300(1991)

Clinical applications of pulmonary artery oximetry (Barker, Steven J.)V1420,22-28(1991)

Clinical evaluation of tumor promotion by CO2 laser (Braun, Robert E.; Liebow, Charles)V1424,138-144(1991)

Clinical experience in applying endoscopic Nd:YAG laser to treat 451 esophagostenotic cases (Wang, Rui-zhong; Wang, Zhen-he; Lu, Kuang-sheng; Yang, Xiao-zhi; Lu, Bo-kao)V1421,203-207(1991)

Clinical experience with an excimer laser angioplasty system (Golobic, Robert A.; Bohley, Thomas K.; Wells, Lisa D.; Sanborn, Timothy A.)V1425,84-92(1991)

Clinical optical dose measurement for PDT: invasive and noninvasive techniques (Bays, Roland; Winterhalter, L.; Funakubo, H.; Monnier, Philippe; Savary, M.; Wagnieres, G.; Braichotte, D.; Chatelain, Andre; van den Bergh, Hubert; Svaasand, Lars O.; Burckhardt, C. W.)V1525,397-408(1991)

CO2 partial matricectomy in the treatment of ingrown toenails (Uppal, Gurvinder S.; Sherk, Henry H.; Black, Johnathan D.; Rhodes, Anthony; Sazy, John; Lane, Gregory J.)V1424,51-52(1991)

Coagulation and precise ablation of biotissues by pulsed sealed-off carbon monoxide laser (Masychev, Victor I.; Alejnikov, Vladislav S.; Klimenko, Vladimir I.)V1427,344-356(1991)

Combined guidance technique using angioscope and fluoroscope images for CO laser angioplasty: in-vivo animal experiment (Arai, Tsunenori; Mizuno, Kyoichi; Sakurada, Masami; Miyamoto, Akira; Arakawa, Koh; Kurita, Akira; Suda, Akira; Kikuchi, Makoto; Nakamura, Haruo; Utsumi, Atsushi; Akai, Yoshiro; Takeuchi, Kiyoshi)V1425,191-195(1991)

Comparative performance of infrared solid state lasers in laser lithotripsy (D'yakonov, G. I.; Konov, Vitaly I.; Mikhailov, V. A.; Nikolaev, D. A.; Pak, S. K.; Shcherbakov, I. A.)V1421,156-162(1991)

Comparative studies on hyperthermia induced by laser light, microwaves, and ultrasonics (Greguss, Pal)V1525,313-324(1991)

Comparative study of gelatin ablation by free-running and Q-switch modes of Er:YAG laser (Konov, Vitaly I.; Kulevsky, Lev A.; Lukashev, Alexei V.; Pashinin, Vladimir P.; Silenok, Alexander S.)V1427,232-242(1991)

Comparison of carbon monoxide and carbon dioxide laser-tissue interaction (Waters, Ruth A.; Thomas, J. M.; Clement, R. M.; Ledger, N. R.)V1427,336-343(1991)

Comparison of the ablation of polymethylmethacrylate by two fiber-optic-compatible infrared lasers (Garino, Jonathan P.; Nazarian, David; Froimson, Mark I.; Grelsamer, Ronald P.; Treat, Michael R.)V1424,43-47(1991)

Comparison of the excimer laser with the erbium yttrium aluminum garnet laser for applications in osteotomy (Li, Zhao-zhang; Van De Merwe, Willem P.; Reinisch, Lou)V1427,152-161(1991)

Comparison of thermal and optical techniques for describing light interaction with vascular grafts, sutures, and thrombus (Obremski, Susan M.; LaMuraglia, Glenn M.; Bruggemann, Ulrich H.; Anderson, R. R.)V1427,327-334(1991)

Computer-aided design and drafting visualization of anatomical structure of the human eye and orbit (Parshall, Robert F.; Sadler, Lewis L.)V1380,200-207(1991)

Computer-aided forensic facial reconstruction (Evenhouse, Raymond J.; Rasmussen, Mary; Sadler, Lewis L.)V1380,147-156(1991)

Computer-assisted surgical planning and automation of laser delivery systems (Zamorano, Lucia J.; Dujovny, Manuel; Dong, Ada; Kadi, A. M.)V1428,59-75(1991)

Contrast-agent-enhanced magnetic resonance imaging: early detection of neoplastic lesions of the CNS (Carvlin, Mark J.; Rosa, Louis; Rajan, Sunder; Francisco, John)V1426,13-21(1991)

Coronary and peripheral angioscopy with carbon dioxide gas and saline in animals (Smits, Pieter C.; Post, Mark J.; Velema, Evelyn; Rienks, Rienk; Borst, Cornelius)V1425,188-190(1991)

Cr,Er:YSGG laser as an instrument for dental surgery (D'yakonov, G. I.; Konov, Vitaly I.; Mikhailov, V. A.; Pak, S. K.; Shcherbakov, I. A.; Ershova, N. I.; Maksimovskiy, Y. V.)V1424,81-86(1991)

Cryosurgical ablation of the prostate (Cohen, Jeffrey K.)V1421,45-45(1991)

Current and future use of lasers in vascular neurosurgery (Chavantes, Maria C.; Zamorano, Lucia J.)V1428,99-127(1991)

Current status of photodynamic therapy for human cancer (Marcus, Stuart L.)V1426,301-310(1991)

Damage induced by pulsed IR laser radiation at transitions between different tissues (Frenz, Martin; Greber, Charlotte M.; Romano, Valerio; Forrer, Martin; Weber, Heinz P.)V1427,9-15(1991)

Decrease in total fluorescence from human arteries with increasing beta-carotene content (Ye, Biqing; Abela, George S.)V1425,45-54(1991)

Degradation of cholesterol crystals in macrophages: the role of phospholipids (Koren, Eugen; Koscec, Mirna; Fugate, Robert D.)V1428,214-223(1991)

Dependence of photodynamic threshold dose on treatment parameters in normal rat liver in vivo (Farrell, Thomas J.; Wilson, Brian C.; Patterson, Michael S.; Chow, Rowena)V1426,146-155(1991)

Design characteristics and point-spread function evaluation of bifocal intraocular lenses (Rol, Pascal O.)V1423,15-19(1991)

Design considerations of a real-time clinical confocal microscope (Masters, Barry R.)V1423,8-14(1991)

Detection of atheroma using Photofrin II and laser-induced fluorescence spectroscopy (Vari, Sandor G.; Papazoglou, Theodore G.; van der Veen, Maurits J.; Papaioannou, Thanassis; Fishbein, Michael C.; Chandra, Mudjianto; Beeder, Clain; Shi, Wei-Qiang; Grundfest, Warren S.)V1426,58-65(1991)

Detection of general anesthetics using a fluorescence-based sensor: incorporation of a single-fiber approach (Abrams, Susan B.; McDonald, Hillary L.; Yager, Paul)V1420,13-21(1991)

Determination of the optical penetration depth in tumors from biopsy samples (Lenz, P.)V1525,183-191(1991)

Development of a novel in-vivo drug/in-vitro light system to investigate mechanisms of cell killing with photodynamic therapy (Hampton, James A.; Selman, Steven H.)V1426,134-145(1991)

Diagnosis of atherosclerotic tissue by resonance fluorescence spectroscopy (Neu, Walter; Haase, Karl K.; Tischler, Christian; Nyga, Ralf; Karsch, Karl R.)V1425,28-36(1991)

Diagnostic and Therapeutic Cardiovascular Interventions (Abela, George S., ed.)V1425(1991)

Differential time-resolved detection of absorbance changes in composite structures (Nossal, Ralph J.; Bonner, Robert F.)V1431,21-28(1991)

Discrete-cosine-transform-based image compression applied to dermatology (Cookson, John P.; Sneiderman, Charles; Rivera, Christopher)V1444,374-378(1991)

Discrimination between urinary tract tissue and urinary stones by fiber-optic-pulsed photothermal radiometry method in vivo (Daidoh, Yuichiro; Arai, Tsunenori; Suda, Akira; Kikuchi, Makoto; Komine, Yukikuni; Murai, Masaru; Nakamura, Hiroshi)V1421,120-123(1991)

Doppler-guided retrograde catheterization system (Frazin, Leon J.; Vonesh, Michael J.; Chandran, Krishnan B.; Khasho, Fouad; Lanza, George M.; Talano, James V.; McPherson, David D.)V1425,207-207(1991)

Dosimetry for lasers and light in dermatology: Monte Carlo simulations of 577nm-pulsed laser penetration into cutaneous vessels (Jacques, Steven L.; Keijzer, Marleen)V1422,2-13(1991)

Dual wavelengths (750/375 nm) laser lithotripsy (Steiner, Rudolf W.; Meier, Thomas H.)V1421,124-128(1991)

Early detection of dysplasia in colon and bladder tissue using laser-induced fluorescence (Rava, Richard P.; Richards-Kortum, Rebecca; Fitzmaurice, Maryann; Cothren, Robert M.; Petras, Robert; Sivak, Michael V.; Levine, Howard)V1426,68-78(1991)

Early diagnosis of lung cancer (Saccomanno, Geno; Bechtel, Joel J.)V1426,2-12(1991)

Effectiveness of porphyrin-like compounds in photodynamic damage of atherosclerotic plaque (Zalessky, Viacheslav N.; Bobrov, Vladimir; Michalkin, Igor; Trunov, Vitaliy)V1426,162-169(1991)

Effect of coagulation on laser light distribution in myocardial tissue (Agah, Ramtin; Sheth, Devang; Motamedi, Massoud E.)V1425,172-179(1991)

Effect of Nd:YAG laser on dentinal bond strength (Dederich, Douglas N.; Tulip, John)V1424,134-137(1991)

Effect of spatial distribution of irradiated sites on injury selectivity in vascular tissue (Cheong, Wai-Fung; Morrison, Paul R.; Trainor, S. W.; Kurban, Amal K.; Tan, Oon T.)V1422,19-26(1991)

Effect of surface boundary on time-resolved reflectance: measurements with a prototype endoscopic catheter (Jacques, Steven L.; Flock, Stephen T.)V1431,12-20(1991)

Effects of He-Ne regional irradiation on 53 cases in the field of pediatric surgery (Guo, Jing-Zhen)V1422,136-139(1991)

Effects of photodynamic treatment on DNA (Oleinick, Nancy L.; Agarwal, Munna L.; Antunez, Antonio R.; Clay, Marian E.; Evans, Helen H.; Harvey, Ella Jo; Rerko, Ronald M.; Xue, Liang-yan)V1427,90-100(1991)

Effects of pressure rise on cw laser ablation of tissue (LeCarpentier, Gerald L.; Motamedi, Massoud E.; Welch, Ashley J.)V1427,273-278(1991)

Efficacy of argon-laser-mediated hot-balloon angioplasty (Sakurada, Masami; Miyamoto, Akira; Mizuno, Kyoichi; Nozaki, Youichi; Tabata, Hirotsugu; Etsuda, Hirokuni; Kurita, Akira; Nakamura, Haruo; Arai, Tsunenori; Suda, Akira; Kikuchi, Makoto; Watanabe, Tamishige; Utsumi, Atsushi; Akai, Yoshiro; Takeuchi, Kiyoshi)V1425,158-164(1991)

Efficacy of photodynamic killing with membrane associated and internalized photosensitizer molecules (Allison, Beth; Jiang, Frank N.; Levy, Julia G.)V1426,200-207(1991)

Elaboration of excimer lasers dosimetry for bone and meniscus cutting and drilling using optical fibers (Jahn, Renate; Dressel, Martin; Neu, Walter; Jungbluth, Karl-Heinz)V1424,23-32(1991)

Electron microscopic study on black pig skin irradiated with pulsed dye laser (504 nm) (Yasuda, Yukio; Tan, Oon T.; Kurban, Amal K.; Tsukada, Sadao)V1422,50-55(1991)

Electrophoretic mobility patterns of collagen following laser welding (Bass, Lawrence S.; Moazami, Nader; Pocsidio, Joanne O.; Oz, Mehmet C.; LoGerfo, Paul; Treat, Michael R.)V1422,123-127(1991)

Enantio-selective optrode for optical isomers of biologically active amines using a new lipophilic aromatic carrier (He, Huarui; Uray, Georg; Wolfbeis, Otto S.)V1510,95-103(1991)

Endoscopic fluorescence detection of early lung cancer (Profio, A. E.; Balchum, Oscar J.; Lam, Stephen)V1426,44-46(1991)

Endoscopic removal of PMMA in hip revision surgery with a CO2 laser (Sazy, John; Kollmer, Charles; Uppal, Gurvinder S.; Lane, Gregory J.; Sherk, Henry H.)V1424,50-50(1991)

Endoscopic YAG laser coagulation for early prostate cancer (McNicholas, Thomas A.; O'Donoghue, Neil)V1421,56-67(1991)

Energy conversion efficiency during optical breakdown (Grad, Ladislav; Diaci, J.; Mozina, Janez)V1525,206-209(1991)

Erbium laser ablation of bone: effect of water content (Walsh, Joseph T.; Hill, D. A.)V1427,27-33(1991)

Evaluation of facial palsy by moire topography (Inokuchi, Ikuo; Kawakami, Shinichiro; Maeta, Manabu; Masuda, Yu)V1429,39-45(1991)

Evaluation of Nile Blue E chalcogen analogs as PDT agents (Foley, James W.; Cincotta, Louis; Cincotta, Anthony H.)V1426,208-215(1991)

Excimer laser in arthroscopic surgery (Koort, Hans J.)V1424,53-59(1991)

Experimental and first clinical results with the alexandrite laser lithotripter (Miller, Kurt; Weber, Hans M.; Rueschoff, Josef; Hautmann, Richard E.)V1421,108-113(1991)

Experimental simulation of holmium laser action on biological tissues (Konov, Vitaly I.; Prokhorov, Alexander M.; Silenok, Alexander S.; Tsarkova, O. G.; Tsvetkov, V. B.; Shcherbakov, I. A.)V1427,220-231(1991)

Fiber fragmentation during laser lithotripsy (Flemming, G.; Brinkmann, Ralf E.; Strunge, Ch.; Engelhardt, R.)V1421,146-152(1991)

Fiber optic biosensors: the situation of the European market (Scheggi, Annamaria V.; Mignani, Anna G.)V1510,40-45(1991)

Flow injection analysis with bioluminescence-based fiber-optic biosensors (Blum, Loic J.; Gautier, Sabine; Coulet, Pierre R.)V1510,46-52(1991)

Fluorescence characteristics of atherosclerotic plaque and malignant tumors (Andersson-Engels, Stefan; Baert, Luc; Berg, Roger; D'Hallewin, Mich A.; Johansson, Jonas; Stenram, U.; Svanberg, Katarina; Svanberg, Sune)V1426,31-43(1991)

Fluorescence detection of tumors: studies on the early diagnosis of microscopic lesions in preclinical and clinical studies (Mang, Thomas S.; McGinnis, Carolyn; Crean, David H.; Khan, S.; Liebow, Charles)V1426,97-110(1991)

Fluorescence spectroscopy of normal and atheromatous human aorta: optimum illumination wavelength (Alexander, Andrew L.; Davenport, Carolyn M.; Gmitro, Arthur F.)V1425,6-15(1991)

Fractal analysis as a means for the quantification of intramandibular trabecular bone loss from dental radiographs (Doyle, Michael D.; Rabin, Harold; Suri, Jasjit S.)V1380,227-235(1991)

Fragmentation methods in laser lithotripsy (Jiang, Zhi X.; Whitehurst, Colin; King, Terence A.)V1421,88-99(1991)

Frequency-doubled alexandrite laser for tissue differentiation in angioplasty (Scheu, M.; Engelhardt, R.)V1425,63-69(1991)

Future Trends in Biomedical Applications of Lasers (Svaasand, Lars O., ed.)V1525(1991)

Grueneisen-stress-induced ablation of biological tissue (Dingus, Ronald S.; Scammon, R. J.)V1427,45-54(1991)

Hearing preservation using CO2 laser for acoustic nerve tumors (Grutsch, James; Heiferman, Kenneth S.; Cerullo, Leonard J.)V1428,136-145(1991)

Helium-neon effects of laser radiation in rats infected with thromboxane B2 (Juri, Hugo; Palma, J. A.; Campana, Vilma; Gavotto, A.; Lapin, R.; Yung, S.; Lillo, J.)V1422,128-135(1991)

High-energy ND:glass laser for oncology (Bouchenkov, Vyatcheslav A.; Utenkov, Boris I.; Zaitsev, V. K.; Bayanov, Valentin I.; Serebryakov, Victor A.)V1427,405-408(1991)

High-sensitivity sensor of gases based on IR tunable diode lasers for human exhalation monitoring (Moskalenko, Konstantin L.; Nadezhdinskii, Alexander I.; Stepanov, Eugene V.)V1426,121-132(1991)

Histochemical identification of malignant and premalignant lesions (Liebow, Charles; Maloney, M. J.)V1426,22-30(1991)

Histological distinction of mechanical and thermal defects produced by nanosecond laser pulses in striated muscle at 1064 nm (Gratzl, Thomas; Dohr, Gottfried; Schmidt-Kloiber, Heinz; Reichel, Erich)V1427,55-62(1991)

Histopathologic assessment of water-dominated photothermal effects produced with laser irradiation (Thomsen, Sharon L.; Cheong, Wai-Fung; Pearce, John A.)V1422,14-18(1991)

HIV detection by in-situ hybridization based on confocal reflected light microscopy (Smith, Louis C.; Jericevic, Zeljko; Cuellar, Roland; Paddock, Stephen W.; Lewis, Dorothy E.)V1428,224-232(1991)

Hollow-tube-guide for UV-power laser beams (Kubo, Uichi; Okada, Kasuyuki; Hashishin, Yuichi)V1420,102-107(1991)

Holmium:YAG and erbium:YAG laser interaction with hard and soft tissue (Charlton, Andrew; Dickinson, Mark R.; King, Terence A.; Freemont, Anthony J.)V1427,189-197(1991)

Holographic interferometric observation of shock wave focusing to extracorporeal shock wave lithotripsy (Takayama, Kazuyoshi; Obara, Tetsuro; Onodera, Osamu)V1358,1180-1190(1991)

Human tooth as an optical device (Altshuler, Grigori B.; Grisimov, Vladimir N.; Ermolaev, Vladimir S.; Vityaz, Irena V.)V1429,95-104(1991)

Hyperthermia treatment of spontaneously occurring oral cavity tumors using a computer-controlled Nd:YAG laser system (Panjehpour, Masoud; Overholt, Bergein F.; Frazier, Donita L.; Klebanow, Edward R.)V1424,179-185(1991)

Hyperthermia treatment using a computer-controlled Nd:YAG laser system in combination with surface cooling (Panjehpour, Masoud; Wilke, August; Frazier, Donita L.; Overholt, Bergein F.)V1427,307-315(1991)

Importance of pulse duration in laser-tissue interactions: a histological study (Kurban, Amal K.; Morrison, Paul R.; Trainor, S. W.; Cheong, Wai-Fung; Yasuda, Yukio; Tan, Oon T.)V1422,43-49(1991)

Influence of helium, oxygen, nitrogen, and room air environment in determining Nd-YAG laser/brain tissue interaction (Chavantes, Maria C.; Vinas, Federico; Zamorano, Lucia J.; Dujovny, Manuel; Dragovic, Ljubisa)V1428,13-22(1991)

Influence of the laser-induced temperature rise in photodynamic therapy (Gottschalk, Wolfgang; Hengst, Joachim; Sroka, Ronald; Unsoeld, Eberhard)V1427,320-326(1991)

Infrared fiber optic sensors: new applications in biology and medicine (Swairjo, Manal; Rothschild, Kenneth J.; Nappi, Bruce; Lane, Alan; Gold, Harris)V1437,60-65(1991)

Infrared tissue ablation: consequences of liquefaction (Zweig, A. D.)V1427,2-8(1991)

In-line power meter for use during laser angioplasty (Smith, Roy E.; Milnes, Peter; Edwards, David H.; Mitchell, David C.; Wood, Richard F.)V1425,116-121(1991)

Integral prism-tipped optical fibers (Friedl, Stephan E.; Kunz, Warren F.; Mathews, Eric D.; Abela, George S.)V1425,134-141(1991)

Interaction of phthalocyanine photodynamic treatment with ionophores and lysosomotrophic agents (Oleinick, Nancy L.; Varnes, Marie E.; Clay, Marian E.; Menegay, Harry J.)V1426,235-243(1991)

Interstitial laser coagulation of the prostate: experimental studies (McNicholas, Thomas A.; Steger, Adrian C.; Bown, Stephen G.; O'Donoghue, Neil)V1421,30-35(1991)

Intracellular uptake and ultrastructural phototoxic effects of sulfonated chlor-aluminum phthalocyanine on bladder tumor cells in vitro (Miller, Kurt; Reich, Ella; Grau, T.)V1426,378-383(1991)

Intraoperative clinical use of low-power laser irradiation following surgical treatment of the tethered spinal cord (Rochkind, Simeone; Alon, M.; Ouaknine, G. E.; Weiss, S.; Avram, J.; Razon, N.; Lubart, Rachel; Friedmann, Harry)V1428,52-58(1991)

Intraoperative endovascular ultrasonography (Eton, Darwin; Ahn, Samuel S.; Baker, J. D.; Pensabene, Joseph; Yeatman, Lawrence S.; Moore, Wesley S.)V1425,182-187(1991)

Intraoperative metastases detection by laser-induced fluorescence spectroscopy (Vari, Sandor G.; Papazoglou, Theodore G.; van der Veen, Maurits J.; Fishbein, Michael C.; Young, J. D.; Chandra, Mudjianto; Papaioannou, Thanassis; Beeder, Clain; Shi, Wei-Qiang; Grundfest, Warren S.)V1426,111-120(1991)

Intravascular ultrasound imaging and intracardiac echocardiography: recent developments and future directions (Pandian, Natesa G.)V1425,198-202(1991)

Investigation of the photodynamic properties of some chlorophyll a derivatives: the effect of doxorubicin on the chlorine e6 photosensitized death of Ehrlich carcinoma cells (Chekulayev, V.; Shevchuk, Igor; Kahru, A.; Mihkelsoo, V. T.; Kallikorm, A. P.)V1426,367-377(1991)

In-vitro ablation of fibrocartilage by XeCl excimer laser (Buchelt, Martin; Papaioannou, Thanassis; Fishbein, Michael C.; Peters, Werner; Beeder, Clain; Grundfest, Warren S.)V1420,249-253(1991)

In-vitro model for evaluation of pulse oximetry (Vegfors, Magnus; Lindberg, Lars-Goran; Lennmarken, Claes; Oberg, Per A.)V1426,79-83(1991)

In-vitro study of the effects of Congo Red on the ablation of atherosclerotic plaque (Beyerbacht, Hugo P.; Aggarwal, Shanti J.; Jansen, E. D.; Welch, Ashley J.)V1427,117-127(1991)

In-vivo optical attenuation in normal rat brain and its implication in PDT (Chen, Qun; Wilson, Brian C.; Patterson, Michael S.; Chopp, Michael; Hetzel, Fred W.)V1426,156-161(1991)

Kinetic models for coagulation processes: determination of rate coefficients in vivo (Pearce, John A.; Cheong, Wai-Fung; Pandit, Kirit; McMurray, Tom J.; Thomsen, Sharon L.)V1422,27-33(1991)

Knowledge-based nursing diagnosis (Hay, Claudette; Hay, D. R.)V1468,314-322(1991)

KTP-532 laser ablation of urethral strictures (Malloy, Terrence P.)V1421,72-72(1991)

KTP-532 laser utilization in endoscopic pelvic lymphadenectomy (Malloy, Terrence P.)V1421,78-78(1991)

Laparoscopically guided bilateral pelvic lymphadenectomy (Gershman, Alex; Danoff, Dudley; Chandra, Mudjianto; Grundfest, Warren S.)V1421,186-188(1991)

Laparoscopic appendectomy (Richards, Kent F.; Christensen, Brent J.)V1421,198-202(1991)

Laparoscopic applications of laser-activated tissue glues (Bass, Lawrence S.; Oz, Mehmet C.; Auteri, Joseph S.; Williams, Mathew R.; Rosen, Jeffrey; Libutti, Steven K.; Eaton, Alexander M.; Lontz, John F.; Nowygrod, Roman; Treat, Michael R.)V1421,164-168(1991)

Laparoscopic use of laser and monopolar electrocautery (Hunter, John G.)V1421,173-183(1991)

Laser angioplasty with lensed fibers and a holmium:YAG laser in iliac artery occlusions (White, Christopher J.; Ramee, Stephen R.; Mesa, Juan E.; Collins, Tyrone J.; Kotmel, Robert F.; Godfrey, Maureen A.)V1425,130-133(1991)

Laser-assisted skin closure at 1.32 microns: the use of a software-driven medical laser system (Dew, Douglas K.; Hsu, Tung M.; Hsu, Long S.; Halpern, Steven J.; Michaels, Charles E.)V1422,111-115(1991)

Laser Doppler flowmetry in neurosurgery (Fasano, Victor A.; Urciuoli, Rosa; Bolognese, Paolo; Fontanella, Marco)V1428,2-12(1991)

Laser effects on fibrin clot response by human meniscal fibrochondrocytes in organ culture (Forman, Scott K.; Oz, Mehmet C.; Wong, Edison; Treat, Michael R.; Kiernan, Howard)V1424,2-6(1991)

Laser-flash photographic studies of Er:YAG laser ablation of water (Jacques, Steven L.; Gofstein, Gary)V1427,63-67(1991); V1525,309-312(1991)

Laser-induced fluorescence spectroscopy of pathologically enlarged prostate gland in vitro (Chandra, Mudjianto; Gershman, Alex; Papazoglou, Theodore G.; Bender, Leon; Danoff, Dudley; Papaioannou, Thanassis; Vari, Sandor G.; Coons, Gregory; Grundfest, Warren S.)V1421,68-71(1991)

Laser-induced shock wave effects on red blood cells (Flotte, Thomas J.; Frisoli, Joan K.; Goetschkes, Margaret; Doukas, Apostolos G.)V1427,36-44(1991)

Laser lithotripsy with a pulsed dye laser: correlation between threshold energy and optical properties (Sterenborg, H. J.; van Swol, Christiaan F.; Biganic, Dane D.; van Gemert, Martin J.)V1427,256-266(1991)

Laser market in the 1990s (Levitt, Morris R.)V1520,1-36(1991)

LAser Microprobe Mass Spectrometry (LAMMS) in dental science: basic principles, instrumentation, and applications (Rechmann, Peter; Tourmann, J. L.; Kaufmann, Raimund)V1424,106-115(1991)

Laser photochemotherapy of psoriasis (Tuchin, Valery V.; Utz, Sergey R.; Barabanov, Alexander J.; Dovzansky, S. I.; Ulyanov, A. N.; Aravin, Vladislav A.; Khomutova, T. G.)V1422,85-96(1991)

Lasers in dentistry (Keller, Ulrich)V1525,282-288(1991)

Lasers in Dermatology and Tissue Welding (Tan, Oon T.; White, Rodney A.; White, John V., eds.)V1422(1991)

Lasers in Orthopedic, Dental, and Veterinary Medicine (Dederich, Douglas N.; O'Brien, Stephen J.; Trent, Ava M.; Wigdor, Harvey A., eds.)V1424(1991)

Lasers in Urology, Laparoscopy, and General Surgery (Watson, Graham M.; Steiner, Rudolf W.; Pietrafitta, Joseph J., eds.)V1421(1991)

Laser surgery for selected small animal soft-tissue conditions (Bartels, Kenneth E.)V1424,164-170(1991)

Laser-Tissue Interaction II (Jacques, Steven L., ed.)V1427(1991)

Lesion-specific laser catheters for angioplasty (Murphy-Chutorian, Douglas)V1420,244-248(1991)

Light absorption characteristics of the human meniscus: applications for laser ablation (Vangsness, C. T.; Huang, Jay; Smith, Chadwick F.)V1424,16-19(1991)

Light dosimetry in vivo in interstitial photodynamic therapy of human tumors (Reynes, Anne M.; Diebold, Simon; Lignon, Dominique; Granjon, Yves; Guillemin, Francois)V1525,177-182(1991)

Local hyperthermia for the treatment of diseases of the prostate (Servadio, Ciro)V1421,6-11(1991)

Local interstitial hyperthermia in malignant brain tumors using a low-power Nd:YAG laser (Bettag, Martin; Ulrich, Frank; Kahn, Thomas; Seitz, R.)V1525,409-411(1991)

Localization of the active site of an enzyme, bacterial luciferase, using two-quantum affinity modification (Benimetskaya, L. Z.; Gitelzon, I. I.; Kozionov, Andrew L.; Novozhilov, S. Y.; Petushkov, V. N.; Rodionova, N. S.; Stockman, M. I.)V1525,242-245(1991)

Low-power laser effects in equine traumatology and postsurgically (Antikas, Theo G.)V1424,186-197(1991)

Lung imaging fluorescence endoscope: development and experimental prototype (Palcic, Branko; Lam, Stephen; MacAulay, Calum; Hung, Jacklyn; Jaggi, Bruno; Radjinia, Massud; Pon, Alfred; Profio, A. E.)V1448,113-117(1991)

Mathematical modeling for laser PUVA treatment of psoriasis (Medvedev, Boris A.; Tuchin, Valery V.; Yaroslavsky, Ilya V.)V1422,73-84(1991)

Mechanical and acoustic analysis in ultrasonic angioplasty (Detwiler, Paul W.; Watkins, James F.; Rose, Eric A.; Ratner, A.; Vu, Louis P.; Severinsky, J. Y.; Rosenschein, Uri)V1425,149-155(1991)

Mechanism of injurious effect of excimer (308 nm) laser on the cell (Nevorotin, Alexey J.; Kallikorm, A. P.; Zeltzer, G. L.; Kull, Mart M.; Mihkelsoo, V. T.)V1427,381-397(1991)

Mechanisms of ultraviolet and mid-infrared tissue ablation (Hibst, Raimund)V1525,162-169(1991)

Medical applications of three-dimensional and four-dimensional laser scanning of facial morphology (Sadler, Lewis L.; Chen, Xiaoming; Figueroa, Alvaro A.; Aduss, Howard)V1380,158-162(1991)

Medical marketplace (Moretti, Michael)V1520,118-131(1991)

Medical prosthetic applications of growth simulations in four-dimensional facial morphology (Sadler, Lewis L.; Chen, Xiaoming; Fyler, Ann)V1380,137-146(1991)

Method for the analysis of the 3-D shape of the face and changes in the shape brought about by facial surgery (Coombes, Anne M.; Linney, Alfred D.; Richards, Robin; Moss, James P.)V1380,180-189(1991)

Microlocalization of Photofrin in neoplastic lesions (Korbelik, Mladen; Krosl, Gorazd; Lam, Stephen; Chaplin, David J.; Palcic, Branko)V1426,172-179(1991)

Monte Carlo simulations and measurements of signals in laser Doppler flowmetry on human skin (Koelink, M. H.; de Mul, F. F.; Greve, Jan; Graaff, Reindert; Dassel, A. C.; Aarnouds, J. G.)V1431,63-72(1991)

Multiphoton photobiologic effects (Hasan, Tayyaba)V1427,70-76(1991)

Naphthalocyanines relevant to the search for second-generation PDT sensitizers (Sounik, James R.; Rihter, Boris D.; Ford, William E.; Rodgers, Michael A.; Kenney, Malcolm E.)V1426,340-349(1991)

Nd:YAG laser irradiation in the treatment of upper tract urothelial tumors (Gaboardi, Franco; Melodia, Tommaso; Bozzola, Andrea; Galli, Luigi)V1421,79-85(1991)

Nd:YAG laser treatment in patients with prostatic adenocarcinoma stage A (Gaboardi, Franco; Dotti, Ernesto; Bozzola, Andrea; Galli, Luigi)V1421,50-55(1991)

New approaches to local destruction of tumors: interstitial laser hyperthermia and photodynamic therapy (Bown, Stephen G.)V1525,325-330(1991)

New concept of a compact multiwavelength solid state laser for laser-induced shock wave lithotripsy (Steiger, Erwin)V1421,140-146(1991)

New device for interactive image-guided surgery (Galloway, Robert L.; Edwards, Charles A.; Thomas, Judith G.; Schreiner, Steven; Maciunas, Robert J.)V1444,9-18(1991)

New results in dosimetry of laser radiation in medical treatment (Beuthan, Jurgen; Hagemann, Roland; Mueller, Gerhard J.; Schaldach, Brita J.; Zur, Ch.)V1420,225-233(1991)

New sensitizers for PDT (Morgan, Alan R.; Garbo, Greta M.; Krivak, T.; Mastroianni, Marta; Petousis, Nikolaos H.; St Clair, T.; Weisenberger, M.; van Lier, Johan E.)V1426,350-355(1991)

Nile Blue derivatives as lysosomotropic photosensitizers (Lin, Chi-Wei; Shulok, Janine R.; Kirley, S. D.; Cincotta, Louis; Foley, James W.)V1426,216-227(1991)

Noninvasive ultrasound-produced volume lesion in prostate (Foster, Richard S.)V1421,47-47(1991)

Novel method for preventing solar ultraviolet-radiation-induced skin cancer (Shi, Weimin; Cui, Ting; Sigel, George H.)V1422,62-72(1991)

Ophthalmic Technologies (Puliafito, Carmen A., ed.)V1423(1991)

Optical and mechanical parameter detection of calcified plaque for laser angioplasty (Stetz, Mark L.; O'Brien, Kenneth M.; Scott, John J.; Baker, Glenn S.; Deckelbaum, Lawrence I.)V1425,55-62(1991)

Optical Fibers in Medicine VI (Katzir, Abraham, ed.)V1420(1991)

Optical Methods for Tumor Treatment and Early Diagnosis: Mechanisms and Techniques (Dougherty, Thomas J., ed.)V1426(1991)

Optical quantification of sodium, potassium, and calcium ions in diluted human plasma based on ion-selective liquid membranes (Spichiger, Ursula E.; Seiler, Kurt; Wang, Kemin; Suter, Gaby; Morf, Werner E.; Simon, Wilhelm)V1510,118-130(1991)

Optical system for control of longitudinal displacement (Dick, Sergei C.; Markhvida, Igor V.; Tanin, Leonid V.)V1454,447-452(1991)

PDT of rat mammary adenocarcinoma in vitro and in a rat dorsal-skin-flap window chamber using Photofrin and chloroaluminum-sulfonated phthalocyanine (Flock, Stephen T.; Jacques, Steven L.; Small, Susan M.; Stern, Scott J.)V1427,77-89(1991)

Peak pressures and temperatures within laser-ablated tissues (Furzikov, Nickolay P.; Dmitriev, A. C.; Lekhtsier, Eugeny N.; Orlov, M. Y.; Semyenov, Alexander D.; Tyurin, Vladimir S.)V1427,288-297(1991)

Percutaneous laser balloon coagulation of accessory pathways (McMath, Linda P.; Schuger, Claudio D.; Crilly, Richard J.; Spears, J. R.)V1425,165-171(1991)

Percutaneous lumbar discectomy using Ho:YAG laser (Black, Johnathan D.; Sherk, Henry H.; Uppal, Gurvinder S.; Sazy, John; Meller, Menachem M.; Rhodes, Anthony; Lane, Gregory J.)V1424,20-22(1991)

Percutaneous peripheral excimer laser angioplasty: immediate success rate and short-term outcome (Visona, Adriana; Liessi, Guido; Bonanome, Andrea; Lusiani, Luigi; Miserocchi, Luigi; Pagnan, Antonio)V1425,75-83(1991)

Perspectives of powerful laser technique for medicine (Konov, Vitaly I.; Prokhorov, Alexander M.; Shcherbakov, I. A.)V1525,250-252(1991)

Pharmacokinetics of Photofrin II distribution in man (Kessel, David; Nseyo, Unyime O.; Schulz, Veronique; Sykes, Elizabeth)V1426,180-187(1991)

Photoablation (Doerschel, Klaus; Mueller, Gerhard J.)V1525,253-279(1991)

Photodynamic therapy in the prophylactic management of bladder cancer (Nseyo, Unyime O.; Lundahl, Scott L.; Merrill, Daniel C.)V1426,287-292(1991)

Photodynamic therapy of malignant brain tumors: supplementary postoperative light delivery by implanted optical fibers: field fractionation (Muller, Paul J.; Wilson, Brian C.)V1426,254-265(1991)

Photodynamic therapy of pet animals with spontaneously occurring head and neck carcinomas (Beck, Elsa R.; Hetzel, Fred W.)V1426,311-315(1991)

Photoproduct formation of endogeneous protoporphyrin and its photodynamic activity (Koenig, Karsten; Schneckenburger, Herbert; Rueck, Angelika C.; Auchter, S.)V1525,412-419(1991)

Photosensitive liposomes as potential drug delivery vehicles for photodynamic therapy (Morgan, Chris G.; Mitchell, A. C.; Chowdhary, R. K.)V1525,391-396(1991)

Photosensitization is required for antiretroviral activity of hypericin (Carpenter, Susan; Tossberg, John; Kraus, George A.)V1426,228-234(1991)

Phthalocyanines as phototherapeutic agents for tumors (Jori, Guilio; Reddi, Elena; Biolo, Roberta; Polo, Laura; Valduga, Giuliana)V1525,367-376(1991)

Plastic hollow fibers employed for CO2 laser power transmission in oral surgery (Calderon, S.; Gannot, Israel; Dror, Jacob; Dahan, Reuben; Croitoru, Nathan I.)V1420,108-115(1991)

Polarimetric optical fiber sensor for biochemical measurements (Heideman, Rene; Blikman, Albert; Koster, Rients; Kooyman, Rob P.; Greve, Jan)V1510,131-137(1991)

Portable and very inexpensive optical fiber sensor for entero-gastric reflux detection (Baldini, Francesco; Falciai, Riccardo; Bechi, Paolo; Cosi, Franco; Bini, Andrea; Milanesi, Francesco)V1510,58-62(1991)

Practical and theoretical considerations on the use of quantitative histology to measure thermal damage of collagen in cardiovascular tissues (Thomsen, Sharon L.)V1425,110-113(1991)

Precise individualized armature for ear reconstruction (Evenhouse, Raymond J.; Chen, Xiaoming)V1380,248-253(1991)

Preliminary experience with laser reinforcement of vascular anastomoses (Oz, Mehmet C.; Bass, Lawrence S.; Williams, Mathew R.; Benvenisty, Alan I.; Hardy, Mark A.; Libutti, Steven K.; Eaton, Alexander M.; Treat, Michael R.; Nowygrod, Roman)V1422,147-150(1991)

Preliminary results of laser-assisted sealing of hand-sewn canine esophageal anastomoses (Auteri, Joseph S.; Oz, Mehmet C.; Sanchez, Juan A.; Bass, Lawrence S.; Jeevanandam, Valluvan; Williams, Mathew R.; Smith, Craig R.; Treat, Michael R.)V1421,182-184(1991)

Preliminary results with sutured colonic anastomoses reinforced with dye-enhanced fibrinogen and a diode laser (Libutti, Steven K.; Williams, Mathew R.; Oz, Mehmet C.; Forde, Kenneth A.; Bass, Lawrence S.; Weinstein, Samuel; Auteri, Joseph S.; Treat, Michael R.; Nowygrod, Roman)V1421,169-172(1991)

Preliminary stress/strain analysis of laser-soldered and -sutured vascular tissue (Ashton, Robert C.; Oz, Mehmet C.; Lontz, John F.; Lemole, Gerald M.)V1422,151-155(1991)

Processing of bioceramic implants with excimer laser (Koort, Hans J.)V1427,173-180(1991)

Promising applications of scanning electron-beam-pumped laser devices in medicine and biology (Katsap, Victor N.; Koshevoy, Alexander V.; Meerovich, Gennady A.; Ulasjuk, Vladimir N.)V1420,259-265(1991)

Promotional effects of CO2 laser on neoplastic lesions in hamsters (Kingsbury, Jeffrey S.; Margarone, Joseph E.; Satchidanand, S.; Liebow, Charles)V1427,363-367(1991)

Pulp blood flow assessment in human teeth by laser Doppler flowmetry (Pettersson, Hans; Oberg, Per A.)V1424,116-119(1991)

Pulsed CO2 laser for intra-articular cartilage vaporization and subchondral bone perforation in horses (Nixon, Alan J.; Roth, Jerry E.; Krook, Lennart)V1424,198-208(1991)

Pulsed lasers in dentistry: sense or nonsense? (Koort, Hans J.; Frentzen, Matthias)V1424,87-98(1991)

Q-switched Nd:glass-laser-induced acoustic pulses in lithotripsy (D'yakonov, G. I.; Mikhailov, V. A.; Pak, S. K.; Shcherbakov, I. A.; Andreev, V. G.; Rudenko, O. V.; Sapozhnikov, A. V.)V1421,153-155(1991)

Radiolabeled red blood cells for the direct measurement of the blood flow kinetics in experimental tumors after photodynamic therapy (Paquette, Benoit; Rousseau, Jacques; Ouellet, Rene; van Lier, Johan E.)V1426,362-366(1991)

Range of modalities for prostate therapy (Watson, Graham M.)V1421,2-5(1991)

Rational anatomical treatment of basal cell carcinoma with photodynamic therapy (Keller, Gregory S.)V1426,266-270(1991)

Recanalization of azoospermia due to a Mullerian duct cyst by Nd:YAG laser (Gaboardi, Franco; Bozzola, Andrea; Zago, Tiziano; Gulf, Gildo M.; Galli, Luigi)V1421,73-77(1991)

Reflective optical sensor system for measurement of intracranial pressure (Zoghi, Behbood; Rastegar, Sohi)V1420,63-71(1991)

Remote 3-D laser topographic mapping with dental application (Altschuler, Bruce R.)V1380,238-247(1991)

Removal of dental filling materials by Er:YAG laser radiation (Hibst, Raimund; Keller, Ulrich)V1424,120-126(1991)

Restenosis after hot-tip laser-balloon angioplasty: histologic evaluation of the samples removed by Simpson atherectomy (Barbieri, Enrico; Tanganelli, Pietro; Taddei, Giuseppe; Perbellini, Antonio; Attino, Vito; Destro, Gianni; Zardini, Piero)V1425,122-127(1991)

Role of various wavelengths in phototherapy (Lubart, Rachel; Wollman, Yoram; Friedmann, Harry; Rochkind, Simeone; Laulicht, Israel)V1422,140-146(1991)

Root resection of endodontically treated teeth by Erbium:YAG laser radiation (Paghdiwala, A. F.)V1424,150-159(1991)

Sampling system for in vivo ultrasound images (Jensen, Jorgen A.; Mathorne, Jan)V1444,221-231(1991)

Selective tumor destruction with photodynamic therapy: exploitation of photodynamic thresholds (Barr, Hugh)V1525,331-340(1991)

Simple analytical model for laser-induced tissue ablation (Kukreja, Lalit M.; Braun, R.; Hess, P.)V1427,243-254(1991)

Site-specific laser modification (cleavage) of oligodeoxynucleotides (Benimetskaya, L. Z.; Bulychev, N. V.; Kozionov, Andrew L.; Koshkin, A. A.; Lebedev, A. V.; Novozhilov, S. Y.; Stockman, M. I.)V1525,210-211(1991)

Six years of transendoscopic Nd:YAG application in large animals (Tate, Lloyd P.; Glasser, Mardi)V1424,209-217(1991)

Skin closure with dye-enhanced laser welding and fibrinogen (Wider, Todd M.; Libutti, Steven K.; Greenwald, Daniel P.; Oz, Mehmet C.; Yager, Jeffrey S.; Treat, Michael R.; Hugo, Norman E.)V1422,56-61(1991)

Small single-sensor for temperature, flow, and pressure measurement (Sun, Mei H.; Kamal, Arvind)V1420,44-52(1991)

Spectral-luminescence analysis as method of tissue states evaluation and laser influence on tissues before and after transplantation (Loshchenov, V. B.; Baryshev, M. V.; Svystushkin, V. M.; Ovchinnikov, U. M.; Babyn, A. V.; Schaldach, Brita J.; Mueller, Gerhard J.)V1420,271-281(1991)

Spectroscopic studies of second-generation sensitizers and their photochemical reactions in liposomes and cells (Ehrenberg, Benjamin; Gross, Eitan; Lavi, Adina; Johnson, Fred M.; Malik, Zvi)V1426,244-251(1991)

Stage III endobronchial squamous cell cancer: survival after Nd:YAG laser combined with photodynamic therapy versus Nd:YAG laser or photodynamic therapy alone (McCaughan, James S.; Barabash, Rostislav D.; Hawley, Philip)V1426,279-287(1991)

Stimulatory effects of Nd:YAG lasers on canine articular cartilage (Lane, Gregory J.; Sherk, Henry H.; Kollmer, Charles; Uppal, Gurvinder S.; Rhodes, Anthony; Sazy, John; Black, Johnathan D.; Lee, Steven)V1424,7-11(1991)

Stone/ureter identification during alexandrite laser lithotripsy (Scheu, M.; Flemming, G.; Engelhardt, R.)V1421,100-107(1991)

Stress analysis in patella by three-dimensional photoelasticity (Chen, Riqi; Zhang, Jianxing; Jiang, Kunsheng)V1554A,407-417(1991)

Studies on the absence of photodynamic mechanism in the normal pancreas (Mang, Thomas S.; Wieman, T. J.; Crean, David H.)V1426,188-199(1991)

Study of bone ablation dynamics with sequenced pulses (Izatt, Joseph A.; Albagli, Douglas; Itzkan, Irving; Feld, Michael S.)V1427,110-116(1991)

Syntheses of porphyrin and chlorin dimers for photodynamic therapy (Pandey, Ravindra K.; Vicente, M. G.; Shiau, Fuu-Yau; Dougherty, Thomas J.; Smith, Kevin M.)V1426,356-361(1991)

Synthetic approaches to long-wavelength photosensitizers for photodynamic therapy (Shiau, Fuu-Yau; Pandey, Ravindra K.; Dougherty, Thomas J.; Smith, Kevin M.)V1426,330-339(1991)

Technique of CO2 laser arthroscopic surgery (Meller, Menachem M.; Sherk, Henry H.; Rhodes, Anthony; Sazy, John; Uppal, Gurvinder S.; Lane, Gregory J.)V1424,60-61(1991)

Temperature distributions in laser-irradiated tissues (Valderrama, Giuseppe L.; Fredin, Leif G.; Berry, Michael J.; Dempsey, B. P.; Harpole, George M.)V1427,200-213(1991)

Temperature increase during in vitro 308 nm excimer laser ablation of porcine aortic tissue (Gijsbers, Geert H.; Sprangers, Rene L.; van den Broecke, Duco G.; van Wieringen, Niek; Brugmans, Marco J.; van Gemert, Martin J.)V1425,94-101(1991)

Temperature variations of reflection, transmission, and fluorescence of the arterial wall (Chambettaz, Francois; Clivaz, Xavier; Marquis-Weible, Fabienne D.; Salathe, R. P.)V1427,134-140(1991)

Theoretical analysis of stone fragmentation rates (Sterenborg, H. J.)V1525,170-176(1991)

Three-Dimensional Bioimaging Systems and Lasers in the Neurosciences (Boggan, James E.; Cerullo, Leonard J.; Smith, Louis C., eds.)V1428(1991)

Time-resolved diffuse reflectance and transmittance studies in tissue simulating phantoms: a comparison between theory and experiment (Madsen, Steen J.; Patterson, Michael S.; Wilson, Brian C.; Park, Young D.; Moulton, J. D.; Jacques, Steven L.; Hefetz, Yaron)V1431,42-51(1991)

Time-resolved fluorescence of normal and atherosclerotic arteries (Pradhan, Asima; Das, Bidyut B.; Liu, C. H.; Alfano, Robert R.; O'Brien, Kenneth M.; Stetz, Mark L.; Scott, John J.; Deckelbaum, Lawrence I.)V1425,2-5(1991)

Time-Resolved Spectroscopy and Imaging of Tissues (Chance, Britton; Katzir, Abraham, eds.)V1431(1991)

Tissue ablation by holmium:YSGG laser pulses through saline and blood (van Leeuwen, Ton G.; van der Veen, Maurits J.; Verdaasdonk, Rudolf M.; Borst, Cornelius)V1427,214-219(1991)

Tissue interactions of ball-tipped and multifiber catheters (Mitchell, David C.; Smith, Roy E.; Walters, Tena K.; Murray, Alan; Wood, Richard F.)V1427,181-188(1991)

Tooth pulp reaction following Er:YAG laser application (Keller, Ulrich; Hibst, Raimund)V1424,127-133(1991)

Toward an explanation of laser-induced stimulation and damage of cell cultures (Friedmann, Harry; Lubart, Rachel; Laulicht, Israel; Rochkind, Simeone)V1427,357-362(1991)

Transendoscopic and freehand use of flexible hollow fibers for carbon dioxide laser surgery in the upper airway of the horse: a preliminary report (Palmer, Scott E.)V1424,218-220(1991)

Transurethral laser prostatectomy using a right-angle laser delivery system (Johnson, Douglas E.; Levinson, A. K.; Greskovich, Frank J.; Cromeens, Douglas M.; Ro, Jae Y.; Costello, Anthony J.; Wishnow, Kenneth I.)V1421,36-41(1991)

Transurethral microwave heating without urethral cooling: theory and experimental results (Petrovich, Zbigniew; Astrahan, Melvin; Baert, Luc)V1421,14-17(1991)

Transurethral microwave hyperthermia for benign prostatic hyperplasia: the Leuven clinical experience (Baert, Luc; Ameye, Filip; Willemen, Patrick; Petrovich, Zbigniew)V1421,18-29(1991)

Transurethral ultrasound-guided laser-induced prostatectomy (Babayan, Richard K.; Roth, Robert A.)V1421,42-44(1991)

Tripyrroledimethine-derived ("texaphyrin"-type) macrocycles: potential photosensitizers which absorb in the far-red spectral region (Sessler, Jonathan L.; Hemmi, Gregory; Maiya, Bhaskar G.; Harriman, Anthony; Judy, Millard M.; Boriak, Richard; Matthews, James L.; Ehrenberg, Benjamin; Malik, Zvi; Nitzan, Yeshayahu; Rueck, Angelika C.)V1426,318-329(1991)

Ultrafast imaging of vascular tissue ablation by an XeCl excimer laser (Neu, Walter; Nyga, Ralf; Tischler, Christian; Haase, Karl K.; Karsch, Karl R.)V1425,37-44(1991)

Use of a 3-D visualization system in the planning and evaluation of facial surgery (Linney, Alfred D.; Moss, James P.; Richards, Robin; Mosse, C. A.; Grindrod, S. R.; Coombes, Anne M.)V1380,190-199(1991)

Use of chloro-aluminum sulfonated phthalocyanine as a photosensitizer in the treatment of malignant tumors in dogs and cats (Peavy, George M.; Klein, Mary K.; Newman, H. C.; Roberts, Walter G.; Berns, Michael W.)V1424,171-178(1991)

UV-fluorescence spectroscopic technique in the diagnosis of breast, ovarian, uterus, and cervix cancer (Das, Bidyut B.; Glassman, W. L. S.; Alfano, Robert R.; Cleary, Joseph; Prudente, R.; Celmer, E.; Lubicz, Stephanie)V1427,368-373(1991)

Vanderbilt Free-Electron Laser Center for Biomedical and Materials Research (Tolk, Norman H.; Brau, Charles A.; Edwards, Glenn S.; Margaritondo, Giorgio; McKinley, J. T.)V1552,7-13(1991)

Vascular parameters from angiographic images (Close, Robert A.; Duckwiler, Gary R.; Vinuela, Fernando; Dion, Jacques E.)V1444,196-203(1991)

Vascular tissue welding of the CO2 laser: limitations (Dalsing, Michael C.; Kruepper, Peter; Cikrit, Dolores F.)V1422,98-102(1991)

Venous occlusion plethysmography based on fiber-optic sensor using the microbending principle (Stenow, Eric; Rohman, H.; Eriksson, L.-E.; Oberg, Per A.)V1420,29-33(1991)

Wavelength selection in laser arthroscopy (Black, Johnathan D.; Sherk, Henry H.; Meller, Menachem M.; Uppal, Gurvinder S.; Divan, James; Sazy, John; Rhodes, Anthony; Lane, Gregory J.)V1424,12-15(1991)

ZD multipurpose neurosurgical image-guided localizing unit: experience in 103 consecutive cases of open stereotaxis (Zamorano, Lucia J.; Dujovny, Manuel)V1428,30-51(1991)

Meteorology—see also atmospheric optics

AcidMODES: a major field study to evaluate regional-scale air pollution models (Ching, Jason K.; Bowne, Norman E.)V1491,360-370(1991)

Automated band selection for multispectral meteorological applications (Westerman, Steven D.; Drake, R. M.; Yool, Stephen R.; Brandley, M.; DeJulio, R.)V1492,263-271(1991)

BEST: a new satellite for a climatology study in the tropics (Orgeret, Marc)V1490,14-22(1991)

ESA Earth observation polar platform program (Rast, Michael; Readings, C. J.)V1490,51-58(1991)

Expert system for fusing weather and doctrinal information used in the intelligence preparation of the battlefield (McWilliams, Gary; Kirby, Steve; Eskridge, Thomas; Newberry, Jeff)V1468,417-428(1991)

German ATMOS program (Puls, Juergen)V1490,2-13(1991)

GLOB(MET)SAT: French proposals for monitoring global change and weather from the polar orbit (Durpaire, Jean-Pierre; Ratier, A.; Dagras, C.)V1490,23-38(1991)

Imager of METEOSAT second generation (Hollier, Pierre A.)V1490,74-81(1991)

Lidar profiles of atmospheric structure properties (Philbrick, Charles R.)V1492,76-84(1991)

METEOSAT second-generation program (Markland, Chris A.)V1490,39-50(1991)

Remote colorimetry and its applications (Sheffer, Dan; Ben-Shalom, Ami; Devir, Adam D.)V1493,232-243(1991)

Remote sensing of coastal environmental hazards (Huh, Oscar K.; Roberts, Harry H.; Rouse, Lawrence J.)V1492,378-386(1991)

ScaRaB Earth radiation budget scanning radiometer (Monge, J.L.; Kandel, Robert S.; Pakhomov, L. A.; Bauche, B.)V1490,84-93(1991)

Use of an expert system to predict thunderstorms and severe weather (Passner, Jeffrey E.; Lee, Robert R.)V1468,2-10(1991)

Metrology—see also microlithography; nondestructive testing; optical testing

0.50 um contact measurement and characterization (Lindsay, Tracy K.; Orvek, Kevin J.; Mumaw, Richard T.)V1464,104-118(1991)

10th Annual Symp on Microlithography (Wiley, James N., ed.)V1496(1991)

Accuracy improvement by mathematical correction method and electro-optical device certification (Bulichev, A.; Voronin, G.; Kaganov, S.; Kuzmin, Vladimir S.; Porjadin, V.; Chibisov, V.)V1500,151-162(1991)

Accuracy/repeatability test for a video photogrammetric measurement (Gustafson, Peter C.)V1526,36-41(1991)

AcidMODES: a major field study to evaluate regional-scale air pollution models (Ching, Jason K.; Bowne, Norman E.)V1491,360-370(1991)

Adaptive control of photolithography (Crisalle, Oscar D.; Soper, Robert A.; Mellichamp, Duncan A.; Seborg, Dale E.)V1464,508-526(1991)

Adaptive process control for a rapid thermal processor (Dilhac, Jean-Marie R.; Ganibal, Christian; Bordeneuve, J.; Dahhou, B.; Amat, L.; Picard, A.)V1393,395-403(1991)

Advanced confocal technique for submicron CD measurements (Rohde, Axel; Saffert, Ralf; Fitch, John)V1464,438-446(1991)

Advanced real-time scanning concept for full dynamics recording, high image quality, and superior measurement accuracy (Lindstrom, Kjell M.; Wallin, Bo)V1488,389-398(1991)

Advanced Techniques for Integrated Circuit Processing (Bondur, James; Turner, Terry R., eds.)V1392(1991)

Aircraft exterior scratch measurement system using machine vision (Sarr, Dennis P.)V1472,177-184(1991)

Algorithm for the generation of look-up range table in 3-D sensing (Su, Xianyu; Zhou, Wen-Sheng)V1332,355-357(1991)

Analysis of laser welding process with the mathematical model GMDH (De Iorio, I.; Sergi, Vincenzo; Tagliaferri, V.)V1397,787-790(1991)

Analysis of temperature distribution and slip in rapid thermal processing (Lee, Hyouk; Yoo, Young-Don; Shin, Hyun-Dong; Earmme, Youn-Young; Kim, Choong-Ki)V1393,404-410(1991)

Analysis of the accuracy of aerostatic slideway made from granite (Li, Jinian)V1533,150-154(1991)

Angle measurement by moire interference technique (Singh, Brahm P.; Chitnis, Vijay T.)V1554B,335-338(1991)

Application of adaptive network theory to dry-etch monitoring and control (Deshmukh, V. G.; Hope, D. A.; Cox, T. I.; Hydes, A. J.)V1392,352-360(1991)

Application of a reduced area electrical test pattern to precise pattern registration measurements (Rominger, James P.)V1496,224-231(1991)

Application of speckle metrology at a nuclear waste repository (Conley, Edgar; Genin, Joseph)V1332,798-801(1991)

Application of thermal wave technology to thickness and grain size monitoring of aluminum films (Opsal, Jon L.)V1596,120-131(1991)

Application of transmission electron detection to x-ray mask calibrations and inspection (Postek, Michael T.; Larrabee, Robert D.; Keery, William J.; Marx, Egon)V1464,35-47(1991)

Applications of diamond-turned null reflectors for generalized aspheric metrology (McCann, James T.)V1332,843-849(1991)

Applications of holographic gratings in x-ray mask metrology (Hansen, Matthew E.)V1396,78-79(1991)

Applications of latent image metrology in microlithography (Adams, Thomas E.)V1464,294-312(1991)

Automated dimensional inspection of cars in crash tests with digital photogrammetry (Beyer, Horst A.)V1526,134-141(1991)

Automated visual inspection for LSI wafer patterns using a derivative-polarity comparison algorithm (Maeda, Shunji; Hiroi, Takashi; Makihira, Hiroshi; Kubota, Hitoshi)V1567,100-109(1991)

Automatic, high-resolution analysis of low-noise fringes (Lassahn, Gordon D.)V1332,690-695(1991)

Automatization of measurement and processing of experimental data in photoelasticity (Zhavoronok, I. V.; Nemchinov, V. V.; Litvin, S. A.; Skanavi, A. M.; Pavlov, V. V.; Evsenev, V. S.)V1554A,371-379(1991)

Basic use of acoustic speckle pattern for metrology and sea waves study (He, Duo-Min; He, Ming-Shia)V1332,808-819(1991)

Beam characterization and measurement of propagation attributes (Sasnett, Michael W.; Johnston, Timothy J.)V1414,21-32(1991)

Beryllium scatter analysis program (Behlau, Jerry L.; Granger, Edward M.; Hannon, John J.; Baumler, Mark; Reilly, James F.)V1530,218-230(1991)

Bidirectional reflectance distribution function raster scan technique for curved samples (McIntosh, Malcolm B.; McNeely, Joseph R.)V1530,263-268(1991)

Black and white cinefilm (ASA 5) applied to dynamic laser speckle and its processing (He, Anzhi; Yu, Gui Ying; Ni, Xiao W.; Fan, Hua Ying)V1554A,747-749(1991)

Business, manufacturing, and system integration issues in cluster tool process control (Richardson, David)V1392,302-314(1991)

Calibration for the SAGE III/EOS instruments (Chu, William P.; McCormick, M. P.; Zawodny, J. M.; McMaster, Leonard R.)V1491,243-250(1991)

Calibration of a CCD camera on a hybrid coordinate measuring machine for industrial metrology (Bruzzone, Elisabetta; Mangili, Fulvia)V1526,96-112(1991)

Capability assessment and comparison of the Nikon 2i, Nikon 3i, and IMS-2000 registration measurement devices (Henderson, Robert K.)V1496,198-216(1991)

Catalog of infrared and cryo-optical properties for selected materials (Heaney, James B.; Alley, Phillip W.; Bradley, Scott E.)V1485,140-159(1991)

Characterization of automatic overlay measurement technique for sub-half-micron devices (Kawai, Akira; Fujiwara, Keiji; Tsujita, Kouichirou; Nagata, Hitoshi)V1464,267-277(1991)

Characterization of hot-isostatic-pressed optical-quality beryllium (Behlau, Jerry L.; Baumler, Mark)V1530,208-217(1991)

Characterization of wavelength offset for optimization of deep-UV stepper performance (Jones, Susan K.; Dudley, Bruce W.; Peters, Charles R.; Kellam, Mark D.; Pavelchek, Edward K.)V1464,546-553(1991)

Charging effects in low-voltage SEM metrology (Monahan, Kevin M.; Benschop, Jozef P.; Harris, Tom A.)V1464,2-9(1991)

Charging phenomena in e-beam metrology (Levy, Dorron; Harris, Karl L.)V1464,413-423(1991)

Clinical measuring system for the form and position errors of circular workpieces using optical fiber sensors (Tan, Jiubin; Qiang, Xifu; Ding, Xuemei)V1572,552-557(1991)

Cluster tool software and hardware architecture (Huntley, Dave)V1392,315-330(1991)

Cn2 estimates in the boundary layer for damp unstable conditions (Tunick, Arnold; Rachele, Henry; Miller, Walter B.)V1487,51-62(1991)

Coastal survey with a multispectral video system (Niedrauer, Terren M.)V1492,240-251(1991)

Comparison of Gauss' and Petermann's formulas for real single-mode fibers by far-field pattern technique (Pospisilova, Marie; Schneiderova, Martina)V1504,287-291(1991)

Comparison of low-scatter-mirror PSD derived from multiple-wavelength BRDFs and WYKO profilometer data (Wong, Wallace K.; Wang, Dexter; Benoit, Robert T.; Barthol, Peter)V1530,86-103(1991)

Computational model of the imaging process in scanning-x microscopy (Gallarda, Harry S.; Jain, Ramesh C.)V1464,459-473(1991)

Control of proximity effects on CD uniformity through the use of process parameters derived from a statistically designed experiment (Christensen, Lorna D.; Bell, Kenneth L.)V1463,504-514(1991)

Correction of inherent scan nonplanarity in the Boeing infrared sensor calibration facility (Chase, Robert P.)V1533,138-149(1991)

Corrections to the Golden Rule (Fearn, Heidi; Lamb, Willis E.)V1497,245-254(1991)

Criteria for accurate cutback attenuation measurements (Haigh, N. R.; Linton, R. S.; Johnson, R.; Grigsby, R.)V1366,259-264(1991)

Critical-angle refractometry: accuracy analysis (Tentori, Diana)V1527,216-224(1991)

Critical dimension control using development end point detection for wafers with multilayer structures (Hagi, Toshio; Okuda, Yoshimitsu; Ohkuma, Tohru)V1464,215-221(1991)

Cross-sectional imaging in SEM: signal formation mechanism and CD measurements (Firstein, Leon A.; Noz, Arthur)V1464,81-88(1991)

Cryoscatter measurements of beryllium (Lippey, Barret; Krone-Schmidt, Wilfried)V1530,150-161(1991)

Defect reduction strategies for process control and yield improvement (Liljegren, Douglas R.)V1392,681-687(1991)

Deformation measurements of the human tympanic membrane under static pressure using automated moire topography (Dirckx, Joris J.; Decraemer, Willem F.)V1429,34-38(1991)

Deformation study of the Klang Gates Dam with multispectral analysis method (Hassan, Azmi; Subari, Mustofa; Som, Zainal A.)V1526,142-156(1991)

Delayed elasticity in Zerodur at room temperature (Pepi, John W.; Golini, Donald)V1533,212-221(1991)

Description and calibration of a fully automated infrared scatterometer (Mainguy, Stephan; Olivier, Michel; Josse, Michel A.; Guidon, Michel)V1530,269-282(1991)

Design considerations for multipurpose bidirectional reflectometers (Neu, John T.; Bressler, Martin)V1530,244-254(1991)

Design criteria of an integrated optics microdisplacement sensor (d'Alessandro, Antonio; De Sario, Marco; D'Orazio, Antonella; Petruzzelli, Vincenzo)V1332,554-562(1991)

Design parameters of an EO sensor (Tanwar, Lakhan S.; Jain, P. C.; Kunzmann, H.)V1332,877-882(1991)

Detection of stratospheric ozone trends by ground-based microwave observations (Connor, Brian J.; Parrish, Alan; Tsou, Jung-Jung)V1491,218-230(1991)

Detection of the object velocity using the time-varying scattered speckles (Okamoto, Takashi; Asakura, Toshimitsu)V1399,192-199(1991)

Determination of thin-film roughness and volume structure parameters from light-scattering investigations (Duparre, Angela; Kassam, Samer)V1530,283-286(1991)

Digital Talbot interferometer (Tam, Siu-Chung; Silva, Donald E.; Wong, H. L.)V1400,38-48(1991)

Displacement analysis of the interior walls of a pipe using panoramic holointerferometry (Gilbert, John A.; Matthys, Donald R.; Hendren, Christelle M.)V1554B,128-134(1991)

Displacement measurement using grating images detected by CCD image sensor (Hane, Kazuhiro; Grover, Chandra P.)V1332,584-590(1991)

Dynamic behavior of internal elements of high-frequency integrated circuits studied by time-resolved optical-beam-induced current (OBIC) method (Bergner, Harald; Hempel, Klaus; Stamm, Uwe)V1362,484-493(1991)

EBES4: mask/reticle writer for the 90's (Chen, George C.)V1496,107-117(1991)

Edge-profile, materials, and protective coating effects on image quality (Doi, Takeshi; Tadros, Karim H.; Kuyel, Birol; Neureuther, Andrew R.)V1464,336-345(1991)

Effective surface PSD for bare hot-isostatic-pressed beryllium mirrors (Vernold, Cynthia L.; Harvey, James E.)V1530,144-149(1991)

Effect of operating points in submicron CD measurements (Dusa, Mircea V.; Jung, Christoph; Jung, Paul; Hogenkamp, Detlef; Roeth, Klaus-Dieter)V1464,447-458(1991)

Effects of the nonvanishing tip size in mechanical profile measurements (Church, Eugene L.; Takacs, Peter Z.)V1332,504-514(1991)

Effects of wafer cooling characteristics after post-exposure bake on critical dimensions (Lauck, Teresa L.; Nomura, Masafumi; Omori, Tsutae; Yoshioka, Kajutoshi)V1464,527-538(1991)

Electrical probe diagnostics for processing discharges (Mantei, Thomas D.)V1392,466-473(1991)

Electron-beam metrology: the European initiative (Jackman, James J.)V1464,71-90(1991)

Emissivity of silicon wafers during rapid thermal processing (Vandenabeele, Peter; Maex, Karen)V1393,316-336(1991)

Enhanced process control of submicron contact definition (Ostrout, Wayne H.; Hunkler, Sean; Ward, Steven D.)V1392,151-164(1991)

Enhancements to the radiometric calibration facility for the Clouds and the Earth's Radiant Energy System instruments (Folkman, Mark A.; Jarecke, Peter J.; Darnton, Lane A.)V1493,255-266(1991)

Estimation of plastic strain by fractal (Dai, YuZhong; Chiang, Fu-Pen)V1332,767-774(1991)

Etch tailoring through flexible end-point detection (Angell, David; Oehrlein, Gottleib S.)V1392,543-550(1991)

Evaluation of a photoresist process for 0.75-micron g-line lithography (Kasahara, Jack S.; Dusa, Mircea V.; Perera, Thiloma)V1463,492-503(1991)

Evaluation of interference patterns: a comparison of methods (Kreis, Thomas M.; Geldmacher, Juergen)V1554B,718-724(1991)

Experimental assessment of 150-mm P/P+ epitaxial silicon wafer flatness for deep-submicron applications (Huff, Howard R.; Weed, Harrison)V1464,278-293(1991)

Experimental characterization of Fresnel zone plate for hard x-ray applications (Lai, Barry P.; Chrzas, John J.; Yun, Wen-Bing; Legnini, Dan; Viccaro, P. J.; Bionta, Richard M.; Skulina, Kenneth M.)V1550,46-49(1991)

Extending electrical measurements to the 0.5 um regime (Troccolo, Patrick M.; Mantalas, Lynda C.; Allen, Richard A.; Linholm, Loren)V1464,90-103(1991)

Fast-injection Langmuir probe for process diagnostic and control (Patrick, Roger; Schoenborn, Philippe; Linder, Stefan; Baltes, Henry P.)V1392,506-513(1991)

Faying surface-gap measurement of aircraft structures for shim fabrication and installation (Sarr, Dennis P.; Jurick, Thomas W.)V1469,506-514(1991)

Fiber optic based vortex shedder flow meter (Chu, Beatrice C.; Newson, Trevor P.; Jackson, David A.)V1504,251-257(1991)

Fiber optic Fabry-Perot sensors for high-speed heat transfer measurements (Kidd, S. R.; Sinha, P. G.; Barton, James S.; Jones, Julian D.)V1504,180-190(1991)

Fiber optic measurement standards (Pollitt, Stuart)V1504,80-87(1991)

Fiber Optic Metrology and Standards (Soares, Oliverio D., ed.)V1504(1991)

Fiber optic pressure and temperature sensor for down-hole applications (Lequime, Michael; Lecot, C.; Jouve, Philippe; Pouleau, J.)V1511,244-249(1991)

Fiber Optic Sensors: Engineering and Applications (Bruinsma, Anastasius J.; Culshaw, Brian, eds.)V1511(1991)

Fiber optic smart structures: structures that see the light (Measures, Raymond M.)V1332,377-398(1991)

Figure of merit for calibration and comparison of linewidth measurement instruments (Hershey, Robert R.; Zavecz, Terrence E.)V1464,22-34(1991)

Fluid-flow-rate metrology: laboratory uncertainties and traceabilities (Mattingly, G. E.)V1392,386-401(1991)

Focal length measurement using diffraction at a grating (Sirohi, Rajpal S.; Kumar, Harish; Jain, Narinder K.)V1332,50-55(1991)

Forecasting optical turbulence strength: effects of macroscale meteorology and aerosols (Sadot, Danny; Kopeika, Norman S.)V1442,325-334(1991)

Formation of heterostructure devices in a multichamber processing environment with in-vacuo surface analysis diagnostics and in-situ process monitoring (Lucovsky, Gerald; Kim, Sang S.; Fitch, J. T.; Wang, Cheng)V1392,605-616(1991)

Fourier transform infrared spectrophotometry for thin film monitors: computer and equipment integration for enhanced capabilities (Cox, J. N.; Sedayao, J.; Shergill, Gurmeet S.; Villasol, R.; Haaland, David M.)V1392,650-659(1991)

French proposal for IEC/TC 86/WG 4 OTDR calibration (Gauthier, Francis)V1504,55-65(1991)

Frequency-derived distributed optical fiber sensing: backscatter analysis (Rogers, Alan J.; Handerek, Vincent A.; Parvaneh, Farhad)V1511,190-200(1991)

Frequency spectrum analysis and assessment of optical surface flaws (Gao, Wenliang; Zhang, Xiao; Yang, Guoguang)V1530,118-128(1991)

Fresnel zone plate moire patterns and its metrological applications (Jaroszewicz, Zbigniew)V1574,154-158(1991)

Fringe-scanning moire system using a servo-controlled grating (Kurokawa, Haruhisa; Ichikawa, Naoki; Yajima, Nobuyuki)V1332,643-654(1991)

Fundamental mechanisms and doping effects in silicon infrared absorption for temperature measurement by infrared transmission (Sturm, James C.; Reaves, Casper M.)V1393,309-315(1991)

Fundamentals of on-line gauging for machine vision (Novini, Amir R.)V1526,2-16(1991)

Geometric measurement of optical fibers with pulse-counting method (Nie, Qiuhua; Nelson, John C.; Fleming, Simon C.)V1332,409-420(1991)

Germanium photodiodes calibration as standards of optical fiber systems power measurements (Campos, Joaquin; Corredera, Pedro; Pons, Alicia A.; Corrons, Antonio)V1504,66-74(1991)

"Golden standard" wafer design for optical stepper characterization (Kemp, Kevin G.; King, Charles F.; Wu, Wei; Stager, Charles)V1464,260-266(1991)

Graduating rules checking up by laser interferometry (Miron, Nicolae; Sporea, Dan G.)V1500,334-338(1991)

Ground-based monitoring of water vapor in the middle atmosphere: the NRL water-vapor millimeter-wave spectrometer (Bevilacqua, Richard M.; Schwartz, Philip R.; Pauls, Thomas A.; Waltman, William B.; Thacker, Dorsey L.)V1491,231-242(1991)

Half-micrometer linewidth metrology (Knight, Stephen E.; Humphrey, Dean; Bowley, Reginald R.; Cogley, Robert M.)V1464,119-126(1991)

Hartmann-Shack sensor as a component in active optical system to improve the depth resolution of the laser tomographic scanner (Liang, Junzhong; Grimm, B.; Goelz, Stefan; Bille, Josef F.)V1542,543-554(1991)

Helium neon laser optics: scattered light measurements and process control (Perilloux, Bruce E.)V1530,255-262(1991)

High-accuracy edge-matching with an extension of the MPGC-matching algorithm (Gruen, Armin; Stallmann, Dirk)V1526,42-55(1991)

High-resolution computer-aided moire (Sciammarella, Cesar A.; Bhat, Gopalakrishna K.)V1554B,162-173(1991)

High-sensitivity interferometric technique for strain measurements (Voloshin, Arkady S.; Bastawros, Adel F.)V1400,50-60(1991)

High-speed polarimetric measurements for fiber-optic communications (Calvani, Riccardo A.; Caponi, Renato; Piglia, Roberto)V1504,258-263(1991)

High-speed stepper setup using a low-voltage SEM (Benschop, Jozef P.; Monahan, Kevin M.; Harris, Tom A.)V1464,62-70(1991)

High-temperature optical sensor for displacement measurement (Ebbeni, Jean P.)V1504,268-272(1991)

Holographic measurement of deformation using carrier fringe and FFT techniques (Quan, C.; Judge, Thomas R.; Bryanston-Cross, Peter J.)V1507,463-475(1991)

Holographic measurement of the angular error of a table moving along a slideway (Matsuda, Kiyofumi; Tenjimbayashi, Koji)V1332,230-235(1991)

Holographic optical profilometer (Hasman, Erez; Davidson, Nir; Friesem, Asher A.)V1442,372-377(1991)

Holography with a single picosecond pulse (Abramson, Nils H.)V1332,224-229(1991)

Holo-interferometric patterns recorded through a panoramic annular lens (Gilbert, John A.; Greguss, Pal; Kransteuber, Amy S.)V1238,412-420(1991)

I-line lithography for highly reproducible fabrication of surface acoustic wave devices (Berek, Stefan; Knauer, Ulrich; Zottl, Helmut)V1463,515-520(1991)

Improvements in sensitivity and discrimination capability of the PD reticle/mask inspection system (Saito, Juichi; Saijo, Y.)V1496,284-301(1991)

Improving submicron CD measurements through optimum operating points (Dusa, Mircea V.; Roeth, Klaus-Dieter; Jung, Christoph)V1496,217-223(1991)

Industrial Vision Metrology (El-Hakim, Sabry F., ed.)V1526(1991)

Information extracting and application for the combining objective speckle and reflection holography (Cao, Zhengyuan; Cheng, Fang)V1332,358-364(1991)

Infrared BRDF measurements of space shuttle tiles (Young, Raymond P.; Wood, Bobby E.; Stewart, P. L.)V1530,335-342(1991)

Infrared/microwave correlation measurements (Norgard, John D.; Metzger, Don W.; Cleary, John C.; Seifert, Michael)V1540,699-708(1991)

Infrared window damage measured by reflective scatter (Bernt, Marvin L.; Stover, John C.)V1530,42-49(1991)

In-line supervisory control in a photolithographic workcell (Ling, Zhi-Min; Leang, Sovarong; Spanos, Costas J.)V1392,660-669(1991)

Innovative Optics and Phase Conjugate Optics (Ahlers, Rolf-Juergen; Tschudi, Theo T., eds.)V1500(1991)

In-plane displacement measurement on rotating components using pulsed laser ESPI (electronic speckle pattern interferometry) (Preater, Richard W.)V1399,164-171(1991)

In-situ film thickness measurements using acoustic techniques (Bhardwaj, Sanjay; Khuri-Yakub, B. T.)V1392,555-562(1991)

In-situ interferometric measurements in a rapid thermal processor (Dilhac, Jean-Marie R.; Ganibal, Christian; Nolhier, N.; Amat, L.)V1393,349-353(1991)

In-situ investigation of the low-pressure MOCVD growth of III-V compounds using reflectance anisotropy measurements (Drevillon, Bernard; Razeghi, Manijeh)V1361,200-212(1991)

In-situ measurement of piston jitter in a ring resonator (Cunningham, Philip R.; Hay, Stephen O.; Francis, Denise M.; Trott, G.E.)V1414,97-129(1991)

In-situ measurements of radicals and particles in a selective silicon oxide etching plasma (Singh, Jyothi)V1392,474-486(1991)

Integrated Circuit Metrology, Inspection, and Process Control V (Arnold, William H., ed.)V1464(1991)

Integrated optics sensor on silicon for the measurement of displacement, force, and refractive index (Ulbers, Gerd)V1506,99-110(1991)

Interfacial adhesive strength measurement in a multilayered two-level metal device structure (Siddiqui, Humayun R.; Ryan, Vivian; Shimer, Julie A.)V1596,139-157(1991)

Interference phenomenon with correlated masks and its application (Grover, Chandra P.; Hane, Kazuhiro)V1332,624-631(1991)

Interferometric fiber optic sensors for use with composite materials (Measures, Raymond M.; Valis, Tomas; Liu, Kexing; Hogg, W. D.; Ferguson, Suzanne M.; Tapanes, Edward)V1332,421-430(1991)

Interferometric measurement of in-plane motion (Hercher, Michael; Wyntjes, Geert J.)V1332,602-612(1991)

Interferometric optical fiber sensors for absolute measurement of displacement and strain (Kersey, Alan D.; Berkoff, Timothy A.; Dandridge, Anthony D.)V1511,40-50(1991)

Interferometric system for the inspection and measurement of the quality of optical fiber ends (Corredera, Pedro; Pons, Alicia A.; Campos, Joaquin; Corrons, Antonio)V1504,281-286(1991)

Interpretation of measured spectral attenuation curves of optical fibers by deconvolution with source spectrum (Hoefle, Wolfgang)V1504,140-146(1991)

Investigation of fiber-reinforced-plastics-based components by means of holographic interferometry (Jueptner, Werner P.; Bischof, Thomas)V1400,69-79(1991)

Investigation of rapid thermal process-induced defects in ion-implanted Czochralski silicon (Yarling, Charles B.; Hahn, Sookap; Hodul, David T.; Suga, Hisaaki; Smith, W. L.)V1393,192-199(1991)

Issues affecting the characterization of integrated optical devices subjected to ionizing radiation (Hickernell, Robert K.; Sanford, N. A.; Christensen, David H.)V1474,138-147(1991)

Laser-based triangulation techniques in optical inspection of industrial structures (Clarke, Timothy A.; Grattan, Kenneth T.; Lindsey, N. E.)V1332,474-486(1991)

Laser Beam Diagnostics (Hindy, Robert N.; Kohanzadeh, Youssef, eds.)V1414(1991)

Laser metrological measurement of transient strain fields in Hopkinson-bar experiments (Vogel, Dietmar; Michel, Bernd; Totzauer, Werner F.; Schreppel, Ulrich; Clos, Rainer)V1554A,262-274(1991)

Laser moire topography for 3-D contour measurement (Matsumoto, Tetsuya; Kitagawa, Yoichi; Adachi, Masaaki; Hayashi, Akihiro)V1332,530-536(1991)

Laser-scanning tomography and related dark-field nanoscopy method (Montgomery, Paul C.; Gall, Pascal; Ardisasmita, Moh S.; Castagne, Michel; Bonnafe, Jacques; Fillard, Jean-Pierre B.)V1332,563-570(1991)

Light scattered by coated paper (Marx, Egon; Song, J. F.; Vorburger, Theodore V.; Lettieri, Thomas R.)V1332,826-834(1991)

Light scattering from gold-coated ground glass and chemically etched surfaces (Ruiz-Cortes, Victor; Mendez, Eugenio R.; Gu, Zu-Han; Maradudin, Alexei A.)V1558,222-232(1991)

Light scatter variations with respect to wafer orientation in GaAs (Brown, Jeff L.)V1530,299-305(1991)

Linearity of coherence probe metrology: simulation and experiment (Davidson, Mark P.; Monahan, Kevin M.; Monteverde, Robert J.)V1464,155-176(1991)

Linewidth measurement comparison between a photometric optical microscope and a scanning electron microscope backed with Monte Carlo trajectory computations (Nunn, John W.; Turner, Nicholas P.)V1464,50-61(1991)

Localization of hot spots in silicon devices with a laser scanning microscope (Bergner, Harald; Krause, A.; Stamm, Uwe)V1361,723-731(1991)

Long-path differential absorption measurements of tropospheric molecules (Harder, Jerald W.; Mount, George H.)V1491,33-42(1991)

Low-coherence optical reflectometry of laser diode waveguides (Boisrobert, Christian Y.; Franzen, Douglas L.; Danielson, Bruce L.; Christensen, David H.)V1474,285-290(1991)

Low-cost fiber optic sensing systems using spatial division multiplexing (Paton, Barry E.)V1332,446-455(1991)

Measurement of atmospheric composition by the ATMOS instrument from Table Mountain Observatory (Gunson, Michael R.; Irion, Fredrick W.)V1491,335-346(1991)

Measurement of color and scattering phenomena of translucent materials (Sjollema, J. I.; den Exter, Ir. T.; Zijp, Jaap R.; Ten Bosch, Jaap J.)V1500,177-188(1991)

Measurement of fluid velocity fields using digital correlation techniques (Matthys, Donald R.; Gilbert, John A.; Puliparambil, Joseph T.)V1332,850-861(1991)

Measurement of interfacial tension by automated video techniques (Deason, Vance A.; Miller, R. L.; Watkins, Arthur D.; Ward, Michael B.; Barrett, K. B.)V1332,868-876(1991)

Measurement of mode field diameter and fiber bending loss (Kiang, Ying J.; Stigliani, Daniel J.)V1366,252-258(1991)

Measurement of nonlinear constants in photosensitive glass optical fibers (Oesterberg, Ulf L.)V1504,107-109(1991)

Measurement of physical parameters with special fibers (Shiota, Alan T.; Inada, Koichi)V1504,90-97(1991)

Measurement of strains by means of electro-optics holography (Sciammarella, Cesar A.; Bhat, Gopalakrishna K.; Albertazzi, Armando A.)V1396,143-154(1991)

Measurement of surface roughness using optical fibre sensor and microcomputer (Fan, Dapeng; Zhang, Honghai; Chen, Jihong; Chen, Riyao)V1572,11-14(1991)

Measurement of the sizes of the semiconductor crystallites in colored glasses through neutron scattering (Banfi, Giampiero P.; Degiorgio, Vittorio; Rennie, A. R.; Righini, Giancarlo C.)V1361,874-880(1991)

Measurement of wave aberrations of intraocular lenses through holographic interferometry (Carretero, L.; Fuentes, Rosa; Fimia-Gil, Antonio)V1507,458-462(1991); V1508,96-100(1991)

Measurements of atmospheric transmittance of CO_2 laser radiation (Aref'ev, Vladimir N.)V1397,827-830(1991)

Measurements of the InxGa1-xAs/GaAs critical layer thickness (Andersson, Thorvald G.; Ekenstedt, M. J.; Kulakovskii, Vladimir D.; Wang, S. M.; Yao, J. Y.)V1361,434-442(1991)

Measurements on the NIST GEC reference cell (Roberts, James R.; Olthoff, James K.; Van Brunt, R. J.; Whetstone, James R.)V1392,428-436(1991)

Measuring films on and below polycrystalline silicon using reflectometry (Engstrom, Herbert L.; Stokowski, Stanley E.)V1464,566-573(1991)

Measuring for thickness distribution of recording layer of PLH (Zhang, Xiao-Chun; Guo, Lu Rong; Guo, Yongkang)V1461,93-96(1991)

Measuring the geometric accuracy of a very accurate, large, drum-type scanner (Wolber, John W.)V1448,175-180(1991)

Melt dynamics and surface deformation in processing with laser radiation (Kreutz, Ernst W.; Pirch, Norbert)V1502,160-176(1991)

Metallization: Performance and Reliability Issues for VLSI and ULSI (Gildenblat, Gennady S.; Schwartz, Gary P., eds.)V1596(1991)

Metrology issues associated with submicron linewidths (Phan, Khoi; Nistler, John L.; Singh, Bhanwar)V1464,424-437(1991)

Microcomputer-based real-time monitoring and control of single-wafer processing (Hauser, John R.; Gyurcsik, Ronald S.)V1392,340-351(1991)

Microlens arrays in integral photography and optical metrology (Davies, Neil A.; McCormick, Malcolm; Lau, Hon W.)V1544,189-198(1991)

Microwave interferometric measurements of process plasma density (Cheah, Chun-Wah; Cecchi, Joseph L.; Stevens, J. L.)V1392,487-497(1991)

Minimum detectable changes in Rayleigh backscatter from distributed fiber sensors (Garside, Brian K.; Park, R. E.)V1588,150-158(1991)

Moire displacement detection by the photoacoustic technique (Hane, Kazuhiro; Watanabe, S.; Goto, T.)V1332,577-583(1991)

Moire measurements using a panoramic annular lens (Gilbert, John A.; Matthys, Donald R.; Lehner, David L.)V1554B,202-209(1991)

Moire-shift interferometer measurements of the shape of human and cat tympanic membrane (Decraemer, Willem F.; Dirckx, Joris J.)V1429,26-33(1991)

Monitoring and control of rf electrical parameters near plasma loads (Rummel, Paul)V1392,411-420(1991)

Monte Carlo modeling of secondary electron signals from heterogeneous specimens with nonplanar surfaces (Russ, John C.; Dudley, Bruce W.; Jones, Susan K.)V1464,10-21(1991)

Multichannel chromatic interferometry: metrology applications (Tribillon, Gilbert M.; Calatroni, Jose; Sandoz, Patrick)V1332,632-642(1991)

Near real-time operation of a centimeter-scale distributed fiber sensing system (Garside, Brian K.)V1332,399-408(1991)

Network for the detection of stratospheric change (Kurylo, Michael J.)V1491,168-174(1991)

Neural net selection of features for defect inspection (Sasaki, Kenji; Casasent, David P.; Natarajan, Sanjay S.)V1384,228-233(1991)

New apparatus and method for fluid composition monitoring and control (Urmson, John)V1392,421-427(1991)

New measurement techniques for modal power distribution in fibers (Golub, Mikhail A.; Sisakian, Iosif N.; Soifer, Victor A.; Uvarov, G. V.)V1366,273-282(1991)

New optical technique for particle sizing and velocimetry (Xie, Gong-Wie; Scott, Peter D.; Shaw, David T.; Zhang, Yi-Mo)V1500,310-321(1991)

New options of holographic metrology (Steinbichler, Hans)V1507,435-445(1991)

New phase measurement for nonmonotonical fringe patterns (Tang, Shou-Hong; Hung, Yau Y.; Zhu, Qiuming)V1332,731-737(1991)

New technique for multiplying the isoclinic fringes (Wen, Mei-Yuan; Liu, Guang T.)V1332,673-675(1991)

New wavefront sensor for metrology of spherical surfaces (Goelz, Stefan; Persoff, Jeffrey J.; Bittner, Groff D.; Liang, Junzhong; Hsueh, Chi-Fu T.; Bille, Josef F.)V1542,502-510(1991)

New Zeiss interferometer (Kuechel, Michael F.)V1332,655-663(1991)

Nomarski viewing system for an optical surface profiler (Bietry, Joseph R.; Auriemma, R. A.; Bristow, Thomas C.; Merritt, Edward)V1332,537-543(1991)

Noncontacting acoustics-based temperature measurement techniques in rapid thermal processing (Lee, Yong J.; Chou, Ching-Hua; Khuri-Yakub, B. T.; Saraswat, Krishna C.)V1393,366-371(1991)

Noncontact optical microtopography (Costa, Manuel Filipe P.; Almeida, Jose B.)V1400,102-107(1991)

Noncontact technique for the measurement of linear displacement using chirped diffraction gratings (Spillman, William B.; Fuhr, Peter L.)V1332,591-601(1991)

Nonimaging optics and the measurement of diffuse reflectance (Hanssen, Leonard M.; Snail, Keith A.)V1528,142-150(1991)

Noninvasive sensors for in-situ process monitoring and control in advanced microelectronics manufacturing (Moslehi, Mehrdad M.)V1393,280-294(1991)

Nonoptical noncoherent imaging in industrial testing (Delecki, Z. A.; Barakat, M. A.)V1526,157-167(1991)

Novel noncontact sensor for surface topography measurements using fiber optics (Butler, Clive; Gregoriou, Gregorios)V1584,282-293(1991)

Numerical investigation of effect of dynamic range and nonlinearity of detector on phase-stepping holographic interferometry (Fang, Qiang; Luo, Xiangyang; Tan, Yushan)V1332,216-222(1991)

Numerical simulation of thick-linewidth measurements by reflected light (Wojcik, Gregory L.; Mould, John; Monteverde, Robert J.; Prochazka, Jaroslav J.; Frank, John R.)V1464,187-203(1991)

Object-oriented strategies for a vision dimensional metrology system (Pizzi, Nicolino J.; El-Hakim, Sabry F.)V1468,296-304(1991)

Observation of stress voids and grain structure in laser-annealed aluminum using focused ion-beam microscopy (Pramanik, Dipankar; Jain, Vivek)V1596,132-138(1991)

Optical 3-D sensing for measurement of bottomhole pattern (Su, Wan-Yong; Su, Xianyu)V1567,680-682(1991)

Optical characterization of InP epitaxial layers on different substrates (Jiao, Kaili L.; Zheng, J. P.; Kwok, Hoi-Sing; Anderson, Wayne A.)V1361,776-783(1991)

Optical Fabrication and Testing (Campbell, Duncan R.; Johnson, Craig W.; Lorenzen, Manfred, eds.)V1400(1991)

Optical fiber measurements and standardization: status and perspectives (Di Vita, P.)V1504,38-46(1991)

Optical figure testing of prototype mirrors for JPL's precision segmented-reflector program (Hochberg, Eric B.)V1542,511-522(1991)

Optical laser intelligent verification expert system (Jones, Robert H.)V1401,86-93(1991)

Optical metrology for integrated circuit fabrication (Chim, Stanley S.; Kino, Gordon S.)V1464,138-144(1991)

Optical polysilicon over oxide thickness measurement (Kaiser, Anne M.)V1464,554-565(1991)

Optical probing of field dependent effects in GaAs photoconductive switches (Donaldson, William R.; Kingsley, Lawrence E.)V1378,226-236(1991)

Optical Scatter: Applications, Measurement, and Theory (Stover, John C., ed.)V1530(1991)

Optical surface microtopography using phase-shifting Nomarski microscope (Shimada, Wataru; Sato, Tadamitu; Yatagai, Toyohiko)V1332,525-529(1991)

Optical Systems in Adverse Environments (Kuok, M. H.; Silva, Donald E.; Tam, Siu-Chung, eds.)V1399(1991)

Optical techniques for determination of normal shock position in supersonic flows for aerospace applications (Adamovsky, Grigory; Eustace, John G.)V1332,750-756(1991)

Optical Testing and Metrology III: Recent Advances in Industrial Optical Inspection (Grover, Chandra P., ed.)V1332(1991)

Optical three-dimensional sensing for measurement of bottomhole pattern (Su, Wan-Yong; Su, Xianyu)V1332,820-823(1991)

Optimal control of positive optical photoresist development (Carroll, Thomas A.; Ramirez, W. F.)V1464,222-231(1991)

Optimal estimation of finish parameters (Church, Eugene L.; Takacs, Peter Z.)V1530,71-85(1991)

OTDR calibration for attenuation measurement (Moeller, Werner; Heitmann, Walter; Reich, M.)V1504,47-54(1991)

Overlay and matching strategy for large-area lithography (Holbrook, David S.; Donaher, J. C.)V1463,475-486(1991)

Overview of the halogen occultation experiment (Russell, James M.)V1491,110-116(1991)

Passive range and azimuth measuring system (Ronning, E.; Fjarlie, E. J.)V1399,178-183(1991)

Pattern recognition approach to trench bottom-width measurement (Chou, Ching-Hua; Berman, John L.; Chim, Stanley S.; Corle, Timothy R.; Xiao, Guoqing; Kino, Gordon S.)V1464,145-154(1991)

Performance evaluation of a vision dimension metrology system (El-Hakim, Sabry F.; Westmore, David B.)V1526,56-67(1991)

Phase-shift mask technology: requirements for e-beam mask lithography (Dunbrack, Steven K.; Muray, Andrew; Sauer, Charles; Lozes, Richard; Nistler, John L.; Arnold, William H.; Kyser, David F.; Minvielle, Anna M.; Preil, Moshe E.; Singh, Bhanwar; Templeton, Michael K.)V1464,314-326(1991)

Phase-stepping technique in holography (Wen, Zheng; Tan, Yushan)V1461,278-285(1991)

Phase space calculation of bend loss in rectangular light pipes (Gleckman, Philip; Ito, John)V1528,163-168(1991)

Photoelastic transducer for high-temperature applications (Redner, Alex S.; Adamovsky, Grigory; Wesson, Laurence N.)V1332,775-782(1991)

Photoreflectance for the in-situ monitoring of semiconductor growth and processing (Pollak, Fred H.)V1361,109-130(1991)

Photoresist dissolution rates: a comparison of puddle, spray, and immersion processes (Robertson, Stewart A.; Stevenson, J. T.; Holwill, Robert J.; Thirsk, Mark; Daraktchiev, Ivan S.; Hansen, Steven G.)V1464,232-244(1991)

Picosecond optoelectronic semiconductor switching and its application (Brueckner, Volkmar; Bergner, Harald; Lenzner, Matthias; Strobel, Reiner)V1362,510-517(1991)

Polarization-maintaining single-mode fibers: measurement and prediction of fundamental characteristics (Sasek, Ladislav; Vohryzek, Jachym)V1504,147-154(1991)

Prediction of Cn2 on the basis of macroscale meteorology including aerosols (Sadot, Danny; Kopeika, Norman S.)V1487,40-50(1991)

Prediction of coarse-aerosol statistics according to weather forecast (Gottlieb, J.; Kopeika, Norman S.)V1487,184-191(1991)

Prediction of thermal-image quality as a function of weather forecast (Shushan, A.; Meninberg, Y.; Levy, I.; Kopeika, Norman S.)V1487,300-311(1991)

Process control sensor development for the automation of single-wafer processors (Hosch, Jimmy W.)V1392,529-534(1991)

Process monitoring and control with fiber optics (Marcus, Michael A.)V1368,191-202(1991)

Process optimization: a case study on the application of Taguchi methods in optical lithography (Arshak, Khalil I.; Murphy, Eamonn; Arshak, A.)V1463,521-533(1991)

Projection moire as a tool for the automated determination of surface topography (Cardenas-Garcia, Jaime F.; Zheng, S.; Shen, F. Z.)V1554B,210-224(1991)

Pyrometer modeling for rapid thermal processing (Wood, Samuel C.; Apte, Pushkar P.; King, Tsu-Jae; Moslehi, Mehrdad M.; Saraswat, Krishna C.)V1393,337-348(1991)

Quantitative evaluation of optical surfaces using an improved Foucault test approach (Vandenberg, Donald E.; Humbel, William D.; Wertheimer, Alan)V1542,534-542(1991)

Radial ion energy measurements in an electron cyclotron resonance reactor (O'Neill, James A.; Holber, William M.; Caughman, John)V1392,516-528(1991)

Radiometric standards in the USSR (Sapritsky, Victor I.)V1493,58-69(1991)

Rapid Thermal and Related Processing Techniques (Moslehi, Mehrdad M.; Singh, Rajendra, eds.)V1393(1991)

Real-time automation of a dry etching system (McCafferty, Robert H.)V1392,331-339(1991)

Real-time, in-situ measurement of film thickness and uniformity during plasma ashing of photoresist (Davies, John T.; Metz, Thomas E.; Savage, Richard N.; Simmons, Horace O.)V1392,551-554(1991)

Real-time monitoring and control of plasma etching (Butler, Stephanie W.; McLaughlin, Kevin J.; Edgar, Thomas F.; Trachtenberg, Isaac)V1392,361-372(1991)

Recent developments in fiber optic and laser sensors for flow, surface vibration, rotation, and velocity measurements (Arik, Engin B.)V1584,202-211(1991)

Refractive-index measurement using moire deflectometry: working conditions (Tentori, Diana; Lopez Famozo, C.)V1535,209-215(1991)

Requirements of a solar diffuser and measurements of some candidate materials (Guzman, Carmen T.; Palmer, James M.; Slater, Philip N.; Bruegge, Carol J.; Miller, Edward A.)V1493,120-131(1991)

Residual stress evaluation using shearography with large shear displacements (Hathaway, Richard B.; Der Hovanesian, Joseph; Hung, Yau Y.)V1554B,725-735(1991)

Resist parameter extraction with graphical user interface in X (Chiu, Anita S.; Ferguson, Richard A.; Doi, Takeshi; Wong, Alfred K.; Tam, Nelson; Neureuther, Andrew R.)V1466,641-652(1991)

Resist tracking: a lithographic diagnostic tool (Takemoto, Cliff H.; Ziger, David H.; Connor, William; Distasio, Romelia)V1464,206-214(1991)

Review of interferogram analysis methods (Malacara Hernandez, Daniel)V1332,678-689(1991)

Review of measurement systems for evaluating thermal expansion homogeneity of Corning Code 7971 ULETM (Hagy, Henry E.)V1533,198-211(1991)

Review of phase-measuring interferometry (Stahl, H. P.)V1332,704-719(1991)

Review of temperature measurements in the semiconductor industry (Anderson, Richard L.)V1392,437-451(1991)

Review of the performances of industrial 1.5-kW and 3-kW CO2 lasers, mobile on gantry or 2-D machines (Aubert, Philippe; Beck, Rasmus)V1502,52-59(1991)

Round-robin comparison of temperature nonuniformity during RTP due to patterned layers (Vandenabeele, Peter; Maex, Karen)V1393,372-394(1991)

RTP temperature uniformity mapping (Keenan, W. A.; Johnson, Walter H.; Hodul, David T.; Mordo, David)V1393,354-365(1991)

Scatter and contamination of a low-scatter mirror (McNeely, Joseph R.; McIntosh, Malcolm B.; Akerman, M. A.)V1530,288-298(1991)

Scatter and roughness measurements on optical surfaces exposed to space (Schmitt, Dirk-Roger; Swoboda, Helmut; Rosteck, Helmut)V1530,104-110(1991)

Selected Papers on Ellipsometry (Azzam, Rasheed M., ed.)VMS27(1991)

Selected Papers on Speckle Metrology (Sirohi, Rajpal S., ed.)VMS35(1991)

Self-aberration-eliminating interferometer for wavefront measurements (Gorelik, Vladimir S.; Kovalenko, Sergey N.; Turukhano, Boris G.)V1507,488-490(1991)

Semiconductor thin-film optical constant determination and thin-film thickness measurement equipment correlation (Kaiser, Anne M.)V1464,386-392(1991)

Semiwafer metrology project (Bennett-Lilley, Marylyn H.; Hiatt, William M.; Lauchlan, Laurie J.; Mantalas, Lynda C.; Rottmann, Hans; Seliger, Mark; Singh, Bhanwar; Yansen, Don E.)V1464,127-136(1991)

Sequential experimentation strategy and response surface methodologies for photoresist process optimization (Flores, Gary E.; Norbury, David H.)V1464,610-627(1991)

Shadow method of scale with a-posteriori increase of measurements sensitivity (Spornik, Nikolai M.; Serenko, M. Y.)V1508,162-169(1991)

Shot noise limited optical measurements at baseband with noisy lasers (Hobbs, Philip C.)V1376,216-221(1991)

Simple and accurate technique to determine substrate indices by the multilayer Brewster angle measurement (Xiang, Feng; Yip, Gar L.)V1583,271-277(1991)

Simultaneous measurement of refractive index and thickness of thin film by polarized reflectances (Kihara, Tami; Yokomori, Kiyoshi)V1332,783-791(1991)

Single-ion spectroscopy (Bergquist, James C.; Wineland, D. J.; Itano, W. M.; Diedrich, F.; Raizen, M. G.; Elsner, Frank)V1435,82-85(1991)

Software concept for the new Zeiss interferometer (Doerband, Bernd; Wiedmann, Wolfgang; Wegmann, Ulrich; Kuebler, C. W.; Freischlad, Klaus R.)V1332,664-672(1991)

Solution for anomalous scattering of bare HIP Be and CVD SiC mirrors (Vernold, Cynthia L.)V1530,130-143(1991)

Some experimental techniques for characterizing photoresists (Spence, Christopher A.; Ferguson, Richard A.)V1466,324-335(1991)

Space-time correlation properties and their applications of dynamic speckles after propagation through an imaging system and double-random modulation (Ma, Shining; Liu, Ying; Du, Fuli)V1554A,645-648(1991)

Spatial-carrier phase-shifting technique of fringe pattern analysis (Kujawinska, Malgorzata; Schmidt, Joanna)V1508,61-67(1991)

Spatial characterization of YAG power laser beam (Grevey, D. F.; Badawi, K. F.)V1502,32-40(1991)

Statistical approach to optimizing advanced low-voltage SEM operation (Apostolakis, Peter J.)V1464,406-412(1991)

Studies on laser dynamic precision measurement of fine-wire diameters (Bao, Liangbi; Chen, Fuyao; Wu, Shixiong; Xu, Jiangtong; Guan, Zhilian)V1332,862-867(1991)

SUB-3D high-resolution pose measurement system (Hudgens, Jeffrey C.; Tesar, Delbert; Sklar, Michael E.)V1387,271-282(1991)

Submerged reflectance measurements as a function of visible wavelength (Giles, John W.; Voss, Kenneth J.)V1537,140-146(1991)

Submicron linewidth measurement using an interferometric optical profiler (Creath, Katherine)V1464,474-483(1991)

Summary of atmospheric chemistry observations from the Antarctic and Arctic aircraft campaigns (Tuck, Adrian F.)V1491,252-272(1991)

Surface defect detection and classification with light scattering (Gebhardt, Michael; Truckenbrodt, Horst; Harnisch, Bernd)V1500,135-143(1991)

Surface field measurement of photoconductive power switches using the electro-optic Kerr effect (Sardesai, Harshad P.; Nunnally, William C.; Williams, Paul F.)V1378,237-248(1991)

Surface microtopography of thin silver films (Costa, Manuel Filipe P.; Almeida, Jose B.)V1332,544-551(1991)

Surface reflection coefficient correction technique for a microdisplacement OFS (Wei, Cailin)V1572,42-46(1991)

Surface roughness measurements of spherical components (Chen, Yi-Sheng; Wang, Wen-Gui)V1530,111-117(1991)

Synchronous phase-extraction technique and its applications (Hung, Yau Y.; Tang, Shou-Hong; Jin, Guofan; Zhu, Qiuming)V1332,738-747(1991)

Systems-oriented survey of noncontact temperature measurement techniques for rapid thermal processing (Peyton, David; Kinoshita, Hiroyuki; Lo, G. Q.; Kwong, Dim-Lee)V1393,295-308(1991)

Techniques for characterization of silicon penetration during DUV surface imaging (Freeman, Peter W.; Bohland, John F.; Pavelchek, Edward K.; Jones, Susan K.; Dudley, Bruce W.; Bobbio, Stephen M.)V1464,377-385(1991)

Temperature and strain sensing using monomode optical fiber (Farahi, Faramarz; Jackson, David A.)V1511,234-243(1991)

Theory and experiment as tools for assessing surface finish in the UV-visible wavelength region (Ingers, Joakim P.; Thibaudeau, Laurent)V1400,178-185(1991)

Theory of electronic projection correlation and its application in measurement of rigid body displacement and rotation (Li, Xide; Tan, Yushan)V1554B,661-668(1991)

Thickness measurement of combined a-Si and Ti films on c-Si using a monochromatic ellipsometer (Yoo, Chue-San; Jans, Jan C.)V1464,393-403(1991)

Three-dimensional automatic precision measurement system by liquid-crystal plate on moire topography (Arai, Yasuhiko; Yekozeki, Shunsuke; Yamada, Tomoharu)V1554B,266-274(1991)

Three-dimensional gauging with stereo computer vision (Wong, Kam W.; Ke, Ying; Lew, Michael; Obaidat, Mohammed T.)V1526,17-26(1991)

Three-dimensional inspection using laser-based dynamic fringe projection (Harvey, David M.; Shaw, Michael M.; Hobson, Clifford A.; Wood, Christopher M.; Atkinson, John T.; Lalor, Michael J.)V1400,86-93(1991)

Three-dimensional interferometric and fiber-optic displacement measuring probe (Liang, Dawei; Fraser Monteiro, L.; Fraser Monteiro, M. L.; Lu, Boyin)V1511,90-97(1991)

Three-dimensional nanoprofiling of semiconductor surfaces (Montgomery, Paul C.; Fillard, Jean-Pierre B.; Tchandjou, N.; Ardisasmita, Moh S.)V1332,515-524(1991)

Two-dimensional micropattern measurement using precision laser beam scanning (Fujita, Hiroo)V1332,456-467(1991)

Two-dimensional surface strain measurement based on a variation of Yamaguchi's laser-speckle strain gauge (Barranger, John P.)V1332,757-766(1991)

Understanding metrology of polysilicon gates through reflectance measurements and simulation (Tadros, Karim H.; Neureuther, Andrew R.; Guerrieri, Roberto)V1464,177-186(1991)

Uniformity characterization of rapid thermal processor thin films (Yarling, Charles B.; Cook, Dawn M.)V1393,411-420(1991)

Use of diffracted light from latent images to improve lithography control (Hickman, Kirt C.; Gaspar, Susan M.; Bishop, Kenneth P.; Naqvi, H. S.; McNeil, John R.; Tipton, Gary D.; Stallard, Brian R.; Draper, B. L.)V1464,245-257(1991)

Using an expert system to interface mainframe computing resources with an interactive video system (Carey, Raymond; Wible, Sheryl F.; Gaynor, Wayne H.; Hendry, Timothy G.)V1464,500-507(1991)

Using dynamic holography for iron fibers with submicron diameter and high velocity (Xie, Gong-Wie; Hua, Lifan; Patel, Sushil; Scott, Peter D.; Shaw, David T.)V1385,132-141(1991)

Using the Atomic Force Microscope to measure submicron dimensions of integrated circuit devices and processes (Rodgers, Mark R.; Monahan, Kevin M.)V1464,358-366(1991)

Van der Lugt optical correlation for the measurement of leak rates of hermetically sealed packages (Fitzpatrick, Colleen M.; Mueller, Edward P.)V1332,185-192(1991)

Variability in thickness measurements using x-ray fluorescence technique (Baltazar, Inmaculada C.; Mena, Manolo G.)V1392,670-680(1991)

Visualization of impingement field of real-rocket-exhausted jets by using moire deflectometry (He, Anzhi; Yan, Da-Peng; Miao, Peng-Cheng; Wang, Hai-Lin)V1554B,429-434(1991)

Wafer alignment based on existing microstructures (Wyntjes, Geert J.; Hercher, Michael)V1464,539-545(1991)

Wavelength and mode-adjustable source for modal characterization of optical fiber components: application to a new alignment method (Pagnoux, Dominique; Blondy, Jean M.; Facq, Paul)V1504,98-106(1991)

Wavelength and temperature dependence of bending loss in monomode optical fibers (Morgan, Russell D.; Jones, Julian D.; Harper, Philip G.; Barton, James S.)V1504,118-124(1991)

Wavelength-based sensor for the measurement of small angles (Depoortere, Marc; Ebbeni, Jean P.)V1504,264-267(1991)

White-light moire phase-measuring interferometry (Stahl, H. P.)V1332,720-730(1991)

Width gauge for hot steel plates by laser-scanning rangefinder (Adachi, Yuzi; Matsui, Kenichi)V1527,225-233(1991)

Microelectronic processing—see also annealing; epitaxy; etching; microlithography

Adaptive control of photolithography (Crisalle, Oscar D.; Soper, Robert A.; Mellichamp, Duncan A.; Seborg, Dale E.)V1464,508-526(1991)

Adaptive process control for a rapid thermal processor (Dilhac, Jean-Marie R.; Ganibal, Christian; Bordeneuve, J.; Dahhou, B.; Amat, L.; Picard, A.)V1393,395-403(1991)

Advanced Techniques for Integrated Circuit Processing (Bondur, James; Turner, Terry R., eds.)V1392(1991)

Analysis of temperature distribution and slip in rapid thermal processing (Lee, Hyouk; Yoo, Young-Don; Shin, Hyun-Dong; Earmme, Youn-Young; Kim, Choong-Ki)V1393,404-410(1991)

Anomalous diffusion phenomena in two-step rapid thermal diffusion of phosphorus (Cho, Byung-Jin; Kim, Choong-Ki)V1393,180-191(1991)

Application of adaptive network theory to dry-etch monitoring and control (Deshmukh, V. G.; Hope, D. A.; Cox, T. I.; Hydes, A. J.)V1392,352-360(1991)

Applications of excimer lasers in microelectronics (Yu, Chang; Sandhu, Gurtej S.; Mathews, V. K.; Doan, Trung T.)V1598,186-197(1991)

Applications of optical emission spectroscopy in plasma manufacturing systems (Gifford, George G.)V1392,454-465(1991)

Basic mechanisms and application of the laser-induced forward transfer for high-Tc superconducting thin film deposition (Fogarassy, Eric)V1394,169-179(1991)

Business, manufacturing, and system integration issues in cluster tool process control (Richardson, David)V1392,302-314(1991)

Characteristics of gate oxide surface material after exposure to magnetron-enhanced reactive ion etching plasma (Webb, Jennifer M.; Amini, Zahra H.)V1392,47-54(1991)

Characterization of microstructure of Si films grown by laser-enhanced photo-CVD using Si2H6 (Lian, S.; Fowler, Burt; Krishnan, S.; Jung, Le-Tien; Li, C.; Banerjee, Sanjay)V1598,98-107(1991)

Characterization of plasma processes with optical emission spectroscopy (Malchow, Douglas S.)V1392,498-505(1991)

Chlorine or bromine chemistry in reactive ion etching Si-trench etching (Rangelow, Iwilo W.; Fichelscher, Andreas)V1392,240-245(1991)

Cleaved surfaces of high Tc films for making SNS structures (Takeuchi, Ichiro; Tsai, Jaw S.; Tsuge, Hisanao; Matsukura, Noritsuga; Miura, Sadahiko; Yoshitake, T.; Kojima, Yoshikatsu; Matsui, Shinji)V1394,96-101(1991)

Cluster tool software and hardware architecture (Huntley, Dave)V1392,315-330(1991)

Contamination and damage of silicon surfaces during magnetron-enhanced reactive ion etching in a single-wafer system (Tan, Swie-In; Colavito, D. B.)V1392,106-118(1991)

Control of oxygen incorporation and lifetime measurement in Si1-xGex epitaxial films grown by rapid thermal chemical vapor deposition (Sturm, James C.; Schwartz, P. V.; Prinz, Erwin J.; Magee, Charles W.)V1393,252-259(1991)

Critical current enhancement in Y1Ba2Cu3O7-y/Y1Ba2(Cu1-xNix)3O7-y heterostructures (Witanachchi, S.; Lee, S. Y.; Song, L. W.; Kao, Yi-Han; Shaw, David T.)V1394,161-168(1991)

Current trends and issues for low-damage dry etching of optoelectronic devices (Hu, Evelyn L.)V1361,512-522(1991)

Defect reduction strategies for process control and yield improvement (Liljegren, Douglas R.)V1392,681-687(1991)

Defect reduction strategies for submicron manufacturing: tools and methodologies (Coleman, Robyn S.; Chitturi, Prasanna R.)V1392,638-649(1991)

Design, fabrication, and performance of an integrated optoelectronic cellular array (Hibbs-Brenner, Mary K.; Mukherjee, Sayan D.; Skogen, J.; Grung, B.; Kalweit, Edith; Bendett, Mark P.)V1563,10-20(1991)

Development of laser bonding as a manufacturing process for inner lead bonding (Hayward, James D.)V1598,164-174(1991)

Dry etching for silylated resist development (Laporte, Philippe; Van den hove, Luc; Melaku, Yosias)V1392,196-207(1991)

Dry etching of high-aspect ratio contact holes (Wiepking, Mark; LeVan, M.; Mayo, Phyllis)V1392,139-150(1991)

Dynamics of the optical parameters of molten silicon during nanosecond laser annealing (Boneberg, J.; Yavas, O.; Mierswa, B.; Leiderer, Paul)V1598,84-90(1991)

Economic impact of single-wafer multiprocessors (Wood, Samuel C.; Saraswat, Krishna C.; Harrison, J. M.)V1393,36-48(1991)

Effect of Cu at Al grain boundaries on electromigration behavior in Al thin films (Frear, Darrel R.; Michael, J. R.; Kim, C.; Romig, A. D.; Morris, J. W.)V1596,72-82(1991)

Effect of laser irradiation on superconducting properties of laser-deposited YBa2Cu3O7 thin films (Singh, Rajiv K.; Bhattacharya, Deepika; Narayan, Jagdish; Jahncke, Catherine; Paesler, Michael A.)V1394,203-213(1991)

Effects of environmental and installation-specific factors on process gas delivery via mass-flow controller with an emphasis on real-time behavior (Gray, David E.; Benjamin, Neil M.; Chapman, Brian N.)V1392,402-410(1991)

Efficient laser cleaning of small particulates using pulsed laser irradiation synchronized with liquid-film deposition (Tam, Andrew C.; Zapka, Werner; Ziemlich, Winfrid)V1598,13-18(1991)

Electrical probe diagnostics for processing discharges (Mantei, Thomas D.)V1392,466-473(1991)

Electromigration in VLSI metallization (Kwok, Thomas Y.)V1596,60-71(1991)

Emissivity of silicon wafers during rapid thermal processing (Vandenabeele, Peter; Maex, Karen)V1393,316-336(1991)

Enhanced etching of InP by cycling with sputter etching and reactive ion etching (Demos, Alexandros T.; Fogler, H. S.; Pang, Stella W.; Elta, Michael E.)V1392,291-297(1991)

Enhanced process control of submicron contact definition (Ostrout, Wayne H.; Hunkler, Sean; Ward, Steven D.)V1392,151-164(1991)

Epitaxial regrowth of silicon on sapphire by rapid isothermal processing (Madarazo, R.; Pedrine, A. G.; Sol, A. A.; Baranauskas, Vitor)V1393,270-277(1991)

Etch tailoring through flexible end-point detection (Angell, David; Oehrlein, Gottlieb S.)V1392,543-550(1991)

Evaluation of low-pressure silicon dry-etch processes with regard to low-substrate degradation (Engelhardt, Manfred)V1392,38-46(1991)

Excimer laser projector for microelectronics applications (Rumsby, Phil T.; Gower, Malcolm C.)V1598,36-45(1991)

Experimental investigations in laser microsoldering (Hartmann, Martin; Bergmann, Hans W.; Kupfer, Roland)V1598,175-185(1991)

Experimental studies of oxygen incorporation during growth of Y-Ba-Cu-O films by pulsed-laser deposition (Gupta, Arunava)V1394,230-230(1991)

Expert system and process optimization techniques for real-time monitoring and control of plasma processes (Cheng, Jie; Qian, Zhaogang; Irani, Keki B.; Etemad, H.; Elta, Michael E.)V1392,373-384(1991)

Fabrication technology of strained layer heterostructure devices (Van Rossum, Marc)V1361,373-382(1991)

Fast-injection Langmuir probe for process diagnostic and control (Patrick, Roger; Schoenborn, Philippe; Linder, Stefan; Baltes, Henry P.)V1392,506-513(1991)

Fluid-flow-rate metrology: laboratory uncertainties and traceabilities (Mattingly, G. E.)V1392,386-401(1991)

Focused ion-beam vacuum lithography of InP with an ultrathin native oxide resist (Wang, Yuh-Lin; Temkin, Henryk; Harriott, Lloyd R.; Hamm, Robert A.)V1392,588-594(1991)

Formation of heterostructure devices in a multichamber processing environment with in-vacuo surface analysis diagnostics and in-situ process monitoring (Lucovsky, Gerald; Kim, Sang S.; Fitch, J. T.; Wang, Cheng)V1392,605-616(1991)

Fourier transform infrared spectrophotometry for thin film monitors: computer and equipment integration for enhanced capabilities (Cox, J. N.; Sedayao, J.; Shergill, Gurmeet S.; Villasol, R.; Haaland, David M.)V1392,650-659(1991)

From VLSI to ULSI: the subhalf micron challenge (Chatterjee, Pallab K.)V1392,2-26(1991); V1393,2-26(1991); V1394,2-26(1991)

Fundamental mechanisms and doping effects in silicon infrared absorption for temperature measurement by infrared transmission (Sturm, James C.; Reaves, Casper M.)V1393,309-315(1991)

Grain-oriented high-Tc superconductors and their applications (Nelson, Jeffrey G.; Neurgaonkar, Ratnakar R.; Crooks, R.; Rhodes, C. G.)V1394,191-195(1991)

Growth and transport properties of Y-Ba-Cu-O/Pr-Ba-Cu-O superlattices (Lowndes, Douglas H.; Norton, D. P.; Budai, J. D.; Christen, D. K.; Klabunde, C. E.; Warmack, R. J.; Pennycook, Stephen J.)V1394,150-160(1991)

Growth of oxide superconducting thin films by plasma-enhanced MOCVD (Kanehori, Keiichi; Sugii, Nobuyuki)V1394,238-243(1991)

High-dose boron implantation and RTP anneal of polysilicon films for shallow junction diffusion sources and interconnects (Raicu, Bruha; Keenan, W. A.; Current, Michael I.; Mordo, David; Brennan, Roger)V1393,161-171(1991)

High-resolution tri-level process by downstream-microwave rf-biased etching (Rangelow, Iwilo W.)V1392,180-184(1991)

High-temperature degradation-free rapid thermal annealing of GaAs and InP (Pearton, Stephen J.; Katz, Avishay; Geva, M.)V1393,150-160(1991)

Honeywell's submicron polysilicon gate process (Chan, Lap S.; Hertog, Craig K.; Youngner, D. W.)V1394,232-239(1991)

IC rewiring by laser microchemistry (Pelous, G.; Guern, Yves; Gobleid, D.; David, J.; Chion, A.; Tonneau, Didier)V1598,149-158(1991)

Improved planarization techniques applied to a low dielectric constant polyimide used in multilevel metal ICs (Chang, Li-Hsin; Goodner, Ray)V1596,34-45(1991)

Improvement in dry etching of tungsten features (Heitzmann, Michel; Laporte, Philippe; Tabouret, Evelyne)V1392,272-279(1991)

Influence of sheath properties on the profile evolution in reactive ion etching processes (Fichelscher, Andreas; Rangelow, Iwilo W.; Stamm, A.)V1392,77-83(1991)

In-line supervisory control in a photolithographic workcell (Ling, Zhi-Min; Leang, Sovarong; Spanos, Costas J.)V1392,660-669(1991)

In-situ film thickness measurements using acoustic techniques (Bhardwaj, Sanjay; Khuri-Yakub, B. T.)V1392,555-562(1991)

In-situ growth of Y1Ba2Cu3O7-x thin films using XeCl excimer and Nd:YAG lasers (Gerri, Mireille; Marine, W.; Mathey, Yves; Sentis, Marc L.; Delaporte, Philippe C.; Fontaine, Bernard L.; Forestier, Bernard M.)V1503,280-291(1991)

In-situ interferometric measurements in a rapid thermal processor (Dilhac, Jean-Marie R.; Ganibal, Christian; Nolhier, N.; Amat, L.)V1393,349-353(1991)

In-situ measurements of radicals and particles in a selective silicon oxide etching plasma (Singh, Jyothi)V1392,474-486(1991)

Instantaneous etch rate measurement of thin transparent films by interferometry for use in an algorithm to control a plasma etcher (Mishurda, Helen L.; Hershkowitz, Noah)V1392,563-569(1991)

Integrated rapid isothermal processing (Singh, Rajendra; Sinha, Sanjai; Thakur, Randhir P.; Hsu, N. J.)V1393,78-89(1991)

Interference effect on laser trimming and layer thickness optimization (Sun, Yunlong)V1598,91-97(1991)

Investigation of rapid thermal process-induced defects in ion-implanted Czochralski silicon (Yarling, Charles B.; Hahn, Sookap; Hodul, David T.; Suga, Hisaaki; Smith, W. L.)V1393,192-199(1991)

Large-scale industrial application for excimer lasers: via-hole-drilling by photoablation (Bachmann, Friedrich G.)V1361,500-511(1991)

Laser drilling vias in GaAs wafers (Riley, Susan; Schick, Larry A.)V1598,118-120(1991)

Laser-formed structures to facilitate TAB bonding (Ledermann, Peter G.; Johnson, Glen W.; Ritter, Mark B.)V1598,160-163(1991)

Laser-induced gold deposition for thin-film circuit repair (Baum, Thomas H.; Comita, Paul B.; Kodas, Toivo T.)V1598,122-131(1991)

Laser-induced metal deposition and laser cutting techniques for fixing IC design errors (Shaver, David C.; Doran, S. P.; Rothschild, Mordechai; Sedlacek, J. H.)V1596,46-50(1991)

Laser processes for repair of thin-film wiring (Wassick, Thomas A.)V1598,141-148(1991)

Laser process for personalization and repair of multichip modules (Mueller, Heinrich G.; Galanakis, Claire T.; Sommerfeldt, Scott C.; Hirsch, Tom J.; Miracky, Robert F.)V1598,132-140(1991)

Laser processing of germanium (Craciun, Valentin; Mihailescu, Ion N.; Luches, Armando; Kiyak, S. G.; Mikhailova, G. N.)V1392,629-634(1991)

Lasers in Microelectronic Manufacturing (Braren, Bodil, ed.)V1598(1991)

LH electron cyclotron resonance plasma source (Kretschmer, K.-H.; Lorenz, Gerhard; Castrischer, G.; Kessler, I.; Baumann, P.)V1392,246-252(1991)

Low-resistivity contacts to silicon using selective RTCVD of germanium (Grider, Douglas T.; Ozturk, Mehmet C.; Wortman, Jim J.; Littlejohn, Michael A.; Zhong, Y.)V1393,229-239(1991)

Low-temperature in-situ dry cleaning process for epitaxial layer multiprocessing (Moslehi, Mehrdad M.)V1393,90-108(1991)

Magnetically enhanced reactive ion etching of submicron silicon trenches (Cooper, Kent; Nguyen, Bich-Yen; Lin, Jung-Hui; Roman, Bernard J.; Tobin, Phil; Ray, Wayne)V1392,253-264(1991)

Measurements on the NIST GEC reference cell (Roberts, James R.; Olthoff, James K.; Van Brunt, R. J.; Whetstone, James R.)V1392,428-436(1991)

Metallization: Performance and Reliability Issues for VLSI and ULSI (Gildenblat, Gennady S.; Schwartz, Gary P., eds.)V1596(1991)

Microcomputer-based real-time monitoring and control of single-wafer processing (Hauser, John R.; Gyurcsik, Ronald S.)V1392,340-351(1991)

Microstructure and superconducting properties of BiSrCaCuO thin films (Wessels, Bruce W.; Zhang, Jiyue; DiMeo, Frank; Richeson, D. S.; Marks, Tobin J.; DeGroot, D. C.; Kannewurf, C. R.)V1394,232-237(1991)

Microwave interferometric measurements of process plasma density (Cheah, Chun-Wah; Cecchi, Joseph L.; Stevens, J. L.)V1392,487-497(1991)

Miniature signal processor for surveillance sensor applications (Jacobi, William J.; Jensen, Preben D.; Teneketges, Nicholas J.; Wadsworth, Leo A.)V1479,111-119(1991)

Mixed application MMIC technologies: progress in combining RF, digital, and photonic circuits (Swirhun, S.; Bendett, Mark P.; Sokolov, Vladimir; Bauhahn, Paul E.; Sullivan, Charles T.; Mactaggart, R.; Mukherjee, Sayan D.; Hibbs-Brenner, Mary K.; Mondal, J. P.)V1475,223-230(1991)

MOCVD of TlBaCaCuO: structure-property relations and progress toward device applications (Hamaguchi, Norihito; Gardiner, R.; Kirlin, Peter S.)V1394,244-254(1991)

Modeling of photochemical vapor deposition of epitaxial silicon using an ArF excimer laser (Fowler, Burt; Lian, S.; Krishnan, S.; Jung, Le-Tien; Li, C.; Banerjee, Sanjay)V1598,108-117(1991)

Monitoring and control of rf electrical parameters near plasma loads (Rummel, Paul)V1392,411-420(1991)

Multichamber rapid thermal processing (Rosser, Paul J.; Moynagh, P.; Affolter, K. B.)V1393,49-66(1991)

Multichamber reactive ion etching processing for III-V optoelectronic devices (Rothman, Mark A.; Thompson, John A.; Armiento, Craig A.)V1392,598-604(1991)

Multiple photo-assisted CVD of thin-film materials for III-V device technology (Nissim, Yves I.; Moison, Jean M.; Houzay, Francoise; Lebland, F.; Licoppe, C.; Bensoussan, M.)V1393,216-228(1991)

New apparatus and method for fluid composition monitoring and control (Urmson, John)V1392,421-427(1991)

Noncontacting acoustics-based temperature measurement techniques in rapid thermal processing (Lee, Yong J.; Chou, Ching-Hua; Khuri-Yakub, B. T.; Saraswat, Krishna C.)V1393,366-371(1991)

Noninvasive sensors for in-situ process monitoring and control in advanced microelectronics manufacturing (Moslehi, Mehrdad M.)V1393,280-294(1991)

Novel selective-plated heatsink, key to compact 2-watt MMIC amplifier (Taylor, Gordon C.; Bechtle, Daniel W.; Jozwiak, Phillip C.; Liu, Shing G.; Camisa, Raymond L.)V1475,103-112(1991)

Optical properties of Li-doped ZnO films (Valentini, Antonio; Quaranta, Fabio; Vasanelli, L.; Piccolo, R.)V1400,164-170(1991)

Optimizing the structural and electrical properties of Ba2YCu3O7-delta (Phillips, Julia M.; Siegal, Michael P.)V1394,186-190(1991)

Oxygen reactive ion etching of polymers: profile evolution and process mechanisms (Pilz, Wolfgang; Janes, Joachim; Muller, Karl P.; Pelka, Joachim)V1392,84-94(1991)

Pattern etching and selective growth of GaAs by in-situ electron-beam lithography using an oxidized thin layer (Akita, K.; Sugimoto, Yoshimasa; Taneya, M.; Hiratani, Y.; Ohki, Y.; Kawanishi, Hidenori; Katayama, Yoshifumi)V1392,576-587(1991)

Performance and reliability of ultrathin reoxidized nitrided oxides fabricated by rapid thermal processing (Joshi, A. B.; Lo, G. Q.; Shih, Dennis K.; Kwong, Dim-Lee)V1393,122-149(1991)

Phase object imaging inside the airy disc (Tychinsky, Vladimir P.)V1392,570-572(1991)

Physical Concepts of Materials for Novel Optoelectronic Device Applications I: Materials Growth and Characterization (Razeghi, Manijeh, ed.)V1361(1991)

Planar SNS Josephson junctions using multilayer Bi system (Setsune, Kentaro; Mizuno, Koichi; Higashino, Hidetaka; Wasa, Kiyotaka)V1394,79-88(1991)

Plasma diagnostics as inputs to the modeling of the oxygen reactive ion etching of multilevel resist structures (Hope, D. A.; Hydes, A. J.; Cox, T. I.; Deshmukh, V. G.)V1392,185-195(1991)

Plasma modeling in microelectronic processing (Meyyappan, Meyya; Govindan, T. R.; Kreskovsky, John P.)V1392,67-76(1991)

Polysilicon etching for nanometer-scale features (Lajzerowicz, Jean; Tedesco, Serge V.; Pierrat, Christophe; Muyard, D.; Taccussel, M. C.; Laporte, Philippe)V1392,222-231(1991)

Precompetitive cooperative research: the culture of the '90s (Holton, William C.)V1392,27-33(1991); V1393,27-33(1991); 1394,27-33(1991)

Process control sensor development for the automation of single-wafer processors (Hosch, Jimmy W.)V1392,529-534(1991)

Progress In High-Temperature Superconducting Transistors and Other Devices (Narayan, Jagdish; Shaw, David T.; Singh, Rajendra, eds.)V1394(1991)

Progress of an advanced diffusion source plasma reactor (Benjamin, Neil M.; Chapman, Brian N.; Boswell, Rod W.)V1392,95-105(1991)

Pulsed-laser deposition of YBa2Cu3O7-x thin films: processing, properties, and performance (Muenchausen, Ross E.; Foltyn, Stephen R.; Wu, Xin D.; Dye, Robert C.; Nogar, Nicholas S.; Carim, A. H.; Heidelbach, F.; Cooke, D. W.; Taber, R. C.; Quinn, Rod K.)V1394,221-229(1991)

Pyrometer modeling for rapid thermal processing (Wood, Samuel C.; Apte, Pushkar P.; King, Tsu-Jae; Moslehi, Mehrdad M.; Saraswat, Krishna C.)V1393,337-348(1991)

Radial ion energy measurements in an electron cyclotron resonance reactor (O'Neill, James A.; Holber, William M.; Caughman, John)V1392,516-528(1991)

Rapid Thermal and Related Processing Techniques (Moslehi, Mehrdad M.; Singh, Rajendra, eds.)V1393(1991)

Rapid thermal annealing of the through-Ta5Si3 film implantation on GaAs (Huang, Fon-Shan; Chen, W. S.; Hsu, Tzu-min)V1393,172-179(1991)

Rapid thermal processing induced defects and gettering effects in silicon (Hartiti, Bouchaib; Muller, Jean-Claude; Siffert, Paul; Vu, Thuong-Quat)V1393,200-206(1991)

Rapid thermal processing in the manufacturing technology of contacts to InP-based photonic devices (Katz, Avishay)V1393,67-77(1991)

Reactive ion etching of deep isolation trenches using sulfur hexafluoride, chlorine, helium, and oxygen (Krawiec, Theresa M.; Giammarco, Nicholas J.)V1392,265-271(1991)

Real-time automation of a dry etching system (McCafferty, Robert H.)V1392,331-339(1991)

Real-time, in-situ measurement of film thickness and uniformity during plasma ashing of photoresist (Davies, John T.; Metz, Thomas E.; Savage, Richard N.; Simmons, Horace O.)V1392,551-554(1991)

Real-time monitoring and control of plasma etching (Butler, Stephanie W.; McLaughlin, Kevin J.; Edgar, Thomas F.; Trachtenberg, Isaac)V1392,361-372(1991)

Reduced thermal budget processing of high-Tc superconducting thin films and related materials by MOCVD (Sinha, Sanjai; Singh, Rajendra; Hsu, N. J.; Ng, J. T.; Chou, P.; Narayan, Jagdish)V1394,266-276(1991)

Reflow soldering of fine-pitch devices using a Nd:YAG laser (Glynn, Thomas J.; Flanagan, Aiden J.; Redfern, R. M.)V1598,200-205(1991)

Research sputter cluster tool (Clarke, Peter J.)V1392,617-624(1991)

Review of temperature measurements in the semiconductor industry (Anderson, Richard L.)V1392,437-451(1991)

Role of buffer layers in the laser-ablated films on metallic substrates (Shaw, David T.; Narumi, E.; Yang, F.; Patel, Sushil)V1394,214-220(1991)

Role of defects in the ablation of wide-bandgap materials (Dickinson, J. T.; Langford, S. C.; Jensen, L. C.)V1598,72-83(1991)

Round-robin comparison of temperature nonuniformity during RTP due to patterned layers (Vandenabeele, Peter; Maex, Karen)V1393,372-394(1991)

RTP-induced defects in silicon studied by positron annihilation technique (Kulkarni, N. M.; Kulkarni, R. N.; Shaligram, Arvind D.)V1393,207-214(1991)

RTP temperature uniformity mapping (Keenan, W. A.; Johnson, Walter H.; Hodul, David T.; Mordo, David)V1393,354-365(1991)

Selective deposition of polycrystalline SixGe1-x by rapid thermal processing (Ozturk, Mehmet C.; Zhong, Y.; Grider, Douglas T.; Sanganeria, M.; Wortman, Jim J.; Littlejohn, Michael A.)V1393,260-269(1991)

Selective low-temperature chemical vapor deposition of copper from new copper(I) compounds (Jain, Ajay; Shin, H. K.; Chi, Kai-Ming; Hampden-Smith, Mark J.; Kodas, Toivo T.; Farkas, Janos; Paffett, M. F.; Farr, J. D.)V1596,23-33(1991)

Selective metal deposition using low-dose focused ion-beam patterning (Kubena, Randall L.; Stratton, F. P.; Mayer, T. M.)V1392,595-597(1991)

Self-aligned synthesis of titanium silicide by multipulse excimer laser irradiation (Craciun, Valentin; Craciun, Doina; Mihailescu, Ion N.; Kuzmichov, A. V.; Konov, Vitaly I.; Uglov, S. A.)V1392,625-628(1991)

Shadow masked growth for the fabrication of photonic integrated circuits (Demeester, Piet M.; Moerman, Ingrid; Zhu, Youcai; van Daele, Peter; Thomson, J.)V1361,1132-1143(1991)

Si-based epitaxial growth by rapid thermal processing chemical vapor deposition (Jung, K. H.; Hsieh, T. Y.; Kwong, Dim-Lee; Spratt, D. B.)V1393,240-251(1991)

Simulation of ion-enhanced dry-etch processes (Pelka, Joachim)V1392,55-66(1991)

Single-crystal silicon trench etching for fabrication of highly integrated circuits (Engelhardt, Manfred)V1392,210-221(1991)

Single-wafer integrated processing as a manufacturing tool using rapid thermal chemical vapor deposition technology (Kermani, Ahmad)V1393,109-119(1991)

Some fundamental issues on metallization in VLSI (Ferry, David K.; Kozicki, M. N.; Raupp, Gregory B.)V1596,2-11(1991)

SPEEDIE: a profile simulator for etching and deposition (McVittie, James P.; Rey, J. C.; Bariya, A. J.; IslamRaja, M. M.; Cheng, L. Y.; Ravi, S.; Saraswat, Krishna C.)V1392,126-138(1991)

Spin-on glasses in the silicon IC: plague or panacea? (Lifshitz, N.; Pinto, Mark R.)V1596,96-105(1991)

Spin-on-glass/phosphosilicate glass etchback planarization process for 1.0 um CMOS technology (Bogle-Rohwer, Elizabeth; Nulty, James E.; Chu, Wileen; Cohen, Andrew)V1392,280-290(1991)

Stress-induced voiding in aluminum alloy metallizations (Sullivan, Timothy D.; Ryan, James G.; Riendeau, J. R.; Bouldin, Dennis P.)V1596,83-95(1991)

Structural and electrical properties of epitaxial YBCO films on Si (Fork, David K.; Barrera, A.; Phillips, Julia M.; Newman, N.; Fenner, David B.; Geballe, T. H.; Connell, G.A. N.; Boyce, James B.)V1394,202-202(1991)

Superconducting YBa2Cu3O7 films on Si and GaAs with conducting indium tin oxide buffer layers (James, Jonathan H.; Kellett, Bruce J.; Gauzzi, Andrea; Dwir, Benjamin; Pavuna, Davor)V1394,45-61(1991)

Systems-oriented survey of noncontact temperature measurement techniques for rapid thermal processing (Peyton, David; Kinoshita, Hiroyuki; Lo, G. Q.; Kwong, Dim-Lee)V1393,295-308(1991)

Technique of assessing contact ohmicity and their relevance to heterostructure devices (Harrison, H. B.; Reeves, Geoffrey K.)V1596,52-59(1991)

Theoretical and practical aspects of real-time Fourier imaging (Grimard, Dennis S.; Terry, Fred L.; Elta, Michael E.)V1392,535-542(1991)

Thin film processing and device fabrication in the Tl-Ca-Ba-Cu-O system (Martens, Jon S.; Ginley, David S.; Zipperian, Thomas E.; Hietala, Vincent M.; Tigges, Chris P.)V1394,140-149(1991)

Thin-film selective multishot ablation at 248 nm (Hunger, Hans E.; Pietsch, H.; Petzoldt, Stefan; Matthias, Eckart)V1598,19-26(1991)

Threshold measurements in laser-assisted particle removal (Lee, Shyan J.; Imen, Kamran; Allen, Susan D.)V1598,2-12(1991)

Unframed via interconnection of nonplanar device structures (Kim, Manjin J.)V1596,12-22(1991)

Uniformity characterization of rapid thermal processor thin films (Yarling, Charles B.; Cook, Dawn M.)V1393,411-420(1991)

Unique symbol for marking and tracking very small semiconductor products (Martin, James P.)V1598,206-220(1991)

Variability in thickness measurements using x-ray fluorescence technique (Baltazar, Inmaculada C.; Mena, Manolo G.)V1392,670-680(1991)

Versatility of metal organic chemical vapor deposition process for fabrication of high-quality YBCO superconducting thin films (Chern, Chyi S.; Zhao, Jing-Fu; Li, Y. Q.; Norris, Peter E.; Kear, B. H.; Gallois, B.)V1394,255-265(1991)

Vertical oxide etching without inducing change in critical dimensions (Nagy, Andrew)V1392,165-179(1991)

YBa2Cu3O7-x/Au/Nb device structures (Hunt, Brian D.; Foote, Marc C.; Bajuk, Louis J.; Vasquez, R. P.)V1394,89-95(1991)

Microlithography—see also beams; lasers, excimer; microelectronic processing; photomasks; photoresists; projection systems; silicon; x rays

0.5-micron deep-UV lithography using a Micrascan-90 step-and-scan exposure tool (Kuyel, Birol; Barrick, Mark W.; Hong, Alexander; Vigil, Joseph)V1463,646-665(1991)

0.50 um contact measurement and characterization (Lindsay, Tracy K.; Orvek, Kevin J.; Mumaw, Richard T.)V1464,104-118(1991)

10th Annual Symp on Microlithography (Wiley, James N., ed.)V1496(1991)

5X reticle fabrication using MEBES multiphase virtual address and AZ5206 resist (Milner, Kathy S.; Chipman, Paul S.)V1496,180-196(1991)

64-Mbit DRAM production with i-line stepper (Shirai, Hisatsugu; Kobayashi, Katsuyoshi; Nakagawa, Kenji)V1463,256-274(1991)

Acid-catalyzed pinacol rearrangement: chemically amplified reverse polarity change (Sooriyakumaran, Ratna; Ito, Hiroshi; Mash, Eugene A.)V1466,419-428(1991)

Adaptive control of photolithography (Crisalle, Oscar D.; Soper, Robert A.; Mellichamp, Duncan A.; Seborg, Dale E.)V1464,508-526(1991)

Advanced confocal technique for submicron CD measurements (Rohde, Axel; Saffert, Ralf; Fitch, John)V1464,438-446(1991)

Advanced lithographic methods for contact patterning on severe topography (Chu, Ron; Greeneich, James S.; Katz, Barton A.; Lin, Hwang-Kuen; Huang, Dong-Tsair)V1465,238-243(1991)

Advanced Techniques for Integrated Circuit Processing (Bondur, James; Turner, Terry R., eds.)V1392(1991)

Advances in Resist Technology and Processing VIII (Ito, Hiroshi, ed.)V1466(1991)

Airborne chemical contamination of a chemically amplified resist (MacDonald, Scott A.; Clecak, Nicholas J.; Wendt, H. R.; Willson, C. G.; Snyder, C. D.; Knors, C. J.; Deyoe, N. B.; Maltabes, John G.; Morrow, James R.; McGuire, Anne E.; Holmes, Steven J.)V1466,2-12(1991)

Al2O3 etch-stop layer for a phase-shifting mask (Hanyu, Isamu; Nunokawa, Mitsuji; Asai, Satoru; Abe, Masayuki)V1463,595-601(1991)

Applicability of dry developable deep-UV lithography to sub-0.5 um processing (Goethals, Anne-Marie; Baik, Ki-Ho; Van den hove, Luc; Tedesco, Serge V.)V1466,604-615(1991)

Application and integration of a focused ion beam circuit repair system (Lange, John A.; Allen, Chris)V1465,50-56(1991)

Application aspects of the Si-CARL bilayer process (Sebald, Michael; Berthold, Joerg; Beyer, Michael; Leuschner, Rainer; Noelscher, Christoph; Scheler, Ulrich; Sezi, Recai; Ahne, Hellmut; Birkle, Siegfried)V1466,227-237(1991)

Application of an electron-beam scattering parameter extraction method for proximity correction in direct-write electron-beam lithography (Weiss, Rudolf M.; Sills, Robert M.)V1465,192-200(1991)

Application of a reduced area electrical test pattern to precise pattern registration measurements (Rominger, James P.)V1496,224-231(1991)

Application of transmission electron detection to x-ray mask calibrations and inspection (Postek, Michael T.; Larrabee, Robert D.; Keery, William J.; Marx, Egon)V1464,35-47(1991)

Applications of holographic gratings in x-ray mask metrology (Hansen, Matthew E.)V1396,78-79(1991)

Applications of latent image metrology in microlithography (Adams, Thomas E.)V1464,294-312(1991)

Argon fluoride excimer laser source for sub-0.25 mm optical lithography (Sandstrom, Richard L.)V1463,610-616(1991)

Astigmatism and field curvature from pin-bars (Kirk, Joseph P.)V1463,282-291(1991)

Automatic mask-to-wafer alignment and gap control using moire interferometry (Chitnis, Vijay T.; Kowsalya, S.; Rashmi, Dr.; Kanjilal, A. K.; Narain, Ram)V1332,613-622(1991)

Bilayer resist system utilizing alkali-developable organosilicon positive photoresist (Nate, Kazuo; Mizushima, Akiko; Sugiyama, Hisashi)V1466,206-210(1991)

Binary and phase-shifting image design for optical lithography (Liu, Yong; Zakhor, Avideh)V1463,382-399(1991)

CAD system for CGHs and laser beam lithography (Yatagai, Toyohiko; Geiser, Martial; Tian, Ronglong; Tian, Xingkang; Onda, Hajime)V1555,8-12(1991)

Capability assessment and comparison of the Nikon 2i, Nikon 3i, and IMS-2000 registration measurement devices (Henderson, Robert K.)V1496,198-216(1991)

Channeling radiation as an x-ray source for angiography, x-ray lithography, molecular structure determination, and elemental analysis (Uberall, Herbert; Faraday, Bruce J.; Maruyama, Xavier K.; Berman, Barry L.)V1552,198-213(1991)

Characterization of automatic overlay measurement technique for sub-half-micron devices (Kawai, Akira; Fujiwara, Keiji; Tsujita, Kouichirou; Nagata, Hitoshi)V1464,267-277(1991)

Characterization of the radiation from a low-energy X-pinch source (Christou, Christos; Choi, Peter)V1552,278-287(1991)

Characterization of wavelength offset for optimization of deep-UV stepper performance (Jones, Susan K.; Dudley, Bruce W.; Peters, Charles R.; Kellam, Mark D.; Pavelchek, Edward K.)V1464,546-553(1991)

Characterizing a surface imaging process in a high-volume DRAM manufacturing production line (Garza, Cesar M.; Catlett, David L.; Jackson, Ricky A.)V1466,616-627(1991)

Charging effects in low-voltage SEM metrology (Monahan, Kevin M.; Benschop, Jozef P.; Harris, Tom A.)V1464,2-9(1991)

Charging phenomena in e-beam metrology (Levy, Dorron; Harris, Karl L.)V1464,413-423(1991)

Chemically amplified negative-tone photoresist for sub-half-micron device and mask fabrication (Conley, Willard E.; Dundatscheck, Robert; Gelorme, Jeffrey D.; Horvat, John; Martino, Ronald M.; Murphy, Elizabeth; Petrosky, Anne; Spinillo, Gary; Stewart, Kevin; Wilbarg, Robert; Wood, Robert L.)V1466,53-66(1991)

Chemically amplified resists for x-ray and e-beam lithography (Berry, Amanda K.; Graziano, Karen A.; Thompson, Stephen D.; Taylor, James W.; Suh, Doowon; Plumb, Dean)V1465,210-220(1991)

Chlorine or bromine chemistry in reactive ion etching Si-trench etching (Rangelow, Iwilo W.; Fichelscher, Andreas)V1392,240-245(1991)

Chromeless phase-shifted masks: a new approach to phase-shifting masks (Toh, Kenny K.; Dao, Giang T.; Singh, Rajeev; Gaw, Henry T.)V1496,27-53(1991)

Comparison of 248-nm line narrowing resonator optics for deep-UV lithography lasers (Kahlert, Hans-Juergen; Rebhan, Ulrich; Lokai, Peter; Basting, Dirk)V1463,604-609(1991)

Comparison of plasma source with synchrotron source in the Center for X-ray Lithography (Guo, Jerry Z.; Cerrina, Franco)V1465,330-337(1991)

Computer simulation of 0.5-micrometer lithography for a 16-megabit DRAM (Maltabes, John G.; Norris, Katherine C.; Writer, Dean)V1463,326-335(1991)

Conjugate twin-shifter for the new phase-shift method to high-resolution lithography (Ohtsuka, Hiroshi; Abe, Kazutoshi; Onodera, Toshio; Kuwahara, Kazuyuki; Taguchi, Takashi)V1463,112-123(1991)

Control of proximity effects on CD uniformity through the use of process parameters derived from a statistically designed experiment (Christensen, Lorna D.; Bell, Kenneth L.)V1463,504-514(1991)

Cost-effective x-ray lithography (Roltsch, Tom J.)V1465,289-307(1991)

Critical dimension control using development end point detection for wafers with multilayer structures (Hagi, Toshio; Okuda, Yoshimitsu; Ohkuma, Tohru)V1464,215-221(1991)

Critical dimension shift resulting from handling time variation in the track coat process (Kulp, John M.)V1466,630-640(1991)

Cross-sectional imaging in SEM: signal formation mechanism and CD measurements (Firstein, Leon A.; Noz, Arthur)V1464,81-88(1991)

Deep-UV diagnostics using continuous tone photoresist (Kirk, Joseph P.; Hibbs, Michael S.)V1463,575-583(1991)

Deep-UV photolithography linewidth variation from reflective substrates (Dunn, Diana D.; Bruce, James A.; Hibbs, Michael S.)V1463,8-15(1991)

Defect repair for gold absorber/silicon membrane x-ray masks (Stewart, Diane K.; Fuchs, Jacob; Grant, Robert A.; Plotnik, Irving)V1465,64-77(1991)

Design and fabrication of soft x-ray photolithography experimental beam line at Beijing National Synchrotron Radiation Laboratory (Zhou, Changxin)V1465,26-33(1991)

Design methodology for dark-field phase-shifted masks (Toh, Kenny K.; Dao, Giang T.; Gaw, Henry T.; Neureuther, Andrew R.; Fredrickson, Larry R.)V1463,402-413(1991)

Design of reflective relay for soft x-ray lithography (Rodgers, John M.; Jewell, · Tanya E.)V1354,330-336(1991)

Design survey of x-ray/XUV projection lithography systems (Shealy, David L.; Viswanathan, Vriddhachalam K.)V1343,229-240(1991)

DESIRE technology with electron-beam resists: fundamentals, experiments, and simulation (Nicolau, Dan V.; Fulga, Florin; Dusa, Mircea V.)V1465,282-288(1991)

Development of electron moire method using a scanning electron microscope (Kishimoto, Satoshi; Egashira, Mitsuru; Shinya, Norio)V1554B,174-180(1991)

Diffractive optical elements for optoelectronic interconnections (Streibl, Norbert)V1574,34-47(1991)

Disk-shaped VUV+O source used as resist asher and resist developer (Hattori, Shuzo; Collins, George J.; Yu, Zenqi; Sugimoto, Dai; Saita, Masahiro)V1463,539-550(1991)

Dissolution inhibition mechanism of ANR photoresists: crosslinking vs. -OH site consumption (Thackeray, James W.; Orsula, George W.; Rajaratnam, Martha M.; Sinta, Roger F.; Herr, Daniel J.; Pavelchek, Edward K.)V1466,39-52(1991)

Dissolution kinetics of high-resolution novolac resists (Itoh, Katsuyuki; Yamanaka, Koji; Nozue, Hiroshi; Kasama, Kunihiko)V1466,485-496(1991)

Dissolution of poly(p-hydroxystyrene): molecular weight effects (Long, Treva; Rodriguez, Ferdinand)V1466,188-198(1991)

Dose control for short exposures in excimer laser lithography (Hollman, Richard F.)V1377,119-125(1991)

DQN photoresist with tetrahydroxydiphenylmethane as ballasting group in PAC (Tzeng, Chao H.; Lin, Dhei-Jhai; Lin, Song S.; Huang, Dong-Tsair; Lin, Hwang-Kuen)V1466,469-476(1991)

Dry development and plasma durability of resists: melt viscosity and self-diffusion effects (Paniez, Patrick J.; Joubert, Olivier P.; Pons, Michel J.; Oberlin, Jean C.; Weill, Andre P.)V1466,583-591(1991)

Dry etching for silylated resist development (Laporte, Philippe; Van den hove, Luc; Melaku, Yosias)V1392,196-207(1991)

Dynamics of a low-energy X-pinch source (Choi, Peter; Christou, Christos; Aliaga, Raul)V1552,270-277(1991)

E-beam data compaction method for large-capacity mask ROM production (Kanemaru, Toyomi; Nakajima, Takashi; Igarashi, Tadanao; Masuda, Rika; Orita, Nobuyuki)V1496,118-123(1991)

EBES4: mask/reticle writer for the 90's (Chen, George C.)V1496,107-117(1991)

Edge-profile, materials, and protective coating effects on image quality (Doi, Takeshi; Tadros, Karim H.; Kuyel, Birol; Neureuther, Andrew R.)V1464,336-345(1991)

Effect of operating points in submicron CD measurements (Dusa, Mircea V.; Jung, Christoph; Jung, Paul; Hogenkamp, Detlef; Roeth, Klaus-Dieter)V1464,447-458(1991)

Effect of sensitizer spatial distribution on dissolution inhibition in novolak/diazonaphthoquinone resists (Rao, Veena; Kosbar, Laura L.; Frank, Curtis W.; Pease, R. F.)V1466,309-323(1991)

Effect of silylation condition on the silylated image in the DESIRE process (Taira, Kazuo; Takahashi, Junichi; Yanagihara, Kenji)V1466,570-582(1991)

Effects of higher order aberrations on the process window (Gortych, Joseph E.; Williamson, David M.)V1463,368-381(1991)

Effects of interfacial modifications on diamond film adhesion (DeNatale, Jeffrey F.; Flintoff, John F.; Harker, Alan B.)V1534,44-48(1991)

Effects of wafer cooling characteristics after post-exposure bake on critical dimensions (Lauck, Teresa L.; Nomura, Masafumi; Omori, Tsutae; Yoshioka, Kajutoshi)V1464,527-538(1991)

Electron beam lithographic fabrication of computer-generated holograms (West, Andrew A.; Smith, Robin W.)V1507,158-167(1991)

Electron-beam lithography for the microfabrication of OEICs (Engel, Herbert; Doeldissen, Walter)V1506,60-64(1991)

Electron-beam metrology: the European initiative (Jackman, James J.)V1464,71-80(1991)

Electron-Beam, X-Ray, and Ion-Beam Submicrometer Lithographies for Manufacturing (Peckerar, Martin, ed.)V1465(1991)

Etch conditions of photolithographic holograms (Guo, Yongkang; Guo, Lu Rong; Zhang, Xiao-Chun)V1461,97-100(1991)

Evaluation of a high-resolution negative-acting electron-beam resist GMC for photomask manufacturing (Chen, Wen-Chih; Novembre, Anthony E.)V1496,266-323(1991)

Evaluation of a photoresist process for 0.75-micron g-line lithography (Kasahara, Jack S.; Dusa, Mircea V.; Perera, Thiloma)V1463,492-503(1991)

Evaluation of phenolic resists for 193 nm surface imaging (Hartney, Mark A.; Johnson, Donald W.; Spencer, Allen C.)V1466,238-247(1991)

Evaluation of poly(p-trimethylsilylstyrene and p-pentamethyldisilylstyrene sulfone) as high-resolution electron-beam resists (Gozdz, Antoni S.; Ono, Hiroshi; Ito, Seiki; Shelburne, John A.; Matsuda, Minoru)V1466,200-205(1991)

Excimer Laser Materials Processing and Beam Delivery Systems (Piwczyk, Bernhard P., ed.)V1377(1991)

Excimer laser photolithography with a 1:1 broadband catadioptric optics (Zhang, Yudong; Lu, Dunwu; Zou, Haixing; Wang, Zhi-Jiang)V1463,456-463(1991)

Excimer lasers for deep-UV lithography (Elliott, David J.; Sengupta, Uday K.)V1377,6-17(1991)

EXCON: a graphics-based experiment-control manager (Khan, Mumit; Anderson, Paul D.; Cerrina, Franco)V1465,315-323(1991)

Experimental and simulation studies of alignment marks (Wong, Alfred K.; Doi, Takeshi; Dunn, Diana D.; Neureuther, Andrew R.)V1463,315-323(1991)

Experimental assessment of 150-mm P/P+ epitaxial silicon wafer flatness for deep-submicron applications (Huff, Howard R.; Weed, Harrison)V1464,278-293(1991)

Expert system for diagnosis/optimization of microlithography process* (Nicolau, Dan V.; Fulga, Florin; Dusa, Mircea V.)V1468,345-351(1991)

Exploration of fabrication techniques for phase-shifting masks (Pfau, Anton K.; Oldham, William G.; Neureuther, Andrew R.)V1463,124-134(1991)

Exposure dose optimization for a positive resist containing polyfunctional photoactive compound (Trefonas, Peter; Mack, Chris A.)V1466,117-131(1991)

Extending electrical measurements to the 0.5 um regime (Troccolo, Patrick M.; Mantalas, Lynda C.; Allen, Richard A.; Linholm, Loren)V1464,90-103(1991)

Fabrication of grooved-glass substrates by phase-mask lithography (Brock, Phillip J.; Levenson, Marc D.; Zavislan, James M.; Lyerla, James R.; Cheng, John C.; Podlogar, Carl V.)V1463,87-100(1991)

Fabrication of light-guiding devices and fiber-coupling structures by the LIGA process (Rogner, Arnd; Ehrfeld, Wolfgang)V1506,80-91(1991)

Fabrication of micro-optical components by laser beam writing in photoresist (Gale, Michael T.; Lang, Graham K.; Raynor, Jeffrey M.; Schuetz, Helmut)V1506,65-70(1991)

Fabrication of multiphase optical elements for weighted array spot generation (McKee, Paul; Wood, David; Dames, Mark P.; Dix, C.)V1461,17-23(1991)

Fabrication of phase-shifting mask (Ishiwata, Naoyuki; Furukawa, Takao)V1463,423-433(1991)

Fabrication of phase structures with continuous and multilevel profile for diffraction optics (Poleshchuk, Alexander G.)V1574,89-100(1991)

Fabrication technology of strained layer heterostructure devices (Van Rossum, Marc)V1361,373-382(1991)

Figure of merit for calibration and comparison of linewidth measurement instruments (Hershey, Robert R.; Zavecz, Terrence E.)V1464,22-34(1991)

Fine undercut control in bilayer PMMA-P(MMA-MAA) resist system for e-beam lithography with submicrometer resolution (Bogdanov, Alexei L.; Andersson, Eva K.)V1465,324-329(1991)

Focus considerations with high-numerical-aperture widefield lenses (Leebrick, David H.)V1463,275-280(1991)

Focused ion beam induced deposition: a review (Melngailis, John)V1465,36-49(1991)

Focused ion-beam vacuum lithography of InP with an ultrathin native oxide resist (Wang, Yuh-Lin; Temkin, Henryk; Harriott, Lloyd R.; Hamm, Robert A.)V1392,588-594(1991)

From VLSI to ULSI: the subhalf micron challenge (Chatterjee, Pallab K.)V1392,2-26(1991); V1393,2-26(1991); V1394,2-26(1991)

Gas puff Z-pinch x-ray source: a new approach (Fisher, Amnon)V1552,252-253(1991)

Generalized characteristic model for lithography: application to negative chemically amplified resists (Ziger, David H.; Mack, Chris A.; Distasio, Romelia)V1466,270-282(1991)

"Golden standard" wafer design for optical stepper characterization (Kemp, Kevin G.; King, Charles F.; Wu, Wei; Stager, Charles)V1464,260-266(1991)

Half-micrometer linewidth metrology (Knight, Stephen E.; Humphrey, Dean; Bowley, Reginald R.; Cogley, Robert M.)V1464,119-126(1991)

Hierarchical proximity effect correction for e-beam direct writing of 64-Mbit DRAM (Misaka, Akio; Hashimoto, Kazuhiko; Kawamoto, M.; Yamashita, H.; Matsuo, Takahiro; Sakashita, Toshihiko; Harafuji, Kenji; Nomura, Noboru)V1465,174-184(1991)

High-average-power narrow-band KrF excimer laser (Wakabayashi, Osamu; Kowaka, Masahiko; Kobayashi, Yukio)V1463,617-628(1991)

Highly accurate pattern generation using acousto-optical deflection (Sandstrom, Torbjorn; Tison, James K.)V1463,629-637(1991)

Highly sensitive microresinoid siloxane resist for EB and deep-UV lithography (Yamazaki, Satomi; Ishida, Shinji; Matsumoto, Hiroshi; Aizaki, Naoaki; Muramoto, Naohiro; Mine, Katsutoshi)V1466,538-545(1991)

High-power x-ray generation using transition radiation (Piestrup, Melvin A.; Boyers, D. G.; Pincus, Cary I.; Li, Qiang; Harris, J. L.; Bergstrom, J. C.; Caplan, H. S.; Silzer, R. M.; Skopik, D. M.; Moran, M. J.; Maruyama, Xavier K.)V1552,214-239(1991)

High-sensitivity and high-dry-etching durability positive-type electron-beam resist (Tamura, Akira; Yonezawa, Masaji; Sato, Mitsuyoshi; Fujimoto, Yoshiaki)V1465,271-281(1991)

High-speed stepper setup using a low-voltage SEM (Benschop, Jozef P.; Monahan, Kevin M.; Harris, Tom A.)V1464,62-70(1991)

Hitachi e-beam lithography tools for advanced applications (Colbran, William V.)V1496,90-96(1991)

Honeywell's submicron polysilicon gate process (Chan, Lap S.; Hertog, Craig K.; Youngner, D. W.)V1392,232-239(1991)

I-line lithography for highly reproducible fabrication of surface acoustic wave devices (Berek, Stefan; Knauer, Ulrich; Zottl, Helmut)V1463,515-520(1991)

Impact of phase masks on deep-UV lithography (Sewell, Harry)V1463,168-179(1991)

Improvement of focus and exposure latitude by the use of phase-shifting masks for DUV applications (Op de Beeck, Maaike; Tokui, Akira; Fujinaga, Masato; Yoshioka, Nobuyuki; Kamon, Kazuya; Hanawa, Tetsuro; Tsukamoto, Katsuhiro)V1463,180-196(1991)

Improvements in 0.5-micron production wafer steppers (Luehrmann, Paul F.; de Mol, Chris G.; van Hout, Frits J.; George, Richard A.; van der Putten, Harrie B.)V1463,434-445(1991)

Improvements in sensitivity and discrimination capability of the PD reticle/mask inspection system (Saito, Juichi; Saijo, Y.)V1496,284-301(1991)

Improving submicron CD measurements through optimum operating points (Dusa, Mircea V.; Roeth, Klaus-Dieter; Jung, Christoph)V1496,217-223(1991)

Improving the performance and usability of a wet-developable DUV resist for sub-500nm lithography (Samarakone, Nandasiri; Van Driessche, Veerle; Jaenen, Patrick; Van den hove, Luc; Ritchie, Douglas R.; Luehrmann, Paul F.)V1463,16-29(1991)

In-line supervisory control in a photolithographic workcell (Ling, Zhi-Min; Leang, Sovarong; Spanos, Costas J.)V1392,660-669(1991)

In-situ monitoring of silylation mechanisms by laser interferometry (Pierrat, Christophe; Paniez, Patrick J.; Martin, P.)V1466,248-256(1991)

Integrated Circuit Metrology, Inspection, and Process Control V (Arnold, William H., ed.)V1464(1991)

Investigation of interlevel proximity effects case of the gate level over LOCOS (Festes, Gilles; Chollet, Jean-Paul E.)V1463,245-255(1991)

Investigation of self-aligned phase-shifting reticles by simulation techniques (Noelscher, Christoph; Mader, Leonhard)V1463,135-150(1991)

Investigation on the effect of electron-beam acceleration voltage and electron-beam sharpness on 0.2-um patterns (Moniwa, Akemi; Okazaki, Shinji)V1465,154-163(1991)

Is phase-shift mask technology production-worthy? (Chen, Mung)V1463,2-5(1991)

Issues in the repair of x-ray masks (Stewart, Diane K.; Doherty, John A.)V1496,247-265(1991)

Laser alignment modeling using rigorous numerical simulations (Wojcik, Gregory L.; Vaughan, David K.; Mould, John; Leon, Francisco A.; Qian, Qi-de; Lutz, Michael A.)V1463,292-303(1991)

Lau imaging (Cartwright, Steven L.)V1396,481-487(1991)

Lens for microlithography (Hsieh, Hung-yu; Wagner, Jerome F.)V1396,467-472(1991)

Linearity of coherence probe metrology: simulation and experiment (Davidson, Mark P.; Monahan, Kevin M.; Monteverde, Robert J.)V1464,155-176(1991)

Linewidth measurement comparison between a photometric optical microscope and a scanning electron microscope backed with Monte Carlo trajectory computations (Nunn, John W.; Turner, Nicholas P.)V1464,50-61(1991)

Low-loss line-narrowed excimer oscillator for projection photolithography: experiments and simulation (Volkov, Gennady S.; Zaroslov, D. Y.)V1503,146-153(1991)

Mechanism of dissolution inhibition of novolak-diazoquinone resist (Furuta, Mitsuhiro; Asaumi, Shingo; Yokota, Akira)V1466,477-484(1991)

Mechanistic studies on the poly(4-tert-butoxycarbonyloxystyrene)/triphenylsulfonium salt photoinitiation process (Hacker, Nigel P.; Welsh, Kevin M.)V1466,384-393(1991)

Metamorphosis of laser writer (Wilson, Michael A.)V1496,156-170(1991)

Metrology issues associated with submicron linewidths (Phan, Khoi; Nistler, John L.; Singh, Bhanwar)V1464,424-437(1991)

Micrograting device using electron-beam lithography (Yamashita, Tsukasa; Aoyama, Shigeru)V1507,81-93(1991)

Mix-and-match lithography for half-micrometer technology (Flack, Warren W.; Dameron, David H.)V1465,164-172(1991)

Modeling of illumination effects on resist profiles in x-ray lithography (Oertel, Heinrich K.; Weiss, M.; Huber, Hans L.; Vladimirsky, Yuli; Maldonado, Juan R.)V1465,244-253(1991)

Modeling phase-shifting masks (Neureuther, Andrew R.)V1496,80-88(1991)

Moire displacement detection by the photoacoustic technique (Hane, Kazuhiro; Watanabe, S.; Goto, T.)V1332,577-583(1991)

Monte Carlo modeling of secondary electron signals from heterogeneous specimens with nonplanar surfaces (Russ, John C.; Dudley, Bruce W.; Jones, Susan K.)V1464,10-21(1991)

Multilayer optics for soft x-ray projection lithography: problems and prospects (Stearns, Daniel G.; Ceglio, Natale M.; Hawryluk, Andrew M.; Rosen, Robert S.; Vernon, Stephen P.)V1465,80-87(1991)

Multiple scattered electron-beam effect in electron-beam lithography (Saitou, Norio; Iwasaki, Teruo; Murai, Fumio)V1465,185-191(1991)

Multispot scanning exposure system for excimer laser stepper (Yoshitake, Yasuhiro; Oshida, Yoshitada; Tanimoto, Tetsuzou; Tanaka, Minoru; Yoshida, Minoru)V1463,678-687(1991)

Mushroom-shaped gates defined by e-beam lithography down to 80-nm gate lengths and fabrication of pseudomorphic HEMTs with a dry-etched gate recess (Huelsmann, Axel; Kaufel, G.; Raynor, Brian; Koehler, Klaus; Schweizer, T.; Braunstein, Juergen; Schlechtweg, M.; Tasker, Paul J.; Jakobus, Theo F.)V1465,201-208(1991)

Nanometer scale focused ion beam vacuum lithography using an ultrathin oxide resist (Harriott, Lloyd R.; Temkin, Henryk; Chu, C. H.; Wang, Yuh-Lin; Hsieh, Y. F.; Hamm, Robert A.; Panish, Morton B.; Wade, H. H.)V1465,57-63(1991)

Narrow (0.1 um to 0.5 um) copper lines for ultra-large-scale integration technology (Shacham-Diamand, Yosef Y.)V1442,11-19(1991)

Negative chemical amplification resist systems based on polyhydroxystyrenes and N-substituted imides or aldehydes (Ito, Hiroshi; Schildknegt, Klaas; Mash, Eugene A.)V1466,408-418(1991)

Negative resist systems using acid-catalyzed pinacol rearrangement reaction in a phenolic resin matrix (Uchino, Shou-ichi; Iwayanagi, Takao; Ueno, Takumi; Hayashi, Nobuaki)V1466,429-435(1991)

New 0.54 aperture i-line wafer stepper with field-by-field leveling combined with global alignment (van den Brink, Martin A.; Katz, Barton A.; Wittekoek, Stefan)V1463,709-724(1991)

New alignment sensors for wafer stepper (Ota, Kazuya; Magome, Nobutaka; Nishi, Kenji)V1463,304-314(1991)

New aqueous base-developable negative-tone photoresist based on furans (Fahey, James T.; Frechet, Jean M.)V1466,67-74(1991)

New family of 1:1 catadioptric broadband deep-UV high-Na lithography lenses (Zhang, Yudong; Zou, Haixing; Wang, Zhi-Jiang)V1463,688-694(1991)

New i-line and deep-UV optical wafer stepper (Unger, Robert; DiSessa, Peter A.)V1463,725-742(1991)

New i-line lens for half-micron lithography (Takahashi, Kazuhiro; Ohta, Masakatsu; Kojima, Toshiyuki; Noguchi, Miyoko)V1463,696-708(1991)

New phase-shifting mask structure for positive resist process (Miyazaki, Junji; Kamon, Kazuya; Yoshioka, Nobuyuki; Matsuda, Shuichi; Fujinaga, Masato; Watakabe, Yaichiro; Nagata, Hitoshi)V1464,327-335(1991)

Nonmetallic acid generators for i-line and g-line chemically amplified resists (Brunsvold, William R.; Montgomery, Warren; Hwang, Bao)V1466,368-376(1991)

Novel acid-hardening positive photoresist technology (Graziano, Karen A.; Thompson, Stephen D.; Winkle, Mark R.)V1466,75-88(1991)

Novel base-generating photoinitiators for deep-UV lithography (Kutal, Charles; Weit, Scott K.; Allen, Robert D.; MacDonald, Scott A.; Willson, C. G.)V1466,362-367(1991)

Novel high-resolution large-field scan-and-repeat projection lithography system (Jain, Kanti)V1463,666-677(1991)

Novel novolak resins using substituted phenols for high-performance positive photoresist (Kajita, Toru; Ota, Toshiyuki; Nemoto, Hiroaki; Yumoto, Yoshiji; Miura, Takao)V1466,161-173(1991)

Novel quinonediazide-sensitized photoresist system for i-line and deep-UV lithography (Fukunaga, Seiki; Kitaori, Tomoyuki; Koyanagi, Hiroo; Umeda, Shin'ichi; Nagasawa, Kohtaro)V1466,446-457(1991)

Novel surface imaging masking technique for high-aspect-ratio dry etching applications (Calabrese, Gary C.; Abali, Livingstone S.; Bohland, John F.; Pavelchek, Edward K.; Sricharoenchaikit, Prasit; Vizvary, Gerald; Bobbio, Stephen M.; Smith, Patrick)V1466,528-537(1991)

Novel toroidal mirror enhances x-ray lithography beamline at the Center for X-ray Lithography (Cole, Richard K.; Cerrina, Franco)V1465,111-121(1991)

Novolak design for high-resolution positive photoresists (IV): tandem-type novolak resin for high-performance positive photoresists (Hanabata, Makoto; Oi, F.; Furuta, Akihiro)V1466,132-140(1991)

Novolak resin design concept for high-resolution positive resists (Noguchi, Tsutomu; Tomita, Hidemi)V1466,149-160(1991)

Onium salt structure/property relationships in poly(4-tert-butyloxycarbonyloxystyrene) deep-UV resists (Schwartzkopf, George; Niazy, Nagla N.; Das, Siddhartha; Surendran, Geetha; Covington, John B.)V1466,26-38(1991)

Optical/Laser Microlithography IV (Pol, Victor, ed.)V1463(1991)

Optical lithography with chromeless phase-shifted masks (Toh, Kenny K.; Dao, Giang T.; Singh, Rajeev; Gaw, Henry T.)V1463,74-86(1991)

Optical metrology for integrated circuit fabrication (Chim, Stanley S.; Kino, Gordon S.)V1464,138-144(1991)

Optimization of an x-ray mask design for use with horizontal and vertical kinematic mounts (Laird, Daniel L.; Engelstad, Roxann L.; Palmer, Shane R.)V1465,134-144(1991)

Optimization of optical properties of resist processes (Brunner, Timothy A.)V1466,297-308(1991)

Optimization of partial coherence for half-micron i-line lithography (Canestrari, Paolo; Degiorgis, Giorgio A.; De Natale, Paolo; Gazzaruso, Lucia; Rivera, Giovanni)V1463,446-455(1991)

Optimum numerical aperture for optical projection microlithography (Lin, Burn J.)V1463,42-53(1991)

Origins of asymmetry in spin-cast films over topography (Manske, Loni M.; Graves, David B.)V1463,414-422(1991)

Overlay and matching strategy for large-area lithography (Holbrook, David S.; Donaher, J. C.)V1463,475-486(1991)

Oxygen plasma etching of silylated resist in top-imaging lithographic process (Dijkstra, Han J.)V1466,592-603(†1991)

Parametric studies and characterization measurements of x-ray lithography mask membranes (Wells, Gregory M.; Chen, Hector T.; Engelstad, Roxann L.; Palmer, Shane R.)V1465,124-133(1991)

Pattern etching and selective growth of GaAs by in-situ electron-beam lithography using an oxidized thin layer (Akita, K.; Sugimoto, Yoshimasa; Taneya, M.; Hiratani, Y.; Ohki, Y.; Kawanishi, Hidenori; Katayama, Yoshifumi)V1392,576-587(1991)

Pattern recognition approach to trench bottom-width measurement (Chou, Ching-Hua; Berman, John L.; Chim, Stanley S.; Corle, Timothy R.; Xiao, Guoqing; Kino, Gordon S.)V1464,145-154(1991)

Pelliclizing technology (Yamauchi, Takashi)V1496,302-314(1991)

Performance appraisal of the ATEQ CORE-2500 in production (Mechtenberg, Monica L.; Watson, Larry J.)V1496,124-155(1991)

Phase-only encoding method for complex wavefronts (Urquhart, Kristopher S.; Lee, Sing H.)V1555,13-22(1991)

Phase-shifting and other challenges in optical mask technology (Lin, Burn J.)V1496,54-79(1991)

Phase-shifting photolithography applicable to real IC patterns (Yanagishita, Yuichiro; Ishiwata, Naoyuki; Tabata, Yasuko; Nakagawa, Kenji; Shigematsu, Kazumasa)V1463,207-217(1991)

Phase-shifting structures for isolated features (Garofalo, Joseph G.; Kostelak, Robert L.; Yang, Tungsheng)V1463,151-166(1991)

Phase-shift mask applications (Buck, Peter D.; Rieger, Michael L.)V1463,218-228(1991)

Phase-shift mask pattern accuracy requirements and inspection technology (Wiley, James N.; Fu, Tao-Yi; Tanaka, Takashi; Takeuchi, Susumu; Aoyama, Satoshi; Miyazaki, Junji; Watakabe, Yaichiro)V1464,346-355(1991)

Phase-shift mask technology: requirements for e-beam mask lithography (Dunbrack, Steven K.; Muray, Andrew; Sauer, Charles; Lozes, Richard; Nistler, John L.; Arnold, William H.; Kyser, David F.; Minvielle, Anna M.; Preil, Moshe E.; Singh, Bhanwar; Templeton, Michael K.)V1464,314-326(1991)

Photolithographic imaging of computer-generated holographic optical elements (Shvartsman, Felix P.; Oren, Moshe)V1555,71-79(1991)

Photomask fabrication utilizing a Philips/Cambridge vector scan e-beam system (McCutchen, William C.)V1496,97-106(1991)

Photoresist bake conditions and their effects on lithography process control (Norbury, David H.; Love, John C.)V1463,558-573(1991)

Photoresist dissolution rates: a comparison of puddle, spray, and immersion processes (Robertson, Stewart A.; Stevenson, J. T.; Holwill, Robert J.; Thirsk, Mark; Daraktchiev, Ivan S.; Hansen, Steven G.)V1464,232-244(1991)

Physical aging of resists: the continual evolution of lithographic material (Paniez, Patrick J.; Weill, Andre P.; Cohendoz, Stephane D.)V1466,336-344(1991)

Pixelgram: an application of electron-beam lithography for the security printing industry (Lee, Robert A.)V1509,48-54(1991)

Planar diffractive optical elements prepared by electron-beam lithography (Urban, Frantisek; Matejka, Frantisek)V1574,58-65(1991)

Polymer effects on the photochemistry of triarylsulfonium salts (McKean, Dennis R.; Allen, Robert D.; Kasai, Paul H.; MacDonald, Scott A.)V1559,214-221(1991)

Polysilanes for microlithography (Wallraff, Gregory M.; Miller, Robert D.; Clecak, Nicholas J.; Baier, M.)V1466,211-217(1991)

Polysilyne resists for 193 nm excimer laser lithography (Kunz, Roderick R.; Bianconi, Patricia A.; Horn, Mark W.; Paladugu, R. R.; Shaver, David C.; Smith, David A.; Freed, Charles A.)V1466,218-226(1991)

Polyvinylphenols protected with tetrahydropyranyl group in chemical amplification positive deep-UV resist systems (Hayashi, Nobuaki; Schlegel, Leo; Ueno, Takumi; Shiraishi, Hiroshi; Iwayanagi, Takao)V1466,377-383(1991)

Preliminary lithographic characteristics of an all-organic chemically amplified resist formulation for single-layer deep-UV lithography (Nalamasu, Omkaram; Reichmanis, Elsa; Cheng, May; Pol, Victor; Kometani, Janet M.; Houlihan, Frank M.; Neenan, Thomas X.; Bohrer, M. P.; Mixon, D. A.; Thompson, Larry F.; Takemoto, Cliff H.)V1466,13-25(1991)

Preparations and properties of novel positive photosensitive polyimides (Hayase, Rumiko H.; Kihara, Naoko; Oyasato, Naohiko; Matake, S.; Oba, Masayuki)V1466,438-445(1991)

Primary research for mechanism of forming PLH (Guo, Lu Rong; Zhang, Xiao-Chun; Guo, Yongkang)V1463,534-538(1991)

Process control capability using a diaphragm photochemical dispense system (Cambria, Terrell D.; Merrow, Scott F.)V1466,670-675(1991)

Process enhancement for a new generation g-line photolithographic system (Ostrout, Wayne H.; Hiatt, William M.; Kozlowski, Alan E.)V1463,54-73(1991)

Process error limitations on binary optics performance (Cox, James A.; Fritz, Bernard S.; Werner, Thomas R.)V1555,80-88(1991)

Process latitude for the chemical amplification resists AZ PF514 and AZ PN114 (Eckes, Charlotte; Pawlowski, Georg; Przybilla, Klaus J.; Meier, Winfried; Madore, Michel; Dammel, Ralph)V1466,394-407(1991)

Process latitude measurements on chemically amplified resists exposed to synchrotron radiation (Babcock, Carl P.; Taylor, James W.; Sullivan, Monroe; Suh, Doowon; Plumb, Dean; Palmer, Shane R.; Berry, Amanda K.; Graziano, Karen A.; Fedynyshyn, Theodore H.)V1466,653-662(1991)

Process latitudes in projection printing (Barouch, Eytan; Hollerbach, Uwe; Orszag, Steven A.; Bradie, Brian D.; Peckerar, Martin)V1465,254-262(1991)

Process optimization: a case study on the application of Taguchi methods in optical lithography (Arshak, Khalil I.; Murphy, Eamonn; Arshak, A.)V1463,521-533(1991)

Progress in DUV resins (Przybilla, Klaus J.; Roeschert, Heinz; Spiess, Walter; Eckes, Charlotte; Chatterjee, Subhankar; Khanna, Dinesh N.; Pawlowski, Georg; Dammel, Ralph)V1466,174-187(1991)

Progress in the study of development-free vapor photolithography (Hong, Xiao-Yin; Liu, Dan; Li, Zhong-Zhe; Xiao, Ji-Quang; Dong, Gui-Rong)V1466,546-557(1991)

Projection direct imaging for high-density interconnection and printed circuit manufacture (Bergstrom, Neil G.)V1390,509-512(1991)

Proximity effect correction on MEBES for 1x mask fabrication: lithography issues and tradeoffs at 0.25 micron (Muray, Andrew; Dean, Robert L.)V1496,171-179(1991)

Ray tracing homogenizing mirrors for synchrotron x-ray lithography (Homer, Michael; Rosser, Roy J.; Speer, R. J.)V1527,134-144(1991)

Recent developments of x-ray lithography in Canada (Chaker, Mohamed; Boily, S.; Ginovker, A.; Jean, A.; Kieffer, J. C.; Mercier, P. P.; Pepin, Henri; Leung, Pak K.; Currie, John F.; Lafontaine, H.)V1465,16-25(1991)

Reduction of the standing wave effect in positive photoresist using an antireflection coating (Mehrotra, R.; Mathur, B. P.; Sharan, Sunil)V1463,487-491(1991)

Reflective optical designs for soft x-ray projection lithography (Jewell, Tanya E.; Thompson, Kevin P.; Rodgers, John M.)V1527,163-173(1991)

Reliability of contrast and dissolution-rate-derived parameters as predictors of photoresist performance (Spragg, Peggy M.; Hurditch, Rodney J.; Toukhy, Medhat A.; Helbert, John N.; Malhotra, Sandeep)V1466,283-296(1991)

Resist design for dry-developed positive working systems in deep-UV and e-beam lithography (Vinet, Francoise; Chevallier, M.; Pierrat, Christophe; Guibert, Jean C.; Rosilio, Charles; Mouanda, B.; Rosilio, A.)V1466,558-569(1991)

Resist parameter extraction with graphical user interface in X (Chiu, Anita S.; Ferguson, Richard A.; Doi, Takeshi; Wong, Alfred K.; Tam, Nelson; Neureuther, Andrew R.)V1466,641-652(1991)

Resist patterning for sub-quarter-micrometer device fabrications (Chiong, Kaolin G.; Hohn, Fritz J.)V1465,221-236(1991)

Resist schemes for soft x-ray lithography (Taylor, Gary N.; Hutton, Richard S.; Windt, David L.; Mansfield, William M.)V1343,258-273(1991)

Resist tracking: a lithographic diagnostic tool (Takemoto, Cliff H.; Ziger, David H.; Connor, William; Distasio, Romelia)V1464,206-214(1991)

Results of photolithographic cluster cells in actual production (Clifford, Sandra; Hayes, Bruce L.; Brade, Richard)V1463,551-557(1991)

Selected performance parameters and functional principles of a new stepper generation (Kliem, Karl-Heinz; Sczepanski, Volker; Michl, Uwe; Hesse, Reiner)V1463,743-751(1991)

Semiwafer metrology project (Bennett-Lilley, Marylyn H.; Hiatt, William M.; Lauchlan, Laurie J.; Mantalas, Lynda C.; Rottmann, Hans; Seliger, Mark; Singh, Bhanwar; Yansen, Don E.)V1464,127-136(1991)

Sequential experimentation strategy and response surface methodologies for photoresist process optimization (Flores, Gary E.; Norbury, David H.)V1464,610-627(1991)

Simulation analysis of deep-UV chemically amplified resist (Ohfuji, Takeshi; Soenosawa, Masanobu; Nozue, Hiroshi; Kasama, Kunihiko)V1463,345-354(1991)

Simulation of an advanced negative i-line photoresist (Barouch, Eytan; Hollerbach, Uwe; Orszag, Steven A.; Allen, Mary T.; Calabrese, Gary C.)V1463,336-344(1991)

Simulation of connected image reversal and DESIRE techniques for submicron lithography (Nicolau, Dan V.; Dusa, Mircea V.; Fulga, Florin)V1466,663-669(1991)

Simulation of low-energy x-ray lithography using a diamond membrane mask (Hasegawa, Shinya; Suzuki, Katsumi)V1465,145-151(1991)

Simulations of bar printing over a MOSFET device using i-line and deep-UV resists (Barouch, Eytan; Hollerbach, Uwe; Orszag, Steven A.; Szmanda, Charles R.; Thackeray, James W.)V1463,464-474(1991)

Single-component chemically amplified resist materials for electron-beam and x-ray lithography (Novembre, Anthony E.; Tai, Woon W.; Kometani, Janet M.; Hanson, James E.; Nalamasu, Omkaram; Taylor, Gary N.; Reichmanis, Elsa; Thompson, Larry F.)V1466,89-99(1991)

Soft x-ray projection lithography: experiments and practical printers (White, Donald L.; Bjorkholm, John E.; Bokor, J.; Eichner, L.; Freeman, Richard R.; Gregus, J. A.; Jewell, Tanya E.; Mansfield, William M.; MacDowell, Alastair A.; Raab, Eric L.; Silfvast, William T.; Szeto, L. H.; Tennant, Donald M.; Waskiewicz, Warren K.; Windt, David L.; Wood, Obert R.)V1343,204-213(1991)

Some experimental techniques for characterizing photoresists (Spence, Christopher A.; Ferguson, Richard A.)V1466,324-335(1991)

Some fundamental issues on metallization in VLSI (Ferry, David K.; Kozicki, M. N.; Raupp, Gregory B.)V1596,2-11(1991)

Sophisticated masks (Pease, R. F.; Owen, Geraint; Browning, Raymond; Hsieh, Robert L.; Lee, Julienne Y.; Maluf, Nadim I.; Berglund, C. N.)V1496,234-238(1991)

Spherical pinch x-ray generator prototype for microlithography (Kawai, Kenji; Panarella, Emilio; Mostacci, D.)V1465,308-314(1991)

Structural effects of DNQ-PAC backbone on resist lithographic properties (Uenishi, Kazuya; Kawabe, Yasumasa; Kokubo, Tadayoshi; Slater, Sydney G.; Blakeney, Andrew J.)V1466,102-116(1991)

Structure of poly(p-hydroxystyrene) film (Toriumi, Minoru; Yanagimachi, Masatoshi; Masuhara, Hiroshi)V1466,458-468(1991)

Studies of dissolution inhibition mechanism of DNQ-novolak resist (II): effect of extended ortho-ortho bond in novolak (Honda, Kenji; Beauchemin, Bernard T.; Fitzgerald, Edward A.; Jeffries, Alfred T.; Tadros, Sobhy P.; Blakeney, Andrew J.; Hurditch, Rodney J.; Tan, Shiro; Sakaguchi, Shinji)V1466,141-148(1991)

Study of silylation mechanisms and kinetics through variations in silylating agent and resin (Dao, T. T.; Spence, Christopher A.; Hess, Dennis W.)V1466,257-268(1991)

Study of the chemically amplifiable resist materials for electron-beam lithography (Koyanagi, Hiroo; Umeda, Shin'ichi; Fukunaga, Seiki; Kitaori, Tomoyuki; Nagasawa, Kohtaro)V1466,346-361(1991)

Study of the relationship between exposure margin and photolithographic process latitude and mask linearity (Hansen, Steven G.; Dao, Giang T.; Gaw, Henry T.; Qian, Qi-de; Spragg, Peggy M.; Hurditch, Rodney J.)V1463,230-244(1991)

Submicrometer lithographic alignment and overlay strategies (Zaidi, Saleem H.; Naqvi, H. S.; Brueck, Steven R.)V1343,245-255(1991)

Submicron linewidth measurement using an interferometric optical profiler (Creath, Katherine)V1464,474-483(1991)

Super-accurate positioning technique using diffracted moire signals (Takada, Yutaka; Uchida, Yoshiyuki; Akao, Yasuo; Yamada, Jun; Hattori, Shuzo)V1332,571-576(1991)

Surface imaging on the basis of phenolic resin: experiments and simulation (Bauch, Lothar; Jagdhold, Ulrich A.; Dreger, Helge H.; Bauer, Joachim J.; Hoeppner, Wolfgang W.; Erzgraeber, Hartmut H.; Mehliss, Georg G.)V1466,510-519(1991)

Synthesis and lithographic evaluation of alternating copolymers of linear and cyclic alkenyl(di)silanes with sulfur dioxide (Gozdz, Antoni S.; Shelburne, John A.)V1466,520-527(1991)

Synthetic holographic beamsplitters for integrated optics (Saarinen, Jyrki V.; Huttunen, Juhani; Vasara, Antti H.; Noponen, Eero; Turunen, Jari P.; Salin, Arto U.)V1555,128-137(1991)

System design considerations for a production-grade, ESR-based x-ray lithography beamline (Kovacs, Stephen; Melore, Dan; Cerrina, Franco; Cole, Richard K.)V1465,88-99(1991)

Techniques for characterization of silicon penetration during DUV surface imaging (Freeman, Peter W.; Bohland, John F.; Pavelchek, Edward K.; Jones, Susan K.; Dudley, Bruce W.; Bobbio, Stephen M.)V1464,377-385(1991)

Technology and chemistry of high-temperature positive resist (Toukhy, Medhat A.; Sarubbi, Thomas R.; Brzozowy, David J.)V1466,497-507(1991)

Three-dimensional simulation of optical lithography (Toh, Kenny K.; Neureuther, Andrew R.)V1463,356-367(1991)

Transparent phase-shifting mask with multistage phase shifter and comb-shaped shifter (Watanabe, Hisashi; Todokoro, Yoshihiro; Hirai, Yoshihiko; Inoue, Morio)V1463,101-110(1991)

Two-layer 1.2-micron pitch multilevel metal demonstrator using resist patterning by surface imaging and dry development (Martin, Brian; Snowden, Ian M.; Mortimer, Simon H.)V1463,584-594(1991)

Ultraprecise scanning technology (Zernike, Frits; Galburt, Daniel N.)V1343,241-244(1991)

Use of antireflective coatings in deep-UV lithography (Sethi, Satyendra A.; Distasio, Romelia; Ziger, David H.; Lamb, James E.; Flaim, Tony)V1463,30-40(1991)

Using the Atomic Force Microscope to measure submicron dimensions of integrated circuit devices and processes (Rodgers, Mark R.; Monahan, Kevin M.)V1464,358-366(1991)

Variable phase-shift mask for deep-submicron optical lithography (Terasawa, Tsuneo; Hasegawa, Norio; Imai, Akira; Tanaka, Toshihiko; Katagiri, Souichi)V1463,197-206(1991)

Video-based alignment system for x-ray lithography (Hughlett, R. E.; Cooper, Keith A.)V1465,100-110(1991)

Wafer alignment based on existing microstructures (Wyntjes, Geert J.; Hercher, Michael)V1464,539-545(1991)

Wet-developed, high-aspect-ratio resist patterns by 20-keV e-beam lithography (Weill, Andre P.; Amblard, Gilles R.; Lalanne, Frederic P.; Panabiere, Jean-Pierre)V1465,264-270(1991)

What IS a phase-shifting mask? (Levenson, Marc D.)V1496,20-26(1991)

X-ray lithography system development at IBM: overview and status (Maldonado, Juan R.)V1465,2-15(1991)

XUV characterization comparison of Mo/Si multilayer coatings (Windt, David L.; Waskiewicz, Warren K.; Kubiak, Glenn D.; Barbee, Troy W.; Watts, Richard N.)V1343,274-282(1991)

XUV free-electron laser-based projection lithography systems (Newnam, Brian E.)V1343,214-228(1991)

XUV resist characterization: studies with a laser plasma source (Kubiak, Glenn D.)V1343,283-291(1991)

Micro-optics—see also lenses

Advances in the computer-aided design of planarized free-space optical circuits: system simulation (Jahns, Juergen; Brumback, Babette A.)V1555,2-7(1991)

Advances in the optical design of miniaturized optical correlators (Gebelein, Rolin J.; Connely, Shawn W.; Foo, Leslie D.)V1564,452-463(1991)

Analysis of the self-oscillation phenomenon of fiber optically addressed silicon microresonators (Rao, Yun-Jiang; Culshaw, Brian)V1506,126-133(1991)

Characterization of micro-optical components fabricated by deep-etch x-ray lithography (Goettert, Jost; Mohr, Jurgen)V1506,170-178(1991)

Design of a GaAs Fresnel lens array for optical signal-processing devices (Armenise, Mario N.; Impagnatiello, Fabrizio; Passaro, Vittorio M.)V1374,86-96(1991)

Detection of DNA sequence symmetries using parallel micro-optical devices (Christens-Barry, William A.; Terry, David H.; Boone, Bradley G.)V1564,177-188(1991)

Electron-beam lithography for the microfabrication of OEICs (Engel, Herbert; Doeldissen, Walter)V1506,60-64(1991)

Electron-beam-written reflection diffractive microlenses for oblique incidence (Shiono, Teruhiro; Ogawa, Hisahito)V1545,232-240(1991)

Experimental characterization of microlenses for WDM transmission systems (Zoboli, Maurizio; Bassi, Paolo)V1506,160-169(1991)

Fabrication and characterization of semiconductor microlens arrays (Diadiuk, Vicky; Liau, Zong-Long; Walpole, James N.)V1354,496-500(1991)

Fabrication and performance of a one-to-one erect imaging microlens array for fax (Bellman, Robert H.; Borrelli, Nicholas F.; Mann, L. G.; Quintal, J. M.)V1544,209-217(1991)

Fabrication of microlenses by laser-assisted chemical etching (Gratrix, Edward J.; Zarowin, Charles B.)V1544,238-243(1991)

Fabrication of micro-optical components by laser beam writing in photoresist (Gale, Michael T.; Lang, Graham K.; Raynor, Jeffrey M.; Schuetz, Helmut)V1506,65-70(1991)

Fast, inexpensive, diffraction-limited cylindrical microlenses (Snyder, James J.; Reichert, Patrick)V1544,146-151(1991)

Focus grating coupler construction optics: theory, design, and tolerancing (Moore, Kenneth E.; Lawrence, George N.)V1354,581-587(1991)

Free-space optical interconnect using microlasers and modulator arrays (Dickinson, Alex G.; Downs, Maralene M.; LaMarche, R. E.; Prise, Michael E.)V1389,503-514(1991)

Grating beam splitting polarizer using multilayer resist method (Aoyama, Shigeru; Yamashita, Tsukasa)V1545,241-250(1991)

High-resolution micro-objective lens with kinoform corrector (Lenkova, Galina A.; Churin, Evgeny G.)V1574,235-242(1991)

High-speed binary optic microlens array in GaAs (Motamedi, M. E.; Southwell, William H.; Anderson, Robert J.; Hale, Leonard G.; Gunning, William J.; Holz, Michael)V1544,33-44(1991)

Holographic microlenses for optical fiber interconnects (Galloway, Peter C.; Dobson, Peter J.)V1365,131-138(1991)

Influence of atomic decay on micromaser operation (Zhu, Shi-Yao)V1497,240-244(1991)

Integrated free-space optics (Jahns, Juergen)V1354,588-592(1991)

Integrated micro-optics for computing and switching applications (Jahns, Juergen)V1544,246-262(1991)

Intl Conf on the Application and Theory of Periodic Structures (Lerner, Jeremy M.; McKinney, Wayne R., eds.)V1545(1991)

Investigation of fringing fields in liquid-crystal devices (Powell, Norman J.; Kelsall, Robert W.; Love, G. D.; Purvis, Alan)V1545,19-30(1991)

Laser ablation of refractive micro-optic lenslet arrays (Bartley, James A.; Goltsos, William)V1544,140-145(1991)

Light coupling characteristics of planar microlens (Oikawa, Masahiro; Nemoto, Hiroyuki; Hamanaka, Kenjiro; Imanishi, Hideki; Kishimoto, Takashi)V1544,226-237(1991)

Liquid-crystal phase modulators for active micro-optic devices (Purvis, Alan; Williams, Geoffrey; Powell, Norman J.; Clark, Michael G.; Wiltshire, Michael C.)V1455,145-149(1991)

Local microholograms recording on the moving photothermoplastic disk (Kutanov, Askar A.; Akaev, Askar A.; Abdrisaev, Baktybek D.; Snimshikov, Igor A.)V1507,94-98(1991)

Low-cost and compact fiber-to-laser coupling with micro-Fresnel lens (Ogata, Shiro; Yoneda, Masahiro; Maeda, Tetsuo; Imanaka, Koichi)V1544,92-100(1991)

Micro-optical elements in holography (von Bally, Gert; Dirksen, D.; Zou, Y.)V1507,66-72(1991)

Micro-optic lens for data storage (Milster, Tom D.; Trusty, Robert M.; Wang, Mark S.; Froehlich, Fred F.; Erwin, J. K.)V1499,286-292(1991)

Micro-Optics II (Scheggi, Annamaria V., ed.)V1506(1991)

Micro-optic studies using photopolymers (Phillips, Nicholas J.; Barnett, Christopher A.)V1544,10-21(1991)

Micrograting device using electron-beam lithography (Yamashita, Tsukasa; Aoyama, Shigeru)V1507,81-93(1991)

Microlens array fabricated in surface relief with high numerical aperture (Lau, Hon W.; Davies, Neil A.; McCormick, Malcolm)V1544,178-188(1991)

Microlens array for modification of SLM devices (Kathman, Alan D.; Temmen, Mark G.; Scott, Miles L.)V1544,58-65(1991)

Microlens array for staring infrared imager (Werner, Thomas R.; Cox, James A.; Swanson, S.; Holz, Michael)V1544,46-57(1991)

Microlens arrays in Europe (Hutley, M. C.)V1544,134-137(1991)

Microlens arrays in integral photography and optical metrology (Davies, Neil A.; McCormick, Malcolm; Lau, Hon W.)V1544,189-198(1991)

Miniature and Micro-Optics: Fabrication and System Applications (Roychoudhuri, Chandrasekhar; Veldkamp, Wilfrid B., eds.)V1544(1991)

Monolithic two-dimensional surface-emitting laser diode arrays with 45 degree micromirrors (Jansen, Michael; Yang, Jane; Ou, Szutsun S.; Sergant, Moshe; Mawst, Luke J.; Rozenbergs, John; Wilcox, Jarka Z.; Botez, Dan)V1418,32-39(1991)

Multiple-beam accessor using microzone plate elements for optoelectronic integrated circuits (Kodate, Kashiko; Kamiya, Takeshi)V1545,251-260(1991)

New approaches to practical guided-wave passive devices based on ion-exchange technologies in glass (Seki, Masafumi; Sato, Shiro; Nakama, Kenichi; Wada, Hiroshi; Hashizume, Hideki; Kobayashi, Shigeru)V1583,184-195(1991)

Optical fiber demultiplexer for telecommunications (Falciai, Riccardo; Scheggi, Annamaria V.; Cosi, Franco; Cao, J. Y.)V1365,38-42(1991)

Optical interconnects: a solution to very high speed integrated circuits and systems (Chen, Ray T.)V1374,162-175(1991)

Overview of micro-optics: past, present, and future (Veldkamp, Wilfrid B.)V1544,287-299(1991)

Performance and measurements of refractive microlens arrays (Feldblum, Avi Y.; Nijander, Casimir R.; Townsend, Wesley P.; Mayer-Costa, Carlos M.)V1544,200-208(1991)

Photon correlation spectroscopy and electrophoretic light scattering using optical fibers (Macfadyen, Allan J.; Jennings, B. R.)V1367,319-328(1991)

Progress in miniature laser systems for space science particle sizing and velocimetry (Brown, Robert G.)V1506,58-59(1991)

Rigorous diffraction theory of binary optical interconnects (Vasara, Antti H.; Noponen, Eero; Turunen, Jari P.; Miller, J. M.; Taghizadeh, Mohammad R.; Tuovinen, Jussi)V1507,224-238(1991)

Rigorous electromagnetic modeling of diffractive optical elements (Johnson, Eric G.; Kathman, Alan D.)V1545,209-216(1991)

Silicon microlenses for enhanced optical coupling to silicon focal planes (Motamedi, M. E.; Griswold, Marsden P.; Knowlden, Robert E.)V1544,22-32(1991)

Single-pixel measurements on LCDs (Jenkins, A. J.)V1506,188-193(1991)

Subminiature package external cavity laser (Fatah, Rebwar M.; Cox, Maurice K.; Bird, David M.; Cameron, Keith H.)V1501,120-128(1991)

Super-compact dual-axis optical scanning unit applying a tortional spring resonator driven by a piezoelectric actuator (Goto, Hiroshi; Imanaka, Koichi)V1544,272-281(1991)

Techniques for integrating 3-D optical systems (Brenner, Karl-Heinz)V1544,263-270(1991)

Testing binary optics: accurate high-precision efficiency measurements of microlens arrays in the visible (Holz, Michael; Stern, Margaret B.; Medeiros, Shirley; Knowlden, Robert E.)V1544,75-89(1991)

Three-dimensional moving-image display by modulated coherent optical fibers: a proposal (Hoshino, Hideshi; Sato, Koki)V1461,227-231(1991)

Two applications for microlens arrays: detector fill-factor improvement and laser diode collimation (D'Amato, Dante P.; Centamore, Robert M.)V1544,166-177(1991)

Vertical cavity lasers for optical interconnects (Jewell, Jack L.; Lee, Yong H.; McCall, Samuel L.; Scherer, Axel; Harbison, James P.; Florez, Leigh T.; Olsson, N. A.; Tucker, Rodney S.; Burrus, Charles A.; Sandroff, Claude J.)V1389,401-407(1991)

Video Hartmann wavefront diagnostic that incorporates a monolithic microlens array (Toeppen, John S.; Bliss, Erlan S.; Long, Theresa W.; Salmon, J. T.)V1544,218-225(1991)

Microscopy

Advanced confocal technique for submicron CD measurements (Rohde, Axel; Saffert, Ralf; Fitch, John)V1464,438-446(1991)

Advanced materials characterization by means of moire techniques (Aymerich, Francesco; Ginesu, Francesco; Priolo, Pierluigi)V1554B,304-314(1991)

Analysis of images of periodic structures obtained by Photon Scanning Tunneling Microscopy (Goudonnet, Jean-Pierre; Salomon, L.; de Fornel, F.; Adam, P.; Bourillot, E.; Neviere, Michel; Guerin, Philippe)V1545,130-139(1991)

Analysis of multidimensional confocal images (Samarabandu, J. K.; Acharya, Raj S.; Edirisinghe, Chandima D.; Cheng, Ping-Chin; Kim, Hyo-Gun; Lin, T. H.; Summers, R. G.; Musial, C. E.)V1450,296-322(1991)

Application of morphological pseudoconvolutions to scanning-tunneling and atomic force microscopy (Dougherty, Edward R.; Weisman, Andrew; Mizes, Howard; Miller, Robert J.)V1567,88-99(1991)

Automatic segmentation of microvessels using textural analysis (Albert, Thomas; O'Connor, Carol; Harris, Patrick D.)V1450,84-89(1991)

Biomedical Image Processing II (Bovik, Alan C.; Howard, Vyvyan, eds.)V1450(1991)

Capabilities of the optical microscope (Bradbury, Savile)V1439,128-134(1991)

Cellular vibration measurement with a noninvasive optical system (Khanna, Shyam M.)V1429,9-20(1991)

Changes in collagen birefringence: a quantitative histologic marker of thermal damage in skin (Thomsen, Sharon L.; Cheong, Wai-Fung; Pearce, John A.)V1422,34-42(1991)

Characterization of surface contaminants using infrared microspectroscopy (Blair, Dianna S.; Ward, Kenneth J.)V1437,76-79(1991)

Characterization of the radiation from a low-energy X-pinch source (Christou, Christos; Choi, Peter)V1552,278-287(1991)

Characterization of wavelength offset for optimization of deep-UV stepper performance (Jones, Susan K.; Dudley, Bruce W.; Peters, Charles R.; Kellam, Mark D.; Pavelchek, Edward K.)V1464,546-553(1991)

Charging effects in low-voltage SEM metrology (Monahan, Kevin M.; Benschop, Jozef P.; Harris, Tom A.)V1464,2-9(1991)

Coherent detection in confocal microscopy (Wilson, Tony)V1439,104-108(1991)

Comparative study of texture measurements for cellular organelle recognition (Grenier, Marie-Claude; Durand, Louis-Gilles; de Guise, J.)V1450,154-169(1991)

Computational model of the imaging process in scanning-x microscopy (Gallarda, Harry S.; Jain, Ramesh C.)V1464,459-473(1991)

Confocal light microscopy of the living in-situ ocular lens: two- and three-dimensional imaging (Masters, Barry R.)V1443,288-293(1991)

Confocal microscopy for the biological and material sciences: principle, applications, limitations (Brakenhoff, G. J.; van der Voort, H. T.; Visscher, Koen)V1439,121-127(1991)

Confocal redox fluorescence microscopy for the evaluation of corneal hypoxia (Masters, Barry R.; Kriete, Andres; Kukulies, Joerg)V1431,218-223(1991)

Continuous TEM observation of diamond nucleus growth by side-view method (Goto, Yasuyuki; Kurihara, Kazuaki; Sawamoto, Yumiko; Kitakohji, Toshisuke)V1534,49-58(1991)

Cross-sectional imaging in SEM: signal formation mechanism and CD measurements (Firstein, Leon A.; Noz, Arthur)V1464,81-88(1991)

Design and analysis of aspherical multilayer imaging x-ray microscope (Shealy, David L.; Jiang, Wu; Hoover, Richard B.)V1343,122-132(1991)

Design and analysis of a water window imaging x-ray microscope (Hoover, Richard B.; Baker, Phillip C.; Shealy, David L.; Brinkley, B. R.; Walker, Arthur B.; Barbee, Troy W.)V1426,84-96(1991)

Design considerations of a real-time clinical confocal microscope (Masters, Barry R.)V1423,8-14(1991)

Determination of electrostatic potentials and charge distributions in bulk and at interfaces by electron microscopy techniques (Hugsted, B.; Gjonnes, K.; Tafto, J.; Gjonnes, Jon; Matsuhata, H.)V1361,751-757(1991)

Development of electron moire method using a scanning electron microscope (Kishimoto, Satoshi; Egashira, Mitsuru; Shinya, Norio)V1554B,174-180(1991)

Development of the water-window imaging x-ray microscope (Hoover, Richard B.; Shealy, David L.; Baker, Phillip C.; Barbee, Troy W.; Walker, Arthur B.)V1435,338-351(1991)

Digital imaging microscopy: the marriage of spectroscopy and the solid state CCD camera (Jovin, Thomas M.; Arndt-Jovin, Donna J.)V1439,109-120(1991)

Digital restoration of scanning electrochemical microscope images (Bartels, Keith A.; Lee, Chongmok; Bovik, Alan C.; Bard, Allen J.)V1450,30-39(1991)

Direct high-spatial-resolution SIMS (secondary ion mass spectrometry) imaging of labeled nucleosides in human chromosomes (Hallegot, Philippe; Girod, C.; LeBeau, M. M.; Levi-Setti, Riccardo)V1396,311-315(1991)

Dynamics of a low-energy X-pinch source (Choi, Peter; Christou, Christos; Aliaga, Raul)V1552,270-277(1991)

Effect of operating points in submicron CD measurements (Dusa, Mircea V.; Jung, Christoph; Jung, Paul; Hogenkamp, Detlef; Roeth, Klaus-Dieter)V1464,447-458(1991)

Experimental characterization of Fresnel zone plate for hard x-ray applications (Lai, Barry P.; Chrzas, John J.; Yun, Wen-Bing; Legnini, Dan; Viccaro, P. J.; Bionta, Richard M.; Skulina, Kenneth M.)V1550,46-49(1991)

First direct measurements of transformation strains in crack-tip zone (Dadkhah, Mahyar S.; Marshall, David B.; Morris, Winfred L.)V1554A,164-175(1991)

Fluorescence imaging in photodynamic therapy (MacRobert, Alexander J.; Phillips, David)V1439,79-87(1991)

Fourier-transform holographic microscope (Haddad, Waleed S.; Cullen, David; Solem, Johndale C.; Longworth, James W.; McPherson, Armon; Boyer, Keith; Rhodes, Charles K.)V1448,81-88(1991)

Hadamard transform Raman imaging (Morris, Michael D.; Govil, Anurag; Liu, Kei-Lee; Sheng, Rong-Sheng)V1439,95-101(1991)

Half-micrometer linewidth metrology (Knight, Stephen E.; Humphrey, Dean; Bowley, Reginald R.; Cogley, Robert M.)V1464,119-126(1991)

High-resolution electron microscopy of diamond film growth defects and their interactions (Shechtman, Dan; Farabaugh, Edward N.; Robins, Lawrence H.; Feldman, Albert; Hutchison, Jerry L.)V1534,26-43(1991)

High-resolution texture analysis for cytology (Hallouche, Farid; Adams, Alan E.; Hinton, Oliver R.)V1445,504-512(1991)

High-speed stepper setup using a low-voltage SEM (Benschop, Jozef P.; Monahan, Kevin M.; Harris, Tom A.)V1464,62-70(1991)

High spatial and temporal resolution in the optical investigation of biological objects (Damm, Tobias; Kempe, M.; Stamm, Uwe; Stolberg, K. P.; Wabnitz, H.)V1403,686-694(1991)

HIV detection by in-situ hybridization based on confocal reflected light microscopy (Smith, Louis C.; Jericevic, Zeljko; Cuellar, Roland; Paddock, Stephen W.; Lewis, Dorothy E.)V1428,224-232(1991)

Image processing utilizing an APL interface (Zmola, Carl; Kapp, Oscar H.)V1396,51-55(1991)

In-situ measurement technique for solution growth in compound semiconductors (Inatomi, Yuko; Kuribayashi, Kazuhiko)V1557,132-139(1991)

In-situ observation of crystal growth in microgravity by high-resolution microscopies (Tsukamoto, Katsuo; Onuma, Kazuo)V1557,112-123(1991)

Instrumentation for simultaneous kinetic imaging of multiple fluorophores in single living cells (Morris, Stephen J.; Beatty, Diane M.; Welling, Larry W.; Wiegmann, Thomas B.)V1428,148-158(1991)

Integrating FTIR microscopy into surface analysis (Church, Jeffrey S.)V1437,80-88(1991)

Interface characterization of XUV multilayer reflectors using HRTEM and x-ray and XUV reflectance (Windt, David L.; Hull, Robert; Waskiewicz, Warren K.; Kortright, Jeffrey B.)V1343,292-308(1991)

Intermolecular contacts within sickle hemoglobin fibers (Watowich, Stanley J.; Gross, Leon J.; Josephs, Robert)V1396,316-323(1991)

Knowledge-driven image analysis of cell structures (Nederlof, Michel A.; Witkin, Andrew; Taylor, D. L.)V1428,233-241(1991)

Laser Doppler measurements of ameboid cytoplasmic streaming and problems of mathematical modeling of intracellular hydrodynamics (Priezzhev, Alexander V.; Proskurin, S. G.; Romanovsky, Yuri M.)V1402,107-113(1991)

Laser scan microscope and infrared laser scan microcope: two important tools for device testing (Ziegler, Eberhard)V1400,108-115(1991)

Linearity of coherence probe metrology: simulation and experiment (Davidson, Mark P.; Monahan, Kevin M.; Monteverde, Robert J.)V1464,155-176(1991)

Linewidth measurement comparison between a photometric optical microscope and a scanning electron microscope backed with Monte Carlo trajectory computations (Nunn, John W.; Turner, Nicholas P.)V1464,50-61(1991)

Localization of hot spots in silicon devices with a laser scanning microscope (Bergner, Harald; Krause, A.; Stamm, Uwe)V1361,723-731(1991)

Manipulation of single-DNA molecules and measurements of their elastic properties under an optical microscope (Bustamante, Carlos; Finzi, Laura; Sebring, Page E.; Smith, Steven B.)V1435,179-187(1991)

Mapping crystal defects with a digital scanning ultramicroscope (Springer, John M.; Silberman, Enrique; Kroes, Roger L.; Reiss, Don)V1557,192-196(1991)

Measurement of light scattering from cells using an inverted infrared optical trap (Wright, William H.; Sonek, Gregory J.; Numajiri, Yasuyuki; Berns, Michael W.)V1427,279-287(1991)

Measurements of teeth using the Reflex Microscope (Teaford, Mark F.)V1380,33-44(1991)

Measuring for thickness distribution of recording layer of PLH (Zhang, Xiao-Chun; Guo, Lu Rong; Guo, Yongkang)V1461,93-96(1991)

Metrology issues associated with submicron linewidths (Phan, Khoi; Nistler, John L.; Singh, Bhanwar)V1464,424-437(1991)

Microlocalization of Photofrin in neoplastic lesions (Korbelik, Mladen; Krosl, Gorazd; Lam, Stephen; Chaplin, David J.; Palcic, Branko)V1426,172-179(1991)

Microscopic feature extraction from optical sections of contracting cardiac muscle cells recorded at high speed (Roos, Kenneth P.; Lake, David S.; Lubell, Bradford A.)V1428,159-168(1991)

Microscopic fluorescence spectroscopy and diagnosis (Schneckenburger, Herbert; Seidlitz, Harold K.; Wessels, Jurina; Strauss, Wolfgang; Rueck, Angelika C.)V1525,91-98(1991)

Monte Carlo modeling of secondary electron signals from heterogeneous specimens with nonplanar surfaces (Russ, John C.; Dudley, Bruce W.; Jones, Susan K.)V1464,10-21(1991)

Networking of an electron microscope laboratory internally and to the internet (Zmola, Carl; Kapp, Oscar H.)V1396,331-334(1991)

New developments in CCD imaging devices for low-level confocal light imaging (Masters, Barry R.)V1428,169-176(1991)

New optical approaches to the quantitative characterization of crystal growth, segregation, and defect formation (Carlson, D. J.; Wargo, Michael J.; Cao, X. Z.; Witt, August F.)V1557,140-146(1991)

Nomarski viewing system for an optical surface profiler (Bietry, Joseph R.; Auriemma, R. A.; Bristow, Thomas C.; Merritt, Edward)V1332,537-543(1991)

Nondestructive investigations of multilayer dielectrical coatings (Shapiro, Alexander G.; Yaminsky, Igor V.)V1362,834-843(1991)

Nonlinear approach to the 3-D reconstruction of microscopic objects (Wu, Xiangchen; Schwarzmann, Peter)V1450,278-285(1991)

Numerical simulation of thick-linewidth measurements by reflected light (Wojcik, Gregory L.; Mould, John; Monteverde, Robert J.; Prochazka, Jaroslav J.; Frank, John R.)V1464,187-203(1991)

Observation of living cells by x-ray microscopy with a laser-plasma x-ray source (Tomie, Toshihisa; Shimizu, Hazime; Majima, T.; Yamada, Mitsuo; Kanayama, Toshihiko; Yano, M.; Kondo, H.)V1552,254-263(1991)

Observation of stress voids and grain structure in laser-annealed aluminum using focused ion-beam microscopy (Pramanik, Dipankar; Jain, Vivek)V1596,132-138(1991)

Optical imaging of cortical activity in the living brain (Ratzlaff, Eugene H.; Grinvald, Amiran)V1439,88-94(1991)

Optical metrology for integrated circuit fabrication (Chim, Stanley S.; Kino, Gordon S.)V1464,138-144(1991)

Optical profilometry using spatial heterodyne interferometric methods (Frankowski, Gottfried)V1500,114-123(1991)

Optical surface microtopography using phase-shifting Nomarski microscope (Shimada, Wataru; Sato, Tadamitu; Yatagai, Toyohiko)V1332,525-529(1991)

Pattern recognition approach to trench bottom-width measurement (Chou, Ching-Hua; Berman, John L.; Chim, Stanley S.; Corle, Timothy R.; Xiao, Guoqing; Kino, Gordon S.)V1464,145-154(1991)

Photoacoustic microscopy by photodeformation applied to the determination of thermal diffusivity (Balageas, Daniel L.; Boscher, Daniel M.; Deom, Alain A.; Enguehard, Francis; Noirot, Laurence)V1467,278-289(1991)

Photon scanning tunneling microscopy (Goudonnet, Jean-Pierre; Salomon, L.; de Fornel, F.; Chabrier, G.; Warmack, R. J.; Ferrell, Trinidad L.)V1400,116-123(1991)

Progress on the subangstrom field emission scanning transmission electron microscope (Ruan, Shengyang; Kapp, Oscar H.)V1396,298-310(1991)

Quantitative microscope for image cytometry (Jaggi, Bruno; Poon, Steven S.; Pontifex, Brian; Fengler, John J.; Marquis, Jacques; Palcic, Branko)V1448,89-97(1991)

Quasi-elastic light scattering spectroscopy of single biological cells under a microscope (Tanaka, Toyoichi; Nishio, Izumi; Peetermans, Joyce; Gorti, Sridhar)V1403,280-287(1991)

Real-time holographic microscope with nonlinear optics (Goldfain, Ervin)V1527,199-209(1991)

Reduced defocus degradation in a system for high-speed three-dimensional digital microscopy (Quesenberry, Laura A.; Morris, V. A.; Neering, Ian R.; Taylor, Stuart R.)V1428,177-190(1991)

Scanning exciton microscopy and single-molecule resolution and detection (Kopelman, Raoul; Tan, Weihong; Lewis, Aaron; Lieberman, Klony)V1435,96-101(1991)

Scanning optical microscopy: a powerful tool in optical recording (Coombs, James H.; Holtslag, A. H.)V1499,6-20(1991)

Semiautomated detection and measurement of glomerular basement membrane from electron micrographs (Ong, Sim-Heng; Giam, S. T.; Jayasooriah, Mr.; Sinniah, R.)V1445,564-573(1991)

Soft x-ray spectro-microscope (Campuzano, Juan C.; Jennings, G.; Beaulaigue, L.; Rodricks, Brian G.; Brizard, Christine M.)V1345,245-254(1991)

Statistical approach to optimizing advanced low-voltage SEM operation (Apostolakis, Peter J.)V1464,406-412(1991)

Surface nitride synthesis by multipulse excimer laser irradiation (D'Anna, Emilia; Leggieri, Gilberto; Luches, Armando; Martino, M.; Perrone, A.; Majni, G.; Mengucci, P.; Drigo, A. V.; Mihailescu, Ion N.)V1503,256-268(1991)

Three-dimensional confocal microscopy of the living cornea and ocular lens (Masters, Barry R.)V1450,286-294(1991)

Three-dimensional image processing method to compensate for depth-dependent light attenuation in images from a confocal microscope (Aslund, Nils R.; Liljeborg, Anders; Oldmixon, E. H.; Ulfsparre, M.)V1450,329-337(1991)

Threshold of structural transition in nematic drops with normal boundary conditions in AC electric field (Bodnar, Vladimir G.; Koval'chuk, Alexandr V.; Lavrentovich, Oleg D.; Pergamenshchik, V. M.; Sergan, V. V.)V1455,61-72(1991)

Transmission electron microscopy, photoluminescence, and capacitance spectroscopy on GaAs/Si grown by metal organic chemical vapor deposition (Bremond, Georges E.; Said, Hicham; Guillot, Gerard; Meddeb, Jaafar; Pitaval, M.; Draidia, Nasser; Azoulay, Rozette)V1361,732-743(1991)

Two-photon excitation in laser scanning fluorescence microscopy (Strickler, James H.; Webb, Watt W.)V1398,107-118(1991)

Use of a cooled CCD camera for confocal light microscopy (Masters, Barry R.)V1448,98-105(1991)

Using the Atomic Force Microscope to measure submicron dimensions of integrated circuit devices and processes (Rodgers, Mark R.; Monahan, Kevin M.)V1464,358-366(1991)

Verification of tracking servo signal simulation from scanning tunneling microscope surface profiles (Karis, Thomas E.; Best, Margaret E.; Logan, John A.; Lyerla, James R.; Lynch, Robert T.; McCormack, R. P.)V1499,366-376(1991)

Visualization of 3-D phase structure in confocal and conventional microscopy (Cogswell, Carol J.; Sheppard, Colin J.)V1450,323-328(1991)

Wavefront dislocations and phase object registering inside the airy disk (Tychinsky, Vladimir P.; Tavrov, Alexander V.)V1500,207-210(1991)

X-ray projection microscopy and cone-beam microtomography (Wang, Ge; Lin, T. H.; Cheng, Ping-Chin; Shinozaki, D. M.; Newberry, S. P.)V1398,180-190(1991)

Microwaves

Accurate and efficient simulation of MMIC layouts (Wu, Doris I.; Chang, David C.)V1475,140-150(1991)

Accurate design of multiport low-noise MMICs up to 20 GHz (Willems, David; Bahl, I.; Griffin, Edward)V1475,55-61(1991)

A/D conversion of microwave signals using a hybrid optical/electronic technique (Bell, John A.; Hamilton, Michael C.; Leep, David A.; Taylor, Henry F.; Lee, Y.-H.)V1476,326-329(1991)

Advances in power MMIC amplifier technology in space communications (Tserng, Hua Q.; Saunier, Paul)V1475,74-85(1991)

Analog optical processing of radio frequency signals (Sullivan, Daniel P.; Weber, Charles L.)V1476,234-245(1991)

Analysis and measurement of the external modulation of modelocked laser diodes (relative noise performance) (Lam, Benson C.; Kellner, Albert L.; Campion, David C.; Costa, Joannes M.; Yu, Paul K.)V1371,36-45(1991)

Application of Kaband MMIC technology for an Orbiter/ACTS communications experiment (Arndt, George D.; Fink, Patrick W.; Leopold, Louis; Bondyopadhyay, Probir; Shaw, Roland)V1475,231-242(1991)

Application of lighter-than-air platforms to law enforcement (Mataloni, Raymond J.)V1479,306-315(1991)

Asymmetric superlattices for microwave detection (Syme, Richard T.; Kelly, Michael J.; Condie, Angus; Dale, Ian)V1362,467-476(1991)

Bragg reflectors: tapered and untapered (Chong, Chae K.; Razeghi, M. M.; McDermott, David B.; Luhmann, Neville C.; Thumm, M.; Pretterebner, Julius)V1407,226-233(1991)

Broadband microwave and millimeter-wave EOMs with ultraflat frequency response (Pan, J. J.; Li, Yi Q.)V1476,22-31(1991)

Characteristics of thin-film-type Josephson junctions using Bi2Sr2CaCu2Ox/Bi2Sr2CuOy/Bi2Sr2CaCu2Oz structure (Mizuno, Koichi; Higashino, Hidetaka; Setsune, Kentaro; Wasa, Kiyotaka)V1477,197-204(1991)

Coherent optical modulation for antenna remoting (Fitzmartin, Daniel J.; Gels, Robert G.; Balboni, Edmund J.)V1476,56-62(1991)

Compact time-delay shifters that are process insensitive (Lesko, Camille; Hill, William A.; Dietrich, Fred; Nelson, William)V1475,330-339(1991)

Comparative study for bolometric and nonbolometric switching elements for microwave phase shifters (Tabib-Azar, Massood; Bhasin, Kul B.; Romanofsky, Robert R.)V1477,85-94(1991)

Comparison of alternative modulation techniques for microwave optical links (Kasemset, Dumrong; Ackerman, Edward I.; Wanuga, Stephen; Herczfeld, Peter R.; Daryoush, Afshin S.)V1371,104-114(1991)

Comparison of negative and positive polarity reflex diode microwave source (Litz, Marc S.; Huttlin, George A.; Lazard, Carl J.; Golden, Jeffry; Pereira, Nino R.; Hahn, Terry D.)V1407,159-166(1991)

Comparison of relativistic magnetron oscillator models for phase-locking studies (Chen, Shien C.)V1407,100-104(1991)

Coplanar waveguide InP-based HEMT MMICs for microwave and millimeter wave applications (Chou, Chia-Shing; Litvin, K.; Larson, Larry E.; Rosenbaum, Steven E.; Nguyen, Loi D.; Mishra, Umesh K.; Lui, M.; Thompson, M.; Ngo, Catherine M.; Melendes, M.)V1475,151-156(1991)

Coplanar waveguide microwave filter of YBa2Cu3O7 (Chew, Wilbert; Riley, A. L.; Rascoe, Daniel L.; Hunt, Brian D.; Foote, Marc C.; Cooley, Thomas W.; Bajuk, Louis J.)V1477,95-100(1991)

Cyclotron resonance and photoluminescence in GaAs in a microwave field (Ashkinadze, B. M.; Bel'Kov, V. V.; Krasinskaya, A. G.)V1361,866-873(1991)

Design and construction of a wideband efficient electro-optic modulator (Harris, Neville W.; Sobolewski, J. M.; Summers, Charles L.; Eng, Richard S.)V1416,59-69(1991)

Design aspects and comparison between high-Tc superconducting coplanar waveguide and microstrip line (Kong, Keon-Shik; Bhasin, Kul B.; Itoh, Tatsuo)V1477,57-65(1991)

Design of a high-power cross-field amplifier at X-band with an internally coupled waveguide (Eppley, Kenneth; Ko, Kwok)V1407,249-259(1991)

Design of frozen-wave and injected-wave microwave generators using optically controlled switches (Nunnally, William C.)V1378,10-21(1991)

Detection of stratospheric ozone trends by ground-based microwave observations (Connor, Brian J.; Parrish, Alan; Tsou, Jung-Jung)V1491,218-230(1991)

Development of the Aurora high-power microwave source (Huttlin, George A.; Conrad, D. B.; Gavnoudias, S.; Judy, Daniel C.; Lazard, Carl J.; Litz, Marc S.; Pereira, Nino R.; Weidenheimer, Douglas M.)V1407,147-158(1991)

Efficiency increase in a traveling wave tube by tapering the phase velocity of the wave (Schachter, Levi; Nation, John A.)V1407,44-56(1991)

Eight-Gb/s QPSK-SCM over a coherent detection optical link (Hill, Paul M.; Olshansky, Robert)V1579,210-220(1991)

Epitaxial Tl2Ba2CaCu2O8 thin films on LaAlO3 and their microwave device properties (Negrete, George V.; Hammond, Robert B.)V1477,36-44(1991)

Excimer laser performance under various microwave excitation conditions (Klingenberg, Hans H.; Gekat, Frank)V1503,140-145(1991)

Excitation of an excimer laser with microwave resonator (Huenermann, Lucia; Meyer, Rudolph; Richter, Franz; Schnase, Alexander)V1503,134-139(1991)

Experimental study of microwave attenuation of ITO-dielectric recombination film (Huang, Guang L.; Zhang, Jun; Peng, Chuan C.)V1519,179-182(1991)

Experiments on the beam breakup instability in long-pulse electron beam transport through cavity systems (Menge, Peter R.; Bosch, Robert A.; Gilgenbach, Ronald M.; Choi, J. J.; Ching, Hong; Spencer, Thomas A.)V1407,578-588(1991)

Frequency-selective devices using a composite multilayer design (Ma, Yushieh; Varadan, Vijay K.; Varadan, Vasundara V.)V1558,132-137(1991)

Frequency up-conversion for the reflectionless propagation of a high-power microwave pulse in a self-generated plasma (Kuo, Spencer P.; Ren, A.; Zhang, Y. S.)V1407,272-280(1991)

GaAs/GaAlAs integrated optoelectronic transmitter for microwave applications (Yap, Daniel; Narayanan, Authi A.; Rosenbaum, Steven E.; Chou, Chia-Shing; Hooper, W. W.; Quen, R. W.; Walden, Robert H.)V1418,471-476(1991)

Generation and sampling of high-repetition-rate/high-frequency electrical waveforms in microstrip circuits by picosecond optoelectronic technique (Lee, Chi H.)V1390,377-387(1991)

Ground-based microwave radiometry of ozone (Kaempfer, Niklaus A.; Bodenmann, P.; Peter, Reto)V1491,314-322(1991)

Ground-based microwave remote sensing of water vapor in the mesosphere and stratosphere (Croskey, Charles L.; Olivero, John J.; Martone, Joseph P.)V1491,323-334(1991)

Growth of high-Tc superconducting thin films for microwave applications (Wu, Xin D.; Foltyn, Stephen R.; Muenchausen, Ross E.; Dye, Robert C.; Cooke, D. W.; Rollett, A. D.; Garcia, A. R.; Nogar, Nicholas S.; Pique, A.; Edwards, R.)V1477,8-14(1991)

High-density circuit approach for low-cost MMIC circuits (Bauhahn, Paul E.; Geddes, John J.)V1475,122-128(1991)

High-dynamic-range mixer using novel balun structure (Bharj, Sarjit; Taylor, Gordon C.; Denlinger, E. J.; Milgazo, H.)V1475,340-349(1991)

High-efficiency dual-band power amplifier for radar applications (Masliah, Denis A.; Cole, Brad; Platzker, Aryeh; Schindler, Manfred)V1475,113-120(1991)

High-gain high-efficiency TWT (traveling wave tube) amplifiers (Nation, John A.; Ivers, J. D.; Kerslick, G.; Shiffler, Donald; Schachter, Levi)V1407,32-43(1991)

High-temperature superconducting Josephson mixers from deliberate grain boundaries in Tl2CaBa2Cu2O8 (Bourne, Lincoln C.; Cardona, A. H.; James, Tim W.; Fleming, J. S.; Forse, R. W.; Hammond, Robert B.; Hong, J. P.; Kim, T. W.; Fetterman, Harold R.)V1477,205-208(1991)

High-temperature superconducting superconductor/normal metal/superconducting devices (Foote, Marc C.; Hunt, Brian D.; Bajuk, Louis J.)V1477,192-196(1991)

High-temperature superconductive microwave technology for space applications (Leonard, Regis F.; Connolly, Denis J.; Bhasin, Kul B.; Warner, Joseph D.; Alterovitz, Samuel A.)V1394,114-125(1991)

Hybrid optical transmitter for microwave communication (Costa, Joannes M.; Lam, Benson C.; Kellner, Albert L.; Campion, David C.; Yu, Paul K.)V1476,74-80(1991)

Induction linac-driven relativistic klystron and cyclotron autoresonance maser experiments (Goodman, Daniel L.; Birx, Daniel L.; Danly, Bruce G.)V1407,217-225(1991)

Infrared/microwave correlation measurements (Norgard, John D.; Metzger, Don W.; Cleary, John C.; Seifert, Michael)V1540,699-708(1991)

InGaAs HEMT MMIC low-noise amplifier and doublers for EHF SATCOM ground terminals (Chow, P. D.; Lester, J.; Huang, P.; Jones, William L.)V1475,42-47(1991)

Injection locking of a long-pulse relativistic magnetron (Chen, Shien C.; Bekefi, George; Temkin, Richard J.)V1407,67-73(1991)

Insertion of emerging GaAs HBT technology in military and communication system applications (McAdam, Bridget A.; Sharma, Arvind K.; Allen, B.; Kintis, M.)V1475,267-274(1991)

Integrated circuit active antenna elements for monolithic implementation (Chang, Kai)V1475,164-174(1991)

Intense Microwave and Particle Beams II (Brandt, Howard E., ed.)V1407(1991)

Investigation of microwave-pumped excimer and rare-gas laser transitions (Klingenberg, Hans H.; Gekat, Frank)V1412,103-114(1991)

Kinetic stability analysis of the extraordinary mode perturbations in a cylindrical magnetron (Uhm, Han S.; Chen, H. C.; Stark, Robert A.)V1407,113-127(1991)

Ku-band radiation in the UNM backward-wave oscillator experiment (Schamiloglu, Edl; Gahl, John M.; McCarthy, G.)V1407,242-248(1991)

Laser link performance improvements with wideband microwave impedance matching (Baldwin, David L.; Sokolov, Vladimir; Bauhahn, Paul E.)V1476,46-55(1991)

Local hyperthermia for the treatment of diseases of the prostate (Servadio, Ciro)V1421,6-11(1991)

Localized waves in complex environments (Ziolkowski, Richard W.; Besieris, Ioannis M.; Shaarawi, Amr M.)V1407,387-397(1991)

Localized wave transmission physics and engineering (Ziolkowski, Richard W.)V1407,375-386(1991)

Long-pulse modulator-driven cyclotron autoresonance maser and free-electron laser experiments at MIT (Danly, Bruce G.; Hartemann, Frederic V.; Chu, T. S.; Menninger, W. L.; Papavaritis, P.; Pendergast, K. D.; Temkin, Richard J.)V1407,192-201(1991)

Low-noise high-yield octave-band feedback amplifiers to 20 GHz (Minot, Katcha; Cochrane, Mike; Nelson, Bradford; Jones, William L.; Streit, Dwight C.; Liu, Po-Hsin P.)V1475,309-313(1991)

Low-noise MMIC performance in Kaband using ion implantation technology (Mondal, J. P.; Contolatis, T.; Geddes, John J.; Swirhun, S.; Sokolov, Vladimir)V1475,314-318(1991)

Magnetically tapered CARM (cyclotron autoresonance maser) for high power (Wang, Qinsong; McDermott, David B.; Luhmann, Neville C.)V1407,209-216(1991)

Magnetic field effects on plasma-filled backward-wave oscillators (Lin, Anthony T.)V1407,234-241(1991)

Microwave characterization of high-Tc superconducting thin films for simulation and realization of planar microelectronic circuits (Carru, J. C.; Mehri, F.; Chauvel, D.; Crosnier, Y.)V1512,232-239(1991)

Microwave control using a high-gain bias-free optoelectronic switch (Freeman, J. L.; Ray, Sankar; West, David L.; Thompson, Alan G.; LaGasse, M. J.)V1476,320-325(1991)

Microwave fiber optic link with DFB lasers (Huff, David B.; Blauvelt, Henry A.)V1371,244-249(1991)

Microwave fiber optic RF/IF link (Pan, J. J.)V1371,195-204(1991)

Microwave high-dynamic-range EO modulators (Pan, J. J.; Garafalo, David A.)V1371,21-35(1991)

Microwave limb sounder experiments for UARS and EOS (Waters, Joe W.)V1491,104-109(1991)

Microwave monolithic integrated circuits for high-data-rate satellite communications (Turner, Elbert L.; Hill, William A.)V1475,248-256(1991)

Microwave-optical study of an As(III) derivative of Eco RI methylase (Maki, August H.; Tsao, Desiree H.)V1432,119-128(1991)

Microwave properties of YB2Cu3O7-x thin films characterized by an open resonator (Zhou, Shi P.; Wu, Ke Q.; Jabbar, A.; Bao, Jia S.; Lou, Wei G.; Ding, Ai L.; Wang, Shu H.)V1519,793-799(1991)

Microwave warming of biological tissue and its control by IR fiber thermometry (Drizlikh, S.; Zur, A.; Moser, Frank; Katzir, Abraham)V1420,53-62(1991)

Mixed application MMIC technologies: progress in combining RF, digital, and photonic circuits (Swirhun, S.; Bendett, Mark P.; Sokolov, Vladimir; Bauhahn, Paul E.; Sullivan, Charles T.; Mactaggart, R.; Mukherjee, Sayan D.; Hibbs-Brenner, Mary K.; Mondal, J. P.)V1475,223-230(1991)

MMIC: a key technology for future communications satellite antennas (Sorbello, Robert M.; Zaghloul, A. I.; Gupta, R. K.; Geller, B. D.; Assal, F. T.; Potukuchi, J. R.)V1475,175-183(1991)

MMIC compatible photodetector design and characterization (Dallabetta, Kyle A.; de La Chapelle, Michael; Lawrence, Robert C.)V1371,116-127(1991)

MMICs for airborne phased arrays (Scalzi, Gary J.; Turtle, John P.; Carr, Paul H.)V1475,2-9(1991)

Models of driven and mutually coupled relativistic magnetrons with nonlinear frequency-shift and growth-saturation effects (Johnston, George L.; Chen, Shien C.; Davidson, Ronald C.; Bekefi, George)V1407,92-99(1991)

Monitoring of tissue temperature during microwave hyperthermia utilizing a fiber optic liquid crystalline microsensor (Domanski, Andrzej W.; Kostrzewa, Stanislaw; Hliniak, Andrzej)V1420,72-80(1991)

Monolithic microwave integrated circuit activities in ESA-ESTEC (Gatti, Giuliano)V1475,10-24(1991)

Monolithic Microwave Integrated Circuits for Sensors, Radar, and Communications Systems (Leonard, Regis F.; Bhasin, Kul B., eds.)V1475(1991)

Monolithic phased arrays: recent advances (Kinzel, Joseph A.)V1475,158-163(1991)

Negative energy cyclotron resonance maser (Lednum, Eugene E.; McDermott, David B.; Lin, Anthony T.; Luhmann, Neville C.)V1407,202-208(1991)

Nonlinear theory of a three-level laser with microwave coupling: numerical calculation (Fearn, Heidi; Lamb, Willis E.; Scully, Marlan O.)V1497,283-290(1991)

Normally incident plane waves on a chiral slab with linear property variations (Lakhtakia, Akhlesh; Varadan, Vijay K.; Varadan, Vasundara V.)V1558,120-126(1991)

Novel selective-plated heatsink, key to compact 2-watt MMIC amplifier (Taylor, Gordon C.; Bechtle, Daniel W.; Jozwiak, Phillip C.; Liu, Shing G.; Camisa, Raymond L.)V1475,103-112(1991)

Operational characteristics of a phase-locked module of relativistic magnetrons (Levine, Jerrold S.; Benford, James N.; Courtney, R.; Harteneck, Bruce D.)V1407,74-82(1991)

Operation of an L-band relativistic magnetron at 100 hz (Smith, Richard R.; Benford, James N.; Cooksey, N. J.; Aiello, Norm; Levine, Jerrold S.; Harteneck, Bruce D.)V1407,83-91(1991)

Optical cross-modulation method for diagnostic of powerful microwave radiation (Kozar, A. V.; Krupenko, S. A.)V1476,305-312(1991)

Optically activated GaAs MMIC switch for microwave and millimeter wave applications (Paolella, Arthur; Madjar, Asher; Herczfeld, Peter R.; Sturzebecher, Dana)V1378,195-202(1991)

Optical techniques for microwave monolithic circuit characterization (Hung, Hing-Loi A.; Li, Ming-Guang; Lee, Chi H.)V1476,276-281(1991)

Optical Technology for Microwave Applications V (Yao, Shi-Kay, ed.)V1476(1991)

Packaging technology for GaAs MMIC (monolithic microwave integrated circuits) modules (Tomimuro, Hisashi)V1390,214-222(1991)

Parametric simulation studies and injection phase locking of relativistic magnetrons (Chen, Chiping; Chan, Hei-Wai; Davidson, Ronald C.)V1407,105-112(1991)

Performance characteristics of Y-Ba-Cu-O microwave superconducting detectors (Shewchun, John; Marsh, P. F.)V1477,115-138(1991)

Performance enhancement in future communications satellites with MMIC technology insertion (Hung, Hing-Loi A.; Mahle, Christoph E.)V1475,212-222(1991)

Performance of stripline resonators using sputtered YBCO films (Mallory, Derek S.; Kadin, Alan M.; Ballentine, Paul H.)V1477,66-76(1991)

Phase-locking simulation of dual magnetrons (Chen, H. C.; Stark, Robert A.; Uhm, Han S.)V1407,139-146(1991)

Phase-shifter technology assessment: prospects and applications (Sokolov, Vladimir)V1475,288-302(1991)

Photoconductive switching for high-power microwave generation (Pocha, Michael D.; Hofer, Wayne W.)V1378,2-9(1991)

Photon-induced charge separation in molecular systems studied by time-resolved microwave conductivity: molecular optoelectric switches (Warman, John M.; Jonker, Stephan A.; de Haas, Matthijs P.; Verhoeven, Jan W.; Paddon-Row, Michael N.)V1559,159-170(1991)

Planar-grating klystron experiment (Xu, Yian-sun; Jackson, Jonathan A.; Price, Edwin P.; Walsh, John E.)V1407,648-652(1991)

Preliminary study of the admittance diagram as a useful tool in the design of stripline components at microwave frequencies (Franck, Charmaine C.; Franck, Jerome B.)V1527,277-290(1991)

Progress toward steady-state high-efficiency vircators (Poulsen, Peter; Pincosy, Phillip A.; Morrison, Jasper J.)V1407,172-182(1991)

Pulsed microwave excitation of rare-gas halide mixtures (Gekat, Frank; Klingenberg, Hans H.)V1411,47-54(1991)

Range of modalities for prostate therapy (Watson, Graham M.)V1421,2-5(1991)

Recirculating fiber optical RF-memory loop in countermeasure systems (Even-Or, Baruch; Lipsky, S.; Markowitz, Raymond; Herczfeld, Peter R.; Daryoush, Afshin S.; Saedi, Reza)V1371,161-169(1991)

Relativistic klystron amplifier I: high-power operation (Friedman, Moshe; Serlin, Victor; Lau, Yue Y.; Krall, Jonathan)V1407,2-7(1991)

Relativistic klystron amplifier II: high-frequency operation (Serlin, Victor; Friedman, Moshe; Lau, Yue Y.; Krall, Jonathan)V1407,8-12(1991)

Relativistic klystron amplifier III: dynamical limiting currents, nonlinear beam loading, and conversion efficiency (Colombant, Denis G.; Lau, Yue Y.; Friedman, Moshe; Krall, Jonathan; Serlin, Victor)V1407,13-22(1991)

Relativistic klystron amplifier IV: simulation studies of a coaxial-geometry RKA (Krall, Jonathan; Friedman, Moshe; Lau, Yue Y.; Serlin, Victor)V1407,23-31(1991)

Relativistic klystron research for future linear colliders (Westenskow, Glen A.; Houck, Timothy L.; Ryne, Robert D.)V1407,496-501(1991)

Remote sensing of precipitation structures using combined microwave radar and radiometric techniques (Vivekanandan, J.; Turk, F. J.; Bringi, Viswanathan N.)V1558,324-338(1991)

Ring resonators for microwave optoelectronics (Gopalakrishnan, G. K.; Fairchild, B. W.; Yeh, C. L.; Park, C. S.; Chang, Kai; Weichold, Mark H.; Taylor, Henry F.)V1476,270-275(1991)

Rugged 20-km fiber optic link for 2-18-GHz communications (Buckley, Robert H.; Lyons, E. R.; Goga, George)V1371,212-222(1991)

Scattering of waves from dense discrete random media: theory and applications in remote sensing (Tsang, Leung; Ding, Kung-Hau; Kong, Jin A.; Winebrenner, Dale P.)V1558,260-268(1991)

Silicon calorimeter for high-power microwave measurements (Lazard, Carl J.; Pereira, Nino R.; Huttlin, George A.; Litz, Marc S.)V1407,167-171(1991)

Simulation of intense microwave pulse propagation in air breakdown environment (Kuo, Spencer P.; Zhang, Y. S.)V1407,260-271(1991)

Simulation studies of the relativistic magnetron (Stark, Robert A.; Chen, H. C.; Uhm, Han S.)V1407,128-138(1991)

Status of high-temperature superconducting analog devices (Talisa, Salvador H.)V1477,78-83(1991)

Submillimeter receiver components using superconducting tunnel junctions (Wengler, Michael J.; Pance, A.; Liu, B.; Dubash, N.; Pance, Gordana; Miller, Ronald E.)V1477,209-220(1991)

Superconducting devices and system insertion (Rachlin, Adam; Babbitt, Richard; Lenzing, Erik; Cadotte, Roland)V1477,101-114(1991)

Superconductivity Applications for Infrared and Microwave Devices II (Heinen, Vernon O.; Bhasin, Kul B., eds.)V1477(1991)

Theoretical modeling of chiral composites (Apparao, R. T.; Varadan, Vasundara V.; Varadan, Vijay K.)V1558,2-13(1991)

Theory and simulation of the HARmonic amPlifier Free-Electron Laser (HARP/FEL) (Gregoire, Daniel J.; Harvey, Robin J.; Levush, Baruch)V1552,118-126(1991)

Transurethral microwave heating without urethral cooling: theory and experimental results (Petrovich, Zbigniew; Astrahan, Melvin; Baert, Luc)V1421,14-17(1991)

Transurethral microwave hyperthermia for benign prostatic hyperplasia: the Leuven clinical experience (Baert, Luc; Ameye, Filip; Willemen, Patrick; Petrovich, Zbigniew)V1421,18-29(1991)

True-time-delay steering of dual-band phased-array antenna using laser-switched optical beam forming networks (Ng, Willie W.; Tangonan, Gregory L.; Walston, Andrew; Newberg, Irwin L.; Lee, Jar J.; Bernstein, Norman P.)V1371,205-211(1991)

Twin traveling wave tube amplifiers driven by a single backward-wave oscillator (Butler, Jennifer M.; Wharton, Charles B.)V1407,57-66(1991)

Ultra-high-frequency InP-based HEMTs for millimeter wave applications (Greiling, Paul T.; Nguyen, Loi D.)V1475,34-41(1991)

Ultralinear electro-optic modulators for microwave fiber-optic communications (Pan, J. J.; Li, Wei Z.; Li, Yi Q.)V1476,32-43(1991)

Variable-gain MMIC module for space application (Palena, Patricia)V1475,91-102(1991)

Variable time delay for RF/microwave signal processing (Toughlian, Edward N.; Zmuda, Henry)V1476,107-121(1991)

Millimeter waves

Accurate and efficient simulation of MMIC layouts (Wu, Doris I.; Chang, David C.)V1475,140-150(1991)

Advances in power MMIC amplifier technology in space communications (Tserng, Hua Q.; Saunier, Paul)V1475,74-85(1991)

Application of super-resolution techniques to passive millimeter-wave images (Gleed, David G.; Lettington, Alan H.)V1567,65-72(1991)

Broadband microwave and millimeter-wave EOMs with ultraflat frequency response (Pan, J. J.; Li, Yi Q.)V1476,22-31(1991)

Comparison of MESFET and HEMT MMIC technologies using a compact Kaband voltage-controlled oscillator (Swirhun, S.; Geddes, John J.; Sokolov, Vladimir; Bosch, D.; Gawronski, M. J.; Anholt, R.)V1475,303-308(1991)

Coplanar SIMMWIC circuits (Luy, Johann-Friedrich; Strohm, Karl M.; Buechler, J.)V1475,129-139(1991)

Coplanar waveguide InP-based HEMT MMICs for microwave and millimeter wave applications (Chou, Chia-Shing; Litvin, K.; Larson, Larry E.; Rosenbaum, Steven E.; Nguyen, Loi D.; Mishra, Umesh K.; Lui, M.; Thompson, M.; Ngo, Catherine M.; Melendes, M.)V1475,151-156(1991)

Effects of radiation on millimeter wave monolithic integrated circuits (Meulenberg, A.; Hung, Hing-Loi A.; Singer, J. L.; Anderson, Wallace T.)V1475,280-285(1991)

Experimental verification of grating theory for surface-emitting structures (Ayekavadi, Raj; Yeh, C. S.; Butler, Jerome K.; Evans, Gary A.; Stabile, Paul J.; Rosen, Arye)V1418,74-85(1991)

Fiber optic link for millimeter-wave communication satellites (Polifko, David M.; Malone, Steven A.; Daryoush, Afshin S.; Kunath, Richard R.)V1476,91-99(1991)

High-dynamic-range mixer using novel balun structure (Bharj, Sarjit; Taylor, Gordon C.; Denlinger, E. J.; Milgazo, H.)V1475,340-349(1991)

High-efficiency Kaband monolithic pseudomorphic HEMT amplifier (Saunier, Paul; Tserng, Hua Q.; Kao, Yung C.)V1475,86-90(1991)

High-temperature superconductivity space experiment: passive millimeter wave devices (Nisenoff, Martin; Gubser, Don U.; Wolf, Stuart A.; Ritter, J. C.; Price, George E.)V1394,104-113(1991)

Insertion of emerging GaAs HBT technology in military and communication system applications (McAdam, Bridget A.; Sharma, Arvind K.; Allen, B.; Kintis, M.)V1475,267-274(1991)

IR/MMW fusion ATR (automatic target recognition) (Thiede, Edwin C.)VIS07,24-35(1991)

Ka-band MMIC array feed development for deep space applications (Cooley, Thomas W.; Riley, A. L.; Crist, Richard A.; Sukamto, Lin; Jamnejad, V.; Rascoe, Daniel L.)V1475,243-247(1991)

Low-noise MMIC performance in Kaband using ion implantation technology (Mondal, J. P.; Contolatis, T.; Geddes, John J.; Swirhun, S.; Sokolov, Vladimir)V1475,314-318(1991)

Millimeter-wave and optoelectronic applications of heterostructure integrated circuits (Pavlidis, Dimitris)V1362,450-466(1991)

Millimeter wave monolithic antenna and receiver arrays for space-based applications (Rebeiz, Gabriel M.; Ulaby, Fawwaz T.)V1475,199-203(1991)

Millimeter wave pseudomorphic HEMT MMIC phased-array components for space communications (Lan, Guey-Liu; Pao, Cheng K.; Wu, Chan-Shin; Mandolia, G.; Hu, M.; Yuan, Steve; Leonard, Regis F.)V1475,184-192(1991)

Millimeter-wave signal generation and control using optical heterodyne techniques and electro-optic devices (Thaniyavarn, Suwat; Abbas, Gregory L.; Dougherty, William A.)V1371,250-251(1991)

Monolithic GaAs integrated circuit millimeter wave imaging sensors (Weinreb, Sander)V1475,25-31(1991)

Monolithic Microwave Integrated Circuits for Sensors, Radar, and Communications Systems (Leonard, Regis F.; Bhasin, Kul B., eds.)V1475(1991)

Monolithic phased arrays: recent advances (Kinzel, Joseph A.)V1475,158-163(1991)

Ninety-four GHz InAlAs/InGaAs/InP HEMT low-noise down-converter (Chow, P. D.; Tan, K.; Streit, Dwight C.; Garske, D.; Liu, Po-Hsin P.; Yen, Huan-chun)V1475,48-54(1991)

Optical figure testing of prototype mirrors for JPL's precision segmented-reflector program (Hochberg, Eric B.)V1542,511-522(1991)

Optically activated GaAs MMIC switch for microwave and millimeter wave applications (Paolella, Arthur; Madjar, Asher; Herczfeld, Peter R.; Sturzebecher, Dana)V1378,195-202(1991)

Performance enhancement in future communications satellites with MMIC technology insertion (Hung, Hing-Loi A.; Mahle, Christoph E.)V1475,212-222(1991)

Phase-shifter technology assessment: prospects and applications (Sokolov, Vladimir)V1475,288-302(1991)

Physical foundations of high-speed thermomagnetic tuning of spatial-temporal structure of far-IR and submillimeter beams (Shvartsburg, Alexandre B.)V1488,28-35(1991)

Quantum-limited quasiparticle mixers at 100 GHz (Mears, Carl A.; Hu, Qing; Richards, Paul L.; Worsham, A.; Prober, Daniel E.; Raisanen, Antti)V1477,221-233(1991)

SADARM status report (DiNardo, Anthony J.)V1479,228-248(1991)

Space-based millimeter wave debris tracking radar (Chang, Kai; Pollock, Michael A.; Skrehot, Michael K.)V1475,257-266(1991)

Synchrotron radiation lasers (Hirshfield, Jay L.; Park, Gun-Sik)V1552,138-146(1991)

System-level integrated circuit development for phased-array antenna applications (Shalkhauser, Kurt A.; Raquet, Charles A.)V1475,204-209(1991)

Ultra-high-frequency GaInAs/InP devices and circuits for millimeter wave application (Greiling, Paul T.)V1361,47-58(1991)

Ultra-high-frequency InP-based HEMTs for millimeter wave applications (Greiling, Paul T.; Nguyen, Loi D.)V1475,34-41(1991)

Mirrors—see also optical design; reflectance; surfaces; telescopes

Absolute phasing of segmented mirrors using the polarization phase sensor (Klumpe, Herbert W.; Lajza-Rooks, Barbara A.; Blum, James D.)V1398,95-106(1991); V1532,230-240(1991)

Absorbing materials in multilayer mirrors (Grebenshikov, Sergey V.)V1500,189-193(1991)

Absorption calorimetry and laser-induced damage threshold measurements of antireflective-coated ZnSe and metal mirrors at 10.6 um (Rahe, Manfred; Oertel, E.; Reinhardt, L.; Ristau, Detleu; Welling, Herbert)V1441,113-126(1991)

Active and Adaptive Optical Systems (Ealey, Mark A., ed.)V1542(1991)

Adaptive optical transfer function modeling (Gaffard, Jean-Paul; Ledanois, Guy)V1542,34-45(1991)

Advanced technology development for high-fluence laser applications (Novaro, Marc; Andre, Michel L.)V1501,183-193(1991)

Aligning diamond-turned optics using visible-light interferometry (Figoski, John W.)V1354,540-546(1991)

Alignment of an aspheric mirror subsystem for an advanced infrared catadioptric system (Tingstad, James S.)V1527,194-198(1991)

Analysis of Optical Structures (O'Shea, Donald C., ed.)V1532(1991)

Analysis of thermal stability of fused optical structure (Powell, William R.)V1532,126-136(1991)

Analytic solutions and numerical results for the optical and radiative properties of V-trough concentrators (Fraidenraich, Naum)V1528,15-30(1991)

Apodized outcouplers for unstable resonators (Budzinski, Christel; Grunwald, Ruediger; Pinz, Ingo; Schaefer, Dieter; Schoennagel, Horst)V1500,264-274(1991)

Applications of diamond-turned null reflectors for generalized aspheric metrology (McCann, James T.)V1332,843-849(1991)

Aspheric testing using null mirrors (Murty, Mantravady V.; Kumar, Vas; von Handorf, Robert J.)V1332,107-114(1991)

Astigmatic unstable resonator with an intracavity deformable mirror (Neal, Daniel R.; McMillin, Pat; Michie, Robert B.)V1542,449-458(1991)

Atmospheric turbulence sensing for a multiconjugate adaptive optics system (Johnston, Dustin C.; Welsh, Byron M.)V1542,76-87(1991)

Beryllium and titanium cost-adjustment report (Owen, John; Ulph, Eric)V1485,128-137(1991)

Beryllium galvanometer mirrors (Weissman, Harold)V1485,13-19(1991)

Bonding skill between the fused-quartz mirror and the composite substrate (Jiang, Shibin; Wang, Huirong; Jiang, Yasi; Wang, Biao; Chen, Menda)V1535,143-147(1991)

Calculation of flux density produced by CPC reflectors on distant targets (Gordon, Jeff M.; Rabl, Ari)V1528,152-162(1991)

Chemical-vapor-deposited silicon and silicon carbide optical substrates for severe environments (Goela, Jitendra S.; Pickering, Michael A.; Taylor, Raymond L.)V1330,25-38(1991)

Compact, low-power precision beam-steering mirror (DeWeerd, Herman)V1454,207-214(1991)

Comparative study of carbon and boron carbide spacing layers inside soft x-ray mirrors (Boher, Pierre; Houdy, Philippe; Kaikati, P.; Barchewitz, Robert J.; Van Ijzendoorn, L. J.; Li, Zhigang; Smith, David J.; Joud, J. C.)V1345,165-179(1991)

Comparison of asymmetrical and symmetrical nonimaging reflectors for east-west circular cylindrical solar receivers (Mills, David R.; Monger, A. G.; Morrison, G. L.)V1528,44-55(1991)

Comparisons of deformable-mirror models and influence functions (Hiddleston, H. R.; Lyman, Dwight D.; Schafer, Eric L.)V1542,20-33(1991)

Computer-aided engineering, manufacturing, and testing of extremely fast steering mirrors (Hubert, Alexis; Hammond, Mark W.)V1532,249-260(1991)

Control of thermally induced porosity for the fabrication of beryllium optics (Moreen, Harry A.)V1485,54-61(1991)

Cryoscatter measurements of beryllium (Lippey, Barret; Krone-Schmidt, Wilfried)V1530,150-161(1991)

Deformable-mirror concept for adaptive optics in space (Kuo, Chin-Po)V1542,420-431(1991)

Deposition- controlled uniformity of multilayer mirrors (Jankowski, Alan F.; Makowiecki, Daniel M.; McKernan, M. A.; Foreman, R. J.; Patterson, R. G.)V1343,32-38(1991)

Design and analysis of a dither mirror control system (Kline-Schoder, Robert J.; Wright, Michael J.)V1489,189-200(1991)

Design and manufacture of an ultralightweight solid deployable reflector (Tremblay, Gary A.; Derby, Eddy A.)V1532,114-123(1991)

Design and optical performance of beryllium assessment mirrors (Thomas, Brigham B.; Maxey, L. C.; Miller, Arthur C.)V1485,20-30(1991)

Design and performance of a small two-axis high-bandwidth steering mirror (Loney, Gregory C.)V1454,198-206(1991)

Design of lightweight beryllium optics, factors effecting producibility, and cost of near-net-shape blanks (Clement, Thomas P.)V1485,31-38(1991)

Design of nonimaging lenses and lens-mirror combinations (Minano, Juan C.; Gonzalez, Juan C.)V1528,104-115(1991)

Design of reflective relay for soft x-ray lithography (Rodgers, John M.; Jewell, Tanya E.)V1354,330-336(1991)

Design survey of x-ray/XUV projection lithography systems (Shealy, David L.; Viswanathan, Vriddhachalam K.)V1343,229-240(1991)

Development and testing of lightweight composite reflector panels (Helms, Richard G.; Porter, Christopher C.; Kuo, Chin-Po; Tsuyuki, Glenn T.)V1532,64-80(1991)

Development of beryllium-mirror turning technology (Arnold, Jones B.)V1485,96-105(1991)

Diffraction-limited Nd:glass and alexandrite lasers using graded reflectivity mirror unstable resonators (Snell, Kevin J.; Duplain, Gaetan; Parent, Andre; Labranche, Bruno; Galarneau, Pierre)V1410,99-106(1991)

Dual ion-beam sputtering: a new coating technology for the fabrication of high-power CO2 laser mirrors (Daugy, Eric; Pointu, Bernard; Villela, Gerard; Vincent, Bernard)V1502,203-212(1991)

Durable, nonchanging, metal-dielectric and all-dielectric mirror coatings (Guenther, Karl H.; Balasubramanian, Kunjithapa; Hu, X. Q.)V1485,240-244(1991)

Dynamic model of deformable adaptive mirror (Glebova, Svetlana N.; Lavrov, Nikolaj A.)V1500,275-280(1991)

Effective surface PSD for bare hot-isostatic-pressed beryllium mirrors (Vernold, Cynthia L.; Harvey, James E.)V1530,144-149(1991)

Efficient, uniform transverse coupling of diode arrays to laser rods using nonimaging optics (Minnigh, Stephen W.; Knights, Mark G.; Avidor, Joel M.; Chicklis, Evan P.)V1528,129-134(1991)

Ellipsoidal mirrors as modular elements in the design of off-axis optical systems (Stavroudis, Orestes N.)V1354,627-646(1991)

Fabrication and characterization of Si-based soft x-ray mirrors (Schmiedeskamp, Bernt; Heidemann, B.; Kleineberg, Ulf; Kloidt, Andreas; Kuehne, Mikhael; Mueller, H.; Mueller, Peter; Nolting, Kerstin; Heinzmann, Ulrich)V1343,64-72(1991)

Fabrication of a fast, aspheric beryllium mirror (Paquin, Roger A.; Gardopee, George J.)V1485,39-45(1991)

Fabrication of a grazing-incidence telescope by grinding and polishing techniques on aluminum (Gallagher, Dennis J.; Cash, Webster C.; Green, James C.)V1343,155-161(1991)

Fabrication of cold light mirror of film projector by direct electron-beam-evaporated TiO2 and SiO2 starting materials in neutral oxygen atmosphere (Zhong, Di S.; Xu, Guang Z.; Liu, Wi)V1519,350-358(1991)

Fabrication of special-purpose optical components (Leuw, David H.)V1442,31-41(1991)

Finite element analysis enhancement of cryogenic testing (Thiem, Clare D.; Norton, Douglas A.)V1532,39-47(1991)

Finite element analysis of deformation in large optics due to space environment radiation (Merzbacher, Celia I.; Friebele, E. J.; Ruller, Jacqueline A.; Matic, P.)V1533,222-228(1991)

Fitting capability of deformable mirror (Jiang, Wen-Han; Ling, Ning; Rao, Xuejun; Shi, Fan)V1542,130-137(1991)

Fixing the Hubble Space Telescope (Crocker, James H.)V1494,2-8(1991)

High-bandwidth control for low-area-density deformable mirrors (How, Jonathan P.; Anderson, Eric H.; Miller, David W.; Hall, Steven R.)V1489,148-162(1991)

High laser-damage threshold and low-cost sol-gel-coated epoxy-replicated mirrors (Floch, Herve G.; Berger, Michel; Novaro, Marc; Thomas, Ian M.)V1441,304-315(1991)

High-power nanosecond pulse iodine laser provided with SBS mirror (Dolgopolov, Y. V.; Kirillov, Gennadi A.; Kochemasov, G. G.; Kulikov, Stanislav M.; Murugov, V. M.; Pevny, S. N.; Sukharev, S. A.)V1412,267-275(1991)

High-precision interferometric testing of spherical mirrors with long radius of curvature (Freischlad, Klaus R.; Kuechel, Michael F.; Wiedmann, Wolfgang; Kaiser, Winfried; Mayer, Max)V1332,8-17(1991)

High-speed, high-resolution image reading technique using multi-area sensors (Uehira, Kazutake)V1448,182-190(1991)

Holographic mirrors laminated into windshields for automotive head-up display and solar protective glazing applications (Beeck, Manfred-Andreas; Frost, Thorsten; Windeln, Wilbert)V1507,394-406(1991)

Imaging characteristics of the development model of the SAX x-ray imaging concentrators (Citterio, Oberto; Conconi, Paolo; Conti, Giancarlo; Mattaini, E.; Santambrogio, E.; Cusumano, G.; Sacco, B.; Braueninger, Heinrich; Burkert, Wolfgang)V1343,145-154(1991)

Imaging performance and tests of soft x-ray telescopes (Spiller, Eberhard; McCorkle, R.; Wilczynski, Janusz S.; Golub, Leon; Nystrom, George U.; Takacs, Peter Z.; Welch, Charles W.)V1343,134-144(1991)

Implementation issues in the control of a flexible mirror testbed (Anderson, Eric H.; How, Jonathan P.)V1542,392-405(1991)

In-line fiber Fabry-Perot interferometer with high-reflectance internal mirrors (Lee, Chung E.; Gibler, William N.; Atkins, Robert A.; Taylor, Henry F.)V1584,396-399(1991)

Interfaces in Mo/Si multilayers (Slaughter, Jon M.; Kearney, Patrick A.; Schulze, Dean W.; Falco, Charles M.; Hills, C. R.; Saloman, Edward B.; Watts, Richard N.)V1343,73-82(1991)

Investigation of unobscured mirror telescope for telecom purposes (Sand, Rolf)V1522,103-110(1991)

IR objective with internal scan mirror (Eisenberg, Shai; Menache, Ram)V1442,133-138(1991)

Johns Hopkins adaptive optics coronagraph (Clampin, Mark; Durrance, Samuel T.; Golimowski, D. A.; Barkhouser, Robert H.)V1542,165-174(1991)

Large active mirror in aluminium (Leblanc, Jean-M.; Rozelot, Jean-Pierre)V1535,122-129(1991)

Latest developments of active optics of the ESO NTT and the implications for the ESO VLT (Noethe, L.; Andreoni, G.; Franza, F.; Giordano, P.; Merkle, Fritz; Wilson, Raymond N.)V1542,293-296(1991)

Lessons learned in recent beryllium-mirror fabrication (Wells, James A.; Lombard, Calvin M.; Sloan, George B.; Moore, Wally W.; Martin, Claude E.)V1485,2-12(1991)

Lightweight composite mirrors: present and future challenges (Brand, Richard A.; Spinar, Karen K.)V1532,57-63(1991)

Lightweight SXA metal matrix composite collimator (Johnson, R. B.; Ahmad, Anees; Hadaway, James B.; Geiger, Alan L.)V1535,136-142(1991)

Lunar liquid-mirror telescopes (Borra, Ermanno F.; Content, R.)V1494,219-227(1991)

Manufacture of fast, aspheric, bare beryllium optics for radiation hard, spaceborne sensor systems (Sweeney, Michael N.)V1485,116-127(1991)

Manufacture of ISO mirrors (Ruch, Eric)V1494,265-278(1991)

Material characterization of beryllium mirrors exhibiting anomalous scatter (Egert, Charles M.)V1530,162-170(1991)

Measuring phase errors of an array or segmented mirror with a single far-field intensity distribution (Tyson, Robert K.)V1542,62-75(1991)

Mechanical and thermal disturbances of the PSR moderate focus-mission structure (Shih, Choon-Foo; Lou, Michael C.)V1532,81-90(1991)

Metallic alternative to glass mirrors (active mirrors in aluminum): a review (Rozelot, Jean-Pierre; Leblanc, Jean-M.)V1494,481-490(1991)

Metal mirror review (Janeczko, Donald J.)VCR38,258-280(1991)

Method of making ultralight primary mirrors (Zito, Richard R.)V1494,491-497(1991)

Mirrors as power filters (Kortright, Jeffrey B.)V1345,38-41(1991)

Mirrors for optical telescopes (Miroshnikov, Mikhail M.)V1533,286-298(1991)

Monolithic two-dimensional surface-emitting laser diode arrays with 45 degree micromirrors (Jansen, Michael; Yang, Jane; Ou, Szutsun S.; Sergant, Moshe; Mawst, Luke J.; Rozenbergs, John; Wilcox, Jarka Z.; Botez, Dan)V1418,32-39(1991)

Multichannel mirror systems for high-speed framing recording (Ushakov, Leonid S.; Ponomaryov, A. M.; Trofimenko, Vladimir V.; Drozhbin, Yu. A.)V1358,447-450(1991)

Multikilowatt transverse-flow CO2 laser with variable reflectivity mirrors (Serri, Laura; Maggi, C.; Garifo, Luciano; De Silvestri, S.; Magni, Vittorio C.; Svelto, Orazio)V1397,469-472(1991)

Multilayer reflectors for the "water window" (Xu, Shi; Evans, Brian L.)V1343,110-121(1991)

Multimode laser beams behaviour through variable reflectivity mirrors (Porras, Miguel A.; Alda, Javier; Bernabeu, Eusebio)V1397,645-648(1991)

Multipulse laser-induced failure prediction for Mo metal mirrors (Becker, Michael F.; Ma, Chun-Chi; Walser, Rodger M.)V1441,174-187(1991)

New configuration of a generator and regenerative amplifier built on three mirrors (Piotrowski, Jan)V1391,272-278(1991)

New development of friction speeding mechanism for high-speed rotating mirror device (Wei, Yan-nian; Jiang, Pei-sheng; Zhang, Zeng-xiang; Wu, Ming-da; Shi, Gao-yi; Ye, Niao-ting)V1358,457-460(1991)

Nonimaging concentrators for diode-pumped slab lasers (Lacovara, Phil; Gleckman, Philip; Holman, Robert L.; Winston, Roland)V1528,135-141(1991)

Nonimaging optics and the measurement of diffuse reflectance (Hanssen, Leonard M.; Snail, Keith A.)V1528,142-150(1991)

Nonimaging Optics: Maximum Efficiency Light Transfer (Winston, Roland; Holman, Robert L., eds.)V1528(1991)

Nonimaging optics: optical design at the thermodynamic limit (Winston, Roland)V1528,2-6(1991)

Nonlinear finite element analysis of the Starlab 80-cm telescope primary-mirror suspension system (Arnold, William R.)V1532,103-113(1991)

Nonrejected earth radiance performance of the visible ultraviolet experiment sensor (Betts, Timothy C.; Dowling, Jerome M.; Friedman, Richard M.)V1479,120-126(1991)

Novel toroidal mirror enhances x-ray lithography beamline at the Center for X-ray Lithography (Cole, Richard K.; Cerrina, Franco)V1465,111-121(1991)

Numerical studies of resonators with on-axis holes in mirrors for FEL applications (Keselbrener, Michel; Ruschin, Shlomo; Lissak, Boaz; Gover, Avraham)V1415,38-47(1991)

Off-axis spherical element telescope with binary optic corrector (Brown, Daniel M.; Kathman, Alan D.)V1555,114-127(1991)

Optical design considerations for next-generation space and lunar telescopes (Korsch, Dietrich G.)V1494,111-118(1991)

Optical performance of an infrared astronomical telescope with a 5-axis secondary mirror (Ettedgui-Atad, Eli; Humphries, Colin M.)V1532,241-248(1991)

Passive Materials for Optical Elements (Wilkerson, Gary W., ed.)V1535(1991)

Performance of multilayer-coated figured optics for soft x-rays near the diffraction limit (Raab, Eric L.; Tennant, Donald M.; Waskiewicz, Warren K.; MacDowell, Alastair A.; Freeman, Richard R.)V1343,104-109(1991)

Plane and concave VUV and soft x-ray multilayered mirrors (Cao, Jianlin; Miao, Tongqun; Qian, Longsheng; Zhu, Xioufang; Li, Futian; Ma, Yueying; Qian, Limin; Chen, Po; Chen, Xingdan)V1345,225-232(1991)

Polarizing optics for the soft x-ray regime: whispering-gallery mirrors and multilayer beamsplitters (Braud, John P.)V1548,69-72(1991)

Prepulse suppression using a self-induced ultrashort pulse plasma mirror (Gold, David; Nathel, Howard; Bolton, Paul R.; White, William E.; Van Woerkom, Linn D.)V1413,41-52(1991)

Primary mirror control system for the Galileo telescope (Bortoletto, Favio; Baruffolo, A.; Bonoli, C.; D'Alessandro, Maurizio; Fantinel, D.; Giudici, G.; Ragazzoni, Roberto; Salvadori, L.; Vanini, P.)V1542,225-235(1991)

Ray tracing homogenizing mirrors for synchrotron x-ray lithography (Homer, Michael; Rosser, Roy J.; Speer, R. J.)V1527,134-144(1991)

Real-time wavefront correction system using a zonal deformable mirror and a Hartmann sensor (Salmon, J. T.; Bliss, Erlan S.; Long, Theresa W.; Orham, Edward L.; Presta, Robert W.; Swift, Charles D.; Ward, Richard S.)V1542,459-467(1991)

Recent developments in nonimaging secondary concentrators for linear receiver solar collectors (Gordon, Jeff M.)V1528,32-43(1991)

Recent developments in production of thin x-ray reflecting foils (Hudec, Rene; Valnicek, Boris; Cervencl, J.; Gerstman, T.; Inneman, Adolf; Nejedly, Pavel; Svatek, Lubomir)V1343,162-163(1991)

Reconstruction of the Hubble Space Telescope mirror figure from out-of-focus stellar images (Roddier, Claude A.; Roddier, Francois J.)V1494,78-84(1991)

Reflective and Refractive Optical Materials for Earth and Space Applications (Riedl, Max J.; Hale, Robert R.; Parsonage, Thomas B., eds.)V1485(1991)

Reflective optical designs for soft x-ray projection lithography (Jewell, Tanya E.; Thompson, Kevin P.; Rodgers, John M.)V1527,163-173(1991)

Reflectors for efficient and uniform distribution of radiation for lighting and infrared based on nonimaging optics (Cai, Wen; Gordon, Jeff M.; Kashin, Peter; Rabl, Ari)V1528,118-128(1991)

Relationship between fluctuation in mirror radius (within one polygon) and the jitter (Horikawa, Hiroshi; Sugisaki, Iwao; Tashiro, Masaru)V1454,46-59(1991)

Relationship between jitter and deformation of mirrors (Horikawa, Hiroshi; Miura, Masayuki; Uchida, Toshiya)V1454,20-32(1991)

Relay Mirror Experiment scoring analysis and the effects of atmospheric turbulence (Sydney, Paul F.; Dillow, Michael A.; Anspach, Joel E.; Kervin, Paul W.; Lee, Terence B.)V1482,196-208(1991)

Robotic-based fabrication system for aspheric reflectors (Zimmerman, Jerrold; Jones, Robert A.; Rupp, Wiktor J.)VCR38,184-192(1991)

Scatter and contamination of a low-scatter mirror (McNeely, Joseph R.; McIntosh, Malcolm B.; Akerman, M. A.)V1530,288-298(1991)

Segmented mirror figure control for a space-based far-IR astronomical telescope (Redding, David C.; Breckenridge, William G.; Lau, Kenneth; Sevaston, George E.; Levine, Bruce M.; Shaklan, Stuart B.)V1489,201-215(1991)

Selected Papers on Holographic and Diffractive Lenses and Mirrors (Stone, Thomas W.; Thompson, Brian J., eds.)VMS34(1991)

Shape control of piezoelectric bimorph mirrors (Burke, Shawn E.; Hubbard, James E.)V1532,207-214(1991)

Simple 180° field-of-view F-theta all-sky camera (Andreic, Zeljko)V1500,293-304(1991)

Sliding control of a single-axis steering mirror (Connors, Bruce P.)V1489,136-147(1991)

Soft X-UV silver silicon multilayer mirrors (Shao, Jian D.; Fan, Zheng X.; Guo, Yong H.; Jin, Lei)V1519,298-301(1991)

Solar astronomy with a 19-segment adaptive mirror (Acton, D. S.; Smithson, Robert C.)V1542,159-164(1991)

Solution for anomalous scattering of bare HIP Be and CVD SiC mirrors (Vernold, Cynthia L.)V1530,130-143(1991)

Space Astronomical Telescopes and Instruments (Bely, Pierre Y.; Breckinridge, James B., eds.)V1494(1991)

Space-based visible all-reflective stray light telescope (Wang, Dexter; Gardner, Leo R.; Wong, Wallace K.; Hadfield, Peter)V1479,57-70(1991)

Specification of precision optical pointing systems (Medbery, James D.; Germann, Lawrence M.)V1489,163-176(1991)

Spectrum investigation and imaging of laser-produced plasma by multilayer x-ray optics (Platonov, Yu. Y.; Salashchenko, N. N.; Shmaenok, L. A.)V1440,188-196(1991)

Surface control techniques for the segmented primary mirror in the large lunar telescope (Gleckler, Anthony D.; Pflibsen, Kent P.; Ulich, Bobby L.; Smith, Duane D.)V1494,454-471(1991)

Surface distortions of a 3.5-meter mirror subjected to thermal variations (Cho, Myung K.; Poczulp, Gary A.)V1532,137-145(1991)

Temperature control of the 3.5-meter WIYN telescope primary mirror (Goble, Larry W.)V1532,161-169(1991)

Thermal analysis of multifacet-mirror ring resonator for XUV free-electron lasers (McVey, Brian D.; Goldstein, John C.; McFarland, Robert D.; Newnam, Brian E.)V1441,457-468(1991)

Thermal and structural analysis of the GOES scan mirror's on-orbit performance (Zurmehly, George E.; Hookman, Robert A.)V1532,170-176(1991)

Thermodynamics of light concentrators (Ries, Harald; Smestad, Greg P.; Winston, Roland)V1528,7-14(1991)

Thick, fine-grained beryllium optical coatings (Murray, Brian W.; Ulph, Eric; Richard, Peter N.)V1485,106-115(1991)

Three materials soft x-ray mirrors: theory and application (Boher, Pierre; Hennet, L.; Houdy, Philippe)V1345,198-212(1991)

Time reversal, enhancement, and suppression of dipole radiation by a phase-conjugate mirror (Bochove, Erik J.)V1497,222-227(1991)

Total internal reflection mirrors fabricated in polymeric optical waveguides via excimer laser ablation (Trewhella, Jeannine M.; Oprysko, Modest M.)V1377,64-72(1991)

Trade-offs in rotary mirror scanner design (Colquhoun, Allan B.; Cowan, Donald W.; Shepherd, Joseph)V1454,12-19(1991)

Two-mirror projection systems for simulating telescopes (Hannan, Paul G.; Davila, Pam M.)V1527,26-36(1991)

Ultraviolet reflector materials for solar detoxification of hazardous waste (Jorgensen, Gary J.; Govindarajan, Rangaprasad)V1536,194-205(1991)

University of Hawaii adaptive optics system: III. Wavefront curvature sensor (Graves, J. E.; McKenna, Daniel)V1542,262-272(1991)

ZnS/Me heat mirror systems (Zhang, Xiao P.; Yu, Shan-qing; Ma, Min W.)V1519,514-520(1991)

Modulation

0.5-GHz cw mode-locked Nd:glass laser (Ling, Junda D.; Yan, Li I.; Liu, YuanQun; Lee, Chi H.; Soos, Jolanta I.)V1454,353-362(1991)

All-silicon Fabry-Perot modulator based on thermo-optic effect (Rendina, Ivo; Cocorullo, Giuseppe)V1583,338-343(1991)

Amplitude-modulated laser-driven fiber-optic RF interferometric strain sensor (Schoenwald, Jeffrey S.)V1418,450-458(1991)

Analog optical processing of radio frequency signals (Sullivan, Daniel P.; Weber, Charles L.)V1476,234-245(1991)

Analysis and measurement of the external modulation of modelocked laser diodes (relative noise performance) (Lam, Benson C.; Kellner, Albert L.; Campion, David C.; Costa, Joannes M.; Yu, Paul K.)V1371,36-45(1991)

Analysis of evanescent coupling in waveguide modulators (Bradley, Joe C.; Kellner, Albert L.)V1476,330-336(1991)

Analysis of the multichannel coherent FSK subcarrier multiplexing system with pilot carrier and phase noise cancelling scheme (Lee, Yang-Hang; Wu, Jingshown; Tsao, Hen-Wai)V1372,140-149(1991)

Analysis on the characteristics of the optical fiber compensation network for the intensity modulation optical fiber sensors (Zhong, Xian-Xin; Li, Jianshu; Fu, Xin; Huang, Shang-Lian)V1572,84-87(1991)

Application of electro-optic modulator in photomechanics (Zhang, Yuan P.)V1554B,669-678(1991)

ASK transmitter for high-bit-rate systems (Schiellerup, Gert; Pedersen, Rune J.)V1372,27-38(1991)

Asymmetric Fabry-Perot modulators for optical signal processing and optical computing applications (Kilcoyne, M. K.; Whitehead, Mark; Coldren, Larry A.)V1389,422-454(1991)

Broadband electromagnetic environment monitoring using semiconductor electroabsorption modulators (Pappert, Stephen A.; Lin, S. C.; Orazi, Richard J.; McLandrich, Matthew N.; Yu, Paul K.; Li, S. T.)V1476,282-293(1991)

Broadband microwave and millimeter-wave EOMs with ultraflat frequency response (Pan, J. J.; Li, Yi Q.)V1476,22-31(1991)

Carrier recovery and filtering in optical BPSK systems using external cavity semiconductor lasers (Pires, Joao J.; Rocha, Jose R.)V1372,118-127(1991)

Cascaded optical modulators: any advantages? (Sierak, Paul)V1398,238-249(1991)

Coherent communication systems research and development at AT&T Bell Laboratories, Solid State Technology Center (Park, Yong K.; Delavaux, Jean-Marc P.; Tench, Robert E.; Cline, Terry W.; Tzeng, Liang D.; Kuo, Chien-yu C.; Wagner, Earl J.; Flores, Carlos F.; Van Eijk, Peter; Pleiss, T. C.; Barski, S.; Owen, B.; Twu, Yih-Jye; Dutta, Niloy K.; Riggs, R. S.; Ogawa, Kinichiro K.)V1372,219-227(1991)

Coherent detection: n-ary PPM versus PCM (Cryan, Robert A.; Unwin, Rodney T.; Massarella, Alistair J.; Sibley, Martin J.; Garrett, Ian)V1372,64-71(1991)

Coherent optical modulation for antenna remoting (Fitzmartin, Daniel J.; Gels, Robert G.; Balboni, Edmund J.)V1476,56-62(1991)

Collinear asymmetrical polymer waveguide modulator (Chen, Ray T.; Sadovnik, Lev S.)V1559,449-460(1991)

Comparison of alternative modulation techniques for microwave optical links (Kasemset, Dumrong; Ackerman, Edward I.; Wanuga, Stephen; Herczfeld, Peter R.; Daryoush, Afshin S.)V1371,104-114(1991)

Design and construction of a wideband efficient electro-optic modulator (Harris, Neville W.; Sobolewski, J. M.; Summers, Charles L.; Eng, Richard S.)V1416,59-69(1991)

Design and fabrication of amplitude modulators for CATV (Mahapatra, Amaresh; Cooper, Ronald F.)V1374,296-299(1991)

Development of wideband 16-channel acousto-optical modulators on the LiNbO3 and TeO2 crystals (Bokov, Lev; Demidov, Anatoly J.; Zadorin, Anatoly; Kushnarev, Igor; Serebrennikov, Leonid J.; Sharangovich, Sergey)V1505,186-198(1991)

Effect of parameter variation on the performance of InGaAsP/InP multiple-quantum-well electroabsorption/electrorefraction modulators (Xiong, F. K.; Zhu, Long D.; Wang, C. W.)V1519,665-669(1991)

Electro-optic illuminating module (Pesl, Ales A.)V1454,299-305(1991)

Electro-optic modulator for high-speed Nd:YAG laser communication (Petsch, Thomas)V1522,72-82(1991)

Electro-optic resonant modulator for coherent optical communication (Robinson, Deborah L.; Chen, Chien-Chung; Hemmati, Hamid)V1417,421-430(1991)

Enumerative modulation coding with arbitrary constraints and postmodulation error correction coding for data storage systems (Mansuripur, Masud)V1499,72-86(1991)

Fabry-Perot etalon as a CO2 laser Q-switch modulator (Walocha, Jerzy)V1391,286-289(1991)

Fine structure characteristics of semiconductor laser diode coupled to an external cavity (Ghiasi, Ali; Gopinath, Anand)V1583,170-174(1991)

Generation of 1-GHz ps optical pulses by direct modulation of a 1.532-um DFB-LD (Nie, Chao-Jiang; Li, Ying; Wu, Fang D.; Chen, Ying-Li; Li, Qi; Hua, Yi-Min)V1579,264-267(1991)

High-current, high-bandwidth laser diode current driver (Copeland, David J.; Zimmerman, Robert K.)V1417,412-420(1991)

High-density waveguide modulator arrays for parallel interconnection (Bristow, Julian P.; Mukherjee, Sayan D.; Khan, M. N.; Hibbs-Brenner, Mary K.; Sullivan, Charles T.; Kalweit, Edith)V1389,535-546(1991)

High-dynamic-range fiber optic link using external modulator diode pumped Nd:YAG lasers (Childs, Richard B.; O'Byrne, Vincent A.)V1371,223-232(1991)

High-dynamic-range, low-noise analog optical links using external modulators: analysis and demonstration (Betts, Gary E.; Johnson, Leonard M.; Cox, Charles H.)V1371,252-257(1991)

High-efficiency vertical-cavity lasers and modulators (Coldren, Larry A.; Corzine, Scott W.; Geels, Randall S.; Gossard, Arthur C.; Law, K. K.; Merz, James L.; Scott, Jeffrey W.; Simes, Robert J.; Yan, Ran H.)V1362,24-37(1991)

High-Frequency Analog Fiber Optic Systems (Sierak, Paul, ed.)V1371(1991)

High-frequency fiber optic phase modulator using piezoelectric polymer coating (Imai, Masaaki M.; Yano, T.; Ohtsuka, Yoshihiro)V1371,13-20(1991)

Hybrid modulation properties of the Epson LCTV (Kirsch, James C.; Loudin, Jeffrey A.; Gregory, Don A.)V1558,432-441(1991)

Indium tin oxide single-mode waveguide modulator (Chen, Ray T.; Robinson, Daniel; Lu, Huey T.; Sadovnik, Lev S.; Ho, Zonh-Zen)V1583,362-374(1991)

InGaAs/AlGaAs vertical optical modulators and sources on a transparent GaAs substrate (Buydens, Luc; Demeester, Piet M.; De Dobbelaere, P.; van Daele, Peter)V1362,50-58(1991)

Inhomogeneous quarter-wave transformers for a waveguide electro-optic modulator (Harris, Neville W.; Wong, D. M.)V1416,86-99(1991)

Instant measurement of phase-characteristic curve (Sadovnik, Lev S.; Rich, Chris C.)V1559,424-428(1991)

Integrated-optical modulators for bandpass analog links (Johnson, Leonard M.; Betts, Gary E.; Roussell, Harold V.)V1371,2-7(1991)

Integrated optics in optical engineering (Popov, Yury V.)V1399,207-213(1991)

Integrated optics intensity modulators in the GaAs/AlGaAs system (Khan, M. A.; Naumaan, Ahmed; Van Hove, James M.)V1396,753-759(1991)

Intercircuit optical interconnects using quantum-well modulators (Goodfellow, Robert C.; Goodwin, Martin J.; Moseley, Andrew J.)V1389,594-599(1991)

LiNbO3-based multichannel electro-optical light modulators (Kiselyov, Boris S.; Mikaelian, Andrei L.; Novoselov, B. A.; Shkitin, Vladimir A.; Arkhontov, L. B.; Evtikhiev, Nickolay N.)V1621,126-137(1991)

Linearization of electro-optic modulators by a cascade coupling of phase-modulating electrodes (Skeie, Halvor; Johnson, Richard V.)V1583,153-164(1991)

Linearized external modulator for analog applications (Trisno, Yudhi S.; Chen, Lian K.; Huber, David R.)V1371,8-12(1991)

Liquid-crystal phase modulator used in DSPI (Kadono, Hirofumi; Toyooka, Satoru)V1554A,628-638(1991)

Lithium niobate proton-exchange technology for phase-amplitude modulators (Varasi, Mauro; Vannucci, Antonello; Signorazzi, Mario)V1583,165-169(1991)

Magneto-optical linear multichannel light modulator for recording of one-dimensional holograms (Abakumov, D. M.)V1621,138-147(1991)

Margin measurements of pulse amplitude modulation channels (Lauffenburger, Jim; Arachtingi, John W.; Robbins, Jamey L.; Kim, Jong)V1499,104-113(1991)

Measurement of biological tissue metabolism using phase modulation spectroscopic technology (Weng, Jian; Zhang, M. Z.; Simons, K.; Chance, Britton)V1431,161-170(1991)

Microwave high-dynamic-range EO modulators (Pan, J. J.; Garafalo, David A.)V1371,21-35(1991)

Minimum polarization modulation: a highly bandwidth efficient coherent optical modulation scheme (Benedetto, Sergio; Kazovsky, Leonid G.; Poggiolini, Pierluigi T.)V1579,112-121(1991)

Modal noise reduction in analog fiber optic links by superposition of high-frequency modulation (Pepeljugoski, Petar K.; Lau, Kam-Yin)V1371,233-243(1991)

Modeling of traveling wave Ti:LiNbO3 waveguide modulators (Rocca, Corrado; Montrosset, Ivo; Gollinucci, Stefano; Ghione, Giovanni)V1372,39-47(1991)

Modulation characteristics of high-power phase-locked arrays of antiguides (Anderson, Eric R.; Jansen, Michael; Botez, Dan; Mawst, Luke J.; Roth, Thomas J.; Yang, Jane)V1417,543-549(1991)

Modulation doping and delta doping of III-V compound semiconductors (Hendriks, Peter; Zwaal, E. A.; Haverkort, Jos E.; Wolter, Joachim H.)V1362,217-227(1991)

Multichannel chromatic interferometry: metrology applications (Tribillon, Gilbert M.; Calatroni, Jose; Sandoz, Patrick)V1332,632-642(1991)

Multimode approach to optical fiber components and sensors (Johnstone, Walter; Thursby, G.; Culshaw, Brian; Murray, S.; Gill, M.; McDonach, Alaster; Moodie, D. G.; Fawcett, G. M.; Stewart, George; McCallion, Kevin J.)V1506,145-149(1991)

Multispectral imaging with frequency-modulated reticles (Sanders, Jeffrey S.; Halford, Carl E.)V1478,52-63(1991)

Multistability, instability, and chaos for intracavity magneto-optic modulation output (Yang, Darang; Song, Ruhua H.; Hu, Zhiping; Le, Shixiao)V1417,440-450(1991)

New configuration of 2 x 2 switching elements using Mach-Zehnder electro-optic modulators (Armenise, Mario N.; Castagnolo, Beniamino; Pesce, Anastasia; Rizzi, Maria L.)V1583,210-220(1991)

New modulation scheme for coherent lightwave communication: Direct-Modulation PSK (Naito, Takao; Chikama, Terumi; Kuwahara, Hideo)V1372,200-207(1991)

New modulation scheme for optical fiber point temperature sensor (Farahi, Faramarz)V1504,237-246(1991)

New technique for characterizing holographic recording materials (Leclere, Philippe; Renotte, Yvon L.; Lion, Yves F.)V1559,298-307(1991)

New type of IR to visible real-time image converter: design and fabrication (Sun, Fang-kui; Yang, Mao-hua; Gao, Shao-hong; Zhao, Shi-jie)V1488,2-5(1991)

Optical A/D conversion based on acousto-optic theta modulation (Li, Yao; Zhang, Yan)V1474,167-173(1991)

Optical carrier modulation by integrated optical devices in lithium niobate (Rasch, Andreas; Buss, Wolfgang; Goering, Rolf; Steinberg, Steffen; Karthe, Wolfgang)V1522,83-92(1991)

Optical cross-modulation method for diagnostic of powerful microwave radiation (Kozar, A. V.; Krupenko, S. A.)V1476,305-312(1991)

Optical decoding and coherent detection of a four-level FSK signal (O'Byrne, Vincent A.; Tatlock, Timothy; Stone, Samuel M.)V1372,88-93(1991)

Optical fiber n-ary PPM: approaching fundamental limits in receiver sensitivity (Cryan, Robert A.; Unwin, Rodney T.)V1579,133-143(1991)

Optical second-harmonic generation by polymer-dispersed liquid-crystal films (Yuan, Haiji J.; Li, Le; Palffy-Muhoray, Peter)V1455,73-83(1991)

Optimization of externally modulated analog optical links (Betts, Gary E.; Johnson, Leonard M.; Cox, Charles H.)V1562,281-302(1991)

Optimization of reflection electro-absorption modulators (Pezeshki, Bardia; Thomas, Dominique; Harris, James S.)V1362,559-565(1991)

Performance comparison of neural network and statistical discriminant processing techniques for automatic modulation recognition (Hill, Peter C.; Orzeszko, Gabriel R.)V1469,329-340(1991)

Performance of on-off modulated lightwave signals with phase noise (Azizoglu, Murat; Humblet, Pierre A.)V1579,168-178(1991)

Photonic computer-aided design tools for high-speed optical modulators (Liu, Pao-Lo)V1371,46-55(1991)

Physical foundations of high-speed thermomagnetic tuning of spatial-temporal structure of far-IR and submillimeter beams (Shvartsburg, Alexandre B.)V1488,28-35(1991)

Pockels' effect in polycrystalline ZnS planar waveguides (Wong, Brian; Kitai, Adrian H.; Jessop, Paul E.)V1398,261-268(1991)

Polarization dependence and uniformity of FLC layers for phase modulation (Biernacki, Paul D.; Brown, Tyler; Freeman, Mark O.)V1455,167-178(1991)

Polarization-modulated coherent optical communication systems (Betti, Silvello; Curti, Franco; De Marchis, Giancarlo; Iannone, Eugenio)V1579,100-111(1991)

Polarization modulation with frequency shift redundancy and frequency modulation with polarization redundancy for POLSK and FSK systems (Marone, Giuseppe; Calvani, Riccardo A.; Caponi, Renato)V1579,122-132(1991)

Polymer-based electro-optic modulators: fabrication and performance (Haas, David R.; Man, Hong-Tai; Teng, Chia-Chi; Chiang, Kophu P.; Yoon, Hyun N.; Findakly, Talal K.)V1371,56-67(1991)

Polymer-dispersed liquid-crystal shutters for IR imaging (McCargar, James W.; Doane, J. W.; West, John L.; Anderson, Thomas W.)V1455,54-60(1991)

Problems and solutions in fiber-optic amplitude-modulated sensors (Brenci, Massimo; Mencaglia, Andrea; Mignani, Anna G.)V1504,212-220(1991)

Quantum-well excitonic devices for optical computing (Singh, Jasprit; Bhattacharya, Pallab K.)V1362,586-597(1991)

Quasirelaxation-free guest-host poled-polymer waveguide modulator: material, technology, and characterization (Levenson, R.; Liang, J.; Toussaere, E.; Carenco, Alain; Zyss, Joseph)V1560,251-261(1991)

Real-time holographic display: improvements using a multichannel acousto-optic modulator and holographic optical elements (St.-Hilaire, Pierre; Benton, Stephen A.; Lucente, Mark; Underkoffler, John S.; Yoshikawa, Hiroshi)V1461,254-261(1991)

Reflection-mode polymeric interference modulators (Yankelevich, Diego; Knoesen, Andre; Eldering, Charles A.; Kowel, Stephen T.)V1560,406-415(1991)

Semiconductor laser transmitters for millimeter-wave fiber-optic links (Pan, J. J.)V1476,63-73(1991)

Silicon modulator for integrated optics (Cocorullo, Giuseppe; Della Corte, Francesco G.; Rendina, Ivo; Cutolo, Antonello)V1374,132-137(1991)

Space-time correlation properties and their applications of dynamic speckles after propagation through an imaging system and double-random modulation (Ma, Shining; Liu, Ying; Du, Fuli)V1554A,645-648(1991)

Symmetric 3 x 3 coupler based demodulator for fiber optic interferometric sensors (Brown, David A.; Cameron, C. B.; Keolian, Robert M.; Gardner, David L.; Garrett, Steven L.)V1584,328-335(1991)

Theoretical analysis of optical phase diversity FSK receivers (Yang, Shien-Chi; Tsao, Hen-Wai; Wu, Jingshown)V1372,128-139(1991)

Ultralinear electro-optic modulators for microwave fiber-optic communications (Pan, J. J.; Li, Wei Z.; Li, Yi Q.)V1476,32-43(1991)

Vectorial photoelectric effect at 2.69 keV (Shaw, Ping-Shine; Hanany, Shaul; Liu, Yee; Church, Eric D.; Fleischman, Judith R.; Kaaret, Philip E.; Novick, Robert; Santangelo, A.)V1548,118-131(1991)

Velocity-matched III-V travelling wave electro-optic modulator (Wang, S. Y.; Tan, Michael T.; Houng, Y. M.)V1371,98-103(1991)

Velocity-matched electro-optic modulator (Bridges, William B.; Sheehy, Finbar T.; Schaffner, James H.)V1371,68-77(1991)

Wide-band analog frequency modulation of optic signals using indirect techniques (Fitzmartin, Daniel J.; Balboni, Edmund J.; Gels, Robert G.)V1371,78-86(1991)

Wideband NLO organic external modulators (Findakly, Talal K.; Teng, Chia-Chi; Walpita, Lak M.)V1476,14-21(1991)

Y-branch optical modulator (Jaeger, Nicolas A.; Lai, Winnie C.)V1583,202-209(1991)

Modulation transfer function—see also optical design

Adaptive optical transfer function modeling (Gaffard, Jean-Paul; Ledanois, Guy)V1542,34-45(1991)

Airborne imaging system performance model (Redus, Wesley D.)V1498,2-23(1991)

Characterization of the atmospheric modulation transfer function using the target contrast characterizer (Watkins, Wendell R.; Billingsley, Daniel R.; Palacios, Fernando R.; Crow, Samuel B.; Jordan, Jay B.)V1486,17-24(1991)

Comparisons of deformable-mirror models and influence functions (Hiddleston, H. R.; Lyman, Dwight D.; Schafer, Eric L.)V1542,20-33(1991)

Determination of FLIR LOS stabilization errors (Pinsky, Howard J.)V1488,334-342(1991)

Effects of phasing on MRT target visibility (Holst, Gerald C.)V1488,90-98(1991)

Evaluation of the 3-D radon transform algorithm for cone beam reconstruction (Grangeat, Pierre; Le Masson, Patrick; Melennec, Pierre; Sire, Pascal)V1445,320-331(1991)

Geometrical and radiometrical signal transfer characteristics of a color CCD camera with 21-million pixels (Lenz, Reimar K.; Lenz, Udo)V1526,123-132(1991)

Imaging model for underwater range-gated imaging systems (Strand, Michael P.)V1537,151-160(1991)

Impact on the medium MTF by model estimation of b (Estep, Leland; Arnone, Robert A.)V1537,89-96(1991)

Light-pulse-induced background in image intensifiers with MCPs (Dashevsky, Boris E.)V1449,25-29(1991)

Measurement of modulation transfer function of desert atmospheres (McDonald, Carlos)V1487,203-219(1991)

Miscellaneous modulation transfer function effects relating to sample summing (Kennedy, Howard V.)V1488,165-176(1991)

MTF characteristics of a Scophony scene projector (Schildwachter, Eric F.; Boreman, Glenn D.)V1488,48-57(1991)

MTF of photolithographic hologram (Guo, Lu Rong; Guo, Yongkang; Zhang, Xiao-Chun)V1392,119-123(1991)

MTF optimization in lens design (Rimmer, Matthew P.; Bruegge, Thomas J.; Kuper, Thomas G.)V1354,83-91(1991)

Numerical calculation of image motion and vibration modulation transfer function (Hadar, Ofer; Fisher, Moshe; Kopeika, Norman S.)V1482,79-91(1991)

Numerical calculation of image motion and vibration modulation transfer functions: a new method (Hadar, Ofer; Dror, Itai; Kopeika, Norman S.)V1533,61-74(1991)

Optimum choice of anamorphic ratio and boost filter parameters for a SPRITE-based infrared sensor (Fredin, Per)V1488,432-442(1991)

Overall atmospheric MTF and aerosol MTF cutoff (Dror, Itai; Kopeika, Norman S.)V1487,192-202(1991)

Reimaging system for evaluating high-resolution charge-coupled-device arrays (Chambers, Robert J.; Warren, David W.; Lawrie, David J.; Lomheim, Terrence S.; Luu, K. T.; Shima, Ralph M.; Schlegel, J. D.)V1488,312-326(1991)

Simulation of partially obscured scenes using the radiosity method (Borel, Christoph C.; Gerstl, Siegfried A.)V1486,271-277(1991)

Sine wave measurements of SPRITE detector MTF (Barnard, Kenneth J.; Boreman, Glenn D.; Plogstedt, Allen E.; Anderson, Barry K.)V1488,426-431(1991)

Space-spectrum resolution function of the imaging spectrometer (Han, Xin Z.; Qiu, Ying)V1479,156-161(1991)

Specifications for image stabilization systems (Hilkert, James M.; Bowen, Max L.; Wang, Joe)V1498,24-38(1991)

Moire—see also interferometry

Advanced materials characterization by means of moire techniques (Aymerich, Francesco; Ginesu, Francesco; Priolo, Pierluigi)V1554B,304-314(1991)

Aero-optics analysis using moire deflectometry (Abushagur, Mustafa A.; Elmanasreh, Ahmed)V1554B,298-302(1991)

Analysis of measurement principle of moire interferometer using Fourier method (Wang, Hai-Lin; Miao, Peng-Cheng; He, Anzhi)V1527,419-422(1991)

Analysis of moire deflectometry by wave optics (Wang, Hai-Lin; Miao, Peng-Cheng; Yan, Da-Peng; He, Anzhi)V1545,268-273(1991)

Analysis of the displacements of cylindrical shells by moire techniques (Laermann, Karl-Hans)V1554B,248-256(1991)

Angle measurement by moire interference technique (Singh, Brahm P.; Chitnis, Vijay T.)V1554B,335-338(1991)

Automated fringe analysis for moire interferometry (Brownell, John B.; Parker, Richard J.)V1554B,481-492(1991)

Automatic mask-to-wafer alignment and gap control using moire interferometry (Chitnis, Vijay T.; Kowsalya, S.; Rashmi, Dr.; Kanjilal, A. K.; Narain, Ram)V1332,613-622(1991)

Biomechanical research of joints: IV. the biohinge of primates (Zhang, Renxiang; Yu, Jie; Lan, Zu-yun; Qu, Wen-ji; Zhang, Hong-zi; Zhang, Kui; Zhang, Liang)V1380,116-121(1991)

Carrier pattern analysis of moire interferometry using Fourier transform (Morimoto, Yoshiharu; Post, Daniel; Gascoigne, Harold E.)V1554B,493-502(1991)

Computer-aided moire strain analysis on thermoplastic models at critical temperature (Barillot, Marc; Jacquot, Pierre M.; Di Chirico, Giuseppe)V1554A,867-878(1991)

Contouring using gratings created on a LCD panel (Asundi, Anand K.; Wong, C. M.)V1400,80-85(1991)

Correspondence in damage phenomena and R-curve behavior in ceramics and geomaterials using moire interferometry (Perry, Kenneth E.; Epstein, Jonathan S.; May, G. B.; Shull, J. E.)V1554A,209-227(1991)

Curvature of bending shells by moire interferometry (Wang, Lingli; Yun, Dazhen)V1554B,436-440(1991)

Deformation measurements of the human tympanic membrane under static pressure using automated moire topography (Dirckx, Joris J.; Decraemer, Willem F.)V1429,34-38(1991)

Design and application of a moire interferometer (He, Anzhi; Wang, Hai-Lin; Yan, Da-Peng; Miao, Peng-Cheng)V1527,334-337(1991)

Design optimization of moire interferometers for rapid 3-D manufacturing inspection (Dubowsky, Steven; Holly, Krisztina J.; Murray, Annie L.; Wander, Joseph M.)V1386,10-20(1991)

Determination of surface quality using moire methods (Keren, Eliezer; Kreske, Kathi; Zac, Yaacov; Livnat, Ami)V1442,266-274(1991)

Determination of thermal stresses in a bimaterial specimen by moire interferometry (Kang, Yilan; Jia, Youquan; Du, Ji)V1554B,514-522(1991)

Development of electron moire method using a scanning electron microscope (Kishimoto, Satoshi; Egashira, Mitsuru; Shinya, Norio)V1554B,174-180(1991)

Diagnostics of arc plasma by moire deflectometry (Gao, Yiqing; Liu, Yupin)V1554B,193-199(1991)

Displacement measurement of track on magnetic tape by Fourier transform grid method (Sogabe, Yasushi; Morimoto, Yoshiharu; Murata, Shigeki)V1554B,289-297(1991)

Displacements and strains in thick-walled composite rings subjected to external pressure using moire interferometry (Gascoigne, Harold E.; Abdallah, Mohamed G.)V1554B,315-322(1991)

Eliminating system errors of a large-aperture and high-sensitivity moire deflector by real-time holography (Yan, Da-Peng; Wang, Hai-Lin; Miao, Peng-Cheng; He, Anzhi)V1527,442-447(1991)

Evaluation of a new electrical resistance shear-strain gauge using moire interferometry (Ifju, Peter G.)V1554B,420-428(1991)

Evaluation of facial palsy by moire topography (Inokuchi, Ikuo; Kawakami, Shinichiro; Maeta, Manabu; Masuda, Yu)V1429,39-45(1991)

Fresnel zone plate moire patterns and its metrological applications (Jaroszewicz, Zbigniew)V1574,154-158(1991)

Fringe analysis in moire interferometry (Asundi, Anand K.)V1554B,472-480(1991)

Fringe-scanning moire system using a servo-controlled grating (Kurokawa, Haruhisa; Ichikawa, Naoki; Yajima, Nobuyuki)V1332,643-654(1991)

Geometrical moire method for displaying directly the stress field of a circular disk (Wang, Ji-Zhong; Wang, Yun-Shan; Yin, Yuan-Cheng; Wu, Rui-Lan)V1554B,188-192(1991)

High-resolution computer-aided moire (Sciammarella, Cesar A.; Bhat, Gopalakrishna K.)V1554B,162-173(1991)

High-sensitivity interferometric technique for strain measurements (Voloshin, Arkady S.; Bastawros, Adel F.)V1400,50-60(1991)

High-speed photography at the Cavendish Laboratory (Field, John E.)V1358,2-17(1991)

High-speed photography of high-resolution moire patterns (Whitworth, Martin B.; Huntley, Jonathan M.; Field, John E.)V1358,677-682(1991)

Holographic in-plane displacement measurement in cracked specimens in plane stress state (Sainov, Ventseslav C.; Simova, Eli S.)V1461,175-183(1991)

Improvement on phase-shifting method precision and application on shadow moire method (Mauvoisin, Gerard; Bremand, Fabrice J.; Lagarde, Alexis)V1554B,181-187(1991)

Information retrieval from ultrasonic (acoustical) holograms by moire principle (Greguss, Pal)V1238,421-427(1991)

Interferometric moire analysis of wood and paper structures (Hyzer, Jim B.; Shih, J.; Rowlands, Robert E.)V1554B,371-382(1991)

Interlaminar deformations on the cylindrical surface of a hole in laminated composites by moire interferometry (Boeman, Raymond G.)V1554B,323-330(1991)

Intersection of crack borders with free surfaces: an engineering interpretation of optical experimental results (Smith, C. W.; Constantinescu, D. M.)V1554A,102-115(1991)

Laser moire topography for 3-D contour measurement (Matsumoto, Tetsuya; Kitagawa, Yoichi; Adachi, Masaaki; Hayashi, Akihiro)V1332,530-536(1991)

Measurement of residual stress during implant resistance welding of plastics (Park, Joon B.; Benatar, Avraham)V1554B,357-370(1991)

Measurement of the dynamic crack-tip displacement field using high-resolution moire photography (Whitworth, Martin B.; Huntley, Jonathan M.; Field, John E.)V1554B,282-288(1991)

Moire-holographic evaluation of the strain fields at the weld toes of welded structures (Pappalettere, Carmine; Trentadue, Bartolo; Monno, Giuseppe)V1554B,99-105(1991)

Moire displacement detection by the photoacoustic technique (Hane, Kazuhiro; Watanabe, S.; Goto, T.)V1332,577-583(1991)

Moire image overlapping method for PCB inspection designator (Chang, Rong-Seng; Hu, Yeu-Jent)V1567,216-219(1991)

Moire interferometry for subdynamic tests in normal light environment (Luo, Zhishan; Luo, Chanzou; Hu, Qinghua; Mu, Zongxue)V1554B,523-528(1991)

Moire interferometry of sticking film using white light (Luo, Zhishan; Zhang, Souyi; Tuo, Sueshang)V1554B,339-342(1991)

Moire measurements using a panoramic annular lens (Gilbert, John A.; Matthys, Donald R.; Lehner, David L.)V1554B,202-209(1991)

Moire pattern and interferogram transform (Lu, Zhen-Wu)V1554B,383-388(1991)

Moire sensor as an automatic feeler gauge (Konwerska-Hrabowska, Joanna; Kryszczynski, Tadeusz; Smolka, M.; Olbrysz, P.; Widomski, L.)V1554B,225-232(1991)

Moire topography with the aid of phase-shift method (Yoshizawa, Toru; Tomisawa, Teiyu)V1554B,441-450(1991)

New method of 3-D shape measurement by moire technique (He, Anzhi; Li, Qun Z.; Miao, Peng-Cheng)V1545,278-281(1991)

New technique for rapid making specimen: gratings for moire interferometry (Jia, Youquan; Kang, Yilan; Du, Ji)V1554B,331-334(1991)

Novel grating methods for optical inspection (Asundi, Anand K.)V1554B,708-715(1991)

Ophthalmic lenses testing by moire deflectometry (Yonte, T.; Quiroga, J.; Alda, J.; Bernabeu, Eusebio)V1554B,233-241(1991)

Phase-shifting hand-held diffraction moire interferometer (Deason, Vance A.; Ward, Michael B.)V1554B,390-398(1991)

Phase-shift moire camera for real-time measurements of three-dimensional shape information (Turney, Jerry L.; Lysogorski, Charles; Gottschalk, Paul G.; Chiu, Arnold H.)V1380,53-63(1991)

Photoelasticity, holography, and moire for strain-stress-state studies of NPS, HPS underground and machinery structures: comparative analysis of physical modeling and numerical methods (Khesin, G. L.)V1554A,313-322(1991)

Polarization approach to high-sensitivity moire interferometry (Salbut, Leszek A.; Patorski, Krzysztof; Kujawinska, Malgorzata)V1554B,451-460(1991)

Production of gratings with desired pitch using the Talbot effect (Li, Wen L.; Zhong, An)V1554B,275-280(1991)

Projection moire as a tool for the automated determination of surface topography (Cardenas-Garcia, Jaime F.; Zheng, S.; Shen, F. Z.)V1554B,210-224(1991)

Projection moire deflectometry for mapping phase objects (Wang, Ming; Ma, Li; Zeng, Jing-gen; Cheng, Qi-Xian; Pan, Chuan K.)V1554B,242-246(1991)

Projection moire using PSALM (Asundi, Anand K.)V1554B,257-265(1991)

Real-time inspection of pavement by moire patterns (Guralnick, Sidney A.; Suen, Eric S.)V1396,664-677(1991)

Real-time optical spatial filtering system with white-light source for displaying color moire (Fang, Cui-Chang; Liao, Yan-Biao; Ma, De-Yuan)V1554A,915-921(1991)

Refractive-index measurement using moire deflectometry: working conditions (Tentori, Diana; Lopez Famozo, C.)V1535,209-215(1991)

Second Intl Conf on Photomechanics and Speckle Metrology: Moire Techniques, Holographic Interferometry, Optical NDT, and Applications to Fluid Mechanics (Chiang, Fu-Pen, ed.)V1554B(1991)

Single-lens moire contouring method (Harding, Kevin G.; Kaltenbacher, Eric; Bieman, Leonard H.)V1385,246-255(1991)

Spatial-carrier phase-shifting technique of fringe pattern analysis (Kujawinska, Malgorzata; Schmidt, Joanna)V1508,61-67(1991)

Spatial phase-shifting techniques of fringe pattern analysis in photomechanics (Kujawinska, Malgorzata; Wojiak, Joanna)V1554B,503-513(1991)

Spatial techniques of fringe pattern analysis in interferometry (Kujawinska, Malgorzata; Spik, Andrzej)V1391,303-312(1991)

Stress-static and strength research of building of NPS and nuclear reactor under power and thermal loads including creep and relaxation influence (Savostjanov, V. N.; Zavalishin, S. I.; Smirnov, S. B.; Morosova, D. V.)V1554A,579-585(1991)

Study of displacement and residual displacement field of an interface crack by moire interferometry (Wang, Y. Y.; Chiang, Fu-Pen; Barsoum, Roshdy S.; Chou, S. T.)V1554B,344-356(1991)

Study of plate vibrations by moire holography (Blanco, J.; Fernandez, J. L.; Doval, A. F.; Lopez, C.; Pino, F.; Perez-Amor, Mariano)V1508,180-190(1991)

Submicrometer lithographic alignment and overlay strategies (Zaidi, Saleem H.; Naqvi, H. S.; Brueck, Steven R.)V1343,245-255(1991)

Super-accurate positioning technique using diffracted moire signals (Takada, Yutaka; Uchida, Yoshiyuki; Akao, Yasuo; Yamada, Jun; Hattori, Shuzo)V1332,571-576(1991)

Thermal strain measurements of solder joints in electronic packaging using moire interferometry (Woychik, Charls G.; Guo, Yi F.)V1554B,461-470(1991)

Three-dimensional automatic precision measurement system by liquid-crystal plate on moire topography (Arai, Yasuhiko; Yekozeki, Shunsuke; Yamada, Tomoharu)V1554B,266-274(1991)

Three-dimensional shape restoration using virtual grating phase detection from deformed grating (Zhou, Shaoxiang; Jiang, Jinyou; Wang, Qimin)V1358,788-792(1991)

Ultrahigh-sensitivity moire interferometry (Han, Bongtae)V1554B,399-411(1991)

Vibration-insensitive moire interferometry system for off-table applications (Guo, Yi F.)V1554B,412-419(1991)

Visualization of impingement field of real-rocket-exhausted jets by using moire deflectometry (He, Anzhi; Yan, Da-Peng; Miao, Peng-Cheng; Wang, Hai-Lin)V1554B,429-434(1991)

Void-crack interaction in aluminum single crystal (Li, X. M.; Chiang, Fu-Pen)V1554A,285-296(1991)

Molecules—see also spectroscopy, molecular

Design of chromophores for nonlinear optical applications (Burland, Donald M.; Rice, J. E.; Downing, J.; Michl, J.)V1560,111-119(1991)

Dielectric relaxation studies of x2 dye containing polystyrene films (Schen, Michael A.; Mopsik, Fred)V1560,315-325(1991)

Electro-optical light modulation in novel azo-dye-substituted poled polymers (Shuto, Yoshito; Amano, Michiyuki; Kaino, Toshikuni)V1560,184-195(1991)

Laser Applications in Life Sciences (Akhmanov, Sergei A.; Poroshina, Marina Y., eds.)V1403(1991)

Model systems for optoelectronic devices based on nonlinear molecular absorption (Speiser, Shammai)V1560,434-442(1991)

Molecular to material design for anomalous-dispersion phase-matched second-harmonic generation (Cahill, Paul A.; Tallant, David R.; Kowalczyk, T. C.; Singer, Kenneth D.)V1560,130-138(1991)

More realistic and efficient algorithm for the drawing of 3-D space-filling molecular models (Wang, Yanqun)V1606,1027-1036(1991)

Nonlinear susceptibilities investigated by the electroabsorption of polymer ion-hemicyanine dye complexes (Nomura, Shintaro; Kobayashi, Takayoshi T.; Nakanishi, Hachiro; Matsuda, Hiro; Okada, Shuji; Tomiyama, Hiromitsu)V1560,272-277(1991)

Photon-induced charge separation in molecular systems studied by time-resolved microwave conductivity: molecular optoelectric switches (Warman, John M.; Jonker, Stephan A.; de Haas, Matthijs P.; Verhoeven, Jan W.; Paddon-Row, Michael N.)V1559,159-170(1991)

Structure/property relationships for molecular second-order nonlinear optics (Marder, Seth R.; Cheng, Lap Tak A.; Tiemann, Bruce G.; Beratan, David N.)V1560,86-97(1991)

Synthesis and nonlinear optical activity of cumulenes (Ermer, Susan P.; Lovejoy, Steven M.; Leung, Doris; Spitzer, Ronnie; Hansen, Glenn A.; Stone, Richard E.)V1560,120-129(1991)

Theoretical analysis of the third-order nonlinear optical properties of linear cyanines and polyenes (Pierce, Brian M.)V1560,148-161(1991)

Theoretical insight into the quadratic nonlinear optical response of organics: derivatives of pyrene and triaminotrinitrobenzene (Bredas, Jean-Luc; Dehu, C.; Meyers, F.; Zyss, Joseph)V1560,98-110(1991)

Third-order optical nonlinearities and femtosecond responses in metallophthalocyanine thin films made by vacuum deposition, molecular beam epitaxy, and spin coating (Wada, Tatsuo; Hosoda, Masahiro; Garito, Anthony F.; Sasabe, Hiroyuki; Terasaki, A.; Kobayashi, Takayoshi T.; Tada, Hiroaki; Koma, Atsushi)V1560,162-171(1991)

Monochromators

Evaluation of the NOAA-11 solar backscatter ultraviolet radiometer, Mod 2 (SBUV/2): inflight calibration (Weiss, Howard; Cebula, Richard P.; Laamann, K.; Hudson, Robert D.)V1493,80-90(1991)

Multilayer monochromator for synchrotron radiation angiography (Baron, Alfred Q.; Barbee, Troy W.; Brown, George S.)V1343,84-94(1991)

Multilayers—see also coatings; interfaces; thin films; x rays

Absorbing materials in multilayer mirrors (Grebenshikov, Sergey V.)V1500,189-193(1991)

Advanced lithographic methods for contact patterning on severe topography (Chu, Ron; Greeneich, James S.; Katz, Barton A.; Lin, Hwang-Kuen; Huang, Dong-Tsair)V1465,238-243(1991)

Advanced X-Ray/EUV Radiation Sources and Applications (Knauer, James P.; Shenoy, Gopal K., eds.)V1345(1991)

Application aspects of the Si-CARL bilayer process (Sebald, Michael; Berthold, Joerg; Beyer, Michael; Leuschner, Rainer; Noelscher, Christoph; Scheler, Ulrich; Sezi, Recai; Ahne, Hellmut; Birkle, Siegfried)V1466,227-237(1991)

Approaches to the construction of intrinsically acentric chromophoric NLO materials: chemical elaboration and resultant properties of self-assembled multilayer structures (Allan, D. S.; Kubota, F.; Marks, Tobin J.; Zhang, T. J.; Lin, W. P.; Wong, George K.)V1560,362-369(1991)

AR layer properties for high-power laser prepared by neutral-solution processing (Wu, Fang F.; Su, Kai L.)V1519,347-349(1991)

Basis and applicaton of evanescent fluorescence measurement (Yuan, Y. F.; Heavens, Oliver S.)V1519,434-439(1991)

Bilayer resist system utilizing alkali-developable organosilicon positive photoresist (Nate, Kazuo; Mizushima, Akiko; Sugiyama, Hisashi)V1466,206-210(1991)

Broadband multilayer coated blazed grating for x-ray wavelengths below 0.6 nm (den Boggende, Antonius J.; Bruijn, Marcel P.; Verhoeven, Jan; Zeijlemaker, H.; Puik, Eric J.; Padmore, Howard A.)V1345,189-197(1991)

Comparative study of carbon and boron carbide spacing layers inside soft x-ray mirrors (Boher, Pierre; Houdy, Philippe; Kaikati, P.; Barchewitz, Robert J.; Van Ijzendoorn, L. J.; Li, Zhigang; Smith, David J.; Joud, J. C.)V1345,165-179(1991)

Critical dimension control using development end point detection for wafers with multilayer structures (Hagi, Toshio; Okuda, Yoshimitsu; Ohkuma, Tohru)V1464,215-221(1991)

Deposition- controlled uniformity of multilayer mirrors (Jankowski, Alan F.; Makowiecki, Daniel M.; McKernan, M. A.; Foreman, R. J.; Patterson, R. G.)V1343,32-38(1991)

Design and analysis of aspherical multilayer imaging x-ray microscope (Shealy, David L.; Jiang, Wu; Hoover, Richard B.)V1343,122-132(1991)

Design considerations for multilayer-coated Schwarzchild objectives for the XUV (Kortright, Jeffrey B.; Underwood, James H.)V1343,95-103(1991)

Design of narrow band XUV and EUV coronagraphs using multilayer optics (Walker, Arthur B.; Allen, Maxwell J.; Barbee, Troy W.; Hoover, Richard B.)V1343,415-427(1991)

Electrical and nonlinear optical properties of zirconium phosphonate multilayer assemblies (Katz, Howard E.; Schilling, M. L.; Ungashe, S.; Putvinski, T. M.; Scheller, G. E.; Chidsey, C. E.; Wilson, William L.)V1560,370-376(1991)

EUV performance of a multilayer-coated high-density toroidal grating (Keski-Kuha, Ritva A.; Thomas, Roger J.; Neupert, Werner M.; Condor, Charles E.; Gum, Jeffrey S.)V1343,566-575(1991)

Evaluation of poly(p-trimethylsilylstyrene and p-pentamethyldisilylstyrene sulfone) as high-resolution electron-beam resists (Gozdz, Antoni S.; Ono, Hiroshi; Ito, Seiki; Shelburne, John A.; Matsuda, Minoru)V1466,200-205(1991)

Fabrication and performance at 1.33 nm of a 0.24-um period multilayer grating (Berrouane, H.; Khan Malek, Chantal; Andre, Jean-Michel; Lesterlin, L.; Ladan, F. R.; Rivoira, R.; Lepetre, Yves; Barchewitz, Robert J.)V1343,428-436(1991)

Fabrication of cold light mirror of film projector by direct electron-beam-evaporated TiO2 and SiO2 starting materials in neutral oxygen atmosphere (Zhong, Di S.; Xu, Guang Z.; Liu, Wi)V1519,350-358(1991)

Fabrication of multilayer Bragg-Fresnel zone plates for the soft x-ray range (Khan Malek, Chantal; Moreno, T.; Guerin, Philippe; Ladan, F. R.; Rivoira, R.; Barchewitz, Robert J.)V1343,56-61(1991)

Fine undercut control in bilayer PMMA-P(MMA-MAA) resist system for e-beam lithography with submicrometer resolution (Bogdanov, Alexei L.; Andersson, Eva K.)V1465,324-329(1991)

Finite element study on indentations into TiN- and multiple TiN/Ti- layers on steel (Wang, H. F.; Wagendristel, A.; Yang, X.; Torzicky, P.; Bangert, H.)V1519,405-410(1991)

Formation of boron nitride and silicon nitride bilayer films by ion-beam-enhanced deposition (Feng, Yi P.; Jiang, Bing Y.; Yang, Gen Q.; Huang, Wei; Zheng, Zhi H.; Liu, Xiang H.; Zou, Shi C.)V1519,440-443(1991)

Frequency-selective devices using a composite multilayer design (Ma, Yushieh; Varadan, Vijay K.; Varadan, Vasundara V.)V1558,132-137(1991)

Grating beam splitting polarizer using multilayer resist method (Aoyama, Shigeru; Yamashita, Tsukasa)V1545,241-250(1991)

Grating efficiency theory versus experimental data in extreme situations (Neviere, Michel; den Boggende, Antonius J.; Padmore, Howard A.; Hollis, K.)V1545,76-87(1991)

Hierarchical proximity effect correction for e-beam direct writing of 64-Mbit DRAM (Misaka, Akio; Hashimoto, Kazuhiko; Kawamoto, M.; Yamashita, H.; Matsuo, Takahiro; Sakashita, Toshihiko; Harafuji, Kenji; Nomura, Noboru)V1465,174-184(1991)

Highly damage-resistant reflectors for 248 nm formed by fluorides multilayers (Izawa, Takao; Yamamura, N.; Uchimura, R.; Hashimoto, I.; Yakuoh, T.; Owadano, Yoshirou; Matsumoto, Y.; Yano, M.)V1441,339-344(1991)

High-quality heavily strained II-VI quantum well (Li, Jie; He, Li; Tang, Wen G.; Shan, W.; Yuan, Shi X.)V1519,660-664(1991)

High-reflective multilayers as narrowband VUV filters (Zukic, Muamer; Torr, Douglas G.)V1485,216-227(1991)

Highly sensitive microresinoid siloxane resist for EB and deep-UV lithography (Yamazaki, Satomi; Ishida, Shinji; Matsumoto, Hiroshi; Aizaki, Naoaki; Muramoto, Naohiro; Mine, Katsutoshi)V1466,538-545(1991)

Imaging performance and tests of soft x-ray telescopes (Spiller, Eberhard; McCorkle, R.; Wilczynski, Janusz S.; Golub, Leon; Nystrom, George U.; Takacs, Peter Z.; Welch, Charles W.)V1343,134-144(1991)

Interface characterization of XUV multilayer reflectors using HRTEM and x-ray and XUV reflectance (Windt, David L.; Hull, Robert; Waskiewicz, Warren K.; Kortright, Jeffrey B.)V1343,292-308(1991)

Interface properties of a-C:H/a-SiOx:H multilayer (Cui, Jing B.; Zhang, Wei P.; Fang, Rong C.; Wang, Chang S.; Zhou, Guien; Wu, Jan X.)V1519,419-422(1991)

Interfaces in Mo/Si multilayers (Slaughter, Jon M.; Kearney, Patrick A.; Schulze, Dean W.; Falco, Charles M.; Hills, C. R.; Saloman, Edward B.; Watts, Richard N.)V1343,73-82(1991)

Investigation of neutral atom and ion emission during laser conditioning of multilayer HfO2-SiO2 coatings (Schildbach, M. A.; Chase, Lloyd L.; Hamza, Alex V.)V1441,287-293(1991)

Laser plasma XUV sources: a role for excimer lasers? (Bijkerk, Fred; Shevelko, A. P.)V1503,380-390(1991)

Lateral-periodicity evaluation of multilayer Bragg reflector surface roughness using x-ray diffraction (Takenaka, Hisataka; Ishii, Yoshikazu)V1345,180-188(1991)

Long-period x-ray standing waves generated by total external reflection (Bedzyk, Michael J.)V1550,151-155(1991)

Manufacturing and reproduction of holographic optical elements in dichromated gelatin films for operation in the infrared (Stojanoff, Christo G.; Tropartz, Stephan; Brasseur, Olivier; Kubitzek, Ruediger)V1485,274-280(1991)

Materials for optimal multilayer coating (Grebenshikov, Sergey V.)V1519,302-307(1991)

Mirror fabrication for full-wafer laser technology (Webb, David J.; Benedict, Melvin K.; Bona, Gian-Luca; Buchmann, Peter L.; Daetwyler, K.; Dietrich, H. P.; Moser, A.; Sasso, G.; Vettiger, Peter; Voegeli, O.)V1418,231-239(1991)

Mirrors as power filters (Kortright, Jeffrey B.)V1345,38-41(1991)

Multilayered optically activated devices based on organic third-order nonlinear optical materials (Swanson, David A.; Altman, Joe C.; Sullivan, Brian J.; Spitzer, Ronnie; Hansen, Glenn A.; Stone, Richard E.)V1560,416-425(1991)

Multilayered superconducting tunnel junctions for use as high-energy-resolution x-ray detectors (Rippert, Edward D.; Song, S. N.; Ketterson, John B.; Ulmer, Melville P.)V1549,283-288(1991)

Multilayer mirrors and filters for imaging the earth's inner magnetosphere (Schulze, Dean W.; Sandel, Bill R.; Broadfoot, A. L.)V1549,319-328(1991)

Multilayer monochromator for synchrotron radiation angiography (Baron, Alfred Q.; Barbee, Troy W.; Brown, George S.)V1343,84-94(1991)

Multilayer OPC for one-shot two-color printer (Sakai, Katsuo)V1458,179-191(1991)

Multilayer optics for soft x-ray projection lithography: problems and prospects (Stearns, Daniel G.; Ceglio, Natale M.; Hawryluk, Andrew M.; Rosen, Robert S.; Vernon, Stephen P.)V1465,80-87(1991)

Multilayer reflectors for the "water window" (Xu, Shi; Evans, Brian L.)V1343,110-121(1991)

Nondestructive investigations of multilayer dielectric coatings (Shapiro, Alexander G.; Yaminsky, Igor V.)V1362,834-843(1991)

Optical characterization of Hg1-xCdxTe/CdTe/GaAs multilayers grown by molecular beam epitaxy (Liu, Wei J.; Liu, Pu L.; Shi, Guo L.; Zhu, Jing-Bing; He, Li; Xie, Qin X.; Yuan, Shi X.)V1519,481-488(1991)

Optimization of quadrilayer structures for various magneto-optical recording materials (He, Ping; McGahan, William A.; Woollam, John A.)V1499,401-411(1991)

Performance of multilayer-coated figured optics for soft x-rays near the diffraction limit (Raab, Eric L.; Tennant, Donald M.; Waskiewicz, Warren K.; MacDowell, Alastair A.; Freeman, Richard R.)V1343,104-109(1991)

Performance of the Multi-Spectral Solar Telescope Array IV: the soft x-ray and extreme ultraviolet filters (Lindblom, Joakim F.; O'Neal, Ray H.; Walker, Arthur B.; Powell, Forbes R.; Barbee, Troy W.; Hoover, Richard B.; Powell, Stephen F.)V1343,544-557(1991)

Photochemistry and photophysics of stilbene and diphenylpolyene surfactants in supported multilayer films (Spooner, Susan P.; Whitten, David G.)V1436,82-91(1991)

Plane and concave VUV and soft x-ray multilayered mirrors (Cao, Jianlin; Miao, Tongqun; Qian, Longsheng; Zhu, Xioufang; Li, Futian; Ma, Yueying; Qian, Limin; Chen, Po; Chen, Xingdan)V1345,225-232(1991)

Polysilanes for microlithography (Wallraff, Gregory M.; Miller, Robert D.; Clecak, Nicholas J.; Baier, M.)V1466,211-217(1991)

Reactive ion-beam-sputtering of fluoride coatings for the UV/VUV range (Schink, Harald; Kolbe, J.; Zimmermann, F.; Ristau, Detleu; Welling, Herbert)V1441,327-338(1991)

Reduction of the standing wave effect in positive photoresist using an antireflection coating (Mehrotra, R.; Mathur, B. P.; Sharan, Sunil)V1463,487-491(1991)

Removal of adsorbed gases with CO2 snow (Zito, Richard R.)V1494,427-433(1991)

Scattering from multilayer coatings: a linear systems model (Harvey, James E.; Lewotsky, Kristin L.)V1530,35-41(1991)

Selection, growth, and characterization of materials for MBE-produced x-ray optics (Kearney, Patrick A.; Slaughter, Jon M.; Falco, Charles M.)V1343,25-31(1991)

Silicon/silicon oxide and silicon/silicon nitride multilayers for XUV optical applications (Boher, Pierre; Houdy, Philippe; Hennet, L.; Delaboudiniere, Jean-Pierre; Kuehne, Mikhael; Mueller, Peter; Li, Zhigang; Smith, David J.)V1343,39-55(1991)

Simulation of local layer-thickness deviation on multilayer diffraction (Guo, S. P.; He, X. C.; Redko, S. V.; Wu, Z. Q.)V1519,400-404(1991)

Soft and hard x-ray reflectivities of multilayers fabricated by alternating-material sputter deposition (Takenaka, Hisataka; Ishii, Yoshikazu; Kinoshita, Hiroo; Kurihara, Kenji)V1345,213-224(1991)

Soft x-ray multilayers fabricated by electron-beam deposition (Sudoh, Masaaki; Yokoyama, Ryouhei; Sumiya, Mitsuo; Yamamoto, Masaki; Yanagihara, Mihiro; Namioka, Takeshi)V1343,14-24(1991)

Soft X-UV silver silicon multilayer mirrors (Shao, Jian D.; Fan, Zheng X.; Guo, Yong H.; Jin, Lei)V1519,298-301(1991)

Solar EUV/FUV line polarimetry: instruments and methods (Hoover, Richard B.; Fineschi, Silvano; Fontenla, Juan; Walker, Arthur B.)V1343,389-403(1991)

Spectrum investigation and imaging of laser-produced plasma by multilayer x-ray optics (Platonov, Yu. Y.; Salashchenko, N. N.; Shmaenok, L. A.)V1440,188-196(1991)

Standing spin wave modes in permalloy-FeCr multilayer films (Chen, H. Y.; Luo, Y. Q.)V1519,761-764(1991)

Structure and optical properties of a-C:H/a-SiOx:H multilayer thin films (Zhang, Wei P.; Cui, Jing B.; Xie, Shan; Song, Yi Z.; Wang, Chang S.; Zhou, Guien; Wu, Jan X.)V1519,23-25(1991)

Study of C-H stretching vibration in hybrid Langmuir-Blodgett/alumina multilayers by infrared spectroscopy (Zheng, Tian S.; Liu, Li Y.; Xing, Zhongjin; Wang, Wen C.; Shen, Yuanhua; Zhang, Zhiming)V1519,339-346(1991)

Study of TEM micrographs of thin-film cross-section replica using spectral analysis (Mei, Ting; Liu, Xu; Tang, Jinfa; Gu, Peifu)V1554A,570-578(1991)

Suppression of columnar-structure formation in Mo-Si layered synthetic microstructures (Niibe, Masahito; Hayashida, Masami; Iizuka, Takashi; Miyake, Akira; Watanabe, Yutaka; Takahashi, Rie; Fukuda, Yasuaki)V1343,2-13(1991)

Synthesis and lithographic evaluation of alternating copolymers of linear and cyclic alkenyl(di)silanes with sulfur dioxide (Gozdz, Antoni S.; Shelburne, John A.)V1466,520-527(1991)

Thickness measurement of combined a-Si and Ti films on c-Si using a monochromatic ellipsometer (Yoo, Chue-San; Jans, Jan C.)V1464,393-403(1991)

Three materials soft x-ray mirrors: theory and application (Boher, Pierre; Hennet, L.; Houdy, Philippe)V1345,198-212(1991)

X-Ray/EUV Optics for Astronomy, Microscopy, Polarimetry, and Projection Lithography (Hoover, Richard B.; Walker, Arthur B., eds.)V1343(1991)

X-ray multilayer-coated reflection gratings: theory and applications (Neviere, Michel; den Boggende, Antonius J.)V1545,116-124(1991)

XUV characterization comparison of Mo/Si multilayer coatings (Windt, David L.; Waskiewicz, Warren K.; Kubiak, Glenn D.; Barbee, Troy W.; Watts, Richard N.)V1343,274-282(1991)

Multiphoton processes—see also spectroscopy

Applications of resonance ionization spectroscopy for semiconductor, environmental and biomedical analysis, and for DNA sequencing (Arlinghaus, Heinrich F.; Spaar, M. T.; Thonnard, N.; McMahon, A. W.; Jacobson, K. B.)V1435,26-35(1991)

Biomedical applications of laser photoionization (Xiong, Xiaoxiong; Moore, Larry J.; Fassett, John D.; O'Haver, Thomas C.)V1435,188-196(1991)

Depth profiling resonance ionization mass spectrometry of electronic materials (Downey, Stephen W.; Emerson, A. B.)V1435,19-25(1991)

Dispersion of n2 in solids (Sheik-Bahae, Mansoor; Hutchings, David C.; Hagan, David J.; Soileau, M. J.; Van Stryland, Eric W.)V1441,430-443(1991)

Dynamics of self-organized x(2) gratings in optical fibers (Kamal, Avais; Terhune, Robert W.; Weinberger, Doreen A.)V1516,137-153(1991)

Effect of polarization on two-color laser-induced damage (Becker, Wilhelm; McIver, John K.; Guenther, Arthur H.)V1441,541-552(1991)

Hole-burning and picosecond time-resolved spectroscopy of isolated molecular clusters (Wittmeyer, Stacey A.; Kaziska, Andrew J.; Shchuka, Maria I.; Topp, Michael R.)V1435,267-278(1991)

Interaction of 1064 nm photons with the Al2O3(1120) surface (Schildbach, M. A.; Hamza, Alex V.)V1441,139-145(1991)

Luminescence and ionization of krypton by multiphotonic excitation near the 3P1 resonant state (Saissac, M.; Berejny, P.; Millet, P.; Yousfi, M.; Salamero, Y.)V1397,739-742(1991)

Measurement of nonlinear constants in photosensitive glass optical fibers (Oesterberg, Ulf L.)V1504,107-109(1991)

Model of second-harmonic generation in glass fibers based on multiphoton ionization interference effects (Anderson, Dana Z.; Mizrahi, Victor; Sipe, John E.)V1516,154-161(1991)

Multiphoton-absorption-induced structural changes in fused silica (Silin, Andrejs R.)V1513,270-273(1991)

Multiphoton excited emission in zinc selenide and other crystals (Prokhorov, Kirill A.; Prokhorov, Alexander M.; Djibladze, Merab I.; Kekelidze, George N.; Gorelik, Vladimir S.)V1501,80-84(1991)

Multiphoton resonance ionization of molecules desorbed from surfaces by ion beams (Winograd, Nicholas; Hrubowchak, D. M.; Ervin, M. H.; Wood, M. C.)V1435,2-11(1991)

New photovoltaic effect in semiconductor junctions n+/p (Dominguez Ferrari, E.; Encinas Sanz, F.; Guerra Perez, J. M.)V1397,725-727(1991)

Novel device for increasing the laser pulse intensity in multiphoton ionization mass spectrometry (Liang, Dawei; Fraser Monteiro, L.; Fraser Monteiro, M. L.)V1501,129-134(1991)

Observation of multiphoton photoemission from a NEA GaAs photocathode (Wang, Liming; Hou, Xun; Cheng, Zhao)V1358,1156-1160(1991)

Optical Methods for Ultrasensitive Detection and Analysis: Techniques and Applications (Fearey, Bryan L., ed.)V1435(1991)

Overview of experimental investigations of laser bandwidth effects in nonlinear optics (Elliott, Daniel S.)V1376,22-33(1991)

Photoemission under three-photon excitation in a NEA GaAs photocathode (Wang, Liming; Hou, Xun; Cheng, Zhao)V1415,120-126(1991)

Photoinductional change of silicate glasses optical parameters at two-photon laser radiation absorption (Glebov, Leonid B.; Efimov, Oleg M.; Mekryukov, A. M.)V1513,274-282(1991)

Photon dynamics in tissue imaging (Chance, Britton; Haselgrove, John C.; Wang, Nai-Guang; Maris, Michael B.; Sevick, Eva M.)V1525,68-82(1991)

Photo resonance excitation and ionization characteristics of atoms by pulsed laser irradiation (Tanazawa, Takeshi; Adachi, Hajime A.; Nakahara, Ktsuhiko; Nittoh, Koichi; Yoshida, Toshifumi; Yoshida, Tadashi; Matsuda, Yasuhiko)V1435,310-321(1991)

Physical models of second-harmonic generation in optical fibers (Lesche, Bernhard)V1516,125-136(1991)

Pulse reshaping and coherent sideband generation effects on multiphoton excitation of polyatomic molecules (Garner, Steven T.; Cantrell, Cyrus D.)V1497,188-196(1991)

Test of Geltman theory of multiple ionization of xenon by intense laser pulses (Bruzzese, Riccardo; Berardi, V.; de Lisio, C.; Solimeno, Salvatore; Spinelli, N.)V1397,735-738(1991)

Two-photon absorption calculations in HgCdTe (Nathan, Vaidya)V1441,553-556(1991)

Two-photon excitation in laser scanning fluorescence microscopy (Strickler, James H.; Webb, Watt W.)V1398,107-118(1991)

Upconversion intensity and multiphonon relaxation of Er3+-doped glasses (Tanabe, Setsuhisa; Hirao, Kazuyuki; Soga, Naohiro; Hanada, Teiichi)V1513,340-348(1991)

Volume optical memory by two-photon interaction (Rentzepis, Peter M.; Dvornikov, Alexander S.)V1563,198-207(1991)

Multiplexing—see also optical communications

4-channel, 662-Mb/s medium-density WDM system with Fabry-Perot laser diodes for subscriber loop applications (Wang, Lon A.; Chapuran, Thomas E.; Menendez, Ronald C.)V1363,85-91(1991)

Analysis of the multichannel coherent FSK subcarrier multiplexing system with pilot carrier and phase noise cancelling scheme (Lee, Yang-Hang; Wu, Jingshown; Tsao, Hen-Wai)V1372,140-149(1991)

Analysis of wavelength division multiplexing technique for optical data storage (Yuk, Tung Ip; Palais, Joseph C.)V1401,130-137(1991)

Behavior of WDM system for intensity modulation (Pierre, Guillaume; Jarret, Bertrand; Brun, Eric)V1511,201-211(1991)

Channel waveguide Mach-Zehnder interferometer for wavelength splitting and combining (Tervonen, Ari; Poyhonen, Pekka; Honkanen, Seppo; Tahkokorpi, Markku T.)V1513,71-75(1991)

Components for Fiber Optic Applications V (Kopera, Paul M., ed.)V1365(1991)

Design and optimization of demultiplexer in ion-exchanged glass waveguides (Mazzola, M.; Montrosset, Ivo; Fincato, Antonello)V1365,2-12(1991)

Design of novel integrated optic devices utilizing depressed index waveguides (Lopez-Amo, Manuel; Menendez-Valdes, Pedro; Sanz, Inmaculada; Muriel, Miguel A.)V1374,74-85(1991)

Free-space optical TDM switch (Goel, Kamal K.; Prucnal, Paul R.; Stacy, John L.; Krol, Mark F.; Johns, Steven T.)V1476,314-319(1991)

Further performance characteristics of a high-sensitivity 64 x 64 element InSb hybrid focal-plane array (Fischer, Robert C.; Martin, Charles J.; Niblack, Curtiss A.; Timlin, Harold A.; Wimmers, James T.)V1494,414-418(1991)

High-power KrF lasers (Key, Michael H.; Bailly-Salins, Rene; Edwards, B.; Harvey, Erol C.; Hirst, Graeme J.; Hooker, Chris J.; Kidd, A. K.; Madraszek, E. M.; Rodgers, P. A.; Ross, Ian N.; Shaw, M. J.; Steyer, M.)V1397,9-17(1991)

High-resolution integrated optic holographic wavelength division multiplexer (Liu, William Y.; Strzelecki, Eva M.; Lin, Freddie S.; Jannson, Tomasz)V1365,20-24(1991)

High-selectivity spectral multiplexers-demultiplexers usable in optical telecommunications obtained from multidielectric coatings at the end of optical fibers (Richier, R.; Amra, Claude)V1504,202-210(1991)

InSb linear multiplexed FPAs for the CRAF/Cassini visible and infrared mapping spectrometer (Niblack, Curtiss A.; Blessinger, Michael A.; Forsthoefel, John J.; Staller, Craig O.; Sobel, Harold R.)V1494,403-413(1991)

Liquid-crystal phase modulators for active micro-optic devices (Purvis, Alan; Williams, Geoffrey; Powell, Norman J.; Clark, Michael G.; Wiltshire, Michael C.)V1455,145-149(1991)

Low- cost fiber optic sensing systems using spatial division multiplexing (Paton, Barry E.)V1332,446-455(1991)

Multiplex and multichannel detection of near-infrared Raman scattering (Chase, Bruce D.)V1439,47-57(1991)

Multiplexed approach for the fiber optic gyro inertial measurement unit (Page, Jerry L.)V1367,93-102(1991)

Multiplex Fabry-Perot interferometer (Snell, Hilary E.; Hays, Paul B.)V1492,403-407(1991)

Multiplexing of remote all-fiber Michelson interferometers with lead insensitivity (Santos, Jose L.; Farahi, Faramarz; Newson, Trevor P.; Leite, Antonio P.; Jackson, David A.)V1511,179-189(1991)

Novel approach to optical frequency synthesis in coherent lightwave systems (Fernando, P. N.; Fake, M.; Seeds, A. J.)V1372,152-163(1991)

Novel high-speed communication system (Dawber, William N.; Hirst, Peter F.; Condon, Brian P.; Maitland, Arthur; Sutton, Phillip)V1476,81-90(1991)

Novel technique for the measurement of coupling length in directional couplers (Cheng, Hsing C.; Ramaswamy, Ramu V.)V1474,291-297(1991)

Optical approach to proximity-operations communications for Space Station Freedom (Marshalek, Robert G.)V1417,53-62(1991)

Optical fiber demultiplexer for telecommunications (Falciai, Riccardo; Scheggi, Annamaria V.; Cosi, Franco; Cao, J. Y.)V1365,38-42(1991)

Optical performance of wavelength division multiplexers made by ion-exchange in glass (Nissim, Carlos; Beguin, Alain; Laborde, Pascale; Lerminiaux, Christian)V1365,13-19(1991)

Optimizing the performance of a frequency-division distributed-optical-fiber sensing system (Leung, Chung-yee; Wu, Jiunn-Shyong; Ho, M. Y.; Chen, Kuang-yi)V1572,566-571(1991)

Optoelectronic devices for fiber-optic sensor interface systems (Hong, C. S.; Hager, Harold E.; Capron, Barbara; Mantz, Joseph L.; Beranek, Mark W.; Huggins, Raymond W.; Chan, Eric Y.; Voitek, Mark; Griffith, David M.; Livezey, Darrell L.; Scharf, Bruce R.)V1418,177-187(1991)

Parallel coherence receiver for quasidistributed optical sensor (Sansonetti, Pierre; Guerin, J. J.; Lequime, Michael; Debrie, J.)V1588,143-149(1991)

Performance analysis of lightwave packet communication networks (Ramaswamy, Raju)V1364,153-162(1991)

Photonics technology for aerospace applications (Figueroa, Luis; Hong, C. S.; Miller, Glen E.; Porter, Charles R.; Smith, David K.)V1418,153-176(1991)

Predetection correlation in a spread-spectrum multiplexing system for fiber optic interferometers (Al-Raweshidy, H. S.; Uttamchandani, Deepak G.)V1367,329-336(1991)

Simulation of parvocellular demultiplexing (Martinez-Uriegas, Eugenio)V1453,300-313(1991)

Spread spectrum technique for passive multiplexing of interferometric optical fiber sensors (Uttamchandani, Deepak G.; Al-Raweshidy, H. S.)V1511,212-219(1991)

Ten-channel single-mode wavelength division demultiplexer in near IR (Chen, Ray T.; Lu, Huey T.; Robinson, Daniel; Wang, Michael R.)V1583,135-142(1991)

Time-division optical microarea networks (Prucnal, Paul R.; Johns, Steven T.; Krol, Mark F.; Stacy, John L.)V1389,462-476(1991)

Transmissive serial interferometric fiber-optic sensor array (Chen, Xiaoguang; Tang, Weizhong; Zhou, Wen)V1572,294-298(1991)

Two-dimensional encoding of images using discrete reticles (Wellfare, Michael R.)V1478,33-40(1991)

Wavelength division and subcarrier system based on Brillouin amplification (Lee, Yang-Hang; Wu, Jingshown; Kao, Ming-Seng; Tsao, Hen-Wai)V1579,155-166(1991)

Wavelength division multiplexers for optical fiber amplifiers (Nagy, Peter A.; Meyer, Tim J.; Tekippe, Vincent J.)V1365,33-37(1991)

Wavelength division multiplexing based on mode-selective coupling (Ouellette, Francois; Duguay, Michel A.)V1365,25-32(1991)

Wavelength division multiplexing of services in a fiber-to-the-home system (Unterleitner, Fred C.)V1363,92-96(1991)

Wavelength-multiplexed fiber optic position encoder for aircraft control systems (Beheim, Glenn; Krasowski, Michael J.; Sotomayor, Jorge L.; Fritsch, Klaus; Flatico, Joseph M.; Bathurst, Richard L.; Eustace, John G.; Anthan, Donald J.)V1369,50-59(1991)

Navigation—see also gyroscopes

Actively controlled multiple-sensor system for feature extraction (Daily, Michael J.; Silberberg, Teresa M.)V1472,85-96(1991)

Airborne Reconnaissance XV (Augustyn, Thomas W.; Henkel, Paul A., eds.)V1538(1991)

ASTRO 1M: a new system for attitude determination in space (Elstner, Christian; Lichtenauer, Gert; Skarus, Waldemar)V1478,150-159(1991)

Autonomous guidance, navigation, and control bridging program plan (McSwain, G. G.; Fernandes, Stan T.; Doane, Kent B.)V1478,228-238(1991)

Autonomous navigation of small co-orbiting satellites using C/A GPS code (Barresi, Giangrande; Soddu, Claudio; Rondinelli, Giuseppe; Caporicci, Lucio; Loria, A.)V1495,246-258(1991)

Characterization of time-division-multiplexed digital optical position transducer (Varshneya, Deepak; Lapierre, A.)V1584,188-201(1991)

Developmental test and evaluation plans for the advanced tactical air reconnaissance system (Minor, John L.; Jenquin, Michael J.)V1538,18-39(1991)

Effects of alternate pictorial pathway displays and stereo 3-D presentation on simulated transport landing approach performance (Busquets, Anthony M.; Parrish, Russell V.; Williams, Steven P.)V1457,91-102(1991)

Hybrid digital/optical ATR system (Goodwin, David B.; Cappiello, Gregory G.; Coppeta, David A.; Govignon, Jacques P.)V1564,536-549(1991)

Image Understanding for Aerospace Applications (Nasr, Hatem N., ed.)V1521(1991)

Improved ring laser gyro navigator (Hadfield, Michael J.; Stiles, Tom; Seidel, David; Miller, William G.; Hensley, David; Wisotsky, Steve; Foote, Michael; Gregory, Rob)V1478,126-144(1991)

Infrared lidar windshear detection for commercial aircraft and the edge technique: a new method for atmospheric wind measurement (Targ, Russell; Bowles, Roland L.; Korb, C. L.; Gentry, Bruce M.; Souilhac, Dominique J.)V1521,144-157(1991)

INS integrated motion analysis for autonomous vehicle navigation (Roberts, Barry; Bazakos, Michael E.)V1521,2-13(1991)

Interframe registration and preprocessing of image sequences (Tajbakhsh, Shahram; Boyce, James F.)V1521,14-22(1991)

Landmark recognition using motion-derived scene structures (Sadjadi, Firooz A.)V1521,98-105(1991)

Linear position sensing by light exchange between two lossy waveguides (Sultan, Michel F.; O'Rourke, Michael J.)V1584,212-219(1991)

Low-cost, low-risk approach to tactical reconnaissance (Beving, James E.; Fishell, Wallace G.)V1538,14-17(1991)

Low-intensity conflict aircraft systems (Henkel, Paul A.)V1538,2-4(1991)

Passive range sensor refinement using texture and segmentation (Sridhar, Banavar; Phatak, Anil; Chatterji, Gano B.)V1478,178-189(1991)

Planetary lander guidance using binary phase-only filters (Reid, Max B.; Hine, Butler P.)V1564,384-394(1991)

Plastic optical fibers for automotive applications (Suganuma, Heiroku; Matsunaga, Tadayo)V1592,12-17(1991)

Reconnaissance mission planning (Fishell, Wallace G.; Fox, Alex J.)V1538,5-13(1991)

Sensing and environment perception for a mobile vehicle (Blais, Francois; Rioux, Marc)V1480,94-101(1991)

Sensors and Sensor Systems for Guidance and Navigation (Wade, Jack; Tuchman, Avi, eds.)V1478(1991)

Sensor system for comet approach and landing (Bonsignori, Roberto; Maresi, Luca)V1478,76-91(1991)

Study of active 3-D terrain mapping for helicopter landings (Velger, Mordekhai; Toker, Gregory)V1478,168-176(1991)

Tactical reconnaissance mission survivability requirements (Lareau, Andy G.; Collins, Ross)V1538,81-98(1991)

Time-optimal maneuver guidance design with sensor line-of-sight constraint (Hartman, Richard D.; Lutze, Frederick H.; Cliff, Eugene M.)V1478,64-75(1991)

Wide-field-of-view star tracker camera (Lewis, Isabella T.; Ledebuhr, Arno G.; Axelrod, Timothy S.; Kordas, Joseph F.; Hills, Robert F.)V1478,2-12(1991)

Neural networks—see also optical computing; pattern recognition

Acquiring rules of selecting cells by using neural network (Yu, He; Zheng, XiangJun; Ye, Yizheng; Wang, LiHong)V1469,412-417(1991)

Adaptive neural methods for multiplexing oriented edges (Marshall, Jonathan A.)V1382,282-291(1991)

Adaptive nonuniformity correction for IR focal-plane arrays using neural networks (Scribner, Dean A.; Sarkady, Kenneth A.; Kruer, Melvin R.; Caulfield, John T.; Hunt, J. D.; Herman, Charles)V1541,100-109(1991)

Adaptive Signal Processing (Haykin, Simon, ed.)V1565(1991)

Adaptive versions of the Ho-Kashyap learning algorithm (Hassoun, Mohamad H.)V1558,459-465(1991)

Analog retina model for detecting moving objects against a moving background (Searfus, Robert M.; Eeckman, Frank H.; Colvin, Michael E.; Axelrod, Timothy S.)V1473,95-101(1991)

Analysis and analog implementation of directionally sensitive shunting inhibitory neural networks (Bouzerdoum, Abdesselam; Nabet, Bahram; Pinter, Robert B.)V1473,29-40(1991)

An Introduction to Biological and Artificial Neural Networks for Pattern Recognition (Kabrisky, Matthew; Rogers, Steven K.)VTT04(1991)

ANN approach for 2-D echocardiographic image processing: application of neocognitron model to LV boundary formation (Wan, Liqun; Li, Dapeng; Wee, William G.; Han, Chia Y.; Porembka, D. T.)V1469,432-440(1991)

ANN-implemented robust vision model (Teng, Chungte; Ligomenides, Panos A.)V1382,74-86(1991)

Application of a multilayer network in image object classification (Tang, Yonghong; Wee, William G.; Han, Chia Y.)V1469,113-120(1991)

Application of back-propagation to the recognition of handwritten digits using morphologically derived features (Hepp, Daniel J.)V1451,228-233(1991)

Application of generalized radial basis functions to the problem of object recognition (Thau, Robert S.)V1469,37-47(1991)

Application of neural networks for the synthesis of binary correlation filters for optical pattern recognition (Mahalanobis, Abhijit; Nadar, Mariappan S.)V1469,292-302(1991)

Application of neural networks in optimization problems: a review (Ashenayi, Kaveh)V1396,285-296(1991)

Application of neural networks to group technology (Caudell, Thomas P.; Smith, Scott D.; Johnson, G. C.; Wunsch, Donald C.)V1469,612-621(1991)

Application of neural networks to range-Doppler imaging (Wu, Xiaoqing; Zhu, Zhaoda)V1569,484-490(1991)

Application of neural network to restoration of signals degraded by a stochastic, shift-variant impulse response function and additive noise (Bilgen, Mehmet; Hung, Hsien-Sen)V1569,260-268(1991)

Application of perceptron to the detecting of particle motion (Li, Jie-gu; Yuan, Qiang)V1469,178-187(1991)

Applications of Artificial Neural Networks II (Rogers, Steven K., ed.)V1469(1991)

Applications of chaotic neurodynamics in pattern recognition (Baird, Bill; Freeman, Walter J.; Eeckman, Frank H.; Yao, Yong)V1469,12-23(1991)

Applications of learning strategies to pattern recognition (Rizki, Mateen M.; Tamburino, Louis A.; Zmuda, Michael A.)V1469,384-391(1991)

Applications of neural networks in experimental high-energy physics (Denby, Bruce)V1469,648-658(1991)

Applications of Optical Engineering: Proceedings of OE/Midwest '90 (Dubiel, Mary K.; Eppinger, Hans E.; Gillespie, Richard E.; Guzik, Rudolph P.; Pearson, James E., eds.)V1396(1991)

Approach to invariant object recognition on grey-level images by exploiting neural network models (Rybak, Ilya A.; Golovan, Alexander V.; Gusakova, Valentina I.)V1469,472-482(1991)

Artificial neural net learns the Sobel operators (and more) (Weller, Scott W.)V1469,219-224(1991)

Artificial neural network and image processing using the Adaptive Solutions' architecture (Baker, Thomas E.)V1452,502-511(1991)

Artificial neural network for supervised learning based on residual analysis (Chan, Keith C.; Vieth, John O.; Wong, Andrew K.)V1469,359-372(1991)

Artificial neural network models for image understanding (Kulkarni, Arun D.; Byars, P.)V1452,512-522(1991)

Artificial neural networks and Abelian harmonic analysis (Rodriguez, Domingo; Pertuz-Campo, Jairo)V1565,492-503(1991)

Artificial neural networks as TV signal processors (Spence, Clay D.; Pearson, John C.; Sverdlove, Ronald)V1469,665-670(1991)

Artificial neural networks for automatic object recognition (Rogers, Steven K.; Ruck, Dennis W.; Kabrisky, Matthew; Tarr, Gregory L.; Oxley, Mark E.)VIS07,231-243(1991)

Artificial neural system approach to IR target identification (Holland, Orgal T.; Tarr, Tomas; Farsaie, Ali; Fuller, James M.)V1469,102-112(1991)

Aspect networks: using multiple views to learn and recognize 3-D objects (Seibert, Michael; Waxman, Allen M.)V1383,10-19(1991)

Aspects of reconfigurable neural networks (Babic, Ranko)V1469,575-580(1991)

Atmospheric propagation effects on pattern recognition by neural networks (Giever, John C.; Hoock, Donald W.)V1486,302-313(1991)

Automatic adjustment of display window (gray level) for MR images using a neural network (Ohhashi, Akinami; Yamada, Shinichi; Haruki, Kazuhito; Hatano, Hisaaki; Fujii, Yumi; Yamaguchi, Koujiro; Ogata, Hakaru)V1444,63-74(1991)

Automatic design of signal processors using neural networks (Menon, Murali M.; Van Allen, Eric J.)V1469,322-328(1991)

Automatic digitization of contour lines for digital map production (Yla-Jaaski, Juha; Yu, Xiaohan)V1472,201-207(1991)

Automatic Object Recognition (Sadjadi, Firooz A., ed.)V1471(1991)

Autowave media and neural networks (Balkarey, Yu. I.; Evtikhov, M. G.; Elinson, Matvey I.)V1621,238-249(1991)

Bidirectional log-polar mapping for invariant object recognition (Mehanian, Couroush; Rak, Steven J.)V1471,200-209(1991)

Cellular-automata-based learning network for pattern recognition (Tzionas, Panagiotis; Tsalides, Ph.; Thanailakis, A.)V1606,269-280(1991)

Cerebellum as a neuronal machine: modern talking (Dunin-Barkowski, W. L.)V1621,250-258(1991)

Challenges of vision theory: self-organization of neural mechanisms for stable steering of object-grouping data in visual motion perception (Marshall, Jonathan A.)V1569,200-215(1991)

Chaotic nature of mesh networks with distributed routing (Rucinski, Andrzej; Drexel, Peter G.; Dziurla, Barbara)V1390,388-398(1991)

Class of learning algorithms for multilayer perceptron (Abbasi, M.; Sayeh, Mohammad R.)V1396,237-242(1991)

Color character recognition method based on a model of human visual processing (Yamaba, Kazuo; Miyake, Yoichi)V1453,290-299(1991)

Combined approach for large-scale pattern recognition with translational, rotational, and scaling invariances (Xu, Qing; Inigo, Rafael M.; McVey, Eugene S.)V1471,378-389(1991)

Combining neural networks and decision trees (Sankar, Ananth; Mammone, Richard J.)V1469,374-383(1991)

Compact optical associative memory (Burns, Thomas J.; Rogers, Steven K.; Kabrisky, Matthew; Vogel, George A.)V1469,208-218(1991)

Compact optical neuro-processors (Paek, Eung Gi; Wullert, John R.; Von Lehman, A.; Patel, J. S.; Martin, R.)V1621,340-350(1991)

Comparative evaluation of neural-based versus conventional segmentors (Daniell, Cindy E.; Kemsley, David; Bouyssounouse, Xavier)V1471,436-451(1991)

Comparative study of model fitting by using neural network and regression (Niu, Aiqun; Li, Dapeng; Wan, Liqun; Wee, William G.; Carrier, Charles W.)V1469,495-505(1991)

Comparison of a multilayered perceptron with standard classification techniques in the presense of noise (Willson, Gregory B.)V1469,351-358(1991)

Comparison of backpropagation neural networks and statistical techniques for analysis of geological features in Landsat imagery (Parikh, Jo Ann; DaPonte, John S.; Damodaran, Meledath; Karageorgiou, Angelos; Podaras, Petros)V1469,526-538(1991)

Comparison of neural network classifiers to quadratic classifiers for sensor fusion (Brown, Joseph R.; Bergondy, Daniel; Archer, Susan J.)V1469,539-543(1991)

Comparison of sinusoidal perceptron with multilayer classical perceptron (Karimi, B.; Baradaran, T.; Ashenayi, Kaveh; Vogh, James)V1396,226-236(1991)

Comparison of techniques for disparate sensor fusion (Huntsberger, Terrance L.)V1383,589-595(1991)

Composite damage assessment employing an optical neural network processor and an embedded fiber-optic sensor array (Grossman, Barry G.; Gao, Xing; Thursby, Michael H.)V1588,64-75(1991)

Computer interpretation of thallium SPECT studies based on neural network analysis (Wang, David C.; Karvelis, K. C.)V1445,574-575(1991)

Connectionist learning systems for control (Baker, Walter L.; Farrell, Jay A.)V1382,181-198(1991)

Connectionist natural language parsing with BrainC (Mueller, Adrian; Zell, Andreas)V1469,188-196(1991)

Constructing attribute classes by example learning: the research of attribute-based knowledge-style pattern recognition (Zhou, Lijia; Song, Hongjun; Zhao, S.)V1469,404-411(1991)

Continuous recognition of sonar targets using neural networks (Venugopal, Kootala P.; Pandya, Abhijit S.; Sudhakar, Raghavan)V1471,44-53(1991)

Contrast, size, and orientation-invariant target detection in infrared imagery (Zhou, Yi-Tong; Crawshaw, Richard D.)V1471,404-411(1991)

Controller implemented by recording the fuzzy rules by backpropagation neural networks (Ying, Xingren; Zeng, Nan)V1469,846-851(1991)

Coping with complexity in the navigation of an autonomous mobile robot (Dodds, David R.)V1388,448-452(1991)

Crystal surface analysis using matrix textural features classified by a probabilistic neural network (Sawyer, Curry R.; Quach, Viet; Nason, Donald; van den Berg, Lodewijk)V1567,254-263(1991)

Design of an analog VLSI chip for a neural network target tracker (Narathong, Chiewcharn; Inigo, Rafael M.)V1452,523-531(1991)

Detection and classification of undersea objects using multilayer perceptrons (Shazeer, Dov J.; Bello, M.)V1469,622-636(1991)

Detection of liver metastisis using the backpropagation algorithm and linear discriminant analysis (DaPonte, John S.; Parikh, Jo Ann; Katz, David A.)V1469,441-450(1991)

Detection of tool wear using multisensor readings defused by artificial neural network (Masory, Oren)V1469,515-525(1991)

Development of a neural network for early detection of renal osteodystrophy (Cheng, Shirley N.; Chan, Heang-Ping; Adler, Ronald; Niklason, Loren T.; Chang, Chair-Li)V1450,90-98(1991)

Development of a smart workstation for use in mammography (Giger, Maryellen L.; Nishikawa, Robert M.; Doi, Kunio; Yin, Fang-Fang; Vyborny, Carl J.; Schmidt, Robert A.; Metz, Charles E.; Wu, Yuzheng; MacMahon, Heber; Yoshimura, Hitoshi)V1445,101-103(1991)

Different types of theories in neurocomputation (Bastida, M. R.; Figueroa-Nazuno, Jesus)V1469,150-156(1991)

DIGNET: a self-organizing neural network for automatic pattern recognition and classification (Thomopoulos, Stelios C.; Bougoulias, Dimitrios K.)V1470,253-262(1991)

Dynamically reconfigurable multiprocessor system for high-order-bidirectional-associative-memory-based image recognition (Wu, Chwan-Hwa; Roland, David A.)V1471,210-221(1991)

EGOLOGY: psychological spatial breakthrough for social redirection—multidisciplinary spatial focus for individuals/humankind (Thompson, Robert A.; Thompson, Louise A.)V1469,451-462(1991)

Efficient recognition with high-order neural networks (Carlson, Rolf; Jeffries, Clark)V1469,684-696(1991)

Electron trapping materials for adaptive learning in photonic neural networks (Farhat, Nabil H.)V1621,310-319(1991)

Failure of outer-product learning to perform higher-order mapping (Kinser, Jason M.)V1541,187-198(1991)

Family of K-winner networks (Wolfe, William J.; Mathis, Donald W.; Anderson, C.; Rothman, Jay; Gottler, Michael; Brady, G.; Walker, R.; Duane, G.; Alaghband, Gita)V1382,240-254(1991)

Fast algorithm for a neocognitron neural network with back-propagation (Qing, Kent P.; Means, Robert W.)V1569,111-120(1991)

Fast-digital multiplication using multizero neural networks (Hu, Chia-Lun J.)V1469,586-591(1991)

Fast identification of images using neural networks (Min, Kwang-Shik; Min, Hisook L.)V1469,129-136(1991)

Fast optoelectronic neurocomputer for character recognition (Zhang, Lin; Robinson, Michael G.; Johnson, Kristina M.)V1469,225-229(1991)

Fault tolerance of optoelectronic neural networks (Ghosh, Anjan K.)V1563,120-120(1991)

Faying surface-gap measurement of aircraft structures for shim fabrication and installation (Sarr, Dennis P.; Jurick, Thomas W.)V1469,506-514(1991)

Feature extractor giving distortion invariant hierarchical feature space (Lampinen, Jouko)V1469,832-842(1991)

Feature trajectory reduction of integrated autoregressive processes based on a multilayer self-organizing neural network (Klose, Joerg; Altena, Oliver)V1565,504-517(1991)

Framework for load apportioning and interactive force control using a Hopfield neural network (Copeland, Bruce R.; Anderson, Joseph N.)V1381,177-188(1991)

Frequency-based pattern recognition using neural networks (Lu, Simon W.)V1569,452-462(1991)

Fusion of multiple-sensor imagery based on target motion characteristics (Tsao, Tien-Ren J.; Libert, John M.)V1470,37-47(1991)

Fuzzy logic and neural networks in artificial intelligence and pattern recognition (Sanchez, Elie)V1569,474-483(1991)

Generalization of the problem of correspondence in long-range motion and the proposal for a solution (Stratton, Norman A.; Vaina, Lucia M.)V1468,176-185(1991)

Generalized neocognitron model for facial recognition (Chen, Su-Shing; Hong, Young-Sik)V1569,463-473(1991)

Generalized neural networks for tactical target image segmentation (Tarr, Gregory L.; Rogers, Steven K.; Kabrisky, Matthew; Oxley, Mark E.; Priddy, Kevin L.)V1469,2-11(1991)

Generalizing from a small set of training exemplars for handwritten digit recognition (Simon, Wayne E.; Carter, Jeffrey R.)V1469,592-601(1991)

Generating good design from bad design: dynamical network approach (Sayeh, Mohammad R.; Ragu, A.)V1396,276-284(1991)

Generation of exploratory schedules in closed loop for enhanced machine learning (Guez, Allon; Ahmad, Ziauddin)V1469,750-755(1991)

Handwritten zip code recognition using an optical radial basis function classifier (Neifeld, Mark A.; Rakshit, S.; Psaltis, Demetri)V1469,250-255(1991)

Hardware implementation of a neural network performing multispectral image fusion (Kagel, Joseph H.)V1469,659-664(1991)

Hetero-association for pattern translation (Yu, Francis T.; Lu, Taiwei; Yang, Xiangyang)V1507,210-221(1991)

Heterogeneous input neuration for network-based object recognition architectures (Gnazzo, John F.)V1569,239-246(1991)

Hierarchical multisensor analysis for robotic exploration (Eberlein, Susan; Yates, Gigi; Majani, Eric)V1388,578-586(1991)

Hierarchical network for clutter and texture modeling (Luttrell, Stephen P.)V1565,518-528(1991)

Hierarchical neural net with pyramid data structures for region labeling of images (Rosten, David P.; Yuen, P.; Hunt, Bobby R.)V1472,118-127(1991)

High-density CCD neurocomputer chip for accurate real-time segmentation of noisy images (Roth, Michael W.; Thompson, K. E.; Kulp, R. L.; Alvarez, E. B.)V1469,25-36(1991)

High-speed optical interconnects for parallel processing and neural networks (Barnes, Nigel; Healey, Peter; McKee, Paul; O'Neill, Alan; Rejman-Greene, Marek A.; Scott, Geoff; Smith, David W.; Webb, Roderick P.; Wood, David)V1389,477-483(1991)

Ho-Kashyap CAAP 1:1 associative processors (Casasent, David P.; Telfer, Brian A.)V1382,158-166(1991)

Holograms in optical computing (Caulfield, H. J.)VIS08,54-61(1991)

Holographic associative memory of biological systems (Gariaev, Peter P.; Chudin, Viktor I.; Komissarov, Gennady G.; Berezin, Andrey A.; Vasiliev, Antoly A.)V1621,280-291(1991)

Holographic associative memory with bipolar features (Wang, Xu-Ming; Mu, Guoguang)V1558,518-528(1991)

Hough transform implementation using an analog associative network (Arrue, Begona C.; Inigo, Rafael M.)V1469,420-431(1991)

How to use optics in neural nets: a perspectus (Casasent, David P.)V1564,630-631(1991)

Hybrid digital/optical ATR system (Goodwin, David B.; Cappiello, Gregory G.; Coppeta, David A.; Govignon, Jacques P.)V1564,536-549(1991)

Hybrid modulation properties of the Epson LCTV (Kirsch, James C.; Loudin, Jeffrey A.; Gregory, Don A.)V1558,432-441(1991)

Image processing based on supervised learning in neural networks (Hasegawa, Akira; Zhang, Wei; Itoh, Kazuyoshi; Ichioka, Yoshiki)V1621,374-379(1991)

Image quality measurements with a neural brightness perception model (Grogan, Timothy A.; Wu, Mei)V1453,16-30(1991)

Image recognition, learning, and control in a cellular automata network (Raghavan, Raghu)V1469,89-101(1991)

Implementing neural-morphological operations using programmable logic (Shih, Frank Y.; Moh, Jenlong)V1382,99-110(1991)

Imposing a temporal structure in neural networks (Gupta, Lalit; Sayeh, Mohammad R.; Upadhye, Anand M.)V1396,266-269(1991)

Improved adaptive resonance theory (Shih, Frank Y.; Moh, Jenlong)V1382,26-36(1991)

Influence of different nonlinearity functions on Perceptron performance (Ashenayi, Kaveh; Vogh, James)V1396,215-225(1991)

Innovative architectural and theoretical considerations yield efficient fuzzy logic controller VLSI design (Basehore, Paul; Yestrebsky, Joseph T.)V1470,190-196(1991)

Integrated circuits with three-dimensional optical interconnections: an element base of neural networks (Gulyaev, Yuri V.; Elinson, Matvey I.; Kopylov, Yuri L.; Perov, Polievkt I.)V1621,84-92(1991)

Integrated neural network system for histological image understanding (Refenes, A. N.; Jain, N.; Alsulaiman, Mansour M.)V1386,62-74(1991)

Intelligent Robots and Computer Vision IX: Neural, Biological, and 3-D Methods (Casasent, David P., ed.)V1382(1991)

Interactive tools for assisting the extraction of cartographic features (Hunt, Bobby R.; Ryan, Thomas W.; Sementilli, P.; DeKruger, D.)V1472,208-218(1991)

Invariant pattern recognition via higher order preprocessing and backprop (Davis, Jon P.; Schmidt, William A.)V1469,804-811(1991)

Iterative neural networks for skeletonization and thinning (Krishnapuram, Raghu J.; Chen, Ling-Fan)V1382,271-281(1991)

Joint space/spatial-frequency representations as preprocessing steps for neural nets; joint recognition of separately learned patterns; results and limitations (Rueff, Manfred; Frankhauser, P.; Dettki, Frank)V1382,255-270(1991)

Knowledge-based system using a neural network (Szabo, Raisa R.; Pandya, Abhijit S.; Szabo, Bela)V1468,794-801(1991)

Knowledge representation by dynamic competitive learning techniques (Racz, Janos; Klotz, Tamas)V1469,778-783(1991)

Landmark-based partial shape recognition by a BAM neural network (Liu, Xianjun; Ansari, Nirwan)V1606,1069-1079(1991)

Large-scale networks via self-organizing hierarchical networks (Smotroff, Ira G.; Friedman, David H.; Connolly, Dennis)V1469,544-550(1991)

LCS: a natural language comprehension system (Trigano, Philippe; Talon, Benedicte; Baltazart, Didier; Demko, Christophe; Newstead, Emma)V1468,866-874(1991)

Learning filter systems with maximum correlation and maximum separation properties (Lenz, Reiner; Osterberg, Mats)V1469,784-795(1991)

Learning in linear feature-discovery networks (Leen, Todd K.)V1565,472-481(1991)

Learning spatially coherent properties of the visual world in connectionist networks (Becker, Suzanna; Hinton, Geoffrey E.)V1569,218-226(1991)

Less interclass disturbance learning for unsupervised neural computing (Liu, Lurng-Kuo; Ligomenides, Panos A.)V1606,496-507(1991)

Linear interconnection architecture in parallel implementation of neural network models (Mostafavi, M. T.)V1396,193-201(1991)

Linear lattice architectures that utilize the central limit for image analysis, Gaussian operators, sine, cosine, Fourier, and Gabor transforms (Ben-Arie, Jezekiel)V1606,823-838(1991)

Linear programming for learning in neural networks (Raghavan, Raghu)V1472,139-148(1991)

Liquid-crystal television optical neural network: architecture, design, and models (Yu, Francis T.)V1455,150-166(1991)

Locally linear neural networks for optical correlators (Gustafson, Steven C.; Little, Gordon R.; Olczak, Eugene G.)V1469,268-274(1991)

Massively parallel image restoration (Menon, Murali M.)V1471,185-190(1991)

Massively parallel implementation of neural network architectures (Omidvar, Omid M.; Wilson, Charles L.)V1452,532-543(1991)

Mean-field stereo correspondence for natural images (Klarquist, William N.; Acton, Scott T.; Ghosh, Joydeep)V1453,321-332(1991)

Millimeter wave radar stationary-target classification using a high-order neural network (Hughen, James H.; Hollon, Kenneth R.)V1469,341-350(1991)

Mobile robot system for the handicapped (Palakal, Mathew J.; Chien, Yung-Ping; Chittajallu, Siva K.; Xue, Qing L.)V1468,456-466(1991)

Modeling inner and outer plexiform retinal processing using nonlinear coupled resistive networks (Andreou, Andreas G.; Boahen, Kwabena A.)V1453,270-281(1991)

Modeling of local neural networks of the visual cortex and applications to image processing (Rybak, Ilya A.; Shevtsova, Natalia A.; Podladchikova, Lubov N.)V1469,737-748(1991)

Multifunctional hybrid neural network: real-time laboratory results (Natarajan, Sanjay S.; Casasent, David P.; Smokelin, John-Scott)V1564,474-488(1991)

Multiple resonant boundary contour system (Lehar, Steve M.; Worth, Andrew J.)V1469,50-62(1991)

Multiple target-to-track association and track estimation system using a neural network (Yee, Mark L.; Casasent, David P.)V1481,418-429(1991)

Multiresolution segmentation of forward-looking IR and SAR imagery using neural networks (Beck, Hal E.; Bergondy, Daniel; Brown, Joseph R.; Sari-Sarraf, Hamed)V1381,600-609(1991)

Multisensor object segmentation using a neural network (Gaughan, Patrick T.; Flachs, Gerald M.; Jordan, Jay B.)V1469,812-819(1991)

Multistage object recognition using dynamical-link graph matching (Flaton, Kenneth A.)V1469,137-148(1991)

Multitask neural network for vision machine systems (Gupta, Madan M.; Knopf, George K.)V1382,60-73(1991)

Multitask neurovision processor with extensive feedback and feedforward connections (Gupta, Madan M.; Knopf, George K.)V1606,482-495(1991)

Natural language parsing in a hybrid connectionist-symbolic architecture (Mueller, Adrian; Zell, Andreas)V1468,875-881(1991)

Negative fuse network (Liu, Shih-Chii; Harris, John G.)V1473,185-193(1991)

Network compensation for missing sensors (Ahumada, Albert J.; Mulligan, Jeffrey B.)V1453,134-146(1991)

Neural control of smart electromagnetic structures (Thursby, Michael H.; Yoo, Kisuck; Grossman, Barry G.)V1588,218-228(1991)

Neural data association (Kim, Kwang H.; Shafai, Bahram)V1481,406-417(1991)

Neural edge detector (Enab, Yehia M.)V1382,292-303(1991)

Neural model for feature matching in stereo vision (Wang, Shengrui; Poussart, Denis; Gagne, Simon)V1382,37-48(1991)

Neural net selection of features for defect inspection (Sasaki, Kenji; Casasent, David P.; Natarajan, Sanjay S.)V1384,228-233(1991)

Neural network adaptive optics for the multiple-mirror telescope (Wizinowich, Peter L.; Lloyd-Hart, Michael; McLeod, Brian A.; Colucci, D'nardo; Dekany, Richard G.; Wittman, David; Angel, J. R.; McCarthy, Donald W.; Hulburd, William G.; Sandler, David G.)V1542,148-158(1991)

Neural-network-aided design for image processing (Vitsnudel, Ilia; Ginosar, Ran; Zeevi, Yehoshua Y.)V1606,1086-1091(1991)

Neural network air-conditioning system for individual comfort (Takemori, Toshikazu; Miyasaka, Nobuji; Hirose, Shozo)V1469,157-165(1991)

Neural network approach for object orientation classification (Yeung, Keith K.; Zakarauskas, Pierre; McCray, Allan G.)V1569,133-146(1991)

Neural network approach to multipath delay estimation (Welstead, Stephen T.; Ward, Michael J.; Keefer, Christopher W.)V1565,482-491(1991)

Neural network approach to power system security (Daneshdoost, Morteza)V1396,270-275(1991)

Neural network architecture for form and motion perception (Grossberg, Stephen)V1469,24-26(1991)

Neural-network-based image processing of human corneal endothelial micrograms (Hasegawa, Akira; Zhang, Wei; Itoh, Kazuyoshi; Ichioka, Yoshiki)V1558,414-421(1991)

Neural-network-based vision processing for autonomous robot guidance (Pomerleau, Dean A.)V1469,121-128(1991)

Neural network design for channel routing (Zargham, Mehdi R.; Sayeh, Mohammad R.)V1396,202-208(1991)

Neural network edge detector (Spreeuwers, Luuk J.)V1451,204-215(1991)

Neural network for improving terrain elevation measurement from stereo images (Jordan, Michael)V1567,179-187(1991)

Neural network for inferring the shape of occluded objects (Citkusev, Ljubomir; Vaina, Lucia M.)V1468,786-793(1991)

Neural network for passive acoustic discrimination between surface and submarine targets (Baran, Robert H.; Coughlin, James P.)V1471,164-176(1991)

Neural network labeling of the Gulf Stream (Lybanon, Matthew; Molinelli, Eugene J.; Muncill, G.; Pepe, Kevin)V1469,637-647(1991)

Neural network modeling of radar backscatter from an ocean surface using chaos theory (Leung, Henry; Haykin, Simon)V1565,279-286(1991)

Neural network model of dynamic form perception: implications of retinal persistence and extraretinal sharpening for the perception of moving boundaries (Ogmen, Haluk)V1606,350-359(1991)

Neural network optimization, components, and design selection (Weller, Scott W.)V1354,371-378(1991)

Neural networks and computers based on in-phase optics (Kovatchev, Methodi; Ilieva, R.)V1621,259-267(1991)

Neural networks and model-based approaches to object identification (Perlovsky, Leonid I.)V1606,1080-1085(1991)

Neural networks application in autonomous path generation for mobile robots (Pourboghrat, Farzad)V1396,243-251(1991)

Neural networks for ATR parameters adaptation (Amehdi, Hossien; Nasr, Hatem N.)V1483,177-184(1991)

Neural networks for Fredholm-type integral equations (Vemuri, V.; Jang, Gyu-Sang)V1469,563-574(1991)

Neural networks for halftoning of color images (Ling, Daniel T.; Just, Dieter)V1452,10-20(1991)

Neural networks for medical image segmentation (Lin, Wei-Chung; Tsao, Chen-Kuo; Chen, Chin-Tu; Feng, Yu-Jen)V1445,376-385(1991)

Neural networks for robot navigation (Pandya, Abhijit S.; Luebbers, Paul G.)V1468,802-811(1991)

Neural networks for smart structures with fiber optic sensors (Sayeh, Mohammad R.; Viswanathan, R.; Dhali, Shirshak K.)V1396,417-429(1991)

Neural networks for the recognition of skilled arm and hand movements (Vaina, Lucia M.; Tuncer, Temel E.)V1468,990-999(1991)

Neural networks implementation on a parallel machine (Wang, Chung Ching; Shirazi, Behrooz)V1396,209-214(1991)

Neural networks in bandwidth compression (Habibi, Ali)V1567,334-340(1991)

Neural network vision integration with learning (Toborg, Scott T.)V1469,77-88(1991)

Neural optoelectronic correlator for pattern recognition (Figue, J.; Refregier, Philippe; Rajbenbach, Henri J.; Huignard, Jean-Pierre)V1564,550-561(1991)

Noise tolerance of adaptive resonance theory neural network for binary pattern recognition (Kim, Yong-Soo; Mitra, Sunanda)V1565,323-330(1991)

Nonlinear dynamics of neuromorphic optical system with spatio-temporal interactions (Vorontsov, Michael A.; Iroshnikov, N. G.)V1621,292-298(1991)

Novel transform for image description and compression with implementation by neural architectures (Ben-Arie, Jezekiel; Rao, K. R.)V1569,367-374(1991)

Oh say, can you see? The physiology of vision (Young, Richard A.)V1453,92-123(1991)

On-focal-plane-array feature extraction using a 3-D artificial neural network (3DANN): Part I (Carson, John C.)V1541,141-144(1991)

On-focal-plane-array feature extraction using 3-D artificial neural network (3DANN): Part II (Carson, John C.)V1541,227-231(1991)

Optical associative memories based on time-delayed four-wave mixing (Belov, M. N.; Manykin, Edward A.)V1621,268-279(1991)

Optical auto- and heteroassociative memory based on a high-order neural network (Kiselyov, Boris S.; Kulakov, Nickolay Y.; Mikaelian, Andrei L.; Shkitin, Vladimir A.)V1621,328-339(1991)

Optical Distributed Inference Network (ODIN) (Bostel, Ashley J.; McOwan, Peter W.; Hall, Trevor J.)V1385,165-172(1991)

Optical/electronical hybrid three-layer neural network for pattern recognition (Zhang, Yanxin; Shen, Jinyuan)V1469,303-307(1991)

Optical engineering for neural networks: an emerging technology (Sayeh, Mohammad R.)V1396,734-743(1991)

Optical higher order double-layer associative memory (Lam, David T.; Carroll, John E.)V1505,104-114(1991)

Optical image segmentor using wavelet filtering techniques as the front-end of a neural network classifier (Veronin, Christopher P.; Rogers, Steven K.; Kabrisky, Matthew; Welsh, Byron M.; Priddy, Kevin L.; Ayer, Kevin W.)V1469,281-291(1991)

Optical implementation of a Hopfield-type neural network by the use of persistent spectral hole-burning media (Ollikainen, Olavi; Rebane, A.)V1621,351-361(1991)

Optical implementation of associative memory based on parallel rank-one interconnections (Jeon, Ho-In; Abushagur, Mustafa A.; Caulfield, H. J.)V1564,522-535(1991)

Optical implementation of neocognitron and its applications to radar signature discrimination (Chao, Tien-Hsin; Stoner, William W.)V1558,505-517(1991)

Optical implementation of winner-take-all models of neural networks (Kobzev, E. F.; Vorontsov, Michael A.)V1402,165-174(1991)

Optical Information Processing Systems and Architectures III (Javidi, Bahram, ed.)V1564(1991)

Optical interconnect and neural network applications based on a simplified holographic N4 recording technique (Lin, Freddie S.; Lu, Taiwei; Kostrzewski, Andrew A.; Chou, Hung)V1558,406-413(1991)

Optical Memory and Neural Networks (Mikaelian, Andrei L., ed.)V1621(1991)

Optical multilayer neural networks (Psaltis, Demetri; Qiao, Yong)V1564,489-494(1991)

Optical neural network: architecture, design, and models (Yu, Francis T.)V1558,390-405(1991)

Optical neural networks: an implementation (Siahmakoun, Azad; Blair, Steven M.; Weiss, Markus R.)V1396,190-192(1991)

Optical neural networks based on liquid-crystal light valves and photorefractive crystals (Owechko, Yuri; Soffer, Bernard H.)V1455,136-144(1991)

Optical neural networks using electron trapping materials (Jutamulia, Suganda; Storti, George M.; Seiderman, William M.)V1558,442-447(1991)

Optical neural network with reconfigurable holographic interconnection (Kirk, Andrew G.; Kendall, G. D.; Imam, H.; Hall, Trevor J.)V1621,320-327(1991)

Optical neurocomputer architectures using spatial light modulators (Robinson, Michael G.; Zhang, Lin; Johnson, Kristina M.)V1469,240-249(1991)

Optical principles of information processing in supercomputer architecture (Burtsev, Vsevolod S.)V1621,380-387(1991)

Optical processing and hybrid neural nets (Casasent, David P.)V1469,256-267(1991)

Optimal generalized weighted-order-statistic filters (Yin, Lin; Astola, Jaakko; Neuvo, Yrjo A.)V1606,431-442(1991)

Optimization neural net for multiple-target data association: real-time optical lab results (Yee, Mark L.; Casasent, David P.)V1469,308-319(1991)

Optimum structure learning algorithms for competitive learning neural network (Uchiyama, Toshio; Sakai, Mitsuhiro; Saito, Tomohide; Nakamura, Taichi)V1451,192-203(1991)

Optoelectronically implemented neural network for early visual processing (Mehanian, Courosh; Aull, Brian F.; Nichols, Kirby B.)V1469,275-280(1991)

Optoelectronic hardware issues for implementation of simulated annealing or Boltzmann machines (Lalanne, Philippe; Chavel, Pierre)V1621,388-401(1991)

Optoelectronic neuron (Pankove, Jacques I.; Radehaus, C.; Borghs, Gustaaf)V1361,620-627(1991)

Optoelectronic neuron arrays (Psaltis, Demetri; Lin, Steven H.)V1562,204-212(1991)

Order-statistic filters on matrices of images (Wilson, Stephen S.)V1451,242-253(1991)

Parallel optical information, concept, and response evolver: POINCARE (Caulfield, H. J.; Caulfield, Kimberly)V1469,232-239(1991)

Parameter estimation for process control with neural networks (Samad, Tariq; Mathur, Anoop)V1469,766-777(1991)

Pattern recognition, attention, and information bottlenecks in the primate visual system (Van Essen, David; Olshausen, B.; Anderson, Clifford H.; Gallant, J. T. L.)V1473,17-28(1991)

Pattern recognition, neural networks, and artificial intelligence (Bezdek, James C.)V1468,924-935(1991)

Pattern recognition using w-orbit finite automata (Liu, Ying; Ma, Hede)V1606,226-240(1991)

Perceptual alternation of ambiguous patterns: a model based on an artificial neural network (Riani, Massimo; Masulli, Francesco; Simonotto, Enrico)V1469,166-177(1991)

Performance comparison of neural network and statistical discriminant processing techniques for automatic modulation recognition (Hill, Peter C.; Orzeszko, Gabriel R.)V1469,329-340(1991)

Performance evaluation of different neural network training algorithms in error control coding (Hussain, Mukhtar; Bedi, Jatinder S.)V1469,697-707(1991)

Performance of a neural-network-based 3-D object recognition system (Rak, Steven J.; Kolodzy, Paul J.)V1471,177-184(1991)

Phonetic-to-acoustic and acoustic-to-phonetic mapping using recurrent neural networks (Kumar, Vinod V.; Krishnamurthy, Ashok K.; Ahalt, Stanley C.)V1469,484-494(1991)

Photoneuron: dynamically reconfigurable information processing control element utilizing embedded-fiber waveguide interconnects (Glista, Andrew S.)V1563,139-155(1991)

Photorefractive devices for optical information processing (Yeh, Pochi A.; Gu, Claire; Hong, John H.)V1562,32-43(1991)

Pixel level data fusion: from algorithm to chip (Mathur, Bimal P.; Wang, H. T.; Liu, Shih-Chii; Koch, Christof; Luo, Jin)V1473,153-160(1991)

Polychromatic neural networks (Yu, Francis T.; Yang, Xiangyang; Gregory, Don A.)V1558,450-458(1991)

Polynomial neural nets for signal and image processing in chaotic backgrounds (Gardner, Sheldon)V1567,451-463(1991)

Polynomial neural network for robot forward and inverse kinematics learning computations (Chen, C. L. P.; McAulay, Alastair D.)V1468,394-405(1991)

Pose determination of spinning satellites using tracks of novel regions (Lee, Andrew J.; Casasent, David P.)V1383,72-83(1991)

Preconditions and prospects for the construction of parallel digital neurocomputers with programmable architecture (Kalyayev, Anatoli V.; Brukhomitsky, Yuri A.; Galuyev, Gennady A.; Chernukhin, Yu. V.)V1621,299-308(1991)

Problems of large neurodynamics system modeling: optical synergetics and neural networks (Vorontsov, Michael A.)V1402,116-144(1991)

Progress and perspectives on optical information processing in Japan (Ishihara, Satoshi)V1389,68-74(1991); V1390,68-74(1991)

Projection methods for evaluation of Hopfield-type CAM models (Berus, Tomasz; Macukow, Bohdan)V1564,562-570(1991)

Prototype neural network pattern recognition testbed (Worrell, Steven W.; Robertson, James A.; Varner, Thomas L.; Garvin, Charles G.)V1382,219-227(1991)

Pseudodeep hologram and its properties (Denisyuk, Yuri N.; Ganzherli, N. M.)V1238,2-12(1991)

Pyramid nets for computer vision (Stinson, Michael C.)V1386,53-61(1991)

Quantization and sampling considerations of computer-generated hologram for optical interconnection (Long, Pin; Hsu, Dahsiung)V1461,270-277(1991)

Quantum holography and neurocomputer architectures (Schempp, Walter)VIS08,62-144(1991)

Random mapping network for tactical target reacquisition after loss of track (Church, Susan D.; Burman, Jerry A.)V1471,192-199(1991)

Real-time motion detection using an analog VLSI zero-crossing chip (Bair, Wyeth; Koch, Christof)V1473,59-65(1991)

Recent developments of the SNNS neural network simulator (Zell, Andreas; Mache, Neils; Sommer, Tilman; Korb, Thomas)V1469,708-718(1991)

Recent research on optical neural networks in Japan (Ishihara, Satoshi; Mori, Masahiko)V1621,362-372(1991)

Recognition of handwritten katakana in a frame using moment invariants based on neural network (Agui, Takeshi; Takahashi, Hiroki; Nakajima, Masayuki; Nagahashi, Hiroshi)V1606,188-197(1991)

Recurrent neural network application to image filtering: 2-D Kalman filtering approach (Swiniarski, Roman W.; Dzielinski, Andrzej; Skoneczny, Slawomir; Butler, Michael P.)V1451,234-241(1991)

Recursion and feedback in image algebra (Ritter, Gerhard X.; Davidson, Jennifer L.)V1406,74-86(1991)

Reed-Solomon encoder/decoder application using a neural network (Hussain, Mukhtar; Bedi, Jatinder S.)V1469,463-471(1991)

Relating binary and continuous problem entropy to backpropagation network architecture (Smith, A.; Dagli, Cihan H.)V1469,551-562(1991)

Relaxation properties and learning paradigms in complex systems (Basti, Gianfranco; Perrone, Antonio; Morgavi, Giovanna)V1469,719-736(1991)

Research and improvement of Vogl's acceleration algorithm (Song, Hongjun; Zhou, Lijia)V1469,581-584(1991)

RETINA (RETinally INspired Architecture project) (Caulfield, H. J.; Wilkins, Nathan A.)V1564,496-503(1991)

Rotation and scale invariant pattern recognition using a multistaged neural network (Minnix, Jay I.; McVey, Eugene S.; Inigo, Rafael M.)V1606,241-251(1991)

Self-organized criticality in neural networks (Makarenko, Vladimir I.; Kirillov, A. B.)V1469,843-845(1991)

Self-organizing leader clustering in a neural network using a fuzzy learning rule (Newton, Scott C.; Mitra, Sunanda)V1565,331-337(1991)

Sensor-knowledge-command fusion paradigm for man-machine systems (Lee, Sukhan; Schenker, Paul S.; Park, Jun S.)V1383,391-402(1991)

Shape discrimination using invariant Fourier representation and a neural network classifier (Wu, Hsien-Huang; Schowengerdt, Robert A.)V1569,147-154(1991)

Shape recognition combining mathematical morphology and neural networks (Schmitt, Michel; Mattioli, Juliette)V1469,392-403(1991)

Signal processing with neural networks: throwing off the yoke of linearity (Hecht-Nielsen, Robert)V1541,146-151(1991)

Signal processing with radial basis function networks using expectation-maximization algorithm clustering (Ukrainec, Andrew; Haykin, Simon)V1565,529-539(1991)

Simulation of a neural network for decentralized detection of a signal in noise (Amirmehrabi, Hamid; Viswanathan, R.)V1396,252-265(1991)

Single-chip imager and feature extractor (Tanner, John E.; Luo, Jin)V1473,76-87(1991)

Source location of acoustic emissions from atmospheric leakage using neural networks (Barga, Roger S.; Friesel, Mark A.; Meador, Jack L.)V1469,602-611(1991)

Space-time system architecture for the neural optical computing (Lo, Yee-Man V.)V1382,199-208(1991)

Spatial light modulators and their applications (Yatagai, Toyohiko; Tian, Ronglong)V1564,691-696(1991)

Speed ranges accommodated by network architectures of elementary velocity estimators (Courellis, Spiridon H.; Marmarelis, Vasilis Z.)V1606,336-349(1991)

Statistical and neural network classifiers in model-based 3-D object recognition (Newton, Scott C.; Nutter, Brian S.; Mitra, Sunanda)V1382,209-218(1991)

Statistical image algebra: a Bayesian approach (Davidson, Jennifer L.; Cressie, N. A.)V1569,288-297(1991)

Stereo vision: a neural network application to constraint satisfaction problem (Mousavi, Madjid S.; Schalkoff, Robert J.)V1382,228-239(1991)

Stochastic and Neural Methods in Signal Processing, Image Processing, and Computer Vision (Chen, Su-Shing, ed.)V1569(1991)

Stochastic neural nets and vision (Fall, Thomas C.)V1468,778-785(1991)

Study of LiNbO3 in optical associative memory (Yuan, Xiao L.; Sayeh, Mohammad R.)V1396,178-187(1991)

Subsampled vector quantization with nonlinear estimation using neural network approach (Sun, Huifang; Manikopoulos, Constantine N.; Hsu, Hwei P.)V1605,214-220(1991)

Supervised color constancy for machine vision (Novak, Carol L.; Shafer, Steven A.)V1453,353-368(1991)

Systolic array architecture of a new VLSI vision chip (Means, Robert W.)V1566,388-393(1991)

Target cuing: a heterogeneous neural network approach (McCauley, Howard M.)V1469,69-76(1991)

Target detection using multilayer feedforward neural networks (Scherf, Alan V.; Scott, Peter A.)V1469,63-68(1991)

Target identification by means of adaptive neural networks in thermal infrared images (Acheroy, Marc P.; Mees, W.)V1569,121-132(1991)

Temperature stability of Bragg grating resonant frequency (Bradley, Eric M.; Rybka, Theodore W.; Yu, Paul K.)V1418,272-278(1991)

Template learning in morphological neural nets (Davidson, Jennifer L.; Sun, K.)V1568,176-187(1991)

Texture classification and neural network methods (Visa, Ari J.)V1469,820-831(1991)

Timbre discrimination of signals with identical pitch using neural networks (Sayegh, Samir I.; Pomalaza-Raez, Carlos A.; Tepper, E.; Beer, B. A.)V1569,100-110(1991)

Time series prediction with a radial basis function neural network (Potts, Michael A.; Broomhead, David S.)V1565,255-266(1991)

Two-dimensional dynamic neural network optical system with simplest types of large-scale interactions (Vorontsov, Michael A.; Zheleznykh, N. I.; Larichev, A. V.)V1402,154-164(1991)

Two-valued neural logic network (Hsu, L. S.; Loe, K. F.; Chan, Sing C.; Teh, H. H.)V1469,197-207(1991)

Unsupervised training of structuring elements (Wilson, Stephen S.)V1568,188-199(1991)

Use of 3 X 3 integrated optic polarizer/splitters for a smart aerospace plane structure (Seshamani, Ramani; Alex, T. K.)V1489,56-64(1991)

Use of magneto-optic spatial light modulators and linear detector arrays in inner-product associative memories (Goff, John R.)V1558,466-475(1991)

Utilizing the central limit theorem for parallel multiple-scale image processing with neural architectures (Ben-Arie, Jezekiel)V1569,227-238(1991)

Video-image-based neural network guidance system with adaptive view-angles for autonomous vehicles (Luebbers, Paul G.; Pandya, Abhijit S.)V1469,756-765(1991)

Visual Communications and Image Processing '91: Image Processing (Tzou, Kou-Hu; Koga, Toshio, eds.)V1606(1991)

Visual Information Processing: From Neurons to Chips (Mathur, Bimal P.; Koch, Christof, eds.)V1473(1991)

VLSI implementable neural networks for target tracking (Himes, Glenn S.; Inigo, Rafael M.; Narathong, Chiewcharn)V1469,671-682(1991)

Weighted-outer-product associative neural network (Ji, Han-Bing)V1606,1060-1068(1991)

What have neural networks to offer statistical pattern processing? (Lowe, David)V1565,460-471(1991)

X-ray inspection utilizing knowledge-based feature isolation with a neural network classifier (Nolan, Adam R.; Hu, Yong-Lin; Wee, William G.)V1472,157-164(1991)

Noise

Adaptive image sequence noise filtering methods (Katsaggelos, Aggelos K.; Kleihorst, Richard P.; Efstratiadis, Serafim N.; Lagendijk, Reginald L.)V1606,716-727(1991)

Adaptive notch filter for removal of coherent noise from infrared scanner data (Jaggi, Sandeep)V1541,134-140(1991)

Analysis and measurement of the external modulation of modelocked laser diodes (relative noise performance) (Lam, Benson C.; Kellner, Albert L.; Campion, David C.; Costa, Joannes M.; Yu, Paul K.)V1371,36-45(1991)

Analysis of upper and lower bounds of the frame noise in linear detector arrays (Jaggi, Sandeep)V1541,152-162(1991)

Application of canonical correlation analysis in detection in presence of spatially correlated noise (Chen, Wei G.; Reilly, James P.; Wong, Kon M.)V1566,464-475(1991)

Artificial neural networks as TV signal processors (Spence, Clay D.; Pearson, John C.; Sverdlove, Ronald)V1469,665-670(1991)

Automatic, high-resolution analysis of low-noise fringes (Lassahn, Gordon D.)V1332,690-695(1991)

Bayesian matching technique for detecting simple objects in heavily noisy environment (Baras, John S.; Frantzeskakis, Emmanuil N.)V1569,341-353(1991)

Beam position noise and other fundamental noise processes that affect optical storage (Levenson, Marc D.)V1376,259-271(1991)

Calculations on the Hanle effect with phase and amplitude fluctuating laser fields (Bergeman, Thomas H.; Ryan, Robert E.)V1376,54-67(1991)

CCD performance model for airborne reconnaissance (Donn, Matthew; Waeber, Bruce)V1538,189-200(1991)

Cerenkov background radiation in imaging detectors (Rosenblatt, Edward I.; Beaver, Edward A.; Cohen, R. D.; Linsky, J. B.; Lyons, R. W.)V1449,72-86(1991)

Characterization and effect of system noise in a differential angle of arrival measurement device (Waldie, Arthur H.; Drexler, James J.; Qualtrough, John A.; Soules, David B.; Eaton, Frank D.; Peterson, William A.; Hines, John R.)V1487,103-108(1991)

Characterizing frequency chirps and phase fluctuations during noisy laser transients (Abraham, Neal B.)V1376,284-293(1991)

Clutter metrics in infrared target acquisition (Tidhar, Gil; Rotman, Stanley R.)V1442,310-324(1991)

Comparison of a multilayered perceptron with standard classification techniques in the presense of noise (Willson, Gregory B.)V1469,351-358(1991)

Comparison of electro-optic diagnostic systems (Hagans, Karla G.; Sargis, Paul G.)V1346,404-408(1991)

Comparison of time and frequency multiplexing techniques in multicomponent FM spectroscopy (Muecke, Robert J.; Werle, Peter W.; Slemr, Franz; Prettl, William)V1433,136-144(1991)

Computer-generated correlated noise images for various statistical distributions (Wenaas, Holly; Weeks, Arthur R.; Myler, Harley R.)V1569,410-421(1991)

Correction method for optical-signal detection-error caused by quantum noise (Fujihashi, Chugo)V1583,298-306(1991)

Crosstalk in direct-detection optical fiber FDMA networks (Hamdy, Walid M.; Humblet, Pierre A.)V1579,184-194(1991)

Design of low-noise wide-dynamic-range GaAs optical preamps (Bayruns, Robert J.; Laverick, Timothy; Scheinberg, Norman; Stofman, Daniel)V1541,83-90(1991)

Detecting spatial and temporal dot patterns in noise (Drum, Bruce)V1453,153-164(1991)

Deterministic and noise-induced phase jumps in the ring laser gyroscope (Chyba, Thomas H.)V1376,132-142(1991)

Deterministic and quantum noise in dye lasers (Raymer, M. G.; Gadomski, Wojciech; Hodges, Steven E.; Adkison, D.)V1376,128-131(1991)

Deterministic fluctuations in an intracavity-coupled solid state laser (Bracikowski, Christopher; Roy, Rajarshi)V1376,103-116(1991)

Diffraction analysis of optical disk readout signal deterioration caused by mark-size fluctuation (Honguh, Yoshinori)V1527,315-321(1991)

Digital halftoning using a blue-noise mask (Mitsa, Theophano; Parker, Kevin J.)V1452,47-56(1991)

Double-beam laser absorption spectroscopy: shot noise-limited performance at baseband with a novel electronic noise canceler (Haller, Kurt L.; Hobbs, Philip C.)V1435,298-309(1991)

Enhanced Schawlow-Townes linewidth in lasers with nonorthogonal transverse eigenmodes (Mussche, Paul L.; Siegman, Anthony E.)V1376,153-163(1991)

Estimation of scene correlation lengths (Futterman, Walter I.; Schweitzer, Eric L.; Newt, J. E.)V1486,127-140(1991)

Evaluation of the effect of noise on subjective image quality (Barten, Peter G.)V1453,2-15(1991)

Excess noise in fiber gyroscope sources (Burns, William K.; Moeller, Robert P.; Dandridge, Anthony D.)V1367,87-92(1991)

Feedback noise in single-mode semiconductor lasers (Lenstra, Daan; Cohen, Julius S.)V1376,245-258(1991)

Fixed-pattern-noise cancellation in linear pyro arrays (Jain, Subhash C.; Malhotra, H. S.; Sarebahi, K. N.; Bist, K. S.)V1488,410-413(1991)

Fluctuations in atomic fluorescence induced by laser noise (Vemuri, Gautam)V1376,34-46(1991)

Focus sensing method with improved pattern noise rejection (Marshall, Daniel R.)V1499,332-339(1991)

General jump model for laser noise: non-Markovian phase and frequency jumps (Levine, Alfred M.; Ozizmir, Ercument; Zaibel, Reuben; Prior, Yehiam)V1376,47-53(1991)

Hierarchical network for clutter and texture modeling (Luttrell, Stephen P.)V1565,518-528(1991)

Hybrid infrared focal-plane signal and noise modeling (Johnson, Jerris F.; Lomheim, Terrence S.)V1541,110-126(1991)

Image noise smoothing based on nonparametric statistics (Chuang, Keh-Shih; Huang, H. K.)V1445,496-503(1991)

Image restoration and identification using the EM and adaptive RAP algorithms (Kuo, Shyh-Shiaw; Mammone, Richard J.)V1606,419-430(1991)

Imaging through a low-light-level Raman amplifier (Duncan, Michael D.; Mahon, R.; Tankersley, Lawrence L.; Reintjes, John F.)V1409,127-134(1991)

InGaAsP/InP distributed-feedback lasers for long-wavelength optical communication systems, lambda=1.55 um: electrical and optical noises study (Orsal, Bernard; Alabedra, Robert; Signoret, P.; Letellier, H.)V1512,112-123(1991)

Laser Noise (Roy, Rajarshi, ed.)V1376(1991)

Linear feature SNR enhancement in radon transform space (Meckley, John R.)V1569,375-385(1991)

Liquid-nitrogen-cooled low-noise radiation pulse detector amplifier (Trojnar, Eugeniusz; Trojanowski, Stanislaw; Czechowicz, Roman; Derwiszynski, Mariusz; Kocyba, Krzysztof)V1391,230-237(1991)

LMMSE restoration of blurred and noisy image sequences (Ozkan, Mehmet K.; Sezan, M. I.; Erdem, A. T.; Tekalp, Ahmet M.)V1606,743-754(1991)

Low-frequency intensity fluctuations in external cavity semiconductor lasers (McInerney, John G.)V1376,236-244(1991)

Methods and applications for intensity stabilization of pulsed and cw lasers from 257 nm to 10.6 microns (Miller, Peter J.)V1376,180-191(1991)

Modal noise reduction in analog fiber optic links by superposition of high-frequency modulation (Pepeljugoski, Petar K.; Lau, Kam-Yin)V1371,233-243(1991)

Model-based morphology (Haralick, Robert M.; Dougherty, Edward R.; Katz, Philip L.)V1472,108-117(1991)

New algorithms for adaptive median filters (Hwang, Humor; Haddad, Richard A.)V1606,400-407(1991)

New approaches to signal-to-noise ratio optimization in background-limited photothermal measurements (Rice, Patrick D.; Thorne, John B.; Bobbitt, Donald R.)V1435,104-113(1991)

Noise and gain performance for an Er3+-doped fiber amplifier pumped at 980 nm or 1480 nm (Vendeltorp-Pommer, Helle; Pedersen, Frands B.; Bjarklev, Anders; Hedegaard Povlsen, Joern)V1373,254-265(1991)

Noise characteristics of rare-earth-doped fiber sources and amplifiers (Morkel, Paul R.; Laming, Richard I.; Cowle, Gregory J.; Payne, David N.)V1373,224-233(1991)

Noise in He:Ne lasers near threshold (Singh, Surendra; Mortazavi, Mansour; Phillips, K. J.; Young, M. R.)V1376,143-152(1991)

Noise in semiconductor lasers and its impact on optical communication systems (Agrawal, Govind P.)V1376,224-235(1991)

Noise mechanisms of high-temperature superconducting infrared detectors (Khalil, Ali E.)V1477,148-158(1991)

Noise on multiple-tau photon correlation data (Schaetzel, Klaus; Peters, Rainer)V1430,109-115(1991)

Noise performance of microchannel plate imaging systems (McCammon, Kent G.; Hagans, Karla G.; Hankla, A.)V1346,398-403(1991)

Noise reduction in heart movies by motion-compensated filtering (Reinen, Tor A.)V1606,755-763(1991)

Noise reduction strategy for hybrid IR focal-plane arrays (Fowler, Albert M.; Gatley, Ian)V1541,127-133(1991)

Noise statistics of ratiometric signal processing systems (Svetkoff, Donald J.; Xydis, Thomas G.; Kilgus, Donald B.)V1385,113-122(1991)

Noise tolerance of adaptive resonance theory neural network for binary pattern recognition (Kim, Yong-Soo; Mitra, Sunanda)V1565,323-330(1991)

Observer detection of image degradation caused by irreversible data compression processes (Chen, Ji; Flynn, Michael J.; Gross, Barry; Spizarny, David)V1444,256-264(1991)

Overview of experimental investigations of laser bandwidth effects in nonlinear optics (Elliott, Daniel S.)V1376,22-33(1991)

Passive laser phase noise suppression technique for fiber interferometers (Kersey, Alan D.; Berkoff, Timothy A.)V1367,310-318(1991)

Perceptual noise measurement of displays (Chakraborty, Dev P.; Pfeiffer, Douglas E.; Brikman, Inna)V1443,183-190(1991)

Performance of on-off modulated lightwave signals with phase noise (Azizoglu, Murat; Humblet, Pierre A.)V1579,168-178(1991)

Performance of Reed-Solomon codes in mulichannel CPFSK coherent optical communications (Wu, Jyh-Horng; Wu, Jingshown)V1579,195-209(1991)

Phase-map unwrapping: a comparison of some traditional methods and a presentation of a new approach (Owner-Petersen, Mette)V1508,73-82(1991)

Photon-noise reduction experiments with a Q-switched Nd:YAG laser (Kumar, Prem; Huang, Jianming; Aytur, Orhan)V1376,192-197(1991)

Physical and psychophysical evaluation of CRT noise performance (Ji, Tinglan; Roehring, Hans; Blume, Hartwig R.; Seeley, George W.; Browne, Michael P.)V1444,136-150(1991)

Polynomial neural nets for signal and image processing in chaotic backgrounds (Gardner, Sheldon)V1567,451-463(1991)

Pulsating instabilities and chaos in Raman lasers (Harrison, Robert G.; Lu, Weiping; Jiad, K.; Uppal, J. S.)V1376,94-102(1991)

QED theory of excess spontaneous emission noise (Milonni, Peter W.)V1376,164-169(1991)

Quantum-limited quasiparticle mixers at 100 GHz (Mears, Carl A.; Hu, Qing; Richards, Paul L.; Worsham, A.; Prober, Daniel E.; Raisanen, Antti)V1477,221-233(1991)

Quantum noise reduction in lasers with nonlinear absorbers (Ritsch, Helmut; Zoller, Peter)V1376,206-215(1991)

Reduction of spectral linewidth and FM noise in semiconductor lasers by application of optical feedback (Li, Hua; Park, J. D.; Seo, Dongsun; Marin, L. D.; McInerney, John G.; Telle, H. R.)V1376,172-179(1991)

Regular and chaotic pulsations of radiation intensity in a CO2 laser with modulated parameters (Bondarenko, Alexander V.; Glova, Alexander F.; Kozlov, Sergei N.; Lebedev, Fedor V.; Likhanskii, Vladimir V.; Napartovich, Anatoly P.; Pis'menny, Vladislav D.; Yaztsev, Vladimir P.)V1376,117-127(1991)

Shot-noise-limited high-power green laser for higher density optical disk systems (Kubota, Shigeo; Oka, Michio)V1499,314-323(1991)

Shot noise limited optical measurements at baseband with noisy lasers (Hobbs, Philip C.)V1376,216-221(1991)

Signal and background models in nonstandard IR systems (Snyder, John L.)V1498,52-63(1991)

Signal processing for low-light-level, high-precision CCD imaging (McCurnin, Thomas W.; Schooley, Larry C.; Sims, Gary R.)V1448,225-236(1991)

Spontaneous emission from free-electron lasers (Schmitt, Mark J.)V1552,107-117(1991)

Spontaneous emission noise reduction of a laser output by extracavity destructive interference (Diels, Jean-Claude; Lai, Ming)V1376,198-205(1991)

Staring sensor MRT measurement and modeling (Mooney, Jonathan M.)V1540,550-564(1991)

Statistics of laser switch-on (San Miguel, Maximino)V1376,272-283(1991)

Stochastic detecting images from strong noise field in visual communications (Cai, De-Fu)V1606,926-933(1991)

Stochastic simulations of light-scattering noise in two-wave mixing: application to artificial Kerr media (McGraw, Robert L.)V1409,135-147(1991)

Strategies for unwrapping noisy interferograms in phase-sampling interferometry (Andrae, P.; Mieth, Ulrike; Osten, Wolfgang)V1508,50-60(1991)

Switching noise in a medium-film copper/polyimide multichip module (Sandborn, Peter A.; Hashemi, Seyed H.; Weigler, William)V1389,177-186(1991)

Terrain and target segmentation using coherent laser radar (Renhorn, Ingmar G.; Letalick, Dietmar; Millnert, Mille)V1480,35-45(1991)

Theory of laser noise (Lax, Melvin)V1376,2-20(1991)

Three-dimensional analysis framework and measurement methodology for imaging system noise (D'Agostino, John A.; Webb, Curtis M.)V1488,110-121(1991)

Toward phase noise reduction in a Nd:YLF laser using electro-optic feedback control (Brown, David L.; Seka, Wolf D.; Letzring, Samuel A.)V1410,209-214(1991)

Use of Fokker-Planck equations for the statistical properties of laser light (Jung, Peter; Risken, H.)V1376,82-93(1991)

Using morphology in document image processing (Concepcion, Vicente P.; Grzech, Matthew P.; D'Amato, Donald P.)V1606,132-140(1991)

Write noise from optical heads with nonachromatic beam expansion prisms (Kay, David B.; Chase, Scott B.; Gage, Edward C.; Silverstein, Barry D.)V1499,281-285(1991)

Nondestructive testing—see also metrology; photomechanics

Application of phase-stepping speckle interferometry to shape and deformation measurement of a 3-D surface (Dobbins, B. N.; He, Shi P.; Kapasi, S.; Wang, Liu Sheng; Button, B. L.; Wu, Xiao-Ping)V1554A,772-780(1991)

Application of visioplasticity to an experimental analysis of the shearing phenomenon (Koga, Nobuhiro; Kudoh, Takeshi; Murakawa, Masao)V1554A,84-92(1991)

Applications of tridimensional heat calibration to a thermographic nondestructive evaluation station (Maldague, Xavier; Fortin, Louis; Picard, J.)V1467,239-251(1991)

Automated measurement of 3-D shapes by a dual-beam digital speckle interferometric technique (Shi, Dahuan; Qin, Jing; Hung, Yau Y.)V1554A,680-689(1991)

Automatic recognition of multidimensional objects buried in layered elastic background media (Ayme-Bellegarda, Eveline J.; Habashy, Tarek M.; Bellegarda, Jerome R.)V1471,18-29(1991)

Broad-range holographic contouring of diffuse surfaces by dual-beam illumination: study of two related techniques (Rastogi, Pramod K.; Pflug, Leopold)V1554B,48-55(1991)

Computer-aided speckle interferometry: Part II—an alternative approach using spectral amplitude and phase information (Chen, Duanjun; Chiang, Fu-Pen; Tan, Yushan; Don, H. S.)V1554A,706-717(1991)

Cone beam for medical imaging and NDE (Smith, Bruce D.)V1450,13-17(1991)

Contouring by DSPI for surface inspection (Paoletti, Domenica; Schirripa Spagnolo, Giuseppe)V1554A,660-667(1991)

Determination of displacements, strain, and rotations from holographic interferometry data using a 2-D fringe order function (Albertazzi, Armando A.)V1554B,64-74(1991)

Determination of the homogeneous degree of deformation of the bearing cap of the motor by reflection holography (Cao, Zhengyuan; Cheng, Fang)V1554B,81-85(1991)

Development of electron moire method using a scanning electron microscope (Kishimoto, Satoshi; Egashira, Mitsuru; Shinya, Norio)V1554B,174-180(1991)

Diagnostics of arc plasma by moire deflectometry (Gao, Yiqing; Liu, Yupin)V1554B,193-199(1991)

Displacement analysis of the interior walls of a pipe using panoramic holointerferometry (Gilbert, John A.; Matthys, Donald R.; Hendren, Christelle M.)V1554A,128-134(1991)

Dynamic two-beam speckle interferometry (Ahmadshahi, Mansour A.; Krishnaswamy, Sridhar; Nemat-Nasser, Siavouche)V1554A,620-627(1991)

Effects of material combinations on the bimaterial fracture behavior (Wang, Wei-Chung; Chen, Jin-Tzaih)V1554A,60-69(1991)

Electronic shearography versus ESPI in nondestructive evaluation (Hung, Yau Y.)V1554B,692-700(1991)

Evaluation of interference patterns: a comparison of methods (Kreis, Thomas M.; Geldmacher, Juergen)V1554B,718-724(1991)

Experimental investigation of dynamic mixed-mode fracture initiation (Lambros, John M.; Mason, James J.; Rosakis, Ares J.)V1554A,70-83(1991)

Flaw dynamics and vibrothermographic-thermoelastic nondestructive evaluation of advanced composite materials (Tenek, Lazarus H.; Henneke, Edmund G.)V1467,252-263(1991)

Holographic devices using photo-induced effect in nondestructive testing techniques (Dovgalenko, George Y.; Onischenko, Yuri I.)V1559,479-486(1991)

Holographic in-plane displacement measurement in cracked specimens in plane stress state (Sainov, Ventseslav C.; Simova, Eli S.)V1461,175-183(1991)

Holographic interferometry analysis of sealed, disposable containers for internal defects (Cohn, Gerald E.; Domanik, Richard A.)V1429,195-206(1991)

Hybrid method to analyze the stress state in piecewise homogeneous two-dimensional objects (Laermann, Karl-Hans)V1554A,143-150(1991)

Industrial applications of spectroscopic imaging (Miller, Richard M.; Birmingham, John J.; Cummins, Philip G.; Singleton, Scott)V1439,66-78(1991)

Influence of fixing stress on the sensitivity of HNDT (Yang, Yu X.; Tan, Yushan)V1554B,768-773(1991)

Interferometric moire analysis of wood and paper structures (Hyzer, Jim B.; Shih, J.; Rowlands, Robert E.)V1554A,371-382(1991)

Interpretation of holographic and shearographic fringes for estimating the size and depth of debonds in laminated plates (Shang, H. M.; Chau, Fook S.; Tay, C. J.; Toh, Siew-Lok)V1554B,680-691(1991)

Laser ultrasonics: generation and detection considerations for improved signal-to-noise ratio (Wagner, James W.; Deaton, John B.; McKie, Andrew D.; Spicer, James B.)V1332,491-501(1991)

Low-temperature deformation measurement of materials by the use of laser speckle photography method (Nakahara, Sumio; Hisada, Shigeyoshi; Fujita, Takeyoshi; Sugihara, Kiyoshi)V1554A,602-609(1991)

Material testing by the laser speckle strain gauge (Yamaguchi, Ichirou; Kobayashi, Koichi)V1554A,240-249(1991)

Measurement of residual stresses in plastic materials by electronic shearography (Long, Kah W.; Hung, Yau Y.; Der Hovanesian, Joseph)V1554A,116-123(1991)

Measurement of the dynamic crack-tip displacement field using high-resolution moire photography (Whitworth, Martin B.; Huntley, Jonathan M.; Field, John E.)V1554B,282-288(1991)

Measurement of thermal deformation of square plate by using holographic interferometry (Kang, Dae I.; Kwon, Young H.; Ko, Yeong-Uk)V1554B,139-147(1991)

Moire-holographic evaluation of the strain fields at the weld toes of welded structures (Pappalettere, Carmine; Trentadue, Bartolo; Monno, Giuseppe)V1554B,99-105(1991)

Moire pattern and interferogram transform (Lu, Zhen-Wu)V1554B,383-388(1991)

New optical approaches to the quantitative characterization of crystal growth, segregation, and defect formation (Carlson, D. J.; Wargo, Michael J.; Cao, X. Z.; Witt, August F.)V1557,140-146(1991)

No-blind-area one-photograph HNDT tire analyzer (Ge, Fang X.; Xiong, Xian M.)V1554B,785-789(1991)

Noncontact lifetime characterization technique for LWIR HgCdTe using transient millimeter-wave reflectance (Schimert, Thomas R.; Tyan, John; Barnes, Scott L.; Kenner, Vern E.; Brouns, Austin J.; Wilson, H. L.)V1484,19-30(1991)

Noncontact measurement of microscopic displacement and vibration by means of fiber optics bundle (Toba, Eiji; Shimosaka, Tetsuya; Shimazu, Hideto)V1584,353-363(1991)

Nondestructive evaluation of turbine blades vibrating in resonant modes (Sciammarella, Cesar A.; Ahmadshahi, Mansour A.)V1554B,743-753(1991)

Nondestructive investigations of multilayer dielectrical coatings (Shapiro, Alexander G.; Yaminsky, Igor V.)V1362,834-843(1991)

Nondestructive testing of printed circuit board by phase-shifting interferometry (Lu, Yue-guang; Jiang, Lingzhen; Zou, Lixun; Zhao, Xia; Sun, Junyong)V1332,287-291(1991)

Ophthalmic lenses testing by moire deflectometry (Yonte, T.; Quiroga, J.; Alda, J.; Bernabeu, Eusebio)V1554B,233-241(1991)

Optical isodyne measurements in fracture mechanics (Pindera, Jerzy T.; Wen, Baicheng)V1554A,196-205(1991)

Optical nondestructive examination for honeycomb structure (Zhu, Jian T.)V1554B,774-784(1991)

Phase-measurement of interferometry techniques for nondestructive testing (Creath, Katherine)V1554B,701-707(1991)

Phase-shifting hand-held diffraction moire interferometer (Deason, Vance A.; Ward, Michael B.)V1554B,390-398(1991)

Phase-stepping technique in holography (Wen, Zheng; Tan, Yushan)V1461,278-285(1991)

Postprocessing of thermograms in infrared nondestructive testing (Vavilov, Vladimir P.; Maldague, Xavier; Saptzin, V. M.)V1540,488-495(1991)

Practical method for automatic detection of brightest ridge line of photomechanical fringe pattern (Ding, Zu-Quan; Yuan, Xun-Hua)V1554A,898-906(1991)

Pulsed holographic recording of very high speed transient events (Steckenrider, John S.; Ehrlich, Michael J.; Wagner, James W.)V1554B,106-112(1991)

Quantitative analysis of contact deformation using holographic interferometry and speckle photography (Joo, Jin W.; Kwon, Young H.; Park, Seung O.)V1554B,113-118(1991)

Recent development in practical optical nondestructive testing (Hung, Yau Y.)V1554A,29-45(1991); V1554B,29-45(1991)

Reducing errors in bending tests of ceramic materials (Furgiuele, Franco M.; Lamberti, Antonio; Pagnotta, Leonard)V1554A,275-284(1991)

Residual stress evaluation using shearography with large shear displacements (Hathaway, Richard B.; Der Hovanesian, Joseph; Hung, Yau Y.)V1554B,725-735(1991)

Scattered-light optical isodynes: basis for 3-D isodyne stress analysis (Pindera, Jerzy T.)V1554A,458-471(1991)

Second Intl Conf on Photomechanics and Speckle Metrology: Moire Techniques, Holographic Interferometry, Optical NDT, and Applications to Fluid Mechanics (Chiang, Fu-Pen, ed.)V1554B(1991)

Second Intl Conf on Photomechanics and Speckle Metrology: Speckle Techniques, Birefringence Methods, and Applications to Solid Mechanics (Chiang, Fu-Pen, ed.)V1554A(1991)

Some applications of a HOE-based desensitized interferometer in materials research (Boone, Pierre M.; Jacquot, Pierre M.)V1554A,512-521(1991)

Source location of acoustic emissions from atmospheric leakage using neural networks (Barga, Roger S.; Friesel, Mark A.; Meador, Jack L.)V1469,602-611(1991)

Speckle measurement for 3-D surface movement (Hilbig, Jens; Ritter, Reinold)V1554A,588-592(1991)

Statistical interferometry based on the statistics of speckle phase (Kadono, Hirofumi; Toyooka, Satoru)V1554A,718-726(1991)

Strain measurements on soft materials application to cloth and papers (Bremand, Fabrice J.; Lagarde, Alexis)V1554B,650-660(1991)

Stress intensity factors of edge-cracked semi-infinite plates under transient thermal loading (Wang, Wei-Chung; Chen, Tsai-Lin; Hwang, Chi-Hung)V1554A,124-135(1991)

Study of displacement and residual displacement field of an interface crack by moire interferometry (Wang, Y. Y.; Chiang, Fu-Pen; Barsoum, Roshdy S.; Chou, S. T.)V1554B,344-356(1991)

Surface contouring using electronic speckle pattern interferometry (Kerr, David; Rodriguez-Vera, Ramon; Mendoza Santoyo, Fernando)V1554A,668-679(1991)

Temperature rise due to dynamic crack growth in Beta-C titanium (Zehnder, Alan T.; Kallivayalil, Jacob A.)V1554A,48-59(1991)

Vibration modal analysis using stroboscopic digital speckle pattern interferometry (Wang, Xizhou; Tan, Yushan)V1554A,907-914(1991)

White-light speckle applied to composites (Guo, Maolin; Zhao, Yin; Wu, Zhongming; Du, Shanyi)V1554A,657-658(1991)

Whole field displacement and strain rosettes by grating objective speckle method (Tu, Meirong; Gielisse, Peter J.; Xu, Wei)V1554A,593-601(1991)

Nonlinear optics—see also four-wave mixing; phase conjugation; second-harmonic generation

7th Mtg in Israel on Optical Engineering (Oron, Moshe; Shladov, Itzhak, eds.)V1442(1991)

Absorptive nonlinear semiconductor amplifiers for fast optical switching (Barnsley, Peter E.; Marshall, Ian W.; Fiddyment, Phillip J.; Robertson, Michael J.)V1378,116-126(1991)

Analysis of Brillouin-enhanced four-wave mixing phase-conjugate systems (Dansereau, Jeffrey P.; Hills, Louis S.; Mani, Siva A.)V1409,67-82(1991)

Applications of chaotic neurodynamics in pattern recognition (Baird, Bill; Freeman, Walter J.; Eeckman, Frank H.; Yao, Yong)V1469,12-23(1991)

Applications of the FEL to NLO spectroscopy of semiconductors (Pidgeon, Carl R.)V1501,178-182(1991)

Approaches to the construction of intrinsically acentric chromophoric NLO materials: chemical elaboration and resultant properties of self-assembled multilayer structures (Allan, D. S.; Kubota, F.; Marks, Tobin J.; Zhang, T. J.; Lin, W. P.; Wong, George K.)V1560,362-369(1991)

Beam-coupling by stimulated Brillouin scattering (Falk, Joel; Chu, Raijun; Kanefsky, Morton; Hua, Xuelei)V1409,83-86(1991)

Bulk darkening of flux-grown KTiOPO4 (Rockafellow, David R.; Teppo, Edward A.; Jacco, John C.)V1561,112-118(1991)

Cavity-QED-enhanced spontaneous emission and lasing in liquid droplets (Campillo, Anthony J.; Eversole, J. D.; Lin, H.-B.; Merritt, C. D.)V1497,78-89(1991)

Chaotic behavior of a Raman ring laser (Englund, John C.)V1497,292-299(1991)

Coherence and optical Kerr nonlinearity (Depoortere, Marc)V1504,133-139(1991)

Coherent coupling of lasers using a photorefractive ring oscillator (Luo, Jhy-Ming)V1409,100-105(1991)

Continuous automatic scanning picosecond optical parametric source using MgO LiNbO3 in the 700-2200 nm (He, Huijuan; Lu, Yutian; Dong, Jingyuan; Zhao, Quingchun)V1409,18-23(1991)

Contributions of pi' bonding to the nonlinear optical properties of inorganic polymers (Ferris, Kim F.; Risser, Steven M.)V1441,510-520(1991)

Corrections to the Golden Rule (Fearn, Heidi; Lamb, Willis E.)V1497,245-254(1991)

Decay of the nonlinear susceptibility components in main-chain functionalized poled polymers (Meyrueix, Remi; LeCompte, J. P.; Tapolsky, Gilles)V1560,454-466(1991)

Degenerate four-wave mixing and optical nonlinearities in quantum-confined CdSe microcrystallites (Park, S. H.; Casey, Michael P.; Falk, Joel)V1409,9-13(1991)

Degenerate four-wave mixing using wave pump beams near the critical angle: two distinct behaviors (Malouin, Christian; Song, Li; Thibault, S.; Denariez-Roberge, Marguerite M.; Lessard, Roger A.)V1559,385-392(1991)

Design of chromophores for nonlinear optical applications (Burland, Donald M.; Rice, J. E.; Downing, J.; Michl, J.)V1560,111-119(1991)

Design of highly soluble extended pi-electron oligomers capable of supporting stabilized delocalized bipolaronic states for NLO applications (Havelka, Kathleen O.; Spangler, Charles W.)V1560,66-74(1991)

Determination of the orientational order parameters <P*2>, <P*4> in a polysilane LB film via polarization-dependent THG (Neher, Dieter; Mittler-Neher, S.; Cha, M.; Stegeman, George I.; Embs, F. W.; Wegner, Gerhard; Miller, Robert D.; Willson, C. G.)V1560,335-343(1991)

Dielectric relaxation studies of x2 dye containing polystyrene films (Schen, Michael A.; Mopsik, Fred)V1560,315-325(1991)

Diffraction effects in stimulated Raman scattering (Scalora, Michael; Haus, Joseph W.)V1497,153-164(1991)

Dispersion of x(3) in fused aromatic ladder polymers and their precursors probed by third-harmonic generation (Meth, Jeffrey S.; Vanherzeele, Herman A.; Jenekhe, Samson A.; Roberts, Michael F.; Agrawal, A. K.; Yang, Chen-Jen)V1560,13-24(1991)

Display nonlinearity in digital image processing for visual communications (Peli, Eli)V1606,508-519(1991)

Dynamic holographic microphasometry (Kozhevnikov, Nikolai M.)V1507,509-516(1991)

Dynamics of stiff macromolecules in concentrated polymer solutions: model of statistical reorientations (Brazhnik, O. D.; Khokhlov, A. R.)V1402,70-77(1991)

Effective electron mass in narrow-band-gap IR materials under different physical conditions (Ghatak, Kamakhya P.; Biswas, Shambhu N.)V1484,149-166(1991)

Electrical and nonlinear optical properties of zirconium phosphonate multilayer assemblies (Katz, Howard E.; Schilling, M. L.; Ungashe, S.; Putvinski, T. M.; Scheller, G. E.; Chidsey, C. E.; Wilson, William L.)V1560,370-376(1991)

Electrochromic property and chemical sensitivity of conducting polymer PAn film (Yuan, Ren K.; Gu, Zhi P.; Yuan, Hong; Yuan, Xue S.; Wang, Yong B.; Liu, Xiang N.; Shen, Xue C.)V1519,831-834(1991)

Electromagnetic field calculations for a tightly focused laser beam incident upon a microdroplet: applications to nonlinear optics (Barton, John P.; Alexander, Dennis R.)V1497,64-77(1991)

Electro-optical light modulation in novel azo-dye-substituted poled polymers (Shuto, Yoshito; Amano, Michiyuki; Kaino, Toshikuni)V1560,184-195(1991)

Electro-optic coefficients in electric-field poled-polymer waveguides (Smith, Barton A.; Herminghaus, Stephan; Swalen, Jerome D.)V1560,400-405(1991)

Excited-state nonlinear optical processes in polymers (Heflin, James R.; Zhou, Qihou L.; Garito, Anthony F.)V1497,398-407(1991)

Experimental measurements and electron correlation theory of third-order nonlinear optical processes in linear chains (Heflin, James R.; Cai, Yongming; Zhou, Qihou L.; Garito, Anthony F.)V1560,2-12(1991)

Experimental simulation analysis of nonlinear problem: investigation into the mechanical and optical behavior of silver chloride of photoplastic material (Yin, Zhi Xiang; Zhang, Shikun; Li, Zong Yan)V1559,487-496(1991)

Extended pi-electron systems incorporating stabilized quinoidal bipolarons: anthracenyl polyene-polycarbonate composites (Nickel, Eric G.; Spangler, Charles W.)V1560,35-43(1991)

Fabrication and nonlinear optical properties of mixed and layered colloidal particles (McDonald, Joseph K.; LaiHing, Kenneth)V1497,367-370(1991)

Fabrication techniques of photopolymer-clad waveguides for nonlinear polymeric modulators (Tumolillo, Thomas A.; Ashley, Paul R.)V1559,65-73(1991)

Ferroelectric microdomain reversal on Y-cut LiNbO3 surfaces (Seibert, Holger; Sohler, Wolfgang)V1362,370-376(1991)

Frequency doubling, absorption, and grating formation in glass fibers: effective defects or defective effects? (Russell, Philip S.; Poyntz-Wright, L. J.; Hand, Duncan P.)V1373,126-139(1991)

Frequency-tunable THG measurements of x(3) between 1-2.1um of organic conjugated-polymer films using an optical parametric oscillator (Gierulski, Alfred; Naarmann, Herbert; Schrof, Wolfgang; Ticktin, Anton)V1560,172-182(1991)

Gain and threshold in noninversion lasers (Scully, Marlan O.; Zhu, Shi-Yao; Narducci, Lorenzo M.; Fearn, Heidi)V1497,264-276(1991)

Gate capacitance of MOS field effect devices of nonlinear optical materials in the presence of a parallel magnetic field (Ghatak, Kamakhya P.; Biswas, Shambhu N.; Banik, S. N.)V1409,240-257(1991)

Generalized Raman gain in nonparabolic semiconductors under strong magnetic field (Ghatak, Kamakhya P.; Ghoshal, Ardhendhu; De, Badal)V1409,178-190(1991)

Generation of ultrashort ultraviolet optical pulses using sum-frequency in LBO crystals (Guo, Ting; Qiu, Peixia; Lin, Fucheng)V1409,24-27(1991)

Ground-based experiments on the growth and characterization of L-arginine phosphate crystals (Rao, S. M.; Cao, C.; Batra, Ashok K.; Lal, Ravindra B.; Mookherji, Tripty K.)V1557,283-292(1991)

Growth and Characterization of Materials for Infrared Detectors and Nonlinear Optical Switches (Longshore, Randolph E.; Baars, Jan W., eds.)V1484(1991)

Growth of thin films of organic nonlinear optical materials by vapor growth processes: an overview and examination of shortfalls (Frazier, Donald O.; Penn, Benjamin G.; Witherow, William K.; Paley, M. S.)V1557,86-97(1991)

Growth, surface passivation, and characterization of CdSe microcrystallites in glass with respect to their application in nonlinear optics (Woggon, Ulrike; Rueckmann, I.; Kornack, J.; Mueller, Matthias; Cesnulevicius, J.; Kolenda, Jonas; Petrauskas, Mendogas)V1362,888-898(1991)

Guest-host polymer fibers and fractal clusters for nonlinear optics (Kuzyk, Mark G.; Andrews, Mark P.; Paek, Un-Chul; Dirk, Carl W.)V1560,44-53(1991)

Guided-wave nonlinear optics in 2-docosylamino-5-nitropyridine Langmuir-Blodgett films (Bosshard, Christian; Kuepfer, Manfred; Floersheimer, M.; Guenter, Peter)V1560,344-352(1991)

Heat capacity of MOS field-effect devices of optical materials in the presence of a strong magnetic field (Ghatak, Kamakhya P.; Biswas, Shambhu N.)V1485,206-214(1991)

High-efficiency tunable mid-infrared generation in KNbO3 (Guyer, Dean R.; Bosenberg, W. R.; Braun, Frank D.)V1409,14-17(1991)

Higher order effect of regular pumping in lasers and masers (Zhu, Shi-Yao; Scully, Marlan O.; Su, Chang)V1497,255-262(1991)

Highest observed second harmonic intensity from a multilayered Langmuir-Blodgett film structure (Ashwell, Geoffrey J.; Dawnay, Emma J.; Kuczynski, Andrzej P.; Martin, Philip J.)V1361,589-598(1991)

High-order nonlinear susceptibilities (Hammond, Richard T.)V1409,148-153(1991)

High-sensitivity photorefractivity in bulk and multiple-quantum-well semiconductors (Partovi, Afshin; Glass, Alastair M.; Feldman, Robert D.)V1561,20-32(1991)

High-Tg nonlinear optical polymer: poly(N-MNA acrylamide) (Herman, Warren N.; Rosen, Warren A.; Sperling, L. H.; Murphy, C. J.; Jain, H.)V1560,206-213(1991)

Imaging through a low-light-level Raman amplifier (Duncan, Michael D.; Mahon, R.; Tankersley, Lawrence L.; Reintjes, John F.)V1409,127-134(1991)

Importance of nonlinear gain in semiconductor lasers (Agrawal, Govind P.; Gray, George R.)V1497,444-455(1991)

Influence of atomic decay on micromaser operation (Zhu, Shi-Yao)V1497,240-244(1991)

Innovative Optics and Phase Conjugate Optics (Ahlers, Rolf-Juergen; Tschudi, Theo T., eds.)V1500(1991)

Inorganic Crystals for Optics, Electro-Optics, and Frequency Conversion (Bordui, Peter F., ed.)V1561(1991)

In-plane poling of doped-polymer films: creation of two asymmetric directions for second-order optical nonlinearity (Berkovic, Garry; Yitzchaik, Shlomo; Krongauz, Valeri)V1560,238-242(1991)

Intelligent laser systems: adaptive compensation of phase distortions in nonlinear system with two-dimensional feedback (Vorontsov, Michael A.; Larichev, A. V.)V1409,260-266(1991)

Intercomparison of homogeneous laser models with transverse effects (Bandy, Donna K.; Hunter, L. W.; Jones, Darlena J.)V1497,142-152(1991)

Intrapulse stimulated Raman scattering and ultrashort solitons in optical fibers (Headley, Clifford; Agrawal, Govind P.; Reardon, A. C.)V1497,197-201(1991)

Ion-exchanged waveguides in semiconductor-doped glasses (Righini, Giancarlo C.; Pelli, Stefano; De Blasi, C.; Fagherazzi, Giuliano; Manno, D.)V1513,105-111(1991)

Ionic conductivity and damage mechanisms in KTiOPO4 crystals (Morris, Patricia A.; Crawford, Michael K.; Roelofs, Mark G.; Bierlein, John D.; Baer, Thomas M.)V1561,104-111(1991)

Laboratory development of a nonlinear optical tracking filter (Block, Kenneth L.; Whitworth, Ernest E.; Bergin, Joseph E.)V1483,62-65(1991)

Langmuir-Blodgett films for second-order nonlinear optics (Penner, Thomas L.; Armstrong, Nancy J.; Willand, Craig S.; Schildkraut, Jay S.; Robello, Douglas R.)V1560,377-386(1991)

Laser-induced phase transitions in liquid crystals and distributed feedback-fluctuations, energy exchange, and instabilities: squeezed polarized states and intensity correlations (Arakelian, Sergei M.; Chilingarian, Yu. S.; Alaverdian, R. B.; Alodjants, A. P.; Drnoian, V. E.; Karaian, A. S.)V1402,175-191(1991)

Linear and nonlinear optical effects in polymer waveguides (Braeuer, Andreas; Bartuch, Ulrike; Zeisberger, M.; Bauer, T.; Dannberg, Peter)V1559,470-478(1991)

Linear and nonlinear optical properties of polymeric Langmuir-Blodgett films (Penner, Thomas L.; Willand, Craig S.; Robello, Douglas R.; Schildkraut, Jay S.; Ulman, Abraham)V1436,169-178(1991)

Linear and nonlinear optical properties of substituted pyrrolo[1,2-a]quinolines (van Hulst, Niek F.; Heesink, Gerard J.; Bolger, Bouwe; Kelderman, E.; Verboom, W.; Reinhoudt, D. N.)V1361,581-588(1991)

Measurement of nonlinear constants in photosensitive glass optical fibers (Oesterberg, Ulf L.)V1504,107-109(1991)

Measurement of nonlinear optical coefficients by phase-matched harmonic generation (Eckardt, Robert C.; Byer, Robert L.)V1561,119-127(1991)

Measurements of second hyperpolarisabilities of diphenylpolyenes by means of phase-conjugate interferometry (Persoons, Andre P.; Van Wonterghem, Bruno M.; Tackx, Peter)V1409,220-229(1991)

Melt-processed calcium aluminate fibers: structural and optical properties (Wallenberger, Frederick T.; Weston, Norman E.; Brown, Sherman D.)V1484,116-124(1991)

Microscopic mechanism of optical nonlinearity in conjugated polymers and other quasi-one-dimensional systems (Mazumdar, Sumit; Guo, Dandan; Dixit, Sham N.)V1436,136-149(1991)

Model systems for optoelectronic devices based on nonlinear molecular absorption (Speiser, Shammai)V1560,434-442(1991)

Molecular to material design for anomalous-dispersion phase-matched second-harmonic generation (Cahill, Paul A.; Tallant, David R.; Kowalczyk, T. C.; Singer, Kenneth D.)V1560,130-138(1991)

Multilayered optically activated devices based on organic third-order nonlinear optical materials (Swanson, David A.; Altman, Joe C.; Sullivan, Brian J.; Spitzer, Ronnie; Hansen, Glenn A.; Stone, Richard E.)V1560,416-425(1991)

New class of mainchain chromophoric nonlinear optical polymers (Lindsay, Geoffrey A.; Stenger-Smith, John D.; Henry, Ronald A.; Hoover, James M.; Kubin, R. F.)V1497,418-422(1991)

New high-performance material in nonlinear optics field, the polymer blend PMMA-EVA: a first investigation (Carbonara, Giuseppe; Mormile, Pasquale; Abbate, G.; Bernini, Umberto; Maddalena, P.; Malinconico, Mario)V1361,688-691(1991)

New organic crystal material for SHG, 2-cyano-3-(3,4-methylene dioxy phenyl)-2-propionic acid ethyl ester (Mori, Yasushi; Sano, Kenji; Todori, Kenji; Kawamonzen, Yosiaki)V1560,310-314(1991)

New polymers with large and stable second-order nonlinear optical effects (Chen, Mai; Yu, Luping; Dalton, Larry R.; Shi, Youngqiang; Steier, William H.)V1409,202-213(1991)

New syndioregic main-chain, nonlinear optical polymers, and their ellipsometric characterization (Lindsay, Geoffrey A.; Nee, Soe-Mie F.; Hoover, James M.; Stenger-Smith, John D.; Henry, Ronald A.; Kubin, R. F.; Seltzer, Michael D.)V1560,443-453(1991)

New thermistor material for thermistor bolometer: material preparation and characterization (Umadevi, P.; Nagendra, C. L.; Thutupalli, G. K.; Mahadevan, K.; Yadgiri, G.)V1484,125-135(1991)

Nondegenerate two-wave mixing in shaped microparticle suspensions (Pizzoferrato, R.; De Spirito, M.; Zammit, Ugo; Marinelli, M.; Rogovin, Dan N.; Scholl, James F.)V1409,192-201(1991)

Nonlinear absorbance effects in bacteriorhodopsin (Rayfield, George W.)V1436,150-159(1991)

Nonlinear absorption of organic compounds in the picosecond laser radiation field (Arutunian, A. H.; Hovanessian, V. A.; Sarkissian, K. A.)V1402,2-6(1991)

Nonlinear approach to the 3-D reconstruction of microscopic objects (Wu, Xiangchen; Schwarzmann, Peter)V1450,278-285(1991)

Nonlinear dynamics of neuromorphic optical system with spatio-temporal interactions (Vorontsov, Michael A.; Iroshnikov, N. G.)V1621,292-298(1991)

Nonlinear electromagnetic field response of high-Tc superconducting microparticle composites (Haus, Joseph W.; Chung-Yau, F.; Bowden, Charles M.)V1497,382-388(1991)

Nonlinear energy transfer between nanosecond pulses in iron-doped InP crystals (Roosen, Gerald)V1362,398-416(1991)

Nonlinearity and lens design (Tatian, Berge)V1354,154-164(1991)

Nonlinear laser interactions with saltwater aerosols (Alexander, Dennis R.; Poulain, D. E.; Schaub, Scott A.; Barton, John P.)V1497,90-97(1991)

Nonlinear model of the optimal tolerance design for a lens system (Zhuang, Song Lin; Qu, Zhijin)V1354,177-179(1991)

Nonlinear optical amplification and oscillation in spherical microdroplets (Kurizki, Gershon; Goldner, E.)V1497,48-62(1991)

Nonlinear optical and piezoelectric behavior of liquid-crystalline elastomers (Hirschmann, Harald; Meier, Wolfgang; Finkelmann, Heino)V1559,27-38(1991)

Nonlinear optical components with liquid crystals (Dumitru, Mihaela A.; Honciuc, Maria; Sterian, Livia)V1500,339-348(1991)

Nonlinear optical effects under laser pulse interaction with tissues (Altshuler, Grigori B.; Belikov, Andrey V.; Erofeev, Andrey V.)V1427,141-150(1991)

Nonlinear optical processes in droplets with single-mode laser excitation (Chang, Richard K.; Chen, Gang; Hill, Steven C.; Barber, Peter W.)V1497,2-13(1991)

Nonlinear optical properties of a fiber with an organic core crystal grown from solution (Ohmi, Toshihiko; Yoshikawa, Nobuo; Sakai, Koji; Koike, Tomoyuki; Umegaki, Shinsuke)V1361,606-612(1991)

Nonlinear optical properties of bacteriorhodopsin: assignment of the third-order polarizability based on two-photon absorption spectroscopy (Birge, Robert R.; Masthay, M. B.; Stuart, Jeffrey A.; Tallent, Jack R.; Zhang, Chian-Fan)V1432,129-140(1991)

Nonlinear optical properties of chalcogenide As-S, As-Se glasses (Andriesh, A.; Bertolotti, Mario; Chumach, V.; Fazio, E.; Ferrari, A.; Liakhou, G.; Sibilia, C.)V1513,137-147(1991)

Nonlinear optical properties of composite materials (Haus, Joseph W.; Inguva, Ramarao)V1497,350-356(1991)

Nonlinear optical properties of germanium diselenide glasses (Haro-Poniatowski, Emmanuel; Guasti, Manuel F.)V1513,86-92(1991)

Nonlinear optical properties of inorganic coordination polymers and organometallic complexes (Thompson, Mark E.; Chiang, William; Myers, Lori K.; Langhoff, Charles A.)V1497,423-429(1991)

Nonlinear optical properties of N-(4-nitro-2-pyridinyl)-phenylalaninol single crystals (Sutter, Kurt; Hulliger, J.; Knoepfle, G.; Saupper, N.; Guenter, Peter)V1560,296-301(1991)

Nonlinear optical properties of new KTiOPO4 isostructures (Phillips, Mark L.; Harrison, William T.; Stucky, Galen D.)V1561,84-92(1991)

Nonlinear optical properties of nipi and hetero nipi superlattices and their application for optoelectronics (Doehler, Gottfried H.)V1361,443-468(1991)

Nonlinear Optical Properties of Organic Materials IV (Singer, Kenneth D., ed.)V1560(1991)

Nonlinear optical properties of phenosafranin-doped substrates (Speiser, Shammai)V1559,238-244(1991)

Nonlinear optical properties of poled polymers (Gadret, G.; Kajzar, Francois; Raimond, P.)V1560,226-237(1991)

Nonlinear optical properties of quantum-well structures and some of their applications (Sung, C. C.)V1497,456-466(1991)

Nonlinear optical properties of suspensions of microparticles: electrostrictive effect and enhanced backscatter (Kang, Chih-Chieh; Fiddy, Michael A.)V1497,372-381(1991)

Nonlinear optical properties of xanthone derivatives (Imanishi, Yasuo; Itoh, Yuzo; Kakuta, Atsushi; Mukoh, Akio)V1361,570-580(1991)

Nonlinear optical susceptibility of CuCl quantum dot glass (Justus, Brian L.; Seaver, Mark E.; Ruller, Jacqueline A.; Campillo, Anthony J.)V1409,2-8(1991)

Nonlinear optical transmission of an integrated optical bent coupler in semiconductor-doped glass (Guntau, Matthias; Possner, Torsten; Braeuer, Andreas; Dannberg, Peter)V1513,112-122(1991)

Nonlinear Optics II (Fisher, Robert A.; Reintjes, John F., eds.)V1409(1991)

Nonlinear Optics and Materials (Cantrell, Cyrus D.; Bowden, Charles M., eds.)V1497(1991)

Nonlinear optics in poled polymers with two-dimensional asymmetry (Berkovic, Garry; Krongauz, Valeri; Yitzchaik, Shlomo)V1442,44-52(1991)

Nonlinear phenomena in semiconductor lasers (Otsuka, Kenju)V1497,432-443(1991)

Nonlinear polymers and devices (Moehlmann, Gustaaf R.; Horsthuis, Winfried H.; Hams, Benno H.)V1512,34-39(1991)

Nonlinear properties of oriented purple membrane films derived from second-harmonic generation under picosecond excitation: prospect of electro-optical measurements of ultrafast photoelectric respon (Sharkov, A. V.; Gillbro, T.)V1403,434-438(1991)

Nonlinear properties of poled-polymer films: SHG and electro-optic measurements (Morichere, D.; Dumont, Michel L.; Levy, Yves; Gadret, G.; Kajzar, Francois)V1560,214-225(1991)

Nonlinear properties of quasi-optic open resonator with a layer of Crv solution in heavy alcohol under the conditions of magnetic resonance (Vertiy, Alexey A.; Gavrilov, Sergey P.)V1362,702-709(1991)

Nonlinear susceptibilities investigated by the electroabsorption of polymer ion-hemicyanine dye complexes (Nomura, Shintaro; Kobayashi, Takayoshi T.; Nakanishi, Hachiro; Matsuda, Hiro; Okada, Shuji; Tomiyama, Hiromitsu)V1560,272-277(1991)

Nonlinear theory of a three-level laser with microwave coupling: numerical calculation (Fearn, Heidi; Lamb, Willis E.; Scully, Marlan O.)V1497,283-290(1991)

Novel linear and ladder polymers from tetraynes for nonlinear optics (Okada, Shuji; Matsuda, Hiro; Masaki, Atsushi; Nakanishi, Hachiro; Hayamizu, Kikuko)V1560,25-34(1991)

Numerical wavefront propagation through inhomogeneous media (Zakeri, Gholam-Ali)V1558,103-112(1991)

One-dimensional rotatory waves in the optical systems with nonlinear large-scale field interactions (Ivanov, Vladimir Y.; Larichev, A. V.; Vorontsov, Michael A.)V1402,145-153(1991)

Optical bistability and signal competition in active cavity with photochromic nonlinearity of bacteriorhodopsin (Taranenko, Victor B.; Vasnetsov, M. V.)V1621,169-179(1991)

Optical frequency shifter based on stimulated Brillouin scattering in birefringent optical fiber (Duffy, Christopher J.; Tatam, Ralph P.)V1511,155-165(1991)

Optical implementation of winner-take-all models of neural networks (Kobzev, E. F.; Vorontsov, Michael A.)V1402,165-174(1991)

Optical nonlinearities due to long-lived electron-hole plasmas (Dawson, Philip; Galbraith, Ian; Kucharska, Alicia I.; Foxon, C. T.)V1362,384-390(1991)

Optical nonlinearities in ZnSe multiple-quantum-wells (Liu, Yudong; Liu, Shutian; Li, Chunfei; Shen, Dezen; Fan, Xi W.; Fan, Guang H.; Chen, Lian C.)V1362,436-447(1991)

Optical nonlinearities of ZnSe thin films (Chen, Lian C.; Zhang, Ji Y.; Fan, Xi W.; Yang, Ai H.; Zheng, Zhu H.)V1519,450-453(1991)

Optically nonlinear polymeric devices (Moehlmann, Gustaaf R.; Horsthuis, Winfried H.; Mertens, Hans W.; Diemeer, Mart B.; Suyten, F. M.; Hendriksen, B.; Duchet, Christian; Fabre, P.; Brot, Christian; Copeland, J. M.; Mellor, J. R.; van Tomme, E.; van Daele, Peter; Baets, Roel G.)V1560,426-433(1991)

Optical processing using photorefractive GaAs and InP (Liu, Duncan T.; Cheng, Li-Jen; Luke, Keung L.)V1409,116-126(1991)

Optimum design of optical array used as pseudoconjugator (Xiao, Guohua; Song, Ruhua H.; Hu, Zhiping; Le, Shixiao)V1409,106-113(1991)

Organic salts with large electro-optic coefficients (Perry, Joseph W.; Marder, Seth R.; Perry, Kelly J.; Sleva, E. T.; Yakymyshyn, Christopher P.; Stewart, Kevin R.; Boden, Eugene P.)V1560,302-309(1991)

Overview of experimental investigations of laser bandwidth effects in nonlinear optics (Elliott, Daniel S.)V1376,22-33(1991)

Overview of stimulated Brillouin scattering in microdroplets (Cantrell, Cyrus D.)V1497,28-47(1991)

Parametric pulse breakup due to population pulsations in three-level systems (DiMarco, Steven F.; Cantrell, Cyrus D.)V1497,178-187(1991)

Phase-conjugate optical preprocessing filter for small-target tracking (Block, Kenneth L.; Whitworth, Ernest E.; Bergin, Joseph E.)V1481,32-34(1991)

Phase-locking and unstability of light waves in Raman-active crystals (Azarenkov, Aleksey N.; Altshuler, Grigori B.; Belashenkov, Nickolay R.; Inochkin, Mickle V.; Karasev, Viatcheslav B.; Kozlov, Sergey A.)V1409,154-164(1991)

Phase-matched second harmonic generations in poled dye-polymer waveguides (Sugihara, Okihiro; Kinoshita, Takeshi; Okabe, M.; Kunioka, S.; Nonaka, Y.; Sasaki, Keisuke)V1361,599-605(1991)

Photochemistry and Photoelectrochemistry of Organic and Inorganic Molecular Thin Films (Frank, Arthur J.; Lawrence, Marcus F.; Ramasesha, S.; Wamser, Carl C., eds.)V1436(1991)

Photoconducting nonlinear optical polymers (Li, Lian; Jeng, Ru J.; Lee, J. Y.; Kumar, Jayant; Tripathy, Sukant K.)V1560,243-250(1991)

Photoemission from quantum-confined structure of nonlinear optical materials (Ghatak, Kamakhya P.; Biswas, Shambhu N.; Ghoshal, Ardhendhu; Biswas, Shambhu N.)V1409,28-57(1991); V1484,136-148(1991)

Photoinduced effects in optical waveguides (Dianov, Evgeni M.; Kazansky, Peter G.; Stepanov, D. Y.)V1516,81-98(1991)

Photo-induced refractive-index changes and birefringence in optically nonlinear polyester (Shi, Youngqiang; Steier, William H.; Yu, Luping; Chen, Mai; Dalton, Larry R.)V1559,118-126(1991)

Photon Statistics and Coherence in Nonlinear Optics (Perina, Jan, ed.)VMS39(1991)

Photorefractive gratings in the organic crystal 2-cyclooctylamino-5-nitropyridine doped with 7,7,8,8-tetracyanoquinodimethane (Sutter, Kurt; Hulliger, J.; Guenter, Peter)V1560,290-295(1991)

Photorefractive two-wave mixing characteristics for image amplification in diffusion-driven media (Gilbreath, G. C.; Clement, Anne E.; Fugera, S. N.; Mizell, Gregory J.)V1409,87-99(1991)

Photorefractivity in doped nonlinear organic polymers (Moerner, William E.; Walsh, C. P.; Scott, J. C.; Ducharme, Stephen P.; Burland, Donald M.; Bjorklund, Gary C.; Twieg, Robert J.)V1560,278-289(1991)

Picosecond orientational dynamics of complex molecules studied by incoherent light three-wave mixing (Apanasevich, P. A.; Kozich, V. P.; Vodchitz, A. I.; Kontsevoy, B. L.)V1403,475-477(1991)

Point defects in KTP and their possible role in laser damage (Scripsick, Michael P.; Edwards, Gary J.; Halliburton, Larry E.; Belt, Roger F.; Kappers, Lawrence A.)V1561,93-103(1991)

Poled polyimides as thermally stable electro-optic polymer (Wu, Jeong W.; Valley, John F.; Stiller, Marc A.; Ermer, Susan P.; Binkley, E. S.; Kenney, John T.; Lipscomb, George F.; Lytel, Richard S.)V1560,196-205(1991)

Preparation and characterization of silver-colloid/polymer-composite nonlinear optical materials (LaPeruta, Richard; Van Wagenen, E. A.; Roche, J. J.; Kitipichai, Prakob; Wnek, Gary E.; Korenowski, G. M.)V1497,357-366(1991)

Problems of large neurodynamics system modeling: optical synergetics and neural networks (Vorontsov, Michael A.)V1402,116-144(1991)

Progress in diode laser upconversion (Dixon, G. J.; Kean, P. N.)V1561,147-150(1991)

Progress in organic third-order nonlinear optical materials (Kuzyk, Mark G.)V1436,160-168(1991)

Pulse reshaping and coherent sideband generation effects on multiphoton excitation of polyatomic molecules (Garner, Steven T.; Cantrell, Cyrus D.)V1497,188-196(1991)

Quantum wells and artificially structured materials for nonlinear optics (Harris, James S.; Fejer, Martin M.)V1361,262-273(1991)

Quasirelaxation-free guest-host poled-polymer waveguide modulator: material, technology, and characterization (Levenson, R.; Liang, J.; Toussaere, E.; Carenco, Alain; Zyss, Joseph)V1560,251-261(1991)

Real-time holographic microscope with nonlinear optics (Goldfain, Ervin)V1527,199-209(1991)

Recording of laser radiation power distribution with the use of nonlinear optical effects (Korchazhkin, S. V.; Krasnova, L. O.)V1358,966-972(1991)

Relationship between conjugation length and third-order nonlinearity in bis-donor substituted diphenyl polyenes (Spangler, Charles W.; Havelka, Kathleen O.; Becker, Mark W.; Kelleher, Tracy A.; Cheng, Lap Tak A.)V1560,139-147(1991)

Resonances and internal electric energy density in droplets (Hill, Steven C.; Barber, Peter W.; Chowdhury, Dipakbin Q.; Khaled, Elsayed-Esam M.; Mazumder, Mohiuddin)V1497,16-27(1991)

Resonant behavior of the temporal response of the photorefractive InP:Fe under dc fields (Abdelghani-Idrissi, Ahmed M.; Ozkul, Cafer; Wolffer, Nicole; Gravey, Philippe; Picoli, Gilbert)V1362,417-427(1991)

Reversible phase transition and third-order nonlinearity of phthalocyanine derivatives (Suda, Yasumasa; Shigehara, Kiyotaka; Yamada, Akira; Matsuda, Hiro; Okada, Shuji; Masaki, Atsushi; Nakanishi, Hachiro)V1560,75-83(1991)

Second-harmonic generation of mode-locked Nd:YAG and Nd:YLF lasers using LiB3O5 (Reed, Murray K.; Tyminski, Jacek K.; Bischel, William K.)V1410,179-184(1991)

Second-order and third-order processes as diagnostic tools for the analysis of organic polymers: an overview (Marowsky, Gerd; Luepke, G.)V1560,328-334(1991)

Selected Papers on Nonlinear Optics (Brandt, Howard E., ed.)VMS32(1991)

Self-action of supremely short light pulses in fibers (Azarenkov, Aleksey N.; Altshuler, Grigori B.; Kozlov, Sergey A.)V1409,166-177(1991)

Semiconductor-doped glass as a nonlinear material (Speit, Burkhard; Remitz, K. E.; Neuroth, Norbert N.)V1361,1128-1131(1991)

Semiconductor-doped glasses: nonlinear and electro-optical properties (Ekimov, A. I.; Kudryavtsev, I. A.; Chepick, D. I.; Shumilov, S. K.)V1513,123-129(1991)

Semiconductor laser amplifiers as all-optical frequency converters (Schunk, Nikolaus; Grosskopf, Gerd; Ludwig, Reinhold; Schnabel, Ronald; Weber, Hans-Georg)V1362,391-397(1991)

Semiconductor laser diode with weak optical feedback: self-coupling effects on P-I characteristics (Milani, Marziale; Mazzoleni, S.; Brivio, Franca)V1474,83-97(1991)

Semiempirical method for calculation of dynamic susceptibilities (Svendsen, Erik N.; Stroyer-Hansen, T.)V1361,1048-1055(1991)

Similarity and difference between degenerate parametric oscillators and two-photon correlated-spontaneous-emission lasers (Zhu, Shi-Yao; Scully, Marlan O.)V1497,277-282(1991)

Sol-gel processed novel multicomponent inorganic oxide: organic polymer composites for nonlinear optics (Zhang, Yue; Cui, Y. P.; Wung, C. J.; Prasad, Paras N.; Burzynski, Ryszard)V1560,264-271(1991)

Spatial frequency filtering on the basis of the nonlinear optics phenomena (Kudryavtseva, Anna D.; Tcherniega, Nicolaii V.; Brekhovskikh, Galina L.; Sokolovskaya, Albina I.)V1385,190-199(1991)

Spectral characteristics of Brillouin-enhanced four-wave mixing for pulsed and CW inputs (Lebow, Paul S.; Ackerman, John R.)V1409,60-66(1991)

Stability analysis of semidiscrete schemes for thermal blooming computation (Ulrich, Peter B.)V1408,192-202(1991)

Statistics of spontaneously generated Raman solitons (Englund, John C.; Bowden, Charles M.)V1497,218-221(1991)

Stimulated Brillouin scatter SIX-code (Litvak, Marvin M.; Wagner, Richard J.)V1415,62-71(1991)

Stimulated Raman diagnostics in diesel droplets (Golombok, Michael)V1497,100-119(1991)

Stimulated Raman scattering with initially nonclassical light (Perina, Jan)V1402,192-192(1991)

Stochastic simulations of light-scattering noise in two-wave mixing: application to artificial Kerr media (McGraw, Robert L.)V1409,135-147(1991)

Structure/property relationships for molecular second-order nonlinear optics (Marder, Seth R.; Cheng, Lap Tak A.; Tiemann, Bruce G.; Beratan, David N.)V1560,86-97(1991)

Study on the surface currents density with second-harmonic generation from silver (Chen, Zhan; Jiang, Hongbing; Zheng, J. B.; Zhang, Zhiming)V1437,103-109(1991)

Superheating phenomena in absorbing microdroplets irradiated by pulsed lasers (Armstrong, Robert L.)V1497,132-140(1991)

Superradiance and exciton dynamics in molecular aggregates (Fidder, Henk; Terpstra, Jacob; Wiersma, Douwe A.)V1403,530-544(1991)

Synthesis and incorporation of ladder polymer subunits in copolyamides, pendant polymers, and composites for enhanced nonlinear optical response (Spangler, Charles W.; Saindon, Michelle L.; Nickel, Eric G.; Sapochak, Linda S.; Polis, David W.; Dalton, Larry R.; Norwood, Robert A.)V1497,408-417(1991)

Synthesis and nonlinear optical activity of cumulenes (Ermer, Susan P.; Lovejoy, Steven M.; Leung, Doris; Spitzer, Ronnie; Hansen, Glenn A.; Stone, Richard E.)V1560,120-129(1991)

Synthesis and nonlinear optical properties of preformed polymers forming Langmuir-Blodgett films (Verbiest, Thierry; Persoons, Andre P.; Samyn, Celest)V1560,353-361(1991)

Synthesis, characterization, and processing of organic nonlinear optical polymers (Druy, Mark A.; Glatkowski, Paul J.)V1409,214-219(1991)

Theoretical analysis of the third-order nonlinear optical properties of linear cyanines and polyenes (Pierce, Brian M.)V1560,148-161(1991)

Theoretical insight into the quadratic nonlinear optical response of organics: derivatives of pyrene and triaminotrinitrobenzene (Bredas, Jean-Luc; Dehu, C.; Meyers, F.; Zyss, Joseph)V1560,98-110(1991)

Theoretical study of optical pockels and Kerr coefficients of polyenes (Albert, I. D.; Ramasesha, S.)V1436,179-189(1991)

Theoretical study of the fifth harmonic generation in organic liquids (Popescu, Ion M.; Puscas, Niculae Tiberiu N.; Sterian, Paul E.; Irimescu, Dorin I.; Podoleanu, Adrian G.)V1361,1041-1047(1991)

Theory and experiments on passive negative feedback pulse control in active/passive mode-locked solid state lasers (Agnesi, Antoniangelo; Fogliani, Manlio F.; Reali, Giancarlo C.; Kubecek, Vaclav)V1415,242-247(1991)

Theory of transient self-focusing (Hsia, Kangmin; Cantrell, Cyrus D.)V1497,166-177(1991)

Third-order nonlinear optical characterization of side-chain copolymers (Norwood, Robert A.; Sounik, James R.; Popolo, J.; Holcomb, D. R.)V1560,54-65(1991)

Third-order optical nonlinearities and femtosecond responses in metallophthalocyanine thin films made by vacuum deposition, molecular beam epitaxy, and spin coating (Wada, Tatsuo; Hosoda, Masahiro; Garito, Anthony F.; Sasabe, Hiroyuki; Terasaki, A.; Kobayashi, Takayoshi T.; Tada, Hiroaki; Koma, Atsushi)V1560,162-171(1991)

Time-resolved Raman spectroscopy from reacting optically levitated microdroplets (Carls, Joseph C.; Brock, James R.)V1497,120-131(1991)

Time reversal, enhancement, and suppression of dipole radiation by a phase-conjugate mirror (Bochove, Erik J.)V1497,222-227(1991)

Transition from homoclinic to heteroclinic chaos in coupled laser arrays (Otsuka, Kenju)V1497,300-312(1991)

Two-color interferometry using a detuned frequency-doubling crystal (Koch, Karl W.; Moore, Gerald T.)V1516,67-74(1991)

Two-dimensional dynamic neural network optical system with simplest types of large-scale interactions (Vorontsov, Michael A.; Zheleznykh, N. I.; Larichev, A. V.)V1402,154-164(1991)

Ultralinear electro-optic modulators for microwave fiber-optic communications (Pan, J. J.; Li, Wei Z.; Li, Yi Q.)V1476,32-43(1991)

Unreasonable effectiveness of laser physics in the life sciences: nonlinear optics in control, diagnostics, and modeling of biophysical processes (Akhmanov, Sergei A.)V1403,2-2(1991)

USSR-CSFR Joint Seminar on Nonlinear Optics in Control, Diagnostics, and Modeling of Biophysical Processes (Akhmanov, Sergei A.; Zadkov, Victor N., eds.)V1402(1991)

Vector beam propagation using Hertz vectors (Milsted, Carl S.; Cantrell, Cyrus D.)V1497,202-215(1991)

Vector squeezed states (Lee, Heun J.; Haus, Joseph W.)V1497,228-239(1991)

Wave-function engineering in Si-Ge microstructures: linear and nonlinear optical response (Jaros, Milan; Turton, Richard M.)V1362,242-253(1991)

What we can learn about second-harmonic generation in germanosilicate glass from the analogous effect in semiconductor-doped glasses (Lawandy, Nabil M.)V1516,99-112(1991)

Wideband NLO organic external modulators (Findakly, Talal K.; Teng, Chia-Chi; Walpita, Lak M.)V1476,14-21(1991)

Object recognition—see also neural networks; pattern recognition

Accelerated detection of image objects and their orientations with distance transforms (Ford, Gary E.; Paglieroni, David W.; Tsujimoto, Eric M.)V1452,244-255(1991)

Adaptive snakes: control of damping and material parameters (Samadani, Ramin)V1570,202-213(1991)

Adaptive surface reconstruction (Terzopoulos, Demetri; Vasilescu, Manuela)V1383,257-264(1991)

Algorithm development and evaluation on the Multifunction Target Acquisition Processor (Haskett, Michael C.; Lidke, Steve L.)VIS07,14-23(1991)

Applications of Digital Image Processing XIV (Tescher, Andrew G., ed.)V1567(1991)

Architecture and performance of a hardware collision-checking accelerator (Bardin, R. K.; Libby, Vibeke)V1566,394-404(1991)

Artificial neural networks for automatic object recognition (Rogers, Steven K.; Ruck, Dennis W.; Kabrisky, Matthew; Tarr, Gregory L.; Oxley, Mark E.)VIS07,231-243(1991)

Aspect networks: using multiple views to learn and recognize 3-D objects (Seibert, Michael; Waxman, Allen M.)V1383,10-19(1991)

ATC (automatic target cueing) algorithm evaluation (Gleason, Jim M.; Sherman, James W.)V1406,169-170(1991)

Automated instrumentation, evaluation, and diagnostics of automatic target recognition systems (Nasr, Hatem N.)VIS07,202-213(1991)

Automatic Object Recognition (Nasr, Hatem N., ed.)VIS07(1991)

Automatic recognition of bone for x-ray bone densitometry (Shepp, Larry A.; Vardi, Y.; Lazewatsky, J.; Libeau, James; Stein, Jay A.)V1452,216-224(1991)

Automatic shape recognition of human limbs to avoid errors due to skin marker shifting in motion analysis (Hatze, Herbert; Baca, Arnold)V1567,264-276(1991)

Autoregressive identification method for partially occluded industrial object recognition (Ionescu, Dan; Damerji, Tayeb)V1406,40-41(1991)

Bayesian matching technique for detecting simple objects in heavily noisy environment (Baras, John S.; Frantzeskakis, Emmanuil N.)V1569,341-353(1991)

Binary object analysis hardware area parameter acceleration (Seitzler, Thomas M.)V1567,25-30(1991)

Challenges of vision theory: self-organization of neural mechanisms for stable steering of object-grouping data in visual motion perception (Marshall, Jonathan A.)V1569,200-215(1991)

Characteristic views and perspective aspect graphs of quadric-surfaced solids (Chen, Shuang; Freeman, Herbert)V1383,2-9(1991)

Clutter metrics in infrared target acquisition (Tidhar, Gil; Rotman, Stanley R.)V1442,310-324(1991)

Coherent digital/optical system for automatic target recognition (Derr, John I.; Ghaffari, Tammy G.)V1406,127-128(1991)

Color space analysis of road detection algorithms (Crisman, Jill D.)V1569,492-506(1991)

Computing motion parameters from sparse multisensor range data for telerobotics (Vemuri, Baba C.; Skofteland, G.)V1383,97-108(1991)

Contour estimation using global shape constraints and local forces (Deng, Keqiang; Wilson, Joseph N.)V1570,227-233(1991)

Deformable surfaces: a free-form shape representation (Delingette, Herve; Hebert, Martial; Ikeuchi, Katsushi)V1570,21-30(1991)

Deformable templates, robust statistics, and Hough transforms (Yuille, Alan L.; Peterson, Carsten; Honda, Ko)V1570,166-174(1991)

Detection of surface-laid minefields using a hierarchical image processing algorithm (McFee, John E.; Russell, Kevin L.; Ito, Mabo R.)V1567,42-52(1991)

Development and use of confidence intervals for automatic target recognition evaluation (Sherman, James W.)VIS07,144-169(1991)

Development of an electronic terrain board as a processor test and evaluation tool (Walters, Clarence P.; Lorenzo, Maximo)VIS07,181-201(1991)

Development of automatic target recognizers for Army applications (Jones, Terry L.)VIS07,4-13(1991)

Direct computation of geometric features from motion disparities and shading (Weinshall, Daphna)V1570,274-285(1991)

Efficient system for 3-D object recognition (Sobh, Tarek M.; Alameldin, Tarek K.)V1383,359-366(1991)

Energy-based segmentation of very sparse range surfaces (Lerner, Mark; Boult, Terrance E.)V1383,277-284(1991)

Extracting features to recognize partially occluded objects (Koch, Mark W.; Ramamurthy, Arjun)V1567,638-649(1991)

Face recognition based on depth maps and surface curvature (Gordon, Gaile G.)V1570,234-247(1991)

Fast one-pass algorithm to label objects and compute their features (Thai, Tan Q.)V1567,533-541(1991)

Feature correspondence in multiple sensor data fusion (Broida, Ted J.)V1383,635-651(1991)

Finding a grasped object's pose using joint angle and torque constraints (Siegel, David M.)V1383,151-165(1991)

From points to surfaces (Fua, Pascal; Sander, Peter T.)V1570,286-296(1991)

Fusion of stereo views: estimating structure and motion using a robust method (Weng, Juyang; Cohen, Paul)V1383,321-332(1991)

Generalized neocognitron model for facial recognition (Chen, Su-Shing; Hong, Young-Sik)V1569,463-473(1991)

Geometric Methods in Computer Vision (Vemuri, Baba C., ed.)V1570(1991)

Global minima via dynamic programming: energy minimizing active contours (Chandran, Sharat; Maejima, Tsukasa; Miyazaki, Sanae)V1570,391-402(1991)

Heterogeneous input neuration for network-based object recognition architectures (Gnazzo, John F.)V1569,239-246(1991)

Hierarchical decomposition and axial representation of shape (Rom, Hillel; Medioni, Gerard)V1570,262-273(1991)

Hierarchical fusion of geometric constraints for image segmentation (Seetharaman, Guna S.; Chu, Chee-Hung H.)V1383,582-588(1991)

Hierarchical multisensor analysis for robotic exploration (Eberlein, Susan; Yates, Gigi; Majani, Eric)V1388,578-586(1991)

Hierarchical target representation for autonomous recognition using distributed sensors (Luo, Ren C.; Kay, Michael G.)V1383,537-544(1991)

Hybrid digital/optical ATR system (Goodwin, David B.; Cappiello, Gregory G.; Coppeta, David A.; Govignon, Jacques P.)V1564,536-549(1991)

Image representation by integrating curvatures and Delaunay triangulations (Wu, Chengke; Mohr, Roger)V1570,362-370(1991)

Image Understanding in the '90s: Building Systems that Work (Mitchell, Brian T., ed.)V1406(1991)

Importance of sensor models to model-based vision applications (Zelnio, Edmund G.)VIS07,112-121(1991)

Invariant feature matching in parameter space with application to line features (Hecker, Y. C.; Bolle, Ruud M.)V1570,298-314(1991)

IR/MMW fusion ATR (automatic target recognition) (Thiede, Edwin C.)VIS07,24-35(1991)

Issues in automatic object recognition: linking geometry and material data to predictive signature codes (Deitz, Paul; Muuss, Michael J.; Davisson, Edwin O.)VIS07,40-56(1991)

Issues in parallelism in object recognition (Bhandarkar, Suchendra M.; Suk, Minsoo)V1384,234-245(1991)

Knowledge- and model-based ATR (automatic target recognition) algorithms adaptation (Nasr, Hatem N.; Bazakos, Michael E.; Sadjadi, Firooz A.; Amehdi, Hossien)VIS07,122-129(1991)

Lessons learned from a commercial module approach to real-time ATR development (Lidke, Steve L.; Haskett, Michael C.)V1406,203-203(1991)

Machine vision applications of image invariants: real-time processing experiments (Payton, Paul M.; Haines, Barry K.; Smedley, Kirk G.; Barrett, Eamon B.)V1406,58-71(1991)

Matching 3-D smooth surfaces with their 2-D projections using 3-D distance maps (Lavallee, Stephane; Szeliski, Richard; Brunie, Lionel)V1570,322-336(1991)

Model-based ATR (automatic target recognition) systems for the military (Tatum, William F.)VIS07,130-139(1991)

Model-based surface classification (Newman, Timothy S.; Flynn, Patrick J.; Jain, Anil K.)V1570,250-261(1991)

Model-based vision: an operational reality? (Mundy, Joseph L.)V1567,124-141(1991)

Model-based vision using geometric hashing (Akerman, Alexander; Patton, Ronald)V1406,30-39(1991)

Modular algorithms for depth characterization of object surfaces with ultrasonic sensors (Brule, Michel J.; Soucy, L.)V1388,432-441(1991)

Morphological segmentation and 3-D rendering of the brain in magnetic resonance imaging (Connor, William H.; Diaz, Pedro J.)V1568,327-334(1991)

Morphologic edge detection in range images (Gupta, Sundeep; Krishnapuram, Raghu J.)V1568,335-346(1991)

Motion estimation without correspondences and object tracking over long time sequences (Goldgof, Dmitry B.; Lee, Hua; Huang, Thomas S.)V1383,109-121(1991)

Multilevel evidence fusion for the recognition of 3-D objects: an overview of computer vision research at IBM/T.J. Watson (Bolle, Ruud M.; Califano, Andrea; Kender, John R.; Kjeldsen, Rick; Mohan, Rakesh)V1383,305-318(1991)

Multispectral and multisensor adaptive automatic object recognition (Sadjadi, Firooz A.)VIS07,218-230(1991)

Neural network approach for object orientation classification (Yeung, Keith K.; Zakarauskas, Pierre; McCray, Allan G.)V1569,133-146(1991)

New method of 3-D object recognition (He, Anzhi; Li, Qun Z.; Miao, Peng-Cheng)V1567,698-702(1991)

Object classification for obstacle avoidance (Regensburger, Uwe; Graefe, Volker)V1388,112-119(1991)

Object detection in real-time (Solder, Ulrich; Graefe, Volker)V1388,104-111(1991)

On seeing spaghetti: a novel self-adjusting seven-parameter Hough space for analyzing flexible extruded objects (Kender, John R.; Kjeldsen, Rick)V1570,315-321(1991)

Optical gray-scale morphology for target detection (Casasent, David P.; Schaefer, Roland H.)V1568,313-326(1991)

Orientation-based differential geometric representations for computer vision applications (Taubes, C. H.; Liang, Ping)V1570,96-102(1991)

Physically based and probabilistic models for computer vision (Szeliski, Richard; Terzopoulos, Demetri)V1570,140-152(1991)

Polynomial approach for morphological operations on 2-D and 3-D images (Bhattacharya, Prabir; Qian, Kai)V1383,530-536(1991)

Polynomial regression analysis for estimating motion from image sequences (Frau, Juan; Llario, Vicenc; Oliver, Gabriel)V1388,329-340(1991)

Primary set of characteristic views for 3-D objects (Chen, Shuang; Freeman, Herbert)V1570,352-361(1991)

Probabilistic modeling of surfaces (Szeliski, Richard)V1570,154-165(1991)

Recognition and positioning of rigid objects using algebraic moment invariants (Taubin, Gabriel; Cooper, David B.)V1570,175-186(1991)

Recognition and tracking of moving objects (El-Konyaly, Sayed H.; Enab, Yehia M.; Soltan, Hesham)V1388,317-328(1991)

Recognition of movement object collision (Chang, Hsiao T.; Sun, Geng-tian; Zhang, Yan)V1388,442-446(1991)

Recognition of partially occluded objects using B-spline representation (Salari, Ezzatollah; Balaji, Sridhar)V1384,115-123(1991)

Recognition, tracking, and pose estimation of arbitrarily shaped 3-D objects in cluttered intensity and range imagery (Gottschalk, Paul G.; Turney, Jerry L.; Chiu, Arnold H.; Mudge, Trevor N.)V1383,84-96(1991)

Recognizing human eyes (Hallinan, Peter W.)V1570,214-226(1991)

Recursive computation of a wire-frame representation of a scene from dynamic stereo using belief functions (Tirumalai, Arun P.; Schunck, Brian G.; Jain, Ramesh C.)V1569,28-42(1991)

Representing the dynamics of the occluding contour (Seales, W. B.; Dyer, Charles R.)V1383,47-58(1991)

Robust statistical method for background extraction in image segmentation (Rodriguez, Arturo A.; Mitchell, O. R.)V1569,182-199(1991)

SCORPIUS: a successful image understanding implementation (Wilkes, Scott C.)V1406,110-121(1991)

Selected Papers on Automatic Object Recognition (Nasr, Hatem N., ed.)VMS41(1991)

Sensor Fusion III: 3-D Perception and Recognition (Schenker, Paul S., ed.)V1383(1991)

Set discrimination analysis tools for grey-level morphological operators (Vogt, Robert C.)V1568,200-211(1991)

Shape discrimination using invariant Fourier representation and a neural network classifier (Wu, Hsien-Huang; Schowengerdt, Robert A.)V1569,147-154(1991)

Shape metrics from curvature-scale space and curvature-tuned smoothing (Dudek, Gregory)V1570,75-85(1991)

Shape representation and nonrigid motion tracking using deformable superquadrics (Metaxas, Dimitri; Terzopoulos, Demetri)V1570,12-20(1991)

Signature prediction models for flir target recognition (Velten, Vincent J.)VIS07,98-107(1991)

Simultaneous object estimation and image reconstruction in a Bayesian setting (Hanson, Kenneth M.)V1452,180-191(1991)

Stereo matching, error detection, and surface reconstruction (Stewart, Charles V.)V1383,285-296(1991)

Stochastic field-based object recognition in computer vision (Zhu, Dongping; Beex, A. A.; Conners, Richard W.)V1569,174-181(1991)

Surface property determination for planetary rovers (Severson, William E.; Douglass, Robert J.; Hennessy, Stephen J.; Boyd, Robert; Anhalt, David J.)V1388,490-501(1991)

Surface reconstruction method using deformable templates (Wang, Yuan-Fang; Wang, Jih-Fang)V1383,265-276(1991)

Surface reconstruction with discontinuities (Lee, David T.; Shiau, Jyh-Jen H.)V1383,297-304(1991)

System transfer modeling for automatic target recognizer evaluations (Clark, Lloyd G.)VIS07,170-180(1991)

Target acquisition model appropriate for dynamically changing scenarios (Rotman, Stanley R.; Gordon, E. S.)V1442,335-346(1991)

Target identification by means of adaptive neural networks in thermal infrared images (Acheroy, Marc P.; Mees, W.)V1569,121-132(1991)

Three-dimensional object recognition using multiple sensors (Hackett, Jay K.; Lavoie, Matt J.; Shah, Mubarak A.)V1383,611-622(1991)

Three-dimensional target recognition from fusion of dense range and intensity images (Ramirez, Manuel; Mitra, Sunanda)V1567,632-637(1991)

Toward tactile sensor-based exploration in a robotic environment (Gadagkar, Hrishikesh P.; Trivedi, Mohan M.)V1383,142-150(1991)

Unified approach to multisensor simulation of target signatures (Stewart, Stephen R.; LaHaie, Ivan J.; Lyons, Jayne T.)VIS07,57-97(1991)

Unsupervised target detection in a single IR image frame (Zhou, Yi-Tong)V1567,502-510(1991)

Vision-based strip inspection hardware for metal production (Seitzler, Thomas M.)V1567,73-76(1991)

Weighted least-squared error method for object localization (Wu, Hsiang-Lung)V1384,90-99(1991)

Workstation recognition using a constrained edge-based Hough transform for mobile robot navigation (Vaughn, David L.; Arkin, Ronald C.)V1383,503-514(1991)

Oceanography

Basic use of acoustic speckle pattern for metrology and sea waves study (He, Duo-Min; He, Ming-Shia)V1332,808-819(1991)

Circular polarization effects in the light scattering from single and suspensions of dinoflagellates (Shapiro, Daniel B.; Hunt, Arlon J.; Quinby-Hunt, Mary S.; Hull, Patricia G.)V1537,30-41(1991)

Computer model for predicting underwater color images (Palowitch, Andrew W.; Jaffe, Jules S.)V1537,128-139(1991)

Cooled focal-plane assembly for ocean color and temperature scanner (Nakayama, Masao; Izawa, Toshiyuki; Fujisada, Hiroyuki; Tange, Yoshio; Miyachi, Yuji; Sato, Ryota; Ishida, Juro; Tanii, Jun)V1490,207-215(1991)

Coupled-dipole approximation: predicting scattering by nonspherical marine organisms (Hull, Patricia G.; Hunt, Arlon J.; Quinby-Hunt, Mary S.; Shapiro, Daniel B.)V1537,21-29(1991)

Design of an imaging spectrometer for observing ocean color (Weng, Zhicheng; Chen, Zhiyong; Cong, Xiaojie)V1527,338-348(1991)

Detector perturbation of ocean radiance measurements (Macdonald, Burns; Helliwell, William S.; Sanborn, James; Voss, Kenneth J.)V1537,104-114(1991)

Development of the marine-aggregated-particle profiling and enumerating rover (Costello, David K.; Carder, Kendall L.; Steward, Robert G.)V1537,161-172(1991)

Dynamic sea image generation (Levesque, Martin P.)V1486,294-300(1991)

Experiment for testing the closure property in ocean optics (Maffione, Robert A.; Honey, Richard C.; Brown, Robert A.)V1537,115-126(1991)

Extension of model of beam spreading in seawater to include dependence on the scattering phase function (Schippnick, Paul F.)V1537,185-193(1991)

General purpose fiber optic hydrophone made of castable epoxy (Garrett, Steven L.; Brown, David A.; Beaton, Brian L.; Wetterskog, Kevin; Serocki, John)V1367,13-29(1991)

High-resolution spectral analysis and modeling of infrared ocean surface radiometric clutter (McGlynn, John D.; Ellis, Kenneth K.; Kryskowski, David)V1486,141-150(1991)

Imaging model for underwater range-gated imaging systems (Strand, Michael P.)V1537,151-160(1991)

Impact on the medium MTF by model estimation of b (Estep, Leland; Arnone, Robert A.)V1537,89-96(1991)

Infrared clutter measurements of marine backgrounds (Schwering, Piet B.)V1486,25-36(1991)

Infrared instrumental complex for remote measurement of ocean surface temperature distribution (Miroshnikov, Mikhail M.; Minyeev, V. N.; Povarkov, V. I.; Samkov, V. M.; Solovyev, V. I.)V1540,496-505(1991)

Instrument for underwater measurement of optical backscatter (Maffione, Robert A.; Dana, David R.; Honey, Richard C.)V1537,173-184(1991)

Integrated oceanographic image understanding system (Lybanon, Matthew; Peckinpaugh, Sarah; Holyer, Ronald J.; Cambridge, Vivian)V1406,180-189(1991)

Isotropic light source for underwater applications (Brown, Robert A.; Honey, Richard C.; Maffione, Robert A.)V1537,147-150(1991)

Laser range-gated underwater imaging including polarization discrimination (Swartz, Barry A.; Cummings, James D.)V1537,42-56(1991)

Laser remote sensing of natural water organics (Babichenko, Sergey M.; Poryvkina, Larisa)V1492,319-323(1991)

Modeling and experiments with a subsea laser radar system (Bjarnar, Morten L.; Klepsvik, John O.; Nilsen, Jan E.)V1537,74-88(1991)

Neural network labeling of the Gulf Stream (Lybanon, Matthew; Molinelli, Eugene J.; Muncill, G.; Pepe, Kevin)V1469,637-647(1991)

New, simple method of extracting temperature of liquid water from Raman scattering (Liu, Zhi-Shen; Ma, Jun; Zhang, Jin-Long; Chen, Wen-Zhong)V1558,306-316(1991)

Ocean color and temperature scanner for ADEOS (Tanii, Jun; Machida, Tsuneo; Ayada, Haruki; Katsuyama, Yoshihiko; Ishida, Juro; Iwasaki, Nobuo; Tange, Yoshio; Miyachi, Yuji; Sato, Ryota)V1490,200-206(1991)

Optical aberrations in underwater holography and their compensation (Watson, John; Kilpatrick, J. M.)V1461,245-253(1991)

Optical power budget and device time-constant considerations in undersea laser-based sensor design (Leatham, James G.)V1537,194-202(1991)

Raman lidar for measuring backscattering in the China Sea (Liu, Zhi-Shen; Zhang, Jin-Long; Chen, Wen-Zhong; Huang, Xiao-Sheng; Ma, Jun)V1558,379-383(1991)

Recent progress in interferometric fiber sensor technology (Kersey, Alan D.)V1367,2-12(1991)

Selected Papers on Multiple Scattering in Plane Parallel Atmospheres and Oceans: Methods (Kattawar, George W., ed.)VMS42(1991)

Shape reconstruction from far-field patterns in a stratified ocean environment (Xu, Yongzhi)V1471,78-86(1991)

Tale of two underwater lenses (Evans, Clinton E.; Doshi, Rekha)V1537,203-214(1991)

Underwater Imaging, Photography, and Visibility (Spinrad, Richard W., ed.)V1537(1991)

Underwater solar light field: analytical model from a WKB evaluation (Tessendorf, Jerry A.)V1537,10-20(1991)

Universal multifractal theory and observations of land and ocean surfaces, and of clouds (Lavallee, Daniel; Lovejoy, Shaun; Schertzer, Daniel)V1558,60-75(1991)

Variation of the point spread function in the Sargasso Sea (Voss, Kenneth J.)V1537,97-103(1991)

Vector and scalar field interpretation (Farrell, Edward J.; Aukrust, Trond; Oberhuber, Josef M.)V1459,136-147(1991)

Vertical ocean reflectance at low altitudes for narrow laser beams (Crittenden, Eugene C.; Rodeback, G. W.; Milne, Edmund A.; Cooper, Alfred W.)V1492,187-197(1991)

Visualizing underwater acoustic matched-field processing (Rosenblum, Lawrence; Kamgar-Parsi, Behzad; Karahalios, Margarida; Heitmeyer, Richard)V1459,274-282(1991)

Wavefront reconstruction of acoustic waves in a variable ocean (Porter, Robert P.; Mourad, Pierre D.; Al-Kurd, Azmi)V1558,91-102(1991)

Ophthalmology—see also medicine

Accurate method for measuring oblique astigmatism and oblique power of ophthalmic lenses (Wihardjo, Erning; Silva, Donald E.)V1529,57-62(1991)

Acoustic effects of Q-switched Nd:YAG on crystalline lens (Porindla, Sridhar N.; Rylander, Henry G.; Welch, Ashley J.)V1427,267-272(1991)

Analysis of an adjustable slit design for correcting astigmatism (Clapham, Terrance N.; D'Arcy, John; Bechtel, Lorne; Glockler, Hermann; Munnerlyn, Charles R.; McDonnell, Peter J.; Garbus, Jenny)V1423,2-7(1991)

Analysis of photon correlation functions obtained from eye lenses in vivo (Van Laethem, Marc; Xia, Jia-zhi; Clauwaert, Julius)V1403,732-742(1991)

Antireflection coating standards of ophthalmic resin lens materials (Porden, Mark)V1529,115-123(1991)

Argon dye photocoagulator for microsurgery of the interior structure of the eye (Wolinski, Wieslaw; Kazmirowski, Antoni; Kesik, Jerzy; Korobowicz, Witold; Spytkowski, Wojciech)V1391,334-340(1991)

Axial and transverse displacement tolerances during excimer laser surgery for myopia (Shimmick, John K.; Munnerlyn, Charles R.; Clapham, Terrance N.; McDonald, Marguerite B.)V1423,140-153(1991)

Biomechanics of the cornea (Buzard, Kurt A.; Hoeltzel, David A.)V1423,70-81(1991)

Biomedical applications of laser technology at Los Alamos (Bigio, Irving J.; Loree, Thomas R.)V1403,776-780(1991)

Calibration issues in the measurement of ocular movement and position using computer image processing (Weeks, Arthur R.; Myler, Harley R.; Jolson, Alfred S.)V1567,77-87(1991)

Cold-laser microsurgery of the retina with a syringe-guided 193 nm excimer laser (Lewis, Aaron; Palanker, Daniel; Hemo, Itzhak; Pe'er, Jacob; Zauberman, Hanan)V1423,98-102(1991)

Color vision enhancement with spectacles (Perrott, Colin M.)V1529,31-36(1991)

Computed estimation of visual acuity after laser refractive keratectomy (Rol, Pascal O.; Parel, Jean-Marie; Hanna, Khalil)V1423,89-93(1991)

Confocal light microscopy of the living in-situ ocular lens: two- and three-dimensional imaging (Masters, Barry R.)V1443,288-293(1991)

Corneal and retinal energy density with various laser beam delivery systems and contact lenses (Dewey, David)V1423,105-116(1991)

Corneal epithelium, visual acuity, and laser refractive keratectomy (Simon, Gabriel; Parel, Jean-Marie; Kervick, Gerard N.; Rol, Pascal O.; Hanna, Khalil; Thompson, Keith P.)V1423,154-156(1991)

Corneal refractive surgery using an ultraviolet (213 nm) solid state laser (Ren, Qiushi; Gailitis, Raymond P.; Thompson, Keith P.; Penney, Carl M.; Lin, J. T.; Waring, George O.)V1423,129-139(1991)

Corneal topography: the dark side of the moon (Bores, Leo D.)V1423,28-39(1991); V1429,217-228(1991)

Correlation tracking: a new technology applied to laser photocoagulation (Forster, Albert A.)V1423,103-104(1991)

Design characteristics and point-spread function evaluation of bifocal intraocular lenses (Rol, Pascal O.)V1423,15-19(1991)

Design considerations of a real-time clinical confocal microscope (Masters, Barry R.)V1423,8-14(1991)

Endocular ophthalmoscope: miniaturization and optical imaging quality (Rol, Pascal O.; Beck, Dominik; Niederer, Peter)V1423,84-88(1991)

Excimer laser cutting of corneal transplants (Tamkivi, Raivo; Schotter, Leo L.; Pakhomova, Tat'yana A.)V1503,375-378(1991)

Extraction of the foveal center and lesion boundary from fundus images (Ishaq, Naseem; Taylor, Kenneth; Steliou, Kypros; Delaney, William)V1381,153-159(1991)

Fourier transform method to determine human corneal endothelial morphology (Masters, Barry R.; Lee, Yimkul; Rhodes, William T.)V1429,82-90(1991)

Fractal patterns in the human retina and their physiological correlates (Masters, Barry R.)V1380,218-226(1991)

Global status of diffraction optics as the basis for an intraocular lens (Isaacson, William B.)V1529,71-83(1991)

Hartmann-Shack sensor as a component in active optical system to improve the depth resolution of the laser tomographic scanner (Liang, Junzhong; Grimm, B.; Goelz, Stefan; Bille, Josef F.)V1542,543-554(1991)

Holographic interferometric analysis of the bovine cornea expansion (Foerster, Werner; Kasprzak, Henryk; von Bally, Gert; Busse, H.)V1429,146-151(1991)

Holographic interferometry of the corneal surface (Friedlander, Miles H.; Mulet, Miguel; Buzard, Kurt A.; Granet, Nicole; Baker, Phillip C.)V1423,62-69(1991); V1429,229-236(1991)

Image quality evaluation of multifocal intraocular lenses (Silberman, Donn M.)V1423,20-28(1991)

Improved plastic molding technology for ophthalmic lens and contact lens (Galic, George J.; Maus, Steven)V1529,13-21(1991)

Improved PMMA single-piece haptic materials (Healy, Donald D.; Wilcox, Christopher D.)V1529,84-93(1991)

Interaction of erbium laser radiation with corneal tissue (Wannop, Neil M.; Charlton, Andrew; Dickinson, Mark R.; King, Terence A.)V1423,163-166(1991)

Interdependence of design, optical evaluation, and visual performance of ophthalmic lenses (Freeman, Michael H.)V1529,2-12(1991)

Laser surgical unit for photoablative and photothermal keratoplasty (Cartlidge, Andy G.; Parel, Jean-Marie; Yokokura, Takashi; Lowery, Joseph A.; Kobayashi, K.; Nose, I.; Lee, William; Simon, Gabriel; Denham, David B.)V1423,167-174(1991)

Mathematical models of laser/tissue interactions for treatment and diagnosis in ophthalmology (Zheltov, Georgi I.; Glazkov, V. N.; Kirkovsky, A. N.; Podol'tsev, A. S.)V1403,752-753(1991)

Measurement of fluorescence spectra and quantum yields of 193 nm ArF laser photoablation of the cornea and synthetic lenticules (Milne, Peter J.; Zika, Rod G.; Parel, Jean-Marie; Denham, David B.; Penney, Carl M.)V1423,122-129(1991)

Measurement of the axial eye length and retinal thickness by laser Doppler interferometry (Hitzenberger, Christoph K.; Fercher, Adolf F.; Juchem, M.)V1423,46-50(1991)

Moulding process for contact lens (Skipper, Richard S.; Shepherd, David W.)V1529,22-30(1991)

Neural-network-based image processing of human corneal endothelial micrograms (Hasegawa, Akira; Zhang, Wei; Itoh, Kazuyoshi; Ichioka, Yoshiki)V1558,414-421(1991)

New results in human eye laser diagnostics (Shubochkin, L. P.; Tuchin, Valery V.)V1403,720-731(1991)

New wavefront sensor for metrology of spherical surfaces (Goelz, Stefan; Persoff, Jeffrey J.; Bittner, Groff D.; Liang, Junzhong; Hsueh, Chi-Fu T.; Bille, Josef F.)V1542,502-510(1991)

Ophthalmic antireflection coatings with same residual reflective colors on ophthalmic optics with different refractive indices (Jin, Tianfeng; Yuan, Youxin)V1529,132-137(1991)

Opthalmic Lens Design and Fabrication (Perrott, Colin M., ed.)V1529(1991)

Ophthalmic lenses testing by moire deflectometry (Yonte, T.; Quiroga, J.; Alda, J.; Bernabeu, Eusebio)V1554B,233-241(1991)

Ophthalmic Technologies (Puliafito, Carmen A., ed.)V1423(1991)

Optical engineering of an excimer laser ophthalmic surgery system (Yoder, Paul R.)V1442,162-171(1991)

Optical properties of diffractive, bifocal intraocular lenses (Larsson, Michael; Beckman, Claes; Nystrom, Alf; Hard, Sverker; Sjostrand, Johan)V1529,63-70(1991)

Photodynamic treatment of lens epithelial cells for cataract surgery (Lingua, Robert W.; Parel, Jean-Marie; Simon, Gabriel; Li, Kam)V1423,58-61(1991)

Photogrammetric measurements of retinal nerve fiber layer thickness along the disc margin and peripapillary region (Takamoto, Takenori; Schwartz, Bernard)V1380,64-74(1991)

Physical mechanism of high-power pulse laser ophthalmology: experimental research (He, Anzhi; Lu, Jian-Feng; Ni, Xiao W.; Li, Yong N.)V1423,117-120(1991)

Plastic photochromic eyewear: a status report (Crano, John C.; Elias, Richard C.)V1529,124-131(1991)

Raman and FT-IR studies of ocular tissues (Ozaki, Yukihiro; Mizuno, Aritake)V1403,710-719(1991)

Ray tracing through progressive ophthalmic lenses (Bourdoncle, Bernard; Chauveau, J. P.; Mercier, Jean-Louis M.)V1354,194-199(1991)

Reactive low-voltage ion plating of optical coatings on ophthalmic lenses (Balasubramanian, Kunjithapa; Richmond, Jeff; Hu, X. Q.; Guenther, Karl H.)V1529,106-114(1991)

Recent progress in artificial vision (Normann, Richard A.)V1423,40-45(1991)

Safety of medical excimer lasers with an emphasis on compressed gases (Sliney, David H.; Clapham, Terrance N.)V1423,157-162(1991)

Sola ASL in Spectralite and polycarbonate aspheric lens designs (Machol, Steven)V1529,45-56(1991)

Sola ASL in Spectralite strikes the perfect balance between cosmetics and optics (Machol, Steven; Modglin, Luan)V1529,38-44(1991)

Spatially resolved water concentration determination in human eye lenses using Raman microspectroscopy (Siebinga, I.; de Mul, F. F.; Vrensen, G. F.; Greve, Jan)V1403,746-748(1991)

Sutureless cataract incision closure using laser-activated tissue glues (Eaton, Alexander M.; Bass, Lawrence S.; Libutti, Steven K.; Schubert, Herman D.; Treat, Michael R.)V1423,52-57(1991)

Three-dimensional confocal microscopy of the living cornea and ocular lens (Masters, Barry R.)V1450,286-294(1991)

Unresolved issues in excimer laser corneal surgery (Trokel, Stephen L.)V1423,94-97(1991)

Use of krypton laser stimulation in the treatment of dry eye syndrome (Kecik, Tadeusz; Switka-Wieclawska, Iwona; Ciszewska, Joanna; Portacha, Lidia)V1391,341-345(1991)

Vibrational spectroscopy in the ophthalmological field (Bertoluzza, Alessandro; Monti, P.; Simoni, R.)V1403,743-745(1991)

Why and how to coat ophthalmic lenses (Guenther, Karl H.)V1529,96-105(1991)

Optical chaos

Applications of chaotic neurodynamics in pattern recognition (Baird, Bill; Freeman, Walter J.; Eeckman, Frank H.; Yao, Yong)V1469,12-23(1991)

Chaotic behavior of a Raman ring laser (Englund, John C.)V1497,292-299(1991)

Chaotic nature of mesh networks with distributed routing (Rucinski, Andrzej; Drexel, Peter G.; Dziurla, Barbara)V1390,388-398(1991)

Neural network modeling of radar backscatter from an ocean surface using chaos theory (Leung, Henry; Haykin, Simon)V1565,279-286(1991)

Numerical study of the onset of chaos in coupled resonators (Rogers, Mark E.; Rought, Nathan W.)V1415,24-37(1991)

Pulsating instabilities and chaos in Raman lasers (Harrison, Robert G.; Lu, Weiping; Jiad, K.; Uppal, J. S.)V1376,94-102(1991)

Regular and chaotic pulsations of radiation intensity in a CO_2 laser with modulated parameters (Bondarenko, Alexander V.; Glova, Alexander F.; Kozlov, Sergei N.; Lebedev, Fedor V.; Likhanskii, Vladimir V.; Napartovich, Anatoly P.; Pis'menny, Vladislav D.; Yaztsev, Vladimir P.)V1376,117-127(1991)

Transition from homoclinic to heteroclinic chaos in coupled laser arrays (Otsuka, Kenju)V1497,300-312(1991)

Optical communications—see also amplifiers; antennas; fiber optic networks; laser, semiconductor; modulation; multiplexing; receivers; satellites; space optics; transmission

1.25-Gb/s wideband LED driver design using active matching techniques (Gershman, Vladimir)V1474,75-82(1991)

1W cw diode-pumped Nd:YAG laser for coherent space communication systems (Johann, Ulrich A.; Seelert, Wolf)V1522,158-168(1991)

4-channel, 662-Mb/s medium-density WDM system with Fabry-Perot laser diodes for subscriber loop applications (Wang, Lon A.; Chapuran, Thomas E.; Menendez, Ronald C.)V1363,85-91(1991)

500 MHz baseband fiber optic transmission system for medical imaging applications (Cheng, Xin; Huang, H. K.)V1364,204-208(1991)

90 degree optical hybrid for coherent receivers (Garreis, Reiner)V1522,210-219(1991)

A 120-Mbit/s QPPM high-power semiconductor transmitter performance and reliability (Greulich, Peter; Hespeler, Bernd; Spatscheck, Thomas)V1417,358-369(1991)

A 39-photon/bit direct-detection receiver at 810 nm, BER = 1x10-6, 60-Mbit/s QPPM (MacGregor, Andrew D.; Dion, Bruno; Noeldeke, Christoph; Duchmann, Olivier)V1417,374-380(1991)

Acquisition and tracking performance measurements for a high-speed area array detector system (Short, Ralph C.; Cosgrove, Michael A.; Clark, David L.; Martino, Anthony J.; Park, Hong-Woo; Seery, Bernard D.)V1417,131-141(1991)

Adaptive semiconductor laser phased arrays for real-time multiple-access communications (Pan, J. J.; Cordeiro, D.)V1476,157-169(1991)

Advanced Fiber Communications Technologies (Kazovsky, Leonid G., ed.)V1579(1991)

Advances in Erbium-doped fiber amplifiers for optical communications (Zyskind, John L.)V1373,80-92(1991)

All-refractive telescope for next-generation inter-satellite communication (Heimbeck, Hans-Jorg)V1354,434-437(1991)

Analysis of the multichannel coherent FSK subcarrier multiplexing system with pilot carrier and phase noise cancelling scheme (Lee, Yang-Hang; Wu, Jingshown; Tsao, Hen-Wai)V1372,140-149(1991)

Application of Kaband MMIC technology for an Orbiter/ACTS communications experiment (Arndt, George D.; Fink, Patrick W.; Leopold, Louis; Bondyopadhyay, Probir; Shaw, Roland)V1475,231-242(1991)

Applications of fiber amplifiers to high-data-rate nonlinear transmission (Spirit, David M.; Brown, Graeme N.; Marshall, Ian W.; Blank, Lutz C.)V1373,197-208(1991)

Applications of optical polymer waveguide devices on future optical communication and signal processing (Keil, Norbert; Strebel, Bernhard N.; Yao, HuiHai; Krauser, Juergen)V1559,278-287(1991)

Applications of optical switches in fiber optic communication networks (Hanson, Daniel; Gosset, Nathalie M.)V1363,48-56(1991)

ASK transmitter for high-bit-rate systems (Schiellerup, Gert; Pedersen, Rune J.)V1372,27-38(1991)

Atmospheric visibility monitoring for planetary optical communications (Cowles, Kelly A.)V1487,272-279(1991)

Back-reflection measurements on the SILEX telescope (Birkl, Reinhard; Manhart, Sigmund)V1522,252-258(1991)

Balanced optical mixer integrated in InGaAlAs/InP for coherent receivers (Caldera, Claudio; De Bernardi, Carlo; Destefanis, Giovanni; Meliga, Marina; Morasca, Salvatore; Rigo, Cesare F.; Stano, Alessandro)V1372,82-87(1991)

Bandwidth requirements for direct detection optical communication receivers with PPM signaling (Davidson, Frederic M.; Sun, Xiaoli; Krainak, Michael A.)V1417,75-88(1991)

Beam-tracker and point-ahead system for optical communications II: servo performance (LaSala, Paul V.; McLaughlin, Chris)V1482,121-137(1991)

BPSK homodyne and DPSK heterodyne receivers for free-space communication with ND:host lasers (Bopp, Matthias; Huether, Gerhard; Spatscheck, Thomas; Specker, Harald; Wiesmann, Theo J.)V1522,199-209(1991)

Bridging issues in DQDB subnetworks (Tantawy, Ahmed N.)V1364,268-276(1991)

Building and campus networks for fiber distributed data interface (McIntosh, Thomas F.)V1364,84-93(1991)

Campus fiber optic enterprise networks (Weeks, Richard A.)V1364,222-227(1991)

Carrier recovery and filtering in optical BPSK systems using external cavity semiconductor lasers (Pires, Joao J.; Rocha, Jose R.)V1372,118-127(1991)

Coarse pointing assembly for the SILEX program, or how to achieve outstanding pointing accuracy with simple hardware associated with consistent control laws (Buvat, Daniel; Muller, Gerard; Peyrot, Patrick)V1417,251-261(1991)

Code Division Multiple Access system candidate for integrated modular avionics (Mendez, Antonio J.; Gagliardi, Robert M.)V1369,67-71(1991)

Coherent communication systems research and development at AT&T Bell Laboratories, Solid State Technology Center (Park, Yong K.; Delavaux, Jean-Marc F.; Tench, Robert E.; Cline, Terry W.; Tzeng, Liang D.; Kuo, Chien-yu C.; Wagner, Earl J.; Flores, Carlos F.; Van Eijk, Peter; Pleiss, T. C.; Barski, S.; Owen, B.; Twu, Yih-Jye; Dutta, Niloy K.; Riggs, R. S.; Ogawa, Kinichiro K.)V1372,219-227(1991)

Coherent detection: n-ary PPM versus PCM (Cryan, Robert A.; Unwin, Rodney T.; Massarella, Alistair J.; Sibley, Martin J.; Garrett, Ian)V1372,64-71(1991)

Coherent Lightwave Communications: Fifth in a Series (Steele, Roger C.; Sunak, Harish R., eds.)V1372(1991)

Coherent optical fiber communications (Meissner, P.)V1522,182-193(1991)

Coherent optical modulation for antenna remoting (Fitzmartin, Daniel J.; Gels, Robert G.; Balboni, Edmund J.)V1476,56-62(1991)

Coherent optical transmission systems with optical amplifiers (Steele, Roger C.; Walker, Nigel G.)V1372,173-187(1991)

Comparative life test of 0.8-um laser diodes for SILEX under NRZ and QPPM modulation (Menke, Bodo; Loeffler, Roland)V1417,316-327(1991)

Comparison of alternative modulation techniques for microwave optical links (Kasemset, Dumrong; Ackerman, Edward I.; Wanuga, Stephen; Herczfeld, Peter R.; Daryoush, Afshin S.)V1371,104-114(1991)

Comparison of optical technologies for a high-data-rate Mars link (Spence, Rodney L.)V1417,550-561(1991)

Comparison of simulation and experimental measurements of avalanche photodiode receiver performance (Mecherle, G. S.; Henderson, Robert J.)V1417,537-542(1991)

Compensation efficiency of an optical adaptive transmitter (Feng, Yue-Zhong; Song, Zhengfang; Gong, Zhi-Ben)V1417,370-372(1991)

Components and applications for high-speed optical analog links (Johnson, Peter T.; Debney, Brian T.; Carter, Andrew C.)V1371,87-97(1991)

Computer-controlled two-segment DFB local oscillator laser tuning for multichannel coherent systems (Johnson, Peter T.; Hankey, Judith; Debney, Brian T.)V1372,188-199(1991)

Correction method for optical-signal detection-error caused by quantum noise (Fujihashi, Chugo)V1583,298-306(1991)

Cost aspects of narrowband and broadband passive optical networks (Jones, J. R.; Sharpe, Randall B.)V1363,106-118(1991)

Crossconnects in a SONET network (Bootman, Steven R.)V1363,142-148(1991)

Current and future activities in the area of optical space communications in Japan (Fujise, Masayuki; Araki, Ken'ichi; Arikawa, Hiroshi; Furuhama, Yoji)V1522,14-26(1991)

CWRUnet: case history of a campus-wide fiber-to-the-desktop network (Neff, Raymond K.; Klingensmith, H. W.; Gumpf, Jeffrey A.; Haigh, Peter J.)V1364,245-256(1991)

Definition of interorbit link optical terminals with diode-pumped Nd:host laser technology (Marini, Andrea E.; Della Torre, Antonio)V1522,222-233(1991)

Design considerations for air-to-air laser communications (Casey, William L.; Doughty, Glenn R.; Marston, Robert K.; Muhonen, John)V1417,89-98(1991)

Design of adaptive optical equalizers for fiber optic communication systems (Ghosh, Anjan K.; Barner, Jim; Paparao, Palacharla; Allen, Susan D.; Imen, Kamran)V1371,170-181(1991)

Design of a diode-pumped Nd:YAG laser communication system (Sontag, Heinz; Johann, Ulrich A.; Pribil, Klaus)V1417,573-587(1991)

Design of a periscopic coarse pointing assembly for optical multiple access (Gatenby, Paul V.; Boereboom, Peter; Grant, Michael A.)V1522,126-134(1991)

Design study for high-performance optical communications satellite terminal with high-power laser diode transmitter (Hildebrand, Ulrich; Seeliger, Reinhard; Smutny, Berry; Sand, Rolf)V1522,50-60(1991)

Diode-pumped Nd:YAG laser transmitter for free-space optical communications (Nava, Enzo; Re Garbagnati, Giuseppe; Garbi, Maurizio; Marchiori, Livio D.; Marini, Andrea E.)V1417,307-315(1991)

Distribution fiber FTTH/FTTC trial results and deployment strategies (Coleman, John D.)V1363,2-12(1991)

DS1 mapping considerations for the synchronous optical network (Cubbage, Robert W.; Littlewood, Paul A.)V1363,163-171(1991)

Dynamic allocation of buffer space in the bridge of two interconnected token rings (Das, Alok K.; Muhuri, K.)V1364,61-69(1991)

Economic factors for a free-space laser communication system (Begley, David L.; Marshalek, Robert G.)V1522,234-242(1991)

Effect of input pulse shape on FSK optical coherent communication system (Shadaram, Mehdi; John, Eugine)V1365,108-115(1991)

Effect of microaccelerations on an optical space communication system (Wittig, Manfred E.)V1522,278-286(1991)

Effect of polarization-mode dispersion on coherent optical distribution systems with shared local oscillator (Kao, Ming-Seng)V1579,221-229(1991)

Effects of atmospheric conditions on the performance of free-space infrared communications (Grotzinger, Timothy L.)V1417,484-495(1991)

Eight-Gb/s QPSK-SCM over a coherent detection optical link (Hill, Paul M.; Olshansky, Robert)V1579,210-220(1991)

Electro-optic modulator for high-speed Nd:YAG laser communication (Petsch, Thomas)V1522,72-82(1991)

Electro-optic resonant modulator for coherent optical communication (Robinson, Deborah L.; Chen, Chien-Chung; Hemmati, Hamid)V1417,421-430(1991)

End-to-end model of a diode-pumped Nd:YAG pulsed laser (Mayer, Richard C.; Dreisewerd, Douglas W.)V1415,248-258(1991)

Equipment development for an atmospheric-transmission measurement campaign (Ruiz, Domingo; Kremer, Paul J.)V1417,212-222(1991)

Erbium-doped fiber amplifiers for future undersea transmission systems (Aspell, Jennifer; Bergano, Neal S.)V1373,188-196(1991)

European SILEX project and other advanced concepts for optical space communications (Oppenhaeuser, Gotthard; Wittig, Manfred E.; Popescu, Alexandru F.)V1522,2-13(1991)

Evolution of fiber-to-the-curb networks toward broadband capabilities (Menendez, Ronald C.; Lu, Kevin W.; Rizzo, Annmarie; Lemberg, Howard L.)V1363,97-105(1991)

Evolution of the DQDB hybrid multiplexing for an integrated service packetized traffic (Gagnaire, A.; Ponsard, Benoit)V1364,277-288(1991)

Examining cable plant bandwidth for FDDI (Hayes, James E.)V1364,115-119(1991)

Experimental implementation of an optical multiple-aperture antenna for space communications (Neubert, Wolfgang M.; Leeb, Walter R.; Scholtz, Arpad L.)V1522,93-102(1991)

Extending HIPPI at 800-mega-bits-per-second over serial links using HOT ROD technology (Annamalai, Kadiresan)V1364,178-189(1991)

Far-field and wavefront characterization of a high-power semiconductor laser for free-space optical communications (Cornwell, Donald M.; Saif, Babak N.)V1417,431-439(1991)

Fault management of a fiber optic LAN (Spencer, Paul E.; Zaharakis, Steven C.; Denton, Richard T.)V1364,228-234(1991)

Fault-tolerant capacity-1 protocol for very fast local networks (Dobosiewicz, Wlodek; Gburzynski, Pawel)V1470,123-133(1991)

FDDI, Campus-Wide, and Metropolitan Area Networks (Annamalai, Kadiresan; Cudworth, Stewart K.; Kasiewicz, Allen B., eds.)V1364(1991)

FDDI components for workstation interconnection (Anderson, Stephen J.; Bulusu, Dutt V.; Racette, James; Scholl, Frederick W.; Zack, Tim; Abbott, Peter G.)V1364,94-100(1991)

FDDI network cabling (Stevens, R. S.)V1364,101-114(1991)

Fiber channel: the next standard peripheral interface and more (Cummings, Roger)V1364,170-177(1991)

Fiber hub in a second-generation ethernet system at Taylor University (Rowan, Paul)V1364,262-266(1991)

Fiber in the loop: an evolution in services and systems (Engineer, Carl P.)V1363,19-29(1991)

Fiber optic link for millimeter-wave communication satellites (Polifko, David M.; Malone, Steven A.; Daryoush, Afshin S.; Kunath, Richard R.)V1476,91-99(1991)

Fiber optic network for mining seismology (Lach, Zbigniew; Zientkiewicz, Jacek K.)V1364,209-220(1991)

Fiber optics in CATV networks (Wolfe, Ronald; Laor, Herzel)V1363,125-132(1991)

Fiber Optics in the Subscriber Loop (Hutcheson, Lynn D.; Kahn, David A., eds.)V1363(1991)

Fiber optics network for the adverse coal mining environment (Zientkiewicz, Jacek K.; Lach, Zbigniew)V1366,45-56(1991)

Free-Space Laser Communication Technologies III (Begley, David L.; Seery, Bernard D., eds.)V1417(1991)

Free-space simulator for laser transmission (Inagaki, Keizo; Nohara, Mitsuo; Araki, Ken'ichi; Fujise, Masayuki; Furuhama, Yoji)V1417,160-169(1991)

Generation of 1-GHz ps optical pulses by direct modulation of a 1.532-um DFB-LD (Nie, Chao-Jiang; Li, Ying; Wu, Fang D.; Chen, Ying-Li; Li, Qi; Hua, Yi-Min)V1579,264-267(1991)

GOPEX: a deep-space optical communications demonstration with the Galileo spacecraft (Wilson, Keith E.; Schwartz, Jon A.; Lesh, James R.)V1417,22-28(1991)

GSFC conceptual design study for an intersatellite optical multiple access communication system (Fox, Neil D.; Maynard, William L.; Clarke, Ernest S.; Bruno, Ronald C.)V1417,452-463(1991)

Heterodyne acquisition and tracking in a free-space diode laser link (Hueber, Martin F.; Scholtz, Arpad L.; Leeb, Walter R.)V1417,233-239(1991)

High-current, high-bandwidth laser diode current driver (Copeland, David J.; Zimmerman, Robert K.)V1417,412-420(1991)

High-Frequency Analog Fiber Optic Systems (Sierak, Paul, ed.)V1371(1991)

High-performance FDDI NIU for streaming voice, video, and data (Bergman, Larry A.; Hartmayer, Ron; Wu, Wennie H.; Cassell, P.; Edgar, G.; Lambert, James L.; Mancini, Richard; Jeng, J.; Pardo, C.; Halloran, Frank; Martinez, James C.)V1364,14-21(1991)

High-power diode-pumped solid state lasers for optical space communications (Koechner, Walter; Burnham, Ralph L.; Kasinski, Jeff; Bournes, Patrick A.; DiBiase, Don; Le, Khoa; Marshall, Larry R.; Hays, Alan D.)V1522,169-179(1991)

High-speed polymer optical fiber networks (Bulusu, Dutt V.; Zack, Tim; Scholl, Frederick W.; Coden, Michael H.; Steele, Robert E.; Miller, Gregory D.; Lynn, Mark A.)V1364,49-60(1991)

Homodyne PSK receivers with laser diode sources (Mecherle, G. S.; Henderson, Robert J.)V1417,99-107(1991)

How to meet intersatellite links mission requirements by an adequate optical terminal design (Duchmann, Olivier; Planche, G.)V1417,30-41(1991)

Hybrid optical transmitter for microwave communication (Costa, Joannes M.; Lam, Benson C.; Kellner, Albert L.; Campion, David C.; Yu, Paul K.)V1476,74-80(1991)

Image processing techniques for aquisition of faint point laser targets in space (Garcia-Prieto, Rafael; Spiero, Francois; Popescu, Alexandru F.)V1522,267-276(1991)

Impact of fiber backscatter on loop video transmission without optical isolator (Das, Santanu K.; Ocenasek, Josef)V1363,172-176(1991)

Implementation of FDDI in the intelligent wiring hub (Tarrant, Peter J.; Truman, Alan K.)V1364,2-6(1991)

InGaAs/InP monolithic photoreceivers for 1.3-1.5 um optical fiber transmission (Scavennec, Andre; Billard, M.; Blanconnier, P.; Caquot, E.; Carer, P.; Giraudet, Louis; Nguyen, L.; Lugiez, F.; Praseuth, Jean-Pierre)V1362,331-337(1991)

Injection chaining of diode-pumped single-frequency ring lasers for free-space communication (Cheng, Emily A.; Kane, Thomas J.; Wallace, Richard W.; Cornwell, Donald M.)V1417,300-306(1991)

Integrated "Byte-to-light" solution for fiber optic data communication (Kubinec, James J.; Somerville, James A.; Chown, David P.; Birch, Martin J.)V1364,130-143(1991)

Integrated optical preamplifier technology for optical signal processing and optical communication systems (Eichen, Elliot G.; Powazinik, William; Meland, Edmund; Bryant, R.; Rideout, William C.; Schlafer, John; Lauer, Robert B.)V1474,260-267(1991)

Integration of a coherent optical receiver with adaptive image rejection capability (Lachs, Gerard; Zaidi, Syed M.; Singh, Amit K.; Henning, Rudolf E.; Trascritti, D.; Kim, H.; Bhattacharya, Pallab K.; Pamulapati, J.; McCleer, P. J.; Haddad, George I.; Peng, S.)V1474,248-259(1991)

Investigation of unobscured mirror telescope for telecom purposes (Sand, Rolf)V1522,103-110(1991)

Jitter considerations for FDDI PMD (Fukuoka, Takashi; Tejika, Yasuhiro; Takada, Hisashi; Takahashi, Hidenori; Hamasaki, Yiji)V1364,40-48(1991)

Laser link performance improvements with wideband microwave impedance matching (Baldwin, David L.; Sokolov, Vladimir; Bauhahn, Paul E.)V1476,46-55(1991)

Laser terminal attitude determination via autonomous star tracking (Chapman, William W.; Fitzmaurice, Michael W.)V1417,277-290(1991)

Life test of high-power Matsushita BTRS laser diodes (Holcomb, Terry L.)V1417,328-337(1991)

Lifetest on a high-power laser diode array transmitter (Greulich, Peter; Hespeler, Bernd; Spatscheck, Thomas)V1522,144-153(1991)

Line coding for high-speed fiber optic transmission systems (Subramanian, K. R.; Dubey, V. K.; Low, J. P.; Tan, L. S.)V1364,190-201(1991)

Liquid crystals for lasercom applications (Tan, Chin; Carlson, Robert T.)V1417,391-401(1991)

Low-back-reflection, low-loss fiber switch (Roberts, Harold A.; Emmons, David J.; Beard, Michael S.; Lu, Liang-ju)V1363,62-69(1991)

Low-data-rate coherent optical link demonstration using frequency-stabilized solid-state lasers (Chen, Chien-Chung; Win, Moe Z.; Marshall, William K.; Lesh, James R.)V1417,170-181(1991)

Low-insertion-loss, high-precision liquid crystal optical phased array (Cassarly, William J.; Ehlert, John C.; Henry, D.)V1417,110-121(1991)

Low-loss L-band microwave fiber optic link for control of a T/R module (Wanuga, Stephen; Ackerman, Edward I.; Kasemset, Dumrong; Hogue, David W.; Chinn, Stephen R.)V1374,97-106(1991)

Management of an adaptable-bit-rate video service in a MAN environment (Marini, Michele; Albanese, Andres)V1364,289-294(1991)

Metropolitan area networks: a corner stone in the broadband era (Ghanem, Adel)V1364,312-319(1991)

Micromachined scanning mirrors for laser beam deflection (Huber, Peter; Gerlach-Meyer, U.)V1522,135-141(1991)

Minimum polarization modulation: a highly bandwidth efficient coherent optical modulation scheme (Benedetto, Sergio; Kazovsky, Leonid G.; Poggiolini, Pierluigi T.)V1579,112-121(1991)

Modeling direct detection and coherent-detection lightwave communication systems that utilize cascaded erbium-doped fiber amplifiers (Joss, Brian T.; Sunak, Harish R.)V1372,94-117(1991)

Modeling of traveling wave Ti:LiNbO3 waveguide modulators (Rocca, Corrado; Montrosset, Ivo; Gollinucci, Stefano; Ghione, Giovanni)V1372,39-47(1991)

Modular FDDI bridge and concentrator (Coden, Michael H.; Bulusu, Dutt V.; Ramsey, Brian; Sztuka, Edward; Morrow, Joel)V1364,22-39(1991)

Modular packaging for FTTC and B-ISDN (Koht, Lowell)V1363,158-162(1991)

Multibit optical sensor networking (Pervez, Anjum)V1511,220-231(1991)

Multichannel optical data link (Ota, Yusuke; Swartz, Robert G.)V1364,146-152(1991)

Multigigabit solitary-wave propagation in both the normal and anomalous dispersion regions of optical fibers (Potasek, M. J.; Tabor, Mark)V1579,232-236(1991)

Multistability, instability, and chaos for intracavity magneto-optic modulation output (Yang, Darang; Song, Ruhua H.; Hu, Zhiping; Le, Shixiao)V1417,440-450(1991)

NASA's flight-technology development program: a 650-Mbit/s laser communications testbed (Hayden, William L.; Fitzmaurice, Michael W.; Nace, Dave; Lokerson, Donald; Minott, Peter O.; Chapman, William W.)V1417,182-199(1991)

NAVSEA gigabit optical MAN prototype history and status (Albanese, Andres; Devetzis, Tasco N.; Ippoliti, A, G.; Karr, Michael A.; Maszczak, M. W.; Dorris, H. N.; Davis, James H.)V1364,320-326(1991)

Nd:host laser-based optical terminal development study for intersatellite links (Marini, Andrea E.; Della Torre, Antonio; Popescu, Alexandru F.)V1417,200-211(1991)

Near-term applications of optical switching in the metropolitan networks (King, F. D.; Tremblay, Yves)V1364,295-303(1991)

Network powering architecture for fiber-to-the-subscriber systems (Pellerin, Sharon J.)V1363,186-190(1991)

New CATV fiber-to-the-subscriber architectures (Kim, Gary)V1363,133-140(1991)

New integrated optic TE/TM splitter made on LiNbO3 isotropic cut (Duchet, Christian; Flaaronning, Nils; Brot, Christian; Sarrabay, Laurence)V1372,72-81(1991)

New modulation scheme for coherent lightwave communication: Direct-Modulation PSK (Naito, Takao; Chikama, Terumi; Kuwahara, Hideo)V1372,200-207(1991)

Noise in semiconductor lasers and its impact on optical communication systems (Agrawal, Govind P.)V1376,224-235(1991)

Nonmechanical steering of laser beams by multiple aperture antennas: tolerance analysis (Neubert, Wolfgang M.; Leeb, Walter R.; Scholtz, Arpad L.)V1417,122-130(1991)

Novel angular discriminator for spatial tracking in free-space laser communications (Fung, Jackie S.)V1417,224-232(1991)

Novel approach to optical frequency synthesis in coherent lightwave systems (Fernando, P. N.; Fake, M.; Seeds, A. J.)V1372,152-163(1991)

Novel high-speed communication system (Dawber, William N.; Hirst, Peter F.; Condon, Brian P.; Maitland, Arthur; Sutton, Phillip)V1476,81-90(1991)

Novel optical-fiber-based conical scan tracking device (Johann, Ulrich A.; Pribil, Klaus; Sontag, Heinz)V1522,243-252(1991)

Optical approach to proximity-operations communications for Space Station Freedom (Marshalek, Robert G.)V1417,53-62(1991)

Optical carrier modulation by integrated optical devices in lithium niobate (Rasch, Andreas; Buss, Wolfgang; Goering, Rolf; Steinberg, Steffen; Karthe, Wolfgang)V1522,83-92(1991)

Optical data communication compel the design of a new class of storage media (Frietman, Edward E.; Dekker, L.; van Nifterick, W.)V1401,19-26(1991)

Optical decoding and coherent detection of a four-level FSK signal (O'Byrne, Vincent A.; Tatlock, Timothy; Stone, Samuel M.)V1372,88-93(1991)

Optical dividers for quadrant avalanche photodiode detectors (Green, Samuel I.)V1417,496-512(1991)

Optical fiber demultiplexer for telecommunications (Falciai, Riccardo; Scheggi, Annamaria V.; Cosi, Franco; Cao, J. Y.)V1365,38-42(1991)

Optical fiber filter comprising a single-coupler fiber ring (or loop) and a double-coupler fiber mirror (Ja, (Yu) Frank H.)V1372,48-61(1991)

Optical fiber n-ary PPM: approaching fundamental limits in receiver sensitivity (Cryan, Robert A.; Unwin, Rodney T.)V1579,133-143(1991)

Optical ISL transmitter design that uses a high-power LD amplifier (Nohara, Mitsuo; Harada, Takashi; Fujise, Masayuki)V1417,338-345(1991)

Optical link demonstration with a lightweight transceiver breadboard (Hemmati, Hamid; Lesh, James R.; Apostolopoulos, John G.; Del Castillo, Hector M.; Martinez, A. S.)V1417,476-483(1991)

Optical phase-locked loop for free-space laser communications with heterodyne detection (Win, Moe Z.; Chen, Chien-Chung; Scholtz, Robert A.)V1417,42-52(1991)

Optical Space Communication II (Franz, Juergen, ed.)V1522(1991)

Optical space-to-ground link availability assessment and diversity requirements (Chapman, William W.; Fitzmaurice, Michael W.)V1417,63-74(1991)

Optical subscriber line transmission system to support an ISDN primary-rate interface (Wataya, Hideo; Tsuchiya, Toshiyuki)V1363,72-84(1991)

Optimum receiver structure and filter design for MPAM optical space communication systems (Al-Ramli, Intesar F.)V1522,111-123(1991)

Options for campus fiber networks (Henderson, Byron B.; Green, Emily N.)V1364,235-244(1991)

Overview of coherent lightwave communications (Steele, Roger C.)V1372,2-3(1991)

Passive components for the subscriber loop (Morrel, William G.)V1363,40-47(1991)

Passive optic solution for an urban rehabilitation topology (Petruziello, David)V1363,30-37(1991)

Performance analysis of direct-detection optical DPSK systems using a dual-detector optical receiver (Pires, Joao J.; Rocha, Jose R.)V1579,144-154(1991)

Performance analysis of lightwave packet communication networks (Ramaswamy, Raju)V1364,153-162(1991)

Performance of a demonstration system for simultaneous laser beacon tracking and low-data-rate optical communications with multiple platforms (Short, Ralph C.; Cosgrove, Michael A.; Clark, David L.; Oleski, Paul J.)V1417,464-475(1991)

Performance of a phased array semiconductor laser source for coherent laser communications (Probst, David K.; Rice, Robert R.)V1417,346-357(1991)

Performance of on-off modulated lightwave signals with phase noise (Azizoglu, Murat; Humblet, Pierre A.)V1579,168-178(1991)

Performance of pseudo-orthogonal codes in temporal, spatial, and spectral code division multiple access systems (Mendez, Antonio J.; Gagliardi, Robert M.)V1364,163-169(1991)

Performance of Reed-Solomon codes in mulichannel CPFSK coherent optical communications (Wu, Jyh-Horng; Wu, Jingshown)V1579,195-209(1991)

Physical-connection compliance testing for FDDI (Baldwin, Christopher)V1364,120-129(1991)

Plan for the development and demonstration of optical communications for deep space (Lesh, James R.; Deutsch, L. J.; Weber, W. J.)V1522,27-35(1991)

Planning for fiber optic use at the University of Massachusetts (Sailer, Donald R.)V1364,257-261(1991)

Pointing, acquisition, and tracking system of the European SILEX program: a major technological step for intersatellite optical communication (Bailly, Michel; Perez, Eric)V1417,142-157(1991)

Polarization-diversity fiber networks (Cullen, Thomas J.)V1372,164-172(1991)

Polarization-modulated coherent optical communication systems (Betti, Silvello; Curti, Franco; De Marchis, Giancarlo; Iannone, Eugenio)V1579,100-111(1991)

Polarization modulation with frequency shift redundancy and frequency modulation with polarization redundancy for POLSK and FSK systems (Marone, Giuseppe; Calvani, Riccardo A.; Caponi, Renato)V1579,122-132(1991)

Potential roles of optical interconnections within broadband switching modules (Lalk, Gail R.; Habiby, Sarry F.; Hartman, Davis H.; Krchnavek, Robert R.; Wilson, Donald K.; Young, Kenneth C.)V1389,386-400(1991)

Prismatic anamorphic beam expanders for free-space optical communications (Jonas, Reginald P.)V1417,402-411(1991)

Proposed one- and two-fiber-to-the-pedestal architectural evolution (Schiffler, Richard A.)V1363,13-18(1991)

Qualification testing of a diode-laser transmitter for free-space coherent communications (Pillsbury, Allen D.; Taylor, John A.)V1417,292-299(1991)

Realization of a coherent optical DPSK (differential phase-shift keying) heterodyne transmission system with 565 MBit/s at 1.064 um (Wandernoth, Bernhard; Franz, Juergen)V1522,194-198(1991)

Realization of heterodyne acquisition and tracking with diode lasers at lambda=1.55 um (Hueber, Martin F.; Leeb, Walter R.; Scholtz, Arpad L.)V1522,259-267(1991)

Recent progress of coherent lightwave systems at Bellcore (Sessa, William B.; Welter, Rudy; Wagner, Richard E.; Maeda, Mari W.)V1372,208-218(1991)

Recirculating fiber optical RF-memory loop in countermeasure systems (Even-Or, Baruch; Lipsky, S.; Markowitz, Raymond; Herczfeld, Peter R.; Daryoush, Afshin S.; Saedi, Reza)V1371,161-169(1991)

Reed-Solomon encoder/decoder application using a neural network (Hussain, Mukhtar; Bedi, Jatinder S.)V1469,463-471(1991)

Reliability analysis of a multiple-laser-diode beacon for intersatellite links (Mauroschat, Andreas)V1417,513-524(1991)

Reliability considerations for fiber optic systems in telecommunications (Shelton, Douglas S.)V1366,2-8(1991)

Research and development of a NYNEX switched multi-megabit data service prototype system (Maman, K. H.; Haines, Robert; Chatterjee, Samir)V1364,304-311(1991)

Rings in a SONET network (Clendening, Steven J.)V1363,149-157(1991)

Rugged 20-km fiber optic link for 2-18-GHz communications (Buckley, Robert H.; Lyons, E. R.; Goga, George)V1371,212-222(1991)

SAFENET II: The Navy's FDDI-based computer network standard (Paige, Jeffrey L.; Howard, Edward A.)V1364,7-13(1991)

Second-generation high-data-rate interorbit link based on diode-pumped Nd:YAG laser technology (Sontag, Heinz; Johann, Ulrich A.; Pribil, Klaus)V1522,61-69(1991)

Selected Papers on Free-Space Laser Communications (Begley, David L., ed.)VMS30(1991)

Semiconductor laser diode with weak optical feedback: self-coupling effects on P-I characteristics (Milani, Marziale; Mazzoleni, S.; Brivio, Franca)V1474,83-97(1991)

Semiconductor lasers for coherent lightwave communcation (Dutta, Niloy K.)V1372,4-12(1991)

Semiconductor laser transmitters for millimeter-wave fiber-optic links (Pan, J. J.)V1476,63-73(1991)

Sensitivity of direct-detection lightwave receivers using optical preamplifiers (Tonguz, Ozan K.; Kazovsky, Leonid G.)V1579,179-183(1991)

Simulation model and on-ground performances validation of the PAT system for the SILEX program (Cossec, Francois R.; Doubrere, Patrick; Perez, Eric)V1417,262-276(1991)

SILEX project: the first European optical intersatellite link experiment (Laurent, Bernard; Duchmann, Olivier)V1417,2-12(1991)

Single-frequency Nd:YAG laser development for space communication (Letterer, Rudolf; Wallmeroth, Klaus)V1522,154-157(1991)

SOLACOS: a diode-pumped Nd:YAG laser breadboard for coherent space communication system verification (Pribil, Klaus; Johann, Ulrich A.; Sontag, Heinz)V1522,36-47(1991)

Solving clock distribution problems in FDDI concentrators (Li, Gabriel)V1364,72-83(1991)

SONET inter-vendor compatibility (Bowmaster, Thomas A.; Cockings, Orville R.; Swanson, Robert A.)V1363,119-124(1991)

Space station laser communication transceiver (Fitzmaurice, Michael W.; Hayden, William L.)V1417,13-21(1991)

Spatial acquisition and tracking for deep-space optical communication packages (Chen, Chien-Chung; Jeganathan, Muthu; Lesh, James R.)V1417,240-250(1991)

Spectral and modulation characteristics of tunable multielectrode DBR lasers (Ferreira, Mario F.; Rocha, Jose R.; Pinto, Joao L.)V1372,14-26(1991)

Spread spectrum technique for passive multiplexing of interferometric optical fiber sensors (Uttamchandani, Deepak G.; Al-Raweshidy, H. S.)V1511,212-219(1991)

Stability considerations in relay lens design for optical communications (Gardam, Allan; Jonas, Reginald P.)V1417,381-390(1991)

Step towards optimal topology of communication networks (Michalewicz, Zbigniew)V1470,112-122(1991)

Substrate-mode holograms for board-to-board two-way communications (Huang, Yang-Tung; Kostuk, Raymond K.)V1474,39-44(1991)

Theoretical analysis of optical phase diversity FSK receivers (Yang, Shien-Chi; Tsao, Hen-Wai; Wu, Jingshown)V1372,128-139(1991)

Two Gbit/s photonic backplane for telephone cards interconnection (Donati, Silvano; Martini, Giuseppe; Francese, Francesco)V1389,665-671(1991)

Ultralinear low-noise amplifier technology for space communications (Watkins, E. T.; Yu, K. K.; Yau, W.; Wu, Chan-Shin; Yuan, Steve)V1475,62-72(1991)

Wavelength division and subcarrier system based on Brillouin amplification (Lee, Yang-Hang; Wu, Jingshown; Kao, Ming-Seng; Tsao, Hen-Wai)V1579,155-166(1991)

Wavelength division multiplexing of services in a fiber-to-the-home system (Unterleitner, Fred C.)V1363,92-96(1991)

Wide-band analog frequency modulation of optic signals using indirect techniques (Fitzmartin, Daniel J.; Balboni, Edmund J.; Gels, Robert G.)V1371,78-86(1991)

Optical computing—see also associative processing; neural networks; optical interconnects; optical processing; parallel processing; pattern recognition; spatial light modulators; switches

Algorithms and architectures for performing Boolean equations using self electro-optic effect devices (Spence, Scott E.; Reyes, Roy M.)V1474,199-207(1991)

Asymmetric Fabry-Perot modulators for optical signal processing and optical computing applications (Kilcoyne, M. K.; Whitehead, Mark; Coldren, Larry A.)V1389,422-454(1991)

Characteristics of a ferroelectric liquid crystal spatial light modulator with a dielectric mirror (Kato, Naoki; Sekura, Rieko; Yamanaka, Junko; Ebihara, Teruo; Yamamoto, Shuhei)V1455,190-205(1991)

Correlation-based optical numeric processors (Casasent, David P.; Woodford, Paul)V1563,112-119(1991)

Design and evaluation of optical switching architectures (Ramesh, S. K.; Smith, Thomas D.)V1474,208-211(1991)

Design, fabrication, and performance of an integrated optoelectronic cellular array (Hibbs-Brenner, Mary K.; Mukherjee, Sayan D.; Skogen, J.; Grung, B.; Kalweit, Edith; Bendett, Mark P.)V1563,10-20(1991)

Designing digital optical computing systems: power distribution and crosstalk (Pratt, Jonathan P.; Heuring, Vincent P.)V1505,124-129(1991)

Design of a Gaussian elimination architecture for the DOC II processor (Murdocca, Miles J.; Levy, Saul Y.)V1563,255-266(1991)

Digital optical computer II (Guilfoyle, Peter S.; Stone, Richard V.)V1563,214-222(1991)

Digital optical computers at Boulder (Jordan, Harry F.)V1505,87-98(1991)

DOC II 32-bit digital optical computer: optoelectronic hardware and software (Stone, Richard V.; Zeise, Frederick F.; Guilfoyle, Peter S.)V1563,267-278(1991)

Efficiency of liquid-crystal light valves as polarization rotators (Collings, Neil; Xue, Wei; Pedrini, G.)V1505,12-19(1991)

Experimental comparison of optical binary phase-only filter and high-pass matched filter correlation (Leib, Kenneth G.; Brandstetter, Robert W.; Drake, Marvin D.; Franks, Glen B.; Siewert, Ronald O.)V1483,140-154(1991)

First programmable digital optical processor: optical cellular logic image processor (Craig, Robert G.; Wherrett, Brian S.; Walker, Andrew C.; McKnight, D. J.; Redmond, Ian R.; Snowdon, John F.; Buller, G. S.; Restall, Edward J.; Wilson, R. A.; Wakelin, S.; McArdle, N.; Meredith, P.; Miller, J. M.; Taghizadeh, Mohammad R.; Mackinnon, G.; Smith, S. D.)V1505,76-86(1991)

Free-space board-to-board optical interconnections (Tsang, Dean Z.)V1563,66-71(1991)

Handwritten zip code recognition using an optical radial basis function classifier (Neifeld, Mark A.; Rakshit, S.; Psaltis, Demetri)V1469,250-255(1991)

High-level design of digital computers using optical logic arrays (Murdocca, Miles J.)V1474,176-187(1991)

High-power laser arrays for optical computing (Zucker, Erik P.; Craig, Richard R.; Mehuys, David G.; Nam, Derek W.; Welch, David F.; Scifres, Donald R.)V1563,223-228(1991)

High-speed 128-element avalanche photodiode array for optical computing applications (Webb, Paul P.; Dion, Bruno)V1563,236-243(1991)

High-speed signal processing architectures using charge-coupled devices (Boddu, Jayabharat; Udpa, Satish S.; Udpa, L.; Chan, Shiu Chuen M.)V1562,251-262(1991)

Holograms in optical computing (Caulfield, H. J.)VIS08,54-61(1991)

Holographic space-variant prism arrays for free-space data permutation in digital optical networks (Kobolla, Harald; Sauer, Frank; Schwider, Johannes; Streibl, Norbert; Voelkel, Reinhard)V1507,175-182(1991)

Hybrid optical array logic system (Kakizaki, Sunao; Miyazaki, Daisuke; Yoshikawa, Eiji; Tanida, Jun; Ichioka, Yoshiki)V1505,199-205(1991)

III-V semiconductor integrated optoelectronics for optical computing (Wada, Osamu)V1362,598-607(1991)

Integration of an edge extraction cellular automaton (Taboury, Jean; Chavel, Pierre)V1505,115-123(1991)

Massive optical interconnections for computer applications (Neff, John A.)V1389,27-38(1991); V1390,27-38(1991)

Matrix optical system for plane-point correlation (Curatu, Eugen O.)V1527,368-375(1991)

Microwatt all-optical switches, array memories, and flip-flops (Wang, Chang H.; Lloyd, Ashley D.; Wherrett, Brian S.)V1505,130-140(1991)

Module for optical parallel logic using bistable optically addressed ferroelectric spatial light modulators (Guibert, L.; Killinger, M.; de Bougrenet de la Tocnaye, Jean-Louis M.)V1505,99-103(1991)

Monte Carlo and diffusion calculations of photon migration in noninfinite highly scattering media (Haselgrove, John C.; Leigh, John S.; Yee, Conway; Wang, Nai-Guang; Maris, Michael B.; Chance, Britton)V1431,30-41(1991)

Motivation for DOC III: 64-bit digital optical computer (Guilfoyle, Peter S.)V1500,14-22(1991); V1503,14-22(1991); 1505,2-10(1991)

Neural networks and computers based on in-phase optics (Kovatchev, Methodi; Ilieva, R.)V1621,259-267(1991)

OLAS: optical logic array structures (Jones, Robert H.; Hadjinicolaou, M. G.; Musgrave, G.)V1401,138-145(1991)

Optical bus protocol for a distributed-shared-memory multiprocessor (Davis, Martin H.; Ramachandran, Umakishore)V1563,176-187(1991)

Optical Enhancements to Computing Technology (Neff, John A., ed.)V1563(1991)

Optical fiber units with ferroelectric liquid crystals for optical computing (Domanski, Andrzej W.; Roszko, Marcin; Sierakowski, Marek W.)V1362,907-915(1991)

Optical interconnects in high-bandwidth computing (Dekker, L.; Frietman, Edward E.)V1505,148-157(1991)

Optical locator for horizon sensing (Fallon, James J.; Selby, Vaughn H.)V1495,268-279(1991)

Optically implementable algorithm for convolution/correlation of long data streams (Zhang, Yan; Li, Yao)V1474,188-198(1991)

Optical neural networks using electron trapping materials (Jutamulia, Suganda; Storti, George M.; Seiderman, William M.)V1558,442-447(1991)

Optical neurocomputer architectures using spatial light modulators (Robinson, Michael G.; Zhang, Lin; Johnson, Kristina M.)V1469,240-249(1991)

Optical parallel set of half adders using spatial light rebroadcasters (McAulay, Alastair D.; Wang, Junqing; Xu, Xin; Zeng, Ming)V1564,685-690(1991)

Optical system for DOC II (Hudyma, Russell M.; Arndt, Thomas D.; Fischer, Robert E.)V1563,244-254(1991)

Optics for Computers: Architectures and Technologies (Lebreton, Guy J., ed.)V1505(1991)

Optoelectronic approach to optical parallel processing based on the photonic parallel memory (Matsuda, Kenichi; Shibata, Jun)V1562,21-29(1991)

Optoelectronic hardware issues for implementation of simulated annealing or Boltzmann machines (Lalanne, Philippe; Chavel, Pierre)V1621,388-401(1991)

Optoelectronic master chip for optical computing (Lang, Robert J.; Kim, Jae H.; Larsson, Anders G.; Nouhi, Akbar; Cody, Jeffrey; Lin, Steven H.; Psaltis, Demetri; Tiberio, Richard C.; Porkolab, Gyorgy A.; Wolf, Edward D.)V1563,2-7(1991)

Optoelectronic neuron (Pankove, Jacques I.; Radehaus, C.; Borghs, Gustaaf)V1361,620-627(1991)

Parallel optical interconnects utilizing VLSI/FLC spatial light modulators (Genco, Sheryl M.)V1563,45-57(1991)

Photocryosar: bistable element for optoelectronic computing (Ryabushkin, Oleg A.; Bader, Vladimir A.)V1505,67-74(1991)

Photon echo as a tool for optical digital processing (Manykin, Edward A.; Chernishev, Nicholas A.)V1505,141-146(1991)

Photorefractive spatial light modulators and their applications to optical computing (Zhao, Mingjun; Li, Yulin; Qin, Yuwen; Wang, Zhao)V1558,529-534(1991)

Polarization-based all-optical bistable element (Domanski, Andrzej W.; Karpierz, Miroslaw A.; Strojewski, Dariusz)V1505,59-66(1991)

Potential digital optical computer III architectures: the next generation (Guilfoyle, Peter S.; Morozov, Valentin N.)V1563,279-283(1991)

Progress and perspectives on optical information processing in Japan (Ishihara, Satoshi)V1389,68-74(1991); V1390,68-74(1991)

Quantization and sampling considerations of computer-generated hologram for optical interconnection (Long, Pin; Hsu, Dahsiung)V1461,270-277(1991)

Quantum holography and neurocomputer architectures (Schempp, Walter)VIS08,62-144(1991)

Quantum-well excitonic devices for optical computing (Singh, Jasprit; Bhattacharya, Pallab K.)V1362,586-597(1991)

Recent research on optical neural networks in Japan (Ishihara, Satoshi; Mori, Masahiko)V1621,362-372(1991)

Relationship of image algebra to the optical processing of signals and imagery (Schmalz, Mark S.; Wilson, Joseph N.)V1474,212-234(1991)

Role for optics in future parallel processing (Rudolph, Larry)V1505,175-185(1991)

Sixty-four channel acousto-optical Bragg cells for optical computing applications (Graves, David W.)V1563,229-235(1991)

Space-time system architecture for the neural optical computing (Lo, Yee-Man V.)V1382,199-208(1991)

Spatial light modulators and their applications (Yatagai, Toyohiko; Tian, Ronglong)V1564,691-696(1991)

Studies on distributed sensing and processing for the control of large flexible spacecraft (Montgomery, Raymond C.; Ghosh, David)V1480,126-137(1991)

Two new developments for optoelectronic bus systems (Jiang, Jie; Kraemer, Udo)V1505,166-174(1991)

Vertical-cavity surface-emitting lasers and arrays for optical interconnect and optical computing applications (Wang, S. C.; Dziura, T. G.; Yang, Y. J.)V1563,27-33(1991)

Optical data storage—see also magneto-optics; optical disks; optical recording

10-mm-thick head mechanism for a stacked optical disk system (Seya, Eiichi; Matsumoto, Kiyoshi; Nihei, Hideki; Ichikawa, Atsushi; Moriyama, Shigeo; Nakamura, Shigeru; Mita, Seiichi)V1499,269-273(1991)

90-mm magneto-optical disk drive with digital servo (Miyoshi, Masahiro; Kawasaki, Tetsuharu; Takahashi, Satoru; Mine, Hironori)V1499,116-119(1991)

Accelerated aging studies for polycarbonate optical disk substrates (Nikles, David E.; Forbes, Charles E.)V1499,39-41(1991)

Analysis of wavelength division multiplexing technique for optical data storage (Yuk, Tung Ip; Palais, Joseph C.)V1401,130-137(1991)

Applications of diamond-like carbon films for write-once optical recording (Armeyev, V. Y.; Arslanbekov, A. H.; Chapliev, N. I.; Konov, Vitaly I.; Ralchenko, V. G.; Strelnitsky, V. E.)V1621,2-10(1991)

Aspects of the application of image and data storage in compact disk formats (Davies, David H.; Clark, David F.)V1401,2-8(1991)

Automated systems for quantitative analysis of holograms (Pryputniewicz, Ryszard J.)VIS08,215-246(1991)

Beam position noise and other fundamental noise processes that affect optical storage (Levenson, Marc D.)V1376,259-271(1991)

Blue laser devices for optical data storage (Lenth, Wilfried; Kozlovsky, William J.; Risk, William P.)V1499,308-313(1991)

Clinical aspects of quality assurance in PACS (Marglin, Stephen I.)V1444,83-86(1991)

Clock generation circuit of a sampled servo optical disk drive (Ozawa, Yasuyuki; Hamaguchi, S.; Moritsugu, Masaharu; Futamata, Akio)V1499,136-142(1991)

Commercial applications for optical data storage (Tas, Jeroen)V1401,10-18(1991)

Compact magneto-optical disk head integrated with chip elements (Yamanaka, Yutaka; Katayama, Ryuichi; Komatsu, Y.; Ishikawa, S.; Itoh, M.; Ono, Yuzo)V1499,263-268(1991)

Compatability test for phase-change erasable and WORM media in a multifunction drive (Ohara, Shunji; Ishida, Takashi; Inokuchi, Chikashi; Furutani, Tadashige; Ishibashi, Kenzo; Kurahashi, Akira; Yoshida, Tomio)V1499,187-194(1991)

Crystalization kinetics in Ge2Sb2Te5 phase-change recording films (Ozawa, Kenji; Ogino, Shinji; Satoh, Yoshikazu; Urushidani, Tatuo; Ueda, Atushi; Deno, Hiroshi; Kawakami, Haruo)V1499,180-186(1991)

Development of a high-performance 86-mm MO disk by using polycarbonate substrate (Ito, Katsunori; Shibata, Yasumasa; Honda, Katsunori; Tsukahara, Makoto; Kojima, Kotaro; Kimura, Masakatsu; Fujisima, Toshihiko; Yoshinaga, Kazuomi; Kanai, Toshitaka; Shimizu, Keijiro)V1499,382-385(1991)

Differential spot-size focus servo (Milster, Tom D.; Wang, Mark S.; Froehlich, Fred F.; Kann, J. L.; Treptau, Jeffrey P.; Erwin, K. E.)V1499,348-353(1991)

Digital map databases in support of avionic display systems (Trenchard, Michael E.; Lohrenz, Maura C.; Rosche, Henry; Wischow, Perry B.)V1456,318-326(1991)

Directions in optical storage (Jipson, Victor B.)V1499,2-4(1991)

Directly overwritable magneto-optical disk with light-power-modulation method using no initializing magnet (Tsutsumi, K.; Nakaki, Y.; Fukami, T.; Tokunaga, T.)V1499,55-61(1991)

Direct overwritable magneto-optical exchange-coupled multilayered disk by laser power modulation recording (Saito, Jun; Akasaka, Hideki)V1499,44-54(1991)

Direct overwriting system by light-intensity modulation using triple-layer disks (Maeda, Fumisada; Arai, Masayuki; Owa, Hideo; Takahashi, Hiroo; Kaneko, Masahiko)V1499,62-69(1991)

Display and applied holography in culture development (Markov, Vladimir B.)VIS08,268-304(1991)

DyFeCo magneto-optical disks with a Ce-SiO2 protective film (Naitou, Kazunori; Numata, Takehiko; Nakashima, Kazuo; Maeda, Miyozo; Koshino, Nagaaki)V1499,386-392(1991)

Effects of focus misregistration on optical disk performance (Finkelstein, Blair I.; Childers, Ed R.)V1499,438-449(1991)

Electric field effect on the persistent hole burning of quinone derivatives (Nishimura, Tetsuya; Yagyu, Eiji; Yoshimura, Motomu; Tsukada, Noriaki; Takeyama, Tetsu)V1436,31-37(1991)

Electron trapping for mass-data-storage memory (Lindmayer, Joseph; Goldsmith, Paul; Gross, Kirk)V1401,103-112(1991)

Enumerative modulation coding with arbitrary constraints and postmodulation error correction coding for data storage systems (Mansuripur, Masud)V1499,72-86(1991)

Erasable compact disk utilizing phase-change material and multipulse recording method (Ohno, Eiji; Nishiuchi, Kenichi; Yamada, Noboru; Akahira, Nobuo)V1499,171-179(1991)

Erasable optical 3D memory using novel electron trapping materials (Jutamulia, Suganda; Storti, George M.; Seiderman, William M.; Lindmayer, Joseph)V1401,113-118(1991)

Extremely durable CD-ROM with a novel structure (Yamaguchi, H.; Tsukamoto, Y.; Watanabe, F.; Sato, A.; Saito, M.; Honda, H.; Murahata, M.; Yanagisawa, M.; Tsuno, Toshio)V1499,29-38(1991)

Fail-safe WORM file system (Ooi, B. C.)V1401,27-34(1991)

Fast-laser-power control for high-density optical disk systems (Satoh, Hiroharu; Kinoshita, Yoshio; Okao, Keiichi; Tanaka, Masahiko; Nagatani, Hiroyuki; Honguh, Yoshinori)V1499,324-329(1991)

Flying head read/write characteristics using a monolithically integrated laser diode/photodiode at a wavelength of 1.3 um (Ukita, Hiroo; Katagiri, Yoshitada; Nakada, Hiroshi)V1499,248-262(1991)

Focus-error detection from far-field image flow (Ishibashi, Hiromichi; Tanaka, Shin-ichi; Moriya, Mitsuro)V1499,340-347(1991)

Focus grating coupler construction optics: theory, design, and tolerancing (Moore, Kenneth E.; Lawrence, George N.)V1354,581-587(1991)

Focus sensing method with improved pattern noise rejection (Marshall, Daniel R.)V1499,332-339(1991)

High-density optical MUSE disk (Tsuchiya, Yoichi; Terasaki, Hitoshi; Ota, Osamu)V1499,450-456(1991)

Highly reliable 7-GB, 1.2- to 2.2-MB/s, 12-inch write-once optical disk (Watanabe, Hitoshi; Koyama, Eiji; Nunomura, T.; Taii, T.; Miura, Michio; Gotoh, Akira; Nishida, T.; Horigome, Shinkichi; Ohta, Norio)V1499,21-28(1991)

High-performance optical disk systems for tactical applications (Haritatos, Fred N.)V1499,196-202(1991)

High-speed large-capacity optical disk using pit-edge recording and MCAV method (Maeda, Takeshi; Saito, Atsushi; Sugiyama, Hisataka; Ojima, Masahiro; Arai, Shinichi; Shigematu, Kazuo)V1499,414-418(1991)

High-speed recording technologies for a magneto-optical disk system (Sukeda, Hirofumi; Tsuchinaga, Hiroyuki; Tanaka, Satoshi; Niihara, Toshio; Nakamura, Shigeru; Mita, Seiichi; Yamada, Yukinori; Ohta, Norio; Fukushima, Mitsugi)V1499,419-425(1991)

High-speed swing arm three-beam optical head (Yak, A. S.; Low, Toh-Siew; Lim, Siak-Piang)V1401,74-81(1991)

Holographic data storage on the photothermoplastic tape carrier (Akaev, Askar A.; Zhumaliev, K. M.; Jamankyzov, N.)V1621,182-193(1991)

Holography (Greguss, Pal; Jeong, Tung H., eds.)VIS08(1991)

Improved phantom for quality control of laser scanner digitizers in PACS (Halpern, Ethan J.; Esser, Peter D.)V1444,104-115(1991)

Interchangeability of optical disks (Shimamoto, Masayoshi; Yamada, Koichi; Watanabe, Isao; Nakajima, Yoshiki; Ito, Osamu; Tanaka, Kunimaro)V1499,393-400(1991)

Investigation of the possibility of creating a multichannel photodetector based on the avalanche MRS-structure (Sadyigov, Z. Y.; Gasanov, A. G.; Yusipov, N. Y.; Golovin, V. M.; Gulanian, Emin H.; Vinokurov, Y. S.; Simonov, A. V.)V1621,158-168(1991)

Laser beam modeling in optical storage systems (Treptau, Jeffrey P.; Milster, Tom D.; Flagello, Donis G.)V1415,317-321(1991)

Light-sensitive organic media for optical discs (Barachevsky, Valery A.; Rot, A. S.; Zaks, I. N.)V1621,33-44(1991)

Lissajous analysis of focus crosstalk in optical disk systems (Grove, Steven L.; Getreuer, Kurt W.; Schell, David L.)V1499,354-359(1991)

Local microholograms recording on the moving photothermoplastic disk (Kutanov, Askar A.; Akaev, Askar A.; Abdrisaev, Baktybek D.; Snimshikov, Igor A.)V1507,94-98(1991)

Magnetically induced super resolution in a novel magneto-optical disk (Aratani, Katsuhisa; Fukumoto, Atsushi; Ohta, Masumi; Kaneko, Masahiko; Watanabe, Kenjirou)V1499,209-215(1991)

Magnetic and electronic properties of Co/Pd superlattices (Victora, Randall H.; MacLaren, J. M.)V1499,378-381(1991)

Magneto-optical reading and writing integrated heads: a way to a multigigabyte multi-rigid-disk drive (Renard, Stephane; Valette, Serge)V1499,238-247(1991)

Magneto-optical signal detection with elliptically polarized light (Nakao, Takeshi; Arimoto, Akira A.; Takahashi, Masahiko)V1499,433-437(1991)

Magneto-optic data storage in the '90s (Funkenbusch, Arnold W.)V1396,699-708(1991)

Margin measurements of pulse amplitude modulation channels (Lauffenburger, Jim; Arachtingi, John W.; Robbins, Jamey L.; Kim, Jong)V1499,104-113(1991)

Micro-optic lens for data storage (Milster, Tom D.; Trusty, Robert M.; Wang, Mark S.; Froehlich, Fred F.; Erwin, J. K.)V1499,286-292(1991)

Multifunction grating for signal detection of optical disk (Gupta, Mool C.; Peng, Song-Tsuen)V1499,303-306(1991)

Multitrack rewritable optical disk system for high-performance applications: 14-inch TODS (Cinelli, Joseph L.; Kozak, Taras)V1499,203-208(1991)

Need for quality assurance related to PACS (Rowberg, Alan H.)V1444,80-82(1991)

New approach to high-density recording on a magneto-optical disk (Fujita, Goro; Urakawa, Yoshinori; Yamagami, Tamotsu; Watanabe, Tetsu)V1499,426-432(1991)

New possibilities of photosensitive glasses for three-dimensional phase hologram recording (Glebov, Leonid B.; Nikonorov, Nikolai V.; Panysheva, Elena I.; Petrovskii, Gurii T.; Savvin, Vladimir V.; Tunimanova, Irina V.; Tsekhomskii, Victor A.)V1621,21-32(1991)

New tracking method for two-beam optical heads using continuously grooved disks (Irie, Mitsuru; Takeshita, Nobuo; Fujita, Teruo; Kime, Kenjiro)V1499,360-365(1991)

Optical associative memories based on time-delayed four-wave mixing (Belov, M. N.; Manykin, Edward A.)V1621,268-279(1991)

Optical associative memory for nontraditional architecture digital computers and database management systems (Burtsev, Vsevolod S.; Fyodorov, Vyatcheslav B.)V1621,215-226(1991)

Optical auto- and heteroassociative memory based on a high-order neural network (Kiselyov, Boris S.; Kulakov, Nickolay Y.; Mikaelian, Andrei L.; Shkitin, Vladimir A.)V1621,328-339(1991)

Optical data communication compel the design of a new class of storage media (Frietman, Edward E.; Dekker, L.; van Nifterick, W.)V1401,19-26(1991)

Optical Data Storage '91 (Burke, James J.; Shull, Thomas A.; Imamura, Nobutake, eds.)V1499(1991)

Optical data storage in photosensitive fibers (Campbell, Robert J.; Kashyap, Raman)V1499,160-164(1991)

Optical Data Storage Technologies (Chua, Soo-Jin; McCallum, John C., eds.)V1401(1991)

Optical fibers for magneto-optical recording (Opsasnick, Michael N.; Stancil, Daniel D.; White, Sean T.; Tsai, Ming-Horn)V1499,276-280(1991)

Optical information recording on vitreous semiconductors with a thermoplastic method of visualization (Panasyuk, L. M.; Forsh, A. A.)V1621,74-82(1991)

Optically coupled three-dimensional memory system (Koyanagi, Mitsumasa)V1390,467-476(1991)

Optical Memory and Neural Networks (Mikaelian, Andrei L., ed.)V1621(1991)

Optical memory in electro-optical crystals (Berezhnoy, Anatoly A.; Popov, Yury V.)V1401,44-49(1991)

Optical memory using localized photoinduced anisotropy in a synthetic dye-polymer (Kuo, Chai-Pei)V1499,148-159(1991)

Optical recording material based on bacteriorhodopsin modified with hydroxylamine (Vsevolodov, N. N.; Dyukova, T. V.; Druzhko, A. B.; Shakhbazyan, V. Y.)V1621,11-20(1991)

Optical recording materials (Savant, Gajendra D.; Jannson, Joanna)V1461,79-90(1991)

Organo-metallic thin films for erasable photochromic laser discs (Hua, Zhong Y.; Chen, G. R.; Wang, Z. H.)V1519,2-7(1991)

Performance comparison of various data codes in Z-CAV optical recording (Lee, Tzuo-chang; Chen, Di)V1499,87-103(1991)

Photothermoplastic molecular heterostructures (Cherkasov, Yuri A.)V1621,62-73(1991)

Polarizing holographic optical elements for optical data storage (Ono, Yuzo)V1555,177-181(1991)

Programmable holographic scanning (Zhengmin, Li; Li, Min; Tang, Jinfa)V1401,66-73(1991)

Quality assurance of PACS systems with laser film digitizers (Esser, Peter D.; Halpern, Ethan J.)V1444,100-103(1991)

Read-write simulation and numerical noise for WORM optical disk and drive (Chao, Shiuh; Yang, Tsong-Yo)V1401,35-43(1991)

Recent advances in optical pickup head with holographic optical elements (Kato, Makoto; Kadowaki, Shin-ichi; Komma, Yoshiaki; Hori, Yoshikazu)V1507,36-44(1991)

Reversible phase-change optical recording by using microcellular GeSbTeCo recording film (Okamine, Shigenori; Terao, Motoyasu; Andoo, Keikichi; Miyauchi, Yasushi)V1499,166-170(1991)

Scanning optical microscopy: a powerful tool in optical recording (Coombs, James H.; Holtslag, A. H.)V1499,6-20(1991)

Selection of image acquisition methods (Donnelly, Joseph J.)V1444,351-356(1991)

Shot-noise-limited high-power green laser for higher density optical disk systems (Kubota, Shigeo; Oka, Michio)V1499,314-323(1991)

Some problems of implementation of the memory system based on optical solitons in fibers (Shcherbakov, A. S.)V1621,204-214(1991)

Study of optical data recording based on photoluminescence effect (Petrov, V. V.; Zymenko, V. I.; Kravetz, V. G.; Polishchuk, E. Y.; Sushko, A. M.)V1621,45-50(1991)

Super resolution in a magneto-optical disk with an active mask (Fukumoto, Atsushi; Aratani, Katsuhisa; Yoshimura, Shunji; Udagawa, Toshiki; Ohta, Masumi; Kaneko, Masahiko)V1499,216-225(1991)

Theoretical background and practical processing for art and technical work in dichromated gelatin holography (Coblijn, Alexander B.)VIS08,305-324(1991)

Thermomagnetic recording on amorphous ferrimagnetic films (Aleksandrov, K. S.; Berman, G. P.; Frolov, G. I.; Seredkin, V. A.)V1621,51-61(1991)

Three-beam CD optical pickup using a holographic optical element (Yoshida, Yoshio; Miyake, Takahiro; Kurata, Yukio; Ishikawa, Toshio)V1401,58-65(1991)

Three-beam overwritable magneto-optic disk drive (Watabe, Akinori; Yamada, Ichiro; Yamamoto, Manabu; Katoh, Kikuji)V1499,226-235(1991)

Tracking method of optical tape recorder using acousto-optic scanning (Narahara, Tatsuya; Kumai, Satoshi; Nakao, Takashi; Ozue, Tadashi)V1499,120-128(1991)

Two-photon 3-D optical memories (Esener, Sadik C.; Rentzepis, Peter M.)V1499,144-147(1991)

Use of FEM modes in time-domain servo simulations (Ernst, Charles H.)V1499,129-135(1991)

Use of rigorous vector coupled-wave theory for designing and tolerancing surface-relief diffractive components for magneto-optical heads (Haggans, Charles W.; Kostuk, Raymond K.)V1499,293-302(1991)

Use of the Society of Motion Picture and Television Engineers test pattern in picture archiving and communication systems (PACS) (Gray, Joel E.)V1444,118-124(1991)

Verification of tracking servo signal simulation from scanning tunneling microscope surface profiles (Karis, Thomas E.; Best, Margaret E.; Logan, John A.; Lyerla, James R.; Lynch, Robert T.; McCormack, R. P.)V1499,366-376(1991)

Visual aspects of picture storage (Brettel, Hans)V1401,50-55(1991)

Volume optical memory by two-photon interaction (Rentzepis, Peter M.; Dvornikov, Alexander S.)V1563,198-207(1991)

Write noise from optical heads with nonachromatic beam expansion prisms (Kay, David B.; Chase, Scott B.; Gage, Edward C.; Silverstein, Barry D.)V1499,281-285(1991)

You can't just plug it in: digital image networks/picture archiving and communication systems installation (Gelish, Anthony)V1444,363-372(1991)

Optical design—see also aberrations; aspherics; coatings; geometrical optics; lenses; modulation transfer function; optical engineering; optical systems; optical transfer function; optomechanical design; standards; stray light; surfaces; telescopes; thin films

1990 International Lens Design Conference lens design problems: the design of a NonLens (Clark, Peter P.; Londono, Carmina)V1354,555-569(1991)

1990 Intl Lens Design Conf (Lawrence, George N., ed.)V1354(1991)

7th Mtg in Israel on Optical Engineering (Oron, Moshe; Shladov, Itzhak, eds.)V1442(1991)

Aberration Theory Made Simple (Mahajan, Virendra N.)VTT06(1991)

Absolute phasing of segmented mirrors using the polarization phase sensor (Klumpe, Herbert W.; Lajza-Rooks, Barbara A.; Blum, James D.)V1398,95-106(1991)

Adaptive optical transfer function modeling (Gaffard, Jean-Paul; Boyer, Corinne)V1483,92-103(1991)

Adaptive optics, transfer loops modeling (Boyer, Corinne; Gaffard, Jean-Paul; Barrat, Jean-Pierre; Lecluse, Yves)V1483,77-91(1991)

Advances in the computer-aided design of planarized free-space optical circuits: system simulation (Jahns, Juergen; Brumback, Babette A.)V1555,2-7(1991)

Advances in the optical design of miniaturized optical correlators (Gebelein, Rolin J.; Connely, Shawn W.; Foo, Leslie D.)V1564,452-463(1991)

Aligning diamond-turned optics using visible-light interferometry (Figoski, John W.)V1354,540-546(1991)

All-reflective four-element zoom telescope: design and analysis (Johnson, R. B.; Hadaway, James B.; Burleson, Thomas A.; Watts, Bob; Park, Ernest D.)V1354,669-675(1991)

All-reflective phased array imaging telescopes (Stuhlinger, Tilman W.)V1354,438-446(1991)

All-refractive telescope for next-generation inter-satellite communication (Heimbeck, Hans-Jorg)V1354,434-437(1991)

Analysis and design of binary gratings for broadband, infrared, low-reflectivity surfaces (Moharam, M. G.)V1485,254-259(1991)

Analysis on image performance of a moisture-absorbed plastic singlet for an optical disk (Tanaka, Yuki; Miyamae, Hiroshi)V1354,395-401(1991)

Analytic solutions and numerical results for the optical and radiative properties of V-trough concentrators (Fraidenraich, Naum)V1528,15-30(1991)

Angle encoding with the folded Brewster interferometer (Young, Niels O.)V1454,235-244(1991)

Athermalized FLIR optics (Rogers, Philip J.)V1354,742-751(1991)

Attempt to develop a zoom-lens-design expert system (Weng, Zhicheng; Chen, Zhiyong; Yang, Yu-Hong; Ren, Tao; Cong, Xiaojie; Yao, Yuchuan; He, Fengling; Li, Yuan-Yuan)V1527,349-356(1991)

Axial and transverse displacement tolerances during excimer laser surgery for myopia (Shimmick, John K.; Munnerlyn, Charles R.; Clapham, Terrance N.; McDonald, Marguerite B.)V1423,140-153(1991)

Binary optics from a ray-tracing point of view (Southwell, William H.)V1354,38-42(1991)

Binary optics in lens design (Kathman, Alan D.; Pitalo, Stephen K.)V1354,297-309(1991)

CAD system for CGHs and laser beam lithography (Yatagai, Toyohiko; Geiser, Martial; Tian, Ronglong; Tian, Xingkang; Onda, Hajime)V1555,8-12(1991)

Calculation method of the Strehl Definition for decentral optical systems (Zhuang, Song Lin; Chen, Huai'an)V1354,252-253(1991)

Calculation of flux density produced by CPC reflectors on distant targets (Gordon, Jeff M.; Rabl, Ari)V1528,152-162(1991)

Calculation of wave aberration in optical systems with holographic optical elements (Yuan, Yanrong)V1354,43-52(1991)

CAN-AM Eastern '90 (Antos, Ronald L.; Krisiloff, Allen J., eds.)V1398(1991)

Case study of elastomeric lens mounts (Fischer, Robert E.)V1533,27-35(1991)

Center for Optics Manufacturing (Pollicove, Harvey M.; Moore, Duncan T.)V1398,90-93(1991)

Color vision enhancement with spectacles (Perrott, Colin M.)V1529,31-36(1991)

Combat vehicle stereo HMD (Rallison, Richard D.; Schicker, Scott R.)V1456,179-190(1991)

Comments on the Seidel aberration theory (Kang, Songgao; Lu, Kaichang; Zhu, Yafei)V1527,376-379(1991)

Compact nonimaging lens with totally internally reflecting facets (Parkyn, William A.; Pelka, David G.)V1528,70-81(1991)

Comparison of asymmetrical and symmetrical nonimaging reflectors for east-west circular cylindrical solar receivers (Mills, David R.; Monger, A. G.; Morrison, G. L.)V1528,44-55(1991)

Computer-aided alignment of a grazing-incidence ring resonator for a visible wavelength free-electron laser (Hudyma, Russell M.; Eigler, Lynne C.)V1354,523-532(1991)

Computing illumination-bundle focusing by lens systems (Forkner, John F.)V1354,210-215(1991)

Concentric systems for adaptation as spectrographs (Mertz, Lawrence N.)V1354,457-459(1991)

Construction of the 16-meter large lunar telescope (Omar, Husam A.)V1494,135-146(1991)

Correction of secondary and higher-order spectrum using special materials (Mercado, Romeo I.)V1535,184-198(1991)

Corrections of aberrations using HOEs in UV and visible imaging systems (Yang, Zishao; Rosenbruch, Klaus J.)V1354,323-327(1991)

Current Developments in Optical Design and Optical Engineering (Fischer, Robert E.; Smith, Warren J., eds.)V1527(1991)

Damage resistant optics for a megajoule solid state laser (Campbell, John H.; Rainer, Frank; Kozlowski, Mark R.; Wolfe, C. R.; Thomas, Ian M.; Milanovich, Fred P.)V1441,444-456(1991)

Debris collision warning sensor telescope design (Brown, Robert J.)V1527,155-162(1991)

Design and analysis of illumination systems (Zochling, Gunter)V1354,617-626(1991)

Design and application of a moire interferometer (He, Anzhi; Wang, Hai-Lin; Yan, Da-Peng; Miao, Peng-Cheng)V1527,334-337(1991)

Design and fabrication of large-format holographic lenses (Stojanoff, Christo G.; Kubitzek, Ruediger; Tropartz, Stephan; Brasseur, Olivier; Froehlich, K.)V1527,48-60(1991)

Design and fabrication of x-ray/EUV optics for photoemission experimental beam line at Hefei National Synchrotron Radiation Lab. (Zhou, Changxin; Sun, Deming)V1345,281-287(1991)

Design and optical performance of beryllium assessment mirrors (Thomas, Brigham B.; Maxey, L. C.; Miller, Arthur C.)V1485,20-30(1991)

Design challenges for the 1990s (Shafer, David R.)V1354,608-616(1991)

Design considerations for multilayer-coated Schwarzchild objectives for the XUV (Kortright, Jeffrey B.; Underwood, James H.)V1343,95-103(1991)

Design of achromatized hybrid diffractive lens systems (Londono, Carmina; Clark, Peter P.)V1354,30-37(1991)

Design of an athermalized three-fields-of-view infrared sensor (Wickholm, David R.)V1488,58-63(1991)

Design of an imaging spectrometer for observing ocean color (Weng, Zhicheng; Chen, Zhiyong; Cong, Xiaojie)V1527,338-348(1991)

Design of an IR non-lens, or how I buried 100 mm of germanium (Aikens, David M.)V1485,183-194(1991)

Design of array systems using shared symmetry (Miao, Cheng-Hsi)V1354,447-456(1991)

Design of CGH for special image formation (Tang, Yaw-Tzong; Lee, Wai-Hon; Chang, Ming-Wen)V1555,194-199(1991)

Design of double-Gauss objective by using concentric lenses (Juang, Jeng-Dang; Chang, Ming-Wen)V1354,273-276(1991)

Design of high-performance aplanatic achromats for the near-ultraviolet waveband (Al-Baho, Tareq I.; Learner, R. C.; Maxwell, Jonathan)V1354,417-428(1991)

Design of lightweight beryllium optics, factors effecting producibility, and cost of near-net-shape blanks (Clement, Thomas P.)V1485,31-38(1991)

Design of narrow band XUV and EUV coronagraphs using multilayer optics (Walker, Arthur B.; Allen, Maxwell J.; Barbee, Troy W.; Hoover, Richard B.)V1343,415-427(1991)

Design of nonimaging lenses and lens-mirror combinations (Minano, Juan C.; Gonzalez, Juan C.)V1528,104-115(1991)

Design of optical systems with HOE by DEMOS program (Gan, Michael A.; Zhdanov, Dmitriy D.; Novoselskiy, Vadim V.; Ustinov, Sergey I.; Fiodorov, Alexander O.; Potyemin, Igor S.; Bezdidko, Sergey N.)V1574,254-260(1991)

Design of reflective relay for soft x-ray lithography (Rodgers, John M.; Jewell, Tanya E.)V1354,330-336(1991)

Design of some achromatic imaging hybrid diffractive-refractive lenses (Twardowski, Patrice J.; Meyrueis, Patrick)V1507,55-65(1991)

Design of two- and three-element diffractive telescopes (Buralli, Dale A.; Morris, G. M.)V1354,292-296(1991)

Design optimization of optical filters for space applications (Annapurna, M. N.; Nagendra, C. L.; Thutupalli, G. K.)V1485,260-271(1991)

Design rules for pseudocolor transmission holographic stereograms (Andrews, John R.)V1507,407-415(1991)

Designs of two-glass apochromats and superachromats (Mercado, Romeo I.)V1354,262-272(1991)

Design survey of x-ray/XUV projection lithography systems (Shealy, David L.; Viswanathan, Vriddhachalam K.)V1343,229-240(1991)

Developments in projection lenses for HDTV (Rudolph, John D.)V1456,15-28(1991)

Differential coating objective (Goldfain, Ervin)V1527,126-133(1991)

Diffraction performance calculations in lens design (Malacara Hernandez, Daniel)V1354,2-14(1991)

Diffractive doublet corrected on-axis at two wavelengths (Farn, Michael W.; Goodman, Joseph W.)V1354,24-29(1991)

Disparity between combiners in a double-combiner head-up display (Cohen, Jonathan; Reichert, Abraham)V1456,250-261(1991)

Easily fabricated wide-angle telescope (Owen, R. C.)V1354,430-433(1991)

Efficient, uniform transverse coupling of diode arrays to laser rods using nonimaging optics (Minnigh, Stephen W.; Knights, Mark G.; Avidor, Joel M.; Chicklis, Evan P.)V1528,129-134(1991)

Electron-beam-written reflection diffractive microlenses for oblique incidence (Shiono, Teruhiro; Ogawa, Hisahito)V1545,232-240(1991)

Elements with long-focal depth and high-lateral resolution (Davidson, Nir; Friesem, Asher A.; Hasman, Erez)V1442,22-25(1991)

Ellipsoidal mirrors as modular elements in the design of off-axis optical systems (Stavroudis, Orestes N.)V1354,627-646(1991)

Enhancement of Conrady's "D-d" method (Gintner, Henry)V1354,97-102(1991)

Exact sine condition in the presence of spherical aberration (Shibuya, Masato)V1354,240-245(1991)

Examples of the topographies of the wavefront-variance merit function at different aberration orders (Johnston, Steve C.)V1354,77-82(1991)

Existence of local minima in lens design (Kidger, Michael J.; Leamy, Paul T.)V1354,69-76(1991)

Expert systems in lens design (Dilworth, Donald C.)V1354,359-370(1991)

Fast, inexpensive, diffraction-limited cylindrical microlenses (Snyder, James J.; Reichert, Patrick)V1544,146-151(1991)

Focus grating coupler construction optics: theory, design, and tolerancing (Moore, Kenneth E.; Lawrence, George N.)V1354,581-587(1991)

Focusing and collimating laser diodes and laser diode arrays (Coetzee, Frans M.; Casasent, David P.)V1544,108-122(1991)

Four-meter lunar engineering telescope (Peacock, Keith; Giannini, Judith A.; Kilgus, Charles C.; Bely, Pierre Y.; May, B. S.; Cooper, Shannon A.; Schlimm, Gerard H.; Sounder, Charles; Ormond, Karen A.; Cheek, Eric A.)V1494,147-159(1991)

Fundamentals of the optical tolerance budget (Smith, Warren J.)V1354,474-481(1991)

Future of global optimization in optical design (Sturlesi, Doron; O'Shea, Donald C.)V1354,54-68(1991)

Gaussian scaling laws for diffraction: top-hat irradiance and Gaussian beam propagation through a paraxial optical train (Townsend, Sallie S.; Cunningham, Philip R.)V1415,154-194(1991)

Global optimization using the y-ybar diagram (Brown, Daniel M.)V1527,19-25(1991)

High-fidelity diagnostic beam sampling of a tunable high-energy laser (Mitchell, John H.; Cohen, Harold L.; Hanley, Stephen T.)V1414,141-150(1991)

High-performance, wide-magnification-range IR zoom telescope with automatic compensation for temperature effects (Shechterman, Mark)V1442,276-285(1991)

High-power laser/optical-fiber-coupling device (Falciai, Riccardo; Pascucci, Tania)V1506,120-125(1991)

High-resolution thermal imager with a field-of-view of 112 degrees (Matsushita, Tadashi; Suzuki, Hiroshi; Wakabayashi, Satoshi; Tajime, Toru)V1488,368-375(1991)

History of infrared optics (Johnson, R. B.; Feng, Chen)VCR38,3-18(1991)

Image plane tilt in optical systems (Sasian, Jose M.)V1527,85-95(1991)

Importance of dispersion tolerances in infrared lens design (Korniski, Ronald J.; Thompson, Kevin P.)V1354,402-407(1991)

Improved plastic molding technology for magneto-optical disk substrate (Galic, George J.)V1396,539-546(1991)

Infrared Optical Design and Fabrication (Hartmann, Rudolf; Smith, Warren J., eds.)VCR38(1991)

Infrared zoom lenses in the eighties and beyond (Mann, Allen)V1540,338-349(1991)

Integrated free-space optics (Jahns, Juergen)V1354,588-592(1991)

Integration of geometrical and physical optics (Lawrence, George N.; Moore, Kenneth E.)V1415,322-329(1991)

Interactive Design and Electro-optic Analysis Liaise (IDEAL) (Alukaidey, Talib A.; Pettiford, Alvin A.)V1390,513-522(1991)

Interdependence of design, optical evaluation, and visual performance of ophthalmic lenses (Freeman, Michael H.)V1529,2-12(1991)

Investigation of process sensitivity for electron-beam evaporation of beryllium (Egert, Charles M.; Schmoyer, D. D.; Nordin, C. W.; Berry, A.)V1485,64-77(1991)

Investigation of unobscured mirror telescope for telecom purposes (Sand, Rolf)V1522,103-110(1991)

IR objective with internal scan mirror (Eisenberg, Shai; Menache, Ram)V1442,133-138(1991)

ISOCAM: a camera for the ISO satellite optical bench development (Auternaud, Danielle)V1488,64-72(1991)

Key technologies for IR zoom lenses: aspherics and athermalization (Nory, Pierre)VCR38,142-152(1991)

Knowledge-based environment for optical system design (Johnson, R. B.)V1354,346-358(1991)

Lens design for the infrared (Fischer, Robert E.)VCR38,19-43(1991)

Lens for microlithography (Hsieh, Hung-yu; Wagner, Jerome F.)V1396,467-472(1991)

Lens with maximum power occupying a given cylindrical volume (Reardon, Patrick J.; Chipman, Russell A.)V1354,234-239(1991)

Linearized ray-trace analysis (Redding, David C.; Breckenridge, William G.)V1354,216-221(1991)

Lunar liquid-mirror telescopes (Borra, Ermanno F.; Content, R.)V1494,219-227(1991)

Manufacture of ISO mirrors (Ruch, Eric)V1494,265-278(1991)

Membrane light modulators: engineering design considerations (O'Mara, Daniel M.; Schiller, Craig M.; Warde, Cardinal)V1527,110-117(1991)

Microlens array for modification of SLM devices (Kathman, Alan D.; Temmen, Mark G.; Scott, Miles L.)V1544,58-65(1991)

Modeling and Simulation of Laser Systems II (Schnurr, Alvin D., ed.)V1415(1991)

Monochromatic quartet: a search for the global optimum (O'Shea, Donald C.)V1354,548-554(1991)

MTF optimization in lens design (Rimmer, Matthew P.; Bruegge, Thomas J.; Kuper, Thomas G.)V1354,83-91(1991)

Multiaperture spectrometer design for the Atmospheric Infrared Sounder (Pagano, Robert J.; Hatch, Marcus R.)V1354,460-471(1991)

Multimode IRST/FLIR design issues (Armstrong, George R.; Oakley, Philip J.; Ranat, Bhadrayu M.)VCR38,120-141(1991)

Narcissus in current generation FLIR systems (Ford, Eric H.; Hasenauer, David M.)VCR38,95-119(1991)

Neural network optimization, components, and design selection (Weller, Scott W.)V1354,371-378(1991)

New developments in the field of chemical infrared fiber sensors (Kellner, Robert A.; Taga, Karim)V1510,232-241(1991)

New family of 1:1 catadioptric broadband deep-UV high-Na lithography lenses (Zhang, Yudong; Zou, Haixing; Wang, Zhi-Jiang)V1463,688-694(1991)

New iterative algorithm for the design of phase-only gratings (Farn, Michael W.)V1555,34-42(1991)

New method of increasing the sensitivity of Schlieren interferometer using two Wollaston prisms and its application to flow field (Yan, Da-Peng; He, Anzhi; Yang, Zu Q.; Zhu, Yi Yun)V1554B,636-640(1991)

New type of large-angle binocular microtelescopes (Lu, Kaichang; Zhu, Yafei; Kang, Songgao)V1527,413-418(1991)

Nonimaging concentrators for diode-pumped slab lasers (Lacovara, Phil; Gleckman, Philip; Holman, Robert L.; Winston, Roland)V1528,135-141(1991)

Nonimaging Optics: Maximum Efficiency Light Transfer (Winston, Roland; Holman, Robert L., eds.)V1528(1991)

Nonimaging optics: optical design at the thermodynamic limit (Winston, Roland)V1528,2-6(1991)

Nonlinearity and lens design (Tatian, Berge)V1354,154-164(1991)

Nonlinear model of the optimal tolerance design for a lens system (Zhuang, Song Lin; Qu, Zhijin)V1354,177-179(1991)

Opthalmic Lens Design and Fabrication (Perrott, Colin M., ed.)V1529(1991)

Optical analysis of segmented aircraft windows (Jones, Mike I.; Jones, Mark S.)V1498,110-127(1991)

Optical analysis of thermal-induced structural distortions (Weinswig, Shepard A.; Hookman, Robert A.)V1527,118-125(1991)

Optical characteristics of sapphire laser scalpels analysed by ray-tracing (Verdaasdonk, Rudolf M.; Borst, Cornelius)V1420,136-140(1991)

Optical design and development of a small barcode scanning module (Wike, Charles K.; Lindacher, Joseph M.)V1398,119-126(1991)

Optical design and optimization with physical optics (Lawrence, George N.; Moore, Kenneth E.)V1354,15-22(1991)

Optical design considerations for next-generation space and lunar telescopes (Korsch, Dietrich G.)V1494,111-118(1991)

Optical design of a high-power fiber-optic coupler (English, R. Edward; Halpin, John M.; House, F. A.; Paris, Robert D.)V1527,174-179(1991)

Optical design of dual-combiner head-up displays (Kirkham, Anthony J.)V1354,310-315(1991)

Optical design of high-aperture aspherical projection lens (Osawa, Atsuo; Fukuda, Kyohei; Hirata, Kouji)V1354,337-343(1991)

Optical design of the moderate-resolution imaging spectrometer-tilt for the Earth Observing System (Maymon, Peter W.)V1492,286-297(1991)

Optical design with physical optics using GLAD (Lawrence, George N.)V1354,126-135(1991)

Optical power budget and device time-constant considerations in undersea laser-based sensor design (Leatham, James G.)V1537,194-202(1991)

Optical system for DOC II (Hudyma, Russell M.; Arndt, Thomas D.; Fischer, Robert E.)V1563,244-254(1991)

Optics for vector scanning (Ehrmann, Jonathan S.)V1454,245-256(1991)

Optimization of athermal systems (Benham, Paul; Kidger, Michael J.)V1354,120-125(1991)

Optimization of lens designer to manufacturer communications (Fischer, Robert E.)V1354,506-522(1991)

Optimization of original lens structure type from optical lens data base (Yang, Daren; Jiang, Huilin; Li, Gongde; Yang, Huamin)V1527,456-461(1991)

Optimization of the optical transfer function (Kidger, Michael J.; Benham, Paul)V1354,92-96(1991)

Optimization of the Seidel image errors by bending of lenses using a 4th-degree merit function (Aurin, Friedrich A.)V1354,180-185(1991)

Optimization using the OSLO and Super-OSLO programs (Sinclair, Douglas C.)V1354,116-119(1991)

Optimum receiver structure and filter design for MPAM optical space communication systems (Al-Ramli, Intesar F.)V1522,111-123(1991)

Optomechanics and Dimensional Stability (Paquin, Roger A.; Vukobratovich, Daniel, eds.)V1533(1991)

Orthogonal cylindrical diffractive lens for parallel readout optical disk system (Urquhart, Kristopher S.; Marchand, Philippe J.; Lee, Sing H.; Esener, Sadik C.)V1555,214-223(1991)

Overview of CODE V development (Harris, Thomas I.)V1354,104-111(1991)

Pacing elements of IR system design (Zissis, George J.)VCR38,44-54(1991)

Panoramic security (Greguss, Pal)V1509,55-66(1991)

Performance limitations of miniature optical correlators (Crandall, Charles M.; Giles, Michael K.; Clark, Natalie)V1564,98-109(1991)

Performance of microstrip proportional counters for x-ray astronomy on spectrum-roentgen-gamma (Budtz-Jorgensen, Carl; Bahnsen, Axel; Christensen, Finn E.; Madsen, M. M.; Olesen, C.; Schnopper, Herbert W.)V1549,429-437(1991)

Photolithographic imaging of computer-generated holographic optical elements (Shvartsman, Felix P.; Oren, Moshe)V1555,71-79(1991)

Photonic computer-aided design tools for high-speed optical modulators (Liu, Pao-Lo)V1371,46-55(1991)

Plastic lens array with the function of forming unit magnification erect image using roof prisms (Takahashi, Eietsu; Tanji, Shigeo; Tanaka, Akira; Hayashi, Yuji)V1527,145-154(1991)

Power of one power (Shenker, Martin)V1354,647-653(1991)

Practical use of generalized simulated annealing optimization on microcomputers (Hearn, Gregory K.)V1354,186-191(1991)

Preliminary study of the admittance diagram as a useful tool in the design of stripline components at microwave frequencies (Franck, Charmaine C.; Franck, Jerome B.)V1527,277-290(1991)

Principles of optimization in lens design developed at the Institute of Optics in Berlin (West) (Kross, Juergen)V1354,165-170(1991)

Prismatic anamorphic beam expanders for free-space optical communications (Jonas, Reginald P.)V1417,402-411(1991)

Problem solving of optical design by R graph with bidirectional search mechanics (Chang, Rong-Seng; Chen, Der-Chin)V1354,379-385(1991)

Progress in binocular design (Seil, Konrad)V1533,48-60(1991)

Quadric surfaces: some derivations and applications (Larkin, Eric W.)V1354,222-231(1991)

Radiation-resistant optical glasses (Marker, Alexander J.; Hayden, Joseph S.; Speit, Burkhard)V1485,160-172(1991)

Rank-down method for automatic lens design (Ooki, Hiroshi)V1354,171-176(1991)

Ray tracing homogenizing mirrors for synchrotron x-ray lithography (Homer, Michael; Rosser, Roy J.; Speer, R. J.)V1527,134-144(1991)

Ray-tracing of optically modified fiber tips for laser angioplasty (Verdaasdonk, Rudolf M.; Borst, Cornelius)V1425,102-109(1991)

Ray tracing through progressive ophthalmic lenses (Bourdoncle, Bernard; Chauveau, J. P.; Mercier, Jean-Louis M.)V1354,194-199(1991)

Recent developments in nonimaging secondary concentrators for linear receiver solar collectors (Gordon, Jeff M.)V1528,32-43(1991)

Reflective optical designs for soft x-ray projection lithography (Jewell, Tanya E.; Thompson, Kevin P.; Rodgers, John M.)V1527,163-173(1991)

Reflectors for efficient and uniform distribution of radiation for lighting and infrared based on nonimaging optics (Cai, Wen; Gordon, Jeff M.; Kashin, Peter; Rabl, Ari)V1528,118-128(1991)

Refractive-index interpolation fit criterion for materials used in optical design (Korniski, Ronald J.)VCR38,193-217(1991)

Reimaging system for evaluating high-resolution charge-coupled-device arrays (Chambers, Robert J.; Warren, David W.; Lawrie, David J.; Lomheim, Terrence S.; Luu, K. T.; Shima, Ralph M.; Schlegel, J. D.)V1488,312-326(1991)

Reluctant glass formers and their applications in lens design (Johnson, R. B.; Feng, Chen; Ethridge, Edwin C.)V1527,2-18(1991); V1535,231-247(1991)

Review of methods for the design of unsymmetrical optical systems (Sasian, Jose M.)V1396,453-466(1991)

Role of aspherics in zoom lens design (Betensky, Ellis I.)V1354,656-662(1991)

Selected Papers on Laser Design (Weichel, Hugo, ed.)VMS29(1991)

Selected Papers on Optical Tolerancing (Wiese, Gary E., ed.)VMS36(1991)

Semiactive telescope for the French PRONAOS submillimetric mission (Duran, Michel; Luquet, Philippe; Buisson, F.; Cousin, B.)V1494,357-376(1991)

Separation of function in the ASAP software package (Johnston, Steve C.; Greynolds, Alan W.)V1354,136-141(1991)

Shared aperture for two beams of different wavelength using the Talbot effect (Hector, Scott D.; Swanson, Gary J.)V1555,200-213(1991)

Silicon microlenses for enhanced optical coupling to silicon focal planes (Motamedi, M. E.; Griswold, Marsden P.; Knowlden, Robert E.)V1544,22-32(1991)

Simple design considerations for binary optical holographic elements (Franck, Jerome B.; Hodgkin, Van A.)V1555,63-70(1991)

Simple method for finding the valid ray-surface intersection (Freeman, David E.)V1354,200-209(1991)

Simple multidimensional quadratic extrapolation method for the correction of specific aberrations in lens systems (Maxwell, Jonathan; Hull, Chris S.)V1354,277-285(1991)

Simplified pupil enlargement technique (Radl, Bruce M.)V1457,314-317(1991)

Simultaneous alignment and multiple surface figure testing of optical system components via wavefront aberration measurement and reverse optimization (Lundgren, Mark A.; Wolfe, William L.)V1354,533-539(1991)

Small-computer program for optical design and analysis written in "C" (Beckmann, Leo H.)V1354,254-261(1991)

Sola ASL in Spectralite and polycarbonate aspheric lens designs (Machol, Steven)V1529,45-56(1991)

Sola ASL in Spectralite strikes the perfect balance between cosmetics and optics (Machol, Steven; Modglin, Luan)V1529,38-44(1991)

Spatial sampling errors for a satellite-borne scanning radiometer (Manalo, Natividad D.; Smith, G. L.)V1493,281-291(1991)

Status and future of polymeric materials in imaging systems (Lytle, John D.)V1354,388-394(1991)

Status and needs of infrared optical property information for optical designers (Wolfe, William L.)V1354,696-741(1991)

Surface contributions of the wave aberrations up to the eighth degree (Aurin, Friedrich A.)V1527,61-72(1991)

SYNOPSYS—a lens design computer program package (Dilworth, Donald C.)V1354,112-115(1991)

Systematic approach to axial gradient lens design (Wang, David Y.; Moore, Duncan T.)V1354,599-605(1991)

Tactical Infrared Systems (Tuttle, Jerry W., ed.)V1498(1991)

Techniques for designing hybrid diffractive optical systems (Brown, Daniel M.; Pitalo, Stephen K.)V1527,73-84(1991)

Tertiary spectrum manipulation in apochromats (Maxwell, Jonathan)V1354,408-416(1991)

Theory of color correction by use of chromatic magnification (Ames, Alan J.)V1354,286-290(1991)

Theory of two-component zoom systems (Oskotsky, Mark L.)V1527,37-47(1991)

Thermal aberration analysis of a laser-diode-pumped Nd:YAG laser (Kubota, Shigeo; Oka, Michio; Kaneda, Yushi; Masuda, Hisashi)V1354,572-580(1991)

Thermal compensation of infrared achromatic objectives with three optical materials (Rayces, Juan L.; Lebich, Lan)V1354,752-759(1991)

Thermodynamics of light concentrators (Ries, Harald; Smestad, Greg P.; Winston, Roland)V1528,7-14(1991)

Tilt-tolerant dual-band interferometer objective lens (Jones, Mike I.)V1498,158-162(1991)

Towards global optimization with adaptive simulated annealing (Forbes, Greg W.; Jones, Andrew E.)V1354,144-153(1991)

Two-dimensional control of radiant intensity by use of nonimaging optics (Kuppenheimer, John D.; Lawson, Robert I.)V1528,93-103(1991)

Two-mirror projection systems for simulating telescopes (Hannan, Paul G.; Davila, Pam M.)V1527,26-36(1991)

Ultra-high-performance zoom lens for the visible waveband (Neil, Iain A.)V1354,684-694(1991)

Using hybrid refractive-diffractive elements in infrared Petzval objectives (Wood, Andrew P.)V1354,316-322(1991)

Very high temperature fiber processing and testing through the use of ultrahigh solar energy concentration (Jacobson, Benjamin A.; Gleckman, Philip; Holman, Robert L.; Sagie, Daniel; Winston, Roland)V1528,82-85(1991)

Ways that designers and fabricators can help each other (Willey, Ronald R.; Durham, Mark E.)V1354,501-505(1991)

Zoom lens design using GRIN materials (Tsuchida, Hirofumi; Aoki, Norihiko; Hyakumura, Kazushi; Yamamoto, Kimiaki)V1354,246-251(1991)

Zoom lenses with a single moving element (Johnson, R. B.; Feng, Chen)V1354,676-683(1991)

Zoom lens with aspherical lens for camcorder (Yatsu, Masahiko; Deguchi, Masaharu; Maruyama, Takesuke)V1354,663-668(1991)

Optical disks—see also magneto-optics; optical data storage; optical recording

10-mm-thick head mechanism for a stacked optical disk system (Seya, Eiichi; Matsumoto, Kiyoshi; Nihei, Hideki; Ichikawa, Atsushi; Moriyama, Shigeo; Nakamura, Shigeru; Mita, Seiichi)V1499,269-273(1991)

Accelerated aging studies for polycarbonate optical disk substrates (Nikles, David E.; Forbes, Charles E.)V1499,39-41(1991)

Compact magneto-optical disk head integrated with chip elements (Yamanaka, Yutaka; Katayama, Ryuichi; Komatsu, Y.; Ishikawa, S.; Itoh, M.; Ono, Yuzo)V1499,263-268(1991)

Design of a motionless head for parallel readout optical disk (Marchand, Philippe J.; Krishnamoorthy, Ashok V.; Ambs, Pierre; Gresser, Julien; Esener, Sadik C.; Lee, Sing H.)V1505,38-49(1991)

Development of a high-performance 86-mm MO disk by using polycarbonate substrate (Ito, Katsunori; Shibata, Yasumasa; Honda, Katsunori; Tsukahara, Makoto; Kojima, Kotaro; Kimura, Masakatsu; Fujisima, Toshihiko; Yoshinaga, Kazuomi; Kanai, Toshitaka; Shimizu, Keijiro)V1499,382-385(1991)

Diffraction analysis of optical disk readout signal deterioration caused by mark-size fluctuation (Honguh, Yoshinori)V1527,315-321(1991)

Directly overwritable magneto-optical disk with light-power-modulation method using no initializing magnet (Tsutsumi, K.; Nakaki, Y.; Fukami, T.; Tokunaga, T.)V1499,55-61(1991)

Direct overwritable magneto-optical exchange-coupled multilayered disk by laser power modulation recording (Saito, Jun; Akasaka, Hideki)V1499,44-54(1991)

Direct overwriting system by light-intensity modulation using triple-layer disks (Maeda, Fumisada; Arai, Masayuki; Owa, Hideo; Takahashi, Hiroo; Kaneko, Masahiko)V1499,62-69(1991)

DyFeCo magneto-optical disks with a Ce-SiO2 protective film (Naitou, Kazunori; Numata, Takehiko; Nakashima, Kazuo; Maeda, Miyozo; Koshino, Nagaaki)V1499,386-392(1991)

Effects of focus misregistration on optical disk performance (Finkelstein, Blair I.; Childers, Ed R.)V1499,438-449(1991)

Erasable compact disk utilizing phase-change material and multipulse recording method (Ohno, Eiji; Nishiuchi, Kenichi; Yamada, Noboru; Akahira, Nobuo)V1499,171-179(1991)

Eraseable optical disk systems for signal processing (Bessette, Oliver E.; Cinelli, Joseph L.)V1474,162-166(1991)

Extremely durable CD-ROM with a novel structure (Yamaguchi, H.; Tsukamoto, Y.; Watanabe, F.; Sato, A.; Saito, M.; Honda, H.; Murahata, M.; Yanagisawa, M.; Tsuno, Toshio)V1499,29-38(1991)

Fabrication of grooved-glass substrates by phase-mask lithography (Brock, Phillip J.; Levenson, Marc D.; Zavislan, James M.; Lyerla, James R.; Cheng, John C.; Podlogar, Carl V.)V1463,87-100(1991)

Fast-laser-power control for high-density optical disk systems (Satoh, Hiroharu; Kinoshita, Yoshio; Okao, Keiichi; Tanaka, Masahiko; Nagatani, Hiroyuki; Honguh, Yoshinori)V1499,324-329(1991)

Flying head read/write characteristics using a monolithically integrated laser diode/photodiode at a wavelength of 1.3 um (Ukita, Hiroo; Katagiri, Yoshitada; Nakada, Hiroshi)V1499,248-262(1991)

Handwritten zip code recognition using an optical radial basis function classifier (Neifeld, Mark A.; Rakshit, S.; Psaltis, Demetri)V1469,250-255(1991)

High-density optical MUSE disk (Tsuchiya, Yoichi; Terasaki, Hitoshi; Ota, Osamu)V1499,450-456(1991)

Highly reliable 7-GB, 1.2- to 2.2-MB/s, 12-inch write-once optical disk (Watanabe, Hitoshi; Koyama, Eiji; Nunomura, T.; Taii, T.; Miura, Michio; Gotoh, Akira; Nishida, T.; Horigome, Shinkichi; Ohta, Norio)V1499,21-28(1991)

High-performance optical disk systems for tactical applications (Haritatos, Fred N.)V1499,196-202(1991)

High-speed large-capacity optical disk using pit-edge recording and MCAV method (Maeda, Takeshi; Saito, Atsushi; Sugiyama, Hisataka; Ojima, Masahiro; Arai, Shinichi; Shigematu, Kazuo)V1499,414-418(1991)

High-speed recording technologies for a magneto-optical disk system (Sukeda, Hirofumi; Tsuchinaga, Hiroyuki; Tanaka, Satoshi; Niihara, Toshio; Nakamura, Shigeru; Mita, Seiichi; Yamada, Yukinori; Ohta, Norio; Fukushima, Mitsugi)V1499,419-425(1991)

Interchangeability of optical disks (Shimamoto, Masayoshi; Yamada, Koichi; Watanabe, Isao; Nakajima, Yoshiki; Ito, Osamu; Tanaka, Kunimaro)V1499,393-400(1991)

Investigation of the possibility of creating a multichannel photodetector based on the avalanche MRS-structure (Sadyigov, Z. Y.; Gasanov, A. G.; Yusipov, N. Y.; Golovin, V. M.; Gulanian, Emin H.; Vinokurov, Y. S.; Simonov, A. V.)V1621,158-168(1991)

Laser beam modeling in optical storage systems (Treptau, Jeffrey P.; Milster, Tom D.; Flagello, Donis G.)V1415,317-321(1991)

Laser disk: a practicum on effective image management (Myers, Richard F.)V1460,50-58(1991)

Light-sensitive organic media for optical discs (Barachevsky, Valery A.; Rot, A. S.; Zaks, I. N.)V1621,33-44(1991)

Magnetically induced super resolution in a novel magneto-optical disk (Aratani, Katsuhisa; Fukumoto, Atsushi; Ohta, Masumi; Kaneko, Masahiko; Watanabe, Kenjirou)V1499,209-215(1991)

Magneto-optical reading and writing integrated heads: a way to a multigigabyte multi-rigid-disk drive (Renard, Stephane; Valette, Serge)V1499,238-247(1991)

Magneto-optical signal detection with elliptically polarized light (Nakao, Takeshi; Arimoto, Akira A.; Takahashi, Masahiko)V1499,433-437(1991)

Magneto-optic data storage in the '90s (Funkenbusch, Arnold W.)V1396,699-708(1991)

Measurement of laser spot quality (Milster, Tom D.; Treptau, Jeffrey P.)V1414,91-96(1991)

Multi-media PACS integrated with HIS/RIS employing magneto-optical disks (Umeda, Tokuo; Inamura, Kiyonari; Inamoto, Kazuo; Kondoh, Hiroshi P.; Kozuka, Takahiro)V1446,199-210(1991)

Multifunction grating for signal detection of optical disk (Gupta, Mool C.; Peng, Song-Tsuen)V1499,303-306(1991)

Multitrack rewritable optical disk system for high-performance applications: 14-inch TODS (Cinelli, Joseph L.; Kozak, Taras)V1499,203-208(1991)

New approach to high-density recording on a magneto-optical disk (Fujita, Goro; Urakawa, Yoshinori; Yamagami, Tamotsu; Watanabe, Tetsu)V1499,426-432(1991)

New tracking method for two-beam optical heads using continuously grooved disks (Irie, Mitsuru; Takeshita, Nobuo; Fujita, Teruo; Kime, Kenjiro)V1499,360-365(1991)

Off-line image exchange between two PACS modules using the "ISAC" magneto-optical disk (Minato, Kotaro; Komori, Masaru; Nakano, Yoshihisa; Yonekura, Yoshiharu; Sasayama, Satoshi; Takahashi, Takashi; Konishi, Junji; Abe, Mituyuki; Sato, Kazuhiro; Hosoba, Minoru)V1446,195-198(1991)

One-dimensional hologram recording in thick photolayer disk (Mikaelian, Andrei L.; Gulanian, Emin H.; Kretlov, Boris S.; Molchanova, L. V.; Semichev, V. A.)V1621,194-203(1991)

Optical Data Storage '91 (Burke, James J.; Shull, Thomas A.; Imamura, Nobutake, eds.)V1499(1991)

Orthogonal cylindrical diffractive lens for parallel readout optical disk system (Urquhart, Kristopher S.; Marchand, Philippe J.; Lee, Sing H.; Esener, Sadik C.)V1555,214-223(1991)

Super resolution in a magneto-optical disk with an active mask (Fukumoto, Atsushi; Aratani, Katsuhisa; Yoshimura, Shunji; Udagawa, Toshiki; Ohta, Masumi; Kaneko, Masahiko)V1499,216-225(1991)

Three-beam overwritable magneto-optic disk drive (Watabe, Akinori; Yamada, Ichiro; Yamamoto, Manabu; Katoh, Kikuji)V1499,226-235(1991)

Verification of tracking servo signal simulation from scanning tunneling microscope surface profiles (Karis, Thomas E.; Best, Margaret E.; Logan, John A.; Lyerla, James R.; Lynch, Robert T.; McCormack, R. P.)V1499,366-376(1991)

Optical engineering—see also optical design; optical fabrication; optical testing; optomechanical design

Analysis of optical measurements of SIF and singularity order in rocket motor geometry (Chang, Che-way)V1554A,250-261(1991)

Applications of Optical Engineering: Proceedings of OE/Midwest '90 (Dubiel, Mary K.; Eppinger, Hans E.; Gillespie, Richard E.; Guzik, Rudolph P.; Pearson, James E., eds.)V1396(1991)

Designing the right visor (Gilboa, Pini)V1456,154-163(1991)

Development and analysis of a simple model for an IR sensor (Ballik, Edward A.; Wan, William)V1488,249-256(1991)

Intersection of crack borders with free surfaces: an engineering interpretation of optical experimental results (Smith, C. W.; Constantinescu, D. M.)V1554A,102-115(1991)

Modular removable precision mechanism for alignment of an FUV spatial heterodyne interferometer (Tom, James L.; Cotton, Daniel M.; Bush, Brett C.; Chung, Ray; Chakrabarti, Supriya)V1549,308-312(1991)

New holographic system for measuring vibration (Cai, Yunliang)V1554B,75-80(1991)

Spatial light modulator on the base of shape memory effect (Antonov, Victor A.; Shelyakov, Alexander V.)V1474,116-123(1991)

Optical fabrication—see also polishing; optical testing; standards

1536 x 1024 CCD image sensor (Wong, Kwok Y.; Torok, Georgia R.; Chang, Win-Chyi; Meisenzahl, Eric J.)V1447,283-287(1991)

Advanced broadband baffle materials (Seals, Roland D.)V1485,78-87(1991)

Advanced processes for the shaping of the Zerodur glass ceramic (Marx, Thomas A.)V1398,94-94(1991)

Advanced processing of the Zerodur R glass ceramic (Marx, Thomas A.)V1535,130-135(1991)

Analysis and performance limits of diamond-turned diffractive lenses for the 3-5 and 8-12 micrometer regions (Riedl, Max J.; McCann, James T.)VCR38,153-163(1991)

Application and machining of Zerodur for optical purposes (Reisert, Norbert)V1400,171-177(1991)

Application of statistical design in materials development and production (Pouskouleli, G.; Wheat, T. A.; Ahmad, A.; Varma, Sudhanshu; Prasad, S. E.)V1590,179-190(1991)

Automation in optics manufacturing (Pollicove, Harvey M.; Moore, Duncan T.)V1354,482-486(1991)

Beryllium and titanium cost-adjustment report (Owen, John; Ulph, Eric)V1485,128-137(1991)

Beryllium galvanometer mirrors (Weissman, Harold)V1485,13-19(1991)

Center for Optics Manufacturing (Pollicove, Harvey M.; Moore, Duncan T.)V1398,90-93(1991)

Computer-Controlled Optical Surfacing (Jones, Robert A., ed.)VMS40(1991)

Control of thermally induced porosity for the fabrication of beryllium optics (Moreen, Harry A.)V1485,54-61(1991)

Cr3+ tunable laser glass (Izumitani, Tetsuro; Zou, Xuolu; Wang, Y.)V1535,150-159(1991)

Creation of aspheric beryllium optical surfaces directly in the hot isostatic pressing consolidation process (Gildner, Donald; Marder, James M.)V1485,46-53(1991)

Design and fabrication of large-format holographic lenses (Stojanoff, Christo G.; Kubitzek, Ruediger; Tropartz, Stephan; Brasseur, Olivier; Froehlich, K.)V1527,48-60(1991)

Design and fabrication of x-ray/EUV optics for photoemission experimental beam line at Hefei National Synchrotron Radiation Lab. (Zhou, Changxin; Sun, Deming)V1345,281-287(1991)

Design and specification of diamond-turned optics (Clark, Robert A.)VCR38,164-183(1991)

Design of lightweight beryllium optics, factors effecting producibility, and cost of near-net-shape blanks (Clement, Thomas P.)V1485,31-38(1991)

Development of beryllium-mirror turning technology (Arnold, Jones B.)V1485,96-105(1991)

Fabrication and characterization of semiconductor microlens arrays (Diadiuk, Vicky; Liau, Zong-Long; Walpole, James N.)V1354,496-500(1991)

Fabrication and performance of a one-to-one erect imaging microlens array for fax (Bellman, Robert H.; Borrelli, Nicholas F.; Mann, L. G.; Quintal, J. M.)V1544,209-217(1991)

Fabrication of a fast, aspheric beryllium mirror (Paquin, Roger A.; Gardopee, George J.)V1485,39-45(1991)

Fabrication of a grazing-incidence telescope by grinding and polishing techniques on aluminum (Gallagher, Dennis J.; Cash, Webster C.; Green, James C.)V1343,155-161(1991)

Fabrication of cold light mirror of film projector by direct electron-beam-evaporated TiO2 and SiO2 starting materials in neutral oxygen atmosphere (Zhong, Di S.; Xu, Guang Z.; Liu, Wi)V1519,350-358(1991)

Fabrication of microlenses by laser-assisted chemical etching (Gratrix, Edward J.; Zarowin, Charles B.)V1544,238-243(1991)

Fabrication of phase structures with continuous and multilevel profile for diffraction optics (Poleshchuk, Alexander G.)V1574,89-100(1991)

Fabrication of special-purpose optical components (Leuw, David H.)V1442,31-41(1991)

Fabrication of toroidal and coma-corrected toroidal diffraction gratings from spherical master gratings using elastically deformable substrates: a progress report (Huber, Martin C.; Timothy, J. G.; Morgan, Jeffrey S.; Lemaitre, Gerard R.; Tondello, Giuseppe; Naletto, Giampiero)V1494,472-480(1991)

Fabrication techniques of photopolymer-clad waveguides for nonlinear polymeric modulators (Tumollillo, Thomas A.; Ashley, Paul R.)V1559,65-73(1991)

Fast, inexpensive, diffraction-limited cylindrical microlenses (Snyder, James J.; Reichert, Patrick)V1544,146-151(1991)

Finish machining of optical components in mass production (Grodnikov, Alexander I.; Korovkin, Vladimir P.)V1400,186-193(1991)

Fundamentals of the optical tolerance budget (Smith, Warren J.)V1354,474-481(1991)

High-speed oscillation free lapping and polishing process for optical lenses (Richter, G.)V1400,158-163(1991)

History of and potential for optical bonding agents in the visible (Magyar, James T.)V1535,55-58(1991)

Holographic characterization of DYE-PVA films studied at 442 nm for optical elements fabrication (Couture, Jean J.)V1559,429-435(1991)

Hubble Space Telescope optics: problems and solutions (Burrows, Christopher J.)V1494,528-533(1991)

Improved plastic molding technology for ophthalmic lens and contact lens (Galic, George J.; Maus, Steven)V1529,13-21(1991)

Infrared Optical Design and Fabrication (Hartmann, Rudolf; Smith, Warren J., eds.)VCR38(1991)

Integrated micro-optics for computing and switching applications (Jahns, Juergen)V1544,246-262(1991)

Investigation of process sensitivity for electron-beam evaporation of beryllium (Egert, Charles M.; Schmoyer, D. D.; Nordin, C. W.; Berry, A.)V1485,64-77(1991)

Large-aperture (80-cm diameter) phase plates for beam smoothing on Nova (Woods, Bruce W.; Thomas, Ian M.; Henesian, Mark A.; Dixit, Sham N.; Powell, Howard T.)V1410,47-54(1991)

Laser ablation of refractive micro-optic lenslet arrays (Bartley, James A.; Goltsos, William)V1544,140-145(1991)

Laser heating and evaporation of glass and glass-borning materials and its application for creating MOC (Veiko, Vadim P.; Yakovlev, Evgeni B.; Frolov, V. V.; Chujko, V. A.; Kromin, A. K.; Abbakumov, M. O.; Shakola, A. T.; Fomichov, P. A.)V1544,152-163(1991)

Lessons learned in recent beryllium-mirror fabrication (Wells, James A.; Lombard, Calvin M.; Sloan, George B.; Moore, Wally W.; Martin, Claude E.)V1485,2-12(1991)

Light-absorbing, lightweight beryllium baffle materials (Murray, Brian W.; Floyd, Dennis R.; Ulph, Eric)V1485,88-95(1991)

Manufacture of fast, aspheric, bare beryllium optics for radiation hard, spacebome sensor systems (Sweeney, Michael N.)V1485,116-127(1991)

Metal mirror review (Janeczko, Donald J.)VCR38,258-280(1991)

Method of making ultralight primary mirrors (Zito, Richard R.)V1494,491-497(1991)

Microlens array fabricated in surface relief with high numerical aperture (Lau, Hon W.; Davies, Neil A.; McCormick, Malcolm)V1544,178-188(1991)

Microlens array for staring infrared imager (Werner, Thomas R.; Cox, James A.; Swanson, S.; Holz, Michael)V1544,46-57(1991)

Miniature and Micro-Optics: Fabrication and System Applications (Roychoudhuri, Chandrasekhar; Veldkamp, Wilfrid B., eds.)V1544(1991)

Moulding process for contact lens (Skipper, Richard S.; Shepherd, David W.)V1529,22-30(1991)

New approach to the simulation of optical manufacturing processes (Oinen, Donald E.; Billow, Nick W.)V1354,487-493(1991)

New glasses for optics and optoelectronics (Morian, Hans F.)V1400,146-157(1991)

New methods for economic production of prisms and lenses (Richter, G.)V1400,11-23(1991)

Optical bonding agents for IR and UV refracting elements (Pellicori, Samuel F.)V1535,48-54(1991)

Optical Fabrication and Testing (Campbell, Duncan R.; Johnson, Craig W.; Lorenzen, Manfred, eds.)V1400(1991)

Optical recording materials (Savant, Gajendra D.; Jannson, Joanna)V1461,79-90(1991)

Optimization of lens designer to manufacturer communications (Fischer, Robert E.)V1354,506-522(1991)

Performance of a variable-line-spaced master reflection grating for use in the reflection grating spectrometer on the x-ray multimirror mission (Bixler, Jay V.; Hailey, Charles J.; Mauche, C. W.; Teague, Peter F.; Thoe, Robert S.; Kahn, Steven M.; Paerels, F. B.)V1549,420-428(1991)

Production of sapphire domes by the growth of near-net-shape single crystals (Biderman, Shlomo; Horowitz, Atara; Einav, Yehezkel; Ben-Amar, Gabi; Gazit, Dan; Stern, Adin; Weiss, Matania)V1535,27-34(1991)

Proportional counter windows for the Bragg crystal spectrometer on AXAF (Markert, Thomas H.; Bauer, J.; Canizares, Claude R.; Isobe, T.; Nenonen, Seppo A.; O'Connor, J.; Schattenburg, Mark L.; Flanagan, Kathryn A.; Zombeck, Martin V.)V1549,408-419(1991)

Recent progress in the growth and characterization of large Ge single crystals for IR optics and microelectronics (Azoulay, Moshe; Gafni, Gabriella; Roth, Michael)V1535,35-45(1991)

Reflective and Refractive Optical Materials for Earth and Space Applications (Riedl, Max J.; Hale, Robert R.; Parsonage, Thomas B., eds.)V1485(1991)

Reimaging system for evaluating high-resolution charge-coupled-device arrays (Chambers, Robert J.; Warren, David W.; Lawrie, David J.; Lomheim, Terrence S.; Luu, K. T.; Shima, Ralph M.; Schlegel, J. D.)V1488,312-326(1991)

Reluctant glass formers and their applications in lens design (Johnson, R. B.; Feng, Chen; Ethridge, Edwin C.)V1527,2-18(1991)

Robotic-based fabrication system for aspheric reflectors (Zimmerman, Jerrold; Jones, Robert A.; Rupp, Wiktor J.)VCR38,184-192(1991)

Scattering properties of ZrF4-based glasses prepared by the gas film levitation technique (Lopez, Adolphe; Baniel, P.; Gall, Pascal; Granier, Jean E.)V1590,191-202(1991)

Selected Papers on Optical Tolerancing (Wiese, Gary E., ed.)VMS36(1991)

Solid matrix Christiansen filters (Milanovic, Zoran; Jacobson, Michael R.; Macleod, H. A.)V1535,160-170(1991)

Techniques for integrating 3-D optical systems (Brenner, Karl-Heinz)V1544,263-270(1991)

Theory and experiment as tools for assessing surface finish in the UV-visible wavelength region (Ingers, Joakim P.; Thibaudeau, Laurent)V1400,178-185(1991)

Thick, fine-grained beryllium optical coatings (Murray, Brian W.; Ulph, Eric Richard, Peter N.)V1485,106-115(1991)

Virtual-phase charge-coupled device image sensors for industrial and scientific applications (Khvilivitzky, A. T.; Berezin, V. Y.; Lazovsky, L. Y.; Tataurschikov, S. S.; Pisarevsky, A. N.; Vydrevich, M. G.; Kossov, V. G.)V1447,184-190(1991)

Ways that designers and fabricators can help each other (Willey, Ronald R.; Durham, Mark E.)V1354,501-505(1991)

Work-induced stress and long-term stability in optically polished silicon (Bender, John W.; Wahl, Roger L.)V1533,264-276(1991)

Optical inspection—see also industrial optics; machine vision; robotics

AI application in shoe industry CAD/CAM (Wang, He-Chen; Lou, Da-li; Xian, Wu; Song, Xiang; Yong, Jiang)V1386,273-276(1991)

Absolute range measurement system for real-time 3-D vision (Wood, Christopher M.; Shaw, Michael M.; Harvey, David M.; Hobson, Clifford A.; Lalor, Michael J.; Atkinson, John T.)V1332,301-313(1991)

Active stereo inspection using computer solids models (Nurre, Joseph H.)V1381,171-176(1991)

Algorithm for quality inspection of characters printed on chip resistors (Numagami, Yasuhiko; Hattori, Yasuyuki; Nakamura, Osamu; Minami, Toshi)V1606,970-979(1991)

Application of fiber optic sensors in pavement maintenance (Shadaram, Mehdi; Solehjou, Amin; Nazarian, Soheil)V1332,487-490(1991)

Application of real-time holographic interferometry in the nondestructive inspection of electronic parts and assemblies (Wood, Craig P.; Trolinger, James D.)V1332,122-131(1991)

Applications of an automated particle detection and identification system in VLSI wafer processing (Hattori, Takeshi; Koyata, Sakuo)V1464,367-376(1991)

Automated approach to the correlation of defect locations to electrical test results to determine yield reducing defects (Slama, M. M.; Patterson, Angela C.)V1464,602-609(1991)

Automated dimensional inspection of cars in crash tests with digital photogrammetry (Beyer, Horst A.)V1526,134-141(1991)

Automated optical grading of timber (Sobey, Peter J.)V1379,168-179(1991)

Automated visual inspection for LSI wafer patterns using a derivative-polarity comparison algorithm (Maeda, Shunji; Hiroi, Takashi; Makihira, Hiroshi; Kubota, Hitoshi)V1567,100-109(1991)

Automated visual inspection of printed wiring boards (Chin, Roland T.; Iverson, Rolf D.)VCR36,93-104(1991)

Automated visual inspection system for IC bonding wires using morphological processing (Tsukahara, Hiroyuki; Nakashima, Masato; Sugawara, Takehisa)V1384,15-26(1991)

Automated wafer inspection in the manufacturing line (Harrigan, Jeanne E.; Stoller, Meryl D.)V1464,596-601(1991)

Automatically inspecting gross features of machined objects using three-dimensional depth data (Marshall, Andrew D.)V1386,243-254(1991)

Automatic inspection of optical fibers (Silberberg, Teresa M.)V1472,150-156(1991)

Automatic inspection system for full-color printed matter (Meguro, Shin-Ichi; Nunotani, Masakatu; Tanimizu, Katsuyuki; Sano, Mutsuo; Ishii, Akira)V1384,27-37(1991)

Computerized detection and identification of the types of defects on crystal blanks (Bow, Sing T.; Chen, Pei)V1396,646-655(1991)

CT image processing for hardwood log inspection (Zhu, Dongping; Conners, Richard W.; Araman, Philip A.)V1567,232-243(1991)

Damage detection in peanut grade samples using chromaticity and luminance (Dowell, Floyd E.; Powell, J. H.)V1379,136-140(1991)

Design optimization of moire interferometers for rapid 3-D manufacturing inspection (Dubowsky, Steven; Holly, Krisztina J.; Murray, Annie L.; Wander, Joseph M.)V1386,10-20(1991)

Digital-signal-processor-based inspection of populated surface-mount technology printed circuit boards (Hartley, David A.; Hobson, Clifford A.; Lilley, F.)V1567,277-282(1991)

Electronic speckle pattern interferometry for 3-D dynamic deformation analysis in industrial environments (Guelker, Gerd; Haack, Olaf; Hinsch, Klaus D.; Hoelscher, Claudia; Kuls, Juergen; Platen, Winfried)V1500,124-134(1991)

Extraction of hierarchical structures from complicated 2-D shapes (Han, Joon H.; Kim, Myung J.; Cho, Kwang J.)V1381,122-129(1991)

Fiber optic damage detection for an aircraft leading edge (Measures, Raymond M.; LeBlanc, Michel; Hogg, W. D.; McEwen, Keith; Park, B. K.)V1332,431-443(1991)

Fiber optic smart structures: structures that see the light (Measures, Raymond M.)V1332,377-398(1991)

Flat plate project (Wijbrans, Klaas C.; Korsten, Maarten J.)V1386,197-205(1991)

Flexible gray-level vision system based on multiple cell-feature description and generalized Hough transform (Sano, Mutsuo; Ishii, Akira; Meguro, Shin-Ichi)V1381,101-110(1991)

Fresh market carrot inspection by machine vision (Howarth, M. S.; Searcy, Stephen W.)V1379,141-150(1991)

Fuzzy logic for fault diagnosis (Comly, James B.; Bonissone, Piero P.; Dausch, Mark E.)V1381,390-400(1991)

High-accuracy capacitive position sensing for low-inertia actuators (Stokes, Brian P.)V1454,223-229(1991)

High-Speed Inspection Architectures, Barcoding, and Character Recognition (Chen, Michael J., ed.)V1384(1991)

Holographic instrumentation for monitoring crystal growth in space (Trolinger, James D.; Lal, Ravindra B.; Batra, Ashok K.)V1332,151-165(1991)

Holographic interferometry analysis of sealed, disposable containers for internal defects (Cohn, Gerald E.; Domanik, Richard A.)V1429,195-206(1991)

Holotag: a novel holographic label (Soares, Oliverio D.; Bernardo, Luis M.; Pinto, M. I.; Morais, F. V.)V1332,166-184(1991)

Image processing methodology for optimizing the quality of corks in the punching process (Davies, Roger; Correia, Bento A.; Carvalho, Fernando D.)V1567,244-253(1991)

Image registration for automated inspection of 2-D electronic circuit patterns (Rodriguez, Arturo A.; Mandeville, Jon R.)V1384,2-14(1991)

Improvements in sensitivity and discrimination capability of the PD reticle/mask inspection system (Saito, Juichi; Saijo, Y.)V1496,284-301(1991)

Improving submicron CD measurements through optimum operating points (Dusa, Mircea V.; Roeth, Klaus-Dieter; Jung, Christoph)V1496,217-223(1991)

Industrial applications of optical fuzzy syntactic pattern recognition (Caulfield, H. J.)V1332,294-300(1991)

Influence of fixing stress on the sensitivity of HNDT (Yang, Yu X.; Tan, Yushan)V1554B,768-773(1991)

In-line wafer inspection using 100-megapixel-per-second digital image processing technology (Dickerson, Gary; Wallace, Rick P.)V1464,584-595(1991)

Inspection of a class of industrial objects using a dense range map and CAD model (Ailisto, Heikki J.; Paakkari, Jussi; Moring, Ilkka)V1384,50-59(1991)

Inspection system using still vision for a rotating laser-textured dull roll (Torao, Akira; Uchida, Hiroyuki; Moriya, Susumu; Ichikawa, Fumihiko; Kataoka, Kenji; Wakui, Tsuneyoshi)V1358,843-850(1991)

Integrated Circuit Metrology, Inspection, and Process Control V (Arnold, William H., ed.)V1464(1991)

Integration of an application accelerator for high-speed inspection (Kille, Knut; Ahlers, Rolf-Juergen; Hager, K.)V1386,222-227(1991)

Intelligent visual inspection of food products (Chan, John P.; Batchelor, Bruce G.; Harris, I. P.; Perry Beng, S. J.)V1386,171-179(1991)

Interferometric analysis for nondestructive product monitoring (Cohn, Gerald E.; Domanik, Richard A.)V1396,131-142(1991)

Interpretation of holographic and shearographic fringes for estimating the size and depth of debonds in laminated plates (Shang, H. M.; Chau, Fook S.; Tay, C. J.; Toh, Siew-Lok)V1554B,680-691(1991)

Laser-based triangulation techniques in optical inspection of industrial structures (Clarke, Timothy A.; Grattan, Kenneth T.; Lindsey, N. E.)V1332,474-486(1991)

Laser scan microscope and infrared laser scan microscope: two important tools for device testing (Ziegler, Eberhard)V1400,108-115(1991)

Machine vision feedback for on-line correction of manufacturing processes: a control formulation (Taylor, Andrew T.; Wang, Paul K.)V1567,220-231(1991)

Machine vision inspection of fluorescent lamps (Bains, Narinder; David, Frank)V1386,232-242(1991)

Machine vision platform requirements for successful implementation and support in the semiconductor assembly manufacturing environment (LeBeau, Christopher J.)V1386,228-231(1991)

Machine vision system for ore sizing (Eichelberger, Christopher L.; Blair, Steven M.; Khorana, Brij M.)V1396,678-686(1991)

Machine Vision Systems Integration in Industry (Batchelor, Bruce G.; Waltz, Frederick M., eds.)V1386(1991)

Mapping of imbedded contaminants in transparent material by optical scatter (Rudberg, Donald A.; Stover, John C.; McGary, Douglas E.)V1530,232-239(1991)

Measurement of interfacial tension by automated video techniques (Deason, Vance A.; Miller, R. L.; Watkins, Arthur D.; Ward, Michael B.; Barrett, K. B.)V1332,868-876(1991)

Moire image overlapping method for PCB inspection designator (Chang, Rong-Seng; Hu, Yeu-Jent)V1567,216-219(1991)

Near real-time operation of a centimeter-scale distributed fiber sensing system (Garside, Brian K.)V1332,399-408(1991)

New scheme for real-time visual inspection of bearing roller (Zhang, Yu; Xian, Wu; Li, Li-Ping; Hall, Ernest L.; Tu, James Z.)V1384,60-65(1991)

New stereo laser triangulation device for specular surface inspection (Samson, Marc; Dufour, Marc L.)V1332,314-322(1991)

Novel grating methods for optical inspection (Asundi, Anand K.)V1554B,708-715(1991)

Off-line machine recognition of forgeries (Randolph, David; Krishnan, Ganapathy)V1386,255-264(1991)

Optical correlator vision system for a manufacturing robot assembly cell (Brandstetter, Robert W.; Fonneland, Nils J.; Zanella, R.; Yearwood, M.)V1385,173-189(1991)

Optical leak detection of oxygen using IR laser diodes (Disimile, Peter J.; Fox, Curtis F.; Toy, Norman)V1492,64-75(1991)

Optical system for automatic inspection of curved surfaces (Livi, S.; Magnani, M.; Pieri, Silvano; Romoli, Andrea)V1500,144-150(1991)

Optics in Agriculture (DeShazer, James A.; Meyer, George E., eds.)V1379(1991)

Pelliclizing technology (Yamauchi, Takashi)V1496,302-314(1991)

Performance evaluation of a vision dimension metrology system (El-Hakim, Sabry F.; Westmore, David B.)V1526,56-67(1991)

Performance of an optoelectronic probe used with coordinate measuring machines (Shams, Iden; Butler, Clive)V1589,120-125(1991)

Phase-shift mask pattern accuracy requirements and inspection technology (Wiley, James N.; Fu, Tao-Yi; Tanaka, Takashi; Takeuchi, Susumu; Aoyama, Satoshi; Miyazaki, Junji; Watakabe, Yaichiro)V1464,346-355(1991)

Photon scanning tunneling microscopy (Goudonnet, Jean-Pierre; Salomon, L.; de Fornel, F.; Chabrier, G.; Warmack, R. J.; Ferrell, Trinidad L.)V1400,116-123(1991)

Process monitoring and control with fiber optics (Marcus, Michael A.)V1368,191-202(1991)

Production quality control problems (Doney, Thomas A.)V1381,9-20(1991)

Proposals for a computer-controlled orbital scanning camera for remote image aquisition (Nwagboso, Christopher O.)V1454,111-122(1991)

Real-time automatic inspection under adverse conditions (Carvalho, Fernando D.; Correia, Fernando C.; Freitas, Jose C.; Rodrigues, F. C.)V1399,130-136(1991)

Real-time classification of wooden boards (Poelzleitner, Wolfgang; Schwingshakl, Gert)V1384,38-49(1991)

Real-time inspection of pavement by moire patterns (Guralnick, Sidney A.; Suen, Eric S.)V1396,664-677(1991)

Real-time quality inspection system for textile industries (Karkanis, S.; Tsoutsou, K.; Vergados, J.; Dimitriadis, Basile D.)V1500,164-170(1991)

Recent development in practical optical nondestructive testing (Hung, Yau Y.)V1554A,29-45(1991); V1554B,29-45(1991)

Remote visual monitoring of seal performance in aircraft jacks using fiber optics (Nwagboso, Christopher O.; Whomes, Terence L.; Davies, P. B.)V1386,30-41(1991)

Research on enhancing signal and SNR in laser/IR inspection of solder joints quality (Xiong, Zhengjun; Cheng, Xuezhong; Liu, Xiande)V1467,410-415(1991)

Rigorous optical theory of the D Sight phenomenon (Reynolds, Rodger L.; Hageniers, Omer L.)V1332,85-96(1991) ·

Spectral signature analysis for industrial inspection (Rauchmiller, Robert F.; Vanderbok, Raymond S.)V1384,100-114(1991)

Study of microbial growth I: by diffraction (Williams, Gareth T.; Bahuguna, Ramendra D.; Arteaga, Humberto; Le Joie, Elaine N.)V1332,802-804(1991)

Study of microbial growth II: by holographic interferomery (Bahuguna, Ramendra D.; Williams, Gareth T.; Pour, I. K.; Raman, R.)V1332,805-807(1991)

Study of oxidization process in real time using speckle correlation (Muramatsu, Mikiya; Guedes, G. H.; Matsuda, Kiyofumi; Barnes, Thomas H.)V1332,792-797(1991)

Submicron linewidth measurement using an interferometric optical profiler (Creath, Katherine)V1464,474-483(1991)

Surface inspection using optical fiber sensor (Abe, Makoto; Ohta, Shigekata; Sawabe, Masaji)V1332,366-376(1991)

Surface roughness effects on light scattered by submicron particles on surfaces (Bawolek, Edward J.; Hirleman, Edwin D.)V1464,574-583(1991)

System calibration and part alignment for inspection of 2-D electronic circuit patterns (Rodriguez, Arturo A.; Mandeville, Jon R.; Wu, Frederick Y.)V1332,25-35(1991)

Thermosense XIII (Baird, George S., ed.)V1467(1991)

Three-dimensional machine vision using line-scan sensors (Godber, Simon X.; Robinson, Max; Evans, Paul)V1526,170-189(1991)

TV holography and image processing in practical use (Lokberg, Ole J.; Ellingsrud, Svein; Vikhagen, Eiolf)V1332,142-150(1991)

Two-dimensional boundary inspection using autoregressive model (Wani, M. A.; Batchelor, Bruce G.)V1384,83-89(1991)

Two stage object identification system in the Delft intelligent assembly cell (Buurman, Johannes; Bierhuizen, David J.)V1386,185-196(1991)

Use of absorption spectroscopy for refined petroleum product discrimination (Short, Michael)V1480,72-79(1991)

Use of high-speed videography to solve a structural vibration problem in overhead cranes (Clayton, Donal L.; Clayton, Richard J.)V1346,33-41(1991)

User interfaces for automated visual inspection systems (Waltz, Frederick M.)VCR36,105-137(1991)

Using computer vision for detecting watercore in apples (Throop, James A.; Rehkugler, Gerald E.; Upchurch, Bruce L.)V1379,124-135(1991)

Van der Lugt optical correlation for the measurement of leak rates of hermetically sealed packages (Fitzpatrick, Colleen M.; Mueller, Edward P.)V1332,185-192(1991)

Vision-based strip inspection hardware for metal production (Seitzler, Thomas M.)V1567,73-76(1991)

Vision methods for inspection of greenhouse poinsettia plants (Meyer, George E.; Troyer, W. W.; Fitzgerald, Jay B.)V1379,99-105(1991)

Vision sensing and processing system for monitoring and control of welding and other high-luminosity processes (Agapakis, John E.; Bolstad, Jon O.)V1385,32-38(1991)

Visual inspection machine for solder joints using tiered illumination (Takagi, Yuuji; Hata, Seiji; Hibi, Susumu; Beutel, Wilhelm)V1386,21-29(1991)

Visual inspection system using multidirectional 3-D imager (Koezuka, Tetsuo; Kakinoki, Yoshikazu; Hashinami, Shinji; Nakashima, Masato)V1332,323-331(1991)

X-ray laminography analysis of ultra-fine-pitch solder connections on ultrathin boards (Adams, John A.)V1464,484-497(1991)

Optical interconnects—see also fiber optic networks; integrated optics; integrated optoelectronics; optical computing; switches; waveguides

Accuracy of CGH encoding schemes for optical data processing (Casasent, David P.; Coetzee, Frans M.; Natarajan, Sanjay S.; Xu, Tianning; Yu, Daming; Liu, Hua-Kuang)V1555,23-33(1991)

Adaptive optical interconnection (Caulfield, H. J.; Schamschula, Marius P.; Verber, Carl M.)V1563,103-111(1991)

Advances in the computer-aided design of planarized free-space optical circuits: system simulation (Jahns, Juergen; Brumback, Babette A.)V1555,2-7(1991)

All-optical interconnection networks (Ghafoor, Arif)V1390,454-466(1991)

Analysis of alignment in optical interconnection systems (Ghosh, Anjan K.; Beech, Russell S.)V1389,630-641(1991)

Applications of electro-optic polymers to optical interconnects (Lytel, Richard S.; Lipscomb, George F.; Kenney, John T.; Ticknor, Anthony J.)V1563,122-138(1991)

Architecture of an integrated computer-aided design system for optoelectronics (Kiamilev, Fouad E.; Fan, J.; Catanzaro, Brian E.; Esener, Sadik C.; Lee, Sing H.)V1390,311-329(1991)

Asymmetric Fabry-Perot modulators for optical signal processing and optical computing applications (Kilcoyne, M. K.; Whitehead, Mark; Coldren, Larry A.)V1389,422-454(1991)

Binary optic interconnects: design, fabrication and limits on implementation (Wong, Vincent V.; Swanson, Gary J.)V1544,123-133(1991)

Capabilities of simple lenses in a free-space perfect shuffle (Miller, Andrew S.; Sawchuk, Alexander A.)V1563,81-92(1991)

Compact PIN-amplifier module for giga bit rates optical interconnection (Suzuki, Tomihiro; Mikamura, Yasuki; Murata, Kazuo; Sekiguchi, Takeshi; Shiga, Nobuo; Murakami, Yasunori)V1389,455-461(1991)

Comparison of optical and electronic 3-dimensional circuits (Stirk, Charles W.; Psaltis, Demetri)V1389,580-593(1991)

Computer-generated holograms fabricated on photopolymer for optical interconnect applications (Oren, Moshe)V1389,527-534(1991)

Current status and future research of the Delft 'supercomputer' project (Frietman, Edward E.; Dekker, L.; van Nifterick, W.; Demeester, Piet M.; van Daele, Peter; Smit, W.)V1390,434-453(1991)

Design and evaluation of optical switching architectures (Ramesh, S. K.; Smith, Thomas D.)V1474,208-211(1991)

Design and fabrication of computer-generated holographic fan-out elements for a matrix/matrix interconnection scheme (Kirk, Andrew G.; Imam, H.; Bird, K.; Hall, Trevor J.)V1574,121-132(1991)

Design of photonic switches for optimizing performances of interconnection networks (Armenise, Mario N.; Castagnolo, Beniamino)V1374,186-197(1991)

Diamond multichip modules (McSheery, Tracy D.)V1563,21-26(1991)

Diffractive optical elements for optoelectronic interconnections (Streibl, Norbert)V1574,34-47(1991)

Dynamically reconfigurable optical interconnect architecture for parallel multiprocessor systems (Girard, Mary M.; Husbands, Charles R.; Antoszewska, Reza)V1563,156-167(1991)

Dynamic holographic interconnects: experimental study using photothermoplastics with improved cycling properties (Gravey, Philippe; Moisan, Jean-Yves)V1507,239-246(1991)

Effects of optoelectronic device characteristics on the performance and design of POEM systems (Esener, Sadik C.; Kiamilev, Fouad E.; Krishnamoorthy, Ashok V.; Marchand, Philippe J.)V1562,11-20(1991)

Electro-optic polymer devices for optical interconnects (Lytel, Richard S.; Binkley, E. S.; Girton, Dexter G.; Kenney, John T.; Lipscomb, George F.; Ticknor, Anthony J.; Van Eck, Timothy E.)V1389,547-558(1991)

Engineering fabrication technology of multiplexed hologram in single-mode optical waveguide (Lin, Freddie S.; Chen, Jenkins C.; Nguyen, Cong T.; Liu, William Y.)V1461,39-50(1991)

Fabrication of computer-generated holograms for the interconnection of single-mode devices (Baettig, Rainer K.; Toms, Dennis J.; Guest, Clark C.)V1563,93-102(1991)

First programmable digital optical processor: optical cellular logic image processor (Craig, Robert G.; Wherrett, Brian S.; Walker, Andrew C.; McKnight, D. J.; Redmond, Ian R.; Snowdon, John F.; Buller, G. S.; Restall, Edward J.; Wilson, R. A.; Wakelin, S.; McArdle, N.; Meredith, P.; Miller, J. M.; Taghizadeh, Mohammad R.; Mackinnon, G.; Smith, S. D.)V1505,76-86(1991)

Free-space board-to-board optical interconnections (Tsang, Dean Z.)V1563,66-71(1991)

Free space cascaded optical logic demonstration (McCormick, Frederick B.; Tooley, Frank A.; Cloonan, Thomas J.; Brubaker, John L.; Lentine, Anthony L.; Morrison, Richard L.; Hinterlong, Steve J.; Herron, Michael J.; Walker, Sonya L.)V1396,508-521(1991)

Free-space optical interconnect using microlasers and modulator arrays (Dickinson, Alex G.; Downs, Maralene M.; LaMarche, R. E.; Prise, Michael E.)V1389,503-514(1991)

Free-space optical TDM switch (Goel, Kamal K.; Prucnal, Paul R.; Stacy, John L.; Krol, Mark F.; Johns, Steven T.)V1476,314-319(1991)

GaAs/GaAlAs integrated optoelectronic transmitter for microwave applications (Yap, Daniel; Narayanan, Authi A.; Rosenbaum, Steven E.; Chou, Chia-Shing; Hooper, W. W.; Quen, R. W.; Walden, Robert H.)V1418,471-476(1991)

Hermaphroditic small tactical connector for single-fiber applications (Darden, Bruce V.; LeFevre, B. G.; Kalomiris, Vasilios E.)V1474,300-308(1991)

High-density waveguide modulator arrays for parallel interconnection (Bristow, Julian P.; Mukherjee, Sayan D.; Khan, M. N.; Hibbs-Brenner, Mary K.; Sullivan, Charles T.; Kalweit, Edith)V1389,535-546(1991)

High-performance GaAs on silicon technology for VLSI, MMICs, and optical interconnects (Christou, Aristos)V1361,354-361(1991)

High-speed board-to-board optical interconnection (Chen, Ray T.; Lu, Huey T.; Robinson, Daniel; Plant, David V.; Fetterman, Harold R.)V1559,110-117(1991)

High-speed optical interconnects for parallel processing and neural networks (Barnes, Nigel; Healey, Peter; McKee, Paul; O'Neill, Alan; Rejman-Greene, Marek A.; Scott, Geoff; Smith, David W.; Webb, Roderick P.; Wood, David)V1389,477-483(1991)

Holographic optical backplane for boards interconnection (Sebillotte-Caron, Claudine)V1389,600-611(1991)

Holographic optical backplane hardware implementation for parallel and distributed processors (Kim, Richard C.; Lin, Freddie S.)V1474,27-38(1991)

Holographic optical elements for free-space clock distribution (Zarschizky, Helmut; Karstensen, Holger; Gerndt, Christian; Klement, Ekkehard; Schneider, Hartmut W.)V1389,484-495(1991)

Holographic optical interconnects for multichip modules (Feldman, Michael R.)V1390,427-433(1991)

Holographic optical switching with photorefractive crystals (Song, Q. W.; Lee, Mowchen C.; Talbot, Peter J.)V1558,143-148(1991)

Improvements in self electro-optic effect devices: toward system implementation (Morgan, Robert A.)V1562,213-227(1991)

Integrated circuits with three-dimensional optical interconnections: an element base of neural networks (Gulyaev, Yuri V.; Elinson, Matvey I.; Kopylov, Yuri L.; Perov, Polievkt I.)V1621,84-92(1991)

Integrated compact matrix-vector multipliers for optical interconnect applications (Lin, Freddie S.; Strzelecki, Eva M.; Liu, William Y.)V1389,642-647(1991)

Integrated micro-optics for computing and switching applications (Jahns, Juergen)V1544,246-262(1991)

Integrated optic components for advanced turbine engine control systems (Emo, Stephen M.; Kinney, Terrance R.; Wong, Ka-Kha)V1374,266-276(1991)

Integrated-optic interconnects and fiber-optic WDM data links based on volume holography (Jannson, Tomasz; Lin, Freddie S.; Moslehi, Behzad M.; Shirk, Kevin W.)V1555,159-176(1991)

Integrated Optics and Optoelectronics II (Wong, Ka-Kha, ed.)V1374(1991)

Integrated packaging of optical backplanes (Jahns, Juergen)V1389,523-526(1991)

Integration of an edge extraction cellular automaton (Taboury, Jean; Chavel, Pierre)V1505,115-123(1991)

Intercircuit optical interconnects using quantum-well modulators (Goodfellow, Robert C.; Goodwin, Martin J.; Moseley, Andrew J.)V1389,594-599(1991)

Interconnection requirements in avionic systems (Vergnolle, Claude; Houssay, Bruno)V1389,648-658(1991)

Massive connectivity and SEEDs (Chirovsky, Leo M.)V1562,228-241(1991)

Massive optical interconnections for computer applications (Neff, John A.)V1389,27-38(1991); V1390,27-38(1991)

Metal vapor gain media based on multiphoton dissociation of organometallics (Samoriski, Brian; Wiedeger, S.; Villarica, M.; Chaiken, Joseph)V1412,12-18(1991)

Methods for comparative analysis of waveform degradation in electrical and optical high-performance interconnections (Merkelo, Henri; McCredie, B. D.; Veatch, M. S.; Quinn, D. L.; Dorneich, M.; Doi, Yutaka)V1390,91-163(1991)

Methylene-blue-sensitized gelatin used for the fabrication of holographic optical elements in the near-infrared (Cappolla, Nadia; Lessard, Roger A.)V1389,612-620(1991)

Microelectronic Interconnects and Packages: Optical and Electrical Technologies (Arjavalingam, Gnanalingam; Pazaris, James, eds.)V1389(1991)

Microelectronic Interconnects and Packages: System and Process Integration (Carruthers, John R.; Tewksbury, Stuart K., eds.)V1390(1991)

Microreflective elements for integrated planar optics interconnects (Sheng, Yunlong; Delisle, Claude; Moreau, Louis; Song, Li; Lessard, Roger A.; Arsenault, Henri H.)V1559,222-228(1991)

Molecular beam epitaxy GaAs on Si: material and devices for optical interconnects (Panayotatos, Paul; Georgakilas, Alexandros; Mourrain, Jean-Loic; Christou, Aristos)V1361,1100-1109(1991)

Motivation for DOC III: 64-bit digital optical computer (Guilfoyle, Peter S.)V1505,2-10(1991)

Multi-Gb/s optical computer interconnect (Sauer, Jon R.)V1579,49-61(1991)

Multimode stripe waveguides for optical interconnections (Maile, Michael; Weidel, Edgar)V1563,188-196(1991)

Multiprocessor optical bus (Kostuk, Raymond K.; Huang, Yang-Tung; Kato, Masayuki)V1389,515-522(1991)

Neural networks and computers based on in-phase optics (Kovatchev, Methodi; Ilieva, R.)V1621,259-267(1991)

New method of recording holographic optical elements applied to optical interconnection in VLSI (Zhao, Feng; Sun, Junyong; Geng, Wanzhen; Jiang, Lingzhen; Hong, Jing)V1461,262-264(1991)

New way of optical interconnection for VLSI (Zhao, Feng; Geng, Wanzhen; Jiang, Lingzhen; Hong, Jing)V1555,241-242(1991)

Novel acousto-optic photonic switch (Wu, Kuang-Yi; Weverka, Robert T.; Wagner, Kelvin H.; Garvin, Charles G.; Roth, Richard S.)V1563,168-175(1991)

Optical approaches to overcome present limitations for interconnection and control in parallel electronic architectures (Maurin, T.; Devos, F.)V1505,158-165(1991)

Optical Enhancements to Computing Technology (Neff, John A., ed.)V1563(1991)

Optical implementation of a Hopfield-type neural network by the use of persistent spectral hole-burning media (Ollikainen, Olavi; Rebane, A.)V1621,351-361(1991)

Optical interconnect and neural network applications based on a simplified holographic N4 recording technique (Lin, Freddie S.; Lu, Taiwei; Kostrzewski, Andrew A.; Chou, Hung)V1558,406-413(1991)

Optical interconnections based on waveguide holograms (Putilin, Andrei N.)V1621,93-101(1991)

Optical interconnects: a solution to very high speed integrated circuits and systems (Chen, Ray T.)V1374,162-175(1991)

Optical interconnects for parallel processing (Guha, Aloke; Bristow, Julian P.; Sullivan, Charles T.; Husain, Anis)V1389,375-385(1991)

Optical interconnects in high-bandwidth computing (Dekker, L.; Frietman, Edward E.)V1505,148-157(1991)

Optically coupled 3-D common memory with GaAs on Si structure (Hirose, Masataka; Takata, H.; Koyanagi, Mitsumasa)V1362,316-322(1991)

Optically powered optoelectronic integrated circuits for optical interconnects (Brown, Julia J.; Gardner, J. T.; Forrest, Stephen R.)V1474,236-242(1991)

Optical neural network with reconfigurable holographic interconnection (Kirk, Andrew G.; Kendall, G. D.; Imam, H.; Hall, Trevor J.)V1621,320-327(1991)

Optical receivers in ECL for 1 GHz parallel links (Wieland, Joerg B.; Melchior, Hans M.)V1389,659-664(1991)

Optical switches based on semiconductor optical amplifiers (Kalman, Robert F.; Dias, Antonio R.; Chau, Kelvin K.; Goodman, Joseph W.)V1563,34-44(1991)

Optical Technology for Signal Processing Systems (Bendett, Mark P., ed.)V1474(1991)

Optics for Computers: Architectures and Technologies (Lebreton, Guy J., ed.)V1505(1991)

Optoelectronic compare-and-exchange switches based on BILED circuits (Mao, Xianjun; Liu, Shutian; Wang, Ruibo)V1563,58-63(1991)

Optomechanics of a free-space photonic switch: the components (Brubaker, John L.; McCormick, Frederick B.; Tooley, Frank A.; Sasian, Jose M.; Cloonan, Thomas J.; Lentine, Anthony L.; Hinterlong, Steve J.; Herron, Michael J.)V1533,88-96(1991)

Optomechanics of a free-space photonic switching fabric: the system (McCormick, Frederick B.; Tooley, Frank A.; Brubaker, John L.; Sasian, Jose M.; Cloonan, Thomas J.; Lentine, Anthony L.; Morrison, Richard L.; Crisci, R. J.; Walker, Sonya L.; Hinterlong, Steve J.; Herron, Michael J.)V1533,97-114(1991)

Overview of optical interconnect technology (Chiu, George; Oprysko, Modest M.)V1389,364-374(1991)

Packaging issues for free-space interconnects at the board level (Kostuk, Raymond K.)V1563,72-80(1991)

Packaging issues for free-space optically interconnected multiprocessors (Ozguz, Volkan H.; Esener, Sadik C.; Lee, Sing H.)V1390,477-488(1991)

Parallel optical interconnects utilizing VLSI/FLC spatial light modulators (Genco, Sheryl M.)V1563,45-57(1991)

Photochemical formation of polymeric optical waveguides and devices for optical interconnection applications (Beeson, Karl W.; Horn, Keith A.; McFarland, Michael J.; Wu, Chengjiu; Yardley, James T.)V1374,176-185(1991)

Photoneuron: dynamically reconfigurable information processing control element utilizing embedded-fiber waveguide interconnects (Glista, Andrew S.)V1563,139-155(1991)

Photonic Multichip Packaging (PMP) using electro-optic organic materials and devices (McDonald, John F.; Vlannes, Nickolas P.; Lu, Toh-Ming; Wnek, Gary E.; Nason, Theodore C.; You, Lu)V1390,286-301(1991)

Photonic switching implementations of 2-input, 2-output switching nodes based on 2-D and 3-D crossover networks (Cloonan, Thomas J.; McCormick, Frederick B.)V1396,488-500(1991)

Pipelined communications on optical busses (Guo, Zicheng; Melhem, Rami G.; Hall, Richard W.; Chiarulli, Donald M.; Levitan, Steven P.)V1390,415-426(1991)

Potential roles of optical interconnections within broadband switching modules (Lalk, Gail R.; Habiby, Sarry F.; Hartman, Davis H.; Krchnavek, Robert R.; Wilson, Donald K.; Young, Kenneth C.)V1389,386-400(1991)

Power economy using point-to-point optical interconnect links (Hartman, Davis H.; Reith, Leslie A.; Habiby, Sarry F.; Lalk, Gail R.; Booth, Bruce L.; Marchegiano, Joseph E.; Hohman, James L.)V1390,368-376(1991)

Progress and perspectives on optical information processing in Japan (Ishihara, Satoshi)V1389,68-74(1991); V1390,68-74(1991)

Propagating beam method simulation of planar multimode optical waveguides (McMullin, James N.)V1389,621-629(1991)

Proposed electro-optic package with bidirectional lensed coupling (Rajasekharan, K.; Michalka, Timothy)V1389,568-579(1991)

Quantization and sampling considerations of computer-generated hologram for optical interconnection (Long, Pin; Hsu, Dahsiung)V1461,270-277(1991)

Restrictions on the geometry of the reference wave in holocoupler devices (Calvo, Maria L.; De Pedraza-Velasco, L.)V1507,288-301(1991)

Rigorous diffraction theory of binary optical interconnects (Vasara, Antti H.; Noponen, Eero; Turunen, Jari P.; Miller, J. M.; Taghizadeh, Mohammad R.; Tuovinen, Jussi)V1507,224-238(1991)

Self-routing interconnection structures using coincident pulse techniques (Chiarulli, Donald M.; Levitan, Steven P.; Melhem, Rami G.)V1390,403-414(1991)

Substrate-mode holograms for board-to-board two-way communications (Huang, Yang-Tung; Kostuk, Raymond K.)V1474,39-44(1991)

Time-division optical microarea networks (Prucnal, Paul R.; Johns, Steven T.; Krol, Mark F.; Stacy, John L.)V1389,462-476(1991)

Two Gbit/s photonic backplane for telephone cards interconnection (Donati, Silvano; Martini, Giuseppe; Francese, Francesco)V1389,665-671(1991)

Unique advantages of optics over electronics for interconnections (Caulfield, H. J.)V1390,399-402(1991)

Vertical 3-D integration of silicon waveguides in a Si-SiO2-Si-SiO2-Si structure (Soref, Richard A.; Namavar, Fereydoon; Cortesi, Elisabetta; Friedman, Lionel R.; Lareau, Richard)V1389,408-421(1991)

Vertical cavity lasers for optical interconnects (Jewell, Jack L.; Lee, Yong H.; McCall, Samuel L.; Scherer, Axel; Harbison, James P.; Florez, Leigh T.; Olsson, N. A.; Tucker, Rodney S.; Burrus, Charles A.; Sandroff, Claude J.)V1389,401-407(1991)

Vertical-cavity surface-emitting lasers and arrays for optical interconnect and optical computing applications (Wang, S. C.; Dziura, T. G.; Yang, Y. J.)V1563,27-33(1991)

Wavelength-sensitive holographic optical interconnects (Lin, Freddie S.; Zaleta, David E.; Jannson, Tomasz)V1474,45-56(1991)

Optical materials—see also coatings; crystals; diamonds; gallium arsenide materials; glass; optical properties; photoconductors; photodetectors; photoresists; polymers; semiconductors; silicon; sol-gels; thin films

1990 Intl Lens Design Conf (Lawrence, George N., ed.)V1354(1991)

Advanced infrared optically black baffle materials (Seals, Roland D.; Egert, Charles M.; Allred, David D.)V1330,164-177(1991)

Advantages of drawing crystal-core fibers in microgravity (Shlichta, Paul J.; Nerad, Bruce A.)V1557,10-23(1991)

Analysis of the Focar-type silver-halide heterogeneous media (Andreeva, Olga V.)V1238,231-234(1991)

Analysis on image performance of a moisture-absorbed plastic singlet for an optical disk (Tanaka, Yuki; Miyamae, Hiroshi)V1354,395-401(1991)

Application of high-rate crystal growth technique to single crystals of nucleic acid bases (Zachova, J.; Shtepanek, J.; Zaitseva, N. P.)V1402,216-222(1991)

Athermalized FLIR optics (Rogers, Philip J.)V1354,742-751(1991)

Beryllium and titanium cost-adjustment report (Owen, John; Ulph, Eric)V1485,128-137(1991)

Beryllium galvanometer mirrors (Weissman, Harold)V1485,13-19(1991)

Bidirectional transmittance distribution function measurements on ZnSe and on ZnS Cleartran (Melozzi, Mauro; Mazzoni, Alessandro; Curti, G.)V1512,178-188(1991)

Bonding and nonequilibrium crystallization of a-C:H/a-Se and a-C:H/KCl (He, Da-Ren; Ji, Xiuyan; Wang, Ruo Bao; Liu, Qihai; Wang, Wangdi; Liu, Maili; Chen, Weizong; Liu, Zhiyuan; Ji, Wanxi; Zhang, Renji)V1362,696-701(1991)

Bonding skill between the fused-quartz mirror and the composite substrate (Jiang, Shibin; Wang, Huirong; Jiang, Yasi; Wang, Biao; Chen, Menda)V1535,143-147(1991)

Catalog of infrared and cryo-optical properties for selected materials (Heaney, James B.; Alley, Phillip W.; Bradley, Scott E.)V1485,140-159(1991)

Characteristic curves and phase-exposition characteristics of holographic photomaterials (Churaev, A. L.; Artyomova, V. V.)V1238,158-165(1991)

Characterization of proton-exchanged and annealed proton-exchanged optical waveguides in z-cut LiNbO3 (Nikolopoulos, John; Yip, Gar L.)V1374,30-36(1991)

Characterization of the photorefractive effect in Ti:LiNbO3 stripe waveguides (Volk, Raimund; Sohler, Wolfgang)V1362,820-826(1991)

Chemical-vapor-deposited silicon and silicon carbide optical substrates for severe environments (Goela, Jitendra S.; Pickering, Michael A.; Taylor, Raymond L.)V1330,25-38(1991)

Color reflection holograms with photopolymer plates (Zhang, Jingfang; Ma, Chunrong; Lang, Hengyuan)V1238,306-310(1991)

Commercial crystal growth in space (Wilcox, William R.)V1557,31-41(1991)

Compact spaceflight solution crystal-growth system (Trolinger, James D.; Lal, Ravindra B.; Vikram, Chandra S.; Witherow, William K.)V1557,250-258(1991)

Comparison of laser-induced damage of optical crystals from the USA and USSR (Soileau, M. J.; Wei, Tai-Huei; Said, Ali A.; Chapliev, N. I.; Garnov, Sergei V.; Epifanov, Alexandre S.)V1441,10-15(1991)

Correction of secondary and higher-order spectrum using special materials (Mercado, Romeo I.)V1535,184-198(1991)

Correlation between the laser-induced breakdown threshold in solids, liquids, and gases (Bettis, Jerry R.)V1441,521-534(1991)

Coupled-mode analysis of dynamic polarization volume holograms (Huang, Tizhi; Wagner, Kelvin H.)V1559,377-384(1991)

Creation of aspheric beryllium optical surfaces directly in the hot isostatic pressing consolidation process (Gildner, Donald; Marder, James M.)V1485,46-53(1991)

Crystal growth by solute diffusion in Earth orbit (Lind, M. D.; Nielsen, K. F.)V1557,259-270(1991)

Crystal Growth in Space and Related Optical Diagnostics (Trolinger, James D.; Lal, Ravindra B., eds.)V1557(1991)

Damage to InAs surface from long-pulse 10.6um laser radiation (Kovalev, Valeri I.)V1441,536-540(1991)

Dark stability of holograms and holographic investigation of slow diffusion in polymers (Veniaminov, Andrei V.)V1238,266-270(1991)

Design and development of a transparent Bridgman furnace (Wells, Mark E.; Groff, Mary B.)V1557,71-77(1991)

Design and development of the Zeolite Crystal Growth Facility (Fiske, Michael R.)V1557,78-85(1991)

Design and specification of diamond-turned optics (Clark, Robert A.)VCR38,164-183(1991)

Designing apochromatic telescope objectives with liquid lenses (Sigler, Robert D.)V1535,89-112(1991)

Design of an IR non-lens, or how I buried 100 mm of germanium (Aikens, David M.)V1485,183-194(1991)

Design of chromophores for nonlinear optical applications (Burland, Donald M.; Rice, J. E.; Downing, J.; Michl, J.)V1560,111-119(1991)

Design of high-performance aplanatic achromats for the near-ultraviolet waveband (Al-Baho, Tareq I.; Learner, R. C.; Maxwell, Jonathan)V1354,417-428(1991)

Design of lightweight beryllium optics, factors effecting producibility, and cost of near-net-shape blanks (Clement, Thomas P.)V1485,31-38(1991)

Development of beryllium-mirror turning technology (Arnold, Jones B.)V1485,96-105(1991)

Diffusion and solubility of ion-implanted Nd and Er in LiNbO3 (Buchal, Christoph; Mohr, S.)V1361,881-892(1991)

Dispersion of n2 in solids (Sheik-Bahae, Mansoor; Hutchings, David C.; Hagan, David J.; Soileau, M. J.; Van Stryland, Eric W.)V1441,430-443(1991)

Display holograms in Du Pont's OmniDex films (Zager, Stephen A.; Weber, Andrew M.)V1461,58-67(1991)

Dynamic holographic interconnects: experimental study using photothermoplastics with improved cycling properties (Gravey, Philippe; Moisan, Jean-Yves)V1507,239-246(1991)

Effective transmission holograms produced on CK-type photoresist (Bulygin, A. R.)V1238,248-252(1991)

Effect of ionizing radiations and thermal treatment on the infrared transmittance of polycrystal CsI (Miley, George H.; Barnouin, O.; Procoli, Alfredo)V1441,16-26(1991)

Effect of the space environment on thermal control coatings (Harada, Yoshiro; Mell, Richard J.; Wilkes, Donald R.)V1330,90-101(1991)

Electro-optic measurements of dye/polymer systems (Wang, Chin H.; Guan, H. W.; Zhang, J. F.)V1559,39-48(1991)

Environments stressful to optical materials in low earth orbit (Musikant, Solomon; Malloy, W. J.)V1330,119-130(1991)

Excimer laser optics (Morozov, N. V.; Sagitov, S. I.; Sergeyev, P. B.)V1441,557-564(1991)

Excited state cross sections for Er-doped glasses (Zemon, Stanley A.; Lambert, Gary M.; Miniscalco, William J.; Davies, Richard W.; Hall, Bruce T.; Folweiler, Robert C.; Wei, Ta-Sheng; Andrews, Leonard J.; Singh, Mahendra P.)V1373,21-32(1991)

Experimental and theoretical studies of second-phase scattering in IR transmitting ZnS-based windows (Chen, William W.; Dunn, Bruce S.; Zhang, Jimin)V1535,199-208(1991)

Fabrication and characterization of microwave-plasma-assisted chemical vapor deposited dielectric coatings (Wood, Roger M.; Greenham, A. C.; Nichols, B. A.; Nourshargh, Noorallah; Lewis, Keith L.)V1441,316-326(1991)

Fabrication and properties of chalcogenide IR diffractive elements (Ewen, Peter J.; Slinger, Christopher W.; Zakery, A.; Zekak, A.; Owen, A. E.)V1512,101-111(1991)

Fabrication of a fast, aspheric beryllium mirror (Paquin, Roger A.; Gardopee, George J.)V1485,39-45(1991)

Fabrication of rare-earth-doped optical fiber (DiGiovanni, David J.)V1373,2-8(1991)

Ferroelectric microdomain reversal on Y-cut LiNbO3 surfaces (Seibert, Holger; Sohler, Wolfgang)V1362,370-376(1991)

Fiber Laser Sources and Amplifiers II (Digonnet, Michel J., ed.)V1373(1991)

Finite element analysis of the transient behavior of optical components under irradiation (Borik, Stefan; Giesen, Adolf)V1441,420-429(1991)

Frequency doubling, absorption, and grating formation in glass fibers: effective defects or defective effects? (Russell, Philip S.; Poyntz-Wright, L. J.; Hand, Duncan P.)V1373,126-139(1991)

Frequency doubling and optical parametric oscillation with potassium niobate (Polzik, Eugene S.; Kimble, H. J.)V1561,143-146(1991)

Gate capacitance of MOS field effect devices of nonlinear optical materials in the presence of a parallel magnetic field (Ghatak, Kamakhya P.; Biswas, Shambhu N.; Banik, S. N.)V1409,240-257(1991)

Ground-based experiments on the growth and characterization of L-arginine phosphate crystals (Rao, S. M.; Cao, C.; Batra, Ashok K.; Lal, Ravindra B.; Mookherji, Tripty K.)V1557,283-292(1991)

Growth and Characterization of Materials for Infrared Detectors and Nonlinear Optical Switches (Longshore, Randolph E.; Baars, Jan W., eds.)V1484(1991)

Growth of thin films of organic nonlinear optical materials by vapor growth processes: an overview and examination of shortfalls (Frazier, Donald O.; Penn, Benjamin G.; Witherow, William K.; Paley, M. S.)V1557,86-97(1991)

Growth, surface passivation, and characterization of CdSe microcrystallites in glass with respect to their application in nonlinear optics (Woggon, Ulrike; Rueckmann, I.; Kornack, J.; Mueller, Matthias; Cesnulevicius, J.; Kolenda, Jonas; Petrauskas, Mendogas)V1362,888-898(1991)

Heat capacity of MOS field-effect devices of optical materials in the presence of a strong magnetic field (Ghatak, Kamakhya P.; Biswas, Shambhu N.)V1485,206-214(1991)

Heterogeneous recording media (Sukhanov, Vitaly I.)V1238,226-230(1991)

High-efficiency reflective holograms: subangstrom spectral selectors (Goncharov, V. F.; Popov, Alexander P.; Veniaminov, Andrei V.)V1238,97-102(1991)

High-emittance surfaces for high-temperature space radiator applications (Banks, Bruce A.; Rutledge, Sharon K.; Hotes, Deborah)V1330,66-77(1991)

Highly sensitive absorption measurements in organic thin films and optical media (Skumanich, Andrew)V1559,267-277(1991)

High-resolution photographic material of Institute of Atomic Energy for holography in intersecting beams (Ryabova, R. V.; Barinova, E. S.; Myatezh, O. V.; Zaborov, A. N.; Romash, E. V.; Chernyi, D. I.)V1238,166-170(1991)

High-sensitive layers of dichromated gelatin for hologram recording by continuous wave and pulsed laser radiation (Kalyashora, L. N.; Michailova, A. G.; Pavlov, A. P.; Rykov, V. S.; Guba, B. S.)V1238,189-194(1991)

Holographic characteristics of IAE and PFG-01 photoplates for colored pulsed holography (Vorzobova, N. D.; Rjabova, R. V.; Schvarzvald, A. I.)V1238,476-477(1991)

Holographic optical elements by dry photopolymer embossing (Shvartsman, Felix P.)V1461,313-320(1991)

Holographic recording in photorefractive media containing bacteriorhodopsin (Kozhevnikov, Nikolai M.; Barmenkov, Yuri O.; Lipovskaya, Margarita J.)V1507,517-524(1991)

Holographic recordings on 2-hydroxyethyl methacrylate and applications of water-immersed holograms (Yacoubian, Araz; Savant, Gajendra D.; Aye, Tin M.)V1559,403-409(1991)

Homogeneous thin film lens on LiNbO3 (Jiang, Pisu; Laybourn, Peter J.; Righini, Giancarlo C.)V1362,899-906(1991)

Importance of dispersion tolerances in infrared lens design (Korniski, Ronald J.; Thompson, Kevin P.)V1354,402-407(1991)

Influence of photon energy on the photoemission from ultrafast electronic materials (Ghatak, Kamakhya P.; Ghoshal, Ardhendu; Bhattacharyya, Sankar; Mondal, Manabendra)V1346,471-489(1991)

Infrared Optical Design and Fabrication (Hartmann, Rudolf; Smith, Warren J., eds.)VCR38(1991)

Infrared refractive-index measurement results for single-crystal and polycrystal germanium (Hilton, Albert R.)V1498,128-137(1991)

Infrared Technology XVII (Andresen, Bjorn F.; Scholl, Marija S.; Spiro, Irving J., eds.)V1540(1991)

Inhomogeneous line broadening of optical transitions in Nd3+ and Er3+ doped preforms and fibers (Briancon, Anne-Marie; Jacquier, Bernard; Gacon, Jean-Claude; Le Sergent, Christian; Marcerou, Jean-Francois)V1373,9-20(1991)

Inorganic Crystals for Optics, Electro-Optics, and Frequency Conversion (Bordui, Peter F., ed.)V1561(1991)

In-situ observation of crystal growth in microgravity by high-resolution microscopies (Tsukamoto, Katsuo; Onuma, Kazuo)V1557,112-123(1991)

Interferometric technique for the concurrent determination of thermo-optic and thermal expansion coefficients (Jewell, John M.; Askins, Charles G.; Aggarwal, Ishwar D.)V1441,38-44(1991)

Investigation of process sensitivity for electron-beam evaporation of beryllium (Egert, Charles M.; Schmoyer, D. D.; Nordin, C. W.; Berry, A.)V1485,64-77(1991)

Investigation of the arsenic sulphide films for relief-phase holograms (Yusupov, I. Y.; Mikhailov, M. D.; Herke, R. R.; Goray, L. I.; Mamedov, S. B.; Yakovuk, O. A.)V1238,240-247(1991)

Ionization potentials effects on CGL/CTL photoconductors (Shi, Xiao D.)V1519,884-889(1991)

Kinetics of formation of holographic structure of a hologram mirror in dichromated gelatin (Kzuzhilin, Yu E.; Mel'nichenko, Yu. B.; Shilov, V. V.)V1238,200-205(1991)

Large active mirror in aluminium (Leblanc, Jean-M.; Rozelot, Jean-Pierre)V1535,122-129(1991)

Large zeolites: why and how to grow in space (Sacco, Albert)V1557,6-9(1991)

Laser-Induced Damage in Optical Materials: 1990 (Bennett, Harold E.; Chase, Lloyd L.; Guenther, Arthur H.; Newman, Brian; Soileau, M. J., eds.)V1441(1991)

Latest developments in crystal growth and characterization of efficient acousto-optic materials (Paradies, C. J.; Glicksman, M. E.; Jones, O. C.; Kim, G. T.; Lin, Jen T.; Gottlieb, Milton S.; Singh, N. B.)V1561,2-5(1991)

Lessons learned in recent beryllium-mirror fabrication (Wells, James A.; Lombard, Calvin M.; Sloan, George B.; Moore, Wally W.; Martin, Claude E.)V1485,2-12(1991)

Light-absorbing, lightweight beryllium baffle materials (Murray, Brian W.; Floyd, Dennis R.; Ulph, Eric)V1485,88-95(1991)

Light-sensitive organic media for optical discs (Barachevsky, Valery A.; Rot, A. S.; Zaks, I. N.)V1621,33-44(1991)

Lightweight SXA metal matrix composite collimator (Johnson, R. B.; Ahmad, Anees; Hadaway, James B.; Geiger, Alan L.)V1535,136-142(1991)

Linear and nonlinear optical properties of substituted pyrrolo[1,2-a]quinolines (van Hulst, Niek F.; Heesink, Gerard J.; Bolger, Bouwe; Kelderman, E.; Verboom, W.; Reinhoudt, D. N.)V1361,581-588(1991)

Long-wave infrared detectors based on III-V materials (Maserjian, Joseph L.)V1540,127-134(1991)

Materials technology for SIRTF (Coulter, Daniel R.; Dolgin, Benjamin P.; Rainen, R.; O'Donnell, Timothy P.)V1540,119-126(1991)

Metal mirror review (Janeczko, Donald J.)VCR38,258-280(1991)

Methylene-blue-sensitized gelatin used for the fabrication of holographic optical elements in the near-infrared (Cappolla, Nadia; Lessard, Roger A.)V1389,612-620(1991)

Modifications of optical properties with ceramic coatings (Besmann, Theodore M.; Abdel-Latif, A. I.)V1330,78-89(1991)

Moisture- and water-induced crack growth in optical materials (Cranmer, David C.; Freiman, Stephen W.; White, Grady S.; Raynes, Alan S.)V1330,152-163(1991)

Molecular to material design for anomalous-dispersion phase-matched second-harmonic generation (Cahill, Paul A.; Tallant, David R.; Kowalczyk, T. C.; Singer, Kenneth D.)V1560,130-138(1991)

Monte Carlo calculations of laser-induced free-electron heating in SiO2 (Arnold, Douglas; Cartier, E.; Fischetti, Massimo V.)V1441,478-487(1991)

Multilayered optically activated devices based on organic third-order nonlinear optical materials (Swanson, David A.; Altman, Joe C.; Sullivan, Brian J.; Spitzer, Ronnie; Hansen, Glenn A.; Stone, Richard E.)V1560,416-425(1991)

NASA microgravity materials science program (Sokolowski, Robert S.)V1557,2-5(1991)

New high-performance material in nonlinear optics field, the polymer blend PMMA-EVA: a first investigation (Carbonara, Giuseppe; Mormile, Pasquale; Abbate, G.; Bernini, Umberto; Maddalena, P.; Malinconico, Mario)V1361,688-691(1991)

New organic crystal material for SHG, 2-cyano-3-(3,4-methylene dioxy phenyl)-2-propionic acid ethyl ester (Mori, Yasushi; Sano, Kenji; Todori, Kenji; Kawamonzen, Yosiaki)V1560,310-314(1991)

New polymers with large and stable second-order nonlinear optical effects (Chen, Mai; Yu, Luping; Dalton, Larry R.; Shi, Youngqiang; Steier, William H.)V1409,202-213(1991)

New thermistor material for thermistor bolometer: material preparation and characterization (Umadevi, P.; Nagendra, C. L.; Thutupalli, G. K.; Mahadevan, K.; Yadgiri, G.)V1485,195-205(1991)

Nonlinear absorbance effects in bacteriorhodopsin (Rayfield, George W.)V1436,150-159(1991)

Nonlinear optical properties of composite materials (Haus, Joseph W.; Inguva, Ramarao)V1497,350-356(1991)

Nonlinear optical properties of inorganic coordination polymers and organometallic complexes (Thompson, Mark E.; Chiang, William; Myers, Lori K.; Langhoff, Charles A.)V1497,423-429(1991)

Nonlinear optical properties of new KTiOPO4 isostructures (Phillips, Mark L.; Harrison, William T.; Stucky, Galen D.)V1561,84-92(1991)

Nonlinear Optical Properties of Organic Materials IV (Singer, Kenneth D., ed.)V1560(1991)

Nonlinear optical properties of xanthone derivatives (Imanishi, Yasuo; Itoh, Yuzo; Kakuta, Atsushi; Mukoh, Akio)V1361,570-580(1991)

Nonlinear Optics and Materials (Cantrell, Cyrus D.; Bowden, Charles M., eds.)V1497(1991)

Novel perfluorinated antireflective and protective coating for KDP and other optical materials (Thomas, Ian M.; Campbell, John H.)V1441,294-303(1991)

O/I-MBE: formation of highly ordered phthalocyanine/semiconductor junctions by molecular-beam epitaxy: photoelectrochemical characterization (Armstrong, Neal R.; Nebesny, Ken W.; Collins, Greg E.; Lee, Paul A.; Chau, Lai K.; Arbour, Claude; Parkinson, Bruce)V1559,18-26(1991)

Optical characteristics of the Teflon AF fluoro-plastic materials (Lowry, Jay H.; Mendlowitz, Joseph S.; Subramanian, N. S.)V1330,142-151(1991)

Optical information recording on vitreous semiconductors with a thermoplastic method of visualization (Panasyuk, L. M.; Forsh, A. A.)V1621,74-82(1991)

Optical materials: evaluation methodology and data base utility (Patty, Charles E.; McMahon, David M.)V1535,13-26(1991)

Optical materials for infrared range of spectrum (Petrovskii, Gurii T.)V1540,401-411(1991)

Optical materials for the infrared (Wolfe, William L.)VCR38,55-68(1991)

Optical materials for use under extreme service conditions (Glebov, Leonid B.; Petrovskii, Gurii T.; Tshavelev, Oleg S.)V1399,200-206(1991)

Optical Materials Technology for Energy Efficiency and Solar Energy Conversion X (Lampert, Carl M.; Granqvist, Claes G., eds.)V1536(1991)

Optical properties of GaAs, GaP, and CVD diamond (Klocek, Paul; Hoggins, James T.; McKenna, T. A.; Trombetta, John M.; Boucher, Maurice W.)V1498,147-157(1991)

Optical properties of KDP crystals grown at high growth rates (Zaitseva, N. P.; Ganikhanov, Ferous S.; Katchalov, O. V.; Efimkov, V. F.; Pastukhov, S. A.; Sobolev, V. B.)V1402,223-230(1991)

Optical recording material based on bacteriorhodopsin modified with hydroxylamine (Vsevolodov, N. N.; Dyukova, T. V.; Druzhko, A. B.; Shakhbazyan, V. Y.)V1621,11-20(1991)

Optical stability of diffuse reflectance materials in space (Hale, Robert R.)V1485,173-182(1991)

Optical Surfaces Resistant to Severe Environments (Musikant, Solomon, ed.)V1330(1991)

Optimization of optical properties of pigmented foils for radiative cooling applications: model calculations (Nilsson, Torbjorn M.; Niklasson, Gunnar A.)V1536,169-182(1991)

Optimization of quadrilayer structures for various magneto-optical recording materials (He, Ping; McGahan, William A.; Woollam, John A.)V1499,401-411(1991)

Optoelectronic Materials and Device Concepts (Razeghi, Manijeh)VPM05(1991)

Overpressure proof testing of large infrared windows for aircraft applications (Pruszynski, Charles J.)V1498,163-170(1991)

Parameters of recording media for 3-D holograms (Mikhailov, I. A.)V1238,140-143(1991)

Particle image velocimetry experiments for the IML-I spaceflight (Trolinger, James D.; Lal, Ravindra B.; Batra, Ashok K.; McIntosh, D.)V1557,98-109(1991)

Passive Materials for Optical Elements (Wilkerson, Gary W., ed.)V1535(1991)

Pattern recognition in autoradiography of optical materials (Ashurov, Mukhsin K.; Gafitullina, Dilyara)V1550,50-54(1991)

Performance and spectroscopic characterization of irradiated Nd:YAG (Rose, Todd S.; Fincher, Curtis L.; Fields, Renny A.)V1561,43-49(1991)

PFG-04 photographic plates based on the nonhardened dichromated gelatin for recordng color reflection holograms (Artemjev, S. V.; Koval, G. I.; Obyknovennaja, I. E.; Cherkasov, A. S.; Shevtsov, M. K.)V1238,206-210(1991)

Photoanisotropic polymeric media and their application in optical devices (Barachevsky, Valery A.)V1559,184-193(1991)

Photochemistry and fluorescence spectroscopy of polymeric materials containing triphenylsulfonium salts (Hacker, Nigel P.; Welsh, Kevin M.)V1559,139-150(1991)

Photochemistry and Photoelectrochemistry of Organic and Inorganic Molecular Thin Films (Frank, Arthur J.; Lawrence, Marcus F.; Ramasesha, S.; Wamser, Carl C., eds.)V1436(1991)

Photoconductivity of high-bandgap materials (Ho, Ping-Tong; Peng, F.; Goldhar, J.; Nolting, Eugene E.; Parsons, C.)V1378,210-216(1991)

Photodegradation of a laser dye in a silica gel matrix (Glab, Wallace L.; Bistransin, Mark; Borst, Walter L.)V1497,389-395(1991)

Photoemission from quantum-confined structure of nonlinear optical materials (Ghatak, Kamakhya P.; Ghoshal, Ardhendhu; Biswas, Shambhu N.)V1409,28-57(1991)

Photopolymer Device Physics, Chemistry, and Applications II (Lessard, Roger A., ed.)V1559(1991)

Photopolymers for optical devices in the USSR (Mikaelian, Andrei L.; Barachevsky, Valery A.)V1559,246-257(1991)

Photorefractive gratings in the organic crystal 2-cyclooctylamino-5-nitropyridine doped with 7,7,8,8-tetracyanoquinodimethane (Sutter, Kurt; Hulliger, J.; Guenter, Peter)V1560,290-295(1991)

Photothermoplastic media with organic and inorganic photosemiconductors for hologram recording (Zaichenko, O. V.; Komarov, Vyacheslav A.; Kuzmin, I. V.)V1238,271-274(1991)

Physicochemistry of holographic recording and processing in dichromated gelatin (Sherstyuk, Valentin P.; Maloletov, Sergei M.; Kondratenko, Nina A.)V1238,211-217(1991)

Physics of multishot laser damage to optical materials (Manenkov, Alexander A.; Nechitailo, V. S.)V1441,392-405(1991)

Point defects in KTP and their possible role in laser damage (Scripsick, Michael P.; Edwards, Gary J.; Halliburton, Larry E.; Belt, Roger F.; Kappers, Lawrence A.)V1561,93-103(1991)

Preparation and characterization of silver-colloid/polymer-composite nonlinear optical materials (LaPeruta, Richard; Van Wagenen, E. A.; Roche, J. J.; Kitipichai, Prakob; Wnek, Gary E.; Korenowski, G. M.)V1497,357-366(1991)

Progress in organic third-order nonlinear optical materials (Kuzyk, Mark G.)V1436,160-168(1991)

Progress in true-color holography (Jeong, Tung H.; Wesly, Edward J.)V1238,298-305(1991)

Properties of volume reflection silver-halide gelatin holograms (Kosobokova, N. L.; Usanov, Yu. Y.; Shevtsov, M. K.)V1238,183-188(1991)

Proportional counter windows for the Bragg crystal spectrometer on AXAF (Markert, Thomas H.; Bauer, J.; Canizares, Claude R.; Isobe, T.; Nenonen, Seppo A.; O'Connor, J.; Schattenburg, Mark L.; Flanagan, Kathryn A.; Zombeck, Martin V.)V1549,408-419(1991)

Proton-diffused channel waveguides on Y-cut LiNbO3 using a self-aligned SiO2-cap diffusion method (Son, Yung-Sung; Lee, Hyung-Jae; Yi, Sang-Yoon; Shin, Sang-Yung)V1374,23-29(1991)

Proton exchange in LiTaO3 with different stoichiometric composition (Savatinova, Ivanka T.; Kuneva, M.; Levi, Zelma; Atuchin, V.; Ziling, K.; Armenise, Mario N.)V1374,37-46(1991)

Pseudocolor reflection hologram properties recorded using monochrome photographic materials (Vanin, V. A.; Vorobjov, S. P.)V1238,324-331(1991)

Pyroelectric linear array IR detectors with CCD multiplexer (Norkus, Volkmar; Neumann, Norbert; Walther, Ludwig; Hofmann, Guenter; Schieferdecker, Jorg; Krauss, Matthias G.; Budzier, Helmut; Hess, Norbert)V1484,98-105(1991)

Radiation effects on various optical components for the Mars Observer Spacecraft (Lowry, Jay H.; Iffrig, C. D.)V1330,132-141(1991)

Recent development of proton-exchanged waveguides and devices in lithium niobate using phosphoric acid (Pun, Edwin Y.)V1374,2-13(1991)

Recording mechanism and postpolymerizing self-amplification of holograms (Gulnazarov, Eduard S.; Smirnova, Tatiana N.; Tikhonov, Evgenij A.)V1238,235-239(1991)

Reflection silver-halide gelatin holograms (Usanov, Yu. Y.; Vavilova, Ye. A.; Kosobokova, N. L.; Shevtsov, M. K.)V1238,178-182(1991)

Reflective and Refractive Optical Materials for Earth and Space Applications (Riedl, Max J.; Hale, Robert R.; Parsonage, Thomas B., eds.)V1485(1991)

Refractive-index interpolation fit criterion for materials used in optical design (Korniski, Ronald J.)VCR38,193-217(1991)

Relationship between conjugation length and third-order nonlinearity in bis-donor substituted diphenyl polyenes (Spangler, Charles W.; Havelka, Kathleen O.; Becker, Mark W.; Kelleher, Tracy A.; Cheng, Lap Tak A.)V1560,139-147(1991)

Relief holograms recording on liquid photopolymerizable layers (Boiko, Yuri B.; Granchak, Vasilij M.; Dilung, Iosiph I.; Solovjev, Vladimir S.; Sisakian, Iosif N.; Soifer, Victor A.)V1238,253-257(1991)

Research of fast stages of latent image formation in holographic photoemulsions influenced by ultrashort radiation pulses (Starobogatov, Igor O.; Nicolaev, S. D.)V1238,153-157(1991)

Role of valence-band excitation in laser ablation of KC1 (Haglund, Richard F.; Tang, Kai; Bunton, P. H.; Wang, Ling-jun)V1441,127-138(1991)

Second harmonic generation of diode laser radiation in KNbO3 (van Hulst, Niek F.; Heesink, Gerard J.; de Leeuw, H.; Bolger, Bouwe)V1362,631-646(1991)

Selected Papers on Laser Design (Weichel, Hugo, ed.)VMS29(1991)

Selective photodetectors: a view from the USSR (Ovsyuk, Victor N.; Svitashev, Konstantin K.)V1540,424-431(1991)

Semiempirical method for calculation of dynamic susceptibilities (Svendsen, Erik N.; Stroyer-Hansen, T.)V1361,1048-1055(1991)

Signal and noise characteristics of space-time light modulators (Komarov, Vyacheslav A.; Melnik, N. E.)V1238,275-285(1991)

Simple high-precision extinction method for measuring refractive index of transparent materials (Nee, Soe-Mie F.; Bennett, Harold E.)V1441,31-37(1991)

Solid state ionics and optical materials technology for energy efficiency, solar energy conversion, and environment control (Lusis, Andrejs R.)V1536,116-124(1991)

Some cases of applying photothermoplastic carriers in holographic interferometry and speckle photography (Anikin, V. I.; Dementiev, I. V.; Zhurminskii, Igor L.; Nasedkhina, N. V.; Panasiuk, L. M.)V1238,286-296(1991)

Some needs for the characterization of the dispersive properties of transmissive-optical materials (Dodge, Marilyn J.)V1535,2-12(1991)

Some principles for formation of self-developing dichromate media (Sherstyuk, Valentin P.; Malov, Alexander N.; Maloletov, Sergei M.; Kalinkin, Vyacheslav V.)V1238,218-223(1991)

Space station atomic-oxygen-resistant coatings (Grieser, James L.; Freeland, Alan W.; Fink, Jeffrey D.; Meinke, Gary E.; Hildreth, Eugene N.)V1330,102-110(1991)

Status and future of polymeric materials in imaging systems (Lytle, John D.)V1354,388-394(1991)

Status and needs of infrared optical property information for optical designers (Wolfe, William L.)V1354,696-741(1991)

Steady-state modeling of large-diameter crystal growth using baffles (Sahai, Viveik; Williamson, John W.; Overfelt, Tony)V1557,60-70(1991)

Study of electronic stage of the Herschel effect in holographic emulsions with different types of chemical sensitization (Mikhailov, V. N.; Grinevitskaya, O. V.; Zagorskaya, Z. A.; Mikhailova, V. I.)V1238,144-152(1991)

Study of optical data recording based on photoluminescence effect (Petrov, V. V.; Zymenko, V. I.; Kravetz, V. G.; Polishchuk, E. Y.; Sushko, A. M.)V1621,45-50(1991)

Study of the structure of reflection holgrams on a dichromate gelatin layer (Meshalkina, M. N.; Smirnov, V. V.)V1238,195-199(1991)

Synthesis, characterization, and processing of organic nonlinear optical polymers (Druy, Mark A.; Glatkowski, Paul J.)V1409,214-219(1991)

Tactical Infrared Systems (Tuttle, Jerry W., ed.)V1498(1991)

Theoretical analysis of the third-order nonlinear optical properties of linear cyanines and polyenes (Pierce, Brian M.)V1560,148-161(1991)

Theoretical insight into the quadratic nonlinear optical response of organics: derivatives of pyrene and triaminotrinitrobenzene (Bredas, Jean-Luc; Dehu, C.; Meyers, F.; Zyss, Joseph)V1560,98-110(1991)

Theoretical study of optical pockels and Kerr coefficients of polyenes (Albert, I. D.; Ramasesha, S.)V1436,179-189(1991)

Theoretical study of the fifth harmonic generation in organic liquids (Popescu, Ion M.; Puscas, Niculae Tiberiu N.; Sterian, Paul E.; Irimescu, Dorin I.; Podoleanu, Adrian G.)V1361,1041-1047(1991)

Thermal compensation of infrared achromatic objectives with three optical materials (Rayces, Juan L.; Lebich, Lan)V1354,752-759(1991)

Thermal stability and microstructure study of WSi0.6/GaAs by XRD and TEM (Zhang, Shu Y.; Tan, Shun; Wang, Chang S.; Zhao, Te X.)V1519,43-46(1991)

Thermoelectric power in fiber optic and laser materials under cross-field configuration (Ghatak, Kamakhya P.)V1584,435-447(1991)

Thermomagnetic recording on amorphous ferrimagnetic films (Aleksandrov, K. S.; Berman, G. P.; Frolov, G. I.; Seredkin, V. A.)V1621,51-61(1991)

Thin film SIMNI material formed by low energy nitrogen implantation and epitaxial growth (Lin, Cheng L.; Li, Jinghua H.; Zou, Shi C.)V1519,104-108(1991)

Three-Dimensional Holography: Science, Culture, Education (Jeong, Tung H.; Markov, Vladimir B., eds.)V1238(1991)

Three-layer material for the registration of colored holograms (Smaev, V. P.; Galpern, A. D.; Kiriencko, Yu. A.)V1238,311-315(1991)

Threshold of structural transition in nematic drops with normal boundary conditions in AC electric field (Bodnar, Vladimir G.; Koval'chuk, Alexandr V.; Lavrentovich, Oleg D.; Pergamenshchik, V. M.; Sergan, V. V.)V1455,61-72(1991)

Time dependence of laser-induced surface breakdown in fused silica at 355 nm in the nanosecond regime (Albagli, Douglas; Izatt, Joseph A.; Hayes, Gary B.; Banish, Bryan; Janes, G. Sargent; Itzkan, Irving; Feld, Michael S.)V1441,146-153(1991)

Time-resolved photon echo and fluorescence anisotropy study of organically doped sol-gel glasses (Chronister, Eric L.; L'Esperance, Drew M.; Pelo, John; Middleton, John; Crowell, Robert A.)V1559,56-64(1991)

Ultra-fine-grain silver-halide photographic materials for holography on the flexible film base (Logak, L. G.; Fassakhova, H. H.; Antonova, N. E.; Minina, L. A.; Gainutdinov, R. K.)V1238,171-176(1991)

Uniformity and transmission of proportional counter window materials for use with AXAF (Flanagan, Kathryn A.; Austin, G. K.; Cobuzzi, J. C.; Goddard, R.; Hughes, John P.; McLaughlin, Edward R.; Podgorski, William A.; Rose, V.; Roy, Adrian G.; Zombeck, Martin V.; Markert, Thomas H.; Bauer, J.; Isobe, T.; Schattenburg, Mark L.)V1549,395-407(1991)

Vacuum outgassing from diffuse-absorptive baffle materials (Egert, Charles M.; Basford, J. A.)V1330,178-185(1991)

Vector squeezed states (Lee, Heun J.; Haus, Joseph W.)V1497,228-239(1991)

Volume holograms in liquid photopolymerizable layers (Boiko, Yuri B.; Granchak, Vasilij M.; Dilung, Iosiph I.; Mironchenko, Vladislav Y.)V1238,258-265(1991)

X-ray diffraction study of GaSb/AlSb strained-layer-superlattices grown on miscut (100) substrates (Macrander, Albert T.; Schwartz, Gary P.; Gualteri, Gregory J.; Gilmer, George)V1550,122-133(1991)

Zn3P2: new material for optoelectronic devices (Misiewicz, Jan; Szatkowski, Jan; Mirowska, N.; Gumienny, Zbigniew; Placzek-Popko, E.)V1561,6-18(1991)

Optical processing—see also associative processing; information processing; neural networks; optical computing; parallel processing; pattern recognition;

Accuracy of CGH encoding schemes for optical data processing (Casasent, David P.; Coetzee, Frans M.; Natarajan, Sanjay S.; Xu, Tianning; Yu, Daming; Liu, Hua-Kuang)V1555,23-33(1991)

Analog optical processing of radio frequency signals (Sullivan, Daniel P.; Weber, Charles L.)V1476,234-245(1991)

Application of optical signal processing: fingerprint identification (Fielding, Kenneth H.; Horner, Joseph L.; Makekau, Charles K.)V1564,224-230(1991)

Bispectral magnitude and phase recovery using a wide bandwidth acousto-optical processor (Kniffen, Stacy K.; Becker, Michael F.; Powers, Edward J.)V1564,617-627(1991)

Building an optical pattern recognizer (Lindberg, Perry C.; Gregory, Don A.)V1470,220-225(1991)

Characteristics of a ferroelectric liquid crystal spatial light modulator with a dielectric mirror (Kato, Naoki; Sekura, Rieko; Yamanaka, Junko; Ebihara, Teruo; Yamamoto, Shuhei)V1455,190-205(1991)

Compact optical associative memory (Burns, Thomas J.; Rogers, Steven K.; Kabrisky, Matthew; Vogel, George A.)V1469,208-218(1991)

Compact optical neuro-processors (Paek, Eung Gi; Wullert, John R.; Von Lehman, A.; Patel, J. S.; Martin, R.)V1621,340-350(1991)

Comparison of optically addressed spatial light modulators (Hudson, Tracy D.; Kirsch, James C.; Gregory, Don A.)V1474,101-111(1991)

Computer-generated holograms for optical data processing (Casasent, David P.)V1544,101-107(1991)

Correlation-based optical numeric processors (Casasent, David P.; Woodford, Paul)V1563,112-119(1991)

Design of a GaAs acousto-optic correlator for real-time processing (Armenise, Mario N.; Impagnatiello, Fabrizio; Passaro, Vittorio M.; Pansini, Evangelista)V1562,160-171(1991)

Devices for Optical Processing (Gookin, Debra M., ed.)V1562(1991)

Effects of optoelectronic device characteristics on the performance and design of POEM systems (Esener, Sadik C.; Kiamilev, Fouad E.; Krishnamoorthy, Ashok V.; Marchand, Philippe J.)V1562,11-20(1991)

Effects of thresholding in multiobject binary joint transform correlation (Javidi, Bahram; Wang, Jianping; Tang, Qing)V1564,212-223(1991)

Electronic interface for high-frame-rate electrically addressed spatial light modulators (Kozaitis, Samuel P.; Kirschner, K.; Kelly, E.; Been, D.; Delgado, J.; Velez, E.; Alkindy, A.; Al-Houra, H.; Ali, F.)V1474,112-115(1991)

Error codes applied to optical algebraic processors (Ellett, Scott A.; Walkup, John F.; Krile, Thomas F.)V1564,634-643(1991)

Experimental comparison of optical binary phase-only filter and high-pass matched filter correlation (Leib, Kenneth G.; Brandstetter, Robert W.; Drake, Marvin D.; Franks, Glen B.; Siewert, Ronald O.)V1483,140-154(1991)

Extraction of features from images using video feedback (Boone, Bradley G.; Shukla, Oodaye B.; Terry, David H.)V1471,390-403(1991)

Fast optoelectronic neurocomputer for character recognition (Zhang, Lin; Robinson, Michael G.; Johnson, Kristina M.)V1469,225-229(1991)

Hermaphroditic small tactical connector for single-fiber applications (Darden, Bruce V.; LeFevre, B. G.; Kalomiris, Vasilios E.)V1474,300-308(1991)

High contrast ratio InxGa1-xAs/GaAs multiple-quantum-well spatial light modulators (Harwit, Alex; Fernandez, R.; Eades, Wendell D.)V1541,38-47(1991)

High-speed signal processing architectures using charge-coupled devices (Boddu, Jayabharat; Udpa, Satish S.; Udpa, L.; Chan, Shiu Chuen M.)V1562,251-262(1991)

Holographic optical backplane hardware implementation for parallel and distributed processors (Kim, Richard C.; Lin, Freddie S.)V1474,27-38(1991)

Hybrid modulation properties of the Epson LCTV (Kirsch, James C.; Loudin, Jeffrey A.; Gregory, Don A.)V1558,432-441(1991)

Hybrid solution for high-speed target acquisition and identification systems (Udomkesmalee, Suraphol; Scholl, Marija S.; Shumate, Michael S.)V1468,81-91(1991)

Image processing based on supervised learning in neural networks (Hasegawa, Akira; Zhang, Wei; Itoh, Kazuyoshi; Ichioka, Yoshiki)V1621,374-379(1991)

Impact of device characteristics on optical processor design (Turpin, Terry M.)V1562,2-10(1991)

Integrated C3I optical processor (Kaminski, Robert L.)V1564,156-164(1991)

Integrated optical preamplifier technology for optical signal processing and optical communication systems (Eichen, Elliot G.; Powazinik, William; Meland, Edmund; Bryant, R.; Rideout, William C.; Schlafer, John; Lauer, Robert B.)V1474,260-267(1991)

Interference effects and the occurrence of blind spots in coherent optical processors (Christie, Simon; Cai, Xian-Yang; Kvasnik, Frank)V1507,202-209(1991)

Liquid-Crystal Devices and Materials (Drzaic, Paul S.; Efron, Uzi, eds.)V1455(1991)

Liquid-crystal devices in planar optics (Armitage, David; Ticknor, Anthony J.)V1455,206-212(1991)

Liquid-crystal-television-based optical-digital processor for measurement of shortening velocity in single rat heart cells (Yelamarty, Rao V.; Yu, Francis T.; Moore, Russell L.; Cheung, Joseph Y.)V1398,170-179(1991)

Locally linear neural networks for optical correlators (Gustafson, Steven C.; Little, Gordon R.; Olczak, Eugene G.)V1469,268-274(1991)

Measurement of radiation-induced attenuation in optical fibers by optical-time-domain reflectometry (Looney, Larry D.; Lyons, Peter B.)V1474,132-137(1991)

Mercurous halides for long time-delay Bragg cells (Brandt, Gerald B.; Singh, N. B.; Gottlieb, Milton S.)V1454,336-343(1991)

Multiple target-to-track association and track estimation system using a neural network (Yee, Mark L.; Casasent, David P.)V1481,418-429(1991)

Noise and discrimination performance of the MINACE optical correlation filter (Ravichandran, Gopalan; Casasent, David P.)V1471,233-248(1991)

Nonlinear dynamics of neuromorphic optical system with spatio-temporal interactions (Vorontsov, Michael A.; Iroshnikov, N. G.)V1621,292-298(1991)

Nonlinear optical processing using phase grating (Sadovnik, Lev S.; Demichovskaya, Olga; Chen, Ray T.)V1545,200-208(1991)

Optical A/D conversion based on acousto-optic theta modulation (Li, Yao; Zhang, Yan)V1474,167-173(1991)

Optical correlation filters for large-class OCR applications (Casasent, David P.; Iyer, Anand K.; Gopalaswamy, Srinivasan)V1470,208-219(1991)

Optical correlator field demonstration (Kirsch, James C.; Gregory, Don A.; Hudson, Tracy D.; Loudin, Jeffrey A.; Crowe, William M.)V1482,69-78(1991)

Optical correlator techniques applied to robotic vision (Hine, Butler P.; Reid, Max B.; Downie, John D.)V1564,416-426(1991)

Optical enhancements of joint-Fourier-transform correlator by image subtraction (Perez, Osvaldo; Karim, Mohammad A.)V1471,255-264(1991)

Optical evaluation of the microchannel spatial light modulator (Duffey, Jason N.; Hudson, Tracy D.; Kirsch, James C.)V1558,422-431(1991)

Optical inference processing techniques for scene analysis (Casasent, David P.)V1564,504-510(1991)

Optical Information Processing Systems and Architectures III (Javidi, Bahram, ed.)V1564(1991)

Optically implementable algorithm for convolution/correlation of long data streams (Zhang, Yan; Li, Yao)V1474,188-198(1991)

Optically powered optoelectronic integrated circuits for optical interconnects (Brown, Julia J.; Gardner, J. T.; Forrest, Stephen R.)V1474,236-242(1991)

Optical neural networks based on liquid-crystal light valves and photorefractive crystals (Owechko, Yuri; Soffer, Bernard H.)V1455,136-144(1991)

Optical principles of information processing in supercomputer architecture (Burtsev, Vsevolod S.)V1621,380-387(1991)

Optical processing and hybrid neural nets (Casasent, David P.)V1469,256-267(1991)

Optimal correlation filters for implementation on deformable mirror devices (Vijaya Kumar, B. V. K.; Carlson, Daniel W.)V1558,476-486(1991)

Optimization neural net for multiple-target data association: real-time optical lab results (Yee, Mark L.; Casasent, David P.)V1469,308-319(1991)

Optoelectronic implementation of filtered-back-projection tomography algorithm (Lu, Tongxin; Udpa, Satish S.; Udpa, L.)V1564,704-713(1991)

Optomechanical M x N fiber-optic matrix switch (Pesavento, Gerry A.)V1474,57-61(1991)

Pacifist's guide to optical computers (Caulfield, H. J.)V1564,632-632(1991)

Parallel optical information, concept, and response evolver: POINCARE (Caulfield, H. J.; Caulfield, Kimberly)V1469,232-239(1991)

Performance of high-dynamic-range CCD arrays with various epilayer structures (Smith, Dale J.; Harrison, Lorna J.; Pellegrino, John M.; Simon, Deborah R.)V1562,242-250(1991)

Photorefractive devices for optical information processing (Yeh, Pochi A.; Gu, Claire; Hong, John H.)V1562,32-43(1991)

Photothermoplastic molecular heterostructures (Cherkasov, Yuri A.)V1621,62-73(1991)

Point target detection, location, and track initiation: initial optical lab results (Carender, Neil H.; Casasent, David P.)V1481,35-48(1991)

Progress in organic third-order nonlinear optical materials (Kuzyk, Mark G.)V1436,160-168(1991)

Realization of infinite-impulse response filters using acousto-optic cells (Ghosh, Anjan K.)V1564,593-601(1991)

Real-time optical processor for increasing resolution beyond the diffraction limit (Dhadwal, Harbans S.; Noel, Eric)V1564,664-673(1991)

Recent research on optical neural networks in Japan (Ishihara, Satoshi; Mori, Masahiko)V1621,362-372(1991)

Relationship of image algebra to the optical processing of signals and imagery (Schmalz, Mark S.; Wilson, Joseph N.)V1474,212-234(1991)

Some problems of implementation of the memory system based on optical solitons in fibers (Shcherbakov, A. S.)V1621,204-214(1991)

Spatial light modulator on the base of shape memory effect (Antonov, Victor A.; Shelyakov, Alexander V.)V1474,116-123(1991)

Staring phased-array radar using photorefractive crystals (Weverka, Robert T.; Wagner, Kelvin H.)V1564,676-684(1991)

Transmissive analogue SLM using a chiral smectic liquid crystal switched by CdSe TFTs (Crossland, William A.; Davey, A. B.; Sparks, Adrian P.; Lee, Michael J.; Wright, S. W.; Judge, C. P.)V1455,264-273(1991)

Two-dimensional spatial light modulators using polarization-sensitive multiple-quantum-well light valve (Jain, Faquir C.; Drake, G.; Chung, C.; Bhattacharjee, K. K.; Cheung, S. K.)V1564,714-722(1991)

Wavelength-sensitive holographic optical interconnects (Lin, Freddie S.; Zaleta, David E.; Jannson, Tomasz)V1474,45-56(1991)

Wide-angular aperture acousto-optic Bragg cell (Weverka, Robert T.; Wagner, Kelvin H.)V1562,66-72(1991)

Optical properties—see also optical materials

Accurate method for neutron fluence control used in improving neutron-transmutation-doped silicon for detectors (Halmagean, Eugenia T.; Lazarovici, Doina N.; Udrea-Spenea, Marian N.)V1484,106-114(1991)

Advanced infrared optically black baffle materials (Seals, Roland D.; Egert, Charles M.; Allred, David D.)V1330,164-177(1991)

Advanced processing of the Zerodur R glass ceramic (Marx, Thomas A.)V1535,130-135(1991)

Angular-selective cermet films produced from a magnetically filtered cathodic arc (Smith, Geoffrey B.; Ng, M. W.; Ditchburn, Robert J.; Martin, Philip J.; Netterfield, Roger P.)V1536,126-137(1991)

Anodic oxides on HgZnTe (Eger, David; Zigelman, Alex)V1484,48-54(1991)

Antireflection coatings of sputter-deposited SnOxFy and SnNxFy (Yin, Zhiqiang; Stjerna, B. A.; Granqvist, Claes G.)V1536,149-157(1991)

Band structure and optical properties of silicon carbide (Gavrilenko, Vladimir I.; Frolov, Sergey I.)V1361,171-182(1991)

Behavior of amorphous semiconductors As2S3 layers after photon, electron, or x-ray exposures (Guttmann, Peter; Danev, Gentsho; Spassova, E.; Babin, Sergei V.)V1361,999-1010(1991)

Beryllium galvanometer mirrors (Weissman, Harold)V1485,13-19(1991)

Bidirectional transmittance distribution function measurements on ZnSe and on ZnS Cleartran (Melozzi, Mauro; Mazzoni, Alessandro; Curti, G.)V1512,178-188(1991)

Bragg grating formation and germanosilicate fiber photosensitivity (Meltz, Gerald; Morey, William W.)V1516,185-199(1991)

Catalog of infrared and cryo-optical properties for selected materials (Heaney, James B.; Alley, Phillip W.; Bradley, Scott E.)V1485,140-159(1991)

Cation intercalation in electrochromic NiOx films (Scarminio, J.; Gorenstein, Annette; Decker, Franco; Passerini, S.; Pileggi, R.; Scrosati, Bruno)V1536,70-80(1991)

Changes in optical density of normal vessel wall and lipid atheromatous plaque after Nd:YAG laser irradiation (Schwarzmaier, Hans-Joachim; Heintzen, Matthias P.; Zumdick, Mathias; Kaufmann, Raimund; Wolbarsht, Myron L.)V1427,128-133(1991)

Characteristic curves and phase-exposition characteristics of holographic photomaterials (Churaev, A. L.; Artyomova, V. V.)V1238,158-165(1991)

Characteristics of domain formation and poling in potassium niobate (Jarman, Richard H.; Johnson, Barry C.)V1561,33-42(1991)

Characterization and switching study of an optically controlled GaAs switch (Stoudt, David C.; Mazzola, Michael S.; Griffiths, Scott F.)V1378,280-285(1991)

Characterization of anodic fluoride films on Hg1-xCdxTe (Esquivias, Ignacio; Dal Colle, M.; Brink, D.; Baars, Jan W.; Bruder, Martin)V1484,55-66(1991)

Characterization of DC-PVA films for holographic recording materials (Leclere, Philippe; Renotte, Yvon L.; Lion, Yves F.)V1507,339-344(1991)

Comparison of photorefractive effects and photogenerated components in polarization-maintaining fibers (Kanellopoulos, S. E.; Guedes Valente, Luiz C.; Handerek, Vincent A.; Rogers, Alan J.)V1516,200-210(1991)

Comparison of thermal and optical techniques for describing light interaction with vascular grafts, sutures, and thrombus (Obremski, Susan M.; LaMuraglia, Glenn M.; Bruggemann, Ulrich H.; Anderson, R. R.)V1427,327-334(1991)

Condensation mechanisms and properties of rf-sputtered a-Si:H (Ligachev, Valery A.; Filikov, V. A.; Gordeev, V. N.)V1519,214-219(1991)

Cr3+ tunable laser glass (Izumitani, Tetsuro; Zou, Xuolu; Wang, Y.)V1535,150-159(1991)

Critical-point phonons of diamond (Klein, Claude A.; Hartnett, Thomas M.; Robinson, Clifford J.)V1534,117-138(1991)

Cryogenic refractive indices of cadmium telluride coatings in wavelength range from 2.5 to 20 um (Feng, Weiting; Yen, Yi X.; Zhu, Cui Y.)V1535,224-230(1991)

Crystal growth and characterization of rare-earth-doped gallates of alkaline earth and lanthanum (Ryba-Romanowski, Witold; Golab, Stanislaw; Berkowski, Marek)V1391,2-5(1991)

Crystallization and optothermal characteristics of germanate glasses (Montenero, Angelo; Gnappi, G.; Bertolotti, Mario; Sibilia, C.; Fazio, E.; Liakhou, G.)V1513,234-242(1991)

CVD diamond as an optical material for adverse environments (Snail, Keith A.)V1330,46-64(1991)

Defect centers and photoinduced self-organization in Ge-doped silica core fiber (Tsai, T. E.; Griscom, David L.)V1516,14-28(1991)

Dependence of optical properties of thermal-evaporated lead telluride films on substrate temperature (Feng, Weiting; Yen, Yi X.; Zhu, Cui Y.)V1519,333-338(1991)

Design and optical performance of beryllium assessment mirrors (Thomas, Brigham B.; Maxey, L. C.; Miller, Arthur C.)V1485,20-30(1991)

Designing apochromatic telescope objectives with liquid lenses (Sigler, Robert D.)V1535,89-112(1991)

Detection by mirage effect of the counter-ion flux between an electrochrome and a liquid electrolyte: application to WO3, Prussian blue, and lutetium diphthalocyanine films (Plichon, V.; Giron, J. C.; Delboulbe, J. P.; Lerbet, F.)V1536,37-47(1991)

Detector perturbation of ocean radiance measurements (Macdonald, Burns; Helliwell, William S.; Sanborn, James; Voss, Kenneth J.)V1537,104-114(1991)

Determination of optical constants of thin film in the soft x-ray region (Guo, Yong H.; Fan, Zheng X.; Bin, Ouyang; Jin, Lei; Shao, Jian D.)V1519,327-332(1991)

Determination of the optical constants of thin chemical-vapor-deposited diamond windows from 0.5 to 6.5 eV (Robins, Lawrence H.; Farabaugh, Edward N.; Feldman, Albert)V1534,105-116(1991)

Device concepts for SiGe optoelectronics (Kasper, Erich; Presting, Hartmut)V1362,302-312(1991)

Diamond Optics IV (Feldman, Albert; Holly, Sandor, eds.)V1534(1991)

Diamond windows for the infrared: fact and fallacy (Klein, Claude A.)VCR38,218-257(1991)

Distorted local environment about Zn and transition metals on the copper sites in YBa2Cu3O7 (Bridges, Frank; Li, Guoguang; Boyce, James B.; Claeson, Tord)V1550,76-84(1991)

Dual-core ion-exchanged glass waveguides (Li, Ming-Jun; Honkanen, Seppo; Wang, Wei-Jian; Leonelli, Richard; Najafi, S. I.)V1513,410-417(1991)

Dynamics of phase-grating formation in optical fibers (An, Sunghyuck; Sipe, John E.)V1516,175-184(1991)

Dynamics of self-organized x(2) gratings in optical fibers (Kamal, Avais; Terhune, Robert W.; Weinberger, Doreen A.)V1516,137-153(1991)

Effect of oxygen on optical properties of yttria thin films (Ying, Xuantong; Feldman, Albert; Farabaugh, Edward N.)V1519,321-326(1991)

Effect of the glass composition on the emission band of erbium-doped active fibers (Cognolato, Livio; Gnazzo, Angelantonio; Sordo, Bruno; Cocito, Giuseppe)V1579,249-256(1991)

Electrical and optical properties of porous glass (Rysiakiewicz-Pasek, Ewa; Marczuk, Krystyna)V1513,283-290(1991)

Electrochromic properties and temperature dependence of chemically deposited Ni(OH)x thin films (Fantini, Marcia C.; Bezerra, George H.; Carvalho, C. R.; Gorenstein, Annette)V1536,81-92(1991)

Electronic and optical properties of silicide/silicon IR detectors (Cabanski, Wolfgang A.; Schulz, Max J.)V1484,81-97(1991)

Electro-optical effects in semiconductor superlattices (Voos, Michel; Voisin, Paul)V1361,416-423(1991)

Experimental simulation analysis of nonlinear problem: investigation into the mechanical and optical behavior of silver chloride of photoplastic material (Yin, Zhi Xiang; Zhang, Shikun; Li, Zong Yan)V1559,487-496(1991)

Experimental study of migration depth for the photons measured at sample surface (Cui, Weijia; Kumar, Chellappa; Chance, Britton)V1431,180-191(1991)

Experimental study of the optical properties of LTCVD SiO2 (Aharoni, Herzl; Swart, Pieter L.)V1442,118-125(1991)

Experiment for testing the closure property in ocean optics (Maffione, Robert A.; Honey, Richard C.; Brown, Robert A.)V1537,115-126(1991)

Extension of model of beam spreading in seawater to include dependence on the scattering phase function (Schippnick, Paul F.)V1537,185-193(1991)

Fabrication and nonlinear optical properties of mixed and layered colloidal particles (McDonald, Joseph K.; LaiHing, Kenneth)V1497,367-370(1991)

Fiber defects in Ge-doped fibers: towards a coherent picture (LaRochelle, Sophie)V1516,55-57(1991)

Finding a single molecule in a haystack: laser spectroscopy of solids from (square root of)N to N=1 (Moerner, William E.; Ambrose, William P.)V1435,244-251(1991)

Finite element analysis of the transient behavior of optical components under irradiation (Borik, Stefan; Giesen, Adolf)V1441,420-429(1991)

Frequency doubling, absorption, and grating formation in glass fibers: effective defects or defective effects? (Russell, Philip S.; Poyntz-Wright, L. J.; Hand, Duncan P.)V1373,126-139(1991)

Frequency doubling in optical fibers: a complex puzzle (Margulis, Walter; Carvalho, Isabel C.; Lesche, Bernhard)V1516,60-66(1991)

Glasses for Optoelectronics II (Righini, Giancarlo C., ed.)V1513(1991)

Growth and properties of GaxIn1-xAs (x<O.47) on InP by MOCVD (Du, MingZe; Yuan, JinShan; Jin, Yixin; Zhou, Tian Ming; Hong, Jiang; Hong, ChunRong; Zhang, BaoLin)V1361,699-705(1991)

High-resolution photographic material of Institute of Atomic Energy for holography in intersecting beams (Ryabova, R. V.; Barinova, E. S.; Myatezh, O. V.; Zaborov, A. N.; Romash, E. V.; Chernyi, D. I.)V1238,166-170(1991)

Human tooth as an optical device (Altshuler, Grigori B.; Grisimov, Vladimir N.; Ermolaev, Vladimir S.; Vityaz, Irena V.)V1429,95-104(1991)

Imaging model for underwater range-gated imaging systems (Strand, Michael P.)V1537,151-160(1991)

Imaging of subsurface regions of random media by remote sensing (Barbour, Randall L.; Graber, Harry L.; Aronson, Raphael; Lubowsky, Jack)V1431,192-203(1991)

Influence of phototransformed molecules on optical properties of finite cholesteric liquid-crystal cell (Pinkevich, Igor P.; Reshetnyak, Victor Y.; Reznikov, Yuriy)V1455,122-133(1991)

Influence of surface passivation on the optical bleaching of CdSe microcrystallites embedded in glass (Rueckmann, I.; Woggon, Ulrike; Kornack, J.; Mueller, Matthias; Cesnulevicius, J.; Kolenda, Jonas; Petrauskas, Mendogas)V1513,78-85(1991)

Influence of the laser-induced temperature rise in photodynamic therapy (Gottschalk, Wolfgang; Hengst, Joachim; Sroka, Ronald; Unsoeld, Eberhard)V1427,320-326(1991)

Infrared reflectivity: a tool for bond investigation in II-VI ternaries (Granger, Rene')V1484,39-46(1991)

Instrument for underwater measurement of optical backscatter (Maffione, Robert A.; Dana, David R.; Honey, Richard C.)V1537,173-184(1991)

Intl Workshop on Photoinduced Self-Organization Effects in Optical Fiber (Ouellette, Francois, ed.)V1516(1991)

Investigation into the characteristics of a-C:H films irradiated by electron beam (Gu, Shu L.; He, Yu L.; Wang, Zhi C.)V1519,175-178(1991)

In-vitro model for evaluation of pulse oximetry (Vegfors, Magnus; Lindberg, Lars-Goran; Lennmarken, Claes; Oberg, Per A.)V1426,79-83(1991)

Ion-assisted deposition of graded index silicon oxynitride coatings (Al-Jumaily, Ghanim A.; Gagliardi, F. J.; McColl, P.; Mizerka, Larry J.)V1441,360-365(1991)

Laser glasses (Lunter, Sergei G.; Dymnikov, Alexander A.; Przhevuskii, Alexander K.; Fedorov, Yurii K.)V1513,349-359(1991)

Laser lithotripsy with a pulsed dye laser: correlation between threshold energy and optical properties (Sterenborg, H. J.; van Swol, Christiaan F.; Biganic, Dane D.; van Gemert, Martin J.)V1427,256-266(1991)

Light scattering properties of new materials for glazing applications (Bergkvist, Mikael; Roos, Arne)V1530,352-362(1991)

Lightweight SXA metal matrix composite collimator (Johnson, R. B.; Ahmad, Anees; Hadaway, James B.; Geiger, Alan L.)V1535,136-142(1991)

Linear and nonlinear optical properties of substituted pyrrolo[1,2-a]quinolines (van Hulst, Niek F.; Heesink, Gerard J.; Bolger, Bouwe; Kelderman, E.; Verboom, W.; Reinhoudt, D. N.)V1361,581-588(1991)

Low-temperature viscosity measurements of infrared transmitting halide glasses (Seddon, Angela B.; Cardoso, A. V.)V1513,255-263(1991)

Melt-processed calcium aluminate fibers: structural and optical properties (Wallenberger, Frederick T.; Weston, Norman E.; Brown, Sherman D.)V1484,116-124(1991)

Melt processing of calcium aluminate fibers with sapphirelike infrared transmission (Wallenberger, Frederick T.; Koutsky, J. A.; Brown, Sherman D.)V1590,72-82(1991)

Method of investigations of the optical properties of anisotropic materials using modulation of light polarization (Gumienny, Zbigniew; Misiewicz, Jan)V1527,462-465(1991)

Microstructures and domain size effects in diamond films characterized by Raman spectroscopy (Nemanich, Robert J.; Bergman, Larry; LeGrice, Yvonne M.; Turner, K. F.; Humphreys, T. P.)V1437,2-12(1991)

Model of second-harmonic generation in glass fibers based on multiphoton ionization interference effects (Anderson, Dana Z.; Mizrahi, Victor; Sipe, John E.)V1516,154-161(1991)

Modifications of optical properties with ceramic coatings (Besmann, Theodore M.; Abdel-Latif, A. I.)V1330,78-89(1991)

New preparation method and properties of diamondlike carbon films (Yu, Bing Kun; Chen, Xao Min)V1534,223-229(1991)

New thermistor material for thermistor bolometer: material preparation and characterization (Umadevi, P.; Nagendra, C. L.; Thutupalli, G. K.; Mahadevan, K.; Yadgiri, G.)V1485,195-205(1991)

Nonlinear optical properties of a fiber with an organic core crystal grown from solution (Ohmi, Toshihiko; Yoshikawa, Nobuo; Sakai, Koji; Koike, Tomoyuki; Umegaki, Shinsuke)V1361,606-612(1991)

Nonlinear optical properties of chalcogenide As-S, As-Se glasses (Andriesh, A.; Bertolotti, Mario; Chumach, V.; Fazio, E.; Ferrari, A.; Liakhou, G.; Sibilia, C.)V1513,137-147(1991)

Nonlinear optical properties of composite materials (Haus, Joseph W.; Inguva, Ramarao)V1497,350-356(1991)

Nonlinear optical properties of germanium diselenide glasses (Haro-Poniatowski, Emmanuel; Guasti, Manuel F.)V1513,86-92(1991)

Nonlinear optical properties of inorganic coordination polymers and organometallic complexes (Thompson, Mark E.; Chiang, William; Myers, Lori K.; Langhoff, Charles A.)V1497,423-429(1991)

Nonlinear optical properties of nipi and hetero nipi superlattices and their application for optoelectronics (Doehler, Gottfried H.)V1361,443-468(1991)

Nonlinear Optical Properties of Organic Materials IV (Singer, Kenneth D., ed.)V1560(1991)

Nonlinear optical properties of poled polymers (Gadret, G.; Kajzar, Francois; Raimond, P.)V1560,226-237(1991)

Nonlinear optical properties of suspensions of microparticles: electrostrictive effect and enhanced backscatter (Kang, Chih-Chieh; Fiddy, Michael A.)V1497,372-381(1991)

Nonlinear optical properties of xanthone derivatives (Imanishi, Yasuo; Itoh, Yuzo; Kakuta, Atsushi; Mukoh, Akio)V1361,570-580(1991)

Nonlinear properties of poled-polymer films: SHG and electro-optic measurements (Morichere, D.; Dumont, Michel L.; Levy, Yves; Gadret, G.; Kajzar, Francois)V1560,214-225(1991)

Novel GaP/InP strained heterostructures: growth, characterization, and technological perspectives (Recio, Miguel; Ruiz, Ana; Melendez, J.; Rodriguez, Jose M.; Armelles, Gaspar; Dotor, M. L.; Briones, Fernando)V1361,469-478(1991)

Optical and environmentally protective coatings for potassium dihydrogen phosphate harmonic converter crystals (Thomas, Ian M.)V1561,70-82(1991)

Optical characteristics of the Teflon AF fluoro-plastic materials (Lowry, Jay H.; Mendlowitz, Joseph S.; Subramanian, N. S.)V1330,142-151(1991)

Optical characterization of photolithographic metal grids (Osmer, Kurt A.; Jones, Mike I.)V1498,138-146(1991)

Optical characterization of solar-selective transmitting coatings (Roos, Arne)V1536,158-168(1991)

Optically induced creation, transformation, and organization of defects and color centers in optical fibers (Russell, Philip S.; Hand, Duncan P.; Chow, Y. T.; Poyntz-Wright, L. J.)V1516,47-54(1991)

Optical performance of angle-dependent light-control glass (Maeda, Koichi; Ishizuka, S.; Tsujino, T.; Yamamoto, H.; Takigawa, Akio)V1536,138-148(1991)

Optical properties: a trip through the glass map (Marker, Alexander J.)V1535,60-77(1991)

Optical properties of amorphous hydrogenated carbon layers (Stenzel, Olaf; Schaarschmidt, Guenther; Roth, Sylvia; Schmidt, Guenther; Scharff, Wolfram)V1534,148-157(1991)

Optical properties of barcode symbols for laser scanning (Quinn, Anna M.; Eastman, Jay M.)V1384,138-144(1991)

Optical properties of DC arc-discharge plasma CVD diamond (Trombetta, John M.; Hoggins, James T.; Klocek, Paul; McKenna, T. A.)V1534,77-88(1991)

Optical properties of GaAs, GaP, and CVD diamond (Klocek, Paul; Hoggins, James T.; McKenna, T. A.; Trombetta, John M.; Boucher, Maurice W.)V1498,147-157(1991)

Optical properties of glass materials obtained by inorganic sol-gel synthesis (Glebov, Leonid B.; Evstropiev, Sergei K.; Petrovskii, Gurii T.; Shashkin, Viktor S.)V1513,224-231(1991)

Optical properties of granular Sn films with coating Al (Wu, Guang M.; Qian, Zheng X.)V1519,315-320(1991)

Optical properties of InGaAs/InP strained quantum wells (Abram, Richard A.; Wood, Andrew C.; Robbins, D. J.)V1361,424-433(1991)

Optical properties of KDP crystals grown at high growth rates (Zaitseva, N. P.; Ganikhanov, Ferous S.; Katchalov, O. V.; Efimkov, V. F.; Pastukhov, S. A.; Sobolev, V. B.)V1402,223-230(1991)

Optical properties of Li-doped ZnO films (Valentini, Antonio; Quaranta, Fabio; Vasanelli, L.; Piccolo, R.)V1400,164-170(1991)

Optical properties of molecular beam epitaxy grown ZnTe epilayers (Kudlek, Gotthard; Presser, Nazmir; Gutowski, Juergen; Mathine, David L.; Kobayashi, Masakazu; Gunshor, Robert L.)V1361,150-158(1991)

Optical properties of oxide films prepared by ion-beam-sputter deposition (Tang, Xue F.; Fan, Zheng X.; Wang, Zhi-Jiang)V1519,96-98(1991)

Optical properties of short-period Si/Ge superlattices grown on (001) Ge studied with photoreflectance (Menczigar, Ulrich; Dahmen, M.; Zachai, Reinhard; Eberl, K.; Abstreiter, Gerhard)V1361,282-292(1991)

Optical properties of some ion-assisted deposited oxides (Andreani, F.; Luridiana, S.; Mao, Shu Z.)V1519,18-22(1991)

Optical properties of ZnS/diamond composites (Xue, L. A.; Noh, T. W.; Sievers, A. J.; Raj, Rishi)V1534,183-196(1991)

Optical property measurements in turbid media using frequency-domain photon migration (Tromberg, Bruce J.; Svaasand, Lars O.; Tsay, Tsong-Tseh; Haskell, Richard C.; Berns, Michael W.)V1525,52-58(1991)

Optical Surfaces Resistant to Severe Environments (Musikant, Solomon, ed.)V1330(1991)

Optimization of optical properties of pigmented foils for radiative cooling applications: model calculations (Nilsson, Torbjorn M.; Niklasson, Gunnar A.)V1536,169-182(1991)

Optimization of optical properties of resist processes (Brunner, Timothy A.)V1466,297-308(1991)

Overpressure proof testing of large infrared windows for aircraft applications (Pruszynski, Charles J.)V1498,163-170(1991)

Phenomenological description of self-organized x(2) grating formation in centrosymmetric doped optical fibers (Chmela, Pavel)V1516,116-124(1991)

Photoacoustic absorption spectrum of some rat and bovine tissues in the ultraviolet-visible range (Bernini, Umberto; Russo, Paolo)V1427,398-404(1991)

Photoconductivity decay method for determining minority carrier lifetime of p-type HgCdTe (Reichman, Joseph)V1484,31-38(1991)

Photodegradation of a laser dye in a silica gel matrix (Glab, Wallace L.; Bistransin, Mark; Borst, Walter L.)V1497,389-395(1991)

Photoinduced effects in optical waveguides (Dianov, Evgeni M.; Kazansky, Peter G.; Stepanov, D. Y.)V1516,81-98(1991)

Photoinduced second-harmonic generation and luminescence of defects in Ge-doped silica fibers (Krol, Denise M.; Atkins, Robert M.; Lemaire, Paul J.)V1516,38-46(1991)

Photoinduced self-organization in optical fiber: some answered and unanswered questions (Ouellette, Francois; Gagnon, Daniel; LaRochelle, Sophie; Poirier, Michel)V1516,2-13(1991)

Photoluminescence and deep-level transient spectroscopy of DX-centers in selectively silicon-doped GaAs-AlAs superlattices (Ababou, Soraya; Benyattou, Taha; Marchand, Jean J.; Mayet, Louis; Guillot, Gerard; Mollot, Francis; Planel, Richard)V1361,706-711(1991)

Photoquenching and characterization studies in a bulk optically controlled GaAs semiconductor switch (Lakdawala, Vishnu K.; Schoenbach, Karl H.; Roush, Randy A.; Barevadia, Gordon R.; Mazzola, Michael S.)V1378,259-270(1991)

Photosensitive germanosilicate preforms and fibers (Williams, Doug L.; Davey, Steven T.; Kashyap, Raman; Armitage, J. R.; Ainslie, B. J.)V1513,158-167(1991)

Photosensitivity in optical fibers: detection, characterization, and application to the fabrication of in-core fiber index gratings (Malo, Bernard; Bilodeau, Francois; Johnson, Derwyn C.; Skinner, Iain M.; Hill, Kenneth O.; Morse, Ted F.; Kilian, Arnd; Reinhart, Larry J.; Oh, Kyunghwan)V1590,83-93(1991)

Physical and optical properties of organically modified silicates doped with laser and NLO dyes (Capozzi, Carol A.; Pye, L. D.)V1513,320-329(1991)

Planar optical waveguides on glasses and glass-ceramic materials (Glebov, Leonid B.; Nikonorov, Nikolai V.; Petrovskii, Gurii T.)V1513,56-70(1991)

Pockels' effect in polycrystalline ZnS planar waveguides (Wong, Brian; Kitai, Adrian H.; Jessop, Paul E.)V1398,261-268(1991)

Polyimide-coated embedded optical fiber sensors (Nath, Dilip K.; Nelson, Gary W.; Griffin, Stephen E.; Harrington, C. T.; He, Yi-Fei; Reinhart, Larry J.; Paine, D. C.; Morse, Ted F.)V1489,17-32(1991)

Preparation and characterization of silver-colloid/polymer-composite nonlinear optical materials (LaPeruta, Richard; Van Wagenen, E. A.; Roche, J. J.; Kitipichai, Prakob; Wnek, Gary E.; Korenowski, G. M.)V1497,357-366(1991)

Principles of phase-resolved optical measurements (Jacques, Steven L.)V1525,143-153(1991)

Progress on the variable reflectivity electrochromic window (Goldner, Ronald B.; Arntz, Floyd O.; Berera, G.; Haas, Terry E.; Wei, G.; Wong, Kwok-keung; Yu, Phillip C.)V1536,63-69(1991)

Properties and processing of the TeX glasses (Zhang, Xhang H.; Ma, Hong-Li; Lucas, Jacques)V1513,209-214(1991)

Properties of CVD diamond for optical applications (Gray, Kevin J.; Lu, Grant)V1534,60-66(1991)

Properties of diamonds with varying isotopic composition (Banholzer, William; Fulghum, Stephen)V1501,163-176(1991)

Properties of optical waves in turbid media (Svaasand, Lars O.; Tromberg, Bruce J.)V1525,41-51(1991)

Radiative transfer in the cloudy atmosphere: modeling radiative transport (Gabriel, Philip; Stephens, Graeme L.; Tsay, Si-Chee)V1558,76-90(1991)

Raman scattering characterization of direct gap Si/Ge superlattices (White, Julian D.; Gell, Michael A.; Fasol, Gerhard; Gibbings, C. J.; Tuppen, C. G.)V1361,293-301(1991)

Rare-earth-doped fluoride glasses for active optical fiber applications (Adam, Jean-Luc; Smektala, Frederic; Denoue, Emmanuel; Lucas, Jacques)V1513,150-157(1991)

Rise time and recovery of GaAs photoconductive semiconductor switches (Zutavern, Fred J.; Loubriel, Guillermo M.; O'Malley, Marty W.; McLaughlin, Dan L.; Helgeson, Wes D.)V1378,271-279(1991)

SHG in fiber: is a high-conversion efficiency possible? (Ouellette, Francois)V1516,113-114(1991)

Soft x-ray, optical, and thermal properties of hard carbon films (Alvey, Mark D.)V1330,39-45(1991)

Solid state ionics and optical materials technology for energy efficiency, solar energy conversion, and environment control (Lusis, Andrejs R.)V1536,116-124(1991)

Some new insights in the physics of quantum-well devices (Tsu, Raphael)V1361,313-324(1991)

Space station atomic-oxygen-resistant coatings (Grieser, James L.; Freeland, Alan W.; Fink, Jeffrey D.; Meinke, Gary E.; Hildreth, Eugene N.)V1330,102-110(1991)

Status and needs of infrared optical property information for optical designers (Wolfe, William L.)V1354,696-741(1991)

Structure and optical properties of a-C:H/a-SiOx:H multilayer thin films (Zhang, Wei P.; Cui, Jing B.; Xie, Shan; Song, Yi Z.; Wang, Chang S.; Zhou, Guien; Wu, Jan X.)V1519,23-25(1991)

Structural and optical properties of chalcogenide glass-ceramics (Boehm, Leah; Assaf, Haim)V1535,171-181(1991)

Structural and optical properties of semiconducting microcrystallite-doped SiO2 glass films prepared by rf-sputtering (Tsunetomo, Keiji; Shimizu, Ryuichiro; Yamamoto, Masaki; Osaka, Yukio)V1513,93-104(1991)

Structure and properties of electrochromic WO3 produced by sol-gel methods (Bell, J. M.; Green, David C.; Patterson, A.; Smith, Geoffrey B.; MacDonald, K. A.; Lee, K. D.; Kirkup, L.; Cullen, J. D.; West, B. O.; Spiccia, L.; Kenny, M. J.; Wielunski, L. S.)V1536,29-36(1991)

Study of properties of a-Si1-xGex:H prepared by SAP-CVD method (Wang, Yi-Ming; Jing, Lian-hua; Pang, Da-wen)V1361,325-335(1991)

Study on the special properties of electrochromic film of a-WO3 (Luo, Zhongkuan)V1489,124-134(1991)

Synthesis and nonlinear optical properties of preformed polymers forming Langmuir-Blodgett films (Verbiest, Thierry; Persoons, Andre P.; Samyn, Celest)V1560,353-361(1991)

Synthesis and properties of sol-gel-derived AgClxBr1-x colloid containing sodium alumo borosilicate glasses (Mennig, Martin; Schmidt, Helmut; Fink, Claudia)V1590,152-159(1991)

Temperature variations of reflection, transmission, and fluorescence of the arterial wall (Chambettaz, Francois; Clivaz, Xavier; Marquis-Weible, Fabienne D.; Salathe, R. P.)V1427,134-140(1991)

Third-order optical nonlinearities and femtosecond responses in metallophthalocyanine thin films made by vacuum deposition, molecular beam epitaxy, and spin coating (Wada, Tatsuo; Hosoda, Masahiro; Garito, Anthony F.; Sasabe, Hiroyuki; Terasaki, A.; Kobayashi, Takayoshi T.; Tada, Hiroaki; Koma, Atsushi)V1560,162-171(1991)

TiNxOy-Cu coatings for low-emissive solar-selective absorbers (Lazarov, M.; Roehle, B.; Eisenhammer, T.; Sizmann, R.)V1536,183-191(1991)

Transparent storage layers for H+ and Li+ ions prepared by sol-gel technique (Valla, Bruno; Tonazzi, Juan C.; Macedo, Marcelo A.; Dall'Antonia, L. H.; Aegerter, Michel A.; Gomes, M. A.; Bulhoes, Luis O.)V1536,48-62(1991)

Two-dimensional model for high-efficiency microgroove silicon solar cells (Sobhan, M. A.; Islam, M. N.)V1536,246-257(1991)

UV spectroscopy of optical fibers and preforms (Williams, Doug L.; Davey, Steven T.; Kashyap, Raman; Armitage, J. R.; Ainslie, B. J.)V1516,29-37(1991)

Vacuum outgassing from diffuse-absorptive baffle materials (Egert, Charles M.; Basford, J. A.)V1330,178-185(1991)

Variation of optical properties of gel-derived VO2 thin films with temperature (Hou, Li S.; Lu, Song W.; Gan, Fuxi)V1519,580-588(1991)

Variation of the point spread function in the Sargasso Sea (Voss, Kenneth J.)V1537,97-103(1991)

Verdet constant and its dispersion in optical glasses (Westenberger, Gerhard; Hoffmann, Hans J.; Jochs, Werner W.; Przybilla, Gudrun)V1535,113-120(1991)

Volume holographic optics recording in photopolymerizable layers (Boiko, Yuri B.)V1507,318-327(1991)

What we can learn about second-harmonic generation in germanosilicate glass from the analogous effect in semiconductor-doped glasses (Lawandy, Nabil M.)V1516,99-112(1991)

Optical recording—see also magneto-optics; optical data storage; optical disks

Acousto-optics in integrated-optic devices for optical recording (Petrov, Dmitry V.; Belostotsky, A. L.; Dolgopolov, V. G.; Leonov, A. S.; Fedjukhin, L. A.)V1374,152-159(1991)

Advanced thin films for optical storage (Gan, Fuxi)V1519,530-538(1991)

Applications of diamond-like carbon films for write-once optical recording (Armeyev, V. Y.; Arslanbekov, A. H.; Chapliev, N. I.; Konov, Vitaly I.; Ralchenko, V. G.; Strelnitsky, V. E.)V1621,2-10(1991)

Automated systems for quantitative analysis of holograms (Pryputniewicz, Ryszard J.)VIS08,215-246(1991)

Bridging the gap between Soviet and Western holography (Phillips, Nicholas J.)VIS08,206-214(1991)

Compatability test for phase-change erasable and WORM media in a multifunction drive (Ohara, Shunji; Ishida, Takashi; Inokuchi, Chikashi; Furutani, Tadashige; Ishibashi, Kenzo; Kurahashi, Akira; Yoshida, Tomio)V1499,187-194(1991)

Crystalization kinetics in Ge2Sb2Te5 phase-change recording films (Ozawa, Kenji; Ogino, Shinji; Satoh, Yoshikazu; Urushidani, Tatuo; Ueda, Atushi; Deno, Hiroshi; Kawakami, Haruo)V1499,180-186(1991)

Directions in optical storage (Jipson, Victor B.)V1499,2-4(1991)

Directly overwritable magneto-optical disk with light-power-modulation method using no initializing magnet (Tsutsumi, K.; Nakaki, Y.; Fukami, T.; Tokunaga, T.)V1499,55-61(1991)

Direct overwritable magneto-optical exchange-coupled multilayered disk by laser power modulation recording (Saito, Jun; Akasaka, Hideki)V1499,44-54(1991)

Direct overwriting system by light-intensity modulation using triple-layer disks (Maeda, Fumisada; Arai, Masayuki; Owa, Hideo; Takahashi, Hiroo; Kaneko, Masahiko)V1499,62-69(1991)

Erasable compact disk utilizing phase-change material and multipulse recording method (Ohno, Eiji; Nishiuchi, Kenichi; Yamada, Noboru; Akahira, Nobuo)V1499,171-179(1991)

Herbarium holographicum (Boone, Pierre M.)VIS08,370-380(1991)

High-capacity optical spatial switch based on reversible holograms (Mikaelian, Andrei L.; Salakhutdinov, Viktor K.; Vsevolodov, N. N.; Dyukova, T. V.)V1621,148-157(1991)

High-speed large-capacity optical disk using pit-edge recording and MCAV method (Maeda, Takeshi; Saito, Atsushi; Sugiyama, Hisataka; Ojima, Masahiro; Arai, Shinichi; Shigematu, Kazuo)V1499,414-418(1991)

High-speed recording technologies for a magneto-optical disk system (Sukeda, Hirofumi; Tsuchinaga, Hiroyuki; Tanaka, Satoshi; Niihara, Toshio; Nakamura, Shigeru; Mita, Seiichi; Yamada, Yukinori; Ohta, Norio; Fukushima, Mitsugi)V1499,419-425(1991)

Holographic data storage on the photothermoplastic tape carrier (Akaev, Askar A.; Zhumaliev, K. M.; Jamankyzov, N.)V1621,182-193(1991)

Holographic portraits (Bjelkhagen, Hans I.)VIS08,347-353(1991)

Holographic recordings on 2-hydroxyethyl methacrylate and applications of water-immersed holograms (Yacoubian, Araz; Savant, Gajendra D.; Aye, Tin M.)V1559,403-409(1991)

Holography (Greguss, Pal; Jeong, Tung H., eds.)VIS08(1991)

Interaction of cracks, waves, and contacts: a photomechanics study (Rossmanith, H. P.)V1554A,2-28(1991); V1554B,2-28(1991)

Langmuir-Blodgett films of tetra-tert-butyl-phenoxy phthalocyanine iron [II] (Luo, Tao; Zhang, Wei Q.; Gan, Fuxi)V1519,826-830(1991)

Laser-induced phase transition in crystal InSb films as used in optical storage (Sun, Yang; Li, Cheng F.; Deng, He; Gan, Fuxi)V1519,554-558(1991)

Light-sensitive organic media for optical discs (Barachevsky, Valery A.; Rot, A. S.; Zaks, I. N.)V1621,33-44(1991)

Magneto-optical linear multichannel light modulator for recording of one-dimensional holograms (Abakumov, D. M.)V1621,138-147(1991)

Metal-ion-doped polymer systems for real-time holographic recording (Lessard, Roger A.; Changkakoti, Rupak; Manivannan, Gurusamy)V1559,438-448(1991)

Multitrack rewritable optical disk system for high-performance applications: 14-inch TODS (Cinelli, Joseph L.; Kozak, Taras)V1499,203-208(1991)

New approach to high-density recording on a magneto-optical disk (Fujita, Goro; Urakawa, Yoshinori; Yamagami, Tamotsu; Watanabe, Tetsu)V1499,426-432(1991)

New method of recording holographic optical elements applied to optical interconnection in VLSI (Zhao, Feng; Sun, Junyong; Geng, Wanzhen; Jiang, Lingzhen; Hong, Jing)V1461,262-264(1991)

New possibilities of photosensitive glasses for three-dimensional phase hologram recording (Glebov, Leonid B.; Nikonorov, Nikolai V.; Panysheva, Elena I.; Petrovskii, Gurii T.; Savvin, Vladimir V.; Tunimanova, Irina V.; Tsekhomskii, Victor A.)V1621,21-32(1991)

New technique for characterizing holographic recording materials (Leclere, Philippe; Renotte, Yvon L.; Lion, Yves F.)V1559,298-307(1991)

Non-gelatin-dichromated holographic film (Guo, Lu Rong; Cheng, Qirei; Wang, Kuoping)V1461,91-92(1991)

One-dimensional hologram recording in thick photolayer disk (Mikaelian, Andrei L.; Gulanian, Emin H.; Kretlov, Boris S.; Molchanova, L. V.; Semichev, V. A.)V1621,194-203(1991)

Optical associative memory for nontraditional architecture digital computers and database management systems (Burtsev, Vsevolod S.; Fyodorov, Vyatcheslav B.)V1621,215-226(1991)

Optical Data Storage '91 (Burke, James J.; Shull, Thomas A.; Imamura, Nobutake, eds.)V1499(1991)

Optical fibers for magneto-optical recording (Opsasnick, Michael N.; Stancil, Daniel D.; White, Sean T.; Tsai, Ming-Horn)V1499,276-280(1991)

Optical information recording on vitreous semiconductors with a thermoplastic method of visualization (Panasyuk, L. M.; Forsh, A. A.)V1621,74-82(1991)

Optical interconnect and neural network applications based on a simplified holographic N4 recording technique (Lin, Freddie S.; Lu, Taiwei; Kostrzewski, Andrew A.; Chou, Hung)V1558,406-413(1991)

Optical Memory and Neural Networks (Mikaelian, Andrei L., ed.)V1621(1991)

Optical memory using localized photoinduced anisotropy in a synthetic dye-polymer (Kuo, Chai-Pei)V1499,148-159(1991)

Optical recording material based on bacteriorhodopsin modified with hydroxylamine (Vsevolodov, N. N.; Dyukova, T. V.; Druzhko, A. B.; Shakhbazyan, V. Y.)V1621,11-20(1991)

Optical recording materials (Savant, Gajendra D.; Jannson, Joanna)V1461,79-90(1991)

Optimization of quadrilayer structures for various magneto-optical recording materials (He, Ping; McGahan, William A.; Woollam, John A.)V1499,401-411(1991)

Optimization of reconstruction geometry for maximum diffraction efficiency in HOE: the influence of recording material (Belendez, A.; Pascual, I.; Fimia-Gil, Antonio)V1574,77-83(1991)

Organic-dye films for write-once optical storage (Zhou, Jian P.; Shu, Ju P.; Xu, Hui J.)V1519,559-564(1991)

Organo-metallic thin film for erasable optical recording medium (Shu, Ju P.; Zhou, Jian P.; Xu, Shi Z.)V1519,565-569(1991)

Overview of a high-performance polygonal scanner subsystem (Rynkowski, Gerald)V1454,102-110(1991)

Performance comparison of various data codes in Z-CAV optical recording (Lee, Tzuo-chang; Chen, Di)V1499,87-103(1991)

Photoanisotropic incoherent-to-coherent conversion using five-wave mixing (Huang, Tizhi; Wagner, Kelvin H.)V1562,44-54(1991)

Photopolymer elements for an optical correlator system (Brandstetter, Robert W.; Fonneland, Nils J.)V1559,308-320(1991)

Photothermoplastic molecular heterostructures (Cherkasov, Yuri A.)V1621,62-73(1991)

Preparation and magneto-optical properties of NdDyFeCoTi amorphous films (Zhang, Si J.; Yang, Xiao Y.; Li, Xiao L.; Zhang, Feng P.)V1519,744-751(1991)

Pulsed holographic recording of very high speed transient events (Steckenrider, John S.; Ehrlich, Michael J.; Wagner, James W.)V1554B,106-112(1991)

Recording and erasing characteristics of GeSbTe-based phase change thin films (Hou, Li S.; Zhu, Chang X.; Gu, Dong H.)V1519,548-553(1991)

Reducing and magnifying holograms (Fargion, Daniele)VIS08,354-359(1991)

Reversible phase-change optical recording by using microcellular GeSbTeCo recording film (Okamine, Shigenori; Terao, Motoyasu; Andoo, Keikichi; Miyauchi, Yasushi)V1499,166-170(1991)

(Sb2Se3)1-x Nix alloy thin films and its application in erasable phase change optical recording (Xue, Song S.; Fan, Zheng X.; Gan, Fuxi)V1519,570-574(1991)

Scanning optical microscopy: a powerful tool in optical recording (Coombs, James H.; Holtslag, A. H.)V1499,6-20(1991)

Shot-noise-limited high-power green laser for higher density optical disk systems (Kubota, Shigeo; Oka, Michio)V1499,314-323(1991)

Study of optical data recording based on photoluminescence effect (Petrov, V. V.; Zymenko, V. I.; Kravetz, V. G.; Polishchuk, E. Y.; Sushko, A. M.)V1621,45-50(1991)

Theoretical background and practical processing for art and technical work in dichromated gelatin holography (Coblijn, Alexander B.)VIS08,305-324(1991)

Thermomagnetic recording on amorphous ferrimagnetic films (Aleksandrov, K. S.; Berman, G. P.; Frolov, G. I.; Seredkin, V. A.)V1621,51-61(1991)

Three-beam overwritable magneto-optic disk drive (Watabe, Akinori; Yamada, Ichiro; Yamamoto, Manabu; Katoh, Kikuji)V1499,226-235(1991)

Tracking method of optical tape recorder using acousto-optic scanning (Narahara, Tatsuya; Kumai, Satoshi; Nakao, Takashi; Ozue, Tadashi)V1499,120-128(1991)

Wavelength shift in DCG holograms (Lelievre, Sylviane; Pawluczyk, Romuald)V1559,288-297(1991)

Write noise from optical heads with nonachromatic beam expansion prisms (Kay, David B.; Chase, Scott B.; Gage, Edward C.; Silverstein, Barry D.)V1499,281-285(1991)

Optical systems—see also imaging systems; optical design; optomechanical design

Active optics system for a 3.5-meter structured mirror (Stepp, Larry M.; Roddier, Nicolas; Dryden, David M.; Cho, Myung K.)V1542,175-185(1991)

Advanced matrix optics and its incidence in laser optics (Wang, Shaomin; Bernabeu, Eusebio; Alda, Javier)V1397,595-602(1991)

Application of the ProtoWare simulation testbed to the design and evaluation of advanced avionics (Bubb, Daniel; Wilson, Leo T.; Stoltz, John R.)V1483,18-28(1991)

Automatic active athermalization of infrared optical systems (Kuerbitz, Gunther)V1540,612-621(1991)

Computational model for the stereoscopic optics of a head-mounted display (Robinett, Warren; Rolland, Jannick P.)V1457,140-160(1991)

Current Developments in Optical Design and Optical Engineering (Fischer, Robert E.; Smith, Warren J., eds.)V1527(1991)

Detection of the object velocity using the time-varying scattered speckles (Okamoto, Takashi; Asakura, Toshimitsu)V1399,192-199(1991)

Development of a high-speed high-precision laser plotter (Tachihara, Satoru; Miyoshi, Tamihiro)V1527,305-314(1991)

Distributed optical fiber sensing (Rogers, Alan J.)V1504,2-24(1991); V1506,2-24(1991); 1507,2-24(1991)

Excimer laser photolithography with a 1:1 broadband catadioptric optics (Zhang, Yudong; Lu, Dunwu; Zou, Haixing; Wang, Zhi-Jiang)V1463,456-463(1991)

High-level PC-based laser system modeling (Taylor, Michael S.)V1415,300-309(1991)

High-power laser/optical-fiber-coupling device (Falciai, Riccardo; Pascucci, Tania)V1506,120-125(1991)

High-resolution thermal imager with a field-of-view of 112 degrees (Matsushita, Tadashi; Suzuki, Hiroshi; Wakabayashi, Satoshi; Tajime, Toru)V1488,368-375(1991)

History of infrared optics (Johnson, R. B.; Feng, Chen)VCR38,3-18(1991)

Holographic center high-mounted stoplight (Smith, Ronald T.)V1461,186-198(1991)

Holographic optical security systems (Fagan, William F.)VIS08,381-386(1991)

Image plane tilt in optical systems (Sasian, Jose M.)V1527,85-95(1991)

Improvements in 0.5-micron production wafer steppers (Luehrmann, Paul F.; de Mol, Chris G.; van Hout, Frits J.; George, Richard A.; van der Putten, Harrie B.)V1463,434-445(1991)

Infrared microanalysis of contaminants at grazing incidence (Reffner, John A.)V1437,89-94(1991)

Large one-step holographic stereogram (Honda, Toshio; Kang, Der-Kuan; Shimura, Kei; Enomoto, H.; Yamaguchi, Masahiro; Ohyama, Nagaaki)V1461,156-166(1991)

Laser beam diagnostics: a conference overview (Sadowski, Thomas J.)V1414,136-140(1991)

Laser surgical unit for photoablative and photothermal keratoplasty (Cartlidge, Andy G.; Parel, Jean-Marie; Yokokura, Takashi; Lowery, Joseph A.; Kobayashi, K.; Nose, I.; Lee, William; Simon, Gabriel; Denham, David B.)V1423,167-174(1991)

Latest advances in CAD data interfacing: a standardization project of ISO/TC 172/SC1 task group "optical database" (Wise, Timothy D.; Wieder, Eckart)V1346,79-85(1991)

Matrix optical system for plane-point correlation (Curatu, Eugen O.)V1527,368-375(1991)

Matrix representation of multimode beam transformation (Alda, Javier; Porras, Miguel A.; Bernabeu, Eusebio)V1527,240-251(1991)

Modeling and Simulation of Laser Systems II (Schnurr, Alvin D., ed.)V1415(1991)

Moving M2 mirror without pointing offset (Ragazzoni, Roberto; Bortoletto, Favio)V1542,236-246(1991)

Narcissus in current generation FLIR systems (Ford, Eric H.; Hasenauer, David M.)VCR38,95-119(1991)

One-dimensional rotatory waves in the optical systems with nonlinear large-scale field interactions (Ivanov, Vladimir Y.; Larichev, A. V.; Vorontsov, Michael A.)V1402,145-153(1991)

Optical system design for a lunar optical interferometer (Colavita, Mark M.; Shao, Michael; Hines, Braden E.; Levine, Bruce M.; Gershman, Robert)V1494,168-181(1991)

Optical system for automatic inspection of curved surfaces (Livi, S.; Magnani, M.; Pieri, Silvano; Romoli, Andrea)V1500,144-150(1991)

Optical system for control of longitudinal displacement (Dick, Sergei C.; Markhvida, Igor V.; Tanin, Leonid V.)V1454,447-452(1991)

Optical Systems in Adverse Environments (Kuok, M. H.; Silva, Donald E.; Tam, Siu-Chung, eds.)V1399(1991)

Passive integrated optics in optical sensor systems (Parriaux, Olivier M.)V1506,111-119(1991)

Passive range and azimuth measuring system (Ronning, E.; Fjarlie, E. J.)V1399,178-183(1991)

Phased-array antenna control by a monolithic photonic integrated circuit (Hietala, Vincent M.; Vawter, Gregory A.; Meyer, W. J.; Kravitz, Stanley H.)V1476,170-175(1991)

Realization of a coherent optical DPSK (differential phase-shift keying) heterodyne transmission system with 565 MBit/s at 1.064 um (Wandernoth, Bernhard; Franz, Juergen)V1522,194-198(1991)

Realization of heterodyne acquisition and tracking with diode lasers at lambda=1.55 um (Hueber, Martin F.; Leeb, Walter R.; Scholtz, Arpad L.)V1522,259-267(1991)

Research and development of optical measurement techniques for aerospace propulsion research: a NASA/Lewis Research Center perspective (Lesco, Daniel J.)V1554B,530-539(1991)

Review of methods for the design of unsymmetrical optical systems (Sasian, Jose M.)V1396,453-466(1991)

Review of nondiffracting Bessel beams (LaPointe, Michael R.)V1527,258-276(1991)

Selected performance parameters and functional principles of a new stepper generation (Kliem, Karl-Heinz; Sczepanski, Volker; Michl, Uwe; Hesse, Reiner)V1463,743-751(1991)

Sensitivity analysis of Navy tactical decision-aid FLIR performance codes (Milne, Edmund A.; Cooper, Alfred W.; Reategui, Rodolfo; Walker, Philip L.)V1486,151-161(1991)

Space-borne beam pointing (Eller, E. D.; LaMont, Douglas V.; Rodden, Jack J.)V1415,2-12(1991)

Specification of precision optical pointing systems (Medbery, James D.; Germann, Lawrence M.)V1489,163-176(1991)

System concepts for a large UV/optical/IR telescope on the moon (Nein, Max E.; Davis, Billy)V1494,98-110(1991)

System design for lunar-based optical and submillimeter interferometers (Gershman, Robert; Mahoney, Michael J.; Rayman, M. D.; Shao, Michael; Snyder, Gerald C.)V1494,160-167(1991)

Three-dimensional display of inside of human body by holographic stereogram (Sato, Koki; Akiyama, Iwaki; Shoji, Hideo; Sumiya, Daigaku; Wada, Tuneyo; Katsuma, Hidetoshi; Itoh, Koichi)V1461,124-131(1991)

Three-dimensional integration of optical systems (Brenner, Karl-Heinz)V1506,94-98(1991)

Two-dimensional dynamic neural network optical system with simplest types of large-scale interactions (Vorontsov, Michael A.; Zheleznykh, N. I.; Larichev, A. V.)V1402,154-164(1991)

University of Hawaii adaptive optics system: I. General approach (Roddier, Francois J.; Graves, J. E.; McKenna, Daniel; Northcott, Malcolm J.)V1542,248-253(1991)

University of Hawaii adaptive optics system: II. Computer simulation (Northcott, Malcolm J.)V1542,254-261(1991)

Upgrade of the LLNL Nova laser for inertial confinement fusion (Murray, John R.; Trenholme, J. B.; Hunt, John T.; Frank, D. N.; Lowdermilk, W. H.; Storm, E.)V1410,28-39(1991)

Optical testing—see also lenses; metrology; mirrors; optical fabrication; surfaces

Absolute interferometric testing of spherical surfaces (Truax, Bruce E.)V1400,61-68(1991)

Absolute measurement of spherical surfaces (Creath, Katherine; Wyant, James C.)V1332,2-7(1991)

Accuracies in FLIR test equipment (Fourier, Ron)V1442,109-117(1991)

Analysis of Optical Structures (O'Shea, Donald C., ed.)V1532(1991)

Analysis of powertrain noise and vibration using interferometry (Griffen, Christopher T.; Fryska, Slawomir T.; Bernier, Paul R.)V1554B,754-766(1991)

Analysis of thermal stability of fused optical structure (Powell, William R.)V1532,126-136(1991)

Aspheric surface testing techniques (Stahl, H. P.)V1332,66-76(1991)

Aspheric testing using null mirrors (Murty, Mantravady V.; Kumar, Vas; von Handorf, Robert J.)V1332,107-114(1991)

Astigmatism and field curvature from pin-bars (Kirk, Joseph P.)V1463,282-291(1991)

Automatic inspection technique for optical surface flaws (Yang, Guoguang; Gao, Wenliang; Cheng, Shangyi)V1332,56-63(1991)

AutoSPEC image evaluation laboratory (Brown, James C.; Webb, Curtis M.; Bell, Paul A.; Washington, Randolph T.; Riordan, Richard J.)V1488,300-311(1991)

Characterization of the dimensional stability of advanced metallic materials using an optical test bench structure (Hsieh, Cheng; O'Donnell, Timothy P.)V1533,240-251(1991)

Chemical-vapor-deposited silicon and silicon carbide optical substrates for severe environments (Goela, Jitendra S.; Pickering, Michael A.; Taylor, Raymond L.)V1330,25-38(1991)

Comparison of Kodak Professional Digital Camera System images to conventional film, still video, and freeze-frame images (Kent, Richard A.; McGlone, John T.; Zoltowski, Norbert W.)V1448,27-44(1991)

Cone array of Fourier lenses for contouring applications (Goldfain, Ervin)V1527,210-215(1991)

Deep-UV diagnostics using continuous tone photoresist (Kirk, Joseph P.; Hibbs, Michael S.)V1463,575-583(1991)

Design, analysis, and testing of a CCD array mounting structure (Sultana, John A.; O'Neill, Mark B.)V1532,27-38(1991)

Design and manufacture of an ultralightweight solid deployable reflector (Tremblay, Gary A.; Derby, Eddy A.)V1532,114-123(1991)

Design and specification of diamond-turned optics (Clark, Robert A.)VCR38,164-183(1991)

Determination of FLIR LOS stabilization errors (Pinsky, Howard J.)V1488,334-342(1991)

Determination of surface quality using moire methods (Keren, Eliezer; Kreske, Kathi; Zac, Yaacov; Livnat, Ami)V1442,266-274(1991)

Development and testing of lightweight composite reflector panels (Helms, Richard G.; Porter, Christopher C.; Kuo, Chin-Po; Tsuyuki, Glenn T.)V1532,64-80(1991)

Diffraction analysis of optical disk readout signal deterioration caused by mark-size fluctuation (Honguh, Yoshinori)V1527,315-321(1991)

Diffraction performance calculations in lens design (Malacara Hernandez, Daniel)V1354,2-14(1991)

Digital Talbot interferometer (Tam, Siu-Chung; Silva, Donald E.; Wong, H. L.)V1400,38-48(1991)

Extending electrical measurements to the 0.5 um regime (Troccolo, Patrick M.; Mantalas, Lynda C.; Allen, Richard A.; Linholm, Loren)V1464,90-103(1991)

Fiber access maintenance leverages (Sinnott, Heather; MacLeod, Wade)V1363,196-200(1991)

Field-portable laser beam diagnostics (Forrest, Gary T.)V1414,55-64(1991)

Finite element analysis enhancement of cryogenic testing (Thiem, Clare D.; Norton, Douglas A.)V1532,39-47(1991)

Finite element analysis of large lenses for the Keck telescope high-resolution echelle spectrograph (Bigelow, Bruce C.)V1532,15-26(1991)

Free-space simulator for laser transmission (Inagaki, Keizo; Nohara, Mitsuo; Araki, Ken'ichi; Fujise, Masayuki; Furuhama, Yoji)V1417,160-169(1991)

Helmet-mounted sight and display testing (Boehm, Hans-Dieter V.; Schreyer, H.; Schranner, R.)V1456,95-123(1991)

High-precision interferometric testing of spherical mirrors with long radius of curvature (Freischlad, Klaus R.; Kuechel, Michael F.; Wiedmann, Wolfgang; Kaiser, Winfried; Mayer, Max)V1332,8-17(1991)

High-spatial-resolution and high-sensitivity interferometric optical-time-domain reflectometer (Kobayashi, Masaru; Noda, Juichi; Takada, Kazumasa; Taylor, Henry F.)V1474,278-284(1991)

Holographic interferometry in corrosion studies of metals: I. Theoretical aspects (Habib, Khaled J.)V1332,193-204(1991)

Holographic interferometry in corrosion studies of metals: II. Applications (Habib, Khaled J.)V1332,205-215(1991)

Holographic optical profilometer (Hasman, Erez; Davidson, Nir; Friesem, Asher A.)V1442,372-377(1991)

Holographic testing canal of adaptive optical systems (Gan, Michael A.; Potyemin, Igor S.)V1507,549-560(1991)

Imaging characteristics of the development model of the SAX x-ray imaging concentrators (Citterio, Oberto; Conconi, Paolo; Conti, Giancarlo; Mattaini, E.; Santambrogio, E.; Cusumano, G.; Sacco, B.; Braueninger, Heinrich; Burkert, Wolfgang)V1343,145-154(1991)

In-flight performance of the Goddard high-resolution spectrograph of the Hubble Space Telescope (Troeltzsch, John R.; Ebbets, Dennis C.; Garner, Harry W.; Tuffli, A.; Breyer, R.; Kinsey, J.; Peck, C.; Lindler, Don; Feggans, J.)V1494,9-15(1991)

Infrared simulation of missile dome heating (Rich, Brian W.)V1540,781-786(1991)

In-line testing for fiber subscriber loop applications (Jiang, Jing-Wen; So, Vincent; Lessard, Michel; Vella, Paul J.)V1363,191-195(1991)

In-situ structural studies of the underpotential deposition of copper onto an iodine-covered platinum surface using x-ray standing waves (Bommarito, G. M.; Acevedo, D.; Rodriguez, J. R.; Abruna, H. D.)V1550,156-170(1991)

Interferometer accuracy and precision (Selberg, Lars A.)V1400,24-32(1991)

Interferometer for testing aspheric surfaces with electron-beam computer-generated holograms (Gemma, Takashi; Hideshima, Masayuki; Taya, Makoto; Watanabe, Nobuko)V1332,77-84(1991)

Interferometer for testing of general aspherics using computer-generated holograms (Arnold, Steven M.; Jain, Anil K.)V1396,473-480(1991)

ISTS array detector test facility (Thomas, Paul J.; Hollinger, Allan B.; Chu, Kan M.; Harron, John W.)V1488,36-47(1991)

Laser beam width, divergence, and propagation factor: status and experience with the draft standard (Fleischer, John M.)V1414,2-11(1991)

Long-term reliability and performance testing of fiber optic sensors for engineering applications (Weinberger, Alex; Weinberger, Ervin)V1367,30-45(1991)

Model-based flaw reconstruction using limited-view x-ray projections and flawless prototype image (Hung, Hsien-Sen; Eray, Mete)V1550,34-45(1991)

Multisensor fusion methodologies compared (Swan, John; Shields, Frank J.)V1483,219-230(1991)

Next generation thermal control coatings (Grieser, James L.; Swisher, Richard L.; Phipps, James A.; Pelleymounter, Douglas R.; Hildreth, Eugene N.)V1330,111-118(1991)

Noncontact optical microtopography (Costa, Manuel Filipe P.; Almeida, Jose B.)V1400,102-107(1991)

Nonlinear finite element analysis of the Starlab 80-cm telescope primary-mirror suspension system (Arnold, William R.)V1532,103-113(1991)

Novel interferometer setup for evaluating the sum of surface contributions to transmitted wavefront distortion (Williamson, Ray)V1527,188-193(1991)

On-line rapid testing of the optical transfer function (Xiang, Caixin; Xiang, Yang)V1527,427-436(1991)

Optical analysis of thermal-induced structural distortions (Weinswig, Shepard A.; Hookman, Robert A.)V1527,118-125(1991)

Optical aspheric surface profiler using phase shift interferometry (Sasaki, Kenji; Ono, Akira)V1332,97-106(1991)

Optical delay tester (Wakana, Shin-ichi; Nagai, Toshiaki; Hama, Soichi; Goto, Yoshiro)V1479,283-290(1991)

Optical Fabrication and Testing (Campbell, Duncan R.; Johnson, Craig W.; Lorenzen, Manfred, eds.)V1400(1991)

Optical figure testing of prototype mirrors for JPL's precision segmented-reflector program (Hochberg, Eric B.)V1542,511-522(1991)

Optical materials: evaluation methodology and data base utility (Patty, Charles E.; McMahon, David M.)V1535,13-26(1991)

Optical performance of an infrared astronomical telescope with a 5-axis secondary mirror (Ettedgui-Atad, Eli; Humphries, Colin M.)V1532,241-248(1991)

Optical phase-conjugate resonators, bistabilities, and applications (Venkateswarlu, Putcha; Dokhanian, Mostafa; Sekhar, Prayaga C.; George, M. C.; Jagannath, H.)V1332,245-266(1991)

Optical profilometry using spatial heterodyne interferometric methods (Frankowski, Gottfried)V1500,114-123(1991)

Optical Testing and Metrology III: Recent Advances in Industrial Optical Inspection (Grover, Chandra P., ed.)V1332(1991)

Optical testing by dynamic holographic interferometry with photorefractive crystals and computer image processing (Vlad, Ionel V.; Popa, Dragos; Petrov, M. P.; Kamshilin, Alexei A.)V1332,236-244(1991)

Optical testing with wavelength scanning interferometer (Okada, Katsuyuki; Tsujiuchi, Jumpei)V1400,33-37(1991)

Optomechanics and Dimensional Stability (Paquin, Roger A.; Vukobratovich, Daniel, eds.)V1533(1991)

Overview of the finite element method in optical systems (Hatheway, Alson E.)V1532,2-14(1991)

Performance and measurements of refractive microlens arrays (Feldblum, Avi Y.; Nijander, Casimir R.; Townsend, Wesley P.; Mayer-Costa, Carlos M.)V1544,200-208(1991)

Performance of the Multi-Spectral Solar Telescope Array III: optical characteristics of the Ritchey-Chretien and Cassegrain Telescopes (Hoover, Richard B.; Baker, Phillip C.; Hadaway, James B.; Johnson, R. B.; Peterson, Cynthia; Gabardi, David R.; Walker, Arthur B.; Lindblom, Joakim F.; DeForest, Craig E.; O'Neal, Ray H.)V1343,189-202(1991)

Phase-conjugate Twyman-Green interferometer for testing conicoidal surfaces (Shukla, Ram P.; Dokhanian, Mostafa; Venkateswarlu, Putcha; George, M. C.)V1332,274-286(1991)

Pulsed electron-beam testing of optical surfaces (Murray, Brian W.; Johnson, Edward A.)V1330,2-24(1991)

Real-time wavefront measurement with lambda/10 fringe spacing for the optical shop (Freischlad, Klaus R.; Kuechel, Michael F.; Schuster, Karl-Heinz; Wegmann, Ulrich; Kaiser, Winfried)V1332,18-24(1991)

Reconstruction of the Hubble Space Telescope mirror figure from out-of-focus stellar images (Roddier, Claude A.; Roddier, Francois J.)V1494,78-84(1991)

Reporting data for arrays with many elements (Coles, Christopher L.; Phillips, Wayne S.; Vincent, John D.)V1488,327-333(1991)

Results from the calibration of the Extreme Ultraviolet Explorer instruments (Welsh, Barry Y.; Jelinsky, Patrick; Vedder, Peter W.; Vallerga, John V.; Finley, David S.; Malina, Roger F.)V1343,166-174(1991)

Rigidity test of large and high-precision instruments (Ma, Pin-Zhong)V1532,177-186(1991)

Rigorous optical theory of the D Sight phenomenon (Reynolds, Rodger L.; Hageniers, Omer L.)V1332,85-96(1991)

Scattering measurements of optical coatings in high-power lasers (Chen, Yi-Sheng)V1332,115-120(1991)

Selected Papers on Interferometry (Hariharan, P., ed.)VMS28(1991)

Selected Papers on Speckle Metrology (Sirohi, Rajpal S., ed.)VMS35(1991)

Set of two 45 - 90 - 45 prisms equivalent to the Fresnel rhomb (Murty, M. V.; Shukla, Ram P.; Apparao, K. V.)V1332,41-49(1991)

Simple test for the 90 degree angle in prisms (Malacara Hernandez, Daniel; Flores, Ricardo)V1332,36-40(1991)

Simultaneous alignment and multiple surface figure testing of optical system components via wavefront aberration measurement and reverse optimization (Lundgren, Mark A.; Wolfe, William L.)V1354,533-539(1991)

Space-borne beam pointing (Eller, E. D.; LaMont, Douglas V.; Rodden, Jack J.)V1415,2-12(1991)

Spatial frequency selective error sensing for space-based, wide field-of-view, multiple-aperture imaging systems (Erteza, Ahmed; Schneeberger, Timothy J.)V1527,182-187(1991)

Statistical properties of intensity fluctuations produced by rough surfaces under the speckle pattern illumination (Yoshimura, Takeaki; Fujiwara, Kazuo; Miyazaki, Eiichi)V1332,835-842(1991)

Studies of structures and phase transitions of Langmuir monolayers using synchrotron radiation (Dutta, Pulak)V1550,134-139(1991)

Surface defect detection and classification with light scattering (Gebhardt, Michael; Truckenbrodt, Horst; Harnisch, Bernd)V1500,135-143(1991)

Surface distortions of a 3.5-meter mirror subjected to thermal variations (Cho, Myung K.; Poczulp, Gary A.)V1532,137-145(1991)

Switching speeds in NCAP displays: dependence on collection angle and wavelength (Reamey, Robert H.; Montoya, Wayne; Wartenberg, Mark)V1455,39-44(1991)

Temperature chamber FLIR and missile test system (Johnson, W. Todd; Lavi, Moshe; Sapir, Eyal)V1488,343-354(1991)

Testing binary optics: accurate high-precision efficiency measurements of microlens arrays in the visible (Holz, Michael; Stern, Margaret B.; Medeiros, Shirley; Knowlden, Robert E.)V1544,75-89(1991)

Three-dimensional analysis framework and measurement methodology for imaging system noise (D'Agostino, John A.; Webb, Curtis M.)V1488,110-121(1991)

Three-dimensional surface inspection using interferometric grating and 2-D FFT-based technique (Hung, Yau Y.; Tang, Shou-Hong; Zhu, Qiuming)V1332,696-703(1991)

White-light transmission holographic interferometry using chromatic corrective filters (Grover, Chandra P.)V1332,132-141(1991)

Work-induced stress and long-term stability in optically polished silicon (Bender, John W.; Wahl, Roger L.)V1533,264-276(1991)

X Rays in Materials Analysis II: Novel Applications and Recent Developments (Mills, Dennis M., ed.)V1550(1991)

Optical transfer function—see also optical design

Optimization of the optical transfer function (Kidger, Michael J.; Benham, Paul)V1354,92-96(1991)

Transer function of a 3-D reflective hologram (Odinokov, S. B.; Poddubnaya, T. E.; Rozhkov, Oleg V.)V1238,103-108(1991)

Optoelectronics—see also integrated optoelectronics

Advanced InGaAs/InP p-type pseudomorphic MODFET (Malzahn, Eric; Heuken, Michael; Gruetzmacher, Dettev; Stollenwerk, M.; Heime, Klaus)V1362,199-204(1991)

All-optical switching of picosecond pulses in GaAs MQW waveguides (Li Kam Wa, Patrick; Park, Choong-Bum; Miller, Alan)V1474,2-7(1991)

Application of fiber optic delay lines and semiconductor optoelectronics to microwave signal processing (Taylor, Henry F.)V1371,150-160(1991)

Application of surface effects externally to alter optical and electronic properties of existing optoelectronic devices: a ten-year update (Hava, Shlomo)V1540,350-358(1991)

Applications of high-Tc superconductors in optoelectronics (Sobolewski, Roman)V1501,14-27(1991); V1502,14-27(1991); 1506,25-38(1991)

Cold to hot electron transition devices (Chang, C. Y.)V1362,978-983(1991)

Comparison of optical and electronic 3-dimensional circuits (Stirk, Charles W.; Psaltis, Demetri)V1389,580-593(1991)

Distributed optical fiber sensing (Rogers, Alan J.)V1504,2-24(1991); V1506,2-24(1991); V1507,2-24(1991)

Electronic materials basic research program managed by the Advanced Technology Directorate of the U.S. Army Strategic Defense Command (Martin, William D.)V1559,10-17(1991)

Frequency response study of traps in III-V compound semiconductors (Kachwalla, Zain)V1361,784-793(1991)

High-resolution MS-type Saticon pick-up tube with optimized electron optical properties (Barden, Raimund; Mertelmeier, Thomas; Traupe, U.)V1449,136-147(1991)

High-speed GaAs metal-semiconductor-metal photodetectors with sub-0.1um finger width and finger spacing (Chou, Stephen Y.; Liu, Yue; Fischer, Paul B.)V1474,243-247(1991)

Holographic optical backplane hardware implementation for parallel and distributed processors (Kim, Richard C.; Lin, Freddie S.)V1474,27-38(1991)

Influence of photon energy on the photoemission from ultrafast electronic materials (Ghatak, Kamakhya P.; Ghoshal, Ardhendhu; Bhattacharyya, Sankar; Mondal, Manabendra)V1346,471-489(1991)

Infrared and Optoelectronic Materials and Devices (Naumaan, Ahmed; Corsi, Carlo; Baixeras, Joseph M.; Kreisler, Alain J., eds.)V1512(1991)

Infrared optical response of superconducting YBaCuO thin films (Sun, Han D.; Zhou, Fang Q.; Zhao, Xing R.; Wang, Lingjie; Yi, Xin J.)V1477,174-177(1991)

Integration of a coherent optical receiver with adaptive image rejection capability (Lachs, Gerard; Zaidi, Syed M.; Singh, Amit K.; Henning, Rudolf E.; Trascritti, D.; Kim, H.; Bhattacharya, Pallab K.; Pamulapati, J.; McCleer, P. J.; Haddad, George I.; Peng, S.)V1474,248-259(1991)

Interplay between photons and superconductors (Gilabert, Alain; Azema, Alain; Roustan, Jean-Claude; Maneval, Jean-Paul)V1477,140-147(1991)

Low-loss waveguides on silicon substrates for photonic circuits (Davis, Richard L.; Lee, Sae H.)V1474,20-26(1991)

Material for future InP-based optoelectronics: InGaAsP versus InGaAlAs (Quillec, Maurice)V1361,34-46(1991)

Microscopic mechanism of optical nonlinearity in conjugated polymers and other quasi-one-dimensional systems (Mazumdar, Sumit; Guo, Dandan; Dixit, Sham N.)V1436,136-149(1991)

Multiple-beam accessor using microzone plate elements for optoelectronic integrated circuits (Kodate, Kashiko; Kamiya, Takeshi)V1545,251-260(1991)

Nonlinear optic in-situ diagnostics of a crystalline film in molecular-beam-epitaxy devices (Krasnov, Victor F.; Musher, Semion L.; Prots, V. I.; Rubenchik, Aleksandr M.; Ryabchenko, Vladimir E.; Stupak, Mikhail F.)V1506,179-187(1991)

Optical and electrical properties of Al-Al2O3-Cu tunnel junctions (Shu, Q. Q.; Tian, X. M.; Chen, X. Y.; Cai, C. Z.; Zheng, K. Q.; Ma, W. G.)V1519,675-679(1991)

Optical/electronical hybrid three-layer neural network for pattern recognition (Zhang, Yanxin; Shen, Jinyuan)V1469,303-307(1991)

Optical implementation of neocognitron and its applications to radar signature discrimination (Chao, Tien-Hsin; Stoner, William W.)V1558,505-517(1991)

Optically powered optoelectronic integrated circuits for optical interconnects (Brown, Julia J.; Gardner, J. T.; Forrest, Stephen R.)V1474,236-242(1991)

Optical Technology for Signal Processing Systems (Bendett, Mark P., ed.)V1474(1991)

Optimal characteristics of rheology and electric field in deformable polymer films of optoelectronic image formation devices (Tarasov, Victor A.; Kuleshov, Nickolay B.; Novoselets, Mikhail K.; Sarkisov, Sergey S.)V1559,331-342(1991)

Optoelectronic Gabor detector for transient signals (Zhang, Yan; Li, Yao; Tolimieri, Richard; Kanterakis, Emmanuel G.; Katz, Al; Lu, X. J.; Caviris, Nicholas P.)V1481,23-31(1991)

Optoelectronic hardware issues for implementation of simulated annealing or Boltzmann machines (Lalanne, Philippe; Chavel, Pierre)V1621,388-401(1991)

Optoelectronic LED-photodiode pairs for moisture and gas sensors in the spectral range 1.8-4.8 um (Yakovlev, Yurii P.; Baranov, Alexej N.; Imenkov, Albert N.; Mikhailova, Maya P.)V1510,170-177(1991)

Optomechanical M x N fiber-optic matrix switch (Pesavento, Gerry A.)V1474,57-61(1991)

Physical Concepts of Materials for Novel Optoelectronic Device Applications I: Materials Growth and Characterization (Razeghi, Manijeh, ed.)V1361(1991)

Physical Concepts of Materials for Novel Optoelectronic Device Applications II: Device Physics and Applications (Razeghi, Manijeh, ed.)V1362(1991)

Picosecond optoelectronic semiconductor switching and its application (Brueckner, Volkmar; Bergner, Harald; Lenzner, Matthias; Strobel, Reiner)V1362,510-517(1991)

P-n heterojunction and Schottky barrier formation between poly(3-methylthiophene) and n-type cadmium sulfide (Frank, Arthur J.; Glenis, Spyridon)V1436,50-57(1991)

Portable and very inexpensive optical fiber sensor for entero-gastric reflux detection (Baldini, Francesco; Falciai, Riccardo; Bechi, Paolo; Cosi, Franco; Bini, Andrea; Milanesi, Francesco)V1510,58-62(1991)

Production considerations necessary to produce large quantities of optoelectronic devices by MOCVD epitaxy (Boldish, Steven I.)V1449,51-64(1991)

Radar-optronic tracking experiment for short- and medium-range aerial combat (Ravat, Christian J.; Mestre, J. P.; Rose, C.; Schorter, M.)V1478,239-246(1991)

Rapid isothermal process technology for optoelectronic applications (Singh, Rajendra)V1418,203-216(1991)

Rapid thermal processing in the manufacturing technology of contacts to InP-based photonic devices (Katz, Avishay)V1393,67-77(1991)

Real-time holography using the high-resolution LCTV-SLM (Hashimoto, Nobuyuki; Morokawa, Shigeru; Kitamura, Kohei)V1461,291-302(1991)

Recent progress on research of materials for optoelectronic device applications in China (Chen, Liang-Hui; Kong, Mei-Ying; Wang, Yi-Ming)V1361,60-73(1991)

Semiconductor laser diode with weak optical feedback: self-coupling effects on P-I characteristics (Milani, Marziale; Mazzoleni, S.; Brivio, Franca)V1474,83-97(1991)

Ultra-high-frequency GaInAs/InP devices and circuits for millimeter wave application (Greiling, Paul T.)V1361,47-58(1991)

Ultrafast measurements of carrier transport optical nonlinearities in a GaAs/AlGaAs MQW SEED device (Park, Choong-Bum; Li Kam Wa, Patrick; Miller, Alan)V1474,8-17(1991)

Optomechanical design—see also alignment; optical design

Absolute phasing of segmented mirrors using the polarization phase sensor (Klumpe, Herbert W.; Lajza-Rooks, Barbara A.; Blum, James D.)V1532,230-240(1991)

Accelerometer-based platform stabilization (Algrain, Marcelo C.)V1482,367-382(1991)

Acquisition, Tracking, and Pointing V (Masten, Michael K.; Stockum, Larry A., eds.)V1482(1991)

Active optics technology: an overview (Ray, Frank B.)V1532,188-206(1991)

Active vibration filtering: application to an optical delay line for stellar interferometer (Koehler, Bertrand; Bourlon, Philippe M.; Dugue, Michel)V1489,177-188(1991)

Advances in optomechanics (Vukobratovich, Daniel)V1396,436-446(1991)

Analysis and testing of a soft actuation system for segmented-reflector articulation and isolation (Jandura, Louise; Agronin, Michael L.)V1542,213-224(1991)

Analysis of an adjustable slit design for correcting astigmatism (Clapham, Terrance N.; D'Arcy, John; Bechtel, Lorne; Glockler, Hermann; Munnerlyn, Charles R.; McDonnell, Peter J.; Garbus, Jenny)V1423,2-7(1991)

Analysis of elastomer lens mountings (Valente, Tina M.; Richard, Ralph M.)V1533,21-26(1991)

Analysis of Optical Structures (O'Shea, Donald C., ed.)V1532(1991)

Analysis of the accuracy of aerostatic slideway made from granite (Li, Jinian)V1533,150-154(1991)

Analysis of thermal stability of fused optical structure (Powell, William R.)V1532,126-136(1991)

Angle encoding with the folded Brewster interferometer (Young, Niels O.)V1454,235-244(1991)

Attacking dimensional instability problems in graphite/epoxy structures (Krumweide, Gary C.; Brand, Richard A.)V1533,252-261(1991)

Automatic-adjusting optical axis for linear CCD scanner (Chang, Rong-Seng; Chen, Der-Chin)V1527,357-360(1991)

Axial stresses with toroidal lens-to-mount interfaces (Yoder, Paul R.)V1533,2-11(1991)

Beam Deflection and Scanning Technologies (Beiser, Leo; Marshall, Gerald F., eds.)V1454(1991)

Bearings for rotary scanners: an overview (Preston, Ralph G.; Colquhoun, Allan B.; Shepherd, Joseph)V1454,124-131(1991)

Case study of elastomeric lens mounts (Fischer, Robert E.)V1533,27-35(1991)

Characterization and enhancement of the damping within composite beams (FitzSimons, Philip M.; Trahan, Daniel J.)V1489,230-242(1991)

Characterization of the dimensional stability of advanced metallic materials using an optical test bench structure (Hsieh, Cheng; O'Donnell, Timothy P.)V1533,240-251(1991)

Coarse pointing assembly for the SILEX program, or how to achieve outstanding pointing accuracy with simple hardware associated with consistent control laws (Buvat, Daniel; Muller, Gerard; Peyrot, Patrick)V1417,251-261(1991)

Combination-matching problems in the layout design of minilaser rangefinder (Wang, Erqi; Song, Dehui)V1400,124-128(1991)

Combination of mechanical athermalization with manual in IR zoom telescope (Chen, Ruiyi; Zheng, Dayue; Zhou, Xiuli; Zhang, Xingde)V1540,724-728(1991)

Come-on-plus project: an upgrade of the come-on adaptive optics prototype system (Gendron, Eric; Cuby, Jean-Gabriel; Rigaut, Francois; Lena, Pierre J.; Fontanella, Jean-Claude; Rousset, Gerard; Gaffard, Jean-Paul; Boyer, Corinne; Richard, Jean-Claude; Vittot, M.; Merkle, Fritz; Hubin, Norbert)V1542,297-307(1991)

Compact high-accuracy Inductosyn-based gimbal control system (Liebst, Brad; Verbanets, William R.; Kimbrell, James E.)V1482,425-438(1991)

Computer-aided engineering, manufacturing, and testing of extremely fast steering mirrors (Hubert, Alexis; Hammond, Mark W.)V1532,249-260(1991)

Considerations in the design of servo amplifiers for high-performance scanning systems (Bukys, Albert)V1454,186-195(1991)

Correction of inherent scan nonplanarity in the Boeing infrared sensor calibration facility (Chase, Robert P.)V1533,138-149(1991)

Delayed elasticity in Zerodur at room temperature (Pepi, John W.; Golini, Donald)V1533,212-221(1991)

Design, analysis, and testing of a CCD array mounting structure (Sultana, John A.; O'Neill, Mark B.)V1532,27-38(1991)

Design and analysis of a dither mirror control system (Kline-Schoder, Robert J.; Wright, Michael J.)V1489,189-200(1991)

Design and manufacture of an ultralightweight solid deployable reflector (Tremblay, Gary A.; Derby, Eddy A.)V1532,114-123(1991)

Design and performance of a small two-axis high-bandwidth steering mirror (Loney, Gregory C.)V1454,198-206(1991)

Design and testing of a cube-corner array for laser ranging (James, William E.; Steel, William H.; Evans, Nelson O.)V1400,129-136(1991)

Design considerations for multipurpose bidirectional reflectometers (Neu, John T.; Bressler, Martin)V1530,244-254(1991)

Design considerations for use of a double-dove prism behind a concentric dome (Gibbons, Robert C.; Cooper, E. E.; Legan, R. G.)V1498,64-69(1991)

Design equations for a polygon laser scanner (Beiser, Leo)V1454,60-66(1991)

Design, fabrication, and integration of holographic dispersive solar concentrator for terrestrial applications (Stojanoff, Christo G.; Kubitzek, Ruediger; Tropartz, Stephan; Froehlich, K.; Brasseur, Olivier)V1536,206-214(1991)

Design of a periscopic coarse pointing assembly for optical multiple access (Gatenby, Paul V.; Boereboom, Peter; Grant, Michael A.)V1522,126-134(1991)

Design of compact IR zoom telescope (Chen, Ruiyi; Zhou, Xiuli; Zhang, Xingde)V1540,717-723(1991)

Deterministic errors in pointing and tracking systems I: identification and correction of static errors (Keitzer, Scott; Kimbrell, James E.; Greenwald, David)V1482,406-414(1991)

Deterministic errors in pointing and tracking systems II: identification and correction of dynamic errors (Kimbrell, James E.; Greenwald, David; Smith, Robert; Kidd, Keith)V1482,415-424(1991)

Development and testing of lightweight composite reflector panels (Helms, Richard G.; Porter, Christopher C.; Kuo, Chin-Po; Tsuyuki, Glenn T.)V1532,64-80(1991)

Development of the water-window imaging x-ray microscope (Hoover, Richard B.; Shealy, David L.; Baker, Phillip C.; Barbee, Troy W.; Walker, Arthur B.)V1435,338-351(1991)

Distributed-parameter estimation for NASA Mini-Mast truss through displacement measurements (Huang, Jen-Kuang; Shen, Ji-Yao; Taylor, Lawrence W.)V1489,266-277(1991)

Double-pulsed TEA CO2 laser (Li, Xiang Ying; Shi, Zhenbang; Sun, Ning)V1412,246-251(1991)

Dynamic behavior and structure optimum of high-speed gear mechanism of high-speed photography apparatus (Wang, Jianshe; Liu, Jian-Hua)V1533,175-184(1991)

Dynamic characteristics of joint-dominated space trusses (Shih, Choon-Foo; Kuo, Chin-Po)V1532,91-102(1991)

Effect of microaccelerations on an optical space communication system (Wittig, Manfred E.)V1522,278-286(1991)

Electronically gated airborne video camera (Sturz, Richard A.)V1538,77-80(1991)

Embedded fiber-optic sensors in large structures (Udd, Eric)V1588,178-181(1991)

Enhanced thematic mapper cold focal plane: design and testing (Yang, Bing T.)V1488,399-409(1991)

Experimental and analytical studies on fixed mask assembly for APS with enhanced cooling (Kuzay, Tuncer M.; Collins, Jeffrey T.; Khounsary, Ali M.; Viccaro, P. J.)V1345,55-70(1991)

Ferrofluid film bearing for enhancement of rotary scanner performance (Cheever, Charles J.; Li, Zhixin; Raj, K.)V1454,139-151(1991)

Filter and window assemblies for high-power insertion device synchrotron radiation sources (Khounsary, Ali M.; Viccaro, P. J.; Kuzay, Tuncer M.)V1345,42-54(1991)

Finite element analysis enhancement of cryogenic testing (Thiem, Clare D.; Norton, Douglas A.)V1532,39-47(1991)

Finite element analysis of deformation in large optics due to space environment radiation (Merzbacher, Celia I.; Friebele, E. J.; Ruller, Jacqueline A.; Matic, P.)V1533,222-228(1991)

Finite element analysis of large lenses for the Keck telescope high-resolution echelle spectrograph (Bigelow, Bruce C.)V1532,15-26(1991)

Fixation method with high-orientation accuracy for optical terminals in space (Bauer, Dietrich; Lober, K.; Seeliger, Reinhard)V1533,277-285(1991)

Generic telescope truss (Pressel, Philip)V1532,50-56(1991)

Geometric error coupling in instrument ball bearings (Kingsbury, Edward P.; Francis, Henry A.)V1454,152-158(1991)

High-bandwidth alignment sensing in active optical systems (Kishner, Stanley J.)V1532,215-229(1991)

High-bandwidth control for low-area-density deformable mirrors (How, Jonathan P.; Anderson, Eric H.; Miller, David W.; Hall, Steven R.)V1489,148-162(1991)

Injection chaining of diode-pumped single-frequency ring lasers for free-space communication (Cheng, Emily A.; Kane, Thomas J.; Wallace, Richard W.; Cornwell, Donald M.)V1417,300-306(1991)

Integrated optics bus access module for intramachine communication (Karioja, Pentti; Tammela, Simo; Hannula, Tapio)V1533,129-137(1991)

Interference fit equations for lens cell design (Richard, Ralph M.; Valente, Tina M.)V1533,12-20(1991)

Large active mirror in aluminium (Leblanc, Jean-M.; Rozelot, Jean-Pierre)V1535,122-129(1991)

Laser designation integration into M-65 turret (Goldmunz, Menachem; Bloomberg, Steve; Neugarten, Michael L.)V1442,149-153(1991)

Laser galvo-angle-encoder with zero added inertia (Hercher, Michael; Wyntjes, Geert J.)V1454,230-234(1991)

Lightweight composite mirrors: present and future challenges (Brand, Richard A.; Spinar, Karen K.)V1532,57-63(1991)

Lightweight SXA metal matrix composite collimator (Johnson, R. B.; Ahmad, Anees; Hadaway, James B.; Geiger, Alan L.)V1535,136-142(1991)

Linear resonant approach to scanning (Confer, Charles L.; Burrer, Gordon J.)V1454,215-222(1991)

Line-of-sight alignment of a multisensor system (Wilk, Shalom; Goldmunz, Menachem; Shahaf, Nachum; Klein, Yitschak; Goldman, Shmuel; Oren, Ehud)V1442,140-148(1991)

Line-of-sight stabilization: sensor blending (Pettit, Christopher J.)V1489,278-287(1991)

Mechanical and thermal disturbances of the PSR moderate focus-mission structure (Shih, Choon-Foo; Lou, Michael C.)V1532,81-90(1991)

Mechanical cooler development program for ASTER (Kawada, Masakuni; Fujisada, Hiroyuki)V1490,299-308(1991)

Microdisplacement positioning system for a diffraction grating ruling engine (Yang, Hou-Min; Wang, Xiaolin; Zhang, Yinxian)V1533,185-192(1991)

Micromachined scanning mirrors for laser beam deflection (Huber, Peter; Gerlach-Meyer, U.)V1522,135-141(1991)

Mirrors for optical telescopes (Miroshnikov, Mikhail M.)V1533,286-298(1991)

Modeling and simulation of friction (Haessig, David A.; Friedland, Bernard)V1482,383-396(1991)

Moderate-resolution imaging spectrometer-tilt baseline concept (Magner, Thomas J.)V1492,272-285(1991)

Modular removable precision mechanism for alignment of an FUV spatial heterodyne interferometer (Tom, James L.; Cotton, Daniel M.; Bush, Brett C.; Chung, Ray; Chakrabarti, Supriya)V1549,308-312(1991)

Multiple-diode laser optomechanical issues (jackson, John E.; Armentrout, Ben A.; Buck, J. P.; Chenoweth, Amos J.; Elliott, G. A.; Fox, Allen M.; Ganley, J. T.; Gray, W. C.; Jett, L. L.; Johnson, Kevin M.; Kelsey, J. F.; Minelli, R. J.; Rose, G. E.; Shepherd, W. J.; Zino, Joseph D.)V1533,75-86(1991)

New design of the illuminating system for transmission film copy (Pesl, Ales A.)V1448,218-224(1991)

Nonlinear finite element analysis of the Starlab 80-cm telescope primary-mirror suspension system (Arnold, William R.)V1532,103-113(1991)

Numerical calculation of image motion and vibration modulation transfer function (Hadar, Ofer; Fisher, Moshe; Kopeika, Norman S.)V1482,79-91(1991)

Numerical calculation of image motion and vibration modulation transfer functions: a new method (Hadar, Ofer; Dror, Itai; Kopeika, Norman S.)V1533,61-74(1991)

Optical design and development of a small barcode scanning module (Wike, Charles K.; Lindacher, Joseph M.)V1398,119-126(1991)

Optical engineering of an excimer laser ophthalmic surgery system (Yoder, Paul R.)V1442,162-171(1991)

Optical link demonstration with a lightweight transceiver breadboard (Hemmati, Hamid; Lesh, James R.; Apostolopoulos, John G.; Del Castillo, Hector M.; Martinez, A. S.)V1417,476-483(1991)

Optical performance of an infrared astronomical telescope with a 5-axis secondary mirror (Ettedgui-Atad, Eli; Humphries, Colin M.)V1532,241-248(1991)

Optical system for DOC II (Hudyma, Russell M.; Arndt, Thomas D.; Fischer, Robert E.)V1563,244-254(1991)

Optics for vector scanning (Ehrmann, Jonathan S.)V1454,245-256(1991)

Optimal control/structure integrated design of a flexible space platform with articulated appendages (Kelkar, Atul G.; Alberts, Thomas E.)V1489,243-253(1991)

Optimization of an x-ray mask design for use with horizontal and vertical kinematic mounts (Laird, Daniel L.; Engelstad, Roxann L.; Palmer, Shane R.)V1465,134-144(1991)

Optomechanics and Dimensional Stability (Paquin, Roger A.; Vukobratovich, Daniel, eds.)V1533(1991)

Optomechanics of a free-space photonic switch: the components (Brubaker, John L.; McCormick, Frederick B.; Tooley, Frank A.; Sasian, Jose M.; Cloonan, Thomas J.; Lentine, Anthony L.; Hinterlong, Steve J.; Herron, Michael J.)V1533,88-96(1991)

Optomechanics of a free-space photonic switching fabric: the system (McCormick, Frederick B.; Tooley, Frank A.; Brubaker, John L.; Sasian, Jose M.; Cloonan, Thomas J.; Lentine, Anthony L.; Morrison, Richard L.; Crisci, R. J.; Walker, Sonya L.; Hinterlong, Steve J.; Herron, Michael J.)V1533,97-114(1991)

Overview of a high-performance polygonal scanner subsystem (Rynkowski, Gerald)V1454,102-110(1991)

Overview of the finite element method in optical systems (Hatheway, Alson E.)V1532,2-14(1991)

Packaging considerations of fiber-optic laser sources (Heikkinen, Veli; Tukkiniemi, Kari; Vahakangas, Jouko; Hannula, Tapio)V1533,115-121(1991)

Penning discharge VUV and soft x-ray source (Cao, Jianlin; Li, Futian; Qian, Limin; Chen, Po; Ma, Yueying; Chen, Xingdan)V1345,71-77(1991)

Progress in binocular design (Seil, Konrad)V1533,48-60(1991)

Progress report on a five-axis fast guiding secondary for the University of Hawaii 2.2-meter telescope (Cavedoni, Charles P.; Graves, J. E.; Pickles, A. J.)V1542,273-282(1991)

Qualification testing of a diode-laser transmitter for free-space coherent communications (Pillsbury, Allen D.; Taylor, John A.)V1417,292-299(1991)

Rapid-cooled lens cell (Stubbs, David M.; Hsu, Ike C.)V1533,36-47(1991)

Relationship between fluctuation in mirror radius (within one polygon) and the jitter (Horikawa, Hiroshi; Sugisaki, Iwao; Tashiro, Masaru)V1454,46-59(1991)

Relationship between jitter and deformation of mirrors (Horikawa, Hiroshi; Miura, Masayuki; Uchida, Toshiya)V1454,20-32(1991)

Review of measurement systems for evaluating thermal expansion homogeneity of Corning Code 7971 ULETM (Hagy, Henry E.)V1533,198-211(1991)

Rigidity test of large and high-precision instruments (Ma, Pin-Zhong)V1532,177-186(1991)

Rigid lightweight optical bench for a spaceborne FUV spatial heterodyne interferometer (Tom, James L.; Cotton, Daniel M.; Bush, Brett C.; Chung, Ray; Chakrabarti, Supriya)V1549,302-307(1991)

Roller chain supports for large optics (Vukobratovich, Daniel; Richard, Ralph M.)V1396,522-534(1991)

Scheme of optical synthetic-aperture telescope (Ma, Pin-Zhong)V1533,163-174(1991)

Segmented mirror figure control for a space-based far-IR astronomical telescope (Redding, David C.; Breckenridge, William G.; Lau, Kenneth; Sevaston, George E.; Levine, Bruce M.; Shaklan, Stuart B.)V1489,201-215(1991)

Self-acting gas bearings for high-speed scanners (Preston, Ralph G.; Robinson, D. H.; Shepherd, Joseph)V1454,132-138(1991)

Sensor line-of-sight stabilization (Cooper, C. J.)V1498,39-51(1991)

Shape control of piezoelectric bimorph mirrors (Burke, Shawn E.; Hubbard, James E.)V1532,207-214(1991)

Single-mode fiber optic rotary joint for aircraft applications (Lewis, Warren H.; Miller, Michael B.)V1369,79-86(1991)

Six degree-of-freedom magnetically suspended fine-steering mirror (Medbery, James D.; Germann, Lawrence M.)V1482,397-405(1991)

Sliding control of a single-axis steering mirror (Connors, Bruce P.)V1489,136-147(1991)

Small-satellite constellation for many uses (Seiberling, Walter E.; Traxler-Lee, Laura A.; Collins, Sean K.)V1495,32-41(1991)

Small-Satellite Technology and Applications (Horais, Brian J., ed.)V1495(1991)

Space Infrared Telescope Facility cryogenic and optical technology (Mason, Peter V.; Kiceniuk, T.; Plamondon, Joseph A.; Petrick, Walt)V1540,88-96(1991)

Spatial acquisition and tracking for deep-space optical communication packages (Chen, Chien-Chung; Jeganathan, Muthu; Lesh, James R.)V1417,240-250(1991)

Specification of precision optical pointing systems (Medbery, James D.; Germann, Lawrence M.)V1489,163-176(1991)

Specifications for image stabilization systems (Hilkert, James M.; Bowen, Max L.; Wang, Joe)V1498,24-38(1991)

Stability analysis of optomechanical components (Kerbis, Esther; Morrison, Richard L.; Cloonan, Thomas J.; Downs, Maralene M.)V1396,447-452(1991)

Stability analysis on 3-axes servo revolution pedestal system (Lu, Eh; Yang, Hongbo; Meng, Qinglai; Han, Rong-jiu)V1533,155-162(1991)

Stability considerations in relay lens design for optical communications (Gardam, Allan; Jonas, Reginald P.)V1417,381-390(1991)

State equalization and resonant control systems (Bigley, William J.)V1482,350-366(1991)

Structures Sensing and Control (Breakwell, John; Varadan, Vijay K., eds.)V1489(1991)

Subminiature package external cavity laser (Fatah, Rebwar M.; Cox, Maurice K.; Bird, David M.; Cameron, Keith H.)V1501,120-128(1991)

Super-high-speed reflex-type moving image camera (Drozhbin, Yu. A.; Trofimenko, Vladimir V.)V1358,454-456(1991)

Surface distortions of a 3.5-meter mirror subjected to thermal variations (Cho, Myung K.; Poczulp, Gary A.)V1532,137-145(1991)

Synthesis and property research of birefringent polymers with predicted optical-mechanical parameters (Askadskij, A. A.; Marshalkovich, A. S.; Latysh, E. G.; Goleneva, L. M.; Pastukhov, A. V.; Sidorova, G. I.)V1554A,426-431(1991)

Telescope enclosure flow visualization (Forbes, Fred F.; Wong, Woon-Yin; Baldwin, Jack; Siegmund, Walter A.; Limmongkol, Siriluk; Comfort, Charles H.)V1532,146-160(1991)

Temperature control of the 3.5-meter WIYN telescope primary mirror (Goble, Larry W.)V1532,161-169(1991)

Theoretical limits of dimensional stability for space structures (Dolgin, Benjamin P.; Moacanin, Jovan; O'Donnell, Timothy P.)V1533,229-239(1991)

Thermal and structural analysis of the GOES scan mirror's on-orbit performance (Zurmehly, George E.; Hookman, Robert A.)V1532,170-176(1991)

Thirty-two-channel LED array spectrometer module with compact optomechanical construction (Malinen, Jouko; Keranen, Heimo; Hannula, Tapio; Hyvarinen, Timo S.)V1533,122-128(1991)

Tilt corrector based on spring-type magnetostrictive actuators (Aksinin, V. I.; Apollonov, V. V.; Chetkin, Sergue A.; Kijko, Vadim V.; Muraviev, S. V.; Vdovin, Gleb V.)V1500,93-104(1991)

Trade-offs in rotary mirror scanner design (Colquhoun, Allan B.; Cowan, Donald W.; Shepherd, Joseph)V1454,12-19(1991)

Tropospheric emission spectrometer for the Earth Observing System (Glavich, Thomas A.; Beer, Reinhard)V1540,148-159(1991)

Ultrafast optical-mechanical camera (Drozhbin, Yu. A.; Trofimenko, Vladimir V.; Chernova, T. I.)V1358,451-453(1991)

University of Hawaii adaptive optics system: II. Computer simulation (Northcott, Malcolm J.)V1542,254-261(1991)

University of Hawaii adaptive optics system: III. Wavefront curvature sensor (Graves, J. E.; McKenna, Daniel)V1542,262-272(1991)

Update of scanner selection, performances, and multiaxis configurations (Montagu, Jean I.)V1454,160-173(1991)

Vibration sensing in flexible structures using a distributed-effect modal domain optical fiber sensor (Reichard, Karl M.; Lindner, Douglas K.; Claus, Richard O.)V1489,218-229(1991)

Visualization of impingement field of real-rocket-exhausted jets by using moire deflectometry (He, Anzhi; Yan, Da-Peng; Miao, Peng-Cheng; Wang, Hai-Lin)V1554A,429-434(1991)

Wave-optic model to determine image quality through supersonic boundary and mixing layers (Lawson, Shelah M.; Clark, Rodney L.; Banish, Michele R.; Crouse, Randy F.)V1488,268-278(1991)

WEBERSAT: a low-cost imaging satellite (Twiggs, Robert J.; Reister, K. R.)V1495,12-18(1991)

Work-induced stress and long-term stability in optically polished silicon (Bender, John W.; Wahl, Roger L.)V1533,264-276(1991)

Parallel processing—see also optical computing; optical processing

All-optical interconnection networks (Ghafoor, Arif)V1390,454-466(1991)

Application of neural networks in optimization problems: a review (Ashenayi, Kaveh)V1396,285-296(1991)

Applications of a minimum sum path algorithm implemented on the connection machine (Rosenfeld, J. P.; Krecker, Donald K.; Hord, R. M.)V1406,147-147(1991)

Applications of the massively parallel machine, the MasPar MP-1, to Earth sciences (Fischer, James R.; Strong, James P.; Dorband, John E.; Tilton, James C.)V1492,229-238(1991)

Architecture for a multiprocessing system based on data flow processing elements in a MAXbus system (Bulsink, Bennie J.; Klok, Frits H.)V1384,215-227(1991)

Artificial neural network and image processing using the Adaptive Solutions' architecture (Baker, Thomas E.)V1452,502-511(1991)

Automated grading of venous beading: an algorithm and parallel implementation (Shen, Zhijiang; Gregson, Peter H.; Cheng, Heng-Da; Kozousek, V.)V1606,632-640(1991)

Cellular-automata-based learning network for pattern recognition (Tzionas, Panagiotis; Tsalides, Ph.; Thanailakis, A.)V1606,269-280(1991)

Class of learning algorithms for multilayer perceptron (Abbasi, M.; Sayeh, Mohammad R.)V1396,237-242(1991)

Comparison of sinusoidal perceptron with multilayer classical perceptron (Karimi, B.; Baradaran, T.; Ashenayi, Kaveh; Vogh, James)V1396,226-236(1991)

Current status and future research of the Delft 'supercomputer' project (Frietman, Edward E.; Dekker, L.; van Nifterick, W.; Demeester, Piet M.; van Daele, Peter; Smit, W.)V1390,434-453(1991)

Data-driven parallel architecture for syntactic pattern recognition (Tseng, Chien-Chao; Hwang, Shu-Yuen)V1384,257-268(1991)

Decomposition and inversion of von Neumann-like convolution operators (Manseur, Zohra Z.; Wilson, David C.)V1568,164-173(1991)

Dynamically reconfigurable optical interconnect architecture for parallel multiprocessor systems (Girard, Mary M.; Husbands, Charles R.; Antoszewska, Reza)V1563,156-167(1991)

Efficient transformation algorithm for 3-D images (Vepsalainen, Ari M.; Rantala, Aarne E.)V1452,64-75(1991)

Experiences with a parallel architecture for image analysis (Evans, Robert H.; Williams, Elmer F.; Brant, Karl)V1406,201-202(1991)

Focal-plane processing algorithm and architecture for laser speckle interferometry (Tomlinson, Harold W.; Weir, Michael P.; Michon, G. J.; Possin, G. E.; Chovan, J.)V1541,178-186(1991)

Fusion of multiple views of multiple reference points using a parallel distributed processing approach (Wolfe, William J.; Magee, Michael)V1383,20-25(1991)

Hadamard transform-based object recognition using an array processor (Celenk, Mehmet; Moiz, Saifuddin)V1468,764-775(1991)

Heterogeneous parallel processor for a model-based vision system (Segal, Andrew C.)V1396,601-614(1991)

High-density packaging and interconnect of massively parallel image processors (Carson, John C.; Indin, Ronald)V1541,232-239(1991)

High-level parallel architecture for a rule-based system (Karne, Ramesh K.; Sood, Arun K.)V1468,938-949(1991)

High-speed optical interconnects for parallel processing and neural networks (Barnes, Nigel; Healey, Peter; McKee, Paul; O'Neill, Alan; Rejman-Greene, Marek A.; Scott, Geoff; Smith, David W.; Webb, Roderick P.; Wood, David)V1389,477-483(1991)

Holographic space-variant prism arrays for free-space data permutation in digital optical networks (Kobolla, Harald; Sauer, Frank; Schwider, Johannes; Streibl, Norbert; Voelkel, Reinhard)V1507,175-182(1991)

Hybrid bipixel structuring element decomposition and Euclidean morphological transforms (Zhou, Ziheng; Venetsanopoulos, Anastasios N.)V1606,309-319(1991)

Hybrid optical array logic system (Kakizaki, Sunao; Miyazaki, Daisuke; Yoshikawa, Eiji; Tanida, Jun; Ichioka, Yoshiki)V1505,199-205(1991)

Image algebra preprocessor for the MasPar parallel computer (Meyer, Trevor E.; Davidson, Jennifer L.)V1568,125-136(1991)

Influence of different nonlinearity functions on Perceptron performance (Ashenayi, Kaveh; Vogh, James)V1396,215-225(1991)

Integrated C3I optical processor (Kaminski, Robert L.)V1564,156-164(1991)

Integration of an edge extraction cellular automaton (Taboury, Jean; Chavel, Pierre)V1505,115-123(1991)

Issues in parallelism in object recognition (Bhandarkar, Suchendra M.; Suk, Minsoo)V1384,234-245(1991)

Linear interconnection architecture in parallel implementation of neural network models (Mostafavi, M. T.)V1396,193-201(1991)

Massively parallel implementation of neural network architectures (Omidvar, Omid M.; Wilson, Charles L.)V1452,532-543(1991)

Massively parallel synthetic-aperture radar autofocus (Mastin, Gary A.; Plimpton, Steven J.; Ghiglia, Dennis C.)V1566,341-352(1991)

Max-polynomials and template decomposition (Li, Dong)V1568,149-156(1991)

MIMD (multiple instruction multiple data) multiprocessor system for real-time image processing (Pirsch, Peter; Jeschke, Hartwig)V1452,544-555(1991)

Miniaturized low-power parallel processor for space applications (Jacobi, William J.; Jensen, Preben D.; Teneketges, Nicholas J.; Wadsworth, Leo A.)V1495,205-213(1991)

Multi-Gb/s optical computer interconnect (Sauer, Jon R.)V1579,49-61(1991)

Neural networks implementation on a parallel machine (Wang, Chung Ching; Shirazi, Behrooz)V1396,209-214(1991)

New parallel algorithms for thinning of binary images (Bhattacharya, Prabir; Lu, Xun)V1468,734-739(1991)

On-focal-plane superconducting signal processing for low- and intermediate-temperature operation (Smetana, Daryl L.; Carson, John C.)V1541,220-226(1991)

Optical approaches to overcome present limitations for interconnection and control in parallel electronic architectures (Maurin, T.; Devos, F.)V1505,158-165(1991)

Optical bus protocol for a distributed-shared-memory multiprocessor (Davis, Martin H.; Ramachandran, Umakishore)V1563,176-187(1991)

Optical Enhancements to Computing Technology (Neff, John A., ed.)V1563(1991)

Optical implementation of associative memory based on parallel rank-one interconnections (Jeon, Ho-In; Abushagur, Mustafa A.; Caulfield, H. J.)V1564,522-535(1991)

Optical interconnects for parallel processing (Guha, Aloke; Bristow, Julian P.; Sullivan, Charles T.; Husain, Anis)V1389,375-385(1991)

Optical interconnects in high-bandwidth computing (Dekker, L.; Frietman, Edward E.)V1505,148-157(1991)

Optically coupled three-dimensional memory system (Koyanagi, Mitsumasa)V1390,467-476(1991)

Optics for Computers: Architectures and Technologies (Lebreton, Guy J., ed.)V1505(1991)

Optoelectronic approach to optical parallel processing based on the photonic parallel memory (Matsuda, Kenichi; Shibata, Jun)V1562,21-29(1991)

Optoelectronic hardware issues for implementation of simulated annealing or Boltzmann machines (Lalanne, Philippe; Chavel, Pierre)V1621,388-401(1991)

Packaging issues for free-space optically interconnected multiprocessors (Ozguz, Volkan H.; Esener, Sadik C.; Lee, Sing H.)V1390,477-488(1991)

Packet-switching algorithm for SIMD computers and its application to parallel computer vision (Maresca, Massimo)V1384,206-214(1991)

Parallel computation of the continuous wavelet transform (Gertner, Izidor; Peskin, Richard L.; Walther, Sandra S.)V1565,414-422(1991)

Parallel constructs for three-dimensional registration on a SIMD (single-instruction stream/multiple-data stream) processor (Morioka, Craig A.; Chan, Kelby K.; Huang, H. K.)V1445,534-538(1991)

Parallel data fusion on a hypercube multiprocessor (Davis, Paul B.; Cate, D.; Abidi, Mongi A.)V1383,515-529(1991)

Parallel DC notch filter (Kwok, Kam-cheung; Chan, Ming-kam)V1567,709-719(1991)

Parallel implementation of some fast adaptive algorithms on a digital signal processor network (Reynaud, Roger; Chebira, Abdennasser; Demoment, Guy)V1566,302-311(1991)

Parallel message-passing architecture for path planning (Tavora, Jose; Lourtie, Pedro M.)V1468,524-535(1991)

Parallel optical interconnects utilizing VLSI/FLC spatial light modulators (Genco, Sheryl M.)V1563,45-57(1991)

Parallel path planning in unknown terrains (Prassler, Erwin E.; Milios, Evangelos E.)V1388,2-13(1991)

Parallel processing approach to transform-based image coding (Normile, James; Wright, Dan; Chu, Ke-Chiang; Yeh, Chia L.)V1452,480-484(1991)

Parallel reduced-instruction-set-computer architecture for real-time symbolic pattern matching (Parson, Dale E.)V1468,960-971(1991)

Parallel uses for serial arithmetic in signal processors (Owens, Robert M.; Irwin, Mary J.)V1566,252-262(1991)

Performance analysis through memory of a proposed parallel architecture for the efficient use of memory in image processing applications (Faruque, Abdullah; Fong, David Y.)V1606,865-877(1991)

Performance of a parallel bispectrum estimation code (Carmona, Edward A.; Matson, Charles L.)V1566,329-340(1991)

Pipelined communications on optical busses (Guo, Zicheng; Melhem, Rami G.; Hall, Richard W.; Chiarulli, Donald M.; Levitan, Steven P.)V1390,415-426(1991)

Preconditions and prospects for the construction of parallel digital neurocomputers with programmable architecture (Kalyayev, Anatoli V.; Brukhomitsky, Yuri A.; Galuyev, Gennady A.; Chernukhin, Yu. V.)V1621,299-308(1991)

Primary mirror control system for the Galileo telescope (Bortoletto, Favio; Baruffolo, A.; Bonoli, C.; D'Alessandro, Maurizio; Fantinel, D.; Giudici, G.; Ragazzoni, Roberto; Salvadori, L.; Vanini, P.)V1542,225-235(1991)

RETINA (RETinally INspired Architecture project) (Caulfield, H. J.; Wilkins, Nathan A.)V1564,496-503(1991)

Role for optics in future parallel processing (Rudolph, Larry)V1505,175-185(1991)

Rotation invariant object classification using fast Fourier transform features (Celenk, Mehmet; Datari, Srinivasa R.)V1468,752-763(1991)

SCORPIUS: lessons learned in managing an image understanding system (Onishi, Randall M.; Bogdanowicz, Julius F.; Watanabe, Miki)V1406,171-178(1991)

Self-routing interconnection structures using coincident pulse techniques (Chiarulli, Donald M.; Levitan, Steven P.; Melhem, Rami G.)V1390,403-414(1991)

Software development tools for implementing vision systems on multiprocessors (Choudhary, Alok; Ranka, Sanjay)V1406,148-161(1991)

Superhigh-definition image processing on a parallel signal processing system (Fujii, Tetsurou; Sawabe, Tomoko; Ohta, Naohisa; Ono, Sadayasu)V1605,339-350(1991)

Survey of parallel architectures used for three image processing algorithms (Smith, Ross)V1396,615-623(1991)

Survey: omnifont-printed character recognition (Tian, Qi; Zhang, Peng; Alexander, Thomas; Kim, Yongmin)V1606,260-268(1991)

Template decomposition and inversion over hexagonally sampled images (Lucas, Dean; Gibson, Laurie)V1568,157-163(1991)

Theories in distributed decision fusion: comparison and generalization (Thomopoulos, Stelios C.)V1383,623-634(1991)

Transputer-based parallel algorithms for automatic object recognition (Bison, Paolo G.; Braggiotti, Alberto; Grinzato, Ermanno G.)V1471,369-377(1991)

Two new developments for optoelectronic bus systems (Jiang, Jie; Kraemer, Udo)V1505,166-174(1991)

Use of heterogeneous distributed memory parallel systems in image processing (Pinfold, Wilfred)V1406,132-137(1991)

Particles—see also aerosols; pollution

Applications of an automated particle detection and identification system in VLSI wafer processing (Hattori, Takeshi; Koyata, Sakuo)V1464,367-376(1991)

Applications of neural networks in experimental high-energy physics (Denby, Bruce)V1469,648-658(1991)

Automated wafer inspection in the manufacturing line (Harrigan, Jeanne E.; Stoller, Meryl D.)V1464,596-601(1991)

Backscattering image resolution as a function of particle density (Rochon, Paul L.; Bissonnette, Daniel)V1530,50-57(1991)

Cavity-QED-enhanced spontaneous emission and lasing in liquid droplets (Campillo, Anthony J.; Eversole, J. D.; Lin, H.-B.; Merritt, C. D.)V1497,78-89(1991)

Determination of the temperature of a single particle heated by a highly concentrated laser beam (Herve, Philippe; Bednarczyk, Sophie; Masclet, Philippe)V1487,387-395(1991)

Double-exposure phase-shifting holography applied to particle velocimetry (Lai, Tianshu; Tan, Yushan)V1554B,580-585(1991)

Dynamic particle holographic instrument (Wang, Guozhi; Feng, San; Wang, Zhengrong; Wang, Shuyan)V1358,73-81(1991)

Effects of proton damage on charge-coupled devices (Janesick, James R.; Soli, George; Elliott, Tom; Collins, Stewart A.)V1447,87-108(1991)

Efficient laser cleaning of small particulates using pulsed laser irradiation synchronized with liquid-film deposition (Tam, Andrew C.; Zapka, Werner; Ziemlich, Winfrid)V1598,13-18(1991)

Electromagnetic field calculations for a tightly focused laser beam incident upon a microdroplet: applications to nonlinear optics (Barton, John P.; Alexander, Dennis R.)V1497,64-77(1991)

Electron trapping and acceleration in a modified elongated betatron (Song, Yuanxu Y.; Fisher, Amnon; Prohaska, Robert M.; Rostoker, Norman)V1407,430-441(1991)

Enhanced backscattering of s- and p-polarized light from particles above a substrate (Greffet, Jean-Jacques; Sentenac, Anne)V1558,288-294(1991)

Fabrication and nonlinear optical properties of mixed and layered colloidal particles (McDonald, Joseph K.; LaiHing, Kenneth)V1497,367-370(1991)

Flow-field velocity measurements for nonisothermal systems (Johnson, Edward J.; Hyer, Paul V.; Culotta, Paul W.; Clark, Ivan O.)V1557,168-179(1991)

Ground-based PIV and numerical flow visualization results from the surface-tension-driven convection experiment (Pline, Alexander D.; Wernet, Mark P.; Hsieh, Kwang-Chung)V1557,222-234(1991)

High-resolution submicron particle sizing by dynamic light scattering (Nicoli, David F.)V1430,19-36(1991)

Improvement of recordable depth of field in far-field holography for analysis of particle size (Lai, Tianshu; Tan, Yushan)V1461,286-290(1991)

Incinerator technology overview (Santoleri, Joseph J.)V1434,2-13(1991)

Large electron accelerators powered by intense relativistic electron beams (Friedman, Moshe; Serlin, Victor; Lau, Yue Y.; Krall, Jonathan)V1407,474-478(1991)

Laser-induced volatilization and ionization of aerosol particles for their mass spectral analysis in real time (Sinha, Mahadeva P.)V1437,150-156(1991)

Laser linac in vacuum: assessment of a high-energy particle accelerator (Moore, Gerald T.; Bochove, Erik J.; Scully, Marlan O.)V1497,328-337(1991)

Laser linac: nondiffractive beam and gas-loading effects (Bochove, Erik J.; Moore, Gerald T.; Scully, Marlan O.; Wodkiewicz, K.)V1497,338-347(1991)

Measurement of particles and drops in combusting flows (Ereaut, Peter R.)V1554B,556-565(1991)

Measuring photon pathlengths by quasielastic light scattering in a multiply scattering medium (Nossal, Ralph J.; Schmitt, Joseph M.)V1430,37-47(1991)

Multiple fiber optic probe for several sensing applications (Dhadwal, Harbans S.; Ansari, Rafat R.)V1584,262-272(1991)

Multiple-scattering effects on pulse propagation through burning particles (Ma, Yushieh; Varadan, Vijay K.; Varadan, Vasundara V.)V1487,220-225(1991)

Negative deuterium ion thermal energy measurements in a volume ion source (Bacal, Marthe; Courteille, C.; Devynck, Pascal; Jones-King, Yolanda D.; Leroy, Renan; Stern, Raul A.)V1407,605-609(1991)

New optical technique for particle sizing and velocimetry (Xie, Gong-Wie; Scott, Peter D.; Shaw, David T.; Zhang, Yi-Mo)V1500,310-321(1991)

Nondegenerate two-wave mixing in shaped microparticle suspensions (Pizzoferrato, R.; De Spirito, M.; Zammit, Ugo; Marinelli, M.; Rogovin, Dan N.; Scholl, James F.)V1409,192-201(1991)

Nonlinear electromagnetic field response of high-Tc superconducting microparticle composites (Haus, Joseph W.; Chung-Yau, F.; Bowden, Charles M.)V1497,382-388(1991)

Nonlinear optical amplification and oscillation in spherical microdroplets (Kurizki, Gershon; Goldner, E.)V1497,48-62(1991)

Nonlinear optical processes in droplets with single-mode laser excitation (Chang, Richard K.; Chen, Gang; Hill, Steven C.; Barber, Peter W.)V1497,2-13(1991)

Nonlinear optical properties of suspensions of microparticles: electrostrictive effect and enhanced backscatter (Kang, Chih-Chieh; Fiddy, Michael A.)V1497,372-381(1991)

Nonlinear Optics and Materials (Cantrell, Cyrus D.; Bowden, Charles M., eds.)V1497(1991)

Numerical and optical evaluation of particle image velocimetry images (Farrell, Patrick V.)V1554B,610-621(1991)

Numerical inversion method for determining aerodynamic effects on particulate exhaust plumes from onboard irradiance data (Cousins, Daniel)V1467,402-409(1991)

Overview of stimulated Brillouin scattering in microdroplets (Cantrell, Cyrus D.)V1497,28-47(1991)

Particle analysis in liquid flow by the registration of elastic light scattering in the condition of laser beam scanning (Dubrovsky, V.; Grinevich, A. E.; Ossin, A. B.)V1403,344-346(1991)

Particle image velocimetry experiments for the IML-I spaceflight (Trolinger, James D.; Lal, Ravindra B.; Batra, Ashok K.; McIntosh, D.)V1557,98-109(1991)

Pelliclizing technology (Yamauchi, Takashi)V1496,302-314(1991)

Phase-locking simulation of dual magnetrons (Chen, H. C.; Stark, Robert A.; Uhm, Han S.)V1407,139-146(1991)

Plasma betatron without gas breakdown (Ishizuka, Hiroshi; Yee, K.; Fisher, Amnon; Rostoker, Norman)V1407,442-455(1991)

Progress in miniature laser systems for space science particle sizing and velocimetry (Brown, Robert G.)V1506,58-59(1991)

Recent developments on the NRL Modified Betatron Accelerator (Golden, Jeffry; Len, Lek K.; Smith, Tab J.; Dialetis, Demos; Marsh, S. J.; Smith, Kevin; Mathew, Joseph; Loschialpo, Peter; Seto, Lloyd; Chang, Jeng-Hsien; Kapetanakos, Christos A.)V1407,418-429(1991)

Reduction of beam breakup growth by cavity cross-couplings in recirculating accelerators (Colombant, Denis G.; Lau, Yue Y.; Chernin, David P.)V1407,484-495(1991)

Relativistic klystron research for future linear colliders (Westenskow, Glen A.; Houck, Timothy L.; Ryne, Robert D.)V1407,496-501(1991)

Removal of small particles from surfaces by pulsed laser irradiation: observations and a mechanism (Kelley, J. D.; Stuff, Michael I.; Hovis, Floyd E.; Linford, Gary J.)V1415,211-219(1991)

Resonances and internal electric energy density in droplets (Hill, Steven C.; Barber, Peter W.; Chowdhury, Dipakbin Q.; Khaled, Elsayed-Esam M.; Mazumder, Mohiuddin)V1497,16-27(1991)

Search for short-lived particles using holography (Brucker, E. B.)V1461,206-214(1991)

Simulations of intracavity laser heating of particles (Linford, Gary J.)V1415,196-210(1991)

Some issues on beam breakup in linear accelerators (Lau, Yue Y.; Colombant, Denis G.)V1407,479-483(1991)

Spherical and nonspherical aerosol and particulate characterization using optical pattern recognition techniques (Marshall, Martin S.; Benner, Robert E.)V1564,121-134(1991)

Status of the proof-of-concept experiment for the spiral line induction accelerator (Bailey, Vernon L.; Corcoran, Patrick; Edighoffer, J. A.; Fockler, J.; Lidestri, Joseph P.; Putnam, Sidney D.; Tiefenback, Michael G.)V1407,400-406(1991)

Stimulated Raman diagnostics in diesel droplets (Golombok, Michael)V1497,100-119(1991)

Superheating phenomena in absorbing microdroplets irradiated by pulsed lasers (Armstrong, Robert L.)V1497,132-140(1991)

Surface roughness effects on light scattered by submicron particles on surfaces (Bawolek, Edward J.; Hirleman, Edwin D.)V1464,574-583(1991)

Tandem betatron accelerator (Keinigs, Rhon K.)V1407,456-466(1991)

Threshold measurements in laser-assisted particle removal (Lee, Shyan J.; Imen, Kamran; Allen, Susan D.)V1598,2-12(1991)

Time-resolved Raman spectroscopy from reacting optically levitated microdroplets (Carls, Joseph C.; Brock, James R.)V1497,120-131(1991)

UCSD high-energy x-ray timing experiment cosmic ray particle anticoincidence detector (Hink, Paul L.; Rothschild, Richard E.; Pelling, Michael R.; MacDonald, Daniel R.; Gruber, Duane E.)V1549,193-202(1991)

Using digital images to measure and discriminate small particles in cotton (Taylor, Robert A.; Godbey, Luther C.)V1379,16-27(1991)

Using dynamic holography for iron fibers with submicron diameter and high velocity (Xie, Gong-Wie; Hua, Lifan; Patel, Sushil; Scott, Peter D.; Shaw, David T.)V1385,132-141(1991)

WEBERSAT: measuring micrometeorite impacts in a polar orbit (Evans, Phillip R.)V1495,149-156(1991)

Pattern recognition—see also correlation; filters; image processing; neural networks; object recognition; optical computing; optical processing

Accuracy of the output peak localization in two-dimensional matched filtering (Chalasinska-Macukow, Katarzyna)V1391,295-302(1991)

Acquiring rules of selecting cells by using neural network (Yu, He; Zheng, XiangJun; Ye, Yizheng; Wang, LiHong)V1469,412-417(1991)

Advanced in-plane rotation-invariant filter results (Ravichandran, Gopalan; Casasent, David P.)V1567,466-479(1991)

Affine-invariant recognition of gray-scale objects by Fourier descriptors (Fenske, Axel; Burkhardt, Hans)V1567,53-64(1991)

Algorithm for quality inspection of characters printed on chip resistors (Numagami, Yasuhiko; Hattori, Yasuyuki; Nakamura, Osamu; Minami, Toshi)V1606,970-979(1991)

Algorithm for statistical classification of radar clutter into one of several categories (Nechval, Nicholas A.)V1470,282-293(1991)

Amplitude-encoded phase-only filters for pattern recognition: influence of the bleaching procedure (Campos, Juan; Janowska-Dmoch, Bozena; Styczynski, K.; Turon, F.; Yzuel, Maria J.; Chalasinska-Macukow, Katarzyna)V1574,141-147(1991)

Analysis of the displacements of cylindrical shells by moire techniques (Laermann, Karl-Hans)V1554B,248-256(1991)

An Introduction to Biological and Artificial Neural Networks for Pattern Recognition (Kabrisky, Matthew; Rogers, Steven K.)VTT04(1991)

ANN approach for 2-D echocardiographic image processing: application of neocognitron model to LV boundary formation (Wan, Liqun; Li, Dapeng; Wee, William G.; Han, Chia Y.; Porembka, D. T.)V1469,432-440(1991)

Anthropomorphic classification using three-dimensional Fourier descriptor (Lee, Nahm S.; Park, Kyung S.)V1450,133-143(1991)

Application of a multilayer network in image object classification (Tang, Yonghong; Wee, William G.; Han, Chia Y.)V1469,113-120(1991)

Application of back-propagation to the recognition of handwritten digits using morphologically derived features (Hepp, Daniel J.)V1451,228-233(1991)

Application of generalized radial basis functions to the problem of object recognition (Thau, Robert S.)V1469,37-47(1991)

Application of neural networks for the synthesis of binary correlation filters for optical pattern recognition (Mahalanobis, Abhijit; Nadar, Mariappan S.)V1469,292-302(1991)

Application of neural networks to group technology (Caudell, Thomas P.; Smith, Scott D.; Johnson, G. C.; Wunsch, Donald C.)V1469,612-621(1991)

Application of perceptron to the detecting of particle motion (Li, Jie-gu; Yuan, Qiang)V1469,178-187(1991)

Application of the phase and amplitude modulating properties of LCTVs (Kirsch, James C.; Loudin, Jeffrey A.; Gregory, Don A.)V1474,90-100(1991)

Applications of Artificial Neural Networks II (Rogers, Steven K., ed.)V1469(1991)

Applications of chaotic neurodynamics in pattern recognition (Baird, Bill; Freeman, Walter J.; Eeckman, Frank H.; Yao, Yong)V1469,12-23(1991)

Applications of learning strategies to pattern recognition (Rizki, Mateen M.; Tamburino, Louis A.; Zmuda, Michael A.)V1469,384-391(1991)

Applications of neural networks in experimental high-energy physics (Denby, Bruce)V1469,648-658(1991)

Applications of the bispectrum in radar signature analysis and target identification (Jouny, Ismail; Garber, Frederick D.; Moses, Randolph L.; Walton, Eric K.)V1471,142-153(1991)

Approach to invariant object recognition on grey-level images by exploiting neural network models (Rybak, Ilya A.; Golovan, Alexander V.; Gusakova, Valentina I.)V1469,472-482(1991)

Artificial neural network for supervised learning based on residual analysis (Chan, Keith C.; Vieth, John O.; Wong, Andrew K.)V1469,359-372(1991)

Artificial neural system approach to IR target identification (Holland, Orgal T.; Tarr, Tomas; Farsaie, Ali; Fuller, James M.)V1469,102-112(1991)

Assumption truth maintenance in model-based ATR algorithm design (Bennett, Laura F.; Johnson, Rubin; Hudson, C. I.)V1470,263-274(1991)

Atmospheric propagation effects on pattern recognition by neural networks (Giever, John C.; Hoock, Donald W.)V1486,302-313(1991)

ATR performance modeling for building multiscenario adaptive systems (Nasr, Hatem N.)V1483,112-117(1991)

Automated calculation of nonadditive measures for object recognition (Tahani, Hossein; Keller, James M.)V1381,379-389(1991)

Automated cyst recognition from x-ray photographs (Nedkova, Rumiana; Delchev, Georgy)V1429,105-107(1991)

Automated registration of terrain range images using surface feature level sets (Wheeler, Frederick W.; Vaz, Richard F.; Cyganski, David)V1606,78-85(1991)

Automated visual inspection for LSI wafer patterns using a derivative-polarity comparison algorithm (Maeda, Shunji; Hiroi, Takashi; Makihira, Hiroshi; Kubota, Hitoshi)V1567,100-109(1991)

Automatic and operator-assisted solid modeling of objects for automatic recognition (Stenstrom, J. R.; Connolly, C. I.)V1470,275-281(1991)

Automatic Object Recognition (Sadjadi, Firooz A., ed.)V1471(1991)

Automatic object recognition: critical issues and current approaches (Sadjadi, Firooz A.)V1471,303-313(1991)

Automatic recognition of multidimensional objects buried in layered elastic background media (Ayme-Bellegarda, Eveline J.; Habashy, Tarek M.; Bellegarda, Jerome R.)V1471,18-29(1991)

Automatic target recognition using acousto-optic image correlator (Molley, Perry A.; Kast, Brian A.)V1471,224-232(1991)

Autonomous navigation of structured city roads (Aubert, Didier; Kluge, Karl; Thorpe, Charles E.)V1388,141-151(1991)

Background characterization using a second-order moment function (Scholl, Marija S.; Udomkesmalee, Suraphol)V1468,92-98(1991)

Best fit ellipse for cell shape analysis (Wali, Rahman; Colef, Michael; Barba, Joseph)V1606,665-674(1991)

Bidirectional log-polar mapping for invariant object recognition (Mehanian, Courosh; Rak, Steven J.)V1471,200-209(1991)

Biomedical structure recognition by successive approximations (Venturi, Giovanni; Dellepiane, Silvana G.; Vernazza, Gianni L.)V1606,217-225(1991)

Bipolar correlations in composite circular harmonic filters (Leclerc, Luc; Sheng, Yunlong; Arsenault, Henri H.)V1564,78-85(1991)

Building an optical pattern recognizer (Lindberg, Perry C.; Gregory, Don A.)V1470,220-225(1991)

Case study of design trade-offs for ternary phase-amplitude filters (Flannery, David L.; Phillips, William E.; Reel, Richard L.)V1564,65-77(1991)

Cellular-automata-based learning network for pattern recognition (Tzionas, Panagiotis; Tsalides, Ph.; Thanailakis, A.)V1606,269-280(1991)

Characteristic pattern matching based on morphology (Zhao, Dongming)V1606,86-96(1991)

Character string detection algorithm using horizontal boundaries, and its application to a part number entry system (Amano, Tomio; Yamashita, Akio; Takahashi, Hiroyasu)V1452,330-339(1991)

Clue derivation and selection activities in a robot vision system (Reihani, Kamran; Thompson, Wiley E.)V1468,305-312(1991)

Color-invariant character recognition and character-background color identification by multichannel matched filter (Campos, Juan; Millan, Maria S.; Yzuel, Maria J.; Ferreira, Carlos)V1564,189-198(1991)

Combined approach for large-scale pattern recognition with translational, rotational, and scaling invariances (Xu, Qing; Inigo, Rafael M.; McVey, Eugene S.)V1471,378-389(1991)

Combining neural networks and decision trees (Sankar, Ananth; Mammone, Richard J.)V1469,374-383(1991)

Compact, one-lens JTC using a transmissive amorphous silicon FLC-SLM (LAPS-SLM) (Haemmerli, Jean-Francois; Iwaki, Tadao; Yamamoto, Shuhei)V1564,275-284(1991)

Comparative study of texture measurements for cellular organelle recognition (Grenier, Marie-Claude; Durand, Louis-Gilles; de Guise, J.)V1450,154-169(1991)

Comparison of a multilayered perceptron with standard classification techniques in the presense of noise (Willson, Gregory B.)V1469,351-358(1991)

Comparison of images on the basis of structural features (Markov, Vladimir B.; Shishkov, Vladimir F.; Voroshnin, A. B.)V1238,118-122(1991)

Comparison of neural network classifiers to quadratic classifiers for sensor fusion (Brown, Joseph R.; Bergondy, Daniel; Archer, Susan J.)V1469,539-543(1991)

Composite image joint transform correlator (Mendlovic, David; Konforti, Naim; Deutsch, Meir; Marom, Emanuel)V1442,182-192(1991)

Computer-aided photorefractive pattern recognition (Sun, Ching-Cherng; Chang, Ming-Wen; Yeh, Smile; Cheng, Nai-Jen)V1564,199-210(1991)

Computer interpretation of thallium SPECT studies based on neural network analysis (Wang, David C.; Karvelis, K. C.)V1445,574-575(1991)

Computerized detection and identification of the types of defects on crystal blanks (Bow, Sing T.; Chen, Pei)V1396,646-655(1991)

Considering multiple-surface hypotheses in a Bayesian hierarchy (LaValle, Steven M.; Hutchinson, Seth A.)V1569,2-15(1991)

Context specification for text recognition in forms (Anderson, Kelly L.; Barrett, William A.)V1384,270-279(1991)

Continuous recognition of sonar targets using neural networks (Venugopal, Kootala P.; Pandya, Abhijit S.; Sudhakar, Raghavan)V1471,44-53(1991)

Contrast, size, and orientation-invariant target detection in infrared imagery (Zhou, Yi-Tong; Crawshaw, Richard D.)V1471,404-411(1991)

Convexity-based method for extracting object parts from 3-D surfaces (Vaina, Lucia M.; Zlateva, Stoyanka D.)V1468,710-719(1991)

Data-driven parallel architecture for syntactic pattern recognition (Tseng, Chien-Chao; Hwang, Shu-Yuen)V1384,257-268(1991)

Data Structures and Target Classification (Libby, Vibeke, ed.)V1470(1991)

Decomposing morphological structure element into neighborhood configurations (Gong, Wei; Shi, Qinyun; Cheng, Minde)V1606,153-164(1991)

Design and testing of three-level optimal correlation filters (Hendrix, Charles D.; Vijaya Kumar, B. V. K.; Stalker, K. T.; Kast, Brian A.; Shori, Raj K.)V1564,2-13(1991)

Detecting spatial and temporal dot patterns in noise (Drum, Bruce)V1453,153-164(1991)

Detection and classification of undersea objects using multilayer perceptrons (Shazeer, Dov J.; Bello, M.)V1469,622-636(1991)

Detection of DNA sequence symmetries using parallel micro-optical devices (Christens-Barry, William A.; Terry, David H.; Boone, Bradley G.)V1564,177-188(1991)

Detection of liver metastisis using the backpropagation algorithm and linear discriminant analysis (DaPonte, John S.; Parikh, Jo Ann; Katz, David A.)V1469,441-450(1991)

Detection of tool wear using multisensor readings defused by artificial neural network (Masory, Oren)V1469,515-525(1991)

Development of the marine-aggregated-particle profiling and enumerating rover (Costello, David K.; Carder, Kendall L.; Steward, Robert G.)V1537,161-172(1991)

DIGNET: a self-organizing neural network for automatic pattern recognition and classification (Thomopoulos, Stelios C.; Bougoulias, Dimitrios K.)V1470,253-262(1991)

Discrete random set models for shape synthesis and analysis (Goutsias, John I.; Wen, Chuanju)V1606,174-185(1991)

Distortion- and intensity-invariant optical correlation filter system (Rahmati, Mohammad; Hassebrook, Laurence G.; Bhushan, M.)V1567,480-489(1991)

Distribution of the pattern spectrum mean for convex base images (Dougherty, Edward R.; Sand, Francis M.)V1451,114-124(1991)

Dynamically reconfigurable multiprocessor system for high-order-bidirectional-associative-memory-based image recognition (Wu, Chwan-Hwa; Roland, David A.)V1471,210-221(1991)

Effectiveness of certain features for optical character recognition (Kovacs, Emoke; Marosi, Istvan)V1384,338-343(1991)

Efficient computation of various types of skeletons (Vincent, Luc M.)V1445,297-311(1991)

EGOLOGY: psychological spatial breakthrough for social redirection—multidisciplinary spatial focus for individuals/humankind (Thompson, Robert A.; Thompson, Louise A.)V1469,451-462(1991)

Elliptical coordinate transformed phase-only filter for shift and scale invariant pattern recognition (Garcia, Javier; Ferreira, Carlos; Szoplik, Tomasz)V1574,133-140(1991)

Error probabilities of minimum-distance classifiers (Poublan, Helene; Castanie, Francis)V1569,329-340(1991)

Estimation of linear parametric models of non-Gaussian discrete random fields (Tugnait, Jitendra K.)V1452,204-215(1991)

Estimation of prospective locations in mature hydrocarbon producing areas (Isaksen, Tron)V1452,270-291(1991)

Evaluation of multiresolution elastic matching using MRI data (Gee, Jim C.; Reivich, Martin; Bilaniuk, L.; Hackney, D.; Zimmerman, R.; Kovacic, Stane; Bajcsy, Ruzena R.)V1445,226-234(1991)

Example of a Bayes network of relations among visual features (Agosta, John M.)V1569,16-27(1991)

Extracting characters from illustration document by relaxation (Takeda, Haruo)V1386,128-134(1991)

Extraction of features from images using video feedback (Boone, Bradley G.; Shukla, Oodaye B.; Terry, David H.)V1471,390-403(1991)

Extraction of human stomach using computational geometry (Aisaka, Kazuo; Arai, Kiyoshi; Tsutsui, Kumiko; Hashizume, Akihide)V1445,312-317(1991)

Fast algorithm for a neocognitron neural network with back-propagation (Qing, Kent P.; Means, Robert W.)V1569,111-120(1991)

Fast algorithm for size analysis of irregular pore areas (Yuan, Li-Ping)V1451,125-136(1991)

Fast holographic correlator for machine vision systems (Nekrasov, Victor V.)V1507,170-174(1991)

Fast identification of images using neural networks (Min, Kwang-Shik; Min, Hisook L.)V1469,129-136(1991)

Fast optoelectronic neurocomputer for character recognition (Zhang, Lin; Robinson, Michael G.; Johnson, Kristina M.)V1469,225-229(1991)

Feature extractor giving distortion invariant hierarchical feature space (Lampinen, Jouko)V1469,832-842(1991)

Formulation and performance evaluation of adaptive, sequential radar-target-recognition algorithms (Snorrason, Ogmundur; Garber, Frederick D.)V1471,116-127(1991)

Fourier cross-correlation and invariance transformation for affine groups (Segman, Joseph)V1606,788-802(1991)

Frequency-based pattern recognition using neural networks (Lu, Simon W.)V1569,452-462(1991)

Fuzzy logic and neural networks in artificial intelligence and pattern recognition (Sanchez, Elie)V1569,474-483(1991)

Generalizing from a small set of training exemplars for handwritten digit recognition (Simon, Wayne E.; Carter, Jeffrey R.)V1469,592-601(1991)

Geometrical and morphological image processing algorithm (Reihani, Kamran; Thompson, Wiley E.)V1452,319-329(1991)

Geometric modeling of noisy image objects (Lipari, Charles A.)V1468,905-917(1991)

Geometric property measurement of convex objects using fuzzy sets (Poelzleitner, Wolfgang)V1381,411-422(1991)

Global approach toward the evaluation of thermal infrared countermeasures (Verlinde, Patrick S.; Proesmans, Marc)V1486,58-65(1991)

Gradient descent techniques for feature detection template generation (Pont, W. F.; Gader, Paul D.)V1568,247-260(1991)

Ground target classification using moving target indicator radar signatures (Yoon, Chun S.)V1470,243-252(1991)

Hetero-association for pattern translation (Yu, Francis T.; Lu, Taiwei; Yang, Xiangyang)V1507,210-221(1991)

Hierarchical network for clutter and texture modeling (Luttrell, Stephen P.)V1565,518-528(1991)

Hierarchical neural net with pyramid data structures for region labeling of images (Rosten, David P.; Yuen, P.; Hunt, Bobby R.)V1472,118-127(1991)

High-Speed Inspection Architectures, Barcoding, and Character Recognition (Chen, Michael J., ed.)V1384(1991)

Ho-Kashyap CAAP 1:1 associative processors (Casasent, David P.; Telfer, Brian A.)V1382,158-166(1991)

Holography, Interferometry, and Optical Pattern Recognition in Biomedicine (Podbielska, Halina, ed.)V1429(1991)

Human face recognition by P-type Fourier descriptor (Aibara, Tsunehiro; Ohue, Kenji; Matsuoka, Yasushi)V1606,198-203(1991)

Hybrid pattern recognition system for the robotic vision (Li, Yulin; Zhao, Mingjun; Zhao, Li)V1385,200-205(1991)

Hybrid solution for high-speed target acquisition and identification systems (Udomkesmalee, Suraphol; Scholl, Marija S.; Shumate, Michael S.)V1468,81-91(1991)

Image Algebra and Morphological Image Processing II (Gader, Paul D.; Dougherty, Edward R., eds.)V1568(1991)

Image analysis applications for grain science (Zayas, Inna Y.; Steele, James L.)V1379,151-161(1991)

Image analysis using attributed fuzzy tournament matching algorithm (Shaout, Adnan)V1381,357-367(1991)

Image Processing Algorithms and Techniques II (Civanlar, Mehmet R.; Mitra, Sanjit K.; Moorhead, Robert J., eds.)V1452(1991)

Image processing to locate corn plants (Jia, Jiancheng; Krutz, Gary W.; Gibson, Harry G.)V1396,656-663(1991)

Image quality, dollars, and very low contrast documents (Weideman, William E.)V1454,382-390(1991)

Image recognition, learning, and control in a cellular automata network (Raghavan, Raghu)V1469,89-101(1991)

Image segmentation with genetic algorithms: a formulation and implementation (Seetharaman, Guna S.; Narasimhan, Anand; Sathe, Anand; Storc, Lisa)V1569,269-273(1991)

Imposing a temporal structure in neural networks (Gupta, Lalit; Sayeh, Mohammad R.; Upadhye, Anand M.)V1396,266-269(1991)

Improved adaptive resonance theory (Shih, Frank Y.; Moh, Jenlong)V1382,26-36(1991)

Industrial applications of optical fuzzy syntactic pattern recognition (Caulfield, H. J.)V1332,294-300(1991)

Influence of input information coding for correlation operations (Maze, Sylvie; Joffre, Pascal; Refregier, Philippe)V1505,20-31(1991)

Initial key word OCR filter results (Casasent, David P.; Iyer, Anand K.; Ravichandran, Gopalan)V1384,324-337(1991)

Integration of an edge extraction cellular automaton (Taboury, Jean; Chavel, Pierre)V1505,115-123(1991)

Intelligent vision process for robot manipulation (Chen, Alexander Y.; Chen, Eugene Y.)V1381,226-239(1991)

Intelligent word-based text recognition (Hoenes, Frank; Bleisinger, Rainer; Dengel, Andreas R.)V1384,305-316(1991)

Interactive tools for assisting the extraction of cartographic features (Hunt, Bobby R.; Ryan, Thomas W.; Sementilli, P.; DeKruger, D.)V1472,208-218(1991)

Interactive tools for extraction of cartographic calibration data from aerial photography (Hunt, Bobby R.; Ryan, Thomas W.; Gifford, E.)V1472,190-200(1991)

Intermodulation effects in pure phase-only correlation method (Chalasinska-Macukow, Katarzyna; Turon, F.; Yzuel, Maria J.; Campos, Juan)V1564,285-293(1991)

Invariant pattern recognition via higher order preprocessing and backprop (Davis, Jon P.; Schmidt, William A.)V1469,804-811(1991)

Inverse filtering technique for the synthesis of distortion-invariant optical correlation filters (Shen, Weisheng; Zhang, Shen; Tao, Chunkan)V1567,691-697(1991)

Inverse scattering problems in the acoustic resonance region of an underwater target (Gaunaurd, Guillermo C.)V1471,30-41(1991)

Japanese document recognition and retrieval system using programmable SIMD processor (Miyahara, Sueharu; Suzuki, Akira; Tada, Shunkichi; Kawatani, Takahiko)V1384,317-323(1991)

Joint space/spatial-frequency representations as preprocessing steps for neural nets; joint recognition of separately learned patterns; results and limitations (Rueff, Manfred; Frankhauser, P.; Dettki, Frank)V1382,255-270(1991)

Labeled object identification for the mobile servicing system on the space station (Zakaria, Marwan F.; Ng, Terence K.)V1386,121-127(1991)

Landmark-based partial shape recognition by a BAM neural network (Liu, Xianjun; Ansari, Nirwan)V1606,1069-1079(1991)

Learnability of min-max pattern classifiers (Yang, Ping-Fai; Maragos, Petros)V1606,294-308(1991)

Learning filter systems with maximum correlation and maximum separation properties (Lenz, Reiner; Osterberg, Mats)V1469,784-795(1991)

Learning object shapes from examples (Shariat, Hormoz)V1567,194-203(1991)

Learning procedure for the recognition of 3-D objects from 2-D images (Bart, Mischa; Buurman, Johannes; Duin, Robert P.)V1381,66-77(1991)

Linear programming for learning in neural networks (Raghavan, Raghu)V1472,139-148(1991)

Linear programming solutions to problems in logical inference and space-variant image restoration (Digumarthi, Ramji V.; Payton, Paul M.; Barrett, Eamon B.)V1472,128-138(1991)

Localized feature selection to maximize discrimination (Duell, Kenneth A.; Freeman, Mark O.)V1564,22-33(1991)

Low-level image segmentation via texture recognition (Patel, Devesh; Stonham, T. J.)V1606,621-629(1991)

Machine verification of traced signatures (Krishnan, Ganapathy; Jones, David E.)V1468,563-572(1991)

Marker-controlled picture segmentation applied to electrical logging images (Rivest, Jean-Francois; Beucher, Serge; Delhomme, J.)V1451,179-190(1991)

Markov random fields for texture classification (Chen, Chaur-Chin)V1569,274-285(1991)

Marriage between digital holography and optical pattern recognition (Wyrowski, Frank; Bernhardt, Michael)V1555,146-153(1991)

Massively parallel implementation of neural network architectures (Omidvar, Omid M.; Wilson, Charles L.)V1452,532-543(1991)

Matched spatial filtering by feature-extracted reference patterns using cross-correlated signals (Kamemaru, Shun-ichi; Yano, Jun-ichi; Itoh, Haruyasu)V1564,143-154(1991)

Mean-field theory for grayscale texture synthesis using Gibbs random fields (Elfadel, Ibrahim M.; Yuille, Alan L.)V1569,248-259(1991)

Microcomputer-based image processing system for CT/MRI scans: II. Expert system (Kwok, John C.; Yu, Peter K.; Cheng, Andrew Y.; Ho, Wai-Chin)V1445,446-455(1991)

Millimeter wave radar stationary-target classification using a high-order neural network (Hughen, James H.; Hollon, Kenneth R.)V1469,341-350(1991)

Model-based automatic target recognition development tools (Nasr, Hatem N.; Amehdi, Hossien)V1471,283-290(1991)

Model-based system for automatic target recognition (Verly, Jacques G.; Delanoy, Richard L.; Dudgeon, Dan E.)V1471,266-282(1991)

Model generation and partial matching of left ventricular boundaries (Tehrani, Saeid; Weymouth, Terry E.; Mancini, G. B.)V1445,434-445(1991)

Model group indexing for recognition (Clemens, David T.; Jacobs, David W.)V1381,30-42(1991)

Modeling of local neural networks of the visual cortex and applications to image processing (Rybak, Ilya A.; Shevtsova, Natalia A.; Podladchikova, Lubov N.)V1469,737-748(1991)

Model of human preattentive visual detection of edge orientation anomalies (Brecher, Virginia H.; Bonner, Raymond; Read, C.)V1473,39-51(1991)

Moire image overlapping method for PCB inspection designator (Chang, Rong-Seng; Hu, Yeu-Jent)V1567,216-219(1991)

Morphological feature-set optimization using the genetic algorithm (Trenkle, John M.; Schlosser, Steve; Vogt, Robert C.)V1568,212-223(1991)

Morphological processing for the analysis of disordered structures (Casasent, David P.; Sturgill, Robert; Schaefer, Roland H.)V1567,683-690(1991)

MSE and hierarchical optical associative processor system (Casasent, David P.; Chien, Sung-Il)V1382,304-310(1991)

Multiple-hypothesis-based multiple-sensor spatial data fusion algorithm (Leung, Dominic S.; Williams, D. S.)V1471,314-325(1991)

Multisensor fusion using the sensor algorithm research expert system (Bullock, Michael E.; Miltonberger, Thomas W.; Reinholdtsen, Paul A.; Wilson, Kathleen)V1471,291-302(1991)

Multitarget detection and estimation parallel algorithm (Krikelis, A.)V1482,307-316(1991)

Neural network architecture for form and motion perception (Grossberg, Stephen)V1469,24-26(1991)

Neural networks for ATR parameters adaptation (Amehdi, Hossien; Nasr, Hatem N.)V1483,177-184(1991)

Neural optoelectronic correlator for pattern recognition (Figue, J.; Refregier, Philippe; Rajbenbach, Henri J.; Huignard, Jean-Pierre)V1564,550-561(1991)

New decision tree algorithm for handwritten numerals recognition using topological features (Impedovo, Sebastiano; Dimauro, Giovanni; Pirlo, Giuseppe)V1384,280-284(1991)

New method for designing face image classifiers using 3-D CG model (Akamatsu, Shigeru; Sasaki, Tsutomu; Masui, Nobuhiko; Fukamachi, Hideo; Suenaga, Yasuhito)V1606,204-216(1991)

New techniques for repertory grid analysis (Liseth, Ole J.; Bezdek, James C.; Ford, Kenneth M.; Adams-Webber, Jack R.)V1468,256-267(1991)

Noise and discrimination performance of the MINACE optical correlation filter (Ravichandran, Gopalan; Casasent, David P.)V1471,233-248(1991)

Noise tolerance of adaptive resonance theory neural network for binary pattern recognition (Kim, Yong-Soo; Mitra, Sunanda)V1565,323-330(1991)

Nonlinear optical flow estimation and segmentation (Geurtz, Alexander M.)V1567,110-121(1991)

Normalization of correlations (Kast, Brian A.; Dickey, Fred M.)V1564,34-42(1991)

Numerical evaluation of the efficiency of camouflage systems in the thermal infrared (Proesmans, Marc; Verlinde, Patrick S.)V1486,102-114(1991)

Object-enhanced optical correlation (Scholl, Marija S.; Shumate, Michael S.; Udomkesmalee, Suraphol)V1564,165-176(1991)

Object recognition using coding schemes (Sadjadi, Firooz A.)V1471,428-434(1991)

Optical correlation filters for large-class OCR applications (Casasent, David P.; Iyer, Anand K.; Gopalaswamy, Srinivasan)V1470,208-219(1991)

Optical correlation filters to locate destination address blocks in OCR (Casasent, David P.; Ravichandran, Gopalan)V1384,344-354(1991)

Optical correlator field test results (Hudson, Tracy D.; Gregory, Don A.; Kirsch, James C.; Loudin, Jeffrey A.; Crowe, William M.)V1564,54-64(1991)

Optical correlators in texture analysis (Honkonen, Veijo; Jaaskelainen, Timo; Parkkinen, Jussi P.)V1564,43-51(1991)

Optical/electronic hybrid three-layer neural network for pattern recognition (Zhang, Yanxin; Shen, Jinyuan)V1469,303-307(1991)

Optical higher order double-layer associative memory (Lam, David T.; Carroll, John E.)V1505,104-114(1991)

Optical image segmentor using wavelet filtering techniques as the front-end of a neural network classifier (Veronin, Christopher P.; Rogers, Steven K.; Kabrisky, Matthew; Welsh, Byron M.; Priddy, Kevin L.; Ayer, Kevin W.)V1469,281-291(1991)

Optical implementation of neocognitron and its applications to radar signature discrimination (Chao, Tien-Hsin; Stoner, William W.)V1558,505-517(1991)

Optical Information Processing Systems and Architectures III (Javidi, Bahram, ed.)V1564(1991)

Optical processing and hybrid neural nets (Casasent, David P.)V1469,256-267(1991)

Optical processing of wire-frame models for object recognition (Kozaitis, Samuel P.; Cofer, Rufus H.)V1471,249-254(1991)

Optimal correlation filters for implementation on deformable mirror devices (Vijaya Kumar, B. V. K.; Carlson, Daniel W.)V1558,476-486(1991)

Optimal distortion-invariant quadratic filters (Gheen, Gregory)V1564,112-120(1991)

Optimum structure learning algorithms for competitive learning neural network (Uchiyama, Toshio; Sakai, Mitsuhiro; Saito, Tomohide; Nakamura, Taichi)V1451,192-203(1991)

Order-statistic filters on matrices of images (Wilson, Stephen S.)V1451,242-253(1991)

Parametric analysis of target/decoy performance (Kerekes, John P.)V1483,155-166(1991)

Parsing algorithm for line-drawing pattern recognition (Wang, Patrick S.; Zhang, Y. Y.)V1384,68-74(1991)

Passive range sensor refinement using texture and segmentation (Sridhar, Banavar; Phatak, Anil; Chatterji, Gano B.)V1478,178-189(1991)

Pattern detection using a modified composite filter with nonlinear joint transform correlator (Vallmitjana, Santiago; Juvells, Ignacio; Carnicer, Arturo; Campos, Juan)V1564,266-274(1991)

Pattern recognition approach to trench bottom-width measurement (Chou, Ching-Hua; Berman, John L.; Chim, Stanley S.; Corle, Timothy R.; Xiao, Guoqing; Kino, Gordon S.)V1464,145-154(1991)

Pattern recognition in pulmonary computerized tomography images using Markovian modeling (Preteux, Francoise; Moubarak, Michel; Grenier, Philippe)V1450,72-83(1991)

Pattern recognition, neural networks, and artificial intelligence (Bezdek, James C.)V1468,924-935(1991)

Pattern recognition using w-orbit finite automata (Liu, Ying; Ma, Hede)V1606,226-240(1991)

Pattern recognition with incoherent light and rotation invariance (Elizur, Eran; Friesem, Asher A.)V1442,230-234(1991)

Pattern spectrum morphology for texture discrimination and object recognition (Lee, Bonita G.; Tom, Victor T.)V1381,80-91(1991)

PC-based hardware implementation of the maximum-likelihood classifier for the shuttle ice detection system (Jaggi, Sandeep)V1452,340-350(1991)

Performance comparison of neural network and statistical discriminant processing techniques for automatic modulation recognition (Hill, Peter C.; Orzeszko, Gabriel R.)V1469,329-340(1991)

Performance limitations of miniature optical correlators (Crandall, Charles M.; Giles, Michael K.; Clark, Natalie)V1564,98-109(1991)

Performance of a neural-network-based 3-D object recognition system (Rak, Steven J.; Kolodzy, Paul J.)V1471,177-184(1991)

Photothermoplastic spatial filters for optical pattern recognition (Isaev, Urkaly T.; Akaev, Askar A.; Kutanov, Askar A.)V1507,198-201(1991)

Position-, scale-, and rotation-invariant photorefractive correlator (Ryan, Vincent; Fielding, Kenneth H.)V1564,86-97(1991)

Possibilities of the fringe pattern learning system VARNA (Nedkova, Rumiana)V1429,214-216(1991)

Prototype neural network pattern recognition testbed (Worrell, Steven W.; Robertson, James A.; Varner, Thomas L.; Garvin, Charles G.)V1382,219-227(1991)

Purely real correlation filters (Mahalanobis, Abhijit; Song, Sewoong)V1564,14-21(1991)

Quality factors of handwritten characters based on human visual perception (Kato, Takahito; Yamada, Mitsuho)V1453,43-50(1991)

Radar image understanding for complex space objects (Hemler, Paul F.)V1381,55-65(1991)

Random mapping network for tactical target reacquisition after loss of track (Church, Susan D.; Burman, Jerry A.)V1471,192-199(1991)

Real-time architecture based on the image processing module family (Kimura, Shigeru; Murakami, Yoshiyuki; Matsuda, Hikaru)V1483,10-17(1991)

Real-time classification of wooden boards (Poelzleitner, Wolfgang; Schwingshakl, Gert)V1384,38-49(1991)

Real-time motion detection using an analog VLSI zero-crossing chip (Bair, Wyeth; Koch, Christof)V1473,59-65(1991)

Recognition criterion for two-dimensional minimum resolvable temperature difference (Kennedy, Howard V.)V1488,196-202(1991)

Recognition in face space (Turk, Matthew A.; Pentland, Alexander P.)V1381,43-54(1991)

Recognition of contacts between objects in the presence of uncertainties (Xiao, Jing)V1470,134-145(1991)

Recognition of handwritten katakana in a frame using moment invariants based on neural network (Agui, Takeshi; Takahashi, Hiroki; Nakajima, Masayuki; Nagahashi, Hiroshi)V1606,188-197(1991)

Representing three-dimensional shapes for visual recognition (Hoffman, Donald)V1445,2-4(1991)

Research and improvement of Vogl's acceleration algorithm (Song, Hongjun; Zhou, Lijia)V1469,581-584(1991)

Residue-producing E-filters and their applications in medical image analysis (Preston, Kendall)V1450,59-70(1991)

Rotation and scale invariant pattern recognition using a multistaged neural network (Minnix, Jay I.; McVey, Eugene S.; Inigo, Rafael M.)V1606,241-251(1991)

Salient contour extraction for target recognition (Rao, Kashi; Liou, James J.)V1482,293-306(1991)

Scanning strategies for target detection (Gertner, Izidor; Zeevi, Yehoshua Y.)V1470,148-166(1991)

Selected Papers on Automatic Object Recognition (Nasr, Hatem N., ed.)VMS41(1991)

Sensor fusion using K-nearest neighbor concepts (Scott, David R.; Flachs, Gerald M.; Gaughan, Patrick T.)V1383,367-378(1991)

Shape recognition combining mathematical morphology and neural networks (Schmitt, Michel; Mattioli, Juliette)V1469,392-403(1991)

Shape recognition in the Fourier domain (Udomkesmalee, Suraphol)V1564,464-472(1991)

Shape registration by morphological operations (Chang, Long-Wen; Tsai, Mong-Jean)V1606,120-131(1991)

Signal and Image Processing Systems Performance Evaluation, Simulation, and Modeling (Nasr, Hatem N.; Bazakos, Michael E., eds.)V1483(1991)

Single-chip imager and feature extractor (Tanner, John E.; Luo, Jin)V1473,76-87(1991)

Slope histogram detection of forged handwritten signatures (Wilkinson, Timothy S.; Goodman, Joseph W.)V1384,293-304(1991)

Soft morphological filters (Koskinen, Lasse; Astola, Jaakko; Neuvo, Yrjo A.)V1568,262-270(1991)

Spherical and nonspherical aerosol and particulate characterization using optical pattern recognition techniques (Marshall, Martin S.; Benner, Robert E.)V1564,121-134(1991)

Statistical and neural network classifiers in model-based 3-D object recognition (Newton, Scott C.; Nutter, Brian S.; Mitra, Sunanda)V1382,209-218(1991)

Statistical approach to model matching (Wells, William M.)V1381,22-29(1991)

Statistical morphology (Yuille, Alan L.; Vincent, Luc M.; Geiger, Davi)V1568,271-282(1991)

Stochastic and Neural Methods in Signal Processing, Image Processing, and Computer Vision (Chen, Su-Shing, ed.)V1569(1991)

Strategies for the color character recognition by optical multichannel correlation (Millan, Maria S.; Yzuel, Maria J.; Campos, Juan; Ferreira, Carlos)V1507,183-197(1991)

Structural identity in visual-perceptual recognition (Ligomenides, Panos A.)V1382,14-25(1991)

Studies in robust approaches to object detection in high-clutter background (Shirvaikar, Mukul V.; Trivedi, Mohan M.)V1468,52-59(1991)

Survey of radar-based target recognition techniques (Cohen, Marvin N.)V1470,233-242(1991)

Survey: omnifont-printed character recognition (Tian, Qi; Zhang, Peng; Alexander, Thomas; Kim, Yongmin)V1606,260-268(1991)

Table recognition for automated document entry system (Kojima, Haruhiko; Akiyama, Teruo)V1384,285-292(1991)

Target acquisition modeling based on human visual system performance (Valeton, J. M.; van Meeteren, Aart)V1486,68-84(1991)

Target cuing: a heterogeneous neural network approach (McCauley, Howard M.)V1469,69-76(1991)

Target detection performance using 3-D laser radar images (Green, Thomas J.; Shapiro, Jeffrey H.; Menon, Murali M.)V1471,328-341(1991)

Target detection using co-occurrence matrix segmentation and its hardware implementation (Auborn, John E.; Fuller, James M.; McCauley, Howard M.)V1482,246-252(1991)

Target detection using multilayer feedforward neural networks (Scherf, Alan V.; Scott, Peter A.)V1469,63-68(1991)

Technique for ground/image truthing using a digital map to reduce the number of required measurements (Der, Sandor Z.; Dome, G. J.; Rusche, Gerald A.)V1483,167-176(1991)

Texture classification and neural network methods (Visa, Ari J.)V1469,820-831(1991)

Texture segmentation in aerial images (Monjoux, E.; Brunet, Gerard; Rudant, J. P.)V1452,310-318(1991)

Theoretical approach to hyperacuity tests based on resolution criteria for two-line images (Mondal, Pronab K.; Calvo, Maria L.; Chevalier, Margarita L.; Lakshminarayanan, Vasudevan)V1429,108-116(1991)

Three-dimensional display of MRI data in neurosurgery: segmentation and rendering aspects (Barillot, Christian; Lachmann, F.; Gibaud, Bernard; Scarabin, Jean-Marie)V1445,54-65(1991)

Three-dimensional object representation by array grammars (Wang, Patrick S.)V1381,210-216(1991)

Three-dimensional orientation from texture using Gabor wavelets (Super, Boaz J.; Bovik, Alan C.)V1606,574-586(1991)

Throughput comparison of optical and digital correlators for automatic target recognition (Huang, Chao H.; Gheen, Gregory; Washwell, Edward R.)V1564,427-438(1991)

Tissue identification in MR images by adaptive cluster analysis (Gutfinger, Dan; Hertzberg, Efrat M.; Tolxdorff, T.; Greensite, F.; Sklansky, Jack)V1445,288-296(1991)

Track-while-image in the presence of background (Mentle, Robert E.; Shapiro, Jeffrey H.)V1471,342-353(1991)

Trade-offs between pseudohexagonal and pseudocubic filters (Preston, Kendall)V1451,75-90(1991)

Training image collection at CECOM's Center for Night Vision and Electro-Optics (Harr, Richard W.)V1483,231-239(1991)

Transputer-based parallel algorithms for automatic object recognition (Bison, Paolo G.; Braggiotti, Alberto; Grinzato, Ermanno G.)V1471,369-377(1991)

Two inverse problems in mathematical morphology (Schmitt, Michel)V1568,283-294(1991)

Two-view vision system for 3-D texture recovery (Yu, Xiaohan; Yla-Jaaski, Juha)V1468,834-842(1991)

Unsupervised training of structuring elements (Wilson, Stephen S.)V1568,188-199(1991)

Use of syntactic recognition with sampled boundary distances (Wang, David T.; Peng, Ming-Chien; Lee, Jueen; Chen, Jyh-Woei; Ng, Peter A.)V1386,206-219(1991)

Using local orientation and hierarchical spatial feature matching for the robust recognition of objects (Seitz, Peter; Lang, Graham K.)V1606,252-259(1991)

Validity criterion for compact and separate fuzzy partitions and its justification (Xie, Xuanli; Beni, Gerardo)V1381,401-410(1991)

Vision-based model of artificial texture perception (Landraud, Anne M.)V1453,314-320(1991)

Visualization of manufacturing process data in N-dimensional spaces: a reanalysis of the data (Fulop, Ann C.; Allen, Donald M.; Deffner, Gerhard)V1459,69-76(1991)

Visual thinking in organizational analysis (Grantham, Charles E.)V1459,77-84(1991)

Warping of a computerized 3-D atlas to match brain image volumes for quantitative neuroanatomical and functional analysis (Evans, Alan C.; Dai, W.; Collins, L.; Neelin, Peter; Marrett, Sean)V1445,236-246(1991)

Phase

Accuracy and high-speed technique for autoprocessing of Young's fringes (Chen, Wenyi; Tan, Yushan)V1554A,879-885(1991)

Al2O3 etch-stop layer for a phase-shifting mask (Hanyu, Isamu; Nunokawa, Mitsuji; Asai, Satoru; Abe, Masayuki)V1463,595-601(1991)

Analysis and experimental performance of reduced-resolution binary phase-only filters (Kozaitis, Samuel P.; Tepedelenlioglu, N.; Foor, Wesley E.)V1564,373-383(1991)

Automated fringe analysis for moire interferometry (Brownell, John B.; Parker, Richard J.)V1554B,481-492(1991)

Binary and phase-shifting image design for optical lithography (Liu, Yong; Zakhor, Avideh)V1463,382-399(1991)

Ceramic phase-shifters for electronically steerable antenna systems (Selmi, Fathi; Ghodgaonkar, Deepak K.; Hughes, Raymond; Varadan, Vasundara V.; Varadan, Vijay K.)V1489,97-107(1991)

Chromeless phase-shifted masks: a new approach to phase-shifting masks (Toh, Kenny K.; Dao, Giang T.; Singh, Rajeev; Gaw, Henry T.)V1496,27-53(1991)

Compact time-delay shifters that are process insensitive (Lesko, Camille; Hill, William A.; Dietrich, Fred; Nelson, William)V1475,330-339(1991)

Comparison of relativistic magnetron oscillator models for phase-locking studies (Chen, Shien C.)V1407,100-104(1991)

Compatability test for phase-change erasable and WORM media in a multifunction drive (Ohara, Shunji; Ishida, Takashi; Inokuchi, Chikashi; Furutani, Tadashige; Ishibashi, Kenzo; Kurahashi, Akira; Yoshida, Tomio)V1499,187-194(1991)

Conjugate twin-shifter for the new phase-shift method to high-resolution lithography (Ohtsuka, Hiroshi; Abe, Kazutoshi; Onodera, Toshio; Kuwahara, Kazuyuki; Taguchi, Takashi)V1463,112-123(1991)

Constructing an optimal binary phase-only filter using a genetic algorithm (Calloway, David L.)V1564,395-402(1991)

Crystalization kinetics in Ge2Sb2Te5 phase-change recording films (Ozawa, Kenji; Ogino, Shinji; Satoh, Yoshikazu; Urushidani, Tatuo; Ueda, Atushi; Deno, Hiroshi; Kawakami, Haruo)V1499,180-186(1991)

Design methodology for dark-field phase-shifted masks (Toh, Kenny K.; Dao, Giang T.; Gaw, Henry T.; Neureuther, Andrew R.; Fredrickson, Larry R.)V1463,402-413(1991)

Digital shearing laser interferometry for heterodyne array phasing (Cederquist, Jack N.; Fienup, James R.; Marron, Joseph C.; Schulz, Timothy J.; Seldin, J. H.)V1416,266-277(1991)

Direct readout of dynamic phase changes in a fiber-optic homodyne interferometer (Jin, Wei; Uttamchandani, Deepak G.; Culshaw, Brian)V1504,125-132(1991)

Double-exposure phase-shifting holography applied to particle velocimetry (Lai, Tianshu; Tan, Yushan)V1554B,580-585(1991)

Dynamic holographic microphasometry (Kozhevnikov, Nikolai M.)V1507,509-516(1991)

Dynamics of phase-grating formation in optical fibers (An, Sunghyuck; Sipe, John E.)V1516,175-184(1991)

Effects of phasing on MRT target visibility (Holst, Gerald C.)V1488,90-98(1991)

Empirical performance of binary phase-only synthetic discriminant functions (Carhart, Gary W.; Draayer, Bret F.; Billings, Paul A.; Giles, Michael K.)V1564,348-362(1991)

Erasable compact disk utilizing phase-change material and multipulse recording method (Ohno, Eiji; Nishiuchi, Kenichi; Yamada, Noboru; Akahira, Nobuo)V1499,171-179(1991)

Estimation of focused laser beam SBS (stimulated Brillouin scatter) threshold dependence on beam shape and phase aberrations (Clendening, Charles W.)V1415,72-78(1991)

Exploration of fabrication techniques for phase-shifting masks (Pfau, Anton K.; Oldham, William G.; Neureuther, Andrew R.)V1463,124-134(1991)

Fabrication of grooved-glass substrates by phase-mask lithography (Brock, Phillip J.; Levenson, Marc D.; Zavislan, James M.; Lyerla, James R.; Cheng, John C.; Podlogar, Carl V.)V1463,87-100(1991)

Fabrication of phase-shifting mask (Ishiwata, Naoyuki; Furukawa, Takao)V1463,423-433(1991)

Fringe analysis in moire interferometry (Asundi, Anand K.)V1554B,472-480(1991)

Impact of phase masks on deep-UV lithography (Sewell, Harry)V1463,168-179(1991)

Improvement of focus and exposure latitude by the use of phase-shifting masks for DUV applications (Op de Beeck, Maaike; Tokui, Akira; Fujinaga, Masato; Yoshioka, Nobuyuki; Kamon, Kazuya; Hanawa, Tetsuro; Tsukamoto, Katsuhiro)V1463,180-196(1991)

Improvement on phase-shifting method precision and application on shadow moire method (Mauvoisin, Gerard; Bremand, Fabrice J.; Lagarde, Alexis)V1554B,181-187(1991)

Increasing the isoplanatic patch size with phase-derivative adaptive optics (Feng, Yue-Zhong; Gong, Zhi-Ben; Song, Zhengfang)V1487,356-360(1991)

Injection locking of a long-pulse relativistic magnetron (Chen, Shien C.; Bekefi, George; Temkin, Richard J.)V1407,67-73(1991)

Instant measurement of phase-characteristic curve (Sadovnik, Lev S.; Rich, Chris C.)V1559,424-428(1991)

Invariant phase-only filters for phase-encoded inputs (Kallman, Robert R.; Goldstein, Dennis H.)V1564,330-347(1991)

Investigation of self-aligned phase-shifting reticles by simulation techniques (Noelscher, Christoph; Mader, Leonhard)V1463,135-150(1991)

Large-aperture (80-cm diameter) phase plates for beam smoothing on Nova (Woods, Bruce W.; Thomas, Ian M.; Henesian, Mark A.; Dixit, Sham N.; Powell, Howard T.)V1410,47-54(1991)

Laser-induced phase transition in crystal InSb films as used in optical storage (Sun, Yang; Li, Cheng F.; Deng, He; Gan, Fuxi)V1519,554-558(1991)

Measurement of the piezoelectric effect in bone using quasiheterodyne holographic interferometry (Ovryn, Benjie; Haacke, E. M.)V1429,172-182(1991)

Measuring phase errors of an array or segmented mirror with a single far-field intensity distribution (Tyson, Robert K.)V1542,62-75(1991)

Models of driven and mutually coupled relativistic magnetrons with nonlinear frequency-shift and growth-saturation effects (Johnston, George L.; Chen, Shien C.; Davidson, Ronald C.; Bekefi, George)V1407,92-99(1991)

Moire topography with the aid of phase-shift method (Yoshizawa, Toru; Tomisawa, Teiyu)V1554B,441-450(1991)

New phase measurement for nonmonotonical fringe patterns (Tang, Shou-Hong; Hung, Yau Y.; Zhu, Qiuming)V1332,731-737(1991)

New phase-shifting mask structure for positive resist process (Miyazaki, Junji; Kamon, Kazuya; Yoshioka, Nobuyuki; Matsuda, Shuichi; Fujinaga, Masato; Watakabe, Yaichiro; Nagata, Hitoshi)V1464,327-335(1991)

Nondestructive testing of printed circuit board by phase-shifting interferometry (Lu, Yue-guang; Jiang, Lingzhen; Zou, Lixun; Zhao, Xia; Sun, Junyong)V1332,287-291(1991)

Novel analog phase tracker for interferometric fiber optic sensor applications (Berkoff, Timothy A.; Kersey, Alan D.; Moeller, Robert P.)V1367,53-58(1991)

Operational characteristics of a phase-locked module of relativistic magnetrons (Levine, Jerrold S.; Benford, James N.; Courtney, R.; Harteneck, Bruce D.)V1407,74-82(1991)

Optical carrier modulation by integrated optical devices in lithium niobate (Rasch, Andreas; Buss, Wolfgang; Goering, Rolf; Steinberg, Steffen; Karthe, Wolfgang)V1522,83-92(1991)

Optical correlation using a phase-only liquid-crystal-over-silicon spatial light modulator (Potter, Duncan J.; Ranshaw, M. J.; Al-Chalabi, Adil O.; Fancey, Norman E.; Sillitto, Richard M.; Vass, David G.)V1564,363-372(1991)

Optical/Laser Microlithography IV (Pol, Victor, ed.)V1463(1991)

Optical lithography with chromeless phase-shifted masks (Toh, Kenny K.; Dao, Giang T.; Singh, Rajeev; Gaw, Henry T.)V1463,74-86(1991)

Parametric simulation studies and injection phase locking of relativistic magnetrons (Chen, Chiping; Chan, Hei-Wai; Davidson, Ronald C.)V1407,105-112(1991)

Performance analysis of direct-detection optical DPSK systems using a dual-detector optical receiver (Pires, Joao J.; Rocha, Jose R.)V1579,144-154(1991)

Performance of pyramidal phase-only filtering of infrared imagery (Kozaitis, Samuel P.; Petrilak, Robert)V1564,403-413(1991)

Phased-array receiver development using high-performance HEMT MMICs (Liu, Louis; Jones, William L.; Carandang, R.; Lam, Wayne W.; Yonaki, J.; Streit, Dwight C.; Kasody, R.)V1475,193-198(1991)

Phase-locking simulation of dual magnetrons (Chen, H. C.; Stark, Robert A.; Uhm, Han S.)V1407,139-146(1991)

Phase-map unwrapping: a comparison of some traditional methods and a presentation of a new approach (Owner-Petersen, Mette)V1508,73-82(1991)

Phase-matched second-harmonic generation of infrared wavelengths in optical fibers (Kashyap, Raman; Davey, Steven T.; Williams, Doug L.)V1516,164-174(1991)

Phase-measurement of interferometry techniques for nondestructive testing (Creath, Katherine)V1554B,701-707(1991)

Phase measuring fiber optic electronic speckle pattern interferometer (Joenathan, Charles; Khorana, Brij M.)V1396,155-163(1991)

Phase-measuring fiber optic ESPI system: phase-step calibration and error sources (Joenathan, Charles; Khorana, Brij M.)V1554B,56-63(1991)

Phase-shifting hand-held diffraction moire interferometer (Deason, Vance A.; Ward, Michael B.)V1554B,390-398(1991)

Phase-shifting photolithography applicable to real IC patterns (Yanagishita, Yuichiro; Ishiwata, Naoyuki; Tabata, Yasuko; Nakagawa, Kenji; Shigematsu, Kazumasa)V1463,207-217(1991)

Phase-shifting structures for isolated features (Garofalo, Joseph G.; Kostelak, Robert L.; Yang, Tungsheng)V1463,151-166(1991)

Phase-shift mask applications (Buck, Peter D.; Rieger, Michael L.)V1463,218-228(1991)

Phase-shift mask pattern accuracy requirements and inspection technology (Wiley, James N.; Fu, Tao-Yi; Tanaka, Takashi; Takeuchi, Susumu; Aoyama, Satoshi; Miyazaki, Junji; Watakabe, Yaichiro)V1464,346-355(1991)

Phase-shift mask technology: requirements for e-beam mask lithography (Dunbrack, Steven K.; Muray, Andrew; Sauer, Charles; Lozes, Richard; Nistler, John L.; Arnold, William H.; Kyser, David F.; Minvielle, Anna M.; Preil, Moshe E.; Singh, Bhanwar; Templeton, Michael K.)V1464,314-326(1991)

Phase stability of a standing-wave free-electron laser (Sharp, William M.; Rangarajan, G.; Sessler, Andrew M.; Wurtele, Jonathan S.)V1407,535-545(1991)

Phase-stepping technique in holography (Wen, Zheng; Tan, Yushan)V1461,278-285(1991)

Planetary lander guidance using binary phase-only filters (Reid, Max B.; Hine, Butler P.)V1564,384-394(1991)

Polarization dependence and uniformity of FLC layers for phase modulation (Biernacki, Paul D.; Brown, Tyler; Freeman, Mark O.)V1455,167-178(1991)

Polarization-state mixing in multiple-beam diffraction and its application to solving the phase problem (Shen, Qun)V1550,27-33(1991)

Projection moire using PSALM (Asundi, Anand K.)V1554B,257-265(1991)

Reversible phase-change optical recording by using microcellular GeSbTeCo recording film (Okamine, Shigenori; Terao, Motoyasu; Andoo, Keikichi; Miyauchi, Yasushi)V1499,166-170(1991)

Review of interferogram analysis methods (Malacara Hernandez, Daniel)V1332,678-689(1991)

Review of phase-measuring interferometry (Stahl, H. P.)V1332,704-719(1991)

Spatial-carrier phase-shifting technique of fringe pattern analysis (Kujawinska, Malgorzata; Schmidt, Joanna)V1508,61-67(1991)

Spatial phase-shifting techniques of fringe pattern analysis in photomechanics (Kujawinska, Malgorzata; Wojiak, Joanna)V1554B,503-513(1991)

Spatial randomization of the scattered optical radiation (Angelsky, Oleg V.; Magun, I. I.; Maksimyak, Peter P.; Perun, T. O.)V1402,231-235(1991)

Strategies for unwrapping noisy interferograms in phase-sampling interferometry (Andrae, P.; Mieth, Ulrike; Osten, Wolfgang)V1508,50-60(1991)

Studies of structures and phase transitions of Langmuir monolayers using synchrotron radiation (Dutta, Pulak)V1550,134-139(1991)

Study of phase transition VO2 thin film (Gao, Jian C.; Lin, Zhi H.; Han, Li Y.)V1519,159-163(1991)

Synchronous phase-extraction technique and its applications (Hung, Yau Y.; Tang, Shou-Hong; Jin, Guofan; Zhu, Qiuming)V1332,738-747(1991)

Testing of PZT shifters for interferometric measurements (Schmit, Joanna; Piatkowski, Tadeusz)V1391,313-317(1991)

Theoretical study of ultrafast dephasing by four-wave mixing (Hoerner, Claudine; Lavoine, J. P.; Villaeys, A. A.)V1362,863-869(1991)

Transparent phase-shifting mask with multistage phase shifter and comb-shaped shifter (Watanabe, Hisashi; Todokoro, Yoshihiro; Hirai, Yoshihiko; Inoue, Morio)V1463,101-110(1991)

Twin traveling wave tube amplifiers driven by a single backward-wave oscillator (Butler, Jennifer M.; Wharton, Charles B.)V1407,57-66(1991)

Variable phase-shift mask for deep-submicron optical lithography (Terasawa, Tsuneo; Hasegawa, Norio; Imai, Akira; Tanaka, Toshihiko; Katagiri, Souichi)V1463,197-206(1991)

White-light moire phase-measuring interferometry (Stahl, H. P.)V1332,720-730(1991)

Phase conjugation—see also nonlinear optics; phase

Advanced matrix optics and its incidence in laser optics (Wang, Shaomin; Bernabeu, Eusebio; Alda, Javier)V1397,595-602(1991)

Amplification of amplitude modulated signals in a self-pumped photorefractive phase conjugator (Petersen, Paul M.; Buchhave, Preben)V1362,582-585(1991)

Analysis of Brillouin-enhanced four-wave mixing phase-conjugate systems (Dansereau, Jeffrey P.; Hills, Louis S.; Mani, Siva A.)V1409,67-82(1991)

CO2 lasers and phase conjugation (Sherstobitov, Vladimir E.; Kalinin, Victor P.; Goryachkin, Dmitriy A.; Romanov, Nikolay A.; Dimakov, Sergey A.; Kuprenyuk, Victor I.)V1415,79-89(1991)

Fidelity of Brillouin amplification with Gaussian input beams (Jones, David C.; Ridley, Kevin D.; Cook, Gary; Scott, Andrew M.)V1409,46-52(1991)

High-fidelity fast-response phase conjugators for visible and ultraviolet applications (O'Key, Michael A.; Osborne, Michael R.)V1500,53-64(1991)

High-fidelity phase conjugation generated by holograms: application to imaging through multimode fibers (Pan, Anpei; Marhic, Michel E.; Epstein, Max)V1396,99-106(1991)

Holographic precompensation for one-way transmission of diffraction-limited beams through diffusing media (Pan, Anpei; Marhic, Michel E.)V1461,8-16(1991)

Image formation through inhomogeneities (Leith, Emmett N.; Chen, Chiaohsiang; Cunha, Andre)V1396,80-84(1991)

Innovative Optics and Phase Conjugate Optics (Ahlers, Rolf-Juergen; Tschudi, Theo T., eds.)V1500(1991)

Large-viewing-angle rainbow hologram by holographic phase conjugation (Fan, Cheng; Jiang, Chaochuan; Guo, Lu Rong)V1461,265-269(1991)

Master oscillator-amplifier Nd:YAG laser with a SBS phase-conjugate mirror (Ayral, Jean-Luc; Montel, J.; Huignard, Jean-Pierre)V1500,81-92(1991)

Measurements of second hyperpolarisabilities of diphenylpolyenes by means of phase-conjugate interferometry (Persoons, Andre P.; Van Wonterghem, Bruno M.; Tackx, Peter)V1409,220-229(1991)

Nonlinear Optics II (Fisher, Robert A.; Reintjes, John F., eds.)V1409(1991)

Optical methods for laser beams control (Mak, Arthur A.; Soms, Leonid N.)V1415,110-119(1991)

Optical phase-conjugate resonators, bistabilities, and applications (Venkateswarlu, Putcha; Dokhanian, Mostafa; Sekhar, Prayaga C.; George, M. C.; Jagannath, H.)V1332,245-266(1991)

Optimum design of optical array used as pseudoconjugator (Xiao, Guohua; Song, Ruhua H.; Hu, Zhiping; Le, Shixiao)V1409,106-113(1991)

Performance tests of a 1500 degree-of-freedom adaptive optics system for atmospheric compensation (Cuellar, Louis; Johnson, Paul A.; Sandler, David G.)V1542,468-476(1991)

Phase-conjugate interferometry by using dye-doped polymer films (Nakagawa, Kazuo; Egami, Chikara; Suzuki, Takayoshi; Fujiwara, Hirofumi)V1332,267-273(1991)

Phase-conjugate Twyman-Green interferometer for testing conicoidal surfaces (Shukla, Ram P.; Dokhanian, Mostafa; Venkateswarlu, Putcha; George, M. C.)V1332,274-286(1991)

Phase-conjugating elements in optical information processing networks (Tschudi, Theo T.; Denz, Cornelia; Kobialka, Torsten)V1500,80-80(1991)

Phase conjugation as a probe for noncentrosymmetry grating formation in organics (Charra, Fabrice; Nunzi, Jean-Michel; Messier, Jean)V1516,211-219(1991)

Phase conjugation by four-wave mixing in nematic liquid crystals (Almeida, Silverio P.; Varamit, Srisuda P.)V1500,34-45(1991)

Phase conjugation in LiF and NaF color center crystals (Basiev, Tasoltan T.; Zverev, Peter G.; Mirov, Sergey B.; Pal, Suranjan)V1500,65-71(1991)

Real-time interferometry by optical phase conjugation in dye-doped film (Fujiwara, Hirofumi)V1507,492-499(1991)

Real-time one-pass distortion correction (Oldekop, Erik; Siahmakoun, Azad)V1396,174-177(1991)

Spectral characteristics of Brillouin-enhanced four-wave mixing for pulsed and CW inputs (Lebow, Paul S.; Ackerman, John R.)V1409,60-66(1991)

Subattomole detection in the condensed phase by nonlinear laser spectroscopy based on degenerate four-wave mixing (Tong, William G.)V1435,90-94(1991)

TEA CO2 laser mirror by degenerate four-wave mixing (Vigroux, Luc M.; Bourdet, Gilbert L.; Cassard, Philippe; Ouhayoun, Michel O.)V1500,74-79(1991)

Ti:Al2O3 laser with an intracavity phase-conjugate mirror (Jankiewicz, Zdzislaw; Szydlak, J.; Skorczakowski, M.; Zendzian, W.; Dvornikov, S. S.; Kondratiuk, N. V.; Skripko, G. A.)V1391,101-104(1991)

Time reversal, enhancement, and suppression of dipole radiation by a phase-conjugate mirror (Bochove, Erik J.)V1497,222-227(1991)

Transient stimulated Raman scattering: theory and experiments of pulse shortening and phase conjugation properties (Agnesi, Antoniangelo; Reali, Giancarlo C.; Kubecek, Vaclav)V1415,104-109(1991)

Photoacoustics—see also acoustics; acousto-optics

Acoustic effects of Q-switched Nd:YAG on crystalline lens (Porindla, Sridhar N.; Rylander, Henry G.; Welch, Ashley J.)V1427,267-272(1991)

Caries selective ablation by pulsed lasers (Hennig, Thomas; Rechmann, Peter; Pilgrim, C.; Schwarzmaier, Hans-Joachim; Kaufmann, Raimund)V1424,99-105(1991)

Comparison of the excimer laser with the erbium yttrium aluminum garnet laser for applications in osteotomy (Li, Zhao-zhang; Van De Merwe, Willem P.; Reinisch, Lou)V1427,152-161(1991)

Determination of SBS-induced damage limits in large fused silica optics for intense, time-varying laser pulses (Kyrazis, Demos T.; Weiland, Timothy L.)V1441,469-477(1991)

Emissions monitoring by infrared photoacoustic spectroscopy (Jalenak, Wayne)V1434,46-54(1991)

Energy conversion efficiency during optical breakdown (Grad, Ladislav; Diaci, J.; Mozina, Janez)V1525,206-209(1991)

Grueneisen-stress-induced ablation of biological tissue (Dingus, Ronald S.; Scammon, R. J.)V1427,45-54(1991)

Histological distinction of mechanical and thermal defects produced by nanosecond laser pulses in striated muscle at 1064 nm (Gratzl, Thomas; Dohr, Gottfried; Schmidt-Kloiber, Heinz; Reichel, Erich)V1427,55-62(1991)

Laser-induced shock wave effects on red blood cells (Flotte, Thomas J.; Frisoli, Joan K.; Goetschkes, Margaret; Doukas, Apostolos G.)V1427,36-44(1991)

Laser lithotripsy with a pulsed dye laser: correlation between threshold energy and optical properties (Sterenborg, H. J.; van Swol, Christiaan F.; Biganic, Dane D.; van Gemert, Martin J.)V1427,256-266(1991)

Moire displacement detection by the photoacoustic technique (Hane, Kazuhiro; Watanabe, S.; Goto, T.)V1332,577-583(1991)

Multishot ablation of thin films: sensitive detection of film/substrate transition by shockwave monitoring (Hunger, Hans E.; Petzoldt, Stefan; Pietsch, H.; Reif, J.; Matthias, Eckart)V1441,283-286(1991)

Narrowband alexandrite laser injection seeded with frequency-dithered diode laser (Schwemmer, Geary K.; Lee, Hyo S.; Prasad, Coorg R.)V1492,52-62(1991)

New concept of a compact multiwavelength solid state laser for laser-induced shock wave lithotripsy (Steiger, Erwin)V1421,140-146(1991)

Optical and mechanical parameter detection of calcified plaque for laser angioplasty (Stetz, Mark L.; O'Brien, Kenneth M.; Scott, John J.; Baker, Glenn S.; Deckelbaum, Lawrence I.)V1425,55-62(1991)

Photoacoustic absorption spectrum of some rat and bovine tissues in the ultraviolet-visible range (Bernini, Umberto; Russo, Paolo)V1427,398-404(1991)

Photoacoustic characterization of surface absorption (Reicher, David W.; Wilson, Scott R.; Kranenberg, C. F.; Raja, M. Y.; McNeil, John R.; Brueck, Steven R.)V1441,106-112(1991)

Photothermal spectroscopy as a sensitive spectroscopic tool (Tam, Andrew C.)V1435,114-127(1991)

Photochemistry—see also chemistry; films; holography; photography

Acid-catalyzed pinacol rearrangement: chemically amplified reverse polarity change (Sooriyakumaran, Ratna; Ito, Hiroshi; Mash, Eugene A.)V1466,419-428(1991)

Amplitude-encoded phase-only filters for pattern recognition: influence of the bleaching procedure (Campos, Juan; Janowska-Dmoch, Bozena; Styczynski, K.; Turon, F.; Yzuel, Maria J.; Chalasinska-Macukow, Katarzyna)V1574,141-147(1991)

Analysis of the Focar-type silver-halide heterogeneous media (Andreeva, Olga V.)V1238,231-234(1991)

Antitumor drugs as photochemotherapeutic agents (Andreoni, Alessandra; Colasanti, Alberto; Kisslinger, Annamaria; Malatesta, Vincenzo; Mastrocinque, Michele; Roberti, Giuseppe)V1525,351-366(1991)

Artificial photosynthesis at octane/water interface in the presence of hydrated chlorophyll a oligomer thin film (Volkov, Alexander G.; Gugeshashvili, M. I.; Kandelaki, M. D.; Markin, V. S.; Zelent, B.; Munger, G.; Leblanc, Roger M.)V1436,68-79(1991)

Asymmetric photopotentials from thin polymeric porphyrin films (Wamser, Carl C.; Senthilathipan, Velu; Li, Wen)V1436,114-124(1991)

Behavior of a thin liquid film under thermal stimulation: theory and applications to infrared interferometry (Ledoyen, Fernand; Lewandowski, Jacques; Cormier, Maurice)V1507,328-338(1991)

Characteristic curves and phase-exposition characteristics of holographic photomaterials (Churaev, A. L.; Artyomova, V. V.)V1238,158-165(1991)

Characterization of DC-PVA films for holographic recording materials (Leclere, Philippe; Renotte, Yvon L.; Lion, Yves F.)V1507,339-344(1991)

Characterization of photobiophysical properties of sensitizers used in photodynamic therapy (Roeder, Beate; Naether, Dirk)V1525,377-384(1991)

Charge separation in functionalized tetrathiafulvalene derivatives (Fox, Marye A.; Pan, Horng-Lon)V1436,2-7(1991)

Collisional effects in laser detection of tropospheric OH (Crosley, David R.)V1433,58-68(1991)

Comparison of dichromated gelatin and Du Pont HRF-700 photopolymer as media for holographic notch filters (Salter, Jeffery L.; Loeffler, Mary F.)V1555,268-278(1991)

Cryogenic limb array etalon spectrometer: experiment description (Roche, Aidan E.; Kumer, John B.)V1491,91-103(1991)

Dark stability of holograms and holographic investigation of slow diffusion in polymers (Veniaminov, Andrei V.)V1238,266-270(1991)

Determination of the optical penetration depth in tumors from biopsy samples (Lenz, P.)V1525,183-191(1991)

Development and experimental investigation of a copying procedure for the reproduction of large-format transmissive holograms (Kubitzek, Ruediger; Froelich, Klaus; Tropartz, Stephan; Stojanoff, Christo G.)V1507,365-372(1991)

Development and investigation of dichromated gelatin film for the fabrication of large-format holograms operating at 400-900 nm (Tropartz, Stephan; Brasseur, Olivier; Kubitzek, Ruediger; Stojanoff, Christo G.)V1507,345-353(1991)

DQN photoresist with tetrahydroxydiphenylmethane as ballasting group in PAC (Tzeng, Chao H.; Lin, Dhei-Jhai; Lin, Song S.; Huang, Dong-Tsair; Lin, Hwang-Kuen)V1466,469-476(1991)

Dry photopolymer embossing: novel photoreplication technology for surface relief holographic optical elements (Shvartsman, Felix P.)V1507,383-391(1991)

Electric field effect on the persistent hole burning of quinone derivatives (Nishimura, Tetsuya; Yagyu, Eiji; Yoshimura, Motomu; Tsukada, Noriaki; Takeyama, Tetsu)V1436,31-37(1991)

Electrochromic properties of poly(pyrrole)/dodecylbenzenesulfonate (Peres, Rosa C.; De Paoli, Marco-Aurelio; Panero, Stefania; Scrosati, Bruno)V1559,151-158(1991)

Electrophotographic properties of thiophene derivatives as charge transport material (Kuroda, Masami; Kawate, K.; Nabeta, Osamu; Furusho, N.)V1458,155-161(1991)

Evaluation of phenolic resists for 193 nm surface imaging (Hartney, Mark A.; Johnson, Donald W.; Spencer, Allen C.)V1466,238-247(1991)

Exposure dose optimization for a positive resist containing polyfunctional photoactive compound (Trefonas, Peter; Mack, Chris A.)V1466,117-131(1991)

Four-dimensional photosensitive materials: applications for shaping, time-domain holographic dissection, and scanning of picosecond pulses (Saari, Peeter M.; Kaarli, Rein K.; Sonajalg, Heiki)V1507,500-508(1991)

Generalized characteristic model for lithography: application to negative chemically amplified resists (Ziger, David H.; Mack, Chris A.; Distasio, Romelia)V1466,270-282(1991)

High-resolution photographic material of Institute of Atomic Energy for holography in intersecting beams (Ryabova, R. V.; Barinova, E. S.; Myatezh, O. V.; Zaborov, A. N.; Romash, E. V.; Chernyi, D. I.)V1238,166-170(1991)

History and status of black and white photographic processing chemicals as effluents (Horn, Richard R.)V1458,69-75(1991)

Hologram: liquid-crystal composites (Ingwall, Richard T.; Adams, Timothy)V1555,279-290(1991)

Holographic transmission elements using improved photopolymer films (Gambogi, William J.; Gerstadt, William A.; Mackara, Steven R.; Weber, Andrew M.)V1555,256-267(1991)

Importance of proton transfer in contact charging (Wollmann, Daphne; Diaz, Art F.)V1458,192-200(1991)

Investigation of the arsenic sulphide films for relief-phase holograms (Yusupov, I. Y.; Mikhailov, M. D.; Herke, R. R.; Goray, L. I.; Mamedov, S. B.; Yakovuk, O. A.)V1238,240-247(1991)

Kinetics of formation of holographic structure of a hologram mirror in dichromated gelatin (Kzuzhilin, Yu E.; Mel'nichenko, Yu. B.; Shilov, V. V.)V1238,200-205(1991)

Laser/light tissue interaction: on the mechanism of optical breakdown (Siomos, Konstadinos)V1525,154-161(1991)

Light distributor for endoscopic photochemotherapy (Lenz, P.)V1525,192-195(1991)

Light dosimetry in vivo in interstitial photodynamic therapy of human tumors (Reynes, Anne M.; Diebold, Simon; Lignon, Dominique; Granjon, Yves; Guillemin, Francois)V1525,177-182(1991)

Light energy conversion with pheophytin a and chlorophyll a monolayers at the optical transparent electrode (Leblanc, Roger M.; Blanchet, P.-F.; Cote, D.; Gugeshashvili, M. I.; Munger, G.; Volkov, Alexander G.)V1436,92-102(1991)

Mechanistic studies on the poly(4-tert-butoxycarbonyloxystyrene)/triphenylsulfonium salt photoinitiation process (Hacker, Nigel P.; Welsh, Kevin M.)V1466,384-393(1991)

Microscopic mechanism of optical nonlinearity in conjugated polymers and other quasi-one-dimensional systems (Mazumdar, Sumit; Guo, Dandan; Dixit, Sham N.)V1436,136-149(1991)

Modified chloroaluminium phthalocyanine: an organic semiconductor with high photoactivity (Dodelet, Jean-Pol; Gastonguay, Louis; Veilleux, George; Saint-Jacques, Robert G.; Cote, Roland; Guay, Daniel; Tourillon, Gerard)V1436,38-49(1991)

Multiphoton photobiologic effects (Hasan, Tayyaba)V1427,70-76(1991)

New developing process for PVCz holograms (Yamagishi, Yasuo; Ishitsuka, Takeshi; Kuramitsu, Yoko; Yoneda, Yasuhiro)V1461,68-72(1991)

New technique for characterizing holographic recording materials (Leclere, Philippe; Renotte, Yvon L.; Lion, Yves F.)V1559,298-307(1991)

Non-gelatin-dichromated holographic film (Guo, Lu Rong; Cheng, Qirei; Wang, Kuoping)V1461,91-92(1991)

Nonlinear optical properties of phenosafranin-doped substrates (Speiser, Shammai)V1559,238-244(1991)

Nonmetallic acid generators for i-line and g-line chemically amplified resists (Brunsvold, William R.; Montgomery, Warren; Hwang, Bao)V1466,368-376(1991)

Novel base-generating photoinitiators for deep-UV lithography (Kutal, Charles; Weit, Scott K.; Allen, Robert D.; MacDonald, Scott A.; Willson, C. G.)V1466,362-367(1991)

O/I-MBE: formation of highly ordered phthalocyanine/semiconductor junctions by molecular-beam epitaxy: photoelectrochemical characterization (Armstrong, Neal R.; Nebesny, Ken W.; Collins, Greg E.; Lee, Paul A.; Chau, Lai K.; Arbour, Claude; Parkinson, Bruce)V1559,18-26(1991)

Optimization of reconstruction geometry for maximum diffraction efficiency in HOE: the influence of recording material (Belendez, A.; Pascual, I.; Fimia-Gil, Antonio)V1574,77-83(1991)

Parameters of recording media for 3-D holograms (Mikhailov, I. A.)V1238,140-143(1991)

PFG-04 photographic plates based on the nonhardened dichromated gelatin for recording color reflection holograms (Artemjev, S. V.; Koval, G. I.; Obyknovennaja, I. E.; Cherkasov, A. S.; Shevtsov, M. K.)V1238,206-210(1991)

Photoanisotropic polymeric media and their application in optical devices (Barachevsky, Valery A.)V1559,184-193(1991)

Photochemical and thermal treatment of dichromated gelatin film for the manufacturing of holographic optical elements for operation in the IR (Stojanoff, Christo G.; Schuette, H.; Brasseur, Olivier; Kubitzek, Ruediger; Tropartz, Stephan)V1559,321-330(1991)

Photochemical delineation of waveguides in polymeric thin films (Beeson, Karl W.; Horn, Keith A.; Lau, Christina; McFarland, Michael J.; Schwind, David; Yardley, James T.)V1559,258-266(1991)

Photochemical formation of polymeric optical waveguides and devices for optical interconnection applications (Beeson, Karl W.; Horn, Keith A.; McFarland, Michael J.; Wu, Chengjiu; Yardley, James T.)V1374,176-185(1991)

Photochemical generation of gradient indices in glass (Mendoza, Edgar A.; Gafney, Harry D.; Morse, David L.)V1378,139-144(1991)

Photochemical hole burning in rigidly coupled polyacenes (Iannone, Mark A.; Salt, Kimberly L.; Scott, Gary W.; Yamashita, Tomihiro)V1559,172-183(1991)

Photochemistry and fluorescence spectroscopy of polymeric materials containing triphenylsulfonium salts (Hacker, Nigel P.; Welsh, Kevin M.)V1559,139-150(1991)

Photochemistry and Photoelectrochemistry of Organic and Inorganic Molecular Thin Films (Frank, Arthur J.; Lawrence, Marcus F.; Ramasesha, S.; Wamser, Carl C., eds.)V1436(1991)

Photochemistry and photophysics of stilbene and diphenylpolyene surfactants in supported multilayer films (Spooner, Susan P.; Whitten, David G.)V1436,82-91(1991)

Photoelectrographic printing with persistently conductive masters based on onium salt acid photogenerators (Bugner, Douglas E.; Fulmer, Gary G.; Riblett, Susan E.)V1458,162-178(1991)

Photo-induced degradation of selected polyimides in the presence of oxygen: a rapid decomposition process (Hoyle, Charles E.; Creed, David; Anzures, Edguardo; Subramanian, P.; Nagarajan, Rajamani)V1559,101-109(1991)

Photophysics of 1,3,5-triaryl-2-pyrazolines (Sahyun, Melville R.; Crooks, G. P.; Sharma, D. K.)V1436,125-133(1991)

Photopolymer Device Physics, Chemistry, and Applications II (Lessard, Roger A., ed.)V1559(1991)

Photoproduct formation of endogeneous protoporphyrin and its photodynamic activity (Koenig, Karsten; Schneckenburger, Herbert; Rueck, Angelika C.; Auchter, S.)V1525,412-419(1991)

Photosensitive liposomes as potential drug delivery vehicles for photodynamic therapy (Morgan, Chris G.; Mitchell, A. C.; Chowdhary, R. K.)V1525,391-396(1991)

Photothermoplastic media with organic and inorganic photosemiconductors for hologram recording (Zaichenko, O. V.; Komarov, Vyacheslav A.; Kuzmin, I. V.)V1238,271-274(1991)

Phthalocyanines as phototherapeutic agents for tumors (Jori, Guilio; Reddi, Elena; Biolo, Roberta; Polo, Laura; Valduga, Giuliana)V1525,367-376(1991)

Physicochemistry of holographic recording and processing in dichromated gelatin (Sherstyuk, Valentin P.; Maloletov, Sergei M.; Kondratenko, Nina A.)V1238,211-217(1991)

P-n heterojunction and Schottky barrier formation between poly(3-methylthiophene) and n-type cadmium sulfide (Frank, Arthur J.; Glenis, Spyridon)V1436,50-57(1991)

Polymer effects on the photochemistry of triarylsulfonium salts (McKean, Dennis R.; Allen, Robert D.; Kasai, Paul H.; MacDonald, Scott A.)V1559,214-221(1991)

Polymerization kinetics of mono- and multifunctional monomers initiated by high-intensity laser pulses: dependence of rate on peak-pulse intensity and chemical structure (Hoyle, Charles E.; Sundell, Per-Erik; Trapp, Martin A.; Kang, Deokman; Sheng, D.; Nagarajan, Rajamani)V1559,202-213(1991)

Printing Technologies for Images, Gray Scale, and Color (Dove, Derek B.; Abe, Takao; Heinzl, Joachim L., eds.)V1458(1991)

Process control capability using a diaphragm photochemical dispense system (Cambria, Terrell D.; Merrow, Scott F.)V1466,670-675(1991)

Properties of volume reflection silver-halide gelatin holograms (Kosobokova, N. L.; Usanov, Yu. Y.; Shevtsov, M. K.)V1238,183-188(1991)

Pulsed photoconductivity of chlorophyll a (Kassi, Hassane; Hotchandani, Surat; Leblanc, Roger M.; Beaudoin, N.; Dery, M.)V1436,58-67(1991)

Recording mechanism and postpolymerizing self-amplification of holograms (Gulnazarov, Eduard S.; Smirnova, Tatiana N.; Tikhonov, Evgenij A.)V1238,235-239(1991)

Reflection holographic optical elements in silver-halide-sensitized gelatin (Pascual, I.; Belendez, A.; Fimia-Gil, Antonio)V1574,72-76(1991)

Reflection silver-halide gelatin holograms (Usanov, Yu. Y.; Vavilova, Ye. A.; Kosobokova, N. L.; Shevtsov, M. K.)V1238,178-182(1991)

Relief holograms recording on liquid photopolymerizable layers (Boiko, Yuri B.; Granchak, Vasilij M.; Dilung, Iosiph I.; Solovjev, Vladimir S.; Sisakian, Iosif N.; Soifer, Victor A.)V1238,253-257(1991)

Research of fast stages of latent image formation in holographic photoemulsions influenced by ultrashort radiation pulses (Starobogatov, Igor O.; Nicolaev, S. D.)V1238,153-157(1991)

Site-specific laser modification (cleavage) of oligodeoxynucleotides (Benimetskaya, L. Z.; Bulychev, N. V.; Kozionov, Andrew L.; Koshkin, A. A.; Lebedev, A. V.; Novozhilov, S. Y.; Stockman, M. I.)V1525,210-211(1991)

Solid state conductivity and photoconductivity studies of an ion-exchange polymer/dye system (Huang, Zhiqing; Ordonez, I.; Ioannidis, Andreas A.; Langford, C. H.; Lawrence, Marcus F.)V1436,103-113(1991)

Solubility, dispersion, and carbon adsorption of a chromium hydroxyazo complex in a toner (Gutierrez, Adolfo R.; Diaz, Art F.)V1458,201-204(1991)

Soluble polyacetylenes derived from the ring-opening metathesis polymerization of substituted cyclooctatetraenes: electrochemical characterization and Schottky barrier devices (Jozefiak, Thomas H.; Sailor, Michael J.; Ginsburg, Eric J.; Gorman, Christopher B.; Lewis, Nathan S.; Grubbs, Robert H.)V1436,8-19(1991)

Some new developments in display holography (Hegedus, Zoltan S.; Hariharan, P.)V1238,480-488(1991)

Some principles for formation of self-developing dichromate media (Sherstyuk, Valentin P.; Malov, Alexander N.; Maloletov, Sergei M.; Kalinkin, Vyacheslav V.)V1238,218-223(1991)

Study of electronic stage of the Herschel effect in holographic emulsions with different types of chemical sensitization (Mikhailov, V. N.; Grinevitskaya, O. V.; Zagorskaya, Z. A.; Mikhailova, V. I.)V1238,144-152(1991)

Theoretical study of optical pockels and Kerr coefficients of polyenes (Albert, I. D.; Ramasesha, S.)V1436,179-189(1991)

Three-Dimensional Holography: Science, Culture, Education (Jeong, Tung H.; Markov, Vladimir B., eds.)V1238(1991)

Triplet-sensitized reactions of some main chain liquid-crystalline polyaryl cinnamates (Subramanian, P.; Creed, David; Hoyle, Charles E.; Venkataram, Krishnan)V1559,461-469(1991)

Ultra-fine-grain silver-halide photographic materials for holography on the flexible film base (Logak, L. G.; Fassakhova, H. H.; Antonova, N. E.; Minina, L. A.; Gainutdinov, R. K.)V1238,171-176(1991)

UV laser-induced photofragmentation and photoionization of dimethylcadmium chemisorbed on silicon (Simonov, Alexander P.; Varakin, Vladimir N.; Panesh, Anatoly M.)V1436,20-30(1991)

Volume holograms in liquid photopolymerizable layers (Boiko, Yuri B.; Granchak, Vasilij M.; Dilung, Iosiph I.; Mironchenko, Vladislav Y.)V1238,258-265(1991)

Volume holographic optics recording in photopolymerizable layers (Boiko, Yuri B.)V1507,318-327(1991)

Wavelength shifting and bandwidth broadening in DCG (Corlatan, Dorina; Schaefer, Martin; Anders, Gerhard)V1507,354-364(1991)

Photoconductors

Adaptive semiconductor laser phased arrays for real-time multiple-access communications (Pan, J. J.; Cordeiro, D.)V1476,157-169(1991)

Characterization and switching study of an optically controlled GaAs switch (Stoudt, David C.; Mazzola, Michael S.; Griffiths, Scott F.)V1378,280-285(1991)

Considerations of the limits and capabilities of light-activated switches (Zucker, Oved S.; Giorgi, David M.; Griffin, Adam; Hargis, David E.; Long, James; McIntyre, Iain A.; Page, Kevin J.; Solone, Paul J.; Wein, Deborah S.)V1378,22-33(1991)

Corbino-capacitance technique for contactless measurements on conducting layers: application to persistent photoconductivity (Hansen, Ole P.; Kristensen, Anders; Bruus, Henrik; Razeghi, Manijeh)V1362,192-198(1991)

Deep-level configuration of GaAs:Si:Cu: a material for a new type of optoelectronic switch (Schoenbach, Karl H.; Schulz, Hans-Joachim; Lakdawala, Vishnu K.; Kimpel, B. M.; Brinkmann, Ralf P.; Germer, Rudolf K.; Barevadia, Gordon R.)V1362,428-435(1991)

Design of frozen-wave and injected-wave microwave generators using optically controlled switches (Nunnally, William C.)V1378,10-21(1991)

Fabrication and performance of CdSe/CdS/ZnS photoconductor for liquid-crystal light valve (Zhuang, Song Lin; Jiang, Yingqui; Qiu, Yinggang; Gu, Lingjuan; Cai, Zhonghua; Chen, Wei)V1558,28-33(1991)

Generation and sampling of high-repetition-rate/high-frequency electrical waveforms in microstrip circuits by picosecond optoelectronic technique (Lee, Chi H.)V1390,377-387(1991)

Generation of stable low-jitter kilovolt amplitude picosecond pulses (Sarkar, Tapan K.; Banerjee, Partha P.)V1378,34-42(1991)

Graded AlxGa1-xAs photoconductive devices for high-efficiency picosecond optoelectronic switching (Morse, Jeffrey D.; Mariella, Raymond P.; Dutton, Robert W.)V1378,55-59(1991)

High-power waveform generation using photoconductive switches (Oicles, Jeffrey A.; Helava, Heikki I.; Grant, Jon R.; Ragle, Larry O.; Wessman, Susan C.)V1378,60-69(1991)

Hot carrier silicon phototransistor (Asmontas, Steponas; Gradauskas, Jonas; Sirmulis, Edmundas)V1512,131-134(1991)

Ionization potentials effects on CGL/CTL photoconductors (Shi, Xiao D.)V1519,884-889(1991)

Multilayer OPC for one-shot two-color printer (Sakai, Katsuo)V1458,179-191(1991)

New type of IR to visible real-time image converter: design and fabrication (Sun, Fang-kui; Yang, Mao-hua; Gao, Shao-hong; Zhao, Shi-jie)V1488,2-5(1991)

Novel doping superlattice-based PbTe-IR detector device (Oswald, Josef; Pippan, Manfred; Tranta, Beate; Bauer, Guenther E.)V1362,534-543(1991)

Optically Activated Switching (Zutavern, Fred J., ed.)V1378(1991)

Optical probing of field dependent effects in GaAs photoconductive switches (Donaldson, William R.; Kingsley, Lawrence E.)V1378,226-236(1991)

Performances of gallium arsenide on silicon substrate photoconductive detectors (Constant, Monique T.; Boussekey, Luc; Decoster, Didier; Vilcot, Jean-Pierre)V1362,156-162(1991)

Photoconducting nonlinear optical polymers (Li, Lian; Jeng, Ru J.; Lee, J. Y.; Kumar, Jayant; Tripathy, Sukant K.)V1560,243-250(1991)

Photoconductivity of bridged polymeric phthalocyanines (Meier, Hans; Albrecht, Wolfgang; Hanack, Michael)V1559,89-100(1991)

Photoconductivity of high-bandgap materials (Ho, Ping-Tong; Peng, F.; Goldhar, J.; Nolting, Eugene E.; Parsons, C.)V1378,210-216(1991)

Photocurrent response to picosecond pulses in semiconductors: application to EL2 in gallium arsenide (Pugnet, Michel; Collet, Jacques; Nardo, Laurent)V1361,96-108(1991)

Photoelectrographic printing with persistently conductive masters based on onium salt acid photogenerators (Bugner, Douglas E.; Fulmer, Gary G.; Riblett, Susan E.)V1458,162-178(1991)

Photoquenching and characterization studies in a bulk optically controlled GaAs semiconductor switch (Lakdawala, Vishnu K.; Schoenbach, Karl H.; Roush, Randy A.; Barevadia, Gordon R.; Mazzola, Michael S.)V1378,259-270(1991)

Picosecond photocurrent measurements of negative differential velocity in GaAs/AlAs superlattices (Minot, Christophe; Le Person, H.; Mollot, Francis; Palmier, Jean F.)V1362,301-308(1991)

Pulsed photoconductivity of chlorophyll a (Kassi, Hassane; Hotchandani, Surat; Leblanc, Roger M.; Beaudoin, N.; Dery, M.)V1436,58-67(1991)

Research on relaxation process of a-Si:H film photoconductivity and the trap effect (Gong, Dao B.)V1519,281-286(1991)

Rise time and recovery of GaAs photoconductive semiconductor switches (Zutavern, Fred J.; Loubriel, Guillermo M.; O'Malley, Marty W.; McLaughlin, Dan L.; Helgeson, Wes D.)V1378,271-279(1991)

Saturnus: the UCLA compact infrared free-electron laser project (Dodd, James W.; Aghamir, F.; Barletta, W. A.; Cline, David B.; Hartman, Steven C.; Katsouleas, Thomas C.; Kolonko, J.; Park, Sanghyun; Pellegrini, Claudio; Terrien, J. C.; Davis, J. G.; Joshi, Chan J.; Luhmann, Neville C.; McDermott, David B.; Ivanchenkov, S. N.; Lachin, Yu Y.; Varfolomeev, A. A.)V1407,467-473(1991)

Solid state conductivity and photoconductivity studies of an ion-exchange polymer/dye system (Huang, Zhiqing; Ordonez, I.; Ioannidis, Andreas A.; Langford, C. H.; Lawrence, Marcus F.)V1436,103-113(1991)

Subnanosecond, high-voltage photoconductive switching in GaAs (Druce, Robert L.; Pocha, Michael D.; Griffin, Kenneth L.; O'Bannon, Jim)V1378,43-54(1991)

Surface field measurement of photoconductive power switches using the electro-optic Kerr effect (Sardesai, Harshad P.; Nunnally, William C.; Williams, Paul F.)V1378,237-248(1991)

Temporal model of optically initiated GaAs avalanche switches (Falk, R. A.; Adams, Jeff C.)V1378,70-81(1991)

Photodetectors—see also detectors

128 x 128 MWIR InSb focal plane and camera system (Parrish, William J.; Blackwell, John D.; Paulson, Robert C.; Arnold, Harold)V1512,68-77(1991)

Absolute measurements of radiation sources spectral brightness and detectors quantum efficiency (Penin, A. N.; Klyshko, D. N.)V1562,143-148(1991)

Accurate method for neutron fluence control used in improving neutron-transmutation-doped silicon for detectors (Halmagean, Eugenia T.; Lazarovici, Doina N.; Udrea-Spenea, Marian N.)V1484,106-114(1991)

Advanced infrared focal-plane arrays (Amingual, Daniel)V1512,40-51(1991)

Advances in multiple-quantum-well IR detectors (Bloss, Walter L.; O'Loughlin, Michael J.; Rosenbluth, Mary)V1541,2-10(1991)

Analysis of neuropeptides using capillary zone electrophoresis with multichannel fluorescence detection (Sweedler, Jonathan V.; Shear, Jason B.; Fishman, Harvey A.; Zare, Richard N.; Scheller, Richard H.)V1439,37-46(1991)

Anodic oxides on HgZnTe (Eger, David; Zigelman, Alex)V1484,48-54(1991)

Application of IR photorecorders based on ionization chambers for fast processes investigation (Egorov, V. V.; Lazarchuk, V. P.; Murugov, V. M.; Sheremetyev, Yu. N.)V1358,984-991(1991)

Application of the NO/O3 chemiluminescence technique to measurements of reactive nitrogen species in the stratosphere (Fahey, David W.)V1433,212-223(1991)

Asymmetric superlattices for microwave detection (Syme, Richard T.; Kelly, Michael J.; Condie, Angus; Dale, Ian)V1362,467-476(1991)

Band-structure dependence of impact ionization: bulk semiconductors, strained Ge/Si alloys, and multiple-quantum-well avalanche photodetectors (Czajkowski, Igor K.; Allam, Jeremy; Adams, Alfred R.)V1362,179-190(1991)

Characteristics of MSM detectors for meander channel CCD imagers on GaAs (Kosel, Peter B.; Bozorgebrahimi, Nercy; Iyer, J.)V1541,48-59(1991)

Characterization of anodic fluoride films on Hg1-xCdxTe (Esquivias, Ignacio; Dal Colle, M.; Brink, D.; Baars, Jan W.; Bruder, Martin)V1484,55-66(1991)

Characterization of picosecond GaAs metal-semiconductor-metal photodetectors (Rosenzweig, Josef; Moglestue, C.; Axmann, A.; Schneider, Joachim J.; Huelsmann, Axel; Lambsdorff, M.; Kuhl, Juergen; Klingenstein, M.; Leier, H.; Forchel, Alfred W.)V1362,168-178(1991)

Devices for generation and detection of subnanosecond IR and FIR radiation pulses (Beregulin, Eugene V.; Ganichev, Sergey D.; Yaroshetskii, Ilya D.; Lang, Peter T.; Schatz, Wolfgan; Renk, Karl F.)V1362,853-862(1991)

Double-resonant tunneling via surface plasmons in a metal-semiconductor system (Sreseli, Olga M.; Belyakov, Ludvig V.; Goryachev, D. N.; Rumyantsev, B. L.; Yaroshetskii, Ilya D.)V1440,326-331(1991)

Effects of the lunar environment on optical telescopes and instruments (Johnson, Charles L.; Dietz, Kurtis L.)V1494,208-218(1991)

Efficiency of a 5V/5-mW power by light power supply for avionics applications (Sherman, Bradley D.; Mendez, Antonio J.; Morookian, John-Michael)V1369,60-66(1991)

Electromagnetically carrier depleted IR photodetector (Djuric, Zoran G.; Piotrowski, Jozef)V1540,622-632(1991)

Electronic and optical properties of silicide/silicon IR detectors (Cabanski, Wolfgang A.; Schulz, Max J.)V1484,81-97(1991)

Enhanced thematic mapper cold focal plane: design and testing (Yang, Bing T.)V1488,399-409(1991)

Free-surface temperature measurement of shock-loaded tin using ultrafast infrared pyrometry (Mondot, Michel; Remiot, Christian)V1558,351-361(1991)

GaInAs PIN photodetectors on semi-insulating substrates (Crawford, Deborah L.; Wey, Y. G.; Bowers, John E.; Hafich, Michael J.; Robinson, Gary Y.)V1371,138-141(1991)

Growth and Characterization of Materials for Infrared Detectors and Nonlinear Optical Switches (Longshore, Randolph E.; Baars, Jan W., eds.)V1484(1991)

High-speed GaAs metal-semiconductor-metal photodetectors with sub-0.1um finger width and finger spacing (Chou, Stephen Y.; Liu, Yue; Fischer, Paul B.)V1474,243-247(1991)

Hot carrier photoeffects in inhomogeneous semiconductors and their applications to light detectors (Amosova, L. P.; Marmur, I. Y.; Oksman, Ya. A.; Ashmontas, S.; Gradauskas, I.; Shirmulis, E.)V1440,406-413(1991)

Hybrid infrared focal-plane signal and noise modeling (Johnson, Jerris F.; Lomheim, Terrence S.)V1541,110-126(1991)

III-V monolithic resonant photoreceiver on silicon substrate for long-wavelength operation (Aboulhouda, S.; Razeghi, Manijeh; Vilcot, Jean-Pierre; Decoster, Didier; Francois, M.; Maricot, S.)V1362,494-498(1991)

Implementing early visual processing in analog VLSI: light adaptation (Mann, James)V1473,128-136(1991)

Infrared and Optoelectronic Materials and Devices (Naumaan, Ahmed; Corsi, Carlo; Baixeras, Joseph M.; Kreisler, Alain J., eds.)V1512(1991)

Infrared detectors from YBaCuO thin films (Zhou, Fang Q.; Sun, Han D.; Zhao, Xing R.; Wang, Lingjie; Yi, Xin J.)V1477,178-181(1991)

Infrared photodetector based on the photofluxonic effect in superconducting thin films (Kadin, Alan M.; Leung, Michael; Smith, Andrew D.; Murduck, J. M.)V1477,156-165(1991)

InGaAs/GaAs interdigitated metal-semiconductor-metal (IMSM) photodetectors operational at 1.3 um grown by molecular beam epitaxy (Elman, Boris S.; Chirravuri, Jagannath; Choudhury, A. N.; Silletti, Andrew; Negri, A. J.; Powers, J.)V1362,610-616(1991)

InGaAs/InP monolithic photoreceivers for 1.3-1.5 um optical fiber transmission (Scavennec, Andre; Billard, M.; Blanconnier, P.; Caquot, E.; Carer, P.; Giraudet, Louis; Nguyen, L.; Lugiez, F.; Praseuth, Jean-Pierre)V1362,331-337(1991)

InP-based quantum-well infrared photodetectors (Gunapala, S. D.; Levine, Barry F.; Ritter, D.; Hamm, Robert A.; Panish, Morton B.)V1541,11-23(1991)

Interplay between photons and superconductors (Gilabert, Alain; Azema, Alain; Roustan, Jean-Claude; Maneval, Jean-Paul)V1477,140-147(1991)

Intrinsic carrier concentration and effective masses in Hg1-xMnxTe (Rogalski, Antoni)V1512,189-194(1991)

Investigation of the possibility of creating a multichannel photodetector based on the avalanche MRS-structure (Sadyigov, Z. Y.; Gasanov, A. G.; Yusipov, N. Y.; Golovin, V. M.; Gulanian, Emin H.; Vinokurov, Y. S.; Simonov, A. V.)V1621,158-168(1991)

Ion implantation and diffusion for electrical junction formation in HgCdTe (Bubulac, Lucia O.)V1484,67-71(1991)

Large-format 1280 x 1024 full-frame CCD image sensor with a lateral-overflow drain and transparent gate electrode (Stevens, Eric G.; Kosman, Steven L.; Cassidy, John C.; Chang, Win-Chyi; Miller, Wesley A.)V1447,274-282(1991)

Liquid-nitrogen-cooled low-noise radiation pulse detector amplifier (Trojnar, Eugeniusz; Trojanowski, Stanislaw; Czechowicz, Roman; Derwiszynski, Mariusz; Kocyba, Krzysztof)V1391,230-237(1991)

Long-pathlength DOAS (differential optical absorption spectrometer) system for the in situ measurement of xylene in indoor air (Biermann, Heinz W.; Green, Martina; Seiber, James N.)V1433,2-7(1991)

Long-wavelength GaAs/AlxGa1-xAs quantum-well infrared photodetectors (Levine, Barry F.; Bethea, Clyde G.; Stayt, J. W.; Glogovsky, K. G.; Leibenguth, R. E.; Gunapala, S. D.; Pei, S. S.; Kuo, Jenn-Ming)V1540,232-237(1991)

Long-wavelength GaAs quantum-well infrared photodetectors (Levine, Barry F.)V1362,163-167(1991)

Magnetoconcentration nonequilibrium IR photodetectors (Piotrowski, Jozef; Djuric, Zoran G.)V1512,84-90(1991)

Measuring device for space-temporal characteristics of technological lasers radiation (Mnatsakanyan, Eduard A.; Andreev, V. G.; Bestalanny, S. I.; Velikoselsky, V.V.; Grinyakin, A.P.; Grinyakin, V.P.; Kolesnikov, S.A.; Rozhnov, U.V.)V1414,130-133(1991)

MMIC compatible photodetector design and characterization (Dallabetta, Kyle A.; de La Chapelle, Michael; Lawrence, Robert C.)V1371,116-127(1991)

Model of a thinned CCD (Blouke, Morley M.)V1439,136-143(1991)

Multialkali photocathodes grown by molecular beam epitaxy technique (Dubovoi, I. A.; Chernikov, A. S.; Prokhorov, Alexander M.; Schelev, Mikhail Y.; Ushakov, V. K.)V1358,134-138(1991)

Novel high-speed dual-wavelength InAlAs/InGaAs graded superlattice Schottky barrier photodiode for 0.8- and 1.3-um detection (Hwang, Kiu C.; Li, Sheng S.; Kao, Yung C.)V1371,128-137(1991)

Observation of multiphoton photoemission from a NEA GaAs photocathode (Wang, Liming; Hou, Xun; Cheng, Zhao)V1358,1156-1160(1991)

On-board calibration device for a wide field-of-view instrument (Krawczyk, Rodolphe; Chessel, Jean-Pierre; Durpaire, Jean-Pierre; Durieux, Alain; Churoux, Pascal; Briottet, Xavier)V1493,2-15(1991)

Optical response in high-temperature superconducting thin films (Thiede, David A.)V1484,72-80(1991)

Performance of high-dynamic-range CCD arrays with various epilayer structures (Smith, Dale J.; Harrison, Lorna J.; Pellegrino, John M.; Simon, Deborah R.)V1562,242-250(1991)

Performances of gallium arsenide on silicon substrate photoconductive detectors (Constant, Monique T.; Boussekey, Luc; Decoster, Didier; Vilcot, Jean-Pierre)V1362,156-162(1991)

Photoconductivity decay method for determining minority carrier lifetime of p-type HgCdTe (Reichman, Joseph)V1484,31-38(1991)

Photodetectors: how to integrate them with microelectronic and optical devices (Decoster, Didier; Vilcot, Jean-Pierre)V1362,959-966(1991)

Preamplifiers of the high-ohmic high-speed photodetector signals (Kovrigin, Yevgeny; Potylitsyn, Yevgeny)V1362,967-975(1991)

Pyroelectric linear array IR detectors with CCD multiplexer (Norkus, Volkmar; Neumann, Norbert; Walther, Ludwig; Hofmann, Guenter; Schieferdecker, Jorg; Krauss, Matthias G.; Budzier, Helmut; Hess, Norbert)V1484,98-105(1991)

Recent progress in device-oriented II-VI research at the University of Wuerzburg (Landwehr, Gottfried; Waag, Andreas; Hofmann, K.; Kallis, N.; Bicknell-Tassius, Robert N.)V1362,282-290(1991)

Review of ITT/EOPD's special purpose photosensitive devices and technologies (Johnson, C. B.)V1396,360-376(1991)

Selective photodetectors: a view from the USSR (Ovsyuk, Victor N.; Svitashev, Konstantin K.)V1540,424-431(1991)

Si:Ga focal-plane arrays for satellite and ground-based telescopes (Mottier, Patrick; Agnese, Patrick; Lagage, Pierre O.)V1494,419-426(1991); V1512,60-67(1991)

Signal, noise, and readout considerations in the development of amorphous silicon photodiode arrays for radiotherapy and diagnostic x-ray imaging (Antonuk, Larry E.; Boudry, J.; Kim, Chung-Won; Longo, M.; Morton, E. J.; Yorkston, J.; Street, Robert A.)V1443,108-119(1991)

Solutions to modeling of imaging IR systems for missile applications: MICOM imaging IR system performance model-90 (Owen, Philip R.; Dawson, James A.; Borg, Eric J.)V1488,122-132(1991)

Submicron thin-film metal-oxide-metal infrared detectors (Wilke, Ingrid; Moix, Dominique; Herrmann, W.; Kneubuhl, F. K.)V1442,2-10(1991)

Transferred electron photocathode with greater than 5% quantum efficiency beyond 1 micron (Costello, Kenneth A.; Davis, Gary A.; Weiss, Robert; Aebi, Verle W.)V1449,40-50(1991)

Ultraviolet light imaging technology and applications (Yokoi, Takane; Suzuki, Kenji; Oba, Koichiro)V1449,30-39(1991)

Veiling glare in the F4111 image intensifier (Acharya, Mukund; Bunch, Robert M.)V1396,377-388(1991)

Vertical ocean reflectance at low altitudes for narrow laser beams (Crittenden, Eugene C.; Rodeback, G. W.; Milne, Edmund A.; Cooper, Alfred W.)V1492,187-197(1991)

Visible/infrared integrated double detector: application to obstacle detection in automotive (Phase 2) (Simonne, John J.; Pham, Vui V.; Esteve, Daniel; Clot, Jean; Mahrane, Achour; Beconne, Jean P.)V1589,139-147(1991)

Photodiodes—see also diodes

A 39-photon/bit direct-detection receiver at 810 nm, BER = 1x10-6, 60-Mbit/s QPPM (MacGregor, Andrew D.; Dion, Bruno; Noeldeke, Christoph; Duchmann, Olivier)V1417,374-380(1991)

Comparative analysis of external factors' influences on the GaP light-emitting p-n-junctions (Rizikov, Igor V.; Svechnikov, Georgy S.; Bulyarsky, Sergey V.; Ambrozevich, Alexander S.)V1362,664-673(1991)

Comparison of simulation and experimental measurements of avalanche photodiode receiver performance (Mecherle, G. S.; Henderson, Robert J.)V1417,537-542(1991)

Computer simulation of staring-array thermal imagers (Bradley, D. J.; Dennis, Peter N.; Baddiley, C. J.; Murphy, Kevin S.; Carpenter, Stephen R.; Wilson, W. G.)V1488,186-195(1991)

Correlation of surface topography and coating damage with changes in the responsivity of silicon PIN photodiodes (Huffaker, Diana L.; Walser, Rodger M.; Becker, Michael F.)V1441,365-380(1991)

Development of a large pixel, spectrally optimized, pinned photodiode/interline CCD detector for the Earth Observing System/Moderate-Resolution Imaging Spectrometer-Tilt Instrument (Ewin, Audrey J.; Jhabvala, Murzy; Shu, Peter K.)V1479,12-20(1991)

Double-resonant tunneling via surface plasmons in layered gratings (Sreseli, Olga M.; Belyakov, Ludvig V.; Goryachev, D. N.; Rumyantsev, B. L.; Yaroshetskii, Ilya D.)V1545,149-158(1991)

Edge effects in silicon photodiode arrays (Kenney, Steven B.; Hirleman, Edwin D.)V1480,82-93(1991)

Equivalent capacitance photodiode determination by an optical method (Podoleanu, Adrian G.; Sterian, Paul E.; Popescu, Ion M.; Puscas, Niculae Tiberiu N.)V1362,721-726(1991)

External factors' influences on AIIIBV light-emitting structures (Svechnikov, Georgy S.; Rizikov, Igor V.)V1362,674-683(1991)

GaAs/GaAlAs superlattice avalanche photodiode at wavelength L = 0.8 um (Kibeya, Saidi; Orsal, Bernard; Alabedra, Robert; Lippens, D.)V1501,97-106(1991)

GaInAs PIN photodetectors on semi-insulating substrates (Crawford, Deborah L.; Wey, Y. G.; Bowers, John E.; Hafich, Michael J.; Robinson, Gary Y.)V1371,138-141(1991)

Germanium photodiodes calibration as standards of optical fiber systems power measurements (Campos, Joaquin; Corredera, Pedro; Pons, Alicia A.; Corrons, Antonio)V1504,66-74(1991)

High-performance a-Si:H TFT-driven linear image sensor and its application to a compact scanner (Miyake, Hiroyuki; Sakai, Kazuhiro; Abe, Tsutomu; Sakai, Yoshihiko; Hotta, Hiroyuki; Sugino, Hajime; Ito, Hisao; Ozawa, Takashi)V1448,150-156(1991)

High-performance InGaAs PIN and APD (avalanche photdiode) detectors for 1.54 um eyesafe rangefinding (Olsen, Gregory H.; Ackley, Donald A.; Hladky, J.; Spadafora, J.; Woodruff, K. M.; Lange, M. J.; Van Orsdel, Brian T.; Forrest, Stephen R.; Liu, Y.)V1419,24-31(1991)

High-speed 128-element avalanche photodiode array for optical computing applications (Webb, Paul P.; Dion, Bruno)V1563,236-243(1991)

Laser link performance improvements with wideband microwave impedance matching (Baldwin, David L.; Sokolov, Vladimir; Bauhahn, Paul E.)V1476,46-55(1991)

Laser photocathode development for high-current electron source (Moustaizis, Stavros D.; Fotakis, Costas; Girardeau-Montaut, Jean-Pierre)V1552,50-56(1991)

Low-temperature operation of silicon drift detectors (Sumner, Timothy J.; Roe, S.; Rochester, G. K.; Hall, G.; Evensen, Per; Avset, B. S.)V1549,265-273(1991)

Multichannel optical data link (Ota, Yusuke; Swartz, Robert G.)V1364,146-152(1991)

New photovoltaic effect in semiconductor junctions n+/p (Dominguez Ferrari, E.; Encinas Sanz, F.; Guerra Perez, J. M.)V1397,725-727(1991)

Novel device for short-pulse generation using optoelectronic feedback (Phelan, Paul J.; Hegarty, John; Elsasser, Wolfgang E.)V1362,623-630(1991)

Novel high-speed dual-wavelength InAlAs/InGaAs graded superlattice Schottky barrier photodiode for 0.8- and 1.3-um detection (Hwang, Kiu C.; Li, Sheng S.; Kao, Yung C.)V1371,128-137(1991)

Optical dividers for quadrant avalanche photodiode detectors (Green, Samuel I.)V1417,496-512(1991)

Optoelectronic LED-photodiode pairs for moisture and gas sensors in the spectral range 1.8-4.8 um (Yakovlev, Yurii P.; Baranov, Alexej N.; Imenkov, Albert N.; Mikhailova, Maya P.)V1510,170-177(1991)

Planar InGaAs APD (avalanche photodiode) for eyesafe laser rangefinding applications (Webb, Paul P.)V1419,17-23(1991)

Process monitoring during CO2 laser cutting (Jorgensen, Henning; Olsen, Flemming O.)V1412,198-208(1991)

Quantum efficiency and crosstalk of an improved backside-illuminated indium antimonide focal plane array (Bloom, I.; Nemirovsky, Yael)V1442,286-297(1991)

Silicon/PVDF integrated double detector: application to obstacle detection in automotive (Simonne, John J.; Pham, Vui V.; Esteve, Daniel; Alaoui-Amine, Mohammed; Bousbiat, Essaid)V1374,107-115(1991)

Temperature-monitored/controlled silicon photodiodes for standardization (Eppeldauer, George)V1479,71-77(1991)

Transferred electron photocathode with greater than 5% quantum efficiency beyond 1 micron (Costello, Kenneth A.; Davis, Gary A.; Weiss, Robert; Aebi, Verle W.)V1449,40-50(1991)

Photogrammetry—see also remote sensing

Accuracy/repeatability test for a video photogrammetric measurement (Gustafson, Peter C.)V1526,36-41(1991)

Analysis of results from high-speed photogrammetry of flow tracers in blast waves (Dewey, John M.; McMillin, Douglas J.)V1358,246-253(1991)

Automated dimensional inspection of cars in crash tests with digital photogrammetry (Beyer, Horst A.)V1526,134-141(1991)

Automated photogrammetric surface reconstruction with structured light (Maas, Hans-Gerd)V1526,70-77(1991)

Calibration procedures for the space vision system experiment (MacLean, Steve G.; Pinkney, H. F.)V1526,113-122(1991)

Deformation study of the Klang Gates Dam with multispectral analysis method (Hassan, Azmi; Subari, Mustofa; Som, Zainal A.)V1526,142-156(1991)

Design of eye movement monitoring system for practical environment (Nakamura, Hiroyuki; Kobayashi, Hitoshi; Taya, Katsuo; Ishigami, Shigenobu)V1456,226-238(1991)

Fast and precise method to extract vanishing points (Coelho, Christopher; Straforini, Marco; Campani, Marco)V1388,398-408(1991)

Generation of synthetic stereo views from digital terrain models and digitized photographs (Bethel, James S.)V1457,49-53(1991)

Industrial Vision Metrology (El-Hakim, Sabry F., ed.)V1526(1991)

Linear resection, intersection, and perspective-independent model matching in photogrammetry: theory (Barrett, Eamon B.; Brill, Michael H.; Haag, Nils N.; Payton, Paul M.)V1567,142-169(1991)

Matching in image/object dual spaces (Zhang, Yaonan)V1526,195-202(1991)

Moire-shift interferometer measurements of the shape of human and cat tympanic membrane (Decraemer, Willem F.; Dirckx, Joris J.)V1429,26-33(1991)

Photogrammetric measurements of retinal nerve fiber layer thickness along the disc margin and peripapillary region (Takamoto, Takenori; Schwartz, Bernard)V1380,64-74(1991)

Problems of photogrammetry of moving target in water (Han, Xin Z.)V1537,215-220(1991)

Reconstructing visible surfaces (Schenk, Anton F.; Toth, Charles)V1526,78-89(1991)

Three-dimensional contour of crack tips using a grating method (Andresen, Klaus; Kamp, B.; Ritter, Reinold)V1554A,93-100(1991)

Three-dimensional machine vision using line-scan sensors (Godber, Simon X.; Robinson, Max; Evans, Paul)V1526,170-189(1991)

X-ray photogrammetry of the hip revisited (Turner-Smith, Alan R.; White, Steven P.; Bulstrode, Christopher)V1380,75-84(1991)

Photography—see also cameras; films; imaging systems; lenses; mirrors; photography, high-speed

A 1.3-megapixel-resolution portable CCD electronic still camera (Jackson, Todd A.; Bell, Cynthia S.)V1448,2-12(1991)

Applications of aerial photography to law enforcement and disaster assessment: a consideration of the state-of-the-art (Cox, William J.; Biache, Andrew)V1479,364-369(1991)

Chemical symmetry: developers that look like bleach agents for holography (Bjelkhagen, Hans I.; Phillips, Nicholas J.; Ce, Wang)V1461,321-328(1991)

Contextual image understanding of airport photographs (Nasr, Hatem N.)V1521,24-33(1991)

Day/night aerial surveillance system for fishery patrol (Uhl, Bernd)V1538,140-147(1991)

Direct-drive film magazines (Lewis, George R.)V1538,167-179(1991)

DSP-based stabilization systems for LOROP cameras (Quinn, James)V1538,150-166(1991)

Evaluation of the effect of noise on subjective image quality (Barten, Peter G.)V1453,2-15(1991)

History and status of black and white photographic processing chemicals as effluents (Horn, Richard R.)V1458,69-75(1991)

Laser beam scanner for uniform halftones (Ando, Toshinori)V1458,128-132(1991)

Lens evaluation for electronic photography (Bell, Cynthia S.)V1448,59-68(1991)

Metal sulfide thin films on glass as solar control, solar absorber, decorative, and photographic coatings (Nair, Padmanabhan K.; Nair, M. T.; Fernandez, A. M.; Garcia, V. M.; Hernandez, A. B.)V1485,228-239(1991)

Microlens arrays in integral photography and optical metrology (Davies, Neil A.; McCormick, Malcolm; Lau, Hon W.)V1544,189-198(1991)

Photoelectrographic printing with persistently conductive masters based on onium salt acid photogenerators (Bugner, Douglas E.; Fulmer, Gary G.; Riblett, Susan E.)V1458,162-178(1991)

Real-time speckle photography using photorefractive Bi12SiO20 crystal (Nakagawa, Kiyoshi; Minemoto, Takumi)V1508,191-200(1991)

Shear-lens photography for holographic stereograms (Molteni, William J.)V1461,132-141(1991)

Simple 180o field-of-view F-theta all-sky camera (Andreic, Zeljko)V1500,293-304(1991)

Speckle photography for investigation of bones supported by different fixing devices (Kasprzak, Henryk; Podbielska, Halina; Pennig, Dietmar)V1429,55-61(1991)

Technology trends in electrophotographic printers (Starkweather, Gary K.)V1458,120-127(1991)

Underwater Imaging, Photography, and Visibility (Spinrad, Richard W., ed.)V1537(1991)

Using digital-scanned aerial photography for wetlands delineation (Anderson, John E.; Roos, Maurits)V1492,252-262(1991)

Photography, high-speed—see also image tubes; intensifiers; photography; streak cameras

10 MHz multichannel image detection and processing system (Damstra, Geert C.; Eenink, A. H.)V1358,644-651(1991)

19th Intl Congress on High-Speed Photography and Photonics (Fuller, Peter W., ed.)V1358(1991)

200,000-frame-per-second drum camera with nanosecond synchronized laser illumination (Briscoe, Dennis)V1346,319-323(1991)

328 MHz synchroscan streak camera (Schelev, Mikhail Y.; Serdyuchenko, Yuri N.; Vaschenko, G. O.)V1358,569-573(1991)

Ablation of hard dental tissues with an ArF-pulsed excimer laser (Neev, Joseph; Raney, Daniel; Whalen, William E.; Fujishige, Jack T.; Ho, Peter D.; McGrann, John V.; Berns, Michael W.)V1427,162-172(1991)

Active mode-locking of external cavity semiconductor laser with 1GHz repetition rate (Wang, Xianhua; Chen, Guofu; Liu, D.; Xu, L.; Ruan, S.)V1358,775-779(1991)

Advanced image intensifier systems for low-light high-speed imaging (Bruno, Theresa L.; Wirth, Allan)V1358,109-116(1991)

Advances in the practical application of schlieren photography in industry (Herbrich, Horst R.)V1358,24-28(1991)

Advantage of simultaneous streak and framing records in the field of detonics (Held, Manfred)V1358,904-913(1991)

Aerodynamic interferograms of explosive field (Miao, Peng-Cheng; Wang, Hai-Lin; Yan, Da-Peng; He, Anzhi)V1554B,641-644(1991)

Aluminum metal combustion in water revealed by high-speed microphotography (Tao, William C.; Frank, Alan M.; Clements, Rochelle E.; Shepherd, Joseph E.)V1346,300-310(1991)

Analysis of pyrotechnic devices by laser-illuminated high-speed photography (Dosser, Larry R.; Stark, Margaret A.)V1346,293-299(1991)

Analysis of results from high-speed photogrammetry of flow tracers in blast waves (Dewey, John M.; McMillin, Douglas J.)V1358,246-253(1991)

Application of a linear picosecond streak camera to the investigation of a 1.55 um mode-locked Er3+ fiber laser (de Souza, Eunezio A.; Cruz, C. H.; Scarparo, Marco A.; Prokhorov, Alexander M.; Postovalov, V. E.; Vorobiev, N. S.; Schelev, Mikhail Y.)V1358,556-560(1991)

Application of cylindrical blast waves to impact studies of materials (Parry, David J.; Stewardson, H. R.; Ahmad, S. H.; Al-Maliky, Noori S.)V1358,1057-1064(1991)

Application of electron-sensitive CCD for taking off the time-dispersed pictures from image tube phosphor screens (Bryukhnevitch, G. I.; Dalinenko, I. N.; Kuz'min, G. A.; Libenson, B. N.; Malyarov, A. V.; Moskalev, B. B.; Postovalov, V. E.; Prokhorov, Alexander M.; Schelev, Mikhail Y.)V1358,739-749(1991)

Application of high-speed infrared emission spectroscopy in reacting flows (Klingenberg, Guenter; Rockstroh, Helmut)V1358,851-858(1991)

Application of high-speed mirror chronograph 3CX-1 to plasma investigations in visible and medium infrared spectrum ranges (Drozhbin, Yu. A.; Zvorykin, V. D.; Polyansky, S. V.; Sychugov, G. V.; Trofimenko, Vladimir V.; Yarova, A. G.)V1358,1029-1034(1991)

Application of high-speed photography in the design of new initiating systems (Steele, Ann F.)V1358,1123-1133(1991); Uyemura, Tsuneyoshi)V1358,351-357(1991)

Application of IR photorecorders based on ionization chambers for fast processes investigation (Egorov, V. V.; Lazarchuk, V. P.; Murugov, V. M.; Sheremetyev, Yu. N.)V1358,984-991(1991)

Application of the half-filter method to the flash radiography using a neutral filter in the range of x-rays (Gerstenmayer, Jean-Louis; Vibert, Patrick)V1346,286-292(1991)

Applications of a soft x-ray streak camera in laser-plasma interaction studies (Tsakiris, George D.)V1358,174-192(1991)

Applications of copper vapor laser lighting in high-speed motion analysis (Hogan, Daniel C.)V1346,324-330(1991)

Applications of high-resolution still video cameras to ballistic imaging (Snyder, Donald R.; Kosel, Frank M.)V1346,216-225(1991)

Applications of synchroscan and dual-sweep streak camera techniques to free-electron laser experiments (Lumpkin, Alex H.)V1552,42-49(1991)

Applications of Z-Plane memory technology to high-frame rate imaging systems (Shanken, Stuart N.; Ludwig, David E.)V1346,210-215(1991)

Appreciation of the Society of Motion Picture and Television Engineers: the originators of the Congress (Lunn, George H.)V1358,1161-1163(1991)

Approximation of the Compton scattered radiation (Burq, Catherine; Vibert, Patrick)V1346,276-285(1991)

Auto-focus video camera system with bag-type lens (Sugiura, Norio; Morita, Shinzo)V1358,442-446(1991)

Automatic film reading system for high-speed photography (Zhou, Renkui; Yu, Chongzhen; Ma, Jiankang; Zhu, Wenkai)V1358,1245-1251(1991)

Avalanche transistor selection for long-term stability in streak camera sweep and pulser applications (Thomas, Stan W.; Griffith, Roger L.; Teruya, Alan T.)V1358,578-588(1991)

Breakdown in pulsed-power semiconductor switches (Germer, Rudolf K.; Mohr, Joachim; Schoenbach, Karl H.)V1358,925-931(1991)

C 850X picosecond high-resolution streak camera (Mens, Alain; Dalmasso, J. M.; Sauneuf, Richard; Verrecchia, R.; Roth, J. M.; Tomasini, F.; Miehe, Joseph A.; Rebuffie, Jean-Claude)V1358,315-328(1991)

Characteristics of sprinklers and water spray mists for fire safety (Jackman, Louise A.; Lavelle, Stephen P.; Nolan, P. F.)V1358,831-842(1991)

Circular encoding in large-area multiexposure holography (Aliaga, R.; Choi, Peter; Chuaqui, Hernan H.)V1358,1257-1264(1991)

Collapsing cavities in reactive and nonreactive media (Bourne, Neil K.; Field, John E.)V1358,1046-1056(1991)

Comparative study of gelatin ablation by free-running and Q-switch modes of Er:YAG laser (Konov, Vitaly I.; Kulevsky, Lev A.; Lukashev, Alexei V.; Pashinin, Vladimir P.; Silenok, Alexander S.)V1427,232-242(1991)

Comparison of electro-optic diagnostic systems (Hagans, Karla G.; Sargis, Paul G.)V1346,404-408(1991)

Conversion of schlieren systems to high-speed interferometers (Anderson, Roland C.; Milton, James E.)V1358,992-1002(1991)

Damage to polymer-coated glass surfaces by small-particle impact (Chaudhri, Mohammad M.; Smith, Alan L.)V1358,683-689(1991)

Deflection evaluation using time-resolved radiography (Fry, David A.; Lucoro, Jacobo P.)V1346,270-275(1991)

Designing and application of solid state lasers for streak cameras calibration (Babushkin, A. V.; Vorobiev, N. S.; Prokhorov, Alexander M.; Schelev, Mikhail Y.)V1346,410-417(1991)

Detection and localization of absorbers in scattering media using frequency-domain principles (Berndt, Klaus W.; Lakowicz, Joseph R.)V1431,149-160(1991)

Determination of burst initiation location and tear propagation velocity during air burst testing of latex condoms (Davidhazy, Andrew)V1358,654-659(1991)

Development of new beamsplitter system for real-time high-speed holographic interferometry (Yamamoto, Yoshitaka)V1358,940-951(1991)

Development of picosecond x-ray framing camera (Chang, Zenghu; Hou, Xun; Zhang, Xiaoqiu; Gong, Meixia; Niu, Lihong; Yong, Hongru; Liu, Xiouqin; Lei, Zhiyuan)V1358,614-618(1991)

Development of subnanosecond framing cameras in IOFAN (Ludikov, V. V.; Prokhorov, Alexander M.; Chevokin, Victor K.)V1346,418-436(1991)

Diagnostic of the reaction behaviour of insensitive high explosives under jet attack (Held, Manfred)V1358,1021-1028(1991)

Digital readout for image converter cameras (Honour, Joseph)V1358,713-718(1991)

Digitize your films without losing resolution (Kallhammer, Jan-Erik O.)V1358,631-636(1991)

Discrimination and classification with Xybion multispectral video systems (Frost, Paul A.)V1358,398-408(1991)

Disk-cathode flash x-ray tube driven by a repetitive type of Blumlein pulser (Sato, Eiichi; Kimura, Shingo; Isobe, Hiroshi; Takahashi, Kei; Tamakawa, Yoshiharu; Yanagisawa, Toru)V1358,146-153(1991)

Dynamic behavior and structure optimum of high-speed gear mechanism of high-speed photography apparatus (Wang, Jianshe; Liu, Jian-Hua)V1533,175-184(1991)

Dynamic particle holographic instrument (Wang, Guozhi; Feng, San; Wang, Zhengrong; Wang, Shuyan)V1358,73-81(1991)

Dynamic photoelasticity applied to crack-branching investigations (Hammami, Slimane; Cottron, M.; Lagarde, Alexis)V1554A,136-142(1991)

Dynamic photoelasticity with a split Hopkinson pressure bar (Morris, David R.; Watson, A. J.)V1358,254-261(1991)

Electro-optical transducer employing liquid crystal target for processing images in real-time scale (Ignatosyan, S. S.; Simonov, V. P.; Stepanov, Boris M.)V1358,100-108(1991)

Enhanced holographic recording capabilities for dynamic applications (Hough, Gary R.; Gustafson, D. M.; Thursby, William R.)V1346,194-199(1991)

Estimation of limit time resolution in image streak camera (Klementyev, V. G.)V1358,1070-1074(1991)

Evaluation of dynamic range for LLNL streak cameras using high-contrast pulses and "pulse podiatry" on the NOVA laser system (Richards, James B.; Weiland, Timothy L.; Prior, John A.)V1346,384-389(1991)

Evolution of high-speed photography and photonics techniques in detonics experiments (Cavailler, Claude)V1358,210-226(1991)

Examination of EOT operation with spherical elements or slit accelerating diaphragm (Korzhenevich, Irina M.; Kolesov, G. V.; Lebedev, V. B.; Petrokovich, O. A.; Feldman, G. G.)V1358,1084-1089(1991)

Experimental investigation of image degradation created by a high-velocity flow field (Couch, Lori L.; Kalin, David A.; McNeal, Terry)V1486,417-423(1991)

Experimental investigation on the flow behavior of liquid aluminum inside pressure-die-casting dies using high-speed photography (Jiang, Xuping; Wu, Guobing)V1358,1237-1244(1991)

Experimental research on the casing-shaped charge (Gao, Er-xin)V1358,1115-1119(1991)

Femtosecond streak tube (Kinoshita, Katsuyuki; Suyama, Motohiro; Inagaki, Yoshinori; Ishihara, Y.; Ito, Masuo)V1358,490-496(1991)

First results on a developmental deflection tube and its associated electronics for streak camera applications (Froehly, Claude; Laucournet, A.; Miehe, Joseph A.; Rebuffie, Jean-Claude; Roth, J. M.; Tomasini, F.)V1358,532-540(1991)

Flash soft radiography: its adaption to the study of breakup mechanisms of liquid jets into a high-density gas (Krehl, Peter; Warken, D.)V1358,162-173(1991)

Focusing of shock waves in water and its observation by the schlieren method (Isuzugawa, Kohji; Horiuchi, Makoto; Okumura, Yoshiyuki)V1358,1003-1010(1991)

Fourth-generation motion analyzer (Balch, Kris S.)V1358,373-397(1991)

Fundamental studies for the high-intensity long-duration flash x-ray generator for biomedical radiography (Sato, Eiichi; Isobe, Hiroshi; Takahashi, Kei; Tamakawa, Yoshiharu; Yanagisawa, Toru)V1358,193-200(1991)

Gatable photonic detector and its image processing (Kume, Hidehiro; Nakamura, Haruhito; Suzuki, Makoto)V1358,1144-1155(1991)

Generation of hard x-ray pulse trains with the help of high-frequency oscillating systems for fast processes recording (Romanovsky, V. F.; Kovalenko, A. N.; Sapozhnikova, T. I.; Tushev, N. R.; Abgaryan, A.)V1358,140-145(1991)

High-density memory packaging technology high-speed imaging applications (Frew, Dean L.)V1346,200-209(1991)

High-dynamic-range image readout system (Mens, Alain; Ducrocq, N.; Mazataud, D.; Mugnier, A.; Eouzan, J. Y.; Heurtaux, J. C.; Tomasini, F.; Mathae, J. C.)V1358,719-731(1991)

Higher-order theory of caustics for fast running cracks under general loadings (Nishioka, Teiichi; Murakami, R.; Matsuo, Seitaro; Ohishi, Y.)V1554A,802-813(1991)

High-intensity soft-flash x-ray generator utilizing a low-vacuum diode (Isobe, Hiroshi; Sato, Eiichi; Shikoda, Arimitsu; Takahashi, Kei; Tamakawa, Yoshiharu; Yanagisawa, Toru)V1358,471-478(1991)

High-precision measurements of the 24-beam UV-OMEGA laser (Jaanimagi, Paul A.; Hestdalen, C.; Kelly, John H.; Seka, Wolf D.)V1358,337-343(1991)

High-speed CCD video camera (Germer, Rudolf K.; Meyer-Ilse, Werner)V1358,346-350(1991)

High-speed laser speckle photography (Huntley, Jonathan M.; Field, John E.)V1554A,756-765(1991)

High-speed microcinematography of aerosols (Lavelle, Stephen P.; Jackman, Louise A.; Nolan, P. F.)V1358,821-830(1991)

High-speed photographic study of a cavitation bubble (Soh, W. K.)V1358,1011-1015(1991)

High-speed photographic study of impact on fibers and woven fabrics (Field, John E.; Sun, Q.)V1358,703-712(1991)

High-speed photography applied for the investigation of the dynamics of free falling evaporating droplets (Guryashkin, L. P.; Stasenko, A. L.)V1358,974-978(1991)

High-speed photography at the Cavendish Laboratory (Field, John E.)V1358,2-17(1991)

High-speed photography of high-resolution moire patterns (Whitworth, Martin B.; Huntley, Jonathan M.; Field, John E.)V1358,677-682(1991)

High-speed readout CCDs (Ball, K.; Burt, D. J.; Smith, Graham W.)V1358,409-420(1991)

High-speed still video photography for ballistic range applications (Speyer, Brian A.)V1358,1215-1221(1991)

High-speed time-resolved holographic interferometer using solid state shutters (Racca, Roberto G.; Dewey, John M.)V1358,932-939(1991)

High-speed video instrumentation system (Gorenflo, Ronald L.; Stockum, Larry A.; Barnett, Brett)V1346,42-53(1991)

Holographic interferometric observation of shock wave focusing to extracorporeal shock wave lithotripsy (Takayama, Kazuyoshi; Obara, Tetsuro; Onodera, Osamu)V1358,1180-1190(1991)

Holographic visualization of hypervelocity explosive events (Cullis, I. C.; Parker, Richard J.; Sewell, Derek)V1358,52-64(1991)

Hubble Space Telescope: mission, history, and systems (Endelman, Lincoln L.)V1358,422-441(1991)

Human movement analysis with image processing in real time (Fauvet, E.; Paindavoine, M.; Cannard, F.)V1358,620-630(1991)

Hycam camera study of the features of a deflagrating munition (Kinsey, Trevor J.; Bussell, T. J.; Chick, M. C.)V1358,914-924(1991)

Image converter streak cameras with super-light-speed scanning (Averin, V. I.; Gus'kova, M. S.; Korzhenevich, Irina M.; Kolesov, G. V.; Lebedev, V. B.; Maranichenko, N. I.; Sobolev, A. A.)V1358,589-602(1991)

Imaging of excimer laser vascular tissue ablation by ultrafast photography (Nyga, Ralf; Neu, Walter; Preisack, M.; Haase, Karl K.; Karsch, Karl R.)V1525,119-123(1991)

Improved image quality from retroreflective screens by spectral smearing (Stevens, Richard F.)V1358,1265-1267(1991)

Improved version of the PIF01 streak image tube (Degtyareva, V. P.; Fedotov, V. I.; Moskalev, B. B.; Postovalov, V. E.; Prokhorov, Alexander M.; Schelev, Mikhail Y.; Soldatov, N. F.)V1358,524-531(1991)

Inspection system using still vision for a rotating laser-textured dull roll (Torao, Akira; Uchida, Hiroyuki; Moriya, Susumu; Ichikawa, Fumihiko; Kataoka, Kenji; Wakui, Tsuneyoshi)V1358,843-850(1991)

Instantaneous measurement of density from double-simultaneous interferograms (Desse, Jean-Michel; Pegneaux, Jean-Claude)V1358,766-774(1991)

Interactions of laser-induced cavitation bubbles with a rigid boundary (Ward, Barry; Emmony, David C.)V1358,1035-1045(1991)

Interference visualization of infrared images (Vlasov, Nikolai G.; Korchazhkin, S. V.; Manikalo, V. V.)V1358,1018-1020(1991)

Interferometric measurements of a high-velocity mixing/shear layer (Peters, Bruce R.; Kalin, David A.)V1486,410-416(1991)

Interferometry, streak photography, and stereo photography of laser-driven miniature flying plates (Paisley, Dennis L.; Montoya, Nelson I.; Stahl, David B.; Garcia, Ismel A.)V1358,760-765(1991)

Investigation of diesel injection jets using high-speed photography and speed holography (Eisfeld, Fritz)V1358,660-671(1991)

Kilohertz range pulsed x-ray generator having a hot cathode triode (Sato, Eiichi; Shikoda, Arimitsu; Isobe, Hiroshi; Tamakawa, Yoshiharu; Yanagisawa, Toru; Honda, Keiji; Yokota, Yoshiharu)V1358,479-487(1991)

Laser generation of Stoneley waves at liquid-solid boundaries (Ward, Barry; Emmony, David C.)V1358,1228-1236(1991)

Laser high-speed photography in cylindrical charge shell burst process (Li, Guozhu; Wu, Wenming; Su, Degong)V1358,1120-1122(1991)

Laser high-speed photography systems used to ammunition measures and tests (Wang, Yuren)V1346,331-337(1991)

Laser light sheet investigation into transonic external aerodynamics (Towers, Catherine E.; Towers, David P.; Judge, Thomas R.; Bryanston-Cross, Peter J.)V1358,952-965(1991)

Laser optical fiber high-speed camera (Xia, Sheng-jie; Yang, Ye-min; Tang, Di-zhu)V1358,43-45(1991)

Laser-produced plasma x-ray diagnostics with an x-ray streak camera at the Iskra-4 plant (Berkovsky, A. G.; Gubanov, Y. I.; Pryanishnikov, Ivan G.; Murugov, V. M.; Petrov, S. I.; Senik, A. V.)V1358,750-755(1991)

Laser-target diagnostics instrumentation system (Bousek, Ronald R.)V1414,175-184(1991)

Long-term flat-field behavior on LLNL streak cameras: preliminary results (Hatch, Jeffrey A.; Montgomery, David S.; Prior, John A.)V1346,371-375(1991)

Low-cost high-quality range camera system (Sewell, Derek)V1358,1209-1214(1991)

Measurement of the dynamic crack-tip displacement field using high-resolution moire photography (Whitworth, Martin B.; Huntley, Jonathan M.; Field, John E.)V1554B,282-288(1991)

Measurement of the shock profiles with streak technique and different detonating arrangements (Held, Manfred)V1346,311-318(1991)

Measurement of triggering instabilities of Imacon 500 streak cameras (Bowley, David J.; Rickett, Ph.; Babushkin, A. V.; Vorobiev, N. S.; Prokhorov, Alexander M.; Schelev, Mikhail Y.)V1358,550-555(1991)

Measurements with a 35-psec gate time microchannel plate camera (Bell, Perry M.; Kilkenny, Joseph D.; Hanks, Roy L.; Landen, Otto L.)V1346,456-464(1991)

Miniaturized semiconductor light source system for Cranz-Schardin applications (Stasicki, Boleslaw; Meier, G. E.)V1358,1222-1227(1991)

Modular equipment for single-frame photography in wide-time and spectral ranges (Bass, V. I.; Gus'kova, M. S.; Lebedev, V. B.; Mikhaylenko, B. M.; Saulevich, S. V.; Seleznev, V. P.; Feldman, G. G.; Chernyshov, N. A.)V1358,1075-1083(1991)

Modular streak camera for laser ranging (Prochazka, Ivan; Hamal, Karel; Schelev, Mikhail Y.; Lozovoi, V. I.; Postovalov, V. E.)V1358,574-577(1991)

Multialkali photocathodes grown by molecular beam epitaxy technique (Dubovoi, I. A.; Chernikov, A. S.; Prokhorov, Alexander M.; Schelev, Mikhail Y.; Ushakov, V. K.)V1358,134-138(1991)

Multichannel mirror systems for high-speed framing recording (Ushakov, Leonid S.; Ponomaryov, A. M.; Trofimenko, Vladimir V.; Drozhbin, Yu. A.)V1358,447-450(1991)

Multiframing image converter camera (Frontov, H. N.; Serdyuchenko, Yuri N.)V1358,311-314(1991)

Multipass holographic interferometry for low-density gas flow analysis (Surget, Jean; Dunet, G.)V1358,65-72(1991)

Multiple-channel correlated double sampling amplifier hybrid to support a 64 parallel output CCD array (Raanes, Chris A.; McNeill, John A.; Cunningham, Andrew P.)V1358,637-643(1991)

New approach to synchroballistic photography (McDowell, Maurice W.; Klee, H. W.; Griffith, Derek J.)V1358,227-236(1991)

New development of friction speeding mechanism for high-speed rotating mirror device (Wei, Yan-nian; Jiang, Pei-sheng; Zhang, Zeng-xiang; Wu, Ming-da; Shi, Gao-yi; Ye, Niao-ting)V1358,457-460(1991)

New electron optic for high-speed single-frame and streak image intensifier tubes (Dashevsky, Boris E.)V1358,561-568(1991)

New framing image tube with high-spatial-resolution (Chang, Zenghu; Hou, Xun; Zhang, Yongfeng; Zhu, Wenhua; Niu, Lihong; Liu, Xiouqin)V1358,541-545(1991)

New generation of copper vapor lasers for high-speed photography (Walder, Brian T.)V1358,811-820(1991)

New image-capturing techniques for a high-speed motion analyzer (Balch, Kris S.)V1346,2-23(1991)

New method for doing flat-field intensity calibrations of multiplexed ITT Streak Cameras (Hugenberg, Keith F.)V1346,390-397(1991)

New rippleflash system for large-area high-intensity lighting in harsh environments (Rendell, John T.)V1358,806-810(1991)

New streak tubes of the P500 series: features and experimental results (Rebuffie, Jean-Claude; Mens, Alain)V1358,511-523(1991)

New streak tube with femtosecond time resolution (Feldman, G. G.; Ilyna, T. A.; Korjenevitch, I. N.; Syrtzev, V. N.)V1358,497-502(1991)

New technologies in lighting systems for high-speed film and photography regarding high-intensity and heat problems (Severon, Burkhard)V1358,1202-1208(1991)

Noise performance of microchannel plate imaging systems (McCammon, Kent G.; Hagans, Karla G.; Hankla, A.)V1346,398-403(1991)

Novel technical advances provide easy solutions to tough motion analysis problems (Brown, Michael J.)V1346,24-32(1991)

Observation of multiphoton photoemission from a NEA GaAs photocathode (Wang, Liming; Hou, Xun; Cheng, Zhao)V1358,1156-1160(1991)

One-dimensional CCD linear array readout device (Borodin, A. M.; Ivanov, K. N.; Naumov, S. K.; Philippov, S. A.; Postovalov, V. E.; Prokhorov, Alexander M.; Stepanov, M. S.; Schelev, Mikhail Y.)V1358,756-758(1991)

One-frame subnanosecond spectroscopy camera (Silkis, E. G.; Titov, V. D.; Feldman, G. G.; Zhilkina, V. M.; Petrokovich, O. A.; Syrtzev, V. N.)V1358,46-49(1991)

Optical diagnostics of line-focused laser-produced plasmas (Lin, Li-Huang; Chen, Shisheng; Jiang, Z. M.; Ge, Wen; Qian, Aidi D.; Bin, Ouyang; Li, Yongchun L.; Kang, Yilan; Xu, Zhizhan)V1346,490-501(1991)

Orthogonal shadowgraphic nanolite stations (Celens, Eduard A.; Chabotier, A.)V1358,1103-1114(1991)

Peculiarities of frame memory design for slow-scan readout system for scientific application (Vysogorets, Mikhail V.; Mitrofanova, Natalya N.; Petrov, Mikhail Y.; Platonov, Valeri N.; Chulkin, Alexey D.)V1358,1066-1069(1991)

Picosecond intensifier gating with a plated webbing cathode underlay (Thomas, Stan W.; Trevino, Jimmy)V1358,84-90(1991)

Picosecond techniques application for definition of nonuniform ingradients inside turbid medium (Vorobiev, N. S.; Serafimovich, O. A.; Smirnov, A. V.)V1358,698-702(1991)

Picosecond x-ray streak camera improvement (Hou, Xun; Zhang, Xiaoqiu; Gong, Meixia; Chang, Zenghu; Lei, Zhiyuan; Yang, Binzhou; Yu, Hongbin; Liu, Xiouqin; Shan, Bin; Gao, Shengshen; Zhao, Wei)V1358,868-873(1991)

Picosecond x-ray streak cameras (Averin, V. I.; Bryukhnevitch, G. I.; Kolesov, G. V.; Lebedev, V. B.; Miller, V. A.; Saulevich, S. V.; Shulika, A. N.)V1358,603-605(1991)

Possibility of keeping color picture in an image converter camera (Zhao, Zongyao)V1358,1252-1256(1991)

Profile-related time resolution for a femtosecond streak tube (Liu, Yueping; Sibbett, Wilson)V1358,503-510(1991)

Projectile velocity and spin rate by image processing of synchro-ballistic photography (Hughett, Paul)V1346,237-248(1991)

Propagation of the spherical short-duration shock wave in a straight tunnel (Ahn, Jae W.; Song, So-Young; Lee, Jun Wung; Yang, Joon Mook)V1358,269-277(1991)

Ratio-telecontrolled strobolume for high-speed photography of depressurized towing tanks (Lin, Wenzheng; Jiang, Aibao; Zhou, Menzhen)V1358,29-36(1991)

Recent advances in gated x-ray imaging at LLNL (Wiedwald, Douglas J.; Bell, Perry M.; Kilkenny, Joseph D.; Bonner, R.; Montgomery, David S.)V1346,449-455(1991)

Recent research and development in electron image tubes/cameras/systems (Prokhorov, Alexander M.; Schelev, Mikhail Y.)V1358,280-289(1991)

Repetitive flash x-ray generator as an energy transfer source utilizing a compact-glass body diode (Shikoda, Arimitsu; Sato, Eiichi; Kimura, Shingo; Isobe, Hiroshi; Takahashi, Kei; Tamakawa, Yoshiharu; Yanagisawa, Toru)V1358,154-161(1991)

Repetitive flash x-ray generator operated at low-dose rates for a medical x-ray television system (Sato, Eiichi; Isobe, Hiroshi; Takahashi, Kei; Tamakawa, Yoshiharu; Yanagisawa, Toru)V1358,462-470(1991)

Repetitive flash x-ray generator utilizing an enclosed-type diode with a ring-shaped graphite cathode (Isobe, Hiroshi; Sato, Eiichi; Kimura, Shingo; Tamakawa, Yoshiharu; Yanagisawa, Toru; Honda, Keiji; Yokota, Yoshiharu)V1358,201-208(1991)

Research on macroeffects and micromechanism of martensite phase transition of shape memory alloys by high-speed photography (Yang, Jie; Wu, Yuehua; Zhou, Yusheng; Uyemura, T.)V1358,672-676(1991)

Selectable one-to-four-port, very high speed 512 x 512 charge-injection device (Zarnowski, Jeffrey J.; Williams, Bryn; Pace, M.; Joyner, M.; Carbone, Joseph; Borman, C.; Arnold, Frank S.; Wadsworth, Mark V.)V1447,191-201(1991)

Simultaneous imaging and interferometric turbule visualization in a high-velocity mixing/shear layer (Kalin, David A.; Saylor, Danny A.; Street, Troy A.)V1358,780-787(1991)

Slapper detonator flyer microphotography with a multiframe Kerr cell and Cranz-Schardin camera (McDaniel, Olin K.)V1358,1164-1179(1991)

Some comparative results of two approaches in computer simulation of electron lenses for streak image tubes (Degtyareva, V. P.; Ivanov, V. Y.; Ignatov, A. M.; Kolesnikov, Sergey V.; Kulikov, Yu. V.; Monastyrski, M. A.; Niu, Hanben; Schelev, Mikhail Y.)V1358,546-548(1991)

Soviet-American image converter cameras "PROSCHEN" (Briscoe, Dennis; Shrivastava, Chinmaya A.; Nebeker, Sidney J.; Hsu, S.; Lozovoi, V. I.; Postovalov, V. E.; Prokhorov, Alexander M.; Schelev, Mikhail Y.; Serdyuchenko, Yuri N.; Vaschenko, G. O.)V1358,329-336(1991)

Space-qualified streak camera for the Geodynamic Laser Ranging System (Johnson, C. B.; Abshire, James B.; Zagwodzki, Thomas W.; Hunkler, L. T.; Letzring, Samuel A.; Jaanimagi, Paul A.)V1346,340-370(1991)

Stable picosecond solid state YA103:Nd3+ laser for streak cameras dynamic evaluation (Babushkin, A. V.)V1358,888-894(1991)

Streak and smear: a definition of terminology (Haddleton, Graham P.)V1358,18-22(1991)

Sub-100 psec x-ray gating cameras for ICF imaging applications (Kilkenny, Joseph D.; Bell, Perry M.; Hammel, Bruce A.; Hanks, Roy L.; Landen, Otto L.; McEwan, Thomas E.; Montgomery, David S.; Turner, R. E.; Wiedwald, Douglas J.; Bradley, David K.)V1358,117-133(1991)

Subnanosecond high-speed framing of prebreakdown phenomena (Pfeiffer, Wolfgang; Stolz, Dieter; Zipfl, P.)V1358,1191-1201(1991)

Subnanosecond intensifier gating using heavy and mesh cathode underlays (Thomas, Stan W.; Shimkunas, Alex R.; Mauger, Philip E.)V1358,91-99(1991)

Super-high-speed reflex-type moving image camera (Drozhbin, Yu. A.; Trofimenko, Vladimir V.)V1358,454-456(1991)

Systems analysis and design for next generation high-speed video systems (Snyder, Donald R.; Rowe, W. J.)V1346,226-236(1991)

Techniques for capturing over 10,000 images/second with intensified imagers (Balch, Kris S.)V1358,358-372(1991)

Temporal and spectral analysis of the synchronization of synchroscan streak cameras (Cunin, B.; Geist, P.; Heisel, Francine; Martz, A.; Miehe, Joseph A.)V1358,606-613(1991)

Temporal fiducial for picosecond streak cameras in laser fusion experiments (Mens, Alain; Gontier, D.; Giraud, P.; Thebault, J. P.)V1358,878-887(1991)

Theoretical and experimental performance evaluations of Picoframe framing cameras (Liu, Yueping; Sibbett, Wilson; Walker, David R.)V1358,300-310(1991)

Thomson parabolic spectrograph with microchannel plate framing camera as register of ionic parabolae (Lebedev, V. B.; Saulevich, S. V.)V1358,874-877(1991)

Time aberrations of combined focusing system of high-speed image converter (Korzhenevich, Irina M.; Kolesov, G. V.)V1358,1090-1095(1991)

Timing between streak cameras with a precision of 10 ps (Lerche, Richard A.)V1346,376-383(1991)

Two-dimensional electron-bombarded CCD readout device for picosecond electron-optical information system (Ivanov, K. N.; Krutikov, N. I.; Naumov, S. K.; Pischelin, E. V.; Semenov, V. A.; Stepanov, M. S.; Postovalov, V. E.; Prokhorov, Alexander M.; Schelev, Mikhail Y.)V1358,732-738(1991)

Two-frequency picosecond laser with electro-optical feedback (Vorobiev, N. S.; Konoplev, O. A.)V1358,895-901(1991)

Ultrafast imaging of vascular tissue ablation by an XeCl excimer laser (Neu, Walter; Nyga, Ralf; Tischler, Christian; Haase, Karl K.; Karsch, Karl R.)V1425,37-44(1991)

Ultrafast optical-mechanical camera (Drozhbin, Yu. A.; Trofimenko, Vladimir V.; Chernova, T. I.)V1358,451-453(1991)

Ultrafast streak camera evaluations of phase noise from an actively stabilized colliding-pulse-mode-locked ring dye laser (Walker, David R.; Sleat, William E.; Evans, J.; Sibbett, Wilson)V1358,860-867(1991)

Ultrahigh- and High-Speed Photography, Videography, Photonics, and Velocimetry '90 (Jaanimagi, Paul A.; Neyer, Barry T.; Shaw, Larry L., eds.)V1346(1991)

ULTRANAC: a new programmable image converter framing camera (Garfield, Brian R.; Riches, Mark J.)V1358,290-299(1991)

Use of an image converter camera for analysis of ballistic resistance of lightweight armor materials (van Bree, J. L.; van Riet, E. J.)V1358,692-697(1991)

Use of high-speed photography in the evaluation of polymer materials (Dear, John P.)V1358,37-42(1991)

Use of high-speed videography to solve a structural vibration problem in overhead cranes (Clayton, Donal L.; Clayton, Richard J.)V1346,33-41(1991)

Visualization study on pool boiling heat transfer (Kamei, Shuya; Hirata, Masaru)V1358,979-983(1991)

Wide-aperture x-ray image converter tubes (Dashevsky, Boris E.; Podvyaznikov, V. A.; Prokhorov, Alexander M.; Chevokin, Victor K.)V1346,437-442(1991)

Wire explosion at reduced pressures (Lee, Eun Soo; Song, So-Young; Jhung, Kyu Soo; Kim, Ung; Lee, Sang-Soo)V1358,262-268(1991)

Workshop on standards for streak camera characterization (Jaanimagi, Paul A.)V1346,504-504(1991)

Photomasks—see also microlithography

10th Annual Symp on Microlithography (Wiley, James N., ed.)V1496(1991)

5X reticle fabrication using MEBES multiphase virtual address and AZ5206 resist (Milner, Kathy S.; Chipman, Paul S.)V1496,180-196(1991)

Capability assessment and comparison of the Nikon 2i, Nikon 3i, and IMS-2000 registration measurement devices (Henderson, Robert K.)V1496,198-216(1991)

Chemically amplified negative-tone photoresist for sub-half-micron device and mask fabrication (Conley, Willard E.; Dundatscheck, Robert; Gelorme, Jeffrey D.; Horvat, John; Martino, Ronald M.; Murphy, Elizabeth; Petrosky, Anne; Spinillo, Gary; Stewart, Kevin; Wilbarg, Robert; Wood, Robert L.)V1466,53-66(1991)

Defect repair for gold absorber/silicon membrane x-ray masks (Stewart, Diane K.; Fuchs, Jacob; Grant, Robert A.; Plotnik, Irving)V1465,64-77(1991)

E-beam data compaction method for large-capacity mask ROM production (Kanemaru, Toyomi; Nakajima, Takashi; Igarashi, Tadanao; Masuda, Rika; Orita, Nobuyuki)V1496,118-123(1991)

Evaluation of a high-resolution negative-acting electron-beam resist GMC for photomask manufacturing (Chen, Wen-Chih; Novembre, Anthony E.)V1496,266-283(1991)

Focused ion beam induced deposition: a review (Melngailis, John)V1465,36-49(1991)

Improvements in sensitivity and discrimination capability of the PD reticle/mask inspection system (Saito, Juichi; Saijo, Y.)V1496,284-301(1991)

Is phase-shift mask technology production-worthy? (Chen, Mung)V1463,2-5(1991)

Issues in the repair of x-ray masks (Stewart, Diane K.; Doherty, John A.)V1496,247-265(1991)

Markets for marking systems (Austin, Patrick D.)V1517,150-175(1991)

Metamorphosis of laser writer (Wilson, Michael A.)V1496,156-170(1991)

Micro-optic studies using photopolymers (Phillips, Nicholas J.; Barnett, Christopher A.)V1544,10-21(1991)

Modeling phase-shifting masks (Neureuther, Andrew R.)V1496,80-88(1991)

Novel method to fabricate corrugation for distributed-feedback lasers using a grating photomask (Okai, Makoto O.; Harada, Tatsuo)V1545,218-224(1991)

Novel surface imaging masking technique for high-aspect-ratio dry etching applications (Calabrese, Gary C.; Abali, Livingstone N.; Bohland, John F.; Pavelchek, Edward K.; Sricharoenchaikit, Prasit; Vizvary, Gerald; Bobbio, Stephen M.; Smith, Patrick)V1466,528-537(1991)

Optimization of an x-ray mask design for use with horizontal and vertical kinematic mounts (Laird, Daniel L.; Engelstad, Roxann L.; Palmer, Shane R.)V1465,134-144(1991)

Parametric studies and characterization measurements of x-ray lithography mask membranes (Wells, Gregory M.; Chen, Hector T.; Engelstad, Roxann L.; Palmer, Shane R.)V1465,124-133(1991)

Pelliclizing technology (Yamauchi, Takashi)V1496,302-314(1991)

Performance appraisal of the ATEQ CORE-2500 in production (Mechtenberg, Monica L.; Watson, Larry J.)V1496,124-155(1991)

Phase-shifting and other challenges in optical mask technology (Lin, Burn J.)V1496,54-79(1991)

Photomask fabrication utilizing a Philips/Cambridge vector scan e-beam system (McCutchen, William C.)V1496,97-106(1991)

Proximity effect correction on MEBES for 1x mask fabrication: lithography issues and tradeoffs at 0.25 micron (Muray, Andrew; Dean, Robert L.)V1496,171-179(1991)

Recent developments of x-ray lithography in Canada (Chaker, Mohamed; Boily, S.; Ginovker, A.; Jean, A.; Kieffer, J. C.; Mercier, P. P.; Pepin, Henri; Leung, Pak K.; Currie, John F.; Lafontaine, H.)V1465,16-25(1991)

Simulation of low-energy x-ray lithography using a diamond membrane mask (Hasegawa, Shinya; Suzuki, Katsumi)V1465,145-151(1991)

Sophisticated masks (Pease, R. F.; Owen, Geraint; Browning, Raymond; Hsieh, Robert L.; Lee, Julienne Y.; Maluf, Nadim I.; Berglund, C. N.)V1496,234-238(1991)

Three-dimensional contour of crack tips using a grating method (Andresen, Klaus; Kamp, B.; Ritter, Reinold)V1554A,93-100(1991)

What IS a phase-shifting mask? (Levenson, Marc D.)V1496,20-26(1991)

Photomechanics—see also metrology; nondestructive testing

Accumulating displacement fields from different steps in laser or white-light speckle methods (Shao, C. A.; King, H. J.; Wang, Yeong-Kang; Chiang, Fu-Pen)V1554A,613-618(1991)

Actual light deflections in regions of crack tips and their influence on measurements in photomechanics (Hecker, Friedrich W.; Pindera, Jerzy T.; Wen, Baicheng)V1554A,151-162(1991)

All-optical data-input device based on fiber optic interferometric strain gauges (Fuerstenau, Norbert; Schmidt, Walter)V1367,357-366(1991)

Analysis of diffracted stress fields around a noncharged borehole with dynamic photoelasticity and gauges (Zhu, Zhenhai; Qu, Guangjian; Yang, Yongqi; Shang, Jian)V1554A,472-481(1991)

Analysis of macro-model composites with Fabry-Perot fiber-optic sensors (Fogg, Brian R.; Miller, William V.; Lesko, John J.; Carman, Gregory P.; Vengsarkar, Ashish M.; Reifsnider, Kenneth L.; Claus, Richard O.)V1588,14-25(1991)

Analysis of optical measurements of SIF and singularity order in rocket motor geometry (Chang, Che-way)V1554A,250-261(1991)

Application of a dynamic photoelasticity technique for the study of two- and three-dimensional structures under seismic, shock, and explosive loads effect (Kostin, Ivan K.; Dvalishvili, V. V.; Ureneva, E. V.; Freishist, N. A.; Fjodorov, A. V.; Dmitriyenko, O. L.)V1554A,418-425(1991)

Application of electro-optic modulator in photomechanics (Zhang, Yuan P.)V1554B,669-678(1991)

Application of high-speed photography to the study of high-strain-rate materials testing (Ruiz, D.; Harding, John; Noble, J. P.; Hillsdon, G. K.)V1358,1134-1143(1991)

Application of photoelastic coating method on elastoplastic stress analysis of rotation disk (Dong, Benhan; Gao, Penfei; Wang, Ju)V1554A,400-406(1991)

Approach for applying holographic interferometry to large deformations and modifications (Schumann, Walter)V1507,526-537(1991)

Automated fringe analysis for moire interferometry (Brownell, John B.; Parker, Richard J.)V1554B,481-492(1991)

Automatization of measurement and processing of experimental data in photoelasticity (Zhavoronok, I. V.; Nemchinov, V. V.; Litvin, S. A.; Skanavi, A. M.; Pavlov, V. V.; Evsenev, V. S.)V1554A,371-379(1991)

Carrier pattern analysis of moire interferometry using Fourier transform (Morimoto, Yoshiharu; Post, Daniel; Gascoigne, Harold E.)V1554B,493-502(1991)

Composite damage assessment employing an optical neural network processor and an embedded fiber-optic sensor array (Grossman, Barry G.; Gao, Xing; Thursby, Michael H.)V1588,64-75(1991)

Computer-aided moire strain analysis on thermoplastic models at critical temperature (Barillot, Marc; Jacquot, Pierre M.; Di Chirico, Giuseppe)V1554A,867-878(1991)

Computerized vibration analysis of hot objects (Lokberg, Ole J.; Rosvold, Geir O.; Malmo, Jan T.; Ellingsrud, Svein)V1508,153-160(1991)

Contouring using gratings created on a LCD panel (Asundi, Anand K.; Wong, C. M.)V1400,80-85(1991)

Cooperative implementation of a high-temperature acoustic sensor (Baldini, S. E.; Nowakowski, Edward; Smith, Herb G.; Friebele, E. J.; Putnam, Martin A.; Rogowski, Robert S.; Melvin, Leland D.; Claus, Richard O.; Tran, Tuan A.; Holben, Milford S.)V1588,125-131(1991)

Curvature of bending shells by moire interferometry (Wang, Lingli; Yun, Dazhen)V1554B,436-440(1991)

Damage detection in woven-composite materials using embedded fiber-optic sensors (Bonniau, Philippe; Chazelas, Jean; Lecuellet, Jerome; Gendre, Francois; Turpin, Marc; Le Pesant, Jean-Pierre; Brevignon, Michele)V1588,52-63(1991)

Damage to polymer-coated glass surfaces by small-particle impact (Chaudhri, Mohammad M.; Smith, Alan L.)V1358,683-689(1991)

Demodulation of a fiber Fabry-Perot strain sensor using white light interferometry (Zuliani, Gary; Hogg, W. D.; Liu, Kexing; Measures, Raymond M.)V1588,308-313(1991)

Determination of burst initiation location and tear propagation velocity during air burst testing of latex condoms (Davidhazy, Andrew)V1358,654-659(1991)

Determination of the adhesive load by holographic interferometry using the results of FEM calculations (Bischof, Thomas; Jueptner, Werner P.)V1508,90-95(1991)

Determination of thermal stresses in a bimaterial specimen by moire interferometry (Kang, Yilan; Jia, Youquan; Du, Ji)V1554B,514-522(1991)

Development of a fiber Fabry-Perot strain gauge (Hogg, W. D.; Janzen, Doug; Valis, Tomas; Measures, Raymond M.)V1588,300-307(1991)

Digital speckle correlation search method and its application (Zhou, Xingeng; Gao, Jianxing)V1554A,886-895(1991)

Displacements and strains in thick-walled composite rings subjected to external pressure using moire interferometry (Gascoigne, Harold E.; Abdallah, Mohamed G.)V1554B,315-322(1991)

Distributed-parameter estimation for NASA Mini-Mast truss through displacement measurements (Huang, Jen-Kuang; Shen, Ji-Yao; Taylor, Lawrence W.)V1489,266-277(1991)

DSPI: a tool for analyzing thermal strain on ceramic and composite materials (Hoefling, Roland; Aswendt, Petra; Totzauer, Werner F.; Jueptner, Werner P.)V1508,135-142(1991)

Dynamic caustic method is applied to fracture (Chen, Zengtao; Guo, Maolin; Wang, Duo; Bi, Xianzhi)V1554A,206-208(1991)

Dynamic photoelasticity applied to crack-branching investigations (Hammami, Slimane; Cottron, M.; Lagarde, Alexis)V1554A,136-142(1991)

Dynamic photoelasticity with a split Hopkinson pressure bar (Morris, David R.; Watson, A. J.)V1358,254-261(1991)

Dynamic photoelastic study on mechanism of short-delay blasting (Yang, Renshu; Yang, Yongqi)V1554A,341-348(1991)

Dynamic two-beam speckle interferometry (Ahmadshahi, Mansour A.; Krishnaswamy, Sridhar; Nemat-Nasser, Siavouche)V1554A,620-627(1991)

Elasto-plastic contact between rollers (Chu, Kunliang; Li, Penghui)V1554A,192-195(1991)

Electronic speckle pattern interferometry for 3-D dynamic deformation analysis in industrial environments (Guelker, Gerd; Haack, Olaf; Hinsch, Klaus D.; Hoelscher, Claudia; Kuls, Juergen; Platen, Winfried)V1500,124-134(1991)

Estimation of plastic strain by fractal (Dai, YuZhong; Chiang, Fu-Pen)V1332,767-774(1991)

Experimental research on optical fibre microbending sensors in on-line measuring of deep-hole drilling bit wear (Yang, Zhiguo; Zhong, Hengyong; Cheng, Jubing; Wang, Youguan)V1572,252-257(1991)

Experimental simulation analysis of nonlinear problem: investigation into the mechanical and optical behavior of silver chloride of photoplastic material (Yin, Zhi Xiang; Zhang, Shikun; Li, Zong Yan)V1559,487-496(1991)

Fabry-Perot fiber-optic sensors in full-scale fatigue testing on an F-15 aircraft (Murphy, Kent A.; Gunther, Michael F.; Vengsarkar, Ashish M.; Claus, Richard O.)V1588,134-142(1991)

Fiber optic multiple sensor for simultaneous measurements of temperature and vibrations (Brenci, Massimo; Mencaglia, Andrea; Mignani, Anna G.; Barbero, V.; Cimbrico, P. L.; Pessino, P.)V1572,318-324(1991)

Fiber optic sensor considerations and developments for smart structures (Measures, Raymond M.)V1588,282-299(1991)

Fiber optic sensor for the study of temperature and structural integrity of PZT: epoxy composite materials (Vishnoi, Gargi; Pillai, P.K. C.; Goel, T. C.)V1572,94-100(1991)

Fiber Optic Smart Structures and Skins IV (Claus, Richard O.; Udd, Eric, eds.)V1588(1991)

Fiber optic technique for simultaneous measurement of strain and temperature variations in composite materials (Michie, W. C.; Culshaw, Brian; Roberts, Scott S.; Davidson, Roger)V1588,342-355(1991)

Four cases of engineering vibration studies using pulsed ESPI (Mendoza Santoyo, Fernando; Shellabear, Michael C.; Tyrer, John R.)V1508,143-152(1991)

Fringe formation in speckle shearing interferometry (Xu, Boqin; Wu, Xiao-Ping)V1554A,789-799(1991)

Geometrical moire method for displaying directly the stress field of a circular disk (Wang, Ji-Zhong; Wang, Yun-Shan; Yin, Yuan-Cheng; Wu, Rui-Lan)V1554B,188-192(1991)

Higher-order theory of caustics for fast running cracks under general loadings (Nishioka, Teiichi; Murakami, R.; Matsuo, Seitaro; Ohishi, Y.)V1554A,802-813(1991)

High-sensitivity interferometric technique for strain measurements (Voloshin, Arkady S.; Bastawros, Adel F.)V1400,50-60(1991)

High-speed holographic interferometric study of the propagation of the electrohydraulic shock wave (Jia, Youquan; Sun, Yongda)V1554B,135-138(1991)

High-speed photographic study of impact on fibers and woven fabrics (Field, John E.; Sun, Q.)V1358,703-712(1991)

High-speed photography of high-resolution moire patterns (Whitworth, Martin B.; Huntley, Jonathan M.; Field, John E.)V1358,677-682(1991)

High-temperature optical sensor for displacement measurement (Ebbeni, Jean P.)V1504,268-272(1991)

Holographic measurement of deformation using carrier fringe and FFT techniques (Quan, C.; Judge, Thomas R.; Bryanston-Cross, Peter J.)V1507,463-475(1991)

Holographic photoelasticity applied to ceramics fracture (Guo, Maolin; Wang, Yigong; Zhao, Yin; Du, Shanyi)V1554A,310-312(1991)

Holography as a tool for widespread industrial applications: analysis and comments (Smigielski, Paul)V1508,38-49(1991)

Improvement of the sensitivity of dynamic white-light speckle method and its application (Fang, Ruhua; Cao, Zhengyuan)V1554A,649-656(1991)

Improvement on phase-shifting method precision and application on shadow moire method (Mauvoisin, Gerard; Bremand, Fabrice J.; Lagarde, Alexis)V1554B,181-187(1991)

Industrial Applications of Holographic and Speckle Measuring Techniques (Jueptner, Werner P., ed.)V1508(1991)

Industrial applications of self-diffraction phenomena in holography on photorefractive crystals (Dovgalenko, George Y.; Yeskov-Soskovetz, Vladimir M.; Cherkashin, G. V.)V1508,110-115(1991)

Influence of the orthotropy of the ductile materials and the stress-assisted diffusion on the caustics (Papadopoulos, George A.)V1554A,826-834(1991)

Information extracting and application for the combining objective speckle and reflection holography (Cao, Zhengyuan; Cheng, Fang)V1332,358-364(1991)

In-plane displacement measurement on rotating components using pulsed laser ESPI (electronic speckle pattern interferometry) (Preater, Richard W.)V1399,164-171(1991)

Integrated photoelasticity for residual stresses in glass specimens of complicated shape (Aben, Hillar K.; Idnurm, S. J.; Josepson, J. I.; Kell, K.-J. E.; Puro, A. E.)V1554A,298-309(1991)

Intelligent composites containing measuring fiber-optic networks for continuous self-diagnosis (Sansonetti, Pierre; Lequime, Michael; Engrand, D.; Guerin, J. J.; Davidson, Roger; Roberts, Scott S.; Fornari, B.; Martinelli, Mario; Escobar Rojo, Priscilla; Gusmeroli, Valeria; Ferdinand, Pierre; Plantey, J.; Crowther, Margaret F.; Culshaw, Brian; Michie, W. C.)V1588,198-209(1991)

Interaction of cracks, waves, and contacts: a photomechanics study (Rossmanith, H. P.)V1554A,2-28(1991); V1554B,2-28(1991)

Interface cracks close to free surfaces: a caustic study (Rossmanith, H. P.; Beer, Rudolf J.; Knasmillner, R. E.)V1554A,850-860(1991)

Interferometric signal processing schemes for the measurement of strain (Berkoff, Timothy A.; Kersey, Alan D.)V1588,169-176(1991)

Interlaminar deformations on the cylindrical surface of a hole in laminated composites by moire interferometry (Boeman, Raymond G.)V1554B,323-330(1991)

Interpretation of optical caustic patterns obtained during unsteady crack growth: an analysis based on a higher-order transient expansion (Liu, Cheng; Rosakis, Ares J.)V1554A,814-825(1991)

Intersection of crack borders with free surfaces: an engineering interpretation of optical experimental results (Smith, C. W.; Constantinescu, D. M.)V1554A,102-115(1991)

Investigation of fiber-reinforced-plastics-based components by means of holographic interferometry (Jueptner, Werner P.; Bischof, Thomas)V1400,69-79(1991)

Investigation of optical fibers as sensors for condition monitoring of composite materials (Nielsen, Peter L.)V1588,229-240(1991)

Laser metrological measurement of transient strain fields in Hopkinson-bar experiments (Vogel, Dietmar; Michel, Bernd; Totzauer, Werner F.; Schreppel, Ulrich; Clos, Rainer)V1554A,262-274(1991)

Lasers applied to photoelastic stress measurements (Lukasiewicz, Stan)V1554A,349-358(1991)

Laser-stimulated thermoelastic stress wave inside a solid medium and measurement of its initial speed (He, Anzhi; Ni, Xiao W.; Lu, Jian-Feng; Wang, Chang Xing; Wang, Hong Y.)V1554A,547-552(1991)

Measurement of residual stress during implant resistance welding of plastics (Park, Joon B.; Benatar, Avraham)V1554B,357-370(1991)

Measurement of small strain of a solid body by two-frequency laser optical fiber sensor (Fang, Yin; Sheng, Kemin)V1572,453-456(1991)

Measurement of strains by means of electro-optics holography (Sciammarella, Cesar A.; Bhat, Gopalakrishna K.; Albertazzi, Armando A.)V1396,143-154(1991)

Mechanical properties of composite materials containing embedded fiber-optic sensors (Roberts, Scott S.; Davidson, Roger)V1588,326-341(1991)

Method for evaluating displacement of objects using the Wigner distribution function (Widjaja, Joewono; Uozumi, Jun; Asakura, Toshimitsu)V1400,94-100(1991)

Method of caustics for anisotropic materials (Rossmanith, H. P.)V1554A,835-849(1991)

Microinteraction of optical fibers embedded in laminated composites (Singh, Hemant; Sirkis, James S.; Dasgupta, Abhijit)V1588,76-85(1991)

Model of an axially strained weakly guiding optical fiber modal pattern (Egalon, Claudio O.; Rogowski, Robert S.)V1588,241-254(1991)

Modern optical method for strain separation in photoplasticity (Zhang, Zuxun)V1554A,482-487(1991)

Multimode fiber-optic Mach-Zehnder interferometric strain sensor (Wang, An; Xie, Haiming)V1572,444-449(1991)

Multipulse polarized-holographic set-up for isodromic fringe registration in the field of running stress waves (Khesin, G. L.; Sakharov, V. N.; Zhavoronok, I. V.; Rottenkolber, Hans; Schorner, Jurgen)V1554B,86-90(1991)

New holographic system for measuring vibration (Cai, Yunliang)V1554B,75-80(1991)

New technique for multiplying the isoclinic fringes (Wen, Mei-Yuan; Liu, Guang T.)V1332,673-675(1991)

New technique for rapid making specimen: gratings for moire interferometry (Jia, Youquan; Kang, Yilan; Du, Ji)V1554A,331-334(1991)

Optical coatings to reduce temperature sensitivity of polarization-maintaining fibers for smart structures and skins (Zhang, Feng; Lit, John W.)V1588,100-109(1991)

Optical fibre interferometer for monitoring tool wear (Zheng, S. X.; McBride, R.; Hale, K. F.; Jones, Barry E.; Barton, James S.; Jones, Julian D.)V1572,359-364(1991)

Optical heterodyne fiber-coil deformation sensor operating in a wide dynamic range (Ohtsuka, Yoshihiro; Nishi, Y.; Sawae, S.; Tanaka, Satoshi)V1572,347-352(1991)

Optical isodyne measurements in fracture mechanics (Pindera, Jerzy T.; Wen, Baicheng)V1554A,196-205(1991)

Optical sensors embedded in composite materials (Bocquet, Jean-Claud; Lecoy, Pierre; Baptiste, Didier)V1588,210-217(1991)

PC-based determination of 3-D deformation using holographic interferometry (Schad, Hanspeter; Schweizer, Edwin; Jodlbauer, Heibert)V1507,446-449(1991)

Phase-strain-temperature model for structurally embedded interferometric optical fiber strain sensors with applications (Sirkis, James S.)V1588,26-43(1991)

Photoelastical mixed-solution method of contact problems of the roll-shape member of limited length (Qu, Zhihao; Jiang, Weixing; Peng, Huihong)V1554A,503-510(1991)

Photoelastic experimental research on the Wan-An lock by oblique incidence method (Wang, Feng; Dai, Fu-long)V1554A,359-370(1991)

Photoelastic investigation of statics, kinetics, and dynamics of crack formation in transparent models and natural structural elements (Taratorin, B. I.; Sakharov, V. N.; Komlev, O. U.; Stcherbakov, V. N.; Starchevsky, A. V.)V1554A,449-456(1991)

Photoelastic investigation of stress waves using models of viscoelastic materials (Dmokhovskij, A. V.; Filippov, I. G.; Skropkin, S. A.; Kobakhidze, T. G.)V1554A,323-330(1991)

Photoelasticity, holography, and moire for strain-stress-state studies of NPS, HPS underground and machinery structures: comparative analysis of physical modeling and numerical methods (Khesin, G. L.)V1554A,313-322(1991)

Photoelastic modeling of linear and nonlinear creep using two- and three-dimensional models (Vardanjan, G. S.; Musatov, L. G.)V1554A,496-502(1991)

Photoelastic sensors for automatic control system of dam safety (Konwerska-Hrabowska, Joanna; Kryszczynski, Tadeusz; Tomaszewicz, Tomasz; Lietz, J.; Mazurkiewicz, Wojciech)V1554A,388-399(1991)

Photoelastic stress analysis of bridge bearings (Allison, Ian M.)V1554A,332-340(1991)

Photoelastic stress investigation in underground large hole in permafrost soil (statics, thermoelasticity, dynamics, photoelastic strain-gauges) (Savostjanov, V. N.; Dvalishvili, V. V.; Sakharov, V. N.; Isajkin, A. S.; Frishter, L.; Starchevsky, A. V.)V1554A,380-386(1991)

Photoelastic study of friction at multipoint contacts (Dally, James W.; Chen, Yung-Mien)V1554A,434-443(1991)

Photosensitivity of selenium-bismuth films with varigap structure (Popov, A.; Mikhalev, N.; Karalyunts, A.; Smirnov, O.; Vasilyeva, N.)V1519,457-462(1991)

Picosecond techniques application for definition of nonuniform ingradients inside turbid medium (Vorobiev, N. S.; Serafimovich, O. A.; Smirnov, A. V.)V1358,698-702(1991)

Practical method for automatic detection of brightest ridge line of photomechanical fringe pattern (Ding, Zu-Quan; Yuan, Xun-Hua)V1554A,898-906(1991)

Projection moire deflectometry for mapping phase objects (Wang, Ming; Ma, Li; Zeng, Jing-gen; Cheng, Qi-Xian; Pan, Chuan K.)V1554B,242-246(1991)

Pulsed holographic and speckle interferometry using Hopkinson loading technique to investigate the dynamical deformation on plates (Han, Lei; Wu, Xiao-Ping; Hu, Shisheng)V1358,793-803(1991)

Range of measurement of computer-aided speckle interferometry (Chen, Duanjun; Li, Shen; Hsu, T. Y.; Chiang, Fu-Pen)V1554A,922-931(1991)

Rapid procedure for obtaining time-average interferograms of vibrating bodies (Rapoport, Eliezer; Bar, Doron; Shiloh, Klara)V1442,383-391(1991)

Real-time optical spatial filtering system with white-light source for displaying color moire (Fang, Cui-Chang; Liao, Yan-Biao; Ma, De-Yuan)V1554A,915-921(1991)

Real-time structural integrity monitoring using a passive quadrature demodulated, localized Michelson optical fiber interferometer capable of simultaneous strain and acoustic emission sensing (Tapanes, Edward)V1588,356-367(1991)

Recent development in practical optical nondestructive testing (Hung, Yau Y.)V1554B,29-45(1991)

Reconstruction of three-dimensional displacement fields by carrier holography (Xu, Zhu; Chen, Ke-long; Wen, Zhen-chu; Yang, Han-Guo; He, Xiao-yuan; Zhang, Bao-he)V1399,172-177(1991)

Research for mechanical properties of rock and clay by laser speckle photography (Wu, Ruiqi)V1554A,690-695(1991)

Research on macroeffects and micromechanism of martensite phase transition of shape memory alloys by high-speed photography (Yang, Jie; Wu, Yuehua; Zhou, Yusheng; Uyemura, T.)V1358,672-676(1991)

Role of adhesion in optical-fiber-based smart composite structures and its implementation in strain analysis for the modeling of an embedded optical fiber (DiFrancia, Celene; Claus, Richard O.; Ward, T. C.)V1588,44-49(1991)

Sapphire fiber interferometer for microdisplacement measurements at high temperatures (Murphy, Kent A.; Fogg, Brian R.; Wang, George Z.; Vengsarkar, Ashish M.; Claus, Richard O.)V1588,117-124(1991)

Scattered-light optical isodynes: basis for 3-D isodyne stress analysis (Pindera, Jerzy T.)V1554A,458-471(1991)

Second Intl Conf on Photomechanics and Speckle Metrology: Speckle Techniques, Birefringence Methods, and Applications to Solid Mechanics (Chiang, Fu-Pen, ed.)V1554A(1991)

Smart civil structures: an overview (Huston, Dryver R.)V1588,182-188(1991)

Solution of optimization problems of perforated and box-shaped structures by photoelasticity and numerical methods (Shvej, E. M.; Latysh, E. G.; Morgunov, A. N.; Mikhalchenko, O. E.; Stchetinin, A. L.; Volokh, K. Y.)V1554A,488-495(1991)

Spatial phase-shifting techniques of fringe pattern analysis in photomechanics (Kujawinska, Malgorzata; Wojiak, Joanna)V1554B,503-513(1991)

Statistical analysis of white-light speckle photography (Fang, Qiang; Tan, Yushan)V1554A,696-704(1991)

Strain measurements on soft materials application to cloth and papers (Bremand, Fabrice J.; Lagarde, Alexis)V1554B,650-660(1991)

Strain sensing using a fiber-optic Bragg grating (Melle, Serge M.; Liu, Kexing; Measures, Raymond M.)V1588,255-263(1991)

Stress analysis in patella by three-dimensional photoelasticity (Chen, Riqi; Zhang, Jianxing; Jiang, Kunsheng)V1554A,407-417(1991)

Stress-static and strength research of building of NPS and nuclear reactor under power and thermal loads including creep and relaxation influence (Savostjanov, V. N.; Zavalishin, S. I.; Smirnov, S. B.; Morosova, D. V.)V1554A,579-585(1991)

Study of displacement and residual displacement field of an interface crack by moire interferometry (Wang, Y. Y.; Chiang, Fu-Pen; Barsoum, Roshdy S.; Chou, S. T.)V1554B,344-356(1991)

Study of electronic shearing speckle technique (Qin, Yuwen; Wang, Jinqi; Ji, Xinhua)V1554A,739-746(1991)

Study of plate vibrations by moire holography (Blanco, J.; Fernandez, J. L.; Doval, A. F.; Lopez, C.; Pino, F.; Perez-Amor, Mariano)V1508,180-190(1991)

Study of stick slip behavior in interface friction using optical fiber pull-out experiment (Tsai, Kun-Hsieh; Kim, Kyung-Suk)V1554A,529-541(1991)

Study of the fracture process using laser holographic interferometry and image analysis (Castro-Montero, Alberto; Shah, S. P.; Bjelkhagen, Hans I.)V1396,122-130(1991)

Study of welding residual stress by means of laser holography interference (Yang, Jiahua)V1554B,155-160(1991)

Thermal plastic metal coatings on optical fiber sensors (Sirkis, James S.; Dasgupta, Abhijit)V1588,88-99(1991)

Thermal strain measurements of solder joints in electronic packaging using moire interferometry (Woychik, Charls G.; Guo, Yi F.)V1554A,461-470(1991)

Two-dimensional surface strain measurement based on a variation of Yamaguchi's laser-speckle strain gauge (Barranger, John P.)V1332,757-766(1991)

Ultrasonic NDE (nondestructive evaluation) for composite materials using embedded fiber optic interferometric sensors (Liu, Kexing; Ferguson, Suzanne M.; Davis, Andrew; McEwen, Keith; Measures, Raymond M.)V1398,206-212(1991)

Use of an image converter camera for analysis of ballistic resistance of lightweight armor materials (van Bree, J. L.; van Riet, E. J.)V1358,692-697(1991)

Use of high-resolution TV camera in photomechanics (Yatagai, Toyohiko; Ino, Tomomi)V1554B,646-649(1991)

Use of high-speed photography in the evaluation of polymer materials (Dear, John P.)V1358,37-42(1991)

Use of scattered-light photoelasticity at crack tips (Ravi-Chandar, K.)V1554A,228-238(1991)

Vibration-insensitive moire interferometry system for off-table applications (Guo, Yi F.)V1554B,412-419(1991)

Vibration sensing in flexible structures using a distributed-effect modal domain optical fiber sensor (Reichard, Karl M.; Lindner, Douglas K.; Claus, Richard O.)V1489,218-229(1991)

Void-crack interaction in aluminum single crystal (Li, X. M.; Chiang, Fu-Pen)V1554A,285-296(1991)

White-light speckle applied to composites (Guo, Maolin; Zhao, Yin; Wu, Zhongming; Du, Shanyi)V1554A,657-658(1991)

Whole-field stress fringe compensation using photoelastic carrier shifting and optical information processing (Zhang, Xi; Wang, Baishi; Li, Yao W.)V1554A,444-448(1991)

Wire explosion at reduced pressures (Lee, Eun Soo; Song, So-Young; Jhung, Kyu Soo; Kim, Ung; Lee, Sang-Soo)V1358,262-268(1991)

Photometry—see also radiometry

Absolute measurements of radiation sources spectral brightness and detectors quantum efficiency (Penin, A. N.; Klyshko, D. N.)V1562,143-148(1991)

Caries selective ablation by pulsed lasers (Hennig, Thomas; Rechmann, Peter; Pilgrim, C.; Schwarzmaier, Hans-Joachim; Kaufmann, Raimund)V1424,99-105(1991)

Design and analysis of illumination systems (Zochling, Gunter)V1354,617-626(1991)

Fiber optic based chemical sensor system for in-situ process measurements using the photothermal effect (Walker, Karl-Heinz; Sontag, Heinz)V1510,212-217(1991)

High-throughput narrowband 83.4-nm self-filtering camera (Zukic, Muamer; Torr, Douglas G.; Torr, Marsha R.)V1549,234-244(1991)

Initial performance of the high-speed photometer (Richards, Evan; Percival, Jeff; Nelson, Matthew; Hatter, Edward; Fitch, John E.; White, Richard L.)V1494,40-48(1991)

Mobile system for measuring retroreflectance of traffic signs (Lumia, John J.)V1385,15-26(1991)

Monolithic integrated-circuit charge amplifier and comparator for MAMA readout (Cole, Edward H.; Smeins, Larry G.)V1549,46-51(1991)

Near-infrared time-resolved spectroscopy and fast scanning spectrophotometry in ischemic human forearm (Ferrari, Marco; De Blasi, Roberto A.; Bruscaglioni, Piero; Barilli, Marco; Carraresi, Luca; Gurioli, G. M.; Quaglia, Enrico; Zaccanti, Giovanni)V1431,276-283(1991)

Noninvasive hemoglobin oxygenation monitor and computed tomography by NIR spectrophotometry (Oda, Ichiro; Ito, Yasunobu; Eda, Hideo; Tamura, Tomomi; Takada, Michinosuke; Abumi, Rentaro; Nagai, Katumi; Nakagawa, Hachiro; Tamura, Masahide)V1431,284-293(1991)

Optical quantification of sodium, potassium, and calcium ions in diluted human plasma based on ion-selective liquid membranes (Spichiger, Ursula E.; Seiler, Kurt; Wang, Kemin; Suter, Gaby; Morf, Werner E.; Simon, Wilhelm)V1510,118-130(1991)

Performance of low-resistance microchannel-plate stacks (Siegmund, Oswald H.; Stock, Joseph)V1549,81-89(1991)

Photometric models in multispectral machine vision (Brill, Michael H.)V1453,369-380(1991)

Quantitative analysis of hemoglobin oxygenation state of rat head by time-resolved photometry using picosecond laser pulse at 1064 nm (Nomura, Yasutomo; Tamura, Mamoru)V1431,102-109(1991)

Temperature-monitored/controlled silicon photodiodes for standardization (Eppeldauer, George)V1479,71-77(1991)

Photorefraction—see also crystals; refraction; spatial light modulators

Amplification of amplitude modulated signals in a self-pumped photorefractive phase conjugator (Petersen, Paul M.; Buchhave, Preben)V1362,582-585(1991)

Analysis of Bragg diffraction in optical memories and optical correlators (Gheen, Gregory)V1564,135-142(1991)

Characterization of the photorefractive effect in Ti:LiNbO3 stripe waveguides (Volk, Raimund; Sohler, Wolfgang)V1362,820-826(1991)

Coherent coupling of lasers using a photorefractive ring oscillator (Luo, Jhy-Ming)V1409,100-105(1991)

Compact optical associative memory (Burns, Thomas J.; Rogers, Steven K.; Kabrisky, Matthew; Vogel, George A.)V1469,208-218(1991)

Comparison of photorefractive effects and photogenerated components in polarization-maintaining fibers (Kanellopoulos, S. E.; Guedes Valente, Luiz C.; Handerek, Vincent A.; Rogers, Alan J.)V1516,200-210(1991)

Computer-aided photorefractive pattern recognition (Sun, Ching-Cherng; Chang, Ming-Wen; Yeh, Smile; Cheng, Nai-Jen)V1564,199-210(1991)

Detection of phase objects in transparent liquids using nonlinear coupling in BaTiO3 crystal (Siahmakoun, Azad; Shen, Xuanguo)V1396,535-538(1991)

High-sensitivity photorefractivity in bulk and multiple-quantum-well semiconductors (Partovi, Afshin; Glass, Alastair M.; Feldman, Robert D.)V1561,20-32(1991)

Holographic devices using photo-induced effect in nondestructive testing techniques (Dovgalenko, George Y.; Onischenko, Yuri I.)V1559,479-486(1991)

Holographic optical switching with photorefractive crystals (Song, Q. W.; Lee, Mowchen C.; Talbot, Peter J.)V1558,143-148(1991)

Holographic recording in photorefractive media containing bacteriorhodopsin (Kozhevnikov, Nikolai M.; Barmenkov, Yuri O.; Lipovskaya, Margarita J.)V1507,517-524(1991)

Industrial applications of self-diffraction phenomena in holography on photorefractive crystals (Dovgalenko, George Y.; Yeskov-Soskovetz, Vladimir M.; Cherkashin, G. V.)V1508,110-115(1991)

Instant measurement of phase-characteristic curve (Sadovnik, Lev S.; Rich, Chris C.)V1559,424-428(1991)

Investigation on phase biological micro-objects with a holographic interferometric microscope on the basis of the photorefractive Bi12TiO20 crystal (Tontchev, Dimitar A.; Zhivkova, Svetla; Miteva, Margarita G.; Grigoriev, Ivo D.; Ivanov, I.)V1429,76-80(1991)

Light-induced volume-phase holograms for cold neutrons (Ibel, Konrad; Matull, Ralph; Rupp, Romano A.; Eschkoetter, Peter; Hehmann, Joerg)V1559,393-402(1991)

Nonlinear energy transfer between nanosecond pulses in iron-doped InP crystals (Roosen, Gerald)V1362,398-416(1991)

Nonlinear optical properties of N-(4-nitro-2-pyridinyl)-phenylalaninol single crystals (Sutter, Kurt; Hulliger, J.; Knoepfle, G.; Saupper, N.; Guenter, Peter)V1560,296-301(1991)

Novel technique for efficient wave mixing in photorefractive materials (Mathey, P.; Launay, Jean C.; Pauliat, G.; Roosen, Gerald)V1500,26-33(1991)

Optical neural networks based on liquid-crystal light valves and photorefractive crystals (Owechko, Yuri; Soffer, Bernard H.)V1455,136-144(1991)

Optical processing using photorefractive GaAs and InP (Liu, Duncan T.; Cheng, Li-Jen; Luke, Keung L.)V1409,116-126(1991)

Optical testing by dynamic holographic interferometry with photorefractive crystals and computer image processing (Vlad, Ionel V.; Popa, Dragos; Petrov, M. P.; Kamshilin, Alexei A.)V1332,236-244(1991)

Photoanisotropic incoherent-to-coherent conversion using five-wave mixing (Huang, Tizhi; Wagner, Kelvin H.)V1562,44-54(1991)

Photoconducting nonlinear optical polymers (Li, Lian; Jeng, Ru J.; Lee, J. Y.; Kumar, Jayant; Tripathy, Sukant K.)V1560,243-250(1991)

Photopolymers for optical devices in the USSR (Mikaelian, Andrei L.; Barachevsky, Valery A.)V1559,246-257(1991)

Photorefractive devices for optical information processing (Yeh, Pochi A.; Gu, Claire; Hong, John H.)V1562,32-43(1991)

Photorefractive gratings in the organic crystal 2-cyclooctylamino-5-nitropyridine doped with 7,7,8,8-tetracyanoquinodimethane (Sutter, Kurt; Hulliger, J.; Guenter, Peter)V1560,290-295(1991)

Photorefractive spatial light modulators and their applications to optical computing (Zhao, Mingjun; Li, Yulin; Qin, Yuwen; Wang, Zhao)V1558,529-534(1991)

Photorefractive two-wave mixing characteristics for image amplification in diffusion-driven media (Gilbreath, G. C.; Clement, Anne E.; Fugera, S. N.; Mizell, Gregory J.)V1409,87-99(1991)

Photorefractivity in doped nonlinear organic polymers (Moerner, William E.; Walsh, C. P.; Scott, J. C.; Ducharme, Stephen P.; Burland, Donald M.; Bjorklund, Gary C.; Twieg, Robert J.)V1560,278-289(1991)

Position-, scale-, and rotation-invariant photorefractive correlator (Ryan, Vincent; Fielding, Kenneth H.)V1564,86-97(1991)

Real-time speckle photography using photorefractive Bi12SiO20 crystal (Nakagawa, Kiyoshi; Minemoto, Takumi)V1508,191-200(1991)

Resonant behavior of the temporal response of the photorefractive InP:Fe under dc fields (Abdelghani-Idrissi, Ahmed M.; Ozkul, Cafer; Wolffer, Nicole; Gravey, Philippe; Picoli, Gilbert)V1362,417-427(1991)

Self-diffraction for active stabilization of interference field for reflection hologram recording (Kalyashov, E. V.; Tyutchev, M. V.)V1238,442-446(1991)

Sol-gel processed novel multicomponent inorganic oxide: organic polymer composites for nonlinear optics (Zhang, Yue; Cui, Y. P.; Wung, C. J.; Prasad, Paras N.; Burzynski, Ryszard)V1560,264-271(1991)

Staring phased-array radar using photorefractive crystals (Weverka, Robert T.; Wagner, Kelvin H.)V1564,676-684(1991)

Study of LiNbO3 in optical associative memory (Yuan, Xiao L.; Sayeh, Mohammad R.)V1396,178-187(1991)

Tunable holographic interferometer using photorefractive crystal (Yang, Guanglu; Siahmakoun, Azad)V1396,552-556(1991)

Photoresists—see also microlithography

10th Annual Symp on Microlithography (Wiley, James N., ed.)V1496(1991)

Acid-catalyzed pinacol rearrangement: chemically amplified reverse polarity change (Sooriyakumaran, Ratna; Ito, Hiroshi; Mash, Eugene A.)V1466,419-428(1991)

Advanced lithographic methods for contact patterning on severe topography (Chu, Ron; Greeneich, James S.; Katz, Barton A.; Lin, Hwang-Kuen; Huang, Dong-Tsair)V1465,238-243(1991)

Advances in Resist Technology and Processing VIII (Ito, Hiroshi, ed.)V1466(1991)

Airborne chemical contamination of a chemically amplified resist (MacDonald, Scott A.; Clecak, Nicholas J.; Wendt, H. R.; Willson, C. G.; Snyder, C. D.; Knors, C. J.; Deyoe, N. B.; Maltabes, John G.; Morrow, James R.; McGuire, Anne E.; Holmes, Steven J.)V1466,2-12(1991)

Applicability of dry developable deep-UV lithography to sub-0.5 um processing (Goethals, Anne-Marie; Baik, Ki-Ho; Van den hove, Luc; Tedesco, Serge V.)V1466,604-615(1991)

Application aspects of the Si-CARL bilayer process (Sebald, Michael; Berthold, Joerg; Beyer, Michael; Leuschner, Rainer; Noelscher, Christoph; Scheler, Ulrich; Sezi, Recai; Ahne, Hellmut; Birkle, Siegfried)V1466,227-237(1991)

Application of an electron-beam scattering parameter extraction method for proximity correction in direct-write electron-beam lithography (Weiss, Rudolf M.; Sills, Robert M.)V1465,192-200(1991)

Bilayer resist system utilizing alkali-developable organosilicon positive photoresist (Nate, Kazuo; Mizushima, Akiko; Sugiyama, Hisashi)V1466,206-210(1991)

Blazing of transmission gratings for astronomical use (Neviere, Michel)V1545,11-18(1991)

Characterizing a surface imaging process in a high-volume DRAM manufacturing production line (Garza, Cesar M.; Catlett, David L.; Jackson, Ricky A.)V1466,616-627(1991)

Chemically amplified negative-tone photoresist for sub-half-micron device and mask fabrication (Conley, Willard E.; Dundatscheck, Robert; Gelorme, Jeffrey D.; Horvat, John; Martino, Ronald M.; Murphy, Elizabeth; Petrosky, Anne; Spinillo, Gary; Stewart, Kevin; Wilbarg, Robert; Wood, Robert L.)V1466,53-66(1991)

Chemically amplified resists for x-ray and e-beam lithography (Berry, Amanda K.; Graziano, Karen A.; Thompson, Stephen D.; Taylor, James W.; Suh, Doowon; Plumb, Dean)V1465,210-220(1991)

Critical dimension shift resulting from handling time variation in the track coat process (Kulp, John M.)V1466,630-640(1991)

DESIRE technology with electron-beam resists: fundamentals, experiments, and simulation (Nicolau, Dan V.; Fulga, Florin; Dusa, Mircea V.)V1465,282-288(1991)

Disk-shaped VUV+O source used as resist asher and resist developer (Hattori, Shuzo; Collins, George J.; Yu, Zenqi; Sugimoto, Dai; Saita, Masahiro)V1463,539-550(1991)

Dissolution inhibition mechanism of ANR photoresists: crosslinking vs. -OH site consumption (Thackeray, James W.; Orsula, George W.; Rajaratnam, Martha M.; Sinta, Roger F.; Herr, Daniel J.; Pavelchek, Edward K.)V1466,39-52(1991)

Dissolution kinetics of high-resolution novolac resists (Itoh, Katsuyuki; Yamanaka, Koji; Nozue, Hiroshi; Kasama, Kunihiko)V1466,485-496(1991)

Dissolution of poly(p-hydroxystyrene): molecular weight effects (Long, Treva; Rodriguez, Ferdinand)V1466,188-198(1991)

DQN photoresist with tetrahydroxydiphenylmethane as ballasting group in PAC (Tzeng, Chao H.; Lin, Dhei-Jhai; Lin, Song S.; Huang, Dong-Tsair; Lin, Hwang-Kuen)V1466,469-476(1991)

Dry development and plasma durability of resists: melt viscosity and self-diffusion effects (Paniez, Patrick J.; Joubert, Olivier P.; Pons, Michel J.; Oberlin, Jean C.; Weill, Andre P.)V1466,583-591(1991)

Dry etching for silylated resist development (Laporte, Philippe; Van den hove, Luc; Melaku, Yosias)V1392,196-207(1991)

Effective transmission holograms produced on CK-type photoresist (Bulygin, A. R.)V1238,248-252(1991)

Effect of sensitizer spatial distribution on dissolution inhibition in novolak/diazonaphthoquinone resists (Rao, Veena; Kosbar, Laura L.; Frank, Curtis W.; Pease, R. F.)V1466,309-323(1991)

Effect of silylation condition on the silylated image in the DESIRE process (Taira, Kazuo; Takahashi, Junichi; Yanagihara, Kenji)V1466,570-582(1991)

Electron-Beam, X-Ray, and Ion-Beam Submicrometer Lithographies for Manufacturing (Peckerar, Martin, ed.)V1465(1991)

Evaluation of a high-resolution negative-acting electron-beam resist GMC for photomask manufacturing (Chen, Wen-Chih; Novembre, Anthony E.)V1496,266-283(1991)

Evaluation of a photoresist process for 0.75-micron g-line lithography (Kasahara, Jack S.; Dusa, Mircea V.; Perera, Thiloma)V1463,492-503(1991)

Evaluation of phenolic resists for 193 nm surface imaging (Hartney, Mark A.; Johnson, Donald W.; Spencer, Allen C.)V1466,238-247(1991)

Evaluation of poly(p-trimethylsilylstyrene and p-pentamethyldisilylstyrene sulfone)s as high-resolution electron-beam resists (Gozdz, Antoni S.; Ono, Hiroshi; Ito, Seiki; Shelburne, John A.; Matsuda, Minoru)V1466,200-205(1991)

Exposure dose optimization for a positive resist containing polyfunctional photoactive compound (Trefonas, Peter; Mack, Chris A.)V1466,117-131(1991)

Fabrication of micro-optical components by laser beam writing in photoresist (Gale, Michael T.; Lang, Graham K.; Raynor, Jeffrey M.; Schuetz, Helmut)V1506,65-70(1991)

Fine undercut control in bilayer PMMA-P(MMA-MAA) resist system for e-beam lithography with submicrometer resolution (Bogdanov, Alexei L.; Andersson, Eva K.)V1465,324-329(1991)

Generalized characteristic model for lithography: application to negative chemically amplified resists (Ziger, David H.; Mack, Chris A.; Distasio, Romelia)V1466,270-282(1991)

Highly sensitive microresinoid siloxane resist for EB and deep-UV lithography (Yamazaki, Satomi; Ishida, Shinji; Matsumoto, Hiroshi; Aizaki, Naoaki; Muramoto, Naohiro; Mine, Katsutoshi)V1466,538-545(1991)

High-sensitive photopolymer for large-size holograms (Ishikawa, Toshiharu; Kuwabara, Y.; Koseki, Kenichi; Yamaoka, Tsuguo)V1461,73-78(1991)

High-sensitivity and high-dry-etching durability positive-type electron-beam resist (Tamura, Akira; Yonezawa, Masaji; Sato, Mitsuyoshi; Fujimoto, Yoshiaki)V1465,271-281(1991)

Impact of phase masks on deep-UV lithography (Sewell, Harry)V1463,168-179(1991)

Improvements in 0.5-micron production wafer steppers (Luehrmann, Paul F.; de Mol, Chris G.; van Hout, Frits J.; George, Richard A.; van der Putten, Harrie B.)V1463,434-445(1991)

Improving the performance and usability of a wet-developable DUV resist for sub-500nm lithography (Samarakone, Nandasiri; Van Driessche, Veerle; Jaenen, Patrick; Van den hove, Luc; Ritchie, Douglas R.; Luehrmann, Paul F.)V1463,16-29(1991)

In-situ monitoring of silylation mechanisms by laser interferometry (Pierrat, Christophe; Paniez, Patrick J.; Martin, P.)V1466,248-256(1991)

Investigation of interlevel proximity effects case of the gate level over LOCOS (Festes, Gilles; Chollet, Jean-Paul E.)V1463,245-255(1991)

Investigation of self-aligned phase-shifting reticles by simulation techniques (Noelscher, Christoph; Mader, Leonhard)V1463,135-150(1991)

Investigation on the effect of electron-beam acceleration voltage and electron-beam sharpness on 0.2-um patterns (Moniwa, Akemi; Okazaki, Shinji)V1465,154-163(1991)

Mechanism of dissolution inhibition of novolak-diazoquinone resist (Furuta, Mitsuhiro; Asaumi, Shingo; Yokota, Akira)V1466,477-484(1991)

Mechanistic studies on the poly(4-tert-butoxycarbonyloxystyrene)/triphenylsulfonium salt photoinitiation process (Hacker, Nigel P.; Welsh, Kevin M.)V1466,384-393(1991)

Microreflective elements for integrated planar optics interconnects (Sheng, Yunlong; Delisle, Claude; Moreau, Louis; Song, Li; Lessard, Roger A.; Arsenault, Henri H.)V1559,222-228(1991)

Modeling of illumination effects on resist profiles in x-ray lithography (Oertel, Heinrich K.; Weiss, M.; Huber, Hans L.; Vladimirsky, Yuli; Maldonado, Juan R.)V1465,244-253(1991)

Nanometer scale focused ion beam vacuum lithography using an ultrathin oxide resist (Harriott, Lloyd R.; Temkin, Henryk; Chu, C. H.; Wang, Yuh-Lin; Hsieh, Y. F.; Hamm, Robert A.; Panish, Morton B.; Wade, H. H.)V1465,57-63(1991)

Nd:YAG laser machining with multilevel resist kinoforms (Ekberg, Mats; Larsson, Michael; Bolle, Aldo; Hard, Sverker)V1527,236-239(1991)

Negative chemical amplification resist systems based on polyhydroxystyrenes and N-substituted imides or aldehydes (Ito, Hiroshi; Schildknegt, Klaas; Mash, Eugene A.)V1466,408-418(1991)

Negative resist systems using acid-catalyzed pinacol rearrangement reaction in a phenolic resin matrix (Uchino, Shou-ichi; Iwayanagi, Takao; Ueno, Takumi; Hayashi, Nobuaki)V1466,429-435(1991)

New aqueous base-developable negative-tone photoresist based on furans (Fahey, James T.; Frechet, Jean M.)V1466,67-74(1991)

New phase-shifting mask structure for positive resist process (Miyazaki, Junji; Kamon, Kazuya; Yoshioka, Nobuyuki; Matsuda, Shuichi; Fujinaga, Masato; Watakabe, Yaichiro; Nagata, Hitoshi)V1464,327-335(1991)

Nonlinear effects in hidden picture amplification and contrast improvement in polymer electron and Roentgenoresist PMMA (Aleksandrov, A. P.; Genkin, Vladimir N.; Myl'nikov, M. Y.; Rukhman, N. V.)V1440,442-453(1991)

Nonmetallic acid generators for i-line and g-line chemically amplified resists (Brunsvold, William R.; Montgomery, Warren; Hwang, Bao)V1466,368-376(1991)

Novel acid-hardening positive photoresist technology (Graziano, Karen A.; Thompson, Stephen D.; Winkle, Mark R.)V1466,75-88(1991)

Novel base-generating photoinitiators for deep-UV lithography (Kutal, Charles; Weit, Scott K.; Allen, Robert D.; MacDonald, Scott A.; Willson, C. G.)V1466,362-367(1991)

Novel novolak resins using substituted phenols for high-performance positive photoresist (Kajita, Toru; Ota, Toshiyuki; Nemoto, Hiroaki; Yumoto, Yoshiji; Miura, Takao)V1466,161-173(1991)

Novel quinonediazide-sensitized photoresist system for i-line and deep-UV lithography (Fukunaga, Seiki; Kitaori, Tomoyuki; Koyanagi, Hiroo; Umeda, Shin'ichi; Nagasawa, Kohtaro)V1466,446-457(1991)

Novel surface imaging masking technique for high-aspect-ratio dry etching applications (Calabrese, Gary C.; Abali, Livingstone N.; Bohland, John F.; Pavelchek, Edward K.; Sricharoenchaikit, Prasit; Vizvary, Gerald; Bobbio, Stephen M.; Smith, Patrick)V1466,528-537(1991)

Novolak design for high-resolution positive photoresists (IV): tandem-type novolak resin for high-performance positive photoresists (Hanabata, Makoto; Oi, F.; Furuta, Akihiro)V1466,132-140(1991)

Novolak resin design concept for high-resolution positive resists (Noguchi, Tsutomu; Tomita, Hidemi)V1466,149-160(1991)

Onium salt structure/property relationships in poly(4-tert-butyloxycarbonyloxystyrene) deep-UV resists (Schwartzkopf, George; Niazy, Nagla N.; Das, Siddhartha; Surendran, Geetha; Covington, John B.)V1466,26-38(1991)

Optical/Laser Microlithography IV (Pol, Victor, ed.)V1463(1991)

Optimal control of positive optical photoresist development (Carroll, Thomas A.; Ramirez, W. F.)V1464,222-231(1991)

Optimization of optical properties of resist processes (Brunner, Timothy A.)V1466,297-308(1991)

Oxygen plasma etching of silylated resist in top-imaging lithographic process (Dijkstra, Han J.)V1466,592-603(1991)

Photo-induced adhesion changes: a technique for patterning lightguide structures (Festl, H. G.; Franke, Hilmar)V1559,410-423(1991)

Photoresist bake conditions and their effects on lithography process control (Norbury, David H.; Love, John C.)V1463,558-573(1991)

Photoresist dissolution rates: a comparison of puddle, spray, and immersion processes (Robertson, Stewart A.; Stevenson, J. T.; Holwill, Robert J.; Thirsk, Mark; Daraktchiev, Ivan S.; Hansen, Steven G.)V1464,232-244(1991)

Physical aging of resists: the continual evolution of lithographic material (Paniez, Patrick J.; Weill, Andre P.; Cohendoz, Stephane D.)V1466,336-344(1991)

Plasma diagnostics as inputs to the modeling of the oxygen reactive ion etching of multilevel resist structures (Hope, D. A.; Hydes, A. J.; Cox, T. I.; Deshmukh, V. G.)V1392,185-195(1991)

Polymer effects on the photochemistry of triarylsulfonium salts (McKean, Dennis R.; Allen, Robert D.; Kasai, Paul H.; MacDonald, Scott A.)V1559,214-221(1991)

Polysilanes for microlithography (Wallraff, Gregory M.; Miller, Robert D.; Clecak, Nicholas J.; Baier, M.)V1466,211-217(1991)

Polysilyne resists for 193 nm excimer laser lithography (Kunz, Roderick R.; Bianconi, Patricia A.; Horn, Mark W.; Paladugu, R. R.; Shaver, David C.; Smith, David A.; Freed, Charles A.)V1466,218-226(1991)

Polyvinylphenols protected with tetrahydropyranyl group in chemical amplification positive deep-UV resist systems (Hayashi, Nobuaki; Schlegel, Leo; Ueno, Takumi; Shiraishi, Hiroshi; Iwayanagi, Takao)V1466,377-383(1991)

Preliminary lithographic characteristics of an all-organic chemically amplified resist formulation for single-layer deep-UV lithography (Nalamasu, Omkaram; Reichmanis, Elsa; Cheng, May; Pol, Victor; Kometani, Janet M.; Houlihan, Frank M.; Neenan, Thomas X.; Bohrer, M. P.; Mixon, D. A.; Thompson, Larry F.; Takemoto, Cliff H.)V1466,13-25(1991)

Preparations and properties of novel positive photosensitive polyimides (Hayase, Rumiko H.; Kihara, Naoko; Oyasato, Naohiko; Matake, S.; Oba, Masayuki)V1466,438-445(1991)

Primary research for mechanism of forming PLH (Guo, Lu Rong; Zhang, Xiao-Chun; Guo, Yongkang)V1463,534-538(1991)

Process control capability using a diaphragm photochemical dispense system (Cambria, Terrell D.; Merrow, Scott F.)V1466,670-675(1991)

Process latitude for the chemical amplification resists AZ PF514 and AZ PN114 (Eckes, Charlotte; Pawlowski, Georg; Przybilla, Klaus J.; Meier, Winfried; Madore, Michel; Dammel, Ralph)V1466,394-407(1991)

Process latitude measurements on chemically amplified resists exposed to synchrotron radiation (Babcock, Carl P.; Taylor, James W.; Sullivan, Monroe; Suh, Doowon; Plumb, Dean; Palmer, Shane R.; Berry, Amanda K.; Graziano, Karen A.; Fedynyshyn, Theodore H.)V1466,653-662(1991)

Process latitudes in projection printing (Barouch, Eytan; Hollerbach, Uwe; Orszag, Steven A.; Bradie, Brian D.; Peckerar, Martin)V1465,254-262(1991)

Progress in DUV resins (Przybilla, Klaus J.; Roeschert, Heinz; Spiess, Walter; Eckes, Charlotte; Chatterjee, Subhankar; Khanna, Dinesh N.; Pawlowski, Georg; Dammel, Ralph)V1466,174-187(1991)

Progress in the study of development-free vapor photolithography (Hong, Xiao-Yin; Liu, Dan; Li, Zhong-Zhe; Xiao, Ji-Quang; Dong, Gui-Rong)V1466,546-557(1991)

Reliability of contrast and dissolution-rate-derived parameters as predictors of photoresist performance (Spragg, Peggy M.; Hurditch, Rodney J.; Toukhy, Medhat A.; Helbert, John N.; Malhotra, Sandeep)V1466,283-296(1991)

Resist design for dry-developed positive working systems in deep-UV and e-beam lithography (Vinet, Francoise; Chevallier, M.; Pierrat, Christophe; Guibert, Jean C.; Rosilio, Charles; Mouanda, B.; Rosilio, A.)V1466,558-569(1991)

Resist parameter extraction with graphical user interface in X (Chiu, Anita S.; Ferguson, Richard A.; Doi, Takeshi; Wong, Alfred K.; Tam, Nelson; Neureuther, Andrew R.)V1466,641-652(1991)

Resist patterning for sub-quarter-micrometer device fabrications (Chiong, Kaolin G.; Hohn, Fritz J.)V1465,221-236(1991)

Resist schemes for soft x-ray lithography (Taylor, Gary N.; Hutton, Richard S.; Windt, David L.; Mansfield, William M.)V1343,258-273(1991)

Resist tracking: a lithographic diagnostic tool (Takemoto, Cliff H.; Ziger, David H.; Connor, William; Distasio, Romelia)V1464,206-214(1991)

Sequential experimentation strategy and response surface methodologies for photoresist process optimization (Flores, Gary E.; Norbury, David H.)V1464,610-627(1991)

Simulation analysis of deep-UV chemically amplified resist (Ohfuji, Takeshi; Soenosawa, Masanobu; Nozue, Hiroshi; Kasama, Kunihiko)V1463,345-354(1991)

Simulation of an advanced negative i-line photoresist (Barouch, Eytan; Hollerbach, Uwe; Orszag, Steven A.; Allen, Mary T.; Calabrese, Gary C.)V1463,336-344(1991)

Simulation of connected image reversal and DESIRE techniques for submicron lithography (Nicolau, Dan V.; Dusa, Mircea V.; Fulga, Florin)V1466,663-669(1991)

Simulations of bar printing over a MOSFET device using i-line and deep-UV resists (Barouch, Eytan; Hollerbach, Uwe; Orszag, Steven A.; Szmanda, Charles R.; Thackeray, James W.)V1463,464-474(1991)

Single-component chemically amplified resist materials for electron-beam and x-ray lithography (Novembre, Anthony E.; Tai, Woon W.; Kometani, Janet M.; Hanson, James E.; Nalamasu, Omkaram; Taylor, Gary N.; Reichmanis, Elsa; Thompson, Larry F.)V1466,89-99(1991)

Some experimental techniques for characterizing photoresists (Spence, Christopher A.; Ferguson, Richard A.)V1466,324-335(1991)

Structural effects of DNQ-PAC backbone on resist lithographic properties (Uenishi, Kazuya; Kawabe, Yasumasa; Kokubo, Tadayoshi; Slater, Sydney G.; Blakeney, Andrew J.)V1466,102-116(1991)

Structure of poly(p-hydroxystyrene) film (Toriumi, Minoru; Yanagimachi, Masatoshi; Masuhara, Hiroshi)V1466,458-468(1991)

Studies of dissolution inhibition mechanism of DNQ-novolak resist (II): effect of extended ortho-ortho bond in novolak (Honda, Kenji; Beauchemin, Bernard T.; Fitzgerald, Edward A.; Jeffries, Alfred T.; Tadros, Sobhy P.; Blakeney, Andrew J.; Hurditch, Rodney J.; Tan, Shiro; Sakaguchi, Shinji)V1466,141-148(1991)

Study of silylation mechanisms and kinetics through variations in silylating agent and resin (Dao, T. T.; Spence, Christopher A.; Hess, Dennis W.)V1466,257-268(1991)

Surface imaging on the basis of phenolic resin: experiments and simulation (Bauch, Lothar; Jagdhold, Ulrich A.; Dreger, Helge H.; Bauer, Joachim J.; Hoeppner, Wolfgang W.; Erzgraeber, Hartmut H.; Mehliss, Georg G.)V1466,510-519(1991)

Synthesis and lithographic evaluation of alternating copolymers of linear and cyclic alkenyl(di)silanes with sulfur dioxide (Gozdz, Antoni S.; Shelburne, John A.)V1466,520-527(1991)

Technology and chemistry of high-temperature positive resist (Toukhy, Medhat A.; Sarubbi, Thomas R.; Brzozowy, David J.)V1466,497-507(1991)

Triplet-sensitized reactions of some main chain liquid-crystalline polyaryl cinnamates (Subramanian, P.; Creed, David; Hoyle, Charles E.; Venkataram, Krishnan)V1559,461-469(1991)

Two-layer 1.2-micron pitch multilevel metal demonstrator using resist patterning by surface imaging and dry development (Martin, Brian; Snowden, Ian M.; Mortimer, Simon H.)V1463,584-594(1991)

Visible-laser-light-induced polymerization: an overview of photosensitive formulations (Fouassier, Jean-Pierre)V1559,76-88(1991)

Wet-developed, high-aspect-ratio resist patterns by 20-keV e-beam lithography (Weill, Andre P.; Amblard, Gilles R.; Lalanne, Frederic P.; Panabiere, Jean-Pierre)V1465,264-270(1991)

XUV resist characterization: studies with a laser plasma source (Kubiak, Glenn D.)V1343,283-291(1991)

Plasmas—see also laser-matter interaction

100-Hz KrF laser plasma x-ray source (Turcu, I. C.; Gower, Malcolm C.; Reason, C. J.; Huntington, P.; Schulz, M.; Michette, Alan G.; Bijkerk, Fred; Louis, Eric; Tallents, Gregory J.; Al-Hadithi, Yas; Batani, D.)V1503,391-405(1991)

Application of high-speed mirror chronograph 3CX-1 to plasma investigations in visible and medium infrared spectrum ranges (Drozhbin, Yu. A.; Zvorykin, V. D.; Polyansky, S. V.; Sychugov, G. V.; Trofimenko, Vladimir V.; Yarova, A. G.)V1358,1029-1034(1991)

Application of spatially resolving holographic interferometry to plasma diagnostics (Neger, Theo; Jaeger, Helmut; Philipp, Harald; Pretzler, Georg; Widmann, Klaus; Woisetschlaeger, Jakob)V1507,476-487(1991)

Applications of a soft x-ray streak camera in laser-plasma interaction studies (Tsakiris, George D.)V1358,174-192(1991)

Applications of laser plasmas in XUV photoabsorption spectroscopy (Kennedy, Eugene T.; Costello, John T.; Mosnier, Jean-Paul)V1503,406-415(1991)

Applications of optical emission spectroscopy in plasma manufacturing systems (Gifford, George G.)V1392,454-465(1991)

Characterization of plasma processes with optical emission spectroscopy (Malchow, Douglas S.)V1392,498-505(1991)

Characterization of the radiation from a low-energy X-pinch source (Christou, Christos; Choi, Peter)V1552,278-287(1991)

Comparison of negative and positive polarity reflex diode microwave source (Litz, Marc S.; Huttlin, George A.; Lazard, Carl J.; Golden, Jeffry; Pereira, Nino R.; Hahn, Terry D.)V1407,159-166(1991)

Comparison of plasma source with synchrotron source in the Center for X-ray Lithography (Guo, Jerry Z.; Cerrina, Franco)V1465,330-337(1991)

Computer modeling of unsteady gas-dynamical and optical phenomena in low-temperature laser plasma (Kanevsky, M. F.; Bolshov, L. A.; Chernov, S. Y.; Vorobjev, V. A.)V1440,154-165(1991)

Continuous inertial confinement fusion for the generation of very intense plasma jets (Winterberg, F.)V1407,322-325(1991)

Deposition of a-Si:H using a supersonically expanding argon plasma (Meeusen, G. J.; Qing, Z.; Wilbers, A. T.; Schram, D. C.)V1519,252-257(1991)

Diagnostics of a DC plasma torch (Russell, Derrek; Taborek, Peter)V1534,14-23(1991)

Diagnostics of arc plasma by moire deflectometry (Gao, Yiqing; Liu, Yupin)V1554B,193-199(1991)

Dynamics of a low-energy X-pinch source (Choi, Peter; Christou, Christos; Aliaga, Raul)V1552,270-277(1991)

Dynamics of near-surface plasma formation and laser absorption waves under the action of microsecond laser radiation with different wavelengths on absorbing condensed media (Min'ko, L. Y.; Chumakov, A. N.)V1440,166-178(1991)

Electrical probe diagnostics for processing discharges (Mantei, Thomas D.)V1392,466-473(1991)

Electron density measurements of laser-induced surface plasma by means of a beam deflection technique (Michaelis, Alexander; Uhlenbusch, Juergen; Vioel, Wolfgang)V1397,709-712(1991)

Experimental investigation of 1.06 um laser interaction with Al target in air (Gang, Yuan)V1415,225-227(1991)

Experimental method for gas kinetic temperature measurements in a thermal plasma (Reynolds, Larry D.; Shaw, C. B.)V1554B,622-631(1991)

Experimental study of the parameters of the laser-induced plasma observed in welding of iron targets with continuous high-power CO2 lasers (Poueyo, Anne; Sabatier, Lilian; Deshors, G.; Fabbro, Remy; de Frutos, Angel M.; Bermejo, Dionisio; Orza, Jose M.)V1502,140-147(1991)

Experiments with SPRITE 12 ps facility (Tallents, Gregory J.; Key, Michael H.; Norreys, P.; Jacoby, J.; Kodama, R.; Tragin, N.; Baldis, Hector A.; Dunn, James; Brown, D.)V1413,70-76(1991)

Fast-injection Langmuir probe for process diagnostic and control (Patrick, Roger; Schoenborn, Philippe; Linder, Stefan; Baltes, Henry P.)V1392,506-513(1991)

Fiber optic sensor for plasma current diagnostics in tokamaks (Kozhevnikov, Nikolai M.; Barmenkov, Yuri O.; Belyakov, V. A.; Medvedev, A. A.; Razdobarin, G. T.)V1584,138-144(1991)

Frequency up-conversion for the reflectionless propagation of a high-power microwave pulse in a self-generated plasma (Kuo, Spencer P.; Ren, A.; Zhang, Y. S.)V1407,272-280(1991)

High-power copper vapor laser development (Aoki, Nobutada; Kimura, Hironobu; Konagai, Chikara; Shirayama, Shimpey; Miyazawa, Tatsuo; Takahashi, Tomoyuki)V1412,2-11(1991)

Hydrodynamic evolution and radiation emission from an impulse-heated solid-density plasma (Lyubomirsky, I.; Durfee, C.; Milchberg, Howard M.)V1413,108-111(1991)

Hydrodynamic evolution of picosecond laser plasmas (Landen, Otto L.; Vu, Brian-Tinh; Stearns, Daniel G.; Alley, W. E.)V1413,120-130(1991)

In-situ measurements of radicals and particles in a selective silicon oxide etching plasma (Singh, Jyothi)V1392,474-486(1991)

Interference detection of plasma of laser field interaction with optical thin films (Ni, Xiao W.; Lu, Jian-Feng; He, Anzhi; Ma, Zi; Zhou, Jiu L.)V1554B,632-635(1991)

Interferometric measurement for the plasma produced by Q-switched laser in air near the surface of an Al Target (Lu, Jian-Feng; Ni, Xiao W.; Wang, Hong Y.; He, Anzhi)V1554B,593-597(1991)

Intracavity spectroscopy measurements of atom and ion densities in near-surface laser plasma (Burakov, V. S.; Lopasov, V. P.; Naumenkov, P. A.; Raikov, S. N.)V1440,270-276(1991)

Ku-band radiation in the UNM backward-wave oscillator experiment (Schamiloglu, Edl; Gahl, John M.; McCarthy, G.)V1407,242-248(1991)

Laboratory soft x-ray source based on high-temperature capillary arc discharge (Belov, Sergei N.; Golubev, Evgeny M.; Vinokurova, Elena G.)V1552,264-267(1991)

Laser plasma XUV sources: a role for excimer lasers? (Bijkerk, Fred; Shevelko, A. P.)V1503,380-390(1991)

Laser-produced continua for studies in the XUV (Carroll, P. K.; O'Sullivan, Gerard D.)V1503,416-427(1991)

Laser-produced plasma x-ray diagnostics with an x-ray streak camera at the Iskra-4 plant (Berkovsky, A. G.; Gubanov, Y. I.; Pryanishnikov, Ivan G.; Murugov, V. M.; Petrov, S. I.; Senik, A. V.)V1358,750-755(1991)

LH electron cyclotron resonance plasma source (Kretschmer, K.-H.; Lorenz, Gerhard; Castrischer, G.; Kessler, I.; Baumann, P.)V1392,246-252(1991)

Liquid droplet supercritical explosion in the field of CO2 laser radiation and influence of plasma chemical reactions on initiation of optical breakdown in air (Budnik, A. P.; Popov, A. G.)V1440,135-145(1991)

L-shell x-ray spectroscopy of laser-produced plasmas in the 1-keV region (Batani, D.; Giulietti, Antonio; Palladino, Libero; Tallents, Gregory J.; Turcu, I. C.)V1503,479-491(1991)

Magnetic field effects on plasma-filled backward-wave oscillators (Lin, Anthony T.)V1407,234-241(1991)

Magnetic fields generated in laser-produced plasmas (Shainoga, I. S.; Shentsev, N. I.; Zakharov, N. S.)V1440,277-290(1991)

Measurements on the NIST GEC reference cell (Roberts, James R.; Olthoff, James K.; Van Brunt, R. J.; Whetstone, James R.)V1392,428-436(1991)

Mechanisms of ionization for gas adjoining to plasma in intense laser beam (Fisher, Vladimir)V1440,179-187(1991)

Microwave interferometric measurements of process plasma density (Cheah, Chun-Wah; Cecchi, Joseph L.; Stevens, J. L.)V1392,487-497(1991)

Multicomponent electric-arc source of metallic plasma (Karpov, D. A.; Nazikov, S. N.)V1519,115-121(1991)

Near-surface plasma initiation model for short laser pulses (Vas'kovsky, Yu. M.; Rovinsky, R. E.; Fedjushin, B. T.; Filatova, S. A.)V1440,241-249(1991)

Negative deuterium ion thermal energy measurements in a volume ion source (Bacal, Marthe; Courteille, C.; Devynck, Pascal; Jones-King, Yolanda D.; Leroy, Renan; Stern, Raul A.)V1407,605-609(1991)

Nonlinear bremsstrahlung in nonequilibrium relativistic beam-plasma systems (Brandt, Howard E.)V1407,326-353(1991)

Nonlinear wakefield generation and relativistic optical guiding of intense laser pulses in plasmas (Esarey, Eric; Sprangle, Phillip; Ting, Antonio C.)V1407,407-417(1991)

Numerical simulation of energy transport mechanisms in high-intensity laser-matter interaction experiments (Ocana, Jose L.)V1502,299-310(1991)

Observation of living cells by x-ray microscopy with a laser-plasma x-ray source (Tomie, Toshihisa; Shimizu, Hazime; Majima, T.; Yamada, Mitsuo; Kanayama, Toshihiko; Yano, M.; Kondo, H.)V1552,254-263(1991)

Opacity studies with laser-produced plasma as an x-ray source (Eidmann, K.; Lanig, E. M.; Schwanda, W.; Sigel, Richard; Tsakiris, George D.)V1502,320-330(1991)

Optical and mechanical parameter detection of calcified plaque for laser angioplasty (Stetz, Mark L.; O'Brien, Kenneth M.; Scott, John J.; Baker, Glenn S.; Deckelbaum, Lawrence I.)V1425,55-62(1991)

Optical diagnostics of line-focused laser-produced plasmas (Lin, Li-Huang; Chen, Shisheng; Jiang, Z. M.; Ge, Wen; Qian, Aidi D.; Bin, Ouyang; Li, Yongchun L.; Kang, Yilan; Xu, Zhizhan)V1346,490-501(1991)

Optical emission spectroscopy of diamond-producing plasmas (Plano, Linda S.)V1437,13-23(1991)

Optically ionized plasma recombination x-ray lasers (Amendt, Peter; Eder, David C.; Wilks, S. C.; Dunning, M. J.; Keane, Christopher J.)V1413,59-69(1991)

Optical Radiation Interaction with Matter (Bonch-Bruevich, Aleksei M.; Konov, Vitaly I.; Libenson, Michail N., eds.)V1440(1991)

Oxygen reactive ion etching of polymers: profile evolution and process mechanisms (Pilz, Wolfgang; Janes, Joachim; Muller, Karl P.; Pelka, Joachim)V1392,84-94(1991)

Phase-locking simulation of dual magnetrons (Chen, H. C.; Stark, Robert A.; Uhm, Han S.)V1407,139-146(1991)

Plasma betatron without gas breakdown (Ishizuka, Hiroshi; Yee, K.; Fisher, Amnon; Rostoker, Norman)V1407,442-455(1991)

Plasma diagnostics as inputs to the modeling of the oxygen reactive ion etching of multilevel resist structures (Hope, D. A.; Hydes, A. J.; Cox, T. I.; Deshmukh, V. G.)V1392,185-195(1991)

Plasma heating by ultrashort laser pulse in the regime of anomalous skin effect (Gamaly, Eugene G.; Kiselev, A. Y.; Tikhonchuk, V. T.)V1413,95-106(1991)

Plasma interaction with powerful laser beams (Giulietti, Antonio; Giulietti, Danilo; Willi, Oswald)V1502,270-283(1991)

Plasma modeling in microelectronic processing (Meyyappan, Meyya; Govindan, T. R.; Kreskovsky, John P.)V1392,67-76(1991)

Plasma motion velocity along laser beam and continuous optical discharge in gas flow (Budnik, A. P.; Gus'kov, K. G.; Raizer, Yu. P.; Surjhikov, S. T.)V1397,721-724(1991)

Plasma parameter determination formed under the influence of CO2 laser radiation on the obstacle in the air using optical methods (Vas'kovsky, Yu. M.; Gordeeva, I. A.; Korenev, A. S.; Rovinsky, R. E.; Cenina, I. S.; Shirokova, I. P.)V1440,229-240(1991)

Plasma parameters in microwave-plasma-assisted chemical vapor deposition of diamond (Weimer, Wayne A.; Cerio, Frank M.; Johnson, Curtis E.)V1534,9-13(1991)

Prepulse suppression using a self-induced ultrashort pulse plasma mirror (Gold, David; Nathel, Howard; Bolton, Paul R.; White, William E.; Van Woerkom, Linn D.)V1413,41-52(1991)

Probe measurements in a CO2 laser plasma (Leys, C.; Sona, P.; Muys, Peter F.)V1397,399-402(1991)

Progress of an advanced diffusion source plasma reactor (Benjamin, Neil M.; Chapman, Brian N.; Boswell, Rod W.)V1392,95-105(1991)

Propagation of plasma beams across the magnetic field (Rahman, Hafiz-ur; Yur, Gung; White, R. S.; Wessel, Frank J.; Song, Joshua J.; Rostoker, Norman)V1407,589-597(1991)

Pump-probe investigation of picosecond laser-gas target interactions (Durfee, C.; Milchberg, Howard M.)V1413,78-80(1991)

Q-switching and pulse shaping with IR lasers (Brinkmann, Ralf E.; Bauer, K.)V1421,134-139(1991)

Reflectivity of stimulated Brillouin scattering in picosecond time scales (Labaune, C.; Rozmus, Wojtek; Baldis, Hector A.; Mounaix, P.; Pesme, Denis; Baton, S.; LaFontaine, Bruno; Villeneuve, D. M.; Enright, G. D.)V1413,138-143(1991)

Research of Cr2O3 thin film deposited by arc discharge plasma deposition as heat-radiation absorbent in electric vacuum devices (Deng, Hong; Wang, Xiang D.; Yuan, Lei)V1519,735-739(1991)

Shock-layer-induced ultraviolet emissions measured by rocket payloads (Caveny, Leonard H.; Mann, David M.)V1479,102-110(1991)

Short-Pulse High-Intensity Lasers and Applications (Baldis, Hector A., ed.)V1413(1991)

Shortwave radiation induced by CO2 laser pulse interaction with aluminum target (Golovin, A. F.; Golub, A. P.; Zemtsov, S. S.; Fedyushin, B. T.)V1440,250-259(1991)

Silicon nitride film formed by NH3 plasma-enhanced thermal nitridation (Gu, Zhi G.; Li, Bing Z.)V1519,247-251(1991)

Simulation of intense microwave pulse propagation in air breakdown environment (Kuo, Spencer P.; Zhang, Y. S.)V1407,260-271(1991)

Soft x-ray emission characteristics from laser plasma sources (Chen, Shisheng; Xu, Zhizhan; Li, Yao-lin; Wang, Xiaofang; Qian, Aidi D.)V1552,288-295(1991)

Spectroscopic determination of the parameters of an iron plasma produced by a CO2 laser (de Frutos, Angel M.; Sabatier, Lilian; Poueyo, Anne; Fabbro, Remy; Bermejo, Dionisio; Orza, Jose M.)V1397,717-720(1991)

Spectroscopic measurements of plasmas temperature and density during high-energy pulsed laser-materials interaction processes (Astic, Dominique; Vigliano, Patrick; Autric, Michel L.; Inglesakis, Georges; Prat, Ch.)V1397,713-716(1991)

Spectrum investigation and imaging of laser-produced plasma by multilayer x-ray optics (Platonov, Yu. Y.; Salashchenko, N. N.; Shmaenok, L. A.)V1440,188-196(1991)

Spherical pinch x-ray generator prototype for microlithography (Kawai, Kenji; Panarella, Emilio; Mostacci, D.)V1465,308-314(1991)

Spontaneous magnetic field diffusion from laser plasma (Bodulinsky, V. K.; Kondratenko, P. S.)V1440,392-396(1991)

Study of x-ray emission from picosecond laser-plasma interaction (Chen, Hong; Chuang, Yung-Ho; Delettrez, J.; Uchida, S.; Meyerhofer, David D.)V1413,112-119(1991)

Subnanosecond high-speed framing of prebreakdown phenomena (Pfeiffer, Wolfgang; Stolz, Dieter; Zipfl, P.)V1358,1191-1201(1991)

Thin-film technology in high-resolution, high-density AC plasma displays (Andreadakis, Nicholas C.)V1456,310-315(1991)

Thomson scattering diagnostics of discharge plasmas in an excimer laser (Yamakoshi, H.; Kato, M.; Kubo, Y.; Enokizono, H.; Uchino, K.; Muraoka, K.; Takahashi, A.; Maeda, Mitsuo)V1397,119-122(1991)

Thresholds of plasma arising under the pulse CO2 laser radiation interaction with an obstacle in air and energetic balance of the process (Babaeva, N. A.; Vas'kovsky, Yu. M.; Zhavoronkov, M. I.; Rovinsky, R. E.; Rjabinkina, V. A.)V1440,260-269(1991)

Time-resolved x-ray absorption spectroscopy apparatus using laser plasma as an x-ray source (Yoda, Osamu; Miyashita, Atsumi; Murakami, Kouichi; Aoki, Sadao; Yamaguchi, Naohiro)V1503,463-466(1991)

Time-space resolved optical study of the plasma produced by laser ablation of Ge: the role of oxygen pressure (Vega, Fidel; Solis, Javier; Afonso, Carmen N.)V1397,807-811(1991)

VUV wall stabilized argon arc discharge source (Li, Futian; Cao, Jianlin; Chen, Po; Qian, Limin; Jin, Lei; Chen, Xingdan)V1345,78-88(1991)

X-ray and optical diagnostics of a 100-fs laser-produced plasma (Geindre, Jean-Paul; Audebert, Patrick; Chenais-Popovics, Claude; Gauthier, Jean-Claude J.; Benattar, Rene; Chambaret, J. P.; Mysyrowicz, Andre; Antonetti, Andre)V1502,311-318(1991)

X-ray evaluation on residual stresses in vapor-deposited hard coatings (Xu, Kewei; Chen, Jin; Gao, Runsheng; He, Jia W.; Zhao, Cheng; Li, Shi Z.)V1519,765-770(1991)

XUV resist characterization: studies with a laser plasma source (Kubiak, Glenn D.)V1343,283-291(1991)

Polarimetry—see also polarization

Bragg crystal polarimeter for the Spectrum-X-Gamma misson (Holley, Jeff; Silver, Eric H.; Ziock, Klaus P.; Novick, Robert; Kaaret, Philip E.; Weisskopf, Martin C.; Elsner, Ronald F.; Beeman, Jeff)V1343,500-511(1991)

Conformational analysis and circular dichroism of bilirubin, the yellow pigment of jaundice (Lightner, David A.; Person, Richard; Peterson, Blake; Puzicha, Gisbert; Pu, Yu-Ming; Bojadziev, Stefan)V1432,2-13(1991)

Cylindrical proportional counter for x-ray polarimetry (Costa, Enrico; Piro, Luigi; Rubini, Alda; Soffitta, Paolo; Massaro, Enrico; Matt, Giorgio; Medici, Gastone; Manzo, Giuseppe; Re, Stefano)V1343,469-476(1991)

Distributed optical fiber sensing (Rogers, Alan J.)V1504,2-24(1991); V1506,2-24(1991); 1507,2-24(1991)

Electronic polarimetric detection system for optical fiber sensor application (Brooking, Nicholas L.; Guedes Valente, Luiz C.; Kawase, Liliana R.; Afonso, Jose A.)V1572,88-93(1991)

Frequency-derived distributed optical fiber sensing: backscatter analysis (Rogers, Alan J.; Handerek, Vincent A.; Parvaneh, Farhad)V1511,190-200(1991)

Further observations of vectorial effects in the x-ray photoemission from caesium iodide (Fraser, George W.; Lees, John E.; Pearson, James F.)V1343,438-456(1991)

Hard x-ray polarimeter utilizing Compton scattering (Sakurai, Hirohisa; Noma, M.; Niizeki, H.)V1343,512-518(1991)

High-speed polarimetric measurements for fiber-optic communications (Calvani, Riccardo A.; Caponi, Renato; Piglia, Roberto)V1504,258-263(1991)

New low-density, high-porosity lithium hydride-beryllium hydride foam: properties and applications to x-ray astronomy (Maienschein, Jon L.; Barry, Patrick E.; McMurphy, Frederick E.; Bowers, John S.)V1343,477-484(1991)

Photoelectric effect from CsI by polarized soft x-rays (Shaw, Ping-Shine; Church, Eric D.; Hanany, Shaul; Liu, Yee; Fleischman, Judith R.; Kaaret, Philip E.; Novick, Robert; Manzo, Giuseppe)V1343,485-499(1991)

Polarimetric monomode optical fibre sensor for monitoring tool wear (Zheng, S. X.; Hale, K. F.; Jones, Barry E.)V1572,268-272(1991)

Polarimetric optical fiber pressure sensor (Li, Luksun; Kerr, Anthony; Giles, Ian P.)V1584,170-177(1991)

Polarimetric optical fiber sensor for biochemical measurements (Heideman, Rene; Blikman, Albert; Koster, Rients; Kooyman, Rob P.; Greve, Jan)V1510,131-137(1991)

Polarimetric segmentation of SAR imagery (Burl, Michael C.; Novak, Leslie M.)V1471,92-115(1991)

Polarimetric sensor strain sensitivity in different thermal operating conditions (De Maria, Letizia; Escobar Rojo, Priscilla; Martinelli, Mario; Pistoni, Natale C.)V1366,304-312(1991)

Polarization sensitivity of x-ray photocathodes in the 60-200eV band (Fraser, George W.; Pain, M. D.; Pearson, James F.; Lees, John E.; Binns, C. R.; Shaw, Ping-Shine; Fleischman, Judith R.)V1548,132-148(1991)

Predicted performance of the lithium scattering and graphite crystal polarimeter for the Spectrum-X-Gamma mission (Weisskopf, Martin C.; Elsner, Ronald F.; Novick, Robert; Kaaret, Philip E.; Silver, Eric H.)V1343,457-468(1991)

Production and Analysis of Polarized X Rays (Siddons, D. P., ed.)V1548(1991)

Solar EUV/FUV line polarimetry: instruments and methods (Hoover, Richard B.; Fineschi, Silvano; Fontenla, Juan; Walker, Arthur B.)V1343,389-403(1991)

Solar EUV/FUV line polarimetry: observational parameters and theoretical considerations (Fineschi, Silvano; Hoover, Richard B.; Fontenla, Juan; Walker, Arthur B.)V1343,376-388(1991)

Status of the stellar x-ray polarimeter for the Spectrum-X-Gamma mission (Kaaret, Philip E.; Novick, Robert; Shaw, Ping-Shine; Hanany, Shaul; Liu, Yee; Fleischman, Judith R.; Sunyaev, Rashid; Lapshov, I.; Weisskopf, Martin C.; Elsner, Ronald F.; Ramsey, Brian D.; Silver, Eric H.; Ziock, Klaus P.; Costa, Enrico; Piro, Luigi; Soffitta, Paolo; Manzo, Giuseppe; Giarrusso, Salvatore; Santangelo, Andrea E.; Scarsi, Livio; Fraser, George W.; Pearson, James F.; Lees, John E.; Perola, G. C.; Massaro, Enrico; Matt, Giorgio)V1548,106-117(1991)

Stress analysis in ion-exchanged waveguides by using a polarimetric technique (Gonella, Francesco; Mazzi, Giulio; Quaranta, Alberto)V1513,425-433(1991)

Theoretical models for stellar x-ray polarization in compact objects (Meszaros, Peter)V1548,13-22(1991)

Use of polarization to separate on-axis scattered and unscattered light in red blood cells (Sardar, Dhiraj K.; Nemati, Babak; Barrera, Frederick J.)V1427,374-380(1991)

XUV polarimeter for undulator radiation measurements (Gluskin, Efim S.; Mattson, J. E.; Bader, Samuel D.; Viccaro, P. J.; Barbee, Troy W.; Brookes, N.; Pitas, Alan A.; Watts, Richard N.)V1548,56-68(1991)

Polarization—see also polarimetry

Absolute phasing of segmented mirrors using the polarization phase sensor (Klumpe, Herbert W.; Lajza-Rooks, Barbara A.; Blum, James D.)V1532,230-240(1991)

Accretion dynamics and polarized x-ray emission of magnetized neutron stars (Arons, Jonathan)V1548,2-12(1991)

Acid-catalyzed pinacol rearrangement: chemically amplified reverse polarity change (Sooriyakumaran, Ratna; Ito, Hiroshi; Mash, Eugene A.)V1466,419-428(1991)

All-fiber pressure sensor up to 100 MPa (Bock, Wojtek J.; Wolinski, Tomasz R.; Domanski, Andrzej W.)V1511,250-254(1991)

Analysis of polarization properties of shallow metallic gratings by an extended Rayleigh-Fano theory (Koike, Masato; Namioka, Takeshi)V1545,88-94(1991)

Applications and characteristics of polished polarization-splitting couplers (Lefevre, Herve C.; Simonpietri, Pascal; Martin, Philippe)V1365,65-73(1991)

Circular intensity differential scattering measurements in the soft x-ray region of the spectrum (~16 EV to 500 EV) (Maestre, Marcos F.; Bustamante, Carlos; Snyder, Patricia A.; Rowe, Ednor M.; Hansen, Roger W.)V1548,179-187(1991)

Circular polarization effects in the light scattering from single and suspensions of dinoflagellates (Shapiro, Daniel B.; Hunt, Arlon J.; Quinby-Hunt, Mary S.; Hull, Patricia G.)V1537,30-41(1991)

Correction method for optical-signal detection-error caused by quantum noise (Fujihashi, Chugo)V1583,298-306(1991)

Coupled-dipole approximation: predicting scattering by nonspherical marine organisms (Hull, Patricia G.; Hunt, Arlon J.; Quinby-Hunt, Mary S.; Shapiro, Daniel B.)V1537,21-29(1991)

Coupled-mode analysis of dynamic polarization volume holograms (Huang, Tizhi; Wagner, Kelvin H.)V1559,377-384(1991)

Defect enhancement of local electric fields in dielectric films (Risser, Steven M.; Ferris, Kim F.)V1441,262-268(1991)

Depolarized fiber optic gyro for future tactical applications (Bramson, Michael D.)V1367,155-160(1991)

Double dispersion from dichromated gelatin volume transmission gratings (Sheat, Dennis E.; Chamberlin, Giles R.; McCartney, David J.)V1461,35-38(1991)

Dual-eigenstate polarization preserving fiber optic sensor (Yu, Dong X.; Storti, George M.)V1584,236-242(1991)

Effect of polarization-mode dispersion on coherent optical distribution systems with shared local oscillator (Kao, Ming-Seng)V1579,221-229(1991)

Effect of polarization on two-color laser-induced damage (Becker, Wilhelm; McIver, John K.; Guenther, Arthur H.)V1441,541-552(1991)

Grating beam splitting polarizer using multilayer resist method (Aoyama, Shigeru; Yamashita, Tsukasa)V1545,241-250(1991)

High-efficiency holograms on the basis of polarization recording by means of beams from opposite directions (Shatalin, Igor D.)V1238,68-73(1991)

High-speed digital ellipsometer for the study of fiber optic sensor systems (Saxena, Indu)V1367,367-373(1991)

Investigation of strain birefringence and wavefront distortion in 001 plates of KD2PO4 (De Yoreo, James J.; Woods, Bruce W.)V1561,50-58(1991)

Laser linac in vacuum: assessment of a high-energy particle accelerator (Moore, Gerald T.; Bochove, Erik J.; Scully, Marlan O.)V1497,328-337(1991)

Laser range-gated underwater imaging including polarization discrimination (Swartz, Barry A.; Cummings, James D.)V1537,42-56(1991)

Linear and nonlinear optical properties of polymeric Langmuir-Blodgett films (Penner, Thomas L.; Willand, Craig S.; Robello, Douglas R.; Schildkraut, Jay S.; Ulman, Abraham)V1436,169-178(1991)

Liquid circular polarizer in laser system (Cesarz, Tadeusz; Klosowicz, Stanislaw; Zmija, Jozef)V1391,244-249(1991)

Low-frequency chiral coatings (Ro, Ru-Yen; Varadan, Vasundara V.; Varadan, Vijay K.)V1489,46-55(1991)

Magnetic circular dichroism measurements of benzene in the 9-eV region with synchrotron radiation (Snyder, Patricia A.; Munger, Robert; Hansen, Roger W.; Rowe, Ednor M.)V1548,188-196(1991)

Measurement of fibre Verdet constant with twist method (Dong, Xiaopeng; Hu, Hao; Qian, Jingren)V1572,56-60(1991)

Measurement of polarization model dispersion and mode-coupling parameter of a polarization-maintaining fiber (Huang, Zhaoming; Wang, Chunhua; Zhang, Jinghua)V1572,140-143(1991)

Measurement of synchrotron beam polarization (Singman, Leif V.; Davis, Brent A.; Holmberg, D. L.; Blake, Richard L.; Hockaday, Robert G.)V1548,80-92(1991)

Method of investigations of the optical properties of anisotropic materials using modulation of light polarization (Gumienny, Zbigniew; Misiewicz, Jan)V1527,462-465(1991)

Minimum polarization modulation: a highly bandwidth efficient coherent optical modulation scheme (Benedetto, Sergio; Kazovsky, Leonid G.; Poggiolini, Pierluigi T.)V1579,112-121(1991)

Modelization of a semi-leaky waveguide: application to a polarizer (Saint-Andre, Francoise; Benech, Pierre; Kevorkian, Antoine P.)V1583,278-288(1991)

Multipulse polarized-holographic set-up for isodromic fringe registration in the field of running stress waves (Khesin, G. L.; Sakharov, V. N.; Zhavoronok, I. V.; Rottenkolber, Hans; Schorner, Jurgen)V1554B,86-90(1991)

New experimental methods and theory of Raman optical activity (Nafie, Laurence A.; Che, Diping; Yu, Gu-Sheng; Freedman, Teresa B.)V1432,37-49(1991)

New lithium niobate Y-junction polarization divider: theoretical study (Conese, Tiziana; De Pascale, Olga; Matteo, Annamaria; Armenise, Mario N.)V1583,249-255(1991)

New technique for multiplying the isoclinic fringes (Wen, Mei-Yuan; Liu, Guang T.)V1332,673-675(1991)

Novel configuration of a two-mode-interference polarization splitter with a buffer layer (Antuofermo, Pasquale; Losacco, Aurora M.; De Pascale, Olga)V1583,143-149(1991)

Novel system for measuring extinction ratio on polarization-maintaining fibers and their devices (Xu, Sen-lu; Sheng, Lie-yi; Zhu, Lie-wei)V1367,303-308(1991)

Photoelastic modeling of linear and nonlinear creep using two- and three-dimensional models (Vardanjan, G. S.; Musatov, L. G.)V1554A,496-502(1991)

Photoelastic stress analysis of bridge bearings (Allison, Ian M.)V1554A,332-340(1991)

Polarization approach to high-sensitivity moire interferometry (Salbut, Leszek A.; Patorski, Krzysztof; Kujawinska, Malgorzata)V1554B,451-460(1991)

Polarization-based all-optical bistable element (Domanski, Andrzej W.; Karpierz, Miroslaw A.; Strojewski, Dariusz)V1505,59-66(1991)

Polarization conversion through the excitation of electromagnetic modes on a grating (Bryan-Brown, G. P.; Elston, S. J.; Sambles, J. R.)V1545,167-178(1991)

Polarization dependence and uniformity of FLC layers for phase modulation (Biernacki, Paul D.; Brown, Tyler; Freeman, Mark O.)V1455,167-178(1991)

Polarization dependence of light scattered from rough surfaces with steep slopes (O'Donnell, Kevin A.; Knotts, Micheal E.)V1558,362-367(1991)

Polarization-dependent EXAFS studies in layered copper oxide superconductors (Mini, Susan M.; Alp, Esen E.; Ramanathan, Mohan; Bommannavar, A.; Hyun, O. B.)V1345,260-269(1991)

Polarization-dependent x-ray spectroscopy of high-Tc superconductors (Heald, Steven M.; Tranquada, John M.)V1550,67-75(1991)

Polarization-diversity fiber networks (Cullen, Thomas J.)V1372,164-172(1991)

Polarization effects in fiber lasers: phenomena, theory, and applications (Lin, Jin T.; Gambling, William A.)V1373,42-53(1991)

Polarization holographic elements (Kakichashvili, Shermazan D.)V1574,101-108(1991)

Polarization-maintaining single-mode fibers: measurement and prediction of fundamental characteristics (Sasek, Ladislav; Vohryzek, Jachym)V1504,147-154(1991)

Polarization-maintaining single-mode fibers with layered core (Sasek, Ladislav; Vohryzek, Jachym; Sochor, Vaclav; Paulicka, Ivan; van Nhac, Nguyen; Franek, Alexandr)V1572,151-156(1991)

Polarization-modulated coherent optical communication systems (Betti, Silvello; Curti, Franco; De Marchis, Giancarlo; Iannone, Eugenio)V1579,100-111(1991)

Polarization modulation with frequency shift redundancy and frequency modulation with polarization redundancy for POLSK and FSK systems (Marone, Giuseppe; Calvani, Riccardo A.; Caponi, Renato)V1579,122-132(1991)

Polarization of emission lines from relativistic accretion disk (Chen, Kaiyou)V1548,23-33(1991)

Polarization sensitivity of x-ray photocathodes in the 60-200eV band (Fraser, George W.; Pain, M. D.; Pearson, James F.; Lees, John E.; Binns, C. R.; Shaw, Ping-Shine; Fleischman, Judith R.)V1548,132-148(1991)

Polarization-state mixing in multiple-beam diffraction and its application to solving the phase problem (Shen, Qun)V1550,27-33(1991)

Polarized nature of synchrotron radiation (Kim, Kwang-je)V1548,73-79(1991)

Polarized x-ray absorption spectroscopy for the study of superconductors and magnetic materials (Ramanathan, Mohan; Alp, Esen E.; Mini, Susan M.; Salem-Sugui, S.; Bommannavar, A.)V1548,168-178(1991)

Polarizing holographic optical elements for optical data storage (Ono, Yuzo)V1555,177-181(1991)

Polarizing optics for the soft x-ray regime: whispering-gallery mirrors and multilayer beamsplitters (Braud, John P.)V1548,69-72(1991)

Polarizing x-ray optics for synchrotron radiation (Hart, Michael)V1548,46-55(1991)

Production and Analysis of Polarized X Rays (Siddons, D. P., ed.)V1548(1991)

Sensitivity of polarization-maintaining fibers to temperature variations (Ruffin, Paul B.; Sung, C. C.)V1478,160-167(1991)

Set of two 45 - 90 - 45 prisms equivalent to the Fresnel rhomb (Murty, M. V.; Shukla, Ram P.; Apparao, K. V.)V1332,41-49(1991)

Silicon nitride single-polarization optical waveguides on silicon substrates (De Brabander, Gregory N.; Boyd, Joseph T.; Jackson, Howard E.)V1583,327-330(1991)

Simultaneous measurement of refractive index and thickness of thin film by polarized reflectances (Kihara, Tami; Yokomori, Kiyoshi)V1332,783-791(1991)

Spectroscopy and polarimetry capabilities of the INTEGRAL imager: Monte Carlo simulation results (Swinyard, Bruce M.; Malaguti, Giuseppe; Caroli, Ezio; Dean, Anthony J.; Di Cocco, Guido)V1548,94-105(1991)

Stokes vectors, Mueller matrices, and polarized scattered light: experimental applications to optical surfaces and all other scatterers (Bickel, William S.; Videen, Gorden W.)V1530,7-14(1991)

Stress-static and strength research of building of NPS and nuclear reactor under power and thermal loads including creep and relaxation influence (Savostjanov, V. N.; Zavalishin, S. I.; Smirnov, S. B.; Morosova, D. V.)V1554A,579-585(1991)

Survey of synchrotron radiation devices producing circular or variable polarization (Kim, Kwang-je)V1345,116-124(1991)

Techniques of production and analysis of polarized synchrotron radiation (Mills, Dennis M.)V1345,125-136(1991)

Theoretical investigation of near-edge phenomena in magnetic systems (Carra, Paolo)V1548,35-44(1991)

Theoretical models for stellar x-ray polarization in compact objects (Meszaros, Peter)V1548,13-22(1991)

Three-dimensional holograms in polarization-sensitive media (Kakichashvili, Shermazan D.; Kilosanidze, Barbara N.)V1238,74-79(1991)

Two-dimensional spatial light modulators using polarization-sensitive multiple-quantum-well light valve (Jain, Faquir C.; Drake, G.; Chung, C.; Bhattacharjee, K. K.; Cheung, S. K.)V1564,714-722(1991)

Two-wave coupled-wave theory of the polarizing properties of volume phase gratings (Tholl, Hans D.)V1555,101-111(1991)

Vectorial photoelectric effect at 2.69 keV (Shaw, Ping-Shine; Hanany, Shaul; Liu, Yee; Church, Eric D.; Fleischman, Judith R.; Kaaret, Philip E.; Novick, Robert; Santangelo, A.)V1548,118-131(1991)

What can we learn that's new and interesting about condensed matter systems using polarized x rays? (Platzman, P. M.)V1548,34-34(1991)

X-ray absorption spectroscopy with polarized synchrotron radiation (Alp, Esen E.; Mini, Susan M.; Ramanathan, Mohan; Hyun, O. B.)V1345,137-145(1991)

Pollution—see also environmental effects; particles

Advances in tunable diode laser technology for atmospheric monitoring applications (Wall, David L.)V1433,94-103(1991)

Analysis of exhaust from clean-fuel vehicles using FTIR spectroscopy (Rieger, Paul L.; Maddox, Christine E.)V1433,290-301(1991)

Applicability of open-path monitors at Superfund sites (Padgett, Joseph; Pritchett, Thomas H.)V1433,352-364(1991)

Application of tunable diode lasers in control of high-pure-material technologies (Nadezhdinskii, Alexander I.; Stepanov, Eugene V.; Kuznetzov, Andrian I.; Devyatykh, Grigory G.; Maximov, G. A.; Khorshev, V. A.; Shapin, S. M.)V1418,487-495(1991)

Auto exhaust gas analysis by FTIR spectroscopy (Herget, William F.; Lowry, Steven R.)V1433,275-289(1991)

Chemical amplifier for peroxy radical measurements based on luminol chemiluminescence (Cantrell, Chris A.; Shetter, Richard E.; Lind, John A.; Gilliland, Curt A.; Calvert, Jack G.)V1433,263-268(1991)

Chemical, Biochemical, and Environmental Fiber Sensors II (Lieberman, Robert A.; Wlodarczyk, Marek T., eds.)V1368(1991)

Collisional effects in laser detection of tropospheric OH (Crosley, David R.)V1433,58-68(1991)

Developing a long-path diode array spectrometer for tropospheric chemistry studies (Lanni, Thomas R.; Demerjian, Kenneth L.)V1433,21-24(1991)

Developing a trial burn plan (Smith, Walter S.; Wong, Tony; Williams, Gary L.; Brintle, David G.)V1434,14-25(1991)

Diode laser spectroscopy of atmospheric pollutants (Nadezhdinskii, Alexander I.; Stepanov, Eugene V.)V1433,202-210(1991)

DOAS (differential optical absorption spectroscopy) urban pollution measurements (Stevens, Robert K.; Vossler, T. L.)V1433,25-35(1991)

Dynamic range limits in field determination of fluorescence using fiber optic sensors (Chudyk, Wayne; Pohlig, Kenneth)V1368,105-114(1991)

Extractive sampling systems for continuous emissions monitors (White, John R.)V1434,104-112(1991)

Fast-response water vapor and carbon dioxide sensor (Kohsiek, W.)V1511,114-119(1991)

Field and laboratory studies of Fourier transform infrared spectroscopy in continuous emissions monitoring applications (Plummer, Grant M.)V1434,78-89(1991)

FTIR: fundamentals and applications in the analysis of dilute vehicle exhaust (Gierczak, Christine A.; Andino, J. M.; Butler, James W.; Heiser, G. A.; Jesion, G.; Korniski, T. J.)V1433,315-328(1991)

Gradient microbore liquid chromatography with dual-wavelength absorbance detection: tunable analyzers for remote chemical monitoring (Sulya, Andrew W.; Moore, Leslie K.; Synovec, Robert E.)V1434,147-158(1991)

History and status of black and white photographic processing chemicals as effluents (Horn, Richard R.)V1458,69-75(1991)

Incinerator technology overview (Santoleri, Joseph J.)V1434,2-13(1991)

Infrared monitoring of combustion (Bates, Stephen C.; Morrison, Philip W.; Solomon, Peter R.)V1434,28-38(1991)

In situ measurement of methyl bromide in indoor air using long-path FTIR spectroscopy (Green, Martina; Seiber, James N.; Biermann, Heinz W.)V1433,270-274(1991)

In situ monitoring for hydrocarbons using fiber optic chemical sensors (Klainer, Stanley M.; Thomas, Johnny R.; Dandge, Dileep K.; Frank, Chet A.; Butler, Marcus S.; Arman, Helen; Goswami, Kisholoy)V1434,119-126(1991)

Kinetic studies of phosgene reduction via in situ Fourier transform infrared analysis (Farquharson, Stuart; Chauvel, J. P.)V1434,135-146(1991)

Laser-induced fluorescence in contaminated soils (Lurk, Paul W.; Cooper, Stafford S.; Malone, Philip G.; Olsen, R. S.; Lieberman, Stephen H.)V1434,114-118(1991)

Long-pathlength DOAS (differential optical absorption spectrometer) system for the in situ measurement of xylene in indoor air (Biermann, Heinz W.; Green, Martina; Seiber, James N.)V1433,2-7(1991)

Low-cost in-soil organic contaminant sensor (Brossia, Charles E.; Wu, Samuel C.)V1368,115-120(1991)

Measurements of nitrous acid, nitrate radicals, formaldehyde, and nitrogen dioxide for the Southern California Air Quality Study by differential optical absorption spectroscopy (Winer, Arthur M.; Biermann, Heinz W.)V1433,44-55(1991)

Multicomponent analysis using established techniques (Dillehay, David L.)V1434,56-66(1991)

Preliminary field demonstration of a fiber optic trichloroethylene sensor (Angel, S. M.; Langry, Kevin; Colston, B. W.; Roe, Jeffrey N.; Daley, Paul F.; Milanovich, Fred P.)V1368,98-104(1991)

Study of nighttime NO3 chemistry by differential optical absorption spectroscopy (Plane, John M.; Nien, Chia-Fu)V1433,8-20(1991)

Use of a Fourier transform spectrometer as a remote sensor at Superfund sites (Russwurm, George M.; Kagann, Robert H.; Simpson, Orman A.; McClenny, William A.)V1433,302-314(1991)

Use of Fourier transform spectroscopy in combustion effluent monitoring (Howe, Gordon S.; McIntosh, Bruce C.)V1434,90-103(1991)

Wedge imaging spectrometer: application to drug and pollution law enforcement (Elerding, George T.; Thunen, John G.; Woody, Loren M.)V1479,380-392(1991)

Polymers

Accurate measurement of thin-polymeric-films index variations: application to elasto-optic effect and to photochromism (Dumont, Michel L.; Morichere, D.; Sekkat, Z.; Levy, Yves)V1559,127-138(1991)

Addressing factors for polymer-dispersed liquid-crystal displays (Margerum, J. D.; Lackner, Anna M.; Erdmann, John H.; Sherman, E.)V1455,27-38(1991)

Advances in Resist Technology and Processing VIII (Ito, Hiroshi, ed.)V1466(1991)

Analysis on image performance of a moisture-absorbed plastic singlet for an optical disk (Tanaka, Yuki; Miyamae, Hiroshi)V1354,395-401(1991)

Anamorphosor for scintillating plastic optical fiber applications (Chiron, Bernard)V1592,158-164(1991)

Applications of electro-optic polymers to optical interconnects (Lytel, Richard S.; Lipscomb, George F.; Kenney, John T.; Ticknor, Anthony J.)V1563,122-138(1991)

Applications of optical polymer waveguide devices on future optical communication and signal processing (Keil, Norbert; Strebel, Bernhard N.; Yao, HuiHai; Krauser, Juergen)V1559,278-287(1991)

Artificial dielectric waveguides from semiconductor-embedded polymers (Grebel, Haim)V1583,355-361(1991)

Bandwidth measurements of polymer optical fibers (Karim, Douglas P.)V1592,31-41(1991)

Bilayer resist system utilizing alkali-developable organosilicon positive photoresist (Nate, Kazuo; Mizushima, Akiko; Sugiyama, Hisashi)V1466,206-210(1991)

Characterization of fluorescent plastic optical fibers for x-ray beam detection (Laguesse, Michel F.; Bourdinaud, Michel J.)V1592,96-107(1991)

Characterization of PMMA-EVA blend via photoacoustic technique (Carbonara, Giuseppe; Mormile, Pasquale; Bernini, Umberto; Russo, Paolo; Malinconico, Mario; Volpe, M. G.)V1361,1038-1040(1991)

Charge buildup in polypropylene thin films (Ding, Hai)V1519,847-856(1991)

Chemically amplified negative-tone photoresist for sub-half-micron device and mask fabrication (Conley, Willard E.; Dundatscheck, Robert; Gelorme, Jeffrey D.; Horvat, John; Martino, Ronald M.; Murphy, Elizabeth; Petrosky, Anne; Spinillo, Gary; Stewart, Kevin; Wilbarg, Robert; Wood, Robert L.)V1466,53-66(1991)

Collinear asymmetrical polymer waveguide modulator (Chen, Ray T.; Sadovnik, Lev S.)V1559,449-460(1991)

Comparative study of the dielectric and optical response of PDLC films (Seekola, Desmond; Kelly, Jack R.)V1455,19-26(1991)

Complex researches optically controlled liquid-crystal spatial-time light modulators on the photoconductivity organic polymer basis (Groznov, Michail A.)V1500,281-289(1991)

Connection system designed for plastic optical fiber local area networks (Cirillo, James R.; Jennings, Kurt L.; Lynn, Mark A.; Messuri, Dominic A.; Steele, Robert E.)V1592,53-59(1991)

Contributions of pi' bonding to the nonlinear optical properties of inorganic polymers (Ferris, Kim F.; Risser, Steven M.)V1441,510-520(1991)

Controlled structuring of polymer surfaces by UV laser irradiation (Bahners, Thomas; Kesting, Wolfgang; Schollmeyer, Eckhard)V1503,206-214(1991)

Cumulative effect and cutting quality improvement of XeCl laser ablation of PMMA (Lou, Qihong; Guo, Hongping)V1598,221-226(1991)

Decay of the nonlinear susceptibility components in main-chain functionalized poled polymers (Meyrueix, Remi; LeCompte, J. P.; Tapolsky, Gilles)V1560,454-466(1991)

Design of highly soluble extended pi-electron oligomers capable of supporting stabilized delocalized bipolaronic states for NLO applications (Havelka, Kathleen O.; Spangler, Charles W.)V1560,66-74(1991)

Determination of the orientational order parameters <P*2>, <P*4> in a polysilane LB film via polarization-dependent THG (Neher, Dieter; Mittler-Neher, S.; Cha, M.; Stegeman, George I.; Embs, F. W.; Wegner, Gerhard; Miller, Robert D.; Willson, C. G.)V1560,335-343(1991)

Development of chemical sensors using plastic optical fiber (Zhou, Quan; Tabacco, Mary B.; Rosenblum, Karl W.)V1592,108-113(1991)

Dielectric relaxation studies of x2 dye containing polystyrene films (Schen, Michael A.; Mopsik, Fred)V1560,315-325(1991)

Diffusion of spherical probes in aqueous systems containing the semiflexible polymer hydroxypropylcellulose (Mustafa, Mazidah B.; Russo, Paul S.)V1430,132-141(1991)

Dispersion of x(3) in fused aromatic ladder polymers and their precursors probed by third-harmonic generation (Meth, Jeffrey S.; Vanherzeele, Herman A.; Jenekhe, Samson A.; Roberts, Michael F.; Agrawal, A. K.; Yang, Chen-Jen)V1560,13-24(1991)

Dissolution kinetics of high-resolution novolac resists (Itoh, Katsuyuki; Yamanaka, Koji; Nozue, Hiroshi; Kasama, Kunihiko)V1466,485-496(1991)

Dissolution of poly(p-hydroxystyrene): molecular weight effects (Long, Treva; Rodriguez, Ferdinand)V1466,188-198(1991)

DQN photoresist with tetrahydroxydiphenylmethane as ballasting group in PAC (Tzeng, Chao H.; Lin, Dhei-Jhai; Lin, Song S.; Huang, Dong-Tsair; Lin, Hwang-Kuen)V1466,469-476(1991)

Droplet-size effects in light scattering from polymer-dispersed liquid-crystal films (Montgomery, G. P.; West, John L.; Tamura-Lis, Winifred)V1455,45-53(1991)

Droplet-size polydispersity in polymer-dispersed liquid-crystal films (Vaz, Nuno A.; Smith, George W.; VanSteenkiste, T. H.; Montgomery, G. P.)V1455,110-122(1991)

Dry development and plasma durability of resists: melt viscosity and self-diffusion effects (Paniez, Patrick J.; Joubert, Olivier P.; Pons, Michel J.; Oberlin, Jean C.; Weill, Andre P.)V1466,583-591(1991)

Dry photopolymer embossing: novel photoreplication technology for surface relief holographic optical elements (Shvartsman, Felix P.)V1507,383-391(1991)

Dye orientation in organic guest host systems on ferroelectric polymers (Osterfeld, Martin; Knabke, Gerhard; Franke, Hilmar)V1559,49-55(1991)

Dynamical structure factor of a solution of charged rod-like polymers in the isotropic phase (Maeda, T.; Doi, Masao)V1403,268-277(1991)

Dynamic light scattering from a side-chain liquid crystalline polymer in a nematic solvent (Devanand, Krisha)V1430,160-164(1991)

Dynamic photoelasticity applied to crack-branching investigations (Hammami, Slimane; Cottron, M.; Lagarde, Alexis)V1554A,136-142(1991)

Dynamics of stiff macromolecules in concentrated polymer solutions: model of statistical reorientations (Brazhnik, O. D.; Khokhlov, A. R.)V1402,70-77(1991)

Dynamics of wormlike chains: theory and computer simulations (Aragon, Sergio R.; Luo, Rolland)V1430,65-84(1991)

Effective temperatures of polymer laser ablation (Furzikov, Nickolay P.)V1503,231-235(1991)

Effect of counterion distribution on the electrostatic component of the persistence length of flexible linear polyions (Klearman, Debbie; Schmitz, Kenneth S.)V1430,236-255(1991)

Effect of polymer mixtures on the performance of PDLC films (Heavin, Scott D.; Fung, Bing M.; Mears, Richard B.; Sluss, James J.; Batchman, Theodore E.)V1455,12-18(1991)

Effect of sensitizer spatial distribution on dissolution inhibition in novolak/diazonaphthoquinone resists (Rao, Veena; Kosbar, Laura L.; Frank, Curtis W.; Pease, R. F.)V1466,309-323(1991)

Electrochromic properties of poly(pyrrole)/dodecylbenzenesulfonate (Peres, Rosa C.; De Paoli, Marco-Aurelio; Panero, Stefania; Scrosati, Bruno)V1559,151-158(1991)

Electrochromic property and chemical sensitivity of conducting polymer PAn film (Yuan, Ren K.; Gu, Zhi P.; Yuan, Hong; Yuan, Xue S.; Wang, Yong B.; Liu, Xiang N.; Shen, Xue C.)V1519,831-834(1991)

Electro-optical light modulation in novel azo-dye-substituted poled polymers (Shuto, Yoshito; Amano, Michiyuki; Kaino, Toshikuni)V1560,184-195(1991)

Electro-optic coefficients in electric-field poled-polymer waveguides (Smith, Barton A.; Herminghaus, Stephan; Swalen, Jerome D.)V1560,400-405(1991)

Electro-optic measurements of dye/polymer systems (Wang, Chin H.; Guan, H. W.; Zhang, J. F.)V1559,39-48(1991)

Electro-optic polymer devices for optical interconnects (Lytel, Richard S.; Binkley, E. S.; Girton, Dexter G.; Kenney, John T.; Lipscomb, George F.; Ticknor, Anthony J.; Van Eck, Timothy E.)V1389,547-558(1991)

Evaluation of phenolic resists for 193 nm surface imaging (Hartney, Mark A.; Johnson, Donald W.; Spencer, Allen C.)V1466,238-247(1991)

Evaluation of polymeric thin film waveguides as chemical sensors (Bowman, Elizabeth M.; Burgess, Lloyd W.)V1368,239-250(1991)

Evaluation of poly(p-trimethylsilylstyrene and p-pentamethyldisilylstyrene sulfone)s as high-resolution electron-beam resists (Gozdz, Antoni S.; Ono, Hiroshi; Ito, Seiki; Shelburne, John A.; Matsuda, Minoru)V1466,200-205(1991)

Excimer laser processing of ceramics and fiber-reinforced polymers assisted by a diagnostic system (Geiger, Manfred; Lutz, Norbert; Biermann, Stephan)V1503,238-248(1991)

Excited-state nonlinear optical processes in polymers (Heflin, James R.; Zhou, Qihou L.; Garito, Anthony F.)V1497,398-407(1991)

Experimental measurements and electron correlation theory of third-order nonlinear optical processes in linear chains (Heflin, James R.; Cai, Yongming; Zhou, Qihou L.; Garito, Anthony F.)V1560,2-12(1991)

Extended pi-electron systems incorporating stabilized quinoidal bipolarons: anthracenyl polyene-polycarbonate composites (Nickel, Eric G.; Spangler, Charles W.)V1560,35-43(1991)

Fabrication techniques of photopolymer-clad waveguides for nonlinear polymeric modulators (Tumolillo, Thomas A.; Ashley, Paul R.)V1559,65-73(1991)

Four-dimensional photosensitive materials: applications for shaping, time-domain holographic dissection, and scanning of picosecond pulses (Saari, Peeter M.; Kaarli, Rein K.; Sonajalg, Heiki)V1507,500-508(1991)

Frequency-tunable THG measurements of x(3) between 1-2.1um of organic conjugated-polymer films using an optical parametric oscillator (Gierulski, Alfred; Naarmann, Herbert; Schrof, Wolfgang; Ticktin, Anton)V1560,172-182(1991)

Graded-index polymer optical fiber by new random copolymerization technique (Koike, Yasuhiro; Hondo, Yukie; Nihei, Eisuke)V1592,62-72(1991)

Guest-host polymer fibers and fractal clusters for nonlinear optics (Kuzyk, Mark G.; Andrews, Mark P.; Paek, Un-Chul; Dirk, Carl W.)V1560,44-53(1991)

High-frequency fiber optic phase modulator using piezoelectric polymer coating (Imai, Masaaki M.; Yano, T.; Ohtsuka, Yoshihiro)V1371,13-20(1991)

Highly efficient plastic optical fluorescent fibers and sensors (Chiron, Bernard)V1592,86-95(1991)

High-sensitive photopolymer for large-size holograms (Ishikawa, Toshiharu; Kuwabara, Y.; Koseki, Kenichi; Yamaoka, Tsuguo)V1461,73-78(1991)

High-speed board-to-board optical interconnection (Chen, Ray T.; Lu, Huey T.; Robinson, Daniel; Plant, David V.; Fetterman, Harold R.)V1559,110-117(1991)

High-Tg nonlinear optical polymer: poly(N-MNA acrylamide) (Herman, Warren N.; Rosen, Warren A.; Sperling, L. H.; Murphy, C. J.; Jain, H.)V1560,206-213(1991)

Holographic optical elements by dry photopolymer embossing (Shvartsman, Felix P.)V1461,313-320(1991)

Identification of modes in dynamic scattering on ternary polymer mixtures (Akcasu, A. Z.)V1430,142-143(1991)

Improved plastic molding technology for magneto-optical disk substrate (Galic, George J.)V1396,539-546(1991)

Improved PMMA single-piece haptic materials (Healy, Donald D.; Wilcox, Christopher D.)V1529,84-93(1991)

Influencing adherence properties of polymers by excimer laser radiation (Breuer, J.; Metev, S.; Sepold, Gerd; Krueger, G.; Hennemann, O. D.)V1503,223-230(1991)

In-plane poling of doped-polymer films: creation of two asymmetric directions for second-order optical nonlinearity (Berkovic, Garry; Yitzchaik, Shlomo; Krongauz, Valeri)V1560,238-242(1991)

In-situ monitoring of silylation mechanisms by laser interferometry (Pierrat, Christophe; Paniez, Patrick J.; Martin, P.)V1466,248-256(1991)

Ion implantation of polymers for electrical conductivity enhancement (Bridwell, Lynn B.; Wang, Y. Q.)V1519,878-883(1991)

Key issues in selecting plastic optical fibers used in novel medical sensors (Kosa, Nadhir B.)V1592,114-121(1991)

KrF laser ablation of polyurethane (Kueper, Stephan; Brannon, James H.)V1598,27-35(1991)

Langmuir-Blodgett films for second-order nonlinear optics (Penner, Thomas L.; Armstrong, Nancy J.; Willand, Craig S.; Schildkraut, Jay S.; Robello, Douglas R.)V1560,377-386(1991)

Langmuir-Blodgett films of tetra-tert-butyl-phenoxy phthalocyanine iron [II] (Luo, Tao; Zhang, Wei Q.; Gan, Fuxi)V1519,826-830(1991)

Laser Doppler diagnostics of vorticity and phenomenological description of the flows of dilute polymer solutions in model tubes (Fedoseeva, E. V.; Polyakova, M. S.)V1403,355-358(1991)

Light budget and optimization strategies for display applications of dichroic nematic droplet/polymer films (Drzaic, Paul S.)V1455,255-263(1991)

Light-induced volume-phase holograms for cold neutrons (Ibel, Konrad; Matull, Ralph; Rupp, Romano A.; Eschkoetter, Peter; Hehmann, Joerg)V1559,393-402(1991)

Light scattering holographic optical elements formation in photopolymerizable layers (Boiko, Yuri B.; Granchak, Vasilij M.)V1507,544-548(1991)

Linear and nonlinear optical effects in polymer waveguides (Braeuer, Andreas; Bartuch, Ulrike; Zeisberger, M.; Bauer, T.; Dannberg, Peter)V1559,470-478(1991)

Liquid-crystal-doped polymers as volume holographic elements (Chen, Geng-Sheng; Brady, David J.)V1562,128-135(1991)

Local area network applications of plastic optical fiber (Cirillo, James R.; Jennings, Kurt L.; Lynn, Mark A.; Messuri, Dominic A.; Steele, Robert E.)V1592,42-52(1991)

Low-loss polymer thin-film optical waveguides (Amleshi, Peerouz M.; Naylor, David L.)V1396,396-403(1991)

Mapping of imbedded contaminants in transparent material by optical scatter (Rudberg, Donald A.; Stover, John C.; McGary, Douglas E.)V1530,232-239(1991)

Mass-producible optical guided-wave devices fabricated by photopolymerization (Hosokawa, Hayami; Horie, Noriyoshi; Yamashita, Tsukasa)V1559,229-237(1991)

Material testing by the laser speckle strain gauge (Yamaguchi, Ichirou; Kobayashi, Koichi)V1554A,240-249(1991)

Mechanism of dissolution inhibition of novolak-diazoquinone resist (Furuta, Mitsuhiro; Asaumi, Shingo; Yokota, Akira)V1466,477-484(1991)

Mechanistic studies on the poly(4-tert-butoxycarbonyloxystyrene)/triphenylsulfonium salt photoinitiation process (Hacker, Nigel P.; Welsh, Kevin M.)V1466,384-393(1991)

Metal-ion-doped polymer systems for real-time holographic recording (Lessard, Roger A.; Changkakoti, Rupak; Manivannan, Gurusamy)V1559,438-448(1991)

Micro-optic studies using photopolymers (Phillips, Nicholas J.; Barnett, Christopher A.)V1544,10-21(1991)

Microscopic mechanism of optical nonlinearity in conjugated polymers and other quasi-one-dimensional systems (Mazumdar, Sumit; Guo, Dandan; Dixit, Sham N.)V1436,136-149(1991)

Molecular anchoring at the droplet wall in PDLC materials (Crawford, Gregory P.; Ondris-Crawford, Renate; Doane, J. W.)V1455,2-11(1991)

Multilayered optically activated devices based on organic third-order nonlinear optical materials (Swanson, David A.; Altman, Joe C.; Sullivan, Brian J.; Spitzer, Ronnie; Hansen, Glenn A.; Stone, Richard E.)V1560,416-425(1991)

Multiline holographic notch filters (Ning, Xiaohui; Masso, Jon D.)V1545,125-129(1991)

Multiple mode and multiple source coupling into polymer thin-film waveguides (Potter, B. L.; Walker, D. S.; Greer, L.; Saavedra, Steven S.; Reichert, William M.)V1368,251-257(1991)

Nd:YAG-laser-based time-domain reflectometry measurements of the intrinsic reflection signature from PMMA fiber splices (Lawson, Christopher M.; Michael, Robert R.; Dressel, Earl M.; Harmony, David W.)V1592,73-83(1991)

Near-UV laser ablation of doped polymers (Ihlemann, Juergen; Bolle, Matthias; Luther, Klaus; Troe, Juergen)V1361,1011-1019(1991)

Negative chemical amplification resist systems based on polyhydroxystyrenes and N-substituted imides or aldehydes (Ito, Hiroshi; Schildknegt, Klaas; Mash, Eugene A.)V1466,408-418(1991)

New aqueous base-developable negative-tone photoresist based on furans (Fahey, James T.; Frechet, Jean M.)V1466,67-74(1991)

New class of mainchain chromophoric nonlinear optical polymers (Lindsay, Geoffrey A.; Stenger-Smith, John D.; Henry, Ronald A.; Hoover, James M.; Kubin, R. F.)V1497,418-422(1991)

New high-performance material in nonlinear optics field, the polymer blend PMMA-EVA: a first investigation (Carbonara, Giuseppe; Mormile, Pasquale; Abbate, G.; Bernini, Umberto; Maddalena, P.; Malinconico, Mario)V1361,688-691(1991)

New polymers with large and stable second-order nonlinear optical effects (Chen, Mai; Yu, Luping; Dalton, Larry R.; Shi, Youngqiang; Steier, William H.)V1409,202-213(1991)

New syndioregic main-chain, nonlinear optical polymers, and their ellipsometric characterization (Lindsay, Geoffrey A.; Nee, Soe-Mie F.; Hoover, James M.; Stenger-Smith, John D.; Henry, Ronald A.; Kubin, R. F.; Seltzer, Michael D.)V1560,443-453(1991)

Nonlinear effects in hidden picture amplification and contrast improvement in polymer electron and Roentgenoresist PMMA (Aleksandrov, A. P.; Genkin, Vladimir N.; Myl'nikov, M. Y.; Rukhman, N. V.)V1440,442-453(1991)

Nonlinear optical properties of inorganic coordination polymers and organometallic complexes (Thompson, Mark E.; Chiang, William; Myers, Lori K.; Langhoff, Charles A.)V1497,423-429(1991)

Nonlinear Optical Properties of Organic Materials IV (Singer, Kenneth D., ed.)V1560(1991)

Nonlinear optical properties of phenosafranin-doped substrates (Speiser, Shammai)V1559,238-244(1991)

Nonlinear optical properties of poled polymers (Gadret, G.; Kajzar, Francois; Raimond, P.)V1560,226-237(1991)

Nonlinear optics in poled polymers with two-dimensional asymmetry (Berkovic, Garry; Krongauz, Valeri; Yitzchaik, Shlomo)V1442,44-52(1991)

Nonlinear polymers and devices (Moehlmann, Gustaaf R.; Horsthuis, Winfried H.; Hams, Benno H.)V1512,34-39(1991)

Nonlinear properties of poled-polymer films: SHG and electro-optic measurements (Morichere, D.; Dumont, Michel L.; Levy, Yves; Gadret, G.; Kajzar, Francois)V1560,214-225(1991)

Nonlinear susceptibilities investigated by the electroabsorption of polymer ion-hemicyanine dye complexes (Nomura, Shintaro; Kobayashi, Takayoshi T.; Nakanishi, Hachiro; Matsuda, Hiro; Okada, Shuji; Tomiyama, Hiromitsu)V1560,272-277(1991)

Novel acid-hardening positive photoresist technology (Graziano, Karen A.; Thompson, Stephen D.; Winkle, Mark R.)V1466,75-88(1991)

Novel base-generating photoinitiators for deep-UV lithography (Kutal, Charles; Weit, Scott K.; Allen, Robert D.; MacDonald, Scott A.; Willson, C. G.)V1466,362-367(1991)

Novel linear and ladder polymers from tetraynes for nonlinear optics (Okada, Shuji; Matsuda, Hiro; Masaki, Atsushi; Nakanishi, Hachiro; Hayamizu, Kikuko)V1560,25-34(1991)

Novel novolak resins using substituted phenols for high-performance positive photoresist (Kajita, Toru; Ota, Toshiyuki; Nemoto, Hiroaki; Yumoto, Yoshiji; Miura, Takao)V1466,161-173(1991)

Novel perfluorinated antireflective and protective coating for KDP and other optical materials (Thomas, Ian M.; Campbell, John H.)V1441,294-303(1991)

Novel plastic image-transmitting fiber (Suzuki, Fumio)V1592,150-157(1991)

Novolak design for high-resolution positive photoresists (IV): tandem-type novolak resin for high-performance positive photoresists (Hanabata, Makoto; Oi, F.; Furuta, Akihiro)V1466,132-140(1991)

Novolak resin design concept for high-resolution positive resists (Noguchi, Tsutomu; Tomita, Hidemi)V1466,149-160(1991)

Optical and environmentally protective coatings for potassium dihydrogen phosphate harmonic converter crystals (Thomas, Ian M.)V1561,70-82(1991)

Optically nonlinear polymeric devices (Moehlmann, Gustaaf R.; Horsthuis, Winfried H.; Mertens, Hans W.; Diemeer, Mart B.; Suyten, F. M.; Hendriksen, B.; Duchet, Christian; Fabre, P.; Brot, Christian; Copeland, J. M.; Mellor, J. R.; van Tomme, E.; van Daele, Peter; Baets, Roel G.)V1560,426-433(1991)

Optical memory using localized photoinduced anisotropy in a synthetic dye-polymer (Kuo, Chai-Pei)V1499,148-159(1991)

Optical waveguides in polymer materials by ion implantation (Frank, Werner F.; Kulisch, Juergen R.; Franke, Hilmar; Rueck, Dorothee M.; Brunner, Stefan; Lessard, Roger A.)V1559,344-353(1991)

Optimal characteristics of rheology and electric field in deformable polymer films of optoelectronic image formation devices (Tarasov, Victor A.; Kuleshov, Nickolay B.; Novoselets, Mikhail K.; Sarkisov, Sergey S.)V1559,331-342(1991)

Organic-dye films for write-once optical storage (Zhou, Jian P.; Shu, Ju P.; Xu, Hui J.)V1519,559-564(1991)

Organic electro-optic devices for optical interconnnection (Lipscomb, George F.; Lytel, Richard S.; Ticknor, Anthony J.; Van Eck, Timothy E.; Girton, Dexter G.; Ermer, Susan P.; Valley, John F.; Kenney, John T.; Binkley, E. S.)V1560,388-399(1991)

Overview of EO polymers for guided-wave devices (Ashley, Paul R.; Tumolillo, Thomas A.)V1583,316-326(1991)

Oxygen reactive ion etching of polymers: profile evolution and process mechanisms (Pilz, Wolfgang; Janes, Joachim; Muller, Karl P.; Pelka, Joachim)V1392,84-94(1991)

Peculiarities of anisotropic photopolymerization in films (Krongauz, Vadim V.; Schmelzer, E. R.)V1559,354-376(1991)

Phase-matched second harmonic generations in poled dye-polymer waveguides (Sugihara, Okihiro; Kinoshita, Takeshi; Okabe, M.; Kunioka, S.; Nonaka, Y.; Sasaki, Keisuke)V1361,599-605(1991)

Photoablative etching of materials for optoelectronic integrated devices (Lemoine, Patrick; Magan, John D.; Blau, Werner)V1377,45-56(1991)

Photoanisotropic polymeric media and their application in optical devices (Barachevsky, Valery A.)V1559,184-193(1991)

Photochemical hole burning in rigidly coupled polyacenes (Iannone, Mark A.; Salt, Kimberly L.; Scott, Gary W.; Yamashita, Tomihiro)V1559,172-183(1991)

Photochemistry and fluorescence spectroscopy of polymeric materials containing triphenylsulfonium salts (Hacker, Nigel P.; Welsh, Kevin M.)V1559,139-150(1991)

Photoconducting nonlinear optical polymers (Li, Lian; Jeng, Ru J.; Lee, J. Y.; Kumar, Jayant; Tripathy, Sukant K.)V1560,243-250(1991)

Photoconductivity of bridged polymeric phthalocyanines (Meier, Hans; Albrecht, Wolfgang; Hanack, Michael)V1559,89-100(1991)

Photoelastic investigation of statics, kinetics, and dynamics of crack formation in transparent models and natural structural elements (Taratorin, B. I.; Sakharov, V. N.; Komlev, O. U.; Stcherbakov, V. N.; Starchevsky, A. V.)V1554A,449-456(1991)

Photoelastic investigation of stress waves using models of viscoelastic materials (Dmokhovskij, A. V.; Filippov, I. G.; Skropkin, S. A.; Kobakhidze, T. G.)V1554A,323-330(1991)

Photo-induced degradation of selected polyimides in the presence of oxygen: a rapid decomposition process (Hoyle, Charles E.; Creed, David; Anzures, Edguardo; Subramanian, P.; Nagarajan, Rajamani)V1559,101-109(1991)

Photo-induced refractive-index changes and birefringence in optically nonlinear polyester (Shi, Youngqiang; Steier, William H.; Yu, Luping; Chen, Mai; Dalton, Larry R.)V1559,118-126(1991)

Photon Correlation Spectroscopy: Multicomponent Systems (Schmitz, Kenneth S., ed.)V1430(1991)

Photonic Multichip Packaging (PMP) using electro-optic organic materials and devices (McDonald, John F.; Vlannes, Nickolas P.; Lu, Toh-Ming; Wnek, Gary E.; Nason, Theodore C.; You, Lu)V1390,286-301(1991)

Photopolymer Device Physics, Chemistry, and Applications II (Lessard, Roger A., ed.)V1559(1991)

Photopolymer elements for an optical correlator system (Brandstetter, Robert W.; Fonneland, Nils J.)V1559,308-320(1991)

Photopolymers for optical devices in the USSR (Mikaelian, Andrei L.; Barachevsky, Valery A.)V1559,246-257(1991)

Photorefractivity in doped nonlinear organic polymers (Moerner, William E.; Walsh, C. P.; Scott, J. C.; Ducharme, Stephen P.; Burland, Donald M.; Bjorklund, Gary C.; Twieg, Robert J.)V1560,278-289(1991)

Physical aging of resists: the continual evolution of lithographic material (Paniez, Patrick J.; Weill, Andre P.; Cohendoz, Stephane D.)V1466,336-344(1991)

Plastic optical fiber applications for lighting of airports and buildings (Jaquet, Patrick J.)V1592,165-172(1991)

Plastic-optical-fiber-based photonic switch (Grimes, Gary J.; Blyler, Lee L.; Larson, Allen L.; Farleigh, Scott E.)V1592,139-149(1991)

Plastic Optical Fibers (Kitazawa, Mototaka; Kreidl, John F.; Steele, Robert E., eds.)V1592(1991)

Plastic optical fibers for automotive applications (Suganuma, Heiroku; Matsunaga, Tadayo)V1592,12-17(1991)

Plastic star coupler (Yuuki, Hayato; Ito, Takeharu; Sugimoto, Tetsuo)V1592,2-11(1991)

Poled polyimides as thermally stable electro-optic polymer (Wu, Jeong W.; Valley, John F.; Stiller, Marc A.; Ermer, Susan P.; Binkley, E. S.; Kenney, John T.; Lipscomb, George F.; Lytel, Richard S.)V1560,196-205(1991)

Poly(bis-alkylthio-acetylen): a dual-mode laser-sensitive material (Baumann, Reinhard; Bargon, Joachim; Roth, Hans-Klaus)V1463,638-643(1991)

Polymer-based electro-optic modulators: fabrication and performance (Haas, David R.; Man, Hong-Tai; Teng, Chia-Chi; Chiang, Kophu P.; Yoon, Hyun N.; Findakly, Talal K.)V1371,56-67(1991)

Polymer-dispersed liquid-crystal shutters for IR imaging (McCargar, James W.; Doane, J. W.; West, John L.; Anderson, Thomas W.)V1455,54-60(1991)

Polymer effects on the photochemistry of triarylsulfonium salts (McKean, Dennis R.; Allen, Robert D.; Kasai, Paul H.; MacDonald, Scott A.)V1559,214-221(1991)

Polymeric optical waveguides for device applications (McFarland, Michael J.; Beeson, Karl W.; Horn, Keith A.; Nahata, Ajay; Wu, Chengjiu; Yardley, James T.)V1583,344-354(1991)

Polymerization kinetics of mono- and multifunctional monomers initiated by high-intensity laser pulses: dependence of rate on peak-pulse intensity and chemical structure (Hoyle, Charles E.; Sundell, Per-Erik; Trapp, Martin A.; Kang, Deokman; Sheng, D.; Nagarajan, Rajamani)V1559,202-213(1991)

Polymer waveguide systems for nonlinear and electro-optic applications (Pantelis, Philip; Hill, Julian R.; Kashyap, Raman)V1559,2-9(1991)

Polysilanes for microlithography (Wallraff, Gregory M.; Miller, Robert D.; Clecak, Nicholas J.; Baier, M.)V1466,211-217(1991)

Polysilyne resists for 193 nm excimer laser lithography (Kunz, Roderick R.; Bianconi, Patricia A.; Horn, Mark W.; Paladugu, R. R.; Shaver, David C.; Smith, David A.; Freed, Charles A.)V1466,218-226(1991)

Polyvinylphenols protected with tetrahydropyranyl group in chemical amplification positive deep-UV resist systems (Hayashi, Nobuaki; Schlegel, Leo; Ueno, Takumi; Shiraishi, Hiroshi; Iwayanagi, Takao)V1466,377-383(1991)

Preparations and properties of novel positive photosensitive polyimides (Hayase, Rumiko H.; Kihara, Naoko; Oyasato, Naohiko; Matake, S.; Oba, Masayuki)V1466,438-445(1991)

Probe diffusion in polymer solutions (Phillies, George D.)V1430,118-131(1991)

Processing of polytetrafluoroethylene with high-power VUV laser radiation (Basting, Dirk; Sowada, Ulrich; Voss, F.; Oesterlin, Peter)V1412,80-83(1991)

Progress in DUV resins (Przybilla, Klaus J.; Roeschert, Heinz; Spiess, Walter; Eckes, Charlotte; Chatterjee, Subhankar; Khanna, Dinesh N.; Pawlowski, Georg; Dammel, Ralph)V1466,174-187(1991)

Prospects for the development and application of plastic optical fibers (Groh, Werner; Kuder, James E.; Theis, Juergen)V1592,20-30(1991)

Quasielastic and electrophoretic light scattering studies of polyelectrolyte-micelle complexes (Rigsbee, Daniel R.; Dubin, Paul L.)V1430,203-215(1991)

Quasirelaxation-free guest-host poled-polymer waveguide modulator: material, technology, and characterization (Levenson, R.; Liang, J.; Toussaere, E.; Carenco, Alain; Zyss, Joseph)V1560,251-261(1991)

Real-time interferometry by optical phase conjugation in dye-doped film (Fujiwara, Hirofumi)V1507,492-499(1991)

Reflection-mode polymeric interference modulators (Yankelevich, Diego; Knoesen, Andre; Eldering, Charles A.; Kowel, Stephen T.)V1560,406-415(1991)

Relationship between conjugation length and third-order nonlinearity in bis-donor substituted diphenyl polyenes (Spangler, Charles W.; Havelka, Kathleen O.; Becker, Mark W.; Kelleher, Tracy A.; Cheng, Lap Tak A.)V1560,139-147(1991)

Resist design for dry-developed positive working systems in deep-UV and e-beam lithography (Vinet, Francoise; Chevallier, M.; Pierrat, Christophe; Guibert, Jean C.; Rosilio, Charles; Mouanda, B.; Rosilio, A.)V1466,558-569(1991)

Role of counterion size and distribution on the electrostatic component to the persistence length of wormlike polyions (Schmitz, Kenneth S.)V1430,216-235(1991)

Scintillating plastic optical fiber radiation detectors in high-energy particle physics (Bross, Alan D.)V1592,122-132(1991)

Second-order and third-order processes as diagnostic tools for the analysis of organic polymers: an overview (Marowsky, Gerd; Luepke, G.)V1560,328-334(1991)

Single-component chemically amplified resist materials for electron-beam and x-ray lithography (Novembre, Anthony E.; Tai, Woon W.; Kometani, Janet M.; Hanson, James E.; Nalamasu, Omkaram; Taylor, Gary N.; Reichmanis, Elsa; Thompson, Larry F.)V1466,89-99(1991)

Sol-gel processed novel multicomponent inorganic oxide: organic polymer composites for nonlinear optics (Zhang, Yue; Cui, Y. P.; Wung, C. J.; Prasad, Paras N.; Burzynski, Ryszard)V1560,264-271(1991)

Soluble polyacetylenes derived from the ring-opening metathesis polymerization of substituted cyclooctatetraenes: electrochemical characterization and Schottky barrier devices (Jozefiak, Thomas H.; Sailor, Michael J.; Ginsburg, Eric J.; Gorman, Christopher B.; Lewis, Nathan S.; Grubbs, Robert H.)V1436,8-19(1991)

Solution of optimization problems of perforated and box-shaped structures by photoelasticity and numerical methods (Shvej, E. M.; Latysh, E. G.; Morgunov, A. N.; Mikhalchenko, O. E.; Stchetinin, A. L.; Volokh, K. Y.)V1554A,488-495(1991)

Splicing plastic optical fibers (Carson, Susan D.; Salazar, Robert A.)V1592,134-138(1991)

Status and future of polymeric materials in imaging systems (Lytle, John D.)V1354,388-394(1991)

Structural effects of DNQ-PAC backbone on resist lithographic properties (Uenishi, Kazuya; Kawabe, Yasumasa; Kokubo, Tadayoshi; Slater, Sydney G.; Blakeney, Andrew J.)V1466,102-116(1991)

Structure of poly(p-hydroxystyrene) film (Toriumi, Minoru; Yanagimachi, Masatoshi; Masuhara, Hiroshi)V1466,458-468(1991)

Studies of dissolution inhibition mechanism of DNQ-novolak resist (II): effect of extended ortho-ortho bond in novolak (Honda, Kenji; Beauchemin, Bernard T.; Fitzgerald, Edward A.; Jeffries, Alfred T.; Tadros, Sobhy P.; Blakeney, Andrew J.; Hurditch, Rodney J.; Tan, Shiro; Sakaguchi, Shinji)V1466,141-148(1991)

Study of photoelectric transformation process at p-PAn/n-Si interface (Yuan, Ren K.; Liu, Yu X.; Yuan, Hong; Wang, Yong B.; Zheng, Xiang Q.; Xu, Jian)V1519,396-399(1991)

Study of silylation mechanisms and kinetics through variations in silylating agent and resin (Dao, T. T.; Spence, Christopher A.; Hess, Dennis W.)V1466,257-268(1991)

Study of the chemically amplifiable resist materials for electron-beam lithography (Koyanagi, Hiroo; Umeda, Shin'ichi; Fukunaga, Seiki; Kitaori, Tomoyuki; Nagasawa, Kohtaro)V1466,346-361(1991)

Switching speeds in NCAP displays: dependence on collection angle and wavelength (Reamey, Robert H.; Montoya, Wayne; Wartenberg, Mark)V1455,39-44(1991)

Synthesis and incorporation of ladder polymer subunits in copolyamides, pendant polymers, and composites for enhanced nonlinear optical response (Spangler, Charles W.; Saindon, Michelle L.; Nickel, Eric G.; Sapochak, Linda S.; Polis, David W.; Dalton, Larry R.; Norwood, Robert A.)V1497,408-417(1991)

Synthesis and lithographic evaluation of alternating copolymers of linear and cyclic alkenyl(di)silanes with sulfur dioxide (Gozdz, Antoni S.; Shelburne, John A.)V1466,520-527(1991)

Synthesis and nonlinear optical activity of cumulenes (Ermer, Susan P.; Lovejoy, Steven M.; Leung, Doris; Spitzer, Ronnie; Hansen, Glenn A.; Stone, Richard E.)V1560,120-129(1991)

Synthesis and nonlinear optical properties of preformed polymers forming Langmuir-Blodgett films (Verbiest, Thierry; Persoons, Andre P.; Samyn, Celest)V1560,353-361(1991)

Synthesis and property research of birefringent polymers with predicted optical-mechanical parameters (Askadskij, A. A.; Marshalkovich, A. S.; Latysh, E. G.; Goleneva, L. M.; Pastukhov, A. V.; Sidorova, G. I.)V1554A,426-431(1991)

Synthesis, characterization, and processing of organic nonlinear optical polymers (Druy, Mark A.; Glatkowski, Paul J.)V1409,214-219(1991)

Technology and chemistry of high-temperature positive resist (Toukhy, Medhat A.; Sarubbi, Thomas R.; Brzozowy, David J.)V1466,497-507(1991)

Theoretical analysis of the third-order nonlinear optical properties of linear cyanines and polyenes (Pierce, Brian M.)V1560,148-161(1991)

Theoretical study of optical pockels and Kerr coefficients of polyenes (Albert, I. D.; Ramasesha, S.)V1436,179-189(1991)

Thermo-activated photoetching of PMMA in direct writing by laser beam (Maslenitsyn, S. F.; Svetovoy, V. B.)V1440,436-441(1991)

Third-order nonlinear optical characterization of side-chain copolymers (Norwood, Robert A.; Sounik, James R.; Popolo, J.; Holcomb, D. R.)V1560,54-65(1991)

Three-dimensional lithography: laser modeling using photopolymers (Heller, Timothy B.)V1454,272-282(1991)

Transmittances of thin polymer films and their suitability as a supportive substrate for a soft x-ray solar filter (Williams, Memorie; Hansen, Evan; Reyes-Mena, Arturo; Allred, David D.)V1549,147-154(1991)

Using photon correlation spectroscopy to study polymer coil internal dynamic behavior (Selser, James C.; Ellis, Albert R.; Schaller, J. K.; McKiernan, M. L.; Devanand, Krisha)V1430,85-88(1991)

Visible-laser-light-induced polymerization: an overview of photosensitive formulations (Fouassier, Jean-Pierre)V1559,76-88(1991)

Waveguiding characteristics in polyimide films with different chemistry of formation (Chakravorty, Kishore K.; Chien, Chung-Ping)V1389,559-567(1991)

Wideband NLO organic external modulators (Findakly, Talal K.; Teng, Chia-Chi; Walpita, Lak M.)V1476,14-21(1991)

Printing

3M's Dry Silver technology: an ideal media for electronic imaging (Morgan, David A.)V1458,62-67(1991)

Advances in color laser printing (Tompkins, Neal)V1458,154-154(1991)

Bidirectional printing method for a thermal ink transfer printer (Nagato, Hitoshi; Ohno, Tadayoshi)V1458,84-91(1991)

Butterfly line scanner: rotary twin reflective deflector that desensitizes scan-line jitter to wobble of the rotational axis (Marshall, Gerald F.; Vettese, Thomas; Carosella, John H.)V1454,37-45(1991)

Color hard copy: a self-tuning color correction algorithm based on a colorimetric model (Petschik, Benno)V1458,108-114(1991)

Color printing technologies (Sahni, Omesh)V1458,4-16(1991)

Comparing laser printing and barcode scanning designs (MacArthur, Thomas D.)V1384,176-184(1991)

Development of a high-speed high-precision laser plotter (Tachihara, Satoru; Miyoshi, Tamihiro)V1527,305-314(1991)

Digital halftoning using a blue-noise mask (Mitsa, Theophano; Parker, Kevin J.)V1452,47-56(1991)

Electrophotographic properties of thiophene derivatives as charge transport material (Kuroda, Masami; Kawate, K.; Nabeta, Osamu; Furusho, N.)V1458,155-161(1991)

Electrostatic screen-through ink jet printing technique (Nakazawa, Akira; Kutami, Michinori; Ozaki, Mitsuo; Suzuki, Shigeharu; Kikuchi, Hideyuki)V1458,115-118(1991)

Eurosprint proofing system (Froehlich, Helmut H.)V1458,51-60(1991)

Gamut mapping computer-generated imagery (Wallace, William E.; Stone, Maureen C.)V1460,20-28(1991)

Genetic algorithm approach to visual model-based halftone pattern design (Chu, Chee-Hung H.; Kottapalli, M. S.)V1606,470-481(1991)

Gray scale and resolution enhancement capabilities of edge emitter imaging stations (Leksell, David; Kun, Zoltan K.; Asars, Juris A.; Phillips, Norman J.; Brandt, Gerald B.; Stringer, J. T.; Matty, T. C.; Gigante, Joseph R.)V1458,133-144(1991)

High-quality image recorder and evaluation (Suzuki, Masayuki; Shiraiwa, Yoshinobu; Minoura, Kazuo)V1454,370-381(1991)

High-resolution high-speed laser recording and printing using low-speed polygons (Razzaghi-Masoud, Mahmoud)V1458,145-153(1991)

High-speed nonsilver lithographic system for laser direct imaging (DoMinh, Thap)V1458,68-68(1991)

History and status of black and white photographic processing chemicals as effluents (Horn, Richard R.)V1458,69-75(1991)

Image Handling and Reproduction Systems Integration (Bender, Walter J.; Plouffe, Wil, eds.)V1460(1991)

Images: from a printer's perspective (Sarkar, N. R.)V1458,42-50(1991)

Laptop page printer realized by thermal transfer technology (Drees, Friedrich-Wilhelm; Pekruhn, Wolfgang)V1458,80-83(1991)

Laser beam scanner for uniform halftones (Ando, Toshinori)V1458,128-132(1991)

Linear LED arrays for film annotation and for high-speed high-resolution printing on ordinary paper (Wareberg, P. G.; Scholes, R.; Taylor, R.)V1538,112-123(1991)

Model-based halftoning (Pappas, Thrasyvoulos N.; Neuhoff, David L.)V1453,244-255(1991)

Multilayer OPC for one-shot two-color printer (Sakai, Katsuo)V1458,179-191(1991)

Normal-contrast lith (Netz, Yoel; Hoffman, Arnold)V1458,61-61(1991)

Novel block segmentation and processing for Chinese-English document (Chien, Bing-Shan; Jeng, Bor S.; Sun, San-Wei; Chang, Gan-How; Shyu, Keh-Haw; Shih, Chun-Hsi)V1606,588-598(1991)

Photoelectrographic printing with persistently conductive masters based on onium salt acid photogenerators (Bugner, Douglas E.; Fulmer, Gary G.; Riblett, Susan E.)V1458,162-178(1991)

Physics and psychophysics of color reproduction (Giorgianni, Edward)V1458,2-3(1991)

Potential usefulness of a video printer for producing secondary images from digitized chest radiographs (Nishikawa, Robert M.; MacMahon, Heber; Doi, Kunio; Bosworth, Eric)V1444,180-189(1991)

Precise method for measuring the edge acuity of electronically (digitally) printed hard-copy images (White, William)V1398,24-28(1991)

Printing Technologies for Images, Gray Scale, and Color (Dove, Derek B.; Abe, Takao; Heinzl, Joachim L., eds.)V1458(1991)

Simulating watercolor by modeling diffusion, pigment, and paper fibers (Small, David)V1460,140-146(1991)

Solubility, dispersion, and carbon adsorption of a chromium hydroxyazo complex in a toner (Gutierrez, Adolfo R.; Diaz, Art F.)V1458,201-204(1991)

Stationary platen 2-axis scanner (Schermer, Mack J.)V1454,257-264(1991)

Study of thermal dye diffusion (Koshizuka, Kunihiro; Abe, Takao)V1458,97-104(1991)

Systems considerations in color printing (Roetling, Paul G.)V1458,17-24(1991)

Technology trends in electrophotographic printers (Starkweather, Gary K.)V1458,120-127(1991)

Thermal dye transfer color hard-copy image stability (Newmiller, Chris)V1458,92-96(1991)

Thermal transfer printing with heat amplification (Aviram, Ari; Shih, Kwang K.; Sachdev, Krishna)V1458,105-107(1991)

Trends in color hard copy (Testan, Peter)V1458,25-28(1991)

Trends in color hard-copy technology in Japan (Abe, Takao)V1458,29-40(1991)

Variable-blocksize transform coding of four-color printed images (Kaup, Andre; Aach, Til)V1605,420-427(1991)

Prisms—see also optical design; refraction

Blazing of transmission gratings for astronomical use (Neviere, Michel)V1545,11-18(1991)

Design considerations for use of a double-dove prism behind a concentric dome (Gibbons, Robert C.; Cooper, E. E.; Legan, R. G.)V1498,64-69(1991)

Holographic space-variant prism arrays for free-space data permutation in digital optical networks (Kobolla, Harald; Sauer, Frank; Schwider, Johannes; Streibl, Norbert; Voelkel, Reinhard)V1507,175-182(1991)

Integral prism-tipped optical fibers (Friedl, Stephan E.; Kunz, Warren F.; Mathews, Eric D.; Abela, George S.)V1425,134-141(1991)

New methods for economic production of prisms and lenses (Richter, G.)V1400,11-23(1991)

Plastic lens array with the function of forming unit magnification erect image using roof prisms (Takahashi, Eietsu; Tanji, Shigeo; Tanaka, Akira; Hayashi, Yuji)V1527,145-154(1991)

Prismatic anamorphic beam expanders for free-space optical communications (Jonas, Reginald P.)V1417,402-411(1991)

Set of two 45 - 90 - 45 prisms equivalent to the Fresnel rhomb (Murty, M. V.; Shukla, Ram P.; Apparao, K. V.)V1332,41-49(1991)

Simple test for the 90 degree angle in prisms (Malacara Hernandez, Daniel; Flores, Ricardo)V1332,36-40(1991)

Study on image-stabilizing reflecting prisms in the case of a finite angular perturbation (Yao, Wu; Lian, Tongshu)V1527,448-455(1991)

Write noise from optical heads with nonachromatic beam expansion prisms (Kay, David B.; Chase, Scott B.; Gage, Edward C.; Silverstein, Barry D.)V1499,281-285(1991)

Projection systems—see also imaging systems; microlithography

Acousto-optic color projection system (Hubin, Thomas)V1454,313-322(1991)

AEDC direct-write scene generation test capabilities (Lowry, Heard S.; Elrod, Parker D.; Johnson, R. J.)V1454,453-464(1991)

Characterization of automatic overlay measurement technique for sub-half-micron devices (Kawai, Akira; Fujiwara, Keiji; Tsujita, Kouichirou; Nagata, Hitoshi)V1464,267-277(1991)

Current HDTV overview in the United States, Japan, and Europe (Cripps, Dale E.)V1456,60-64(1991)

Development of a large-screen high-definition laser video projection system (Clynick, Tony)V1456,51-57(1991)

Development of an IR projector using a deformable mirror device (Lanteigne, David J.)V1486,376-379(1991)

Developments in projection lenses for HDTV (Rudolph, John D.)V1456,15-28(1991)

Dynamic infrared scene projection technology (Mobley, Scottie B.)V1486,325-332(1991)

Effects of higher order aberrations on the process window (Gortych, Joseph E.; Williamson, David M.)V1463,368-381(1991)

Excimer laser projector for microelectronics applications (Rumsby, Phil T.; Gower, Malcolm C.)V1598,36-45(1991)

"Golden standard" wafer design for optical stepper characterization (Kemp, Kevin G.; King, Charles F.; Wu, Wei; Stager, Charles)V1464,260-266(1991)

High-definition projection television (Browning, Iben)V1456,48-50(1991)

Large Screen Projection, Avionic, and Helmet-Mounted Displays (Assenheim, Harry M.; Flasck, Richard A.; Lippert, Thomas M.; Bentz, Jerry, eds.)V1456(1991)

Laser-based display technology development at the Naval Ocean Systems Center (Phillips, Thomas E.; Trias, John A.; Lasher, Mark E.; Poirier, Peter M.; Dahlke, Weldon J.; Robinson, Waldo R.)V1454,290-298(1991)

MTF characteristics of a Scophony scene projector (Schildwachter, Eric F.; Boreman, Glenn D.)V1488,48-57(1991)

New 0.54 aperture i-line wafer stepper with field-by-field leveling combined with global alignment (van den Brink, Martin A.; Katz, Barton A.; Wittekoek, Stefan)V1463,709-724(1991)

Novel high-resolution large-field scan-and-repeat projection lithography system (Jain, Kanti)V1463,666-677(1991)

Optical design of high-aperture aspherical projection lens (Osawa, Atsuo; Fukuda, Kyohei; Hirata, Kouji)V1354,337-343(1991)

Performance of NCAP projection displays (Jones, Philip J.; Tomita, Akira; Wartenberg, Mark)V1456,6-14(1991)

Projection screens for high-definition television (Kirkpatrick, Michael D.; Mihalakis, George M.)V1456,40-47(1991)

Reflective optical designs for soft x-ray projection lithography (Jewell, Tanya E.; Thompson, Kevin P.; Rodgers, John M.)V1527,163-173(1991)

Self-acting gas bearings for high-speed scanners (Preston, Ralph G.; Robinson, D. H.; Shepherd, Joseph)V1454,132-138(1991)

Stacked STN LCDs for true-color projection systems (Gulick, Paul E.; Conner, Arlie R.)V1456,76-82(1991)

Two-mirror projection systems for simulating telescopes (Hannan, Paul G.; Davila, Pam M.)V1527,26-36(1991)

Propagation—see also atmospheric optics; beams; turbulence

Aerosol models for optical and IR propagation in the marine atmospheric boundary layer (de Leeuw, Gerrit)V1487,130-159(1991)

Analysis of lightwave propagation in the bent waveguide by the Galerkin Method (Maruta, Akihiro; Matsuhara, Masanori)V1583,307-313(1991)

Anisoplanatism and the use of laser guide stars (Goad, Larry E.)V1542,100-109(1991)

Atmospheric continuum absorption models (Delaye, Corinne T.; Thomas, Michael E.)V1487,291-298(1991)

Atmospheric propagation effects on pattern recognition by neural networks (Giever, John C.; Hoock, Donald W.)V1486,302-313(1991)

Atmospheric visibility monitoring for planetary optical communications (Cowles, Kelly A.)V1487,272-279(1991)

Beam characterization and measurement of propagation attributes (Sasnett, Michael W.; Johnston, Timothy J.)V1414,21-32(1991)

Beam quality in laser amplifiers (Martinez-Herrero, R.; Mejias, P. M.)V1397,623-626(1991)

Comparison of laser radar transmittance for the five atmospheric models (Au, Robert H.)V1487,280-290(1991)

Compensation efficiency of an optical adaptive transmitter (Feng, Yue-Zhong; Song, Zhengfang; Gong, Zhi-Ben)V1417,370-372(1991)

Correlation between the aerosol profiles measurements, the meteorological conditions, and the atmospheric IR transmission in a mediterranean marine atmosphere (Tanguy, Mireille; Bonhommet, Herve; Autric, Michel L.; Vigliano, Patrick)V1487,172-183(1991)

Correlation of bifrequency beam propagating in a folded turbulent path (Song, Zhengfang; Ma, Jun)V1487,382-386(1991)

Diffraction effects in directed radiation beams (Hafizi, Bahman; Sprangle, Phillip)V1407,316-321(1991)

Dynamics of laser light transmission losses in aerosols (Malicka, Marianna; Parma, Ludvik)V1391,181-189(1991)

Effective properties of electromagnetic wave propagation in some composite media (Artola, Michel; Cessenat, Michel)V1558,14-21(1991)

Effective refractive indices of three-phase optical coatings (Ma, Yushieh; Varadan, Vijay K.; Varadan, Vasundara V.)V1558,138-142(1991)

Effects of thin and subvisible cirrus on HEL far-field intensity calculations at various wavelengths (Harada, Larrene K.)V1408,28-40(1991)

Empirical modeling of laser propagation effects in intermediate turbulence (Tofsted, David H.)V1487,372-381(1991)

Equipment development for an atmospheric-transmission measurement campaign (Ruiz, Domingo; Kremer, Paul J.)V1417,212-222(1991)

Errors inherent in the restoration of imagery acquired through remotely sensed refractive interfaces and scattering media (Schmalz, Mark S.)V1479,183-198(1991)

Evaluation of the Navy Oceanic Vertical Aerosol Model using lidar and PMS particle-size spectrometers (Jensen, Douglas R.)V1487,160-171(1991)

Finite aperture effects on intensity fluctuations of laser radiation in a turbulent atmosphere (Mazar, Reuven; Bronshtein, Alexander)V1487,361-371(1991)

Fluctuation variation of a CO2 laser pulse intensity during its interaction with a cloud (Almaev, R. K.; Semenov, L. P.; Slesarev, A. G.; Volkovitsky, O. A.)V1397,831-834(1991)

Focusing infrared laser beams on targets in space without using adaptive optics (McKechnie, Thomas S.)V1408,119-135(1991)

Forward-looking IR and lidar atmospheric propagation in the infrared field program (Koenig, George G.; Bissonnette, Luc R.)V1487,240-249(1991)

Free propagation of high-order moments of laser beam intensity distribution (Sanchez, Miguel; Hernandez Neira, Jose Luis; Delgado, J.; Calvo, G.)V1397,635-638(1991)

Frequency up-conversion for the reflectionless propagation of a high-power microwave pulse in a self-generated plasma (Kuo, Spencer P.; Ren, A.; Zhang, Y. S.)V1407,272-280(1991)

Functional reconstruction predictions of uplink whole beam Strehl ratios in the presence of thermal blooming (Enguehard, S.; Hatfield, Brian)V1408,186-191(1991)

Gaussian scaling laws for diffraction: top-hat irradiance and Gaussian beam propagation through a paraxial optical train (Townsend, Sallie S.; Cunningham, Philip R.)V1415,154-194(1991)

GRAND: a 4-D wave optics code for atmospheric laser propagation (Mehta, Naresh C.)V1487,398-409(1991)

Ground-to-space multiline propagation at 1.3 um (Crawford, Douglas P.; Harada, Larrene K.)V1408,167-177(1991)

Helmholtz beam propagation by the method of Lanczos reduction (Fleck, Joseph A.)V1583,228-239(1991)

Higher-order theory of caustics for fast running cracks under general loadings (Nishioka, Teiichi; Murakami, R.; Matsuo, Seitaro; Ohishi, Y.)V1554A,802-813(1991)

Interpretation of optical caustic patterns obtained during unsteady crack growth: an analysis based on a higher-order transient expansion (Liu, Cheng; Rosakis, Ares J.)V1554A,814-825(1991)

Laser beam propagation through inhomogeneous amplifying media (Martinez-Herrero, R.; Mejias, P. M.)V1397,619-622(1991)

Laser beam width, divergence, and propagation factor: status and experience with the draft standard (Fleischer, John M.)V1414,2-11(1991)

Lidar evaluation of the propagation environment (Uthe, Edward E.; Livingston, John M.)V1487,228-239(1991)

Localized waves in complex environments (Ziolkowski, Richard W.; Besieris, Ioannis M.; Shaarawi, Amr M.)V1407,387-397(1991)

Localized wave transmission physics and engineering (Ziolkowski, Richard W.)V1407,375-386(1991)

Measurements of atmospheric transmittance of CO2 laser radiation (Aref'ev, Vladimir N.)V1397,827-830(1991)

Modeling turbulent transport in laser beam propagation (Wallace, James)V1408,19-27(1991)

Multiple-scattering effects on pulse propagation through burning particles (Ma, Yushieh; Varadan, Vijay K.; Varadan, Vasundara V.)V1487,220-225(1991)

Multiple scattering of laser light in dense aerosol (Parma, Ludvik; Malicka, Marianna)V1391,190-198(1991)

Multiscattered lidar returns from atmospheric aerosols (Hutt, Daniel L.; Bissonnette, Luc R.; Durand, Louis-Gilles)V1487,250-261(1991)

Nonlinear wakefield generation and relativistic optical guiding of intense laser pulses in plasmas (Esarey, Eric; Sprangle, Phillip; Ting, Antonio C.)V1407,407-417(1991)

Normally incident plane waves on a chiral slab with linear property variations (Lakhtakia, Akhlesh; Varadan, Vijay K.; Varadan, Vasundara V.)V1558,120-126(1991)

Numerical experiments in propagation with wind velocity fluctuation (Carlson, Lawrence W.)V1408,203-211(1991)

Numerical wavefront propagation through inhomogeneous media (Zakeri, Gholam-Ali)V1558,103-112(1991)

Observations of uplink and retroreflected scintillation in the Relay Mirror Experiment (Lightsey, Paul A.; Anspach, Joel E.; Sydney, Paul F.)V1482,209-222(1991)

Optical quality of a combined aerodynamic window (Du, Keming; Franek, Joachim; Loosen, Peter; Zefferer, H.; Shen, Junquan)V1397,639-643(1991)

Optical space-to-ground link availability assessment and diversity requirements (Chapman, William W.; Fitzmaurice, Michael W.)V1417,63-74(1991)

Parametric pulse breakup due to population pulsations in three-level systems (DiMarco, Steven F.; Cantrell, Cyrus D.)V1497,178-187(1991)

Path integral approach to thermal blooming (Enguehard, S.; Hatfield, Brian)V1408,178-185(1991)

PCI (phase compensation instability) and minishear (Fried, David L.; Szeto, Roque K.)V1408,150-166(1991)

Propagation Engineering: Fourth in a Series (Bissonnette, Luc R.; Miller, Walter B., eds.)V1487(1991)

Propagation invariance of laser beam parameters through optical systems (Martinez-Herrero, R.; Mejias, P. M.; Hernandez Neira, Jose Luis; Sanchez, Miguel)V1397,627-630(1991)

Propagation of High-Energy Laser Beams Through the Earth's Atmosphere II (Ulrich, Peter B.; Wilson, LeRoy E., eds.)V1408(1991)

Pulse propagation in random media (Ishimaru, Akira)V1558,127-129(1991)

Pulse reshaping and coherent sideband generation effects on multiphoton excitation of polyatomic molecules (Garner, Steven T.; Cantrell, Cyrus D.)V1497,188-196(1991)

Realistic wind effects on turbulence and thermal blooming compensation (Long, Jerry E.; Hills, Louis S.; Gebhardt, Frederick G.)V1408,58-71(1991)

Review of nondiffracting Bessel beams (LaPointe, Michael R.)V1527,258-276(1991)

Review of the physics of small-scale thermal blooming in uplink propagation (Enguehard, S.; Hatfield, Brian)V1415,128-137(1991)

Short-pulse electromagnetics for sensing applications (Felsen, Leopold B.; Vecchi, G.; Carin, L.; Bertoni, H. L.)V1471,154-162(1991)

Simple beam-propagation measurements on ion lasers (Guggenhiemer, Steven; Wright, David L.)V1414,12-20(1991)

Simulation and design of integrated optical waveguide devices by the BPM (Yip, Gar L.)V1583,240-248(1991)

Simulation of intense microwave pulse propagation in air breakdown environment (Kuo, Spencer P.; Zhang, Y. S.)V1407,260-271(1991)

Spatial characterization of high-power multimode laser beams (Serna, Julio; Martinez-Herrero, R.; Mejias, P. M.)V1397,631-634(1991)

Stability analysis of semidiscrete schemes for thermal blooming computation (Ulrich, Peter B.)V1408,192-202(1991)

Theoretical modeling of chiral composites (Apparao, R. T.; Varadan, Vasundara V.; Varadan, Vijay K.)V1558,2-13(1991)

Theory of transient self-focusing (Hsia, Kangmin; Cantrell, Cyrus D.)V1497,166-177(1991)

Thermal blooming critical power and adaptive optics correction for the ground-based laser (Smith, David C.; Townsend, Sallie S.)V1408,112-118(1991)

Time development of acoustic bullets, wave-zone form of focus wave modes, and other solutions of the acoustic equations (Moses, Harry E.; Prosser, Reese T.)V1407,354-374(1991)

Turbulence at the inner scale (Tatarskii, V. I.)V1408,2-9(1991)

Ultrashort pulse propagation in visible semiconductor diode laser amplifiers (Tatum, Jim A.; MacFarlane, Duncan L.)V1497,320-325(1991)

Use of adaptive optics for minimizing atmospheric distortion of optical waves (Lukin, Vladimir P.)V1408,86-95(1991)

Variable wind direction effects on thermal blooming correction (Hills, Louis S.; Long, Jerry E.; Gebhardt, Frederick G.)V1408,41-57(1991)

Vector beam propagation method based on finite-difference (Huang, Weiping W.; Xu, Chenglin; Chu, S. T.; Chaudhuri, Sujeet K.)V1583,268-270(1991)

Vector beam propagation using Hertz vectors (Milsted, Carl S.; Cantrell, Cyrus D.)V1497,202-215(1991)

Vector squeezed states (Lee, Heun J.; Haus, Joseph W.)V1497,228-239(1991)

Visible extinction measurements in rain and snow using a forward-scatter meter (Hutt, Daniel L.; Oman, James)V1487,312-323(1991)

Wave-optics analysis of fast-beam focusing (Shih, Chun-Ching)V1415,150-153(1991)

Wide bandwidth spectral measurements of atmospheric tilt turbulence (Tiszauer, Detlev H.; Smith, Richard C.)V1408,72-83(1991)

Wide-spectral-range transmissometer used for fog measurements (Turner, Vernon; Trowbridge, Christian A.)V1487,262-271(1991)

Pulses

Amplified quantum fluctuation as a mechanism for generating ultrashort pulses in semiconductor lasers (Yuan, Ruixi; Taylor, Henry F.)V1497,313-319(1991)

Beam control and power conditioning of GEKKO glass laser system (Nakatsuka, Masahiro; Jitsuno, Takahisa; Kanabe, Tadashi; Urushihara, Shinji; Miyanaga, Noriaki; Nakai, Sadao S.)V1411,108-115(1991)

Characteristics of pulsed nuclear-reactor-pumped flowing gas lasers (Neuman, William A.; Fincke, James R.)V1411,28-40(1991)

Comparison of negative and positive polarity reflex diode microwave source (Litz, Marc S.; Huttlin, George A.; Lazard, Carl J.; Golden, Jeffry; Pereira, Nino R.; Hahn, Terry D.)V1407,159-166(1991)

Computational and experimental progress on laser-activated gas avalanche switches for broadband, high-power electromagnetic pulse generation (Mayhall, David J.; Yee, Jick H.; Villa, Francesco)V1378,101-114(1991)

Development of solid state pulse power supply for copper vapor laser (Fujii, Takashi; Nemoto, Koshichi; Ishikawa, Rikio; Hayashi, Kazuo; Noda, Etsuo)V1412,50-57(1991)

Dispersion of picosecond pulses propagating on microstrip interconnections on semiconductor integrated-circuit substrates (Pasik, Michael F.; Cangellaris, Andreas C.; Prince, John L.)V1389,297-301(1991)

Display holography with pulsed ruby laser: recording and copying (Kliot-Dashinskaya, I. M.; Mikhailova, V. I.; Paltsev, G. P.; Strigun, V. L.)V1238,465-469(1991)

Effect of input pulse shape on FSK optical coherent communication system (Shadaram, Mehdi; John, Eugine)V1365,108-115(1991)

Effect of magnetic quantization on the pulse power output of spontaneous emission for II-VI and IV-VI lasers operated at low temperature (Ghatak, Kamakhya P.; Mitra, Bhaswati; Biswas, Shambhu N.)V1411,137-148(1991)

Electrical characteristics of lossy interconnections for high-performance computer applications (Deutsch, Alina; Kopcsay, Gerard V.; Ranieri, V. A.; Cataldo, J. K.; Galligan, E. A.; Graham, W. S.; McGouey, R. P.; Nunes, S. L.; Paraszczak, Jurij R.; Ritsko, J. J.; Serino, R. J.; Shih, D. Y.; Wilczynski, Janusz S.)V1389,161-176(1991)

Evaluation of electrolytic capacitors for high-peak current pulse duty (Harris, Kevin; McDuff, G. G.; Burkes, Tom R.)V1411,87-99(1991)

Frequency up-conversion for the reflectionless propagation of a high-power microwave pulse in a self-generated plasma (Kuo, Spencer P.; Ren, A.; Zhang, Y. S.)V1407,272-280(1991)

GaAs opto-thyristor for pulsed power applications (Hur, Jung H.; Hadizad, Peyman; Zhao, Hanmin; Hummel, Steven R.; Dapkus, P. D.; Fetterman, Harold R.; Gundersen, Martin A.)V1378,95-100(1991)

Generation of 1-GHz ps optical pulses by direct modulation of a 1.532-um DFB-LD (Nie, Chao-Jiang; Li, Ying; Wu, Fang D.; Chen, Ying-Li; Li, Qi; Hua, Yi-Min)V1579,264-267(1991)

Generation of hard x-ray pulse trains with the help of high-frequency oscillating systems for fast processes recording (Romanovsky, V. F.; Kovalenko, A. N.; Sapozhnikova, T. I.; Tushev, N. R.; Abgaryan, A. A.)V1358,140-145(1991)

Generation of stable low-jitter kilovolt amplitude picosecond pulses (Sarkar, Tapan K.; Banerjee, Partha P.)V1378,34-42(1991)

High-voltage picosecond pulse generation using avalanche diodes (McEwan, Thomas E.; Hanks, Roy L.)V1346,465-470(1991)

Holographic characteristics of IAE and PFG-01 photoplates for colored pulsed holography (Vorzobova, N. D.; Rjabova, R. V.; Schvarzvald, A. I.)V1238,476-477(1991)

Holography with a single picosecond pulse (Abramson, Nils H.)V1332,224-229(1991)

Intrapulse stimulated Raman scattering and ultrashort solitons in optical fibers (Headley, Clifford; Agrawal, Govind P.; Reardon, A. C.)V1497,197-201(1991)

Localized waves in complex environments (Ziolkowski, Richard W.; Besieris, Ioannis M.; Shaarawi, Amr M.)V1407,387-397(1991)

Localized wave transmission physics and engineering (Ziolkowski, Richard W.)V1407,375-386(1991)

Long-pulse electron gun for laser applications (Bayless, John R.; Burkhart, Craig P.)V1411,42-46(1991)

Magnetic pulse compression in the prepulse circuit for a 1 kW, 1kHz XeCl excimer laser (Ekelmans, G. B.; van Goor, Frederik A.; Trentelman, M.; Witteman, Wilhelmus J.)V1397,569-572(1991)

Magnetic pulse compressor for high-power excimer discharge laser pumping (Vizir, V. A.; Kiryukhin, Yu. B.; Manylov, V. I.; Nosenko, S. P.; Sychev, S. P.; Chervyakov, V. V.; Shubkin, N. G.)V1411,63-68(1991)

Methods and applications for intensity stabilization of pulsed and cw lasers from 257 nm to 10.6 microns (Miller, Peter J.)V1376,180-191(1991)

Modeling of pumping kinetics of an iodine photodissociation laser with long pumping pulse (Rohlena, Karel; Beranek, J.; Masek, Karel)V1415,259-268(1991)

Modern holographic studio (Bryskin, V. Z.; Krylov, Vitaly N.; Staselko, D. I.)V1238,448-451(1991)

Modifications to achieve subnanosecond jitter from standard commercial excimer lasers (Rust, Kenneth R.)V1411,80-86(1991)

Modifications to the LP-140 pulsed CO2 laser for lidar use (Jaenisch, Holger M.; Spiers, Gary D.)V1411,127-136(1991)

Monochromatic and two-color recording of holographic portraits with the use of pulsed lasers (Vorzobova, N. D.; Sizov, V. N.; Rjabova, R. V.)V1238,462-464(1991)

Multiple-spectral structure of the 578.2 nm line for copper vapor laser (Wang, Yongjiang; Pan, Bailiang; Ding, Xiande; Qian, Yujun; Shi, Shuyi)V1412,60-66(1991)

Nd-doped lasers with widely variable pulsewidths (Lin, Li-Huang; Ge, Wen; Kang, Yilan; Chen, Shisheng; Bin, Ouyang; Li, Shiying; Wang, Shijie)V1410,65-71(1991)

Newly developed excitation circuit for kHz pulsed lasers (Yasuoka, Koichi; Ishii, Akira; Tamagawa, Tohru; Ohshima, Iwao)V1412,32-37(1991)

Nonlinear energy transfer between nanosecond pulses in iron-doped InP crystals (Roosen, Gerald)V1362,398-416(1991)

Nonlinear wakefield generation and relativistic optical guiding of intense laser pulses in plasmas (Esarey, Eric; Sprangle, Phillip; Ting, Antonio C.)V1407,407-417(1991)

Novel device for increasing the laser pulse intensity in multiphoton ionization mass spectrometry (Liang, Dawei; Fraser Monteiro, L.; Fraser Monteiro, M. L.)V1501,129-134(1991)

Novel device for short-pulse generation using optoelectronic feedback (Phelan, Paul J.; Hegarty, John; Elsasser, Wolfgang E.)V1362,623-630(1991)

Observation of power gain in an inductive pulsed power system with an optically activated semiconductor closing and opening switch (Kung, Chun C.; Funk, Eric E.; Chauchard, Eve A.; Rhee, M. J.; Lee, Chi H.; Yan, Li I.)V1378,250-258(1991)

Optical coherent transients induced by time-delayed fluctuating pulses (Finkelstein, Vladimir; Berman, Paul R.)V1376,68-79(1991)

Optimization of pulsed laser power supply system (Alci, Mustafa; Yilbas, Bekir S.; Danisman, Kenan; Ciftlikli, Cebrail; Altuner, Mehmet)V1411,100-106(1991)

Parametric pulse breakup due to population pulsations in three-level systems (DiMarco, Steven F.; Cantrell, Cyrus D.)V1497,178-187(1991)

Performance characteristics of a discharge-pumped XeCl laser driven by a pulse forming network containing nonlinear ferroelectric capacitors (Fairlie, S. A.; Smith, Paul W.)V1411,56-62(1991)

Pockels cell driver with high repetition rate (Brzezinski, Ryszard)V1391,279-285(1991)

Pulse compression of KrF laser radiation by stimulated scattering (Alimpiev, Sergei S.; Bukreev, Viacheslav S.; Kusakin, Vladimir I.; Likhansky, Sergey V.; Obidin, Alexey Z.; Vartapetov, Serge K.; Veselovsky, Igor A.)V1503,154-158(1991)

Pulsed hologram recording by nanosecond and subnanosecond small-dimensioned laser system (Bespalov, V. G.; Dikasov, A. B.)V1238,470-475(1991)

Pulsed laser system for recording large-scale color hologram (Bespalov, V. G.; Krylov, Vitaly N.; Sizov, V. N.)V1238,457-461(1991)

Pulsed microwave excitation of rare-gas halide mixtures (Gekat, Frank; Klingenberg, Hans H.)V1411,47-54(1991)

Pulse formation and characteristics of the cw mode-locked titanium-doped sapphire laser (Zschocke, Wolfgang; Stamm, Uwe; Heumann, Ernst; Ledig, Mario; Guenzel, Uwe; Kvapil, Jiri; Koselja, Michal P.; Kubelka, Jiri)V1501,49-56(1991)

Pulse Power for Lasers III (McDuff, G. G., ed.)V1411(1991)

Pulse reshaping and coherent sideband generation effects on multiphoton excitation of polyatomic molecules (Garner, Steven T.; Cantrell, Cyrus D.)V1497,188-196(1991)

Radio frequency pulse compression experiments at SLAC (Farkas, Zoltan D.; Lavine, T. L.; Menegat, A.; Miller, Roger H.; Nantista, C.; Spalek, G.; Wilson, P. B.)V1407,502-511(1991)

Recent developments on the NRL Modified Betatron Accelerator (Golden, Jeffry; Len, Lek K.; Smith, Tab J.; Dialetis, Demos; Marsh, S. J.; Smith, Kevin; Mathew, Joseph; Loschialpo, Peter; Seto, Lloyd; Chang, Jeng-Hsien; Kapetanakos, Christos A.)V1407,418-429(1991)

Selected Papers on Ultrafast Laser Technology (Gosnell, Timothy R.; Taylor, Antoinette J., eds.)VMS44(1991)

Self-action of supremely short light pulses in fibers (Azarenkov, Aleksey N.; Altshuler, Grigori B.; Kozlov, Sergey A.)V1409,166-177(1991)

Shaping of the second-harmonic pulse temporal profile (Patron, Zbigniew)V1391,259-263(1991)

Short-pulse electromagnetics for sensing applications (Felsen, Leopold B.; Vecchi, G.; Carin, L.; Bertoni, H. L.)V1471,154-162(1991)

Simulation of intense microwave pulse propagation in air breakdown environment (Kuo, Spencer P.; Zhang, Y. S.)V1407,260-271(1991)

Termination for minimal reflection of high-speed pulse propagation along multiple-coupled microstrip lines (Kuo, Jen-Tsai; Tzuang, Ching-Kuang C.)V1389,156-160(1991)

Thyratron-switched, L-C inverter, prepulse-sustainer, laser discharge circuit (Pacala, Thomas J.; Tranis, Art; Laudenslager, James B.; Kinley, Fred G.)V1411,69-79(1991)

Transient stimulated Raman scattering: theory and experiments of pulse shortening and phase conjugation properties (Agnesi, Antoniangelo; Reali, Giancarlo C.; Kubecek, Vaclav)V1415,104-109(1991)

Ultrashort pulse propagation in visible semiconductor diode laser amplifiers (Tatum, Jim A.; MacFarlane, Duncan L.)V1497,320-325(1991)

Uniform energy discharge for pulsed lasers (Koschmann, Eric C.)V1411,118-126(1991)

Wideband bistatic radar signal processing using a coherent detection architecture (Belcher, Melvin L.; Garmon, Jeff P.)V1476,224-233(1991)

Q-switching

Environmental testing of a Q-switched Nd:YLF laser and a Nd:YAG ring laser (Robinson, Deborah L.)V1417,562-572(1991)

Fabry-Perot etalon as a CO2 laser Q-switch modulator (Walocha, Jerzy)V1391,286-289(1991)

High repetition rate Q-switched erbium glass lasers (Hamlin, Scott J.; Myers, John D.; Myers, Michael J.)V1419,100-106(1991)

Numerical model of Q-switched solid state laser (Altshuler, Grigori B.; Kargin, Igor U.; Khloponin, Leonid V.; Khramov, Valery)V1415,269-280(1991)

Passive Q-switching of eyesafe Er:glass lasers (Denker, Boris I.; Maksimova, G. V.; Osiko, Vyacheslav V.; Prokhorov, Alexander M.; Sverchkov, Sergey E.; Sverchkov, Yuri E.; Horvath, Zoltan G.)V1419,50-54(1991)

Pockels cell driver with high repetition rate (Brzezinski, Ryszard)V1391,279-285(1991)

Positive-branch unstable resonator for Nd:YAG laser with Q-switching (Marczak, Jan; Rycyk, Antoni; Sarzynski, Antoni)V1391,48-51(1991)

Unstable resonator with a super-Gaussian dielectric mirror for Nd:YAG Q-switched laser (Firak, Jozef; Marczak, Jan; Sarzynski, Antoni)V1391,42-47(1991)

Quantum wells—see also excitons; gallium arsenide materials; integrated optoelectronics; lasers, semiconductor

Advanced InGaAs/InP p-type pseudomorphic MODFET (Malzahn, Eric; Heuken, Michael; Gruetzmacher, Dettev; Stollenwerk, M.; Heime, Klaus)V1362,199-204(1991)

Advances in multiple-quantum-well IR detectors (Bloss, Walter L.; O'Loughlin, Michael J.; Rosenbluth, Mary)V1541,2-10(1991)

All-optical switching of picosecond pulses in GaAs MQW waveguides (Li Kam Wa, Patrick; Park, Choong-Bum; Miller, Alan)V1474,2-7(1991)

Attenuation of leaky waves in GaAs/AlGaAs MQW waveguides formed on a GaAs substrate (Kubica, Jacek M.)V1506,134-139(1991)

Band-structure dependence of impact ionization: bulk semiconductors, strained Ge/Si alloys, and multiple-quantum-well avalanche photodetectors (Czajkowski, Igor K.; Allam, Jeremy; Adams, Alfred R.)V1362,179-190(1991)

Behaviour of a single quantum-well under deep level transient spectroscopy (DLTS) measurement: a new theoretical model (Letartre, Xavier; Stievenard, Didier; Barbier, E.)V1362,778-789(1991)

Bias dependence of the hole tunneling time in AlAs/GaAs resonant tunneling structures (Van Hoof, Chris A.; Goovaerts, Etienne; Borghs, Gustaaf)V1362,291-300(1991)

Characterization of GRIN-SCH-SQW amplifiers (Haake, John M.; Zediker, Mark S.; Balestra, Chet L.; Krebs, Danny J.; Levy, Joseph L.)V1418,298-308(1991)

Degenerate four-wave mixing and optical nonlinearities in quantum-confined CdSe microcrystallites (Park, S. H.; Casey, Michael P.; Falk, Joel)V1409,9-13(1991)

Differentiation of the nonradiative recombination properties of the two interfaces of molecular beam epitaxy grown GaAs-GaAlAs quantum wells (Sermage, Bernard; Gerard, Jean M.; Bergomi, Lorenzo; Marzin, Jean Y.)V1361,131-135(1991)

Effective electron mass in narrow-band-gap IR materials under different physical conditions (Ghatak, Kamakhya P.; Biswas, Shambhu N.)V1484,149-166(1991)

Effect of parameter variation on the performance of InGaAsP/InP multiple-quantum-well electroabsorption/electrorefraction modulators (Xiong, F. K.; Zhu, Long D.; Wang, C. W.)V1519,665-669(1991)

Efficient optical waveguide modulation based on Wannier-Stark localization in a InGaAs-InAlAs superlattice (Bigan, Erwan; Allovon, Michel; Carre, Madeleine; Carenco, Alain; Voisin, Paul)V1362,553-558(1991)

Energy and momentum relaxation of electrons in GaAs quantum-wells: effect of nondrifting hot phonons and interface roughness (Gupta, Rita P.; Balkan, N.; Ridley, Brian K.; Emeny, M. T.)V1362,798-803(1991)

Energy levels of GaAs/A1xGa1-xAs double-barrier quantum wells (Chen, Yong; Neu, G.; Deparis, C.; Massies, J.)V1361,860-865(1991)

Epitaxial growth and photoluminescence investigations of InP/InAs quantum well grown by hydride vapor phase epitaxy (Banvillet, Henri; Gil, E.; Vasson, A. M.; Cadoret, R.; Tabata, A.; Benyattou, Taha; Guillot, Gerard)V1361,972-979(1991)

Evaluation of diode laser failure mechanisms and factors influencing reliability (Baumann, John A.; Shepard, Allan H.; Waters, Robert G.; Yellen, Steven L.; Harding, Charleton M.; Serreze, Harvey B.)V1418,328-337(1991)

Excitonic photoabsorption study of AlGaAs/GaAs multiple-quantum-well grown by low-pressure MOCVD (Kwon, O'Dae; Lee, Seung-Won; Choi, Woong-Lim; Kim, Kwang-Il; Jeong, Yoon-Ha)V1361,802-808(1991)

Experimental determination of recombination mechanisms in strained and unstrained quantum-well lasers (Rideout, William C.)V1418,99-107(1991)

Far-IR magneto-emission study of the quantum-hall state and breakdown of the quantum-hall effect (Raymond, Andre; Chaubet, C.; Razeghi, Manijeh)V1362,275-281(1991)

High contrast ratio InxGa1-xAs/GaAs multiple-quantum-well spatial light modulators (Harwit, Alex; Fernandez, R.; Eades, Wendell D.)V1541,38-47(1991)

High-efficiency vertical-cavity lasers and modulators (Coldren, Larry A.; Corzine, Scott W.; Geels, Randall S.; Gossard, Arthur C.; Law, K. K.; Merz, James L.; Scott, Jeffrey W.; Simes, Robert J.; Yan, Ran H.)V1362,24-37(1991)

High-power single-element pseudomorphic InGaAs/GaAs/AlGaAs single-quantum-well lasers for pumping Er-doped fiber amplifiers (Larsson, Anders G.; Forouhar, Siamak; Cody, Jeffrey; Lang, Robert J.; Andrekson, Peter A.)V1418,292-297(1991)

High-quality heavily strained II-VI quantum well (Li, Jie; He, Li; Tang, Wen G.; Shan, W.; Yuan, Shi X.)V1519,660-664(1991)

High-sensitivity photorefractivity in bulk and multiple-quantum-well semiconductors (Partovi, Afshin; Glass, Alastair M.; Feldman, Robert D.)V1561,20-32(1991)

Hot carrier relaxation in bulk InGaAs and quantum-wells (Gregory, Andrew; Phillips, Richard T.; Majumder, Fariduddin A.)V1362,268-274(1991)

Hot electron instabilities and light emission in GaAs quantum wells (Balkan, N.; Ridley, Brian K.)V1361,927-934(1991)

Improvements in self electro-optic effect devices: toward system implementation (Morgan, Robert A.)V1562,213-227(1991)

Influence of the p-type doping of the InP cladding layer on the threshold current density in 1.5 um QW lasers (Sermage, Bernard; Blez, M.; Kazmierski, Christophe; Ougazzaden, A.; Mircea, Andrei; Bouley, J. C.)V1362,617-622(1991)

InGaAs/InP distributed feedback quantum-well lasers (Temkin, Henryk; Logan, Ralph A.; Tanbun-Ek, Tawaee)V1418,88-98(1991)

InP-based quantum-well infrared photodetectors (Gunapala, S. D.; Levine, Barry F.; Ritter, D.; Hamm, Robert A.; Panish, Morton B.)V1541,11-23(1991)

Intercircuit optical interconnects using quantum-well modulators (Goodfellow, Robert C.; Goodwin, Martin J.; Moseley, Andrew J.)V1389,594-599(1991)

Investigation of uniform deposition of GaInAsP quantum wells by MOCVD (Puetz, Norbert; Miner, Carla J.; Hingston, G.; Moore, Chris J.; Watt, B.; Hillier, Glen)V1361,692-698(1991)

Long-wavelength GaAs/AlxGa1-xAs quantum-well infrared photodetectors (Levine, Barry F.; Bethea, Clyde G.; Stayt, J. W.; Glogovsky, K. G.; Leibenguth, R. E.; Gunapala, S. D.; Pei, S. S.; Kuo, Jenn-Ming)V1540,232-237(1991)

Long-wavelength GaAs quantum-well infrared photodetectors (Levine, Barry F.)V1362,163-167(1991)

Low-substrate temperature molecular beam epitaxy growth and thermal stability of strained InGaAs/GaAs single-quantum-wells (Elman, Boris S.; Koteles, Emil S.; Melman, Paul; Rothman, Mark A.)V1361,362-372(1991)

Magnetocapacitance and photoluminescence spectroscopy studies of charge storage, bistability, and energy relaxation effects in resonant tunneling devices (Eaves, Lawrence; Hayes, David; Leadbeater, M. L.; Simmonds, P. E.; Skolnick, Maurice S.)V1362,520-533(1991)

Massive connectivity and SEEDs (Chirovsky, Leo M.)V1562,228-241(1991)

Nonlinear optical gain in InGaAs/InGaAsP quantum-wells (Rosenzweig, M.; Moehrle, M.; Dueser, H.; Tischel, M.; Heitz, R.; Hoffmann, Axel)V1362,876-887(1991)

Nonlinear optical properties of nipi and hetero nipi superlattices and their application for optoelectronics (Doehler, Gottfried H.)V1361,443-468(1991)

Nonlinear optical properties of quantum-well structures and some of their applications (Sung, C. C.)V1497,456-466(1991)

Nonlinear optical susceptibility of CuCl quantum dot glass (Justus, Brian L.; Seaver, Mark E.; Ruller, Jacqueline A.; Campillo, Anthony J.)V1409,2-8(1991)

Numerical solution of the one-dimensional Schrodinger equation: application to heterostructures and superlattices (Bottacchi, Stefano)V1362,727-749(1991)

Observation of tunneling emission from a single-quantum-well using deep-level transient spectroscopy (Letartre, Xavier; Stievenard, Didier)V1361,1144-1155(1991)

Optical nonlinearities due to long-lived electron-hole plasmas (Dawson, Philip; Galbraith, Ian; Kucharska, Alicia I.; Foxon, C. T.)V1362,384-390(1991)

Optical nonlinearities in ZnSe multiple-quantum-wells (Liu, Yudong; Liu, Shutian; Li, Chunfei; Shen, Dezen; Fan, Xi W.; Fan, Guang H.; Chen, Lian C.)V1362,436-447(1991)

Optical properties of InGaAs/InP strained quantum wells (Abram, Richard A.; Wood, Andrew C.; Robbins, D. J.)V1361,424-433(1991)

Optimization of reflection electro-absorption modulators (Pezeshki, Bardia; Thomas, Dominique; Harris, James S.)V1362,559-565(1991)

Optimization of strained layer InGaAs/GaAs quantum-well lasers (Fu, Richard J.; Hong, C. S.; Chan, Eric Y.; Booher, Dan J.; Figueroa, Luis)V1418,108-115(1991)

Optoelectronically implemented neural network for early visual processing (Mehanian, Courosh; Aull, Brian F.; Nichols, Kirby B.)V1469,275-280(1991)

Photoemission from quantum-confined structure of nonlinear optical materials (Ghatak, Kamakhya P.; Biswas, Shambhu N.; Ghoshal, Ardhendhu; Biswas, Shambhu N.)V1409,28-57(1991); V1484,136-148(1991)

Predicting diode laser performance (Lim, G.; Park, Youngsoh; Zmudzinski, C. A.; Zory, Peter S.; Miller, L. M.; Cockerill, T. M.; Coleman, James J.; Hong, C. S.; Figueroa, Luis)V1418,123-131(1991)

Quantum-well devices for optics in digital systems (Miller, David A.)V1389,496-502(1991)

Quantum-well excitonic devices for optical computing (Singh, Jasprit; Bhattacharya, Pallab K.)V1362,586-597(1991)

Quantum wells and artificially structured materials for nonlinear optics (Harris, James S.; Fejer, Martin M.)V1361,262-273(1991)

Quasi-three-dimensional electron systems and superlattices in wide parabolic wells: fabrication and physics (Shayegan, Mansour)V1362,228-239(1991)

Radiative processes in quantum-confined structures (Merz, James L.; Holtz, Per O.)V1361,76-88(1991)

Refractive index of multiple-quantum-well waveguides subject to impurity induced disordering using boron and fluorine (Hansen, Stein I.; Marsh, John H.; Roberts, John S.; Jeynes, C.)V1362,361-369(1991)

Regular doping structures: a Si-based, quantum-well infrared detector (Koch, J. F.)V1362,544-552(1991)

Some new insights in the physics of quantum-well devices (Tsu, Raphael)V1361,313-324(1991)

Spectral hole burning of strongly confined CdSe quantum dots (Spiegelberg, Christine; Henneberger, Fritz; Puls, J.)V1362,951-958(1991)

Strained quantum-well leaky-mode diode laser arrays (Shiau, T. H.; Sun, Shang-zhu; Schaus, Christian F.; Zheng, Kang; Hadley, G. R.; Hohimer, John P.)V1418,116-122(1991)

Strained semiconductors and heterostructures: synthesis and applications (Bhattacharya, Pallab K.; Singh, Jasprit)V1361,394-405(1991)

Structure optimization of selectively doped heterojunctions: evidences for a magnetically induced Wigner solidification (Etienne, Bernard; Paris, E.; Dorin, C.; Thierry-Mieg, V.; Williams, F. I.; Glattly, D. C.; Deville, G.; Andrei, E. Y.; Probst, O.)V1362,256-267(1991)

Study of GaAs/AlGaAs quantum-well structures grown by MOVPE using tertiarybutylarsine (Lee, Hyung G.; Kim, HyungJun; Park, S. H.; Langer, Dietrich W.)V1361,893-900(1991)

Study of novel oscillations in degenerate GaAs/A1GaAs quantum-wells using electro-optic voltage probing (Tsui, Ernest S.; Vickers, Anthony J.)V1362,804-810(1991)

Theory of optical-phonon interactions in a rectangular quantum wire (Stroscio, Michael A.; Kim, K. W.; Littlejohn, Michael A.)V1362,566-579(1991)

Thermoelectric power in fiber optic and laser materials under cross-field configuration (Ghatak, Kamakhya P.)V1584,435-447(1991)

Time-resolved luminescence experiments on modulation n-doped GaAs quantum wells (Lopez, C.; Meseguer, Francisco; Sanchez-Dehesa, Jose; Ruehle, Wolfgang W.; Ploog, Klaus H.)V1361,89-95(1991)

Tunneling recombination of carriers at type-II interface in GaInAsSb-GaSb heterostructures (Titkov, A. N.; Yakovlev, Yurii P.; Baranov, Alexej N.; Cheban, V. N.)V1361,669-673(1991)

Two-dimensional spatial light modulators using polarization-sensitive multiple-quantum-well light valve (Jain, Faquir C.; Drake, G.; Chung, C.; Bhattacharjee, K. K.; Cheung, S. K.)V1564,714-722(1991)

Ultrafast measurements of carrier transport optical nonlinearities in a GaAs/AlGaAs MQW SEED device (Park, Choong-Bum; Li Kam Wa, Patrick; Miller, Alan)V1474,8-17(1991)

Radar—see also lidar

Advanced Signal Processing Algorithms, Architectures, and Implementations II (Luk, Franklin T., ed.)V1566(1991)

Algorithm for statistical classification of radar clutter into one of several categories (Nechval, Nicholas A.)V1470,282-293(1991)

Analog optical processing of radio frequency signals (Sullivan, Daniel P.; Weber, Charles L.)V1476,234-245(1991)

Applications of the bispectrum in radar signature analysis and target identification (Jouny, Ismail; Garber, Frederick D.; Moses, Randolph L.; Walton, Eric K.)V1471,142-153(1991)

Asynchronous data fusion for target tracking with a multitasking radar and optical sensor (Blair, William D.; Rice, Theodore R.; Alouani, Ali T.; Xia, P.)V1482,234-245(1991)

BEST: a new satellite for a climatology study in the tropics (Orgeret, Marc)V1490,14-22(1991)

Broadband electromagnetic environment monitoring using semiconductor electroabsorption modulators (Pappert, Stephen A.; Lin, S. C.; Orazi, Richard J.; McLandrich, Matthew N.; Yu, Paul K.; Li, S. T.)V1476,282-293(1991)

Compact low-power acousto-optic range-Doppler-angle processor for a pulsed-Doppler radar (Pape, Dennis R.; Vlannes, Nickolas P.; Patel, Dharmesh P.; Phuvan, Sonlinh)V1476,201-213(1991)

Comparison of a multilayered perceptron with standard classification techniques in the presense of noise (Willson, Gregory B.)V1469,351-358(1991)

Comparison of synthetic-aperture radar autofocus techniques: phase gradient versus subaperture (Calloway, Terry M.; Jakowatz, Charles V.; Thompson, Paul A.; Eichel, Paul H.)V1566,353-364(1991)

Complex phase error and motion estimation in synthetic aperture radar imaging (Soumekh, Mehrdad; Yang, H.)V1452,104-113(1991)

Detection of contraband brought into the United States by aircraft and other transportation methods: a changing problem (Bruder, Joseph A.; Greneker, E. F.; Nathanson, F. E.; Henneberger, T. C.)V1479,316-321(1991)

Direct optical phase shifter for phased-array systems (Vawter, Gregory A.; Hietala, Vincent M.; Kravitz, Stanley H.; Meyer, W. J.)V1476,102-106(1991)

Effect of resampling on the MUSIC algorithm when applied to inverse-synthetic-aperture radar (Nash, Graeme)V1566,476-482(1991)

Efficient use of data structures for digital monopulse feature extraction (McEachern, Robert; Eckhardt, Andrew J.; Nauda, Alexander)V1470,226-232(1991)

Feature discrimination using multiband classification techniques (Ivey, Jim; Fairchild, Scott; Peterson, James R.; Stahl, Charles G.)V1567,170-178(1991)

Formulation and performance evaluation of adaptive, sequential radar-target-recognition algorithms (Snorrason, Ogmundur; Garber, Frederick D.)V1471,116-127(1991)

Ground target classification using moving target indicator radar signatures (Yoon, Chun S.)V1470,243-252(1991)

High-efficiency dual-band power amplifier for radar applications (Masliah, Denis A.; Cole, Brad; Platzker, Aryeh; Schindler, Manfred)V1475,113-120(1991)

High-power diode-pumped solid state lasers for optical space communications (Koechner, Walter; Burnham, Ralph L.; Kasinski, Jeff; Bournes, Patrick A.; DiBiase, Don; Le, Khoa; Marshall, Larry R.; Hays, Alan D.)V1522,169-179(1991)

High-resolution airborne multisensor system (Prutzer, Steven; Biron, David G.; Quist, Theodore M.)V1480,46-61(1991)

Importance of sensor models to model-based vision applications (Zelnio, Edmund G.)VIS07,112-121(1991)

Infrared tellurium two-dimensional acousto-optic processor for synthetic aperture radar (Souilhac, Dominique J.; Billerey, Dominique)V1521,158-174(1991)

Integrated processor architecture for multisensor signal processing (Nasburg, Robert E.; Stillman, Steve M.; Nguyen, M. T.)V1481,84-95(1991)

IR laser-light backscattering by an arbitrarily shaped dielectric object with rough surface (Wu, Zhensen; Cheng, Denghui)V1558,251-257(1991)

Issues in automatic object recognition: linking geometry and material data to predictive signature codes (Deitz, Paul; Muuss, Michael J.; Davisson, Edwin O.)VIS07,40-56(1991)

Laser Radar VI (Becherer, Richard J., ed.)V1416(1991)

Lasers and electro-optic technology in natural resource management (Greer, Jerry D.)V1396,342-352(1991)

Ll-filters in CFAR (constant false-alarm rate) detection (Mahmood Reza, Syed; Willett, Peter K.)V1451,298-308(1991)

Low-noise MMIC performance in Kaband using ion implantation technology (Mondal, J. P.; Contolatis, T.; Geddes, John J.; Swirhun, S.; Sokolov, Vladimir)V1475,314-318(1991)

Massively parallel synthetic-aperture radar autofocus (Mastin, Gary A.; Plimpton, Steven J.; Ghiglia, Dennis C.)V1566,341-352(1991)

Millimeter wave radar stationary-target classification using a high-order neural network (Hughen, James H.; Hollon, Kenneth R.)V1469,341-350(1991)

Modeling and experiments with a subsea laser radar system (Bjarnar, Morten L.; Klepsvik, John O.; Nilsen, Jan E.)V1537,74-88(1991)

Multifrequency beamspace Root-MUSIC: an experimental evaluation (Zoltowski, Michael D.; Kautz, Gregory M.; Silverstein, Seth D.)V1566,452-463(1991)

Naval Research Laboratory flex processor for radar signal processing (Alter, James J.; Evins, James B.; Letellier, J. P.)V1566,296-301(1991)

Neural data association (Kim, Kwang H.; Shafai, Bahram)V1481,406-417(1991)

Neural network modeling of radar backscatter from an ocean surface using chaos theory (Leung, Henry; Haykin, Simon)V1565,279-286(1991)

New inverse synthetic aperture radar algorithm for translational motion compensation (Bocker, Richard P.; Henderson, Thomas B.; Jones, Scott A.; Frieden, B. R.)V1569,298-310(1991)

New track-to-track association logic for almost identical multiple sensors (Malakian, Kourken; Vidmar, Anthony)V1481,315-328(1991)

Polarimetric segmentation of SAR imagery (Burl, Michael C.; Novak, Leslie M.)V1471,92-115(1991)

Quasioptical MESFET VCOs (Bundy, Scott; Mader, Tom; Popovic, Zoya; Ellinson, Reinold; Hjelme, Dag R.; Surette, Mark R.; Yadlowski, Michael; Mickelson, Alan R.)V1475,319-329(1991)

Radar-optronic tracking experiment for short- and medium-range aerial combat (Ravat, Christian J.; Mestre, J. P.; Rose, C.; Schorter, M.)V1478,239-246(1991)

Recirculating fiber optical RF-memory loop in countermeasure systems (Even-Or, Baruch; Lipsky, S.; Markowitz, Raymond; Herczfeld, Peter R.; Daryoush, Afshin S.; Saedi, Reza)V1371,161-169(1991)

Relationships between autofocus methods for SAR and self-survey techniques for SONAR (Wahl, Daniel E.; Jakowatz, Charles V.; Ghiglia, Dennis C.; Eichel, Paul H.)V1567,32-40(1991)

Remote sensing of precipitation structures using combined microwave radar and radiometric techniques (Vivekanandan, J.; Turk, F. J.; Bringi, Viswanathan N.)V1558,324-338(1991)

Robust CFAR detection using order statistic processors for Weibull-distributed clutter (Nagle, Daniel T.; Saniie, Jafar)V1481,49-63(1991)

Some analytical and statistical properties of Fisher information (Frieden, B. R.)V1569,311-316(1991)

Space-based millimeter wave debris tracking radar (Chang, Kai; Pollock, Michael A.; Skrehot, Michael K.)V1475,257-266(1991)

Staring phased-array radar using photorefractive crystals (Weverka, Robert T.; Wagner, Kelvin H.)V1564,676-684(1991)

Survey of radar-based target recognition techniques (Cohen, Marvin N.)V1470,233-242(1991)

Synthetic aperture radar of JERS-1 (Ono, Makoto; Nemoto, Yoshiaki)V1490,184-190(1991)

Target identification by means of impulse radar (Abrahamsson, S.; Brusmark, B.; Gaunaurd, Guillermo C.; Strifors, Hans C.)V1471,130-141(1991)

Terrain and target segmentation using coherent laser radar (Renhorn, Ingmar G.; Letalick, Dietmar; Millnert, Mille)V1480,35-45(1991)

Three-dimensional morphology for target detection (Patterson, Tim J.)V1471,358-368(1991)

TREIS: a concept for a user-affordable, user-friendly radar satellite system for tropical forest monitoring (Raney, R. K.; Specter, Christine N.)V1492,298-306(1991)

Unified approach to multisensor simulation of target signatures (Stewart, Stephen R.; LaHaie, Ivan J.; Lyons, Jayne T.)VIS07,57-97(1991)

Wideband bistatic radar signal processing using a coherent detection architecture (Belcher, Melvin L.; Garmon, Jeff P.)V1476,224-233(1991)

Radiance

Detector perturbation of ocean radiance measurements (Macdonald, Burns; Helliwell, William S.; Sanborn, James; Voss, Kenneth J.)V1537,104-114(1991)

Superradiance and exciton dynamics in molecular aggregates (Fidder, Henk; Terpstra, Jacob; Wiersma, Douwe A.)V1403,530-544(1991)

Underwater solar light field: analytical model from a WKB evaluation (Tessendorf, Jerry A.)V1537,10-20(1991)

Variation of the point spread function in the Sargasso Sea (Voss, Kenneth J.)V1537,97-103(1991)

Radiation—see also gamma rays; infrared; lasers; microwaves; millimeter waves; radiology; synchrotron radiation; ultraviolet; x rays

Accretion dynamics and polarized x-ray emission of magnetized neutron stars (Arons, Jonathan)V1548,2-12(1991)

Analytical studies of large closed-loop arrays (Fikioris, George; Freeman, D. K.; King, Ronold W.; Shen, Hao-Ming; Wu, Tai T.)V1407,295-305(1991)

Automatic counting of chromosome fragments for the determination of radiation dose (Smith, Warren E.; Leung, Billy C.; Leary, James F.)V1398,142-150(1991)

Balloon-born investigations of total and aerosol attenuation continuous spectra in the stratosphere (Mirzoeva, Larisa A.; Kiseleva, Margaret S.; Reshetnikova, Irina N.)V1540,444-449(1991)

Calculation of flux density produced by CPC reflectors on distant targets (Gordon, Jeff M.; Rabl, Ari)V1528,152-162(1991)

Cerenkov background radiation in imaging detectors (Rosenblatt, Edward I.; Beaver, Edward A.; Cohen, R. D.; Linsky, J. B.; Lyons, R. W.)V1449,72-86(1991)

Defect-induced stabilization of Fermi level in bulk silicon and at the silicon-metal interface (Iwanowski, Ryszard J.; Tatarkiewicz, Jakub J.)V1361,765-775(1991)

Diffraction effects in directed radiation beams (Hafizi, Bahman; Sprangle, Phillip)V1407,316-321(1991)

Discrete-angle radiative transfer in a multifractal medium (Davis, Anthony B.; Lovejoy, Shaun; Schertzer, Daniel)V1558,37-59(1991)

Displacement damage in Si imagers for space applications (Dale, Cheryl J.; Marshall, Paul W.)V1447,70-86(1991)

Effects of proton damage on charge-coupled devices (Janesick, James R.; Soli, George; Elliott, Tom; Collins, Stewart A.)V1447,87-108(1991)

Effects of radiation on millimeter wave monolithic integrated circuits (Meulenberg, A.; Hung, Hing-Loi A.; Singer, J. L.; Anderson, Wallace T.)V1475,280-285(1991)

Experimental study of electromagnetic missiles from a hyperboloidal lens (Shen, Hao-Ming; Wu, Tai T.; Myers, John M.)V1407,286-294(1991)

Experimental study of the resonance of a circular array (Shen, Hao-Ming)V1407,306-315(1991)

FIR optical cavity oscillation is observed with the AT&T Bell Laboratories free-electron laser (Shaw, Earl D.; Chichester, Robert J.; La Porta, A.)V1552,14-23(1991)

Foundation, excavation, and radiation-shielding concepts for a 16-m large lunar telescope (Chua, Koon M.; Johnson, Stewart W.)V1494,119-134(1991)

Helium-neon effects of laser radiation in rats infected with thromboxane B2 (Juri, Hugo; Palma, J. A.; Campana, Vilma; Gavotto, A.; Lapin, R.; Yung, S.; Lillo, J.)V1422,128-135(1991)

Influence of dose rate on radiation-induced loss in optical fibers (Henschel, Henning; Koehn, Otmar; Schmidt, Hans U.)V1399,49-63(1991)

Interframe registration and preprocessing of image sequences (Tajbakhsh, Shahram; Boyce, James F.)V1521,14-22(1991)

Intraoperative clinical use of low-power laser irradiation following surgical treatment of the tethered spinal cord (Rochkind, Simeone; Alon, M.; Ouaknine, G. E.; Weiss, S.; Avram, J.; Razon, N.; Lubart, Rachel; Friedmann, Harry)V1428,52-58(1991)

Ion implantation of polymers for electrical conductivity enhancement (Bridwell, Lynn B.; Wang, Y. Q.)V1519,878-883(1991)

Issues affecting the characterization of integrated optical devices subjected to ionizing radiation (Hickernell, Robert K.; Sanford, N. A.; Christensen, David H.)V1474,138-147(1991)

Laser-based flow cytometric analysis of genotoxicity of humans exposed to ionizing radiation during the Chernobyl accident (Jensen, Ronald H.; Bigbee, William L.; Langlois, Richard G.; Grant, Stephen G.; Pleshanov, Pavel G.; Chirkov, Andre A.; Pilinskaya, Maria A.)V1403,372-380(1991)

LiTaO3 and LiNbO3:Ti responses to ionizing radiation (Padden, Richard J.; Taylor, Edward W.; Sanchez, Anthony D.; Berry, J. N.; Chapman, S. P.; DeWalt, Steve A.; Wong, Ka-Kha)V1474,148-159(1991)

Luminescence molulation for the characterization of radiation damage within scintillator material (Bayer, Eberhard G.)V1361,195-199(1991)

Measurement of radiation-induced attenuation in optical fibers by optical-time-domain reflectometry (Looney, Larry D.; Lyons, Peter B.)V1474,132-137(1991)

Method and device that prevent target sensors from being radiation overexposed in the presence of a nuclear blast (Holubowicz, Kazimierz S.)V1456,274-285(1991)

Neutron transmutation doping of silicon in the VVR-S type nuclear reactor (Halmagean, Eugenia T.)V1362,984-991(1991)

New 2/3-inch MF image pick-up tubes for HDTV camera (Kobayashi, Akira; Ishikawa, Masayoshi; Suzuki, Takayoshi; Ikeya, Morihiro; Shimomoto, Yasuharu)V1449,148-156(1991)

New sensitizers for PDT (Morgan, Alan R.; Garbo, Greta M.; Krivak, T.; Mastroianni, Marta; Petousis, Nikolaos H.; St Clair, T.; Weisenberger, M.; van Lier, Johan E.)V1426,350-355(1991)

Nonlinear bremsstrahlung in nonequilibrium relativistic beam-plasma systems (Brandt, Howard E.)V1407,326-353(1991)

Nonlinear optical effects under laser pulse interaction with tissues (Altshuler, Grigori B.; Belikov, Andrey V.; Erofeev, Andrey V.)V1427,141-150(1991)

Nonselective thermal detectors of radiation (Pankratov, Nickolai A.)V1540,432-443(1991)

Notch and large-area CCD imagers (Bredthauer, Richard A.; Pinter, Jeff H.; Janesick, James R.; Robinson, Lloyd B.)V1447,310-315(1991)

Optical fiber radiation damage measurements (Ediriweera, Sanath R.; Kvasnik, Frank)V1399,64-75(1991)

Optical sensor system for Japanese Earth resources satellite 1 (Hino, Hideo; Takei, Mitsuru; Ono, Hiromi; Nagura, Riichi; Narimatsu, Yoshito; Hiramatsu, Masaru; Harada, Hisashi; Ogikubo, Kazuhiro)V1490,166-176(1991)

Polarization and directionality of the Earth's reflectances: the POLDER instrument (Lorsignol, Jean; Hollier, Pierre A.; Deshayes, Jean-Pierre)V1490,155-163(1991)

Polarization of emission lines from relativistic accretion disk (Chen, Kaiyou)V1548,23-33(1991)

Polarized nature of synchrotron radiation (Kim, Kwang-je)V1548,73-79(1991)

Radiation concerns for the Solar-A soft x-ray telescope (Acton, Loren W.; Morrison, Mons D.; Janesick, James R.; Elliott, Tom)V1447,123-139(1991)

Radiation effects on various optical components for the Mars Observer Spacecraft (Lowry, Jay H.; Iffrig, C. D.)V1330,132-141(1991)

Radiation-induced crosstalk in guided-wave devices (Taylor, Edward W.; Padden, Richard J.; Sanchez, Anthony D.; Chapman, S. P.; Berry, J. N.; DeWalt, Steve A.)V1474,126-131(1991)

Radiation-resistant optical glasses (Marker, Alexander J.; Hayden, Joseph S.; Speit, Burkhard)V1485,160-172(1991)

Radiative transfer in the cloudy atmosphere: modeling radiative transport (Gabriel, Philip; Stephens, Graeme L.; Tsay, Si-Chee)V1558,76-90(1991)

Radiometric standards in the USSR (Sapritsky, Victor I.)V1493,58-69(1991)

Remote sensing of precipitation structures using combined microwave radar and radiometric techniques (Vivekanandan, J.; Turk, F. J.; Bringi, Viswanathan N.)V1558,324-338(1991)

ScaRaB Earth radiation budget scanning radiometer (Monge, J.L.; Kandel, Robert S.; Pakhomov, L. A.; Bauche, B.)V1490,84-93(1991)

Scintillating plastic optical fiber radiation detectors in high-energy particle physics (Bross, Alan D.)V1592,122-132(1991)

Short-Wavelength Radiation Sources (Sprangle, Phillip, ed.)V1552(1991)

Spectroscopical studies of the ionizing-radiation-induced damage in optical fibers (Ediriweera, Sanath R.; Kvasnik, Frank)V1504,110-117(1991)

Spontaneous emission from free-electron lasers (Schmitt, Mark J.)V1552,107-117(1991)

Stereotactic multibeam radiation therapy system in a PACS environment (Fresne, Francoise; Le Gall, G.; Barillot, Christian; Gibaud, Bernard; Manens, J. P.; Toumoulin, Christine; Lemoine, D.; Chenal, C.; Scarabin, Jean-Marie)V1444,26-36(1991)

Stimulated relativistic harmonic generation from intense laser interactions with beams and plasmas (Sprangle, Phillip; Esarey, Eric)V1552,147-153(1991)

Submicron structures—promising filters in EUV: a review (Gruntman, Michael A.)V1549,385-394(1991)

Transition radiation in foil stack and x-ray laser (Yan, Zu Q.; Ruan, Ke F.)V1519,183-191(1991)

Two-dimensional control of radiant intensity by use of nonimaging optics (Kuppenheimer, John D.; Lawson, Robert I.)V1528,93-103(1991)

Vanderbilt Free-Electron Laser Center for Biomedical and Materials Research (Tolk, Norman H.; Brau, Charles A.; Edwards, Glenn S.; Margaritondo, Giorgio; McKinley, J. T.)V1552,7-13(1991)

XUV polarimeter for undulator radiation measurements (Gluskin, Efim S.; Mattson, J. E.; Bader, Samuel D.; Viccaro, P. J.; Barbee, Troy W.; Brookes, N.; Pitas, Alan A.; Watts, Richard N.)V1548,56-68(1991)

Radiology—see also medical imaging; x rays

Automated cyst recognition from x-ray photographs (Nedkova, Rumiana; Delchev, Georgy)V1429,105-107(1991)

Automated labeling of coronary arterial tree segments in angiographic projection data (Dumay, Adrie C.; Gerbrands, Jan J.; Geest, Rob J.; Verbruggen, Patricia E.; Reiber, J. H.)V1445,38-46(1991)

Automated techniques for quality assurance of radiological image modalities (Goodenough, David J.; Atkins, Frank B.; Dyer, Stephen M.)V1444,87-99(1991)

Automatic counting of chromosome fragments for the determination of radiation dose (Smith, Warren E.; Leung, Billy C.; Leary, James F.)V1398,142-150(1991)

Brightness and contrast adjustments for different tissue densities in digital chest radiographs (McNitt-Gray, Michael F.; Taira, Ricky K.; Eldredge, Sandra L.; Razavi, Mahmood)V1445,468-478(1991)

Clinical aspects of quality assurance in PACS (Marglin, Stephen I.)V1444,83-86(1991)

Clinical aspects of the Mayo/IBM PACS project (Forbes, Glenn S.; Morin, Richard L.; Pavlicek, William)V1446,318-322(1991)

Clinical evaluation of a 2K x 2K workstation for primary diagnosis in pediatric radiology (Razavi, Mahmood; Sayre, James W.; Simons, Margaret A.; Hamedaninia, Azar; Boechat, Maria I.; Hall, Theodore R.; Kangarloo, Hooshang; Taira, Ricky K.; Chuang, Keh-Shih; Kashifian, Payam)V1446,24-34(1991)

Clinical experience with a stereoscopic image workstation (Henri, Christopher J.; Collins, D. L.; Pike, G. B.; Olivier, A.; Peters, Terence M.)V1444,306-317(1991)

Compact open-architecture computed radiography system (Huang, H. K.; Lim, Art J.; Kangarloo, Hooshang; Eldredge, Sandra L.; Loloyan, Mansur; Chuang, Keh-Shih)V1443,198-202(1991)

Comparison of morphological and conventional edge detectors in medical imaging applications (Kaabi, Lotfi; Loloyan, Mansur; Huang, H. K.)V1445,11-23(1991)

Correlation between the detection and interpretation of image features (Fuhrman, Carl R.; King, Jill L.; Obuchowski, Nancy A.; Rockette, Howard E.; Sashin, Donald; Harris, Kathleen M.; Gur, David)V1446,422-429(1991)

Correlation model for a class of medical images (Zhang, Ya-Qin; Loew, Murray H.; Pickholtz, Raymond L.)V1445,367-373(1991)

Correlations between time required for radiological diagnoses, readers' performance, display environments, and difficulty of cases (Gur, David; Rockette, Howard E.; Sumkin, Jules H.; Hoy, Ronald J.; Feist, John H.; Thaete, F. L.; King, Jill L.; Slasky, B. S.; Miketic, Linda M.; Straub, William H.)V1446,284-288(1991)

Definition and evaluation of the data-link layer of PACnet (Alsafadi, Yasser H.; Martinez, Ralph; Sanders, William H.)V1446,129-140(1991)

Design considerations for a high-resolution film scanner for teleradiology applications (Kocher, Thomas E.; Whiting, Bruce R.)V1446,459-464(1991)

Detailed description of the Mayo/IBM PACS (Gehring, Dale G.; Persons, Kenneth R.; Rothman, Melvyn L.; Salutz, James R.; Morin, Richard L.)V1446,248-252(1991)

Diagnostic report acquisition unit for the Mayo/IBM PACS project (Brooks, Everett G.; Rothman, Melvyn L.)V1446,217-226(1991)

Digital film library implementation (Kishore, Sheel; Khalsa, Satjeet S.; Seshadri, Sridhar B.; Arenson, Ronald L.)V1446,188-194(1991)

Digital radiology with solid state linear x-ray detectors (Munier, Bernard; Prieur-Drevon, P.; Chabbal, Jean)V1447,44-55(1991)

Digital replication of chest radiographs without altering diagnostic observer performance (Flynn, Michael J.; Davies, Eric; Spizarny, David; Beute, Gordon; Peterson, Ed; Eyler, William R.; Gross, Barry; Chen, Ji)V1444,172-179(1991)

Direct digital image transfer gateway (Mun, In K.; Kim, Y. S.; Mun, Seong Ki)V1444,232-237(1991)

Display systems for medical imaging (Erdekian, Vahram V.; Trombetta, Steven P.)V1444,151-158(1991)

Distributing the server function in a multiring PAC system (Lynne, Kenton J.)V1446,177-187(1991)

DRILL: a standardized radiology-teaching knowledge base (Rundle, Debra A.; Evers, K.; Seshadri, Sridhar B.; Arenson, Ronald L.)V1446,405-413(1991)

Dynamic spatial reconstructor: a high-speed, stop-action, 3-D, digital radiographic imager of moving internal organs and blood (Jorgensen, Steven M.; Whitlock, S. V.; Thomas, Paul J.; Roessler, R. W.; Ritman, Erik L.)V1346,180-191(1991)

Effect of experimental design on sample size (Rockette, Howard E.; Obuchowski, Nancy A.; Gur, David; Good, Walter F.)V1446,276-283(1991)

Effects of transmission errors on medical images (Pronios, Nikolaos-John B.; Yovanof, Gregory S.)V1446,108-128(1991)

Equalization radiography with radiation quality modulation (Geluk, Ronald J.; Vlasbloem, Hugo)V1443,143-152(1991)

European activities towards a hospital-integrated PACS based on open systems (Kouwenberg, Jef M.; Ottes, Fenno P.; Bakker, Albert R.)V1446,357-361(1991)

Evaluation of a moving slit technique for mammography (Rosenthal, Marc S.; Sashin, Donald; Herron, John M.; Maitz, Glenn S.; Boyer, Joseph W.; Gur, David)V1443,132-142(1991)

Evaluation of total workstation CT interpretation quality: a single-screen pilot study (Beard, David V.; Perry, John R.; Muller, K.; Misra, Ram B.; Brown, P.; Hemminger, Bradley M.; Johnston, R. E.; Mauro, M.; Jaques, P. F.; Schiebler, M.)V1446,52-58(1991)

Folder management on a multimodality PACS display station (Feingold, Eric; Seshadri, Sridhar B.; Arenson, Ronald L.)V1446,211-216(1991)

Full-frame entropy coding for radiological image compression (Lo, Shih-Chung B.; Krasner, Brian H.; Mun, Seong Ki; Horii, Steven C.)V1444,265-271(1991)

Fundamental studies for the high-intensity long-duration flash x-ray generator for biomedical radiography (Sato, Eiichi; Isobe, Hiroshi; Takahashi, Kei; Tamakawa, Yoshiharu; Yanagisawa, Toru)V1358,193-200(1991)

High-resolution teleradiology applications within the hospital (Jost, R. G.; Blaine, G. J.; Kocher, Thomas E.; Muka, Edward; Whiting, Bruce R.)V1446,2-9(1991)

Hospital-integrated PACS at the University Hospital of Geneva (Ratib, Osman M.; Ligier, Yves; Hochstrasser, Denis; Scherrer, Jean-Raoul)V1446,330-340(1991)

Image acquisition unit for the Mayo/IBM PACS project (Reardon, Frank J.; Salutz, James R.)V1446,481-491(1991)

Image and modality control issues in the objective evaluation of manipulation techniques for digital chest images (Rehm, Kelly; Seeley, George W.; Dallas, William J.)V1445,24-35(1991)

Image delivery performance of a CT/MR PACS module applied in neuroradiology (Lou, Shyh-Liang; Loloyan, Mansur; Weinberg, Wolfram S.; Valentino, Daniel J.; Lufkin, Robert B.; Hanafee, William; Bentson, John R.; Jabour, Bradly; Huang, H. K.)V1446,302-311(1991)

Image-display optimization using clinical history (Nodine, Calvin F.; Brikman, Inna; Kundel, Harold L.; Douglas, A.; Seshadri, Sridhar B.; Arenson, Ronald L.)V1444,56-62(1991)

Improved real-time volumetric ultrasonic imaging system (Pavy, Henry G.; Smith, Stephen W.; von Ramm, Olaf T.)V1443,54-61(1991)

Initial experiences with PACS in a clinical and research environment (Honeyman, Janice C.; Staab, Edward V.; Frost, Meryll M.)V1446,362-368(1991)

Interactive graphics system for locating plunge electrodes in cardiac MRI images (Laxer, Cary; Johnson, G. A.; Kavanagh, Katherine M.; Simpson, Edward V.; Ideker, Raymond E.; Smith, William M.)V1444,190-195(1991)

Interactive image processing in swallowing research (Dengel, Gail A.; Robbins, JoAnne; Rosenbek, John C.)V1445,88-94(1991)

Investigation of new filtering schemes for computerized detection of lung nodules (Yoshimura, Hitoshi; Giger, Maryellen L.; Matsumoto, Tsuneo; Doi, Kunio; MacMahon, Heber; Montner, Steven M.)V1445,47-51(1991)

Iso-precision scaling of digitized mammograms to facilitate image analysis (Karssemeijer, Nico; van Erning, Leon J.)V1445,166-177(1991)

Management system for a PACS network in a hospital environment (Mattheus, Rudy A.; Temmerman, Yvan; Verhellen, P.; Osteaux, Michel)V1446,341-351(1991)

Medical diagnostic imaging support systems for military medicine (Goeringer, Fred)V1444,340-350(1991)

Medical Imaging V: Image Capture, Formatting, and Display (Kim, Yongmin, ed.)V1444(1991)

Medical Imaging V: Image Physics (Schneider, Roger H., ed.)V1443(1991)

Medical Imaging V: Image Processing (Loew, Murray H., ed.)V1445(1991)

Medical Imaging V: PACS Design and Evaluation (Jost, R. G., ed.)V1446(1991)

Multichannel fiber optic broadband video communication system for monitoring CT/MR examinations (Huang, H. K.; Kangarloo, Hooshang; Tecotzky, Raymond H.; Cheng, Xin; Vanderweit, Don)V1444,214-220(1991)

Multicomputer performance evaluation tool and its application to the Mayo/IBM image archival system (Pavicic, Mark J.; Ding, Yingjai)V1446,370-378(1991)

Multi-media PACS integrated with HIS/RIS employing magneto-optical disks (Umeda, Tokuo; Inamura, Kiyonari; Inamoto, Kazuo; Kondoh, Hiroshi P.; Kozuka, Takahiro)V1446,199-210(1991)

Multiple communication networks for a radiological PACS (Wong, Albert W.; Stewart, Brent K.; Lou, Shyh-Liang; Chan, Kelby K.; Huang, H. K.)V1446,73-80(1991)

Need for quality assurance related to PACS (Rowberg, Alan H.)V1444,80-82(1991)

Object-oriented model for medical image database (Aubry, Florent; Bizais, Yves; Gibaud, Bernard; Forte, Anne-Marie; Chameroy, Virginie; Di Paola, Robert; Scarabin, Jean-Marie)V1446,168-176(1991)

Off-line image exchange between two PACS modules using the "ISAC" magneto-optical disk (Minato, Kotaro; Komori, Masaru; Nakano, Yoshihisa; Yonekura, Yoshiharu; Sasayama, Satoshi; Takahashi, Takashi; Konishi, Junji; Abe, Mituyuki; Sato, Kazuhiro; Hosoba, Minoru)V1446,195-198(1991)

One-year clinical experience with a fully digitized nuclear medicine department: organizational and economical aspects (Anema, P. C.; de Graaf, C. N.; Wilmink, J. B.; Hall, David; Hoekstra, A.; van Rijk, P. P.; Van Isselt, J. W.; Viergever, Max A.)V1446,352-356(1991)

Operational infrastructure for a clinical PACS (Boehme, Johannes M.; Chimiak, William J.; Choplin, Robert H.; Maynard, C. D.)V1446,312-317(1991)

Optimization and evaluation of an image intensifier TV system for digital chest imaging (Angelhed, Jan-Erik; Mansson, Lars G.; Kheddache, Susanne)V1444,159-170(1991)

PACS and teleradiology for on-call support of abdominal imaging (Horii, Steven C.; Garra, Brian S.; Mun, Seong Ki; Zeman, Robert K.; Levine, Betty A.; Fielding, Robert)V1446,10-15(1991)

PACS: "Back to the Future" (Carey, Bruce; Seshadri, Sridhar B.; Arenson, Ronald L.)V1446,414-419(1991)

PACS for GU radiology (Hayrapetian, Alek; Barbaric, Zoran L.; Weinberg, Wolfram S.; Chan, Kelby K.; Loloyan, Mansur; Taira, Ricky K.; Huang, H. K.)V1446,243-247(1991)

PACS modeling and development: requirements versus reality (Seshadri, Sridhar B.; Kishore, Sheel; Khalsa, Satjeet S.; Feingold, Eric; Arenson, Ronald L.)V1446,388-395(1991)

PACS reading time comparision: the workstation versus alternator for ultrasound (Horii, Steven C.; Garra, Brian S.; Mun, Seong Ki; Singer, Jon; Zeman, Robert K.; Levine, Betty A.; Fielding, Robert; Lo, Ben)V1446,475-480(1991)

Performance characteristics of an ultrafast network for PACS (Stewart, Brent K.; Lou, Shyh-Liang; Wong, Albert W.; Chan, Kelby K.; Huang, H. K.)V1446,141-153(1991)

Physical and psychophysical evaluation of CRT noise performance (Ji, Tinglan; Roehring, Hans; Blume, Hartwig R.; Seeley, George W.; Browne, Michael P.)V1444,136-150(1991)

Potential usefulness of a video printer for producing secondary images from digitized chest radiographs (Nishikawa, Robert M.; MacMahon, Heber; Doi, Kunio; Bosworth, Eric)V1444,180-189(1991)

Prefetching: PACS image management optimization using HIS/RIS information (Lodder, Herman; van Poppel, Bas M.; Bakker, Albert R.)V1446,227-233(1991)

Preliminary results of a PACS implementation (Kishore, Sheel; Khalsa, Satjeet S.; Seshadri, Sridhar B.; Arenson, Ronald L.)V1446,236-242(1991)

Present status and future directions of the Mayo/IBM PACS project (Morin, Richard L.; Forbes, Glenn S.; Gehring, Dale G.; Salutz, James R.; Pavlicek, William)V1446,436-441(1991)

Qualitative approach to medical image databases (Bizais, Yves; Gibaud, Bernard; Forte, Anne-Marie; Aubry, Florent; Di Paola, Robert; Scarabin, Jean-Marie)V1446,156-167(1991)

Quality assurance of PACS systems with laser film digitizers (Esser, Peter D.; Halpern, Ethan J.)V1444,100-103(1991)

RadGSP: a medical image display and user interface for UWGSP3 (Yee, David K.; Lee, Woobin; Kim, Dong-Lok; Haass, Clark D.; Rowberg, Alan H.; Kim, Yongmin)V1444,292-305(1991)

Radiologists' confidence in detecting abnormalities on chest images and their subjective judgments of image quality (King, Jill L.; Gur, David; Rockette, Howard E.; Curtin, Hugh D.; Obuchowski, Nancy A.; Thaete, F. L.; Britton, Cynthia A.; Metz, Charles E.)V1446,268-275(1991)

Radiology workstation for mammography: preliminary observations, eyetracker studies, and design (Beard, David V.; Johnston, R. E.; Pisano, E.; Hemminger, Bradley M.; Pizer, Stephen M.)V1446,289-296(1991)

Rapid display of radiographic images (Cox, Jerome R.; Moore, Stephen M.; Whitman, Robert A.; Blaine, G. J.; Jost, R. G.; Karlsson, L. M.; Monsees, Thomas L.; Hassen, Gregory L.; David, Timothy C.)V1446,40-51(1991)

Reliability issues in PACS (Taira, Ricky K.; Chan, Kelby K.; Stewart, Brent K.; Weinberg, Wolfram S.)V1446,451-458(1991)

Removing vertical lines generated when x-ray images are digitized (Oyama, Yoshiro; Tani, Yuichiro; Shigemura, Naoshi; Abe, Toshio; Matsuda, Koyo; Kubota, Shigeto; Inami, Takashi)V1444,413-423(1991)

Repetitive flash x-ray generator operated at low-dose rates for a medical x-ray television system (Sato, Eiichi; Isobe, Hiroshi; Takahashi, Kei; Tamakawa, Yoshiharu; Yanagisawa, Toru)V1358,462-470(1991)

Selection of image acquisition methods (Donnelly, Joseph J.)V1444,351-356(1991)

Signa Tutor: results and future directions (Rundle, Debra A.; Watson, Carolyn K.; Seshadri, Sridhar B.; Wehrli, Felix W.)V1446,379-387(1991)

Solid models for CT/MR image display: accuracy and utility in surgical planning (Mankovich, Nicholas J.; Yue, Alvin; Ammirati, Mario; Kioumehr, Farhad; Turner, Scott)V1444,2-8(1991)

Subjective evaluation of image enhancements in improving the visibility of pathology in chest radiographs (Plessis, Brigitte; Goldberg, Morris; Belanger, Garry; Hickey, Nancy M.)V1445,539-554(1991)

Sunset: a hardware-oriented algorithm for lossless compression of gray-scale images (Langdon, Glen G.)V1444,272-282(1991)

Surface definition technique for clinical imaging (Liao, Wen-gen; Simovsky, Ilya; Li, Andrew; Kramer, David M.; Kaufman, Leon; Rhodes, Michael L.)V1444,47-55(1991)

Technical and clinical evaluations of a 2048 x 2048-matrix digital radiography system for gastrointestinal examinations (Ogura, Toshihiro; Masuda, Yukihisa; Fujita, Hiroshi; Inoue, Nobuo; Yonekura, Fukuo; Miyagi, Yoshihiro; Takatsu, Kazuaki; Akahira, Katsuyoshi; Tsuruta, Shigehiko; Kamiya, Masami; Takahashi, Fumitaka; Oda, Kazuyuki; Ikeda, Shigeyuki; Koike, Kouichi)V1443,153-157(1991)

Teleradiology in the local environment (Staab, Edward V.; Honeyman, Janice C.; Frost, Meryll M.; Bidgood, W. D.)V1446,16-22(1991)

Temperature and fading effects of fiber-optic dosimeters for radiotherapy (Bueker, Harald; Gripp, S.; Haesing, Friedrich W.)V1572,410-418(1991)

Tissue identification in MR images by adaptive cluster analysis (Gutfinger, Dan; Hertzberg, Efrat M.; Tolxdorff, T.; Greensite, F.; Sklansky, Jack)V1445,288-296(1991)

Vascular parameters from angiographic images (Close, Robert A.; Duckwiler, Gary R.; Vinuela, Fernando; Dion, Jacques E.)V1444,196-203(1991)

Visualization and volumetric compression (Chan, Kelby K.; Lau, Christina C.; Chuang, Keh-Shih; Morioka, Craig A.)V1444,250-255(1991)

X-window-based 2K display workstation (Weinberg, Wolfram S.; Hayrapetian, Alek; Cho, Paul S.; Valentino, Daniel J.; Taira, Ricky K.; Huang, H. K.)V1446,35-39(1991)

Radiometry—see also photometry

Advanced visible and near-IR radiometer for ADEOS (Iwasaki, Nobuo; Tange, Yoshio; Miyachi, Yuji; Inoue, Kouichi; Kadowaki, Tomoko; Tanaka, Hirokazu; Michioka, Hidekazu)V1490,216-221(1991)

Airborne and spaceborne thermal multispectral remote sensing (Watanabe, Hiroshi; Sano, Masaharu; Mills, F.; Chang, Sheng-Huei; Masuda, Shoichi)V1490,317-323(1991)

ASTER calibration concept (Ono, Akira; Sakuma, Fumihiro)V1490,285-298(1991)

Calibrated intercepts for solar radiometers used in remote sensor calibration (Gellman, David I.; Biggar, Stuart F.; Slater, Philip N.; Bruegge, Carol J.)V1493,175-180(1991)

Calibration for the medium-resolution imaging spectrometer (Baudin, Gilles; Chessel, Jean-Pierre; Cutter, Mike A.; Lobb, Daniel R.)V1493,16-27(1991)

Calibration of Passive Remote Observing Optical and Microwave Instrumentation (Guenther, Bruce W., ed.)V1493(1991)

Characterization of tropospheric methane through space-based remote sensing (Ashcroft, Peter)V1491,48-55(1991)

CiNeRaMa model: a useful tool for detection range estimate (Talmore, Eli T.)V1442,362-371(1991)

Clouds as calibration targets for AVHRR reflected-solar channels: results from a two-year study at NOAA/NESDIS (Abel, Peter)V1493,195-206(1991)

Cooling system for short-wave infrared radiometer of JERS-1 optical sensor (Ohmori, Yasuhiro; Arakawa, Isao; Nakai, Akira; Tsubosaka, Kazuyoshi)V1490,177-183(1991)

Cryogenic radiometers and intensity-stabilized lasers for EOS radiometric calibrations (Foukal, Peter; Hoyt, Clifford C.; Jauniskis, L.)V1493,72-79(1991)

Design considerations for infrared imaging radiometers (Fraedrich, Douglas S.)V1486,2-7(1991)

Development and analysis of a simple model for an IR sensor (Ballik, Edward A.; Wan, William)V1488,249-256(1991)

Different aspects of backgrounds in various spectral bands (Ben-Shalom, Ami; Devir, Adam D.; Ribak, Erez N.; Talmore, Eli T.; Balfour, L. S.; Brandman, N.)V1486,238-257(1991)

Enhancements to the radiometric calibration facility for the Clouds and the Earth's Radiant Energy System instruments (Folkman, Mark A.; Jarecke, Peter J.; Darnton, Lane A.)V1493,255-266(1991)

Evaluation of the NOAA-11 solar backscatter ultraviolet radiometer, Mod 2 (SBUV/2): inflight calibration (Weiss, Howard; Cebula, Richard P.; Laamann, K.; Hudson, Robert D.)V1493,80-90(1991)

Flight solar calibrations using the mirror attenuator mosaic: low-scattering mirror (Lee, Robert B.)V1493,267-280(1991)

Ground-based microwave radiometry of ozone (Kaempfer, Niklaus A.; Bodenmann, P.; Peter, Reto)V1491,314-322(1991)

High radiometric performance CCD for the third-generation stratospheric aerosol and gas experiment (Delamere, W. A.; Baer, James W.; Ebben, Thomas H.; Flores, James S.; Kleiman, Gary; Blouke, Morley M.; McCormick, M. P.)V1447,204-213(1991)

High-resolution multichannel mm-wave radiometer for the detection of stratospheric ClO (Gerber, Louis; Kaempfer, Niklaus A.)V1491,211-217(1991)

High-resolution spectral analysis and modeling of infrared ocean surface radiometric clutter (McGlynn, John D.; Ellis, Kenneth K.; Kryskowski, David)V1486,141-150(1991)

Hollow waveguides for sensor applications (Saggese, Steven J.; Harrington, James A.; Sigel, George H.)V1368,2-14(1991)

In-flight calibration of a helicopter-mounted Daedalus multispectral scanner (Balick, Lee K.; Golanics, Charles J.; Shines, Janet E.; Biggar, Stuart F.; Slater, Philip N.)V1493,215-223(1991)

Infrared background measurements at White Sands Missile Range, NM (Troyer, David E.; Fouse, Timothy; Murdaugh, William O.; Zammit, Michael G.; Rogers, Stephen B.; Skrzypczak, J. A.; Colley, Charles B.; Taczak, William J.)V1486,396-409(1991)

In-line power meter for use during laser angioplasty (Smith, Roy E.; Milnes, Peter; Edwards, David H.; Mitchell, David C.; Wood, Richard F.)V1425,116-121(1991)

Intensified multispectral imaging measuring in the spatial, frequency, and time domains with a single instrument (Kennedy, Benjamin J.)V1346,68-74(1991)

Large integrating sphere of prelaunch calibration system for Japanese Earth Resources Satellite optical sensors (Suzuki, Naoshi; Narimatsu, Yoshito; Nagura, Riichi; Sakuma, Fumihiro; Ono, Akira)V1493,48-57(1991)

Laser-target diagnostics instrumentation system (Bousek, Ronald R.)V1414,175-184(1991)

Measurement of tropospheric carbon monoxide using gas filter radiometers (Reichle, Henry G.)V1491,15-25(1991)

Monolithic GaAs integrated circuit millimeter wave imaging sensors (Weinreb, Sander)V1475,25-31(1991)

Optical stability of diffuse reflectance materials in space (Hale, Robert R.)V1485,173-182(1991)

Overview of ASTER design concept (Fujisada, Hiroyuki; Ono, Akira)V1490,244-254(1991)

Overview of the halogen occultation experiment (Russell, James M.)V1491,110-116(1991)

Pattern recognition in autoradiography of optical materials (Ashurov, Mukhsin K.; Gafitullina, Dilyara)V1550,50-54(1991)

Prelaunch calibration system for optical sensors of Japanese Earth Resources Satellite (Sakuma, Fumihiro; Ono, Akira)V1493,37-47(1991)

Pressure modulator infrared radiometer (PMIRR) optical system alignment and performance (Chrisp, Michael P.; Macenka, Steve A.)V1540,213-218(1991)

Quantitative evaluation of errors in remote measurements using a thermal imager (Engel, Michael Y.; Balfour, L. S.)V1442,298-307(1991)

Radiometric calibration of an airborne multispectral scanner (Markham, Brian L.; Ahmad, Suraiya P.; Jackson, Ray D.; Moran, M. S.; Biggar, Stuart F.; Gellman, David I.; Slater, Philip N.)V1493,207-214(1991)

Radiometric calibration of space-borne Fourier transform infrared-emission spectrometers: proposed scenario for European Space Agency's Michelson interferometer for passive atmospheric sounding (Lamarre, Daniel; Giroux, Jean)V1493,28-36(1991)

Radiometric calibration of SPOT 2 HRV: a comparison of three methods (Biggar, Stuart F.; Dinguirard, Magdeleine C.; Gellman, David I.; Henry, Patrice; Jackson, Ray D.; Moran, M. S.; Slater, Philip N.)V1493,155-162(1991)

Radiometric calibration plan for the Clouds and the Earth's Radiant Energy System scanning instruments (Jarecke, Peter J.; Folkman, Mark A.; Darnton, Lane A.)V1493,244-254(1991)

Radiometric stability of the shuttle-borne solar backscatter ultraviolet spectrometer (Cebula, Richard P.; Hilsenrath, Ernest; Kelly, Thomas J.; Batluck, Georgiann R.)V1493,91-99(1991)

Radiometric standards in the USSR (Sapritsky, Victor I.)V1493,58-69(1991)

Radiometric versus thermometric calibration of IR test systems: which is best? (Richardson, Philip I.)V1488,80-88(1991)

Ratioing radiometer for use with a solar diffuser (Palmer, James M.; Slater, Philip N.)V1493,106-117(1991)

Reflectance stability analysis of Spectralon diffuse calibration panels (Bruegge, Carol J.; Stiegman, Albert E.; Coulter, Daniel R.; Hale, Robert R.; Diner, David J.; Springsteen, Arthur W.)V1493,132-142(1991)

Results of calibrations of the NOAA-11 AVHRR made by reference to calibrated SPOT imagery at White Sands, N.M. (Nianzeng, Che; Grant, Barbara G.; Flittner, David E.; Slater, Philip N.; Biggar, Stuart F.; Jackson, Ray D.; Moran, M. S.)V1493,182-194(1991)

Short-wavelength infrared subsystem design status of ASTER (Akasaka, Akira; Ono, Makoto; Sakurai, Yasushi; Hayashida, Bun)V1490,269-277(1991)

Solar-diffuser panel and ratioing radiometer approach to satellite sensor on-board calibration (Slater, Philip N.; Palmer, James M.)V1493,100-105(1991)

Spatial sampling errors for a satellite-borne scanning radiometer (Manalo, Natividad D.; Smith, G. L.)V1493,281-291(1991)

State-of-the-art transfer radiometer for testing and calibration of FLIR test equipment (Kopolovich, Zvi; Naor, Yoram; Cabib, Dario; Johnson, W. Todd; Sapir, Eyal)V1540,565-577(1991)

Stray light effects on calibrations using a solar diffuser (Palmer, James M.)V1493,143-154(1991)

Surface and aerosol models for use in radiative transfer codes (Hart, Quinn J.)V1493,163-174(1991)

Technique for improving the calibration of large-area sphere sources (Walker, James H.; Cromer, Chris L.; McLean, James T.)V1493,224-230(1991)

Temperature distributions in laser-irradiated tissues (Valderrama, Giuseppe L.; Fredin, Leif G.; Berry, Michael J.; Dempsey, B. P.; Harpole, George M.)V1427,200-213(1991)

Temperature-monitored/controlled silicon photodiodes for standardization (Eppeldauer, George)V1479,71-77(1991)

Thermal infrared subsystem design status of ASTER (Aoki, Yutaka; Ohmae, Hirokazu; Kitamura, Shin-ichi)V1490,278-284(1991)

Time-resolved infrared radiometry of multilayer organic coatings using surface and subsurface heating (Maclachlan Spicer, J. W.; Kerns, W. D.; Aamodt, Leonard C.; Murphy, John C.)V1467,311-321(1991)

Transfer function considerations for the CERES (cloud's and earth's radiant energy system) scanning radiometer (Manalo, Natividad D.; Smith, G. L.; Barkstrom, Bruce R.)V1521,106-116(1991)

Transient radiometric measurements with a PtSi IR camera (Konopka, Wayne L.; Soel, Michael A.; Celentano, A.; Calia, V.)V1488,355-365(1991)

Update: high-speed/high-volume radiometric testing of Z-technology focal planes (Johnson, Jerome L.)V1541,210-219(1991)

Visible and near-infrared subsystem and common signal processor design status of ASTER (Takahashi, Fumiho; Hiramatsu, Masaru; Watanabe, Fumito; Narimatsu, Yoshito; Nagura, Riichi)V1490,255-268(1991)

VUV wall stabilized argon arc discharge source (Li, Futian; Cao, Jianlin; Chen, Po; Qian, Limin; Jin, Lei; Chen, Xingdan)V1345,78-88(1991)

Raman effect—see also scattering; spectroscopy

Advanced approaches in luminescence and Raman spectroscopy (Vo-Dinh, Tuan)V1435,197-202(1991)

Chaotic behavior of a Raman ring laser (Englund, John C.)V1497,292-299(1991)

Compact Raman instrumentation for process and environmental monitoring (Carrabba, Michael M.; Spencer, Kevin M.; Rauh, R. D.)V1434,127-134(1991)

Determination of gasoline fuel properties by Raman spectroscopy (Clarke, Richard H.; Chung, W. M.; Wang, Q.; De Jesus, Stephen T.)V1437,198-204(1991)

Diffraction effects in stimulated Raman scattering (Scalora, Michael; Haus, Joseph W.)V1497,153-164(1991)

Fiber optic sensor probe for in-situ surface-enhanced Raman monitoring (Vo-Dinh, Tuan; Stokes, D. L.; Li, Ying-Sing; Miller, Gordon H.)V1368,203-209(1991)

Frequency up-conversion of a discharge-pumped molecular fluorine laser by stimulated Raman scattering in H2 (Kakehata, Masayuki; Hashimoto, Etsu; Kannari, Fumihiko; Obara, Minoru)V1397,185-189(1991)

Gain and threshold in noninversion lasers (Scully, Marlan O.; Zhu, Shi-Yao; Narducci, Lorenzo M.; Fearn, Heidi)V1497,264-276(1991)

Generalized Raman gain in nonparabolic semiconductors under strong magnetic field (Ghatak, Kamakhya P.; Ghoshal, Ardhendhu; De, Badal)V1409,178-190(1991)

High-power KrF lasers (Key, Michael H.; Bailly-Salins, Rene; Edwards, B.; Harvey, Erol C.; Hirst, Graeme J.; Hooker, Chris J.; Kidd, A. K.; Madraszek, E. M.; Rodgers, P. A.; Ross, Ian N.; Shaw, M. J.; Steyer, M.)V1397,9-17(1991)

High-temperature Raman scattering behavior in diamond (Herchen, Harald; Cappelli, Mark A.)V1534,158-168(1991)

Holographic nonlinear Raman spectroscopy of large molecules of biological importance (Ivanov, Anatoliy A.; Koroteev, Nikolai I.; Fishman, A. I.)V1429,132-144(1991); V1432,141-153(1991)

Microstructures and domain size effects in diamond films characterized by Raman spectroscopy (Nemanich, Robert J.; Bergman, Larry; LeGrice, Yvonne M.; Turner, K. F.; Humphreys, T. P.)V1437,2-12(1991)

New experimental methods and theory of Raman optical activity (Nafie, Laurence A.; Che, Diping; Yu, Gu-Sheng; Freedman, Teresa B.)V1432,37-49(1991)

New problems of femtosecond time-domain CARS (coherent antistokes Raman spectroscopy) of large molecules (Kolomoitsev, D. V.; Nikitin, S. Y.)V1402,31-43(1991)

Optical measurements of electrodynamically levitated microparticles (Davis, E. J.)V1435,216-242(1991)

Pericyclic photochemical ring-opening reactions are complete in picoseconds: a time-resolved UV resonance Raman study (Reid, Philip J.; Doig, Stephen J.; Mathies, Richard A.)V1432,172-183(1991)

Phase-locking and unstability of light waves in Raman-active crystals (Azarenkov, Aleksey N.; Altshuler, Grigori B.; Belashenkov, Nickolay R.; Inochkin, Mickle V.; Karasev, Viatcheslav B.; Kozlov, Sergey A.)V1409,154-164(1991)

Physical effects in time-domain CARS (coherent antistokes Raman spectroscopy) of molecular gases (Kolomoitsev, D. V.; Nikitin, S. Y.)V1402,11-30(1991)

Picosecond time-resolved resonance Raman spectroscopy of bacteriorhodopsin: structure and kinetics of the J, K, and KL intermediates (Doig, Stephen J.; Reid, Philip J.; Mathies, Richard A.)V1432,184-196(1991)

Pulsating instabilities and chaos in Raman lasers (Harrison, Robert G.; Lu, Weiping; Jiad, K.; Uppal, J. S.)V1376,94-102(1991)

Raman scattering characterization of direct gap Si/Ge superlattices (White, Julian D.; Gell, Michael A.; Fasol, Gerhard; Gibbings, C. J.; Tuppen, C. G.)V1361,293-301(1991)

Raman scattering determination of nonpersistent optical control of electron density in a heterojunction (Richards, David R.; Fasol, Gerhard; Ploog, Klaus H.)V1361,246-254(1991)

Raman study of icosahedral C60 (Sinha, Kislay; Menendez, Jose; Adams, G. B.; Page, J. B.; Sankey, Otto F.; Lamb, Lowell; Huffman, Donald R.)V1437,32-35(1991)

Resonance Raman scattering from the primary electron donor in photosynthetic reaction centers from Rhodobacter sphaeroides (Bocian, David F.)V1432,166-171(1991)

Spectroscopic properties of the potentiometric probe merocyanine-540 in solutions and liposomes (Ehrenberg, Benjamin; Pevzner, Eliyahu)V1432,154-163(1991)

Statistics of spontaneously generated Raman solitons (Englund, John C.; Bowden, Charles M.)V1497,218-221(1991)

Stimulated Raman diagnostics in diesel droplets (Golombok, Michael)V1497,100-119(1991)

Superradiant Raman free-electron lasers (Tsui, King H.)V1407,281-284(1991)

Symmetrization analysis of lattice-vibrational modes and study of Raman-IR spectra for B-BaB2O4 (Hong, Shuili)V1437,194-197(1991)

Theory and simulation of Raman scattering in intense short-pulse laser-plasma interactions (Wilks, S. C.; Kruer, William L.; Langdon, A. B.; Amendt, Peter; Eder, David C.; Keane, Christopher J.)V1413,131-137(1991)

Time development of AlGaAs single-quantum-well laser facet temperature on route to catastrophical breakdown (Tang, Wade C.; Rosen, Hal J.; Vettiger, Peter; Webb, David J.)V1418,338-342(1991)

Time-resolved Raman spectroscopy from reacting optically levitated microdroplets (Carls, Joseph C.; Brock, James R.)V1497,120-131(1991)

Transient stimulated Raman scattering: theory and experiments of pulse shortening and phase conjugation properties (Agnesi, Antoniangelo; Reali, Giancarlo C.; Kubecek, Vaclav)V1415,104-109(1991)

Ultrafast and not-so-fast dynamics of cytochrome oxidase: the ligand shuttle and its possible functional significance (Woodruff, William H.; Dyer, R. B.; Einarsdottir, Oloef; Peterson, Kristen A.; Stoutland, Page O.; Bagley, K. A.; Palmer, Graham; Schoonover, J. R.; Kliger, David S.; Goldbeck, Robert A.; Dawes, T. D.; Martin, Jean-Louis; Lambry, J.-C.; Atherton, Stephen J.; Hubig, Stefan M.)V1432,205-210(1991)

Vibrational Raman characterization of hard-carbon and diamond films (Ager, Joel W.; Veirs, D. K.; Marchon, Bruno; Cho, Namhee; Rosenblatt, Gern M.)V1437,24-31(1991)

Receivers—see also optical communications

90 degree optical hybrid for coherent receivers (Garreis, Reiner)V1522,210-219(1991)

A 39-photon/bit direct-detection receiver at 810 nm, BER = 1x10-6, 60-Mbit/s QPPM (MacGregor, Andrew D.; Dion, Bruno; Noeldeke, Christoph; Duchmann, Olivier)V1417,374-380(1991)

Analysis of the multichannel coherent FSK subcarrier multiplexing system with pilot carrier and phase noise cancelling scheme (Lee, Yang-Hang; Wu, Jingshown; Tsao, Hen-Wai)V1372,140-149(1991)

Balanced optical mixer integrated in InGaAlAs/InP for coherent receivers (Caldera, Claudio; De Bernardi, Carlo; Destefanis, Giovanni; Meliga, Marina; Morasca, Salvatore; Rigo, Cesare F.; Stano, Alessandro)V1372,82-87(1991)

BPSK homodyne and DPSK heterodyne receivers for free-space communication with ND:host lasers (Bopp, Matthias; Huether, Gerhard; Spatscheck, Thomas; Specker, Harald; Wiesmann, Theo J.)V1522,199-209(1991)

Carrier recovery and filtering in optical BPSK systems using external cavity semiconductor lasers (Pires, Joao J.; Rocha, Jose R.)V1372,118-127(1991)

Coherent Lightwave Communications: Fifth in a Series (Steele, Roger C.; Sunak, Harish R., eds.)V1372(1991)

Comparison of simulation and experimental measurements of avalanche photodiode receiver performance (Mecherle, G. S.; Henderson, Robert J.)V1417,537-542(1991)

Computer-controlled two-segment DFB local oscillator laser tuning for multichannel coherent systems (Johnson, Peter T.; Hankey, Judith; Debney, Brian T.)V1372,188-199(1991)

Correction method for optical-signal detection-error caused by quantum noise (Fujihashi, Chugo)V1583,298-306(1991)

Design of low-noise wide-dynamic-range GaAs optical preamps (Bayruns, Robert J.; Laverick, Timothy; Scheinberg, Norman; Stofman, Daniel)V1541,83-90(1991)

Design optimization of three-stage GaAs monolithic optical amplifier using SPICE (Yadav, M. S.; Dumka, D. C.; Ramola, Ramesh C.; Johri, Subodh; Kothari, Harshad S.; Singh, Babu R.)V1362,811-819(1991)

High-dynamic-range mixer using novel balun structure (Bharj, Sarjit; Taylor, Gordon C.; Denlinger, E. J.; Milgazo, H.)V1475,340-349(1991)

Homodyne PSK receivers with laser diode sources (Mecherle, G. S.; Henderson, Robert J.)V1417,99-107(1991)

Interferometric acousto-optic receiver results (Gill, E. T.; Tsui, J. B.)V1476,190-200(1991)

Low-data-rate coherent optical link demonstration using frequency-stabilized solid-state lasers (Chen, Chien-Chung; Win, Moe Z.; Marshall, William K.; Lesh, James R.)V1417,170-181(1991)

Method to find the transimpedance gain of optical receivers using measured S-parameters (Saad, Ricardo E.; Souza, Rui F.)V1371,142-148(1991)

Microwave monolithic integrated circuits for high-data-rate satellite communications (Turner, Elbert L.; Hill, William A.)V1475,248-256(1991)

Millimeter wave monolithic antenna and receiver arrays for space-based applications (Rebeiz, Gabriel M.; Ulaby, Fawwaz T.)V1475,199-203(1991)

Optical fiber n-ary PPM: approaching fundamental limits in receiver sensitivity (Cryan, Robert A.; Unwin, Rodney T.)V1579,133-143(1991)

Optical receivers in ECL for 1GHz parallel links (Wieland, Joerg B.; Melchior, Hans M.)V1389,659-664(1991)

Optimum receiver structure and filter design for MPAM optical space communication systems (Al-Ramli, Intesar F.)V1522,111-123(1991)

Parallel coherence receiver for quasidistributed optical sensor (Sansonetti, Pierre; Guerin, J. J.; Lequime, Michael; Debrie, J.)V1588,143-149(1991)

Performance analysis of direct-detection optical DPSK systems using a dual-detector optical receiver (Pires, Joao J.; Rocha, Jose R.)V1579,144-154(1991)

Performance of Reed-Solomon codes in mulichannel CPFSK coherent optical communications (Wu, Jyh-Horng; Wu, Jingshown)V1579,195-209(1991)

Photosensitized receptor inactivation with He-Ne laser: preliminary results (Arber, Simon; Rymer, William Z.; Crumrine, David)V1428,23-29(1991)

Receivers for eyesafe laser rangefinders: an overview (Crawford, Ian D.)V1419,9-16(1991)

Recent progress of coherent lightwave systems at Bellcore (Sessa, William B.; Welter, Rudy; Wagner, Richard E.; Maeda, Mari W.)V1372,208-218(1991)

SAW real-time Doppler analysis (Martin, Tom A.)V1416,52-58(1991)

Sensitivity of direct-detection lightwave receivers using optical preamplifiers (Tonguz, Ozan K.; Kazovsky, Leonid G.)V1579,179-183(1991)

Submillimeter receiver components using superconducting tunnel junctions (Wengler, Michael J.; Pance, A.; Liu, B.; Dubash, N.; Pance, Gordana; Miller, Ronald E.)V1477,209-220(1991)

Theoretical analysis of optical phase diversity FSK receivers (Yang, Shien-Chi; Tsao, Hen-Wai; Wu, Jingshown)V1372,128-139(1991)

Wideband acousto-optic spectrometer (Chang, I. C.)V1476,257-268(1991)

Reflectance—see also mirrors; surfaces

Analysis and design of binary gratings for broadband, infrared, low-reflectivity surfaces (Moharam, M. G.)V1485,254-259(1991)

Atmospheric code sensitivity to uncertainties in aerosol optical depth characteristics (Teillet, Philippe M.; Fedosejevs, Gunar; Ahern, Francis J.; Gauthier, Robert P.; Sirois, J.)V1492,213-223(1991)

Back-reflection measurements on the SILEX telexcope (Birkl, Reinhard; Manhart, Sigmund)V1522,252-258(1991)

Bragg reflectors: tapered and untapered (Chong, Chae K.; Razeghi, M. M.; McDermott, David B.; Luhmann, Neville C.; Thumm, M.; Pretterebner, Julius)V1407,226-233(1991)

Broadband, antireflection coating designs for large-aperture infrared windows (Balasubramanian, Kunjithapa; Le, Tam V.; Guenther, Karl H.; Kumar, Vas)V1485,245-253(1991)

Color correction using principle components (Trussell, Henry J.; Vrhel, Michael J.)V1452,2-9(1991)

Compact nonimaging lens with totally internally reflecting facets (Parkyn, William A.; Pelka, David G.)V1528,70-81(1991)

Deep-UV photolithography linewidth variation from reflective substrates (Dunn, Diana D.; Bruce, James A.; Hibbs, Michael S.)V1463,8-15(1991)

Design considerations for multipurpose bidirectional reflectometers (Neu, John T.; Bressler, Martin)V1530,244-254(1991)

Dry/wet wire reflectivities at 10.6 um (Young, Donald S.; Osche, Gregory R.; Fisher, Kirk L.; Lok, Y. F.)V1416,221-228(1991)

Experimental study of the laser retroreflection of various surfaces (Liu, Wen-Qing; Jiang, Rong-Xi; Wang, Ya-Ping; Xia, Yu-Xing)V1530,240-243(1991)

Flight solar calibrations using the mirror attenuator mosaic: low-scattering mirror (Lee, Robert B.)V1493,267-280(1991)

Forest decline model development with LANDSAT TM, SPOT, and DEM DATA (Brockhaus, John A.; Campbell, Michael V.; Khorram, Siamak; Bruck, Robert I.; Stallings, Casson)V1492,200-205(1991)

French proposal for IEC/TC 86/WG 4 OTDR calibration (Gauthier, Francis)V1504,55-65(1991)

Generalization of Bragg reflector geometry: application to (Ga,Al)As - (Ca,Sr)F2 reflectors (Fontaine, Chantal; Requena, Philippe; Munoz-Yague, Antonio)V1362,59-66(1991)

Hydrodynamic evolution of picosecond laser plasmas (Landen, Otto L.; Vu, Brian-Tinh; Stearns, Daniel G.; Alley, W. E.)V1413,120-130(1991)

Improvement of specular reflection pyrometer (Wen, Lin Ying; Hua, Yun)V1367,300-302(1991)

Infrared reflectivity: a tool for bond investigation in II-VI ternaries (Granger, Rene')V1484,39-46(1991)

Infrared window damage measured by reflective scatter (Bernt, Marvin L.; Stover, John C.)V1530,42-49(1991)

In-situ investigation of the low-pressure MOCVD growth of III-V compounds using reflectance anisotropy measurements (Drevillon, Bernard; Razeghi, Manijeh)V1361,200-212(1991)

In-situ structural studies of the underpotential deposition of copper onto an iodine-covered platinum surface using x-ray standing waves (Bommarito, G. M.; Acevedo, D.; Rodriguez, J. R.; Abruna, H. D.)V1550,156-170(1991)

Interface characterization of XUV multilayer reflectors using HRTEM and x-ray and XUV reflectance (Windt, David L.; Hull, Robert; Waskiewicz, Warren K.; Kortright, Jeffrey B.)V1343,292-308(1991)

Investigation of interlevel proximity effects case of the gate level over LOCOS (Festes, Gilles; Chollet, Jean-Paul E.)V1463,245-255(1991)

Lateral-periodicity evaluation of multilayer Bragg reflector surface roughness using x-ray diffraction (Takenaka, Hisataka; Ishii, Yoshikazu)V1345,180-188(1991)

Long-period x-ray standing waves generated by total external reflection (Bedzyk, Michael J.)V1550,151-155(1991)

Low-coherence optical reflectometry of laser diode waveguides (Boisrobert, Christian Y.; Franzen, Douglas L.; Danielson, Bruce L.; Christensen, David H.)V1474,285-290(1991)

Measurement of radiation-induced attenuation in optical fibers by optical-time-domain reflectometry (Looney, Larry D.; Lyons, Peter B.)V1474,132-137(1991)

Measuring films on and below polycrystalline silicon using reflectometry (Engstrom, Herbert L.; Stokowski, Stanley E.)V1464,566-573(1991)

Metal sulfide thin films on glass as solar control, solar absorber, decorative, and photographic coatings (Nair, Padmanabhan K.; Nair, M. T.; Fernandez, A. M.; Garcia, V. M.; Hernandez, A. B.)V1485,228-239(1991)

Modeling the distribution of optical radiation in diffusely reflecting materials (Birth, Gerald S.)V1379,81-88(1991)

Nd:YAG-laser-based time-domain reflectometry measurements of the intrinsic reflection signature from PMMA fiber splices (Lawson, Christopher M.; Michael, Robert R.; Dressel, Earl M.; Harmony, David W.)V1592,73-83(1991)

Near real-time operation of a centimeter-scale distributed fiber sensing system (Garside, Brian K.)V1332,399-408(1991)

New formulation of reflection coefficient in propagation theory for imaging inhomogeneous media and for nonuniform waveguides (Berger, Henry; Del Bosque-Izaguirre, Delma)V1521,117-130(1991)

Noncontact lifetime characterization technique for LWIR HgCdTe using transient millimeter-wave reflectance (Schimert, Thomas R.; Tyan, John; Barnes, Scott L.; Kenner, Vern E.; Brouns, Austin J.; Wilson, H. L.)V1484,19-30(1991)

Nonimaging optics and the measurement of diffuse reflectance (Hanssen, Leonard M.; Snail, Keith A.)V1528,142-150(1991)

Optical diagnostics of mercuric iodide crystal growth (Burger, Arnold; Morgan, S. H.; Silberman, Enrique; Nason, Donald)V1557,245-249(1991)

Optical Scatter: Applications, Measurement, and Theory (Stover, John C., ed.)V1530(1991)

Optical time-domain reflectometry performance enhancement using erbium-doped fiber amplifiers (Keeble, Peter J.)V1366,39-44(1991)

OTDR calibration for attenuation measurement (Moeller, Werner; Heitmann, Walter; Reich, M.)V1504,47-54(1991)

Radiometric calibration of SPOT 2 HRV: a comparison of three methods (Biggar, Stuart F.; Dinguirard, Magdeleine C.; Gellman, David I.; Henry, Patrice; Jackson, Ray D.; Moran, M. S.; Slater, Philip N.)V1493,155-162(1991)

Reduction of spectral linewidth and FM noise in semiconductor lasers by application of optical feedback (Li, Hua; Park, J. D.; Seo, Dongsun; Marin, L. D.; McInerney, John G.; Telle, H. R.)V1376,172-179(1991)

Reflectance at the red edge as a sensitive indicator of the damage of trees and its correlation to the state of the photosynthetic system (Ruth, Bernhard)V1521,131-142(1991)

Reflectance stability analysis of Spectralon diffuse calibration panels (Bruegge, Carol J.; Stiegman, Albert E.; Coulter, Daniel R.; Hale, Robert R.; Diner, David J.; Springsteen, Arthur W.)V1493,132-142(1991)

Reflection spectrum of multiple chirped gratings (Shellan, Jeffrey B.; Yeh, Pochi A.)V1545,179-188(1991)

Reflectivity of stimulated Brillouin scattering in picosecond time scales (Labaune, C.; Rozmus, Wojtek; Baldis, Hector A.; Mounaix, P.; Pesme, Denis; Baton, S.; LaFontaine, Bruno; Villeneuve, D. M.; Enright, G. D.)V1413,138-143(1991)

Requirements of a solar diffuser and measurements of some candidate materials (Guzman, Carmen T.; Palmer, James M.; Slater, Philip N.; Bruegge, Carol J.; Miller, Edward A.)V1493,120-131(1991)

Scaling properties of optical reflectance from quasiperiodic superlattices (Wu, Xiang; Yao, He S.; Feng, Wei G.)V1519,625-631(1991)

Scattering parameters of volume phase hologram: Bragg's approach (Gubanov, V. A.; Kiselyov, Boris S.)V1621,102-113(1991)

Self-reflection and self-transmission of pulsed radiation by laser-evaporated media (Furzikov, Nickolay P.)V1415,228-239(1991)

Simultaneous measurement of refractive index and thickness of thin film by polarized reflectances (Kihara, Tami; Yokomori, Kiyoshi)V1332,783-791(1991)

Soft and hard x-ray reflectivities of multilayers fabricated by alternating-material sputter deposition (Takenaka, Hisataka; Ishii, Yoshikazu; Kinoshita, Hiroo; Kurihara, Kenji)V1345,213-224(1991)

Stray-light implications of scratch/dig specifications (Lewis, Isabella T.; Ledebuhr, Arno G.; Bernt, Marvin L.)V1530,22-34(1991)

Structural design of the large deployable reflector (Satter, Celeste M.; Lou, Michael C.)V1494,279-300(1991)

Submerged reflectance measurements as a function of visible wavelength (Giles, John W.; Voss, Kenneth J.)V1537,140-146(1991)

Time-resolved diffuse reflectance and transmittance studies in tissue simulating phantoms: a comparison between theory and experiment (Madsen, Steen J.; Patterson, Michael S.; Wilson, Brian C.; Park, Young D.; Moulton, J. D.; Jacques, Steven L.; Hefetz, Yaron)V1431,42-51(1991)

Time-resolved reflectance spectroscopy (Jacques, Steven L.; Flock, Stephen T.)V1525,35-40(1991)

Ultrahigh resolution OTDR using streak camera technology (Sawaki, Akihiro; Miwa, Mitsuharu; Roehrenbeck, Paul W.)V1366,324-331(1991)

Vertical ocean reflectance at low altitudes for narrow laser beams (Crittenden, Eugene C.; Rodeback, G. W.; Milne, Edmund A.; Cooper, Alfred W.)V1492,187-197(1991)

Wave interactions with continuous fractal layers (Kim, Yun J.; Jaggard, Dwight L.)V1558,113-119(1991)

XUV characterization comparison of Mo/Si multilayer coatings (Windt, David L.; Waskiewicz, Warren K.; Kubiak, Glenn D.; Barbee, Troy W.; Watts, Richard N.)V1343,274-282(1991)

Reflectors—see mirrors; telescopes

Refraction—see also photorefraction; refractive index

Critical-angle refractometry: accuracy analysis (Tentori, Diana)V1527,216-224(1991)

Optical fiber refractometer and its application in the sugar industry (Ma, Junxian; Yang, Shuwen)V1572,377-381(1991)

Some needs for the characterization of the dispersive properties of transmissive-optical materials (Dodge, Marilyn J.)V1535,2-12(1991)

Refractive index—see also gradient-index optics; lenses; optical design; optical properties; refraction

Accurate measurement of thin-polymeric-films index variations: application to elasto-optic effect and to photochromism (Dumont, Michel L.; Morichere, D.; Sekkat, Z.; Levy, Yves)V1559,127-138(1991)

Analysis of moire deflectometry by wave optics (Wang, Hai-Lin; Miao, Peng-Cheng; Yan, Da-Peng; He, Anzhi)V1545,268-273(1991)

Cn2 estimates in the boundary layer for damp unstable conditions (Tunick, Arnold; Rachele, Henry; Miller, Walter B.)V1487,51-62(1991)

Computed estimation of visual acuity after laser refractive keratectomy (Rol, Pascal O.; Parel, Jean-Marie; Hanna, Khalil)V1423,89-93(1991)

Correlation between the laser-induced breakdown threshold in solids, liquids, and gases (Bettis, Jerry R.)V1441,521-534(1991)

Cryogenic refractive indices of cadmium telluride coatings in wavelength range from 2.5 to 20 um (Feng, Weiting; Yen, Yi X.; Zhu, Cui Y.)V1535,224-230(1991)

Determination of thickness and refractal index of HgCdMnTe/CdMnTe VPE films by IR transmission spectrum (Chen, Wei M.; Ma, Ke J.; Yu, Zhen Z.; Ji, Hua M.)V1519,521-524(1991)

Diagnostics of arc plasma by moire deflectometry (Gao, Yiqing; Liu, Yupin)V1554B,193-199(1991)

Effective refractive indices of three-phase optical coatings (Ma, Yushieh; Varadan, Vijay K.; Varadan, Vasundara V.)V1558,138-142(1991)

Experimental characterization of microlenses for WDM transmission systems (Zoboli, Maurizio; Bassi, Paolo)V1506,160-169(1991)

Expressions for the spherical-wave-structure function based on a bump spectrum model for the index of refraction (Richardson, Christina E.; Andrews, Larry C.)V1487,19-30(1991)

Influence of phototransformed molecules on optical properties of finite cholesteric liquid-crystal cell (Pinkevich, Igor P.; Reshetnyak, Victor Y.; Reznikov, Yuriy)V1455,122-133(1991)

Infrared refractive-index measurement results for single-crystal and polycrystal germanium (Hilton, Albert R.)V1498,128-137(1991)

Integrated optics sensor on silicon for the measurement of displacement, force, and refractive index (Ulbers, Gerd)V1506,99-110(1991)

Interaction of exposure time and system noise with angle-of-arrival measurements (Eaton, Frank D.; Peterson, William A.; Hines, John R.; Waldie, Arthur H.; Drexler, James J.; Qualtrough, John A.; Soules, David B.)V1487,84-90(1991)

New look at bump spectra models for temperature and refractive-index fluctuations (Sisterman, Elizabeth A.; Andrews, Larry C.)V1487,345-355(1991)

New method of increasing the sensitivity of Schlieren interferometer using two Wollaston prisms and its application to flow field (Yan, Da-Peng; He, Anzhi; Yang, Zu Q.; Zhu, Yi Yun)V1554B,636-640(1991)

Novel approach for the refractive index gradient measurement in microliter volumes using fiber-optic technology (Synovec, Robert E.; Renn, Curtiss N.)V1435,128-139(1991)

Photochemical delineation of waveguides in polymeric thin films (Beeson, Karl W.; Horn, Keith A.; Lau, Christina; McFarland, Michael J.; Schwind, David; Yardley, James T.)V1559,258-266(1991)

Photo-induced refractive-index changes and birefringence in optically nonlinear polyester (Shi, Youngqiang; Steier, William H.; Yu, Luping; Chen, Mai; Dalton, Larry R.)V1559,118-126(1991)

Prediction of Cn2 on the basis of macroscale meteorology including aerosols (Sadot, Danny; Kopeika, Norman S.)V1487,40-50(1991)

Preparation of Pb1-xGexTe crystal with high refractive index for IR coating (Zhang, Su Y.; Xu, Bu Y.; Zhang, Feng S.; Yan, Yixun)V1519,508-513(1991)

Refractive-index interpolation fit criterion for materials used in optical design (Korniski, Ronald J.)VCR38,193-217(1991)

Refractive index measurement capabilities at the National Institute of Standards and Technology (Dodge, Marilyn J.)V1441,56-60(1991)

Refractive-index measurement using moire deflectometry: working conditions (Tentori, Diana; Lopez Famozo, C.)V1535,209-215(1991)

Refractive index of multiple-quantum-well waveguides subject to impurity induced disordering using boron and fluorine (Hansen, Stein I.; Marsh, John H.; Roberts, John S.; Jeynes, C.)V1362,361-369(1991)

Refractive properties of TGS aqueous solution for two-color interferometry (Vikram, Chandra S.; Witherow, William K.; Trolinger, James D.)V1557,197-201(1991)

Sensing refractive-turbulence profiles (Cn2) using wavefront phase measurements from multiple reference sources (Welsh, Byron M.)V1487,91-102(1991)

Simple high-precision extinction method for measuring refractive index of transparent materials (Nee, Soe-Mie F.; Bennett, Harold E.)V1441,31-37(1991)

U-shaped fiber-optic refractive-index sensor and its applications (Takeo, Takashi; Hattori, Hajime)V1544,282-286(1991)

Remote sensing—see also satellites; sensors; telescopes

Advanced visible and near-IR radiometer for ADEOS (Iwasaki, Nobuo; Tange, Yoshio; Miyachi, Yuji; Inoue, Kouichi; Kadowaki, Tomoko; Tanaka, Hirokazu; Michioka, Hidekazu)V1490,216-221(1991)

Aerospace remote sensing monitoring of inland water quality (Gitelson, Anatoly A.)V1492,307-318(1991)

Airborne and spaceborne thermal multispectral remote sensing (Watanabe, Hiroshi; Sano, Masaharu; Mills, F.; Chang, Sheng-Huei; Masuda, Shoichi)V1490,317-323(1991)

Airborne lidar elastic scattering, fluorescent scattering, and differential absorption observations (Uthe, Edward E.)V1479,393-402(1991)

Airborne lidar measurements of ozone and aerosols in the summertime Arctic troposphere (Browell, Edward V.)V1491,7-14(1991)

Airborne Reconnaissance XV (Augustyn, Thomas W.; Henkel, Paul A., eds.)V1538(1991)

Analysis of vegetation stress and damage from images of the high-resolution airborne pushbroom image spectrograph compact airborne spectrographic imager (Mueksch, Michaela C.)V1399,157-161(1991)

Angle-only tracking and prediction of boost vehicle position (Tsai, Ming-Jer; Rogal, Fannie A.)V1481,281-291(1991)

Applicability of open-path monitors at Superfund sites (Padgett, Joseph; Pritchett, Thomas H.)V1433,352-364(1991)

Application of backscatter absorption gas imaging to the detection of chemicals related to drug production (Kulp, Thomas J.; Garvis, Darrel G.; Kennedy, Randall B.; McRae, Thomas G.)V1479,352-363(1991)

Application of IR staring arrays to space surveillance (Cantella, Michael J.; Ide, M. H.; O'Donnell, P. J.; Tsaur, Bor-Yeu)V1540,634-652(1991)

Application of MHT to group-to-object tracking (Kovacich, Michael; Casaletto, Tom; Lutjens, William; McIntyre, David; Ansell, Ralph; VanDyk, Ed)V1481,357-370(1991)

Applications of aerial photography to law enforcement and disaster assessment: a consideration of the state-of-the-art (Cox, William J.; Biache, Andrew)V1479,364-369(1991)

Applications of the massively parallel machine, the MasPar MP-1, to Earth sciences (Fischer, James R.; Strong, James P.; Dorband, John E.; Tilton, James C.)V1492,229-238(1991)

ASTER calibration concept (Ono, Akira; Sakuma, Fumihiro)V1490,285-298(1991)

ATHENA: a high-resolution wide-area coverage commercial remote sensing system (Claybaugh, William R.; Megill, L. R.)V1495,81-94(1991)

ATLID: the first preoperational ATmospheric LIDar for the European polar platform (Lange, Robert; Endemann, Martin J.; Reiland, Werner; Krawczyk, Rodolphe; Hofer, Bruno)V1492,24-37(1991)

Atmospheric code sensitivity to uncertainties in aerosol optical depth characteristics (Teillet, Philippe M.; Fedosejevs, Gunar; Ahern, Francis J.; Gauthier, Robert P.; Sirois, J.)V1492,213-223(1991)

Automated band selection for multispectral meteorological applications (Westerman, Steven D.; Drake, R. M.; Yool, Stephen R.; Brandley, M.; DeJulio, R.)V1492,263-271(1991)

Automatic reconstruction of buildings from aerial imagery (Sinha, Saravajit S.)V1468,698-709(1991)

Buried object remote detection technology for law enforcement (Del Grande, Nancy K.; Clark, Greg A.; Durbin, Philip F.; Fields, David J.; Hernandez, Jose E.; Sherwood, Robert J.)V1479,335-351(1991)

Calibrated intercepts for solar radiometers used in remote sensor calibration (Gellman, David I.; Biggar, Stuart F.; Slater, Philip N.; Bruegge, Carol J.)V1493,175-180(1991)

Calibration of Passive Remote Observing Optical and Microwave Instrumentation (Guenther, Bruce W., ed.)V1493(1991)

CCD star sensors for Indian remote sensing satellites (Alex, T. K.; Rao, V. K.)V1478,101-105(1991)

Characterization of tropospheric methane through space-based remote sensing (Ashcroft, Peter)V1491,48-55(1991)

Chemical amplifier for peroxy radical measurements based on luminol chemiluminescence (Cantrell, Chris A.; Shetter, Richard E.; Lind, John A.; Gilliland, Curt A.; Calvert, Jack G.)V1433,263-268(1991)

Clustering algorithms for a PC-based hardware implementation of the unsupervised classifier for the shuttle ice detection system (Jaggi, Sandeep)V1451,289-297(1991)

Coastal survey with a multispectral video system (Niedrauer, Terren M.)V1492,240-251(1991)

Coherent laser radar for target classification (Kranz, Wolfgang)V1479,270-274(1991)

Commercial remote sensing small-satellite feasibility study (Birk, Ronald J.; Tompkins, Jim M.; Burns, Gregory S.)V1495,2-11(1991)

Comparison of backpropagation neural networks and statistical techniques for analysis of geological features in Landsat imagery (Parikh, Jo Ann; DaPonte, John S.; Damodaran, Meledath; Karageorgiou, Angelos; Podaras, Petros)V1469,526-538(1991)

Computer-aided performance evaluation system for the on-board data compression system in HIRIS (Qian, Shen-en; Wang, Ruqin; Li, Shuqiu; Dai, Yisong)V1483,196-206(1991)

Concept for the subresolution measurement of earthquake strain fields using SPOT panchromatic imagery (Crippen, Robert E.; Blom, Ronald G.)V1492,370-377(1991)

Cooled focal-plane assembly for ocean color and temperature scanner (Nakayama, Masao; Izawa, Toshiyuki; Fujisada, Hiroyuki; Tange, Yoshio; Miyachi, Yuji; Sato, Ryota; Ishida, Juro; Tanii, Jun)V1490,207-215(1991)

Cooling system for short-wave infrared radiometer of JERS-1 optical sensor (Ohmori, Yasuhiro; Arakawa, Isao; Nakai, Akira; Tsubosaka, Kazuyoshi)V1490,177-183(1991)

Cryogenic radiometers and intensity-stabilized lasers for EOS radiometric calibrations (Foukal, Peter; Hoyt, Clifford C.; Jauniskis, L.)V1493,72-79(1991)

DARPA initiatives in small-satellite technologies (Bonometti, Robert J.; Wheatley, Alvis A.; Flynn, Lin; Nicastri, Edward; Sudol, R.)V1495,166-176(1991)

Design and testing of data fusion systems for the U.S. Customs Service drug interdiction program (Stoltz, John R.; Cole, Donald C.)V1479,423-434(1991)

Design considerations for EOS direct broadcast (Vermillion, Charles H.; Chan, Paul H.)V1492,224-228(1991)

Detection of contraband brought into the United States by aircraft and other transportation methods: a changing problem (Bruder, Joseph A.; Greneker, E. F.; Nathanson, F. E.; Henneberger, T. C.)V1479,316-321(1991)

Detection of stratospheric ozone trends by ground-based microwave observations (Connor, Brian J.; Parrish, Alan; Tsou, Jung-Jung)V1491,218-230(1991)

Developmental test and evaluation plans for the advanced tactical air reconnaissance system (Minor, John L.; Jenquin, Michael J.)V1538,18-39(1991)

Development of 1- and 2-um coherent Doppler lidars for atmospheric sensing (Chan, Kin-Pui; Killinger, Dennis K.)V1492,111-114(1991)

Development of an airborne excimer-based UV-DIAL for monitoring ozone and sulfur dioxide in the lower troposphere (Bristow, Michael P.; Diebel, D. E.; Bundy, Donald H.; Edmonds, Curtis M.; Turner, Ruldopha M.; McElroy, James L.)V1491,68-74(1991)

Different aspects of backgrounds in various spectral bands (Ben-Shalom, Ami; Devir, Adam D.; Ribak, Erez N.; Talmore, Eli T.; Balfour, L. S.; Brandman, N.)V1486,238-257(1991)

Differential time-resolved detection of absorbance changes in composite structures (Nossal, Ralph J.; Bonner, Robert F.)V1431,21-28(1991)

Discussion on spectral sampling of imaging spectrometer (Han, Xin Z.)V1538,99-102(1991)

Eagle-class small satellite for LEO applications (O'Neil, Jason; Goralczyk, Steven M.)V1495,72-80(1991)

Earth and Atmospheric Remote Sensing (Curran, Robert J.; Smith, James A.; Watson, Ken, eds.)V1492(1991)

Earth Observing System (Wilson, Stan; Dozier, Jeff)V1491,117-124(1991)

Edge technique: a new method for atmospheric wind measurements with lidar (Korb, C. L.; Gentry, Bruce M.)V1416,177-182(1991)

EGOLOGY: psychological spatial breakthrough for social redirection—multidisciplinary spatial focus for individuals/humankind (Thompson, Robert A.; Thompson, Louise A.)V1469,451-462(1991)

Enhanced thematic mapper cold focal plane: design and testing (Yang, Bing T.)V1488,399-409(1991)

Errors inherent in the restoration of imagery acquired through remotely sensed refractive interfaces and scattering media (Schmalz, Mark S.)V1479,183-198(1991)

ESA Earth observation polar platform program (Rast, Michael; Readings, C. J.)V1490,51-58(1991)

Estimation of scene correlation lengths (Futterman, Walter I.; Schweitzer, Eric L.; Newt, J. E.)V1486,127-140(1991)

Evaluation of the NOAA-11 solar backscatter ultraviolet radiometer, Mod 2 (SBUV/2): inflight calibration (Weiss, Howard; Cebula, Richard P.; Laamann, K.; Hudson, Robert D.)V1493,80-90(1991)

Firefly system concept (Nichols, Joseph D.)V1540,202-206(1991)

Fluorescent imaging (Thompson, Jill C.)V1482,253-257(1991)

Forest decline model development with LANDSAT TM, SPOT, and DEM DATA (Brockhaus, John A.; Campbell, Michael V.; Khorram, Siamak; Bruck, Robert I.; Stallings, Casson)V1492,200-205(1991)

Future European and Japanese Remote-Sensing Sensors and Programs (Slater, Philip N., ed.)V1490(1991)

Future technologies for lidar/DIAL remote sensing (Allario, Frank; Barnes, Norman P.; Storm, Mark E.)V1492,92-110(1991)

German ATMOS program (Puls, Juergen)V1490,2-13(1991)

Global ozone monitoring by occultation of stars (Bertaux, J. L.)V1490,133-145(1991)

GLOB(MET)SAT: French proposals for monitoring global change and weather from the polar orbit (Durpaire, Jean-Pierre; Ratier, A.; Dagras, C.)V1490,23-38(1991)

Ground-based lidar for long-term and network measurements of ozone (McDermid, I. S.; Schmoe, Martha S.; Walsh, T. D.)V1491,175-181(1991)

Ground-based microwave remote sensing of water vapor in the mesosphere and stratosphere (Croskey, Charles L.; Olivero, John J.; Martone, Joseph P.)V1491,323-334(1991)

High-resolution multichannel mm-wave radiometer for the detection of stratospheric ClO (Gerber, Louis; Kaempfer, Niklaus A.)V1491,211-217(1991)

Imager of METEOSAT second generation (Hollier, Pierre A.)V1490,74-81(1991)

Image Understanding for Aerospace Applications (Nasr, Hatem N., ed.)V1521(1991)

Imaging of subsurface regions of random media by remote sensing (Barbour, Randall L.; Graber, Harry L.; Aronson, Raphael; Lubowsky, Jack)V1431,52-62(1991); V1431,192-203(1991)

Improvement in detection of small wildfires (Sleigh, William J.)V1540,207-212(1991)

In-flight calibration of a helicopter-mounted Daedalus multispectral scanner (Balick, Lee K.; Golanics, Charles J.; Shines, Janet E.; Biggar, Stuart F.; Slater, Philip N.)V1493,215-223(1991)

Infrared lidars for atmospheric remote sensing (Menzies, Robert T.)V1540,160-163(1991)

Infrared lidar windshear detection for commercial aircraft and the edge technique: a new method for atmospheric wind measurement (Targ, Russell; Bowles, Roland L.; Korb, C. L.; Gentry, Bruce M.; Souilhac, Dominique J.)V1521,144-157(1991)

Infrared monitoring of combustion (Bates, Stephen C.; Morrison, Philip W.; Solomon, Peter R.)V1434,28-38(1991)

Infrared spectrometer for ground-based profiling of atmospheric temperature and humidity (Shaw, Joseph A.; Churnside, James H.; Westwater, Edward R.)V1540,681-686(1991)

Integration of diverse remote sensing data sets for geologic mapping and resource exploration (Kruse, Fred A.; Dietz, John B.)V1492,326-337(1991)

Intensified multispectral imaging measuring in the spatial, frequency, and time domains with a single instrument (Kennedy, Benjamin J.)V1346,68-74(1991)

Interferometric monitor for greenhouse gasses for ADEOS (Tsuno, Katsuhiko; Kameda, Yoshihiko; Kondoh, Kayoko; Hirai, Shoichi)V1490,222-232(1991)

Japanese mission overview of JERS and ASTER programs (Yamaguchi, Yasushi; Tsu, Hiroji; Sato, Isao)V1490,324-334(1991)

Large integrating sphere of prelaunch calibration system for Japanese Earth Resources Satellite optical sensors (Suzuki, Naoshi; Narimatsu, Yoshito; Nagura, Riichi; Sakuma, Fumihiro; Ono, Akira)V1493,48-57(1991)

Laser remote sensing of natural water organics (Babichenko, Sergey M.; Poryvkina, Larisa)V1492,319-323(1991)

Lasers and electro-optic technology in natural resource management (Greer, Jerry D.)V1396,342-352(1991)

Lidar profiles of atmospheric structure properties (Philbrick, Charles R.)V1492,76-84(1991)

Low-cost, low-risk approach to tactical reconnaissance (Beving, James E.; Fishell, Wallace G.)V1538,14-17(1991)

Low-cost spacecraft buses for remote sensing applications (Harvey, Edwin L.; Cullen, Robert M.)V1495,134-145(1991)

Low-cost space platforms for detection and tracking technologies (Cullen, Robert M.)V1479,295-305(1991)

Low-intensity conflict aircraft systems (Henkel, Paul A.)V1538,2-4(1991)

Measurement of tropospheric carbon monoxide using gas filter radiometers (Reichle, Henry G.)V1491,15-25(1991)

Measuring tropospheric ozone using differential absorption lidar technique (Proffitt, Michael H.; Langford, A. O.)V1491,2-6(1991)

Mechanical cooler development program for ASTER (Kawada, Masakuni; Fujisada, Hiroyuki)V1490,299-308(1991)

METEOSAT second-generation program (Markland, Chris A.)V1490,39-50(1991)

Midinfrared backscatter spectra of selected agricultural crops (Narayanan, Ram M.; Green, Steven E.; Alexander, Dennis R.)V1379,116-122(1991)

Millimeter wave monolithic antenna and receiver arrays for space-based applications (Rebeiz, Gabriel M.; Ulaby, Fawwaz T.)V1475,199-203(1991)

Miniature signal processor for surveillance sensor applications (Jacobi, William J.; Jensen, Preben D.; Teneketges, Nicholas J.; Wadsworth, Leo A.)V1479,111-119(1991)

Mission overview of ADEOS program (Iwasaki, Nobuo; Hara, Norikazu; Kajii, Makoto; Tange, Yoshio; Miyachi, Yuji; Sato, Ryota; Inoue, Kouichi)V1490,192-199(1991)

Mission study overview of Japanese polar orbiting platform program (Moriyama, Takashi; Nakayama, Kimihiko; Homma, M.; Haruyama, Yukio)V1490,310-316(1991)

Model "T" satellite series: small satellites designed for scientific and commercial use (McMillen, Donald V.)V1495,95-102(1991)

Moire sensor as an automatic feeler gauge (Konwerska-Hrabowska, Joanna; Kryszczynski, Tadeusz; Smolka, M.; Olbrysz, P.; Widomski, L.)V1554B,225-232(1991)

MOMS-02/Spacelab D-2: a high-resolution multispectral stereo scanner for the second German Spacelab mission (Ackermann, F.; Bodechtel, Joh; Lanzl, Franz; Meissner, D.; Seige, Peter; Winkenbach, H.; Zilger, Johannes)V1490,94-101(1991)

Multiorder etalon sounder for vertical temperature profiling: technique and performance analysis (Wang, Jin-Xue; Hays, Paul B.; Drayson, S. R.)V1492,391-402(1991)

Multiplex Fabry-Perot interferometer (Snell, Hilary E.; Hays, Paul B.)V1492,403-407(1991)

Multisensor image processing (de Salabert, Arturo; Pike, T. K.; Sawyer, F. G.; Jones-Parry, I. H.; Rye, A. J.; Oddy, C. J.; Johnson, D. G.; Mason, D.; Wielogorski, A. L.; Plassard, T.; Serpico, Sebastiano B.; Hindley, N.)V1521,74-88(1991)

NASA's Geostationary Earth Observatory and its optical instruments (Koczor, Ronald J.)V1527,98-109(1991)

NASA's program in lidar remote sensing (Theon, John S.; Vaughan, William W.; Browell, Edward V.; Jones, William D.; McCormick, M. P.; Melfi, Samuel H.; Menzies, Robert T.; Schwemmer, Geary K.; Spinhirne, James D.)V1492,2-23(1991)

Neural network for passive acoustic discrimination between surface and submarine targets (Baran, Robert H.; Coughlin, James P.)V1471,164-176(1991)

Neural network labeling of the Gulf Stream (Lybanon, Matthew; Molinelli, Eugene J.; Muncill, G.; Pepe, Kevin)V1469,637-647(1991)

Neural networks for Fredholm-type integral equations (Vemuri, V.; Jang, Gyu-Sang)V1469,563-574(1991)

New concept for a highly compact imaging Fourier transform spectrometer (Simeoni, Denis)V1479,127-138(1991)

New concepts in remote sensing and geolocation (Seastone, A. J.)V1495,228-239(1991)

New formulation of reflection coefficient in propagation theory for imaging inhomogeneous media and for nonuniform waveguides (Berger, Henry; Del Bosque-Izaguirre, Delma)V1521,117-130(1991)

New spectroscopic instrumentation for measurement of stratospheric trace species by remote sensing of scattered skylight (Mount, George H.; Jakoubek, Roger O.; Sanders, Ryan W.; Harder, Jerald W.; Solomon, Susan; Winkler, Richard; Thompson, Thomas; Harrop, Walter)V1491,188-193(1991)

O3, NO2, NO3, SO2, and aerosol measurements in Beijing (Xue, Qing-yu; Guo, Song; Zhao, Xue-peng; Nieu, Jian-guo; Zhang, Yi-ping)V1491,75-82(1991)

Ocean color and temperature scanner for ADEOS (Tanii, Jun; Machida, Tsuneo; Ayada, Haruki; Katsuyama, Yoshihiko; Ishida, Juro; Iwasaki, Nobuo; Tange, Yoshio; Miyachi, Yuji; Sato, Ryota)V1490,200-206(1991)

On-board calibration device for a wide field-of-view instrument (Krawczyk, Rodolphe; Chessel, Jean-Pierre; Durpaire, Jean-Pierre; Durieux, Alain; Churoux, Pascal; Briottet, Xavier)V1493,2-15(1991)

Optical design of the moderate-resolution imaging spectrometer-tilt for the Earth Observing System (Maymon, Peter W.)V1492,286-297(1991)

Optical leak detection of oxygen using IR laser diodes (Disimile, Peter J.; Fox, Curtis F.; Toy, Norman)V1492,64-75(1991)

Optical locator for horizon sensing (Fallon, James J.; Selby, Vaughn H.)V1495,268-279(1991)

Optical mapping instrument (Bagot, K. H.)V1490,126-132(1991)

Optical reflectance sensor for detecting plants (Shropshire, Geoffrey J.; Von Bargen, Kenneth; Mortensen, David A.)V1379,222-235(1991)

Optical sensor system for Japanese Earth resources satellite 1 (Hino, Hideo; Takei, Mitsuru; Ono, Hiromi; Nagura, Riichi; Narimatsu, Yoshito; Hiramatsu, Masaru; Harada, Hisashi; Ogikubo, Kazuhiro)V1490,166-176(1991)

Overview of U.S. Fishery requirements (Springer, Steven C.; McLean, Craig)V1479,372-379(1991)

Pegasus air-launched space booster payload interfaces and processing procedures for small optical payloads (Mosier, Marty R.; Harris, Gary N.; Whitmeyer, Charlie)V1495,177-192(1991)

POPS: parallel opportunistic photointerpretation system (Howard, Michael D.)V1471,422-427(1991)

Potential of tunable lasers for optimized dual-color laser ranging (Lund, Glenn I.; Gaignebet, Jean)V1492,166-175(1991)

Process monitoring during CO2 laser cutting (Jorgensen, Henning; Olsen, Flemming O.)V1412,198-208(1991)

Radiometric calibration plan for the Clouds and the Earth's Radiant Energy System scanning instruments (Jarecke, Peter J.; Folkman, Mark A.; Darnton, Lane A.)V1493,244-254(1991)

Radiometric stability of the shuttle-borne solar backscatter ultraviolet spectrometer (Cebula, Richard P.; Hilsenrath, Ernest; Kelly, Thomas J.; Batluck, Georgiann R.)V1493,91-99(1991)

Recent lidar measurements of stratospheric ozone and temperature within the Network for the Detection of Stratospheric Change (McGee, Thomas J.; Ferrare, Richard; Butler, James J.; Frost, Robert L.; Gross, Michael; Margitan, James)V1491,182-187(1991)

Reconnaissance and imaging sensor test facilities at Eglin Air Force Base (Pratt, Stephen R.; Tucker, Robert)V1538,40-45(1991)

Remote colorimetry and its applications (Sheffer, Dan; Ben-Shalom, Ami; Devir, Adam D.)V1493,232-243(1991)

Remote Sensing of Atmospheric Chemistry (McElroy, James L.; McNeal, Robert J., eds.)V1491(1991)

Remote sensing of coastal environmental hazards (Huh, Oscar K.; Roberts, Harry H.; Rouse, Lawrence J.)V1492,378-386(1991)

Remote sensing of volcanic ash hazards to aircraft (Rose, William I.; Schneider, David J.)V1492,387-390(1991)

Remote spectral fingerprinting for law enforcement (Huguenin, Robert L.; Tahmoush, Donald J.)V1479,403-411(1991)

Remote spectral identification of surface aggregates by thermal imaging techniques: progress report (Scholen, Douglas E.; Clerke, William H.; Burns, Gregory S.)V1492,358-369(1991)

Role of orbital observations in detecting and monitoring geological hazards: prospects for the future (Pieri, David C.)V1492,410-417(1991)

Scattering of waves from dense discrete random media: theory and applications in remote sensing (Tsang, Leung; Ding, Kung-Hau; Kong, Jin A.; Winebrenner, Dale P.)V1558,260-268(1991)

SCORPIUS: final report (Onishi, Randall M.)V1472,56-65(1991)

Sensors and Sensor Integration (Dean, Peter D., ed.)V1480(1991)

Short-wavelength infrared subsystem design status of ASTER (Akasaka, Akira; Ono, Makoto; Sakurai, Yasushi; Hayashida, Bun)V1490,269-277(1991)

Simulation of partially obscured scenes using the radiosity method (Borel, Christoph C.; Gerstl, Siegfried A.)V1486,271-277(1991)

Small satellites: current legal issues in remote sensing (Stern, Jill A.)V1495,42-51(1991)

Small satellites for water cycle experiments (Rondinelli, Giuseppe; Di Girolamo, Sergio; Barresi, Giangrande)V1495,19-31(1991)

Small-Satellite Technology and Applications (Horais, Brian J., ed.)V1495(1991)

Small-target acquisition and typing by AASAP (Huguenin, Robert L.; Tahmoush, Donald J.)V1481,64-72(1991)

Space-based sensing of atmospheric conditions over data-void regions (Behunek, Jan L.; Vonder Haar, Thomas H.)V1479,93-100(1991)

Space-based visible surveillance experiment (Dyjak, Charles P.; Harrison, David C.)V1479,42-56(1991)

Spatial sampling errors for a satellite-borne scanning radiometer (Manalo, Natividad D.; Smith, G. L.)V1493,281-291(1991)

Spectral stratigraphy (Lang, Harold R.)V1492,351-357(1991)

SPOT 4 HRVIR instrument and future high-resolution stereo instruments (Fratter, C.; Reulet, Jean-Francois; Jouan, Jacky)V1490,59-73(1991)

SPOT system and defence applications (Bernard, Christian)V1521,66-73(1991)

Static horizon sensor for remote sensing satellite (Kamalakar, J. A.; Jain, Yashwant K.; Laxmiprasad, A. S.; Shashikala, M.)V1478,92-100(1991)

Studies on distributed sensing and processing for the control of large flexible spacecraft (Montgomery, Raymond C.; Ghosh, David)V1480,126-137(1991)

Summary of atmospheric chemistry observations from the Antarctic and Arctic aircraft campaigns (Tuck, Adrian F.)V1491,252-272(1991)

Surface and aerosol models for use in radiative transfer codes (Hart, Quinn J.)V1493,163-174(1991)

Surveillance Technologies (Gowrinathan, Sankaran; Mataloni, Raymond J.; Schwartz, Stanley J., eds.)V1479(1991)

Synthetic aperture radar of JERS-1 (Ono, Makoto; Nemoto, Yoshiaki)V1490,184-190(1991)

Target lifetimes in natural resource management (Greer, Jerry D.)V1538,69-76(1991)

Technique for measuring atmospheric effects on image metrics (Crow, Samuel B.; Watkins, Wendell R.; Palacios, Fernando R.; Billingsley, Daniel R.)V1486,333-344(1991)

TECHSTARS: small, smart space systems (Higbee, Terry A.)V1495,103-114(1991)

Territorial analysis by fusion of LANDSAT and SAR data (Vernazza, Gianni L.; Dambra, Carlo; Parizzi, Francesco; Roli, Fabio; Serpico, Sebastiano B.)V1492,206-212(1991)

Thermal and radiometric modeling of terrain backgrounds (Conant, John A.; Hummel, John R.)V1486,217-230(1991)

Thermal infrared imagery from the Geoscan Mk II scanner and its calibration: two case histories from Nevada—Ludwig Skarn (Yerington District) & Virginia City (Lyon, Ronald J.; Honey, Frank R.)V1492,339-350(1991)

Thermal infrared subsystem design status of ASTER (Aoki, Yutaka; Ohmae, Hirokazu; Kitamura, Shin-ichi)V1490,278-284(1991)

TREIS: a concept for a user-affordable, user-friendly radar satellite system for tropical forest monitoring (Raney, R. K.; Specter, Christine N.)V1492,298-306(1991)

Uniform field laser illuminator for remote sensing (Di Benedetto, John A.; Capelle, Gene; Lutz, Stephen S.)V1492,115-125(1991)

Use of a Fourier transform spectrometer as a remote sensor at Superfund sites (Russwurm, George M.; Kagann, Robert H.; Simpson, Orman A.; McClenny, William A.)V1433,302-314(1991)

Use of a multibeam transmitter for significant improvement in signal-dynamic-range reduction and near-range coverage for incoherent lidar systems (Zhao, Yanzeng; Hardesty, R. M.; Post, Madison J.)V1492,85-90(1991)

Use of satellite data to determine the distribution of ozone in the troposphere (Fishman, Jack; Watson, Catherine E.; Brackett, Vincent G.; Fakhruzzaman, Khan; Veiga, Robert)V1491,348-359(1991)

Use of spline expansions and regularization in the unfolding of data from spaceborne sensors (Fisher, Thornton R.; Perez, Joseph D.)V1479,212-225(1991)

User interface development for semiautomated imagery exploitation (O'Connor, R. P.; Bohling, Edward H.)V1472,26-37(1991)

Vertical ocean reflectance at low altitudes for narrow laser beams (Crittenden, Eugene C.; Rodeback, G. W.; Milne, Edmund A.; Cooper, Alfred W.)V1492,187-197(1991)

Visible and near-infrared subsystem and common signal processor design status of ASTER (Takahashi, Fumiho; Hiramatsu, Masaru; Watanabe, Fumito; Narimatsu, Yoshito; Nagura, Riichi)V1490,255-268(1991)

Wave interactions with continuous fractal layers (Kim, Yun J.; Jaggard, Dwight L.)V1558,113-119(1991)

Resonators—see also lasers

10.6-um TEM00 beam transmission characteristics of a hollow circular cross-section multimode waveguide (Jenkins, R. M.; Devereux, R. W.)V1512,135-142(1991)

Advanced matrix optics and its incidence in laser optics (Wang, Shaomin; Bernabeu, Eusebio; Alda, Javier)V1397,595-602(1991)

Analytical studies of large closed-loop arrays (Fikioris, George; Freeman, D. K.; King, Ronold W.; Shen, Hao-Ming; Wu, Tai T.)V1407,295-305(1991)

Apodized outcouplers for unstable resonators (Budzinski, Christel; Grunwald, Ruediger; Pinz, Ingo; Schaefer, Dieter; Schoennagel, Horst)V1500,264-274(1991)

Bragg reflectors: tapered and untapered (Chong, Chae K.; Razeghi, M. M.; McDermott, David B.; Luhmann, Neville C.; Thumm, M.; Pretterebner, Julius)V1407,226-233(1991)

Brightness enhancement of solid state laser oscillators in single-mode lasing using novel inside-resonator optical elements with radially variable transmission (Lukishova, Svetlana G.; Mendez, Nestor R.; Ter-Mikirtychev, Valery V.; Tulajkova, Tamara V.)V1527,380-391(1991)

Characterization, modeling, and design of dielectric resonators based on thin ceramic tape (Morris, Jacqueline H.; Belopolsky, Yakov)V1389,236-248(1991)

Characterization of a high-critical-temperature superconducting thin film by the ring resonator method (Pyee, Maurice; Meisse, Pascal; Baudrand, Henry; Chaubet, Michel)V1512,240-248(1991)

Characterization of high-Tc coplanar transmission lines and resonators (Kessler, Jochen; Dill, Roland; Russer, Peter)V1477,45-56(1991)

Compact probe for all-fiber optically addressed silicon cantilever microresonators (Rao, Yun-Jiang; Uttamchandani, Deepak G.; Culshaw, Brian)V1572,287-292(1991)

Comparison of 248-nm line narrowing resonator optics for deep-UV lithography lasers (Kahlert, Hans-Juergen; Rebhan, Ulrich; Lokai, Peter; Basting, Dirk)V1463,604-609(1991)

Comparison of welding results with stable and unstable resonators (Franek, Joachim; Du, Keming; Pflueger, Silke; Imhoff, Ralf; Loosen, Peter)V1397,791-795(1991)

Computer-aided alignment of a grazing-incidence ring resonator for a visible wavelength free-electron laser (Hudyma, Russell M.; Eigler, Lynne C.)V1354,523-532(1991)

Diffraction-limited Nd:glass and alexandrite lasers using graded reflectivity mirror unstable resonators (Snell, Kevin J.; Duplain, Gaetan; Parent, Andre; Labranche, Bruno; Galarneau, Pierre)V1410,99-106(1991)

Enhanced Schawlow-Townes linewidth in lasers with nonorthogonal transverse eigenmodes (Mussche, Paul L.; Siegman, Anthony E.)V1376,153-163(1991)

Excitation of an excimer laser with microwave resonator (Huenermann, Lucia; Meyer, Rudolph; Richter, Franz; Schnase, Alexander)V1503,134-139(1991)

Experimental study of the resonance of a circular array (Shen, Hao-Ming)V1407,306-315(1991)

Fabrication of unstable resonator diode lasers (Largent, Craig C.; Gallant, David J.; Yang, Jane; Allen, Michael S.; Jansen, Michael)V1418,40-45(1991)

Focal plane intensity distribution of copper vapor laser with different unstable resonators (Nikonchuk, Michael O.; Polyakov, Igor V.)V1412,72-78(1991)

Frequency stability of a solid state mode-locked laser system (Simpson, Thomas B.; Doft, F.; Malley, Michael M.; Sutton, George W.; Day, Timothy)V1410,133-140(1991)

I-line lithography for highly reproducible fabrication of surface acoustic wave devices (Berek, Stefan; Knauer, Ulrich; Zottl, Helmut)V1463,515-520(1991)

In-situ measurement of piston jitter in a ring resonator (Cunningham, Philip R.; Hay, Stephen O.; Francis, Denise M.; Trott, G.E.)V1414,97-129(1991)

Investigations of cumulative beam breakup in radio-frequency linacs (Bohn, Courtlandt L.; Delayen, Jean R.)V1407,566-577(1991)

Master oscillator-amplifier Nd:YAG laser with a SBS phase-conjugate mirror (Ayral, Jean-Luc; Montel, J.; Huignard, Jean-Pierre)V1500,81-92(1991)

Mathematical simulation of composite optical systems loaded with active medium (Apollonova, O. V.; Elkin, Nickolai N.; Korjov, M. Y.; Korotkov, V. A.; Likhanskii, Vladimir V.; Napartovich, Anatoly P.; Troshchiev, V. E.)V1501,108-119(1991)

Matrix representation of multimode beam transformation (Alda, Javier; Porras, Miguel A.; Bernabeu, Eusebio)V1527,240-251(1991)

Microwave characterization of high-Tc superconducting thin films for simulation and realization of planar microelectronic circuits (Carru, J. C.; Mehri, F.; Chauvel, D.; Crosnier, Y.)V1512,232-239(1991)

Mode analysis of an unstable resonator with an internal aperture (Rabczuk, G.)V1391,267-271(1991)

Modified off-axis unstable resonator for copper vapor laser (Lando, Mordechai; Belker, D.; Lerrer, A.; Lotem, Haim; Dikman, A.; Bialolanker, Gabriel; Lavi, S.; Gabay, Shimon)V1412,19-26(1991)

Multimode laser beams behaviour through variable reflectivity mirrors (Porras, Miguel A.; Alda, Javier; Bernabeu, Eusebio)V1397,645-648(1991)

Multipass unstable negative branch resonator with a spatial filter for a transverse-flow CO2 laser (Rabczuk, G.)V1397,387-389(1991)

New approach to creation of stable and unstable active resonators for high-power solid state lasers (Apollonov, V. V.; Chetkin, Sergue A.; Kislov, V. I.; Vdovin, Gleb V.)V1502,83-94(1991)

Nonlinear properties of quasi-optic open resonator with a layer of Crv solution in heavy alcohol under the conditions of magnetic resonance (Vertiy, Alexey A.; Gavrilov, Sergey P.)V1362,702-709(1991)

Numerical studies of resonators with on-axis holes in mirrors for FEL applications (Keselbrener, Michel; Ruschin, Shlomo; Lissàk, Boaz; Gover, Avraham)V1415,38-47(1991)

Numerical study of the onset of chaos in coupled resonators (Rogers, Mark E.; Rought, Nathan W.)V1415,24-37(1991)

Optically self-excited miniature fixed-beam resonator sensor (Gu, Lizhong; Ma, Jiancheng; Wang, Jiazhen)V1572,450-452(1991)

Optical phase-conjugate resonators, bistabilities, and applications (Venkateswarlu, Putcha; Dokhanian, Mostafa; Sekhar, Prayaga C.; George, M. C.; Jagannath, H.)V1332,245-266(1991)

Performance of stripline resonators using sputtered YBCO films (Mallory, Derek S.; Kadin, Alan M.; Ballentine, Paul H.)V1477,66-76(1991)

Planar-grating klystron experiment (Xu, Yian-sun; Jackson, Jonathan A.; Price, Edwin P.; Walsh, John E.)V1407,648-652(1991)

Pointing stability of copper vapor laser with novel off-axis unstable resonator (Lando, Mordechai; Belker, D.; Lerrer, A.; Lotem, Haim; Dikman, A.; Bialolanker, Gabriel; Lavi, S.; Gabay, Shimon)V1442,172-180(1991)

Positive-branch unstable resonator for Nd:YAG laser with Q-switching (Marczak, Jan; Rycyk, Antoni; Sarzynski, Antoni)V1391,48-51(1991)

Reduction of spectral linewidth and FM noise in semiconductor lasers by application of optical feedback (Li, Hua; Park, J. D.; Seo, Dongsun; Marin, L. D.; McInerney, John G.; Telle, H. R.)V1376,172-179(1991)

Resonators for coaxial slow-flow CO2 lasers (Habich, Uwe; Bauer, Axel; Loosen, Peter; Plum, Heinz-Dieter)V1397,383-386(1991)

Ring resonators for microwave optoelectronics (Gopalakrishnan, G. K.; Fairchild, B. W.; Yeh, C. L.; Park, C. S.; Chang, Kai; Weichold, Mark H.; Taylor, Henry F.)V1476,270-275(1991)

Selected Papers on Laser Design (Weichel, Hugo, ed.)VMS29(1991)

TEA CO2 laser mirror by degenerate four-wave mixing (Vigroux, Luc M.; Bourdet, Gilbert L.; Cassard, Philippe; Ouhayoun, Michel O.)V1500,74-79(1991)

Threshold current density of InGaAsP/InP surface-emitting laser diodes with hemispherical resonator (Jing, Xing-Liang; Zhang, Yong-Tao; Chen, Yi-Xin)V1418,434-441(1991)

Two-dimensional periodic structures in solid state laser resonator (Okulov, Alexey Y.)V1410,221-232(1991)

Unstable resonator with a super-Gaussian dielectric mirror for Nd:YAG Q-switched laser (Firak, Jozef; Marczak, Jan; Sarzynski, Antoni)V1391,42-47(1991)

Robotics—see also artificial intelligence; computer vision; expert systems; industrial optics; machine vision; optical inspection; object recognition; pattern recognition

Achieving a balance between autonomy and tele-operation in specifying plans for a planetary rover (Lyons, Damian M.; Allton, Judith H.)V1387,124-133(1991)

Adaptive gross motion control: a case study (Leahy, Michael B.; Whalen, P. V.; Lamont, Gary B.)V1387,148-158(1991)

Ad-hoc and derived parking curves (Lyon, Douglas)V1388,39-49(1991)

Algorithmic approaches to optimal route planning (Mitchell, Joseph S.)V1388,248-259(1991)

Analysis of cooperative robot manipulators on a mobile platform (Murphy, Steve H.; Wen, John T.; Saridis, George N.)V1387,14-25(1991)

Analysis of terrain-map-matching using multisensing techniques for applications to autonomous vehicle navigation (Page, Lance A.; Shen, Chi N.)V1383,471-482(1991)

Application of H oo control design techniques to improve dynamics of dexterous manipulation (Chapel, Jim D.; Su, Renjeng)V1387,284-295(1991)

Applications of Artificial Intelligence IX (Trivedi, Mohan M., ed.)V1468(1991)

Atomic temporal interval relations in branching time: calculation and application (Anger, Frank D.; Ladkin, Peter B.; Rodriguez, Rita V.)V1468,122-136(1991)

Automated assembly system for large space structures (Wil!, Ralph W.; Rhodes, Marvin D.)V1387,60-71(1991)

Automation of vehicle processing at Space Station Freedom (Vargo, Rick C.; Sklar, Michael E.; Wegerif, Daniel G.)V1387,72-81(1991)

Autonomous navigation in a dynamic environment (Davies, Henry C.; Kayaalp, Ali E.; Moezzi, Saied)V1388,165-175(1991)

Autonomous navigation of structured city roads (Aubert, Didier; Kluge, Karl; Thorpe, Charles E.)V1388,141-151(1991)

Building and maintaining a local on-orbit reference frame model (Viggh, Herbert E.)V1387,224-236(1991)

CLIPS implementation of a knowledge-based distributed control of an autonomous mobile robot (Bou-Ghannam, Akram A.; Doty, Keith L.)V1468,504-515(1991)

Clue derivation and selection activities in a robot vision system (Reihani, Kamran; Thompson, Wiley E.)V1468,305-312(1991)

Comparison of techniques for disparate sensor fusion (Huntsberger, Terrance L.)V1383,589-595(1991)

Complexity of computing reachable workspaces for redundant manipulators (Alameldin, Tarek K.; Palis, Michael A.; Rajasekaran, Sanguthevar; Badler, Norman I.)V1381,217-225(1991)

Computing motion parameters from sparse multisensor range data for telerobotics (Vemuri, Baba C.; Skofteland, G.)V1383,97-108(1991)

Control of flexible, kinematically redundant robot manipulators (Nguyen, Luong A.; Walker, Ian D.; De Figueiredo, Rui; deFigueiredo, Rui J.)V1387,296-312(1991)

Control scheme for sensor fusion for navigation of autonomous mobile robots (Murphy, Robin R.)V1383,436-447(1991)

Cooperative Intelligent Robotics in Space (Stoney, William E.; deFigueiredo, Rui J., eds.)V1387(1991)

Coordinating sensing and local navigation (Slack, Marc G.)V1383,459-470(1991)

Coping with complexity in the navigation of an autonomous mobile robot (Dodds, David R.)V1388,448-452(1991)

Design and control of ultralight manipulators for interplanetary exploration (Byler, Eric A.)V1387,313-327(1991)

Design and testing of a nonreactive, fingertip, tactile display for interaction with remote environments (Patrick, Nicholas J.; Sheridan, Thomas B.; Massimino, Michael J.; Marcus, Beth A.)V1387,215-224(1991)

Design/implementation architecture for complex multirobot systems (Herd, James T.; Duffy, Neil D.; Philip, Gary P.; Davidson, Alan C.; Eccles, N. J.)V1387,194-201(1991)

Detecting difficult roads and intersections without map knowledge for robot vehicle navigation (Crisman, Jill D.; Thorpe, Charles E.)V1388,152-164(1991)

Determination of flint wheel orientation for the automated assembly of lighters (Safabakhsh, Reza)V1472,185-189(1991)

Distributed architecture for intelligent robotics (Gouveia, Feliz A.; Barthes, Jean-Paul A.; Oliveira, Eugenio C.)V1468,516-523(1991)

Efficient method for computing the force distribution of a three-fingered grasp (Walker, Ian D.; Cheatham, John B.; Chen, Yu-Che)V1387,256-270(1991)

Environment for simulation and animation of sensor-based robots (Chen, ChuXin; Trivedi, Mohan M.; Bidlack, Clint R.; Lassiter, Terrell N.)V1468,354-366(1991)

Environment model for mobile robots indoor navigation (Roth-Tabak, Yuval; Weymouth, Terry E.)V1388,453-463(1991)

Experimental testbed for cooperative robotic manipulators (Desrochers, Alan A.)V1387,2-13(1991)

Experiments in real-time visual control (Griswold, Norman C.; Kehtarnavaz, Nasser)V1388,342-349(1991)

Experiments in tele-operator and autonomous control of space robotic vehicles (Alexander, Harold L.)V1388,560-565(1991)

Exploiting known topologies to navigate with low-computation sensing (Miller, David P.; Gat, Erann)V1383,425-435(1991)

Eye-slaved pointing system for tele-operator control (Razdan, Rikki; Kielar, Alan)V1388,361-371(1991)

Fast algorithm for obtaining dense depth maps for high-speed navigation (Khalili, Payman; Jain, Ramesh C.)V1388,210-221(1991)

Finding a grasped object's pose using joint angle and torque constraints (Siegel, David M.)V1383,151-165(1991)

Flat plate project (Wijbrans, Klaas C.; Korsten, Maarten J.)V1386,197-205(1991)

Formalization and implementation of topological visual navigation in two dimensions (Kender, John R.; Park, Il-Pyung; Yang, David)V1388,476-489(1991)

Framework for load apportioning and interactive force control using a Hopfield neural network (Copeland, Bruce R.; Anderson, Joseph N.)V1381,177-188(1991)

From object structure to object function (Zlateva, Stoyanka D.; Vaina, Lucia M.)V1468,379-393(1991)

Fusing human and machine skills for remote robotic operations (Schenker, Paul S.; Kim, Won S.; Venema, Steven; Bejczy, Antal K.)V1383,202-223(1991)

Fuzzy logic: principles, applications, and perspectives (Zadeh, Lotfi A.)V1468,582-582(1991)

Grasp-oriented sensing and control (Grupen, Roderic A.; Weiss, Richard S.; Oskard, David N.)V1383,189-201(1991)

Gripper for truss structure assembly (Kelley, Robert B.; Tsai, Jodi; Bethel, Jeff; Peiffer, John)V1387,38-46(1991)

Hand-eye coordination for grasping moving objects (Allen, Peter K.; Yoshimi, Billibon; Timcenko, Alexander; Michelman, Paul)V1383,176-188(1991)

HelpMate autonomous mobile robot navigation system (King, Steven J.; Weiman, Carl F.)V1388,190-198(1991)

Heuristic search approach for mobile robot trap recovery (Zhao, Yilin; BeMent, Spencer L.)V1388,122-130(1991)

Hierarchical fusion of geometric constraints for image segmentation (Seetharaman, Guna S.; Chu, Chee-Hung H.)V1383,582-588(1991)

Hierarchical multisensor analysis for robotic exploration (Eberlein, Susan; Yates, Gigi; Majani, Eric)V1388,578-586(1991)

Hierarchical planner for space truss assembly (Mathur, Rajive K.; Sanderson, Arthur C.)V1387,47-57(1991)

Hierarchical terrain representations for off-road navigation (Gowdy, Jay W.; Stentz, Anthony; Hebert, Martial)V1388,131-140(1991)

High-speed sensor-based systems for mobile robotics (Kayaalp, Ali E.; Moezzi, Saied; Davies, Henry C.)V1406,98-109(1991)

Hybrid navigational control scheme for autonomous platforms (Holland, John; Everett, Hobart R.; Gilbreath, Gary A.)V1388,291-298(1991)

Impact of uncertain terrain models on the weighted region problem (Mobasseri, Bijan J.)V1388,270-277(1991)

Implementation and control of a 3 degree-of-freedom, force-reflecting manual controller (Kim, Whee-Kuk; Bevill, Pat; Tesar, Delbert)V1387,392-406(1991)

Implementation of a 3-D laser imager-based robot navigation system with location identification (Boltinghouse, Susan T.; Burke, James; Ho, Daniel)V1388,14-29(1991)

Innovative architectural and theoretical considerations yield efficient fuzzy logic controller VLSI design (Basehore, Paul; Yestrebsky, Joseph T.)V1470,190-196(1991)

Integrated mobile robot control (Amidi, Omead; Thorpe, Charles E.)V1388,504-523(1991)

Integrated vision system for object identification and localization using 3-D geometrical models (Bidlack, Clint R.; Trivedi, Mohan M.)V1468,270-280(1991)

Integrating acoustical and optical sensory data for mobile robots (Wang, Gang)V1468,479-482(1991)

Integration of a computer vision system with an IBM 7535 robot (Gonzalez, Orlando; Johnson, Carroll; Starks, Scott A.)V1381,284-291(1991)

Intelligent grasp planning strategy for robotic hands (Walker, Ian D.; Cheatham, John B.; Chen, Yu-Che)V1468,974-989(1991)

Intelligent piloting tools for control of an autonomous mobile robot (Malotaux, Eric; Alimenti, Rodolphe; Bogaert, Marc; Gaspart, Pierre)V1388,372-383(1991)

Intelligent vision process for robot manipulation (Chen, Alexander Y.; Chen, Eugene Y.)V1381,226-239(1991)

Issues in mobile robotics: the unmanned ground vehicle program tele-operated vehicle (Aviles, Walter A.; Hughes, T. W.; Everett, Hobart R.; Umeda, A. Y.; Martin, Stephen W.; Koyamatsu, A. H.; Solorzano, M.; Laird, Robin T.; McArthur, S. P.)V1388,587-597(1991)

Knowledge-based process planning and line design in robotized assembly (Delchambre, Alain)V1468,367-378(1991)

LANELOK: an algorithm for extending the lane sensing operating range to 100-feet (Kenue, Surender K.)V1388,222-233(1991)

LaneLok: an improved Hough transform algorithm for lane sensing using strategic search methods (Kenue, Surender K.; Wybo, David R.)V1468,538-550(1991)

Least-squares-based data fusion strategies and robotic applications (Eason, Richard O.; Kamata, Seiichiro)V1383,566-573(1991)

Low-cost, low-risk approach to tactical reconnaissance (Beving, James E.; Fishell, Wallace G.)V1538,14-17(1991)

Method for robot path adaptation using scalar sensor data (Cullen, Christopher P.)V1388,62-71(1991)

Minimum jerk trajectory planning for robotic manipulators (Kyriakopoulos, Konstantinos J.; Saridis, George N.)V1387,159-164(1991)

Mobile Robots V (Chun, Wendell H.; Wolfe, William J., eds.)V1388(1991)

Mobile robot system for the handicapped (Palakal, Mathew J.; Chien, Yung-Ping; Chittajallu, Siva K.; Xue, Qing L.)V1468,456-466(1991)

Model-based task planning system for a space laboratory environment (Chi, Sung-Do; Zeigler, Bernard P.; Cellier, Francois E.)V1387,182-193(1991)

Modeling and simulation of friction (Haessig, David A.; Friedland, Bernard)V1482,383-396(1991)

Model reference adaptive control of flexible robots in the presence of sudden load changes (Steinvorth, Rodrigo; Kaufman, Howard; Neat, Gregory W.)V1387,136-147(1991)

Modular robotic architecture (Smurlo, Richard; Laird, Robin T.)V1388,566-577(1991)

Multiagent collaboration for experimental calibration of an autonomous mobile robot (Vachon, Bertrand; Berge-Cherfaoui, Veronique)V1468,483-492(1991)

Multiple target tracking system (Lu, Simon W.)V1388,299-305(1991)

Multiple unfused passive sensors for operating in busy indoor environments (Konishi, Mashide; Brooks, Rodney A.)V1383,448-458(1991)

Neural networks application in autonomous path generation for mobile robots (Pourboghrat, Farzad)V1396,243-251(1991)

Neural networks for robot navigation (Pandya, Abhijit S.; Luebbers, Paul G.)V1468,802-811(1991)

Neural networks for the recognition of skilled arm and hand movements (Vaina, Lucia M.; Tuncer, Temel E.)V1468,990-999(1991)

New robot slip sensor using optical fibre and its application (Chen, Jinjiang)V1572,284-286(1991)

Nondeterministic approaches in data fusion: a review (Abdulghafour, Muhamad; Goddard, J.; Abidi, Mongi A.)V1383,596-610(1991)

Object classification for obstacle avoidance (Regensburger, Uwe; Graefe, Volker)V1388,112-119(1991)

Object detection in real-time (Solder, Ulrich; Graefe, Volker)V1388,104-111(1991)

Operator-coached machine vision for space telerobotics (Bon, Bruce; Wilcox, Brian H.; Litwin, Todd; Gennery, Donald B.)V1387,337-342(1991)

Operator/system communication: an optimizing decision tool (Sobh, Tarek M.; Alameldin, Tarek K.)V1388,524-535(1991)

Optical target location using machine vision in space robotics tasks (Sklair, Cheryl W.; Gatrell, Lance B.; Hoff, William A.; Magee, Michael)V1387,380-391(1991)

Optimal control/structure integrated design of a flexible space platform with articulated appendages (Kelkar, Atul G.; Alberts, Thomas E.)V1489,243-253(1991)

Parallel message-passing architecture for path planning (Tavora, Jose; Lourtie, Pedro M.)V1468,524-535(1991)

Parallel path planning in unknown terrains (Prassler, Erwin E.; Milios, Evangelos E.)V1388,2-13(1991)

Path planning algorithm for a mobile robot* (Fan, Kuo-Chin; Lui, Po-Chang)V1468,1010-1021(1991)

Plan-behavior interaction in autonomous navigation (Lim, William Y.; Eilbert, James L.)V1388,464-475(1991)

Polynomial neural network for robot forward and inverse kinematics learning computations (Chen, C. L. P.; McAulay, Alastair D.)V1468,394-405(1991)

Polynomial regression analysis for estimating motion from image sequences (Frau, Juan; Llario, Vicenc; Oliver, Gabriel)V1388,329-340(1991)

Pose determination of spinning satellites using tracks of novel regions (Lee, Andrew J.; Casasent, David P.)V1383,72-83(1991)

Positioning method using polarization-detecting optical sensor for precision robot systems (Sakai, Masao; Nagayama, Akira; Sasakura, Kunihiko)V1385,8-14(1991)

Range image-based object detection and localization for HERMIES III mobile robot (Sluder, John C.; Bidlack, Clint R.; Abidi, Mongi A.; Trivedi, Mohan M.; Jones, Judson P.; Sweeney, Frank J.)V1468,642-652(1991)

Ray-following model of sonar range sensing (Wilkes, David R.; Dudek, Gregory; Jenkin, Michael R.; Milios, Evangelos E.)V1388,536-542(1991)

Real-time map building for fast mobile robot obstacle avoidance (Borenstein, Johann; Koren, Yoram)V1388,74-81(1991)

Recent advances in the development and transfer of machine vision technologies for space (deFigueiredo, Rui J.; Pendleton, Thomas W.)V1387,330-336(1991)

Recognition and tracking of moving objects (El-Konyaly, Sayed H.; Enab, Yehia M.; Soltan, Hesham)V1388,317-328(1991)

Recognition of contacts between objects in the presence of uncertainties (Xiao, Jing)V1470,134-145(1991)

Recognition of movement object collision (Chang, Hsiao T.; Sun, Geng-tian; Zhang, Yan)V1388,442-446(1991)

Recovering epipolar geometry in 3-D vision systems (Schenk, Anton F.; Toth, Charles)V1457,66-73(1991)

Remote driving: one eye or two (Bryant, Keith; Ince, Ilhan)V1457,120-132(1991)

Research sputter cluster tool (Clarke, Peter J.)V1392,617-624(1991)

Research state-of-the-art of mobile robots in China (Wu, Lin; Zhao, Jinglun; Zhang, Peng; Li, Shiqing)V1388,598-601(1991)

Robotics in near-earth space (Card, Michael E.)V1387,101-108(1991)

Robot location densities (Malik, Raashid; Polkowski, Edward T.)V1388,280-290(1991)

Robot self-location based on corner detection (Malik, Raashid; Polkowski, Edward T.)V1388,306-316(1991)

Robot vision system for obstacle avoidance planning (Attolico, Giovanni; Caponetti, Laura; Chiaradia, Maria T.; Distante, Arcangelo)V1388,50-61(1991)

Role of computer graphics in space telerobotics: preview and predictive displays (Bejczy, Antal K.; Venema, Steven; Kim, Won S.)V1387,365-377(1991)

Safe motion planning for mobile agents: a model of reactive planning for multiple mobile agents (Fujimura, Kikuo)V1388,260-269(1991)

Scheme for sensory data integration (Shen, Helen C.; Basir, O. A.)V1383,403-408(1991)

Sensing and environment perception for a mobile vehicle (Blais, Francois; Rioux, Marc)V1480,94-101(1991)

Sensor-based identification of control parameters for intelligent gripping (Wood, Hugh C.; Vaidyanathan, C. S.)V1387,245-254(1991)

Sensor Fusion III: 3-D Perception and Recognition (Schenker, Paul S., ed.)V1383(1991)

Sensor fusion at different levels of data abstraction (Chen, Su-Shing)V1383,574-581(1991)

Sensor-knowledge-command fusion paradigm for man-machine systems (Lee, Sukhan; Schenker, Paul S.; Park, Jun S.)V1383,391-402(1991)

Simulator for developing mobile robot control systems (Roning, Juha J.; Riekki, Jukka P.; Kemppainen, Seppo)V1388,350-360(1991)

Situation assessment for space telerobotics (Bruno, Guy; Morgenthaler, Matthew K.)V1387,352-358(1991)

Space roles for robots (Cliff, Rodger A.)V1387,98-100(1991)

Stability evaluation of the PUMA-560 robot arm under model mismatch (Larsson, T.; Perev, K.; Valavanis, Kimon P.; Gardner, S.)V1387,165-168(1991)

Standard control language for two different robotic manipulators (Chen, Robert S.; Malstrom, Eric M.; Parker, Sandra C.)V1381,189-200(1991)

State estimation for distributed systems with sensing delay (Alexander, Harold L.)V1470,103-111(1991)

Stereoscopic versus orthogonal view displays for performance of a remote manipulation task (Spain, Edward H.; Holzhausen, Klause-Peter)V1457,103-110(1991)

Stereotactic multibeam radiation therapy system in a PACS environment (Fresne, Francoise; Le Gall, G.; Barillot, Christian; Gibaud, Bernard; Manens, J. P.; Toumoulin, Christine; Lemoine, D.; Chenal, C.; Scarabin, Jean-Marie)V1444,26-36(1991)

Stereovision and color segmentation for autonomous navigation (Sung, Eric)V1388,176-187(1991)

Stereo vision for planetary rovers: stochastic modeling to near-real-time implementation (Matthies, Larry H.)V1570,187-200(1991)

SUB-3D high-resolution pose measurement system (Hudgens, Jeffrey C.; Tesar, Delbert; Sklar, Michael E.)V1387,271-282(1991)

Subsumption architecture control system for space proximity maneuvering (Viggh, Herbert E.)V1387,202-214(1991)

Surface property determination for planetary rovers (Severson, William E.; Douglass, Robert J.; Hennessy, Stephen J.; Boyd, Robert; Anhalt, David J.)V1388,490-501(1991)

Task decomposition, distribution, and localization for intelligent robot coordination (Kountouris, Vasilios G.; Stephanou, Harry E.)V1387,169-180(1991)

Telepresence for planetary exploration (McGreevy, Michael W.; Stoker, Carol R.)V1387,110-123(1991)

Telerobotic capabilities for space operations (Akin, David L.)V1387,359-364(1991)

Terrain acquisition algorithm for an autonomous mobile robot with finite-range sensors (Smith, John M.; Choo, Chang Y.; Nasrabadi, Nasser M.)V1468,493-501(1991)

Terrain adaptive footfall placement using real-time range images (Dodds, David R.)V1388,543-548(1991)

Terrain classification in navigation of an autonomous mobile robot (Dodds, David R.)V1388,82-89(1991)

Testbed for tele-autonomous operation of multiarmed robotic servicers in space (Morgenthaler, Matthew K.; Bruno, Guy; Spofford, John R.; Greunke, Roy G.; Gatrell, Lance B.)V1387,82-95(1991)

Test of a vision-based autonomous space station robotic task (Castellano, Anthony R.; Hwang, Vincent S.; Stoney, William E.)V1387,343-350(1991)

Theories in distributed decision fusion: comparison and generalization (Thomopoulos, Stelios C.)V1383,623-634(1991)

Three-dimensional position determination from motion (Nashman, Marilyn; Chaconas, Karen)V1383,166-175(1991)

Towards a general formula for analogical learning leading to more autonomous systems (Cooke, Daniel E.; Patterson, Dan W.; Starks, Scott A.)V1381,299-305(1991)

Towards a versatile control system for mobile robots (Noreils, Fabrice R.)V1388,384-396(1991)

Towards integrated autonomous systems (Jain, Ramesh C.; Roth-Tabak, Yuval)V1468,188-201(1991)

Toward tactile sensor-based exploration in a robotic environment (Gadagkar, Hrishikesh P.; Trivedi, Mohan M.)V1383,142-150(1991)

Using fixation for direct recovery of motion and shape in the general case (Taalebinezhaad, M. A.)V1388,199-209(1991)

Using real-time stereopsis for mobile robot control (Bonasso, R. P.; Nishihara, H. K.)V1387,237-244(1991)

Using robust statistics for sensor fusion (McKendall, Raymond; Mintz, Max)V1383,547-565(1991)

Validation of vision-based obstacle detection algorithms for low-altitude helicopter flight (Suorsa, Raymond E.; Sridhar, Banavar)V1388,90-103(1991)

Vehicle path planning via dual-world representations (Peck, Alex N.; Breul, Harry T.)V1388,30-38(1991)

Video-image-based neural network guidance system with adaptive view-angles for autonomous vehicles (Luebbers, Paul G.; Pandya, Abhijit S.)V1469,756-765(1991)

Visual surveillance system based on spatio-temporal model of moving objects in industrial workroom environments (Motamed, Cina; Schmitt, Alain)V1606,961-969(1991)

Wheeled planetary rover testbed (Price, R. S.; Chun, Wendell H.; Hammond, Mark W.; Hubert, Alexis)V1388,550-559(1991)

Why mobile robots need a spatial memory (Haber, Ralph N.; Haber, Lyn)V1383,411-424(1991)

Workstation recognition using a constrained edge-based Hough transform for mobile robot navigation (Vaughn, David L.; Arkin, Ronald C.)V1383,503-514(1991)

Satellites—see also astronomy; remote sensing; space optics

A 120-Mbit/s QPPM high-power semiconductor transmitter performance and reliability (Greulich, Peter; Hespeler, Bernd; Spatscheck, Thomas)V1417,358-369(1991)

Acquisition, Tracking, and Pointing V (Masten, Michael K.; Stockum, Larry A., eds.)V1482(1991)

Advanced visible and near-IR radiometer for ADEOS (Iwasaki, Nobuo; Tange, Yoshio; Miyachi, Yuji; Inoue, Kouichi; Kadowaki, Tomoko; Tanaka, Hirokazu; Michioka, Hidekazu)V1490,216-221(1991)

Advanced X-ray Astrophysics Facility science instruments (Winkler, Carl E.; Dailey, Carroll C.; Cumings, Nesbitt P.)V1494,301-313(1991)

Application results for an augmented video tracker (Pierce, Bill)V1482,182-195(1991)

Applications of laser ranging to ocean, ice, and land topography (Degnan, John J.)V1492,176-186(1991)

Arithmetic coding model for compression of LANDSAT images (Perez, Arnulfo; Kamata, Seiichiro; Kawaguchi, Eiji)V1605,879-884(1991)

ASTER calibration concept (Ono, Akira; Sakuma, Fumihiro)V1490,285-298(1991)

ATHENA: a high-resolution wide-area coverage commercial remote sensing system (Claybaugh, William R.; Megill, L. R.)V1495,81-94(1991)

ATLID: the first preoperational ATmospheric LIDar for the European polar platform (Lange, Robert; Endemann, Martin J.; Reiland, Werner; Krawczyk, Rodolphe; Hofer, Bruno)V1492,24-37(1991)

Augmented tracking and acquisition system for GBL satellite illumination (Brodsky, Aaron; Goodrich, Alan; Lawson, David G.; Holm, Richard W.)V1482,159-169(1991)

Automated band selection for multispectral meteorological applications (Westerman, Steven D.; Drake, R. M.; Yool, Stephen R.; Brandley, M.; DeJulio, R.)V1492,263-271(1991)

Autonomous navigation of small co-orbiting satellites using C/A GPS code (Barresi, Giangrande; Soddu, Claudio; Rondinelli, Giuseppe; Caporicci, Lucio; Loria, A.)V1495,246-258(1991)

CCD star sensors for Indian remote sensing satellites (Alex, T. K.; Rao, V. K.)V1478,101-105(1991)

Circular streak camera application for satellite laser ranging (Prochazka, Ivan; Hamal, Karel; Kirchner, G.; Schelev, Mikhail Y.; Postovalov, V. E.)V1449,116-120(1991)

Coarse pointing assembly for the SILEX program, or how to achieve outstanding pointing accuracy with simple hardware associated with consistent control laws (Buvat, Daniel; Muller, Gerard; Peyrot, Patrick)V1417,251-261(1991)

Commercial remote sensing small-satellite feasibility study (Birk, Ronald J.; Tompkins, Jim M.; Burns, Gregory S.)V1495,2-11(1991)

Comparison of optical technologies for a high-data-rate Mars link (Spence, Rodney L.)V1417,550-561(1991)

Complementary experiments for tether dynamics analysis (Wingo, Dennis R.; Bankston, Cheryl D.)V1495,123-133(1991)

Cooled focal-plane assembly for ocean color and temperature scanner (Nakayama, Masao; Izawa, Toshiyuki; Fujisada, Hiroyuki; Tange, Yoshio; Miyachi, Yuji; Sato, Ryota; Ishida, Juro; Tanii, Jun)V1490,207-215(1991)

Cryogenic limb array etalon spectrometer: experiment description (Roche, Aidan E.; Kumer, John B.)V1491,91-103(1991)

Current and future activities in the area of optical space communications in Japan (Fujise, Masayuki; Araki, Ken'ichi; Arikawa, Hiroshi; Furuhama, Yoji)V1522,14-26(1991)

DARPA initiatives in small-satellite technologies (Bonometti, Robert J.; Wheatley, Alvis A.; Flynn, Lin; Nicastri, Edward; Sudol, R.)V1495,166-176(1991)

Definition of interorbit link optical terminals with diode-pumped Nd:host laser technology (Marini, Andrea E.; Della Torre, Antonio)V1522,222-233(1991)

Design and analysis of the closed-loop pointing system of a scientific satellite (Heyler, Gene A.; Garlick, Dean S.; Yionoulis, Steve M.)V1481,198-208(1991)

Design and development of the Zeolite Crystal Growth Facility (Fiske, Michael R.)V1557,78-85(1991)

Design considerations for EOS direct broadcast (Vermillion, Charles H.; Chan, Paul H.)V1492,224-228(1991)

Design of a diode-pumped Nd:YAG laser communication system (Sontag, Heinz; Johann, Ulrich A.; Pribil, Klaus)V1417,573-587(1991)

Design of a periscopic coarse pointing assembly for optical multiple access (Gatenby, Paul V.; Boereboom, Peter; Grant, Michael A.)V1522,126-134(1991)

Design study for high-performance optical communications satellite terminal with high-power laser diode transmitter (Hildebrand, Ulrich; Seeliger, Reinhard; Smutny, Berry; Sand, Rolf)V1522,50-60(1991)

Diode-laser-based range sensor (Seshamani, Ramani)V1471,354-356(1991)

Discrete piezoelectric sensors and actuators for active control of two-dimensional spacecraft components (Bayer, Janice I.; Varadan, Vasundara V.; Varadan, Vijay K.)V1480,102-114(1991)

Eagle-class small satellite for LEO applications (O'Neil, Jason; Goralczyk, Steven M.)V1495,72-80(1991)

Earth Observing System (Wilson, Stan; Dozier, Jeff)V1491,117-124(1991)

Effect of microaccelerations on an optical space communication system (Wittig, Manfred E.)V1522,278-286(1991)

Effects of base motion on space-based precision laser tracking in the Relay Mirror Experiment (Anspach, Joel E.; Sydney, Paul F.; Hendry, Gregg)V1482,170-181(1991)

European SILEX project and other advanced concepts for optical space communications (Oppenhaeuser, Gotthard; Wittig, Manfred E.; Popescu, Alexandru F.)V1522,2-13(1991)

Eyes in the skies: building satellites for education (Hansen, Verne W.; Summers, Robert A.; Clapp, William G.)V1495,115-122(1991)

Fiber optic link for millimeter-wave communication satellites (Polifko, David M.; Malone, Steven A.; Daryoush, Afshin S.; Kunath, Richard R.)V1476,91-99(1991)

Forest decline model development with LANDSAT TM, SPOT, and DEM DATA (Brockhaus, John A.; Campbell, Michael V.; Khorram, Siamak; Bruck, Robert I.; Stallings, Casson)V1492,200-205(1991)

Free-Space Laser Communication Technologies III (Begley, David L.; Seery, Bernard D., eds.)V1417(1991)

Future European and Japanese Remote-Sensing Sensors and Programs (Slater, Philip N., ed.)V1490(1991)

GaAs monolithic RF modules for SARSAT distress beacons (Cauley, Mike A.)V1475,275-279(1991)

German ATMOS program (Puls, Juergen)V1490,2-13(1991)

GLOB(MET)SAT: French proposals for monitoring global change and weather from the polar orbit (Durpaire, Jean-Pierre; Ratier, A.; Dagras, C.)V1490,23-38(1991)

Global ozone monitoring by occultation of stars (Bertaux, J. L.)V1490,133-145(1991)

GOPEX: a deep-space optical communications demonstration with the Galileo spacecraft (Wilson, Keith E.; Schwartz, Jon A.; Lesh, James R.)V1417,22-28(1991)

Ground systems and operations concepts for the Space Infrared Telescope Facility (Miller, Richard B.)V1540,38-46(1991)

GSFC conceptual design study for an intersatellite optical multiple access communication system (Fox, Neil D.; Maynard, William L.; Clarke, Ernest S.; Bruno, Ronald C.)V1417,452-463(1991)

Heterodyne acquisition and tracking in a free-space diode laser link (Hueber, Martin F.; Scholtz, Arpad L.; Leeb, Walter R.)V1417,233-239(1991)

High-power diode-pumped solid state lasers for optical space communications (Koechner, Walter; Burnham, Ralph L.; Kasinski, Jeff; Bournes, Patrick A.; DiBiase, Don; Le, Khoa; Marshall, Larry R.; Hays, Alan D.)V1522,169-179(1991)

High-resolution imaging interferometer (Ballangrud, Ase; Jaeger, Tycho C.; Wang, Gunnar)V1521,89-96(1991)

How to meet intersatellite links mission requirements by an adequate optical terminal design (Duchmann, Olivier; Planche, G.)V1417,30-41(1991)

Imager of METEOSAT second generation (Hollier, Pierre A.)V1490,74-81(1991)

Imaging capabilities of small satellites: Indian experience (Alex, T. K.)V1495,52-58(1991)

Improved limb atmospheric spectrometer and retroreflector in-space for ADEOS (Sasano, Yasuhiro; Asada, Kazuya; Sugimoto, Nobuo; Yokota, Tatsuya; Suzuki, Makoto; Minato, Atsushi; Matsuzaki, Akiyoshi; Akimoto, Hajime)V1490,233-242(1991)

Infrared Space Observatory optical subsystem (Singer, Christian; Massoni, Jean A.; Mossbacher, Bernard; Cinotti, Ciro)V1494,255-264(1991)

Intelligent variable-resolution laser scanner for the space vision system (Blais, Francois; Rioux, Marc; MacLean, Steve G.)V1482,473-479(1991)

Japanese mission overview of JERS and ASTER programs (Yamaguchi, Yasushi; Tsu, Hiroji; Sato, Isao)V1490,324-334(1991)

Labeled object identification for the mobile servicing system on the space station (Zakaria, Marwan F.; Ng, Terence K.)V1386,121-127(1991)

Large integrating sphere of prelaunch calibration system for Japanese Earth Resources Satellite optical sensors (Suzuki, Naoshi; Narimatsu, Yoshito; Nagura, Riichi; Sakuma, Fumihiro; Ono, Akira)V1493,48-57(1991)

Laser terminal attitude determination via autonomous star tracking (Chapman, William W.; Fitzmaurice, Michael W.)V1417,277-290(1991)

Low-cost spacecraft buses for remote sensing applications (Harvey, Edwin L.; Cullen, Robert M.)V1495,134-145(1991)

Low-cost space flight for attached payloads (Perkins, Frederick W.)V1495,157-163(1991)

Low-cost space platforms for detection and tracking technologies (Cullen, Robert M.)V1479,295-305(1991)

Low-insertion-loss, high-precision liquid crystal optical phased array (Cassarly, William J.; Ehlert, John C.; Henry, D.)V1417,110-121(1991)

Materials technology for SIRTF (Coulter, Daniel R.; Dolgin, Benjamin P.; Rainen, R.; O'Donnell, Timothy P.)V1540,119-126(1991)

Mechanical cooler development program for ASTER (Kawada, Masakuni; Fujisada, Hiroyuki)V1490,299-308(1991)

Medium-resolution imaging spectrometer (Baudin, Gilles; Bessudo, Richard; Cutter, Mike A.; Lobb, Daniel R.; Bezy, Jean L.)V1490,102-113(1991)

METEOSAT second-generation program (Markland, Chris A.)V1490,39-50(1991)

Microwave limb sounder experiments for UARS and EOS (Waters, Joe W.)V1491,104-109(1991)

Microwave monolithic integrated circuits for high-data-rate satellite communications (Turner, Elbert L.; Hill, William A.)V1475,248-256(1991)

Miniaturized low-power parallel processor for space applications (Jacobi, William J.; Jensen, Preben D.; Teneketges, Nicholas J.; Wadsworth, Leo A.)V1495,205-213(1991)

Mission design for the Space Infrared Telescope Facility (Kwok, Johnny H.; Osmolovsky, Michael J.)V1540,27-37(1991)

Mission overview of ADEOS program (Iwasaki, Nobuo; Hara, Norikazu; Kajii, Makoto; Tange, Yoshio; Miyachi, Yuji; Sato, Ryota; Inoue, Kouichi)V1490,192-199(1991)

Mission study overview of Japanese polar orbiting platform program (Moriyama, Takashi; Nakayama, Kimihiko; Homma, M.; Haruyama, Yukio)V1490,310-316(1991)

MMIC: a key technology for future communications satellite antennas (Sorbello, Robert M.; Zaghloul, A. I.; Gupta, R. K.; Geller, B. D.; Assal, F. T.; Potukuchi, J. R.)V1475,175-183(1991)

Model "T" satellite series: small satellites designed for scientific and commercial use (McMillen, Donald V.)V1495,95-102(1991)

Moderate-resolution imaging spectrometer-tilt baseline concept (Magner, Thomas J.)V1492,272-285(1991)

Multimission sensor for the RESERVES small-satellite program (Kilston, Steven; Kilston, Vera M.; Utsch, Thomas F.)V1495,193-204(1991)

NASA's flight-technology development program: a 650-Mbit/s laser communications testbed (Hayden, William L.; Fitzmaurice, Michael W.; Nace, Dave; Lokerson, Donald; Minott, Peter O.; Chapman, William W.)V1417,182-199(1991)

Nd:host laser-based optical terminal development study for intersatellite links (Marini, Andrea E.; Della Torre, Antonio; Popescu, Alexandru F.)V1417,200-211(1991)

New concepts in remote sensing and geolocation (Seastone, A. J.)V1495,228-239(1991)

New generation control system for ultra-low-jitter satellite tracking (Verbanets, William R.; Greenwald, David)V1482,112-120(1991)

Novel angular discriminator for spatial tracking in free-space laser communications (Fung, Jackie S.)V1417,224-232(1991)

Objectives for the Space Infrared Telescope Facility (Spehalski, Richard J.; Werner, Michael J.)V1540,2-14(1991)

Ocean color and temperature scanner for ADEOS (Tanii, Jun; Machida, Tsuneo; Ayada, Haruki; Katsuyama, Yoshihiko; Ishida, Juro; Iwasaki, Nobuo; Tange, Yoshio; Miyachi, Yuji; Sato, Ryota)V1490,200-206(1991)

Optical approach to proximity-operations communications for Space Station Freedom (Marshalek, Robert G.)V1417,53-62(1991)

Optical filters for the wind imaging interferometer (Sellar, R. G.; Gault, William A.; Karp, Christopher K.)V1479,140-155(1991)

Optical ISL transmitter design that uses a high-power LD amplifier (Nohara, Mitsuo; Harada, Takashi; Fujise, Masayuki)V1417,338-345(1991)

Optical sensor system for Japanese Earth resources satellite 1 (Hino, Hideo; Takei, Mitsuru; Ono, Hiromi; Nagura, Riichi; Narimatsu, Yoshito; Hiramatsu, Masaru; Harada, Hisashi; Ogikubo, Kazuhiro)V1490,166-176(1991)

Optical Space Communication II (Franz, Juergen, ed.)V1522(1991)

Optical space-to-ground link availability assessment and diversity requirements (Chapman, William W.; Fitzmaurice, Michael W.)V1417,63-74(1991)

Overview of ASTER design concept (Fujisada, Hiroyuki; Ono, Akira)V1490,244-254(1991)

Overview of the halogen occultation experiment (Russell, James M.)V1491,110-116(1991)

Overview of U.S. Fishery requirements (Springer, Steven C.; McLean, Craig)V1479,372-379(1991)

Passive-sensor data fusion (Kolitz, Stephan E.)V1481,329-340(1991)

Pegasus air-launched space booster payload interfaces and processing procedures for small optical payloads (Mosier, Marty R.; Harris, Gary N.; Whitmeyer, Charlie)V1495,177-192(1991)

Performance enhancement in future communications satellites with MMIC technology insertion (Hung, Hing-Loi A.; Mahle, Christoph E.)V1475,212-222(1991)

Pointing, acquisition, and tracking system of the European SILEX program: a major technological step for intersatellite optical communication (Bailly, Michel; Perez, Eric)V1417,142-157(1991)

Pose determination of spinning satellites using tracks of novel regions (Lee, Andrew J.; Casasent, David P.)V1383,72-83(1991)

Prelaunch calibration system for optical sensors of Japanese Earth Resources Satellite (Sakuma, Fumihiro; Ono, Akira)V1493,37-47(1991)

Ratioing radiometer for use with a solar diffuser (Palmer, James M.; Slater, Philip N.)V1493,106-117(1991)

Relay Mirror Experiment overview: a GBL pointing and tracking demonstration (Dierks, Jeffrey S.; Ross, Susan E.; Brodsky, Aaron; Kervin, Paul W.; Holm, Richard W.)V1482,146-158(1991)

Reliability analysis of a multiple-laser-diode beacon for intersatellite links (Mauroschat, Andreas)V1417,513-524(1991)

Remote sensing of coastal environmental hazards (Huh, Oscar K.; Roberts, Harry H.; Rouse, Lawrence J.)V1492,378-386(1991)

Remote sensing of volcanic ash hazards to aircraft (Rose, William I.; Schneider, David J.)V1492,387-390(1991)

Role of orbital observations in detecting and monitoring geological hazards: prospects for the future (Pieri, David C.)V1492,410-417(1991)

Satellite-borne laser for adaptive optics reference (Greenaway, Alan H.)V1494,386-393(1991)

Scanning infrared earth sensor for INSAT-II (Alex, T. K.; Kamalakar, J. A.)V1478,106-111(1991)

Scientific data compression for space: a modified block truncation coding algorithm (Lu, Wei-Wei; Gough, M. P.; Davies, Peter N.)V1470,197-205(1991)

Second-generation high-data-rate interorbit link based on diode-pumped Nd:YAG laser technology (Sontag, Heinz; Johann, Ulrich A.; Pribil, Klaus)V1522,61-69(1991)

Short-wavelength infrared subsystem design status of ASTER (Akasaka, Akira; Ono, Makoto; Sakurai, Yasushi; Hayashida, Bun)V1490,269-277(1991)

Signal processor for space-based visible sensing (Anderson, James C.; Downs, G. S.; Trepagnier, Pierre C.)V1479,78-92(1991)

SILEX project: the first European optical intersatellite link experiment (Laurent, Bernard; Duchmann, Olivier)V1417,2-12(1991)

Silicon x-ray array detector on spectrum-x-gamma satellite (Sipila, Heikki; Huttunen, Pekka; Kamarainen, Veikko J.; Vilhu, Osmi; Kurki, Jouko; Leppelmeier, Gilbert W.; Taylor, Ivor; Niemela, Arto; Laegsgaard, Erik; Sunyaev, Rashid)V1549,246-255(1991)

Simulation model and on-ground performances validation of the PAT system for the SILEX program (Cossec, Francois R.; Doubrere, Patrick; Perez, Eric)V1417,262-276(1991)

SIRTF focal-plane technologies (Capps, Richard W.; Bothwell, Mary)V1540,47-50(1991)

Small-capacity low-cost (Ni-H2) design concept for commercial, military, and higher volume aerospace applications (Wheeler, James R.; Cook, William D.; Smith, Ron)V1495,280-285(1991)

Small-satellite constellation for many uses (Seiberling, Walter E.; Traxler-Lee, Laura A.; Collins, Sean K.)V1495,32-41(1991)

Small satellites: current legal issues in remote sensing (Stern, Jill A.)V1495,42-51(1991)

Small-satellite sensors for multispectral space surveillance (Kostishack, Daniel F.)V1495,214-227(1991)

Small satellites for water cycle experiments (Rondinelli, Giuseppe; Di Girolamo, Sergio; Barresi, Giangrande)V1495,19-31(1991)

Small-Satellite Technology and Applications (Horais, Brian J., ed.)V1495(1991)

SPOT 4 HRVIR instrument and future high-resolution stereo instruments (Fratter, C.; Reulet, Jean-Francois; Jouan, Jacky)V1490,59-73(1991)

SPOT system and defence applications (Bernard, Christian)V1521,66-73(1991)

Solid state magnetic azimuth sensor for small satellites (Rouse, Gordon F.; Stauffer, Donald R.; French, Howard B.)V1495,240-245(1991)

Space-based sensing of atmospheric conditions over data-void regions (Behunek, Jan L.; Vonder Haar, Thomas H.)V1479,93-100(1991)

Space experiments using small satellites (Schor, Matthew J.)V1495,146-148(1991)

Space Infrared Telescope Facility structural design requirements (MacNeal, Paul D.; Lou, Michael C.; Chen, Gun-Shing)V1540,68-85(1991)

Space Infrared Telescope Facility telescope overview (Schember, Helene R.; Manhart, Paul K.; Guiar, Cecilia N.; Stevens, James H.)V1540,51-62(1991)

Space station laser communication transceiver (Fitzmaurice, Michael W.; Hayden, William L.)V1417,13-21(1991)

Static horizon sensor for remote sensing satellite (Kamalakar, J. A.; Jain, Yashwant K.; Laxmiprasad, A. S.; Shashikala, M.)V1478,92-100(1991)

Station-keeping strategy for multiple-spacecraft interferometry (DeCou, Anthony B.)V1494,440-451(1991)

Stratospheric aerosol and gas experiment III: aerosol and trace gas measurements for the Earth Observing System (McCormick, M. P.; Chu, William P.; Zawodny, J. M.; Mauldin, Lemuel E.; McMaster, Leonard R.)V1491,125-141(1991)

Structural design considerations for the Space Infrared Telescope Facility (MacNeal, Paul D.; Lou, Michael C.)V1494,236-254(1991)

Structural design of the large deployable reflector (Satter, Celeste M.; Lou, Michael C.)V1494,279-300(1991)

Surveillance Technologies (Gowrinathan, Sankaran; Mataloni, Raymond J.; Schwartz, Stanley J., eds.)V1479(1991)

Synthetic aperture radar of JERS-1 (Ono, Makoto; Nemoto, Yoshiaki)V1490,184-190(1991)

TECHSTARS: small, smart space systems (Higbee, Terry A.)V1495,103-114(1991)

Territorial analysis by fusion of LANDSAT and SAR data (Vernazza, Gianni L.; Dambra, Carlo; Parizzi, Francesco; Roli, Fabio; Serpico, Sebastiano B.)V1492,206-212(1991)

Thermal analysis of a small expendable tether satellite package (Randorf, Jeffrey A.)V1495,259-267(1991)

Thermal infrared subsystem design status of ASTER (Aoki, Yutaka; Ohmae, Hirokazu; Kitamura, Shin-ichi)V1490,278-284(1991)

Thermal systems analysis for the Space Infrared Telescope Facility dewar (Bhandari, Pradeep; Petrick, Stanley W.; Schember, Helene R.)V1540,97-108(1991)

TREIS: a concept for a user-affordable, user-friendly radar satellite system for tropical forest monitoring (Raney, R. K.; Specter, Christine N.)V1492,298-306(1991)

Upper-atmosphere research satellite: an overview (McNeal, Robert J.)V1491,84-90(1991)

Use of satellite data to determine the distribution of ozone in the troposphere (Fishman, Jack; Watson, Catherine E.; Brackett, Vincent G.; Fakhruzzaman, Khan; Veiga, Robert)V1491,348-359(1991)

Variable-gain MMIC module for space application (Palena, Patricia)V1475,91-102(1991)

Visible and near-infrared subsystem and common signal processor design status of ASTER (Takahashi, Fumiho; Hiramatsu, Masaru; Watanabe, Fumito; Narimatsu, Yoshito; Nagura, Riichi)V1490,255-268(1991)

WEBERSAT: a low-cost imaging satellite (Twiggs, Robert J.; Reister, K. R.)V1495,12-18(1991)

WEBERSAT: data analysis and dynamic behavior (Smith, Jay L.)V1495,59-70(1991)

WEBERSAT: measuring micrometeorite impacts in a polar orbit (Evans, Phillip R.)V1495,149-156(1991)

Scanning—see also lasers, applications

0.5-micron deep-UV lithography using a Micrascan-90 step-and-scan exposure tool (Kuyel, Birol; Barrick, Mark W.; Hong, Alexander; Vigil, Joseph)V1463,646-665(1991)

Adaptive notch filter for removal of coherent noise from infrared scanner data (Jaggi, Sandeep)V1541,134-140(1991)

Advanced real-time scanning concept for full dynamics recording, high image quality, and superior measurement accuracy (Lindstrom, Kjell M.; Wallin, Bo)V1488,389-398(1991)

AEDC direct-write scene generation test capabilities (Lowry, Heard S.; Elrod, Parker D.; Johnson, R. J.)V1454,453-464(1991)

Airborne and spaceborne thermal multispectral remote sensing (Watanabe, Hiroshi; Sano, Masaharu; Mills, F.; Chang, Sheng-Huei; Masuda, Shoichi)V1490,317-323(1991)

Analysis of barcode digitization techniques (Boles, John A.; Hems, Randall K.)V1384,195-204(1991)

Analysis of one-dimensional barcode (Wang, Ynjiun P.; Pavlidis, Theo; Swartz, Jerome)V1384,145-160(1991)

Analysis of the flying light spot experiment on SPRITE detector (Gu, Bo-qi; Feng, Wen-qing)V1488,443-446(1991)

Angle encoding with the folded Brewster interferometer (Young, Niels O.)V1454,235-244(1991)

Automatic-adjusting optical axis for linear CCD scanner (Chang, Rong-Seng; Chen, Der-Chin)V1527,357-360(1991)

Automatic film reading system for high-speed photography (Zhou, Renkui; Yu, Chongzhen; Ma, Jiankang; Zhu, Wenkai)V1358,1245-1251(1991)

Beam Deflection and Scanning Technologies (Beiser, Leo; Marshall, Gerald F., eds.)V1454(1991)

Bearings for rotary scanners: an overview (Preston, Ralph G.; Colquhoun, Allan B.; Shepherd, Joseph)V1454,124-131(1991)

Butterfly line scanner: rotary twin reflective deflector that desensitizes scan-line jitter to wobble of the rotational axis (Marshall, Gerald F.; Vettese, Thomas; Carosella, John H.)V1454,37-45(1991)

Camera and Input Scanner Systems (Chang, Win-Chyi; Milch, James R., eds.)V1448(1991)

Capabilities of the optical microscope (Bradbury, Savile)V1439,128-134(1991)

Colorimetric calibration for scanners and media (Hung, Po-Chieh)V1448,164-174(1991)

Compact, low-power precision beam-steering mirror (DeWeerd, Herman)V1454,207-214(1991)

Comparing laser printing and barcode scanning designs (MacArthur, Thomas D.)V1384,176-184(1991)

Considerations in the design of servo amplifiers for high-performance scanning systems (Bukys, Albert)V1454,186-195(1991)

Cooled focal-plane assembly for ocean color and temperature scanner (Nakayama, Masao; Izawa, Toshiyuki; Fujisada, Hiroyuki; Tange, Yoshio; Miyachi, Yuji; Sato, Ryota; Ishida, Juro; Tanii, Jun)V1490,207-215(1991)

Coordinate measuring system for 2-D scanners (Bukatin, Vladimir V.)V1454,283-288(1991)

Design and performance of a small two-axis high-bandwidth steering mirror (Loney, Gregory C.)V1454,198-206(1991)

Design considerations for a high-resolution film scanner for teleradiology applications (Kocher, Thomas E.; Whiting, Bruce R.)V1446,459-464(1991)

Design considerations for use of a double-dove prism behind a concentric dome (Gibbons, Robert C.; Cooper, E. E.; Legan, R. G.)V1498,64-69(1991)

Design equations for a polygon laser scanner (Beiser, Leo)V1454,60-66(1991)

Development of a precision high-speed flying spot position detector (Jodoin, Ronald E.; Loce, Robert P.; Nowak, William J.; Costanza, Daniel W.)V1398,61-70(1991)

Diffraction analysis of beams for barcode scanning (Eastman, Jay M.; Quinn, Anna M.)V1384,185-194(1991)

Dimension and lacunarity measurement of IR images using Hilbert scanning (Moghaddam, Baback; Hintz, Kenneth J.; Stewart, Clayton V.)V1486,115-126(1991)

Electronic f-theta correction for hologon deflector systems (Whitman, Tony; Araghi, Mehdi N.)V1454,426-433(1991)

Fast access to reduced-resolution subsamples of high-resolution images (Isaacson, Joel S.)V1460,80-91(1991)

Ferrofluid film bearing for enhancement of rotary scanner performance (Cheever, Charles J.; Li, Zhixin; Raj, K.)V1454,139-151(1991)

Firefly system concept (Nichols, Joseph D.)V1540,202-206(1991)

Frequency modulation of printed gratings as a protection against copying (Spannenburg, S.)V1509,88-104(1991)

Geometric error coupling in instrument ball bearings (Kingsbury, Edward P.; Francis, Henry A.)V1454,152-158(1991)

Geometric measurement of optical fibers with pulse-counting method (Nie, Qiuhua; Nelson, John C.; Fleming, Simon C.)V1332,409-420(1991)

Hexagonal sampling and filtering for target detection with a scanning E-O sensor (Sperling, I.; Drummond, Oliver E.; Reed, Irving S.)V1481,2-11(1991)

High-accuracy capacitive position sensing for low-inertia actuators (Stokes, Brian P.)V1454,223-229(1991)

High-density two-dimensional barcode (Wang, Ynjiun P.; Pavlidis, Theo; Swartz, Jerome)V1384,169-175(1991)

Highly accurate pattern generation using acousto-optical deflection (Sandstrom, Torbjorn; Tison, James K.)V1463,629-637(1991)

High-performance 400-DPI A4-size contact image sensor module for scanner and G4 fax applications (Yeh, Long-Ching; Wu, Way-Chen; Tang, Ru-Shyah; Tsai, Yong-Song; Chiang, Eugene; Chiao, Pat; Chu, Chu-Lin; Wang, Weng-Lyang)V1527,361-367(1991)

High-performance a-Si:H TFT-driven linear image sensor and its application to a compact scanner (Miyake, Hiroyuki; Sakai, Kazuhiro; Abe, Tsutomu; Sakai, Yoshihiko; Hotta, Hiroyuki; Sugino, Hajime; Ito, Hisao; Ozawa, Takashi)V1448,150-156(1991)

High-quality image recorder and evaluation (Suzuki, Masayuki; Shiraiwa, Yoshinobu; Minoura, Kazuo)V1454,370-381(1991)

High-resolution high-speed laser recording and printing using low-speed polygons (Razzaghi-Masoud, Mahmoud)V1458,145-153(1991)

High-resolution teleradiology applications within the hospital (Jost, R. G.; Blaine, G. J.; Kocher, Thomas E.; Muka, Edward; Whiting, Bruce R.)V1446,2-9(1991)

High-Speed Inspection Architectures, Barcoding, and Character Recognition (Chen, Michael J., ed.)V1384(1991)

Hilbert scanning arithmetic coding for multispectral image compression (Perez, Arnulfo; Kamata, Seiichiro; Kawaguchi, Eiji)V1567,354-361(1991)

Holographic deflectors for graphic arts applications: an overview (Kramer, Charles J.)V1454,68-100(1991)

Holographic disk scanner for active infrared sensors (Kawauchi, Yoshikazu; Toyoda, Ryuuichi; Kimura, Minoru; Kawata, Koichi)V1555,224-227(1991)

Image converter streak cameras with super-light-speed scanning (Averin, V. I.; Gus'kova, M. S.; Korzhenevich, Irina M.; Kolesov, G. V.; Lebedev, V. B.; Maranichenko, N. I.; Sobolev, A. A.)V1358,589-602(1991)

Image quality, dollars, and very low contrast documents (Weideman, William E.)V1454,382-390(1991)

Improvement in detection of small wildfires (Sleigh, William J.)V1540,207-212(1991)

In-flight calibration of a helicopter-mounted Daedalus multispectral scanner (Balick, Lee K.; Golanics, Charles J.; Shines, Janet E.; Biggar, Stuart F.; Slater, Philip N.)V1493,215-223(1991)

Integration of diverse remote sensing data sets for geologic mapping and resource exploration (Kruse, Fred A.; Dietz, John B.)V1492,326-337(1991)

Intelligent variable-resolution laser scanner for the space vision system (Blais, Francois; Rioux, Marc; MacLean, Steve G.)V1482,473-479(1991)

Investigation of higher-order diffraction in a one-crystal 2-D scanner (Melamed, Nathan T.; Gottlieb, Milton S.)V1454,306-312(1991)

Laser-based display technology development at the Naval Ocean Systems Center (Phillips, Thomas E.; Trias, John A.; Lasher, Mark E.; Poirier, Peter M.; Dahlke, Weldon J.; Robinson, Waldo R.)V1454,290-298(1991)

Laser galvo-angle-encoder with zero added inertia (Hercher, Michael; Wyntjes, Geert J.)V1454,230-234(1991)

Linear resonant approach to scanning (Confer, Charles L.; Burrer, Gordon J.)V1454,215-222(1991)

Localization of hot spots in silicon devices with a laser scanning microscope (Bergner, Harald; Krause, A.; Stamm, Uwe)V1361,723-731(1991)

Measuring the geometric accuracy of a very accurate, large, drum-type scanner (Wolber, John W.)V1448,175-180(1991)

Medical applications of three-dimensional and four-dimensional laser scanning of facial morphology (Sadler, Lewis L.; Chen, Xiaoming; Figueroa, Alvaro A.; Aduss, Howard)V1380,158-162(1991)

Micromachined scanning mirrors for laser beam deflection (Huber, Peter; Gerlach-Meyer, U.)V1522,135-141(1991)

MOMS-02/Spacelab D-2: a high-resolution multispectral stereo scanner for the second German Spacelab mission (Ackermann, F.; Bodechtel, Joh; Lanzl, Franz; Meissner, D.; Seige, Peter; Winkenbach, H.; Zilger, Johannes)V1490,94-101(1991)

Monogon laser scanner with no line wobble (Beiser, Leo)V1454,33-36(1991)

Monte Carlo calculation of light distribution in an integrating cavity illuminator (Kaplan, Martin)V1448,206-217(1991)

Multispot scanning exposure system for excimer laser stepper (Yoshitake, Yasuhiro; Oshida, Yoshitada; Tanimoto, Tetsuzou; Tanaka, Minoru; Yoshida, Minoru)V1463,678-687(1991)

Narcissus in current generation FLIR systems (Ford, Eric H.; Hasenauer, David M.)VCR38,95-119(1991)

N-dimensional Hilbert scanning and its application to data compression (Perez, Arnulfo; Kamata, Seiichiro; Kawaguchi, Eiji)V1452,430-441(1991)

Novel high-resolution large-field scan-and-repeat projection lithography system (Jain, Kanti)V1463,666-677(1991)

Novel optical-fiber-based conical scan tracking device (Johann, Ulrich A.; Pribil, Klaus; Sontag, Heinz)V1522,243-252(1991)

Ocean color and temperature scanner for ADEOS (Tanii, Jun; Machida, Tsuneo; Ayada, Haruki; Katsuyama, Yoshihiko; Ishida, Juro; Iwasaki, Nobuo; Tange, Yoshio; Miyachi, Yuji; Sato, Ryota)V1490,200-206(1991)

On-line acquisition of CT and MRI studies from multiple scanners (Weinberg, Wolfram S.; Loloyan, Mansur; Chan, Kelby K.)V1446,430-435(1991)

Optical design and development of a small barcode scanning module (Wike, Charles K.; Lindacher, Joseph M.)V1398,119-126(1991)

Optical locator for horizon sensing (Fallon, James J.; Selby, Vaughn H.)V1495,268-279(1991)

Optical properties of barcode symbols for laser scanning (Quinn, Anna M.; Eastman, Jay M.)V1384,138-144(1991)

Optical scanning system for a CCD telecine for HDTV (Kurtz, Andrew F.; Kessler, David)V1448,191-205(1991)

Optics for vector scanning (Ehrmann, Jonathan S.)V1454,245-256(1991)

Overview of a high-performance polygonal scanner subsystem (Rynkowski, Gerald)V1454,102-110(1991)

Particle analysis in liquid flow by the registration of elastic light scattering in the condition of laser beam scanning (Dubrovsky, V.; Grinevich, A. E.; Ossin, A. B.)V1403,344-346(1991)

Programmable holographic scanning (Zhengmin, Li; Li, Min; Tang, Jinfa)V1401,66-73(1991)

Proposals for a computer-controlled orbital scanning camera for remote image aquisition (Nwagboso, Christopher O.)V1454,111-122(1991)

Radiometric calibration of an airborne multispectral scanner (Markham, Brian L.; Ahmad, Suraiya P.; Jackson, Ray D.; Moran, M. S.; Biggar, Stuart F.; Gellman, David I.; Slater, Philip N.)V1493,207-214(1991)

Radiometric calibration plan for the Clouds and the Earth's Radiant Energy System scanning instruments (Jarecke, Peter J.; Folkman, Mark A.; Darnton, Lane A.)V1493,244-254(1991)

Range data analysis using cross-stripe structured-light system (Kim, Whoi-Yul)V1385,216-228(1991)

Real-time optical scanning system for measurement of chest volume changes during anesthesia (Duffy, Neil D.; Drummond, Gordon D.; McGowan, Steve; Dessesard, Pascal)V1380,46-52(1991)

Relationship between fluctuation in mirror radius (within one polygon) and the jitter (Horikawa, Hiroshi; Sugisaki, Iwao; Tashiro, Masaru)V1454,46-59(1991)

Relationship between jitter and deformation of mirrors (Horikawa, Hiroshi; Miura, Masayuki; Uchida, Toshiya)V1454,20-32(1991)

Remote spectral identification of surface aggregates by thermal imaging techniques: progress report (Scholen, Douglas E.; Clerke, William H.; Burns, Gregory S.)V1492,358-369(1991)

Scanner analyzer target (Simonis, Roland)V1454,364-369(1991)

Scanning exciton microscopy and single-molecule resolution and detection (Kopelman, Raoul; Tan, Weihong; Lewis, Aaron; Lieberman, Klony)V1435,96-101(1991)

Scanning imaging absorption spectrometer for atmospheric chartography (Burrows, John P.; Chance, Kelly V.)V1490,146-154(1991)

Scanning infrared earth sensor for INSAT-II (Alex, T. K.; Kamalakar, J. A.)V1478,106-111(1991)

Scanning method of receiving high-effective reflective holograms on bichromate gelatin (Imedadze, Theodore S.; Kakichashvili, Shermazan D.)V1238,439-441(1991)

Scanning optical microscopy: a powerful tool in optical recording (Coombs, James H.; Holtslag, A. H.)V1499,6-20(1991)

Scanning strategies for target detection (Gertner, Izidor; Zeevi, Yehoshua Y.)V1470,148-166(1991)

ScaRaB Earth radiation budget scanning radiometer (Monge, J.L.; Kandel, Robert S.; Pakhomov, L. A.; Bauche, B.)V1490,84-93(1991)

Self-acting gas bearings for high-speed scanners (Preston, Ralph G.; Robinson, D. H.; Shepherd, Joseph)V1454,132-138(1991)

Simultaneous patient translation during CT scanning (Crawford, Carl R.; King, Kevin F.)V1443,203-213(1991)

Six degree-of-freedom magnetically suspended fine-steering mirror (Medbery, James D.; Germann, Lawrence M.)V1482,397-405(1991)

Software calibration of scan system distortions (Weisz, John R.)V1454,265-271(1991)

Spatial sampling errors for a satellite-borne scanning radiometer (Manalo, Natividad D.; Smith, G. L.)V1493,281-291(1991)

Standards for oscillatory scanners (Ludwiszewski, Alan P.)V1454,174-185(1991)

Stationary platen 2-axis scanner (Schermer, Mack J.)V1454,257-264(1991)

Study of active 3-D terrain mapping for helicopter landings (Velger, Mordekhai; Toker, Gregory)V1478,168-176(1991)

Super-compact dual-axis optical scanning unit applying a tortional spring resonator driven by a piezoelectric actuator (Goto, Hiroshi; Imanaka, Koichi)V1544,272-281(1991)

Temporal and spectral analysis of the synchronization of synchroscan streak cameras (Cunin, B.; Geist, P.; Heisel, Francine; Martz, A.; Miehe, Joseph A.)V1358,606-613(1991)

Thermal infrared imagery from the Geoscan Mk II scanner and its calibration: two case histories from Nevada—Ludwig Skarn (Yerington District) & Virginia City (Lyon, Ronald J.; Honey, Frank R.)V1492,339-350(1991)

Thin family: a new barcode concept (Allais, David C.)V1384,161-168(1991)

Three-dimensional lithography: laser modeling using photopolymers (Heller, Timothy B.)V1454,272-282(1991)

Tracking method of optical tape recorder using acousto-optic scanning (Narahara, Tatsuya; Kumai, Satoshi; Nakao, Takashi; Ozue, Tadashi)V1499,120-128(1991)

Trade-offs in rotary mirror scanner design (Colquhoun, Allan B.; Cowan, Donald W.; Shepherd, Joseph)V1454,12-19(1991)

Two-dimensional encoding of images using discrete reticles (Wellfare, Michael R.)V1478,33-40(1991)

Two-dimensional micropattern measurement using precision laser beam scanning (Fujita, Hiroo)V1332,456-467(1991)

Ultra-high scanning holographic deflector unit (Kramer, Charles J.; Szalowski, Rafal; Watkins, Mark)V1454,434-446(1991)

Underwater laser scanning system (Austin, Roswell W.; Duntley, Seibert Q.; Ensminger, Richard L.; Petzold, Theodore J.; Smith, Raymond C.)V1537,57-73(1991)

Update of scanner selection, performances, and multiaxis configurations (Montagu, Jean I.)V1454,160-173(1991)

Using digital-scanned aerial photography for wetlands delineation (Anderson, John E.; Roos, Maurits)V1492,252-262(1991)

Velocity profiling in linear and rotational systems (Crabtree, Daniel L.)V1482,458-472(1991)

Vision sensor system for the environment of disaster (Kawauchi, Yoshikazu; Kawata, Koichi)V1507,538-543(1991)

Width gauge for hot steel plates by laser-scanning rangefinder (Adachi, Yuzi; Matsui, Kenichi)V1527,225-233(1991)

Scattering—see also Raman effect; surfaces

Airborne lidar elastic scattering, fluorescent scattering, and differential absorption observations (Uthe, Edward E.)V1479,393-402(1991)

Analysis of absorption, scattering, and hemoglobin saturation using phase-modulation spectroscopy (Sevick, Eva M.; Weng, Jian; Maris, Michael B.; Chance, Britton)V1431,264-275(1991)

Analysis of multiple-multipole scattering by time-resolved spectroscopy and spectrometry (Frank, Klaus; Hoeper, J.; Zuendorf, J.; Tauschek, D.; Kessler, Manfred; Wiesner, J.; Wokaun, Alexander J.)V1431,2-11(1991)

Analysis of thin-film losses from guided wave attenuations with photothermal deflection technique (Liu, Xu; Tang, Jinfa; Pelletier, Emile P.)V1554A,558-569(1991)

Anisotropic polarizability and diffusion of protein in water solutions studied by laser light scattering (Petrova, G. P.; Petrusevich, Yu. M.; Borisov, B. A.)V1403,387-389(1991)

Anomalous scattering from optical surfaces with roughness and permittivity perturbations (Elson, John M.)V1530,196-207(1991)

Application of a dynamic photoelasticity technique for the study of two- and three-dimensional structures under seismic, shock, and explosive loads effect (Kostin, Ivan K.; Dvalishvili, V. V.; Ureneva, E. V.; Freishist, N. A.; Fjodorov, A. V.; Dmitriyenko, O. L.)V1554A,418-425(1991)

Application of an electron-beam scattering parameter extraction method for proximity correction in direct-write electron-beam lithography (Weiss, Rudolf M.; Sills, Robert M.)V1465,192-200(1991)

Approximation of the Compton scattered radiation (Burq, Catherine; Vibert, Patrick)V1346,276-285(1991)

Automatic recognition of multidimensional objects buried in layered elastic background media (Ayme-Bellegarda, Eveline J.; Habashy, Tarek M.; Bellegarda, Jerome R.)V1471,18-29(1991)

Backscattering image resolution as a function of particle density (Rochon, Paul L.; Bissonnette, Daniel)V1530,50-57(1991)

Beam-coupling by stimulated Brillouin scattering (Falk, Joel; Chu, Raijun; Kanefsky, Morton; Hua, Xuelei)V1409,83-86(1991)

Beryllium scatter analysis program (Behlau, Jerry L.; Granger, Edward M.; Hannon, John J.; Baumler, Mark; Reilly, James F.)V1530,218-230(1991)

Bidirectional reflectance distribution function raster scan technique for curved samples (McIntosh, Malcolm B.; McNeely, Joseph R.)V1530,263-268(1991)

Bragg crystal polarimeter for the Spectrum-X-Gamma misson (Holley, Jeff; Silver, Eric H.; Ziock, Klaus P.; Novick, Robert; Kaaret, Philip E.; Weisskopf, Martin C.; Elsner, Ronald F.; Beeman, Jeff)V1343,500-511(1991)

Brownian dynamics simulation of polarized light scattering from wormlike chains (Allison, Stuart A.)V1430,50-64(1991)

Certain results of the interaction of laser light with various media (Drobnik, Antoni)V1391,211-214(1991)

Characterization of a vesicle distribution in equilibrium with larger aggregates by accurate static and dynamic laser light scattering measurements (Cantu, Laura; Corti, Mario; Lago, Paolo; Musolino, Mario)V1430,144-159(1991)

Characterization of hot-isostatic-pressed optical-quality beryllium (Behlau, Jerry L.; Baumler, Mark)V1530,208-217(1991)

Characterization of motion of probes and networks in gels by laser light scattering under sample rotation (Kobayasi, Syoyu)V1403,296-305(1991)

Circular intensity differential scattering measurements in the soft x-ray region of the spectrum (~16 EV to 500 EV) (Maestre, Marcos F.; Bustamante, Carlos; Snyder, Patricia A.; Rowe, Ednor M.; Hansen, Roger W.)V1548,179-187(1991)

Circular polarization effects in the light scattering from single and suspensions of dinoflagellates (Shapiro, Daniel B.; Hunt, Arlon J.; Quinby-Hunt, Mary S.; Hull, Patricia G.)V1537,30-41(1991)

Coherence in single and multiple scattering of light from randomly rough surfaces (Gu, Zu-Han; Maradudin, Alexei A.; Mendez, Eugenio R.)V1530,60-70(1991)

Comparison of low-scatter-mirror PSD derived from multiple-wavelength BRDFs and WYKO profilometer data (Wong, Wallace K.; Wang, Dexter; Benoit, Robert T.; Barthol, Peter)V1530,86-103(1991)

Control of thermally induced porosity for the fabrication of beryllium optics (Moreen, Harry A.)V1485,54-61(1991)

Coupled-dipole approximation: predicting scattering by nonspherical marine organisms (Hull, Patricia G.; Hunt, Arlon J.; Quinby-Hunt, Mary S.; Shapiro, Daniel B.)V1537,21-29(1991)

Critical frequencies for large-scale resonance signatures from elastic bodies (Werby, Michael F.; Gaunaurd, Guillermo C.)V1471,2-17(1991)

Cryoscatter measurements of beryllium (Lippey, Barret; Krone-Schmidt, Wilfried)V1530,150-161(1991)

Cylindrical proportional counter for x-ray polarimetry (Costa, Enrico; Piro, Luigi; Rubini, Alda; Soffitta, Paolo; Massaro, Enrico; Matt, Giorgio; Medici, Gastone; Manzo, Giuseppe; Re, Stefano)V1343,469-476(1991)

Data analysis methods for near-infrared spectroscopy of tissue: problems in determining the relative cytochrome aa3 concentration (Cope, Mark; van der Zee, Pieter; Essenpreis, Matthias; Arridge, Simon R.; Delpy, David T.)V1431,251-262(1991)

Description and calibration of a fully automated infrared scatterometer (Mainguy, Stephan; Olivier, Michel; Josse, Michel A.; Guidon, Michel)V1530,269-282(1991)

Design considerations for multipurpose bidirectional reflectometers (Neu, John T.; Bressler, Martin)V1530,244-254(1991)

Detection and localization of absorbers in scattering media using frequency-domain principles (Berndt, Klaus W.; Lakowicz, Joseph R.)V1431,149-160(1991)

Detection of the object velocity using the time-varying scattered speckles (Okamoto, Takashi; Asakura, Toshimitsu)V1399,192-199(1991)

Determination of thin-film roughness and volume structure parameters from light-scattering investigations (Duparre, Angela; Kassam, Samer)V1530,283-286(1991)

Development of the marine-aggregated-particle profiling and enumerating rover (Costello, David K.; Carder, Kendall L.; Steward, Robert G.)V1537,161-172(1991)

Diagnostics of functional state of blood by registration of light scattering intensity variations due to the reversible aggregation of red blood cells (Tukhvatulin, R. T.; Vaulin, P. P.)V1403,390-391(1991)

Diffraction effects in stimulated Raman scattering (Scalora, Michael; Haus, Joseph W.)V1497,153-164(1991)

Diffusing wave spectroscopy studies of gelling systems (Horne, David S.)V1430,166-180(1991)

Diffusion of spherical probes in aqueous systems containing the semiflexible polymer hydroxypropylcellulose (Mustafa, Mazidah B.; Russo, Paul S.)V1430,132-141(1991)

Dimension and time effects caused by nonlocal scattering of laser radiation from a rough metal surface (Dolgina, A. N.; Kovalev, A. A.; Kondratenko, P. S.)V1440,342-353(1991)

Discrete-angle radiative transfer in a multifractal medium (Davis, Anthony B.; Lovejoy, Shaun; Schertzer, Daniel)V1558,37-59(1991)

Droplet-size effects in light scattering from polymer-dispersed liquid-crystal films (Montgomery, G. P.; West, John L.; Tamura-Lis, Winifred)V1455,45-53(1991)

Dynamical structure factor of a solution of charged rod-like polymers in the isotropic phase (Maeda, T.; Doi, Masao)V1403,268-277(1991)

Dynamic light scattering from a side-chain liquid crystalline polymer in a nematic solvent (Devanand, Krisha)V1430,160-164(1991)

Dynamic light scattering from strongly interacting multicomponent systems: salt-free polyelectrolyte solutions (Sedlak, Marian; Amis, Eric J.; Konak, Cestmir; Stepanek, Petr)V1430,191-202(1991)

Dynamic light scattering studies of resorcinol formaldehyde gels as precursors of organic aerogels (Cotts, Patricia M.; Pekala, Rick)V1430,181-190(1991)

Dynamics of wormlike chains: theory and computer simulations (Aragon, Sergio R.; Luo, Rolland)V1430,65-84(1991)

Effective properties of electromagnetic wave propagation in some composite media (Artola, Michel; Cessenat, Michel)V1558,14-21(1991)

Effective refractive indices of three-phase optical coatings (Ma, Yushieh; Varadan, Vijay K.; Varadan, Vasundara V.)V1558,138-142(1991)

Effective surface PSD for bare hot-isostatic-pressed beryllium mirrors (Vernold, Cynthia L.; Harvey, James E.)V1530,144-149(1991)

Effect of counterion distribution on the electrostatic component of the persistence length of flexible linear polyions (Klearman, Debbie; Schmitz, Kenneth S.)V1430,236-255(1991)

Effects of atmospheric conditions on the performance of free-space infrared communications (Grotzinger, Timothy L.)V1417,484-495(1991)

Elasticity of biomembranes studied by dynamic light scattering (Fujime, Satoru; Miyamoto, Shigeaki)V1403,306-315(1991)

Electromagnetic field calculations for a tightly focused laser beam incident upon a microdroplet: applications to nonlinear optics (Barton, John P.; Alexander, Dennis R.)V1497,64-77(1991)

Electromagnetic scattering from a finite cylinder with complex permittivity (Murphy, Robert A.; Christodoulou, Christos G.; Phillips, Ronald L.)V1558,295-305(1991)

Enhanced backscattering of s- and p-polarized light from particles above a substrate (Greffet, Jean-Jacques; Sentenac, Anne)V1558,288-294(1991)

Errors inherent in the restoration of imagery acquired through remotely sensed refractive interfaces and scattering media (Schmalz, Mark S.)V1479,183-198(1991)

Estimation of focused laser beam SBS (stimulated Brillouin scatter) threshold dependence on beam shape and phase aberrations (Clendening, Charles W.)V1415,72-78(1991)

Evaluation of parameters in stimulated backward Brillouin scattering (de Oliveira, C. A.; Jen, Cheng K.)V1590,101-106(1991)

Experimental and theoretical studies of second-phase scattering in IR transmitting ZnS-based windows (Chen, William W.; Dunn, Bruce S.; Zhang, Jimin)V1535,199-208(1991)

Experimental investigations of the autocorrelation function for inhomogeneous scatterers (Singh, Brahm P.; Chopra, S.)V1558,317-321(1991)

Experimental study of the laser retroreflection of various surfaces (Liu, Wen-Qing; Jiang, Rong-Xi; Wang, Ya-Ping; Xia, Yu-Xing)V1530,240-243(1991)

Experiment for testing the closure property in ocean optics (Maffione, Robert A.; Honey, Richard C.; Brown, Robert A.)V1537,115-126(1991)

Extension of model of beam spreading in seawater to include dependence on the scattering phase function (Schippnick, Paul F.)V1537,185-193(1991)

Fidelity of Brillouin amplification with Gaussian input beams (Jones, David C.; Ridley, Kevin D.; Cook, Gary; Scott, Andrew M.)V1500,46-52(1991)

Frequency-derived distributed optical fiber sensing: backscatter analysis (Rogers, Alan J.; Handerek, Vincent A.; Parvaneh, Farhad)V1511,190-200(1991)

Frequency spectra of erythrocyte membrane flickering measured by laser light scattering (Bek, A. M.; Kononenko, Vadim L.)V1403,384-386(1991)

Frequency spectrum analysis and assessment of optical surface flaws (Gao, Wenliang; Zhang, Xiao; Yang, Guoguang)V1530,118-128(1991)

Frequency up-conversion of a discharge-pumped molecular fluorine laser by stimulated Raman scattering in H2 (Kakehata, Masaru; Hashimoto, Etsu; Kannari, Fumihiko; Obara, Minoru)V1397,185-189(1991)

Grating line shape characterization using scatterometry (Bishop, Kenneth P.; Gaspar, Susan M.; Milner, Lisa-Michelle; Naqvi, H. S.; McNeil, John R.)V1545,64-73(1991)

Hard x-ray polarimeter utilizing Compton scattering (Sakurai, Hirohisa; Noma, M.; Niizeki, H.)V1343,512-518(1991)

Helium neon laser optics: scattered light measurements and process control (Perilloux, Bruce E.)V1530,255-262(1991)

Hierarchical proximity effect correction for e-beam direct writing of 64-Mbit DRAM (Misaka, Akio; Hashimoto, Kazuhiko; Kawamoto, M.; Yamashita, H.; Matsuo, Takahiro; Sakashita, Toshihiko; Harafuji, Kenji; Nomura, Noboru)V1465,174-184(1991)

High-fidelity fast-response phase conjugators for visible and ultraviolet applications (O'Key, Michael A.; Osborne, Michael R.)V1500,53-64(1991)

High-resolution submicron particle sizing by dynamic light scattering (Nicoli, David F.)V1430,19-36(1991)

High-spatial-resolution and high-sensitivity interferometric optical-time-domain reflectometer (Kobayashi, Masaru; Noda, Juichi; Takada, Kazumasa; Taylor, Henry F.)V1474,278-284(1991)

High-temperature Raman scattering behavior in diamond (Herchen, Harald; Cappelli, Mark A.)V1534,158-168(1991)

Identification of modes in dynamic scattering on ternary polymer mixtures (Akcasu, A. Z.)V1430,142-143(1991)

IR laser-light backscattering by an arbitrarily shaped dielectric object with rough surface (Wu, Zhensen; Cheng, Denghui)V1558,251-257(1991)

Imaging of subsurface regions of random media by remote sensing (Barbour, Randall L.; Graber, Harry L.; Aronson, Raphael; Lubowsky, Jack)V1431,52-62(1991); V1431,192-203(1991)

Impact on the medium MTF by model estimation of b (Estep, Leland; Arnone, Robert A.)V1537,89-96(1991)

Improved formalism for rough-surface scattering of acoustic and electromagnetic waves (Milder, D. M.)V1558,213-221(1991)

Indirect illumination to reduce veiling luminance in seawater (Wells, Willard H.)V1537,2-9(1991)

Infrared BRDF measurements of space shuttle tiles (Young, Raymond P.; Wood, Bobby E.; Stewart, P. L.)V1530,335-342(1991)

Infrared window damage measured by reflective scatter (Bernt, Marvin L.; Stover, John C.)V1530,42-49(1991)

In-situ surface x-ray scattering of metal monolayers adsorbed at solid-liquid interfaces (Toney, Michael F.; Gordon, Joseph G.; Melroy, Owen R.)V1550,140-150(1991)

Instrument for underwater measurement of optical backscatter (Maffione, Robert A.; Dana, David R.; Honey, Richard C.)V1537,173-184(1991)

Intrapulse stimulated Raman scattering and ultrashort solitons in optical fibers (Headley, Clifford; Agrawal, Govind P.; Reardon, A. C.)V1497,197-201(1991)

Inverse scattering problems in the acoustic resonance region of an underwater target (Gaunaurd, Guillermo C.)V1471,30-41(1991)

Investigation of vesicle-capsular plague antigen complex formation by elastic laser radiation scattering (Guseva, N. P.; Maximova, I. L.; Romanov, S. V.; Shubochkin, L. P.; Tatarintsev, S. V.)V1403,332-334(1991)

Investigations of the phenomenon of light scattering on particles of atmospheric aerosols (Drobnik, Antoni)V1391,204-210(1991)

Laser light scattering studies of biological gels (Burne, P. M.; Sellen, D. B.)V1403,288-295(1991)

Lidar backscatter calculations for solid-sphere and layered-sphere aerosols (Youmans, Douglas G.)V1416,151-162(1991)

Light scattered by coated paper (Marx, Egon; Song, J. F.; Vorburger, Theodore V.; Lettieri, Thomas R.)V1332,826-834(1991)

Light scattering from binary optics (Ricks, Douglas W.; Ajmera, Ramesh)V1555,89-100(1991)

Light scattering from gold-coated ground glass and chemically etched surfaces (Ruiz-Cortes, Victor; Mendez, Eugenio R.; Gu, Zu-Han; Maradudin, Alexei A.)V1558,222-232(1991)

Light scattering holographic optical elements formation in photopolymerizable layers (Boiko, Yuri B.; Granchak, Vasilij M.)V1507,544-548(1991)

Light scattering matrices of a densely packed binary system of hard spheres (Maksimova, I. L.; Shubochkin, L. P.)V1403,749-751(1991)

Light scattering properties of new materials for glazing applications (Bergkvist, Mikael; Roos, Arne)V1530,352-362(1991)

Light scatter variations with respect to wafer orientation in GaAs (Brown, Jeff L.)V1530,299-305(1991)

Mapping of imbedded contaminants in transparent material by optical scatter (Rudberg, Donald A.; Stover, John C.; McGary, Douglas E.)V1530,232-239(1991)

Material characterization of beryllium mirrors exhibiting anomalous scatter (Egert, Charles M.)V1530,162-170(1991)

Measurement of color and scattering phenomena of translucent materials (Sjollema, J. I.; den Exter, Ir. T.; Zijp, Jaap R.; Ten Bosch, Jaap J.)V1500,177-188(1991)

Measurement of light scattering from cells using an inverted infrared optical trap (Wright, William H.; Sonek, Gregory J.; Numajiri, Yasuyuki; Berns, Michael W.)V1427,279-287(1991)

Measurement of the sizes of the semiconductor crystallites in colored glasses through neutron scattering (Banfi, Giampiero P.; Degiorgio, Vittorio; Rennie, A. R.; Righini, Giancarlo C.)V1361,874-880(1991)

Measurements of lightness: dependence on the position of a white in the field of view (McCann, John J.; Savoy, Robert L.)V1453,402-411(1991)

Measuring photon pathlengths by quasielastic light scattering in a multiply scattering medium (Nossal, Ralph J.; Schmitt, Joseph M.)V1430,37-47(1991)

Minimum detectable changes in Rayleigh backscatter from distributed fiber sensors (Garside, Brian K.; Park, R. E.)V1588,150-158(1991)

Modeling the distribution of optical radiation in diffusely reflecting materials (Birth, Gerald S.)V1379,81-88(1991)

Monte Carlo and diffusion calculations of photon migration in noninfinite highly scattering media (Haselgrove, John C.; Leigh, John S.; Yee, Conway; Wang, Nai-Guang; Maris, Michael B.; Chance, Britton)V1431,30-41(1991)

Multiple quasi-elastic circular-polarized light scattering by brownian moving aspherical particles (Korolevich, A. N.; Khairullina, A. Y.)V1403,364-371(1991)

Multiple scattered electron-beam effect in electron-beam lithography (Saitou, Norio; Iwasaki, Teruo; Murai, Fumio)V1465,185-191(1991)

Multiple-scattering effects on pulse propagation through burning particles (Ma, Yushieh; Varadan, Vijay K.; Varadan, Vasundara V.)V1487,220-225(1991)

Multiple scattering of laser light in dense aerosol (Parma, Ludvik; Malicka, Marianna)V1391,190-198(1991)

Multiplex and multichannel detection of near-infrared Raman scattering (Chase, Bruce D.)V1439,47-57(1991)

Multistage XeCl excimer system "Cactus" and some investigations of stimulated scattering in liquids (Karpov, V. I.; Korobkin, V. V.; Naboichenko, A. K.; Dolgolenko, D. A.)V1503,492-502(1991)

New experimental methods and theory of Raman optical activity (Nafie, Laurence A.; Che, Diping; Yu, Gu-Sheng; Freedman, Teresa B.)V1432,37-49(1991)

New low-density, high-porosity lithium hydride-beryllium hydride foam: properties and applications to x-ray astronomy (Maienschein, Jon L.; Barry, Patrick E.; McMurphy, Frederick E.; Bowers, John S.)V1343,477-484(1991)

New optical technique for particle sizing and velocimetry (Xie, Gong-Wie; Scott, Peter D.; Shaw, David T.; Zhang, Yi-Mo)V1500,310-321(1991)

New, simple method of extracting temperature of liquid water from Raman scattering (Liu, Zhi-Shen; Ma, Jun; Zhang, Jin-Long; Chen, Wen-Zhong)V1558,306-316(1991)

NIR/CCD Raman spectroscopy: second battle of a revolution (McCreery, Richard L.)V1439,25-36(1991)

Nonlinear optical amplification and oscillation in spherical microdroplets (Kurizki, Gershon; Goldner, E.)V1497,48-62(1991)

Nonlinear optical effects under laser pulse interaction with tissues (Altshuler, Grigori B.; Belikov, Andrey V.; Erofeev, Andrey V.)V1427,141-150(1991)

Nonlinear optical processes in droplets with single-mode laser excitation (Chang, Richard K.; Chen, Gang; Hill, Steven C.; Barber, Peter W.)V1497,2-13(1991)

Nonrejected earth radiance performance of the visible ultraviolet experiment sensor (Betts, Timothy C.; Dowling, Jerome M.; Friedman, Richard M.)V1479,120-126(1991)

Normally incident plane waves on a chiral slab with linear property variations (Lakhtakia, Akhlesh; Varadan, Vijay K.; Varadan, Vasundara V.)V1558,120-126(1991)

Optical correlation studies of biological objects (Angelsky, Oleg V.; Maksimyak, Peter P.)V1403,667-673(1991)

Optical methods for laser beams control (Mak, Arthur A.; Soms, Leonid N.)V1415,110-119(1991)

Optical scatter: an overview (Stover, John C.)V1530,2-6(1991)

Optical Scatter: Applications, Measurement, and Theory (Stover, John C., ed.)V1530(1991)

Optical velocity sensor for air data applications (Smart, Anthony E.)V1480,62-71(1991)

Optimal estimation of finish parameters (Church, Eugene L.; Takacs, Peter Z.)V1530,71-85(1991)

Optimal surface characteristics for instruments for use in laser neurosurgery (Heiferman, Kenneth S.; Cramer, K. E.; Walsh, Joseph T.)V1428,128-134(1991)

Overview of stimulated Brillouin scattering in microdroplets (Cantrell, Cyrus D.)V1497,28-47(1991)

Particle analysis in liquid flow by the registration of elastic light scattering in the condition of laser beam scanning (Dubrovsky, V.; Grinevich, A. E.; Ossin, A. B.)V1403,344-346(1991)

Performance of NCAP projection displays (Jones, Philip J.; Tomita, Akira; Wartenberg, Mark)V1456,6-14(1991)

Photon correlation spectroscopy and electrophoretic light scattering using optical fibers (Macfadyen, Allan J.; Jennings, B. R.)V1367,319-328(1991)

Photon Correlation Spectroscopy: Multicomponent Systems (Schmitz, Kenneth S., ed.)V1430(1991)

Photon correlation spectroscopy: technique and instrumentation (Thomas, John C.)V1430,2-18(1991)

Plasma interaction with powerful laser beams (Giulietti, Antonio; Giulietti, Danilo; Willi, Oswald)V1502,270-283(1991)

Polarization dependence of light scattered from rough surfaces with steep slopes (O'Donnell, Kevin A.; Knotts, Micheal E.)V1558,362-367(1991)

Possibility of associative reconstruction of reflection hologram in phase-conjugate interferometer (Lazaruk, A. M.; Rubanov, Alexander S.; Serebryakova, L. M.)V1621,114-124(1991)

Possibility of resolution of internal macromolecular relaxation by dynamic light scattering (Timchenko, A. A.; Griko, N. B.; Serdyuk, I. N.)V1403,340-343(1991)

Predicted performance of the lithium scattering and graphite crystal polarimeter for the Spectrum-X-Gamma mission (Weisskopf, Martin C.; Elsner, Ronald F.; Novick, Robert; Kaaret, Philip E.; Silver, Eric H.)V1343,457-468(1991)

Predicting diode laser performance (Lim, G.; Park, Youngsoh; Zmudzinski, C. A.; Zory, Peter S.; Miller, L. M.; Cockerill, T. M.; Coleman, James J.; Hong, C. S.; Figueroa, Luis)V1418,123-131(1991)

Probe diffusion in polymer solutions (Phillies, George D.)V1430,118-131(1991)

Pulse compression of KrF laser radiation by stimulated scattering (Alimpiev, Sergei S.; Bukreev, Viacheslav S.; Kusakin, Vladimir I.; Likhansky, Sergey V.; Obidin, Alexey Z.; Vartapetov, Serge K.; Veselovsky, Igor A.)V1503,154-158(1991)

Pulse propagation in random media (Ishimaru, Akira)V1558,127-129(1991)

Quasielastic and electrophoretic light scattering studies of polyelectrolyte-micelle complexes (Rigsbee, Daniel R.; Dubin, Paul L.)V1430,203-215(1991)

Quasi-elastic light scattering spectroscopy of single biological cells under a microscope (Tanaka, Toyoichi; Nishio, Izumi; Peetermans, Joyce; Gorti, Sridhar)V1403,280-287(1991)

Raman lidar for measuring backscattering in the China Sea (Liu, Zhi-Shen; Zhang, Jin-Long; Chen, Wen-Zhong; Huang, Xiao-Sheng; Ma, Jun)V1558,379-383(1991)

Raman scattering characterization of direct gap Si/Ge superlattices (White, Julian D.; Gell, Michael A.; Fasol, Gerhard; Gibbings, C. J.; Tuppen, C. G.)V1361,293-301(1991)

Raman scattering determination of nonpersistent optical control of electron density in a heterojunction (Richards, David R.; Fasol, Gerhard; Ploog, Klaus H.)V1361,246-254(1991)

Remote sensing of precipitation structures using combined microwave radar and radiometric techniques (Vivekanandan, J.; Turk, F. J.; Bringi, Viswanathan N.)V1558,324-338(1991)

Resonances and internal electric energy density in droplets (Hill, Steven C.; Barber, Peter W.; Chowdhury, Dipakbin Q.; Khaled, Elsayed-Esam M.; Mazumder, Mohiuddin)V1497,16-27(1991)

Role of counterion size and distribution on the electrostatic component to the persistence length of wormlike polyions (Schmitz, Kenneth S.)V1430,216-235(1991)

Role of the surface height correlation function in the enhanced backscattering of light from random metallic surfaces (Maradudin, Alexei A.; Michel, T.)V1558,233-250(1991)

Rough-interface scattering without plane waves (Berman, David H.)V1558,191-201(1991)

Scatter and contamination of a low-scatter mirror (McNeely, Joseph R.; McIntosh, Malcolm B.; Akerman, M. A.)V1530,288-298(1991)

Scatter and roughness measurements on optical surfaces exposed to space (Schmitt, Dirk-Roger; Swoboda, Helmut; Rosteck, Helmut)V1530,104-110(1991)

Scattering and thermal emission from spatially inhomogeneous atmospheric rain and cloud (Jin, Ya-Qiu)V1487,324-332(1991)

Scattering contribution to the error budget of an emissive IR calibration sphere (Chalupa, John; Cobb, W. K.; Murdock, Tom L.)V1530,343-351(1991)

Scattering from multilayer coatings: a linear systems model (Harvey, James E.; Lewotsky, Kristin L.)V1530,35-41(1991)

Scattering from objects near a rough surface (Rino, Charles L.; Ngo, Hoc D.)V1558,339-350(1991)

Scattering from slightly rough crystal surfaces (Church, Eugene L.)V1530,171-184(1991)

Scattering from very rough metallic and dielectric surfaces and enhanced backscattering (Ishimaru, Akira; Chen, Jei S.)V1558,182-190(1991)

Scattering in paper coatings (Hyvarinen, Timo S.; Sumen, Juha)V1530,325-334(1991)

Scattering liquid crystal in optical attenuator applications (Karppinen, Arto; Kopola, Harri K.; Myllyla, Risto A.)V1455,179-189(1991)

Scattering measurements of optical coatings in high-power lasers (Chen, Yi-Sheng)V1332,115-120(1991)

Scattering of waves from dense discrete random media: theory and applications in remote sensing (Tsang, Leung; Ding, Kung-Hau; Kong, Jin A.; Winebrenner, Dale P.)V1558,260-268(1991)

Scattering parameters of volume phase hologram: Bragg's approach (Gubanov, V. A.; Kiselyov, Boris S.)V1621,102-113(1991)

Scattering properties of ZrF4-based glasses prepared by the gas film levitation technique (Lopez, Adolphe; Baniel, P.; Gall, Pascal; Granier, Jean E.)V1590,191-202(1991)

Scatter properties of gratings at ultraviolet and visible wavelengths (Hoose, John F.; Olson, Jeffrey)V1545,160-166(1991)

Selected Papers on Multiple Scattering in Plane Parallel Atmospheres and Oceans: Methods (Kattawar, George W., ed.)VMS42(1991)

SERS used to study the effect of Langmuir-Blodgett spacer layers on metal surface (Yu, Bing Kun; Li, Yu; Wang, Yingting)V1530,363-369(1991)

Shape reconstruction from far-field patterns in a stratified ocean environment (Xu, Yongzhi)V1471,78-86(1991)

Short-pulse electromagnetics for sensing applications (Felsen, Leopold B.; Vecchi, G.; Carin, L.; Bertoni, H. L.)V1471,154-162(1991)

Simulation of vertical profiles of extinction and backscatter coefficients in very low stratus clouds and subcloud regions (Kilmer, Neal H.; Rachele, Henry)V1487,109-122(1991)

Sinusoidal surfaces as standards for BRDF instruments (Marx, Egon; Lettieri, Thomas R.; Vorburger, Theodore V.; McIntosh, Malcolm B.)V1530,15-21(1991)

Slow-motion acquisition of laser beam profiles after propagation through gun blast (Kay, Armin V.)V1486,8-16(1991)

Small-angle scattering measurement (Gu, Zu-Han; Dummer, Richard S.)V1558,368-378(1991)

Soft x-ray resonant magnetic scattering study of thin films and multilayers (Kao, Chi-Chang; Johnson, Erik D.; Hastings, Jerome B.; Siddons, D. P.; Vettier, C.)V1548,149-157(1991)

Solution for anomalous scattering of bare HIP Be and CVD SiC mirrors (Vernold, Cynthia L.)V1530,130-143(1991)

Some computational results for rough-surface scattering (DeSanto, John A.; Wombell, Richard J.)V1558,202-212(1991)

Space-based visible all-reflective stray light telescope (Wang, Dexter; Gardner, Leo R.; Wong, Wallace K.; Hadfield, Peter)V1479,57-70(1991)

Sparse random distribution of noninteracting small chiral spheres in a chiral host medium (Lakhtakia, Akhlesh; Varadan, Vijay K.; Varadan, Vasundara V.)V1558,22-27(1991)

Spatial randomization of the scattered optical radiation (Angelsky, Oleg V.; Magun, I. I.; Maksimyak, Peter P.; Perun, T. O.)V1402,231-235(1991)

Statistical properties of intensity fluctuations produced by rough surfaces under the speckle pattern illumination (Yoshimura, Takeaki; Fujiwara, Kazuo; Miyazaki, Eiichi)V1332,835-842(1991)

Statistics of spontaneously generated Raman solitons (Englund, John C.; Bowden, Charles M.)V1497,218-221(1991)

Stimulated Brillouin scatter SIX-code (Litvak, Marvin M.; Wagner, Richard J.)V1415,62-71(1991)

Stochastic simulations of light-scattering noise in two-wave mixing: application to artificial Kerr media (McGraw, Robert L.)V1409,135-147(1991)

Stokes vectors, Mueller matrices, and polarized scattered light: experimental applications to optical surfaces and all other scatterers (Bickel, William S.; Videen, Gorden W.)V1530,7-14(1991)

Stray-light implications of scratch/dig specifications (Lewis, Isabella T.; Ledebuhr, Arno G.; Bernt, Marvin L.)V1530,22-34(1991)

Stray-light reduction in a WFOV star tracker lens (Lewis, Isabella T.; Ledebuhr, Arno G.; Axelrod, Timothy S.; Ruddell, Scott A.)V1530,306-324(1991)

Study and characterization of three-dimensional angular distributions of acoustic scattering from spheroidal targets (George, Jacob; Werby, Michael F.)V1471,66-77(1991)

Study of anomalous scatter characteristics (Stover, John C.; Bernt, Marvin L.; Henning, Timothy D.)V1530,185-195(1991)

Surface defect detection and classification with light scattering (Gebhardt, Michael; Truckenbrodt, Horst; Harnisch, Bernd)V1500,135-143(1991)

Surface roughness effects on light scattered by submicron particles on surfaces (Bawolek, Edward J.; Hirleman, Edwin D.)V1464,574-583(1991)

Surface roughness measurements of spherical components (Chen, Yi-Sheng; Wang, Wen-Gui)V1530,111-117(1991)

Symmetrization analysis of lattice-vibrational modes and study of Raman-IR spectra for B-BaB2O4 (Hong, Shuili)V1437,194-197(1991)

Target identification by means of impulse radar (Abrahamsson, S.; Brusmark, B.; Gaunaurd, Guillermo C.; Strifors, Hans C.)V1471,130-141(1991)

Target shape and material composition from resonance echoes of submerged elongated elastic targets (Dean, Cleon E.; Werby, Michael F.)V1471,54-65(1991)

Theoretical investigation of near-edge phenomena in magnetic systems (Carra, Paolo)V1548,35-44(1991)

Theoretical modeling of chiral composites (Apparao, R. T.; Varadan, Vasundara V.; Varadan, Vijay K.)V1558,2-13(1991)

Theory and experiment as tools for assessing surface finish in the UV-visible wavelength region (Ingers, Joakim P.; Thibaudeau, Laurent)V1400,178-185(1991)

Theory and simulation of Raman scattering in intense short-pulse laser-plasma interactions (Wilks, S. C.; Kruer, William L.; Langdon, A. B.; Amendt, Peter; Eder, David C.; Keane, Christopher J.)V1413,131-137(1991)

Thomson scattering diagnostics of discharge plasmas in an excimer laser (Yamakoshi, H.; Kato, M.; Kubo, Y.; Enokizono, H.; Uchino, K.; Muraoka, K.; Takahashi, A.; Maeda, Mitsuo)V1397,119-122(1991)

Time-resolved techniques: an overview (Larson, Bennett C.; Tischler, J. Z.)V1345,90-100(1991)

Time-resolved x-ray scattering studies using CCD detectors (Clarke, Roy; Dos Passos, Waldemar; Lowe, Walter P.; Rodricks, Brian G.; Brizard, Christine M.)V1345,101-114(1991)

Transitions in model membranes (Earnshaw, J. C.; Winch, P. J.)V1403,316-325(1991)

Tumor detection using time-resolved light transillumination (Berg, Roger; Andersson-Engels, Stefan; Jarlman, Olof; Svanberg, Sune)V1525,59-67(1991)

Use of polarization to separate on-axis scattered and unscattered light in red blood cells (Sardar, Dhiraj K.; Nemati, Babak; Barrera, Frederick J.)V1427,374-380(1991)

Use of scattered-light photoelasticity at crack tips (Ravi-Chandar, K.)V1554A,228-238(1991)

Using photon correlation spectroscopy to study polymer coil internal dynamic behavior (Selser, James C.; Ellis, Albert R.; Schaller, J. K.; McKiernan, M. L.; Devanand, Krisha)V1430,85-88(1991)

Variation of the point spread function in the Sargasso Sea (Voss, Kenneth J.)V1537,97-103(1991)

Visible extinction measurements in rain and snow using a forward-scatter meter (Hutt, Daniel L.; Oman, James)V1487,312-323(1991)

Wave Propagation and Scattering in Varied Media II (Varadan, Vijay K., ed.)V1558(1991)

Zero-angle scattering of light in oriented organic liquids: classical and quantum states for both linear and nonlinear scattering (Arakelian, Sergei M.; Chilingarian, Yu. S.)V1403,326-331(1991)

Second-harmonic generation—see also nonlinear optics

Approaches to the construction of intrinsically acentric chromophoric NLO materials: chemical elaboration and resultant properties of self-assembled multilayer structures (Allan, D. S.; Kubota, F.; Marks, Tobin J.; Zhang, T. J.; Lin, W. P.; Wong, George K.)V1560,362-369(1991)

Defect centers and photoinduced self-organization in Ge-doped silica core fiber (Tsai, T. E.; Griscom, David L.)V1516,14-28(1991)

Design of chromophores for nonlinear optical applications (Burland, Donald M.; Rice, J. E.; Downing, J.; Michl, J.)V1560,111-119(1991)

Electrical and nonlinear optical properties of zirconium phosphonate multilayer assemblies (Katz, Howard E.; Schilling, M. L.; Ungashe, S.; Putvinski, T. M.; Scheller, G. E.; Chidsey, C. E.; Wilson, William L.)V1560,370-376(1991)

Ferroelectric microdomain reversal on Y-cut LiNbO3 surfaces (Seibert, Holger; Sohler, Wolfgang)V1362,370-376(1991)

Frequency doubling in optical fibers: a complex puzzle (Margulis, Walter; Carvalho, Isabel C.; Lesche, Bernhard)V1516,60-66(1991)

Highest observed second harmonic intensity from a multilayered Langmuir-Blodgett film structure (Ashwell, Geoffrey J.; Dawnay, Emma J.; Kuczynski, Andrzej P.; Martin, Philip J.)V1361,589-598(1991)

High-Tg nonlinear optical polymer: poly(N-MNA acrylamide) (Herman, Warren N.; Rosen, Warren A.; Sperling, L. H.; Murphy, C. J.; Jain, H.)V1560,206-213(1991)

In-plane poling of doped-polymer films: creation of two asymmetric directions for second-order optical nonlinearity (Berkovic, Garry; Yitzchaik, Shlomo; Krongauz, Valeri)V1560,238-242(1991)

Integrated optical devices for high-data-rate serial-to-parallel conversion (Verber, Carl M.; Kenan, Richard P.; Tan, Ronson K.; Bao, Y.)V1374,68-73(1991)

Intl Workshop on Photoinduced Self-Organization Effects in Optical Fiber (Ouellette, Francois, ed.)V1516(1991)

Langmuir-Blodgett films for second-order nonlinear optics (Penner, Thomas L.; Armstrong, Nancy J.; Willand, Craig S.; Schildkraut, Jay S.; Robello, Douglas R.)V1560,377-386(1991)

Laser boresighting by second-harmonic generation (Adel, Michael E.; Buckwald, Bob A.; Cabib, Dario)V1442,68-80(1991)

Laser-induced diffusion and second-harmonic generation in glasses (Aksenov, V. P.)V1440,377-383(1991)

Measurement of nonlinear optical coefficients by phase-matched harmonic generation (Eckardt, Robert C.; Byer, Robert L.)V1561,119-127(1991)

Model of second-harmonic generation in glass fibers based on multiphoton ionization interference effects (Anderson, Dana Z.; Mizrahi, Victor; Sipe, John E.)V1516,154-161(1991)

Molecular to material design for anomalous-dispersion phase-matched second-harmonic generation (Cahill, Paul A.; Tallant, David R.; Kowalczyk, T. C.; Singer, Kenneth D.)V1560,130-138(1991)

New organic crystal material for SHG, 2-cyano-3-(3,4-methylene dioxy phenyl)-2-propionic acid ethyl ester (Mori, Yasushi; Sano, Kenji; Todori, Kenji; Kawamonzen, Yosiaki)V1560,310-314(1991)

Nonlinear optical and piezoelectric behavior of liquid-crystalline elastomers (Hirschmann, Harald; Meier, Wolfgang; Finkelmann, Heino)V1559,27-38(1991)

Nonlinear optical properties of a fiber with an organic core crystal grown from solution (Ohmi, Toshihiko; Yoshikawa, Nobuo; Sakai, Koji; Koike, Tomoyuki; Umegaki, Shinsuke)V1361,606-612(1991)

Nonlinear optical properties of N-(4-nitro-2-pyridinyl)-phenylalaninol single crystals (Sutter, Kurt; Hulliger, J.; Knoepfle, G.; Saupper, N.; Guenter, Peter)V1560,296-301(1991)

Nonlinear optical properties of poled polymers (Gadret, G.; Kajzar, Francois; Raimond, P.)V1560,226-237(1991)

Nonlinear optic frequency converters with lowered sensibility to spectral width of laser radiation (Volosov, Vladimir D.)V1500,105-110(1991)

Nonlinear optic in-situ diagnostics of a crystalline film in molecular-beam-epitaxy devices (Krasnov, Victor F.; Musher, Semion L.; Prots, V. I.; Rubenchik, Aleksandr M.; Ryabchenko, Vladimir E.; Stupak, Mikhail F.)V1506,179-187(1991)

Optical second-harmonic generation by polymer-dispersed liquid-crystal films (Yuan, Haiji J.; Li, Le; Palffy-Muhoray, Peter)V1455,73-83(1991)

Organic salts with large electro-optic coefficients (Perry, Joseph W.; Marder, Seth R.; Perry, Kelly J.; Sleva, E. T.; Yakymyshyn, Christopher P.; Stewart, Kevin R.; Boden, Eugene P.)V1560,302-309(1991)

Phase conjugation as a probe for noncentrosymmetry grating formation in organics (Charra, Fabrice; Nunzi, Jean-Michel; Messier, Jean)V1516,211-219(1991)

Phase-matched second-harmonic generation of infrared wavelengths in optical fibers (Kashyap, Raman; Davey, Steven T.; Williams, Doug L.)V1516,164-174(1991)

Phase-matched second harmonic generations in poled dye-polymer waveguides (Sugihara, Okihiro; Kinoshita, Takeshi; Okabe, M.; Kunioka, S.; Nonaka, Y.; Sasaki, Keisuke)V1361,599-605(1991)

Phenomenological description of self-organized x(2) grating formation in centrosymmetric doped optical fibers (Chmela, Pavel)V1516,116-124(1991)

Photoinduced second-harmonic generation and luminescence of defects in Ge-doped silica fibers (Krol, Denise M.; Atkins, Robert M.; Lemaire, Paul J.)V1516,38-46(1991)

Physical models of second-harmonic generation in optical fibers (Lesche, Bernhard)V1516,125-136(1991)

Quantum wells and artificially structured materials for nonlinear optics (Harris, James S.; Fejer, Martin M.)V1361,262-273(1991)

Quasi-phase-matched frequency conversion in lithium niobate and lithium tantalate waveguides (Lim, Eric J.; Matsumoto, S.; Bortz, M. L.; Fejer, Martin M.)V1561,135-142(1991)

Second-harmonic generation as an in situ diagnostic for corrosion (Klenerman, David; Spowage, K.; Walpole, B.)V1437,95-102(1991)

Second harmonic generation of diode laser radiation in KNbO3 (van Hulst, Niek F.; Heesink, Gerard J.; de Leeuw, H.; Bolger, Bouwe)V1362,631-646(1991)

Second-harmonic generation of mode-locked Nd:YAG and Nd:YLF lasers using LiB3O5 (Reed, Murray K.; Tyminski, Jacek K.; Bischel, William K.)V1410,179-184(1991)

Second-order and third-order processes as diagnostic tools for the analysis of organic polymers: an overview (Marowsky, Gerd; Luepke, G.)V1560,328-334(1991)

Selected Papers on Nonlinear Optics (Brandt, Howard E., ed.)VMS32(1991)

Shaping of the second-harmonic pulse temporal profile (Patron, Zbigniew)V1391,259-263(1991)

Sol-gel processed novel multicomponent inorganic oxide: organic polymer composites for nonlinear optics (Zhang, Yue; Cui, Y. P.; Wung, C. J.; Prasad, Paras N.; Burzynski, Ryszard)V1560,264-271(1991)

Structure/property relationships for molecular second-order nonlinear optics (Marder, Seth R.; Cheng, Lap Tak A.; Tiemann, Bruce G.; Beratan, David N.)V1560,86-97(1991)

Study on the surface currents density with second-harmonic generation from silver (Chen, Zhan; Jiang, Hongbing; Zheng, J. B.; Zhang, Zhiming)V1437,103-109(1991)

Surface normal SHG in PLZT thin film waveguides (Zou, Lian C.; Malloy, Kevin J.; Wu, A. Y.)V1519,707-711(1991)

Test of photovoltaic model of photoinduced second-harmonic generation in optical fibers (Dianov, Evgeni M.; Kazansky, Peter G.; Krautschik, Christof G.; Stepanov, D. Y.)V1516,75-80(1991)

Two-color interferometry using a detuned frequency-doubling crystal (Koch, Karl W.; Moore, Gerald T.)V1516,67-74(1991)

UV spectroscopy of optical fibers and preforms (Williams, Doug L.; Davey, Steven T.; Kashyap, Raman; Armitage, J. R.; Ainslie, B. J.)V1516,29-37(1991)

What we can learn about second-harmonic generation in germanosilicate glass from the analogous effect in semiconductor-doped glasses (Lawandy, Nabil M.)V1516,99-112(1991)

Semiconductors—see also gallium arsenide materials; microelectronic processing; silicon

128 x 128 MWIR InSb focal plane and camera system (Parrish, William J.; Blackwell, John D.; Paulson, Robert C.; Arnold, Harold)V1512,68-77(1991)

Absorptive nonlinear semiconductor amplifiers for fast optical switching (Barnsley, Peter E.; Marshall, Ian W.; Fiddyment, Phillip J.; Robertson, Michael J.)V1378,116-126(1991)

Advanced infrared detector materials (Mullin, John B.)V1512,144-154(1991)

Advanced infrared focal-plane arrays (Amingual, Daniel)V1512,40-51(1991)

Amorphous silicon periodic and quasiperiodic superlattices (Chen, Kun J.; Du, Jia F.; Li, Zhi F.; Xu, Jun; Jiang, Jian G.; Feng, Duan; Fritzsche, Hellmut)V1519,632-639(1991)

Analysis and control of semiconductor crystal growth with reflectance-difference spectroscopy and spectroellipsometry (Aspnes, David E.)V1361,551-561(1991)

Anisotropic local melting of semiconductors under light pulse irradiation (Fattakhov, Yakh'ya V.; Khaibullin, Ildus B.; Bayazitov, Rustem M.)V1440,16-23(1991)

Anodic oxides on HgZnTe (Eger, David; Zigelman, Alex)V1484,48-54(1991)

Application of AlGaAs/GaAs superlattice for negative-differential-resistance transistor (Liu, W. C.; Lour, W. S.; Sun, C. Y.; Lee, Y. S.; Guo, D. F.)V1519,670-674(1991)

Application of surface effects externally to alter optical and electronic properties of existing optoelectronic devices: a ten-year update (Hava, Shlomo)V1540,350-358(1991)

Applications of GaAs grade-period doping superlattice for negative-differential-resistance device (Liu, W. C.; Sun, C. Y.; Lour, W. S.; Guo, D. F.; Lee, Y. S.)V1519,640-644(1991)

Applications of resonance ionization spectroscopy for semiconductor, environmental and biomedical analysis, and for DNA sequencing (Arlinghaus, Heinrich F.; Spaar, M. T.; Thonnard, N.; McMahon, A. W.; Jacobson, K. B.)V1435,26-35(1991)

Applications of the FEL to NLO spectroscopy of semiconductors (Pidgeon, Carl R.)V1501,178-182(1991)

Atomic scale simulation of the growth of CdTe layers on GaAs (Rouhani, Mehdi D.; Laroussi, M.; Gue, A. M.; Esteve, Daniel)V1361,954-962(1991)

Average energy gap of AIBIIIC2VI optoelectronic materials (Kumar, V.; Chandra, Dinesh)V1361,809-811(1991)

Band structure and optical properties of silicon carbide (Gavrilenko, Vladimir I.; Frolov, Sergey I.)V1361,171-182(1991)

Band-structure dependence of impact ionization: bulk semiconductors, strained Ge/Si alloys, and multiple-quantum-well avalanche photodetectors (Czajkowski, Igor K.; Allam, Jeremy; Adams, Alfred R.)V1362,179-190(1991)

Behavior of amorphous semiconductors As2S3 layers after photon, electron, or x-ray exposures (Guttmann, Peter; Danev, Gentsho; Spassova, E.; Babin, Sergei V.)V1361,999-1010(1991)

Characterization of anodic fluoride films on Hg1-xCdxTe (Esquivias, Ignacio; Dal Colle, M.; Brink, D.; Baars, Jan W.; Bruder, Martin)V1484,55-66(1991)

Charge separation in functionalized tetrathiafulvalene derivatives (Fox, Marye A.; Pan, Horng-Lon)V1436,2-7(1991)

Close-spaced vapor transport of II-VI semiconductors (Perrier, Gerard)V1536,258-267(1991)

Comparative analysis of external factors' influences on the GaP light-emitting p-n-junctions (Rizikov, Igor V.; Svechnikov, Georgy S.; Bulyarsky, Sergey V.; Ambrozevich, Alexander S.)V1362,664-673(1991)

Comparative study for bolometric and nonbolometric switching elements for microwave phase shifters (Tabib-Azar, Massood; Bhasin, Kul B.; Romanofsky, Robert R.)V1477,85-94(1991)

Competitiveness (Minihan, Charles E.)V1496,2-18(1991)

Considerations of the limits and capabilities of light-activated switches (Zucker, Oved S.; Giorgi, David M.; Griffin, Adam; Hargis, David E.; Long, James; McIntyre, Iain A.; Page, Kevin J.; Solone, Paul J.; Wein, Deborah S.)V1378,22-33(1991)

Continuous flow manufacturing (Bowers, George H.)V1496,239-246(1991)

Crystallization and optothermal characteristics of germanate glasses (Montenero, Angelo; Gnappi, G.; Bertolotti, Mario; Sibilia, C.; Fazio, E.; Liakhou, G.)V1513,234-242(1991)

Current state of gas-source molecular beam epitaxy for growth of optoelectronic materials (Pessa, Markus; Hakkarainen, T.; Keskinen, Jari; Rakennus, K.; Salokatve, A.; Zhang, G.; Asonen, Harry M.)V1361,529-542(1991)

Current transport in charge injection devices (Wu, Chao-Wen; Lin, Hao-Hsiung)V1362,768-777(1991)

Current-voltage characteristics of resonant tunneling diodes (Sen, Susanta; Nag, B. R.; Midday, S.)V1362,750-759(1991)

Deep levels in III-V compounds, heterostructures, and superlattices (Bourgoin, J. C.; Feng, S. L.; von Bardeleben, H. J.)V1361,184-194(1991)

Degenerate four-wave mixing and optical nonlinearities in quantum-confined CdSe microcrystallites (Park, S. H.; Casey, Michael P.; Falk, Joel)V1409,9-13(1991)

Depth profiling resonance ionization mass spectrometry of electronic materials (Downey, Stephen W.; Emerson, A. B.)V1435,19-25(1991)

Design of frozen-wave and injected-wave microwave generators using optically controlled switches (Nunnally, William C.)V1378,10-21(1991)

Design of low-noise wide-dynamic-range GaAs optical preamps (Bayruns, Robert J.; Laverick, Timothy; Scheinberg, Norman; Stofman, Daniel)V1541,83-90(1991)

Determination of electrostatic potentials and charge distributions in bulk and at interfaces by electron microscopy techniques (Hugsted, B.; Gjonnes, K.; Tafto, J.; Gjonnes, Jon; Matsuhata, H.)V1361,751-757(1991)

Determination of thickness and refractal index of HgCdMnTe/CdMnTe VPE films by IR transmission spectrum (Chen, Wei M.; Ma, Ke J.; Yu, Zhen Z.; Ji, Hua M.)V1519,521-524(1991)

Developments in precursors for II-VI semiconductors (Mullin, John B.; Cole-Hamilton, D. J.; McQueen, A. E.; Hails, J. E.)V1361,1116-1127(1991)

Devices for generation and detection of subnanosecond IR and FIR radiation pulses (Beregulin, Eugene V.; Ganichev, Sergey D.; Yaroshetskii, Ilya D.; Lang, Peter T.; Schatz, Wolfgan; Renk, Karl F.)V1362,853-862(1991)

Dispersion of picosecond pulses propagating on microstrip interconnections on semiconductor integrated-circuit substrates (Pasik, Michael F.; Cangellaris, Andreas C.; Prince, John L.)V1389,297-301(1991)

Double-resonant tunneling via surface plasmons in a metal-semiconductor system (Sreseli, Olga M.; Belyakov, Ludvig V.; Goryachev, D. N.; Rumyantsev, B. L.; Yaroshetskii, Ilya D.)V1440,326-331(1991)

Effective electron mass in narrow-band-gap IR materials under different physical conditions (Ghatak, Kamakhya P.; Biswas, Shambhu N.)V1484,149-166(1991)

Effects of TMSb/TEGa ratios on epilayer properties of gallium antimonide grown by low-pressure MOCVD (Wu, T. S.; Su, Yan K.; Juang, F. S.; Li, N. Y.; Gan, K. J.)V1361,23-33(1991)

Electrical and optical properties of As- and Li-doped ZnSe films (Hingerl, Kurt; Lilja, J.; Toivonen, M.; Pessa, Markus; Jantsch, Wolfgang; As, D. J.; Rothemund, W.; Juza, P.; Sitter, Helmut)V1361,943-953(1991)

Electronic properties of mercury-based type-III superlattices (Guldner, Yves; Manasses, J.)V1361,657-668(1991)

Electronic properties of Si-doped nipi structures in GaAs (Fong, C. Y.; Gallup, R. F.; Nelson, J. S.)V1361,479-488(1991)

Electronic structure of Ge(001) 2x1 by different angle-resolved photoemission techniques: EDC, CFS and CIS (Kipp, Lutz; Manzke, Recardo; Skibowski, Michael)V1361,794-801(1991)

Electro-optical effects in semiconductor superlattices (Voos, Michel; Voisin, Paul)V1361,416-423(1991)

Emission of the 1.54um Er-related peaks by impact excitation of Er atoms in InP and its characteristics (Isshiki, Hideo; Kobayashi, Hitoshi; Yugo, Shigemi; Saito, Riichiro; Kimura, Tadamasa; Ikoma, Toshiaki)V1361,223-227(1991)

Equilibrium and nonequilibrium properties of semiconductors with multiply ionizable deep centers (Brinkmann, Ralf P.; Schoenbach, Karl H.; Schulz, Hans-Joachim)V1361,274-280(1991)

Exciton spectroscopy of semiconductor materials used in laser elements (Nasibov, Alexander S.; Markov, L. S.; Fedorov, D. L.; Shapkin, P. V.; Korostelin, Y. V.; Machintsev, G. A.)V1361,901-908(1991)

EXCON: a graphics-based experiment-control manager (Khan, Mumit; Anderson, Paul D.; Cerrina, Franco)V1465,315-323(1991)

External factors' influences on AIIIBV light-emitting structures (Svechnikov, Georgy S.; Rizikov, Igor V.)V1362,674-683(1991)

Fabrication and characterization of semiconductor microlens arrays (Diadiuk, Vicky; Liau, Zong-Long; Walpole, James N.)V1354,496-500(1991)

Fabrication technology of strained layer heterostructure devices (Van Rossum, Marc)V1361,373-382(1991)

Flight experiment to investigate microgravity effects on solidification phenomena of selected materials (Maag, Carl R.; Hansen, Patricia A.)V1557,24-30(1991)

Frequency response study of traps in III-V compound semiconductors (Kachwalla, Zain)V1361,784-793(1991)

Generalized Raman gain in nonparabolic semiconductors under strong magnetic field (Ghatak, Kamakhya P.; Ghoshal, Ardhendhu; De, Badal)V1409,178-190(1991)

Generalized transport model for heterojunction: a computer modeling approach (Bellomi, Giovanni; Bottacchi, Stefano)V1362,760-767(1991)

Growth and characterization of semiconducting Fe-Si2 thin layers on Si(111) (Rizzi, Angela; Moritz, Heiko; Lueth, Hans)V1361,827-833(1991)

Growth and characterization of ultrathin SimGen strained-layer superlattices (Presting, Hartmut; Jaros, Milan; Abstreiter, Gerhard)V1512,250-277(1991)

Growth and characterization of ZnSe and ZnTe grown on GaAs by hot-wall epitaxy (Hingerl, Kurt; Pesek, Andreas; Sitter, Helmut; Krost, Alois; Zahn, Dietrich R.; Richter, W.; Kudlek, Gotthard; Gutowski, Juergen)V1361,383-393(1991)

Growth and properties of GaxIn1-xAs (x<O.47) on InP by MOCVD (Du, MingZe; Yuan, JinShan; Jin, Yixin; Zhou, Tian Ming; Hong, Jiang; Hong, ChunRong; Zhang, BaoLin)V1361,699-705(1991)

Growth by liquid phase epitaxy and characterization of GaInAsSb and InAsSbP alloys for mid-infrared applications (2-3 um) (Tournie, Eric; Lazzari, J. L.; Mani, Habib; Pitard, F.; Alibert, Claude L.; Joullie, Andre F.)V1361,641-656(1991)

Growth dynamics of lattice-matched and strained layer III-V compounds in molecular beam epitaxy (Joyce, Bruce A.; Zhang, J.; Foxon, C. T.; Vvedensky, D. D.; Shitara, T.; Myers-Beaghton, A. K.)V1361,13-22(1991)

Growth of CdTe-CdMnTe heterostructures by molecular beam epitaxy (Bicknell-Tassius, Robert N.)V1484,11-18(1991)

Growth of ZnSe-ZnTe strained-layer supperlattices by atmospheric pressure MOCVD on transparent substrate CaF2 (111) (Pan, Chuan K.; Jiang, F. Y.; Fan, Guang H.; Ma, Y. Z.; Fan, Xi W.)V1519,645-651(1991)

Heat capacity of MOS field-effect devices of optical materials in the presence of a strong magnetic field (Ghatak, Kamakhya P.; Biswas, Shambhu N.)V1485,206-214(1991)

Heteroepitaxial growth of InP and GaInAs on GaAs substrates using nonhydride sources (Chu, Shirley S.; Chu, Ting L.; Yoo, C. H.; Smith, G. L.)V1361,1020-1025(1991)

Heteroepitaxy of II-VI and IV-VI semiconductors on Si substrates (Zogg, Hans; Tiwari, A. N.; Blunier, Stefan; Maissen, Clau; Masek, Jiri)V1361,406-413(1991)

High-efficiency UV and blue emitting devices prepared by MOVPE and low-energy electron-beam irradiation treatment (Akasaki, Isamu; Amano, H.)V1361,138-149(1991)

High-performance metal/SiO2/InSb capacitor fabricated by photoenhanced chemical vapor deposition (Sun, Tai-Ping; Lee, Si-Chen; Liu, Kou-Chen; Pang, Yen-Ming; Yang, Sheng-Jenn)V1361,1033-1037(1991)

High-power switching with electron-beam-controlled semiconductors (Brinkmann, Ralf P.; Schoenbach, Karl H.; Roush, Randy A.; Stoudt, David C.; Lakdawala, Vishnu K.; Gerdin, Glenn A.)V1378,203-208(1991)

High-power waveform generation using photoconductive switches (Oicles, Jeffrey A.; Helava, Heikki I.; Grant, Jon R.; Ragle, Larry O.; Wessman, Susan C.)V1378,60-69(1991)

High-sensitivity photorefractivity in bulk and multiple-quantum-well semiconductors (Partovi, Afshin; Glass, Alastair M.; Feldman, Robert D.)V1561,20-32(1991)

Holographic diffraction gratings on the base of chalcogenide semiconductors (Indutnyi, I. Z.; Robur, I.; Romanenko, Peter F.; Stronski, Alexander V.)V1555,243-253(1991)

Hot carrier photoeffects in inhomogeneous semiconductors and their applications to light detectors (Amosova, L. P.; Marmur, I. Y.; Oksman, Ya. A.; Ashmontas, S.; Gradauskas, I.; Shirmulis, E.)V1440,406-413(1991)

III-V semiconductor integrated optoelectronics for optical computing (Wada, Osamu)V1362,598-607(1991)

Incorporation of As into HgCdTe grown by MOCVD (He, Jin; Yu, Zhen Z.; Ma, Ke J.; Jia, Pei M.; Yang, Jian R.; Shen, Shou Z.; Chen, Wei M.; Yang, Ji M.)V1519,499-507(1991)

Indium tin oxide single-mode waveguide modulator (Chen, Ray T.; Robinson, Daniel; Lu, Huey T.; Sadovnik, Lev S.; Ho, Zonh-Zen)V1583,362-374(1991)

Influence of a high-power laser beam on electrophysical and photoelectrical properties of epitaxial films of CdxHg1-xTe (x is approximately equal to 0.2) (Szeregij, E. M.; Ugrin, J. O.; Virt, I. S.; Abeynayake, C.; Kuzma, Marian)V1391,199-203(1991)

Influence of defects on dynamics of semiconductors (Ge, Si, GaAs) heating by laser radiation (Moin, M.)V1440,2-7(1991)

Influence of surface passivation on the optical bleaching of CdSe microcrystallites embedded in glass (Rueckmann, I.; Woggon, Ulrike; Kornack, J.; Mueller, Matthias; Cesnulevicius, J.; Kolenda, Jonas; Petrauskas, Mendogas)V1513,78-85(1991)

Infrared and Optoelectronic Materials and Devices (Naumaan, Ahmed; Corsi, Carlo; Baixeras, Joseph M.; Kreisler, Alain J., eds.)V1512(1991)

Infrared reflectivity: a tool for bond investigation in II-VI ternaries (Granger, Rene')V1484,39-46(1991)

In-situ investigation of the low-pressure MOCVD growth of III-V compounds using reflectance anisotropy measurements (Drevillon, Bernard; Razeghi, Manijeh)V1361,200-212(1991)

Interaction of hydrogen at InP(100) surfaces before and after ion bombardment (Allinger, Thomas; Persch, V.; Schaefer, Juergen A.; Meng, Y.; De, H.; Anderson, J.; Lapeyre, G. J.)V1361,935-942(1991)

Interface demarcation in Bridgman-Stockbarger crystal growth of II-VI compounds (Gillies, Donald C.; Lehoczky, S. L.; Szofran, Frank R.; Su, Ching-Hua; Larson, David J.)V1484,2-10(1991)

Intl Conf on Thin Film Physics and Applications (Zhou, Shixun; Wang, Yonglin, eds.)V1519(1991)

Intrinsic carrier concentration and effective masses in Hg1-xMnxTe (Rogalski, Antoni)V1512,189-194(1991)

Investigation of hot electron emission in MOS structure under avalanche conditions (Solonko, Alexander G.)V1435,360-365(1991)

Ion-exchanged waveguides in semiconductor-doped glasses (Righini, Giancarlo C.; Pelli, Stefano; De Blasi, C.; Fagherazzi, Giuliano; Manno, D.)V1513,105-111(1991)

Ion implantation and diffusion for electrical junction formation in HgCdTe (Bubulac, Lucia O.)V1484,67-71(1991)

Kinetic properties of adsorbed particles photostimulated migration upon the surface of ionic-covalent-type semiconductors (Kluyev, V. G.; Kushnir, M. A.; Latyshev, A. N.; Voloshina, T. V.)V1440,303-308(1991)

Laser-induced etched grating on InP for integrated optical circuit elements (Grebel, Haim; Pien, P.)V1583,331-337(1991)

Laser processing of germanium (Craciun, Valentin; Mihailescu, Ion N.; Luches, Armando; Kiyak, S. G.; Mikhailova, G. N.)V1392,629-634(1991)

Laser-scanning tomography and related dark-field nanoscopy method (Montgomery, Paul C.; Gall, Pascal; Ardisasmita, Moh S.; Castagne, Michel; Bonnafe, Jacques; Fillard, Jean-Pierre B.)V1332,563-570(1991)

Lasers in Microelectronic Manufacturing (Braren, Bodil, ed.)V1598(1991)

Lattice-mismatched elemental and compound semiconductor heterostructures for 2-D and 3-D applications (Lee, El-Hang)V1362,499-509(1991)

Light energy conversion with pheophytin a and chlorophyll a monolayers at the optical transparent electrode (Leblanc, Roger M.; Blanchet, P.-F.; Cote, D.; Gugeshashvili, M. I.; Munger, G.; Volkov, Alexander G.)V1436,92-102(1991)

Low-loss waveguides on silicon substrates for photonic circuits (Davis, Richard L.; Lee, Sae H.)V1474,20-26(1991)

Low-power optically addressed spatial light modulators using MBE-grown III-V structures (Maserjian, Joseph L.; Larsson, Anders G.)V1562,85-92(1991)

Material for future InP-based optoelectronics: InGaAsP versus InGaAlAs (Quillec, Maurice)V1361,34-46(1991)

Measurement of the sizes of the semiconductor crystallites in colored glasses through neutron scattering (Banfi, Giampiero P.; Degiorgio, Vittorio; Rennie, A. R.; Righini, Giancarlo C.)V1361,874-880(1991)

Metal-organic molecular beam epitaxy of II-VI materials (Summers, Christopher J.; Wagner, Brent K.; Benz, Rudolph G.; Rajavel, D.)V1512,170-176(1991)

Microscopic origin of the shallow-deep transition of impurity levels in III-V and II-VI semiconductors (Chadi, D. J.)V1361,228-231(1991)

Modeling of InP metal organic chemical vapor deposition (Black, Linda R.; Clark, Ivan O.; Kui, Jianming; Jesser, William A.)V1557,54-59(1991)

Modified chloroaluminium phthalocyanine: an organic semiconductor with high photoactivity (Dodelet, Jean-Pol; Gastonguay, Louis; Veilleux, George; Saint-Jacques, Robert G.; Cote, Roland; Guay, Daniel; Tourillon, Gerard)V1436,38-49(1991)

Modulation doping and delta doping of III-V compound semiconductors (Hendriks, Peter; Zwaal, E. A.; Haverkort, Jos E.; Wolter, Joachim H.)V1362,217-227(1991)

Molecular beam epitaxial growth of ZnSe-ZnS strained-layer superlattices (Shen, Ai D.; Cui, Jie; Wang, Hai L.; Wang, Zhi-Jiang)V1519,656-659(1991)

Molecular beam epitaxy of CdTe and HgCdTe on large-area Si(100) (Sporken, R.; Lange, M. D.; Faurie, Jean-Pierre)V1512,155-163(1991)

Monolithic epitaxial IV-VI compound IR-sensor arrays on Si substrates for the SWIR, MWIR and LWIR range (Zogg, Hans; Masek, Jiri; Maissen, Clau; Hoshino, Taizo J.; Blunier, Stefan; Tiwari, A. N.)V1361,1079-1086(1991)

MOVPE technology in device applications for telecommunication (Moss, Rodney H.)V1361,1170-1181(1991)

Multilayer InSb diodes grown by molecular beam epitaxy for near-ambient temperature operation (Ashley, Timothy; Dean, A. B.; Elliott, Charles T.; Houlton, M. R.; McConville, C. F.; Tarry, H. A.; Whitehouse, Colin R.)V1361,238-244(1991)

Mushroom-shaped gates defined by e-beam lithography down to 80-nm gate lengths and fabrication of pseudomorphic HEMTs with a dry-etched gate recess (Huelsmann, Axel; Kaufel, G.; Raynor, Brian; Koehler, Klaus; Schweizer, T.; Braunstein, Juergen; Schlechtweg, M.; Tasker, Paul J.; Jakobus, Theo F.)V1465,201-208(1991)

New approaches to ultrasensitive magnetic resonance (Bowers, C. R.; Buratto, S. K.; Carson, Paul; Cho, H. M.; Hwang, J. Y.; Mueller, L.; Pizarro, P. J.; Shykind, David; Weitekamp, Daniel P.)V1435,36-50(1991)

New concept for multiwafer production of highly uniform III-V layers for optoelectronic applications by MOVPE (Heyen, Meino)V1362,146-153(1991)

New materials for high-performance III-V ICs and OEICs: an industrial approach (Martin, Gerard M.; Frijlink, P. M.)V1362,67-74(1991)

New semiconductor material A1xInAsySb/InAs: LPE synthesis and properties (Charykov, N. A.; Litvak, Alexandr M.; Moiseev, K. D.; Yakovlev, Yurii P.)V1512,198-203(1991)

Noncontact lifetime characterization technique for LWIR HgCdTe using transient millimeter-wave reflectance (Schimert, Thomas R.; Tyan, John; Barnes, Scott L.; Kenner, Vern E.; Brouns, Austin J.; Wilson, H. L.)V1484,19-30(1991)

Nonlinear energy transfer between nanosecond pulses in iron-doped InP crystals (Roosen, Gerald)V1362,398-416(1991)

Nonlinear optical properties of chalcogenide As-S, As-Se glasses (Andriesh, A.; Bertolotti, Mario; Chumach, V.; Fazio, E.; Ferrari, A.; Liakhou, G.; Sibilia, C.)V1513,137-147(1991)

Nonlinear optical properties of germanium diselenide glasses (Haro-Poniatowski, Emmanuel; Guasti, Manuel F.)V1513,86-92(1991)

Nonlinear optical properties of nipi and hetero nipi superlattices and their application for optoelectronics (Doehler, Gottfried H.)V1361,443-468(1991)

Nonlinear optical transmission of an integrated optical bent coupler in semiconductor-doped glass (Guntau, Matthias; Possner, Torsten; Braeuer, Andreas; Dannberg, Peter)V1513,112-122(1991)

Nonstoichiometry effect on mercury thiogallate luminescence (Yelisseyev, Alexander P.; Sinyakova, Elena F.)V1512,204-212(1991)

Novel doping superlattice-based PbTe-IR detector device (Oswald, Josef; Pippan, Manfred; Tranta, Beate; Bauer, Guenther E.)V1362,534-543(1991)

Novel GaP/InP strained heterostructures: growth, characterization,and technological perspectives (Recio, Miguel; Ruiz, Ana; Melendez, J.; Rodriguez, Jose M.; Armelles, Gaspar; Dotor, M. L.; Briones, Fernando)V1361,469-478(1991)

Novel narrow-gap semiconductor systems (Stradling, R. A.)V1361,630-640(1991)

Numerical solution of the one-dimensional Schrodinger equation: application to heterostructures and superlattices (Bottacchi, Stefano)V1362,727-749(1991)

Optical and electrical properties of Al-Al2O3-Cu tunnel junctions (Shu, Q. Q.; Tian, X. M.; Chen, X. Y.; Cai, C. Z.; Zheng, K. Q.; Ma, W. G.)V1519,675-679(1991)

Optical characterization of Hg1-xCdxTe/CdTe/GaAs multilayers grown by molecular beam epitaxy (Liu, Wei J.; Liu, Pu L.; Shi, Guo L.; Zhu, Jing-Bing; He, Li; Xie, Qin X.; Yuan, Shi X.)V1519,481-488(1991)

Optical characterization of InP epitaxial layers on different substrates (Jiao, Kaili L.; Zheng, J. P.; Kwok, Hoi-Sing; Anderson, Wayne A.)V1361,776-783(1991)

Optical cross-modulation method for diagnostic of powerful microwave radiation (Kozar, A. V.; Krupenko, S. A.)V1476,305-312(1991)

Optical investigation of microcrystals in glasses (Ferrara, M.; Lugara, M.; Moro, C.; Cingolani, R.; De Blasi, C.; Manno, D.; Righini, Giancarlo C.)V1513,130-136(1991)

Optically Activated Switching (Zutavern, Fred J., ed.)V1378(1991)

Optically coupled 3-D common memory with GaAs on Si structure (Hirose, Masataka; Takata, H.; Koyanagi, Mitsumasa)V1362,316-322(1991)

Optical nonlinearities in ZnSe multiple-quantum-wells (Liu, Yudong; Liu, Shutian; Li, Chunfei; Shen, Dezen; Fan, Xi W.; Fan, Guang H.; Chen, Lian C.)V1362,436-447(1991)

Optical probing of field dependent effects in GaAs photoconductive switches (Donaldson, William R.; Kingsley, Lawrence E.)V1378,226-236(1991)

Optical processing using photorefractive GaAs and InP (Liu, Duncan T.; Cheng, Li-Jen; Luke, Keung L.)V1409,116-126(1991)

Optical properties of molecular beam epitaxy grown ZnTe epilayers (Kudlek, Gotthard; Presser, Nazmir; Gutowski, Juergen; Mathine, David L.; Kobayashi, Masakazu; Gunshor, Robert L.)V1361,150-158(1991)

Optoelectronic Materials and Device Concepts (Razeghi, Manijeh)VPM05(1991)

Optoelectronic neuron (Pankove, Jacques I.; Radehaus, C.; Borghs, Gustaaf)V1361,620-627(1991)

ORION semiconductor optical detectors: research and development (Khryapov, V. T.; Ponomarenko, Vladimir P.; Butkevitch, V. G.; Taubkin, I. I.; Stafeev, V. I.; Popov, S. A.; Osipov, V. V.)V1540,412-423(1991)

Phonons in PbTe and PbTe:Cr strained layers (Baleva, Mitra I.; Momchilova, Maia M.)V1361,712-722(1991)

Photoconductive switching for high-power microwave generation (Pocha, Michael D.; Hofer, Wayne W.)V1378,2-9(1991)

Photoconductivity decay method for determining minority carrier lifetime of p-type HgCdTe (Reichman, Joseph)V1484,31-38(1991)

Photoconductivity of bridged polymeric phthalocyanines (Meier, Hans; Albrecht, Wolfgang; Hanack, Michael)V1559,89-100(1991)

Photoconductivity of high-bandgap materials (Ho, Ping-Tong; Peng, F.; Goldhar, J.; Nolting, Eugene E.; Parsons, C.)V1378,210-216(1991)

Photoelectrochemical characteristics of slurry-coated CdSeTe films (Murali, K. R.; Subramanian, V.; Rangarajan, N.; Lakshmanan, A. S.; Rangarajan, S. K.)V1536,289-295(1991)

Photoelectrochemical etching of n-InP producing antireflecting structures for solar cells (Soltz, David; Cescato, Lucila H.; Decker, Franco)V1536,268-276(1991)

Photoemission from periodic structure of graded superlattices under magnetic field (Ghatak, Kamakhya P.)V1545,282-293(1991)

Photoemission from quantum-confined structure of nonlinear optical materials (Ghatak, Kamakhya P.; Biswas, Shambhu N.)V1484,136-148(1991)

Photoluminescence and surface photovoltaic spectra of strained InP on GaAs by MOCVD (Zhuang, Weihua; Chen, Chao; Teng, Da; Yu, Jin-zhong; Li, Yu Z.)V1361,980-986(1991)

Photonics technology for aerospace applications (Figueroa, Luis; Hong, C. S.; Miller, Glen E.; Porter, Charles R.; Smith, David K.)V1418,153-176(1991)

Photoreflectance for the in-situ monitoring of semiconductor growth and processing (Pollak, Fred H.)V1361,109-130(1991)

Photovoltaic HgCdTe MWIR-detector arrays on (100)CdZnTe/(100)GaAs grown by hot-wall-beam epitaxy (Gresslehner, Karl-Heinz; Schirz, W.; Humenberger, Josef; Sitter, Helmut; Andorfer, J.; Lischka, Klaus)V1361,1087-1093(1991)

Physical Concepts of Materials for Novel Optoelectronic Device Applications I: Materials Growth and Characterization (Razeghi, Manijeh, ed.)V1361(1991)

Physical Concepts of Materials for Novel Optoelectronic Device Applications II: Device Physics and Applications (Razeghi, Manijeh, ed.)V1362(1991)

P-n heterojunction and Schottky barrier formation between poly(3-methylthiophene) and n-type cadmium sulfide (Frank, Arthur J.; Glenis, Spyridon)V1436,50-57(1991)

Preparation, electrochemical, photoelectrochemical, and solid state characteristics of In-incorporated TiO2 thin films for solar energy applications (Badawy, Waheed A.; El-Giar, Emad M.)V1536,277-288(1991)

Properties of ZnSe/ZnS grown by MOVPE on a rotating substrate (Soellner, Joerg; Heuken, Michael; Heime, Klaus)V1361,963-971(1991)

Prospects for hybrid electronics (Singh, Rajendra; Sinha, Sanjai; Krueger, D. J.)V1394,62-67(1991)

Quantum efficiency and crosstalk of an improved backside-illuminated indium antimonide focal plane array (Bloom, I.; Nemirovsky, Yael)V1442,286-297(1991)

Raman spectra of ZnSe-ZnTe strained-layer superlattice (Cui, Jie; Wang, Hai L.; Gan, Fuxi)V1519,652-655(1991)

Reactive ion etching of InP and its optical assessment (MacLeod, Roderick W.; Sotomayor Torres, Clivia M.; Razeghi, Manijeh; Stanley, C. R.; Wilkinson, Chris D.)V1361,562-567(1991)

Real-time x-ray studies of semiconductor device structures (Clarke, Roy; Dos Passos, Waldemar; Chan, Yi-Jen; Pavlidis, Dimitris; Lowe, Walter P.; Rodricks, Brian G.; Brizard, Christine M.)V1361,2-12(1991)

Recent progress in device-oriented II-VI research at the University of Wuerzburg (Landwehr, Gottfried; Waag, Andreas; Hofmann, K.; Kallis, N.; Bicknell-Tassius, Robert N.)V1362,282-290(1991)

Recent progress on research of materials for optoelectronic device applications in China (Chen, Liang-Hui; Kong, Mei-Ying; Wang, Yi-Ming)V1361,60-73(1991)

Regular doping structures: a Si-based, quantum-well infrared detector (Koch, J. F.)V1362,544-552(1991)

Relaxation-rate of phonon momentum in semiconductors (Gupta, Rita P.; Ridley, Brian K.)V1362,790-797(1991)

Resonant behavior of the temporal response of the photorefractive InP:Fe under dc fields (Abdelghani-Idrissi, Ahmed M.; Ozkul, Cafer; Wolffer, Nicole; Gravey, Philippe; Picoli, Gilbert)V1362,417-427(1991)

Scaling properties of optical reflectance from quasiperiodic superlattices (Wu, Xiang; Yao, He S.; Feng, Wei G.)V1519,625-631(1991)

Selected performance parameters and functional principles of a new stepper generation (Kliem, Karl-Heinz; Sczepanski, Volker; Michl, Uwe; Hesse, Reiner)V1463,743-751(1991)

Selectively grown InxGa1-xAs and InxGa1-xP structures: locally resolved stoichiometry determination by Raman spectroscopy (Geurts, Jean; Finders, J.; Kayser, O.; Opitz, B.; Maassen, M.; Westphalen, R.; Balk, P.)V1361,744-750(1991)

Self-aligned synthesis of titanium silicide by multipulse excimer laser irradiation (Craciun, Valentin; Craciun, Doina; Mihailescu, Ion N.; Kuzmichov, A. V.; Konov, Vitaly I.; Uglov, S. A.)V1392,625-628(1991)

Semiconductor-doped glass as a nonlinear material (Speit, Burkhard; Remitz, K. E.; Neuroth, Norbert N.)V1361,1128-1131(1991)

Semiconductor-doped glasses: nonlinear and electro-optical properties (Ekimov, A. I.; Kudryavtsev, I. A.; Chepick, D. I.; Shumilov, S. K.)V1513,123-129(1991)

Semiconductor thin-film optical constant determination and thin-film thickness measurement equipment correlation (Kaiser, Anne M.)V1464,386-392(1991)

Semiconductor waveguides for optical switching (Laval, Suzanne)V1362,82-92(1991)

Sensitivity and selectivity enhancement in semiconductor gas sensors (Egashira, Makoto; Shimizu, Yasuhiro)V1519,467-476(1991)

Shadow masked growth for the fabrication of photonic integrated circuits (Demeester, Piet M.; Moerman, Ingrid; Zhu, Youcai; van Daele, Peter; Thomson, J.)V1361,1132-1143(1991)

Silicon calorimeter for high-power microwave measurements (Lazard, Carl J.; Pereira, Nino R.; Huttlin, George A.; Litz, Marc S.)V1407,167-171(1991)

Silicon carbide layers produced by rapid thermal chemical vapor deposition (Ruddell, F. H.; McNeill, D.; Armstrong, Brian M.; Gamble, Harold S.)V1361,159-170(1991)

Solid state conductivity and photoconductivity studies of an ion-exchange polymer/dye system (Huang, Zhiqing; Ordonez, I.; Ioannidis, Andreas A.; Langford, C. H.; Lawrence, Marcus F.)V1436,103-113(1991)

Soluble polyacetylenes derived from the ring-opening metathesis polymerization of substituted cyclooctatetraenes: electrochemical characterization and Schottky barrier devices (Jozefiak, Thomas H.; Sailor, Michael J.; Ginsburg, Eric J.; Gorman, Christopher B.; Lewis, Nathan S.; Grubbs, Robert H.)V1436,8-19(1991)

Spectral hole burning of strongly confined CdSe quantum dots (Spiegelberg, Christine; Henneberger, Fritz; Puls, J.)V1362,951-958(1991)

Spectrum of surface electromagnetic waves in CdxHg1-xTe crystals at 0.3 K < T < 77 K (Vertiy, Alexey A.; Beletskii, N. N.; Gorbatyuk, I. N.; Ivanchenko, I. V.; Popenko, N. A.; Rarenko, I. M.; Tarapov, Sergey I.)V1361,1070-1078(1991)

Stabilization of CdxHg1-xTe heterointerfaces (Clifton, Paul A.; Brown, Paul D.)V1361,1063-1069(1991)

Strained semiconductors and heterostructures: synthesis and applications (Bhattacharya, Pallab K.; Singh, Jasprit)V1361,394-405(1991)

Structural and optical properties of semiconducting microcrystallite-doped SiO2 glass films prepared by rf-sputtering (Tsunetomo, Keiji; Shimizu, Ryuichiro; Yamamoto, Masaki; Osaka, Yukio)V1513,93-104(1991)

Studies of correlation of molecular structure under preparation conditions for noncrystalline selenium thin films with aid of computer simulation (Popov, A.; Vasiljeva, Natalja V.)V1519,37-42(1991)

Studies of InSb metal oxide semiconductor structure fabricated by photo-CVD using Si2H6 and N2O (Huang, C. J.; Su, Yan K.; Leu, R. L.)V1519,70-73(1991)

Study of p-ZnTe/n-CdTe thin film heterojunction (Wu, Ping)V1519,477-480(1991)

Superconducting YBa2Cu3O7 films on Si and GaAs with conducting indium tin oxide buffer layers (James, Jonathan H.; Kellett, Bruce J.; Gauzzi, Andrea; Dwir, Benjamin; Pavuna, Davor)V1394,45-61(1991)

Superconductor/semiconductor structure and its application to superconducting devices (Nishino, Toshikazu; Hatano, Mutsuko; Hasegawa, Haruhiro)V1394,36-44(1991)

Temporal model of optically initiated GaAs avalanche switches (Falk, R. A.; Adams, Jeff C.)V1378,70-81(1991)

Three-dimensional nanoprofiling of semiconductor surfaces (Montgomery, Paul C.; Fillard, Jean-Pierre B.; Tchandjou, N.; Ardisasmita, Moh S.)V1332,515-524(1991)

Time-addressing of coherence-tuned optical fibre sensors based on a multimode laser diode (Santos, Jose L.; Jackson, David A.)V1572,325-330(1991)

Transfer matrix approach to the design of InP/InGaAsP ARROW structures (Kubica, Jacek M.; Domanski, Andrzej W.)V1391,24-31(1991)

Transport time and single-particle relaxation time in two-dimensional semiconductors (Gold, Alfred)V1362,309-313(1991)

Tunable cyclotron-resonance laser in germanium (Kremser, Christian; Unterrainer, Karl; Gornik, Erich; Strasser, G.; Pidgeon, Carl R.)V1501,69-79(1991)

Tunneling recombination of carriers at type-II interface in GaInAsSb-GaSb heterostructures (Titkov, A. N.; Yakovlev, Yurii P.; Baranov, Alexej N.; Cheban, V. N.)V1361,669-673(1991)

Type-II heterojunctions in GaSb-InAs solid solutions: physics and applications (Mikhailova, Maya P.; Baranov, Alexej N.; Imenkov, Albert N.; Yakovlev, Yurii P.)V1361,674-685(1991)

Ultra-high-frequency GaInAs/InP devices and circuits for millimeter wave application (Greiling, Paul T.)V1361,47-58(1991)

Using an expert system to interface mainframe computing resources with an interactive video system (Carey, Raymond; Wible, Sheryl F.; Gaynor, Wayne H.; Hendry, Timothy G.)V1464,500-507(1991)

Wave-function engineering in Si-Ge microstructures: linear and nonlinear optical response (Jaros, Milan; Turton, Richard M.)V1362,242-253(1991)

Wave-vector-dependent magneto-optics in semiconductors (De Salvo, Edmondo; Girlanda, Raffaello)V1362,870-875(1991)

Sensor fusion—see sensors

Sensors—see also detectors; fiber optic sensors; photodetectors; remote sensing

1536 x 1024 CCD image sensor (Wong, Kwok Y.; Torok, Georgia R.; Chang, Win-Chyi; Meisenzahl, Eric J.)V1447,283-287(1991)

640 x 480 MOS PtSi IR sensor (Sauer, Donald J.; Shallcross, Frank V.; Hsueh, Fu-Lung; Meray, G. M.; Levine, Peter A.; Gilmartin, Harvey R.; Villani, Thomas S.; Esposito, Benjamin J.; Tower, John R.)V1540,285-296(1991)

Accuracies in FLIR test equipment (Fourier, Ron)V1442,109-117(1991)

Acquisition, Tracking, and Pointing V (Masten, Michael K.; Stockum, Larry A., eds.)V1482(1991)

Actively controlled multiple-sensor system for feature extraction (Daily, Michael J.; Silberberg, Teresa M.)V1472,85-96(1991)

Adaptive control of propellant slosh for launch vehicles (Adler, James M.; Lee, Michael S.; Saugen, John D.)V1480,11-22(1991)

Adaptive detection of subpixel targets using multiband frame sequences (Stocker, Alan D.; Yu, Xiaoli; Winter, Edwin M.; Hoff, Lawrence E.)V1481,156-169(1991)

Adaptive selection of sensors based on individual performances in a multisensor environment (Parra-Loera, Ramon; Thompson, Wiley E.; Salvi, Ajit P.)V1470,30-36(1991)

Advanced broadband baffle materials (Seals, Roland D.)V1485,78-87(1991)

Airborne electro-optical sensor: performance predictions and design considerations (Mishra, R. K.; Pillai, A. M.; Sheshadri, M. R.; Sarma, C. G.)V1482,138-145(1991)

Airborne imaging system performance model (Redus, Wesley D.)V1498,2-23(1991)

Airborne infrared and visible sensors used for law enforcement and drug interdiction (Aikens, David M.; Young, William R.)V1479,435-444(1991)

Airborne seeker evaluation and test system (Jollie, William B.)V1482,92-103(1991)

Airborne tunable diode laser sensor for high-precision concentration and flux measurements of carbon monoxide and methane (Sachse, Glen W.; Collins, Jim E.; Hill, G. F.; Wade, L. O.; Burney, Lewis G.; Ritter, J. A.)V1433,157-166(1991)

Algorithmic sensor failure detection on passive antenna arrays (Chun, Joohwan; Luk, Franklin T.)V1566,483-492(1991)

Amorphous silicon thin film x-ray sensor (Wei, Guang P.)V1519,225-233(1991)

Analog CCD processors for image filtering (Yang, Woodward)V1473,114-127(1991)

Analog retina model for detecting moving objects against a moving background (Searfus, Robert M.; Eeckman, Frank H.; Colvin, Michael E.; Axelrod, Timothy S.)V1473,95-101(1991)

Application of backscatter absorption gas imaging to the detection of chemicals related to drug production (Kulp, Thomas J.; Garvis, Darrel G.; Kennedy, Randall B.; McRae, Thomas G.)V1479,352-363(1991)

Application of Dempster-Shafer theory to a novel control scheme for sensor fusion (Murphy, Robin R.)V1569,55-68(1991)

Application of lighter-than-air platforms to law enforcement (Mataloni, Raymond J.)V1479,306-315(1991)

Application of MHT to group-to-object tracking (Kovacich, Michael; Casaletto, Tom; Lutjens, William; McIntyre, David; Ansell, Ralph; VanDyk, Ed)V1481,357-370(1991)

Application of the ProtoWare simulation testbed to the design and evaluation of advanced avionics (Bubb, Daniel; Wilson, Leo T.; Stoltz, John R.)V1483,18-28(1991)

Application results for an augmented video tracker (Pierce, Bill)V1482,182-195(1991)

a-Si:H TFT-driven high-gray-scale contact image sensor with a ground-mesh-type multiplex circuit (Kobayashi, Kenichi; Abe, Tsutomu; Miyake, Hiroyuki; Kashimura, Hirotsugu; Ozawa, Takashi; Hamano, Toshihisa; Fennell, Leonard E.; Turner, William D.; Weisfield, Richard L.)V1448,157-163(1991)

ASTRO 1M: a new system for attitude determination in space (Elstner, Christian; Lichtenauer, Gert; Skarus, Waldemar)V1478,150-159(1991)

Asynchronous data fusion for target tracking with a multitasking radar and optical sensor (Blair, William D.; Rice, Theodore R.; Alouani, Ali T.; Xia, P.)V1482,234-245(1991)

Automated characterization of Z-technology sensor modules (Gilcrest, Andrew S.)V1541,240-249(1991)

Automatic Object Recognition (Nasr, Hatem N., ed.)VIS07(1991)

Biological basis for space-variant sensor design II: implications for VLSI sensor design (Rojer, Alan S.; Schwartz, Eric L.)V1386,44-52(1991)

Calibration for the SAGE III/EOS instruments (Chu, William P.; McCormick, M. P.; Zawodny, J. M.; McMaster, Leonard R.)V1491,243-250(1991)

Calibration of EOS multispectral imaging sensors and solar irradiance variability (Mecherikunnel, Ann)V1493,292-302(1991)

CCD performance model (Dial, O. E.)V1479,2-11(1991)

CCD performance model for airborne reconnaissance (Donn, Matthew; Waeber, Bruce)V1538,189-200(1991)

CCD star sensors for Indian remote sensing satellites (Alex, T. K.; Rao, V. K.)V1478,101-105(1991)

Characteristics of a dynamic holographic sensor for shape control of a large reflector (Welch, Sharon S.; Cox, David E.)V1480,2-10(1991)

Charge-Coupled Devices and Solid State Optical Sensors II (Blouke, Morley M., ed.)V1447(1991)

Chemo-optical microsensing systems (Lambeck, Paul V.)V1511,100-113(1991)

CiNeRaMa model: a useful tool for detection range estimate (Talmore, Eli T.)V1442,362-371(1991)

Coherent launch-site atmospheric wind sounder (Targ, Russell; Hawley, James G.; Otto, Robert G.; Kavaya, Michael J.)V1478,211-227(1991)

Colorimetric characterization of CCD sensors by spectrophotometry (Daligault, Laurence; Glasser, Jean)V1512,124-130(1991)

Comparison of analog and digital strategies for automatic vibration control of lightweight space structures (Hong, Suk-Yoon; Varadan, Vasundara V.; Varadan, Vijay K.)V1489,75-83(1991)

Comparison of mono- and stereo-camera systems for autonomous vehicle tracking (Kehtarnavaz, Nasser; Griswold, Norman C.; Eem, J. K.)V1468,467-478(1991)

Comparison of neural network classifiers to quadratic classifiers for sensor fusion (Brown, Joseph R.; Bergondy, Daniel; Archer, Susan J.)V1469,539-543(1991)

Control scheme for sensor fusion for navigation of autonomous mobile robots (Murphy, Robin R.)V1383,436-447(1991)

Conversion of sensor data for real-time scene generation (Libby, Vibeke; Bardin, R. K.)V1470,59-64(1991)

Cooling system for short-wave infrared radiometer of JERS-1 optical sensor (Ohmori, Yasuhiro; Arakawa, Isao; Nakai, Akira; Tsubosaka, Kazuyoshi)V1490,177-183(1991)

Coordinating sensing and local navigation (Slack, Marc G.)V1383,459-470(1991)

Correction of inherent scan nonplanarity in the Boeing infrared sensor calibration facility (Chase, Robert P.)V1533,138-149(1991)

Data Structures and Target Classification (Libby, Vibeke, ed.)V1470(1991)

Design criteria of an integrated optics microdisplacement sensor (d'Alessandro, Antonio; De Sario, Marco; D'Orazio, Antonella; Petruzzelli, Vincenzo)V1332,554-562(1991)

Design of an athermalized three-fields-of-view infrared sensor (Wickholm, David R.)V1488,58-63(1991)

Design parameters of an EO sensor (Tanwar, Lakhan S.; Jain, P. C.; Kunzmann, H.)V1332,877-882(1991)

Detection of tool wear using multisensor readings defused by artificial neural network (Masory, Oren)V1469,515-525(1991)

Developing a trial burn plan (Smith, Walter S.; Wong, Tony; Williams, Gary L.; Brintle, David G.)V1434,14-25(1991)

Development of a low-pass far-infrared filter for lunar observer horizon sensor application (Mobasser, Sohrab; Horwitz, Larry S.; Griffith, O'Dale)V1540,764-774(1991)

Development of automatic target recognizers for Army applications (Jones, Terry L.)VIS07,4-13(1991)

Diode-laser-based range sensor (Seshamani, Ramani)V1471,354-356(1991)

Direct view thermal imager (Reinhold, Ralph R.)V1447,251-262(1991)

Discrete piezoelectric sensors and actuators for active control of two-dimensional spacecraft components (Bayer, Janice I.; Varadan, Vasundara V.; Varadan, Vijay K.)V1480,102-114(1991)

Displacement measurement using grating images detected by CCD image sensor (Hane, Kazuhiro; Grover, Chandra P.)V1332,584-590(1991)

Distributed-effect optical fiber sensors for trusses and plates (Reichard, Karl M.; Lindner, Douglas K.)V1480,115-125(1991)

Effects of deformable mirror/wavefront sensor separation in laser beam trains (Schafer, Eric L.; Lyman, Dwight D.)V1415,310-316(1991)

Enantio-selective optrode for optical isomers of biologically active amines using a new lipophilic aromatic carrier (He, Huarui; Uray, Georg; Wolfbeis, Otto S.)V1510,95-103(1991)

Environmental Sensing and Combustion Diagnostics (Santoleri, Joseph J., ed.)V1434(1991)

Environment for simulation and animation of sensor-based robots (Chen, ChuXin; Trivedi, Mohan M.; Bidlack, Clint R.; Lassiter, Terrell N.)V1468,354-366(1991)

Error analysis of combined stereo/optical-flow passive ranging (Barniv, Yair)V1479,259-267(1991)

Excess carrier profile in a moving Gaussian spot illumination semiconductor panel and its use as a laser sensor (Rahnavard, Mohammad H.; Bakhtazad, A.)V1385,123-130(1991)

Exploiting known topologies to navigate with low-computation sensing (Miller, David P.; Gat, Erann)V1383,425-435(1991)

Extractive sampling systems for continuous emissions monitors (White, John R.)V1434,104-112(1991)

Feature correspondence in multiple sensor data fusion (Broida, Ted J.)V1383,635-651(1991)

Fiber coupled heterodyne interferometric displacement sensor (Nerheim, Noble M.)V1542,523-533(1991)

Fiber optic based multiprobe system for intraoperative monitoring of brain functions (Mayevsky, Avraham; Flamm, E. S.; Pennie, W.; Chance, Britton)V1431,303-313(1991)

Fiber Optic Sensors: Engineering and Applications (Bruinsma, Anastasius J.; Culshaw, Brian, eds.)V1511(1991)

Finding a grasped object's pose using joint angle and torque constraints (Siegel, David M.)V1383,151-165(1991)

Flight test integration and evaluation of the LANTIRN system on the F-15E (Presuhn, Gary G.; Zeis, Joseph E.)V1479,249-258(1991)

Focal-plane architectures and signal processing (Jayadev, T. S.)V1541,163-166(1991)

Frequency-modulated reticle trackers in the pupil plane (Taylor, James S.; Krapels, Keith A.)V1478,41-49(1991)

Functional row of pyroelectric sensors for infrared devices used in ecological monitoring (Savinykh, Viktor P.; Glushko, A. A.)V1540,450-454(1991)

Fusion of multiple-sensor imagery based on target motion characteristics (Tsao, Tien-Ren J.; Libert, John M.)V1470,37-47(1991)

Fusion or confusion: knowledge or nonsense? (Rothman, Peter L.; Denton, Richard V.)V1470,2-12(1991)

Gas sensor based on an integrated optical interferometer (Brandenburg, Albrecht; Edelhaeuser, Rainer; Hutter, Frank)V1510,148-159(1991)

Generic model for line-of-sight analysis and calibration (Afik, Zvi; Shammas, A.; Schwartz, Roni; Gal, Eli)V1442,392-398(1991)

Global modeling approach for multisensor problems (Chung, Yi-Nung; Emre, Erol; Gustafson, Donald L.)V1481,306-314(1991)

Hartmann-Shack wavefront sensor using a binary optic lenslet array (Kwo, Deborah P.; Damas, George; Zmek, William P.; Haller, Mitch)V1544,66-74(1991)

High-accuracy capacitive position sensing for low-inertia actuators (Stokes, Brian P.)V1454,223-229(1991)

High-fill-factor monolithic infrared image sensor (Kimata, Masafumi; Yutani, Naoki; Yagi, Hirofumi; Nakanishi, Junji; Tsubouchi, Natsuro; Seto, Toshiki)V1540,238-249(1991)

High-performance 400-DPI A4-size contact image sensor module for scanner and G4 fax applications (Yeh, Long-Ching; Wu, Way-Chen; Tang, Ru-Shyah; Tsai, Yong-Song; Chiang, Eugene; Chiao, Pat; Chu, Chu-Lin; Wang, Weng-Lyang)V1527,361-367(1991)

High-performance a-Si:H TFT-driven linear image sensor and its application to a compact scanner (Miyake, Hiroyuki; Sakai, Kazuhiro; Abe, Tsutomu; Sakai, Yoshihiko; Hotta, Hiroyuki; Sugino, Hajime; Ito, Hisao; Ozawa, Takashi)V1448,150-156(1991)

High-resolution airborne multisensor system (Prutzer, Steven; Biron, David G.; Quist, Theodore M.)V1480,46-61(1991)

High-spatial-resolution FLIR (Tucker, Christopher J.; Mitchell, Robert J.)V1498,92-98(1991)

High-Tc bolometer developments for planetary missions (Brasunas, John C.; Lakew, Brook)V1477,166-173(1991)

Hollow and dielectric waveguides for infrared spectroscopic applications (Saggese, Steven J.; Harrington, James A.; Sigel, George H.)V1437,44-53(1991)

Holographic disk scanner for active infrared sensors (Kawauchi, Yoshikazu; Toyoda, Ryuuichi; Kimura, Minoru; Kawata, Koichi)V1555,224-227(1991)

Holographic optical elements as laser irradiation sensor components (Leib, Kenneth G.; Pernick, Benjamin J.)V1532,261-270(1991)

Impact of tactical maneuvers on EO sensor imagery (Hanson, David S.)V1538,48-63(1991)

Importance of sensor models to model-based vision applications (Zelnio, Edmund G.)VIS07,112-121(1991)

Improved limb atmospheric spectrometer and retroreflector in-space for ADEOS (Sasano, Yasuhiro; Asada, Kazuya; Sugimoto, Nobuo; Yokota, Tatsuya; Suzuki, Makoto; Minato, Atsushi; Matsuzaki, Akiyoshi; Akimoto, Hajime)V1490,233-242(1991)

Improved ring laser gyro navigator (Hadfield, Michael J.; Stiles, Tom; Seidel, David; Miller, William G.; Hensley, David; Wisotsky, Steve; Foote, Michael; Gregory, Rob)V1478,126-144(1991)

Indirect spectroscopic detection of singlet oxygen during photodynamic therapy (Tromberg, Bruce J.; Dvornikov, Tatiana; Berns, Michael W.)V1427,101-108(1991)

Infrared Sensors: Detectors, Electronics, and Signal Processing (Jayadev, T. S., ed.)V1541(1991)

Instrumentation concepts for multiplexed Bragg grating sensors (Giesler, Leslie E.; Dunphy, James R.; Morey, William W.; Meltz, Gerald; Glenn, William H.)V1480,138-142(1991)

Instrument for underwater measurement of optical backscatter (Maffione, Robert A.; Dana, David R.; Honey, Richard C.)V1537,173-184(1991)

Integrated optics temperature sensor (d'Alessandro, Antonio; De Sario, Marco; D'Orazio, Antonella; Petruzzelli, Vincenzo)V1399,184-191(1991)

Integrating acoustical and optical sensory data for mobile robots (Wang, Gang)V1468,479-482(1991)

Interconnection requirements in avionic systems (Vergnolle, Claude; Houssay, Bruno)V1389,648-658(1991)

Interferometric optical fiber sensors for absolute measurement of displacement and strain (Kersey, Alan D.; Berkoff, Timothy A.; Dandridge, Anthony D.)V1511,40-50(1991)

IR/MMW fusion ATR (automatic target recognition) (Thiede, Edwin C.)VIS07,24-35(1991)

Langmuir-Blodgett films of immunoglobulin G and direct immunochemical sensing (Turko, Illarion V.; Pikuleva, Irene A.; Yurkevich, Igor S.; Chashchin, Vadim L.)V1510,53-56(1991)

Large-format 1280 x 1024 full-frame CCD image sensor with a lateral-overflow drain and transparent gate electrode (Stevens, Eric G.; Kosman, Steven L.; Cassidy, John C.; Chang, Win-Chyi; Miller, Wesley A.)V1447,274-282(1991)

Laser applications for multisensor systems (Smolka, Greg L.; Strother, George T.; Ott, Carl)V1498,70-80(1991)

Laser beam diagnostics: a conference overview (Sadowski, Thomas J.)V1414,136-140(1991)

Laser radar based on diode lasers (Levenstein, Harold)V1416,30-43(1991)

Lens evaluation for electronic photography (Bell, Cynthia S.)V1448,59-68(1991)

Lessons learned in recent beryllium-mirror fabrication (Wells, James A.; Lombard, Calvin M.; Sloan, George B.; Moore, Wally W.; Martin, Claude E.)V1485,2-12(1991)

Lidar for expendable launch vehicles (Lee, Michael S.)V1480,23-34(1991)

Line-of-sight alignment of a multisensor system (Wilk, Shalom; Goldmunz, Menachem; Shahaf, Nachum; Klein, Yitschak; Goldman, Shmuel; Oren, Ehud)V1442,140-148(1991)

Line-of-sight stabilization: sensor blending (Pettit, Christopher J.)V1489,278-287(1991)

Linear array camera interface techniques (DeLuca, Dan)V1396,558-565(1991)

Local and remote track-file registration using minimum description length (Kenefic, Richard J.)V1481,430-439(1991)

Low-cost space platforms for detection and tracking technologies (Cullen, Robert M.)V1479,295-305(1991)

Manufacture of fast, aspheric, bare beryllium optics for radiation hard, spaceborne sensor systems (Sweeney, Michael N.)V1485,116-127(1991)

Method and device that prevent target sensors from being radiation overexposed in the presence of a nuclear blast (Holubowicz, Kazimierz S.)V1456,274-285(1991)

Miniature laser Doppler anemometer for sensor concepts (Damp, Stephan)V1418,459-470(1991)

Mission overview of ADEOS program (Iwasaki, Nobuo; Hara, Norikazu; Kajii, Makoto; Tange, Yoshio; Miyachi, Yuji; Sato, Ryota; Inoue, Kouichi)V1490,192-199(1991)

Mission study overview of Japanese polar orbiting platform program (Moriyama, Takashi; Nakayama, Kimihiko; Homma, M.; Haruyama, Yukio)V1490,310-316(1991)

Mission verification systems for FMS applications (Flaherty, Marty)V1538,64-68(1991)

MITAS: multisensor imaging technology for airborne surveillance (Thomas, John)V1470,65-74(1991)

Mobile robot system for the handicapped (Palakal, Mathew J.; Chien, Yung-Ping; Chittajallu, Siva K.; Xue, Qing L.)V1468,456-466(1991)

Model-based ATR (automatic target recognition) systems for the military (Tatum, William F.)VIS07,130-139(1991)

Multiagent collaboration for experimental calibration of an autonomous mobile robot (Vachon, Bertrand; Berge-Cherfaoui, Veronique)V1468,483-492(1991)

Multimission sensor for the RESERVES small-satellite program (Kilston, Steven; Kilston, Vera M.; Utsch, Thomas F.)V1495,193-204(1991)

Multiple-hypothesis-based multiple-sensor spatial data fusion algorithm (Leung, Dominic S.; Williams, D. S.)V1471,314-325(1991)

Multiple-sensor cueing using a heuristic search (David, Philip)V1468,1000-1009(1991)

Multiple unfused passive sensors for operating in busy indoor environments (Konishi, Mashide; Brooks, Rodney A.)V1383,448-458(1991)

Multiplexed mid-wavelength IR long linear photoconductive focal-plane arrays (Kreider, James F.; Preis, Mark K.; Roberts, Peter C.; Owen, Larry D.; Scott, Walter M.; Walmsley, Charles F.; Quin, Alan)V1488,376-388(1991)

Multisensor analysis tool (Gerlach, Francis W.; Cook, Daniel B.)V1488,134-143(1991)

Multisensor approach in linear-Gaussian estimation of 3-D angular motion (Algrain, Marcelo C.; Saniie, Jafar)V1478,201-210(1991)

Multisensor fusion using the sensor algorithm research expert system (Bullock, Michael E.; Miltonberger, Thomas W.; Reinholdtsen, Paul A.; Wilson, Kathleen)V1471,291-302(1991)

Multispectral and multisensor adaptive automatic object recognition (Sadjadi, Firooz A.)VIS07,218-230(1991)

Multispectral imaging with frequency-modulated reticles (Sanders, Jeffrey S.; Halford, Carl E.)V1478,52-63(1991)

Multiwindow method for spectrum estimation and sinusoid detection in an array environment (Onn, Ruth; Steinhardt, Allan O.)V1566,427-438(1991)

Neural network vision integration with learning (Toborg, Scott T.)V1469,77-88(1991)

New estimation architecture for multisensor data fusion (Covino, Joseph M.; Griffiths, Barry E.)V1478,114-125(1991)

New method for sensor data fusion in machine vision (Wang, Yuan-Fang)V1570,31-42(1991)

New track-to-track association logic for almost identical multiple sensors (Malakian, Kourken; Vidmar, Anthony)V1481,315-328(1991)

Noninvasive sensors for in-situ process monitoring and control in advanced microelectronics manufacturing (Moslehi, Mehrdad M.)V1393,280-294(1991)

Notch and large-area CCD imagers (Bredthauer, Richard A.; Pinter, Jeff H.; Janesick, James R.; Robinson, Lloyd B.)V1447,310-315(1991)

On-focal-plane superconducting signal processing for low- and intermediate-temperature operation (Smetana, Daryl L.; Carson, John C.)V1541,220-226(1991)

Optical analysis of segmented aircraft windows (Jones, Mike I.; Jones, Mark S.)V1498,110-127(1991)

Optical locator for horizon sensing (Fallon, James J.; Selby, Vaughn H.)V1495,268-279(1991)

Optical power budget and device time-constant considerations in undersea laser-based sensor design (Leatham, James G.)V1537,194-202(1991)

Optical quantification of sodium, potassium, and calcium ions in diluted human plasma based on ion-selective liquid membranes (Spichiger, Ursula E.; Seiler, Kurt; Wang, Kemin; Suter, Gaby; Morf, Werner E.; Simon, Wilhelm)V1510,118-130(1991)

Optical techniques for determination of normal shock position in supersonic flows for aerospace applications (Adamovsky, Grigory; Eustace, John G.)V1332,750-756(1991)

Optical velocity sensor for air data applications (Smart, Anthony E.)V1480,62-71(1991)

Optimization of a gimbal-scanned infrared seeker (Williams, Elmer F.; Evans, Robert H.; Brant, Karl; Stockum, Larry A.)V1482,104-111(1991)

Optimum choice of anamorphic ratio and boost filter parameters for a SPRITE-based infrared sensor (Fredin, Per)V1488,432-442(1991)

Optoelectronic LED-photodiode pairs for moisture and gas sensors in the spectral range 1.8-4.8 um (Yakovlev, Yurii P.; Baranov, Alexej N.; Imenkov, Albert N.; Mikhailova, Maya P.)V1510,170-177(1991)

Parametric analysis of target/decoy performance (Kerekes, John P.)V1483,155-166(1991)

Passive and active sensors for autonomous space applications (Tchoryk, Peter; Gleichman, Kurt W.; Carmer, Dwayne C.; Morita, Yuji; Trichel, Milton; Gilbert, R. K.)V1479,164-182(1991)

Passive integrated optics in optical sensor systems (Parriaux, Olivier M.)V1506,111-119(1991)

Passive range sensor refinement using texture and segmentation (Sridhar, Banavar; Phatak, Anil; Chatterji, Gano B.)V1478,178-189(1991)

Passive-sensor data fusion (Kolitz, Stephan E.)V1481,329-340(1991)

Performance evaluation methods for multiple-target-tracking algorithms (Fridling, Barry E.; Drummond, Oliver E.)V1481,371-383(1991)

Photoelastic transducer for high-temperature applications (Redner, Alex S.; Adamovsky, Grigory; Wesson, Laurence N.)V1332,775-782(1991)

Positioning method using polarization-detecting optical sensor for precision robot systems (Sakai, Masao; Nagayama, Akira; Sasakura, Kunihiko)V1385,8-14(1991)

Prelaunch calibration system for optical sensors of Japanese Earth Resources Satellite (Sakuma, Fumihiro; Ono, Akira)V1493,37-47(1991)

Preparation of tin oxide and insulating oxide thin films for multilayered gas sensors (Feng, Chang D.; Shimizu, Yasuhiro; Egashira, Makoto)V1519,8-13(1991)

Pseudo K-means approach to the multisensor multitarget tracking problem (Thompson, Wiley E.; Parra-Loera, Ramon; Tao, Chin-Wang)V1470,48-58(1991)

Radar-optronic tracking experiment for short- and medium-range aerial combat (Ravat, Christian J.; Mestre, J. P.; Rose, C.; Schorter, M.)V1478,239-246(1991)

Radiometric calibration of SPOT 2 HRV: a comparison of three methods (Biggar, Stuart F.; Dinguirard, Magdeleine C.; Gellman, David I.; Henry, Patrice; Jackson, Ray D.; Moran, M. S.; Slater, Philip N.)V1493,155-162(1991)

Ratioing radiometer for use with a solar diffuser (Palmer, James M.; Slater, Philip N.)V1493,106-117(1991)

Reconnaissance and imaging sensor test facilities at Eglin Air Force Base (Pratt, Stephen R.; Tucker, Robert)V1538,40-45(1991)

Relative performance studies for focal-plane arrays (Murphy, Kevin S.; Bradley, D. J.; Dennis, Peter N.)V1488,178-185(1991)

Requirements of a solar diffuser and measurements of some candidate materials (Guzman, Carmen T.; Palmer, James M.; Slater, Philip N.; Bruegge, Carol J.; Miller, Edward A.)V1493,120-131(1991)

Results of calibrations of the NOAA-11 AVHRR made by reference to calibrated SPOT imagery at White Sands, N.M. (Nianzeng, Che; Grant, Barbara G.; Flittner, David E.; Slater, Philip N.; Biggar, Stuart F.; Jackson, Ray D.; Moran, M. S.)V1493,182-194(1991)

Scanning infrared earth sensor for INSAT-II (Alex, T. K.; Kamalakar, J. A.)V1478,106-111(1991)

Scattering contribution to the error budget of an emissive IR calibration sphere (Chalupa, John; Cobb, W. K.; Murdock, Tom L.)V1530,343-351(1991)

Scientific results from the Hubble Space Telescope fine-guidance sensors (Taff, Laurence G.)V1494,66-77(1991)

Sensing and environment perception for a mobile vehicle (Blais, Francois; Rioux, Marc)V1480,94-101(1991)

Sensitivity and selectivity enhancement in semiconductor gas sensors (Egashira, Makoto; Shimizu, Yasuhiro)V1519,467-476(1991)

Sensor Fusion III: 3-D Perception and Recognition (Schenker, Paul S., ed.)V1383(1991)

Sensor fusion approach to optimization for human perception: an observer-optimized tricolor IR target locating sensor (Miller, Walter E.)V1482,224-233(1991)

Sensors and Sensor Integration (Dean, Peter D., ed.)V1480(1991)

Sensors and Sensor Systems for Guidance and Navigation (Wade, Jack; Tuchman, Avi, eds.)V1478(1991)

Sensor system for comet approach and landing (Bonsignori, Roberto; Maresi, Luca)V1478,76-91(1991)

Signal and Data Processing of Small Targets 1991 (Drummond, Oliver E., ed.)V1481(1991)

Signature prediction models for flir target recognition (Velten, Vincent J.)VIS07,98-107(1991)

Simulation of scenes, sensors, and tracker for space-based acquisition, tracking, and pointing experiments (DeYoung, David B.)V1415,13-21(1991)

Simulation study to characterize thermal infrared sensor false alarms (Sabol, Bruce M.; Mixon, Harold D.)V1486,258-270(1991)

Simultaneous active/passive IR vehicle detection (Baum, Jerrold E.; Rak, Steven J.)V1416,209-220(1991)

Small-satellite sensors for multispectral space surveillance (Kostishack, Daniel F.)V1495,214-227(1991)

Small-Satellite Technology and Applications (Horais, Brian J., ed.)V1495(1991)

Smart sensors (Corsi, Carlo)V1512,52-59(1991)

Solar-diffuser panel and ratioing radiometer approach to satellite sensor on-board calibration (Slater, Philip N.; Palmer, James M.)V1493,100-105(1991)

Solid state magnetic azimuth sensor for small satellites (Rouse, Gordon F.; Stauffer, Donald R.; French, Howard B.)V1495,240-245(1991)

Space-based millimeter wave debris tracking radar (Chang, Kai; Pollock, Michael A.; Skrehot, Michael K.)V1475,257-266(1991)

Space Infrared Telescope Facility science instruments overview (Bothwell, Mary)V1540,15-26(1991)

Specification of precision optical pointing systems (Medbery, James D.; Germann, Lawrence M.)V1489,163-176(1991)

Staring sensor MRT measurement and modeling (Mooney, Jonathan M.)V1540,550-564(1991)

State estimation for distributed systems with sensing delay (Alexander, Harold L.)V1470,103-111(1991)

Static horizon sensor for remote sensing satellite (Kamalakar, J. A.; Jain, Yashwant K.; Laxmiprasad, A. S.; Shashikala, M.)V1478,92-100(1991)

Stray-light reduction in a WFOV star tracker lens (Lewis, Isabella T.; Ledebuhr, Arno G.; Axelrod, Timothy S.; Ruddell, Scott A.)V1530,306-324(1991)

Studies on distributed sensing and processing for the control of large flexible spacecraft (Montgomery, Raymond C.; Ghosh, David)V1480,126-137(1991)

Study of active 3-D terrain mapping for helicopter landings (Velger, Mordekhai; Toker, Gregory)V1478,168-176(1991)

Survey of multisensor data fusion systems (Linn, Robert J.; Hall, David L.; Llinas, James)V1470,13-29(1991)

System considerations for detection and tracking of small targets using passive sensors (DeBell, David A.)V1481,180-186(1991)

Tactical Infrared Systems (Tuttle, Jerry W., ed.)V1498(1991)

Target lifetimes in natural resource management (Greer, Jerry D.)V1538,69-76(1991)

Technology trends for high-performance windows (Askinazi, Joel)V1498,100-109(1991)

Terrain acquisition algorithm for an autonomous mobile robot with finite-range sensors (Smith, John M.; Choo, Chang Y.; Nasrabadi, Nasser M.)V1468,493-501(1991)

Three-dimensional imaging using TDI CCD sensors (Fenster, Aaron; Holdsworth, David W.; Drangova, Maria)V1447,28-33(1991)

Three-dimensional object recognition using multiple sensors (Hackett, Jay K.; Lavoie, Matt J.; Shah, Mubarak A.)V1383,611-622(1991)

Time-optimal maneuver guidance design with sensor line-of-sight constraint (Hartman, Richard D.; Lutze, Frederick H.; Cliff, Eugene M.)V1478,64-75(1991)

Toward tactile sensor-based exploration in a robotic environment (Gadagkar, Hrishikesh P.; Trivedi, Mohan M.)V1383,142-150(1991)

Ultraviolet-laser-induced fluorescence imaging sensor (Thompson, Jill C.)V1479,412-414(1991)

Unified approach to multisensor simulation of target signatures (Stewart, Stephen R.; LaHaie, Ivan J.; Lyons, Jayne T.)VIS07,57-97(1991)

Update on the C2NVEO FLIR90 and ACQUIRE sensor performance models (Scott, Luke B.; Tomkinson, David M.)V1488,99-109(1991)

Use of anatomical knowledge to register 3-D blood vessel data derived from DSA with MR images (Hill, Derek L.; Hawkes, David J.; Hardingham, Charles R.)V1445,348-357(1991)

Using robust statistics for sensor fusion (McKendall, Raymond; Mintz, Max)V1383,547-565(1991)

Video Hartmann wavefront diagnostic that incorporates a monolithic microlens array (Toeppen, John S.; Bliss, Erlan S.; Long, Theresa W.; Salmon, J. T.)V1544,218-225(1991)

Virtual-phase charge-coupled device image sensors for industrial and scientific applications (Khvilivitzky, A. T.; Berezin, V. Y.; Lazovsky, L. Y.; Tataurschikov, S. S.; Pisarevsky, A. N.; Vydrevich, M. G.; Kossov, V. G.)V1447,184-190(1991)

Vision sensor system for the environment of disaster (Kawauchi, Yoshikazu; Kawata, Koichi)V1507,538-543(1991)

Wavefront curvature sensing and compensation methods in adaptive optics (Roddier, Francois J.)V1487,123-128(1991)

Why mobile robots need a spatial memory (Haber, Ralph N.; Haber, Lyn)V1383,411-424(1991)

Signal processing—see also architectures; optical processing

Accurate fast Hankel matrix solver (Bojanczyk, Adam W.; Lee, Tong J.; Luk, Franklin T.)V1566,74-83(1991)

Acousto-optic architectures for multidimensional phased-array antenna processing (Riza, Nabeel A.)V1476,144-156(1991)

Acousto-optic estimation of autocorrelation and spectra using triple correlations and bispectra (Sadler, Brian M.; Giannakis, Georgios B.; Smith, Dale L.)V1476,246-256(1991)

Adaptive beamforming using recursive eigenstructure updating with subspace constraint (Yu, Kai-Bor)V1565,288-295(1991)

Adaptive chirplet: an adaptive generalized wavelet-like transform (Mann, Steven; Haykin, Simon)V1565,402-413(1991)

Adaptive deconvolution based on spectral decomposition (Ahlen, Anders; Sternad, Mikael)V1565,130-142(1991)

Adaptive detection of subpixel targets using multiband frame sequences (Stocker, Alan D.; Yu, Xiaoli; Winter, Edwin M.; Hoff, Lawrence E.)V1481,156-169(1991)

Adaptive filters and blind equalizers for mixed-phase channels (Tugnait, Jitendra K.)V1565,209-220(1991)

Adaptive Signal Processing (Haykin, Simon, ed.)V1565(1991)

A/D conversion of microwave signals using a hybrid optical/electronic technique (Bell, John A.; Hamilton, Michael C.; Leep, David A.; Taylor, Henry F.; Lee, Y.-H.)V1476,326-329(1991)

Advanced multichip module packaging and interconnect issues for GaAs signal processors operating above 1 GHz clock rates (Gilbert, Barry K.; Thompson, R.; Fokken, G.; McNeff, W.; Prentice, Jeffrey A.; Rowlands, David O.; Staniszewski, A.; Walters, W.; Zahn, S.; Pan, George W.)V1390,235-248(1991)

Advanced Signal Processing Algorithms, Architectures, and Implementations II (Luk, Franklin T., ed.)V1566(1991)

Algorithm for statistical classification of radar clutter into one of several categories (Nechval, Nicholas A.)V1470,282-293(1991)

Algorithmic sensor failure detection on passive antenna arrays (Chun, Joohwan; Luk, Franklin T.)V1566,483-492(1991)

Algorithms and architectures for implementing large-velocity filter banks (Stocker, Alan D.; Jensen, Preben D.)V1481,140-155(1991)

Angle-only tracking and prediction of boost vehicle position (Tsai, Ming-Jer; Rogal, Fannie A.)V1481,281-291(1991)

Application of adaptive filters to the problem of reducing microphony in arrays of pyroelectric infrared detectors (Carmichael, I. C.; White, Paul R.)V1541,167-177(1991)

Application of canonical correlation analysis in detection in presence of spatially correlated noise (Chen, Wei G.; Reilly, James P.; Wong, Kon M.)V1566,464-475(1991)

Application of fiber optic delay lines and semiconductor optoelectronics to microwave signal processing (Taylor, Henry F.)V1371,150-160(1991)

Application of MHT to group-to-object tracking (Kovacich, Michael; Casaletto, Tom; Lutjens, William; McIntyre, David; Ansell, Ralph; VanDyk, Ed)V1481,357-370(1991)

Application of neural network to restoration of signals degraded by a stochastic, shift-variant impulse response function and additive noise (Bilgen, Mehmet; Hung, Hsien-Sen)V1569,260-268(1991)

Applications of optical polymer waveguide devices on future optical communication and signal processing (Keil, Norbert; Strebel, Bernhard N.; Yao, HuiHai; Krauser, Juergen)V1559,278-287(1991)

Applications of the bispectrum in radar signature analysis and target identification (Jouny, Ismail; Garber, Frederick D.; Moses, Randolph L.; Walton, Eric K.)V1471,142-153(1991)

Architecture and performance of a hardware collision-checking accelerator (Bardin, R. K.; Libby, Vibeke)V1566,394-404(1991)

Architecture for adaptive eigenstructure decomposition based on systolic QRD (Erlich, Simha; Yao, Kung)V1565,47-56(1991)

Arithmetic processor design for the T9000 transputer (Knowles, Simon C.)V1566,230-243(1991)

Arithmetic unit based on a high-speed multiplier with a redundant-binary addition tree (Takagi, Naofumi)V1566,244-251(1991)

Artificial neural networks and Abelian harmonic analysis (Rodriguez, Domingo; Pertuz-Campo, Jairo)V1565,492-503(1991)

Automatic design of signal processors using neural networks (Menon, Murali M.; Van Allen, Eric J.)V1469,322-328(1991)

Backward consistency concept and a new decomposition of the error propagation dynamics in RLS algorithms (Slock, Dirk T.)V1565,14-24(1991)

Bayesian iterative method for blind deconvolution (Neri, Alessandro; Scarano, Gaetano; Jacovitti, Giovanni)V1565,196-208(1991)

Bidirectional log-polar mapping for invariant object recognition (Mehanian, Courosh; Rak, Steven J.)V1471,200-209(1991)

Blind equalization (Proakis, John G.; Nikias, Chrysostomos L.)V1565,76-87(1991)

Blind equalization and deconvolution (Bellini, Sandro)V1565,88-101(1991)

Blind equalization based on cepstra of power spectrum and tricoherence (Bessios, Anthony G.; Nikias, Chrysostomos L.)V1565,166-177(1991)

Compact low-power acousto-optic range-Doppler-angle processor for a pulsed-Doppler radar (Pape, Dennis R.; Vlannes, Nickolas P.; Patel, Dharmesh P.; Phuvan, Sonlinh)V1476,201-213(1991)

Comparative performance study of several blind equalization algorithms (Shynk, John J.; Gooch, Richard P.; Krishnamurthy, Giridhar; Chan, Christina K.)V1565,102-117(1991)

Comparison of synthetic-aperture radar autofocus techniques: phase gradient versus subaperture (Calloway, Terry M.; Jakowatz, Charles V.; Thompson, Paul A.; Eichel, Paul H.)V1566,353-364(1991)

Comparison of three efficient-detail-synthesis methods for modeling using under-sampled data (Iannarilli, Frank J.; Wohlers, Martin R.)V1486,314-324(1991)

Comparison of two kernels for the modified Wigner distribution function (Griffin, Chintana; Nuttall, Albert H.)V1566,439-451(1991)

Compensated digital readout family (Ludwig, David E.; Skow, Michael)V1541,73-82(1991)

Computation and meaning of Gabor coefficients (An, Myoung H.; Conner, Michael; Tolimieri, Richard; Orr, Richard S.)V1565,383-401(1991)

Computer-aided performance evaluation system for the on-board data compression system in HIRIS (Qian, Shen-en; Wang, Ruqin; Li, Shuqiu; Dai, Yisong)V1483,196-206(1991)

Cone beam for medical imaging and NDE (Smith, Bruce D.)V1450,13-17(1991)

CORDIC processor architectures (Boehme, Johann F.; Timmermann, D.; Hahn, H.; Hosticka, Bedrich J.)V1566,208-219(1991)

Cramer-Rao bound for multiple-target tracking (Daum, Frederick E.)V1481,341-344(1991)

Design and performance analysis of optoelectronic adaptive infinite-impulse response filters (Ghosh, Anjan K.)V1565,69-73(1991)

Design, fabrication, and testing of a 7-bit binary fiber optic delay line (Goutzoulis, Anastasios P.; Davies, D. K.)V1371,182-194(1991)

Design of a GaAs Fresnel lens array for optical signal-processing devices (Armenise, Mario N.; Impagnatiello, Fabrizio; Passaro, Vittorio M.)V1374,86-96(1991)

Design of M-band filter banks based on wavelet transform (Yaou, Ming-Haw; Chang, Wen-Thong)V1605,149-159(1991)

Detection and tracking of small targets in persistence (Toumodge, Shawn S.)V1481,221-232(1991)

Detection of moving subpixel targets in infrared clutter with space-time filtering (Braunreiter, Dennis C.; Banh, Nam D.)V1481,73-83(1991)

Devices for Optical Processing (Gookin, Debra M., ed.)V1562(1991)

Digital-signal-processor-based inspection of populated surface-mount technology printed circuit boards (Hartley, David A.; Hobson, Clifford A.; Lilley, F.)V1567,277-282(1991)

Dimensionality of signal sets (Orr, Richard S.)V1565,435-446(1991)

DSP array for real-time adaptive sidelobe cancellation (Rorabaugh, Terry L.; Vaccaro, John J.; Grace, Kevin H.; Pauer, Eric K.)V1566,312-322(1991)

DSP-based stabilization systems for LOROP cameras (Quinn, James)V1538,150-166(1991)

Dual-band optical system for IR multicolor signal processing (Kimball, Paulette R.; Fraser, James C.; Johnson, Jeffrey P.; Siegel, Andrew M.)V1540,687-698(1991)

Dynamic end-to-end model testbed for IR detection algorithms (Iannarilli, Frank J.; Wohlers, Martin R.)V1483,66-76(1991)

Effect of resampling on the MUSIC algorithm when applied to inverse-synthetic-aperture radar (Nash, Graeme)V1566,476-482(1991)

Effects of constellation shaping on blind equalization (Zervas, E.; Proakis, John G.; Eyuboglu, Vedat)V1565,178-187(1991)

Efficient computation of densely sampled wavelet transforms (Jones, Douglas L.; Baraniuk, Richard G.)V1566,202-206(1991)

Efficient use of data structures for digital monopulse feature extraction (McEachern, Robert; Eckhardt, Andrew J.; Nauda, Alexander)V1470,226-232(1991)

End-to-end model for detection performance evaluation against scenario-specific targets (Iannarilli, Frank J.; Wohlers, Martin R.)V1488,226-236(1991)

End-to-end scenario-generating model for IRST performance analysis (Iannarilli, Frank J.; Wohlers, Martin R.)V1481,187-197(1991)

Eraseable optical disk systems for signal processing (Bessette, Oliver E.; Cinelli, Joseph L.)V1474,162-166(1991)

Error analysis in unnormalized floating point arithmetic (Barlow, Jesse L.; Zaccone, Richard J.)V1566,286-294(1991)

Error probabilities of minimum-distance classifiers (Poublan, Helene; Castanie, Francis)V1569,329-340(1991)

Evaluation of image tracker algorithms (Marshall, William C.)V1483,207-218(1991)

Existence of cross terms in the wavelet transform (Kadambe, Shubha; Boudreaux-Bartels, G. F.)V1565,423-434(1991)

Existing gap between theory and application of blind equalization (Ding, Zhi; Johnson, C. R.)V1565,154-165(1991)

Extension of Rader's algorithm for high-speed multidimensional autocorrelation (Rinaldo, R.; Bernardini, Riccardo; Cortelazzo, Guido M.)V1606,773-787(1991)

Fast algorithm and architecture for constrained adaptive sidelobe cancellation (Games, Richard A.; Eastman, Willard L.; Sousa, Michael J.)V1566,323-328(1991)

Fast recursive-least-squares algorithms for Toeplitz matrices (Qiao, Sanzheng)V1566,47-58(1991)

Fast RLS adaptive algorithms and Chandrasekhar equations (Demoment, Guy; Reynaud, Roger)V1565,357-367(1991)

Feature trajectory reduction of integrated autoregressive processes based on a multilayer self-organizing neural network (Klose, Joerg; Altena, Oliver)V1565,504-517(1991)

Fiber optic delay lines for wideband signal processing (Taylor, Henry F.; Gweon, S.; Fang, S. P.; Lee, Chung E.)V1562,264-275(1991)

Finite-precision error analysis of a QR-decomposition-based lattice predictor (Syed, Mushtaq A.; Mathews, V. J.)V1565,25-34(1991)

Flow-control mechanism for distributed systems (Maitan, Jacek)V1470,88-97(1991)

Focal-plane architectures and signal processing (Jayadev, T. S.)V1541,163-166(1991)

Formulation and performance evaluation of adaptive, sequential radar-target-recognition algorithms (Snorrason, Ogmundur; Garber, Frederick D.)V1471,116-127(1991)

Fractional Brownian motion and its fractal dimension estimation (Zhang, Peng; Martinez, Andrew B.; Barad, Herbert S.)V1569,398-409(1991)

Future directions in focal-plane signal processing for spaceborne scientific imagers (Fossum, Eric R.)V1541,62-67(1991)

General approach for obtaining joint representations in signal analysis and an application to scale (Cohen, Leon)V1566,109-133(1991)

Generic modular imaging IR signal processor (Auborn, John E.; Harris, William R.)V1483,2-9(1991)

Global modeling approach for multisensor problems (Chung, Yi-Nung; Emre, Erol; Gustafson, Donald L.)V1481,306-314(1991)

Ground target classification using moving target indicator radar signatures (Yoon, Chun S.)V1470,243-252(1991)

Guided-wave magneto-optic and acousto-optic Bragg cells for RF signal processing (Tsai, Chen S.)V1562,55-65(1991)

Hexagonal sampling and filtering for target detection with a scanning E-O sensor (Sperling, I.; Drummond, Oliver E.; Reed, Irving S.)V1481,2-11(1991)

Hierarchical network for clutter and texture modeling (Luttrell, Stephen P.)V1565,518-528(1991)

High-Frequency Analog Fiber Optic Systems (Sierak, Paul, ed.)V1371(1991)

High-resolution spectral analysis and modeling of infrared ocean surface radiometric clutter (McGlynn, John D.; Ellis, Kenneth K.; Kryskowski, David)V1486,141-150(1991)

Impact of tactical maneuvers on EO sensor imagery (Hanson, David S.)V1538,48-63(1991)

Implementation of an angle-only tracking filter (Allen, Ross R.; Blackman, Samuel S.)V1481,292-303(1991)

Incremental model for target maneuver estimation (Chang, Wen-Thong; Lin, Shao-An)V1481,242-253(1991)

Information-theoretic approach to optimal quantization (Lorenzo, Maximo; Der, Sandor Z.; Moulton, Joseph R.)V1483,118-137(1991)

Infrared Sensors: Detectors, Electronics, and Signal Processing (Jayadev, T. S., ed.)V1541(1991)

In-line Fabry-Perot interferometric temperature sensor with digital signal processing (Yeh, Yunhae; Lee, J. H.; Lee, Chung E.; Taylor, Henry F.)V1584,72-78(1991)

Instantaneous amplitude and frequency estimation: performance bounds and applications to source localization (Arun, K. S.; Liang, R. M.)V1566,157-166(1991)

Instantaneous quantities and uncertainty concepts for signal-dependent time-frequency distributions (Jones, Graeme; Boashash, Boualem)V1566,167-178(1991)

Integrated processor architecture for multisensor signal processing (Nasburg, Robert E.; Stillman, Steve M.; Nguyen, M. T.)V1481,84-95(1991)

Intensity edge detection with stack filters (Yoo, Jisang; Bouman, Charles A.; Delp, Edward J.; Coyle, Edward J.)V1451,58-69(1991)

Interferometric acousto-optic receiver results (Gill, E. T.; Tsui, J. B.)V1476,190-200(1991)

Interferometric signal processing schemes for the measurement of strain (Berkoff, Timothy A.; Kersey, Alan D.)V1588,169-176(1991)

Linear modeling algorithm for tracking time-varying signals (Bachnak, Rafic A.)V1481,12-22(1991)

Ll-filters in CFAR (constant false-alarm rate) detection (Mahmood Reza, Syed; Willett, Peter K.)V1451,298-308(1991)

Local and remote track-file registration using minimum description length (Kenefic, Richard J.)V1481,430-439(1991)

Massively parallel implementation for real-time Gabor decomposition (Dufaux, Frederic; Ebrahimi, Touradj; Kunt, Murat)V1606,851-864(1991)

Massively parallel synthetic-aperture radar autofocus (Mastin, Gary A.; Plimpton, Steven J.; Ghiglia, Dennis C.)V1566,341-352(1991)

Matrix reformulation of the Gabor transform (Balart, Rogelio)V1565,447-457(1991)

Maximum-likelihood blind equalization (Ghosh, Monisha; Weber, Charles L.)V1565,188-195(1991)

Maximum likelihood estimation of differential delay and differential Doppler (Greene, Herbert G.; MacMullan, Jay)V1470,98-102(1991)

Midcourse multitarget tracking using continuous representation (Zak, Michail; Toomarian, Nikzad)V1481,386-397(1991)

Miniature signal processor for surveillance sensor applications (Jacobi, William J.; Jensen, Preben D.; Teneketges, Nicholas J.; Wadsworth, Leo A.)V1479,111-119(1991)

Miniaturized low-power parallel processor for space applications (Jacobi, William J.; Jensen, Preben D.; Teneketges, Nicholas J.; Wadsworth, Leo A.)V1495,205-213(1991)

Minimum-phase LU factorization preconditioner for Toeplitz matrices (Ku, Ta-Kang; Kuo, C.-C. J.)V1566,59-73(1991)

Model-based analysis of 3-D spatial-temporal IR clutter suppression filtering (Chan, David S.)V1481,117-128(1991)

Modeling ultrasound speckle formation and its dependence on imaging system's response (Rao, Navalgund A.; Zhu, Hui)V1443,81-95(1991)

Modular implementations of linearly constrained beamformers (Liu, Tsung-Ching; Van Veen, Barry D.)V1566,419-426(1991)

Moving-point-target tracking in low SNR (Shen, Zhen-kang; Mao, Xuguang; Jin, Yiping)V1481,233-240(1991)

Multifrequency beamspace Root-MUSIC: an experimental evaluation (Zoltowski, Michael D.; Kautz, Gregory M.; Silverstein, Seth D.)V1566,452-463(1991)

Multimission sensor for the RESERVES small-satellite program (Kilston, Steven; Kilston, Vera M.; Utsch, Thomas F.)V1495,193-204(1991)

Multiple-target tracking in a cluttered environment and intelligent track record (Tomasini, Bernard; Cassassolles, Emmanuel; Poyet, Patrice; Maynard de Lavalette, Guy M.; Siffredi, Brigitte)V1468,60-71(1991)

Multiple-target tracking using the SME filter with polar coordinate measurements (Sastry, C. R.; Kamen, Edward W.)V1481,261-280(1991)

Multisensor fusion methodologies compared (Swan, John; Shields, Frank J.)V1483,219-230(1991)

Multitarget tracking and multidimensional assignment problems (Poore, Aubrey B.; Rijavec, Nenad)V1481,345-356(1991)

Multiwindow method for spectrum estimation and sinusoid detection in an array environment (Onn, Ruth; Steinhardt, Allan O.)V1566,427-438(1991)

Naval Research Laboratory flex processor for radar signal processing (Alter, James J.; Evins, James B.; Letellier, J. P.)V1566,296-301(1991)

Neural network approach to multipath delay estimation (Welstead, Stephen T.; Ward, Michael J.; Keefer, Christopher W.)V1565,482-491(1991)

Neural network for passive acoustic discrimination between surface and submarine targets (Baran, Robert H.; Coughlin, James P.)V1471,164-176(1991)

Neural network modeling of radar backscatter from an ocean surface using chaos theory (Leung, Henry; Haykin, Simon)V1565,279-286(1991)

Neural networks for ATR parameters adaptation (Amehdi, Hossien; Nasr, Hatem N.)V1483,177-184(1991)

New address-generation-unit architecture for video signal processing (Kitagaki, Kazukuni; Oto, Takeshi; Demura, Tatsuhiko; Araki, Yoshitsugu; Takada, Tomoji)V1606,891-900(1991)

New algorithms for on-line computation of elementary functions (Kla, Sylvanus; Mazenc, Christophe; Merrheim, Xavier; Muller, Jean-Michel M.)V1566,275-285(1991)

Noise statistics of ratiometric signal processing systems (Svetkoff, Donald J.; Xydis, Thomas G.; Kilgus, Donald B.)V1385,113-122(1991)

Noniterative subspace updating (DeGroat, Ronald D.; Dowling, Eric M.)V1566,376-387(1991)

Nonlinear Image Processing II (Arce, Gonzalo R.; Boncelet, Charles G.; Dougherty, Edward R., eds.)V1451(1991)

Nonlinear prediction and the Wiener process (Nisbet, K. C.; McLaughlin, S.; Mulgrew, B.)V1565,244-254(1991)

Nonlinear regression for signal processing (Restrepo, Alfredo)V1451,258-268(1991)

Novel regular-array ASIC architecture for 2-D ROS sorting (van Swaaij, Michael F.; Catthoor, Francky V.; De Man, Hugo J.)V1606,901-910(1991)

Novel signal processing scheme for ruby-fluorescence-based fiber-optic temperature sensor (Zhang, Zhiyi; Grattan, Kenneth T.; Palmer, Andrew W.)V1511,264-274(1991)

Novel transform for image description and compression with implementation by neural architectures (Ben-Arie, Jezekiel; Rao, K. R.)V1569,367-374(1991)

Numerical stability issues in fast least-squares adaptive algorithms (Regalia, Phillip A.)V1565,2-13(1991)

Observable-based parametrizations (Felsen, Leopold B.)V1471,88-91(1991)

On-focal-plane superconducting signal processing for low- and intermediate-temperature operation (Smetana, Daryl L.; Carson, John C.)V1541,220-226(1991)

On-line arithmetic for recurrence problems (Ercegovac, Milos D.)V1566,263-274(1991)

Optical frequency shifter based on stimulated Brillouin scattering in birefringent optical fiber (Duffy, Christopher J.; Tatam, Ralph P.)V1511,155-165(1991)

Optical Hartley-transform-based adaptive filter (Abushagur, Mustafa A.; Berinato, Robert J.)V1564,602-609(1991)

Optical sampling and demultiplexing applied to A/D conversion (Bell, John A.; Hamilton, Michael C.; Leep, David A.)V1562,276-280(1991)

Optical Technology for Signal Processing Systems (Bendett, Mark P., ed.)V1474(1991)

Optoelectronic Gabor detector for transient signals (Zhang, Yan; Li, Yao; Tolimieri, Richard; Kanterakis, Emmanuel G.; Katz, Al; Lu, X. J.; Caviris, Nicholas P.)V1481,23-31(1991)

Optoelectronic sampling receiver for time-division multiplexed signal processing applications (Yaseen, Mohammed; Walker, Stuart D.)V1562,319-326(1991)

Orthogonal polynomials, Hankel matrices, and the Lanczos algorithm (Boley, Daniel L.)V1566,84-95(1991)

Parallel computation of the continuous wavelet transform (Gertner, Izidor; Peskin, Richard L.; Walther, Sandra S.)V1565,414-422(1991)

Parallel implementation of some fast adaptive algorithms on a digital signal processor network (Reynaud, Roger; Chebira, Abdennasser; Demoment, Guy)V1566,302-311(1991)

Parallel uses for serial arithmetic in signal processors (Owens, Robert M.; Irwin, Mary J.)V1566,252-262(1991)

Performance comparison of neural network and statistical discriminant processing techniques for automatic modulation recognition (Hill, Peter C.; Orzeszko, Gabriel R.)V1469,329-340(1991)

Performance evaluation methods for multiple-target-tracking algorithms (Fridling, Barry E.; Drummond, Oliver E.)V1481,371-383(1991)

Performance evaluation of different neural network training algorithms in error control coding (Hussain, Mukhtar; Bedi, Jatinder S.)V1469,697-707(1991)

Performance of a parallel bispectrum estimation code (Carmona, Edward A.; Matson, Charles L.)V1566,329-340(1991)

Polarimetric segmentation of SAR imagery (Burl, Michael C.; Novak, Leslie M.)V1471,92-115(1991)

Polynomial neural nets for signal and image processing in chaotic backgrounds (Gardner, Sheldon)V1567,451-463(1991)

Programmable processor for multidimensional digital signal processing (Abdelrazik, Mohamed B.)V1606,812-822(1991)

Pseudoinverse matrix methods for signal reconstruction from partial data (Feichtinger, Hans G.)V1606,766-772(1991)

Pyramid median filtering by block threshold decomposition (Zhou, Hongbing; Zeng, Bing; Neuvo, Yrjo A.)V1451,2-12(1991)

Quantifying predictability for applications in signal separation (Taylor, William W.)V1565,267-278(1991)

Rapid enhancement and compression of image data (Balram, Nikhil; Moura, Jose M.)V1606,374-385(1991)

Real-time video signal processing by generalized DDA and control memories: three-dimensional rotation and mapping (Hama, Hiromitsu; Yamashita, Kazumi)V1606,878-890(1991)

Recent results in adaptive array detection (Kalson, S. Z.)V1566,406-418(1991)

Recursive total-least-squares adaptive filtering (Dowling, Eric M.; DeGroat, Ronald D.)V1565,35-46(1991)

Reed-Solomon encoder/decoder application using a neural network (Hussain, Mukhtar; Bedi, Jatinder S.)V1469,463-471(1991)

Resolution advantages of quadratic signal processing (Atlas, Les E.; Fang, Jing; Loughlin, Patrick; Music, Wayne)V1566,134-143(1991)

Restoration of subpixel detail using the regularized pseudoinverse of the imaging operator (Abbiss, John B.; Brames, Bryan J.)V1566,365-375(1991)

Robust CFAR detection using order statistic processors for Weibull-distributed clutter (Nagle, Daniel T.; Saniie, Jafar)V1481,49-63(1991)

SADARM status report (DiNardo, Anthony J.)V1479,228-248(1991)

Sampling system for in vivo ultrasound images (Jensen, Jorgen A.; Mathorne, Jan)V1444,221-231(1991)

SAW real-time Doppler analysis (Martin, Tom A.)V1416,52-58(1991)

Schur RLS adaptive filtering using systolic arrays (Strobach, Peter)V1565,307-322(1991)

Self-organizing leader clustering in a neural network using a fuzzy learning rule (Newton, Scott C.; Mitra, Sunanda)V1565,331-337(1991)

Signal and Data Processing of Small Targets 1991 (Drummond, Oliver E., ed.)V1481(1991)

Signal and Image Processing Systems Performance Evaluation, Simulation, and Modeling (Nasr, Hatem N.; Bazakos, Michael E., eds.)V1483(1991)

Signal enhancement in noise- and clutter-corrupted images using adaptive predictive filtering techniques (Soni, Tarun; Zeidler, James R.; Rao, Bhaskar D.; Ku, Walter H.)V1565,338-344(1991)

Signal processing for nonlinear systems (Broomhead, David S.)V1565,228-243(1991)

Signal processing with neural networks: throwing off the yoke of linearity (Hecht-Nielsen, Robert)V1541,146-151(1991)

Signal processing with radial basis function networks using expectation-maximization algorithm clustering (Ukrainec, Andrew; Haykin, Simon)V1565,529-539(1991)

Signal processor for space-based visible sensing (Anderson, James C.; Downs, G. S.; Trepagnier, Pierre C.)V1479,78-92(1991)

Simulation of infrared backgrounds using two-dimensional models (Cadzow, James A.; Wilkes, D. M.; Peters, Richard A.; Li, Xingkang; Patel, Jamshed N.)V1486,352-363(1991)

Small-target acquisition and typing by AASAP (Huguenin, Robert L.; Tahmoush, Donald J.)V1481,64-72(1991)

Smart sensors (Corsi, Carlo)V1512,52-59(1991)

Solution of the Yule-Walker equations (Gohberg, I.; Koltracht, Israel; Xiao, Tongsan D.)V1566,14-22(1991)

Some fast Toeplitz least-squares algorithms (Nagy, James G.; Plemmons, Robert J.)V1566,35-46(1991)

Some properties of the two-dimensional pseudomedian filter (Schulze, Mark A.; Pearce, John A.)V1451,48-57(1991)

Source location of acoustic emissions from atmospheric leakage using neural networks (Barga, Roger S.; Friesel, Mark A.; Meador, Jack L.)V1469,602-611(1991)

Stability of Bareiss algorithm (Bojanczyk, Adam W.; Brent, Richard P.; de Hoog, F. R.)V1566,23-34(1991)

STARCON: a reconfigurable fieldable signal processing system (Brandt, S. A.; Budenske, John)V1406,122-126(1991)

Statistical initial orbit determination (Taff, Laurence G.; Belkin, Barry; Schweiter, G. A.; Sommar, K.)V1481,440-448(1991)

Statistical signal processing in Raman spectroscopy of biological samples (Praus, Petr; Stepanek, Josef)V1403,76-84(1991)

Stop-and-go sign algorithms for blind equalization (Hatzinakos, Dimitrios)V1565,118-129(1991)

Studies on distributed sensing and processing for the control of large flexible spacecraft (Montgomery, Raymond C.; Ghosh, David)V1480,126-137(1991)

Substrate-mode holograms for board-to-board two-way communications (Huang, Yang-Tung; Kostuk, Raymond K.)V1474,39-44(1991)

Super-exponential methods for blind deconvolution (Shalvi, Ofir; Weinstein, Ehud)V1565,143-152(1991)

Surface-acoustic-wave acousto-optic devices for wide-bandwidth signal processing and switching applications (Garvin, Charles G.; Sadler, Brian M.)V1562,303-318(1991)

Surveillance test bed for SDIO (Wesley, Michael; Osterheld, Robert; Kyser, Jeff; Farr, Michele; Vandergriff, Linda J.)V1481,209-220(1991)

Survey of multisensor data fusion systems (Linn, Robert J.; Hall, David L.; Llinas, James)V1470,13-29(1991)

Survey of radar-based target recognition techniques (Cohen, Marvin N.)V1470,233-242(1991)

Systematic treatment of order-recursive least-squares algorithms (Ling, Fuyun; Proakis, John G.; Zhao, Ke.)V1565,296-306(1991)

Systolic adaptive beamforming: from theory to practice (Hargrave, Philip)V1566,2-11(1991)

Systolic array architecture of a new VLSI vision chip (Means, Robert W.)V1566,388-393(1991)

Target identification by means of impulse radar (Abrahamsson, S.; Brusmark, B.; Gaunaurd, Guillermo C.; Strifors, Hans C.)V1471,130-141(1991)

TeO2 slow surface acoustic wave Bragg cell (Yao, Shi-Kay)V1476,214-221(1991)

Timbre discrimination of signals with identical pitch using neural networks (Sayegh, Samir I.; Pomalaza-Raez, Carlos A.; Tepper, E.; Beer, B. A.)V1569,100-110(1991)

Time-derivative adaptive silicon photoreceptor array (Delbruck, Tobi; Mead, Carver A.)V1541,92-99(1991)

Time-frequency distributions for propulsion-system diagnostics (Griffin, Michael F.; Tulpule, Sharayu)V1566,179-189(1991)

Time series prediction with a radial basis function neural network (Potts, Michael A.; Broomhead, David S.)V1565,255-266(1991)

Time-varying higher order spectra (Boashash, Boualem; O'Shea, Peter)V1566,98-108(1991)

Time-varying system identification via explicit filtering of the parameter estimates (Bellegarda, Jerome R.)V1566,190-201(1991)

Two-dimensional signal deconvolution: design issues related to a novel multisensor-based approach (Sidiropoulos, N. D.; Baras, John S.; Berenstein, C. A.)V1569,356-366(1991)

Ultrafast measurements of carrier transport optical nonlinearities in a GaAs/AlGaAs MQW SEED device (Park, Choong-Bum; Li Kam Wa, Patrick; Miller, Alan)V1474,8-17(1991)

Uncertainty, information, and time-frequency distributions (Williams, William J.; Brown, Mark L.; Hero, Alfred O.)V1566,144-156(1991)

Use of threshold decomposition theory to derive basic properties of median filters (Hawley, Robert W.; Gallagher, Neal C.)V1451,91-100(1991)

Using sound to extract meaning from complex data (Scaletti, Carla; Craig, Alan B.)V1459,207-219(1991)

Variable time delay for RF/microwave signal processing (Toughlian, Edward N.; Zmuda, Henry)V1476,107-121(1991)

Visible and near-infrared subsystem and common signal processor design status of ASTER (Takahashi, Fumiho; Hiramatsu, Masaru; Watanabe, Fumito; Narimatsu, Yoshito; Nagura, Riichi)V1490,255-268(1991)

VLSI processor for high-performance arithmetic computations (McQuillan, S. E.; McCanny, J. V.)V1566,220-229(1991)

Wavelets and adaptive signal processing (Resnikoff, Howard L.)V1565,370-382(1991)

What have neural networks to offer statistical pattern processing? (Lowe, David)V1565,460-471(1991)

Wideband bistatic radar signal processing using a coherent detection architecture (Belcher, Melvin L.; Garmon, Jeff P.)V1476,224-233(1991)

Zak transform as an adaptive tool (Tolimieri, Richard; Conner, Michael)V1565,345-356(1991)

Silicon—see also microelectronic processing; semiconductors

Accurate method for neutron fluence control used in improving neutron-transmutation-doped silicon for detectors (Halmagean, Eugenia T.; Lazarovici, Doina N.; Udrea-Spenea, Marian N.)V1484,106-114(1991)

Amorphous silicon periodic and quasiperiodic superlattices (Chen, Kun J.; Du, Jia F.; Li, Zhi F.; Xu, Jun; Jiang, Jian G.; Feng, Duan; Fritzsche, Hellmut)V1519,632-639(1991)

Amorphous silicon thin film x-ray sensor (Wei, Guang P.)V1519,225-233(1991)

Analysis of the self-oscillation phenomenon of fiber optically addressed silicon microresonators (Rao, Yun-Jiang; Culshaw, Brian)V1506,126-133(1991)

Applicability of dry developable deep-UV lithography to sub-0.5 um processing (Goethals, Anne-Marie; Baik, Ki-Ho; Van den hove, Luc; Tedesco, Serge V.)V1466,604-615(1991)

a-Si:H TFT-driven high-gray-scale contact image sensor with a ground-mesh-type multiplex circuit (Kobayashi, Kenichi; Abe, Tsutomu; Miyake, Hiroyuki; Kashimura, Hirotsugu; Ozawa, Takashi; Hamano, Toshihisa; Fennell, Leonard E.; Turner, William D.; Weisfield, Richard L.)V1448,157-163(1991)

Characterization of microstructure of Si films grown by laser-enhanced photo-CVD using Si2H6 (Lian, S.; Fowler, Burt; Krishnan, S.; Jung, Le-Tien; Li, C.; Banerjee, Sanjay)V1598,98-107(1991)

Charge characteristics of thin rapid-thermal-nitrided SiOxNy film in MIS structure (Chen, Pu S.; Yang, Jing)V1519,258-262(1991)

Computer modeling of the dynamics of nanosecond laser annealing of amorphous thin silicon layers (Zhvavyi, S.; Sadovskaya, O.)V1440,8-15(1991)

Condensation mechanisms and properties of rf-sputtered a-Si:H (Ligachev, Valery A.; Filikov, V. A.; Gordeev, V. N.)V1519,214-219(1991)

Coplanar SIMMWIC circuits (Luy, Johann-Freidrich; Strohm, Karl M.; Buechler, J.)V1475,129-139(1991)

Crystallization of hydrogenated amorphous silicon film and its fractal structure (Lin, Hong Y.; Yang, Dao M.; Li, Ying X.)V1519,210-213(1991)

Data transmission at 1.3 um using silicon spatial light modulator (Xiao, Xiaodong; Goel, Kamal K.; Sturm, James C.; Schwartz, P. V.)V1476,301-304(1991)

Defect-induced stabilization of Fermi level in bulk silicon and at the silicon-metal interface (Iwanowski, Ryszard J.; Tatarkiewicz, Jakub J.)V1361,765-775(1991)

Density of localized states in glow-discharge a-Si1-x Cx:H (Oktu, Ozcau; Usala, Sandro; Adriaenssens, Guy J.; Tolunay, H.; Eray, A.)V1361,812-818(1991)

Device concepts for SiGe optoelectronics (Kasper, Erich; Presting, Hartmut)V1361,302-312(1991)

Dislocation mechanism of periodical relief formation in laser-irradiated silicon (Shandybina, Galina)V1440,384-389(1991)

Effect of silylation condition on the silylated image in the DESIRE process (Taira, Kazuo; Takahashi, Junichi; Yanagihara, Kenji)V1466,570-582(1991)

Electronic and optical properties of silicide/silicon IR detectors (Cabanski, Wolfgang A.; Schulz, Max J.)V1484,81-97(1991)

Epitaxial growth of the semiconducting silicide FeSi2 on silicon (Chevrier, Joel S.; Thanh, V. L.; Derrien, J.)V1512,278-288(1991)

Formation and electronic properties of epitaxial erbium silicide (Nguyen, Tan T.; Veuillen, J. Y.)V1512,289-298(1991)

Formation of SiO2 film on plastic substrate by liquid-phase-deposition method (Kitaoka, Masaki; Honda, Hisao; Yoshida, Harunobu; Takigawa, Akio; Kawahara, Hideo)V1519,109-114(1991)

GaN single-crystal films on silicon substrates grown by MOVPE (Nagatomo, Takao; Ochiai, Ichiro; Ookoshi, Shigeo; Omoto, Osamu)V1519,90-95(1991)

Heteroepitaxy of II-VI and IV-VI semiconductors on Si substrates (Zogg, Hans; Tiwari, A. N.; Blunier, Stefan; Maissen, Clau; Masek, Jiri)V1361,406-413(1991)

High-field electron trapping and detrapping characteristics in thin SiOxNy films (Yang, Bing L.; Liu, Bai Y.; Cheng, Y. C.; Wong, H.)V1519,241-246(1991)

Highly sensitive microresinoid siloxane resist for EB and deep-UV lithography (Yamazaki, Satomi; Ishida, Shinji; Matsumoto, Hiroshi; Aizaki, Naoaki; Muramoto, Naohiro; Mine, Katsutoshi)V1466,538-545(1991)

High-performance a-Si:H TFT-driven linear image sensor and its application to a compact scanner (Miyake, Hiroyuki; Sakai, Kazuhiro; Abe, Tsutomu; Sakai, Yoshihiko; Hotta, Hiroyuki; Sugino, Hajime; Ito, Hisao; Ozawa, Takashi)V1448,150-156(1991)

Hot carrier silicon phototransistor (Asmontas, Steponas; Gradauskas, Jonas; Sirmulis, Edmundas)V1512,131-134(1991)

Integrated optical channel waveguides in silicon using SiGe alloys (Splett, Armin O.; Schmidtchen, Joachim; Schueppert, B.; Petermann, Klaus)V1362,827-833(1991)

Ion beam modification of glasses (Mazzoldi, Paolo; Carnera, Alberto; Caccavale, F.; Granozzi, G.; Bertoncello, R.; Battaglin, Giancarlo; Boscolo-Boscoletto, A.; Polato, P.)V1513,182-197(1991)

Laser-induced thermal desorption studies of surface reaction kinetics (George, Steven M.; Coon, P. A.; Gupta, P.; Wise, M. L.)V1437,157-165(1991)

Low-temperature operation of silicon drift detectors (Sumner, Timothy J.; Roe, S.; Rochester, G. K.; Hall, G.; Evensen, Per; Avset, B. S.)V1549,265-273(1991)

Measuring films on and below polycrystalline silicon using reflectometry (Engstrom, Herbert L.; Stokowski, Stanley E.)V1464,566-573(1991)

Micromachined scanning mirrors for laser beam deflection (Huber, Peter; Gerlach-Meyer, U.)V1522,135-141(1991)

Modeling of photochemical vapor deposition of epitaxial silicon using an ArF excimer laser (Fowler, Burt; Lian, S.; Krishnan, S.; Jung, Le-Tien; Li, C.; Banerjee, Sanjay)V1598,108-117(1991)

Neutron transmutation doping of silicon in the VVR-S type nuclear reactor (Halmagean, Eugenia T.)V1362,984-991(1991)

New deposition system for the preparation of doped a-Si:H (Wu, Zhao P.; Chen, Ru G.; Wang, Yonglin)V1519,194-198(1991)

Optical polysilicon over oxide thickness measurement (Kaiser, Anne M.)V1464,554-565(1991)

Optical properties of short-period Si/Ge superlattices grown on (001) Ge studied with photoreflectance (Menczigar, Ulrich; Dahmen, M.; Zachai, Reinhard; Eberl, K.; Abstreiter, Gerhard)V1361,282-292(1991)

Oxygen plasma etching of silylated resist in top-imaging lithographic process (Dijkstra, Han J.)V1466,592-603(1991)

Peculiarities of the formation of periodic structures on silicon under millisecond laser radiation (Kokin, A. N.; Libenson, Michail N.; Minaev, Sergei M.)V1440,338-341(1991)

Preparation of SiO2 film utilizing equilibrium reaction in aqueous solution (Kawahara, Hideo; Sakai, Y.; Goda, Takuji; Hishinuma, Akihiro; Takemura, Kazuo)V1513,198-203(1991)

Progress in the study of development-free vapor photolithography (Hong, Xiao-Yin; Liu, Dan; Li, Zhong-Zhe; Xiao, Ji-Quang; Dong, Gui-Rong)V1466,546-557(1991)

Raman scattering characterization of direct gap Si/Ge superlattices (White, Julian D.; Gell, Michael A.; Fasol, Gerhard; Gibbings, C. J.; Tuppen, C. G.)V1361,293-301(1991)

Recent developments of x-ray lithography in Canada (Chaker, Mohamed; Boily, S.; Ginovker, A.; Jean, A.; Kieffer, J. C.; Mercier, P. P.; Pepin, Henri; Leung, Pak K.; Currie, John F.; Lafontaine, H.)V1465,16-25(1991)

Recent progress in artificial vision (Normann, Richard A.)V1423,40-45(1991)

Recent progress in Si thin film technology for solar cells (Kuwano, Yukinori; Nakano, Shoichi; Tsuda, Shinya)V1519,200-209(1991)

Research on relaxation process of a-Si:H film photoconductivity and the trap effect (Gong, Dao B.)V1519,281-286(1991)

Resist design for dry-developed positive working systems in deep-UV and e-beam lithography (Vinet, Francoise; Chevallier, M.; Pierrat, Christophe; Guibert, Jean C.; Rosilio, Charles; Mouanda, B.; Rosilio, A.)V1466,558-569(1991)

Resonant IR laser-induced diffusion of oxygen in silicon (Artsimovich, M. V.; Baranov, A. N.; Krivov, V. V.; Kudriavtsev, E. M.; Lotkova, E. N.; Makeev, B. H.; Mogilnik, I. F.; Pavlovich, V. N.; Romanuk, B. N.; Soroka, V. I.; Tokarevski, V. V.; Zotov, S. D.)V1397,729-733(1991)

Resonant tunneling in microcrystalline silicon quantum box diode (Tsu, Raphael; Ye, Qui-Yi; Nicollian, Edward H.)V1361,232-235(1991)

Si:Ga focal-plane arrays for satellite and ground-based telescopes (Mottier, Patrick; Agnese, Patrick; Lagage, Pierre O.)V1512,60-67(1991)

Silicon-on-silicon microsystem in plastic packages (Novak, Agneta; Glaes, Anders; Blom, Claes; Hentzell, Hans; Hodges, Charles; Kalhur, Farzeen)V1389,80-86(1991)

Silicon nitride film formed by NH3 plasma-enhanced thermal nitridation (Gu, Zhi G.; Li, Bing Z.)V1519,247-251(1991)

Simulation of connected image reversal and DESIRE techniques for submicron lithography (Nicolau, Dan V.; Dusa, Mircea V.; Fulga, Florin)V1466,663-669(1991)

Soft X-UV silver silicon multilayer mirrors (Shao, Jian D.; Fan, Zheng X.; Guo, Yong H.; Jin, Lei)V1519,298-301(1991)

Some new insights in the physics of quantum-well devices (Tsu, Raphael)V1361,313-324(1991)

SPAT studies of near-surface defects in silicon induced by BF2+ and F++B+ implantation (Li, Xiao Q.; Lin, Cheng L.; Zou, Shi C.; Weng, Hei M.; Han, Xue D.)V1519,14-17(1991)

Studies of InSb metal oxide semiconductor structure fabricated by photo-CVD using Si2H6 and N2O (Huang, C. J.; Su, Yan K.; Leu, R. L.)V1519,70-73(1991)

Study of properties of a-Si1-xGex:H prepared by SAP-CVD method (Wang, Yi-Ming; Jing, Lian-hua; Pang, Da-wen)V1361,325-335(1991)

Study of silylation mechanisms and kinetics through variations in silylating agent and resin (Dao, T. T.; Spence, Christopher A.; Hess, Dennis W.)V1466,257-268(1991)

Study of structural imperfections in epitaxial beta-SiC layers by method of x-ray differential diffractometry (Baranov, Igor M.; Kutt, R. N.; Nikitina, Irina P.)V1361,1110-1115(1991)

Study of the microstructures in Ar+ laser crystallized films of a-Si:H for active layer of thin film transistors (Huang, Xin F.; Zhang, Xiang D.; Zhu, Wei Y.; Chen, Ying Y.)V1519,220-224(1991)

Study on the high-field current transport mechanisms in thin SiOxNy films (Yang, Bing L.; Liu, Bai Y.; Chen, D. N.; Cheng, Y. C.; Wong, H.)V1519,269-274(1991)

Surface flashover of silicon (Williams, Paul F.; Peterkin, Frank E.; Ridolfi, Tim; Buresh, L. L.; Hankla, B. J.)V1378,217-225(1991)

Surface imaging on the basis of phenolic resin: experiments and simulation (Bauch, Lothar; Jagdhold, Ulrich A.; Dreger, Helge H.; Bauer, Joachim J.; Hoeppner, Wolfgang W.; Erzgraeber, Hartmut H.; Mehliss, Georg G.)V1466,510-519(1991)

Technique for improving the calibration of large-area sphere sources (Walker, James H.; Cromer, Chris L.; McLean, James T.)V1493,224-230(1991)

Temperature instability in silicon-based microheating device (Shie, Jin-Shown; Lian, Jiunn-Long)V1362,655-663(1991)

Thermal stabilities of a-Si:H films and its application to thyristor elements (Sun, Yue Z.; Chen, Chun X.; Xie, Qi Y.; Yin, Chen Z.; He, Yu L.)V1519,234-240(1991)

Time-derivative adaptive silicon photoreceptor array (Delbruck, Tobi; Mead, Carver A.)V1541,92-99(1991)

Two-dimensional model for high-efficiency microgroove silicon solar cells (Sobhan, M. A.; Islam, M. N.)V1536,246-257(1991)

Understanding metrology of polysilicon gates through reflectance measurements and simulation (Tadros, Karim H.; Neureuther, Andrew R.; Guerrieri, Roberto)V1464,177-186(1991)

Vertical 3-D integration of silicon waveguides in a Si-SiO2-Si-SiO2-Si structure (Soref, Richard A.; Namavar, Fereydoon; Cortesi, Elisabetta; Friedman, Lionel R.; Lareau, Richard)V1389,408-421(1991)

Wave-function engineering in Si-Ge microstructures: linear and nonlinear optical response (Jaros, Milan; Turton, Richard M.)V1362,242-253(1991)

Work-induced stress and long-term stability in optically polished silicon (Bender, John W.; Wahl, Roger L.)V1533,264-276(1991)

X-ray lithography system development at IBM: overview and status (Maldonado, Juan R.)V1465,2-15(1991)

Solar energy—see also chromogenics; coatings

Analytic solutions and numerical results for the optical and radiative properties of V-trough concentrators (Fraidenraich, Naum)V1528,15-30(1991)

Angular confining cavities for photovoltaics (Minano, Juan C.; Luque, Antonio)V1528,58-69(1991)

Application of nonimaging optical concentrators to infrared energy detection (Ning, Xiaohui)V1528,88-92(1991)

Asymmetric photopotentials from thin polymeric porphyrin films (Wamser, Carl C.; Senthilathipan, Velu; Li, Wen)V1436,114-124(1991)

Beam quality measurements and improvement in solar-pumped laser systems (Bernstein, Hana; Thompson, George A.; Yogev, Amnon; Oron, Moshe)V1442,81-88(1991)

Calibrated intercepts for solar radiometers used in remote sensor calibration (Gellman, David I.; Biggar, Stuart F.; Slater, Philip N.; Bruegge, Carol J.)V1493,175-180(1991)

Calibration of EOS multispectral imaging sensors and solar irradiance variability (Mecherikunnel, Ann)V1493,292-302(1991)

Charge separation in functionalized tetrathiafulvalene derivatives (Fox, Marye A.; Pan, Horng-Lon)V1436,2-7(1991)

Close-spaced vapor transport of II-VI semiconductors (Perrier, Gerard)V1536,258-267(1991)

Clouds as calibration targets for AVHRR reflected-solar channels: results from a two-year study at NOAA/NESDIS (Abel, Peter)V1493,195-206(1991)

Compact nonimaging lens with totally internally reflecting facets (Parkyn, William A.; Pelka, David G.)V1528,70-81(1991)

Comparison of asymmetrical and symmetrical nonimaging reflectors for east-west circular cylindrical solar receivers (Mills, David R.; Monger, A. G.; Morrison, G. L.)V1528,44-55(1991)

Design, fabrication, and integration of holographic dispersive solar concentrator for terrestrial applications (Stojanoff, Christo G.; Kubitzek, Ruediger; Tropartz, Stephan; Froehlich, K.; Brasseur, Olivier)V1536,206-214(1991)

Design of nonimaging lenses and lens-mirror combinations (Minano, Juan C.; Gonzalez, Juan C.)V1528,104-115(1991)

Droplet-size effects in light scattering from polymer-dispersed liquid-crystal films (Montgomery, G. P.; West, John L.; Tamura-Lis, Winifred)V1455,45-53(1991)

Electrodeposited nickel-cobalt thin films for photothermal conversion of solar energy (Karuppiah, N.; John, S.; Natarajan, Sanjay S.; Sivan, V.)V1536,215-221(1991)

Fiber optic lighting system for plant production (St. George, Dennis R.; Feddes, John J.)V1379,69-80(1991)

Flight solar calibrations using the mirror attenuator mosaic: low-scattering mirror (Lee, Robert B.)V1493,267-280(1991)

Infrared techniques applied to large solar arrays: a ten-year update (Hodor, James R.; Decker, Herman J.; Barney, Jesus J.)V1540,331-337(1991)

Luminescence and chemical potential of solar cells (Smestad, Greg P.; Ries, Harald)V1536,234-245(1991)

Nonimaging Optics: Maximum Efficiency Light Transfer (Winston, Roland; Holman, Robert L., eds.)V1528(1991)

Nonimaging optics: optical design at the thermodynamic limit (Winston, Roland)V1528,2-6(1991)

Optical characterization of solar-selective transmitting coatings (Roos, Arne)V1536,158-168(1991)

Optical Materials Technology for Energy Efficiency and Solar Energy Conversion X (Lampert, Carl M.; Granqvist, Claes G., eds.)V1536(1991)

Optical performance of angle-dependent light-control glass (Maeda, Koichi; Ishizuka, S.; Tsujino, T.; Yamamoto, H.; Takigawa, Akio)V1536,138-148(1991)

Optimization of optical properties of pigmented foils for radiative cooling applications: model calculations (Nilsson, Torbjorn M.; Niklasson, Gunnar A.)V1536,169-182(1991)

Photoelectrochemical characteristics of slurry-coated CdSeTe films (Murali, K. R.; Subramanian, V.; Rangarajan, N.; Lakshmanan, A. S.; Rangarajan, S. K.)V1536,289-295(1991)

Photoelectrochemical etching of n-InP producing antireflecting structures for solar cells (Soltz, David; Cescato, Lucila H.; Decker, Franco)V1536,268-276(1991)

P-n heterojunction and Schottky barrier formation between poly(3-methylthiophene) and n-type cadmium sulfide (Frank, Arthur J.; Glenis, Spyridon)V1436,50-57(1991)

Preparation, electrochemical, photoelectrochemical, and solid state characteristics of In-incorporated TiO2 thin films for solar energy applications (Badawy, Waheed A.; El-Giar, Emad M.)V1536,277-288(1991)

Pulsed photoconductivity of chlorophyll a (Kassi, Hassane; Hotchandani, Surat; Leblanc, Roger M.; Beaudoin, N.; Dery, M.)V1436,58-67(1991)

Ratioing radiometer for use with a solar diffuser (Palmer, James M.; Slater, Philip N.)V1493,106-117(1991)

Recent developments in nonimaging secondary concentrators for linear receiver solar collectors (Gordon, Jeff M.)V1528,32-43(1991)

Recent progress in Si thin film technology for solar cells (Kuwano, Yukinori; Nakano, Shoichi; Tsuda, Shinya)V1519,200-209(1991)

Solar-diffuser panel and ratioing radiometer approach to satellite sensor on-board calibration (Slater, Philip N.; Palmer, James M.)V1493,100-105(1991)

Solar-powered blackbody-pumped lasers (Christiansen, Walter H.; Sirota, J. M.)V1397,821-825(1991)

Solid state conductivity and photoconductivity studies of an ion-exchange polymer/dye system (Huang, Zhiqing; Ordonez, I.; Ioannidis, Andreas A.; Langford, C. H.; Lawrence, Marcus F.)V1436,103-113(1991)

Solid state ionics and optical materials technology for energy efficiency, solar energy conversion, and environment control (Lusis, Andrejs R.)V1536,116-124(1991)

Stray light effects on calibrations using a solar diffuser (Palmer, James M.)V1493,143-154(1991)

Theory of solar-powered, cavityless waveguide laser (Mel'nikov, Lev Y.; Kochelap, Viatcheslav A.; Izmailov, I. A.)V1501,144-149(1991)

Thermodynamics of light concentrators (Ries, Harald; Smestad, Greg P.; Winston, Roland)V1528,7-14(1991)

TiNxOy-Cu coatings for low-emissive solar-selective absorbers (Lazarov, M.; Roehle, B.; Eisenhammer, T.; Sizmann, R.)V1536,183-191(1991)

Two-dimensional control of radiant intensity by use of nonimaging optics (Kuppenheimer, John D.; Lawson, Robert I.)V1528,93-103(1991)

Two-dimensional model for high-efficiency microgroove silicon solar cells (Sobhan, M. A.; Islam, M. N.)V1536,246-257(1991)

Ultraviolet reflector materials for solar detoxification of hazardous waste (Jorgensen, Gary J.; Govindarajan, Rangaprasad)V1536,194-205(1991)

Very high temperature fiber processing and testing through the use of ultrahigh solar energy concentration (Jacobson, Benjamin A.; Gleckman, Philip; Holman, Robert L.; Sagie, Daniel; Winston, Roland)V1528,82-85(1991)

Sol-gels—see also coatings

Active glasses prepared by the sol-gel method including islands of CdS or silver (Reisfeld, Renata; Minti, Harry; Eyal, Marek)V1513,360-367(1991)

Advanced technology development for high-fluence laser applications (Novaro, Marc; Andre, Michel L.)V1501,183-193(1991)

Development of optical waveguides by sol-gel techniques for laser patterning (Schmidt, Helmut; Krug, Herbert; Kasemann, Reiner; Tiefensee, Frank)V1590,36-43(1991)

Glasses including quantum dots of cadmium sulfide, silver, and laser dyes (Reisfeld, Renata; Eyal, Marek; Chernyak, Valery; Jorgensen, Christian K.)V1590,215-228(1991)

Gradient-index fiber-optic preforms by a sol-gel method (Banash, Mark A.; Caldwell, J. B.; Che, Tessie M.; Mininni, Robert M.; Soskey, Paul R.; Warden, Victor N.; Pope, Edward J.)V1590,8-13(1991)

Influence of processing variables on the optical properties of SiO2-TiO2 planar waveguides (Weisenbach, Lori; Zelinski, Brian J.; O'Kelly, John; Morreale, Jeanne; Roncone, Ronald L.; Burke, James J.)V1590,50-58(1991)

IR-spectroscopical investigations on the glass structure of porous and sintered compacts of colloidal silica gels (Clasen, Rolf; Hornfeck, M.; Theiss, W.)V1513,243-254(1991)

Low-temperature ion exchange of dried gels for potential waveguide fabrication in glasses (Risen, William M.; Morse, Ted F.; Tsagaropoulos, George)V1590,44-49(1991)

Optical properties of glass materials obtained by inorganic sol-gel synthesis (Glebov, Leonid B.; Evstropiev, Sergei K.; Petrovskii, Gurii T.; Shashkin, Viktor S.)V1513,224-231(1991)

Planar and strip optical waveguides by sol-gel method and laser densification (Guglielmi, Massimo; Colombo, Paolo; Mancinelli Degli Esposti, Luca; Righini, Giancarlo C.; Pelli, Stefano)V1513,44-49(1991)

Sol-gel derived BaTiO3 thin films (Xiang, Xiao L.; Hou, Li S.; Gan, Fuxi)V1519,712-716(1991)

Sol-gel overview: transparent, microporous silica, its synthesis and characterization (Klein, Lisa C.)V1590,2-7(1991)

Structure and properties of electrochromic WO3 produced by sol-gel methods (Bell, J. M.; Green, David C.; Patterson, A.; Smith, Geoffrey B.; MacDonald, K. A.; Lee, K. D.; Kirkup, L.; Cullen, J. D.; West, B. O.; Spiccia, L.; Kenny, M. J.; Wielunski, L. S.)V1536,29-36(1991)

Submolecular Glass Chemistry and Physics (Bray, Phillip; Kreidl, Norbert J., eds.)V1590(1991)

Supported sol-gel thin-film glasses embodying laser dyes II: three-layered waveguide assemblies (Haruvy, Yair; Heller, Adam; Webber, Stephen E.)V1590,59-70(1991)

Synthesis and properties of sol-gel-derived AgClxBr1-x colloid containing sodium alumo borosilicate glasses (Mennig, Martin; Schmidt, Helmut; Fink, Claudia)V1590,152-159(1991)

Transparent storage layers for H+ and Li+ ions prepared by sol-gel technique (Valla, Bruno; Tonazzi, Juan C.; Macedo, Marcelo A.; Dall'Antonia, L. H.; Aegerter, Michel A.; Gomes, M. A.; Bulhoes, Luis O.)V1536,48-62(1991)

Waveguide formation by laser irradiation of sol-gel coatings (Zaugg, Thomas C.; Fabes, Brian D.; Weisenbach, Lori; Zelinski, Brian J.)V1590,26-35(1991)

Sources—see also radiation; sun; synchrotron radiation

10.6 um laser frequency stabilization system with two optical circuits (Hu, Yu; Li, Xian; Zhu, Dayong; Ye, Naiqun; Pen, Shengyang)V1409,230-239(1991)

Disk-cathode flash x-ray tube driven by a repetitive type of Blumlein pulser (Sato, Eiichi; Kimura, Shingo; Isobe, Hiroshi; Takahashi, Kei; Tamakawa, Yoshiharu; Yanagisawa, Toru)V1358,146-153(1991)

Efficient, uniform transverse coupling of diode arrays to laser rods using nonimaging optics (Minnigh, Stephen W.; Knights, Mark G.; Avidor, Joel M.; Chicklis, Evan P.)V1528,129-134(1991)

Flash soft radiography: its adaption to the study of breakup mechanisms of liquid jets into a high-density gas (Krehl, Peter; Warken, D.)V1358,162-173(1991)

Flexible object-centered illuminator (Uber, Gordon T.)V1385,2-7(1991)

High-repetition-rate x-ray preionization source (van Goor, Frederik A.)V1397,563-568(1991)

Isotropic light source for underwater applications (Brown, Robert A.; Honey, Richard C.; Maffione, Robert A.)V1537,147-150(1991)

Miniaturized semiconductor light source system for Cranz-Schardin applications (Stasicki, Boleslaw; Meier, G. E.)V1358,1222-1227(1991)

Mobile system for measuring retroreflectance of traffic signs (Lumia, John J.)V1385,15-26(1991)

Monte Carlo calculation of light distribution in an integrating cavity illuminator (Kaplan, Martin)V1448,206-217(1991)

New rippleflash system for large-area high-intensity lighting in harsh environments (Rendell, John T.)V1358,806-810(1991)

New technologies in lighting systems for high-speed film and photography regarding high-intensity and heat problems (Severon, Burkhard)V1358,1202-1208(1991)

Penning discharge VUV and soft x-ray source (Cao, Jianlin; Li, Futian; Qian, Limin; Chen, Po; Ma, Yueying; Chen, Xingdan)V1345,71-77(1991)

Photoinitiation of chemical lasers using radiation from a cylindral surface discharge (Beverly, R. E.)V1397,581-584(1991)

Plastic optical fiber applications for lighting of airports and buildings (Jaquet, Patrick J.)V1592,165-172(1991)

Repetitive flash x-ray generator as an energy transfer source utilizing a compact-glass body diode (Shikoda, Arimitsu; Sato, Eiichi; Kimura, Shingo; Isobe, Hiroshi; Takahashi, Kei; Tamakawa, Yoshiharu; Yanagisawa, Toru)V1358,154-161(1991)

Repetitive flash x-ray generator utilizing an enclosed-type diode with a ring-shaped graphite cathode (Isobe, Hiroshi; Sato, Eiichi; Kimura, Shingo; Tamakawa, Yoshiharu; Yanagisawa, Toru; Honda, Keiji; Yokota, Yoshiharu)V1358,201-208(1991)

Scanning exciton microscopy and single-molecule resolution and detection (Kopelman, Raoul; Tan, Weihong; Lewis, Aaron; Lieberman, Klony)V1435,96-101(1991)

Solar-powered blackbody-pumped lasers (Christiansen, Walter H.; Sirota, J. M.)V1397,821-825(1991)

Test results on pulsed cesium amalgam flashlamps for solid state laser pumping (Witting, Harald L.)V1410,90-98(1991)

Uniform energy discharge for pulsed lasers (Koschmann, Eric C.)V1411,118-126(1991)

VUV wall stabilized argon arc discharge source (Li, Futian; Cao, Jianlin; Chen, Po; Qian, Limin; Jin, Lei; Chen, Xingdan)V1345,78-88(1991)

Space optics—see also astronomy; robotics; satellites; telescopes

A 39-photon/bit direct-detection receiver at 810 nm, BER = 1x10-6, 60-Mbit/s QPPM (MacGregor, Andrew D.; Dion, Bruno; Noeldeke, Christoph; Duchmann, Olivier)V1417,374-380(1991)

Achieving a balance between autonomy and tele-operation in specifying plans for a planetary rover (Lyons, Damian M.; Allton, Judith H.)V1387,124-133(1991)

Acquisition and tracking performance measurements for a high-speed area array detector system (Short, Ralph C.; Cosgrove, Michael A.; Clark, David L.; Martino, Anthony J.; Park, Hong-Woo; Seery, Bernard D.)V1417,131-141(1991)

Active vibration filtering: application to an optical delay line for stellar interferometer (Koehler, Bertrand; Bourlon, Philippe M.; Dugue, Michel)V1489,177-188(1991)

Adaptive gross motion control: a case study (Leahy, Michael B.; Whalen, P. V.; Lamont, Gary B.)V1387,148-158(1991)

Adaptive structures technology programs for space-based optical systems (Betros, Robert S.; Bronowicki, Allen J.; Manning, Raymond A.)V1542,406-419(1991)

Advanced Si IR detectors using molecular beam epitaxy (Lin, TrueLon; Jones, E. W.; George, T.; Ksendzov, Alexander; Huberman, M. L.)V1540,135-139(1991)

Advanced X-ray Astrophysics Facility science instruments (Winkler, Carl E.; Dailey, Carroll C.; Cumings, Nesbitt P.)V1494,301-313(1991)

Analysis and testing of a soft actuation system for segmented-reflector articulation and isolation (Jandura, Louise; Agronin, Michael L.)V1542,213-224(1991)

Analysis of cooperative robot manipulators on a mobile platform (Murphy, Steve H.; Wen, John T.; Saridis, George N.)V1387,14-25(1991)

Angular positioning mechanism for the ultraviolet coronagraph spectrometer (Ostaszewski, Miroslaw A.; Guy, Larry J.)V1482,13-25(1991)

Application of laser radar to autonomous spacecraft landing (Gleichman, Kurt W.; Tchoryk, Peter; Sampson, Robert E.)V1416,286-294(1991)

ASTRO 1M: a new system for attitude determination in space (Elstner, Christian; Lichtenauer, Gert; Skarus, Waldemar)V1478,150-159(1991)

ATHENA: a high-resolution wide-area coverage commercial remote sensing system (Claybaugh, William R.; Megill, L. R.)V1495,81-94(1991)

Automated assembly system for large space structures (Will, Ralph W.; Rhodes, Marvin D.)V1387,60-71(1991)

Automation of vehicle processing at Space Station Freedom (Vargo, Rick C.; Sklar, Michael E.; Wegerif, Daniel G.)V1387,72-81(1991)

Autonomous guidance, navigation, and control bridging program plan (McSwain, G. G.; Fernandes, Stan T.; Doane, Kent B.)V1478,228-238(1991)

Bandwidth requirements for direct detection optical communication receivers with PPM signaling (Davidson, Frederic M.; Sun, Xiaoli; Krainak, Michael A.)V1417,75-88(1991)

Bandwidth, throughput, and information capacity of fiber optic networks for space systems (Choudry, Amar)V1369,121-125(1991)

Building and maintaining a local on-orbit reference frame model (Viggh, Herbert E.)V1387,224-236(1991)

Calibration for the medium-resolution imaging spectrometer (Baudin, Gilles; Chessel, Jean-Pierre; Cutter, Mike A.; Lobb, Daniel R.)V1493,16-27(1991)

Calibration of EOS multispectral imaging sensors and solar irradiance variability (Mecherikunnel, Ann)V1493,292-302(1991)

Calibration of Passive Remote Observing Optical and Microwave Instrumentation (Guenther, Bruce W., ed.)V1493(1991)

Calibration procedures for the space vision system experiment (MacLean, Steve G.; Pinkney, H. F.)V1526,113-122(1991)

Comparative life test of 0.8-um laser diodes for SILEX under NRZ and QPPM modulation (Menke, Bodo; Loeffler, Roland)V1417,316-327(1991)

Comparison of optical technologies for a high-data-rate Mars link (Spence, Rodney L.)V1417,550-561(1991)

Comparison of the angular resolution limit and SNR of the Hubble Space Telescope and the large ground-based telescopes (Souilhac, Dominique J.; Billerey, Dominique)V1494,503-526(1991)

Complementary experiments for tether dynamics analysis (Wingo, Dennis R.; Bankston, Cheryl D.)V1495,123-133(1991)

Construction of the 16-meter large lunar telescope (Omar, Husam A.)V1494,135-146(1991)

Cooperative Intelligent Robotics in Space (Stoney, William E.; deFigueiredo, Rui J., eds.)V1387(1991)

Current and future activities in the area of optical space communications in Japan (Fujise, Masayuki; Araki, Ken'ichi; Arikawa, Hiroshi; Furuhama, Yoji)V1522,14-26(1991)

Current instrument status of the airborne visible/infrared imaging spectrometer (AVIRIS) (Eastwood, Michael L.; Sarture, Charles M.; Chrien, Thomas G.; Green, Robert O.; Porter, Wallace M.)V1540,164-175(1991)

DARPA initiatives in small-satellite technologies (Bonometti, Robert J.; Wheatley, Alvis A.; Flynn, Lin; Nicastri, Edward; Sudol, R.)V1495,166-176(1991)

Definition of interorbit link optical terminals with diode-pumped Nd:host laser technology (Marini, Andrea E.; Della Torre, Antonio)V1522,222-233(1991)

Deformable-mirror concept for adaptive optics in space (Kuo, Chin-Po)V1542,420-431(1991)

Design and control of ultralight manipulators for interplanetary exploration (Byler, Eric A.)V1387,313-327(1991)

Design and performance of an automatic gain control system for the high-energy x-ray timing experiment (Pelling, Michael R.; Rothschild, Richard E.; MacDonald, Daniel R.; Hertel, Robert H.; Nishiie, Edward S.)V1549,134-146(1991)

Design and testing of a nonreactive, fingertip, tactile display for interaction with remote environments (Patrick, Nicholas J.; Sheridan, Thomas B.; Massimino, Michael J.; Marcus, Beth A.)V1387,215-222(1991)

Design of a diode-pumped Nd:YAG laser communication system (Sontag, Heinz; Johann, Ulrich A.; Pribil, Klaus)V1417,573-587(1991)

Design of a periscopic coarse pointing assembly for optical multiple access (Gatenby, Paul V.; Boereboom, Peter; Grant, Michael A.)V1522,126-134(1991)

Design optimization of optical filters for space applications (Annapurna, M. N.; Nagendra, C. L.; Thutupalli, G. K.)V1485,260-271(1991)

Design study for high-performance optical communications satellite terminal with high-power laser diode transmitter (Hildebrand, Ulrich; Seeliger, Reinhard; Smutny, Berry; Sand, Rolf)V1522,50-60(1991)

Development and test of the Starlab control system (LaMont, Douglas V.; Mar, Lim O.; Rodden, Jack J.)V1482,2-12(1991)

Diode-laser-based range sensor (Seshamani, Ramani)V1471,354-356(1991)

Diode-pumped Nd:YAG laser transmitter for free-space optical communications (Nava, Enzo; Re Garbagnati, Giuseppe; Garbi, Maurizio; Marchiori, Livio D.; Marini, Andrea E.)V1417,307-315(1991)

Displacement damage in Si imagers for space applications (Dale, Cheryl J.; Marshall, Paul W.)V1447,70-86(1991)

Distributed-parameter estimation for NASA Mini-Mast truss through displacement measurements (Huang, Jen-Kuang; Shen, Ji-Yao; Taylor, Lawrence W.)V1489,266-277(1991)

Dynamic characteristics of joint-dominated space trusses (Shih, Choon-Foo; Kuo, Chin-Po)V1532,91-102(1991)

Economic factors for a free-space laser communication system (Begley, David L.; Marshalek, Robert G.)V1522,234-242(1991)

Effect of microaccelerations on an optical space communication system (Wittig, Manfred E.)V1522,278-286(1991)

Effects of the lunar environment on optical telescopes and instruments (Johnson, Charles L.; Dietz, Kurtis L.)V1494,208-218(1991)

Enhancements to the radiometric calibration facility for the Clouds and the Earth's Radiant Energy System instruments (Folkman, Mark A.; Jarecke, Peter J.; Darnton, Lane A.)V1493,255-266(1991)

Environmental testing of a Q-switched Nd:YLF laser and a Nd:YAG ring laser (Robinson, Deborah L.)V1417,562-572(1991)

Environments stressful to optical materials in low earth orbit (Musikant, Solomon; Malloy, W. J.)V1330,119-130(1991)

Equipment development for an atmospheric-transmission measurement campaign (Ruiz, Domingo; Kremer, Paul J.)V1417,212-222(1991)

European SILEX project and other advanced concepts for optical space communications (Oppenhaeuser, Gotthard; Wittig, Manfred E.; Popescu, Alexandru F.)V1522,2-13(1991)

Evaluation of a CCD star camera at Table Mountain Observatory (Strikwerda, Thomas E.; Fisher, H. L.; Frank, L. J.; Kilgus, Charles C.; Gray, Connie B.; Barnes, Donald L.)V1478,13-23(1991)

Experimental testbed for cooperative robotic manipulators (Desrochers, Alan A.)V1387,2-13(1991)

Experiments in tele-operator and autonomous control of space robotic vehicles (Alexander, Harold L.)V1388,560-565(1991)

Fabrication of toroidal and coma-corrected toroidal diffraction gratings from spherical master gratings using elastically deformable substrates: a progress report (Huber, Martin C.; Timothy, J. G.; Morgan, Jeffrey S.; Lemaitre, Gerard R.; Tondello, Giuseppe; Naletto, Giampiero)V1494,472-480(1991)

Far-field and wavefront characterization of a high-power semiconductor laser for free-space optical communications (Cornwell, Donald M.; Saif, Babak N.)V1417,431-439(1991)

Fiber optics in liquid propellant rocket engine environments (Delcher, Ray C.; Dinnsen, Doug K.; Barkhoudarian, S.)V1369,114-120(1991)

Filled-arm Fizeau telescope (Synnott, Stephen P.)V1494,334-343(1991)

Finite element analysis of deformation in large optics due to space environment radiation (Merzbacher, Celia I.; Friebele, E. J.; Ruller, Jacqueline A.; Matic, P.)V1533,222-228(1991)

Fixation method with high-orientation accuracy for optical terminals in space (Bauer, Dietrich; Lober, K.; Seeliger, Reinhard)V1533,277-285(1991)

Fixing the Hubble Space Telescope (Crocker, James H.)V1494,2-8(1991)

Foundation, excavation, and radiation-shielding concepts for a 16-m large lunar telescope (Chua, Koon M.; Johnson, Stewart W.)V1494,119-134(1991)

Four-meter lunar engineering telescope (Peacock, Keith; Giannini, Judith A.; Kilgus, Charles C.; Bely, Pierre Y.; May, B. S.; Cooper, Shannon A.; Schlimm, Gerard H.; Sounder, Charles; Ormond, Karen A.; Cheek, Eric A.)V1494,147-159(1991)

Free-Space Laser Communication Technologies III (Begley, David L.; Seery, Bernard D., eds.)V1417(1991)

Free-space simulator for laser transmission (Inagaki, Keizo; Nohara, Mitsuo; Araki, Ken'ichi; Fujise, Masayuki; Furuhama, Yoji)V1417,160-169(1991)

Future directions in focal-plane signal processing for spaceborne scientific imagers (Fossum, Eric R.)V1541,62-67(1991)

Future technologies for lidar/DIAL remote sensing (Allario, Frank; Barnes, Norman P.; Storm, Mark E.)V1492,92-110(1991)

Geoscience laser ranging system design and performance predictions (Anderson, Kent L.)V1492,142-152(1991)

Goldhelox: a project to view the x-ray sun (Fair, Melody)V1549,182-192(1991)

GOPEX: a deep-space optical communications demonstration with the Galileo spacecraft (Wilson, Keith E.; Schwartz, Jon A.; Lesh, James R.)V1417,22-28(1991)

Gripper for truss structure assembly (Kelley, Robert B.; Tsai, Jodi; Bethel, Jeff; Peiffer, John)V1387,38-46(1991)

GSFC conceptual design study for an intersatellite optical multiple access communication system (Fox, Neil D.; Maynard, William L.; Clarke, Ernest S.; Bruno, Ronald C.)V1417,452-463(1991)

Hierarchical multisensor analysis for robotic exploration (Eberlein, Susan; Yates, Gigi; Majani, Eric)V1388,578-586(1991)

Hierarchical planner for space truss assembly (Mathur, Rajive K.; Sanderson, Arthur C.)V1387,47-57(1991)

High-bandwidth alignment sensing in active optical systems (Kishner, Stanley J.)V1532,215-229(1991)

High-current, high-bandwidth laser diode current driver (Copeland, David J.; Zimmerman, Robert K.)V1417,412-420(1991)

High-resolution, two-dimensional imaging, microchannel-plate detector for use on a sounding rocket experiment (Bush, Brett C.; Cotton, Daniel M.; Siegmund, Oswald H.; Chakrabarti, Supriya; Harris, Walter; Clarke, John T.)V1549,290-301(1991)

High-temperature superconductive microwave technology for space applications (Leonard, Regis F.; Connolly, Denis J.; Bhasin, Kul B.; Warner, Joseph D.; Alterovitz, Samuel A.)V1394,114-125(1991)

High-temperature superconductivity space experiment: passive millimeter wave devices (Nisenoff, Martin; Gubser, Don U.; Wolf, Stuart A.; Ritter, J. C.; Price, George E.)V1394,104-113(1991)

HOLIDDO: an interferometer for space experiments (Mary, Joel; Bernard, Yves; Lefaucheux, Francoise; Gonzalez, Francois)V1557,147-155(1991)

How to meet intersatellite links mission requirements by an adequate optical terminal design (Duchmann, Olivier; Planche, G.)V1417,30-41(1991)

HST image processing: an overview of algorithms for image restoration (Gonsalves, Robert A.; Nisenson, Peter)V1567,294-307(1991)

HST image processing: how does it work and what are the problems? (White, Richard L.; Hanisch, Robert J.)V1567,308-316(1991)

HST phase retrieval: a parameter estimation (Lyon, Richard G.; Miller, Peter E.; Gruszczak, Anthony)V1567,317-326(1991)

Hubble extra-solar planet interferometer (Shao, Michael)V1494,347-356(1991)

Hubble Space Telescope: mission, history, and systems (Endelman, Lincoln L.)V1358,422-441(1991)

Hubble Space Telescope optics: problems and solutions (Burrows, Christopher J.)V1494,528-533(1991)

Hubble Space Telescope optics status (Burrows, Christopher J.)V1567,284-293(1991)

Hybrid fiber optic/electrical network for launch vehicles (Clark, Timothy E.; Curran, Mark E.)V1369,98-106(1991)

Imaging gas scintillation proportional counters for ASTRO-D (Ohashi, T.; Makishima, K.; Ishida, M.; Tsuru, T.; Tashiro, M.; Mihara, Teruyoshi; Kohmura, Y.; Inoue, Hiroyuki)V1549,9-19(1991)

Imaging pulse-counting detector systems for space ultraviolet astrophysics missions (Timothy, J. G.)V1494,394-402(1991)

In-flight performance of the faint object camera of the Hubble Space Telescope (Greenfield, Perry E.; Paresce, Francesco; Baxter, David; Hodge, P.; Hook, R.; Jakobsen, P.; Jedrzejewski, Robert; Nota, Anatonella; Sparks, W. B.; Towers, Nigel M.; Laurance, R. J.; Macchetto, F.)V1494,16-39(1991)

In-flight performance of the Goddard high-resolution spectrograph of the Hubble Space Telescope (Troeltzsch, John R.; Ebbets, Dennis C.; Garner, Harry W.; Tuffli, A.; Breyer, R.; Kinsey, J.; Peck, C.; Lindler, Don; Feggans, J.)V1494,9-15(1991)

Infrared BRDF measurements of space shuttle tiles (Young, Raymond P.; Wood, Bobby E.; Stewart, P. L.)V1530,335-342(1991)

Infrared Space Observatory optical subsystem (Singer, Christian; Massoni, Jean A.; Mossbacher, Bernard; Cinotti, Ciro)V1494,255-264(1991)

Initial performance of the high-speed photometer (Richards, Evan; Percival, Jeff; Nelson, Matthew; Hatter, Edward; Fitch, John E.; White, Richard L.)V1494,40-48(1991)

Injection chaining of diode-pumped single-frequency ring lasers for free-space communication (Cheng, Emily A.; Kane, Thomas J.; Wallace, Richard W.; Cornwell, Donald M.)V1417,300-306(1991)

ISOCAM: a camera for the ISO satellite optical bench development (Auternaud, Danielle)V1488,64-72(1991)

Issues in mobile robotics: the unmanned ground vehicle program tele-operated vehicle (Aviles, Walter A.; Hughes, T. W.; Everett, Hobart R.; Umeda, A. Y.; Martin, Stephen W.; Koyamatsu, A. H.; Solorzano, M.; Laird, Robin T.; McArthur, S. P.)V1388,587-597(1991)

Laser terminal attitude determination via autonomous star tracking (Chapman, William W.; Fitzmaurice, Michael W.)V1417,277-290(1991)

Laser transmitter for lidar in-space technology experiment (Chang, John H.; Cimolino, Marc C.; Petros, Mulugeta)V1492,43-46(1991)

Life test of high-power Matsushita BTRS laser diodes (Holcomb, Terry L.)V1417,328-337(1991)

Line-of-sight stabilization: sensor blending (Pettit, Christopher J.)V1489,278-287(1991)

Long-wave infrared detectors based on III-V materials (Maserjian, Joseph L.)V1540,127-134(1991)

Lunar dust: implications for astronomical observatories (Johnson, Stewart W.; Chua, Koon M.; Burns, Jack O.; Slane, Frederic A.)V1494,194-207(1991)

Lunar liquid-mirror telescopes (Borra, Ermanno F.; Content, R.)V1494,219-227(1991)

Manufacture of ISO mirrors (Ruch, Eric)V1494,265-278(1991)

Medium-resolution imaging spectrometer (Baudin, Gilles; Bessudo, Richard; Cutter, Mike A.; Lobb, Daniel R.; Bezy, Jean L.)V1490,102-113(1991)

Metallic alternative to glass mirrors (active mirrors in aluminum): a review (Rozelot, Jean-Pierre; Leblanc, Jean-M.)V1494,481-490(1991)

Method of making ultralight primary mirrors (Zito, Richard R.)V1494,491-497(1991)

Microchannel-plate detectors for space-based astronomy (Crocker, James H.; Cox, Colin R.; Ray, Knute A.; Sen, Amit)V1494,434-439(1991)

Minimum jerk trajectory planning for robotic manipulators (Kyriakopoulos, Konstantinos J.; Saridis, George N.)V1387,159-164(1991)

Model-based task planning system for a space laboratory environment (Chi, Sung-Do; Zeigler, Bernard P.; Cellier, Francois E.)V1387,182-193(1991)

Moderate-resolution imaging spectrometer-tilt baseline concept (Magner, Thomas J.)V1492,272-285(1991)

Monolithic microwave integrated circuit activities in ESA-ESTEC (Gatti, Giuliano)V1475,10-24(1991)

Multimission sensor for the RESERVES small-satellite program (Kilston, Steven; Kilston, Vera M.; Utsch, Thomas F.)V1495,193-204(1991)

Multiple degree-of-freedom tracking for attitude control of an experimental system on tether-stabilized platform (Angrilli, Francesco; Baglioni, Pietro; Bianchini, Gianandrea; Da Forno, R.; Fanti, Giullo; Mozzi, Massimo)V1482,26-39(1991)

Narrowband optical interference filters (Cotton, John M.; Casey, William L.)V1417,525-536(1991)

NASA's flight-technology development program: a 650-Mbit/s laser communications testbed (Hayden, William L.; Fitzmaurice, Michael W.; Nace, Dave; Lokerson, Donald; Minott, Peter O.; Chapman, William W.)V1417,182-199(1991)

NASA's Geostationary Earth Observatory and its optical instruments (Koczor, Ronald J.)V1527,98-109(1991)

Nd:host laser-based optical terminal development study for intersatellite links (Marini, Andrea E.; Della Torre, Antonio; Popescu, Alexandru F.)V1417,200-211(1991)

Next-generation space telescope: a large UV-IR successor to HST (Illingworth, Garth)V1494,86-97(1991)

Nonrejected earth radiance performance of the visible ultraviolet experiment sensor (Betts, Timothy C.; Dowling, Jerome M.; Friedman, Richard M.)V1479,120-126(1991)

Operator-coached machine vision for space telerobotics (Bon, Bruce; Wilcox, Brian H.; Litwin, Todd; Gennery, Donald B.)V1387,337-342(1991)

Optical approach to proximity-operations communications for Space Station Freedom (Marshalek, Robert G.)V1417,53-62(1991)

Optical design considerations for next-generation space and lunar telescopes (Korsch, Dietrich G.)V1494,111-118(1991)

Optical link demonstration with a lightweight transceiver breadboard (Hemmati, Hamid; Lesh, James R.; Apostolopoulos, John G.; Del Castillo, Hector M.; Martinez, A. S.)V1417,476-483(1991)

Optical mapping instrument (Bagot, K. H.)V1490,126-132(1991)

Optical phase-locked loop for free-space laser communications with heterodyne detection (Win, Moe Z.; Chen, Chien-Chung; Scholtz, Robert A.)V1417,42-52(1991)

Optical Space Communication II (Franz, Juergen, ed.)V1522(1991)

Optical stability of diffuse reflectance materials in space (Hale, Robert R.)V1485,173-182(1991)

Optical system design for a lunar optical interferometer (Colavita, Mark M.; Shao, Michael; Hines, Braden E.; Levine, Bruce M.; Gershman, Robert)V1494,168-181(1991)

Optical target location using machine vision in space robotics tasks (Sklair, Cheryl W.; Gatrell, Lance B.; Hoff, William A.; Magee, Michael)V1387,380-391(1991)

Optimal control/structure integrated design of a flexible space platform with articulated appendages (Kelkar, Atul G.; Alberts, Thomas E.)V1489,243-253(1991)

ORFEUS alignment concept (Graue, Roland; Kampf, Dirk; Rippel, Harald; Witte, G.)V1494,377-385(1991)

Overview of fiber optics in the natural space environment (Barnes, Charles E.; Dorsky, Leonard; Johnston, Alan R.; Bergman, Larry A.; Stassinopoulos, E.)V1366,9-16(1991)

Passive and active sensors for autonomous space applications (Tchoryk, Peter; Gleichman, Kurt W.; Carmer, Dwayne C.; Morita, Yuji; Trichel, Milton; Gilbert, R. K.)V1479,164-182(1991)

PC-based hardware implementation of the maximum-likelihood classifier for the shuttle ice detection system (Jaggi, Sandeep)V1452,340-350(1991)

Performance and spectroscopic characterization of irradiated Nd:YAG (Rose, Todd S.; Fincher, Curtis L.; Fields, Renny A.)V1561,43-49(1991)

Performance of a demonstration system for simultaneous laser beacon tracking and low-data-rate optical communications with multiple platforms (Short, Ralph C.; Cosgrove, Michael A.; Clark, David L.; Oleski, Paul J.)V1417,464-475(1991)

Phase retrieval for the Hubble Space Telescope using iterative propagation algorithms (Fienup, James R.)V1567,327-332(1991)

Plan for the development and demonstration of optical communications for deep space (Lesh, James R.; Deutsch, L. J.; Weber, W. J.)V1522,27-35(1991)

Pointing, acquisition, and tracking system of the European SILEX program: a major technological step for intersatellite optical communication (Bailly, Michel; Perez, Eric)V1417,142-157(1991)

Pressure modulator infrared radiometer (PMIRR) optical system alignment and performance (Chrisp, Michael P.; Macenka, Steve A.)V1540,213-218(1991)

Prismatic anamorphic beam expanders for free-space optical communications (Jonas, Reginald P.)V1417,402-411(1991)

Progress in miniature laser systems for space science particle sizing and velocimetry (Brown, Robert G.)V1506,58-59(1991)

Progress with PN-CCDs for the XMM satellite mission (Braeuninger, Heinrich; Hauff, D.; Lechner, P.; Lutz, G.; Kink, W.; Meidinger, Norbert; Metzner, G.; Predehl, Peter; Reppin, C.; Strueder, Lothar; Truemper, Joachim; Kendziorra, E.; Staubert, R.; Radeka, V.; Rehak, P.; Rescia, S.; Bertuccio, G.; Gatti, E.; Longoni, Antonio; Sampietro, Marco; Findeis, N.; Holl, P.; Kemmer, J.; von Zanthier, C.)V1549,330-339(1991)

Qualification testing of a diode-laser transmitter for free-space coherent communications (Pillsbury, Allen D.; Taylor, John A.)V1417,292-299(1991)

Radiation effects on various optical components for the Mars Observer Spacecraft (Lowry, Jay H.; Iffrig, C. D.)V1330,132-141(1991)

Realization of a coherent optical DPSK (differential phase-shift keying) heterodyne transmission system with 565 MBit/s at 1.064 um (Wandernoth, Bernhard; Franz, Juergen)V1522,194-198(1991)

Recent advances in the development and transfer of machine vision technologies for space (deFigueiredo, Rui J.; Pendleton, Thomas W.)V1387,330-336(1991)

Reconstruction of the Hubble Space Telescope mirror figure from out-of-focus stellar images (Roddier, Claude A.; Roddier, Francois J.)V1494,78-84(1991)

Reliability analysis of a multiple-laser-diode beacon for intersatellite links (Mauroschat, Andreas)V1417,513-524(1991)

Robotics in near-earth space (Card, Michael E.)V1387,101-108(1991)

Robust image coding with a model of adaptive retinal processing (Narayanswamy, Ramkumar; Alter-Gartenberg, Rachel; Huck, Friedrich O.)V1385,93-103(1991)

Role of computer graphics in space telerobotics: preview and predictive displays (Bejczy, Antal K.; Venema, Steven; Kim, Won S.)V1387,365-377(1991)

SALSA: a synthesis array for lunar submillimeter astronomy (Mahoney, Michael J.; Marsh, Kenneth A.)V1494,182-193(1991)

Scientific results from the Hubble Space Telescope fine-guidance sensors (Taff, Laurence G.)V1494,66-77(1991)

Second-generation high-data-rate interorbit link based on diode-pumped Nd:YAG laser technology (Sontag, Heinz; Johann, Ulrich A.; Pribil, Klaus)V1522,61-69(1991)

Semiactive telescope for the French PRONAOS submillimetric mission (Duran, Michel; Luquet, Philippe; Buisson, F.; Cousin, B.)V1494,357-376(1991)

Sensors and Sensor Systems for Guidance and Navigation (Wade, Jack; Tuchman, Avi, eds.)V1478(1991)

Sensor system for comet approach and landing (Bonsignori, Roberto; Maresi, Luca)V1478,76-91(1991)

Shock-layer-induced ultraviolet emissions measured by rocket payloads (Caveny, Leonard H.; Mann, David M.)V1479,102-110(1991)

SILEX project: the first European optical intersatellite link experiment (Laurent, Bernard; Duchmann, Olivier)V1417,2-10(1991)

Simulation model and on-ground performances validation of the PAT system for the SILEX program (Cossec, Francois R.; Doubrere, Patrick; Perez, Eric)V1417,262-276(1991)

Situation assessment for space telerobotics (Bruno, Guy; Morgenthaler, Matthew K.)V1387,352-358(1991)

SOLACOS: a diode-pumped Nd:YAG laser breadboard for coherent space communication system verification (Pribil, Klaus; Johann, Ulrich A.; Sontag, Heinz)V1522,36-47(1991)

Solid state lasers for planetary exploration (Greene, Ben; Taubman, Matthew; Watts, Jeffrey W.; Gaither, Gary L.)V1492,126-139(1991)

Space Astronomical Telescopes and Instruments (Bely, Pierre Y.; Breckinridge, James B., eds.)V1494(1991)

Space-based visible all-reflective stray light telescope (Wang, Dexter; Gardner, Leo R.; Wong, Wallace K.; Hadfield, Peter)V1479,57-70(1991)

Space-based visible surveillance experiment (Dyjak, Charles P.; Harrison, David C.)V1479,42-56(1991)

Space-borne beam pointing (Eller, E. D.; LaMont, Douglas V.; Rodden, Jack J.)V1415,2-12(1991)

Space-qualified laser transmitter for lidar applications (Chang, John H.; Reithmaier, Karl D.)V1492,38-42(1991)

Space roles for robots (Cliff, Rodger A.)V1387,98-100(1991)

Space-spectrum resolution function of the imaging spectrometer (Han, Xin Z.; Qiu, Ying)V1479,156-161(1991)

Space station laser communication transceiver (Fitzmaurice, Michael W.; Hayden, William L.)V1417,13-21(1991)

Space telescope imaging spectrograph 2048 CCD and its characteristics (Delamere, W. A.; Ebben, Thomas H.; Murata-Seawalt, Debbie; Blouke, Morley M.; Reed, Richard; Woodgate, Bruce E.)V1479,21-30(1991)

Spatial acquisition and tracking for deep-space optical communication packages (Chen, Chien-Chung; Jeganathan, Muthu; Lesh, James R.)V1417,240-250(1991)

Stability considerations in relay lens design for optical communications (Gardam, Allan; Jonas, Reginald P.)V1417,381-390(1991)

Static horizon sensor for remote sensing satellite (Kamalakar, J. A.; Jain, Yashwant K.; Laxmiprasad, A. S.; Shashikala, M.)V1478,92-100(1991)

Station-keeping strategy for multiple-spacecraft interferometry (DeCou, Anthony B.)V1494,440-451(1991)

Statistical initial orbit determination (Taff, Laurence G.; Belkin, Barry; Schweiter, G. A.; Sommar, K.)V1481,440-448(1991)

Stereo vision for planetary rovers: stochastic modeling to near-real-time implementation (Matthies, Larry H.)V1570,187-200(1991)

Stratospheric wind infrared limb sounder (Rider, David M.; McCleese, Daniel J.)V1540,142-147(1991)

Structural design considerations for the Space Infrared Telescope Facility (MacNeal, Paul D.; Lou, Michael C.)V1494,236-254(1991)

Structural design of the large deployable reflector (Satter, Celeste M.; Lou, Michael C.)V1494,279-300(1991)

Subsumption architecture control system for space proximity maneuvering (Viggh, Herbert E.)V1387,202-214(1991)

Surface control techniques for the segmented primary mirror in the large lunar telescope (Gleckler, Anthony D.; Pflibsen, Kent P.; Ulich, Bobby L.; Smith, Duane D.)V1494,454-471(1991)

Surface property determination for planetary rovers (Severson, William E.; Douglass, Robert J.; Hennessy, Stephen J.; Boyd, Robert; Anhalt, David J.)V1388,490-501(1991)

Surveillance Technologies (Gowrinathan, Sankaran; Mataloni, Raymond J.; Schwartz, Stanley J., eds.)V1479(1991)

System concepts for a large UV/optical/IR telescope on the moon (Nein, Max E.; Davis, Billy)V1494,98-110(1991)

System design for lunar-based optical and submillimeter interferometers (Gershman, Robert; Mahoney, Michael J.; Rayman, M. D.; Shao, Michael; Snyder, Gerald C.)V1494,160-167(1991)

Telepresence for planetary exploration (McGreevy, Michael W.; Stoker, Carol R.)V1387,110-123(1991)

Telerobotic capabilities for space operations (Akin, David L.)V1387,359-364(1991)

Testbed for tele-autonomous operation of multiarmed robotic servicers in space (Morgenthaler, Matthew K.; Bruno, Guy; Spofford, John R.; Greunke, Roy G.; Gatrell, Lance B.)V1387,82-95(1991)

Test of a vision-based autonomous space station robotic task (Castellano, Anthony R.; Hwang, Vincent S.; Stoney, William E.)V1387,343-350(1991)

Theoretical limits of dimensional stability for space structures (Dolgin, Benjamin P.; Moacanin, Jovan; O'Donnell, Timothy P.)V1533,229-239(1991)

Thermal and structural analysis of the GOES scan mirror's on-orbit performance (Zurmehly, George E.; Hookman, Robert A.)V1532,170-176(1991)

Three-dimensional vision: requirements and applications in a space environment (Noseworthy, J. R.; Gerhardt, Lester A.)V1387,26-37(1991)

Toward an ESA strategy for optical interferometry outside the atmosphere: recommendations by the ESA space interferometry study team (Noordam, Jan E.)V1494,344-346(1991)

Toward the optimization of passive ground targets in spaceborne laser ranging (Lund, Glenn I.; Renault, Herve)V1492,153-165(1991)

Tropospheric emission spectrometer for the Earth Observing System (Glavich, Thomas A.; Beer, Reinhard)V1540,148-159(1991)

Use of spline expansions and regularization in the unfolding of data from spaceborne sensors (Fisher, Thornton R.; Perez, Joseph D.)V1479,212-225(1991)

Wheeled planetary rover testbed (Price, R. S.; Chun, Wendell H.; Hammond, Mark W.; Hubert, Alexis)V1388,550-559(1991)

Spatial light modulators—see also acousto-optics; crystals, liquid; optical computing; photorefraction;

Advanced magneto-optic spatial light modulator device development (Ross, William E.; Lambeth, David N.)V1562,93-102(1991)

Analysis and experimental performance of reduced-resolution binary phase-only filters (Kozaitis, Samuel P.; Tepedelenlioglu, N.; Foor, Wesley E.)V1564,373-383(1991)

Application of the phase and amplitude modulating properties of LCTVs (Kirsch, James C.; Loudin, Jeffrey A.; Gregory, Don A.)V1474,90-100(1991)

Characteristics of a ferroelectric liquid crystal spatial light modulator with a dielectric mirror (Kato, Naoki; Sekura, Rieko; Yamanaka, Junko; Ebihara, Teruo; Yamamoto, Shuhei)V1455,190-205(1991)

Charge-coupled-device-addressed liquid-crystal light valve: an update (Efron, Uzi; Byles, W. R.; Goodwin, Norman W.; Forber, Richard A.; Sayyah, Keyvan; Wu, Chiung S.; Welkowsky, Murray S.)V1455,237-247(1991)

Compact, one-lens JTC using a transmissive amorphous silicon FLC-SLM (LAPS-SLM) (Haemmerli, Jean-Francois; Iwaki, Tadao; Yamamoto, Shuhei)V1564,275-284(1991)

Comparison of optically addressed spatial light modulators (Hudson, Tracy D.; Kirsch, James C.; Gregory, Don A.)V1474,101-111(1991)

Complex researches optically controlled liquid-crystal spatial-time light modulators on the photoconductivity organic polymer basis (Groznov, Michail A.)V1500,281-289(1991)

Computer-generated holograms optimized for illumination with partially coherent light using silicon backplane spatial light modulators as the recording device (O'Brien, Dominic C.; Mears, Robert J.)V1505,32-37(1991)

Correction of magneto-optic device phase errors in optical correlators through filter design modifications (Downie, John D.; Reid, Max B.; Hine, Butler P.)V1564,308-319(1991)

Data transmission at 1.3 um using silicon spatial light modulator (Xiao, Xiaodong; Goel, Kamal K.; Sturm, James C.; Schwartz, P. V.)V1476,301-304(1991)

Development of an IR projector using a deformable mirror device (Lanteigne, David J.)V1486,376-379(1991)

Devices for Optical Processing (Gookin, Debra M., ed.)V1562(1991)

Efficiency of liquid-crystal light valves as polarization rotators (Collings, Neil; Xue, Wei; Pedrini, G.)V1505,12-19(1991)

Electron-beam-addressed lithium niobate spatial light modulator (Hillman, Robert L.; Melnik, George A.; Tsakiris, Todd N.; Leard, Francis L.; Jurgilewicz, Robert P.; Warde, Cardinal)V1562,136-142(1991)

Electron-beam-addressed membrane light modulator for IR scene projection (Horsky, Thomas N.; Schiller, Craig M.; Genetti, George J.; O'Mara, Daniel M.; Hamnett, Whitney S.; Warde, Cardinal)V1540,527-532(1991)

Subject Index

Electronic interface for high-frame-rate electrically addressed spatial light modulators (Kozaitis, Samuel P.; Kirschner, K.; Kelly, E.; Been, D.; Delgado, J.; Velez, E.; Alkindy, A.; Al-Houra, H.; Ali, F.)V1474,112-115(1991)

Electro-optical transducer employing liquid crystal target for processing images in real-time scale (Ignatosyan, S. S.; Simonov, V. P.; Stepanov, Boris M.)V1358,100-108(1991)

Fabrication and performance of CdSe/CdS/ZnS photoconductor for liquid-crystal light valve (Zhuang, Song Lin; Jiang, Yingqui; Qiu, Yinggang; Gu, Lingjuan; Cai, Zhonghua; Chen, Wei)V1558,28-33(1991)

Fourier processing with binary spatial light modulators (Hossack, William J.; Vass, David G.; Underwood, Ian)V1564,697-702(1991)

Full-complex modulation with two one-parameter SLMs (Juday, Richard D.; Florence, James M.)V1558,499-504(1991)

Full-complex spatial filtering with a phase mostly DMD (Florence, James M.; Juday, Richard D.)V1558,487-498(1991)

High contrast ratio InxGa1-xAs/GaAs multiple-quantum-well spatial light modulators (Harwit, Alex; Fernandez, R.; Eades, Wendell D.)V1541,38-47(1991)

High-performance spatial light modulator (Underwood, Ian; Vass, David G.; Sillitto, Richard M.; Bradford, George; Fancey, Norman E.; Al-Chalabi, Adil O.; Birch, Martin J.; Crossland, William A.; Sparks, Adrian P.; Latham, Steve C.)V1562,107-115(1991)

High-resolution display using a laser-addressed ferroelectric liquid-crystal light valve (Nakajima, Hajime; Kisaki, Jyunko; Tahata, Shin; Horikawa, Tsuyoshi; Nishi, Kazuro)V1456,29-39(1991)

High-speed signal processing architectures using charge-coupled devices (Boddu, Jayabharat; Udpa, Satish S.; Udpa, L.; Chan, Shiu Chuen M.)V1562,251-262(1991)

Impact of device characteristics on optical processor design (Turpin, Terry M.)V1562,2-10(1991)

Influence of input information coding for correlation operations (Maze, Sylvie; Joffre, Pascal; Refregier, Philippe)V1505,20-31(1991)

Joint transform correlator using nonlinear ferroelectric liquid-crystal spatial light modulator (Kohler, A.; Fracasso, B.; Ambs, Pierre; de Bougrenet de la Tocnaye, Jean-Louis M.)V1564,236-243(1991)

Liquid-Crystal Devices and Materials (Drzaic, Paul S.; Efron, Uzi, eds.)V1455(1991)

Liquid-crystal devices in planar optics (Armitage, David; Ticknor, Anthony J.)V1455,206-212(1991)

Liquid-crystal-television-based optical-digital processor for measurement of shortening velocity in single rat heart cells (Yelamarty, Rao V.; Yu, Francis T.; Moore, Russell L.; Cheung, Joseph Y.)V1398,170-179(1991)

Low-power optically addressed spatial light modulators using MBE-grown III-V structures (Maserjian, Joseph L.; Larsson, Anders G.)V1562,85-92(1991)

Membrane light modulators: engineering design considerations (O'Mara, Daniel M.; Schiller, Craig M.; Warde, Cardinal)V1527,110-117(1991)

Microlens array for modification of SLM devices (Kathman, Alan D.; Temmen, Mark G.; Scott, Miles L.)V1544,58-65(1991)

Module for optical parallel logic using bistable optically addressed ferroelectric spatial light modulators (Guibert, L.; Killinger, M.; de Bougrenet de la Tocnaye, Jean-Louis M.)V1505,99-103(1991)

Optical correlation using a phase-only liquid-crystal-over-silicon spatial light modulator (Potter, Duncan J.; Ranshaw, M. J.; Al-Chalabi, Adil O.; Fancey, Norman E.; Sillitto, Richard M.; Vass, David G.)V1564,363-372(1991)

Optical evaluation of the microchannel spatial light modulator (Duffey, Jason N.; Hudson, Tracy D.; Kirsch, James C.)V1558,422-431(1991)

Optically addressed spatial light modulator with nipin aSi:H layers and bistable ferroelectric liquid crystal (Cambon, P.; Killinger, M.; de Bougrenet de la Tocnaye, Jean-Louis M.)V1562,116-125(1991)

Optical neural networks using electron trapping materials (Jutamulia, Suganda; Storti, George M.; Seiderman, William M.)V1558,442-447(1991)

Optical neurocomputer architectures using spatial light modulators (Robinson, Michael G.; Zhang, Lin; Johnson, Kristina M.)V1469,240-249(1991)

Optical parallel set of half adders using spatial light rebroadcasters (McAulay, Alastair D.; Wang, Junqing; Xu, Xin; Zeng, Ming)V1564,685-690(1991)

Optical thresholding with a liquid crystal light valve (Shariv, I.; Friesem, Asher A.)V1442,258-263(1991)

Optics for Computers: Architectures and Technologies (Lebreton, Guy J., ed.)V1505(1991)

Optimal correlation filters for implementation on deformable mirror devices (Vijaya Kumar, B. V. K.; Carlson, Daniel W.)V1558,476-486(1991)

Parallel optical interconnects utilizing VLSI/FLC spatial light modulators (Genco, Sheryl M.)V1563,45-57(1991)

Photorefractive devices for optical information processing (Yeh, Pochi A.; Gu, Claire; Hong, John H.)V1562,32-43(1991)

Photorefractive spatial light modulators and their applications to optical computing (Zhao, Mingjun; Li, Yulin; Qin, Yuwen; Wang, Zhao)V1558,529-534(1991)

Programmable holographic scanning (Zhengmin, Li; Li, Min; Tang, Jinfa)V1401,66-73(1991)

Real-time holography using the high-resolution LCTV-SLM (Hashimoto, Nobuyuki; Morokawa, Shigeru; Kitamura, Kohei)V1461,291-302(1991)

Recent advances in surface plasmon spatial light modulators (Caldwell, Martin E.; Yeatman, Eric M.)V1505,50-58(1991)

Rotation- and scale-invariant joint transform correlator using FLC-SLMs (Mitsuoka, Yasuyuki; Iwaki, Tadao; Yamamoto, Shuhei)V1564,244-252(1991)

Schottky diode silicon liquid-crystal light valve (Sayyah, Keyvan; Efron, Uzi; Forber, Richard A.; Goodwin, Norman W.; Reif, Philip G.)V1455,249-254(1991)

Signal and noise characteristics of space-time light modulators (Komarov, Vyacheslav A.; Melnik, N. E.)V1238,275-285(1991)

Simulating arbitrary response curves with available response curves for SLMs (Caulfield, H. J.; Reardon, Janine; Javidi, Bahram)V1562,103-106(1991)

Spatial light modulator on the base of shape memory effect (Antonov, Victor A.; Shelyakov, Alexander V.)V1474,116-123(1991)

Spatial light modulators and their applications (Yatagai, Toyohiko; Tian, Ronglong)V1564,691-696(1991)

Tradeoffs in the design and operation of optically addressed spatial light modulators (Perlmutter, S. H.; Doroski, D.; Landreth, Bruce; Gabor, A. M.; Barbier, Pierre R.; Moddel, Garrett)V1562,74-84(1991)

Transmissive analogue SLM using a chiral smectic liquid crystal switched by CdSe TFTs (Crossland, William A.; Davey, A. B.; Sparks, Adrian P.; Lee, Michael J.; Wright, S. W.; Judge, C. P.)V1455,264-273(1991)

Two-dimensional spatial light modulators using polarization-sensitive multiple-quantum-well light valve (Jain, Faquir C.; Drake, G.; Chung, C.; Bhattacharjee, K. K.; Cheung, S. K.)V1564,714-722(1991)

Use of magneto-optic spatial light modulators and linear detector arrays in inner-product associative memories (Goff, John R.)V1558,466-475(1991)

Using liquid-crystal TVs in Vander Lugt optical correlators (Clark, Natalie; Crandall, Charles M.; Giles, Michael K.)V1564,439-451(1991)

Ways of the high-speed increasing of magneto-optical spatial light modulators (Randoshkin, Vladimir V.)V1469,796-803(1991)

Speckle

Accumulating displacement fields from different steps in laser or white-light speckle methods (Shao, C. A.; King, H. J.; Wang, Yeong-Kang; Chiang, Fu-Pen)V1554A,613-618(1991)

Accuracy and high-speed technique for autoprocessing of Young's fringes (Chen, Wenyi; Tan, Yushan)V1554A,879-885(1991)

Adaptive filtering and enhancement method of speckle pattern (Yang, Yibing; Wang, Houshu)V1554A,781-788(1991)

Angular sensitivity of holograms with a reference speckle wave (Darsky, Alexei M.; Markov, Vladimir B.)V1238,54-61(1991)

Application of phase-stepping speckle interferometry to shape and deformation measurement of a 3-D surface (Dobbins, B. N.; He, Shi P.; Kapasi, S.; Wang, Liu Sheng; Button, B. L.; Wu, Xiao-Ping)V1554A,772-780(1991)

Application of speckle metrology at a nuclear waste repository (Conley, Edgar; Genin, Joseph)V1332,798-801(1991)

Application of visioplasticity to an experimental analysis of the shearing phenomenon (Koga, Nobuhiro; Kudoh, Takeshi; Murakawa, Masao)V1554A,84-92(1991)

Automated measurement of 3-D shapes by a dual-beam digital speckle interferometric technique (Shi, Dahuan; Qin, Jing; Hung, Yau Y.)V1554A,680-689(1991)

Automatic data processing of speckle fringe pattern (Dupre, Jean-Christophe; Lagarde, Alexis)V1554A,766-771(1991)

Automatic heterodyning of fiber-optic speckle pattern interferometry (Valera Robles, Jesus D.; Harvey, David; Jones, Julian D.)V1508,170-179(1991)

Basic use of acoustic speckle pattern for metrology and sea waves study (He, Duo-Min; He, Ming-Shia)V1332,808-819(1991)

Biospeckle phenomena and their applications to blood-flow measurements (Aizu, Yoshihisa; Asakura, Toshimitsu)V1431,239-250(1991)

Black and white cinefilm (ASA 5) applied to dynamic laser speckle and its processing (He, Anzhi; Yu, Gui Ying; Ni, Xiao W.; Fan, Hua Ying)V1554A,747-749(1991)

Computer-aided speckle interferometry: Part II—an alternative approach using spectral amplitude and phase information (Chen, Duanjun; Chiang, Fu-Pen; Tan, Yushan; Don, H. S.)V1554A,706-717(1991)

Computerized vibration analysis of hot objects (Lokberg, Ole J.; Rosvold, Geir O.; Malmo, Jan T.; Ellingsrud, Svein)V1508,153-160(1991)

Contouring by DSPI for surface inspection (Paoletti, Domenica; Schirripa Spagnolo, Giuseppe)V1554A,660-667(1991)

Correlation of speckle pattern generated by TEM10 laser mode (Grzegorzewski, Bronislaw; Kowalczyk, M.; Mallek, Janusz)V1391,290-294(1991)

Defocused white light speckle method for object contouring (Asundi, Anand K.)V1385,239-245(1991)

Detection of the object velocity using the time-varying scattered speckles (Okamoto, Takashi; Asakura, Toshimitsu)V1399,192-199(1991)

Digital speckle correlation search method and its application (Zhou, Xingeng; Gao, Jianxing)V1554A,886-895(1991)

DSPI: a tool for analyzing thermal strain on ceramic and composite materials (Hoefling, Roland; Aswendt, Petra; Totzauer, Werner F.; Jueptner, Werner P.)V1508,135-142(1991)

Dynamic caustic method is applied to fracture (Chen, Zengtao; Guo, Maolin; Wang, Duo; Bi, Xianzhi)V1554A,206-208(1991)

Dynamic two-beam speckle interferometry (Ahmadshahi, Mansour A.; Krishnaswamy, Sridhar; Nemat-Nasser, Siavouche)V1554A,620-627(1991)

Electronic shearography versus ESPI in nondestructive evaluation (Hung, Yau Y.)V1554B,692-700(1991)

Electronic speckle pattern interferometry for 3-D dynamic deformation analysis in industrial environments (Guelker, Gerd; Haack, Olaf; Hinsch, Klaus D.; Hoelscher, Claudia; Kuls, Juergen; Platen, Winfried)V1500,124-134(1991)

Experimental investigation of speckle-size distribution (Markhvida, Igor V.; Tanin, Leonid V.; Drobot, Igor L.)V1508,128-134(1991)

Four cases of engineering vibration studies using pulsed ESPI (Mendoza Santoyo, Fernando; Shellabear, Michael C.; Tyrer, John R.)V1508,143-152(1991)

Fringe formation in speckle shearing interferometry (Xu, Boqin; Wu, Xiao-Ping)V1554A,789-799(1991)

Fringe quality in pulsed TV-holography (Spooren, Rudie)V1508,118-127(1991)

High-speed laser speckle photography (Huntley, Jonathan M.; Field, John E.)V1554A,756-765(1991)

High-speed photography at the Cavendish Laboratory (Field, John E.)V1358,2-17(1991)

Illumination coherence effects in laser-speckle imaging (Voelz, David G.; Idell, Paul S.; Bush, Keith A.)V1416,260-265(1991)

Improvement of the sensitivity of dynamic white-light speckle method and its application (Fang, Ruhua; Cao, Zhengyuan)V1554A,649-656(1991)

Industrial Applications of Holographic and Speckle Measuring Techniques (Jueptner, Werner P., ed.)V1508(1991)

Information capacity of holograms with reference speckle wave (Darsky, Alexei M.; Markov, Vladimir B.)V1509,36-46(1991)

Information extracting and application for the combining objective speckle and reflection holography (Cao, Zhengyuan; Cheng, Fang)V1332,358-364(1991)

In-plane ESPI measurements in hostile environments (Preater, Richard W.; Swain, Robin C.)V1554A,727-738(1991)

Laser diagnostic techniques in a resonant incinerator (Cadou, Christopher P.; Logan, Pamela; Karagozian, Ann; Marchant, Roy; Smith, Owen I.)V1434,67-77(1991)

Laser metrological measurement of transient strain fields in Hopkinson-bar experiments (Vogel, Dietmar; Michel, Bernd; Totzauer, Werner F.; Schreppel, Ulrich; Clos, Rainer)V1554A,262-274(1991)

Laser speckle and its temporal variability: the implications for biomedical holography (Briers, J. D.)V1429,48-54(1991)

Laser speckle and optical fiber sensors for micromovements monitoring in biotissues (Tuchin, Valery V.; Ampilogov, Andrey V.; Bogoroditsky, Alexander G.; Rabinovich, Emmanuil M.; Ryabukho, Vladimir P.; Ul'yanov, Sergey S.; V'yushkin, Maksim E.)V1420,81-92(1991); V1429,62-73(1991)

L-filter design using the gradient search algorithm (Roy, Sumit)V1451,254-256(1991)

Liquid-crystal phase modulator used in DSPI (Kadono, Hirofumi; Toyooka, Satoru)V1554A,628-638(1991)

Low-temperature deformation measurement of materials by the use of laser speckle photography method (Nakahara, Sumio; Hisada, Shigeyoshi; Fujita, Takeyoshi; Sugihara, Kiyoshi)V1554A,602-609(1991)

Material testing by the laser speckle strain gauge (Yamaguchi, Ichirou; Kobayashi, Koichi)V1554A,240-249(1991)

Method for evaluating displacement of objects using the Wigner distribution function (Widjaja, Joewono; Uozumi, Jun; Asakura, Toshimitsu)V1400,94-100(1991)

Modeling ultrasound speckle formation and its dependence on imaging system's response (Rao, Navalgund A.; Zhu, Hui)V1443,81-95(1991)

Mode-locked Nd:YLF laser for precision range-Doppler imaging (Simpson, Thomas B.; Malley, Michael M.; Sutton, George W.; Doft, F.)V1416,2-9(1991)

Modern optical method for strain separation in photoplasticity (Zhang, Zuxun)V1554A,482-487(1991)

Multiple-channel sensing with fiber specklegrams (Wu, Shudong; Yin, Shizhuo; Rajan, Sumati; Yu, Francis T.)V1584,415-424(1991)

Numerical and optical evaluation of particle image velocimetry images (Farrell, Patrick V.)V1554B,610-621(1991)

Optical system for control of longitudinal displacement (Dick, Sergei C.; Markhvida, Igor V.; Tanin, Leonid V.)V1454,447-452(1991)

Optimal algorithms for holographic image reconstruction of an object with a glint using nonuniform illumination (Nowakowski, Jerzy; Wlodawski, Mitchell)V1416,229-240(1991)

Phase measuring fiber optic electronic speckle pattern interferometer (Joenathan, Charles; Khorana, Brij M.)V1396,155-163(1991)

Phase-measuring fiber optic ESPI system: phase-step calibration and error sources (Joenathan, Charles; Khorana, Brij M.)V1554B,56-63(1991)

Phenomenon of fringe patterns from speckle photogram and interferometry within a film (Bo, Weiyun; Wang, Rupeng)V1554A,750-754(1991)

Pulsed holographic and speckle interferometry using Hopkinson loading technique to investigate the dynamical deformation on plates (Han, Lei; Wu, Xiao-Ping; Hu, Shisheng)V1358,793-803(1991)

Quantitative analysis of contact deformation using holographic interferometry and speckle photography (Joo, Jin W.; Kwon, Young H.; Park, Seung O.)V1554B,113-118(1991)

Range of measurement of computer-aided speckle interferometry (Chen, Duanjun; Li, Shen; Hsu, T. Y.; Chiang, Fu-Pen)V1554A,922-931(1991)

Real-time nonlinear optical correlator in speckle metrology (Ogiwara, Akifumi; Ohtsubo, Junji)V1564,294-305(1991)

Real-time optical spatial filtering system with white-light source for displaying color moire (Fang, Cui-Chang; Liao, Yan-Biao; Ma, De-Yuan)V1554A,915-921(1991)

Real-time speckle photography using photorefractive Bi12SiO20 crystal (Nakagawa, Kiyoshi; Minemoto, Takumi)V1508,191-200(1991)

Research for mechanical properties of rock and clay by laser speckle photography (Wu, Ruiqi)V1554A,690-695(1991)

Residual stress evaluation using shearography with large shear displacements (Hathaway, Richard B.; Der Hovanesian, Joseph; Hung, Yau Y.)V1554B,725-735(1991)

Resolution limits for high-resolution imaging lidar (Idell, Paul S.)V1416,250-259(1991)

Second Intl Conf on Photomechanics and Speckle Metrology: Speckle Techniques, Birefringence Methods, and Applications to Solid Mechanics (Chiang, Fu-Pen, ed.)V1554A(1991)

Selected Papers on Speckle Metrology (Sirohi, Rajpal S., ed.)VMS35(1991)

Simple technique for 3-D displacements measurement using synthesis of holographic interferometry and speckle photography (Kwon, Young H.; Park, Seung O.; Park, B. C.)V1554A,639-644(1991)

Some cases of applying photothermoplastic carriers in holographic interferometry and speckle photography (Anikin, V. I.; Dementiev, I. V.; Zhurminskii, Igor L.; Nasedkhina, N. V.; Panasiuk, L. M.)V1238,286-296(1991)

Space-time correlation properties and their applications of dynamic speckles after propagation through an imaging system and double-random modulation (Ma, Shining; Liu, Ying; Du, Fuli)V1554A,645-648(1991)

Spatial techniques of fringe pattern analysis in interferometry (Kujawinska, Malgorzata; Spik, Andrzej)V1391,303-312(1991)

Speckle measurement for 3-D surface movement (Hilbig, Jens; Ritter, Reinold)V1554A,588-592(1991)

Speckle photography for investigation of bones supported by different fixing devices (Kasprzak, Henryk; Podbielska, Halina; Pennig, Dietmar)V1429,55-61(1991)

Statistical analysis of white-light speckle photography (Fang, Qiang; Tan, Yushan)V1554A,696-704(1991)

Statistical interferometry based on the statistics of speckle phase (Kadono, Hirofumi; Toyooka, Satoru)V1554A,718-726(1991)

Statistical properties of intensity fluctuations produced by rough surfaces under the speckle pattern illumination (Yoshimura, Takeaki; Fujiwara, Kazuo; Miyazaki, Eiichi)V1332,835-842(1991)

Study of electronic shearing speckle technique (Qin, Yuwen; Wang, Jinqi; Ji, Xinhua)V1554A,739-746(1991)

Study of measuring skin blood flow using speckle counting (Liu, Ying; Ma, Shining; Du, Fuli; Peng, Xiang; Ye, Shenhua)V1554A,610-612(1991)

Study of oxidization process in real time using speckle correlation (Muramatsu, Mikiya; Guedes, G. H.; Matsuda, Kiyofumi; Barnes, Thomas H.)V1332,792-797(1991)

Surface contouring using electronic speckle pattern interferometry (Kerr, David; Rodriguez-Vera, Ramon; Mendoza Santoyo, Fernando)V1554A,668-679(1991)

Two-dimensional surface strain measurement based on a variation of Yamaguchi's laser-speckle strain gauge (Barranger, John P.)V1332,757-766(1991)

Use of laser diodes and monomode optical fiber in electronic speckle pattern interferometry (Atcha, Hashim; Tatam, Ralph P.)V1584,425-434(1991)

Vibration modal analysis using stroboscopic digital speckle pattern interferometry (Wang, Xizhou; Tan, Yushan)V1554A,907-914(1991)

White-light speckle applied to composites (Guo, Maolin; Zhao, Yin; Wu, Zhongming; Du, Shanyi)V1554A,657-658(1991)

Whole field displacement and strain rosettes by grating objective speckle method (Tu, Meirong; Gielisse, Peter J.; Xu, Wei)V1554A,593-601(1991)

Spectroscopy—see also Raman effect; spectroscopy, atomic; spectroscopy, molecular; spectrum analysis

4096 x 4096 pixel CCD mosaic imager for astronomical applications (Geary, John C.; Luppino, Gerard A.; Bredthauer, Richard A.; Hlivak, Robert J.; Robinson, Lloyd B.)V1447,264-273(1991)

Advanced portable four-wavelength NIR analyzer for rapid chemical composition analysis (Malinen, Jouko; Hyvarinen, Timo S.)V1510,204-209(1991)

Advanced X-ray Astrophysics Facility science instruments (Winkler, Carl E.; Dailey, Carroll C.; Cumings, Nesbitt P.)V1494,301-313(1991)

Advances in analytical chemistry (Arendale, William F.; Congo, Richard T.; Nielsen, Bruce J.)V1434,159-170(1991)

Advances in R&D in near-infrared spectroscopy in Japan (Kawano, Sumio; Iwamoto, Mutsuo)V1379,2-9(1991)

Advances in tunable diode laser technology for atmospheric monitoring applications (Wall, David L.)V1433,94-103(1991)

Aerodyne research mobile infrared methane monitor (McManus, J. B.; Kebabian, Paul L.; Kolb, Charles E.)V1433,330-339(1991)

Aircraft laser infrared absorption spectrometer (ALIAS) for polar ozone studies (Webster, Chris R.; May, R. D.)V1540,187-194(1991)

Air quality monitoring with the differential optical absorption spectrometer (Stevens, Robert K.; Conner, Teri L.)V1491,56-67(1991)

Analysis and control of semiconductor crystal growth with reflectance-difference spectroscopy and spectroellipsometry (Aspnes, David E.)V1361,551-561(1991)

Analysis of absorption, scattering, and hemoglobin saturation using phase-modulation spectroscopy (Sevick, Eva M.; Weng, Jian; Maris, Michael B.; Chance, Britton)V1431,264-275(1991)

Analysis of exhaust from clean-fuel vehicles using FTIR spectroscopy (Rieger, Paul L.; Maddox, Christine E.)V1433,290-301(1991)

Analysis of multiple-multipole scattering by time-resolved spectroscopy and spectrometry (Frank, Klaus; Hoeper, J.; Zuendorf, J.; Tauschek, D.; Kessler, Manfred; Wiesner, J.; Wokaun, Alexander J.)V1431,2-11(1991)

Analysis of photon correlation functions obtained from eye lenses in vivo (Van Laethem, Marc; Xia, Jia-zhi; Clauwaert, Julius)V1403,732-742(1991)

Analysis of silage composition by near-infrared reflectance spectroscopy (Reeves, James B.; Blosser, Timothy H.; Colenbrander, V. F.)V1379,28-38(1991)

Analytical Raman and atomic spectroscopies using charge-coupled-device detection (Bilhorn, Robert B.; Ferris, Nancy S.)V1439,15-24(1991)

Application of FM spectroscopy in atmospheric trace gas monitoring: a study of some factors influencing the instrument design (Werle, Peter W.; Josek, K.; Slemr, Franz)V1433,128-135(1991)

Application of high-speed infrared emission spectroscopy in reacting flows (Klingenberg, Guenter; Rockstroh, Helmut)V1358,851-858(1991)

Application of low-noise CID imagers in scientific instrumentation cameras (Carbone, Joseph; Hutton, J.; Arnold, Frank S.; Zarnowski, Jeffrey J.; VanGorden, Steve; Pilon, Michael J.; Wadsworth, Mark V.)V1447,229-242(1991)

Application of tunable diode lasers in control of high-pure-material technologies (Nadezhdinskii, Alexander I.; Stepanov, Eugene V.; Kuznetzov, Andrian I.; Devyatykh, Grigory G.; Maximov, G. A.; Khorshev, V. A.; Shapin, S. M.)V1418,487-495(1991)

Applications of an automated particle detection and identification system in VLSI wafer processing (Hattori, Takeshi; Koyata, Sakuo)V1464,367-376(1991)

Applications of charge-coupled and charge-injection devices in analytical spectroscopy (Denton, M. B.)V1447,2-11(1991)

Applications of laser plasmas in XUV photoabsorption spectroscopy (Kennedy, Eugene T.; Costello, John T.; Mosnier, Jean-Paul)V1503,406-415(1991)

Applications of optical emission spectroscopy in plasma manufacturing systems (Gifford, George G.)V1392,454-465(1991)

Applications of the FEL to NLO spectroscopy of semiconductors (Pidgeon, Carl R.)V1501,178-182(1991)

Applied Spectroscopy in Material Science (Saperstein, David D., ed.)V1437(1991)

Array detectors in astronomy (Lesser, Michael P.)V1439,144-151(1991)

Auto exhaust gas analysis by FTIR spectroscopy (Herget, William F.; Lowry, Steven R.)V1433,275-289(1991)

Automated band selection for multispectral meteorological applications (Westerman, Steven D.; Drake, R. M.; Yool, Stephen R.; Brandley, M.; DeJulio, R.)V1492,263-271(1991)

Background correction in multiharmonic Fourier transform fluorescence lifetime spectroscopy (Swift, Kerry M.; Mitchell, George W.)V1431,171-178(1991)

Biochemical measurement of bilirubin with an evanescent wave optical sensor (Poscio, Patrick; Depeursinge, Ch.; Emery, Y.; Parriaux, Olivier M.; Voirin, G.)V1510,112-117(1991)

Biomedical applications of laser photoionization (Xiong, Xiaoxiong; Moore, Larry J.; Fassett, John D.; O'Haver, Thomas C.)V1435,188-196(1991)

Biomedical applications of laser technology at Los Alamos (Bigio, Irving J.; Loree, Thomas R.)V1403,776-780(1991)

Biospeckle phenomena and their applications to blood-flow measurements (Aizu, Yoshihisa; Asakura, Toshimitsu)V1431,239-250(1991)

Calibration for the medium-resolution imaging spectrometer (Baudin, Gilles; Chessel, Jean-Pierre; Cutter, Mike A.; Lobb, Daniel R.)V1493,16-27(1991)

Characterization of plasma processes with optical emission spectroscopy (Malchow, Douglas S.)V1392,498-505(1991)

Characterization of PMMA-EVA blend via photoacoustic technique (Carbonara, Giuseppe; Mormile, Pasquale; Bernini, Umberto; Russo, Paolo; Malinconico, Mario; Volpe, M. G.)V1361,1038-1040(1991)

Characterization of the fluorescence lifetimes of the ionic species found in aqueous solutions of hematoporphyrin IX as a function of pH (Nadeau, Pierre; Pottier, R.; Szabo, Arthur G.; Brault, Daniel; Vever-Bizet, C.)V1398,151-161(1991)

Chemical analysis of human urinary and renal calculi by Raman laser fiber-optics method (Hong, Nguyen T.; Phat, Darith; Plaza, Pascal; Daudon, Michel; Dao, Nguyen Q.)V1525,132-142(1991)

Combined x-ray absorption spectroscopy and x-ray powder diffraction (Dent, Andrew J.; Derbyshire, Gareth E.; Greaves, G. N.; Ramsdale, Christine A.; Couves, J. W.; Jones, Richard; Catlow, C. R.; Thomas, John M.)V1550,97-107(1991)

Compact Raman instrumentation for process and environmental monitoring (Carrabba, Michael M.; Spencer, Kevin M.; Rauh, R. D.)V1434,127-134(1991)

Comparison of time and frequency multiplexing techniques in multicomponent FM spectroscopy (Muecke, Robert J.; Werle, Peter W.; Slemr, Franz; Prettl, William)V1433,136-144(1991)

Cryogenic limb array etalon spectrometer: experiment description (Roche, Aidan E.; Kumer, John B.)V1491,91-103(1991)

Current instrument status of the airborne visible/infrared imaging spectrometer (AVIRIS) (Eastwood, Michael L.; Sarture, Charles M.; Chrien, Thomas G.; Green, Robert O.; Porter, Wallace M.)V1540,164-175(1991)

Data analysis methods for near-infrared spectroscopy of tissue: problems in determining the relative cytochrome aa3 concentration (Cope, Mark; van der Zee, Pieter; Essenpreis, Matthias; Arridge, Simon R.; Delpy, David T.)V1431,251-262(1991)

Design and performance of a PtSi spectroscopic infrared array and detector head (Cizdziel, Philip J.)V1488,6-27(1991)

Detecting dilution-narrowed systems (Jaffe, Steven M.; Yen, William M.)V1435,252-257(1991)

Detection and localization of absorbers in scattering media using frequency-domain principles (Berndt, Klaus W.; Lakowicz, Joseph R.)V1431,149-160(1991)

Detection of atheroma using Photofrin II and laser-induced fluorescence spectroscopy (Vari, Sandor G.; Papazoglou, Theodore G.; van der Veen, Maurits J.; Papaioannou, Thanassis; Fishbein, Michael C.; Chandra, Mudjianto; Beeder, Clain; Shi, Wei-Qiang; Grundfest, Warren S.)V1426,58-65(1991)

Determination of inhomogeneous trace absorption by using exponential expansion of the absorption Pade-approximant (Dobrego, Kirill V.)V1433,365-374(1991)

Developing a long-path diode array spectrometer for tropospheric chemistry studies (Lanni, Thomas R.; Demerjian, Kenneth L.)V1433,21-24(1991)

Diagnosis of atherosclerotic tissue by resonance fluorescence spectroscopy (Neu, Walter; Haase, Karl K.; Tischler, Christian; Nyga, Ralf; Karsch, Karl R.)V1425,28-36(1991)

Diagnostics of a compact UV-preionized XeCl laser with BCl3 halogen donor (Peet, Viktor E.; Treshchalov, Alexei B.; Slivinskij, E. V.)V1412,138-148(1991)

Differential time-resolved detection of absorbance changes in composite structures (Nossal, Ralph J.; Bonner, Robert F.)V1431,21-28(1991)

Diffusing wave spectroscopy studies of gelling systems (Horne, David S.)V1430,166-180(1991)

Diffusion of intensity modulated near-infrared light in turbid media (Fishkin, Joshua B.; Gratton, Enrico; vandeVen, Martin J.; Mantulin, William W.)V1431,122-135(1991)

Digital imaging microscopy: the marriage of spectroscopy and the solid state CCD camera (Jovin, Thomas M.; Arndt-Jovin, Donna J.)V1439,109-120(1991)

Diode laser spectroscopy of atmospheric pollutants (Nadezhdinskii, Alexander I.; Stepanov, Eugene V.)V1433,202-210(1991)

Discussion on spectral sampling of imaging spectrometer (Han, Xin Z.)V1538,99-102(1991)

DOAS (differential optical absorption spectroscopy) urban pollution measurements (Stevens, Robert K.; Vossler, T. L.)V1433,25-35(1991)

Double-beam laser absorption spectroscopy: shot noise-limited performance at baseband with a novel electronic noise canceler (Haller, Kurt L.; Hobbs, Philip C.)V1435,298-309(1991)

Double-mode CO2 laser with complex cavity for ultrasensitive sub-Doppler spectroscopy (Kurochkin, Vadim Y.; Petrovsky, Victor N.; Protsenko, Evgeniy D.; Rurukin, Alexander N.; Golovchenko, Anotoly M.)V1435,322-330(1991)

Dynamic light scattering from strongly interacting multicomponent systems: salt-free polyelectrolyte solutions (Sedlak, Marian; Amis, Eric J.; Konak, Cestmir; Stepanek, Petr)V1430,191-202(1991)

Early detection of dysplasia in colon and bladder tissue using laser-induced fluorescence (Rava, Richard P.; Richards-Kortum, Rebecca; Fitzmaurice, Maryann; Cothren, Robert M.; Petras, Robert; Sivak, Michael V.; Levine, Howard)V1426,68-78(1991)

Effect of surface boundary on time-resolved reflectance: measurements with a prototype endoscopic catheter (Jacques, Steven L.; Flock, Stephen T.)V1431,12-20(1991)

Electronic structure of Ge(001) 2x1 by different angle-resolved photoemission techniques: EDC, CFS and CIS (Kipp, Lutz; Manzke, Recardo; Skibowski, Michael)V1361,794-801(1991)

Emissions monitoring by infrared photoacoustic spectroscopy (Jalenak, Wayne)V1434,46-54(1991)

Environmental Sensing and Combustion Diagnostics (Santoleri, Joseph J., ed.)V1434(1991)

Excimer-laser-induced fluorescence spectroscopy of human arteries during laser ablation (Abel, B.; Hippler, Horst; Koerber, B.; Morguet, A.; Neu, Walter)V1525,110-118(1991)

Exciton spectroscopy of semiconductor materials used in laser elements (Nasibov, Alexander S.; Markov, L. S.; Fedorov, D. L.; Shapkin, P. V.; Korostelin, Y. V.; Machintsev, G. A.)V1361,901-908(1991)

Experimental method for gas kinetic temperature measurements in a thermal plasma (Reynolds, Larry D.; Shaw, C. B.)V1554B,622-631(1991)

Fabrication of toroidal and coma-corrected toroidal diffraction gratings from spherical master gratings using elastically deformable substrates: a progress report (Huber, Martin C.; Timothy, J. G.; Morgan, Jeffrey S.; Lemaitre, Gerard R.; Tondello, Giuseppe; Naletto, Giampiero)V1494,472-480(1991)

Faint object spectrograph early performance (Harms, Richard J.; Fitch, John E.)V1494,49-65(1991)

Far-IR emission spectroscopy on electron-beam-irradiated AlGaAs/GaAs heterostructures (Diessel, Edgar; Sigg, Hans; von Klitzing, Klaus)V1361,1094-1099(1991)

Far-IR studies of moderately doped molecular beam epitaxy grown GaAs on Si(100) (Morley, Stefan; Eickhoff, Thomas; Zahn, Dietrich R.; Richter, W.; Woolf, D.; Westwood, D. I.; Williams, R. H.)V1361,213-222(1991)

Fiber optic measurement of intracellular pH in intact rat liver using pH-sensitive dyes (Felberbauer, Franz; Graf, Juerg)V1510,63-71(1991)

Fiber optic remote Fourier transform infrared spectroscopy (Druy, Mark A.; Glatkowski, Paul J.; Stevenson, William A.)V1584,48-52(1991)

Fiber optic sensor probe for in-situ surface-enhanced Raman monitoring (Vo-Dinh, Tuan; Stokes, D. L.; Li, Ying-Sing; Miller, Gordon H.)V1368,203-209(1991)

Fiber optic thermometer using Fourier transform spectroscopy (Beheim, Glenn; Sotomayor, Jorge L.; Flatico, Joseph M.; Azar, Massood T.)V1584,64-71(1991)

Field and laboratory studies of Fourier transform infrared spectroscopy in continuous emissions monitoring applications (Plummer, Grant M.)V1434,78-89(1991)

FIR lasers as local oscillators in submillimeter astronomy (Roeser, Hans-Peter; van der Wal, Peter)V1501,194-197(1991)

Fluorescence line-narrowing spectroscopy in the study of chemical carcinogenesis (Jankowiak, Ryszard; Jeong, H.; Small, Gerald J.)V1435,203-213(1991)

Fluorescence spectroscopy of normal and atheromatous human aorta: optimum illumination wavelength (Alexander, Andrew L.; Davenport, Carolyn M.; Gmitro, Arthur F.)V1425,6-15(1991)

Fourier analysis of near-infrared spectra (McClure, W. F.)V1379,45-51(1991)

Frequency-domain fluorescence spectroscopy: instrumentation and applications to the biosciences (Lakowicz, Joseph R.; Gryczynski, Ignacy; Malak, Henryk; Johnson, Michael L.; Laczko, Gabor; Wiczk, Wieslaw M.; Szmacinski, Henryk; Kusba, Jozef)V1435,142-160(1991)

Frequency-domain measurements of changes of optical pathlength during spreading depression in a rodent brain model (Maris, Michael B.; Mayevsky, Avraham; Sevick, Eva M.; Chance, Britton)V1431,136-148(1991)

Frequency-modulation absorption spectroscopy for trace species detection: theoretical and experimental comparison among methods (Silver, Joel A.; Bomse, David S.; Stanton, Alan C.)V1435,64-71(1991)

Frequency modulation spectroscopy for chemical sensing of the environment (Cooper, David E.; Riris, Haris; van der Laan, Jan E.)V1433,120-127(1991)

Frequency up-conversion in Pr3+ doped fibers (Gomes, Anderson S.; de Araujo, Cid B.; Moraes, E. S.; Opalinska, M. M.; Gouveia-Neto, A. S.)V1579,257-263(1991)

FTIR: fundamentals and applications in the analysis of dilute vehicle exhaust (Gierczak, Christine A.; Andino, J. M.; Butler, James W.; Heiser, G. A.; Jesion, G.; Korniski, T. J.)V1433,315-328(1991)

Gaseous incinerator emissions analysis by FTIR (Fourier transform infrared) spectroscopy (Herget, William F.; Demirgian, Jack)V1434,39-45(1991)

Global ozone monitoring by occultation of stars (Bertaux, J. L.)V1490,133-145(1991)

Ground-based monitoring of water vapor in the middle atmosphere: the NRL water-vapor millimeter-wave spectrometer (Bevilacqua, Richard M.; Schwartz, Philip R.; Pauls, Thomas A.; Waltman, William B.; Thacker, Dorsey L.)V1491,231-242(1991)

Growth and characterization of ultrathin SimGen strained-layer superlattices (Presting, Hartmut; Jaros, Milan; Abstreiter, Gerhard)V1512,250-277(1991)

Growth and characterization of ZnSe and ZnTe grown on GaAs by hot-wall epitaxy (Hingerl, Kurt; Pesek, Andreas; Sitter, Helmut; Krost, Alois; Zahn, Dietrich R.; Richter, W.; Kudlek, Gotthard; Gutowski, Juergen)V1361,383-393(1991)

Guided-wave nonlinear spectroscopy of silica glasses (Aleksandrov, I. V.; Nesterova, Z. V.; Petrovskii, Gurii T.)V1513,309-312(1991)

High-dynamic-range MCP structures (Slater, David C.; Timothy, J. G.)V1549,68-80(1991)

Highly sensitive absorption measurements in organic thin films and optical media (Skumanich, Andrew)V1559,267-277(1991)

High-power x-ray generation using transition radiation (Piestrup, Melvin A.; Boyers, D. G.; Pincus, Cary I.; Li, Qiang; Harris, J. L.; Bergstrom, J. C.; Caplan, H. S.; Silzer, R. M.; Skopik, D. M.; Moran, M. J.; Maruyama, Xavier K.)V1552,214-239(1991)

High-resolution studies of atmospheric IR emission spectra (Murcray, Frank J.; Murcray, F. H.; Goldman, Aaron; Blatherwick, Ronald D.; Murcray, David J.)V1491,282-287(1991)

High-resolution submicron particle sizing by dynamic light scattering (Nicoli, David F.)V1430,19-36(1991)

High-sensitivity sensor of gases based on IR tunable diode lasers for human exhalation monitoring (Moskalenko, Konstantin L.; Nadezhdinskii, Alexander I.; Stepanov, Eugene V.)V1426,121-132(1991)

High-temperature Raman spectra of nioboborate glass melts (Yang, Quanzu; Wang, Zhongcai; Wang, Shizhuo)V1513,264-269(1991)

Hole-burning spectroscopy of phthalocyanine Langmuir-Blodget films (Adamec, F.; Ambroz, M.; Dian, J.; Vacha, M.; Hala, J.; Balog, P.; Brynda, E.)V1402,82-84(1991)

Hollow and dielectric waveguides for infrared spectroscopic applications (Saggese, Steven J.; Harrington, James A.; Sigel, George H.)V1437,44-53(1991)

Hollow waveguides for sensor applications (Saggese, Steven J.; Harrington, James A.; Sigel, George H.)V1368,2-14(1991)

Imaging spectroscope with an optical recombination system (Li, Wenchong; Ma, Chunhua)V1428,242-248(1991)

Improved limb atmospheric spectrometer and retroreflector in-space for ADEOS (Sasano, Yasuhiro; Asada, Kazuya; Sugimoto, Nobuo; Yokota, Tatsuya; Suzuki, Makoto; Minato, Atsushi; Matsuzaki, Akiyoshi; Akimoto, Hajime)V1490,233-242(1991)

Indirect spectroscopic detection of singlet oxygen during photodynamic therapy (Tromberg, Bruce J.; Dvornikov, Tatiana; Berns, Michael W.)V1427,101-108(1991)

Industrial applications of spectroscopic imaging (Miller, Richard M.; Birmingham, John J.; Cummins, Philip G.; Singleton, Scott)V1439,66-78(1991)

In-flight performance of the Goddard high-resolution spectrograph of the Hubble Space Telescope (Troeltzsch, John R.; Ebbets, Dennis C.; Garner, Harry W.; Tuffli, A.; Breyer, R.; Kinsey, J.; Peck, C.; Lindler, Don; Feggans, J.)V1494,9-15(1991)

Infrared focal-plane design for the comet rendezvous/asteroid flyby and Cassini visible and infrared mapping spectrometers (Staller, Craig O.; Niblack, Curtiss A.; Evans, Thomas G.; Blessinger, Michael A.; Westrick, Anthony)V1540,219-230(1991)

Infrared spectrometer for ground-based profiling of atmospheric temperature and humidity (Shaw, Joseph A.; Churnside, James H.; Westwater, Edward R.)V1540,681-686(1991)

InSb linear multiplexed FPAs for the CRAF/Cassini visible and infrared mapping spectrometer (Niblack, Curtiss A.; Blessinger, Michael A.; Forsthoefel, John J.; Staller, Craig O.; Sobel, Harold R.)V1494,403-413(1991)

In-situ CARS detection of H2 in the CVD of Si3N4 (Hay, Stephen O.; Veltri, R. D.; Lee, W. Y.; Roman, Ward C.)V1435,352-358(1991)

In situ measurement of methyl bromide in indoor air using long-path FTIR spectroscopy (Green, Martina; Seiber, James N.; Biermann, Heinz W.)V1433,270-274(1991)

Instrument design and test results of the new all-reflection spatial heterodyne spectrometer (Bush, Brett C.; Cotton, Daniel M.; Vickers, James S.; Chakrabarti, Supriya)V1549,376-384(1991)

Intl Conf on Scientific Optical Imaging (Denton, M. B., ed.)V1439(1991)

Intracavity spectroscopy measurements of atom and ion densities in near-surface laser plasma (Burakov, V. S.; Lopasov, V. P.; Naumenkov, P. A.; Raikov, S. N.)V1440,270-276(1991)

Intraoperative metastases detection by laser-induced fluorescence spectroscopy (Vari, Sandor G.; Papazoglou, Theodore G.; van der Veen, Maurits J.; Fishbein, Michael C.; Young, J. D.; Chandra, Mudjianto; Papaioannou, Thanassis; Beeder, Clain; Shi, Wei-Qiang; Grundfest, Warren S.)V1426,111-120(1991)

Inverse scattering problems in the acoustic resonance region of an underwater target (Gaunaurd, Guillermo C.)V1471,30-41(1991)

IR reflectance spectroscopy and AES investigation of titanium ion-beam-doped silica (Belostotsky, Vladimir I.; Solinov, Vladimir F.)V1513,313-318(1991)

IR-spectroscopical investigations on the glass structure of porous and sintered compacts of colloidal silica gels (Clasen, Rolf; Hornfeck, M.; Theiss, W.)V1513,243-254(1991)

Kinetic studies of phosgene reduction via in situ Fourier transform infrared analysis (Farquharson, Stuart; Chauvel, J. P.)V1434,135-146(1991)

Laboratory soft x-ray source based on high-temperature capillary arc discharge (Belov, Sergei N.; Golubev, Evgeny M.; Vinokurova, Elena G.)V1552,264-267(1991)

Laser Applications in Life Sciences (Akhmanov, Sergei A.; Poroshina, Marina Y., eds.)V1403(1991)

Laser desorption jet cooling spectroscopy (de Vries, Mattanjah S.; Hunziker, Heinrich E.; Meijer, Gerard; Wendt, H. R.)V1437,129-137(1991)

Laser-excited fluorescence and fluorescence probes for diagnosing bulk damage in cable insulation (Ordonez, Ishmael D.; Crafton, J.; Murdock, R. H.; Hatfield, Lynn L.; Menzel, E. R.)V1437,184-193(1991)

Laser femtosecond resonant ionization and mapping of biomolecules (Letokhov, Vladilen S.)V1403,4-4(1991)

Laser glasses (Lunter, Sergei G.; Dymnikov, Alexander A.; Przhevuskii, Alexander K.; Fedorov, Yurii K.)V1513,349-359(1991)

Laser-induced fluorescence spectroscopy of pathologically enlarged prostate gland in vitro (Chandra, Mudjianto; Gershman, Alex; Papazoglou, Theodore G.; Bender, Leon; Danoff, Dudley; Papaioannou, Thanassis; Vari, Sandor G.; Coons, Gregory; Grundfest, Warren S.)V1421,68-71(1991)

Linear and nonlinear optical properties of polymeric Langmuir-Blodgett films (Penner, Thomas L.; Willand, Craig S.; Robello, Douglas R.; Schildkraut, Jay S.; Ulman, Abraham)V1436,169-178(1991)

Lippmann volume holographic filters for Rayleigh line rejection in Raman spectroscopy (Rich, Chris C.; Cook, David M.)V1461,2-7(1991)

Long-path differential absorption measurements of tropospheric molecules (Harder, Jerald W.; Mount, George H.)V1491,33-42(1991)

Long-path intracavity laser for the measurement of atmospheric trace gases (McManus, J. B.; Kolb, Charles E.)V1433,340-351(1991)

Long-pathlength DOAS (differential optical absorption spectrometer) system for the in situ measurement of xylene in indoor air (Biermann, Heinz W.; Green, Martina; Seiber, James N.)V1433,2-7(1991)

L-shell x-ray spectroscopy of laser-produced plasmas in the 1-keV region (Batani, D.; Giulietti, Antonio; Palladino, Libero; Tallents, Gregory J.; Turcu, I. C.)V1503,479-491(1991)

Magnetic circular dichroism measurements of benzene in the 9-eV region with synchrotron radiation (Snyder, Patricia A.; Munger, Robert; Hansen, Roger W.; Rowe, Ednor M.)V1548,188-196(1991)

Magnetic circular dichroism studies with soft x-rays (Tjeng, L. H.; Rudolf, P.; Meigs, G.; Sette, Francesco; Chen, Chien-Te; Idzerda, Y. U.)V1548,160-167(1991)

Matrix-assisted laser desorption by Fourier transform mass spectrometry (Nuwaysir, Lydia M.; Wilkins, Charles L.)V1437,112-123(1991)

Measurement of atmospheric composition by the ATMOS instrument from Table Mountain Observatory (Gunson, Michael R.; Irion, Fredrick W.)V1491,335-346(1991)

Measurement of Atmospheric Gases (Schiff, Harold I., ed.)V1433(1991)

Measurement of biological tissue metabolism using phase modulation spectroscopic technology (Weng, Jian; Zhang, M. Z.; Simons, K.; Chance, Britton)V1431,161-170(1991)

Measurement of the tropospheric hydroxyl radical by long-path absorption (Mount, George H.)V1491,26-32(1991)

Measurement of triplet optical densities of organic compounds by means of CW laser excitation (Pavlopoulos, Theodore G.)V1437,168-183(1991)

Measurements of nitrous acid, nitrate radicals, formaldehyde, and nitrogen dioxide for the Southern California Air Quality Study by differential optical absorption spectroscopy (Winer, Arthur M.; Biermann, Heinz W.)V1433,44-55(1991)

Measuring photon pathlengths by quasielastic light scattering in a multiply scattering medium (Nossal, Ralph J.; Schmitt, Joseph M.)V1430,37-47(1991)

Medical applications of holography (von Bally, Gert)V1525,2-8(1991)

Medium-resolution imaging spectrometer (Baudin, Gilles; Bessudo, Richard; Cutter, Mike A.; Lobb, Daniel R.; Bezy, Jean L.)V1490,102-113(1991)

Michelson interferometer for passive atmosphere sounding (Posselt, Winfried)V1490,114-125(1991)

Microchannel-plate detectors for space-based astronomy (Crocker, James H.; Cox, Colin R.; Ray, Knute A.; Sen, Amit)V1494,434-439(1991)

Microscopic fluorescence spectroscopy and diagnosis (Schneckenburger, Herbert; Seidlitz, Harold K.; Wessels, Jurina; Strauss, Wolfgang; Rueck, Angelika C.)V1525,91-98(1991)

Microwave limb sounder experiments for UARS and EOS (Waters, Joe W.)V1491,104-109(1991)

Moderate-resolution imaging spectrometer-tilt baseline concept (Magner, Thomas J.)V1492,272-285(1991)

Modular removable precision mechanism for alignment of an FUV spatial heterodyne interferometer (Tom, James L.; Cotton, Daniel M.; Bush, Brett C.; Chung, Ray; Chakrabarti, Supriya)V1549,308-312(1991)

Moisture influence on near-infrared prediction of wheat hardness (Windham, William R.; Gaines, Charles S.; Leffler, Richard G.)V1379,39-44(1991)

Monte Carlo and diffusion calculations of photon migration in noninfinite highly scattering media (Haselgrove, John C.; Leigh, John S.; Yee, Conway; Wang, Nai-Guang; Maris, Michael B.; Chance, Britton)V1431,30-41(1991)

Multiorder etalon sounder for vertical temperature profiling: technique and performance analysis (Wang, Jin-Xue; Hays, Paul B.; Drayson, S. R.)V1492,391-402(1991)

Multiple-spectral structure of the 578.2 nm line for copper vapor laser (Wang, Yongjiang; Pan, Bailiang; Ding, Xiande; Qian, Yujun; Shi, Shuyi)V1412,60-66(1991)

Nb tunnel junctions as x-ray spectrometers (Rando, Nicola; Peacock, Anthony J.; Foden, Clare; van Dordrecht, Axel; Engelhardt, Ralph; Lumley, John M.; Pereira, Carl)V1549,340-356(1991)

Near-infrared imaging in vivo: imaging of Hb oxygenation in living tissues (Araki, Ryuichiro; Nashimoto, Ichiro)V1431,321-332(1991)

Near-infrared optical monitoring of cardiac oxygen sufficiency through thoracic wall without open-chest surgery (Kakihana, Yasuyuki; Tamura, Mamoru)V1431,314-321(1991)

Near-infrared time-resolved spectroscopy and fast scanning spectrophotometry in ischemic human forearm (Ferrari, Marco; De Blasi, Roberto A.; Bruscaglioni, Piero; Barilli, Marco; Carraresi, Luca; Gurioli, G. M.; Quaglia, Enrico; Zaccanti, Giovanni)V1431,276-283(1991)

New approaches to signal-to-noise ratio optimization in background-limited photothermal measurements (Rice, Patrick D.; Thorne, John B.; Bobbitt, Donald R.)V1435,104-113(1991)

New approaches to ultrasensitive magnetic resonance (Bowers, C. R.; Buratto, S. K.; Carson, Paul; Cho, H. M.; Hwang, J. Y.; Mueller, L.; Pizarro, P. J.; Shykind, David; Weitekamp, Daniel P.)V1435,36-50(1991)

New method for tissue indentification: resonance fluorescence spectroscopy (Neu, Walter)V1525,124-131(1991)

New results in dosimetry of laser radiation in medical treatment (Beuthan, Jurgen; Hagemann, Roland; Mueller, Gerhard J.; Schaldach, Brita J.; Zur, Ch.)V1420,225-233(1991)

New spectroscopic instrumentation for measurement of stratospheric trace species by remote sensing of scattered skylight (Mount, George H.; Jakoubek, Roger O.; Sanders, Ryan W.; Harder, Jerald W.; Solomon, Susan; Winkler, Richard; Thompson, Thomas; Harrop, Walter)V1491,188-193(1991)

NIR/CCD Raman spectroscopy: second battle of a revolution (McCreery, Richard J.)V1439,25-36(1991)

Noise on multiple-tau photon correlation data (Schaetzel, Klaus; Peters, Rainer)V1430,109-115(1991)

Noninvasive hemoglobin oxygenation monitor and computed tomography by NIR spectrophotometry (Oda, Ichiro; Ito, Yasunobu; Eda, Hideo; Tamura, Tomomi; Takada, Michinosuke; Abumi, Rentaro; Nagai, Katumi; Nakagawa, Hachiro; Tamura, Masahide)V1431,284-293(1991)

Noninvasive measurement of regional cerebrovascular oxygen saturation in humans using optical spectroscopy (McCormick, Patrick W.; Stewart, Melville; Lewis, Gary)V1431,294-302(1991)

Nonlinear laser interactions with saltwater aerosols (Alexander, Dennis R.; Poulain, D. E.; Schaub, Scott A.; Barton, John P.)V1497,90-97(1991)

Novel GaP/InP strained heterostructures: growth, characterization,and technological perspectives (Recio, Miguel; Ruiz, Ana; Melendez, J.; Rodriguez, Jose M.; Armelles, Gaspar; Dotor, M. L.; Briones, Fernando)V1361,469-478(1991)

Observation of tunneling emission from a single-quantum-well using deep-level transient spectroscopy (Letartre, Xavier; Stievenard, Didier)V1361,1144-1155(1991)

One-frame subnanosecond spectroscopy camera (Silkis, E. G.; Titov, V. D.; Feldman, G. G.; Zhilkina, V. M.; Petrokovich, O. A.; Syrtzev, V. N.)V1358,46-49(1991)

Optical design of the moderate-resolution imaging spectrometer-tilt for the Earth Observing System (Maymon, Peter W.)V1492,286-297(1991)

Optical diagnostics of mercuric iodide crystal growth (Burger, Arnold; Morgan, S. H.; Silberman, Enrique; Nason, Donald)V1557,245-249(1991)

Optical measurements of electrodynamically levitated microparticles (Davis, E. J.)V1435,216-242(1991)

Optical Methods for Ultrasensitive Detection and Analysis: Techniques and Applications (Fearey, Bryan L., ed.)V1435(1991)

Optical sensor system for Japanese Earth resources satellite 1 (Hino, Hideo; Takei, Mitsuru; Ono, Hiromi; Nagura, Riichi; Narimatsu, Yoshito; Hiramatsu, Masaru; Harada, Hisashi; Ogikubo, Kazuhiro)V1490,166-176(1991)

Optics for tunable diode laser spectrometers (Riedel, Wolfgang J.)V1433,179-189(1991)

Optimization of affecting parameters in relation to pulsed CO2 laser design (Danisman, Kenan; Yilbas, Bekir S.; Altuner, Mehmet; Ciftlikli, Cebrail)V1412,218-226(1991)

ORFEUS alignment concept (Graue, Roland; Kampf, Dirk; Rippel, Harald; Witte, G.)V1494,377-385(1991)

Overview of the spectroscopy of the atmosphere using far-infrared emission experiment (Russell, James M.)V1491,142-150(1991)

Performance characteristics of the imaging MAMA detector systems for SOHO, STIS, and FUSE/Lyman (Timothy, J. G.)V1549,221-233(1991)

Performance of a variable-line-spaced master reflection grating for use in the reflection grating spectrometer on the x-ray multimirror mission (Bixler, Jay V.; Hailey, Charles J.; Mauche, C. W.; Teague, Peter F.; Thoe, Robert S.; Kahn, Steven M.; Paerels, F. B.)V1549,420-428(1991)

Performance of low-resistance microchannel-plate stacks (Siegmund, Oswald H.; Stock, Joseph)V1549,81-89(1991)

Phonons in PbTe and PbTe:Cr strained layers (Baleva, Mitra I.; Momchilova, Maia M.)V1361,712-722(1991)

Photochemistry and fluorescence spectroscopy of polymeric materials containing triphenylsulfonium salts (Hacker, Nigel P.; Welsh, Kevin M.)V1559,139-150(1991)

Photoluminescence and deep-level transient spectroscopy of DX-centers in selectively silicon-doped GaAs-AlAs superlattices (Ababou, Soraya; Benyattou, Taha; Marchand, Jean J.; Mayet, Louis; Guillot, Gerard; Mollot, Francis; Planel, Richard)V1361,706-711(1991)

Photon correlation spectroscopic studies of filamentous actin networks (Newman, Jay E.; San Biagio, Pier L.; Schick, Kenneth L.)V1430,89-108(1991)

Photon correlation spectroscopy and electrophoretic light scattering using optical fibers (Macfadyen, Allan J.; Jennings, B. R.)V1367,319-328(1991)

Photon Correlation Spectroscopy: Multicomponent Systems (Schmitz, Kenneth S., ed.)V1430(1991)

Photon correlation spectroscopy of chromaffin granules lysis (Ermakov, Yu. A.; Engel, J.; Donath, E.)V1403,338-339(1991)

Photon correlation spectroscopy: technique and instrumentation (Thomas, John C.)V1430,2-18(1991)

Photon dynamics in tissue imaging (Chance, Britton; Haselgrove, John C.; Wang, Nai-Guang; Maris, Michael B.; Sevick, Eva M.)V1525,68-82(1991)

Photon migration in a model of the head measured using time- and frequency-domain techniques: potentials of spectroscopy and imaging (Sevick, Eva M.; Chance, Britton)V1431,84-96(1991)

Photothermal displacement spectroscopy of optical coatings (Su, Xing; Fan, Zheng X.)V1519,80-84(1991)

Photothermal spectroscopy as a sensitive spectroscopic tool (Tam, Andrew C.)V1435,114-127(1991)

Polarization-dependent x-ray spectroscopy of high-Tc superconductors (Heald, Steven M.; Tranquada, John M.)V1550,67-75(1991)

Polarized x-ray absorption spectroscopy for the study of superconductors and magnetic materials (Ramanathan, Mohan; Alp, Esen E.; Mini, Susan M.; Salem-Sugui, S.; Bommannavar, A.)V1548,168-178(1991)

Probe diffusion in polymer solutions (Phillies, George D.)V1430,118-131(1991)

Progress in fiber-remote gas correlation spectrometry (Dakin, John P.; Edwards, Henry O.)V1510,160-169(1991)

Proportional counter windows for the Bragg crystal spectrometer on AXAF (Markert, Thomas H.; Bauer, J.; Canizares, Claude R.; Isobe, T.; Nenonen, Seppo A.; O'Connor, J.; Schattenburg, Mark L.; Flanagan, Kathryn A.; Zombeck, Martin V.)V1549,408-419(1991)

Pulsed photothermal spectroscopy applied to lanthanide and actinide speciation (Berg, John M.; Morris, David E.; Clark, David L.; Tait, C. D.; Woodruff, William H.; Van Der Sluys, William G.)V1435,331-337(1991)

Quantitative analysis at the molecular level of laser/neural tissue interactions using a liposome model system (VanderMeulen, David L.; Misra, Prabhakar; Michael, Jason; Spears, Kenneth G.; Khoka, Mustafa)V1428,91-98(1991)

Radiometric calibration of space-borne Fourier transform infrared-emission spectrometers: proposed scenario for European Space Agency's Michelson interferometer for passive atmospheric sounding (Lamarre, Daniel; Giroux, Jean)V1493,28-36(1991)

Radiometric stability of the shuttle-borne solar backscatter ultraviolet spectrometer (Cebula, Richard P.; Hilsenrath, Ernest; Kelly, Thomas J.; Batluck, Georgiann R.)V1493,91-99(1991)

Raman and FT-IR studies of ocular tissues (Ozaki, Yukihiro; Mizuno, Aritake)V1403,710-719(1991)

Recent developments using GaAs as an x-ray detector (Sumner, Timothy J.; Grant, S. M.; Bewick, A.; Li, J. P.; Spooner, N. J.; Smith, K.; Beaumont, Steven P.)V1549,256-264(1991)

Recent progress in intrinsic fiber optic chemical sensing (Lieberman, Robert A.)V1368,15-24(1991)

Remote Raman spectroscopic imaging of human artery wall (Phat, Darith; Vuong, Phat N.; Plaza, Pascal; Cheilan, Francis; Dao, Nguyen Q.)V1525,196-205(1991)

Remote Raman spectroscopy using diode lasers and fiber-optic probes (Angel, S. M.; Myrick, Michael L.; Vess, Thomas M.)V1435,72-81(1991)

Remote sensing of volcanic ash hazards to aircraft (Rose, William I.; Schneider, David J.)V1492,387-390(1991)

Research and development of optical measurement techniques for aerospace propulsion research: a NASA/Lewis Research Center perspective (Lesco, Daniel J.)V1554B,530-539(1991)

Retrieval and molecule sensitivity studies for the global ozone monitoring experiment and the scanning imaging absorption spectrometer for atmospheric chartography (Chance, Kelly V.; Burrows, John P.; Schneider, Wolfgang)V1491,151-165(1991)

Revolutionary impact of today's array detector technology on chemical analysis (Radspinner, David A.; Fields, Robert E.; Earle, Colin W.; Denton, M. B.)V1439,2-14(1991)

Rigid lightweight optical bench for a spaceborne FUV spatial heterodyne interferometer (Tom, James L.; Cotton, Daniel M.; Bush, Brett C.; Chung, Ray; Chakrabarti, Supriya)V1549,302-307(1991)

Scanning imaging absorption spectrometer for atmospheric chartography (Burrows, John P.; Chance, Kelly V.)V1490,146-154(1991)

Scientific laser trends (Messenger, Heather)V1520,58-92(1991)

Selected Papers on Interferometry (Hariharan, P., ed.)VMS28(1991)

Selectively grown InxGa1-xAs and InxGa1-xP structures: locally resolved stoichiometry determination by Raman spectroscopy (Geurts, Jean; Finders, J.; Kayser, O.; Opitz, B.; Maassen, M.; Westphalen, R.; Balk, P.)V1361,744-750(1991)

Semiconductor lasers in analytical chemistry (Patonay, Gabor; Antoine, Miquel D.; Boyer, A. E.)V1435,52-63(1991)

Silicon x-ray array detector on spectrum-x-gamma satellite (Sipila, Heikki; Huttunen, Pekka; Kamarainen, Veikko J.; Vilhu, Osmi; Kurki, Jouko; Leppelmeier, Gilbert W.; Taylor, Ivor; Niemela, Arto; Laegsgaard, Erik; Sunyaev, Rashid)V1549,246-255(1991)

Simulation of time-resolved optical-CT imaging (Yamada, Yukio; Hasegawa, Yasuo)V1431,73-82(1991)

Single-ion spectroscopy (Bergquist, James C.; Wineland, D. J.; Itano, W. M.; Diedrich, F.; Raizen, M. G.; Elsner, Frank)V1435,82-85(1991)

Small-satellite sensors for multispectral space surveillance (Kostishack, Daniel F.)V1495,214-227(1991)

Soft x-ray spectro-microscope (Campuzano, Juan C.; Jennings, G.; Beaulaigue, L.; Rodricks, Brian G.; Brizard, Christine M.)V1345,245-254(1991)

Solid state lasers with passive mode-locking and negative feedback for picosecond spectroscopy (Danelius, R.; Grigonis, R.; Piskarskas, Algis S.; Podenas, D.; Sirutkaitis, V.)V1402,198-208(1991)

Space telescope imaging spectrograph 2048 CCD and its characteristics (Delamere, W. A.; Ebben, Thomas H.; Murata-Seawalt, Debbie; Blouke, Morley M.; Reed, Richard; Woodgate, Bruce E.)V1447,288-297(1991)

Spatially resolved water concentration determination in human eye lenses using Raman microspectroscopy (Siebinga, I.; de Mul, F. F.; Vrensen, G. F.; Greve, Jan)V1403,746-748(1991)

Spectral stratigraphy (Lang, Harold R.)V1492,351-357(1991)

Spectroscopical studies of the ionizing-radiation-induced damage in optical fibers (Ediriweera, Sanath R.; Kvasnik, Frank)V1504,110-117(1991)

Spectroscopic characteristics of Eu-doped aluminosilicate optical fiber preform (Oh, Kyunghwan; Morse, Ted F.; Reinhart, Larry J.; Kilian, Arnd)V1590,94-100(1991)

Spectroscopic determination of the parameters of an iron plasma produced by a CO2 laser (de Frutos, Angel M.; Sabatier, Lilian; Poueyo, Anne; Fabbro, Remy; Bermejo, Dionisio; Orza, Jose M.)V1397,717-720(1991)

Spectroscopic measurements of plasmas temperature and density during high-energy pulsed laser-materials interaction processes (Astic, Dominique; Vigliano, Patrick; Autric, Michel L.; Inglesakis, Georges; Prat, Ch.)V1397,713-716(1991)

Spectroscopic observations of CO2, CH4, N2O, and CO from Kitt Peak, 1979-1990 (Livingston, William C.; Wallace, Lloyd V.)V1491,43-47(1991)

Spectroscopic properties of Er3+-doped glasses for the realization of active waveguides by ion-exchange technique (Cognolato, Livio; De Bernardi, Carlo; Ferraris, M.; Gnazzo, Angelantonio; Morasca, Salvatore; Scarano, Domenica)V1513,368-377(1991)

Spectroscopic studies of second-generation sensitizers and their photochemical reactions in liposomes and cells (Ehrenberg, Benjamin; Gross, Eitan; Lavi, Adina; Johnson, Fred M.; Malik, Zvi)V1426,244-251(1991)

Spectroscopy and polarimetry capabilities of the INTEGRAL imager: Monte Carlo simulation results (Swinyard, Bruce M.; Malaguti, Giuseppe; Caroli, Ezio; Dean, Anthony J.; Di Cocco, Guido)V1548,94-105(1991)

Stimulated Raman diagnostics in diesel droplets (Golombok, Michael)V1497,100-119(1991)

Stoichiometry of laser-deposited Bi-Sr-Ca-Cu-O films on silicon and mass spectrometric investigations of superconductors (Becker, J. S.; Lorenz, M.; Dietze, H.-J.)V1598,227-238(1991)

Stone/ureter identification during alexandrite laser lithotripsy (Scheu, M.; Flemming, G.; Engelhardt, R.)V1421,100-107(1991)

Stratospheric spectroscopy with the far-infrared spectrometer: overview and recent results (Traub, Wesley A.; Chance, Kelly V.; Johnson, David G.; Jucks, Kenneth W.)V1491,298-307(1991)

Structural investigations of the (Si1-x,Gex)O2 single-crystal thin films by x-ray photoelectron spectroscopy (Sorokina, Svetlana; Dikov, Juriy)V1519,128-133(1991)

Studies of yeast cell oxygenation and energetics by laser fluorometry of reduced nicotinamide adenine dinucleotide (Pan, Fu-shih; Chen, Stephen; Mintzer, Robert; Chen, Chin-Tu; Schumacker, Paul)V1396,5-8(1991)

Study of C-H stretching vibration in hybrid Langmuir-Blodgett/alumina multilayers by infrared spectroscopy (Zheng, Tian S.; Liu, Li Y.; Xing, Zhongjin; Wang, Wen C.; Shen, Yuanhua; Zhang, Zhiming)V1519,339-346(1991)

Study of nighttime NO3 chemistry by differential optical absorption spectroscopy (Plane, John M.; Nien, Chia-Fu)V1433,8-20(1991)

Study of the optical features of YBa2Cu3O7-x films by SEW spectroscopy (Vaicikauskas, V.; Maldutis, Evaldas K.; Kajokas, R.)V1440,357-363(1991)

Study of x-ray emission from picosecond laser-plasma interaction (Chen, Hong; Chuang, Yung-Ho; Delettrez, J.; Uchida, S.; Meyerhofer, David D.)V1413,112-119(1991)

Study on Hadamard transform imaging spectroscopy (Wu, Ji-Zong; Deng, Jia-cheng; Chen, Ben-zhi)V1399,122-129(1991)

Subattomole detection in the condensed phase by nonlinear laser spectroscopy based on degenerate four-wave mixing (Tong, William G.)V1435,90-94(1991)

System for evaluation of trace gas concentration in the atmosphere based on the differential optical absorption spectroscopy technique (Hallstadius, Hans; Uneus, Leif; Wallin, Suante)V1433,36-43(1991)

Time-gated fluorescence spectroscopy and imaging of porphyrins and phthalocyanines (Cubeddu, Rinaldo; Canti, Gianfranco L.; Taroni, Paola; Valentini, G.)V1525,17-25(1991)

Time-resolved diffuse reflectance and transmittance studies in tissue simulating phantoms: a comparison between theory and experiment (Madsen, Steen J.; Patterson, Michael S.; Wilson, Brian C.; Park, Young D.; Moulton, J. D.; Jacques, Steven L.; Hefetz, Yaron)V1431,42-51(1991)

Time-resolved fluorescence of normal and atherosclerotic arteries (Pradhan, Asima; Das, Bidyut B.; Liu, C. H.; Alfano, Robert R.; O'Brien, Kenneth M.; Stetz, Mark L.; Scott, John J.; Deckelbaum, Lawrence I.)V1425,2-5(1991)

Time-resolved polarization luminescence spectroscopy of hematoporphyrin in liposomes (Chernyaeva, E. B.; Golubeva, N. A.; Koroteev, Nikolai I.; Lobanov, O. V.; Vardanyan, A. G.)V1402,7-10(1991)

Time-resolved Raman spectroscopy from reacting optically levitated microdroplets (Carls, Joseph C.; Brock, James R.)V1497,120-131(1991)

Time-resolved reflectance spectroscopy (Jacques, Steven L.; Flock, Stephen T.)V1525,35-40(1991)

Time-Resolved Spectroscopy and Imaging of Tissues (Chance, Britton; Katzir, Abraham, eds.)V1431(1991)

Time-resolved transillumination for medical diagnostics (Berg, Roger; Andersson-Engels, Stefan; Jarlman, Olof; Svanberg, Sune)V1431,110-119(1991)

Time-resolved x-ray absorption spectroscopy apparatus using laser plasma as an x-ray source (Yoda, Osamu; Miyashita, Atsumi; Murakami, Kouichi; Aoki, Sadao; Yamaguchi, Naohiro)V1503,463-466(1991)

Transient and persistent hole-burning of photosystem II preparations (Hayes, John M.; Tang, D.; Jankowiak, Ryszard; Small, Gerald J.)V1435,258-266(1991)

Transient of electrostatic potential at GaAs/AlAs heterointerfaces characterized by x-ray photoemission spectroscopy (Hirakawa, Kazuhiko; Hashimoto, Y.; Ikoma, Toshiaki)V1361,255-261(1991)

Transmission electron microscopy, photoluminescence, and capacitance spectroscopy on GaAs/Si grown by metal organic chemical vapor deposition (Bremond, Georges E.; Said, Hicham; Guillot, Gerard; Meddeb, Jaafar; Pitaval, M.; Draidia, Nasser; Azoulay, Rozette)V1361,732-743(1991)

Tropospheric emission spectrometer for the Earth Observing System (Glavich, Thomas A.; Beer, Reinhard)V1540,148-159(1991)

Tunable diode laser spectrometer for high-precision concentration and ratio measurements of long-lived atmospheric gases (Fried, Alan; Drummond, James R.; Henry, Bruce; Fox, Jack)V1433,145-156(1991)

Tunable diode laser systems for trace gas monitoring (Mackay, Gervase I.; Karecki, David R.; Schiff, Harold I.)V1433,104-119(1991)

Tunneling spectroscopy at nanometer scale in molecular beam epitaxy grown (Al)GaAs multilayers (Albrektsen, O.; Koenraad, Paul; Salemink, Huub W.)V1361,338-342(1991)

Use of absorption spectroscopy for refined petroleum product discrimination (Short, Michael)V1480,72-79(1991)

Use of admittance spectroscopy to probe the DX-centers in AlGaAs (Subramanian, S.; Chakravarty, S.; Anand, S.; Arora, B. M.)V1362,205-216(1991)

Use of Fourier transform spectroscopy in combustion effluent monitoring (Howe, Gordon S.; McIntosh, Bruce C.)V1434,90-103(1991)

Use of intelligent multichannel analyzer for study of biological objects (Karimov, M. G.)V1402,209-215(1991)

Using photon correlation spectroscopy to study polymer coil internal dynamic behavior (Selser, James C.; Ellis, Albert R.; Schaller, J. K.; McKiernan, M. L.; Devanand, Krisha)V1430,85-88(1991)

UV-fluorescence spectroscopic technique in the diagnosis of breast, ovarian, uterus, and cervix cancer (Das, Bidyut B.; Glassman, W. L. S.; Alfano, Robert R.; Cleary, Joseph; Prudente, R.; Celmer, E.; Lubicz, Stephanie)V1427,368-373(1991)

Vibrational Raman characterization of hard-carbon and diamond films (Ager, Joel W.; Veirs, D. K.; Marchon, Bruno; Cho, Namhee; Rosenblatt, Gern M.)V1437,24-31(1991)

Vibrational spectroscopy in the ophthalmological field (Bertoluzza, Alessandro; Monti, P.; Simoni, R.)V1403,743-745(1991)

VUV/photofragmentation laser-induced fluorescence sensor for the measurement of atmospheric ammonia (Sandholm, Scott T.; Bradshaw, John D.)V1433,69-80(1991)

X-ray absorption fine structure of systems in the anharmonic limit (Mustre de Leon, Jose; Conradson, Steven D.; Batistic, I.; Bishop, A. R.; Raistrick, Ian D.; Jackson, W. E.; Brown, George S.)V1550,85-96(1991)

X-ray absorption spectroscopy: how is it done? what can it tell us? (Hayes, Tim)V1550,56-66(1991)

X-ray absorption spectroscopy with polarized synchrotron radiation (Alp, Esen E.; Mini, Susan M.; Ramanathan, Mohan; Hyun, O. B.)V1345,137-145(1991)

Spectroscopy, atomic—see also spectroscopy

Applications of resonance ionization spectroscopy for semiconductor, environmental and biomedical analysis, and for DNA sequencing (Arlinghaus, Heinrich F.; Spaar, M. T.; Thonnard, N.; McMahon, A. W.; Jacobson, K. B.)V1435,26-35(1991)

Bistability of the Sn donor in AlxGa1-xAs and GaAs under pressure studied by Mossbauer spectroscopy (Gibart, Pierre; Williamson, Don L.)V1362,938-950(1991)

Calculations on the Hanle effect with phase and amplitude fluctuating laser fields (Bergeman, Thomas H.; Ryan, Robert E.)V1376,54-67(1991)

Depth profiling resonance ionization mass spectrometry of electronic materials (Downey, Stephen W.; Emerson, A. B.)V1435,19-25(1991)

Detection with a charge-coupled device in atomic emission spectroscopy (Bilhorn, Robert B.)V1448,74-80(1991)

Fluctuations in atomic fluorescence induced by laser noise (Vemuri, Gautam)V1376,34-46(1991)

High-efficiency resonance ionization mass spectrometric analysis by external laser cavity enhancement techniques (Johnson, Stephen G.; Rios, E. L.; Miller, Charles M.; Fearey, Bryan L.)V1435,292-297(1991)

Integrating FTIR microscopy into surface analysis (Church, Jeffrey S.)V1437,80-88(1991)

Measurement of trace isotopes by photon burst mass spectrometry (Fairbank, William M.; Hansen, C. S.; LaBelle, R. D.; Pan, X. J.; Chamberlin, E. P.; Fearey, Bryan L.; Gritzo, R. E.; Keller, Richard A.; Miller, Charles M.; Oona, H.)V1435,86-89(1991)

Photo resonance excitation and ionization characteristics of atoms by pulsed laser irradiation (Tanazawa, Takeshi; Adachi, Hajime A.; Nakahara, Ktsuhiko; Nittoh, Koichi; Yoshida, Toshifumi; Yoshida, Tadashi; Matsuda, Yasuhiko)V1435,310-321(1991)

Resonant and nonresonant ionization in sputtered initiated laser ionization spectrometry (Havrilla, George J.; Nicholas, Mark; Bryan, Scott R.; Pruett, J. G.)V1435,12-18(1991)

Theory and molecular dynamics simulation of one-photon electronic excitation of multiatomic molecules (Grishanin, B. A.; Vachev, V. D.; Zadkov, Victor N.)V1402,44-52(1991)

Spectroscopy, molecular—see also spectroscopy

Adaptation of coherent Raman methods for the investigation of biological samples in vivo (Lau, A.; Pfeiffer, M.; Werncke, W.)V1403,212-220(1991)

Advanced approaches in luminescence and Raman spectroscopy (Vo-Dinh, Tuan)V1435,197-202(1991)

Advances in polarized fluorescence depletion measurement of cell membrane protein rotation (Barisas, B. G.; Rahman, N. A.; Londo, T. R.; Herman, J. R.; Roess, Debrah A.)V1432,52-63(1991)

Advances in structural studies of viruses by Raman spectroscopy (Towse, Stacy A.; Benevides, James M.; Thomas, George J.)V1403,6-14(1991)

Applied Spectroscopy in Material Science (Saperstein, David D., ed.)V1437(1991)

Biomolecular Spectroscopy II (Birge, Robert R.; Nafie, Laurence A., eds.)V1432(1991)

Characterization of fluorescence-labeled DNA by time-resolved fluorescence spectroscopy (Seidel, Claus; Rittinger, K.; Cortes, J.; Goody, R. S.; Koellner, Malte; Wolfrum, Juergen M.; Greulich, Karl O.)V1432,105-116(1991)

Characterization of surface contaminants using infrared microspectroscopy (Blair, Dianna S.; Ward, Kenneth J.)V1437,76-79(1991)

Chlorine sensing by optical techniques (Momin, S. A.; Narayanaswamy, R.)V1510,180-186(1991)

Coherent holographic Raman spectroscopy of molecules (Ivanov, Anatoliy A.; Koroteev, Nikolai I.; Fishman, A. I.)V1403,174-184(1991)

Conformational analysis and circular dichroism of bilirubin, the yellow pigment of jaundice (Lightner, David A.; Person, Richard; Peterson, Blake; Puzicha, Gisbert; Pu, Yu-Ming; Bojadziev, Stefan)V1432,2-13(1991)

Conformational analysis of organic molecules in liquids with polarization-sensitive coherent anti-Stokes Raman scattering spectroscopy (Ivanov, Anatoliy A.; Koroteev, Nikolai I.; Mironov, S. F.; Fishman, A. I.)V1403,243-245(1991)

Cytochrome c at charged interfaces studied by resonance Raman and surface-enhanced resonance Raman spectroscopy (Hildebrandt, Peter)V1403,102-111(1991)

Dehydrogenase enzyme/coenzyme/substrate interactions (Hester, R. E.; Austin, J. C.)V1403,15-21(1991)

Determination of gasoline fuel properties by Raman spectroscopy (Clarke, Richard H.; Chung, W. M.; Wang, Q.; De Jesus, Stephen T.)V1437,198-204(1991)

Differential Raman spectroscopic study of the interaction of nickel (II) cation with adenine nucleotides (Mojzes, Peter)V1403,167-171(1991)

Direct measurement of vibrational energy relaxation in photoexcited deoxyhemoglobin using picosecond Raman spectroscopy (Hopkins, John B.; Xu, Xiaobing; Lingle, Robert; Zhu, Huiping; Yu, Soo-chang)V1432,221-226(1991)

Drug-target interactions on a single living cell: an approach by optical microspectroscopy (Manfait, Michel; Morjani, Hamid; Millot, Jean-Marc; Debal, Vincent; Angiboust, Jean-Francois; Nabiev, I. R.)V1403,695-707(1991)

Dynamics of ultrafast photoprocesses in Zn-octaethylporphyrin and Zn-octaethylporphin pi-monoanions (Chirvony, V. S.; Sinyakov, G. N.; Gadonas, R.; Krasauskas, V.; Pelakauskas, A.)V1403,504-506(1991)

Dynamic vs static bending rigidities for DNA and M13 virus (Schurr, J. M.; Song, Lu; Kim, Ug-Sung)V1403,248-257(1991)

Effect of alternating magnetic fields on the properties of water systems (Berezin, M. V.; Levshin, L. V.; Saletsky, A. M.)V1403,335-337(1991)

Excimer formation and singlet-singlet energy transfer of organoluminophores in the premicellar and micellar-polyelectrolytes solutions (Bisenbaev, A. K.; Levshin, L. V.; Saletsky, A. M.)V1403,606-610(1991)

Excitation energy relaxation in model aggregates of photosynthetic pigments upon picosecond laser excitation (Chirvony, V. S.; Zenkevich, E. I.; Gadonas, R.; Krasauskas, V.; Pelakauskas, A.)V1403,638-640(1991)

Feasibility of luminescence-eliminated anti-Stokes Raman spectroscopy (Ishibashi, Taka-aki; Hamaguchi, Hiro-o)V1403,555-562(1991)

Femtosecond infrared spectroscopy and molecular dynamics (Hochstrasser, Robin M.)V1403,3-3(1991)

Femtosecond processes in allophycocyanin trimers (Khoroshilov, E. V.; Kryukov, I. V.; Kryukov, P. G.; Sharkov, A. V.; Gillbro, T.)V1403,431-433(1991)

Femtosecond spectroscopy of acidified and neutral bacteriorhodopsin (Kobayashi, Takayoshi T.; Terauchi, Mamoru; Kouyama, Tsutomu; Yoshizawa, Masayuki; Taiji, Makoto)V1403,407-416(1991)

Finding a single molecule in a haystack: laser spectroscopy of solids from (square root of)N to N=1 (Moerner, William E.; Ambrose, William P.)V1435,244-251(1991)

Fluorescence spectrochronography of protein intramolecular dynamics (Chikishev, A. Y.; Ladokhin, Alexey S.; Shkurinov, A. P.)V1403,448-456(1991)

Four-photon spectroscopy of excited molecules: relaxation pathways and rates from high-excited states of 1,4-diphenylbutadiene molecules (Bogdanov, V. L.; Kulya, S. V.; Spiro, A. G.)V1403,470-474(1991)

Fractal dynamics of fluorescence energy transfer in biomembranes (Dewey, T. G.)V1432,64-75(1991)

Generation of free radicals in high-intensity laser photolysis of organic microcyclic compounds: time-resolved spectroscopy and EPR study (Angelov, D.; Gantchev, Ts.; Grabner, G.; Getoff, N.; Keskinova, E.; Shopova, Maria)V1403,572-574(1991)

High-power UV laser photolysis of nucleosides: final product analysis (Angelov, D.; Berger, M.; Cadet, J.; Ballini, Jean-Pierre; Keskinova, E.; Vigny, Paul)V1403,575-577(1991)

Hole-burning and picosecond time-resolved spectroscopy of isolated molecular clusters (Wittmeyer, Stacey A.; Kaziska, Andrew J.; Shchuka, Maria I.; Topp, Michael R.)V1435,267-278(1991)

Holographic nonlinear Raman spectroscopy of large molecules of biological importance (Ivanov, Anatoliy A.; Koroteev, Nikolai I.; Fishman, A. I.)V1429,132-144(1991); V1432,141-153(1991)

Infrared microanalysis of contaminants at grazing incidence (Reffner, John A.)V1437,89-94(1991)

In-situ characterization of resin chemistry with infrared transmitting optical fibers and infrared spectroscopy (Druy, Mark A.; Glatkowski, Paul J.; Stevenson, William A.)V1437,66-74(1991)

Integrating FTIR microscopy into surface analysis (Church, Jeffrey S.)V1437,80-88(1991)

Inter- and intramolecular processes in metalloporphyrins: study by transient absorption and resonance Raman and coherent anti-Stokes Raman scattering (Apanasevich, P. A.; Chirvony, V. S.; Kruglik, S. G.; Kvach, V. V.; Orlovich, V. A.)V1403,195-211(1991)

Intercalation between antitumor anthracyclines and DNA as probed by resonance and surface-enhanced Raman spectroscopy (Smulevich, G.; Mantini, A. R.; Casu, M.; Marzocchi, M. P.)V1403,125-127(1991)

Interpretation of the resonance Raman spectra of hemoproteins and model compounds (Solovyov, K. N.; Gladkov, L. L.)V1403,132-133(1991)

Intracellular location, picosecond kinetics, and light-induced reactions of photosensitizing porphyrins (Schneckenburger, Herbert; Seidlitz, Harold K.; Wessels, Jurina; Rueck, Angelika C.)V1403,646-652(1991)

Intramolecular processes of excitation energy redistribution in metalloporphyrins: examination by transient absorption and resonance Raman and coherent anti-Stokes Raman scattering spectroscopies (Apanasevich, P. A.; Chyrvony, V. S.; Kruglik, S. G.; Kvach, V. V.; Orlovich, V. A.)V1403,240-242(1991)

Investigation of single biological cells and chromosomes by confocal Raman microspectroscopy (Puppels, G. J.; Otto, C.; de Mul, F. F.; Greve, Jan)V1403,146-146(1991)

In vivo energy transfer studies in photosynthetic systems by subpicosecond timing (Shreve, A. P.; Trautman, Jay K.; Owens, T. G.; Frank, Harry A.; Albricht, Andreas C.)V1403,394-399(1991)

Laser investigation of molecular dynamic processes of the relaxational spectra formation (Neporent, B. S.)V1403,600-605(1991)

Laser spectroscopy of carotenoids in plant bio-objects (Kozlova, T. G.; Lobacheva, M. I.; Pravdin, A. B.; Romakina, M. Y.; Sinichkin, Yury P.; Tuchin, Valery V.)V1403,159-160(1991)

Laser spectroscopy of proton dynamics in hydrogen bonds (Hochstrasser, Robin M.; Oppenlaender, A.; Pierre, M.; Rambaud, C.; Silbey, R.; Skinner, J. L.; Trommsdorff, H. P.; Vial, J.-C.)V1403,221-229(1991)

Microstructures and domain size effects in diamond films characterized by Raman spectroscopy (Nemanich, Robert J.; Bergman, Larry; LeGrice, Yvonne M.; Turner, K. F.; Humphreys, T. P.)V1437,2-12(1991)

Molecular aggregates of quinuclidine and chlorophyll a (Korppi-Tommola, Jouko E.; Hakkarainen, Aulis; Helenius, Vesa F.)V1403,457-465(1991)

Molecular spectroscopy of biological molecules: Raman, NMR, and CD study of monophosphate dinucleosides at different degrees of protonation (Bertoluzza, Alessandro; Fagnano, C.; Morelli, M. A.; Tosi, M. R.; Tugnoli, V.)V1403,150-152(1991)

Multiphoton resonance ionization of molecules desorbed from surfaces by ion beams (Winograd, Nicholas; Hrubowchak, D. M.; Ervin, M. H.; Wood, M. C.)V1435,2-11(1991)

Nanosecond time-resolved natural and magnetic circular dichroism spectroscopy of protein dynamics (Goldbeck, Robert A.; Bjorling, Sophie; Kliger, David S.)V1432,14-27(1991)

New experimental methods and theory of Raman optical activity (Nafie, Laurence A.; Che, Diping; Yu, Gu-Sheng; Freedman, Teresa B.)V1432,37-49(1991)

New problems of femtosecond time-domain CARS (coherent antistokes Raman spectroscopy) of large molecules (Kolomoitsev, D. V.; Nikitin, S. Y.)V1402,31-43(1991)

Nonlinear optical properties of bacteriorhodopsin: assignment of the third-order polarizability based on two-photon absorption spectroscopy (Birge, Robert R.; Masthay, M. B.; Stuart, Jeffrey A.; Tallent, Jack R.; Zhang, Chian-Fan)V1432,129-140(1991)

Nonresonant background suppression in coherent anti-Stokes Raman scattering spectra of dissolved molecules (Lucassen, G. W.; Scholten, T. A.; de Boey, W. P.; de Mul, F. F.; Greve, Jan)V1403,185-194(1991)

Novel device for increasing the laser pulse intensity in multiphoton ionization mass spectrometry (Liang, Dawei; Fraser Monteiro, L.; Fraser Monteiro, M. L.)V1501,129-134(1991)

Optical emission spectroscopy of diamond-producing plasmas (Plano, Linda S.)V1437,13-23(1991)

Optical studies of nitrogen-doped amorphous carbon: laboratory and interstellar investigations (Kaufman, James H.; Metin, Serhat; Saperstein, David D.)V1437,36-41(1991)

Pericyclic photochemical ring-opening reactions are complete in picoseconds: a time-resolved UV resonance Raman study (Reid, Philip J.; Doig, Stephen J.; Mathies, Richard A.)V1432,172-183(1991)

Photophysics of 1,3,5-triaryl-2-pyrazolines (Sahyun, Melville R.; Crooks, G. P.; Sharma, D. K.)V1436,125-133(1991)

Physical effects in time-domain CARS (coherent antistokes Raman spectroscopy) of molecular gases (Kolomoitsev, D. V.; Nikitin, S. Y.)V1402,11-30(1991)

Picosecond absorption and circular dichroism studies of proteins (Simon, John D.; Xie, Xiaoliang; Dunn, Robert C.)V1432,211-220(1991)

Picosecond kinetics and Sn S1 absorption spectra of retinoids and carotenoids (Bondarev, S. L.; Tikhomirov, S. A.; Bachilo, S. M.)V1403,497-499(1991)

Picosecond reaction dynamics in photosynthetic and proton pumping systems: picosecond time-resolved Raman spectroscopy of electronic and vibrationally excited states (Atkinson, George H.)V1403,50-58(1991)

Picosecond time-resolved fluorescence spectroscopy of phytochrome and stentorin (Song, Pill-Soon)V1403,590-599(1991)

Picosecond time-resolved resonance Raman spectroscopy of bacteriorhodopsin: structure and kinetics of the J, K, and KL intermediates (Doig, Stephen J.; Reid, Philip J.; Mathies, Richard A.)V1432,184-196(1991)

Polynuclear membranes as a substrate for obtaining surface-enhanced Raman scattering films (Oleynikov, V. A.; Sokolov, K. V.; Hodorchenko, P. V.; Nabiev, I. R.)V1403,164-166(1991)

Raman and FT-IR characterization of biologically relevant Langmuir-Blodgett films (Katayama, Norihisa; Fukui, Masahiko; Ozaki, Yukihiro; Araki, Toshinari; Yokoi, Seiichi; Iriyama, Keiji)V1403,147-149(1991)

Raman and multichannel Raman spectroscopy of biological systems (Bertoluzza, Alessandro; Caramazza, R.; Fagnano, C.)V1403,40-49(1991)

Raman spectroscopy of biological molecules: uncharged phospholipid/ polyamine interactions in the presence of bivalent cations (Bertoluzza, Alessandro; Bonora, S.; Fini, G.; Morelli, M. A.)V1403,153-155(1991)

Raman study of icosahedral C60 (Sinha, Kislay; Menendez, Jose; Adams, G. B.; Page, J. B.; Sankey, Otto F.; Lamb, Lowell; Huffman, Donald R.)V1437,32-35(1991)

Recent developments by femtosecond spectroscopy in biological ultrafast free radical reactions (Gauduel, Yann; Pommeret, Stanislas; Yamada, Noelle; Antonetti, Andre)V1403,417-426(1991)

Resonance Raman scattering from the primary electron donor in photosynthetic reaction centers from Rhodobacter sphaeroides (Bocian, David F.)V1432,166-171(1991)

Resonance Raman spectra of hematoporphyrin derivative (Golubeva, N. G.; Wang, Litszin)V1403,134-138(1991)

Resonance Raman spectra of transient species of a respiration enzyme detected with an artificial cardiovascular system and Raman/absorption simultaneous measurement system (Kitagawa, Teizo; Ogura, Takashi)V1403,563-571(1991)

Resonance Raman studies of photosynthetic membrane proteins (Lutz, Marc; Mattioli, Tony; Moenne-Loccoz, Pierre; Zhou, Qing; Robert, Bruno)V1403,59-65(1991)

Resonance Raman studies of the peptide bond: implications for the geometry of the electronic-excited state and the nature of the vibronic linewidth (Hudson, Bruce S.)V1403,27-36(1991)

Resonance Raman study of Zn-porphyrin pi-anions (Gurinovich, G. P.; Kruglik, S. G.; Kvach, V. V.; Terekhov, S. N.)V1403,139-141(1991)

Rotational diffusion of receptors for epidermal growth factor measured by time-resolved phosphorescence depolarization (Zidovetzki, Raphael; Johnson, David A.; Arndt-Jovin, Donna J.; Jovin, Thomas M.)V1432,76-81(1991)

Sequence dependence of the length of the B to Z junctions in DNA (Peticolas, Warner L.; Dai, Z.; Thomas, G. A.)V1403,22-26(1991)

Solution conformation of biomolecules from infrared vibrational circular dichroism spectroscopy (Diem, Max)V1432,28-36(1991)

Spectral properties of the polaron model of a protein (Balabaev, N. K.; Lakhno, V. D.)V1403,478-486(1991)

Spectroscopic properties of the potentiometric probe merocyanine-540 in solutions and liposomes (Ehrenberg, Benjamin; Pevzner, Eliyahu)V1432,154-163(1991)

Spectroscopy of proteins on surfaces: implications for protein orientation and protein/protein interactions (Cotton, Therese M.; Rospendowski, Bernard; Schlegel, Vicki; Uphaus, Robert A.; Wang, Danli L.; Eng, Lars H.; Stankovich, Marion T.)V1403,93-101(1991)

Statistical signal processing in Raman spectroscopy of biological samples (Praus, Petr; Stepanek, Josef)V1403,76-84(1991)

Structure and dynamics of the active site of peroxidases as revealed by resonance Raman spectroscopy (Smulevich, G.; English, A. M.; Spiro, T. G.)V1403,440-447(1991)

Studies of the excited states of biological systems using UV-excited resonance Raman and picosecond transient Raman spectroscopy (Gustafson, Terry L.; Iwata, Koichi; Weaver, William L.; Huston, Lisa A.; Benson, Ronda L.)V1403,545-554(1991)

Subpicosecond electron transfer in reaction centers of photosynthetic bacteria (Shuvalov, V. A.; Ganago, A. O.; Shkuropatov, A. Y.; Klevanik, A. V.)V1403,400-406(1991)

Superradiance and exciton dynamics in molecular aggregates (Fidder, Henk; Terpstra, Jacob; Wiersma, Douwe A.)V1403,530-544(1991)

Surface-enhanced hyper-Raman and near-IR FT-Raman studies of biomolecules (Yu, Nai-Teng; Nie, Shuming)V1403,112-124(1991)

Surface-enhanced Raman spectroscopy in the structural studies of biomolecules: the state of the art (Nabiev, I. R.; Sokolov, K. V.; Efremov, R. G.; Chumanov, G. D.)V1403,85-92(1991)

Surface-enhanced Raman spectroscopy of adenosine and 5'AMP: evolution in time (Sanches-Cortes, S.; Garcia-Ramos, J. V.)V1403,142-145(1991)

Surface-enhanced resonance hyper-Raman spectra of bacteriorhodopsin adsorbed on silver colloids (Baranov, A. V.; Nabiev, I. R.)V1403,128-131(1991)

Symmetrization analysis of lattice-vibrational modes and study of Raman-IR spectra for B-BaB2O4 (Hong, Shuili)V1437,194-197(1991)

Theory of non-Condon femtosecond quantum beats in electronic transitions of dye molecules (Mazurenko, Yuri T.; Smirnov, V. V.)V1403,466-469(1991)

Time resolution of events in an enzyme's active site at 4 K and 300 K using resonance Raman spectroscopy (Carey, Paul R.; Kim, Munsok; Tonge, Peter J.)V1403,37-39(1991)

Time-resolved infrared studies of the dynamics of ligand binding to cytochrome c oxidase (Dyer, R. B.; Peterson, Kristen A.; Stoutland, Page O.; Einarsdottir, Oloef; Woodruff, William H.)V1432,197-204(1991)

Ultrafast and not-so-fast dynamics of cytochrome oxidase: the ligand shuttle and its possible functional significance (Woodruff, William H.; Dyer, R. B.; Einarsdottir, Oloef; Peterson, Kristen A.; Stoutland, Page O.; Bagley, K. A.; Palmer, Graham; Schoonover, J. R.; Kliger, David S.; Goldbeck, Robert A.; Dawes, T. D.; Martin, Jean-Louis; Lambry, J.-C.; Atherton, Stephen J.; Hubig, Stefan M.)V1432,205-210(1991)

Use of fluorescence spectroscopy to elucidate structural features of the nicotinic acetylcholine receptor (Johnson, David A.; Valenzuela, C. F.)V1432,82-90(1991)

UV resonance Raman spectroscopic study of Trp residues in a hydrophobic environment (Efremov, R. G.; Feofanov, A. V.; Nabiev, I. R.)V1403,161-163(1991)

Vibrational Raman optical activity of biological molecules (Barron, L. D.; Gargaro, A. R.; Hecht, L.; Wen, Z. Q.; Hug, W.)V1403,66-75(1991)

Vibrational spectroscopy of biological molecules: halocompound/nucleic acid component interactions (Bottura, Giorgio; Filippetti, P.; Tinti, A.)V1403,156-158(1991)

Spectrum analysis—see also spectroscopy

Acousto-optic estimation of autocorrelation and spectra using triple correlations and bispectra (Sadler, Brian M.; Giannakis, Georgios B.; Smith, Dale J.)V1476,246-256(1991)

Adsorption of bromine in CuBr laser (Wang, Yongjiang; Shi, Shuyi; Qian, Yujun; Pan, Bailiang; Ding, Xiande; Guo, Qiang; Chen, Lideng; Fan, Ruixing)V1412,67-71(1991)

Aerospace remote sensing monitoring of inland water quality (Gitelson, Anatoly A.)V1492,307-318(1991)

Analysis of atmospheric trace constituents from high-resolution infrared balloon-borne and ground-based solar absorption spectra (Goldman, Aaron; Murcray, Frank J.; Rinsland, C. P.; Blatherwick, Ronald D.; Murcray, F. H.; Murcray, David G.)V1491,194-202(1991)

Analysis of molecular adsorbates by laser-induced thermal desorption (McIver, Robert T.; Hemminger, John C.; Parker, D.; Li, Y.; Land, Donald P.; Pettiette-Hall, C. L.)V1437,124-128(1991)

Analysis of vegetation stress and damage from images of the high-resolution airborne pushbroom image spectrograph compact airborne spectrographic imager (Mueksch, Michaela C.)V1399,157-161(1991)

Angular positioning mechanism for the ultraviolet coronagraph spectrometer (Ostaszewski, Miroslaw A.; Guy, Larry J.)V1482,13-25(1991)

Clinical applications of pulmonary artery oximetry (Barker, Steven J.)V1420,22-28(1991)

Colorimetric characterization of CCD sensors by spectrophotometry (Daligault, Laurence; Glasser, Jean)V1512,124-130(1991)

Computer-aided performance evaluation system for the on-board data compression system in HIRIS (Qian, Shen-en; Wang, Ruqin; Li, Shuqiu; Dai, Yisong)V1483,196-206(1991)

Concentric systems for adaptation as spectrographs (Mertz, Lawrence N.)V1354,457-459(1991)

Design and analysis of the reflection grating arrays for the X-Ray Multi-Mirror Mission (Atkinson, Dennis P.; Bixler, Jay V.; Geraghty, Paul; Hailey, Charles J.; Klingmann, Jeffrey L.; Montesanti, Richard C.; Kahn, Steven M.; Paerels, F. B.)V1343,530-541(1991)

Design of an imaging spectrometer for observing ocean color (Weng, Zhicheng; Chen, Zhiyong; Cong, Xiaojie)V1527,338-348(1991)

Design of spherical varied line-space gratings for a high-resolution EUV spectrometer (Harada, Tatsuo; Kita, Toshiaki; Bowyer, C. S.; Hurwitz, Mark)V1545,2-10(1991)

Determination of the altitude of the nitric acid layer from very high resolution, ground-based IR solar spectra (Blatherwick, Ronald D.; Murcray, Frank J.; Murcray, David G.; Locker, M. H.)V1491,203-210(1991)

Deviated-plane varied-line-space grating spectrograph (Edelstein, Jerry)V1545,145-148(1991)

Evolution of an excimer laser gas mix (Boardman, A. D.; Hodgson, Elizabeth M.; Spence, A. J.; Richardson, A. D.; Richardson, M. B.)V1503,160-166(1991)

Feature selection technique for classification of hyperspectral AVIRIS data (Shen, Sylvia S.; Trang, Bonnie Y.)V1567,188-193(1991)

Fourier transform infrared spectrophotometry for thin film monitors: computer and equipment integration for enhanced capabilities (Cox, J. N.; Sedayao, J.; Shergill, Gurmeet S.; Villasol, R.; Haaland, David M.)V1392,650-659(1991)

Hadamard transform Raman imaging (Morris, Michael D.; Govil, Anurag; Liu, Kei-Lee; Sheng, Rong-Sheng)V1439,95-101(1991)

High-efficiency resonance ionization mass spectrometric analysis by external laser cavity enhancement techniques (Johnson, Stephen G.; Rios, E. L.; Miller, Charles M.; Fearey, Bryan L.)V1435,292-297(1991)

High-resolution stigmatic EUV spectroheliometer for studies of the fine scale structure of the solar chromosphere, transition region, and corona (Timothy, J. G.; Berger, Thomas E.; Morgan, Jeffrey S.; Walker, Arthur B.; Bhattacharyya, Jagadish C.; Jain, Surendra K.; Saxena, Ajay K.; Huber, Martin C.; Tondello, Giuseppe; Naletto, Giampiero)V1343,350-358(1991)

High-resolution studies of atmospheric IR emission spectra (Murcray, Frank J.; Murcray, F. H.; Goldman, Aaron; Blatherwick, Ronald D.; Murcray, David G.)V1491,282-287(1991)

Hubble Space Telescope: mission, history, and systems (Endelman, Lincoln L.)V1358,422-441(1991)

Importance of proton transfer in contact charging (Wollmann, Daphne; Diaz, Art F.)V1458,192-200(1991)

Influence of gas composition on XeCl laser performance (Jursich, Gregory M.; Von Drusek, William A.; Mulderink, Ken; Olchowka, V.; Reid, John; Brimacombe, Robert K.)V1412,115-122(1991)

Instrumentation to measure the near-IR spectrum of small fruits (Jaenisch, Holger M.; Niedzwiecki, Abraham J.; Cernosek, John D.; Johnson, R. B.; Seeley, John S.; Dull, Gerald G.; Leffler, Richard G.)V1379,162-167(1991)

Interpretation of measured spectral attenuation curves of optical fibers by deconvolution with source spectrum (Hoefle, Wolfgang)V1504,140-146(1991)

Laser-induced volatilization and ionization of aerosol particles for their mass spectral analysis in real time (Sinha, Mahadeva P.)V1437,150-156(1991)

LAser Microprobe Mass Spectrometry (LAMMS) in dental science: basic principles, instrumentation, and applications (Rechmann, Peter; Tourmann, J. L.; Kaufmann, Raimund)V1424,106-115(1991)

Laser probe mass spectrometry (Campana, Joseph E.)V1437,138-149(1991)

Laser remote sensing of natural water organics (Babichenko, Sergey M.; Poryvkina, Larisa)V1492,319-323(1991)

Local spectrum analysis of medical images (Baskurt, Atilla; Peyrin, Francoise; Min, Zhu-Yue; Goutte, Robert)V1445,485-495(1991)

Matched-filter algorithm for subpixel spectral detection in hyperspectral image data (Borough, Howard C.)V1541,199-208(1991)

Measurement of trace isotopes by photon burst mass spectrometry (Fairbank, William M.; Hansen, C. S.; LaBelle, R. D.; Pan, X. J.; Chamberlin, E. P.; Fearey, Bryan L.; Gritzo, R. E.; Keller, Richard A.; Miller, Charles M.; Oona, H.)V1435,86-89(1991)

Measurements of the atmospheric turbulence spectrum and intermittency using laser scintillation (Frehlich, Rod G.)V1487,10-18(1991)

Midinfrared backscatter spectra of selected agricultural crops (Narayanan, Ram M.; Green, Steven E.; Alexander, Dennis R.)V1379,116-122(1991)

Multiaperture spectrometer design for the Atmospheric Infrared Sounder (Pagano, Robert J.; Hatch, Marcus R.)V1354,460-471(1991)

Multichannel chromatic interferometry: metrology applications (Tribillon, Gilbert M.; Calatroni, Jose; Sandoz, Patrick)V1332,632-642(1991)

New concept for a highly compact imaging Fourier transform spectrometer (Simeoni, Denis)V1479,127-138(1991)

New look at bump spectra models for temperature and refractive-index fluctuations (Sisterman, Elizabeth A.; Andrews, Larry C.)V1487,345-355(1991)

Nondestructive determination of the solids content of horticultural products (Birth, Gerald S.; Dull, Gerald G.; Leffler, Richard G.)V1379,10-15(1991)

Novel monolithic chip-integrated color spectrometer: the distributed-wavelength filter component (Holm-Kennedy, James W.; Tsang, Koon Wing; Sze, Wah Wai; Jiang, Fenglai; Yang, Datong)V1527,322-331(1991)

Optical reflectance sensor for detecting plants (Shropshire, Geoffrey J.; Von Bargen, Kenneth; Mortensen, David A.)V1379,222-235(1991)

Optics in Agriculture (DeShazer, James A.; Meyer, George E., eds.)V1379(1991)

Optimal estimation of finish parameters (Church, Eugene L.; Takacs, Peter Z.)V1530,71-85(1991)

Optoelectronic devices for fiber-optic sensor interface systems (Hong, C. S.; Hager, Harold E.; Capron, Barbara; Mantz, Joseph L.; Beranek, Mark W.; Huggins, Raymond W.; Chan, Eric Y.; Voitek, Mark; Griffith, David M.; Livezey, Darrell L.; Scharf, Bruce R.)V1418,177-187(1991)

ORFEUS: orbiting and retrievable far and extreme ultraviolet spectrometer (Rippel, Harald; Kampf, Dirk; Graue, Roland)V1343,520-529(1991)

Overview of ASTER design concept (Fujisada, Hiroyuki; Ono, Akira)V1490,244-254(1991)

Reflectance anisotropy spectrometer for real-time crystal growth investigations (Acher, O.; Benferhat, Ramdane; Drevillon, Bernard; Razeghi, Manijeh)V1361,1156-1163(1991)

Reflection spectrum of multiple chirped gratings (Shellan, Jeffrey B.; Yeh, Pochi A.)V1545,179-188(1991)

Remote spectral fingerprinting for law enforcement (Huguenin, Robert L.; Tahmoush, Donald J.)V1479,403-411(1991)

Resonant and nonresonant ionization in sputtered initiated laser ionization spectrometry (Havrilla, George J.; Nicholas, Mark; Bryan, Scott R.; Pruett, J. G.)V1435,12-18(1991)

Results from the calibration of the Extreme Ultraviolet Explorer instruments (Welsh, Barry Y.; Jelinsky, Patrick; Vedder, Peter W.; Vallerga, John V.; Finley, David S.; Malina, Roger F.)V1343,166-174(1991)

Shock-layer-induced ultraviolet emissions measured by rocket payloads (Caveny, Leonard H.; Mann, David M.)V1479,102-110(1991)

Small-target acquisition and typing by AASAP (Huguenin, Robert L.; Tahmoush, Donald J.)V1481,64-72(1991)

Space-spectrum resolution function of the imaging spectrometer (Han, Xin Z.; Qiu, Ying)V1479,156-161(1991)

Space telescope imaging spectrograph 2048 CCD and its characteristics (Delamere, W. A.; Ebben, Thomas H.; Murata-Seawalt, Debbie; Blouke, Morley M.; Reed, Richard; Woodgate, Bruce E.)V1479,21-30(1991)

Spectral and time-resolved measurements of pollutants on water surface by an XeCl laser fluorosensor (Barbini, Roberto; Fantoni, Roberta; Palucci, Antonio; Ribezzo, Sergio; van der Steen, Hendricus J.)V1503,363-374(1991)

Spectral hole burning of strongly confined CdSe quantum dots (Spiegelberg, Christine; Henneberger, Fritz; Puls, J.)V1362,951-958(1991)

Spectral signature analysis for industrial inspection (Rauchmiller, Robert F.; Vanderbok, Raymond S.)V1384,100-114(1991)

Spectrum of surface electromagnetic waves in CdxHg1-xTe crystals at 0.3 K < T < 77 K (Vertiy, Alexey A.; Beletskii, N. N.; Gorbatyuk, I. N.; Ivanchenko, I. V.; Popenko, N. A.; Rarenko, I. M.; Tarapov, Sergey I.)V1361,1070-1078(1991)

Strain measurements on soft materials application to cloth and papers (Bremand, Fabrice J.; Lagarde, Alexis)V1554B,650-660(1991)

Stratospheric ozone concentration profiles from Spacelab-1 solar occultation infrared absorption spectra (De Maziere, Martine M.; Camy-Peyret, C.; Lippens, C.; Papineau, N.)V1491,288-297(1991)

Study of TEM micrographs of thin-film cross-section replica using spectral analysis (Mei, Ting; Liu, Xu; Tang, Jinfa; Gu, Peifu)V1554A,570-578(1991)

Thirty-two-channel LED array spectrometer module with compact optomechanical construction (Malinen, Jouko; Keranen, Heimo; Hannula, Tapio; Hyvarinen, Timo S.)V1533,122-128(1991)

Thomson parabolic spectrograph with microchannel plate framing camera as register of ionic parabolae (Lebedev, V. B.; Saulevich, S. V.)V1358,874-877(1991)

Time-integrating optical raster spectrum analyzer using semiconductor laser as input modulator (Larkin, Alexander I.; Matveev, Alexander; Mironov, Yury)V1621,414-423(1991)

Two-frequency correlation in a turbulent atmosphere (Mazar, Reuven; Rozental, Mark)V1487,31-39(1991)

Ultra-High-Resolution XUV Spectroheliograph II: predicted performance (Walker, Arthur B.; Lindblom, Joakim F.; Timothy, J. G.; Allen, Maxwell J.; DeForest, Craig E.; Kankelborg, Charles C.; O'Neal, Ray H.; Paris, Elizabeth S.; Willis, Thomas D.; Barbee, Troy W.; Hoover, Richard B.)V1343,319-333(1991)

Use of spline expansions and regularization in the unfolding of data from spaceborne sensors (Fisher, Thornton R.; Perez, Joseph D.)V1479,212-225(1991)

Vision methods for inspection of greenhouse poinsettia plants (Meyer, George E.; Troyer, W. W.; Fitzgerald, Jay B.)V1379,99-105(1991)

Wedge imaging spectrometer: application to drug and pollution law enforcement (Elerding, George T.; Thunen, John G.; Woody, Loren M.)V1479,380-392(1991)

Wideband acousto-optic spectrometer (Chang, I. C.)V1476,257-268(1991)

Wide-spectral-range transmissometer used for fog measurements (Turner, Vernon; Trowbridge, Christian A.)V1487,262-271(1991)

XMM space telescope: development plan for the lightweight replicated x-ray gratings (Montesanti, Richard C.; Atkinson, Dennis P.; Edwards, David F.; Klingmann, Jeffrey L.)V1343,558-565(1991)

Standards

Adaptive perceptual quantization for video compression (Puri, Atul; Aravind, R.)V1605,297-300(1991)

Aerospace resource document: fiber optic interconnection hardware (Little, William R.)V1589,20-23(1991)

Antireflection coating standards of ophthalmic resin lens materials (Porden, Mark)V1529,115-123(1991)

Are optical fiber sensors intrinsically, inherently, or relatively safe? (McGeehin, Peter)V1504,75-79(1991)

Arithmetic processor design for the T9000 transputer (Knowles, Simon C.)V1566,230-243(1991)

Automated techniques for quality assurance of radiological image modalities (Goodenough, David J.; Atkins, Frank B.; Dyer, Stephen M.)V1444,87-99(1991)

Bridging issues in DQDB subnetworks (Tantawy, Ahmed N.)V1364,268-276(1991)

Building and campus networks for fiber distributed data interface (McIntosh, Thomas F.)V1364,84-93(1991)

Building an infrastructure for system integration: machine vision standards (Bieman, Leonard H.; Peyton, James A.)VCR36,3-19(1991)

Campus fiber optic enterprise networks (Weeks, Richard A.)V1364,222-227(1991)

Color standards for electronic imaging (McDowell, David Q.)VCR37,40-53(1991)

Comparison of image compression techniques for high quality based on properties of visual perception (Algazi, V. R.; Reed, Todd R.)V1567,589-598(1991)

Compensated digital readout family (Ludwig, David E.; Skow, Michael)V1541,73-82(1991)

Crossconnects in a SONET network (Bootman, Steven R.)V1363,142-148(1991)

Development and implementation of MIS-36477 laser damage certification of designator optical components (Mordaunt, David W.; Nieuwsma, Daniel E.)V1441,27-30(1991)

Digital Image Compression Techniques (Jones, Paul W.; Rabbani, Majid)VTT07(1991)

Discussion of the standard practice for the location of wet insulation in roofing systems using infrared imaging (ASTM C1153-90) (Sopko, Victor)V1467,83-89(1991)

DS1 mapping considerations for the synchronous optical network (Cubbage, Robert W.; Littlewood, Paul A.)V1363,163-171(1991)

Effect of standards on new equipment design by new international standards and industry restraints (Endelman, Lincoln L.)V1346,90-92(1991)

Evolution of the DQDB hybrid multiplexing for an integrated service packetized traffic (Gagnaire, A.; Ponsard, Benoit)V1364,277-288(1991)

Evolving JPEG color data compression standard (Mitchell, Joan L.; Pennebaker, William B.)VCR37,68-97(1991)

Extending HIPPI at 800-mega-bits-per-second over serial links using HOT ROD technology (Annamalai, Kadiresan)V1364,178-189(1991)

FDDI, Campus-Wide, and Metropolitan Area Networks (Annamalai, Kadiresan; Cudworth, Stewart K.; Kasiewicz, Allen B., eds.)V1364(1991)

FDDI network cabling (Stevens, R. S.)V1364,101-114(1991)

Fiber channel: the next standard peripheral interface and more (Cummings, Roger)V1364,170-177(1991)

Fiber optic measurement standards (Pollitt, Stuart)V1504,80-87(1991)

Fiber Optic Metrology and Standards (Soares, Oliverio D., ed.)V1504(1991)

Fiber Optics in the Subscriber Loop (Hutcheson, Lynn D.; Kahn, David A., eds.)V1363(1991)

French proposal for IEC/TC 86/WG 4 OTDR calibration (Gauthier, Francis)V1504,55-65(1991)

Frequency spectrum analysis and assessment of optical surface flaws (Gao, Wenliang; Zhang, Xiao; Yang, Guoguang)V1530,118-128(1991)

Germanium photodiodes calibration as standards of optical fiber systems power measurements (Campos, Joaquin; Corredera, Pedro; Pons, Alicia A.; Corrons, Antonio)V1504,66-74(1991)

High-bandwidth recording in a hostile environment (Darg, David A.; Kikuchi, Akira)V1538,124-131(1991)

High-density memory packaging technology high-speed imaging applications (Frew, Dean L.)V1346,200-209(1991)

High-performance FDDI NIU for streaming voice, video, and data (Bergman, Larry A.; Hartmayer, Ron; Wu, Wennie H.; Cassell, P.; Edgar, G.; Lambert, James L.; Mancini, Richard; Jeng, J.; Pardo, C.; Halloran, Frank; Martinez, James C.)V1364,14-21(1991)

Image acquisition: quality control for document image scanners (Gershbock, Richard)VCR37,20-39(1991)

Imagery technology database (Courtot, Marilyn E.; Nier, Michael)VCR37,221-246(1991)

Implementation of FDDI in the intelligent wiring hub (Tarrant, Peter J.; Truman, Alan K.)V1364,2-6(1991)

Integrated "Byte-to-light" solution for fiber optic data communication (Kubinec, James J.; Somerville, James A.; Chown, David P.; Birch, Martin J.)V1364,130-143(1991)

Interchangeability of optical disks (Shimamoto, Masayoshi; Yamada, Koichi; Watanabe, Isao; Nakajima, Yoshiki; Ito, Osamu; Tanaka, Kunimaro)V1499,393-400(1991)

Jitter considerations for FDDI PMD (Fukuoka, Takashi; Tejika, Yasuhiro; Takada, Hisashi; Takahashi, Hidenori; Hamasaki, Yiji)V1364,40-48(1991)

Laser beam width, divergence, and propagation factor: status and experience with the draft standard (Fleischer, John M.)V1414,2-11(1991)

Latest advances in CAD data interfacing: a standardization project of ISO/TC 172/SC1 task group "optical database" (Wise, Timothy D.; Wieder, Eckart)V1346,79-85(1991)

Long-term fiber reliability (Saifi, Mansoor A.)V1366,58-70(1991)

Long-term reliability and performance testing of fiber optic sensors for engineering applications (Weinberger, Alex; Weinberger, Ervin)V1367,30-45(1991)

Margin measurements of pulse amplitude modulation channels (Lauffenburger, Jim; Arachtingi, John W.; Robbins, Jamey L.; Kim, Jong)V1499,104-113(1991)

Measurement of Poisson's ratio of nonmetallic materials by laser holographic interferometry (Zhu, Jian T.)V1554B,148-154(1991)

Metropolitan area networks: a corner stone in the broadband era (Ghanem, Adel)V1364,312-319(1991)

Mother's recipe is not a standard (Shell, Forney L.)V1346,86-89(1991)

MPEG (moving pictures expert group) video compression algorithm: a review (Le Gall, Didier J.)V1452,444-457(1991)

Multimedia courseware in an open-systems environment: a DoD strategy (Welsch, Lawrence A.)VCR37,207-220(1991)

NAVSEA gigabit optical MAN prototype history and status (Albanese, Andres; Devetzis, Tasco N.; Ippoliti, A, G.; Karr, Michael A.; Maszczak, M. W.; Dorris, H. N.; Davis, James H.)V1364,320-326(1991)

New subband scheme for super-HDTV coding (Tanimoto, Masayuki; Yamada, Akio; Naito, Yoichi)V1605,394-405(1991)

On enhancing the performance of the ACR-NEMA protocol (Maydell, Ursula M.; Hassanein, Hossam S.; Deng, Shuang)V1446,81-92(1991)

Optical fiber measurements and standardization: status and perspectives (Di Vita, P.)V1504,38-46(1991)

Optical subscriber line transmission system to support an ISDN primary-rate interface (Wataya, Hideo; Tsuchiya, Toshiyuki)V1363,72-84(1991)

Options for campus fiber networks (Henderson, Byron B.; Green, Emily N.)V1364,235-244(1991)

OTDR calibration for attenuation measurement (Moeller, Werner; Heitmann, Walter; Reich, M.)V1504,47-54(1991)

Performance tests of large CCDs (Robinson, Lloyd B.; Brown, William E.; Gilmore, Kirk; Stover, Richard J.; Wei, Mingzhi; Geary, John C.)V1447,214-228(1991)

Physical-connection compliance testing for FDDI (Baldwin, Christopher)V1364,120-129(1991)

Picture archiving and communications systems protocol based on ISO-OSI standard (Martinez, Ralph; Nam, Jiseung; Dallas, William J.; Osada, Masakazu; McNeill, Kevin M.; Ozeki, Takeshi; Komatsu, Ken-ichi)V1446,100-107(1991)

Pod-mounted MIL-STD-2179B recorder (Kessler, William D.; Abeille, Pierre; Sulzer, Jean-Francois)V1538,104-111(1991)

Preliminary review of imaging standards (Ren, Victor; Hatfield, Donald J.; Deacutis, Martin)VCR37,54-67(1991)

Process latitude measurements on chemically amplified resists exposed to synchrotron radiation (Babcock, Carl P.; Taylor, James W.; Sullivan, Monroe; Suh, Doowon; Plumb, Dean; Palmer, Shane R.; Berry, Amanda K.; Graziano, Karen A.; Fedynyshyn, Theodore H.)V1466,653-662(1991)

Quality assurance from a manufacturer's standpoint (Anderson, William J.)V1444,128-133(1991)

Radiometric standards in the USSR (Sapritsky, Victor I.)V1493,58-69(1991)

Reduction of blocking artifacts using motion-compensated spatial-temporal filtering (Chu, Frank J.; Yeh, Chia L.)V1452,38-46(1991)

Refractive index measurement capabilities at the National Institute of Standards and Technology (Dodge, Marilyn J.)V1441,56-60(1991)

Reliability of contrast and dissolution-rate-derived parameters as predictors of photoresist performance (Spragg, Peggy M.; Hurditch, Rodney J.; Toukhy, Medhat A.; Helbert, John N.; Malhotra, Sandeep)V1466,283-296(1991)

Research and development of a NYNEX switched multi-megabit data service prototype system (Maman, K. H.; Haines, Robert; Chatterjee, Samir)V1364,304-311(1991)

Rings in a SONET network (Clendening, Steven J.)V1363,149-157(1991)

Round table discussion on standards (Endelman, Lincoln L.)V1346,93-94(1991)

SAFENET II: The Navy's FDDI-based computer network standard (Paige, Jeffrey L.; Howard, Edward A.)V1364,7-13(1991)

Scaled discrete cosine transform algorithms for JPEG and MPEG implementations on fused multiply/add architectures (Feig, Ephraim; Linzer, Elliot)V1452,458-467(1991)

Simulation of free-electron lasers in the presence of measured magnetic field errors of the undulator (Marable, William P.; Tang, Cha-Mei; Esarey, Eric)V1552,80-86(1991)

Sinusoidal surfaces as standards for BRDF instruments (Marx, Egon; Lettieri, Thomas R.; Vorburger, Theodore V.; McIntosh, Malcolm B.)V1530,15-21(1991)

Some experimental techniques for characterizing photoresists (Spence, Christopher A.; Ferguson, Richard A.)V1466,324-335(1991)

SONET inter-vendor compatibility (Bowmaster, Thomas A.; Cockings, Orville R.; Swanson, Robert A.)V1363,119-124(1991)

Standardization efforts for the preservation of electronic imagery (Adelstein, Peter Z.; Storm, William D.)VCR37,159-179(1991)

Standardization of image quality measurements of medical x-ray image intensifier systems (Sandrik, John M.)VCR37,180-206(1991)

Standards for electronic imaging for facsimile systems (Urban, Stephen J.)VCR37,113-145(1991)

Standards for electronic imaging for graphic arts systems (Dunn, S. T.; Dunn, Patrice M.)VCR37,98-112(1991)

Standards for Electronic Imaging Systems (Nier, Michael; Courtot, Marilyn E., eds.)VCR37(1991)

Standards for flat panel display systems (Greeson, James C.)VCR37,146-158(1991)

Standards for image input devices: review and forecast (Gilblom, David L.)VCR37,3-19(1991)

Standards for oscillatory scanners (Ludwiszewski, Alan P.)V1454,174-185(1991)

Status and importance of optical standards (Parks, Robert E.)V1346,76-78(1991)

Stray-light implications of scratch/dig specifications (Lewis, Isabella T.; Ledebuhr, Arno G.; Bernt, Marvin L.)V1530,22-34(1991)

Study of binary image compression using universal coding (Nakano, Yasuhiko; Chiba, Hirotaka; Okada, Yoshiyuki; Yoshida, Shigeru; Mori, Masahiro)V1605,874-878(1991)

Summary of color definition activity in the graphic arts (McDowell, David Q.)V1460,29-35(1991)

Superhigh-definition image communication: an application perspective (Kohli, Jagdish C.)V1605,351-361(1991)

Three-dimensional tolerance verification for computer vision systems (Griffin, Paul M.; Taboada, John)V1381,292-298(1991)

Visual factors and image analysis in the encoding of high-quality still images (Algazi, V. R.; Reed, Todd R.; Ford, Gary E.; Estes, Robert R.)V1605,329-338(1991)

Windowed motion compensation (Watanabe, Hiroshi; Singhal, Sharad)V1605,582-589(1991)

Workshop on standards for streak camera characterization (Jaanimagi, Paul A.)V1346,504-504(1991)

Statistical optics

Adaptive filtering and enhancement method of speckle pattern (Yang, Yibing; Wang, Houshu)V1554A,781-788(1991)

Basic manufacturability interval (Billings, Daniel A.)V1468,434-445(1991)

Characterization of simulated and open-air atmospheric turbulence (Razdan, Anil K.; Singh, Brahm P.; Chopra, S.; Modi, M. B.)V1558,384-388(1991)

Coherence and optical Kerr nonlinearity (Depoortere, Marc)V1504,133-139(1991)

Development of high resolution statistically nonstationary infrared earthlimb radiance scenes (Strugala, Lisa A.; Newt, J. E.; Futterman, Walter I.; Schweitzer, Eric L.; Herman, Bruce J.; Sears, Robert D.)V1486,176-187(1991)

Focusing infrared laser beams on targets in space without using adaptive optics (McKechnie, Thomas S.)V1408,119-135(1991)

Noise on multiple-tau photon correlation data (Schaetzel, Klaus; Peters, Rainer)V1430,109-115(1991)

Phenomenon of fringe patterns from speckle photogram and interferometry within a film (Bo, Weiyun; Wang, Rupeng)V1554A,750-754(1991)

Propagation of variances in belief networks (Neapolitan, Richard E.)V1468,333-344(1991)

Role of the surface height correlation function in the enhanced backscattering of light from random metallic surfaces (Maradudin, Alexei A.; Michel, T.)V1558,233-250(1991)

Small-angle scattering measurement (Gu, Zu-Han; Dummer, Richard S.)V1558,368-378(1991)

Statistical analysis of white-light speckle photography (Fang, Qiang; Tan, Yushan)V1554A,696-704(1991)

Statistical interferometry based on the statistics of speckle phase (Kadono, Hirofumi; Toyooka, Satoru)V1554A,718-726(1991)

Wave Propagation and Scattering in Varied Media II (Varadan, Vijay K., ed.)V1558(1991)

Stray light—see also optical design

Advanced infrared optically black baffle materials (Seals, Roland D.; Egert, Charles M.; Allred, David D.)V1330,164-177(1991)

Monte Carlo calculation of light distribution in an integrating cavity illuminator (Kaplan, Martin)V1448,206-217(1991)

Nonrejected earth radiance performance of the visible ultraviolet experiment sensor (Betts, Timothy C.; Dowling, Jerome M.; Friedman, Richard M.)V1479,120-126(1991)

SIRTF stray light analysis (Elliott, David G.; St. Clair Dinger, Ann)V1540,63-67(1991)

Stray light effects on calibrations using a solar diffuser (Palmer, James M.)V1493,143-154(1991)

Stray light issues for background-limited infrared telescope operation (Scholl, Marija S.; Scholl, James W.)V1540,109-118(1991)

Stray-light reduction in a WFOV star tracker lens (Lewis, Isabella T.; Ledebuhr, Arno G.; Axelrod, Timothy S.; Ruddell, Scott A.)V1530,306-324(1991)

Vacuum outgassing from diffuse-absorptive baffle materials (Egert, Charles M.; Basford, J. A.)V1330,178-185(1991)

Streak cameras—see also image tubes; intensifiers; photography, high-speed

19th Intl Congress on High-Speed Photography and Photonics (Fuller, Peter W., ed.)V1358(1991)

328 MHz synchroscan streak camera (Schelev, Mikhail Y.; Serdyuchenko, Yuri N.; Vaschenko, G. O.)V1358,569-573(1991)

Advantage of simultaneous streak and framing records in the field of detonics (Held, Manfred)V1358,904-913(1991)

Application of a linear picosecond streak camera to the investigation of a 1.55 um mode-locked Er3+ fiber laser (de Souza, Eunezio A.; Cruz, C. H.; Scarparo, Marco A.; Prokhorov, Alexander M.; Postovalov, V. E.; Vorobiev, N. S.; Schelev, Mikhail Y.)V1358,556-560(1991)

Applications of a soft x-ray streak camera in laser-plasma interaction studies (Tsakiris, George D.)V1358,174-192(1991)

Applications of synchroscan and dual-sweep streak camera techniques to free-electron laser experiments (Lumpkin, Alex H.)V1552,42-49(1991)

Avalanche transistor selection for long-term stability in streak camera sweep and pulser applications (Thomas, Stan W.; Griffith, Roger L.; Teruya, Alan T.)V1358,578-588(1991)

C 850X picosecond high-resolution streak camera (Mens, Alain; Dalmasso, J. M.; Sauneuf, Richard; Verrecchia, R.; Roth, J. M.; Tomasini, F.; Miehe, Joseph A.; Rebuffie, Jean-Claude)V1358,315-328(1991)

Circular streak camera application for satellite laser ranging (Prochazka, Ivan; Hamal, Karel; Kirchner, G.; Schelev, Mikhail Y.; Postovalov, V. E.)V1449,116-120(1991)

Comparison of electro-optic diagnostic systems (Hagans, Karla G.; Sargis, Paul G.)V1346,404-408(1991)

Design and operational characteristics of a PV 001 image tube incorporated with EB CCD readout (Bryukhnevitch, G. I.; Dalinenko, I. N.; Ivanov, K. N.; Kaidalov, S. A.; Kuz'min, G. A.; Moskalev, B. B.; Naumov, S. K.; Pischelin, E. V.; Postovalov, V. E.; Prokhorov, Alexander M.; Schelev, Mikhail Y.)V1449,109-115(1991)

Designing and application of solid state lasers for streak cameras calibration (Babushkin, A. V.; Vorobiev, N. S.; Prokhorov, Alexander M.; Schelev, Mikhail Y.)V1346,410-417(1991)

Estimation of limit time resolution in image streak camera (Klementyev, V. G.)V1358,1070-1074(1991)

Evaluation of dynamic range for LLNL streak cameras using high-contrast pulses and "pulse podiatry" on the NOVA laser system (Richards, James B.; Weiland, Timothy L.; Prior, John A.)V1346,384-389(1991)

Femtosecond streak tube (Kinoshita, Katsuyuki; Suyama, Motohiro; Inagaki, Yoshinori; Ishihara, Y.; Ito, Masuo)V1358,490-496(1991)

First results on a developmental deflection tube and its associated electronics for streak camera applications (Froehly, Claude; Laucournet, A.; Miehe, Joseph A.; Rebuffie, Jean-Claude; Roth, J. M.; Tomasini, F.)V1358,532-540(1991)

High-dynamic-range image readout system (Mens, Alain; Ducrocq, N.; Mazataud, D.; Mugnier, A.; Eouzan, J. Y.; Heurtaux, J. C.; Tomasini, F.; Mathae, J. C.)V1358,719-731(1991)

High-precision measurements of the 24-beam UV-OMEGA laser (Jaanimagi, Paul A.; Hestdalen, C.; Kelly, John H.; Seka, Wolf D.)V1358,337-343(1991)

Image converter streak cameras with super-light-speed scanning (Averin, V. I.; Gus'kova, M. S.; Korzhenevich, Irina M.; Kolesov, G. V.; Lebedev, V. B.; Maranichenko, N. I.; Sobolev, A. A.)V1358,589-602(1991)

Improved version of the PIF01 streak image tube (Degtyareva, V. P.; Fedotov, V. I.; Moskalev, B. B.; Postovalov, V. E.; Prokhorov, Alexander M.; Schelev, Mikhail Y.; Soldatov, N. F.)V1358,524-531(1991)

Interferometry, streak photography, and stereo photography of laser-driven miniature flying plates (Paisley, Dennis L.; Montoya, Nelson I.; Stahl, David B.; Garcia, Ismel A.)V1358,760-765(1991)

Laser-produced plasma x-ray diagnostics with an x-ray streak camera at the Iskra-4 plant (Berkovsky, A. G.; Gubanov, Y. I.; Pryanishnikov, Ivan G.; Murugov, V. M.; Petrov, S. I.; Senik, A. V.)V1358,750-755(1991)

Long-term flat-field behavior on LLNL streak cameras: preliminary results (Hatch, Jeffrey A.; Montgomery, David S.; Prior, John A.)V1346,371-375(1991)

Measurement of triggering instabilities of Imacon 500 streak cameras (Bowley, David J.; Rickett, Ph.; Babushkin, A. V.; Vorobiev, N. S.; Prokhorov, Alexander M.; Schelev, Mikhail Y.)V1358,550-555(1991)

Modular streak camera for laser ranging (Prochazka, Ivan; Hamal, Karel; Schelev, Mikhail Y.; Lozovoi, V. I.; Postovalov, V. E.)V1358,574-577(1991)

Multichannel mirror systems for high-speed framing recording (Ushakov, Leonid S.; Ponomaryov, A. M.; Trofimenko, Vladimir V.; Drozhbin, Yu. A.)V1358,447-450(1991)

Multiframing image converter camera (Frontov, H. N.; Serdyuchenko, Yuri N.)V1358,311-314(1991)

New electron optic for high-speed single-frame and streak image intensifier tubes (Dashevsky, Boris E.)V1358,561-568(1991)

New method for doing flat-field intensity calibrations of multiplexed ITT Streak Cameras (Hugenberg, Keith F.)V1346,390-397(1991)

New streak tubes of the P500 series: features and experimental results (Rebuffie, Jean-Claude; Mens, Alain)V1358,511-523(1991)

New streak tube with femtosecond time resolution (Feldman, G. G.; Ilyna, T. A.; Korjenevitch, I. N.; Syrtzev, V. N.)V1358,497-502(1991)

One-dimensional CCD linear array readout device (Borodin, A. M.; Ivanov, K. N.; Naumov, S. K.; Philippov, S. A.; Postovalov, V. E.; Prokhorov, Alexander M.; Stepanov, M. S.; Schelev, Mikhail Y.)V1358,756-758(1991)

Picosecond techniques application for definition of nonuniform ingradients inside turbid medium (Vorobiev, N. S.; Serafimovich, O. A.; Smirnov, A. V.)V1358,698-702(1991)

Picosecond x-ray streak camera improvement (Hou, Xun; Zhang, Xiaoqiu; Gong, Meixia; Chang, Zenghu; Lei, Zhiyuan; Yang, Binzhou; Yu, Hongbin; Liu, Xiouqin; Shan, Bin; Gao, Shengshen; Zhao, Wei)V1358,868-873(1991)

Picosecond x-ray streak cameras (Averin, V. I.; Bryukhnevitch, G. I.; Kolesov, G. V.; Lebedev, V. B.; Miller, V. A.; Saulevich, S. V.; Shulika, A. N.)V1358,603-605(1991)

Profile-related time resolution for a femtosecond streak tube (Liu, Yueping; Sibbett, Wilson)V1358,503-510(1991)

Quantitative analysis of hemoglobin oxygenation state of rat head by time-resolved photometry using picosecond laser pulse at 1064 nm (Nomura, Yasutomo; Tamura, Mamoru)V1431,102-109(1991)

Recent research and development in electron image tubes/cameras/systems (Prokhorov, Alexander M.; Schelev, Mikhail Y.)V1358,280-289(1991)

Some comparative results of two approaches in computer simulation of electron lenses for streak image tubes (Degtyareva, V. P.; Ivanov, V. Y.; Ignatov, A. M.; Kolesnikov, Sergey V.; Kulikov, Yu. V.; Monastyrski, M. A.; Niu, Hanben; Schelev, Mikhail Y.)V1358,546-548(1991)

Soviet-American image converter cameras "PROSCHEN" (Briscoe, Dennis; Shrivastava, Chinmaya A.; Nebeker, Sidney J.; Hsu, S.; Lozovoi, V. I.; Postovalov, V. E.; Prokhorov, Alexander M.; Schelev, Mikhail Y.; Serdyuchenko, Yuri N.; Vaschenko, G. O.)V1358,329-336(1991)

Space-qualified streak camera for the Geodynamic Laser Ranging System (Johnson, C. B.; Abshire, James B.; Zagwodzki, Thomas W.; Hunkler, L. T.; Letzring, Samuel A.; Jaanimagi, Paul A.)V1346,340-370(1991)

Stable picosecond solid state YA103:Nd3+ laser for streak cameras dynamic evaluation (Babushkin, A. V.)V1358,888-894(1991)

Streak and smear: a definition of terminology (Haddleton, Graham P.)V1358,18-22(1991)

Streak camera phosphors: response to ultrashort excitation (Jaanimagi, Paul A.; Hestdalen, C.)V1346,443-448(1991)

Temporal and spectral analysis of the synchronization of synchroscan streak cameras (Cunin, B.; Geist, P.; Heisel, Francine; Martz, A.; Miehe, Joseph A.)V1358,606-613(1991)

Temporal fiducial for picosecond streak cameras in laser fusion experiments (Mens, Alain; Gontier, D.; Giraud, P.; Thebault, J. P.)V1358,878-887(1991)

Theoretical and experimental performance evaluations of Picoframe framing cameras (Liu, Yueping; Sibbett, Wilson; Walker, David R.)V1358,300-310(1991)

Timing between streak cameras with a precision of 10 ps (Lerche, Richard A.)V1346,376-383(1991)

Transferred electron photocathode with greater than 5% quantum efficiency beyond 1 micron (Costello, Kenneth A.; Davis, Gary A.; Weiss, Robert; Aebi, Verle W.)V1449,40-50(1991)

Two-frequency picosecond laser with electro-optical feedback (Vorobiev, N. S.; Konoplev, O. A.)V1358,895-901(1991)

Ultrafast streak camera evaluations of phase noise from an actively stabilized colliding-pulse-mode-locked ring dye laser (Walker, David R.; Sleat, William E.; Evans, J.; Sibbett, Wilson)V1358,860-867(1991)

Ultrahigh resolution OTDR using streak camera technology (Sawaki, Akihiro; Miwa, Mitsuharu; Roehrenbeck, Paul W.)V1366,324-331(1991)

ULTRANAC: a new programmable image converter framing camera (Garfield, Brian R.; Riches, Mark J.)V1358,290-299(1991)

Wide-aperture x-ray image converter tubes (Dashevsky, Boris E.; Podvyaznikov, V. A.; Prokhorov, Alexander M.; Chevokin, Victor K.)V1346,437-442(1991)

Workshop on standards for streak camera characterization (Jaanimagi, Paul A.)V1346,504-504(1991)

Sun—see also astronomy; sources

Active sun telescope array (Walker, Arthur B.; Timothy, J. G.; Barbee, Troy W.; Hoover, Richard B.)V1343,334-347(1991)

Design of narrow band XUV and EUV coronagraphs using multilayer optics (Walker, Arthur B.; Allen, Maxwell J.; Barbee, Troy W.; Hoover, Richard B.)V1343,415-427(1991)

Goldhelox: a project to view the x-ray sun (Fair, Melody)V1549,182-192(1991)

High-resolution stigmatic EUV spectroheliometer for studies of the fine scale structure of the solar chromosphere, transition region, and corona (Timothy, J. G.; Berger, Thomas E.; Morgan, Jeffrey S.; Walker, Arthur B.; Bhattacharyya, Jagadish C.; Jain, Surendra K.; Saxena, Ajay K.; Huber, Martin E.; Tondello, Giuseppe; Naletto, Giampiero)V1343,350-358(1991)

Novel method for preventing solar ultraviolet-radiation-induced skin cancer (Shi, Weimin; Cui, Ting; Sigel, George H.)V1422,62-72(1991)

Performance of the Multi-Spectral Solar Telescope Array III: optical characteristics of the Ritchey-Chretien and Cassegrain Telescopes (Hoover, Richard B.; Baker, Phillip C.; Hadaway, James B.; Johnson, R. B.; Peterson, Cynthia; Gabardi, David R.; Walker, Arthur B.; Lindblom, Joakim F.; DeForest, Craig E.; O'Neal, Ray H.)V1343,189-202(1991)

Performance of the Multi-Spectral Solar Telescope Array V: temperature diagnostic response to the optically thin solar plasma (DeForest, Craig E.; Kankelborg, Charles C.; Allen, Maxwell J.; Paris, Elizabeth S.; Willis, Thomas D.; Lindblom, Joakim F.; O'Neal, Ray H.; Walker, Arthur B.; Barbee, Troy W.; Hoover, Richard B.; Barbee, Troy W.; Gluskin, Efim S.)V1343,404-414(1991)

Performance of the Multi-Spectral Solar Telescope Array VI: performance and characteristics of the photographic films (Hoover, Richard B.; Walker, Arthur B.; DeForest, Craig E.; Allen, Maxwell J.; Lindblom, Joakim F.)V1343,175-188(1991)

Proposed conversion of the McMath Telescope to 4.0-meter aperture for solar observations in the IR (Livingston, William C.)V1494,498-502(1991)

SOHO space satellite: UV instrumentation (Poland, Arthur I.; Domingo, Vicente)V1343,310-318(1991)

Solar EUV/FUV line polarimetry: instruments and methods (Hoover, Richard B.; Fineschi, Silvano; Fontenla, Juan; Walker, Arthur B.)V1343,389-403(1991)

Solar EUV/FUV line polarimetry: observational parameters and theoretical considerations (Fineschi, Silvano; Hoover, Richard B.; Fontenla, Juan; Walker, Arthur B.)V1343,376-388(1991)

Spectroscopic observations of CO2, CH4, N2O, and CO from Kitt Peak, 1979-1990 (Livingston, William C.; Wallace, Lloyd V.)V1491,43-47(1991)

Survey of hard x-ray imaging concepts currently proposed for viewing solar flares (Campbell, Jonathan W.; Davis, John M.; Emslie, A. G.)V1343,359-375(1991)

Ultra-High-Resolution XUV Spectroheliograph II: predicted performance (Walker, Arthur B.; Lindblom, Joakim F.; Timothy, J. G.; Allen, Maxwell J.; DeForest, Craig E.; Kankelborg, Charles C.; O'Neal, Ray H.; Paris, Elizabeth S.; Willis, Thomas D.; Barbee, Troy W.; Hoover, Richard B.)V1343,319-333(1991)

Wide bandwidth spectral measurements of atmospheric tilt turbulence (Tiszauer, Detlev H.; Smith, Richard C.)V1408,72-83(1991)

Superconductors—see also thin films

Aging and surface instability in high-Tc superconductors (Larkins, Grover L.; Jones, W. K.; Lu, Q.; Levay, C.; Albaijes, D.)V1477,26-33(1991)

Anomalous optical response of YBa2Cu3O7-x thin films during superconducting transitions (Xi, Xiaoxing; Venkatesan, T.; Etemad, Shahab; Hemmick, D.; Li, Q.)V1477,20-25(1991)

Application of rf superconductivity to high-brightness and high-gradient ion beam accelerators (Delayen, Jean R.; Bohn, Courtlandt L.; Roche, C. T.)V1407,524-534(1991)

Applications of high-Tc superconductors in optoelectronics (Sobolewski, Roman)V1501,14-27(1991); V1502,14-27(1991); 1506,25-38(1991)

Basic mechanisms and application of the laser-induced forward transfer for high-Tc superconducting thin film deposition (Fogarassy, Eric)V1394,169-179(1991)

Bi(Pb)-Sr-Ca-Cu-O superconducting films prepared by chemical spray deposition (Li, Chang J.; Liu, Li M.; Yao, Qi)V1519,779-787(1991)

Characteristics of thin-film-type Josephson junctions using Bi2Sr2CaCu2Ox/Bi2Sr2CuOy/Bi2Sr2CaCu2Oz structure (Mizuno, Koichi; Higashino, Hidetaka; Setsune, Kentaro; Wasa, Kiyotaka)V1477,197-204(1991)

Characterization and preparation of high-Tc YBa2Cu3O7-x thin films on Si with conducting indium oxide as a buffer layer (Zhang, Z. J.; Luo, Wei A.; Zeng, Y. Y.; Yang, N. P.; Cai, Y. M.; Shen, X. L.; Chen, H. S.; Hua, Zhong Y.)V1519,790-792(1991)

Characterization of a high-critical-temperature superconducting thin film by the ring resonator method (Pyee, Maurice; Meisse, Pascal; Baudrand, Henry; Chaubet, Michel)V1512,240-248(1991)

Characterization of high-Tc coplanar transmission lines and resonators (Kessler, Jochen; Dill, Roland; Russer, Peter)V1477,45-56(1991)

Cleaved surfaces of high Tc films for making SNS structures (Takeuchi, Ichiro; Tsai, Jaw S.; Tsuge, Hisanao; Matsukura, Noritsuga; Miura, Sadahiko; Yoshitake, T.; Kojima, Yoshikatsu; Matsui, Shinji)V1394,96-101(1991)

Comparative study for bolometric and nonbolometric switching elements for microwave phase shifters (Tabib-Azar, Massood; Bhasin, Kul B.; Romanofsky, Robert R.)V1477,85-94(1991)

Coplanar waveguide microwave filter of YBa2Cu3O7 (Chew, Wilbert; Riley, A. L.; Rascoe, Daniel L.; Hunt, Brian D.; Foote, Marc C.; Cooley, Thomas W.; Bajuk, Louis J.)V1477,95-100(1991)

Critical current enhancement in Y1Ba2Cu3O7-y/Y1Ba2(Cu1-xNix)3O7-y heterostructures (Witanachchi, S.; Lee, S. Y.; Song, L. W.; Kao, Yi-Han; Shaw, David T.)V1394,161-168(1991)

Design aspects and comparison between high-Tc superconducting coplanar waveguide and microstrip line (Kong, Keon-Shik; Bhasin, Kul B.; Itoh, Tatsuo)V1477,57-65(1991)

Distorted local environment about Zn and transition metals on the copper sites in YBa2Cu3O7 (Bridges, Frank; Li, Guoguang; Boyce, James B.; Claeson, Tord)V1550,76-84(1991)

Dynamical properties of superconducting superlattices (Andronov, Alexander A.; Genkin, G.; Ghinovker, M.; Kurin, V.; Nefedov, I.; Okomelkov, A.; Shereshevsky, I.)V1362,684-695(1991)

Effect of laser irradiation on superconducting properties of laser-deposited YBa2Cu3O7 thin films (Singh, Rajiv K.; Bhattacharya, Deepika; Narayan, Jagdish; Jahncke, Catherine; Paesler, Michael A.)V1394,203-213(1991)

Ellipsometric studies of the optical anisotropy of GdBa2Cu3O7-x epitaxial films (Liu, Ansheng; Keller, Ole)V1512,226-231(1991)

Epitaxial films YBa2Cu3O7-delta(jc(78K)>106A/cm2) on sapphire and SrTiO3: peculiarities and differences in conditions of film growth and properties (Predtechensky, M.; Smal, A.; Varlamov, Yu.)V1477,234-241(1991)

Epitaxial Tl2Ba2CaCu2O8 thin films on LaAlO3 and their microwave device properties (Negrete, George V.; Hammond, Robert B.)V1477,36-44(1991)

Experimental studies of oxygen incorporation during growth of Y-Ba-Cu-O films by pulsed-laser deposition (Gupta, Arunava)V1394,230-230(1991)

Grain-oriented high-Tc superconductors and their applications (Nelson, Jeffrey G.; Neurgaonkar, Ratnakar R.; Crooks, R.; Rhodes, C. G.)V1394,191-195(1991)

Growth and properties of YBCO thin films by metal-organic chemical vapor deposition and plasma-enhanced MOCVD (Zhao, Jing-Fu; Li, Y. Q.; Chern, Chyi S.; Huang, W.; Norris, Peter E.; Gallois, B.; Kear, B. H.; Lu, P.; Kulesha, G. A.; Cosandey, F.)V1362,135-143(1991)

Growth and transport properties of Y-Ba-Cu-O/Pr-Ba-Cu-O superlattices (Lowndes, Douglas H.; Norton, D. P.; Budai, J. D.; Christen, D. K.; Klabunde, C. E.; Warmack, R. J.; Pennycook, Stephen J.)V1394,150-160(1991)

Growth of high-Tc superconducting thin films for microwave applications (Wu, Xin D.; Foltyn, Stephen R.; Muenchausen, Ross E.; Dye, Robert C.; Cooke, D. W.; Rollett, A. D.; Garcia, A. R.; Nogar, Nicholas S.; Pique, A.; Edwards, R.)V1477,8-14(1991)

Growth of oxide superconducting thin films by plasma-enhanced MOCVD (Kanehori, Keiichi; Sugii, Nobuyuki)V1394,238-243(1991)

High-Tc bolometer developments for planetary missions (Brasunas, John C.; Lakew, Brook)V1477,166-173(1991)

High-Tc superconducting infrared bolometric detector (Cole, Barry E.)V1477,126-138(1991)

High-Tc superconductors, physics, and applications (Bok, Julien)V1362,94-101(1991)

High-temperature superconducting Josephson junction devices (Simon, Randy W.)V1394,70-78(1991)

High-temperature superconducting Josephson mixers from deliberate grain boundaries in Tl2CaBa2Cu2O8 (Bourne, Lincoln C.; Cardona, A. H.; James, Tim W.; Fleming, J. S.; Forse, R. W.; Hammond, Robert B.; Hong, J. P.; Kim, T. W.; Fetterman, Harold R.)V1477,205-208(1991)

High-temperature superconducting superconductor/normal metal/ superconducting devices (Foote, Marc C.; Hunt, Brian D.; Bajuk, Louis J.)V1477,192-196(1991)

High-temperature superconductive microwave technology for space applications (Leonard, Regis F.; Connolly, Denis J.; Bhasin, Kul B.; Warner, Joseph D.; Alterovitz, Samuel A.)V1394,114-125(1991)

High-temperature superconductivity space experiment: passive millimeter wave devices (Nisenoff, Martin; Gubser, Don U.; Wolf, Stuart A.; Ritter, J. C.; Price, George E.)V1394,104-113(1991)

High-temperature superconductor junction technology (Simon, Randy W.)V1477,184-191(1991)

HTC microbolometer for far-infrared detection (Barholm-Hansen, Claus; Levinsen, Mogens T.)V1512,218-225(1991)

Infrared detectors from YBaCuO thin films (Zhou, Fang Q.; Sun, Han D.; Zhao, Xing R.; Wang, Lingjie; Yi, Xin J.)V1477,178-181(1991)

Infrared optical response of superconducting YBaCuO thin films (Sun, Han D.; Zhou, Fang Q.; Zhao, Xing R.; Wang, Lingjie; Yi, Xin J.)V1477,174-177(1991)

Infrared photodetector based on the photofluxonic effect in superconducting thin films (Kadin, Alan M.; Leung, Michael; Smith, Andrew D.; Murduck, J. M.)V1477,156-165(1991)

In-situ growth of Y1Ba2Cu3O7-x thin films using XeCl excimer and Nd:YAG lasers (Gerri, Mireille; Marine, W.; Mathey, Yves; Sentis, Marc L.; Delaporte, Philippe C.; Fontaine, Bernard L.; Forestier, Bernard M.)V1503,280-291(1991)

In-situ laser preparation of high-Tc superconducting thin film at 450-550 degree C (An, Cheng W.; Fan, Yong C.; Lu, Dong S.; Li, Zai Q.)V1519,818-821(1991)

In-situ low-temperature and epitaxial growth of high-Tc superconducting films using oxygen-discharge-assisted laser ablation method (Fan, Yong C.; An, Cheng W.; Lu, Dong S.; Li, Zai Q.)V1519,775-778(1991)

Interplay between photons and superconductors (Gilabert, Alain; Azema, Alain; Roustan, Jean-Claude; Maneval, Jean-Paul)V1477,140-147(1991)

Investigation of high-temperature superconductors under the effect of picosecond laser pulses (Agranat, M. B.; Ashitkov, S. I.; Granovsky, A. B.; Kuznetsov, V. I.)V1440,397-400(1991)

Ion-beam-sputtering deposition and etching of high-Tc YBCO superconducting thin films (Zhao, Xing R.; Hao, Jian H.; Zhou, Fang Q.; Sun, Han D.; Wang, Lingjie; Yi, Xin J.)V1519,772-774(1991)

Ion implantation of polymers for electrical conductivity enhancement (Bridwell, Lynn B.; Wang, Y. Q.)V1519,878-883(1991)

Laser-triggered superconducting opening switch (Dhali, Shirshak K.; Mohsin, Mohammad)V1396,353-359(1991)

Loss of heat relaxation in superconducting materials near Tc (Ermolaeva, Tatiana T.; Emelin, S. E.)V1361,1164-1165(1991)

Loss of heat relaxation in the superconducting and nonsuperconducting ceramics near Tc (Ermolaeva, Tatiana T.)V1361,1166-1167(1991)

Luminescence studies in the process of preparation high-Tc superconducting films with excimer laser ablation (Fan, Yong C.; An, Cheng W.; Lu, Dong S.; Li, Zai Q.)V1519,813-817(1991)

Metal-superconductor-insulator transitions in oxide materials (Liang, W. Y.)V1362,127-134(1991)

Microstructure and superconducting properties of BiSrCaCuO thin films (Wessels, Bruce W.; Zhang, Jiyue; DiMeo, Frank; Richeson, D. S.; Marks, Tobin J.; DeGroot, D. C.; Kannewurf, C. R.)V1394,232-237(1991)

Microstructure and superconducting properties of Y1Ba2Cu3O7-delta thin films grown by rapid thermal annealing (RTA) as a function of crystalline structure of zirconia substrates (Hosseini Teherani, Ferechteh; Kreisler, Alain J.; Baixeras, Joseph M.)V1362,921-929(1991)

Microwave characterization of high-Tc superconducting thin films for simulation and realization of planar microelectronic circuits (Carru, J. C.; Mehri, F.; Chauvel, D.; Crosnier, Y.)V1512,232-239(1991)

Microwave properties of YB2Cu3O7-x thin films characterized by an open resonator (Zhou, Shi P.; Wu, Ke Q.; Jabbar, A.; Bao, Jia S.; Lou, Wei G.; Ding, Ai L.; Wang, Shu H.)V1519,793-799(1991)

MOCVD of TlBaCaCuO: structure-property relations and progress toward device applications (Hamaguchi, Norihito; Gardiner, R.; Kirlin, Peter S.)V1394,244-254(1991)

Multilayered superconducting tunnel junctions for use as high-energy-resolution x-ray detectors (Rippert, Edward D.; Song, S. N.; Ketterson, John B.; Ulmer, Melville P.)V1549,283-288(1991)

Nb tunnel junctions as x-ray spectrometers (Rando, Nicola; Peacock, Anthony J.; Foden, Clare; van Dordrecht, Axel; Engelhardt, Ralph; Lumley, John M.; Pereira, Carl)V1549,340-356(1991)

New way of making a lunar telescope (Chen, Peter C.)V1494,228-233(1991)

Noise mechanisms of high-temperature superconducting infrared detectors (Khalil, Ali E.)V1477,148-158(1991)

Nonlinear electromagnetic field response of high-Tc superconducting microparticle composites (Haus, Joseph W.; Chung-Yau, F.; Bowden, Charles M.)V1497,382-388(1991)

On-focal-plane superconducting signal processing for low- and intermediate-temperature operation (Smetana, Daryl L.; Carson, John C.)V1541,220-226(1991)

Optical response in high-temperature superconducting thin films (Thiede, David A.)V1484,72-80(1991)

Optimizing the structural and electrical properties of Ba2YCu3O7-delta (Phillips, Julia M.; Siegal, Michael P.)V1394,186-190(1991)

Performance characteristics of Y-Ba-Cu-O microwave superconducting detectors (Shewchun, John; Marsh, P. F.)V1477,115-138(1991)

Performance of stripline resonators using sputtered YBCO films (Mallory, Derek S.; Kadin, Alan M.; Ballentine, Paul H.)V1477,66-76(1991)

Planar SNS Josephson junctions using multilayer Bi system (Setsune, Kentaro; Mizuno, Koichi; Higashino, Hidetaka; Wasa, Kiyotaka)V1394,79-88(1991)

Polarization-dependent EXAFS studies in layered copper oxide superconductors (Mini, Susan M.; Alp, Esen E.; Ramanathan, Mohan; Bommannavar, A.; Hyun, O. B.)V1345,260-269(1991)

Polarization-dependent x-ray spectroscopy of high-Tc superconductors (Heald, Steven M.; Tranquada, John M.)V1550,67-75(1991)

Possible enhancement in bolometric response using free-standing film of YBa2Cu3Ox (Ng, Hon K.; Kilibarda, S.)V1477,15-19(1991)

Preparation and O+ implantation of Y-Ba-Cu-O superconducting thin films by sputtering and RTA process (Fan, Xiang J.; Guo, Huai X.; Jiang, Chang Z.; Pen, You G.; Liu, Chang; Pen, Zhi L.; Li, Hong T.)V1519,808-812(1991)

Preparation of superconducting Y-Ba-Cu-O thin films by rf magnetron sputtering (Huang, Zong T.; Li, Guo Z.; Zeng, Guang L.; Huang, Jiang P.; Xiong, Guilan)V1519,788-789(1991)

Preparation of thin superconducting YBCO films by ion-beam mixing (Fan, Xiang J.; Pen, You G.; Guo, Huai X.; Li, Hong T.; Liu, Chang; Jiang, Chang Z.; Pen, Zhi L.)V1519,805-807(1991)

Progress In High-Temperature Superconducting Transistors and Other Devices (Narayan, Jagdish; Shaw, David T.; Singh, Rajendra, eds.)V1394(1991)

Progress toward thin-film-based tape conductors (Berdahl, Paul H.; Russo, Rick)V1394,180-185(1991)

Prospects for hybrid electronics (Singh, Rajendra; Sinha, Sanjai; Krueger, D. J.)V1394,62-67(1991)

Pulsed-laser deposition of YBa2Cu3O7-x thin films: processing, properties, and performance (Muenchausen, Ross E.; Foltyn, Stephen R.; Wu, Xin D.; Dye, Robert C.; Nogar, Nicholas S.; Carim, A. H.; Heidelbach, F.; Cooke, D. W.; Taber, R. C.; Quinn, Rod K.)V1394,221-229(1991)

Quantum-limited quasiparticle mixers at 100 GHz (Mears, Carl A.; Hu, Qing; Richards, Paul L.; Worsham, A.; Prober, Daniel E.; Raisanen, Antti)V1477,221-233(1991)

Reduced thermal budget processing of high-Tc superconducting thin films and related materials by MOCVD (Sinha, Sanjai; Singh, Rajendra; Hsu, N. J.; Ng, J. T.; Chou, P.; Narayan, Jagdish)V1394,266-276(1991)

Role of buffer layers in the laser-ablated films on metallic substrates (Shaw, David T.; Narumi, E.; Yang, F.; Patel, Sushil)V1394,214-220(1991)

Speed of optically-controlled superconducting devices (Kwok, Hoi-Sing; Shi, Lei; Zheng, J. P.; Dong, S. Y.; Pang, Y.; Prasad, Paras N.)V1394,196-200(1991)

Status of high-temperature superconducting analog devices (Talisa, Salvador H.)V1477,78-83(1991)

Stoichiometry of laser-deposited Bi-Sr-Ca-Cu-O films on silicon and mass spectrometric investigations of superconductors (Becker, J. S.; Lorenz, M.; Dietze, H.-J.)V1598,227-238(1991)

Structural and electrical properties of epitaxial YBCO films on Si (Fork, David K.; Barrera, A.; Phillips, Julia M.; Newman, N.; Fenner, David B.; Geballe, T. H.; Connell, G.A. N.; Boyce, James B.)V1394,202-202(1991)

Study of the optical features of YBa2Cu3O7-x films by SEW spectroscopy (Vaicikauskas, V.; Maldutis, Evaldas K.; Kajokas, R.)V1440,357-363(1991)

Submillimeter receiver components using superconducting tunnel junctions (Wengler, Michael J.; Pance, A.; Liu, B.; Dubash, N.; Pance, Gordana; Miller, Ronald E.)V1477,209-220(1991)

Superconducting bolometers: high-Tc and low-Tc (Richards, Paul L.)V1477,2-6(1991)

Superconducting devices and system insertion (Rachlin, Adam; Babbitt, Richard; Lenzing, Erik; Cadotte, Roland)V1477,101-114(1991)

Superconducting YBa2Cu3O7 films for novel optoelectronic device structures (Pavuna, Davor; Dwir, Benjamin; Gauzzi, Andrea; James, Jonathan H.; Kellett, Bruce J.)V1362,102-116(1991)

Superconducting YBa2Cu3O7 films on Si and GaAs with conducting indium tin oxide buffer layers (James, Jonathan H.; Kellett, Bruce J.; Gauzzi, Andrea; Dwir, Benjamin; Pavuna, Davor)V1394,45-61(1991)

Superconductivity and structural changes of Ar ion-implanted YBa2Cu3O7-x thin films (Li, Yi J.; Ren, Cong X.; Chen, Guo L.; Chen, Jian M.; Zou, Shi C.)V1519,800-804(1991)

Superconductivity Applications for Infrared and Microwave Devices II (Heinen, Vernon O.; Bhasin, Kul B., eds.)V1477(1991)

Superconductor/semiconductor structure and its application to superconducting devices (Nishino, Toshikazu; Hatano, Mutsuko; Hasegawa, Haruhiro)V1394,36-44(1991)

Tensor resistivity and tensor susceptibility of oriented high-Tc superconducting film YBaCuO (Hou, Bi H.; Qi, Zhen Z.)V1519,822-823(1991)

Thin film processing and device fabrication in the Tl-Ca-Ba-Cu-O system (Martens, Jon S.; Ginley, David S.; Zipperian, Thomas E.; Hietala, Vincent M.; Tigges, Chris P.)V1394,140-149(1991)

Thin films of YBaCuO for electronic applications (Baixeras, Joseph M.; Hosseini Teherani, Ferechteh; Kreisler, Alain J.; Straboni, Alain; Barla, Kathy)V1362,117-126(1991)

Versatility of metal organic chemical vapor deposition process for fabrication of high-quality YBCO superconducting thin films (Chern, Chyi S.; Zhao, Jing-Fu; Li, Y. Q.; Norris, Peter E.; Kear, B. H.; Gallois, B.)V1394,255-265(1991)

YBa2Cu3O7-x/Au/Nb device structures (Hunt, Brian D.; Foote, Marc C.; Bajuk, Louis J.; Vasquez, R. P.)V1394,89-95(1991)

Surfaces—see also mirrors; optical testing; scattering

About investigation of absolute surface relief by holographic method (Rachkovsky, Leonid I.; Tanin, Leonid V.; Rubanov, Alexander S.)V1461,232-240(1991)

Absolute interferometric testing of spherical surfaces (Truax, Bruce E.)V1400,61-68(1991)

Absolute measurement of spherical surfaces (Creath, Katherine; Wyant, James C.)V1332,2-7(1991)

Absolute range measurement system for real-time 3-D vision (Wood, Christopher M.; Shaw, Michael M.; Harvey, David M.; Hobson, Clifford A.; Lalor, Michael J.; Atkinson, John T.)V1332,301-313(1991)

Advanced broadband baffle materials (Seals, Roland D.)V1485,78-87(1991)

Aging and surface instability in high-Tc superconductors (Larkins, Grover L.; Jones, W. K.; Lu, Q.; Levay, C.; Albaijes, D.)V1477,26-33(1991)

Anomalous scattering from optical surfaces with roughness and permittivity perturbations (Elson, John M.)V1530,196-207(1991)

Applicability of dry developable deep-UV lithography to sub-0.5 um processing (Goethals, Anne-Marie; Baik, Ki-Ho; Van den hove, Luc; Tedesco, Serge V.)V1466,604-615(1991)

Application of surface effects externally to alter optical and electronic properties of existing optoelectronic devices: a ten-year update (Hava, Shlomo)V1540,350-358(1991)

Applications of diamond-turned null reflectors for generalized aspheric metrology (McCann, James T.)V1332,843-849(1991)

Aspheric surface testing techniques (Stahl, H. P.)V1332,66-76(1991)

Automated photogrammetric surface reconstruction with structured light (Maas, Hans-Gerd)V1526,70-77(1991)

Automatic inspection technique for optical surface flaws (Yang, Guoguang; Gao, Wenliang; Cheng, Shangyi)V1332,56-63(1991)

Beryllium scatter analysis program (Behlau, Jerry L.; Granger, Edward M.; Hannon, John J.; Baumler, Mark; Reilly, James F.)V1530,218-230(1991)

Bidirectional reflectance distribution function raster scan technique for curved samples (McIntosh, Malcolm B.; McNeely, Joseph R.)V1530,263-268(1991)

Characterization of hot-isostatic-pressed optical-quality beryllium (Behlau, Jerry L.; Baumler, Mark)V1530,208-217(1991)

Characterization of surface contaminants using infrared microspectroscopy (Blair, Dianna S.; Ward, Kenneth J.)V1437,76-79(1991)

Characterizing a surface imaging process in a high-volume DRAM manufacturing production line (Garza, Cesar M.; Catlett, David L.; Jackson, Ricky A.)V1466,616-627(1991)

Closed-form onset threshold analysis of defect-driven surface and bulk laser damage (O'Connell, Robert M.)V1441,406-419(1991)

Coherence in single and multiple scattering of light from randomly rough surfaces (Gu, Zu-Han; Maradudin, Alexei A.; Mendez, Eugenio R.)V1530,60-70(1991)

Color and Grassmann-Cayley coordinates of shape (Petrov, A. P.)V1453,342-352(1991)

Comparison of low-scatter-mirror PSD derived from multiple-wavelength BRDFs and WYKO profilometer data (Wong, Wallace K.; Wang, Dexter; Benoit, Robert T.; Barthol, Peter)V1530,86-103(1991)

Computer-Controlled Optical Surfacing (Jones, Robert A., ed.)VMS40(1991)

Cone array of Fourier lenses for contouring applications (Goldfain, Ervin)V1527,210-215(1991)

Constructing topologically connected surfaces for the comprehensive analysis of 3-D medical structures (Kalvin, Alan D.; Cutting, Court B.; Haddad, Betsy; Noz, Marilyn E.)V1445,247-258(1991)

Contour estimation using global shape constraints and local forces (Deng, Keqiang; Wilson, Joseph N.)V1570,227-233(1991)

Convexity-based method for extracting object parts from 3-D surfaces (Vaina, Lucia M.; Zlateva, Stoyanka D.)V1468,710-719(1991)

Correlation of surface topography and coating damage with changes in the responsivity of silicon PIN photodiodes (Huffaker, Diana L.; Walser, Rodger M.; Becker, Michael F.)V1441,365-380(1991)

Creation of aspheric beryllium optical surfaces directly in the hot isostatic pressing consolidation process (Gildner, Donald; Marder, James M.)V1485,46-53(1991)

Deformable surfaces: a free-form shape representation (Delingette, Herve; Hebert, Martial; Ikeuchi, Katsushi)V1570,21-30(1991)

Design and specification of diamond-turned optics (Clark, Robert A.)VCR38,164-183(1991)

Design parameters of an EO sensor (Tanwar, Lakhan S.; Jain, P. C.; Kunzmann, H.)V1332,877-882(1991)

Determination of surface quality using moire methods (Keren, Eliezer; Kreske, Kathi; Zac, Yaacov; Livnat, Ami)V1442,266-274(1991)

Determination of thin-film roughness and volume structure parameters from light-scattering investigations (Duparre, Angela; Kassam, Samer)V1530,283-286(1991)

Development of beryllium-mirror turning technology (Arnold, Jones B.)V1485,96-105(1991)

Differential properties from adaptive thin-plate splines (Sinha, Saravajit S.)V1570,64-74(1991)

Diffractive properties of surface-relief microstructures (Qu, Dong-Ning; Burge, Ronald E.; Yuan, X.)V1506,152-159(1991)

Dimension and time effects caused by nonlocal scattering of laser radiation from a rough metal surface (Dolgina, A. N.; Kovalev, A. A.; Kondratenko, P. S.)V1440,342-353(1991)

Direct method for reconstructing shape from shading (Oliensis, John; Dupuis, Paul)V1570,116-128(1991)

Effective surface PSD for bare hot-isostatic-pressed beryllium mirrors (Vernold, Cynthia L.; Harvey, James E.)V1530,144-149(1991)

Effect of silylation condition on the silylated image in the DESIRE process (Taira, Kazuo; Takahashi, Junichi; Yanagihara, Kenji)V1466,570-582(1991)

Effects of the nonvanishing tip size in mechanical profile measurements (Church, Eugene L.; Takacs, Peter Z.)V1332,504-514(1991)

Efficient visual representation and reconstruction from generalized curvature measures (Barth, Erhardt; Caelli, Terry M.; Zetzsche, Christoph)V1570,86-95(1991)

Energy bands of graphite and CsC8-GIC, and CO physisorption on graphite basal surface (Yang, Yong; Guo, Jian Q.; Lu, Dong)V1519,444-448(1991)

Energy functions for regularization algorithms (Delingette, Herve; Hebert, Martial; Ikeuchi, Katsushi)V1570,104-115(1991)

Estimation of plastic strain by fractal (Dai, YuZhong; Chiang, Fu-Pen)V1332,767-774(1991)

Experimental study of the laser retroreflection of various surfaces (Liu, Wen-Qing; Jiang, Rong-Xi; Wang, Ya-Ping; Xia, Yu-Xing)V1530,240-243(1991)

Formation of large-scale relief on a target surface under multiple-pulsed action of laser radiation (Brailovsky, A. B.; Gaponov, Sergey V.; Dorofeev, I. A.; Lutschin, V. I.; Semenov, V. E.)V1440,84-89(1991)

Frequency spectrum analysis and assessment of optical surface flaws (Gao, Wenliang; Zhang, Xiao; Yang, Guoguang)V1530,118-128(1991)

From points to surfaces (Fua, Pascal; Sander, Peter T.)V1570,286-296(1991)

From voxel to curvature (Monga, Olivier; Ayache, Nicholas; Sander, Peter T.)V1570,382-390(1991)

Geometric Methods in Computer Vision (Vemuri, Baba C., ed.)V1570(1991)

Global minima via dynamic programming: energy minimizing active contours (Chandran, Sharat; Maejima, Tsukasa; Miyazaki, Sanae)V1570,391-402(1991)

Heterophase isotopic SF6 molecule separation in the surface electromagnetic wavefield (Bordo, V. G.; Ershov, I. A.; Kravchenko, V. A.; Petrov, Yu. N.)V1440,364-369(1991)

High-emittance surfaces for high-temperature space radiator applications (Banks, Bruce A.; Rutledge, Sharon K.; Hotes, Deborah)V1330,66-77(1991)

High-precision interferometric testing of spherical mirrors with long radius of curvature (Freischlad, Klaus R.; Kuechel, Michael F.; Wiedmann, Wolfgang; Kaiser, Winfried; Mayer, Max)V1332,8-17(1991)

Holographic optical profilometer (Hasman, Erez; Davidson, Nir; Friesem, Asher A.)V1442,372-377(1991)

Holographic soundfield visualization for nondestructive testing of hot surfaces (Crostack, Horst-Artur; Meyer, E. H.; Pohl, Klaus-Juergen)V1508,101-109(1991)

Improved formalism for rough-surface scattering of acoustic and electromagnetic waves (Milder, D. M.)V1558,213-221(1991)

Influence of diamond turning and surface cleaning processes on the degradation of KDP crystal surfaces (Kozlowski, Mark R.; Thomas, Ian M.; Edwards, Gary; Stanion, Ken; Fuchs, Baruch A.; Latanich, L.)V1561,59-69(1991)

Influence of phototransformed molecules on optical properties of finite cholesteric liquid-crystal cell (Pinkevich, Igor P.; Reshetnyak, Victor Y.; Reznikov, Yuriy)V1455,122-133(1991)

Infrared microanalysis of contaminants at grazing incidence (Reffner, John A.)V1437,89-94(1991)

Integrating FTIR microscopy into surface analysis (Church, Jeffrey S.)V1437,80-88(1991)

Interaction of 1064 nm photons with the Al2O3(1120) surface (Schildbach, M. A.; Hamza, Alex V.)V1441,139-145(1991)

Interferometer for testing aspheric surfaces with electron-beam computer-generated holograms (Gemma, Takashi; Hideshima, Masayuki; Taya, Makoto; Watanabe, Nobuko)V1332,77-84(1991)

Ion beam milling of fused silica for window fabrication (Wilson, Scott R.; Reicher, David W.; Kranenberg, C. F.; McNeil, John R.; White, Patricia L.; Martin, Peter M.; McCready, David E.)V1441,82-86(1991)

IR laser-light backscattering by an arbitrarily shaped dielectric object with rough surface (Wu, Zhensen; Cheng, Denghui)V1558,251-257(1991)

Kinetics of surface ordering: Pb on Ni(001) (Eng, Peter J.; Stephens, Peter; Tse, Teddy)V1550,110-121(1991)

Laser-induced generation of surface periodic structures resulting from the waveguide mode interaction (Bazakutsa, P. V.; Maslennikov, V. L.; Sychugov, V. A.; Yakovlev, V. A.)V1440,370-376(1991)

Laser-induced thermal desorption studies of surface reaction kinetics (George, Steven M.; Coon, P. A.; Gupta, P.; Wise, M. L.)V1437,157-165(1991)

Laser moire topography for 3-D contour measurement (Matsumoto, Tetsuya; Kitagawa, Yoichi; Adachi, Masaaki; Hayashi, Akihiro)V1332,530-536(1991)

Light-absorbing, lightweight beryllium baffle materials (Murray, Brian W.; Floyd, Dennis R.; Ulph, Eric)V1485,88-95(1991)

Light-induced polishing of evaporating surface (Tokarev, Vladimir N.; Konov, Vitaly I.)V1503,269-278(1991)

Light scattering from gold-coated ground glass and chemically etched surfaces (Ruiz-Cortes, Victor; Mendez, Eugenio R.; Gu, Zu-Han; Maradudin, Alexei A.)V1558,222-232(1991)

Linearized ray-trace analysis (Redding, David C.; Breckenridge, William G.)V1354,216-221(1991)

Matching 3-D smooth surfaces with their 2-D projections using 3-D distance maps (Lavallee, Stephane; Szeliski, Richard; Brunie, Lionel)V1570,322-336(1991)

Measurement of SEW phase velocity by optical heterodyning method (Libenson, Michail N.; Makin, Vladimir S.; Trubaev, Vladimir V.)V1440,354-356(1991)

Measurement of surface roughness using optical fibre sensor and microcomputer (Fan, Dapeng; Zhang, Honghai; Chen, Jihong; Chen, Riyao)V1572,11-14(1991)

Model-based surface classification (Newman, Timothy S.; Flynn, Patrick J.; Jain, Anil K.)V1570,250-261(1991)

Multipulse laser-induced failure prediction for Mo metal mirrors (Becker, Michael F.; Ma, Chun-Chi; Walser, Rodger M.)V1441,174-187(1991)

New approaches to ultrasensitive magnetic resonance (Bowers, C. R.; Buratto, S. K.; Carson, Paul; Cho, H. M.; Hwang, J. Y.; Mueller, L.; Pizarro, P. J.; Shykind, David; Weitekamp, Daniel P.)V1435,36-50(1991)

New multisource contouring method (Rachkovsky, Leonid I.; Drobot, Igor L.; Tanin, Leonid V.)V1507,450-457(1991)

New stereo laser triangulation device for specular surface inspection (Samson, Marc; Dufour, Marc L.)V1332,314-322(1991)

Nomarski viewing system for an optical surface profiler (Bietry, Joseph R.; Auriemma, R. A.; Bristow, Thomas C.; Merritt, Edward)V1332,537-543(1991)

Noncontact optical microtopography (Costa, Manuel Filipe P.; Almeida, Jose B.)V1400,102-107(1991)

Novel noncontact sensor for surface topography measurements using fiber optics (Butler, Clive; Gregoriou, Gregorios)V1584,282-293(1991)

Optical 3-D sensing for measurement of bottomhole pattern (Su, Wan-Yong; Su, Xianyu)V1567,680-682(1991)

Optical aspheric surface profiler using phase shift interferometry (Sasaki, Kenji; Ono, Akira)V1332,97-106(1991)

Optical scatter: an overview (Stover, John C.)V1530,2-6(1991)

Optical Scatter: Applications, Measurement, and Theory (Stover, John C., ed.)V1530(1991)

Optical surface microtopography using phase-shifting Nomarski microscope (Shimada, Wataru; Sato, Tadamitu; Yatagai, Toyohiko)V1332,525-529(1991)

Optical system for automatic inspection of curved surfaces (Livi, S.; Magnani, M.; Pieri, Silvano; Romoli, Andrea)V1500,144-150(1991)

Optical Testing and Metrology III: Recent Advances in Industrial Optical Inspection (Grover, Chandra P., ed.)V1332(1991)

Optical three-dimensional sensing for measurement of bottomhole pattern (Su, Wan-Yong; Su, Xianyu)V1332,820-823(1991)

Optimal estimation of finish parameters (Church, Eugene L.; Takacs, Peter Z.)V1530,71-85(1991)

Orientation-based differential geometric representations for computer vision applications (Taubes, C. H.; Liang, Ping)V1570,96-102(1991)

Phase-conjugate Twyman-Green interferometer for testing conicoidal surfaces (Shukla, Ram P.; Dokhanian, Mostafa; Venkateswarlu, Putcha; George, M. C.)V1332,274-286(1991)

Photoacoustic characterization of surface absorption (Reicher, David W.; Wilson, Scott R.; Kranenberg, C. F.; Raja, M. Y.; McNeil, John R.; Brueck, Steven R.)V1441,106-112(1991)

Photon scanning tunneling microscopy (Goudonnet, Jean-Pierre; Salomon, L.; de Fornel, F.; Chabrier, G.; Warmack, R. J.; Ferrell, Trinidad L.)V1400,116-123(1991)

Polarization dependence of light scattered from rough surfaces with steep slopes (O'Donnell, Kevin A.; Knotts, Micheal E.)V1558,362-367(1991)

Probabilistic modeling of surfaces (Szeliski, Richard)V1570,154-165(1991)

Processing of bioceramic implants with excimer laser (Koort, Hans J.)V1427,173-180(1991)

Pulsed electron-beam testing of optical surfaces (Murray, Brian W.; Johnson, Edward A.)V1330,2-24(1991)

Quadric surfaces: some derivations and applications (Larkin, Eric W.)V1354,222-231(1991)

Raman lidar for measuring backscattering in the China Sea (Liu, Zhi-Shen; Zhang, Jin-Long; Chen, Wen-Zhong; Huang, Xiao-Sheng; Ma, Jun)V1558,379-383(1991)

Reconstructing visible surfaces (Schenk, Anton F.; Toth, Charles)V1526,78-89(1991)

Removal of small particles from surfaces by pulsed laser irradiation: observations and a mechanism (Kelley, J. D.; Stuff, Michael I.; Hovis, Floyd E.; Linford, Gary J.)V1415,211-219(1991)

Resonant and nonresonant ionization in sputtered initiated laser ionization spectrometry (Havrilla, George J.; Nicholas, Mark; Bryan, Scott R.; Pruett, J. G.)V1435,12-18(1991)

Review of interferogram analysis methods (Malacara Hernandez, Daniel)V1332,678-689(1991)

Rigorous optical theory of the D Sight phenomenon (Reynolds, Rodger L.; Hageniers, Omer L.)V1332,85-96(1991)

Role of the surface height correlation function in the enhanced backscattering of light from random metallic surfaces (Maradudin, Alexei A.; Michel, T.)V1558,233-250(1991)

Scatter and roughness measurements on optical surfaces exposed to space (Schmitt, Dirk-Roger; Swoboda, Helmut; Rosteck, Helmut)V1530,104-110(1991)

Scattering from multilayer coatings: a linear systems model (Harvey, James E.; Lewotsky, Kristin L.)V1530,35-41(1991)

Scattering from objects near a rough surface (Rino, Charles L.; Ngo, Hoc D.)V1558,339-350(1991)

Scattering from slightly rough crystal surfaces (Church, Eugene L.)V1530,171-184(1991)

Scattering from very rough metallic and dielectric surfaces and enhanced backscattering (Ishimaru, Akira; Chen, Jei S.)V1558,182-190(1991)

Second-harmonic generation as an in situ diagnostic for corrosion (Klenerman, David; Spowage, K.; Walpole, B.)V1437,95-102(1991)

Selected Papers on Optical Tolerancing (Wiese, Gary E., ed.)VMS36(1991)

Self-reflection and self-transmission of pulsed radiation by laser-evaporated media (Furzikov, Nickolay P.)V1415,228-239(1991)

SERS used to study the effect of Langmuir-Blodgett spacer layers on metal surface (Yu, Bing Kun; Li, Yu; Wang, Yingting)V1530,363-369(1991)

Shape-from-focus: surface reconstruction of hybrid surfaces (Chen, Su-Shing; Tang, Wu-bin; Xu, Jian-hua)V1569,446-450(1991)

Shape metrics from curvature-scale space and curvature-tuned smoothing (Dudek, Gregory)V1570,75-85(1991)

Simple method for finding the valid ray-surface intersection (Freeman, David E.)V1354,200-209(1991)

Simulation of the process of laser beam machining based on the model of surface vaporization (Stankiewicz, Maria; Zachorowski, Jan)V1391,174-180(1991)

Simultaneous alignment and multiple surface figure testing of optical system components via wavefront aberration measurement and reverse optimization (Lundgren, Mark A.; Wolfe, William L.)V1354,533-539(1991)

Sinusoidal surfaces as standards for BRDF instruments (Marx, Egon; Lettieri, Thomas R.; Vorburger, Theodore V.; McIntosh, Malcolm B.)V1530,15-21(1991)

Some computational results for rough-surface scattering (DeSanto, John A.; Wombell, Richard J.)V1558,202-212(1991)

Spatial and temporal surface interpolation using wavelet bases (Pentland, Alexander P.)V1570,43-62(1991)

Spatio-temporal curvature measures for flow-field analysis (Zetzsche, Christoph; Barth, Erhardt; Berkmann, J.)V1570,337-350(1991)

Statistical properties of intensity fluctuations produced by rough surfaces under the speckle pattern illumination (Yoshimura, Takeaki; Fujiwara, Kazuo; Miyazaki, Eiichi)V1332,835-842(1991)

Stokes vectors, Mueller matrices, and polarized scattered light: experimental applications to optical surfaces and all other scatterers (Bickel, William S.; Videen, Gorden W.)V1530,7-14(1991)

Strategies for tunable frequency-selective surfaces (Lakhtakia, Akhlesh)V1489,108-111(1991)

Structure of very thin metal film (Huang, Yong L.; Jiang, Ping)V1519,142-145(1991)

Study of anomalous scatter characteristics (Stover, John C.; Bernt, Marvin L.; Henning, Timothy D.)V1530,185-195(1991)

Study of PEO on LTI carbon surfaces by ellipsometry and tribometry (Wang, Jinyu; Stroup, Eric; Wang, Xing F.; Andrade, Joseph D.)V1519,835-841(1991)

Study on the surface currents density with second-harmonic generation from silver (Chen, Zhan; Jiang, Hongbing; Zheng, J. B.; Zhang, Zhiming)V1437,103-109(1991)

Subsurface polishing damage of fused silica: nature and effect on laser damage of coated surfaces (Tesar, Aleta A.; Brown, Norman J.; Taylor, John R.; Stolz, Christopher J.)V1441,154-172(1991)

Surface defect detection and classification with light scattering (Gebhardt, Michael; Truckenbrodt, Horst; Harnisch, Bernd)V1500,135-143(1991)

Surface inspection using optical fiber sensor (Abe, Makoto; Ohta, Shigekata; Sawabe, Masaji)V1332,366-376(1991)

Surface layers of metals alloyed with a pulsed laser (Pawlak, Ryszard; Raczynski, Tomasz)V1391,170-173(1991)

Surface microtopography of thin silver films (Costa, Manuel Filipe P.; Almeida, Jose B.)V1332,544-551(1991)

Surface reflection coefficient correction technique for a microdisplacement OFS (Wei, Cailin)V1572,42-46(1991)

Surface roughness effects on light scattered by submicron particles on surfaces (Bawolek, Edward J.; Hirleman, Edwin D.)V1464,574-583(1991)

Surface roughness measurements of spherical components (Chen, Yi-Sheng; Wang, Wen-Gui)V1530,111-117(1991)

Surface structures formation by pulse heating of metals in oxidized environment (Bazhenov, V. V.; Libenson, Michail N.; Makin, Vladimir S.; Trubaev, Vladimir V.)V1440,332-337(1991)

Theory and experiment as tools for assessing surface finish in the UV-visible wavelength region (Ingers, Joakim P.; Thibaudeau, Laurent)V1400,178-185(1991)

Thermocapillary mechanism of laser pulse alloying of a metal surface layer (Kostrubiec, Franciszek)V1391,224-227(1991)

Thin films of solid electrolytes and studies of their surface (Pan, Xiao R.; Gu, Zhi X.)V1519,85-89(1991)

Three-dimensional nanoprofiling of semiconductor surfaces (Montgomery, Paul C.; Fillard, Jean-Pierre B.; Tchandjou, N.; Ardisasmita, Moh S.)V1332,515-524(1991)

Three-dimensional surface inspection using interferometric grating and 2-D FFT-based technique (Hung, Yau Y.; Tang, Shou-Hong; Zhu, Qiuming)V1332,696-703(1991)

Total internal reflection studies of a ferroelectric liquid crystal/anisotropic solid interface (Zhuang, Zhiming; Clark, Noel A.; Meadows, Michael R.)V1455,105-109(1991)

UV laser-induced photofragmentation and photoionization of dimethylcadmium chemisorbed on silicon (Simonov, Alexander P.; Varakin, Vladimir N.; Panesh, Anatoly M.)V1436,20-30(1991)

White-light moire phase-measuring interferometry (Stahl, H. P.)V1332,720-730(1991)

Switches—see also optical interconnects

34-Mb/s TDM photonic switching system (Yuan, Weitao; Zha, Kaide; Guo, Yili; Zhou, Bing-Kun)V1572,78-83(1991)

Absorptive nonlinear semiconductor amplifiers for fast optical switching (Barnsley, Peter E.; Marshall, Ian W.; Fiddyment, Phillip J.; Robertson, Michael J.)V1378,116-126(1991)

Active matrix LCDs driven by two- and three-terminal switches: a comparison (den Boer, Willem; Yaniv, Zvi)V1455,248-248(1991)

All-optical interconnection networks (Ghafoor, Arif)V1390,454-466(1991)

All-optical switching of picosecond pulses in GaAs MQW waveguides (Li Kam Wa, Patrick; Park, Choong-Bum; Miller, Alan)V1474,2-7(1991)

All-optical Ti:LiNbO3 waveguide switch (d'Alessandro, Antonio; De Sario, Marco; D'Orazio, Antonella; Petruzzelli, Vincenzo)V1378,127-138(1991)

Applications and testing of an opto-mechanical switch (Thevenot, Clarel; Newhouse, Mark A.; Annunziata, Frank A.)V1398,250-260(1991)

Applications of optical switches in fiber optic communication networks (Hanson, Daniel; Gosset, Nathalie M.)V1363,48-56(1991)

Bistable optical switching in GaAs multiple-quantum-well epitaxial etalons (Oudar, Jean-Louis; Sfez, B. G.; Kuszelewicz, Robert)V1361,490-498(1991)

Breakdown in pulsed-power semiconductor switches (Germer, Rudolf K.; Mohr, Joachim; Schoenbach, Karl H.)V1358,925-931(1991)

Characterization and switching study of an optically controlled GaAs switch (Stoudt, David C.; Mazzola, Michael S.; Griffiths, Scott F.)V1378,280-285(1991)

Computational and experimental progress on laser-activated gas avalanche switches for broadband, high-power electromagnetic pulse generation (Mayhall, David J.; Yee, Jick H.; Villa, Francesco)V1378,101-114(1991)

Considerations of the limits and capabilities of light-activated switches (Zucker, Oved S.; Giorgi, David M.; Griffin, Adam; Hargis, David E.; Long, James; McIntyre, Iain A.; Page, Kevin J.; Solone, Paul J.; Wein, Deborah S.)V1378,22-33(1991)

Cost-effective optical switch matrix for microwave phased array (Pan, J. J.; Chau, Seung L.; Li, Wei Z.; Grove, Charles H.)V1476,133-142(1991)

Deep-level configuration of GaAs:Si:Cu: a material for a new type of optoelectronic switch (Schoenbach, Karl H.; Schulz, Hans-Joachim; Lakdawala, Vishnu K.; Kimpel, B. M.; Brinkmann, Ralf P.; Germer, Rudolf K.; Barevadia, Gordon R.)V1362,428-435(1991)

Design and evaluation of optical switching architectures (Ramesh, S. K.; Smith, Thomas D.)V1474,208-211(1991)

Design of frozen-wave and injected-wave microwave generators using optically controlled switches (Nunnally, William C.)V1378,10-21(1991)

Design of photonic switches for optimizing performances of interconnection networks (Armenise, Mario N.; Castagnolo, Beniamino)V1374,186-197(1991)

Electronic/photonic inversion channel technology for optoelectronic ICs and photonic switching (Taylor, Geoff W.; Cooke, Paul W.; Kiely, Philip A.; Claisse, Paul R.; Sargood, Stephen K.; Doctor, D. P.; Vang, T.; Evaldsson, P.; Daryanani, Sonu J.)V1476,2-13(1991)

Free space cascaded optical logic demonstration (McCormick, Frederick B.; Tooley, Frank A.; Cloonan, Thomas J.; Brubaker, John L.; Lentine, Anthony L.; Morrison, Richard L.; Hinterlong, Steve J.; Herron, Michael J.; Walker, Sonya L.)V1396,508-521(1991)

Free-space optical TDM switch (Goel, Kamal K.; Prucnal, Paul R.; Stacy, John L.; Krol, Mark F.; Johns, Steven T.)V1476,314-319(1991)

GaAs opto-thyristor for pulsed power applications (Hur, Jung H.; Hadizad, Peyman; Zhao, Hanmin; Hummel, Steven R.; Dapkus, P. D.; Fetterman, Harold R.; Gundersen, Martin A.)V1378,95-100(1991)

Gating techniques for imaging digicon tubes (Swanberg, Norman; Urbach, Michael K.; Ginaven, Robert O.)V1346,249-267(1991)

Generation of stable low-jitter kilovolt amplitude picosecond pulses (Sarkar, Tapan K.; Banerjee, Partha P.)V1378,34-42(1991)

Generic applications for Si and GaAs optical switching devices utilizing semiconductor lasers as an optical source (Rosen, Arye; Stabile, Paul J.; Zutavern, Fred J.; Loubriel, Guillermo M.)V1378,187-194(1991)

Graded AlxGa1-xAs photoconductive devices for high-efficiency picosecond optoelectronic switching (Morse, Jeffrey D.; Mariella, Raymond P.; Dutton, Robert W.)V1378,55-59(1991)

Heat monitoring by fiber-optic microswitches (Mencaglia, Andrea; Brenci, Massimo; Falciai, Riccardo; Guzzi, D.; Pascucci, Tania)V1506,140-144(1991)

High-capacity optical spatial switch based on reversible holograms (Mikaelian, Andrei L.; Salakhutdinov, Viktor K.; Vsevolodov, N. N.; Dyukova, T. V.)V1621,148-157(1991)

High on-off ratio, ultrafast optical switch for optically controlled phased array (Pan, J. J.; Li, Wei Z.)V1476,122-132(1991)

High-power/high-pulse repetition frequency (PRF) pulse generation using a laser-diode-activated photoconductive GaAs switch (Kim, A. H.; Zeto, Robert J.; Youmans, Robert J.; Kondek, Christine D.; Weiner, Maurice; Lalevic, Bogoliub)V1378,173-178(1991)

High-power switching with electron-beam-controlled semiconductors (Brinkmann, Ralf P.; Schoenbach, Karl H.; Roush, Randy A.; Stoudt, David C.; Lakdawala, Vishnu K.; Gerdin, Glenn A.)V1378,203-208(1991)

High-power waveform generation using photoconductive switches (Oicles, Jeffrey A.; Helava, Heikki I.; Grant, Jon R.; Ragle, Larry O.; Wessman, Susan C.)V1378,60-69(1991)

High-repetition-rate pseudospark switches for pulsed high-power lasers (Bickel, P.; Christiansen, Jens; Frank, Klaus; Goertler, Andreas; Hartmann, Werner; Kozlik, Claudius; Wiesneth, Peter)V1503,167-175(1991)

Image analysis for diagnostics in photonic switching (Morrison, Richard L.)V1396,568-574(1991)

Integrated compact matrix-vector multipliers for optical interconnect applications (Lin, Freddie S.; Strzelecki, Eva M.; Liu, William Y.)V1389,642-647(1991)

Laser-triggered superconducting opening switch (Dhali, Shirshak K.; Mohsin, Mohammad)V1396,353-359(1991)

Low-back-reflection, low-loss fiber switch (Roberts, Harold A.; Emmons, David J.; Beard, Michael S.; Lu, Liang-ju)V1363,62-69(1991)

Magneto-optical linear multichannel light modulator for recording of one-dimensional holograms (Abakumov, D. M.)V1621,138-147(1991)

Microwatt all-optical switches, array memories, and flip-flops (Wang, Chang H.; Lloyd, Ashley D.; Wherrett, Brian S.)V1505,130-140(1991)

Microwave control using a high-gain bias-free optoelectronic switch (Freeman, J. L.; Ray, Sankar; West, David L.; Thompson, Alan G.; LaGasse, M. J.)V1476,320-325(1991)

Modeling of a long-pulse high-efficiency XeCl laser with double-discharge and fast magnetic switch (Kobhio, M. N.; Fontaine, Bernard L.; Hueber, Jean-Marc; Delaporte, Philippe C.; Forestier, Bernard M.; Sentis, Marc L.)V1503,88-97(1991)

Model systems for optoelectronic devices based on nonlinear molecular absorption (Speiser, Shammai)V1560,434-442(1991)

Multi-Gb/s optical computer interconnect (Sauer, Jon R.)V1579,49-61(1991)

Multiple-mode reconfigurable electro-optic switching network for optical fiber sensor array (Chen, Ray T.; Wang, Michael R.; Jannson, Tomasz; Baumbick, Robert J.)V1374,223-236(1991)

Multiplexed approach for the fiber optic gyro inertial measurement unit (Page, Jerry L.)V1367,93-102(1991)

Multiwavelength optical switch based on the acousto-optic tunable filter (Liew, Soung C.)V1363,57-61(1991)

Near-term applications of optical switching in the metropolitan networks (King, F. D.; Tremblay, Yves)V1364,295-303(1991)

New configuration of 2 x 2 switching elements using Mach-Zehnder electro-optic modulators (Armenise, Mario N.; Castagnolo, Beniamino; Pesce, Anastasia; Rizzi, Maria L.)V1583,210-220(1991)

New lithium niobate Y-junction polarization divider: theoretical study (Conese, Tiziana; De Pascale, Olga; Matteo, Annamaria; Armenise, Mario N.)V1583,249-255(1991)

Novel acousto-optic photonic switch (Wu, Kuang-Yi; Weverka, Robert T.; Wagner, Kelvin H.; Garvin, Charles G.; Roth, Richard S.)V1563,168-175(1991)

Observation of power gain in an inductive pulsed power system with an optically activated semiconductor closing and opening switch (Kung, Chun C.; Funk, Eric E.; Chauchard, Eve A.; Rhee, M. J.; Lee, Chi H.; Yan, Li I.)V1378,250-258(1991)

OLAS: optical logic array structures (Jones, Robert H.; Hadjinicolaou, M. G.; Musgrave, G.)V1401,138-145(1991)

Optically activated GaAs MMIC switch for microwave and millimeter wave applications (Paolella, Arthur; Madjar, Asher; Herczfeld, Peter R.; Sturzebecher, Dana)V1378,195-202(1991)

Optically Activated Switching (Zutavern, Fred J., ed.)V1378(1991)

Optically nonlinear polymeric devices (Moehlmann, Gustaaf R.; Horsthuis, Winfried H.; Mertens, Hans W.; Diemeer, Mart B.; Suyten, F. M.; Hendriksen, B.; Duchet, Christian; Fabre, P.; Brot, Christian; Copeland, J. M.; Mellor, J. R.; van Tomme, E.; van Daele, Peter; Baets, Roel G.)V1560,426-433(1991)

Optically triggered GaAs thyristor switches: integrated structures for environmental hardening (Carson, Richard F.; Weaver, Harry T.; Hughes, Robert C.; Zipperian, Thomas E.; Brennan, Thomas M.; Hammons, B. E.)V1378,84-94(1991)

Optical probing of field dependent effects in GaAs photoconductive switches (Donaldson, William R.; Kingsley, Lawrence E.)V1378,226-236(1991)

Optical requirements for light-activated switches (McIntyre, Iain A.; Giorgi, David M.; Hargis, David E.; Zucker, Oved S.)V1378,162-172(1991)

Optical switches based on semiconductor optical amplifiers (Kalman, Robert F.; Dias, Antonio R.; Chau, Kelvin K.; Goodman, Joseph W.)V1563,34-44(1991)

Optoelectronic approach to optical parallel processing based on the photonic parallel memory (Matsuda, Kenichi; Shibata, Jun)V1562,21-29(1991)

Optoelectronic compare-and-exchange switches based on BILED circuits (Mao, Xianjun; Liu, Shutian; Wang, Ruibo)V1563,58-63(1991)

Optomechanical M x N fiber-optic matrix switch (Pesavento, Gerry A.)V1474,57-61(1991)

Optomechanics of a free-space photonic switching fabric: the system (McCormick, Frederick B.; Tooley, Frank A.; Brubaker, John L.; Sasian, Jose M.; Cloonan, Thomas J.; Lentine, Anthony L.; Morrison, Richard L.; Crisci, R. J.; Walker, Sonya L.; Hinterlong, Steve J.; Herron, Michael J.)V1533,97-114(1991)

Optomechanics of a free-space photonic switch: the components (Brubaker, John L.; McCormick, Frederick B.; Tooley, Frank A.; Sasian, Jose M.; Cloonan, Thomas J.; Lentine, Anthony L.; Hinterlong, Steve J.; Herron, Michael J.)V1533,88-96(1991)

Photoconductive switching for high-power microwave generation (Pocha, Michael D.; Hofer, Wayne W.)V1378,2-9(1991)

Photoconductivity of high-bandgap materials (Ho, Ping-Tong; Peng, F.; Goldhar, J.; Nolting, Eugene E.; Parsons, C.)V1378,210-216(1991)

Photonic switching implementations of 2-input, 2-output switching nodes based on 2-D and 3-D crossover networks (Cloonan, Thomas J.; McCormick, Frederick B.)V1396,488-500(1991)

Photon-induced charge separation in molecular systems studied by time-resolved microwave conductivity: molecular optoelectronic switches (Warman, John M.; Jonker, Stephan A.; de Haas, Matthijs P.; Verhoeven, Jan W.; Paddon-Row, Michael N.)V1559,159-170(1991)

Photoquenching and characterization studies in a bulk optically controlled GaAs semiconductor switch (Lakdawala, Vishnu K.; Schoenbach, Karl H.; Roush, Randy A.; Barevadia, Gordon R.; Mazzola, Michael S.)V1378,259-270(1991)

Picosecond optoelectronic semiconductor switching and its application (Brueckner, Volkmar; Bergner, Harald; Lenzner, Matthias; Strobel, Reiner)V1362,510-517(1991)

Plastic-optical-fiber-based photonic switch (Grimes, Gary J.; Blyler, Lee L.; Larson, Allen L.; Farleigh, Scott E.)V1592,139-149(1991)

Potential roles of optical interconnections within broadband switching modules (Lalk, Gail R.; Habiby, Sarry F.; Hartman, Davis H.; Krchnavek, Robert R.; Wilson, Donald K.; Young, Kenneth C.)V1389,386-400(1991)

Quantum-well devices for optics in digital systems (Miller, David A.)V1389,496-502(1991)

Radiation effects on dynamical behavior of LiNbO3 switching devices (Kanofsky, Alvin S.; Minford, William J.; Watson, James E.)V1374,59-66(1991)

Reconfiguration phase in rearrangeable multihop lightwave networks (Acampora, Anthony S.; Labourdette, Jean-Francois P.)V1579,30-39(1991)

Rise time and recovery of GaAs photoconductive semiconductor switches (Zutavern, Fred J.; Loubriel, Guillermo M.; O'Malley, Marty W.; McLaughlin, Dan L.; Helgeson, Wes D.)V1378,271-279(1991)

Scattering liquid crystal in optical attenuator applications (Karppinen, Arto; Kopola, Harri K.; Myllyla, Risto A.)V1455,179-189(1991)

Semiconductor waveguides for optical switching (Laval, Suzanne)V1362,82-92(1991)

Spatial dynamics of picosecond CO2 laser pulses produced by optical switching in Ge (Pogorelsky, Igor V.; Fisher, A. S.; Veligdan, James T.; Russell, P.)V1413,21-31(1991)

Subnanosecond, high-voltage photoconductive switching in GaAs (Druce, Robert L.; Pocha, Michael D.; Griffin, Kenneth L.; O'Bannon, Jim)V1378,43-54(1991)

Surface field measurement of photoconductive power switches using the electro-optic Kerr effect (Sardesai, Harshad P.; Nunnally, William C.; Williams, Paul F.)V1378,237-248(1991)

Surface flashover of silicon (Williams, Paul F.; Peterkin, Frank E.; Ridolfi, Tim; Buresh, L. L.; Hankla, B. J.)V1378,217-225(1991)

Switching noise in a medium-film copper/polyimide multichip module (Sandborn, Peter A.; Hashemi, Seyed H.; Weigler, William)V1389,177-186(1991)

Symmetry properties of reverse-delta-beta directional couplers (Smith, Terrance L.; Misemer, David K.; Attanasio, Daniel V.; Crow, Gretchen L.; Smith, Wiley K.; Swierczek, Mary J.; Watson, James E.)V1512,92-100(1991)

Temporal model of optically initiated GaAs avalanche switches (Falk, R. A.; Adams, Jeff C.)V1378,70-81(1991)

Triggering GaAs lock-on switches with laser diode arrays (Loubriel, Guillermo M.; Buttram, Malcolm T.; Helgeson, Wes D.; McLaughlin, Dan L.; O'Malley, Marty W.; Zutavern, Fred J.; Rosen, Arye; Stabile, Paul J.)V1378,179-186(1991)

Two-dimensional edge- and surface-emitting semiconductor laser arrays for optically activated switching (Evans, Gary A.; Rosen, Arye; Stabile, Paul J.; Bour, David P.; Carlson, Nils W.; Connolly, John C.)V1378,146-161(1991)

Y-branch optical modulator (Jaeger, Nicolas A.; Lai, Winnie C.)V1583,202-209(1991)

Synchrotron radiation—see also radiation; sources; x rays

8-GeV synchrotron radiation facility at Nishi-Harima (Kamitsubo, Hiromichi)V1345,16-25(1991)

Advanced X-Ray/EUV Radiation Sources and Applications (Knauer, James P.; Shenoy, Gopal K., eds.)V1345(1991)

Applications of powder diffraction in materials science using synchrotron radiation (Hart, Michael)V1550,11-17(1991)

Channeling radiation as an x-ray source for angiography, x-ray lithography, molecular structure determination, and elemental analysis (Uberall, Herbert; Faraday, Bruce J.; Maruyama, Xavier K.; Berman, Barry L.)V1552,198-213(1991)

Characterization of micro-optical components fabricated by deep-etch x-ray lithography (Goettert, Jost; Mohr, Jurgen)V1506,170-178(1991)

Comparison of plasma source with synchrotron source in the Center for X-ray Lithography (Guo, Jerry Z.; Cerrina, Franco)V1465,330-337(1991)

Cost-effective x-ray lithography (Roltsch, Tom J.)V1465,289-307(1991)

Defect repair for gold absorber/silicon membrane x-ray masks (Stewart, Diane K.; Fuchs, Jacob; Grant, Robert A.; Plotnik, Irving)V1465,64-77(1991)

Design and fabrication of soft x-ray photolithography experimental beam line at Beijing National Synchrotron Radiation Laboratory (Zhou, Changxin)V1465,26-33(1991)

Design and fabrication of x-ray/EUV optics for photoemission experimental beam line at Hefei National Synchrotron Radiation Lab. (Zhou, Changxin; Sun, Deming)V1345,281-287(1991)

Experimental and analytical studies on fixed mask assembly for APS with enhanced cooling (Kuzay, Tuncer M.; Collins, Jeffrey T.; Khounsary, Ali M.; Viccaro, P. J.)V1345,55-70(1991)

Experimental characterization of Fresnel zone plate for hard x-ray applications (Lai, Barry P.; Chrzas, John J.; Yun, Wen-Bing; Legnini, Dan; Viccaro, P. J.; Bionta, Richard M.; Skulina, Kenneth M.)V1550,46-49(1991)

Filter and window assemblies for high-power insertion device synchrotron radiation sources (Khounsary, Ali M.; Viccaro, P. J.; Kuzay, Tuncer M.)V1345,42-54(1991)

Magnetic circular dichroism measurements of benzene in the 9-eV region with synchrotron radiation (Snyder, Patricia A.; Munger, Robert; Hansen, Roger W.; Rowe, Ednor M.)V1548,188-196(1991)

Magnetic circular dichroism studies with soft x-rays (Tjeng, L. H.; Rudolf, P.; Meigs, G.; Sette, Francesco; Chen, Chien-Te; Idzerda, Y. U.)V1548,160-167(1991)

Measurement of synchrotron beam polarization (Singman, Leif V.; Davis, Brent A.; Holmberg, D. L.; Blake, Richard L.; Hockaday, Robert G.)V1548,80-92(1991)

Mirrors as power filters (Kortright, Jeffrey B.)V1345,38-41(1991)

Modeling of illumination effects on resist profiles in x-ray lithography (Oertel, Heinrich K.; Weiss, Martin; Huber, Hans L.; Vladimirsky, Yuli; Maldonado, Juan R.)V1465,244-253(1991)

Modified chloroaluminium phthalocyanine: an organic semiconductor with high photoactivity (Dodelet, Jean-Pol; Gastonguay, Louis; Veilleux, George; Saint-Jacques, Robert G.; Cote, Roland; Guay, Daniel; Tourillon, Gerard)V1436,38-49(1991)

Multilayer monochromator for synchrotron radiation angiography (Baron, Alfred Q.; Barbee, Troy W.; Brown, George S.)V1343,84-94(1991)

Novel toroidal mirror enhances x-ray lithography beamline at the Center for X-ray Lithography (Cole, Richard K.; Cerrina, Franco)V1465,111-121(1991)

Polarization sensitivity of x-ray photocathodes in the 60-200eV band (Fraser, George W.; Pain, M. D.; Pearson, James F.; Lees, John E.; Binns, C. R.; Shaw, Ping-Shine; Fleischman, Judith R.)V1548,132-148(1991)

Polarized nature of synchrotron radiation (Kim, Kwang-je)V1548,73-79(1991)

Polarized x-ray absorption spectroscopy for the study of superconductors and magnetic materials (Ramanathan, Mohan; Alp, Esen E.; Mini, Susan M.; Salem-Sugui, S.; Bommannavar, A.)V1548,168-178(1991)

Polarizing x-ray optics for synchrotron radiation (Hart, Michael)V1548,46-55(1991)

Power distribution from insertion device x-ray sources (Viccaro, P. J.)V1345,28-37(1991)

Process latitude measurements on chemically amplified resists exposed to synchrotron radiation (Babcock, Carl P.; Taylor, James W.; Sullivan, Monroe; Suh, Doowon; Plumb, Dean; Palmer, Shane R.; Berry, Amanda K.; Graziano, Karen A.; Fedynyshyn, Theodore H.)V1466,653-662(1991)

Production and Analysis of Polarized X Rays (Siddons, D. P., ed.)V1548(1991)

Ray tracing homogenizing mirrors for synchrotron x-ray lithography (Homer, Michael; Rosser, Roy J.; Speer, R. J.)V1527,134-144(1991)

Real-time x-ray studies of semiconductor device structures (Clarke, Roy; Dos Passos, Waldemar; Chan, Yi-Jen; Pavlidis, Dimitris; Lowe, Walter P.; Rodricks, Brian G.; Brizard, Christine M.)V1361,2-12(1991)

Role of valence-band excitation in laser ablation of KC1 (Haglund, Richard F.; Tang, Kai; Bunton, P. H.; Wang, Ling-jun)V1441,127-138(1991)

Simulation of low-energy x-ray lithography using a diamond membrane mask (Hasegawa, Shinya; Suzuki, Katsumi)V1465,145-151(1991)

Soft x-ray resonant magnetic scattering study of thin films and multilayers (Kao, Chi-Chang; Johnson, Erik D.; Hastings, Jerome B.; Siddons, D. P.; Vettier, C.)V1548,149-157(1991)

Status of the Advanced Light Source (Marx, Jay N.)V1345,2-10(1991)

Status report on the Advanced Photon Source, summer 1990 (Moncton, David E.)V1345,11-15(1991)

Studies of structures and phase transitions of Langmuir monolayers using synchrotron radiation (Dutta, Pulak)V1550,134-139(1991)

Study of gas scintillation proportional counter physics using synchrotron radiation (Bavdaz, Marcos; Favata, Fabio; Smith, Alan; Parmar, A. N.)V1549,35-44(1991)

Survey of synchrotron radiation devices producing circular or variable polarization (Kim, Kwang-je)V1345,116-124(1991)

Synchrotron radiation lasers (Hirshfield, Jay L.; Park, Gun-Sik)V1552,138-146(1991)

System design considerations for a production-grade, ESR-based x-ray lithography beamline (Kovacs, Stephen; Melore, Dan; Cerrina, Franco; Cole, Richard K.)V1465,88-99(1991)

Techniques of production and analysis of polarized synchrotron radiation (Mills, Dennis M.)V1345,125-136(1991)

Time-resolved techniques: an overview (Larson, Bennett C.; Tischler, J. Z.)V1345,90-100(1991)

X-ray absorption spectroscopy with polarized synchrotron radiation (Alp, Esen E.; Mini, Susan M.; Ramanathan, Mohan; Hyun, O. B.)V1345,137-145(1991)

X-ray detector for time-resolved studies (Rodricks, Brian G.; Brizard, Christine M.; Clarke, Roy; Lowe, Walter P.)V1550,18-26(1991)

X-ray diffraction from materials under extreme pressures (Brister, Keith)V1550,2-10(1991)

X-ray lithography system development at IBM: overview and status (Maldonado, Juan R.)V1465,2-15(1991)

XUV polarimeter for undulator radiation measurements (Gluskin, Efim S.; Mattson, J. E.; Bader, Samuel D.; Viccaro, P. J.; Barbee, Troy W.; Brookes, N.; Pitas, Alan A.; Watts, Richard N.)V1548,56-68(1991)

Synthetic apertures—see also radar

Infrared tellurium two-dimensional acousto-optic processor for synthetic aperture radar (Souilhac, Dominique J.; Billerey, Dominique)V1521,158-174(1991)

Japanese mission overview of JERS and ASTER programs (Yamaguchi, Yasushi; Tsu, Hiroji; Sato, Isao)V1490,324-334(1991)

Proposed conversion of the McMath Telescope to 4.0-meter aperture for solar observations in the IR (Livingston, William C.)V1494,498-502(1991)

SALSA: a synthesis array for lunar submillimeter astronomy (Mahoney, Michael J.; Marsh, Kenneth A.)V1494,182-193(1991)

Scheme of optical synthetic-aperture telescope (Ma, Pin-Zhong)V1533,163-174(1991)

Synthetic aperture radar of JERS-1 (Ono, Makoto; Nemoto, Yoshiaki)V1490,184-190(1991)

Telescopes—see also astonomy; mirrors; satellites; space optics

1990 Intl Lens Design Conf (Lawrence, George N., ed.)V1354(1991)

Active optics system for a 3.5-meter structured mirror (Stepp, Larry M.; Roddier, Nicolas; Dryden, David M.; Cho, Myung K.)V1542,175-185(1991)

Active optics technology: an overview (Ray, Frank B.)V1532,188-206(1991)

Active sun telescope array (Walker, Arthur B.; Timothy, J. G.; Barbee, Troy W.; Hoover, Richard B.)V1343,334-347(1991)

Adaptive optics: a progress review (Hardy, John W.)V1542,2-17(1991)

Adaptive optics for the European very large telescope (Merkle, Fritz; Hubin, Norbert)V1542,283-292(1991)

Adaptive optics system tests at the ESO 3.6-m telescope (Merkle, Fritz; Gehring, G.; Rigaut, Francois; Lena, Pierre J.; Rousset, Gerard; Fontanella, Jean-Claude; Gaffard, Jean-Paul)V1542,308-318(1991)

Adaptive optics using curvature sensing (Forbes, Fred F.; Roddier, Nicolas)V1542,140-147(1991)

Advanced X-ray Astrophysics Facility science instruments (Winkler, Carl E.; Dailey, Carroll C.; Cumings, Nesbitt P.)V1494,301-313(1991)

Alignment and focus control of a telescope using image sharpening (Jones, Peter A.)V1542,194-204(1991)

All-reflective four-element zoom telescope: design and analysis (Johnson, R. B.; Hadaway, James B.; Burleson, Thomas A.; Watts, Bob; Park, Ernest D.)V1354,669-675(1991)

All-reflective phased array imaging telescopes (Stuhlinger, Tilman W.)V1354,438-446(1991)

All-refractive telescope for next-generation inter-satellite communication (Heimbeck, Hans-Jorg)V1354,434-437(1991)

Analysis and testing of a soft actuation system for segmented-reflector articulation and isolation (Jandura, Louise; Agronin, Michael L.)V1542,213-224(1991)

Analysis of Optical Structures (O'Shea, Donald C., ed.)V1532(1991)

Angular positioning mechanism for the ultraviolet coronagraph spectrometer (Ostaszewski, Miroslaw A.; Guy, Larry J.)V1482,13-25(1991)

Aspheric testing using null mirrors (Murty, Mantravady V.; Kumar, Vas; von Handorf, Robert J.)V1332,107-114(1991)

Back-reflection measurements on the SILEX telexcope (Birkl, Reinhard; Manhart, Sigmund)V1522,252-258(1991)

Beam shaping system (MacAndrew, J. A.; Humphries, Mark R.; Welford, W. T.; Golby, John A.; Dickinson, P. H.; Wheeler, J. R.)V1500,172-176(1991)

Beam-tracker and point-ahead system for optical communications II: servo performance (LaSala, Paul V.; McLaughlin, Chris)V1482,121-137(1991)

Bragg crystal polarimeter for the Spectrum-X-Gamma misson (Holley, Jeff; Silver, Eric H.; Ziock, Klaus P.; Novick, Robert; Kaaret, Philip E.; Weisskopf, Martin C.; Elsner, Ronald F.; Beeman, Jeff)V1343,500-511(1991)

CCD camera for an autoguider (Schempp, William V.)V1448,129-133(1991)

Characteristics of a dynamic holographic sensor for shape control of a large reflector (Welch, Sharon S.; Cox, David E.)V1480,2-10(1991)

Coarse pointing assembly for the SILEX program, or how to achieve outstanding pointing accuracy with simple hardware associated with consistent control laws (Buvat, Daniel; Muller, Gerard; Peyrot, Patrick)V1417,251-261(1991)

Combination of mechanical athermalization with manual in IR zoom telescope (Chen, Ruiyi; Zheng, Dayue; Zhou, Xiuli; Zhang, Xingde)V1540,724-728(1991)

Comparison of the angular resolution limit and SNR of the Hubble Space Telescope and the large ground-based telescopes (Souilhac, Dominique J.; Billerey, Dominique)V1494,503-526(1991)

Construction of the 16-meter large lunar telescope (Omar, Husam A.)V1494,135-146(1991)

Current and future activities in the area of optical space communications in Japan (Fujise, Masayuki; Araki, Ken'ichi; Arikawa, Hiroshi; Furuhama, Yoji)V1522,14-26(1991)

Cylindrical proportional counter for x-ray polarimetry (Costa, Enrico; Piro, Luigi; Rubini, Alda; Soffitta, Paolo; Massaro, Enrico; Matt, Giorgio; Medici, Gastone; Manzo, Giuseppe; Re, Stefano)V1343,469-476(1991)

Debris collision warning sensor telescope design (Brown, Robert J.)V1527,155-162(1991)

Design, analysis, and testing of a CCD array mounting structure (Sultana, John A.; O'Neill, Mark B.)V1532,27-38(1991)

Design and analysis of the reflection grating arrays for the X-Ray Multi-Mirror Mission (Atkinson, Dennis P.; Bixler, Jay V.; Geraghty, Paul; Hailey, Charles J.; Klingmann, Jeffrey L.; Montesanti, Richard C.; Kahn, Steven M.; Paerels, F. B.)V1343,530-541(1991)

Designing apochromatic telescope objectives with liquid lenses (Sigler, Robert D.)V1535,89-112(1991)

Design of array systems using shared symmetry (Miao, Cheng-Hsi)V1354,447-456(1991)

Design of compact IR zoom telescope (Chen, Ruiyi; Zhou, Xiuli; Zhang, Xingde)V1540,717-723(1991)

Design of two- and three-element diffractive telescopes (Buralli, Dale A.; Morris, G. M.)V1354,292-296(1991)

Easily fabricated wide-angle telescope (Owen, R. C.)V1354,430-433(1991)

Effects of the lunar environment on optical telescopes and instruments (Johnson, Charles L.; Dietz, Kurtis L.)V1494,208-218(1991)

EUV performance of a multilayer-coated high-density toroidal grating (Keski-Kuha, Ritva A.; Thomas, Roger J.; Neupert, Werner M.; Condor, Charles E.; Gum, Jeffrey S.)V1343,566-575(1991)

Fabrication of a grazing-incidence telescope by grinding and polishing techniques on aluminum (Gallagher, Dennis J.; Cash, Webster C.; Green, James C.)V1343,155-161(1991)

Faint object spectrograph early performance (Harms, Richard J.; Fitch, John E.)V1494,49-65(1991)

Filled-arm Fizeau telescope (Synnott, Stephen P.)V1494,334-343(1991)

Finite element analysis of large lenses for the Keck telescope high-resolution echelle spectrograph (Bigelow, Bruce C.)V1532,15-26(1991)

Fixing the Hubble Space Telescope (Crocker, James H.)V1494,2-8(1991)

Foundation, excavation, and radiation-shielding concepts for a 16-m large lunar telescope (Chua, Koon M.; Johnson, Stewart W.)V1494,119-134(1991)

Four-meter lunar engineering telescope (Peacock, Keith; Giannini, Judith A.; Kilgus, Charles C.; Bely, Pierre Y.; May, B. S.; Cooper, Shannon A.; Schlimm, Gerard H.; Sounder, Charles; Ormond, Karen A.; Cheek, Eric A.)V1494,147-159(1991)

Generic telescope truss (Pressel, Philip)V1532,50-56(1991)

GOPEX: a deep-space optical communications demonstration with the Galileo spacecraft (Wilson, Keith E.; Schwartz, Jon A.; Lesh, James R.)V1417,22-28(1991)

Ground systems and operations concepts for the Space Infrared Telescope Facility (Miller, Richard B.)V1540,38-46(1991)

Hard x-ray polarimeter utilizing Compton scattering (Sakurai, Hirohisa; Noma, M.; Niizeki, H.)V1343,512-518(1991)

High-performance, wide-magnification-range IR zoom telescope with automatic compensation for temperature effects (Shechterman, Mark)V1442,276-285(1991)

High-resolution imaging with multilayer soft x-ray, EUV, and FUV telescopes of modest aperture and cost (Walker, Arthur B.; Lindblom, Joakim F.; Timothy, J. G.; Hoover, Richard B.; Barbee, Troy W.; Baker, Phillip C.; Powell, Forbes R.)V1494,320-333(1991)

High-resolution stigmatic EUV spectroheliometer for studies of the fine scale structure of the solar chromosphere, transition region, and corona (Timothy, J. G.; Berger, Thomas E.; Morgan, Jeffrey S.; Walker, Arthur B.; Bhattacharyya, Jagadish C.; Jain, Surendra K.; Saxena, Ajay K.; Huber, Martin C.; Tondello, Giuseppe; Naletto, Giampiero)V1343,350-358(1991)

HST image processing: an overview of algorithms for image restoration (Gonsalves, Robert A.; Nisenson, Peter)V1567,294-307(1991)

HST image processing: how does it work and what are the problems? (White, Richard L.; Hanisch, Robert J.)V1567,308-316(1991)

HST phase retrieval: a parameter estimation (Lyon, Richard G.; Miller, Peter E.; Gruszczak, Anthony)V1567,317-326(1991)

Hubble Space Telescope: mission, history, and systems (Endelman, Lincoln L.)V1358,422-441(1991)

Hubble Space Telescope optics: problems and solutions (Burrows, Christopher J.)V1494,528-533(1991)

Hubble Space Telescope optics status (Burrows, Christopher J.)V1567,284-293(1991)

Imaging characteristics of the development model of the SAX x-ray imaging concentrators (Citterio, Oberto; Conconi, Paolo; Conti, Giancarlo; Mattaini, E.; Santambrogio, E.; Cusumano, G.; Sacco, B.; Braueninger, Heinrich; Burkert, Wolfgang)V1343,145-154(1991)

Imaging performance analysis of adaptive optical telescopes using laser guide stars (Welsh, Byron M.)V1542,88-99(1991)

Imaging performance and tests of soft x-ray telescopes (Spiller, Eberhard; McCorkle, R.; Wilczynski, Janusz S.; Golub, Leon; Nystrom, George U.; Takacs, Peter Z.; Welch, Charles W.)V1343,134-144(1991)

In-flight performance of the faint object camera of the Hubble Space Telescope (Greenfield, Perry E.; Paresce, Francesco; Baxter, David; Hodge, P.; Hook, R.; Jakobsen, P.; Jedrzejewski, Robert; Nota, Anatonella; Sparks, W. B.; Towers, Nigel M.; Laurance, R. J.; Macchetto, F.)V1494,16-39(1991)

In-flight performance of the Goddard high-resolution spectrograph of the Hubble Space Telescope (Troeltzsch, John R.; Ebbets, Dennis C.; Garner, Harry W.; Tuffli, A.; Breyer, R.; Kinsey, J.; Peck, C.; Lindler, Don; Feggans, J.)V1494,9-15(1991)

Infrared Space Observatory optical subsystem (Singer, Christian; Massoni, Jean A.; Mossbacher, Bernard; Cinotti, Ciro)V1494,255-264(1991)

Initial performance of the high-speed photometer (Richards, Evan; Percival, Jeff; Nelson, Matthew; Hatter, Edward; Fitch, John E.; White, Richard L.)V1494,40-48(1991)

Investigation of unobscured mirror telescope for telecom purposes (Sand, Rolf)V1522,103-110(1991)

Latest developments of active optics of the ESO NTT and the implications for the ESO VLT (Noethe, L.; Andreoni, G.; Franza, F.; Giordano, P.; Merkle, Fritz; Wilson, Raymond N.)V1542,293-296(1991)

Launch area theodolite system (Bradley, Lester M.; Corriveau, John P.; Tindal, Nan E.)V1482,48-60(1991)

Lunar liquid-mirror telescopes (Borra, Ermanno F.; Content, R.)V1494,219-227(1991)

Manufacture of ISO mirrors (Ruch, Eric)V1494,265-278(1991)

MARTINI: system operation and astronomical performance (Doel, A. P.; Dunlop, C. N.; Major, J. V.; Myers, Richard M.; Sharples, R. M.)V1542,319-326(1991)

Materials technology for SIRTF (Coulter, Daniel R.; Dolgin, Benjamin P.; Rainen, R.; O'Donnell, Timothy P.)V1540,119-126(1991)

Metallic alternative to glass mirrors (active mirrors in aluminum): a review (Rozelot, Jean-Pierre; Leblanc, Jean-M.)V1494,481-490(1991)

Method of making ultralight primary mirrors (Zito, Richard R.)V1494,491-497(1991)

Mirrors for optical telescopes (Miroshnikov, Mikhail M.)V1533,286-298(1991)

Mission design for the Space Infrared Telescope Facility (Kwok, Johnny H.; Osmolovsky, Michael G.)V1540,27-37(1991)

Moving M2 mirror without pointing offset (Ragazzoni, Roberto; Bortoletto, Favio)V1542,236-246(1991)

Multiple degree-of-freedom tracking for attitude control of an experimental system on tether-stabilized platform (Angrilli, Francesco; Baglioni, Pietro; Bianchini, Gianandrea; Da Forno, R.; Fanti, Giullo; Mozzi, Massimo)V1482,26-39(1991)

Need for active structures in future large IR and sub-mm telescopes (Rapp, Donald)V1542,328-358(1991)

Neural network adaptive optics for the multiple-mirror telescope (Wizinowich, Peter L.; Lloyd-Hart, Michael; McLeod, Brian A.; Colucci, D'nardo; Dekany, Richard G.; Wittman, David; Angel, J. R.; McCarthy, Donald W.; Hulburd, William G.; Sandler, David G.)V1542,148-158(1991)

New type of large-angle binocular microtelescopes (Lu, Kaichang; Zhu, Yafei; Kang, Songgao)V1527,413-418(1991)

New way of making a lunar telescope (Chen, Peter C.)V1494,228-233(1991)

Next-generation space telescope: a large UV-IR successor to HST (Illingworth, Garth)V1494,86-97(1991)

Nonlinear finite element analysis of the Starlab 80-cm telescope primary-mirror suspension system (Arnold, William R.)V1532,103-113(1991)

Objectives for the Space Infrared Telescope Facility (Spehalski, Richard J.; Werner, Michael J.)V1540,2-14(1991)

Off-axis spherical element telescope with binary optic corrector (Brown, Daniel M.; Kathman, Alan D.)V1555,114-127(1991)

Optical design considerations for next-generation space and lunar telescopes (Korsch, Dietrich G.)V1494,111-118(1991)

Optical design of an off-axis low-distortion UV telescope (Richardson, E. H.)V1494,314-319(1991)

Optical performance of an infrared astronomical telescope with a 5-axis secondary mirror (Ettedgui-Atad, Eli; Humphries, Colin M.)V1532,241-248(1991)

ORFEUS alignment concept (Graue, Roland; Kampf, Dirk; Rippel, Harald; Witte, G.)V1494,377-385(1991)

ORFEUS: orbiting and retrievable far and extreme ultraviolet spectrometer (Rippel, Harald; Kampf, Dirk; Graue, Roland)V1343,520-529(1991)

Performance of the Multi-Spectral Solar Telescope Array III: optical characteristics of the Ritchey-Chretien and Cassegrain Telescopes (Hoover, Richard B.; Baker, Phillip C.; Hadaway, James B.; Johnson, R. B.; Peterson, Cynthia; Gabardi, David R.; Walker, Arthur B.; Lindblom, Joakim F.; DeForest, Craig E.; O'Neal, Ray H.)V1343,189-202(1991)

Performance of the Multi-Spectral Solar Telescope Array IV: the soft x-ray and extreme ultraviolet filters (Lindblom, Joakim F.; O'Neal, Ray H.; Walker, Arthur B.; Powell, Forbes R.; Barbee, Troy W.; Hoover, Richard B.; Powell, Stephen F.)V1343,544-557(1991)

Performance of the Multi-Spectral Solar Telescope Array V: temperature diagnostic response to the optically thin solar plasma (DeForest, Craig E.; Kankelborg, Charles C.; Allen, Maxwell J.; Paris, Elizabeth S.; Willis, Thomas D.; Lindblom, Joakim F.; O'Neal, Ray H.; Walker, Arthur B.; Barbee, Troy W.; Hoover, Richard B.; Barbee, Troy W.; Gluskin, Efim S.)V1343,404-414(1991)

Performance of the Multi-Spectral Solar Telescope Array VI: performance and characteristics of the photographic films (Hoover, Richard B.; Walker, Arthur B.; DeForest, Craig E.; Allen, Maxwell J.; Lindblom, Joakim F.)V1343,175-188(1991)

Phase retrieval for the Hubble Space Telescope using iterative propagation algorithms (Fienup, James R.)V1567,327-332(1991)

Photoelectric effect from CsI by polarized soft x-rays (Shaw, Ping-Shine; Church, Eric D.; Hanany, Shaul; Liu, Yee; Fleischman, Judith R.; Kaaret, Philip E.; Novick, Robert; Manzo, Giuseppe)V1343,485-499(1991)

Power of one power (Shenker, Martin)V1354,647-653(1991)

Predicted performance of the lithium scattering and graphite crystal polarimeter for the Spectrum-X-Gamma mission (Weisskopf, Martin C.; Elsner, Ronald F.; Novick, Robert; Kaaret, Philip E.; Silver, Eric H.)V1343,457-468(1991)

Primary mirror control system for the Galileo telescope (Bortoletto, Favio; Baruffolo, A.; Bonoli, C.; D'Alessandro, Maurizio; Fantinel, D.; Giudici, G.; Ragazzoni, Roberto; Salvadori, L.; Vanini, P.)V1542,225-235(1991)

Proposed conversion of the McMath Telescope to 4.0-meter aperture for solar observations in the IR (Livingston, William C.)V1494,498-502(1991)

Radiation concerns for the Solar-A soft x-ray telescope (Acton, Loren W.; Morrison, Mons D.; Janesick, James R.; Elliott, Tom)V1447,123-139(1991)

Reconstruction of the Hubble Space Telescope mirror figure from out-of-focus stellar images (Roddier, Claude A.; Roddier, Francois J.)V1494,78-84(1991)

Relay Mirror Experiment scoring analysis and the effects of atmospheric turbulence (Sydney, Paul F.; Dillow, Michael A.; Anspach, Joel E.; Kervin, Paul W.; Lee, Terence B.)V1482,196-208(1991)

Removal of adsorbed gases with CO2 snow (Zito, Richard R.)V1494,427-433(1991)

Results from the calibration of the Extreme Ultraviolet Explorer instruments (Welsh, Barry Y.; Jelinsky, Patrick; Vedder, Peter W.; Vallerga, John V.; Finley, David S.; Malina, Roger F.)V1343,166-174(1991)

Rigidity test of large and high-precision instruments (Ma, Pin-Zhong)V1532,177-186(1991)

Scheme of optical synthetic-aperture telescope (Ma, Pin-Zhong)V1533,163-174(1991)

Scientific results from the Hubble Space Telescope fine-guidance sensors (Taff, Laurence G.)V1494,66-77(1991)

Segmented mirror figure control for a space-based far-IR astronomical telescope (Redding, David C.; Breckenridge, William G.; Lau, Kenneth; Sevaston, George E.; Levine, Bruce M.; Shaklan, Stuart B.)V1489,201-215(1991)

Semiactive telescope for the French PRONAOS submillimetric mission (Duran, Michel; Luquet, Philippe; Buisson, F.; Cousin, B.)V1494,357-376(1991)

Si:Ga focal-plane arrays for satellite and ground-based telescopes (Mottier, Patrick; Agnese, Patrick; Lagage, Pierre O.)V1494,419-426(1991)

SIRTF focal-plane technologies (Capps, Richard W.; Bothwell, Mary)V1540,47-50(1991)

SIRTF stray light analysis (Elliott, David G.; St. Clair Dinger, Ann)V1540,63-67(1991)

SOHO space satellite: UV instrumentation (Poland, Arthur I.; Domingo, Vicente)V1343,310-318(1991)

Solar astronomy with a 19-segment adaptive mirror (Acton, D. S.; Smithson, Robert C.)V1542,159-164(1991)

Space Astronomical Telescopes and Instruments (Bely, Pierre Y.; Breckinridge, James B., eds.)V1494(1991)

Space-based visible all-reflective stray light telescope (Wang, Dexter; Gardner, Leo R.; Wong, Wallace K.; Hadfield, Peter)V1479,57-70(1991)

Space Infrared Telescope Facility cryogenic and optical technology (Mason, Peter V.; Kiceniuk, T.; Plamondon, Joseph A.; Petrick, Walt)V1540,88-96(1991)

Space Infrared Telescope Facility science instruments overview (Bothwell, Mary)V1540,15-26(1991)

Space Infrared Telescope Facility structural design requirements (MacNeal, Paul D.; Lou, Michael C.; Chen, Gun-Shing)V1540,68-85(1991)

Space Infrared Telescope Facility telescope overview (Schember, Helene R.; Manhart, Paul K.; Guiar, Cecilia N.; Stevens, James H.)V1540,51-62(1991)

Space telescope imaging spectrograph 2048 CCD and its characteristics (Delamere, W. A.; Ebben, Thomas H.; Murata-Seawalt, Debbie; Blouke, Morley M.; Reed, Richard; Woodgate, Bruce E.)V1447,288-297(1991)

Spatial frequency selective error sensing for space-based, wide field-of-view, multiple-aperture imaging systems (Erteza, Ahmed; Schneeberger, Timothy J.)V1527,182-187(1991)

Station-keeping strategy for multiple-spacecraft interferometry (DeCou, Anthony B.)V1494,440-451(1991)

Stray light issues for background-limited infrared telescope operation (Scholl, Marija S.; Scholl, James W.)V1540,109-118(1991)

Structural design considerations for the Space Infrared Telescope Facility (MacNeal, Paul D.; Lou, Michael C.)V1494,236-254(1991)

Surface control techniques for the segmented primary mirror in the large lunar telescope (Gleckler, Anthony D.; Pflibsen, Kent P.; Ulich, Bobby L.; Smith, Duane D.)V1494,454-471(1991)

Survey of hard x-ray imaging concepts currently proposed for viewing solar flares (Campbell, Jonathan W.; Davis, John M.; Emslie, A. G.)V1343,359-375(1991)

System concepts for a large UV/optical/IR telescope on the moon (Nein, Max E.; Davis, Billy)V1494,98-110(1991)

Telescope enclosure flow visualization (Forbes, Fred F.; Wong, Woon-Yin; Baldwin, Jack; Siegmund, Walter A.; Limmongkol, Siriluk; Comfort, Charles H.)V1532,146-160(1991)

Temperature control of the 3.5-meter WIYN telescope primary mirror (Goble, Larry W.)V1532,161-169(1991)

Thermal systems analysis for the Space Infrared Telescope Facility dewar (Bhandari, Pradeep; Petrick, Stanley W.; Schember, Helene R.)V1540,97-108(1991)

Two-mirror projection systems for simulating telescopes (Hannan, Paul G.; Davila, Pam M.)V1527,26-36(1991)

Ultra-High-Resolution XUV Spectroheliograph II: predicted performance (Walker, Arthur B.; Lindblom, Joakim F.; Timothy, J. G.; Allen, Maxwell J.; DeForest, Craig E.; Kankelborg, Charles C.; O'Neal, Ray H.; Paris, Elizabeth S.; Willis, Thomas D.; Barbee, Troy W.; Hoover, Richard B.)V1343,319-333(1991)

University of Hawaii adaptive optics system: I. General approach (Roddier, Francois J.; Graves, J. E.; McKenna, Daniel; Northcott, Malcolm J.)V1542,248-253(1991)

Virtual-phase charge-coupled device imaging system for astronomy (Khvilivitzky, A. T.; Zuev, A. G.; Rybakov, M. I.; Kiryan, G. V.; Berezin, V. Y.)V1447,64-68(1991)

XMM space telescope: development plan for the lightweight replicated x-ray gratings (Montesanti, Richard C.; Atkinson, Dennis P.; Edwards, David F.; Klingmann, Jeffrey L.)V1343,558-565(1991)

X-Ray/EUV Optics for Astronomy, Microscopy, Polarimetry, and Projection Lithography (Hoover, Richard B.; Walker, Arthur B., eds.)V1343(1991)

Television—see also cameras; imaging systems; video

140-Mbit/s HDTV coding using subband and hybrid techniques (Amor, Hamed; Wietzke, Joachim)V1567,578-588(1991)

3-D TV: joined identification of global motion parameters for stereoscopic sequence coding (Tamtaoui, Ahmed; Labit, Claude)V1605,720-731(1991)

3-DTV research and development in Europe (Sand, Ruediger)V1457,76-84(1991)

45-Mbps multichannel TV coding system (Matsumoto, Shuichi; Hamada, Takahiro; Saito, Masahiro; Murakami, Hitomi)V1605,37-46(1991)

Acousto-optic color projection system (Hubin, Thomas)V1454,313-322(1991)

Artificial neural networks as TV signal processors (Spence, Clay D.; Pearson, John C.; Sverdlove, Ronald)V1469,665-670(1991)

Compact zoom lens for stereoscopic television (Scheiwiller, Peter M.; Murphy, S. P.; Dumbreck, Andrew A.)V1457,2-8(1991)

Computerized vibration analysis of hot objects (Lokberg, Ole J.; Rosvold, Geir O.; Malmo, Jan T.; Ellingsrud, Svein)V1508,153-160(1991)

Current HDTV overview in the United States, Japan, and Europe (Cripps, Dale E.)V1456,60-64(1991)

Design and fabrication of amplitude modulators for CATV (Mahapatra, Amaresh; Cooper, Ronald F.)V1374,296-299(1991)

Developments in projection lenses for HDTV (Rudolph, John D.)V1456,15-28(1991)

Digital picture processing for the transmission of HDTV: the progress in Europe (Le Pannerer, Yves-Marie; Tourtier, Philippe)V1567,556-565(1991)

Estimation of three-dimensional motion in a 3-DTV image sequence (Dugelay, Jean-Luc)V1605,688-696(1991)

Evaluation of the effect of noise on subjective image quality (Barten, Peter G.)V1453,2-15(1991)

Fiber optics in CATV networks (Wolfe, Ronald; Laor, Herzel)V1363,125-132(1991)

Fringe quality in pulsed TV-holography (Spooren, Rudie)V1508,118-127(1991)

High-definition projection television (Browning, Iben)V1456,48-50(1991)

High-precision digital charge-coupled device TV system (Vishnevsky, G. I.; Ioffe, S. A.; Berezin, V. Y.; Rybakov, M. I.; Mikhaylov, A. V.; Belyaev, L. V.)V1448,69-72(1991)

Liquid-crystal television optical neural network: architecture, design, and models (Yu, Francis T.)V1455,150-166(1991)

Luminance asymmetry in stereo TV images (Beldie, Ion P.; Kost, Bernd)V1457,242-247(1991)

Model-based coding of facial images based on facial muscle motion through isodensity maps (So, Ikken; Nakamura, Osamu; Minami, Toshi)V1605,263-272(1991)

Motion affine models identification and application to television image coding (Sanson, Henri)V1605,570-581(1991)

Motion-compensated subsampling of HDTV (Belfor, Ricardo A.; Lagendijk, Reginald L.; Biemond, Jan)V1605,274-284(1991)

Motion compensation by block matching and vector postprocessing in subband coding of TV signals at 15 Mbit/s (Lallauret, Fabrice; Barba, Dominique)V1605,26-36(1991)

New CATV fiber-to-the-subscriber architectures (Kim, Gary)V1363,133-140(1991)

Object extraction method for image synthesis (Inoue, Seiki)V1606,43-54(1991)

Optical enhancements of joint-Fourier-transform correlator by image subtraction (Perez, Osvaldo; Karim, Mohammad A.)V1471,255-264(1991)

Optical scanning system for a CCD telecine for HDTV (Kurtz, Andrew F.; Kessler, David)V1448,191-205(1991)

"Perfect" displays and "perfect" image compression in space and time (Klein, Stanley A.; Carney, Thom)V1453,190-205(1991)

Performance of an HDTV codec adopting transform and motion compensation techniques (Barbero, Marzio; Cucchi, Silvio; Muratori, Mario)V1567,566-577(1991)

Performance of NCAP projection displays (Jones, Philip J.; Tomita, Akira; Wartenberg, Mark)V1456,6-14(1991)

Policy issues affecting telephone company opportunities in CATV and deployment of fiber optic cable (Keane, William K.)V1363,184-185(1991)

Present state of HDTV coding in Japan and future prospect (Murakami, Hitomi)V1567,544-555(1991)

Projection screens for high-definition television (Kirkpatrick, Michael D.; Mihalakis, George M.)V1456,40-47(1991)

Real-time nonlinear optical correlator in speckle metrology (Ogiwara, Akifumi; Ohtsubo, Junji)V1564,294-305(1991)

Single-fiber wideband transmission systems for HDTV production applications (Cheng, Xin; Levin, Paul A.; Nguyen, Hat)V1363,177-181(1991)

Static and dynamic spatial resolution in image coding: an investigation of eye movements (Stelmach, Lew B.; Tam, Wa J.; Hearty, Paul J.)V1453,147-152(1991)

Stereoscopic Displays and Applications II (Merritt, John O.; Fisher, Scott S., eds.)V1457(1991)

Stereoscopic video and the quest for virtual reality: an annotated bibliography of selected topics (Starks, Michael R.)V1457,327-342(1991)

Superhigh-definition image communication: an application perspective (Kohli, Jagdish C.)V1605,351-361(1991)

Surface contouring using TV holography (Atcha, Hashim; Tatam, Ralph P.; Buckberry, Clive H.; Davies, Jeremy C.; Jones, Julian D.)V1504,221-232(1991)

Use of flicker-free television products for stereoscopic display applications (Woods, Andrew J.; Docherty, Tom; Koch, Rolf)V1457,322-326(1991)

Use of high-resolution TV camera in photomechanics (Yatagai, Toyohiko; Ino, Tomomi)V1554B,646-649(1991)

Using liquid-crystal TVs in Vander Lugt optical correlators (Clark, Natalie; Crandall, Charles M.; Giles, Michael K.)V1564,439-451(1991)

Temperature—see also thermal effects

Accurate temperature measurement in thermography: an overview of relevant features, parameters, and definitions (Hamrelius, Torbjorn)V1467,448-457(1991)

Adaptive process control for a rapid thermal processor (Dilhac, Jean-Marie R.; Ganibal, Christian; Bordeneuve, J.; Dahhou, B.; Amat, L.; Picard, A.)V1393,395-403(1991)

All-fiber temperature sensing system by using polarization-maintaining fiber coupler as beamsplitter (Fang, Yin; Rong, Jian; Sheng, Kemin; Qin, Pingling)V1572,197-200(1991)

Analysis of temperature distribution and slip in rapid thermal processing (Lee, Hyouk; Yoo, Young-Don; Shin, Hyun-Dong; Earmme, Youn-Young; Kim, Choong-Ki)V1393,404-410(1991)

Application of fiber optic thermometry to the monitoring of winding temperatures in medium- and large-power transformers (Wickersheim, Kenneth A.)V1584,3-14(1991)

Automatic active athermalization of infrared optical systems (Kuerbitz, Gunther)V1540,612-621(1991)

Bare fiber temperature sensor (Soares, Edmundo A.; Dantas, Tarcisio M.)V1367,261-265(1991)

Basic principle and system research of self-emission fiber optic temperature sensor (Nie, Chao-Jiang; Yu, Jing-ming; Wu, Fang D.; Gao, Yu-ping)V1584,87-93(1991)

Blackbody radiators for field calibration (Cross, Edward F.; Mauritz, F. L.; Bixler, H. A.; Kaegi, E. M.; Wiemokly, Gary D.)V1540,756-763(1991)

Buried object remote detection technology for law enforcement (Del Grande, Nancy K.; Clark, Greg A.; Durbin, Philip F.; Fields, David J.; Hernandez, Jose E.; Sherwood, Robert J.)V1479,335-351(1991)

Buried-steam-line temperature and heat loss calculation (MacDavid, Jacob H.)V1467,11-17(1991)

Comparison of some algorithms commonly used in IR pyrometry: a computer simulation (Barani, Gianni; Tofani, Alessandro)V1467,458-468(1991)

Computer-aided moire strain analysis on thermoplastic models at critical temperature (Barillot, Marc; Jacquot, Pierre M.; Di Chirico, Giuseppe)V1554A,867-878(1991)

Cryoscatter measurements of beryllium (Lippey, Barret; Krone-Schmidt, Wilfried)V1530,150-161(1991)

Crystal growth by solute diffusion in Earth orbit (Lind, M. D.; Nielsen, K. F.)V1557,259-270(1991)

Dependence of optical properties of thermal-evaporated lead telluride films on substrate temperature (Feng, Weiting; Yen, Yi X.; Zhu, Cui Y.)V1519,333-338(1991)

Determination of the temperature of a single particle heated by a highly concentrated laser beam (Herve, Philippe; Bednarczyk, Sophie; Masclet, Philippe)V1487,387-395(1991)

Experimental method for gas kinetic temperature measurements in a thermal plasma (Reynolds, Larry D.; Shaw, C. B.)V1554B,622-631(1991)

Fiber Optic and Laser Sensors IX (DePaula, Ramon P.; Udd, Eric, eds.)V1584(1991)

Fiber optic based miniature high-temperature probe exploiting coherence-tuned signal recovery via multimode laser diode illumination (Gerges, Awad S.; Jackson, David A.)V1504,233-236(1991)

Fiber optic Fabry-Perot sensors for high-speed heat transfer measurements (Kidd, S. R.; Sinha, P. G.; Barton, James S.; Jones, Julian D.)V1504,180-190(1991)

Fiber optic high-temperature sensor for applications in iron and steel industries (Hao, Tianyou; Zhou, Feng-Shen; Xie, Xiou-Qioun; Hu, Ji-Wu; Wang, Wei-Yen)V1584,32-38(1991)

Fiber optic liquid crystalline microsensor for temperature measurement in high magnetic field (Domanski, Andrzej W.; Kostrzewa, Stanislaw)V1510,72-77(1991)

Fiber optic multiple sensor for simultaneous measurements of temperature and vibrations (Brenci, Massimo; Mencaglia, Andrea; Mignani, Anna G.; Barbero, V.; Cimbrico, P. L.; Pessino, P.)V1572,318-324(1991)

Fiber optic pressure and temperature sensor for down-hole applications (Lequime, Michael; Lecot, C.; Jouve, Philippe; Pouleau, J.)V1511,244-249(1991)

Fiber optic sensor applied to measure high temperature under high-pressure condition (Xiao, Wen; Chen, Yaosheng; Gao, Wei; Xue, Mingqiu)V1572,170-174(1991)

Fiber optic sensor for simultaneous measurement of strain and temperature (Vengsarkar, Ashish M.; Michie, W. C.; Jankovic, Lilja; Culshaw, Brian; Claus, Richard O.)V1367,249-260(1991)

Fiber optic sensor for the study of temperature and structural integrity of PZT: epoxy composite materials (Vishnoi, Gargi; Pillai, P.K. C.; Goel, T. C.)V1572,94-100(1991)

Fiber optic sensors for heat transfer studies (Farahi, Faramarz; Jones, Julian D.; Jackson, David A.)V1584,53-61(1991)

Fiber optic technique for simultaneous measurement of strain and temperature variations in composite materials (Michie, W. C.; Culshaw, Brian; Roberts, Scott S.; Davidson, Roger)V1588,342-355(1991)

Fiber optic temperature probe system for inner body (Liu, Bo; Deng, Xingzhong; Cao, Wei; Cheng, Xianping; Xie, Tuqiang; Zhong, Zugen)V1572,211-215(1991)

Fiber optic temperature sensor based on bend losses (Liu, Rui-Fu; Xi, Xiao-chun; Li, Wei-min; Tian, Da-chao)V1572,180-184(1991)

Fiber optic temperature sensor for aerospace applications (Jensen, Stephen C.; Tilstra, Shelle D.; Barnabo, Geoffrey A.; Thomas, David C.; Phillips, Richard W.)V1369,87-95(1991)

Fiber optic thermometer using Fourier transform spectroscopy (Beheim, Glenn; Sotomayor, Jorge L.; Flatico, Joseph M.; Azar, Massood T.)V1584,64-71(1991)

Fiber sensor design for turbine engines (Tobin, Kenneth W.; Beshears, David L.; Turley, W. D.; Lewis, Wilfred; Noel, Bruce W.)V1584,23-31(1991)

Field testing of a fiber optic rotor temperature monitor for power generators (Brown, Stewart K.; Mannik, Len)V1584,15-22(1991)

Free-surface temperature measurement of shock-loaded tin using ultrafast infrared pyrometry (Mondot, Michel; Remiot, Christian)V1558,351-361(1991)

Fundamental mechanisms and doping effects in silicon infrared absorption for temperature measurement by infrared transmission (Sturm, James C.; Reaves, Casper M.)V1393,309-315(1991)

Heat monitoring by fiber-optic microswitches (Mencaglia, Andrea; Brenci, Massimo; Falciai, Riccardo; Guzzi, D.; Pascucci, Tania)V1506,140-144(1991)

High-temperature optical sensor for displacement measurement (Ebbeni, Jean P.)V1504,268-272(1991)

High-temperature Raman spectra of nioboborate glass melts (Yang, Quanzu; Wang, Zhongcai; Wang, Shizhuo)V1513,264-269(1991)

HTC microbolometer for far-infrared detection (Barholm-Hansen, Claus; Levinsen, Mogens T.)V1512,218-225(1991)

Hybride fiber-optic temperature sensors on the base of LiNbO3 and LiNbO3:Ti waveguides (Goering, Rolf)V1511,275-280(1991)

Improvement of specular reflection pyrometer (Wen, Lin Ying; Hua, Yun)V1367,300-302(1991)

Influence of atmospheric conditions on two-color temperature discrimination (Mallory, William R.)V1540,365-369(1991)

Infrared instrumental complex for remote measurement of ocean surface temperature distribution (Miroshnikov, Mikhail M.; Minyeev, V. N.; Povarkov, V. I.; Samkov, V. M.; Solovyev, V. I.)V1540,496-505(1991)

Infrared simulation of missile dome heating (Rich, Brian W.)V1540,781-786(1991)

In-line Fabry-Perot interferometric temperature sensor with digital signal processing (Yeh, Yunhae; Lee, J. H.; Lee, Chung E.; Taylor, Henry F.)V1584,72-78(1991)

Integrated optics temperature sensor (d'Alessandro, Antonio; De Sario, Marco; D'Orazio, Antonella; Petruzzelli, Vincenzo)V1399,184-191(1991)

Interferometric technique for the concurrent determination of thermo-optic and thermal expansion coefficients (Jewell, John M.; Askins, Charles G.; Aggarwal, Ishwar D.)V1441,38-44(1991)

Laser beam deflection: a method to investigate convection in vapor growth experiments (Lenski, Harald; Braun, Michael)V1557,124-131(1991)

Laser Raman measurements of dielectric coatings as a function of temperature (Exarhos, Gregory J.; Hess, Nancy J.; Ryan, Samantha)V1441,190-199(1991)

Liquid-nitrogen-cooled low-noise radiation pulse detector amplifier (Trojnar, Eugeniusz; Trojanowski, Stanislaw; Czechowicz, Roman; Derwiszynski, Mariusz; Kocyba, Krzysztof)V1391,230-237(1991)

Low-temperature deformation measurement of materials by the use of laser speckle photography method (Nakahara, Sumio; Hisada, Shigeyoshi; Fujita, Takeyoshi; Sugihara, Kiyoshi)V1554A,602-609(1991)

Low-temperature viscosity measurements of infrared transmitting halide glasses (Seddon, Angela B.; Cardoso, A. V.)V1513,255-263(1991)

Measurement of the temperature field in confined jet impingement using phase-stepping video holography (Dobbins, B. N.; He, Shi P.; Jambunathan, K.; Kapasi, S.; Wang, Liu Sheng; Button, B. L.)V1554B,586-592(1991)

Microbend pressure sensor for high-temperature environments (Majercak, David; Sernas, Valentinas; Polymeropoulos, Constantine E.; Sigel, George H.)V1584,162-169(1991)

Multimode fiber-optic temperature sensor system based on dual-wavelength difference absorption principle (Zhang, Zaixuan; Lin, Dan; Fang, Xiao; Jing, Shangzhong)V1572,201-204(1991)

Near-infrared fiber optic temperature sensor (Schoen, Christian; Sharma, Shiv K.; Seki, Arthur; Angel, S. M.)V1584,79-86(1991)

New look at bump spectra models for temperature and refractive-index fluctuations (Sisterman, Elizabeth A.; Andrews, Larry C.)V1487,345-355(1991)

New modulation scheme for optical fiber point temperature sensor (Farahi, Faramarz)V1504,237-246(1991)

New, simple method of extracting temperature of liquid water from Raman scattering (Liu, Zhi-Shen; Ma, Jun; Zhang, Jin-Long; Chen, Wen-Zhong)V1558,306-316(1991)

Noncontacting acoustics-based temperature measurement techniques in rapid thermal processing (Lee, Yong J.; Chou, Ching-Hua; Khuri-Yakub, B. T.; Saraswat, Krishna C.)V1393,366-371(1991)

Noninvasive sensors for in-situ process monitoring and control in advanced microelectronics manufacturing (Moslehi, Mehrdad M.)V1393,280-294(1991)

Novel signal processing scheme for ruby-fluorescence-based fiber-optic temperature sensor (Zhang, Zhiyi; Grattan, Kenneth T.; Palmer, Andrew W.)V1511,264-274(1991)

Optical fiber sensor for temperature measurement from 600 to 1900 C in gas turbine engines (Tregay, George W.; Calabrese, Paul R.; Kaplin, Peter L.; Finney, Mark J.)V1589,38-47(1991)

Optically powered sensor system using conventional electrical sensors (Nieuwkoop, E.; Kapsenberg, Th.; Steenvoorden, G. K.; Bruinsma, Anastasius J.)V1511,255-263(1991)

Optically powered thermistor with optical fiber link (Shi, Jinshan; Wang, Yutian)V1572,175-179(1991)

Peak pressures and temperatures within laser-ablated tissues (Furzikov, Nickolay P.; Dmitriev, A. C.; Lekhtsier, Eugeny N.; Orlov, M. Y.; Semyenov, Alexander D.; Tyurin, Vladimir S.)V1427,288-297(1991)

Performance of the Multi-Spectral Solar Telescope Array V: temperature diagnostic response to the optically thin solar plasma (DeForest, Craig E.; Kankelborg, Charles C.; Allen, Maxwell J.; Paris, Elizabeth S.; Willis, Thomas D.; Lindblom, Joakim F.; O'Neal, Ray H.; Walker, Arthur B.; Barbee, Troy W.; Hoover, Richard B.; Barbee, Troy W.; Gluskin, Efim S.)V1343,404-414(1991)

Photochemical hole burning in rigidly coupled polyacenes (Iannone, Mark A.; Salt, Kimberly L.; Scott, Gary W.; Yamashita, Tomihiro)V1559,172-183(1991)

Photoelastic stress investigation in underground large hole in permafrost soil (statics, thermoelasticity, dynamics, photoelastic strain-gauges) (Savostjanov, V. N.; Dvalishvili, V. V.; Sakharov, V. N.; Isajkin, A. S.; Frishter, L.; Starchevsky, A. V.)V1554A,380-386(1991)

Projection moire deflectometry for mapping phase objects (Wang, Ming; Ma, Li; Zeng, Jing-gen; Cheng, Qi-Xian; Pan, Chuan K.)V1554B,242-246(1991)

Properties of liquid-nitrogen-cooled electronic elements (Trojnar, Eugeniusz; Trojanowski, Stanislaw; Czechowicz, Roman; Derwiszynski, Mariusz; Kocyba, Krzysztof)V1391,238-243(1991)

Pyrometer modeling for rapid thermal processing (Wood, Samuel C.; Apte, Pushkar P.; King, Tsu-Jae; Moslehi, Mehrdad M.; Saraswat, Krishna C.)V1393,337-348(1991)

Rapid Thermal and Related Processing Techniques (Moslehi, Mehrdad M.; Singh, Rajendra, eds.)V1393(1991)

Real-time temperature measurement on PCB:s, hybrids, and microchips (Wallin, Bo)V1467,180-187(1991)

Reflective fiber temperature sensor using a bimetallic transducer (Liu, Rui-Fu; Jiang, G.; Jiang, H. Y.; Xi, Xiao-chun; Song, W. S.)V1572,189-191(1991)

Refractive properties of TGS aqueous solution for two-color interferometry (Vikram, Chandra S.; Witherow, William K.; Trolinger, James D.)V1557,197-201(1991)

Remote temperature sensing of a pulsed thermionic cathode (Del Grande, J. M.)V1467,427-437(1991)

Research of optical fiber pyrometer sensor for gas-making furnace (Deng, Xingzhong; Zhong, Zugen; Cheng, Xianping; Liu, Bo; Cao, Wei)V1572,220-223(1991)

Research of signal transmission and processing of fiber-optic pyrometer (Cheng, Xianping; Deng, Xingzhong; Xie, Tuqiang; Cao, Wei; Liu, Bo)V1572,216-219(1991)

Research on the temperature of thin film under ion beam bombarding (Zhao, Yun F.; Sun, Zhu Z.; Pang, Shi J.)V1519,411-414(1991)

Review of temperature measurements in the semiconductor industry (Anderson, Richard L.)V1392,437-451(1991)

Round-robin comparison of temperature nonuniformity during RTP due to patterned layers (Vandenabeele, Peter; Maex, Karen)V1393,372-394(1991)

RTP temperature uniformity mapping (Keenan, W. A.; Johnson, Walter H.; Hodul, David T.; Mordo, David)V1393,354-365(1991)

Sensitivity of polarization-maintaining fibers to temperature variations (Ruffin, Paul B.; Sung, C. C.)V1478,160-167(1991)

Silicon calorimeter for high-power microwave measurements (Lazard, Carl J.; Pereira, Nino R.; Huttlin, George A.; Litz, Marc S.)V1407,167-171(1991)

Strain sensing using a fiber-optic Bragg grating (Melle, Serge M.; Liu, Kexing; Measures, Raymond M.)V1588,255-263(1991)

Study on pyrometer with double Y-type optical fibers (Li, Zhiquan; Shi, Jinshan; Wang, Yutian)V1572,185-188(1991)

Surface temperature and shear stress measurement using liquid crystals (Toy, Norman; Savory, Eric; Disimile, Peter J.)V1489,112-123(1991)

Surveying and damping heat loss from machines with high surface temperatures: thermography as a tool (Perch-Nielsen, Thomas; Paulsen, Otto; Drivsholm, Christian)V1467,169-179(1991)

Systems-oriented survey of noncontact temperature measurement techniques for rapid thermal processing (Peyton, David; Kinoshita, Hiroyuki; Lo, G. Q.; Kwong, Dim-Lee)V1393,295-308(1991)

Temperature and strain sensing using monomode optical fiber (Farahi, Faramarz; Jackson, David A.)V1511,234-243(1991)

Temperature compensation of a highly birefringent optical fiber current sensor (McStay, Daniel; Chu, W. W.; Rogers, Alan J.)V1584,118-123(1991)

Temperature control of the 3.5-meter WIYN telescope primary mirror (Goble, Larry W.)V1532,161-169(1991)

Temperature dependence of resistance of diamond film synthesized by microwave plasma CVD (Yang, Bang C.; Gou, Li; Jia, Yu M.; Ran, Jun G.; Zheng, Chang Q.; Tang, Xia)V1519,864-865(1991)

Temperature instability in silicon-based microheating device (Shie, Jin-Shown; Lian, Jiunn-Long)V1362,655-663(1991)

Temperature rise due to dynamic crack growth in Beta-C titanium (Zehnder, Alan T.; Kallivayalil, Jacob A.)V1554A,48-59(1991)

Thermal sensing of fireball plumes (Toossi, Reza)V1467,384-393(1991)

Thermosense XIII (Baird, George S., ed.)V1467(1991)

Time characteristics in HTS rare-earth-doped optical fiber at high temperature (Yang, Yang; Li, Naiji; Wang, Huiwen)V1572,205-210(1991)

Two-colour ratio pyrometer with optical fiber (Wang, Yutian; Shi, Jinshan; Li, Zhiquan)V1572,192-196(1991)

Two-mode elliptical-core fiber sensors for measurement of strain and temperature (Wang, Anbo; Wang, Zhiguang; Vengsarkar, Ashish M.; Claus, Richard O.)V1584,294-303(1991)

Universal equation for IR thermometer and its applications (Liu, Jian; Bao, Xue-Cheng; Zhang, Cai-Gen; Zhang, You-Wen)V1540,744-755(1991)

UV optical fiber distributed temperature sensor (Paton, Andrew T.; Scott, Chris J.)V1367,274-281(1991)

Variation of optical properties of gel-derived VO2 thin films with temperature (Hou, Li S.; Lu, Song W.; Gan, Fuxi)V1519,580-588(1991)

Wavelength and temperature dependence of bending loss in monomode optical fibers (Morgan, Russell D.; Jones, Julian D.; Harper, Philip G.; Barton, James S.)V1504,118-124(1991)

Thermal blooming—see also atmospheric optics

Analytical Raman and atomic spectroscopies using charge-coupled-device detection (Bilhorn, Robert B.; Ferris, Nancy S.)V1439,15-24(1991)

Effects of thin and subvisible cirrus on HEL far-field intensity calculations at various wavelengths (Harada, Larrene K.)V1408,28-40(1991)

Functional reconstruction predictions of uplink whole beam Strehl ratios in the presence of thermal blooming (Enguehard, S.; Hatfield, Brian)V1408,186-191(1991)

GRAND: a 4-D wave optics code for atmospheric laser propagation (Mehta, Naresh C.)V1487,398-409(1991)

Modeling turbulent transport in laser beam propagation (Wallace, James)V1408,19-27(1991)

Path integral approach to thermal blooming (Enguehard, S.; Hatfield, Brian)V1408,178-185(1991)

Realistic wind effects on turbulence and thermal blooming compensation (Long, Jerry E.; Hills, Louis S.; Gebhardt, Frederick G.)V1408,58-71(1991)

Review of the physics of small-scale thermal blooming in uplink propagation (Enguehard, S.; Hatfield, Brian)V1415,128-137(1991)

Stability analysis of semidiscrete schemes for thermal blooming computation (Ulrich, Peter B.)V1408,192-202(1991)

Thermal blooming critical power and adaptive optics correction for the ground-based laser (Smith, David C.; Townsend, Sallie S.)V1408,112-118(1991)

Use of adaptive optics for minimizing atmospheric distortion of optical waves (Lukin, Vladimir P.)V1408,86-95(1991)

Variable wind direction effects on thermal blooming correction (Hills, Louis S.; Long, Jerry E.; Gebhardt, Frederick G.)V1408,41-57(1991)

Thermal effects—see also temperature

Adaptive process control for a rapid thermal processor (Dilhac, Jean-Marie R.; Ganibal, Christian; Bordeneuve, J.; Dahhou, B.; Amat, L.; Picard, A.)V1393,395-403(1991)

All-silicon Fabry-Perot modulator based on thermo-optic effect (Rendina, Ivo; Cocorullo, Giuseppe)V1583,338-343(1991)

Analysis of kinetic rate modeling of thermal damage in laser-irradiated tissue (Rastegar, Sohi; Glenn, T.)V1427,300-306(1991)

Analysis of molecular adsorbates by laser-induced thermal desorption (McIver, Robert T.; Hemminger, John C.; Parker, D.; Li, Y.; Land, Donald P.; Pettiette-Hall, C. L.)V1437,124-128(1991)

Analysis of temperature distribution and slip in rapid thermal processing (Lee, Hyouk; Yoo, Young-Don; Shin, Hyun-Dong; Earmme, Youn-Young; Kim, Choong-Ki)V1393,404-410(1991)

Analysis of thermal stability of fused optical structure (Powell, William R.)V1532,126-136(1991)

Anisotropic local melting of semiconductors under light pulse irradiation (Fattakhov, Yakh'ya V.; Khaibullin, Ildus B.; Bayazitov, Rustem M.)V1440,16-23(1991)

Anomalous diffusion phenomena in two-step rapid thermal diffusion of phosphorus (Cho, Byung-Jin; Kim, Choong-Ki)V1393,180-191(1991)

Application issues of fiber-optic sensors in aircraft structures (Lu, Zhuo J.; Blaha, Franz A.)V1588,276-281(1991)

Athermalization of IR optical systems (Rogers, Philip J.)VCR38,69-94(1991)

Beam quality measurements and improvement in solar-pumped laser systems (Bernstein, Hana; Thompson, George A.; Yogev, Amnon; Oron, Moshe)V1442,81-88(1991)

Changes in collagen birefringence: a quantitative histologic marker of thermal damage in skin (Thomsen, Sharon L.; Cheong, Wai-Fung; Pearce, John A.)V1422,34-42(1991)

Characteristic features of melting surface of tin produced with laser pulse (Rozniakowski, Kazimierz)V1391,215-223(1991)

Comparison of carbon monoxide and carbon dioxide laser-tissue interaction (Waters, Ruth A.; Thomas, J. M.; Clement, R. M.; Ledger, N. R.)V1427,336-343(1991)

Control of oxygen incorporation and lifetime measurement in Si1-xGex epitaxial films grown by rapid thermal chemical vapor deposition (Sturm, James C.; Schwartz, P. V.; Prinz, Erwin J.; Magee, Charles W.)V1393,252-259(1991)

Control of thermally induced porosity for the fabrication of beryllium optics (Moreen, Harry A.)V1485,54-61(1991)

Cooperative implementation of a high-temperature acoustic sensor (Baldini, S. E.; Nowakowski, Edward; Smith, Herb G.; Friebele, E. J.; Putnam, Martin A.; Rogowski, Robert S.; Melvin, Leland D.; Claus, Richard O.; Tran, Tuan A.; Holben, Milford S.)V1588,125-131(1991)

Critical dimension shift resulting from handling time variation in the track coat process (Kulp, John M.)V1466,630-640(1991)

Cryogenic refractive indices of cadmium telluride coatings in wavelength range from 2.5 to 20 um (Feng, Weiting; Yen, Yi X.; Zhu, Cui Y.)V1535,224-230(1991)

Cryosurgical ablation of the prostate (Cohen, Jeffrey K.)V1421,45-45(1991)

Damage induced by pulsed IR laser radiation at transitions between different tissues (Frenz, Martin; Greber, Charlotte M.; Romano, Valerio; Forrer, Martin; Weber, Heinz P.)V1427,9-15(1991)

Design and development of a transparent Bridgman furnace (Wells, Mark E.; Groff, Mary B.)V1557,71-77(1991)

Design of an athermalized three-fields-of-view infrared sensor (Wickholm, David R.)V1488,58-63(1991)

Determination of thermal stresses in a bimaterial specimen by moire interferometry (Kang, Yilan; Jia, Youquan; Du, Ji)V1554A,514-522(1991)

Dissolution inhibition mechanism of ANR photoresists: crosslinking vs. -OH site consumption (Thackeray, James W.; Orsula, George W.; Rajaratnam, Martha M.; Sinta, Roger F.; Herr, Daniel J.; Pavelchek, Edward K.)V1466,39-52(1991)

Dissolution kinetics of high-resolution novolac resists (Itoh, Katsuyuki; Yamanaka, Koji; Nozue, Hiroshi; Kasama, Kunihiko)V1466,485-496(1991)

Effect of coagulation on laser light distribution in myocardial tissue (Agah, Ramtin; Sheth, Devang; Motamedi, Massoud E.)V1425,172-179(1991)

Effect of sensitizer spatial distribution on dissolution inhibition in novolak/diazonaphthoquinone resists (Rao, Veena; Kosbar, Laura L.; Frank, Curtis W.; Pease, R. F.)V1466,309-323(1991)

Effect of the space environment on thermal control coatings (Harada, Yoshiro; Mell, Richard J.; Wilkes, Donald R.)V1330,90-101(1991)

Effects of film thickness on the thermoelectric behavior of pyrolytic ZnO thin film (Ambia, M. G.; Islam, M. N.; Hakim, M. O.)V1536,222-232(1991)

Effects of pressure rise on cw laser ablation of tissue (LeCarpentier, Gerald L.; Motamedi, Massoud E.; Welch, Ashley J.)V1427,273-278(1991)

Efficacy of argon-laser-mediated hot-balloon angioplasty (Sakurada, Masami; Miyamoto, Akira; Mizuno, Kyoichi; Nozaki, Youichi; Tabata, Hirotsugu; Etsuda, Hirokuni; Kurita, Akira; Nakamura, Haruo; Arai, Tsunenori; Suda, Akira; Kikuchi, Makoto; Watanabe, Tamishige; Utsumi, Atsushi; Akai, Yoshiro; Takeuchi, Kiyoshi)V1425,158-164(1991)

Electron microscopic study on black pig skin irradiated with pulsed dye laser (504 nm) (Yasuda, Yukio; Tan, Oon T.; Kurban, Amal K.; Tsukada, Sadao)V1422,50-55(1991)

Emissivity of silicon wafers during rapid thermal processing (Vandenabeele, Peter; Maex, Karen)V1393,316-336(1991)

Endoscopic YAG laser coagulation for early prostate cancer (McNicholas, Thomas A.; O'Donoghue, Neil)V1421,56-67(1991)

Epitaxial regrowth of silicon on sapphire by rapid isothermal processing (Madarazo, R.; Pedrine, A. G.; Sol, A. A.; Baranauskas, Vitor)V1393,270-277(1991)

Experimental and analytical studies on fixed mask assembly for APS with enhanced cooling (Kuzay, Tuncer M.; Collins, Jeffrey T.; Khounsary, Ali M.; Viccaro, P. J.)V1345,55-70(1991)

Fiber optic based chemical sensor system for in-situ process measurements using the photothermal effect (Walker, Karl-Heinz; Sontag, Heinz)V1510,212-217(1991)

Finite element analysis enhancement of cryogenic testing (Thiem, Clare D.; Norton, Douglas A.)V1532,39-47(1991)

Formation of optical elements by photothermo-induced crystallization of glass (Glebov, Leonid B.; Nikonorov, Nikolai V.; Petrovskii, Gurii T.; Kharchenko, Mikhail V.)V1440,24-35(1991)

Fundamental mechanisms and doping effects in silicon infrared absorption for temperature measurement by infrared transmission (Sturm, James C.; Reaves, Casper M.)V1393,309-315(1991)

Heat capacity of MOS field-effect devices of optical materials in the presence of a strong magnetic field (Ghatak, Kamakhya P.; Biswas, Shambhu N.)V1485,206-214(1991)

High-density packaging and interconnect of massively parallel image processors (Carson, John C.; Indin, Ronald)V1541,232-239(1991)

High-dose boron implantation and RTP anneal of polysilicon films for shallow junction diffusion sources and interconnects (Raicu, Bruha; Keenan, W. A.; Current, Michael I.; Mordo, David; Brennan, Roger)V1393,161-171(1991)

High-emittance surfaces for high-temperature space radiator applications (Banks, Bruce A.; Rutledge, Sharon K.; Hotes, Deborah)V1330,66-77(1991)

High-performance, wide-magnification-range IR zoom telescope with automatic compensation for temperature effects (Shechterman, Mark)V1442,276-285(1991)

High-temperature degradation-free rapid thermal annealing of GaAs and InP (Pearton, Stephen J.; Katz, Avishay; Geva, M.)V1393,150-160(1991)

Histopathologic assessment of water-dominated photothermal effects produced with laser irradiation (Thomsen, Sharon L.; Cheong, Wai-Fung; Pearce, John A.)V1422,14-18(1991)

Hyperthermia treatment using a computer-controlled Nd:YAG laser system in combination with surface cooling (Panjehpour, Masoud; Wilke, August; Frazier, Donita L.; Overholt, Bergein F.)V1427,307-315(1991)

Influence of laser irradiation on the kinetics of oxide layer growth and their optical characteristics (Uglov, A.; Krivonogov, Yu.)V1440,310-320(1991)

Influence of the laser-induced temperature rise in photodynamic therapy (Gottschalk, Wolfgang; Hengst, Joachim; Sroka, Ronald; Unsoeld, Eberhard)V1427,320-326(1991)

Infrared tissue ablation: consequences of liquefaction (Zweig, A. D.)V1427,2-8(1991)

In-situ interferometric measurements in a rapid thermal processor (Dilhac, Jean-Marie R.; Ganibal, Christian; Nolhier, N.; Amat, L.)V1393,349-353(1991)

Integrated rapid isothermal processing (Singh, Rajendra; Sinha, Sanjai; Thakur, Randhir P.; Hsu, N. J.)V1393,78-89(1991)

Interface demarcation in Bridgman-Stockbarger crystal growth of II-VI compounds (Gillies, Donald C.; Lehoczky, S. L.; Szofran, Frank R.; Su, Ching-Hua; Larson, David J.)V1484,2-10(1991)

Interference measurement of thermal effect induced by optical pumping on YAG crystal rod (Luo, Bikai; Ni, Xiao W.; Zhang, Qi; He, Anzhi)V1554A,542-546(1991)

Interstitial laser coagulation of the prostate: experimental studies (McNicholas, Thomas A.; Steger, Adrian C.; Bown, Stephen G.; O'Donoghue, Neil)V1421,30-35(1991)

Investigation of rapid thermal process-induced defects in ion-implanted Czochralski silicon (Yarling, Charles B.; Hahn, Sookap; Hodul, David T.; Suga, Hisaaki; Smith, W. L.)V1393,192-199(1991)

Ion-exchange strengthening of high-average-power phosphate laser glass (Lee, Huai-Chuan; Meissner, Helmuth E.)V1441,87-103(1991)

Kinetic models for coagulation processes: determination of rate coefficients in vivo (Pearce, John A.; Cheong, Wai-Fung; Pandit, Kirit; McMurray, Tom J.; Thomsen, Sharon L.)V1422,27-33(1991)

Laser neural tissue interactions using bilayer membrane models (VanderMeulen, David L.; Khoka, Mustafa; Spears, Kenneth G.)V1428,84-90(1991)

Laser-stimulated thermoelastic stress wave inside a solid medium and measurement of its initial speed (He, Anzhi; Ni, Xiao W.; Lu, Jian-Feng; Wang, Chang Xing; Wang, Hong Y.)V1554A,547-552(1991)

Liquid as a deformed crystal: the model of a liquid structure (Yakovlev, Evgeni B.)V1440,36-49(1991)

Local hyperthermia for the treatment of diseases of the prostate (Servadio, Ciro)V1421,6-11(1991)

Low-resistivity contacts to silicon using selective RTCVD of germanium (Grider, Douglas T.; Ozturk, Mehmet C.; Wortman, Jim J.; Littlejohn, Michael A.; Zhong, Y.)V1393,229-239(1991)

Low-temperature in-situ dry cleaning process for epitaxial layer multiprocessing (Moslehi, Mehrdad M.)V1393,90-108(1991)

Measurement of temperature distributions after pulsed IR radiation impact in biological tissue models with fluorescent thin films (Romano, Valerio; Greber, Charlotte M.; Frenz, Martin; Forrer, Martin; Weber, Heinz P.)V1427,16-26(1991)

Measurement of thermal deformation of square plate by using holographic interferometry (Kang, Dae I.; Kwon, Young H.; Ko, Yeong-Uk)V1554B,139-147(1991)

Measuring device for space-temporal characteristics of technological lasers radiation (Mnatsakanyan, Eduard A.; Andreev, V. G.; Bestalanny, S. I.; Velikoselsky, V.V.; Grinyakin, A.P.; Grinyakin, V.P.; Kolesnikov, S.A.; Rozhnov, U.V.)V1414,130-133(1991)

Mechanical and thermal disturbances of the PSR moderate focus-mission structure (Shih, Choon-Foo; Lou, Michael C.)V1532,81-90(1991)

Melt dynamics and surface deformation in processing with laser radiation (Kreutz, Ernst W.; Pirch, Norbert)V1502,160-176(1991)

Microwave warming of biological tissue and its control by IR fiber thermometry (Drizlikh, S.; Zur, A.; Moser, Frank; Katzir, Abraham)V1420,53-62(1991)

Monitoring of tissue temperature during microwave hyperthermia utilizing a fiber optic liquid crystalline microsensor (Domanski, Andrzej W.; Kostrzewa, Stanislaw; Hliniak, Andrzej)V1420,72-80(1991)

Multichamber rapid thermal processing (Rosser, Paul J.; Moynagh, P.; Affolter, K. B.)V1393,49-66(1991)

Multiple photo-assisted CVD of thin-film materials for III-V device technology (Nissim, Yves I.; Moison, Jean M.; Houzay, Francoise; Lebland, F.; Licoppe, C.; Bensoussan, M.)V1393,216-228(1991)

Negative resist systems using acid-catalyzed pinacol rearrangement reaction in a phenolic resin matrix (Uchino, Shou-ichi; Iwayanagi, Takao; Ueno, Takumi; Hayashi, Nobuaki)V1466,429-435(1991)

New approaches to signal-to-noise ratio optimization in background-limited photothermal measurements (Rice, Patrick D.; Thorne, John B.; Bobbitt, Donald R.)V1435,104-113(1991)

New thermistor material for thermistor bolometer: material preparation and characterization (Umadevi, P.; Nagendra, C. L.; Thutupalli, G. K.; Mahadevan, K.; Yadgiri, G.)V1484,125-135(1991); V1485,195-205(1991)

Next generation thermal control coatings (Grieser, James L.; Swisher, Richard L.; Phipps, James A.; Pelleymounter, Douglas R.; Hildreth, Eugene N.)V1330,111-118(1991)

Noncontacting acoustics-based temperature measurement techniques in rapid thermal processing (Lee, Yong J.; Chou, Ching-Hua; Khuri-Yakub, B. T.; Saraswat, Krishna C.)V1393,366-371(1991)

Optical analysis of thermal-induced structural distortions (Weinswig, Shepard A.; Hookman, Robert A.)V1527,118-125(1991)

Optical coatings to reduce temperature sensitivity of polarization-maintaining fibers for smart structures and skins (Zhang, Feng; Lit, John W.)V1588,100-109(1991)

Optical Radiation Interaction with Matter (Bonch-Bruevich, Aleksei M.; Konov, Vitaly I.; Libenson, Michail N., eds.)V1440(1991)

Peculiarities of metal surface heating by repetitively pulsed CO2 laser radiation with the blowing (Anisimov, N. R.; Buzykin, O. G.; Zayakin, A. A.; Makarov, V. V.; Makashev, N. K.; Nosachev, L. V.; Frolov, I. P.)V1440,206-210(1991)

Performance and reliability of ultrathin reoxidized nitrided oxides fabricated by rapid thermal processing (Joshi, A. B.; Lo, G. Q.; Shih, Dennis K.; Kwong, Dim-Lee)V1393,122-149(1991)

Performance of milliKelvin Si bolometers as x-ray and exotic particle detectors (Zammit, C. C.; Sumner, Timothy J.; Lea, M. J.; Fozooni, P.; Hepburn, I. D.)V1549,274-282(1991)

Photoablation using the Holmium:YAG laser: a laboratory and clinical study (Waidhauser, Erich; Markmiller, U.; Enders, S.; Hessel, Stefan F.; Ulrich, Frank; Beck, Oskar J.; Feld, Michael S.)V1428,75-83(1991)

Photoelastic transducer for high-temperature applications (Redner, Alex S.; Adamovsky, Grigory; Wesson, Laurence N.)V1332,775-782(1991)

Photothermal spectroscopy as a sensitive spectroscopic tool (Tam, Andrew C.)V1435,114-127(1991)

Physical aging of resists: the continual evolution of lithographic material (Paniez, Patrick J.; Weill, Andre P.; Cohendoz, Stephane D.)V1466,336-344(1991)

Positron emission tomography of laser-induced interstitial hyperthermia in cerebral gliomas (Ulrich, Frank; Bettag, Martin; Langen, K. J.)V1428,135-135(1991)

Practical and theoretical considerations on the use of quantitative histology to measure thermal damage of collagen in cardiovascular tissues (Thomsen, Sharon L.)V1425,110-113(1991)

Preliminary evaluation of collagen as a component in the thermally induced 'weld' (Lemole, G. M.; Anderson, R. R.; DeCoste, Sue)V1422,116-122(1991)

Process latitude for the chemical amplification resists AZ PF514 and AZ PN114 (Eckes, Charlotte; Pawlowski, Georg; Przybilla, Klaus J.; Meier, Winfried; Madore, Michel; Dammel, Ralph)V1466,394-407(1991)

Pulsed CO2 laser-material interaction: mechanical coupling and reflected and scattered radiation (Prat, Ch.; Autric, Michel L.; Inglesakis, Georges; Astic, Dominique)V1397,701-704(1991)

Pulsed photothermal spectroscopy applied to lanthanide and actinide speciation (Berg, John M.; Morris, David E.; Clark, David L.; Tait, C. D.; Woodruff, William H.; Van Der Sluys, William G.)V1435,331-337(1991)

Pyrometer modeling for rapid thermal processing (Wood, Samuel C.; Apte, Pushkar P.; King, Tsu-Jae; Moslehi, Mehrdad M.; Saraswat, Krishna C.)V1393,337-348(1991)

Quantitative analysis at the molecular level of laser/neural tissue interactions using a liposome model system (VanderMeulen, David L.; Misra, Prabhakar; Michael, Jason; Spears, Kenneth G.; Khoka, Mustafa)V1428,91-98(1991)

Radiometric versus thermometric calibration of IR test systems: which is best? (Richardson, Philip I.)V1488,80-88(1991)

Rapid isothermal process technology for optoelectronic applications (Singh, Rajendra)V1418,203-216(1991)

Rapid Thermal and Related Processing Techniques (Moslehi, Mehrdad M.; Singh, Rajendra, eds.)V1393(1991)

Rapid thermal annealing of the through-Ta5Si3 film implantation on GaAs (Huang, Fon-Shan; Chen, W. S.; Hsu, Tzu-min)V1393,172-179(1991)

Rapid thermal processing induced defects and gettering effects in silicon (Hartiti, Bouchaib; Muller, Jean-Claude; Siffert, Paul; Vu, Thuong-Quat)V1393,200-206(1991)

Rapid thermal processing in the manufacturing technology of contacts to InP-based photonic devices (Katz, Avishay)V1393,67-77(1991)

Reduced thermal budget processing of high-Tc superconducting thin films and related materials by MOCVD (Sinha, Sanjai; Singh, Rajendra; Hsu, N. J.; Ng, J. T.; Chou, P.; Narayan, Jagdish)V1394,266-276(1991)

Removal of small particles from surfaces by pulsed laser irradiation: observations and a mechanism (Kelley, J. D.; Stuff, Michael I.; Hovis, Floyd E.; Linford, Gary J.)V1415,211-219(1991)

Review of measurement systems for evaluating thermal expansion homogeneity of Corning Code 7971 ULETM (Hagy, Henry E.)V1533,198-211(1991)

Rigid lightweight optical bench for a spaceborne FUV spatial heterodyne interferometer (Tom, James L.; Cotton, Daniel M.; Bush, Brett C.; Chung, Ray; Chakrabarti, Supriya)V1549,302-307(1991)

Round-robin comparison of temperature nonuniformity during RTP due to patterned layers (Vandenabeele, Peter; Maex, Karen)V1393,372-394(1991)

RTP-induced defects in silicon studied by positron annihilation technique (Kulkarni, N. M.; Kulkarni, R. N.; Shaligram, Arvind D.)V1393,207-214(1991)

RTP temperature uniformity mapping (Keenan, W. A.; Johnson, Walter H.; Hodul, David T.; Mordo, David)V1393,354-365(1991)

Sapphire fiber interferometer for microdisplacement measurements at high temperatures (Murphy, Kent A.; Fogg, Brian R.; Wang, George Z.; Vengsarkar, Ashish M.; Claus, Richard O.)V1588,117-124(1991)

Scattering and thermal emission from spatially inhomogeneous atmospheric rain and cloud (Jin, Ya-Qiu)V1487,324-332(1991)

Selective deposition of polycrystalline SixGe1-x by rapid thermal processing (Ozturk, Mehmet C.; Zhong, Y.; Grider, Douglas T.; Sanganeria, M.; Wortman, Jim J.; Littlejohn, Michael A.)V1393,260-269(1991)

Si-based epitaxial growth by rapid thermal processing chemical vapor deposition (Jung, K. H.; Hsieh, T. Y.; Kwong, Dim-Lee; Spratt, D. B.)V1393,240-251(1991)

Simulations of intracavity laser heating of particles (Linford, Gary J.)V1415,196-210(1991)

Single-wafer integrated processing as a manufacturing tool using rapid thermal chemical vapor deposition technology (Kermani, Ahmad)V1393,109-119(1991)

Slow thermodeformation of metals with fast laser heating (Baloshin, Yu. A.; Yurevich, V. I.; Sud'enkov, Yu. V.)V1440,71-77(1991)

Some aspects of the application of the laser flash method for investigations on the thermal diffusivity of porous materials (Drobnik, Antoni; Rozniakowski, Kazimierz; Wojtatowicz, Tomasz W.)V1391,361-369(1991)

Sound generation by thermocavitation-induced cw laser in solutions (Rastopov, S. F.; Sukhodolsky, A. T.)V1440,127-134(1991)

Stress intensity factors of edge-cracked semi-infinite plates under transient thermal loading (Wang, Wei-Chung; Chen, Tsai-Lin; Hwang, Chi-Hung)V1554A,124-135(1991)

Superheating phenomena in absorbing microdroplets irradiated by pulsed lasers (Armstrong, Robert L.)V1497,132-140(1991)

Surface distortions of a 3.5-meter mirror subjected to thermal variations (Cho, Myung K.; Poczulp, Gary A.)V1532,137-145(1991)

Systems-oriented survey of noncontact temperature measurement techniques for rapid thermal processing (Peyton, David; Kinoshita, Hiroyuki; Lo, G. Q.; Kwong, Dim-Lee)V1393,295-308(1991)

Technology and chemistry of high-temperature positive resist (Toukhy, Medhat A.; Sarubbi, Thomas R.; Brzozowy, David J.)V1466,497-507(1991)

Temperature distributions in laser-irradiated tissues (Valderrama, Giuseppe L.; Fredin, Leif G.; Berry, Michael J.; Dempsey, B. P.; Harpole, George M.)V1427,200-213(1991)

Temperature increase during in vitro 308 nm excimer laser ablation of porcine aortic tissue (Gijsbers, Geert H.; Sprangers, Rene L.; van den Broecke, Duco G.; van Wieringen, Niek; Brugmans, Marco J.; van Gemert, Martin J.)V1425,94-101(1991)

Temperature instability in silicon-based microheating device (Shie, Jin-Shown; Lian, Jiunn-Long)V1362,655-663(1991)

Temperature response of biological tissues to nonablative pulsed CO2 laser irradiation (van Gemert, Martin J.; Brugmans, Marco J.; Gijsbers, Geert H.; Kemper, J.; van der Meulen, F. W.; Nijdam, D. C.)V1427,316-319(1991)

Temperature variations of reflection, transmission, and fluorescence of the arterial wall (Chambettaz, Francois; Clivaz, Xavier; Marquis-Weible, Fabienne D.; Salathe, R. P.)V1427,134-140(1991)

Thermal analysis of a small expendable tether satellite package (Randorf, Jeffrey A.)V1495,259-267(1991)

Thermal and structural analysis of the GOES scan mirror's on-orbit performance (Zurmehly, George E.; Hookman, Robert A.)V1532,170-176(1991)

Thermal effects in diode-laser-pumped monolithic Nd:glass lasers (Schmitt, Randal L.; Spence, Paul A.; Scerbak, David G.)V1410,55-64(1991)

Thermal lensing and frequency chirp in a heated CdTe modulator crystal and its effects on laser radar performance (Eng, Richard S.; Kachelmyer, Alan L.; Harris, Neville W.)V1416,70-85(1991)

Thermal lensing effect in fast-axial flow CO2 lasers (Moissl, M.; Paul, R.; Breining, K.; Giesen, Adolf; Huegel, Helmut)V1397,395-398(1991)

Thermal plastic metal coatings on optical fiber sensors (Sirkis, James S.; Dasgupta, Abhijit)V1588,88-99(1991)

Thermal process of laser-induced damage in optical thin films (Fan, Zheng X.; Wu, Zou L.; Shi, Zeng R.)V1519,359-364(1991)

Thermal regulation applied to CO2 laser self-quenching of complex geometry workpieces (Bataille, F.; Kechemair, Didier; Pawlovski, C.; Houdjal, R.)V1397,839-842(1991)

Thermal relaxation of tellurium-halide-based glasses (Ma, Hong-Li; Zhang, Xhang H.; Lucas, Jacques)V1590,146-151(1991)

Thermal stabilities of a-Si:H films and its application to thyristor elements (Sun, Yue Z.; Chen, Chun X.; Xie, Qi Y.; Yin, Chen Z.; He, Yu L.)V1519,234-240(1991)

Thermal stability and microstructure study of WSi0.6/GaAs by XRD and TEM (Zhang, Shu Y.; Tan, Shun; Wang, Chang S.; Zhao, Te X.)V1519,43-46(1991)

Thermal strain measurements of solder joints in electronic packaging using moire interferometry (Woychik, Charls G.; Guo, Yi F.)V1554B,461-470(1991)

Thermal transport properties of optical thin films (Swimm, Randall T.)V1441,45-55(1991)

Thermocapillary mechanism of laser pulse alloying of a metal surface layer (Kostrubiec, Franciszek)V1391,224-227(1991)

Time development of AlGaAs single-quantum-well laser facet temperature on route to catastrophical breakdown (Tang, Wade C.; Rosen, Hal J.; Vettiger, Peter; Webb, David J.)V1418,338-342(1991)

Transurethral microwave heating without urethral cooling: theory and experimental results (Petrovich, Zbigniew; Astrahan, Melvin; Baert, Luc)V1421,14-17(1991)

Transurethral microwave hyperthermia for benign prostatic hyperplasia: the Leuven clinical experience (Baert, Luc; Ameye, Filip; Willemen, Patrick; Petrovich, Zbigniew)V1421,18-29(1991)

Uniformity characterization of rapid thermal processor thin films (Yarling, Charles B.; Cook, Dawn M.)V1393,411-420(1991)

Velocity measurements in molten pools during high-power laser interaction with metals (Caillibotte, Georges; Kechemair, Didier; Sabatier, Lilian)V1412,209-211(1991)

Wavelength selection in laser arthroscopy (Black, Johnathan D.; Sherk, Henry H.; Meller, Menachem M.; Uppal, Gurvinder S.; Divan, James; Sazy, John; Rhodes, Anthony; Lane, Gregory J.)V1424,12-15(1991)

Thermal imaging—see also imaging systems; infrared

640 x 480 MOS PtSi IR sensor (Sauer, Donald J.; Shallcross, Frank V.; Hsueh, Fu-Lung; Meray, G. M.; Levine, Peter A.; Gilmartin, Harvey R.; Villani, Thomas S.; Esposito, Benjamin J.; Tower, John R.)V1540,296-296(1991)

Ablation of hard dental tissues with an ArF-pulsed excimer laser (Neev, Joseph; Raney, Daniel; Whalen, William E.; Fujishige, Jack T.; Ho, Peter D.; McGrann, John V.; Berns, Michael W.)V1427,162-172(1991)

Accurate temperature measurement in thermography: an overview of relevant features, parameters, and definitions (Hamrelius, Torbjorn)V1467,448-457(1991)

Advanced real-time scanning concept for full dynamics recording, high image quality, and superior measurement accuracy (Lindstrom, Kjell M.; Wallin, Bo)V1488,389-398(1991)

Airborne infrared and visible sensors used for law enforcement and drug interdiction (Aikens, David M.; Young, William R.)V1479,435-444(1991)

Analysis of the flying light spot experiment on SPRITE detector (Gu, Bo-qi; Feng, Wen-qing)V1488,443-446(1991)

Answer to the dynamic (fretting effect) and static (oxide) behavior of electric contact surfaces: based on a five-year infrared thermographic study (Paez-Leon, Cristobal J.; Patino, Antonio R.; Aguillon, Luis)V1467,188-194(1991)

Applications and development of IR techniques for building research in Finland (Kaasinen, Harri I.; Kauppi, Ari; Nykanen, Esa)V1467,90-98(1991)

Applications of diffractive optics to uncooled infrared imagers (Cox, James A.)V1540,606-611(1991)

Applications of tridimensional heat calibration to a thermographic nondestructive evaluation station (Maldague, Xavier; Fortin, Louis; Picard, J.)V1467,239-251(1991)

Assessment of the optimum operating conditions for 2-D focal-plane-array systems (Bourne, Robert W.; Jefferys, E. A.; Murphy, Kevin S.)V1488,73-79(1991)

AutoSPEC image evaluation laboratory (Brown, James C.; Webb, Curtis M.; Bell, Paul A.; Washington, Randolph T.; Riordan, Richard J.)V1488,300-311(1991)

Behavioral observations in thermal imaging of the big brown bat: Eptesicus fuscus (Kirkwood, James J.; Cartwright, Anne)V1467,369-371(1991)

Bidirectional printing method for a thermal ink transfer printer (Nagato, Hitoshi; Ohno, Tadayoshi)V1458,84-91(1991)

Buried-steam-line temperature and heat loss calculation (MacDavid, Jacob H.)V1467,11-17(1991)

Chemistry of the Konica Dry Color System (Suda, Yoshihiko; Ohbayashi, Keiji; Onodera, Kaoru)V1458,76-78(1991)

Clustering algorithms for a PC-based hardware implementation of the unsupervised classifier for the shuttle ice detection system (Jaggi, Sandeep)V1451,289-297(1991)

Color printing technologies (Sahni, Omesh)V1458,4-16(1991)

Comparison of some algorithms commonly used in IR pyrometry: a computer simulation (Barani, Gianni; Tofani, Alessandro)V1467,458-468(1991)

Computer simulation of staring-array thermal imagers (Bradley, D. J.; Dennis, Peter N.; Baddiley, C. J.; Murphy, Kevin S.; Carpenter, Stephen R.; Wilson, W. G.)V1488,186-195(1991)

Correlation between images in the long-wave infrared and short-wave infrared of natural ground terrain (Agassi, Eyal; Wilner, Kalman; Ben-Yosef, Nissim)V1442,126-132(1991)

Corrosion evaluation of coated sheet metal by means of thermography and image analysis (Jernberg, Per)V1467,295-302(1991)

Critical look at AlGaAs/GaAs multiple-quantum-well infrared detectors for thermal imaging applications (Adams, Frank W.; Cuff, K. F.; Gal, George; Harwit, Alex; Whitney, Raymond L.)V1541,24-37(1991)

Design and performance of a 486 x 640 pixel platinum silicide IR imaging system (Clark, David L.; Berry, Joseph R.; Compagna, Gary L.; Cosgrove, Michael A.; Furman, Geoffrey G.; Heydweiller, James R.; Honickman, Harris; Rehberg, Raymond A.; Sorlie, Paul H.; Nelson, Edward T.)V1540,303-311(1991)

Detection of citrus freeze damage with natural color and color IR video systems (Blazquez, Carlos H.)V1467,394-401(1991)

Development of a fire and forget imaging infrared seeker missile simulation (Hall, Charles S.; Alongi, Robert E.; Fortner, Russ L.; Fraser, Laurie K.)V1483,29-38(1991)

Direct view thermal imager (Reinhold, Ralph R.)V1447,251-262(1991)

Discussion of the standard practice for the location of wet insulation in roofing systems using infrared imaging (ASTM C1153-90) (Sopko, Victor)V1467,83-89(1991)

Effects of phasing on MRT target visibility (Holst, Gerald C.)V1488,90-98(1991)

Electron-beam-addressed membrane light modulator for IR scene projection (Horsky, Thomas N.; Schiller, Craig M.; Genetti, George J.; O'Mara, Daniel M.; Hamnett, Whitney S.; Warde, Cardinal)V1540,527-532(1991)

Evaluation of the IR signature of dynamic air targets (Porta, Paola M.)V1540,508-518(1991)

Experimental simulation of IR signatures (Mulero, Manuel A.; Barreiros, Manuel A.)V1540,519-526(1991)

Field documentation and client presentation of IR inspections on new masonry structures (McMullan, Phillip C.)V1467,66-74(1991)

Flaw dynamics and vibrothermographic-thermoelastic nondestructive evaluation of advanced composite materials (Tenek, Lazarus H.; Henneke, Edmund G.)V1467,252-263(1991)

Functional row of pyroelectric sensors for infrared devices used in ecological monitoring (Savinykh, Viktor P.; Glushko, A. A.)V1540,450-454(1991)

General principles of constructing the nuclei of nonlinear IR-image transformations (Nesteruk, Vsevolod F.)V1540,468-476(1991)

Generation of realistic IR images of tactical targets in obscured environments (Greenleaf, William G.; Siniard, Sheri M.; Tait, Mary B.)V1486,364-375(1991)

Global approach toward the evaluation of thermal infrared countermeasures (Verlinde, Patrick S.; Proesmans, Marc)V1486,58-65(1991)

High-fill-factor monolithic infrared image sensor (Kimata, Masafumi; Yutani, Naoki; Yagi, Hirofumi; Nakanishi, Junji; Tsubouchi, Natsuro; Seto, Toshiki)V1540,238-249(1991)

High-performance FLIR testing using reflective-target technology (McHugh, Stephen W.)V1540,775-780(1991)

High-performance InSb 256 x 256 infrared camera (Blackwell, John D.; Parrish, William J.; Kincaid, Glen T.)V1479,324-334(1991)

High-performance IR thermography system based on Class II Thermal Imaging Common Modules (Bell, Ian G.)V1467,438-447(1991)

High-resolution thermal imager with a field-of-view of 112 degrees (Matsushita, Tadashi; Suzuki, Hiroshi; Wakabayashi, Satoshi; Tajime, Toru)V1488,368-375(1991)

High-spatial-resolution FLIR (Tucker, Christopher J.; Mitchell, Robert J.)V1498,92-98(1991)

Image display and background analysis with the Naval Postgraduate School infrared search and target designation system (Cooper, Alfred W.; Lentz, William J.; Baca, Michael J.; Bernier, J. D.)V1486,47-57(1991)

Images of turbulent, absorbing-emitting atmospheres and their application to windshear detection (Watt, David W.; Philbrick, Daniel A.)V1467,357-368(1991)

Improved IR image generator for real-time scene simulation (Keller, Catherine E.; Stenger, Anthony J.; Bernstein, Uri)V1486,278-285(1991)

Incorporation of time-dependent thermodynamic models and radiation propagation models into IR 3-D synthetic image generation models (Schott, John R.; Raqueno, Rolando; Salvaggio, Carl; Kraus, Eugene J.)V1540,533-549(1991)

Infrared background measurements at White Sands Missile Range, NM (Troyer, David E.; Fouse, Timothy; Murdaugh, William O.; Zammit, Michael G.; Rogers, Stephen B.; Skrzypczak, J. A.; Colley, Charles B.; Taczak, William J.)V1486,396-409(1991)

Infrared detection of moist areas in monumental buildings based on thermal inertia analysis (Grinzato, Ermanno G.; Mazzoldi, Andrea)V1467,75-82(1991)

Infrared Imaging Systems: Design, Analysis, Modeling and Testing II (Holst, Gerald C., ed.)V1488(1991)

Infrared Technology XVII (Andresen, Bjorn F.; Scholl, Marija S.; Spiro, Irving J., eds.)V1540(1991)

Infrared thermal-wave studies of coatings and composites (Favro, Lawrence D.; Ahmed, Tasdiq; Crowther, D. J.; Jin, Huijia J.; Kuo, Pao K.; Thomas, Robert L.; Wang, X.)V1467,290-294(1991)

Infrared thermographic analysis of snow ski tracks (Roberts, Charles C.)V1467,207-218(1991)

Infrared-thermography-based pipeline leak detection systems (Weil, Gary J.; Graf, Richard J.)V1467,18-33(1991)

Integrating thermography into the Palisades Nuclear Plant's electrical predictive maintenance program (Ridley, W. C.)V1467,51-58(1991)

Introducing multiple-dynamic-windows in thermal imaging (Bales, Maurice; Boulton, Herbert)V1467,195-206(1991)

IR CCD staring imaging system (Zhou, Qibo)V1540,677-680(1991)

Is it worth it?— statistics of corporate-based IR program results (Johnson, Peter F.)V1467,47-50(1991)

Laptop page printer realized by thermal transfer technology (Drees, Friedrich-Wilhelm; Pekruhn, Wolfgang)V1458,80-83(1991)

Lightweight surveillance FLIR (Fawcett, James M.)V1498,82-91(1991)

Long-wavelength GexSi1-x/Si heterojunction infrared detectors and focal-plane arrays (Tsaur, Bor-Yeu; Chen, C. K.; Marino, S. A.)V1540,580-595(1991)

Low-cost high-performance InSb 256 x 256 infrared camera (Parrish, William J.; Blackwell, John D.; Kincaid, Glen T.; Paulson, Robert C.)V1540,274-284(1991)

Measurement of point spread function of thermal imager (Ryu, Zee Man)V1467,469-474(1991)

Miscellaneous modulation transfer function effects relating to sample summing (Kennedy, Howard V.)V1488,165-176(1991)

Multiorder etalon sounder for vertical temperature profiling: technique and performance analysis (Wang, Jin-Xue; Hays, Paul B.; Drayson, S. R.)V1492,391-402(1991)

Multiplexed mid-wavelength IR long linear photoconductive focal-plane arrays (Kreider, James F.; Preis, Mark K.; Roberts, Peter C.; Owen, Larry D.; Scott, Walter M.; Walmsley, Charles F.; Quin, Alan)V1488,376-388(1991)

New method of target acquisition in the presence of clutter (Tidhar, Gil; Rotman, Stanley R.)V1486,188-199(1991)

Numerical evaluation of the efficiency of camouflage systems in the thermal infrared (Proesmans, Marc; Verlinde, Patrick S.)V1486,102-114(1991)

Numerical inversion method for determining aerodynamic effects on particulate exhaust plumes from onboard irradiance data (Cousins, Daniel)V1467,402-409(1991)

Optical image transformation including IR region for information conformity of their formation and perception processes based on Fibonacci polynomials and series (Miroshnikov, Mikhail M.; Nesteruk, Vsevolod F.)V1540,477-487(1991)

Optimum choice of anamorphic ratio and boost filter parameters for a SPRITE-based infrared sensor (Fredin, Per)V1488,432-442(1991)

Percutaneous laser balloon coagulation of accessory pathways (McMath, Linda P.; Schuger, Claudio D.; Crilly, Richard J.; Spears, J. R.)V1425,165-171(1991)

Performance of infrared systems under field conditions (Chrzanowski, Krzysztof)V1512,78-83(1991)

Photoacoustic microscopy by photodeformation applied to the determination of thermal diffusivity (Balageas, Daniel L.; Boscher, Daniel M.; Deom, Alain A.; Enguehard, Francis; Noirot, Laurence)V1467,278-289(1991)

Postprocessing of thermograms in infrared nondestructive testing (Vavilov, Vladimir P.; Maldague, Xavier; Saptzin, V. M.)V1540,488-495(1991)

Practical method for automatic detection of brightest ridge line of photomechanical fringe pattern (Ding, Zu-Quan; Yuan, Xun-Hua)V1554A,898-906(1991)

Predicting electronic component lifetime using thermography (Moy, Richard Q.; Vargas, Raymund; Eubanks, Charles)V1467,154-160(1991)

Prediction of thermal-image quality as a function of weather forecast (Shushan, A.; Meninberg, Y.; Levy, I.; Kopeika, Norman S.)V1487,300-311(1991)

Programmable command interpreter to automate image processing of IR thermography (Hughett, Paul)V1467,416-426(1991)

Qualitative and quantitative evaluation of moisture in thermal insulation by using thermography (Vavilov, Vladimir P.; Ivanov, A. I.; Sengulye, A. A.)V1467,230-233(1991)

Quantitative evaluation of cavities and inclusions in solids using IR thermography (Madrid, Angel)V1467,322-336(1991)

Quantitative evaluation of errors in remote measurements using a thermal imager (Engel, Michael Y.; Balfour, L. S.)V1442,298-307(1991)

Quantitative measurement of thermal parameters over large areas using pulse-video thermography (Hobbs, Chris P.; Kenway-Jackson, Damian; Milne, James M.)V1467,264-277(1991)

Quantitative thermal gradient imaging of biological surfaces (Swanson, Curtis J.; Wingard, Christopher J.)V1467,372-383(1991)

Real-time quantitative imaging for semiconductor crystal growth, control, and characterization (Wargo, Michael J.)V1557,271-282(1991)

Real-time temperature measurement on PCB:s, hybrids, and microchips (Wallin, Bo)V1467,180-187(1991)

Recognition criterion for two-dimensional minimum resolvable temperature difference (Kennedy, Howard V.)V1488,196-202(1991)

Remote spectral identification of surface aggregates by thermal imaging techniques: progress report (Scholen, Douglas E.; Clerke, William H.; Burns, Gregory S.)V1492,358-369(1991)

Remote temperature sensing of a pulsed thermionic cathode (Del Grande, J. M.)V1467,427-437(1991)

Research on enhancing signal and SNR in laser/IR inspection of solder joints quality (Xiong, Zhengjun; Cheng, Xuezhong; Liu, Xiande)V1467,410-415(1991)

Sensor fusion approach to optimization for human perception: an observer-optimized tricolor IR target locating sensor (Miller, Walter E.)V1482,224-233(1991)

Ship signature measurements for tactical decision-aid input (Cooper, Alfred W.; Milne, Edmund A.; Crittenden, Eugene C.; Walker, Philip L.; Moore, E.; Lentz, William J.)V1486,37-46(1991)

Simulation of partially obscured scenes using the radiosity method (Borel, Christoph C.; Gerstl, Siegfried A.)V1486,271-277(1991)

Simulation of sampling effects in FPAs (Cook, Thomas H.; Hall, Charles S.; Smith, Frederick G.; Rogne, Timothy J.)V1488,214-225(1991)

Simulation study to characterize thermal infrared sensor false alarms (Sabol, Bruce M.; Mixon, Harold D.)V1486,258-270(1991)

Solutions to modeling of imaging IR systems for missile applications: MICOM imaging IR system performance model-90 (Owen, Philip R.; Dawson, James A.; Borg, Eric J.)V1488,122-132(1991)

So now what?— things to do if your IR program stops producing results (Lucier, Ronald D.)V1467,59-62(1991)

Soviet IR imagers and their applications: short state of the art (Vavilov, Vladimir P.)V1540,460-465(1991)

Spectral stratigraphy (Lang, Harold R.)V1492,351-357(1991)

SPRITE detector characterization through impulse response testing (Anderson, Barry K.; Boreman, Glenn D.; Barnard, Kenneth J.; Plogstedt, Allen E.)V1488,416-425(1991)

Status report on thermographer certification (Baird, George S.; Mack, Russell T.)V1467,63-63(1991)

Studies on defocus in thermal imaging systems (Venkateswara Rao, B.)V1399,145-156(1991)

Study of thermal dye diffusion (Koshizuka, Kunihiro; Abe, Takao)V1458,97-104(1991)

Supervision of self-heating in peat stockpiles by aerial thermography (Tervo, Matti; Kauppinen, Timo)V1467,161-168(1991)

Surveying and damping heat loss from machines with high surface temperatures: thermography as a tool (Perch-Nielsen, Thomas; Paulsen, Otto; Drivsholm, Christian)V1467,169-179(1991)

Surveying the elements of successful infrared predictive maintenance programs (Snell, John R.; Spring, Robert W.)V1467,2-10(1991)

Target acquisition modeling based on human visual system performance (Valeton, J. M.; van Meeteren, Aart)V1486,68-84(1991)

Target identification by means of adaptive neural networks in thermal infrared images (Acheroy, Marc P.; Mees, W.)V1569,121-132(1991)

Technology trends for high-performance windows (Askinazi, Joel)V1498,100-109(1991)

Temperature chamber FLIR and missile test system (Johnson, W. Todd; Lavi, Moshe; Sapir, Eyal)V1488,343-354(1991)

Thermal analysis of masonry block buildings during construction (Allen, Lee R.; Semanovich, Sharon A.)V1467,99-103(1991)

Thermal analysis of the bottle forming process (Wilson, Jeannie S.)V1467,219-228(1991)

Thermal and radiometric modeling of terrain backgrounds (Conant, John A.; Hummel, John R.)V1486,217-230(1991)

Thermal diagnostics for monitoring welding parameters in real time (Fuchs, Elizabeth A.; Mahin, K. W.; Ortega, A. R.; Bertram, L. A.; Williams, Dean R.; Pomplun, Alan R.)V1467,136-149(1991)

Thermal dye transfer color hard-copy image stability (Newmiller, Chris)V1458,92-96(1991)

Thermal infrared imagery from the Geoscan Mk II scanner and its calibration: two case histories from Nevada—Ludwig Skarn (Yerington District) & Virginia City (Lyon, Ronald J.; Honey, Frank R.)V1492,339-350(1991)

Thermal model for real-time textured IR background simulation (Bernstein, Uri; Keller, Catherine E.)V1486,345-351(1991)

Thermal sensing of fireball plumes (Toossi, Reza)V1467,384-393(1991)

Thermal signature training for military observers (LaFollette, Robert; Horger, John D.)V1488,289-299(1991)

Thermal transfer printing with heat amplification (Aviram, Ari; Shih, Kwang K.; Sachdev, Krishna)V1458,105-107(1991)

Thermographic analysis of the anisotropy in the thermal conductivity of composite materials (Burleigh, Douglas D.; De La Torre, William)V1467,303-310(1991)

Thermographic monitoring of lubricated couplings (Wurzbach, Richard N.)V1467,41-46(1991)

Thermography and complementary method: a tool for cost-effective measures in retrofitting buildings (Lyberg, Mats D.; Ljungberg, Sven-Ake)V1467,104-115(1991)

Thermosense XIII (Baird, George S., ed.)V1467(1991)

Three-dimensional analysis framework and measurement methodology for imaging system noise (D'Agostino, John A.; Webb, Curtis M.)V1488,110-121(1991)

Time-resolved infrared radiometry of multilayer organic coatings using surface and subsurface heating (Maclachlan Spicer, J. W.; Kerns, W. D.; Aamodt, Leonard C.; Murphy, John C.)V1467,311-321(1991)

Time-resolved videothermography at above-frame-rate frequencies (Shepard, Steven M.; Sass, David T.; Imirowicz, Thomas P.; Meng, A.)V1467,234-238(1991)

Trends in color hard-copy technology in Japan (Abe, Takao)V1458,29-40(1991)

Twenty-five years of aerodynamic research with IR imaging (Gartenberg, Ehud; Roberts, A. S.)V1467,338-356(1991)

Update on the C2NVEO FLIR90 and ACQUIRE sensor performance models (Scott, Luke B.; Tomkinson, David M.)V1488,99-109(1991)

Using IR thermography as a manufacturing tool to analyze and repair defects in printed circuit boards (Fike, Daniel K.)V1467,150-153(1991)

Utility gains through infrared predictive maintenance (Black, James E.)V1467,34-40(1991)

Variable emissivity plates under a three-dimensional sky background (Meitzler, Thomas J.; Gonda, Teresa G.; Jones, Jack C.; Reynolds, William R.)V1486,380-389(1991)

What is MRT and how do I get one? (Hoover, Carl W.; Webb, Curtis M.)V1488,280-288(1991)

Thin films—see also coatings; optical materials

Ablation of ITO and TO films from glass substrates (Meringdal, Frode; Slinde, Harald)V1503,292-298(1991)

Accurate measurement of thin-polymeric-films index variations: application to elasto-optic effect and to photochromism (Dumont, Michel L.; Morichere, D.; Sekkat, Z.; Levy, Yves)V1559,127-138(1991)

Advanced thin films for optical storage (Gan, Fuxi)V1519,530-538(1991)

Al2O3 etch-stop layer for a phase-shifting mask (Hanyu, Isamu; Nunokawa, Mitsuji; Asai, Satoru; Abe, Masayuki)V1463,595-601(1991)

Amorphous silicon thin film x-ray sensor (Wei, Guang P.)V1519,225-233(1991)

Analysis of thin-film losses from guided wave attenuations with photothermal deflection technique (Liu, Xu; Tang, Jinfa; Pelletier, Emile P.)V1554A,558-569(1991)

Angular-selective cermet films produced from a magnetically filtered cathodic arc (Smith, Geoffrey B.; Ng, M. W.; Ditchburn, Robert J.; Martin, Philip J.; Netterfield, Roger P.)V1536,126-137(1991)

Anisotropic electrical and optical behavior and preferential orientation in early transition metal tellurides thin films (Mathey, Yves; Pailharey, Daniel; Gerri, Mireille; Bonnot, Anne-Marie; Sorbier, J. P.)V1361,909-916(1991)

Anomalous optical response of YBa2Cu3O7-x thin films during superconducting transitions (Xi, Xiaoxing; Venkatesan, T.; Etemad, Shahab; Hemmick, D.; Li, Q.)V1477,20-25(1991)

Antireflection coatings of sputter-deposited SnOxFy and SnNxFy (Yin, Zhiqiang; Stjerna, B. A.; Granqvist, Claes G.)V1536,149-157(1991)

Application of thermal wave technology to thickness and grain size monitoring of aluminum films (Opsal, Jon L.)V1596,120-131(1991)

Application of YBa2Cu3O7-x thin film in high-Tc semiconducting infrared detector (Zhou, Bing; Chen, Ju X.; Shi, Bao A.; Wu, Ru J.; Gong, Shuxing)V1519,454-456(1991)

Applications of pulsed photothermal deflection technique in the study of laser-induced damage in optical coatings (Wu, Zhouling; Reichling, M.; Fan, Zheng X.; Wang, Zhi-Jiang)V1441,214-227(1991)

Approaches to the construction of intrinsically acentric chromophoric NLO materials: chemical elaboration and resultant properties of self-assembled multilayer structures (Allan, D. S.; Kubota, F.; Marks, Tobin J.; Zhang, T. J.; Lin, W. P.; Wong, George K.)V1560,362-369(1991)

Artificial photosynthesis at octane/water interface in the presence of hydrated chlorophyll a oligomer thin film (Volkov, Alexander G.; Gugeshashvili, M. I.; Kandelaki, M. D.; Markin, V. S.; Zelent, B.; Munger, G.; Leblanc, Roger M.)V1436,68-79(1991)

Asymmetric photopotentials from thin polymeric porphyrin films (Wamser, Carl C.; Senthilathipan, Velu; Li, Wen)V1436,114-124(1991)

Atomic layer growth of zinc oxide and zinc sulphide (Sanders, Brian W.; Kitai, Adrian H.)V1398,81-87(1991)

Basis and applicaton of evanescent fluorescence measurement (Yuan, Y. F.; Heavens, Oliver S.)V1519,434-439(1991)

Bi(Pb)-Sr-Ca-Cu-O superconducting films prepared by chemical spray deposition (Li, Chang J.; Liu, Li M.; Yao, Qi)V1519,779-787(1991)

Boron depth profiles in a-Si1-xCx:H(B) films after thermal annealing (Liao, Chang G.; Zheng, Zhi H.; Wang, Yong Q.; Yang, Sheng S.)V1519,152-155(1991)

Cation intercalation in electrochromic NiOx films (Scarminio, J.; Gorenstein, Annette; Decker, Franco; Passerini, S.; Pileggi, R.; Scrosati, Bruno)V1536,70-80(1991)

Characteristics of thin-film-type Josephson junctions using Bi2Sr2CaCu2Ox/Bi2Sr2CuOy/Bi2Sr2CaCu2Oz structure (Mizuno, Koichi; Higashino, Hidetaka; Setsune, Kentaro; Wasa, Kiyotaka)V1477,197-204(1991)

Characterization and preparation of high-Tc YBa2Cu3O7-x thin films on Si with conducting indium oxide as a buffer layer (Zhang, Z. J.; Luo, Wei A.; Zeng, Y. Y.; Yang, N. P.; Cai, Y. M.; Shen, X. L.; Chen, H. S.; Hua, Zhong Y.)V1519,790-792(1991)

Characterization of anodic fluoride films on Hg1-xCdxTe (Esquivias, Ignacio; Dal Colle, M.; Brink, D.; Baars, Jan W.; Bruder, Martin)V1484,55-66(1991)

Characterization of GaAs thin films grown by molecular beam epitaxy on Si-on-insulator (Zhu, Wen H.; Lin, Cheng L.; Yu, Yue H.; Li, Aizhen; Zou, Shi C.; Hemment, Peter L.)V1519,423-427(1991)

Characterization of high-Tc coplanar transmission lines and resonators (Kessler, Jochen; Dill, Roland; Russer, Peter)V1477,45-56(1991)

Charge buildup in polypropylene thin films (Ding, Hai)V1519,847-856(1991)

Charge characteristics of thin rapid-thermal-nitrided SiOxNy film in MIS structure (Chen, Pu S.; Yang, Jing)V1519,258-262(1991)

Chemo-optical microsensing systems (Lambeck, Paul V.)V1511,100-113(1991)

Comparative study of carbon and boron carbide spacing layers inside soft x-ray mirrors (Boher, Pierre; Houdy, Philippe; Kaikati, P.; Barchewitz, Robert J.; Van Ijzendoorn, L. J.; Li, Zhigang; Smith, David J.; Joud, J. C.)V1345,165-179(1991)

Condensation mechanisms and properties of rf-sputtered a-Si:H (Ligachev, Valery A.; Filikov, V. A.; Gordeev, V. N.)V1519,214-219(1991)

Corona charged polychlorotrifluorotylene film electrets, and its charge storage and transport (Xia, Zhong F.; Jiang, Jian)V1519,866-871(1991)

Corrosion of ZnSe by alternating high voltage in a saline solution (King, Joseph A.)V1535,216-223(1991)

Cryogenic refractive indices of cadmium telluride coatings in wavelength range from 2.5 to 20 um (Feng, Weiting; Yen, Yi X.; Zhu, Cui Y.)V1535,224-230(1991)

Crystalization kinetics in Ge2Sb2Te5 phase-change recording films (Ozawa, Kenji; Ogino, Shinji; Satoh, Yoshikazu; Urushidani, Tatuo; Ueda, Atushi; Deno, Hiroshi; Kawakami, Haruo)V1499,180-186(1991)

Crystallization of hydrogenated amorphous silicon film and its fractal structure (Lin, Hong Y.; Yang, Dao M.; Li, Ying X.)V1519,210-213(1991)

CVD diamond as an optical material for adverse environments (Snail, Keith A.)V1330,46-64(1991)

Defect enhancement of local electric fields in dielectric films (Risser, Steven M.; Ferris, Kim F.)V1441,262-268(1991)

Defects in SIPOS film studied by ESR (Wang, Yun Z.; Pan, Yao L.)V1519,860-863(1991)

Dependence of optical properties of thermal-evaporated lead telluride films on substrate temperature (Feng, Weiting; Yen, Yi X.; Zhu, Cui Y.)V1519,333-338(1991)

Depopulation kinetics of electron traps in thin oxynitride films (Wong, H.; Cheng, Y. C.; Yang, Bing L.; Liu, Bai Y.)V1519,494-498(1991)

Deposition of a-Si:H using a supersonically expanding argon plasma (Meeusen, G. J.; Qing, Z.; Wilbers, A. T.; Schram, D. C.)V1519,252-257(1991)

Deposition of silica coatings on Incoloy 800H substrates using a high-power laser (Fellowes, Fiona C.; Steen, William M.)V1502,213-222(1991)

Detection by mirage effect of the counter-ion flux between an electrochrome and a liquid electrolyte: application to WO3, Prussian blue, and lutetium diphthalocyanine films (Plichon, V.; Giron, J. C.; Delboulbe, J. P.; Lerbet, F.)V1536,37-47(1991)

Determination of optical constants of thin film in the soft x-ray region (Guo, Yong H.; Fan, Zheng X.; Bin, Ouyang; Jin, Lei; Shao, Jian D.)V1519,327-332(1991)

Determination of the orientational order parameters <P*2>, <P*4> in a polysilane LB film via polarization-dependent THG (Neher, Dieter; Mittler-Neher, S.; Cha, M.; Stegeman, George I.; Embs, F. W.; Wegner, Gerhard; Miller, Robert D.; Willson, C. G.)V1560,335-343(1991)

Determination of thickness and refractal index of HgCdMnTe/CdMnTe VPE films by IR transmission spectrum (Chen, Wei M.; Ma, Ke J.; Yu, Zhen Z.; Ji, Hua M.)V1519,521-524(1991)

Determination of thin-film roughness and volume structure parameters from light-scattering investigations (Duparre, Angela; Kassam, Samer)V1530,283-286(1991)

Development of laminated nickel/manganese oxide and nickel/niobium oxide electrochromic devices (Ma, Y. P.; Yu, Phillip C.; Lampert, Carl M.)V1536,93-103(1991)

Diamond-like carbon thin films prepared by rf-plasma CVD (Jiang, Jie; Liu, Chen Z.)V1519,717-724(1991)

Diamond Optics IV (Feldman, Albert; Holly, Sandor, eds.)V1534(1991)

Diffraction by one-dimensional or two-dimensional periodic arrays of conducting plates (Petit, Roger; Bouchitte, G.; Tayeb, Gerard; Zolla, F.)V1545,31-41(1991)

Dispersion of nonlinear magnetostatic surface waves on thin films (Boardman, A. D.; Nikitov, S. A.; Wang, Qi; Bao, Jia S.; Cai, Ying S.; Shen, Janice)V1519,609-615(1991)

Dual ion-beam sputtering: a new coating technology for the fabrication of high-power CO2 laser mirrors (Daugy, Eric; Pointu, Bernard; Villela, Gerard; Vincent, Bernard)V1502,203-212(1991)

Durable, nonchanging, metal-dielectric and all-dielectric mirror coatings (Guenther, Karl H.; Balasubramanian, Kunjithapa; Hu, X. Q.)V1485,240-244(1991)

DyFeCo magneto-optical disks with a Ce-SiO2 protective film (Naitou, Kazunori; Numata, Takehiko; Nakashima, Kazuo; Maeda, Miyozo; Koshino, Nagaaki)V1499,386-392(1991)

Effect of Cu at Al grain boundaries on electromigration behavior in Al thin films (Frear, Darrel R.; Michael, J. R.; Kim, C.; Romig, A. D.; Morris, J. W.)V1596,72-82(1991)

Effect of oxygen on optical properties of yttria thin films (Ying, Xuantong; Feldman, Albert; Farabaugh, Edward N.)V1519,321-326(1991)

Effects of film thickness on the thermoelectric behavior of pyrolytic ZnO thin film (Ambia, M. G.; Islam, M. N.; Hakim, M. O.)V1536,222-232(1991)

Electrical resistance-strain characteristics and structure of amorphous Ni-Si-B thin films (Cheng, Xian A.; Gu, Qi H.; Chen, Bing Y.)V1519,33-36(1991)

Electrochromic materials for smart window applications (Ashrit, Pandurang V.; Bader, G.; Girouard, Fernand E.; Truong, Vo-Van)V1401,119-129(1991)

Electrochromic properties and temperature dependence of chemically deposited Ni(OH)x thin films (Fantini, Marcia C.; Bezerra, George H.; Carvalho, C. R.; Gorenstein, Annette)V1536,81-92(1991)

Electrochromic property and chemical sensitivity of conducting polymer PAn film (Yuan, Ren K.; Gu, Zhi P.; Yuan, Hong; Yuan, Xue S.; Wang, Yong B.; Liu, Xiang N.; Shen, Xue C.)V1519,831-834(1991)

Electrochromism in cobalt oxyhydroxide thin films (Gorenstein, Annette; Polo Da Fonseca, C. N.; Torresi, R. M.)V1536,104-115(1991)

Electrodeposited nickel-cobalt thin films for photothermal conversion of solar energy (Karuppiah, N.; John, S.; Natarajan, Sanjay S.; Sivan, V.)V1536,215-221(1991)

Electromigration physical modeling of failure in thin film structures (Lloyd, James R.)V1596,106-117(1991)

Epitaxial films YBa2Cu3O7-delta(jc(78K)>106A/cm2) on sapphire and SrTiO3: peculiarities and differences in conditions of film growth and properties (Predtechensky, M.; Smal, A.; Varlamov, Yu.)V1477,234-241(1991)

Epitaxial Tl2Ba2CaCu2O8 thin films on LaAlO3 and their microwave device properties (Negrete, George V.; Hammond, Robert B.)V1477,36-44(1991)

Establishment of new criterion aiding the control of antireflection coating semiconductor diodes (Lu, Yu C.; Li, Da Y.; Chen, Jian G.; Luo, Bin)V1519,463-466(1991)

Evaluation of polymeric thin film waveguides as chemical sensors (Bowman, Elizabeth M.; Burgess, Lloyd W.)V1368,239-250(1991)

Excimer laser deposition and characterization of tin and tin-oxide films (Borsella, E.; De Padova, P.; Larciprete, Rosanna)V1503,312-320(1991)

Excimer laser processing of diamond-like films (Ageev, Vladimir P.; Glushko, T. N.; Dorfman, V. F.; Kuzmichov, A. V.; Pypkin, B. N.)V1503,453-462(1991)

Experimental study of microwave attenuation of ITO-dielectric recombination film (Huang, Guang L.; Zhang, Jun; Peng, Chuan C.)V1519,179-182(1991)

Experimental study of the optical properties of LTCVD SiO2 (Aharoni, Herzl; Swart, Pieter L.)V1442,118-125(1991)

Fatigue-resistant coating of SiO2 glass (Tomozawa, Minoru; Han, Won-Taek; Davis, Kenneth M.)V1590,160-167(1991)

Fiber optic displacement sensor for measurement of thin film thickness (Wang, Jianhua)V1572,264-267(1991)

Fiber optic remote Fourier transform infrared spectroscopy (Druy, Mark A.; Glatkowski, Paul J.; Stevenson, William A.)V1584,48-52(1991)

Finite element study on indentations into TiN- and multiple TiN/Ti- layers on steel (Wang, H. F.; Wagendristel, A.; Yang, X.; Torzicky, P.; Bangert, H.)V1519,405-410(1991)

Formation of boron nitride and silicon nitride bilayer films by ion-beam-enhanced deposition (Feng, Yi P.; Jiang, Bing Y.; Yang, Gen Q.; Huang, Wei; Zheng, Zhi H.; Liu, Xiang H.; Zou, Shi C.)V1519,440-443(1991)

Formation of SiO2 film on plastic substrate by liquid-phase-deposition method (Kitaoka, Masaki; Honda, Hisao; Yoshida, Harunobu; Takigawa, Akio; Kawahara, Hideo)V1519,109-114(1991)

Formation of titanium nitride films by Xe+ ion-beam-enhanced deposition in a N2 gas environment (Wang, Xi; Yang, Gen Q.; Liu, Xiang H.; Zheng, Zhi H.; Huang, Wei; Zhou, Zu Y.; Zou, Shi C.)V1519,740-743(1991)

Fractal simulation of aggregation in magnetic thin film (Wang, Yi P.; Wang, Yu J.)V1519,605-608(1991)

Frequency-tunable THG measurements of x(3) between 1-2.1um of organic conjugated-polymer films using an optical parametric oscillator (Gierulski, Alfred; Naarmann, Herbert; Schrof, Wolfgang; Ticktin, Anton)V1560,172-182(1991)

GaN single-crystal films on silicon substrates grown by MOVPE (Nagatomo, Takao; Ochiai, Ichiro; Ookoshi, Shigeo; Omoto, Osamu)V1519,90-95(1991)

Gray scale and resolution enhancement capabilities of edge emitter imaging stations (Leksell, David; Kun, Zoltan K.; Asars, Juris A.; Phillips, Norman J.; Brandt, Gerald B.; Stringer, J. T.; Matty, T. C.; Gigante, Joseph R.)V1458,133-144(1991)

Growth mechanism of orientated PLZT thin films sputtered on glass substrate (Zhang, Rui T.; Ge, Ming; Luo, Wei G.)V1519,757-760(1991)

Growth of high-Tc superconducting thin films for microwave applications (Wu, Xin D.; Foltyn, Stephen R.; Muenchausen, Ross E.; Dye, Robert C.; Cooke, D. W.; Rollett, A. D.; Garcia, A. R.; Nogar, Nicholas S.; Pique, A.; Edwards, R.)V1477,8-14(1991)

Growth of PbTiO3 films by photo-MOCVD (Shimizu, Masaru; Katayama, Takuma; Fujimoto, Masashi; Shiosaki, Tadashi)V1519,122-127(1991)

Growth of rf-sputtered selenium thin films (Yuan, Xiang L.; Min, Szuk W.; Fang, Zhi Y.; Yu, Da W.; Qi, Lei)V1519,167-171(1991)

Growth of thin films of organic nonlinear optical materials by vapor growth processes: an overview and examination of shortfalls (Frazier, Donald O.; Penn, Benjamin G.; Witherow, William K.; Paley, M. S.)V1557,86-97(1991)

Guided-wave nonlinear optics in 2-docosylamino-5-nitropyridine Langmuir-Blodgett films (Bosshard, Christian; Kuepfer, Manfred; Floersheimer, M.; Guenter, Peter)V1560,344-352(1991)

Hard carbon coating on ZnS (Azran, A.; Elfersi, U.; Greenfield, Ephraim)V1442,54-57(1991)

Heating and damage of thin metal films under the conditions of disturbed equilibrium (Minaeva, E. M.; Libenson, Michail N.)V1440,63-70(1991)

Highest observed second harmonic intensity from a multilayered Langmuir-Blodgett film structure (Ashwell, Geoffrey J.; Dawnay, Emma J.; Kuczynski, Andrzej P.; Martin, Philip J.)V1361,589-598(1991)

High-field electron trapping and detrapping characteristics in thin SiOxNy films (Yang, Bing L.; Liu, Bai Y.; Cheng, Y. C.; Wong, H.)V1519,241-246(1991)

Highly sensitive absorption measurements in organic thin films and optical media (Skumanich, Andrew)V1559,267-277(1991)

High-reflective multilayers as narrowband VUV filters (Zukic, Muamer; Torr, Douglas G.)V1485,216-227(1991)

High-selectivity spectral multiplexers-demultiplexers usable in optical telecommunications obtained from multidielectric coatings at the end of optical fibers (Richier, R.; Amra, Claude)V1504,202-210(1991)

High-temperature superconducting Josephson mixers from deliberate grain boundaries in Tl2CaBa2Cu2O8 (Bourne, Lincoln C.; Cardona, A. H.; James, Tim W.; Fleming, J. S.; Forse, R. W.; Hammond, Robert B.; Hong, J. P.; Kim, T. W.; Fetterman, Harold R.)V1477,205-208(1991)

Hole-burning spectroscopy of phthalocyanine Langmuir-Blodget films (Adamec, F.; Ambroz, M.; Dian, J.; Vacha, M.; Hala, J.; Balog, P.; Brynda, E.)V1402,82-84(1991)

Holographic characterization of DYE-PVA films studied at 442 nm for optical elements fabrication (Couture, Jean J.)V1559,429-435(1991)

Homogeneous thin film lens on LiNbO3 (Jiang, Pisu; Laybourn, Peter J.; Righini, Giancarlo C.)V1362,899-906(1991)

Humidity dependence of ceramic substrate electroluminescent devices (Young, Richard; Kitai, Adrian H.)V1398,71-80(1991)

Incorporation of As into HgCdTe grown by MOCVD (He, Jin; Yu, Zhen Z.; Ma, Ke J.; Jia, Pei M.; Yang, Jian R.; Shen, Shou Z.; Chen, Wei M.; Yang, Ji M.)V1519,499-507(1991)

Indirect spectroscopic detection of singlet oxygen during photodynamic therapy (Tromberg, Bruce J.; Dvornikov, Tatiana; Berns, Michael W.)V1427,101-108(1991)

Infrared photodetector based on the photofluxonic effect in superconducting thin films (Kadin, Alan M.; Leung, Michael; Smith, Andrew D.; Murduck, J. M.)V1477,156-165(1991)

Infrared transparent conductive coatings deposited by activated reactive evaporation (Marcovitch, Orna; Zipin, Hedva; Klein, Z.; Lubezky, I.)V1442,58-58(1991)

In-situ laser preparation of high-Tc superconducting thin film at 450-550 degree C (An, Cheng W.; Fan, Yong C.; Lu, Dong S.; Li, Zai Q.)V1519,818-821(1991)

In-situ low-temperature and epitaxial growth of high-Tc superconducting films using oxygen-discharge-assisted laser ablation method (Fan, Yong C.; An, Cheng W.; Lu, Dong S.; Li, Zai Q.)V1519,775-778(1991)

Interaction between the excitons and electrons in ZnSe1-xSx epilayer under high excitation (Guan, Z. P.; Zheng, Zhu H.; Zhang, J. H.; Lu, Y. M.; Fan, Guang H.; Fan, Xi W.)V1519,26-32(1991)

Interface stress at thin film semiconductor heterostructures (Nishino, Taneo)V1519,382-390(1991)

Interfacial adhesive strength measurement in a multilayered two-level metal device structure (Siddiqui, Humayun R.; Ryan, Vivian; Shimer, Julie A.)V1596,139-157(1991)

Interference detection of plasma of laser field interaction with optical thin films (Ni, Xiao W.; Lu, Jian-Feng; He, Anzhi; Ma, Zi; Zhou, Jiu L.)V1554B,632-635(1991)

Interplay between photons and superconductors (Gilabert, Alain; Azema, Alain; Roustan, Jean-Claude; Maneval, Jean-Paul)V1477,140-147(1991)

Intl Conf on Thin Film Physics and Applications (Zhou, Shixun; Wang, Yonglin, eds.)V1519(1991)

Investigation into the characteristics of a-C:H films irradiated by electron beam (Gu, Shu L.; He, Yu L.; Wang, Zhi C.)V1519,175-178(1991)

Investigation of defects in HgCdTe epi-films grown from Te solutions (Wang, Yue; Tang, Zhi J.; Zhuang, Wei S.; He, Jing F.)V1519,428-433(1991)

Investigation of Ti-Al-Mo-V alloy nitride coatings by ARC technique (Wang, Ren; Yang, Guang Y.; Wu, Bei X.; Fu, Bao W.; Zhan, Yun C.; Zhang, Yun H.)V1519,146-151(1991)

Investigation on the anomalous structure of the nanocrystal Ti and Zr films (Shi, W.; Kong, J.; Shen, H.; Du, G.; Yao, W.; Qi, Zhen Z.)V1519,138-141(1991)

Investigation on the DLC films prepared by dual-ion-beam-sputtering deposition (Wang, Tianmin; Wang, Weijie; Liu, Guidng; Huang, Liangpu; Luo, Chuntai; Liu, Dingquan; Xu, Ming; Yang, Yimin)V1519,890-900(1991)

Ion-beam-sputtering deposition and etching of high-Tc YBCO superconducting thin films (Zhao, Xing R.; Hao, Jian H.; Zhou, Fang Q.; Sun, Han D.; Wang, Lingjie; Yi, Xin J.)V1519,772-774(1991)

Ion implantation of diamond-like carbon films (Xiang, Jin Z.; Zheng, Zhi H.; Liao, Chang G.; Xiong, Jing; Wang, Yong Q.; Zhang, Fang Q.)V1519,683-687(1991)

Kinetics of surface ordering: Pb on Ni(001) (Eng, Peter J.; Stephens, Peter; Tse, Teddy)V1550,110-121(1991)

Langmuir-Blodgett films for second-order nonlinear optics (Penner, Thomas L.; Armstrong, Nancy J.; Willand, Craig S.; Schildkraut, Jay S.; Robello, Douglas R.)V1560,377-386(1991)

Langmuir-Blodgett films of immunoglobulin G and direct immunochemical sensing (Turko, Illarion V.; Pikuleva, Irene A.; Yurkevich, Igor S.; Chashchin, Vadim L.)V1510,53-56(1991)

Langmuir-Blodgett films of tetra-tert-butyl-phenoxy phthalocyanine iron [II] (Luo, Tao; Zhang, Wei Q.; Gan, Fuxi)V1519,826-830(1991)

Large-scale production of a broadband antireflection coating on ophthalmic lenses with ion-assisted deposition (Andreani, F.; Luridiana, S.; Mao, Shu Z.)V1519,63-69(1991)

Laser-assisted deposition of thin films onto transparent substrates from liquid-phase organometallic precursor: iron acetylacetonate (Shafeev, George A.; Laude, Lucien D.)V1503,321-329(1991)

Laser conditioning and electronic defects of HfO2 and SiO2 thin films (Kozlowski, Mark R.; Staggs, Michael C.; Rainer, Frank; Stathis, J. H.)V1441,269-282(1991)

Laser-induced damage of diamond films (Read, Harold E.; Merker, M.; Gurtman, G. A.; Wilson, Russell S.)V1441,345-359(1991)

Laser-induced optical effects in light-sensitive complexes (Kotov, Gennady A.; Filippov, N.; Shandybina, Galina)V1440,321-324(1991)

Laser-induced phase transition in crystal InSb films as used in optical storage (Sun, Yang; Li, Cheng F.; Deng, He; Gan, Fuxi)V1519,554-558(1991)

Laser Raman measurements of dielectric coatings as a function of temperature (Exarhos, Gregory J.; Hess, Nancy J.; Ryan, Samantha)V1441,190-199(1991)

Lateral-periodicity evaluation of multilayer Bragg reflector surface roughness using x-ray diffraction (Takenaka, Hisataka; Ishii, Yoshikazu)V1345,180-188(1991)

Light energy conversion with pheophytin a and chlorophyll a monolayers at the optical transparent electrode (Leblanc, Roger M.; Blanchet, P.-F.; Cote, D.; Gugeshashvili, M. I.; Munger, G.; Volkov, Alexander G.)V1436,92-102(1991)

Light scattering holographic optical elements formation in photopolymerizable layers (Boiko, Yuri B.; Granchak, Vasilij M.)V1507,544-548(1991)

Linear and nonlinear optical properties of polymeric Langmuir-Blodgett films (Penner, Thomas L.; Willand, Craig S.; Robello, Douglas R.; Schildkraut, Jay S.; Ulman, Abraham)V1436,169-178(1991)

Long-period x-ray standing waves generated by total external reflection (Bedzyk, Michael J.)V1550,151-155(1991)

Low-energy hydrodynamic mechanism of laser destruction of thin films (Sarnakov, S. M.)V1440,112-114(1991)

Low-loss polymer thin-film optical waveguides (Amleshi, Peerouz M.; Naylor, David L.)V1396,396-403(1991)

Luminescence studies in the process of preparation high-Tc superconducting films with excimer laser ablation (Fan, Yong C.; An, Cheng W.; Lu, Dong S.; Li, Zai Q.)V1519,813-817(1991)

Magnetically induced super resolution in a novel magneto-optical disk (Aratani, Katsuhisa; Fukumoto, Atsushi; Ohta, Masumi; Kaneko, Masahiko; Watanabe, Kenjirou)V1499,209-215(1991)

Magnetic and electronic properties of Co/Pd superlattices (Victora, Randall H.; MacLaren, J. M.)V1499,378-381(1991)

Magnetic steering of energy flow of linear and nonlinear magnetostatic waves in ferrimagnetic films (Wang, Qi; Bao, Jia S.; Boardman, A. D.)V1519,589-596(1991)

Magneto-optical Kerr rotation of thin ferromagnetic films (Zhai, H. R.; Xu, Y. B.; Lu, M.; Miao, Y. Z.; Hogue, K. L.; Naik, H. M.; Ahamd, M.; Dunifer, G. L.)V1519,575-579(1991)

Magneto-optical studies of n-type Hg0.622Cd0.378Te grown by molecular beam epitaxy (Liu, Wei J.; Liu, Pu L.; Shi, Guo L.; Zhu, Jing-Bing; Yuan, Shi X.; Xie, Qin X.; He, Li)V1519,415-418(1991)

Measuring films on and below polycrystalline silicon using reflectometry (Engstrom, Herbert L.; Stokowski, Stanley E.)V1464,566-573(1991)

Metal sulfide thin films on glass as solar control, solar absorber, decorative, and photographic coatings (Nair, Padmanabhan K.; Nair, M. T.; Fernandez, A. M.; Garcia, V. M.; Hernandez, A. B.)V1485,228-239(1991)

Microstructure of titanium oxide thin films (Fan, Ru Y.; Lu, Yue M.; Song, Xiang Y.)V1519,134-137(1991)

Microwave properties of YB2Cu3O7-x thin films characterized by an open resonator (Zhou, Shi P.; Wu, Ke Q.; Jabbar, A.; Bao, Jia S.; Lou, Wei G.; Ding, Ai L.; Wang, Shu H.)V1519,793-799(1991)

Modeling of InP metal organic chemical vapor deposition (Black, Linda R.; Clark, Ivan O.; Kui, Jianming; Jesser, William A.)V1557,54-59(1991)

Model of a thin-film optical fiber fluorosensor (Egalon, Claudio O.; Rogowski, Robert S.)V1368,134-149(1991)

Modified chloroaluminium phthalocyanine: an organic semiconductor with high photoactivity (Dodelet, Jean-Pol; Gastonguay, Louis; Veilleux, George; Saint-Jacques, Robert G.; Cote, Roland; Guay, Daniel; Tourillon, Gerard)V1436,38-49(1991)

Morphology and laser damage studies by atomic force microscopy of e-beam evaporation deposited antireflection and high-reflection coatings (Tesar, Aleta A.; Balooch, M.; Shotts, K. W.; Siekhaus, Wigbert J.)V1441,228-236(1991)

Multicomponent electric-arc source of metallic plasma (Karpov, D. A.; Nazikov, S. N.)V1519,115-121(1991)

Multiple mode and multiple source coupling into polymer thin-film waveguides (Potter, B. L.; Walker, D. S.; Greer, L.; Saavedra, Steven S.; Reichert, William M.)V1368,251-257(1991)

Multiple-pulse laser damage to thin film optical coating (Li, Zhong Y.; Li, Cheng F.; Guo, Ju P.)V1519,374-379(1991)

Multishot ablation of thin films: sensitive detection of film/substrate transition by shockwave monitoring (Hunger, Hans E.; Petzoldt, Stefan; Pietsch, H.; Reif, J.; Matthias, Eckart)V1441,283-286(1991)

Multistable magnetostatic waves in thin film (Boardman, A. D.; Nikitov, S. A.; Wang, Qi; Bao, Jia S.; Cai, Ying S.)V1519,597-604(1991)

Narrowband optical interference filters (Cotton, John M.; Casey, William L.)V1417,525-536(1991)

New approach to colored thin film electroluminescence (Xu, Xu R.; Lei, Gang; Xu, Zheng; Shen, Meng Y.)V1519,525-528(1991)

New approach to obtain uniform thickness ZnS thin film interference filters (Mei, Yuan S.; Shang, Shi X.; Shan, Jin A.; Sun, Jian G.)V1519,370-373(1991)

New coating technology and ion source (Yan, Yi S.)V1519,192-193(1991)

New definition of laser damage threshold of thin film (Ni, Xiao W.; Lu, Jian-Feng; He, Anzhi; Ma, Zi; Zhou, Jiu L.)V1527,437-441(1991)

New definition of laser damage threshold of thin film: optical breakdown threshold (Ni, Xiao W.; Lu, Jian-Feng; He, Anzhi; Ma, Zi; Zhou, Jiu L.)V1519,365-369(1991)

New deposition system for the preparation of doped a-Si:H (Wu, Zhao P.; Chen, Ru G.; Wang, Yonglin)V1519,194-198(1991)

New ion-beam sources and their applications to thin film physics (Wei, David T.; Kaufman, Harold R.)V1519,47-55(1991)

New preparation method and properties of diamondlike carbon films (Yu, Bing Kun; Chen, Xao Min)V1534,223-229(1991)

New results on proton nuclear magnetic resonance of a-SiN:H films (Wang, Ji S.; Xu, Chun F.)V1519,857-859(1991)

New thin-film transistor structure and its processing method for liquid-crystal displays (Kuo, Yue)V1456,288-299(1991)

Nonlinear optic in-situ diagnostics of a crystalline film in molecular-beam-epitaxy devices (Krasnov, Victor F.; Musher, Semion L.; Prots, V. I.; Rubenchik, Aleksandr M.; Ryabchenko, Vladimir E.; Stupak, Mikhail F.)V1506,179-187(1991)

Novel cathodic arc plasma PVD system with column target for the deposition of TiN film and other metallic films (Liu, Wei Y.; Li, Yu M.; Cui, Zhan J.; He, Tian X.)V1519,172-174(1991)

Novel quinonediazide-sensitized photoresist system for i-line and deep-UV lithography (Fukunaga, Seiki; Kitaori, Tomoyuki; Koyanagi, Hiroo; Umeda, Shin'ichi; Nagasawa, Kohtaro)V1466,446-457(1991)

Ohmic and Schottky contacts to GaSb (Wu, T. S.; Su, Yan K.; Juang, F. S.; Li, N. Y.; Gan, K. J.)V1519,263-268(1991)

O/I-MBE: formation of highly ordered phthalocyanine/semiconductor junctions by molecular-beam epitaxy: photoelectrochemical characterization (Armstrong, Neal R.; Nebesny, Ken W.; Collins, Greg E.; Lee, Paul A.; Chau, Lai K.; Arbour, Claude; Parkinson, Bruce)V1559,18-26(1991)

Optical characterization of damage resistant kilolayer rugate filters (Elder, Melanie L.; Jancaitis, Kenneth S.; Milam, David; Campbell, John H.)V1441,237-246(1991)

Optical characterization of photolithographic metal grids (Osmer, Kurt A.; Jones, Mike I.)V1498,138-146(1991)

Optical characterization of solar-selective transmitting coatings (Roos, Arne)V1536,158-168(1991)

Optical detector prepared by high-Tc superconducting thin film (Wang, Lingjie; Zhou, Fang Q.; Zhao, Xing R.; Sun, Han D.; Yi, Xin J.)V1540,738-741(1991)

Optical emission spectroscopy in diamond-like carbon film deposition by glow discharge (Zhang, Wei P.; Chen, Jing; Fang, Rong C.; Hu, Ke L.)V1519,680-682(1991)

Optical Materials Technology for Energy Efficiency and Solar Energy Conversion X (Lampert, Carl M.; Granqvist, Claes G., eds.)V1536(1991)

Optical nonlinearities of ZnSe thin films (Chen, Lian C.; Zhang, Ji Y.; Fan, Xi W.; Yang, Ai H.; Zheng, Zhu H.)V1519,450-453(1991)

Optical properties of amorphous hydrogenated carbon layers (Stenzel, Olaf; Schaarschmidt, Guenther; Roth, Sylvia; Schmidt, Guenther; Scharff, Wolfram)V1534,148-157(1991)

Optical properties of granular Sn films with coating Al (Wu, Guang M.; Qian, Zheng X.)V1519,315-320(1991)

Optical properties of Li-doped ZnO films (Valentini, Antonio; Quaranta, Fabio; Vasanelli, L.; Piccolo, R.)V1400,164-170(1991)

Optical properties of oxide films prepared by ion-beam-sputter deposition (Tang, Xue F.; Fan, Zheng X.; Wang, Zhi-Jiang)V1519,96-98(1991)

Optical properties of some ion-assisted deposited oxides (Andreani, F.; Luridiana, S.; Mao, Shu Z.)V1519,18-22(1991)

Optical studies of nitrogen-doped amorphous carbon: laboratory and interstellar investigations (Kaufman, James H.; Metin, Serhat; Saperstein, David D.)V1437,36-41(1991)

Optical thin film devices (Mao, Shu Z.)V1519,288-297(1991)

Optical transition characteristic energies of amorphous and polycrystalline tin oxide films (Muhamad, M. R.; Majid, W. H.; Ariffin, Z.)V1519,872-877(1991)

Optics in adverse environments (Macleod, H. A.)V1399,2-6(1991)

Optimization of optical properties of resist processes (Brunner, Timothy A.)V1466,297-308(1991)

Organic-dye films for write-once optical storage (Zhou, Jian P.; Shu, Ju P.; Xu, Hui J.)V1519,559-564(1991)

Organic electro-optic devices for optical interconnnection (Lipscomb, George F.; Lytel, Richard S.; Ticknor, Anthony J.; Van Eck, Timothy E.; Girton, Dexter G.; Ermer, Susan P.; Valley, John F.; Kenney, John T.; Binkley, E. S.)V1560,388-399(1991)

Organo-metallic thin film for erasable optical recording medium (Shu, Ju P.; Zhou, Jian P.; Xu, Shi Z.)V1519,565-569(1991)

Organo-metallic thin films for erasable photochromatic laser discs (Hua, Zhong Y.; Chen, G. R.; Wang, Z. H.)V1519,2-7(1991)

Parametric studies and characterization measurements of x-ray lithography mask membranes (Wells, Gregory M.; Chen, Hector T.; Engelstad, Roxann L.; Palmer, Shane R.)V1465,124-133(1991)

Performance of stripline resonators using sputtered YBCO films (Mallory, Derek S.; Kadin, Alan M.; Ballentine, Paul H.)V1477,66-76(1991)

Photoacoustic characterization of surface absorption (Reicher, David W.; Wilson, Scott R.; Kranenberg, C. F.; Raja, M. Y.; McNeil, John R.; Brueck, Steven R.)V1441,106-112(1991)

Photochemical delineation of waveguides in polymeric thin films (Beeson, Karl W.; Horn, Keith A.; Lau, Christina; McFarland, Michael J.; Schwind, David; Yardley, James T.)V1559,258-266(1991)

Photochemistry and Photoelectrochemistry of Organic and Inorganic Molecular Thin Films (Frank, Arthur J.; Lawrence, Marcus F.; Ramasesha, S.; Wamser, Carl C., eds.)V1436(1991)

Photochemistry and photophysics of stilbene and diphenylpolyene surfactants in supported multilayer films (Spooner, Susan P.; Whitten, David G.)V1436,82-91(1991)

Photoelectrochemical characteristics of slurry-coated CdSeTe films (Murali, K. R.; Subramanian, V.; Rangarajan, N.; Lakshmanan, A. S.; Rangarajan, S. K.)V1536,289-295(1991)

Photo-induced adhesion changes: a technique for patterning lightguide structures (Festl, H. G.; Franke, Hilmar)V1559,410-423(1991)

Photo-induced degradation of selected polyimides in the presence of oxygen: a rapid decomposition process (Hoyle, Charles E.; Creed, David; Anzures, Edguardo; Subramanian, P.; Nagarajan, Rajamani)V1559,101-109(1991)

Photosensitivity of selenium-bismuth films with varigap structure (Popov, A.; Mikhalev, N.; Karalyunts, A.; Smirnov, O.; Vasilyeva, N.)V1519,457-462(1991)

Photothermal displacement spectroscopy of optical coatings (Su, Xing; Fan, Zheng X.)V1519,80-84(1991)

Pockels' effect in polycrystalline ZnS planar waveguides (Wong, Brian; Kitai, Adrian H.; Jessop, Paul E.)V1398,261-268(1991)

Poled polyimides as thermally stable electro-optic polymer (Wu, Jeong W.; Valley, John F.; Stiller, Marc A.; Ermer, Susan P.; Binkley, E. S.; Kenney, John T.; Lipscomb, George F.; Lytel, Richard S.)V1560,196-205(1991)

Polynuclear membranes as a substrate for obtaining surface-enhanced Raman scattering films (Oleynikov, V. A.; Sokolov, K. V.; Hodorchenko, P. V.; Nabiev, I. R.)V1403,164-166(1991)

Possible enhancement in bolometric response using free-standing film of YBa2Cu3Ox (Ng, Hon K.; Kilibarda, S.)V1477,15-19(1991)

Preferential growth in ion-beam-enhanced deposition of Ti(C,N) films (Chen, You S.; Sun, Yi L.; Zhang, Fu M.; Mou, Hai C.)V1519,56-62(1991)

Preliminary study of the admittance diagram as a useful tool in the design of stripline components at microwave frequencies (Franck, Charmaine C.; Franck, Jerome B.)V1527,277-290(1991)

Preparation and magneto-optical properties of NdDyFeCoTi amorphous films (Zhang, Si J.; Yang, Xiao Y.; Li, Xiao L.; Zhang, Feng P.)V1519,744-751(1991)

Preparation and O+ implantation of Y-Ba-Cu-O superconducting thin films by sputtering and RTA process (Fan, Xiang J.; Guo, Huai X.; Jiang, Chang Z.; Pen, You G.; Liu, Chang; Pen, Zhi L.; Li, Hong T.)V1519,808-812(1991)

Preparation, electrochemical, photoelectrochemical, and solid state characteristics of In-incorporated TiO2 thin films for solar energy applications (Badawy, Waheed A.; El-Giar, Emad M.)V1536,277-288(1991)

Preparation of Pb1-xGexTe crystal with high refractive index for IR coating (Zhang, Su Y.; Xu, Bu Y.; Zhang, Feng S.; Yan, Yixun)V1519,508-513(1991)

Preparation of PbTiO3 thin film by dc single-target magnetron sputtering (Yang, Bang C.; Wang, Ju Y.; Jia, Yu M.; Huang, Yong L.)V1519,725-728(1991)

Preparation of superconducting Y-Ba-Cu-O thin films by rf magnetron sputtering (Huang, Zong T.; Li, Guo Z.; Zeng, Guang L.; Huang, Jiang P.; Xiong, Guilan)V1519,788-789(1991)

Preparation of thin superconducting YBCO films by ion-beam mixing (Fan, Xiang J.; Pen, You G.; Guo, Huai X.; Li, Hong T.; Liu, Chang; Jiang, Chang Z.; Pen, Zhi L.)V1519,805-807(1991)

Preparation of tin oxide and insulating oxide thin films for multilayered gas sensors (Feng, Chang D.; Shimizu, Yasuhiro; Egashira, Makoto)V1519,8-13(1991)

Progress on the variable reflectivity electrochromic window (Goldner, Ronald B.; Arntz, Floyd O.; Berera, G.; Haas, Terry E.; Wei, G.; Wong, Kwok-keung; Yu, Phillip C.)V1536,63-69(1991)

Properties and applications of ferroelectric and piezoelectric thin films (Shiosaki, Tadashi)V1519,694-703(1991)

Pulsed-laser deposition of oxides over large areas (Greer, James A.; Van Hook, H. J.)V1377,79-90(1991)

Pulsed photoconductivity of chlorophyll a (Kassi, Hassane; Hotchandani, Surat; Leblanc, Roger M.; Beaudoin, N.; Dery, M.)V1436,58-67(1991)

Raman and FT-IR characterization of biologically relevant Langmuir-Blodgett films (Katayama, Norihisa; Fukui, Masahiko; Ozaki, Yukihiro; Araki, Toshinari; Yokoi, Seiichi; Iriyama, Keiji)V1403,147-149(1991)

Raman study of icosahedral C60 (Sinha, Kislay; Menendez, Jose; Adams, G. B.; Page, J. B.; Sankey, Otto F.; Lamb, Lowell; Huffman, Donald R.)V1437,32-35(1991)

Reactive ion-beam-sputtering of fluoride coatings for the UV/VUV range (Schink, Harald; Kolbe, J.; Zimmermann, F.; Ristau, Detleu; Welling, Herbert)V1441,327-338(1991)

Recent progress in Si thin film technology for solar cells (Kuwano, Yukinori; Nakano, Shoichi; Tsuda, Shinya)V1519,200-209(1991)

Recording and erasing characteristics of GeSbTe-based phase change thin films (Hou, Li S.; Zhu, Chang X.; Gu, Dong H.)V1519,548-553(1991)

Research of Cr2O3 thin film deposited by arc discharge plasma deposition as heat-radiation absorbent in electric vacuum devices (Deng, Hong; Wang, Xiang D.; Yuan, Lei)V1519,735-739(1991)

Research on influence of substrate to crystal growth by ion beams (Yin, Dong; Guan, Wen X.; Sun, Shu Z.; Zhang, Zhao A.)V1519,164-166(1991)

Research on relaxation process of a-Si:H film photoconductivity and the trap effect (Gong, Dao B.)V1519,281-286(1991)

Research on the temperature of thin film under ion beam bombarding (Zhao, Yun F.; Sun, Zhu Z.; Pang, Shi J.)V1519,411-414(1991)

Reversible phase-change optical recording by using microcellular GeSbTeCo recording film (Okamine, Shigenori; Terao, Motoyasu; Andoo, Keikichi; Miyauchi, Yasushi)V1499,166-170(1991)

Reversible phase transition and third-order nonlinearity of phthalocyanine derivatives (Suda, Yasumasa; Shigehara, Kiyotaka; Yamada, Akira; Matsuda, Hiro; Okada, Shuji; Masaki, Atsushi; Nakanishi, Hachiro)V1560,75-83(1991)

(Sb2Se3)1-x Nix alloy thin films and its application in erasable phase change optical recording (Xue, Song S.; Fan, Zheng X.; Gan, Fuxi)V1519,570-574(1991)

Scaling properties of optical reflectance from quasiperiodic superlattices (Wu, Xiang; Yao, He S.; Feng, Wei G.)V1519,625-631(1991)

Selected Papers on Ellipsometry (Azzam, Rasheed M., ed.)VMS27(1991)

Semiconductor thin-film optical constant determination and thin-film thickness measurement equipment correlation (Kaiser, Anne M.)V1464,386-392(1991)

Silicon nitride film formed by NH3 plasma-enhanced thermal nitridation (Gu, Zhi G.; Li, Bing Z.)V1519,247-251(1991)

Simulation of local layer-thickness deviation on multilayer diffraction (Guo, S. P.; He, X. C.; Redko, S. V.; Wu, Z. Q.)V1519,400-404(1991)

Simultaneous measurement of refractive index and thickness of thin film by polarized reflectances (Kihara, Tami; Yokomori, Kiyoshi)V1332,783-791(1991)

Single crystallinity and oxygen diffusion in high-quality YBa2Cu3O7-delta films (Wu, Zi L.; Wei, M. Z.; Chen, Y. X.; Ren, Cong X.; Zhang, J. H.)V1519,618-624(1991)

Small polaron conduction in amorphous CoMnNiO thin film (Tan, Hui; Tao, Ming D.; Gin, Dong; Han, Ying; Lin, Cheng L.)V1519,752-756(1991)

Soft x-ray, optical, and thermal properties of hard carbon films (Alvey, Mark D.)V1330,39-45(1991)

Soft x-ray windows for position-sensitive proportional counters (Viitanen, Veli-Pekka; Nenonen, Seppo A.; Sipila, Heikki; Mutikainen, Risto)V1549,28-34(1991)

Sol-gel derived BaTiO3 thin films (Xiang, Xiao L.; Hou, Li S.; Gan, Fuxi)V1519,712-716(1991)

Soluble polyacetylenes derived from the ring-opening metathesis polymerization of substituted cyclooctatetraenes: electrochemical characterization and Schottky barrier devices (Jozefiak, Thomas H.; Sailor, Michael J.; Ginsburg, Eric J.; Gorman, Christopher B.; Lewis, Nathan S.; Grubbs, Robert H.)V1436,8-19(1991)

Specific properties of ferroelectric thin films (Liu, Wei G.)V1519,704-706(1991)

Sputtering of silicate glasses (Kai, Teruhiko; Takebe, Hiromichi; Morinaga, Kenji)V1519,99-103(1991)

Standing spin wave modes in permalloy-FeCr multilayer films (Chen, H. Y.; Luo, Y. Q.)V1519,761-764(1991)

Structural investigations of the (Si1-x,Gex)O2 single-crystal thin films by x-ray photoelectron spectroscopy (Sorokina, Svetlana; Dikov, Juriy)V1519,128-133(1991)

Structure and optical properties of a-C:H/a-SiOx:H multilayer thin films (Zhang, Wei P.; Cui, Jing B.; Xie, Shan; Song, Yi Z.; Wang, Chang S.; Zhou, Guien; Wu, Jan X.)V1519,23-25(1991)

Structure and properties of electrochromic WO3 produced by sol-gel methods (Bell, J. M.; Green, David C.; Patterson, A.; Smith, Geoffrey B.; MacDonald, K. A.; Lee, K. D.; Kirkup, L.; Cullen, J. D.; West, B. O.; Spiccia, L.; Kenny, M. J.; Wielunski, L. S.)V1536,29-36(1991)

Structure of poly(p-hydroxystyrene) film (Toriumi, Minoru; Yanagimachi, Masatoshi; Masuhara, Hiroshi)V1466,458-468(1991)

Structure of very thin metal film (Huang, Yong L.; Jiang, Ping)V1519,142-145(1991)

Studies of correlation of molecular structure under preparation conditions for noncrystalline selenium thin films with aid of computer simulation (Popov, A.; Vasiljeva, Natalja V.)V1519,37-42(1991)

Study of C-H stretching vibration in hybrid Langmuir-Blodgett/alumina multilayers by infrared spectroscopy (Zheng, Tian S.; Liu, Li Y.; Xing, Zhongjin; Wang, Wen C.; Shen, Yuanhua; Zhang, Zhiming)V1519,339-346(1991)

Study of crystal structure of vacuum-evaporated Ag-Cu thin film (Sun, Da M.)V1519,688-691(1991)

Study of HgCdTe/CdTe interface structure grown by metal-organic chemical vapor deposition (Ma, Ke J.; Yu, Zhen Z.; Yanh, Jian R.; Shen, Shou Z.; He, Jin; Chen, Wei M.; Song, Xiang Y.)V1519,489-493(1991)

Study of PEO on LTI carbon surfaces by ellipsometry and tribometry (Wang, Jinyu; Stroup, Eric; Wang, Xing F.; Andrade, Joseph D.)V1519,835-841(1991)

Study of phase transition VO2 thin film (Gao, Jian C.; Lin, Zhi H.; Han, Li Y.)V1519,159-163(1991)

Study of p-ZnTe/n-CdTe thin film heterojunction (Wu, Ping)V1519,477-480(1991)

Study of TEM micrographs of thin-film cross-section replica using spectral analysis (Mei, Ting; Liu, Xu; Tang, Jinfa; Gu, Peifu)V1554A,570-578(1991)

Study of the microstructures in Ar+ laser crystallized films of a-Si:H for active layer of thin film transistors (Huang, Xin F.; Zhang, Xiang D.; Zhu, Wei Y.; Chen, Ying Y.)V1519,220-224(1991)

Study on different proportion W-Ti (C) binary alloy carbide thin film (Zhang, Yun H.; Wu, Bei X.; Yang, Guang Y.; Wang, Ren)V1519,729-734(1991)

Study on the high-field current transport mechanisms in thin SiOxNy films (Yang, Bing L.; Liu, Bai Y.; Chen, D. N.; Cheng, Y. C.; Wong, H.)V1519,269-274(1991)

Study on the mechanism of ZnS antireflecting coating with high strength (Yu, Ju X.; Tang, Jia T.)V1519,308-314(1991)

Study on the special properties of electrochromic film of a-WO3 (Luo, Zhongkuan)V1489,124-134(1991)

Submillimeter receiver components using superconducting tunnel junctions (Wengler, Michael J.; Pance, A.; Liu, B.; Dubash, N.; Pance, Gordana; Miller, Ronald E.)V1477,209-220(1991)

Superconducting devices and system insertion (Rachlin, Adam; Babbitt, Richard; Lenzing, Erik; Cadotte, Roland)V1477,101-114(1991)

Superconductivity and structural changes of Ar ion-implanted YBa2Cu3O7-x thin films (Li, Yi J.; Ren, Cong X.; Chen, Guo L.; Chen, Jian M.; Zou, Shi C.)V1519,800-804(1991)

Superconductivity Applications for Infrared and Microwave Devices II (Heinen, Vernon O.; Bhasin, Kul B., eds.)V1477(1991)

Surface microtopography of thin silver films (Costa, Manuel Filipe P.; Almeida, Jose B.)V1332,544-551(1991)

Surface normal SHG in PLZT thin film waveguides (Zou, Lian C.; Malloy, Kevin J.; Wu, A. Y.)V1519,707-711(1991)

Synthesis and nonlinear optical properties of preformed polymers forming Langmuir-Blodgett films (Verbiest, Thierry; Persoons, Andre P.; Samyn, Celest)V1560,353-361(1991)

Ta/Al alloy thin film medium-power attenuator (Yang, Bang C.; Jia, Yu M.)V1519,156-158(1991)

Temperature dependence of resistance of diamond film synthesized by microwave plasma CVD (Yang, Bang C.; Gou, Li; Jia, Yu M.; Ran, Jun G.; Zheng, Chang Q.; Tang, Xia)V1519,864-865(1991)

Tensor resistivity and tensor susceptibility of oriented high-Tc superconducting film YBaCuO (Hou, Bi H.; Qi, Zhen Z.)V1519,822-823(1991)

Thermal process of laser-induced damage in optical thin films (Fan, Zheng X.; Wu, Zou L.; Shi, Zeng R.)V1519,359-364(1991)

Thermal stabilities of a-Si:H films and its application to thyristor elements (Sun, Yue Z.; Chen, Chun X.; Xie, Qi Y.; Yin, Chen Z.; He, Yu L.)V1519,234-240(1991)

Thermal stress modeling for diamond-coated optical windows (Klein, Claude A.)V1441,488-509(1991)

Thermal transport properties of optical thin films (Swimm, Randall T.)V1441,45-55(1991)

Thermoelectric voltage in slant-angle-deposited metallic films (Verechshagin, I. I.; Oksman, Ya. A.)V1440,401-405(1991)

Thickness measurement of combined a-Si and Ti films on c-Si using a monochromatic ellipsometer (Yoo, Chue-San; Jans, Jan C.)V1464,393-403(1991)

Thin film fabrication of stabilized zirconia for solid oxide fuel cells (Setoguchi, Toshihiko; Eguchi, Koichi; Arai, Hiromichi)V1519,74-79(1991)

Thin film magnetic recording (Chen, Yi-Xin)V1519,539-547(1991)

Thin-film selective multishot ablation at 248 nm (Hunger, Hans E.; Pietsch, H.; Petzoldt, Stefan; Matthias, Eckart)V1598,19-26(1991)

Thin film SIMNI material formed by low energy nitrogen implantation and epitaxial growth (Lin, Cheng L.; Li, Jinghua H.; Zou, Shi C.)V1519,104-108(1991)

Thin films of solid electrolytes and studies of their surface (Pan, Xiao R.; Gu, Zhi X.)V1519,85-89(1991)

Thin-film technology in high-resolution, high-density AC plasma displays (Andreadakis, Nicholas C.)V1456,310-315(1991)

Third-order optical nonlinearities and femtosecond responses in metallophthalocyanine thin films made by vacuum deposition, molecular beam epitaxy, and spin coating (Wada, Tatsuo; Hosoda, Masahiro; Garito, Anthony F.; Sasabe, Hiroyuki; Terasaki, A.; Kobayashi, Takayoshi T.; Tada, Hiroaki; Koma, Atsushi)V1560,162-171(1991)

TiNxOy-Cu coatings for low-emissive solar-selective absorbers (Lazarov, M.; Roehle, B.; Eisenhammer, T.; Sizmann, R.)V1536,183-191(1991)

Transparent storage layers for H+ and Li+ ions prepared by sol-gel technique (Valla, Bruno; Tonazzi, Juan C.; Macedo, Marcelo A.; Dall'Antonia, L. H.; Aegerter, Michel A.; Gomes, M. A.; Bulhoes, Luis O.)V1536,48-62(1991)

Triplet-sensitized reactions of some main chain liquid-crystalline polyaryl cinnamates (Subramanian, P.; Creed, David; Hoyle, Charles E.; Venkataram, Krishnan)V1559,461-469(1991)

Understanding metal-dielectric-metal absorption interference filters using lumped circuit theory and transmission line theory (Pastor, Rickey G.)V1396,501-507(1991)

Understanding of the abnormal wavelength effect of overcoats (Wu, Zhouling; Reichling, M.; Fan, Zheng X.; Wang, Zhi-Jiang)V1441,200-213(1991)

Uniformity characterization of rapid thermal processor thin films (Yarling, Charles B.; Cook, Dawn M.)V1393,411-420(1991)

UV laser-induced photofragmentation and photoionization of dimethylcadmium chemisorbed on silicon (Simonov, Alexander P.; Varakin, Vladimir N.; Panesh, Anatoly M.)V1436,20-30(1991)

Variation of optical properties of gel-derived VO2 thin films with temperature (Hou, Li S.; Lu, Song W.; Gan, Fuxi)V1519,580-588(1991)

Ways of the high-speed increasing of magneto-optical spatial light modulators (Randoshkin, Vladimir V.)V1469,796-803(1991)

X-ray diffraction study of GaSb/AlSb strained-layer-superlattices grown on miscut (100) substrates (Macrander, Albert T.; Schwartz, Gary P.; Gualteri, Gregory J.; Gilmer, George)V1550,122-133(1991)

X Rays in Materials Analysis II: Novel Applications and Recent Developments (Mills, Dennis M., ed.)V1550(1991)

ZnS/Me heat mirror systems (Zhang, Xiao P.; Yu, Shan-qing; Ma, Min W.)V1519,514-520(1991)

ZnS:Mn thin film electroluminescent display devices using hafnium dioxide as insulating layer (Hsu, C. T.; Li, J. W.; Liu, C. S.; Su, Yan K.; Wu, T. S.; Yokoyama, M.)V1519,391-395(1991)

Three dimensions—see also displays; object recognition; vision

3-D camera based on differential optical absorbance (Houde, Regis; Laurendeau, Denis; Poussart, Denis)V1332,343-354(1991)

3-D TV: joined identification of global motion parameters for stereoscopic sequence coding (Tamtaoui, Ahmed; Labit, Claude)V1605,720-731(1991)

3-DTV research and development in Europe (Sand, Ruediger)V1457,76-84(1991)

Absolute range measurement system for real-time 3-D vision (Wood, Christopher M.; Shaw, Michael M.; Harvey, David M.; Hobson, Clifford A.; Lalor, Michael J.; Atkinson, John T.)V1332,301-313(1991)

Adaptive surface reconstruction (Terzopoulos, Demetri; Vasilescu, Manuela)V1383,257-264(1991)

Algorithm for the generation of look-up range table in 3-D sensing (Su, Xianyu; Zhou, Wen-Sheng)V1332,355-357(1991)

Allowable delay time of images with motion parallax and high-speed image generation (Satoh, Takanori; Tomono, Akira; Kishino, Fumio)V1606,1014-1021(1991)

Analysis and representation of complex structures in separated flows (Helman, James L.; Hesselink, Lambertus B.)V1459,88-96(1991)

Analysis of optical measurements of SIF and singularity order in rocket motor geometry (Chang, Che-way)V1554A,250-261(1991)

Application of a discrete-space representation to three-dimensional medical imaging (Toennies, Klaus D.; Tronnier, Uwe)V1444,19-25(1991)

Application of holographic interferometry to a three-dimensional flow field (Doerr, Stephen E.)V1554A,544-555(1991)

Application of phase-stepping speckle interferometry to shape and deformation measurement of a 3-D surface (Dobbins, B. N.; He, Shi P.; Kapasi, S.; Wang, Liu Sheng; Button, B. L.; Wu, Xiao-Ping)V1554A,772-780(1991)

Aspect networks: using multiple views to learn and recognize 3-D objects (Seibert, Michael; Waxman, Allen M.)V1383,10-19(1991)

Automated measurement of 3-D shapes by a dual-beam digital speckle interferometric technique (Shi, Dahuan; Qin, Jing; Hung, Yau Y.)V1554A,680-689(1991)

Automated three-dimensional registration of medical images (Neiw, Han-Min; Chen, Chin-Tu; Lin, Wei-Chung; Pelizzari, Charles A.)V1445,259-264(1991)

Automatic 3-D reconstruction of vascular geometry from two orthogonal angiographic image sequences (Close, Robert A.)V1445,513-522(1991)

Automatic acquisition of movement information by a knowledge-based recognition approach (Bae, Kyongtae T.; Altschuler, Martin D.)V1380,108-115(1991)

Automatic analysis system for three-dimensional angiograms (Higgins, William E.; Spyra, Wolfgang J.; Karwoski, Ronald A.; Ritman, Erik L.)V1445,276-286(1991)

Autonomous navigation in a dynamic environment (Davies, Henry C.; Kayaalp, Ali E.; Moezzi, Saied)V1388,165-175(1991)

Autostereoscopic (3-D without glasses) display for personal computer applications (Eichenlaub, Jesse B.)V1398,48-51(1991)

Basic principles of stereographic software development (Hodges, Larry F.)V1457,9-17(1991)

Biostereometric Technology and Applications (Herron, Robin E., ed.)V1380(1991)

Brain surface maps from 3-D medical images (Lu, Jiuhuai; Hansen, Eric W.; Gazzaniga, Michael S.)V1459,117-124(1991)

Broad-range holographic contouring of diffuse surfaces by dual-beam illumination: study of two related techniques (Rastogi, Pramod K.; Pflug, Leopold)V1554B,48-55(1991)

Characteristic views and perspective aspect graphs of quadric-surfaced solids (Chen, Shuang; Freeman, Herbert)V1383,2-9(1991)

Clinical experience with a stereoscopic image workstation (Henri, Christopher J.; Collins, D. L.; Pike, G. B.; Olivier, A.; Peters, Terence M.)V1444,306-317(1991)

Color-coding reproduction of 3-D object with rainbow holography (Fan, Cheng; Jiang, Chaochuan; Guo, Lu Rong)V1461,51-55(1991)

Color-encoded depth: an image enhancement tool (Bieman, Leonard H.)V1385,229-238(1991)

Color quantization aspects in stereopsis (Hebbar, Prashant D.; McAllister, David F.)V1457,233-241(1991)

Compact motion representation based on global features for semantic image sequence coding (Labit, Claude; Nicolas, Henri)V1605,697-708(1991)

Compact zoom lens for stereoscopic television (Scheiwiller, Peter M.; Murphy, S. P.; Dumbreck, Andrew A.)V1457,2-8(1991)

Comparison of 3-D display formats for CAD applications (McWhorter, Shane W.; Hodges, Larry F.; Rodriguez, Walter E.)V1457,85-90(1991)

Comparison of stereoscopic cursors for the interactive manipulation of B-splines (Barham, Paul T.; McAllister, David F.)V1457,18-26(1991)

Comparison of three-dimensional surface rendering techniques (Thomas, Judith G.; Galloway, Robert L.; Edwards, Charles A.; Haden, Gerald L.; Maciunas, Robert J.)V1444,379-388(1991)

Complexity of computing reachable workspaces for redundant manipulators (Alameldin, Tarek K.; Palis, Michael A.; Rajasekaran, Sanguthevar; Badler, Norman I.)V1381,217-225(1991)

Computational model for the stereoscopic optics of a head-mounted display (Robinett, Warren; Rolland, Jannick P.)V1457,140-160(1991)

Computer-aided design and drafting visualization of anatomical structure of the human eye and orbit (Parshall, Robert F.; Sadler, Lewis L.)V1380,200-207(1991)

Computer-aided forensic facial reconstruction (Evenhouse, Raymond J.; Rasmussen, Mary; Sadler, Lewis L.)V1380,147-156(1991)

Computer-generated holograms of linear segments (Navarro, Maria T.; Egozcue, Juan J.; Fimia-Gil, Antonio)V1507,142-148(1991)

Computing image flow and scene depth: an estimation-theoretic fusion-based framework (Singh, Ajit)V1383,122-140(1991)

Confocal light microscopy of the living in-situ ocular lens: two- and three-dimensional imaging (Masters, Barry R.)V1443,288-293(1991)

Confocal microscopy for the biological and material sciences: principle, applications, limitations (Brakenhoff, G. J.; van der Voort, H. T.; Visscher, Koen)V1439,121-127(1991)

Constructing topologically connected surfaces for the comprehensive analysis of 3-D medical structures (Kalvin, Alan D.; Cutting, Court B.; Haddad, Betsy; Noz, Marilyn E.)V1445,247-258(1991)

Convexity-based method for extracting object parts from 3-D surfaces (Vaina, Lucia M.; Zlateva, Stoyanka D.)V1468,710-719(1991)

Defocused white light speckle method for object contouring (Asundi, Anand K.)V1385,239-245(1991)

Dense-depth map from multiple views (Attolico, Giovanni; Caponetti, Laura; Chiaradia, Maria T.; Distante, Arcangelo; Stella, Ettore)V1383,34-46(1991)

Dense stereo correspondence using color (Jordan, John R.; Bovik, Alan C.)V1382,111-122(1991)

Depth cueing for visual search and cursor positioning (Reinhart, William F.)V1457,221-232(1991)

Depth determination using complex logarithmic mapping (Bartlett, Sandra L.; Jain, Ramesh C.)V1382,3-13(1991)

Design rules for pseudocolor transmission holographic stereograms (Andrews, John R.)V1507,407-415(1991)

Detection of unresolved target tracks in infrared imagery (Rajala, Sarah A.; Nolte, Loren W.; Aanstoos, James V.)V1606,360-371(1991)

Development of a stereoscopic three-dimensional drawing application (Carver, Donald E.; McAllister, David F.)V1457,54-65(1991)

Development of criteria to compare model-based texture analysis methods (Soh, Young-Sung; Murthy, S. N.; Huntsberger, Terrance L.)V1381,561-573(1991)

Dynamic integration of visual cues for position estimation (Das, Subhodev; Ahuja, Narendra)V1382,341-352(1991)

Dynamic range data acquisition and pose estimation for 3-D regular objects (Marszalec, Janusz A.; Heikkila, Tapio A.; Jarviluoma, Markku)V1382,443-452(1991)

Effect of viewing distance and disparity on perceived depth (Gooding, Linda; Miller, Michael E.; Moore, Jana; Kim, Seong-Han)V1457,259-266(1991)

Effects of alternate pictorial pathway displays and stereo 3-D presentation on simulated transport landing approach performance (Busquets, Anthony M.; Parrish, Russell V.; Williams, Steven P.)V1457,91-102(1991)

Efficient system for 3-D object recognition (Sobh, Tarek M.; Alameldin, Tarek K.)V1383,359-366(1991)

Efficient transformation algorithm for 3-D images (Vepsalainen, Ari M.; Rantala, Aarne E.)V1452,64-75(1991)

Electro-optical autostereoscopic displays using large cylindrical lenses (Hattori, Tomohiko)V1457,283-289(1991)

Elements of real-space imaging: a proposed taxonomy (Naimark, Michael)V1457,169-179(1991)

Energy-based segmentation of very sparse range surfaces (Lerner, Mark; Boult, Terrance E.)V1383,277-284(1991)

Error analysis on target localization from two projection images (Lee, Byung-Uk; Adler, John R.; Binford, Thomas O.)V1380,96-107(1991)

Estimation of three-dimensional motion in a 3-DTV image sequence (Dugelay, Jean-Luc)V1605,688-696(1991)

Evaluation of a pose estimation algorithm using single perspective view (Chandra, T.; Abidi, Mongi A.)V1382,409-426(1991)

Evaluation of interference fringe pattern on spatially curved objects (Laermann, Karl-Hans)V1554A,522-528(1991)

Experiments with perceptual grouping (Shiu, Yiu C.)V1381,130-141(1991)

Exploiting geometric relationships for object modeling and recognition (Walker, Ellen L.)V1382,353-363(1991)

Feature correspondence in multiple sensor data fusion (Broida, Ted J.)V1383,635-651(1991)

Fusion of multiple views of multiple reference points using a parallel distributed processing approach (Wolfe, William J.; Magee, Michael)V1383,20-25(1991)

Fusion of stereo views: estimating structure and motion using a robust method (Weng, Juyang; Cohen, Paul)V1383,321-332(1991)

Generation of synthetic stereo views from digital terrain models and digitized photographs (Bethel, James S.)V1457,49-53(1991)

Geometric Methods in Computer Vision (Vemuri, Baba C., ed.)V1570(1991)

Grasp-oriented sensing and control (Grupen, Roderic A.; Weiss, Richard S.; Oskard, David N.)V1383,189-201(1991)

Hand-eye coordination for grasping moving objects (Allen, Peter K.; Yoshimi, Billibon; Timcenko, Alexander; Michelman, Paul)V1383,176-188(1991)

Hierarchical target representation for autonomous recognition using distributed sensors (Luo, Ren C.; Kay, Michael G.)V1383,537-544(1991)

High-resolution fully 3-D mapping of human surfaces by laser array camera and data representations (Bae, Kyongtae T.; Altschuler, Martin D.)V1380,171-178(1991)

High-speed integrated rendering algorithm for interpreting multiple-variable 3-D data (Miyazawa, Tatsuo)V1459,36-47(1991)

Image quality metrics for volumetric laser displays (Williams, Rodney D.; Donohoo, Daniel)V1457,210-220(1991)

Implementation of a 3-D stereovision system for the production of customized orthotic accessories (Daher, Reinhard; McAdam, Wylie; Pizey, Gordon)V1526,90-93(1991)

Incorporation of time-dependent thermodynamic models and radiation propagation models into IR 3-D synthetic image generation models (Schott, John R.; Raqueno, Rolando; Salvaggio, Carl; Kraus, Eugene J.)V1540,533-549(1991)

Inspection of a class of industrial objects using a dense range map and CAD model (Ailisto, Heikki J.; Paakkari, Jussi; Moring, Ilkka)V1384,50-59(1991)

Integrated vision system for object identification and localization using 3-D geometrical models (Bidlack, Clint R.; Trivedi, Mohan M.)V1468,270-280(1991)

Intensity interpolation for branching in reconstructing three-dimensional objects from serial cross-sections (Liang, Cheng-Chung; Chen, Chin-Tu; Lin, Wei-Chung)V1445,456-467(1991)

Interaction of objects in a virtual environment: a two-point paradigm (Bryson, Steve T.)V1457,180-187(1991)

Interactive graphics system for multivariate data display (Becker, Richard A.; Cleveland, William S.; Shyu, William M.; Wilks, Allan R.)V1459,48-56(1991)

Interpolation of stereo data using Lagrangian polynomials (Bachnak, Rafic A.; Yamout, Jihad S.)V1457,27-36(1991)

Invariant reconstruction of 3-D curves and surfaces (Stevenson, Robert L.; Delp, Edward J.)V1382,364-375(1991)

Investigation of methods of combining functional evidence for 3-D object recognition (Stark, Louise; Hall, Lawrence O.; Bowyer, Kevin W.)V1381,334-345(1991)

Knowledge-based direct 3-D texture segmentation system for confocal microscopic images (Lang, Zhengping; Zhang, Zhen; Scarberry, Randell E.; Shao, Weimin; Sun, Xu-Mei)V1468,826-833(1991)

Learning procedure for the recognition of 3-D objects from 2-D images (Bart, Mischa; Buurman, Johannes; Duin, Robert P.)V1381,66-77(1991)

Luminance asymmetry in stereo TV images (Beldie, Ion P.; Kost, Bernd)V1457,242-247(1991)

Machine vision applications of image invariants: real-time processing experiments (Payton, Paul M.; Haines, Barry K.; Smedley, Kirk G.; Barrett, Eamon B.)V1406,58-71(1991)

Matching in image/object dual spaces (Zhang, Yaonan)V1526,195-202(1991)

Mean-field stereo correspondence for natural images (Klarquist, William N.; Acton, Scott T.; Ghosh, Joydeep)V1453,321-332(1991)

Measurement of fractal dimension using 3-D technique (Chuang, Keh-Shih; Valentino, Daniel J.; Huang, H. K.)V1445,341-347(1991)

Medical applications of three-dimensional and four-dimensional laser scanning of facial morphology (Sadler, Lewis L.; Chen, Xiaoming; Figueroa, Alvaro A.; Aduss, Howard)V1380,158-162(1991)

Method for the analysis of the 3-D shape of the face and changes in the shape brought about by facial surgery (Coombes, Anne M.; Linney, Alfred D.; Richards, Robin; Moss, James P.)V1380,180-189(1991)

Model-based flaw reconstruction using limited-view x-ray projections and flawless prototype image (Hung, Hsien-Sen; Eray, Mete)V1550,34-45(1991)

Model-based vision using geometric hashing (Akerman, Alexander; Patton, Ronald)V1406,30-39(1991)

Modeling and experiments with a subsea laser radar system (Bjarnar, Morten L.; Klepsvik, John O.; Nilsen, Jan E.)V1537,74-88(1991)

Modeling and visualization of scattered volumetric data (Nielson, Gregory M.; Dierks, Tim)V1459,22-33(1991)

Modeling nonhomogeneous 3-D objects for thermal and visual image synthesis (Karthik, Sankaran; Nandhakumar, N.; Aggarwal, Jake K.)V1468,686-697(1991)

Model of the left ventricle 3-D global motion: application to MRI data (Friboulet, Denis; Magnin, Isabelle E.; Mathieu, Christophe; Revel, D.; Amiel, Michel)V1445,106-117(1991)

Moire topography with the aid of phase-shift method (Yoshizawa, Toru; Tomisawa, Teiyu)V1554B,441-450(1991)

Monocular passive range sensing using defocus information (Prasad, K. V.; Mammone, Richard J.)V1385,280-291(1991)

More realistic and efficient algorithm for the drawing of 3-D space-filling molecular models (Wang, Yanqun)V1606,1027-1036(1991)

Motion analysis for visually-guided navigation (Hildreth, Ellen C.)V1382,167-180(1991)

Motion estimation without correspondences and object tracking over long time sequences (Goldgof, Dmitry B.; Lee, Hua; Huang, Thomas S.)V1383,109-121(1991)

Multilevel evidence fusion for the recognition of 3-D objects: an overview of computer vision research at IBM/T.J. Watson (Bolle, Ruud M.; Califano, Andrea; Kender, John R.; Kjeldsen, Rick; Mohan, Rakesh)V1383,305-318(1991)

Multiresolution range-guided stereo matching (Tate, Kevin J.; Li, Ze-Nian)V1383,491-502(1991)

Multisensor approach in linear-Gaussian estimation of 3-D angular motion (Algrain, Marcelo C.; Saniie, Jafar)V1478,201-210(1991)

Natural pixel decomposition for interferometric tomography (Cha, Dong J.; Cha, Soyoung S.)V1554B,600-609(1991)

Neural model for feature matching in stereo vision (Wang, Shengrui; Poussart, Denis; Gagne, Simon)V1382,37-48(1991)

Neural network for improving terrain elevation measurement from stereo images (Jordan, Michael)V1567,179-187(1991)

New cooperative edge linking (Bonnin, Patrick; Zavidovique, Bertrand)V1381,142-152(1991)

New method for constructing 3-D liver from CT images (Sun, Yung-Nien; Chen, Jiann-Jone; Lin, Xi-Zhang; Mao, Chi-Wu)V1606,653-664(1991)

New method for designing face image classifiers using 3-D CG model (Akamatsu, Shigeru; Sasaki, Tsutomu; Masui, Nobuhiko; Fukamachi, Hideo; Suenaga, Yasuhito)V1606,204-216(1991)

New method of 3-D shape measurement by moire technique (He, Anzhi; Li, Qun Z.; Miao, Peng-Cheng)V1545,278-281(1991)

Object-oriented data management for interactive visual analysis of three-dimensional fluid-flow models (Walther, Sandra S.; Peskin, Richard L.)V1459,232-243(1991)

Object-oriented strategies for a vision dimensional metrology system (Pizzi, Nicolino J.; El-Hakim, Sabry F.)V1468,296-304(1991)

Octree optimization (Globus, Al)V1459,2-10(1991)

On-focal-plane-array feature extraction using a 3-D artificial neural network (3DANN): Part I (Carson, John C.)V1541,141-144(1991)

On-focal-plane-array feature extraction using 3-D artificial neural network (3DANN): Part II (Carson, John C.)V1541,227-231(1991)

Optical 3-D sensing for measurement of bottomhole pattern (Su, Wan-Yong; Su, Xianyu)V1567,680-682(1991)

Optical three-dimensional sensing for measurement of bottomhole pattern (Su, Wan-Yong; Su, Xianyu)V1332,820-823(1991)

Optics, Illumination, and Image Sensing for Machine Vision V (Harding, Kevin G.; Svetkoff, Donald J.; Uber, Gordon T.; Wittels, Norman, eds.)V1385(1991)

Parallel algorithm for volumetric segmentation (Liou, Shih-Ping; Jain, Ramesh C.)V1381,447-458(1991)

Parallel constructs for three-dimensional registration on a SIMD (single-instruction stream/multiple-data stream) processor (Morioka, Craig A.; Chan, Kelby K.; Huang, H. K.)V1445,534-538(1991)

Perceptual training with cues for hazard detection in off-road driving (Merritt, John O.; CuQlock-Knopp, V. G.)V1457,133-138(1991)

Phase-shift moire camera for real-time measurements of three-dimensional shape information (Turney, Jerry L.; Lysogorski, Charles; Gottschalk, Paul G.; Chiu, Arnold H.)V1380,53-63(1991)

Photoelastical mixed-solution method of contact problems of the roll-shape member of limited length (Qu, Zhihao; Jiang, Weixing; Peng, Huihong)V1554A,503-510(1991)

Photogrammetric measurements of retinal nerve fiber layer thickness along the disc margin and peripapillary region (Takamoto, Takenori; Schwartz, Bernard)V1380,64-74(1991)

Polynomial approach for morphological operations on 2-D and 3-D images (Bhattacharya, Prabir; Qian, Kai)V1383,530-536(1991)

Positioning accuracy of a virtual stereographic pointer in a real stereoscopic video world (Drascic, David; Milgram, Paul)V1457,302-313(1991)

Practical low-cost stereo head-mounted display (Pausch, Randy; Dwivedi, Pramod; Long, Allan C.)V1457,198-208(1991)

Preliminary tests of maximum likelihood image reconstruction method on 3-D real data and some practical considerations for the data corrections (Liu, Yi-Hwa; Holmes, Timothy J.; Koshy, Matthew)V1428,191-199(1991)

Probabilistic modeling of surfaces (Szeliski, Richard)V1570,154-165(1991)

Progress in autostereoscopic display technology at Dimension Technologies Inc. (Eichenlaub, Jesse B.)V1457,290-301(1991)

Psychophysical estimation of the human depth combination rule (Landy, Michael S.; Maloney, Laurence T.; Young, Mark J.)V1383,247-254(1991)

Qualitative three-dimensional shape from stereo (Wildes, Richard P.)V1382,453-463(1991)

Quantitative analysis of three-dimensional landmark coordinate data (Richtsmeier, Joan T.)V1380,12-23(1991)

Radiative tetrahedral lattices (Driver, Jesse W.; Buckalew, Chris)V1459,109-116(1991)

Range data analysis using cross-stripe structured-light system (Kim, Whoi-Yul)V1385,216-228(1991)

Range data from stereo images of edge points (Lim, Hong-Seh)V1382,434-442(1991)

Real-time optical scanning system for measurement of chest volume changes during anesthesia (Duffy, Neil J.; Drummond, Gordon D.; McGowan, Steve; Dessesard, Pascal)V1380,46-52(1991)

Recent applications of biostereometrics in research and education (Herron, Robin E.)V1380,2-5(1991)

Recognition, tracking, and pose estimation of arbitrarily shaped 3-D objects in cluttered intensity and range imagery (Gottschalk, Paul G.; Turney, Jerry L.; Chiu, Arnold H.; Mudge, Trevor N.)V1383,84-96(1991)

Reconstruction of three-dimensional displacement fields by carrier holography (Xu, Zhu; Chen, Ke-long; Wen, Zhen-chu; Yang, Han-Guo; He, Xiao-yuan; Zhang, Bao-he)V1399,172-177(1991)

Recovering 3-D translation of a rigid surface by a binocular observer using moments (Al-Hudaithi, Aziz; Udpa, Satish S.)V1567,490-501(1991)

Recovering epipolar geometry in 3-D vision systems (Schenk, Anton F.; Toth, Charles)V1457,66-73(1991)

Recursive computation of a wire-frame representation of a scene from dynamic stereo using belief functions (Tirumalai, Arun P.; Schunck, Brian G.; Jain, Ramesh C.)V1569,28-42(1991)

Reduced defocus degradation in a system for high-speed three-dimensional digital microscopy (Quesenberry, Laura A.; Morris, V. A.; Neering, Ian R.; Taylor, Stuart R.)V1428,177-190(1991)

Remote 3-D laser topographic mapping with dental application (Altschuler, Bruce R.)V1380,238-247(1991)

Remote driving: one eye or two (Bryant, Keith; Ince, Ilhan)V1457,120-132(1991)

Representing the dynamics of the occluding contour (Seales, W. B.; Dyer, Charles R.)V1383,47-58(1991)

Restoration of distorted depth maps calculated from stereo sequences (Damour, Kevin; Kaufman, Howard)V1452,78-89(1991)

Reversible image data compression based on HINT (hierarchical interpolation) decorrelation and arithmetic coding (Roos, Paul; Viergever, Max A.)V1444,283-290(1991)

Robust self-calibration and evidential reasoning for building environment maps (Tirumalai, Arun P.; Schunck, Brian G.; Jain, Ramesh C.)V1383,345-358(1991)

Rule-based system to reconstruct 3-D tree structure from two views (Liu, Iching; Sun, Ying)V1606,67-77(1991)

Scene description: an iterative approach (Mulgaonkar, Prasanna G.; Decurtins, Jeff; Cowan, Cregg K.)V1382,320-330(1991)

Segmentation using range data and structured light (Hu, Gongzhu)V1381,482-489(1991)

Selection devices for field-sequential stereoscopic displays: a brief history (Lipton, Lenny)V1457,274-282(1991)

Sensor fusion using K-nearest neighbor concepts (Scott, David R.; Flachs, Gerald M.; Gaughan, Patrick T.)V1383,367-378(1991)

Shape-from-X: psychophysics and computation (Buelthoff, Heinrich H.; Yuille, Alan L.)V1383,235-246(1991)

Shape perception from binocular disparity and structure-from-motion (Tittle, James S.; Braunstein, Myron L.)V1383,225-234(1991)

Shape reconstruction and object recognition using angles in an image (Fukada, Youji)V1381,111-121(1991)

Simple technique for 3-D displacements measurement using synthesis of holographic interferometry and speckle photography (Kwon, Young H.; Park, Seung O.; Park, B. C.)V1554A,639-644(1991)

Simplified pupil enlargement technique (Radl, Bruce M.)V1457,314-317(1991)

Simultaneous graphics and multislice raster image display for interactive image-guided surgery (Edwards, Charles A.; Galloway, Robert L.; Thomas, Judith G.; Schreiner, Steven; Maciunas, Robert J.)V1444,38-46(1991)

Single-lens moire contouring method (Harding, Kevin G.; Kaltenbacher, Eric; Bieman, Leonard H.)V1385,246-255(1991)

Slice plane generation for three-dimensional image viewing using multiprocessing (Ho, Bruce K.; Ma, Marco; Chuang, Keh-Shih)V1445,95-100(1991)

Solid models for CT/MR image display: accuracy and utility in surgical planning (Mankovich, Nicholas J.; Yue, Alvin; Ammirati, Mario; Kioumehr, Farhad; Turner, Scott)V1444,2-8(1991)

Some effects on depth-position and course-prediction judgments in 2-D and 3-D displays (Miller, Robert H.; Beaton, Robert J.)V1457,248-258(1991)

Spatial frequency analysis for the computer-generated holography of 3-D objects (Tommasi, Tullio; Bianco, Bruno)V1507,136-141(1991)

Statistical and neural network classifiers in model-based 3-D object recognition (Newton, Scott C.; Nutter, Brian S.; Mitra, Sunanda)V1382,209-218(1991)

Stereo matching, error detection, and surface reconstruction (Stewart, Charles V.)V1383,285-296(1991)

Stereoscopic Displays and Applications II (Merritt, John O.; Fisher, Scott S., eds.)V1457(1991)

Stereoscopic ray tracing of implicitly defined surfaces (Devarajan, Ravinder; McAllister, David F.)V1457,37-48(1991)

Stereoscopic versus orthogonal view displays for performance of a remote manipulation task (Spain, Edward H.; Holzhausen, Klause-Peter)V1457,103-110(1991)

Stereoscopic video and the quest for virtual reality: an annotated bibliography of selected topics (Starks, Michael R.)V1457,327-342(1991)

Stereotactic multibeam radiation therapy system in a PACS environment (Fresne, Francoise; Le Gall, G.; Barillot, Christian; Gibaud, Bernard; Manens, J. P.; Toumoulin, Christine; Lemoine, D.; Chenal, C.; Scarabin, Jean-Marie)V1444,26-36(1991)

Stereo vision: a neural network application to constraint satisfaction problem (Mousavi, Madjid S.; Schalkoff, Robert J.)V1382,228-239(1991)

Stereo vision for planetary rovers: stochastic modeling to near-real-time implementation (Matthies, Larry H.)V1570,187-200(1991)

Storing and managing three-dimensional digital medical image information (Chapman, Michael A.; Denby, N.)V1526,190-194(1991)

Structure and motion of entire polyhedra (Sobh, Tarek M.; Alameldin, Tarek K.)V1388,425-431(1991)

Structure of a scene from two and three projections (Tommasi, Tullio)V1383,26-33(1991)

Study and characterization of three-dimensional angular distributions of acoustic scattering from spheroidal targets (George, Jacob; Werby, Michael F.)V1471,66-77(1991)

Study of active 3-D terrain mapping for helicopter landings (Velger, Mordekhai; Toker, Gregory)V1478,168-176(1991)

Surface definition technique for clinical imaging (Liao, Wen-gen; Simovsky, Ilya; Li, Andrew; Kramer, David M.; Kaufman, Leon; Rhodes, Michael L.)V1444,47-55(1991)

Surface digitizing of anatomical subjects with DIGIBOT-4 (Koch, Stephen; Koch, Eric)V1380,163-170(1991)

Surface reconstruction method using deformable templates (Wang, Yuan-Fang; Wang, Jih-Fang)V1383,265-276(1991)

Surface reconstruction with discontinuities (Lee, David T.; Shiau, Jyh-Jen H.)V1383,297-304(1991)

Synthesized holograms in medicine and industry (Tsujiuchi, Jumpei)VIS08,326-334(1991)

Techniques for integrating 3-D optical systems (Brenner, Karl-Heinz)V1544,263-270(1991)

Teleoperator performance with virtual window display (Cole, Robert E.; Merritt, John O.; Coleman, Richard; Ikehara, Curtis)V1457,111-119(1991)

Three-dimensional automatic precision measurement system by liquid-crystal plate on moire topography (Arai, Yasuhiko; Yekozeki, Shunsuke; Yamada, Tomoharu)V1554B,266-274(1991)

Three-dimensional color Doppler imaging of the carotid artery (Picot, Paul A.; Rickey, Daniel W.; Mitchell, J. R.; Rankin, Richard N.; Fenster, Aaron)V1444,206-213(1991)

Three-dimensional CT image segmentation by volume growing (Zhu, Dongping; Conners, Richard W.; Araman, Philip A.)V1606,685-696(1991)

Three-dimensional description of symmetric objects from range images (Alvertos, Nicolas; D'Cunha, Ivan)V1382,388-396(1991)

Three-dimensional face model reproduction method using multiview images (Nagashima, Yoshio; Agawa, Hiroshi; Kishino, Fumio)V1606,566-573(1991)

Three-dimensional fibre-optic position sensor (Yang, Q.; Butler, Clive)V1572,558-563(1991)

Three-dimensional gauging with stereo computer vision (Wong, Kam W.; Ke, Ying; Lew, Michael; Obaidat, Mohammed T.)V1526,17-26(1991)

Three-dimensional imaging laparoscope (Jones, Edwin R.; McLaurin, A. P.; Mason, J. L.)V1457,318-321(1991)

Three-dimensional imaging using TDI CCD sensors (Fenster, Aaron; Holdsworth, David W.; Drangova, Maria)V1447,28-33(1991)

Three-dimensional inspection using laser-based dynamic fringe projection (Harvey, David M.; Shaw, Michael M.; Hobson, Clifford A.; Wood, Christopher M.; Atkinson, John T.; Lalor, Michael J.)V1400,86-93(1991)

Three-dimensional integration of optical systems (Brenner, Karl-Heinz)V1506,94-98(1991)

Three-dimensional lithography: laser modeling using photopolymers (Heller, Timothy B.)V1454,272-282(1991)

Three-dimensional machine vision using line-scan sensors (Godber, Simon X.; Robinson, Max; Evans, Paul)V1526,170-189(1991)

Three-dimensional magnetic resonance imaging of the head (Keeler, Elaine K.; Oyen, Ordean J.)V1380,24-32(1991)

Three-dimensional motion analysis and structure recovering by multistage Hough transform (Nakajima, Shigeyoshi; Zhou, Mingyong; Hama, Hiromitsu; Yamashita, Kazumi)V1605,709-719(1991)

Three-dimensional moving-image display by modulated coherent optical fibers: a proposal (Hoshino, Hideshi; Sato, Koki)V1461,227-231(1991)

Three-dimensional object recognition using multiple sensors (Hackett, Jay K.; Lavoie, Matt J.; Shah, Mubarak A.)V1383,611-622(1991)

Three-dimensional object representation by array grammars (Wang, Patrick S.)V1381,210-216(1991)

Three-dimensional orientation from texture using Gabor wavelets (Super, Boaz J.; Bovik, Alan C.)V1606,574-586(1991)

Three-dimensional position determination from motion (Nashman, Marilyn; Chaconas, Karen)V1383,166-175(1991)

Three-dimensional reconstruction from cone beam projections (Ohishi, Satoru; Yamaguchi, Masahiro; Ohyama, Nagaaki; Honda, Toshio)V1443,280-285(1991)

Three-dimensional reconstruction from optical flow using temporal integration (Rangachar, Ramesh M.; Hong, Tsai-Hong; Herman, Martin; Luck, Randall L.)V1382,331-340(1991)

Three-dimensional reconstruction of liver from 2-D tomographic slices (Chou, Jin-Shin; Chen, Chin-Tu; Giger, Maryellen L.; Kahn, Charles E.; Bae, Kyongtae T.; Lin, Wei-Chung)V1396,45-50(1991)

Three-dimensional reconstruction using virtual planes and horopters (Grosso, Enrico; Sandini, Giulio; Tistarelli, Massimo)V1570,371-381(1991)

Three-dimensional scene interpretation through information fusion (Shen, Sylvia S.)V1382,427-433(1991)

Three-dimensional scene reconstruction using optimal information fusion (Hong, Lang)V1383,333-344(1991)

Three-dimensional shape restoration using virtual grating phase detection from deformed grating (Zhou, Shaoxiang; Jiang, Jinyou; Wang, Qimin)V1358,788-792(1991)

Three-dimensional simulation of optical lithography (Toh, Kenny K.; Neureuther, Andrew R.)V1463,356-367(1991)

Three-dimensional subband decompositions for hierarchical video coding (Bosveld, Frank; Lagendijk, Reginald L.; Biemond, Jan)V1605,769-780(1991)

Three-dimensional tolerance verification for computer vision systems (Griffin, Paul M.; Taboada, John)V1381,292-298(1991)

Three-dimensional vision and figure-ground separation by visual cortex (Grossberg, Stephen)V1382,2-2(1991)

Three-dimensional vision: requirements and applications in a space environment (Noseworthy, J. R.; Gerhardt, Lester A.)V1387,26-37(1991)

Three-dimensional visualization and quantification of evolving amorphous objects (Silver, Deborah E.; Zabusky, Norman J.)V1459,97-108(1991)

Topographic mapping for stereo and motion processing (Mallot, Hanspeter A.; Zielke, Thomas; Storjohann, Kai; von Seelen, Werner)V1382,397-408(1991)

Triangulating between parallel splitting contours using a simplicial algorithm (Miranda, Rick; McCracken, Thomas O.; Fedde, Chris)V1380,210-217(1991)

Use of a 3-D visualization system in the planning and evaluation of facial surgery (Linney, Alfred D.; Moss, James P.; Richards, Robin; Mosse, C. A.; Grindrod, S. R.; Coombes, Anne M.)V1380,190-199(1991)

Use of anatomical knowledge to register 3-D blood vessel data derived from DSA with MR images (Hill, Derek L.; Hawkes, David J.; Hardingham, Charles R.)V1445,348-357(1991)

Use of flicker-free television products for stereoscopic display applications (Woods, Andrew J.; Docherty, Tom; Koch, Rolf)V1457,322-326(1991)

Use of scattered-light photoelasticity at crack tips (Ravi-Chandar, K.)V1554A,228-238(1991)

User benefits of visualization with 3-D stereoscopic displays (Wichansky, Anna M.)V1457,267-271(1991)

Variable emissivity plates under a three-dimensional sky background (Meitzler, Thomas J.; Gonda, Teresa G.; Jones, Jack C.; Reynolds, William R.)V1486,380-389(1991)

Vesalius project: interactive computers in anatomical instruction (McCracken, Thomas O.; Roper, Stephen D.; Spurgeon, Thomas L.)V1380,6-10(1991)

Virtual environment for the exploration of three-dimensional steady flows (Bryson, Steve T.; Levit, Creon C.)V1457,161-168(1991)

Virtual environment system for simulation of leg surgery (Pieper, Steve; Delp, Scott; Rosen, Joseph; Fisher, Scott S.)V1457,188-197(1991)

Visual inspection system using multidirectional 3-D imager (Koezuka, Tetsuo; Kakinoki, Yoshikazu; Hashinami, Shinji; Nakashima, Masato)V1332,323-331(1991)

Visualization and volumetric compression (Chan, Kelby K.; Lau, Christina C.; Chuang, Keh-Shih; Morioka, Craig A.)V1444,250-255(1991)

Visualization of liver in 3-D (Chen, Chin-Tu; Chou, Jin-Shin; Giger, Maryellen L.; Kahn, Charles E.; Bae, Kyongtae T.; Lin, Wei-Chung)V1444,75-77(1991)

Visualization of manufacturing process data in N-dimensional spaces: a reanalysis of the data (Fulop, Ann C.; Allen, Donald M.; Deffner, Gerhard)V1459,69-76(1991)

X-ray photogrammetry of the hip revisited (Turner-Smith, Alan R.; White, Steven P.; Bulstrode, Christopher)V1380,75-84(1991)

Tomography—see also medical imaging

Adaptive projection technique for CT images refinement (Kuo, Shyh-Shiaw; Mammone, Richard J.)V1606,641-652(1991)

Aero-optics analysis using moire deflectometry (Abushagur, Mustafa A.; Elmanasreh, Ahmed)V1554B,298-302(1991)

Automated techniques for quality assurance of radiological image modalities (Goodenough, David J.; Atkins, Frank B.; Dyer, Stephen M.)V1444,87-99(1991)

Automated three-dimensional registration of medical images (Neiw, Han-Min; Chen, Chin-Tu; Lin, Wei-Chung; Pelizzari, Charles A.)V1445,259-264(1991)

Automatic adjustment of display window (gray level) for MR images using a neural network (Ohhashi, Akinami; Yamada, Shinichi; Haruki, Kazuhito; Hatano, Hisaaki; Fujii, Yumi; Yamaguchi, Koujiro; Ogata, Hakaru)V1444,63-74(1991)

Clinical image-intensifier-based volume CT imager for angiography (Ning, Ruola; Barsotti, John B.; Kido, Daniel K.; Kruger, Robert A.)V1443,236-249(1991)

Comparison of three-dimensional surface rendering techniques (Thomas, Judith G.; Galloway, Robert L.; Edwards, Charles A.; Haden, Gerald L.; Maciunas, Robert J.)V1444,379-388(1991)

Composite PET and MRI for accurate localization and metabolic modeling: a very useful tool for research and clinic (Bidaut, Luc M.)V1445,66-77(1991)

Cone beam for medical imaging and NDE (Smith, Bruce D.)V1450,13-17(1991)

CT image processing for hardwood log inspection (Zhu, Dongping; Conners, Richard W.; Araman, Philip A.)V1567,232-243(1991)

CT imaging with an image intensifier: using a radiation therapy simulator as a CT scanner (Silver, Michael D.; Nishiki, Masayuki; Tochimura, Katsumi; Arita, Masataka; Drawert, Bruce M.; Judd, Thomas C.)V1443,250-260(1991)

Detection and visualization of porosity in industrial CT scans of aluminum die castings (Andrews, Lee T.; Klingler, Joseph W.; Schindler, Jeffery A.; Begemann, Michael S.; Farron, Donald; Vaughan, Bobbi; Riggs, Bud; Cestaro, John)V1459,125-135(1991)

Development of a low-cost computed tomography image processing system (Hughes, Simon H.; Slocum, Robert E.)V1396,575-581(1991)

Development of computer-aided functions in clinical neurosurgery with PACS (Mukasa, Minoru; Aoki, Makoto; Satoh, Minoru; Kowada, Masayoshi; Kikuchi, K.)V1446,253-265(1991)

Diffuse tomography (Gruenbaum, F. A.; Kohn, Philip D.; Latham, Geoff A.; Singer, Jay R.; Zubelli, Jorge P.)V1431,232-238(1991)

Dynamic spatial reconstructor: a high-speed, stop-action, 3-D, digital radiographic imager of moving internal organs and blood (Jorgensen, Steven M.; Whitlock, S. V.; Thomas, Paul J.; Roessler, R. W.; Ritman, Erik L.)V1346,180-191(1991)

Hartmann-Shack sensor as a component in active optical system to improve the depth resolution of the laser tomographic scanner (Liang, Junzhong; Grimm, B.; Goelz, Stefan; Bille, Josef F.)V1542,543-554(1991)

High-precision video frame grabber for computed tomography (Drawert, Bruce M.; Slocum, Robert E.)V1396,566-567(1991)

Holographic-coordinate-transform-based system for direct Fourier tomographic reconstruction (Huang, Qiang; Freeman, Mark O.)V1564,644-655(1991)

Incorporation of structural CT and MR images in PET image reconstruction (Chen, Chin-Tu; Ouyang, Xiaolong; Ordonez, Caesar; Hu, Xiaoping; Wong, Wing H.; Metz, Charles E.)V1445,222-225(1991)

Integrated photoelasticity for residual stresses in glass specimens of complicated shape (Aben, Hillar K.; Idnurm, S. J.; Josepson, J. I.; Kell, K.-J. E.; Puro, A. E.)V1554A,298-309(1991)

Laser-scanning tomography and related dark-field nanoscopy method (Montgomery, Paul C.; Gall, Pascal; Ardisasmita, Moh S.; Castagne, Michel; Bonnafe, Jacques; Fillard, Jean-Pierre B.)V1332,563-570(1991)

Low-cost computed tomography system using an image intensifier (Hoeft, Gregory L.; Hughes, Simon H.; Slocum, Robert E.)V1396,638-645(1991)

MAP image reconstruction using intensity and line processes for emission tomography data (Yan, Xiao-Hong; Leahy, Richard M.)V1452,158-169(1991)

Mapping crystal defects with a digital scanning ultramicroscope (Springer, John M.; Silberman, Enrique; Kroes, Roger L.; Reiss, Don)V1557,192-196(1991)

Medical Imaging V: Image Physics (Schneider, Roger H., ed.)V1443(1991)

Model-based flaw reconstruction using limited-view x-ray projections and flawless prototype image (Hung, Hsien-Sen; Eray, Mete)V1550,34-45(1991)

Natural pixel decomposition for interferometric tomography (Cha, Dong J.; Cha, Soyoung S.)V1554B,600-609(1991)

Near-infrared imaging in vivo: imaging of Hb oxygenation in living tissues (Araki, Ryuichiro; Nashimoto, Ichiro)V1431,321-332(1991)

New algorithm and an efficient parallel implementation of the expectation maximization technique in PET (positron emission tomography) imaging (Buyukkoc, Cagatay; Persiano, G.)V1452,170-179(1991)

New method for constructing 3-D liver from CT images (Sun, Yung-Nien; Chen, Jiann-Jone; Lin, Xi-Zhang; Mao, Chi-Wu)V1606,653-664(1991)

Noninvasive hemoglobin oxygenation monitor and computed tomography by NIR spectrophotometry (Oda, Ichiro; Ito, Yasunobu; Eda, Hideo; Tamura, Tomomi; Takada, Michinosuke; Abumi, Rentaro; Nagai, Katumi; Nakagawa, Hachiro; Tamura, Masahide)V1431,284-293(1991)

Optical computer-assisted tomography realized by coherent detection imaging incorporating laser heterodyne method for biomedical applications (Inaba, Humio; Toida, Masahiro; Ichimura, Tsutomu)V1399,108-115(1991)

Optical tomography by heterodyne holographic interferometry (Vukicevic, Dalibor; Neger, Theo; Jaeger, Helmut; Woisetschlaeger, Jakob; Philipp, Harald)VIS08,160-193(1991)

Optoelectronic implementation of filtered-back-projection tomography algorithm (Lu, Tongxin; Udpa, Satish S.; Udpa, L.)V1564,704-713(1991)

Pattern recognition in pulmonary computerized tomography images using Markovian modeling (Preteux, Francoise; Moubarak, Michel; Grenier, Philippe)V1450,72-83(1991)

PET reconstruction using multisensor fusion techniques (Acharya, Raj S.; Gai, Nevile)V1445,207-221(1991)

Positron emission tomography of laser-induced interstitial hyperthermia in cerebral gliomas (Ulrich, Frank; Bettag, Martin; Langen, K. J.)V1428,135-135(1991)

Resolution enhancement of CT images (Kuo, Shyh-Shiaw; Mammone, Richard J.)V1450,18-29(1991)

Simulated annealing image reconstruction for an x-ray coded source tomograph (El Alaoui, Mohsine; Magnin, Isabelle E.; Amiel, Michel)V1569,80-87(1991)

Simulation of time-resolved optical-CT imaging (Yamada, Yukio; Hasegawa, Yasuo)V1431,73-82(1991)

Simultaneous patient translation during CT scanning (Crawford, Carl R.; King, Kevin F.)V1443,203-213(1991)

Spline-based tomographic reconstruction method (Guedon, Jean-Pierre; Bizais, Yves)V1443,214-225(1991)

Three-dimensional CT image segmentation by volume growing (Zhu, Dongping; Conners, Richard W.; Araman, Philip A.)V1606,685-696(1991)

Three-dimensional reconstruction from cone beam projection by a block iterative technique (Peyrin, Francoise; Goutte, Robert; Amiel, Michel)V1443,268-279(1991)

Three-dimensional reconstruction from cone beam projections (Ohishi, Satoru; Yamaguchi, Masahiro; Ohyama, Nagaaki; Honda, Toshio)V1443,280-285(1991)

Three-dimensional reconstruction of liver from 2-D tomographic slices (Chou, Jin-Shin; Chen, Chin-Tu; Giger, Maryellen L.; Kahn, Charles E.; Bae, Kyongtae T.; Lin, Wei-Chung)V1396,45-50(1991)

Tomographic imaging using picosecond pulses of light (Hebden, Jeremy C.; Kruger, Robert A.; Wong, K. S.)V1443,294-300(1991)

Use of cross-validation as a stopping rule in emission tomography image reconstruction (Coakley, Kevin J.; Llacer, Jorge)V1443,226-233(1991)

Using correlated CT images in compensation for attenuation in PET (positron emission tomography) image reconstruction (Yu, Xiaolin; Chen, Chin-Tu; Bartlett, R.; Pelizzari, Charles A.; Ordonez, Caesar)V1396,56-58(1991)

Visualization of liver in 3-D (Chen, Chin-Tu; Chou, Jin-Shin; Giger, Maryellen L.; Kahn, Charles E.; Bae, Kyongtae T.; Lin, Wei-Chung)V1444,75-77(1991)

X-ray projection microscopy and cone-beam microtomography (Wang, Ge; Lin, T. H.; Cheng, Ping-Chin; Shinozaki, D. M.; Newberry, S. P.)V1398,180-190(1991)

Tracking/ranging

Acquisition and tracking performance measurements for a high-speed area array detector system (Short, Ralph C.; Cosgrove, Michael A.; Clark, David L.; Martino, Anthony J.; Park, Hong-Woo; Seery, Bernard D.)V1417,131-141(1991)

Acquisition, Tracking, and Pointing V (Masten, Michael K.; Stockum, Larry A., eds.)V1482(1991)

Adaptive detection of subpixel targets using multiband frame sequences (Stocker, Alan D.; Yu, Xiaoli; Winter, Edwin M.; Hoff, Lawrence E.)V1481,156-169(1991)

Adaptive selection of sensors based on individual performances in a multisensor environment (Parra-Loera, Ramon; Thompson, Wiley E.; Salvi, Ajit P.)V1470,30-36(1991)

Airborne electro-optical sensor: performance predictions and design considerations (Mishra, R. K.; Pillai, A. M.; Sheshadri, M. R.; Sarma, C. G.)V1482,138-145(1991)

Airborne Reconnaissance XV (Augustyn, Thomas W.; Henkel, Paul A., eds.)V1538(1991)

Airborne seeker evaluation and test system (Jollie, William B.)V1482,92-103(1991)

Algorithm for statistical classification of radar clutter into one of several categories (Nechval, Nicholas A.)V1470,282-293(1991)

Algorithms and architectures for implementing large-velocity filter banks (Stocker, Alan D.; Jensen, Preben D.)V1481,140-155(1991)

Analytic approach to centroid performance analysis (Schultz, Kenneth I.)V1416,199-208(1991)

Angle-only tracking and prediction of boost vehicle position (Tsai, Ming-Jer; Rogal, Fannie A.)V1481,281-291(1991)

Angular positioning mechanism for the ultraviolet coronagraph spectrometer (Ostaszewski, Miroslaw A.; Guy, Larry J.)V1482,13-25(1991)

Application of MHT to group-to-object tracking (Kovacich, Michael; Casaletto, Tom; Lutjens, William; McIntyre, David; Ansell, Ralph; VanDyk, Ed)V1481,357-370(1991)

Application of neural networks to range-Doppler imaging (Wu, Xiaoqing; Zhu, Zhaoda)V1569,484-490(1991)

Application of perceptron to the detecting of particle motion (Li, Jie-gu; Yuan, Qiang)V1469,178-187(1991)

Application of the ProtoWare simulation testbed to the design and evaluation of advanced avionics (Bubb, Daniel; Wilson, Leo T.; Stoltz, John R.)V1483,18-28(1991)

Application results for an augmented video tracker (Pierce, Bill)V1482,182-195(1991)

Applications of laser ranging to ocean, ice, and land topography (Degnan, John J.)V1492,176-186(1991)

Approximation-based video tracking system (Deng, Keqiang; Wilson, Joseph N.)V1568,304-312(1991)

Asynchronous data fusion for target tracking with a multitasking radar and optical sensor (Blair, William D.; Rice, Theodore R.; Alouani, Ali T.; Xia, P.)V1482,234-245(1991)

ATR performance modeling for building multiscenario adaptive systems (Nasr, Hatem N.)V1483,112-117(1991)

Augmented tracking and acquisition system for GBL satellite illumination (Brodsky, Aaron; Goodrich, Alan; Lawson, David G.; Holm, Richard W.)V1482,159-169(1991)

Automatic Object Recognition (Sadjadi, Firooz A., ed.)V1471(1991)

Automatic target detection for surveillance (Ramesh, Nagarajan; Sethi, Ishwar K.; Cheung, Huey)V1468,72-80(1991)

Bayesian estimation of smooth object motion using data from direction-sensitive velocity sensors (Fong, David Y.; Pomalaza-Raez, Carlos A.)V1569,156-161(1991)

Beam-tracker and point-ahead system for optical communications II: servo performance (LaSala, Paul V.; McLaughlin, Chris)V1482,121-137(1991)

Bounds on the performance of optimal four-dimensional filters for detection of low-contrast IR point targets (Wohlers, Martin R.)V1481,129-139(1991)

Butterfly line scanner: rotary twin reflective deflector that desensitizes scan-line jitter to wobble of the rotational axis (Marshall, Gerald F.; Vettese, Thomas; Carosella, John H.)V1454,37-45(1991)

Carbon dioxide eyesafe laser rangefinders (Powell, Richard K.; Berdanier, Barry N.; McKay, James)V1419,126-140(1991)

Centroid tracking of range-Doppler images (Kachelmyer, Alan L.; Nordquist, David P.)V1416,184-198(1991)

Challenges of vision theory: self-organization of neural mechanisms for stable steering of object-grouping data in visual motion perception (Marshall, Jonathan A.)V1569,200-215(1991)

CiNeRaMa model: a useful tool for detection range estimate (Talmore, Eli T.)V1442,362-371(1991)

Circular streak camera application for satellite laser ranging (Prochazka, Ivan; Hamal, Karel; Kirchner, G.; Schelev, Mikhail Y.; Postovalov, V. E.)V1449,116-120(1991)

Combination-matching problems in the layout design of minilaser rangefinder (Wang, Erqi; Song, Dehui)V1400,124-128(1991)

Compact high-accuracy Inductosyn-based gimbal control system (Liebst, Brad; Verbanets, William R.; Kimbrell, James E.)V1482,425-438(1991)

Computing motion parameters from sparse multisensor range data for telerobotics (Vemuri, Baba C.; Skofteland, G.)V1383,97-108(1991)

Control of night vision pilotage systems (Heaton, Mark W.; Ewing, William S.)V1482,444-457(1991)

Conversion of sensor data for real-time scene generation (Libby, Vibeke; Bardin, R. K.)V1470,59-64(1991)

Correlation tracking: a new technology applied to laser photocoagulation (Forster, Albert A.)V1423,103-104(1991)

Cramer-Rao bound for multiple-target tracking (Daum, Frederick E.)V1481,341-344(1991)

Data Structures and Target Classification (Libby, Vibeke, ed.)V1470(1991)

Design and analysis of the closed-loop pointing system of a scientific satellite (Heyler, Gene A.; Garlick, Dean S.; Yionoulis, Steve M.)V1481,198-208(1991)

Design and testing of a cube-corner array for laser ranging (James, William E.; Steel, William H.; Evans, Nelson O.)V1400,129-136(1991)

Design and testing of data fusion systems for the U.S. Customs Service drug interdiction program (Stoltz, John R.; Cole, Donald C.)V1479,423-434(1991)

Design of an analog VLSI chip for a neural network target tracker (Narathong, Chiewcharn; Inigo, Rafael M.)V1452,523-531(1991)

Detection and tracking of small targets in persistence (Toumodge, Shawn S.)V1481,221-232(1991)

Detection of contraband brought into the United States by aircraft and other transportation methods: a changing problem (Bruder, Joseph A.; Greneker, E. F.; Nathanson, F. E.; Henneberger, T. C.)V1479,316-321(1991)

Detection of moving subpixel targets in infrared clutter with space-time filtering (Braunreiter, Dennis C.; Banh, Nam D.)V1481,73-83(1991)

Detection of targets in terrain clutter by using multispectral infrared image processing (Hoff, Lawrence E.; Evans, John R.; Bunney, Laura)V1481,98-109(1991)

Deterministic errors in pointing and tracking systems I: identification and correction of static errors (Keitzer, Scott; Kimbrell, James E.; Greenwald, David)V1482,406-414(1991)

Deterministic errors in pointing and tracking systems II: identification and correction of dynamic errors (Kimbrell, James E.; Greenwald, David; Smith, Robert; Kidd, Keith)V1482,415-424(1991)

Development and test of the Starlab control system (LaMont, Douglas V.; Mar, Lim O.; Rodden, Jack J.)V1482,2-12(1991)

Development of a fire and forget imaging infrared seeker missile simulation (Hall, Charles S.; Alongi, Robert E.; Fortner, Russ L.; Fraser, Laurie K.)V1483,29-38(1991)

Development of automatic target recognizers for Army applications (Jones, Terry L.)VIS07,4-13(1991)

Diode-laser-based range sensor (Seshamani, Ramani)V1471,354-356(1991)

Discrete frequency-versus-radius reticle trackers (Driggers, Ronald G.)V1478,24-32(1991)

Dynamic end-to-end model testbed for IR detection algorithms (Iannarilli, Frank J.; Wohlers, Martin R.)V1483,66-76(1991)

Effects of base motion on space-based precision laser tracking in the Relay Mirror Experiment (Anspach, Joel E.; Sydney, Paul F.; Hendry, Gregg)V1482,170-181(1991)

End-to-end model for detection performance evaluation against scenario-specific targets (Iannarilli, Frank J.; Wohlers, Martin R.)V1488,226-236(1991)

End-to-end scenario-generating model for IRST performance analysis (Iannarilli, Frank J.; Wohlers, Martin R.)V1481,187-197(1991)

Error analysis of combined stereo/optical-flow passive ranging (Barniv, Yair)V1479,259-267(1991)

European SILEX project and other advanced concepts for optical space communications (Oppenhaeuser, Gotthard; Wittig, Manfred E.; Popescu, Alexandru F.)V1522,2-13(1991)

Evaluation of image tracker algorithms (Marshall, William C.)V1483,207-218(1991)

Evaluation of the IR signature of dynamic air targets (Porta, Paola M.)V1540,508-518(1991)

Eyesafe diode laser rangefinder technology (Perger, Andreas; Metz, Jurgen; Tiedeke, J.; Rille, Eduard P.)V1419,75-83(1991)

Eyesafe high-pulse-rate laser progress at Hughes (Stultz, Robert D.; Nieuwsma, Daniel E.; Gregor, Eduard)V1419,64-74(1991)

Eyesafe Lasers: Components, Systems, and Applications (Johnson, Anthony M., ed.)V1419(1991)

Fast algorithm for image-based ranging (Menon, P. K.; Chatterji, Gano B.; Sridhar, Banavar)V1478,190-200(1991)

Flight test integration and evaluation of the LANTIRN system on the F-15E (Presuhn, Gary G.; Zeis, Joseph E.)V1479,249-258(1991)

Focusing on targets through exclusion (Mueller, Walter; Olson, James; Martin, Andrew; Hinchman, John H.)V1472,66-75(1991)

Frequency-modulated reticle trackers in the pupil plane (Taylor, James S.; Krapels, Keith A.)V1478,41-49(1991)

Fusing human and machine skills for remote robotic operations (Schenker, Paul S.; Kim, Won S.; Venema, Steven; Bejczy, Antal K.)V1383,202-223(1991)

Fusion of stereo views: estimating structure and motion using a robust method (Weng, Juyang; Cohen, Paul)V1383,321-332(1991)

Fusion or confusion: knowledge or nonsense? (Rothman, Peter L.; Denton, Richard V.)V1470,2-12(1991)

Fuzzy logic approach to multitarget tracking in clutter (Priebe, Russell; Jones, Richard A.)V1482,265-274(1991)

GaAs monolithic RF modules for SARSAT distress beacons (Cauley, Mike A.)V1475,275-279(1991)

Geoscience laser ranging system design and performance predictions (Anderson, Kent L.)V1492,142-152(1991)

Global modeling approach for multisensor problems (Chung, Yi-Nung; Emre, Erol; Gustafson, Donald L.)V1481,306-314(1991)

Hand-eye coordination for grasping moving objects (Allen, Peter K.; Yoshimi, Billibon; Timcenko, Alexander; Michelman, Paul)V1383,176-188(1991)

Heterodyne acquisition and tracking in a free-space diode laser link (Hueber, Martin F.; Scholtz, Arpad L.; Leeb, Walter R.)V1417,233-239(1991)

Hexagonal sampling and filtering for target detection with a scanning E-O sensor (Sperling, I.; Drummond, Oliver E.; Reed, Irving S.)V1481,2-11(1991)

High-accuracy target tracking algorithm based on deviation vector of the local window grey center (Shen, Zhen-kang; Zhao, Mingsheng)V1482,325-336(1991)

High-accuracy tracking algorithm based on iterating tracking window center towards the target's center (Shen, Zhen-kang; Han, Zhenduo; Zhao, Mingsheng)V1482,337-347(1991)

High-performance InGaAs PIN and APD (avalanche photdiode) detectors for 1.54 um eyesafe rangefinding (Olsen, Gregory H.; Ackley, Donald A.; Hladky, J.; Spadafora, J.; Woodruff, K. M.; Lange, M. J.; Van Orsdel, Brian T.; Forrest, Stephen R.; Liu, Y.)V1419,24-31(1991)

High-repetition-rate eyesafe rangefinders (Corcoran, Vincent J.)V1419,160-169(1991)

High-speed fine-motion tracking of some parts of a target (Casals, Alicia; Amat, Josep; Quesada, Jose L.; Sanchez, Luis)V1482,317-324(1991)

High-speed sensor-based systems for mobile robotics (Kayaalp, Ali E.; Moezzi, Saied; Davies, Henry C.)V1406,98-109(1991)

High-speed short-range laser rangefinder (Gielen, Robert M.; Slegtenhorst, Ronald P.)V1419,153-159(1991)

Holographic image reconstruction from interferograms of laser-illuminated complex targets (Wlodawski, Mitchell; Nowakowski, Jerzy)V1416,241-249(1991)

Hough transform computer-generated holograms: new output format (Carender, Neil H.; Casasent, David P.; Coetzee, Frans M.; Yu, Daming)V1555,182-193(1991)

Image display and background analysis with the Naval Postgraduate School infrared search and target designation system (Cooper, Alfred W.; Lentz, William J.; Baca, Michael J.; Bernier, J. D.)V1486,47-57(1991)

Image processing techniques for aquisition of faint point laser targets in space (Garcia-Prieto, Rafael; Spiero, Francois; Popescu, Alexandru F.)V1522,267-276(1991)

Imaging autotracker technology for guided missile systems (Hammon, Ricky K.; Helton, Monte K.)V1482,258-264(1991)

Implementation of an angle-only tracking filter (Allen, Ross R.; Blackman, Samuel S.)V1481,292-303(1991)

Incremental model for target maneuver estimation (Chang, Wen-Thong; Lin, Shao-An)V1481,242-253(1991)

Integrated image processing and tracker performance prediction workstation (Schneeberger, Timothy J.; McIntire, Harold D.)V1567,2-8(1991)

Integrated processor architecture for multisensor signal processing (Nasburg, Robert E.; Stillman, Steve M.; Nguyen, M. T.)V1481,84-95(1991)

Intelligent variable-resolution laser scanner for the space vision system (Blais, Francois; Rioux, Marc; MacLean, Steve G.)V1482,473-479(1991)

Issues in automatic object recognition: linking geometry and material data to predictive signature codes (Deitz, Paul; Muuss, Michael J.; Davisson, Edwin O.)VIS07,40-56(1991)

Kalman-based computation of optical flow fields (Viswanath, Harsha C.; Jones, Richard A.)V1482,275-284(1991)

Laboratory development of a nonlinear optical tracking filter (Block, Kenneth L.; Whitworth, Ernest E.; Bergin, Joseph E.)V1483,62-65(1991)

Laser applications for multisensor systems (Smolka, Greg L.; Strother, George T.; Ott, Carl)V1498,70-80(1991)

Laser designation integration into M-65 turret (Goldmunz, Menachem; Bloomberg, Steve; Neugarten, Michael L.)V1442,149-153(1991)

Laser terminal attitude determination via autonomous star tracking (Chapman, William W.; Fitzmaurice, Michael W.)V1417,277-290(1991)

Laser Tracker II: Sandia National Laboratories second-generation laser tracking system (Patrick, Duane L.)V1482,61-68(1991)

Launch area theodolite system (Bradley, Lester M.; Corriveau, John P.; Tindal, Nan E.)V1482,48-60(1991)

Linear modeling algorithm for tracking time-varying signals (Bachnak, Rafic A.)V1481,12-22(1991)

Local and remote track-file registration using minimum description length (Kenefic, Richard J.)V1481,430-439(1991)

Low-cost real-time hardware in the loop FCS performance evaluation (Cifarelli, Salvatore; Magrini, Sandro)V1482,480-490(1991)

Low-cost space platforms for detection and tracking technologies (Cullen, Robert M.)V1479,295-305(1991)

Low-intensity conflict aircraft systems (Henkel, Paul A.)V1538,2-4(1991)

Low-power optical correlator for sonar range finding (Turk, Harris; Leepa, Douglas C.; Snyder, Robert F.; Soos, Jolanta I.; Bronstein, Sam)V1454,344-352(1991)

Matched-filter algorithm for subpixel spectral detection in hyperspectral image data (Borough, Howard C.)V1541,199-208(1991)

Maximum likelihood estimation of affine-modeled image motion (Shaltaf, Samir J.; Namazi, Nader M.)V1567,609-620(1991)

Microprocessor-based laser range finder (Rao, M. K.; Tam, Siu-Chung)V1399,116-121(1991)

Midcourse multitarget tracking using continuous representation (Zak, Michial; Toomarian, Nikzad)V1481,386-397(1991)

MITAS: multisensor imaging technology for airborne surveillance (Thomas, John)V1470,65-74(1991)

Modeling and simulation of friction (Haessig, David A.; Friedland, Bernard)V1482,383-396(1991)

Modular streak camera for laser ranging (Prochazka, Ivan; Hamal, Karel; Schelev, Mikhail Y.; Lozovoi, V. I.; Postovalov, V. E.)V1358,574-577(1991)

Monocular passive range sensing using defocus information (Prasad, K. V.; Mammone, Richard J.)V1385,280-291(1991)

Morphologic edge detection in range images (Gupta, Sundeep; Krishnapuram, Raghu J.)V1568,335-346(1991)

Motion estimation without correspondences and object tracking over long time sequences (Goldgof, Dmitry B.; Lee, Hua; Huang, Thomas S.)V1383,109-121(1991)

Moving-point-target tracking in low SNR (Shen, Zhen-kang; Mao, Xuguang; Jin, Yiping)V1481,233-240(1991)

Multiple-hypothesis-based multiple-sensor spatial data fusion algorithm (Leung, Dominic S.; Williams, D. S.)V1471,314-325(1991)

Multiple-target tracking in a cluttered environment and intelligent track record (Tomasini, Bernard; Cassassolles, Emmanuel; Poyet, Patrice; Maynard de Lavalette, Guy M.; Siffredi, Brigitte)V1468,60-71(1991)

Multiple-target tracking using the SME filter with polar coordinate measurements (Sastry, C. R.; Kamen, Edward W.)V1481,261-280(1991)

Multiple degree-of-freedom tracking for attitude control of an experimental system on tether-stabilized platform (Angrilli, Francesco; Baglioni, Pietro; Bianchini, Gianandrea; Da Forno, R.; Fanti, Giullo; Mozzi, Massimo)V1482,26-39(1991)

Multiple target-to-track association and track estimation system using a neural network (Yee, Mark L.; Casasent, David P.)V1481,418-429(1991)

Multiple target tracking system (Lu, Simon W.)V1388,299-305(1991)

Multiresolution range-guided stereo matching (Tate, Kevin J.; Li, Ze-Nian)V1383,491-502(1991)

Multisensor approach in linear-Gaussian estimation of 3-D angular motion (Algrain, Marcelo C.; Saniie, Jafar)V1478,201-210(1991)

Multitarget adaptive gate tracker with linear prediction (Liu, Zhili)V1482,285-292(1991)

Multitarget detection and estimation parallel algorithm (Krikelis, A.)V1482,307-316(1991)

Multitarget tracking and multidimensional assignment problems (Poore, Aubrey B.; Rijavec, Nenad)V1481,345-356(1991)

Neural data association (Kim, Kwang H.; Shafai, Bahram)V1481,406-417(1991)

New concepts in remote sensing and geolocation (Seastone, A. J.)V1495,228-239(1991)

New estimation architecture for multisensor data fusion (Covino, Joseph M.; Griffiths, Barry E.)V1478,114-125(1991)

New generation control system for ultra-low-jitter satellite tracking (Verbanets, William R.; Greenwald, David)V1482,112-120(1991)

New track-to-track association logic for almost identical multiple sensors (Malakian, Kourken; Vidmar, Anthony)V1481,315-328(1991)

Novel angular discriminator for spatial tracking in free-space laser communications (Fung, Jackie S.)V1417,224-232(1991)

Novel optical-fiber-based conical scan tracking device (Johann, Ulrich A.; Pribil, Klaus; Sontag, Heinz)V1522,243-252(1991)

Optical correlator field demonstration (Kirsch, James C.; Gregory, Don A.; Hudson, Tracy D.; Loudin, Jeffrey A.; Crowe, William M.)V1482,69-78(1991)

Optical dividers for quadrant avalanche photodiode detectors (Green, Samuel I.)V1417,496-512(1991)

Optical flow techniques for moving target detection (Russo, Paul; Markandey, Vishal; Bui, Trung H.; Shrode, David)V1383,60-71(1991)

Optical target location using machine vision in space robotics tasks (Sklair, Cheryl W.; Gatrell, Lance B.; Hoff, William A.; Magee, Michael)V1387,380-391(1991)

Optic flow: multiple instantaneous rigid motions (Zhuang, Xinhua; Wang, Tao; Zhang, Peng)V1569,434-445(1991)

Optimal algorithms for holographic image reconstruction of an object with a glint using nonuniform illumination (Nowakowski, Jerzy; Wlodawski, Mitchell)V1416,229-240(1991)

Optimization neural net for multiple-target data association: real-time optical lab results (Yee, Mark L.; Casasent, David P.)V1469,308-319(1991)

Optimization of a gimbal-scanned infrared seeker (Williams, Elmer F.; Evans, Robert H.; Brant, Karl; Stockum, Larry A.)V1482,104-111(1991)

Overview of U.S. Fishery requirements (Springer, Steven C.; McLean, Craig)V1479,372-379(1991)

Parallel rule inferencing for automatic target recognition (Pacelli, Jean L.; Geyer, Steve L.; Ramsey, Timothy S.)V1472,76-84(1991)

Passive range and azimuth measuring system (Ronning, E.; Fjarlie, E. J.)V1399,178-183(1991)

Passive range sensor refinement using texture and segmentation (Sridhar, Banavar; Phatak, Anil; Chatterji, Gano B.)V1478,178-189(1991)

Passive-sensor data fusion (Kolitz, Stephan E.)V1481,329-340(1991)

Performance evaluation methods for multiple-target-tracking algorithms (Fridling, Barry E.; Drummond, Oliver E.)V1481,371-383(1991)

Performance of a demonstration system for simultaneous laser beacon tracking and low-data-rate optical communications with multiple platforms (Short, Ralph C.; Cosgrove, Michael A.; Clark, David L.; Oleski, Paul J.)V1417,464-475(1991)

Phase-conjugate optical preprocessing filter for small-target tracking (Block, Kenneth L.; Whitworth, Ernest E.; Bergin, Joseph E.)V1481,32-34(1991)

Pioneer unmanned air vehicle accomplishments during Operation Desert Storm (Christner, James H.)V1538,201-207(1991)

Planar InGaAs APD (avalanche photodiode) for eyesafe laser rangefinding applications (Webb, Paul P.)V1419,17-23(1991)

Pointing, acquisition, and tracking system of the European SILEX program: a major technological step for intersatellite optical communication (Bailly, Michel; Perez, Eric)V1417,142-157(1991)

Point target detection, location, and track initiation: initial optical lab results (Carender, Neil H.; Casasent, David P.)V1481,35-48(1991)

Polynomial regression analysis for estimating motion from image sequences (Frau, Juan; Llario, Vicenc; Oliver, Gabriel)V1388,329-340(1991)

Pose determination of spinning satellites using tracks of novel regions (Lee, Andrew J.; Casasent, David P.)V1383,72-83(1991)

Potential of tunable lasers for optimized dual-color laser ranging (Lund, Glenn I.; Gaignebet, Jean)V1492,166-175(1991)

Precision tracking of small target in IR systems (Lu, Huanzhang; Sun, Zhong-kang)V1481,398-405(1991)

Predictive control for 4-D guidance (Ilie, Stiharu-Alexe)V1482,491-501(1991)

Problems of photogrammetry of moving target in water (Han, Xin Z.)V1537,215-220(1991)

Pseudo K-means approach to the multisensor multitarget tracking problem (Thompson, Wiley E.; Parra-Loera, Ramon; Tao, Chin-Wang)V1470,48-58(1991)

Radar-optronic tracking experiment for short- and medium-range aerial combat (Ravat, Christian J.; Mestre, J. P.; Rose, C.; Schorter, M.)V1478,239-246(1991)

Random mapping network for tactical target reacquisition after loss of track (Church, Susan D.; Burman, Jerry A.)V1471,192-199(1991)

Range data analysis using cross-stripe structured-light system (Kim, Whoi-Yul)V1385,216-228(1991)

Range image-based object detection and localization for HERMIES III mobile robot (Sluder, John C.; Bidlack, Clint R.; Abidi, Mongi A.; Trivedi, Mohan M.; Jones, Judson P.; Sweeney, Frank J.)V1468,642-652(1991)

Ray-following model of sonar range sensing (Wilkes, David R.; Dudek, Gregory; Jenkin, Michael R.; Milios, Evangelos E.)V1388,536-542(1991)

Realization of heterodyne acquisition and tracking with diode lasers at lambda=1.55 um (Hueber, Martin F.; Leeb, Walter R.; Scholtz, Arpad L.)V1522,259-267(1991)

Real-time architecture based on the image processing module family (Kimura, Shigeru; Murakami, Yoshiyuki; Matsuda, Hikaru)V1483,10-17(1991)

Receivers for eyesafe laser rangefinders: an overview (Crawford, Ian D.)V1419,9-16(1991)

Recognition and tracking of moving objects (El-Konyaly, Sayed H.; Enab, Yehia M.; Soltan, Hesham)V1388,317-328(1991)

Recognition, tracking, and pose estimation of arbitrarily shaped 3-D objects in cluttered intensity and range imagery (Gottschalk, Paul G.; Turney, Jerry L.; Chiu, Arnold H.; Mudge, Trevor N.)V1383,84-96(1991)

Reconnaissance and imaging sensor test facilities at Eglin Air Force Base (Pratt, Stephen R.; Tucker, Robert)V1538,40-45(1991)

Reconnaissance mission planning (Fishell, Wallace G.; Fox, Alex J.)V1538,5-13(1991)

Recovering absolute depth and motion of multiple objects from intensity images (Jiang, Fan; Schunck, Brian G.)V1569,162-173(1991)

Relay Mirror Experiment overview: a GBL pointing and tracking demonstration (Dierks, Jeffrey S.; Ross, Susan E.; Brodsky, Aaron; Kervin, Paul W.; Holm, Richard W.)V1482,146-158(1991)

Robot location densities (Malik, Raashid; Polkowski, Edward T.)V1388,280-290(1991)

Robot self-location based on corner detection (Malik, Raashid; Polkowski, Edward T.)V1388,306-316(1991)

Scanning strategies for target detection (Gertner, Izidor; Zeevi, Yehoshua Y.)V1470,148-166(1991)

Selected Papers on Free-Space Laser Communications (Begley, David L., ed.)VMS30(1991)

Sensing and environment perception for a mobile vehicle (Blais, Francois; Rioux, Marc)V1480,94-101(1991)

Sensitivity analysis of Navy tactical decision-aid FLIR performance codes (Milne, Edmund A.; Cooper, Alfred W.; Reategui, Rodolfo; Walker, Philip L.)V1486,151-161(1991)

Sensor fusion approach to optimization for human perception: an observer-optimized tricolor IR target locating sensor (Miller, Walter E.)V1482,224-233(1991)

Sensor system for comet approach and landing (Bonsignori, Roberto; Maresi, Luca)V1478,76-91(1991)

Sensors and Sensor Systems for Guidance and Navigation (Wade, Jack; Tuchman, Avi, eds.)V1478(1991)

Signal and Data Processing of Small Targets 1991 (Drummond, Oliver E., ed.)V1481(1991)

Signal and Image Processing Systems Performance Evaluation, Simulation, and Modeling (Nasr, Hatem N.; Bazakos, Michael E., eds.)V1483(1991)

Simulation model and on-ground performances validation of the PAT system for the SILEX program (Cossec, Francois R.; Doubrere, Patrick; Perez, Eric)V1417,262-276(1991)

Simulation of scenes, sensors, and tracker for space-based acquisition, tracking, and pointing experiments (DeYoung, David B.)V1415,13-21(1991)

SIRE (sight-integrated ranging equipment): an eyesafe laser rangefinder for armored vehicle fire control systems (Keeter, Howard S.; Gudmundson, Glen A.; Woodall, Milton A.)V1419,84-93(1991)

Six degree-of-freedom magnetically suspended fine-steering mirror (Medbery, James D.; Germann, Lawrence M.)V1482,397-405(1991)

Sled tracking system (Downey, George A.; Fountain, H. W.; Riding, Thomas J.; Eggleston, James; Hopkins, Michael; Adams, Billy)V1482,40-47(1991)

Solid state lasers for planetary exploration (Greene, Ben; Taubman, Matthew; Watts, Jeffrey W.; Gaither, Gary L.)V1492,126-139(1991)

Space-based millimeter wave debris tracking radar (Chang, Kai; Pollock, Michael A.; Skrehot, Michael K.)V1475,257-266(1991)

Space-borne beam pointing (Eller, E. D.; LaMont, Douglas V.; Rodden, Jack J.)V1415,2-12(1991)

Spatial acquisition and tracking for deep-space optical communication packages (Chen, Chien-Chung; Jeganathan, Muthu; Lesh, James R.)V1417,240-250(1991)

State equalization and resonant control systems (Bigley, William J.)V1482,350-366(1991)

Statistical initial orbit determination (Taff, Laurence G.; Belkin, Barry; Schweiter, G. A.; Sommar, K.)V1481,440-448(1991)

Surveillance test bed for SDIO (Wesley, Michael; Osterheld, Robert; Kyser, Jeff; Farr, Michele; Vandergriff, Linda J.)V1481,209-220(1991)

Survey of helmet tracking technologies (Ferrin, Frank J.)V1456,86-94(1991)

Survey of multisensor data fusion systems (Linn, Robert J.; Hall, David L.; Llinas, James)V1470,13-29(1991)

Synthesis of a maneuver detector and adaptive gain tracking filter (Gardner, Kenneth R.; Kasky, Thomas J.)V1481,254-260(1991)

System considerations for detection and tracking of small targets using passive sensors (DeBell, David A.)V1481,180-186(1991)

Tactical reconnaissance mission survivability requirements (Lareau, Andy G.; Collins, Ross)V1538,81-98(1991)

Target detection using co-occurrence matrix segmentation and its hardware implementation (Auborn, John E.; Fuller, James M.; McCauley, Howard M.)V1482,246-252(1991)

Target lifetimes in natural resource management (Greer, Jerry D.)V1538,69-76(1991)

Terrain adaptive footfall placement using real-time range images (Dodds, David R.)V1388,543-548(1991)

Terrain and target segmentation using coherent laser radar (Renhorn, Ingmar G.; Letalick, Dietmar; Millnert, Mille)V1480,35-45(1991)

Three-dimensional inspection using laser-based dynamic fringe projection (Harvey, David M.; Shaw, Michael M.; Hobson, Clifford A.; Wood, Christopher M.; Atkinson, John T.; Lalor, Michael J.)V1400,86-93(1991)

Three-dimensional morphology for target detection (Patterson, Tim J.)V1471,358-368(1991)

Three-dimensional position determination from motion (Nashman, Marilyn; Chaconas, Karen)V1383,166-175(1991)

Time-optimal maneuver guidance design with sensor line-of-sight constraint (Hartman, Richard D.; Lutze, Frederick H.; Cliff, Eugene M.)V1478,64-75(1991)

Toward the optimization of passive ground targets in spaceborne laser ranging (Lund, Glenn I.; Renault, Herve)V1492,153-165(1991)

Track-while-image in the presence of background (Mentle, Robert E.; Shapiro, Jeffrey H.)V1471,342-353(1991)

VLSI implementable neural networks for target tracking (Himes, Glenn S.; Inigo, Rafael M.; Narathong, Chiewcharn)V1469,671-682(1991)

Wide-field-of-view star tracker camera (Lewis, Isabella T.; Ledebuhr, Arno G.; Axelrod, Timothy S.; Kordas, Joseph F.; Hills, Robert F.)V1478,2-12(1991)

Transforms—see also Fourier transforms

Accelerated detection of image objects and their orientations with distance transforms (Ford, Gary E.; Paglieroni, David W.; Tsujimoto, Eric M.)V1452,244-255(1991)

Adaptive chirplet: an adaptive generalized wavelet-like transform (Mann, Steven; Haykin, Simon)V1565,402-413(1991)

Application of optical signal processing: fingerprint identification (Fielding, Kenneth H.; Horner, Joseph L.; Makekau, Charles K.)V1564,224-230(1991)

Compact joint transform correlators in planar-integrated packages (Ghosh, Anjan K.)V1564,231-235(1991)

Compact, one-lens JTC using a transmissive amorphous silicon FLC-SLM (LAPS-SLM) (Haemmerli, Jean-Francois; Iwaki, Tadao; Yamamoto, Shuhei)V1564,275-284(1991)

Computation and meaning of Gabor coefficients (An, Myoung H.; Conner, Michael; Tolimieri, Richard; Orr, Richard S.)V1565,383-401(1991)

Connectivity-preserving morphological image transformations (Bloomberg, Dan S.)V1606,320-334(1991)

Deformable templates, robust statistics, and Hough transforms (Yuille, Alan L.; Peterson, Carsten; Honda, Ko)V1570,166-174(1991)

Dimensionality of signal sets (Orr, Richard S.)V1565,435-446(1991)

Effects of M-transform for bit-error resilement in the adaptive DCT coding (Yamane, Nobumoto; Morikawa, Yoshitaka; Hamada, Hiroshi)V1605,679-686(1991)

Effects of thresholding in multiobject binary joint transform correlation (Javidi, Bahram; Wang, Jianping; Tang, Qing)V1564,212-223(1991)

Efficient computation of densely sampled wavelet transforms (Jones, Douglas L.; Baraniuk, Richard G.)V1566,202-206(1991)

Existence of cross terms in the wavelet transform (Kadambe, Shubha; Boudreaux-Bartels, G. F.)V1565,423-434(1991)

Hadamard transform-based object recognition using an array processor (Celenk, Mehmet; Moiz, Saifuddin)V1468,764-775(1991)

Hough transform computer-generated holograms: new output format (Carender, Neil H.; Casasent, David P.; Coetzee, Frans M.; Yu, Daming)V1555,182-193(1991)

Hough transform implementation using an analog associative network (Arrue, Begona C.; Inigo, Rafael M.)V1469,420-431(1991)

Image representation by group theoretic approach (Segman, Joseph; Zeevi, Yehoshua Y.)V1606,97-109(1991)

Joint transform correlator using nonlinear ferroelectric liquid-crystal spatial light modulator (Kohler, A.; Fracasso, B.; Ambs, Pierre; de Bougrenet de la Tocnaye, Jean-Louis M.)V1564,236-243(1991)

Linear feature SNR enhancement in radon transform space (Meckley, John R.)V1569,375-385(1991)

Matrix reformulation of the Gabor transform (Balart, Rogelio)V1565,447-457(1991)

Novel transform for image description and compression with implementation by neural architectures (Ben-Arie, Jezekiel; Rao, K. R.)V1569,367-374(1991)

Optical Hartley-transform-based adaptive filter (Abushagur, Mustafa A.; Berinato, Robert J.)V1564,602-609(1991)

Parallel computation of the continuous wavelet transform (Gertner, Izidor; Peskin, Richard L.; Walther, Sandra S.)V1565,414-422(1991)

Parallel processing approach to transform-based image coding (Normile, James; Wright, Dan; Chu, Ke-Chiang; Yeh, Chia L.)V1452,480-484(1991)

Pattern detection using a modified composite filter with nonlinear joint transform correlator (Vallmitjana, Santiago; Juvells, Ignacio; Carnicer, Arturo; Campos, Juan)V1564,266-274(1991)

Quantization analysis of the binary joint transform correlator in the presence of nonlinear compression (Javidi, Bahram; Wang, Jianping)V1564,254-265(1991)

Real-time nonlinear optical correlator in speckle metrology (Ogiwara, Akifumi; Ohtsubo, Junji)V1564,294-305(1991)

Recovery for a real image from its Hartley transform modulus only (Dong, Bizhen; Gu, Benyuan; Yang, Guo-Zhen)V1429,117-126(1991)

Rotation- and scale-invariant joint transform correlator using FLC-SLMs (Mitsuoka, Yasuyuki; Iwaki, Tadao; Yamamoto, Shuhei)V1564,244-252(1991)

Spatial and temporal surface interpolation using wavelet bases (Pentland, Alexander P.)V1570,43-62(1991)

Study on Hadamard transform imaging spectroscopy (Wu, Ji-Zong; Deng, Jia-cheng; Chen, Ben-zhi)V1399,122-129(1991)

Texture discrimination using wavelets (Carter, Patricia H.)V1567,432-438(1991)

Three-dimensional motion analysis and structure recovering by multistage Hough transform (Nakajima, Shigeyoshi; Zhou, Mingyong; Hama, Hiromitsu; Yamashita, Kazumi)V1605,709-719(1991)

Utilizing the central limit theorem for parallel multiple-scale image processing with neural architectures (Ben-Arie, Jezekiel)V1569,227-238(1991)

Wavelets and adaptive signal processing (Resnikoff, Howard L.)V1565,370-382(1991)

Zak transform as an adaptive tool (Tolimieri, Richard; Conner, Michael)V1565,345-356(1991)

Transmission—see also optical communications

10.6-um TEM00 beam transmission characteristics of a hollow circular cross-section multimode waveguide (Jenkins, R. M.; Devereux, R. W.)V1512,135-142(1991)

45-Mbps multichannel TV coding system (Matsumoto, Shuichi; Hamada, Takahiro; Saito, Masahiro; Murakami, Hitomi)V1605,37-46(1991)

500 MHz baseband fiber optic transmission system for medical imaging applications (Cheng, Xin; Huang, H. K.)V1364,204-208(1991)

A 120-Mbit/s QPPM high-power semiconductor transmitter performance and reliability (Greulich, Peter; Hespeler, Bernd; Spatscheck, Thomas)V1417,358-369(1991)

Advanced Fiber Communications Technologies (Kazovsky, Leonid G., ed.)V1579(1991)

Analysis and modeling of uniformly- and nonuniformly-coupled lossy lines for interconnections and packaging in hybrid and monolithic circuits (Orhanovic, Neven; Hayden, Leonard A.; Tripathi, Vijai K.)V1389,273-284(1991)

Analysis of optimum-frame-rate in low-bit-rate video coding (Takishima, Yasuhiro; Wada, M.; Murakami, Hitomi)V1605,635-645(1991)

Analysis of skin effect loss in high-frequency interconnects with finite metallization thickness (Kiang, Jean-Fu)V1389,340-351(1991)

ASK transmitter for high-bit-rate systems (Schiellerup, Gert; Pedersen, Rune J.)V1372,27-38(1991)

Atmospheric laser-transmission tables simply generated (Mallory, William R.)V1540,359-364(1991)

Broadband, antireflection coating designs for large-aperture infrared windows (Balasubramanian, Kunjithapa; Le, Tam V.; Guenther, Karl H.; Kumar, Vas)V1485,245-253(1991)

CAE tools for verifying high-performance digital systems (Rubin, Lawrence M.)V1390,336-358(1991)

Coherent communication systems research and development at AT&T Bell Laboratories, Solid State Technology Center (Park, Yong K.; Delavaux, Jean-Marc P.; Tench, Robert E.; Cline, Terry W.; Tzeng, Liang D.; Kuo, Chien-yu C.; Wagner, Earl J.; Flores, Carlos F.; Van Eijk, Peter; Pleiss, T. C.; Barski, S.; Owen, B.; Twu, Yih-Jye; Dutta, Niloy K.; Riggs, R. S.; Ogawa, Kinichiro K.)V1372,219-227(1991)

Coherent Lightwave Communications: Fifth in a Series (Steele, Roger C.; Sunak, Harish R., eds.)V1372(1991)

Comparison of photorefractive effects and photogenerated components in polarization-maintaining fibers (Kanellopoulos, S. E.; Guedes Valente, Luiz C.; Handerek, Vincent A.; Rogers, Alan J.)V1516,200-210(1991)

Design optimization of optical filters for space applications (Annapurna, M. N.; Nagendra, C. L.; Thutupalli, G. K.)V1485,260-271(1991)

Design study for high-performance optical communications satellite terminal with high-power laser diode transmitter (Hildebrand, Ulrich; Seeliger, Reinhard; Smutny, Berry; Sand, Rolf)V1522,50-60(1991)

Determination of thickness and refractal index of HgCdMnTe/CdMnTe VPE films by IR transmission spectrum (Chen, Wei M.; Ma, Ke J.; Yu, Zhen Z.; Ji, Hua M.)V1519,521-524(1991)

Digital video codec for medium bitrate transmission (Ebrahimi, Touradj; Dufaux, Frederic; Moccagatta, Iole; Campbell, T. G.; Kunt, Murat)V1605,2-15(1991)

Diode-pumped Nd:YAG laser transmitter for free-space optical communications (Nava, Enzo; Re Garbagnati, Giuseppe; Garbi, Maurizio; Marchiori, Livio D.; Marini, Andrea E.)V1417,307-315(1991)

Discrete frequency-versus-radius reticle trackers (Driggers, Ronald G.)V1478,24-32(1991)

Effects of conductor losses on cross-talk in multilevel-coupled VLSI interconnections (van Deventer, T. E.; Katehi, Linda P.; Cangellaris, Andreas C.)V1389,285-296(1991)

Eight-Gb/s QPSK-SCM over a coherent detection optical link (Hill, Paul M.; Olshansky, Robert)V1579,210-220(1991)

Electrical characteristics of lossy interconnections for high-performance computer applications (Deutsch, Alina; Kopcsay, Gerard V.; Ranieri, V. A.; Cataldo, J. K.; Galligan, E. A.; Graham, W. S.; McGouey, R. P.; Nunes, S. L.; Paraszczak, Jurij R.; Ritsko, J. J.; Serino, R. J.; Shih, D. Y.; Wilczynski, Janusz S.)V1389,161-176(1991)

Experimental study of chiral composites (Ro, Ru-Yen; Varadan, Vasundara V.; Varadan, Vijay K.)V1558,269-287(1991)

Frequency domain evaluation of the accuracy of lumped element models for RLC transmission lines (Delbare, William; Dhaene, Tom; Vanhauwermeiren, Luc; De Zutter, Daniel)V1389,257-272(1991)

Frequency-selective devices using a composite multilayer design (Ma, Yushieh; Varadan, Vijay K.; Varadan, Vasundara V.)V1558,132-137(1991)

GaAs monolithic RF modules for SARSAT distress beacons (Cauley, Mike A.)V1475,275-279(1991)

High-power transmission through step-index multimode fibers (Setchell, Robert E.; Meeks, Kent D.; Trott, Wayne M.; Klingsporn, Paul E.; Berry, Dante M.)V1441,61-70(1991)

Hybrid optical transmitter for microwave communication (Costa, Joannes M.; Lam, Benson C.; Kellner, Albert L.; Campion, David C.; Yu, Paul K.)V1476,74-80(1991)

Laplacian pyramid coding of prediction error images (Stiller, Christoph; Lappe, Dirk)V1605,47-57(1991)

Laptop image transmission equipment (Mocenter, Michael M.)V1538,132-139(1991)

Laser transmitter for lidar in-space technology experiment (Chang, John H.; Cimolino, Marc C.; Petros, Mulugeta)V1492,43-46(1991)

Lifetest on a high-power laser diode array transmitter (Greulich, Peter; Hespeler, Bernd; Spatscheck, Thomas)V1522,144-153(1991)

Luminance asymmetry in stereo TV images (Beldie, Ion P.; Kost, Bernd)V1457,242-247(1991)

Metal sulfide thin films on glass as solar control, solar absorber, decorative, and photographic coatings (Nair, Padmanabhan K.; Nair, M. T.; Fernandez, A. M.; Garcia, V. M.; Hernandez, A. B.)V1485,228-239(1991)

Minimum polarization modulation: a highly bandwidth efficient coherent optical modulation scheme (Benedetto, Sergio; Kazovsky, Leonid G.; Poggiolini, Pierluigi T.)V1579,112-121(1991)

Motion video coding for packet-switching networks: an integrated approach (Gilge, Michael; Gusella, Riccardo)V1605,592-603(1991)

Multigigabit solitary-wave propagation in both the normal and anomalous dispersion regions of optical fibers (Potasek, M. J.; Tabor, Mark)V1579,232-236(1991)

Optical fibre system with CCLID for the transmission of two-dimensional images (Li, Shaohui; Wen, Shengping; Zhang, Zhipeng; Lu, Feizhen)V1572,543-547(1991)

Optical ISL transmitter design that uses a high-power LD amplifier (Nohara, Mitsuo; Harada, Takashi; Fujise, Masayuki)V1417,338-345(1991)

Optical leak detection of oxygen using IR laser diodes (Disimile, Peter J.; Fox, Curtis F.; Toy, Norman)V1492,64-75(1991)

Performance analysis of direct-detection optical DPSK systems using a dual-detector optical receiver (Pires, Joao J.; Rocha, Jose R.)V1579,144-154(1991)

Photoinduced self-organization in optical fiber: some answered and unanswered questions (Ouellette, Francois; Gagnon, Daniel; LaRochelle, Sophie; Poirier, Michel)V1516,2-13(1991)

Polarization-modulated coherent optical communication systems (Betti, Silvello; Curti, Franco; De Marchis, Giancarlo; Iannone, Eugenio)V1579,100-111(1991)

Polarization modulation with frequency shift redundancy and frequency modulation with polarization redundancy for POLSK and FSK systems (Marone, Giuseppe; Calvani, Riccardo A.; Caponi, Renato)V1579,122-132(1991)

Principles of phase-resolved optical measurements (Jacques, Steven L.)V1525,143-153(1991)

Qualification testing of a diode-laser transmitter for free-space coherent communications (Pillsbury, Allen D.; Taylor, John A.)V1417,292-299(1991)

Radiation-resistant optical glasses (Marker, Alexander J.; Hayden, Joseph S.; Speit, Burkhard)V1485,160-172(1991)

Realistic model for battlefield fire plume simulation (Bruce, Dorothy)V1486,231-236(1991)

Realization of a coherent optical DPSK (differential phase-shift keying) heterodyne transmission system with 565 MBit/s at 1.064 um (Wandernoth, Bernhard; Franz, Juergen)V1522,194-198(1991)

Recent progress of coherent lightwave systems at Bellcore (Sessa, William B.; Welter, Rudy; Wagner, Richard E.; Maeda, Mari W.)V1372,208-218(1991)

Scattering parameters of volume phase hologram: Bragg's approach (Gubanov, V. A.; Kiselyov, Boris S.)V1621,102-113(1991)

Secret transmission method of character data in motion picture communication (Tanaka, Kiyoshi; Nakamura, Yasuhiro; Matsui, Kineo)V1605,646-649(1991)

Single-fiber wideband transmission systems for HDTV production applications (Cheng, Xin; Levin, Paul A.; Nguyen, Hat)V1363,177-181(1991)

Solid state lasers for planetary exploration (Greene, Ben; Taubman, Matthew; Watts, Jeffrey W.; Gaither, Gary L.)V1492,126-139(1991)

Space-qualified laser transmitter for lidar applications (Chang, John H.; Reithmaier, Karl D.)V1492,38-42(1991)

Spectral and modulation characteristics of tunable multielectrode DBR lasers (Ferreira, Mario F.; Rocha, Jose R.; Pinto, Joao L.)V1372,14-26(1991)

Stability analysis of semidiscrete schemes for thermal blooming computation (Ulrich, Peter B.)V1408,192-202(1991)

Stochastic detecting images from strong noise field in visual communications (Cai, De-Fu)V1606,926-933(1991)

System interconnect issues for subnanosecond signal transmission (Moresco, Larry L.)V1390,202-213(1991)

Teleradiology in the local environment (Staab, Edward V.; Honeyman, Janice C.; Frost, Meryll M.; Bidgood, W. D.)V1446,16-22(1991)

Termination for minimal reflection of high-speed pulse propagation along multiple-coupled microstrip lines (Kuo, Jen-Tsai; Tzuang, Ching-Kuang C.)V1389,156-160(1991)

Time-resolved diffuse reflectance and transmittance studies in tissue simulating phantoms: a comparison between theory and experiment (Madsen, Steen J.; Patterson, Michael S.; Wilson, Brian C.; Park, Young D.; Moulton, J. D.; Jacques, Steven L.; Hefetz, Yaron)V1431,42-51(1991)

Transmission characteristics of multimode optical fiber with multisplices (Das, Alok K.; Mandal, Anup K.)V1365,144-155(1991)

Transmission of high-speed voltage waves at the junction of three transmission lines (Sakagami, Iwata; Kaji, Akihiro)V1389,329-339(1991)

Transmission of the motion of a walker by model-based image coding (Kimoto, Tadahiko; Yasuda, Yasuhiko)V1605,253-262(1991)

Two-layer pyramid image coding scheme for interworking of video services in ATM (Sikora, Thomas; Tan, T. K.; Pang, Khee K.)V1605,624-634(1991)

Uniformity and transmission of proportional counter window materials for use with AXAF (Flanagan, Kathryn A.; Austin, G. K.; Cobuzzi, J. C.; Goddard, R.; Hughes, John P.; McLaughlin, Edward R.; Podgorski, William A.; Rose, V.; Roy, Adrian G.; Zombeck, Martin V.; Markert, Thomas H.; Bauer, J.; Isobe, T.; Schattenburg, Mark L.)V1549,395-407(1991)

Use of a multibeam transmitter for significant improvement in signal-dynamic-range reduction and near-range coverage for incoherent lidar systems (Zhao, Yanzeng; Hardesty, R. M.; Post, Madison J.)V1492,85-90(1991)

Variable-bit-rate HDTV coding algorithm for ATM environments for B-ISDN (Kinoshita, Taizo; Nakahashi, Tomoko; Takizawa, Masaaki)V1605,604-613(1991)

Wavelength division and subcarrier system based on Brillouin amplification (Lee, Yang-Hang; Wu, Jingshown; Kao, Ming-Seng; Tsao, Hen-Wai)V1579,155-166(1991)

Wide-band analog frequency modulation of optic signals using indirect techniques (Fitzmartin, Daniel J.; Balboni, Edmund J.; Gels, Robert G.)V1371,78-86(1991)

Wide-spectral-range transmissometer used for fog measurements (Turner, Vernon; Trowbridge, Christian A.)V1487,262-271(1991)

ZnS/Me heat mirror systems (Zhang, Xiao P.; Yu, Shan-qing; Ma, Min W.)V1519,514-520(1991)

Turbulence—see also atmospheric optics

Blur identification and image restoration with the expectation-maximization algorithm (Durack, Donald L.)V1487,72-83(1991)

Characterization and effect of system noise in a differential angle of arrival measurement device (Waldie, Arthur H.; Drexler, James J.; Qualtrough, John A.; Soules, David B.; Eaton, Frank D.; Peterson, William A.; Hines, John R.)V1487,103-108(1991)

Characterization of simulated and open-air atmospheric turbulence (Razdan, Anil K.; Singh, Brahm P.; Chopra, S.; Modi, M. B.)V1558,384-388(1991)

Cn2 estimates in the boundary layer for damp unstable conditions (Tunick, Arnold; Rachele, Henry; Miller, Walter B.)V1487,51-62(1991)

Correlation of bifrequency beam propagating in a folded turbulent path (Song, Zhengfang; Ma, Jun)V1487,382-386(1991)

Double-passage imaging through turbulence (Dainty, J. C.; Mavroidis, Theo; Solomon, Christopher J.)V1487,2-9(1991)

Empirical modeling of laser propagation effects in intermediate turbulence (Tofsted, David H.)V1487,372-381(1991)

Experimental investigation of image degradation created by a high-velocity flow field (Couch, Lori L.; Kalin, David A.; McNeal, Terry)V1486,417-423(1991)

Expressions for the spherical-wave-structure function based on a bump spectrum model for the index of refraction (Richardson, Christina E.; Andrews, Larry C.)V1487,19-30(1991)

Finite aperture effects on intensity fluctuations of laser radiation in a turbulent atmosphere (Mazar, Reuven; Bronshtein, Alexander)V1487,361-371(1991)

Forecasting optical turbulence strength: effects of macroscale meteorology and aerosols (Sadot, Danny; Kopeika, Norman S.)V1442,325-334(1991)

GRAND: a 4-D wave optics code for atmospheric laser propagation (Mehta, Naresh C.)V1487,398-409(1991)

Ground-to-space multiline propagation at 1.3 um (Crawford, Douglas P.; Harada, Larrene K.)V1408,167-177(1991)

High-resolution astronomical observations using deconvolution from wavefront sensing (Michau, Vincent; Marais, T.; Laurent, Jean; Primot, Jerome; Fontanella, Jean-Claude; Tallon, M.; Fuensalida, Jesus J.)V1487,64-71(1991)

Images of turbulent, absorbing-emitting atmospheres and their application to windshear detection (Watt, David W.; Philbrick, Daniel A.)V1467,357-368(1991)

Increasing the isoplanatic patch size with phase-derivative adaptive optics (Feng, Yue-Zhong; Gong, Zhi-Ben; Song, Zhengfang)V1487,356-360(1991)

Interaction of exposure time and system noise with angle-of-arrival measurements (Eaton, Frank D.; Peterson, William A.; Hines, John R.; Waldie, Arthur H.; Drexler, James J.; Qualtrough, John A.; Soules, David B.)V1487,84-90(1991)

Interferometric measurements of a high-velocity mixing/shear layer (Peters, Bruce R.; Kalin, David A.)V1486,410-416(1991)

Measurement of modulation transfer function of desert atmospheres (McDonald, Carlos)V1487,203-219(1991)

Measurements of the atmospheric turbulence spectrum and intermittency using laser scintillation (Frehlich, Rod G.)V1487,10-18(1991)

Modeling and simulation of systems imaging through atmospheric turbulence (Caponi, Maria Z.)V1415,138-149(1991)

Modeling time-dependent obscuration for simulated imaging of dust and smoke clouds (Hoock, Donald W.)V1486,164-175(1991)

Modeling turbulent transport in laser beam propagation (Wallace, James)V1408,19-27(1991)

New look at bump spectra models for temperature and refractive-index fluctuations (Sisterman, Elizabeth A.; Andrews, Larry C.)V1487,345-355(1991)

Observations of uplink and retroreflected scintillation in the Relay Mirror Experiment (Lightsey, Paul A.; Anspach, Joel E.; Sydney, Paul F.)V1482,209-222(1991)

Prediction of Cn2 on the basis of macroscale meteorology including aerosols (Sadot, Danny; Kopeika, Norman S.)V1487,40-50(1991)

Propagation Engineering: Fourth in a Series (Bissonnette, Luc R.; Miller, Walter B., eds.)V1487(1991)

Propagation of High-Energy Laser Beams Through the Earth's Atmosphere II (Ulrich, Peter B.; Wilson, LeRoy E., eds.)V1408(1991)

Pulse propagation in random media (Ishimaru, Akira)V1558,127-129(1991)

Realistic wind effects on turbulence and thermal blooming compensation (Long, Jerry E.; Hills, Louis S.; Gebhardt, Frederick G.)V1408,58-71(1991)

Relay Mirror Experiment scoring analysis and the effects of atmospheric turbulence (Sydney, Paul F.; Dillow, Michael A.; Anspach, Joel E.; Kervin, Paul W.; Lee, Terence B.)V1482,196-208(1991)

Remote alignment of adaptive optical systems with far-field optimization (Mehta, Naresh C.)V1408,96-111(1991)

Sensing refractive-turbulence profiles (Cn2) using wavefront phase measurements from multiple reference sources (Welsh, Byron M.)V1487,91-102(1991)

Turbulence at the inner scale (Tatarskii, V. I.)V1408,2-9(1991)

Two-frequency correlation in a turbulent atmosphere (Mazar, Reuven; Rozental, Mark)V1487,31-39(1991)

Use of adaptive optics for minimizing atmospheric distortion of optical waves (Lukin, Vladimir P.)V1408,86-95(1991)

Wavefront curvature sensing and compensation methods in adaptive optics (Roddier, Francois J.)V1487,123-128(1991)

Wide bandwidth spectral measurements of atmospheric tilt turbulence (Tiszauer, Detlev H.; Smith, Richard C.)V1408,72-83(1991)

Ultrafast phenomena—see also lasers; photography; high-speed; streak cameras

19th Intl Congress on High-Speed Photography and Photonics (Fuller, Peter W., ed.)V1358(1991)

200,000-frame-per-second drum camera with nanosecond synchronized laser illumination (Briscoe, Dennis)V1346,319-323(1991)

Advanced concepts of electron-beam-pumped excimer lasers (Tittel, Frank K.; Canarelli, P.; Dane, C. B.; Hofmann, Thomas; Sauerbrey, Roland A.; Sharp, Tracy E.; Szabo, Gabor; Wilson, William L.; Wisoff, P. J.; Yamaguchi, Shigeru)V1397,21-29(1991)

All-optical Ti:LiNbO3 waveguide switch (d'Alessandro, Antonio; De Sario, Marco; D'Orazio, Antonella; Petruzzelli, Vincenzo)V1378,127-138(1991)

Amplified quantum fluctuation as a mechanism for generating ultrashort pulses in semiconductor lasers (Yuan, Ruixi; Taylor, Henry F.)V1497,313-319(1991)

Analysis of pyrotechnic devices by laser-illuminated high-speed photography (Dosser, Larry R.; Stark, Margaret A.)V1346,293-299(1991)

Application of Fabry-Perot velocimetry to hypervelocity impact experiments (Chau, Henry H.; Osher, John E.)V1346,103-112(1991)

Applications of copper vapor laser lighting in high-speed motion analysis (Hogan, Daniel C.)V1346,324-330(1991)

Ballistic imaging of biomedical samples using picosecond optical Kerr gate (Wang, LeMing; Liu, Y.; Ho, Ping-Pei; Alfano, Robert R.)V1431,97-101(1991)

Characteristics of amplification of ultrashort laser pulses in excimer media (Kannari, Fumihiko; Obara, Minoru)V1397,85-89(1991)

Characterization of a subpicosecond XeF (C->A) excimer laser (Hofmann, Thomas; Sharp, Tracy E.; Dane, C. B.; Wisoff, P. J.; Wilson, William L.; Tittel, Frank K.; Szabo, Gabor)V1412,84-90(1991)

Computational and experimental progress on laser-activated gas avalanche switches for broadband, high-power electromagnetic pulse generation (Mayhall, David J.; Yee, Jick H.; Villa, Francesco)V1378,101-114(1991)

Continuous automatic scanning picosecond optical parametric source using MgO LiNbO3 in the 700-2200 nm (He, Huijuan; Lu, Yutian; Dong, Jingyuan; Zhao, Quingchun)V1409,18-23(1991)

Designing and application of solid state lasers for streak cameras calibration (Babushkin, A. V.; Vorobiev, N. S.; Prokhorov, Alexander M.; Schelev, Mikhail Y.)V1346,410-417(1991)

Development of subnanosecond framing cameras in IOFAN (Ludikov, V. V.; Prokhorov, Alexander M.; Chevokin, Victor K.)V1346,418-436(1991)

Direct measurement of vibrational energy relaxation in photoexcited deoxyhemoglobin using picosecond Raman spectroscopy (Hopkins, John B.; Xu, Xiaobing; Lingle, Robert; Zhu, Huiping; Yu, Soo-chang)V1432,221-226(1991)

Dynamics of ultrafast photoprocesses in Zn-octaethylporphyrin and Zn-octaethylphlorin pi-monoanions (Chirvony, V. S.; Sinyakov, G. N.; Gadonas, R.; Krasauskas, V.; Pelakauskas, A.)V1403,504-506(1991)

Effect of magnetic quantization on the pulse power output of spontaneous emission for II-VI and IV-VI lasers operated at low temperature (Ghatak, Kamakhya P.; Mitra, Bhaswati; Biswas, Shambhu N.)V1411,137-148(1991)

Electronic system for generation of laser pulses by self-injection method (Brzezinski, Ryszard; Piotrowski, Jan)V1391,127-134(1991)

Enhanced holographic recording capabilities for dynamic applications (Hough, Gary R.; Gustafson, D. M.; Thursby, William R.)V1346,194-199(1991)

Evaluation of dynamic range for LLNL streak cameras using high-contrast pulses and "pulse podiatry" on the NOVA laser system (Richards, James B.; Weiland, Timothy L.; Prior, John A.)V1346,384-389(1991)

Excitation energy relaxation in model aggregates of photosynthetic pigments upon picosecond laser excitation (Chirvony, V. S.; Zenkevich, E. I.; Gadonas, R.; Krasauskas, V.; Pelakauskas, A.)V1403,638-640(1991)

Experiments with SPRITE 12 ps facility (Tallents, Gregory J.; Key, Michael H.; Norreys, P.; Jacoby, J.; Kodama, R.; Tragin, N.; Baldis, Hector A.; Dunn, James; Brown, D.)V1413,70-76(1991)

Femtosecond processes in allophycocyanin trimers (Khoroshilov, E. V.; Kryukov, I. V.; Kryukov, P. G.; Sharkov, A. V.; Gillbro, T.)V1403,431-433(1991)

Femtosecond spectroscopy of acidified and neutral bacteriorhodopsin (Kobayashi, Takayoshi T.; Terauchi, Mamoru; Kouyama, Tsutomu; Yoshizawa, Masayuki; Taiji, Makoto)V1403,407-416(1991)

Four-dimensional photosensitive materials: applications for shaping, time-domain holographic dissection, and scanning of picosecond pulses (Saari, Peeter M.; Kaarli, Rein K.; Sonajalg, Heiki)V1507,500-508(1991)

Frequency stability of a solid state mode-locked laser system (Simpson, Thomas B.; Doft, F.; Malley, Michael M.; Sutton, George W.; Day, Timothy)V1410,133-140(1991)

GaAs opto-thyristor for pulsed power applications (Hur, Jung H.; Hadizad, Peyman; Zhao, Hanmin; Hummel, Steven R.; Dapkus, P. D.; Fetterman, Harold R.; Gundersen, Martin A.)V1378,95-100(1991)

Generation of stable low-jitter kilovolt amplitude picosecond pulses (Sarkar, Tapan K.; Banerjee, Partha P.)V1378,34-42(1991)

Generation of ultrashort high-average-power passively mode-locked pulses from a Nd:YLF laser with a nonlinear external-coupled cavity at high-repetition rates (Chee, Joseph K.; Kong, Mo-Nga; Liu, Jia-ming)V1413,14-20(1991)

Generation of ultrashort ultraviolet optical pulses using sum-frequency in LBO crystals (Guo, Ting; Qiu, Peixia; Lin, Fucheng)V1409,24-27(1991)

Graded AlxGa1-xAs photoconductive devices for high-efficiency picosecond optoelectronic switching (Morse, Jeffrey D.; Mariella, Raymond P.; Dutton, Robert W.)V1378,55-59(1991)

High-power/high-pulse repetition frequency (PRF) pulse generation using a laser-diode-activated photoconductive GaAs switch (Kim, A. H.; Zeto, Robert J.; Youmans, Robert J.; Kondek, Christine D.; Weiner, Maurice; Lalevic, Bogoliub)V1378,173-178(1991)

High-power nanosecond pulse iodine laser provided with SBS mirror (Dolgopolov, Y. V.; Kirillov, Gennadi A.; Kochemasov, G. G.; Kulikov, Stanislav M.; Murugov, V. M.; Pevny, S. N.; Sukharev, S. A.)V1412,267-275(1991)

High-power waveform generation using photoconductive switches (Oicles, Jeffrey A.; Helava, Heikki I.; Grant, Jon R.; Ragle, Larry O.; Wessman, Susan C.)V1378,60-69(1991)

High-voltage picosecond pulse generation using avalanche diodes (McEwan, Thomas E.; Hanks, Roy L.)V1346,465-470(1991)

Holography with a single picosecond pulse (Abramson, Nils H.)V1332,224-229(1991)

Hydrodynamic evolution and radiation emission from an impulse-heated solid-density plasma (Lyubomirsky, I.; Durfee, C.; Milchberg, Howard M.)V1413,108-111(1991)

Hydrodynamic evolution of picosecond laser plasmas (Landen, Otto L.; Vu, Brian-Tinh; Stearns, Daniel G.; Alley, W. E.)V1413,120-130(1991)

Influence of photon energy on the photoemission from ultrafast electronic materials (Ghatak, Kamakhya P.; Ghoshal, Ardhendhu; Bhattacharyya, Sankar; Mondal, Manabendra)V1346,471-489(1991)

Instantaneous velocity field measurement of objects in coaxial rotation using digital image velocimetry (Cho, Young-Chung; Park, Hong-Woo)V1346,160-171(1991)

In vivo energy transfer studies in photosynthetic systems by subpicosecond timing (Shreve, A. P.; Trautman, Jay K.; Owens, T. G.; Frank, Harry A.; Albricht, Andreas C.)V1403,394-399(1991)

Ion yields from strong optical field ionization experiments using 100-femtosecond laser pulses (Fittinghoff, David N.; Bolton, Paul R.; Chang, Britton; Van Woerkom, Linn D.; White, William E.)V1413,81-88(1991)

Laser high-speed photography systems used to ammunition measures and tests (Wang, Yuren)V1346,331-337(1991)

Laser photocathode development for high-current electron source (Moustaizis, Stavros D.; Fotakis, Costas; Girardeau-Montaut, Jean-Pierre)V1552,50-56(1991)

Line-imaging Fabry-Perot interferometer (Mathews, Allen R.; Warnes, Richard H.; Hemsing, Willard F.; Whittemore, Gerald R.)V1346,122-132(1991)

Measurements with a 35-psec gate time microchannel plate camera (Bell, Perry M.; Kilkenny, Joseph D.; Hanks, Roy L.; Landen, Otto L.)V1346,456-464(1991)

Modeling of electron density produced by femtosecond laser on metallic photocathodes (Girardeau-Montaut, Jean-Pierre; Girardeau-Montaut, Claire)V1502,331-335(1991)

Modeling the absorption of intense, short laser pulses in steep density gradients (Alley, W. E.)V1413,89-94(1991)

Modeling the pedestal in a chirped-pulse-amplification laser (Chuang, Yung-Ho; Peatross, J.; Meyerhofer, David D.)V1413,32-40(1991)

Molecular dynamics of stilbene molecule under laser excitation (Vachev, V. D.; Zadkov, Victor N.)V1403,487-496(1991)

Multigigabit solitary-wave propagation in both the normal and anomalous dispersion regions of optical fibers (Potasek, M. J.; Tabor, Mark)V1579,232-236(1991)

New configuration of a generator and regenerative amplifier built on three mirrors (Piotrowski, Jan)V1391,272-278(1991)

New problems of femtosecond time-domain CARS (coherent antistokes Raman spectroscopy) of large molecules (Kolomoitsev, D. V.; Nikitin, S. Y.)V1402,31-43(1991)

Nonlinear properties of oriented purple membrane films derived from second-harmonic generation under picosecond excitation: prospect of electro-optical measurements of ultrafast photoelectric respon (Sharkov, A. V.; Gillbro, T.)V1403,434-438(1991)

One-frame subnanosecond spectroscopy camera (Silkis, E. G.; Titov, V. D.; Feldman, G. G.; Zhilkina, V. M.; Petrokovich, O. A.; Syrtzev, V. N.)V1358,46-49(1991)

Optical coherent transients induced by time-delayed fluctuating pulses (Finkelstein, Vladimir; Berman, Paul R.)V1376,68-79(1991)

Optically ionized plasma recombination x-ray lasers (Amendt, Peter; Eder, David C.; Wilks, S. C.; Dunning, M. J.; Keane, Christopher J.)V1413,59-69(1991)

Pericyclic photochemical ring-opening reactions are complete in picoseconds: a time-resolved UV resonance Raman study (Reid, Philip J.; Doig, Stephen J.; Mathies, Richard A.)V1432,172-183(1991)

Photophysics of 1,3,5-triaryl-2-pyrazolines (Sahyun, Melville R.; Crooks, G. P.; Sharma, D. K.)V1436,125-133(1991)

Pico- and femtosecond pulses in the UV and XUV (Dinev, S.; Dreischuh, A.)V1403,427-430(1991)

Picosecond absorption and circular dichroism studies of proteins (Simon, John D.; Xie, Xiaoliang; Dunn, Robert C.)V1432,211-220(1991)

Picosecond kinetics and Sn S1 absorption spectra of retinoids and carotenoids (Bondarev, S. L.; Tikhomirov, S. A.; Bachilo, S. M.)V1403,497-499(1991)

Picosecond optoelectronic semiconductor switching and its application (Brueckner, Volkmar; Bergner, Harald; Lenzner, Matthias; Strobel, Reiner)V1362,510-517(1991)

Picosecond orientational dynamics of complex molecules studied by incoherent light three-wave mixing (Apanasevich, P. A.; Kozich, V. P.; Vodchitz, A. I.; Kontsevoy, B. L.)V1403,475-477(1991)

Plasma heating by ultrashort laser pulse in the regime of anomalous skin effect (Gamaly, Eugene G.; Kiselev, A. Y.; Tikhonchuk, V. T.)V1413,95-106(1991)

Possibility of short-pulses generation in excimer lasers by self-injection (Badziak, Jan; Dubicki, Adam; Piotrowski, Jan)V1391,117-126(1991)

Prepulse suppression using a self-induced ultrashort pulse plasma mirror (Gold, David; Nathel, Howard; Bolton, Paul R.; White, William E.; Van Woerkom, Linn D.)V1413,41-52(1991)

Provitamin D photoisomerization kinetics upon picosecond laser irradiation: role of previtamin conformational nonequilibrium (Terenetskaya, I. P.; Repeyev, Yu A.)V1403,500-503(1991)

Pulsed hologram recording by nanosecond and subnanosecond small-dimensioned laser system (Bespalov, V. G.; Dikasov, A. B.)V1238,470-475(1991)

Pump-probe investigation of picosecond laser-gas target interactions (Durfee, C.; Milchberg, Howard M.)V1413,78-80(1991)

Quantitative analysis of hemoglobin oxygenation state of rat head by time-resolved photometry using picosecond laser pulse at 1064 nm (Nomura, Yasutomo; Tamura, Mamoru)V1431,102-109(1991)

Recent advances in excimer laser technology at Los Alamos (Bigio, Irving J.; Czuchlewski, Stephen J.; McCown, Andrew W.; Taylor, Antoinette J.)V1397,47-53(1991)

Recent developments by femtosecond spectroscopy in biological ultrafast free radical reactions (Gauduel, Yann; Pommeret, Stanislas; Yamada, Noelle; Antonetti, Andre)V1403,417-426(1991)

Reflectivity of stimulated Brillouin scattering in picosecond time scales (Labaune, C.; Rozmus, Wojtek; Baldis, Hector A.; Mounaix, P.; Pesme, Denis; Baton, S.; LaFontaine, Bruno; Villeneuve, D. M.; Enright, G. D.)V1413,138-143(1991)

Research of fast stages of latent image formation in holographic photoemulsions influenced by ultrashort radiation pulses (Starobogatov, Igor O.; Nicolaev, S. D.)V1238,153-157(1991)

Selected Papers on Ultrafast Laser Technology (Gosnell, Timothy R.; Taylor, Antoinette J., eds.)VMS44(1991)

Self-action of supremely short light pulses in fibers (Azarenkov, Aleksey N.; Altshuler, Grigori B.; Kozlov, Sergey A.)V1409,166-177(1991)

Short-Pulse High-Intensity Lasers and Applications (Baldis, Hector A., ed.)V1413(1991)

Short pulse self-focusing (Strickland, Donna; Corkum, Paul B.)V1413,54-58(1991)

Simplified VISAR system (Sweatt, William C.; Stanton, Philip L.; Crump, O. B.)V1346,151-159(1991)

Solid state lasers with passive mode-locking and negative feedback for picosecond spectroscopy (Danelius, R.; Grigonis, R.; Piskarskas, Algis S.; Podenas, D.; Sirutkaitis, V.)V1402,198-208(1991)

Space-qualified streak camera for the Geodynamic Laser Ranging System (Johnson, C. B.; Abshire, James B.; Zagwodzki, Thomas W.; Hunkler, L. T.; Letzring, Samuel A.; Jaanimagi, Paul A.)V1346,340-370(1991)

Spatial dynamics of picosecond CO_2 laser pulses produced by optical switching in Ge (Pogorelsky, Igor V.; Fisher, A. S.; Veligdan, James T.; Russell, P.)V1413,21-31(1991)

Speed of optically-controlled superconducting devices (Kwok, Hoi-Sing; Shi, Lei; Zheng, J. P.; Dong, S. Y.; Pang, Y.; Prasad, Paras N.)V1394,196-200(1991)

Stanford picosecond FEL center (Swent, Richard L.; Schwettman, H. A.; Smith, T. I.)V1552,24-35(1991)

Streak camera phosphors: response to ultrashort excitation (Jaanimagi, Paul A.; Hestdalen, C.)V1346,443-448(1991)

Striped Fabry-Perots: improved efficiency for velocimetry (McMillan, Charles F.; Steinmetz, Lloyd L.)V1346,113-121(1991)

Studies of the excited states of biological systems using UV-excited resonance Raman and picosecond transient Raman spectroscopy (Gustafson, Terry L.; Iwata, Koichi; Weaver, William L.; Huston, Lisa A.; Benson, Ronda L.)V1403,545-554(1991)

Study of x-ray emission from picosecond laser-plasma interaction (Chen, Hong; Chuang, Yung-Ho; Delettrez, J.; Uchida, S.; Meyerhofer, David D.)V1413,112-119(1991)

Subnanosecond, high-voltage photoconductive switching in GaAs (Druce, Robert L.; Pocha, Michael D.; Griffin, Kenneth L.; O'Bannon, Jim)V1378,43-54(1991)

Subpicosecond electron transfer in reaction centers of photosynthetic bacteria (Shuvalov, V. A.; Ganago, A. O.; Shkuropatov, A. Y.; Klevanik, A. V.)V1403,400-406(1991)

Theoretical study of ultrafast dephasing by four-wave mixing (Hoerner, Claudine; Lavoine, J. P.; Villaeys, A. A.)V1362,863-869(1991)

Theory and simulation of Raman scattering in intense short-pulse laser-plasma interactions (Wilks, S. C.; Kruer, William L.; Langdon, A. B.; Amendt, Peter; Eder, David C.; Keane, Christopher J.)V1413,131-137(1991)

Third-order nonlinear optical characterization of side-chain copolymers (Norwood, Robert A.; Sounik, James R.; Popolo, J.; Holcomb, D. R.)V1560,54-65(1991)

Three-dimensional holography of nonstationary waves (Mazurenko, Yuri T.)V1238,85-96(1991)

Time-resolved infrared studies of the dynamics of ligand binding to cytochrome c oxidase (Dyer, R. B.; Peterson, Kristen A.; Stoutland, Page O.; Einarsdottir, Oloef; Woodruff, William H.)V1432,197-204(1991)

Time-resolved techniques: an overview (Larson, Bennett C.; Tischler, J. Z.)V1345,90-100(1991)

Time-resolved x-ray scattering studies using CCD detectors (Clarke, Roy; Dos Passos, Waldemar; Lowe, Walter P.; Rodricks, Brian G.; Brizard, Christine M.)V1345,101-114(1991)

Timing between streak cameras with a precision of 10 ps (Lerche, Richard A.)V1346,376-383(1991)

Ultrafast and not-so-fast dynamics of cytochrome oxidase: the ligand shuttle and its possible functional significance (Woodruff, William H.; Dyer, R. B.; Einarsdottir, Oloef; Peterson, Kristen A.; Stoutland, Page O.; Bagley, K. A.; Palmer, Graham; Schoonover, J. R.; Kliger, David S.; Goldbeck, Robert A.; Dawes, T. D.; Martin, Jean-Louis; Lambry, J.-C.; Atherton, Stephen J.; Hubig, Stefan M.)V1432,205-210(1991)

Ultrafast measurements of carrier transport optical nonlinearities in a GaAs/AlGaAs MQW SEED device (Park, Choong-Bum; Li Kam Wa, Patrick; Miller, Alan)V1474,8-17(1991)

Ultrafast pulse generation in fiber lasers (Kafka, James D.; Baer, Thomas M.)V1373,140-149(1991)

Ultrahigh- and High-Speed Photography, Videography, Photonics, and Velocimetry '90 (Jaanimagi, Paul A.; Neyer, Barry T.; Shaw, Larry L., eds.)V1346(1991)

Ultrashort pulse generation in solid state lasers (Fujimoto, James G.; Schulz, Peter A.; Fan, Tso Y.)V1413,2-13(1991)

Uses of Fabry-Perot velocimeters in studies of high explosives detonation (Breithaupt, R. D.; Tarver, Craig M.)V1346,96-102(1991)

Velocity interferometry of miniature flyer plates with subnanosecond time resolution (Paisley, Dennis L.; Montoya, Nelson I.; Stahl, David B.; Garcia, Ismel A.; Hemsing, Willard F.)V1346,172-178(1991)

VISAR: displacement-mode data reduction (Hemsing, Willard F.)V1346,141-150(1991)

VISAR: line-imaging interferometer (Hemsing, Willard F.; Mathews, Allen R.; Warnes, Richard H.; Whittemore, Gerald R.)V1346,133-140(1991)

X-ray and optical diagnostics of a 100-fs laser-produced plasma (Geindre, Jean-Paul; Audebert, Patrick; Chenais-Popovics, Claude; Gauthier, Jean-Claude J.; Benattar, Rene; Chambaret, J. P.; Mysyrowicz, Andre; Antonetti, Andre)V1502,311-318(1991)

Ultraviolet—see also lasers, excimer; microlithography

0.5-micron deep-UV lithography using a Micrascan-90 step-and-scan exposure tool (Kuyel, Birol; Barrick, Mark W.; Hong, Alexander; Vigil, Joseph)V1463,646-665(1991)

Advanced X-Ray/EUV Radiation Sources and Applications (Knauer, James P.; Shenoy, Gopal K., eds.)V1345(1991)

Applications of laser plasmas in XUV photoabsorption spectroscopy (Kennedy, Eugene T.; Costello, John T.; Mosnier, Jean-Paul)V1503,406-415(1991)

Clinical experience with an excimer laser angioplasty system (Golobic, Robert A.; Bohley, Thomas K.; Wells, Lisa D.; Sanborn, Timothy A.)V1425,84-92(1991)

Comparison of 248-nm line narrowing resonator optics for deep-UV lithography lasers (Kahlert, Hans-Juergen; Rebhan, Ulrich; Lokai, Peter; Basting, Dirk)V1463,604-609(1991)

Computer simulation of 0.5-micrometer lithography for a 16-megabit DRAM (Maltabes, John G.; Norris, Katherine C.; Writer, Dean)V1463,326-335(1991)

Deep-UV diagnostics using continuous tone photoresist (Kirk, Joseph P.; Hibbs, Michael S.)V1463,575-583(1991)

Deep-UV photolithography linewidth variation from reflective substrates (Dunn, Diana D.; Bruce, James A.; Hibbs, Michael S.)V1463,8-15(1991)

Design of high-performance aplanatic achromats for the near-ultraviolet waveband (Al-Baho, Tareq I.; Learner, R. C.; Maxwell, Jonathan)V1354,417-428(1991)

Design of narrow band XUV and EUV coronagraphs using multilayer optics (Walker, Arthur B.; Allen, Maxwell J.; Barbee, Troy W.; Hoover, Richard B.)V1343,415-425(1991)

Disk-shaped VUV+O source used as resist asher and resist developer (Hattori, Shuzo; Collins, George J.; Yu, Zenqi; Sugimoto, Dai; Saita, Masahiro)V1463,539-550(1991)

Dose control for short exposures in excimer laser lithography (Hollman, Richard F.)V1377,119-125(1991)

EUV performance of a multilayer-coated high-density toroidal grating (Keski-Kuha, Ritva A.; Thomas, Roger J.; Neupert, Werner M.; Condor, Charles E.; Gum, Jeffrey S.)V1343,566-575(1991)

EUV, X-Ray, and Gamma-Ray Instrumentation for Astronomy II (Siegmund, Oswald H.; Rothschild, Richard E., eds.)V1549(1991)

Excimer Lasers and Applications III (Letardi, Tommaso; Laude, Lucien D., eds.)V1503(1991)

Excimer lasers for deep-UV lithography (Elliott, David J.; Sengupta, Uday K.)V1377,6-17(1991)

Field-portable laser beam diagnostics (Forrest, Gary T.)V1414,55-64(1991)

Fused silica fibers for the delivery of high-power UV radiation (Artjushenko, Vjacheslav G.; Konov, Vitaly I.; Pashinin, Vladimir P.; Silenok, Alexander S.; Blinov, Leonid M.; Solomatin, A. M.; Shilov, I. P.; Volodko, V. V.; Mueller, Gerhard J.; Schaldach, Brita J.; Ulrich, R.; Neuberger, Wolfgang)V1420,149-156(1991)

Generation of ultrashort ultraviolet optical pulses using sum-frequency in LBO crystals (Guo, Ting; Qiu, Peixia; Lin, Fucheng)V1409,24-27(1991)

High-resolution imaging with multilayer soft x-ray, EUV, and FUV telescopes of modest aperture and cost (Walker, Arthur B.; Lindblom, Joakim F.; Timothy, J. G.; Hoover, Richard B.; Barbee, Troy W.; Baker, Phillip C.; Powell, Forbes R.)V1494,320-333(1991)

High-resolution stigmatic EUV spectroheliometer for studies of the fine scale structure of the solar chromosphere, transition region, and corona (Timothy, J. G.; Berger, Thomas E.; Morgan, Jeffrey S.; Walker, Arthur B.; Bhattacharyya, Jagadish C.; Jain, Surendra K.; Saxena, Ajay K.; Huber, Martin C.; Tondello, Giuseppe; Naletto, Giampiero)V1343,350-358(1991)

High-resolution, two-dimensional imaging, microchannel-plate detector for use on a sounding rocket experiment (Bush, Brett C.; Cotton, Daniel M.; Siegmund, Oswald H.; Chakrabarti, Supriya; Harris, Walter; Clarke, John T.)V1549,290-301(1991)

High-throughput narrowband 83.4-nm self-filtering camera (Zukic, Muamer; Torr, Douglas G.; Torr, Marsha R.)V1549,234-244(1991)

Imaging pulse-counting detector systems for space ultraviolet astrophysics missions (Timothy, J. G.)V1494,394-402(1991)

Impact of phase masks on deep-UV lithography (Sewell, Harry)V1463,168-179(1991)

Improvement of focus and exposure latitude by the use of phase-shifting masks for DUV applications (Op de Beeck, Maaike; Tokui, Akira; Fujinaga, Masato; Yoshioka, Nobuyuki; Kamon, Kazuya; Hanawa, Tetsuro; Tsukamoto, Katsuhiro)V1463,180-196(1991)

Improving the performance and usability of a wet-developable DUV resist for sub-500nm lithography (Samarakone, Nandasiri; Van Driessche, Veerle; Jaenen, Patrick; Van den hove, Luc; Ritchie, Douglas R.; Luehrmann, Paul F.)V1463,16-29(1991)

Interface characterization of XUV multilayer reflectors using HRTEM and x-ray and XUV reflectance (Windt, David L.; Hull, Robert; Waskiewicz, Warren K.; Kortright, Jeffrey B.)V1343,292-308(1991)

Laser photochemotherapy of psoriasis (Tuchin, Valery V.; Utz, Sergey R.; Barabanov, Alexander J.; Dovzansky, S. I.; Ulyanov, A. N.; Aravin, Vladislav A.; Khomutova, T. G.)V1422,85-96(1991)

Laser plasma XUV sources: a role for excimer lasers? (Bijkerk, Fred; Shevelko, A. P.)V1503,380-390(1991)

Laser-produced continua for studies in the XUV (Carroll, P. K.; O'Sullivan, Gerard D.)V1503,416-427(1991)

Luminescence molulation for the characterization of radiation damage within scintillator material (Bayer, Eberhard G.)V1361,195-199(1991)

Mathematical modeling for laser PUVA treatment of psoriasis (Medvedev, Boris A.; Tuchin, Valery V.; Yaroslavsky, Ilya V.)V1422,73-84(1991)

New i-line and deep-UV optical wafer stepper (Unger, Robert; DiSessa, Peter A.)V1463,725-742(1991)

Novel method for preventing solar ultraviolet-radiation-induced skin cancer (Shi, Weimin; Cui, Ting; Sigel, George H.)V1422,62-72(1991)

Optical design of an off-axis low-distortion UV telescope (Richardson, E. H.)V1494,314-319(1991)

Optical fibers for UV applications (Fabian, Heinz; Grzesik, Ulrich; Woerner, K.-H.; Klein, Karl-Friedrich)V1513,168-173(1991)

ORFEUS: orbiting and retrievable far and extreme ultraviolet spectrometer (Rippel, Harald; Kampf, Dirk; Graue, Roland)V1343,520-529(1991)

ORFEUS alignment concept (Graue, Roland; Kampf, Dirk; Rippel, Harald; Witte, G.)V1494,377-385(1991)

Output power stabilization of a XeCl excimer laser by HCl gas injection (Ogura, Satoshi; Kawakubo, Yukio; Sasaki, Kouji; Kubota, Yoshiyuki; Miki, Atsushi)V1412,123-128(1991)

Penning discharge VUV and soft x-ray source (Cao, Jianlin; Li, Futian; Qian, Limin; Chen, Po; Ma, Yueying; Chen, Xingdan)V1345,71-77(1991)

Performance of the Multi-Spectral Solar Telescope Array IV: the soft x-ray and extreme ultraviolet filters (Lindblom, Joakim F.; O'Neal, Ray H.; Walker, Arthur B.; Powell, Forbes R.; Barbee, Troy W.; Hoover, Richard B.; Powell, Stephen F.)V1343,544-557(1991)

Performance of the Multi-Spectral Solar Telescope Array V: temperature diagnostic response to the optically thin solar plasma (DeForest, Craig E.; Kankelborg, Charles C.; Allen, Maxwell J.; Paris, Elizabeth S.; Willis, Thomas D.; Lindblom, Joakim F.; O'Neal, Ray H.; Walker, Arthur B.; Barbee, Troy W.; Hoover, Richard B.; Barbee, Troy W.; Gluskin, Efim S.)V1343,404-414(1991)

Plane and concave VUV and soft x-ray multilayered mirrors (Cao, Jianlin; Miao, Tongqun; Qian, Longsheng; Zhu, Xioufang; Li, Futian; Ma, Yueying; Qian, Limin; Chen, Po; Chen, Xingdan)V1345,225-232(1991)

Radiometric stability of the shuttle-borne solar backscatter ultraviolet spectrometer (Cebula, Richard P.; Hilsenrath, Ernest; Kelly, Thomas J.; Batluck, Georgiann R.)V1493,91-99(1991)

Results from the calibration of the Extreme Ultraviolet Explorer instruments (Welsh, Barry Y.; Jelinsky, Patrick; Vedder, Peter W.; Vallerga, John V.; Finley, David S.; Malina, Roger F.)V1343,166-174(1991)

SHG in fiber: is a high-conversion efficiency possible? (Ouellette, Francois)V1516,113-114(1991)

Shock-layer-induced ultraviolet emissions measured by rocket payloads (Caveny, Leonard H.; Mann, David M.)V1479,102-110(1991)

Simplified model of the back surface of a charge-coupled device (Blouke, Morley M.; Delamere, W. A.; Womack, G.)V1447,142-155(1991)

Simulation analysis of deep-UV chemically amplified resist (Ohfuji, Takeshi; Soenosawa, Masanobu; Nozue, Hiroshi; Kasama, Kunihiko)V1463,345-354(1991)

Simulations of bar printing over a MOSFET device using i-line and deep-UV resists (Barouch, Eytan; Hollerbach, Uwe; Orszag, Steven A.; Szmanda, Charles R.; Thackeray, James W.)V1463,464-474(1991)

SOHO space satellite: UV instrumentation (Poland, Arthur I.; Domingo, Vicente)V1343,310-318(1991)

Solar EUV/FUV line polarimetry: instruments and methods (Hoover, Richard B.; Fineschi, Silvano; Fontenla, Juan; Walker, Arthur B.)V1343,389-403(1991)

Solar EUV/FUV line polarimetry: observational parameters and theoretical considerations (Fineschi, Silvano; Hoover, Richard B.; Fontenla, Juan; Walker, Arthur B.)V1343,376-388(1991)

Status of the Advanced Light Source (Marx, Jay N.)V1345,2-10(1991)

Submicron structures—promising filters in EUV: a review (Gruntman, Michael A.)V1549,385-394(1991)

Ultra-High-Resolution XUV Spectroheliograph II: predicted performance (Walker, Arthur B.; Lindblom, Joakim F.; Timothy, J. G.; Allen, Maxwell J.; DeForest, Craig E.; Kankelborg, Charles C.; O'Neal, Ray H.; Paris, Elizabeth S.; Willis, Thomas D.; Barbee, Troy W.; Hoover, Richard B.)V1343,319-333(1991)

Ultraprecise scanning technology (Zernike, Frits; Galburt, Daniel N.)V1343,241-244(1991)

Ultraviolet light imaging technology and applications (Yokoi, Takane; Suzuki, Kenji; Oba, Koichiro)V1449,30-39(1991)

Ultraviolet reflector materials for solar detoxification of hazardous waste (Jorgensen, Gary J.; Govindarajan, Rangaprasad)V1536,194-205(1991)

Use of antireflective coatings in deep-UV lithography (Sethi, Satyendra A.; Distasio, Romelia; Ziger, David H.; Lamb, James E.; Flaim, Tony)V1463,30-40(1991)

UV-VIS solid state excimer laser: XeF in crystalline argon (Zerza, Gerald; Knopp, F.; Kometer, R.; Sliwinski, G.; Schwentner, N.)V1410,202-208(1991)

VUV wall stabilized argon arc discharge source (Li, Futian; Cao, Jianlin; Chen, Po; Qian, Limin; Jin, Lei; Chen, Xingdan)V1345,78-88(1991)

Wet-developed, high-aspect-ratio resist patterns by 20-keV e-beam lithography (Weill, Andre P.; Amblard, Gilles R.; Lalanne, Frederic P.; Panabiere, Jean-Pierre)V1465,264-270(1991)

X-Ray/EUV Optics for Astronomy, Microscopy, Polarimetry, and Projection Lithography (Hoover, Richard B.; Walker, Arthur B., eds.)V1343(1991)

XUV characterization comparison of Mo/Si multilayer coatings (Windt, David L.; Waskiewicz, Warren K.; Kubiak, Glenn D.; Barbee, Troy W.; Watts, Richard N.)V1343,274-282(1991)

XUV free-electron laser-based projection lithography systems (Newnam, Brian E.)V1343,214-228(1991)

XUV resist characterization: studies with a laser plasma source (Kubiak, Glenn D.)V1343,283-291(1991)

Velocimetry—see also Doppler effect; flows

Application of Fabry-Perot velocimetry to hypervelocity impact experiments (Chau, Henry H.; Osher, John E.)V1346,103-112(1991)

Applications of copper vapor laser lighting in high-speed motion analysis (Hogan, Daniel C.)V1346,324-330(1991)

Automatic heterodyning of fiber-optic speckle pattern interferometry (Valera Robles, Jesus D.; Harvey, David; Jones, Julian D.)V1508,170-179(1991)

Biospeckle phenomena and their applications to blood-flow measurements (Aizu, Yoshihisa; Asakura, Toshimitsu)V1431,239-250(1991)

Depth dependent laser Doppler perfusion measurements: theory and instrumentation (Koelink, M. H.; de Mul, F. F.; Greve, Jan; Graaff, Reindert; Dassel, A. C.; Aarnouds, J. G.)V1403,347-349(1991)

Development, manufacturing, and integration of holographic optical elements for laser Doppler velocimetry applications (Stojanoff, Christo G.; Tholl, Hans D.; Luebbers, Hubertus; Windeln, Wilbert)V1507,426-434(1991)

Development of a laser Doppler system for measurement of velocity fields in PVT crystal growth systems (Jones, O. C.; Glicksman, M. E.; Lin, Jen T.; Kim, G. T.; Singh, N. B.)V1557,202-208(1991)

Double-exposure phase-shifting holography applied to particle velocimetry (Lai, Tianshu; Tan, Yushan)V1554B,580-585(1991)

Edge technique: a new method for atmospheric wind measurements with lidar (Korb, C. L.; Gentry, Bruce M.)V1416,177-182(1991)

Flow-field velocity measurements for nonisothermal systems (Johnson, Edward J.; Hyer, Paul V.; Culotta, Paul W.; Clark, Ivan O.)V1557,168-179(1991)

Ground-based PIV and numerical flow visualization results from the surface-tension-driven convection experiment (Pline, Alexander D.; Wernet, Mark P.; Hsieh, Kwang-Chung)V1557,222-234(1991)

Improvement in spatial resolution of a forward-scatter laser Doppler velocimeter (Mozumdar, Subir; Bond, Robert L.)V1584,254-261(1991)

Instantaneous velocity field measurement of objects in coaxial rotation using digital image velocimetry (Cho, Young-Chung; Park, Hong-Woo)V1346,160-171(1991)

In-vivo blood flow velocity measurements using the self-mixing effect in a fiber-coupled semiconductor laser (Koelink, M. H.; Slot, M.; de Mul, F. F.; Greve, Jan; Graaff, Reindert; Dassel, A. C.; Aarnouds, J. G.)V1511,120-128(1991)

Laser velocimetry applications (Soreide, David C.; McGarvey, John A.)V1416,280-285(1991)

Line-imaging Fabry-Perot interferometer (Mathews, Allen R.; Warnes, Richard H.; Hemsing, Willard F.; Whittemore, Gerald R.)V1346,122-132(1991)

Measurement of fluid velocity fields using digital correlation techniques (Matthys, Donald R.; Gilbert, John A.; Puliparambil, Joseph T.)V1332,850-861(1991)

Miniature laser Doppler anemometer for sensor concepts (Damp, Stephan)V1418,459-470(1991)

Monte Carlo simulations and measurements of signals in laser Doppler flowmetry on human skin (Koelink, M. H.; de Mul, F. F.; Greve, Jan; Graaff, Reindert; Dassel, A. C.; Aarnouds, J. G.)V1431,63-72(1991)

New optical technique for particle sizing and velocimetry (Xie, Gong-Wie; Scott, Peter D.; Shaw, David T.; Zhang, Yi-Mo)V1500,310-321(1991)

Numerical and optical evaluation of particle image velocimetry images (Farrell, Patrick V.)V1554B,610-621(1991)

Optical velocity sensor for air data applications (Smart, Anthony E.)V1480,62-71(1991)

Particle image velocimetry experiments for the IML-I spaceflight (Trolinger, James D.; Lal, Ravindra B.; Batra, Ashok K.; McIntosh, D.)V1557,98-109(1991)

Progress in miniature laser systems for space science particle sizing and velocimetry (Brown, Robert G.)V1506,58-59(1991)

Projectile velocity and spin rate by image processing of synchro-ballistic photography (Hughett, Paul)V1346,237-248(1991)

Pulsed holographic recording of very high speed transient events (Steckenrider, John S.; Ehrlich, Michael J.; Wagner, James W.)V1554B,106-112(1991)

Research and development of optical measurement techniques for aerospace propulsion research: a NASA/Lewis Research Center perspective (Lesco, Daniel J.)V1554B,530-539(1991)

Simplified VISAR system (Sweatt, William C.; Stanton, Philip L.; Crump, O. B.)V1346,151-159(1991)

Striped Fabry-Perots: improved efficiency for velocimetry (McMillan, Charles F.; Steinmetz, Lloyd L.)V1346,113-121(1991)

Uses of Fabry-Perot velocimeters in studies of high explosives detonation (Breithaupt, R. D.; Tarver, Craig M.)V1346,96-102(1991)

Velocity interferometry of miniature flyer plates with subnanosecond time resolution (Paisley, Dennis L.; Montoya, Nelson I.; Stahl, David B.; Garcia, Ismel A.; Hemsing, Willard F.)V1346,172-178(1991)

VISAR: displacement-mode data reduction (Hemsing, Willard F.)V1346,141-150(1991)

VISAR: line-imaging interferometer (Hemsing, Willard F.; Mathews, Allen R.; Warnes, Richard H.; Whittemore, Gerald R.)V1346,133-140(1991)

Video—see also cameras; television

Accuracy/repeatability test for a video photogrammetric measurement (Gustafson, Peter C.)V1526,36-41(1991)

Adaptive perceptual quantization for video compression (Puri, Atul; Aravind, R.)V1605,297-300(1991)

Analysis of optimum-frame-rate in low-bit-rate video coding (Takishima, Yasuhiro; Wada, M.; Murakami, Hitomi)V1605,635-645(1991)

Application of one-dimensional high-speed video camera system to motion analysis (Yokoyama, Naoki; Uyemura, Tsuneyoshi)V1358,351-357(1991)

Application results for an augmented video tracker (Pierce, Bill)V1482,182-195(1991)

Applications of high-resolution still video cameras to ballistic imaging (Snyder, Donald R.; Kosel, Frank M.)V1346,216-225(1991)

Applications of Z-Plane memory technology to high-frame rate imaging systems (Shanken, Stuart N.; Ludwig, David E.)V1346,210-215(1991)

Approximation-based video tracking system (Deng, Keqiang; Wilson, Joseph N.)V1568,304-312(1991)

Auto-focus video camera system with bag-type lens (Sugiura, Norio; Morita, Shinzo)V1358,442-446(1991)

Automatic shape recognition of human limbs to avoid errors due to skin marker shifting in motion analysis (Hatze, Herbert; Baca, Arnold)V1567,264-276(1991)

Bayesian approach to segmentation of temporal dynamics in video data (Jones, Coleen T.; Sauer, Ken D.)V1605,522-533(1991)

Blink comparison techniques applied to medical images (Craine, Eric R.; Craine, Brian L.)V1444,389-399(1991)

Block-adaptive quantization of multiple-frame motion field (Lavagetto, Fabio; Leonardi, Riccardo)V1605,534-545(1991)

Camera zoom/pan estimation and compensation for video compression (Tse, Yi-tong; Baker, Richard L.)V1452,468-479(1991)

Characteristics of sprinklers and water spray mists for fire safety (Jackman, Louise A.; Lavelle, Stephen P.; Nolan, P. F.)V1358,831-842(1991)

Coastal survey with a multispectral video system (Niedrauer, Terren M.)V1492,240-251(1991)

Coding of digital TV by motion-compensated Gabor decomposition (Dufaux, Frederic; Ebrahimi, Touradj; Geurtz, Alexander M.; Kunt, Murat)V1567,362-379(1991)

Coding of motion vectors for motion-compensated predictive/interpolative video coder (Chen, Cheng-Tie; Jeng, Fure-Ching)V1605,812-821(1991)

Compact motion representation based on global features for semantic image sequence coding (Labit, Claude; Nicolas, Henri)V1605,697-708(1991)

Comparison of directionally based and nondirectionally based subband image coders (Bamberger, Roberto H.; Smith, Mark J.)V1605,757-768(1991)

Comparison of image compression techniques for high quality based on properties of visual perception (Algazi, V. R.; Reed, Todd R.)V1567,589-598(1991)

Computer animation method for simulating polymer flow for injection-molded parts (Perry, Meg W.; Rumbaugh, Richard C.; Frost, David P.)V1459,155-156(1991)

Construction of efficient variable-length codes with clear synchronizing codewords for digital video applications (Lei, Shaw-Min)V1605,863-873(1991)

Convergence of video and computing (Carlson, Curtis R.)V1472,2-5(1991)

Design of parallel multiresolution filter banks by simulated annealing (Li, Wei; Basso, Andrea; Popat, Ashok C.; Nicoulin, Andre; Kunt, Murat)V1605,124-136(1991)

Detection of citrus freeze damage with natural color and color IR video systems (Blazquez, Carlos H.)V1467,394-401(1991)

Development of a large-screen high-definition laser video projection system (Clynick, Tony)V1456,51-57(1991)

Digital video codec for medium bitrate transmission (Ebrahimi, Touradj; Dufaux, Frederic; Moccagatta, Iole; Campbell, T. G.; Kunt, Murat)V1605,2-15(1991)

Digitize your films without losing resolution (Kallhammer, Jan-Erik O.)V1358,631-636(1991)

Discrimination and classification with Xybion multispectral video systems (Frost, Paul A.)V1358,398-408(1991)

Distribution fiber FTTH/FTTC trial results and deployment strategies (Coleman, John D.)V1363,2-12(1991)

Edge-based block matching technique for video motion estimation (Han, Richard Y.; Zakhor, Avideh)V1452,395-408(1991)

Edge-based subband image coding technique for encoding the upper-frequency bands (Mohsenian, Nader; Nasrabadi, Nasser M.)V1605,781-792(1991)

Electronically gated airborne video camera (Sturz, Richard A.)V1538,77-80(1991)

Entropy coding for wavelet transform of image and its application for motion picture coding (Ohta, Mutsumi; Yano, Mitsuharu; Nishitani, Takao)V1605,456-466(1991)

Estimation and prediction of object-oriented segmentation for video predictive coding (Brofferio, Sergio C.; Comunale, Domenico; Tubaro, Stefano)V1605,500-510(1991)

Evolution of fiber-to-the-curb networks toward broadband capabilities (Menendez, Ronald C.; Lu, Kevin W.; Rizzo, Annmarie; Lemberg, Howard L.)V1363,97-105(1991)

Experiments in holographic video imaging (Benton, Stephen A.)VIS08,247-267(1991)

Extraction of features from images using video feedback (Boone, Bradley G.; Shukla, Oodaye B.; Terry, David H.)V1471,390-403(1991)

Eyes in the skies: building satellites for education (Hansen, Verne W.; Summers, Robert A.; Clapp, William G.)V1495,115-122(1991)

Fiber in the loop: an evolution in services and systems (Engineer, Carl P.)V1363,19-29(1991)

Fourth-generation motion analyzer (Balch, Kris S.)V1358,373-397(1991)

HDTV compression with vector quantization of transform coefficients (Wu, Siu W.; Gersho, Allen)V1605,73-84(1991)

Hierarchical block motion estimation for video subband coding (Jeon, Joon-hyeon; Hahm, Cheul-hee; Kim, Jae-Kyoon)V1605,954-962(1991)

Hierarchical motion-compensated deinterlacing (Woods, John W.; Han, Soo-Chul)V1605,805-810(1991)

Hierarchical motion-compensated interframe DPCM algorithm for low-bit-rate coding (Xie, Kan; Van Eycken, Luc; Oosterlinck, Andre J.)V1567,380-389(1991)

High-density optical MUSE disk (Tsuchiya, Yoichi; Terasaki, Hitoshi; Ota, Osamu)V1499,450-456(1991)

High-dynamic-range image readout system (Mens, Alain; Ducrocq, N.; Mazataud, D.; Mugnier, A.; Eouzan, J. Y.; Heurtaux, J. C.; Tomasini, F.; Mathae, J. C.)V1358,719-731(1991)

Highly efficient entropy coding of multilevel images using a modified arithmetic code (Chen, Yan-Ping; Yasuda, Yasuhiko)V1605,822-831(1991)

High-resolution CCD still/video and still-still/video systems (Kee, Richard C.)V1448,13-20(1991)

High-speed CCD video camera (Germer, Rudolf K.; Meyer-Ilse, Werner)V1358,346-350(1991)

High-speed electronic memory video recording (Thomas, Don L.)V1448,140-147(1991)

High-speed programmable digitizer for real-time video compression experiments (Cox, Norman R.)V1605,906-915(1991)

High-speed still video photography for ballistic range applications (Speyer, Brian A.)V1358,1215-1221(1991)

High-speed two-dimensional pyramid image coding method and its implementation (Sahinoglou, Haralambos; Cabrera, Sergio D.)V1605,793-804(1991)

High-speed video instrumentation system (Gorenflo, Ronald L.; Stockum, Larry A.; Barnett, Brett)V1346,42-53(1991)

Human facial motion modeling, analysis, and synthesis for video compression (Huang, Thomas S.; Reddy, Subhash C.; Aizawa, Kiyoharu)V1605,234-241(1991)

Hybrid coder for image sequences using detailed motion estimates (Nickel, Michael; Husoy, John H.)V1605,963-971(1991)

Image compression for digital video tape recording with high-speed playback capability (Wu, Siu W.; Gersho, Allen)V1452,352-363(1991)

Image processing system for brain and neural tissue (Sun, Bingrong; Xu, Jiafang)V1606,1022-1026(1991)

Imaging in digestive videoendoscopy (Guadagni, Stefano; Nadeau, Theodore R.; Lombardi, Loreto; Pistoia, Francesco; Pistoia, Maria A.)V1420,178-182(1991)

Impact of fiber backscatter on loop video transmission without optical isolator (Das, Santanu K.; Ocenasek, Josef)V1363,172-176(1991)

Intensified multispectral imaging measuring in the spatial, frequency, and time domains with a single instrument (Kennedy, Benjamin J.)V1346,68-74(1991)

Interactive analysis of transient field data (Dickinson, Robert R.)V1459,166-176(1991)

Iterative motion estimation method using triangular patches for motion compensation (Nakaya, Yuichiro; Harashima, Hiroshi)V1605,546-557(1991)

Lapped orthogonal transform for motion-compensated video compression (Lynch, William E.; Reibman, Amy R.)V1605,285-296(1991)

Launch area theodolite system (Bradley, Lester M.; Corriveau, John P.; Tindal, Nan E.)V1482,48-60(1991)

Low light level imaging systems application considerations and calculations (Caudle, Dennis E.)V1346,54-63(1991)

Management of an adaptable-bit-rate video service in a MAN environment (Marini, Michele; Albanese, Andres)V1364,289-294(1991)

Maximum likelihood estimation of affine-modeled image motion (Shaltaf, Samir J.; Namazi, Nader M.)V1567,609-620(1991)

Method for removing background regions from moving images (Fujimoto, Tsuyoshi; Shoman, Mineo; Hase, Masahiko)V1606,599-606(1991)

Method to convert image resolution using M-band-extended QMF banks (Kawashima, Masahisa; Tominaga, Hideyoshi)V1605,107-111(1991)

MITAS: multisensor imaging technology for airborne surveillance (Thomas, John)V1470,65-74(1991)

Motion-compensated priority discrete cosine transform coding of image sequences (Efstratiadis, Serafim N.; Huang, Yunming G.; Xiong, Z.; Galatsanos, Nikolas P.; Katsaggelos, Aggelos K.)V1605,16-25(1991)

Motion-compensated wavelet transform coding for color video compression (Zhang, Ya-Qin; Zafar, Sohail)V1605,301-316(1991)

Motion field estimation for complex scenes (Driessen, Johannes N.; Biemond, Jan)V1605,511-521(1991)

Motion video coding for packet-switching networks: an integrated approach (Gilge, Michael; Gusella, Riccardo)V1605,592-603(1991)

Near-real-time biplanar fluoroscopic tracking system for the video tumor fighter (Lawson, Michael A.; Wika, Kevin G.; Gillies, George T.; Ritter, Rogers C.)V1445,265-275(1991)

New address-generation-unit architecture for video signal processing (Kitagaki, Kazukuni; Oto, Takeshi; Demura, Tatsuhiko; Araki, Yoshitsugu; Takada, Tomoji)V1606,891-900(1991)

New image-capturing techniques for a high-speed motion analyzer (Balch, Kris S.)V1346,2-23(1991)

New method for identifying features of an image on a digital video display (Doyle, Michael D.)V1380,86-95(1991)

Nonuniform image motion estimation in transformed-domain (Namazi, Nader M.; Lipp, John I.)V1567,659-669(1991)

Novel technical advances provide easy solutions to tough motion analysis problems (Brown, Michael J.)V1346,24-32(1991)

Optical subscriber line transmission system to support an ISDN primary-rate interface (Wataya, Hideo; Tsuchiya, Toshiyuki)V1363,72-84(1991)

Perceptual training with cues for hazard detection in off-road driving (Merritt, John O.; CuQlock-Knopp, V. G.)V1457,133-138(1991)

Performance evaluation of subband coding and optimization of its filter coefficients (Katto, Jiro; Yasuda, Yasuhiko)V1605,95-106(1991)

Positioning accuracy of a virtual stereographic pointer in a real stereoscopic video world (Drascic, David; Milgram, Paul)V1457,302-313(1991)

Postprocessing of video sequence using motion-dependent median filters (Lee, Ching-Long; Jeng, Bor S.; Ju, Rong-Hauh; Huang, Huang-Cheng; Kan, Kou-Sou; Huang, Jei-Shyong; Liu, Tsann-Shyong)V1606,728-734(1991)

Potential usefulness of a video printer for producing secondary images from digitized chest radiographs (Nishikawa, Robert M.; MacMahon, Heber; Doi, Kunio; Bosworth, Eric)V1444,180-189(1991)

Practical videography (Sturz, Richard A.)V1346,64-67(1991)

Process of videotape making: presentation design, software, and hardware (Dickinson, Robert R.; Brady, Dan R.; Bennison, Tim; Burns, Thomas; Pines, Sheldon)V1459,178-189(1991)

Proposed one- and two-fiber-to-the-pedestal architectural evolution (Schiffler, Richard A.)V1363,13-18(1991)

Quality assessment of video image capture systems (Rowberg, Alan H.; Lian, Jing)V1444,125-127(1991)

Quantitative measurement of thermal parameters over large areas using pulse-video thermography (Hobbs, Chris P.; Kenway-Jackson, Damian; Milne, James M.)V1467,264-277(1991)

Real-time video signal processing by generalized DDA and control memories: three-dimensional rotation and mapping (Hama, Hiromitsu; Yamashita, Kazumi)V1606,878-890(1991)

Reconstruction of quincunx-coded image sequences using vector median (Oistamo, Kai; Neuvo, Yrjo A.)V1606,735-742(1991)

Remote driving: one eye or two (Bryant, Keith; Ince, Ilhan)V1457,120-132(1991)

Restoration of spatially variant motion blurs in sequential imagery (Trussell, Henry J.; Fogel, Sergei)V1452,139-145(1991)

Secret transmission method of character data in motion picture communication (Tanaka, Kiyoshi; Nakamura, Yasuhiro; Matsui, Kineo)V1605,646-649(1991)

Signal extension and noncausal filtering for subband coding of images (Martucci, Stephen A.)V1605,137-148(1991)

Signal loss recovery in DCT-based image and video codecs (Wang, Yao; Zhu, Qin-Fan)V1605,667-678(1991)

Statistically optimized PR-QMF design (Caglar, Hakan; Liu, Yipeng; Akansu, Ali N.)V1605,86-94(1991)

Stereoscopic holographic cinematography (Albe, Felix; Smigielski, Paul)V1358,1098-1102(1991)

Stereoscopic video and the quest for virtual reality: an annotated bibliography of selected topics (Starks, Michael R.)V1457,327-342(1991)

Subband coding of video using energy-adaptive arithmetic coding and statistical feedback-free rate control (Popat, Ashok C.; Nicoulin, Andre; Basso, Andrea; Li, Wei; Kunt, Murat)V1605,940-953(1991)

Subband decomposition procedure for quincunx sampling grids (Kim, Chai W.; Ansari, Rashid)V1605,112-123(1991)

System for making scientific videotapes (Appino, Perry A.; Farrell, Edward J.)V1459,157-165(1991)

Systems analysis and design for next generation high-speed video systems (Snyder, Donald R.; Rowe, W. J.)V1346,226-236(1991)

Techniques for capturing over 10,000 images/second with intensified imagers (Balch, Kris S.)V1358,358-372(1991)

Temporal projection for motion estimation and motion compensating interpolation (Robert, Philippe)V1605,558-569(1991)

Three-dimensional motion analysis and structure recovering by multistage Hough transform (Nakajima, Shigeyoshi; Zhou, Mingyong; Hama, Hiromitsu; Yamashita, Kazumi)V1605,709-719(1991)

Three-dimensional subband decompositions for hierarchical video coding (Bosveld, Frank; Lagendijk, Reginald L.; Biemond, Jan)V1605,769-780(1991)

Time-resolved videothermography at above-frame-rate frequencies (Shepard, Steven M.; Sass, David T.; Imirowicz, Thomas P.; Meng, A.)V1467,234-238(1991)

Transmission of the motion of a walker by model-based image coding (Kimoto, Tadahiko; Yasuda, Yasuhiko)V1605,253-262(1991)

TV holography and image processing in practical use (Lokberg, Ole J.; Ellingsrud, Svein; Vikhagen, Eiolf)V1332,142-150(1991)

Ultrahigh- and High-Speed Photography, Videography, Photonics, and Velocimetry '90 (Jaanimagi, Paul A.; Neyer, Barry T.; Shaw, Larry L., eds.)V1346(1991)

Use of a human visual model in subband coding of color video signal with adaptive chrominance signal vector quantization (Barba, Dominique; Hanen, Jose)V1605,408-419(1991)

Use of color, color infrared, black and white films, and video systems in detecting health, stress, and disease in vegetation (Blazquez, Carlos H.)V1379,106-115(1991)

Use of high-speed videography to solve a structural vibration problem in overhead cranes (Clayton, Donal L.; Clayton, Richard J.)V1346,33-41(1991)

Use of the Society of Motion Picture and Television Engineers test pattern in picture archiving and communication systems (PACS) (Gray, Joel E.)V1444,118-124(1991)

Using sound to extract meaning from complex data (Scaletti, Carla; Craig, Alan B.)V1459,207-219(1991)

Variable-bit-rate HDTV coding algorithm for ATM environments for B-ISDN (Kinoshita, Taizo; Nakahashi, Tomoko; Takizawa, Masaaki)V1605,604-613(1991)

Velocity measurements in molten pools during high-power laser interaction with metals (Caillibotte, Georges; Kechemair, Didier; Sabatier, Lilian)V1412,209-211(1991)

Video-based alignment system for x-ray lithography (Hughlett, R. E.; Cooper, Keith A.)V1465,100-110(1991)

Video browsing using brightness data (Otsuji, Kiyotaka; Tonomura, Yoshinobu; Ohba, Yuji)V1606,980-989(1991)

Video compression algorithm with adaptive bit allocation and quantization (Viscito, Eric; Gonzales, Cesar A.)V1605,58-72(1991)

Visual Communications and Image Processing '91: Visual Communication (Tzou, Kou-Hu; Koga, Toshio, eds.)V1605(1991)

Visual factors and image analysis in the encoding of high-quality still images (Algazi, V. R.; Reed, Todd R.; Ford, Gary E.; Estes, Robert R.)V1605,329-338(1991)

VLSI implementation of a buffer, universal quantizer, and frame-rate-control processor (Uwabu, H.; Kakii, Eiji; Lacombe, R.; Maruyama, Masanori; Fujiwara, Hiroshi)V1605,928-937(1991)

Windowed motion compensation (Watanabe, Hiroshi; Singhal, Sharad)V1605,582-589(1991)

Wind tunnel model aircraft attitude and motion analysis (Mostafavi, Hassan)V1483,104-111(1991)

Visibility—see also atmospheric optics

Accumulating displacement fields from different steps in laser or white-light speckle methods (Shao, C. A.; King, H. J.; Wang, Yeong-Kang; Chiang, Fu-Pen)V1554A,613-618(1991)

Atmospheric visibility monitoring for planetary optical communications (Cowles, Kelly A.)V1487,272-279(1991)

Battlefield training in impaired visibility (Gammarino, Rudolph R.; Surhigh, James W.)V1419,115-125(1991)

Computer model for predicting underwater color images (Palowitch, Andrew W.; Jaffe, Jules S.)V1537,128-139(1991)

Coupled-dipole approximation: predicting scattering by nonspherical marine organisms (Hull, Patricia G.; Hunt, Arlon J.; Quinby-Hunt, Mary S.; Shapiro, Daniel B.)V1537,21-29(1991)

Day/night aerial surveillance system for fishery patrol (Uhl, Bernd)V1538,140-147(1991)

Isotropic light source for underwater applications (Brown, Robert A.; Honey, Richard C.; Maffione, Robert A.)V1537,147-150(1991)

Optical technique for the compensation of the temperature-dependent Verdet constant in Faraday rotation magnetometers (Hamid, Sohail; Tatam, Ralph P.)V1511,78-89(1991)

Possibility of liquid crystal display panels for a space-saving PACS workstation (Komori, Masaru; Minato, Kotaro; Takahashi, Takashi; Nakano, Yoshihisa; Sakurai, Tsunetaro)V1444,334-337(1991)

Quantization of color image components in the DCT domain (Peterson, Heidi A.; Peng, Hui; Morgan, J. H.; Pennebaker, William B.)V1453,210-222(1991)

Submerged reflectance measurements as a function of visible wavelength (Giles, John W.; Voss, Kenneth J.)V1537,140-146(1991)

Target acquisition model appropriate for dynamically changing scenarios (Rotman, Stanley R.; Gordon, E. S.)V1442,335-346(1991)

Underwater Imaging, Photography, and Visibility (Spinrad, Richard W., ed.)V1537(1991)

Visual characteristics of LED display pushbuttons for avionic applications (Vanni, Paolo; Isoldi, Felice)V1456,300-309(1991)

Vision—see also biology; color; three dimensions; visualization

Analog retina model for detecting moving objects against a moving background (Searfus, Robert M.; Eeckman, Frank H.; Colvin, Michael E.; Axelrod, Timothy S.)V1473,95-101(1991)

Analysis and analog implementation of directionally sensitive shunting inhibitory neural networks (Bouzerdoum, Abdesselam; Nabet, Bahram; Pinter, Robert B.)V1473,29-40(1991)

Apparent contrast and surface color in complex scenes (Arend, Lawrence)V1453,412-421(1991)

Approach to invariant object recognition on grey-level images by exploiting neural network models (Rybak, Ilya A.; Golovan, Alexander V.; Gusakova, Valentina I.)V1469,472-482(1991)

Biological basis for space-variant sensor design I: parameters of monkey and human spatial vision (Rojer, Alan S.; Schwartz, Eric L.)V1382,132-144(1991)

Biological basis for space-variant sensor design II: implications for VLSI sensor design (Rojer, Alan S.; Schwartz, Eric L.)V1386,44-52(1991)

B-transformation and Fibonacci-transformations of optical images for the information conformity of their perception (Miroshnikov, Mikhail M.; Nesteruk, Vsevolod F.)V1500,322-333(1991)

Challenges of vision theory: self-organization of neural mechanisms for stable steering of object-grouping data in visual motion perception (Marshall, Jonathan A.)V1569,200-215(1991)

Clue derivation and selection activities in a robot vision system (Reihani, Kamran; Thompson, Wiley E.)V1468,305-312(1991)

Color and Grassmann-Cayley coordinates of shape (Petrov, A. P.)V1453,342-352(1991)

Colorimetry, normal human vision, and visual display (Thornton, William A.)V1456,219-225(1991)

Color image enhancement with spectacles (Perrott, Colin M.)V1529,31-36(1991)

Comparison of 2-D planar and 3-D perspective display formats in multidimensional data visualization (Merwin, David H.; Wickens, Christopher D.)V1456,211-218(1991)

Comparison of image compression techniques for high quality based on properties of visual perception (Algazi, V. R.; Reed, Todd R.)V1567,589-598(1991)

Computational model of an integrated vision system (Uttal, William; Shepherd, Thomas; Lovell, Robb E.; Dayanand, Sriram)V1453,258-269(1991)

Corneal topography: the dark side of the moon (Bores, Leo D.)V1423,28-39(1991); V1429,217-228(1991)

Depth determination using complex logarithmic mapping (Bartlett, Sandra L.; Jain, Ramesh C.)V1382,3-13(1991)

Design of eye movement monitoring system for practical environment (Nakamura, Hiroyuki; Kobayashi, Hitoshi; Taya, Katsuo; Ishigami, Shigenobu)V1456,226-238(1991)

Detecting spatial and temporal dot patterns in noise (Drum, Bruce)V1453,153-164(1991)

Determining range information from self-motion: the template model (Sobey, Peter J.)V1382,123-131(1991)

Ecological approach to partial binocular overlap (Melzer, James E.; Moffitt, Kirk W.)V1456,124-131(1991)

Efficient visual representation and reconstruction from generalized curvature measures (Barth, Erhardt; Caelli, Terry M.; Zetzsche, Christoph)V1570,86-95(1991)

Electronic interface for high-frame-rate electrically addressed spatial light modulators (Kozaitis, Samuel P.; Kirschner, K.; Kelly, E.; Been, D.; Delgado, J.; Velez, E.; Alkindy, A.; Al-Houra, H.; Ali, F.)V1474,112-115(1991)

Eye-slaved pointing system for tele-operator control (Razdan, Rikki; Kielar, Alan)V1388,361-371(1991)

Finding distinctive colored regions in images (Syeda, Tanveer F.)V1381,574-581(1991)

Generalization of the problem of correspondence in long-range motion and the proposal for a solution (Stratton, Norman A.; Vaina, Lucia M.)V1468,176-185(1991)

Helmet-mounted sight and display testing (Boehm, Hans-Dieter V.; Schreyer, H.; Schranner, R.)V1456,95-123(1991)

Human Vision, Visual Processing, and Digital Display II (Rogowitz, Bernice E.; Brill, Michael H.; Allebach, Jan P., eds.)V1453(1991)

Human visual performance model for crewstation design (Larimer, James O.; Prevost, Michael P.; Arditi, Aries R.; Azueta, Steven; Bergen, James R.; Lubin, Jeffrey)V1456,196-210(1991)

Image quality evaluation of multifocal intraocular lenses (Silberman, Donn M.)V1423,20-28(1991)

Implementing early vision algorithms in analog hardware: an overview (Koch, Christof)V1473,2-16(1991)

Importance of phosphor persistence characteristics in reducing visual distress symptoms in VDT users (Hayosh, Thomas D.)V1454,399-405(1991)

Indirect illumination to reduce veiling luminance in seawater (Wells, Willard H.)V1537,2-9(1991)

Intelligent Robots and Computer Vision IX: Neural, Biological, and 3-D Methods (Casasent, David P., ed.)V1382(1991)

LaneLok: an improved Hough transform algorithm for lane sensing using strategic search methods (Kenue, Surender K.; Wybo, David R.)V1468,538-550(1991)

Large and small color differences: predicting them from hue scaling (Chan, Hoover; Abramov, Israel; Gordon, James)V1453,381-389(1991)

Large Screen Projection, Avionic, and Helmet-Mounted Displays (Assenheim, Harry M.; Flasck, Richard A.; Lippert, Thomas M.; Bentz, Jerry, eds.)V1456(1991)

Mathematical theories of shape: do they model perception? (Mumford, David)V1570,2-10(1991)

Mean-field stereo correspondence for natural images (Klarquist, William N.; Acton, Scott T.; Ghosh, Joydeep)V1453,321-332(1991)

Measurements of lightness: dependence on the position of a white in the field of view (McCann, John J.; Savoy, Robert L.)V1453,402-411(1991)

Minimum resolution for human face detection and identification (Samal, Ashok)V1453,81-89(1991)

Modeling inner and outer plexiform retinal processing using nonlinear coupled resistive networks (Andreou, Andreas G.; Boahen, Kwabena A.)V1453,270-281(1991)

Modeling of local neural networks of the visual cortex and applications to image processing (Rybak, Ilya A.; Shevtsova, Natalia A.; Podladchikova, Lubov N.)V1469,737-748(1991)

Model of human preattentive visual detection of edge orientation anomalies (Brecher, Virginia H.; Bonner, Raymond; Read, C.)V1473,39-51(1991)

Motion analysis for visually-guided navigation (Hildreth, Ellen C.)V1382,167-180(1991)

Multitask neural network for vision machine systems (Gupta, Madan M.; Knopf, George K.)V1382,60-73(1991)

Network compensation for missing sensors (Ahumada, Albert J.; Mulligan, Jeffrey B.)V1453,134-146(1991)

Neural edge detector (Enab, Yehia M.)V1382,292-303(1991)

Neural network for inferring the shape of occluded objects (Citkusev, Ljubomir; Vaina, Lucia M.)V1468,786-793(1991)

New human vision system model for spatio-temporal image signals (Matsui, Toshikazu; Hirahara, Shuzo)V1453,282-289(1991)

New perspectives on image enhancement for the visually impaired (Peli, Eli)V1382,49-59(1991)

Objective evaluation of the feeling of depth in 2-D or 3-D images using the convergence angle of the eyes (Yamada, Mitsuho; Hiruma, Nobuyuki; Hoshino, Haruo)V1453,51-57(1991)

Observer performance in dynamic displays: effect of frame rate on visual signal detection in noisy images (Whiting, James S.; Honig, David A.; Carterette, Edward; Eigler, Neal)V1453,165-175(1991)

Oh say, can you see? The physiology of vision (Young, Richard A.)V1453,92-123(1991)

Packing geometry of human cone photoreceptors: variations with eccentricity and evidence for local anisotropy (Sloan, Kenneth R.; Curcio, Christine A.)V1453,124-133(1991)

Pattern recognition, attention, and information bottlenecks in the primate visual system (Van Essen, David; Olshausen, B.; Anderson, Clifford H.; Gallant, J. T. L.)V1473,17-28(1991)

Perceiving the coherent movements of spatially separated features (Mowafy, Lyn; Lappin, Joseph S.)V1453,177-187(1991)

Perceptual noise measurement of displays (Chakraborty, Dev P.; Pfeiffer, Douglas E.; Brikman, Inna)V1443,183-190(1991)

"Perfect" displays and "perfect" image compression in space and time (Klein, Stanley A.; Carney, Thom)V1453,190-205(1991)

Psychophysical estimation of the human depth combination rule (Landy, Michael S.; Maloney, Laurence T.; Young, Mark J.)V1383,247-254(1991)

Quality factors of handwritten characters based on human visual perception (Kato, Takahito; Yamada, Mitsuho)V1453,43-50(1991)

Real-time motion detection using an analog VLSI zero-crossing chip (Bair, Wyeth; Koch, Christof)V1473,59-65(1991)

Recent progress in artificial vision (Normann, Richard A.)V1423,40-45(1991)

Receptive fields and the theory of discriminant operators (Gupta, Madan M.; Hungenahally, Suresh K.)V1382,87-98(1991)

Representing three-dimensional shapes for visual recognition (Hoffman, Donald)V1445,2-4(1991)

RETINA (RETinally INspired Architecture project) (Caulfield, H. J.; Wilkins, Nathan A.)V1564,496-503(1991)

Selective edge detection based on harmonic oscillator wave functions (Kawakami, Hajimu)V1468,156-166(1991)

Self-organized criticality in neural networks (Makarenko, Vladimir I.; Kirillov, A. B.)V1469,843-845(1991)

Shape-from-X: psychophysics and computation (Buelthoff, Heinrich H.; Yuille, Alan L.)V1383,235-246(1991)

Shape perception from binocular disparity and structure-from-motion (Tittle, James S.; Braunstein, Myron L.)V1383,225-234(1991)

Silicon retina with adaptive photoreceptors (Mahowald, Misha A.)V1473,52-58(1991)

Simulation of parvocellular demultiplexing (Martinez-Uriegas, Eugenio)V1453,300-313(1991)

Static and dynamic spatial resolution in image coding: an investigation of eye movements (Stelmach, Lew B.; Tam, Wa J.; Hearty, Paul J.)V1453,147-152(1991)

Structural identity in visual-perceptual recognition (Ligomenides, Panos A.)V1382,14-25(1991)

Target acquisition modeling based on human visual system performance (Valeton, J. M.; van Meeteren, Aart)V1486,68-84(1991)

Theoretical approach to hyperacuity tests based on resolution criteria for two-line images (Mondal, Pronab K.; Calvo, Maria L.; Chevalier, Margarita L.; Lakshminarayanan, Vasudevan)V1429,108-116(1991)

Three-dimensional vision and figure-ground separation by visual cortex (Grossberg, Stephen)V1382,2-2(1991)

Visual aspects of picture storage (Brettel, Hans)V1401,50-55(1991)

Visual field information in nap-of-the-earth flight by teleoperated helmet-mounted displays (Grunwald, Arthur J.; Kohn, S.; Merhav, S. J.)V1456,132-153(1991)

Visual Information Processing: From Neurons to Chips (Mathur, Bimal P.; Koch, Christof, eds.)V1473(1991)

Visualization of image from 2-D strings using visual reasoning (Li, Xiao-Rong; Chang, Shi-Kuo)V1468,720-731(1991)

Visual motion detection: emulation of retinal peripheral visual field (Gupta, Madan M.; Digney, Bruce L.)V1381,346-356(1991)

Visual processing, transformability of primaries, and visual efficiency of display devices (Thornton, William A.)V1453,390-401(1991)

Visual communications—see also image processing; television; video

45-Mbps multichannel TV coding system (Matsumoto, Shuichi; Hamada, Takahiro; Saito, Masahiro; Murakami, Hitomi)V1605,37-46(1991)

Cheops: a modular processor for scalable video coding (Bove, V. M.; Watlington, John)V1605,886-893(1991)

Classified transform coding of images using two-channel conjugate vector quantization (Nam, J. Y.; Rao, K. R.)V1605,202-213(1991)

Classified vector quantizer based on minimum-distance partitioning (Kim, Dong S.; Lee, Sang U.)V1605,190-201(1991)

Coding of motion vectors for motion-compensated predictive/interpolative video coder (Chen, Cheng-Tie; Jeng, Fure-Ching)V1605,812-821(1991)

Comparison of mono- and stereo-camera systems for autonomous vehicle tracking (Kehtarnavaz, Nasser; Griswold, Norman C.; Eem, J. K.)V1468,467-478(1991)

Design of M-band filter banks based on wavelet transform (Yaou, Ming-Haw; Chang, Wen-Thong)V1605,149-159(1991)

Design of parallel multiresolution filter banks by simulated annealing (Li, Wei; Basso, Andrea; Popat, Ashok C.; Nicoulin, Andre; Kunt, Murat)V1605,124-136(1991)

Digital video codec for medium bitrate transmission (Ebrahimi, Touradj; Dufaux, Frederic; Moccagatta, Iole; Campbell, T. G.; Kunt, Murat)V1605,2-15(1991)

Display nonlinearity in digital image processing for visual communications (Peli, Eli)V1606,508-519(1991)

Divergence as a measure of visual recognition: bias and errors caused by small samples (Lau, Manhot; Okagaki, Takashi)V1606,705-713(1991)

Electro-optic illuminating module (Pesl, Ales A.)V1454,299-305(1991)

Estimation and prediction of object-oriented segmentation for video predictive coding (Brofferio, Sergio C.; Comunale, Domenico; Tubaro, Stefano)V1605,500-510(1991)

Extension of Rader's algorithm for high-speed multidimensional autocorrelation (Rinaldo, R.; Bernardini, Riccardo; Cortelazzo, Guido M.)V1606,773-787(1991)

HDTV compression with vector quantization of transform coefficients (Wu, Siu W.; Gersho, Allen)V1605,73-84(1991)

Highly efficient entropy coding of multilevel images using a modified arithmetic code (Chen, Yan-Ping; Yasuda, Yasuhiko)V1605,822-831(1991)

High-speed hardware architecture for high-definition videotex system (Maruyama, Mitsuru; Sakamoto, Hideki; Ishibashi, Yutaka; Nishimura, Kazutoshi)V1605,916-927(1991)

Image vector quantization with block-adaptive scalar prediction (Gupta, Smita; Gersho, Allen)V1605,179-189(1991)

ISDN audio color-graphics teleconferencing system (Oyaizu, Ikuro; Tanaka, Kiyoto; Yamaguchi, Toshikazu; Miyabo, Katsuaki; Takahashi, Mamoru)V1606,990-1001(1991)

Laplacian pyramid coding of prediction error images (Stiller, Christoph; Lappe, Dirk)V1605,47-57(1991)

Method for removing background regions from moving images (Fujimoto, Tsuyoshi; Shoman, Mineo; Hase, Masahiko)V1606,599-606(1991)

Method to convert image resolution using M-band-extended QMF banks (Kawashima, Masahisa; Tominaga, Hideyoshi)V1605,107-111(1991)

Model for packet image communication in a centralized distribution system (Torbey, Habib H.; Zhang, Zhensheng)V1605,650-666(1991)

Motion-compensated priority discrete cosine transform coding of image sequences (Efstratiadis, Serafim N.; Huang, Yunming G.; Xiong, Z.; Galatsanos, Nikolas P.; Katsaggelos, Aggelos K.)V1605,16-25(1991)

Motion compensation by block matching and vector postprocessing in subband coding of TV signals at 15 Mbit/s (Lallauret, Fabrice; Barba, Dominique)V1605,26-36(1991)

Motion field estimation for complex scenes (Driessen, Johannes N.; Biemond, Jan)V1605,511-521(1991)

Performance evaluation of subband coding and optimization of its filter coefficients (Katto, Jiro; Yasuda, Yasuhiko)V1605,95-106(1991)

Probabilistic model for quadtree representation of binary images (Chou, Chun-Hsien; Chu, Chih-Peng)V1605,832-843(1991)

Pseudoinverse matrix methods for signal reconstruction from partial data (Feichtinger, Hans G.)V1606,766-772(1991)

Statistically optimized PR-QMF design (Caglar, Hakan; Liu, Yipeng; Akansu, Ali N.)V1605,86-94(1991)

Stochastic detecting images from strong noise field in visual communications (Cai, De-Fu)V1606,926-933(1991)

Subband decomposition procedure for quincunx sampling grids (Kim, Chai W.; Ansari, Rashid)V1605,112-123(1991)

Subband video-coding algorithm and its feasibility on a transputer video coder (Brofferio, Sergio C.; Marcozzi, Elena; Mori, Luigi; Raveglia, Dalmazio)V1605,894-905(1991)

Three-dimensional face model reproduction method using multiview images (Nagashima, Yoshio; Agawa, Hiroshi; Kishino, Fumio)V1606,566-573(1991)

Transparency and blur as selective cues for complex visual information (Colby, Grace; Scholl, Laura)V1460,114-125(1991)

Tree-structured vector quantization with input-weighted distortion measures (Cosman, Pamela C.; Oehler, Karen; Heaton, Amanda A.; Gray, Robert M.)V1605,162-171(1991)

Two-layer pyramid image coding scheme for interworking of video services in ATM (Sikora, Thomas; Tan, T. K.; Pang, Khee K.)V1605,624-634(1991)

Vector quantization of image pyramids with the ECPNN algorithm (de Garrido, Diego P.; Pearlman, William A.; Finamore, Weiler A.)V1605,221-232(1991)

Video compression algorithm with adaptive bit allocation and quantization (Viscito, Eric; Gonzales, Cesar A.)V1605,58-72(1991)

Visual Communications and Image Processing '91: Image Processing (Tzou, Kou-Hu; Koga, Toshio, eds.)V1606(1991)

Visual Communications and Image Processing '91: Visual Communication (Tzou, Kou-Hu; Koga, Toshio, eds.)V1605(1991)

Visualization—see also vision

Adaptive isosurface generation in a distortion-rate framework (Ning, Paul C.; Hesselink, Lambertus B.)V1459,11-21(1991)

Analysis and representation of complex structures in separated flows (Helman, James L.; Hesselink, Lambertus B.)V1459,88-96(1991)

Biostereometric Technology and Applications (Herron, Robin E., ed.)V1380(1991)

Brain surface maps from 3-D medical images (Lu, Jiuhuai; Hansen, Eric W.; Gazzaniga, Michael S.)V1459,117-124(1991)

Collaborative processing to extract myocardium from a sequence of two-dimensional echocardiograms (Revankar, Shriram; Sher, David B.; Rosenthal, Steven)V1459,268-273(1991)

Comparison of 2-D planar and 3-D perspective display formats in multidimensional data visualization (Merwin, David H.; Wickens, Christopher D.)V1456,211-218(1991)

Computer-aided design and drafting visualization of anatomical structure of the human eye and orbit (Parshall, Robert F.; Sadler, Lewis L.)V1380,200-207(1991)

Computer-aided forensic facial reconstruction (Evenhouse, Raymond J.; Rasmussen, Mary; Sadler, Lewis L.)V1380,147-156(1991)

Computer animation method for simulating polymer flow for injection-molded parts (Perry, Meg W.; Rumbaugh, Richard C.; Frost, David P.)V1459,155-156(1991)

Constructing topologically connected surfaces for the comprehensive analysis of 3-D medical structures (Kalvin, Alan D.; Cutting, Court B.; Haddad, Betsy; Noz, Marilyn E.)V1445,247-258(1991)

Detection and visualization of porosity in industrial CT scans of aluminum die castings (Andrews, Lee T.; Klingler, Joseph W.; Schindler, Jeffery A.; Begeman, Michael S.; Farron, Donald; Vaughan, Bobbi; Riggs, Bud; Cestaro, John)V1459,125-135(1991)

Development of image processing techniques for applications in flow visualization and analysis (Disimile, Peter J.; Shoe, Bridget; Toy, Norman; Savory, Eric; Tahouri, Bahman)V1489,66-74(1991)

Efficient extraction of local myocardial motion with optical flow and a resolution hierarchy (Srikantan, Geetha; Sher, David B.; Newberger, Ed)V1459,258-267(1991)

Extracting Meaning from Complex Data: Processing, Display, Interaction II (Farrell, Edward J., ed.)V1459(1991)

Fast finite-state codebook design algorithm for vector quantization (Chang, Ruey-Feng; Chen, Wen-Tsuen)V1605,172-178(1991)

Global geometric, sound, and color controls for iconographic displays of scientific data (Smith, Stuart; Grinstein, Georges G.; Pickett, Ronald M.)V1459,192-206(1991)

High-resolution fully 3-D mapping of human surfaces by laser array camera and data representations (Bae, Kyongtae T.; Altschuler, Martin D.)V1380,171-178(1991)

High-speed integrated rendering algorithm for interpreting multiple-variable 3-D data (Miyazawa, Tatsuo)V1459,36-47(1991)

Holographic soundfield visualization for nondestructive testing of hot surfaces (Crostack, Horst-Artur; Meyer, E. H.; Pohl, Klaus-Juergen)V1508,101-109(1991)

Image resampling in remote sensing and image visualization applications (Trainer, Thomas J.; Sun, Fang-Kuo)V1567,650-658(1991)

Interactive analysis of transient field data (Dickinson, Robert R.)V1459,166-176(1991)

Interactive graphics system for multivariate data display (Becker, Richard A.; Cleveland, William S.; Shyu, William M.; Wilks, Allan R.)V1459,48-56(1991)

Medical prosthetic applications of growth simulations in four-dimensional facial morphology (Sadler, Lewis L.; Chen, Xiaoming; Fyler, Ann)V1380,137-146(1991)

Method for the analysis of the 3-D shape of the face and changes in the shape brought about by facial surgery (Coombes, Anne M.; Linney, Alfred D.; Richards, Robin; Moss, James P.)V1380,180-189(1991)

Modeling and visualization of scattered volumetric data (Nielson, Gregory M.; Dierks, Tim)V1459,22-33(1991)

Network visualization: user interface issues (Becker, Richard A.; Eick, Stephen G.; Miller, Eileen O.; Wilks, Allan R.)V1459,150-154(1991)

Object-oriented data management for interactive visual analysis of three-dimensional fluid-flow models (Walther, Sandra S.; Peskin, Richard L.)V1459,232-243(1991)

Octree optimization (Globus, Al)V1459,2-10(1991)

Precise individualized armature for ear reconstruction (Evenhouse, Raymond J.; Chen, Xiaoming)V1380,248-253(1991)

Process of videotape making: presentation design, software, and hardware (Dickinson, Robert R.; Brady, Dan R.; Bennison, Tim; Burns, Thomas; Pines, Sheldon)V1459,178-189(1991)

Project DaVinci (Winarsky, Norman; Alexander, Joanna R.)V1459,67-68(1991)

Radiative tetrahedral lattices (Driver, Jesse W.; Buckalew, Chris)V1459,109-116(1991)

Subsampled vector quantization with nonlinear estimation using neural network approach (Sun, Huifang; Manikopoulos, Constantine N.; Hsu, Hwei P.)V1605,214-220(1991)

Surface digitizing of anatomical subjects with DIGIBOT-4 (Koch, Stephen; Koch, Eric)V1380,163-170(1991)

Surface temperature and shear stress measurement using liquid crystals (Toy, Norman; Savory, Eric; Disimile, Peter J.)V1489,112-123(1991)

System for making scientific videotapes (Appino, Perry A.; Farrell, Edward J.)V1459,157-165(1991)

Telescope enclosure flow visualization (Forbes, Fred F.; Wong, Woon-Yin; Baldwin, Jack; Siegmund, Walter A.; Limmongkol, Siriluk; Comfort, Charles H.)V1532,146-160(1991)

Three-dimensional visualization and quantification of evolving amorphous objects (Silver, Deborah E.; Zabusky, Norman J.)V1459,97-108(1991)

Use of a 3-D visualization system in the planning and evaluation of facial surgery (Linney, Alfred D.; Moss, James P.; Richards, Robin; Mosse, C. A.; Grindrod, S. R.; Coombes, Anne M.)V1380,190-199(1991)

Validation of vision-based obstacle detection algorithms for low-altitude helicopter flight (Suorsa, Raymond E.; Sridhar, Banavar)V1388,90-103(1991)

Vector and scalar field interpretation (Farrell, Edward J.; Aukrust, Trond; Oberhuber, Josef M.)V1459,136-147(1991)

Virtual environment technology (Zeltzer, David L.)V1459,86-86(1991)

Visualization and comparison of simulation results in computational fluid dynamics (Felger, Wolfgang; Astheimer, Peter)V1459,222-231(1991)

Visualization of manufacturing process data in N-dimensional spaces: a reanalysis of the data (Fulop, Ann C.; Allen, Donald M.; Deffner, Gerhard)V1459,69-76(1991)

Visualization tool for human-machine interface designers (Prevost, Michael P.; Banda, Carolyn P.)V1459,58-66(1991)

Visualizing underwater acoustic matched-field processing (Rosenblum, Lawrence; Kamgar-Parsi, Behzad; Karahalios, Margarida; Heitmeyer, Richard)V1459,274-282(1991)

Visual thinking in organizational analysis (Grantham, Charles E.)V1459,77-84(1991)

Visual workbench for analyzing the behavior of dynamical systems (Cahoon, Peter)V1459,244-253(1991)

Water

Aerospace remote sensing monitoring of inland water quality (Gitelson, Anatoly A.)V1492,307-318(1991)

Close-spaced vapor transport of II-VI semiconductors (Perrier, Gerard)V1536,258-267(1991)

Effect of alternating magnetic fields on the properties of water systems (Berezin, M. V.; Levshin, L. V.; Saletsky, A. M.)V1403,335-337(1991)

Fiber optic sensor for nitrates in water (MacCraith, Brian D.; Maxwell, J.)V1510,195-203(1991)

Ground-based microwave remote sensing of water vapor in the mesosphere and stratosphere (Croskey, Charles L.; Olivero, John J.; Martone, Joseph P.)V1491,323-334(1991)

Ground-based monitoring of water vapor in the middle atmosphere: the NRL water-vapor millimeter-wave spectrometer (Bevilacqua, Richard M.; Schwartz, Philip R.; Pauls, Thomas A.; Waltman, William B.; Thacker, Dorsey L.)V1491,231-242(1991)

Indirect illumination to reduce veiling luminance in seawater (Wells, Willard H.)V1537,2-9(1991)

Interferential diagnosis of self-focusing of Q-switched YAG laser in liquid (Lu, Jian-Feng; Wang, Chang Xing; Miao, Peng-Cheng; Ni, Xiao W.; He, Anzhi)V1415,220-224(1991)

Laser-flash photographic studies of Er:YAG laser ablation of water (Jacques, Steven L.; Gofstein, Gary)V1427,63-67(1991)

Liquid droplet supercritical explosion in the field of CO2 laser radiation and influence of plasma chemical reactions on initiation of optical breakdown in air (Budnik, A. P.; Popov, A. G.)V1440,135-145(1991)

On-line optical determination of water in ethanol (Kessler, Manfred A.)V1510,218-223(1991)

Problems of photogrammetry of moving target in water (Han, Xin Z.)V1537,215-220(1991)

Small satellites for water cycle experiments (Rondinelli, Giuseppe; Di Girolamo, Sergio; Barresi, Giangrande)V1495,19-31(1991)

Submerged reflectance measurements as a function of visible wavelength (Giles, John W.; Voss, Kenneth J.)V1537,140-146(1991)

Underwater laser scanning system (Austin, Roswell W.; Duntley, Seibert Q.; Ensminger, Richard L.; Petzold, Theodore J.; Smith, Raymond C.)V1537,57-73(1991)

Underwater solar light field: analytical model from a WKB evaluation (Tessendorf, Jerry A.)V1537,10-20(1991)

Water continuum in the 15- to 25-um spectral region: evidence for H2O2 in the atmosphere (Devir, Adam D.; Neumann, M.; Lipson, Steven G.; Oppenheim, Uri P.)V1442,347-359(1991)

Wavefronts

Active and Adaptive Optical Systems (Ealey, Mark A., ed.)V1542(1991)

Adaptive optics system tests at the ESO 3.6-m telescope (Merkle, Fritz; Gehring, G.; Rigaut, Francois; Lena, Pierre J.; Rousset, Gerard; Fontanella, Jean-Claude; Gaffard, Jean-Paul)V1542,308-318(1991)

Adaptive optics, transfer loops modeling (Boyer, Corinne; Gaffard, Jean-Paul)V1542,46-61(1991)

Adaptive optics using curvature sensing (Forbes, Fred F.; Roddier, Nicolas)V1542,140-147(1991)

Algorithms for wavefront reconstruction out of curvature sensing data (Roddier, Nicolas)V1542,120-129(1991)

Analysis of diffracted stress fields around a noncharged borehole with dynamic photoelasticity and gauges (Zhu, Zhenhai; Qu, Guangjian; Yang, Yongqi; Shang, Jian)V1554A,472-481(1991)

Anisoplanatism and the use of laser guide stars (Goad, Larry E.)V1542,100-109(1991)

Atmospheric turbulence sensing for a multiconjugate adaptive optics system (Johnston, Dustin C.; Welsh, Byron M.)V1542,76-87(1991)

Come-on-plus project: an upgrade of the come-on adaptive optics prototype system (Gendron, Eric; Cuby, Jean-Gabriel; Rigaut, Francois; Lena, Pierre J.; Fontanella, Jean-Claude; Rousset, Gerard; Gaffard, Jean-Paul; Boyer, Corinne; Richard, Jean-Claude; Vittot, M.; Merkle, Fritz; Hubin, Norbert)V1542,297-307(1991)

Effective properties of electromagnetic wave propagation in some composite media (Artola, Michel; Cessenat, Michel)V1558,14-21(1991)

Expressions for the spherical-wave-structure function based on a bump spectrum model for the index of refraction (Richardson, Christina E.; Andrews, Larry C.)V1487,19-30(1991)

Fitting capability of deformable mirror (Jiang, Wen-Han; Ling, Ning; Rao, Xuejun; Shi, Fan)V1542,130-137(1991)

Hartmann-Shack wavefront sensor using a binary optic lenslet array (Kwo, Deborah P.; Damas, George; Zmek, William P.; Haller, Mitch)V1544,66-74(1991)

High-energy laser wavefront sensors (Geary, Joseph M.)V1414,66-79(1991)

High-resolution astronomical observations using deconvolution from wavefront sensing (Michau, Vincent; Marais, T.; Laurent, Jean; Primot, Jerome; Fontanella, Jean-Claude; Tallon, M.; Fuensalida, Jesus J.)V1487,64-71(1991)

Integration of geometrical and physical optics (Lawrence, George N.; Moore, Kenneth E.)V1415,322-329(1991)

Investigation of strain birefringence and wavefront distortion in 001 plates of KD2PO4 (De Yoreo, James J.; Woods, Bruce W.)V1561,50-58(1991)

Johns Hopkins adaptive optics coronagraph (Clampin, Mark; Durrance, Samuel T.; Golimowski, D. A.; Barkhouser, Robert H.)V1542,165-174(1991)

Measuring phase errors of an array or segmented mirror with a single far-field intensity distribution (Tyson, Robert K.)V1542,62-75(1991)

New wavefront sensor for metrology of spherical surfaces (Goelz, Stefan; Persoff, Jeffrey J.; Bittner, Groff D.; Liang, Junzhong; Hsueh, Chi-Fu T.; Bille, Josef F.)V1542,502-510(1991)

Numerical wavefront propagation through inhomogeneous media (Zakeri, Gholam-Ali)V1558,103-112(1991)

Progress report on a five-axis fast guiding secondary for the University of Hawaii 2.2-meter telescope (Cavedoni, Charles P.; Graves, J. E.; Pickles, A. J.)V1542,273-282(1991)

Quantitative evaluation of optical surfaces using an improved Foucault test approach (Vandenberg, Donald E.; Humbel, William D.; Wertheimer, Alan)V1542,534-542(1991)

Quantum holography (Granik, Alex T.; Caulfield, H. J.)VIS08,33-38(1991)

Quantum holography and neurocomputer architectures (Schempp, Walter)VIS08,62-144(1991)

Real-time wavefront correction system using a zonal deformable mirror and a Hartmann sensor (Salmon, J. T.; Bliss, Erlan S.; Long, Theresa W.; Orham, Edward L.; Presta, Robert W.; Swift, Charles D.; Ward, Richard S.)V1542,459-467(1991)

Satellite-borne laser for adaptive optics reference (Greenaway, Alan H.)V1494,386-393(1991)

Self-aberration-eliminating interferometer for wavefront measurements (Gorelik, Vladimir S.; Kovalenko, Sergey N.; Turukhano, Boris G.)V1507,488-490(1991)

Self-referencing Mach-Zehnder interferometer as a laser system diagnostic (Feldman, Mark; Mockler, Daniel J.; English, R. Edward; Byrd, Jerry L.; Salmon, J. T.)V1542,490-501(1991)

Sensing refractive-turbulence profiles (Cn2) using wavefront phase measurements from multiple reference sources (Welsh, Byron M.)V1487,91-102(1991)

Simulation of optical diagnostics for crystal growth: models and results (Banish, Michele R.; Clark, Rodney L.; Kathman, Alan D.; Lawson, Shelah M.)V1557,209-221(1991)

Spatial frequency analysis for the computer-generated holography of 3-D objects (Tommasi, Tullio; Bianco, Bruno)V1507,136-141(1991)

University of Hawaii adaptive optics system: III. Wavefront curvature sensor (Graves, J. E.; McKenna, Daniel)V1542,262-272(1991)

Video Hartmann wavefront diagnostic that incorporates a monolithic microlens array (Toeppen, John S.; Bliss, Erlan S.; Long, Theresa W.; Salmon, J. T.)V1544,218-225(1991)

Wavefront control model of a beam control experiment (Cielinski, Amy J.)V1542,434-448(1991)

Wavefront curvature sensing and compensation methods in adaptive optics (Roddier, Francois J.)V1487,123-128(1991)

Wavefront dislocations and phase object registering inside the airy disk (Tychinsky, Vladimir P.; Tavrov, Alexander V.)V1500,207-210(1991)

Wavefront reconstruction of acoustic waves in a variable ocean (Porter, Robert P.; Mourad, Pierre D.; Al-Kurd, Azmi)V1558,91-102(1991)

Wave-optics analysis of fast-beam focusing (Shih, Chun-Ching)V1415,150-153(1991)

Whenever two beams interfere, one fringe equals one wave in the plane of interference, always (Williamson, Ray)V1527,252-257(1991)

Waveguides—see also fiber optics; guided waves; integrated optics; integrated optoelectronics

10.6-um TEM00 beam transmission characteristics of a hollow circular cross-section multimode waveguide (Jenkins, R. M.; Devereux, R. W.)V1512,135-142(1991)

Absorption, fluorescence, and stimulated emission in Ti-diffused Er:LiNbO3 waveguides (Brinkmann, R.; Sohler, Wolfgang; Suche, Hubertus)V1362,377-382(1991)

Accurate measurement of thin-polymeric-films index variations: application to elasto-optic effect and to photochromism (Dumont, Michel L.; Morichere, D.; Sekkat, Z.; Levy, Yves)V1559,127-138(1991)

Accurate modeling of the index profile in annealed proton-exchanged LiNbO3 waveguides (Nikolopoulos, John; Yip, Gar L.)V1583,71-82(1991)

Achromatization of optical waveguide components (Spaulding, Kevin E.; Morris, G. M.)V1507,45-54(1991)

Acoustic mode analysis of multilayered structures for the design of acousto-optic devices (Armenise, Mario N.; Matteo, Annamaria; Passaro, Vittorio M.)V1583,256-267(1991)

All-optical switching of picosecond pulses in GaAs MQW waveguides (Li Kam Wa, Patrick; Park, Choong-Bum; Miller, Alan)V1474,2-7(1991)

All-optical Ti:LiNbO3 waveguide switch (d'Alessandro, Antonio; De Sario, Marco; D'Orazio, Antonella; Petruzzelli, Vincenzo)V1378,127-138(1991)

Analysis of evanescent coupling in waveguide modulators (Bradley, Joe C.; Kellner, Albert L.)V1476,330-336(1991)

Analysis of lightwave propagation in the bent waveguide by the Galerkin Method (Maruta, Akihiro; Matsuhara, Masanori)V1583,307-313(1991)

Analysis of the refractive-index profile in ion-exchanged waveguides (Righini, Giancarlo C.; Pelli, Stefano; Saracini, R.; Battaglin, Giancarlo; Scaglione, Antonio)V1513,418-424(1991)

Annealing properties of proton-exchanged waveguides in LiNbO3 fabricated using stearic acid (Pun, Edwin Y.; Loi, K. K.; Zhao, S.; Chung, P. S.)V1583,102-108(1991)

Applications of optical polymer waveguide devices on future optical communication and signal processing (Keil, Norbert; Strebel, Bernhard N.; Yao, HuiHai; Krauser, Juergen)V1559,278-287(1991)

Artificial dielectric waveguides from semiconductor-embedded polymers (Grebel, Haim)V1583,355-361(1991)

Attenuation of leaky waves in GaAs/AlGaAs MQW waveguides formed on a GaAs substrate (Kubica, Jacek M.)V1506,134-139(1991)

Blackbody-pumped CO2 lasers using Gaussian and waveguide cavities (Chang, Jim J.; Christiansen, Walter H.)V1412,150-163(1991)

Blue laser devices for optical data storage (Lenth, Wilfried; Kozlovsky, William J.; Risk, William P.)V1499,308-313(1991)

Buried-glass waveguides by ion exchange through ionic barrier (Li, Ming-Jun; Honkanen, Seppo; Wang, Wei-Jian; Najafi, S. I.; Tervonen, Ari; Poyhonen, Pekka)V1506,52-57(1991)

Buried-ridge-stripe lasers monolithically integrated with butt-coupled passive waveguides for OEIC (Remiens, D.; Hornung, V.; Rose, B.; Robein, D.)V1362,323-330(1991)

Cavityless dielectric-waveguide-mode generation in a weakly amplifying gaseous medium (Mel'nikov, Lev Y.; Kochelap, Viatcheslav A.; Izmailov, I. A.)V1397,603-610(1991)

Channel waveguide Mach-Zehnder interferometer for wavelength splitting and combining (Tervonen, Ari; Poyhonen, Pekka; Honkanen, Seppo; Tahkokorpi, Markku T.)V1513,71-75(1991)

Characterization, modeling, and design optimization of integrated optical waveguide devices in glass (Yip, Gar L.)V1513,26-36(1991)

Characterization of high-Tc coplanar transmission lines and resonators (Kessler, Jochen; Dill, Roland; Russer, Peter)V1477,45-56(1991)

Characterization of ion-exchange waveguide made with diluted KNO3 (Kishioka, Kiyoshi)V1583,19-26(1991)

Characterization of micro-optical components fabricated by deep-etch x-ray lithography (Goettert, Jost; Mohr, Jurgen)V1506,170-178(1991)

Characterization of planar optical waveguides by K+ ion exchange in glass at 1.152 and 1.523 um (Yip, Gar L.; Kishioka, Kiyoshi; Xiang, Feng; Chen, J. Y.)V1583,14-18(1991)

Characterization of proton-exchanged and annealed proton-exchanged optical waveguides in z-cut LiNbO3 (Nikolopoulos, John; Yip, Gar L.)V1374,30-36(1991)

Characterization of the photorefractive effect in Ti:LiNbO3 stripe waveguides (Volk, Raimund; Sohler, Wolfgang)V1362,820-826(1991)

Collinear asymmetrical polymer waveguide modulator (Chen, Ray T.; Sadovnik, Lev S.)V1559,449-460(1991)

Compact far-infrared free-electron laser (Ride, Sally K.; Golightly, W.)V1552,128-137(1991)

Comparative study for bolometric and nonbolometric switching elements for microwave phase shifters (Tabib-Azar, Massood; Bhasin, Kul B.; Romanofsky, Robert R.)V1477,85-94(1991)

Coplanar waveguide InP-based HEMT MMICs for microwave and millimeter wave applications (Chou, Chia-Shing; Litvin, K.; Larson, Larry E.; Rosenbaum, Steven E.; Nguyen, Loi D.; Mishra, Umesh K.; Lui, M.; Thompson, M.; Ngo, Catherine M.; Melendes, M.)V1475,151-156(1991)

Coplanar waveguide microwave filter of YBa2Cu3O7 (Chew, Wilbert; Riley, A. L.; Rascoe, Daniel L.; Hunt, Brian D.; Foote, Marc C.; Cooley, Thomas W.; Bajuk, Louis J.)V1477,95-100(1991)

Design and evaluation of optical switching architectures (Ramesh, S. K.; Smith, Thomas D.)V1474,208-211(1991)

Design and optimization of demultiplexer in ion-exchanged glass waveguides (Mazzola, M.; Montrosset, Ivo; Fincato, Antonello)V1365,2-12(1991)

Design aspects and comparison between high-Tc superconducting coplanar waveguide and microstrip line (Kong, Keon-Shik; Bhasin, Kul B.; Itoh, Tatsuo)V1477,57-65(1991)

Design of a high-power cross-field amplifier at X-band with an internally coupled waveguide (Eppley, Kenneth; Ko, Kwok)V1407,249-259(1991)

Design of novel integrated optic devices utilizing depressed index waveguides (Lopez-Amo, Manuel; Menendez-Valdes, Pedro; Sanz, Inmaculada; Muriel, Miguel A.)V1374,74-85(1991)

Design of photonic switches for optimizing performances of interconnection networks (Armenise, Mario N.; Castagnolo, Beniamino)V1374,186-197(1991)

Development of optical waveguides by sol-gel techniques for laser patterning (Schmidt, Helmut; Krug, Herbert; Kasemann, Reiner; Tiefensee, Frank)V1590,36-43(1991)

Diffusion and solubility of ion-implanted Nd and Er in LiNbO3 (Buchal, Christoph; Mohr, S.)V1361,881-892(1991)

Dual-core ion-exchanged glass waveguides (Li, Ming-Jun; Honkanen, Seppo; Wang, Wei-Jian; Leonelli, Richard; Najafi, S. I.)V1513,410-417(1991)

Dye orientation in organic guest host systems on ferroelectric polymers (Osterfeld, Martin; Knabke, Gerhard; Franke, Hilmar)V1559,49-55(1991)

Effective laser with active element rectangular geometry (Danilov, Alexander A.; Nikirui, Ernest Y.; Osiko, Vyacheslav V.; Polushkin, Valery G.; Sorokin, Svjatoslav N.; Timoshechkin, M. I.)V1362,916-920(1991)

Efficient optical waveguide modulation based on Wannier-Stark localization in a InGaAs-InAlAs superlattice (Bigan, Erwan; Allovon, Michel; Carre, Madeleine; Carenco, Alain; Voisin, Paul)V1362,553-558(1991)

Electro-optic coefficients in electric-field poled-polymer waveguides (Smith, Barton A.; Herminghaus, Stephan; Swalen, Jerome D.)V1560,400-405(1991)

Engineering fabrication technology of multiplexed hologram in single-mode optical waveguide (Lin, Freddie S.; Chen, Jenkins C.; Nguyen, Cong T.; Liu, William Y.)V1461,39-50(1991)

Excimer laser fabrication of waveguide devices (Stiller, Marc A.)V1377,73-78(1991)

Excimer laser micromachining for passive fiber coupling to polymeric waveguide devices (Booth, Bruce L.; Hohman, James L.; Keating, Kenneth B.; Marchegiano, Joseph E.; Witman, Sandy L.)V1377,57-63(1991)

Experimental and theoretical investigation of surface- and bulk-induced attenuation in solution-deposited waveguides (Roncone, Ronald L.; Burke, James J.; Weisenbach, Lori; Zelinski, Brian J.)V1590,14-25(1991)

Experimental studies of proton-exchanged waveguides in lithium niobate using toluic acid (Pun, Edwin Y.; Loi, K. K.; Chung, P. S.)V1583,64-70(1991)

Fabrication of large multimode glass waveguides by dry silver ion exchange in vacuum (Tammela, Simo; Pohjonen, Harri; Honkanen, Seppo; Tervonen, Ari)V1583,37-42(1991)

Fabrication of light-guiding devices and fiber-coupling structures by the LIGA process (Rogner, Arnd; Ehrfeld, Wolfgang)V1506,80-91(1991)

Fabrication techniques of photopolymer-clad waveguides for nonlinear polymeric modulators (Tumolillo, Thomas A.; Ashley, Paul R.)V1559,65-73(1991)

Ferroelectric microdomain reversal on Y-cut LiNbO3 surfaces (Seibert, Holger; Sohler, Wolfgang)V1362,370-376(1991)

Fully proton-exchanged lithium niobate waveguides with grating (Zhang, H.; Li, Ming-Jun; Najafi, S. I.; Schwelb, Otto)V1583,83-89(1991)

Glasses for Optoelectronics II (Righini, Giancarlo C., ed.)V1513(1991)

Glass waveguides by ion exchange with ionic masking (Wang, Wei-Jian; Li, Ming-Jun; Honkanen, Seppo; Najafi, S. I.; Tervonen, Ari)V1513,434-440(1991)

Grating splitter for glass waveguide (Jin, Guoliang; Shen, Ronggui; Ying, Zaisheng)V1513,50-55(1991)

Guided-wave nonlinear optics in 2-docosylamino-5-nitropyridine Langmuir-Blodgett films (Bosshard, Christian; Kuepfer, Manfred; Floersheimer, M.; Guenter, Peter)V1560,344-352(1991)

Helmholtz beam propagation by the method of Lanczos reduction (Fleck, Joseph A.)V1583,228-239(1991)

High-density waveguide modulator arrays for parallel interconnection (Bristow, Julian P.; Mukherjee, Sayan D.; Khan, M. N.; Hibbs-Brenner, Mary K.; Sullivan, Charles T.; Kalweit, Edith)V1389,535-546(1991)

High-efficient fiber-to-stripe waveguide coupler (Domanski, Andrzej W.; Roszko, Marcin; Sierakowski, Marek W.)V1362,844-852(1991)

High-resolution integrated optic holographic wavelength division multiplexer (Liu, William Y.; Strzelecki, Eva M.; Lin, Freddie S.; Jannson, Tomasz)V1365,20-24(1991)

High-silica cascaded three-waveguide couplers for wideband filtering by flame hydrolysis on Si (Barbarossa, Giovanni; Laybourn, Peter J.)V1583,122-128(1991)

High-silica low-loss three-waveguide couplers on Si by flame hydrolysis deposition (Barbarossa, Giovanni; Laybourn, Peter J.)V1513,37-43(1991)

High-speed board-to-board optical interconnection (Chen, Ray T.; Lu, Huey T.; Robinson, Daniel; Plant, David V.; Fetterman, Harold R.)V1559,110-117(1991)

H:LiNbO3 optical waveguides made from pyrophosphoric acid (Ziling, C. C.; Pokrovskii, L.; Terpugov, N. V.; Kuneva, M.; Savatinova, Ivanka T.; Armenise, Mario N.)V1583,90-101(1991)

Hollow and dielectric waveguides for infrared spectroscopic applications (Saggese, Steven J.; Harrington, James A.; Sigel, George H.)V1437,44-53(1991)

Hollow curved Al2O3 waveguides for CO2 laser surgery (Gregory, Christopher J.; Harrington, James A.; Altkorn, Robert I.; Haidle, Rudy H.; Helenowski, Tomasz)V1420,169-175(1991)

Hollow-tube-guide for UV-power laser beams (Kubo, Uichi; Okada, Kasuyuki; Hashishin, Yuichi)V1420,102-107(1991)

Homogeneous thin film lens on LiNbO3 (Jiang, Pisu; Laybourn, Peter J.; Righini, Giancarlo C.)V1362,899-906(1991)

Indium tin oxide single-mode waveguide modulator (Chen, Ray T.; Robinson, Daniel; Lu, Huey T.; Sadovnik, Lev S.; Ho, Zonh-Zen)V1583,362-374(1991)

Influence of processing variables on the optical properties of SiO2-TiO2 planar waveguides (Weisenbach, Lori; Zelinski, Brian J.; O'Kelly, John; Morreale, Jeanne; Roncone, Ronald L.; Burke, James J.)V1590,50-58(1991)

Infrared cables and catheters for medical applications (Artjushenko, Vjacheslav G.; Ivchenko, N.; Konov, Vitaly I.; Kryukov, A. P.; Krupchitsky, Vladimir P.; Kuznetcov, R.; Lerman, A. A.; Litvinenko, E. G.; Nabatov, A. O.; Plotnichenko, V. G.; Prokhorov, Alexander M.; Pylnov, I. L.; Tsibulya, Andrew B.; Vojtsekhovsky, V. V.; Ashraf, N.; Neuberger, Wolfgang; Moran, Kelly B.; Mueller, Gerhard J.; Schaldach, Brita J.)V1420,157-168(1991)

Integrated compact matrix-vector multipliers for optical interconnect applications (Lin, Freddie S.; Strzelecki, Eva M.; Liu, William Y.)V1389,642-647(1991)

Integrated optical channel waveguides in silicon using SiGe alloys (Splett, Armin O.; Schmidtchen, Joachim; Schueppert, B.; Petermann, Klaus)V1362,827-833(1991)

Integrated Optical Circuits (Wong, Ka-Kha, ed.)V1583(1991)

Integrated optical device in a fiber gyroscope (Rasch, Andreas; Goering, Rolf; Karthe, Wolfgang; Schroeter, Siegmund; Ecke, Wolfgang; Schwotzer, Guenter; Willsch, Reinhardt)V1511,149-154(1991)

Integrated optical devices for high-data-rate serial-to-parallel conversion (Verber, Carl M.; Kenan, Richard P.; Tan, Ronson K.; Bao, Y.)V1374,68-73(1991)

Integrated optical devices with silicon oxynitride prepared by plasma-enhanced chemical vapor deposition (PECVD) on Si and GaAs substrates (Peters, Dethard; Mueller, Joerg)V1362,338-349(1991)

Integrated optical Mach-Zehnder interferometers in glass (Lefebvre, P.; Vahid-Shahidi, A.; Albert, Jacques; Najafi, S. I.)V1583,221-225(1991)

Integrated-optic interconnects and fiber-optic WDM data links based on volume holography (Jannson, Tomasz; Lin, Freddie S.; Moslehi, Behzad M.; Shirk, Kevin W.)V1555,159-176(1991)

Integrated Optics and Optoelectronics II (Wong, Ka-Kha, ed.)V1374(1991)

Integrated optics intensity modulators in the GaAs/AlGaAs system (Khan, M. A.; Naumaan, Ahmed; Van Hove, James M.)V1396,753-759(1991)

Interconnection requirements in avionic systems (Vergnolle, Claude; Houssay, Bruno)V1389,648-658(1991)

Ion-exchanged waveguides: current status (Srivastava, Ramakant; Ramaswamy, Ramu V.)V1583,2-13(1991)

Ion-exchanged waveguides in glass: simulation and experiment (Wolf, Barbara; Fabricius, Norbert; Foss, Wolfgang; Dorsel, Andreas N.)V1506,40-51(1991)

Ion-exchanged waveguides in semiconductor-doped glasses (Righini, Giancarlo C.; Pelli, Stefano; De Blasi, C.; Fagherazzi, Giuliano; Manno, D.)V1513,105-111(1991)

Ionic conductivity and damage mechanisms in KTiOPO4 crystals (Morris, Patricia A.; Crawford, Michael K.; Roelofs, Mark G.; Bierlein, John D.; Baer, Thomas M.)V1561,104-111(1991)

Issues affecting the characterization of integrated optical devices subjected to ionizing radiation (Hickernell, Robert K.; Sanford, N. A.; Christensen, David H.)V1474,138-147(1991)

K-ion-exchange waveguide directional coupler sensor (Chen, Zheng; Dai, Ji Zhi)V1572,129-131(1991)

KTP waveguides for frequency upconversion of strained-layer InGaAs laser diodes (Risk, William P.; Nadler, Ch. K.)V1561,130-134(1991)

Langmuir-Blodgett films of immunoglobulin G and direct immunochemical sensing (Turko, Illarion V.; Pikuleva, Irene A.; Yurkevich, Igor S.; Chashchin, Vadim L.)V1510,53-56(1991)

Laser-induced generation of surface periodic structures resulting from the waveguide mode interaction (Bazakutsa, P. V.; Maslennikov, V. L.; Sychugov, V. A.; Yakovlev, V. A.)V1440,370-376(1991)

Length-minimization design considerations in photonic integrated circuits incorporating directional couplers (Boyd, Joseph T.; Radens, Carl J.; Kauffman, Michael T.)V1583,138-143(1991)

Linear and nonlinear optical effects in polymer waveguides (Braeuer, Andreas; Bartuch, Ulrike; Zeisberger, M.; Bauer, T.; Dannberg, Peter)V1559,470-478(1991)

Linearization of electro-optic modulators by a cascade coupling of phase-modulating electrodes (Skeie, Halvor; Johnson, Richard V.)V1583,153-164(1991)

LiTaO3 and LiNbO3:Ti responses to ionizing radiation (Padden, Richard J.; Taylor, Edward W.; Sanchez, Anthony D.; Berry, J. N.; Chapman, S. P.; DeWalt, Steve A.; Wong, Ka-Kha)V1474,148-159(1991)

Lithium niobate proton-exchange technology for phase-amplitude modulators (Varasi, Mauro; Vannucci, Antonello; Signorazzi, Mario)V1583,165-169(1991)

Lithium niobate waveguide devices: present performance and future applications (Bulmer, Catherine H.)V1583,176-183(1991)

Low-loss polymer thin-film optical waveguides (Amleshi, Peerouz M.; Naylor, David L.)V1396,396-403(1991)

Low-loss waveguides on silicon substrates for photonic circuits (Davis, Richard L.; Lee, Sae H.)V1474,20-26(1991)

Low-loss Y-couplers for fiber optic gyro applications (Page, Jerry L.)V1374,287-293(1991)

Low-temperature ion exchange of dried gels for potential waveguide fabrication in glasses (Risen, William M.; Morse, Ted F.; Tsagaropoulos, George)V1590,44-49(1991)

Mass-producible optical guided-wave devices fabricated by photopolymerization (Hosokawa, Hayami; Horie, Noriyoshi; Yamashita, Tsukasa)V1559,229-237(1991)

Micromachined structure for coupling optical fibers to integrated optical waveguides (Goel, Sanjay; Naylor, David L.)V1396,404-410(1991)

Micro-Optics II (Scheggi, Annamaria V., ed.)V1506(1991)

Mode conversion in Y-branch waveguides (Chan, Hau P.; Chung, P. S.; Pun, Edwin Y.)V1583,129-134(1991)

Modeling of acousto-optic interaction in multilayer guiding structures (Armenise, Mario N.; Matteo, Annamaria; Passaro, Vittorio M.)V1583,289-297(1991)

Modeling of traveling wave Ti:LiNbO3 waveguide modulators (Rocca, Corrado; Montrosset, Ivo; Gollinucci, Stefano; Ghione, Giovanni)V1372,39-47(1991)

Modelization of a semi-leaky waveguide: application to a polarizer (Saint-Andre, Francoise; Benech, Pierre; Kevorkian, Antoine P.)V1583,278-288(1991)

Multimode stripe waveguides for optical interconnections (Maile, Michael; Weidel, Edgar)V1563,188-196(1991)

New approaches to practical guided-wave passive devices based on ion-exchange technologies in glass (Seki, Masafumi; Sato, Shiro; Nakama, Kenichi; Wada, Hiroshi; Hashizume, Hideki; Kobayashi, Shigeru)V1583,184-195(1991)

New configuration of 2 x 2 switching elements using Mach-Zehnder electro-optic modulators (Armenise, Mario N.; Castagnolo, Beniamino; Pesce, Anastasia; Rizzi, Maria L.)V1583,210-220(1991)

New formulation of reflection coefficient in propagation theory for imaging inhomogeneous media and for nonuniform waveguides (Berger, Henry; Del Bosque-Izaguirre, Delma)V1521,117-130(1991)

New integrated optic TE/TM splitter made on LiNbO3 isotropic cut (Duchet, Christian; Flaaronning, Nils; Brot, Christian; Sarrabay, Laurence)V1372,72-81(1991)

New lithium niobate Y-junction polarization divider: theoretical study (Conese, Tiziana; De Pascale, Olga; Matteo, Annamaria; Armenise, Mario N.)V1583,249-255(1991)

New syndioregic main-chain, nonlinear optical polymers, and their ellipsometric characterization (Lindsay, Geoffrey A.; Nee, Soe-Mie F.; Hoover, James M.; Stenger-Smith, John D.; Henry, Ronald A.; Kubin, R. F.; Seltzer, Michael D.)V1560,443-453(1991)

Nondestructive investigations of multilayer dielectrical coatings (Shapiro, Alexander G.; Yaminsky, Igor V.)V1362,834-843(1991)

Nonlinear optical transmission of an integrated optical bent coupler in semiconductor-doped glass (Guntau, Matthias; Possner, Torsten; Braeuer, Andreas; Dannberg, Peter)V1513,112-122(1991)

Nonlinear polymers and devices (Moehlmann, Gustaaf R.; Horsthuis, Winfried H.; Hams, Benno H.)V1512,34-39(1991)

Novel configuration of a two-mode-interference polarization splitter with a buffer layer (Antuofermo, Pasquale; Losacco, Aurora M.; De Pascale, Olga)V1583,143-149(1991)

Novel technique for analysis and design of diffused Ti:LiNbO3 optical planar waveguides (Lopez-Higuera, Jose M.; Lopez-Amo, Manuel; Muriel, Miguel A.)V1374,144-151(1991)

Novel technique for the measurement of coupling length in directional couplers (Cheng, Hsing C.; Ramaswamy, Ramu V.)V1474,291-297(1991)

Optical bus protocol for a distributed-shared-memory multiprocessor (Davis, Martin H.; Ramachandran, Umakishore)V1563,176-187(1991)

Optical cross-modulation method for diagnostic of powerful microwave radiation (Kozar, A. V.; Krupenko, S. A.)V1476,305-312(1991)

Optical damage threshold of ion-exchanged glass waveguides at 1.06 um (Albert, Jacques; Wang, Wei-Jian; Najafi, S. I.)V1583,27-31(1991)

Optical interconnections based on waveguide holograms (Putilin, Andrei N.)V1621,93-101(1991)

Optical interconnects: a solution to very high speed integrated circuits and systems (Chen, Ray T.)V1374,162-175(1991)

Optical performance of wavelength division multiplexers made by ion-exchange in glass (Nissim, Carlos; Beguin, Alain; Laborde, Pascale; Lerminiaux, Christian)V1365,13-19(1991)

Optical sampling and demultiplexing applied to A/D conversion (Bell, John A.; Hamilton, Michael C.; Leep, David A.)V1562,276-280(1991)

Optical waveguides in polymer materials by ion implantation (Frank, Werner F.; Kulisch, Juergen R.; Franke, Hilmar; Rueck, Dorothee M.; Brunner, Stefan; Lessard, Roger A.)V1559,344-353(1991)

Optimization of externally modulated analog optical links (Betts, Gary E.; Johnson, Leonard M.; Cox, Charles H.)V1562,281-302(1991)

Overview of EO polymers for guided-wave devices (Ashley, Paul R.; Tumolillo, Thomas A.)V1583,316-326(1991)

Passive optical silica-on-silicon waveguide components (McGoldrick, Elizabeth; Hubbard, Steven D.; Maxwell, Graeme D.; Thomas, N.)V1374,118-125(1991)

Perturbation theory for optical bistability of prism and grating couplers and comparison with rigorous method (Akhouayri, Hassan; Vincent, Patrick; Neviere, Michel)V1545,140-144(1991)

Phase-matched second harmonic generations in poled dye-polymer waveguides (Sugihara, Okihiro; Kinoshita, Takeshi; Okabe, M.; Kunioka, S.; Nonaka, Y.; Sasaki, Keisuke)V1361,599-605(1991)

Phase space calculation of bend loss in rectangular light pipes (Gleckman, Philip; Ito, John)V1528,163-168(1991)

Photochemical delineation of waveguides in polymeric thin films (Beeson, Karl W.; Horn, Keith A.; Lau, Christina; McFarland, Michael J.; Schwind, David; Yardley, James T.)V1559,258-266(1991)

Photochemical formation of polymeric optical waveguides and devices for optical interconnection applications (Beeson, Karl W.; Horn, Keith A.; McFarland, Michael J.; Wu, Chengjiu; Yardley, James T.)V1374,176-185(1991)

Photodetectors: how to integrate them with microelectronic and optical devices (Decoster, Didier; Vilcot, Jean-Pierre)V1362,959-966(1991)

Photo-induced adhesion changes: a technique for patterning lightguide structures (Festl, H. G.; Franke, Hilmar)V1559,410-423(1991)

Photoinduced effects in optical waveguides (Dianov, Evgeni M.; Kazansky, Peter G.; Stepanov, D. Y.)V1516,81-98(1991)

Photolithographic processing of integrated optic devices in glasses (Mendoza, Edgar A.; Gafney, Harry D.; Morse, David L.)V1583,43-51(1991)

Photonic Multichip Packaging (PMP) using electro-optic organic materials and devices (McDonald, John F.; Vlannes, Nickolas P.; Lu, Toh-Ming; Wnek, Gary E.; Nason, Theodore C.; You, Lu)V1390,286-301(1991)

Planar and strip optical waveguides by sol-gel method and laser densification (Guglielmi, Massimo; Colombo, Paolo; Mancinelli Degli Esposti, Luca; Righini, Giancarlo C.; Pelli, Stefano)V1513,44-49(1991)

Planar optical waveguides on glasses and glass-ceramic materials (Glebov, Leonid B.; Nikonorov, Nikolai V.; Petrovskii, Gurii T.)V1513,56-70(1991)

Pockels' effect in polycrystalline ZnS planar waveguides (Wong, Brian; Kitai, Adrian H.; Jessop, Paul E.)V1398,261-268(1991)

Polymer-based electro-optic modulators: fabrication and performance (Haas, David R.; Man, Hong-Tai; Teng, Chia-Chi; Chiang, Kophu P.; Yoon, Hyun N.; Findakly, Talal K.)V1371,56-67(1991)

Polymeric optical waveguides for device applications (McFarland, Michael J.; Beeson, Karl W.; Horn, Keith A.; Nahata, Ajay; Wu, Chengjiu; Yardley, James T.)V1583,344-354(1991)

Polymer waveguide systems for nonlinear and electro-optic applications (Pantelis, Philip; Hill, Julian R.; Kashyap, Raman)V1559,2-9(1991)

Power economy using point-to-point optical interconnect links (Hartman, Davis H.; Reith, Leslie A.; Habiby, Sarry F.; Lalk, Gail R.; Booth, Bruce L.; Marchegiano, Joseph E.; Hohman, James L.)V1390,368-376(1991)

Propagating beam method simulation of planar multimode optical waveguides (McMullin, James N.)V1389,621-629(1991)

Propagation and diffusion characteristics of optical waveguides made by electric field-assisted k+-ion exchange in glass (Noutsios, Peter C.; Yip, Gar L.; Kishioka, Kiyoshi)V1374,14-22(1991)

Proposed electro-optic package with bidirectional lensed coupling (Rajasekharan, K.; Michalka, Timothy)V1389,568-579(1991)

Proton-diffused channel waveguides on Y-cut LiNbO3 using a self-aligned SiO2-cap diffusion method (Son, Yung-Sung; Lee, Hyung-Jae; Yi, Sang-Yoon; Shin, Sang-Yung)V1374,23-29(1991)

Proton exchange in LiTaO3 with different stoichiometric composition (Savatinova, Ivanka T.; Kuneva, M.; Levi, Zelma; Atuchin, V.; Ziling, K.; Armenise, Mario N.)V1374,37-46(1991)

Proton exchange: past, present, and future (Jackel, Janet L.)V1583,54-63(1991)

Proton-exchange X-cut lithium tantalate fiber optic gyro chips (Wong, Ka-Kha; Killian, Kevin M.; Dimitrov-Kuhl, K. P.; Long, Margaret; Fleming, J. T.; van de Vaart, Herman)V1374,278-286(1991)

Quantum wells and artificially structured materials for nonlinear optics (Harris, James S.; Fejer, Martin M.)V1361,262-273(1991)

Quasi-phase-matched frequency conversion in lithium niobate and lithium tantalate waveguides (Lim, Eric J.; Matsumoto, S.; Bortz, M. L.; Fejer, Martin M.)V1561,135-142(1991)

Quasirelaxation-free guest-host poled-polymer waveguide modulator: material, technology, and characterization (Levenson, R.; Liang, J.; Toussaere, E.; Carenco, Alain; Zyss, Joseph)V1560,251-261(1991)

Radiation effects on dynamical behavior of LiNbO3 switching devices (Kanofsky, Alvin S.; Minford, William J.; Watson, James E.)V1374,59-66(1991)

Radiation-induced crosstalk in guided-wave devices (Taylor, Edward W.; Padden, Richard J.; Sanchez, Anthony D.; Chapman, S. P.; Berry, J. N.; DeWalt, Steve A.)V1474,126-131(1991)

Rare-earth-doped LiNbO3 waveguide amplifiers and lasers (Sohler, Wolfgang)V1583,110-121(1991)

Recent development of proton-exchanged waveguides and devices in lithium niobate using phosphoric acid (Pun, Edwin Y.)V1374,2-13(1991)

Refractive index of multiple-quantum-well waveguides subject to impurity induced disordering using boron and fluorine (Hansen, Stein I.; Marsh, John H.; Roberts, John S.; Jeynes, C.)V1362,361-369(1991)

Semiconductor waveguides for optical switching (Laval, Suzanne)V1362,82-92(1991)

Silica optical integrated devices (Kobayashi, Soichi; Sumida, Shin S.; Miyashita, Tadashi M.)V1374,300-303(1991)

Silicon nitride single-polarization optical waveguides on silicon substrates (De Brabander, Gregory N.; Boyd, Joseph T.; Jackson, Howard E.)V1583,327-330(1991)

Simulation and design of integrated optical waveguide devices by the BPM (Yip, Gar L.)V1583,240-248(1991)

Single-mode fibers to single-mode waveguides coupling with minimum Fresnel back-reflection (Sneh, Anat; Ruschin, Shlomo; Marom, Emanuel)V1442,252-257(1991)

Small-signal gain for parabolic profile beams in free-electron lasers (Elliott, C. J.)V1552,175-181(1991)

Spectrometer on a chip: InP-based integrated grating spectrograph for wavelength-multiplexed optical processing (Soole, Julian B.; Scherer, Axel; LeBlanc, Herve P.; Andreadakis, Nicholas C.; Bhat, Rajaram; Koza, M. A.)V1474,268-276(1991)

Spectroscopic properties of Er3+-doped glasses for the realization of active waveguides by ion-exchange technique (Cognolato, Livio; De Bernardi, Carlo; Ferraris, M.; Gnazzo, Angelantonio; Morasca, Salvatore; Scarano, Domenica)V1513,368-377(1991)

Stress analysis in ion-exchanged waveguides by using a polarimetric technique (Gonella, Francesco; Mazzi, Giulio; Quaranta, Alberto)V1513,425-433(1991)

Stripe waveguides by Cs+- and K+-exchange in neodymium-doped soda silicate glasses for laser application (Possner, Torsten; Ehrt, Doris; Sargsjan, Geworg; Unger, Clemens)V1513,378-385(1991)

Study using nuclear techniques of waveguides produced by electromigration processes (Battaglin, Giancarlo; De Marchi, Giovanna; Losacco, Aurora M.; Mazzoldi, Paolo; Miotello, Antonio; Quaranta, Alberto; Valentini, Antonio)V1513,441-450(1991)

Supported sol-gel thin-film glasses embodying laser dyes II: three-layered waveguide assemblies (Haruvy, Yair; Heller, Adam; Webber, Stephen E.)V1590,59-70(1991)

Surface normal SHG in PLZT thin film waveguides (Zou, Lian C.; Malloy, Kevin J.; Wu, A. Y.)V1519,707-711(1991)

Technique for measuring stress-induced birefringence (Heidel, Jeffrey R.; Zediker, Mark S.)V1418,240-247(1991)

Ten-channel single-mode wavelength division demultiplexer in near IR (Chen, Ray T.; Lu, Huey T.; Robinson, Daniel; Wang, Michael R.)V1583,135-142(1991)

Theory of solar-powered, cavityless waveguide laser (Mel'nikov, Lev Y.; Kochelap, Viatcheslav A.; Izmailov, I. A.)V1501,144-149(1991)

Total internal reflection mirrors fabricated in polymeric optical waveguides via excimer laser ablation (Trewhella, Jeannine M.; Oprysko, Modest M.)V1377,64-72(1991)

Transendoscopic and freehand use of flexible hollow fibers for carbon dioxide laser surgery in the upper airway of the horse: a preliminary report (Palmer, Scott E.)V1424,218-220(1991)

Transfer matrix approach to the design of InP/InGaAsP ARROW structures (Kubica, Jacek M.; Domanski, Andrzej W.)V1391,24-31(1991)

Tunability of cascaded grating that is used in distributed feedback laser (Rahnavard, Mohammad H.; Moheimany, O. R.; Abiri-Jahromi, H.)V1367,374-381(1991)

Two Gbit/s photonic backplane for telephone cards interconnection (Donati, Silvano; Martini, Giuseppe; Francese, Francesco)V1389,665-671(1991)

UV waveguide gas laser for biological and medical diagnostic methods (Kushlevsky, S. V.; Patrin, V. V.; Provorov, Alexander S.; Salmin, V. V.)V1403,799-800(1991)

Vector beam propagation method based on finite-difference (Huang, Weiping W.; Xu, Chenglin; Chu, S. T.; Chaudhuri, Sujeet K.)V1583,268-270(1991)

Velocity-matched electro-optic modulator (Bridges, William B.; Sheehy, Finbar T.; Schaffner, James H.)V1371,68-77(1991)

Vertical 3-D integration of silicon waveguides in a Si-SiO2-Si-SiO2-Si structure (Soref, Richard A.; Namavar, Fereydoon; Cortesi, Elisabetta; Friedman, Lionel R.; Lareau, Richard)V1389,408-421(1991)

Waveguide formation by laser irradiation of sol-gel coatings (Zaugg, Thomas C.; Fabes, Brian D.; Weisenbach, Lori; Zelinski, Brian J.)V1590,26-35(1991)

Waveguide hologram star couplers (Caulfield, H. J.; Johnson, R. B.; Huang, Qiang)V1555,154-158(1991)

Waveguiding characteristics in polyimide films with different chemistry of formation (Chakravorty, Kishore K.; Chien, Chung-Ping)V1389,559-567(1991)

Wavelength division multiplexing based on mode-selective coupling (Ouellette, Francois; Duguay, Michel A.)V1365,25-32(1991)

WDS measurement of thallium-diffused glass waveguides and numerical simulation: a comparison (Bourhis, Jean-Francois; Guerin, Philippe; Teral, Stephane; Haux, Denys; Di Maggio, Michel)V1583,383-388(1991)

Wideband NLO organic external modulators (Findakly, Talal K.; Teng, Chia-Chi; Walpita, Lak M.)V1476,14-21(1991)

Y-branch optical modulator (Jaeger, Nicolas A.; Lai, Winnie C.)V1583,202-209(1991)

X rays—see also astronomy; medical imaging; radiology; synchrotron radiation

100-Hz KrF laser plasma x-ray source (Turcu, I. C.; Gower, Malcolm C.; Reason, C. J.; Huntington, P.; Schulz, M.; Michette, Alan G.; Bijkerk, Fred; Louis, Eric; Tallents, Gregory J.; Al-Hadithi, Yas; Batani, D.)V1503,391-405(1991)

19th Intl Congress on High-Speed Photography and Photonics (Fuller, Peter W., ed.)V1358(1991)

8-GeV synchrotron radiation facility at Nishi-Harima (Kamitsubo, Hiromichi)V1345,16-25(1991)

Accretion dynamics and polarized x-ray emission of magnetized neutron stars (Arons, Jonathan)V1548,2-12(1991)

Active sun telescope array (Walker, Arthur B.; Timothy, J. G.; Barbee, Troy W.; Hoover, Richard B.)V1343,334-347(1991)

Adaptive coding method of x-ray mammograms (Baskurt, Atilla; Magnin, Isabelle E.; Bremond, Alain; Charvet, Pierre Y.)V1444,240-249(1991)

Advanced X-Ray/EUV Radiation Sources and Applications (Knauer, James P.; Shenoy, Gopal K., eds.)V1345(1991)

Amorphous silicon thin film x-ray sensor (Wei, Guang P.)V1519,225-233(1991)

Application of the half-filter method to the flash radiography using a neutral filter in the range of x-rays (Gerstenmayer, Jean-Louis; Vibert, Patrick)V1346,286-292(1991)

Application of transmission electron detection to x-ray mask calibrations and inspection (Postek, Michael T.; Larrabee, Robert D.; Keery, William J.; Marx, Egon)V1464,35-47(1991)

Applications of a soft x-ray streak camera in laser-plasma interaction studies (Tsakiris, George D.)V1358,174-192(1991)

Applications of powder diffraction in materials science using synchrotron radiation (Hart, Michael)V1550,11-17(1991)

Approximation of the Compton scattered radiation (Burq, Catherine; Vibert, Patrick)V1346,276-285(1991)

Automated cyst recognition from x-ray photographs (Nedkova, Rumiana; Delchev, Georgy)V1429,105-107(1991)

Automatic recognition of bone for x-ray bone densitometry (Shepp, Larry A.; Vardi, Y.; Lazewatsky, J.; Libeau, James; Stein, Jay A.)V1452,216-224(1991)

Bayesian signal reconstruction from Fourier transform magnitude and x-ray crystallography (Doerschuk, Peter C.)V1569,70-79(1991)

Bragg crystal polarimeter for the Spectrum-X-Gamma misson (Holley, Jeff; Silver, Eric H.; Ziock, Klaus P.; Novick, Robert; Kaaret, Philip E.; Weisskopf, Martin C.; Elsner, Ronald F.; Beeman, Jeff)V1343,500-511(1991)

Broadband multilayer coated blazed grating for x-ray wavelengths below 0.6 nm (den Boggende, Antonius J.; Bruijn, Marcel P.; Verhoeven, Jan; Zeijlemaker, H.; Puik, Eric J.; Padmore, Howard A.)V1345,189-197(1991)

CCD focal-plane imaging detector for the JET-X instrument on spectrum R-G (Wells, Alan A.; Castelli, C. M.; Holland, Andrew D.; McCarthy, Kieran J.; Spragg, J. E.; Whitford, C. H.)V1549,357-373(1991)

Channeling radiation as an x-ray source for angiography, x-ray lithography, molecular structure determination, and elemental analysis (Uberall, Herbert; Faraday, Bruce J.; Maruyama, Xavier K.; Berman, Barry L.)V1552,198-213(1991)

Characteristics of a high-pressure gas proportional counter filled with xenon (Sakurai, Hirohisa; Ramsey, Brian D.)V1549,20-27(1991)

Characterization of fluorescent plastic optical fibers for x-ray beam detection (Laguesse, Michel F.; Bourdinaud, Michel J.)V1592,96-107(1991)

Characterization of the Bridgman crystal growth process by radiographic imaging (Fripp, Archibald L.; Debnam, W. J.; Woodell, G. W.; Berry, R. F.; Simchick, Richard T.; Sorokach, S. K.; Barber, Patrick G.)V1557,236-244(1991)

Characterization of the radiation from a low-energy X-pinch source (Christou, Christos; Choi, Peter)V1552,278-287(1991)

Chemically amplified resists for x-ray and e-beam lithography (Berry, Amanda K.; Graziano, Karen A.; Thompson, Stephen D.; Taylor, James W.; Suh, Doowon; Plumb, Dean)V1465,210-220(1991)

Circular intensity differential scattering measurements in the soft x-ray region of the spectrum (~16 EV to 500 EV) (Maestre, Marcos F.; Bustamante, Carlos; Snyder, Patricia A.; Rowe, Ednor M.; Hansen, Roger W.)V1548,179-187(1991)

Combined x-ray absorption spectroscopy and x-ray powder diffraction (Dent, Andrew J.; Derbyshire, Gareth E.; Greaves, G. N.; Ramsdale, Christine A.; Couves, J. W.; Jones, Richard; Catlow, C. R.; Thomas, John M.)V1550,97-107(1991)

Compact open-architecture computed radiography system (Huang, H. K.; Lim, Art J.; Kangarloo, Hooshang; Eldredge, Sandra L.; Loloyan, Mansur; Chuang, Keh-Shih)V1443,198-202(1991)

Comparative study of carbon and boron carbide spacing layers inside soft x-ray mirrors (Boher, Pierre; Houdy, Philippe; Kaikati, P.; Barchewitz, Robert J.; Van Ijzendoorn, L. J.; Li, Zhigang; Smith, David J.; Joud, J. C.)V1345,165-179(1991)

Cost-effective x-ray lithography (Roltsch, Tom J.)V1465,289-307(1991)

Crystal growth by solute diffusion in Earth orbit (Lind, M. D.; Nielsen, K. F.)V1557,259-270(1991)

CT imaging with an image intensifier: using a radiation therapy simulator as a CT scanner (Silver, Michael D.; Nishiki, Masayuki; Tochimura, Katsumi; Arita, Masataka; Drawert, Bruce M.; Judd, Thomas C.)V1443,250-260(1991)

Cylindrical proportional counter for x-ray polarimetry (Costa, Enrico; Piro, Luigi; Rubini, Alda; Soffitta, Paolo; Massaro, Enrico; Matt, Giorgio; Medici, Gastone; Manzo, Giuseppe; Re, Stefano)V1343,469-476(1991)

Dead-time effects in microchannel-plate imaging detectors (Zombeck, Martin V.; Fraser, George W.)V1549,90-100(1991)

Defect repair for gold absorber/silicon membrane x-ray masks (Stewart, Diane K.; Fuchs, Jacob; Grant, Robert A.; Plotnik, Irving)V1465,64-77(1991)

Deflection evaluation using time-resolved radiography (Fry, David A.; Lucoro, Jacobo P.)V1346,270-275(1991)

Deposition- controlled uniformity of multilayer mirrors (Jankowski, Alan F.; Makowiecki, Daniel M.; McKernan, M. A.; Foreman, R. J.; Patterson, R. G.)V1343,32-38(1991)

Design and analysis of aspherical multilayer imaging x-ray microscope (Shealy, David L.; Jiang, Wu; Hoover, Richard B.)V1343,122-132(1991)

Design and analysis of a water window imaging x-ray microscope (Hoover, Richard B.; Baker, Phillip C.; Shealy, David L.; Brinkley, B. R.; Walker, Arthur B.; Barbee, Troy W.)V1426,84-96(1991)

Design and analysis of the reflection grating arrays for the X-Ray Multi-Mirror Mission (Atkinson, Dennis P.; Bixler, Jay V.; Geraghty, Paul; Hailey, Charles J.; Klingmann, Jeffrey L.; Montesanti, Richard C.; Kahn, Steven M.; Paerels, F. B.)V1343,530-541(1991)

Design and fabrication of soft x-ray photolithography experimental beam line at Beijing National Synchrotron Radiation Laboratory (Zhou, Changxin)V1465,26-33(1991)

Design and fabrication of x-ray/EUV optics for photoemission experimental beam line at Hefei National Synchrotron Radiation Lab. (Zhou, Changxin; Sun, Deming)V1345,281-287(1991)

Design and performance of an automatic gain control system for the high-energy x-ray timing experiment (Pelling, Michael R.; Rothschild, Richard E.; MacDonald, Daniel R.; Hertel, Robert H.; Nishiie, Edward S.)V1549,134-146(1991)

Design considerations for multilayer-coated Schwarzschild objectives for the XUV (Kortright, Jeffrey B.; Underwood, James H.)V1343,95-103(1991)

Design of reflective relay for soft x-ray lithography (Rodgers, John M.; Jewell, Tanya E.)V1354,330-336(1991)

Design survey of x-ray/XUV projection lithography systems (Shealy, David L.; Viswanathan, Vriddhachalam K.)V1343,229-240(1991)

Determination of optical constants of thin film in the soft x-ray region (Guo, Yong H.; Fan, Zheng X.; Bin, Ouyang; Jin, Lei; Shao, Jian D.)V1519,327-332(1991)

Development of an XUV-IR free-electron laser user facility for scientific research and industrial applications (Newnam, Brian E.; Warren, Roger W.; Conradson, Steven D.; Goldstein, John C.; McVey, Brian D.; Schmitt, Mark J.; Elliott, C. J.; Burns, M. J.; Carlsten, Bruce E.; Chan, Kwok-Chi D.; Johnson, W. J.; Wang, Tai-San; Sheffield, Richard L.; Meier, Karl L.; Olsher, R. H.; Scott, Marion L.; Griggs, J. E.)V1552,154-174(1991)

Development of a synchrotron CCD-based area detector for structural biology (Kalata, Kenneth; Phillips, Walter C.; Stanton, Martin J.; Li, Youli)V1345,270-280(1991)

Development of picosecond x-ray framing camera (Chang, Zenghu; Hou, Xun; Zhang, Xiaoqiu; Gong, Meixia; Niu, Lihong; Yong, Hongru; Liu, Xiouqin; Lei, Zhiyuan)V1358,614-618(1991)

Development of the water-window imaging x-ray microscope (Hoover, Richard B.; Shealy, David L.; Baker, Phillip C.; Barbee, Troy W.; Walker, Arthur B.)V1435,338-351(1991)

Diffractive optics for x rays: the state of the art (Michette, Alan G.)V1574,8-21(1991)

Digital radiology with solid state linear x-ray detectors (Munier, Bernard; Prieur-Drevon, P.; Chabbal, Jean)V1447,44-55(1991)

Disk-cathode flash x-ray tube driven by a repetitive type of Blumlein pulser (Sato, Eiichi; Kimura, Shingo; Isobe, Hiroshi; Takahashi, Kei; Tamakawa, Yoshiharu; Yanagisawa, Toru)V1358,146-153(1991)

Distorted local environment about Zn and transition metals on the copper sites in YBa2Cu3O7 (Bridges, Frank; Li, Guoguang; Boyce, James B.; Claeson, Tord)V1550,76-84(1991)

Dynamics of a low-energy X-pinch source (Choi, Peter; Christou, Christos; Aliaga, Raul)V1552,270-277(1991)

Dynamic spatial reconstructor: a high-speed, stop-action, 3-D, digital radiographic imager of moving internal organs and blood (Jorgensen, Steven M.; Whitlock, S. V.; Thomas, Paul J.; Roessler, R. W.; Ritman, Erik L.)V1346,180-191(1991)

Electron-Beam, X-Ray, and Ion-Beam Submicrometer Lithographies for Manufacturing (Peckerar, Martin, ed.)V1465(1991)

Enhancement of dental x-ray images by two-channel image processing (Mitra, Sanjit K.; Yu, Tian-Hu)V1445,156-165(1991)

Equalization radiography with radiation quality modulation (Geluk, Ronald J.; Vlasbloem, Hugo)V1443,143-152(1991)

Error analysis on target localization from two projection images (Lee, Byung-Uk; Adler, John R.; Binford, Thomas O.)V1380,96-107(1991)

EUV, X-Ray, and Gamma-Ray Instrumentation for Astronomy II (Siegmund, Oswald H.; Rothschild, Richard E., eds.)V1549(1991)

Evaluation of a moving slit technique for mammography (Rosenthal, Marc S.; Sashin, Donald; Herron, John M.; Maitz, Glenn S.; Boyer, Joseph W.; Gur, David)V1443,132-142(1991)

Excimer laser with sealed x-ray preionizer (Atjezhev, Vladimir V.; Belov, Sergey R.; Bukreev, Viacheslav S.; Vartapetov, Serge K.; Zhukov, Alexander N.; Zigenshin, Ilnur T.; Prokhorov, Alexander M.; Soldatkin, Alexey E.; Stepanov, Yuri D.)V1503,197-199(1991)

Experimental and analytical studies on fixed mask assembly for APS with enhanced cooling (Kuzay, Tuncer M.; Collins, Jeffrey T.; Khounsary, Ali M.; Viccaro, P. J.)V1345,55-70(1991)

Experimental characterization of Fresnel zone plate for hard x-ray applications (Lai, Barry P.; Chrzas, John J.; Yun, Wen-Bing; Legnini, Dan; Viccaro, P. J.; Bionta, Richard M.; Skulina, Kenneth M.)V1550,46-49(1991)

Fabrication and characterization of Si-based soft x-ray mirrors (Schmiedeskamp, Bernt; Heidemann, B.; Kleineberg, Ulf; Kloidt, Andreas; Kuehne, Mikhael; Mueller, H.; Mueller, Peter; Nolting, Kerstin; Heinzmann, Ulrich)V1343,64-72(1991)

Fabrication and performance at 1.33 nm of a 0.24-um period multilayer grating (Berrouane, H.; Khan Malek, Chantal; Andre, Jean-Michel; Lesterlin, L.; Ladan, F. R.; Rivoira, R.; Lepetre, Yves; Barchewitz, Robert J.)V1343,428-436(1991)

Fabrication of a grazing-incidence telescope by grinding and polishing techniques on aluminum (Gallagher, Dennis J.; Cash, Webster C.; Green, James C.)V1343,155-161(1991)

Fabrication of multilayer Bragg-Fresnel zone plates for the soft x-ray range (Khan Malek, Chantal; Moreno, T.; Guerin, Philippe; Ladan, F. R.; Rivoira, R.; Barchewitz, Robert J.)V1343,56-61(1991)

Fiber optic interferometric x-ray dosimeter (Barone, Fabrizio; Bernini, Umberto; Conti, M.; Del Guerra, Alberto; Di Fiore, Luciano; Maddalena, P.; Milano, L.; Russo, G.; Russo, Paolo)V1584,304-307(1991)

Filter and window assemblies for high-power insertion device synchrotron radiation sources (Khounsary, Ali M.; Viccaro, P. J.; Kuzay, Tuncer M.)V1345,42-54(1991)

Finite thickness effect of a zone plate on focusing hard x-rays (Yun, Wen-Bing; Chrzas, John J.; Viccaro, P. J.)V1345,146-164(1991)

Flash soft radiography: its adaption to the study of breakup mechanisms of liquid jets into a high-density gas (Krehl, Peter; Warken, D.)V1358,162-173(1991)

Fundamental studies for the high-intensity long-duration flash x-ray generator for biomedical radiography (Sato, Eiichi; Isobe, Hiroshi; Takahashi, Kei; Tamakawa, Yoshiharu; Yanagisawa, Toru)V1358,193-200(1991)

Further observations of vectorial effects in the x-ray photoemission from caesium iodide (Fraser, George W.; Lees, John E.; Pearson, James F.)V1343,438-456(1991)

Gas puff Z-pinch x-ray source: a new approach (Fisher, Amnon)V1552,252-253(1991)

Generation of hard x-ray pulse trains with the help of high-frequency oscillating systems for fast processes recording (Romanovsky, V. F.; Kovalenko, A. N.; Sapozhnikova, T. I.; Tushev, N. R.; Abgaryan, A. A.)V1358,140-145(1991)

Goldhelox: a project to view the x-ray sun (Fair, Melody)V1549,182-192(1991)

Hard x-ray imaging via crystal diffraction: first results of reflectivity measurements (Frontera, Filippo; De Chiara, P.; Gambaccini, M.; Landini, G.; Pasqualini, G.)V1549,113-119(1991)

Hard x-ray polarimeter utilizing Compton scattering (Sakurai, Hirohisa; Noma, M.; Niizeki, H.)V1343,512-518(1991)

High-dynamic-range image readout system (Mens, Alain; Ducrocq, N.; Mazataud, D.; Mugnier, A.; Eouzan, J. Y.; Heurtaux, J. C.; Tomasini, F.; Mathae, J. C.)V1358,719-731(1991)

High-energy x-ray diffraction (Freund, Andreas K.)V1345,234-244(1991)

High-intensity soft-flash x-ray generator utilizing a low-vacuum diode (Isobe, Hiroshi; Sato, Eiichi; Shikoda, Arimitsu; Takahashi, Kei; Tamakawa, Yoshiharu; Yanagisawa, Toru)V1358,471-478(1991)

High-power x-ray generation using transition radiation (Piestrup, Melvin A.; Boyers, D. G.; Pincus, Cary I.; Li, Qiang; Harris, J. L.; Bergstrom, J. C.; Caplan, H. S.; Silzer, R. M.; Skopik, D. M.; Moran, M. J.; Maruyama, Xavier K.)V1552,214-239(1991)

High-repetition-rate x-ray preionization source (van Goor, Frederik A.)V1397,563-568(1991)

High-resolution synchrotron x-radiation diffraction imaging of crystals grown in microgravity and closely related terrestrial crystals (Steiner, Bruce W.; Dobbyn, Ronald C.; Black, David; Burdette, Harold; Kuriyama, Masao; Spal, Richard; van den Berg, Lodewijk; Fripp, Archibald L.; Simchick, Richard T.; Lal, Ravindra B.; Batra, Ashok K.; Matthiesen, David; Ditchek, Brian M.)V1557,156-167(1991)

Imaging characteristics of the development model of the SAX x-ray imaging concentrators (Citterio, Oberto; Conconi, Paolo; Conti, Giancarlo; Mattaini, E.; Santambrogio, E.; Cusumano, G.; Sacco, B.; Braueninger, Heinrich; Burkert, Wolfgang)V1343,145-154(1991)

Imaging gas scintillation proportional counters for ASTRO-D (Ohashi, T.; Makishima, K.; Ishida, M.; Tsuru, T.; Tashiro, M.; Mihara, Teruyoshi; Kohmura, Y.; Inoue, Hiroyuki)V1549,9-19(1991)

Imaging performance and tests of soft x-ray telescopes (Spiller, Eberhard; McCorkle, R.; Wilczynski, Janusz S.; Golub, Leon; Nystrom, George U.; Takacs, Peter Z.; Welch, Charles W.)V1343,134-144(1991)

Imaging the sun in hard x-rays: spatial and rotating modulation collimators (Campbell, Jonathan W.; Davis, John M.; Emslie, A. G.)V1549,155-179(1991)

Improved phantom for quality control of laser scanner digitizers in PACS (Halpern, Ethan J.; Esser, Peter D.)V1444,104-115(1991)

In-situ structural studies of the underpotential deposition of copper onto an iodine-covered platinum surface using x-ray standing waves (Bommarito, G. M.; Acevedo, D.; Rodriguez, J. R.; Abruna, H. D.)V1550,156-170(1991)

In-situ surface x-ray scattering of metal monolayers adsorbed at solid-liquid interfaces (Toney, Michael F.; Gordon, Joseph G.; Melroy, Owen R.)V1550,140-150(1991)

Interfaces in Mo/Si multilayers (Slaughter, Jon M.; Kearney, Patrick A.; Schulze, Dean W.; Falco, Charles M.; Hills, C. R.; Saloman, Edward B.; Watts, Richard N.)V1343,73-82(1991)

Issues in the repair of x-ray masks (Stewart, Diane K.; Doherty, John A.)V1496,247-265(1991)

Kilohertz range pulsed x-ray generator having a hot cathode triode (Sato, Eiichi; Shikoda, Arimitsu; Isobe, Hiroshi; Tamakawa, Yoshiharu; Yanagisawa, Toru; Honda, Keiji; Yokota, Yoshiharu)V1358,479-487(1991)

Kinetics of surface ordering: Pb on Ni(001) (Eng, Peter J.; Stephens, Peter; Tse, Teddy)V1550,110-121(1991)

Laboratory soft x-ray source based on high-temperature capillary arc discharge (Belov, Sergei N.; Golubev, Evgeny M.; Vinokurova, Elena G.)V1552,264-267(1991)

Large-aperture CCD x-ray detector for protein crystallography using a fiber-optic taper (Strauss, Michael G.; Westbrook, Edwin M.; Naday, Istvan; Coleman, T. A.; Westbrook, M. L.; Travis, D. J.; Sweet, Robert M.; Pflugrath, J. W.; Stanton, Martin J.)V1447,12-27(1991)

Laser-produced plasma x-ray diagnostics with an x-ray streak camera at the Iskra-4 plant (Berkovsky, A. G.; Gubanov, Y. I.; Pryanishnikov, Ivan G.; Murugov, V. M.; Petrov, S. I.; Senik, A. V.)V1358,750-755(1991)

Lateral-periodicity evaluation of multilayer Bragg reflector surface roughness using x-ray diffraction (Takenaka, Hisataka; Ishii, Yoshikazu)V1345,180-188(1991)

Long-period x-ray standing waves generated by total external reflection (Bedzyk, Michael J.)V1550,151-155(1991)

Low-dose magnetic-field-immune biplanar fluoroscopy for neurosurgery (Ramos, P. A.; Lawson, Michael A.; Wika, Kevin G.; Allison, Stephen W.; Quate, E. G.; Molloy, J. A.; Ritter, Rogers C.; Gillies, George T.)V1443,160-170(1991)

Low-temperature operation of silicon drift detectors (Sumner, Timothy J.; Roe, S.; Rochester, G. K.; Hall, G.; Evensen, Per; Avset, B. S.)V1549,265-273(1991)

L-shell x-ray spectroscopy of laser-produced plasmas in the 1-keV region (Batani, D.; Giulietti, Antonio; Palladino, Libero; Tallents, Gregory J.; Turcu, I. C.)V1503,479-491(1991)

Luminescence molulation for the characterization of radiation damage within scintillator material (Bayer, Eberhard G.)V1361,195-199(1991)

Magnetic circular dichroism studies with soft x-rays (Tjeng, L. H.; Rudolf, P.; Meigs, G.; Sette, Francesco; Chen, Chien-Te; Idzerda, Y. U.)V1548,160-167(1991)

MCP image intensifier in the 100-KeV to 1-MeV x-ray range (Veaux, Jacqueline; Cavailler, Claude; Gex, Jean-Pierre; Hauducoeur, Alain; Hivernage, M.)V1449,13-24(1991)

Measurement of synchrotron beam polarization (Singman, Leif V.; Davis, Brent A.; Holmberg, D. L.; Blake, Richard L.; Hockaday, Robert G.)V1548,80-92(1991)

Measurements with a 35-psec gate time microchannel plate camera (Bell, Perry M.; Kilkenny, Joseph D.; Hanks, Roy L.; Landen, Otto L.)V1346,456-464(1991)

Mirrors as power filters (Kortright, Jeffrey B.)V1345,38-41(1991)

Model-based flaw reconstruction using limited-view x-ray projections and flawless prototype image (Hung, Hsien-Sen; Eray, Mete)V1550,34-45(1991)

Model-based vision system for automatic recognition of structures in dental radiographs (Acharya, Raj S.; Samarabandu, J. K.; Hausmann, E.; Allen, K. A.)V1450,170-177(1991)

Modeling of illumination effects on resist profiles in x-ray lithography (Oertel, Heinrich K.; Weiss, M.; Huber, Hans L.; Vladimirsky, Yuli; Maldonado, Juan R.)V1465,244-253(1991)

Multilayered superconducting tunnel junctions for use as high-energy-resolution x-ray detectors (Rippert, Edward D.; Song, S. N.; Ketterson, John B.; Ulmer, Melville P.)V1549,283-288(1991)

Multilayer optics for soft x-ray projection lithography: problems and prospects (Stearns, Daniel G.; Ceglio, Natale M.; Hawryluk, Andrew M.; Rosen, Robert S.; Vernon, Stephen P.)V1465,80-87(1991)

Multilayer reflectors for the "water window" (Xu, Shi; Evans, Brian L.)V1343,110-121(1991)

Nb tunnel junctions as x-ray spectrometers (Rando, Nicola; Peacock, Anthony J.; Foden, Clare; van Dordrecht, Axel; Engelhardt, Ralph; Lumley, John M.; Pereira, Carl)V1549,340-356(1991)

New low-density, high-porosity lithium hydride-beryllium hydride foam: properties and applications to x-ray astronomy (Maienschein, Jon L.; Barry, Patrick E.; McMurphy, Frederick E.; Bowers, John S.)V1343,477-484(1991)

Novel toroidal mirror enhances x-ray lithography beamline at the Center for X-ray Lithography (Cole, Richard K.; Cerrina, Franco)V1465,111-121(1991)

Observation of living cells by x-ray microscopy with a laser-plasma x-ray source (Tomie, Toshihisa; Shimizu, Hazime; Majima, T.; Yamada, Mitsuo; Kanayama, Toshihiko; Yano, M.; Kondo, H.)V1552,254-263(1991)

One-dimensional proportional counters for x-ray all-sky monitors (Matsuoka, Masaru; Yamauchi, Masamitsu; Nakamura, Haruo; Kondo, M.; Kawai, N.; Yoshida, A.; Imai, Tohru)V1549,2-8(1991)

Opacity studies with laser-produced plasma as an x-ray source (Eidmann, K.; Lanig, E. M.; Schwanda, W.; Sigel, Richard; Tsakiris, George D.)V1502,320-330(1991)

Optical diagnostics of line-focused laser-produced plasmas (Lin, Li-Huang; Chen, Shisheng; Jiang, Z. M.; Ge, Wen; Qian, Aidi D.; Bin, Ouyang; Li, Yongchun L.; Kang, Yilan; Xu, Zhizhan)V1346,490-501(1991)

Optimization of an x-ray mask design for use with horizontal and vertical kinematic mounts (Laird, Daniel L.; Engelstad, Roxann L.; Palmer, Shane R.)V1465,134-144(1991)

Parametric studies and characterization measurements of x-ray lithography mask membranes (Wells, Gregory M.; Chen, Hector T.; Engelstad, Roxann L.; Palmer, Shane R.)V1465,124-133(1991)

Parametric study of small-volume long-pulse x-ray preionized XeCl laser with double-discharge and fast magnetic switch (Hueber, Jean-Marc; Fontaine, Bernard L.; Kobhio, M. N.; Delaporte, Philippe C.; Forestier, Bernard M.; Sentis, Marc L.)V1503,62-70(1991)

Penning discharge VUV and soft x-ray source (Cao, Jianlin; Li, Futian; Qian, Limin; Chen, Po; Ma, Yueying; Chen, Xingdan)V1345,71-77(1991)

Performance of a variable-line-spaced master reflection grating for use in the reflection grating spectrometer on the x-ray multimirror mission (Bixler, Jay V.; Hailey, Charles J.; Mauche, C. W.; Teague, Peter F.; Thoe, Robert S.; Kahn, Steven M.; Paerels, F. B.)V1549,420-428(1991)

Performance of microstrip proportional counters for x-ray astronomy on spectrum-roentgen-gamma (Budtz-Jorgensen, Carl; Bahnsen, Axel; Christensen, Finn E.; Madsen, M. M.; Olesen, C.; Schnopper, Herbert W.)V1549,429-437(1991)

Performance of milliKelvin Si bolometers as x-ray and exotic particle detectors (Zammit, C. C.; Sumner, Timothy J.; Lea, M. J.; Fozooni, P.; Hepburn, I. D.)V1549,274-282(1991)

Performance of multilayer-coated figured optics for soft x-rays near the diffraction limit (Raab, Eric L.; Tennant, Donald M.; Waskiewicz, Warren K.; MacDowell, Alastair A.; Freeman, Richard R.)V1343,104-109(1991)

Performance of the Multi-Spectral Solar Telescope Array III: optical characteristics of the Ritchey-Chretien and Cassegrain Telescopes (Hoover, Richard B.; Baker, Phillip C.; Hadaway, James B.; Johnson, R. B.; Peterson, Cynthia; Gabardi, David R.; Walker, Arthur B.; Lindblom, Joakim F.; DeForest, Craig E.; O'Neal, Ray H.)V1343,189-202(1991)

Performance of the Multi-Spectral Solar Telescope Array VI: performance and characteristics of the photographic films (Hoover, Richard B.; Walker, Arthur B.; DeForest, Craig E.; Allen, Maxwell J.; Lindblom, Joakim F.)V1343,175-188(1991)

Photoelectric effect from CsI by polarized soft x-rays (Shaw, Ping-Shine; Church, Eric D.; Hanany, Shaul; Liu, Yee; Fleischman, Judith R.; Kaaret, Philip E.; Novick, Robert; Manzo, Giuseppe)V1343,485-499(1991)

Picosecond x-ray streak camera improvement (Hou, Xun; Zhang, Xiaoqiu; Gong, Meixia; Chang, Zenghu; Lei, Zhiyuan; Yang, Binzhou; Yu, Hongbin; Liu, Xiouqin; Shan, Bin; Gao, Shengshen; Zhao, Wei)V1358,868-873(1991)

Picosecond x-ray streak cameras (Averin, V. I.; Bryukhnevitch, G. I.; Kolesov, G. V.; Lebedev, V. B.; Miller, V. A.; Saulevich, S. V.; Shulika, A. N.)V1358,603-605(1991)

Plane and concave VUV and soft x-ray multilayered mirrors (Cao, Jianlin; Miao, Tongqun; Qian, Longsheng; Zhu, Xioufang; Li, Futian; Ma, Yueying; Qian, Limin; Chen, Po; Chen, Xingdan)V1345,225-232(1991)

Polarization-dependent EXAFS studies in layered copper oxide superconductors (Mini, Susan M.; Alp, Esen E.; Ramanathan, Mohan; Bommannavar, A.; Hyun, O. B.)V1345,260-269(1991)

Polarization-dependent x-ray spectroscopy of high-Tc superconductors (Heald, Steven M.; Tranquada, John M.)V1550,67-75(1991)

Polarization of emission lines from relativistic accretion disk (Chen, Kaiyou)V1548,23-33(1991)

Polarization sensitivity of x-ray photocathodes in the 60-200eV band (Fraser, George W.; Pain, M. D.; Pearson, James F.; Lees, John E.; Binns, C. R.; Shaw, Ping-Shine; Fleischman, Judith R.)V1548,132-148(1991)

Polarized nature of synchrotron radiation (Kim, Kwang-je)V1548,73-79(1991)

Polarizing optics for the soft x-ray regime: whispering-gallery mirrors and multilayer beamsplitters (Braud, John P.)V1548,69-72(1991)

Polarizing x-ray optics for synchrotron radiation (Hart, Michael)V1548,46-55(1991)

Power distribution from insertion device x-ray sources (Viccaro, P. J.)V1345,28-37(1991)

Predicted performance of the lithium scattering and graphite crystal polarimeter for the Spectrum-X-Gamma mission (Weisskopf, Martin C.; Elsner, Ronald F.; Novick, Robert; Kaaret, Philip E.; Silver, Eric H.)V1343,457-468(1991)

Production and Analysis of Polarized X Rays (Siddons, D. P., ed.)V1548(1991)

Production of x-rays by the interaction of charged particle beams with periodic structures and crystalline materials (Rule, Donald W.; Fiorito, Ralph B.; Piestrup, Melvin A.; Gary, Charles K.; Maruyama, Xavier K.)V1552,240-251(1991)

Progress with PN-CCDs for the XMM satellite mission (Braueninger, Heinrich; Hauff, D.; Lechner, P.; Lutz, G.; Kink, W.; Meidinger, Norbert; Metzner, G.; Predehl, Peter; Reppin, C.; Strueder, Lothar; Truemper, Joachim; Kendziorra, E.; Staubert, R.; Radeka, V.; Rehak, P.; Rescia, S.; Bertuccio, G.; Gatti, E.; Longoni, Antonio; Sampietro, Marco; Findeis, N.; Holl, P.; Kemmer, J.; von Zanthier, C.)V1549,330-339(1991)

Proportional counter windows for the Bragg crystal spectrometer on AXAF (Markert, Thomas H.; Bauer, J.; Canizares, Claude R.; Isobe, T.; Nenonen, Seppo A.; O'Connor, J.; Schattenburg, Mark L.; Flanagan, Kathryn A.; Zombeck, Martin V.)V1549,408-419(1991)

Real-time x-ray studies of semiconductor device structures (Clarke, Roy; Dos Passos, Waldemar; Chan, Yi-Jen; Pavlidis, Dimitris; Lowe, Walter P.; Rodricks, Brian G.; Brizard, Christine M.)V1361,2-12(1991)

Recent advances in gated x-ray imaging at LLNL (Wiedwald, Douglas J.; Bell, Perry M.; Kilkenny, Joseph D.; Bonner, R.; Montgomery, David S.)V1346,449-455(1991)

Recent developments in production of thin x-ray reflecting foils (Hudec, Rene; Valnicek, Boris; Cervencl, J.; Gerstman, T.; Inneman, Adolf; Nejedly, Pavel; Svatek, Lubomir)V1343,162-163(1991)

Recent developments using GaAs as an x-ray detector (Sumner, Timothy J.; Grant, S. M.; Bewick, A.; Li, J. P.; Spooner, N. J.; Smith, K.; Beaumont, Steven P.)V1549,256-264(1991)

Removing vertical lines generated when x-ray images are digitized (Oyama, Yoshiro; Tani, Yuichiro; Shigemura, Naoshi; Abe, Toshio; Matsuda, Koyo; Kubota, Shigeto; Inami, Takashi)V1444,413-423(1991)

Repetitive flash x-ray generator as an energy transfer source utilizing a compact-glass body diode (Shikoda, Arimitsu; Sato, Eiichi; Kimura, Shingo; Isobe, Hiroshi; Takahashi, Kei; Tamakawa, Yoshiharu; Yanagisawa, Toru)V1358,154-161(1991)

Repetitive flash x-ray generator operated at low-dose rates for a medical x-ray television system (Sato, Eiichi; Isobe, Hiroshi; Takahashi, Kei; Tamakawa, Yoshiharu; Yanagisawa, Toru)V1358,462-470(1991)

Repetitive flash x-ray generator utilizing an enclosed-type diode with a ring-shaped graphite cathode (Isobe, Hiroshi; Sato, Eiichi; Kimura, Shingo; Tamakawa, Yoshiharu; Yanagisawa, Toru; Honda, Keiji; Yokota, Yoshiharu)V1358,201-208(1991)

Resist schemes for soft x-ray lithography (Taylor, Gary N.; Hutton, Richard S.; Windt, David L.; Mansfield, William M.)V1343,258-273(1991)

Reversible compression of industrial radiographs using multiresolution decorrelation (Chen, Keshi; Ramabadran, Tenkasi V.)V1567,397-401(1991)

Selection, growth, and characterization of materials for MBE-produced x-ray optics (Kearney, Patrick A.; Slaughter, Jon M.; Falco, Charles M.)V1343,25-31(1991)

Semiautomatic x-ray inspection system (Amladi, Nandan G.; Finegan, Michael K.; Wee, William G.)V1472,165-176(1991)

Short-Wavelength Radiation Sources (Sprangle, Phillip, ed.)V1552(1991)

Signal-to-noise performance in cesium iodide x-ray fluorescent screens (Hillen, Walter; Eckenbach, W.; Quadflieg, P.; Zaengel, Thomas T.)V1443,120-131(1991)

Silicon/silicon oxide and silicon/silicon nitride multilayers for XUV optical applications (Boher, Pierre; Houdy, Philippe; Hennet, L.; Delaboudiniere, Jean-Pierre; Kuehne, Mikhael; Mueller, Peter; Li, Zhigang; Smith, David J.)V1343,39-55(1991)

Silicon x-ray array detector on spectrum-x-gamma satellite (Sipila, Heikki; Huttunen, Pekka; Kamarainen, Veikko J.; Vilhu, Osmi; Kurki, Jouko; Leppelmeier, Gilbert W.; Taylor, Ivor; Niemela, Arto; Laegsgaard, Erik; Sunyaev, Rashid)V1549,246-255(1991)

Simulation of low-energy x-ray lithography using a diamond membrane mask (Hasegawa, Shinya; Suzuki, Katsumi)V1465,145-151(1991)

Soft and hard x-ray reflectivities of multilayers fabricated by alternating-material sputter deposition (Takenaka, Hisataka; Ishii, Yoshikazu; Kinoshita, Hiroo; Kurihara, Kenji)V1345,213-224(1991)

Soft x-ray emission characteristics from laser plasma sources (Chen, Shisheng; Xu, Zhizhan; Li, Yao-lin; Wang, Xiaofang; Qian, Aidi D.)V1552,288-295(1991)

Soft x-ray multilayers fabricated by electron-beam deposition (Sudoh, Masaaki; Yokoyama, Ryouhei; Sumiya, Mitsuo; Yamamoto, Masaki; Yanagihara, Mihiro; Namioka, Takeshi)V1343,14-24(1991)

Soft x-ray projection lithography: experiments and practical printers (White, Donald L.; Bjorkholm, John E.; Bokor, J.; Eichner, L.; Freeman, Richard R.; Gregus, J. A.; Jewell, Tanya E.; Mansfield, William M.; MacDowell, Alastair A.; Raab, Eric L.; Silfvast, William T.; Szeto, L. H.; Tennant, Donald M.; Waskiewicz, Warren K.; Windt, David L.; Wood, Obert R.)V1343,204-213(1991)

Soft x-ray resonant magnetic scattering study of thin films and multilayers (Kao, Chi-Chang; Johnson, Erik D.; Hastings, Jerome B.; Siddons, D. P.; Vettier, C.)V1548,149-157(1991)

Soft x-ray spectro-microscope (Campuzano, Juan C.; Jennings, G.; Beaulaigue, L.; Rodricks, Brian G.; Brizard, Christine M.)V1345,245-254(1991)

Soft x-ray windows for position-sensitive proportional counters (Viitanen, Veli-Pekka; Nenonen, Seppo A.; Sipila, Heikki; Mutikainen, Risto)V1549,28-34(1991)

Soft X-UV silver silicon multilayer mirrors (Shao, Jian D.; Fan, Zheng X.; Guo, Yong H.; Jin, Lei)V1519,298-301(1991)

Spectrum investigation and imaging of laser-produced plasma by multilayer x-ray optics (Platonov, Yu. Y.; Salashchenko, N. N.; Shmaenok, L. A.)V1440,188-196(1991)

Spherical pinch x-ray generator prototype for microlithography (Kawai, Kenji; Panarella, Emilio; Mostacci, D.)V1465,308-314(1991)

Standardization of image quality measurements of medical x-ray image intensifier systems (Sandrik, John M.)VCR37,180-206(1991)

Status of the Advanced Light Source (Marx, Jay N.)V1345,2-10(1991)

Status of the stellar x-ray polarimeter for the Spectrum-X-Gamma mission (Kaaret, Philip E.; Novick, Robert; Shaw, Ping-Shine; Hanany, Shaul; Liu, Yee; Fleischman, Judith R.; Sunyaev, Rashid; Lapshov, I.; Weisskopf, Martin C.; Elsner, Ronald F.; Ramsey, Brian D.; Silver, Eric H.; Ziock, Klaus P.; Costa, Enrico; Piro, Luigi; Soffitta, Paolo; Manzo, Giuseppe; Giarrusso, Salvatore; Santangelo, Andrea E.; Scarsi, Livio; Fraser, George W.; Pearson, James F.; Lees, John E.; Perola, G. C.; Massaro, Enrico; Matt, Giorgio)V1548,106-117(1991)

Status report on the Advanced Photon Source, summer 1990 (Moncton, David E.)V1345,11-15(1991)

Study of structural imperfections in epitaxial beta-SiC layers by method of x-ray differential diffractometry (Baranov, Igor M.; Kutt, R. N.; Nikitina, Irina P.)V1361,1110-1115(1991)

Study of x-ray emission from picosecond laser-plasma interaction (Chen, Hong; Chuang, Yung-Ho; Delettrez, J.; Uchida, S.; Meyerhofer, David D.)V1413,112-119(1991)

Sub-100 psec x-ray gating cameras for ICF imaging applications (Kilkenny, Joseph D.; Bell, Perry M.; Hammel, Bruce A.; Hanks, Roy L.; Landen, Otto L.; McEwan, Thomas E.; Montgomery, David S.; Turner, R. E.; Wiedwald, Douglas J.; Bradley, David K.)V1358,117-133(1991)

Suppression of columnar-structure formation in Mo-Si layered synthetic microstructures (Niibe, Masahito; Hayashida, Masami; Iizuka, Takashi; Miyake, Akira; Watanabe, Yutaka; Takahashi, Rie; Fukuda, Yasuaki)V1343,2-13(1991)

Survey of hard x-ray imaging concepts currently proposed for viewing solar flares (Campbell, Jonathan W.; Davis, John M.; Emslie, A. G.)V1343,359-375(1991)

Survey of synchrotron radiation devices producing circular or variable polarization (Kim, Kwang-je)V1345,116-124(1991)

System design considerations for a production-grade, ESR-based x-ray lithography beamline (Kovacs, Stephen; Melore, Dan; Cerrina, Franco; Cole, Richard K.)V1465,88-99(1991)

Technical and clinical evaluations of a 2048 x 2048-matrix digital radiography system for gastrointestinal examinations (Ogura, Toshihiro; Masuda, Yukihisa; Fujita, Hiroshi; Inoue, Nobuo; Yonekura, Fukuo; Miyagi, Yoshihiro; Takatsu, Kazuaki; Akahira, Katsuyoshi; Tsuruta, Shigehiko; Kamiya, Masami; Takahashi, Fumitaka; Oda, Kazuyuki; Ikeda, Shigeyuki; Koike, Kouichi)V1443,153-157(1991)

Techniques of production and analysis of polarized synchrotron radiation (Mills, Dennis M.)V1345,125-136(1991)

Theoretical investigation of near-edge phenomena in magnetic systems (Carra, Paolo)V1548,35-44(1991)

Theoretical models for stellar x-ray polarization in compact objects (Meszaros, Peter)V1548,13-22(1991)

Three materials soft x-ray mirrors: theory and application (Boher, Pierre; Hennet, L.; Houdy, Philippe)V1345,198-212(1991)

Time-resolved techniques: an overview (Larson, Bennett C.; Tischler, J. Z.)V1345,90-100(1991)

Time-resolved x-ray absorption spectroscopy apparatus using laser plasma as an x-ray source (Yoda, Osamu; Miyashita, Atsumi; Murakami, Kouichi; Aoki, Sadao; Yamaguchi, Naohiro)V1503,463-466(1991)

Time-resolved x-ray scattering studies using CCD detectors (Clarke, Roy; Dos Passos, Waldemar; Lowe, Walter P.; Rodricks, Brian G.; Brizard, Christine M.)V1345,101-114(1991)

Transmittances of thin polymer films and their suitability as a supportive substrate for a soft x-ray solar filter (Williams, Memorie; Hansen, Evan; Reyes-Mena, Arturo; Allred, David D.)V1549,147-154(1991)

UCSD high-energy x-ray timing experiment cosmic ray particle anticoincidence detector (Hink, Paul L.; Rothschild, Richard E.; Pelling, Michael R.; MacDonald, Daniel R.; Gruber, Duane E.)V1549,193-202(1991)

UCSD high-energy x-ray timing experiment magnetic shield design and test results (Rothschild, Richard E.; Pelling, Michael R.; Hink, Paul L.)V1549,120-133(1991)

Uniformity and transmission of proportional counter window materials for use with AXAF (Flanagan, Kathryn A.; Austin, G. K.; Cobuzzi, J. C.; Goddard, R.; Hughes, John P.; McLaughlin, Edward R.; Podgorski, William A.; Rose, V.; Roy, Adrian G.; Zombeck, Martin V.; Markert, Thomas H.; Bauer, J.; Isobe, T.; Schattenburg, Mark L.)V1549,395-407(1991)

Vectorial photoelectric effect at 2.69 keV (Shaw, Ping-Shine; Hanany, Shaul; Liu, Yee; Church, Eric D.; Fleischman, Judith R.; Kaaret, Philip E.; Novick, Robert; Santangelo, A.)V1548,118-131(1991)

Video-based alignment system for x-ray lithography (Hughlett, R. E.; Cooper, Keith A.)V1465,100-110(1991)

What can we learn that's new and interesting about condensed matter systems using polarized x rays? (Platzman, P. M.)V1548,34-34(1991)

Wide-aperture x-ray image converter tubes (Dashevsky, Boris E.; Podvyaznikov, V. A.; Prokhorov, Alexander M.; Chevokin, Victor K.)V1346,437-442(1991)

XMM space telescope: development plan for the lightweight replicated x-ray gratings (Montesanti, Richard C.; Atkinson, Dennis P.; Edwards, David F.; Klingmann, Jeffrey L.)V1343,558-565(1991)

X-ray absorption fine structure of systems in the anharmonic limit (Mustre de Leon, Jose; Conradson, Steven D.; Batistic, I.; Bishop, A. R.; Raistrick, Ian D.; Jackson, W. E.; Brown, George S.)V1550,85-96(1991)

X-ray absorption spectroscopy: how is it done? what can it tell us? (Hayes, Tim)V1550,56-66(1991)

X-ray absorption spectroscopy with polarized synchrotron radiation (Alp, Esen E.; Mini, Susan M.; Ramanathan, Mohan; Hyun, O. B.)V1345,137-145(1991)

X-ray detector for time-resolved studies (Rodricks, Brian G.; Brizard, Christine M.; Clarke, Roy; Lowe, Walter P.)V1550,18-26(1991)

X-ray diffraction from materials under extreme pressures (Brister, Keith)V1550,2-10(1991)

X-ray diffraction study of GaSb/AlSb strained-layer-superlattices grown on miscut (100) substrates (Macrander, Albert T.; Schwartz, Gary P.; Gualteri, Gregory J.; Gilmer, George)V1550,122-133(1991)

X-Ray/EUV Optics for Astronomy, Microscopy, Polarimetry, and Projection Lithography (Hoover, Richard B.; Walker, Arthur B., eds.)V1343(1991)

X-ray holography for sequencing DNA (Yorkey, Thomas J.; Brase, James M.; Trebes, James E.; Lane, Stephen M.; Gray, Joe W.)V1345,255-259(1991)

X-ray interferometric observatory (Martin, Christopher)V1549,203-220(1991)

X-ray laminography analysis of ultra-fine-pitch solder connections on ultrathin boards (Adams, John A.)V1464,484-497(1991)

X-ray lithography system development at IBM: overview and status (Maldonado, Juan R.)V1465,2-15(1991)

X-ray multilayer-coated reflection gratings: theory and applications (Neviere, Michel; den Boggende, Antonius J.)V1545,116-124(1991)

X-ray photogrammetry of the hip revisited (Turner-Smith, Alan R.; White, Steven P.; Bulstrode, Christopher)V1380,75-84(1991)

X-ray preionization for high-repetition-rate discharge excimer lasers (van Goor, Frederik A.; Witteman, Wilhelmus J.)V1412,91-102(1991)

X Rays in Materials Analysis II: Novel Applications and Recent Developments (Mills, Dennis M., ed.)V1550(1991)

XUV polarimeter for undulator radiation measurements (Gluskin, Efim S.; Mattson, J. E.; Bader, Samuel D.; Viccaro, P. J.; Barbee, Troy W.; Brookes, N.; Pitas, Alan A.; Watts, Richard N.)V1548,56-68(1991)

Author/Editor Index

Aach, Til See Kaup, Andre: V1605,420-427(1991)

Aamodt, Leonard C. See Maclachlan Spicer, J. W.: V1467,311-321(1991)

Aanstoos, James V. See Rajala, Sarah A.: V1606,360-371(1991)

Aarnouds, J. G.
See Koelink, M. H.: V1403,347-349(1991)
See Koelink, M. H.: V1511,120-128(1991)
See Koelink, M. H.: V1431,63-72(1991)

Aase, Sven O.
; Ramstad, Tor A.: Some fundamental experiments in subband coding of images,V1605,734-744(1991)

Ababou, Soraya
; Benyattou, Taha; Marchand, Jean J.; Mayet, Louis; Guillot, Gerard; Mollot, Francis; Planel, Richard: Photoluminescence and deep-level transient spectroscopy of DX-centers in selectively silicon-doped GaAs-AlAs superlattices,V1361,706-711(1991)

Abakumov, D. M.
: Magneto-optical linear multichannel light modulator for recording of one-dimensional holograms,V1621,138-147(1991)

Abali, Livingstone N. See Calabrese, Gary C.: V1466,528-537(1991)

Abbakumov, M. O. See Veiko, Vadim P.: V1544,152-163(1991)

Abbas, Gregory L. See Thaniyavarn, Suwat: V1371,250-251(1991)

Abbasi, M.
; Sayeh, Mohammad R.: Class of learning algorithms for multilayer perceptron,V1396,237-242(1991)

Abbate, G. See Carbonara, Giuseppe: V1361,688-691(1991)

Abbiss, John B.
; Brames, Bryan J.: Restoration of subpixel detail using the regularized pseudoinverse of the imaging operator,V1566,365-375(1991)

Abbott, Mark A.
; Messner, Richard A.: Use of coordinate mapping as a method for image data reduction,V1381,272-282(1991)

Abbott, Peter G. See Anderson, Stephen J.: V1364,94-100(1991)

Abdallah, Mohamed G. See Gascoigne, Harold E.: V1554B,315-322(1991)

Abdel-Latif, A. I. See Besmann, Theodore M.: V1330,78-89(1991)

Abdelghani-Idrissi, Ahmed M.
; Ozkul, Cafer; Wolffer, Nicole; Gravey, Philippe; Picoli, Gilbert: Resonant behavior of the temporal response of the photorefractive InP:Fe under dc fields,V1362,417-427(1991)

Abdelrazik, Mohamed B.
: Programmable processor for multidimensional digital signal processing,V1606,812-822(1991)

Abdrisaev, Baktybek D. See Kutanov, Askar A.: V1507,94-98(1991)

Abdulghafour, Muhamad
; Goddard, J.; Abidi, Mongi A.: Nondeterministic approaches in data fusion: a review,V1383,596-610(1991)

Abe, Kazutoshi See Ohtsuka, Hiroshi: V1463,112-123(1991)

Abe, Koichi See Jen, Cheng K.: V1590,107-119(1991)

Abe, Makoto
; Ohta, Shigekata; Sawabe, Masaji: Surface inspection using optical fiber sensor,V1332,366-376(1991)

Abe, Masayuki See Hanyu, Isamu: V1463,595-601(1991)

Abe, Mituyuki See Minato, Kotaro: V1446,195-198(1991)

Abe, Takao
: Trends in color hard-copy technology in Japan,V1458,29-40(1991)
See Dove, Derek B.: V1458(1991)
See Koshizuka, Kunihiro: V1458,97-104(1991)

Abe, Toshio See Oyama, Yoshiro: V1444,413-423(1991)

Abe, Tsutomu
See Kobayashi, Kenichi: V1448,157-163(1991)
See Miyake, Hiroyuki: V1448,150-156(1991)

Abeille, Pierre See Kessler, William D.: V1538,104-111(1991)

Abel, B.
; Hippler, Horst; Koerber, B.; Morguet, A.; Neu, Walter: Excimer-laser-induced fluorescence spectroscopy of human arteries during laser ablation,V1525,110-118(1991)

Abel, Peter
: Clouds as calibration targets for AVHRR reflected-solar channels: results from a two-year study at NOAA/NESDIS,V1493,195-206(1991)

Abela, George S.
ed.: *Diagnostic and Therapeutic Cardiovascular Interventions*,V1425(1991)
See Friedl, Stephan E.: V1425,134-141(1991)
See Ye, Biqing: V1425,45-54(1991)

Abele, C. C. See Bunis, Jenifer L.: V1377,30-36(1991)

Aben, Hillar K.
; Idnurm, S. J.; Josepson, J. I.; Kell, K.-J. E.; Puro, A. E.: Integrated photoelasticity for residual stresses in glass specimens of complicated shape,V1554A,298-309(1991)

Abeynayake, C. See Szeregij, E. M.: V1391,199-203(1991)

Abgaryan, A. A. See Romanovsky, V. F.: V1358,140-145(1991)

Abidi, Mongi A.
See Abdulghafour, Muhamad: V1383,596-610(1991)
See Chandra, T.: V1382,409-426(1991)
See Davis, Paul B.: V1383,515-529(1991)
See Sluder, John C.: V1468,642-652(1991)

Abiri-Jahromi, H.
See Rahnavard, Mohammad H.: V1562,327-337(1991)
See Rahnavard, Mohammad H.: V1367,374-381(1991)

Aboulhouda, S.
; Razeghi, Manijeh; Vilcot, Jean-Pierre; Decoster, Didier; Francois, M.; Maricot, S.: III-V monolithic resonant photoreceiver on silicon substrate for long-wavelength operation,V1362,494-498(1991)

Abraham, C. See Jensen, Robert K.: V1380,124-136(1991)

Abraham, Neal B.
: Characterizing frequency chirps and phase fluctuations during noisy laser transients,V1376,284-293(1991)

Abrahamsson, S.
; Brusmark, B.; Gaunaurd, Guillermo C.; Strifors, Hans C.: Target identification by means of impulse radar,V1471,130-141(1991)

Abram, Richard A.
; Wood, Andrew C.; Robbins, D. J.: Optical properties of InGaAs/InP strained quantum wells,V1361,424-433(1991)

Abramov, Israel See Chan, Hoover: V1453,381-389(1991)

Abrams, Susan B.
; McDonald, Hillary L.; Yager, Paul: Detection of general anesthetics using a fluorescence-based sensor: incorporation of a single-fiber approach,V1420,13-21(1991)

Abramson, Nils H.
: Holography and relativity,VIS08,2-32(1991)
: Holography with a single picosecond pulse,V1332,224-229(1991)

Abreu, M. A.
See Freitas, Jose C.: V1399,16-23(1991)
See Freitas, Jose C.: V1399,42-48(1991)

Abrukov, Victor S.
; Ilyin, Stanislav V.: Analysis of dynamic characteristics of nonstationary gas streams using interferometry techniques,V1554B,540-543(1991)

Abruna, H. D. See Bommarito, G. M.: V1550,156-170(1991)

Abshire, James B. See Johnson, C. B.: V1346,340-370(1991)

Abstreiter, Gerhard
See Menczigar, Ulrich: V1361,282-292(1991)
See Presting, Hartmut: V1512,250-277(1991)

Abumi, Rentaro See Oda, Ichiro: V1431,284-293(1991)

Abushagur, Mustafa A.
; Berinato, Robert J.: Optical Hartley-transform-based adaptive filter,V1564,602-609(1991)
; Elmanasreh, Ahmed: Aero-optics analysis using moire deflectometry,V1554B,298-302(1991)
See Jeon, Ho-In: V1564,522-535(1991)

Acampora, Anthony S.
; Labourdette, Jean-Francois P.: Reconfiguration phase in rearrangeable multihop lightwave networks,V1579,30-39(1991)
See Gidron, Rafael: V1579,40-48(1991)

Acebal, Robert
; Dansereau, Jeffrey P.; Jones, R.; Malins, Robert J.; Schreiber, H.; Smith, W.; Taylor, S.; Duncan, William A.; Patterson, S.: Reactive-flow modeling of the H/NF2/BiF reaction system,V1397,191-196(1991)

Acevedo, D. See Bommarito, G. M.: V1550,156-170(1991)

Acharya, Mukund
; Bunch, Robert M.: Veiling glare in the F4111 image intensifier,V1396,377-388(1991)

Acharya, Raj S.
; Gai, Nevile: PET reconstruction using multisensor fusion techniques,V1445,207-221(1991)
; Samarabandu, J. K.; Hausmann, E.; Allen, K. A.: Model-based vision system for automatic recognition of structures in dental radiographs,V1450,170-177(1991)
See Morales, Aldo W.: V1452,258-269(1991)
See Samarabandu, J. K.: V1450,296-322(1991)

Acher, O.
; Benferhat, Ramdane; Drevillon, Bernard; Razeghi, Manijeh: Reflectance anisotropy spectrometer for real-time crystal growth investigations,V1361,1156-1163(1991)

Acheroy, Marc P.
; Mees, W.: Target identification by means of adaptive neural networks in thermal infrared images,V1569,121-132(1991)

Ackaert, A. See Pollentier, Ivan K.: V1361,1056-1062(1991)

Ackerman, Edward I.
See Kasemset, Dumrong: V1371,104-114(1991)
See Wanuga, Stephen: V1374,97-106(1991)

Ackerman, John R. See Lebow, Paul S.: V1409,60-66(1991)

Ackermann, F.
; Bodechtel, Joh; Lanzl, Franz; Meissner, D.; Seige, Peter; Winkenbach, H.; Zilger, Johannes: MOMS-02/Spacelab D-2: a high-resolution multispectral stereo scanner for the second German Spacelab mission,V1490,94-101(1991)

Ackley, Donald A. See Olsen, Gregory H.: V1419,24-31(1991)

Acosta, E.
; Gomez-Reino, Carlos; Gonzalez, R. M.: Design of an anamorphic gradient-index lens to correct astigmatism in diode lasers,V1401,82-85(1991)

Acton, D. S.
; Smithson, Robert C.: Solar astronomy with a 19-segment adaptive mirror,V1542,159-164(1991)

Acton, Loren W.
; Morrison, Mons D.; Janesick, James R.; Elliott, Tom: Radiation concerns for the Solar-A soft x-ray telescope,V1447,123-139(1991)

Acton, Scott T. See Klarquist, William N.: V1453,321-332(1991)

Adachi, Hajime A. See Tanazawa, Takeshi: V1435,310-321(1991)

Adachi, Masaaki See Matsumoto, Tetsuya: V1332,530-536(1991)

Adachi, Yuzi
; Matsui, Kenichi: Width gauge for hot steel plates by laser-scanning rangefinder,V1527,225-233(1991)

Adam, Jean-Luc
; Smektala, Frederic; Denoue, Emmanuel; Lucas, Jacques: Rare-earth-doped fluoride glasses for active optical fiber applications,V1513,150-157(1991)

Adam, P. See Goudonnet, Jean-Pierre: V1545,130-139(1991)

Adamec, F.
; Ambroz, M.; Dian, J.; Vacha, M.; Hala, J.; Balog, P.; Brynda, E.: Hole-burning spectroscopy of phthalocyanine Langmuir-Blodget films,V1402,82-84(1991)

Adamiak, S.
; Bylica, A.; Kuzma, Marian: Structures of laser hardening of high-speed steel SW7M,V1391,382-386(1991)

Adamovsky, Grigory
; Eustace, John G.: Optical techniques for determination of normal shock position in supersonic flows for aerospace applications,V1332,750-756(1991)
See Eustace, John G.: V1584,320-327(1991)
See Redner, Alex S.: V1332,775-782(1991)

Adams, Alan E. See Hallouche, Farid: V1445,504-512(1991)

Adams, Alfred R. See Czajkowski, Igor K.: V1362,179-190(1991)

Adams, Billy See Downey, George A.: V1482,40-47(1991)

Adams, Bruce See Schietinger, Chuck W.: V1366,284-293(1991)

Adams, Frank W.
; Cuff, K. F.; Gal, George; Harwit, Alex; Whitney, Raymond L.: Critical look at AlGaAs/GaAs multiple-quantum-well infrared detectors for thermal imaging applications,V1541,24-37(1991)

Adams, G. B. See Sinha, Kislay: V1437,32-35(1991)

Adams, Jeff C. See Falk, R. A.: V1378,70-81(1991)

Adams, John A.
: X-ray laminography analysis of ultra-fine-pitch solder connections on ultrathin boards,V1464,484-497(1991)

Adams, Stephen D. See Williams, E. J.: V1379,236-245(1991)

Adams, Thomas E.
: Applications of latent image metrology in microlithography,V1464,294-312(1991)

Adams, Timothy See Ingwall, Richard T.: V1555,279-290(1991)

Adams-Webber, Jack R. See Liseth, Ole J.: V1468,256-267(1991)

Adel, Michael E.
; Buckwald, Bob A.; Cabib, Dario: Laser boresighting by second-harmonic generation,V1442,68-80(1991)

Adelstein, Peter Z.
; Storm, William D.: Standardization efforts for the preservation of electronic imagery,VCR37,159-179(1991)

Adema, G. M. See Rinne, Glenn A.: V1389,110-121(1991)

Adiseshan, Prakash
; Faber, Tracy L.: Classification of tissue-types by combining relaxation labeling with edge detection,V1445,128-132(1991)

Adkison, D. See Raymer, M. G.: V1376,128-131(1991)

Adler, James M.
; Lee, Michael S.; Saugen, John D.: Adaptive control of propellant slosh for launch vehicles,V1480,11-22(1991)

Adler, John R. See Lee, Byung-Uk: V1380,96-107(1991)

Adler, Michael S.
: GE high-density interconnect: a solution to the system interconnect problem,V1390,504-508(1991)

Adler, Richard A. See Humphries, Stanley: V1407,512-523(1991)

Adler, Ronald See Cheng, Shirley N.: V1450,90-98(1991)

Adriaenssens, Guy J. See Oktu, Ozcau: V1361,812-818(1991)

Aduss, Howard See Sadler, Lewis L.: V1380,158-162(1991)

Aebi, Verle W. See Costello, Kenneth A.: V1449,40-50(1991)

Aegerter, Michel A. See Valla, Bruno: V1536,48-62(1991)

Affolter, K. B. See Rosser, Paul J.: V1393,49-66(1991)

Afik, Zvi
; Shammas, A.; Schwartz, Roni; Gal, Eli: Generic model for line-of-sight analysis and calibration,V1442,392-398(1991)
See Meidan, Moshe: V1540,729-737(1991)

Afonso, Carmen N. See Vega, Fidel: V1397,807-811(1991)

Afonso, Jose A.
See Brooking, Nicholas L.: V1572,88-93(1991)
See Guedes Valente, Luiz C.: V1584,96-102(1991)

Agah, Ramtin
; Sheth, Devang; Motamedi, Massoud E.: Effect of coagulation on laser light distribution in myocardial tissue,V1425,172-179(1991)

Agapakis, John E.
; Bolstad, Jon O.: Vision sensing and processing system for monitoring and control of welding and other high-luminosity processes,V1385,32-38(1991)

Agarwal, Munna L. See Oleinick, Nancy L.: V1427,90-100(1991)

Agarwal, Rajiv See Jiang, Ching-Long: V1418,261-271(1991)

Agassi, Eyal
; Wilner, Kalman; Ben-Yosef, Nissim: Correlation between images in the long-wave infrared and short-wave infrared of natural ground terrain,V1442,126-132(1991)

Agawa, Hiroshi See Nagashima, Yoshio: V1606,566-573(1991)

Ageev, L. A. See Miloslavsky, V. K.: V1440,90-96(1991)

Ageev, Vladimir P.
; Glushko, T. N.; Dorfman, V. F.; Kuzmichov, A. V.; Pypkin, B. N.: Excimer laser processing of diamond-like films,V1503,453-462(1991)

Ager, Joel W.
; Veirs, D. K.; Marchon, Bruno; Cho, Namhee; Rosenblatt, Gern M.: Vibrational Raman characterization of hard-carbon and diamond films,V1437,24-31(1991)

Agerskov, Carsten See Dougherty, Edward R.: V1606,141-152(1991)

Aggarwal, Anil K.
; Kaura, Sushil K.; Chhachhia, D. P.: Recent advances in white-light display holography at CSIO, Chandigarh,V1238,18-27(1991)

Aggarwal, Ishwar D. See Jewell, John M.: V1441,38-44(1991)

Aggarwal, Jake K. See Karthik, Sankaran: V1468,686-697(1991)

Aggarwal, Shanti J. See Beyerbacht, Hugo P.: V1427,117-127(1991)

Aghamir, F. See Dodd, James W.: V1407,467-473(1991)

Agnese, Patrick See Mottier, Patrick: V1494,419-426(1991); 1512,60-67(1991)

Agnesi, Antoniangelo
; Fogliani, Manlio F.; Reali, Giancarlo C.; Kubecek, Vaclav: Theory and experiments on passive negative feedback pulse control in active/passive mode-locked solid state lasers,V1415,242-247(1991)
; Reali, Giancarlo C.; Kubecek, Vaclav: Transient stimulated Raman scattering: theory and experiments of pulse shortening and phase conjugation properties,V1415,104-109(1991)

Agosta, John M.
: Example of a Bayes network of relations among visual features,V1569,16-27(1991)

Agranat, M. B.
; Ashitkov, S. I.; Granovsky, A. B.; Kuznetsov, V. I.: Investigation of high-temperature superconductors under the effect of picosecond laser pulses,V1440,397-400(1991)

Agrawal, A. K. See Meth, Jeffrey S.: V1560,13-24(1991)

Agrawal, Govind P.
: Noise in semiconductor lasers and its impact on optical communication systems,V1376,224-235(1991)
; Gray, George R.: Importance of nonlinear gain in semiconductor lasers,V1497,444-455(1991)
See Headley, Clifford: V1497,197-201(1991)

Agronin, Michael L. See Jandura, Louise: V1542,213-224(1991)

Agui, Takeshi
; Nagae, Takanori; Nakajima, Masayuki: Digital halftoning using a generalized Peano scan,V1606,912-916(1991)
; Takahashi, Hiroki; Nakajima, Masayuki; Nagahashi, Hiroshi: Recognition of handwritten katakana in a frame using moment invariants based on neural network,V1606,188-197(1991)

Aguillon, Luis See Paez-Leon, Cristobal J.: V1467,188-194(1991)

Ahalt, Stanley C. See Kumar, Vinod V.: V1469,484-494(1991)

Ahamd, M. See Zhai, H. R.: V1519,575-579(1991)

Aharon, O.
; Elior, A.; Herskowitz, M.; Lebiush, E.; Rosenwaks, Salman: Experiment and modeling of O2(1 delta) generation in a bubble-column-type reactor for chemically pumped iodine lasers,V1397,251-255(1991)

Aharoni, Herzl
; Swart, Pieter L.: Experimental study of the optical properties of LTCVD SiO2,V1442,118-125(1991)

Ahern, Francis J. See Teillet, Philippe M.: V1492,213-223(1991)

Ahlen, Anders
; Sternad, Mikael: Adaptive deconvolution based on spectral decomposition,V1565,130-142(1991)

Ahlers, Rolf-Juergen
: Case studies in machine vision integration,VCR36,56-63(1991)
; Tschudi, Theo T.; eds.: *Innovative Optics and Phase Conjugate Optics,*V1500(1991)
See Kille, Knut: V1386,76-83(1991)
See Kille, Knut: V1386,222-227(1991)

Ahlquist, Gregory C.
: Vision for automated imagery exploitation,V1381,2-8(1991)

Ahmad, A. See Pouskouleli, G.: V1590,179-190(1991)

Ahmad, Anees See Johnson, R. B.: V1535,136-142(1991)

Ahmad, S. H. See Parry, David J.: V1358,1057-1064(1991)

Ahmad, Suraiya P. See Markham, Brian L.: V1493,207-214(1991)

Ahmad, Ziauddin See Guez, Allon: V1469,750-755(1991)

Ahmadshahi, Mansour A.
; Krishnaswamy, Sridhar; Nemat-Nasser, Siavouche: Dynamic two-beam speckle interferometry,V1554A,620-627(1991)
See Sciammarella, Cesar A.: V1554B,743-753(1991)

Ahmed, Tasdiq See Favro, Lawrence D.: V1467,290-294(1991)

Ahn, Hong-Young
; Tou, Julius T.: Segmentation via fusion of edge and needle map,V1468,896-904(1991)

Ahn, Jae W.
; Song, So-Young; Lee, Jun Wung; Yang, Joon Mook: Propagation of the spherical short-duration shock wave in a straight tunnel,V1358,269-277(1991)

Ahn, Samuel S. See Eton, Darwin: V1425,182-187(1991)

Ahne, Hellmut See Sebald, Michael: V1466,227-237(1991)

Ahuja, Narendra See Das, Subhodev: V1382,341-352(1991)

Ahumada, Albert J.
; Mulligan, Jeffrey B.: Network compensation for missing sensors,V1453,134-146(1991)

Aibara, Tsunehiro
; Ohue, Kenji; Matsuoka, Yasushi: Human face recognition by P-type Fourier descriptor,V1606,198-203(1991)

Aiello, Norm See Smith, Richard R.: V1407,83-91(1991)

Aiga, Masao
; Omura, Etsuji E.: Long-wavelength lasers and detectors fabricated on InP/GaAs superheteroepitaxial wafer,V1418,217-222(1991)

Aikens, David M.
: Design of an IR non-lens, or how I buried 100 mm of germanium,V1485,183-194(1991)
; Young, William R.: Airborne infrared and visible sensors used for law enforcement and drug interdiction,V1479,435-444(1991)

Ailisto, Heikki J.
; Paakkari, Jussi; Moring, Ilkka: Inspection of a class of industrial objects using a dense range map and CAD model,V1384,50-59(1991)

Ainslie, B. J.
See Massicott, Jennifer F.: V1373,93-102(1991)
See Williams, Doug L.: V1513,158-167(1991)
See Williams, Doug L.: V1516,29-37(1991)

Aisaka, Kazuo
; Arai, Kiyoshi; Tsutsui, Kumiko; Hashizume, Akihide: Extraction of human stomach using computational geometry,V1445,312-317(1991)

Aizaki, Naoaki See Yamazaki, Satomi: V1466,538-545(1991)

Aizawa, Kiyoharu
; Komatsu, Takashi; Saito, Takahiro: Acquisition of very high resolution images using stereo cameras,V1605,318-328(1991)
See Huang, Thomas S.: V1605,234-241(1991)

Aizenberg, G. E. See Rotman, Stanley R.: V1442,205-215(1991)

Aizu, Yoshihisa
; Asakura, Toshimitsu: Biospeckle phenomena and their applications to blood-flow measurements,V1431,239-250(1991)

Ajmera, Ramesh See Ricks, Douglas W.: V1555,89-100(1991)

Akaev, Askar A.
; Zhumaliev, K. M.; Jamankyzov, N.: Holographic data storage on the photothermoplastic tape carrier,V1621,182-193(1991)
See Isaev, Urkaly T.: V1507,198-201(1991)
See Kutanov, Askar A.: V1507,94-98(1991)

Akahira, Katsuyoshi See Ogura, Toshihiro: V1443,153-157(1991)

Akahira, Nobuo See Ohno, Eiji: V1499,171-179(1991)

Akai, Yoshiro
See Arai, Tsunenori: V1425,191-195(1991)
See Sakurada, Masami: V1425,158-164(1991)

Akamatsu, Shigeru
; Sasaki, Tsutomu; Masui, Nobuhiko; Fukamachi, Hideo; Suenaga, Yasuhito: New method for designing face image classifiers using 3-D CG model,V1606,204-216(1991)

Akansu, Ali N. See Caglar, Hakan: V1605,86-94(1991)

Akao, Yasuo See Takada, Yutaka: V1332,571-576(1991)

Akasaka, Akira
; Ono, Makoto; Sakurai, Yasushi; Hayashida, Bun: Short-wavelength infrared subsystem design status of ASTER,V1490,269-277(1991)

Akasaka, Hideki See Saito, Jun: V1499,44-54(1991)

Akasaki, Isamu
; Amano, H.: High-efficiency UV and blue emitting devices prepared by MOVPE and low-energy electron-beam irradiation treatment,V1361,138-149(1991)

Akcasu, A. Z.
: Identification of modes in dynamic scattering on ternary polymer mixtures,V1430,142-143(1991)

Akerman, Alexander
; Patton, Ronald: Model-based vision using geometric hashing,V1406,30-39(1991)

Akerman, M. A. See McNeely, Joseph R.: V1530,288-298(1991)

Akhmanov, Sergei A.
: Unreasonable effectiveness of laser physics in the life sciences: nonlinear optics in control, diagnostics, and modeling of biophysical processes,V1403,2-2(1991)
; Poroshina, Marina Y.; eds.: *Laser Applications in Life Sciences,*V1403(1991)
; Zadkov, Victor N.; eds.: *USSR-CSFR Joint Seminar on Nonlinear Optics in Control, Diagnostics, and Modeling of Biophysical Processes,* V1402(1991)
See Gordienko, Vyacheslav M.: V1416,102-114(1991)
See Pogosov, Gregory A.: V1416,115-124(1991)

Akhouayri, Hassan
; Vincent, Patrick; Neviere, Michel: Perturbation theory for optical bistability of prism and grating couplers and comparison with rigorous method,V1545,140-144(1991)

Akimoto, Hajime See Sasano, Yasuhiro: V1490,233-242(1991)

Akin, David L.
: Telerobotic capabilities for space operations,V1387,359-364(1991)

Akita, K.
; Sugimoto, Yoshimasa; Taneya, M.; Hiratani, Y.; Ohki, Y.; Kawanishi, Hidenori; Katayama, Yoshifumi: Pattern etching and selective growth of GaAs by in-situ electron-beam lithography using an oxidized thin layer,V1392,576-587(1991)

Akiyama, Iwaki See Sato, Koki: V1461,124-131(1991)

Akiyama, Mamoru
See Iyoda, Mitsuhiro: V1415,342-349(1991)
See Iyoda, Mitsuhiro: V1397,457-460(1991)

Akiyama, Teruo See Kojima, Haruhiko: V1384,285-292(1991)

Akkapeddi, Prasad R.
; Macomber, Steven H.: Surface-emitting, distributed-feedback laser as a source for laser radar,V1416,44-49(1991)

Aksenov, V. P.
: Laser-induced diffusion and second-harmonic generation in glasses,V1440,377-383(1991)

Aksinin, V. I.
; Apollonov, V. V.; Chetkin, Sergue A.; Kijko, Vadim V.; Muraviev, S. V.; Vdovin, Gleb V.: Tilt corrector based on spring-type magnetostrictive actuators,V1500,93-104(1991)

Al-Baho, Tareq I.
; Learner, R. C.; Maxwell, Jonathan: Design of high-performance aplanatic achromats for the near-ultraviolet waveband,V1354,417-428(1991)

Al-Chalabi, Adil O.
See Potter, Duncan J.: V1564,363-372(1991)
See Underwood, Ian: V1562,107-115(1991)

Al-Hadithi, Yas See Turcu, I. C.: V1503,391-405(1991)

Al-Houra, H. See Kozaitis, Samuel P.: V1474,112-115(1991)

Al-Hudaithi, Aziz
; Udpa, Satish S.: Recovering 3-D translation of a rigid surface by a binocular observer using moments,V1567,490-501(1991)

Al-Hujazi, Ezzet H.
; Sood, Arun K.: Integration of edge- and region-based techniques for range image segmentation,V1381,589-599(1991)

Al-Jumaily, Ghanim A.
; Gagliardi, F. J.; McColl, P.; Mizerka, Larry J.: Ion-assisted deposition of graded index silicon oxynitride coatings,V1441,360-365(1991)

Al-Kurd, Azmi See Porter, Robert P.: V1558,91-102(1991)

Al-Maliky, Noori S. See Parry, David J.: V1358,1057-1064(1991)

Al-Ramli, Intesar F.
: Optimum receiver structure and filter design for MPAM optical space communication systems,V1522,111-123(1991)

Al-Raweshidy, H. S.
; Uttamchandani, Deepak G.: Predetection correlation in a spread-spectrum multiplexing system for fiber optic interferometers,V1367,329-336(1991)
See Uttamchandani, Deepak G.: V1511,212-219(1991)

Alabedra, Robert
See Kibeya, Saidi: V1501,97-106(1991)
See Orsal, Bernard: V1512,112-123(1991)

Alaghband, Gita See Wolfe, William J.: V1382,240-254(1991)

Alaluf, M. See Gannot, Israel: V1442,156-161(1991)

Alameldin, Tarek K.
; Badler, Norman I.; Sobh, Tarek M.: Hybrid system for computing reachable workspaces for redundant manipulators,V1386,112-120(1991)
; Palis, Michael A.; Rajasekaran, Sanguthevar; Badler, Norman I.: Complexity of computing reachable workspaces for redundant manipulators,V1381,217-225(1991)
See Sobh, Tarek M.: V1383,359-366(1991)
See Sobh, Tarek M.: V1388,524-535(1991)
See Sobh, Tarek M.: V1388,425-431(1991)

Alaoui-Amine, Mohammed See Simonne, John J.: V1374,107-115(1991)

Alaverdian, R. B. See Arakelian, Sergei M.: V1402,175-191(1991)

Alavie, A. T. See Caimi, Frank M.: V1589,90-99(1991)

Albagli, Douglas
; Izatt, Joseph A.; Hayes, Gary B.; Banish, Bryan; Janes, G. Sargent; Itzkan, Irving; Feld, Michael S.: Time dependence of laser-induced surface breakdown in fused silica at 355 nm in the nanosecond regime,V1441,146-153(1991)
See Izatt, Joseph A.: V1427,110-116(1991)

Albaijes, D. See Larkins, Grover L.: V1477,26-33(1991)

Albanese, Andres
; Devetzis, Tasco N.; Ippoliti, A, G.; Karr, Michael A.; Maszczak, M. W.; Dorris, H. N.; Davis, James H.: NAVSEA gigabit optical MAN prototype history and status,V1364,320-326(1991)
See Marini, Michele: V1364,289-294(1991)

Albe, Felix
; Smigielski, Paul: Stereoscopic holographic cinematography,V1358,1098-1102(1991)

Albert, I. D.
; Ramasesha, S.: Theoretical study of optical pockels and Kerr coefficients of polyenes,V1436,179-189(1991)

Albert, Jacques
; Wang, Wei-Jian; Najafi, S. I.: Optical damage threshold of ion-exchanged glass waveguides at 1.06 um,V1583,27-31(1991)
See Lefebvre, P.: V1583,221-225(1991)
See Najafi, S. I.: V1583,32-36(1991)

Albert, Thomas
; O'Connor, Carol; Harris, Patrick D.: Automatic segmentation of microvessels using textural analysis,V1450,84-89(1991)

Albertazzi, Armando A.
: Determination of displacements, strain, and rotations from holographic interferometry data using a 2-D fringe order function,V1554B,64-74(1991)
See Sciammarella, Cesar A.: V1429,183-194(1991)
See Sciammarella, Cesar A.: V1396,143-154(1991)

Alberts, Thomas E. See Kelkar, Atul G.: V1489,243-253(1991)

Albrecht, Wolfgang See Meier, Hans: V1559,89-100(1991)

Albrektsen, O.
; Koenraad, Paul; Salemink, Huub W.: Tunneling spectroscopy at nanometer scale in molecular beam epitaxy grown (Al)GaAs multilayers,V1361,338-342(1991)

Albricht, Andreas C. See Shreve, A. P.: V1403,394-399(1991)

Alci, Mustafa
; Yilbas, Bekir S.; Danisman, Kenan; Ciftlikli, Cebrail; Altuner, Mehmet: Optimization of pulsed laser power supply system,V1411,100-106(1991)

Alcoz, J. J. See Lee, Chung E.: V1588,110-116(1991)

Alda, J. See Yonte, T.: V1554B,233-241(1991)

Alda, Javier
; Porras, Miguel A.; Bernabeu, Eusebio: Matrix representation of multimode beam transformation,V1527,240-251(1991)
See Porras, Miguel A.: V1397,645-648(1991)
See Wang, Shaomin: V1397,595-602(1991)

Alejnikov, Vladislav S.
; Masychev, Victor I.; Karpetscki, V. V.: Chemical problems of high-power sealed-off carbon monoxide lasers,V1412,173-184(1991)
See Masychev, Victor I.: V1412,227-235(1991)
See Masychev, Victor I.: V1427,344-356(1991)

Aleksandrov, A. P.
; Genkin, Vladimir N.; Myl'nikov, M. Y.; Rukhman, N. V.: Nonlinear effects in hidden picture amplification and contrast improvement in polymer electron and Roentgenoresist PMMA,V1440,442-453(1991)

Aleksandrov, I. V.
; Nesterova, Z. V.; Petrovskii, Gurii T.: Guided-wave nonlinear spectroscopy of silica glasses,V1513,309-312(1991)

Aleksandrov, K. S.
; Berman, G. P.; Frolov, G. I.; Seredkin, V. A.: Thermomagnetic recording on amorphous ferrimagnetic films,V1621,51-61(1991)

Alex, T. K.
: Imaging capabilities of small satellites: Indian experience,V1495,52-58(1991)
; Kamalakar, J. A.: Scanning infrared earth sensor for INSAT-II,V1478,106-111(1991)
; Rao, V. K.: CCD star sensors for Indian remote sensing satellites,V1478,101-105(1991)
See Seshamani, Ramani: V1489,56-64(1991)

Alexander, Andrew L.
; Davenport, Carolyn M.; Gmitro, Arthur F.: Fluorescence spectroscopy of normal and atheromatous human aorta: optimum illumination wavelength,V1425,6-15(1991)
See Davenport, Carolyn M.: V1425,16-27(1991)

Alexander, Dennis R.
; Poulain, D. E.; Schaub, Scott A.; Barton, John P.: Nonlinear laser interactions with saltwater aerosols,V1497,90-97(1991)
See Barton, John P.: V1497,64-77(1991)
See Narayanan, Ram M.: V1379,116-122(1991)

Alexander, Harold L.
: Experiments in tele-operator and autonomous control of space robotic vehicles,V1388,560-565(1991)
: State estimation for distributed systems with sensing delay,V1470,103-111(1991)

Alexander, Joanna R. See Winarsky, Norman: V1459,67-68(1991)

Alexander, Thomas
See Gove, Robert J.: V1444,318-333(1991)
See Tian, Qi: V1606,260-268(1991)

Alfano, Robert R.
See Das, Bidyut B.: V1427,368-373(1991)
See Pradhan, Asima: V1425,2-5(1991)
See Wang, LeMing: V1431,97-101(1991)

Algazi, V. R.
; Maurincomme, Eric; Ford, Gary E.: Error free and transparent coding of images using approximations by splines,V1452,364-370(1991)
; Reed, Todd R.: Comparison of image compression techniques for high quality based on properties of visual perception,V1567,589-598(1991)
; Reed, Todd R.; Ford, Gary E.; Estes, Robert R.: Visual factors and image analysis in the encoding of high-quality still images,V1605,329-338(1991)

Algrain, Marcelo C.
: Accelerometer-based platform stabilization,V1482,367-382(1991)
; Saniie, Jafar: Multisensor approach in linear-Gaussian estimation of 3-D angular motion,V1478,201-210(1991)

Ali, F. See Kozaitis, Samuel P.: V1474,112-115(1991)

Ali, Mir A.
; Meldrum, Gerald L.; Krieger, Jeffry M.: Glass requirements for encapsulating metallurgical diodes,V1513,215-223(1991)

Ali-Yahia, Tahar
; Dana, Michel: High-performance CAM-based Prolog execution scheme,V1468,950-959(1991)

Aliaga, R.
; Choi, Peter; Chuaqui, Hernan H.: Circular encoding in large-area multiexposure holography,V1358,1257-1264(1991)

Aliaga, Raul See Choi, Peter: V1552,270-277(1991)

Alibert, Claude L. See Tournie, Eric: V1361,641-656(1991)

Alimenti, Rodolphe See Malotaux, Eric: V1388,372-383(1991)

Alimpiev, Sergei S.
; Bukreev, Viacheslav S.; Kusakin, Vladimir I.; Likhansky, Sergey V.; Obidin, Alexey Z.; Vartapetov, Serge K.; Veselovsky, Igor A.: Pulse compression of KrF laser radiation by stimulated scattering,V1503,154-158(1991)

Alkindy, A. See Kozaitis, Samuel P.: V1474,112-115(1991)

Allais, David C.
: Thin family: a new barcode concept,V1384,161-168(1991)

Allam, Jeremy See Czajkowski, Igor K.: V1362,179-190(1991)

Allan, D. S.
; Kubota, F.; Marks, Tobin J.; Zhang, T. J.; Lin, W. P.; Wong, George K.: Approaches to the construction of intrinsically acentric chromophoric NLO materials: chemical elaboration and resultant properties of self-assembled multilayer structures,V1560,362-369(1991)

Allario, Frank
; Barnes, Norman P.; Storm, Mark E.: Future technologies for lidar/DIAL remote sensing,V1492,92-110(1991)

Allebach, Jan P.
See Balasubramanian, Raja: V1453,58-69(1991)
See Rogowitz, Bernice E.: V1453(1991)

Allen, B. See McAdam, Bridget A.: V1475,267-274(1991)

Allen, C. K. See Guharay, Samar K.: V1407,610-619(1991)

Allen, Chris See Lange, John A.: V1465,50-56(1991)

Allen, Donald M. See Fulop, Ann C.: V1459,69-76(1991)

Allen, K. A. See Acharya, Raj S.: V1450,170-177(1991)

Allen, Lee R.
; Semanovich, Sharon A.: Thermal analysis of masonry block buildings during construction,V1467,99-103(1991)

Allen, Mark G. See Davis, Steven J.: V1377,113-118(1991)

Allen, Mary T. See Barouch, Eytan: V1463,336-344(1991)

Allen, Maxwell J.
See DeForest, Craig E.: V1343,404-414(1991)
See Hoover, Richard B.: V1343,175-188(1991)
See Walker, Arthur B.: V1343,415-427(1991)
See Walker, Arthur B.: V1343,319-333(1991)

Allen, Michael S. See Largent, Craig C.: V1418,40-45(1991)

Allen, Peter K.
; Yoshimi, Billibon; Timcenko, Alexander; Michelman, Paul: Hand-eye coordination for grasping moving objects,V1383,176-188(1991)

Allen, Richard A. See Troccolo, Patrick M.: V1464,90-103(1991)

Allen, Robert D.
See Kutal, Charles: V1466,362-367(1991)
See McKean, Dennis R.: V1559,214-221(1991)

Allen, Ross R.
; Blackman, Samuel S.: Implementation of an angle-only tracking filter,V1481,292-303(1991)

Allen, Susan D.
See Ghosh, Anjan K.: V1371,170-181(1991)
See Imen, Kamran: V1365,60-64(1991)
See Lee, Shyan J.: V1598,2-12(1991)

Alley, Phillip W. See Heaney, James B.: V1485,140-159(1991)

Alley, W. E.
: Modeling the absorption of intense, short laser pulses in steep density gradients,V1413,89-94(1991)
See Landen, Otto L.: V1413,120-130(1991)

Allinger, Thomas
; Persch, V.; Schaefer, Juergen A.; Meng, Y.; De, H.; Anderson, J.; Lapeyre, G. J.: Interaction of hydrogen at InP(100) surfaces before and after ion bombardment,V1361,935-942(1991)
See Schaefer, Juergen A.: V1361,1026-1032(1991)

Allison, Beth
; Jiang, Frank N.; Levy, Julia G.: Efficacy of photodynamic killing with membrane associated and internalized photosensitizer molecules,V1426,200-207(1991)

Allison, Ian M.
: Photoelastic stress analysis of bridge bearings,V1554A,332-340(1991)

Allison, Stephen W. See Ramos, P. A.: V1443,160-170(1991)

Allison, Stuart A.
: Brownian dynamics simulation of polarized light scattering from wormlike chains,V1430,50-64(1991)

Alloncle, Anne P.
; Viernes, Jacques; Dufresne, Daniel; Clement, X.; Guerin, Jean M.; Testud, P.: Study of the interaction of a high-power laser radiation and a transparent liquid,V1397,675-678(1991)

Allovon, Michel See Bigan, Erwan: V1362,553-558(1991)

Allred, David D.
See Seals, Roland D.: V1330,164-177(1991)
See Williams, Memorie: V1549,147-154(1991)

Allton, Judith H. See Lyons, Damian M.: V1387,124-133(1991)

Almaev, R. K.
; Semenov, L. P.; Slesarev, A. G.; Volkovitsky, O. A.: Fluctuation variation of a CO_2 laser pulse intensity during its interaction with a cloud,V1397,831-834(1991)

Almeida, Jose B.
See Costa, Manuel Filipe P.: V1400,102-107(1991)
See Costa, Manuel Filipe P.: V1332,544-551(1991)

Almeida, Silverio P.
; Varamit, Srisuda P.: Phase conjugation by four-wave mixing in nematic liquid crystals,V1500,34-45(1991)

Alodjants, A. P. See Arakelian, Sergei M.: V1402,175-191(1991)

Alon, M. See Rochkind, Simeone: V1428,52-58(1991)

Alongi, Robert E. See Hall, Charles S.: V1483,29-38(1991)

Alouani, Ali T. See Blair, William D.: V1482,234-245(1991)

Alp, Esen E.
; Mini, Susan M.; Ramanathan, Mohan; Hyun, O. B.: X-ray absorption spectroscopy with polarized synchrotron radiation,V1345,137-145(1991)
See Mini, Susan M.: V1345,260-269(1991)
See Ramanathan, Mohan: V1548,168-178(1991)

Alperin, Noam
; Hoffmann, Kenneth R.; Doi, Kunio: Automated extraction of vascular information from angiographic images using a vessel-tracking algorithm, V1396,27-31(1991)

Alsafadi, Yasser H.
; Martinez, Ralph; Sanders, William H.: Definition and evaluation of the data-link layer of PACnet, V1446,129-140(1991)

Alsulaiman, Mansour M. See Refenes, A. N.: V1386,62-74(1991)

Altena, Oliver See Klose, Joerg: V1565,504-517(1991)

Alter, James J.
; Evins, James B.; Letellier, J. P.: Naval Research Laboratory flex processor for radar signal processing, V1566,296-301(1991)

Alter-Gartenberg, Rachel
; Fales, Carl L.; Huck, Friedrich O.; Rahman, Zia-ur; Reichenbach, Stephen E.: Multiresponse imaging system design for improved resolution, V1605,745-756(1991)
See Narayanswamy, Ramkumar: V1385,93-103(1991)
See Rahman, Zia-ur: V1488,237-248(1991)
See Reichenbach, Stephen E.: V1569,422-433(1991)

Alterovitz, Samuel A. See Leonard, Regis F.: V1394,114-125(1991)

Altkorn, Robert I. See Gregory, Christopher C.: V1420,169-175(1991)

Altman, Joe C. See Swanson, David A.: V1560,416-425(1991)

Altschuler, Bruce R.
: Remote 3-D laser topographic mapping with dental application, V1380,238-247(1991)

Altschuler, Martin D.
See Bae, Kyongtae T.: V1380,108-115(1991)
See Bae, Kyongtae T.: V1380,171-178(1991)

Altshuler, Grigori B.
; Belashenkov, Nickolay R.; Karasev, Viatcheslav B.; Okishev, Andrey V.: New effects in sound generation in organic dye solutions, V1440,116-126(1991)
; Belikov, Andrey V.; Erofeev, Andrey V.: Nonlinear optical effects under laser pulse interaction with tissues, V1427,141-150(1991)
; Grisimov, Vladimir N.; Ermolaev, Vladimir S.; Vityaz, Irena V.: Human tooth as an optical device, V1429,95-104(1991)
; Kargin, Igor U.; Khloponin, Leonid V.; Khramov, Valery: Numerical model of Q-switched solid state laser, V1415,269-280(1991)
See Azarenkov, Aleksey N.: V1409,154-164(1991)
See Azarenkov, Aleksey N.: V1409,166-177(1991)

Altuner, Mehmet
See Alci, Mustafa: V1411,100-106(1991)
See Danisman, Kenan: V1412,218-226(1991)

Alukaidey, Talib A.
; Pettiford, Alvin A.: Interactive Design and Electro-optic Analysis Liaise (IDEAL), V1390,513-522(1991)

Alvarez, E. B. See Roth, Michael W.: V1469,25-36(1991)

Alvertos, Nicolas
; D'Cunha, Ivan: Three-dimensional description of symmetric objects from range images, V1382,388-396(1991)

Alvey, Mark D.
: Soft x-ray, optical, and thermal properties of hard carbon films, V1330,39-45(1991)

Amano, H. See Akasaki, Isamu: V1361,138-149(1991)

Amano, Michiyuki See Shuto, Yoshito: V1560,184-195(1991)

Amano, Tomio
; Yamashita, Akio; Takahashi, Hiroyasu: Character string detection algorithm using horizontal boundaries, and its application to a part number entry system, V1452,330-339(1991)

Amantea, Robert See Evans, Gary A.: V1418,406-413(1991)

Amat, Josep See Casals, Alicia: V1482,317-324(1991)

Amat, L.
See Dilhac, Jean-Marie R.: V1393,395-403(1991)
See Dilhac, Jean-Marie R.: V1393,349-353(1991)

Ambia, M. G.
; Islam, M. N.; Hakim, M. O.: Effects of film thickness on the thermoelectric behavior of pyrolytic ZnO thin film, V1536,222-232(1991)

Amblard, Gilles R. See Weill, Andre P.: V1465,264-270(1991)

Ambree, P. See Gruska, Bernd: V1361,758-764(1991)

Ambrose, William P. See Moerner, William E.: V1435,244-251(1991)

Ambroz, M. See Adamec, F.: V1402,82-84(1991)

Ambrozevich, Alexander S. See Rizikov, Igor V.: V1362,664-673(1991)

Ambs, Pierre
See Kohler, A.: V1564,236-243(1991)
See Marchand, Philippe J.: V1505,38-49(1991)

Amehdi, Hossien
; Nasr, Hatem N.: Neural networks for ATR parameters adaptation, V1483,177-184(1991)
See Nasr, Hatem N.: VIS07,122-129(1991)
See Nasr, Hatem N.: V1471,283-290(1991)

Amendt, Peter
; Eder, David C.; Wilks, S. C.; Dunning, M. J.; Keane, Christopher J.: Optically ionized plasma recombination x-ray lasers, V1413,59-69(1991)
See Wilks, S. C.: V1413,131-137(1991)

Ames, Alan J.
: Theory of color correction by use of chromatic magnification, V1354,286-290(1991)

Ameye, Filip See Baert, Luc: V1421,18-29(1991)

Amidi, Omead
; Thorpe, Charles E.: Integrated mobile robot control, V1388,504-523(1991)

Amiel, Michel
See El Alaoui, Mohsine: V1569,80-87(1991)
See Friboulet, Denis: V1445,106-117(1991)
See Peyrin, Francoise: V1443,268-279(1991)

Amingual, Daniel
: Advanced infrared focal-plane arrays, V1512,40-51(1991)

Amini, Zahra H. See Webb, Jennifer M.: V1392,47-54(1991)

Amirmehrabi, Hamid
; Viswanathan, R.: Simulation of a neural network for decentralized detection of a signal in noise, V1396,252-265(1991)

Amis, Eric J. See Sedlak, Marian: V1430,191-202(1991)

Amit, M. See Chuchem, D.: V1397,277-281(1991)

Amladi, Nandan G.
; Finegan, Michael K.; Wee, William G.: Semiautomatic x-ray inspection system, V1472,165-176(1991)

Amleshi, Peerouz M.
; Naylor, David L.: Low-loss polymer thin-film optical waveguides, V1396,396-403(1991)

Ammirati, Mario See Mankovich, Nicholas J.: V1444,2-8(1991)

Amor, Hamed
; Wietzke, Joachim: 140-Mbit/s HDTV coding using subband and hybrid techniques, V1567,578-588(1991)

Amos, Lynn G. See Vethanayagam, Thirukumar K.: V1366,343-350(1991)

Amosova, L. P.
; Marmur, I. Y.; Oksman, Ya. A.; Ashmontas, S.; Gradauskas, I.; Shirmulis, E.: Hot carrier photoeffects in inhomogeneous semiconductors and their applications to light detectors, V1440,406-413(1991)

Ampilogov, Andrey V. See Tuchin, Valery V.: V1420,81-92(1991); 1429,62-73(1991)

Amra, Claude See Richier, R.: V1504,202-210(1991)

An, Cheng W.
; Fan, Yong C.; Lu, Dong S.; Li, Zai Q.: In-situ laser preparation of high-Tc superconducting thin film at 450-550 degree C, V1519,818-821(1991)
See Fan, Yong C.: V1519,775-778(1991)
See Fan, Yong C.: V1519,813-817(1991)

An, Myoung H.
; Conner, Michael; Tolimieri, Richard; Orr, Richard S.: Computation and meaning of Gabor coefficients, V1565,383-401(1991)

An, Sunghyuck
; Sipe, John E.: Dynamics of phase-grating formation in optical fibers, V1516,175-184(1991)

Anand, S. See Subramanian, S.: V1362,205-216(1991)

Anbar, Michael
: Objective assessment of clinical computerized thermal images, V1445,479-484(1991)

Anders, Gerhard See Corlatan, Dorina: V1507,354-364(1991)

Anders, K. See Ploetz, F.: V1510,224-230(1991)

Anderson, Barry K.
; Boreman, Glenn D.; Barnard, Kenneth J.; Plogstedt, Allen E.: SPRITE detector characterization through impulse response testing, V1488,416-425(1991)
See Barnard, Kenneth J.: V1488,426-431(1991)

Anderson, C. See Wolfe, William J.: V1382,240-254(1991)

Anderson, Clifford H. See Van Essen, David: V1473,17-28(1991)

Anderson, Dana Z.
: Mizrahi, Victor; Sipe, John E.: Model of second-harmonic generation in glass fibers based on multiphoton ionization interference effects,V1516,154-161(1991)

Anderson, Eric H.
; How, Jonathan P.: Implementation issues in the control of a flexible mirror testbed,V1542,392-405(1991)
See How, Jonathan P.: V1489,148-162(1991)

Anderson, Eric R.
; Jansen, Michael; Botez, Dan; Mawst, Luke J.; Roth, Thomas J.; Yang, Jane: Modulation characteristics of high-power phase-locked arrays of antiguides,V1417,543-549(1991)
See Botez, Dan: V1474,64-74(1991)
See Mawst, Luke J.: V1418,353-357(1991)

Anderson, Forrest L.
: Three-dimensional real-time ultrasonic imaging using ellipsoidal backprojection,V1443,62-80(1991)

Anderson, J. See Allinger, Thomas: V1361,935-942(1991)

Anderson, James C.
; Downs, G. S.; Trepagnier, Pierre C.: Signal processor for space-based visible sensing,V1479,78-92(1991)

Anderson, John E.
; Roos, Maurits: Using digital-scanned aerial photography for wetlands delineation,V1492,252-262(1991)

Anderson, Joseph N. See Copeland, Bruce R.: V1381,177-188(1991)

Anderson, Kelly L.
; Barrett, William A.: Context specification for text recognition in forms,V1384,270-279(1991)

Anderson, Kent L.
: Geoscience laser ranging system design and performance predictions,V1492,142-152(1991)

Anderson, Mary P.
; Brown, David G.; Loew, Murray H.: Evaluation of medical image compression by Gabor elementary functions,V1444,407-412(1991)

Anderson, Paul D. See Khan, Mumit: V1465,315-323(1991)

Anderson, R. R.
See Lemole, G. M.: V1422,116-122(1991)
See Obremski, Susan M.: V1427,327-334(1991)

Anderson, Richard L.
: Review of temperature measurements in the semiconductor industry,V1392,437-451(1991)

Anderson, Robert J. See Motamedi, M. E.: V1544,33-44(1991)

Anderson, Roland C.
; Milton, James E.: Conversion of schlieren systems to high-speed interferometers,V1358,992-1002(1991)

Anderson, Stephen J.
; Bulusu, Dutt V.; Racette, James; Scholl, Frederick W.; Zack, Tim; Abbott, Peter G.: FDDI components for workstation interconnection,V1364,94-100(1991)

Anderson, Stuart M.
; Zahniser, Mark S.: Open-path tunable diode laser absorption for eddy correlation flux measurements of atmospheric trace gases,V1433,167-178(1991)

Anderson, Thomas W. See McCargar, James W.: V1455,54-60(1991)

Anderson, Wallace T. See Meulenberg, A.: V1475,280-285(1991)

Anderson, Wayne A. See Jiao, Kaili L.: V1361,776-783(1991)

Anderson, William J.
: Quality assurance from a manufacturer's standpoint,V1444,128-133(1991)

Anderson, William P. See Dasgupta, Samhita: V1374,211-222(1991)

Anderson, William T. See Kilmer, Joyce P.: V1366,85-91(1991)

Andersson, Eva K. See Bogdanov, Alexei L.: V1465,324-329(1991)

Andersson, Thorvald G.
; Ekenstedt, M. J.; Kulakovskii, Vladimir D.; Wang, S. M.; Yao, J. Y.: Measurements of the InxGa1-xAs/GaAs critical layer thickness,V1361,434-442(1991)

Andersson-Engels, Stefan
; Baert, Luc; Berg, Roger; D'Hallewin, Mich A.; Johansson, Jonas; Stenram, U.; Svanberg, Katarina; Svanberg, Sune: Fluorescence characteristics of atherosclerotic plaque and malignant tumors,V1426,31-43(1991)
See Berg, Roger: V1431,110-119(1991)
See Berg, Roger: V1525,59-67(1991)

Andino, J. M. See Gierczak, Christine A.: V1433,315-328(1991)

Ando, Toshinori
: Laser beam scanner for uniform halftones,V1458,128-132(1991)

Andonovic, Ivan See Stewart, George: V1368,230-238(1991)

Andoo, Keikichi See Okamine, Shigenori: V1499,166-170(1991)

Andorfer, J. See Gresslehner, Karl-Heinz: V1361,1087-1093(1991)

Andrade, Joseph D. See Wang, Jinyu: V1519,835-841(1991)

Andrae, P.
; Mieth, Ulrike; Osten, Wolfgang: Strategies for unwrapping noisy interferograms in phase-sampling interferometry,V1508,50-60(1991)

Andre, Jean-Claude See Saatdjian, E.: V1397,535-538(1991)

Andre, Jean-Michel See Berrouane, H.: V1343,428-436(1991)

Andre, Michel L.
; Coutant, Jacques; Dautray, Robert; Decroisette, Michel; Lompre, Louis A.; Naudy, Michel; Manus, Claude; Mainfray, Gerard L.; Migus, Arnold; Normand, Didier; Sauteret, Christian; Watteau, Jean P.: Interaction of ultra-bright lasers with matter: program of the French Commissariat a l'Energie Atomique,V1502,286-298(1991)
See Fleurot, Noel A.: V1502,230-241(1991)
See Novaro, Marc: V1501,183-193(1991)

Andreadakis, Nicholas C.
: Thin-film technology in high-resolution, high-density AC plasma displays,V1456,310-315(1991)
See Soole, Julian B.: V1474,268-276(1991)

Andreani, F.
; Luridiana, S.; Mao, Shu Z.: Large-scale production of a broadband antireflection coating on ophthalmic lenses with ion-assisted deposition,V1519,63-69(1991)
; Luridiana, S.; Mao, Shu Z.: Optical properties of some ion-assisted deposited oxides,V1519,18-22(1991)

Andreev, V. G.
See D'yakonov, G. I.: V1421,153-155(1991)
See Mnatsakanyan, Eduard A.: V1414,130-133(1991)

Andreeva, Olga V.
: Analysis of the Focar-type silver-halide heterogeneous media,V1238,231-234(1991)

Andrei, E. Y. See Etienne, Bernard: V1362,256-267(1991)

Andreic, Zeljko
: Simple 180o field-of-view F-theta all-sky camera,V1500,293-304(1991)

Andrekson, Peter A. See Larsson, Anders G.: V1418,292-297(1991)

Andreoni, Alessandra
; Colasanti, Alberto; Kisslinger, Annamaria; Malatesta, Vincenzo; Mastrocinque, Michele; Roberti, Giuseppe: Antitumor drugs as photochemotherapeutic agents,V1525,351-366(1991)

Andreoni, G. See Noethe, L.: V1542,293-296(1991)

Andreou, Andreas G.
; Boahen, Kwabena A.: Modeling inner and outer plexiform retinal processing using nonlinear coupled resistive networks,V1453,270-281(1991)

Andresen, Bjorn F.
; Scholl, Marija S.; Spiro, Irving J.; eds.: *Infrared Technology XVII*,V1540(1991)

Andresen, Klaus
; Kamp, B.; Ritter, Reinold: Three-dimensional contour of crack tips using a grating method,V1554A,93-100(1991)

Andress, Keith M.
: Hierarchical Dempster-Shafer evidential reasoning for image interpretation,V1569,43-54(1991)
; Wilson, David L.: Optimization of morphological structuring elements for angiogram enhancement,V1445,6-10(1991)

Andrews, John R.
: Design rules for pseudocolor transmission holographic stereograms,V1507,407-415(1991)
; Stinehour, Judith E.; Lean, Meng H.; Potyondy, David O.; Wawrzynek, Paul A.; Ingraffea, Anthony R.; Rainsdon, Michael D.: Holographic display of computer simulations,V1461,110-123(1991)

Andrews, Larry C.
See Richardson, Christina E.: V1487,19-30(1991)
See Sisterman, Elizabeth A.: V1487,345-355(1991)

Andrews, Lee T.
; Klingler, Joseph W.; Schindler, Jeffery A.; Begeman, Michael S.; Farron, Donald; Vaughan, Bobbi; Riggs, Bud; Cestaro, John: Detection and visualization of porosity in industrial CT scans of aluminum die castings,V1459,125-135(1991)

Andrews, Leonard J. See Zemon, Stanley A.: V1373,21-32(1991)

Andrews, Mark P. See Kuzyk, Mark G.: V1560,44-53(1991)

Andricacos, P. C. See Krongelb, Sol: V1389,249-256(1991)

Andriesh, A.
; Bertolotti, Mario; Chumach, V.; Fazio, E.; Ferrari, A.; Liakhou, G.; Sibilia, C.: Nonlinear optical properties of chalcogenide As-S, As-Se glasses,V1513,137-147(1991)

Andronov, Alexander A.
; Genkin, G.; Ghinovker, M.; Kurin, V.; Nefedov, I.; Okomelkov, A.; Shereshevsky, I.: Dynamical properties of superconducting superlattices,V1362,684-695(1991)

Anema, P. C.
; de Graaf, C. N.; Wilmink, J. B.; Hall, David; Hoekstra, A.; van Rijk, P. P.; Van Isselt, J. W.; Viergever, Max A.: One-year clinical experience with a fully digitized nuclear medicine department: organizational and economical aspects,V1446,352-356(1991)

Angel, J. R. See Wizinowich, Peter L.: V1542,148-158(1991)

Angel, S. M.
; Langry, Kevin; Colston, B. W.; Roe, Jeffrey N.; Daley, Paul F.; Milanovich, Fred P.: Preliminary field demonstration of a fiber optic trichloroethylene sensor,V1368,98-104(1991)
; Myrick, Michael L.; Vess, Thomas M.: Remote Raman spectroscopy using diode lasers and fiber-optic probes,V1435,72-81(1991)
See Schoen, Christian: V1584,79-86(1991)

Angelhed, Jan-Erik
; Mansson, Lars G.; Kheddache, Susanne: Optimization and evaluation of an image intensifier TV system for digital chest imaging,V1444,159-170(1991)

Angell, David
; Oehrlein, Gottlieb S.: Etch tailoring through flexible end-point detection,V1392,543-550(1991)

Angelov, D.
; Berger, M.; Cadet, J.; Ballini, Jean-Pierre; Keskinova, E.; Vigny, Paul: High-power UV laser photolysis of nucleosides: final product analysis,V1403,575-577(1991)
; Dimitrov, S.; Keskinova, E.; Pashev, I.; Russanova, V.; Stefanovsky, Yu.: Picosecond laser cross: linking histones to DNA in chromatin: implication in studying histone/DNA interactions,V1403,230-239(1991)
; Gantchev, Ts.; Grabner, G.; Getoff, N.; Keskinova, E.; Shopova, Maria: Generation of free radicals in high-intensity laser photolysis of organic microcyclic compounds: time-resolved spectroscopy and EPR study,V1403,572-574(1991)

Angelsky, Oleg V.
; Magun, I. I.; Maksimyak, Peter P.; Perun, T. O.: Spatial randomization of the scattered optical radiation,V1402,231-235(1991)
; Maksimyak, Peter P.: Optical correlation studies of biological objects,V1403,667-673(1991)

Anger, Frank D.
; Ladkin, Peter B.; Rodriguez, Rita V.: Atomic temporal interval relations in branching time: calculation and application,V1468,122-136(1991)

Angiboust, Jean-Francois See Manfait, Michel: V1403,695-707(1991)

Angrilli, Francesco
; Baglioni, Pietro; Bianchini, Gianandrea; Da Forno, R.; Fanti, Giullo; Mozzi, Massimo: Multiple degree-of-freedom tracking for attitude control of an experimental system on tether-stabilized platform,V1482,26-39(1991)

Anhalt, David J. See Severson, William E.: V1388,490-501(1991)

Anholt, R. See Swirhun, S.: V1475,303-308(1991)

Anikin, V. I.
; Dementiev, I. V.; Zhurminskii, Igor L.; Nasedkhina, N. V.; Panasiuk, L. M.: Some cases of applying photothermoplastic carriers in holographic interferometry and speckle photography,V1238,286-296(1991)

Anisimov, N. R.
; Buzykin, O. G.; Zayakin, A. A.; Makarov, V. V.; Makashev, N. K.; Nosachev, L. V.; Frolov, I. P.: Peculiarities of metal surface heating by repetitively pulsed CO2 laser radiation with the blowing,V1440,206-210(1991)

Anliker, M. See Muser, Markus H.: V1448,106-112(1991)

Annamalai, Kadiresan
: Extending HIPPI at 800-mega-bits-per-second over serial links using HOT ROD technology,V1364,178-189(1991)
; Cudworth, Stewart K.; Kasiewicz, Allen B.; eds.: *FDDI, Campus-Wide, and Metropolitan Area Networks*,V1364(1991)

Annapurna, M. N.
; Nagendra, C. L.; Thutupalli, G. K.: Design optimization of optical filters for space applications,V1485,260-271(1991)

Annunziata, Frank A. See Thevenot, Clarel: V1398,250-260(1991)

Ansari, Nirwan
; Huang, Kuo-Wei: Nonparametric dominant point detection,V1606,31-42(1991)
See Chang, Yiher: V1606,839-850(1991)
See Liu, Xianjun: V1606,1069-1079(1991)

Ansari, Rafat R. See Dhadwal, Harbans S.: V1584,262-272(1991)

Ansari, Rashid See Kim, Chai W.: V1605,112-123(1991)

Ansell, Ralph See Kovacich, Michael: V1481,357-370(1991)

Anspach, Joel E.
; Sydney, Paul F.; Hendry, Gregg: Effects of base motion on space-based precision laser tracking in the Relay Mirror Experiment,V1482,170-181(1991)
See Lightsey, Paul A.: V1482,209-222(1991)
See Sydney, Paul F.: V1482,196-208(1991)

Anthan, Donald J. See Beheim, Glenn: V1369,50-59(1991)

Antikas, Theo G.
: Low-power laser effects in equine traumatology and postsurgically,V1424,186-197(1991)

Antipenko, Boris M. See Bouchenkov, Vyatcheslav A.: V1410,185-188(1991); 1427,409-412(1991)

Antoine, Miquel D. See Patonay, Gabor: V1435,52-63(1991)

Antonetti, Andre
See Gauduel, Yann: V1403,417-426(1991)
See Geindre, Jean-Paul: V1502,311-318(1991)

Antonov, Victor A.
; Shelyakov, Alexander V.: Spatial light modulator on the base of shape memory effect,V1474,116-123(1991)

Antonova, N. E. See Logak, L. G.: V1238,171-176(1991)

Antonuk, Larry E.
; Boudry, J.; Kim, Chung-Won; Longo, M.; Morton, E. J.; Yorkston, J.; Street, Robert A.: Signal, noise, and readout considerations in the development of amorphous silicon photodiode arrays for radiotherapy and diagnostic x-ray imaging,V1443,108-119(1991)

Antos, Ronald L.
; Krisiloff, Allen J.; eds.: *CAN-AM Eastern '90*,V1398(1991)

Antoszewska, Reza See Girard, Mary M.: V1563,156-167(1991)

Antunez, Antonio R. See Oleinick, Nancy L.: V1427,90-100(1991)

Antuofermo, Pasquale
; Losacco, Aurora M.; De Pascale, Olga: Novel configuration of a two-mode-interference polarization splitter with a buffer layer,V1583,143-149(1991)

Antyuhov, V.
; Bondarenko, Alexander V.; Glova, Alexander F.; Golubenzev, A. A.; Danshikov, E.; Kachurin, O. R.; Lebedev, Fedor V.; Likhanskii, Vladimir V.; Napartovich, Anatoly P.; Pis'menny, Vladislav D.; Yarzev, V.; Yaztsev, Vladimir P.: Coupled CO2 lasers,V1397,355-365(1991); 1415,48-59(1991)

Anufrik, S. S.
; Znosko, K. F.; Kurgansky, A. D.: XeCl laser with LC-circuit excitation research,V1391,87-92(1991)

Anzures, Edguardo See Hoyle, Charles E.: V1559,101-109(1991)

Aoki, Makoto See Mukasa, Minoru: V1446,253-265(1991)

Aoki, Nobutada
; Kimura, Hironobu; Konagai, Chikara; Shirayama, Shimpey; Miyazawa, Tatsuo; Takahashi, Tomoyuki: High-power copper vapor laser development,V1412,2-11(1991)

Aoki, Norihiko See Tsuchida, Hirofumi: V1354,246-251(1991)

Aoki, Sadao See Yoda, Osamu: V1503,463-466(1991)

Aoki, Yutaka
; Ohmae, Hirokazu; Kitamura, Shin-ichi: Thermal infrared subsystem design status of ASTER,V1490,278-284(1991)

Aoyama, Satoshi See Wiley, James N.: V1464,346-355(1991)

Aoyama, Shigeru
; Yamashita, Tsukasa: Grating beam splitting polarizer using multilayer resist method,V1545,241-250(1991)
See Yamashita, Tsukasa: V1507,81-93(1991)

Apanasevich, P. A.
; Chirvony, V. S.; Kruglik, S. G.; Kvach, V. V.; Orlovich, V. A.: Inter- and intramolecular processes in metalloporphyrins: study by transient absorption and resonance Raman and coherent anti-Stokes Raman scattering,V1403,195-211(1991)
; Chyrvony, V. S.; Kruglik, S. G.; Kvach, V. V.; Orlovich, V. A.: Intramolecular processes of excitation energy redistribution in metalloporphyrins: examination by transient absorption and resonance Raman and coherent anti-Stokes Raman scattering spectroscopies,V1403,240-242(1991)
; Kozich, V. P.; Vodchitz, A. I.; Kontsevoy, B. L.: Picosecond orientational dynamics of complex molecules studied by incoherent light three-wave mixing,V1403,475-477(1991)

Apollonov, V. V.
; Chetkin, Sergue A.; Kislov, V. I.; Vdovin, Gleb V.: New approach to creation of stable and unstable active resonators for high-power solid state lasers,V1502,83-94(1991)
See Aksinin, V. I.: V1500,93-104(1991)

Apollonova, O. V.
; Elkin, Nickolai N.; Korjov, M. Y.; Korotkov, V. A.; Likhanskii, Vladimir V.; Napartovich, Anatoly P.; Troshchiev, V. E.: Mathematical simulation of composite optical systems loaded with active medium,V1501,108-119(1991)

Apostolakis, Peter J.
: Statistical approach to optimizing advanced low-voltage SEM operation,V1464,406-412(1991)

Apostolopoulos, John G. See Hemmati, Hamid: V1417,476-483(1991)

Apparao, K. V. See Murty, M. V.: V1332,41-49(1991)

Apparao, R. T.
; Varadan, Vasundara V.; Varadan, Vijay K.: Theoretical modeling of chiral composites,V1558,2-13(1991)

Appelbaum, Ami
See Cheng, Wood-Hi: V1418,279-283(1991)
See Jiang, Ching-Long: V1418,261-271(1991)

Appino, Perry A.
; Farrell, Edward J.: System for making scientific videotapes,V1459,157-165(1991)

Apte, Pushkar P. See Wood, Samuel C.: V1393,337-348(1991)

Arabnia, Hamid R.
; Chen, Ching-Yi: Remote media vision-based computer input device,V1606,917-925(1991)

Arachtingi, John W. See Lauffenburger, Jim: V1499,104-113(1991)

Araghi, Mehdi N. See Whitman, Tony: V1454,426-433(1991)

Aragon, Sergio R.
; Luo, Rolland: Dynamics of wormlike chains: theory and computer simulations,V1430,65-84(1991)

Aragones, J. M. See Velarde, G.: V1502,242-257(1991)

Arai, Hiromichi See Setoguchi, Toshihiko: V1519,74-79(1991)

Arai, Kiyoshi See Aisaka, Kazuo: V1445,312-317(1991)

Arai, Masayuki See Maeda, Fumisada: V1499,62-69(1991)

Arai, S. See Endo, Masamori: V1397,267-270(1991)

Arai, Shinichi See Maeda, Takeshi: V1499,414-418(1991)

Arai, Tsunenori
; Mizuno, Kyoichi; Sakurada, Masami; Miyamoto, Akira; Arakawa, Koh; Kurita, Akira; Suda, Akira; Kikuchi, Makoto; Nakamura, Haruo; Utsumi, Atsushi; Akai, Yoshiro; Takeuchi, Kiyoshi: Combined guidance technique using angioscope and fluoroscope images for CO laser angioplasty: in-vivo animal experiment,V1425,191-195(1991)
See Daidoh, Yuichiro: V1421,120-123(1991)
See Sakurada, Masami: V1425,158-164(1991)

Arai, Yasuhiko
; Yekozeki, Shunsuke; Yamada, Tomoharu: Three-dimensional automatic precision measurement system by liquid-crystal plate on moire topography,V1554B,266-274(1991)

Arakawa, Isao See Ohmori, Yasuhiro: V1490,177-183(1991)

Arakawa, Koh See Arai, Tsunenori: V1425,191-195(1991)

Arakelian, Sergei M.
; Chilingarian, Yu. S.; Alaverdian, R. B.; Alodjants, A. P.; Drnoian, V. E.; Karaian, A. S.: Laser-induced phase transitions in liquid crystals and distributed feedback-fluctuations, energy exchange, and instabilities: squeezed polarized states and intensity correlations,V1402,175-191(1991)
; Chilingarian, Yu. S.: Zero-angle scattering of light in oriented organic liquids: classical and quantum states for both linear and nonlinear scattering,V1403,326-331(1991)

Araki, Ken'ichi
See Fujise, Masayuki: V1522,14-26(1991)
See Inagaki, Keizo: V1417,160-169(1991)

Araki, Ryuichiro
; Nashimoto, Ichiro: Near-infrared imaging in vivo: imaging of Hb oxygenation in living tissues,V1431,321-332(1991)

Araki, Toshinari See Katayama, Norihisa: V1403,147-149(1991)

Araki, Yoshitsugu See Kitagaki, Kazukuni: V1606,891-900(1991)

Araman, Philip A.
See Zhu, Dongping: V1567,232-243(1991)
See Zhu, Dongping: V1606,685-696(1991)

Arata, Louis K.
; Dhawan, Atam P.; Thomas, Stephen R.: Model-based labeling, analysis, and three-dimensional visualization from two-dimensional medical images,V1446,465-474(1991)

Aratani, Katsuhisa
; Fukumoto, Atsushi; Ohta, Masumi; Kaneko, Masahiko; Watanabe, Kenjirou: Magnetically induced super resolution in a novel magneto-optical disk,V1499,209-215(1991)
See Fukumoto, Atsushi: V1499,216-225(1991)

Araujo, Roger J.
; Borrelli, Nicholas F.: Optical effects induced in oxide glasses by irradiation,V1590,138-145(1991)

Aravin, Vladislav A. See Tuchin, Valery V.: V1422,85-96(1991)

Aravind, R. See Puri, Atul: V1605,297-300(1991)

Arber, Simon
; Rymer, William Z.; Crumrine, David: Photosensitized receptor inactivation with He-Ne laser: preliminary results,V1428,23-29(1991)

Arbour, Claude See Armstrong, Neal R.: V1559,18-26(1991)

Arbus, R. See Leporcq, B.: V1397,153-156(1991)

Arce, Gonzalo R.
; Boncelet, Charles G.; Dougherty, Edward R.; eds.: *Nonlinear Image Processing II*,V1451(1991)
See Boncelet, Charles G.: V1451,70-74(1991)

Archer, Susan J. See Brown, Joseph R.: V1469,539-543(1991)

Ardisasmita, Moh S.
See Montgomery, Paul C.: V1332,563-570(1991)
See Montgomery, Paul C.: V1332,515-524(1991)

Arditi, Aries R. See Larimer, James O.: V1456,196-210(1991)

Arditty, Herve J. See Lefevre, Herve C.: V1367,72-80(1991)

Aref'ev, Vladimir N.
: Measurements of atmospheric transmittance of CO2 laser radiation,V1397,827-830(1991)

Arend, Lawrence
: Apparent contrast and surface color in complex scenes,V1453,412-421(1991)

Arendale, William F.
; Congo, Richard T.; Nielsen, Bruce J.: Advances in analytical chemistry,V1434,159-170(1991)

Arenson, Ronald L.
See Carey, Bruce: V1446,414-419(1991)
See Feingold, Eric: V1446,211-216(1991)
See Kishore, Sheel: V1446,188-194(1991)
See Kishore, Sheel: V1446,236-242(1991)
See Kundel, Harold L.: V1446,297-300(1991)
See Nodine, Calvin F.: V1444,56-62(1991)
See Rundle, Debra A.: V1446,405-413(1991)
See Seshadri, Sridhar B.: V1446,388-395(1991)

Aretz, H. T. See Gregory, Kenton W.: V1425,217-225(1991)

Aridgides, Athanasios
; Fernandez, Manuel F.; Randolph, D.; Ferris, D.: Adaptive 4-D IR clutter suppression filtering technique,V1481,110-116(1991)
See Fernandez, Manuel F.: V1481,172-179(1991)

Ariffin, Z. See Muhamad, M. R.: V1519,872-877(1991)

Arik, Engin B.
: Recent developments in fiber optic and laser sensors for flow, surface vibration, rotation, and velocity measurements,V1584,202-211(1991)

Arikawa, Hiroshi See Fujise, Masayuki: V1522,14-26(1991)

Arimoto, Akira A. See Nakao, Takeshi: V1499,433-437(1991)

Arita, Masataka See Silver, Michael D.: V1443,250-260(1991)

Arjavalingam, Gnanalingam
; Pazaris, James; eds.: *Microelectronic Interconnects and Packages: Optical and Electrical Technologies,*V1389(1991)

Arkhontov, L. B. See Kiselyov, Boris S.: V1621,126-137(1991)

Arkin, Ronald C. See Vaughn, David L.: V1383,503-514(1991)

Arlinghaus, Heinrich F.
; Spaar, M. T.; Thonnard, N.; McMahon, A. W.; Jacobson, K. B.: Applications of resonance ionization spectroscopy for semiconductor, environmental and biomedical analysis, and for DNA sequencing,V1435,26-35(1991)

Arman, Helen See Klainer, Stanley M.: V1434,119-126(1991)

Armelles, Gaspar See Recio, Miguel: V1361,469-478(1991)

Armenise, Mario N.
; Castagnolo, Beniamino: Design of photonic switches for optimizing performances of interconnection networks,V1374,186-197(1991)
; Castagnolo, Beniamino; Pesce, Anastasia; Rizzi, Maria L.: New configuration of 2 x 2 switching elements using Mach-Zehnder electro-optic modulators,V1583,210-220(1991)
; Impagnatiello, Fabrizio; Passaro, Vittorio M.: Design of a GaAs Fresnel lens array for optical signal-processing devices,V1374,86-96(1991)
; Impagnatiello, Fabrizio; Passaro, Vittorio M.; Pansini, Evangelista: Design of a GaAs acousto-optic correlator for real-time processing,V1562,160-171(1991)
; Matteo, Annamaria; Passaro, Vittorio M.: Acoustic mode analysis of multilayered structures for the design of acousto-optic devices,V1583,256-267(1991)
; Matteo, Annamaria; Passaro, Vittorio M.: Modeling of acousto-optic interaction in multilayer guiding structures,V1583,289-297(1991)
See Conese, Tiziana: V1583,249-255(1991)
See Savatinova, Ivanka T.: V1374,37-46(1991)
See Ziling, C. C.: V1583,90-101(1991)

Armentrout, Ben A. See Jackson, John E.: V1533,75-86(1991)

Armeyev, V. Y.
; Arslanbekov, A. H.; Chapliev, N. I.; Konov, Vitaly I.; Ralchenko, V. G.; Strelnitsky, V. E.: Applications of diamond-like carbon films for write-once optical recording,V1621,2-10(1991)

Armiento, Craig A. See Rothman, Mark A.: V1392,598-604(1991)

Armitage, David
; Ticknor, Anthony J.: Liquid-crystal devices in planar optics,V1455,206-212(1991)

Armitage, J. R.
See Williams, Doug L.: V1513,158-167(1991)
See Williams, Doug L.: V1516,29-37(1991)

Armstrong, Brian M. See Ruddell, F. H.: V1361,159-170(1991)

Armstrong, George R.
; Oakley, Philip J.; Ranat, Bhadrayu M.: Multimode IRST/FLIR design issues,VCR38,120-141(1991)

Armstrong, Nancy J. See Penner, Thomas L.: V1560,377-386(1991)

Armstrong, Neal R.
; Nebesny, Ken W.; Collins, Greg E.; Lee, Paul A.; Chau, Lai K.; Arbour, Claude; Parkinson, Bruce: O/I-MBE: formation of highly ordered phthalocyanine/semiconductor junctions by molecular-beam epitaxy: photoelectrochemical characterization,V1559,18-26(1991)

Armstrong, Robert L.
: Superheating phenomena in absorbing microdroplets irradiated by pulsed lasers,V1497,132-140(1991)

Arndt, George D.
; Fink, Patrick W.; Leopold, Louis; Bondyopadhyay, Probir; Shaw, Roland: Application of Kaband MMIC technology for an Orbiter/ACTS communications experiment,V1475,231-242(1991)

Arndt, Thomas D. See Hudyma, Russell M.: V1563,244-254(1991)

Arndt-Jovin, Donna J.
See Jovin, Thomas M.: V1439,109-120(1991)
See Zidovetzki, Raphael: V1432,76-81(1991)

Arnold, Douglas
; Cartier, E.; Fischetti, Massimo V.: Monte Carlo calculations of laser-induced free-electron heating in SiO2,V1441,478-487(1991)

Arnold, Frank S.
See Carbone, Joseph: V1447,229-242(1991)
See Zarnowski, Jeffrey J.: V1447,191-201(1991)

Arnold, Harold See Parrish, William J.: V1512,68-77(1991)

Arnold, Jones B.
: Development of beryllium-mirror turning technology,V1485,96-105(1991)

Arnold, Steven M.
; Jain, Anil K.: Interferometer for testing of general aspherics using computer-generated holograms,V1396,473-480(1991)

Arnold, William H.
ed.: *Integrated Circuit Metrology, Inspection, and Process Control V,*V1464(1991)
See Dunbrack, Steven K.: V1464,314-326(1991)

Arnold, William R.
: Nonlinear finite element analysis of the Starlab 80-cm telescope primary-mirror suspension system,V1532,103-113(1991)

Arnon, S. See Rotman, Stanley R.: V1442,194-204(1991)

Arnone, Robert A. See Estep, Leland: V1537,89-96(1991)

Arntz, Floyd O. See Goldner, Ronald B.: V1536,63-69(1991)

Arons, Jonathan
: Accretion dynamics and polarized x-ray emission of magnetized neutron stars,V1548,2-12(1991)

Aronson, Raphael See Barbour, Randall L.: V1431,52-62(1991); 1431,192-203(1991)

Arora, B. M. See Subramanian, S.: V1362,205-216(1991)

Arridge, Simon R.
; van der Zee, Pieter; Cope, Mark; Delpy, David T.: Reconstruction methods for infrared absorption imaging,V1431,204-215(1991)
See Cope, Mark: V1431,251-262(1991)

Arrott, Matthew See Chen, Chang W.: V1450,231-242(1991)

Arrue, Begona C.
; Inigo, Rafael M.: Hough transform implementation using an analog associative network,V1469,420-431(1991)

Arsenault, Henri H.
See Gagne, Philippe: V1564,656-663(1991)
See Leclerc, Luc: V1564,78-85(1991)
See Sheng, Yunlong: V1559,222-228(1991)
See Sheng, Yunlong: V1564,320-329(1991)

Arshak, A. See Arshak, Khalil I.: V1463,521-533(1991)

Arshak, Khalil I.
; Murphy, Eamonn; Arshak, A.: Process optimization: a case study on the application of Taguchi methods in optical lithography,V1463,521-533(1991)

Arslanbekov, A. H. See Armeyev, V. Y.: V1621,2-10(1991)

Arteaga, Humberto See Williams, Gareth T.: V1332,802-804(1991)

Artemjev, S. V.
; Koval, G. I.; Obyknovennaja, I. E.; Cherkasov, A. S.; Shevtsov, M. K.: PFG-04 photographic plates based on the nonhardened dichromated gelatin for recordng color reflection holograms,V1238,206-210(1991)

Artjushenko, Vjacheslav G.
; Ivchenko, N.; Konov, Vitaly I.; Kryukov, A. P.; Krupchitsky, Vladimir P.; Kuznetcov, R.; Lerman, A. A.; Litvinenko, E. G.; Nabatov, A. O.; Plotnichenko, V. G.; Prokhorov, Alexander M.; Pylnov, I. L.; Tsibulya, Andrew B.; Vojtsekhovsky, V. V.; Ashraf, N.; Neuberger, Wolfgang; Moran, Kelly B.; Mueller, Gerhard J.; Schaldach, Brita J.: Infrared cables and catheters for medical applications,V1420,157-168(1991)
; Konov, Vitaly I.; Pashinin, Vladimir P.; Silenok, Alexander S.; Blinov, Leonid M.; Solomatin, A. M.; Shilov, I. P.; Volodko, V. V.; Mueller, Gerhard J.; Schaldach, Brita J.; Ulrich, R.; Neuberger, Wolfgang: Fused silica fibers for the delivery of high-power UV radiation,V1420,149-156(1991)
; Konov, Vitaly I.; Pashinin, Vladimir P.; Silenok, Alexander S.; Mueller, Gerhard J.; Schaldach, Brita J.; Ulrich, R.: Mechanism of excimer-laser-induced absorption in fused silica fibers,V1420,176-176(1991)
; Konov, Vitaly I.; Konstantinov, N. Y.; Pashinin, Vladimir P.; Silenok, Alexander S.; Mueller, Gerhard J.; Schaldach, Brita J.; Ulrich, R.; Neuberger, Wolfgang; Castro, Jose L.: Mechanisms of UV-laser-induced absorption in fused silica fibers,V1590,131-136(1991)

Artola, Michel
; Cessenat, Michel: Effective properties of electromagnetic wave propagation in some composite media,V1558,14-21(1991)

Artsimovich, M. V.
; Baranov, A. N.; Krivov, V. V.; Kudriavtsev, E. M.; Lotkova, E. N.; Makeev, B. H.; Mogilnik, I. F.; Pavlovich, V. N.; Romanuk, B. N.; Soroka, V. I.; Tokarevski, V. V.; Zotov, S. D.: Resonant IR laser-induced diffusion of oxygen in silicon,V1397,729-733(1991)

Artyomova, V. V. See Churaev, A. L.: V1238,158-165(1991)

Arun, K. S.
; Liang, R. M.: Instantaneous amplitude and frequency estimation: performance bounds and applications to source localization,V1566,157-166(1991)

Arutunian, A. H.

; Hovanessian, V. A.; Sarkissian, K. A.: Laser microphotolysis of biological objects with the application of UV solid state lasers,V1402,102-106(1991)

; Hovanessian, V. A.; Sarkissian, K. A.: Nonlinear absorption of organic compounds in the picosecond laser radiation field,V1402,2-6(1991)

; Hovanessian, V. A.; Sarkissian, K. A.: Study of aggregation phenomenon of hematoporphyrin derivative by laser microphotolysis,V1403,585-587(1991)

As, D. J. See Hingerl, Kurt: V1361,943-953(1991)

Asada, Kazuya See Sasano, Yasuhiro: V1490,233-242(1991)

Asada, Osamu See Matsushiro, Nobuhito: V1452,21-26(1991)

Asai, Satoru See Hanyu, Isamu: V1463,595-601(1991)

Asakura, Toshimitsu

See Aizu, Yoshihisa: V1431,239-250(1991)

See Okamoto, Takashi: V1399,192-199(1991)

See Widjaja, Joewono: V1400,94-100(1991)

Asano, J. See Minemura, Tetsuroh: V1361,344-353(1991)

Asars, Juris A. See Leksell, David: V1458,133-144(1991)

Asaumi, Shingo See Furuta, Mitsuhiro: V1466,477-484(1991)

Ashcroft, Peter

: Characterization of tropospheric methane through space-based remote sensing,V1491,48-55(1991)

Ashenayi, Kaveh

: Application of neural networks in optimization problems: a review,V1396,285-296(1991)

; Vogh, James: Influence of different nonlinearity functions on Perceptron performance,V1396,215-225(1991)

See Karimi, B.: V1396,226-236(1991)

Ashidate, Shu-ichi

; Takashima, Toshiaki; Kannari, Fumihiko; Obara, Minoru: Theoretical study of the high-power pulsed vibrational overtone HF chemical laser,V1397,283-286(1991)

Ashitkov, S. I. See Agranat, M. B.: V1440,397-400(1991)

Ashkinadze, B. M.

; Bel'Kov, V. V.; Krasinskaya, A. G.: Cyclotron resonance and photoluminescence in GaAs in a microwave field,V1361,866-873(1991)

Ashley, Paul R.

; Tumolillo, Thomas A.: Overview of EO polymers for guided-wave devices,V1583,316-326(1991)

See Tumolillo, Thomas A.: V1559,65-73(1991)

Ashley, Timothy

; Dean, A. B.; Elliott, Charles T.; Houlton, M. R.; McConville, C. F.; Tarry, H. A.; Whitehouse, Colin R.: Multilayer InSb diodes grown by molecular beam epitaxy for near-ambient temperature operation,V1361,238-244(1991)

Ashmontas, S. See Amosova, L. P.: V1440,406-413(1991)

Ashraf, N. See Artjushenko, Vjacheslav G.: V1420,157-168(1991)

Ashrit, Pandurang V.

; Bader, G.; Girouard, Fernand E.; Truong, Vo-Van: Electrochromic materials for smart window applications,V1401,119-129(1991)

Ashton, Robert C.

; Oz, Mehmet C.; Lontz, John F.; Lemole, Gerald M.: Preliminary stress/strain analysis of laser-soldered and -sutured vascular tissue,V1422,151-155(1991)

Ashurov, Mukhsin K.

; Gafitullina, Dilyara: Pattern recognition in autoradiography of optical materials,V1550,50-54(1991)

Ashwell, Geoffrey J.

; Dawnay, Emma J.; Kuczynski, Andrzej P.; Martin, Philip J.: Highest observed second harmonic intensity from a multilayered Langmuir-Blodgett film structure,V1361,589-598(1991)

Askadskij, A. A.

; Marshalkovich, A. S.; Latysh, E. G.; Goleneva, L. M.; Pastukhov, A. V.; Sidorova, G. I.: Synthesis and property research of birefringent polymers with predicted optical-mechanical parameters,V1554A,426-431(1991)

Askinazi, Joel

: Technology trends for high-performance windows,V1498,100-109(1991)

Askins, Charles G. See Jewell, John M.: V1441,38-44(1991)

Aslund, Nils R.

; Liljeborg, Anders; Oldmixon, E. H.; Ulfsparre, M.: Three-dimensional image processing method to compensate for depth-dependent light attenuation in images from a confocal microscope,V1450,329-337(1991)

Asmontas, Steponas

; Gradauskas, Jonas; Sirmulis, Edmundas: Hot carrier silicon phototransistor,V1512,131-134(1991)

Asom, Moses T. See Morgan, Robert A.: V1562,149-159(1991)

Asonen, Harry M. See Pessa, Markus: V1361,529-542(1991)

Aspell, Jennifer

; Bergano, Neal S.: Erbium-doped fiber amplifiers for future undersea transmission systems,V1373,188-196(1991)

Aspnes, David E.

: Analysis and control of semiconductor crystal growth with reflectance-difference spectroscopy and spectroellipsometry,V1361,551-561(1991)

Assaf, Haim See Boehm, Leah: V1535,171-181(1991)

Assal, F. T. See Sorbello, Robert M.: V1475,175-183(1991)

Assaleh, Khaled T.

; Zeevi, Yehoshua Y.; Gertner, Izidor: Realization of the Zak-Gabor representation of images,V1606,532-540(1991)

Assenheim, Harry M.

; Flasck, Richard A.; Lippert, Thomas M.; Bentz, Jerry; eds.: *Large Screen Projection, Avionic, and Helmet-Mounted Displays,*V1456(1991)

Astheimer, Peter See Felger, Wolfgang: V1459,222-231(1991)

Astic, Dominique

; Vigliano, Patrick; Autric, Michel L.; Inglesakis, Georges; Prat, Ch.: Spectroscopic measurements of plasmas temperature and density during high-energy pulsed laser-materials interaction processes,V1397,713-716(1991)

See Prat, Ch.: V1397,701-704(1991)

Astola, Jaakko

See Koskinen, Lasse: V1451,102-113(1991)

See Koskinen, Lasse: V1568,262-270(1991)

See Yin, Lin: V1451,216-227(1991)

See Yin, Lin: V1606,431-442(1991)

Astrahan, Melvin See Petrovich, Zbigniew: V1421,14-17(1991)

Asundi, Anand K.

: Defocused white light speckle method for object contouring,V1385,239-245(1991)

: Fringe analysis in moire interferometry,V1554B,472-480(1991)

: Novel grating methods for optical inspection,V1554B,708-715(1991)

: Projection moire using PSALM,V1554B,257-265(1991)

; Wong, C. M.: Contouring using gratings created on a LCD panel,V1400,80-85(1991)

Aswendt, Petra See Hoefling, Roland: V1508,135-142(1991)

Atack, Douglas See Stationwala, Mustafa I.: V1358,237-245(1991)

Atcha, Hashim

; Tatam, Ralph P.; Buckberry, Clive H.; Davies, Jeremy C.; Jones, Julian D.: Surface contouring using TV holography,V1504,221-232(1991)

; Tatam, Ralph P.: Use of laser diodes and monomode optical fiber in electronic speckle pattern interferometry,V1584,425-434(1991)

Atherton, Stephen J. See Woodruff, William H.: V1432,205-210(1991)

Atjezhev, Vladimir V.

; Belov, Sergey R.; Bukreev, Viacheslav S.; Vartapetov, Serge K.; Zhukov, Alexander N.; Ziganshin, Ilnur T.; Prokhorov, Alexander M.; Soldatkin, Alexey E.; Stepanov, Yuri D.: Excimer laser with sealed x-ray preionizer,V1503,197-199(1991)

Atkins, Frank B. See Goodenough, David J.: V1444,87-99(1991)

Atkins, Robert A.

See Lee, Chung E.: V1584,396-399(1991)

See Lee, Chung E.: V1588,110-116(1991)

Atkins, Robert M. See Krol, Denise M.: V1516,38-46(1991)

Atkinson, Dennis P.

; Bixler, Jay V.; Geraghty, Paul; Hailey, Charles J.; Klingmann, Jeffrey L.; Montesanti, Richard C.; Kahn, Steven M.; Paerels, F. B.: Design and analysis of the reflection grating arrays for the X-Ray Multi-Mirror Mission,V1343,530-541(1991)

See Montesanti, Richard C.: V1343,558-565(1991)

Atkinson, George H.

: Picosecond reaction dynamics in photosynthetic and proton pumping systems: picosecond time-resolved Raman spectroscopy of electronic and vibrationally excited states,V1403,50-58(1991)

Atkinson, John T.

See Harvey, David M.: V1400,86-93(1991)

See Wood, Christopher M.: V1332,301-313(1991)

Atlas, Les E.

; Fang, Jing; Loughlin, Patrick; Music, Wayne: Resolution advantages of quadratic signal processing,V1566,134-143(1991)

Atsuta, Toshio See Fujii, H.: V1397,213-220(1991)

Attanasio, Daniel V. See Smith, Terrance L.: V1512,92-100(1991)

Attino, Vito See Barbieri, Enrico: V1425,122-127(1991)

Attolico, Giovanni
; Caponetti, Laura; Chiaradia, Maria T.; Distante, Arcangelo; Stella, Ettore: Dense-depth map from multiple views,V1383,34-46(1991)
; Caponetti, Laura; Chiaradia, Maria T.; Distante, Arcangelo: Robot vision system for obstacle avoidance planning,V1388,50-61(1991)

Atuchin, V. See Savatinova, Ivanka T.: V1374,37-46(1991)

Au, Robert H.
: Atmospheric effects on laser systems,V1399,8-15(1991)
: Comparison of laser radar transmittance for the five atmospheric models,V1487,280-290(1991)

Aubert, Didier
; Kluge, Karl; Thorpe, Charles E.: Autonomous navigation of structured city roads,V1388,141-151(1991)

Aubert, Philippe
; Beck, Rasmus: Review of the performances of industrial 1.5-kW and 3-kW CO2 lasers, mobile on gantry or 2-D machines,V1502,52-59(1991)

Auborn, John E.
; Fuller, James M.; McCauley, Howard M.: Target detection using co-occurrence matrix segmentation and its hardware implementation,V1482,246-252(1991)
; Harris, William R.: Generic modular imaging IR signal processor,V1483,2-9(1991)
See McCauley, Howard M.: V1479,416-422(1991)

Aubrecht, I. See Miler, Miroslav: V1574,22-33(1991)

Aubry, Florent
; Bizais, Yves; Gibaud, Bernard; Forte, Anne-Marie; Chameroy, Virginie; Di Paola, Robert; Scarabin, Jean-Marie: Object-oriented model for medical image database,V1446,168-176(1991)
See Badaoui, Said: V1567,31-31(1991)
See Bizais, Yves: V1446,156-167(1991)

Auchter, S. See Koenig, Karsten: V1525,412-419(1991)

Audebert, Patrick See Geindre, Jean-Paul: V1502,311-318(1991)

Auge, Jacques C. See Marcerou, Jean-Francois: V1373,168-186(1991)

Augustyn, Thomas W.
; Henkel, Paul A.; eds.: *Airborne Reconnaissance XV*,V1538(1991)

Aukrust, Trond See Farrell, Edward J.: V1459,136-147(1991)

Aull, Brian F. See Mehanian, Courosh: V1469,275-280(1991)

Aumann, Hartmut H.
: Atmospheric infrared sounder on the Earth Observing System: in-orbit spectral calibration,V1540,176-186(1991)

Auriemma, R. A. See Bietry, Joseph R.: V1332,537-543(1991)

Aurin, Friedrich A.
: Optimization of the Seidel image errors by bending of lenses using a 4th-degree merit function,V1354,180-185(1991)
: Surface contributions of the wave aberrations up to the eighth degree,V1527,61-72(1991)

Austin, G. K. See Flanagan, Kathryn A.: V1549,395-407(1991)

Austin, J. C. See Hester, R. E.: V1403,15-21(1991)

Austin, Patrick D.
: Markets for marking systems,V1517,150-175(1991)

Austin, Roswell W.
; Duntley, Seibert Q.; Ensminger, Richard L.; Petzold, Theodore J.; Smith, Raymond C.: Underwater laser scanning system,V1537,57-73(1991)

Auteri, Joseph S.
; Oz, Mehmet C.; Sanchez, Juan A.; Bass, Lawrence S.; Jeevanandam, Valluvan; Williams, Mathew R.; Smith, Craig R.; Treat, Michael R.: Preliminary results of laser-assisted sealing of hand-sewn canine esophageal anastomoses,V1421,182-184(1991)
See Bass, Lawrence S.: V1421,164-168(1991)
See Libutti, Steven K.: V1421,169-172(1991)

Auternaud, Danielle
: ISOCAM: a camera for the ISO satellite optical bench development,V1488,64-72(1991)

Autric, Michel L.
See Astic, Dominique: V1397,713-716(1991)
See Prat, Ch.: V1397,701-704(1991)
See Tanguy, Mireille: V1487,172-183(1991)

Avdoshin, A. P. See Fakhrutdinov, I. H.: V1399,98-106(1991)

Avent, R. R.
; Conners, Richard W.: Automatic method for inspecting plywood shear samples,V1468,281-295(1991)

Averin, V. I.
; Bryukhnevitch, G. I.; Kolesov, G. V.; Lebedev, V. B.; Miller, V. A.; Saulevich, S. V.; Shulika, A. N.: Picosecond x-ray streak cameras,V1358,603-605(1991)
; Gus'kova, M. S.; Korzhenevich, Irina M.; Kolesov, G. V.; Lebedev, V. B.; Maranichenko, N. I.; Sobolev, A. A.: Image converter streak cameras with super-light-speed scanning,V1358,589-602(1991)

Avidor, Joel M. See Minnigh, Stephen W.: V1528,129-134(1991)

Aviles, Walter A.
; Hughes, T. W.; Everett, Hobart R.; Umeda, A. Y.; Martin, Stephen W.; Koyamatsu, A. H.; Solorzano, M.; Laird, Robin T.; McArthur, S. P.: Issues in mobile robotics: the unmanned ground vehicle program tele-operated vehicle,V1388,587-597(1991)

Aviram, Ari
; Shih, Kwang K.; Sachdev, Krishna: Thermal transfer printing with heat amplification,V1458,105-107(1991)

Avizonis, Petras V. See Bohn, Willy L.: V1397,235-238(1991)

Avram, J. See Rochkind, Simeone: V1428,52-58(1991)

Avset, B. S. See Sumner, Timothy J.: V1549,265-273(1991)

Axelrod, Timothy S.
See Lewis, Isabella T.: V1530,306-324(1991)
See Lewis, Isabella T.: V1478,2-12(1991)
See Searfus, Robert M.: V1473,95-101(1991)

Axmann, A. See Rosenzweig, Josef: V1362,168-178(1991)

Ayache, Nicholas See Monga, Olivier: V1570,382-390(1991)

Ayada, Haruki See Tanii, Jun: V1490,200-206(1991)

Aydin, M. See Hora, Heinrich: V1502,258-269(1991)

Aye, Tin M. See Yacoubian, Araz: V1559,403-409(1991)

Ayekavadi, Raj
; Yeh, C. S.; Butler, Jerome K.; Evans, Gary A.; Stabile, Paul J.; Rosen, Arye: Experimental verification of grating theory for surface-emitting structures,V1418,74-85(1991)

Ayer, Kevin W. See Veronin, Christopher P.: V1469,281-291(1991)

Ayme-Bellegarda, Eveline J.
; Habashy, Tarek M.; Bellegarda, Jerome R.: Automatic recognition of multidimensional objects buried in layered elastic background media,V1471,18-29(1991)

Aymerich, Francesco
; Ginesu, Francesco; Priolo, Pierluigi: Advanced materials characterization by means of moire techniques,V1554B,304-314(1991)

Ayral, Jean-Luc
; Montel, J.; Huignard, Jean-Pierre: Master oscillator-amplifier Nd:YAG laser with a SBS phase-conjugate mirror,V1500,81-92(1991)

Aytur, Orhan See Kumar, Prem: V1376,192-197(1991)

Azar, Massood T. See Beheim, Glenn: V1584,64-71(1991)

Azarenkov, Aleksey N.
; Altshuler, Grigori B.; Belashenkov, Nickolay R.; Inochkin, Mickle V.; Karasev, Viatcheslav B.; Kozlov, Sergey A.: Phase-locking and unstability of light waves in Raman-active crystals,V1409,154-164(1991)
; Altshuler, Grigori B.; Kozlov, Sergey A.: Self-action of supremely short light pulses in fibers,V1409,166-177(1991)

Azema, Alain See Gilabert, Alain: V1477,140-147(1991)

Azizoglu, Murat
; Humblet, Pierre A.: Performance of on-off modulated lightwave signals with phase noise,V1579,168-178(1991)

Azoulay, Moshe
; Gafni, Gabriella; Roth, Michael: Recent progress in the growth and characterization of large Ge single crystals for IR optics and microelectronics,V1535,35-45(1991)

Azoulay, Rozette See Bremond, Georges E.: V1361,732-743(1991)

Azran, A.
; Elfersi, U.; Greenfield, Ephraim: Hard carbon coating on ZnS,V1442,54-57(1991)

Azueta, Steven See Larimer, James O.: V1456,196-210(1991)

Azzam, Rasheed M.
ed.: *Selected Papers on Ellipsometry*,VMS27(1991)

Baars, Jan W.
See Esquivias, Ignacio: V1484,55-66(1991)
See Longshore, Randolph E.: V1484(1991)

Babaeva, N. A.
; Vas'kovsky, Yu. M.; Zhavoronkov, M. I.; Rovinsky, R. E.; Rjabinkina, V. A.: Thresholds of plasma arising under the pulse CO2 laser radiation interaction with an obstacle in air and energetic balance of the process,V1440,260-269(1991)

Babayan, Richard K.
; Roth, Robert A.: Transurethral ultrasound-guided laser-induced prostatectomy,V1421,42-44(1991)

Babbitt, Richard See Rachlin, Adam: V1477,101-114(1991)

Babcock, Carl P.
; Taylor, James W.; Sullivan, Monroe; Suh, Doowon; Plumb, Dean; Palmer, Shane R.; Berry, Amanda K.; Graziano, Karen A.; Fedynyshyn, Theodore H.: Process latitude measurements on chemically amplified resists exposed to synchrotron radiation,V1466,653-662(1991)

Babcock, Steven See Humphries, Stanley: V1407,512-523(1991)

Babeau, A. See Twieg, Robert J.: V1455,86-96(1991)

Babic, Ranko
: Aspects of reconfigurable neural networks,V1469,575-580(1991)

Babichenko, Sergey M.
; Poryvkina, Larisa: Laser remote sensing of natural water organics,V1492,319-323(1991)

Babin, Sergei A.
; Kuklin, A. E.: Comparison of high-current discharges with axial and transverse gas flow for UV ion lasers,V1397,589-592(1991)

Babin, Sergei V. See Guttmann, Peter: V1361,999-1010(1991)

Babushkin, A. V.
: Stable picosecond solid state YA103:Nd3+ laser for streak cameras dynamic evaluation,V1358,888-894(1991)
; Vorobiev, N. S.; Prokhorov, Alexander M.; Schelev, Mikhail Y.: Designing and application of solid state lasers for streak cameras calibration,V1346,410-417(1991)
See Bowley, David J.: V1358,550-555(1991)

Babyn, A. V. See Loshchenov, V. B.: V1420,271-281(1991)

Baca, Arnold See Hatze, Herbert: V1567,264-276(1991)

Baca, Michael J. See Cooper, Alfred W.: V1486,47-57(1991)

Bacal, Marthe
; Courteille, C.; Devynck, Pascal; Jones-King, Yolanda D.; Leroy, Renan; Stern, Raul A.: Negative deuterium ion thermal energy measurements in a volume ion source,V1407,605-609(1991)

Bacci, Mauro See Baldini, Francesco: V1368,210-217(1991)

Bach, Bernhard W. See Cotton, Daniel M.: V1549,313-318(1991)

Bachilo, S. M. See Bondarev, S. L.: V1403,497-499(1991)

Bachmann, Friedrich G.
: Large-scale industrial application for excimer lasers: via-hole-drilling by photoablation,V1361,500-511(1991); 1377,18-29(1991)

Bachnak, Rafic A.
: Linear modeling algorithm for tracking time-varying signals,V1481,12-22(1991)
; Yamout, Jihad S.: Interpolation of stereo data using Lagrangian polynomials,V1457,27-36(1991)

Bacis, Roger
; Bonnet, Jean C.; Bouvier, A. J.; Churassy, S.; Grozet, P.; Erba, B.; Georges, Eric; Jouvet, C.; Lamarre, J.; Louvet, Y.; Nota, M.; Pigache, Daniel R.; Ross, A. J.; Setra, M.: New emission spectra from chemically excited oxygen and potentiality as a visible chemical laser,V1397,173-176(1991)

Bacon, Fredrick See Krohn, David A.: V1420,126-135(1991)

Badaoui, Said
; Aubry, Florent: MIMS: a medical image management system,V1567,31-31(1991)

Badawi, K. F. See Grevey, D. F.: V1502,32-40(1991)

Badawy, Waheed A.
; El-Giar, Emad M.: Preparation, electrochemical, photoelectrochemical, and solid state characteristics of In-incorporated TiO2 thin films for solar energy applications,V1536,277-288(1991)

Baddiley, C. J. See Bradley, D. J.: V1488,186-195(1991)

Bader, G. See Ashrit, Pandurang V.: V1401,119-129(1991)

Bader, Samuel D. See Gluskin, Efim S.: V1548,56-68(1991)

Bader, Vladimir A.
See Ryabushkin, Oleg A.: V1362,75-79(1991)
See Ryabushkin, Oleg A.: V1505,67-74(1991)

Badler, Norman I.
See Alameldin, Tarek K.: V1381,217-225(1991)
See Alameldin, Tarek K.: V1386,112-120(1991)

Badziak, Jan
; Drazek, Wieslaw; Dubicki, Adam; Perlinski, Leszek: Influence of gas composition and pressure on the pulse duration of electron beam controlled discharge XeCl laser,V1397,81-84(1991)
; Dubicki, Adam; Piotrowski, Jan: Possibility of short-pulses generation in excimer lasers by self-injection,V1391,117-126(1991)
; Dzwigalski, Zygmunt: Influence of voltage pulse parameters on the beam current density distribution of large-aperture electron gun,V1391,250-253(1991)

Badzian, Andrzej R. See Zhu, Wei: V1534,230-242(1991)

Bae, Kyongtae T.
; Altschuler, Martin D.: Automatic acquisition of movement information by a knowledge-based recognition approach,V1380,108-115(1991)
; Altschuler, Martin D.: High-resolution fully 3-D mapping of human surfaces by laser array camera and data representations,V1380,171-178(1991)
See Chen, Chin-Tu: V1444,75-77(1991)
See Chou, Jin-Shin: V1396,45-50(1991)

Baer, C. See Meidan, Moshe: V1540,729-737(1991)

Baer, James W.
See Delamere, W. A.: V1479,31-40(1991)
See Delamere, W. A.: V1447,204-213(1991)

Baer, Thomas M.
See Kafka, James D.: V1373,140-149(1991)
See Morris, Patricia A.: V1561,104-111(1991)

Baert, Luc
; Ameye, Filip; Willemen, Patrick; Petrovich, Zbigniew: Transurethral microwave hyperthermia for benign prostatic hyperplasia: the Leuven clinical experience,V1421,18-29(1991)
; Berg, Roger; van Damme, B.; D'Hallewin, Mich A.; Johansson, Jonas; Svanberg, Katarina; Svanberg, Sune: Clinical fluorescence diagnosis of human bladder carcinoma following low-dose photofrin injection,V1525,385-390(1991)
See Andersson-Engels, Stefan: V1426,31-43(1991)
See Petrovich, Zbigniew: V1421,14-17(1991)

Baets, Roel G. See Moehlmann, Gustaaf R.: V1560,426-433(1991)

Baettig, Rainer K.
; Toms, Dennis J.; Guest, Clark C.: Fabrication of computer-generated holograms for the interconnection of single-mode devices,V1563,93-102(1991)

Bagdasarov, V. H. See Manenkov, Alexander A.: V1420,254-258(1991)

Bagg, John C.
; Ohlhaber, Jack: Fiber optics in a broadband bus architecture,V1369,126-137(1991)

Bagley, K. A. See Woodruff, William H.: V1432,205-210(1991)

Bagley, Steven C.
; Kopec, Gary E.: Applications of text-image editing,V1460,71-79(1991)

Baglioni, Pietro See Angrilli, Francesco: V1482,26-39(1991)

Bagot, K. H.
: Optical mapping instrument,V1490,126-132(1991)

Bahl, I. See Willems, David: V1475,55-61(1991)

Bahners, Thomas
; Kesting, Wolfgang; Schollmeyer, Eckhard: Controlled structuring of polymer surfaces by UV laser irradiation,V1503,206-214(1991)

Bahnsen, Axel See Budtz-Jorgensen, Carl: V1549,429-437(1991)

Bahram-pour, A. R.
; Mehdizadeh, E.; Bolorizadeh, M. A.; Shojaey, M.: Precise uncoupling theory to study gain of gas-dynamic laser,V1397,539-542(1991)

Bahuguna, Ramendra D.
; Williams, Gareth T.; Pour, I. K.; Raman, R.: Study of microbial growth II: by holographic interferomery,V1332,805-807(1991)
See Williams, Gareth T.: V1332,802-804(1991)

Baier, M. See Wallraff, Gregory M.: V1466,211-217(1991)

Baik, Ki-Ho See Goethals, Anne-Marie: V1466,604-615(1991)

Bailey, Robert B. See Kozlowski, Lester J.: V1540,250-261(1991)

Bailey, Vernon L.
; Corcoran, Patrick; Edighoffer, J. A.; Fockler, J.; Lidestri, Joseph P.; Putnam, Sidney D.; Tiefenback, Michael G.: Status of the proof-of-concept experiment for the spiral line induction accelerator,V1407,400-406(1991)

Bailly, Michel
; Perez, Eric: Pointing, acquisition, and tracking system of the European SILEX program: a major technological step for intersatellite optical communication,V1417,142-157(1991)

Bailly-Salins, Rene See Key, Michael H.: V1397,9-17(1991)

Bains, Narinder
; David, Frank: Machine vision inspection of fluorescent lamps,V1386,232-242(1991)

Bair, Wyeth
; Koch, Christof: Real-time motion detection using an analog VLSI zero-crossing chip,V1473,59-65(1991)

Baird, Bill
; Freeman, Walter J.; Eeckman, Frank H.; Yao, Yong: Applications of chaotic neurodynamics in pattern recognition,V1469,12-23(1991)

Baird, George S.
ed.: *Thermosense XIII,*V1467(1991)
; Mack, Russell T.: Status report on thermographer certification,V1467,63-63(1991)

Baixeras, Joseph M.
; Hosseini Teherani, Ferechteh; Kreisler, Alain J.; Straboni, Alain; Barla, Kathy: Thin films of YBaCuO for electronic applications,V1362,117-126(1991)
See Hosseini Teherani, Ferechteh: V1362,921-929(1991)
See Naumaan, Ahmed: V1512(1991)

Bajcsy, Ruzena R. See Gee, Jim C.: V1445,226-234(1991)

Bajkov, A. V. See Shestakov, S. D.: V1440,423-435(1991)

Bajuk, Louis J.
See Chew, Wilbert: V1477,95-100(1991)
See Foote, Marc C.: V1477,192-196(1991)
See Hunt, Brian D.: V1394,89-95(1991)

Baker, Glenn S. See Stetz, Mark L.: V1425,55-62(1991)

Baker, Howard J.
; Laidler, Ian: Spatial and electrical characteristics of capacitively ballasted rf laser discharges,V1397,545-548(1991)

Baker, J. D. See Eton, Darwin: V1425,182-187(1991)

Baker, Mark E. See Pomeroy, Robert S.: V1439,60-65(1991)

Baker, Phillip C.
See Friedlander, Miles H.: V1423,62-69(1991); 1429,229-236(1991)
See Hoover, Richard B.: V1426,84-96(1991)
See Hoover, Richard B.: V1435,338-351(1991)
See Hoover, Richard B.: V1343,189-202(1991)
See Walker, Arthur B.: V1494,320-333(1991)

Baker, R. A. See Henshall, Gordon D.: V1418,286-291(1991)

Baker, Richard L. See Tse, Yi-tong: V1452,468-479(1991)

Baker, Thomas E.
: Artificial neural network and image processing using the Adaptive Solutions' architecture,V1452,502-511(1991)

Baker, Walter L.
; Farrell, Jay A.: Connectionist learning systems for control,V1382,181-198(1991)

Bakhtazad, A.
See Rahnavard, Mohammad H.: V1385,123-130(1991)
See Rahnavard, Mohammad H.: V1562,327-337(1991)

Bakker, Albert R.
See Kouwenberg, Jef M.: V1446,357-361(1991)
See Lodder, Herman: V1446,227-233(1991)
See Stut, W. J.: V1446,396-404(1991)
See van Gennip, Elisabeth M.: V1446,442-450(1991)

Bala, Krishna
; Stern, Thomas E.: Topologies for linear lightwave networks,V1579,62-73(1991)

Balabaev, N. K.
; Lakhno, V. D.: Spectral properties of the polaron model of a protein,V1403,478-486(1991)
; Lemak, A. S.; Fushman, D. A.; Mironova, Yu. V.: Simulation of spin label behavior on a model surface,V1402,53-69(1991)

Balageas, Daniel L.
; Boscher, Daniel M.; Deom, Alain A.; Enguehard, Francis; Noirot, Laurence: Photoacoustic microscopy by photodeformation applied to the determination of thermal diffusivity,V1467,278-289(1991)

Balaji, Sridhar See Salari, Ezzatollah: V1384,115-123(1991)

Balart, Rogelio
: Matrix reformulation of the Gabor transform,V1565,447-457(1991)

Balasubramanian, Kunjithapa
; Le, Tam V.; Guenther, Karl H.; Kumar, Vas: Broadband, antireflection coating designs for large-aperture infrared windows,V1485,245-253(1991)

; Richmond, Jeff; Hu, X. Q.; Guenther, Karl H.: Reactive low-voltage ion plating of optical coatings on ophthalmic lenses,V1529,106-114(1991)
See Guenther, Karl H.: V1485,240-244(1991)

Balasubramanian, Raja
; Allebach, Jan P.: New approach to palette selection for color images,V1453,58-69(1991)

Balboni, Edmund J.
See Fitzmartin, Daniel J.: V1476,56-62(1991)
See Fitzmartin, Daniel J.: V1371,78-86(1991)

Balch, Kris S.
: Fourth-generation motion analyzer,V1358,373-397(1991)
: New image-capturing techniques for a high-speed motion analyzer,V1346,2-23(1991)
: Techniques for capturing over 10,000 images/second with intensified imagers,V1358,358-372(1991)

Balchum, Oscar J. See Profio, A. E.: V1426,44-46(1991)

Baldini, Francesco
: Recent progress in fiber optic pH sensing,V1368,184-190(1991)
; Bacci, Mauro; Bracci, Susanna: Analysis of acid-base indicators covalently bound on glass supports,V1368,210-217(1991)
; Falciai, Riccardo; Bechi, Paolo; Cosi, Franco; Bini, Andrea; Milanesi, Francesco: Portable and very inexpensive optical fiber sensor for enterogastric reflux detection,V1510,58-62(1991)
See Falciai, Riccardo: V1572,424-427(1991)

Baldini, S. E.
; Nowakowski, Edward; Smith, Herb G.; Friebele, E. J.; Putnam, Martin A.; Rogowski, Robert S.; Melvin, Leland D.; Claus, Richard O.; Tran, Tuan A.; Holben, Milford S.: Cooperative implementation of a high-temperature acoustic sensor,V1588,125-131(1991)

Baldis, Hector A.
ed.: *Short-Pulse High-Intensity Lasers and Applications,*V1413(1991)
See Labaune, C.: V1413,138-143(1991)
See Tallents, Gregory J.: V1413,70-76(1991)

Baldwin, Christopher
: Physical-connection compliance testing for FDDI,V1364,120-129(1991)

Baldwin, David L.
; Sokolov, Vladimir; Bauhahn, Paul E.: Laser link performance improvements with wideband microwave impedance matching,V1476,46-55(1991)

Baldwin, Jack See Forbes, Fred F.: V1532,146-160(1991)

Bales, Maurice
; Boulton, Herbert: Introducing multiple-dynamic-windows in thermal imaging,V1467,195-206(1991)

Balestra, Chet L.
See Haake, John M.: V1418,298-308(1991)
See Udd, Eric: V1418,134-152(1991)

Baleva, Mitra I.
; Momchilova, Maia M.: Phonons in PbTe and PbTe:Cr strained layers,V1361,712-722(1991)

Balfour, L. S.
See Ben-Shalom, Ami: V1486,238-257(1991)
See Engel, Michael Y.: V1442,298-307(1991)

Balick, Lee K.
; Golanics, Charles J.; Shines, Janet E.; Biggar, Stuart F.; Slater, Philip N.: In-flight calibration of a helicopter-mounted Daedalus multispectral scanner,V1493,215-223(1991)

Balk, P. See Geurts, Jean: V1361,744-750(1991)

Balkan, N.
; Ridley, Brian K.: Hot electron instabilities and light emission in GaAs quantum wells,V1361,927-934(1991)
See Gupta, Rita P.: V1362,798-803(1991)

Balkarey, Yu. I.
; Evtikhov, M. G.; Elinson, Matvey I.: Autowave media and neural networks,V1621,238-249(1991)

Ball, K.
; Burt, D. J.; Smith, Graham W.: High-speed readout CCDs,V1358,409-420(1991)

Ballangrud, Ase
; Jaeger, Tycho C.; Wang, Gunnar: High-resolution imaging interferometer,V1521,89-96(1991)

Ballentine, Paul H. See Mallory, Derek S.: V1477,66-76(1991)

Ballik, Edward A.
; Wan, William: Development and analysis of a simple model for an IR sensor,V1488,249-256(1991)

Ballini, Jean-Pierre See Angelov, D.: V1403,575-577(1991)

Balog, P. See Adamec, F.: V1402,82-84(1991)

Balooch, M. See Tesar, Aleta A.: V1441,228-236(1991)

Baloshin, Yu. A.
; Yurevich, V. I.; Sud'enkov, Yu. V.: Slow thermodeformation of metals with fast laser heating,V1440,71-77(1991)

Balram, Nikhil
; Moura, Jose M.: Rapid enhancement and compression of image data,V1606,374-385(1991)

Baltazar, Inmaculada C.
; Mena, Manolo G.: Variability in thickness measurements using x-ray fluorescence technique,V1392,670-680(1991)

Baltazart, Didier See Trigano, Philippe: V1468,866-874(1991)

Baltes, Henry P. See Patrick, Roger: V1392,506-513(1991)

Bamberger, Roberto H.
; Smith, Mark J.: Comparison of directionally based and nondirectionally based subband image coders,V1605,757-768(1991)

Ban, Vladimir S. See Olsen, Gregory H.: V1540,596-605(1991)

Banash, Mark A.
; Caldwell, J. B.; Che, Tessie M.; Mininni, Robert M.; Soskey, Paul R.; Warden, Victor N.; Pope, Edward J.: Gradient-index fiber-optic preforms by a sol-gel method,V1590,8-13(1991)

Banda, Carolyn P. See Prevost, Michael P.: V1459,58-66(1991)

Bandy, Donna K.
; Hunter, L. W.; Jones, Darlena J.: Intercomparison of homogeneous laser models with transverse effects,V1497,142-152(1991)

Banerjee, Partha P. See Sarkar, Tapan K.: V1378,34-42(1991)

Banerjee, Pranab K. See Singh, Anjali: V1366,184-190(1991)

Banerjee, Sanjay
See Fowler, Burt: V1598,108-117(1991)
See Lian, S.: V1598,98-107(1991)

Banfi, Giampiero P.
; Degiorgio, Vittorio; Rennie, A. R.; Righini, Giancarlo C.: Measurement of the sizes of the semiconductor crystallites in colored glasses through neutron scattering,V1361,874-880(1991)

Bangert, H. See Wang, H. F.: V1519,405-410(1991)

Banh, Nam D. See Braunreiter, Dennis C.: V1481,73-83(1991)

Banholzer, William
; Fulghum, Stephen: Properties of diamonds with varying isotopic composition,V1501,163-176(1991)

Baniel, P. See Lopez, Adolphe: V1590,191-202(1991)

Banik, S. N. See Ghatak, Kamakhya P.: V1409,240-257(1991)

Banish, Bryan See Albagli, Douglas: V1441,146-153(1991)

Banish, Michele R.
; Clark, Rodney L.; Kathman, Alan D.; Lawson, Shelah M.: Simulation of optical diagnostics for crystal growth: models and results,V1557,209-221(1991)
See Lawson, Shelah M.: V1488,268-278(1991)

Banks, Bruce A.
; Rutledge, Sharon K.; Hotes, Deborah: High-emittance surfaces for high-temperature space radiator applications,V1330,66-77(1991)

Bankston, Cheryl D. See Wingo, Dennis R.: V1495,123-133(1991)

Banvillet, Henri
; Gil, E.; Vasson, A. M.; Cadoret, R.; Tabata, A.; Benyattou, Taha; Guillot, Gerard: Epitaxial growth and photoluminescence investigations of InP/InAs quantum well grown by hydride vapor phase epitaxy,V1361,972-979(1991)

Banyasz, Istvan
: Resolution problems in holography,V1574,282-293(1991)

Bao, Jia S.
See Boardman, A. D.: V1519,609-615(1991)
See Boardman, A. D.: V1519,597-604(1991)
See Wang, Qi: V1519,589-596(1991)
See Zhou, Shi P.: V1519,793-799(1991)

Bao, Liangbi
; Chen, Fuyao; Wu, Shixiong; Xu, Jiangtong; Guan, Zhilian: Studies on laser dynamic precision measurement of fine-wire diameters,V1332,862-867(1991)

Bao, Xiu Y. See Zhao, Shu L.: V1519,275-280(1991)

Bao, Xue-Cheng See Liu, Jian: V1540,744-755(1991)

Bao, Y. See Verber, Carl M.: V1374,68-73(1991)

Baptiste, Didier See Bocquet, Jean-Claud: V1588,210-217(1991)

Bar, Doron See Rapoport, Eliezer: V1442,383-391(1991)

Bar, I.
; Ben-Porat, T.; Cohen, A.; Heflinger, Dov; Miron, G.; Tzuk, Y.; Rosenwaks, Salman: Studies of short-wavelength chemical lasers: enhanced emission of Pb atoms following detonation of lead azide via a supersonic nozzle,V1397,169-172(1991)

Bar-Joseph, Dan
: 2.014-micron Cr;Tm:YAG: optimization of doping concentration for flash lamp operation,V1410,142-147(1991)

Barabanov, Alexander J. See Tuchin, Valery V.: V1422,85-96(1991)

Barabash, Rostislav D. See McCaughan, James S.: V1426,279-287(1991)

Barachevsky, Valery A.
: Photoanisotropic polymeric media and their application in optical devices,V1559,184-193(1991)
; Rot, A. S.; Zaks, I. N.: Light-sensitive organic media for optical discs,V1621,33-44(1991)
See Mikaelian, Andrei L.: V1559,246-257(1991)

Barad, Herbert S. See Zhang, Peng: V1569,398-409(1991)

Baradaran, T. See Karimi, B.: V1396,226-236(1991)

Barakat, M. A. See Delecki, Z. A.: V1526,157-167(1991)

Baran, Robert H.
; Coughlin, James P.: Neural network for passive acoustic discrimination between surface and submarine targets,V1471,164-176(1991)

Baranauskas, Vitor See Madarazo, R.: V1393,270-277(1991)

Barani, Gianni
; Tofani, Alessandro: Comparison of some algorithms commonly used in IR pyrometry: a computer simulation,V1467,458-468(1991)

Baraniuk, Richard G. See Jones, Douglas L.: V1566,202-206(1991)

Baranov, A. N. See Artsimovich, M. V.: V1397,729-733(1991)

Baranov, A. V.
; Nabiev, I. R.: Surface-enhanced resonance hyper-Raman spectra of bacteriorhodopsin adsorbed on silver colloids,V1403,128-131(1991)

Baranov, Alexej N.
See Mikhailova, Maya P.: V1361,674-685(1991)
See Titkov, A. N.: V1361,669-673(1991)
See Yakovlev, Yurii P.: V1510,170-177(1991)

Baranov, D. V.
; Zolotov, Evgeny M.; Pelekhaty, V. M.; Tavlykaev, R. F.: Phase-bias tuning and extinction ratio improvement of Mach-Zehnder interferometer,V1583,389-394(1991)

Baranov, Igor M.
; Kutt, R. N.; Nikitina, Irina P.: Study of structural imperfections in epitaxial beta-SiC layers by method of x-ray differential diffractometry,V1361,1110-1115(1991)

Baras, John S.
; Frantzeskakis, Emmanuil N.: Bayesian matching technique for detecting simple objects in heavily noisy environment,V1569,341-353(1991)
See Sidiropoulos, N. D.: V1569,356-366(1991)

Barba, Dominique
; Hanen, Jose: Use of a human visual model in subband coding of color video signal with adaptive chrominance signal vector quantization,V1605,408-419(1991)
See Lallauret, Fabrice: V1605,26-36(1991)

Barba, Joseph See Wali, Rahman: V1606,665-674(1991)

Barbaric, Zoran L. See Hayrapetian, Alek: V1446,243-247(1991)

Barbarossa, Giovanni
; Laybourn, Peter J.: High-silica cascaded three-waveguide couplers for wideband filtering by flame hydrolysis on Si,V1583,122-128(1991)
; Laybourn, Peter J.: High-silica low-loss three-waveguide couplers on Si by flame hydrolysis deposition,V1513,37-43(1991)

Barbee, Troy W.
See Baron, Alfred Q.: V1343,84-94(1991)
See DeForest, Craig E.: V1343,404-414(1991); 1343,404-414(1991)
See Gluskin, Efim S.: V1548,56-68(1991)
See Hoover, Richard B.: V1426,84-96(1991)
See Hoover, Richard B.: V1435,338-351(1991)
See Lindblom, Joakim F.: V1343,544-557(1991)
See Walker, Arthur B.: V1343,334-347(1991)
See Walker, Arthur B.: V1343,415-427(1991)
See Walker, Arthur B.: V1494,320-333(1991)
See Walker, Arthur B.: V1343,319-333(1991)
See Windt, David L.: V1343,274-282(1991)

Barber, Patrick G. See Fripp, Archibald L.: V1557,236-244(1991)

Barber, Peter W.
    See Chang, Richard K.: V1497,2-13(1991)
    See Hill, Steven C.: V1497,16-27(1991)

Barbero, Marzio
    ; Cucchi, Silvio; Muratori, Mario: Performance of an HDTV codec adopting
        transform and motion compensation techniques,V1567,566-577(1991)

Barbero, V. See Brenci, Massimo: V1572,318-324(1991)

Barbier, E. See Letartre, Xavier: V1362,778-789(1991)

Barbier, Pierre R. See Perlmutter, S. H.: V1562,74-84(1991)

Barbieri, Enrico
    ; Tanganelli, Pietro; Taddei, Giuseppe; Perbellini, Antonio; Attino, Vito;
        Destro, Gianni; Zardini, Piero: Restenosis after hot-tip laser-balloon
        angioplasty: histologic evaluation of the samples removed by Simpson
        atherectomy,V1425,122-127(1991)

Barbieri, Fabio See Ferrario, Mario: V1450,108-117(1991)

Barbini, Roberto
    ; Fantoni, Roberta; Palucci, Antonio; Ribezzo, Sergio; van der Steen,
        Hendricus J.: Spectral and time-resolved measurements of pollutants on
        water surface by an XeCl laser fluorosensor,V1503,363-374(1991)

Barbour, Randall L.
    ; Graber, Harry L.; Aronson, Raphael; Lubowsky, Jack: Imaging of subsurface
        regions of random media by remote sensing,V1431,52-62(1991);
        1431,192-203(1991)

Barchewitz, Robert J.
    See Berrouane, H.: V1343,428-436(1991)
    See Boher, Pierre: V1345,165-179(1991)
    See Khan Malek, Chantal: V1343,56-61(1991)

Bard, Allen J. See Bartels, Keith A.: V1450,30-39(1991)

Barden, Raimund
    ; Mertelmeier, Thomas; Traupe, U.: High-resolution MS-type Saticon pick-up
        tube with optimized electron optical properties,V1449,136-147(1991)

Bardin, R. K.
    ; Libby, Vibeke: Architecture and performance of a hardware collision-
        checking accelerator,V1566,394-404(1991)
    See Libby, Vibeke: V1470,59-64(1991)

Bardon, Didier
    : Adaptive typography for dynamic mapping environments,V1460,126-
        139(1991)

Barel, Alain R. See Schrever, Koen: V1572,107-112(1991)

Barevadia, Gordon R.
    See Lakdawala, Vishnu K.: V1378,259-270(1991)
    See Schoenbach, Karl H.: V1362,428-435(1991)

Barga, Roger S.
    ; Friesel, Mark A.; Meador, Jack L.: Source location of acoustic emissions
        from atmospheric leakage using neural networks,V1469,602-611(1991)

Bargon, Joachim See Baumann, Reinhard: V1463,638-643(1991)

Barham, Paul T.
    ; McAllister, David F.: Comparison of stereoscopic cursors for the interactive
        manipulation of B-splines,V1457,18-26(1991)

Barholm-Hansen, Claus
    ; Levinsen, Mogens T.: HTC microbolometer for far-infrared
        detection,V1512,218-225(1991)

Barilli, Marco See Ferrari, Marco: V1431,276-283(1991)

Barillot, Christian
    ; Lachmann, F.; Gibaud, Bernard; Scarabin, Jean-Marie: Three-dimensional
        display of MRI data in neurosurgery: segmentation and rendering
        aspects,V1445,54-65(1991)
    See Fresne, Francoise: V1444,26-36(1991)

Barillot, Marc
    ; Jacquot, Pierre M.; Di Chirico, Giuseppe: Computer-aided moire strain
        analysis on thermoplastic models at critical temperature,V1554A,867-
        878(1991)

Barinova, E. S. See Ryabova, R. V.: V1238,166-170(1991)

Barisas, B. G.
    ; Rahman, N. A.; Londo, T. R.; Herman, J. R.; Roess, Debrah A.: Advances in
        polarized fluorescence depletion measurement of cell membrane protein
        rotation,V1432,52-63(1991)

Bariya, A. J. See McVittie, James P.: V1392,126-138(1991)

Barker, Steven J.
    : Clinical applications of pulmonary artery oximetry,V1420,22-28(1991)

Barkhoudarian, S. See Delcher, Ray C.: V1369,114-120(1991)

Barkhouser, Robert H. See Clampin, Mark: V1542,165-174(1991)

Barkstrom, Bruce R. See Manalo, Natividad D.: V1521,106-116(1991)

Barla, Kathy See Baixeras, Joseph M.: V1362,117-126(1991)

Barletta, W. A. See Dodd, James W.: V1407,467-473(1991)

Barlow, Jesse L.
    ; Zaccone, Richard J.: Error analysis in unnormalized floating point
        arithmetic,V1566,286-294(1991)

Barmashenko, B. D.
    ; Kochelap, Viatcheslav A.: Possibility of long population inversion in active
        media for IR chemical lasers,V1397,303-307(1991)

Barmenkov, Yuri O.
    See Kozhevnikov, Nikolai M.: V1584,387-395(1991)
    See Kozhevnikov, Nikolai M.: V1584,138-144(1991)
    See Kozhevnikov, Nikolai M.: V1507,517-524(1991)

Barnabo, Geoffrey A. See Jensen, Stephen C.: V1369,87-95(1991)

Barnard, Kenneth J.
    ; Boreman, Glenn D.; Plogstedt, Allen E.; Anderson, Barry K.: Sine wave
        measurements of SPRITE detector MTF,V1488,426-431(1991)
    See Anderson, Barry K.: V1488,416-425(1991)

Barnard, Steven M.
    ; Walt, David R.: Antibody-based fiber optic sensors for environmental and
        process-control applications,V1368,86-92(1991)

Barnault, B.
    ; Barraud, Roger; Forestier, L.; Georges, Eric; Louvet, Y.; Mouthon, Alain;
        Ory, M.; Pigache, Daniel R.: High-power chemical oxygen-iodine
        laser,V1397,231-234(1991)
    ; Joly, V.; Pigache, Daniel R.: Population of high-vibrational levels of the
        iodine ground state in its dissociation process by singlet
        oxygen,V1397,257-260(1991)

Barner, Jim See Ghosh, Anjan K.: V1371,170-181(1991)

Barnes, Charles E.
    ; Dorsky, Leonard; Johnston, Alan R.; Bergman, Larry A.; Stassinopoulos, E.:
        Overview of fiber optics in the natural space environment,V1366,9-
        16(1991)

Barnes, Christopher See Kossentini, Faouzi: V1452,383-394(1991)

Barnes, Donald L. See Strikwerda, Thomas E.: V1478,13-23(1991)

Barnes, Nigel
    ; Healey, Peter; McKee, Paul; O'Neill, Alan; Rejman-Greene, Marek A.;
        Scott, Geoff; Smith, David W.; Webb, Roderick P.; Wood, David: High-
        speed optical interconnects for parallel processing and neural
        networks,V1389,477-483(1991)

Barnes, Norman P. See Allario, Frank: V1492,92-110(1991)

Barnes, Scott L. See Schimert, Thomas R.: V1484,19-30(1991)

Barnes, Thomas H. See Muramatsu, Mikiya: V1332,792-797(1991)

Barnett, Brett See Gorenflo, Ronald L.: V1346,42-53(1991)

Barnett, Christopher A. See Phillips, Nicholas J.: V1544,10-21(1991)

Barney, Jesus J. See Hodor, James R.: V1540,331-337(1991)

Barniv, Yair
    : Error analysis of combined stereo/optical-flow passive ranging,V1479,259-
        267(1991)

Barnouin, O. See Miley, George H.: V1441,16-26(1991)

Barnsley, Peter E.
    ; Marshall, Ian W.; Fiddyment, Phillip J.; Robertson, Michael J.: Absorptive
        nonlinear semiconductor amplifiers for fast optical switching,V1378,116-
        126(1991)

Baron, Alfred Q.
    ; Barbee, Troy W.; Brown, George S.: Multilayer monochromator for
        synchrotron radiation angiography,V1343,84-94(1991)

Barone, Fabrizio
    ; Bernini, Umberto; Conti, M.; Del Guerra, Alberto; Di Fiore, Luciano;
        Maddalena, L.; Milano, L.; Russo, G.; Russo, Paolo: Fiber optic
        interferometric x-ray dosimeter,V1584,304-307(1991)

Baronnet, Jean M. See Sontag, Andre: V1397,287-290(1991)

Barouch, Eytan
    ; Hollerbach, Uwe; Orszag, Steven A.; Bradie, Brian D.; Peckerar, Martin:
        Process latitudes in projection printing,V1465,254-262(1991)
    ; Hollerbach, Uwe; Orszag, Steven A.; Allen, Mary T.; Calabrese, Gary C.:
        Simulation of an advanced negative i-line photoresist,V1463,336-
        344(1991)
    ; Hollerbach, Uwe; Orszag, Steven A.; Szmanda, Charles R.; Thackeray,
        James W.: Simulations of bar printing over a MOSFET device using i-line
        and deep-UV resists,V1463,464-474(1991)

Barr, Donald E. See Seraphim, Donald: V1389,39-54(1991); 1390,39-54(1991)

Barr, Hugh
: Selective tumor destruction with photodynamic therapy: exploitation of photodynamic thresholds, V1525,331-340(1991)

Barranger, John P.
: Two-dimensional surface strain measurement based on a variation of Yamaguchi's laser-speckle strain gauge, V1332,757-766(1991)

Barrat, Jean-Pierre See Boyer, Corinne: V1483,77-91(1991)

Barraud, Roger
See Barnault, B.: V1397,231-234(1991)
See Georges, Eric: V1397,243-246(1991)

Barreiros, Manuel A. See Mulero, Manuel A.: V1540,519-526(1991)

Barrera, A. See Fork, David K.: V1394,202-202(1991)

Barrera, Frederick J. See Sardar, Dhiraj K.: V1427,374-380(1991)

Barresi, Giangrande
; Soddu, Claudio; Rondinelli, Giuseppe; Caporicci, Lucio; Loria, A.: Autonomous navigation of small co-orbiting satellites using C/A GPS code, V1495,246-258(1991)
See Rondinelli, Giuseppe: V1495,19-31(1991)

Barrett, Eamon B.
; Brill, Michael H.; Haag, Nils N.; Payton, Paul M.: Linear resection, intersection, and perspective-independent model matching in photogrammetry: theory, V1567,142-169(1991)
; Pearson, James J.; eds.: *Image Understanding and the Man-Machine Interface III*, V1472(1991)
See Digumarthi, Ramji V.: V1472,128-138(1991)
See Payton, Paul M.: V1406,58-71(1991)

Barrett, K. B. See Deason, Vance A.: V1332,868-876(1991)

Barrett, William A. See Anderson, Kelly L.: V1384,270-279(1991)

Barrick, Mark W. See Kuyel, Birol: V1463,646-665(1991)

Barron, L. D.
; Gargaro, A. R.; Hecht, L.; Wen, Z. Q.; Hug, W.: Vibrational Raman optical activity of biological molecules, V1403,66-75(1991)

Barry, Patrick E. See Maienschein, Jon L.: V1343,477-484(1991)

Barsalou, Norman See Longbotham, Harold G.: V1451,36-47(1991)

Barski, S. See Park, Yong K.: V1372,219-227(1991)

Barsotti, John B. See Ning, Ruola: V1443,236-249(1991)

Barsoum, Roshdy S. See Wang, Y. Y.: V1554B,344-356(1991)

Bart, Mischa
; Buurman, Johannes; Duin, Robert P.: Learning procedure for the recognition of 3-D objects from 2-D images, V1381,66-77(1991)

Bartels, Keith A.
; Lee, Chongmok; Bovik, Alan C.; Bard, Allen J.: Digital restoration of scanning electrochemical microscope images, V1450,30-39(1991)

Bartels, Kenneth E.
: Laser surgery for selected small animal soft-tissue conditions, V1424,164-170(1991)

Barten, Peter G.
: Evaluation of the effect of noise on subjective image quality, V1453,2-15(1991)

Barth, Erhardt
; Caelli, Terry M.; Zetzsche, Christoph: Efficient visual representation and reconstruction from generalized curvature measures, V1570,86-95(1991)
See Zetzsche, Christoph: V1570,337-350(1991)

Barthes, Jean-Paul A. See Gouveia, Feliz A.: V1468,516-523(1991)

Barthol, Peter See Wong, Wallace K.: V1530,86-103(1991)

Barticevic, Zdenka See Pacheco, Monica: V1361,819-826(1991)

Bartlett, R. See Yu, Xiaolin: V1396,56-58(1991)

Bartlett, Sandra L.
; Jain, Ramesh C.: Depth determination using complex logarithmic mapping, V1382,3-13(1991)

Bartlett, Steven C.
; Farahi, Faramarz; Jackson, David A.: Common-path optical fiber heterodyne interferometric current sensor, V1504,247-250(1991)

Bartley, James A.
; Goltsos, William: Laser ablation of refractive micro-optic lenslet arrays, V1544,140-145(1991)

Bartolini, R. See Ciocci, Franco: V1501,154-162(1991)

Barton, James S.
See Kidd, S. R.: V1504,180-190(1991)
See Morgan, Russell D.: V1504,118-124(1991)
See Zheng, S. X.: V1572,359-364(1991)

Barton, John P.
; Alexander, Dennis R.: Electromagnetic field calculations for a tightly focused laser beam incident upon a microdroplet: applications to nonlinear optics, V1497,64-77(1991)
See Alexander, Dennis R.: V1497,90-97(1991)

Bartuch, Ulrike See Braeuer, Andreas: V1559,470-478(1991)

Baruffolo, A. See Bortoletto, Favio: V1542,225-235(1991)

Baryshev, M. V. See Loshchenov, V. B.: V1420,271-281(1991)

Basart, John P. See Chackalackal, Mathew S.: V1568,347-356(1991)

Basehore, Paul
; Yestrebsky, Joseph T.: Innovative architectural and theoretical considerations yield efficient fuzzy logic controller VLSI design, V1470,190-196(1991)

Basford, J. A. See Egert, Charles M.: V1330,178-185(1991)

Basiev, Tasoltan T.
; Zverev, Peter G.; Mirov, Sergey B.; Pal, Suranjan: Phase conjugation in LiF and NaF color center crystals, V1500,65-71(1991)

Basir, O. A. See Shen, Helen C.: V1383,403-408(1991)

Baskurt, Atilla
; Magnin, Isabelle E.; Bremond, Alain; Charvet, Pierre Y.: Adaptive coding method of x-ray mammograms, V1444,240-249(1991)
; Peyrin, Francoise; Min, Zhu-Yue; Goutte, Robert: Local spectrum analysis of medical images, V1445,485-495(1991)

Bass, Lawrence S.
; Moazami, Nader; Pocsidio, Joanne O.; Oz, Mehmet C.; LoGerfo, Paul; Treat, Michael R.: Electrophoretic mobility patterns of collagen following laser welding, V1422,123-127(1991)
; Oz, Mehmet C.; Auteri, Joseph S.; Williams, Mathew R.; Rosen, Jeffrey; Libutti, Steven K.; Eaton, Alexander M.; Lontz, John F.; Nowygrod, Roman; Treat, Michael R.: Laparoscopic applications of laser-activated tissue glues, V1421,164-168(1991)
See Auteri, Joseph S.: V1421,182-184(1991)
See Eaton, Alexander M.: V1423,52-57(1991)
See Libutti, Steven K.: V1421,169-172(1991)
See Oz, Mehmet C.: V1422,147-150(1991)

Bass, V. I.
; Gus'kova, M. S.; Lebedev, V. B.; Mikhaylenko, B. M.; Saulevich, S. V.; Seleznev, V. P.; Feldman, G. G.; Chernyshov, N. A.: Modular equipment for single-frame photography in wide-time and spectral ranges, V1358,1075-1083(1991)

Bassi, Paolo See Zoboli, Maurizio: V1506,160-169(1991)

Bassiouni, Mostafa See Kasparis, Takis: V1521,46-54(1991)

Basso, Andrea
See Li, Wei: V1605,124-136(1991)
See Popat, Ashok C.: V1605,940-953(1991)

Bastawros, Adel F. See Voloshin, Arkady S.: V1400,50-60(1991)

Basti, Gianfranco
; Perrone, Antonio; Morgavi, Giovanna: Relaxation properties and learning paradigms in complex systems, V1469,719-736(1991)

Bastiaens, H. M.
; Peters, Peter J.; Witteman, Wilhelmus J.: Measurement of the gain in a XeF (C-A) laser pumped by a coaxial e-beam, V1397,77-80(1991)

Bastida, M. R.
; Figueroa-Nazuno, Jesus: Different types of theories in neurocomputation, V1469,150-156(1991)

Basting, Dirk
; Sowada, Ulrich; Voss, F.; Oesterlin, Peter: Processing of polytetrafluoroethylene with high-power VUV laser radiation, V1412,80-83(1991)
See Kahlert, Hans-Juergen: V1463,604-609(1991)
See Mueller-Horsche, Elmar: V1503,28-39(1991)

Bataille, F.
; Kechemair, Didier; Houdjal, R.: Real-time actuating of laser power and scanning velocity for thermal regulation during laser hardening, V1502,135-139(1991)
; Kechemair, Didier; Pawlowski, C.; Houdjal, R.: Thermal regulation applied to CO2 laser self-quenching of complex geometry workpieces, V1397,839-842(1991)

Batani, D.
; Giulietti, Antonio; Palladino, Libero; Tallents, Gregory J.; Turcu, I. C.: L-shell x-ray spectroscopy of laser-produced plasmas in the 1-keV region, V1503,479-491(1991)
See Turcu, I. C.: V1503,391-405(1991)

Batchelor, Bruce G.
: Tools for designing industrial vision systems,VCR36,138-166(1991)
; Waltz, Frederick M.; eds.: *Machine Vision Systems Integration*,VCR36(1991)
; Waltz, Frederick M.; eds.: *Machine Vision Systems Integration in Industry*,V1386(1991)
See Chan, John P.: V1386,163-170(1991)
See Chan, John P.: V1386,171-179(1991)
See Lambert, Robin A.: V1381,582-588(1991)
See Wani, M. A.: V1384,83-89(1991)

Batchman, Theodore E. See Heavin, Scott D.: V1455,12-18(1991)

Bates, Stephen C.
; Morrison, Philip W.; Solomon, Peter R.: Infrared monitoring of combustion,V1434,28-38(1991)

Bathurst, Richard L. See Beheim, Glenn: V1369,50-59(1991)

Batistic, I. See Mustre de Leon, Jose: V1550,85-96(1991)

Batluck, Georgiann R. See Cebula, Richard P.: V1493,91-99(1991)

Baton, S. See Labaune, C.: V1413,138-143(1991)

Batra, Ashok K.
See Rao, S. M.: V1557,283-292(1991)
See Steiner, Bruce W.: V1557,156-167(1991)
See Trolinger, James D.: V1332,151-165(1991)
See Trolinger, James D.: V1557,98-109(1991)

Battaglin, Giancarlo
; De Marchi, Giovanna; Losacco, Aurora M.; Mazzoldi, Paolo; Miotello, Antonio; Quaranta, Alberto; Valentini, Antonio: Study using nuclear techniques of waveguides produced by electromigration processes,V1513,441-450(1991)
See Mazzoldi, Paolo: V1513,182-197(1991)
See Righini, Giancarlo C.: V1513,418-424(1991)

Bauch, Lothar
; Jagdhold, Ulrich A.; Dreger, Helge H.; Bauer, Joachim J.; Hoeppner, Wolfgang W.; Erzgraeber, Hartmut H.; Mehliss, Georg G.: Surface imaging on the basis of phenolic resin: experiments and simulation,V1466,510-519(1991)

Bauche, B. See Monge, J.L.: V1490,84-93(1991)

Baudin, Gilles
; Bessudo, Richard; Cutter, Mike A.; Lobb, Daniel R.; Bezy, Jean L.: Medium-resolution imaging spectrometer,V1490,102-113(1991)
; Chessel, Jean-Pierre; Cutter, Mike A.; Lobb, Daniel R.: Calibration for the medium-resolution imaging spectrometer,V1493,16-27(1991)

Baudrand, Henry See Pyee, Maurice: V1512,240-248(1991)

Bauer, Axel See Habich, Uwe: V1397,383-386(1991)

Bauer, Dietrich
; Lober, K.; Seeliger, Reinhard: Fixation method with high-orientation accuracy for optical terminals in space,V1533,277-285(1991)

Bauer, Guenther E. See Oswald, Josef: V1362,534-543(1991)

Bauer, J.
See Flanagan, Kathryn A.: V1549,395-407(1991)
See Markert, Thomas H.: V1549,408-419(1991)

Bauer, Joachim J. See Bauch, Lothar: V1466,510-519(1991)

Bauer, K. See Brinkmann, Ralf E.: V1421,134-139(1991)

Bauer, T. See Braeuer, Andreas: V1559,470-478(1991)

Bauhahn, Paul E.
; Geddes, John J.: High-density circuit approach for low-cost MMIC circuits,V1475,122-128(1991)
See Baldwin, David L.: V1476,46-55(1991)
See Swirhun, S.: V1475,223-230(1991)

Baum, Jerrold E.
; Rak, Steven J.: Simultaneous active/passive IR vehicle detection,V1416,209-220(1991)

Baum, Thomas H.
; Comita, Paul B.; Kodas, Toivo T.: Laser-induced gold deposition for thin-film circuit repair,V1598,122-131(1991)

Baumann, John A.
; Shepard, Allan H.; Waters, Robert G.; Yellen, Steven L.; Harding, Charleton M.; Serreze, Harvey B.: Evaluation of diode laser failure mechanisms and factors influencing reliability,V1418,328-337(1991)

Baumann, P. See Kretschmer, K.-H.: V1392,246-252(1991)

Baumann, Reinhard
; Bargon, Joachim; Roth, Hans-Klaus: Poly(bis-alkylthio-acetylen): a dual-mode laser-sensitive material,V1463,638-643(1991)

Baumbick, Robert J.
: Potential for integrated optical circuits in advanced aircraft with fiber optic control and monitoring systems,V1374,238-250(1991)
: Review of the Fiber Optic Control System Integration program,V1589,12-19(1991)
See Chen, Ray T.: V1374,223-236(1991)

Baumgartner, R.
; Heil, P.; Jocham, D.; Kriegmair, M.; Stepp, Herbert; Unsoeld, Eberhard: Improved instrumentation for photodynamic fluorescence detection of cancer,V1525,246-248(1991)

Baumler, Mark
See Behlau, Jerry L.: V1530,218-230(1991)
See Behlau, Jerry L.: V1530,208-217(1991)

Baur, Charles
; Beer, Simon: High-speed vision system based on computer graphics models,V1385,85-92(1991)

Bauser, Elizabeth See Rinker, Michael: V1362,14-23(1991)

Bavarian, Behnam
See Vaezi, Matt M.: V1606,803-809(1991)
See Vaezi, Matt M.: V1452,57-63(1991)

Bavdaz, Marcos
; Favata, Fabio; Smith, Alan; Parmar, A. N.: Study of gas scintillation proportional counter physics using synchrotron radiation,V1549,35-44(1991)

Bawolek, Edward J.
; Hirleman, Edwin D.: Surface roughness effects on light scattered by submicron particles on surfaces,V1464,574-583(1991)

Baxter, David See Greenfield, Perry E.: V1494,16-39(1991)

Baxter, Lisa C.
; Coggins, James M.: Supervised pixel classification using a feature space derived from an artificial visual system,V1381,459-469(1991)

Bayanov, Valentin I.
See Bouchenkov, Vyatcheslav A.: V1427,405-408(1991)
See Bouchenkov, Vyatcheslav A.: V1410,244-247(1991)

Bayazitov, Rustem M. See Fattakhov, Yakh'ya V.: V1440,16-23(1991)

Bayer, Eberhard G.
: Luminescence molulation for the characterization of radiation damage within scintillator material,V1361,195-199(1991)

Bayer, Janice I.
; Varadan, Vasundara V.; Varadan, Vijay K.: Discrete piezoelectric sensors and actuators for active control of two-dimensional spacecraft components,V1480,102-114(1991)

Bayless, John R.
; Burkhart, Craig P.: Long-pulse electron gun for laser applications,V1411,42-46(1991)

Bayruns, Robert J.
; Laverick, Timothy; Scheinberg, Norman; Stofman, Daniel: Design of low-noise wide-dynamic-range GaAs optical preamps,V1541,83-90(1991)

Bays, Roland
; Winterhalter, L.; Funakubo, H.; Monnier, Philippe; Savary, M.; Wagnieres, G.; Braichotte, D.; Chatelain, Andre; van den Bergh, Hubert; Svaasand, Lars O.; Burckhardt, C. W.: Clinical optical dose measurement for PDT: invasive and noninvasive techniques,V1525,397-408(1991)
See Braichotte, D.: V1525,212-218(1991)

Bazakos, Michael E.
See Nasr, Hatem N.: VIS07,122-129(1991)
See Nasr, Hatem N.: V1483(1991)
See Roberts, Barry: V1521,2-13(1991)

Bazakutsa, P. V.
; Maslennikov, V. L.; Sychugov, V. A.; Yakovlev, V. A.: Laser-induced generation of surface periodic structures resulting from the waveguide mode interaction,V1440,370-376(1991)

Bazhenov, V. V.
; Libenson, Michail N.; Makin, Vladimir S.; Trubaev, Vladimir V.: Surface structures formation by pulse heating of metals in oxidized environment,V1440,332-337(1991)

Bazhenov, V. Y.
; Burykin, N. M.; Soskin, M. S.; Taranenko, Victor B.; Vasnetsov, M. V.: Holographic spectral selectors and filters based on phase gratings and planar waveguides,V1574,148-153(1991)

Beales, M. S. See Johnston, James S.: V1589,126-132(1991)

Beard, David V.
; Johnston, R. E.; Pisano, E.; Hemminger, Bradley M.; Pizer, Stephen M.: Radiology workstation for mammography: preliminary observations, eyetracker studies, and design, V1446, 289-296(1991)
; Perry, John R.; Muller, K.; Misra, Ram B.; Brown, P.; Hemminger, Bradley M.; Johnston, R. E.; Mauro, M.; Jaques, P. F.; Schiebler, M.: Evaluation of total workstation CT interpretation quality: a single-screen pilot study, V1446, 52-58(1991)

Beard, Michael S. See Roberts, Harold A.: V1363, 62-69(1991)

Beaton, Brian L. See Garrett, Steven L.: V1367, 13-29(1991)

Beaton, Robert J. See Miller, Robert H.: V1457, 248-258(1991)

Beatty, Diane M. See Morris, Stephen J.: V1428, 148-158(1991)

Beauchemin, Bernard T. See Honda, Kenji: V1466, 141-148(1991)

Beaudoin, N. See Kassi, Hassane: V1436, 58-67(1991)

Beaulaigue, L. See Campuzano, Juan C.: V1345, 245-254(1991)

Beaumont, Steven P. See Sumner, Timothy J.: V1549, 256-264(1991)

Beaver, Edward A. See Rosenblatt, Edward I.: V1449, 72-86(1991)

Becherer, Richard J.
ed.: *Laser Radar VI*, V1416(1991)
; Kahan, Lloyd R.: Ground-based CW atmospheric Doppler performance modeling, V1416, 306-313(1991)

Bechi, Paolo
See Baldini, Francesco: V1510, 58-62(1991)
See Falciai, Riccardo: V1572, 424-427(1991)

Bechtel, Joel J. See Saccomanno, Geno: V1426, 2-12(1991)

Bechtel, Lorne See Clapham, Terrance N.: V1423, 2-7(1991)

Bechtle, Daniel W. See Taylor, Gordon C.: V1475, 103-112(1991)

Beck, Dominik See Rol, Pascal O.: V1423, 84-88(1991)

Beck, Elsa R.
; Hetzel, Fred W.: Photodynamic therapy of pet animals with spontaneously occurring head and neck carcinomas, V1426, 311-315(1991)

Beck, Hal E.
; Bergondy, Daniel; Brown, Joseph R.; Sari-Sarraf, Hamed: Multiresolution segmentation of forward-looking IR and SAR imagery using neural networks, V1381, 600-609(1991)

Beck, M.
; Berger, Peter; Dausinger, Friedrich; Huegel, Helmut: Aspects of keyhole/melt interaction in high-speed laser welding, V1397, 769-774(1991)

Beck, Oskar J. See Waidhauser, Erich: V1428, 75-83(1991)

Beck, Rasmus See Aubert, Philippe: V1502, 52-59(1991)

Beck, Robert N.
: Imaging science in the 1990s, V1396, 688-695(1991)

Becker, J. S.
; Lorenz, M.; Dietze, H.-J.: Stoichiometry of laser-deposited Bi-Sr-Ca-Cu-O films on silicon and mass spectrometric investigations of superconductors, V1598, 227-238(1991)

Becker, Mark W. See Spangler, Charles W.: V1560, 139-147(1991)

Becker, Michael F.
; Ma, Chun-Chi; Walser, Rodger M.: Multipulse laser-induced failure prediction for Mo metal mirrors, V1441, 174-187(1991)
See Huffaker, Diana L.: V1441, 365-380(1991)
See Kniffen, Stacy K.: V1564, 617-627(1991)

Becker, Richard A.
; Cleveland, William S.; Shyu, William M.; Wilks, Allan R.: Interactive graphics system for multivariate data display, V1459, 48-56(1991)
; Eick, Stephen G.; Miller, Eileen O.; Wilks, Allan R.: Network visualization: user interface issues, V1459, 150-154(1991)

Becker, Suzanna
; Hinton, Geoffrey E.: Learning spatially coherent properties of the visual world in connectionist networks, V1569, 218-226(1991)

Becker, Wilhelm
; McIver, John K.; Guenther, Arthur H.: Effect of polarization on two-color laser-induced damage, V1441, 541-552(1991)

Beckman, Claes See Larsson, Michael: V1529, 63-70(1991)

Beckmann, Leo H.
: Small-computer program for optical design and analysis written in "C", V1354, 254-261(1991)

Beconne, Jean P. See Simonne, John J.: V1589, 139-147(1991)

Bedi, Jatinder S.
See Hussain, Mukhtar: V1469, 697-707(1991)
See Hussain, Mukhtar: V1469, 463-471(1991)

Bednarczyk, Sophie See Herve, Philippe: V1487, 387-395(1991)

Bedzyk, Michael J.
: Long-period x-ray standing waves generated by total external reflection, V1550, 151-155(1991)

Beech, Russell S. See Ghosh, Anjan K.: V1389, 630-641(1991)

Beeck, Manfred-Andreas
; Frost, Thorsten; Windeln, Wilbert: Holographic mirrors laminated into windshields for automotive head-up display and solar protective glazing applications, V1507, 394-406(1991)

Beeder, Clain
See Buchelt, Martin: V1420, 249-253(1991)
See Papaioannou, Thanassis: V1420, 203-211(1991)
See Vari, Sandor G.: V1426, 58-65(1991)
See Vari, Sandor G.: V1426, 111-120(1991)

Beeman, Jeff See Holley, Jeff: V1343, 500-511(1991)

Been, D. See Kozaitis, Samuel P.: V1474, 112-115(1991)

Beer, B. A. See Sayegh, Samir I.: V1569, 100-110(1991)

Beer, Reinhard See Glavich, Thomas A.: V1540, 148-159(1991)

Beer, Rudolf J. See Rossmanith, H. P.: V1554A, 850-860(1991)

Beer, Simon See Baur, Charles: V1385, 85-92(1991)

Beeson, Karl W.
; Horn, Keith A.; Lau, Christina; McFarland, Michael J.; Schwind, David; Yardley, James T.: Photochemical delineation of waveguides in polymeric thin films, V1559, 258-266(1991)
; Horn, Keith A.; McFarland, Michael J.; Wu, Chengjiu; Yardley, James T.: Photochemical formation of polymeric optical waveguides and devices for optical interconnection applications, V1374, 176-185(1991)
See McFarland, Michael J.: V1583, 344-354(1991)

Beex, A. A. See Zhu, Dongping: V1569, 174-181(1991)

Begeman, Michael S. See Andrews, Lee T.: V1459, 125-135(1991)

Begleiter, Eric
: Edible holography: the application of holographic techniques to food processing, V1461, 102-109(1991)

Begley, David L.
ed.: *Selected Papers on Free-Space Laser Communications*, VMS30(1991)
; Marshalek, Robert G.: Economic factors for a free-space laser communication system, V1522, 234-242(1991)
; Seery, Bernard D.; eds.: *Free-Space Laser Communication Technologies III*, V1417(1991)

Beguin, Alain See Nissim, Carlos: V1365, 13-19(1991)

Beheim, Glenn
; Krasowski, Michael J.; Sotomayor, Jorge L.; Fritsch, Klaus; Flatico, Joseph M.; Bathurst, Richard L.; Eustace, John G.; Anthan, Donald J.: Wavelength-multiplexed fiber optic position encoder for aircraft control systems, V1369, 50-59(1991)
; Sotomayor, Jorge L.; Flatico, Joseph M.; Azar, Massood T.: Fiber optic thermometer using Fourier transform spectroscopy, V1584, 64-71(1991)

Behlau, Jerry L.
; Baumler, Mark: Characterization of hot-isostatic-pressed optical-quality beryllium, V1530, 208-217(1991)
; Granger, Edward M.; Hannon, John J.; Baumler, Mark; Reilly, James F.: Beryllium scatter analysis program, V1530, 218-230(1991)

Behunek, Jan L.
; Vonder Haar, Thomas H.: Space-based sensing of atmospheric conditions over data-void regions, V1479, 93-100(1991)

Beiser, Leo
: Design equations for a polygon laser scanner, V1454, 60-66(1991)
: Monogon laser scanner with no line wobble, V1454, 33-36(1991)
; Marshall, Gerald F.; eds.: *Beam Deflection and Scanning Technologies*, V1454(1991)

Bejczy, Antal K.
; Venema, Steven; Kim, Won S.: Role of computer graphics in space telerobotics: preview and predictive displays, V1387, 365-377(1991)
See Schenker, Paul S.: V1383, 202-223(1991)

Bek, A. M.
; Kononenko, Vadim L.: Frequency spectra of erythrocyte membrane flickering measured by laser light scattering, V1403, 384-386(1991)

Bekefi, George
See Chen, Shien C.: V1407, 67-73(1991)
See Johnston, George L.: V1407, 92-99(1991)

Bel'Kov, V. V. See Ashkinadze, B. M.: V1361, 866-873(1991)

Belanger, Garry See Plessis, Brigitte: V1445, 539-554(1991)

Belashenkov, Nickolay R.
See Altshuler, Grigori B.: V1440,116-126(1991)
See Azarenkov, Aleksey N.: V1409,154-164(1991)

Belcher, Melvin L.
; Garmon, Jeff P.: Wideband bistatic radar signal processing using a coherent detection architecture,V1476,224-233(1991)

Beldie, Ion P.
; Kost, Bernd: Luminance asymmetry in stereo TV images,V1457,242-247(1991)

Belendez, A.
; Pascual, I.; Fimia-Gil, Antonio: Effective holographic grating model to analyze thick holograms,V1507,268-276(1991)
; Pascual, I.; Fimia-Gil, Antonio: Optimization of reconstruction geometry for maximum diffraction efficiency in HOE: the influence of recording material,V1574,77-83(1991)
See Pascual, I.: V1507,373-378(1991)
See Pascual, I.: V1574,72-76(1991)

Beletskii, N. N. See Vertiy, Alexey A.: V1361,1070-1078(1991)

Belfor, Ricardo A.
; Lagendijk, Reginald L.; Biemond, Jan: Motion-compensated subsampling of HDTV,V1605,274-284(1991)

Belforte, David A.
: Industrial laser market,V1520,93-117(1991)
: Job shop market,V1517,176-196(1991)
; Levitt, Morris R.; eds.: *The Marketplace for Industrial Lasers*,V1517(1991)

Belikov, Andrey V. See Altshuler, Grigori B.: V1427,141-150(1991)

Belker, D.
See Lando, Mordechai: V1412,19-26(1991)
See Lando, Mordechai: V1442,172-180(1991)

Belkin, Barry See Taff, Laurence G.: V1481,440-448(1991)

Bell, Cynthia S.
: Lens evaluation for electronic photography,V1448,59-68(1991)
See Jackson, Todd A.: V1448,2-12(1991)

Bell, Ian G.
: High-performance IR thermography system based on Class II Thermal Imaging Common Modules,V1467,438-447(1991)

Bell, J. M.
; Green, David C.; Patterson, A.; Smith, Geoffrey B.; MacDonald, K. A.; Lee, K. D.; Kirkup, L.; Cullen, J. D.; West, B. O.; Spiccia, L.; Kenny, M. J.; Wielunski, L. S.: Structure and properties of electrochromic WO3 produced by sol-gel methods,V1536,29-36(1991)

Bell, John A.
; Hamilton, Michael C.; Leep, David A.; Taylor, Henry F.; Lee, Y.-H.: A/D conversion of microwave signals using a hybrid optical/electronic technique,V1476,326-329(1991)
; Hamilton, Michael C.; Leep, David A.: Optical sampling and demultiplexing applied to A/D conversion,V1562,276-280(1991)

Bell, Kenneth L. See Christensen, Lorna D.: V1463,504-514(1991)

Bell, Paul A. See Brown, James C.: V1488,300-311(1991)

Bell, Perry M.
; Kilkenny, Joseph D.; Hanks, Roy L.; Landen, Otto L.: Measurements with a 35-psec gate time microchannel plate camera,V1346,456-464(1991)
See Kilkenny, Joseph D.: V1358,117-133(1991)
See Wiedwald, Douglas J.: V1346,449-455(1991)

Bellegarda, Jerome R.
: Time-varying system identification via explicit filtering of the parameter estimates,V1566,190-201(1991)
See Ayme-Bellegarda, Eveline J.: V1471,18-29(1991)

Bellini, Sandro
: Blind equalization and deconvolution,V1565,88-101(1991)

Bellis, J. G. See Fordham, John L.: V1449,87-98(1991)

Bellman, Robert H.
; Borrelli, Nicholas F.; Mann, L. G.; Quintal, J. M.: Fabrication and performance of a one-to-one erect imaging microlens array for fax,V1544,209-217(1991)

Bello, M. See Shazeer, Dov J.: V1469,622-636(1991)

Bellomi, Giovanni
; Bottacchi, Stefano: Generalized transport model for heterojunction: a computer modeling approach,V1362,760-767(1991)

Belopolsky, Yakov See Morris, Jacqueline H.: V1389,236-248(1991)

Belostotsky, A. L. See Petrov, Dmitry V.: V1374,152-159(1991)

Belostotsky, Vladimir I.
; Solinov, Vladimir F.: IR reflectance spectroscopy and AES investigation of titanium ion-beam-doped silica,V1513,313-318(1991)

Belov, M. N.
; Manykin, Edward A.: Optical associative memories based on time-delayed four-wave mixing,V1621,268-279(1991)

Belov, Sergei N.
; Golubev, Evgeny M.; Vinokurova, Elena G.: Laboratory soft x-ray source based on high-temperature capillary arc discharge,V1552,264-267(1991)
; Vangonen, Albert I.; Levina, Olga V.; Puhov, Anatoly M.: Photodestruction of organic compounds exposed to pulsed VUV irradiation,V1503,503-509(1991)

Belov, Sergey R. See Atjezhev, Vladimir V.: V1503,197-199(1991)

Belt, Roger F. See Scripsick, Michael P.: V1561,93-103(1991)

Bely, Pierre Y.
; Breckinridge, James B.; eds.: *Space Astronomical Telescopes and Instruments*,V1494(1991)
See Peacock, Keith: V1494,147-159(1991)

Belyaev, L. V. See Vishnevsky, G. I.: V1448,69-72(1991)

Belyakov, Ludvig V.
See Sreseli, Olga M.: V1440,326-331(1991)
See Sreseli, Olga M.: V1545,149-158(1991)

Belyakov, V. A. See Kozhevnikov, Nikolai M.: V1584,138-144(1991)

BeMent, Spencer L. See Zhao, Yilin: V1388,122-130(1991)

Ben-Amar, Gabi See Biderman, Shlomo: V1535,27-34(1991)

Ben-Arie, Jezekiel
: Linear lattice architectures that utilize the central limit for image analysis, Gaussian operators, sine, cosine, Fourier, and Gabor transforms,V1606,823-838(1991)
: Utilizing the central limit theorem for parallel multiple-scale image processing with neural architectures,V1569,227-238(1991)
; Huddleston, James: Grouping and forming quantitative descriptions of image features by a novel parallel algorithm,V1606,2-19(1991)
; Rao, K. R.: Novel transform for image description and compression with implementation by neural architectures,V1569,367-374(1991)

Ben-Kish, A. See Shafir, Ehud: V1442,236-241(1991)

Ben-Porat, T. See Bar, I.: V1397,169-172(1991)

Ben-Shalom, Ami
; Devir, Adam D.; Ribak, Erez N.; Talmore, Eli T.; Balfour, L. S.; Brandman, N.: Different aspects of backgrounds in various spectral bands,V1486,238-257(1991)
See Sheffer, Dan: V1493,232-243(1991)

Ben-Yosef, Nissim See Agassi, Eyal: V1442,126-132(1991)

Benaim, George See Mattioli, Stefano: V1421,114-119(1991)

Benamati, Brian L. See Parulski, Kenneth A.: V1448,45-58(1991)

Benatar, Avraham See Park, Joon B.: V1554B,357-370(1991)

Benattar, Rene See Geindre, Jean-Paul: V1502,311-318(1991)

Bender, E. See Johnson, C. B.: V1449,2-12(1991)

Bender, John W.
; Wahl, Roger L.: Work-induced stress and long-term stability in optically polished silicon,V1533,264-276(1991)

Bender, Leon See Chandra, Mudjianto: V1421,68-71(1991)

Bender, Walter J.
; Plouffe, Wil; eds.: *Image Handling and Reproduction Systems Integration*,V1460(1991)
; Rosenberg, Charles: Image enhancement using nonuniform sampling,V1460,59-70(1991)
See Jacobson, Nathaniel: V1453,70-80(1991)

Bendett, Mark P.
ed.: *Optical Technology for Signal Processing Systems*,V1474(1991)
See Hibbs-Brenner, Mary K.: V1563,10-20(1991)
See Swirhun, S.: V1475,223-230(1991)

Benech, Pierre See Saint-Andre, Francoise: V1583,278-288(1991)

Benedetto, Sergio
; Kazovsky, Leonid G.; Poggiolini, Pierluigi T.: Minimum polarization modulation: a highly bandwidth efficient coherent optical modulation scheme,V1579,112-121(1991)

Benedict, Melvin K. See Webb, David J.: V1418,231-239(1991)

Benevides, James M. See Towse, Stacy A.: V1403,6-14(1991)

Benferhat, Ramdane See Acher, O.: V1361,1156-1163(1991)

Benford, James N.
See Levine, Jerrold S.: V1407,74-82(1991)
See Smith, Richard R.: V1407,83-91(1991)

Benham, Paul
; Kidger, Michael J.: Optimization of athermal systems,V1354,120-125(1991)
See Kidger, Michael J.: V1354,92-96(1991)

Beni, Gerardo See Xie, Xuanli: V1381,401-410(1991)

Benimetskaya, L. Z.
; Bulychev, N. V.; Kozionov, Andrew L.; Koshkin, A. A.; Lebedev, A. V.;
Novozhilov, S. Y.; Stockman, M. I.: Site-specific laser modification
(cleavage) of oligodeoxynucleotides,V1525,210-211(1991)
; Gitelzon, I. I.; Kozionov, Andrew L.; Novozhilov, S. Y.; Petushkov, V. N.;
Rodionova, N. S.; Stockman, M. I.: Localization of the active site of an
enzyme, bacterial luciferase, using two-quantum affinity
modification,V1525,242-245(1991)

Benjamin, Neil M.
; Chapman, Brian N.; Boswell, Rod W.: Progress of an advanced diffusion
source plasma reactor,V1392,95-105(1991)
See Gray, David E.: V1392,402-410(1991)

Benner, Robert E. See Marshall, Martin S.: V1564,121-134(1991)

Bennett, Harold E.
; Chase, Lloyd L.; Guenther, Arthur H.; Newman, Brian; Soileau, M. J.; eds.:
Laser-Induced Damage in Optical Materials: 1990,V1441(1991)
See Nee, Soe-Mie F.: V1441,31-37(1991)

Bennett, Laura F.
; Johnson, Rubin; Hudson, C. I.: Assumption truth maintenance in model-
based ATR algorithm design,V1470,263-274(1991)

Bennett-Lilley, Marylyn H.
; Hiatt, William M.; Lauchlan, Laurie J.; Mantalas, Lynda C.; Rottmann, Hans;
Seliger, Mark; Singh, Bhanwar; Yansen, Don E.: Semiwafer metrology
project,V1464,127-136(1991)

Bennison, Tim See Dickinson, Robert R.: V1459,178-189(1991)

Benoit, Robert T. See Wong, Wallace K.: V1530,86-103(1991)

Benschop, Jozef P.
; Monahan, Kevin M.; Harris, Tom A.: High-speed stepper setup using a low-
voltage SEM,V1464,62-70(1991)
See Monahan, Kevin M.: V1464,2-9(1991)

Benson, Ronda L. See Gustafson, Terry L.: V1403,545-554(1991)

Benson, Stephen V. See Litvinenko, Vladimir N.: V1552,2-6(1991)

Bensoussan, M. See Nissim, Yves I.: V1393,216-228(1991)

Benton, Stephen A.
: Experiments in holographic video imaging,VIS08,247-267(1991)
ed.: *Practical Holography V*,V1461(1991)
See Farmer, William J.: V1461,215-226(1991)
See Halle, Michael W.: V1461,142-155(1991)
See St.-Hilaire, Pierre: V1461,254-261(1991)

Bentson, John R. See Lou, Shyh-Liang: V1446,302-311(1991)

Bentz, Jerry See Assenheim, Harry M.: V1456(1991)

Benvenisty, Alan I. See Oz, Mehmet C.: V1422,147-150(1991)

Benyattou, Taha
See Ababou, Soraya: V1361,706-711(1991)
See Banvillet, Henri: V1361,972-979(1991)

Benz, Rudolph G. See Summers, Christopher J.: V1512,170-176(1991)

Beranek, J. See Rohlena, Karel: V1415,259-268(1991)

Beranek, Mark W. See Hong, C. S.: V1418,177-187(1991)

Berardi, V. See Bruzzese, Riccardo: V1397,735-738(1991)

Beratan, David N. See Marder, Seth R.: V1560,86-97(1991)

Berdahl, Paul H.
; Russo, Rick: Progress toward thin-film-based tape conductors,V1394,180-
185(1991)

Berdanier, Barry N. See Powell, Richard K.: V1419,126-140(1991)

Beregulin, Eugene V.
; Ganichev, Sergey D.; Yaroshetskii, Ilya D.; Lang, Peter T.; Schatz, Wolfgan;
Renk, Karl F.: Devices for generation and detection of subnanosecond IR
and FIR radiation pulses,V1362,853-862(1991)

Berejny, P. See Saissac, M.: V1397,739-742(1991)

Berek, Stefan
; Knauer, Ulrich; Zottl, Helmut: I-line lithography for highly reproducible
fabrication of surface acoustic wave devices,V1463,515-520(1991)

Berenstein, C. A. See Sidiropoulos, N. D.: V1569,356-366(1991)

Berera, G. See Goldner, Ronald B.: V1536,63-69(1991)

Beretta, Stefano
; Cairoli, Massimo; Viardi, Marzia: Optimum design of phase gratings for
diffractive optical elements obtained by thin-film deposition,V1544,2-
9(1991)

Berezhnoy, Anatoly A.
; Popov, Yury V.: Optical memory in electro-optical crystals,V1401,44-
49(1991)

Berezin, Andrey A. See Gariaev, Peter P.: V1621,280-291(1991)

Berezin, Boris See Motenko, Boris: V1399,78-81(1991)

Berezin, Juri D. See Bouchenkov, Vyatcheslav A.: V1410,185-188(1991)

Berezin, M. V.
; Levshin, L. V.; Saletsky, A. M.: Effect of alternating magnetic fields on the
properties of water systems,V1403,335-337(1991)

Berezin, U. D. See Bouchenkov, Vyatcheslav A.: V1427,409-412(1991)

Berezin, V. Y.
See Khvilivitzky, A. T.: V1447,184-190(1991)
See Khvilivitzky, A. T.: V1447,64-68(1991)
See Vishnevsky, G. I.: V1447,34-43(1991)
See Vishnevsky, G. I.: V1448,69-72(1991)

Berezinskaya, Aleksandra M.
; Dukhovniy, Anatoliy M.: Diffuse radiation unsteady transformation by
thermal dynamic holograms,V1238,80-84(1991)

Berg, John M.
; Morris, David E.; Clark, David L.; Tait, C. D.; Woodruff, William H.; Van
Der Sluys, William G.: Pulsed photothermal spectroscopy applied to
lanthanide and actinide speciation,V1435,331-337(1991)

Berg, Roger
; Andersson-Engels, Stefan; Jarlman, Olof; Svanberg, Sune: Time-resolved
transillumination for medical diagnostics,V1431,110-119(1991)
; Andersson-Engels, Stefan; Jarlman, Olof; Svanberg, Sune: Tumor detection
using time-resolved light transillumination,V1525,59-67(1991)
See Andersson-Engels, Stefan: V1426,31-43(1991)
See Baert, Luc: V1525,385-390(1991)

Bergano, Neal S. See Aspell, Jennifer: V1373,188-196(1991)

Berge-Cherfaoui, Veronique See Vachon, Bertrand: V1468,483-492(1991)

Bergeman, Thomas H.
; Ryan, Robert E.: Calculations on the Hanle effect with phase and amplitude
fluctuating laser fields,V1376,54-67(1991)

Bergen, James R. See Larimer, James O.: V1456,196-210(1991)

Berger, Henry
; Del Bosque-Izaguirre, Delma: New formulation of reflection coefficient in
propagation theory for imaging inhomogeneous media and for
nonuniform waveguides,V1521,117-130(1991)

Berger, M. See Angelov, D.: V1403,575-577(1991)

Berger, Michel See Floch, Herve G.: V1441,304-315(1991)

Berger, Peter
See Beck, M.: V1397,769-774(1991)
See Holzwarth, Achim: V1503,98-109(1991)

Berger, R. M.
; Chmelir, M.; Reedy, Herman E.; Chambers, Jack P.: Very high reflective all-
dielectric coatings for high-power CO2 lasers,V1397,611-618(1991)

Berger, Thomas E. See Timothy, J. G.: V1343,350-358(1991)

Bergin, Joseph E.
See Block, Kenneth L.: V1483,62-65(1991)
See Block, Kenneth L.: V1481,32-34(1991)

Bergkvist, Mikael
; Roos, Arne: Light scattering properties of new materials for glazing
applications,V1530,352-362(1991)

Berglund, C. N. See Pease, R. F.: V1496,234-238(1991)

Bergman, Larry See Nemanich, Robert J.: V1437,2-12(1991)

Bergman, Larry A.
; Hartmayer, Ron; Wu, Wennie H.; Cassell, P.; Edgar, G.; Lambert, James L.;
Mancini, Richard; Jeng, J.; Pardo, C.; Halloran, Frank; Martinez, James
C.: High-performance FDDI NIU for streaming voice, video, and
data,V1364,14-21(1991)
See Barnes, Charles E.: V1366,9-16(1991)

Bergmann, Hans W.
See Hartmann, Martin: V1598,175-185(1991)
See Kupfer, Roland: V1598,46-60(1991)
See Schubert, Emil: V1503,299-309(1991)

Bergner, Harald
; Hempel, Klaus; Stamm, Uwe: Dynamic behavior of internal elements of
high-frequency integrated circuits studied by time-resolved optical-beam-
induced current (OBIC) method,V1362,484-493(1991)
; Krause, A.; Stamm, Uwe: Localization of hot spots in silicon devices with a
laser scanning microscope,V1361,723-731(1991)
See Brueckner, Volkmar: V1362,510-517(1991)

Bergomi, Lorenzo See Sermage, Bernard: V1361,131-135(1991)

Bergondy, Daniel
See Beck, Hal E.: V1381,600-609(1991)
See Brown, Joseph R.: V1469,539-543(1991)

Bergquist, James C.
; Wineland, D. J.; Itano, W. M.; Diedrich, F.; Raizen, M. G.; Elsner, Frank: Single-ion spectroscopy,V1435,82-85(1991)

Bergstrom, J. C. See Piestrup, Melvin A.: V1552,214-239(1991)

Bergstrom, Neil G.
: Projection direct imaging for high-density interconnection and printed circuit manufacture,V1390,509-512(1991)

Berik, Irina K. See Vill, Arnold A.: V1503,110-114(1991)

Berinato, Robert J. See Abushagur, Mustafa A.: V1564,602-609(1991)

Berkhahn, Glenn
: Sheet-metal cutting market,V1517,117-135(1991)

Berkmann, J. See Zetzsche, Christoph: V1570,337-350(1991)

Berkoff, Timothy A.
; Kersey, Alan D.: Interferometric signal processing schemes for the measurement of strain,V1588,169-176(1991)
; Kersey, Alan D.; Moeller, Robert P.: Novel analog phase tracker for interferometric fiber optic sensor applications,V1367,53-58(1991)
See Kersey, Alan D.: V1511,40-50(1991)
See Kersey, Alan D.: V1367,310-318(1991)

Berkovic, Garry
; Krongauz, Valeri; Yitzchaik, Shlomo: Nonlinear optics in poled polymers with two-dimensional asymmetry,V1442,44-52(1991)
; Yitzchaik, Shlomo; Krongauz, Valeri: In-plane poling of doped-polymer films: creation of two asymmetric directions for second-order optical nonlinearity,V1560,238-242(1991)

Berkovsky, A. G.
; Gubanov, Y. I.; Pryanishnikov, Ivan G.; Murugov, V. M.; Petrov, S. I.; Senik, A. V.: Laser-produced plasma x-ray diagnostics with an x-ray streak camera at the Iskra-4 plant,V1358,750-755(1991)

Berkowski, Marek See Ryba-Romanowski, Witold: V1391,2-5(1991)

Berman, Barry L. See Uberall, Herbert: V1552,198-213(1991)

Berman, David H.
: Rough-interface scattering without plane waves,V1558,191-201(1991)

Berman, G. P. See Aleksandrov, K. S.: V1621,51-61(1991)

Berman, John L. See Chou, Ching-Hua: V1464,145-154(1991)

Berman, Paul R. See Finkelstein, Vladimir: V1376,68-79(1991)

Berman, Richard J.
; Burgess, Lloyd W.: Flow optrodes for chemical analysis,V1368,25-35(1991)

Bermejo, Dionisio
See de Frutos, Angel M.: V1397,717-720(1991)
See Poueyo, Anne: V1502,140-147(1991)

Bernabeu, Eusebio
See Alda, Javier: V1527,240-251(1991)
See Porras, Miguel A.: V1397,645-648(1991)
See Wang, Shaomin: V1397,595-602(1991)
See Yonte, T.: V1554B,233-241(1991)

Bernard, Christian
: SPOT system and defence applications,V1521,66-73(1991)

Bernard, Yves See Mary, Joel: V1557,147-155(1991)

Bernardini, Riccardo See Rinaldo, R.: V1606,773-787(1991)

Bernardo, Luis M. See Soares, Oliverio D.: V1332,166-184(1991)

Berndt, Klaus W.
; Lakowicz, Joseph R.: Detection and localization of absorbers in scattering media using frequency-domain principles,V1431,149-160(1991)

Bernhardt, Michael See Wyrowski, Frank: V1555,146-153(1991)

Bernier, J. D. See Cooper, Alfred W.: V1486,47-57(1991)

Bernier, Paul R. See Griffen, Christopher T.: V1554B,754-766(1991)

Bernini, Umberto
; Russo, Paolo: Photoacoustic absorption spectrum of some rat and bovine tissues in the ultraviolet-visible range,V1427,398-404(1991)
See Barone, Fabrizio: V1584,304-307(1991)
See Carbonara, Giuseppe: V1361,1038-1040(1991)
See Carbonara, Giuseppe: V1361,688-691(1991)

Berns, Michael W.
See Gottfried, Varda: V1442,218-229(1991)
See Kimel, Sol: V1525,341-350(1991)
See Neev, Joseph: V1427,162-172(1991)
See Peavy, George M.: V1424,171-178(1991)

See Tromberg, Bruce J.: V1427,101-108(1991)
See Tromberg, Bruce J.: V1525,52-58(1991)
See Wright, William H.: V1427,279-287(1991)

Bernstein, Hana
; Thompson, George A.; Yogev, Amnon; Oron, Moshe: Beam quality measurements and improvement in solar-pumped laser systems,V1442,81-88(1991)

Bernstein, Norman P. See Ng, Willie W.: V1371,205-211(1991)

Bernstein, Uri
; Keller, Catherine E.: Thermal model for real-time textured IR background simulation,V1486,345-351(1991)
See Keller, Catherine E.: V1486,278-285(1991)

Bernt, Marvin L.
; Stover, John C.: Infrared window damage measured by reflective scatter,V1530,42-49(1991)
See Lewis, Isabella T.: V1530,22-34(1991)
See Stover, John C.: V1530,185-195(1991)

Berrouane, H.
; Khan Malek, Chantal; Andre, Jean-Michel; Lesterlin, L.; Ladan, F. R.; Rivoira, R.; Lepetre, Yves; Barchewitz, Robert J.: Fabrication and performance at 1.33 nm of a 0.24-um period multilayer grating,V1343,428-436(1991)

Berry, A. See Egert, Charles M.: V1485,64-77(1991)

Berry, Amanda K.
; Graziano, Karen A.; Thompson, Stephen D.; Taylor, James W.; Suh, Doowon; Plumb, Dean: Chemically amplified resists for x-ray and e-beam lithography,V1465,210-220(1991)
See Babcock, Carl P.: V1466,653-662(1991)

Berry, Dante M. See Setchell, Robert E.: V1441,61-70(1991)

Berry, J. N.
See Padden, Richard J.: V1474,148-159(1991)
See Taylor, Edward W.: V1474,126-131(1991)

Berry, Joseph R. See Clark, David L.: V1540,303-311(1991)

Berry, Michael J. See Valderrama, Giuseppe L.: V1427,200-213(1991)

Berry, R. F. See Fripp, Archibald L.: V1557,236-244(1991)

Bersenev, V. I. See Gordienko, Vyacheslav M.: V1416,102-114(1991)

Bertaux, J. L.
: Global ozone monitoring by occultation of stars,V1490,133-145(1991)

Berthold, Joerg See Sebald, Michael: V1466,227-237(1991)

Berthold, John W.
: Field test results on fiber optic pressure transmitter system,V1584,39-47(1991)
See Kidwell, J. J.: V1367,192-196(1991)

Bertolotti, Mario
See Andriesh, A.: V1513,137-147(1991)
See Montenero, Angelo: V1513,234-242(1991)

Bertoluzza, Alessandro
; Bonora, S.; Fini, G.; Morelli, M. A.: Raman spectroscopy of biological molecules: uncharged phospholipid/polyamine interactions in the presence of bivalent cations,V1403,153-155(1991)
; Caramazza, R.; Fagnano, C.: Raman and multichannel Raman spectroscopy of biological systems,V1403,40-49(1991)
; Fagnano, C.; Morelli, M. A.; Tosi, M. R.; Tugnoli, V.: Molecular spectroscopy of biological molecules: Raman, NMR, and CD study of monophosphate dinucleosides at different degrees of protonation,V1403,150-152(1991)
; Monti, P.; Simoni, R.: Vibrational spectroscopy in the ophthalmological field,V1403,743-745(1991)

Bertoncello, R. See Mazzoldi, Paolo: V1513,182-197(1991)

Bertoni, H. L. See Felsen, Leopold B.: V1471,154-162(1991)

Bertram, L. A. See Fuchs, Elizabeth A.: V1467,136-149(1991)

Bertuccio, G. See Braueninger, Heinrich: V1549,330-339(1991)

Berus, Tomasz
; Macukow, Bohdan: Projection methods for evaluation of Hopfield-type CAM models,V1564,562-570(1991)

Berwick, Michael
; Pannell, Christopher N.; Russell, Philip S.; Jackson, David A.: Demonstration of birefringent optical fiber frequency shifter employing torsional acoustic waves,V1584,364-373(1991)
; Pannell, Christopher N.; Russell, Philip S.; Jackson, David A.: Demonstration of birefringent optical fibre frequency shifter employing torsional acoustic waves,V1572,157-162(1991)

Beshears, David L. See Tobin, Kenneth W.: V1584,23-31(1991)

Besieris, Ioannis M. See Ziolkowski, Richard W.: V1407,387-397(1991)

Besmann, Theodore M.
; Abdel-Latif, A. I.: Modifications of optical properties with ceramic coatings,V1330,78-89(1991)

Bespalov, V. G.
; Dikasov, A. B.: Pulsed hologram recording by nanosecond and subnanosecond small-dimensioned laser system,V1238,470-475(1991)
; Krylov, Vitaly N.; Sizov, V. N.: Pulsed laser system for recording large-scale color hologram,V1238,457-461(1991)

Bessette, Oliver E.
; Cinelli, Joseph L.: Eraseable optical disk systems for signal processing,V1474,162-166(1991)

Bessho, Yasuyuki See Yamaguchi, Takao: V1418,363-371(1991)

Bessios, Anthony G.
; Nikias, Chrysostomos L.: Blind equalization based on cepstra of power spectrum and tricoherence,V1565,166-177(1991)

Bessudo, Richard See Baudin, Gilles: V1490,102-113(1991)

Best, Margaret E. See Karis, Thomas E.: V1499,366-376(1991)

Bestalanny, S. I. See Mnatsakanyan, Eduard A.: V1414,130-133(1991)

Betensky, Ellis I.
: Role of aspherics in zoom lens design,V1354,656-662(1991)

Beth, Mark-Udo
; Hall, Thomas; Mayerhofer, Wilhelm: Optimization of discharge parameters of an e-beam sustained repetitively pulsed CO2 laser,V1397,577-580(1991)

Bethea, Clyde G. See Levine, Barry F.: V1540,232-237(1991)

Bethel, James S.
: Generation of synthetic stereo views from digital terrain models and digitized photographs,V1457,49-53(1991)

Bethel, Jeff See Kelley, Robert B.: V1387,38-46(1991)

Betros, Robert S.
; Bronowicki, Allen J.; Manning, Raymond A.: Adaptive structures technology programs for space-based optical systems,V1542,406-419(1991)

Bettag, Martin
; Ulrich, Frank; Kahn, Thomas; Seitz, R.: Local interstitial hyperthermia in malignant brain tumors using a low-power Nd:YAG laser,V1525,409-411(1991)
See Ulrich, Frank: V1428,135-135(1991)

Betterton, K. See Twieg, Robert J.: V1455,86-96(1991)

Betti, Silvello
; Curti, Franco; De Marchis, Giancarlo; Iannone, Eugenio: Polarization-modulated coherent optical communication systems,V1579,100-111(1991)

Bettis, Jerry R.
: Correlation between the laser-induced breakdown threshold in solids, liquids, and gases,V1441,521-534(1991)

Betts, Gary E.
; Johnson, Leonard M.; Cox, Charles H.: High-dynamic-range, low-noise analog optical links using external modulators: analysis and demonstration,V1371,252-257(1991)
; Johnson, Leonard M.; Cox, Charles H.: Optimization of externally modulated analog optical links,V1562,281-302(1991)
See Johnson, Leonard M.: V1371,2-7(1991)

Betts, Timothy C.
; Dowling, Jerome M.; Friedman, Richard M.: Nonrejected earth radiance performance of the visible ultraviolet experiment sensor,V1479,120-126(1991)

Beucher, Serge See Rivest, Jean-Francois: V1451,179-190(1991)

Beurle, R. L. See Daemi, M. F.: V1567,621-631(1991)

Beute, Gordon See Flynn, Michael J.: V1444,172-179(1991)

Beutel, Wilhelm See Takagi, Yuuji: V1386,21-29(1991)

Beuthan, Jurgen
; Hagemann, Roland; Mueller, Gerhard J.; Schaldach, Brita J.; Zur, Ch.: New results in dosimetry of laser radiation in medical treatment,V1420,225-233(1991)
; Mueller, Gerhard J.; Schaldach, Brita J.; Zur, Ch.: Fiber design for interstitial laser treatment,V1420,234-241(1991)

Beverly, R. E.
: Photoinitiation of chemical lasers using radiation from a cylindrical surface discharge,V1397,581-584(1991)

Bevilacqua, Richard M.
; Schwartz, Philip R.; Pauls, Thomas A.; Waltman, William B.; Thacker, Dorsey L.: Ground-based monitoring of water vapor in the middle atmosphere: the NRL water-vapor millimeter-wave spectrometer,V1491,231-242(1991)

Bevill, Pat See Kim, Whee-Kuk: V1387,392-406(1991)

Beving, James E.
; Fishell, Wallace G.: Low-cost, low-risk approach to tactical reconnaissance,V1538,14-17(1991)

Bewick, A. See Sumner, Timothy J.: V1549,256-264(1991)

Beyer, Eckhard See Quenzer, A.: V1397,753-759(1991)

Beyer, Horst A.
: Automated dimensional inspection of cars in crash tests with digital photogrammetry,V1526,134-141(1991)

Beyer, Michael See Sebald, Michael: V1466,227-237(1991)

Beyerbacht, Hugo P.
; Aggarwal, Shanti J.; Jansen, E. D.; Welch, Ashley J.: In-vitro study of the effects of Congo Red on the ablation of atherosclerotic plaque,V1427,117-127(1991)

Bezdek, James C.
: Pattern recognition, neural networks, and artificial intelligence,V1468,924-935(1991)
See Liseth, Ole J.: V1468,256-267(1991)

Bezdidko, Sergey N.
: New approach for optimization of optical systems,V1574,250-253(1991)
See Gan, Michael A.: V1574,254-260(1991)

Bezerra, George H. See Fantini, Marcia C.: V1536,81-92(1991)

Bezy, Jean L. See Baudin, Gilles: V1490,102-113(1991)

Bhagvati, Chakravarthy
; Marineau, Peter; Skolnick, Michael M.; Sternberg, Stanley: Morphological algorithms for modeling Gaussian image features,V1606,112-119(1991)

Bhandari, Pradeep
; Petrick, Stanley W.; Schember, Helene R.: Thermal systems analysis for the Space Infrared Telescope Facility dewar,V1540,97-108(1991)

Bhandarkar, Suchendra M.
; Suk, Minsoo: Issues in parallelism in object recognition,V1384,234-245(1991)

Bhardwaj, Sanjay
; Khuri-Yakub, B. T.: In-situ film thickness measurements using acoustic techniques,V1392,555-562(1991)

Bharj, Sarjit
; Taylor, Gordon C.; Denlinger, E. J.; Milgazo, H.: High-dynamic-range mixer using novel balun structure,V1475,340-349(1991)

Bhasin, Kul B.
See Heinen, Vernon O.: V1477(1991)
See Kong, Keon-Shik: V1477,57-65(1991)
See Leonard, Regis F.: V1394,114-125(1991)
See Leonard, Regis F.: V1475(1991)
See Tabib-Azar, Massood: V1477,85-94(1991)

Bhat, Gopalakrishna K.
See Sciammarella, Cesar A.: V1429,183-194(1991)
See Sciammarella, Cesar A.: V1554B,162-173(1991)
See Sciammarella, Cesar A.: V1396,143-154(1991)

Bhat, Rajaram See Soole, Julian B.: V1474,268-276(1991)

Bhattacharjee, K. K. See Jain, Faquir C.: V1564,714-722(1991)

Bhattacharya, Deepika See Singh, Rajiv K.: V1394,203-213(1991)

Bhattacharya, Pallab K.
; Singh, Jasprit: Strained semiconductors and heterostructures: synthesis and applications,V1361,394-405(1991)
See Lachs, Gerard: V1474,248-259(1991)
See Singh, Jasprit: V1362,586-597(1991)

Bhattacharya, Prabir
; Lu, Xun: New parallel algorithms for thinning of binary images,V1468,734-739(1991)
; Qian, Kai: Geometric property measurement of binary images by polynomial representation,V1468,918-922(1991)
; Qian, Kai: Polynomial approach for morphological operations on 2-D and 3-D images,V1383,530-536(1991)

Bhattacharyya, Jagadish C. See Timothy, J. G.: V1343,350-358(1991)

Bhattacharyya, Sankar See Ghatak, Kamakhya P.: V1346,471-489(1991)

Bhullar, B. S. See Boiarski, Anthony A.: V1368,264-272(1991)

Bhushan, M. See Rahmati, Mohammad: V1567,480-489(1991)

Bi, Xianzhi See Chen, Zengtao: V1554A,206-208(1991)

Biache, Andrew See Cox, William J.: V1479,364-369(1991)

Bialolanker, Gabriel
    See Lando, Mordechai: V1412,19-26(1991)
    See Lando, Mordechai: V1442,172-180(1991)

Bian, Buming
    ; Wittels, Norman: Accurate image simulation by hemisphere
        projection,V1453,333-340(1991)

Bian, Hong See Chen, Yu-Feng: V1572,113-117(1991)

Bianchini, Gianandrea See Angrilli, Francesco: V1482,26-39(1991)

Bianco, Bruno See Tommasi, Tullio: V1507,136-141(1991)

Bianconi, Patricia A. See Kunz, Roderick R.: V1466,218-226(1991)

Bibby, Y. W. See Larson, Donald C.: V1572,517-522(1991)

Bibby, Yu Wang See Brewer, Donald R.: V1396,430-434(1991)

Bibeau, Camille See Henesian, Mark A.: V1415,90-103(1991)

Bickel, P.
    ; Christiansen, Jens; Frank, Klaus; Goertler, Andreas; Hartmann, Werner;
        Kozlik, Claudius; Wiesneth, Peter: High-repetition-rate pseudospark
        switches for pulsed high-power lasers,V1503,167-175(1991)

Bickel, William S.
    ; Videen, Gorden W.: Stokes vectors, Mueller matrices, and polarized
        scattered light: experimental applications to optical surfaces and all other
        scatterers,V1530,7-14(1991)

Bickley, Harmon See Myers, M. B.: V1461,242-244(1991)

Bicknell-Tassius, Robert N.
    : Growth of CdTe-CdMnTe heterostructures by molecular beam
        epitaxy,V1484,11-18(1991)
    See Landwehr, Gottfried: V1362,282-290(1991)

Bidaut, Luc M.
    : Composite PET and MRI for accurate localization and metabolic modeling:
        a very useful tool for research and clinic,V1445,66-77(1991)

Biderman, Shlomo
    ; Horowitz, Atara; Einav, Yehezkel; Ben-Amar, Gabi; Gazit, Dan; Stern, Adin;
        Weiss, Matania: Production of sapphire domes by the growth of near-net-
        shape single crystals,V1535,27-34(1991)

Bidgood, W. D. See Staab, Edward V.: V1446,16-22(1991)

Bidlack, Clint R.
    ; Trivedi, Mohan M.: Integrated vision system for object identification and
        localization using 3-D geometrical models,V1468,270-280(1991)
    See Chen, ChuXin: V1468,354-366(1991)
    See Sluder, John C.: V1468,642-652(1991)

Bielak, Leslaw See Heimrath, Adam E.: V1366,265-272(1991)

Bieman, Leonard H.
    : Color-encoded depth: an image enhancement tool,V1385,229-238(1991)
    ; Peyton, James A.: Building an infrastructure for system integration: machine
        vision standards,VCR36,3-19(1991)
    See Harding, Kevin G.: V1385,246-255(1991)

Biemond, Jan
    See Belfor, Ricardo A.: V1605,274-284(1991)
    See Bosveld, Frank: V1605,769-780(1991)
    See Driessen, Johannes N.: V1605,511-521(1991)

Bierhuizen, David J. See Buurman, Johannes: V1386,185-196(1991)

Bierlein, John D. See Morris, Patricia A.: V1561,104-111(1991)

Biermann, Heinz W.
    ; Green, Martina; Seiber, James N.: Long-pathlength DOAS (differential
        optical absorption spectrometer) system for the in situ measurement of
        xylene in indoor air,V1433,2-7(1991)
    See Green, Martina: V1433,270-274(1991)
    See Winer, Arthur M.: V1433,44-55(1991)

Biermann, Stephan
    ; Geiger, Manfred: Integration of diagnostics in high-power laser systems for
        optimization of laser material processing,V1415,330-341(1991)
    See Geiger, Manfred: V1503,238-248(1991)

Biernacki, Paul D.
    ; Brown, Tyler; Freeman, Mark O.: Polarization dependence and uniformity
        of FLC layers for phase modulation,V1455,167-178(1991)

Bietry, Joseph R.
    ; Auriemma, R. A.; Bristow, Thomas C.; Merritt, Edward: Nomarski viewing
        system for an optical surface profiler,V1332,537-543(1991)

Bigan, Erwan
    ; Allovon, Michel; Carre, Madeleine; Carenco, Alain; Voisin, Paul: Efficient
        optical waveguide modulation based on Wannier-Stark localization in a
        InGaAs-InAlAs superlattice,V1362,553-558(1991)

Biganic, Dane D. See Sterenborg, H. J.: V1427,256-266(1991)

Bigbee, William L. See Jensen, Ronald H.: V1403,372-380(1991)

Bigelow, Bruce C.
    : Finite element analysis of large lenses for the Keck telescope high-resolution
        echelle spectrograph,V1532,15-26(1991)

Bigelow, Russell N. See Ingold, Joseph P.: V1589,83-89(1991)

Biggar, Stuart F.
    ; Dinguirard, Magdeleine C.; Gellman, David I.; Henry, Patrice; Jackson, Ray
        D.; Moran, M. S.; Slater, Philip N.: Radiometric calibration of SPOT 2
        HRV: a comparison of three methods,V1493,155-162(1991)
    See Balick, Lee K.: V1493,215-223(1991)
    See Gellman, David I.: V1493,175-180(1991)
    See Markham, Brian L.: V1493,207-214(1991)
    See Nianzeng, Che: V1493,182-194(1991)

Bigio, Irving J.
    ; Czuchlewski, Stephen J.; McCown, Andrew W.; Taylor, Antoinette J.:
        Recent advances in excimer laser technology at Los Alamos,V1397,47-
        53(1991)
    ; Loree, Thomas R.: Biomedical applications of laser technology at Los
        Alamos,V1403,776-780(1991)

Bigley, William J.
    : State equalization and resonant control systems,V1482,350-366(1991)

Bijkerk, Fred
    ; Shevelko, A. P.: Laser plasma XUV sources: a role for excimer
        lasers?,V1503,380-390(1991)
    See Turcu, I. C.: V1503,391-405(1991)

Bilaniuk, L. See Gee, Jim C.: V1445,226-234(1991)

Bilbro, Griff L.
    : General method for accelerating simulated annealing algorithms for
        Bayesian image restoration,V1569,88-98(1991)

Bilgen, Mehmet
    ; Hung, Hsien-Sen: Application of neural network to restoration of signals
        degraded by a stochastic, shift-variant impulse response function and
        additive noise,V1569,260-268(1991)

Bilhorn, Robert B.
    : Detection with a charge-coupled device in atomic emission
        spectroscopy,V1448,74-80(1991)
    ; Ferris, Nancy S.: Analytical Raman and atomic spectroscopies using charge-
        coupled-device detection,V1439,15-24(1991)

Billard, M. See Scavennec, Andre: V1362,331-337(1991)

Bille, Josef F.
    See Goelz, Stefan: V1542,502-510(1991)
    See Liang, Junzhong: V1542,543-554(1991)

Biller, M. See Bussard, Anne B.: V1366,380-386(1991)

Billerey, Dominique
    See Souilhac, Dominique J.: V1494,503-526(1991)
    See Souilhac, Dominique J.: V1521,158-174(1991)

Billings, Daniel A.
    : Basic manufacturability interval,V1468,434-445(1991)

Billings, Paul A. See Carhart, Gary W.: V1564,348-362(1991)

Billingsley, Daniel R.
    See Crow, Samuel B.: V1486,333-344(1991)
    See Watkins, Wendell R.: V1486,17-24(1991)

Billon, Jean P.
    ; Fabre, Edouard; eds.: *Industrial and Scientific Uses of High-Power
        Lasers*,V1502(1991)

Billow, Nick W. See Oinen, Donald E.: V1354,487-493(1991)

Bilodeau, Francois See Malo, Bernard: V1590,83-93(1991)

Bin, Ouyang
    See Guo, Yong H.: V1519,327-332(1991)
    See Li, Shiying: V1410,215-220(1991)
    See Lin, Li-Huang: V1410,65-71(1991)
    See Lin, Li-Huang: V1346,490-501(1991)

Binford, Thomas O. See Lee, Byung-Uk: V1380,96-107(1991)

Bini, Andrea
    See Baldini, Francesco: V1510,58-62(1991)
    See Falciai, Riccardo: V1572,424-427(1991)

Binkley, E. S.
    See Lipscomb, George F.: V1560,388-399(1991)
    See Lytel, Richard S.: V1389,547-558(1991)
    See Wu, Jeong W.: V1560,196-205(1991)

Binns, C. R. See Fraser, George W.: V1548,132-148(1991)

Biolo, Roberta See Jori, Guilio: V1525,367-376(1991)

Bionta, Richard M. See Lai, Barry P.: V1550,46-49(1991)

Birch, Martin J.
See Kubinec, James J.: V1364,130-143(1991)
See Underwood, Ian: V1562,107-115(1991)

Bird, A. J. See Di Cocco, Guido: V1549,102-112(1991)

Bird, David M. See Fatah, Rebwar M.: V1501,120-128(1991)

Bird, K. See Kirk, Andrew G.: V1574,121-132(1991)

Bird, Victor J. See Wesson, Laurence N.: V1367,204-213(1991)

Birge, Robert R.
; Masthay, M. B.; Stuart, Jeffrey A.; Tallent, Jack R.; Zhang, Chian-Fan:
Nonlinear optical properties of bacteriorhodopsin: assignment of the
third-order polarizability based on two-photon absorption
spectroscopy,V1432,129-140(1991)
; Nafie, Laurence A.; eds.: *Biomolecular Spectroscopy II*,V1432(1991)

Birk, Ronald J.
; Tompkins, Jim M.; Burns, Gregory S.: Commercial remote sensing small-
satellite feasibility study,V1495,2-11(1991)

Birkl, Reinhard
; Manhart, Sigmund: Back-reflection measurements on the SILEX
telexcope,V1522,252-258(1991)

Birkle, Siegfried See Sebald, Michael: V1466,227-237(1991)

Birmingham, John J. See Miller, Richard M.: V1439,66-78(1991)

Birnbaum, Milton See Pollack, S. A.: V1410,156-164(1991)

Biron, David G. See Prutzer, Steven: V1480,46-61(1991)

Birth, Gerald S.
: Modeling the distribution of optical radiation in diffusely reflecting
materials,V1379,81-88(1991)
; Dull, Gerald G.; Leffler, Richard G.: Nondestructive determination of the
solids content of horticultural products,V1379,10-15(1991)

Birx, Daniel L. See Goodman, Daniel L.: V1407,217-225(1991)

Bischel, William K. See Reed, Murray K.: V1410,179-184(1991)

Bischof, Thomas
; Jueptner, Werner P.: Determination of the adhesive load by holographic
interferometry using the results of FEM calculations,V1508,90-95(1991)
See Jueptner, Werner P.: V1400,69-79(1991)

Bisenbaev, A. K.
; Levshin, L. V.; Saletsky, A. M.: Excimer formation and singlet-singlet energy
transfer of organoluminophores in the premicellar and micellar-
polyelectrolytes solutions,V1403,606-610(1991)

Bishop, A. R. See Mustre de Leon, Jose: V1550,85-96(1991)

Bishop, Kenneth P.
; Gaspar, Susan M.; Milner, Lisa-Michelle; Naqvi, H. S.; McNeil, John R.:
Grating line shape characterization using scatterometry,V1545,64-
73(1991)
See Hickman, Kirt C.: V1464,245-257(1991)

Bismar, Hisham See Xue, Kefu: V1606,675-684(1991)

Bison, Paolo G.
; Braggiotti, Alberto; Grinzato, Ermanno G.: Transputer-based parallel
algorithms for automatic object recognition,V1471,369-377(1991)

Bissonnette, Daniel See Rochon, Paul L.: V1530,50-57(1991)

Bissonnette, Luc R.
: Blurring effect of aerosols on imaging systems,V1487,333-344(1991)
; Miller, Walter B.; eds.: *Propagation Engineering: Fourth in a
Series*,V1487(1991)
See Hutt, Daniel L.: V1487,250-261(1991)
See Koenig, George G.: V1487,240-249(1991)

Bist, K. S. See Jain, Subhash C.: V1488,410-413(1991)

Bistransin, Mark See Glab, Wallace L.: V1497,389-395(1991)

Biswas, Dipak R.
: Strength and fatigue of optical fibers at different temperatures,V1366,71-
76(1991)

Biswas, Gautam See Kaul, Neeraj: V1468,204-215(1991)

Biswas, Shambhu N.
See Ghatak, Kamakhya P.: V1484,149-166(1991)
See Ghatak, Kamakhya P.: V1415,281-297(1991)
See Ghatak, Kamakhya P.: V1411,137-148(1991)
See Ghatak, Kamakhya P.: V1409,240-257(1991)
See Ghatak, Kamakhya P.: V1485,206-214(1991)
See Ghatak, Kamakhya P.: V1484,136-148(1991)

Bittner, Groff D. See Goelz, Stefan: V1542,502-510(1991)

Bixler, H. A. See Cross, Edward F.: V1540,756-763(1991)

Bixler, Jay V.
; Hailey, Charles J.; Mauche, C. W.; Teague, Peter F.; Thoe, Robert S.; Kahn,
Steven M.; Paerels, F. B.: Performance of a variable-line-spaced master
reflection grating for use in the reflection grating spectrometer on the x-
ray multimirror mission,V1549,420-428(1991)
See Atkinson, Dennis P.: V1343,530-541(1991)

Bizais, Yves
; Gibaud, Bernard; Forte, Anne-Marie; Aubry, Florent; Di Paola, Robert;
Scarabin, Jean-Marie: Qualitative approach to medical image
databases,V1446,156-167(1991)
See Aubry, Florent: V1446,168-176(1991)
See Forte, Anne-Marie: V1445,409-420(1991)
See Guedon, Jean-Pierre: V1443,214-225(1991)

Bjarklev, Anders See Vendeltorp-Pommer, Helle: V1373,254-265(1991)

Bjarnar, Morten L.
; Klepsvik, John O.; Nilsen, Jan E.: Modeling and experiments with a subsea
laser radar system,V1537,74-88(1991)

Bjelkhagen, Hans I.
: Holographic high-resolution endoscopic image recording,V1396,93-
98(1991)
: Holographic portraits,VIS08,347-353(1991)
; Phillips, Nicholas J.; Ce, Wang: Chemical symmetry: developers that look
like bleach agents for holography,V1461,321-328(1991)
See Castro-Montero, Alberto: V1396,122-130(1991)

Bjorkholm, John E. See White, Donald L.: V1343,204-213(1991)

Bjorklund, Gary C. See Moerner, William E.: V1560,278-289(1991)

Bjorling, Sophie See Goldbeck, Robert A.: V1432,14-27(1991)

Blaakman, Andre See Hornak, Joseph P.: V1445,523-533(1991)

Black, David See Steiner, Bruce W.: V1557,156-167(1991)

Black, James E.
: Utility gains through infrared predictive maintenance,V1467,34-40(1991)

Black, Johnathan D.
; Sherk, Henry H.; Uppal, Gurvinder S.; Sazy, John; Meller, Menachem M.;
Rhodes, Anthony; Lane, Gregory J.: Percutaneous lumbar discectomy
using Ho:YAG laser,V1424,20-22(1991)
; Sherk, Henry H.; Meller, Menachem M.; Uppal, Gurvinder S.; Divan, James;
Sazy, John; Rhodes, Anthony; Lane, Gregory J.: Wavelength selection in
laser arthroscopy,V1424,12-15(1991)
See Lane, Gregory J.: V1424,7-11(1991)
See Uppal, Gurvinder S.: V1424,51-52(1991)

Black, Linda R.
; Clark, Ivan O.; Kui, Jianming; Jesser, William A.: Modeling of InP metal
organic chemical vapor deposition,V1557,54-59(1991)

Blackman, Samuel S. See Allen, Ross R.: V1481,292-303(1991)

Blackwell, John D.
; Parrish, William J.; Kincaid, Glen T.: High-performance InSb 256 x 256
infrared camera,V1479,324-334(1991)
See Parrish, William J.: V1512,68-77(1991)
See Parrish, William J.: V1540,274-284(1991)

Blackwell, Luther
: Naval fiber optic system development program,V1589,69-82(1991)

Blackwell, Richard J. See Zediker, Mark S.: V1418,309-315(1991)

Blackwood, Gary H.
; Jacques, Robert; Miller, David W.: MIT multipoint alignment testbed:
technology development for optical interferometry,V1542,371-391(1991)

Blaha, Franz A. See Lu, Zhuo J.: V1588,276-281(1991)

Blaine, G. J.
See Cox, Jerome R.: V1446,40-51(1991)
See Jost, R. G.: V1446,2-9(1991)

Blair, Dianna S.
; Ward, Kenneth J.: Characterization of surface contaminants using infrared
microspectroscopy,V1437,76-79(1991)

Blair, Steven M.
See Eichelberger, Christopher L.: V1396,678-686(1991)
See Siahmakoun, Azad: V1396,190-192(1991)

Blair, William D.
; Rice, Theodore R.; Alouani, Ali T.; Xia, P.: Asynchronous data fusion for
target tracking with a multitasking radar and optical sensor,V1482,234-
245(1991)

Blais, Francois
; Rioux, Marc; MacLean, Steve G.: Intelligent variable-resolution laser
scanner for the space vision system,V1482,473-479(1991)
; Rioux, Marc: Sensing and environment perception for a mobile
vehicle,V1480,94-101(1991)

Blake, James N.
: Magnetic field sensitivity of depolarized fiber optic gyros,V1367,81-86(1991)

Blake, Richard L. See Singman, Leif V.: V1548,80-92(1991)

Blakeney, Andrew J.
See Honda, Kenji: V1466,141-148(1991)
See Uenishi, Kazuya: V1466,102-116(1991)

Blanchet, P.-F. See Leblanc, Roger M.: V1436,92-102(1991)

Blanco, J.
; Fernandez, J. L.; Doval, A. F.; Lopez, C.; Pino, F.; Perez-Amor, Mariano: Study of plate vibrations by moire holography,V1508,180-190(1991)

Blanconnier, P. See Scavennec, Andre: V1362,331-337(1991)

Blank, Lutz C. See Spirit, David M.: V1373,197-208(1991)

Blank, R.
; Friesem, Asher A.: Does a Fourier holographic lens exist?,V1442,26-30(1991)

Blaszczak, Zdzislaw
: Optical Kerr effect in chain carbonyl compounds,V1403,509-511(1991)
: Temperature dependence of optical Kerr effect in aromatic ethers,V1391,146-155(1991)
; Gauden, Pawel: Optical Kerr effect in liquid and gaseous carbon dioxide,V1391,156-163(1991)

Blatherwick, Ronald D.
; Murcray, Frank J.; Murcray, David G.; Locker, M. H.: Determination of the altitude of the nitric acid layer from very high resolution, ground-based IR solar spectra,V1491,203-210(1991)
See Goldman, Aaron: V1491,194-202(1991)
See Murcray, Frank J.: V1491,282-287(1991)

Blau, Werner See Lemoine, Patrick: V1377,45-56(1991)

Blauvelt, Henry A. See Huff, David B.: V1371,244-249(1991)

Blazquez, Carlos H.
: Detection of citrus freeze damage with natural color and color IR video systems,V1467,394-401(1991)
: Use of color, color infrared, black and white films, and video systems in detecting health, stress, and disease in vegetation,V1379,106-115(1991)

Bleha, William P. See Shields, Steven E.: V1455,225-236(1991)

Bleisinger, Rainer See Hoenes, Frank: V1384,305-316(1991)

Blennemann, Heinrich C.
; Pease, R. F.: Novel microstructures for low-distortion chip-to-chip interconnects,V1389,215-235(1991)

Blessinger, Michael A.
See Niblack, Curtiss A.: V1494,403-413(1991)
See Staller, Craig O.: V1540,219-230(1991)

Blez, M. See Sermage, Bernard: V1362,617-622(1991)

Blikman, Albert See Heideman, Rene: V1510,131-137(1991)

Blinov, Leonid M. See Artjushenko, Vjacheslav G.: V1420,149-156(1991)

Bliss, Erlan S.
See Salmon, J. T.: V1542,459-467(1991)
See Toeppen, John S.: V1544,218-225(1991)

Block, Kenneth L.
; Whitworth, Ernest E.; Bergin, Joseph E.: Laboratory development of a nonlinear optical tracking filter,V1483,62-65(1991)
; Whitworth, Ernest E.; Bergin, Joseph E.: Phase-conjugate optical preprocessing filter for small-target tracking,V1481,32-34(1991)

Blocki, Narcyz
; Daszkiewicz, Marek; Galas, Jacek: Computer-driven Fourier diffractometer,V1562,172-183(1991)

Blom, Claes See Novak, Agneta: V1389,80-86(1991)

Blom, Ronald G. See Crippen, Robert E.: V1492,370-377(1991)

Blommel, Fred P.
; Dennis, Peter N.; Bradley, D. J.: Effects of microscan operation on staring infrared sensor imagery,V1540,653-664(1991)

Blondy, Jean M. See Pagnoux, Dominique: V1504,98-106(1991)

Bloom, I.
; Nemirovsky, Yael: Quantum efficiency and crosstalk of an improved backside-illuminated indium antimonide focal plane array,V1442,286-297(1991)

Bloomberg, Dan S.
: Connectivity-preserving morphological image transformations,V1606,320-334(1991)
: Image analysis using threshold reduction,V1568,38-52(1991)

Bloomberg, Steve See Goldmunz, Menachem: V1442,149-153(1991)

Bloss, Walter L.
; O'Loughlin, Michael J.; Rosenbluth, Mary: Advances in multiple-quantum-well IR detectors,V1541,2-10(1991)

Blosser, Timothy H. See Reeves, James B.: V1379,28-38(1991)

Blouke, Morley M.
ed.: *Charge-Coupled Devices and Solid State Optical Sensors II*,V1447(1991)
: Model of a thinned CCD,V1439,136-143(1991)
; Delamere, W. A.; Womack, G.: Simplified model of the back surface of a charge-coupled device,V1447,142-155(1991)
See Delamere, W. A.: V1479,31-40(1991)
See Delamere, W. A.: V1447,204-213(1991)
See Delamere, W. A.: V1447,288-297(1991); 1479,21-30(1991)
See Marsh, Harry H.: V1447,298-309(1991)

Blow, Victor O.
; Giewont, Kenneth J.: Top-side electroluminescence: a failure analysis technique to view electroluminescence along a laser channel,V1366,107-111(1991)

Blum, James D. See Klumpe, Herbert W.: V1398,95-106(1991); 1532,230-240(1991)

Blum, Loic J.
; Gautier, Sabine; Coulet, Pierre R.: Flow injection analysis with bioluminescence-based fiber-optic biosensors,V1510,46-52(1991)

Blume, Hartwig R. See Ji, Tinglan: V1444,136-150(1991)

Blunier, Stefan
See Zogg, Hans: V1361,406-413(1991)
See Zogg, Hans: V1361,1079-1086(1991)

Blyler, Lee L. See Grimes, Gary J.: V1592,139-149(1991)

Bo, Weiyun
; Wang, Rupeng: Phenomenon of fringe patterns from speckle photogram and interferometry within a film,V1554A,750-754(1991)

Boahen, Kwabena A. See Andreou, Andreas G.: V1453,270-281(1991)

Boardman, A. D.
; Hodgson, Elizabeth M.; Spence, A. J.; Richardson, A. D.; Richardson, M. B.: Evolution of an excimer laser gas mix,V1503,160-166(1991)
; Nikitov, S. A.; Wang, Qi; Bao, Jia S.; Cai, Ying S.; Shen, Janice: Dispersion of nonlinear magnetostatic surface waves on thin films,V1519,609-615(1991)
; Nikitov, S. A.; Wang, Qi; Bao, Jia S.; Cai, Ying S.: Multistable magnetostatic waves in thin film,V1519,597-604(1991)
See Wang, Qi: V1519,589-596(1991)

Boashash, Boualem
; O'Shea, Peter: Time-varying higher order spectra,V1566,98-108(1991)
See Jones, Graeme: V1566,167-178(1991)

Bobbio, Stephen M.
See Calabrese, Gary C.: V1466,528-537(1991)
See Freeman, Peter W.: V1464,377-385(1991)

Bobbitt, Donald R. See Rice, Patrick D.: V1435,104-113(1991)

Bobilkov, G. P.
; Genkin, Vladimir N.: Detonation mechanism of breakdowns in dielectrics,V1440,98-104(1991)

Bobrov, Vladimir See Zalessky, Viacheslav N.: V1426,162-169(1991)

Bochove, Erik J.
: Time reversal, enhancement, and suppression of dipole radiation by a phase-conjugate mirror,V1497,222-227(1991)
; Moore, Gerald T.; Scully, Marlan O.; Wodkiewicz, K.: Laser linac: nondiffractive beam and gas-loading effects,V1497,338-347(1991)
See Moore, Gerald T.: V1497,328-337(1991)

Bocian, David F.
: Resonance Raman scattering from the primary electron donor in photosynthetic reaction centers from Rhodobacter sphaeroides,V1432,166-171(1991)

Bock, Wojtek J.
; Wolinski, Tomasz R.; Domanski, Andrzej W.: All-fiber pressure sensor up to 100 MPa,V1511,250-254(1991)
; Wolinski, Tomasz R.; Fontaine, Marie: High-hydrostatic-pressure sensor using elliptical-core optical fibers,V1584,157-161(1991)
See Wolinski, Tomasz R.: V1511,281-288(1991)

Bocker, Richard P.
; Henderson, Thomas B.; Jones, Scott A.; Frieden, B. R.: New inverse synthetic aperture radar algorithm for translational motion compensation,V1569,298-310(1991)

Bocquet, Jean-Claud
; Lecoy, Pierre; Baptiste, Didier: Optical sensors embedded in composite materials,V1588,210-217(1991)

Boddu, Jayabharat
; Udpa, Satish S.; Udpa, L.; Chan, Shiu Chuen M.: High-speed signal processing architectures using charge-coupled devices,V1562,251-262(1991)

Bodechtel, Joh See Ackermann, F.: V1490,94-101(1991)

Boden, Eugene P. See Perry, Joseph W.: V1560,302-309(1991)

Bodenheimer, Joseph S. See Eisenbach, Shlomo: V1442,242-251(1991)

Bodenmann, P. See Kaempfer, Niklaus A.: V1491,314-322(1991)

Bodnar, Vladimir G.
; Koval'chuk, Alexandr V.; Lavrentovich, Oleg D.; Pergamenshchik, V. M.; Sergan, V. V.: Threshold of structural transition in nematic drops with normal boundary conditions in AC electric field,V1455,61-72(1991)

Bodulinsky, V. K.
; Kondratenko, P. S.: Spontaneous magnetic field diffusion from laser plasma,V1440,392-396(1991)

Boechat, Alvaro A. See Su, Daoning: V1502,41-51(1991)

Boechat, Maria I. See Razavi, Mahmood: V1446,24-34(1991)

Boehm, Hans-Dieter V.
; Schreyer, H.; Schranner, R.: Helmet-mounted sight and display testing,V1456,95-123(1991)

Boehm, Leah
; Assaf, Haim: Structural and optical properties of chalcogenide glass-ceramics,V1535,171-181(1991)

Boehme, Johannes M.
; Chimiak, William J.; Choplin, Robert H.; Maynard, C. D.: Operational infrastructure for a clinical PACS,V1446,312-317(1991)

Boehme, Johann F.
; Timmermann, D.; Hahn, H.; Hosticka, Bedrich J.: CORDIC processor architectures,V1566,208-219(1991)

Boeman, Raymond G.
: Interlaminar deformations on the cylindrical surface of a hole in laminated composites by moire interferometry,V1554B,323-330(1991)

Boereboom, Peter See Gatenby, Paul V.: V1522,126-134(1991)

Bogaert, Marc See Malotaux, Eric: V1388,372-383(1991)

Bogdanov, Alexei L.
; Andersson, Eva K.: Fine undercut control in bilayer PMMA-P(MMA-MAA) resist system for e-beam lithography with submicrometer resolution,V1465,324-329(1991)

Bogdanov, V. L.
; Kulya, S. V.; Spiro, A. G.: Four-photon spectroscopy of excited molecules: relaxation pathways and rates from high-excited states of 1,4-diphenylbutadiene molecules,V1403,470-474(1991)

Bogdanowicz, Julius F. See Onishi, Randall M.: V1406,171-178(1991)

Boggan, James E.
; Cerullo, Leonard J.; Smith, Louis C.; eds.: *Three-Dimensional Bioimaging Systems and Lasers in the Neurosciences*,V1428(1991)

Bogle-Rohwer, Elizabeth
; Nulty, James E.; Chu, Wileen; Cohen, Andrew: Spin-on-glass/phosphosilicate glass etchback planarization process for 1.0 um CMOS technology,V1392,280-290(1991)

Bogoroditsky, Alexander G. See Tuchin, Valery V.: V1420,81-92(1991); 1429,62-73(1991)

Boher, Pierre
; Hennet, L.; Houdy, Philippe: Three materials soft x-ray mirrors: theory and application,V1345,198-212(1991)
; Houdy, Philippe; Kaikati, P.; Barchewitz, Robert J.; Van Ijzendoorn, L. J.; Li, Zhigang; Smith, David J.; Joud, J. C.: Comparative study of carbon and boron carbide spacing layers inside soft x-ray mirrors,V1345,165-179(1991)
; Houdy, Philippe; Hennet, L.; Delaboudiniere, Jean-Pierre; Kuehne, Mikhael; Mueller, Peter; Li, Zhigang; Smith, David J.: Silicon/silicon oxide and silicon/silicon nitride multilayers for XUV optical applications,V1343,39-55(1991)

Bohland, John F.
See Calabrese, Gary C.: V1466,528-537(1991)
See Freeman, Peter W.: V1464,377-385(1991)

Bohley, Thomas K. See Golobic, Robert A.: V1425,84-92(1991)

Bohling, Edward H.
; O'Connor, R. P.: Computer vision systems: integration of software architectures,V1406,164-168(1991)
See O'Connor, R. P.: V1472,26-37(1991)

Bohn, Courtlandt L.
; Delayen, Jean R.: Investigations of cumulative beam breakup in radio-frequency linacs,V1407,566-577(1991)
See Delayen, Jean R.: V1407,524-534(1991)

Bohn, Willy L.
; Truesdell, Keith; Latham, William P.; Avizonis, Petras V.: Small-signal gain in the oxygen-iodine laser,V1397,235-238(1991)

Bohrer, M. P. See Nalamasu, Omkaram: V1466,13-25(1991)

Boiarski, Anthony A.
; Ridgway, Richard W.; Miller, Larry S.; Bhullar, B. S.: Integrated optic device for biochemical sensing,V1368,264-272(1991)

Boiko, Yuri B.
: Volume holographic optics recording in photopolymerizable layers,V1507,318-327(1991)
; Granchak, Vasilij M.: Light scattering holographic optical elements formation in photopolymerizable layers,V1507,544-548(1991)
; Granchak, Vasilij M.; Dilung, Iosiph I.; Solovjev, Vladimir S.; Sisakian, Iosif N.; Soifer, Victor A.: Relief holograms recording on liquid photopolymerizable layers,V1238,253-257(1991)
; Granchak, Vasilij M.; Dilung, Iosiph I.; Mironchenko, Vladislav Y.: Volume holograms in liquid photopolymerizable layers,V1238,258-265(1991)

Boily, S. See Chaker, Mohamed: V1465,16-25(1991)

Boisde, Gilbert
; Sebille, Bernard: Co-immobilization of several dyes on optodes for pH measurements,V1510,80-94(1991)

Boisrobert, Christian Y.
; Franzen, Douglas L.; Danielson, Bruce L.; Christensen, David H.: Low-coherence optical reflectometry of laser diode waveguides,V1474,285-290(1991)

Bojadziev, Stefan See Lightner, David A.: V1432,2-13(1991)

Bojanczyk, Adam W.
; Brent, Richard P.; de Hoog, F. R.: Stability of Bareiss algorithm,V1566,23-34(1991)
; Lee, Tong J.; Luk, Franklin T.: Accurate fast Hankel matrix solver,V1566,74-83(1991)

Bok, Julien
: High-Tc superconductors, physics, and applications,V1362,94-101(1991)

Bokor, J. See White, Donald L.: V1343,204-213(1991)

Bokov, Lev
; Demidov, Anatoly J.; Zadorin, Anatoly; Kushnarev, Igor; Serebrennikov, Leonid J.; Sharangovich, Sergey: Development of wideband 16-channel acousto-optical modulators on the LiNbO3 and TeO2 crystals,V1505,186-198(1991)

Bokun, Leszek J.
; Domanski, Andrzej W.: Fiber optic photoplethysmograph,V1420,93-99(1991)

Boldish, Steven I.
: Production considerations necessary to produce large quantities of optoelectronic devices by MOCVD epitaxy,V1449,51-64(1991)

Boles, John A.
; Hems, Randall K.: Analysis of barcode digitization techniques,V1384,195-204(1991)

Boley, Daniel L.
: Orthogonal polynomials, Hankel matrices, and the Lanczos algorithm,V1566,84-95(1991)

Bolger, Bouwe
See van Hulst, Niek F.: V1361,581-588(1991)
See van Hulst, Niek F.: V1362,631-646(1991)

Bollanti, Sarah
; Di Lazzaro, Paolo; Flora, Francesco; Fu, Shufen; Giordano, Gualtiero; Letardi, Tommaso; Lisi, Nicola; Schina, Giovanni; Zheng, Cheng-En: Excimer laser development at the ENEA Frascati Centre: discharge instabilities study,V1503,80-87(1991)
; Di Lazzaro, Paolo; Flora, Francesco; Giordano, Gualtiero; Letardi, Tommaso; Lisi, Nicola; Schina, Giovanni; Zheng, Cheng-En: Large volume XeCl laser with longitudinal gas flow: experimental results and theoretical analysis,V1397,97-102(1991)

Bolle, Aldo See Ekberg, Mats: V1527,236-239(1991)

Bolle, Matthias See Ihlemann, Juergen: V1361,1011-1019(1991)

Bolle, Ruud M.
; Califano, Andrea; Kender, John R.; Kjeldsen, Rick; Mohan, Rakesh: Multilevel evidence fusion for the recognition of 3-D objects: an overview of computer vision research at IBM/T.J. Watson,V1383,305-318(1991)
See Hecker, Y. C.: V1570,298-314(1991)

Bolognese, Paolo See Fasano, Victor A.: V1428,2-12(1991)

Bolorizadeh, M. A. See Bahram-pour, A. R.: V1397,539-542(1991)

Bolshov, L. A. See Kanevsky, M. F.: V1440,154-165(1991)

Bolstad, Jon O. See Agapakis, John E.: V1385,32-38(1991)

Boltinghouse, Susan T.
; Burke, James; Ho, Daniel: Implementation of a 3-D laser imager-based robot navigation system with location identification,V1388,14-29(1991)

Bolton, Paul R.
See Fittinghoff, David N.: V1413,81-88(1991)
See Gold, David: V1413,41-52(1991)

Bom, Nicolaas See Gussenhoven, Elma J.: V1425,203-206(1991)

Bommannavar, A.
See Mini, Susan M.: V1345,260-269(1991)
See Ramanathan, Mohan: V1548,168-178(1991)

Bommarito, G. M.
; Acevedo, D.; Rodriguez, J. R.; Abruna, H. D.: In-situ structural studies of the underpotential deposition of copper onto an iodine-covered platinum surface using x-ray standing waves,V1550,156-170(1991)

Bomse, David S. See Silver, Joel A.: V1435,64-71(1991)

Bon, Bruce
; Wilcox, Brian H.; Litwin, Todd; Gennery, Donald B.: Operator-coached machine vision for space telerobotics,V1387,337-342(1991)

Bona, Gian-Luca See Webb, David J.: V1418,231-239(1991)

Bonanome, Andrea See Visona, Adriana: V1425,75-83(1991)

Bonasso, R. P.
; Nishihara, H. K.: Using real-time stereopsis for mobile robot control,V1387,237-244(1991)

Bonavito, N. L.
; Dorband, John E.; Busse, Tim: Maximum entropy method applied to deblurring images on a MasPar MP-1 computer,V1406,138-146(1991)

Boncelet, Charles G.
; Hardie, Russell C.; Hakami, M. R.; Arce, Gonzalo R.: LUM filters for smoothing and sharpening,V1451,70-74(1991)
See Arce, Gonzalo R.: V1451(1991)

Bonch-Bruevich, Aleksei M.
; Konov, Vitaly I.; Libenson, Michail N.; eds.: *Optical Radiation Interaction with Matter*,V1440(1991)

Bond, Robert L. See Mozumdar, Subir: V1584,254-261(1991)

Bondarenko, Alexander V.
; Glova, Alexander F.; Kozlov, Sergei N.; Lebedev, Fedor V.; Likhanskii, Vladimir V.; Napartovich, Anatoly P.; Pis'menny, Vladislav D.; Yaztsev, Vladimir P.: Regular and chaotic pulsations of radiation intensity in a CO2 laser with modulated parameters,V1376,117-127(1991)
See Antyuhov, V.: V1397,355-365(1991); 1415,48-59(1991)

Bondarev, S. L.
; Tikhomirov, S. A.; Bachilo, S. M.: Picosecond kinetics and Sn S1 absorption spectra of retinoids and carotenoids,V1403,497-499(1991)

Bondur, James
; Turner, Terry R.; eds.: *Advanced Techniques for Integrated Circuit Processing*,V1392(1991)

Bondyopadhyay, Probir See Arndt, George D.: V1475,231-242(1991)

Bone, David A. See Fordham, John L.: V1449,87-98(1991)

Boneberg, J.
; Yavas, O.; Mierswa, B.; Leiderer, Paul: Dynamics of the optical parameters of molten silicon during nanosecond laser annealing,V1598,84-90(1991)

Bonhommet, Herve See Tanguy, Mireille: V1487,172-183(1991)

Bonissone, Piero P. See Comly, James B.: V1381,390-400(1991)

Bonnafe, Jacques See Montgomery, Paul C.: V1332,563-570(1991)

Bonnell, Lee J. See Jen, Cheng K.: V1590,107-119(1991)

Bonner, R. See Wiedwald, Douglas J.: V1346,449-455(1991)

Bonner, Raymond See Brecher, Virginia H.: V1473,39-51(1991)

Bonner, Robert F.
See Nossal, Ralph J.: V1431,21-28(1991)
See Pinto, Joseph F.: V1420,242-243(1991)

Bonnet, Jean C.
See Bacis, Roger: V1397,173-176(1991)
See Godard, Bruno: V1503,71-77(1991)
See Godard, Bruno: V1397,59-62(1991)

Bonniau, Philippe
; Chazelas, Jean; Lecuellet, Jerome; Gendre, Francois; Turpin, Marc; Le Pesant, Jean-Pierre; Brevignon, Michele: Damage detection in woven-composite materials using embedded fiber-optic sensors,V1588,52-63(1991)

Bonnin, Patrick
; Zavidovique, Bertrand: New cooperative edge linking,V1381,142-152(1991)

Bonnot, Anne-Marie See Mathey, Yves: V1361,909-916(1991)

Bonoli, C. See Bortoletto, Favio: V1542,225-235(1991)

Bonometti, Robert J.
; Wheatley, Alvis A.; Flynn, Lin; Nicastri, Edward; Sudol, R.: DARPA initiatives in small-satellite technologies,V1495,166-176(1991)

Bonora, S. See Bertoluzza, Alessandro: V1403,153-155(1991)

Bonsignori, Roberto
; Maresi, Luca: Sensor system for comet approach and landing,V1478,76-91(1991)

Booher, Dan J. See Fu, Richard J.: V1418,108-115(1991)

Boone, Bradley G.
; Shukla, Oodaye B.; Terry, David H.: Extraction of features from images using video feedback,V1471,390-403(1991)
See Christens-Barry, William A.: V1564,177-188(1991)

Boone, Pierre M.
: Herbarium holographicum,VIS08,370-380(1991)
; Jacquot, Pierre M.: Some applications of a HOE-based desensitized interferometer in materials research,V1554A,512-521(1991)

Booth, Bruce L.
; Hohman, James L.; Keating, Kenneth B.; Marchegiano, Joseph E.; Witman, Sandy L.: Excimer laser micromachining for passive fiber coupling to polymeric waveguide devices,V1377,57-63(1991)
See Hartman, Davis H.: V1390,368-376(1991)

Bootman, Steven R.
: Crossconnects in a SONET network,V1363,142-148(1991)

Booysen, Andre
; Spammer, Stephanus J.; Swart, Pieter L.: Ratiometric fiber optic sensor utilizing a fused biconically tapered coupler,V1584,273-279(1991)

Bopp, Matthias
; Huether, Gerhard; Spatscheck, Thomas; Specker, Harald; Wiesmann, Theo J.: BPSK homodyne and DPSK heterodyne receivers for free-space communication with ND:host lasers,V1522,199-209(1991)

Bordeneuve, J. See Dilhac, Jean-Marie R.: V1393,395-403(1991)

Bordo, V. G.
; Ershov, I. A.; Kravchenko, V. A.; Petrov, Yu. N.: Heterophase isotopic SF6 molecule separation in the surface electromagnetic wavefield,V1440,364-369(1991)

Bordui, Peter F.
ed.: *Inorganic Crystals for Optics, Electro-Optics, and Frequency Conversion*,V1561(1991)

Borel, Christoph C.
; Gerstl, Siegfried A.: Simulation of partially obscured scenes using the radiosity method,V1486,271-277(1991)

Boreman, Glenn D.
See Anderson, Barry K.: V1488,416-425(1991)
See Barnard, Kenneth J.: V1488,426-431(1991)
See Schildwachter, Eric F.: V1488,48-57(1991)

Borenstein, Johann
; Koren, Yoram: Real-time map building for fast mobile robot obstacle avoidance,V1388,74-81(1991)

Bores, Leo D.
: Corneal topography: the dark side of the moon,V1423,28-39(1991); 1429,217-228(1991)

Borg, Eric J. See Owen, Philip R.: V1488,122-132(1991)

Borghs, Gustaaf
See Pankove, Jacques I.: V1361,620-627(1991)
See Van Hoof, Chris A.: V1362,291-300(1991)

Boriak, Richard See Sessler, Jonathan L.: V1426,318-329(1991)

Borik, Stefan
; Giesen, Adolf: Finite element analysis of the transient behavior of optical components under irradiation,V1441,420-429(1991)

Borisevich, Nikolai A.
; Zamkovets, A. D.; Ponyavina, A. N.: High-contrast composite infrared filters,V1500,222-231(1991)

Borisov, B. A. See Petrova, G. P.: V1403,387-389(1991)

Borisov, V. M.
; Khristoforov, O. B.; Kirykhin, Yu. B.; Kuznetsov, S. G.; Stepanov, Yu. Y.; Vinokhodov, A. Y.: Kilowatt-range high-repetition-rate excimer lasers,V1503,40-47(1991)

Borman, C. See Zarnowski, Jeffrey J.: V1447,191-201(1991)

Born, I. See Kreider, Gregory: V1381,242-249(1991)

Borodin, A. M.
; Ivanov, K. N.; Naumov, S. K.; Philippov, S. A.; Postovalov, V. E.; Prokhorov, Alexander M.; Stepanov, M. S.; Schelev, Mikhail Y.: One-dimensional CCD linear array readout device,V1358,756-758(1991)

Borough, Howard C.
: Matched-filter algorithm for subpixel spectral detection in hyperspectral image data,V1541,199-208(1991)

Borra, Ermanno F.
; Content, R.: Lunar liquid-mirror telescopes,V1494,219-227(1991)

Borrelli, Nicholas F.
See Araujo, Roger J.: V1590,138-145(1991)
See Bellman, Robert H.: V1544,209-217(1991)

Borsella, E.
; De Padova, P.; Larciprete, Rosanna: Excimer laser deposition and characterization of tin and tin-oxide films,V1503,312-320(1991)

Borst, Cornelius
See Smits, Pieter C.: V1425,188-190(1991)
See van Leeuwen, Ton G.: V1427,214-219(1991)
See Verdaasdonk, Rudolf M.: V1420,136-140(1991)
See Verdaasdonk, Rudolf M.: V1425,102-109(1991)

Borst, Walter L. See Glab, Wallace L.: V1497,389-395(1991)

Borsuk, Pavel S. See Potyrailo, Radislav A.: V1572,434-438(1991)

Bortoletto, Favio
; Baruffolo, A.; Bonoli, C.; D'Alessandro, Maurizio; Fantinel, D.; Giudici, G.; Ragazzoni, Roberto; Salvadori, L.; Vanini, P.: Primary mirror control system for the Galileo telescope,V1542,225-235(1991)
See Ragazzoni, Roberto: V1542,236-246(1991)

Bortolini, James R. See Coyle, Richard J.: V1412,129-137(1991)

Bortz, M. L. See Lim, Eric J.: V1561,135-142(1991)

Bosch, D. See Swirhun, S.: V1475,303-308(1991)

Bosch, Robert A. See Menge, Peter R.: V1407,578-588(1991)

Boscher, Daniel M. See Balageas, Daniel L.: V1467,278-289(1991)

Boscolo-Boscoletto, A. See Mazzoldi, Paolo: V1513,182-197(1991)

Bosenberg, W. R. See Guyer, Dean R.: V1409,14-17(1991)

Bosshard, Christian
; Kuepfer, Manfred; Floersheimer, M.; Guenter, Peter: Guided-wave nonlinear optics in 2-docosylamino-5-nitropyridine Langmuir-Blodgett films,V1560,344-352(1991)

Bossler, Franklin B. See LaViolette, Kerry D.: V1398,213-218(1991)

Bostel, Ashley J.
; McOwan, Peter W.; Hall, Trevor J.: Optical Distributed Inference Network (ODIN),V1385,165-172(1991)

Bosveld, Frank
; Lagendijk, Reginald L.; Biemond, Jan: Three-dimensional subband decompositions for hierarchical video coding,V1605,769-780(1991)

Boswell, Rod W. See Benjamin, Neil M.: V1392,95-105(1991)

Bosworth, Eric See Nishikawa, Robert M.: V1444,180-189(1991)

Bosyi, O. N.
; Efimov, Oleg M.; Mekryukov, A. M.: Investigation of optical parameters of silicate glasses in case of before-threshold effect of laser radiation,V1440,57-62(1991)

Botez, Dan
; Mawst, Luke J.; Jansen, Michael; Anderson, Eric R.; Ou, Szutsun S.; Sergant, Moshe; Peterson, Gary L.; Roth, Thomas J.; Rozenbergs, John: High-power coherent diode lasers,V1474,64-74(1991)
See Anderson, Eric R.: V1417,543-549(1991)
See Jansen, Michael: V1418,32-39(1991)
See Mawst, Luke J.: V1418,353-357(1991)

Bothwell, Mary
: Space Infrared Telescope Facility science instruments overview,V1540,15-26(1991)
See Capps, Richard W.: V1540,47-50(1991)

Botma, H.
; Peters, Peter J.; Witteman, Wilhelmus J.: Efficient e-beam sustained Ar:Xe laser,V1397,573-576(1991)

Bottacchi, Stefano
: Numerical solution of the one-dimensional Schrodinger equation: application to heterostructures and superlattices,V1362,727-749(1991)
See Bellomi, Giovanni: V1362,760-767(1991)

Bottura, Giorgio
; Filippetti, P.; Tinti, A.: Vibrational spectroscopy of biological molecules: halocompound/nucleic acid component interactions,V1403,156-158(1991)

Bou-Ghannam, Akram A.
; Doty, Keith L.: CLIPS implementation of a knowledge-based distributed control of an autonomous mobile robot,V1468,504-515(1991)

Bouchenkov, Vyatcheslav A.
; Utenkov, Boris I.; Antipenko, Boris M.; Berezin, Juri D.; Berezin, U. D.; Malinin, Boris G.; Serebryakov, Victor A.: 10-W Ho laser for surgery,V1410,185-188(1991); 1427,409-412(1991)
; Utenkov, Boris I.; Zaitsev, V. K.; Bayanov, Valentin I.; Serebryakov, Victor A.: High-energy ND:glass laser for oncology,V1427,405-408(1991)
; Utenkov, Boris I.; Zajtsev, Viktor K.; Bayanov, Valentin I.; Serebryakov, Victor A.: High-energy Nd:glass laser for oncology,V1410,244-247(1991)

Boucher, Maurice W. See Klocek, Paul: V1498,147-157(1991)

Bouchitte, G. See Petit, Roger: V1545,31-41(1991)

Boudenne, Jean-Michel See Koenig, Michel: V1502,338-342(1991)

Boudot, C.
; Vastra, I.: Influence of the distance between the welding head and the surface material of the weld shape,V1502,177-189(1991)
; Vastra, I.: Study of different optical fibers,V1502,72-82(1991)
See Vastra, I.: V1502,190-202(1991)

Boudreaux-Bartels, G. F. See Kadambe, Shubha: V1565,423-434(1991)

Boudry, J. See Antonuk, Larry E.: V1443,108-119(1991)

Boufflet, Jean-Paul See Trigano, Philippe: V1468,408-416(1991)

Bougas, V.
; Kalymnios, Demetrios: Plastic fiber couplers using simple polishing techniques,V1504,298-302(1991)

Bougoulias, Dimitrios K. See Thomopoulos, Stelios C.: V1470,253-262(1991)

Boulard, B.
; Jacoboni, Charles: Vapor phase deposition of transition metal fluoride glasses,V1513,204-208(1991)

Bouldin, Dennis P. See Sullivan, Timothy D.: V1596,83-95(1991)

Bouley, J. C. See Sermage, Bernard: V1362,617-622(1991)

Boult, Terrance E. See Lerner, Mark: V1383,277-284(1991)

Boulton, Herbert See Bales, Maurice: V1467,195-206(1991)

Bouman, Charles A. See Yoo, Jisang: V1451,58-69(1991)

Bour, David P.
See Coleman, James J.: V1418,318-327(1991)
See Evans, Gary A.: V1418,406-413(1991)
See Evans, Gary A.: V1378,146-161(1991)

Bourda, C.
; Puig, Thierry T.; Decamps, B.; Condat, M.: Deformation of a gamma/gamma WASPALOY after laser shock,V1502,148-159(1991)

Bourdet, Gilbert L. See Vigroux, Luc M.: V1500,74-79(1991)

Bourdinaud, Michel J. See Laguesse, Michel F.: V1592,96-107(1991)

Bourdoncle, Bernard
; Chauveau, J. P.; Mercier, Jean-Louis M.: Ray tracing through progressive ophthalmic lenses,V1354,194-199(1991)

Bourgoin, J. C.
; Feng, S. L.; von Bardeleben, H. J.: Deep levels in III-V compounds, heterostructures, and superlattices,V1361,184-194(1991)

Bourhis, Jean-Francois
; Guerin, Philippe; Teral, Stephane; Haux, Denys; Di Maggio, Michel: WDS measurement of thallium-diffused glass waveguides and numerical simulation: a comparison,V1583,383-388(1991)

Bourillot, E. See Goudonnet, Jean-Pierre: V1545,130-139(1991)

Bourlon, Philippe M. See Koehler, Bertrand: V1489,177-188(1991)

Bourne, Lincoln C.
; Cardona, A. H.; James, Tim W.; Fleming, J. S.; Forse, R. W.; Hammond, Robert B.; Hong, J. P.; Kim, T. W.; Fetterman, Harold R.: High-temperature superconducting Josephson mixers from deliberate grain boundaries in Tl2CaBa2Cu2O8,V1477,205-208(1991)

Bourne, Neil K.
; Field, John E.: Collapsing cavities in reactive and nonreactive media,V1358,1046-1056(1991)

Bourne, Robert W.
; Jefferys, E. A.; Murphy, Kevin S.: Assessment of the optimum operating conditions for 2-D focal-plane-array systems,V1488,73-79(1991)

Bournes, Patrick A. See Koechner, Walter: V1522,169-179(1991)

Bournot, Philippe See Paolacci, Sylvie: V1397,705-708(1991)

Bousbiat, Essaid See Simonne, John J.: V1374,107-115(1991)

Bousek, Ronald R.
: Laser-target diagnostics instrumentation system,V1414,175-184(1991)

Boussekey, Luc See Constant, Monique T.: V1362,156-162(1991)

Bousselet, Philippe See Marcerou, Jean-Francois: V1373,168-186(1991)

Boutinaud, P.
; Parent, C.; Le Flem, Gille; Moine, Bernard; Pedrini, Christian; Duloisy, E.: Fluorescence properties of Cu+ ion in borate and phosphate glasses,V1590,168-178(1991)

Bouvier, A. J. See Bacis, Roger: V1397,173-176(1991)

Bouyssounouse, Xavier See Daniell, Cindy E.: V1471,436-451(1991)

Bouzerdoum, Abdesselam
; Nabet, Bahram; Pinter, Robert B.: Analysis and analog implementation of directionally sensitive shunting inhibitory neural networks,V1473,29-40(1991)

Bovard, Bertrand G.
; Zhao, Tianji; Macleod, H. A.: Smooth diamond films by reactive ion-beam polishing,V1534,216-222(1991)

Bove, V. M.
; Watlington, John: Cheops: a modular processor for scalable video coding,V1605,886-893(1991)

Bovik, Alan C.
; Howard, Vyvyan; eds.: *Biomedical Image Processing II*,V1450(1991)
See Bartels, Keith A.: V1450,30-39(1991)
See Jordan, John R.: V1382,111-122(1991)
See Silsbee, Peter L.: V1452,409-419(1991)
See Super, Boaz J.: V1606,574-586(1991)

Bow, Sing T.
; Chen, Pei: Computerized detection and identification of the types of defects on crystal blanks,V1396,646-655(1991)

Bowden, Charles M.
See Cantrell, Cyrus D.: V1497(1991)
See Englund, John C.: V1497,218-221(1991)
See Haus, Joseph W.: V1497,382-388(1991)

Bowen, Max L. See Hilkert, James M.: V1498,24-38(1991)

Bowers, C. R.
; Buratto, S. K.; Carson, Paul; Cho, H. M.; Hwang, J. Y.; Mueller, L.; Pizarro, P. J.; Shykind, David; Weitekamp, Daniel P.: New approaches to ultrasensitive magnetic resonance,V1435,36-50(1991)

Bowers, George H.
: Continuous flow manufacturing,V1496,239-246(1991)

Bowers, John E. See Crawford, Deborah L.: V1371,138-141(1991)

Bowers, John S. See Maienschein, Jon L.: V1343,477-484(1991)

Bowles, Roland L.
See Targ, Russell: V1521,144-157(1991)
See Targ, Russell: V1416,130-138(1991)

Bowley, David J.
; Rickett, Ph.; Babushkin, A. V.; Vorobiev, N. S.; Prokhorov, Alexander M.; Schelev, Mikhail Y.: Measurement of triggering instabilities of Imacon 500 streak cameras,V1358,550-555(1991)

Bowley, Reginald R. See Knight, Stephen E.: V1464,119-126(1991)

Bowman, Elizabeth M.
; Burgess, Lloyd W.: Evaluation of polymeric thin film waveguides as chemical sensors,V1368,239-250(1991)

Bowmaster, Thomas A.
; Cockings, Orville R.; Swanson, Robert A.: SONET inter-vendor compatibility,V1363,119-124(1991)

Bown, Stephen G.
: New approaches to local destruction of tumors: interstitial laser hyperthermia and photodynamic therapy,V1525,325-330(1991)
See McNicholas, Thomas A.: V1421,30-35(1991)

Bowne, Norman E. See Ching, Jason K.: V1491,360-370(1991)

Bowyer, C. S. See Harada, Tatsuo: V1545,2-10(1991)

Bowyer, Kevin W.
See Stark, Louise: V1381,334-345(1991)
See Wilkins, Belinda: V1468,662-673(1991)

Boxler, Lawrence H.
; Brown, Glen; Western, Arthur B.: Practical use of reference mirror rotation in holographic interferometry,V1396,85-92(1991)
See Brown, Glen: V1396,164-173(1991)

Boyce, James B.
See Bridges, Frank: V1550,76-84(1991)
See Fork, David K.: V1394,202-202(1991)

Boyce, James F. See Tajbakhsh, Shahram: V1521,14-22(1991)

Boyd, Joseph T.
; Radens, Carl J.; Kauffman, Michael T.: Length-minimization design considerations in photonic integrated circuits incorporating directional couplers,V1374,138-143(1991)
See De Brabander, Gregory N.: V1583,327-330(1991)

Boyd, Robert See Severson, William E.: V1388,490-501(1991)

Boyer, A. E. See Patonay, Gabor: V1435,52-63(1991)

Boyer, Corinne
; Gaffard, Jean-Paul; Barrat, Jean-Pierre; Lecluse, Yves: Adaptive optics, transfer loops modeling,V1483,77-91(1991); 1542,46-61(1991)
See Gaffard, Jean-Paul: V1483,92-103(1991)
See Gendron, Eric: V1542,297-307(1991)

Boyer, James D.
; Mauro, Billie R.; Sanders, Virgil E.: Coating development for high-energy KrF excimer lasers,V1377,92-98(1991)
; Mauro, Billie R.; Sanders, Virgil E.: Development of damage resistant optics for KrF excimer lasers,V1441,255-261(1991)

Boyer, Joseph W. See Rosenthal, Marc S.: V1443,132-142(1991)

Boyer, Keith See Haddad, Waleed S.: V1448,81-88(1991)

Boyers, D. G. See Piestrup, Melvin A.: V1552,214-239(1991)

Boyle, William J.
; Grattan, Kenneth T.; Palmer, Andrew W.; Meggitt, Beverley T.: Application of low-coherence optical fiber Doppler anemometry to fluid-flow measurement: optical system considerations,V1511,51-56(1991)
See Weir, Kenneth: V1584,220-225(1991)

Bozorgebrahimi, Nercy See Kosel, Peter B.: V1541,48-59(1991)

Bozzola, Andrea
See Gaboardi, Franco: V1421,79-85(1991)
See Gaboardi, Franco: V1421,50-55(1991)
See Gaboardi, Franco: V1421,73-77(1991)

Bozzolo, Nora G. See Hartz, William G.: V1398,52-60(1991)

Bracci, Susanna See Baldini, Francesco: V1368,210-217(1991)

Bracikowski, Christopher
; Roy, Rajarshi: Deterministic fluctuations in an intracavity-coupled solid state laser,V1376,103-116(1991)

Brackett, Vincent G. See Fishman, Jack: V1491,348-359(1991)

Bradbury, Savile
: Capabilities of the optical microscope,V1439,128-134(1991)

Brade, Richard See Clifford, Sandra: V1463,551-557(1991)

Bradford, Elaine See Turner, Robert E.: V1377,99-106(1991)

Bradford, George See Underwood, Ian: V1562,107-115(1991)

Bradie, Brian D. See Barouch, Eytan: V1465,254-262(1991)

Bradley, D. J.
; Dennis, Peter N.; Baddiley, C. J.; Murphy, Kevin S.; Carpenter, Stephen R.; Wilson, W. G.: Computer simulation of staring-array thermal imagers,V1488,186-195(1991)
See Blommel, Fred P.: V1540,653-664(1991)
See Murphy, Kevin S.: V1488,178-185(1991)

Bradley, David K. See Kilkenny, Joseph D.: V1358,117-133(1991)

Bradley, Eric M.
; Rybka, Theodore W.; Yu, Paul K.: Temperature stability of Bragg grating resonant frequency,V1418,272-278(1991)

Bradley, Joe C.
; Kellner, Albert L.: Analysis of evanescent coupling in waveguide modulators,V1476,330-336(1991)

Bradley, Lester M.
; Corriveau, John P.; Tindal, Nan E.: Launch area theodolite system,V1482,48-60(1991)

Bradley, Scott E. See Heaney, James B.: V1485,140-159(1991)

Bradshaw, John D.
; van Dijk, Cornelius A.: 2.9 micron laser source for use in the two-photon/laser-induced fluorescence detection of atmospheric OH,V1433,81-91(1991)
See Sandholm, Scott T.: V1433,69-80(1991)

Brady, Dan R. See Dickinson, Robert R.: V1459,178-189(1991)

Brady, David J. See Chen, Geng-Sheng: V1562,128-135(1991)

Brady, G. See Wolfe, William J.: V1382,240-254(1991)

Braeuer, Andreas
; Bartuch, Ulrike; Zeisberger, M.; Bauer, T.; Dannberg, Peter: Linear and nonlinear optical effects in polymer waveguides,V1559,470-478(1991)
See Guntau, Matthias: V1513,112-122(1991)

Braggins, Donald W.
: Shortage of system integrators,VCR36,46-55(1991)

Braggiotti, Alberto See Bison, Paolo G.: V1471,369-377(1991)

Braichotte, D.
; Wagnieres, G.; Monnier, Philippe; Savary, M.; Bays, Roland; van den Bergh, Hubert; Chatelain, Andre: Endoscopic tissue autofluorescence measurements in the upper aerodigestive tract and the bronchi,V1525,212-218(1991)
See Bays, Roland: V1525,397-408(1991)
See Wagnieres, G.: V1525,219-236(1991)

Brailean, James C.
; Giger, Maryellen L.; Chen, Chin-Tu; Sullivan, Barry J.: Quantitative performance evaluation of the EM algorithm applied to radiographic images,V1450,40-46(1991)

Brailovsky, A. B.
; Gaponov, Sergey V.; Dorofeev, I. A.; Lutschin, V. I.; Semenov, V. E.: Formation of large-scale relief on a target surface under multiple-pulsed action of laser radiation,V1440,84-89(1991)

Brakenhoff, G. J.
; van der Voort, H. T.; Visscher, Koen: Confocal microscopy for the biological and material sciences: principle, applications, limitations,V1439,121-127(1991)

Brames, Bryan J. See Abbiss, John B.: V1566,365-375(1991)

Bramson, Michael D.
: Depolarized fiber optic gyro for future tactical applications,V1367,155-160(1991)

Brand, Richard A.
; Spinar, Karen K.: Lightweight composite mirrors: present and future challenges,V1532,57-63(1991)
See Krumweide, Gary C.: V1533,252-261(1991)

Brandenburg, Albrecht
; Edelhaeuser, Rainer; Hutter, Frank: Gas sensor based on an integrated optical interferometer,V1510,148-159(1991)

Brandley, M. See Westerman, Steven D.: V1492,263-271(1991)

Brandman, N. See Ben-Shalom, Ami: V1486,238-257(1991)

Brandstetter, Robert W.
; Fonneland, Nils J.; Zanella, R.; Yearwood, M.: Optical correlator vision system for a manufacturing robot assembly cell,V1385,173-189(1991)
; Fonneland, Nils J.: Photopolymer elements for an optical correlator system,V1559,308-320(1991)
See Leib, Kenneth G.: V1483,140-154(1991)

Brandt, Gerald B.
; Singh, N. B.; Gottlieb, Milton S.: Mercurous halides for long time-delay Bragg cells,V1454,336-343(1991)
See Leksell, David: V1458,133-144(1991)

Brandt, Howard E.
ed.: *Intense Microwave and Particle Beams II,*V1407(1991)
: Nonlinear bremsstrahlung in nonequilibrium relativistic beam-plasma systems,V1407,326-353(1991)
ed.: *Selected Papers on Nonlinear Optics,*VMS32(1991)

Brandt, S. A.
; Budenske, John: STARCON: a reconfigurable fieldable signal processing system,V1406,122-126(1991)

Brannon, James H. See Kueper, Stephan: V1598,27-35(1991)

Brant, Karl
See Evans, Robert H.: V1406,201-202(1991)
See Williams, Elmer F.: V1482,104-111(1991)

Branzalov, Peter P.
: TE two-frequency pulsed laser at 0.337 um and 10.6 um with a plasma cathode,V1412,236-245(1991)

Braren, Bodil
ed.: *Lasers in Microelectronic Manufacturing,*V1598(1991)

Brase, James M. See Yorkey, Thomas J.: V1345,255-259(1991)

Brasseur, Olivier
See Stojanoff, Christo G.: V1536,206-214(1991)
See Stojanoff, Christo G.: V1527,48-60(1991)
See Stojanoff, Christo G.: V1485,274-280(1991)
See Stojanoff, Christo G.: V1559,321-330(1991)
See Tropartz, Stephan: V1507,345-353(1991)

Brasunas, John C.
; Lakew, Brook: High-Tc bolometer developments for planetary missions,V1477,166-173(1991)

Brau, Charles A. See Tolk, Norman H.: V1552,7-13(1991)

Braud, John P.
: Polarizing optics for the soft x-ray regime: whispering-gallery mirrors and multilayer beamsplitters,V1548,69-72(1991)

Braueninger, Heinrich
; Hauff, D.; Lechner, P.; Lutz, G.; Kink, W.; Meidinger, Norbert; Metzner, G.; Predehl, Peter; Reppin, C.; Strueder, Lothar; Truemper, Joachim; Kendziorra, E.; Staubert, R.; Radeka, V.; Rehak, P.; Rescia, S.; Bertuccio, G.; Gatti, E.; Longoni, Antonio; Sampietro, Marco; Findeis, N.; Holl, P.; Kemmer, J.; von Zanthier, C.: Progress with PN-CCDs for the XMM satellite mission,V1549,330-339(1991)
See Citterio, Oberto: V1343,145-154(1991)

Brault, Daniel See Nadeau, Pierre: V1398,151-161(1991)

Braun, Frank D. See Guyer, Dean R.: V1409,14-17(1991)

Braun, Michael See Lenski, Harald: V1557,124-131(1991)

Braun, R. See Kukreja, Lalit M.: V1427,243-254(1991)

Braun, Robert E.
; Liebow, Charles: Clinical evaluation of tumor promotion by CO2 laser,V1424,138-144(1991)

Braunreiter, Dennis C.
; Banh, Nam D.: Detection of moving subpixel targets in infrared clutter with space-time filtering,V1481,73-83(1991)

Braunstein, Juergen See Huelsmann, Axel: V1465,201-208(1991)

Braunstein, Myron L. See Tittle, James S.: V1383,225-234(1991)

Bray, Phillip
; Kreidl, Norbert J.; eds.: *Submolecular Glass Chemistry and Physics,*V1590(1991)

Braze, Bill See Tanaka, Satoru C.: V1448,21-26(1991)

Brazhnik, O. D.
; Khokhlov, A. R.: Dynamics of stiff macromolecules in concentrated polymer solutions: model of statistical reorientations,V1402,70-77(1991)

Breakwell, John
; Varadan, Vijay K.; eds.: *Structures Sensing and Control,*V1489(1991)

Brecher, Virginia H.
; Bonner, Raymond; Read, C.: Model of human preattentive visual detection of edge orientation anomalies,V1473,39-51(1991)

Breckenridge, William G.
See Redding, David C.: V1354,216-221(1991)
See Redding, David C.: V1489,201-215(1991)

Breckinridge, James B. See Bely, Pierre Y.: V1494(1991)

Bredas, Jean-Luc
; Dehu, C.; Meyers, F.; Zyss, Joseph: Theoretical insight into the quadratic nonlinear optical response of organics: derivatives of pyrene and triaminotrinitrobenzene,V1560,98-110(1991)

Bredthauer, Richard A.
; Pinter, Jeff H.; Janesick, James R.; Robinson, Lloyd B.: Notch and large-area CCD imagers,V1447,310-315(1991)
See Geary, John C.: V1447,264-273(1991)

Breining, K. See Moissl, M.: V1397,395-398(1991)

Breithaupt, R. D.
; Tarver, Craig M.: Uses of Fabry-Perot velocimeters in studies of high explosives detonation,V1346,96-102(1991)

Brekhovskikh, Galina L. See Kudryavtseva, Anna D.: V1385,190-199(1991)

Bremand, Fabrice J.
; Lagarde, Alexis: Strain measurements on soft materials application to cloth and papers,V1554B,650-660(1991)
See Mauvoisin, Gerard: V1554B,181-187(1991)

Bremond, Alain See Baskurt, Atilla: V1444,240-249(1991)

Bremond, Georges E.
; Said, Hicham; Guillot, Gerard; Meddeb, Jaafar; Pitaval, M.; Draidia, Nasser; Azoulay, Rozette: Transmission electron microscopy, photoluminescence, and capacitance spectroscopy on GaAs/Si grown by metal organic chemical vapor deposition,V1361,732-743(1991)

Brenci, Massimo
; Mencaglia, Andrea; Mignani, Anna G.; Barbero, V.; Cimbrico, P. L.; Pessino, P.: Fiber optic multiple sensor for simultaneous measurements of temperature and vibrations,V1572,318-324(1991)
; Mencaglia, Andrea; Mignani, Anna G.: Problems and solutions in fiber-optic amplitude-modulated sensors,V1504,212-220(1991)
See Mencaglia, Andrea: V1506,140-144(1991)

Brennan, Roger See Raicu, Bruha: V1393,161-171(1991)

Brennan, Thomas M. See Carson, Richard F.: V1378,84-94(1991)

Brenner, Karl-Heinz
: Techniques for integrating 3-D optical systems,V1544,263-270(1991)
: Three-dimensional integration of optical systems,V1506,94-98(1991)

Brent, Richard P. See Bojanczyk, Adam W.: V1566,23-34(1991)

Bresciani, Paolo
: Logical account of a terminological tool,V1468,245-255(1991)

Bressler, Martin See Neu, John T.: V1530,244-254(1991)

Bretagne, J. See Capitelli, Mario: V1503,126-131(1991)

Brettel, Hans
: Visual aspects of picture storage,V1401,50-55(1991)

Breuer, J.
; Metev, S.; Sepold, Gerd; Krueger, G.; Hennemann, O. D.: Influencing
adherence properties of polymers by excimer laser radiation,V1503,223-
230(1991)

Breul, Harry T. See Peck, Alex N.: V1388,30-38(1991)

Brevignon, Michele See Bonniau, Philippe: V1588,52-63(1991)

Brewer, Donald R.
; Joenathan, Charles; Bibby, Yu Wang; Khorana, Brij M.: Development of an
interferometric fiber optic sensor using diode laser,V1396,430-434(1991)

Brewer, Joe E.
; French, Larry E.: Wafer scale integration modular packaging,V1390,164-
174(1991)

Breyer, R. See Troeltzsch, John R.: V1494,9-15(1991)

Briancon, Anne-Marie
; Jacquier, Bernard; Gacon, Jean-Claude; Le Sergent, Christian; Marcerou,
Jean-Francois: Inhomogeneous line broadening of optical transitions in
Nd3+ and Er3+ doped preforms and fibers,V1373,9-20(1991)

Bridges, Frank
; Li, Guoguang; Boyce, James B.; Claeson, Tord: Distorted local environment
about Zn and transition metals on the copper sites in
YBa2Cu3O7,V1550,76-84(1991)

Bridges, William B.
; Sheehy, Finbar T.; Schaffner, James H.: Velocity-matched electro-optic
modulator,V1371,68-77(1991)

Bridwell, Lynn B.
; Wang, Y. Q.: Ion implantation of polymers for electrical conductivity
enhancement,V1519,878-883(1991)

Brierley, Michael C. See France, Paul W.: V1373,33-39(1991)

Briers, J. D.
: Laser speckle and its temporal variability: the implications for biomedical
holography,V1429,48-54(1991)

Brikman, Inna
See Chakraborty, Dev P.: V1443,183-190(1991)
See Nodine, Calvin F.: V1444,56-62(1991)

Brill, Michael H.
: Photometric models in multispectral machine vision,V1453,369-380(1991)
See Barrett, Eamon B.: V1567,142-169(1991)
See Rogowitz, Bernice E.: V1453(1991)

Brimacombe, Robert K. See Jursich, Gregory M.: V1412,115-122(1991)

Bringi, Viswanathan N. See Vivekanandan, J.: V1558,324-338(1991)

Brink, D. See Esquivias, Ignacio: V1484,55-66(1991)

Brinkley, B. R. See Hoover, Richard B.: V1426,84-96(1991)

Brinkley, James F.
: Semiautomatic medical image segmentation using knowledge of anatomic
shape,V1445,78-87(1991)

Brinkmann, R.
; Sohler, Wolfgang; Suche, Hubertus: Absorption, fluorescence, and
stimulated emission in Ti-diffused Er:LiNbO3 waveguides,V1362,377-
382(1991)

Brinkmann, Ralf E.
; Bauer, K.: Q-switching and pulse shaping with IR lasers,V1421,134-
139(1991)
See Flemming, G.: V1421,146-152(1991)

Brinkmann, Ralf P.
; Schoenbach, Karl H.; Schulz, Hans-Joachim: Equilibrium and
nonequilibrium properties of semiconductors with multiply ionizable
deep centers,V1361,274-280(1991)
; Schoenbach, Karl H.; Roush, Randy A.; Stoudt, David C.; Lakdawala,
Vishnu K.; Gerdin, Glenn A.: High-power switching with electron-beam-
controlled semiconductors,V1378,203-208(1991)
See Schoenbach, Karl H.: V1362,428-435(1991)
See Schulz, Hans-Joachim: V1361,834-847(1991)

Brinkop, Axel
; Laudwein, Norbert: Knowledge-based system for configuring mixing-
machines,V1468,227-234(1991)

Brintle, David G. See Smith, Walter S.: V1434,14-25(1991)

Briones, Fernando See Recio, Miguel: V1361,469-478(1991)

Briottet, Xavier See Krawczyk, Rodolphe: V1493,2-15(1991)

Briscoe, Dennis
: 200,000-frame-per-second drum camera with nanosecond synchronized
laser illumination,V1346,319-323(1991)
; Shrivastava, Chinmaya A.; Nebeker, Sidney J.; Hsu, S.; Lozovoi, V. I.;
Postovalov, V. E.; Prokhorov, Alexander M.; Schelev, Mikhail Y.;
Serdyuchenko, Yuri N.; Vaschenko, G. O.: Soviet-American image
converter cameras "PROSCHEN",V1358,329-336(1991)

Brister, Keith
: X-ray diffraction from materials under extreme pressures,V1550,2-10(1991)

Bristow, Julian P.
; Mukherjee, Sayan D.; Khan, M. N.; Hibbs-Brenner, Mary K.; Sullivan,
Charles T.; Kalweit, Edith: High-density waveguide modulator arrays for
parallel interconnection,V1389,535-546(1991)
See Guha, Aloke: V1389,375-385(1991)

Bristow, Michael P.
; Diebel, D. E.; Bundy, Donald H.; Edmonds, Curtis M.; Turner, Ruldopha M.;
McElroy, James L.: Development of an airborne excimer-based UV-DIAL
for monitoring ozone and sulfur dioxide in the lower
troposphere,V1491,68-74(1991)

Bristow, Thomas C. See Bietry, Joseph R.: V1332,537-543(1991).

Britton, Cynthia A. See King, Jill L.: V1446,268-275(1991)

Brivio, Franca See Milani, Marziale: V1474,83-97(1991)

Brizard, Christine M.
See Campuzano, Juan C.: V1345,245-254(1991)
See Clarke, Roy: V1361,2-12(1991)
See Clarke, Roy: V1345,101-114(1991)
See Rodricks, Brian G.: V1550,18-26(1991)

Broadfoot, A. L. See Schulze, Dean W.: V1549,319-328(1991)

Brock, James R. See Carls, Joseph C.: V1497,120-131(1991)

Brock, Phillip J.
; Levenson, Marc D.; Zavislan, James M.; Lyerla, James R.; Cheng, John C.;
Podlogar, Carl V.: Fabrication of grooved-glass substrates by phase-mask
lithography,V1463,87-100(1991)

Brockhaus, John A.
; Campbell, Michael V.; Khorram, Siamak; Bruck, Robert I.; Stallings, Casson:
Forest decline model development with LANDSAT TM, SPOT, and DEM
DATA,V1492,200-205(1991)

Brocklesby, Bill See Tropper, Anne C.: V1373,152-157(1991)

Brodsky, Aaron
; Goodrich, Alan; Lawson, David G.; Holm, Richard W.: Augmented tracking
and acquisition system for GBL satellite illumination,V1482,159-
169(1991)
See Dierks, Jeffrey S.: V1482,146-158(1991)

Brofferio, Sergio C.
; Comunale, Domenico; Tubaro, Stefano: Estimation and prediction of object-
oriented segmentation for video predictive coding,V1605,500-510(1991)
; Marcozzi, Elena; Mori, Luigi; Raveglia, Dalmazio: Subband video-coding
algorithm and its feasibility on a transputer video coder,V1605,894-
905(1991)

Broida, Ted J.
: Feature correspondence in multiple sensor data fusion,V1383,635-
651(1991)

Brongers, J. D.
: Valuable security features in a competitive banking environment: does
security attract criminals instead of customers?,V1509,105-112(1991)

Bronowicki, Allen J. See Betros, Robert S.: V1542,406-419(1991)

Bronshtein, Alexander See Mazar, Reuven: V1487,361-371(1991)

Bronstein, Sam See Turk, Harris: V1454,344-352(1991)

Brookes, N. See Gluskin, Efim S.: V1548,56-68(1991)

Brooking, Nicholas L.
; Guedes Valente, Luiz C.; Kawase, Liliana R.; Afonso, Jose A.: Electronic
polarimetric detection system for optical fiber sensor
application,V1572,88-93(1991)

Brooks, Everett G.
; Rothman, Melvyn L.: Diagnostic report acquisition unit for the Mayo/IBM
PACS project,V1446,217-226(1991)

Brooks, Jeff W.
: Focal-plane image processing using acoustic charge transport technology,V1541,68-72(1991)

Brooks, Rodney A. See Konishi, Mashide: V1383,448-458(1991)

Broomhead, David S.
: Signal processing for nonlinear systems,V1565,228-243(1991)
See Potts, Michael A.: V1565,255-266(1991)

Bross, Alan D.
: Scintillating plastic optical fiber radiation detectors in high-energy particle physics,V1592,122-132(1991)

Brossia, Charles E.
; Wu, Samuel C.: Low-cost in-soil organic contaminant sensor,V1368,115-120(1991)

Brot, Christian
See Duchet, Christian: V1372,72-81(1991)
See Moehlmann, Gustaaf R.: V1560,426-433(1991)

Brouns, Austin J. See Schimert, Thomas R.: V1484,19-30(1991)

Browell, Edward V.
: Airborne lidar measurements of ozone and aerosols in the summertime Arctic troposphere,V1491,7-14(1991)
: Airborne lidar observations of ozone and aerosols in the wintertime Arctic stratosphere,V1491,273-281(1991)
See Ponsardin, Patrick: V1492,47-51(1991)
See Theon, John S.: V1492,2-23(1991)

Brown, D.
See Tallents, Gregory J.: V1413,70-76(1991)
See Taylor, Roderick S.: V1420,183-192(1991)

Brown, Daniel M.
: Global optimization using the y-ybar diagram,V1527,19-25(1991)
; Kathman, Alan D.: Off-axis spherical element telescope with binary optic corrector,V1555,114-127(1991)
; Pitalo, Stephen K.: Techniques for designing hybrid diffractive optical systems,V1527,73-84(1991)

Brown, David A.
; Cameron, C. B.; Keolian, Robert M.; Gardner, David L.; Garrett, Steven L.: Symmetric 3 x 3 coupler based demodulator for fiber optic interferometric sensors,V1584,328-335(1991)
; Garrett, Steven L.; Conte, D. V.; Smith, R. C.; Rothenberg, E.; Young, M.; Rissberger, Ed: Fiber optic interferometric ellipsoidal shell hydrophone,V1369,2-8(1991)
; Garrett, Steven L.: Interferometric fiber optic accelerometer,V1367,282-288(1991)
See Garrett, Steven L.: V1367,13-29(1991)

Brown, David G. See Anderson, Mary P.: V1444,407-412(1991)

Brown, David L.
; Seka, Wolf D.; Letzring, Samuel A.: Toward phase noise reduction in a Nd:YLF laser using electro-optic feedback control,V1410,209-214(1991)

Brown, Gair D.
; Ingold, Joseph P.; Spence, Scott E.; Paxton, Jack G.: High-impact shock testing of fiber optic components,V1366,351-360(1991)
; Ingold, Joseph P.; Paxton, Jack G.: Long-term performance of fiber optic cable plant in Navy ships,V1589,58-68(1991)

Brown, George S.
See Baron, Alfred Q.: V1343,84-94(1991)
See Mustre de Leon, Jose: V1550,85-96(1991)

Brown, Glen
; Boxler, Lawrence H.; Chun, Patrick K.; Western, Arthur B.: Study of human cardiac cycle using holographic interferometry,V1396,164-173(1991)
See Boxler, Lawrence H.: V1396,85-92(1991)

Brown, Graeme N. See Spirit, David M.: V1373,197-208(1991)

Brown, James C.
; Webb, Curtis M.; Bell, Paul A.; Washington, Randolph T.; Riordan, Richard J.: AutoSPEC image evaluation laboratory,V1488,300-311(1991)

Brown, Jeff L.
: Light scatter variations with respect to wafer orientation in GaAs,V1530,299-305(1991)

Brown, Joseph R.
; Bergondy, Daniel; Archer, Susan J.: Comparison of neural network classifiers to quadratic classifiers for sensor fusion,V1469,539-543(1991)
See Beck, Hal E.: V1381,600-609(1991)

Brown, Julia J.
; Gardner, J. T.; Forrest, Stephen R.: Optically powered optoelectronic integrated circuits for optical interconnects,V1474,236-242(1991)

Brown, K. G. See Upchurch, Billy T.: V1416,21-29(1991)

Brown, Mark L. See Williams, William J.: V1566,144-156(1991)

Brown, Michael J.
: Novel technical advances provide easy solutions to tough motion analysis problems,V1346,24-32(1991)

Brown, Neil H.
; Wood, Hugh C.; Wilson, James N.: Image analysis for vision-based agricultural vehicle guidance,V1379,54-68(1991)

Brown, Norman J. See Tesar, Aleta A.: V1441,154-172(1991)

Brown, P. See Beard, David V.: V1446,52-58(1991)

Brown, Paul D. See Clifton, Paul A.: V1361,1063-1069(1991)

Brown, Robert A.
; Honey, Richard C.; Maffione, Robert A.: Isotropic light source for underwater applications,V1537,147-150(1991)
See Maffione, Robert A.: V1537,115-126(1991)

Brown, Robert G.
: Progress in miniature laser systems for space science particle sizing and velocimetry,V1506,58-59(1991)

Brown, Robert J.
: Debris collision warning sensor telescope design,V1527,155-162(1991)

Brown, Sherman D.
See Wallenberger, Frederick T.: V1484,116-124(1991)
See Wallenberger, Frederick T.: V1590,72-82(1991)

Brown, Simon
: Mathematical modeling and standardization in holography,V1509,79-86(1991)

Brown, Stewart K.
; Mannik, Len: Field testing of a fiber optic rotor temperature monitor for power generators,V1584,15-22(1991)

Brown, Tyler See Biernacki, Paul D.: V1455,167-178(1991)

Brown, William E. See Robinson, Lloyd B.: V1447,214-228(1991)

Browne, Mark A.
; Jolleys, Glenn D.; Joyner, David J.: Comparison of surface and volume presentation of multislice biomedical images,V1450,338-349(1991)

Browne, Michael P. See Ji, Tinglan: V1444,136-150(1991)

Brownell, John B.
; Parker, Richard J.: Automated fringe analysis for moire interferometry,V1554B,481-492(1991)

Browning, Iben
: High-definition projection television,V1456,48-50(1991)

Browning, Raymond See Pease, R. F.: V1496,234-238(1991)

Brubaker, John L.
; McCormick, Frederick B.; Tooley, Frank A.; Sasian, Jose M.; Cloonan, Thomas J.; Lentine, Anthony L.; Hinterlong, Steve J.; Herron, Michael J.: Optomechanics of a free-space photonic switch: the components,V1533,88-96(1991)
See McCormick, Frederick B.: V1396,508-521(1991)
See McCormick, Frederick B.: V1533,97-114(1991)

Bruce, Dorothy
: Realistic model for battlefield fire plume simulation,V1486,231-236(1991)

Bruce, James A. See Dunn, Diana D.: V1463,8-15(1991)

Bruck, Robert I. See Brockhaus, John A.: V1492,200-205(1991)

Brucker, E. B.
: Search for short-lived particles using holography,V1461,206-214(1991)

Bruder, Joseph A.
; Greneker, E. F.; Nathanson, F. E.; Henneberger, T. C.: Detection of contraband brought into the United States by aircraft and other transportation methods: a changing problem,V1479,316-321(1991)

Bruder, Martin See Esquivias, Ignacio: V1484,55-66(1991)

Brueck, Steven R.
See Reicher, David W.: V1441,106-112(1991)
See Zaidi, Saleem H.: V1343,245-255(1991)

Brueckner, Volkmar
; Bergner, Harald; Lenzner, Matthias; Strobel, Reiner: Picosecond optoelectronic semiconductor switching and its application,V1362,510-517(1991)

Bruegge, Carol J.
; Stiegman, Albert E.; Coulter, Daniel R.; Hale, Robert R.; Diner, David J.; Springsteen, Arthur W.: Reflectance stability analysis of Spectralon diffuse calibration panels,V1493,132-142(1991)
See Gellman, David I.: V1493,175-180(1991)
See Guzman, Carmen T.: V1493,120-131(1991)

Bruegge, Thomas J. See Rimmer, Matthew P.: V1354,83-91(1991)

Bruggemann, Ulrich H. See Obremski, Susan M.: V1427,327-334(1991)

Brugmans, Marco J.
See Gijsbers, Geert H.: V1425,94-101(1991)
See van Gemert, Martin J.: V1427,316-319(1991)

Bruijn, Marcel P. See den Boggende, Antonius J.: V1345,189-197(1991)

Bruinsma, Anastasius J.
; Culshaw, Brian; eds.: *Fiber Optic Sensors: Engineering and Applications,*V1511(1991)
See Nieuwkoop, E.: V1511,255-263(1991)

Brukhomitsky, Yuri A. See Kalyayev, Anatoli V.: V1621,299-308(1991)

Brule, Michel J.
; Soucy, L.: Modular algorithms for depth characterization of object surfaces with ultrasonic sensors,V1388,432-441(1991)

Brumback, Babette A. See Jahns, Juergen: V1555,2-7(1991)

Brun, Eric See Pierre, Guillaume: V1511,201-211(1991)

Brun, Philip See Dong, LiXin: V1445,178-187(1991)

Brunet, Gerard See Monjoux, E.: V1452,310-318(1991)

Brunet, Henri
; Mabru, Michel; Rocca Serra, J.; Vannier, C.: Pulsed HF chemical laser using a VUV phototriggered discharge,V1397,273-276(1991)
See Prigent, Pascale: V1397,197-201(1991)
See Voignier, Francois: V1397,297-301(1991)

Brunie, Lionel See Lavallee, Stephane: V1570,322-336(1991)

Brunner, Stefan See Frank, Werner F.: V1559,344-353(1991)

Brunner, Timothy A.
: Optimization of optical properties of resist processes,V1466,297-308(1991)

Bruno, Alain See Kpalma, Kidiyo: V1606,55-66(1991)

Bruno, Guy
; Morgenthaler, Matthew K.: Situation assessment for space telerobotics,V1387,352-358(1991)
See Morgenthaler, Matthew K.: V1387,82-95(1991)

Bruno, Ronald C. See Fox, Neil D.: V1417,452-463(1991)

Bruno, Theresa L.
; Wirth, Allan: Advanced image intensifier systems for low-light high-speed imaging,V1358,109-116(1991)

Brunsvold, William R.
; Montgomery, Warren; Hwang, Bao: Nonmetallic acid generators for i-line and g-line chemically amplified resists,V1466,368-376(1991)

Bruscaglioni, Piero See Ferrari, Marco: V1431,276-283(1991)

Brusmark, B. See Abrahamsson, S.: V1471,130-141(1991)

Bruus, Henrik See Hansen, Ole P.: V1362,192-198(1991)

Bruzzese, Riccardo
; Berardi, V.; de Lisio, C.; Solimeno, Salvatore; Spinelli, N.: Test of Geltman theory of multiple ionization of xenon by intense laser pulses,V1397,735-738(1991)

Bruzzone, Elisabetta
; Mangili, Fulvia: Calibration of a CCD camera on a hybrid coordinate measuring machine for industrial metrology,V1526,96-112(1991)

Bryan, Scott R. See Havrilla, George J.: V1435,12-18(1991)

Bryan-Brown, G. P.
; Elston, S. J.; Sambles, J. R.: Polarization conversion through the excitation of electromagnetic modes on a grating,V1545,167-178(1991)

Bryanston-Cross, Peter J.
See Quan, C.: V1507,463-475(1991)
See Towers, Catherine E.: V1358,952-965(1991)

Bryant, Keith
; Ince, Ilhan: Remote driving: one eye or two,V1457,120-132(1991)

Bryant, R. See Eichen, Elliot G.: V1474,260-267(1991)

Brylowska, Irena
; Paprocki, K.: Influence of laser pulse annealing on the depth distribution of Sb recoil atoms in Si,V1391,164-169(1991)

Brynda, E. See Adamec, F.: V1402,82-84(1991)

Bryskin, V. Z.
; Krylov, Vitaly N.; Staselko, D. I.: Modern holographic studio,V1238,448-451(1991)
; Prostev, A.: Holographic art,V1238,368-369(1991)

Bryson, Steve T.
: Interaction of objects in a virtual environment: a two-point paradigm,V1457,180-187(1991)
; Levit, Creon C.: Virtual environment for the exploration of three-dimensional steady flows,V1457,161-168(1991)

Bryukhnevitch, G. I.
; Dalinenko, I. N.; Kuz'min, G. A.; Libenson, B. N.; Malyarov, A. V.; Moskalev, B. B.; Postovalov, V. E.; Prokhorov, Alexander M.; Schelev, Mikhail Y.: Application of electron-sensitive CCD for taking off the time-dispersed pictures from image tube phosphor screens,V1358,739-749(1991)
; Dalinenko, I. N.; Ivanov, K. N.; Kaidalov, S. A.; Kuz'min, G. A.; Moskalev, B. B.; Naumov, S. K.; Pischelin, E. V.; Postovalov, V. E.; Prokhorov, Alexander M.; Schelev, Mikhail Y.: Design and operational characteristics of a PV 001 image tube incorporated with EB CCD readout,V1449,109-115(1991)
See Averin, V. I.: V1358,603-605(1991)

Brzezinski, Ryszard
: Pockels cell driver with high repetition rate,V1391,279-285(1991)
; Piotrowski, Jan: Electronic system for generation of laser pulses by self-injection method,V1391,127-134(1991)

Brzozowy, David J. See Toukhy, Medhat A.: V1466,497-507(1991)

Bubb, Daniel
; Wilson, Leo T.; Stoltz, John R.: Application of the ProtoWare simulation testbed to the design and evaluation of advanced avionics,V1483,18-28(1991)

Bubulac, Lucia O.
: Ion implantation and diffusion for electrical junction formation in HgCdTe,V1484,67-71(1991)

Buchal, Christoph
; Mohr, S.: Diffusion and solubility of ion-implanted Nd and Er in LiNbO3,V1361,881-892(1991)

Buchegger, F. See Wagnieres, G.: V1525,219-236(1991)

Buchelt, Martin
; Papaioannou, Thanassis; Fishbein, Michael C.; Peters, Werner; Beeder, Clain; Grundfest, Warren S.: In-vitro ablation of fibrocartilage by XeCl excimer laser,V1420,249-253(1991)

Buchhave, Preben See Petersen, Paul M.: V1362,582-585(1991)

Buchmann, Peter L. See Webb, David J.: V1418,231-239(1991)

Bucholtz, Frank
; Dagenais, Dominique M.; Koo, Kee P.; Vohra, Sandeep T.: Recent developments in fiber optic magnetostrictive sensors,V1367,226-235(1991)

Bucholz, Roger C.
: Electromagnetic enviromental effects on shipboard fiber optic installations,V1369,19-23(1991)

Buck, J. P. See Jackson, John E.: V1533,75-86(1991)

Buck, Peter D.
; Rieger, Michael L.: Phase-shift mask applications,V1463,218-228(1991)

Buckalew, Chris See Driver, Jesse W.: V1459,109-116(1991)

Buckberry, Clive H. See Atcha, Hashim: V1504,221-232(1991)

Buckley, L. M. See Kochevar, Irene E.: V1403,756-763(1991)

Buckley, Robert H.
; Lyons, E. R.; Goga, George: Rugged 20-km fiber optic link for 2-18-GHz communications,V1371,212-222(1991)

Buckwald, Bob A. See Adel, Michael E.: V1442,68-80(1991)

Buczek, Harthmuth
; Mayor, J. M.; Regnault, P.: DOE design and manufacture at CSEM,V1574,48-57(1991)

Budai, J. D. See Lowndes, Douglas H.: V1394,150-160(1991)

Budenske, John See Brandt, S. A.: V1406,122-126(1991)

Budnik, A. P.
; Gus'kov, K. G.; Raizer, Yu. P.; Surjhikov, S. T.: Plasma motion velocity along laser beam and continuous optical discharge in gas flow,V1397,721-724(1991)
; Popov, A. G.: Liquid droplet supercritical explosion in the field of CO2 laser radiation and influence of plasma chemical reactions on initiation of optical breakdown in air,V1440,135-145(1991)

Budtz-Jorgensen, Carl
; Bahnsen, Axel; Christensen, Finn E.; Madsen, M. M.; Olesen, C.; Schnopper, Herbert W.: Performance of microstrip proportional counters for x-ray astronomy on spectrum-roentgen-gamma,V1549,429-437(1991)

Budzban, Gregory M.
; DeCatrel, John M.: Markov random fields on a SIMD machine for global region labelling,V1470,175-182(1991)

Budzier, Helmut See Norkus, Volkmar: V1484,98-105(1991)

Budzinski, Christel
; Grunwald, Ruediger; Pinz, Ingo; Schaefer, Dieter; Schoennagel, Horst: Apodized outcouplers for unstable resonators,V1500,264-274(1991)

Buechler, J. See Luy, Johann-Freidrich: V1475,129-139(1991)

Bueckmann, F. See Ploetz, F.: V1510,224-230(1991)

Bueker, Harald
; Gripp, S.; Haesing, Friedrich W.: Temperature and fading effects of fiber-optic dosimeters for radiotherapy,V1572,410-418(1991)

Buelthoff, Heinrich H.
; Yuille, Alan L.: Shape-from-X: psychophysics and computation,V1383,235-246(1991)

Bugner, Douglas E.
; Fulmer, Gary G.; Riblett, Susan E.: Photoelectrographic printing with persistently conductive masters based on onium salt acid photogenerators,V1458,162-178(1991)

Bugrov, N. V.
; Zakharov, N. S.: Fusion of diamond phases of graphite in laser shock waves,V1440,416-422(1991)

Bui, Snow T.
See Kelly, John H.: V1410,40-46(1991)
See Skeldon, Mark D.: V1410,116-124(1991)

Bui, Trung H. See Russo, Paul: V1383,60-71(1991)

Buisson, F. See Duran, Michel: V1494,357-376(1991)

Bukatin, Vladimir V.
: Coordinate measuring system for 2-D scanners,V1454,283-288(1991)

Bukreev, Viacheslav S.
See Alimpiev, Sergei S.: V1503,154-158(1991)
See Atjezhev, Vladimir V.: V1503,197-199(1991)

Bukys, Albert
: Considerations in the design of servo amplifiers for high-performance scanning systems,V1454,186-195(1991)

Bulhoes, Luis O. See Valla, Bruno: V1536,48-62(1991)

Bulichev, A.
; Voronin, G.; Kaganov, S.; Kuzmin, Vladimir S.; Porjadin, V.; Chibisov, V.: Accuracy improvement by mathematical correction method and electro-optical device certification,V1500,151-162(1991)

Buller, G. S. See Craig, Robert G.: V1505,76-86(1991)

Bullock, Michael E.
; Miltonberger, Thomas W.; Reinholdtsen, Paul A.; Wilson, Kathleen: Multisensor fusion using the sensor algorithm research expert system,V1471,291-302(1991)

Bulmer, Catherine H.
: Lithium niobate waveguide devices: present performance and future applications,V1583,176-183(1991)

Bulsink, Bennie J.
; Klok, Frits H.: Architecture for a multiprocessing system based on data flow processing elements in a MAXbus system,V1384,215-227(1991)

Bulstrode, Christopher See Turner-Smith, Alan R.: V1380,75-84(1991)

Bulusu, Dutt V.
; Zack, Tim; Scholl, Frederick W.; Coden, Michael H.; Steele, Robert E.; Miller, Gregory D.; Lynn, Mark A.: High-speed polymer optical fiber networks,V1364,49-60(1991)
See Anderson, Stephen J.: V1364,94-100(1991)
See Coden, Michael H.: V1364,22-39(1991)

Bulyarsky, Sergey V. See Rizikov, Igor V.: V1362,664-673(1991)

Bulychev, N. V. See Benimetskaya, L. Z.: V1525,210-211(1991)

Bulygin, A. R.
: Effective transmission holograms produced on CK-type photoresist,V1238,248-252(1991)
: Production of reflection relief holograms with asymmetric shape of slits profile and the examination of their spectral characteristics,V1238,129-133(1991)

Bumpas, Stanley E. See Pitts, John H.: V1441,71-81(1991)

Bunch, Robert M.
; Caughey, Joseph P.: Angularly-polished optical fiber tips,V1396,411-416(1991)
See Acharya, Mukund: V1396,377-388(1991)

Bundy, Donald H. See Bristow, Michael P.: V1491,68-74(1991)

Bundy, Scott
; Mader, Tom; Popovic, Zoya; Ellinson, Reinold; Hjelme, Dag R.; Surette, Mark R.; Yadlowski, Michael; Mickelson, Alan R.: Quasioptical MESFET VCOs,V1475,319-329(1991)

Bunis, Jenifer L.
; Abele, C. C.; Campbell, James D.; Caudle, George F.: Producing a uniform excimer laser beam for materials processing applications,V1377,30-36(1991)

Bunney, Laura See Hoff, Lawrence E.: V1481,98-109(1991)

Bunton, P. H. See Haglund, Richard F.: V1441,127-138(1991)

Burakov, V. S.
; Lopasov, V. P.; Naumenkov, P. A.; Raikov, S. N.: Intracavity spectroscopy measurements of atom and ion densities in near-surface laser plasma,V1440,270-276(1991)

Buralli, Dale A.
; Morris, G. M.: Design of two- and three-element diffractive telescopes,V1354,292-296(1991)

Buratto, S. K. See Bowers, C. R.: V1435,36-50(1991)

Burckhardt, C. W. See Bays, Roland: V1525,397-408(1991)

Burdette, Harold See Steiner, Bruce W.: V1557,156-167(1991)

Buresh, L. L. See Williams, Paul F.: V1378,217-225(1991)

Burge, Ronald E.
See McOwan, Peter W.: V1384,75-82(1991)
See Qu, Dong-Ning: V1506,152-159(1991)

Burger, Arnold
; Morgan, S. H.; Silberman, Enrique; Nason, Donald: Optical diagnostics of mercuric iodide crystal growth,V1557,245-249(1991)

Burger, Robert J.
; Greenberg, David A.: Fiber array optics for electronic imaging,V1449,174-185(1991)
See Cook, Lee M.: V1449,186-192(1991)

Burgess, Lloyd W.
: Overview of planar waveguide techniques for chemical sensing,V1368,224-229(1991)
See Berman, Richard J.: V1368,25-35(1991)
See Bowman, Elizabeth M.: V1368,239-250(1991)

Burke, Barry E. See Huang, Chin M.: V1447,156-164(1991)

Burke, James See Boltinghouse, Susan T.: V1388,14-29(1991)

Burke, James J.
; Shull, Thomas A.; Imamura, Nobutake; eds.: *Optical Data Storage '91*,V1499(1991)
See Roncone, Ronald L.: V1590,14-25(1991)
See Weisenbach, Lori: V1590,50-58(1991)

Burke, Shawn E.
; Hubbard, James E.: Shape control of piezoelectric bimorph mirrors,V1532,207-214(1991)

Burkert, Wolfgang See Citterio, Oberto: V1343,145-154(1991)

Burkes, Tom R. See Harris, Kevin: V1411,87-99(1991)

Burkhardt, Hans See Fenske, Axel: V1567,53-64(1991)

Burkhart, Craig P. See Bayless, John R.: V1411,42-46(1991)

Burl, Michael C.
; Novak, Leslie M.: Polarimetric segmentation of SAR imagery,V1471,92-115(1991)

Burland, Donald M.
; Rice, J. E.; Downing, J.; Michl, J.: Design of chromophores for nonlinear optical applications,V1560,111-119(1991)
See Moerner, William E.: V1560,278-289(1991)

Burleigh, Douglas D.
; De La Torre, William: Thermographic analysis of the anisotropy in the thermal conductivity of composite materials,V1467,303-310(1991)

Burleson, Thomas A. See Johnson, R. B.: V1354,669-675(1991)

Burman, Jerry A. See Church, Susan D.: V1471,192-199(1991)

Burne, P. M.
; Sellen, D. B.: Laser light scattering studies of biological gels,V1403,288-295(1991)

Burney, Lewis G. See Sachse, Glen W.: V1433,157-166(1991)

Burnham, B. See Litvinenko, Vladimir N.: V1552,2-6(1991)

Burnham, Ralph L.
See Koechner, Walter: V1522,169-179(1991)
See Marshall, Larry R.: V1419,141-152(1991)

Burns, Gregory S.
See Birk, Ronald J.: V1495,2-11(1991)
See Scholen, Douglas E.: V1492,358-369(1991)

Burns, Jack O. See Johnson, Stewart W.: V1494,194-207(1991)

Burns, M. J. See Newnam, Brian E.: V1552,154-174(1991)

Burns, Peter D.
: Image signal modulation and noise analysis of CRT displays,V1454,392-398(1991)

Burns, Thomas See Dickinson, Robert R.: V1459,178-189(1991)

Burns, Thomas J.
; Rogers, Steven K.; Kabrisky, Matthew; Vogel, George A.: Compact optical associative memory,V1469,208-218(1991)

Burns, William K.
; Moeller, Robert P.; Dandridge, Anthony D.: Excess noise in fiber gyroscope sources,V1367,87-92(1991)

Burq, Catherine
; Vibert, Patrick: Approximation of the Compton scattered radiation,V1346,276-285(1991)

Burrer, Gordon J. See Confer, Charles L.: V1454,215-222(1991)

Burrows, Christopher J.
: Hubble Space Telescope optics: problems and solutions,V1494,528-533(1991)
: Hubble Space Telescope optics status,V1567,284-293(1991)

Burrows, John P.
; Chance, Kelly V.: Scanning imaging absorption spectrometer for atmospheric chartography,V1490,146-154(1991)
See Chance, Kelly V.: V1491,151-165(1991)

Burrus, Charles A. See Jewell, Jack L.: V1389,401-407(1991)

Burt, D. J. See Ball, K.: V1358,409-420(1991)

Burtsev, Vsevolod S.
: Optical principles of information processing in supercomputer architecture,V1621,380-387(1991)
; Fyodorov, Vyatcheslav B.: Optical associative memory for nontraditional architecture digital computers and database management systems,V1621,215-226(1991)

Burykin, N. M. See Bazhenov, V. Y.: V1574,148-153(1991)

Burzynski, Ryszard See Zhang, Yue: V1560,264-271(1991)

Bush, Brett C.
; Cotton, Daniel M.; Siegmund, Oswald H.; Chakrabarti, Supriya; Harris, Walter; Clarke, John T.: High-resolution, two-dimensional imaging, microchannel-plate detector for use on a sounding rocket experiment,V1549,290-301(1991)
; Cotton, Daniel M.; Vickers, James S.; Chakrabarti, Supriya: Instrument design and test results of the new all-reflection spatial heterodyne spectrometer,V1549,376-384(1991)
See Cotton, Daniel M.: V1549,313-318(1991)
See Tom, James L.: V1549,308-312(1991)
See Tom, James L.: V1549,302-307(1991)

Bush, Keith A. See Voelz, David G.: V1416,260-265(1991)

Bush, Simon P.
; Jackson, David A.: Dual-channel current sensor capable of simultaneously measuring two currents,V1584,103-109(1991)

Busquets, Anthony M.
; Parrish, Russell V.; Williams, Steven P.: Effects of alternate pictorial pathway displays and stereo 3-D presentation on simulated transport landing approach performance,V1457,91-102(1991)

Buss, Wolfgang See Rasch, Andreas: V1522,83-92(1991)

Bussard, Anne B.
; Biller, M.; Serfas, D. A.: Multichannel data acquistion system for assessing reliability of fiber optic couplers,V1366,380-386(1991)

Busse, H. See Foerster, Werner: V1429,146-151(1991)

Busse, Tim See Bonavito, N. L.: V1406,138-146(1991)

Bussell, T. J. See Kinsey, Trevor J.: V1358,914-924(1991)

Bustamante, Carlos
; Finzi, Laura; Sebring, Page E.; Smith, Steven B.: Manipulation of single-DNA molecules and measurements of their elastic properties under an optical microscope,V1435,179-187(1991)
See Maestre, Marcos F.: V1548,179-187(1991)

Butkevitch, V. G. See Khryapov, V. T.: V1540,412-423(1991)

Butkovsky, A. V.
: Convective evaporation of water aerosol droplet irradiated by CO2 laser,V1440,146-152(1991)

Butler, Clive
; Gregoriou, Gregorios: Novel noncontact sensor for surface topography measurements using fiber optics,V1584,282-293(1991)
See Shams, Iden: V1589,120-125(1991)
See Yang, Q.: V1572,558-563(1991)

Butler, James E. See Hickey, Carolyn F.: V1534,67-76(1991)

Butler, James J. See McGee, Thomas J.: V1491,182-187(1991)

Butler, James W. See Gierczak, Christine A.: V1433,315-328(1991)

Butler, Jennifer M.
; Wharton, Charles B.: Twin traveling wave tube amplifiers driven by a single backward-wave oscillator,V1407,57-66(1991)

Butler, Jerome K.
See Ayekavadi, Raj: V1418,74-85(1991)
See Evans, Gary A.: V1418,406-413(1991)

Butler, Marcus S. See Klainer, Stanley M.: V1434,119-126(1991)

Butler, Michael A.
: Optical fiber interferometric sensors for chemical detection,V1368,46-54(1991)

Butler, Michael P. See Swiniarski, Roman W.: V1451,234-241(1991)

Butler, Stephanie W.
; McLaughlin, Kevin J.; Edgar, Thomas F.; Trachtenberg, Isaac: Real-time monitoring and control of plasma etching,V1392,361-372(1991)

Button, B. L.
See Dobbins, B. N.: V1554A,772-780(1991)
See Dobbins, B. N.: V1554B,586-592(1991)

Buttram, Malcolm T. See Loubriel, Guillermo M.: V1378,179-186(1991)

Buurman, Johannes
; Bierhuizen, David J.: Two stage object identification system in the Delft intelligent assembly cell,V1386,185-196(1991)
See Bart, Mischa: V1381,66-77(1991)

Buus, Jens
: <author>Single Frequency Semiconductor Lasers,VTT05(1991)

Buvat, Daniel
; Muller, Gerard; Peyrot, Patrick: Coarse pointing assembly for the SILEX program, or how to achieve outstanding pointing accuracy with simple hardware associated with consistent control laws,V1417,251-261(1991)

Buydens, Luc
; Demeester, Piet M.; De Dobbelaere, P.; van Daele, Peter: InGaAs/AlGaAs vertical optical modulators and sources on a transparent GaAs substrate,V1362,50-58(1991)
See Demeester, Piet M.: V1361,987-998(1991)
See Pollentier, Ivan K.: V1361,1056-1062(1991)

Buyukkoc, Cagatay
; Persiano, G.: New algorithm and an efficient parallel implementation of the expectation maximization technique in PET (positron emission tomography) imaging,V1452,170-179(1991)

Buzard, Kurt A.
; Hoeltzel, David A.: Biomechanics of the cornea,V1423,70-81(1991)
See Friedlander, Miles H.: V1423,62-69(1991); 1429,229-236(1991)

Buzykin, O. G. See Anisimov, N. R.: V1440,206-210(1991)

Byars, P. See Kulkarni, Arun D.: V1452,512-522(1991)

Byer, Robert L. See Eckardt, Robert C.: V1561,119-127(1991)

Byler, Eric A.
: Design and control of ultralight manipulators for interplanetary exploration,V1387,313-327(1991)

Byles, W. R. See Efron, Uzi: V1455,237-247(1991)

Bylica, A. See Adamiak, S.: V1391,382-386(1991)

Byrd, Jerry L. See Feldman, Mark: V1542,490-501(1991)

Byrom, Ernest See Chan, Cheuk L.: V1450,208-217(1991)

Cabanski, Wolfgang A.
; Schulz, Max J.: Electronic and optical properties of silicide/silicon IR detectors,V1484,81-97(1991)

Cabato, Nellie L. See Wesson, Laurence N.: V1367,204-213(1991)

Cabib, Dario
See Adel, Michael E.: V1442,68-80(1991)
See Kopolovich, Zvi: V1540,565-577(1991)

Cabrera, Sergio D. See Sahinoglou, Haralambos: V1605,793-804(1991)

Caccavale, F. See Mazzoldi, Paolo: V1513,182-197(1991)

Cadet, J. See Angelov, D.: V1403,575-577(1991)

Cadilhac, M. See Tayeb, Gerard: V1545,95-105(1991)

Cadoret, R. See Banvillet, Henri: V1361,972-979(1991)

Cadotte, Roland See Rachlin, Adam: V1477,101-114(1991)

Cadou, Christopher P.
; Logan, Pamela; Karagozian, Ann; Marchant, Roy; Smith, Owen I.: Laser diagnostic techniques in a resonant incinerator,V1434,67-77(1991)

Cadzow, James A.
; Wilkes, D. M.; Peters, Richard A.; Li, Xingkang; Patel, Jamshed N.: Simulation of infrared backgrounds using two-dimensional models,V1486,352-363(1991)

Caelli, Terry M. See Barth, Erhardt: V1570,86-95(1991)

Caglar, Hakan
; Liu, Yipeng; Akansu, Ali N.: Statistically optimized PR-QMF design,V1605,86-94(1991)

Cahill, Paul A.
; Tallant, David R.; Kowalczyk, T. C.; Singer, Kenneth D.: Molecular to material design for anomalous-dispersion phase-matched second-harmonic generation,V1560,130-138(1991)

Cahoon, Peter
: Visual workbench for analyzing the behavior of dynamical systems,V1459,244-253(1991)

Cai, C. Z. See Shu, Q. Q.: V1519,675-679(1991)

Cai, De-Fu
: Stochastic detecting images from strong noise field in visual communications,V1606,926-933(1991)

Cai, Hai-Tao
; Chen, Zhen-Pei: New method of adjusting color of pseudocolor encoding image,V1567,703-708(1991)

Cai, Jine See Cui, DaFu: V1572,386-391(1991)

Cai, Tiequan See Guo, Lu Rong: V1555,293-296(1991)

Cai, Wen
; Gordon, Jeff M.; Kashin, Peter; Rabl, Ari: Reflectors for efficient and uniform distribution of radiation for lighting and infrared based on nonimaging optics,V1528,118-128(1991)

Cai, Xian-Yang See Christie, Simon: V1507,202-209(1991)

Cai, Y. M. See Zhang, Z. J.: V1519,790-792(1991)

Cai, Ying S.
See Boardman, A. D.: V1519,609-615(1991)
See Boardman, A. D.: V1519,597-604(1991)

Cai, Yongming See Heflin, James R.: V1560,2-12(1991)

Cai, Yunliang
: New holographic system for measuring vibration,V1554B,75-80(1991)

Cai, Zhonghua See Zhuang, Song Lin: V1558,28-33(1991)

Caillibotte, Georges
; Kechemair, Didier; Sabatier, Lilian: Experiments on convection in laser-melted pools,V1502,117-122(1991)
; Kechemair, Didier; Sabatier, Lilian: Velocity measurements in molten pools during high-power laser interaction with metals,V1412,209-211(1991)

Caimi, Frank M.
; Neely, Jerry; Grossman, Barry G.; Alavie, A. T.: Use of fiber optic communications and control for a tethered undersea vehicle,V1589,90-99(1991)

Caird, John A. See Laumann, Curt W.: V1414,151-160(1991)

Cairoli, Massimo See Beretta, Stefano: V1544,2-9(1991)

Calabrese, Gary C.
; Abali, Livingstone N.; Bohland, John F.; Pavelchek, Edward K.; Sricharoenchaikit, Prasit; Vizvary, Gerald; Bobbio, Stephen M.; Smith, Patrick: Novel surface imaging masking technique for high-aspect-ratio dry etching applications,V1466,528-537(1991)
See Barouch, Eytan: V1463,336-344(1991)

Calabrese, Paul R. See Tregay, George W.: V1589,38-47(1991)

Calatroni, Jose See Tribillon, Gilbert M.: V1332,632-642(1991)

Caldera, Claudio
; De Bernardi, Carlo; Destefanis, Giovanni; Meliga, Marina; Morasca, Salvatore; Rigo, Cesare F.; Stano, Alessandro: Balanced optical mixer integrated in InGaAlAs/InP for coherent receivers,V1372,82-87(1991)

Calderon, S.
; Gannot, Israel; Dror, Jacob; Dahan, Reuben; Croitoru, Nathan I.: Plastic hollow fibers employed for CO2 laser power transmission in oral surgery,V1420,108-115(1991)

Caldwell, J. B.
: Optical design with Wood lenses,V1354,593-598(1991)
See Banash, Mark A.: V1590,8-13(1991)

Caldwell, Martin E.
; Yeatman, Eric M.: Recent advances in surface plasmon spatial light modulators,V1505,50-58(1991)

Calia, V. See Konopka, Wayne L.: V1488,355-365(1991)

Califano, Andrea See Bolle, Ruud M.: V1383,305-318(1991)

Calloway, David L.
: Constructing an optimal binary phase-only filter using a genetic algorithm,V1564,395-402(1991)

Calloway, Terry M.
; Jakowatz, Charles V.; Thompson, Paul A.; Eichel, Paul H.: Comparison of synthetic-aperture radar autofocus techniques: phase gradient versus subaperture,V1566,353-364(1991)

Calmes, J.-M. See Wagnieres, G.: V1525,219-236(1991)

Calvani, Riccardo A.
; Caponi, Renato; Piglia, Roberto: High-speed polarimetric measurements for fiber-optic communications,V1504,258-263(1991)
See Marone, Giuseppe: V1579,122-132(1991)

Calvert, Jack G. See Cantrell, Chris A.: V1433,263-268(1991)

Calvo, G.
See Sanchez, Miguel: V1397,635-638(1991)
See Trelles, Mario A.: V1403,781-798(1991)

Calvo, Maria L.
; De Pedraza-Velasco, L.: Restrictions on the geometry of the reference wave in holocoupler devices,V1507,288-301(1991)
See Mondal, Pronab K.: V1429,108-116(1991)

Camara, Carmen See Moreno-Bondi, Maria C.: V1368,157-164(1991)

Cambon, P.
; Killinger, M.; de Bougrenet de la Tocnaye, Jean-Louis M.: Optically addressed spatial light modulator with nipin aSi:H layers and bistable ferroelectric liquid crystal,V1562,116-125(1991)

Cambria, Terrell D.
; Merrow, Scott F.: Process control capability using a diaphragm photochemical dispense system,V1466,670-675(1991)

Cambridge, Vivian See Lybanon, Matthew: V1406,180-189(1991)

Cameron, C. B. See Brown, David A.: V1584,328-335(1991)

Cameron, Keith H. See Fatah, Rebwar M.: V1501,120-128(1991)

Camisa, Raymond L. See Taylor, Gordon C.: V1475,103-112(1991)

Campana, Joseph E.
: Laser probe mass spectrometry,V1437,138-149(1991)

Campana, Vilma See Juri, Hugo: V1422,128-135(1991)

Campani, Marco
; Straforini, Marco; Verri, Alessandro: First-order differential technique for optical flow,V1388,409-414(1991)
See Coelho, Christopher: V1388,398-408(1991)

Campbell, Duncan R.
; Johnson, Craig W.; Lorenzen, Manfred; eds.: *Optical Fabrication and Testing*,V1400(1991)

Campbell, James D. See Bunis, Jenifer L.: V1377,30-36(1991)

Campbell, Joe C. See Lee, W. D.: V1365,96-101(1991)

Campbell, John H.
; Rainer, Frank; Kozlowski, Mark R.; Wolfe, C. R.; Thomas, Ian M.; Milanovich, Fred P.: Damage resistant optics for a megajoule solid state laser,V1441,444-456(1991)
See Elder, Melanie L.: V1441,237-246(1991)
See Thomas, Ian M.: V1441,294-303(1991)

Campbell, Jonathan W.
; Davis, John M.; Emslie, A. G.: Imaging the sun in hard x-rays: spatial and rotating modulation collimators,V1549,155-179(1991)
; Davis, John M.; Emslie, A. G.: Survey of hard x-ray imaging concepts currently proposed for viewing solar flares,V1343,359-375(1991)

Campbell, Michael V. See Brockhaus, John A.: V1492,200-205(1991)

Campbell, Robert J.
; Kashyap, Raman: Optical data storage in photosensitive fibers,V1499,160-164(1991)

Campbell, T. G. See Ebrahimi, Touradj: V1605,2-15(1991)

Campillo, Anthony J.
; Eversole, J. D.; Lin, H.-B.; Merritt, C. D.: Cavity-QED-enhanced spontaneous emission and lasing in liquid droplets,V1497,78-89(1991)
See Justus, Brian L.: V1409,2-8(1991)

Campion, David C.
See Costa, Joannes M.: V1476,74-80(1991)
See Lam, Benson C.: V1371,36-45(1991)

Campos, Joaquin
; Corredera, Pedro; Pons, Alicia A.; Corrons, Antonio: Germanium photodiodes calibration as standards of optical fiber systems power measurements,V1504,66-74(1991)
See Corredera, Pedro: V1504,281-286(1991)

Campos, Juan
; Janowska-Dmoch, Bozena; Styczynski, K.; Turon, F.; Yzuel, Maria J.; Chalasinska-Macukow, Katarzyna: Amplitude-encoded phase-only filters for pattern recognition: influence of the bleaching procedure,V1574,141-147(1991)
; Millan, Maria S.; Yzuel, Maria J.; Ferreira, Carlos: Color-invariant character recognition and character-background color identification by multichannel matched filter,V1564,189-198(1991)
See Chalasinska-Macukow, Katarzyna: V1564,285-293(1991)
See Millan, Maria S.: V1507,183-197(1991)
See Vallmitjana, Santiago: V1564,266-274(1991)

Campuzano, Juan C.
; Jennings, G.; Beaulaigue, L.; Rodricks, Brian G.; Brizard, Christine M.: Soft x-ray spectro-microscope,V1345,245-254(1991)

Camy-Peyret, C. See De Maziere, Martine M.: V1491,288-297(1991)

Canarelli, P. See Tittel, Frank K.: V1397,21-29(1991)

Canestrari, Paolo
; Degiorgis, Giorgio A.; De Natale, Paolo; Gazzaruso, Lucia; Rivera, Giovanni: Optimization of partial coherence for half-micron i-line lithography,V1463,446-455(1991)

Cangellaris, Andreas C.
See Pasik, Michael F.: V1389,297-301(1991)
See Prince, John L.: V1390,271-285(1991)
See van Deventer, T. E.: V1389,285-296(1991)

Canizares, Claude R. See Markert, Thomas H.: V1549,408-419(1991)

Cannard, F. See Fauvet, E.: V1358,620-630(1991)

Cantella, Michael J.
; Ide, M. H.; O'Donnell, P. J.; Tsaur, Bor-Yeu: Application of IR staring arrays to space surveillance,V1540,634-652(1991)

Cantello, Maichi See Del Bello, Umberto: V1502,104-116(1991)

Canti, Gianfranco L. See Cubeddu, Rinaldo: V1525,17-25(1991)

Cantrell, Chris A.
; Shetter, Richard E.; Lind, John A.; Gilliland, Curt A.; Calvert, Jack G.: Chemical amplifier for peroxy radical measurements based on luminol chemiluminescence,V1433,263-268(1991)

Cantrell, Cyrus D.
: Overview of stimulated Brillouin scattering in microdroplets,V1497,28-47(1991)
; Bowden, Charles M.; eds.: *Nonlinear Optics and Materials*,V1497(1991)
See DiMarco, Steven F.: V1497,178-187(1991)
See Garner, Steven T.: V1497,188-196(1991)
See Hsia, Kangmin: V1497,166-177(1991)
See Milsted, Carl S.: V1497,202-215(1991)

Cantu, Laura
; Corti, Mario; Lago, Paolo; Musolino, Mario: Characterization of a vesicle distribution in equilibrium with larger aggregates by accurate static and dynamic laser light scattering measurements,V1430,144-159(1991)

Cao, C. See Rao, S. M.: V1557,283-292(1991)

Cao, J. Y. See Falciai, Riccardo: V1365,38-42(1991)

Cao, Jianlin
; Li, Futian; Qian, Limin; Chen, Po; Ma, Yueying; Chen, Xingdan: Penning discharge VUV and soft x-ray source,V1345,71-77(1991)
; Miao, Tongqun; Qian, Longsheng; Zhu, Xioufang; Li, Futian; Ma, Yueying; Qian, Limin; Chen, Po; Chen, Xingdan: Plane and concave VUV and soft x-ray multilayered mirrors,V1345,225-232(1991)
See Li, Futian: V1345,78-88(1991)

Cao, Qiang See Cui, DaFu: V1572,386-391(1991)

Cao, Wei
See Cheng, Xianping: V1572,216-219(1991)
See Deng, Xingzhong: V1572,220-223(1991)
See Liu, Bo: V1572,211-215(1991)

Cao, X. Z. See Carlson, D. J.: V1557,140-146(1991)

Cao, Zheng-Ping
; Huang, Yue-Huai: Research on optical fiber colorimeter,V1572,38-41(1991)

Cao, Zhengyuan
; Cheng, Fang: Determination of the homogeneous degree of deformation of the bearing cap of the motor by reflection holography,V1554B,81-85(1991)
; Cheng, Fang: Information extracting and application for the combining objective speckle and reflection holography,V1332,358-364(1991)
See Fang, Ruhua: V1554A,649-656(1991)

Capelle, Gene See Di Benedetto, John A.: V1492,115-125(1991)

Capitelli, Mario
; Gorse, Claudine; Longo, Savino; Bretagne, J.; Estocq, Emmanuel: Importance of nonequilibrium vibrational kinetics of HCl in XeCl laser modeling,V1503,126-131(1991)

Caplan, H. S. See Piestrup, Melvin A.: V1552,214-239(1991)

Caponetti, Laura
See Attolico, Giovanni: V1383,34-46(1991)
See Attolico, Giovanni: V1388,50-61(1991)

Caponi, Maria Z.
: Modeling and simulation of systems imaging through atmospheric turbulence,V1415,138-149(1991)

Caponi, Renato
See Calvani, Riccardo A.: V1504,258-263(1991)
See Marone, Giuseppe: V1579,122-132(1991)

Caporicci, Lucio See Barresi, Giangrande: V1495,246-258(1991)

Capozzi, Carol A.
; Pye, L. D.: Physical and optical properties of organically modified silicates doped with laser and NLO dyes,V1513,320-329(1991)

Cappelletti, John D. See Gerson, Nahum D.: V1406,129-129(1991)

Cappelli, Mark A. See Herchen, Harald: V1534,158-168(1991)

Cappiello, Gregory G. See Goodwin, David B.: V1564,536-549(1991)

Cappolla, Nadia
; Lessard, Roger A.: Methylene-blue-sensitized gelatin used for the fabrication of holographic optical elements in the near-infrared,V1389,612-620(1991)

Capps, Richard W.
; Bothwell, Mary: SIRTF focal-plane technologies,V1540,47-50(1991)

Capron, Barbara See Hong, C. S.: V1418,177-187(1991)

Caquot, E. See Scavennec, Andre: V1362,331-337(1991)

Caramazza, R. See Bertoluzza, Alessandro: V1403,40-49(1991)

Carandang, R. See Liu, Louis: V1475,193-198(1991)

Carbonara, Giuseppe
; Mormile, Pasquale; Bernini, Umberto; Russo, Paolo; Malinconico, Mario; Volpe, M. G.: Characterization of PMMA-EVA blend via photoacoustic technique,V1361,1038-1040(1991)
; Mormile, Pasquale; Abbate, G.; Bernini, Umberto; Maddalena, P.; Malinconico, Mario: New high-performance material in nonlinear optics field, the polymer blend PMMA-EVA: a first investigation,V1361,688-691(1991)

Carbone, Joseph
; Hutton, J.; Arnold, Frank S.; Zarnowski, Jeffrey J.; VanGorden, Steve; Pilon, Michael J.; Wadsworth, Mark V.: Application of low-noise CID imagers in scientific instrumentation cameras,V1447,229-242(1991)
See Zarnowski, Jeffrey J.: V1447,191-201(1991)

Card, Michael E.
: Robotics in near-earth space,V1387,101-108(1991)

Cardarelli, Donato See Kaiser, Todd J.: V1367,121-126(1991)

Cardenas-Garcia, Jaime F.
; Zheng, S.; Shen, F. Z.: Projection moire as a tool for the automated determination of surface topography,V1554B,210-224(1991)

Carder, Kendall L. See Costello, David K.: V1537,161-172(1991)

Cardona, A. H. See Bourne, Lincoln C.: V1477,205-208(1991)

Cardoso, A. V. See Seddon, Angela B.: V1513,255-263(1991)

Carduner, Keith R.
; Colvin, A. D.; Leong, D. Y.; Schuetzle, Dennis; Mackay, Gervase I.: Application of tunable diode laser spectroscopy to the real-time analysis of engine oil economy,V1433,190-201(1991)

Carenco, Alain
See Bigan, Erwan: V1362,553-558(1991)
See Levenson, R.: V1560,251-261(1991)

Carender, Neil H.
; Casasent, David P.; Coetzee, Frans M.; Yu, Daming: Hough transform computer-generated holograms: new output format,V1555,182-193(1991)
; Casasent, David P.: Point target detection, location, and track initiation: initial optical lab results,V1481,35-48(1991)

Carer, P. See Scavennec, Andre: V1362,331-337(1991)

Caressa, J. P. See Saatdjian, E.: V1397,535-538(1991)

Carey, Bruce
; Seshadri, Sridhar B.; Arenson, Ronald L.: PACS: "Back to the Future",V1446,414-419(1991)

Carey, Paul R.
; Kim, Munsok; Tonge, Peter J.: Time resolution of events in an enzyme's active site at 4 K and 300 K using resonance Raman spectroscopy,V1403,37-39(1991)

Carey, Raymond
; Wible, Sheryl F.; Gaynor, Wayne H.; Hendry, Timothy G.: Using an expert system to interface mainframe computing resources with an interactive video system,V1464,500-507(1991)

Carhart, Gary W.
; Draayer, Bret F.; Billings, Paul A.; Giles, Michael K.: Empirical performance of binary phase-only synthetic discriminant functions,V1564,348-362(1991)

Carim, A. H. See Muenchausen, Ross E.: V1394,221-229(1991)

Carin, L. See Felsen, Leopold B.: V1471,154-162(1991)

Carls, Joseph C.
; Brock, James R.: Time-resolved Raman spectroscopy from reacting optically levitated microdroplets,V1497,120-131(1991)

Carlson, Curtis R.
: Convergence of video and computing,V1472,2-5(1991)

Carlson, D. J.
; Wargo, Michael J.; Cao, X. Z.; Witt, August F.: New optical approaches to the quantitative characterization of crystal growth, segregation, and defect formation,V1557,140-146(1991)

Carlson, Daniel W. See Vijaya Kumar, B. V. K.: V1558,476-486(1991)

Carlson, Lawrence W.
: Numerical experiments in propagation with wind velocity fluctuation,V1408,203-211(1991)

Carlson, Nils W.
See Evans, Gary A.: V1418,406-413(1991)
See Evans, Gary A.: V1378,146-161(1991)

Carlson, Robert T. See Tan, Chin: V1417,391-401(1991)

Carlson, Rolf
; Jeffries, Clark: Efficient recognition with high-order neural networks,V1469,684-696(1991)

Carlsten, Bruce E.
See Newnam, Brian E.: V1552,154-174(1991)
See Thode, Lester E.: V1552,87-106(1991)

Carman, Gregory P. See Fogg, Brian R.: V1588,14-25(1991)

Carmer, Dwayne C.
See Reiley, Michael F.: V1416,295-303(1991)
See Tchoryk, Peter: V1479,164-182(1991)

Carmichael, I. C.
; White, Paul R.: Application of adaptive filters to the problem of reducing microphony in arrays of pyroelectric infrared detectors,V1541,167-177(1991)

Carmona, Edward A.
; Matson, Charles L.: Performance of a parallel bispectrum estimation code,V1566,329-340(1991)

Carnera, Alberto See Mazzoldi, Paolo: V1513,182-197(1991)

Carney, Thom See Klein, Stanley A.: V1453,190-205(1991)

Carnicer, Arturo See Vallmitjana, Santiago: V1564,266-274(1991)

Caroli, Ezio See Swinyard, Bruce M.: V1548,94-105(1991)

Carome, Edward F.
; Kubulins, Vilnis E.; Flanagan, Roger L.; Shamray-Bertaud, Patricia: Intensity-type fiber optic electric current sensor,V1584,110-117(1991)
; Kubulins, Vilnis E.; Flanagan, Roger L.: Lower cost fiber optic vibration sensors,V1589,133-138(1991)
See Eustace, John G.: V1584,320-327(1991)

Carosella, John H. See Marshall, Gerald F.: V1454,37-45(1991)

Carpenter, Stephen R. See Bradley, D. J.: V1488,186-195(1991)

Carpenter, Susan
; Tossberg, John; Kraus, George A.: Photosensitization is required for antiretroviral activity of hypericin,V1426,228-234(1991)

Carr, Paul H. See Scalzi, Gary J.: V1475,2-9(1991)

Carra, Paolo
: Theoretical investigation of near-edge phenomena in magnetic systems,V1548,35-44(1991)

Carrabba, Michael M.
; Spencer, Kevin M.; Rauh, R. D.: Compact Raman instrumentation for process and environmental monitoring,V1434,127-134(1991)

Carraresi, Luca See Ferrari, Marco: V1431,276-283(1991)

Carre, Madeleine See Bigan, Erwan: V1362,553-558(1991)

Carretero, L.
; Fuentes, Rosa; Fimia-Gil, Antonio: Measurement of wave aberrations of intraocular lenses through holographic interferometry,V1507,458-462(1991); 1508,96-100(1991)

Carrier, Charles W. See Niu, Aiqun: V1469,495-505(1991)

Carroll, David L. See Thompson, Laird A.: V1542,110-119(1991)

Carroll, John E. See Lam, David T.: V1505,104-114(1991)

Carroll, P. K.
; O'Sullivan, Gerard D.: Laser-produced continua for studies in the XUV,V1503,416-427(1991)

Carroll, Thomas A.
; Ramirez, W. F.: Optimal control of positive optical photoresist development,V1464,222-231(1991)

Carru, J. C.
; Mehri, F.; Chauvel, D.; Crosnier, Y.: Microwave characterization of high-Tc superconducting thin films for simulation and realization of planar microelectronic circuits,V1512,232-239(1991)

Carruthers, John R.
; Tewksbury, Stuart K.; eds.: *Microelectronic Interconnects and Packages: System and Process Integration*,V1390(1991)

Carson, John C.
: On-focal-plane-array feature extraction using 3-D artificial neural network (3DANN): Part II,V1541,227-231(1991)
: On-focal-plane-array feature extraction using a 3-D artificial neural network (3DANN): Part I,V1541,141-144(1991)
; Indin, Ronald: High-density packaging and interconnect of massively parallel image processors,V1541,232-239(1991)
See Smetana, Daryl L.: V1541,220-226(1991)

Carson, Paul See Bowers, C. R.: V1435,36-50(1991)

Carson, Richard F.
; Weaver, Harry T.; Hughes, Robert C.; Zipperian, Thomas E.; Brennan, Thomas M.; Hammons, B. E.: Optically triggered GaAs thyristor switches: integrated structures for environmental hardening,V1378,84-94(1991)

Carson, Susan D.
; Salazar, Robert A.: Splicing plastic optical fibers,V1592,134-138(1991)

Carter, Andrew C. See Johnson, Peter T.: V1371,87-97(1991)

Carter, Jeffrey R. See Simon, Wayne E.: V1469,592-601(1991)

Carter, John N. See Smart, Richard G.: V1373,158-165(1991)

Carter, Patricia H.
: Texture discrimination using wavelets,V1567,432-438(1991)

Carter, T. See Di Cocco, Guido: V1549,102-112(1991)

Carterette, Edward See Whiting, James S.: V1453,165-175(1991)

Cartier, E. See Arnold, Douglas: V1441,478-487(1991)

Cartlidge, Andy G.
; Parel, Jean-Marie; Yokokura, Takashi; Lowery, Joseph A.; Kobayashi, K.; Nose, I.; Lee, William; Simon, Gabriel; Denham, David B.: Laser surgical unit for photoablative and photothermal keratoplasty,V1423,167-174(1991)

Cartwright, Anne See Kirkwood, James J.: V1467,369-371(1991)

Cartwright, Steven L.
: Lau imaging,V1396,481-487(1991)

Carvalho, C. R. See Fantini, Marcia C.: V1536,81-92(1991)

Carvalho, Fernando D.
; Correia, Fernando C.; Freitas, Jose C.; Rodrigues, F. C.: Real-time automatic inspection under adverse conditions,V1399,130-136(1991)
See Correia, Bento A.: V1567,15-24(1991)
See Davies, Roger: V1459,283-291(1991)
See Davies, Roger: V1567,244-253(1991)
See Freitas, Jose C.: V1399,16-23(1991)
See Freitas, Jose C.: V1399,42-48(1991)
See Pais, Cassiano P.: V1451,282-288(1991)
See Rodrigues, F. C.: V1399,90-97(1991)

Carvalho, Isabel C. See Margulis, Walter: V1516,60-66(1991)

Carver, Donald E.
; McAllister, David F.: Development of a stereoscopic three-dimensional drawing application,V1457,54-65(1991)

Carvlin, Mark J.
; Rosa, Louis; Rajan, Sunder; Francisco, John: Contrast-agent-enhanced magnetic resonance imaging: early detection of neoplastic lesions of the CNS,V1426,13-21(1991)

Casaletto, Tom See Kovacich, Michael: V1481,357-370(1991)

Casals, Alicia
; Amat, Josep; Quesada, Jose L.; Sanchez, Luis: High-speed fine-motion tracking of some parts of a target,V1482,317-324(1991)

Casasent, David P.
: Computer-generated holograms for optical data processing,V1544,101-107(1991)
: How to use optics in neural nets: a perspectus,V1564,630-631(1991)
ed.: *Intelligent Robots and Computer Vision IX: Algorithms and Techniques*,V1381(1991)
ed.: *Intelligent Robots and Computer Vision IX: Neural, Biological, and 3-D Methods*,V1382(1991)
: Optical inference processing techniques for scene analysis,V1564,504-510(1991)
: Optical processing and hybrid neural nets,V1469,256-267(1991)
; Chien, Sung-II: MSE and hierarchical optical associative processor system,V1382,304-310(1991)
; Coetzee, Frans M.; Natarajan, Sanjay S.; Xu, Tianning; Yu, Daming; Liu, Hua-Kuang: Accuracy of CGH encoding schemes for optical data processing,V1555,23-33(1991)
; Iyer, Anand K.; Ravichandran, Gopalan: Initial key word OCR filter results,V1384,324-337(1991)
; Iyer, Anand K.; Gopalaswamy, Srinivasan: Optical correlation filters for large-class OCR applications,V1470,208-219(1991)
; Ravichandran, Gopalan: Optical correlation filters to locate destination address blocks in OCR,V1384,344-354(1991)
; Schaefer, Roland H.; Kokaj, Jahja O.: Morphological processing to reduce shading and illumination effects,V1385,152-164(1991)
; Schaefer, Roland H.: Optical gray-scale morphology for target detection,V1568,313-326(1991)
; Sturgill, Robert; Schaefer, Roland H.: Morphological processing for the analysis of disordered structures,V1567,683-690(1991)
; Telfer, Brian A.: Ho-Kashyap CAAP 1:1 associative processors,V1382,158-166(1991)
; Woodford, Paul: Correlation-based optical numeric processors,V1563,112-119(1991)
See Carender, Neil H.: V1555,182-193(1991)
See Carender, Neil H.: V1481,35-48(1991)
See Coetzee, Frans M.: V1544,108-122(1991)
See Lee, Andrew J.: V1383,72-83(1991)
See Natarajan, Sanjay S.: V1564,474-488(1991)
See Ravichandran, Gopalan: V1567,466-479(1991)
See Ravichandran, Gopalan: V1471,233-248(1991)
See Sasaki, Kenji: V1384,228-233(1991)
See Yee, Mark L.: V1481,418-429(1991)
See Yee, Mark L.: V1469,308-319(1991)

Casey, Curtis J.
; Melzer, James E.: Part-task training with a helmet-integrated display simulator system,V1456,175-178(1991)

Casey, Michael P. See Park, S. H.: V1409,9-13(1991)

Casey, William L.
; Doughty, Glenn R.; Marston, Robert K.; Muhonen, John: Design considerations for air-to-air laser communications,V1417,89-98(1991)
See Cotton, John M.: V1417,525-536(1991)

Cash, Webster C. See Gallagher, Dennis J.: V1343,155-161(1991)

Cason, Charles See Horton, T. E.: V1416,10-20(1991)

Cassard, Philippe See Vigroux, Luc M.: V1500,74-79(1991)

Cassarly, William J.
; Ehlert, John C.; Henry, D.: Low-insertion-loss, high-precision liquid crystal optical phased array,V1417,110-121(1991)

Cassassolles, Emmanuel See Tomasini, Bernard: V1468,60-71(1991)

Cassell, P. See Bergman, Larry A.: V1364,14-21(1991)

Cassidy, John C. See Stevens, Eric G.: V1447,274-282(1991)

Castagne, Michel See Montgomery, Paul C.: V1332,563-570(1991)

Castagnolo, Beniamino
See Armenise, Mario N.: V1374,186-197(1991)
See Armenise, Mario N.: V1583,210-220(1991)

Castanho, Jose Eduardo C. See Tozzi, Clesio L.: V1384,124-132(1991)

Castanie, Francis See Poublan, Helene: V1569,329-340(1991)

Castellano, Anthony R.
; Hwang, Vincent S.; Stoney, William E.: Test of a vision-based autonomous space station robotic task,V1387,343-350(1991)

Castelli, C. M. See Wells, Alan A.: V1549,357-373(1991)

Castle, Richard See Thompson, Laird A.: V1542,110-119(1991)

Castrischer, G. See Kretschmer, K.-H.: V1392,246-252(1991)

Castro, Jose L. See Artjushenko, Vjacheslav G.: V1590,131-136(1991)

Castro, Peter
; Gittings, J.; Choi, Yauho J.: TDI camera for industrial applications,V1448,134-139(1991)

Castro-Montero, Alberto
; Shah, S. P.; Bjelkhagen, Hans I.: Study of the fracture process using laser holographic interferometry and image analysis,V1396,122-130(1991)

Casu, M. See Smulevich, G.: V1403,125-127(1991)

Cataldo, J. K. See Deutsch, Alina: V1389,161-176(1991)

Catanzaro, Brian E. See Kiamilev, Fouad E.: V1390,311-329(1991)

Cate, D. See Davis, Paul B.: V1383,515-529(1991)

Catlett, David L. See Garza, Cesar M.: V1466,616-627(1991)

Catlow, C. R. See Dent, Andrew J.: V1550,97-107(1991)

Catthoor, Francky V. See van Swaaij, Michael F.: V1606,901-910(1991)

Caudell, Thomas P.
; Smith, Scott D.; Johnson, G. C.; Wunsch, Donald C.: Application of neural networks to group technology,V1469,612-621(1991)

Caudle, Dennis E.
: Low light level imaging systems application considerations and calculations,V1346,54-63(1991)

Caudle, George F. See Bunis, Jenifer L.: V1377,30-36(1991)

Caughey, Joseph P. See Bunch, Robert M.: V1396,411-416(1991)

Caughman, John See O'Neill, James A.: V1392,516-528(1991)

Cauley, Mike A.
: GaAs monolithic RF modules for SARSAT distress beacons,V1475,275-279(1991)

Caulfield, H. J.
: Holograms in optical computing,VIS08,54-61(1991)
: Industrial applications of optical fuzzy syntactic pattern recognition,V1332,294-300(1991)
: Pacifist's guide to optical computers,V1564,632-632(1991)
: Unique advantages of optics over electronics for interconnections,V1390,399-402(1991)
; Caulfield, Kimberly: Parallel optical information, concept, and response evolver: POINCARE,V1469,232-239(1991)
; Johnson, R. B.; Huang, Qiang: Waveguide hologram star couplers,V1555,154-158(1991)
; Reardon, Janine; Javidi, Bahram: Simulating arbitrary response curves with available response curves for SLMs,V1562,103-106(1991)
; Schamschula, Marius P.; Verber, Carl M.: Adaptive optical interconnection,V1563,103-111(1991)
; Wilkins, Nathan A.: RETINA (RETinally INspired Architecture project),V1564,496-503(1991)
See Granik, Alex T.: VIS08,33-38(1991)
See Huang, Qiang: V1461,303-312(1991)
See Jeon, Ho-In: V1564,522-535(1991)

Caulfield, John T. See Scribner, Dean A.: V1541,100-109(1991)

Caulfield, Kimberly See Caulfield, H. J.: V1469,232-239(1991)

Cavailler, Claude
: Evolution of high-speed photography and photonics techniques in detonics experiments,V1358,210-226(1991)
See Veaux, Jacqueline: V1449,13-24(1991)

Cavaye, Douglas See Tabbara, Marwan R.: V1425,208-216(1991)

Cavedoni, Charles P.
; Graves, J. E.; Pickles, A. J.: Progress report on a five-axis fast guiding secondary for the University of Hawaii 2.2-meter telescope,V1542,273-282(1991)

Caveny, Leonard H.
; Mann, David M.: Shock-layer-induced ultraviolet emissions measured by rocket payloads,V1479,102-110(1991)

Caviris, Nicholas P. See Zhang, Yan: V1481,23-31(1991)

Cawthon, Michael A. See Smith, Donald V.: V1444,357-362(1991)

Ce, Wang See Bjelkhagen, Hans I.: V1461,321-328(1991)

Cebula, Richard P.
; Hilsenrath, Ernest; Kelly, Thomas J.; Batluck, Georgiann R.: Radiometric stability of the shuttle-borne solar backscatter ultraviolet spectrometer,V1493,91-99(1991)
See Weiss, Howard: V1493,80-90(1991)

Cecchi, Joseph L. See Cheah, Chun-Wah: V1392,487-497(1991)

Cederquist, Jack N.
; Fienup, James R.; Marron, Joseph C.; Schulz, Timothy J.; Seldin, J. H.: Digital shearing laser interferometry for heterodyne array phasing,V1416,266-277(1991)

Ceglio, Natale M. See Stearns, Daniel G.: V1465,80-87(1991)

Celenk, Mehmet
; Datari, Srinivasa R.: Rotation invariant object classification using fast Fourier transform features,V1468,752-763(1991)
; Moiz, Saifuddin: Hadamard transform-based object recognition using an array processor,V1468,764-775(1991)

Celens, Eduard A.
; Chabotier, A.: Orthogonal shadowgraphic nanolite stations,V1358,1103-1114(1991)

Celentano, A. See Konopka, Wayne L.: V1488,355-365(1991)

Cellier, Francois E. See Chi, Sung-Do: V1387,182-193(1991)

Celmer, E. See Das, Bidyut B.: V1427,368-373(1991)

Cenina, I. S. See Vas'kovsky, Yu. M.: V1440,229-240(1991)

Centamore, Robert M. See D'Amato, Dante P.: V1544,166-177(1991)

Cerio, Frank M. See Weimer, Wayne A.: V1534,9-13(1991)

Cernosek, John D. See Jaenisch, Holger M.: V1379,162-167(1991)

Cerny, C. L. See Chu, Ting L.: V1361,523-528(1991)

Cerrina, Franco
See Cole, Richard K.: V1465,111-121(1991)
See Guo, Jerry Z.: V1465,330-337(1991)
See Khan, Mumit: V1465,315-323(1991)
See Kovacs, Stephen: V1465,88-99(1991)

Cerullo, Leonard J.
See Boggan, James E.: V1428(1991)
See Grutsch, James: V1428,136-145(1991)

Cervencl, J. See Hudec, Rene: V1343,162-163(1991)

Cesarz, Tadeusz
; Klosowicz, Stanislaw; Zmija, Jozef: Liquid circular polarizer in laser system,V1391,244-249(1991)

Cescato, Lucila H. See Soltz, David: V1536,268-276(1991)

Cesnulevicius, J.
See Rueckmann, I.: V1513,78-85(1991)
See Woggon, Ulrike: V1362,888-898(1991)

Cessenat, Michel See Artola, Michel: V1558,14-21(1991)

Cestaro, John See Andrews, Lee T.: V1459,125-135(1991)

Cha, Dong J.
; Cha, Soyoung S.: Natural pixel decomposition for interferometric tomography,V1554B,600-609(1991)

Cha, M. See Neher, Dieter: V1560,335-343(1991)

Cha, Soyoung S.
See Cha, Dong J.: V1554B,600-609(1991)
See Slepicka, James S.: V1554B,574-579(1991)

Chabbal, Jean See Munier, Bernard: V1447,44-55(1991)

Chabotier, A. See Celens, Eduard A.: V1358,1103-1114(1991)

Chabrier, G. See Goudonnet, Jean-Pierre: V1400,116-123(1991)

Chackalackal, Mathew S.
; Basart, John P.: Application of mathematical morphology to the automated determination of microstructural characteristics of composites,V1568,347-356(1991)

Chaconas, Karen See Nashman, Marilyn: V1383,166-175(1991)

Chadi, D. J.
: Microscopic origin of the shallow-deep transition of impurity levels in III-V and II-VI semiconductors,V1361,228-231(1991)

Chaiken, Joseph See Samoriski, Brian: V1412,12-18(1991)

Chaker, Mohamed
; Boily, S.; Ginovker, A.; Jean, A.; Kieffer, J. C.; Mercier, P. P.; Pepin, Henri; Leung, Pak K.; Currie, John F.; Lafontaine, H.: Recent developments of x-ray lithography in Canada,V1465,16-25(1991)

Chakrabarti, Supriya
See Bush, Brett C.: V1549,290-301(1991)
See Bush, Brett C.: V1549,376-384(1991)
See Cotton, Daniel M.: V1549,313-318(1991)
See Tom, James L.: V1549,308-312(1991)
See Tom, James L.: V1549,302-307(1991)

Chakraborty, Dev P.
; Pfeiffer, Douglas E.; Brikman, Inna: Perceptual noise measurement of displays,V1443,183-190(1991)

Chakravarty, S. See Subramanian, S.: V1362,205-216(1991)

Chakravorty, Kishore K.
; Chien, Chung-Ping: Waveguiding characteristics in polyimide films with different chemistry of formation,V1389,559-567(1991)

Chalasinska-Macukow, Katarzyna
: Accuracy of the output peak localization in two-dimensional matched filtering,V1391,295-302(1991)
; Turon, F.; Yzuel, Maria J.; Campos, Juan: Intermodulation effects in pure phase-only correlation method,V1564,285-293(1991)
See Campos, Juan: V1574,141-147(1991)

Chalupa, John
; Cobb, W. K.; Murdock, Tom L.: Scattering contribution to the error budget of an emissive IR calibration sphere,V1530,343-351(1991)

Chambaret, J. P. See Geindre, Jean-Paul: V1502,311-318(1991)

Chamberlin, E. P. See Fairbank, William M.: V1435,86-89(1991)

Chamberlin, Giles R. See Sheat, Dennis E.: V1461,35-38(1991)

Chambers, Jack P. See Berger, R. M.: V1397,611-618(1991)

Chambers, Robert J.
; Warren, David W.; Lawrie, David J.; Lomheim, Terrence S.; Luu, K. T.; Shima, Ralph M.; Schlegel, J. D.: Reimaging system for evaluating high-resolution charge-coupled-device arrays,V1488,312-326(1991)

Chambettaz, Francois
; Clivaz, Xavier; Marquis-Weible, Fabienne D.; Salathe, R. P.: Temperature variations of reflection, transmission, and fluorescence of the arterial wall,V1427,134-140(1991)

Chameroy, Virginie See Aubry, Florent: V1446,168-176(1991)

Chan, Andrew K. See Gong, Xue-Mei: V1418,422-433(1991)

Chan, Cheuk L.
; Sullivan, Barry J.; Sahakian, Alan V.; Katsaggelos, Aggelos K.; Frohlich, Thomas; Byrom, Ernest: Spatiotemporal filtering of digital angiographic image sequences corrupted by quantum mottle,V1450,208-217(1991)

Chan, Christina K. See Shynk, John J.: V1565,102-117(1991)

Chan, David S.
: Model-based analysis of 3-D spatial-temporal IR clutter suppression filtering,V1481,117-128(1991)

Chan, Eric Y.
See Fu, Richard J.: V1418,108-115(1991)
See Hong, C. S.: V1418,177-187(1991)

Chan, Hau P.
; Chung, P. S.; Pun, Edwin Y.: Mode conversion in Y-branch waveguides,V1583,129-134(1991)

Chan, Heang-Ping See Cheng, Shirley N.: V1450,90-98(1991)

Chan, Hei-Wai See Chen, Chiping: V1407,105-112(1991)

Chan, Hoover
; Abramov, Israel; Gordon, James: Large and small color differences: predicting them from hue scaling,V1453,381-389(1991)

Chan, John P.
; Batchelor, Bruce G.: Integrating vision and AI in an image processing workstation,V1386,163-170(1991)
; Batchelor, Bruce G.; Harris, I. P.; Perry Beng, S. J.: Intelligent visual inspection of food products,V1386,171-179(1991)

Chan, Keith C.
; Vieth, John O.; Wong, Andrew K.: Artificial neural network for supervised learning based on residual analysis,V1469,359-372(1991)

Chan, Kelby K.
; Lau, Christina C.; Chuang, Keh-Shih; Morioka, Craig A.: Visualization and volumetric compression,V1444,250-255(1991)
See Hayrapetian, Alek: V1446,243-247(1991)
See Morioka, Craig A.: V1445,534-538(1991)
See Stewart, Brent K.: V1446,141-153(1991)
See Taira, Ricky K.: V1446,451-458(1991)
See Weinberg, Wolfram S.: V1446,430-435(1991)
See Wong, Albert W.: V1446,73-80(1991)

Chan, Kin-Pui
; Killinger, Dennis K.: Development of 1- and 2-um coherent Doppler lidars for atmospheric sensing,V1492,111-114(1991)
See Killinger, Dennis K.: V1416,125-128(1991)

Chan, Kwok-Chi D.
; Meier, Karl L.; Nguyen, Dinh C.; Sheffield, Richard L.; Wang, Tai-San; Warren, Roger W.; Wilson, William; Young, Lloyd M.: Compact free-electron laser at the Los Alamos National Laboratory,V1552,69-78(1991)
See Newnam, Brian E.: V1552,154-174(1991)
See Thode, Lester E.: V1552,87-106(1991)

Chan, Lap S.
; Hertog, Craig K.; Youngner, D. W.: Honeywell's submicron polysilicon gate process,V1392,232-239(1991)

Chan, Ming-kam See Kwok, Kam-cheung: V1567,709-719(1991)

Chan, Paul H. See Vermillion, Charles H.: V1492,224-228(1991)

Chan, Shiu Chuen M. See Boddu, Jayabharat: V1562,251-262(1991)

Chan, Sing C. See Hsu, L. S.: V1469,197-207(1991)

Chan, W. K.
; Liu, C. Y.; Wong, Y. W.: Applications of laser techniques in fluid mechanics,V1399,82-89(1991)

Chan, Yi-Jen See Clarke, Roy: V1361,2-12(1991)

Chance, Britton
; Haselgrove, John C.; Wang, Nai-Guang; Maris, Michael B.; Sevick, Eva M.: Photon dynamics in tissue imaging,V1525,68-82(1991)
; Katzir, Abraham; eds.: *Time-Resolved Spectroscopy and Imaging of Tissues*,V1431(1991)
See Cui, Weijia: V1431,180-191(1991)
See Haselgrove, John C.: V1431,30-41(1991)
See Maris, Michael B.: V1431,136-148(1991)
See Mayevsky, Avraham: V1431,303-313(1991)
See Sevick, Eva M.: V1431,264-275(1991)
See Sevick, Eva M.: V1431,84-96(1991)
See Weng, Jian: V1431,161-170(1991)

Chance, Kelly V.
; Burrows, John P.; Schneider, Wolfgang: Retrieval and molecule sensitivity studies for the global ozone monitoring experiment and the scanning imaging absorption spectrometer for atmospheric chartography,V1491,151-165(1991)
See Burrows, John P.: V1490,146-154(1991)
See Traub, Wesley A.: V1491,298-307(1991)

Chandan, Harish C. See Yuce, Hakan H.: V1366,120-128(1991)

Chandra, Dinesh See Kumar, V.: V1361,809-811(1991)

Chandra, Kambhamettu See Goldgof, Dmitry B.: V1450,264-276(1991)

Chandra, Mudjianto
; Gershman, Alex; Papazoglou, Theodore G.; Bender, Leon; Danoff, Dudley; Papaioannou, Thanassis; Vari, Sandor G.; Coons, Gregory; Grundfest, Warren S.: Laser-induced fluorescence spectroscopy of pathologically enlarged prostate gland in vitro,V1421,68-71(1991)
See Gershman, Alex: V1421,186-188(1991)
See Papaioannou, Thanassis: V1420,203-211(1991)
See Vari, Sandor G.: V1426,58-65(1991)
See Vari, Sandor G.: V1426,111-120(1991)

Chandra, Ramesh
; Rusinek, Henry: Tissue volume determinations from brain MRI images: a phantom study,V1445,133-144(1991)

Chandra, S.
; Lee, Yung-Cheng: Quick prototyping center for hybrid-wafer-scale integration (HWSI) multichip modules,V1390,548-559(1991)

Chandra, T.
; Abidi, Mongi A.: Evaluation of a pose estimation algorithm using single perspective view,V1382,409-426(1991)

Chandran, Krishnan B. See Frazin, Leon J.: V1425,207-207(1991)

Chandran, Sharat
; Maejima, Tsukasa; Miyazaki, Sanae: Global minima via dynamic programming: energy minimizing active contours,V1570,391-402(1991)

Chandrasekaran, K. See Sehgal, Chandra M.: V1425,226-233(1991)

Chandrasekhar, N. S.
; Sundari, V.; Vengal, Jacob V.; Rao, P. N.: Computer-Aided System Interconnect Design (CASID) in multipackage environment,V1390,523-536(1991)

Chang, Britton See Fittinghoff, David N.: V1413,81-88(1991)

Chang, C. Y.
: Cold to hot electron transition devices,V1362,978-983(1991)

Chang, Chair-Li See Cheng, Shirley N.: V1450,90-98(1991)

Chang, Che-way
: Analysis of optical measurements of SIF and singularity order in rocket motor geometry,V1554A,250-261(1991)

Chang, Chia-Yuan See Fan, Kuo-Chin: V1468,674-684(1991)

Chang, Ching T. See Sun, C. K.: V1476,294-300(1991)

Chang, David B. See Pollack, S. A.: V1410,156-164(1991)

Chang, David C. See Wu, Doris I.: V1475,140-150(1991)

Chang, Gan-How See Chien, Bing-Shan: V1606,588-598(1991)

Chang, Hsiao T.
; Sun, Geng-tian; Zhang, Yan: Recognition of movement object collision,V1388,442-446(1991)

Chang, I. C.
: Wideband acousto-optic spectrometer,V1476,257-268(1991)

Chang, Jeng-Hsien See Golden, Jeffry: V1407,418-429(1991)

Chang, Jim J.
; Christiansen, Walter H.: Blackbody-pumped CO2 lasers using Gaussian and waveguide cavities,V1412,150-163(1991)

Chang, Jin W.
: Active pattern recognition based on attributes and reasoning,V1468,137-146(1991)

Chang, John H.
; Cimolino, Marc C.; Petros, Mulugeta: Laser transmitter for lidar in-space technology experiment,V1492,43-46(1991)
; Reithmaier, Karl D.: Space-qualified laser transmitter for lidar applications,V1492,38-42(1991)

Chang, Kai
: Integrated circuit active antenna elements for monolithic implementation,V1475,164-174(1991)
; Pollock, Michael A.; Skrehot, Michael K.: Space-based millimeter wave debris tracking radar,V1475,257-266(1991)
See Gopalakrishnan, G. K.: V1476,270-275(1991)

Chang, Li-Hsin
; Goodner, Ray: Improved planarization techniques applied to a low dielectric constant polyimide used in multilevel metal ICs,V1596,34-45(1991)

Chang, Long-Wen
; Tsai, Mong-Jean: Shape registration by morphological operations,V1606,120-131(1991)

Chang, Meitung See Chi, Jifu: V1572,74-77(1991)

Chang, Ming-Wen
See Juang, Jeng-Dang: V1354,273-276(1991)
See Sun, Ching-Cherng: V1564,199-210(1991)
See Tang, Yaw-Tzong: V1555,194-199(1991)

Chang, Po-Rong
; Yeh, Bao-Fuh: Retina-like image acquisition system with wide-range light adaptation,V1606,456-469(1991)

Chang, Richard K.
; Chen, Gang; Hill, Steven C.; Barber, Peter W.: Nonlinear optical processes in droplets with single-mode laser excitation,V1497,2-13(1991)

Chang, Robert S.
; Sengupta, Sonnath; Shaw, Leslie B.; Djeu, Nick: Fabrication of laser materials by laser-heated pedestal growth,V1410,125-132(1991)

Chang, Rong-Seng
; Chen, Der-Chin: Automatic-adjusting optical axis for linear CCD scanner,V1527,357-360(1991)
; Chen, Der-Chin: Problem solving of optical design by R graph with bidirectional search mechanics,V1354,379-385(1991)
; Hu, Yeu-Jent: Moire image overlapping method for PCB inspection designator,V1567,216-219(1991)

Chang, Ruey-Feng
; Chen, Wen-Tsuen: Fast finite-state codebook design algorithm for vector quantization,V1605,172-178(1991)

Chang, Sheng-Huei See Watanabe, Hiroshi: V1490,317-323(1991)

Chang, Shi-Kuo
See Leu, Fang Y.: V1468,620-631(1991)
See Li, Xiao-Rong: V1468,720-731(1991)

Chang, Wen-Thong
; Lin, Shao-An: Incremental model for target maneuver estimation,V1481,242-253(1991)
See Yaou, Ming-Haw: V1605,149-159(1991)

Chang, Win-Chyi
; Milch, James R.; eds.: *Camera and Input Scanner Systems*,V1448(1991)
See Stevens, Eric G.: V1447,274-282(1991)
See Wong, Kwok Y.: V1447,283-287(1991)

Chang, Yiher
; Ansari, Nirwan: Practical VLSI realization of morphological operations,V1606,839-850(1991)

Chang, Zenghu
; Hou, Xun; Zhang, Xiaoqiu; Gong, Meixia; Niu, Lihong; Yong, Hongru; Liu, Xiouqin; Lei, Zhiyuan: Development of picosecond x-ray framing camera,V1358,614-618(1991)
; Hou, Xun; Zhang, Yongfeng; Zhu, Wenhua; Niu, Lihong; Liu, Xiouqin: New framing image tube with high-spatial-resolution,V1358,541-545(1991)
See Hou, Xun: V1358,868-873(1991)

Changkakoti, Rupak See Lessard, Roger A.: V1559,438-448(1991)

Chao, Shiuh
; Yang, Tsong-Yo: Read-write simulation and numerical noise for WORM optical disk and drive,V1401,35-43(1991)

Chao, Tien-Hsin
; Stoner, William W.: Optical implementation of neocognitron and its applications to radar signature discrimination,V1558,505-517(1991)

Chapel, Jim D.
; Su, Renjeng: Application of H oo control design techniques to improve dynamics of dexterous manipulation,V1387,284-295(1991)

Chapelle, Walter E.
: Optical components for future fiber optic systems,V1396,389-395(1991)

Chapliev, N. I.
See Armeyev, V. Y.: V1621,2-10(1991)
See Soileau, M. J.: V1441,10-15(1991)

Chaplin, David J. See Korbelik, Mladen: V1426,172-179(1991)

Chapman, Brian N.
See Benjamin, Neil M.: V1392,95-105(1991)
See Gray, David E.: V1392,402-410(1991)

Chapman, Michael A.
; Denby, N.: Storing and managing three-dimensional digital medical image information,V1526,190-194(1991)

Chapman, S. P.
See Padden, Richard J.: V1474,148-159(1991)
See Taylor, Edward W.: V1474,126-131(1991)

Chapman, William W.
; Fitzmaurice, Michael W.: Laser terminal attitude determination via autonomous star tracking,V1417,277-290(1991)
; Fitzmaurice, Michael W.: Optical space-to-ground link availability assessment and diversity requirements,V1417,63-74(1991)
See Hayden, William L.: V1417,182-199(1991)

Chapuis, G. See Wagnieres, G.: V1525,219-236(1991)

Chapuran, Thomas E. See Wang, Lon A.: V1363,85-91(1991)

Charlton, Andrew
; Dickinson, Mark R.; King, Terence A.; Freemont, Anthony J.: Holmium:YAG and erbium:YAG laser interaction with hard and soft tissue,V1427,189-197(1991)
See Wannop, Neil M.: V1423,163-166(1991)

Charra, Fabrice
; Nunzi, Jean-Michel; Messier, Jean: Phase conjugation as a probe for noncentrosymmetry grating formation in organics,V1516,211-219(1991)

Charvet, Pierre Y. See Baskurt, Atilla: V1444,240-249(1991)

Charykov, N. A.
; Litvak, Alexandr M.; Moiseev, K. D.; Yakovlev, Yurii P.: New semiconductor material A1xInAsySb/InAs: LPE synthesis and properties,V1512,198-203(1991)

Chase, Bruce D.
: Multiplex and multichannel detection of near-infrared Raman scattering,V1439,47-57(1991)

Chase, Lloyd L.
See Bennett, Harold E.: V1441(1991)
See Schildbach, M. A.: V1441,287-293(1991)

Chase, Robert P.
: Correction of inherent scan nonplanarity in the Boeing infrared sensor calibration facility,V1533,138-149(1991)

Chase, Scott B. See Kay, David B.: V1499,281-285(1991)

Chashchin, Vadim L.
See Turko, Illarion V.: V1572,419-423(1991)
See Turko, Illarion V.: V1510,53-56(1991)

Chatelain, Andre
See Bays, Roland: V1525,397-408(1991)
See Braichotte, D.: V1525,212-218(1991)
See Wagnieres, G.: V1525,219-236(1991)

Chatterjee, Pallab K.
: From VLSI to ULSI: the subhalf micron challenge,V1393,2-26(1991); 1394,2-26(1991)
: From VLSI to USLI: the subhalf micron challenge,V1392,2-26(1991)

Chatterjee, Samir See Maman, K. H.: V1364,304-311(1991)

Chatterjee, Subhankar See Przybilla, Klaus J.: V1466,174-187(1991)

Chatterji, Gano B.
See Menon, P. K.: V1478,190-200(1991)
See Sridhar, Banavar: V1478,178-189(1991)

Chau, Fook S. See Shang, H. M.: V1554B,680-691(1991)

Chau, Henry H.
; Osher, John E.: Application of Fabry-Perot velocimetry to hypervelocity impact experiments,V1346,103-112(1991)

Chau, Kelvin K. See Kalman, Robert F.: V1563,34-44(1991)

Chau, Lai K. See Armstrong, Neal R.: V1559,18-26(1991)

Chau, Seung L. See Pan, J. J.: V1476,133-142(1991)

Chaubet, C. See Raymond, Andre: V1362,275-281(1991)

Chaubet, Michel See Pyee, Maurice: V1512,240-248(1991)

Chauchard, Eve A. See Kung, Chun C.: V1378,250-258(1991)

Chaudhri, Mohammad M.
; Smith, Alan L.: Damage to polymer-coated glass surfaces by small-particle impact,V1358,683-689(1991)

Chaudhuri, Sujeet K. See Huang, Weiping W.: V1583,268-270(1991)

Chauveau, J. P. See Bourdoncle, Bernard: V1354,194-199(1991)

Chauvel, D. See Carru, J. C.: V1512,232-239(1991)

Chauvel, J. P. See Farquharson, Stuart: V1434,135-146(1991)

Chavantes, Maria C.
; Vinas, Federico; Zamorano, Lucia J.; Dujovny, Manuel; Dragovic, Ljubisa: Influence of helium, oxygen, nitrogen, and room air environment in determining Nd-YAG laser/brain tissue interaction,V1428,13-22(1991)
; Zamorano, Lucia J.: Current and future use of lasers in vascular neurosurgery,V1428,99-127(1991)

Chavel, Pierre
See Lalanne, Philippe: V1621,388-401(1991)
See Taboury, Jean: V1505,115-123(1991)

Chazelas, Jean See Bonniau, Philippe: V1588,52-63(1991)

Che, Diping See Nafie, Laurence A.: V1432,37-49(1991)

Che, Tessie M. See Banash, Mark A.: V1590,8-13(1991)

Cheah, Chun-Wah
; Cecchi, Joseph L.; Stevens, J. L.: Microwave interferometric measurements of process plasma density,V1392,487-497(1991)

Cheatham, John B.
See Walker, Ian D.: V1387,256-270(1991)
See Walker, Ian D.: V1468,974-989(1991)

Cheban, V. N. See Titkov, A. N.: V1361,669-673(1991)

Chebira, Abdennasser See Reynaud, Roger: V1566,302-311(1991)

Chee, Joseph K.
; Kong, Mo-Nga; Liu, Jia-ming: Generation of ultrashort high-average-power passively mode-locked pulses from a Nd:YLF laser with a nonlinear external-coupled cavity at high-repetition rates,V1413,14-20(1991)

Cheek, Eric A. See Peacock, Keith: V1494,147-159(1991)

Cheever, Charles J.
; Li, Zhixin; Raj, K.: Ferrofluid film bearing for enhancement of rotary scanner performance,V1454,139-151(1991)

Cheilan, Francis See Phat, Darith: V1525,196-205(1991)

Chekulayev, V.
; Shevchuk, Igor; Kahru, A.; Mihkelsoo, V. T.; Kallikorm, A. P.: Investigation of the photodynamic properties of some chlorophyll a derivatives: the effect of doxorubicin on the chlorine e6 photosensitized death of Ehrlich carcinoma cells,V1426,367-377(1991)

Chen, Alexander Y.
; Chen, Eugene Y.: Intelligent vision process for robot manipulation,V1381,226-239(1991)

Chen, Ben-zhi See Wu, Ji-Zong: V1399,122-129(1991)

Chen, Bing Y. See Cheng, Xian A.: V1519,33-36(1991)

Chen, C. K. See Tsaur, Bor-Yeu: V1540,580-595(1991)

Chen, C. L. P.
; McAulay, Alastair D.: Polynomial neural network for robot forward and inverse kinematics learning computations,V1468,394-405(1991)

Chen, Chang W.
; Huang, Thomas S.; Arrott, Matthew: Analysis and visualization of heart motion,V1450,231-242(1991)

Chen, Chao See Zhuang, Weihua: V1361,980-986(1991)

Chen, Chaur-Chin
: Markov random fields for texture classification,V1569,274-285(1991)

Chen, Cheng-Tie
; Jeng, Fure-Ching: Coding of motion vectors for motion-compensated predictive/interpolative video coder,V1605,812-821(1991)

Chen, Chiaohsiang See Leith, Emmett N.: V1396,80-84(1991)

Chen, Chien-Chung
; Jeganathan, Muthu; Lesh, James R.: Spatial acquisition and tracking for deep-space optical communication packages,V1417,240-250(1991)
; Win, Moe Z.; Marshall, William K.; Lesh, James R.: Low-data-rate coherent optical link demonstration using frequency-stabilized solid-state lasers,V1417,170-181(1991)
See Robinson, Deborah L.: V1417,421-430(1991)
See Win, Moe Z.: V1417,42-52(1991)

Chen, Chien-Te See Tjeng, L. H.: V1548,160-167(1991)

Chen, Chin-Tu
; Chou, Jin-Shin; Giger, Maryellen L.; Kahn, Charles E.; Bae, Kyongtae T.; Lin, Wei-Chung: Visualization of liver in 3-D,V1444,75-77(1991)
; Ouyang, Xiaolong; Ordonez, Caesar; Hu, Xiaoping; Wong, Wing H.; Metz, Charles E.: Incorporation of structural CT and MR images in PET image reconstruction,V1445,222-225(1991)
See Brailean, James C.: V1450,40-46(1991)
See Chen, Shiuh-Yung: V1445,386-397(1991)
See Chou, Jin-Shin: V1396,45-50(1991)
See Liang, Cheng-Chung: V1445,456-467(1991)
See Lin, Wei-Chung: V1445,376-385(1991)
See Neiw, Han-Min: V1445,259-264(1991)
See Pan, Fu-shih: V1396,5-8(1991)
See Yu, Xiaolin: V1396,56-58(1991)

Chen, Ching-Yi See Arabnia, Hamid R.: V1606,917-925(1991)

Chen, Chiping
; Chan, Hei-Wai; Davidson, Ronald C.: Parametric simulation studies and injection phase locking of relativistic magnetrons,V1407,105-112(1991)
; Wurtele, Jonathan S.: Theory of multimode interactions in cyclotron autoresonance maser amplifiers,V1407,183-191(1991)

Chen, Chung-Wen
; Huang, Jen-Kuang: Estimation of optimal Kalman filter gain from nonoptimal filter residuals,V1489,254-265(1991)

Chen, Chun X. See Sun, Yue Z.: V1519,234-240(1991)

Chen, ChuXin
; Trivedi, Mohan M.; Bidlack, Clint R.; Lassiter, Terrell N.: Environment for simulation and animation of sensor-based robots,V1468,354-366(1991)

Chen, D. N. See Yang, Bing L.: V1519,269-274(1991)

Chen, Da Yong
; Swerdlow, H.; Harke, H.; Zhang, Jian Z.; Dovichi, Norman J.: Single-color laser-induced fluorescence detection and capillary gel electrophoresis for DNA sequencing,V1435,161-167(1991)

Chen, Der-Chin
See Chang, Rong-Seng: V1527,357-360(1991)
See Chang, Rong-Seng: V1354,379-385(1991)

Chen, Di See Lee, Tzuo-chang: V1499,87-103(1991)

Chen, Duanjun
; Chiang, Fu-Pen; Tan, Yushan; Don, H. S.: Computer-aided speckle interferometry: Part II—an alternative approach using spectral amplitude and phase information,V1554A,706-717(1991)
; Li, Shen; Hsu, T. Y.; Chiang, Fu-Pen: Range of measurement of computer-aided speckle interferometry,V1554A,922-931(1991)

Chen, Eugene Y. See Chen, Alexander Y.: V1381,226-239(1991)

Chen, Fuyao See Bao, Liangbi: V1332,862-867(1991)

Chen, G. R. See Hua, Zhong Y.: V1519,2-7(1991)

Chen, Gang See Chang, Richard K.: V1497,2-13(1991)

Chen, Geng-Sheng
; Brady, David J.: Liquid-crystal-doped polymers as volume holographic elements,V1562,128-135(1991)

Chen, George C.
: EBES4: mask/reticle writer for the 90's,V1496,107-117(1991)

Chen, Guang-Hua See Jiang, Xiang-Liu: V1534,207-213(1991)

Chen, Gun-Shing See MacNeal, Paul D.: V1540,68-85(1991)

Chen, Guofu See Wang, Xianhua: V1358,775-779(1991)

Chen, Guo L. See Li, Yi J.: V1519,800-804(1991)

Chen, H. C.
; Stark, Robert A.; Uhm, Han S.: Phase-locking simulation of dual magnetrons,V1407,139-146(1991)
See Stark, Robert A.: V1407,128-138(1991)
See Uhm, Han S.: V1407,113-127(1991)

Chen, H. S. See Zhang, Z. J.: V1519,790-792(1991)

Chen, H. Y.
; Luo, Y. Q.: Standing spin wave modes in permalloy-FeCr multilayer films,V1519,761-764(1991)

Chen, Hector T. See Wells, Gregory M.: V1465,124-133(1991)

Chen, Hong
; Chuang, Yung-Ho; Delettrez, J.; Uchida, S.; Meyerhofer, David D.: Study of x-ray emission from picosecond laser-plasma interaction,V1413,112-119(1991)

Chen, Huai'an See Zhuang, Song Lin: V1354,252-253(1991)

Chen, Huihuang See Zhao, Mingsheng: V1471,464-473(1991)

Chen, J. Y. See Yip, Gar L.: V1583,14-18(1991)

Chen, Jei S. See Ishimaru, Akira: V1558,182-190(1991)

Chen, Jenkins C. See Lin, Freddie S.: V1461,39-50(1991)

Chen, Ji
; Flynn, Michael J.; Gross, Barry; Spizarny, David: Observer detection of image degradation caused by irreversible data compression processes,V1444,256-264(1991)
See Flynn, Michael J.: V1444,172-179(1991)

Chen, Jian G. See Lu, Yu C.: V1519,463-466(1991)

Chen, Jian M. See Li, Yi J.: V1519,800-804(1991)

Chen, Jiann-Jone See Sun, Yung-Nien: V1606,653-664(1991)

Chen, Jihong See Fan, Dapeng: V1572,11-14(1991)

Chen, Jin See Xu, Kewei: V1519,765-770(1991)

Chen, Jin-Tzaih See Wang, Wei-Chung: V1554A,60-69(1991)

Chen, Jing See Zhang, Wei P.: V1519,680-682(1991)

Chen, Jinjiang
: New robot slip sensor using optical fibre and its application,V1572,284-286(1991)

Chen, Ju X. See Zhou, Bing: V1519,454-456(1991)

Chen, Jyh-Woei See Wang, David T.: V1386,206-219(1991)

Chen, Kaiyou
: Polarization of emission lines from relativistic accretion disk,V1548,23-33(1991)

Chen, Ke-long
; Xu, Zhu; Wen, Zhen-chu; Chen, Yu-zhuo: Reconstruction algorithm of displacement field by using holographic image,V1385,206-213(1991)
See Xu, Zhu: V1399,172-177(1991)

Chen, Keshi
; Ramabadran, Tenkasi V.: Reversible compression of industrial radiographs using multiresolution decorrelation,V1567,397-401(1991)

Chen, Kuan-Ren
; Dawson, John M.: Ion-ripple laser as an advanced coherent radiation source,V1552,185-196(1991)

Chen, Kuang-yi See Leung, Chung-yee: V1572,566-571(1991)

Chen, Kun J.
; Du, Jia F.; Li, Zhi F.; Xu, Jun; Jiang, Jian G.; Feng, Duan; Fritzsche, Hellmut: Amorphous silicon periodic and quasiperiodic superlattices,V1519,632-639(1991)

Chen, Lian C.
; Zhang, Ji Y.; Fan, Xi W.; Yang, Ai H.; Zheng, Zhu H.: Optical nonlinearities of ZnSe thin films,V1519,450-453(1991)
See Liu, Yudong: V1362,436-447(1991)

Chen, Liang-Hui
; Kong, Mei-Ying; Wang, Yi-Ming: Recent progress on research of materials for optoelectronic device applications in China,V1361,60-73(1991)

Chen, Lian K. See Trisno, Yudhi S.: V1371,8-12(1991)

Chen, Lideng See Wang, Yongjiang: V1412,67-71(1991)

Chen, Ling-Fan See Krishnapuram, Raghu J.: V1382,271-281(1991)

Chen, Mai
; Yu, Luping; Dalton, Larry R.; Shi, Youngqiang; Steier, William H.: New polymers with large and stable second-order nonlinear optical effects,V1409,202-213(1991)
See Shi, Youngqiang: V1559,118-126(1991)

Chen, Mei-Xia
: Excess-loss dependence of field shape of cladding modes in single-mode fibers,V1504,274-280(1991)

Chen, Menda See Jiang, Shibin: V1535,143-147(1991)

Chen, Michael J.
ed.: *High-Speed Inspection Architectures, Barcoding, and Character Recognition,*V1384(1991)

Chen, Mung
: Is phase-shift mask technology production-worthy?,V1463,2-5(1991)

Chen, Pei See Bow, Sing T.: V1396,646-655(1991)

Chen, Peter C.
: New way of making a lunar telescope,V1494,228-233(1991)

Chen, Po
See Cao, Jianlin: V1345,71-77(1991)
See Cao, Jianlin: V1345,225-232(1991)
See Li, Futian: V1345,78-88(1991)

Chen, Pu S.
; Yang, Jing: Charge characteristics of thin rapid-thermal-nitrided SiOxNy film in MIS structure,V1519,258-262(1991)

Chen, Qianmei See Chen, Xiaobao: V1572,226-229(1991)

Chen, Qing See Kasparis, Takis: V1521,46-54(1991)

Chen, Qun
; Wilson, Brian C.; Patterson, Michael S.; Chopp, Michael; Hetzel, Fred W.: In-vivo optical attenuation in normal rat brain and its implication in PDT,V1426,156-161(1991)

Chen, Ray T.
: Optical interconnects: a solution to very high speed integrated circuits and systems,V1374,162-175(1991)
; Lu, Huey T.; Robinson, Daniel; Plant, David V.; Fetterman, Harold R.: High-speed board-to-board optical interconnection,V1559,110-117(1991)
; Lu, Huey T.; Robinson, Daniel; Wang, Michael R.: Ten-channel single-mode wavelength division demultiplexer in near IR,V1583,135-142(1991)
; Robinson, Daniel; Lu, Huey T.; Sadovnik, Lev S.; Ho, Zonh-Zen: Indium tin oxide single-mode waveguide modulator,V1583,362-374(1991)
; Sadovnik, Lev S.: Collinear asymmetrical polymer waveguide modulator,V1559,449-460(1991)
; Wang, Michael R.; Jannson, Tomasz; Baumbick, Robert J.: Multiple-mode reconfigurable electro-optic switching network for optical fiber sensor array,V1374,223-236(1991)
See Manasson, Vladimir A.: V1559,194-201(1991)
See Sadovnik, Lev S.: V1545,200-208(1991)

Chen, Riqi
; Zhang, Jianxing; Jiang, Kunsheng: Stress analysis in patella by three-dimensional photoelasticity,V1554A,407-417(1991)

Chen, Riyao See Fan, Dapeng: V1572,11-14(1991)

Chen, Robert S.
; Malstrom, Eric M.; Parker, Sandra C.: Standard control language for two different robotic manipulators,V1381,189-200(1991)

Chen, Rongsheng See Yu, Tong: V1572,469-471(1991); 1584,135-137(1991)

Chen, Ru G. See Wu, Zhao P.: V1519,194-198(1991)

Chen, Ruiyi
; Zheng, Dayue; Zhou, Xiuli; Zhang, Xingde: Combination of mechanical athermalization with manual in IR zoom telescope,V1540,724-728(1991)
; Zhou, Xiuli; Zhang, Xingde: Design of compact IR zoom telescope,V1540,717-723(1991)

Chen, Shang-Liang
; Li, L.; Modern, P. J.; Steen, William M.: In-process laser beam position sensing,V1502,123-134(1991)

Chen, Shien C.
: Comparison of relativistic magnetron oscillator models for phase-locking studies,V1407,100-104(1991)
; Bekefi, George; Temkin, Richard J.: Injection locking of a long-pulse relativistic magnetron,V1407,67-73(1991)
See Johnston, George L.: V1407,92-99(1991)

Chen, Shiping
; Rogers, Alan J.; Meggitt, Beverley T.: Large-dynamic-range elecronically scanned "white-light" interferometer with optical fiber Young's structure,V1504,191-201(1991)
; Rogers, Alan J.; Meggitt, Beverley T.: Large dynamic range electronically scanned "white-light" interferometer with optical fiber Young's structure,V1511,67-77(1991)

Chen, Shisheng
; Xu, Zhizhan; Li, Yao-lin; Wang, Xiaofang; Qian, Aidi D.: Soft x-ray emission characteristics from laser plasma sources,V1552,288-295(1991)
See Li, Shiying: V1410,215-220(1991)
See Lin, Li-Huang: V1410,65-71(1991)
See Lin, Li-Huang: V1346,490-501(1991)

Chen, Shiuh-Yung
; Lin, Wei-Chung; Chen, Chin-Tu: Medical image understanding system based on Dempster-Shafer reasoning,V1445,386-397(1991)

Chen, Shouliu See Sun, Dexing: V1572,508-513(1991)

Chen, Shuang
; Freeman, Herbert: Characteristic views and perspective aspect graphs of quadric-surfaced solids,V1383,2-9(1991)
; Freeman, Herbert: Primary set of characteristic views for 3-D objects,V1570,352-361(1991)

Chen, Stephen See Pan, Fu-shih: V1396,5-8(1991)

Chen, Su-Shing
: Sensor fusion at different levels of data abstraction,V1383,574-581(1991)
ed.: *Stochastic and Neural Methods in Signal Processing, Image Processing, and Computer Vision*,V1569(1991)
; Hong, Young-Sik: Generalized neocognitron model for facial recognition,V1569,463-473(1991)
; Tang, Wu-bin; Xu, Jian-hua: Shape-from-focus: surface reconstruction of hybrid surfaces,V1569,446-450(1991)

Chen, Terry Y.
; Ju, M. S.; Lee, C. Y.; Lo, M. P.: Computer-controlled pulse laser system for dynamic holography,V1554B,92-98(1991)

Chen, Tsai-Lin See Wang, Wei-Chung: V1554A,124-135(1991)

Chen, W. S. See Huang, Fon-Shan: V1393,172-179(1991)

Chen, Wei
; Gong, Yiming; Wang, Ruli; Xu, Yayong: Novel method for the computer analysis of cryomicroscopic images,V1450,198-205(1991)
See Zhuang, Song Lin: V1558,28-33(1991)
See Zhuang, Song Lin: V1558,149-153(1991)

Chen, Wei G.
; Reilly, James P.; Wong, Kon M.: Application of canonical correlation analysis in detection in presence of spatially correlated noise,V1566,464-475(1991)

Chen, Wei M.
; Ma, Ke J.; Yu, Zhen Z.; Ji, Hua M.: Determination of thickness and refractal index of HgCdMnTe/CdMnTe VPE films by IR transmission spectrum,V1519,521-524(1991)
See He, Jin: V1519,499-507(1991)
See Ma, Ke J.: V1519,489-493(1991)

Chen, Weizong See He, Da-Ren: V1362,696-701(1991)

Chen, Wen-Chih
; Novembre, Anthony E.: Evaluation of a high-resolution negative-acting electron-beam resist GMC for photomask manufacturing,V1496,266-283(1991)

Chen, Wen-Tsuen See Chang, Ruey-Feng: V1605,172-178(1991)

Chen, Wen-Zhong
See Liu, Zhi-Shen: V1558,306-316(1991)
See Liu, Zhi-Shen: V1558,379-383(1991)

Chen, Wenyi
; Tan, Yushan: Accuracy and high-speed technique for autoprocessing of Young's fringes,V1554A,879-885(1991)

Chen, William W.
; Dunn, Bruce S.; Zhang, Jimin: Experimental and theoretical studies of second-phase scattering in IR transmitting ZnS-based windows,V1535,199-208(1991)

Chen, X. Y. See Shu, Q. Q.: V1519,675-679(1991)

Chen, Xao Min See Yu, Bing Kun: V1534,223-229(1991)

Chen, Xiaobao
; Chen, Qianmei: Optical fiber pressure transducer with improved sensitivity and linearity,V1572,226-229(1991)

Chen, Xiaoguang
: Novel fiber-optic interferometer with high sensitivity and common-mode compensation,V1572,332-336(1991)
; Tang, Weizhong; Zhou, Wen: Transmissive serial interferometric fiber-optic sensor array,V1572,294-298(1991)

Chen, Xiaoming
See Evenhouse, Raymond J.: V1380,248-253(1991)
See Sadler, Lewis L.: V1380,158-162(1991)
See Sadler, Lewis L.: V1380,137-146(1991)

Chen, Xingdan
See Cao, Jianlin: V1345,71-77(1991)
See Cao, Jianlin: V1345,225-232(1991)
See Li, Futian: V1345,78-88(1991)

Chen, Y. X. See Wu, Zi L.: V1519,618-624(1991)

Chen, Yan-Ping
; Yasuda, Yasuhiko: Highly efficient entropy coding of multilevel images using a modified arithmetic code,V1605,822-831(1991)

Chen, Yaosheng
  ; Xiao, Wen; Xue, Mingqiu: Application of the plastic optical fiber in fiber-optic sensors,V1572,124-128(1991)
  See Luo, Nan: V1572,2-4(1991)
  See Xiao, Wen: V1572,170-174(1991)

Chen, Yi
  : Image analysis applied to black ice detection,V1468,551-562(1991)

Chen, Yi-Sheng
  : Scattering measurements of optical coatings in high-power lasers,V1332,115-120(1991)
  ; Wang, Wen-Gui: Surface roughness measurements of spherical components,V1530,111-117(1991)

Chen, Yi-Xin
  : Thin film magnetic recording,V1519,539-547(1991)
  See Jing, Xing-Liang: V1418,434-441(1991)

Chen, Yidong See Dougherty, Edward R.: V1606,141-152(1991)

Chen, Ying-Li See Nie, Chao-Jiang: V1579,264-267(1991)

Chen, Ying Y. See Huang, Xin F.: V1519,220-224(1991)

Chen, Yong
  ; Neu, G.; Deparis, C.; Massies, J.: Energy levels of GaAs/A1xGa1-xAs double-barrier quantum wells,V1361,860-865(1991)

Chen, You S.
  ; Sun, Yi L.; Zhang, Fu M.; Mou, Hai C.: Preferential growth in ion-beam-enhanced deposition of Ti(C,N) films,V1519,56-62(1991)

Chen, Yu-bao See Lu, Xiaoming: V1572,304-307(1991)

Chen, Yu-Che
  See Walker, Ian D.: V1387,256-270(1991)
  See Walker, Ian D.: V1468,974-989(1991)

Chen, Yu-Feng
  ; Liang, Yue-kun; Bian, Hong: High stability optical fiber sensor system,V1572,113-117(1991)

Chen, Yu-zhuo See Chen, Ke-long: V1385,206-213(1991)

Chen, Yud-Ren See McDonald, Timothy P.: V1379,89-98(1991)

Chen, Yuhai
  ; Jia, Youquan: Two-dimensional flow quantitative visualization by the hybrid method of flow birefringence and boundary integration,V1554B,566-572(1991)

Chen, Yung-Mien See Dally, James W.: V1554A,434-443(1991)

Chen, Zengtao
  ; Guo, Maolin; Wang, Duo; Bi, Xianzhi: Dynamic caustic method is applied to fracture,V1554A,206-208(1991)

Chen, Zhan
  ; Jiang, Hongbing; Zheng, J. B.; Zhang, Zhiming: Study on the surface currents density with second-harmonic generation from silver,V1437,103-109(1991)

Chen, Zhen-Pei See Cai, Hai-Tao: V1567,703-708(1991)

Chen, Zheng
  ; Dai, Ji Zhi: K-ion-exchange waveguide directional coupler sensor,V1572,129-131(1991)

Chen, Zhiyong
  See Weng, Zhicheng: V1527,349-356(1991)
  See Weng, Zhicheng: V1527,338-348(1991)

Chenais-Popovics, Claude See Geindre, Jean-Paul: V1502,311-318(1991)

Chenal, C. See Fresne, Francoise: V1444,26-36(1991)

Cheng, Andrew Y.
  ; Ho, Wai-Chin; Kwok, John C.; Yu, Peter K.: Microcomputer-based image processing system for CT/MRI scans: hardware configuration and software capacity,V1444,400-406(1991)
  See Kwok, John C.: V1445,446-455(1991)

Cheng, Denghui See Wu, Zhensen: V1558,251-257(1991)

Cheng, Emily A.
  ; Kane, Thomas J.; Wallace, Richard W.; Cornwell, Donald M.: Injection chaining of diode-pumped single-frequency ring lasers for free-space communication,V1417,300-306(1991)

Cheng, Fang
  See Cao, Zhengyuan: V1554B,81-85(1991)
  See Cao, Zhengyuan: V1332,358-364(1991)

Cheng, Fulin
  ; Venetsanopoulos, Anastasios N.: Adaptive morphological filter for image processing,V1483,49-59(1991)

Cheng, Heng-Da See Shen, Zhijiang: V1606,632-640(1991)

Cheng, Hsing C.
  ; Ramaswamy, Ramu V.: Novel technique for the measurement of coupling length in directional couplers,V1474,291-297(1991)

Cheng, Jie
  ; Qian, Zhaogang; Irani, Keki B.; Etemad, H.; Elta, Michael E.: Expert system and process optimization techniques for real-time monitoring and control of plasma processes,V1392,373-384(1991)

Cheng, John C. See Brock, Phillip J.: V1463,87-100(1991)

Cheng, Jubing See Yang, Zhiguo: V1572,252-257(1991)

Cheng, L. Y. See McVittie, James P.: V1392,126-138(1991)

Cheng, Lap Tak A.
  See Marder, Seth R.: V1560,86-97(1991)
  See Spangler, Charles W.: V1560,139-147(1991)

Cheng, Li-Jen See Liu, Duncan T.: V1409,116-126(1991)

Cheng, May See Nalamasu, Omkaram: V1466,13-25(1991)

Cheng, Minde See Gong, Wei: V1606,153-164(1991)

Cheng, Nai-Jen See Sun, Ching-Cherng: V1564,199-210(1991)

Cheng, Ping-Chin
  See Samarabandu, J. K.: V1450,296-322(1991)
  See Wang, Ge: V1398,180-190(1991)

Cheng, Qi-Xian See Wang, Ming: V1554B,242-246(1991)

Cheng, Qirei
  See Guo, Lu Rong: V1461,91-92(1991)
  See Guo, Lu Rong: V1555,291-292(1991)

Cheng, Shangyi See Yang, Guoguang: V1332,56-63(1991)

Cheng, Shirley N.
  : Computer simulation of magnetic resonance images using fractal-grown brain slices,V1396,9-14(1991)
  ; Chan, Heang-Ping; Adler, Ronald; Niklason, Loren T.; Chang, Chair-Li: Development of a neural network for early detection of renal osteodystrophy,V1450,90-98(1991)

Cheng, Wood-Hi
  ; Appelbaum, Ami; Huang, Rong-Ting; Renner, Daniel; Cioffi, Ken R.: High-frequency 1.3 um InGaAsP semi-insulating buried crescent lasers for analog applications,V1418,279-283(1991)

Cheng, Xian A.
  ; Gu, Qi H.; Chen, Bing Y.: Electrical resistance-strain characteristics and structure of amorphous Ni-Si-B thin films,V1519,33-36(1991)

Cheng, Xianping
  ; Deng, Xingzhong; Xie, Tuqiang; Cao, Wei; Liu, Bo: Research of signal transmission and processing of fiber-optic pyrometer,V1572,216-219(1991)
  See Deng, Xingzhong: V1572,220-223(1991)
  See Liu, Bo: V1572,211-215(1991)

Cheng, Xiaoxue See Guo, Lu Rong: V1555,300-303(1991)

Cheng, Xin
  ; Huang, H. K.: 500 MHz baseband fiber optic transmission system for medical imaging applications,V1364,204-208(1991)
  ; Levin, Paul A.; Nguyen, Hat: Single-fiber wideband transmission systems for HDTV production applications,V1363,177-181(1991)
  See Huang, H. K.: V1444,214-220(1991)

Cheng, Xuezhong See Xiong, Zhengjun: V1467,410-415(1991)

Cheng, Y. C.
  See Wong, H.: V1519,494-498(1991)
  See Yang, Bing L.: V1519,241-246(1991)
  See Yang, Bing L.: V1519,269-274(1991)

Cheng, Yong-Qing See Liu, Ke: V1567,720-728(1991)

Cheng, Yuqi
  ; Zou, Kun; Shan, Xuekang: All-optical-fiber- and general-halogen-lamp-based remote measuring system for CH4,V1572,392-395(1991)
  See Zou, Kun: V1572,472-476(1991)

Cheng, Zhao
  See Wang, Liming: V1358,1156-1160(1991)
  See Wang, Liming: V1415,120-126(1991)

Chenoweth, Amos J. See Jackson, John E.: V1533,75-86(1991)

Cheong, Wai-Fung
  ; Morrison, Paul R.; Trainor, S. W.; Kurban, Amal K.; Tan, Oon T.: Effect of spatial distribution of irradiated sites on injury selectivity in vascular tissue,V1422,19-26(1991)
  See Kurban, Amal K.: V1422,43-49(1991)
  See Pearce, John A.: V1422,27-33(1991)
  See Thomsen, Sharon L.: V1422,34-42(1991)
  See Thomsen, Sharon L.: V1422,14-18(1991)

Chepick, D. I. See Ekimov, A. I.: V1513,123-129(1991)

Cherkashin, G. V. See Dovgalenko, George Y.: V1508,110-115(1991)

Cherkasov, A. S. See Artemjev, S. V.: V1238,206-210(1991)

Cherkasov, Yuri A.
: Photothermoplastic molecular heterostructures,V1621,62-73(1991)

Chern, Chyi S.
; Zhao, Jing-Fu; Li, Y. Q.; Norris, Peter E.; Kear, B. H.; Gallois, B.: Versatility of metal organic chemical vapor deposition process for fabrication of high-quality YBCO superconducting thin films,V1394,255-265(1991)
See Zhao, Jing-Fu: V1362,135-143(1991)

Cherng, Chung-Pin
; Osinski, Marek A.: Fundamental array mode operation of semiconductor laser arrays using external spatial filtering,V1418,372-385(1991)

Chernikov, A. S. See Dubovoi, I. A.: V1358,134-138(1991)

Chernin, David P.
See Colombant, Denis G.: V1407,484-495(1991)
See Lau, Yue Y.: V1407,546-552(1991)

Chernishev, Nicholas A. See Manykin, Edward A.: V1505,141-146(1991)

Chernousova, I. V. See Gaiduk, Mark I.: V1403,674-675(1991)

Chernov, Boris C.
: Characteristic wave theory for volume holographic gratings with arbitrarily slanted fringes,V1507,302-309(1991)
: Comparison of rigorous and approximate methods of analyzing holographic gratings diffraction,V1238,44-53(1991)

Chernov, S. Y. See Kanevsky, M. F.: V1440,154-165(1991)

Chernova, T. I. See Drozhbin, Yu. A.: V1358,451-453(1991)

Chernukhin, Yu. V. See Kalyayev, Anatoli V.: V1621,299-308(1991)

Cherny, V. V. See Ermakov, Yu. A.: V1403,278-279(1991)

Chernyaeva, E. B.
; Golubeva, N. A.; Koroteev, Nikolai I.; Lobanov, O. V.; Vardanyan, A. G.: Time-resolved polarization luminescence spectroscopy of hematoporphyrin in liposomes,V1402,7-10(1991)
See Hianik, T.: V1402,85-88(1991)

Chernyak, Valery See Reisfeld, Renata: V1590,215-228(1991)

Chernyi, D. I. See Ryabova, R. V.: V1238,166-170(1991)

Chernyshov, N. A. See Bass, V. I.: V1358,1075-1083(1991)

Chervyakov, V. V. See Vizir, V. A.: V1411,63-68(1991)

Chessel, Jean-Pierre
See Baudin, Gilles: V1493,16-27(1991)
See Krawczyk, Rodolphe: V1493,2-15(1991)

Chetkin, Sergue A.
See Aksinin, V. I.: V1500,93-104(1991)
See Apollonov, V. V.: V1502,83-94(1991)

Cheung, Huey See Ramesh, Nagarajan: V1468,72-80(1991)

Cheung, Joseph Y. See Yelamarty, Rao V.: V1398,170-179(1991)

Cheung, S. K. See Jain, Faquir C.: V1564,714-722(1991)

Chevalier, Margarita L. See Mondal, Pronab K.: V1429,108-116(1991)

Chevallier, M. See Vinet, Francoise: V1466,558-569(1991)

Chevokin, Victor K.
See Dashevsky, Boris E.: V1346,437-442(1991)
See Ludikov, V. V.: V1346,418-436(1991)

Chevrier, Joel S.
; Thanh, V. L.; Derrien, J.: Epitaxial growth of the semiconducting silicide FeSi2 on silicon,V1512,278-288(1991)

Chew, Wilbert
; Riley, A. L.; Rascoe, Daniel L.; Hunt, Brian D.; Foote, Marc C.; Cooley, Thomas W.; Bajuk, Louis J.: Coplanar waveguide microwave filter of YBa2Cu3O7,V1477,95-100(1991)

Chhachhia, D. P. See Aggarwal, Anil K.: V1238,18-27(1991)

Chi, Jifu
; Chang, Meitung: Signal processing in fiber-optic interferometer with FM light sources,V1572,74-77(1991)

Chi, Kai-Ming See Jain, Ajay: V1596,23-33(1991)

Chi, Rongsheng
See Lu, Xiaoming: V1572,248-251(1991)
See Lu, Xiaoming: V1572,304-307(1991)

Chi, Sung-Do
; Zeigler, Bernard P.; Cellier, Francois E.: Model-based task planning system for a space laboratory environment,V1387,182-193(1991)

Chiang, Eugene See Yeh, Long-Ching: V1527,361-367(1991)

Chiang, Fu-Pen
ed.: *Second Intl Conf on Photomechanics and Speckle Metrology: Moire Techniques, Holographic Interferometry, Optical NDT, and Applications to Fluid Mechanics*,V1554B(1991)
ed.: *Second Intl Conf on Photomechanics and Speckle Metrology: Speckle Techniques, Birefringence Methods, and Applications to Solid Mechanics*,V1554A(1991)
See Chen, Duanjun: V1554A,706-717(1991)
See Chen, Duanjun: V1554A,922-931(1991)
See Dai, YuZhong: V1332,767-774(1991)
See Li, X. M.: V1554A,285-296(1991)
See Shao, C. A.: V1554A,613-618(1991)
See Wang, Y. Y.: V1554B,344-356(1991)

Chiang, John Y.
; Sullivan, Barry J.: Registration of medical images by coincident bit counting,V1396,15-26(1991)

Chiang, Kophu P. See Haas, David R.: V1371,56-67(1991)

Chiang, William See Thompson, Mark E.: V1497,423-429(1991)

Chiao, Pat See Yeh, Long-Ching: V1527,361-367(1991)

Chiaradia, Maria T.
See Attolico, Giovanni: V1383,34-46(1991)
See Attolico, Giovanni: V1388,50-61(1991)

Chiarulli, Donald M.
; Levitan, Steven P.; Melhem, Rami G.: Self-routing interconnection structures using coincident pulse techniques,V1390,403-414(1991)
See Guo, Zicheng: V1390,415-426(1991)

Chiba, Hirotaka See Nakano, Yasuhiko: V1605,874-878(1991)

Chibisov, V. See Bulichev, A.: V1500,151-162(1991)

Chichester, Robert J. See Shaw, Earl D.: V1552,14-23(1991)

Chick, M. C. See Kinsey, Trevor J.: V1358,914-924(1991)

Chicklis, Evan P. See Minnigh, Stephen W.: V1528,129-134(1991)

Chidsey, C. E. See Katz, Howard E.: V1560,370-376(1991)

Chien, Bing-Shan
; Jeng, Bor S.; Sun, San-Wei; Chang, Gan-How; Shyu, Keh-Haw; Shih, Chun-Hsi: Novel block segmentation and processing for Chinese-English document,V1606,588-598(1991)

Chien, Chung-Ping See Chakravorty, Kishore K.: V1389,559-567(1991)

Chien, Sung-Il See Casasent, David P.: V1382,304-310(1991)

Chien, Yung-Ping See Palakal, Mathew J.: V1468,456-466(1991)

Chikama, Terumi See Naito, Takao: V1372,200-207(1991)

Chikishev, A. Y.
; Khurgin, Yu. I.; Romanovsky, Yuri M.; Shidlovskaya, E. G.: Cluster model of protein molecule,V1403,517-521(1991)
; Ladokhin, Alexey S.; Shkurinov, A. P.: Fluorescence spectrochronography of protein intramolecular dynamics,V1403,448-456(1991)

Childers, Ed R. See Finkelstein, Blair I.: V1499,438-449(1991)

Childs, Richard B.
; O'Byrne, Vincent A.: High-dynamic-range fiber optic link using external modulator diode pumped Nd:YAG lasers,V1371,223-232(1991)

Chilingarian, Yu. S.
See Arakelian, Sergei M.: V1402,175-191(1991)
See Arakelian, Sergei M.: V1403,326-331(1991)

Chim, Stanley S.
; Kino, Gordon S.: Optical metrology for integrated circuit fabrication,V1464,138-144(1991)
See Chou, Ching-Hua: V1464,145-154(1991)

Chimiak, William J.
; Williams, Rodney C.: Using the ACR/NEMA standard with TCP/IP and Ethernet,V1446,93-99(1991)
See Boehme, Johannes M.: V1446,312-317(1991)

Chin, Gene H.
; Cordova, Amado; Goldner, Eric L.: Extended environmental performance of attitude and heading reference grade fiber optic rotation sensors,V1367,107-120(1991)

Chin, Hubert H.
: Intelligent information system: for automation of airborne early warning crew decision processes,V1468,235-244(1991)

Chin, Roland T.
; Iverson, Rolf D.: Automated visual inspection of printed wiring boards,VCR36,93-104(1991)

Ching, Hong See Menge, Peter R.: V1407,578-588(1991)

Ching, Jason K.
: ; Bowne, Norman E.: AcidMODES: a major field study to evaluate regional-scale air pollution models,V1491,360-370(1991)

Chinn, Stephen R. See Wanuga, Stephen: V1374,97-106(1991)

Chion, A. See Pelous, G.: V1598,149-158(1991)

Chiong, Kaolin G.
: ; Hohn, Fritz J.: Resist patterning for sub-quarter-micrometer device fabrications,V1465,221-236(1991)

Chipman, Paul S. See Milner, Kathy S.: V1496,180-196(1991)

Chipman, Russell A. See Reardon, Patrick J.: V1354,234-239(1991)

Chirkov, Andre A. See Jensen, Ronald H.: V1403,372-380(1991)

Chiron, Bernard
: : Anamorphosor for scintillating plastic optical fiber applications,V1592,158-164(1991)
: : Highly efficient plastic optical fluorescent fibers and sensors,V1592,86-95(1991)

Chirovsky, Leo M.
: : Massive connectivity and SEEDs,V1562,228-241(1991)
: See Morgan, Robert A.: V1562,149-159(1991)

Chirravuri, Jagannath See Elman, Boris S.: V1362,610-616(1991)

Chirvony, V. S.
: ; Sinyakov, G. N.; Gadonas, R.; Krasauskas, V.; Pelakauskas, A.: Dynamics of ultrafast photoprocesses in Zn-octaethylporphyrin and Zn-octaethylphlorin pi-monoanions,V1403,504-506(1991)
: ; Zenkevich, E. I.; Gadonas, R.; Krasauskas, V.; Pelakauskas, A.: Excitation energy relaxation in model aggregates of photosynthetic pigments upon picosecond laser excitation,V1403,638-640(1991)
: See Apanasevich, P. A.: V1403,195-211(1991)

Chitnis, Vijay T.
: ; Kowsalya, S.; Rashmi, Dr.; Kanjilal, A. K.; Narain, Ram: Automatic mask-to-wafer alignment and gap control using moire interferometry,V1332,613-622(1991)
: See Singh, Brahm P.: V1554B,335-338(1991)

Chittajallu, Siva K. See Palakal, Mathew J.: V1468,456-466(1991)

Chitturi, Prasanna R. See Coleman, Robyn S.: V1392,638-649(1991)

Chiu, Anita S.
: ; Ferguson, Richard A.; Doi, Takeshi; Wong, Alfred K.; Tam, Nelson; Neureuther, Andrew R.: Resist parameter extraction with graphical user interface in X,V1466,641-652(1991)

Chiu, Arnold H.
: See Gottschalk, Paul G.: V1383,84-96(1991)
: See Turney, Jerry L.: V1380,53-63(1991)

Chiu, George
: ; Oprysko, Modest M.: Overview of optical interconnect technology,V1389,364-374(1991)

Chmela, Pavel
: : Phenomenological description of self-organized x(2) grating formation in centrosymmetric doped optical fibers,V1516,116-124(1991)

Chmelir, M. See Berger, R. M.: V1397,611-618(1991)

Cho, Byung-Jin
: ; Kim, Choong-Ki: Anomalous diffusion phenomena in two-step rapid thermal diffusion of phosphorus,V1393,180-191(1991)

Cho, H. M. See Bowers, C. R.: V1435,36-50(1991)

Cho, Kwang J. See Han, Joon H.: V1381,122-129(1991)

Cho, Kyugon See Shemlon, Stephen: V1381,470-481(1991)

Cho, Myung K.
: ; Poczulp, Gary A.: Surface distortions of a 3.5-meter mirror subjected to thermal variations,V1532,137-145(1991)
: See Stepp, Larry M.: V1542,175-185(1991)

Cho, Namhee See Ager, Joel W.: V1437,24-31(1991)

Cho, Paul S. See Weinberg, Wolfram S.: V1446,35-39(1991)

Cho, Young-Chung
: ; Park, Hong-Woo: Instantaneous velocity field measurement of objects in coaxial rotation using digital image velocimetry,V1346,160-171(1991)

Choi, Boo-Yeon See Lee, Choo-Hie: V1397,91-95(1991)

Choi, Hong K. See Evans, Gary A.: V1418,406-413(1991)

Choi, J. J. See Menge, Peter R.: V1407,578-588(1991)

Choi, Peter
: ; Christou, Christos; Aliaga, Raul: Dynamics of a low-energy X-pinch source,V1552,270-277(1991)
: See Aliaga, R.: V1358,1257-1264(1991)
: See Christou, Christos: V1552,278-287(1991)

Choi, Woong-Lim See Kwon, O'Dae: V1361,802-808(1991)

Choi, Yauho J. See Castro, Peter: V1448,134-139(1991)

Chollet, Jean-Paul E. See Festes, Gilles: V1463,245-255(1991)

Chong, Baoxin See Zhang, Zhipeng: V1572,464-468(1991)

Chong, Chae K.
: ; Razeghi, M. M.; McDermott, David B.; Luhmann, Neville C.; Thumm, M.; Pretterebner, Julius: Bragg reflectors: tapered and untapered,V1407,226-233(1991)

Chong, Tow C. See Chua, Soo-Jin: V1401,96-102(1991)

Choo, Chang Y. See Smith, John M.: V1468,493-501(1991)

Choplin, Robert H. See Boehme, Johannes M.: V1446,312-317(1991)

Chopp, Michael See Chen, Qun: V1426,156-161(1991)

Chopra, S.
: See Razdan, Anil K.: V1558,384-388(1991)
: See Singh, Brahm P.: V1558,317-321(1991)

Choquette, Steven J.
: ; Locascio-Brown, Laurie; Durst, Richard A.: Planar waveguide optical immunosensors,V1368,258-263(1991)

Chorvat, Dusan
: ; Hianik, T.: Current state of research and technical application of lasers in Slovakia: their special use in biophysical research at Comenius University in Bratislava,V1402,194-197(1991)
: ; Shvec, Peter: Dynamics of components of biomembranes,V1403,659-666(1991)
: ; Shvec, Peter; Kvasnichka, P.; Shipocz, Tibor; Jarkovska, B.: Lateral mobility of biological membrane components,V1402,89-92(1991)
: See Shvec, Peter: V1402,78-81(1991)

Chorvath, B. See Shvec, Peter: V1403,635-637(1991)

Chou, Chia-Shing
: ; Litvin, K.; Larson, Larry E.; Rosenbaum, Steven E.; Nguyen, Loi D.; Mishra, Umesh K.; Lui, M.; Thompson, M.; Ngo, Catherine M.; Melendes, M.: Coplanar waveguide InP-based HEMT MMICs for microwave and millimeter wave applications,V1475,151-156(1991)
: See Yap, Daniel: V1418,471-476(1991)

Chou, Ching-Hua
: ; Berman, John L.; Chim, Stanley S.; Corle, Timothy R.; Xiao, Guoqing; Kino, Gordon S.: Pattern recognition approach to trench bottom-width measurement,V1464,145-154(1991)
: See Lee, Yong J.: V1393,366-371(1991)

Chou, Chun-Hsien
: ; Chu, Chih-Peng: Probabilistic model for quadtree representation of binary images,V1605,832-843(1991)

Chou, Hung See Lin, Freddie J.: V1558,406-413(1991)

Chou, Jin-Shin
: ; Chen, Chin-Tu; Giger, Maryellen L.; Kahn, Charles E.; Bae, Kyongtae T.; Lin, Wei-Chung: Three-dimensional reconstruction of liver from 2-D tomographic slices,V1396,45-50(1991)
: See Chen, Chin-Tu: V1444,75-77(1991)

Chou, P. See Sinha, Sanjai: V1394,266-276(1991)

Chou, S. T. See Wang, Y. Y.: V1554B,344-356(1991)

Chou, Stephen Y.
: ; Liu, Yue; Fischer, Paul B.: High-speed GaAs metal-semiconductor-metal photodetectors with sub-0.1um finger width and finger spacing,V1474,243-247(1991)

Choudhary, Alok
: ; Ranka, Sanjay: Software development tools for implementing vision systems on multiprocessors,V1406,148-161(1991)

Choudhury, A. N. See Elman, Boris S.: V1362,610-616(1991)

Choudry, Amar
: : Bandwidth, throughput, and information capacity of fiber optic networks for space systems,V1369,121-125(1991)

Chovan, J. See Tomlinson, Harold W.: V1541,178-186(1991)

Chow, P. D.
: ; Lester, J.; Huang, P.; Jones, William L.: InGaAs HEMT MMIC low-noise amplifier and doublers for EHF SATCOM ground terminals,V1475,42-47(1991)
: ; Tan, K.; Streit, Dwight C.; Garske, D.; Liu, Po-Hsin P.; Yen, Huan-chun: Ninety-four GHz InAlAs/InGaAs/InP HEMT low-noise down-converter,V1475,48-54(1991)

Chow, Rowena See Farrell, Thomas J.: V1426,146-155(1991)

Chow, Y. T. See Russell, Philip S.: V1516,47-54(1991)

Chowdhary, R. K. See Morgan, Chris G.: V1525,391-396(1991)

Chowdhury, Dipakbin Q. See Hill, Steven C.: V1497,16-27(1991)

Chown, David P. See Kubinec, James J.: V1364,130-143(1991)

Chrien, Thomas G. See Eastwood, Michael L.: V1540,164-175(1991)

Chrisp, Michael P.
; Macenka, Steve A.: Pressure modulator infrared radiometer (PMIRR) optical system alignment and performance,V1540,213-218(1991)

Christakis, Anne-Marie
: Le Musee de l'Holographie de Paris and its activities: 1980-1990,V1238,348-350(1991)

Christen, D. K. See Lowndes, Douglas H.: V1394,150-160(1991)

Christens-Barry, William A.
; Terry, David H.; Boone, Bradley G.: Detection of DNA sequence symmetries using parallel micro-optical devices,V1564,177-188(1991)

Christensen, Brent J. See Richards, Kent F.: V1421,198-202(1991)

Christensen, David H.
See Boisrobert, Christian Y.: V1474,285-290(1991)
See Hickernell, Robert K.: V1474,138-147(1991)

Christensen, Finn E. See Budtz-Jorgensen, Carl: V1549,429-437(1991)

Christensen, Lorna D.
; Bell, Kenneth L.: Control of proximity effects on CD uniformity through the use of process parameters derived from a statistically designed experiment,V1463,504-514(1991)

Christenson, Eric See Tucker, Michael R.: V1367,289-299(1991)

Christiansen, Jens See Bickel, P.: V1503,167-175(1991)

Christiansen, Walter H.
; Sirota, J. M.: Solar-powered blackbody-pumped lasers,V1397,821-825(1991)
See Chang, Jim J.: V1412,150-163(1991)

Christie, Phillip
; Styer, Stephen B.: Fractal description of computer interconnections,V1390,359-367(1991)

Christie, Simon
; Cai, Xian-Yang; Kvasnik, Frank: Interference effects and the occurrence of blind spots in coherent optical processors,V1507,202-209(1991)

Christner, James H.
: Pioneer unmanned air vehicle accomplishments during Operation Desert Storm,V1538,201-207(1991)

Christodoulou, Christos G. See Murphy, Robert A.: V1558,295-305(1991)

Christou, Aristos
: High-performance GaAs on silicon technology for VLSI, MMICs, and optical interconnects,V1361,354-361(1991)
See Panayotatos, Paul: V1361,1100-1109(1991)

Christou, Christos
; Choi, Peter: Characterization of the radiation from a low-energy X-pinch source,V1552,278-287(1991)
See Choi, Peter: V1552,270-277(1991)

Chronister, Eric L.
; L'Esperance, Drew M.; Pelo, John; Middleton, John; Crowell, Robert A.: Time-resolved photon echo and fluorescence anisotropy study of organically doped sol-gel glasses,V1559,56-64(1991)

Chrzanowski, Krzysztof
: Performance of infrared systems under field conditions,V1512,78-83(1991)

Chrzas, John J.
See Lai, Barry P.: V1550,46-49(1991)
See Yun, Wen-Bing: V1345,146-164(1991)

Chu, Beatrice C.
; Newson, Trevor P.; Jackson, David A.: Fiber optic based vortex shedder flow meter,V1504,251-257(1991)

Chu, Benjamin
; Wang, Zhulun: Electrophoretic mobility and conformational changes of DNA in agarose gels,V1403,258-267(1991)

Chu, C. H. See Harriott, Lloyd R.: V1465,57-63(1991)

Chu, Chee-Hung H.
; Kottapalli, M. S.: Genetic algorithm approach to visual model-based halftone pattern design,V1606,470-481(1991)
See Seetharaman, Guna S.: V1383,582-588(1991)

Chu, Chih-Peng See Chou, Chun-Hsien: V1605,832-843(1991)

Chu, Chu-Lin See Yeh, Long-Ching: V1527,361-367(1991)

Chu, Frank J.
; Yeh, Chia L.: Reduction of blocking artifacts using motion-compensated spatial-temporal filtering,V1452,38-46(1991)

Chu, Kan M. See Thomas, Paul J.: V1488,36-47(1991)

Chu, Ke-Chiang See Normile, James: V1452,480-484(1991)

Chu, Kunliang
; Li, Penghui: Elasto-plastic contact between rollers,V1554A,192-195(1991)

Chu, Raijun See Falk, Joel: V1409,83-86(1991)

Chu, Ron
; Greeneich, James S.; Katz, Barton A.; Lin, Hwang-Kuen; Huang, Dong-Tsair: Advanced lithographic methods for contact patterning on severe topography,V1465,238-243(1991)

Chu, S. T. See Huang, Weiping W.: V1583,268-270(1991)

Chu, Shirley S.
; Chu, Ting L.; Yoo, C. H.; Smith, G. L.: Heteroepitaxial growth of InP and GaInAs on GaAs substrates using nonhydride sources,V1361,1020-1025(1991)
See Chu, Ting L.: V1361,523-528(1991)

Chu, T. S. See Danly, Bruce G.: V1407,192-201(1991)

Chu, Ting L.
; Chu, Shirley S.; Green, Richard F.; Cerny, C. L.: Epitaxial growth of gallium arsenide from elemental arsenic,V1361,523-528(1991)
See Chu, Shirley S.: V1361,1020-1025(1991)

Chu, W. W.
; McStay, Daniel; Rogers, Alan J.: Miniature HiBi current sensor,V1572,523-527(1991)
See McStay, Daniel: V1584,118-123(1991)

Chu, Wileen See Bogle-Rohwer, Elizabeth: V1392,280-290(1991)

Chu, William P.
; McCormick, M. P.; Zawodny, J. M.; McMaster, Leonard R.: Calibration for the SAGE III/EOS instruments,V1491,243-250(1991)
See McCormick, M. P.: V1491,125-141(1991)

Chua, Koon M.
; Johnson, Stewart W.: Foundation, excavation, and radiation-shielding concepts for a 16-m large lunar telescope,V1494,119-134(1991)
See Johnson, Stewart W.: V1494,194-207(1991)

Chua, Soo-Jin
; Leow, S. K.; Ng, T. B.; Chong, Tow C.; Kanhere, R.: Stable broad near-field (single-lateral mode) semiconductor laser,V1401,96-102(1991)
; McCallum, John C.; eds.: *Optical Data Storage Technologies,*V1401(1991)

Chuang, Keh-Shih
; Huang, H. K.: Image noise smoothing based on nonparametric statistics,V1445,496-503(1991)
; Valentino, Daniel J.; Huang, H. K.: Measurement of fractal dimension using 3-D technique,V1445,341-347(1991)
See Chan, Kelby K.: V1444,250-255(1991)
See Ho, Bruce K.: V1445,95-100(1991)
See Huang, H. K.: V1443,198-202(1991)
See Razavi, Mahmood: V1446,24-34(1991)

Chuang, Yung-Ho
; Peatross, J.; Meyerhofer, David D.: Modeling the pedestal in a chirped-pulse-amplification laser,V1413,32-40(1991)
See Chen, Hong: V1413,112-119(1991)

Chuaqui, Hernan H. See Aliaga, R.: V1358,1257-1264(1991)

Chuchem, D.
; Kalisky, Yehoshua Y.; Amit, M.; Smilanski, Israel: Design and performance of an atmospheric pressure HF chemical laser,V1397,277-281(1991)

Chudin, Viktor I. See Gariaev, Peter P.: V1621,280-291(1991)

Chudyk, Wayne
; Pohlig, Kenneth: Dynamic range limits in field determination of fluorescence using fiber optic sensors,V1368,105-114(1991)

Chujko, V. A. See Veiko, Vadim P.: V1544,152-163(1991)

Chulkin, Alexey D. See Vysogorets, Mikhail V.: V1358,1066-1069(1991)

Chumach, V. See Andriesh, A.: V1513,137-147(1991)

Chumakov, A. N. See Min'ko, L. Y.: V1440,166-178(1991)

Chumanov, G. D. See Nabiev, I. R.: V1403,85-92(1991)

Chun, Joohwan
; Luk, Franklin T.: Algorithmic sensor failure detection on passive antenna arrays,V1566,483-492(1991)

Chun, Patrick K. See Brown, Glen: V1396,164-173(1991)

Chun, Wendell H.
; Wolfe, William J.; eds.: *Mobile Robots V,*V1388(1991)
See Price, R. S.: V1388,550-559(1991)

Chung, C. See Jain, Faquir C.: V1564,714-722(1991)

Chung, Kwangsue
; Tou, Julius T.: Knowledge-based approach to fault diagnosis and control in distributed process environments,V1468,323-332(1991)

Chung, P. S.
See Chan, Hau P.: V1583,129-134(1991)
See Pun, Edwin Y.: V1583,102-108(1991)
See Pun, Edwin Y.: V1583,64-70(1991)

Chung, Ray
See Tom, James L.: V1549,308-312(1991)
See Tom, James L.: V1549,302-307(1991)

Chung, W. M. See Clarke, Richard H.: V1437,198-204(1991)

Chung, Yi-Nung
; Emre, Erol; Gustafson, Donald L.: Global modeling approach for
multisensor problems,V1481,306-314(1991)

Chung-Yau, F. See Haus, Joseph W.: V1497,382-388(1991)

Churaev, A. L.
; Artyomova, V. V.: Characteristic curves and phase-exposition characteristics
of holographic photomaterials,V1238,158-165(1991)

Churassy, S. See Bacis, Roger: V1397,173-176(1991)

Church, Eric D.
See Shaw, Ping-Shine: V1343,485-499(1991)
See Shaw, Ping-Shine: V1548,118-131(1991)

Church, Eugene L.
: Scattering from slightly rough crystal surfaces,V1530,171-184(1991)
; Takacs, Peter Z.: Effects of the nonvanishing tip size in mechanical profile
measurements,V1332,504-514(1991)
; Takacs, Peter Z.: Optimal estimation of finish parameters,V1530,71-
85(1991)

Church, Jeffrey S.
: Integrating FTIR microscopy into surface analysis,V1437,80-88(1991)

Church, Susan D.
; Burman, Jerry A.: Random mapping network for tactical target reacquisition
after loss of track,V1471,192-199(1991)

Churin, Evgeny G.
See Hossfeld, Jens: V1574,159-166(1991)
See Lenkova, Galina A.: V1574,235-242(1991)

Churnside, James H. See Shaw, Joseph A.: V1540,681-686(1991)

Churoux, Pascal See Krawczyk, Rodolphe: V1493,2-15(1991)

Chyba, Thomas H.
: Deterministic and noise-induced phase jumps in the ring laser
gyroscope,V1376,132-142(1991)

Chyrvony, V. S. See Apanasevich, P. A.: V1403,240-242(1991)

Cielinski, Amy J.
: Wavefront control model of a beam control experiment,V1542,434-
448(1991)

Cifarelli, Salvatore
; Magrini, Sandro: Low-cost real-time hardware in the loop FCS performance
evaluation,V1482,480-490(1991)

Ciftlikli, Cebrail
See Alci, Mustafa: V1411,100-106(1991)
See Danisman, Kenan: V1412,218-226(1991)

Cikrit, Dolores F. See Dalsing, Michael C.: V1422,98-102(1991)

Cimbrico, P. L. See Brenci, Massimo: V1572,318-324(1991)

Cimolino, Marc C. See Chang, John H.: V1492,43-46(1991)

Cincotta, Anthony H. See Foley, James W.: V1426,208-215(1991)

Cincotta, Louis
See Foley, James W.: V1426,208-215(1991)
See Lin, Chi-Wei: V1426,216-227(1991)

Cindrich, Ivan
; Lee, Sing H.; eds.: *Computer and Optically Generated Holographic Optics;
4th in a Series,*V1555(1991)

Cinelli, Joseph L.
; Kozak, Taras: Multitrack rewritable optical disk system for high-performance
applications: 14-inch TODS,V1499,203-208(1991)
See Bessette, Oliver E.: V1474,162-166(1991)

Cingolani, R. See Ferrara, M.: V1513,130-136(1991)

Cinotti, Ciro See Singer, Christian: V1494,255-264(1991)

Ciocci, Franco
; Bartolini, R.; Dattoli, Giuseppe; Dipace, A.; Doria, Andrea; Gallerano, Gian
P.; Kimmitt, Maurice F.; Messina, G.; Renieri, Alberto; Sabia, E.; Walsh,
John E.: Compact, free-electron laser devices,V1501,154-162(1991)

Cioffi, Ken R. See Cheng, Wood-Hi: V1418,279-283(1991)

Cipriani, Francois D.
See Morvan, D.: V1397,689-692(1991)
See Roux, Agnes: V1397,693-696(1991)

Cirillo, James R.
; Jennings, Kurt L.; Lynn, Mark A.; Messuri, Dominic A.; Steele, Robert E.:
Connection system designed for plastic optical fiber local area
networks,V1592,53-59(1991)
; Jennings, Kurt L.; Lynn, Mark A.; Messuri, Dominic A.; Steele, Robert E.:
Local area network applications of plastic optical fiber,V1592,42-
52(1991)

Ciszewska, Joanna See Kecik, Tadeusz: V1391,341-345(1991)

Citkusev, Ljubomir
; Vaina, Lucia M.: Neural network for inferring the shape of occluded
objects,V1468,786-793(1991)

Citterio, Oberto
; Conconi, Paolo; Conti, Giancarlo; Mattaini, E.; Santambrogio, E.;
Cusumano, G.; Sacco, B.; Braueninger, Heinrich; Burkert, Wolfgang:
Imaging characteristics of the development model of the SAX x-ray
imaging concentrators,V1343,145-154(1991)

Civanlar, Mehmet R.
; Mitra, Sanjit K.; Moorhead, Robert J.; eds.: *Image Processing Algorithms
and Techniques II,*V1452(1991)

Cizdziel, Philip J.
: Design and performance of a PtSi spectroscopic infrared array and detector
head,V1488,6-27(1991)

Claeson, Tord See Bridges, Frank: V1550,76-84(1991)

Claeys, Cor L. See Kreider, Gregory: V1381,242-249(1991)

Claisse, Paul R. See Taylor, Geoff W.: V1476,2-13(1991)

Clampin, Mark
; Durrance, Samuel T.; Golimowski, D. A.; Barkhouser, Robert H.: Johns
Hopkins adaptive optics coronagraph,V1542,165-174(1991)

Clapham, Terrance N.
; D'Arcy, John; Bechtel, Lorne; Glockler, Hermann; Munnerlyn, Charles R.;
McDonnell, Peter J.; Garbus, Jenny: Analysis of an adjustable slit design
for correcting astigmatism,V1423,2-7(1991)
See Shimmick, John K.: V1423,140-153(1991)
See Sliney, David H.: V1423,157-162(1991)

Clapp, William G. See Hansen, Verne W.: V1495,115-122(1991)

Clark, David F. See Davies, David H.: V1401,2-8(1991)

Clark, David L.
; Berry, Joseph R.; Compagna, Gary L.; Cosgrove, Michael A.; Furman,
Geoffrey G.; Heydweiller, James R.; Honickman, Harris; Rehberg,
Raymond A.; Sorlie, Paul H.; Nelson, Edward T.: Design and
performance of a 486 x 640 pixel platinum silicide IR imaging
system,V1540,303-311(1991)
See Berg, John M.: V1435,331-337(1991)
See Short, Ralph C.: V1417,131-141(1991)
See Short, Ralph C.: V1417,464-475(1991)

Clark, Douglas F. See Stewart, George: V1368,230-238(1991)

Clark, Greg A. See Del Grande, Nancy K.: V1479,335-351(1991)

Clark, Ivan O.
See Black, Linda R.: V1557,54-59(1991)
See Johnson, Edward J.: V1557,168-179(1991)

Clark, Lloyd G.
: System transfer modeling for automatic target recognizer
evaluations,VIS07,170-180(1991)

Clark, Michael G. See Purvis, Alan: V1455,145-149(1991)

Clark, Natalie
; Crandall, Charles M.; Giles, Michael K.: Using liquid-crystal TVs in Vander
Lugt optical correlators,V1564,439-451(1991)
See Crandall, Charles M.: V1564,98-109(1991)

Clark, Noel A. See Zhuang, Zhiming: V1455,105-109(1991)

Clark, Peter P.
; Londono, Carmina: 1990 International Lens Design Conference lens design
problems: the design of a NonLens,V1354,555-569(1991)
See Londono, Carmina: V1354,30-37(1991)

Clark, Robert A.
: Design and specification of diamond-turned optics,VCR38,164-183(1991)

Clark, Rodney L.
See Banish, Michele R.: V1557,209-221(1991)
See Johnson-Cole, Helen: V1488,203-211(1991)
See Lawson, Shelah M.: V1488,268-278(1991)

Clark, Stuart E. See Emmony, David C.: V1397,651-659(1991)

Clark, Timothy E.
; Curran, Mark E.: Hybrid fiber optic/electrical network for launch
vehicles,V1369,98-106(1991)
See Udd, Eric: V1418,134-152(1991)

Clark, Wally T.
; Darrow, K. W.; Skipper, M. C.: Comparison of transient analog data fiber optic links,V1371,258-265(1991)

Clarke, Ernest S. See Fox, Neil D.: V1417,452-463(1991)

Clarke, John T. See Bush, Brett C.: V1549,290-301(1991)

Clarke, Peter J.
: Research sputter cluster tool,V1392,617-624(1991)

Clarke, Richard H.
; Chung, W. M.; Wang, Q.; De Jesus, Stephen T.: Determination of gasoline fuel properties by Raman spectroscopy,V1437,198-204(1991)

Clarke, Roy
; Dos Passos, Waldemar; Chan, Yi-Jen; Pavlidis, Dimitris; Lowe, Walter P.; Rodricks, Brian G.; Brizard, Christine M.: Real-time x-ray studies of semiconductor device structures,V1361,2-12(1991)
; Dos Passos, Waldemar; Lowe, Walter P.; Rodricks, Brian G.; Brizard, Christine M.: Time-resolved x-ray scattering studies using CCD detectors,V1345,101-114(1991)
See Rodricks, Brian G.: V1550,18-26(1991)

Clarke, Timothy A.
; Grattan, Kenneth T.; Lindsey, N. E.: Laser-based triangulation techniques in optical inspection of industrial structures,V1332,474-486(1991)

Claro, Francisco See Pacheco, Monica: V1361,819-826(1991)

Clasen, Rolf
; Hornfeck, M.; Theiss, W.: IR-spectroscopical investigations on the glass structure of porous and sintered compacts of colloidal silica gels,V1513,243-254(1991)

Claus, Richard O.
; Murphy, Kent A.; Fogg, Brian R.; Sun, David; Vengsarkar, Ashish M.: Photofluidics for integrating fiber sensors with high-authority mechanical actuators,V1588,159-168(1991)
; Udd, Eric; eds.: *Fiber Optic Smart Structures and Skins IV*,V1588(1991)
See Baldini, S. E.: V1588,125-131(1991)
See DiFrancia, Celene: V1588,44-49(1991)
See Fogg, Brian R.: V1588,14-25(1991)
See Murphy, Kent A.: V1588,134-142(1991)
See Murphy, Kent A.: V1588,117-124(1991)
See Nader-Rezvani, Navid: V1584,405-414(1991)
See Reichard, Karl M.: V1489,218-229(1991)
See Shaw, J. K.: V1367,337-346(1991)
See Tran, Tuan A.: V1584,178-186(1991)
See Vengsarkar, Ashish M.: V1367,249-260(1991)
See Vengsarkar, Ashish M.: V1588,2-13(1991)
See Wang, Anbo: V1584,294-303(1991)

Clauwaert, Julius See Van Laethem, Marc: V1403,732-742(1991)

Clay, Marian E.
See Oleinick, Nancy L.: V1427,90-100(1991)
See Oleinick, Nancy L.: V1426,235-243(1991)

Claybaugh, William R.
; Megill, L. R.: ATHENA: a high-resolution wide-area coverage commercial remote sensing system,V1495,81-94(1991)

Clayton, Donal L.
; Clayton, Richard J.: Use of high-speed videography to solve a structural vibration problem in overhead cranes,V1346,33-41(1991)

Clayton, Richard J. See Clayton, Donal L.: V1346,33-41(1991)

Cleary, John C. See Norgard, John D.: V1540,699-708(1991)

Cleary, Joseph See Das, Bidyut B.: V1427,368-373(1991)

Clecak, Nicholas J.
See MacDonald, Scott A.: V1466,2-12(1991)
See Wallraff, Gregory M.: V1466,211-217(1991)

Clemens, David T.
; Jacobs, David W.: Model group indexing for recognition,V1381,30-42(1991)

Clement, Anne E. See Gilbreath, G. C.: V1409,87-99(1991)

Clement, Dieter See Watkins, Wendell R.: V1486(1991)

Clement, R. M. See Waters, Ruth A.: V1427,336-343(1991)

Clement, Thomas P.
: Design of lightweight beryllium optics, factors effecting producibility, and cost of near-net-shape blanks,V1485,31-38(1991)

Clement, X. See Alloncle, Anne P.: V1397,675-678(1991)

Clements, Rochelle E. See Tao, William C.: V1346,300-310(1991)

Clendening, Charles W.
: Estimation of focused laser beam SBS (stimulated Brillouin scatter) threshold dependence on beam shape and phase aberrations,V1415,72-78(1991)

Clendening, Steven J.
: Rings in a SONET network,V1363,149-157(1991)

Clerke, William H. See Scholen, Douglas E.: V1492,358-369(1991)

Cleveland, William S. See Becker, Richard A.: V1459,48-56(1991)

Cliff, Eugene M. See Hartman, Richard D.: V1478,64-75(1991)

Cliff, Rodger A.
: Space roles for robots,V1387,98-100(1991)

Clifford, Sandra
; Hayes, Bruce L.; Brade, Richard: Results of photolithographic cluster cells in actual production,V1463,551-557(1991)

Clifton, Paul A.
; Brown, Paul D.: Stabilization of CdxHg1-xTe heterointerfaces,V1361,1063-1069(1991)

Cline, David B. See Dodd, James W.: V1407,467-473(1991)

Cline, Terry W. See Park, Yong K.: V1372,219-227(1991)

Clivaz, Xavier See Chambettaz, Francois: V1427,134-140(1991)

Cloonan, Thomas J.
; McCormick, Frederick B.: Photonic switching implementations of 2-input, 2-output switching nodes based on 2-D and 3-D crossover networks,V1396,488-500(1991)
See Brubaker, John L.: V1533,88-96(1991)
See Kerbis, Esther: V1396,447-452(1991)
See McCormick, Frederick B.: V1396,508-521(1991)
See McCormick, Frederick B.: V1533,97-114(1991)

Clos, Rainer See Vogel, Dietmar: V1554A,262-274(1991)

Close, Robert A.
: Automatic 3-D reconstruction of vascular geometry from two orthogonal angiographic image sequences,V1445,513-522(1991)
; Duckwiler, Gary R.; Vinuela, Fernando; Dion, Jacques E.: Vascular parameters from angiographic images,V1444,196-203(1991)

Clot, Jean See Simonne, John J.: V1589,139-147(1991)

Clynick, Tony
: Development of a large-screen high-definition laser video projection system,V1456,51-57(1991)

Coakley, Kevin J.
; Llacer, Jorge: Use of cross-validation as a stopping rule in emission tomography image reconstruction,V1443,226-233(1991)

Cobb, S. H. See Gole, James L.: V1397,125-135(1991)

Cobb, W. K. See Chalupa, John: V1530,343-351(1991)

Coblijn, Alexander B.
: Holographically protected printwork,V1509,73-78(1991)
: Theoretical background and practical processing for art and technical work in dichromated gelatin holography,VIS08,305-324(1991)

Cobuzzi, J. C. See Flanagan, Kathryn A.: V1549,395-407(1991)

Cochrane, Mike See Minot, Katcha: V1475,309-313(1991)

Cocito, Guiseppe See Cognolato, Livio: V1579,249-256(1991)

Cockerill, T. M. See Lim, G.: V1418,123-131(1991)

Cockings, Orville R. See Bowmaster, Thomas A.: V1363,119-124(1991)

Cocorullo, Giuseppe
; Della Corte, Francesco G.; Rendina, Ivo; Cutolo, Antonello: Silicon modulator for integrated optics,V1374,132-137(1991)
See Rendina, Ivo: V1583,338-343(1991)

Coden, Michael H.
; Bulusu, Dutt V.; Ramsey, Brian; Sztuka, Edward; Morrow, Joel: Modular FDDI bridge and concentrator,V1364,22-39(1991)
See Bulusu, Dutt V.: V1364,49-60(1991)

Cody, Jeffrey
See Lang, Robert J.: V1563,2-7(1991)
See Larsson, Anders G.: V1418,292-297(1991)

Coelho, Christopher
; Straforini, Marco; Campani, Marco: Fast and precise method to extract vanishing points,V1388,398-408(1991)

Coetzee, Frans M.
; Casasent, David P.: Focusing and collimating laser diodes and laser diode arrays,V1544,108-122(1991)
See Carender, Neil H.: V1555,182-193(1991)
See Casasent, David P.: V1555,23-33(1991)

Cofer, Rufus H. See Kozaitis, Samuel P.: V1471,249-254(1991)

Coffield, Patrick C.
: Electro-optical image processing architecture for implementing image algebra operations,V1568,137-148(1991)

Coggins, James M. See Baxter, Lisa C.: V1381,459-469(1991)

Coghlan, Gregory A. See Eustace, John G.: V1584,320-327(1991)

Cogley, Robert M. See Knight, Stephen E.: V1464,119-126(1991)

Cognolato, Livio
; De Bernardi, Carlo; Ferraris, M.; Gnazzo, Angelantonio; Morasca, Salvatore; Scarano, Domenica: Spectroscopic properties of Er3+-doped glasses for the realization of active waveguides by ion-exchange technique,V1513,368-377(1991)
; Gnazzo, Angelantonio; Sordo, Bruno; Cocito, Guiseppe: Effect of the glass composition on the emission band of erbium-doped active fibers,V1579,249-256(1991)

Cogswell, Carol J.
; Sheppard, Colin J.: Visualization of 3-D phase structure in confocal and conventional microscopy,V1450,323-328(1991)

Cohen, A. See Bar, I.: V1397,169-172(1991)

Cohen, Andrew See Bogle-Rohwer, Elizabeth: V1392,280-290(1991)

Cohen, Harold L. See Mitchell, John H.: V1414,141-150(1991)

Cohen, Jeffrey K.
: Cryosurgical ablation of the prostate,V1421,45-45(1991)

Cohen, Jonathan
; Reichert, Abraham: Disparity between combiners in a double-combiner head-up display,V1456,250-261(1991)

Cohen, Julius S. See Lenstra, Daan: V1376,245-258(1991)

Cohen, Leon
: General approach for obtaining joint representations in signal analysis and an application to scale,V1566,109-133(1991)

Cohen, Marvin N.
: Survey of radar-based target recognition techniques,V1470,233-242(1991)

Cohen, Paul See Weng, Juyang: V1383,321-332(1991)

Cohen, R. D. See Rosenblatt, Edward I.: V1449,72-86(1991)

Cohendoz, Stephane D. See Paniez, Patrick J.: V1466,336-344(1991)

Cohn, Brian See Lipson, David: V1368,36-43(1991)

Cohn, Gerald E.
; Domanik, Richard A.: Holographic interferometry analysis of sealed, disposable containers for internal defects,V1429,195-206(1991)
; Domanik, Richard A.: Interferometric analysis for nondestructive product monitoring,V1396,131-142(1991)

Colasanti, Alberto See Andreoni, Alessandra: V1525,351-366(1991)

Colavita, Mark M.
; Hines, Braden E.; Shao, Michael; Klose, George J.; Gibson, B. V.: Prototype high-speed optical delay line for stellar interferometry,V1542,205-212(1991)
; Shao, Michael; Hines, Braden E.; Levine, Bruce M.; Gershman, Robert: Optical system design for a lunar optical interferometer,V1494,168-181(1991)

Colavito, D. B. See Tan, Swie-In: V1392,106-118(1991)

Colbran, William V.
: Hitachi e-beam lithography tools for advanced applications,V1496,90-96(1991)

Colby, Grace
; Scholl, Laura: Transparency and blur as selective cues for complex visual information,V1460,114-125(1991)

Coldren, Larry A.
; Corzine, Scott W.; Geels, Randall S.; Gossard, Arthur C.; Law, K. K.; Merz, James L.; Scott, Jeffrey W.; Simes, Robert J.; Yan, Ran H.: High-efficiency vertical-cavity lasers and modulators,V1362,24-37(1991)
See Geels, Randall S.: V1418,46-56(1991)
See Kilcoyne, M. K.: V1389,422-454(1991)

Cole, Barry E.
: High-Tc superconducting infrared bolometric detector,V1394,126-138(1991)

Cole, Brad See Masliah, Denis A.: V1475,113-120(1991)

Cole, Donald C. See Stoltz, John R.: V1479,423-434(1991)

Cole, Edward H.
; Smeins, Larry G.: Monolithic integrated-circuit charge amplifier and comparator for MAMA readout,V1549,46-51(1991)
See Smeins, Larry G.: V1549,59-65(1991)

Cole, George R. See Van Metre, Richard: V1369,9-18(1991)

Cole, Richard K.
; Cerrina, Franco: Novel toroidal mirror enhances x-ray lithography beamline at the Center for X-ray Lithography,V1465,111-121(1991)
See Kovacs, Stephen: V1465,88-99(1991)

Cole, Robert E.
; Merritt, John O.; Coleman, Richard; Ikehara, Curtis: Teleoperator performance with virtual window display,V1457,111-119(1991)

Cole-Hamilton, D. J. See Mullin, John B.: V1361,1116-1127(1991)

Colef, Michael See Wali, Rahman: V1606,665-674(1991)

Coleman, James J.
; Waters, Robert G.; Bour, David P.: InGaAs-GaAs strained layer lasers: physics and reliability,V1418,318-327(1991)
See Lim, G.: V1418,123-131(1991)

Coleman, John D.
: Distribution fiber FTTH/FTTC trial results and deployment strategies,V1363,2-12(1991)

Coleman, Richard See Cole, Robert E.: V1457,111-119(1991)

Coleman, Robyn S.
; Chitturi, Prasanna R.: Defect reduction strategies for submicron manufacturing: tools and methodologies,V1392,638-649(1991)

Coleman, T. A. See Strauss, Michael G.: V1447,12-27(1991)

Colenbrander, V. F. See Reeves, James B.: V1379,28-38(1991)

Coles, Christopher L.
; Phillips, Wayne S.; Vincent, John D.: Reporting data for arrays with many elements,V1488,327-333(1991)

Collet, Jacques See Pugnet, Michel: V1361,96-108(1991)

Colley, Charles B. See Troyer, David E.: V1486,396-409(1991)

Collings, Neil
; Xue, Wei; Pedrini, G.: Efficiency of liquid-crystal light valves as polarization rotators,V1505,12-19(1991)

Collins, D. L. See Henri, Christopher J.: V1444,306-317(1991)

Collins, George J. See Hattori, Shuzo: V1463,539-550(1991)

Collins, Greg E. See Armstrong, Neal R.: V1559,18-26(1991)

Collins, Jeffrey T. See Kuzay, Tuncer M.: V1345,55-70(1991)

Collins, Jim E. See Sachse, Glen W.: V1433,157-166(1991)

Collins, L. See Evans, Alan C.: V1445,236-246(1991)

Collins, Ross See Lareau, Andy G.: V1538,81-98(1991)

Collins, Sean K. See Seiberling, Walter E.: V1495,32-41(1991)

Collins, Stewart A. See Janesick, James R.: V1447,87-108(1991)

Collins, Tyrone J.
See Ramee, Stephen R.: V1420,199-202(1991)
See White, Christopher J.: V1425,130-133(1991)

Colombant, Denis G.
; Lau, Yue Y.; Chernin, David P.: Reduction of beam breakup growth by cavity cross-couplings in recirculating accelerators,V1407,484-495(1991)
; Lau, Yue Y.; Friedman, Moshe; Krall, Jonathan; Serlin, Victor: Relativistic klystron amplifier III: dynamical limiting currents, nonlinear beam loading, and conversion efficiency,V1407,13-22(1991)
See Lau, Yue Y.: V1407,635-646(1991); 1552,182-184(1991)
See Lau, Yue Y.: V1407,546-552(1991)
See Lau, Yue Y.: V1407,479-483(1991)

Colombo, Paolo See Guglielmi, Massimo: V1513,44-49(1991)

Colquhoun, Allan B.
; Cowan, Donald W.; Shepherd, Joseph: Trade-offs in rotary mirror scanner design,V1454,12-19(1991)
See Preston, Ralph G.: V1454,124-131(1991)

Colston, B. W. See Angel, S. M.: V1368,98-104(1991)

Colucci, D'nardo See Wizinowich, Peter L.: V1542,148-158(1991)

Colvin, A. D. See Carduner, Keith R.: V1433,190-201(1991)

Colvin, Michael E. See Searfus, Robert M.: V1473,95-101(1991)

Combis, P. See David, Jean: V1397,697-700(1991)

Comfort, Charles H. See Forbes, Fred F.: V1532,146-160(1991)

Comita, Paul B. See Baum, Thomas H.: V1598,122-131(1991)

Comly, James B.
; Bonissone, Piero P.; Dausch, Mark E.: Fuzzy logic for fault diagnosis,V1381,390-400(1991)

Compagna, Gary L. See Clark, David L.: V1540,303-311(1991)

Comunale, Domenico See Brofferio, Sergio C.: V1605,500-510(1991)

Conant, John A.
; Hummel, John R.: Thermal and radiometric modeling of terrain backgrounds,V1486,217-230(1991)

Concepcion, Vicente P.
; Grzech, Matthew P.; D'Amato, Donald P.: Using morphology in document image processing,V1606,132-140(1991)

Conconi, Paolo See Citterio, Oberto: V1343,145-154(1991)

Condat, M. See Bourda, C.: V1502,148-159(1991)

Condie, Angus See Syme, Richard T.: V1362,467-476(1991)

Condon, Brian P. See Dawber, William N.: V1476,81-90(1991)

Condor, Charles E. See Keski-Kuha, Ritva A.: V1343,566-575(1991)

Conese, Tiziana
; De Pascale, Olga; Matteo, Annamaria; Armenise, Mario N.: New lithium niobate Y-junction polarization divider: theoretical study,V1583,249-255(1991)

Confer, Charles L.
; Burrer, Gordon J.: Linear resonant approach to scanning,V1454,215-222(1991)

Cong, Xiaojie
See Weng, Zhicheng: V1527,349-356(1991)
See Weng, Zhicheng: V1527,338-348(1991)

Congo, Richard T. See Arendale, William F.: V1434,159-170(1991)

Conley, Edgar
; Genin, Joseph: Application of speckle metrology at a nuclear waste repository,V1332,798-801(1991)

Conley, Willard E.
; Dundatscheck, Robert; Gelorme, Jeffrey D.; Horvat, John; Martino, Ronald M.; Murphy, Elizabeth; Petrosky, Anne; Spinillo, Gary; Stewart, Kevin; Wilbarg, Robert; Wood, Robert L.: Chemically amplified negative-tone photoresist for sub-half-micron device and mask fabrication,V1466,53-66(1991)

Connell, G.A. N. See Fork, David K.: V1394,202-202(1991)

Connelly, James M.
; Vijaya Kumar, B. V. K.; Molley, Perry A.; Stalker, K. T.; Kast, Brian A.: Design and testing of space-domain minimum-average correlation energy filters for 2-D acousto-optic correlators,V1564,572-592(1991)

Connelly, William G. See Gates, James L.: V1540,262-273(1991)

Connely, Shawn W. See Gebelein, Rolin J.: V1564,452-463(1991)

Conner, Arlie R. See Gulick, Paul E.: V1456,76-82(1991)

Conner, Michael
See An, Myoung H.: V1565,383-401(1991)
See Tolimieri, Richard: V1565,345-356(1991)

Conner, Teri L. See Stevens, Robert K.: V1491,56-67(1991)

Conners, Richard W.
See Avent, R. R.: V1468,281-295(1991)
See Zhu, Dongping: V1567,232-243(1991)
See Zhu, Dongping: V1569,174-181(1991)
See Zhu, Dongping: V1606,685-696(1991)

Connolly, C. I. See Stenstrom, J. R.: V1470,275-281(1991)

Connolly, Denis J. See Leonard, Regis F.: V1394,114-125(1991)

Connolly, Dennis See Smotroff, Ira G.: V1469,544-550(1991)

Connolly, John C. See Evans, Gary A.: V1378,146-161(1991)

Connor, Brian J.
; Parrish, Alan; Tsou, Jung-Jung: Detection of stratospheric ozone trends by ground-based microwave observations,V1491,218-230(1991)

Connor, William See Takemoto, Cliff H.: V1464,206-214(1991)

Connor, William H.
; Diaz, Pedro J.: Morphological segmentation and 3-D rendering of the brain in magnetic resonance imaging,V1568,327-334(1991)

Connors, Bruce P.
: Sliding control of a single-axis steering mirror,V1489,136-147(1991)

Conrad, D. B. See Huttlin, George A.: V1407,147-158(1991)

Conradson, Steven D.
See Mustre de Leon, Jose: V1550,85-96(1991)
See Newnam, Brian E.: V1552,154-174(1991)

Constant, Monique T.
; Boussekey, Luc; Decoster, Didier; Vilcot, Jean-Pierre: Performances of gallium arsenide on silicon substrate photoconductive detectors,V1362,156-162(1991)

Constantinescu, D. M. See Smith, C. W.: V1554A,102-115(1991)

Conte, D. V. See Brown, David A.: V1369,2-8(1991)

Content, R. See Borra, Ermanno F.: V1494,219-227(1991)

Conti, Giancarlo See Citterio, Oberto: V1343,145-154(1991)

Conti, M. See Barone, Fabrizio: V1584,304-307(1991)

Contolatis, T. See Mondal, J. P.: V1475,314-318(1991)

Cook, Daniel B. See Gerlach, Francis W.: V1488,134-143(1991)

Cook, David M. See Rich, Chris C.: V1461,2-7(1991)

Cook, Dawn M. See Yarling, Charles B.: V1393,411-420(1991)

Cook, Gary See Jones, David C.: V1500,46-52(1991)

Cook, Lee M.
; Burger, Robert J.: Fabrication and optical performance of fractal fiber optics,V1449,186-192(1991)

Cook, Thomas H.
; Hall, Charles S.; Smith, Frederick G.; Rogne, Timothy J.: Simulation of sampling effects in FPAs,V1488,214-225(1991)

Cook, William D. See Wheeler, James R.: V1495,280-285(1991)

Cooke, D. W.
See Muenchausen, Ross E.: V1394,221-229(1991)
See Wu, Xin D.: V1477,8-14(1991)

Cooke, Daniel E.
; Patterson, Dan W.; Starks, Scott A.: Towards a general formula for analogical learning leading to more autonomous systems,V1381,299-305(1991)

Cooke, Paul W. See Taylor, Geoff W.: V1476,2-13(1991)

Cooksey, N. J. See Smith, Richard R.: V1407,83-91(1991)

Cookson, John P.
; Sneiderman, Charles; Rivera, Christopher: Discrete-cosine-transform-based image compression applied to dermatology,V1444,374-378(1991)

Cooley, Thomas W.
; Riley, A. L.; Crist, Richard A.; Sukamto, Lin; Jamnejad, V.; Rascoe, Daniel L.: Ka-band MMIC array feed development for deep space applications,V1475,243-247(1991)
See Chew, Wilbert: V1477,95-100(1991)

Coombes, Anne M.
; Linney, Alfred D.; Richards, Robin; Moss, James P.: Method for the analysis of the 3-D shape of the face and changes in the shape brought about by facial surgery,V1380,180-189(1991)
See Linney, Alfred D.: V1380,190-199(1991)

Coombs, James H.
; Holtslag, A. H.: Scanning optical microscopy: a powerful tool in optical recording,V1499,6-20(1991)

Coon, P. A. See George, Steven M.: V1437,157-165(1991)

Coons, Gregory See Chandra, Mudjianto: V1421,68-71(1991)

Cooper, Alfred W.
; Lentz, William J.; Baca, Michael J.; Bernier, J. D.: Image display and background analysis with the Naval Postgraduate School infrared search and target designation system,V1486,47-57(1991)
; Milne, Edmund A.; Crittenden, Eugene C.; Walker, Philip L.; Moore, E.; Lentz, William J.: Ship signature measurements for tactical decision-aid input,V1486,37-46(1991)
See Crittenden, Eugene C.: V1492,187-197(1991)
See Milne, Edmund A.: V1486,151-161(1991)

Cooper, C. J.
: Sensor line-of-sight stabilization,V1498,39-51(1991)

Cooper, David B. See Taubin, Gabriel: V1570,175-186(1991)

Cooper, David E.
; Riris, Haris; van der Laan, Jan E.: Frequency modulation spectroscopy for chemical sensing of the environment,V1433,120-127(1991)

Cooper, Donald E. See Kozlowski, Lester J.: V1540,250-261(1991)

Cooper, E. E. See Gibbons, Robert C.: V1498,64-69(1991)

Cooper, Keith A. See Hughlett, R. E.: V1465,100-110(1991)

Cooper, Kent
; Nguyen, Bich-Yen; Lin, Jung-Hui; Roman, Bernard J.; Tobin, Phil; Ray, Wayne: Magnetically enhanced reactive ion etching of submicron silicon trenches,V1392,253-264(1991)

Cooper, Richard K. See Thode, Lester E.: V1552,87-106(1991)

Cooper, Ronald F. See Mahapatra, Amaresh: V1374,296-299(1991)

Cooper, Shannon A. See Peacock, Keith: V1494,147-159(1991)

Cooper, Stafford S. See Lurk, Paul W.: V1434,114-118(1991)

Cope, Mark
; van der Zee, Pieter; Essenpreis, Matthias; Arridge, Simon R.; Delpy, David T.: Data analysis methods for near-infrared spectroscopy of tissue: problems in determining the relative cytochrome aa3 concentration,V1431,251-262(1991)
See Arridge, Simon R.: V1431,204-215(1991)

Copeland, Bruce R.
; Anderson, Joseph N.: Framework for load apportioning and interactive force control using a Hopfield neural network,V1381,177-188(1991)

Copeland, David J.
; Zimmerman, Robert K.: High-current, high-bandwidth laser diode current driver,V1417,412-420(1991)

Copeland, J. M. See Moehlmann, Gustaaf R.: V1560,426-433(1991)

Coppeta, David A. See Goodwin, David B.: V1564,536-549(1991)

Corba, M. See Di Cocco, Guido: V1549,102-112(1991)

Corcoran, Patrick See Bailey, Vernon L.: V1407,400-406(1991)

Corcoran, Vincent J.
: High-repetition-rate eyesafe rangefinders,V1419,160-169(1991)

Cordeiro, D. See Pan, J. J.: V1476,157-169(1991)

Cordova, Amado See Chin, Gene H.: V1367,107-120(1991)

Corkum, Paul B. See Strickland, Donna: V1413,54-58(1991)

Corlatan, Dorina
; Schaefer, Martin; Anders, Gerhard: Wavelength shifting and bandwidth broadening in DCG,V1507,354-364(1991)

Corle, Timothy R. See Chou, Ching-Hua: V1464,145-154(1991)

Cormier, Maurice See Ledoyen, Fernand: V1507,328-338(1991)

Cornwell, Donald M.
; Saif, Babak N.: Far-field and wavefront characterization of a high-power semiconductor laser for free-space optical communications,V1417,431-439(1991)
See Cheng, Emily A.: V1417,300-306(1991)

Corredera, Pedro
; Pons, Alicia A.; Campos, Joaquin; Corrons, Antonio: Interferometric system for the inspection and measurement of the quality of optical fiber ends,V1504,281-286(1991)
See Campos, Joaquin: V1504,66-74(1991)

Correia, Bento A.
; Davies, Roger; Carvalho, Fernando D.; Rodrigues, F. C.: Toros: an image processing system for measuring consignments of wood,V1567,15-24(1991)
See Davies, Roger: V1459,283-291(1991)
See Davies, Roger: V1567,244-253(1991)

Correia, Fernando C. See Carvalho, Fernando D.: V1399,130-136(1991)

Corriveau, John P. See Bradley, Lester M.: V1482,48-60(1991)

Corrons, Antonio
See Campos, Joaquin: V1504,66-74(1991)
See Corredera, Pedro: V1504,281-286(1991)

Corsi, Carlo
: Smart sensors,V1512,52-59(1991)
See Naumaan, Ahmed: V1512(1991)

Cortelazzo, Guido M. See Rinaldo, R.: V1606,773-787(1991)

Cortes, J. See Seidel, Claus: V1432,105-116(1991)

Cortesi, Elisabetta See Soref, Richard A.: V1389,408-421(1991)

Corti, Mario See Cantu, Laura: V1430,144-159(1991)

Corzine, Scott W.
See Coldren, Larry A.: V1362,24-37(1991)
See Geels, Randall S.: V1418,46-56(1991)

Cosandey, F. See Zhao, Jing-Fu: V1362,135-143(1991)

Cosgrove, Michael A.
See Clark, David L.: V1540,303-311(1991)
See Short, Ralph C.: V1417,131-141(1991)
See Short, Ralph C.: V1417,464-475(1991)

Cosi, Franco
See Baldini, Francesco: V1510,58-62(1991)
See Falciai, Riccardo: V1572,424-427(1991)
See Falciai, Riccardo: V1365,38-42(1991)

Cosman, Pamela C.
; Oehler, Karen; Heaton, Amanda A.; Gray, Robert M.: Tree-structured vector quantization with input-weighted distortion measures,V1605,162-171(1991)

Cossec, Francois R.
; Doubrere, Patrick; Perez, Eric: Simulation model and on-ground performances validation of the PAT system for the SILEX program,V1417,262-276(1991)

Costa, Enrico
; Piro, Luigi; Rubini, Alda; Soffitta, Paolo; Massaro, Enrico; Matt, Giorgio; Medici, Gastone; Manzo, Giuseppe; Re, Stefano: Cylindrical proportional counter for x-ray polarimetry,V1343,469-476(1991)
See Kaaret, Philip E.: V1548,106-117(1991)

Costa, Joannes M.
; Lam, Benson C.; Kellner, Albert L.; Campion, David C.; Yu, Paul K.: Hybrid optical transmitter for microwave communication,V1476,74-80(1991)
See Lam, Benson C.: V1371,36-45(1991)

Costa, Manuel Filipe P.
; Almeida, Jose B.: Noncontact optical microtopography,V1400,102-107(1991)
; Almeida, Jose B.: Surface microtopography of thin silver films,V1332,544-551(1991)

Costanza, Daniel W. See Jodoin, Ronald E.: V1398,61-70(1991)

Costello, Anthony J. See Johnson, Douglas E.: V1421,36-41(1991)

Costello, David K.
; Carder, Kendall L.; Steward, Robert G.: Development of the marine-aggregated-particle profiling and enumerating rover,V1537,161-172(1991)

Costello, John T. See Kennedy, Eugene T.: V1503,406-415(1991)

Costello, Kenneth A.
; Davis, Gary A.; Weiss, Robert; Aebi, Verle W.: Transferred electron photocathode with greater than 5% quantum efficiency beyond 1 micron,V1449,40-50(1991)

Cote, D. See Leblanc, Roger M.: V1436,92-102(1991)

Cote, Roland See Dodelet, Jean-Pol: V1436,38-49(1991)

Cothren, Robert M. See Rava, Richard P.: V1426,68-78(1991)

Cotton, Daniel M.
; Bach, Bernhard W.; Bush, Brett C.; Chakrabarti, Supriya: V-groove diffraction grating for use in an FUV spatial heterodyne interferometer,V1549,313-318(1991)
See Bush, Brett C.: V1549,290-301(1991)
See Bush, Brett C.: V1549,376-384(1991)
See Tom, James L.: V1549,308-312(1991)
See Tom, James L.: V1549,302-307(1991)

Cotton, John M.
; Casey, William L.: Narrowband optical interference filters,V1417,525-536(1991)

Cotton, Therese M.
; Rospendowski, Bernard; Schlegel, Vicki; Uphaus, Robert A.; Wang, Danli L.; Eng, Lars H.; Stankovich, Marion T.: Spectroscopy of proteins on surfaces: implications for protein orientation and protein/protein interactions,V1403,93-101(1991)

Cottron, M. See Hammami, Slimane: V1554A,136-142(1991)

Cotts, Patricia M.
; Pekala, Rick: Dynamic light scattering studies of resorcinol formaldehyde gels as precursors of organic aerogels,V1430,181-190(1991)

Couch, Lori L.
; Kalin, David A.; McNeal, Terry: Experimental investigation of image degradation created by a high-velocity flow field,V1486,417-423(1991)

Coughlin, James P. See Baran, Robert H.: V1471,164-176(1991)

Coulet, Pierre R. See Blum, Loic J.: V1510,46-52(1991)

Coulter, Daniel R.
; Dolgin, Benjamin P.; Rainen, R.; O'Donnell, Timothy P.: Materials technology for SIRTF,V1540,119-126(1991)
See Bruegge, Carol J.: V1493,132-142(1991)

Courellis, Spiridon H.
; Marmarelis, Vasilis Z.: Speed ranges accommodated by network architectures of elementary velocity estimators,V1606,336-349(1991)

Courteille, C. See Bacal, Marthe: V1407,605-609(1991)

Courtney, R. See Levine, Jerrold S.: V1407,74-82(1991)

Courtot, Marilyn E.
; Nier, Michael: Imagery technology database,VCR37,221-246(1991)
See Nier, Michael: VCR37(1991)

Cousin, B. See Duran, Michel: V1494,357-376(1991)

Cousins, Daniel
: Numerical inversion method for determining aerodynamic effects on particulate exhaust plumes from onboard irradiance data,V1467,402-409(1991)

Coutant, Jacques See Andre, Michel L.: V1502,286-298(1991)

Couture, Jean J.
: Holographic characterization of DYE-PVA films studied at 442 nm for optical elements fabrication,V1559,429-435(1991)

Couves, J. W. See Dent, Andrew J.: V1550,97-107(1991)

Covelli, L.
; De Iorio, I.; Tagliaferri, V.: Surface absorptance in CO2 laser steel processing,V1397,797-802(1991)

Cover, John R. See Schaeffer, A. R.: V1447,165-176(1991)

Covington, John B. See Schwartzkopf, George: V1466,26-38(1991)

Covino, Joseph M.
; Griffiths, Barry E.: New estimation architecture for multisensor data fusion,V1478,114-125(1991)

Cowan, Cregg K. See Mulgaonkar, Prasanna G.: V1382,320-330(1991)

Cowan, Donald W. See Colquhoun, Allan B.: V1454,12-19(1991)

Cowle, Gregory J.
; Reekie, Laurence; Morkel, Paul R.; Payne, David N.: Narrow linewidth fiber laser sources,V1373,54-65(1991)
See Morkel, Paul R.: V1373,224-233(1991)

Cowles, Kelly A.
: Atmospheric visibility monitoring for planetary optical communications,V1487,272-279(1991)

Cox, Charles H.
See Betts, Gary E.: V1371,252-257(1991)
See Betts, Gary E.: V1562,281-302(1991)

Cox, Colin R. See Crocker, James H.: V1494,434-439(1991)

Cox, David E. See Welch, Sharon S.: V1480,2-10(1991)

Cox, J. N.
; Sedayao, J.; Shergill, Gurmeet S.; Villasol, R.; Haaland, David M.: Fourier transform infrared spectrophotometry for thin film monitors: computer and equipment integration for enhanced capabilities,V1392,650-659(1991)

Cox, James A.
: Applications of diffractive optics to uncooled infrared imagers,V1540,606-611(1991)
; Fritz, Bernard S.; Werner, Thomas R.: Process-dependent kinoform performance,V1507,100-109(1991)
; Fritz, Bernard S.; Werner, Thomas R.: Process error limitations on binary optics performance,V1555,80-88(1991)
See Dobson, David C.: V1545,106-113(1991)
See Werner, Thomas R.: V1544,46-57(1991)

Cox, Jerome R.
; Moore, Stephen M.; Whitman, Robert A.; Blaine, G. J.; Jost, R. G.; Karlsson, L. M.; Monsees, Thomas L.; Hassen, Gregory L.; David, Timothy C.: Rapid display of radiographic images,V1446,40-51(1991)

Cox, Maurice K. See Fatah, Rebwar M.: V1501,120-128(1991)

Cox, Norman R.
: High-speed programmable digitizer for real-time video compression experiments,V1605,906-915(1991)

Cox, T. I.
See Deshmukh, V. G.: V1392,352-360(1991)
See Hope, D. A.: V1392,185-195(1991)

Cox, William J.
; Biache, Andrew: Applications of aerial photography to law enforcement and disaster assessment: a consideration of the state-of-the-art,V1479,364-369(1991)

Coyle, Edward J. See Yoo, Jisang: V1451,58-69(1991)

Coyle, Richard J.
; Serafino, Anthony J.; Grimes, Gary J.; Bortolini, James R.: Excimer laser machining of optical fiber taps,V1412,129-137(1991)

Crabtree, Daniel L.
: Velocity profiling in linear and rotational systems,V1482,458-472(1991)

Craciun, Doina See Craciun, Valentin: V1392,625-628(1991)

Craciun, Valentin
; Craciun, Doina; Mihailescu, Ion N.; Kuzmichov, A. V.; Konov, Vitaly I.; Uglov, S. A.: Self-aligned synthesis of titanium silicide by multipulse excimer laser irradiation,V1392,625-628(1991)
; Mihailescu, Ion N.; Luches, Armando; Kiyak, S. G.; Mikhailova, G. N.: Laser processing of germanium,V1392,629-634(1991)

Crafton, J. See Ordonez, Ishmael D.: V1437,184-193(1991)

Craig, Alan B. See Scaletti, Carla: V1459,207-219(1991)

Craig, Richard R. See Zucker, Erik P.: V1563,223-228(1991)

Craig, Robert G.
; Wherrett, Brian S.; Walker, Andrew C.; McKnight, D. J.; Redmond, Ian R.; Snowdon, John F.; Buller, G. S.; Restall, Edward J.; Wilson, R. A.; Wakelin, S.; McArdle, N.; Meredith, P.; Miller, J. M.; Taghizadeh, Mohammad R.; Mackinnon, G.; Smith, S. D.: First programmable digital optical processor: optical cellular logic image processor,V1505,76-86(1991)

Craig-Ryan, S. P. See Massicott, Jennifer F.: V1373,93-102(1991)

Craine, Brian L. See Craine, Eric R.: V1444,389-399(1991)

Craine, Eric R.
; Craine, Brian L.: Blink comparison techniques applied to medical images,V1444,389-399(1991)

Cramer, K. E. See Heiferman, Kenneth S.: V1428,128-134(1991)

Crandall, Charles M.
; Giles, Michael K.; Clark, Natalie: Performance limitations of miniature optical correlators,V1564,98-109(1991)
See Clark, Natalie: V1564,439-451(1991)

Cranmer, David C.
; Freiman, Stephen W.; White, Grady S.; Raynes, Alan S.: Moisture- and water-induced crack growth in optical materials,V1330,152-163(1991)

Crannage, R. P.
; Johnson, Daniel E.; Dorko, Ernest A.: Surface quenching of singlet delta oxygen,V1397,261-265(1991)

Crano, John C.
; Elias, Richard C.: Plastic photochromic eyewear: a status report,V1529,124-131(1991)

Crase, Robert J.
: Effects of polishing materials on the laser damage threshold of optical coatings,V1441,381-389(1991)

Crawford, Carl R.
; King, Kevin F.: Simultaneous patient translation during CT scanning,V1443,203-213(1991)

Crawford, Deborah L.
; Wey, Y. G.; Bowers, John E.; Hafich, Michael J.; Robinson, Gary Y.: GaInAs PIN photodetectors on semi-insulating substrates,V1371,138-141(1991)

Crawford, Douglas P.
; Harada, Larrene K.: Ground-to-space multiline propagation at 1.3 um,V1408,167-177(1991)

Crawford, Gregory P.
; Ondris-Crawford, Renate; Doane, J. W.: Molecular anchoring at the droplet wall in PDLC materials,V1455,2-11(1991)

Crawford, Ian D.
: Receivers for eyesafe laser rangefinders: an overview,V1419,9-16(1991)

Crawford, Michael K. See Morris, Patricia A.: V1561,104-111(1991)

Crawshaw, Richard D. See Zhou, Yi-Tong: V1471,404-411(1991)

Crean, David H.
See Mang, Thomas S.: V1426,97-110(1991)
See Mang, Thomas S.: V1426,188-199(1991)

Creath, Katherine
: Phase-measurement of interferometry techniques for nondestructive testing,V1554B,701-707(1991)
: Submicron linewidth measurement using an interferometric optical profiler,V1464,474-483(1991)
; Wyant, James C.: Absolute measurement of spherical surfaces,V1332,2-7(1991)

Creed, David
See Hoyle, Charles E.: V1559,101-109(1991)
See Subramanian, P.: V1559,461-469(1991)

Cremona, M. See Mattioli, Stefano: V1421,114-119(1991)

Cressie, N. A. See Davidson, Jennifer L.: V1569,288-297(1991)

Crilly, Richard J. See McMath, Linda P.: V1425,165-171(1991)

Crippen, Robert E.
; Blom, Ronald G.: Concept for the subresolution measurement of earthquake strain fields using SPOT panchromatic imagery,V1492,370-377(1991)

Cripps, Dale E.
: Current HDTV overview in the United States, Japan, and Europe,V1456,60-64(1991)

Crisalle, Oscar D.
; Soper, Robert A.; Mellichamp, Duncan A.; Seborg, Dale E.: Adaptive control of photolithography,V1464,508-526(1991)

Crisci, R. J. See McCormick, Frederick B.: V1533,97-114(1991)

Crisman, Jill D.
: Color space analysis of road detection algorithms,V1569,492-506(1991)
; Thorpe, Charles E.: Detecting difficult roads and intersections without map knowledge for robot vehicle navigation,V1388,152-164(1991)

Crist, Richard A. See Cooley, Thomas W.: V1475,243-247(1991)

Crittenden, Eugene C.
; Rodeback, G. W.; Milne, Edmund A.; Cooper, Alfred W.: Vertical ocean reflectance at low altitudes for narrow laser beams,V1492,187-197(1991)
See Cooper, Alfred W.: V1486,37-46(1991)

Crocker, James H.
: Fixing the Hubble Space Telescope,V1494,2-8(1991)
; Cox, Colin R.; Ray, Knute A.; Sen, Amit: Microchannel-plate detectors for space-based astronomy,V1494,434-439(1991)

Croitoru, Nathan I.
See Calderon, S.: V1420,108-115(1991)
See Gannot, Israel: V1442,156-161(1991)

Cromeens, Douglas M. See Johnson, Douglas E.: V1421,36-41(1991)

Cromer, Chris L. See Walker, James H.: V1493,224-230(1991)

Crooks, G. P. See Sahyun, Melville R.: V1436,125-133(1991)

Crooks, R. See Nelson, Jeffrey G.: V1394,191-195(1991)

Croskey, Charles L.
; Olivero, John J.; Martone, Joseph P.: Ground-based microwave remote sensing of water vapor in the mesosphere and stratosphere,V1491,323-334(1991)

Crosley, David R.
: Collisional effects in laser detection of tropospheric OH,V1433,58-68(1991)

Crosnier, Y. See Carru, J. C.: V1512,232-239(1991)

Cross, Edward F.
; Mauritz, F. L.; Bixler, H. A.; Kaegi, E. M.; Wiemokly, Gary D.: Blackbody radiators for field calibration,V1540,756-763(1991)

Cross, P. H. See Gorton, E. K.: V1397,291-295(1991)

Cross, P. S. See Yaeli, Joseph: V1442,378-382(1991)

Crossland, William A.
; Davey, A. B.; Sparks, Adrian P.; Lee, Michael J.; Wright, S. W.; Judge, C. P.: Transmissive analogue SLM using a chiral smectic liquid crystal switched by CdSe TFTs,V1455,264-273(1991)
See Underwood, Ian: V1562,107-115(1991)

Crostack, Horst-Artur
; Meyer, E. H.; Pohl, Klaus-Juergen: Holographic soundfield visualization for nondestructive testing of hot surfaces,V1508,101-109(1991)

Croteau, R. E. See Palmieri, Francesco: V1451,24-35(1991)

Crouse, Randy F. See Lawson, Shelah M.: V1488,268-278(1991)

Crow, Gretchen L. See Smith, Terrance L.: V1512,92-100(1991)

Crow, Samuel B.
; Watkins, Wendell R.; Palacios, Fernando R.; Billingsley, Daniel R.: Technique for measuring atmospheric effects on image metrics,V1486,333-344(1991)
See Watkins, Wendell R.: V1486,17-24(1991)

Crowe, William M.
See Hudson, Tracy D.: V1564,54-64(1991)
See Kirsch, James C.: V1482,69-78(1991)

Crowell, Robert A. See Chronister, Eric L.: V1559,56-64(1991)

Crowther, D. J. See Favro, Lawrence D.: V1467,290-294(1991)

Crowther, Margaret F. See Sansonetti, Pierre: V1588,198-209(1991)

Crump, O. B. See Sweatt, William C.: V1346,151-159(1991)

Crumrine, David See Arber, Simon: V1428,23-29(1991)

Cruz, C. H. See de Souza, Eunezio A.: V1358,556-560(1991)

Cryan, Robert A.
; Unwin, Rodney T.; Massarella, Alistair J.; Sibley, Martin J.; Garrett, Ian: Coherent detection: n-ary PPM versus PCM,V1372,64-71(1991)
; Unwin, Rodney T.: Optical fiber n-ary PPM: approaching fundamental limits in receiver sensitivity,V1579,133-143(1991)

Csorba, Illes P.
ed.: *Electron Image Tubes and Image Intensifiers II*,V1449(1991)

Cubbage, Robert W.
; Littlewood, Paul A.: DS1 mapping considerations for the synchronous optical network,V1363,163-171(1991)

Cubeddu, Rinaldo
; Canti, Gianfranco L.; Taroni, Paola; Valentini, G.: Time-gated fluorescence spectroscopy and imaging of porphyrins and phthalocyanines,V1525,17-25(1991)

Cuby, Jean-Gabriel See Gendron, Eric: V1542,297-307(1991)

Cucalon, Antoine See Matthews, James E.: V1366,206-214(1991)

Cucchi, Silvio See Barbero, Marzio: V1567,566-577(1991)

Cudworth, Stewart K. See Annamalai, Kadiresan: V1364(1991)

Cuellar, Enrique
See Kennedy, Michael T.: V1366,167-176(1991)
See Roberts, Daniel R.: V1366,129-135(1991)

Cuellar, Louis
; Johnson, Paul A.; Sandler, David G.: Performance tests of a 1500 degree-of-freedom adaptive optics system for atmospheric compensation,V1542,468-476(1991)

Cuellar, Roland See Smith, Louis C.: V1428,224-232(1991)

Cuff, K. F. See Adams, Frank W.: V1541,24-37(1991)

Cui, DaFu
; Cao, Qiang; Han, JingHong; Cai, Jine; Li, YaTing; Zhu, ZeMin; Fan, Jie; Gao, Ning: Optical fibre PH sensor based on immobilized indicator,V1572,386-391(1991)

Cui, Jie
; Wang, Hai L.; Gan, Fuxi: Raman spectra of ZnSe-ZnTe strained-layer superlattice,V1519,652-655(1991)
See Shen, Ai D.: V1519,656-659(1991)

Cui, Jing B.
; Zhang, Wei P.; Fang, Rong C.; Wang, Chang S.; Zhou, Guien; Wu, Jan X.: Interface properties of a-C:H/a-SiOx:H multilayer,V1519,419-422(1991)
See Zhang, Wei P.: V1519,23-25(1991)

Cui, T. J. See Zhuang, Qi: V1397,157-160(1991)

Cui, Ting See Shi, Weimin: V1422,62-72(1991)

Cui, Weijia
; Kumar, Chellappa; Chance, Britton: Experimental study of migration depth for the photons measured at sample surface,V1431,180-191(1991)

Cui, Y. P. See Zhang, Yue: V1560,264-271(1991)

Cui, Zhan J. See Liu, Wei Y.: V1519,172-174(1991)

Cullen, Christopher P.
: Method for robot path adaptation using scalar sensor data,V1388,62-71(1991)

Cullen, David See Haddad, Waleed S.: V1448,81-88(1991)

Cullen, J. D. See Bell, J. M.: V1536,29-36(1991)

Cullen, Robert M.
: Low-cost space platforms for detection and tracking technologies,V1479,295-305(1991)
See Harvey, Edwin L.: V1495,134-145(1991)

Cullen, Thomas J.
: Polarization-diversity fiber networks,V1372,164-172(1991)

Cullis, I. C.
; Parker, Richard J.; Sewell, Derek: Holographic visualization of hypervelocity explosive events,V1358,52-64(1991)

Culotta, Paul W. See Johnson, Edward J.: V1557,168-179(1991)

Culshaw, Brian
: Fiber optic sensor networks,V1511,168-178(1991)
; Liao, Yan-Biao; eds.: *Intl Conf on Optical Fibre Sensors in China*,V1572(1991)
See Bruinsma, Anastasius J.: V1511(1991)
See Jin, Wei: V1504,125-132(1991)
See Johnstone, Walter: V1506,145-149(1991)
See Michie, W. C.: V1588,342-355(1991)
See Rao, Yun-Jiang: V1506,126-133(1991)
See Rao, Yun-Jiang: V1572,287-292(1991)
See Sansonetti, Pierre: V1588,198-209(1991)
See Stewart, George: V1368,230-238(1991)
See Vengsarkar, Ashish M.: V1367,249-260(1991)

Cumings, Nesbitt P. See Winkler, Carl E.: V1494,301-313(1991)

Cummings, James D. See Swartz, Barry A.: V1537,42-56(1991)

Cummings, Roger
: Fiber channel: the next standard peripheral interface and more,V1364,170-177(1991)

Cummins, Philip G. See Miller, Richard M.: V1439,66-78(1991)

Cunha, Andre See Leith, Emmett N.: V1396,80-84(1991)

Cunin, B.
; Geist, P.; Heisel, Francine; Martz, A.; Miehe, Joseph A.: Temporal and spectral analysis of the synchronization of synchroscan streak cameras,V1358,606-613(1991)

Cunningham, Andrew P. See Raanes, Chris A.: V1358,637-643(1991)

Cunningham, Philip R.
; Hay, Stephen O.; Francis, Denise M.; Trott, G.E.: In-situ measurement of piston jitter in a ring resonator,V1414,97-129(1991)
See Suter, Kevin J.: V1414,33-65(1991)
See Townsend, Sallie S.: V1415,154-194(1991)

Cuomo, Frank W. See He, Gang: V1584,152-156(1991)

CuQlock-Knopp, V. G. See Merritt, John O.: V1457,133-138(1991)

Curatu, Eugen O.
  : Matrix optical system for plane-point correlation,V1527,368-375(1991)

Curcio, Christine A. See Sloan, Kenneth R.: V1453,124-133(1991)

Curran, Mark E.
  See Clark, Timothy E.: V1369,98-106(1991)
  See Van Metre, Richard: V1369,9-18(1991)

Curran, Robert J.
  ; Smith, James A.; Watson, Ken; eds.: *Earth and Atmospheric Remote
    Sensing*,V1492(1991)

Current, Michael I. See Raicu, Bruha: V1393,161-171(1991)

Currie, John F. See Chaker, Mohamed: V1465,16-25(1991)

Currin, Michael S. See Sanders, Jeffrey S.: V1488,144-155(1991)

Curti, Franco See Betti, Silvello: V1579,100-111(1991)

Curti, G. See Melozzi, Mauro: V1512,178-188(1991)

Curtin, Hugh D. See King, Jill L.: V1446,268-275(1991)

Cusumano, G. See Citterio, Oberto: V1343,145-154(1991)

Cutolo, Antonello See Cocorullo, Giuseppe: V1374,132-137(1991)

Cutter, Mike A.
  See Baudin, Gilles: V1493,16-27(1991)
  See Baudin, Gilles: V1490,102-113(1991)

Cutting, Court B. See Kalvin, Alan D.: V1445,247-258(1991)

Cyganski, David See Wheeler, Frederick W.: V1606,78-85(1991)

Czajkowski, Igor K.
  ; Allam, Jeremy; Adams, Alfred R.: Band-structure dependence of impact
    ionization: bulk semiconductors, strained Ge/Si alloys, and multiple-
    quantum-well avalanche photodetectors,V1362,179-190(1991)

Czechowicz, Roman
  ; Kopczynski, Krzysztof: Optical pumping of a solid state laser with a high-
    frequency train of pumping pulsed,V1391,52-60(1991)
  See Trojnar, Eugeniusz: V1391,230-237(1991)
  See Trojnar, Eugeniusz: V1391,238-243(1991)

Czuchlewski, Stephen J. See Bigio, Irving J.: V1397,47-53(1991)

D'Agostino, John A.
  ; Webb, Curtis M.: Three-dimensional analysis framework and measurement
    methodology for imaging system noise,V1488,110-121(1991)

d'Alessandro, Antonio
  ; De Sario, Marco; D'Orazio, Antonella; Petruzzelli, Vincenzo: All-optical
    Ti:LiNbO3 waveguide switch,V1378,127-138(1991)
  ; De Sario, Marco; D'Orazio, Antonella; Petruzzelli, Vincenzo: Design
    criteria of an integrated optics microdisplacement sensor,V1332,554-
    562(1991)
  ; De Sario, Marco; D'Orazio, Antonella; Petruzzelli, Vincenzo: Integrated
    optics displacement sensor,V1366,313-323(1991)
  ; De Sario, Marco; D'Orazio, Antonella; Petruzzelli, Vincenzo: Integrated
    optics temperature sensor,V1399,184-191(1991)

D'Alessandro, Maurizio See Bortoletto, Favio: V1542,225-235(1991)

D'Amato, Dante P.
  ; Centamore, Robert M.: Two applications for microlens arrays: detector fill-
    factor improvement and laser diode collimation,V1544,166-177(1991)

D'Amato, Donald P. See Concepcion, Vicente P.: V1606,132-140(1991)

D'Anna, Emilia
  ; Leggieri, Gilberto; Luches, Armando; Martino, M.; Perrone, A.; Majni, G.;
    Mengucci, P.; Drigo, A. V.; Mihailescu, Ion N.: Surface nitride synthesis
    by multipulse excimer laser irradiation,V1503,256-268(1991)

D'Arcy, John See Clapham, Terrance N.: V1423,2-7(1991)

D'Cunha, Ivan See Alvertos, Nicolas: V1382,388-396(1991)

D'Hallewin, Mich A.
  See Andersson-Engels, Stefan: V1426,31-43(1991)
  See Baert, Luc: V1525,385-390(1991)

D'Luna, Lionel J. See Parulski, Kenneth A.: V1448,45-58(1991)

D'Orazio, Antonella
  See d'Alessandro, Antonio: V1378,127-138(1991)
  See d'Alessandro, Antonio: V1332,554-562(1991)
  See d'Alessandro, Antonio: V1366,313-323(1991)
  See d'Alessandro, Antonio: V1399,184-191(1991)

D'Orazio, Tiziana See Distante, Arcangelo: V1381,513-523(1991)

D'yakonov, G. I.
  ; Konov, Vitaly I.; Mikhailov, V. A.; Nikolaev, D. A.; Pak, S. K.; Shcherbakov,
    I. A.: Comparative performance of infrared solid state lasers in laser
    lithotripsy,V1421,156-162(1991)
  ; Konov, Vitaly I.; Mikhailov, V. A.; Pak, S. K.; Shcherbakov, I. A.; Ershova, N.
    I.; Maksimovskiy, Y. V.: Cr,Er:YSGG laser as an instrument for dental
    surgery,V1424,81-86(1991)

  ; Mikhailov, V. A.; Pak, S. K.; Shcherbakov, I. A.; Andreev, V. G.; Rudenko,
    O. V.; Sapozhnikov, A. V.: Q-switched Nd:glass-laser-induced acoustic
    pulses in lithotripsy,V1421,153-155(1991)

D'yakov, Alexander See Gurvich, Alexander S.: V1408,10-18(1991)

Dadkhah, Mahyar S.
  ; Marshall, David B.; Morris, Winfred L.: First direct measurements of
    transformation strains in crack-tip zone,V1554A,164-175(1991)

Daemi, M. F.
  ; Beurle, R. L.: Assessment of the information content of patterns: an
    algorithm,V1567,621-631(1991)
  See Naylor, David C.: V1567,522-532(1991)

Daendliker, Rene See Herzig, Hans-Peter: V1507,247-255(1991)

Daetwyler, K. See Webb, David J.: V1418,231-239(1991)

Da Forno, R. See Angrilli, Francesco: V1482,26-39(1991)

Dagenais, Dominique M. See Bucholtz, Frank: V1367,226-235(1991)

Dagli, Cihan H. See Smith, A.: V1469,551-562(1991)

Dagras, C. See Durpaire, Jean-Pierre: V1490,23-38(1991)

Dahan, Reuben
  See Calderon, S.: V1420,108-115(1991)
  See Gannot, Israel: V1442,156-161(1991)

Dahe, Liu
  ; Liang, Zhujian; Tang, Weiguo: Unsymmetrical spectrum of reflective
    hologram grating,V1507,310-315(1991)

Daher, Reinhard
  ; McAdam, Wylie; Pizey, Gordon: Implementation of a 3-D stereovision
    system for the production of customized orthotic accessories,V1526,90-
    93(1991)

Dahhou, B. See Dilhac, Jean-Marie R.: V1393,395-403(1991)

Dahlke, Weldon J. See Phillips, Thomas E.: V1454,290-298(1991)

Dahmen, M. See Menczigar, Ulrich: V1361,282-292(1991)

Dai, Chao M. See Guo, Lu Rong: V1555,293-296(1991)

Dai, Fu-long See Wang, Feng: V1554A,359-370(1991)

Dai, Ji Zhi See Chen, Zheng: V1572,129-131(1991)

Dai, W. See Evans, Alan C.: V1445,236-246(1991)

Dai, Yisong See Qian, Shen-en: V1483,196-206(1991)

Dai, YuZhong
  ; Chiang, Fu-Pen: Estimation of plastic strain by fractal,V1332,767-774(1991)

Dai, Z. See Peticolas, Warner L.: V1403,22-26(1991)

Daidoh, Yuichiro
  ; Arai, Tsunenori; Suda, Akira; Kikuchi, Makoto; Komine, Yukikuni; Murai,
    Masaru; Nakamura, Hiroshi: Discrimination between urinary tract tissue
    and urinary stones by fiber-optic-pulsed photothermal radiometry method
    in vivo,V1421,120-123(1991)

Dailey, Carroll C. See Winkler, Carl E.: V1494,301-313(1991)

Daily, Michael J.
  : Parameter studies for Markov random field models of early
    vision,V1473,138-152(1991)
  ; Silberberg, Teresa M.: Actively controlled multiple-sensor system for feature
    extraction,V1472,85-96(1991)

Dainty, J. C.
  ; Mavroidis, Theo; Solomon, Christopher J.: Double-passage imaging through
    turbulence,V1487,2-9(1991)

Dakin, John P.
  ; Edwards, Henry O.: Progress in fiber-remote gas correlation
    spectrometry,V1510,160-169(1991)

Dakss, Mark L.
  ; Miniscalco, William J.: Large-signal model and signal/noise ratio analysis for
    Nd3+-doped fiber amplifiers at 1.3 um,V1373,111-124(1991)

Dal Colle, M. See Esquivias, Ignacio: V1484,55-66(1991)

Dal Degan, Nevaino
  ; Lancini, R.; Migliorati, Pierangelo; Pozzi, S.: Color handling in the image
    retrieval system Imagine,V1606,934-940(1991)

Dale, Cheryl J.
  ; Marshall, Paul W.: Displacement damage in Si imagers for space
    applications,V1447,70-86(1991)

Dale, Ian See Syme, Richard T.: V1362,467-476(1991)

Daley, Paul F. See Angel, S. M.: V1368,98-104(1991)

Daligault, Laurence
  ; Glasser, Jean: Colorimetric characterization of CCD sensors by
    spectrophotometry,V1512,124-130(1991)

Dalinenko, I. N.
; Kuz'min, G. A.; Malyarov, A. V.; Prokhorov, Alexander M.; Schelev, Mikhail Y.: Thinned backside-illuminated cooled CCDs for UV and VUV applications,V1449,167-172(1991)
See Bryukhnevitch, G. I.: V1358,739-749(1991)
See Bryukhnevitch, G. I.: V1449,109-115(1991)

Dall'Antonia, L. H. See Valla, Bruno: V1536,48-62(1991)

Dallabetta, Kyle A.
; de La Chapelle, Michael; Lawrence, Robert C.: MMIC compatible photodetector design and characterization,V1371,116-127(1991)

Dallas, William J.
See Martinez, Ralph: V1446,100-107(1991)
See Rehm, Kelly: V1445,24-35(1991)

Dally, James W.
; Chen, Yung-Mien: Photoelastic study of friction at multipoint contacts,V1554A,434-443(1991)

Dalmasso, J. M. See Mens, Alain: V1358,315-328(1991)

Dalsing, Michael C.
; Kruepper, Peter; Cikrit, Dolores F.: Vascular tissue welding of the CO2 laser: limitations,V1422,98-102(1991)

Dalton, Larry R.
See Chen, Mai: V1409,202-213(1991)
See Shi, Youngqiang: V1559,118-126(1991)
See Spangler, Charles W.: V1497,408-417(1991)

Daly, John G.
: Mid-infrared laser applications,V1419,94-99(1991)

Damas, George See Kwo, Deborah P.: V1544,66-74(1991)

Dambra, Carlo See Vernazza, Gianni L.: V1492,206-212(1991)

Damerji, Tayeb See Ionescu, Dan: V1406,40-41(1991)

Dameron, David H. See Flack, Warren W.: V1465,164-172(1991)

Dames, Mark P. See McKee, Paul: V1461,17-23(1991)

Damm, Tobias
; Kempe, M.; Stamm, Uwe; Stolberg, K. P.; Wabnitz, H.: High spatial and temporal resolution in the optical investigation of biological objects,V1403,686-694(1991)

Dammel, Ralph
See Eckes, Charlotte: V1466,394-407(1991)
See Przybilla, Klaus J.: V1466,174-187(1991)

Damodaran, Meledath See Parikh, Jo Ann: V1469,526-538(1991)

Damour, Kevin
; Kaufman, Howard: Restoration of distorted depth maps calculated from stereo sequences,V1452,78-89(1991)

Damp, Stephan
: Miniature laser Doppler anemometer for sensor concepts,V1418,459-470(1991)

Damstra, Geert C.
; Eenink, A. H.: 10 MHz multichannel image detection and processing system,V1358,644-651(1991)

Dana, David R. See Maffione, Robert A.: V1537,173-184(1991)

Dana, Michel See Ali-Yahia, Tahar: V1468,950-959(1991)

Dandge, Dileep K. See Klainer, Stanley M.: V1434,119-126(1991)

Dandridge, Anthony D.
See Burns, William K.: V1367,87-92(1991)
See Kersey, Alan D.: V1511,40-50(1991)

Dane, C. B.
See Hofmann, Thomas: V1412,84-90(1991)
See Tittel, Frank K.: V1397,21-29(1991)

Danelius, R.
; Grigonis, R.; Piskarskas, Algis S.; Podenas, D.; Sirutkaitis, V.: Solid state lasers with passive mode-locking and negative feedback for picosecond spectroscopy,V1402,198-208(1991)

Daneshdoost, Morteza
: Neural network approach to power system security,V1396,270-275(1991)

Danev, Gentsho See Guttmann, Peter: V1361,999-1010(1991)

Daniel, Hani S.
; Moore, Douglas R.: Single-mode MxN star couplers fabricated using fused biconical taper techniques,V1365,53-59(1991)

Daniell, Cindy E.
; Kemsley, David; Bouyssounouse, Xavier: Comparative evaluation of neural-based versus conventional segmentors,V1471,436-451(1991)

Danielson, Bruce L. See Boisrobert, Christian Y.: V1474,285-290(1991)

Danilov, Alexander A.
; Nikirui, Ernest Y.; Osiko, Vyacheslav V.; Polushkin, Valery G.; Sorokin, Svjatoslav N.; Timoshechkin, M. I.: Effective laser with active element rectangular geometry,V1362,916-920(1991)
; Sorokin, Svjatoslav N.: Spontaneous emission amplification and the utmost energy characteristics of solid state lasers on the base of optically dense media,V1362,647-654(1991)

Danisman, Kenan
; Yilbas, Bekir S.; Altuner, Mehmet; Ciftlikli, Cebrail: Optimization of affecting parameters in relation to pulsed CO2 laser design,V1412,218-226(1991)
See Alci, Mustafa: V1411,100-106(1991)

Danly, Bruce G.
; Hartemann, Frederic V.; Chu, T. S.; Menninger, W. L.; Papavaritis, P.; Pendergast, K. D.; Temkin, Richard J.: Long-pulse modulator-driven cyclotron autoresonance maser and free-electron laser experiments at MIT,V1407,192-201(1991)
See Goodman, Daniel L.: V1407,217-225(1991)

Dannberg, Peter
See Braeuer, Andreas: V1559,470-478(1991)
See Guntau, Matthias: V1513,112-122(1991)

Danoff, Dudley
See Chandra, Mudjianto: V1421,68-71(1991)
See Gershman, Alex: V1421,186-188(1991)

Dansereau, Jeffrey P.
; Hills, Louis S.; Mani, Siva A.: Analysis of Brillouin-enhanced four-wave mixing phase-conjugate systems,V1409,67-82(1991)
See Acebal, Robert: V1397,191-196(1991)

Danshikov, E. See Antyuhov, V.: V1397,355-365(1991); 1415,48-59(1991)

Dantas, Tarcisio M. See Soares, Edmundo A.: V1367,261-265(1991)

Dao, Giang T.
See Hansen, Steven G.: V1463,230-244(1991)
See Toh, Kenny K.: V1496,27-53(1991)
See Toh, Kenny K.: V1463,402-413(1991)
See Toh, Kenny K.: V1463,74-86(1991)

Dao, Nguyen Q.
See Hong, Nguyen T.: V1525,132-142(1991)
See Phat, Darith: V1525,196-205(1991)

Dao, T. T.
; Spence, Christopher A.; Hess, Dennis W.: Study of silylation mechanisms and kinetics through variations in silylating agent and resin,V1466,257-268(1991)

Dapkus, P. D. See Hur, Jung H.: V1378,95-100(1991)

DaPonte, John S.
; Parikh, Jo Ann; Katz, David A.: Detection of liver metastisis using the backpropagation algorithm and linear discriminant analysis,V1469,441-450(1991)
See Parikh, Jo Ann: V1469,526-538(1991)

Daraktchiev, Ivan S. See Robertson, Stewart A.: V1464,232-244(1991)

Darden, Bruce V.
; LeFevre, B. G.; Kalomiris, Vasilios E.: Hermaphroditic small tactical connector for single-fiber applications,V1474,300-308(1991)

Darg, David A.
; Kikuchi, Akira: High-bandwidth recording in a hostile environment,V1538,124-131(1991)

Dario, Paolo See Kreider, Gregory: V1381,242-249(1991)

Darnton, Lane A.
See Folkman, Mark A.: V1493,255-266(1991)
See Jarecke, Peter J.: V1493,244-254(1991)

Darrow, K. W. See Clark, Wally T.: V1371,258-265(1991)

Darsky, Alexei M.
; Markov, Vladimir B.: Angular sensitivity of holograms with a reference speckle wave,V1238,54-61(1991)
; Markov, Vladimir B.: Information capacity of holograms with reference speckle wave,V1509,36-46(1991)

Daryanani, Sonu L. See Taylor, Geoff W.: V1476,2-13(1991)

Daryoush, Afshin S.
See Even-Or, Baruch: V1371,161-169(1991)
See Kasemset, Dumrong: V1371,104-114(1991)
See Polifko, David M.: V1476,91-99(1991)

Das, Alok K.
; Mandal, Anup K.; Pandit, Malay K.: Effect of liquid on partially removed cladding SM fiber and its application to sensors,V1572,572-580(1991)
; Mandal, Anup K.: Transmission characteristics of multimode optical fiber with multisplices,V1365,144-155(1991)
; Muhuri, K.: Dynamic allocation of buffer space in the bridge of two interconnected token rings,V1364,61-69(1991)
; Pandit, Malay K.: Analysis and modeling of low-loss fused fiber couplers,V1365,74-85(1991)

Das, Bidyut B.
; Glassman, W. L. S.; Alfano, Robert R.; Cleary, Joseph; Prudente, R.; Celmer, E.; Lubicz, Stephanie: UV-fluorescence spectroscopic technique in the diagnosis of breast, ovarian, uterus, and cervix cancer,V1427,368-373(1991)
See Pradhan, Asima: V1425,2-5(1991)

Das, Santanu K.
; Ocenasek, Josef: Impact of fiber backscatter on loop video transmission without optical isolator,V1363,172-176(1991)

Das, Siddhartha See Schwartzkopf, George: V1466,26-38(1991)

Das, Subhodev
; Ahuja, Narendra: Dynamic integration of visual cues for position estimation,V1382,341-352(1991)

Das, Sujata See Flickner, Myron D.: V1568,113-124(1991)

Dasgupta, Abhijit
See Singh, Hemant: V1588,76-85(1991)
See Sirkis, James S.: V1588,88-99(1991)

Dasgupta, Samhita
; Poppel, Gary L.; Anderson, William P.: Fiber optic controls for aircraft engines: issues and implications,V1374,211-222(1991)

Dashevsky, Boris E.
: Light-pulse-induced background in image intensifiers with MCPs,V1449,25-29(1991)
: Method of estimating light loadings power and repetition frequency effects on the image quality of gated image intensifier tubes,V1449,65-70(1991)
: New electron optic for high-speed single-frame and streak image intensifier tubes,V1358,561-568(1991)
; Podvyaznikov, V. A.; Prokhorov, Alexander M.; Chevokin, Victor K.: Wide-aperture x-ray image converter tubes,V1346,437-442(1991)

Dassel, A. C.
See Koelink, M. H.: V1403,347-349(1991)
See Koelink, M. H.: V1511,120-128(1991)
See Koelink, M. H.: V1431,63-72(1991)

Daszkiewicz, Marek
: Optical diffractometry with directionally variable incident light wave,V1562,184-191(1991)
See Blocki, Narcyz: V1562,172-183(1991)

Datari, Srinivasa R. See Celenk, Mehmet: V1468,752-763(1991)

Dattoli, Giuseppe See Ciocci, Franco: V1501,154-162(1991)

Daudon, Michel See Hong, Nguyen T.: V1525,132-142(1991)

Daugy, Eric
; Pointu, Bernard; Villela, Gerard; Vincent, Bernard: Dual ion-beam sputtering: a new coating technology for the fabrication of high-power CO2 laser mirrors,V1502,203-212(1991)

Daum, Frederick E.
: Cramer-Rao bound for multiple-target tracking,V1481,341-344(1991)

Daurelio, Giuseppe
; Dionoro, G.; Memola Capece Minutolo, F.: Laser welding of INCONEL 600,V1397,783-786(1991)

Dausch, Mark E. See Comly, James B.: V1381,390-400(1991)

Dausinger, Friedrich
See Beck, M.: V1397,769-774(1991)
See Ream, Stanley L.: V1601(1991)

Dautray, Robert See Andre, Michel L.: V1502,286-298(1991)

Dave, Rajesh N.
; Patel, Kalpesh J.: Fuzzy ellipsoidal shell clustering algorithm and detection of elliptical shapes,V1381,320-333(1991)

Davenport, Carolyn M.
; Alexander, Andrew L.; Gmitro, Arthur F.: Optimal fluorescence imaging of atherosclerotic human tissue,V1425,16-27(1991)
See Alexander, Andrew L.: V1425,6-15(1991)

Davey, A. B. See Crossland, William A.: V1455,264-273(1991)

Davey, Steven T.
See Kashyap, Raman: V1516,164-174(1991)
See Williams, Doug L.: V1513,158-167(1991)
See Williams, Doug L.: V1516,29-37(1991)

David, Frank See Bains, Narinder: V1386,232-242(1991)

David, J. See Pelous, G.: V1598,149-158(1991)

David, Jean
; Wettling, J. C.; Combis, P.; Nierat, G.; Rostaing, M.: Quartz gauge and ballistic pendulum measurements of the mechanical impulse imparted to a target by a laser pulse,V1397,697-700(1991)

David, Philip
: Multiple-sensor cueing using a heuristic search,V1468,1000-1009(1991)

David, Timothy C. See Cox, Jerome R.: V1446,40-51(1991)

Davidhazy, Andrew
: Determination of burst initiation location and tear propagation velocity during air burst testing of latex condoms,V1358,654-659(1991)

Davidson, Alan C. See Herd, James T.: V1387,194-201(1991)

Davidson, Frederic M.
; Sun, Xiaoli; Krainak, Michael A.: Bandwidth requirements for direct detection optical communication receivers with PPM signaling,V1417,75-88(1991)

Davidson, Jennifer L.
; Cressie, N. A.: Statistical image algebra: a Bayesian approach,V1569,288-297(1991)
; Sun, K.: Template learning in morphological neural nets,V1568,176-187(1991)
See Meyer, Trevor E.: V1568,125-136(1991)
See Ritter, Gerhard X.: V1406,74-86(1991)

Davidson, Mark P.
; Monahan, Kevin M.; Monteverde, Robert J.: Linearity of coherence probe metrology: simulation and experiment,V1464,155-176(1991)

Davidson, Nir
; Friesem, Asher A.; Hasman, Erez: Elements with long-focal depth and high-lateral resolution,V1442,22-25(1991)
See Hasman, Erez: V1442,372-377(1991)

Davidson, Roger
See Michie, W. C.: V1588,342-355(1991)
See Roberts, Scott S.: V1588,326-341(1991)
See Sansonetti, Pierre: V1588,198-209(1991)

Davidson, Ronald C.
See Chen, Chiping: V1407,105-112(1991)
See Johnston, George L.: V1407,92-99(1991)

Davidson, Scott E. See Wigginton, Stewart C.: V1390,560-567(1991)

Davies, D. K. See Goutzoulis, Anastasios P.: V1371,182-194(1991)

Davies, David H.
; Clark, David F.: Aspects of the application of image and data storage in compact disk formats,V1401,2-8(1991)

Davies, Eric See Flynn, Michael J.: V1444,172-179(1991)

Davies, Henry C.
; Kayaalp, Ali E.; Moezzi, Saied: Autonomous navigation in a dynamic environment,V1388,165-175(1991)
See Kayaalp, Ali E.: V1406,98-109(1991)

Davies, Jeremy C. See Atcha, Hashim: V1504,221-232(1991)

Davies, John T.
; Metz, Thomas E.; Savage, Richard N.; Simmons, Horace O.: Real-time, in-situ measurement of film thickness and uniformity during plasma ashing of photoresist,V1392,551-554(1991)

Davies, Neil A.
; McCormick, Malcolm; Lau, Hon W.: Microlens arrays in integral photography and optical metrology,V1544,189-198(1991)
See Lau, Hon W.: V1544,178-188(1991)

Davies, P. B. See Nwagboso, Christopher O.: V1386,30-41(1991)

Davies, Peter N. See Lu, Wei-Wei: V1470,197-205(1991)

Davies, Richard W. See Zemon, Stanley A.: V1373,21-32(1991)

Davies, Roger
; Correia, Bento A.; Carvalho, Fernando D.; Rodrigues, F. C.: BRICORK: an automatic machine with image processing for the production of corks,V1459,283-291(1991)
; Correia, Bento A.; Carvalho, Fernando D.: Image processing methodology for optimizing the quality of corks in the punching process,V1567,244-253(1991)
See Correia, Bento A.: V1567,15-24(1991)

Davila, Pam M. See Hannan, Paul G.: V1527,26-36(1991)

Davinson, Ian
: Use of optical sensors and signal processing in gas turbine engines,V1374,251-265(1991)

Davis, Andrew
; Ohn, Myo M.; Liu, Kexing; Measures, Raymond M.: Composite cure monitoring with embedded optical fiber sensors,V1489,33-43(1991)
; Ohn, Myo M.; Liu, Kexing; Measures, Raymond M.: Study of an opto-ultrasonic technique for cure monitoring,V1588,264-274(1991)
See Liu, Kexing: V1398,206-212(1991)

Davis, Anthony B.
; Lovejoy, Shaun; Schertzer, Daniel: Discrete-angle radiative transfer in a multifractal medium,V1558,37-59(1991)

Davis, Billy See Nein, Max E.: V1494,98-110(1991)

Davis, Brent A. See Singman, Leif V.: V1548,80-92(1991)

Davis, Darryl N.
; Taylor, Christopher J.: Blackboard architecture for medical image interpretation,V1445,421-432(1991)

Davis, E. J.
: Optical measurements of electrodynamically levitated microparticles,V1435,216-242(1991)

Davis, Gary A. See Costello, Kenneth A.: V1449,40-50(1991)

Davis, J. G. See Dodd, James W.: V1407,467-473(1991)

Davis, James H. See Albanese, Andres: V1364,320-326(1991)

Davis, John M.
See Campbell, Jonathan W.: V1549,155-179(1991)
See Campbell, Jonathan W.: V1343,359-375(1991)

Davis, Jon P.
; Schmidt, William A.: Invariant pattern recognition via higher order preprocessing and backprop,V1469,804-811(1991)

Davis, Kenneth M. See Tomozawa, Minoru: V1590,160-167(1991)

Davis, Lloyd M. See Soper, Steven A.: V1435,168-178(1991)

Davis, Martin H.
; Ramachandran, Umakishore: Optical bus protocol for a distributed-shared-memory multiprocessor,V1563,176-187(1991)

Davis, Paul B.
; Cate, D.; Abidi, Mongi A.: Parallel data fusion on a hypercube multiprocessor,V1383,515-529(1991)

Davis, Richard L.
; Lee, Sae H.: Low-loss waveguides on silicon substrates for photonic circuits,V1474,20-26(1991)

Davis, Steven J.
; Allen, Mark G.: Applications of excimer lasers to combustion research,V1377,113-118(1991)

Davisson, Edwin O. See Deitz, Paul: VIS07,40-56(1991)

Dawber, William N.
; Hirst, Peter F.; Condon, Brian P.; Maitland, Arthur; Sutton, Phillip: Novel high-speed communication system,V1476,81-90(1991)

Dawes, T. D. See Woodruff, William H.: V1432,205-210(1991)

Dawnay, Emma J. See Ashwell, Geoffrey J.: V1361,589-598(1991)

Dawson, James A. See Owen, Philip R.: V1488,122-132(1991)

Dawson, John M. See Chen, Kuan-Ren: V1552,185-196(1991)

Dawson, Philip
; Galbraith, Ian; Kucharska, Alicia I.; Foxon, C. T.: Optical nonlinearities due to long-lived electron-hole plasmas,V1362,384-390(1991)

Day, Gordon W. See Deeter, Merritt N.: V1367,243-248(1991)

Day, Timothy See Simpson, Thomas B.: V1410,133-140(1991)

Dayanand, Sriram See Uttal, William: V1453,258-269(1991)

Dayhoff, Ruth E.
; Maloney, Daniel L.; Kuzmak, Peter; Shepard, Barclay M.: Experiences with a comprehensive hospital information system that incorporates image management capabilities,V1446,323-329(1991)

Daykhovsky, Leon See Papaioannou, Thanassis: V1420,203-211(1991)

De, Badal See Ghatak, Kamakhya P.: V1409,178-190(1991)

De, H. See Allinger, Thomas: V1361,935-942(1991)

Deacutis, Martin See Ren, Victor: VCR37,54-67(1991)

Dean, A. B. See Ashley, Timothy: V1361,238-244(1991)

Dean, Anthony J.
See Di Cocco, Guido: V1549,102-112(1991)
See Swinyard, Bruce M.: V1548,94-105(1991)

Dean, Cleon E.
; Werby, Michael F.: Target shape and material composition from resonance echoes of submerged elongated elastic targets,V1471,54-65(1991)

Dean, Peter D.
ed.: Sensors and Sensor Integration,V1480(1991)

Dean, Robert L. See Muray, Andrew: V1496,171-179(1991)

DeAngelis, Franco E.
: Laser-generated 3-D prototypes,V1598,61-70(1991)

Dear, John P.
: Use of high-speed photography in the evaluation of polymer materials,V1358,37-42(1991)

de Araujo, Cid B. See Gomes, Anderson S.: V1579,257-263(1991)

Deason, Vance A.
; Miller, R. L.; Watkins, Arthur D.; Ward, Michael B.; Barrett, K. B.: Measurement of interfacial tension by automated video techniques,V1332,868-876(1991)
; Ward, Michael B.: Phase-shifting hand-held diffraction moire interferometer,V1554B,390-398(1991)

Deaton, John B. See Wagner, James W.: V1332,491-501(1991)

Debal, Vincent See Manfait, Michel: V1403,695-707(1991)

DeBell, David A.
: System considerations for detection and tracking of small targets using passive sensors,V1481,180-186(1991)

De Bernardi, Carlo
See Caldera, Claudio: V1372,82-87(1991)
See Cognolato, Livio: V1513,368-377(1991)

De Blasi, C.
See Ferrara, M.: V1513,130-136(1991)
See Righini, Giancarlo C.: V1513,105-111(1991)

De Blasi, Roberto A. See Ferrari, Marco: V1431,276-283(1991)

Debnam, W. J. See Fripp, Archibald L.: V1557,236-244(1991)

Debney, Brian T.
See Johnson, Peter T.: V1371,87-97(1991)
See Johnson, Peter T.: V1372,188-199(1991)

de Boey, W. P. See Lucassen, G. W.: V1403,185-194(1991)

de Bougrenet de la Tocnaye, Jean-Louis M.
See Cambon, P.: V1562,116-125(1991)
See Guibert, L.: V1505,99-103(1991)
See Kohler, A.: V1564,236-243(1991)

De Brabander, Gregory N.
; Boyd, Joseph T.; Jackson, Howard E.: Silicon nitride single-polarization optical waveguides on silicon substrates,V1583,327-330(1991)

Debrie, J. See Sansonetti, Pierre: V1588,143-149(1991)

Debusschere, Ingrid See Kreider, Gregory: V1381,242-249(1991)

Decamps, B. See Bourda, C.: V1502,148-159(1991)

DeCatrel, John M. See Budzban, Gregory M.: V1470,175-182(1991)

De Chiara, P. See Frontera, Filippo: V1549,113-119(1991)

Deckelbaum, Lawrence I.
See Pradhan, Asima: V1425,2-5(1991)
See Stetz, Mark L.: V1425,55-62(1991)

Decker, Franco
See Scarminio, J.: V1536,70-80(1991)
See Soltz, David: V1536,268-276(1991)

Decker, Herman J. See Hodor, James R.: V1540,331-337(1991)

DeCoste, Sue See Lemole, G. M.: V1422,116-122(1991)

Decoster, Didier
; Vilcot, Jean-Pierre: Photodetectors: how to integrate them with microelectronic and optical devices,V1362,959-966(1991)
See Aboulhouda, S.: V1362,494-498(1991)
See Constant, Monique T.: V1362,156-162(1991)

DeCou, Anthony B.
: Station-keeping strategy for multiple-spacecraft interferometry,V1494,440-451(1991)

Decraemer, Willem F.
; Dirckx, Joris J.: Moire-shift interferometer measurements of the shape of human and cat tympanic membrane,V1429,26-33(1991)
See Dirckx, Joris J.: V1429,34-38(1991)

Decroisette, Michel See Andre, Michel L.: V1502,286-298(1991)

Decurtins, Jeff See Mulgaonkar, Prasanna G.: V1382,320-330(1991)

Dederich, Douglas N.
; O'Brien, Stephen J.; Trent, Ava M.; Wigdor, Harvey A.; eds.: *Lasers in Orthopedic, Dental, and Veterinary Medicine*,V1424(1991)
; Tulip, John: Effect of Nd:YAG laser on dentinal bond strength,V1424,134-137(1991)

De Dobbelaere, P.
See Buydens, Luc: V1362,50-58(1991)
See Pollentier, Ivan K.: V1361,1056-1062(1991)

Deeter, Merritt N.
; Rose, Allen H.; Day, Gordon W.: Faraday-effect magnetic field sensors based on substituted iron garnets,V1367,243-248(1991)

Deffner, Gerhard See Fulop, Ann C.: V1459,69-76(1991)

De Figueiredo, Rui See Nguyen, Luong A.: V1387,296-312(1991)

deFigueiredo, Rui J.
; Pendleton, Thomas W.: Recent advances in the development and transfer of machine vision technologies for space,V1387,330-336(1991)
See Nguyen, Luong A.: V1387,296-312(1991)
See Stoney, William E.: V1387(1991)

DeForest, Craig E.
; Kankelborg, Charles C.; Allen, Maxwell J.; Paris, Elizabeth S.; Willis, Thomas D.; Lindblom, Joakim F.; O'Neal, Ray H.; Walker, Arthur B.; Barbee, Troy W.; Hoover, Richard B.; Barbee, Troy W.; Gluskin, Efim S.: Performance of the Multi-Spectral Solar Telescope Array V: temperature diagnostic response to the optically thin solar plasma,V1343,404-414(1991)
See Hoover, Richard B.: V1343,189-202(1991)
See Hoover, Richard B.: V1343,175-188(1991)
See Walker, Arthur B.: V1343,319-333(1991)

de Fornel, F.
See Goudonnet, Jean-Pierre: V1545,130-139(1991)
See Goudonnet, Jean-Pierre: V1400,116-123(1991)

Defosse, Yves
; Renotte, Yvon L.; Lion, Yves F.: Calculation of diffraction efficiencies for spherical and cylindrical holographic lenses,V1507,277-287(1991)

de Frutos, Angel M.
; Sabatier, Lilian; Poueyo, Anne; Fabbro, Remy; Bermejo, Dionisio; Orza, Jose M.: Spectroscopic determination of the parameters of an iron plasma produced by a CO2 laser,V1397,717-720(1991)
See Poueyo, Anne: V1502,140-147(1991)

de Garrido, Diego P.
; Pearlman, William A.; Finamore, Weiler A.: Vector quantization of image pyramids with the ECPNN algorithm,V1605,221-232(1991)

Degiorgio, Vittorio See Banfi, Giampiero P.: V1361,874-880(1991)

Degiorgis, Giorgio A. See Canestrari, Paolo: V1463,446-455(1991)

Degnan, John J.
: Applications of laser ranging to ocean, ice, and land topography,V1492,176-186(1991)

de Graaf, C. N. See Anema, P. C.: V1446,352-356(1991)

DeGroat, Ronald D.
; Dowling, Eric M.: Noniterative subspace updating,V1566,376-387(1991)
See Dowling, Eric M.: V1565,35-46(1991)

DeGroot, D. C. See Wessels, Bruce W.: V1394,232-237(1991)

Degtyareva, V. P.
; Fedotov, V. I.; Moskalev, B. B.; Postovalov, V. E.; Prokhorov, Alexander M.; Schelev, Mikhail Y.; Soldatov, N. F.: Improved version of the PIF01 streak image tube,V1358,524-531(1991)
; Ivanov, V. Y.; Ignatov, A. M.; Kolesnikov, Sergey V.; Kulikov, Yu. V.; Monastyrski, M. A.; Niu, Hanben; Schelev, Mikhail Y.: Some comparative results of two approaches in computer simulation of electron lenses for streak image tubes,V1358,546-548(1991)

Deguchi, Masaharu See Yatsu, Masahiko: V1354,663-668(1991)

Deguchi, Masahiro See Nishimura, Kazuhito: V1534,199-206(1991)

de Guise, J. See Grenier, Marie-Claude: V1450,154-169(1991)

de Haas, Matthijs P. See Warman, John M.: V1559,159-170(1991)

de Hoog, F. R. See Bojanczyk, Adam W.: V1566,23-34(1991)

Dehu, C. See Bredas, Jean-Luc: V1560,98-110(1991)

De Iorio, I.
; Sergi, Vincenzo; Tagliaferri, V.: Analysis of laser welding process with the mathematical model GMDH,V1397,787-790(1991)
See Covelli, L.: V1397,797-802(1991)

Deitz, Paul
; Muuss, Michael J.; Davisson, Edwin O.: Issues in automatic object recognition: linking geometry and material data to predictive signature codes,VIS07,40-56(1991)

De Jesus, Stephen T. See Clarke, Richard H.: V1437,198-204(1991)

DeJulio, R. See Westerman, Steven D.: V1492,263-271(1991)

Dekany, Richard G. See Wizinowich, Peter L.: V1542,148-158(1991)

Dekker, L.
; Frietman, Edward E.: Optical interconnects in high-bandwidth computing,V1505,148-157(1991)
See Frietman, Edward E.: V1390,434-453(1991)
See Frietman, Edward E.: V1401,19-26(1991)

DeKruger, D. See Hunt, Bobby R.: V1472,208-218(1991)

Delaboudiniere, Jean-Pierre See Boher, Pierre: V1343,39-55(1991)

de La Chapelle, Michael See Dallabetta, Kyle A.: V1371,116-127(1991)

Delamere, W. A.
; Baer, James W.; Ebben, Thomas H.; Flores, James S.; Kleiman, Gary; Blouke, Morley M.; McCormick, M. P.: High-radiometric-performance CCD for the third-generation stratospheric aerosol and gas experiment,V1479,31-40(1991)
; Baer, James W.; Ebben, Thomas H.; Flores, James S.; Kleiman, Gary; Blouke, Morley M.; McCormick, M. P.: High radiometric performance CCD for the third-generation stratospheric aerosol and gas experiment,V1447,204-213(1991)
; Ebben, Thomas H.; Murata-Seawalt, Debbie; Blouke, Morley M.; Reed, Richard; Woodgate, Bruce E.: Space telescope imaging spectrograph 2048 CCD and its characteristics,V1447,288-297(1991); 1479,21-30(1991)
See Blouke, Morley M.: V1447,142-155(1991)

Delaney, William See Ishaq, Naseem: V1381,153-159(1991)

Delanoy, Richard L. See Verly, Jacques G.: V1471,266-282(1991)

Delaporte, Philippe C.
; Fontaine, Bernard L.; Forestier, Bernard M.; Sentis, Marc L.: Fast-flow gas-dynamic effects in high-pulse repetition-rate excimer lasers,V1397,485-492(1991)
See Gerri, Mireille: V1503,280-291(1991)
See Hueber, Jean-Marc: V1503,62-70(1991)
See Kobhio, M. N.: V1397,555-558(1991)
See Kobhio, M. N.: V1503,88-97(1991)

De La Torre, William See Burleigh, Douglas D.: V1467,303-310(1991)

Delavaux, Jean-Marc P. See Park, Yong K.: V1372,219-227(1991)

Delay, P. See Giannini, Jean-Pierre: V1366,215-222(1991)

Delaye, Corinne T.
; Thomas, Michael E.: Atmospheric continuum absorption models,V1487,291-298(1991)

Delayen, Jean R.
; Bohn, Courtlandt L.; Roche, C. T.: Application of rf superconductivity to high-brightness and high-gradient ion beam accelerators,V1407,524-534(1991)
See Bohn, Courtlandt L.: V1407,566-577(1991)

Delbare, William
; Dhaene, Tom; Vanhauwermeiren, Luc; De Zutter, Daniel: Frequency domain evaluation of the accuracy of lumped element models for RLC transmission lines,V1389,257-272(1991)

Del Bello, Umberto
; Rivela, Cristina; Cantello, Maichi; Penasa, Mauro: Energy balance in high-power CO2 laser welding,V1502,104-116(1991)

Del Bosque-Izaguirre, Delma See Berger, Henry: V1521,117-130(1991)

Delboulbe, J. P. See Plichon, V.: V1536,37-47(1991)

Delbruck, Tobi
; Mead, Carver A.: Time-derivative adaptive silicon photoreceptor array,V1541,92-99(1991)

Del Castillo, Hector M. See Hemmati, Hamid: V1417,476-483(1991)

del Castillo, Maria D.
; Kumpel, Daniel M.: Learning by comparison: improving the task planning capability,V1468,596-607(1991)

Delchambre, Alain
: Knowledge-based process planning and line design in robotized assembly,V1468,367-378(1991)

Delcher, Ray C.
; Dinnsen, Doug K.; Barkhoudarian, S.: Fiber optics in liquid propellant rocket engine environments,V1369,114-120(1991)

Delchev, Georgy See Nedkova, Rumiana: V1429,105-107(1991)

Delecki, Z. A.
; Barakat, M. A.: Nonoptical noncoherent imaging in industrial testing,V1526,157-167(1991)

de Leeuw, Gerrit
: Aerosol models for optical and IR propagation in the marine atmospheric boundary layer,V1487,130-159(1991)

de Leeuw, H. See van Hulst, Niek F.: V1362,631-646(1991)

Delettrez, J. See Chen, Hong: V1413,112-119(1991)

Delgado, J.
See Kozaitis, Samuel P.: V1474,112-115(1991)
See Sanchez, Miguel: V1397,635-638(1991)

Del Grande, J. M.
: Remote temperature sensing of a pulsed thermionic cathode,V1467,427-437(1991)

Del Grande, Nancy K.
; Clark, Greg A.; Durbin, Philip F.; Fields, David J.; Hernandez, Jose E.; Sherwood, Robert J.: Buried object remote detection technology for law enforcement,V1479,335-351(1991)

Del Guerra, Alberto See Barone, Fabrizio: V1584,304-307(1991)

Delhomme, J. See Rivest, Jean-Francois: V1451,179-190(1991)

Delingette, Herve
; Hebert, Martial; Ikeuchi, Katsushi: Deformable surfaces: a free-form shape representation,V1570,21-30(1991)
; Hebert, Martial; Ikeuchi, Katsushi: Energy functions for regularization algorithms,V1570,104-115(1991)

de Lisio, C. See Bruzzese, Riccardo: V1397,735-738(1991)

Delisle, Claude See Sheng, Yunlong: V1559,222-228(1991)

Della Corte, Francesco G. See Cocorullo, Giuseppe: V1374,132-137(1991)

Della Torre, Antonio
See Marini, Andrea E.: V1522,222-233(1991)
See Marini, Andrea E.: V1417,200-211(1991)

Dellepiane, Silvana G. See Venturi, Giovanni: V1606,217-225(1991)

de Lourdes Quinta, Maria
; Freitas, Jose C.; Rodrigues, F. C.; Silva, Jorge A.: Fungal testing of diode laser collimators,V1399,24-29(1991)

Delp, Edward J.
See Stevenson, Robert L.: V1382,364-375(1991)
See Yoo, Jisang: V1451,58-69(1991)

Delp, Scott See Pieper, Steve: V1457,188-197(1991)

Delpy, David T.
See Arridge, Simon R.: V1431,204-215(1991)
See Cope, Mark: V1431,251-262(1991)

DeLuca, Dan
: Linear array camera interface techniques,V1396,558-565(1991)

De Man, Hugo J. See van Swaaij, Michael F.: V1606,901-910(1991)

De Marchi, Giovanna See Battaglin, Giancarlo: V1513,441-450(1991)

De Marchis, Giancarlo See Betti, Silvello: V1579,100-111(1991)

De Marco, Frank P. See Rainer, Frank: V1441,247-254(1991)

De Maria, Letizia
; Escobar Rojo, Priscilla; Martinelli, Mario; Pistoni, Natale C.: Polarimetric sensor strain sensitivity in different thermal operating conditions,V1366,304-312(1991)

De Maziere, Martine M.
; Camy-Peyret, C.; Lippens, C.; Papineau, N.: Stratospheric ozone concentration profiles from Spacelab-1 solar occultation infrared absorption spectra,V1491,288-297(1991)

Dembinski, David T. See Johnson, Dean R.: V1367,140-154(1991)

Demeester, Piet M.
; Moerman, Ingrid; Zhu, Youcai; van Daele, Peter; Thomson, J.: Shadow masked growth for the fabrication of photonic integrated circuits,V1361,1132-1143(1991)
; Pollentier, Ivan K.; Buydens, Luc; van Daele, Peter: Novel optoelectronic devices and integrated circuits using epitaxial lift-off,V1361,987-998(1991)
See Buydens, Luc: V1362,50-58(1991)
See Frietman, Edward E.: V1390,434-453(1991)
See Pollentier, Ivan K.: V1361,1056-1062(1991)

Dementiev, I. V. See Anikin, V. I.: V1238,286-296(1991)

Demerjian, Kenneth L. See Lanni, Thomas R.: V1433,21-24(1991)

De Micheli, Marc See Lallier, Eric: V1506,71-79(1991)

Demichelis, Francesca
; Giachello, G.; Pirri, C. F.; Tagliaferro, Alberto: New mixed sputtering-plasma CVD technique for the deposition of diamondlike films,V1534,140-147(1991)

Demichovskaya, Olga See Sadovnik, Lev S.: V1545,200-208(1991)

Demidov, Anatoly J. See Bokov, Lev: V1505,186-198(1991)

Demirgian, Jack See Herget, William F.: V1434,39-45(1991)

Demiryont, Hulya
: Review on electrochromic devices for automotive glazing,V1536,2-28(1991)

Demko, Christophe See Trigano, Philippe: V1468,866-874(1991)

de Mol, Chris G. See Luehrmann, Paul F.: V1463,434-445(1991)

Demoment, Guy
; Reynaud, Roger: Fast RLS adaptive algorithms and Chandrasekhar equations,V1565,357-367(1991)
See Le Besnerais, Guy: V1569,386-395(1991)
See Reynaud, Roger: V1566,302-311(1991)

Demos, Alexandros T.
; Fogler, H. S.; Pang, Stella W.; Elta, Michael E.: Enhanced etching of InP by cycling with sputter etching and reactive ion etching,V1392,291-297(1991)

Dempsey, B. P. See Valderrama, Giuseppe L.: V1427,200-213(1991)

de Mul, F. F.
See Koelink, M. H.: V1403,347-349(1991)
See Koelink, M. H.: V1511,120-128(1991)
See Koelink, M. H.: V1431,63-72(1991)
See Lucassen, G. W.: V1403,185-194(1991)
See Puppels, G. J.: V1403,146-146(1991)
See Siebinga, I.: V1403,746-748(1991)

Demura, Tatsuhiko See Kitagaki, Kazukuni: V1606,891-900(1991)

Denariez-Roberge, Marguerite M. See Malouin, Christian: V1559,385-392(1991)

DeNatale, Jeffrey F.
; Flintoff, John F.; Harker, Alan B.: Effects of interfacial modifications on diamond film adhesion,V1534,44-48(1991)
See Harker, Alan B.: V1534,2-8(1991)

De Natale, Paolo See Canestrari, Paolo: V1463,446-455(1991)

Denber, Michel J.
: Applications of cellular logic image processing,V1398,29-38(1991)

den Boer, Willem
; Yaniv, Zvi: Active matrix LCDs driven by two- and three-terminal switches: a comparison,V1455,248-248(1991)

den Boggende, Antonius J.
; Bruijn, Marcel P.; Verhoeven, Jan; Zeijlemaker, H.; Puik, Eric J.; Padmore, Howard A.: Broadband multilayer coated blazed grating for x-ray wavelengths below 0.6 nm,V1345,189-197(1991)
See Neviere, Michel: V1545,76-87(1991)
See Neviere, Michel: V1545,116-124(1991)

Denby, Bruce
: Applications of neural networks in experimental high-energy physics,V1469,648-658(1991)

Denby, N. See Chapman, Michael A.: V1526,190-194(1991)

den Exter, Ir. T. See Sjollema, J. I.: V1500,177-188(1991)

Deng, He See Sun, Yang: V1519,554-558(1991)

Deng, Hong
; Wang, Xiang D.; Yuan, Lei: Research of Cr2O3 thin film deposited by arc discharge plasma deposition as heat-radiation absorbent in electric vacuum devices,V1519,735-739(1991)

Deng, Jia-cheng See Wu, Ji-Zong: V1399,122-129(1991)

Deng, Keqiang
; Wilson, Joseph N.: Approximation-based video tracking system,V1568,304-312(1991)
; Wilson, Joseph N.: Contour estimation using global shape constraints and local forces,V1570,227-233(1991)

Deng, Shuang See Maydell, Ursula M.: V1446,81-92(1991)

Deng, Xingzhong
; Zhong, Zugen; Cheng, Xianping; Liu, Bo; Cao, Wei: Research of optical fiber pyrometic sensor for gas-making furnace,V1572,220-223(1991)
See Cheng, Xianping: V1572,216-219(1991)
See Liu, Bo: V1572,211-215(1991)

Deng, Zesheng See Qi, GuangXue: V1572,536-538(1991)

Dengel, Andreas R.
: Object-oriented representation of image space by puzzletrees,V1606,20-30(1991)
See Hoenes, Frank: V1384,305-316(1991)

Dengel, Gail A.
; Robbins, JoAnne; Rosenbek, John C.: Interactive image processing in swallowing research,V1445,88-94(1991)

Denham, David B.
See Cartlidge, Andy G.: V1423,167-174(1991)
See Milne, Peter J.: V1423,122-129(1991)

Denisov, N. N. See Manenkov, Alexander A.: V1420,254-258(1991)

Denisyuk, Yuri N.
; Ganzherli, N. M.: Pseudodeep hologram and its properties,V1238,2-12(1991)

Denker, Boris I.
; Maksimova, G. V.; Osiko, Vyacheslav V.; Prokhorov, Alexander M.; Sverchkov, Sergey E.; Sverchkov, Yuri E.; Horvath, Zoltan G.: Passive Q-switching of eyesafe Er:glass lasers,V1419,50-54(1991)

Denlinger, E. J. See Bharj, Sarjit: V1475,340-349(1991)

Dennis, Peter N.
See Blommel, Fred P.: V1540,653-664(1991)
See Bradley, D. J.: V1488,186-195(1991)
See Murphy, Kevin S.: V1488,178-185(1991)

Deno, Hiroshi See Ozawa, Kenji: V1499,180-186(1991)

Denoue, Emmanuel See Adam, Jean-Luc: V1513,150-157(1991)

Dent, Andrew J.
; Derbyshire, Gareth E.; Greaves, G. N.; Ramsdale, Christine A.; Couves, J. W.; Jones, Richard; Catlow, C. R.; Thomas, John M.: Combined x-ray absorption spectroscopy and x-ray powder diffraction,V1550,97-107(1991)

Dente, Gregory C.
; Walter, Robert F.; Gardner, Daniel C.: Laser-induced medium perturbation in pulsed CO2 lasers,V1397,403-408(1991)

Denton, M. B.
: Applications of charge-coupled and charge-injection devices in analytical spectroscopy,V1447,2-11(1991)
ed.: *Intl Conf on Scientific Optical Imaging*,V1439(1991)
See Pomeroy, Robert S.: V1439,60-65(1991)
See Radspinner, David A.: V1439,2-14(1991)

Denton, Richard T. See Spencer, Paul E.: V1364,228-234(1991)

Denton, Richard V. See Rothman, Peter L.: V1470,2-12(1991)

Denz, Cornelia See Tschudi, Theo T.: V1500,80-80(1991)

de Oliveira, C. A.
; Jen, Cheng K.: Evaluation of parameters in stimulated backward Brillouin scattering,V1590,101-106(1991)

Deom, Alain A. See Balageas, Daniel L.: V1467,278-289(1991)

De Padova, P. See Borsella, E.: V1503,312-320(1991)

De Paoli, Marco-Aurelio See Peres, Rosa C.: V1559,151-158(1991)

Deparis, C. See Chen, Yong: V1361,860-865(1991)

De Pascale, Olga
See Antuofermo, Pasquale: V1583,143-149(1991)
See Conese, Tiziana: V1583,249-255(1991)

DePaula, Ramon P.
; Udd, Eric; eds.: *Fiber Optic and Laser Sensors IX*,V1584(1991)
; Udd, Eric; eds.: *Fiber Optic and Laser Sensors VIII*,V1367(1991)
See Minford, William J.: V1367,46-52(1991)

De Pedraza-Velasco, L. See Calvo, Maria L.: V1507,288-301(1991)

Depeursinge, Ch.
See Poscio, Patrick: V1510,112-117(1991)
See Wagnieres, G.: V1525,219-236(1991)

Depoortere, Marc
: Coherence and optical Kerr nonlinearity,V1504,133-139(1991)
; Ebbeni, Jean P.: Wavelength-based sensor for the measurement of small angles,V1504,264-267(1991)

Der, Sandor Z.
; Dome, G. J.; Rusche, Gerald A.: Technique for ground/image truthing using a digital map to reduce the number of required measurements,V1483,167-176(1991)
See Lorenzo, Maximo: V1483,118-137(1991)

Derby, Eddy A. See Tremblay, Gary A.: V1532,114-123(1991)

Derbyshire, Gareth E. See Dent, Andrew J.: V1550,97-107(1991)

Der Hovanesian, Joseph
See Hathaway, Richard B.: V1554B,725-735(1991)
See Long, Kah W.: V1554A,116-123(1991)

de Ridder, Huib
: Subjective evaluation of scale-space image coding,V1453,31-42(1991)

Derksen, Grant See Taylor, Geoff L.: V1526,27-34(1991)

Derr, John I.
; Ghaffari, Tammy G.: Coherent digital/optical system for automatic target recognition,V1406,127-128(1991)

Derrien, J. See Chevrier, Joel S.: V1512,278-288(1991)

Derwiszynski, Mariusz
See Trojnar, Eugeniusz: V1391,230-237(1991)
See Trojnar, Eugeniusz: V1391,238-243(1991)

Dery, M. See Kassi, Hassane: V1436,58-67(1991)

de Salabert, Arturo
; Pike, T. K.; Sawyer, F. G.; Jones-Parry, I. H.; Rye, A. J.; Oddy, C. J.; Johnson, D. G.; Mason, D.; Wielogorski, A. L.; Plassard, T.; Serpico, Sebastiano B.; Hindley, N.: Multisensor image processing,V1521,74-88(1991)

De Salvo, Edmondo
; Girlanda, Raffaello: Wave-vector-dependent magneto-optics in semiconductors,V1362,870-875(1991)

DeSanto, John A.
; Wombell, Richard J.: Some computational results for rough-surface scattering,V1558,202-212(1991)

De Sario, Marco
See d'Alessandro, Antonio: V1378,127-138(1991)
See d'Alessandro, Antonio: V1332,554-562(1991)
See d'Alessandro, Antonio: V1366,313-323(1991)
See d'Alessandro, Antonio: V1399,184-191(1991)

Deshayes, Jean-Pierre See Lorsignol, Jean: V1490,155-163(1991)

DeShazer, James A.
; Meyer, George E.; eds.: *Optics in Agriculture*,V1379(1991)

Deshmukh, V. G.
; Hope, D. A.; Cox, T. I.; Hydes, A. J.: Application of adaptive network theory to dry-etch monitoring and control,V1392,352-360(1991)
See Hope, D. A.: V1392,185-195(1991)

Deshors, G. See Poueyo, Anne: V1502,140-147(1991)

Deshpande, Ujwal A.
; Howell, Gelston; Shamouilian, Shamouil: High-density interconnect technology for VAX 9000 system,V1390,489-501(1991)

De Silvestri, S. See Serri, Laura: V1397,469-472(1991)

de Souza, Eunezio A.
; Cruz, C. H.; Scarparo, Marco A.; Prokhorov, Alexander M.; Postovalov, V. E.; Vorobiev, N. S.; Schelev, Mikhail Y.: Application of a linear picosecond streak camera to the investigation of a 1.55 um mode-locked Er3+ fiber laser,V1358,556-560(1991)

De Spirito, M. See Pizzoferrato, R.: V1409,192-201(1991)

Desrochers, Alan A.
: Experimental testbed for cooperative robotic manipulators,V1387,2-13(1991)

Desse, Jean-Michel
; Pegneaux, Jean-Claude: Instantaneous measurement of density from double-simultaneous interferograms,V1358,766-774(1991)

Dessesard, Pascal See Duffy, Neil D.: V1380,46-52(1991)

Dessy, Raymond E. See Petersen, James V.: V1368,61-72(1991)

Destefanis, Giovanni See Caldera, Claudio: V1372,82-87(1991)

Destrade, C. See Twieg, Robert J.: V1455,86-96(1991)

Destro, Gianni See Barbieri, Enrico: V1425,122-127(1991)

Dettki, Frank See Rueff, Manfred: V1382,255-270(1991)

Detwiler, Paul W.
; Watkins, James F.; Rose, Eric A.; Ratner, A.; Vu, Louis P.; Severinsky, J. Y.; Rosenschein, Uri: Mechanical and acoustic analysis in ultrasonic angioplasty,V1425,149-155(1991)

Deutsch, Alina
; Kopcsay, Gerard V.; Ranieri, V. A.; Cataldo, J. K.; Galligan, E. A.; Graham, W. S.; McGouey, R. P.; Nunes, S. L.; Paraszczak, Jurij R.; Ritsko, J. J.; Serino, R. J.; Shih, D. Y.; Wilczynski, Janusz S.: Electrical characteristics of lossy interconnections for high-performance computer applications,V1389,161-176(1991)

Deutsch, L. J. See Lesh, James R.: V1522,27-35(1991)

Deutsch, Meir See Mendlovic, David: V1442,182-192(1991)

Devanand, Krisha
: Dynamic light scattering from a side-chain liquid crystalline polymer in a nematic solvent,V1430,160-164(1991)
See Selser, James C.: V1430,85-88(1991)

Devarajan, Ravinder
; McAllister, David F.: Stereoscopic ray tracing of implicitly defined surfaces,V1457,37-48(1991)

Devereux, R. W. See Jenkins, R. M.: V1512,135-142(1991)

Devetzis, Tasco N. See Albanese, Andres: V1364,320-326(1991)

De Vilder, Jan See Schrever, Koen: V1572,107-112(1991)

Deville, G. See Etienne, Bernard: V1362,256-267(1991)

Devir, Adam D.
; Neumann, M.; Lipson, Steven G.; Oppenheim, Uri P.: Water continuum in the 15- to 25-um spectral region: evidence for H2O2 in the atmosphere,V1442,347-359(1991)
See Ben-Shalom, Ami: V1486,238-257(1991)
See Sheffer, Dan: V1493,232-243(1991)

DeVito, Anthony
See Hart, Patrick W.: V1366,334-342(1991)
See Kilmer, Joyce P.: V1366,85-91(1991)

Devos, F. See Maurin, T.: V1505,158-165(1991)

de Vries, Mattanjah S.
; Hunziker, Heinrich E.; Meijer, Gerard; Wendt, H. R.: Laser desorption jet cooling spectroscopy,V1437,129-137(1991)

Devyatykh, Grigory G. See Nadezhdinskii, Alexander I.: V1418,487-495(1991)

Devynck, Pascal See Bacal, Marthe: V1407,605-609(1991)

Dew, Douglas K.
; Hsu, Tung M.; Hsu, Long S.; Halpern, Steven J.; Michaels, Charles E.: Laser-assisted skin closure at 1.32 microns: the use of a software-driven medical laser system,V1422,111-115(1991)

DeWalt, Steve A.
See Padden, Richard J.: V1474,148-159(1991)
See Taylor, Edward W.: V1474,126-131(1991)

DeWeerd, Herman
: Compact, low-power precision beam-steering mirror,V1454,207-214(1991)

Dewey, David
: Corneal and retinal energy density with various laser beam delivery systems and contact lenses,V1423,105-116(1991)

Dewey, John M.
; McMillin, Douglas J.: Analysis of results from high-speed photogrammetry of flow tracers in blast waves,V1358,246-253(1991)
See Racca, Roberto G.: V1358,932-939(1991)

Dewey, T. G.
: Fractal dynamics of fluorescence energy transfer in biomembranes,V1432,64-75(1991)

Deyoe, N. B. See MacDonald, Scott A.: V1466,2-12(1991)

De Yoreo, James J.
; Woods, Bruce W.: Investigation of strain birefringence and wavefront distortion in 001 plates of KD2PO4,V1561,50-58(1991)

DeYoung, David B.
: Simulation of scenes, sensors, and tracker for space-based acquisition, tracking, and pointing experiments,V1415,13-21(1991)

Dezenberg, George J. See Horton, T. E.: V1416,10-20(1991)

De Zutter, Daniel See Delbare, William: V1389,257-272(1991)

Dhadwal, Harbans S.
; Ansari, Rafat R.: Multiple fiber optic probe for several sensing applications,V1584,262-272(1991)
; Noel, Eric: Real-time optical processor for increasing resolution beyond the diffraction limit,V1564,664-673(1991)

Dhaene, Tom See Delbare, William: V1389,257-272(1991)

Dhali, Shirshak K.
; Mohsin, Mohammad: Laser-triggered superconducting opening switch,V1396,353-359(1991)
See Sayeh, Mohammad R.: V1396,417-429(1991)

Dhawan, Atam P. See Arata, Louis K.: V1446,465-474(1991)

Diaci, J. See Grad, Ladislav: V1525,206-209(1991)

Diadiuk, Vicky
; Liau, Zong-Long; Walpole, James N.: Fabrication and characterization of semiconductor microlens arrays,V1354,496-500(1991)

Dial, O. E.
: CCD performance model,V1479,2-11(1991)

Dialetis, Demos See Golden, Jeffry: V1407,418-429(1991)

Dian, J. See Adamec, F.: V1402,82-84(1991)

Dianov, Evgeni M.
; Kazansky, Peter G.; Stepanov, D. Y.: Photoinduced effects in optical waveguides,V1516,81-98(1991)
; Kazansky, Peter G.; Krautschik, Christof G.; Stepanov, D. Y.: Test of photovoltaic model of photoinduced second-harmonic generation in optical fibers,V1516,75-80(1991)

Dias, Antonio R. See Kalman, Robert F.: V1563,34-44(1991)

Diaz, Art F.
See Gutierrez, Adolfo R.: V1458,201-204(1991)
See Wollmann, Daphne: V1458,192-200(1991)

Diaz, Pedro J. See Connor, William H.: V1568,327-334(1991)

Dibble, R. W.
; Ketterle, Wolfgang; Fourgette, D. C.; eds.: *ICALEO '90 Optical Methods in Flow and Particle Diagnostics*,V1602(1991)

Di Benedetto, John A.
; Capelle, Gene; Lutz, Stephen S.: Uniform field laser illuminator for remote sensing,V1492,115-125(1991)

DiBiase, Don See Koechner, Walter: V1522,169-179(1991)

Di Chirico, Giuseppe See Barillot, Marc: V1554A,867-878(1991)

Dick, Sergei C.
; Markhvida, Igor V.; Tanin, Leonid V.: Optical system for control of longitudinal displacement,V1454,447-452(1991)

Dickerson, Gary
; Wallace, Rick P.: In-line wafer inspection using 100-megapixel-per-second digital image processing technology,V1464,584-595(1991)

Dickerson, Stephen L.
; Lee, Kok-Meng; Lee, Eun Ho; Single, Thomas; Li, Da-ren: Intelligent material handling: use of vision,V1381,201-209(1991)

Dickey, Fred M. See Kast, Brian A.: V1564,34-42(1991)

Dickinson, Alex G.
; Downs, Maralene M.; LaMarche, R. E.; Prise, Michael E.: Free-space optical interconnect using microlasers and modulator arrays,V1389,503-514(1991)

Dickinson, J. T.
; Langford, S. C.; Jensen, L. C.: Role of defects in the ablation of wide-bandgap materials,V1598,72-83(1991)

Dickinson, Mark R.
See Charlton, Andrew: V1427,189-197(1991)
See Wannop, Neil M.: V1423,163-166(1991)

Dickinson, P. H. See MacAndrew, J. A.: V1500,172-176(1991)

Dickinson, Robert R.
: Interactive analysis of transient field data,V1459,166-176(1991)
; Brady, Dan R.; Bennison, Tim; Burns, Thomas; Pines, Sheldon: Process of videotape making: presentation design, software, and hardware,V1459,178-189(1991)

Di Cocco, Guido
; Labanti, Claudio; Malaguti, Giuseppe; Rossi, Elio; Schiavone, Filomena; Spizzichino, A.; Traci, A.; Bird, A. J.; Carter, T.; Dean, Anthony J.; Gomm, A. J.; Grant, K. J.; Corba, M.; Quadrini, E.; Rossi, M.; Villa, G. E.; Swinyard, Bruce M.: Imager for gamma-ray astronomy: balloon prototype,V1549,102-112(1991)
See Swinyard, Bruce M.: V1548,94-105(1991)

Diebel, D. E. See Bristow, Michael P.: V1491,68-74(1991)

Diebold, Simon See Reynes, Anne M.: V1525,177-182(1991)

Diedrich, F. See Bergquist, James C.: V1435,82-85(1991)

Diels, Jean-Claude
; Lai, Ming: Spontaneous emission noise reduction of a laser output by extracavity destructive interference,V1376,198-205(1991)

Diem, Max
: Solution conformation of biomolecules from infrared vibrational circular dichroism spectroscopy,V1432,28-36(1991)

Diemeer, Mart B. See Moehlmann, Gustaaf R.: V1560,426-433(1991)

Dierks, Jeffrey S.
; Ross, Susan E.; Brodsky, Aaron; Kervin, Paul W.; Holm, Richard W.: Relay Mirror Experiment overview: a GBL pointing and tracking demonstration,V1482,146-158(1991)

Dierks, Tim See Nielson, Gregory M.: V1459,22-33(1991)

Diessel, Edgar
; Sigg, Hans; von Klitzing, Klaus: Far-IR emission spectroscopy on electron-beam-irradiated A1GaAs/GaAs heterostructures,V1361,1094-1099(1991)

Dietel, W.
; Dorn, P.; Zenk, W.; Zielinski, M.: Laser-induced fluorescence of biological tissue,V1403,653-658(1991)

Dietrich, Fred See Lesko, Camille: V1475,330-339(1991)

Dietrich, H. P. See Webb, David J.: V1418,231-239(1991)

Dietz, Alvin See Turner, Robert E.: V1377,99-106(1991)

Dietz, John B. See Kruse, Fred A.: V1492,326-337(1991)

Dietz, Kurtis L. See Johnson, Charles L.: V1494,208-218(1991)

Dietze, H.-J. See Becker, J. S.: V1598,227-238(1991)

Di Fiore, Luciano See Barone, Fabrizio: V1584,304-307(1991)

DiFrancia, Celene
; Claus, Richard O.; Ward, T. C.: Role of adhesion in optical-fiber-based smart composite structures and its implementation in strain analysis for the modeling of an embedded optical fiber,V1588,44-49(1991)

DiGiovanni, David J.
: Fabrication of rare-earth-doped optical fiber,V1373,2-8(1991)

Di Girolamo, Sergio See Rondinelli, Giuseppe: V1495,19-31(1991)

Digney, Bruce L. See Gupta, Madan M.: V1381,346-356(1991)

Digonnet, Michel J.
ed.: *Fiber Laser Sources and Amplifiers II*,V1373(1991)
See Kalman, Robert F.: V1373,209-222(1991)
See Wysocki, Paul F.: V1373,66-77(1991)
See Wysocki, Paul F.: V1373,234-245(1991)

Digumarthi, Ramji V.
; Mehta, Naresh C.: Simplex optimization method for adaptive optics system alignment,V1408,136-147(1991)
; Payton, Paul M.; Barrett, Eamon B.: Linear programming solutions to problems in logical inference and space-variant image restoration,V1472,128-138(1991)

Dijaili, Sol P. See Sauer, Jon R.: V1579,84-97(1991)

Dijkstra, Han J.
: Oxygen plasma etching of silylated resist in top-imaging lithographic process,V1466,592-603(1991)

Dikasov, A. B. See Bespalov, V. G.: V1238,470-475(1991)

Dikman, A.
See Lando, Mordechai: V1412,19-26(1991)
See Lando, Mordechai: V1442,172-180(1991)

Dikov, Juriy See Sorokina, Svetlana: V1519,128-133(1991)

Di Lazzaro, Paolo
See Bollanti, Sarah: V1503,80-87(1991)
See Bollanti, Sarah: V1397,97-102(1991)

Dilger, Werner
: Computer-aided acquisition of design knowledge,V1468,584-595(1991)

Dilhac, Jean-Marie R.
; Ganibal, Christian; Bordeneuve, J.; Dahhou, B.; Amat, L.; Picard, A.: Adaptive process control for a rapid thermal processor,V1393,395-403(1991)
; Ganibal, Christian; Nolhier, N.; Amat, L.: In-situ interferometric measurements in a rapid thermal processor,V1393,349-353(1991)

Dill, Roland See Kessler, Jochen: V1477,45-56(1991)

Dillehay, David L.
: Multicomponent analysis using established techniques,V1434,56-66(1991)

Dillow, Michael A. See Sydney, Paul F.: V1482,196-208(1991)

Dilung, Iosiph I.
See Boiko, Yuri B.: V1238,253-257(1991)
See Boiko, Yuri B.: V1238,258-265(1991)

Dilworth, Donald C.
: Expert systems in lens design,V1354,359-370(1991)
: SYNOPSYS—a lens design computer program package,V1354,112-115(1991)

Di Maggio, Michel See Bourhis, Jean-Francois: V1583,383-388(1991)

Dimakov, Sergey A. See Sherstobitov, Vladimir E.: V1415,79-89(1991)

DiMarco, Steven F.
; Cantrell, Cyrus D.: Parametric pulse breakup due to population pulsations in three-level systems,V1497,178-187(1991)

Dimauro, Giovanni See Impedovo, Sebastiano: V1384,280-284(1991)

DiMeo, Frank See Wessels, Bruce W.: V1394,232-237(1991)

Dimitriadis, Basile D. See Karkanis, S.: V1500,164-170(1991)

Dimitrov, S. See Angelov, D.: V1403,230-239(1991)

Dimitrov-Kuhl, K. P. See Wong, Ka-Kha: V1374,278-286(1991)

DiNardo, Anthony J.
: SADARM status report,V1479,228-248(1991)

Diner, David J. See Bruegge, Carol J.: V1493,132-142(1991)

Dinev, S.
; Dreischuh, A.: Pico- and femtosecond pulses in the UV and XUV,V1403,427-430(1991)

Ding, Ai L. See Zhou, Shi P.: V1519,793-799(1991)

Ding, Hai
: Charge buildup in polypropylene thin films,V1519,847-856(1991)

Ding, Kung-Hau See Tsang, Leung: V1558,260-268(1991)

Ding, Xiande
See Wang, Yongjiang: V1412,67-71(1991)
See Wang, Yongjiang: V1412,60-66(1991)

Ding, Xuemei See Tan, Jiubin: V1572,552-557(1991)

Ding, Yingjai
See Pavicic, Mark J.: V1446,370-378(1991)
See Persons, Kenneth R.: V1446,60-72(1991)

Ding, Zhi
; Johnson, C. R.: Existing gap between theory and application of blind equalization,V1565,154-165(1991)

Ding, Zu-Quan
; Yuan, Xun-Hua: Practical method for automatic detection of brightest ridge line of photomechanical fringe pattern,V1554A,898-906(1991)

Dinguirard, Magdeleine C. See Biggar, Stuart F.: V1493,155-162(1991)

Dingus, Ronald S.
; Scammon, R. J.: Grueneisen-stress-induced ablation of biological tissue,V1427,45-54(1991)

Dinnsen, Doug K. See Delcher, Ray C.: V1369,114-120(1991)

Dion, Bruno
See MacGregor, Andrew D.: V1417,374-380(1991)
See Webb, Paul P.: V1563,236-243(1991)

Dion, Jacques E. See Close, Robert A.: V1444,196-203(1991)

Dionoro, G. See Daurelio, Giuseppe: V1397,783-786(1991)

Dipace, A. See Ciocci, Franco: V1501,154-162(1991)

Di Paola, Robert
See Aubry, Florent: V1446,168-176(1991)
See Bizais, Yves: V1446,156-167(1991)

DiPietro, Richard See Twieg, Robert J.: V1455,86-96(1991)

Dirckx, Joris J.
; Decraemer, Willem F.: Deformation measurements of the human tympanic membrane under static pressure using automated moire topography,V1429,34-38(1991)
See Decraemer, Willem F.: V1429,26-33(1991)

Dirk, Carl W. See Kuzyk, Mark G.: V1560,44-53(1991)

Dirksen, D. See von Bally, Gert: V1507,66-72(1991)

DiSessa, Peter A. See Unger, Robert: V1463,725-742(1991)

Disimile, Peter J.
; Fox, Curtis F.; Toy, Norman: Optical leak detection of oxygen using IR laser diodes,V1492,64-75(1991)
; Shoe, Bridget; Toy, Norman; Savory, Eric; Tahouri, Bahman: Development of image processing techniques for applications in flow visualization and analysis,V1489,66-74(1991)
; Shoe, Bridget; Toy, Norman: Health monitoring of rocket engines using image processing,V1483,39-48(1991)
See Shoe, Bridget: V1521,34-45(1991)
See Toy, Norman: V1489,112-123(1991)

Distante, Arcangelo
; D'Orazio, Tiziana; Stella, Ettore: Segmentation of orientation maps by an integration of edge- and region-based methods,V1381,513-523(1991)
See Attolico, Giovanni: V1383,34-46(1991)
See Attolico, Giovanni: V1388,50-61(1991)

Distasio, Romelia
See Sethi, Satyendra A.: V1463,30-40(1991)
See Takemoto, Cliff H.: V1464,206-214(1991)
See Ziger, David H.: V1466,270-282(1991)

Ditchburn, Robert J. See Smith, Geoffrey B.: V1536,126-137(1991)

Ditchek, Brian M. See Steiner, Bruce W.: V1557,156-167(1991)

Divan, James See Black, Johnathan D.: V1424,12-15(1991)

Di Vita, P.
: Optical fiber measurements and standardization: status and perspectives,V1504,38-46(1991)

Dix, C. See McKee, Paul: V1461,17-23(1991)

Dixit, Sham N.
See Mazumdar, Sumit: V1436,136-149(1991)
See Woods, Bruce W.: V1410,47-54(1991)

Dixon, G. J.
; Kean, P. N.: Progress in diode laser upconversion,V1561,147-150(1991)

Djeu, Nick See Chang, Robert S.: V1410,125-132(1991)

Djibladze, Merab I. See Prokhorov, Kirill A.: V1501,80-84(1991)

Djuric, Zoran G.
; Piotrowski, Jozef: Electromagnetically carrier depleted IR photodetector,V1540,622-632(1991)
See Piotrowski, Jozef: V1512,84-90(1991)

Dmitriev, A. C. See Furzikov, Nickolay P.: V1427,288-297(1991)

Dmitriyenko, O. L. See Kostin, Ivan K.: V1554A,418-425(1991)

Dmokhovskij, A. V.
; Filippov, I. G.; Skropkin, S. A.; Kobakhidze, T. G.: Photoelastic investigation of stress waves using models of viscoelastic materials,V1554A,323-330(1991)

Doan, Trung T. See Yu, Chang: V1598,186-197(1991)

Doane, J. W.
See Crawford, Gregory P.: V1455,2-11(1991)
See McCargar, James W.: V1455,54-60(1991)

Doane, Kent B. See McSwain, G. G.: V1478,228-238(1991)

Dobbins, B. N.
; He, Shi P.; Kapasi, S.; Wang, Liu Sheng; Button, B. L.; Wu, Xiao-Ping: Application of phase-stepping speckle interferometry to shape and deformation measurement of a 3-D surface,V1554A,772-780(1991)
; He, Shi P.; Jambunathan, K.; Kapasi, S.; Wang, Liu Sheng; Button, B. L.: Measurement of the temperature field in confined jet impingement using phase-stepping video holography,V1554B,586-592(1991)

Dobbyn, Ronald C. See Steiner, Bruce W.: V1557,156-167(1991)

Dobosiewicz, Wlodek
; Gburzynski, Pawel: Fault-tolerant capacity-1 protocol for very fast local networks,V1470,123-133(1991)

Dobrego, Kirill V.
: Determination of inhomogeneous trace absorption by using exponential expansion of the absorption Pade-approximant,V1433,365-374(1991)

Dobson, David C.
; Cox, James A.: Integral equation method for biperiodic diffraction structures,V1545,106-113(1991)

Dobson, Peter J. See Galloway, Peter C.: V1365,131-138(1991)

Docherty, Tom See Woods, Andrew J.: V1457,322-326(1991)

Doctor, D. P. See Taylor, Geoff W.: V1476,2-13(1991)

Dodd, James W.
; Aghamir, F.; Barletta, W. A.; Cline, David B.; Hartman, Steven C.; Katsouleas, Thomas C.; Kolonko, J.; Park, Sanghyun; Pellegrini, Claudio; Terrien, J. C.; Davis, J. G.; Joshi, Chan J.; Luhmann, Neville C.; McDermott, David B.; Ivanchenkov, S. N.; Lachin, Yu Y.; Varfolomeev, A. A.: Saturnus: the UCLA compact infrared free-electron laser project,V1407,467-473(1991)

Dodds, David R.
: Coping with complexity in the navigation of an autonomous mobile robot,V1388,448-452(1991)
: Terrain adaptive footfall placement using real-time range images,V1388,543-549(1991)
: Terrain classification in navigation of an autonomous mobile robot,V1388,82-89(1991)

Dodelet, Jean-Pol
; Gastonguay, Louis; Veilleux, George; Saint-Jacques, Robert G.; Cote, Roland; Guay, Daniel; Tourillon, Gerard: Modified chloroaluminium phthalocyanine: an organic semiconductor with high photoactivity,V1436,38-49(1991)

Dodge, Marilyn J.
: Refractive index measurement capabilities at the National Institute of Standards and Technology,V1441,56-60(1991)
: Some needs for the characterization of the dispersive properties of transmissive-optical materials,V1535,2-12(1991)

Doehler, Gottfried H.
: Nonlinear optical properties of nipi and hetero nipi superlattices and their application for optoelectronics,V1361,443-468(1991)

Doel, A. P.
; Dunlop, C. N.; Major, J. V.; Myers, Richard M.; Sharples, R. M.: MARTINI: system operation and astronomical performance,V1542,319-326(1991)

Doeldissen, Walter See Engel, Herbert: V1506,60-64(1991)

Doerband, Bernd
; Wiedmann, Wolfgang; Wegmann, Ulrich; Kuebler, C. W.; Freischlad, Klaus R.: Software concept for the new Zeiss interferometer,V1332,664-672(1991)

Doerr, Stephen E.
: Application of holographic interferometry to a three-dimensional flow field,V1554B,544-555(1991)

Doerschel, Klaus
; Mueller, Gerhard J.: Photoablation,V1525,253-279(1991)

Doerschuk, Peter C.
: Bayesian signal reconstruction from Fourier transform magnitude and x-ray crystallography,V1569,70-79(1991)

Doft, F.
See Simpson, Thomas B.: V1410,133-140(1991)
See Simpson, Thomas B.: V1416,2-9(1991)

Doherty, John A. See Stewart, Diane K.: V1496,247-265(1991)

Doherty, Peter E.
; Sims, Gary R.: Advanced imaging system for high-precision, high-resolution CCD imaging,V1448,118-128(1991)

Dohr, Gottfried See Gratzl, Thomas: V1427,55-62(1991)

Doi, Kunio
See Alperin, Noam: V1396,27-31(1991)
See Giger, Maryellen L.: V1445,101-103(1991)
See Nishikawa, Robert M.: V1444,180-189(1991)
See Yin, Fang-Fang: V1396,2-4(1991)
See Yoshimura, Hitoshi: V1445,47-51(1991)

Doi, Masao See Maeda, T.: V1403,268-277(1991)

Doi, Takeshi
; Tadros, Karim H.; Kuyel, Birol; Neureuther, Andrew R.: Edge-profile, materials, and protective coating effects on image quality,V1464,336-345(1991)
See Chiu, Anita S.: V1466,641-652(1991)
See Wong, Alfred K.: V1463,315-323(1991)

Doi, Yutaka See Merkelo, Henri: V1390,91-163(1991)

Doig, Stephen J.
; Reid, Philip J.; Mathies, Richard A.: Picosecond time-resolved resonance Raman spectroscopy of bacteriorhodopsin: structure and kinetics of the J, K, and KL intermediates,V1432,184-196(1991)
See Reid, Philip J.: V1432,172-183(1991)

Dokhanian, Mostafa
See Shukla, Ram P.: V1332,274-286(1991)
See Venkateswarlu, Putcha: V1332,245-266(1991)

Dolgin, Benjamin P.
; Moacanin, Jovan; O'Donnell, Timothy P.: Theoretical limits of dimensional stability for space structures,V1533,229-239(1991)
See Coulter, Daniel R.: V1540,119-126(1991)

Dolgina, A. N.
; Kovalev, A. A.; Kondratenko, P. S.: Dimension and time effects caused by nonlocal scattering of laser radiation from a rough metal surface,V1440,342-353(1991)

Dolgolenko, D. A. See Karpov, V. I.: V1503,492-502(1991)

Dolgopolov, V. G. See Petrov, Dmitry V.: V1374,152-159(1991)

Dolgopolov, Y. V.
; Kirillov, Gennadi A.; Kochemasov, G. G.; Kulikov, Stanislav M.; Murugov, V. M.; Pevny, S. N.; Sukharev, S. A.: High-power nanosecond pulse iodine laser provided with SBS mirror,V1412,267-275(1991)

Domanik, Richard A.
See Cohn, Gerald E.: V1429,195-206(1991)
See Cohn, Gerald E.: V1396,131-142(1991)

Domanski, Andrzej W.
; Karpierz, Miroslaw A.; Strojewski, Dariusz: Polarization-based all-optical bistable element,V1505,59-66(1991)
; Kostrzewa, Stanislaw: Fiber optic liquid crystalline microsensor for temperature measurement in high magnetic field,V1510,72-77(1991)
; Kostrzewa, Stanislaw; Hliniak, Andrzej: Monitoring of tissue temperature during microwave hyperthermia utilizing a fiber optic liquid crystalline microsensor,V1420,72-80(1991)
; Roszko, Marcin; Sierakowski, Marek W.: High-efficient fiber-to-stripe waveguide coupler,V1362,844-852(1991)
; Roszko, Marcin; Sierakowski, Marek W.: Optical fiber units with ferroelectric liquid crystals for optical computing,V1362,907-915(1991)
See Bock, Wojtek J.: V1511,250-254(1991)
See Bokun, Leszek J.: V1420,93-99(1991)
See Kubica, Jacek M.: V1391,24-31(1991)

Dome, G. J. See Der, Sandor Z.: V1483,167-176(1991)

Domingo, Concepcion
; Orza, Jose M.; eds.: *8th Intl Symp on Gas Flow and Chemical Lasers*,V1397(1991)

Domingo, Vicente See Poland, Arthur I.: V1343,310-318(1991)

Dominguez Ferrari, E.
; Encinas Sanz, F.; Guerra Perez, J. M.: New photovoltaic effect in semiconductor junctions n+/p,V1397,725-727(1991)

DoMinh, Thap
: High-speed nonsilver lithographic system for laser direct imaging,V1458,68-68(1991)

Don, H. S. See Chen, Duanjun: V1554A,706-717(1991)

Donaher, J. C. See Holbrook, David S.: V1463,475-486(1991)

Donaldson, William R.
; Kingsley, Lawrence E.: Optical probing of field dependent effects in GaAs photoconductive switches,V1378,226-236(1991)

Donath, E. See Ermakov, Yu. A.: V1403,338-339(1991)

Donati, Silvano
; Martini, Giuseppe; Francese, Francesco: Two Gbit/s photonic backplane for telephone cards interconnection,V1389,665-671(1991)

Doney, Thomas A.
: Production quality control problems,V1381,9-20(1991)

Dong, Ada See Zamorano, Lucia J.: V1428,59-75(1991)

Dong, Benhan
; Gao, Penfei; Wang, Ju: Application of photoelastic coating method on elastoplastic stress analysis of rotation disk,V1554A,400-406(1991)

Dong, Bizhen
; Gu, Benyuan; Yang, Guo-Zhen: Recovery for a real image from its Hartley transform modulus only,V1429,117-126(1991)

Dong, Gui-Rong See Hong, Xiao-Yin: V1466,546-557(1991)

Dong, Jingyuan See He, Huijuan: V1409,18-23(1991)

Dong, Linjun See Qian, Anping: V1572,144-147(1991)

Dong, LiXin
; Pelle, Gabriel; Brun, Philip; Unser, Michael A.: Model-based boundary detection in echocardiography using dynamic programming technique,V1445,178-187(1991)

Dong, S. Y. See Kwok, Hoi-Sing: V1394,196-200(1991)

Dong, Xiaopeng
; Hu, Hao; Qian, Jingren: Measurement of fibre Verdet constant with twist method,V1572,56-60(1991)

Donn, Matthew
; Waeber, Bruce: CCD performance model for airborne reconnaissance,V1538,189-200(1991)

Donnelly, Joseph J.
: Selection of image acquisition methods,V1444,351-356(1991)

Donohoo, Daniel See Williams, Rodney D.: V1457,210-220(1991)

Doran, S. P. See Shaver, David C.: V1596,46-50(1991)

Dorband, John E.
See Bonavito, N. L.: V1406,138-146(1991)
See Fischer, James R.: V1492,229-238(1991)

Dorfman, V. F. See Ageev, Vladimir P.: V1503,453-462(1991)

Doria, Andrea See Ciocci, Franco: V1501,154-162(1991)

Dorin, C. See Etienne, Bernard: V1362,256-267(1991)

Dorko, Ernest A. See Crannage, R. P.: V1397,261-265(1991)

Dorn, P. See Dietel, W.: V1403,653-658(1991)

Dorneich, M. See Merkelo, Henri: V1390,91-163(1991)

Dorofeev, I. A. See Brailovsky, A. B.: V1440,84-89(1991)

Doroschenko, V. M.
; Kudriavtsev, N. N.; Sukhov, A. M.: Chemical gas-dynamic mixing CO2 laser pumped by the reactions between N2O and CO,V1397,503-511(1991)

Doroski, D. See Perlmutter, S. H.: V1562,74-84(1991)

Dorris, H. N. See Albanese, Andres: V1364,320-326(1991)

Dorsch, Friedhelm
: High-resolution spectral characterization of high-power laser diodes,V1418,477-486(1991)

Dorsel, Andreas N. See Wolf, Barbara: V1506,40-51(1991)

Dorsky, Leonard See Barnes, Charles E.: V1366,9-16(1991)

Doshi, Rekha See Evans, Clinton E.: V1537,203-214(1991)

Doskolovich, Leonid L.
See Golub, Mikhail A.: V1500,194-206(1991)
See Golub, Mikhail A.: V1500,211-221(1991)

Dos Passos, Waldemar
See Clarke, Roy: V1361,2-12(1991)
See Clarke, Roy: V1345,101-114(1991)

Dosser, Larry R.
; Stark, Margaret A.: Analysis of pyrotechnic devices by laser-illuminated high-speed photography,V1346,293-299(1991)

Dostenko, Alexander V. See Kuchinskii, Sergei A.: V1513,297-308(1991)

Dotor, M. L. See Recio, Miguel: V1361,469-478(1991)

Dotti, Ernesto See Gaboardi, Franco: V1421,50-55(1991)

Doty, Keith L. See Bou-Ghannam, Akram A.: V1468,504-515(1991)

Doubrere, Patrick See Cossec, Francois R.: V1417,262-276(1991)

Dougal, John See Jeffs, Brian D.: V1567,511-521(1991)

Dougherty, Edward R.
; Haralick, Robert M.: Hole spectrum: model-based optimization of morphological filters,V1568,224-232(1991)
; Haralick, Robert M.; Chen, Yidong; Li, Bo; Agerskov, Carsten; Jacobi, Ulrik; Sloth, Poul H.: Morphological pattern-spectra-based Tau-opening optimization,V1606,141-152(1991)
; Mathew, A.; Swarnakar, Vivek: Conditional-expectation-based implementation of the optimal mean-square binary morphological filter,V1451,137-147(1991)
; Sand, Francis M.: Distribution of the pattern spectrum mean for convex base images,V1451,114-124(1991)
; Weisman, Andrew; Mizes, Howard; Miller, Robert J.: Application of morphological pseudoconvolutions to scanning-tunneling and atomic force microscopy,V1567,88-99(1991)
See Arce, Gonzalo R.: V1451(1991)
See Gader, Paul D.: V1568(1991)
See Haralick, Robert M.: V1472,108-117(1991)
See Loce, Robert P.: V1568,233-246(1991)

Dougherty, Thomas J.
ed.: *Optical Methods for Tumor Treatment and Early Diagnosis: Mechanisms and Techniques,*V1426(1991)
See Pandey, Ravindra K.: V1426,356-361(1991)
See Shiau, Fuu-Yau: V1426,330-339(1991)

Dougherty, William A. See Thaniyavarn, Suwat: V1371,250-251(1991)

Doughty, Glenn R. See Casey, William L.: V1417,89-98(1991)

Doughty, J. R. See Gole, James L.: V1397,125-135(1991)

Douglas, A. See Nodine, Calvin F.: V1444,56-62(1991)

Douglass, Robert J. See Severson, William E.: V1388,490-501(1991)

Doukas, Apostolos G. See Flotte, Thomas J.: V1427,36-44(1991)

Doval, A. F. See Blanco, J.: V1508,180-190(1991)

Dove, Derek B.
; Abe, Takao; Heinzl, Joachim L.; eds.: *Printing Technologies for Images, Gray Scale, and Color,*V1458(1991)

Dovgalenko, George Y.
; Onischenko, Yuri I.: Holographic devices using photo-induced effect in nondestructive testing techniques,V1559,479-486(1991)
; Yeskov-Soskovetz, Vladimir M.; Cherkashin, G. V.: Industrial applications of self-diffraction phenomena in holography on photorefractive crystals,V1508,110-115(1991)

Dovichi, Norman J. See Chen, Da Yong: V1435,161-167(1991)

Dovzansky, S. I. See Tuchin, Valery V.: V1422,85-96(1991)

Dowell, Floyd E.
; Powell, J. H.: Damage detection in peanut grade samples using chromaticity and luminance,V1379,136-140(1991)

Dowling, Eric M.
; DeGroat, Ronald D.: Recursive total-least-squares adaptive filtering,V1565,35-46(1991)
See DeGroat, Ronald D.: V1566,376-387(1991)

Dowling, Jerome M. See Betts, Timothy C.: V1479,120-126(1991)

Downey, George A.
; Fountain, H. W.; Riding, Thomas J.; Eggleston, James; Hopkins, Michael; Adams, Billy: Sled tracking system,V1482,40-47(1991)

Downey, Stephen W.
; Emerson, A. B.: Depth profiling resonance ionization mass spectrometry of electronic materials,V1435,19-25(1991)

Downie, John D.
; Reid, Max B.; Hine, Butler P.: Correction of magneto-optic device phase errors in optical correlators through filter design modifications,V1564,308-319(1991)
See Hine, Butler P.: V1564,416-426(1991)

Downing, J. See Burland, Donald M.: V1560,111-119(1991)

Downs, G. S. See Anderson, James C.: V1479,78-92(1991)

Downs, Maralene M.
See Dickinson, Alex G.: V1389,503-514(1991)
See Kerbis, Esther: V1396,447-452(1991)

Doyle, Michael D.
: New method for identifying features of an image on a digital video display,V1380,86-95(1991)
; Rabin, Harold; Suri, Jasjit S.: Fractal analysis as a means for the quantification of intramandibular trabecular bone loss from dental radiographs,V1380,227-235(1991)

Dozier, Jeff See Wilson, Stan: V1491,117-124(1991)

Draayer, Bret F. See Carhart, Gary W.: V1564,348-362(1991)

Drabik, Timothy J.
: Devices for optoelectronic integrated systems,V1562,194-203(1991)

Dragovic, Ljubisa See Chavantes, Maria C.: V1428,13-22(1991)

Draidia, Nasser See Bremond, Georges E.: V1361,732-743(1991)

Drake, G. See Jain, Faquir C.: V1564,714-722(1991)

Drake, Marvin D. See Leib, Kenneth G.: V1483,140-154(1991)

Drake, R. M. See Westerman, Steven D.: V1492,263-271(1991)

Drangova, Maria See Fenster, Aaron: V1447,28-33(1991)

Draper, B. L. See Hickman, Kirt C.: V1464,245-257(1991)

Drascic, David
; Milgram, Paul: Positioning accuracy of a virtual stereographic pointer in a real stereoscopic video world,V1457,302-313(1991)

Drawert, Bruce M.
; Slocum, Robert E.: High-precision video frame grabber for computed tomography,V1396,566-567(1991)
See Silver, Michael D.: V1443,250-260(1991)

Drayson, S. R. See Wang, Jin-Xue: V1492,391-402(1991)

Drazek, Wieslaw See Badziak, Jan: V1397,81-84(1991)

Drees, Friedrich-Wilhelm
; Pekruhn, Wolfgang: Laptop page printer realized by thermal transfer technology,V1458,80-83(1991)

Dreger, Helge H. See Bauch, Lothar: V1466,510-519(1991)

Dreischuh, A. See Dinev, S.: V1403,427-430(1991)

Dreisewerd, Douglas W. See Mayer, Richard C.: V1415,248-258(1991)

Dreizen, Howard M. See Huang, Yunming G.: V1396,624-637(1991)

Dressel, Earl M. See Lawson, Christopher M.: V1592,73-83(1991)

Dressel, Martin See Jahn, Renate: V1424,23-32(1991)

Drevillon, Bernard
; Razeghi, Manijeh: In-situ investigation of the low-pressure MOCVD growth of III-V compounds using reflectance anisotropy measurements,V1361,200-212(1991)
See Acher, O.: V1361,1156-1163(1991)

Drexel, Peter G. See Rucinski, Andrzej: V1390,388-398(1991)

Drexler, James J.
See Eaton, Frank D.: V1487,84-90(1991)
See Waldie, Arthur H.: V1487,103-108(1991)

Dreyer, Donald R.
; Saikkonen, Stuart L.; Hanson, Thomas A.; Linchuck, Barry A.: Impact of relative humidity on mechanical test results for optical fiber,V1366,372-379(1991)
See Vethanayagam, Thirukumar K.: V1366,343-350(1991)

Dreyer, Keith J.
; Simko, Joseph; Held, A. C.: Quantitative analysis of cardiac imaging using expert systems,V1445,398-408(1991)

Driemeyer, D. See Zediker, Mark S.: V1418,309-315(1991)

Driessen, Johannes N.
; Biemond, Jan: Motion field estimation for complex scenes,V1605,511-521(1991)

Driggers, Ronald G.
: Discrete frequency-versus-radius reticle trackers,V1478,24-32(1991)

Drigo, A. V. See D'Anna, Emilia: V1503,256-268(1991)

Driver, Jesse W.
; Buckalew, Chris: Radiative tetrahedral lattices,V1459,109-116(1991)

Drivsholm, Christian See Perch-Nielsen, Thomas: V1467,169-179(1991)

Drizlikh, S.
; Zur, A.; Moser, Frank; Katzir, Abraham: Microwave warming of biological tissue and its control by IR fiber thermometry,V1420,53-62(1991)

Drnoian, V. E. See Arakelian, Sergei M.: V1402,175-191(1991)

Drobnik, Antoni
: Certain results of the interaction of laser light with various media,V1391,211-214(1991)
: Investigations of the phenomenon of light scattering on particles of atmospheric aerosols,V1391,204-210(1991)
; Pieszynski, Krzysztof: Accuracy of aerosol size measurements and the spectrum of applied laser light,V1391,378-381(1991)
; Rozniakowski, Kazimierz; Wojtatowicz, Tomasz W.: Some aspects of the application of the laser flash method for investigations on the thermal diffusivity of porous materials,V1391,361-369(1991)

Drobot, Igor L.
See Markhvida, Igor V.: V1508,128-134(1991)
See Rachkovsky, Leonid I.: V1507,450-457(1991)

Dror, Itai
; Kopeika, Norman S.: Overall atmospheric MTF and aerosol MTF cutoff,V1487,192-202(1991)
See Hadar, Ofer: V1533,61-74(1991)

Dror, Jacob
See Calderon, S.: V1420,108-115(1991)
See Gannot, Israel: V1442,156-161(1991)

Drozhbin, Yu. A.
; Trofimenko, Vladimir V.: Super-high-speed reflex-type moving image camera,V1358,454-456(1991)
; Trofimenko, Vladimir V.; Chernova, T. I.: Ultrafast optical-mechanical camera,V1358,451-453(1991)
; Zvorykin, V. D.; Polyansky, S. V.; Sychugov, G. V.; Trofimenko, Vladimir V.; Yarova, A. G.: Application of high-speed mirror chronograph 3CX-1 to plasma investigations in visible and medium infrared spectrum ranges,V1358,1029-1034(1991)
See Ushakov, Leonid S.: V1358,447-450(1991)

Druce, Robert L.
; Pocha, Michael D.; Griffin, Kenneth L.; O'Bannon, Jim: Subnanosecond, high-voltage photoconductive switching in GaAs,V1378,43-54(1991)

Drum, Bruce
: Detecting spatial and temporal dot patterns in noise,V1453,153-164(1991)

Drummond, Gordon D. See Duffy, Neil D.: V1380,46-52(1991)

Drummond, James R. See Fried, Alan: V1433,145-156(1991)

Drummond, John W.
; Mackay, Gervase I.; Schiff, Harold I.: Measurement of peroxyacetyl nitrate, NO2, and NOx by using a gas chromatograph with a luminol-based detector,V1433,242-252(1991)
; Topham, L. A.; Mackay, Gervase I.; Schiff, Harold I.: Use of chemiluminescence techniques in portable, lightweight, highly sensitive instruments for measuring NO2, NOx, and O3,V1433,224-231(1991)

Drummond, Oliver E.
ed.: *Signal and Data Processing of Small Targets 1991*,V1481(1991)
See Fridling, Barry E.: V1481,371-383(1991)
See Sperling, I.: V1481,2-11(1991)

Druy, Mark A.
; Glatkowski, Paul J.; Stevenson, William A.: Fiber optic remote Fourier transform infrared spectroscopy,V1584,48-52(1991)
; Glatkowski, Paul J.; Stevenson, William A.: In-situ characterization of resin chemistry with infrared transmitting optical fibers and infrared spectroscopy,V1437,66-74(1991)
; Glatkowski, Paul J.: Synthesis, characterization, and processing of organic nonlinear optical polymers,V1409,214-219(1991)

Druzhko, A. B. See Vsevolodov, N. N.: V1621,11-20(1991)

Dryden, David M. See Stepp, Larry M.: V1542,175-185(1991)

Drzaic, Paul S.
: Light budget and optimization strategies for display applications of dichroic nematic droplet/polymer films,V1455,255-263(1991)
; Efron, Uzi; eds.: *Liquid-Crystal Devices and Materials*,V1455(1991)

Du, Chongwu See Pan, Yingtain: V1572,477-482(1991)

Du, Fuli
See Liu, Ying: V1554A,610-612(1991)
See Ma, Shining: V1554A,645-648(1991)

Du, G.
See Shi, W.: V1519,138-141(1991)
See Yang, Y. J.: V1418,414-421(1991)

Du, Ji
See Jia, Youquan: V1554B,331-334(1991)
See Kang, Yilan: V1554B,514-522(1991)

Du, Jia F. See Chen, Kun J.: V1519,632-639(1991)

Du, Keming
; Franek, Joachim; Loosen, Peter; Zefferer, H.; Shen, Junquan: Optical quality of a combined aerodynamic window,V1397,639-643(1991)
See Franek, Joachim: V1397,791-795(1991)

Du, MingZe
; Yuan, JinShan; Jin, Yixin; Zhou, Tian Ming; Hong, Jiang; Hong, ChunRong; Zhang, BaoLin: Growth and properties of GaxIn1-xAs (x<O.47) on InP by MOCVD,V1361,699-705(1991)

Du, Shanyi
See Guo, Maolin: V1554A,310-312(1991)
See Guo, Maolin: V1554A,657-658(1991)

Du, Shu-qin See Zhang, Yue-qing: V1418,444-447(1991)

Duane, G. See Wolfe, William J.: V1382,240-254(1991)

DuanMu, Qingduo See Qiu, Guanming: V1513,396-407(1991)

Dubant, Olivier See Wang, Gang: V1468,11-15(1991)

Dubash, N. See Wengler, Michael J.: V1477,209-220(1991)

Dube, George
ed.: *Solid State Lasers II*,V1410(1991)

Dubey, V. K. See Subramanian, K. R.: V1364,190-201(1991)

Dubicki, Adam
See Badziak, Jan: V1397,81-84(1991)
See Badziak, Jan: V1391,117-126(1991)

Dubiel, Mary K.
; Eppinger, Hans E.; Gillespie, Richard E.; Guzik, Rudolph P.; Pearson, James E.; eds.: *Applications of Optical Engineering: Proceedings of OE/ Midwest '90*,V1396(1991)

Dubik, Boguslawa
; Zajac, Marek: Sphero-chromatic aberration correction of single holo-lens used as a spectral device,V1574,227-234(1991)

Dubin, Paul L. See Rigsbee, Daniel R.: V1430,203-215(1991)

Dubovoi, I. A.
; Chernikov, A. S.; Prokhorov, Alexander M.; Schelev, Mikhail Y.; Ushakov, V. K.: Multialkali photocathodes grown by molecular beam epitaxy technique,V1358,134-138(1991)

Dubowsky, Steven
; Holly, Krisztina J.; Murray, Annie L.; Wander, Joseph M.: Design optimization of moire interferometers for rapid 3-D manufacturing inspection,V1386,10-20(1991)

Dubrovsky, V.
; Grinevich, A. E.; Ossin, A. B.: Particle analysis in liquid flow by the registration of elastic light scattering in the condition of laser beam scanning,V1403,344-346(1991)

Ducharme, Stephen P. See Moerner, William E.: V1560,278-289(1991)

Duchet, Christian
; Flaaronning, Nils; Brot, Christian; Sarrabay, Laurence: New integrated optic TE/TM splitter made on LiNbO3 isotropic cut,V1372,72-81(1991)
See Moehlmann, Gustaaf R.: V1560,426-433(1991)

Duchmann, Olivier
; Planche, G.: How to meet intersatellite links mission requirements by an adequate optical terminal design,V1417,30-41(1991)
See Laurent, Bernard: V1417,2-12(1991)
See MacGregor, Andrew D.: V1417,374-380(1991)

Duckwiler, Gary R. See Close, Robert A.: V1444,196-203(1991)

Ducrocq, N. See Mens, Alain: V1358,719-731(1991)

Dudderar, Thomas D. See Gilbert, John A.: VIS08,146-159(1991)

Dudek, Gregory
: Shape metrics from curvature-scale space and curvature-tuned smoothing,V1570,75-85(1991)
See Wilkes, David R.: V1388,536-542(1991)

Dudgeon, Dan E. See Verly, Jacques G.: V1471,266-282(1991)

Dudley, Bruce W.
See Freeman, Peter W.: V1464,377-385(1991)
See Jones, Susan K.: V1464,546-553(1991)
See Russ, John C.: V1464,10-21(1991)

Duell, Kenneth A.
; Freeman, Mark O.: Localized feature selection to maximize discrimination,V1564,22-33(1991)

Dueser, H. See Rosenzweig, M.: V1362,876-887(1991)

Dufaux, Frederic
; Ebrahimi, Touradj; Geurtz, Alexander M.; Kunt, Murat: Coding of digital TV by motion-compensated Gabor decomposition,V1567,362-379(1991)
; Ebrahimi, Touradj; Kunt, Murat: Massively parallel implementation for real-time Gabor decomposition,V1606,851-864(1991)
See Ebrahimi, Touradj: V1605,2-15(1991)

Duffey, Jason N.
; Hudson, Tracy D.; Kirsch, James C.: Optical evaluation of the microchannel spatial light modulator,V1558,422-431(1991)

Duffy, Christopher J.
; Tatam, Ralph P.: Optical frequency shifter based on stimulated Brillouin scattering in birefringent optical fiber,V1511,155-165(1991)

Duffy, Neil D.
; Drummond, Gordon D.; McGowan, Steve; Dessesard, Pascal: Real-time optical scanning system for measurement of chest volume changes durng anesthesia,V1380,46-52(1991)
See Herd, James T.: V1387,194-201(1991)

Dufour, Marc L. See Samson, Marc: V1332,314-322(1991)

Dufresne, Daniel
See Alloncle, Anne P.: V1397,675-678(1991)
See Morvan, D.: V1397,689-692(1991)

Dugelay, Jean-Luc
: Estimation of three-dimensional motion in a 3-DTV image sequence,V1605,688-696(1991)

Duguay, Michel A. See Ouellette, Francois: V1365,25-32(1991)

Dugue, Michel See Koehler, Bertrand: V1489,177-188(1991)

Duin, Robert P. See Bart, Mischa: V1381,66-77(1991)

Dujovny, Manuel
See Chavantes, Maria C.: V1428,13-22(1991)
See Zamorano, Lucia J.: V1428,59-75(1991)
See Zamorano, Lucia J.: V1428,30-51(1991)

Dukhovniy, Anatoliy M. See Berezinskaya, Aleksandra M.: V1238,80-84(1991)

Dukovic, John O. See Krongelb, Sol: V1389,249-256(1991)

Dull, Gerald G.
See Birth, Gerald S.: V1379,10-15(1991)
See Jaenisch, Holger M.: V1379,162-167(1991)

Duloisy, E. See Boutinaud, P.: V1590,168-178(1991)

Dumay, Adrie C.
; Gerbrands, Jan J.; Geest, Rob J.; Verbruggen, Patricia E.; Reiber, J. H.: Automated labeling of coronary arterial tree segments in angiographic projection data,V1445,38-46(1991)

Dumbreck, Andrew A. See Scheiwiller, Peter M.: V1457,2-8(1991)

Dumitru, Mihaela A.
; Honciuc, Maria; Sterian, Livia: Nonlinear optical components with liquid crystals,V1500,339-348(1991)

Dumka, D. C. See Yadav, M. S.: V1362,811-819(1991)

Dummer, Richard S. See Gu, Zu-Han: V1558,368-378(1991)

Dumont, Michel L.
; Morichere, D.; Sekkat, Z.; Levy, Yves: Accurate measurement of thin-polymeric-films index variations: application to elasto-optic effect and to photochromism,V1559,127-138(1991)
See Morichere, D.: V1560,214-225(1991)

Dunbrack, Steven K.
; Muray, Andrew; Sauer, Charles; Lozes, Richard; Nistler, John L.; Arnold, William H.; Kyser, David F.; Minvielle, Anna M.; Preil, Moshe E.; Singh, Bhanwar; Templeton, Michael K.: Phase-shift mask technology: requirements for e-beam mask lithography,V1464,314-326(1991)

Duncan, Michael D.
; Mahon, R.; Tankersley, Lawrence L.; Reintjes, John F.: Imaging through a low-light-level Raman amplifier,V1409,127-134(1991)

Duncan, William A. See Acebal, Robert: V1397,191-196(1991)

Dundatscheck, Robert See Conley, Willard E.: V1466,53-66(1991)

Dunet, G. See Surget, Jean: V1358,65-72(1991)

Dunifer, G. L. See Zhai, H. R.: V1519,575-579(1991)

Dunin-Barkowski, W. L.
: Cerebellum as a neuronal machine: modern talking,V1621,250-258(1991)

Dunlop, C. N. See Doel, A. P.: V1542,319-326(1991)

Dunn, Bruce S. See Chen, William W.: V1535,199-208(1991)

Dunn, D. A. See Kochevar, Irene E.: V1403,756-763(1991)

Dunn, Dennis F.
; Higgins, William E.; Maida, Anthony; Wakeley, Joseph: Texture boundary classification using Gabor elementary functions,V1606,541-552(1991)

Dunn, Diana D.
; Bruce, James A.; Hibbs, Michael S.: Deep-UV photolithography linewidth variation from reflective substrates,V1463,8-15(1991)
See Wong, Alfred K.: V1463,315-323(1991)

Dunn, James See Tallents, Gregory J.: V1413,70-76(1991)

Dunn, Patrice M. See Dunn, S. T.: VCR37,98-112(1991)

Dunn, Richard B. See Rimmele, Thomas: V1542,186-193(1991)

Dunn, Robert C. See Simon, John D.: V1432,211-220(1991)

Dunn, S. T.
; Dunn, Patrice M.: Standards for electronic imaging for graphic arts systems,VCR37,98-112(1991)

Dunn, Stanley M.
See Keizer, Richard L.: V1406,88-97(1991)
See Shemlon, Stephen: V1381,470-481(1991)

Dunning, M. J. See Amendt, Peter: V1413,59-69(1991)

Dunphy, James R. See Giesler, Leslie E.: V1480,138-142(1991)

Duntley, Seibert Q. See Austin, Roswell W.: V1537,57-73(1991)

Duparre, Angela
; Kassam, Samer: Determination of thin-film roughness and volume structure parameters from light-scattering investigations,V1530,283-286(1991)

Duplain, Gaetan See Snell, Kevin J.: V1410,99-106(1991)

Dupre, Jean-Christophe
; Lagarde, Alexis: Automatic data processing of speckle fringe pattern,V1554A,766-771(1991)

Dupuis, Paul See Oliensis, John: V1570,116-128(1991)

Durack, Donald L.
: Blur identification and image restoration with the expectation-maximization algorithm,V1487,72-83(1991)

Duran, Michel
; Luquet, Philippe; Buisson, F.; Cousin, B.: Semiactive telescope for the French PRONAOS submillimetric mission,V1494,357-376(1991)

Durand, Louis-Gilles
See Grenier, Marie-Claude: V1450,154-169(1991)
See Hutt, Daniel L.: V1487,250-261(1991)

Durbin, Philip F. See Del Grande, Nancy K.: V1479,335-351(1991)

Durfee, C.
; Milchberg, Howard M.: Pump-probe investigation of picosecond laser-gas target interactions,V1413,78-80(1991)
See Lyubomirsky, I.: V1413,108-111(1991)

Durham, Mark E. See Willey, Ronald R.: V1354,501-505(1991)

Durieux, Alain See Krawczyk, Rodolphe: V1493,2-15(1991)

Durpaire, Jean-Pierre
; Ratier, A.; Dagras, C.: GLOB(MET)SAT: French proposals for monitoring global change and weather from the polar orbit,V1490,23-38(1991)
See Krawczyk, Rodolphe: V1493,2-15(1991)

Durrance, Samuel T. See Clampin, Mark: V1542,165-174(1991)

Durst, Richard A. See Choquette, Steven J.: V1368,258-263(1991)

Dusa, Mircea V.
; Jung, Christoph; Jung, Paul; Hogenkamp, Detlef; Roeth, Klaus-Dieter: Effect of operating points in submicron CD measurements,V1464,447-458(1991)
; Roeth, Klaus-Dieter; Jung, Christoph: Improving submicron CD measurements through optimum operating points,V1496,217-223(1991)
See Kasahara, Jack S.: V1463,492-503(1991)
See Nicolau, Dan V.: V1465,282-288(1991)
See Nicolau, Dan V.: V1468,345-351(1991)
See Nicolau, Dan V.: V1466,663-669(1991)

Dutta, Niloy K.
: Semiconductor lasers for coherent lightwave communcation,V1372,4-12(1991)
See Park, Yong K.: V1372,219-227(1991)

Dutta, Pulak
: Studies of structures and phase transitions of Langmuir monolayers using synchrotron radiation,V1550,134-139(1991)

Dutton, Robert W. See Morse, Jeffrey D.: V1378,55-59(1991)

Duval, A. See Giannini, Jean-Pierre: V1366,215-222(1991)

Duveneck, G.
; Ehrat, M.; Widmer, H. M.: Fiber optic evanescent wave biosensor,V1510,138-145(1991)

Dvalishvili, V. V.
See Kostin, Ivan K.: V1554A,418-425(1991)
See Savostjanov, V. N.: V1554A,380-386(1991)

Dvornikov, Alexander S. See Rentzepis, Peter M.: V1563,198-207(1991)

Dvornikov, S. S. See Jankiewicz, Zdzislaw: V1391,101-104(1991)

Dvornikov, Tatiana See Tromberg, Bruce J.: V1427,101-108(1991)

Dvoryankin, A. N.
: Continuous chemical lasers of visible region,V1397,145-152(1991)
; Kulagin, Yu. A.; Kudryavtcev, N. Y.: Influence of mixing on the characteristics of the oxygen-iodine laser,V1397,247-250(1991)
; Makarov, V. N.: Chemical generation of electronically excited nitrogen N2(A3 sigma +u) and lasers on electronic transitions,V1397,177-180(1991)

Dwir, Benjamin
See James, Jonathan H.: V1394,45-61(1991)
See Pavuna, Davor: V1362,102-116(1991)

Dwivedi, Pramod See Pausch, Randy: V1457,198-208(1991)

Dwyer, John L. See Trautwein, Charles M.: V1492,338-338(1991)

Dye, James E. See Walega, James G.: V1433,232-241(1991)

Dye, Robert C.
See Muenchausen, Ross E.: V1394,221-229(1991)
See Wu, Xin D.: V1477,8-14(1991)

Dyer, Charles R. See Seales, W. B.: V1383,47-58(1991)

Dyer, R. B.
; Peterson, Kristen A.; Stoutland, Page O.; Einarsdottir, Oloef; Woodruff, William H.: Time-resolved infrared studies of the dynamics of ligand binding to cytochrome c oxidase,V1432,197-204(1991)
See Woodruff, William H.: V1432,205-210(1991)

Dyer, Stephen M. See Goodenough, David J.: V1444,87-99(1991)

Dyjak, Charles P.
; Harrison, David C.: Space-based visible surveillance experiment,V1479,42-56(1991)

Dymnikov, Alexander A. See Lunter, Sergei G.: V1513,349-359(1991)

Dyott, Richard B.
: Emerging technology in fiber optic sensors,V1396,709-717(1991)
; Huang, Yung Y.; Jannush, D. A.; Morrison, Steve A.: Low-drift fiber-optic gyro for earth-rate applications,V1482,439-443(1991)

Dyukova, T. V.
See Mikaelian, Andrei L.: V1621,148-157(1991)
See Vsevolodov, N. N.: V1621,11-20(1991)

Dzenis, J. See Purinsh, Juris: V1525,289-308(1991)

Dzielinski, Andrzej See Swiniarski, Roman W.: V1451,234-241(1991)

Dziura, T. G.
See Wang, S. C.: V1563,27-33(1991)
See Yang, Y. J.: V1418,414-421(1991)

Dziurla, Barbara See Rucinski, Andrzej: V1390,388-398(1991)

Dzwigalski, Zygmunt See Badziak, Jan: V1391,250-253(1991)

Eades, Wendell D. See Harwit, Alex: V1541,38-47(1991)

Ealey, Mark A.
ed.: *Active and Adaptive Optical Systems*,V1542(1991)

Earle, Colin W. See Radspinner, David A.: V1439,2-14(1991)

Earmme, Youn-Young See Lee, Hyouk: V1393,404-410(1991)

Earnshaw, J. C.
; Winch, P. J.: Transitions in model membranes,V1403,316-325(1991)

Eason, Richard O.
; Kamata, Seiichiro: Least-squares-based data fusion strategies and robotic applications,V1383,566-573(1991)

Easson, William J. See She, K.: V1572,581-587(1991)

Eastman, Jay M.
; Quinn, Anna M.: Diffraction analysis of beams for barcode scanning,V1384,185-194(1991)
See Quinn, Anna M.: V1384,138-144(1991)

Eastman, Willard L. See Games, Richard A.: V1566,323-328(1991)

Eastwood, Michael L.
; Sarture, Charles M.; Chrien, Thomas G.; Green, Robert O.; Porter, Wallace M.: Current instrument status of the airborne visible/infrared imaging spectrometer (AVIRIS),V1540,164-175(1991)

Eaton, Alexander M.
; Bass, Lawrence S.; Libutti, Steven K.; Schubert, Herman D.; Treat, Michael R.: Sutureless cataract incision closure using laser-activated tissue glues,V1423,52-57(1991)
See Bass, Lawrence S.: V1421,164-168(1991)
See Oz, Mehmet C.: V1422,147-150(1991)

Eaton, Frank D.
; Peterson, William A.; Hines, John R.; Waldie, Arthur H.; Drexler, James J.; Qualtrough, John A.; Soules, David B.: Interaction of exposure time and system noise with angle-of-arrival measurements,V1487,84-90(1991)
See Waldie, Arthur H.: V1487,103-108(1991)

Eaves, Lawrence
  ; Hayes, David; Leadbeater, M. L.; Simmonds, P. E.; Skolnick, Maurice S.: Magnetocapacitance and photoluminescence spectroscopy studies of charge storage, bistability, and energy relaxation effects in resonant tunneling devices,V1362,520-533(1991)

Ebben, Thomas H.
  See Delamere, W. A.: V1479,31-40(1991)
  See Delamere, W. A.: V1447,204-213(1991)
  See Delamere, W. A.: V1447,288-297(1991); 1479,21-30(1991)

Ebbeni, Jean P.
  : High-temperature optical sensor for displacement measurement,V1504,268-272(1991)
  See Depoortere, Marc: V1504,264-267(1991)

Ebbets, Dennis C. See Troeltzsch, John R.: V1494,9-15(1991)

Eberhardt, F. See Ploetz, F.: V1510,224-230(1991)

Eberl, K. See Menczigar, Ulrich: V1361,282-292(1991)

Eberlein, Susan
  ; Yates, Gigi; Majani, Eric: Hierarchical multisensor analysis for robotic exploration,V1388,578-586(1991)

Ebihara, Teruo See Kato, Naoki: V1455,190-205(1991)

Ebina, K. See Yokozawa, T.: V1397,513-517(1991)

Ebrahimi, Touradj
  ; Dufaux, Frederic; Moccagatta, Iole; Campbell, T. G.; Kunt, Murat: Digital video codec for medium bitrate transmission,V1605,2-15(1991)
  See Dufaux, Frederic: V1567,362-379(1991)
  See Dufaux, Frederic: V1606,851-864(1991)

Eccles, N. J. See Herd, James T.: V1387,194-201(1991)

Eckardt, Robert C.
  ; Byer, Robert L.: Measurement of nonlinear optical coefficients by phase-matched harmonic generation,V1561,119-127(1991)

Ecke, Wolfgang
  ; Schroeter, Siegmund; Schwotzer, Guenter; Willsch, Reinhardt: All-fiber closed-loop gyroscope with self-calibration,V1511,57-66(1991)
  See Rasch, Andreas: V1511,149-154(1991)

Eckenbach, W. See Hillen, Walter: V1443,120-131(1991)

Eckes, Charlotte
  ; Pawlowski, Georg; Przybilla, Klaus J.; Meier, Winfried; Madore, Michel; Dammel, Ralph: Process latitude for the chemical amplification resists AZ PF514 and AZ PN114,V1466,394-407(1991)
  See Przybilla, Klaus J.: V1466,174-187(1991)

Eckhardt, Andrew J. See McEachern, Robert: V1470,226-232(1991)

Eda, Hideo See Oda, Ichiro: V1431,284-293(1991)

Edelhaeuser, Rainer See Brandenburg, Albrecht: V1510,148-159(1991)

Edelstein, Daniel C.
  : Three-dimensional capacitance modeling of advanced multilayer interconnection technologies,V1389,352-360(1991)

Edelstein, Jerry
  : Deviated-plane varied-line-space grating spectrograph,V1545,145-148(1991)

Eder, David C.
  See Amendt, Peter: V1413,59-69(1991)
  See Wilks, S. C.: V1413,131-137(1991)

Edgar, G. See Bergman, Larry A.: V1364,14-21(1991)

Edgar, Thomas F. See Butler, Stephanie W.: V1392,361-372(1991)

Edighoffer, J. A. See Bailey, Vernon L.: V1407,400-406(1991)

Edirisinghe, Chandima D. See Samarabandu, J. K.: V1450,296-322(1991)

Ediriweera, Sanath R.
  ; Kvasnik, Frank: Optical fiber radiation damage measurements,V1399,64-75(1991)
  ; Kvasnik, Frank: Spectroscopical studies of the ionizing-radiation-induced damage in optical fibers,V1504,110-117(1991)

Edmonds, Curtis M. See Bristow, Michael P.: V1491,68-74(1991)

Edwards, B. See Key, Michael H.: V1397,9-17(1991)

Edwards, Charles A.
  ; Galloway, Robert L.; Thomas, Judith G.; Schreiner, Steven; Maciunas, Robert J.: Simultaneous graphics and multislice raster image display for interactive image-guided surgery,V1444,38-46(1991)
  See Galloway, Robert L.: V1444,9-18(1991)
  See Thomas, Judith G.: V1444,379-388(1991)

Edwards, David F. See Montesanti, Richard C.: V1343,558-565(1991)

Edwards, David H. See Smith, Roy E.: V1425,116-121(1991)

Edwards, Gary See Kozlowski, Mark R.: V1561,59-69(1991)

Edwards, Gary J. See Scripsick, Michael P.: V1561,93-103(1991)

Edwards, Glenn S. See Tolk, Norman H.: V1552,7-13(1991)

Edwards, Henry O. See Dakin, John P.: V1510,160-169(1991)

Edwards, R. See Wu, Xin D.: V1477,8-14(1991)

Eeckman, Frank H.
  See Baird, Bill: V1469,12-23(1991)
  See Searfus, Robert M.: V1473,95-101(1991)

Eem, J. K. See Kehtarnavaz, Nasser: V1468,467-478(1991)

Eenink, A. H. See Damstra, Geert C.: V1358,644-651(1991)

Efimkov, V. F. See Zaitseva, N. P.: V1402,223-230(1991)

Efimov, Oleg M.
  See Bosyi, O. N.: V1440,57-62(1991)
  See Glebov, Leonid B.: V1513,274-282(1991)
  See Glebov, Leonid B.: V1440,50-56(1991)

Efremov, R. G.
  ; Feofanov, A. V.; Nabiev, I. R.: UV resonance Raman spectroscopic study of Trp residues in a hydrophobic environment,V1403,161-163(1991)
  See Nabiev, I. R.: V1403,85-92(1991)

Efron, Uzi
  ; Byles, W. R.; Goodwin, Norman W.; Forber, Richard A.; Sayyah, Keyvan; Wu, Chiung S.; Welkowsky, Murray S.: Charge-coupled-device-addressed liquid-crystal light valve: an update,V1455,237-247(1991)
  See Drzaic, Paul S.: V1455(1991)
  See Sayyah, Keyvan: V1455,249-254(1991)

Efstratiadis, Serafim N.
  ; Huang, Yunming G.; Xiong, Z.; Galatsanos, Nikolas P.; Katsaggelos, Aggelos K.: Motion-compensated priority discrete cosine transform coding of image sequences,V1605,16-25(1991)
  See Katsaggelos, Aggelos K.: V1606,716-727(1991)
  See Kwak, J. Y.: V1396,32-44(1991)

Efthimiopoulos, Tom
  : Efficient population of low-vibrational-number electronic states of excimer molecules: the argon dimer,V1503,430-437(1991)

Egalon, Claudio O.
  ; Rogowski, Robert S.; Tai, Alan C.: Excitation efficiency of an optical fiber core source,V1489,9-16(1991)
  ; Rogowski, Robert S.: Model of an axially strained weakly guiding optical fiber modal pattern,V1588,241-254(1991)
  ; Rogowski, Robert S.: Model of a thin-film optical fiber fluorosensor,V1368,134-149(1991)

Egami, Chikara See Nakagawa, Kazuo: V1332,267-273(1991)

Egashira, Makoto
  ; Shimizu, Yasuhiro: Sensitivity and selectivity enhancement in semiconductor gas sensors,V1519,467-476(1991)
  See Feng, Chang D.: V1519,8-13(1991)

Egashira, Mitsuru See Kishimoto, Satoshi: V1554B,174-180(1991)

Eger, David
  ; Zigelman, Alex: Anodic oxides on HgZnTe,V1484,48-54(1991)

Egert, Charles M.
  : Material characterization of beryllium mirrors exhibiting anomalous scatter,V1530,162-170(1991)
  ; Basford, J. A.: Vacuum outgassing from diffuse-absorptive baffle materials,V1330,178-185(1991)
  ; Schmoyer, D. D.; Nordin, C. W.; Berry, A.: Investigation of process sensitivity for electron-beam evaporation of beryllium,V1485,64-77(1991)
  See Seals, Roland D.: V1330,164-177(1991)

Eggleston, James See Downey, George A.: V1482,40-47(1991)

Egorov, S. Y.
  ; Krasnovsky, A. A.: Laser-induced luminescence of singlet molecular oxygen: generation by drugs and pigments of biological importance,V1403,611-621(1991)

Egorov, V. V.
  ; Lazarchuk, V. P.; Murugov, V. M.; Sheremetyev, Yu. N.: Application of IR photorecorders based on ionization chambers for fast processes investigation,V1358,984-991(1991)

Egozcue, Juan J.
  See Fimia-Gil, Antonio: V1507,153-157(1991)
  See Navarro, Maria T.: V1507,142-148(1991)

Eguchi, Koichi See Setoguchi, Toshihiko: V1519,74-79(1991)

Ehbets, Peter See Herzig, Hans-Peter: V1507,247-255(1991)

Ehlers, Sandy L. See Morgan, Robert E.: V1588,189-197(1991)

Ehlert, John C. See Cassarly, William J.: V1417,110-121(1991)

Ehrat, M. See Duveneck, G.: V1510,138-145(1991)

Ehrenberg, Benjamin
; Gross, Eitan; Lavi, Adina; Johnson, Fred M.; Malik, Zvi: Spectroscopic studies of second-generation sensitizers and their photochemical reactions in liposomes and cells,V1426,244-251(1991)
; Pevzner, Eliyahu: Spectroscopic properties of the potentiometric probe merocyanine-540 in solutions and liposomes,V1432,154-163(1991)
See Sessler, Jonathan L.: V1426,318-329(1991)

Ehrfeld, Wolfgang See Rogner, Arnd: V1506,80-91(1991)

Ehrlich, Michael J. See Steckenrider, John S.: V1554B,106-112(1991)

Ehrlich, Robert B. See Henesian, Mark A.: V1415,90-103(1991)

Ehrmann, Jonathan S.
: Optics for vector scanning,V1454,245-256(1991)

Ehrt, Doris See Possner, Torsten: V1513,378-385(1991)

Eichel, Paul H.
See Calloway, Terry M.: V1566,353-364(1991)
See Wahl, Daniel E.: V1567,32-40(1991)

Eichelberger, Christopher L.
; Blair, Steven M.; Khorana, Brij M.: Machine vision system for ore sizing,V1396,678-686(1991)

Eichen, Elliot G.
; Powazinik, William; Meland, Edmund; Bryant, R.; Rideout, William C.; Schlafer, John; Lauer, Robert B.: Integrated optical preamplifier technology for optical signal processing and optical communication systems,V1474,260-267(1991)

Eichenlaub, Jesse B.
: Autostereoscopic (3-D without glasses) display for personal computer applications,V1398,48-51(1991)
: Progress in autostereoscopic display technology at Dimension Technologies Inc.,V1457,290-301(1991)

Eichner, L. See White, Donald L.: V1343,204-213(1991)

Eick, Stephen G. See Becker, Richard A.: V1459,150-154(1991)

Eickhoff, Thomas See Morley, Stefan: V1361,213-222(1991)

Eid, El-Sayed I. See Kemeny, Sabrina E.: V1447,243-250(1991)

Eidelman, Shmuel
; Grossmann, William; Friedman, Aharon: Nonlinear signal processing using integration of fluid dynamics equations,V1567,439-450(1991)

Eidmann, K.
; Lanig, E. M.; Schwanda, W.; Sigel, Richard; Tsakiris, George D.: Opacity studies with laser-produced plasma as an x-ray source,V1502,320-330(1991)

Eigler, Lynne C. See Hudyma, Russell M.: V1354,523-532(1991)

Eigler, Neal See Whiting, James S.: V1453,165-175(1991)

Eilbert, James L. See Lim, William Y.: V1388,464-475(1991)

Einarsdottir, Oloef
See Dyer, R. B.: V1432,197-204(1991)
See Woodruff, William H.: V1432,205-210(1991)

Einav, Shmuel See Kwiat, Doron: V1443,2-28(1991)

Einav, Yehezkel See Biderman, Shlomo: V1535,27-34(1991)

Eisenbach, Shlomo
; Bodenheimer, Joseph S.: Research on a curved optical fiber,V1442,242-251(1991)

Eisenberg, Shai
; Menache, Ram: IR objective with internal scan mirror,V1442,133-138(1991)

Eisenhammer, T. See Lazarov, M.: V1536,183-191(1991)

Eisfeld, Fritz
: Investigation of diesel injection jets using high-speed photography and speed holography,V1358,660-671(1991)

Ekberg, Mats
; Larsson, Michael; Bolle, Aldo; Hard, Sverker: Nd:YAG laser machining with multilevel resist kinoforms,V1527,236-239(1991)

Ekelmans, G. B.
; van Goor, Frederik A.; Trentelman, M.; Witteman, Wilhelmus J.: Magnetic pulse compression in the prepulse circuit for a 1 kW, 1kHz XeCl excimer laser,V1397,569-572(1991)
See Trentelman, M.: V1397,115-118(1991)
See Witteman, Wilhelmus J.: V1397,37-45(1991)

Ekenstedt, M. J. See Andersson, Thorvald G.: V1361,434-442(1991)

Ekimov, A. I.
; Kudryavtsev, I. A.; Chepick, D. I.; Shumilov, S. K.: Semiconductor-doped glasses: nonlinear and electro-optical properties,V1513,123-129(1991)

El-Giar, Emad M. See Badawy, Waheed A.: V1536,277-288(1991)

El-Hakim, Sabry F.
ed.: *Industrial Vision Metrology*,V1526(1991)
; Westmore, David B.: Performance evaluation of a vision dimension metrology system,V1526,56-67(1991)
See Pizzi, Nicolino J.: V1468,296-304(1991)

El-Konyaly, Sayed H.
; Enab, Yehia M.; Soltan, Hesham: Recognition and tracking of moving objects,V1388,317-328(1991)

El Alaoui, Mohsine
; Magnin, Isabelle E.; Amiel, Michel: Simulated annealing image reconstruction for an x-ray coded source tomograph,V1569,80-87(1991)

Elby, Stuart D. See Gidron, Rafael: V1579,40-48(1991)

Elder, Melanie L.
; Jancaitis, Kenneth S.; Milam, David; Campbell, John H.: Optical characterization of damage resistant kilolayer rugate filters,V1441,237-246(1991)

Eldering, Charles A. See Yankelevich, Diego: V1560,406-415(1991)

Eldredge, Sandra L.
See Huang, H. K.: V1443,198-202(1991)
See McNitt-Gray, Michael F.: V1445,468-478(1991)

Elenkrig, Boris B. See Semenov, Alexandr T.: V1584,348-352(1991)

Elerding, George T.
; Thunen, John G.; Woody, Loren M.: Wedge imaging spectrometer: application to drug and pollution law enforcement,V1479,380-392(1991)

Elfadel, Ibrahim M.
; Picard, Rosalind W.: Miscibility matrices explain the behavior of gray-scale textures generated by Gibbs random fields,V1381,524-535(1991)
; Yuille, Alan L.: Mean-field theory for grayscale texture synthesis using Gibbs random fields,V1569,248-259(1991)

Elfersi, U. See Azran, A.: V1442,54-57(1991)

Elias, Richard C. See Crano, John C.: V1529,124-131(1991)

Elinson, Matvey I.
See Balkarey, Yu. I.: V1621,238-249(1991)
See Gulyaev, Yuri V.: V1621,84-92(1991)

Elior, A. See Aharon, O.: V1397,251-255(1991)

Elizur, Eran
; Friesem, Asher A.: Pattern recognition with incoherent light and rotation invariance,V1442,230-234(1991)

Elkin, Nickolai N. See Apollonova, O. V.: V1501,108-119(1991)

Elksninsh, N. See Purinsh, Juris: V1525,289-308(1991)

Eller, E. D.
; LaMont, Douglas V.; Rodden, Jack J.: Space-borne beam pointing,V1415,2-12(1991)

Ellett, Scott A.
; Walkup, John F.; Krile, Thomas F.: Error codes applied to optical algebraic processors,V1564,634-643(1991)

Ellingsrud, Svein
; Lokberg, Ole J.; Pedersen, Hans M.: Recording and analysis of (high-frequency) sinusoidal vibrations using computerized TV-holography,V1399,30-41(1991)
See Lokberg, Ole J.: V1508,153-160(1991)
See Lokberg, Ole J.: V1332,142-150(1991)

Ellinson, Reinold See Bundy, Scott: V1475,319-329(1991)

Elliott, C. J.
: Small-signal gain for parabolic profile beams in free-electron lasers,V1552,175-181(1991)
See Newnam, Brian E.: V1552,154-174(1991)
See Thode, Lester E.: V1552,87-106(1991)

Elliott, Charles T. See Ashley, Timothy: V1361,238-244(1991)

Elliott, Daniel S.
: Overview of experimental investigations of laser bandwidth effects in nonlinear optics,V1376,22-33(1991)

Elliott, David G.
; St. Clair Dinger, Ann: SIRTF stray light analysis,V1540,63-67(1991)

Elliott, David J.
; Sengupta, Uday K.: Excimer lasers for deep-UV lithography,V1377,6-17(1991)

Elliott, G. A. See Jackson, John E.: V1533,75-86(1991)

Elliott, Tom
See Acton, Loren W.: V1447,123-139(1991)
See Janesick, James R.: V1447,87-108(1991)

Ellis, Albert R. See Selser, James C.: V1430,85-88(1991)

Ellis, Kenneth K. See McGlynn, John D.: V1486,141-150(1991)

Elman, Boris S.
; Chirravuri, Jagannath; Choudhury, A. N.; Silletti, Andrew; Negri, A. J.; Powers, J.: InGaAs/GaAs interdigitated metal-semiconductor-metal (IMSM) photodetectors operational at 1.3 um grown by molecular beam epitaxy,V1362,610-616(1991)
; Koteles, Emil S.; Melman, Paul; Rothman, Mark A.: Low-substrate temperature molecular beam epitaxy growth and thermal stability of strained InGaAs/GaAs single-quantum-wells,V1361,362-372(1991)

Elmanasreh, Ahmed See Abushagur, Mustafa A.: V1554B,298-302(1991)

Elrod, Parker D. See Lowry, Heard S.: V1454,453-464(1991)

Elsasser, Wolfgang E. See Phelan, Paul J.: V1362,623-630(1991)

Elsner, Frank See Bergquist, James C.: V1435,82-85(1991)

Elsner, Ronald F.
See Holley, Jeff: V1343,500-511(1991)
See Kaaret, Philip E.: V1548,106-117(1991)
See Weisskopf, Martin C.: V1343,457-468(1991)

Elson, John M.
: Anomalous scattering from optical surfaces with roughness and permittivity perturbations,V1530,196-207(1991)

Elstner, Christian
; Lichtenauer, Gert; Skarus, Waldemar: ASTRO 1M: a new system for attitude determination in space,V1478,150-159(1991)

Elston, S. J. See Bryan-Brown, G. P.: V1545,167-178(1991)

Elta, Michael E.
See Cheng, Jie: V1392,373-384(1991)
See Demos, Alexandros T.: V1392,291-297(1991)
See Grimard, Dennis S.: V1392,535-542(1991)

Embs, F. W. See Neher, Dieter: V1560,335-343(1991)

Emelin, S. E. See Ermolaeva, Tatiana T.: V1361,1164-1165(1991)

Emeny, M. T. See Gupta, Rita P.: V1362,798-803(1991)

Emerson, A. B. See Downey, Stephen W.: V1435,19-25(1991)

Emery, Y. See Poscio, Patrick: V1510,112-117(1991)

Emmons, David J. See Roberts, Harold A.: V1363,62-69(1991)

Emmony, David C.
; Clark, Stuart E.; Kerr, Noel C.; Omar, Basil A.: Laser interaction with solids,V1397,651-659(1991)
See Ward, Barry: V1358,1035-1045(1991)
See Ward, Barry: V1358,1228-1236(1991)

Emo, Stephen M.
; Kinney, Terrance R.; Wong, Ka-Kha: Integrated optic components for advanced turbine engine control systems,V1374,266-276(1991)

Emre, Erol See Chung, Yi-Nung: V1481,306-314(1991)

Emslie, A. G.
See Campbell, Jonathan W.: V1549,155-179(1991)
See Campbell, Jonathan W.: V1343,359-375(1991)

Enab, Yehia M.
: Neural edge detector,V1382,292-303(1991)
See El-Konyaly, Sayed H.: V1388,317-328(1991)

Encinas Sanz, F. See Dominguez Ferrari, E.: V1397,725-727(1991)

Endelman, Lincoln L.
: Effect of standards on new equipment design by new international standards and industry restraints,V1346,90-92(1991)
: Hubble Space Telescope: mission, history, and systems,V1358,422-441(1991)
: Round table discussion on standards,V1346,93-94(1991)

Endemann, Martin J. See Lange, Robert: V1492,24-37(1991)

Enders, S. See Waidhauser, Erich: V1428,75-83(1991)

Endo, Masamori
; Arai, S.; Yamashita, T.; Uchiyama, Taro: 30-Torr pulsed singlet oxygen generator,V1397,267-270(1991)

Endriz, John G. See Willing, Steven L.: V1418,358-362(1991)

Eng, Lars H. See Cotton, Therese M.: V1403,93-101(1991)

Eng, Peter J.
; Stephens, Peter; Tse, Teddy: Kinetics of surface ordering: Pb on Ni(001),V1550,110-121(1991)

Eng, Richard S.
; Kachelmyer, Alan L.; Harris, Neville W.: Thermal lensing and frequency chirp in a heated CdTe modulator crystal and its effects on laser radar performance,V1416,70-85(1991)
See Harris, Neville W.: V1416,59-69(1991)

Engel, Herbert
; Doeldissen, Walter: Electron-beam lithography for the microfabrication of OEICs,V1506,60-64(1991)

Engel, J. See Ermakov, Yu. A.: V1403,338-339(1991)

Engel, Michael Y.
; Balfour, L. S.: Quantitative evaluation of errors in remote measurements using a thermal imager,V1442,298-307(1991)

Engelhardt, Manfred
: Evaluation of low-pressure silicon dry-etch processes with regard to low-substrate degradation,V1392,38-46(1991)
: Single-crystal silicon trench etching for fabrication of highly integrated circuits,V1392,210-221(1991)

Engelhardt, R.
See Flemming, G.: V1421,146-152(1991)
See Scheu, M.: V1425,63-69(1991)
See Scheu, M.: V1421,100-107(1991)

Engelhardt, Ralph See Rando, Nicola: V1549,340-356(1991)

Engelstad, Roxann L.
See Laird, Daniel L.: V1465,134-144(1991)
See Wells, Gregory M.: V1465,124-133(1991)

Engemann, Detlef
; Faymonville, Rudolf; Felten, Rainer; Frenzl, Otto: Infrared detector arrays with integrating cryogenic read-out electronics,V1362,710-720(1991)

Engineer, Carl P.
: Fiber in the loop: an evolution in services and systems,V1363,19-29(1991)

English, A. M. See Smulevich, G.: V1403,440-447(1991)

English, R. Edward
; Halpin, John M.; House, F. A.; Paris, Robert D.: Optical design of a high-power fiber-optic coupler,V1527,174-179(1991)
See Feldman, Mark: V1542,490-501(1991)

Englund, John C.
: Chaotic behavior of a Raman ring laser,V1497,292-299(1991)
; Bowden, Charles M.: Statistics of spontaneously generated Raman solitons,V1497,218-221(1991)

Engman, Per See Johansson, Micael: V1538,180-188(1991)

Engrand, D. See Sansonetti, Pierre: V1588,198-209(1991)

Engstrom, Herbert L.
; Stokowski, Stanley E.: Measuring films on and below polycrystalline silicon using reflectometry,V1464,566-573(1991)

Enguehard, Francis See Balageas, Daniel L.: V1467,278-289(1991)

Enguehard, S.
; Hatfield, Brian: Functional reconstruction predictions of uplink whole beam Strehl ratios in the presence of thermal blooming,V1408,186-191(1991)
; Hatfield, Brian: Path integral approach to thermal blooming,V1408,178-185(1991)
; Hatfield, Brian: Review of the physics of small-scale thermal blooming in uplink propagation,V1415,128-137(1991)

Enokizono, H. See Yamakoshi, H.: V1397,119-122(1991)

Enomoto, H. See Honda, Toshio: V1461,156-166(1991)

Enomoto, Hajime See Kamoshida, Minoru: V1606,951-960(1991)

Enright, G. D. See Labaune, C.: V1413,138-143(1991)

Ensminger, Richard L. See Austin, Roswell W.: V1537,57-73(1991)

Eouzan, J. Y. See Mens, Alain: V1358,719-731(1991)

Epifanov, Alexandre S. See Soileau, M. J.: V1441,10-15(1991)

Eppeldauer, George
: Temperature-monitored/controlled silicon photodiodes for standardization,V1479,71-77(1991)

Eppinger, Hans E. See Dubiel, Mary K.: V1396(1991)

Eppley, Kenneth
; Ko, Kwok: Design of a high-power cross-field amplifier at X-band with an internally coupled waveguide,V1407,249-259(1991)

Epstein, Jonathan S. See Perry, Kenneth E.: V1554A,209-227(1991)

Epstein, Max See Pan, Anpei: V1396,99-106(1991)

Eray, A. See Oktu, Ozcau: V1361,812-818(1991)

Eray, Mete See Hung, Hsien-Sen: V1550,34-45(1991)

Erba, B. See Bacis, Roger: V1397,173-176(1991)

Ercegovac, Milos D.
: On-line arithmetic for recurrence problems,V1566,263-274(1991)

Erdekian, Vahram V.
; Trombetta, Steven P.: Display systems for medical imaging,V1444,151-158(1991)

Erdem, A. T. See Ozkan, Mehmet K.: V1606,743-754(1991)

Erdmann, John H. See Margerum, J. D.: V1455,27-38(1991)

Ereaut, Peter R.
: Measurement of particles and drops in combusting flows,V1554B,556-565(1991)

Eriksson, L.-E. See Stenow, Eric: V1420,29-33(1991)

Erlich, Simha
; Yao, Kung: Architecture for adaptive eigenstructure decomposition based on systolic QRD,V1565,47-56(1991)

Ermakov, Boris A. See Motenko, Boris: V1399,78-81(1991)

Ermakov, Yu. A.
; Cherny, V. V.: Determination of association parameters for the adsorption of mono- and divalent cations at the lipid membrane surface,V1403,278-279(1991)
; Engel, J.; Donath, E.: Photon correlation spectroscopy of chromaffin granules lysis,V1403,338-339(1991)

Ermer, Susan P.
; Lovejoy, Steven M.; Leung, Doris; Spitzer, Ronnie; Hansen, Glenn A.; Stone, Richard E.: Synthesis and nonlinear optical activity of cumulenes,V1560,120-129(1991)
See Lipscomb, George F.: V1560,388-399(1991)
See Wu, Jeong W.: V1560,196-205(1991)

Ermolaev, Vladimir S. See Altshuler, Grigori B.: V1429,95-104(1991)

Ermolaeva, Tatiana T.
: Loss of heat relaxation in the superconducting and nonsuperconducting ceramics near Tc,V1361,1166-1167(1991)
; Emelin, S. E.: Loss of heat relaxation in superconducting materials near Tc,V1361,1164-1165(1991)

Ernst, Charles H.
: Use of FEM modes in time-domain servo simulations,V1499,129-135(1991)

Erofeev, Andrey V. See Altshuler, Grigori B.: V1427,141-150(1991)

Erokhovets, Valerii K.
; Larchenko, Yu. V.; Leonov, A. M.; Tkachenko, Vadim V.: Laser image recording on a metal/polymer medium,V1621,227-236(1991)

Ershov, I. A. See Bordo, V. G.: V1440,364-369(1991)

Ershova, N. I. See D'yakonov, G. I.: V1424,81-86(1991)

Erteza, Ahmed
; Schneeberger, Timothy J.: Spatial frequency selective error sensing for space-based, wide field-of-view, multiple-aperture imaging systems,V1527,182-187(1991)

Ervin, M. H. See Winograd, Nicholas: V1435,2-11(1991)

Erwin, J. K. See Milster, Tom D.: V1499,286-292(1991)

Erwin, K. E. See Milster, Tom D.: V1499,348-353(1991)

Erzgraeber, Hartmut H. See Bauch, Lothar: V1466,510-519(1991)

Esarey, Eric
; Sprangle, Phillip; Ting, Antonio C.: Nonlinear wakefield generation and relativistic optical guiding of intense laser pulses in plasmas,V1407,407-417(1991)
See Marable, William P.: V1552,80-86(1991)
See Sprangle, Phillip: V1552,147-153(1991)

Eschkoetter, Peter See Ibel, Konrad: V1559,393-402(1991)

Escobar Rojo, Priscilla
See De Maria, Letizia: V1366,304-312(1991)
See Sansonetti, Pierre: V1588,198-209(1991)

Esener, Sadik C.
; Kiamilev, Fouad E.; Krishnamoorthy, Ashok V.; Marchand, Philippe J.: Effects of optoelectronic device characteristics on the performance and design of POEM systems,V1562,11-20(1991)
; Rentzepis, Peter M.: Two-photon 3-D optical memories,V1499,144-147(1991)
See Kiamilev, Fouad E.: V1390,311-329(1991)
See Marchand, Philippe J.: V1505,38-49(1991)
See Ozguz, Volkan H.: V1390,477-488(1991)
See Urquhart, Kristopher S.: V1555,214-223(1991)

Eskridge, Thomas See McWilliams, Gary: V1468,417-428(1991)

Esposito, Benjamin J. See Sauer, Donald J.: V1540,285-296(1991)

Esquivias, Ignacio
; Dal Colle, M.; Brink, D.; Baars, Jan W.; Bruder, Martin: Characterization of anodic fluoride films on Hg1-xCdxTe,V1484,55-66(1991)

Essenpreis, Matthias See Cope, Mark: V1431,251-262(1991)

Esser, Peter D.
; Halpern, Ethan J.: Quality assurance of PACS systems with laser film digitizers,V1444,100-103(1991)
See Halpern, Ethan J.: V1444,104-115(1991)

Estep, Leland
; Arnone, Robert A.: Impact on the medium MTF by model estimation of b,V1537,89-96(1991)

Esterowitz, Leon
See Pinto, Joseph F.: V1420,242-243(1991)
See Pinto, Joseph F.: V1410,175-178(1991)
See Quarles, Gregory J.: V1410,165-174(1991)
See Stoneman, Robert C.: V1410,148-155(1991)

Estes, Robert R. See Algazi, V. R.: V1605,329-338(1991)

Esteve, Daniel
See Rouhani, Mehdi D.: V1361,954-962(1991)
See Simonne, John J.: V1374,107-115(1991)
See Simonne, John J.: V1589,139-147(1991)

Estocq, Emmanuel
See Capitelli, Mario: V1503,126-131(1991)
See Godard, Bruno: V1503,71-77(1991)
See Godard, Bruno: V1397,59-62(1991)

Estraillier, P. See Fleurot, Noel A.: V1502,230-241(1991)

Etemad, H. See Cheng, Jie: V1392,373-384(1991)

Etemad, Shahab See Xi, Xiaoxing: V1477,20-25(1991)

Ethridge, Edwin C. See Johnson, R. B.: V1527,2-18(1991); 1535,231-247(1991)

Etienne, Bernard
; Paris, E.; Dorin, C.; Thierry-Mieg, V.; Williams, F. I.; Glattly, D. C.; Deville, G.; Andrei, E. Y.; Probst, O.: Structure optimization of selectively doped heterojunctions: evidences for a magnetically induced Wigner solidification,V1362,256-267(1991)

Eton, Darwin
; Ahn, Samuel S.; Baker, J. D.; Pensabene, Joseph; Yeatman, Lawrence S.; Moore, Wesley S.: Intraoperative endovascular ultrasonography,V1425,182-187(1991)

Etsuda, Hirokuni See Sakurada, Masami: V1425,158-164(1991)

Ettedgui-Atad, Eli
; Humphries, Colin M.: Optical performance of an infrared astronomical telescope with a 5-axis secondary mirror,V1532,241-248(1991)

Eubanks, Charles See Moy, Richard Q.: V1467,154-160(1991)

Eustace, John G.
; Coghlan, Gregory A.; Yorka, Christian M.; Carome, Edward F.; Adamovsky, Grigory: Fiber optic sensing technique employing rf-modulated interferometry,V1584,320-327(1991)
See Adamovsky, Grigory: V1332,750-756(1991)
See Beheim, Glenn: V1369,50-59(1991)

Evaldsson, P. See Taylor, Geoff W.: V1476,2-13(1991)

Evans, Alan C.
; Dai, W.; Collins, L.; Neelin, Peter; Marrett, Sean: Warping of a computerized 3-D atlas to match brain image volumes for quantitative neuroanatomical and functional analysis,V1445,236-246(1991)

Evans, Brian L. See Xu, Shi: V1343,110-121(1991)

Evans, Chris J. See Polvani, Robert S.: V1441,173-173(1991)

Evans, Clinton E.
; Doshi, Rekha: Tale of two underwater lenses,V1537,203-214(1991)

Evans, Gary A.
; Carlson, Nils W.; Bour, David P.; Liew, So K.; Amantea, Robert; Wang, Christine A.; Choi, Hong K.; Walpole, James N.; Butler, Jerome K.; Ferguson, W. E.: Low-threshold grating surface-emitting laser arrays,V1418,406-413(1991)
; Rosen, Arye; Stabile, Paul J.; Bour, David P.; Carlson, Nils W.; Connolly, John C.: Two-dimensional edge- and surface-emitting semiconductor laser arrays for optically activated switching,V1378,146-161(1991)
See Ayekavadi, Raj: V1418,74-85(1991)

Evans, Helen H. See Oleinick, Nancy L.: V1427,90-100(1991)

Evans, J. See Walker, David R.: V1358,860-867(1991)

Evans, John R. See Hoff, Lawrence E.: V1481,98-109(1991)

Evans, Nelson O. See James, William E.: V1400,129-136(1991)

Evans, Paul See Godber, Simon X.: V1526,170-189(1991)

Evans, Phillip R.
: WEBERSAT: measuring micrometeorite impacts in a polar orbit,V1495,149-156(1991)

Evans, Robert H.
; Williams, Elmer F.; Brant, Karl: Experiences with a parallel architecture for image analysis,V1406,201-202(1991)
See Williams, Elmer F.: V1482,104-111(1991)

Evans, Thomas G. See Staller, Craig O.: V1540,219-230(1991)

Even-Or, Baruch
; Lipsky, S.; Markowitz, Raymond; Herczfeld, Peter R.; Daryoush, Afshin S.; Saedi, Reza: Recirculating fiber optical RF-memory loop in countermeasure systems,V1371,161-169(1991)

Evenhouse, Raymond J.
; Chen, Xiaoming: Precise individualized armature for ear reconstruction,V1380,248-253(1991)
; Rasmussen, Mary; Sadler, Lewis L.: Computer-aided forensic facial reconstruction,V1380,147-156(1991)

Evensen, Per See Sumner, Timothy J.: V1549,265-273(1991)

Everett, Hobart R.
See Aviles, Walter A.: V1388,587-597(1991)
See Holland, John: V1388,291-298(1991)

Evers, K. See Rundle, Debra A.: V1446,405-413(1991)

Eversole, J. D. See Campillo, Anthony J.: V1497,78-89(1991)

Evins, James B. See Alter, James J.: V1566,296-301(1991)

Evsenev, V. S. See Zhavoronok, I. V.: V1554A,371-379(1991)

Evstropiev, Sergei K. See Glebov, Leonid B.: V1513,224-231(1991)

Evtikhiev, Nickolay N. See Kiselyov, Boris S.: V1621,126-137(1991)

Evtikhov, M. G. See Balkarey, Yu. I.: V1621,238-249(1991)

Ewen, Peter J.
; Slinger, Christopher W.; Zakery, A.; Zekak, A.; Owen, A. E.: Fabrication and properties of chalcogenide IR diffractive elements,V1512,101-111(1991)

Ewin, Audrey J.
; Jhabvala, Murzy; Shu, Peter K.: Development of a large pixel, spectrally optimized, pinned photodiode/interline CCD detector for the Earth Observing System/Moderate-Resolution Imaging Spectrometer-Tilt Instrument,V1479,12-20(1991)

Ewing, William S. See Heaton, Mark W.: V1482,444-457(1991)

Exarhos, Gregory J.
; Hess, Nancy J.; Ryan, Samantha: Laser Raman measurements of dielectric coatings as a function of temperature,V1441,190-199(1991)

Eyal, Marek
See Reisfeld, Renata: V1513,360-367(1991)
See Reisfeld, Renata: V1590,215-228(1991)
See Rotman, Stanley R.: V1442,194-204(1991)

Eyler, William R. See Flynn, Michael J.: V1444,172-179(1991)

Eyuboglu, Vedat See Zervas, E.: V1565,178-187(1991)

Ezekiel, Shaoul See Smith, S. P.: V1367,103-106(1991)

Fabbro, Remy
See de Frutos, Angel M.: V1397,717-720(1991)
See Poueyo, Anne: V1502,140-147(1991)

Faber, Tracy L. See Adiseshan, Prakash: V1445,128-132(1991)

Fabes, Brian D. See Zaugg, Thomas C.: V1590,26-35(1991)

Fabian, Heinz
; Grzesik, Ulrich; Woerner, K.-H.; Klein, Karl-Friedrich: Optical fibers for UV applications,V1513,168-173(1991)

Fabre, Edouard
See Billon, Jean P.: V1502(1991)
See Koenig, Michel: V1502,338-342(1991)

Fabre, P. See Moehlmann, Gustaaf R.: V1560,426-433(1991)

Fabricius, Norbert See Wolf, Barbara: V1506,40-51(1991)

Facq, Paul See Pagnoux, Dominique: V1504,98-106(1991)

Fagan, William F.
ed.: *Holographic Optical Security Systems,*V1509(1991)
: Holographic optical security systems,VIS08,381-386(1991)

Fagherazzi, Giuliano See Righini, Giancarlo C.: V1513,105-111(1991)

Fagnano, C.
See Bertoluzza, Alessandro: V1403,150-152(1991)
See Bertoluzza, Alessandro: V1403,40-49(1991)

Fahey, David W.
: Application of the NO/O3 chemiluminescence technique to measurements of reactive nitrogen species in the stratosphere,V1433,212-223(1991)

Fahey, James T.
; Frechet, Jean M.: New aqueous base-developable negative-tone photoresist based on furans,V1466,67-74(1991)

Fair, Melody
: Goldhelox: a project to view the x-ray sun,V1549,182-192(1991)

Fairbank, William M.
; Hansen, C. S.; LaBelle, R. D.; Pan, X. J.; Chamberlin, E. P.; Fearey, Bryan L.; Gritzo, R. E.; Keller, Richard A.; Miller, Charles M.; Oona, H.: Measurement of trace isotopes by photon burst mass spectrometry,V1435,86-89(1991)

Fairchild, B. W. See Gopalakrishnan, G. K.: V1476,270-275(1991)

Fairchild, Scott See Ivey, Jim: V1567,170-178(1991)

Fairfield, Frederick R. See Soper, Steven A.: V1435,168-178(1991)

Fairlie, S. A.
; Smith, Paul W.: Performance characteristics of a discharge-pumped XeCl laser driven by a pulse forming network containing nonlinear ferroelectric capacitors,V1411,56-62(1991)

Fake, M. See Fernando, P. N.: V1372,152-163(1991)

Fakhrutdinov, I. H.
; Avdoshin, A. P.; Moshin, J. N.; Poltavsky, V. V.: Autonomous mobile laser complex,V1399,98-106(1991)

Fakhruzzaman, Khan See Fishman, Jack: V1491,348-359(1991)

Falciai, Riccardo
; Baldini, Francesco; Bechi, Paolo; Cosi, Franco; Bini, Andrea: Bile optical fiber sensor: the method for entero-gastric reflux detection,V1572,424-427(1991)
; Pascucci, Tania: High-power laser/optical-fiber-coupling device,V1506,120-125(1991)
; Scheggi, Annamaria V.; Cosi, Franco; Cao, J. Y.: Optical fiber demultiplexer for telecommunications,V1365,38-42(1991)
See Baldini, Francesco: V1510,58-62(1991)
See Mencaglia, Andrea: V1506,140-144(1991)

Falco, Charles M.
See Kearney, Patrick A.: V1343,25-31(1991)
See Slaughter, Jon M.: V1343,73-82(1991)

Fales, Carl L.
See Alter-Gartenberg, Rachel: V1605,745-756(1991)
See Rahman, Zia-ur: V1488,237-248(1991)

Falk, Joel
; Chu, Raijun; Kanefsky, Morton; Hua, Xuelei: Beam-coupling by stimulated Brillouin scattering,V1409,83-86(1991)
See Park, S. H.: V1409,9-13(1991)

Falk, R. A.
; Adams, Jeff C.: Temporal model of optically initiated GaAs avalanche switches,V1378,70-81(1991)

Fall, Thomas C.
: Stochastic neural nets and vision,V1468,778-785(1991)

Fallon, James J.
; Selby, Vaughn H.: Optical locator for horizon sensing,V1495,268-279(1991)

Fan, Cheng
; Jiang, Chaochuan; Guo, Lu Rong: Color-coding reproduction of 3-D object with rainbow holography,V1461,51-55(1991)
; Jiang, Chaochuan; Guo, Lu Rong: Large-viewing-angle rainbow hologram by holographic phase conjugation,V1461,265-269(1991)

Fan, Dapeng
; Zhang, Honghai; Chen, Jihong; Chen, Riyao: Measurement of surface roughness using optical fibre sensor and microcomputer,V1572,11-14(1991)

Fan, Guang H.
See Guan, Z. P.: V1519,26-32(1991)
See Liu, Yudong: V1362,436-447(1991)
See Pan, Chuan K.: V1519,645-651(1991)

Fan, Hua Ying See He, Anzhi: V1554A,747-749(1991)

Fan, J. See Kiamilev, Fouad E.: V1390,311-329(1991)

Fan, Jie See Cui, DaFu: V1572,386-391(1991)

Fan, Kuo-Chin
; Chang, Chia-Yuan: Object segmentation algorithm for use in recognizing 3-D partially occluded objects*,V1468,674-684(1991)
; Lui, Po-Chang: Path planning algorithm for a mobile robot*,V1468,1010-1021(1991)

Fan, Ruixing See Wang, Yongjiang: V1412,67-71(1991)

Fan, Ru Y.
   ; Lu, Yue M.; Song, Xiang Y.: Microstructure of titanium oxide thin
      films,V1519,134-137(1991)

Fan, Tso Y. See Fujimoto, James G.: V1413,2-13(1991)

Fan, Xiang J.
   ; Guo, Huai X.; Jiang, Chang Z.; Pen, You G.; Liu, Chang; Pen, Zhi L.; Li,
      Hong T.: Preparation and O+ implantation of Y-Ba-Cu-O
      superconducting thin films by sputtering and RTA process,V1519,808-
      812(1991)
   ; Pen, You G.; Guo, Huai X.; Li, Hong T.; Liu, Chang; Jiang, Chang Z.; Pen,
      Zhi L.: Preparation of thin superconducting YBCO films by ion-beam
      mixing,V1519,805-807(1991)

Fan, Xinrui See Xiang, Tingyuan: V1572,372-376(1991)

Fan, Xi W.
   See Chen, Lian C.: V1519,450-453(1991)
   See Guan, Z. P.: V1519,26-32(1991)
   See Liu, Yudong: V1362,436-447(1991)
   See Pan, Chuan K.: V1519,645-651(1991)

Fan, Ya F. See Zhao, Shu L.: V1519,275-280(1991)

Fan, Yong C.
   ; An, Cheng W.; Lu, Dong S.; Li, Zai Q.: In-situ low-temperature and epitaxial
      growth of high-Tc superconducting films using oxygen-discharge-assisted
      laser ablation method,V1519,775-778(1991)
   ; An, Cheng W.; Lu, Dong S.; Li, Zai Q.: Luminescence studies in the process
      of preparation high-Tc superconducting films with excimer laser
      ablation,V1519,813-817(1991)
   See An, Cheng W.: V1519,818-821(1991)

Fan, Zheng X.
   ; Wu, Zou L.; Shi, Zeng R.: Thermal process of laser-induced damage in
      optical thin films,V1519,359-364(1991)
   See Guo, Yong H.: V1519,327-332(1991)
   See Shao, Jian D.: V1519,298-301(1991)
   See Su, Xing: V1519,80-84(1991)
   See Tang, Xue F.: V1519,96-98(1991)
   See Wu, Zhouling: V1441,214-227(1991)
   See Wu, Zhouling: V1441,200-213(1991)
   See Xue, Song S.: V1519,570-574(1991)

Fancey, Norman E.
   See Potter, Duncan J.: V1564,363-372(1991)
   See Underwood, Ian: V1562,107-115(1991)

Fang, Cui-Chang
   ; Liao, Yan-Biao; Ma, De-Yuan: Real-time optical spatial filtering system with
      white-light source for displaying color moire,V1554A,915-921(1991)

Fang, Jing See Atlas, Les E.: V1566,134-143(1991)

Fang, Qiang
   ; Luo, Xiangyang; Tan, Yushan: Numerical investigation of effect of dynamic
      range and nonlinearity of detector on phase-stepping holographic
      interferometry,V1332,216-222(1991)
   ; Tan, Yushan: Statistical analysis of white-light speckle
      photography,V1554A,696-704(1991)

Fang, Rong C.
   See Cui, Jing B.: V1519,419-422(1991)
   See Zhang, Wei P.: V1519,680-682(1991)

Fang, Ruhua
   ; Cao, Zhengyuan: Improvement of the sensitivity of dynamic white-light
      speckle method and its application,V1554A,649-656(1991)

Fang, S. P. See Taylor, Henry F.: V1562,264-275(1991)

Fang, X. J.
   ; Wang, A. B.; Se, H.; Jin, X. D.; Jang, T.; Lin, J. X.: Stability studies of optical
      fiber pressure sensor,V1572,279-283(1991)

Fang, Xian-chen
   ; Guo, Jian: Dynamical and real-time measurement of the fringe visibility of
      optical fiber interferometer,V1572,52-55(1991)

Fang, Xiao See Zhang, Zaixuan: V1572,201-204(1991)

Fang, Yin
   ; Rong, Jian; Sheng, Kemin; Qin, Pingling: All-fiber temperature sensing
      system by using polarization-maintaining fiber coupler as
      beamsplitter,V1572,197-200(1991)
   ; Sheng, Kemin: Measurement of small strain of a solid body by two-
      frequency laser optical fiber sensor,V1572,453-456(1991)

Fang, Zhen-he
   ; Huang, Shao-ming; Shen, Yu-qing: Closed-loop fiber-optic
      gyroscope,V1572,342-346(1991)

Fang, Zhi Y. See Yuan, Xiang L.: V1519,167-171(1991)

Fanti, Giullo See Angrilli, Francesco: V1482,26-39(1991)

Fantinel, D. See Bortoletto, Favio: V1542,225-235(1991)

Fantini, Marcia C.
   ; Bezerra, George H.; Carvalho, C. R.; Gorenstein, Annette: Electrochromic
      properties and temperature dependence of chemically deposited
      Ni(OH)x thin films,V1536,81-92(1991)

Fantoni, Roberta See Barbini, Roberto: V1503,363-374(1991)

Farabaugh, Edward N.
   See Robins, Lawrence H.: V1534,105-116(1991)
   See Shechtman, Dan: V1534,26-43(1991)
   See Ying, Xuantong: V1519,321-326(1991)

Faraday, Bruce J. See Uberall, Herbert: V1552,198-213(1991)

Farahi, Faramarz
   : New modulation scheme for optical fiber point temperature
      sensor,V1504,237-246(1991)
   ; Jackson, David A.: Temperature and strain sensing using monomode optical
      fiber,V1511,234-243(1991)
   ; Jones, Julian D.; Jackson, David A.: Fiber optic sensors for heat transfer
      studies,V1584,53-61(1991)
   See Bartlett, Steven C.: V1504,247-250(1991)
   See Santos, Jose L.: V1511,179-189(1991)

Fargion, Daniele
   : Reduced size holography,V1238,428-438(1991)
   : Reducing and magnifying holograms,VIS08,354-359(1991)

Farhat, Nabil H.
   : Electron trapping materials for adaptive learning in photonic neural
      networks,V1621,310-319(1991)

Farkas, Janos See Jain, Ajay: V1596,23-33(1991)

Farkas, Zoltan D.
   ; Lavine, T. L.; Menegat, A.; Miller, Roger H.; Nantista, C.; Spalek, G.;
      Wilson, P. B.: Radio frequency pulse compression experiments at
      SLAC,V1407,502-511(1991)

Farleigh, Scott E. See Grimes, Gary J.: V1592,139-149(1991)

Farmer, William J.
   ; Benton, Stephen A.; Klug, Michael A.: Application of the edge-lit format to
      holographic stereograms,V1461,215-226(1991)

Farn, Michael W.
   : New iterative algorithm for the design of phase-only gratings,V1555,34-
      42(1991)
   ; Goodman, Joseph W.: Diffractive doublet corrected on-axis at two
      wavelengths,V1354,24-29(1991)

Farnworth, Chuck
   : Need for laser beam diagnostics,V1441,2-8(1991)

Farquharson, Stuart
   ; Chauvel, J. P.: Kinetic studies of phosgene reduction via in situ Fourier
      transform infrared analysis,V1434,135-146(1991)

Farr, J. D. See Jain, Ajay: V1596,23-33(1991)

Farr, Michele See Wesley, Michael: V1481,209-220(1991)

Farrell, Edward J.
   ed.: Extracting Meaning from Complex Data: Processing, Display,
      Interaction II,V1459(1991)
   ; Aukrust, Trond; Oberhuber, Josef M.: Vector and scalar field
      interpretation,V1459,136-147(1991)
   See Appino, Perry A.: V1459,157-165(1991)

Farrell, Jay A. See Baker, Walter L.: V1382,181-198(1991)

Farrell, Patrick V.
   : Numerical and optical evaluation of particle image velocimetry
      images,V1554B,610-621(1991)

Farrell, Thomas J.
   ; Wilson, Brian C.; Patterson, Michael S.; Chow, Rowena: Dependence of
      photodynamic threshold dose on treatment parameters in normal rat liver
      in vivo,V1426,146-155(1991)

Farron, Donald See Andrews, Lee T.: V1459,125-135(1991)

Farsaie, Ali See Holland, Orgal T.: V1469,102-112(1991)

Faruque, Abdullah
   ; Fong, David Y.: Performance analysis through memory of a proposed
      parallel architecture for the efficient use of memory in image processing
      applications,V1606,865-877(1991)

Fasano, Victor A.
   ; Urciuoli, Rosa; Bolognese, Paolo; Fontanella, Marco: Laser Doppler
      flowmetry in neurosurgery,V1428,2-12(1991)

Fasol, Gerhard
    See Richards, David R.: V1361,246-254(1991)
    See White, Julian D.: V1361,293-301(1991)

Fassakhova, H. H. See Logak, L. G.: V1238,171-176(1991)

Fassett, John D. See Xiong, Xiaoxiong: V1435,188-196(1991)

Fatah, Rebwar M.
    ; Cox, Maurice K.; Bird, David M.; Cameron, Keith H.: Subminiature package
        external cavity laser,V1501,120-128(1991)

Fattakhov, Yakh'ya V.
    ; Khaibullin, Ildus B.; Bayazitov, Rustem M.: Anisotropic local melting of
        semiconductors under light pulse irradiation,V1440,16-23(1991)

Faurie, Jean-Pierre See Sporken, R.: V1512,155-163(1991)

Fauvet, E.
    ; Paindavoine, M.; Cannard, F.: Human movement analysis with image
        processing in real time,V1358,620-630(1991)

Favata, Fabio See Bavdaz, Marcos: V1549,35-44(1991)

Favro, Lawrence D.
    ; Ahmed, Tasdiq; Crowther, D. J.; Jin, Huijia J.; Kuo, Pao K.; Thomas, Robert
        L.; Wang, X.: Infrared thermal-wave studies of coatings and
        composites,V1467,290-294(1991)

Fawcett, G. M. See Johnstone, Walter: V1506,145-149(1991)

Fawcett, James M.
    : Lightweight surveillance FLIR,V1498,82-91(1991)

Faymonville, Rudolf See Engemann, Detlef: V1362,710-720(1991)

Fazio, E.
    See Andriesh, A.: V1513,137-147(1991)
    See Montenero, Angelo: V1513,234-242(1991)

Fealy, Stephen V. See O'Brien, Stephen J.: V1424,62-75(1991)

Fearey, Bryan L.
    ed.: *Optical Methods for Ultrasensitive Detection and Analysis: Techniques
        and Applications*,V1435(1991)
    See Fairbank, William M.: V1435,86-89(1991)
    See Johnson, Stephen G.: V1435,292-297(1991)

Fearn, Heidi
    ; Lamb, Willis E.: Corrections to the Golden Rule,V1497,245-254(1991)
    ; Lamb, Willis E.; Scully, Marlan O.: Nonlinear theory of a three-level laser
        with microwave coupling: numerical calculation,V1497,283-290(1991)
    See Scully, Marlan O.: V1497,264-276(1991)

Featherstone, John D.
    ; Zhang, S. H.; Shariati, M.; McCormack, S. M.: Carbon dioxide laser effects
        on caries-like lesions of dental enamel,V1424,145-149(1991)

Fedde, Chris See Miranda, Rick: V1380,210-217(1991)

Feddes, John J. See St. George, Dennis R.: V1379,69-80(1991)

Fediay, Sergey G.
    ; Kuznetzov, Alexsey V.: New method for detection of blood coagulation
        using fiber-optic sensor,V1420,41-43(1991)

Fedjukhin, L. A. See Petrov, Dmitry V.: V1374,152-159(1991)

Fedjushin, B. T. See Vas'kovsky, Yu. M.: V1440,241-249(1991)

Fedorov, D. L. See Nasibov, Alexander S.: V1361,901-908(1991)

Fedorov, Yurii K. See Lunter, Sergei G.: V1513,349-359(1991)

Fedoseeva, E. V.
    ; Polyakova, M. S.: Laser Doppler diagnostics of vorticity and
        phenomenological description of the flows of dilute polymer solutions in
        model tubes,V1403,355-358(1991)

Fedosejevs, Gunar See Teillet, Philippe M.: V1492,213-223(1991)

Fedotov, V. I. See Degtyareva, V. P.: V1358,524-531(1991)

Fedynyshyn, Theodore H. See Babcock, Carl P.: V1466,653-662(1991)

Fedyushin, B. T. See Golovin, A. F.: V1440,250-259(1991)

Feggans, J. See Troeltzsch, John R.: V1494,9-15(1991)

Feichtinger, Hans G.
    : Pseudoinverse matrix methods for signal reconstruction from partial
        data,V1606,766-772(1991)

Feig, Ephraim
    ; Linzer, Elliot: Scaled discrete cosine transform algorithms for JPEG and
        MPEG implementations on fused multiply/add architectures,V1452,458-
        467(1991)

Feinberg, Rick
    ed.: *Current Overviews in Optical Science and Engineering II*,VAT02(1991)

Feingold, Eric
    ; Seshadri, Sridhar B.; Arenson, Ronald L.: Folder management on a
        multimodality PACS display station,V1446,211-216(1991)
    See Seshadri, Sridhar B.: V1446,388-395(1991)

Feist, John H. See Gur, David: V1446,284-288(1991)

Feit, Zeev
    ; Kostyk, D.; Woods, R. J.; Mak, Paul S.: PbEuSeTe/Pb1-xSnxTe buried
        heterostructure diode lasers grown by molecular beam
        epitaxy,V1512,164-169(1991)

Fejer, Martin M.
    See Harris, James S.: V1361,262-273(1991)
    See Lim, Eric J.: V1561,135-142(1991)

Felberbauer, Franz
    ; Graf, Juerg: Fiber optic measurement of intracellular pH in intact rat liver
        using pH-sensitive dyes,V1510,63-71(1991)

Feld, Michael S.
    See Albagli, Douglas: V1441,146-153(1991)
    See Izatt, Joseph A.: V1427,110-116(1991)
    See Waidhauser, Erich: V1428,75-83(1991)

Feldblum, Avi Y.
    ; Nijander, Casimir R.; Townsend, Wesley P.; Mayer-Costa, Carlos M.:
        Performance and measurements of refractive microlens
        arrays,V1544,200-208(1991)

Feldman, Albert
    ; Holly, Sandor; eds.: [$BFS]Diamond Optics IV[$BFE],V1534(1991)
    See Robins, Lawrence H.: V1534,105-116(1991)
    See Shechtman, Dan: V1534,26-43(1991)
    See Ying, Xuantong: V1519,321-326(1991)

Feldman, G. G.
    ; Ilyna, T. A.; Korjenevitch, I. N.; Syrtzev, V. N.: New streak tube with
        femtosecond time resolution,V1358,497-502(1991)
    See Bass, V. I.: V1358,1075-1083(1991)
    See Korzhenevich, Irina M.: V1358,1084-1089(1991)
    See Silkis, E. G.: V1358,46-49(1991)

Feldman, Mark
    ; Mockler, Daniel J.; English, R. Edward; Byrd, Jerry L.; Salmon, J. T.: Self-
        referencing Mach-Zehnder interferometer as a laser system
        diagnostic,V1542,490-501(1991)

Feldman, Michael R.
    : Holographic optical interconnects for multichip modules,V1390,427-
        433(1991)

Feldman, Robert D. See Partovi, Afshin: V1561,20-32(1991)

Feldman, Uri See Jacobson, Nathaniel: V1453,70-80(1991)

Felger, Wolfgang
    ; Astheimer, Peter: Visualization and comparison of simulation results in
        computational fluid dynamics,V1459,222-231(1991)

Fellowes, Fiona C.
    ; Steen, William M.: Deposition of silica coatings on Incoloy 800H substrates
        using a high-power laser,V1502,213-222(1991)

Felsen, Leopold B.
    : Observable-based parametrizations,V1471,88-91(1991)
    ; Vecchi, G.; Carin, L.; Bertoni, H. L.: Short-pulse electromagnetics for
        sensing applications,V1471,154-162(1991)

Felten, Rainer See Engemann, Detlef: V1362,710-720(1991)

Felus, Y. See Rotman, Stanley R.: V1442,194-204(1991)

Feng, Chang D.
    ; Shimizu, Yasuhiro; Egashira, Makoto: Preparation of tin oxide and insulating
        oxide thin films for multilayered gas sensors,V1519,8-13(1991)

Feng, Chen
    See Johnson, R. B.: VCR38,3-18(1991)
    See Johnson, R. B.: V1527,2-18(1991); 1535,231-247(1991)
    See Johnson, R. B.: V1354,676-683(1991)

Feng, Duan See Chen, Kun J.: V1519,632-639(1991)

Feng, Hao See Sang, Fengting: V1412,252-257(1991)

Feng, Qiang See Jeong, Tung H.: V1396,60-70(1991)

Feng, S. L. See Bourgoin, J. C.: V1361,184-194(1991)

Feng, San See Wang, Guozhi: V1358,73-81(1991)

Feng, Wei G. See Wu, Xiang: V1519,625-631(1991)

Feng, Weiting
    ; Yen, Yi X.; Zhu, Cui Y.: Cryogenic refractive indices of cadmium telluride
        coatings in wavelength range from 2.5 to 20 um,V1535,224-230(1991)
    ; Yen, Yi X.; Zhu, Cui Y.: Dependence of optical properties of thermal-
        evaporated lead telluride films on substrate temperature,V1519,333-
        338(1991)

Feng, Wen-qing See Gu, Bo-qi: V1488,443-446(1991)

Feng, Yi P.
; Jiang, Bing Y.; Yang, Gen Q.; Huang, Wei; Zheng, Zhi H.; Liu, Xiang H.; Zou, Shi C.: Formation of boron nitride and silicon nitride bilayer films by ion-beam-enhanced deposition,V1519,440-443(1991)

Feng, Yu-Jen See Lin, Wei-Chung: V1445,376-385(1991)

Feng, Yue-Zhong
; Gong, Zhi-Ben; Song, Zhengfang: Increasing the isoplanatic patch size with phase-derivative adaptive optics,V1487,356-360(1991)
; Song, Zhengfang; Gong, Zhi-Ben: Compensation efficiency of an optical adaptive transmitter,V1417,370-372(1991)

Fengler, John J. See Jaggi, Bruno: V1448,89-97(1991)

Fennell, Leonard E. See Kobayashi, Kenichi: V1448,157-163(1991)

Fenner, David B. See Fork, David K.: V1394,202-202(1991)

Fenske, Axel
; Burkhardt, Hans: Affine-invariant recognition of gray-scale objects by Fourier descriptors,V1567,53-64(1991)

Fenster, Aaron
; Holdsworth, David W.; Drangova, Maria: Three-dimensional imaging using TDI CCD sensors,V1447,28-33(1991)
See Picot, Paul A.: V1444,206-213(1991)

Feofanov, A. V. See Efremov, R. G.: V1403,161-163(1991)

Fercher, Adolf F. See Hitzenberger, Christoph K.: V1423,46-50(1991); 1429,21-25(1991)

Ferdinand, Pierre See Sansonetti, Pierre: V1588,198-209(1991)

Ferguson, Richard A.
See Chiu, Anita S.: V1466,641-652(1991)
See Spence, Christopher A.: V1466,324-335(1991)

Ferguson, Suzanne M.
See Liu, Kexing: V1398,206-212(1991)
See Measures, Raymond M.: V1489,86-96(1991)
See Measures, Raymond M.: V1332,421-430(1991)

Ferguson, W. E. See Evans, Gary A.: V1418,406-413(1991)

Fernandes, Stan T. See McSwain, G. G.: V1478,228-238(1991)

Fernandez, A. M. See Nair, Padmanabhan K.: V1485,228-239(1991)

Fernandez, J. L. See Blanco, J.: V1508,180-190(1991)

Fernandez, Manuel F.
; Aridgides, Athanasios; Randolph, D.; Ferris, D.: Optimal subpixel-level IR frame-to-frame registration,V1481,172-179(1991)
See Aridgides, Athanasios: V1481,110-116(1991)

Fernandez, R.
See Harwit, Alex: V1541,38-47(1991)
See Yang, Y. J.: V1418,414-421(1991)

Fernando, P. N.
; Fake, M.; Seeds, A. J.: Novel approach to optical frequency synthesis in coherent lightwave systems,V1372,152-163(1991)

Ferrara, M.
; Lugara, M.; Moro, C.; Cingolani, R.; De Blasi, C.; Manno, D.; Righini, Giancarlo C.: Optical investigation of microcrystals in glasses,V1513,130-136(1991)

Ferrare, Richard See McGee, Thomas J.: V1491,182-187(1991)

Ferrari, A. See Andriesh, A.: V1513,137-147(1991)

Ferrari, Marco
; De Blasi, Roberto A.; Bruscaglioni, Piero; Barilli, Marco; Carraresi, Luca; Gurioli, G. M.; Quaglia, Enrico; Zaccanti, Giovanni: Near-infrared time-resolved spectroscopy and fast scanning spectrophotometry in ischemic human forearm,V1431,276-283(1991)

Ferrario, Angelo See Mattioli, Stefano: V1421,114-119(1991)

Ferrario, Mario
; Barbieri, Fabio: Computerized system for clinical diagnosis of melanoma,V1450,108-117(1991)

Ferraris, M. See Cognolato, Livio: V1513,368-377(1991)

Ferreira, Carlos
See Campos, Juan: V1564,189-198(1991)
See Garcia, Javier: V1574,133-140(1991)
See Millan, Maria S.: V1507,183-197(1991)

Ferreira, Mario F.
; Rocha, Jose R.; Pinto, Joao L.: Spectral and modulation characteristics of tunable multielectrode DBR lasers,V1372,14-26(1991)

Ferrell, Trinidad L. See Goudonnet, Jean-Pierre: V1400,116-123(1991)

Ferrin, Frank J.
: Survey of helmet tracking technologies,V1456,86-94(1991)

Ferris, D.
See Aridgides, Athanasios: V1481,110-116(1991)
See Fernandez, Manuel F.: V1481,172-179(1991)

Ferris, Kim F.
; Risser, Steven M.: Contributions of pi' bonding to the nonlinear optical properties of inorganic polymers,V1441,510-520(1991)
See Risser, Steven M.: V1441,262-268(1991)

Ferris, Nancy S. See Bilhorn, Robert B.: V1439,15-24(1991)

Ferry, David K.
; Kozicki, M. N.; Raupp, Gregory B.: Some fundamental issues on metallization in VLSI,V1596,2-11(1991)

Fesler, Kenneth A. See Wysocki, Paul F.: V1373,234-245(1991)

Festes, Gilles
; Chollet, Jean-Paul E.: Investigation of interlevel proximity effects case of the gate level over LOCOS,V1463,245-255(1991)

Festl, H. G.
; Franke, Hilmar: Photo-induced adhesion changes: a technique for patterning lightguide structures,V1559,410-423(1991)

Fetterman, Harold R.
See Bourne, Lincoln C.: V1477,205-208(1991)
See Chen, Ray T.: V1559,110-117(1991)
See Hur, Jung H.: V1378,95-100(1991)

Feuerstein, Robert J.
; Januar, Indra; Mickelson, Alan R.; Sauer, Jon R.: Wavelength dependence of proton-exchanged LiNbO3 integrated optic directional couplers from 1.5um - 1.65um,V1583,196-201(1991)

Fevrier, Herve A. See Marcerou, Jean-Francois: V1373,168-186(1991)

Fews, Peter See Koenig, Michel: V1502,338-342(1991)

Fiala, Pavel
; Jerie, Tomas: Aplanatic holographic systems,V1574,179-187(1991)

Fichelscher, Andreas
; Rangelow, Iwilo W.; Stamm, A.: Influence of sheath properties on the profile evolution in reactive ion etching processes,V1392,77-83(1991)
See Rangelow, Iwilo W.: V1392,240-245(1991)

Fidder, Henk
; Terpstra, Jacob; Wiersma, Douwe A.: Superradiance and exciton dynamics in molecular aggregates,V1403,530-544(1991)

Fiddy, Michael A. See Kang, Chih-Chieh: V1497,372-381(1991)

Fiddyment, Phillip J. See Barnsley, Peter E.: V1378,116-126(1991)

Fidorra, F. See Tischel, M.: V1361,917-926(1991)

Field, John E.
: High-speed photography at the Cavendish Laboratory,V1358,2-17(1991)
; Sun, Q.: High-speed photographic study of impact on fibers and woven fabrics,V1358,703-712(1991)
See Bourne, Neil K.: V1358,1046-1056(1991)
See Huntley, Jonathan M.: V1554A,756-765(1991)
See Whitworth, Martin B.: V1358,677-682(1991)
See Whitworth, Martin B.: V1554B,282-288(1991)

Fielden, John
; Kwong, Henry Y.; Wilbrink, Jacob: Reconstructing MR images from incomplete Fourier data using the maximum entropy method,V1445,145-154(1991)

Fielding, Kenneth H.
; Horner, Joseph L.; Makekau, Charles K.: Application of optical signal processing: fingerprint identification,V1564,224-230(1991)
See Ryan, Vincent: V1564,86-97(1991)

Fielding, Robert
See Horii, Steven C.: V1446,10-15(1991)
See Horii, Steven C.: V1446,475-480(1991)

Fields, David J. See Del Grande, Nancy K.: V1479,335-351(1991)

Fields, Renny A. See Rose, Todd S.: V1561,43-49(1991)

Fields, Robert E. See Radspinner, David A.: V1439,2-14(1991)

Fienup, James R.
: Phase retrieval for the Hubble Space Telescope using iterative propagation algorithms,V1567,327-332(1991)
See Cederquist, Jack N.: V1416,266-277(1991)

Fieret, Jim
; Green, J. M.; Heath, R.; O'Key, Michael A.; Osborne, Michael R.; Osbourn, S. J.; Taylor, Arthur F.; Winfield, R. J.: Review of the multikilohertz performance of the CHIRP laser and components,V1503,53-61(1991)

Figoski, John W.
: Aligning diamond-turned optics using visible-light interferometry,V1354,540-546(1991)

Figue, J.
; Refregier, Philippe; Rajbenbach, Henri J.; Huignard, Jean-Pierre: Neural optoelectronic correlator for pattern recognition,V1564,550-561(1991)

Figueroa, Alvaro A. See Sadler, Lewis L.: V1380,158-162(1991)

Figueroa, Luis
; Hong, C. S.; Miller, Glen E.; Porter, Charles R.; Smith, David K.: Photonics technology for aerospace applications,V1418,153-176(1991)
See Fu, Richard J.: V1418,108-115(1991)
See Lim, G.: V1418,123-131(1991)

Figueroa-Nazuno, Jesus See Bastida, M. R.: V1469,150-156(1991)

Fike, Daniel K.
: Using IR thermography as a manufacturing tool to analyze and repair defects in printed circuit boards,V1467,150-153(1991)

Fikioris, George
; Freeman, D. K.; King, Ronold W.; Shen, Hao-Ming; Wu, Tai T.: Analytical studies of large closed-loop arrays,V1407,295-305(1991)

Filatova, S. A. See Vas'kovsky, Yu. M.: V1440,241-249(1991)

Filikov, V. A. See Ligachev, Valery A.: V1519,214-219(1991)

Filippetti, P. See Bottura, Giorgio: V1403,156-158(1991)

Filippov, I. G. See Dmokhovskij, A. V.: V1554A,323-330(1991)

Filippov, N. See Kotov, Gennady A.: V1440,321-324(1991)

Fillard, Jean-Pierre B.
See Montgomery, Paul C.: V1332,563-570(1991)
See Montgomery, Paul C.: V1332,515-524(1991)

Fimia-Gil, Antonio
; Navarro, Maria T.; Egozcue, Juan J.: Computer-generated holograms of diffused objects,V1507,153-157(1991)
See Belendez, A.: V1507,268-276(1991)
See Belendez, A.: V1574,77-83(1991)
See Carretero, L.: V1507,458-462(1991); 1508,96-100(1991)
See Navarro, Maria T.: V1507,142-148(1991)
See Pascual, I.: V1507,373-378(1991)
See Pascual, I.: V1574,72-76(1991)

Finamore, Weiler A. See de Garrido, Diego P.: V1605,221-232(1991)

Fincato, Antonello See Mazzola, M.: V1365,2-12(1991)

Fincher, Curtis L. See Rose, Todd S.: V1561,43-49(1991)

Fincke, James R. See Neuman, William A.: V1411,28-40(1991)

Findakly, Talal K.
; Teng, Chia-Chi; Walpita, Lak M.: Wideband NLO organic external modulators,V1476,14-21(1991)
See Haas, David R.: V1371,56-67(1991)

Findeis, N. See Braueninger, Heinrich: V1549,330-339(1991)

Finders, J. See Geurts, Jean: V1361,744-750(1991)

Finegan, Michael K. See Amladi, Nandan G.: V1472,165-176(1991)

Fineschi, Silvano
; Hoover, Richard B.; Fontenla, Juan; Walker, Arthur B.: Solar EUV/FUV line polarimetry: observational parameters and theoretical considerations,V1343,376-388(1991)
See Hoover, Richard B.: V1343,389-403(1991)

Fini, G. See Bertoluzza, Alessandro: V1403,153-155(1991)

Fink, Claudia See Mennig, Martin: V1590,152-159(1991)

Fink, Jeffrey D. See Grieser, James L.: V1330,102-110(1991)

Fink, Patrick W. See Arndt, George D.: V1475,231-242(1991)

Finkelmann, Heino See Hirschmann, Harald: V1559,27-38(1991)

Finkelstein, Blair I.
; Childers, Ed R.: Effects of focus misregistration on optical disk performance,V1499,438-449(1991)

Finkelstein, Vladimir
; Berman, Paul R.: Optical coherent transients induced by time-delayed fluctuating pulses,V1376,68-79(1991)

Finley, David S. See Welsh, Barry Y.: V1343,166-174(1991)

Finney, Mark J. See Tregay, George W.: V1589,38-47(1991)

Finzi, Laura See Bustamante, Carlos: V1435,179-187(1991)

Fiodorov, Alexander O. See Gan, Michael A.: V1574,254-260(1991)

Fiorito, Ralph B. See Rule, Donald W.: V1552,240-251(1991)

Firak, Jozef
; Marczak, Jan; Sarzynski, Antoni: Unstable resonator with a super-Gaussian dielectric mirror for Nd:YAG Q-switched laser,V1391,42-47(1991)

Firby, R. J.
: Planning, acting, and sensor fusion,V1383,483-489(1991)

Firsov, N. N.
; Priezzhev, Alexander V.; Stepanian, A. S.: Laser nephelometry of erythrocytes in shear flows,V1403,350-354(1991)

Firstein, Leon A.
; Noz, Arthur: Cross-sectional imaging in SEM: signal formation mechanism and CD measurements,V1464,81-88(1991)

Fischer, James R.
; Strong, James P.; Dorband, John E.; Tilton, James C.: Applications of the massively parallel machine, the MasPar MP-1, to Earth sciences,V1492,229-238(1991)

Fischer, Paul B. See Chou, Stephen Y.: V1474,243-247(1991)

Fischer, Robert C.
; Martin, Charles J.; Niblack, Curtiss A.; Timlin, Harold A.; Wimmers, James T.: Further performance characteristics of a high-sensitivity 64 x 64 element InSb hybrid focal-plane array,V1494,414-418(1991)

Fischer, Robert E.
: Case study of elastomeric lens mounts,V1533,27-35(1991)
: Lens design for the infrared,VCR38,19-43(1991)
: Optimization of lens designer to manufacturer communications,V1354,506-522(1991)
; Smith, Warren J.; eds.: [$BFS]Current Developments in Optical Design and Optical Engineering[$BFE],V1527(1991)
; Thomas, Michael J.; Hudyma, Russell M.: Optical glass selection using computerized data base,V1535,78-88(1991)
See Hudyma, Russell M.: V1563,244-254(1991)
See Lipson, David: V1368,36-43(1991)

Fischetti, Massimo V. See Arnold, Douglas: V1441,478-487(1991)

Fishbein, Michael C.
See Buchelt, Martin: V1420,249-253(1991)
See Vari, Sandor G.: V1424,33-42(1991)
See Vari, Sandor G.: V1426,58-65(1991)
See Vari, Sandor G.: V1426,111-120(1991)

Fishell, Wallace G.
; Fox, Alex J.: Reconnaissance mission planning,V1538,5-13(1991)
See Beving, James E.: V1538,14-17(1991)

Fisher, A. S. See Pogorelsky, Igor V.: V1413,21-31(1991)

Fisher, Amnon
: Gas puff Z-pinch x-ray source: a new approach,V1552,252-253(1991)
See Ishizuka, Hiroshi: V1407,442-455(1991)
See Prohaska, Robert M.: V1407,598-604(1991)
See Song, Yuanxu Y.: V1407,430-441(1991)

Fisher, H. L. See Strikwerda, Thomas E.: V1478,13-23(1991)

Fisher, Kirk L. See Young, Donald S.: V1416,221-228(1991)

Fisher, Moshe See Hadar, Ofer: V1482,79-91(1991)

Fisher, Robert A.
; Reintjes, John F.; eds.: [$BFS]Nonlinear Optics II[$BFE],V1409(1991)

Fisher, Scott S.
See Merritt, John O.: V1457(1991)
See Pieper, Steve: V1457,188-197(1991)

Fisher, Thornton R.
; Perez, Joseph D.: Use of spline expansions and regularization in the unfolding of data from spaceborne sensors,V1479,212-225(1991)

Fisher, Vladimir
: Mechanisms of ionization for gas adjoining to plasma in intense laser beam,V1440,179-187(1991)

Fishkin, Joshua B.
; Gratton, Enrico; vandeVen, Martin J.; Mantulin, William W.: Diffusion of intensity modulated near-infrared light in turbid media,V1431,122-135(1991)

Fishman, A. I.
See Ivanov, Anatoliy A.: V1403,174-184(1991)
See Ivanov, Anatoliy A.: V1403,243-245(1991)
See Ivanov, Anatoliy A.: V1429,132-144(1991); 1432,141-153(1991)

Fishman, Harvey A. See Sweedler, Jonathan V.: V1439,37-46(1991)

Fishman, Jack
; Watson, Catherine E.; Brackett, Vincent G.; Fakhruzzaman, Khan; Veiga, Robert: Use of satellite data to determine the distribution of ozone in the troposphere,V1491,348-359(1991)

Fiske, Michael R.
: Design and development of the Zeolite Crystal Growth Facility,V1557,78-85(1991)

Fitch, J. T. See Lucovsky, Gerald: V1392,605-616(1991)

Fitch, John See Rohde, Axel: V1464,438-446(1991)

Fitch, John E.
See Harms, Richard J.: V1494,49-65(1991)
See Richards, Evan: V1494,40-48(1991)

Fittinghoff, David N.
; Bolton, Paul R.; Chang, Britton; Van Woerkom, Linn D.; White, William E.: Ion yields from strong optical field ionization experiments using 100-femtosecond laser pulses,V1413,81-88(1991)

Fitzgerald, Edward A. See Honda, Kenji: V1466,141-148(1991)

Fitzgerald, Jay B. See Meyer, George E.: V1379,99-105(1991)

Fitzmartin, Daniel J.
; Balboni, Edmund J.; Gels, Robert G.: Wide-band analog frequency modulation of optic signals using indirect techniques,V1371,78-86(1991)
; Gels, Robert G.; Balboni, Edmund J.: Coherent optical modulation for antenna remoting,V1476,56-62(1991)

Fitzmaurice, Maryann See Rava, Richard P.: V1426,68-78(1991)

Fitzmaurice, Michael W.
; Hayden, William L.: Space station laser communication transceiver,V1417,13-21(1991)
See Chapman, William W.: V1417,277-290(1991)
See Chapman, William W.: V1417,63-74(1991)
See Hayden, William L.: V1417,182-199(1991)

Fitzpatrick, Colleen M.
; Mueller, Edward P.: Van der Lugt optical correlation for the measurement of leak rates of hermetically sealed packages,V1332,185-192(1991)

FitzSimons, Philip M.
; Trahan, Daniel J.: Characterization and enhancement of the damping within composite beams,V1489,230-242(1991)

Fjarlie, E. J. See Ronning, E.: V1399,178-183(1991)

Fjodorov, A. V. See Kostin, Ivan K.: V1554A,418-425(1991)

Flaaronning, Nils See Duchet, Christian: V1372,72-81(1991)

Flachs, Gerald M.
See Gaughan, Patrick T.: V1469,812-819(1991)
See Scott, David R.: V1383,367-378(1991)

Flack, Warren W.
; Dameron, David H.: Mix-and-match lithography for half-micrometer technology,V1465,164-172(1991)

Flagello, Donis G. See Treptau, Jeffrey P.: V1415,317-321(1991)

Flaherty, Marty
: Mission verification systems for FMS applications,V1538,64-68(1991)

Flaim, Tony See Sethi, Satyendra A.: V1463,30-40(1991)

Flamm, E. S. See Mayevsky, Avraham: V1431,303-313(1991)

Flanagan, Aiden J. See Glynn, Thomas J.: V1598,200-205(1991)

Flanagan, Kathryn A.
; Austin, G. K.; Cobuzzi, J. C.; Goddard, R.; Hughes, John P.; McLaughlin, Edward R.; Podgorski, William A.; Rose, V.; Roy, Adrian G.; Zombeck, Martin V.; Markert, Thomas H.; Bauer, J.; Isobe, T.; Schattenburg, Mark L.: Uniformity and transmission of proportional counter window materials for use with AXAF,V1549,395-407(1991)
See Markert, Thomas H.: V1549,408-419(1991)

Flanagan, Roger L.
See Carome, Edward F.: V1584,110-117(1991)
See Carome, Edward F.: V1589,133-138(1991)

Flannery, David L.
; Phillips, William E.; Reel, Richard L.: Case study of design trade-offs for ternary phase-amplitude filters,V1564,65-77(1991)

Flasck, Richard A. See Assenheim, Harry M.: V1456(1991)

Flatico, Joseph M.
See Beheim, Glenn: V1584,64-71(1991)
See Beheim, Glenn: V1369,50-59(1991)

Flaton, Kenneth A.
: Multistage object recognition using dynamical-link graph matching,V1469,137-148(1991)

Fleck, Joseph A.
: Helmholtz beam propagation by the method of Lanczos reduction,V1583,228-239(1991)

Fleischer, John M.
: Laser beam width, divergence, and propagation factor: status and experience with the draft standard,V1414,2-11(1991)

Fleischman, Judith R.
See Fraser, George W.: V1548,132-148(1991)
See Kaaret, Philip E.: V1548,106-117(1991)
See Shaw, Ping-Shine: V1343,485-499(1991)
See Shaw, Ping-Shine: V1548,118-131(1991)

Fleming, J. S. See Bourne, Lincoln C.: V1477,205-208(1991)

Fleming, J. T. See Wong, Ka-Kha: V1374,278-286(1991)

Fleming, Simon C. See Nie, Qiuhua: V1332,409-420(1991)

Flemming, G.
; Brinkmann, Ralf E.; Strunge, Ch.; Engelhardt, R.: Fiber fragmentation during laser lithotripsy,V1421,146-152(1991)
See Scheu, M.: V1421,100-107(1991)

Fletcher, P. See Jensen, Robert K.: V1380,124-136(1991)

Fleurot, Noel A.
; Andre, Michel L.; Estraillier, P.; Friart, Daniel; Gouedard, C.; Rouyer, C.; Thebault, J. P.; Thiell, Gaston; Veron, Didier: Output pulse and energy capabilities of the PHEBUS laser facility,V1502,230-241(1991)

Flickner, Myron D.
; Lavin, Mark A.; Das, Sujata: Object-oriented language for image and vision execution,V1568,113-124(1991)

Flintoff, John F. See DeNatale, Jeffrey F.: V1534,44-48(1991)

Flittner, David E. See Nianzeng, Che: V1493,182-194(1991)

Floch, Herve G.
; Berger, Michel; Novaro, Marc; Thomas, Ian M.: High laser-damage threshold and low-cost sol-gel-coated epoxy-replicated mirrors,V1441,304-315(1991)

Flock, Stephen T.
; Jacques, Steven L.; Small, Susan M.; Stern, Scott J.: PDT of rat mammary adenocarcinoma in vitro and in a rat dorsal-skin-flap window chamber using Photofrin and chloroaluminum-sulfonated phthalocyanine,V1427,77-89(1991)
See Jacques, Steven L.: V1431,12-20(1991)
See Jacques, Steven L.: V1525,35-40(1991)

Floeder, Steven P.
; Waltz, Frederick M.: Toward low-cost real-time EPLD-based machine vision workstations and target systems,V1386,90-101(1991)

Floersheimer, M. See Bosshard, Christian: V1560,344-352(1991)

Flora, Francesco
See Bollanti, Sarah: V1503,80-87(1991)
See Bollanti, Sarah: V1397,97-102(1991)

Florence, James M.
; Juday, Richard D.: Full-complex spatial filtering with a phase mostly DMD,V1558,487-498(1991)
See Juday, Richard D.: V1558,499-504(1991)

Flores, Carlos F. See Park, Yong K.: V1372,219-227(1991)

Flores, Gary E.
; Norbury, David H.: Sequential experimentation strategy and response surface methodologies for photoresist process optimization,V1464,610-627(1991)

Flores, James S.
See Delamere, W. A.: V1479,31-40(1991)
See Delamere, W. A.: V1447,204-213(1991)

Flores, Ricardo See Malacara Hernandez, Daniel: V1332,36-40(1991)

Florez, Leigh T. See Jewell, Jack L.: V1389,401-407(1991)

Flotte, Thomas J.
; Frisoli, Joan K.; Goetschkes, Margaret; Doukas, Apostolos G.: Laser-induced shock wave effects on red blood cells,V1427,36-44(1991)

Floyd, Dennis R. See Murray, Brian W.: V1485,88-95(1991)

Flynn, Lin See Bonometti, Robert J.: V1495,166-176(1991)

Flynn, Michael J.
; Davies, Eric; Spizarny, David; Beute, Gordon; Peterson, Ed; Eyler, William R.; Gross, Barry; Chen, Ji: Digital replication of chest radiographs without altering diagnostic observer performance,V1444,172-179(1991)
See Chen, Ji: V1444,256-264(1991)

Flynn, Patrick J. See Newman, Timothy S.: V1570,250-261(1991)

Focht, Marlin W. See Morgan, Robert A.: V1562,149-159(1991)

Fockler, J. See Bailey, Vernon L.: V1407,400-406(1991)

Foden, Clare See Rando, Nicola: V1549,340-356(1991)

Foerster, Werner
; Kasprzak, Henryk; von Bally, Gert; Busse, H.: Holographic interferometric analysis of the bovine cornea expansion,V1429,146-151(1991)

Fogarassy, Eric
: Basic mechanisms and application of the laser-induced forward transfer for high-Tc superconducting thin film deposition,V1394,169-179(1991)

Fogel, Sergei See Trussell, Henry J.: V1452,139-145(1991)

Fogg, Brian R.
; Miller, William V.; Lesko, John J.; Carman, Gregory P.; Vengsarkar, Ashish M.; Reifsnider, Kenneth L.; Claus, Richard O.: Analysis of macro-model composites with Fabry-Perot fiber-optic sensors,V1588,14-25(1991)
See Claus, Richard O.: V1588,159-168(1991)
See Murphy, Kent A.: V1588,117-124(1991)

Fogler, H. S. See Demos, Alexandros T.: V1392,291-297(1991)

Fogliani, Manlio F. See Agnesi, Antoniangelo: V1415,242-247(1991)

Fokken, G. See Gilbert, Barry K.: V1390,235-248(1991)

Foley, James W.
; Cincotta, Louis; Cincotta, Anthony H.: Evaluation of Nile Blue E chalcogen analogs as PDT agents,V1426,208-215(1991)
See Lin, Chi-Wei: V1426,216-227(1991)

Folkman, Mark A.
; Jarecke, Peter J.; Darnton, Lane A.: Enhancements to the radiometric calibration facility for the Clouds and the Earth's Radiant Energy System instruments,V1493,255-266(1991)
See Jarecke, Peter J.: V1493,244-254(1991)

Folli, S. See Wagnieres, G.: V1525,219-236(1991)

Foltyn, Stephen R.
See Muenchausen, Ross E.: V1394,221-229(1991)
See Wu, Xin D.: V1477,8-14(1991)

Folweiler, Robert C. See Zemon, Stanley A.: V1373,21-32(1991)

Fomichov, P. A. See Veiko, Vadim P.: V1544,152-163(1991)

Fong, C. Y.
; Gallup, R. F.; Nelson, J. S.: Electronic properties of Si-doped nipi structures in GaAs,V1361,479-488(1991)

Fong, David Y.
: Analysis and simulation of an inhibitive directional selective unit for computer vision,V1606,941-950(1991)
; Pomalaza-Raez, Carlos A.: Bayesian estimation of smooth object motion using data from direction-sensitive velocity sensors,V1569,156-161(1991)
See Faruque, Abdullah: V1606,865-877(1991)

Fonneland, Nils J.
See Brandstetter, Robert W.: V1385,173-189(1991)
See Brandstetter, Robert W.: V1559,308-320(1991)

Fontaine, Bernard L.
See Delaporte, Philippe C.: V1397,485-492(1991)
See Gerri, Mireille: V1503,280-291(1991)
See Hueber, Jean-Marc: V1503,62-70(1991)
See Kobhio, M. N.: V1397,555-558(1991)
See Kobhio, M. N.: V1503,88-97(1991)

Fontaine, Chantal
; Requena, Philippe; Munoz-Yague, Antonio: Generalization of Bragg reflector geometry: application to (Ga,Al)As - (Ca,Sr)F2 reflectors,V1362,59-66(1991)

Fontaine, Marie See Bock, Wojtek J.: V1584,157-161(1991)

Fontanella, Jean-Claude
See Gendron, Eric: V1542,297-307(1991)
See Merkle, Fritz: V1542,308-318(1991)
See Michau, Vincent: V1487,64-71(1991)

Fontanella, Marco See Fasano, Victor A.: V1428,2-12(1991)

Fontenla, Juan
See Fineschi, Silvano: V1343,376-388(1991)
See Hoover, Richard B.: V1343,389-403(1991)

Fontolliet, Ch. See Wagnieres, G.: V1525,219-236(1991)

Foo, Leslie D. See Gebelein, Rolin J.: V1564,452-463(1991)

Foor, Wesley E. See Kozaitis, Samuel P.: V1564,373-383(1991)

Foote, Marc C.
; Hunt, Brian D.; Bajuk, Louis J.: High-temperature superconducting superconductor/normal metal/superconducting devices,V1477,192-196(1991)
See Chew, Wilbert: V1477,95-100(1991)
See Hunt, Brian D.: V1394,89-95(1991)

Foote, Michael See Hadfield, Michael J.: V1478,126-144(1991)

Forber, Richard A.
See Efron, Uzi: V1455,237-247(1991)
See Sayyah, Keyvan: V1455,249-254(1991)

Forbes, Charles E. See Nikles, David E.: V1499,39-41(1991)

Forbes, Fred F.
; Roddier, Nicolas: Adaptive optics using curvature sensing,V1542,140-147(1991)
; Wong, Woon-Yin; Baldwin, Jack; Siegmund, Walter A.; Limmongkol, Siriluk; Comfort, Charles H.: Telescope enclosure flow visualization,V1532,146-160(1991)

Forbes, Glenn S.
; Morin, Richard L.; Pavlicek, William: Clinical aspects of the Mayo/IBM PACS project,V1446,318-322(1991)
See Morin, Richard L.: V1446,436-441(1991)

Forbes, Greg W.
; Jones, Andrew E.: Towards global optimization with adaptive simulated annealing,V1354,144-153(1991)

Forchel, Alfred W. See Rosenzweig, Josef: V1362,168-178(1991)

Ford, Eric H.
; Hasenauer, David M.: Narcissus in current generation FLIR systems,VCR38,95-119(1991)

Ford, Gary E.
; Paglieroni, David W.; Tsujimoto, Eric M.: Accelerated detection of image objects and their orientations with distance transforms,V1452,244-255(1991)
See Algazi, V. R.: V1452,364-370(1991)
See Algazi, V. R.: V1605,329-338(1991)

Ford, Kenneth M. See Liseth, Ole J.: V1468,256-267(1991)

Ford, William E. See Sounik, James R.: V1426,340-349(1991)

Forde, Kenneth A. See Libutti, Steven K.: V1421,169-172(1991)

Fordham, John L.
; Bellis, J. G.; Bone, David A.; Norton, Timothy J.: MIC photon counting detector,V1449,87-98(1991)

Foreman, R. J. See Jankowski, Alan F.: V1343,32-38(1991)

Foresi, James S. See Zediker, Mark S.: V1418,309-315(1991)

Forestier, Bernard M.
See Delaporte, Philippe C.: V1397,485-492(1991)
See Gerri, Mireille: V1503,280-291(1991)
See Hueber, Jean-Marc: V1503,62-70(1991)
See Kobhio, M. N.: V1397,555-558(1991)
See Kobhio, M. N.: V1503,88-97(1991)
See Zeitoun, David: V1397,585-588(1991)

Forestier, L. See Barnault, B.: V1397,231-234(1991)

Fork, David K.
; Barrera, A.; Phillips, Julia M.; Newman, N.; Fenner, David B.; Geballe, T. H.; Connell, G.A. N.; Boyce, James B.: Structural and electrical properties of epitaxial YBCO films on Si,V1394,202-202(1991)

Forkner, John F.
: Computing illumination-bundle focusing by lens systems,V1354,210-215(1991)

Forman, Scott K.
; Oz, Mehmet C.; Wong, Edison; Treat, Michael R.; Kiernan, Howard: Laser effects on fibrin clot response by human meniscal fibrochondrocytes in organ culture,V1424,2-6(1991)

Fornari, B. See Sansonetti, Pierre: V1588,198-209(1991)

Forouhar, Siamak See Larsson, Anders G.: V1418,292-297(1991)

Forrer, Martin
See Frenz, Martin: V1427,9-15(1991)
See Romano, Valerio: V1427,16-26(1991)

Forrest, Gary T.
: Emerging laser technologies,V1520,37-57(1991)
: Field-portable laser beam diagnostics,V1414,55-64(1991)
; Levitt, Morris R.; eds.: [$BFS]The Laser Marketplace 1991[$BFE],V1520(1991)

Forrest, Stephen R.
See Brown, Julia J.: V1474,236-242(1991)
See Olsen, Gregory H.: V1419,24-31(1991)

Forse, R. W. See Bourne, Lincoln C.: V1477,205-208(1991)

Forsh, A. A. See Panasyuk, L. M.: V1621,74-82(1991)

Forster, Albert A.
: Correlation tracking: a new technology applied to laser photocoagulation,V1423,103-104(1991)

Forsthoefel, John J. See Niblack, Curtiss A.: V1494,403-413(1991)

Forsyth, William B.
; Lewis, H. G.: Fusion of human vision system with mid-range IR image processing displays,V1472,18-25(1991)

Forte, Anne-Marie
; Bizais, Yves: Analyzing and interpreting pulmonary tomoscintigraphy
    sequences: realization and perspectives,V1445,409-420(1991)
See Aubry, Florent: V1446,168-176(1991)
See Bizais, Yves: V1446,156-167(1991)

Fortin, Louis See Maldague, Xavier: V1467,239-251(1991)

Fortino, Dennis J.
: Market for high-power CO2 lasers,V1517,61-84(1991)

Fortner, Russ L. See Hall, Charles S.: V1483,29-38(1991)

Fosdick, Jerilyn J.
: Essential for success: people as part of the system,VCR36,34-45(1991)

Foss, Wolfgang See Wolf, Barbara: V1506,40-51(1991)

Fossum, Eric R.
: Future directions in focal-plane signal processing for spaceborne scientific
    imagers,V1541,62-67(1991)
; Song, Jong I.; Rossi, David V.: Two-dimensional electron gas charge-
    coupled devices,V1447,202-203(1991)
See Kemeny, Sabrina E.: V1447,243-250(1991)

Foster, Richard S.
: Noninvasive ultrasound-produced volume lesion in prostate,V1421,47-
    47(1991)

Fotakis, Costas
See Hontzopoulos, Elias I.: V1397,761-768(1991)
See Moustaizis, Stavros D.: V1552,50-56(1991)

Fouassier, Jean-Pierre
: Visible-laser-light-induced polymerization: an overview of photosensitive
    formulations,V1559,76-88(1991)

Foukal, Peter
; Hoyt, Clifford C.; Jauniskis, L.: Cryogenic radiometers and intensity-
    stabilized lasers for EOS radiometric calibrations,V1493,72-79(1991)

Fountain, H. W. See Downey, George A.: V1482,40-47(1991)

Fourgette, D. C. See Dibble, R. W.: V1602(1991)

Fourier, Ron
: Accuracies in FLIR test equipment,V1442,109-117(1991)

Fouse, Timothy See Troyer, David E.: V1486,396-409(1991)

Fowler, Albert M.
; Gatley, Ian: Noise reduction strategy for hybrid IR focal-plane
    arrays,V1541,127-133(1991)

Fowler, Burt
; Lian, S.; Krishnan, S.; Jung, Le-Tien; Li, C.; Banerjee, Sanjay: Modeling of
    photochemical vapor deposition of epitaxial silicon using an ArF excimer
    laser,V1598,108-117(1991)
See Lian, S.: V1598,98-107(1991)

Fox, Alex J. See Fishell, Wallace G.: V1538,5-13(1991)

Fox, Allen M. See Jackson, John E.: V1533,75-86(1991)

Fox, Curtis F. See Disimile, Peter J.: V1492,64-75(1991)

Fox, Jack See Fried, Alan: V1433,145-156(1991)

Fox, Marye A.
; Pan, Horng-Lon: Charge separation in functionalized tetrathiafulvalene
    derivatives,V1436,2-7(1991)

Fox, Neil D.
; Maynard, William L.; Clarke, Ernest S.; Bruno, Ronald C.: GSFC conceptual
    design study for an intersatellite optical multiple access communication
    system,V1417,452-463(1991)

Foxon, C. T.
See Dawson, Philip: V1362,384-390(1991)
See Joyce, Bruce A.: V1361,13-22(1991)

Fozooni, P. See Zammit, C. C.: V1549,274-282(1991)

Fracasso, B. See Kohler, A.: V1564,236-243(1991)

Fraedrich, Douglas S.
: Design considerations for infrared imaging radiometers,V1486,2-7(1991)

Fraidenraich, Naum
: Analytic solutions and numerical results for the optical and radiative
    properties of V-trough concentrators,V1528,15-30(1991)

France, Paul W.
; Brierley, Michael C.: Progress in fluoride fiber lasers and
    amplifiers,V1373,33-39(1991)

Francese, Francesco See Donati, Silvano: V1389,665-671(1991)

Francis, Denise M. See Cunningham, Philip R.: V1414,97-129(1991)

Francis, Henry A. See Kingsbury, Edward P.: V1454,152-158(1991)

Francisco, John See Carvlin, Mark J.: V1426,13-21(1991)

Franck, Charmaine C.
; Franck, Jerome B.: Preliminary study of the admittance diagram as a useful
    tool in the design of stripline components at microwave
    frequencies,V1527,277-290(1991)

Franck, Jerome B.
; Hodgkin, Van A.: Simple design considerations for binary optical
    holographic elements,V1555,63-70(1991)
See Franck, Charmaine C.: V1527,277-290(1991)

Francois, M. See Aboulhouda, S.: V1362,494-498(1991)

Francos, Joseph M.
; Meiri, A. Z.; Porat, Boaz: Modeling of the texture structural components
    using 2-D deterministic random fields,V1606,553-565(1991)

Franek, Alexandr See Sasek, Ladislav: V1572,151-156(1991)

Franek, Joachim
; Du, Keming; Pflueger, Silke; Imhoff, Ralf; Loosen, Peter: Comparison of
    welding results with stable and unstable resonators,V1397,791-
    795(1991)
See Du, Keming: V1397,639-643(1991)

Frank, Alan M. See Tao, William C.: V1346,300-310(1991)

Frank, Arthur J.
; Glenis, Spyridon: P-n heterojunction and Schottky barrier formation
    between poly(3-methylthiophene) and n-type cadmium
    sulfide,V1436,50-57(1991)
; Lawrence, Marcus F.; Ramasesha, S.; Wamser, Carl C.; eds.:
    *Photochemistry and Photoelectrochemistry of Organic and Inorganic
    Molecular Thin Films*,V1436(1991)

Frank, Chet A. See Klainer, Stanley M.: V1434,119-126(1991)

Frank, Curtis W. See Rao, Veena: V1466,309-323(1991)

Frank, D. N. See Murray, John R.: V1410,28-39(1991)

Frank, Harry A. See Shreve, A. P.: V1403,394-399(1991)

Frank, John R. See Wojcik, Gregory L.: V1464,187-203(1991)

Frank, Klaus
; Hoeper, J.; Zuendorf, J.; Tauschek, D.; Kessler, Manfred; Wiesner, J.;
    Wokaun, Alexander J.: Analysis of multiple-multipole scattering by time-
    resolved spectroscopy and spectrometry,V1431,2-11(1991)
See Bickel, P.: V1503,167-175(1991)

Frank, L. J. See Strikwerda, Thomas E.: V1478,13-23(1991)

Frank, Werner F.
; Kulisch, Juergen R.; Franke, Hilmar; Rueck, Dorothee M.; Brunner, Stefan;
    Lessard, Roger A.: Optical waveguides in polymer materials by ion
    implantation,V1559,344-353(1991)

Franke, Hilmar
See Festl, H. G.: V1559,410-423(1991)
See Frank, Werner F.: V1559,344-353(1991)
See Osterfeld, Martin: V1559,49-55(1991)

Frankhauser, P. See Rueff, Manfred: V1382,255-270(1991)

Franklin, T. D. See Gates, James L.: V1540,262-273(1991)

Frankowski, Gottfried
: Optical profilometry using spatial heterodyne interferometric
    methods,V1500,114-123(1991)

Franks, Glen B. See Leib, Kenneth G.: V1483,140-154(1991)

Franks, James K.
: What is eye safe?,V1419,2-8(1991)

Frantzeskakis, Emmanuil N. See Baras, John S.: V1569,341-353(1991)

Franz, Juergen
ed.: *Optical Space Communication II*,V1522(1991)
See Wandernoth, Bernhard: V1522,194-198(1991)

Franza, F. See Noethe, L.: V1542,293-296(1991)

Franzen, Douglas L. See Boisrobert, Christian Y.: V1474,285-290(1991)

Fraser, George W.
; Lees, John E.; Pearson, James F.: Further observations of vectorial effects in
    the x-ray photoemission from caesium iodide,V1343,438-456(1991)
; Pain, M. D.; Pearson, James F.; Lees, John E.; Binns, C. R.; Shaw, Ping-
    Shine; Fleischman, Judith R.: Polarization sensitivity of x-ray
    photocathodes in the 60-200eV band,V1548,132-148(1991)
See Kaaret, Philip E.: V1548,106-117(1991)
See Zombeck, Martin V.: V1549,90-100(1991)

Fraser, James C. See Kimball, Paulette R.: V1540,687-698(1991)

Fraser, Laurie K. See Hall, Charles S.: V1483,29-38(1991)

Fraser Monteiro, L.
See Liang, Dawei: V1501,129-134(1991)
See Liang, Dawei: V1511,90-97(1991)

Fraser Monteiro, M. L.
See Liang, Dawei: V1501,129-134(1991)
See Liang, Dawei: V1511,90-97(1991)

Fratter, C.
; Reulet, Jean-Francois; Jouan, Jacky: SPOT 4 HRVIR instrument and future high-resolution stereo instruments,V1490,59-73(1991)

Frau, Juan
; Llario, Vicenc; Oliver, Gabriel: Polynomial regression analysis for estimating motion from image sequences,V1388,329-340(1991)

Frazier, Donald O.
; Penn, Benjamin G.; Witherow, William K.; Paley, M. S.: Growth of thin films of organic nonlinear optical materials by vapor growth processes: an overview and examination of shortfalls,V1557,86-97(1991)

Frazier, Donita L.
See Panjehpour, Masoud: V1424,179-185(1991)
See Panjehpour, Masoud: V1427,307-315(1991)

Frazier, John C. See Kennett, Ruth D.: V1501,57-68(1991)

Frazin, Leon J.
; Vonesh, Michael J.; Chandran, Krishnan B.; Khasho, Fouad; Lanza, George M.; Talano, James V.; McPherson, David D.: Doppler-guided retrograde catheterization system,V1425,207-207(1991)

Frear, Darrel R.
; Michael, J. R.; Kim, C.; Romig, A. D.; Morris, J. W.: Effect of Cu at Al grain boundaries on electromigration behavior in Al thin films,V1596,72-82(1991)

Frechet, Jean M. See Fahey, James T.: V1466,67-74(1991)

Fredin, Leif G. See Valderrama, Giuseppe L.: V1427,200-213(1991)

Fredin, Per
: Optimum choice of anamorphic ratio and boost filter parameters for a SPRITE-based infrared sensor,V1488,432-442(1991)

Fredricks, Ronald J.
; Johnson, Dean R.; Sabri, Sehbaz H.; Yu, Ming H.: Performance comparison of various low-cost multimode fiber optic rotation rate sensor designs,V1367,127-139(1991)
See Johnson, Dean R.: V1367,140-154(1991)

Fredrickson, Larry R. See Toh, Kenny K.: V1463,402-413(1991)

Freed, Charles A. See Kunz, Roderick R.: V1466,218-226(1991)

Freedman, Teresa B. See Nafie, Laurence A.: V1432,37-49(1991)

Freeland, Alan W. See Grieser, James L.: V1330,102-110(1991)

Freeman, D. K. See Fikioris, George: V1407,295-305(1991)

Freeman, David E.
: Simple method for finding the valid ray-surface intersection,V1354,200-209(1991)

Freeman, Herbert
See Chen, Shuang: V1383,2-9(1991)
See Chen, Shuang: V1570,352-361(1991)

Freeman, J. L.
; Ray, Sankar; West, David L.; Thompson, Alan G.; LaGasse, M. J.: Microwave control using a high-gain bias-free optoelectronic switch,V1476,320-325(1991)

Freeman, Mark O.
See Biernacki, Paul D.: V1455,167-178(1991)
See Duell, Kenneth A.: V1564,22-33(1991)
See Huang, Qiang: V1564,644-655(1991)

Freeman, Michael H.
: Interdependence of design, optical evaluation, and visual performance of ophthalmic lenses,V1529,2-12(1991)

Freeman, Peter W.
; Bohland, John F.; Pavelchek, Edward K.; Jones, Susan K.; Dudley, Bruce W.; Bobbio, Stephen M.: Techniques for characterization of silicon penetration during DUV surface imaging,V1464,377-385(1991)

Freeman, Richard R.
See Raab, Eric L.: V1343,104-109(1991)
See White, Donald L.: V1343,204-213(1991)

Freeman, Walter J. See Baird, Bill: V1469,12-23(1991)

Freemont, Anthony J. See Charlton, Andrew: V1427,189-197(1991)

Frehlich, Rod G.
: Measurements of the atmospheric turbulence spectrum and intermittency using laser scintillation,V1487,10-18(1991)

Freiman, Stephen W. See Cranmer, David C.: V1330,152-163(1991)

Freimanis, R. See Purinsh, Juris: V1525,289-308(1991)

Freischlad, Klaus R.
; Kuechel, Michael F.; Wiedmann, Wolfgang; Kaiser, Winfried; Mayer, Max: High-precision interferometric testing of spherical mirrors with long radius of curvature,V1332,8-17(1991)
; Kuechel, Michael F.; Schuster, Karl-Heinz; Wegmann, Ulrich; Kaiser, Winfried: Real-time wavefront measurement with lambda/10 fringe spacing for the optical shop,V1332,18-24(1991)
See Doerband, Bernd: V1332,664-672(1991)

Freishist, N. A. See Kostin, Ivan K.: V1554A,418-425(1991)

Freisinger, Bernhard
; Pauls, Markus; Schaefer, Johannes H.; Uhlenbusch, Juergen: High-power CO_2 laser excited by 2.45 GHz microwave discharges,V1397,311-318(1991)

Freitas, Jose C.
; Abreu, M. A.; Rodrigues, F. C.; Carvalho, Fernando D.: Misalignments of airborne laser beams due to mechanical vibrations,V1399,42-48(1991)
; Carvalho, Fernando D.; Rodrigues, F. C.; Abreu, M. A.; Marcal, Joao P.: Far-field pattern of laser diodes as function of the relative atmospheric humidity,V1399,16-23(1991)
See Carvalho, Fernando D.: V1399,130-136(1991)
See de Lourdes Quinta, Maria: V1399,24-29(1991)
See Rodrigues, F. C.: V1399,90-97(1991)

French, Howard B. See Rouse, Gordon F.: V1495,240-245(1991)

French, Larry E. See Brewer, Joe E.: V1390,164-174(1991)

Frentzen, Matthias See Koort, Hans J.: V1424,87-98(1991)

Frenz, Martin
; Greber, Charlotte M.; Romano, Valerio; Forrer, Martin; Weber, Heinz P.: Damage induced by pulsed IR laser radiation at transitions between different tissues,V1427,9-15(1991)
See Romano, Valerio: V1427,16-26(1991)

Frenzl, Otto See Engemann, Detlef: V1362,710-720(1991)

Fresne, Francoise
; Le Gall, G.; Barillot, Christian; Gibaud, Bernard; Manens, J. P.; Toumoulin, Christine; Lemoine, D.; Chenal, C.; Scarabin, Jean-Marie: Stereotactic multibeam radiation therapy system in a PACS environment,V1444,26-36(1991)

Freund, Andreas K.
: High-energy x-ray diffraction,V1345,234-244(1991)

Freund, Harold G. See Sharma, Minoti: V1435,280-291(1991)

Frew, Dean L.
: High-density memory packaging technology high-speed imaging applications,V1346,200-209(1991)

Friart, Daniel See Fleurot, Noel A.: V1502,230-241(1991)

Friboulet, Denis
; Magnin, Isabelle E.; Mathieu, Christophe; Revel, D.; Amiel, Michel: Model of the left ventricle 3-D global motion: application to MRI data,V1445,106-117(1991)

Fridling, Barry E.
; Drummond, Oliver E.: Performance evaluation methods for multiple-target-tracking algorithms,V1481,371-383(1991)

Friebele, E. J.
See Baldini, S. E.: V1588,125-131(1991)
See Merzbacher, Celia I.: V1533,222-228(1991)

Fried, Alan
; Drummond, James R.; Henry, Bruce; Fox, Jack: Tunable diode laser spectrometer for high-precision concentration and ratio measurements of long-lived atmospheric gases,V1433,145-156(1991)

Fried, David L.
; Szeto, Roque K.: Measurement of wind velocity spread: signal-to-noise ratio for heterodyne detection of laser backscatter from aerosol,V1416,163-176(1991)
; Szeto, Roque K.: PCI (phase compensation instability) and minishear,V1408,150-166(1991)

Frieden, B. R.
: Some analytical and statistical properties of Fisher information,V1569,311-316(1991)
See Bocker, Richard P.: V1569,298-310(1991)

Friedenberg, Abraham
: Optimization of point source detection,V1442,60-65(1991)

Friedl, Stephan E.
; Kunz, Warren F.; Mathews, Eric D.; Abela, George S.: Integral prism-tipped optical fibers,V1425,134-141(1991)

Friedland, Bernard See Haessig, David A.: V1482,383-396(1991)

Friedlander, Miles H.
; Mulet, Miguel; Buzard, Kurt A.; Granet, Nicole; Baker, Phillip C.: Holographic interferometry of the corneal surface,V1423,62-69(1991); 1429,229-236(1991)

Friedman, Aharon See Eidelman, Shmuel: V1567,439-450(1991)

Friedman, David H. See Smotroff, Ira G.: V1469,544-550(1991)

Friedman, Lionel R. See Soref, Richard A.: V1389,408-421(1991)

Friedman, Moshe
; Serlin, Victor; Lau, Yue Y.; Krall, Jonathan: Large electron accelerators powered by intense relativistic electron beams,V1407,474-478(1991)
; Serlin, Victor; Lau, Yue Y.; Krall, Jonathan: Relativistic klystron amplifier I: high-power operation,V1407,2-7(1991)
See Colombant, Denis G.: V1407,13-22(1991)
See Krall, Jonathan: V1407,23-31(1991)
See Serlin, Victor: V1407,8-12(1991)

Friedman, Richard M. See Betts, Timothy C.: V1479,120-126(1991)

Friedmann, Harry
; Lubart, Rachel; Laulicht, Israel; Rochkind, Simeone: Toward an explanation of laser-induced stimulation and damage of cell cultures,V1427,357-362(1991)
See Lubart, Rachel: V1422,140-146(1991)
See Rochkind, Simeone: V1428,52-58(1991)

Friedrich, E. A. See Kohl, M.: V1525,26-34(1991)

Friesel, Mark A. See Barga, Roger S.: V1469,602-611(1991)

Friesem, Asher A.
See Blank, R.: V1442,26-30(1991)
See Davidson, Nir: V1442,22-25(1991)
See Elizur, Eran: V1442,230-234(1991)
See Hasman, Erez: V1442,372-377(1991)
See Kinrot, O.: V1442,106-108(1991)
See Shariv, I.: V1442,258-263(1991)

Frietman, Edward E.
; Dekker, L.; van Nifterick, W.; Demeester, Piet M.; van Daele, Peter; Smit, W.: Current status and future research of the Delft 'supercomputer' project,V1390,434-453(1991)
; Dekker, L.; van Nifterick, W.: Optical data communication compel the design of a new class of storage media,V1401,19-26(1991)
See Dekker, L.: V1505,148-157(1991)

Frijlink, P. M. See Martin, Gerard M.: V1362,67-74(1991)

Fripp, Archibald L.
; Debnam, W. J.; Woodell, G. W.; Berry, R. F.; Simchick, Richard T.; Sorokach, S. K.; Barber, Patrick G.: Characterization of the Bridgman crystal growth process by radiographic imaging,V1557,236-244(1991)
See Steiner, Bruce W.: V1557,156-167(1991)

Frishter, L. See Savostjanov, V. N.: V1554A,380-386(1991)

Frisoli, Joan K. See Flotte, Thomas J.: V1427,36-44(1991)

Fritsch, Klaus See Beheim, Glenn: V1369,50-59(1991)

Fritz, Bernard S.
See Cox, James A.: V1507,100-109(1991)
See Cox, James A.: V1555,80-88(1991)

Fritzsche, Hellmut See Chen, Kun J.: V1519,632-639(1991)

Froehlich, Fred F.
See Milster, Tom D.: V1499,348-353(1991)
See Milster, Tom D.: V1499,286-292(1991)

Froehlich, Helmut H.
: Eurosprint proofing system,V1458,51-60(1991)

Froehlich, K.
See Stojanoff, Christo G.: V1536,206-214(1991)
See Stojanoff, Christo G.: V1527,48-60(1991)

Froehly, Claude
; Laucournet, A.; Miehe, Joseph A.; Rebuffie, Jean-Claude; Roth, J. M.; Tomasini, F.: First results on a developmental deflection tube and its associated electronics for streak camera applications,V1358,532-540(1991)

Froelich, Klaus See Kubitzek, Ruediger: V1507,365-372(1991)

Frohlich, Thomas
See Chan, Cheuk L.: V1450,208-217(1991)
See Kwak, J. Y.: V1396,32-44(1991)

Froimson, Mark I. See Garino, Jonathan P.: V1424,43-47(1991)

Frolov, G. I. See Aleksandrov, K. S.: V1621,51-61(1991)

Frolov, I. P. See Anisimov, N. R.: V1440,206-210(1991)

Frolov, K.
; Ionin, A. A.; Kelner, M.; Sinitsin, D. V.; Suchkov, A. F.; Zhivukcin, I.: High-power electron beam controlled discharge N2O laser,V1397,461-468(1991)

Frolov, Sergey I. See Gavrilenko, Vladimir I.: V1361,171-182(1991)

Frolov, V. V. See Veiko, Vadim P.: V1544,152-163(1991)

Frontera, Filippo
; De Chiara, P.; Gambaccini, M.; Landini, G.; Pasqualini, G.: Hard x-ray imaging via crystal diffraction: first results of reflectivity measurements,V1549,113-119(1991)

Frontov, H. N.
; Serdyuchenko, Yuri N.: Multiframing image converter camera,V1358,311-314(1991)

Frost, David P. See Perry, Meg W.: V1459,155-156(1991)

Frost, Meryll M.
See Honeyman, Janice C.: V1446,362-368(1991)
See Staab, Edward V.: V1446,16-22(1991)

Frost, Paul A.
: Discrimination and classification with Xybion multispectral video systems,V1358,398-408(1991)

Frost, Robert L. See McGee, Thomas J.: V1491,182-187(1991)

Frost, Thorsten See Beeck, Manfred-Andreas: V1507,394-406(1991)

Fry, David A.
; Lucoro, Jacobo P.: Deflection evaluation using time-resolved radiography,V1346,270-275(1991)

Fryska, Slawomir T. See Griffen, Christopher T.: V1554B,754-766(1991)

Fu, Bao W. See Wang, Ren: V1519,146-151(1991)

Fu, Huimin See Xue, Kefu: V1606,675-684(1991)

Fu, Richard J.
; Hong, C. S.; Chan, Eric Y.; Booher, Dan J.; Figueroa, Luis: Optimization of strained layer InGaAs/GaAs quantum-well lasers,V1418,108-115(1991)

Fu, Shufen See Bollanti, Sarah: V1503,80-87(1991)

Fu, Tao-Yi See Wiley, James N.: V1464,346-355(1991)

Fu, Xin See Zhong, Xian-Xin: V1572,84-87(1991)

Fua, Pascal
; Sander, Peter T.: From points to surfaces,V1570,286-296(1991)

Fuchs, Baruch A. See Kozlowski, Mark R.: V1561,59-69(1991)

Fuchs, Elizabeth A.
; Mahin, K. W.; Ortega, A. R.; Bertram, L. A.; Williams, Dean R.; Pomplun, Alan R.: Thermal diagnostics for monitoring welding parameters in real time,V1467,136-149(1991)

Fuchs, Jacob See Stewart, Diane K.: V1465,64-77(1991)

Fuensalida, Jesus J. See Michau, Vincent: V1487,64-71(1991)

Fuentes, Rosa See Carretero, L.: V1507,458-462(1991); 1508,96-100(1991)

Fuerstenau, Norbert
; Schmidt, Walter: All-optical data-input device based on fiber optic interferometric strain gauges,V1367,357-366(1991)

Fugate, Robert D. See Koren, Eugen: V1428,214-223(1991)

Fugera, S. N. See Gilbreath, G. C.: V1409,87-99(1991)

Fuhr, Peter L. See Spillman, William B.: V1332,591-601(1991)

Fuhrman, Carl R.
; King, Jill L.; Obuchowski, Nancy A.; Rockette, Howard E.; Sashin, Donald; Harris, Kathleen M.; Gur, David: Correlation between the detection and interpretation of image features,V1446,422-429(1991)

Fujihashi, Chugo
: Correction method for optical-signal detection-error caused by quantum noise,V1583,298-306(1991)

Fujii, H.
; Iizuka, Masahiro; Muro, Mikio; Kuchiki, Hirotsuna; Atsuta, Toshio: Development of chemical oxygen-iodine laser for industrial application,V1397,213-220(1991)

Fujii, Takashi
; Nemoto, Koshichi; Ishikawa, Rikio; Hayashi, Kazuo; Noda, Etsuo: Development of solid state pulse power supply for copper vapor laser,V1412,50-57(1991)

Fujii, Tetsurou
; Sawabe, Tomoko; Ohta, Naohisa; Ono, Sadayasu: Superhigh-definition image processing on a parallel signal processing system,V1605,339-350(1991)

Fujii, Yumi See Ohhashi, Akinami: V1444,63-74(1991)

Fujime, Satoru
; Miyamoto, Shigeaki: Elasticity of biomembranes studied by dynamic light scattering,V1403,306-315(1991)

Fujimoto, James G.
; Schulz, Peter A.; Fan, Tso Y.: Ultrashort pulse generation in solid state lasers,V1413,2-13(1991)

Fujimoto, Masashi See Shimizu, Masaru: V1519,122-127(1991)

Fujimoto, R. See Kashiwabara, S.: V1397,803-806(1991)

Fujimoto, Tsuyoshi
; Shoman, Mineo; Hase, Masahiko: Method for removing background regions from moving images,V1606,599-606(1991)

Fujimoto, Yoshiaki See Tamura, Akira: V1465,271-281(1991)

Fujimura, Kikuo
: Safe motion planning for mobile agents: a model of reactive planning for multiple mobile agents,V1388,260-269(1991)

Fujinaga, Masato
See Miyazaki, Junji: V1464,327-335(1991)
See Op de Beeck, Maaike: V1463,180-196(1991)

Fujioka, Tomoo See Ream, Stanley L.: V1601(1991)

Fujisada, Hiroyuki
: High-sensitive thermal video camera with self-scanned 128 InSb linear array,V1540,665-676(1991)
; Ono, Akira: Overview of ASTER design concept,V1490,244-254(1991)
See Kawada, Masakuni: V1490,299-308(1991)
See Nakayama, Masao: V1490,207-215(1991)

Fujise, Masayuki
; Araki, Ken'ichi; Arikawa, Hiroshi; Furuhama, Yoji: Current and future activities in the area of optical space communications in Japan,V1522,14-26(1991)
See Inagaki, Keizo: V1417,160-169(1991)
See Nohara, Mitsuo: V1417,338-345(1991)

Fujishige, Jack T. See Neev, Joseph: V1427,162-172(1991)

Fujisima, Toshihiko See Ito, Katsunori: V1499,382-385(1991)

Fujita, Goro
; Urakawa, Yoshinori; Yamagami, Tamotsu; Watanabe, Tetsu: New approach to high-density recording on a magneto-optical disk,V1499,426-432(1991)

Fujita, Hiroo
: Two-dimensional micropattern measurement using precision laser beam scanning,V1332,456-467(1991)

Fujita, Hiroshi See Ogura, Toshihiro: V1443,153-157(1991)

Fujita, Takeyoshi See Nakahara, Sumio: V1554A,602-609(1991)

Fujita, Teruo See Irie, Mitsuru: V1499,360-365(1991)

Fujiwara, Hirofumi
: Real-time interferometry by optical phase conjugation in dye-doped film,V1507,492-499(1991)
See Nakagawa, Kazuo: V1332,267-273(1991)

Fujiwara, Hiroshi See Uwabu, H.: V1605,928-937(1991)

Fujiwara, Kazuo See Yoshimura, Takeaki: V1332,835-842(1991)

Fujiwara, Keiji See Kawai, Akira: V1464,267-277(1991)

Fukada, Youji
: Shape reconstruction and object recognition using angles in an image,V1381,111-121(1991)

Fukamachi, Hideo See Akamatsu, Shigeru: V1606,204-216(1991)

Fukami, T. See Tsutsumi, K.: V1499,55-61(1991)

Fukuda, Kyohei See Osawa, Atsuo: V1354,337-343(1991)

Fukuda, Yasuaki See Niibe, Masahito: V1343,2-13(1991)

Fukui, Masahiko See Katayama, Norihisa: V1403,147-149(1991)

Fukumoto, Atsushi
; Aratani, Katsuhisa; Yoshimura, Shunji; Udagawa, Toshiki; Ohta, Masumi; Kaneko, Masahiko: Super resolution in a magneto-optical disk with an active mask,V1499,216-225(1991)
See Aratani, Katsuhisa: V1499,209-215(1991)

Fukunaga, Seiki
; Kitaori, Tomoyuki; Koyanagi, Hiroo; Umeda, Shin'ichi; Nagasawa, Kohtaro: Novel quinonediazide-sensitized photoresist system for i-line and deep-UV lithography,V1466,446-457(1991)
See Koyanagi, Hiroo: V1466,346-361(1991)

Fukuoka, Takashi
; Tejika, Yasuhiro; Takada, Hisashi; Takahashi, Hidenori; Hamasaki, Yiji: Jitter considerations for FDDI PMD,V1364,40-48(1991)

Fukushima, Mitsugi See Sukeda, Hirofumi: V1499,419-425(1991)

Fulga, Florin
See Nicolau, Dan V.: V1465,282-288(1991)
See Nicolau, Dan V.: V1468,345-351(1991)
See Nicolau, Dan V.: V1466,663-669(1991)

Fulghum, Stephen See Banholzer, William: V1501,163-176(1991)

Fuller, James M.
See Auborn, John E.: V1482,246-252(1991)
See Holland, Orgal T.: V1469,102-112(1991)

Fuller, Peter W.
ed.: *19th Intl Congress on High-Speed Photography and Photonics*,V1358(1991)

Fulmer, Gary G. See Bugner, Douglas E.: V1458,162-178(1991)

Fulop, Ann C.
; Allen, Donald M.; Deffner, Gerhard: Visualization of manufacturing process data in N-dimensional spaces: a reanalysis of the data,V1459,69-76(1991)

Fulton, Robert E.
; Hughes, Joseph L.; Scott, Waymond R.; Umeagukwu, Charles; Yeh, Chao-Pin: Multidisciplinary analysis and design of printed wiring boards,V1389,144-155(1991)
See Yeh, Chao-Pin: V1389,187-198(1991)

Funakubo, H. See Bays, Roland: V1525,397-408(1991)

Fung, Bing M. See Heavin, Scott D.: V1455,12-18(1991)

Fung, Jackie S.
: Novel angular discriminator for spatial tracking in free-space laser communications,V1417,224-232(1991)

Funk, Eric E. See Kung, Chun C.: V1378,250-258(1991)

Funkenbusch, Arnold W.
: Magneto-optic data storage in the '90s,V1396,699-708(1991)

Furgiuele, Franco M.
; Lamberti, Antonio; Pagnotta, Leonard: Reducing errors in bending tests of ceramic materials,V1554A,275-284(1991)

Furman, Geoffrey G. See Clark, David L.: V1540,303-311(1991)

Furuhama, Yoji
See Fujise, Masayuki: V1522,14-26(1991)
See Inagaki, Keizo: V1417,160-169(1991)

Furukawa, Takao See Ishiwata, Naoyuki: V1463,423-433(1991)

Furusho, N. See Kuroda, Masami: V1458,155-161(1991)

Furuta, Akihiro See Hanabata, Makoto: V1466,132-140(1991)

Furuta, Mitsuhiro
; Asaumi, Shingo; Yokota, Akira: Mechanism of dissolution inhibition of novolak-diazoquinone resist,V1466,477-484(1991)

Furutani, Tadashige See Ohara, Shunji: V1499,187-194(1991)

Furzikov, Nickolay P.
: Effective temperatures of polymer laser ablation,V1503,231-235(1991)
: Physical processes of laser tissue ablation,V1403,764-775(1991)
: Self-reflection and self-transmission of pulsed radiation by laser-evaporated media,V1415,228-239(1991)
; Dmitriev, A. C.; Lekhtsier, Eugeny N.; Orlov, M. Y.; Semyenov, Alexander D.; Tyurin, Vladimir S.: Peak pressures and temperatures within laser-ablated tissues,V1427,288-297(1991)

Fushman, D. A. See Balabaev, N. K.: V1402,53-69(1991)

Futamata, Akio See Ozawa, Yasuyuki: V1499,136-142(1991)

Futterman, Walter I.
; Schweitzer, Eric L.; Newt, J. E.: Estimation of scene correlation lengths,V1486,127-140(1991)
See Strugala, Lisa A.: V1486,176-187(1991)

Fuzessy, Zoltan
; Gyi'mesi, Ferenc: Difference holographic interferometry: an overview,VIS08,194-204(1991)

Fyler, Ann See Sadler, Lewis L.: V1380,137-146(1991)

Fyodorov, Vyatcheslav B. See Burtsev, Vsevolod S.: V1621,215-226(1991)

Gabardi, David R. See Hoover, Richard B.: V1343,189-202(1991)

Gabay, Amnon See Yatsiv, Shaul: V1397,319-329(1991)

Gabay, Shimon
See Lando, Mordechai: V1412,19-26(1991)
See Lando, Mordechai: V1442,172-180(1991)

Gabbouj, Moncef See Zeng, Bing: V1606,443-454(1991)

Gaboardi, Franco
; Bozzola, Andrea; Zago, Tiziano; Gulfi, Gildo M.; Galli, Luigi: Recanalization of azoospermia due to a Mullerian duct cyst by Nd:YAG laser,V1421,73-77(1991)
; Dotti, Ernesto; Bozzola, Andrea; Galli, Luigi: Nd:YAG laser treatment in patients with prostatic adenocarcinoma stage A,V1421,50-55(1991)
; Melodia, Tommaso; Bozzola, Andrea; Galli, Luigi: Nd:YAG laser irradiation in the treatment of upper tract urothelial tumors,V1421,79-85(1991)

Gabor, A. M. See Perlmutter, S. H.: V1562,74-84(1991)

Gabric, R. See Hilton, Peter J.: V1385,27-31(1991)

Gabriel, Philip
; Stephens, Graeme L.; Tsay, Si-Chee: Radiative transfer in the cloudy atmosphere: modeling radiative transport,V1558,76-90(1991)

Gacon, Jean-Claude See Briancon, Anne-Marie: V1373,9-20(1991)

Gadagkar, Hrishikesh P.
; Trivedi, Mohan M.: Toward tactile sensor-based exploration in a robotic environment,V1383,142-150(1991)

Gader, Paul D.
; Dougherty, Edward R.; eds.: [$BFS]Image Algebra and Morphological Image Processing II[$BFE],V1568(1991)
See Pont, W. F.: V1568,247-260(1991)

Gadomski, Wojciech See Raymer, M. G.: V1376,128-131(1991)

Gadonas, R.
See Chirvony, V. S.: V1403,504-506(1991)
See Chirvony, V. S.: V1403,638-640(1991)

Gadret, G.
; Kajzar, Francois; Raimond, P.: Nonlinear optical properties of poled polymers,V1560,226-237(1991)
See Morichere, D.: V1560,214-225(1991)

Gaffard, Jean-Paul
; Boyer, Corinne: Adaptive optical transfer function modeling,V1483,92-103(1991)
; Ledanois, Guy: Adaptive optical transfer function modeling,V1542,34-45(1991)
See Boyer, Corinne: V1483,77-91(1991); 1542,46-61(1991)
See Gendron, Eric: V1542,297-307(1991)
See Merkle, Fritz: V1542,308-318(1991)

Gafitullina, Dilyara See Ashurov, Mukhsin K.: V1550,50-54(1991)

Gafney, Harry D.
See Mendoza, Edgar A.: V1378,139-144(1991)
See Mendoza, Edgar A.: V1583,43-51(1991)

Gafni, Gabriella See Azoulay, Moshe: V1535,35-45(1991)

Gage, Edward C. See Kay, David B.: V1499,281-285(1991)

Gagliardi, F. J. See Al-Jumaily, Ghanim A.: V1441,360-365(1991)

Gagliardi, Robert M.
See Mendez, Antonio J.: V1369,67-71(1991)
See Mendez, Antonio J.: V1364,163-169(1991)

Gagnaire, A.
; Ponsard, Benoit: Evolution of the DQDB hybrid multiplexing for an integrated service packetized traffic,V1364,277-288(1991)

Gagne, Philippe
; Arsenault, Henri H.: Using information from multiple low-resolution images to increase resolution,V1564,656-663(1991)

Gagne, Simon See Wang, Shengrui: V1382,37-48(1991)

Gagnon, Daniel See Ouellette, Francois: V1516,2-13(1991)

Gahl, John M. See Schamiloglu, Edl: V1407,242-248(1991)

Gai, Nevile See Acharya, Raj S.: V1445,207-221(1991)

Gaiduk, Mark I.
; Grigoryants, V. V.; Chernousova, I. V.; Menenkov, V. D.: Study of biological objects in the reflected light with the help of an analogous fiber optic biophotometer,V1403,674-675(1991)

Gaignebet, Jean See Lund, Glenn I.: V1492,166-175(1991)

Gailitis, Raymond P. See Ren, Qiushi: V1423,129-139(1991)

Gaines, Charles S. See Windham, William R.: V1379,39-44(1991)

Gainutdinov, R. K. See Logak, L. G.: V1238,171-176(1991)

Gaither, Gary L. See Greene, Ben: V1492,126-139(1991)

Gajda, Danuta See Wolinski, Wieslaw: V1391(1991)

Gajda, Jerzy K.
; Niesterowicz, Andrzej: Magneto-optical optical fiber switch,V1391,329-331(1991)
See Wolinski, Wieslaw: V1391(1991)

Gakamsky, D. M.
; Goldin, A. A.; Petrov, E. P.; Rubinov, A. N.: Analysis of fluorescence kinetics and the computing of the decay time distribution,V1403,641-643(1991)

Gal, Eli
See Afik, Zvi: V1442,392-398(1991)
See Meidan, Moshe: V1540,729-737(1991)

Gal, George See Adams, Frank W.: V1541,24-37(1991)

Galanakis, Claire T. See Mueller, Heinrich G.: V1598,132-140(1991)

Galarneau, Pierre See Snell, Kevin J.: V1410,99-106(1991)

Galas, Jacek See Blocki, Narcyz: V1562,172-183(1991)

Galatsanos, Nikolas P.
; Katsaggelos, Aggelos K.: Properties of different estimates of the regularizing parameter for the least-squares image restoration problem,V1396,590-600(1991)
See Efstratiadis, Serafim N.: V1605,16-25(1991)

Galbraith, Ian See Dawson, Philip: V1362,384-390(1991)

Galburt, Daniel N. See Zernike, Frits: V1343,241-244(1991)

Gale, Michael T.
; Lang, Graham K.; Raynor, Jeffrey M.; Schuetz, Helmut: Fabrication of micro-optical components by laser beam writing in photoresist,V1506,65-70(1991)

Galiberti, Sandra See Viligiardi, Riccardo: V1425,72-74(1991)

Galic, George J.
: Improved plastic molding technology for magneto-optical disk substrate,V1396,539-546(1991)
; Maus, Steven: Improved plastic molding technology for ophthalmic lens and contact lens,V1529,13-21(1991)

Galkin, S. L. See Ignatyev, Alexander V.: V1584,336-345(1991)

Gall, Pascal
See Lopez, Adolphe: V1590,191-202(1991)
See Montgomery, Paul C.: V1332,563-570(1991)

Gallagher, Dennis J.
; Cash, Webster C.; Green, James C.: Fabrication of a grazing-incidence telescope by grinding and polishing techniques on aluminum,V1343,155-161(1991)

Gallagher, Neal C.
: Binary optics in the '90s,V1396,722-733(1991)
See Hawley, Robert W.: V1451,91-100(1991)

Gallant, David J. See Largent, Craig C.: V1418,40-45(1991)

Gallant, J. T. L. See Van Essen, David: V1473,17-28(1991)

Gallarda, Harry S.
; Jain, Ramesh C.: Computational model of the imaging process in scanning-x microscopy,V1464,459-473(1991)

Gallerano, Gian P. See Ciocci, Franco: V1501,154-162(1991)

Galli, Luigi
See Gaboardi, Franco: V1421,79-85(1991)
See Gaboardi, Franco: V1421,50-55(1991)
See Gaboardi, Franco: V1421,73-77(1991)

Galligan, E. A. See Deutsch, Alina: V1389,161-176(1991)

Gallois, B.
See Chern, Chyi S.: V1394,255-265(1991)
See Zhao, Jing-Fu: V1362,135-143(1991)

Galloway, Peter C.
; Dobson, Peter J.: Holographic microlenses for optical fiber interconnects,V1365,131-138(1991)

Galloway, Robert L.
; Edwards, Charles A.; Thomas, Judith G.; Schreiner, Steven; Maciunas, Robert J.: New device for interactive image-guided surgery,V1444,9-18(1991)
See Edwards, Charles A.: V1444,38-46(1991)
See Thomas, Judith G.: V1444,379-388(1991)

Gallup, R. F. See Fong, C. Y.: V1361,479-488(1991)

Galpern, A. D.
; Smaev, V. P.; Paramonov, A. A.; Kiriencko, Yu. A.: Relief-phase colored hologram registration,V1238,320-323(1991)
See Smaev, V. P.: V1238,311-315(1991)

Galuyev, Gennady A. See Kalyayev, Anatoli V.: V1621,299-308(1991)

Gamaly, Eugene G.
; Kiselev, A. Y.; Tikhonchuk, V. T.: Plasma heating by ultrashort laser pulse in the regime of anomalous skin effect,V1413,95-106(1991)

Gambaccini, M. See Frontera, Filippo: V1549,113-119(1991)

Gamble, Harold S. See Ruddell, F. H.: V1361,159-170(1991)

Gambling, William A. See Lin, Jin T.: V1373,42-53(1991)

Gambogi, William J.
; Gerstadt, William A.; Mackara, Steven R.; Weber, Andrew M.: Holographic transmission elements using improved photopolymer films,V1555,256-267(1991)

Games, Richard A.
; Eastman, Willard L.; Sousa, Michael J.: Fast algorithm and architecture for constrained adaptive sidelobe cancellation,V1566,323-328(1991)

Gammarino, Rudolph R.
; Surhigh, James W.: Battlefield training in impaired visibility,V1419,115-125(1991)

Gan, Fuxi
: Advanced thin films for optical storage,V1519,530-538(1991)
See Cui, Jie: V1519,652-655(1991)
See Hou, Li S.: V1519,580-588(1991)
See Luo, Tao: V1519,826-830(1991)
See Sun, Yang: V1519,554-558(1991)
See Xiang, Xiao L.: V1519,712-716(1991)
See Xue, Song S.: V1519,570-574(1991)

Gan, K. J.
See Wu, T. S.: V1361,23-33(1991)
See Wu, T. S.: V1519,263-268(1991)

Gan, Michael A.
; Potyemin, Igor S.; Poszinskaja, Irina I.: Apo-tele lenses with kinoform elements,V1507,116-125(1991)
; Potyemin, Igor S.; Perveev, Anatoly F.: High-speed apo-lens with kinoform element,V1574,243-249(1991)
; Potyemin, Igor S.: Holographic testing canal of adaptive optical systems,V1507,549-560(1991)
; Zhdanov, Dmitriy D.; Novoselskiy, Vadim V.; Ustinov, Sergey I.; Fiodorov, Alexander O.; Potyemin, Igor S.; Bezdidko, Sergey N.: Design of optical systems with HOE by DEMOS program,V1574,254-260(1991)

Gan, Qing
; Miyahara, Makoto; Kotani, Kazunori: Characteristic analysis of color information based on (R,G,B)-> (H,V,C) color space transformation,V1605,374-381(1991)

Ganago, A. O. See Shuvalov, V. A.: V1403,400-406(1991)

Gang, Yuan
: Experimental investigation of 1.06 um laser interaction with Al target in air,V1415,225-227(1991)

Ganibal, Christian
See Dilhac, Jean-Marie R.: V1393,395-403(1991)
See Dilhac, Jean-Marie R.: V1393,349-353(1991)

Ganichev, Sergey D. See Beregulin, Eugene V.: V1362,853-862(1991)

Ganikhanov, Ferous S. See Zaitseva, N. P.: V1402,223-230(1991)

Ganley, J. T. See Jackson, John E.: V1533,75-86(1991)

Gannot, Israel
; Dror, Jacob; Dahan, Reuben; Alaluf, M.; Croitoru, Nathan I.: Characterization and uses of plastic hollow fibers for CO2 laser energy transmission,V1442,156-161(1991)
See Calderon, S.: V1420,108-115(1991)

Gantchev, Ts. See Angelov, D.: V1403,572-574(1991)

Ganzherli, N. M. See Denisyuk, Yuri N.: V1238,2-12(1991)

Gao, Er-xin
: Experimental research on the casing-shaped charge,V1358,1115-1119(1991)

Gao, Hangjun See Zhang, Zhongxian: V1572,5-10(1991)

Gao, Jian C.
; Lin, Zhi H.; Han, Li Y.: Study of phase transition VO2 thin film,V1519,159-163(1991)

Gao, Jianxing See Zhou, Xingeng: V1554A,886-895(1991)

Gao, Ning See Cui, DaFu: V1572,386-391(1991)

Gao, Penfei See Dong, Benhan: V1554A,400-406(1991)

Gao, Runsheng See Xu, Kewei: V1519,765-770(1991)

Gao, Shao-hong See Sun, Fang-kui: V1488,2-5(1991)

Gao, Shengshen See Hou, Xun: V1358,868-873(1991)

Gao, Wei See Xiao, Wen: V1572,170-174(1991)

Gao, Wenliang
; Zhang, Xiao; Yang, Guoguang: Frequency spectrum analysis and assessment of optical surface flaws,V1530,118-128(1991)
See Yang, Guoguang: V1332,56-63(1991)

Gao, Xing See Grossman, Barry G.: V1588,64-75(1991)

Gao, Yiqing
; Liu, Yupin: Diagnostics of arc plasma by moire deflectometry,V1554B,193-199(1991)

Gao, Yu-ping See Nie, Chao-Jiang: V1584,87-93(1991)

Gaponov, Sergey V. See Brailovsky, A. B.: V1440,84-89(1991)

Garafalo, David A. See Pan, J. J.: V1371,21-35(1991)

Garber, Frederick D.
See Jouny, Ismail: V1471,142-153(1991)
See Snorrason, Ogmundur: V1471,116-127(1991)

Garbi, Maurizio See Nava, Enzo: V1417,307-315(1991)

Garbo, Greta M. See Morgan, Alan R.: V1426,350-355(1991)

Garbus, Jenny See Clapham, Terrance N.: V1423,2-7(1991)

Garbuzov, Dmitriy Z.
; Goncharov, S. E.; Il'in, Y. V.; Mikhailov, A. V.; Ovchinnikov, Alexander V.; Pikhtin, N. A.; Tarasov, I. S.: 1-W cw separate confinement InGaAsP/InP (lamda = 1.3 um) laser diodes and their coupling with optical fibers,V1418,386-393(1991)

Garcia, A. R. See Wu, Xin D.: V1477,8-14(1991)

Garcia, Ismel A.
See Paisley, Dennis L.: V1358,760-765(1991)
See Paisley, Dennis L.: V1346,172-178(1991)

Garcia, Javier
; Ferreira, Carlos; Szoplik, Tomasz: Elliptical coordinate transformed phase-only filter for shift and scale invariant pattern recognition,V1574,133-140(1991)

Garcia, V. M. See Nair, Padmanabhan K.: V1485,228-239(1991)

Garcia-Prieto, Rafael
; Spiero, Francois; Popescu, Alexandru F.: Image processing techniques for aquisition of faint point laser targets in space,V1522,267-276(1991)

Garcia-Ramos, J. V. See Sanches-Cortes, S.: V1403,142-145(1991)

Gardam, Allan
; Jonas, Reginald P.: Stability considerations in relay lens design for optical communications,V1417,381-390(1991)

Gardiner, Peter T.
: Activities at the Smart Structures Research Institute,V1588,314-324(1991)
: Fiber optic rotary position sensors for vehicle and propulsion controls,V1374,200-210(1991)

Gardiner, R. See Hamaguchi, Norihito: V1394,244-254(1991)

Gardner, Daniel C. See Dente, Gregory C.: V1397,403-408(1991)

Gardner, David L. See Brown, David A.: V1584,328-335(1991)

Gardner, J. T. See Brown, Julia J.: V1474,236-242(1991)

Gardner, Kenneth R.
; Kasky, Thomas J.: Synthesis of a maneuver detector and adaptive gain tracking filter,V1481,254-260(1991)

Gardner, Leo R. See Wang, Dexter: V1479,57-70(1991)

Gardner, S. See Larsson, T.: V1387,165-168(1991)

Gardner, Sheldon
: Polynomial neural nets for signal and image processing in chaotic backgrounds,V1567,451-463(1991)

Gardner, Steven D. See Upchurch, Billy T.: V1416,21-29(1991)

Gardopee, George J. See Paquin, Roger A.: V1485,39-45(1991)

Garfield, Brian R.
; Riches, Mark J.: ULTRANAC: a new programmable image converter framing camera,V1358,290-299(1991)

Gargaro, A. R. See Barron, L. D.: V1403,66-75(1991)

Gariaev, Peter P.
; Chudin, Viktor I.; Komissarov, Gennady G.; Berezin, Andrey A.; Vasiliev, Antoly A.: Holographic associative memory of biological systems,V1621,280-291(1991)

Garifo, Luciano See Serri, Laura: V1397,469-472(1991)

Garino, A. See Morvan, D.: V1397,689-692(1991)

Garino, Jonathan P.
; Nazarian, David; Froimson, Mark I.; Grelsamer, Ronald P.; Treat, Michael R.: Comparison of the ablation of polymethylmethacrylate by two fiber-optic-compatible infrared lasers,V1424,43-47(1991)

Garito, Anthony F.
See Heflin, James R.: V1497,398-407(1991)
See Heflin, James R.: V1560,2-12(1991)
See Wada, Tatsuo: V1560,162-171(1991)

Garlick, Dean S. See Heyler, Gene A.: V1481,198-208(1991)

Garmon, Jeff P. See Belcher, Melvin L.: V1476,224-233(1991)

Garner, Harry W. See Troeltzsch, John R.: V1494,9-15(1991)

Garner, Steven T.
; Cantrell, Cyrus D.: Pulse reshaping and coherent sideband generation effects on multiphoton excitation of polyatomic molecules,V1497,188-196(1991)

Garnov, Sergei V. See Soileau, M. J.: V1441,10-15(1991)

Garofalo, Joseph G.
; Kostelak, Robert L.; Yang, Tungsheng: Phase-shifting structures for isolated features,V1463,151-166(1991)

Garra, Brian S.
See Horii, Steven C.: V1446,10-15(1991)
See Horii, Steven C.: V1446,475-480(1991)

Garreis, Reiner
: 90 degree optical hybrid for coherent receivers,V1522,210-219(1991)

Garrett, Ian See Cryan, Robert A.: V1372,64-71(1991)

Garrett, Steven L.
; Brown, David A.; Beaton, Brian L.; Wetterskog, Kevin; Serocki, John: General purpose fiber optic hydrophone made of castable epoxy,V1367,13-29(1991)
See Brown, David A.: V1369,2-8(1991)
See Brown, David A.: V1367,282-288(1991)
See Brown, David A.: V1584,328-335(1991)

Garsha, I. See Purinsh, Juris: V1525,289-308(1991)

Garside, Brian K.
: Near real-time operation of a centimeter-scale distributed fiber sensing system,V1332,399-408(1991)
; Park, R. E.: Minimum detectable changes in Rayleigh backscatter from distributed fiber sensors,V1588,150-158(1991)

Garske, D. See Chow, P. D.: V1475,48-54(1991)

Gartenberg, Ehud
; Roberts, A. S.: Twenty-five years of aerodynamic research with IR imaging,V1467,338-356(1991)

Garvin, Charles G.
; Sadler, Brian M.: Surface-acoustic-wave acousto-optic devices for wide-bandwidth signal processing and switching applications,V1562,303-318(1991)
See Worrell, Steven W.: V1382,219-227(1991)
See Wu, Kuang-Yi: V1563,168-175(1991)

Garvis, Darrel G. See Kulp, Thomas J.: V1479,352-363(1991)

Gary, Charles K. See Rule, Donald W.: V1552,240-251(1991)

Garza, Cesar M.
; Catlett, David L.; Jackson, Ricky A.: Characterizing a surface imaging process in a high-volume DRAM manufacturing production line,V1466,616-627(1991)

Gasanov, A. G. See Sadyigov, Z. Y.: V1621,158-168(1991)

Gascoigne, Harold E.
; Abdallah, Mohamed G.: Displacements and strains in thick-walled composite rings subjected to external pressure using moire interferometry,V1554B,315-322(1991)
See Morimoto, Yoshiharu: V1554B,493-502(1991)

Gaspar, Susan M.
See Bishop, Kenneth P.: V1545,64-73(1991)
See Hickman, Kirt C.: V1464,245-257(1991)

Gaspart, Pierre See Malotaux, Eric: V1388,372-383(1991)

Gastonguay, Louis See Dodelet, Jean-Pol: V1436,38-49(1991)

Gat, Erann See Miller, David P.: V1383,425-435(1991)

Gatenby, Paul V.
; Boereboom, Peter; Grant, Michael A.: Design of a periscopic coarse pointing assembly for optical multiple access,V1522,126-134(1991)

Gates, James L.
; Connelly, William G.; Franklin, T. D.; Mills, Robert E.; Price, Frederick W.; Wittwer, Timothy Y.: 488 x 640-element hybrid platinum silicide Schottky focal-plane array,V1540,262-273(1991)

Gatley, Ian See Fowler, Albert M.: V1541,127-133(1991)

Gatrell, Lance B.
See Morgenthaler, Matthew K.: V1387,82-95(1991)
See Sklair, Cheryl W.: V1387,380-391(1991)

Gatti, E. See Braueninger, Heinrich: V1549,330-339(1991)

Gatti, Giuliano
: Monolithic microwave integrated circuit activities in ESA-ESTEC,V1475,10-24(1991)

Gauch, John M.
: Image contrast enhancement via blurred weighted adaptive histogram equalization,V1606,386-399(1991)
See Zheng, Joe: V1606,1037-1047(1991)

Gauden, Pawel See Blaszczak, Zdzislaw: V1391,156-163(1991)

Gauduel, Yann
; Pommeret, Stanislas; Yamada, Noelle; Antonetti, Andre: Recent developments by femtosecond spectroscopy in biological ultrafast free radical reactions,V1403,417-426(1991)

Gaughan, Patrick T.
; Flachs, Gerald M.; Jordan, Jay B.: Multisensor object segmentation using a neural network,V1469,812-819(1991)
See Scott, David R.: V1383,367-378(1991)

Gault, William A. See Sellar, R. G.: V1479,140-155(1991)

Gaunaurd, Guillermo C.
: Inverse scattering problems in the acoustic resonance region of an underwater target,V1471,30-41(1991)
See Abrahamsson, S.: V1471,130-141(1991)
See Werby, Michael F.: V1471,2-17(1991)

Gauthier, Francis
: French proposal for IEC/TC 86/WG 4 OTDR calibration,V1504,55-65(1991)

Gauthier, Jean-Claude J. See Geindre, Jean-Paul: V1502,311-318(1991)

Gauthier, Robert P. See Teillet, Philippe M.: V1492,213-223(1991)

Gautier, Sabine See Blum, Loic J.: V1510,46-52(1991)

Gauzzi, Andrea
See James, Jonathan H.: V1394,45-61(1991)
See Pavuna, Davor: V1362,102-116(1991)

Gavnoudias, S. See Huttlin, George A.: V1407,147-158(1991)

Gavotto, A. See Juri, Hugo: V1422,128-135(1991)

Gavrilenko, Vladimir I.
; Frolov, Sergey I.: Band structure and optical properties of silicon carbide,V1361,171-182(1991)

Gavrilov, Sergey P. See Vertiy, Alexey A.: V1362,702-709(1991)

Gaw, Henry T.
See Hansen, Steven G.: V1463,230-244(1991)
See Toh, Kenny K.: V1496,27-53(1991)
See Toh, Kenny K.: V1463,402-413(1991)
See Toh, Kenny K.: V1463,74-86(1991)

Gawronski, M. J. See Swirhun, S.: V1475,303-308(1991)

Gaynor, Wayne H. See Carey, Raymond: V1464,500-507(1991)

Gazit, Dan See Biderman, Shlomo: V1535,27-34(1991)

Gazzaniga, Michael S. See Lu, Jiuhuai: V1459,117-124(1991)

Gazzaruso, Lucia See Canestrari, Paolo: V1463,446-455(1991)

Gburzynski, Pawel See Dobosiewicz, Wlodek: V1470,123-133(1991)

Ge, Fang X.
; Xiong, Xian M.: No-blind-area one-photograph HNDT tire analyzer,V1554B,785-789(1991)

Ge, Ming See Zhang, Rui T.: V1519,757-760(1991)

Ge, Wen
See Lin, Li-Huang: V1410,65-71(1991)
See Lin, Li-Huang: V1346,490-501(1991)

Geary, John C.
: Large-format CCDs for astronomical applications,V1439,159-168(1991)
; Luppino, Gerard A.; Bredthauer, Richard A.; Hlivak, Robert J.; Robinson, Lloyd B.: 4096 x 4096 pixel CCD mosaic imager for astronomical applications,V1447,264-273(1991)
See Robinson, Lloyd B.: V1447,214-228(1991)

Geary, Joseph M.
: High-energy laser wavefront sensors,V1414,66-79(1991)

Geballe, T. H. See Fork, David K.: V1394,202-202(1991)

Gebelein, Rolin J.
; Connely, Shawn W.; Foo, Leslie D.: Advances in the optical design of miniaturized optical correlators,V1564,452-463(1991)

Gebhardt, Frederick G.
See Hills, Louis S.: V1408,41-57(1991)
See Long, Jerry E.: V1408,58-71(1991)

Gebhardt, Michael
; Truckenbrodt, Horst; Harnisch, Bernd: Surface defect detection and classification with light scattering,V1500,135-143(1991)

Geddes, John J.
See Bauhahn, Paul E.: V1475,122-128(1991)
See Mondal, J. P.: V1475,314-318(1991)
See Swirhun, S.: V1475,303-308(1991)

Gedrat, Olaf See Toenshoff, Hans K.: V1377,38-44(1991)

Gee, Jim C.
; Reivich, Martin; Bilaniuk, L.; Hackney, D.; Zimmerman, R.; Kovacic, Stane; Bajcsy, Ruzena R.: Evaluation of multiresolution elastic matching using MRI data, V1445,226-234(1991)

Gee, Shirley J.
: FLIPS: Friendly Lisp Image Processing System, V1472,38-45(1991)
: SCORPIUS: lessons learned in developing a successful image understanding system, V1406,190-200(1991)

Geelhaar, Thomas See Wand, Michael D.: V1455,97-104(1991)

Geels, Randall S.
; Corzine, Scott W.; Coldren, Larry A.: Vertical-cavity surface-emitters for optoelectronic integration, V1418,46-56(1991)
See Coldren, Larry A.: V1362,24-37(1991)

Geers, Rony See Van der Stuyft, Emmanuel: V1379,189-200(1991)

Geest, Rob J. See Dumay, Adrie C.: V1445,38-46(1991)

Gehring, Dale G.
; Persons, Kenneth R.; Rothman, Melvyn L.; Salutz, James R.; Morin, Richard L.: Detailed description of the Mayo/IBM PACS, V1446,248-252(1991)
See Morin, Richard L.: V1446,436-441(1991)
See Persons, Kenneth R.: V1446,60-72(1991)

Gehring, G. See Merkle, Fritz: V1542,308-318(1991)

Geiger, Alan L. See Johnson, R. B.: V1535,136-142(1991)

Geiger, Davi See Yuille, Alan L.: V1568,271-282(1991)

Geiger, Manfred
; Lutz, Norbert; Biermann, Stephan: Excimer laser processing of ceramics and fiber-reinforced polymers assisted by a diagnostic system, V1503,238-248(1991)
See Biermann, Stephan: V1415,330-341(1991)

Geindre, Jean-Paul
; Audebert, Patrick; Chenais-Popovics, Claude; Gauthier, Jean-Claude J.; Benattar, Rene; Chambaret, J. P.; Mysyrowicz, Andre; Antonetti, Andre: X-ray and optical diagnostics of a 100-fs laser-produced plasma, V1502,311-318(1991)

Geiser, Martial See Yatagai, Toyohiko: V1555,8-12(1991)

Geist, P. See Cunin, B.: V1358,606-613(1991)

Gekat, Frank
; Klingenberg, Hans H.: Pulsed microwave excitation of rare-gas halide mixtures, V1411,47-54(1991)
See Klingenberg, Hans H.: V1503,140-145(1991)
See Klingenberg, Hans H.: V1412,103-114(1991)

Geldmacher, Juergen See Kreis, Thomas M.: V1554B,718-724(1991)

Gelish, Anthony
: You can't just plug it in: digital image networks/picture archiving and communication systems installation, V1444,363-372(1991)

Gell, Michael A. See White, Julian D.: V1361,293-301(1991)

Geller, B. D. See Sorbello, Robert M.: V1475,175-183(1991)

Gellman, David I.
; Biggar, Stuart F.; Slater, Philip N.; Bruegge, Carol J.: Calibrated intercepts for solar radiometers used in remote sensor calibration, V1493,175-180(1991)
See Biggar, Stuart F.: V1493,155-162(1991)
See Markham, Brian L.: V1493,207-214(1991)

Gelorme, Jeffrey D. See Conley, Willard E.: V1466,53-66(1991)

Gels, Robert G.
See Fitzmartin, Daniel J.: V1476,56-62(1991)
See Fitzmartin, Daniel J.: V1371,78-86(1991)

Geluk, Ronald J.
; Vlasbloem, Hugo: Equalization radiography with radiation quality modulation, V1443,143-152(1991)

Gemma, Takashi
; Hideshima, Masayuki; Taya, Makoto; Watanabe, Nobuko: Interferometer for testing aspheric surfaces with electron-beam computer-generated holograms, V1332,77-84(1991)

Genco, Sheryl M.
: Parallel optical interconnects utilizing VLSI/FLC spatial light modulators, V1563,45-57(1991)

Gendre, Francois See Bonniau, Philippe: V1588,52-63(1991)

Gendron, Eric
; Cuby, Jean-Gabriel; Rigaut, Francois; Lena, Pierre J.; Fontanella, Jean-Claude; Rousset, Gerard; Gaffard, Jean-Paul; Boyer, Corinne; Richard, Jean-Claude; Vittot, M.; Merkle, Fritz; Hubin, Norbert: Come-on-plus project: an upgrade of the come-on adaptive optics prototype system, V1542,297-307(1991)

Genetti, George J. See Horsky, Thomas N.: V1540,527-532(1991)

Geng, Wanzhen
See Zhao, Feng: V1555,297-299(1991)
See Zhao, Feng: V1461,262-264(1991)
See Zhao, Feng: V1555,241-242(1991)

Genin, Joseph See Conley, Edgar: V1332,798-801(1991)

Genkin, G. See Andronov, Alexander A.: V1362,684-695(1991)

Genkin, Vladimir N.
See Aleksandrov, A. P.: V1440,442-453(1991)
See Bobilkov, G. P.: V1440,98-104(1991)

Gennert, Michael A.
: Shape from shading with circular symmetry, V1385,256-258(1991)
; Ren, Biao; Yuille, Alan L.: Stereo matching by energy function minimization, V1385,268-279(1991)

Gennery, Donald B. See Bon, Bruce: V1387,337-342(1991)

Gentry, Bruce M.
See Korb, C. L.: V1416,177-182(1991)
See Targ, Russell: V1521,144-157(1991)

Genuario, Ralph
; Koert, Peter: High-power uhf rectenna for energy recovery in the HCRF (high-current radio frequency) system, V1407,553-565(1991)

Georgakilas, Alexandros See Panayotatos, Paul: V1361,1100-1109(1991)

George, Jacob
; Werby, Michael F.: Study and characterization of three-dimensional angular distributions of acoustic scattering from spheroidal targets, V1471,66-77(1991)

George, John S.
; Lewis, Paul S.; Ranken, D. M.; Kaplan, L.; Wood, C. C.: Anatomical constraints for neuromagnetic source models, V1443,37-51(1991)

George, M. C.
See Shukla, Ram P.: V1332,274-286(1991)
See Venkateswarlu, Putcha: V1332,245-266(1991)

George, Richard A. See Luehrmann, Paul F.: V1463,434-445(1991)

George, Steven M.
; Coon, P. A.; Gupta, P.; Wise, M. L.: Laser-induced thermal desorption studies of surface reaction kinetics, V1437,157-165(1991)

George, T. See Lin, TrueLon: V1540,135-139(1991)

Georges, Eric
; Barraud, Roger; Mouthon, Alain: Chemical oxygen-iodine laser: flow diagnostics and overall qualification, V1397,243-246(1991)
See Bacis, Roger: V1397,173-176(1991)
See Barnault, B.: V1397,231-234(1991)

Georges, John B. See Gidron, Rafael: V1579,40-48(1991)

Georgiades, C. See Persephonis, Peter: V1503,185-196(1991)

Geraghty, Paul See Atkinson, Dennis P.: V1343,530-541(1991)

Gerard, Jean M. See Sermage, Bernard: V1361,131-135(1991)

Gerber, Louis
; Kaempfer, Niklaus A.: High-resolution multichannel mm-wave radiometer for the detection of stratospheric ClO, V1491,211-217(1991)

Gerbrands, Jan J. See Dumay, Adrie C.: V1445,38-46(1991)

Gerdin, Glenn A. See Brinkmann, Ralf P.: V1378,203-208(1991)

Gerdt, David W. See Henning, Michael R.: V1420,34-40(1991)

Gerges, Awad S.
; Jackson, David A.: Fiber optic based miniature high-temperature probe exploiting coherence-tuned signal recovery via multimode laser diode illumination, V1504,233-236(1991)
; Newson, Trevor P.; Jackson, David A.: Interferometric fiber-optic sensing using a multimode laser diode source, V1504,176-179(1991)

Gerhardt, Harald See Mann, Klaus R.: V1503,176-184(1991)

Gerhardt, Lester A. See Noseworthy, J. R.: V1387,26-37(1991)

Gerla, Mario See Kovacevic, Milan: V1579,74-83(1991)

Gerlach, Francis W.
; Cook, Daniel B.: Multisensor analysis tool, V1488,134-143(1991)

Gerlach-Meyer, U. See Huber, Peter: V1522,135-141(1991)

Germann, Lawrence M.
See Medbery, James D.: V1482,397-405(1991)
See Medbery, James D.: V1489,163-176(1991)

Germer, Rudolf K.
; Meyer-Ilse, Werner: High-speed CCD video camera, V1358,346-350(1991)
; Mohr, Joachim; Schoenbach, Karl H.: Breakdown in pulsed-power semiconductor switches, V1358,925-931(1991)
See Schoenbach, Karl H.: V1362,428-435(1991)

Gerndt, Christian See Zarschizky, Helmut: V1389,484-495(1991)

Gerri, Mireille
; Marine, W.; Mathey, Yves; Sentis, Marc L.; Delaporte, Philippe C.; Fontaine, Bernard L.; Forestier, Bernard M.: In-situ growth of Y1Ba2Cu3O7-x thin films using XeCl excimer and Nd:YAG lasers,V1503,280-291(1991)
See Mathey, Yves: V1361,909-916(1991)

Gershbock, Richard
: Image acquisition: quality control for document image scanners,VCR37,20-39(1991)

Gershman, Alex
; Danoff, Dudley; Chandra, Mudjianto; Grundfest, Warren S.: Laparoscopically guided bilateral pelvic lymphadenectomy,V1421,186-188(1991)
See Chandra, Mudjianto: V1421,68-71(1991)
See Papaioannou, Thanassis: V1420,203-211(1991)

Gershman, Robert
; Mahoney, Michael J.; Rayman, M. D.; Shao, Michael; Snyder, Gerald C.: System design for lunar-based optical and submillimeter interferometers,V1494,160-167(1991)
See Colavita, Mark M.: V1494,168-181(1991)

Gershman, Vladimir
: 1.25-Gb/s wideband LED driver design using active matching techniques,V1474,75-82(1991)

Gersho, Allen
See Gupta, Smita: V1605,179-189(1991)
See Wu, Siu W.: V1605,487-498(1991)
See Wu, Siu W.: V1605,73-84(1991)
See Wu, Siu W.: V1452,352-363(1991)

Gerson, Nahum D.
; Cappelletti, John D.; Hinds, Stuart C.; Glenn, Marcus E.: Image understanding, visualization, registration, and data fusion of biomedical brain images,V1406,129-129(1991)

Gerstadt, William A. See Gambogi, William J.: V1555,256-267(1991)

Gerstenmayer, Jean-Louis
; Vibert, Patrick: Application of the half-filter method to the flash radiography using a neutral filter in the range of x-rays,V1346,286-292(1991)

Gerstl, Siegfried A. See Borel, Christoph C.: V1486,271-277(1991)

Gerstman, T. See Hudec, Rene: V1343,162-163(1991)

Gertner, E. R. See Kozlowski, Lester J.: V1540,250-261(1991)

Gertner, Izidor
; Peskin, Richard L.; Walther, Sandra S.: Parallel computation of the continuous wavelet transform,V1565,414-422(1991)
; Zeevi, Yehoshua Y.: Scanning strategies for target detection,V1470,148-166(1991)
See Assaleh, Khaled T.: V1606,532-540(1991)

Getoff, N. See Angelov, D.: V1403,572-574(1991)

Getreuer, Kurt W. See Grove, Steven L.: V1499,354-359(1991)

Geurts, Jean
; Finders, J.; Kayser, O.; Opitz, B.; Maassen, M.; Westphalen, R.; Balk, P.: Selectively grown InxGa1-xAs and InxGa1-xP structures: locally resolved stoichiometry determination by Raman spectroscopy,V1361,744-750(1991)

Geurtz, Alexander M.
: Nonlinear optical flow estimation and segmentation,V1567,110-121(1991)
See Dufaux, Frederic: V1567,362-379(1991)

Geva, M. See Pearton, Stephen J.: V1393,150-160(1991)

Gex, Jean-Pierre See Veaux, Jacqueline: V1449,13-24(1991)

Geyer, Steve L. See Pacelli, Jean L.: V1472,76-84(1991)

Ghaffari, Tammy G. See Derr, John I.: V1406,127-128(1991)

Ghafoor, Arif
: All-optical interconnection networks,V1390,454-466(1991)

Ghanem, Adel
: Metropolitan area networks: a corner stone in the broadband era,V1364,312-319(1991)

Ghatak, Kamakhya P.
: Photoemission from periodic structure of graded superlattices under magnetic field,V1545,282-293(1991)
: Thermoelectric power in fiber optic and laser materials under cross-field configuration,V1584,459-447(1991)
; Biswas, Shambhu N.: Effective electron mass in narrow-band-gap IR materials under different physical conditions,V1484,149-166(1991)
; Biswas, Shambhu N.; Banik, S. N.: Gate capacitance of MOS field effect devices of nonlinear optical materials in the presence of a parallel magnetic field,V1409,240-257(1991)

; Biswas, Shambhu N.: Heat capacity of MOS field-effect devices of optical materials in the presence of a strong magnetic field,V1485,206-214(1991)
; Biswas, Shambhu N.: Photoemission from quantum-confined structure of nonlinear optical materials,V1484,136-148(1991)
; Ghoshal, Ardhendhu; De, Badal: Generalized Raman gain in nonparabolic semiconductors under strong magnetic field,V1409,178-190(1991)
; Ghoshal, Ardhendhu; Bhattacharyya, Sankar; Mondal, Manabendra: Influence of photon energy on the photoemission from ultrafast electronic materials,V1346,471-489(1991)
; Ghoshal, Ardhendhu; Biswas, Shambhu N.: Photoemission from quantum-confined structure of nonlinear optical materials,V1409,28-57(1991)
; Mitra, Bhaswati; Biswas, Shambhu N.: Effect of a longitudinal magnetic field on the diffusion coefficient of the minority carriers in solid state junction lasers,V1415,281-297(1991)
; Mitra, Bhaswati; Biswas, Shambhu N.: Effect of magnetic quantization on the pulse power output of spontaneous emission for II-VI and IV-VI lasers operated at low temperature,V1411,137-148(1991)

Gheen, Gregory
: Analysis of Bragg diffraction in optical memories and optical correlators,V1564,135-142(1991)
: Optimal distortion-invariant quadratic filters,V1564,112-120(1991)
See Huang, Chao H.: V1564,427-438(1991)

Ghiasi, Ali
; Gopinath, Anand: Fine structure characteristics of semiconductor laser diode coupled to an external cavity,V1583,170-174(1991)

Ghiglia, Dennis C.
See Mastin, Gary A.: V1566,341-352(1991)
See Wahl, Daniel E.: V1567,32-40(1991)

Ghinovker, M. See Andronov, Alexander A.: V1362,684-695(1991)

Ghione, Giovanni See Rocca, Corrado: V1372,39-47(1991)

Ghodgaonkar, Deepak K. See Selmi, Fathi: V1489,97-107(1991)

Ghosh, Anjan K.
: Compact joint transform correlators in planar-integrated packages,V1564,231-235(1991)
: Design and performance analysis of optoelectronic adaptive infinite-impulse response filters,V1565,69-73(1991)
: Fault tolerance of optoelectronic neural networks,V1563,120-120(1991)
: Realization of infinite-impulse response filters using acousto-optic cells,V1564,593-601(1991)
; Barner, Jim; Paparao, Palacharla; Allen, Susan D.; Imen, Kamran: Design of adaptive optical equalizers for fiber optic communication systems,V1371,170-181(1991)
; Beech, Russell S.: Analysis of alignment in optical interconnection systems,V1389,630-641(1991)
See Imen, Kamran: V1365,60-64(1991)

Ghosh, David See Montgomery, Raymond C.: V1480,126-137(1991)

Ghosh, Joydeep See Klarquist, William N.: V1453,321-332(1991)

Ghosh, Monisha
; Weber, Charles L.: Maximum-likelihood blind equalization,V1565,188-195(1991)

Ghoshal, Ardhendhu
See Ghatak, Kamakhya P.: V1409,178-190(1991)
See Ghatak, Kamakhya P.: V1346,471-489(1991)
See Ghatak, Kamakhya P.: V1409,28-57(1991)

Giachello, G. See Demichelis, Francesca: V1534,140-147(1991)

Giam, S. T. See Ong, Sim-Heng: V1445,564-573(1991)

Giammarco, Nicholas J. See Krawiec, Theresa M.: V1392,265-271(1991)

Giannacopoulos, A. See Hontzopoulos, Elias I.: V1397,761-768(1991)

Giannakis, Georgios B. See Sadler, Brian M.: V1476,246-256(1991)

Giannetas, B. See Persephonis, Peter: V1503,185-196(1991)

Giannini, Jean-Pierre
; Delay, P.; Duval, A.; Wlodkiewiez, A.: Comparison of performances on analog fiber optic links equipped with either noncontact or physical-contact connectors,V1366,215-222(1991)

Giannini, Judith A. See Peacock, Keith: V1494,147-159(1991)

Giardina, Charles R. See Laplante, Phillip A.: V1568,295-302(1991)

Giarrusso, Salvatore See Kaaret, Philip E.: V1548,106-117(1991)

Gibart, Pierre
; Williamson, Don L.: Bistability of the Sn donor in AlxGa1-xAs and GaAs under pressure studied by Mossbauer spectroscopy,V1362,938-950(1991)

Gibaud, Bernard
See Aubry, Florent: V1446,168-176(1991)
See Barillot, Christian: V1445,54-65(1991)
See Bizais, Yves: V1446,156-167(1991)
See Fresne, Francoise: V1444,26-36(1991)

Gibbings, C. J. See White, Julian D.: V1361,293-301(1991)

Gibbons, Robert C.
; Cooper, E. E.; Legan, R. G.: Design considerations for use of a double-dove prism behind a concentric dome,V1498,64-69(1991)

Gibler, William N.
See Lee, Chung E.: V1584,396-399(1991)
See Lee, Chung E.: V1588,110-116(1991)

Gibney, Mary A. See O'Brien, Stephen J.: V1424,62-75(1991)

Gibson, B. V. See Colavita, Mark M.: V1542,205-212(1991)

Gibson, Harry G.
See Jia, Jiancheng: V1379,246-253(1991)
See Jia, Jiancheng: V1396,656-663(1991)

Gibson, Laurie See Lucas, Dean: V1568,157-163(1991)

Gidron, Rafael
; Elby, Stuart D.; Acampora, Anthony S.; Georges, John B.; Lau, Kam-Yin: TeraNet: a multigigabit-per-second hybrid circuit/packet-switched lightwave network,V1579,40-48(1991)

Gielen, Robert M.
; Slegtenhorst, Ronald P.: High-speed short-range laser rangefinder,V1419,153-159(1991)

Gielisse, Peter J. See Tu, Meirong: V1554A,593-601(1991)

Gierczak, Christine A.
; Andino, J. M.; Butler, James W.; Heiser, G. A.; Jesion, G.; Korniski, T. J.: FTIR: fundamentals and applications in the analysis of dilute vehicle exhaust,V1433,315-328(1991)

Gierulski, Alfred
; Naarmann, Herbert; Schrof, Wolfgang; Ticktin, Anton: Frequency-tunable THG measurements of x(3) between 1-2.1um of organic conjugated-polymer films using an optical parametric oscillator,V1560,172-182(1991)

Gieschen, Nikolaus
; Rocks, Manfred; Olivier, Lutz: Novel integrated acousto-optical LiNbO3 device for application in single-laser self-heterodyne systems,V1579,237-248(1991)

Giesen, Adolf
See Borik, Stefan: V1441,420-429(1991)
See Moissl, M.: V1397,395-398(1991)

Giesler, Leslie E.
; Dunphy, James R.; Morey, William W.; Meltz, Gerald; Glenn, William H.: Instrumentation concepts for multiplexed Bragg grating sensors,V1480,138-142(1991)

Giever, John C.
; Hoock, Donald W.: Atmospheric propagation effects on pattern recognition by neural networks,V1486,302-313(1991)

Giewont, Kenneth J. See Blow, Victor O.: V1366,107-111(1991)

Gifford, E. See Hunt, Bobby R.: V1472,190-200(1991)

Gifford, George G.
: Applications of optical emission spectroscopy in plasma manufacturing systems,V1392,454-465(1991)

Gigante, Joseph R. See Leksell, David: V1458,133-144(1991)

Giger, Maryellen L.
; Nishikawa, Robert M.; Doi, Kunio; Yin, Fang-Fang; Vyborny, Carl J.; Schmidt, Robert A.; Metz, Charles E.; Wu, Yuzheng; MacMahon, Heber; Yoshimura, Hitoshi: Development of a smart workstation for use in mammography,V1445,101-103(1991)
See Brailean, James C.: V1450,40-46(1991)
See Chen, Chin-Tu: V1444,75-77(1991)
See Chou, Jin-Shin: V1396,45-50(1991)
See Yin, Fang-Fang: V1396,2-4(1991)
See Yoshimura, Hitoshi: V1445,47-51(1991)

Gijsbers, Geert H.
; Sprangers, Rene L.; van den Broecke, Duco G.; van Wieringen, Niek; Brugmans, Marco J.; van Gemert, Martin J.: Temperature increase during in vitro 308 nm excimer laser ablation of porcine aortic tissue,V1425,94-101(1991)
See van Gemert, Martin J.: V1427,316-319(1991)

Gil, E. See Banvillet, Henri: V1361,972-979(1991)

Gilabert, Alain
; Azema, Alain; Roustan, Jean-Claude; Maneval, Jean-Paul: Interplay between photons and superconductors,V1477,140-147(1991)

Gilbert, Barry K.
; Thompson, R.; Fokken, G.; McNeff, W.; Prentice, Jeffrey A.; Rowlands, David O.; Staniszewski, A.; Walters, W.; Zahn, S.; Pan, George W.: Advanced multichip module packaging and interconnect issues for GaAs signal processors operating above 1 GHz clock rates,V1390,235-248(1991)

Gilbert, John A.
; Dudderar, Thomas D.: Uses of fiber optics to enhance and extend the capabilities of holographic interferometry,VIS08,146-159(1991)
; Greguss, Pal; Kransteuber, Amy S.: Holo-interferometric patterns recorded through a panoramic annular lens,V1238,412-420(1991)
; Matthys, Donald R.; Hendren, Christelle M.: Displacement analysis of the interior walls of a pipe using panoramic holointerferometry,V1554B,128-134(1991)
; Matthys, Donald R.; Lehner, David L.: Moire measurements using a panoramic annular lens,V1554B,202-209(1991)
See Matthys, Donald R.: V1554B,736-742(1991)
See Matthys, Donald R.: V1332,850-861(1991)

Gilbert, R. K. See Tchoryk, Peter: V1479,164-182(1991)

Gilblom, David L.
: Standards for image input devices: review and forecast,VCR37,3-19(1991)

Gilboa, Pini
: Designing the right visor,V1456,154-163(1991)

Gilbreath, G. C.
; Clement, Anne E.; Fugera, S. N.; Mizell, Gregory J.: Photorefractive two-wave mixing characteristics for image amplification in diffusion-driven media,V1409,87-99(1991)

Gilbreath, Gary A. See Holland, John: V1388,291-298(1991)

Gilcrest, Andrew S.
: Automated characterization of Z-technology sensor modules,V1541,240-249(1991)

Gildenblat, Gennady S.
; Schwartz, Gary P.; eds.: [$BFS]Metallization: Performance and Reliability Issues for VLSI and ULSI[$BFE],V1596(1991)

Gildner, Donald
; Marder, James M.: Creation of aspheric beryllium optical surfaces directly in the hot isostatic pressing consolidation process,V1485,46-53(1991)

Giles, Ian P. See Li, Luksun: V1584,170-177(1991)

Giles, John W.
; Voss, Kenneth J.: Submerged reflectance measurements as a function of visible wavelength,V1537,140-146(1991)

Giles, Michael K.
See Carhart, Gary W.: V1564,348-362(1991)
See Clark, Natalie: V1564,439-451(1991)
See Crandall, Charles M.: V1564,98-109(1991)

Gilge, Michael
; Gusella, Riccardo: Motion video coding for packet-switching networks: an integrated approach,V1605,592-603(1991)

Gilgenbach, Ronald M. See Menge, Peter R.: V1407,578-588(1991)

Gill, E. T.
; Tsui, J. B.: Interferometric acousto-optic receiver results,V1476,190-200(1991)

Gill, M. See Johnstone, Walter: V1506,145-149(1991)

Gillbro, T.
See Khoroshilov, E. V.: V1403,431-433(1991)
See Sharkov, A. V.: V1403,434-438(1991)

Gillespie, Richard E. See Dubiel, Mary K.: V1396(1991)

Gillies, Donald C.
; Lehoczky, S. L.; Szofran, Frank R.; Su, Ching-Hua; Larson, David J.: Interface demarcation in Bridgman-Stockbarger crystal growth of II-VI compounds,V1484,2-10(1991)

Gillies, George T.
See Lawson, Michael A.: V1445,265-275(1991)
See Ramos, P. A.: V1443,160-170(1991)

Gilliland, Curt A. See Cantrell, Chris A.: V1433,263-268(1991)

Gilmartin, Harvey R. See Sauer, Donald J.: V1540,285-296(1991)

Gilmer, George See Macrander, Albert T.: V1550,122-133(1991)

Gilmore, Kirk See Robinson, Lloyd B.: V1447,214-228(1991)

Gilo, Mordechai
; Rabinovitch, Kopel: Color- and intensity-balanced multichannel optical beamsplitter,V1442,90-104(1991)

Gin, Dong See Tan, Hui: V1519,752-756(1991)

Ginaven, Robert O. See Swanberg, Norman: V1346,249-267(1991)

Ginesu, Francesco See Aymerich, Francesco: V1554B,304-314(1991)

Ginley, David S. See Martens, Jon S.: V1394,140-149(1991)

Ginosar, Ran See Vitsnudel, Ilia: V1606,1086-1091(1991)

Ginovker, A. See Chaker, Mohamed: V1465,16-25(1991)

Ginsburg, Eric J. See Jozefiak, Thomas H.: V1436,8-19(1991)

Gintner, Henry
: Enhancement of Conrady's "D-d" method,V1354,97-102(1991)

Giordano, Gualtiero
See Bollanti, Sarah: V1503,80-87(1991)
See Bollanti, Sarah: V1397,97-102(1991)

Giordano, P. See Noethe, L.: V1542,293-296(1991)

Giorgi, David M.
See McIntyre, Iain A.: V1378,162-172(1991)
See Zucker, Oved S.: V1378,22-33(1991)

Giorgianni, Edward
: Physics and psychophysics of color reproduction,V1458,2-3(1991)

Giovannini, Hugues See Lequime, Michael: V1367,236-242(1991)

Girard, Mary M.
; Husbands, Charles R.; Antoszewska, Reza: Dynamically reconfigurable optical interconnect architecture for parallel multiprocessor systems,V1563,156-167(1991)

Girardeau-Montaut, Claire See Girardeau-Montaut, Jean-Pierre: V1502,331-335(1991)

Girardeau-Montaut, Jean-Pierre
; Girardeau-Montaut, Claire: Modeling of electron density produced by femtosecond laser on metallic photocathodes,V1502,331-335(1991)
See Moustaizis, Stavros D.: V1552,50-56(1991)

Giraud, P. See Mens, Alain: V1358,878-887(1991)

Giraudet, Louis See Scavennec, Andre: V1362,331-337(1991)

Girlanda, Raffaello See De Salvo, Edmondo: V1362,870-875(1991)

Girod, C. See Hallegot, Philippe: V1396,311-315(1991)

Giron, J. C. See Plichon, V.: V1536,37-47(1991)

Girouard, Fernand E. See Ashrit, Pandurang V.: V1401,119-129(1991)

Giroux, Jean See Lamarre, Daniel: V1493,28-36(1991)

Girton, Dexter G.
See Lipscomb, George F.: V1560,388-399(1991)
See Lytel, Richard S.: V1389,547-558(1991)

Gitelson, Anatoly A.
: Aerospace remote sensing monitoring of inland water quality,V1492,307-318(1991)

Gitelzon, I. I. See Benimetskaya, L. Z.: V1525,242-245(1991)

Gitomer, Steven J. See Thode, Lester E.: V1552,87-106(1991)

Gittings, J. See Castro, Peter: V1448,134-139(1991)

Giudici, G. See Bortoletto, Favio: V1542,225-235(1991)

Giulietti, Antonio
; Giulietti, Danilo; Willi, Oswald: Plasma interaction with powerful laser beams,V1502,270-283(1991)
See Batani, D.: V1503,479-491(1991)

Giulietti, Danilo See Giulietti, Antonio: V1502,270-283(1991)

Givel, J.-C. See Wagnieres, G.: V1525,219-236(1991)

Gjonnes, Jon See Hugsted, B.: V1361,751-757(1991)

Gjonnes, K. See Hugsted, B.: V1361,751-757(1991)

Glab, Wallace L.
; Bistransin, Mark; Borst, Walter L.: Photodegradation of a laser dye in a silica gel matrix,V1497,389-395(1991)

Gladkov, L. L. See Solovyov, K. N.: V1403,132-133(1991)

Gladysz, D. See Taylor, Roderick S.: V1420,183-192(1991)

Glaes, Anders See Novak, Agneta: V1389,80-86(1991)

Glass, Alastair M. See Partovi, Afshin: V1561,20-32(1991)

Glasser, Jean See Daligault, Laurence: V1512,124-130(1991)

Glasser, Mardi See Tate, Lloyd P.: V1424,209-217(1991)

Glassman, W. L. S. See Das, Bidyut B.: V1427,368-373(1991)

Glatkowski, Paul J.
See Druy, Mark A.: V1584,48-52(1991)
See Druy, Mark A.: V1437,66-74(1991)
See Druy, Mark A.: V1409,214-219(1991)

Glattly, D. C. See Etienne, Bernard: V1362,256-267(1991)

Glavich, Thomas A.
; Beer, Reinhard: Tropospheric emission spectrometer for the Earth Observing System,V1540,148-159(1991)

Glazkov, V. N. See Zheltov, Georgi I.: V1403,752-753(1991)

Gleason, Jim M.
; Sherman, James W.: ATC (automatic target cueing) algorithm evaluation,V1406,169-170(1991)

Gleason, Shaun S.
; Hunt, Martin A.; Jatko, W. B.: Subpixel measurement of image features based on paraboloid surface fit,V1386,135-144(1991)

Glebov, Leonid B.
; Efimov, Oleg M.; Mekryukov, A. M.: Photoinductional change of silicate glasses optical parameters at two-photon laser radiation absorption,V1513,274-282(1991)
; Efimov, Oleg M.: Study of regularities and the mechanism of intrinsic optical breakdown of glasses,V1440,50-56(1991)
; Evstropiev, Sergei K.; Petrovskii, Gurii T.; Shashkin, Viktor S.: Optical properties of glass materials obtained by inorganic sol-gel synthesis,V1513,224-231(1991)
; Nikonorov, Nikolai V.; Petrovskii, Gurii T.; Kharchenko, Mikhail V.: Formation of optical elements by photothermo-induced crystallization of glass,V1440,24-35(1991)
; Nikonorov, Nikolai V.; Panysheva, Elena I.; Petrovskii, Gurii T.; Savvin, Vladimir V.; Tunimanova, Irina V.; Tsekhomskii, Victor A.: New possibilities of photosensitive glasses for three-dimensional phase hologram recording,V1621,21-32(1991)
; Nikonorov, Nikolai V.; Petrovskii, Gurii T.: Planar optical waveguides on glasses and glass-ceramic materials,V1513,56-70(1991)
; Petrovskii, Gurii T.; Tshavelev, Oleg S.: Optical materials for use under extreme service conditions,V1399,200-206(1991)

Glebova, Svetlana N.
; Lavrov, Nikolaj A.: Dynamic model of deformable adaptive mirror,V1500,275-280(1991)

Gleckler, Anthony D.
; Pflibsen, Kent P.; Ulich, Bobby L.; Smith, Duane D.: Surface control techniques for the segmented primary mirror in the large lunar telescope,V1494,454-471(1991)

Gleckman, Philip
; Ito, John: Phase space calculation of bend loss in rectangular light pipes,V1528,163-168(1991)
See Jacobson, Benjamin A.: V1528,82-85(1991)
See Lacovara, Phil: V1528,135-141(1991)

Gleed, David G.
; Lettington, Alan H.: Application of super-resolution techniques to passive millimeter-wave images,V1567,65-72(1991)

Gleichman, Kurt W.
; Tchoryk, Peter; Sampson, Robert E.: Application of laser radar to autonomous spacecraft landing,V1416,286-294(1991)
See Tchoryk, Peter: V1479,164-182(1991)

Glenis, Spyridon See Frank, Arthur J.: V1436,50-57(1991)

Glenn, Marcus E. See Gerson, Nahum D.: V1406,129-129(1991)

Glenn, T. See Rastegar, Sohi: V1427,300-306(1991)

Glenn, William H. See Giesler, Leslie E.: V1480,138-142(1991)

Glicksman, M. E.
See Jones, O. C.: V1557,202-208(1991)
See Paradies, C. J.: V1561,2-5(1991)

Glista, Andrew S.
: Novel fiber optic coupler/repeater for military systems,V1369,24-34(1991)
: Photoneuron: dynamically reconfigurable information processing control element utilizing embedded-fiber waveguide interconnects,V1563,139-155(1991)

Globus, Al
: Octree optimization,V1459,2-10(1991)

Glockler, Hermann See Clapham, Terrance N.: V1423,2-7(1991)

Glogovsky, K. G. See Levine, Barry F.: V1540,232-237(1991)

Glomb, Walter L.
: Fiber optic position transducers for aircraft controls,V1367,162-164(1991)

Glova, Alexander F.
See Antyuhov, V.: V1397,355-365(1991); 1415,48-59(1991)
See Bondarenko, Alexander V.: V1376,117-127(1991)

Glushko, A. A. See Savinykh, Viktor P.: V1540,450-454(1991)

Glushko, T. N. See Ageev, Vladimir P.: V1503,453-462(1991)

Gluskin, Efim S.
; Mattson, J. E.; Bader, Samuel D.; Viccaro, P. J.; Barbee, Troy W.; Brookes, N.; Pitas, Alan A.; Watts, Richard N.: XUV polarimeter for undulator radiation measurements, V1548,56-68(1991)
See DeForest, Craig E.: V1343,404-414(1991)

Glynn, Thomas J.
; Flanagan, Aiden J.; Redfern, R. M.: Reflow soldering of fine-pitch devices using a Nd:YAG laser, V1598,200-205(1991)

Gmitro, Arthur F.
See Alexander, Andrew L.: V1425,6-15(1991)
See Davenport, Carolyn M.: V1425,16-27(1991)

Gnappi, G. See Montenero, Angelo: V1513,234-242(1991)

Gnazzo, Angelantonio
See Cognolato, Livio: V1579,249-256(1991)
See Cognolato, Livio: V1513,368-377(1991)

Gnazzo, John F.
: Heterogeneous input neuration for network-based object recognition architectures, V1569,239-246(1991)

Goad, Joseph H.
; Rinsland, Pamela L.; Kist, Edward H.; Irick, Steven C.: Wavemeter for tuning solid state lasers, V1410,107-115(1991)

Goad, Larry E.
: Anisoplanatism and the use of laser guide stars, V1542,100-109(1991)

Goble, Larry W.
: Temperature control of the 3.5-meter WIYN telescope primary mirror, V1532,161-169(1991)

Gobleid, D. See Pelous, G.: V1598,149-158(1991)

Goda, Takuji See Kawahara, Hideo: V1513,198-203(1991)

Godard, Bruno
; Estocq, Emmanuel; Joulain, Franck; Murer, Pierre; Stehle, Marc X.; Bonnet, Jean C.; Pigache, Daniel R.: Parametric study of a high-average-power XeCl laser, V1503,71-77(1991)
; Estocq, Emmanuel; Stehle, Marc X.; Bonnet, Jean C.; Pigache, Daniel R.: Study of high-average-power excimer laser with circulation loop, V1397,59-62(1991)

Godber, Simon X.
; Robinson, Max; Evans, Paul: Three-dimensional machine vision using line-scan sensors, V1526,170-189(1991)

Godbey, Luther C. See Taylor, Robert A.: V1379,16-27(1991)

Goddard, J. See Abdulghafour, Muhamad: V1383,596-610(1991)

Goddard, R. See Flanagan, Kathryn A.: V1549,395-407(1991)

Godfrey, Maureen A. See White, Christopher J.: V1425,130-133(1991)

Goedseels, Vic See Van der Stuyft, Emmanuel: V1379,189-200(1991)

Goel, Kamal K.
; Prucnal, Paul R.; Stacy, John L.; Krol, Mark F.; Johns, Steven T.: Free-space optical TDM switch, V1476,314-319(1991)
See Xiao, Xiaodong: V1476,301-304(1991)

Goel, Sanjay
; Naylor, David L.: Micromachined structure for coupling optical fibers to integrated optical waveguides, V1396,404-410(1991)

Goel, T. C. See Vishnoi, Gargi: V1572,94-100(1991)

Goela, Jitendra S.
; Pickering, Michael A.; Taylor, Raymond L.: Chemical-vapor-deposited silicon and silicon carbide optical substrates for severe environments, V1330,25-38(1991)

Goelz, Stefan
; Persoff, Jeffrey J.; Bittner, Groff D.; Liang, Junzhong; Hsueh, Chi-Fu T.; Bille, Josef F.: New wavefront sensor for metrology of spherical surfaces, V1542,502-510(1991)
See Liang, Junzhong: V1542,543-554(1991)

Goering, Rolf
: Hybride fiber-optic temperature sensors on the base of LiNbO3 and LiNbO3:Ti waveguides, V1511,275-280(1991)
See Rasch, Andreas: V1511,149-154(1991)
See Rasch, Andreas: V1522,83-92(1991)

Goeringer, Fred
: Medical diagnostic imaging support systems for military medicine, V1444,340-350(1991)

Goertler, Andreas See Bickel, P.: V1503,167-175(1991)

Goethals, Anne-Marie
; Baik, Ki-Ho; Van den hove, Luc; Tedesco, Serge V.: Applicability of dry developable deep-UV lithography to sub-0.5 um processing, V1466,604-615(1991)

Goetschkes, Margaret See Flotte, Thomas J.: V1427,36-44(1991)

Goettert, Jost
; Mohr, Jurgen: Characterization of micro-optical components fabricated by deep-etch x-ray lithography, V1506,170-178(1991)

Goff, John R.
: Use of magneto-optic spatial light modulators and linear detector arrays in inner-product associative memories, V1558,466-475(1991)

Gofstein, Gary See Jacques, Steven L.: V1427,63-67(1991); 1525,309-312(1991)

Goga, George See Buckley, Robert H.: V1371,212-222(1991)

Gohberg, I.
; Koltracht, Israel; Xiao, Tongsan D.: Solution of the Yule-Walker equations, V1566,14-22(1991)

Gokay, Cem
: Industrial applications of metal vapor lasers, V1412,28-31(1991)

Golab, Stanislaw See Ryba-Romanowski, Witold: V1391,2-5(1991)

Golanics, Charles J. See Balick, Lee K.: V1493,215-223(1991)

Golby, John A. See MacAndrew, J. A.: V1500,172-176(1991)

Gold, Alfred
: Transport time and single-particle relaxation time in two-dimensional semiconductors, V1362,309-313(1991)

Gold, David
; Nathel, Howard; Bolton, Paul R.; White, William E.; Van Woerkom, Linn D.: Prepulse suppression using a self-induced ultrashort pulse plasma mirror, V1413,41-52(1991)

Gold, Harris See Swairjo, Manal: V1437,60-65(1991)

Goldbeck, Robert A.
; Bjorling, Sophie; Kliger, David S.: Nanosecond time-resolved natural and magnetic circular dichroism spectroscopy of protein dynamics, V1432,14-27(1991)
See Woodruff, William H.: V1432,205-210(1991)

Goldberg, Morris See Plessis, Brigitte: V1445,539-554(1991)

Golden, Jeffry
; Len, Lek K.; Smith, Tab J.; Dialetis, Demos; Marsh, S. J.; Smith, Kevin; Mathew, Joseph; Loschialpo, Peter; Seto, Lloyd; Chang, Jeng-Hsien; Kapetanakos, Christos A.: Recent developments on the NRL Modified Betatron Accelerator, V1407,418-429(1991)
See Litz, Marc S.: V1407,159-166(1991)

Goldfain, Ervin
: Cone array of Fourier lenses for contouring applications, V1527,210-215(1991)
: Differential coating objective, V1527,126-133(1991)
: Real-time holographic microscope with nonlinear optics, V1527,199-209(1991)

Goldgof, Dmitry B.
; Chandra, Kambhamettu: Application of the nonrigid shape matching algorithm to volumetric cardiac images, V1450,264-276(1991)
; Lee, Hua; Huang, Thomas S.: Motion estimation without correspondences and object tracking over long time sequences, V1383,109-121(1991)
See Mishra, Sanjoy K.: V1450,218-230(1991)
See Wilkins, Belinda: V1468,662-673(1991)

Goldhar, J. See Ho, Ping-Tong: V1378,210-216(1991)

Goldin, A. A. See Gakamsky, D. M.: V1403,641-643(1991)

Golding, Douglas J. See Seka, Wolf D.: V1398,162-169(1991)

Goldman, Aaron
; Murcray, Frank J.; Rinsland, C. P.; Blatherwick, Ronald D.; Murcray, F. H.; Murcray, David G.: Analysis of atmospheric trace constituents from high-resolution infrared balloon-borne and ground-based solar absorption spectra, V1491,194-202(1991)
See Murcray, Frank J.: V1491,282-287(1991)

Goldman, Shmuel See Wilk, Shalom: V1442,140-148(1991)

Goldmann, A. See Schaefer, Juergen A.: V1361,1026-1032(1991)

Goldmunz, Menachem
; Bloomberg, Steve; Neugarten, Michael L.: Laser designation integration into M-65 turret, V1442,149-153(1991)
See Wilk, Shalom: V1442,140-148(1991)

Goldner, E. See Kurizki, Gershon: V1497,48-62(1991)

Goldner, Eric L. See Chin, Gene H.: V1367,107-120(1991)

Goldner, Ronald B.
; Arntz, Floyd O.; Berera, G.; Haas, Terry E.; Wei, G.; Wong, Kwok-keung; Yu, Phillip C.: Progress on the variable reflectivity electrochromic window, V1536,63-69(1991)

Goldsmith, Paul See Lindmayer, Joseph: V1401,103-112(1991)

Goldstein, Dennis H. See Kallman, Robert R.: V1564,330-347(1991)

Goldstein, John C.
See McVey, Brian D.: V1441,457-468(1991)
See Newnam, Brian E.: V1552,154-174(1991)
See Thode, Lester E.: V1552,87-106(1991)

Gole, James L.
; Woodward, J. R.; Cobb, S. H.; Shen, KangKang; Doughty, J. R.: Chemically driven pulsed and continuous visible laser amplifiers and oscillators,V1397,125-135(1991)

Goleneva, L. M. See Askadskij, A. A.: V1554A,426-431(1991)

Golightly, W. See Ride, Sally K.: V1552,128-137(1991)

Golimowski, D. A. See Clampin, Mark: V1542,165-174(1991)

Golini, Donald See Pepi, John W.: V1533,212-221(1991)

Gollinucci, Stefano See Rocca, Corrado: V1372,39-47(1991)

Golobic, Robert A.
; Bohley, Thomas K.; Wells, Lisa D.; Sanborn, Timothy A.: Clinical experience with an excimer laser angioplasty system,V1425,84-92(1991)

Golombok, Michael
: Stimulated Raman diagnostics in diesel droplets,V1497,100-119(1991)

Golovan, Alexander V. See Rybak, Ilya A.: V1469,472-482(1991)

Golovchenko, Anotoly M. See Kurochkin, Vadim Y.: V1435,322-330(1991)

Golovin, A. F.
; Golub, A. P.; Zemtsov, S. S.; Fedyushin, B. T.: Shortwave radiation induced by CO_2 laser pulse interaction with aluminum target,V1440,250-259(1991)

Golovin, V. M. See Sadyigov, Z. Y.: V1621,158-168(1991)

Goltsos, William See Bartley, James A.: V1544,140-145(1991)

Golub, A. P. See Golovin, A. F.: V1440,250-259(1991)

Golub, Leon See Spiller, Eberhard: V1343,134-144(1991)

Golub, Mikhail A.
; Doskolovich, Leonid L.; Kazanskiy, Nikolay L.; Kharitonov, Sergey I.; Orlova, Natalia G.; Sisakian, Iosif N.; Soifer, Victor A.: Computational experiment for computer-generated optical elements,V1500,194-206(1991)
; Doskolovich, Leonid L.; Kazanskiy, Nikolay L.; Kharitonov, Sergey I.; Sisakian, Iosif N.; Soifer, Victor A.: Focusators at letters diffraction design,V1500,211-221(1991)
; Sisakian, Iosif N.; Soifer, Victor A.: Computer-generated optical elements for fiber's mode selection and launching,V1365,156-165(1991)
; Sisakian, Iosif N.; Soifer, Victor A.; Uvarov, G. V.: New measurement techniques for modal power distribution in fibers,V1366,273-282(1991)
; Sisakyan, Iosiph N.; Soifer, Victor A.; Uvarov, G. V.: Mode-selective fiber sensors operating with computer-generated optical elements,V1572,101-106(1991)

Golubenzev, A. A. See Antyuhov, V.: V1397,355-365(1991); 1415,48-59(1991)

Golubev, Evgeny M. See Belov, Sergei N.: V1552,264-267(1991)

Golubeva, N. A. See Chernyaeva, E. B.: V1402,7-10(1991)

Golubeva, N. G.
; Wang, Litszin: Resonance Raman spectra of hematoporphyrin derivative,V1403,134-138(1991)

Golubkov, Sergei P. See Potyrailo, Radislav A.: V1572,434-438(1991)

Gomes, Anderson S.
; de Araujo, Cid B.; Moraes, E. S.; Opalinska, M. M.; Gouveia-Neto, A. S.: Frequency up-conversion in Pr3+ doped fibers,V1579,257-263(1991)

Gomes, M. A. See Valla, Bruno: V1536,48-62(1991)

Gomez-Reino, Carlos
; Linares, Jesus: GRIN fiber lens connectors,V1332,468-473(1991)
See Acosta, E.: V1401,82-85(1991)

Gomm, A. J. See Di Cocco, Guido: V1549,102-112(1991)

Goncharov, S. E. See Garbuzov, Dmitriy Z.: V1418,386-393(1991)

Goncharov, V. F.
; Popov, Alexander P.; Veniaminov, Andrei V.: High-efficiency reflective holograms: subangstrom spectral selectors,V1238,97-102(1991)

Gonda, Teresa G. See Meitzler, Thomas J.: V1486,380-389(1991)

Gonella, Francesco
; Mazzi, Giulio; Quaranta, Alberto: Stress analysis in ion-exchanged waveguides by using a polarimetric technique,V1513,425-433(1991)

Gong, Dao B.
: Research on relaxation process of a-Si:H film photoconductivity and the trap effect,V1519,281-286(1991)

Gong, Leiguang
; Kulikowski, Casimir A.; Mezrich, Reuben S.: Automatic segmentation of brain images: selection of region extraction methods,V1450,144-153(1991)

Gong, Meixia
See Chang, Zenghu: V1358,614-618(1991)
See Hou, Xun: V1358,868-873(1991)

Gong, Shuxing See Zhou, Bing: V1519,454-456(1991)

Gong, Wei
; Shi, Qinyun; Cheng, Minde: Decomposing morphological structure element into neighborhood configurations,V1606,153-164(1991)

Gong, Xue-Mei
; Chan, Andrew K.; Taylor, Henry F.: Lateral-mode discrimination in surface-emitting DBR lasers with cylindrical symmetry,V1418,422-433(1991)

Gong, Yiming See Chen, Wei: V1450,198-205(1991)

Gong, Zhi-Ben
See Feng, Yue-Zhong: V1417,370-372(1991)
See Feng, Yue-Zhong: V1487,356-360(1991)

Gonsalves, Robert A.
; Nisenson, Peter: HST image processing: an overview of algorithms for image restoration,V1567,294-307(1991)

Gontier, D. See Mens, Alain: V1358,878-887(1991)

Gonzales, Cesar A. See Viscito, Eric: V1605,58-72(1991)

Gonzalez, Francois See Mary, Joel: V1557,147-155(1991)

Gonzalez, Juan C. See Minano, Juan C.: V1528,104-115(1991)

Gonzalez, Marcelo S.
; Mena, Manolo G.: Flow behavior of thermoset molding compound in conventional PDIP molds,V1390,568-579(1991)

Gonzalez, Orlando
; Johnson, Carroll; Starks, Scott A.: Integration of a computer vision system with an IBM 7535 robot,V1381,284-291(1991)

Gonzalez, R. M. See Acosta, E.: V1401,82-85(1991)

Gooch, Richard P. See Shynk, John J.: V1565,102-117(1991)

Good, Walter F. See Rockette, Howard E.: V1446,276-283(1991)

Goodenough, David J.
; Atkins, Frank B.; Dyer, Stephen M.: Automated techniques for quality assurance of radiological image modalities,V1444,87-99(1991)

Goodfellow, Robert C.
; Goodwin, Martin J.; Moseley, Andrew J.: Intercircuit optical interconnects using quantum-well modulators,V1389,594-599(1991)

Gooding, Linda
; Miller, Michael E.; Moore, Jana; Kim, Seong-Han: Effect of viewing distance and disparity on perceived depth,V1457,259-266(1991)

Goodman, Daniel L.
; Birx, Daniel L.; Danly, Bruce G.: Induction linac-driven relativistic klystron and cyclotron autoresonance maser experiments,V1407,217-225(1991)

Goodman, Joseph W.
See Farn, Michael W.: V1354,24-29(1991)
See Kalman, Robert F.: V1563,34-44(1991)
See Wilkinson, Timothy S.: V1384,293-304(1991)

Goodner, Ray See Chang, Li-Hsin: V1596,34-45(1991)

Goodrich, Alan See Brodsky, Aaron: V1482,159-169(1991)

Goodwin, David B.
; Cappiello, Gregory G.; Coppeta, David A.; Govignon, Jacques P.: Hybrid digital/optical ATR system,V1564,536-549(1991)

Goodwin, Martin J. See Goodfellow, Robert C.: V1389,594-599(1991)

Goodwin, Norman W.
See Efron, Uzi: V1455,237-247(1991)
See Sayyah, Keyvan: V1455,249-254(1991)

Goody, R. S. See Seidel, Claus: V1432,105-116(1991)

Gookin, Debra M.
ed.: *Devices for Optical Processing*,V1562(1991)

Goovaerts, Etienne See Van Hoof, Chris A.: V1362,291-300(1991)

Gopalakrishnan, G. K.
; Fairchild, B. W.; Yeh, C. L.; Park, C. S.; Chang, Kai; Weichold, Mark H.; Taylor, Henry F.: Ring resonators for microwave optoelectronics,V1476,270-275(1991)

Gopalaswamy, Srinivasan See Casasent, David P.: V1470,208-219(1991)

Gopinath, Anand See Ghiasi, Ali: V1583,170-174(1991)

Goralczyk, Steven M. See O'Neil, Jason: V1495,72-80(1991)

Goray, L. I. See Yusupov, I. Y.: V1238,240-247(1991)

Gorbatyuk, I. N. See Vertiy, Alexey A.: V1361,1070-1078(1991)

Gorbunov, A. V.
: Liquid dendrites growth at laser-induced melting in a NaCl volume,V1440,78-82(1991)

Gordeev, V. N. See Ligachev, Valery A.: V1519,214-219(1991)

Gordeeva, I. A. See Vas'kovsky, Yu. M.: V1440,229-240(1991)

Gordienko, Vyacheslav M.
; Akhmanov, Sergei A.; Bersenev, V. I.; Kosovsky, L. A.; Kurochkin, Nikolai N.; Priezzhev, Alexander V.; Pogosov, Gregory A.; Putivskii, Yu. Y.: Infrared coherent lidar systems for wind velocity measurements,V1416,102-114(1991)
See Pogosov, Gregory A.: V1416,115-124(1991)

Gordon, E. S.
See Rotman, Stanley R.: V1442,205-215(1991)
See Rotman, Stanley R.: V1442,335-346(1991)

Gordon, Gaile G.
: Face recognition based on depth maps and surface curvature,V1570,234-247(1991)

Gordon, James See Chan, Hoover: V1453,381-389(1991)

Gordon, Jeff M.
: Recent developments in nonimaging secondary concentrators for linear receiver solar collectors,V1528,32-43(1991)
; Rabl, Ari: Calculation of flux density produced by CPC reflectors on distant targets,V1528,152-162(1991)
See Cai, Wen: V1528,118-128(1991)

Gordon, Joseph G. See Toney, Michael F.: V1550,140-150(1991)

Gorelik, Vladimir S.
; Kovalenko, Sergey N.; Turukhano, Boris G.: Active stabilization of interferometers by two-frequency phase modulation,V1507,379-382(1991)
; Kovalenko, Sergey N.; Turukhano, Boris G.: Self-aberration-eliminating interferometer for wavefront measurements,V1507,488-490(1991)
See Prokhorov, Kirill A.: V1501,80-84(1991)
See Turukhano, Boris G.: V1500,290-292(1991)

Gorenflo, Ronald L.
; Stockum, Larry A.; Barnett, Brett: High-speed video instrumentation system,V1346,42-53(1991)

Gorenstein, Annette
; Polo Da Fonseca, C. N.; Torresi, R. M.: Electrochromism in cobalt oxyhydroxide thin films,V1536,104-115(1991)
See Fantini, Marcia C.: V1536,81-92(1991)
See Scarminio, J.: V1536,70-80(1991)

Gorman, Christopher B. See Jozefiak, Thomas H.: V1436,8-19(1991)

Gornik, Erich
; Koeck, A.; Thanner, C.; Korte, Lutz: Surface plasmon enhanced light emission in GaAs/AlGaAs light emitting diodes,V1362,1-13(1991)
See Kremser, Christian: V1501,69-79(1991)

Gorse, Claudine See Capitelli, Mario: V1503,126-131(1991)

Gorti, Sridhar See Tanaka, Toyoichi: V1403,280-287(1991)

Gorton, E. K.
; Parcell, E. W.; Cross, P. H.: Pulse shape effects in a twin cell DF laser,V1397,291-295(1991)

Gortych, Joseph E.
; Williamson, David M.: Effects of higher order aberrations on the process window,V1463,368-381(1991)

Goryachev, D. N.
See Sreseli, Olga M.: V1440,326-331(1991)
See Sreseli, Olga M.: V1545,149-158(1991)

Goryachkin, Dmitriy A. See Sherstobitov, Vladimir E.: V1415,79-89(1991)

Gosnell, Timothy R.
; Taylor, Antoinette J.; eds.: [$BFS]Selected Papers on Ultrafast Laser Technology[$BFE],VMS44(1991)

Gossard, Arthur C. See Coldren, Larry A.: V1362,24-37(1991)

Gosset, Nathalie M. See Hanson, Daniel: V1363,48-56(1991)

Goswami, Kisholoy See Klainer, Stanley M.: V1434,119-126(1991)

Goto, Hiroshi
; Imanaka, Koichi: Super-compact dual-axis optical scanning unit applying a tortional spring resonator driven by a piezoelectric actuator,V1544,272-281(1991)

Goto, T.
See Hane, Kazuhiro: V1332,577-583(1991)
See Ishikawa, Ken: V1397,55-58(1991)

Goto, Yasuyuki
; Kurihara, Kazuaki; Sawamoto, Yumiko; Kitakohji, Toshisuke: Continuous TEM observation of diamond nucleus growth by side-view method,V1534,49-58(1991)

Goto, Yoshiro See Wakana, Shin-ichi: V1479,283-290(1991)

Gotoh, Akira See Watanabe, Hitoshi: V1499,21-28(1991)

Gottfried, Varda
; Lindenbaum, Ella S.; Kimel, Sol; Hammer-Wilson, Marie J.; Berns, Michael W.: Laser photodynamic therapy of cancer: the chorioallantoic membrane model for measuring damage to blood vessels in-vivo,V1442,218-229(1991)
See Kimel, Sol: V1525,341-350(1991)

Gottler, Michael See Wolfe, William J.: V1382,240-254(1991)

Gottlieb, J.
; Kopeika, Norman S.: Prediction of coarse-aerosol statistics according to weather forecast,V1487,184-191(1991)

Gottlieb, Milton S.
See Brandt, Gerald B.: V1454,336-343(1991)
See Melamed, Nathan T.: V1454,306-312(1991)
See Paradies, C. J.: V1561,2-5(1991)

Gottschalk, Paul G.
; Turney, Jerry L.; Chiu, Arnold H.; Mudge, Trevor N.: Recognition, tracking, and pose estimation of arbitrarily shaped 3-D objects in cluttered intensity and range imagery,V1383,84-96(1991)
See Turney, Jerry L.: V1380,53-63(1991)

Gottschalk, Wolfgang
; Hengst, Joachim; Sroka, Ronald; Unsoeld, Eberhard: Influence of the laser-induced temperature rise in photodynamic therapy,V1427,320-326(1991)

Gotz, J. See Matejec, Vlastimil: V1513,174-179(1991)

Gou, Li See Yang, Bang C.: V1519,864-865(1991)

Goudonnet, Jean-Pierre
; Salomon, L.; de Fornel, F.; Adam, P.; Bourillot, E.; Neviere, Michel; Guerin, Philippe: Analysis of images of periodic structures obtained by Photon Scanning Tunneling Microscopy,V1545,130-139(1991)
; Salomon, L.; de Fornel, F.; Chabrier, G.; Warmack, R. J.; Ferrell, Trinidad L.: Photon scanning tunneling microscopy,V1400,116-123(1991)

Gouedard, C. See Fleurot, Noel A.: V1502,230-241(1991)

Gough, David W. See Lake, Stephen P.: V1486,286-293(1991)

Gough, M. P. See Lu, Wei-Wei: V1470,197-205(1991)

Goutsias, John I.
; Wen, Chuanju: Discrete random set models for shape synthesis and analysis,V1606,174-185(1991)

Goutte, Robert
See Baskurt, Atilla: V1445,485-495(1991)
See Peyrin, Francoise: V1443,268-279(1991)

Goutzoulis, Anastasios P.
; Davies, D. K.: Design, fabrication, and testing of a 7-bit binary fiber optic delay line,V1371,182-194(1991)

Gouveia, Feliz A.
; Barthes, Jean-Paul A.; Oliveira, Eugenio C.: Distributed architecture for intelligent robotics,V1468,516-523(1991)

Gouveia-Neto, A. S. See Gomes, Anderson S.: V1579,257-263(1991)

Gove, Robert J.
; Lee, Woobin; Kim, Yongmin; Alexander, Thomas: Image computing requirements for the 1990s: from multimedia to medicine,V1444,318-333(1991)

Gover, Avraham See Keselbrener, Michel: V1415,38-47(1991)

Govignon, Jacques P. See Goodwin, David B.: V1564,536-549(1991)

Govil, Anurag See Morris, Michael D.: V1439,95-101(1991)

Govindan, T. R. See Meyyappan, Meyya: V1392,67-76(1991)

Govindarajan, Rangaprasad See Jorgensen, Gary J.: V1536,194-205(1991)

Gowdy, Jay W.
; Stentz, Anthony; Hebert, Martial: Hierarchical terrain representations for off-road navigation,V1388,131-140(1991)

Gower, Malcolm C.
See Rumsby, Phil T.: V1598,36-45(1991)
See Turcu, I. C.: V1503,391-405(1991)

Gowrinathan, Sankaran
; Mataloni, Raymond J.; Schwartz, Stanley J.; eds.: [$BFS]Surveillance Technologies[$BFE],V1479(1991)

Gozdz, Antoni S.
; Ono, Hiroshi; Ito, Seiki; Shelburne, John A.; Matsuda, Minoru: Evaluation of poly(p-trimethylsilylstyrene and p-pentamethyldisilylstyrene sulfone)s as high-resolution electron-beam resists,V1466,200-205(1991)
; Shelburne, John A.: Synthesis and lithographic evaluation of alternating copolymers of linear and cyclic alkenyl(di)silanes with sulfur dioxide,V1466,520-527(1991)

Graaff, Reindert
See Koelink, M. H.: V1403,347-349(1991)
See Koelink, M. H.: V1511,120-128(1991)
See Koelink, M. H.: V1431,63-72(1991)

Graber, Harry L. See Barbour, Randall L.: V1431,52-62(1991); 1431,192-203(1991)

Grabner, G. See Angelov, D.: V1403,572-574(1991)

Grace, Kevin H. See Rorabaugh, Terry L.: V1566,312-322(1991)

Grad, Ladislav
; Diaci, J.; Mozina, Janez: Energy conversion efficiency during optical breakdown,V1525,206-209(1991)

Gradauskas, I. See Amosova, L. P.: V1440,406-413(1991)

Gradauskas, Jonas See Asmontas, Steponas: V1512,131-134(1991)

Graefe, Volker
See Regensburger, Uwe: V1388,112-119(1991)
See Solder, Ulrich: V1388,104-111(1991)

Graf, Juerg See Felberbauer, Franz: V1510,63-71(1991)

Graf, Richard J. See Weil, Gary J.: V1467,18-33(1991)

Graham, W. S. See Deutsch, Alina: V1389,161-176(1991)

Grahek, Frank E. See Walega, James G.: V1433,232-241(1991)

Granchak, Vasilij M.
See Boiko, Yuri B.: V1507,544-548(1991)
See Boiko, Yuri B.: V1238,253-257(1991)
See Boiko, Yuri B.: V1238,258-265(1991)

Granet, Nicole See Friedlander, Miles H.: V1423,62-69(1991); 1429,229-236(1991)

Grangeat, Pierre
; Le Masson, Patrick; Melennec, Pierre; Sire, Pascal: Evaluation of the 3-D radon transform algorithm for cone beam reconstruction,V1445,320-331(1991)

Granger, Edward M. See Behlau, Jerry L.: V1530,218-230(1991)

Granger, Rene'
: Infrared reflectivity: a tool for bond investigation in II-VI ternaries,V1484,39-46(1991)

Granier, Jean E. See Lopez, Adolphe: V1590,191-202(1991)

Granik, Alex T.
; Caulfield, H. J.: Quantum holography,VIS08,33-38(1991)

Granjon, Yves See Reynes, Anne M.: V1525,177-182(1991)

Granovsky, A. B. See Agranat, M. B.: V1440,397-400(1991)

Granozzi, G. See Mazzoldi, Paolo: V1513,182-197(1991)

Granqvist, Claes G.
See Lampert, Carl M.: V1536(1991)
See Yin, Zhiqiang: V1536,149-157(1991)

Grant, Barbara G. See Nianzeng, Che: V1493,182-194(1991)

Grant, Jon R. See Oicles, Jeffrey A.: V1378,60-69(1991)

Grant, K. J. See Di Cocco, Guido: V1549,102-112(1991)

Grant, Michael A. See Gatenby, Paul V.: V1522,126-134(1991)

Grant, Robert A. See Stewart, Diane K.: V1465,64-77(1991)

Grant, S. M. See Sumner, Timothy J.: V1549,256-264(1991)

Grant, Stephen G. See Jensen, Ronald H.: V1403,372-380(1991)

Grantham, Charles E.
: Visual thinking in organizational analysis,V1459,77-84(1991)

Graschew, Georgi See Kohl, M.: V1525,26-34(1991)

Gratrix, Edward J.
; Zarowin, Charles B.: Fabrication of microlenses by laser-assisted chemical etching,V1544,238-243(1991)

Grattan, Kenneth T.
See Boyle, William J.: V1511,51-56(1991)
See Clarke, Timothy A.: V1332,474-486(1991)
See Ning, Yanong N.: V1367,347-356(1991)
See Weir, Kenneth: V1584,220-225(1991)
See Zhang, Zhiyi: V1511,264-274(1991)

Gratton, Enrico See Fishkin, Joshua B.: V1431,122-135(1991)

Gratzl, Thomas
; Dohr, Gottfried; Schmidt-Kloiber, Heinz; Reichel, Erich: Histological distinction of mechanical and thermal defects produced by nanosecond laser pulses in striated muscle at 1064 nm,V1427,55-62(1991)

Grau, T. See Miller, Kurt: V1426,378-383(1991)

Graue, Roland
; Kampf, Dirk; Rippel, Harald; Witte, G.: ORFEUS alignment concept,V1494,377-385(1991)
See Rippel, Harald: V1343,520-529(1991)

Gravert, D. See Twieg, Robert J.: V1455,86-96(1991)

Graves, David B. See Manske, Loni M.: V1463,414-422(1991)

Graves, David W.
: Sixty-four channel acousto-optical Bragg cells for optical computing applications,V1563,229-235(1991)

Graves, J. E.
; McKenna, Daniel: University of Hawaii adaptive optics system: III. Wavefront curvature sensor,V1542,262-272(1991)
See Cavedoni, Charles P.: V1542,273-282(1991)
See Roddier, Francois J.: V1542,248-253(1991)

Gravey, Philippe
; Moisan, Jean-Yves: Dynamic holographic interconnects: experimental study using photothermoplastics with improved cycling properties,V1507,239-246(1991)
See Abdelghani-Idrissi, Ahmed M.: V1362,417-427(1991)

Gray, Connie B. See Strikwerda, Thomas E.: V1478,13-23(1991)

Gray, David E.
; Benjamin, Neil M.; Chapman, Brian N.: Effects of environmental and installation-specific factors on process gas delivery via mass-flow controller with an emphasis on real-time behavior,V1392,402-410(1991)

Gray, George R. See Agrawal, Govind P.: V1497,444-455(1991)

Gray, Joel E.
: Use of the Society of Motion Picture and Television Engineers test pattern in picture archiving and communication systems (PACS),V1444,118-124(1991)

Gray, Joe W. See Yorkey, Thomas J.: V1345,255-259(1991)

Gray, Kevin J.
; Lu, Grant: Properties of CVD diamond for optical applications,V1534,60-66(1991)

Gray, Robert M. See Cosman, Pamela C.: V1605,162-171(1991)

Gray, W. C. See Jackson, John E.: V1533,75-86(1991)

Graziano, Karen A.
; Thompson, Stephen D.; Winkle, Mark R.: Novel acid-hardening positive photoresist technology,V1466,75-88(1991)
See Babcock, Carl P.: V1466,653-662(1991)
See Berry, Amanda K.: V1465,210-220(1991)

Greated, C. A. See She, K.: V1572,581-587(1991)

Greaves, G. N. See Dent, Andrew J.: V1550,97-107(1991)

Grebel, Haim
: Artificial dielectric waveguides from semiconductor-embedded polymers,V1583,355-361(1991)
; Pien, P.: Laser-induced etched grating on InP for integrated optical circuit elements,V1583,331-337(1991)

Grebenshikov, Sergey V.
: Absorbing materials in multilayer mirrors,V1500,189-193(1991)
: Materials for optimal multilayer coating,V1519,302-307(1991)

Greber, Charlotte M.
See Frenz, Martin: V1427,9-15(1991)
See Romano, Valerio: V1427,16-26(1991)

Greby, Daniel F. See van Gilse, Jan: V1414,45-54(1991)

Green, Christian A. See Hayes, Todd R.: V1418,190-202(1991)

Green, David C. See Bell, J. M.: V1536,29-36(1991)

Green, Emily N. See Henderson, Byron B.: V1364,235-244(1991)

Green, J. M. See Fieret, Jim: V1503,53-61(1991)

Green, James C. See Gallagher, Dennis J.: V1343,155-161(1991)

Green, Martina
; Seiber, James N.; Biermann, Heinz W.: In situ measurement of methyl bromide in indoor air using long-path FTIR spectroscopy,V1433,270-274(1991)
See Biermann, Heinz W.: V1433,2-7(1991)

Green, Richard F. See Chu, Ting L.: V1361,523-528(1991)

Green, Robert O. See Eastwood, Michael L.: V1540,164-175(1991)

Green, Samuel I.
: Optical dividers for quadrant avalanche photodiode detectors,V1417,496-512(1991)

Green, Steven E. See Narayanan, Ram M.: V1379,116-122(1991)

Green, Thomas J.
; Shapiro, Jeffrey H.; Menon, Murali M.: Target detection performance using 3-D laser radar images,V1471,328-341(1991)

Greenaway, Alan H.
: Satellite-borne laser for adaptive optics reference,V1494,386-393(1991)

Greenberg, David A. See Burger, Robert J.: V1449,174-185(1991)

Greenberg, Michael R. See Rainer, Frank: V1441,247-254(1991)

Greene, Ben
; Taubman, Matthew; Watts, Jeffrey W.; Gaither, Gary L.: Solid state lasers for planetary exploration,V1492,126-139(1991)

Greene, Herbert G.
; MacMullan, Jay: Maximum likelihood estimation of differential delay and differential Doppler,V1470,98-102(1991)

Greeneich, James S. See Chu, Ron: V1465,238-243(1991)

Greenfield, Ephraim See Azran, A.: V1442,54-57(1991)

Greenfield, Perry E.
; Paresce, Francesco; Baxter, David; Hodge, P.; Hook, R.; Jakobsen, P.; Jedrzejewski, Robert; Nota, Antonella; Sparks, W. B.; Towers, Nigel M.; Laurance, R. J.; Macchetto, F.: In-flight performance of the faint object camera of the Hubble Space Telescope,V1494,16-39(1991)

Greenham, A. C. See Wood, Roger M.: V1441,316-326(1991)

Greenleaf, William G.
; Siniard, Sheri M.; Tait, Mary B.: Generation of realistic IR images of tactical targets in obscured environments,V1486,364-375(1991)

Greensite, F. See Gutfinger, Dan: V1445,288-296(1991)

Greenwald, Daniel P. See Wider, Todd M.: V1422,56-61(1991)

Greenwald, David
See Keitzer, Scott: V1482,406-414(1991)
See Kimbrell, James E.: V1482,415-424(1991)
See Verbanets, William R.: V1482,112-120(1991)

Greenwell, Roger A.
; Paul, Dilip K.; eds.: *Fiber Optics Reliability: Benign and Adverse Environments IV*,V1366(1991)
See Karbassiyoon, Kamran: V1366,178-183(1991)

Greer, James A.
; Van Hook, H. J.: Pulsed-laser deposition of oxides over large areas,V1377,79-90(1991)

Greer, Jerry D.
: Lasers and electro-optic technology in natural resource management,V1396,342-352(1991)
: Target lifetimes in natural resource management,V1538,69-76(1991)

Greer, L. See Potter, B. L.: V1368,251-257(1991)

Greeson, James C.
: Standards for flat panel display systems,VCR37,146-158(1991)

Greffet, Jean-Jacques
; Sentenac, Anne: Enhanced backscattering of s- and p-polarized light from particles above a substrate,V1558,288-294(1991)

Gregg, Richard E. See Gregory, Kenton W.: V1425,217-225(1991)

Gregoire, Daniel J.
; Harvey, Robin J.; Levush, Baruch: Theory and simulation of the HARmonic amPlifier Free-Electron Laser (HARP/FEL),V1552,118-126(1991)

Gregor, Eduard See Stultz, Robert D.: V1419,64-74(1991)

Gregoriou, Gregorios See Butler, Clive: V1584,282-293(1991)

Gregory, Andrew
; Phillips, Richard T.; Majumder, Fariduddin A.: Hot carrier relaxation in bulk InGaAs and quantum-wells,V1362,268-274(1991)

Gregory, Christopher C.
; Harrington, James A.; Altkorn, Robert I.; Haidle, Rudy H.; Helenowski, Tomasz: Hollow curved Al2O3 waveguides for CO2 laser surgery,V1420,169-175(1991)

Gregory, Don A.
See Hudson, Tracy D.: V1474,101-111(1991)
See Hudson, Tracy D.: V1564,54-64(1991)
See Kirsch, James C.: V1474,90-100(1991)
See Kirsch, James C.: V1558,432-441(1991)
See Kirsch, James C.: V1482,69-78(1991)
See Lindberg, Perry C.: V1470,220-225(1991)
See Yu, Francis T.: V1558,450-458(1991)

Gregory, J. A. See Huang, Chin M.: V1447,156-164(1991)

Gregory, Kenton W.
; Aretz, H. T.; Martinelli, Michael A.; Ledet, Earl G.; Hatch, G. F.; Gregg, Richard E.; Sedlacek, Tomas; Haase, W. C.: Intraluminal laser atherectomy with ultrasound and electromagnetic guidance,V1425,217-225(1991)

Gregory, Rob See Hadfield, Michael J.: V1478,126-144(1991)

Gregson, Peter H. See Shen, Zhijiang: V1606,632-640(1991)

Gregus, J. A. See White, Donald L.: V1343,204-213(1991)

Greguss, Pal
: Acoustical and some other nonconventional holography,VIS08,387-401(1991)
: Comparative studies on hyperthermia induced by laser light, microwaves, and ultrasonics,V1525,313-324(1991)
: Information retrieval from ultrasonic (acoustical) holograms by moire principle,V1238,421-427(1991)
: Panoramic security,V1509,55-66(1991)
; Jeong, Tung H.; eds.: *Holography*,VIS08(1991)
See Gilbert, John A.: V1238,412-420(1991)

Greiling, Paul T.
: Ultra-high-frequency GaInAs/InP devices and circuits for millimeter wave application,V1361,47-58(1991)
; Nguyen, Loi D.: Ultra-high-frequency InP-based HEMTs for millimeter wave applications,V1475,34-41(1991)

Grelsamer, Ronald P. See Garino, Jonathan P.: V1424,43-47(1991)

Greneker, E. F. See Bruder, Joseph A.: V1479,316-321(1991)

Grenier, Marie-Claude
; Durand, Louis-Gilles; de Guise, J.: Comparative study of texture measurements for cellular organelle recognition,V1450,154-169(1991)

Grenier, Philippe See Preteux, Francoise: V1450,72-83(1991)

Greskovich, Frank J. See Johnson, Douglas E.: V1421,36-41(1991)

Gresser, Julien See Marchand, Philippe J.: V1505,38-49(1991)

Gresslehner, Karl-Heinz
; Schirz, W.; Humenberger, Josef; Sitter, Helmut; Andorfer, J.; Lischka, Klaus: Photovoltaic HgCdTe MWIR-detector arrays on (100)CdZnTe/(100)GaAs grown by hot-wall-beam epitaxy,V1361,1087-1093(1991)

Greulich, Karl O.
See Hitzler, Hermine: V1503,355-362(1991)
See Seidel, Claus: V1432,105-116(1991)

Greulich, Peter
; Hespeler, Bernd; Spatscheck, Thomas: A 120-Mbit/s QPPM high-power semiconductor transmitter performance and reliability,V1417,358-369(1991)
; Hespeler, Bernd; Spatscheck, Thomas: Lifetest on a high-power laser diode array transmitter,V1522,144-153(1991)

Greunke, Roy G. See Morgenthaler, Matthew K.: V1387,82-95(1991)

Greve, Jan
See Heideman, Rene: V1510,131-137(1991)
See Koelink, M. H.: V1403,347-349(1991)
See Koelink, M. H.: V1511,120-128(1991)
See Koelink, M. H.: V1431,63-72(1991)
See Lucassen, G. W.: V1403,185-194(1991)
See Puppels, G. J.: V1403,146-146(1991)
See Siebinga, I.: V1403,746-748(1991)

Grevey, D. F.
; Badawi, K. F.: Spatial characterization of YAG power laser beam,V1502,32-40(1991)

Greynolds, Alan W. See Johnston, Steve C.: V1354,136-141(1991)

Grezes-Besset, C. See Lallier, Eric: V1506,71-79(1991)

Grider, Douglas T.
; Ozturk, Mehmet C.; Wortman, Jim J.; Littlejohn, Michael A.; Zhong, Y.: Low-resistivity contacts to silicon using selective RTCVD of germanium,V1393,229-239(1991)
See Ozturk, Mehmet C.: V1393,260-269(1991)

Griebsch, Juergen See Holzwarth, Achim: V1503,98-109(1991)

Grieser, James L.
; Freeland, Alan W.; Fink, Jeffrey D.; Meinke, Gary E.; Hildreth, Eugene N.: Space station atomic-oxygen-resistant coatings,V1330,102-110(1991)
; Swisher, Richard L.; Phipps, James A.; Pelleymounter, Douglas R.; Hildreth, Eugene N.: Next generation thermal control coatings,V1330,111-118(1991)

Griffen, Christopher T.
; Fryska, Slawomir T.; Bernier, Paul R.: Analysis of powertrain noise and vibration using interferometry,V1554B,754-766(1991)

Griffin, Adam See Zucker, Oved S.: V1378,22-33(1991)

Griffin, Chintana
; Nuttall, Albert H.: Comparison of two kernels for the modified Wigner
distribution function,V1566,439-451(1991)

Griffin, Edward See Willems, David: V1475,55-61(1991)

Griffin, Kenneth L. See Druce, Robert L.: V1378,43-54(1991)

Griffin, Michael F.
; Tulpule, Sharayu: Time-frequency distributions for propulsion-system
diagnostics,V1566,179-189(1991)

Griffin, Neal L.
: Scheduler's assistant: a tool for intelligent scheduling,V1468,110-121(1991)

Griffin, Paul M.
; Taboada, John: Three-dimensional tolerance verification for computer
vision systems,V1381,292-298(1991)

Griffin, Stephen E. See Nath, Dilip K.: V1489,17-32(1991)

Griffith, David M. See Hong, C. S.: V1418,177-187(1991)

Griffith, Derek J. See McDowell, Maurice W.: V1358,227-236(1991)

Griffith, O'Dale See Mobasser, Sohrab: V1540,764-774(1991)

Griffith, Roger L. See Thomas, Stan W.: V1358,578-588(1991)

Griffiths, Barry E. See Covino, Joseph M.: V1478,114-125(1991)

Griffiths, Scott F. See Stoudt, David C.: V1378,280-285(1991)

Griggs, J. E. See Newnam, Brian E.: V1552,154-174(1991)

Grigonis, R. See Danelius, R.: V1402,198-208(1991)

Grigor'ev, N. N.
; Kudykina, T. A.; Tomchuk, P. M.: Laser-induced destruction of solids due to
photo-excited carriers recombination,V1440,105-111(1991)

Grigor'yants, Vil V. See Matejec, Vlastimil: V1513,174-179(1991)

Grigorenko, A. N. See Nikitin, Petr I.: V1584,124-134(1991)

Grigoriev, Ivo D. See Tontchev, Dimitar A.: V1429,76-80(1991)

Grigoryants, V. V. See Gaiduk, Mark I.: V1403,674-675(1991)

Grigsby, R. See Haigh, N. R.: V1366,259-264(1991)

Griko, N. B. See Timchenko, A. A.: V1403,340-343(1991)

Grimard, Dennis S.
; Terry, Fred L.; Elta, Michael E.: Theoretical and practical aspects of real-time
Fourier imaging,V1392,535-542(1991)

Grimes, Gary J.
; Blyler, Lee L.; Larson, Allen L.; Farleigh, Scott E.: Plastic-optical-fiber-based
photonic switch,V1592,139-149(1991)
See Coyle, Richard J.: V1412,129-137(1991)

Grimm, B. See Liang, Junzhong: V1542,543-554(1991)

Grindrod, S. R. See Linney, Alfred D.: V1380,190-199(1991)

Grinevich, A. E. See Dubrovsky, V.: V1403,344-346(1991)

Grinevitskaya, O. V. See Mikhailov, V. N.: V1238,144-152(1991)

Grinstein, Georges G. See Smith, Stuart: V1459,192-206(1991)

Grinvald, Amiran See Ratzlaff, Eugene H.: V1439,88-94(1991)

Grinyakin, A.P. See Mnatsakanyan, Eduard A.: V1414,130-133(1991)

Grinyakin, V.P. See Mnatsakanyan, Eduard A.: V1414,130-133(1991)

Grinzato, Ermanno G.
; Mazzoldi, Andrea: Infrared detection of moist areas in monumental
buildings based on thermal inertia analysis,V1467,75-82(1991)
See Bison, Paolo G.: V1471,369-377(1991)

Gripp, S. See Bueker, Harald: V1572,410-418(1991)

Griscom, David L. See Tsai, T. E.: V1516,14-28(1991)

Grishanin, B. A.
; Vachev, V. D.; Zadkov, Victor N.: Theory and molecular dynamics
simulation of one-photon electronic excitation of multiatomic
molecules,V1402,44-52(1991)

Grisimov, Vladimir N. See Altshuler, Grigori B.: V1429,95-104(1991)

Griswold, Marsden P. See Motamedi, M. E.: V1544,22-32(1991)

Griswold, Norman C.
; Kehtarnavaz, Nasser: Experiments in real-time visual control,V1388,342-
349(1991)
See Kehtarnavaz, Nasser: V1468,467-478(1991)

Gritzo, R. E. See Fairbank, William M.: V1435,86-89(1991)

Grodnikov, Alexander I.
; Korovkin, Vladimir P.: Finish machining of optical components in mass
production,V1400,186-193(1991)

Groenewegen, L. P. See Stut, W. J.: V1446,396-404(1991)

Groff, Mary B. See Wells, Mark E.: V1557,71-77(1991)

Grogan, Timothy A.
; Wu, Mei: Image quality measurements with a neural brightness perception
model,V1453,16-30(1991)

Groh, Werner
; Kuder, James E.; Theis, Juergen: Prospects for the development and
application of plastic optical fibers,V1592,20-30(1991)

Groke, Karl See Reichel, Erich: V1421,129-133(1991)

Gross, Barry
See Chen, Ji: V1444,256-264(1991)
See Flynn, Michael J.: V1444,172-179(1991)

Gross, Eitan See Ehrenberg, Benjamin: V1426,244-251(1991)

Gross, Kirk See Lindmayer, Joseph: V1401,103-112(1991)

Gross, Leon J. See Watowich, Stanley J.: V1396,316-323(1991)

Gross, Michael See McGee, Thomas J.: V1491,182-187(1991)

Grossberg, Stephen
: Neural network architecture for form and motion perception,V1469,24-
26(1991)
: Three-dimensional vision and figure-ground separation by visual
cortex,V1382,2-2(1991)

Grosskopf, Gerd See Schunk, Nikolaus: V1362,391-397(1991)

Grossman, Barry G.
; Gao, Xing; Thursby, Michael H.: Composite damage assessment employing
an optical neural network processor and an embedded fiber-optic sensor
array,V1588,64-75(1991)
See Caimi, Frank M.: V1589,90-99(1991)
See Thursby, Michael H.: V1588,218-228(1991)

Grossmann, Benoist E. See Ponsardin, Patrick: V1492,47-51(1991)

Grossmann, William See Eidelman, Shmuel: V1567,439-450(1991)

Grosso, Enrico
; Sandini, Giulio; Tistarelli, Massimo: Three-dimensional reconstruction using
virtual planes and horopters,V1570,371-381(1991)

Grotzinger, Timothy L.
: Effects of atmospheric conditions on the performance of free-space infrared
communications,V1417,484-495(1991)

Grove, Charles H. See Pan, J. J.: V1476,133-142(1991)

Grove, Steven L.
; Getreuer, Kurt W.; Schell, David L.: Lissajous analysis of focus crosstalk in
optical disk systems,V1499,354-359(1991)

Grover, Chandra P.
ed.: Optical Testing and Metrology III: Recent Advances in Industrial
Optical Inspection,V1332(1991)
: White-light transmission holographic interferometry using chromatic
corrective filters,V1332,132-141(1991)
; Hane, Kazuhiro: Interference phenomenon with correlated masks and its
application,V1332,624-631(1991)
See Hane, Kazuhiro: V1332,584-590(1991)

Grozet, P. See Bacis, Roger: V1397,173-176(1991)

Groznov, Michail A.
: Complex researches optically controlled liquid-crystal spatial-time light
modulators on the photoconductivity organic polymer basis,V1500,281-
289(1991)

Grubbs, Robert H. See Jozefiak, Thomas H.: V1436,8-19(1991)

Gruber, Duane E. See Hink, Paul L.: V1549,193-202(1991)

Gruen, Armin
; Stallmann, Dirk: High-accuracy edge-matching with an extension of the
MPGC-matching algorithm,V1526,42-55(1991)

Gruenbaum, F. A.
; Kohn, Philip D.; Latham, Geoff A.; Singer, Jay R.; Zubelli, Jorge P.: Diffuse
tomography,V1431,232-238(1991)

Gruenewald, Maria M.
; Hinchman, John H.: Production environment implementation of the stereo
extraction of cartographic features using computer vision and knowledge
base systems in DMA's digital production system,V1468,843-852(1991)

Gruetzmacher, Dettev See Malzahn, Eric: V1362,199-204(1991)

Grundfest, Warren S.
See Buchelt, Martin: V1420,249-253(1991)
See Chandra, Mudjianto: V1421,68-71(1991)
See Gershman, Alex: V1421,186-188(1991)
See Papaioannou, Thanassis: V1420,203-211(1991)
See Vari, Sandor G.: V1424,33-42(1991)
See Vari, Sandor G.: V1426,58-65(1991)
See Vari, Sandor G.: V1426,111-120(1991)

Grung, B. See Hibbs-Brenner, Mary K.: V1563,10-20(1991)

Gruntman, Michael A.
: Submicron structures—promising filters in EUV: a review,V1549,385-394(1991)

Grunwald, Arthur J.
; Kohn, S.; Merhav, S. J.: Visual field information in nap-of-the-earth flight by teleoperated helmet-mounted displays,V1456,132-153(1991)

Grunwald, Ruediger See Budzinski, Christel: V1500,264-274(1991)

Grupen, Roderic A.
; Weiss, Richard S.; Oskard, David N.: Grasp-oriented sensing and control,V1383,189-201(1991)

Gruszczak, Anthony See Lyon, Richard G.: V1567,317-326(1991)

Gruska, Bernd
; Ambree, P.; Wandel, K.; Wielsch, U.: High-conducting p+-InGaAs toplayers produced by simultaneous diffusion of Zn and Cd,V1361,758-764(1991)

Grutsch, James
; Heiferman, Kenneth S.; Cerullo, Leonard J.: Hearing preservation using CO2 laser for acoustic nerve tumors,V1428,136-145(1991)

Gruver, William A. See Hu, Yong-Lin: V1468,653-661(1991)

Gryczynski, Ignacy See Lakowicz, Joseph R.: V1435,142-160(1991)

Grzech, Matthew P. See Concepcion, Vicente P.: V1606,132-140(1991)

Grzegorzewski, Bronislaw
; Kowalczyk, M.; Mallek, Janusz: Correlation of speckle pattern generated by TEM10 laser mode,V1391,290-294(1991)

Grzesik, Ulrich See Fabian, Heinz: V1513,168-173(1991)

Gu, Benyuan See Dong, Bizhen: V1429,117-126(1991)

Gu, Bo-qi
; Feng, Wen-qing: Analysis of the flying light spot experiment on SPRITE detector,V1488,443-446(1991)

Gu, Claire See Yeh, Pochi A.: V1562,32-43(1991)

Gu, Dong H. See Hou, Li S.: V1519,548-553(1991)

Gu, Lingjuan See Zhuang, Song Lin: V1558,28-33(1991)

Gu, Lizhong
; Ma, Jiancheng; Wang, Jiazhen: Optically self-excited miniature fixed-beam resonator sensor,V1572,450-452(1991)

Gu, Peifu See Mei, Ting: V1554A,570-578(1991)

Gu, Qi H. See Cheng, Xian A.: V1519,33-36(1991)

Gu, Shenghua See Li, Naiji: V1572,47-51(1991)

Gu, Shu L.
; He, Yu L.; Wang, Zhi C.: Investigation into the characteristics of a-C:H films irradiated by electron beam,V1519,175-178(1991)

Gu, Zhi G.
; Li, Bing Z.: Silicon nitride film formed by NH3 plasma-enhanced thermal nitridation,V1519,247-251(1991)

Gu, Zhi P. See Yuan, Ren K.: V1519,831-834(1991)

Gu, Zhi X. See Pan, Xiao R.: V1519,85-89(1991)

Gu, Zu-Han
; Dummer, Richard S.: Small-angle scattering measurement,V1558,368-378(1991)
; Maradudin, Alexei A.; Mendez, Eugenio R.: Coherence in single and multiple scattering of light from randomly rough surfaces,V1530,60-70(1991)
See Ruiz-Cortes, Victor: V1558,222-232(1991)

Guadagni, Stefano
; Nadeau, Theodore R.; Lombardi, Loreto; Pistoia, Francesco; Pistoia, Maria A.: Imaging in digestive videoendoscopy,V1420,178-182(1991)

Gualteri, Gregory J. See Macrander, Albert T.: V1550,122-133(1991)

Guan, Genzhi
: Development of an optical fiber and photoelectric coupling V/F converter for 5.4-MV impulse generator,V1572,487-491(1991)

Guan, H. W. See Wang, Chin H.: V1559,39-48(1991)

Guan, Wen X. See Yin, Dong: V1519,164-166(1991)

Guan, Z. P.
; Zheng, Zhu H.; Zhang, J. H.; Lu, Y. M.; Fan, Guang H.; Fan, Xi W.: Interaction between the excitons and electrons in ZnSe1-xSx epilayer under high excitation,V1519,26-32(1991)

Guan, Zhilian See Bao, Liangbi: V1332,862-867(1991)

Guasti, Manuel F. See Haro-Poniatowski, Emmanuel: V1513,86-92(1991)

Guay, Daniel See Dodelet, Jean-Pol: V1436,38-49(1991)

Guba, B. S. See Kalyashora, L. N.: V1238,189-194(1991)

Gubanov, V. A.
; Kiselyov, Boris S.: Scattering parameters of volume phase hologram: Bragg's approach,V1621,102-113(1991)

Gubanov, Y. I. See Berkovsky, A. G.: V1358,750-755(1991)

Gubser, Don U. See Nisenoff, Martin: V1394,104-113(1991)

Gudmundson, Glen A. See Keeter, Howard S.: V1419,84-93(1991)

Gue, A. M. See Rouhani, Mehdi D.: V1361,954-962(1991)

Guedes, G. H. See Muramatsu, Mikiya: V1332,792-797(1991)

Guedes Valente, Luiz C.
; Kawase, Liliana R.; Afonso, Jose A.; Kalinowski, Hypolito J.: Electronic-digital detection system for an optical fiber current sensor,V1584,96-102(1991)
See Brooking, Nicholas L.: V1572,88-93(1991)
See Kanellopoulos, S. E.: V1516,200-210(1991)

Guedon, Jean-Pierre
; Bizais, Yves: Spline-based tomographic reconstruction method,V1443,214-225(1991)

Guelker, Gerd
; Haack, Olaf; Hinsch, Klaus D.; Hoelscher, Claudia; Kuls, Juergen; Platen, Winfried: Electronic speckle pattern interferometry for 3-D dynamic deformation analysis in industrial environments,V1500,124-134(1991)

Guenter, Peter
See Bosshard, Christian: V1560,344-352(1991)
See Sutter, Kurt: V1560,296-301(1991)
See Sutter, Kurt: V1560,290-295(1991)

Guenther, Arthur H.
See Becker, Wilhelm: V1441,541-552(1991)
See Bennett, Harold E.: V1441(1991)

Guenther, Bruce W.
ed.: *Calibration of Passive Remote Observing Optical and Microwave Instrumentation*,V1493(1991)

Guenther, Karl H.
: Why and how to coat ophthalmic lenses,V1529,96-105(1991)
; Balasubramanian, Kunjithapa; Hu, X. Q.: Durable, nonchanging, metal-dielectric and all-dielectric mirror coatings,V1485,240-244(1991)
See Balasubramanian, Kunjithapa: V1485,245-253(1991)
See Balasubramanian, Kunjithapa: V1529,106-114(1991)

Guenzel, Uwe See Zschocke, Wolfgang: V1501,49-56(1991)

Guerin, J. J.
See Sansonetti, Pierre: V1588,198-209(1991)
See Sansonetti, Pierre: V1588,143-149(1991)

Guerin, Jean M. See Alloncle, Anne P.: V1397,675-678(1991)

Guerin, Philippe
See Bourhis, Jean-Francois: V1583,383-388(1991)
See Goudonnet, Jean-Pierre: V1545,130-139(1991)
See Khan Malek, Chantal: V1343,56-61(1991)

Guern, Yves See Pelous, G.: V1598,149-158(1991)

Guerra, J. M. See Martin, E.: V1397,835-838(1991)

Guerra Perez, J. M. See Dominguez Ferrari, E.: V1397,725-727(1991)

Guerrieri, Roberto See Tadros, Karim H.: V1464,177-186(1991)

Guest, Clark C. See Baettig, Rainer K.: V1563,93-102(1991)

Guez, Allon
; Ahmad, Ziauddin: Generation of exploratory schedules in closed loop for enhanced machine learning,V1469,750-755(1991)

Gugeshashvili, M. I.
See Leblanc, Roger M.: V1436,92-102(1991)
See Volkov, Alexander G.: V1436,68-79(1991)

Guggenhiemer, Steven
; Wright, David L.: Simple beam-propagation measurements on ion lasers,V1414,12-20(1991)

Guglielmi, Massimo
; Colombo, Paolo; Mancinelli Degli Esposti, Luca; Righini, Giancarlo C.; Pelli, Stefano: Planar and strip optical waveguides by sol-gel method and laser densification,V1513,44-49(1991)

Guha, Aloke
; Bristow, Julian P.; Sullivan, Charles T.; Husain, Anis: Optical interconnects for parallel processing,V1389,375-385(1991)

Guharay, Samar K.
; Allen, C. K.; Reiser, Martin P.: Electrostatic focusing and RFQ (radio frequency quadrupole) matching system for a low-energy H-beam,V1407,610-619(1991)

Guiar, Cecilia N. See Schember, Helene R.: V1540,51-62(1991)

Guibert, Jean C. See Vinet, Francoise: V1466,558-569(1991)

Guibert, L.
; Killinger, M.; de Bougrenet de la Tocnaye, Jean-Louis M.: Module for optical parallel logic using bistable optically addressed ferroelectric spatial light modulators,V1505,99-103(1991)

Guidon, Michel See Mainguy, Stephan: V1530,269-282(1991)

Guilfoyle, Peter S.
: Motivation for DOC III: 64-bit digital optical computer,V1500,14-22(1991); 1503,14-22(1991); 1505,2-10(1991)
; Morozov, Valentin N.: Potential digital optical computer III architectures: the next generation,V1563,279-283(1991)
; Stone, Richard V.: Digital optical computer II,V1563,214-222(1991)
See Stone, Richard V.: V1563,267-278(1991)

Guillemin, Francois See Reynes, Anne M.: V1525,177-182(1991)

Guillot, Gerard
See Ababou, Soraya: V1361,706-711(1991)
See Banvillet, Henri: V1361,972-979(1991)
See Bremond, Georges E.: V1361,732-743(1991)

Gulanian, Emin H.
See Mikaelian, Andrei L.: V1621,194-203(1991)
See Sadyigov, Z. Y.: V1621,158-168(1991)

Guldner, Yves
; Manasses, J.: Electronic properties of mercury-based type-III superlattices,V1361,657-668(1991)

Gulfi, Gildo M. See Gaboardi, Franco: V1421,73-77(1991)

Gulick, Paul E.
; Conner, Arlie R.: Stacked STN LCDs for true-color projection systems,V1456,76-82(1991)

Gullberg, Grant T. See Zeng, Gengsheng L.: V1445,332-340(1991)

Gulnazarov, Eduard S.
; Smirnova, Tatiana N.; Tikhonov, Evgenij A.: Recording mechanism and postpolymerizing self-amplification of holograms,V1238,235-239(1991)

Gulyaev, Yuri V.
; Elinson, Matvey I.; Kopylov, Yuri L.; Perov, Polievkt I.: Integrated circuits with three-dimensional optical interconnections: an element base of neural networks,V1621,84-92(1991)

Gum, Jeffrey S. See Keski-Kuha, Ritva A.: V1343,566-575(1991)

Gumienny, Zbigniew
; Misiewicz, Jan: Method of investigations of the optical properties of anisotropic materials using modulation of light polarization,V1527,462-465(1991)
See Misiewicz, Jan: V1561,6-18(1991)

Gumpf, Jeffrey A. See Neff, Raymond K.: V1364,245-256(1991)

Gunapala, S. D.
; Levine, Barry F.; Ritter, D.; Hamm, Robert A.; Panish, Morton B.: InP-based quantum-well infrared photodetectors,V1541,11-23(1991)
See Levine, Barry F.: V1540,232-237(1991)

Gundersen, Martin A. See Hur, Jung H.: V1378,95-100(1991)

Gunning, William J. See Motamedi, M. E.: V1544,33-44(1991)

Gunsay, Metin See Jeffs, Brian D.: V1567,511-521(1991)

Gunshor, Robert L. See Kudlek, Gotthard: V1361,150-158(1991)

Gunson, Michael R.
; Irion, Fredrick W.: Measurement of atmospheric composition by the ATMOS instrument from Table Mountain Observatory,V1491,335-346(1991)

Guntau, Matthias
; Possner, Torsten; Braeuer, Andreas; Dannberg, Peter: Nonlinear optical transmission of an integrated optical bent coupler in semiconductor-doped glass,V1513,112-122(1991)

Gunther, Michael F.
See Murphy, Kent A.: V1588,134-142(1991)
See Vengsarkar, Ashish M.: V1588,2-13(1991)

Guo, D. F.
See Liu, W. C.: V1519,670-674(1991)
See Liu, W. C.: V1519,640-644(1991)

Guo, Dandan See Mazumdar, Sumit: V1436,136-149(1991)

Guo, Hongping See Lou, Qihong: V1598,221-226(1991)

Guo, Huai X.
See Fan, Xiang J.: V1519,808-812(1991)
See Fan, Xiang J.: V1519,805-807(1991)

Guo, Jerry Z.
; Cerrina, Franco: Comparison of plasma source with synchrotron source in the Center for X-ray Lithography,V1465,330-337(1991)

Guo, Jian See Fang, Xian-chen: V1572,52-55(1991)

Guo, Jian Q. See Yang, Yong: V1519,444-448(1991)

Guo, Jing-Zhen
: Effects of He-Ne regional irradiation on 53 cases in the field of pediatric surgery,V1422,136-139(1991)

Guo, Ju P. See Li, Zhong Y.: V1519,374-379(1991)

Guo, Liping See Wang, Guomei: V1590,229-236(1991)

Guo, Lu Rong
; Cheng, Qirei; Wang, Kuoping: Non-gelatin-dichromated holographic film,V1461,91-92(1991)
; Cheng, Xiaoxue; Guo, Yongkang; Hsu, Ping: New method of making HOE by copying CGH on NGD,V1555,300-303(1991)
; Dai, Chao M.; Guo, Yongkang; Cai, Tiequan: Antihumidity dichromated gelatin holographic recording material,V1555,293-296(1991)
; Guo, Yongkang; Zhang, Xiao-Chun: MTF of photolithographic hologram,V1392,119-123(1991)
; Wang, Kuoping; Cheng, Qirei: Real-time holographic recording material: NGD,V1555,291-292(1991)
; Zhang, Xiao-Chun; Guo, Yongkang: Primary research for mechanism of forming PLH,V1463,534-538(1991)
See Fan, Cheng: V1461,51-55(1991)
See Fan, Cheng: V1461,265-269(1991)
See Guo, Yongkang: V1461,97-100(1991)
See Zhang, Xiao-Chun: V1461,93-96(1991)

Guo, Maolin
; Wang, Yigong; Zhao, Yin; Du, Shanyi: Holographic photoelasticity applied to ceramics fracture,V1554A,310-312(1991)
; Zhao, Yin; Wu, Zhongming; Du, Shanyi: White-light speckle applied to composites,V1554A,657-658(1991)
See Chen, Zengtao: V1554A,206-208(1991)

Guo, Qiang See Wang, Yongjiang: V1412,67-71(1991)

Guo, S. P.
; He, X. C.; Redko, S. V.; Wu, Z. Q.: Simulation of local layer-thickness deviation on multilayer diffraction,V1519,400-404(1991)

Guo, Song See Xue, Qing-yu: V1491,75-82(1991)

Guo, Ting
; Qiu, Peixia; Lin, Fucheng: Generation of ultrashort ultraviolet optical pulses using sum-frequency in LBO crystals,V1409,24-27(1991)

Guo, Yi F.
: Vibration-insensitive moire interferometry system for off-table applications,V1554B,412-419(1991)
See Woychik, Charls G.: V1554B,461-470(1991)

Guo, Yili See Yuan, Weitao: V1572,78-83(1991)

Guo, Yong H.
; Fan, Zheng X.; Bin, Ouyang; Jin, Lei; Shao, Jian D.: Determination of optical constants of thin film in the soft x-ray region,V1519,327-332(1991)
See Shao, Jian D.: V1519,298-301(1991)

Guo, Yongkang
; Guo, Lu Rong; Zhang, Xiao-Chun: Etch conditions of photolithographic holograms,V1461,97-100(1991)
See Guo, Lu Rong: V1555,293-296(1991)
See Guo, Lu Rong: V1392,119-123(1991)
See Guo, Lu Rong: V1555,300-303(1991)
See Guo, Lu Rong: V1463,534-538(1991)
See Zhang, Xiao-Chun: V1461,93-96(1991)

Guo, Zicheng
; Melhem, Rami G.; Hall, Richard W.; Chiarulli, Donald M.; Levitan, Steven P.: Pipelined communications on optical busses,V1390,415-426(1991)

Gupta, Arunava
: Experimental studies of oxygen incorporation during growth of Y-Ba-Cu-O films by pulsed-laser deposition,V1394,230-230(1991)

Gupta, Lalit
; Sayeh, Mohammad R.; Upadhye, Anand M.: Imposing a temporal structure in neural networks,V1396,266-269(1991)

Gupta, Madan M.
; Digney, Bruce L.: Visual motion detection: emulation of retinal peripheral visual field,V1381,346-356(1991)
; Hungenahally, Suresh K.: Receptive fields and the theory of discriminant operators,V1382,87-98(1991)
; Knopf, George K.: Multitask neural network for vision machine systems,V1382,60-73(1991)
; Knopf, George K.: Multitask neurovision processor with extensive feedback and feedforward connections,V1606,482-495(1991)

Gupta, Mool C.
; Peng, Song-Tsuen: Multifunction grating for signal detection of optical disk,V1499,303-306(1991)

Gupta, P. See George, Steven M.: V1437,157-165(1991)

Gupta, R. K. See Sorbello, Robert M.: V1475,175-183(1991)

Gupta, Rita P.
; Balkan, N.; Ridley, Brian K.; Emeny, M. T.: Energy and momentum relaxation of electrons in GaAs quantum-wells: effect of nondrifting hot phonons and interface roughness,V1362,798-803(1991)
; Ridley, Brian K.: Relaxation-rate of phonon momentum in semiconductors,V1362,790-797(1991)

Gupta, Smita
; Gersho, Allen: Image vector quantization with block-adaptive scalar prediction,V1605,179-189(1991)

Gupta, Sundeep
; Krishnapuram, Raghu J.: Morphologic edge detection in range images,V1568,335-346(1991)

Gur, David
; Rockette, Howard E.; Sumkin, Jules H.; Hoy, Ronald J.; Feist, John H.; Thaete, F. L.; King, Jill L.; Slasky, B. S.; Miketic, Linda M.; Straub, William H.: Correlations between time required for radiological diagnoses, readers' performance, display environments, and difficulty of cases,V1446,284-288(1991)
See Fuhrman, Carl R.: V1446,422-429(1991)
See King, Jill L.: V1446,268-275(1991)
See Rockette, Howard E.: V1446,276-283(1991)
See Rosenthal, Marc S.: V1443,132-142(1991)

Guralnick, Sidney A.
; Suen, Eric S.: Real-time inspection of pavement by moire patterns,V1396,664-677(1991)

Gurinovich, G. P.
; Kruglik, S. G.; Kvach, V. V.; Terekhov, S. N.: Resonance Raman study of Zn-porphyrin pi-anions,V1403,139-141(1991)

Gurioli, G. M. See Ferrari, Marco: V1431,276-283(1991)

Gurtman, G. A. See Read, Harold E.: V1441,345-359(1991)

Gurvich, Alexander S.
; Myakinin, Vladimir A.; Vorob'ev, Valerii V.; D'yakov, Alexander; Pokasov, Vladimir; Pryanichnikov, Victor: Sound generation by laser radiation in air: application to restoration of energy distribution in laser beams,V1408,10-18(1991)

Guryashkin, L. P.
; Stasenko, A. L.: High-speed photography applied for the investigation of the dynamics of free falling evaporating droplets,V1358,974-978(1991)

Gus'kov, K. G. See Budnik, A. P.: V1397,721-724(1991)

Gus'kova, M. S.
See Averin, V. I.: V1358,589-602(1991)
See Bass, V. I.: V1358,1075-1083(1991)

Gusakova, Valentina I. See Rybak, Ilya A.: V1469,472-482(1991)

Gusella, Riccardo See Gilge, Michael: V1605,592-603(1991)

Guseva, N. P.
; Maximova, I. L.; Romanov, S. V.; Shubochkin, L. P.; Tatarintsev, S. V.: Investigation of vesicle-capsular plague antigen complex formation by elastic laser radiation scattering,V1403,332-334(1991)

Gusmeroli, Valeria See Sansonetti, Pierre: V1588,198-209(1991)

Gussenhoven, Elma J.
; Bom, Nicolaas; Li, Wenguang; van Urk, Hero; Pietermann, Herman; van Suylen, Robert J.; Salem, H. K.: Hypoechoic media: a landmark for intravascular ultrasonic imaging,V1425,203-206(1991)

Gustafson, D. M. See Hough, Gary R.: V1346,194-199(1991)

Gustafson, Donald L. See Chung, Yi-Nung: V1481,306-314(1991)

Gustafson, Peter C.
: Accuracy/repeatability test for a video photogrammetric measurement,V1526,36-41(1991)

Gustafson, Steven C.
; Little, Gordon R.; Olczak, Eugene G.: Locally linear neural networks for optical correlators,V1469,268-274(1991)

Gustafson, Terry L.
; Iwata, Koichi; Weaver, William L.; Huston, Lisa A.; Benson, Ronda L.: Studies of the excited states of biological systems using UV-excited resonance Raman and picosecond transient Raman spectroscopy,V1403,545-554(1991)

Gutfinger, Dan
; Hertzberg, Efrat M.; Tolxdorff, T.; Greensite, F.; Sklansky, Jack: Tissue identification in MR images by adaptive cluster analysis,V1445,288-296(1991)

Guth, Gregory O. See Morgan, Robert A.: V1562,149-159(1991)

Gutierrez, Adolfo R.
; Diaz, Art F.: Solubility, dispersion, and carbon adsorption of a chromium hydroxyazo complex in a toner,V1458,201-204(1991)

Gutierrez da Costa, Henrique S. See Tozzi, Clesio L.: V1384,124-132(1991)

Gutowski, Juergen
See Hingerl, Kurt: V1361,383-393(1991)
See Kudlek, Gotthard: V1361,150-158(1991)

Guttmann, Peter
; Danev, Gentsho; Spassova, E.; Babin, Sergei V.: Behavior of amorphous semiconductors As2S3 layers after photon, electron, or x-ray exposures,V1361,999-1010(1991)

Guy, Larry J. See Ostaszewski, Miroslaw A.: V1482,13-25(1991)

Guyer, Dean R.
; Bosenberg, W. R.; Braun, Frank D.: High-efficiency tunable mid-infrared generation in KNbO3,V1409,14-17(1991)

Guzik, Rudolph P. See Dubiel, Mary K.: V1396(1991)

Guzman, Carmen T.
; Palmer, James M.; Slater, Philip N.; Bruegge, Carol J.; Miller, Edward A.: Requirements of a solar diffuser and measurements of some candidate materials,V1493,120-131(1991)

Guzzi, D. See Mencaglia, Andrea: V1506,140-144(1991)

Gweon, S. See Taylor, Henry F.: V1562,264-275(1991)

Gyi'mesi, Ferenc See Fuzessy, Zoltan: VIS08,194-204(1991)

Gyurcsik, Ronald S. See Hauser, John R.: V1392,340-351(1991)

Haack, Olaf See Guelker, Gerd: V1500,124-134(1991)

Haacke, E. M. See Ovryn, Benjie: V1429,172-182(1991)

Haag, Nils N. See Barrett, Eamon B.: V1567,142-169(1991)

Haake, John M.
; Zediker, Mark S.; Balestra, Chet L.; Krebs, Danny J.; Levy, Joseph L.: Characterization of GRIN-SCH-SQW amplifiers,V1418,298-308(1991)
See Zediker, Mark S.: V1418,309-315(1991)

Haaland, David M. See Cox, J. N.: V1392,650-659(1991)

Haas, David R.
; Man, Hong-Tai; Teng, Chia-Chi; Chiang, Kophu P.; Yoon, Hyun N.; Findakly, Talal K.: Polymer-based electro-optic modulators: fabrication and performance,V1371,56-67(1991)

Haas, Terry E. See Goldner, Ronald B.: V1536,63-69(1991)

Haase, Karl K.
See Neu, Walter: V1425,28-36(1991)
See Neu, Walter: V1425,37-44(1991)
See Nyga, Ralf: V1525,119-123(1991)

Haase, W. C. See Gregory, Kenton W.: V1425,217-225(1991)

Haass, Clark D. See Yee, David K.: V1444,292-305(1991)

Habashy, Tarek M. See Ayme-Bellegarda, Eveline J.: V1471,18-29(1991)

Haber, Lyn See Haber, Ralph N.: V1383,411-424(1991)

Haber, Ralph N.
; Haber, Lyn: Why mobile robots need a spatial memory,V1383,411-424(1991)

Habib, Khaled J.
: Holographic interferometry in corrosion studies of metals: I. Theoretical aspects,V1332,193-204(1991)
: Holographic interferometry in corrosion studies of metals: II. Applications,V1332,205-215(1991)

Habibi, Ali
: Neural networks in bandwidth compression,V1567,334-340(1991)

Habiby, Sarry F.
See Hartman, Davis H.: V1390,368-376(1991)
See Lalk, Gail R.: V1389,386-400(1991)

Habich, Uwe
; Bauer, Axel; Loosen, Peter; Plum, Heinz-Dieter: Resonators for coaxial slow-flow CO2 lasers,V1397,383-386(1991)

Hachijin, Michio See Hashimoto, Takashi: V1397,519-522(1991)

Hacker, Nigel P.
; Welsh, Kevin M.: Mechanistic studies on the poly(4-tert-butoxycarbonyloxystyrene)/triphenylsulfonium salt photoinitiation process,V1466,384-393(1991)
; Welsh, Kevin M.: Photochemistry and fluorescence spectroscopy of polymeric materials containing triphenylsulfonium salts,V1559,139-150(1991)

Hackett, Jay K.
; Lavoie, Matt J.; Shah, Mubarak A.: Three-dimensional object recognition using multiple sensors,V1383,611-622(1991)

Hackney, D. See Gee, Jim C.: V1445,226-234(1991)

Hadar, Ofer
; Dror, Itai; Kopeika, Norman S.: Numerical calculation of image motion and vibration modulation transfer functions: a new method,V1533,61-74(1991)
; Fisher, Moshe; Kopeika, Norman S.: Numerical calculation of image motion and vibration modulation transfer function,V1482,79-91(1991)

Hadaway, James B.
See Hoover, Richard B.: V1343,189-202(1991)
See Johnson, R. B.: V1354,669-675(1991)
See Johnson, R. B.: V1535,136-142(1991)

Haddad, Betsy See Kalvin, Alan D.: V1445,247-258(1991)

Haddad, George I. See Lachs, Gerard: V1474,248-259(1991)

Haddad, Richard A. See Hwang, Humor: V1606,400-407(1991)

Haddad, Waleed S.
; Cullen, David; Solem, Johndale C.; Longworth, James W.; McPherson, Armon; Boyer, Keith; Rhodes, Charles K.: Fourier-transform holographic microscope,V1448,81-88(1991)

Haddleton, Graham P.
: Streak and smear: a definition of terminology,V1358,18-22(1991)

Haden, Gerald L. See Thomas, Judith G.: V1444,379-388(1991)

Hadfield, Michael J.
; Stiles, Tom; Seidel, David; Miller, William G.; Hensley, David; Wisotsky, Steve; Foote, Michael; Gregory, Rob: Improved ring laser gyro navigator,V1478,126-144(1991)

Hadfield, Peter See Wang, Dexter: V1479,57-70(1991)

Hadizad, Peyman See Hur, Jung H.: V1378,95-100(1991)

Hadjifotiou, A. See Henshall, Gordon D.: V1418,286-291(1991)

Hadjinicolaou, M. G. See Jones, Robert H.: V1401,138-145(1991)

Hadley, G. R. See Shiau, T. H.: V1418,116-122(1991)

Haemmerli, Jean-Francois
; Iwaki, Tadao; Yamamoto, Shuhei: Compact, one-lens JTC using a transmissive amorphous silicon FLC-SLM (LAPS-SLM),V1564,275-284(1991)

Haese-Coat, Veronique See Kpalma, Kidiyo: V1606,55-66(1991)

Haesing, Friedrich W. See Bueker, Harald: V1572,410-418(1991)

Haessig, David A.
; Friedland, Bernard: Modeling and simulation of friction,V1482,383-396(1991)

Hafich, Michael J. See Crawford, Deborah L.: V1371,138-141(1991)

Hafizi, Bahman
; Sprangle, Phillip: Diffraction effects in directed radiation beams,V1407,316-321(1991)

Hagan, David J. See Sheik-Bahae, Mansoor: V1441,430-443(1991)

Hagans, Karla G.
; Sargis, Paul G.: Comparison of electro-optic diagnostic systems,V1346,404-408(1991)
See McCammon, Kent G.: V1346,398-403(1991)

Hagemann, Roland See Beuthan, Jurgen: V1420,225-233(1991)

Hageniers, Omer L. See Reynolds, Rodger L.: V1332,85-96(1991)

Hager, Gregory D.
: Deciding not to decide using resource-bounded sensing,V1383,379-390(1991)

Hager, Harold E. See Hong, C. S.: V1418,177-187(1991)

Hager, K. See Kille, Knut: V1386,222-227(1991)

Haggans, Charles W.
; Kostuk, Raymond K.: Use of rigorous vector coupled-wave theory for designing and tolerancing surface-relief diffractive components for magneto-optical heads,V1499,293-302(1991)

Hagge, John K.
: State-of-the-art multichip modules for avionics,V1390,175-194(1991)

Hagi, Toshio
; Okuda, Yoshimitsu; Ohkuma, Tohru: Critical dimension control using development end point detection for wafers with multilayer structures,V1464,215-221(1991)

Haglund, Richard F.
; Tang, Kai; Bunton, P. H.; Wang, Ling-jun: Role of valence-band excitation in laser ablation of KC1,V1441,127-138(1991)

Hagy, Henry E.
: Review of measurement systems for evaluating thermal expansion homogeneity of Corning Code 7971 ULETM,V1533,198-211(1991)

Hahm, Cheul-hee See Jeon, Joon-hyeon: V1605,954-962(1991)

Hahn, H. See Boehme, Johann F.: V1566,208-219(1991)

Hahn, Sookap See Yarling, Charles B.: V1393,192-199(1991)

Hahn, Terry D. See Litz, Marc S.: V1407,159-166(1991)

Haidle, Rudy H. See Gregory, Christopher C.: V1420,169-175(1991)

Haigh, N. R.
; Linton, R. S.; Johnson, R.; Grigsby, R.: Criteria for accurate cutback attenuation measurements,V1366,259-264(1991)

Haigh, Peter J. See Neff, Raymond K.: V1364,245-256(1991)

Hailey, Charles J.
See Atkinson, Dennis P.: V1343,530-541(1991)
See Bixler, Jay V.: V1549,420-428(1991)

Hails, J. E. See Mullin, John B.: V1361,1116-1127(1991)

Haines, Barry K. See Payton, Paul M.: V1406,58-71(1991)

Haines, Robert See Maman, K. H.: V1364,304-311(1991)

Hakami, M. R. See Boncelet, Charles G.: V1451,70-74(1991)

Hakim, M. O. See Ambia, M. G.: V1536,222-232(1991)

Hakkarainen, Aulis See Korppi-Tommola, Jouko E.: V1403,457-465(1991)

Hakkarainen, J. M.
; Little, James J.; Lee, Hae-Seung; Wyatt, John L.: Interaction of algorithm and implementation for analog VLSI stereo vision,V1473,173-184(1991)

Hakkarainen, T. See Pessa, Markus: V1361,529-542(1991)

Hala, J. See Adamec, F.: V1402,82-84(1991)

Hale, K. F.
See Zheng, S. X.: V1572,359-364(1991)
See Zheng, S. X.: V1572,268-272(1991)

Hale, Leonard G. See Motamedi, M. E.: V1544,33-44(1991)

Hale, Robert R.
: Optical stability of diffuse reflectance materials in space,V1485,173-182(1991)
See Bruegge, Carol J.: V1493,132-142(1991)
See Riedl, Max J.: V1485(1991)

Halford, Carl E. See Sanders, Jeffrey S.: V1478,52-63(1991)

Hall, Bruce T. See Zemon, Stanley A.: V1373,21-32(1991)

Hall, Charles S.
; Alongi, Robert E.; Fortner, Russ L.; Fraser, Laurie K.: Development of a fire and forget imaging infrared seeker missile simulation,V1483,29-38(1991)
See Cook, Thomas H.: V1488,214-225(1991)

Hall, David See Anema, P. C.: V1446,352-356(1991)

Hall, David L. See Linn, Robert J.: V1470,13-29(1991)

Hall, Ernest L.
; Shell, Richard; Slutzky, Gale D.: Intelligent packaging and material handling,V1381,162-170(1991)
See Zhang, Yu: V1384,60-65(1991)

Hall, G. See Sumner, Timothy J.: V1549,265-273(1991)

Hall, Lawrence O. See Stark, Louise: V1381,334-345(1991)

Hall, Richard W. See Guo, Zicheng: V1390,415-426(1991)

Hall, Steven R. See How, Jonathan P.: V1489,148-162(1991)

Hall, Theodore R. See Razavi, Mahmood: V1446,24-34(1991)

Hall, Thomas See Beth, Mark-Udo: V1397,577-580(1991)

Hall, Trevor J.
See Bostel, Ashley J.: V1385,165-172(1991)
See Kirk, Andrew G.: V1574,121-132(1991)
See Kirk, Andrew G.: V1621,320-327(1991)

Halle, Michael W.
; Benton, Stephen A.; Klug, Michael A.; Underkoffler, John S.: Ultragram: a generalized holographic stereogram,V1461,142-155(1991)

Hallegot, Philippe
; Girod, C.; LeBeau, M. M.; Levi-Setti, Riccardo: Direct high-spatial-resolution SIMS (secondary ion mass spectrometry) imaging of labeled nucleosides in human chromosomes,V1396,311-315(1991)

Haller, Kurt L.
; Hobbs, Philip C.: Double-beam laser absorption spectroscopy: shot noise-limited performance at baseband with a novel electronic noise canceler,V1435,298-309(1991)

Haller, Mitch See Kwo, Deborah P.: V1544,66-74(1991)

Halliburton, Larry E. See Scripsick, Michael P.: V1561,93-103(1991)

Hallinan, Peter W.
: Recognizing human eyes,V1570,214-226(1991)

Halloran, Frank See Bergman, Larry A.: V1364,14-21(1991)

Hallouche, Farid
; Adams, Alan E.; Hinton, Oliver R.: High-resolution texture analysis for cytology,V1445,504-512(1991)

Hallstadius, Hans
; Uneus, Leif; Wallin, Suante: System for evaluation of trace gas concentration in the atmosphere based on the differential optical absorption spectroscopy technique,V1433,36-43(1991)

Halmagean, Eugenia T.
: Neutron transmutation doping of silicon in the VVR-S type nuclear reactor,V1362,984-991(1991)
; Lazarovici, Doina N.; Udrea-Spenea, Marian N.: Accurate method for neutron fluence control used in improving neutron-transmutation-doped silicon for detectors,V1484,106-114(1991)

Halpern, Ethan J.
; Esser, Peter D.: Improved phantom for quality control of laser scanner digitizers in PACS,V1444,104-115(1991)
See Esser, Peter D.: V1444,100-103(1991)

Halpern, Steven J. See Dew, Douglas K.: V1422,111-115(1991)

Halpin, John M. See English, R. Edward: V1527,174-179(1991)

Hama, Hiromitsu
; Yamashita, Kazumi: Real-time video signal processing by generalized DDA and control memories: three-dimensional rotation and mapping,V1606,878-890(1991)
See Nakajima, Shigeyoshi: V1605,709-719(1991)

Hama, Soichi See Wakana, Shin-ichi: V1479,283-290(1991)

Hamada, Hiroshi
See Morikawa, Yoshitaka: V1605,445-455(1991)
See Yamane, Nobumoto: V1605,679-686(1991)

Hamada, Takahiro See Matsumoto, Shuichi: V1605,37-46(1991)

Hamaguchi, Hiro-o See Ishibashi, Taka-aki: V1403,555-562(1991)

Hamaguchi, Norihito
; Gardiner, R.; Kirlin, Peter S.: MOCVD of TlBaCaCuO: structure-property relations and progress toward device applications,V1394,244-254(1991)

Hamaguchi, S. See Ozawa, Yasuyuki: V1499,136-142(1991)

Hamal, Karel
See Prochazka, Ivan: V1449,116-120(1991)
See Prochazka, Ivan: V1358,574-577(1991)

Hamalainen, Rauno M.
See Silvennoinen, Raimo: V1574,84-88(1991)
See Silvennoinen, Raimo: V1574,261-265(1991)

Hamanaka, Kenjiro See Oikawa, Masahiro: V1544,226-237(1991)

Hamano, Toshihisa See Kobayashi, Kenichi: V1448,157-163(1991)

Hamano, Yasunori See Kanazawa, Hirotaka: V1397,445-448(1991)

Hamasaki, Yiji See Fukuoka, Takashi: V1364,40-48(1991)

Hamdy, Walid M.
; Humblet, Pierre A.: Crosstalk in direct-detection optical fiber FDMA networks,V1579,184-194(1991)

Hamedaninia, Azar See Razavi, Mahmood: V1446,24-34(1991)

Hamid, Sohail
; Tatam, Ralph P.: Optical technique for the compensation of the temperature-dependent Verdet constant in Faraday rotation magnetometers,V1511,78-89(1991)

Hamilton, Michael C.
See Bell, John A.: V1476,326-329(1991)
See Bell, John A.: V1562,276-280(1991)

Hamlin, Scott J.
; Myers, John D.; Myers, Michael J.: High repetition rate Q-switched erbium glass lasers,V1419,100-106(1991)

Hamm, Robert A.
See Gunapala, S. D.: V1541,11-23(1991)
See Harriott, Lloyd R.: V1465,57-63(1991)
See Wang, Yuh-Lin: V1392,588-594(1991)

Hammami, Slimane
; Cottron, M.; Lagarde, Alexis: Dynamic photoelasticity applied to crack-branching investigations,V1554A,136-142(1991)

Hammel, Bruce A. See Kilkenny, Joseph D.: V1358,117-133(1991)

Hammer-Wilson, Marie J.
See Gottfried, Varda: V1442,218-229(1991)
See Kimel, Sol: V1525,341-350(1991)

Hammerling, Peter See Koenig, Michel: V1502,338-342(1991)

Hammon, Ricky K.
; Helton, Monte K.: Imaging autotracker technology for guided missile systems,V1482,258-264(1991)

Hammond, Mark L. See Soper, Steven A.: V1435,168-178(1991)

Hammond, Mark W.
See Hubert, Alexis: V1532,249-260(1991)
See Price, R. S.: V1388,550-559(1991)

Hammond, Richard T.
: High-order nonlinear susceptibilities,V1409,148-153(1991)

Hammond, Robert B.
See Bourne, Lincoln C.: V1477,205-208(1991)
See Negrete, George V.: V1477,36-44(1991)

Hammons, B. E. See Carson, Richard F.: V1378,84-94(1991)

Hamnett, Whitney S. See Horsky, Thomas N.: V1540,527-532(1991)

Hampden-Smith, Mark J. See Jain, Ajay: V1596,23-33(1991)

Hampton, D. S. See Rossing, Thomas D.: V1396,108-121(1991)

Hampton, James A.
; Selman, Steven H.: Development of a novel in-vivo drug/in-vitro light system to investigate mechanisms of cell killing with photodynamic therapy,V1426,134-145(1991)

Hamrelius, Torbjorn
: Accurate temperature measurement in thermography: an overview of relevant features, parameters, and definitions,V1467,448-457(1991)

Hams, Benno H. See Moehlmann, Gustaaf R.: V1512,34-39(1991)

Hamza, Alex V.
See Schildbach, M. A.: V1441,139-145(1991)
See Schildbach, M. A.: V1441,287-293(1991)

Han, Bongtae
: Ultrahigh-sensitivity moire interferometry,V1554B,399-411(1991)

Han, Chia Y.
See Hu, Yong-Lin: V1468,653-661(1991)
See Jiang, Dareng: V1468,16-25(1991)
See Krovvidy, Srinivas: V1468,216-226(1991)
See Tang, Yonghong: V1469,113-120(1991)
See Wan, Liqun: V1469,432-440(1991)

Han, JingHong See Cui, DaFu: V1572,386-391(1991)

Han, Joon H.
; Kim, Myung J.; Cho, Kwang J.: Extraction of hierarchical structures from complicated 2-D shapes,V1381,122-129(1991)

Han, Lei
; Wu, Xiao-Ping; Hu, Shisheng: Pulsed holographic and speckle interferometry using Hopkinson loading technique to investigate the dynamical deformation on plates,V1358,793-803(1991)

Han, Li Y. See Gao, Jian C.: V1519,159-163(1991)

Han, Richard Y.
; Zakhor, Avideh: Edge-based block matching technique for video motion estimation,V1452,395-408(1991)

Han, Rong-jiu See Lu, Eh: V1533,155-162(1991)

Han, Soo-Chul See Woods, John W.: V1605,805-810(1991)

Han, Won-Taek See Tomozawa, Minoru: V1590,160-167(1991)

Han, Xin Z.
: Discussion on spectral sampling of imaging spectrometer,V1538,99-102(1991)
: Problems of photogrammetry of moving target in water,V1537,215-220(1991)
; Qiu, Ying: Space-spectrum resolution function of the imaging spectrometer,V1479,156-161(1991)

Han, Xue D. See Li, Xiao Q.: V1519,14-17(1991)

Han, Ying See Tan, Hui: V1519,752-756(1991)

Han, Zhao-Jin
: New transducer for displacement measurement,V1555,304-308(1991)

Han, Zhenduo See Shen, Zhen-kang: V1482,337-347(1991)

Hanabata, Makoto
; Oi, F.; Furuta, Akihiro: Novolak design for high-resolution positive photoresists (IV): tandem-type novolak resin for high-performance positive photoresists,V1466,132-140(1991)

Hanack, Michael See Meier, Hans: V1559,89-100(1991)

Hanada, Teiichi See Tanabe, Setsuhisa: V1513,340-348(1991)

Hanafee, William See Lou, Shyh-Liang: V1446,302-311(1991)

Hanany, Shaul
See Kaaret, Philip E.: V1548,106-117(1991)
See Shaw, Ping-Shine: V1343,485-499(1991)
See Shaw, Ping-Shine: V1548,118-131(1991)

Hanawa, Tetsuro See Op de Beeck, Maaike: V1463,180-196(1991)

Hand, Duncan P.
See Russell, Philip S.: V1373,126-139(1991)
See Russell, Philip S.: V1516,47-54(1991)

Handerek, Vincent A.
See Kanellopoulos, S. E.: V1516,200-210(1991)
See Rogers, Alan J.: V1511,190-200(1991)

Hane, Kazuhiro
; Grover, Chandra P.: Displacement measurement using grating images detected by CCD image sensor,V1332,584-590(1991)
; Watanabe, S.; Goto, T.: Moire displacement detection by the photoacoustic technique,V1332,577-583(1991)
See Grover, Chandra P.: V1332,624-631(1991)

Hanen, Jose See Barba, Dominique: V1605,408-419(1991)

Hanisch, Robert J. See White, Richard L.: V1567,308-316(1991)

Hankey, Judith See Johnson, Peter T.: V1372,188-199(1991)

Hankla, A. See McCammon, Kent G.: V1346,398-403(1991)

Hankla, B. J. See Williams, Paul F.: V1378,217-225(1991)

Hanks, Roy L.
See Bell, Perry M.: V1346,456-464(1991)
See Kilkenny, Joseph D.: V1358,117-133(1991)
See McEwan, Thomas E.: V1346,465-470(1991)

Hanley, Stephen T. See Mitchell, John H.: V1414,141-150(1991)

Hanna, David C.
See Smart, Richard G.: V1373,158-165(1991)
See Tropper, Anne C.: V1373,152-157(1991)

Hanna, Khalil
See Rol, Pascal O.: V1423,89-93(1991)
See Simon, Gabriel: V1423,154-156(1991)

Hannan, Paul G.
; Davila, Pam M.: Two-mirror projection systems for simulating telescopes,V1527,26-36(1991)

Hannon, John J. See Behlau, Jerry L.: V1530,218-230(1991)

Hannula, Tapio
See Heikkinen, Veli: V1533,115-121(1991)
See Karioja, Pentti: V1533,129-137(1991)
See Malinen, Jouko: V1533,122-128(1991)

Hansen, C. S. See Fairbank, William M.: V1435,86-89(1991)

Hansen, Eric W. See Lu, Jiuhuai: V1459,117-124(1991)

Hansen, Evan See Williams, Memorie: V1549,147-154(1991)

Hansen, Glenn A.
See Ermer, Susan P.: V1560,120-129(1991)
See Swanson, David A.: V1560,416-425(1991)

Hansen, Matthew E.
: Applications of holographic gratings in x-ray mask metrology,V1396,78-79(1991)

Hansen, Ole P.
; Kristensen, Anders; Bruus, Henrik; Razeghi, Manijeh: Corbino-capacitance technique for contactless measurements on conducting layers: application to persistent photoconductivity,V1362,192-198(1991)

Hansen, Patricia A. See Maag, Carl R.: V1557,24-30(1991)

Hansen, Roger W.
See Maestre, Marcos F.: V1548,179-187(1991)
See Snyder, Patricia A.: V1548,188-196(1991)

Hansen, Stein I.
; Marsh, John H.; Roberts, John S.; Jeynes, C.: Refractive index of multiple-quantum-well waveguides subject to impurity induced disordering using boron and fluorine,V1362,361-369(1991)

Hansen, Steven G.
; Dao, Giang T.; Gaw, Henry T.; Qian, Qi-de; Spragg, Peggy M.; Hurditch, Rodney J.: Study of the relationship between exposure margin and photolithographic process latitude and mask linearity,V1463,230-244(1991)
See Robertson, Stewart A.: V1464,232-244(1991)

Hansen, Verne W.
; Summers, Robert A.; Clapp, William G.: Eyes in the skies: building satellites for education,V1495,115-122(1991)

Hanson, Daniel
; Gosset, Nathalie M.: Applications of optical switches in fiber optic communication networks,V1363,48-56(1991)

Hanson, David S.
: Impact of tactical maneuvers on EO sensor imagery,V1538,48-63(1991)

Hanson, James E. See Novembre, Anthony E.: V1466,89-99(1991)

Hanson, Kenneth M.
: Simultaneous object estimation and image reconstruction in a Bayesian setting,V1452,180-191(1991)
See Myers, Kyle J.: V1443,172-182(1991)

Hanson, Thomas A. See Dreyer, Donald R.: V1366,372-379(1991)

Hanssen, Leonard M.
; Snail, Keith A.: Nonimaging optics and the measurement of diffuse reflectance,V1528,142-150(1991)

Hanyu, Isamu
; Nunokawa, Mitsuji; Asai, Satoru; Abe, Masayuki: Al2O3 etch-stop layer for a phase-shifting mask,V1463,595-601(1991)

Hao, De-Fu
: Error transfer function for grating interferometer,V1545,261-265(1991)

Hao, Jian H. See Zhao, Xing R.: V1519,772-774(1991)

Hao, Tianyou
; Zhou, Feng-Shen; Xie, Xiou-Qioun; Hu, Ji-Wu; Wang, Wei-Yen: Fiber optic high-temperature sensor for applications in iron and steel industries,V1584,32-38(1991)

Hara, Hiroshi
See Hashimoto, Takashi: V1397,519-522(1991)
See Yokozawa, T.: V1397,513-517(1991)

Hara, Norikazu See Iwasaki, Nobuo: V1490,192-199(1991)

Harada, Hisashi See Hino, Hideo: V1490,166-176(1991)

Harada, Larrene K.
: Effects of thin and subvisible cirrus on HEL far-field intensity calculations at various wavelengths,V1408,28-40(1991)
See Crawford, Douglas P.: V1408,167-177(1991)

Harada, Takashi See Nohara, Mitsuo: V1417,338-345(1991)

Harada, Tatsuo
; Kita, Toshiaki; Bowyer, C. S.; Hurwitz, Mark: Design of spherical varied line-space gratings for a high-resolution EUV spectrometer,V1545,2-10(1991)
See Okai, Makoto O.: V1545,218-224(1991)

Harada, Yoshiro
; Mell, Richard J.; Wilkes, Donald R.: Effect of the space environment on thermal control coatings,V1330,90-101(1991)

Harafuji, Kenji See Misaka, Akio: V1465,174-184(1991)

Haralick, Robert M.
; Dougherty, Edward R.; Katz, Philip L.: Model-based morphology,V1472,108-117(1991)
; Ramesh, Visvanathan: Performance characterization of vision algorithms,V1406,2-16(1991)
See Dougherty, Edward R.: V1568,224-232(1991)
See Dougherty, Edward R.: V1606,141-152(1991)
See Kanungo, Tapas: V1385,104-112(1991)

Harashima, Hiroshi See Nakaya, Yuichiro: V1605,546-557(1991)

Harbison, James P. See Jewell, Jack L.: V1389,401-407(1991)

Hard, Sverker
See Ekberg, Mats: V1527,236-239(1991)
See Larsson, Michael: V1529,63-70(1991)

Harder, Jerald W.
; Mount, George H.: Long-path differential absorption measurements of tropospheric molecules,V1491,33-42(1991)
See Mount, George H.: V1491,188-193(1991)

Hardesty, R. M. See Zhao, Yanzeng: V1492,85-90(1991)

Hardie, Russell C. See Boncelet, Charles G.: V1451,70-74(1991)

Harding, Charleton M. See Baumann, John A.: V1418,328-337(1991)

Harding, John See Ruiz, D.: V1358,1134-1143(1991)

Harding, Kevin G.
; Kaltenbacher, Eric; Bieman, Leonard H.: Single-lens moire contouring method,V1385,246-255(1991)
; Svetkoff, Donald J.; Uber, Gordon T.; Wittels, Norman; eds.: *Optics, Illumination, and Image Sensing for Machine Vision V,*V1385(1991)
See Uber, Gordon T.: VCR36,20-33(1991)

Hardingham, Charles R. See Hill, Derek L.: V1445,348-357(1991)

Hardy, Amos See Mehuys, David G.: V1418,57-63(1991)

Hardy, John W.
: Adaptive optics: a progress review,V1542,2-17(1991)

Hardy, Mark A. See Oz, Mehmet C.: V1422,147-150(1991)

Harger, Carol A. See Soper, Steven A.: V1435,168-178(1991)

Hargis, David E.
See McIntyre, Iain A.: V1378,162-172(1991)
See Zucker, Oved S.: V1378,22-33(1991)

Hargrave, Philip
: Systolic adaptive beamforming: from theory to practice,V1566,2-11(1991)

Hariharan, P.
: Interferometry with laser diodes,V1400,2-10(1991)
ed.: *Selected Papers on Interferometry,*VMS28(1991)
See Hegedus, Zoltan S.: V1238,480-488(1991)

Haritatos, Fred N.
: High-performance optical disk systems for tactical applications,V1499,196-202(1991)

Harke, H. See Chen, Da Yong: V1435,161-167(1991)

Harker, Alan B.
; DeNatale, Jeffrey F.: Pressure effects in the microwave plasma growth of polycrystalline diamond,V1534,2-8(1991)
See DeNatale, Jeffrey F.: V1534,44-48(1991)

Harmony, David W. See Lawson, Christopher M.: V1592,73-83(1991)

Harms, Richard J.
; Fitch, John E.: Faint object spectrograph early performance,V1494,49-65(1991)

Harnagel, Gary L. See Willing, Steven L.: V1418,358-362(1991)

Harnisch, Bernd See Gebhardt, Michael: V1500,135-143(1991)

Haro-Poniatowski, Emmanuel
; Guasti, Manuel F.: Nonlinear optical properties of germanium diselenide glasses,V1513,86-92(1991)

Harper, Philip G. See Morgan, Russell D.: V1504,118-124(1991)

Harpole, George M. See Valderrama, Giuseppe L.: V1427,200-213(1991)

Harr, Richard W.
: Training image collection at CECOM's Center for Night Vision and Electro-Optics,V1483,231-239(1991)

Harrigan, Jeanne E.
; Stoller, Meryl D.: Automated wafer inspection in the manufacturing line,V1464,596-601(1991)

Harriman, Anthony See Sessler, Jonathan L.: V1426,318-329(1991)

Harrington, C. T. See Nath, Dilip K.: V1489,17-32(1991)

Harrington, James A.
See Gregory, Christopher C.: V1420,169-175(1991)
See Saggese, Steven J.: V1437,44-53(1991)
See Saggese, Steven J.: V1368,2-14(1991)

Harriott, Lloyd R.
; Temkin, Henryk; Chu, C. H.; Wang, Yuh-Lin; Hsieh, Y. F.; Hamm, Robert A.; Panish, Morton B.; Wade, H. H.: Nanometer scale focused ion beam vacuum lithography using an ultrathin oxide resist,V1465,57-63(1991)
See Wang, Yuh-Lin: V1392,588-594(1991)

Harris, David B. See Rose, Evan A.: V1411,15-27(1991)

Harris, David M.
: Future trends in the medical laser industry,V1396,696-698(1991)

Harris, Gary N. See Mosier, Marty R.: V1495,177-192(1991)

Harris, I. P. See Chan, John P.: V1386,171-179(1991)

Harris, J. L. See Piestrup, Melvin A.: V1552,214-239(1991)

Harris, James S.
; Fejer, Martin M.: Quantum wells and artificially structured materials for nonlinear optics,V1361,262-273(1991)
See Pezeshki, Bardia: V1362,559-565(1991)

Harris, John G.
: Continuous-time segmentation networks,V1473,161-172(1991)
See Liu, Shih-Chii: V1473,185-193(1991)

Harris, Karl L. See Levy, Dorron: V1464,413-423(1991)

Harris, Kathleen M. See Fuhrman, Carl R.: V1446,422-429(1991)

Harris, Kevin
; McDuff, G. G.; Burkes, Tom R.: Evaluation of electrolytic capacitors for high-peak current pulse duty,V1411,87-99(1991)

Harris, Neville W.
; Sobolewski, J. M.; Summers, Charles L.; Eng, Richard S.: Design and construction of a wideband efficient electro-optic modulator,V1416,59-69(1991)
; Wong, D. M.: Inhomogeneous quarter-wave transformers for a waveguide electro-optic modulator,V1416,86-99(1991)
See Eng, Richard S.: V1416,70-85(1991)

Harris, Patrick D. See Albert, Thomas: V1450,84-89(1991)

Harris, Thomas I.
: Overview of CODE V development,V1354,104-111(1991)

Harris, Tom A.
See Benschop, Jozef P.: V1464,62-70(1991)
See Monahan, Kevin M.: V1464,2-9(1991)

Harris, Walter See Bush, Brett C.: V1549,290-301(1991)

Harris, William R.
: Infrared systems design from an operational requirement using a hypercard-based program,V1488,156-164(1991)
See Auborn, John E.: V1483,2-9(1991)

Harrison, David C. See Dyjak, Charles P.: V1479,42-56(1991)

Harrison, H. B.
; Reeves, Geoffrey K.: Technique of assessing contact ohmicity and their relevance to heterostructure devices,V1596,52-59(1991)

Harrison, J. M. See Wood, Samuel C.: V1393,36-48(1991)

Harrison, Lorna J. See Smith, Dale J.: V1562,242-250(1991)

Harrison, Robert G.
; Lu, Weiping; Jiad, K.; Uppal, J. S.: Pulsating instabilities and chaos in Raman lasers,V1376,94-102(1991)

Harrison, William T. See Phillips, Mark L.: V1561,84-92(1991)

Harron, John W. See Thomas, Paul J.: V1488,36-47(1991)

Harrop, Walter See Mount, George H.: V1491,188-193(1991)

Hart, Michael
: Applications of powder diffraction in materials science using synchrotron radiation,V1550,11-17(1991)
: Polarizing x-ray optics for synchrotron radiation,V1548,46-55(1991)

Hart, Patrick W.
; Tucker, Russ; Yuce, Hakan H.; Varachi, John P.; Wieczorek, Casey J.; DeVito, Anthony: Effect of earthquake motion on the mechanical reliability of optical cables,V1366,334-342(1991)

Hart, Quinn J.
: Surface and aerosol models for use in radiative transfer codes,V1493,163-174(1991)

Hartemann, Frederic V. See Danly, Bruce G.: V1407,192-201(1991)

Harteneck, Bruce D.
See Levine, Jerrold S.: V1407,74-82(1991)
See Smith, Richard R.: V1407,83-91(1991)

Harting, William L. See Wigginton, Stewart C.: V1390,560-567(1991)

Hartiti, Bouchaib
; Muller, Jean-Claude; Siffert, Paul; Vu, Thuong-Quat: Rapid thermal processing induced defects and gettering effects in silicon,V1393,200-206(1991)

Hartley, David A.
; Hobson, Clifford A.; Lilley, F.: Digital-signal-processor-based inspection of populated surface-mount technology printed circuit boards,V1567,277-282(1991)

Hartley, Neil See Sage, Maurice G.: V1390,302-310(1991)

Hartman, Davis H.
; Reith, Leslie A.; Habiby, Sarry F.; Lalk, Gail R.; Booth, Bruce L.; Marchegiano, Joseph E.; Hohman, James L.: Power economy using point-to-point optical interconnect links,V1390,368-376(1991)
See Lalk, Gail R.: V1389,386-400(1991)

Hartman, Richard D.
; Lutze, Frederick H.; Cliff, Eugene M.: Time-optimal maneuver guidance design with sensor line-of-sight constraint,V1478,64-75(1991)

Hartman, Steven C. See Dodd, James W.: V1407,467-473(1991)

Hartmann, F. X. See Rotman, Stanley R.: V1442,194-204(1991)

Hartmann, Martin
; Bergmann, Hans W.; Kupfer, Roland: Experimental investigations in laser microsoldering,V1598,175-185(1991)

Hartmann, Rudolf
; Smith, Warren J.; eds.: *Infrared Optical Design and Fabrication*, VCR38(1991)

Hartmann, Werner See Bickel, P.: V1503,167-175(1991)

Hartmayer, Ron See Bergman, Larry A.: V1364,14-21(1991)

Hartnett, Thomas M. See Klein, Claude A.: V1534,117-138(1991)

Hartney, Mark A.
; Johnson, Donald W.; Spencer, Allen C.: Evaluation of phenolic resists for 193 nm surface imaging, V1466,238-247(1991)

Hartz, William G.
; Bozzolo, Nora G.; Lewis, Catherine C.; Pestak, Christopher J.: Design of an automated imaging system for use in a space experiment, V1398,52-60(1991)

Haruki, Kazuhito See Ohhashi, Akinami: V1444,63-74(1991)

Haruta, K. See Nagai, Haruhiko: V1397,31-36(1991)

Haruvy, Yair
; Heller, Adam; Webber, Stephen E.: Supported sol-gel thin-film glasses embodying laser dyes II: three-layered waveguide assemblies, V1590,59-70(1991)

Haruyama, Yukio See Moriyama, Takashi: V1490,310-316(1991)

Harvey, David See Valera Robles, Jesus D.: V1508,170-179(1991)

Harvey, David M.
; Shaw, Michael M.; Hobson, Clifford A.; Wood, Christopher M.; Atkinson, John T.; Lalor, Michael J.: Three-dimensional inspection using laser-based dynamic fringe projection, V1400,86-93(1991)
See Wood, Christopher M.: V1332,301-313(1991)

Harvey, Edwin L.
; Cullen, Robert M.: Low-cost spacecraft buses for remote sensing applications, V1495,134-145(1991)

Harvey, Ella Jo See Oleinick, Nancy L.: V1427,90-100(1991)

Harvey, Erol C. See Key, Michael H.: V1397,9-17(1991)

Harvey, James E.
; Lewotsky, Kristin L.: Scattering from multilayer coatings: a linear systems model, V1530,35-41(1991)
See Vernold, Cynthia L.: V1530,144-149(1991)

Harvey, Robin J. See Gregoire, Daniel J.: V1552,118-126(1991)

Harwit, Alex
; Fernandez, R.; Eades, Wendell D.: High contrast ratio InxGa1-xAs/GaAs multiple-quantum-well spatial light modulators, V1541,38-47(1991)
See Adams, Frank W.: V1541,24-37(1991)

Hasan, Tayyaba
: Multiphoton photobiologic effects, V1427,70-76(1991)

Hase, Masahiko See Fujimoto, Tsuyoshi: V1606,599-606(1991)

Hasegawa, Akira
; Zhang, Wei; Itoh, Kazuyoshi; Ichioka, Yoshiki: Image processing based on supervised learning in neural networks, V1621,374-379(1991)
; Zhang, Wei; Itoh, Kazuyoshi; Ichioka, Yoshiki: Neural-network-based image processing of human corneal endothelial micrograms, V1558,414-421(1991)

Hasegawa, Haruhiro See Nishino, Toshikazu: V1394,36-44(1991)

Hasegawa, Norio See Terasawa, Tsuneo: V1463,197-206(1991)

Hasegawa, Shinya
; Suzuki, Katsumi: Simulation of low-energy x-ray lithography using a diamond membrane mask, V1465,145-151(1991)

Hasegawa, Yasuo See Yamada, Yukio: V1431,73-82(1991)

Haselgrove, John C.
; Leigh, John S.; Yee, Conway; Wang, Nai-Guang; Maris, Michael B.; Chance, Britton: Monte Carlo and diffusion calculations of photon migration in noninfinite highly scattering media, V1431,30-41(1991)
See Chance, Britton: V1525,68-82(1991)

Hasenauer, David M. See Ford, Eric H.: VCR38,95-119(1991)

Hashemi, Seyed H. See Sandborn, Peter A.: V1389,177-186(1991)

Hashimoto, Etsu See Kakehata, Masayuki: V1397,185-189(1991)

Hashimoto, I. See Izawa, Takao: V1441,339-344(1991)

Hashimoto, Kazuhiko See Misaka, Akio: V1465,174-184(1991)

Hashimoto, Nobuyuki
; Morokawa, Shigeru; Kitamura, Kohei: Real-time holography using the high-resolution LCTV-SLM, V1461,291-302(1991)

Hashimoto, Takashi
; Nakano, Susumu; Hachijin, Michio; Komatsu, Katsuhiko; Hara, Hiroshi: Characteristics of a downstream-mixing CO2 gas-dynamic laser, V1397,519-522(1991)

Hashimoto, Y. See Hirakawa, Kazuhiko: V1361,255-261(1991)

Hashinami, Shinji See Koezuka, Tetsuo: V1332,323-331(1991)

Hashishin, Yuichi See Kubo, Uichi: V1420,102-107(1991)

Hashizume, Akihide See Aisaka, Kazuo: V1445,312-317(1991)

Hashizume, Hideki See Seki, Masafumi: V1583,184-195(1991)

Haskell, Richard C. See Tromberg, Bruce J.: V1525,52-58(1991)

Haskett, Michael C.
; Lidke, Steve L.: Algorithm development and evaluation on the Multifunction Target Acquisition Processor, VIS07,14-23(1991)
See Lidke, Steve L.: V1406,203-203(1991)

Hasman, Erez
; Davidson, Nir; Friesem, Asher A.: Holographic optical profilometer, V1442,372-377(1991)
See Davidson, Nir: V1442,22-25(1991)

Hassan, Azmi
; Subari, Mustofa; Som, Zainal A.: Deformation study of the Klang Gates Dam with multispectral analysis method, V1526,142-156(1991)

Hassanein, Hossam S. See Maydell, Ursula M.: V1446,81-92(1991)

Hassebrook, Laurence G. See Rahmati, Mohammad: V1567,480-489(1991)

Hassen, Gregory L. See Cox, Jerome R.: V1446,40-51(1991)

Hassoun, Mohamad H.
: Adaptive versions of the Ho-Kashyap learning algorithm, V1558,459-465(1991)

Hastings, Jerome B. See Kao, Chi-Chang: V1548,149-157(1991)

Hata, Seiji See Takagi, Yuuji: V1386,21-29(1991)

Hatakoshi, Gen-ichi See Ishikawa, Masayuki M.: V1418,344-352(1991)

Hatanaka, Hidekazu
; Kawahara, Nobuo; Midorikawa, Katsumi; Tashiro, Hideo; Obara, Minoru: High-repetition-rate, multijoule transversely excited atmospheric CO2 laser, V1397,379-382(1991)

Hatano, Hisaaki See Ohhashi, Akinami: V1444,63-74(1991)

Hatano, Mutsuko See Nishino, Toshikazu: V1394,36-44(1991)

Hatch, G. F. See Gregory, Kenton W.: V1425,217-225(1991)

Hatch, Jeffrey A.
; Montgomery, David S.; Prior, John A.: Long-term flat-field behavior on LLNL streak cameras: preliminary results, V1346,371-375(1991)

Hatch, Marcus R. See Pagano, Robert J.: V1354,460-471(1991)

Hatfield, Brian
See Enguehard, S.: V1408,186-191(1991)
See Enguehard, S.: V1408,178-185(1991)
See Enguehard, S.: V1415,128-137(1991)

Hatfield, Donald J. See Ren, Victor: VCR37,54-67(1991)

Hatfield, Lynn L. See Ordonez, Ishmael D.: V1437,184-193(1991)

Hathaway, Richard B.
; Der Hovanesian, Joseph; Hung, Yau Y.: Residual stress evaluation using shearography with large shear displacements, V1554B,725-735(1991)

Hatheway, Alson E.
: Overview of the finite element method in optical systems, V1532,2-14(1991)

Hatter, Edward See Richards, Evan: V1494,40-48(1991)

Hattori, Hajime See Takeo, Takashi: V1544,282-286(1991)

Hattori, Shuzo
; Collins, George J.; Yu, Zenqi; Sugimoto, Dai; Saita, Masahiro: Disk-shaped VUV+O source used as resist asher and resist developer, V1463,539-550(1991)
See Takada, Yutaka: V1332,571-576(1991)

Hattori, Takeshi
; Koyata, Sakuo: Applications of an automated particle detection and identification system in VLSI wafer processing, V1464,367-376(1991)

Hattori, Tomohiko
: Electro-optical autostereoscopic displays using large cylindrical lenses, V1457,283-289(1991)

Hattori, Yasuyuki See Numagami, Yasuhiko: V1606,970-979(1991)

Hatze, Herbert
; Baca, Arnold: Automatic shape recognition of human limbs to avoid errors due to skin marker shifting in motion analysis, V1567,264-276(1991)

Hatzinakos, Dimitrios
: Stop-and-go sign algorithms for blind equalization, V1565,118-129(1991)

Hauducoeur, Alain See Veaux, Jacqueline: V1449,13-24(1991)

Hauff, D. See Braueninger, Heinrich: V1549,330-339(1991)

Haus, Joseph W.
: ; Chung-Yau, F.; Bowden, Charles M.: Nonlinear electromagnetic field response of high-Tc superconducting microparticle composites,V1497,382-388(1991)
: ; Inguva, Ramarao: Nonlinear optical properties of composite materials,V1497,350-356(1991)
: See Lee, Heun J.: V1497,228-239(1991)
: See Scalora, Michael: V1497,153-164(1991)

Hauser, John R.
: ; Gyurcsik, Ronald S.: Microcomputer-based real-time monitoring and control of single-wafer processing,V1392,340-351(1991)

Hauser, Neal A.
: ; Mitchell, Harvey B.: Efficient odd max quantizer for use in transform image coding,V1605,428-433(1991)

Hausmann, E. See Acharya, Raj S.: V1450,170-177(1991)

Hautmann, Richard E. See Miller, Kurt: V1421,108-113(1991)

Haux, Denys See Bourhis, Jean-Francois: V1583,383-388(1991)

Hava, Shlomo
: : Application of surface effects externally to alter optical and electronic properties of existing optoelectronic devices: a ten-year update,V1540,350-358(1991)

Havelka, Kathleen O.
: ; Spangler, Charles W.: Design of highly soluble extended pi-electron oligomers capable of supporting stabilized delocalized bipolaronic states for NLO applications,V1560,66-74(1991)
: See Spangler, Charles W.: V1560,139-147(1991)

Haverkort, Jos E. See Hendriks, Peter: V1362,217-227(1991)

Havrilla, George J.
: ; Nicholas, Mark; Bryan, Scott R.; Pruett, J. G.: Resonant and nonresHaywardonant ionization in sputtered initiated laser ionization spectrometry,V1435,12-18(1991)

Hawkes, David J. See Hill, Derek L.: V1445,348-357(1991)

Hawley, James G. See Targ, Russell: V1478,211-227(1991)

Hawley, Philip See McCaughan, James S.: V1426,279-287(1991)

Hawley, Robert W.
: ; Gallagher, Neal C.: Use of threshold decomposition theory to derive basic properties of median filters,V1451,91-100(1991)

Hawryluk, Andrew M. See Stearns, Daniel G.: V1465,80-87(1991)

Hay, Claudette
: ; Hay, D. R.: Knowledge-based nursing diagnosis,V1468,314-322(1991)

Hay, D. R. See Hay, Claudette: V1468,314-322(1991)

Hay, Stephen O.
: ; Veltri, R. D.; Lee, W. Y.; Roman, Ward C.: In-situ CARS detection of H2 in the CVD of Si3N4,V1435,352-358(1991)
: See Cunningham, Philip R.: V1414,97-129(1991)

Hayakawa, Yoshiaki
: ; Kurokawa, Akihiro: Outlook of fiber-optic gyroscope,V1572,353-358(1991)
: See Kurokawa, Akihiro: V1504,156-164(1991)

Hayamizu, Kikuko See Okada, Shuji: V1560,25-34(1991)

Hayase, Rumiko H.
: ; Kihara, Naoko; Oyasato, Naohiko; Matake, S.; Oba, Masayuki: Preparations and properties of novel positive photosensitive polyimides,V1466,438-445(1991)

Hayashi, Akihiro See Matsumoto, Tetsuya: V1332,530-536(1991)

Hayashi, Kazuo See Fujii, Takashi: V1412,50-57(1991)

Hayashi, Kenkichi See Watanabe, Mikio: V1452,27-36(1991)

Hayashi, Nobuaki
: ; Schlegel, Leo; Ueno, Takumi; Shiraishi, Hiroshi; Iwayanagi, Takao: Polyvinylphenols protected with tetrahydropyranyl group in chemical amplification positive deep-UV resist systems,V1466,377-383(1991)
: See Uchino, Shou-ichi: V1466,429-435(1991)

Hayashi, Yuji See Takahashi, Eietsu: V1527,145-154(1991)

Hayashida, Bun See Akasaka, Akira: V1490,269-277(1991)

Hayashida, Masami See Niibe, Masahito: V1343,2-13(1991)

Hayden, Joseph S. See Marker, Alexander J.: V1485,160-172(1991)

Hayden, Leonard A.
: ; Jong, Jyh-Ming; Rettig, John B.; Tripathi, Vijai K.: Measurements and characterization of multiple-coupled interconnection lines in hybrid and monolithic integrated circuits,V1389,205-214(1991)
: See Orhanovic, Neven: V1389,273-284(1991)

Hayden, William L.
: ; Fitzmaurice, Michael W.; Nace, Dave; Lokerson, Donald; Minott, Peter O.; Chapman, William W.: NASA's flight-technology development program: a 650-Mbit/s laser communications testbed,V1417,182-199(1991)
: See Fitzmaurice, Michael W.: V1417,13-21(1991)

Hayes, Bruce L. See Clifford, Sandra: V1463,551-557(1991)

Hayes, David See Eaves, Lawrence: V1362,520-533(1991)

Hayes, Gary B. See Albagli, Douglas: V1441,146-153(1991)

Hayes, James E.
: : Examining cable plant bandwidth for FDDI,V1364,115-119(1991)

Hayes, Jeffrey A.
: ; Sullivan, Barry J.: Comparison of detection systems in time-of-flight transillumination imaging,V1396,324-330(1991)

Hayes, John M.
: ; Tang, D.; Jankowiak, Ryszard; Small, Gerald J.: Transient and persistent hole-burning of photosystem II preparations,V1435,258-266(1991)

Hayes, Raymond See Marsh, Harry H.: V1447,298-309(1991)

Hayes, Tim
: : X-ray absorption spectroscopy: how is it done? what can it tell us?,V1550,56-66(1991)

Hayes, Todd R.
: ; Kim, Sung J.; Green, Christian A.: Applications of dry etching to InP-based laser fabrication,V1418,190-202(1991)

Haykin, Simon
: ed.: *Adaptive Signal Processing*,V1565(1991)
: See Leung, Henry: V1565,279-286(1991)
: See Mann, Steven: V1565,402-413(1991)
: See Ukrainec, Andrew: V1565,529-539(1991)

Hayosh, Thomas D.
: : Importance of phosphor persistence characteristics in reducing visual distress symptoms in VDT users,V1454,399-405(1991)

Hayrapetian, Alek
: ; Barbaric, Zoran L.; Weinberg, Wolfram S.; Chan, Kelby K.; Loloyan, Mansur; Taira, Ricky K.; Huang, H. K.: PACS for GU radiology,V1446,243-247(1991)
: See Weinberg, Wolfram S.: V1446,35-39(1991)

Hays, Alan D.
: See Koechner, Walter: V1522,169-179(1991)
: See Marshall, Larry R.: V1419,141-152(1991)

Hays, Paul B.
: See Snell, Hilary E.: V1492,403-407(1991)
: See Wang, Jin-Xue: V1492,391-402(1991)

Hayward, James D.
: : Development of laser bonding as a manufacturing process for inner lead bonding,V1598,164-174(1991)

He, Anzhi
: ; Li, Qun Z.; Miao, Peng-Cheng: New method of 3-D object recognition,V1567,698-702(1991)
: ; Li, Qun Z.; Miao, Peng-Cheng: New method of 3-D shape measurement by moire technique,V1545,278-281(1991)
: ; Lu, Jian-Feng; Ni, Xiao W.; Li, Yong N.: Physical mechanism of high-power pulse laser ophthalmology: experimental research,V1423,117-120(1991)
: ; Ni, Xiao W.; Lu, Jian-Feng; Wang, Chang Xing; Wang, Hong Y.: Laser-stimulated thermoelastic stress wave inside a solid medium and measurement of its initial speed,V1554A,547-552(1991)
: ; Wang, Hai-Lin; Yan, Da-Peng; Miao, Peng-Cheng: Design and application of a moire interferometer,V1527,334-337(1991)
: ; Wang, Hai-Lin; Miao, Peng-Cheng; Yan, Da-Peng: Large-aperture high-accuracy lateral shearing interferometer utilizing a Twyman-Green interferometer,V1527,423-426(1991)
: ; Yan, Da-Peng; Miao, Peng-Cheng; Wang, Hai-Lin: Visualization of impingement field of real-rocket-exhausted jets by using moire deflectometry,V1554B,429-434(1991)
: ; Yu, Gui Ying; Ni, Xiao W.; Fan, Hua Ying: Black and white cinefilm (ASA 5) applied to dynamic laser speckle and its processing,V1554A,747-749(1991)
: ; Zhang, Li; Zhang, Jiajun: New optoelectronic implementation of 2-D associative memory,V1563,208-212(1991)
: See Lu, Jian-Feng: V1415,220-224(1991)
: See Lu, Jian-Feng: V1554B,593-597(1991)
: See Luo, Bikai: V1554A,542-546(1991)
: See Miao, Peng-Cheng: V1554B,641-644(1991)
: See Ni, Xiao W.: V1554B,632-635(1991)
: See Ni, Xiao W.: V1527,437-441(1991)

See Ni, Xiao W.: V1519,365-369(1991)
See Wang, Hai-Lin: V1527,419-422(1991)
See Wang, Hai-Lin: V1545,268-273(1991)
See Wang, Hai-Lin: V1545,274-277(1991)
See Yan, Da-Peng: V1527,442-447(1991)
See Yan, Da-Peng: V1554B,636-640(1991)

He, Da-Ren
; Ji, Xiuyan; Wang, Ruo Bao; Liu, Qihai; Wang, Wangdi; Liu, Maili; Chen, Weizong; Liu, Zhiyuan; Ji, Wanxi; Zhang, Renji: Bonding and nonequilibrium crystallization of a-C:H/a-Se and a-C:H/KCl,V1362,696-701(1991)

He, Duo-Min
; He, Ming-Shia: Basic use of acoustic speckle pattern for metrology and sea waves study,V1332,808-819(1991)

He, Fengling See Weng, Zhicheng: V1527,349-356(1991)

He, Gang
; Cuomo, Frank W.; Zuckerwar, Allan: Diaphragm size and sensitivity for fiber optic pressure sensors,V1584,152-156(1991)

He, Huarui
; Uray, Georg; Wolfbeis, Otto S.: Enantio-selective optode for the B-blocker propranolol,V1368,175-180(1991)
; Uray, Georg; Wolfbeis, Otto S.: Enantio-selective optode for optical isomers of biologically active amines using a new lipophilic aromatic carrier,V1510,95-103(1991)
; Wolfbeis, Otto S.: Fluorescence-based optrodes for alkali ions based on the use of ion carriers and lipophilic acid/base indicators,V1368,165-171(1991)

He, Huijuan
; Lu, Yutian; Dong, Jingyuan; Zhao, Quingchun: Continuous automatic scanning picosecond optical parametic source using MgO LiNbO3 in the 700-2200 nm,V1409,18-23(1991)

He, Jia W. See Xu, Kewei: V1519,765-770(1991)

He, Jin
; Yu, Zhen Z.; Ma, Ke J.; Jia, Pei M.; Yang, Jian R.; Shen, Shou Z.; Chen, Wei M.; Yang, Ji M.: Incorporation of As into HgCdTe grown by MOCVD,V1519,499-507(1991)
See Ma, Ke J.: V1519,489-493(1991)

He, Jing F. See Wang, Yue: V1519,428-433(1991)

He, Li
See Li, Jie: V1519,660-664(1991)
See Liu, Wei J.: V1519,415-418(1991)
See Liu, Wei J.: V1519,481-488(1991)

He, Ming-Shia See He, Duo-Min: V1332,808-819(1991)

He, Ping
; McGahan, William A.; Woollam, John A.: Optimization of quadrilayer structures for various magneto-optical recording materials,V1499,401-411(1991)
See Xue, Kefu: V1606,675-684(1991)

He, Q. See Lallier, Eric: V1506,71-79(1991)

He, Shi P.
See Dobbins, B. N.: V1554A,772-780(1991)
See Dobbins, B. N.: V1554B,586-592(1991)

He, Tian X. See Liu, Wei Y.: V1519,172-174(1991)

He, X. C. See Guo, S. P.: V1519,400-404(1991)

He, Xiao-yuan See Xu, Zhu: V1399,172-177(1991)

He, Yi-Fei See Nath, Dilip K.: V1489,17-32(1991)

He, Yu L.
See Gu, Shu L.: V1519,175-178(1991)
See Sun, Yue Z.: V1519,234-240(1991)

Headley, Clifford
; Agrawal, Govind P.; Reardon, A. C.: Intrapulse stimulated Raman scattering and ultrashort solitons in optical fibers,V1497,197-201(1991)

Heald, Steven M.
; Tranquada, John M.: Polarization-dependent x-ray spectroscopy of high-Tc superconductors,V1550,67-75(1991)

Healey, Glenn
: Using color to segment images of 3-D scenes,V1468,814-825(1991)
See Vaezi, Matt M.: V1606,803-809(1991)
See Vaezi, Matt M.: V1452,57-63(1991)

Healey, Peter See Barnes, Nigel: V1389,477-483(1991)

Healey, Rebecca J. See Sharma, Ramesh D.: V1540,314-320(1991)

Healy, Donald D.
; Wilcox, Christopher D.: Improved PMMA single-piece haptic materials,V1529,84-93(1991)

Heaney, James B.
; Alley, Phillip W.; Bradley, Scott E.: Catalog of infrared and cryo-optical properties for selected materials,V1485,140-159(1991)

Hearn, Gregory K.
: Practical use of generalized simulated annealing optimization on microcomputers,V1354,186-191(1991)

Hearty, Paul J. See Stelmach, Lew B.: V1453,147-152(1991)

Heath, R. See Fieret, Jim: V1503,53-61(1991)

Heatherly, Lee See Shaw, Robert W.: V1534,170-174(1991)

Heaton, Amanda A. See Cosman, Pamela C.: V1605,162-171(1991)

Heaton, Mark W.
; Ewing, William S.: Control of night vision pilotage systems,V1482,444-457(1991)

Heavens, Oliver S. See Yuan, Y. F.: V1519,434-439(1991)

Heavin, Scott D.
; Fung, Bing M.; Mears, Richard B.; Sluss, James J.; Batchman, Theodore E.: Effect of polymer mixtures on the performance of PDLC films,V1455,12-18(1991)

Hebbar, Prashant D.
; McAllister, David F.: Color quantization aspects in stereopsis,V1457,233-241(1991)

Hebden, Jeremy C.
; Kruger, Robert A.: Time-of-flight breast imaging system: spatial resolution performance,V1431,225-231(1991)
; Kruger, Robert A.; Wong, K. S.: Tomographic imaging using picosecond pulses of light,V1443,294-300(1991)

Hebert, Martial
See Delingette, Herve: V1570,21-30(1991)
See Delingette, Herve: V1570,104-115(1991)
See Gowdy, Jay W.: V1388,131-140(1991)

Hecht, L. See Barron, L. D.: V1403,66-75(1991)

Hecht-Nielsen, Robert
: Signal processing with neural networks: throwing off the yoke of linearity,V1541,146-151(1991)

Hecker, Friedrich W.
; Pindera, Jerzy T.; Wen, Baicheng: Actual light deflections in regions of crack tips and their influence on measurements in photomechanics,V1554A,151-162(1991)

Hecker, Y. C.
; Bolle, Ruud M.: Invariant feature matching in parameter space with application to line features,V1570,298-314(1991)

Hector, Scott D.
; Swanson, Gary J.: Shared aperture for two beams of different wavelength using the Talbot effect,V1555,200-213(1991)

Hedegaard Povlsen, Joern See Vendeltorp-Pommer, Helle: V1373,254-265(1991)

Heesink, Gerard J.
See van Hulst, Niek F.: V1361,581-588(1991)
See van Hulst, Niek F.: V1362,631-646(1991)

Hefetz, Yaron
See Kochevar, Irene E.: V1403,756-763(1991)
See Madsen, Steen J.: V1431,42-51(1991)

Heflin, James R.
; Cai, Yongming; Zhou, Qihou L.; Garito, Anthony F.: Experimental measurements and electron correlation theory of third-order nonlinear optical processes in linear chains,V1560,2-12(1991)
; Zhou, Qihou L.; Garito, Anthony F.: Excited-state nonlinear optical processes in polymers,V1497,398-407(1991)

Heflinger, Dov See Bar, I.: V1397,169-172(1991)

Hegarty, John See Phelan, Paul J.: V1362,623-630(1991)

Hegedus, Zoltan S.
; Hariharan, P.: Some new developments in display holography,V1238,480-488(1991)

Hehmann, Joerg See Ibel, Konrad: V1559,393-402(1991)

Heidel, Jeffrey R.
; Zediker, Mark S.: Technique for measuring stress-induced birefringence,V1418,240-247(1991)
See Zediker, Mark S.: V1418,309-315(1991)

Heidelbach, F. See Muenchausen, Ross E.: V1394,221-229(1991)

Heideman, Rene
; Blikman, Albert; Koster, Rients; Kooyman, Rob P.; Greve, Jan: Polarimetric optical fiber sensor for biochemical measurements,V1510,131-137(1991)

Heidemann, B. See Schmiedeskamp, Bernt: V1343,64-72(1991)

Heiferman, Kenneth S.
; Cramer, K. E.; Walsh, Joseph T.: Optimal surface characteristics for instruments for use in laser neurosurgery,V1428,128-134(1991)
See Grutsch, James: V1428,136-145(1991)

Heikes, Brian G.
; Miller, William L.; Lee, Meehye: Hydrogen peroxide and organic peroxides in the marine environment,V1433,253-262(1991)

Heikkila, Tapio A. See Marszalec, Janusz A.: V1382,443-452(1991)

Heikkinen, Veli
; Tukkiniemi, Kari; Vahakangas, Jouko; Hannula, Tapio: Packaging considerations of fiber-optic laser sources,V1533,115-121(1991)

Heil, P. See Baumgartner, R.: V1525,246-248(1991)

Heimbeck, Hans-Jorg
: All-refractive telescope for next-generation inter-satellite communication,V1354,434-437(1991)

Heime, Klaus
See Malzahn, Eric: V1362,199-204(1991)
See Soellner, Joerg: V1361,963-971(1991)

Heimrath, Adam E.
; Bielak, Leslaw: Random imperfections of multimode fiber and the signal dispersion,V1366,265-272(1991)

Heinen, Vernon O.Heideman
; Bhasin, Kul B.; eds.: *Superconductivity Applications for Infrared and Microwave Devices II,*V1477(1991)

Heintzen, Matthias P. See Schwarzmaier, Hans-Joachim: V1427,128-133(1991)

Heinzl, Joachim L. See Dove, Derek B.: V1458(1991)

Heinzmann, Ulrich See Schmiedeskamp, Bernt: V1343,64-72(1991)

Heisel, Francine See Cunin, B.: V1358,606-613(1991)

Heiser, G. A. See Gierczak, Christine A.: V1433,315-328(1991)

Heitmann, Walter See Moeller, Werner: V1504,47-54(1991)

Heitmeyer, Richard See Rosenblum, Lawrence: V1459,274-282(1991)

Heitz, R. See Rosenzweig, M.: V1362,876-887(1991)

Heitzmann, Michel
; Laporte, Philippe; Tabouret, Evelyne: Improvement in dry etching of tungsten features,V1392,272-279(1991)

Hekalo, A. V. See Shestakov, S. D.: V1440,423-435(1991)

Helava, Heikki I. See Oicles, Jeffrey A.: V1378,60-69(1991)

Helbert, John N. See Spragg, Peggy M.: V1466,283-296(1991)

Held, A. C. See Dreyer, Keith J.: V1445,398-408(1991)

Held, Manfred
: Advantage of simultaneous streak and framing records in the field of detonics,V1358,904-913(1991)
: Diagnostic of the reaction behaviour of insensitive high explosives under jet attack,V1358,1021-1028(1991)
: Measurement of the shock profiles with streak technique and different detonating arrangements,V1346,311-318(1991)

Helenius, Vesa F. See Korppi-Tommola, Jouko E.: V1403,457-465(1991)

Helenowski, Tomasz See Gregory, Christopher C.: V1420,169-175(1991)

Helgeson, Wes D.
See Loubriel, Guillermo M.: V1378,179-186(1991)
See Zutavern, Fred J.: V1378,271-279(1991)

Heller, Adam See Haruvy, Yair: V1590,59-70(1991)

Heller, Timothy B.
: Three-dimensional lithography: laser modeling using photopolymers,V1454,272-282(1991)

Helliwell, William S. See Macdonald, Burns: V1537,104-114(1991)

Helman, James L.
; Hesselink, Lambertus B.: Analysis and representation of complex structures in separated flows,V1459,88-96(1991)

Helms, Richard G.
; Porter, Christopher C.; Kuo, Chin-Po; Tsuyuki, Glenn T.: Development and testing of lightweight composite reflector panels,V1532,64-80(1991)

Helton, Monte K. See Hammon, Ricky K.: V1482,258-264(1991)

Hemler, Paul F.
: Radar image understanding for complex space objects,V1381,55-65(1991)

Hemmati, Hamid
; Lesh, James R.; Apostolopoulos, John G.; Del Castillo, Hector M.; Martinez, A. S.: Optical link demonstration with a lightweight transceiver breadboard,V1417,476-483(1991)
See Robinson, Deborah L.: V1417,421-430(1991)

Hemment, Peter L. See Zhu, Wen H.: V1519,423-427(1991)

Hemmer, Philip R. See Kane, Jonathan S.: V1564,511-520(1991)

Hemmi, Gregory See Sessler, Jonathan L.: V1426,318-329(1991)

Hemmick, D. See Xi, Xiaoxing: V1477,20-25(1991)

Hemminger, Bradley M.
See Beard, David V.: V1446,52-58(1991)
See Beard, David V.: V1446,289-296(1991)

Hemminger, John C. See McIver, Robert T.: V1437,124-128(1991)

Hemo, Itzhak See Lewis, Aaron: V1423,98-102(1991)

Hempel, Klaus See Bergner, Harald: V1362,484-493(1991)

Hems, Randall K. See Boles, John A.: V1384,195-204(1991)

Hemsing, Willard F.
: VISAR: displacement-mode data reduction,V1346,141-150(1991)
; Mathews, Allen R.; Warnes, Richard H.; Whittemore, Gerald R.: VISAR: line-imaging interferometer,V1346,133-140(1991)
See Mathews, Allen R.: V1346,122-132(1991)
See Paisley, Dennis L.: V1346,172-178(1991)

Henderson, Byron B.
; Green, Emily N.: Options for campus fiber networks,V1364,235-244(1991)

Henderson, Robert J.
See Mecherle, G. S.: V1417,537-542(1991)
See Mecherle, G. S.: V1417,99-107(1991)

Henderson, Robert K.
: Capability assessment and comparison of the Nikon 2i, Nikon 3i, and IMS-2000 registration measurement devices,V1496,198-216(1991)

Henderson, Thomas B. See Bocker, Richard P.: V1569,298-310(1991)

Hendren, Christelle M. See Gilbert, John A.: V1554B,128-134(1991)

Hendriks, Peter
; Zwaal, E. A.; Haverkort, Jos E.; Wolter, Joachim H.: Modulation doping and delta doping of III-V compound semiconductors,V1362,217-227(1991)

Hendriksen, B. See Moehlmann, Gustaaf R.: V1560,426-433(1991)

Hendrix, Charles D.
; Vijaya Kumar, B. V. K.; Stalker, K. T.; Kast, Brian A.; Shori, Raj K.: Design and testing of three-level optimal correlation filters,V1564,2-13(1991)

Hendry, Gregg See Anspach, Joel E.: V1482,170-181(1991)

Hendry, Timothy G. See Carey, Raymond: V1464,500-507(1991)

Henesian, Mark A.
; Wegner, Paul J.; Speck, David R.; Bibeau, Camille; Ehrlich, Robert B.; Laumann, Curt W.; Lawson, Janice K.; Weiland, Timothy L.: Modeling of large-aperture third-harmonic frequency conversion of high-power Nd:glass laser systems,V1415,90-103(1991)
See Wegner, Paul J.: V1414,162-174(1991)
See Woods, Bruce W.: V1410,47-54(1991)

Hengst, Joachim See Gottschalk, Wolfgang: V1427,320-326(1991)

Henkel, Paul A.
: Low-intensity conflict aircraft systems,V1538,2-4(1991)
See Augustyn, Thomas W.: V1538(1991)

Henneberger, Fritz See Spiegelberg, Christine: V1362,951-958(1991)

Henneberger, T. C. See Bruder, Joseph A.: V1479,316-321(1991)

Henneke, Edmund G. See Tenek, Lazarus H.: V1467,252-263(1991)

Hennemann, O. D. See Breuer, J.: V1503,223-230(1991)

Hennessy, Stephen J. See Severson, William E.: V1388,490-501(1991)

Hennet, L.
See Boher, Pierre: V1343,39-55(1991)
See Boher, Pierre: V1345,198-212(1991)

Hennig, Thomas
; Rechmann, Peter; Pilgrim, C.; Schwarzmaier, Hans-Joachim; Kaufmann, Raimund: Caries selective ablation by pulsed lasers,V1424,99-105(1991)

Henning, Michael R.
; Gerdt, David W.; Spraggins, Thomas: Using a fiber-optic pulse sensor in magnetic resonance imaging,V1420,34-40(1991)

Henning, Rudolf E. See Lachs, Gerard: V1474,248-259(1991)

Henning, Timothy D. See Stover, John C.: V1530,185-195(1991)

Henri, Christopher J.
; Collins, D. L.; Pike, G. B.; Olivier, A.; Peters, Terence M.: Clinical experience with a stereoscopic image workstation,V1444,306-317(1991)

Henry, Bruce See Fried, Alan: V1433,145-156(1991)

Henry, D. See Cassarly, William J.: V1417,110-121(1991)

Henry, Patrice See Biggar, Stuart F.: V1493,155-162(1991)

Henry, Ronald A.
See Lindsay, Geoffrey A.: V1497,418-422(1991)
See Lindsay, Geoffrey A.: V1560,443-453(1991)

Henschel, Henning
; Koehn, Otmar; Schmidt, Hans U.: Influence of dose rate on radiation-induced loss in optical fibers,V1399,49-63(1991)

Henshall, Gordon D.
; Hadjifotiou, A.; Baker, R. A.; Warbrick, K. J.: Advances in laser pump sources for erbium-doped fiber amplifiers,V1418,286-291(1991)

Hensley, David See Hadfield, Michael J.: V1478,126-144(1991)

Henson, Michael A.
; Petro, David: Approach to real-time contrast enhancement for airborne reconnaissance applications,V1396,582-589(1991)

Hentzell, Hans See Novak, Agneta: V1389,80-86(1991)

Hepburn, I. D. See Zammit, C. C.: V1549,274-282(1991)

Hepp, Daniel J.
: Application of back-propagation to the recognition of handwritten digits using morphologically derived features,V1451,228-233(1991)

Herbelin, John M.
: Progress toward the demonstration of a visible (blue) chemical laser,V1397,161-167(1991)

Herbrich, Horst R.
: Advances in the practical application of schlieren photography in industry,V1358,24-28(1991)

Herchen, Harald
; Cappelli, Mark A.: High-temperature Raman scattering behavior in diamond,V1534,158-168(1991)

Hercher, Michael
; Wyntjes, Geert J.: Interferometric measurement of in-plane motion,V1332,602-612(1991)
; Wyntjes, Geert J.: Laser galvo-angle-encoder with zero added inertia,V1454,230-234(1991)
See Wyntjes, Geert J.: V1464,539-545(1991)

Herczfeld, Peter R.
See Even-Or, Baruch: V1371,161-169(1991)
See Kasemset, Dumrong: V1371,104-114(1991)
See Paolella, Arthur: V1378,195-202(1991)

Herd, James T.
; Duffy, Neil D.; Philip, Gary P.; Davidson, Alan C.; Eccles, N. J.: Design/implementation architecture for complex multirobot systems,V1387,194-201(1991)

Herget, William F.
; Demirgian, Jack: Gaseous incinerator emissions analysis by FTIR (Fourier transform infrared) spectroscopy,V1434,39-45(1991)
; Lowry, Steven R.: Auto exhaust gas analysis by FTIR spectroscopy,V1433,275-289(1991)

Herke, R. R. See Yusupov, I. Y.: V1238,240-247(1991)

Herman, Bruce J. See Strugala, Lisa A.: V1486,176-187(1991)

Herman, Charles See Scribner, Dean A.: V1541,100-109(1991)

Herman, J. R. See Barisas, B. G.: V1432,52-63(1991)

Herman, Martin
See Rangachar, Ramesh M.: V1382,376-385(1991)
See Rangachar, Ramesh M.: V1382,331-340(1991)

Herman, Warren N.
; Rosen, Warren A.; Sperling, L. H.; Murphy, C. J.; Jain, H.: High-Tg nonlinear optical polymer: poly(N-MNA acrylamide),V1560,206-213(1991)

Herminghaus, Stephan See Smith, Barton A.: V1560,400-405(1991)

Hernandez, A. B. See Nair, Padmanabhan K.: V1485,228-239(1991)

Hernandez, Jose E. See Del Grande, Nancy K.: V1479,335-351(1991)

Hernandez Neira, Jose Luis
See Martinez-Herrero, R.: V1397,627-630(1991)
See Sanchez, Miguel: V1397,635-638(1991)

Hero, Alfred O. See Williams, William J.: V1566,144-156(1991)

Herr, Daniel J. See Thackeray, James W.: V1466,39-52(1991)

Herrmann, Sandy See Zediker, Mark S.: V1418,309-315(1991)

Herrmann, W. See Wilke, Ingrid: V1442,2-10(1991)

Herron, John M. See Rosenthal, Marc S.: V1443,132-142(1991)

Herron, Michael J.
See Brubaker, John L.: V1533,88-96(1991)
See McCormick, Frederick B.: V1396,508-521(1991)
See McCormick, Frederick B.: V1533,97-114(1991)

Herron, Robin E.
ed.: *Biostereometric Technology and Applications,*V1380(1991)
: Recent applications of biostereometrics in research and education,V1380,2-5(1991)

Hershey, Robert R.
; Zavecz, Terrence E.: Figure of merit for calibration and comparison of linewidth measurement instruments,V1464,22-34(1991)

Hershkowitz, Noah See Mishurda, Helen L.: V1392,563-569(1991)

Herskowitz, Gerald J.
; Mezhoudi, Mohcene: Automated fiber optic moisture sensor system,V1368,55-60(1991)

Herskowitz, M. See Aharon, O.: V1397,251-255(1991)

Hertel, Robert H. See Pelling, Michael R.: V1549,134-146(1991)

Hertog, Craig K. See Chan, Lap S.: V1392,232-239(1991)

Hertzberg, Efrat M. See Gutfinger, Dan: V1445,288-296(1991)

Herve, Philippe
; Bednarczyk, Sophie; Masclet, Philippe: Determination of the temperature of a single particle heated by a highly concentrated laser beam,V1487,387-395(1991)

Herzig, Hans-Peter
; Ehbets, Peter; Prongue, Damien; Daendliker, Rene: Fan-out elements by multiple-beam recording in volume holograms,V1507,247-255(1991)

Hespeler, Bernd
See Greulich, Peter: V1417,358-369(1991)
See Greulich, Peter: V1522,144-153(1991)

Hess, Dennis W. See Dao, T. T.: V1466,257-268(1991)

Hess, Nancy J. See Exarhos, Gregory J.: V1441,190-199(1991)

Hess, Norbert See Norkus, Volkmar: V1484,98-105(1991)

Hess, P. See Kukreja, Lalit M.: V1427,243-254(1991)

Hesse, Reiner See Kliem, Karl-Heinz: V1463,743-751(1991)

Hessel, Stefan F. See Waidhauser, Erich: V1428,75-83(1991)

Hesselink, Lambertus B.
See Helman, James L.: V1459,88-96(1991)
See Ning, Paul C.: V1459,11-21(1991)

Hestdalen, C.
See Jaanimagi, Paul A.: V1358,337-343(1991)
See Jaanimagi, Paul A.: V1346,443-448(1991)

Hester, R. E.
; Austin, J. C.: Dehydrogenase enzyme/coenzyme/substrate interactions,V1403,15-21(1991)

Hetzel, Fred W.
See Beck, Elsa R.: V1426,311-315(1991)
See Chen, Qun: V1426,156-161(1991)

Heuken, Michael
See Malzahn, Eric: V1362,199-204(1991)
See Soellner, Joerg: V1361,963-971(1991)

Heumann, Ernst See Zschocke, Wolfgang: V1501,49-56(1991)

Heuring, Vincent P. See Pratt, Jonathan P.: V1505,124-129(1991)

Heurtaux, J. C. See Mens, Alain: V1358,719-731(1991)

Heydweiller, James R. See Clark, David L.: V1540,303-311(1991)

Heyen, Meino
: New concept for multiwafer production of highly uniform III-V layers for optoelectronic applications by MOVPE,V1362,146-153(1991)

Heyler, Gene A.
; Garlick, Dean S.; Yionoulis, Steve M.: Design and analysis of the closed-loop pointing system of a scientific satellite,V1481,198-208(1991)

Hianik, T.
; Kavechansky, J.; Piknova, Barbora: Effects of incorporation and functioning of integral proteins on the physical characteristics of lipid bilayers,V1402,93-96(1991)
; Masarikova, D.; Zhorina, L. V.; Poroshina, Marina Y.; Chernyaeva, E. B.: Influence of hematoporphyrin on the mechanical properties of the lipid bilayer membranes,V1402,85-88(1991)
See Chorvat, Dusan: V1402,194-197(1991)

Hiatt, William M.
See Bennett-Lilley, Marylyn H.: V1464,127-136(1991)
See Ostrout, Wayne H.: V1463,54-73(1991)

Hibbs, Michael S.
See Dunn, Diana D.: V1463,8-15(1991)
See Kirk, Joseph P.: V1463,575-583(1991)

Hibbs-Brenner, Mary K.
; Mukherjee, Sayan D.; Skogen, J.; Grung, B.; Kalweit, Edith; Bendett, Mark
P.: Design, fabrication, and performance of an integrated optoelectronic
cellular array,V1563,10-20(1991)
See Bristow, Julian P.: V1389,535-546(1991)
See Swirhun, S.: V1475,223-230(1991)

Hibi, Susumu See Takagi, Yuuji: V1386,21-29(1991)

Hibst, Raimund
: Mechanisms of ultraviolet and mid-infrared tissue ablation,V1525,162-
169(1991)
; Keller, Ulrich: Removal of dental filling materials by Er:YAG laser
radiation,V1424,120-126(1991)
See Keller, Ulrich: V1424,127-133(1991)

Hickernell, Robert K.
; Sanford, N. A.; Christensen, David H.: Issues affecting the characterization
of integrated optical devices subjected to ionizing radiation,V1474,138-
147(1991)

Hickey, Carolyn F.
; Thorpe, Thomas P.; Morrish, Arthur A.; Butler, James E.; Vold, C.; Snail,
Keith A.: Polishing of filament-assisted CVD diamond films,V1534,67-
76(1991)

Hickey, Nancy M. See Plessis, Brigitte: V1445,539-554(1991)

Hickman, Kirt C.
; Gaspar, Susan M.; Bishop, Kenneth P.; Naqvi, H. S.; McNeil, John R.;
Tipton, Gary D.; Stallard, Brian R.; Draper, B. L.: Use of diffracted light
from latent images to improve lithography control,V1464,245-257(1991)

Hiddleston, H. R.
; Lyman, Dwight D.; Schafer, Eric L.: Comparisons of deformable-mirror
models and influence functions,V1542,20-33(1991)

Hideshima, Masayuki See Gemma, Takashi: V1332,77-84(1991)

Hietala, Vincent M.
; Vawter, Gregory A.; Meyer, W. J.; Kravitz, Stanley H.: Phased-array antenna
control by a monolithic photonic integrated circuit,V1476,170-175(1991)
See Martens, Jon S.: V1394,140-149(1991)
See Vawter, Gregory A.: V1476,102-106(1991)

Higashino, Hidetaka
See Mizuno, Koichi: V1477,197-204(1991)
See Setsune, Kentaro: V1394,79-88(1991)

Higbee, Terry A.
: TECHSTARS: small, smart space systems,V1495,103-114(1991)

Higdon, Noah S. See Ponsardin, Patrick: V1492,47-51(1991)

Higgins, William E.
; Spyra, Wolfgang J.; Karwoski, Ronald A.; Ritman, Erik L.: Automatic analysis
system for three-dimensional angiograms,V1445,276-286(1991)
See Dunn, Dennis F.: V1606,541-552(1991)

Higginson, Lyall A. See Taylor, Roderick S.: V1420,183-192(1991)

Higuchi, Hirofumi See Saito, Takahiro: V1605,382-393(1991)

Hikida, N. See Kasuya, Koichi: V1397,67-70(1991)

Hilbig, Jens
; Ritter, Reinold: Speckle measurement for 3-D surface
movement,V1554A,588-592(1991)

Hildebrand, Ulrich
; Seeliger, Reinhard; Smutny, Berry; Sand, Rolf: Design study for high-
performance optical communications satellite terminal with high-power
laser diode transmitter,V1522,50-60(1991)

Hildebrandt, Peter
: Cytochrome c at charged interfaces studied by resonance Raman and
surface-enhanced resonance Raman spectroscopy,V1403,102-111(1991)

Hildreth, Ellen C.
: Motion analysis for visually-guided navigation,V1382,167-180(1991)

Hildreth, Eugene N.
See Grieser, James L.: V1330,111-118(1991)
See Grieser, James L.: V1330,102-110(1991)

Hilkert, James M.
; Bowen, Max L.; Wang, Joe: Specifications for image stabilization
systems,V1498,24-38(1991)

Hill, D. A. See Walsh, Joseph T.: V1427,27-33(1991)

Hill, Derek L.
; Hawkes, David J.; Hardingham, Charles R.: Use of anatomical knowledge to
register 3-D blood vessel data derived from DSA with MR
images,V1445,348-357(1991)

Hill, G. F. See Sachse, Glen W.: V1433,157-166(1991)

Hill, Julian R. See Pantelis, Philip: V1559,2-9(1991)

Hill, Kenneth O. See Malo, Bernard: V1590,83-93(1991)

Hill, Paul M.
; Olshansky, Robert: Eight-Gb/s QPSK-SCM over a coherent detection optical
link,V1579,210-220(1991)

Hill, Peter C.
; Orzeszko, Gabriel R.: Performance comparison of neural network and
statistical discriminant processing techniques for automatic modulation
recognition,V1469,329-340(1991)

Hill, Steven C.
; Barber, Peter W.; Chowdhury, Dipakbin Q.; Khaled, Elsayed-Esam M.;
Mazumder, Mohiuddin: Resonances and internal electric energy density
in droplets,V1497,16-27(1991)
See Chang, Richard K.: V1497,2-13(1991)

Hill, William A.
See Lesko, Camille: V1475,330-339(1991)
See Turner, Elbert L.: V1475,248-256(1991)

Hillen, Walter
; Eckenbach, W.; Quadflieg, P.; Zaengel, Thomas T.: Signal-to-noise
performance in cesium iodide x-ray fluorescent screens,V1443,120-
131(1991)

Hillenkamp, Franz See Kochevar, Irene E.: V1403,756-763(1991)

Hillier, Glen See Puetz, Norbert: V1361,692-698(1991)

Hillman, Robert L.
; Melnik, George A.; Tsakiris, Todd N.; Leard, Francis L.; Jurgilewicz, Robert
P.; Warde, Cardinal: Electron-beam-addressed lithium niobate spatial
light modulator,V1562,136-142(1991)

Hills, C. R. See Slaughter, Jon M.: V1343,73-82(1991)

Hills, Louis S.
; Long, Jerry E.; Gebhardt, Frederick G.: Variable wind direction effects on
thermal blooming correction,V1408,41-57(1991)
See Dansereau, Jeffrey P.: V1409,67-82(1991)
See Long, Jerry E.: V1408,58-71(1991)

Hills, Peter C.
; Samson, Peter J.: Ignition risks of fiber optic systems,V1589,110-119(1991)

Hills, Robert F. See Lewis, Isabella T.: V1478,2-12(1991)

Hillsdon, G. K. See Ruiz, D.: V1358,1134-1143(1991)

Hilsenrath, Ernest See Cebula, Richard P.: V1493,91-99(1991)

Hilton, Albert R.
: Infrared refractive-index measurement results for single-crystal and
polycrystal germanium,V1498,128-137(1991)
: Infrared transmitting glasses and fibers for chemical analysis,V1437,54-
59(1991)

Hilton, Peter J.
; Gabric, R.; Walternberg, P. T.: Laser Image Detection System (LIDS): laser-
based imaging,V1385,27-31(1991)

Himes, Glenn S.
; Inigo, Rafael M.; Narathong, Chiewcharn: VLSI implementable neural
networks for target tracking,V1469,671-682(1991)

Hinchman, John H.
See Gruenewald, Maria M.: V1468,843-852(1991)
See Mueller, Walter: V1472,66-75(1991)

Hindley, N. See de Salabert, Arturo: V1521,74-88(1991)

Hinds, Stuart C. See Gerson, Nahum D.: V1406,129-129(1991)

Hindy, Robert N.
; Kohanzadeh, Youssef; eds.: *Laser Beam Diagnostics,*V1414(1991)

Hine, Butler P.
; Reid, Max B.; Downie, John D.: Optical correlator techniques applied to
robotic vision,V1564,416-426(1991)
See Downie, John D.: V1564,308-319(1991)
See Reid, Max B.: V1564,384-394(1991)

Hines, Braden E.
See Colavita, Mark M.: V1494,168-181(1991)
See Colavita, Mark M.: V1542,205-212(1991)

Hines, John R.
See Eaton, Frank D.: V1487,84-90(1991)
See Waldie, Arthur H.: V1487,103-108(1991)

Hingerl, Kurt
; Lilja, J.; Toivonen, M.; Pessa, Markus; Jantsch, Wolfgang; As, D. J.; Rothemund, W.; Juza, P.; Sitter, Helmut: Electrical and optical properties of As- and Li-doped ZnSe films,V1361,943-953(1991)
; Pesek, Andreas; Sitter, Helmut; Krost, Alois; Zahn, Dietrich R.; Richter, W.; Kudlek, Gotthard; Gutowski, Juergen: Growth and characterization of ZnSe and ZnTe grown on GaAs by hot-wall epitaxy,V1361,383-393(1991)

Hingston, G. See Puetz, Norbert: V1361,692-698(1991)

Hink, Paul L.
; Rothschild, Richard E.; Pelling, Michael R.; MacDonald, Daniel R.; Gruber, Duane E.: UCSD high-energy x-ray timing experiment cosmic ray particle anticoincidence detector,V1549,193-202(1991)
See Rothschild, Richard E.: V1549,120-133(1991)

Hino, Hideo
; Takei, Mitsuru; Ono, Hiromi; Nagura, Riichi; Narimatsu, Yoshito; Hiramatsu, Masaru; Harada, Hisashi; Ogikubo, Kazuhiro: Optical sensor system for Japanese Earth resources satellite 1,V1490,166-176(1991)

Hinsch, Klaus D. See Guelker, Gerd: V1500,124-134(1991)

Hinterlong, Steve J.
See Brubaker, John L.: V1533,88-96(1991)
See McCormick, Frederick B.: V1396,508-521(1991)
See McCormick, Frederick B.: V1533,97-114(1991)

Hinton, Geoffrey E. See Becker, Suzanna: V1569,218-226(1991)

Hinton, Oliver R. See Hallouche, Farid: V1445,504-512(1991)

Hintz, Kenneth J.
See Moghaddam, Baback: V1486,115-126(1991)
See Moghaddam, Baback: V1406,42-57(1991)
See Moghaddam, Baback: V1471,414-421(1991)

Hippler, Horst See Abel, B.: V1525,110-118(1991)

Hirahara, Shuzo See Matsui, Toshikazu: V1453,282-289(1991)

Hirai, Shoichi See Tsuno, Katsuhiko: V1490,222-232(1991)

Hirai, Yoshihiko See Watanabe, Hisashi: V1463,101-110(1991)

Hirakawa, Kazuhiko
; Hashimoto, Y.; Ikoma, Toshiaki: Transient of electrostatic potential at GaAs/AlAs heterointerfaces characterized by x-ray photoemission spectroscopy,V1361,255-261(1991)

Hiraki, Akio
: New diamond activities at Osaka University,V1534,198-198(1991)
See Nishimura, Kazuhito: V1534,199-206(1991)

Hiramatsu, Masaru
See Hino, Hideo: V1490,166-176(1991)
See Takahashi, Fumiho: V1490,255-268(1991)

Hirao, Kazuyuki See Tanabe, Setsuhisa: V1513,340-348(1991)

Hirao, Takashi See Nishimura, Kazuhito: V1534,199-206(1991)

Hirata, Kouji See Osawa, Atsuo: V1354,337-343(1991)

Hirata, Masaru See Kamei, Shuya: V1358,979-983(1991)

Hiratani, Y. See Akita, K.: V1392,576-587(1991)

Hirleman, Edwin D.
See Bawolek, Edward J.: V1464,574-583(1991)
See Kenney, Steven B.: V1480,82-93(1991)

Hiroi, Takashi See Maeda, Shunji: V1567,100-109(1991)

Hirose, Masataka
; Takata, H.; Koyanagi, Mitsumasa: Optically coupled 3-D common memory with GaAs on Si structure,V1362,316-322(1991)

Hirose, Shozo See Takemori, Toshikazu: V1469,157-165(1991)

Hirsch, Tom J. See Mueller, Heinrich G.: V1598,132-140(1991)

Hirschmann, Harald
; Meier, Wolfgang; Finkelmann, Heino: Nonlinear optical and piezoelectric behavior of liquid-crystalline elastomers,V1559,27-38(1991)

Hirshfield, Jay L.
; Park, Gun-Sik: Synchrotron radiation lasers,V1552,138-146(1991)

Hirst, Graeme J. See Key, Michael H.: V1397,9-17(1991)

Hirst, Peter F. See Dawber, William N.: V1476,81-90(1991)

Hiruma, Nobuyuki See Yamada, Mitsuho: V1453,51-57(1991)

Hisada, Shigeyoshi See Nakahara, Sumio: V1554A,602-609(1991)

Hishinuma, Akihiro See Kawahara, Hideo: V1513,198-203(1991)

Hitzenberger, Christoph K.
; Fercher, Adolf F.; Juchem, M.: Measurement of the axial eye length and retinal thickness by laser Doppler interferometry,V1423,46-50(1991); 1429,21-25(1991)

Hitzler, Hermine
; Leclerc, Norbert; Pfleiderer, Christoph; Wolfrum, Juergen M.; Greulich, Karl O.; Klein, Karl-Friedrich: Bundle of tapered fibers for the transmission of high-power excimer laser pulses,V1503,355-362(1991)

Hivernage, M. See Veaux, Jacqueline: V1449,13-24(1991)

Hjelme, Dag R. See Bundy, Scott: V1475,319-329(1991)

Hladky, J. See Olsen, Gregory H.: V1419,24-31(1991)

Hliniak, Andrzej See Domanski, Andrzej W.: V1420,72-80(1991)

Hlivak, Robert J. See Geary, John C.: V1447,264-273(1991)

Ho, Bruce K.
; Ma, Marco; Chuang, Keh-Shih: Slice plane generation for three-dimensional image viewing using multiprocessing,V1445,95-100(1991)

Ho, Daniel See Boltinghouse, Susan T.: V1388,14-29(1991)

Ho, Huey-Chin C. See Young, Eddie H.: V1476,178-189(1991)

Ho, M. Y. See Leung, Chung-yee: V1572,566-571(1991)

Ho, Peter D. See Neev, Joseph: V1427,162-172(1991)

Ho, Ping-Pei See Wang, LeMing: V1431,97-101(1991)

Ho, Ping-Tong
; Peng, F.; Goldhar, J.; Nolting, Eugene E.; Parsons, C.: Photoconductivity of high-bandgap materials,V1378,210-216(1991)
See Lau, Yue Y.: V1407,546-552(1991)

Ho, Wai-Chin
See Cheng, Andrew Y.: V1444,400-406(1991)
See Kwok, John C.: V1445,446-455(1991)

Ho, Zonh-Zen See Chen, Ray T.: V1583,362-374(1991)

Hobbs, Chris P.
; Kenway-Jackson, Damian; Milne, James M.: Quantitative measurement of thermal parameters over large areas using pulse-video thermography,V1467,264-277(1991)

Hobbs, Philip C.
: Shot noise limited optical measurements at baseband with noisy lasers,V1376,216-221(1991)
See Haller, Kurt L.: V1435,298-309(1991)

Hobson, Clifford A.
See Hartley, David A.: V1567,277-282(1991)
See Harvey, David M.: V1400,86-93(1991)
See Wood, Christopher M.: V1332,301-313(1991)

Hochberg, Eric B.
: Optical figure testing of prototype mirrors for JPL's precision segmented-reflector program,V1542,511-522(1991)

Hochstrasser, Denis See Ratib, Osman M.: V1446,330-340(1991)

Hochstrasser, Robin M.
: Femtosecond infrared spectroscopy and molecular dynamics,V1403,3-3(1991)
; Oppenlaender, A.; Pierre, M.; Rambaud, C.; Silbey, R.; Skinner, J. L.; Trommsdorff, H. P.; Vial, J.-C.: Laser spectroscopy of proton dynamics in hydrogen bonds,V1403,221-229(1991)

Hockaday, Robert G. See Singman, Leif V.: V1548,80-92(1991)

Hodge, P. See Greenfield, Perry E.: V1494,16-39(1991)

Hodges, Charles See Novak, Agneta: V1389,80-86(1991)

Hodges, Larry F.
: Basic principles of stereographic software development,V1457,9-17(1991)
See McWhorter, Shane W.: V1457,85-90(1991)

Hodges, Steven E. See Raymer, M. G.: V1376,128-131(1991)

Hodgkin, Van A. See Franck, Jerome B.: V1555,63-70(1991)

Hodgson, Elizabeth M. See Boardman, A. D.: V1503,160-166(1991)

Hodor, James R.
; Decker, Herman J.; Barney, Jesus J.: Infrared techniques applied to large solar arrays: a ten-year update,V1540,331-337(1991)

Hodorchenko, P. V. See Oleynikov, V. A.: V1403,164-166(1991)

Hodul, David T.
See Keenan, W. A.: V1393,354-365(1991)
See Yarling, Charles B.: V1393,192-199(1991)

Hoefle, Wolfgang
: Interpretation of measured spectral attenuation curves of optical fibers by deconvolution with source spectrum,V1504,140-146(1991)

Hoefling, Roland
; Aswendt, Petra; Totzauer, Werner F.; Jueptner, Werner P.: DSPI: a tool for analyzing thermal strain on ceramic and composite materials, V1508,135-142(1991)

Hoeft, Gregory L.
; Hughes, Simon H.; Slocum, Robert E.: Low-cost computed tomography system using an image intensifier, V1396,638-645(1991)

Hoekstra, A. See Anema, P. C.: V1446,352-356(1991)

Hoelscher, Claudia See Guelker, Gerd: V1500,124-134(1991)

Hoeltzel, David A. See Buzard, Kurt A.: V1423,70-81(1991)

Hoenes, Frank
; Bleisinger, Rainer; Dengel, Andreas R.: Intelligent word-based text recognition, V1384,305-316(1991)

Hoeper, J. See Frank, Klaus: V1431,2-11(1991)

Hoeppner, Wolfgang W. See Bauch, Lothar: V1466,510-519(1991)

Hoerner, Claudine
; Lavoine, J. P.; Villaeys, A. A.: Theoretical study of ultrafast dephasing by four-wave mixing, V1362,863-869(1991)

Hofer, Bruno See Lange, Robert: V1492,24-37(1991)

Hofer, Wayne W. See Pocha, Michael D.: V1378,2-9(1991)

Hoff, Frederick G. See Spillman, William B.: V1367,197-203(1991)

Hoff, Lawrence E.
; Evans, John R.; Bunney, Laura: Detection of targets in terrain clutter by using multispectral infrared image processing, V1481,98-109(1991)
See Stocker, Alan D.: V1481,156-169(1991)

Hoff, William A. See Sklair, Cheryl W.: V1387,380-391(1991)

Hoffman, Alan
; Randall, David: High-performance 256 x 256 InSb FPA for astronomy, V1540,297-302(1991)

Hoffman, Arnold See Netz, Yoel: V1458,61-61(1991)

Hoffman, Donald
: Representing three-dimensional shapes for visual recognition, V1445,2-4(1991)

Hoffmann, Axel
See Rosenzweig, M.: V1362,876-887(1991)
See Tischel, M.: V1361,917-926(1991)

Hoffmann, Hans J. See Westenberger, Gerhard: V1535,113-120(1991)

Hoffmann, Kenneth R. See Alperin, Noam: V1396,27-31(1991)

Hoflund, Gar B. See Upchurch, Billy T.: V1416,21-29(1991)

Hofmann, Guenter See Norkus, Volkmar: V1484,98-105(1991)

Hofmann, K. See Landwehr, Gottfried: V1362,282-290(1991)

Hofmann, Thomas
; Sharp, Tracy E.; Dane, C. B.; Wisoff, P. J.; Wilson, William L.; Tittel, Frank K.; Szabo, Gabor: Characterization of a subpicosecond XeF (C->A) excimer laser, V1412,84-90(1991)
See Tittel, Frank K.: V1397,21-29(1991)

Hogan, Daniel C.
: Applications of copper vapor laser lighting in high-speed motion analysis, V1346,324-330(1991)

Hogenkamp, Detlef See Dusa, Mircea V.: V1464,447-458(1991)

Hogg, W. D.
; Janzen, Doug; Valis, Tomas; Measures, Raymond M.: Development of a fiber Fabry-Perot strain gauge, V1588,300-307(1991)
See Measures, Raymond M.: V1332,431-443(1991)
See Measures, Raymond M.: V1332,421-430(1991)
See Zuliani, Gary: V1588,308-313(1991)

Hoggins, James T.
See Klocek, Paul: V1498,147-157(1991)
See Trombetta, John M.: V1534,77-88(1991)

Hogue, David W. See Wanuga, Stephen: V1374,97-106(1991)

Hogue, K. L. See Zhai, H. R.: V1519,575-579(1991)

Hohimer, John P. See Shiau, T. H.: V1418,116-122(1991)

Hohman, James L.
See Booth, Bruce L.: V1377,57-63(1991)
See Hartman, Davis H.: V1390,368-376(1991)

Hohn, Fritz J. See Chiong, Kaolin G.: V1465,221-236(1991)

Holben, Milford S. See Baldini, S. E.: V1588,125-131(1991)

Holber, William M. See O'Neill, James A.: V1392,516-528(1991)

Holbrook, David S.
; Donaher, J. C.: Overlay and matching strategy for large-area lithography, V1463,475-486(1991)

Holcomb, D. R. See Norwood, Robert A.: V1560,54-65(1991)

Holcomb, Terry L.
: Life test of high-power Matsushita BTRS laser diodes, V1417,328-337(1991)

Holdsworth, David W. See Fenster, Aaron: V1447,28-33(1991)

Holl, P. See Braueninger, Heinrich: V1549,330-339(1991)

Holland, Andrew D. See Wells, Alan A.: V1549,357-373(1991)

Holland, John
; Everett, Hobart R.; Gilbreath, Gary A.: Hybrid navigational control scheme for autonomous platforms, V1388,291-298(1991)

Holland, Orgal T.
; Tarr, Tomas; Farsaie, Ali; Fuller, James M.: Artificial neural system approach to IR target identification, V1469,102-112(1991)

Hollerbach, Uwe
See Barouch, Eytan: V1465,254-262(1991)
See Barouch, Eytan: V1463,336-344(1991)
See Barouch, Eytan: V1463,464-474(1991)

Holley, Jeff
; Silver, Eric H.; Ziock, Klaus P.; Novick, Robert; Kaaret, Philip E.; Weisskopf, Martin C.; Elsner, Ronald F.; Beeman, Jeff: Bragg crystal polarimeter for the Spectrum-X-Gamma misson, V1343,500-511(1991)

Hollier, Pierre A.
: Imager of METEOSAT second generation, V1490,74-81(1991)
See Lorsignol, Jean: V1490,155-163(1991)

Hollinger, Allan B. See Thomas, Paul J.: V1488,36-47(1991)

Hollis, K. See Neviere, Michel: V1545,76-87(1991)

Hollman, Richard F.
: Dose control for short exposures in excimer laser lithography, V1377,119-125(1991)

Hollon, Kenneth R. See Hughen, James H.: V1469,341-350(1991)

Holly, Krisztina J. See Dubowsky, Steven: V1386,10-20(1991)

Holly, Sandor See Feldman, Albert: V1534(1991)

Holm, Richard W.
See Brodsky, Aaron: V1482,159-169(1991)
See Dierks, Jeffrey S.: V1482,146-158(1991)

Holm-Kennedy, James W.
; Tsang, Koon Wing; Sze, Wah Wai; Jiang, Fenglai; Yang, Datong: Novel monolithic chip-integrated color spectrometer: the distributed-wavelength filter component, V1527,322-331(1991)

Holman, Robert L.
See Jacobson, Benjamin A.: V1528,82-85(1991)
See Lacovara, Phil: V1528,135-141(1991)
See Winston, Roland: V1528(1991)

Holmberg, D. L. See Singman, Leif V.: V1548,80-92(1991)

Holmes, Steven J. See MacDonald, Scott A.: V1466,2-12(1991)

Holmes, Timothy J. See Liu, Yi-Hwa: V1428,191-199(1991)

Holmstrom, Roger P.
; Meland, Edmund; Powazinik, William: New structure and method for fabricating InP/InGaAsP buried heterostructure semiconductor lasers, V1418,223-230(1991)

Holst, Gerald C.
: Effects of phasing on MRT target visibility, V1488,90-98(1991)
ed.: *Infrared Imaging Systems: Design, Analysis, Modeling and Testing II*, V1488(1991)

Holton, William C.
: Precompetitive cooperative research: the culture of the '90s, V1392,27-33(1991); 1393,27-33(1991); 1394,27-33(1991)

Holtslag, A. H. See Coombs, James H.: V1499,6-20(1991)

Holtz, Per O. See Merz, James L.: V1361,76-88(1991)

Holubowicz, Kazimierz S.
: Method and device that prevent target sensors from being radiation overexposed in the presence of a nuclear blast, V1456,274-285(1991)

Holwill, Robert J. See Robertson, Stewart A.: V1464,232-244(1991)

Holyer, Ronald J. See Lybanon, Matthew: V1406,180-189(1991)

Holz, Michael
; Stern, Margaret B.; Medeiros, Shirley; Knowlden, Robert E.: Testing binary optics: accurate high-precision efficiency measurements of microlens arrays in the visible, V1544,75-89(1991)
See Motamedi, M. E.: V1544,33-44(1991)
See Werner, Thomas R.: V1544,46-57(1991)

Holzhausen, Klause-Peter See Spain, Edward H.: V1457,103-110(1991)

Holzwarth, Achim
; Griebsch, Juergen; Berger, Peter: Theoretical and experimental investigations on pressure-wave reflections and attenuation in high-power excimer lasers,V1503,98-109(1991)

Homer, Michael
; Rosser, Roy J.; Speer, R. J.: Ray tracing homogenizing mirrors for synchrotron x-ray lithography,V1527,134-144(1991)

Homma, M. See Moriyama, Takashi: V1490,310-316(1991)

Honciuc, Maria See Dumitru, Mihaela A.: V1500,339-348(1991)

Honda, H. See Yamaguchi, H.: V1499,29-38(1991)

Honda, Hisao See Kitaoka, Masaki: V1519,109-114(1991)

Honda, Katsunori See Ito, Katsunori: V1499,382-385(1991)

Honda, Keiji
See Isobe, Hiroshi: V1358,201-208(1991)
See Sato, Eiichi: V1358,479-487(1991)

Honda, Kenji
; Beauchemin, Bernard T.; Fitzgerald, Edward A.; Jeffries, Alfred T.; Tadros, Sobhy P.; Blakeney, Andrew J.; Hurditch, Rodney J.; Tan, Shiro; Sakaguchi, Shinji: Studies of dissolution inhibition mechanism of DNQ-novolak resist (II): effect of extended ortho-ortho bond in novolak,V1466,141-148(1991)

Honda, Ko See Yuille, Alan L.: V1570,166-174(1991)

Honda, Tatsuro
; Matsui, Kenichi: Coating thickness gauge,V1540,709-716(1991)

Honda, Toshio
; Kang, Der-Kuan; Shimura, Kei; Enomoto, H.; Yamaguchi, Masahiro; Ohyama, Nagaaki: Large one-step holographic stereogram,V1461,156-166(1991)
See Ohishi, Satoru: V1443,280-285(1991)

Hondo, Yukie See Koike, Yasuhiro: V1592,62-72(1991)

Honey, Frank R. See Lyon, Ronald J.: V1492,339-350(1991)

Honey, Richard C.
See Brown, Robert A.: V1537,147-150(1991)
See Maffione, Robert A.: V1537,115-126(1991)
See Maffione, Robert A.: V1537,173-184(1991)

Honeyman, Janice C.
; Staab, Edward V.; Frost, Meryll M.: Initial experiences with PACS in a clinical and research environment,V1446,362-368(1991)
See Staab, Edward V.: V1446,16-22(1991)

Hong, Alexander See Kuyel, Birol: V1463,646-665(1991)

Hong, C. S.
; Hager, Harold E.; Capron, Barbara; Mantz, Joseph L.; Beranek, Mark W.; Huggins, Raymond W.; Chan, Eric Y.; Voitek, Mark; Griffith, David M.; Livezey, Darrell L.; Scharf, Bruce R.: Optoelectronic devices for fiber-optic sensor interface systems,V1418,177-187(1991)
See Figueroa, Luis: V1418,153-176(1991)
See Fu, Richard J.: V1418,108-115(1991)
See Lim, G.: V1418,123-131(1991)

Hong, ChunRong See Du, MingZe: V1361,699-705(1991)

Hong, J. P. See Bourne, Lincoln C.: V1477,205-208(1991)

Hong, Jiang See Du, MingZe: V1361,699-705(1991)

Hong, Jing
See Zhao, Feng: V1555,297-299(1991)
See Zhao, Feng: V1461,262-264(1991)
See Zhao, Feng: V1555,241-242(1991)

Hong, John H. See Yeh, Pochi A.: V1562,32-43(1991)

Hong, Lang
: Three-dimensional scene reconstruction using optimal information fusion,V1383,333-344(1991)

Hong, Nguyen T.
; Phat, Darith; Plaza, Pascal; Daudon, Michel; Dao, Nguyen Q.: Chemical analysis of human urinary and renal calculi by Raman laser fiber-optics method,V1525,132-142(1991)

Hong, Shuili
: Symmetrization analysis of lattice-vibrational modes and study of Raman-IR spectra for B-BaB2O4,V1437,194-197(1991)

Hong, Suk-Yoon
; Varadan, Vasundara V.; Varadan, Vijay K.: Comparison of analog and digital strategies for automatic vibration control of lightweight space structures,V1489,75-83(1991)

Hong, Tsai-Hong
See Rangachar, Ramesh M.: V1382,376-385(1991)
See Rangachar, Ramesh M.: V1382,331-340(1991)

Hong, Xiao-Yin
; Liu, Dan; Li, Zhong-Zhe; Xiao, Ji-Quang; Dong, Gui-Rong: Progress in the study of development-free vapor photolithography,V1466,546-557(1991)

Hong, Young-Sik See Chen, Su-Shing: V1569,463-473(1991)

Hongu, Hitoshi
; Karasaki, Hidehiko: New CO2 laser equipped with high-peak pulse power and high-speed drilling process,V1501,198-204(1991)

Honguh, Yoshinori
: Diffraction analysis of optical disk readout signal deterioration caused by mark-size fluctuation,V1527,315-321(1991)
See Satoh, Hiroharu: V1499,324-329(1991)

Honickman, Harris See Clark, David L.: V1540,303-311(1991)

Honig, David A. See Whiting, James S.: V1453,165-175(1991)

Honkanen, Seppo
See Li, Ming-Jun: V1506,52-57(1991)
See Li, Ming-Jun: V1513,410-417(1991)
See Najafi, S. I.: V1583,32-36(1991)
See Tammela, Simo: V1583,37-42(1991)
See Tervonen, Ari: V1513,71-75(1991)
See Wang, Wei-Jian: V1513,434-440(1991)

Honkonen, Veijo
; Jaaskelainen, Timo; Parkkinen, Jussi P.: Optical correlators in texture analysis,V1564,43-51(1991)

Honour, Joseph
: Digital readout for image converter cameras,V1358,713-718(1991)

Honrubia, J. J. See Velarde, G.: V1502,242-257(1991)

Hontzopoulos, Elias I.
; Zervaki, A.; Zergioti, Y.; Hourdakis, G.; Raptakis, E.; Giannacopoulos, A.; Fotakis, Costas: Excimer laser ceramic and metal surface alloying applications,V1397,761-768(1991)
See Hourdakis, G.: V1503,249-255(1991)
See Kollia, Z.: V1503,215-222(1991)

Hoock, Donald W.
: Modeling time-dependent obscuration for simulated imaging of dust and smoke clouds,V1486,164-175(1991)
See Giever, John C.: V1486,302-313(1991)

Hook, R. See Greenfield, Perry E.: V1494,16-39(1991)

Hooker, Chris J. See Key, Michael H.: V1397,9-17(1991)

Hookman, Robert A.
See Weinswig, Shepard A.: V1527,118-125(1991)
See Zurmehly, George E.: V1532,170-176(1991)

Hooper, W. W. See Yap, Daniel: V1418,471-476(1991)

Hoose, John F.
; Loewen, Erwin G.: Anomaly reduction in gratings,V1545,189-199(1991)
; Olson, Jeffrey: Scatter properties of gratings at ultraviolet and visible wavelengths,V1545,160-166(1991)

Hoover, Carl W.
; Webb, Curtis M.: What is MRT and how do I get one?,V1488,280-288(1991)

Hoover, James M.
See Lindsay, Geoffrey A.: V1497,418-422(1991)
See Lindsay, Geoffrey A.: V1560,443-453(1991)

Hoover, Richard B.
; Baker, Phillip C.; Shealy, David L.; Brinkley, B. R.; Walker, Arthur B.; Barbee, Troy W.: Design and analysis of a water window imaging x-ray microscope,V1426,84-96(1991)
; Baker, Phillip C.; Hadaway, James B.; Johnson, R. B.; Peterson, Cynthia; Gabardi, David R.; Walker, Arthur B.; Lindblom, Joakim F.; DeForest, Craig E.; O'Neal, Ray H.: Performance of the Multi-Spectral Solar Telescope Array III: optical characteristics of the Ritchey-Chretien and Cassegrain Telescopes,V1343,189-202(1991)
; Fineschi, Silvano; Fontenla, Juan; Walker, Arthur B.: Solar EUV/FUV line polarimetry: instruments and methods,V1343,389-403(1991)
; Shealy, David L.; Baker, Phillip C.; Barbee, Troy W.; Walker, Arthur B.: Development of the water-window imaging x-ray microscope,V1435,338-351(1991)
; Walker, Arthur B.; DeForest, Craig E.; Allen, Maxwell J.; Lindblom, Joakim F.: Performance of the Multi-Spectral Solar Telescope Array VI: performance and characteristics of the photographic films,V1343,175-188(1991)
; Walker, Arthur B.; eds.: *X-Ray/EUV Optics for Astronomy, Microscopy, Polarimetry, and Projection Lithography*,V1343(1991)
See DeForest, Craig E.: V1343,404-414(1991)
See Fineschi, Silvano: V1343,376-388(1991)

See Lindblom, Joakim F.: V1343,544-557(1991)
See Shealy, David L.: V1343,122-132(1991)
See Walker, Arthur B.: V1343,334-347(1991)
See Walker, Arthur B.: V1343,415-427(1991)
See Walker, Arthur B.: V1494,320-333(1991)
See Walker, Arthur B.: V1343,319-333(1991)

Hope, D. A.
; Hydes, A. J.; Cox, T. I.; Deshmukh, V. G.: Plasma diagnostics as inputs to the modeling of the oxygen reactive ion etching of multilevel resist structures,V1392,185-195(1991)
See Deshmukh, V. G.: V1392,352-360(1991)

Hopkins, John B.
; Xu, Xiaobing; Lingle, Robert; Zhu, Huiping; Yu, Soo-chang: Direct measurement of vibrational energy relaxation in photoexcited deoxyhemoglobin using picosecond Raman spectroscopy,V1432,221-226(1991)

Hopkins, Michael See Downey, George A.: V1482,40-47(1991)

Hopwood, Anthony I.
: New holographic overlays,V1509,26-35(1991)

Hora, Heinrich
; Aydin, M.; Kasotakis, G.; Stening, R. L.: Laser application for fusion using volume ignition and smoothing by suppression of pulsation,V1502,258-269(1991)

Horais, Brian J.
ed.: *Small-Satellite Technology and Applications*,V1495(1991)

Hord, R. M. See Rosenfeld, J. P.: V1406,147-147(1991)

Horger, John D. See LaFollette, Robert: V1488,289-299(1991)

Hori, Yoshikazu See Kato, Makoto: V1507,36-44(1991)

Horie, Noriyoshi See Hosokawa, Hayami: V1559,229-237(1991)

Horigome, Shinkichi See Watanabe, Hitoshi: V1499,21-28(1991)

Horii, Steven C.
; Garra, Brian S.; Mun, Seong Ki; Zeman, Robert K.; Levine, Betty A.; Fielding, Robert: PACS and teleradiology for on-call support of abdominal imaging,V1446,10-15(1991)
; Garra, Brian S.; Mun, Seong Ki; Singer, Jon; Zeman, Robert K.; Levine, Betty A.; Fielding, Robert; Lo, Ben: PACS reading time comparision: the workstation versus alternator for ultrasound,V1446,475-480(1991)
See Lo, Shih-Chung B.: V1444,265-271(1991)

Horikawa, Hiroshi
; Miura, Masayuki; Uchida, Toshiya: Relationship between jitter and deformation of mirrors,V1454,20-32(1991)
; Sugisaki, Iwao; Tashiro, Masaru: Relationship between fluctuation in mirror radius (within one polygon) and the jitter,V1454,46-59(1991)

Horikawa, Tsuyoshi See Nakajima, Hajime: V1456,29-39(1991)

Horioka, K. See Kasuya, Koichi: V1397,67-70(1991)

Horita, Yuukou
; Miyahara, Makoto: Image segmentation based on ULCS color difference,V1606,607-620(1991)

Horiuchi, Makoto See Isuzugawa, Kohji: V1358,1003-1010(1991)

Horn, Berthold K. See Standley, David L.: V1473,194-201(1991)

Horn, Keith A.
See Beeson, Karl W.: V1559,258-266(1991)
See Beeson, Karl W.: V1374,176-185(1991)
See McFarland, Michael J.: V1583,344-354(1991)

Horn, Mark W. See Kunz, Roderick R.: V1466,218-226(1991)

Horn, Richard R.
: History and status of black and white photographic processing chemicals as effluents,V1458,69-75(1991)

Hornak, Joseph P.
; Blaakman, Andre; Rubens, Deborah; Totterman, Saara: Multispectral image segmentation of breast pathology,V1445,523-533(1991)

Horne, David S.
: Diffusing wave spectroscopy studies of gelling systems,V1430,166-180(1991)

Horner, Joseph L. See Fielding, Kenneth H.: V1564,224-230(1991)

Hornfeck, M. See Clasen, Rolf: V1513,243-254(1991)

Hornung, V. See Remiens, D.: V1362,323-330(1991)

Horowitz, Atara See Biderman, Shlomo: V1535,27-34(1991)

Horowitz, Bradley See Pentland, Alexander P.: V1605,467-474(1991)

Horsky, Thomas N.
; Schiller, Craig M.; Genetti, George J.; O'Mara, Daniel M.; Hamnett, Whitney S.; Warde, Cardinal: Electron-beam-addressed membrane light modulator for IR scene projection,V1540,527-532(1991)

Horsthuis, Winfried H.
See Moehlmann, Gustaaf R.: V1512,34-39(1991)
See Moehlmann, Gustaaf R.: V1560,426-433(1991)

Horton, R. L. See Laumann, Curt W.: V1414,151-160(1991)

Horton, T. E.
: Fluid-dynamic perturbations in gas lasers,V1397,549-554(1991)
; Cason, Charles; Dezenberg, George J.: Optomechanical considerations for stable lasers,V1416,10-20(1991)

Horvat, John See Conley, Willard E.: V1466,53-66(1991)

Horvath, Zoltan G. See Denker, Boris I.: V1419,50-54(1991)

Horwitz, Larry S. See Mobasser, Sohrab: V1540,764-774(1991)

Hosch, Jimmy W.
: Process control sensor development for the automation of single-wafer processors,V1392,529-534(1991)

Hoshino, Haruo See Yamada, Mitsuho: V1453,51-57(1991)

Hoshino, Hideshi
; Sato, Koki: Three-dimensional moving-image display by modulated coherent optical fibers: a proposal,V1461,227-231(1991)

Hoshino, Taizo J. See Zogg, Hans: V1361,1079-1086(1991)

Hosoba, Minoru See Minato, Kotaro: V1446,195-198(1991)

Hosoda, Masahiro See Wada, Tatsuo: V1560,162-171(1991)

Hosokawa, Hayami
; Horie, Noriyoshi; Yamashita, Tsukasa: Mass-producible optical guided-wave devices fabricated by photopolymerization,V1559,229-237(1991)

Hossack, William J.
; Vass, David G.; Underwood, Ian: Fourier processing with binary spatial light modulators,V1564,697-702(1991)

Hosseini Teherani, Ferechteh
; Kreisler, Alain J.; Baixeras, Joseph M.: Microstructure and superconducting properties of Y1Ba2Cu3O7-delta thin films grown by rapid thermal annealing (RTA) as a function of crystalline structure of zirconia substrates,V1362,921-929(1991)
See Baixeras, Joseph M.: V1362,117-126(1991)

Hossfeld, Jens
; Jaeger, Erwin; Tschudi, Theo T.; Churin, Evgeny G.; Koronkevich, Voldemar P.: Rectangular focus spots with uniform intensity profile formed by computer-generated holograms,V1574,159-166(1991)

Hosticka, Bedrich J. See Boehme, Johann F.: V1566,208-219(1991)

Hotchandani, Surat See Kassi, Hassane: V1436,58-67(1991)

Hotes, Deborah See Banks, Bruce A.: V1330,66-77(1991)

Hotta, Hiroyuki See Miyake, Hiroyuki: V1448,150-156(1991)

Hou, Bi H.
; Qi, Zhen Z.: Tensor resistivity and tensor susceptibility of oriented high-Tc superconducting film YBaCuO,V1519,822-823(1991)

Hou, Hsieh S.
: Recursive scaled DCT,V1567,402-412(1991)

Hou, Li S.
; Lu, Song W.; Gan, Fuxi: Variation of optical properties of gel-derived VO2 thin films with temperature,V1519,580-588(1991)
; Zhu, Chang X.; Gu, Dong H.: Recording and erasing characteristics of GeSbTe-based phase change thin films,V1519,548-553(1991)
See Xiang, Xiao L.: V1519,712-716(1991)

Hou, Xun
; Zhang, Xiaoqiu; Gong, Meixia; Chang, Zenghu; Lei, Zhiyuan; Yang, Binzhou; Yu, Hongbin; Liu, Xiouqin; Shan, Bin; Gao, Shengshen; Zhao, Wei: Picosecond x-ray streak camera improvement,V1358,868-873(1991)
See Chang, Zenghu: V1358,614-618(1991)
See Chang, Zenghu: V1358,541-545(1991)
See Wang, Liming: V1358,1156-1160(1991)
See Wang, Liming: V1415,120-126(1991)

Houck, Timothy L. See Westenskow, Glen A.: V1407,496-501(1991)

Houde, Regis
; Laurendeau, Denis; Poussart, Denis: 3-D camera based on differential optical absorbance,V1332,343-354(1991)

Houdjal, R.
See Bataille, F.: V1502,135-139(1991)
See Bataille, F.: V1397,839-842(1991)

Houdy, Philippe
See Boher, Pierre: V1345,165-179(1991)
See Boher, Pierre: V1343,39-55(1991)
See Boher, Pierre: V1345,198-212(1991)

Hough, Gary R.
; Gustafson, D. M.; Thursby, William R.: Enhanced holographic recording capabilities for dynamic applications,V1346,194-199(1991)

Hough, Stewart E.
; Stanley, Pamela S.: Militarized infrared touch panels,V1456,240-249(1991)

Houlihan, Frank M. See Nalamasu, Omkaram: V1466,13-25(1991)

Houlton, M. R. See Ashley, Timothy: V1361,238-244(1991)

Houng, Y. M. See Wang, S. Y.: V1371,98-103(1991)

Hourdakis, G.
; Hontzopoulos, Elias I.; Tsetsekou, A.; Zampetakis, Th.; Stournaras, C.: Excimer laser surface treatment of ceramics,V1503,249-255(1991)
See Hontzopoulos, Elias I.: V1397,761-768(1991)

House, F. A. See English, R. Edward: V1527,174-179(1991)

Houssay, Bruno See Vergnolle, Claude: V1389,648-658(1991)

Houzay, Francoise See Nissim, Yves I.: V1393,216-228(1991)

Hovanessian, V. A.
See Arutunian, A. H.: V1402,102-106(1991)
See Arutunian, A. H.: V1402,2-6(1991)
See Arutunian, A. H.: V1403,585-587(1991)

Hovis, Floyd E. See Kelley, J. D.: V1415,211-219(1991)

How, Jonathan P.
; Anderson, Eric H.; Miller, David W.; Hall, Steven R.: High-bandwidth control for low-area-density deformable mirrors,V1489,148-162(1991)
See Anderson, Eric H.: V1542,392-405(1991)

Howard, Edward A. See Paige, Jeffrey L.: V1364,7-13(1991)

Howard, Michael D.
: POPS: parallel opportunistic photointerpretation system,V1471,422-427(1991)

Howard, Vyvyan See Bovik, Alan C.: V1450(1991)

Howarth, M. S.
; Searcy, Stephen W.: Fresh market carrot inspection by machine vision,V1379,141-150(1991)

Howe, Gordon S.
; McIntosh, Bruce C.: Use of Fourier transform spectroscopy in combustion effluent monitoring,V1434,90-103(1991)

Howell, Gelston See Deshpande, Ujwal A.: V1390,489-501(1991)

Hoy, Ronald J. See Gur, David: V1446,284-288(1991)

Hoyle, Charles E.
; Creed, David; Anzures, Edguardo; Subramanian, P.; Nagarajan, Rajamani: Photo-induced degradation of selected polyimides in the presence of oxygen: a rapid decomposition process,V1559,101-109(1991)
; Sundell, Per-Erik; Trapp, Martin A.; Kang, Deokman; Sheng, D.; Nagarajan, Rajamani: Polymerization kinetics of mono- and multifunctional monomers initiated by high-intensity laser pulses: dependence of rate on peak-pulse intensity and chemical structure,V1559,202-213(1991)
See Subramanian, P.: V1559,461-469(1991)

Hoyt, Clifford C. See Foukal, Peter: V1493,72-79(1991)

Hrubowchak, D. M. See Winograd, Nicholas: V1435,2-11(1991)

Hsia, Kangmin
; Cantrell, Cyrus D.: Theory of transient self-focusing,V1497,166-177(1991)

Hsieh, Cheng
; O'Donnell, Timothy P.: Characterization of the dimensional stability of advanced metallic materials using an optical test bench structure,V1533,240-251(1991)

Hsieh, Hung-yu
; Wagner, Jerome F.: Lens for microlithography,V1396,467-472(1991)

Hsieh, Kwang-Chung See Pline, Alexander D.: V1557,222-234(1991)

Hsieh, Robert L. See Pease, R. F.: V1496,234-238(1991)

Hsieh, T. Y. See Jung, K. H.: V1393,240-251(1991)

Hsieh, Y. F. See Harriott, Lloyd R.: V1465,57-63(1991)

Hsu, C. T.
; Li, J. W.; Liu, C. S.; Su, Yan K.; Wu, T. S.; Yokoyama, M.: ZnS:Mn thin film electroluminescent display devices using hafnium dioxide as insulating layer,V1519,391-395(1991)

Hsu, Dahsiung
; Jiao, Jiangzhong; Tao, Huiying; Long, Pin: Recent developments on holography in China,V1238,13-17(1991)
See Long, Pin: V1461,270-277(1991)
See Zou, Yunlu: V1238,452-456(1991)

Hsu, Hwei P. See Sun, Huifang: V1605,214-220(1991)

Hsu, Ike C. See Stubbs, David M.: V1533,36-47(1991)

Hsu, L. S.
; Loe, K. F.; Chan, Sing C.; Teh, H. H.: Two-valued neural logic network,V1469,197-207(1991)

Hsu, Long S. See Dew, Douglas K.: V1422,111-115(1991)

Hsu, N. J.
See Singh, Rajendra: V1393,78-89(1991)
See Sinha, Sanjai: V1394,266-276(1991)

Hsu, Ping See Guo, Lu Rong: V1555,300-303(1991)

Hsu, S. See Briscoe, Dennis: V1358,329-336(1991)

Hsu, T. Y. See Chen, Duanjun: V1554A,922-931(1991)

Hsu, Tung M. See Dew, Douglas K.: V1422,111-115(1991)

Hsu, Tzu-min See Huang, Fon-Shan: V1393,172-179(1991)

Hsueh, Chi-Fu T. See Goelz, Stefan: V1542,502-510(1991)

Hsueh, Fu-Lung See Sauer, Donald J.: V1540,285-296(1991)

Hu, Chia-Lun J.
: Fast-digital multiplication using multizero neural networks,V1469,586-591(1991)

Hu, Evelyn L.
: Current trends and issues for low-damage dry etching of optoelectronic devices,V1361,512-522(1991)

Hu, Gongzhu
: Segmentation using range data and structured light,V1381,482-489(1991)

Hu, Hao See Dong, Xiaopeng: V1572,56-60(1991)

Hu, Ji-Wu See Hao, Tianyou: V1584,32-38(1991)

Hu, Ke L. See Zhang, Wei P.: V1519,680-682(1991)

Hu, M. See Lan, Guey-Liu: V1475,184-192(1991)

Hu, Qing See Mears, Carl A.: V1477,221-233(1991)

Hu, Qinghua See Luo, Zhishan: V1554B,523-528(1991)

Hu, Shichuang
; Ye, Miaoyuan; Qu, Gen: FFT measuring method for magneto-optical ac current measurement,V1572,492-496(1991)
See Ye, Miaoyuan: V1572,483-486(1991)

Hu, Shisheng See Han, Lei: V1358,793-803(1991)

Hu, X. Q.
See Balasubramanian, Kunjithapa: V1529,106-114(1991)
See Guenther, Karl H.: V1485,240-244(1991)

Hu, Xiaoping See Chen, Chin-Tu: V1445,222-225(1991)

Hu, Yeu-Jent See Chang, Rong-Seng: V1567,216-219(1991)

Hu, Yong-Lin
; Wee, William G.; Gruver, William A.; Han, Chia Y.: Computer vision system for automated inspection of molded plastic print wheels,V1468,653-661(1991)
See Nolan, Adam R.: V1472,157-164(1991)

Hu, Yu
; Li, Xian; Zhu, Dayong; Ye, Naiqun; Pen, Shengyang: 10.6 um laser frequency stabilization system with two optical circuits,V1409,230-239(1991)

Hu, Zhiping
See Xiao, Guohua: V1409,106-113(1991)
See Yang, Darang: V1417,440-450(1991)

Hua, Lifan
; Xie, Gong-Wie; Shaw, David T.; Scott, Peter D.: Resolution enhancement in digital in-line holography,V1385,142-151(1991)
See Xie, Gong-Wie: V1385,132-141(1991)

Hua, Xuelei See Falk, Joel: V1409,83-86(1991)

Hua, Yi-Min See Nie, Chao-Jiang: V1579,264-267(1991)

Hua, Yun See Wen, Lin Ying: V1367,300-302(1991)

Hua, Zhong Y.
; Chen, G. R.; Wang, Z. H.: Organo-metallic thin films for erasable photochromic laser discs,V1519,2-7(1991)
See Zhang, Z. J.: V1519,790-792(1991)

Huang, C. J.
; Su, Yan K.; Leu, R. L.: Studies of InSb metal oxide semiconductor structure fabricated by photo-CVD using Si2H6 and N2O,V1519,70-73(1991)

Huang, Chao H.
; Gheen, Gregory; Washwell, Edward R.: Throughput comparison of optical and digital correlators for automatic target recognition,V1564,427-438(1991)

Huang, Chin M.
; Kosicki, Bernard B.; Theriault, Joseph R.; Gregory, J. A.; Burke, Barry E.; Johnson, Brett W.; Hurley, Edward T.: Quantum efficiency model for p+-doped back-illuminated CCD imager,V1447,156-164(1991)

Huang, Dong-Tsair
   See Chu, Ron: V1465,238-243(1991)
   See Tzeng, Chao H.: V1466,469-476(1991)

Huang, Fon-Shan
   ; Chen, W. S.; Hsu, Tzu-min: Rapid thermal annealing of the through-Ta5Si3 film implantation on GaAs,V1393,172-179(1991)

Huang, Guang L.
   ; Zhang, Jun; Peng, Chuan C.: Experimental study of microwave attenuation of ITO-dielectric recombination film,V1519,179-182(1991)

Huang, H. K.
   ; Kangarloo, Hooshang; Tecotzky, Raymond H.; Cheng, Xin; Vanderweit, Don: Multichannel fiber optic broadband video communication system for monitoring CT/MR examinations,V1444,214-220(1991)
   ; Lim, Art J.; Kangarloo, Hooshang; Eldredge, Sandra L.; Loloyan, Mansur; Chuang, Keh-Shih: Compact open-architecture computed radiography system,V1443,198-202(1991)
   See Cheng, Xin: V1364,204-208(1991)
   See Chuang, Keh-Shih: V1445,496-503(1991)
   See Chuang, Keh-Shih: V1445,341-347(1991)
   See Hayrapetian, Alek: V1446,243-247(1991)
   See Kaabi, Lotfi: V1445,11-23(1991)
   See Lou, Shyh-Liang: V1446,302-311(1991)
   See Morioka, Craig A.: V1445,534-538(1991)
   See Stewart, Brent K.: V1446,141-153(1991)
   See Weinberg, Wolfram S.: V1446,35-39(1991)
   See Wong, Albert W.: V1446,73-80(1991)

Huang, Huang-Cheng See Lee, Ching-Long: V1606,728-734(1991)

Huang, Jay See Vangsness, C. T.: V1424,16-19(1991)

Huang, Jei-Shyong See Lee, Ching-Long: V1606,728-734(1991)

Huang, Jen-Kuang
   ; Shen, Ji-Yao; Taylor, Lawrence W.: Distributed-parameter estimation for NASA Mini-Mast truss through displacement measurements,V1489,266-277(1991)
   See Chen, Chung-Wen: V1489,254-265(1991)

Huang, Jiang P. See Huang, Zong T.: V1519,788-789(1991)

Huang, Jianmin See Yu, Jie: V1420,266-270(1991)

Huang, Jianming See Kumar, Prem: V1376,192-197(1991)

Huang, Kuo-Wei See Ansari, Nirwan: V1606,31-42(1991)

Huang, Liangpu See Wang, Tianmin: V1519,890-900(1991)

Huang, Nanmin See Xiang, Tingyuan: V1572,372-376(1991)

Huang, P. See Chow, P. D.: V1475,42-47(1991)

Huang, Qiang
   ; Caulfield, H. J.: Waveguide holography and its applications,V1461,303-312(1991)
   ; Freeman, Mark O.: Holographic-coordinate-transform-based system for direct Fourier tomographic reconstruction,V1564,644-655(1991)
   See Caulfield, H. J.: V1555,154-158(1991)

Huang, Rong-Ting
   See Cheng, Wood-Hi: V1418,279-283(1991)
   See Jiang, Ching-Long: V1418,261-271(1991)

Huang, Shang-Lian
   See Luo, Fei: V1367,221-224(1991)
   See Zhong, Xian-Xin: V1572,84-87(1991)

Huang, Shao-ming See Fang, Zhen-he: V1572,342-346(1991)

Huang, Thomas S.
   ; Reddy, Subhash C.; Aizawa, Kiyoharu: Human facial motion modeling, analysis, and synthesis for video compression,V1605,234-241(1991)
   See Chen, Chang W.: V1450,231-242(1991)
   See Goldgof, Dmitry B.: V1383,109-121(1991)
   See Palaniappan, K.: V1450,186-197(1991)

Huang, Tizhi
   ; Wagner, Kelvin H.: Coupled-mode analysis of dynamic polarization volume holograms,V1559,377-384(1991)
   ; Wagner, Kelvin H.: Photoanisotropic incoherent-to-coherent conversion using five-wave mixing,V1562,44-54(1991)

Huang, W. See Zhao, Jing-Fu: V1362,135-143(1991)

Huang, Wei
   See Feng, Yi P.: V1519,440-443(1991)
   See Wang, Xi: V1519,740-743(1991)

Huang, Wei-Xu
   ; Wing, Omar: Distortion characteristic of transient signals through bend discontinuity of high-speed integrated curcuits,V1389,199-204(1991)

Huang, Weiping W.
   ; Xu, Chenglin; Chu, S. T.; Chaudhuri, Sujeet K.: Vector beam propagation method based on finite-difference,V1583,268-270(1991)

Huang, Xiao-Sheng See Liu, Zhi-Shen: V1558,379-383(1991)

Huang, Xin F.
   ; Zhang, Xiang D.; Zhu, Wei Y.; Chen, Ying Y.: Study of the microstructures in Ar+ laser crystallized films of a-Si:H for active layer of thin film transistors,V1519,220-224(1991)

Huang, Yang-Tung
   ; Kostuk, Raymond K.: Substrate-mode holograms for board-to-board two-way communications,V1474,39-44(1991)
   See Kostuk, Raymond K.: V1389,515-522(1991)

Huang, Yong L.
   ; Jiang, Ping: Structure of very thin metal film,V1519,142-145(1991)
   See Yang, Bang C.: V1519,725-728(1991)

Huang, Yue-Huai
   ; Mu, Lemin: Study on an optical fibre detector in fabric edge control,V1572,539-542(1991)
   See Cao, Zheng-Ping: V1572,38-41(1991)

Huang, Yung Y. See Dyott, Richard B.: V1482,439-443(1991)

Huang, Yunming G.
   ; Dreizen, Howard M.: Prioritized DCT (discrete cosine transform) image coding,V1396,624-637(1991)
   See Efstratiadis, Serafim N.: V1605,16-25(1991)

Huang, Zhaoming
   ; Wang, Chunhua; Zhang, Jinghua: Measurement of polarization model dispersion and mode-coupling parameter of a polarization-maintaining fiber,V1572,140-143(1991)
   See Zhang, Jinghua: V1572,69-73(1991)

Huang, Zhiqing
   ; Ordonez, I.; Ioannidis, Andreas A.; Langford, C. H.; Lawrence, Marcus F.: Solid state conductivity and photoconductivity studies of an ion-exchange polymer/dye system,V1436,103-113(1991)

Huang, Zong T.
   ; Li, Guo Z.; Zeng, Guang L.; Huang, Jiang P.; Xiong, Guilan: Preparation of superconducting Y-Ba-Cu-O thin films by rf magnetron sputtering,V1519,788-789(1991)

Hubbard, James E. See Burke, Shawn E.: V1532,207-214(1991)

Hubbard, Steven D. See McGoldrick, Elizabeth: V1374,118-125(1991)

Hubel, Paul M.
   : Recent advances in color reflection holography,V1461,167-174(1991)

Huber, David R. See Trisno, Yudhi S.: V1371,8-12(1991)

Huber, Hans L. See Oertel, Heinrich K.: V1465,244-253(1991)

Huber, Martin C.
   ; Timothy, J. G.; Morgan, Jeffrey S.; Lemaitre, Gerard R.; Tondello, Giuseppe; Naletto, Giampiero: Fabrication of toroidal and coma-corrected toroidal diffraction gratings from spherical master gratings using elastically deformable substrates: a progress report,V1494,472-480(1991)
   See Timothy, J. G.: V1343,350-358(1991)

Huber, Peter
   ; Gerlach-Meyer, U.: Micromachined scanning mirrors for laser beam deflection,V1522,135-141(1991)

Huberman, M. L. See Lin, TrueLon: V1540,135-139(1991)

Hubert, Alexis
   ; Hammond, Mark W.: Computer-aided engineering, manufacturing, and testing of extremely fast steering mirrors,V1532,249-260(1991)
   See Price, R. S.: V1388,550-559(1991)

Hubig, Stefan M. See Woodruff, William H.: V1432,205-210(1991)

Hubin, Norbert
   See Gendron, Eric: V1542,297-307(1991)
   See Merkle, Fritz: V1542,283-292(1991)

Hubin, Thomas
   : Acousto-optic color projection system,V1454,313-322(1991)

Huck, Friedrich O.
   See Alter-Gartenberg, Rachel: V1605,745-756(1991)
   See Narayanswamy, Ramkumar: V1385,93-103(1991)
   See Rahman, Zia-ur: V1488,237-248(1991)

Huddleston, James See Ben-Arie, Jezekiel: V1606,2-19(1991)

Hudec, Rene
   ; Valnicek, Boris; Cervencl, J.; Gerstman, T.; Inneman, Adolf; Nejedly, Pavel; Svatek, Lubomir: Recent developments in production of thin x-ray reflecting foils,V1343,162-163(1991)

Hudgens, Jeffrey C.
; Tesar, Delbert; Sklar, Michael E.: SUB-3D high-resolution pose measurement system,V1387,271-282(1991)

Hudson, Bruce S.
: Resonance Raman studies of the peptide bond: implications for the geometry of the electronic-excited state and the nature of the vibronic linewidth,V1403,27-36(1991)

Hudson, C. I. See Bennett, Laura F.: V1470,263-274(1991)

Hudson, Robert D. See Weiss, Howard: V1493,80-90(1991)

Hudson, Tracy D.
; Gregory, Don A.; Kirsch, James C.; Loudin, Jeffrey A.; Crowe, William M.: Optical correlator field test results,V1564,54-64(1991)
; Kirsch, James C.; Gregory, Don A.: Comparison of optically addressed spatial light modulators,V1474,101-111(1991)
See Duffey, Jason N.: V1558,422-431(1991)
See Kirsch, James C.: V1482,69-78(1991)

Hudyma, Russell M.
; Arndt, Thomas D.; Fischer, Robert E.: Optical system for DOC II,V1563,244-254(1991)
; Eigler, Lynne C.: Computer-aided alignment of a grazing-incidence ring resonator for a visible wavelength free-electron laser,V1354,523-532(1991)
See Fischer, Robert E.: V1535,78-88(1991)

Hueber, Jean-Marc
; Fontaine, Bernard L.; Kobhio, M. N.; Delaporte, Philippe C.; Forestier, Bernard M.; Sentis, Marc L.: Parametric study of small-volume long-pulse x-ray preionized XeCl laser with double-discharge and fast magnetic switch,V1503,62-70(1991)
See Kobhio, M. N.: V1397,555-558(1991)**Hudgens**
See Kobhio, M. N.: V1503,88-97(1991)

Hueber, Martin F.
; Leeb, Walter R.; Scholtz, Arpad L.: Realization of heterodyne acquisition and tracking with diode lasers at lambda=1.55 um,V1522,259-267(1991)
; Scholtz, Arpad L.; Leeb, Walter R.: Heterodyne acquisition and tracking in a free-space diode laser link,V1417,233-239(1991)

Huegel, Helmut
See Beck, M.: V1397,769-774(1991)
See Moissl, M.: V1397,395-398(1991)

Huelsmann, Axel
; Kaufel, G.; Raynor, Brian; Koehler, Klaus; Schweizer, T.; Braunstein, Juergen; Schlechtweg, M.; Tasker, Paul J.; Jakobus, Theo F.: Mushroom-shaped gates defined by e-beam lithography down to 80-nm gate lengths and fabrication of pseudomorphic HEMTs with a dry-etched gate recess,V1465,201-208(1991)
See Rosenzweig, Josef: V1362,168-178(1991)

Huenermann, Lucia
; Meyer, Rudolph; Richter, Franz; Schnase, Alexander: Excitation of an excimer laser with microwave resonator,V1503,134-139(1991)

Hueter, David C. See Kwak, J. Y.: V1396,32-44(1991)

Huether, Gerhard See Bopp, Matthias: V1522,199-209(1991)

Huff, David B.
; Blauvelt, Henry A.: Microwave fiber optic link with DFB lasers,V1371,244-249(1991)

Huff, Howard R.
; Weed, Harrison: Experimental assessment of 150-mm P/P+ epitaxial silicon wafer flatness for deep-submicron applications,V1464,278-293(1991)

Huffaker, Diana L.
; Walser, Rodger M.; Becker, Michael F.: Correlation of surface topography and coating damage with changes in the responsivity of silicon PIN photodiodes,V1441,365-380(1991)

Huffman, Donald R. See Sinha, Kislay: V1437,32-35(1991)

Hug, W. See Barron, L. D.: V1403,66-75(1991)

Hugenberg, Keith F.
: New method for doing flat-field intensity calibrations of multiplexed ITT Streak Cameras,V1346,390-397(1991)

Hugenschmidt, Manfred See Paolacci, Sylvie: V1397,705-708(1991)

Huggins, Raymond W.
: Multi-analog track fiber-coupled position sensor,V1367,174-180(1991)
See Hong, C. S.: V1418,177-187(1991)

Hughen, James H.
; Hollon, Kenneth R.: Millimeter wave radar stationary-target classification using a high-order neural network,V1469,341-350(1991)

Hughes, John P. See Flanagan, Kathryn A.: V1549,395-407(1991)

Hughes, Joseph L.
; Pahlajrai, Prem: Effects of packaging and interconnect technology on testability of printed wiring boards,V1389,87-97(1991)
See Fulton, Robert E.: V1389,144-155(1991)

Hughes, Raymond See Selmi, Fathi: V1489,97-107(1991)

Hughes, Robert C. See Carson, Richard F.: V1378,84-94(1991)

Hughes, Simon H.
; Slocum, Robert E.: Development of a low-cost computed tomography image processing system,V1396,575-581(1991)
See Hoeft, Gregory L.: V1396,638-645(1991)

Hughes, T. W. See Aviles, Walter A.: V1388,587-597(1991)

Hughett, Paul
: Programmable command interpreter to automate image processing of IR thermography,V1467,416-426(1991)
: Projectile velocity and spin rate by image processing of synchro-ballistic photography,V1346,237-248(1991)

Hughlett, R. E.
; Cooper, Keith A.: Video-based alignment system for x-ray lithography,V1465,100-110(1991)

Hugo, Norman E. See Wider, Todd M.: V1422,56-61(1991)

Hugsted, B.
; Gjonnes, K.; Tafto, J.; Gjonnes, Jon; Matsuhata, H.: Determination of electrostatic potentials and charge distributions in bulk and at interfaces by electron microscopy techniques,V1361,751-757(1991)

Huguenin, Robert L.
; Tahmoush, Donald J.: Remote spectral fingerprinting for law enforcement,V1479,403-411(1991)
; Tahmoush, Donald J.: Small-target acquisition and typing by AASAP,V1481,64-72(1991)

Huh, Oscar K.
; Roberts, Harry H.; Rouse, Lawrence J.: Remote sensing of coastal environmental hazards,V1492,378-386(1991)

Huignard, Jean-Pierre
See Ayral, Jean-Luc: V1500,81-92(1991)
See Figue, J.: V1564,550-561(1991)

Hulburd, William G. See Wizinowich, Peter L.: V1542,148-158(1991)

Hull, Chris S. See Maxwell, Jonathan: V1354,277-285(1991)

Hull, Patricia G.
; Hunt, Arlon J.; Quinby-Hunt, Mary S.; Shapiro, Daniel B.: Coupled-dipole approximation: predicting scattering by nonspherical marine organisms,V1537,21-29(1991)
See Shapiro, Daniel B.: V1537,30-41(1991)

Hull, R. J. See Reilly, James P.: V1397,339-354(1991)

Hull, Robert See Windt, David L.: V1343,292-308(1991)

Hulliger, J.
See Sutter, Kurt: V1560,296-301(1991)
See Sutter, Kurt: V1560,290-295(1991)

Humbel, William D. See Vandenberg, Donald E.: V1542,534-542(1991)

Humblet, Pierre A.
See Azizoglu, Murat: V1579,168-178(1991)
See Hamdy, Walid M.: V1579,184-194(1991)

Humenberger, Josef See Gresslehner, Karl-Heinz: V1361,1087-1093(1991)

Hummel, John R. See Conant, John A.: V1486,217-230(1991)

Hummel, Steven R. See Hur, Jung H.: V1378,95-100(1991)

Humphrey, Dean See Knight, Stephen E.: V1464,119-126(1991)

Humphreys, T. P. See Nemanich, Robert J.: V1437,2-12(1991)

Humphries, Colin M. See Ettedgui-Atad, Eli: V1532,241-248(1991)

Humphries, Mark R. See MacAndrew, J. A.: V1500,172-176(1991)

Humphries, Stanley
; Babcock, Steven; Wilson, J. M.; Adler, Richard A.: Scanning beam switch experiment for intense rf power generation,V1407,512-523(1991)

Hunakova, L. See Shvec, Peter: V1403,635-637(1991)

Hung, Hing-Loi A.
; Li, Ming-Guang; Lee, Chi H.: Optical techniques for microwave monolithic circuit characterization,V1476,276-281(1991)
; Mahle, Christoph E.: Performance enhancement in future communications satellites with MMIC technology insertion,V1475,212-222(1991)
See Meulenberg, A.: V1475,280-285(1991)

Hung, Hsien-Sen
; Eray, Mete: Model-based flaw reconstruction using limited-view x-ray projections and flawless prototype image,V1550,34-45(1991)
See Bilgen, Mehmet: V1569,260-268(1991)

Hung, Jacklyn See Palcic, Branko: V1448,113-117(1991)

Hung, Po-Chieh
: Colorimetric calibration for scanners and media,V1448,164-174(1991)

Hung, Yau Y.
: Electronic shearography versus ESPI in nondestructive evaluation,V1554B,692-700(1991)
: Recent development in practical optical nondestructive testing,V1554A,29-45(1991); 1554B,29-45(1991)
; Tang, Shou-Hong; Jin, Guofan; Zhu, Qiuming: Synchronous phase-extraction technique and its applications,V1332,738-747(1991)
; Tang, Shou-Hong; Zhu, Qiuming: Three-dimensional surface inspection using interferometric grating and 2-D FFT-based technique,V1332,696-703(1991)
; Zhu, Qiuming; Shi, Dahuan; Tang, Shou-Hong: Real-time edge extraction by active defocusing,V1332,332-342(1991)
See Hathaway, Richard B.: V1554B,725-735(1991)
See Long, Kah W.: V1554A,116-123(1991)
See Shi, Dahuan: V1554A,680-689(1991)
See Tang, Shou-Hong: V1332,731-737(1991)

Hungenahally, Suresh K. See Gupta, Madan M.: V1382,87-98(1991)

Hunger, Hans E.
; Petzoldt, Stefan; Pietsch, H.; Reif, J.; Matthias, Eckart: Multishot ablation of thin films: sensitive detection of film/substrate transition by shockwave monitoring,V1441,283-286(1991)
; Pietsch, H.; Petzoldt, Stefan; Matthias, Eckart: Thin-film selective multishot ablation at 248 nm,V1598,19-26(1991)

Hunkler, L. T. See Johnson, C. B.: V1346,340-370(1991)

Hunkler, Sean See Ostrout, Wayne H.: V1392,151-164(1991)

Hunt, Arlon J.
See Hull, Patricia G.: V1537,21-29(1991)
See Shapiro, Daniel B.: V1537,30-41(1991)

Hunt, Bobby R.
: Imagery super-resolution: emerging prospects,V1567,600-608(1991)
; Ryan, Thomas W.; Sementilli, P.; DeKruger, D.: Interactive tools for assisting the extraction of cartographic features,V1472,208-218(1991)
; Ryan, Thomas W.; Gifford, E.: Interactive tools for extraction of cartographic calibration data from aerial photography,V1472,190-200(1991)
See Rosten, David P.: V1472,118-127(1991)

Hunt, Brian D.
; Foote, Marc C.; Bajuk, Louis J.; Vasquez, R. P.: YBa2Cu3O7-x/Au/Nb device structures,V1394,89-95(1991)
See Chew, Wilbert: V1477,95-100(1991)
See Foote, Marc C.: V1477,192-196(1991)

Hunt, J. D. See Scribner, Dean A.: V1541,100-109(1991)

Hunt, John T.
: High-peak-power Nd:glass laser facilities for end users,V1410,2-14(1991)
See Murray, John R.: V1410,28-39(1991)
See Rainer, Frank: V1441,247-254(1991)

Hunt, Martin A. See Gleason, Shaun S.: V1386,135-144(1991)

Hunter, John G.
: Laparoscopic use of laser and monopolar electrocautery,V1421,173-183(1991)

Hunter, L. W. See Bandy, Donna K.: V1497,142-152(1991)

Huntington, P. See Turcu, I. C.: V1503,391-405(1991)

Huntley, Dave
: Cluster tool software and hardware architecture,V1392,315-330(1991)

Huntley, Jonathan M.
; Field, John E.: High-speed laser speckle photography,V1554A,756-765(1991)
See Whitworth, Martin B.: V1358,677-682(1991)
See Whitworth, Martin B.: V1554B,282-288(1991)

Huntsberger, Terrance L.
: Comparison of techniques for disparate sensor fusion,V1383,589-595(1991)
See Soh, Young-Sung: V1381,561-573(1991)

Hunziker, Heinrich E. See de Vries, Mattanjah S.: V1437,129-137(1991)

Hur, Jung H.
; Hadizad, Peyman; Zhao, Hanmin; Hummel, Steven R.; Dapkus, P. D.; Fetterman, Harold R.; Gundersen, Martin A.: GaAs opto-thyristor for pulsed power applications,V1378,95-100(1991)

Hurditch, Rodney J.
See Hansen, Steven G.: V1463,230-244(1991)
See Honda, Kenji: V1466,141-148(1991)
See Spragg, Peggy M.: V1466,283-296(1991)

Hurley, Edward T. See Huang, Chin M.: V1447,156-164(1991)

Hurwitz, Mark See Harada, Tatsuo: V1545,2-10(1991)

Husain, Anis See Guha, Aloke: V1389,375-385(1991)

Husbands, Charles R. See Girard, Mary M.: V1563,156-167(1991)

Husoy, John H. See Nickel, Michael: V1605,963-971(1991)

Hussain, Mukhtar
; Bedi, Jatinder S.: Performance evaluation of different neural network training algorithms in error control coding,V1469,697-707(1991)
; Bedi, Jatinder S.: Reed-Solomon encoder/decoder application using a neural network,V1469,463-471(1991)

Huston, Dryver R.
: Smart civil structures: an overview,V1588,182-188(1991)

Huston, Lisa A. See Gustafson, Terry L.: V1403,545-554(1991)

Hutcheson, Lynn D.
; Kahn, David A.; eds.: *Fiber Optics in the Subscriber Loop*,V1363(1991)

Hutchings, David C. See Sheik-Bahae, Mansoor: V1441,430-443(1991)

Hutchinson, Seth A. See LaValle, Steven M.: V1569,2-15(1991)

Hutchison, Jerry L. See Shechtman, Dan: V1534,26-43(1991)

Hutley, M. C.
: Blazed zone plates for the 10-um spectral region,V1574,2-7(1991)
: Microlens arrays in Europe,V1544,134-137(1991)
: Relative merits of bulk and surface relief diffracting components,V1574,294-302(1991)

Hutt, Daniel L.
; Bissonnette, Luc R.; Durand, Louis-Gilles: Multiscattered lidar returns from atmospheric aerosols,V1487,250-261(1991)
; Oman, James: Visible extinction measurements in rain and snow using a forward-scatter meter,V1487,312-323(1991)

Hutter, Frank See Brandenburg, Albrecht: V1510,148-159(1991)

Huttlin, George A.
; Conrad, D. B.; Gavnoudias, S.; Judy, Daniel C.; Lazard, Carl J.; Litz, Marc S.; Pereira, Nino R.; Weidenheimer, Douglas M.: Development of the Aurora high-power microwave source,V1407,147-158(1991)
See Lazard, Carl J.: V1407,167-171(1991)
See Litz, Marc S.: V1407,159-166(1991)

Hutton, J. See Carbone, Joseph: V1447,229-242(1991)

Hutton, Richard S. See Taylor, Gary N.: V1343,258-273(1991)

Huttunen, Juhani See Saarinen, Jyrki V.: V1555,128-137(1991)

Huttunen, Pekka See Sipila, Heikki: V1549,246-255(1991)

Hwang, Bao See Brunsvold, William R.: V1466,368-376(1991)

Hwang, Chi-Hung See Wang, Wei-Chung: V1554A,124-135(1991)

Hwang, Humor
; Haddad, Richard A.: New algorithms for adaptive median filters,V1606,400-407(1991)

Hwang, J. Y. See Bowers, C. R.: V1435,36-50(1991)

Hwang, Kiu C.
; Li, Sheng S.; Kao, Yung C.: Novel high-speed dual-wavelength InAlAs/InGaAs graded superlattice Schottky barrier photodiode for 0.8- and 1.3-um detection,V1371,128-137(1991)

Hwang, Lih-Tyng
; Turlik, Iwona: Skin effect in high-speed ULSI/VLSI packages,V1390,249-260(1991)
See Rinne, Glenn A.: V1389,110-121(1991)

Hwang, Shu-Yuen
: State-space search as high-level control for machine vision,V1386,145-156(1991)
See Tseng, Chien-Chao: V1384,257-268(1991)

Hwang, Vincent S. See Castellano, Anthony R.: V1387,343-350(1991)

Hyakumura, Kazushi See Tsuchida, Hirofumi: V1354,246-251(1991)

Hydes, A. J.
See Deshmukh, V. G.: V1392,352-360(1991)
See Hope, D. A.: V1392,185-195(1991)

Hyer, Paul V. See Johnson, Edward J.: V1557,168-179(1991)

Hyun, O. B.
See Alp, Esen E.: V1345,137-145(1991)
See Mini, Susan M.: V1345,260-269(1991)

Hyvarinen, Timo S.
; Sumen, Juha: Scattering in paper coatings,V1530,325-334(1991)
See Malinen, Jouko: V1510,204-209(1991)
See Malinen, Jouko: V1533,122-128(1991)

Hyzer, Jim B.
; Shih, J.; Rowlands, Robert E.: Interferometric moire analysis of wood and paper structures,V1554B,371-382(1991)

Iannarilli, Frank J.
; Wohlers, Martin R.: Comparison of three efficient-detail-synthesis methods for modeling using under-sampled data,V1486,314-324(1991)
; Wohlers, Martin R.: Dynamic end-to-end model testbed for IR detection algorithms,V1483,66-76(1991)
; Wohlers, Martin R.: End-to-end model for detection performance evaluation against scenario-specific targets,V1488,226-236(1991)
; Wohlers, Martin R.: End-to-end scenario-generating model for IRST performance analysis,V1481,187-197(1991)

Iannone, Eugenio See Betti, Silvelo: V1579,100-111(1991)

Iannone, Mark A.
; Salt, Kimberly L.; Scott, Gary W.; Yamashita, Tomihiro: Photochemical hole burning in rigidly coupled polyacenes,V1559,172-183(1991)

Ibarra, J. See Ojeda-Castaneda, Jorge: V1500,252-255(1991)

Ibel, Konrad
; Matull, Ralph; Rupp, Romano A.; Eschkoetter, Peter; Hehmann, Joerg: Light-induced volume-phase holograms for cold neutrons,V1559,393-402(1991)

Ichikawa, Atsushi See Seya, Eiichi: V1499,269-273(1991)

Ichikawa, Fumihiko See Torao, Akira: V1358,843-850(1991)

Ichikawa, Naoki See Kurokawa, Haruhisa: V1332,643-654(1991)

Ichimura, Tsutomu See Inaba, Humio: V1399,108-115(1991)

Ichioka, Yoshiki
See Hasegawa, Akira: V1621,374-379(1991)
See Hasegawa, Akira: V1558,414-421(1991)
See Kakizaki, Sunao: V1505,199-205(1991)

Ide, M. H. See Cantella, Michael J.: V1540,634-652(1991)

Ideker, Raymond E. See Laxer, Cary: V1444,190-195(1991)

Idell, Paul S.
: Resolution limits for high-resolution imaging lidar,V1416,250-259(1991)
See Voelz, David G.: V1416,260-265(1991)

Ideno, S. See Kasuya, Koichi: V1397,67-70(1991)

Idnurm, S. J. See Aben, Hillar K.: V1554A,298-309(1991)

Idzerda, Y. U. See Tjeng, L. H.: V1548,160-167(1991)

Iffrig, C. D. See Lowry, Jay H.: V1330,132-141(1991)

Ifju, Peter G.
: Evaluation of a new electrical resistance shear-strain gauge using moire interferometry,V1554B,420-428(1991)

Iga, Kenichi See Tamanuki, Takemasa: V1361,614-617(1991)

Igarashi, Tadanao See Kanemaru, Toyomi: V1496,118-123(1991)

Ignatosyan, S. S.
; Simonov, V. P.; Stepanov, Boris M.: Electro-optical transducer employing liquid crystal target for processing images in real-time scale,V1358,100-108(1991)

Ignatov, A. M. See Degtyareva, V. P.: V1358,546-548(1991)

Ignatyev, Alexander V.
; Galkin, S. L.; Nikolaev, V. A.; Strigalev, V. E.: Fiber optic interferometric sensors using multimode fibers,V1584,336-345(1991)

Igoshin, Valery I.
; Pichugin, Sergei: Short-pulsed H2-F2 amplifier initiated by optical discharge,V1501,150-152(1991)

Ihlemann, Juergen
; Bolle, Matthias; Luther, Klaus; Troe, Juergen: Near-UV laser ablation of doped polymers,V1361,1011-1019(1991)

Iizuka, Masahiro See Fujii, H.: V1397,213-220(1991)

Iizuka, Takashi See Niibe, Masahito: V1343,2-13(1991)

Ikebata, Sigeki See Sasakawa, Koichi: V1386,265-272(1991)

Ikeda, Shigeyuki See Ogura, Toshihiro: V1443,153-157(1991)

Ikegami, Tetsuhiko
; Nakahara, Motohiro: Optical fiber amplifiers,V1362,350-360(1991)

Ikehara, Curtis See Cole, Robert E.: V1457,111-119(1991)

Ikeuchi, Katsushi
See Delingette, Herve: V1570,21-30(1991)
See Delingette, Herve: V1570,104-115(1991)

Ikeya, Morihiro See Kobayashi, Akira: V1449,148-156(1991)

Ikoma, Toshiaki
See Hirakawa, Kazuhiko: V1361,255-261(1991)
See Isshiki, Hideo: V1361,223-227(1991)

Il'in, Y. V. See Garbuzov, Dmitriy Z.: V1418,386-393(1991)

Ilie, Stiharu-Alexe
: Predictive control for 4-D guidance,V1482,491-501(1991)

Ilieva, R. See Kovatchev, Methodi: V1621,259-267(1991)

Illenyi, Andras
; Jessel, M.: Holophonics: a spread-out of the basic ideas on holography into audio-acoustics,VIS08,39-52(1991)

Illingworth, Garth
: Next-generation space telescope: a large UV-IR successor to HST,V1494,86-97(1991)

Ilyin, Stanislav V. See Abrukov, Victor S.: V1554B,540-543(1991)

Ilyna, T. A. See Feldman, G. G.: V1358,497-502(1991)

Imai, Akira See Terasawa, Tsuneo: V1463,197-206(1991)

Imai, Masaaki M.
; Yano, T.; Ohtsuka, Yoshihiro: High-frequency fiber optic phase modulator using piezoelectric polymer coating,V1371,13-20(1991)

Imai, Tohru See Matsuoka, Masaru: V1549,2-8(1991)

Imam, H.
See Kirk, Andrew G.: V1574,121-132(1991)
See Kirk, Andrew G.: V1621,320-327(1991)

Imamura, Nobutake See Burke, James J.: V1499(1991)

Imanaka, Koichi
See Goto, Hiroshi: V1544,272-281(1991)
See Ogata, Shiro: V1544,92-100(1991)

Imanishi, Hideki See Oikawa, Masahiro: V1544,226-237(1991)

Imanishi, Yasuo
; Itoh, Yuzo; Kakuta, Atsushi; Mukoh, Akio: Nonlinear optical properties of xanthone derivatives,V1361,570-580(1991)

Imatake, Shigenori
See Kuribayashi, Shizuma: V1397,439-443(1991)
See Noda, Osama: V1397,427-432(1991)
See Sato, Shunichi: V1397,421-425(1991)
See Sato, Shunichi: V1397,433-437(1991)

Imbert, Michel P. See Tarabelli, D.: V1397,523-526(1991)

Imedadze, Theodore S.
; Kakichashvili, Shermazan D.: Scanning method of receiving high-effective reflective holograms on bichromate gelatin,V1238,439-441(1991)

Imen, Kamran
; Lee, Changhun H.; Yang, Y. Y.; Allen, Susan D.; Ghosh, Anjan K.: Laser-fabricated fiber optical taps for interconnects and optical data processing devices,V1365,60-64(1991)
See Ghosh, Anjan K.: V1371,170-181(1991)
See Lee, Shyan J.: V1598,2-12(1991)

Imenkov, Albert N.
See Mikhailova, Maya P.: V1361,674-685(1991)
See Yakovlev, Yurii P.: V1510,170-177(1991)

Imhoff, Ralf See Franek, Joachim: V1397,791-795(1991)

Imirowicz, Thomas P. See Shepard, Steven M.: V1467,234-238(1991)

Impagnatiello, Fabrizio
See Armenise, Mario N.: V1374,86-96(1991)
See Armenise, Mario N.: V1562,160-171(1991)

Impedovo, Sebastiano
; Dimauro, Giovanni; Pirlo, Giuseppe: New decision tree algorithm for handwritten numerals recognition using topological features,V1384,280-284(1991)

Inaba, Humio
; Toida, Masahiro; Ichimura, Tsutomu: Optical computer-assisted tomography realized by coherent detection imaging incorporating laser heterodyne method for biomedical applications,V1399,108-115(1991)

Inada, Koichi
: Special optical fibers for sensors,V1572,163-168(1991)
See Shiota, Alan T.: V1504,90-97(1991)

Inagaki, Keizo
; Nohara, Mitsuo; Araki, Ken'ichi; Fujise, Masayuki; Furuhama, Yoji: Free-space simulator for laser transmission,V1417,160-169(1991)

Inagaki, Yoshinori See Kinoshita, Katsuyuki: V1358,490-496(1991)

Inami, Takashi See Oyama, Yoshiro: V1444,413-423(1991)

Inamoto, Kazuo See Umeda, Tokuo: V1446,199-210(1991)

Inamura, Kiyonari See Umeda, Tokuo: V1446,199-210(1991)

Inaoka, Noriko
; Suzuki, Hideo; Mori, Masaki; Takabatake, Hirotsugu; Suzuki, Akira: Three-dimensional reconstruction of pulmonary blood vessels by using anatomical knowledge base,V1450,2-12(1991)
See Suzuki, Hideo: V1450,99-107(1991)

Inatomi, Yuko
; Kuribayashi, Kazuhiko: In-situ measurement technique for solution growth in compound semiconductors,V1557,132-139(1991)

Ince, Ilhan See Bryant, Keith: V1457,120-132(1991)

Indin, Ronald See Carson, John C.: V1541,232-239(1991)

Indutnyi, I. Z.
; Robur, I.; Romanenko, Peter F.; Stronski, Alexander V.: Holographic diffraction gratings on the base of chalcogenide semiconductors,V1555,243-253(1991)

Ingers, Joakim P.
; Thibaudeau, Laurent: Theory and experiment as tools for assessing surface finish in the UV-visible wavelength region,V1400,178-185(1991)

Inglesakis, Georges
See Astic, Dominique: V1397,713-716(1991)
See Prat, Ch.: V1397,701-704(1991)

Ingold, Joseph P.
; Sun, Mei H.; Bigelow, Russell N.: Fiber optic pressure sensor,V1589,83-89(1991)
See Brown, Gair D.: V1366,351-360(1991)
See Brown, Gair D.: V1589,58-68(1991)

Ingraffea, Anthony R. See Andrews, John R.: V1461,110-123(1991)

Inguva, Ramarao See Haus, Joseph W.: V1497,350-356(1991)

Ingwall, Richard T.
; Adams, Timothy: Hologram: liquid-crystal composites,V1555,279-290(1991)

Inigo, Rafael M.
See Arrue, Begona C.: V1469,420-431(1991)
See Himes, Glenn S.: V1469,671-682(1991)
See Minnix, Jay I.: V1606,241-251(1991)
See Narathong, Chiewcharn: V1452,523-531(1991)
See Xu, Qing: V1471,378-389(1991)

Inneman, Adolf See Hudec, Rene: V1343,162-163(1991)

Ino, Tomomi See Yatagai, Toyohiko: V1554B,646-649(1991)

Inochkin, Mickle V. See Azarenkov, Aleksey N.: V1409,154-164(1991)

Inokuchi, Chikashi See Ohara, Shunji: V1499,187-194(1991)

Inokuchi, Ikuo
; Kawakami, Shinichiro; Maeta, Manabu; Masuda, Yu: Evaluation of facial palsy by moire topography,V1429,39-45(1991)

Inoue, Hiroyuki See Ohashi, T.: V1549,9-19(1991)

Inoue, Kouichi
See Iwasaki, Nobuo: V1490,216-221(1991)
See Iwasaki, Nobuo: V1490,192-199(1991)

Inoue, M. See Nagai, Haruhiko: V1397,31-36(1991)

Inoue, Morio See Watanabe, Hisashi: V1463,101-110(1991)

Inoue, Nobuo See Ogura, Toshihiro: V1443,153-157(1991)

Inoue, Seiki
: Object extraction method for image synthesis,V1606,43-54(1991)

Inoue, Yasuaki See Yamaguchi, Takao: V1418,363-371(1991)

Ioannidis, Andreas A. See Huang, Zhiqing: V1436,103-113(1991)

Ioffe, S. A. See Vishnevsky, G. I.: V1448,69-72(1991)

Ionescu, Dan
; Damerji, Tayeb: Autoregressive identification method for partially occluded industrial object recognition,V1406,40-41(1991)

Ionin, A. A.
: High-power pulsed and repetitively pulsed electron-beam-controlled discharge CO laser systems,V1502,95-102(1991)
; Kotkov, A. A.; Minkovsky, M. G.; Sinitsin, D. V.: Supersonic electron beam controlled discharge CO laser,V1397,453-456(1991)
See Frolov, K.: V1397,461-468(1991)

Ippoliti, A, G. See Albanese, Andres: V1364,320-326(1991)

Irani, Keki B. See Cheng, Jie: V1392,373-384(1991)

Irick, Steven C. See Goad, Joseph H.: V1410,107-115(1991)

Irie, Mitsuru
; Takeshita, Nobuo; Fujita, Teruo; Kime, Kenjiro: New tracking method for two-beam optical heads using continuously grooved disks,V1499,360-365(1991)

Irimescu, Dorin I. See Popescu, Ion M.: V1361,1041-1047(1991)

Irion, Fredrick W. See Gunson, Michael R.: V1491,335-346(1991)

Iriyama, Keiji See Katayama, Norihisa: V1403,147-149(1991)

Iroshnikov, N. G. See Vorontsov, Michael A.: V1621,292-298(1991)

Irwin, Mary J. See Owens, Robert M.: V1566,252-262(1991)

Isaacson, Joel S.
: Fast access to reduced-resolution subsamples of high-resolution images,V1460,80-91(1991)

Isaacson, William B.
: Global status of diffraction optics as the basis for an intraocular lens,V1529,71-83(1991)

Isaev, Urkaly T.
; Akaev, Askar A.; Kutanov, Askar A.: Photothermoplastic spatial filters for optical pattern recognition,V1507,198-201(1991)

Isajkin, A. S. See Savostjanov, V. N.: V1554A,380-386(1991)

Isaksen, Tron
: Estimation of prospective locations in mature hydrocarbon producing areas,V1452,270-291(1991)

Ishaq, Naseem
; Taylor, Kenneth; Steliou, Kypros; Delaney, William: Extraction of the foveal center and lesion boundary from fundus images,V1381,153-159(1991)

Ishibashi, Hiromichi
; Tanaka, Shin-ichi; Moriya, Mitsuro: Focus-error detection from far-field image flow,V1499,340-347(1991)

Ishibashi, Kenzo See Ohara, Shunji: V1499,187-194(1991)

Ishibashi, Satoshi
; Kishino, Fumio: Color/texture analysis and synthesis for model-based human image coding,V1605,242-252(1991)

Ishibashi, Taka-aki
; Hamaguchi, Hiro-o: Feasibility of luminescence-eliminated anti-Stokes Raman spectroscopy,V1403,555-562(1991)

Ishibashi, Yutaka See Maruyama, Mitsuru: V1605,916-927(1991)

Ishida, Juro
See Nakayama, Masao: V1490,207-215(1991)
See Tanii, Jun: V1490,200-206(1991)

Ishida, M. See Ohashi, T.: V1549,9-19(1991)

Ishida, Shinji See Yamazaki, Satomi: V1466,538-545(1991)

Ishida, Takashi See Ohara, Shunji: V1499,187-194(1991)

Ishigami, Shigenobu See Nakamura, Hiroyuki: V1456,226-238(1991)

Ishihara, Satoshi
: Progress and perspectives on optical information processing in Japan,V1389,68-74(1991); 1390,68-74(1991)
; Mori, Masahiko: Recent research on optical neural networks in Japan,V1621,362-372(1991)

Ishihara, Y. See Kinoshita, Katsuyuki: V1358,490-496(1991)

Ishii, Akira
See Meguro, Shin-Ichi: V1384,27-37(1991)
See Sano, Mutsuo: V1381,101-110(1991)
See Yasuoka, Koichi: V1412,32-37(1991)

Ishii, Masanori See Suzuki, Shinzoh: V1374,126-131(1991)

Ishii, Yoshikazu
See Takenaka, Hisataka: V1345,180-188(1991)
See Takenaka, Hisataka: V1345,213-224(1991)

Ishikawa, Ken
; Takagi, S.; Okamoto, N.; Kakizaki, K.; Sato, S.; Goto, T.: 2.5 kHz high-repetition-rate XeCl excimer laser,V1397,55-58(1991)

Ishikawa, Masayoshi See Kobayashi, Akira: V1449,148-156(1991)

Ishikawa, Masayuki M.
; Itaya, Kazuhiko; Okajima, Masaki; Hatakoshi, Gen-ichi: High-power visible semiconductor lasers,V1418,344-352(1991)

Ishikawa, Rikio See Fujii, Takashi: V1412,50-57(1991)

Ishikawa, S. See Yamanaka, Yutaka: V1499,263-268(1991)

Ishikawa, Toshiharu
; Kuwabara, Y.; Koseki, Kenichi; Yamaoka, Tsuguo: High-sensitive photopolymer for large-size holograms,V1461,73-78(1991)

Ishikawa, Toshio See Yoshida, Yoshio: V1401,58-65(1991)

Ishimaru, Akira
: Pulse propagation in random media,V1558,127-129(1991)
; Chen, Jei S.: Scattering from very rough metallic and dielectric surfaces and enhanced backscattering,V1558,182-190(1991)

Ishitsuka, Takeshi See Yamagishi, Yasuo: V1461,68-72(1991)

Ishiwata, Naoyuki
; Furukawa, Takao: Fabrication of phase-shifting mask,V1463,423-433(1991)
See Yanagishita, Yuichiro: V1463,207-217(1991)

Ishizuka, Hiroshi
; Yee, K.; Fisher, Amnon; Rostoker, Norman: Plasma betatron without gas breakdown,V1407,442-455(1991)

Ishizuka, S. See Maeda, Koichi: V1536,138-148(1991)

Islam, M. N.
See Ambia, M. G.: V1536,222-232(1991)
See Sobhan, M. A.: V1536,246-257(1991)

Islam, Mohammed N. See Sauer, Jon R.:V1579,84-97(1991)

IslamRaja, M. M. See McVittie, James P.:V1392,126-138(1991)

Isobe, Hiroshi
; Sato, Eiichi; Shikoda, Arimitsu; Takahashi, Kei; Tamakawa, Yoshiharu; Yanagisawa, Toru: High-intensity soft-flash x-ray generator utilizing a low-vacuum diode,V1358,471-478(1991)
; Sato, Eiichi; Kimura, Shingo; Tamakawa, Yoshiharu; Yanagisawa, Toru; Honda, Keiji; Yokota, Yoshiharu: Repetitive flash x-ray generator utilizing an enclosed-type diode with a ring-shaped graphite cathode,V1358,201-208(1991)
See Sato, Eiichi: V1358,146-153(1991)
See Sato, Eiichi: V1358,193-200(1991)
See Sato, Eiichi: V1358,479-487(1991)
See Sato, Eiichi: V1358,462-470(1991)
See Shikoda, Arimitsu: V1358,154-161(1991)

Isobe, T.
See Flanagan, Kathryn A.: V1549,395-407(1991)
See Markert, Thomas H.: V1549,408-419(1991)

Isogai, Fumihiko See Sasakawa, Koichi: V1386,265-272(1991)

Isoldi, Felice See Vanni, Paolo: V1456,300-309(1991)

Isshiki, Hideo
; Kobayashi, Hitoshi; Yugo, Shigemi; Saito, Riichiro; Kimura, Tadamasa; Ikoma, Toshiaki: Emission of the 1.54um Er-related peaks by impact excitation of Er atoms in InP and its characteristics,V1361,223-227(1991)

Isuzugawa, Kohji
; Horiuchi, Makoto; Okumura, Yoshiyuki: Focusing of shock waves in water and its observation by the schlieren method,V1358,1003-1010(1991)

Itano, W. M. See Bergquist, James C.: V1435,82-85(1991)

Itaya, Kazuhiko See Ishikawa, Masayuki M.: V1418,344-352(1991)

Ito, Hiroshi
ed.: *Advances in Resist Technology and Processing VIII*,V1466(1991)
; Schildknegt, Klaas; Mash, Eugene A.: Negative chemical amplification resist systems based on polyhydroxystyrenes and N-substituted imides or aldehydes,V1466,408-418(1991)
See Sooriyakumaran, Ratna: V1466,419-428(1991)

Ito, Hisao See Miyake, Hiroyuki: V1448,150-156(1991)

Ito, John See Gleckman, Philip: V1528,163-168(1991)

Ito, Katsunori
; Shibata, Yasumasa; Honda, Katsunori; Tsukahara, Makoto; Kojima, Kotaro; Kimura, Masakatsu; Fujisima, Toshihiko; Yoshinaga, Kazuomi; Kanai, Toshitaka; Shimizu, Keijiro: Development of a high-performance 86-mm MO disk by using polycarbonate substrate,V1499,382-385(1991)

Ito, Kenji See Watanabe, Mikio: V1452,27-36(1991)

Ito, Mabo R. See McFee, John E.: V1567,42-52(1991)

Ito, Masuo See Kinoshita, Katsuyuki: V1358,490-496(1991)

Ito, Osamu See Shimamoto, Masayoshi: V1499,393-400(1991)

Ito, Seiki See Gozdz, Antoni S.: V1466,200-205(1991)

Ito, Takeharu See Yuuki, Hayato: V1592,2-11(1991)

Ito, Yasunobu See Oda, Ichiro: V1431,284-293(1991)

Itoh, Haruyasu See Kamemaru, Shun-ichi: V1564,143-154(1991)

Itoh, Katsuyuki
; Yamanaka, Koji; Nozue, Hiroshi; Kasama, Kunihiko: Dissolution kinetics of high-resolution novolac resists,V1466,485-496(1991)

Itoh, Kazuyoshi
See Hasegawa, Akira: V1621,374-379(1991)
See Hasegawa, Akira: V1558,414-421(1991)

Itoh, Koichi See Sato, Koki: V1461,124-131(1991)

Itoh, M. See Yamanaka, Yutaka: V1499,263-268(1991)

Itoh, Tatsuo See Kong, Keon-Shik: V1477,57-65(1991)

Itoh, Yuzo See Imanishi, Yasuo: V1361,570-580(1991)

Itzkan, Irving
See Albagli, Douglas: V1441,146-153(1991)
See Izatt, Joseph A.: V1427,110-116(1991)

Ivanchenko, I. V. See Vertiy, Alexey A.: V1361,1070-1078(1991)

Ivanchenkov, S. N. See Dodd, James W.: V1407,467-473(1991)

Ivanov, A. I. See Vavilov, Vladimir P.: V1467,230-233(1991)

Ivanov, Anatoliy A.
; Koroteev, Nikolai I.; Fishman, A. I.: Coherent holographic Raman spectroscopy of molecules,V1403,174-184(1991)
; Koroteev, Nikolai I.; Mironov, S. F.; Fishman, A. I.: Conformational analysis of organic molecules in liquids with polarization-sensitive coherent anti-Stokes Raman scattering spectroscopy,V1403,243-245(1991)
; Koroteev, Nikolai I.; Fishman, A. I.: Holographic nonlinear Raman spectroscopy of large molecules of biological importance,V1429,132-144(1991); 1432,141-153(1991)

Ivanov, G. A. See Matejec, Vlastimil: V1513,174-179(1991)

Ivanov, I. See Tontchev, Dimitar A.: V1429,76-80(1991)

Ivanov, K. N.
; Krutikov, N. I.; Naumov, S. K.; Pischelin, E. V.; Semenov, V. A.; Stepanov, M. S.; Postovalov, V. E.; Prokhorov, Alexander M.; Schelev, Mikhail Y.: Two-dimensional electron-bombarded CCD readout device for picosecond electron-optical information system,V1358,732-738(1991)
See Borodin, A. M.: V1358,756-758(1991)
See Bryukhnevitch, G. I.: V1449,109-115(1991)

Ivanov, U. V.
; Pravilov, A. M.; Smirnova, L. G.; Vilesov, A. F.: Kinetics and yield of Xe excitation as third body in the process of N(4S) atom recombination,V1397,181-184(1991)

Ivanov, V. Y. See Degtyareva, V. P.: V1358,546-548(1991)

Ivanov, Vladimir Y.
; Larichev, A. V.; Vorontsov, Michael A.: One-dimensional rotatory waves in the optical systems with nonlinear large-scale field interactions,V1402,145-153(1991)

Ivchenko, N. See Artjushenko, Vjacheslav G.: V1420,157-168(1991)

Ivers, J. D. See Nation, John A.: V1407,32-43(1991)

Iverson, Rolf D. See Chin, Roland T.: VCR36,93-104(1991)

Ivey, Jim
; Fairchild, Scott; Peterson, James R.; Stahl, Charles G.: Feature discrimination using multiband classification techniques,V1567,170-178(1991)

Iwahashi, Masahiro
; Masuda, Shun-ichi: Compaction of color images with arithmetic coding,V1605,844-850(1991)

Iwaki, Tadao
See Haemmerli, Jean-Francois: V1564,275-284(1991)
See Mitsuoka, Yasuyuki: V1564,244-252(1991)

Iwamoto, Mutsuo See Kawano, Sumio: V1379,2-9(1991)

Iwanejko, Leszek
; Pokora, Ludwik J.: Investigations of an excimer laser working with a four-component gaseous mixture He-Kr:Xe-HCl,V1391,105-108(1991)
; Pokora, Ludwik J.; Wolinski, Wieslaw: Prototype of an excimer laser for microprocessing,V1391,98-100(1991)

Iwanowski, Ryszard J.
; Tatarkiewicz, Jakub J.: Defect-induced stabilization of Fermi level in bulk silicon and at the silicon-metal interface,V1361,765-775(1991)

Iwasaki, Nobuo
; Hara, Norikazu; Kajii, Makoto; Tange, Yoshio; Miyachi, Yuji; Sato, Ryota; Inoue, Kouichi: Mission overview of ADEOS program,V1490,192-199(1991)
; Tange, Yoshio; Miyachi, Yuji; Inoue, Kouichi; Kadowaki, Tomoko; Tanaka, Hirokazu; Michioka, Hidekazu: Advanced visible and near-IR radiometer for ADEOS,V1490,216-221(1991)
See Tanii, Jun: V1490,200-206(1991)

Iwasaki, Teruo See Saitou, Norio: V1465,185-191(1991)

Iwata, Fujio See Takahashi, Susumu: V1461,199-205(1991)

Iwata, Koichi See Gustafson, Terry L.: V1403,545-554(1991)

Iwayanagi, Takao
See Hayashi, Nobuaki: V1466,377-383(1991)
See Uchino, Shou-ichi: V1466,429-435(1991)

Iyer, Anand K.
See Casasent, David P.: V1384,324-337(1991)
See Casasent, David P.: V1470,208-219(1991)

Iyer, J. See Kosel, Peter B.: V1541,48-59(1991)

Iyoda, Mitsuhiro
; Murota, Tomoya; Akiyama, Mamoru; Sato, Shunichi: One-dimensional and two-dimensional computer models of industrial CO laser,V1415,342-349(1991)
; Murota, Tomoya; Akiyama, Mamoru; Sato, Shunichi: Two-dimensional computer modeling of discharge-excited CO gas flow,V1397,457-460(1991)

Izatt, Joseph A.
; Albagli, Douglas; Itzkan, Irving; Feld, Michael S.: Study of bone ablation dynamics with sequenced pulses,V1427,110-116(1991)
See Albagli, Douglas: V1441,146-153(1991)

Izawa, Takao
; Yamamura, N.; Uchimura, R.; Hashimoto, I.; Yakuoh, T.; Owadano, Yoshirou; Matsumoto, Y.; Yano, M.: Highly damage-resistant reflectors for 248 nm formed by fluorides multilayers,V1441,339-344(1991)

Izawa, Toshiyuki See Nakayama, Masao: V1490,207-215(1991)

Izmailov, I. A.
See Mel'nikov, Lev Y.: V1397,603-610(1991)
See Mel'nikov, Lev Y.: V1501,144-149(1991)

Izuchi, N. See Kojima, Arata: V1429,162-171(1991)

Izumitani, Tetsuro
; Zou, Xuolu; Wang, Y.: Cr3+ tunable laser glass,V1535,150-159(1991)

Ja, (Yu) Frank H.
: Optical fiber filter comprising a single-coupler fiber ring (or loop) and a double-coupler fiber mirror,V1372,48-61(1991)

Jaanimagi, Paul A.
: Workshop on standards for streak camera characterization,V1346,504-504(1991)
; Hestdalen, C.; Kelly, John H.; Seka, Wolf D.: High-precision measurements of the 24-beam UV-OMEGA laser,V1358,337-343(1991)
; Hestdalen, C.: Streak camera phosphors: response to ultrashort excitation,V1346,443-448(1991)
; Neyer, Barry T.; Shaw, Larry L.; eds.: *Ultrahigh- and High-Speed Photography, Videography, Photonics, and Velocimetry '90*,V1346(1991)
See Johnson, C. B.: V1346,340-370(1991)

Jaaskelainen, Timo
; Kuittinen, Markku: Inverse grating diffraction problems,V1574,272-281(1991)
See Honkonen, Veijo: V1564,43-51(1991)
See Kuittinen, Markku: V1507,258-267(1991)

Jabbar, A. See Zhou, Shi P.: V1519,793-799(1991)

Jabczynski, Jan K.
: Bessel beam generation: theory and application proposal,V1391,254-258(1991)
; Mindak, Marek K.: Investigations of the focal shift of the high-power cw YAG laser beam,V1391,109-116(1991)

Jabour, Bradly See Lou, Shyh-Liang: V1446,302-311(1991)

Jacco, John C. See Rockafellow, David R.: V1561,112-118(1991)

Jackel, Janet L.
: Proton exchange: past, present, and future,V1583,54-63(1991)

Jackman, James J.
: Electron-beam metrology: the European initiative,V1464,71-80(1991)

Jackman, Louise A.
; Lavelle, Stephen P.; Nolan, P. F.: Characteristics of sprinklers and water spray mists for fire safety,V1358,831-842(1991)
See Lavelle, Stephen P.: V1358,821-830(1991)

Jackson, David A.
See Bartlett, Steven C.: V1504,247-250(1991)
See Berwick, Michael: V1584,364-373(1991)
See Berwick, Michael: V1572,157-162(1991)
See Bush, Simon P.: V1584,103-109(1991)
See Chu, Beatrice C.: V1504,251-257(1991)
See Farahi, Faramarz: V1584,53-61(1991)
See Farahi, Faramarz: V1511,234-243(1991)
See Gerges, Awad S.: V1504,233-236(1991)
See Gerges, Awad S.: V1504,176-179(1991)
See Santos, Jose L.: V1511,179-189(1991)
See Santos, Jose L.: V1572,325-330(1991)

Jackson, Howard E. See De Brabander, Gregory N.: V1583,327-330(1991)

Jackson, John E.
; Armentrout, Ben A.; Buck, J. P.; Chenoweth, Amos J.; Elliott, G. A.; Fox, Allen M.; Ganley, J. T.; Gray, W. C.; Jett, L. L.; Johnson, Kevin M.; Kelsey, J. F.; Minelli, R. J.; Rose, G. E.; Shepherd, W. J.; Zino, Joseph D.: Multiple-diode laser optomechanical issues,V1533,75-86(1991)

Jackson, Jonathan A. See Xu, Yian-sun: V1407,648-652(1991)

Jackson, Ray D.
See Biggar, Stuart F.: V1493,155-162(1991)
See Markham, Brian L.: V1493,207-214(1991)
See Nianzeng, Che: V1493,182-194(1991)

Jackson, Ricky A. See Garza, Cesar M.: V1466,616-627(1991)

Jackson, Todd A.
; Bell, Cynthia S.: A 1.3-megapixel-resolution portable CCD electronic still camera,V1448,2-12(1991)

Jackson, W. E. See Mustre de Leon, Jose: V1550,85-96(1991)

Jacobi, Ulrik See Dougherty, Edward R.: V1606,141-152(1991)

Jacobi, William J.
; Jensen, Preben D.; Teneketges, Nicholas J.; Wadsworth, Leo A.: Miniature signal processor for surveillance sensor applications,V1479,111-119(1991)
; Jensen, Preben D.; Teneketges, Nicholas J.; Wadsworth, Leo A.: Miniaturized low-power parallel processor for space applications,V1495,205-213(1991)

Jacoboni, Charles See Boulard, B.: V1513,204-208(1991)

Jacobs, David W. See Clemens, David T.: V1381,30-42(1991)

Jacobson, Benjamin A.
; Gleckman, Philip; Holman, Robert L.; Sagie, Daniel; Winston, Roland: Very high temperature fiber processing and testing through the use of ultrahigh solar energy concentration,V1528,82-85(1991)

Jacobson, K. B. See Arlinghaus, Heinrich F.: V1435,26-35(1991)

Jacobson, Michael R. See Milanovic, Zoran: V1535,160-170(1991)

Jacobson, Nathaniel
; Bender, Walter J.; Feldman, Uri: Alignment and amplification as determinants of expressive color,V1453,70-80(1991)

Jacoby, J. See Tallents, Gregory J.: V1413,70-76(1991)

Jacovitti, Giovanni See Neri, Alessandro: V1565,196-208(1991)

Jacques, Robert See Blackwood, Gary H.: V1542,371-391(1991)

Jacques, Steven L.
ed.: *Laser-Tissue Interaction II*,V1427(1991)
: Principles of phase-resolved optical measurements,V1525,143-153(1991)
; Flock, Stephen T.: Effect of surface boundary on time-resolved reflectance: measurements with a prototype endoscopic catheter,V1431,12-20(1991)
; Flock, Stephen T.: Time-resolved reflectance spectroscopy,V1525,35-40(1991)
; Gofstein, Gary: Laser-flash photographic studies of Er:YAG laser ablation of water,V1427,63-67(1991); 1525,309-312(1991)
; Keijzer, Marleen: Dosimetry for lasers and light in dermatology: Monte Carlo simulations of 577nm-pulsed laser penetration into cutaneous vessels,V1422,2-13(1991)
See Flock, Stephen T.: V1427,77-89(1991)
See Madsen, Steen J.: V1431,42-51(1991)

Jacquier, Bernard See Briancon, Anne-Marie: V1373,9-20(1991)

Jacquot, Pierre M.
See Barillot, Marc: V1554A,867-878(1991)
See Boone, Pierre M.: V1554A,512-521(1991)

Jaeger, Erwin See Hossfeld, Jens: V1574,159-166(1991)

Jaeger, Helmut
See Neger, Theo: V1507,476-487(1991)
See Vukicevic, Dalibor: VIS08,160-193(1991)

Jaeger, Nicolas A.
; Lai, Winnie C.: Y-branch optical modulator,V1583,202-209(1991)

Jaeger, Tycho C. See Ballangrud, Ase: V1521,89-96(1991)

Jaenen, Patrick See Samarakone, Nandasiri: V1463,16-29(1991)

Jaenisch, Holger M.
; Niedzwiecki, Abraham J.; Cernosek, John D.; Johnson, R. B.; Seeley, John S.; Dull, Gerald G.; Leffler, Richard G.: Instrumentation to measure the near-IR spectrum of small fruits,V1379,162-167(1991)
; Spiers, Gary D.: Modifications to the LP-140 pulsed CO2 laser for lidar use,V1411,127-136(1991)

Jaffe, Jules S. See Palowitch, Andrew W.: V1537,128-139(1991)

Jaffe, Steven M.
; Yen, William M.: Detecting dilution-narrowed systems,V1435,252-257(1991)

Jagannath, H. See Venkateswarlu, Putcha: V1332,245-266(1991)

Jagdhold, Ulrich A. See Bauch, Lothar: V1466,510-519(1991)

Jaggard, Dwight L. See Kim, Yun J.: V1558,113-119(1991)

Jaggi, Bruno
   ; Poon, Steven S.; Pontifex, Brian; Fengler, John J.; Marquis, Jacques; Palcic, Branko: Quantitative microscope for image cytometry,V1448,89-97(1991)
   See Palcic, Branko: V1448,113-117(1991)

Jaggi, Sandeep
   : Adaptive notch filter for removal of coherent noise from infrared scanner data,V1541,134-140(1991)
   : Analysis of upper and lower bounds of the frame noise in linear detector arrays,V1541,152-162(1991)
   : Clustering algorithms for a PC-based hardware implementation of the unsupervised classifier for the shuttle ice detection system,V1451,289-297(1991)
   : PC-based hardware implementation of the maximum-likelihood classifier for the shuttle ice detection system,V1452,340-350(1991)

Jahn, Renate
   ; Dressel, Martin; Neu, Walter; Jungbluth, Karl-Heinz: Elaboration of excimer lasers dosimetry for bone and meniscus cutting and drilling using optical fibers,V1424,23-32(1991)

Jahncke, Catherine See Singh, Rajiv K.: V1394,203-213(1991)

Jahns, Juergen
   : Integrated free-space optics,V1354,588-592(1991)
   : Integrated micro-optics for computing and switching applications,V1544,246-262(1991)
   : Integrated packaging of optical backplanes,V1389,523-526(1991)
   ; Brumback, Babette A.: Advances in the computer-aided design of planarized free-space optical circuits: system simulation,V1555,2-7(1991)

Jain, Ajay
   ; Shin, H. K.; Chi, Kai-Ming; Hampden-Smith, Mark J.; Kodas, Toivo T.; Farkas, Janos; Paffett, M. F.; Farr, J. D.: Selective low-temperature chemical vapor deposition of copper from new copper(I) compounds,V1596,23-33(1991)

Jain, Anil K.
   See Arnold, Steven M.: V1396,473-480(1991)
   See Newman, Timothy S.: V1570,250-261(1991)

Jain, Ashit See Ramee, Stephen R.: V1420,199-202(1991)

Jain, Faquir C.
   ; Drake, G.; Chung, C.; Bhattacharjee, K. K.; Cheung, S. K.: Two-dimensional spatial light modulators using polarization-sensitive multiple-quantum-well light valve,V1564,714-722(1991)

Jain, H. See Herman, Warren N.: V1560,206-213(1991)

Jain, Kanti
   : Novel high-resolution large-field scan-and-repeat projection lithography system,V1463,666-677(1991)

Jain, N. See Refenes, A. N.: V1386,62-74(1991)

Jain, Narinder K. See Sirohi, Rajpal S.: V1332,50-55(1991)

Jain, P. C. See Tanwar, Lakhan S.: V1332,877-882(1991)

Jain, Ramesh C.
   ; Roth-Tabak, Yuval: Towards integrated autonomous systems,V1468,188-201(1991)
   See Bartlett, Sandra L.: V1382,3-13(1991)
   See Gallarda, Harry S.: V1464,459-473(1991)
   See Khalili, Payman: V1388,210-221(1991)
   See Liou, Shih-Ping: V1381,447-458(1991)
   See Tirumalai, Arun P.: V1569,28-42(1991)
   See Tirumalai, Arun P.: V1383,345-358(1991)

Jain, Subhash C.
   ; Malhotra, H. S.; Sarebahi, K. N.; Bist, K. S.: Fixed-pattern-noise cancellation in linear pyro arrays,V1488,410-413(1991)

Jain, Surendra K. See Timothy, J. G.: V1343,350-358(1991)

Jain, Vivek See Pramanik, Dipankar: V1596,132-138(1991)

Jain, Yashwant K. See Kamalakar, J. A.: V1478,92-100(1991)

Jaisimha, Mysore Y. See Kanungo, Tapas: V1385,104-112(1991)

Jakobsen, P. See Greenfield, Perry E.: V1494,16-39(1991)

Jakobus, Theo F. See Huelsmann, Axel: V1465,201-208(1991)

Jakoubek, Roger O. See Mount, George H.: V1491,188-193(1991)

Jakowatz, Charles V.
   See Calloway, Terry M.: V1566,353-364(1991)
   See Wahl, Daniel E.: V1567,32-40(1991)

Jalenak, Wayne
   : Emissions monitoring by infrared photoacoustic spectroscopy,V1434,46-54(1991)

Jamankyzov, N. See Akaev, Askar A.: V1621,182-193(1991)

Jambunathan, K. See Dobbins, B. N.: V1554B,586-592(1991)

James, Jonathan H.
   ; Kellett, Bruce J.; Gauzzi, Andrea; Dwir, Benjamin; Pavuna, Davor: Superconducting YBa2Cu3O7 films on Si and GaAs with conducting indium tin oxide buffer layers,V1394,45-61(1991)
   See Pavuna, Davor: V1362,102-116(1991)

James, Tim W. See Bourne, Lincoln C.: V1477,205-208(1991)

James, William E.
   ; Steel, William H.; Evans, Nelson O.: Design and testing of a cube-corner array for laser ranging,V1400,129-136(1991)

Jamnejad, V. See Cooley, Thomas W.: V1475,243-247(1991)

Jancaitis, Kenneth S. See Elder, Melanie L.: V1441,237-246(1991)

Jandura, Louise
   ; Agronin, Michael L.: Analysis and testing of a soft actuation system for segmented-reflector articulation and isolation,V1542,213-224(1991)

Janeczko, Donald J.
   : Metal mirror review,VCR38,258-280(1991)

Janes, G. Sargent See Albagli, Douglas: V1441,146-153(1991)

Janes, Joachim See Pilz, Wolfgang: V1392,84-94(1991)

Janesick, James R.
   ; Soli, George; Elliott, Tom; Collins, Stewart A.: Effects of proton damage on charge-coupled devices,V1447,87-108(1991)
   See Acton, Loren W.: V1447,123-139(1991)
   See Bredthauer, Richard A.: V1447,310-315(1991)

Jang, Gyu-Sang See Vemuri, V.: V1469,563-574(1991)

Jang, T. See Fang, X. J.: V1572,279-283(1991)

Janicijevic, Lj.
   ; Jonoska, M.: Phase-circular hologram as a laser beam splitter,V1574,167-178(1991)

Jankiewicz, Zdzislaw
   ; Szydlak, J.; Skorczakowski, M.; Zendzian, W.; Dvornikov, S. S.; Kondratiuk, N. V.; Skripko, G. A.: Ti:Al2O3 laser with an intracavity phase-conjugate mirror,V1391,101-104(1991)

Jankovic, Lilja See Vengsarkar, Ashish M.: V1367,249-260(1991)

Jankowiak, Ryszard
   ; Jeong, H.; Small, Gerald J.: Fluorescence line-narrowing spectroscopy in the study of chemical carcinogenesis,V1435,203-213(1991)
   See Hayes, John M.: V1435,258-266(1991)

Jankowski, Alan F.
   ; Makowiecki, Daniel M.; McKernan, M. A.; Foreman, R. J.; Patterson, R. G.: Deposition- controlled uniformity of multilayer mirrors,V1343,32-38(1991)

Jannson, Joanna See Savant, Gajendra D.: V1461,79-90(1991)

Jannson, Tomasz
   ; Lin, Freddie S.; Moslehi, Behzad M.; Shirk, Kevin W.: Integrated-optic interconnects and fiber-optic WDM data links based on volume holography,V1555,159-176(1991)
   ; Rich, Chris C.; Sadovnik, Lev S.: Wavelength-dispersive and filtering applications of volume holographic optical elements,V1545,42-63(1991)
   See Chen, Ray T.: V1374,223-236(1991)
   See Lin, Freddie S.: V1474,45-56(1991)
   See Liu, William Y.: V1365,20-24(1991)

Jannush, D. A. See Dyott, Richard B.: V1482,439-443(1991)

Janowska-Dmoch, Bozena See Campos, Juan: V1574,141-147(1991)

Jans, Jan C. See Yoo, Chue-San: V1464,393-403(1991)

Jansen, E. D. See Beyerbacht, Hugo P.: V1427,117-127(1991)

Jansen, Michael
   ; Yang, Jane; Ou, Szutsun S.; Sergant, Moshe; Mawst, Luke J.; Rozenbergs, John; Wilcox, Jarka Z.; Botez, Dan: Monolithic two-dimensional surface-emitting laser diode arrays with 45 degree micromirrors,V1418,32-39(1991)
   See Anderson, Eric R.: V1417,543-549(1991)
   See Botez, Dan: V1474,64-74(1991)
   See Largent, Craig C.: V1418,40-45(1991)
   See Mawst, Luke J.: V1418,353-357(1991)

Jantsch, Wolfgang See Hingerl, Kurt: V1361,943-953(1991)

Januar, Indra See Feuerstein, Robert J.: V1583,196-201(1991)

Janulewicz, K. A. See Wolinski, Wieslaw: V1397,391-393(1991)

Janzen, Doug See Hogg, W. D.: V1588,300-307(1991)

Jaquenoud, Laurent See Salembier, Philippe: V1568,26-37(1991)

Jaques, P. F. See Beard, David V.: V1446,52-58(1991)

Jaquet, Patrick J.
: Plastic optical fiber applications for lighting of airports and buildings, V1592,165-172(1991)

Jarecke, Peter J.
; Folkman, Mark A.; Darnton, Lane A.: Radiometric calibration plan for the Clouds and the Earth's Radiant Energy System scanning instruments, V1493,244-254(1991)
See Folkman, Mark A.: V1493,255-266(1991)

Jarkovska, B. See Chorvat, Dusan: V1402,89-92(1991)

Jarlman, Olof
See Berg, Roger: V1431,110-119(1991)
See Berg, Roger: V1525,59-67(1991)

Jarman, Richard H.
; Johnson, Barry C.: Characteristics of domain formation and poling in potassium niobate, V1561,33-42(1991)

Jaros, Milan
; Turton, Richard M.: Wave-function engineering in Si-Ge microstructures: linear and nonlinear optical response, V1362,242-253(1991)
See Presting, Hartmut: V1512,250-277(1991)

Jaroszewicz, Zbigniew
: Fresnel zone plate moire patterns and its metrological applications, V1574,154-158(1991)
; Kolodziejczyk, Andrzej: Computer-generated diffractive elements focusing light into arbitrary line segments, V1555,236-240(1991)

Jarret, Bertrand See Pierre, Guillaume: V1511,201-211(1991)

Jarviluoma, Markku See Marszalec, Janusz A.: V1382,443-452(1991)

Jatko, W. B. See Gleason, Shaun S.: V1386,135-144(1991)

Jauniskis, L. See Foukal, Peter: V1493,72-79(1991)

Javidi, Bahram
ed.: *Optical Information Processing Systems and Architectures III*, V1564(1991)
; Wang, Jianping; Tang, Qing: Effects of thresholding in multiobject binary joint transform correlation, V1564,212-223(1991)
; Wang, Jianping: Quantization analysis of the binary joint transform correlator in the presence of nonlinear compression, V1564,254-265(1991)
See Caulfield, H. J.: V1562,103-106(1991)

Jayadev, T. S.
: Focal-plane architectures and signal processing, V1541,163-166(1991)
ed.: *Infrared Sensors: Detectors, Electronics, and Signal Processing*, V1541(1991)

Jayasooriah, Mr. See Ong, Sim-Heng: V1445,564-573(1991)

Jean, A. See Chaker, Mohamed: V1465,16-25(1991)

Jedrzejewski, Robert See Greenfield, Perry E.: V1494,16-39(1991)

Jeevanandam, Valluvan See Auteri, Joseph S.: V1421,182-184(1991)

Jefferys, E. A. See Bourne, Robert W.: V1488,73-79(1991)

Jeffries, Alfred T. See Honda, Kenji: V1466,141-148(1991)

Jeffries, Clark See Carlson, Rolf: V1469,684-696(1991)

Jeffs, Brian D.
; Gunsay, Metin; Dougal, John: Resolution enhancement of blurred star field images by maximally sparse restoration, V1567,511-521(1991)

Jeganathan, Muthu See Chen, Chien-Chung: V1417,240-250(1991)

Jelinsky, Patrick See Welsh, Barry Y.: V1343,166-174(1991)

Jen, Cheng K.
; Neron, C.; Shang, Alain; Abe, Koichi; Bonnell, Lee J.; Kushibiki, J.; Saravanos, C.: Acoustic characterization of optical fiber glasses, V1590,107-119(1991)
See de Oliveira, C. A.: V1590,101-106(1991)

Jenekhe, Samson A. See Meth, Jeffrey S.: V1560,13-24(1991)

Jeng, Bor S.
See Chien, Bing-Shan: V1606,588-598(1991)
See Lee, Ching-Long: V1606,728-734(1991)

Jeng, Fure-Ching See Chen, Cheng-Tie: V1605,812-821(1991)

Jeng, J. See Bergman, Larry A.: V1364,14-21(1991)

Jeng, Ru J. See Li, Lian: V1560,243-250(1991)

Jenkin, Michael R. See Wilkes, David R.: V1388,536-542(1991)

Jenkins, A. J.
: Single-pixel measurements on LCDs, V1506,188-193(1991)

Jenkins, R. M.
; Devereux, R. W.: 10.6-um TEM00 beam transmission characteristics of a hollow circular cross-section multimode waveguide, V1512,135-142(1991)

Jennings, B. R. See Macfadyen, Allan J.: V1367,319-328(1991)

Jennings, G. See Campuzano, Juan C.: V1345,245-254(1991)

Jennings, Kurt L.
; Miller, Gregory D.: Loss modeling of amodal fiber-to-fiber interconnects, V1369,36-42(1991)
See Cirillo, James R.: V1592,53-59(1991)
See Cirillo, James R.: V1592,42-52(1991)

Jenquin, Michael J. See Minor, John L.: V1538,18-39(1991)

Jensen, Douglas R.
: Evaluation of the Navy Oceanic Vertical Aerosol Model using lidar and PMS particle-size spectrometers, V1487,160-171(1991)

Jensen, Jorgen A.
; Mathorne, Jan: Sampling system for in vivo ultrasound images, V1444,221-231(1991)

Jensen, L. C. See Dickinson, J. T.: V1598,72-83(1991)

Jensen, Preben D.
See Jacobi, William J.: V1479,111-119(1991)
See Jacobi, William J.: V1495,205-213(1991)
See Stocker, Alan D.: V1481,140-155(1991)

Jensen, Robert K.
; Fletcher, P.; Abraham, C.: Body shape changes in the elderly and the influence of density assumptions on segment inertia parameters, V1380,124-136(1991)

Jensen, Ronald H.
; Bigbee, William L.; Langlois, Richard G.; Grant, Stephen G.; Pleshanov, Pavel G.; Chirkov, Andre A.; Pilinskaya, Maria A.: Laser-based flow cytometric analysis of genotoxicity of humans exposed to ionizing radiation during the Chernobyl accident, V1403,372-380(1991)

Jensen, Stephen C.
; Tilstra, Shelle D.; Barnabo, Geoffrey A.; Thomas, David C.; Phillips, Richard W.: Fiber optic temperature sensor for aerospace applications, V1369,87-95(1991)

Jeon, Ho-In
; Abushagur, Mustafa A.; Caulfield, H. J.: Optical implementation of associative memory based on parallel rank-one interconnections, V1564,522-535(1991)

Jeon, Joon-hyeon
; Hahm, Cheul-hee; Kim, Jae-Kyoon: Hierarchical block motion estimation for video subband coding, V1605,954-962(1991)

Jeong, H. See Jankowiak, Ryszard: V1435,203-213(1991)

Jeong, Tung H.
: Holography and education, VIS08,360-369(1991)
: Holography and education in the United States, V1238,351-354(1991)
: Holography in the '90s, V1396,718-721(1991)
; Feng, Qiang; Wesly, Edward J.; Qu, Zhi-Min: Coherent length and holography, V1396,60-70(1991)
; Markov, Vladimir B.; eds.: *Three-Dimensional Holography: Science, Culture, Education*, V1238(1991)
; Wesly, Edward J.: Progress in true-color holography, V1238,298-305(1991)
See Greguss, Pal: VIS08(1991)

Jeong, Yoon-Ha See Kwon, O'Dae: V1361,802-808(1991)

Jericevic, Zeljko
; McGavran, Loris; Smith, Louis C.: Digital imaging of Giemsa-banded human chromosomes: eigenanalysis and the Fourier phase reconstruction, V1428,200-213(1991)
See Smith, Louis C.: V1428,224-232(1991)

Jerie, Tomas See Fiala, Pavel: V1574,179-187(1991)

Jernberg, Per
: Corrosion evaluation of coated sheet metal by means of thermography and image analysis, V1467,295-302(1991)

Jeschke, Hartwig See Pirsch, Peter: V1452,544-555(1991)

Jesion, G. See Gierczak, Christine A.: V1433,315-328(1991)

Jessel, M. See Illenyi, Andras: VIS08,39-52(1991)

Jesser, William A. See Black, Linda R.: V1557,54-59(1991)

Jessop, Paul E. See Wong, Brian: V1398,261-268(1991)

Jett, James H. See Soper, Steven A.: V1435,168-178(1991)

Jett, L. L. See Jackson, John E.: V1533,75-86(1991)

Jetter, Heinz-Leonard
: Design considerations for high-power industrial excimer lasers,V1503,48-52(1991)

Jewell, Jack L.
; Lee, Yong H.; McCall, Samuel L.; Scherer, Axel; Harbison, James P.; Florez, Leigh T.; Olsson, N. A.; Tucker, Rodney S.; Burrus, Charles A.; Sandroff, Claude J.: Vertical cavity lasers for optical interconnects,V1389,401-407(1991)
See Morgan, Robert A.: V1562,149-159(1991)

Jewell, John M.
; Askins, Charles G.; Aggarwal, Ishwar D.: Interferometric technique for the concurrent determination of thermo-optic and thermal expansion coefficients,V1441,38-44(1991)

Jewell, Tanya E.
; Thompson, Kevin P.; Rodgers, John M.: Reflective optical designs for soft x-ray projection lithography,V1527,163-173(1991)
See Rodgers, John M.: V1354,330-336(1991)
See White, Donald L.: V1343,204-213(1991)

Jeynes, C. See Hansen, Stein I.: V1362,361-369(1991)

Jhabvala, Murzy See Ewin, Audrey J.: V1479,12-20(1991)

Jhung, Kyu Soo See Lee, Eun Soo: V1358,262-268(1991)

Ji, Han-Bing
: Weighted-outer-product associative neural network,V1606,1060-1068(1991)

Ji, Hua M. See Chen, Wei M.: V1519,521-524(1991)

Ji, Tinglan
; Roehring, Hans; Blume, Hartwig R.; Seeley, George W.; Browne, Michael P.: Physical and psychophysical evaluation of CRT noise performance,V1444,136-150(1991)

Ji, Wanxi See He, Da-Ren: V1362,696-701(1991)

Ji, Xinhua See Qin, Yuwen: V1554A,739-746(1991)

Ji, Xiuyan See He, Da-Ren: V1362,696-701(1991)

Jia, Jiancheng
; Krutz, Gary W.; Gibson, Harry G.: Corn plant locating by image processing,V1379,246-253(1991)
; Krutz, Gary W.; Gibson, Harry G.: Image processing to locate corn plants,V1396,656-663(1991)

Jia, Pei M. See He, Jin: V1519,499-507(1991)

Jia, Youquan
; Kang, Yilan; Du, Ji: New technique for rapid making specimen: gratings for moire interferometry,V1554B,331-334(1991)
; Sun, Yongda: High-speed holographic interferometric study of the propagation of the electrohydraulic shock wave,V1554B,135-138(1991)
See Chen, Yuhai: V1554B,566-572(1991)
See Kang, Yilan: V1554B,514-522(1991)

Jia, Yu M.
See Yang, Bang C.: V1519,725-728(1991)
See Yang, Bang C.: V1519,156-158(1991)
See Yang, Bang C.: V1519,864-865(1991)

Jiad, K. See Harrison, Robert G.: V1376,94-102(1991)

Jiang, Aibao See Lin, Wenzheng: V1358,29-36(1991)

Jiang, Bing Y. See Feng, Yi P.: V1519,440-443(1991)

Jiang, Chang Z.
See Fan, Xiang J.: V1519,808-812(1991)
See Fan, Xiang J.: V1519,805-807(1991)

Jiang, Chaochuan
See Fan, Cheng: V1461,51-55(1991)
See Fan, Cheng: V1461,265-269(1991)

Jiang, Ching-Long
; Agarwal, Rajiv; Kuwamoto, Hide; Huang, Rong-Ting; Appelbaum, Ami; Renner, Daniel; Su, Chin B.: Optimization of grating depth and layer thicknesses for DFB lasers,V1418,261-271(1991)

Jiang, Dareng
; Han, Chia Y.; Wee, William G.: Prototype expert system for preventive control in power plants,V1468,16-25(1991)

Jiang, Desheng
; Ye, Qizheng; Zhang, Shengpei; Li, Faxian: Parallel network for optical fiber sensors,V1572,313-317(1991)

Jiang, F. Y. See Pan, Chuan K.: V1519,645-651(1991)

Jiang, Fan
; Schunck, Brian G.: Recovering absolute depth and motion of multiple objects from intensity images,V1569,162-173(1991)

Jiang, Fenglai See Holm-Kennedy, James W.: V1527,322-331(1991)

Jiang, Frank N. See Allison, Beth: V1426,200-207(1991)

Jiang, G.
See Liu, Rui-Fu: V1572,189-191(1991)
See Wang, Yutian: V1572,230-234(1991)

Jiang, H. Y. See Liu, Rui-Fu: V1572,189-191(1991)

Jiang, Hong-Tao
See Shi, Yi-Wei: V1572,308-312(1991)
See Wang, Yao-Cai: V1572,32-37(1991)

Jiang, Hongbing See Chen, Zhan: V1437,103-109(1991)

Jiang, Huilin See Yang, Daren: V1527,456-461(1991)

Jiang, Jian See Xia, Zhong F.: V1519,866-871(1991)

Jiang, Jian G. See Chen, Kun J.: V1519,632-639(1991)

Jiang, Jie
; Kraemer, Udo: Two new developments for optoelectronic bus systems,V1505,166-174(1991)
; Liu, Chen Z.: Diamond-like carbon thin films prepared by rf-plasma CVD,V1519,717-724(1991)

Jiang, Jing-Wen
; So, Vincent; Lessard, Michel; Vella, Paul J.: In-line testing for fiber subscriber loop applications,V1363,191-195(1991)

Jiang, Jinyou See Zhou, Shaoxiang: V1358,788-792(1991)

Jiang, Kunsheng See Chen, Riqi: V1554A,407-417(1991)

Jiang, Lingzhen
See Lu, Yue-guang: V1332,287-291(1991)
See Zhao, Feng: V1555,297-299(1991)
See Zhao, Feng: V1461,262-264(1991)
See Zhao, Feng: V1555,241-242(1991)

Jiang, Pei-sheng See Wei, Yan-nian: V1358,457-460(1991)

Jiang, Ping See Huang, Yong L.: V1519,142-145(1991)

Jiang, Pisu
; Laybourn, Peter J.; Righini, Giancarlo C.: Homogeneous thin film lens on LiNbO3,V1362,899-906(1991)

Jiang, Rong-Xi See Liu, Wen-Qing: V1530,240-243(1991)

Jiang, Shibin
; Wang, Huirong; Jiang, Yasi; Wang, Biao; Chen, Menda: Bonding skill between the fused-quartz mirror and the composite substrate,V1535,143-147(1991)

Jiang, Weixing See Qu, Zhihao: V1554A,503-510(1991)

Jiang, Wen-Han
; Ling, Ning; Rao, Xuejun; Shi, Fan: Fitting capability of deformable mirror,V1542,130-137(1991)

Jiang, Wu See Shealy, David L.: V1343,122-132(1991)

Jiang, Xiang-Liu
; Zhang, Fang Q.; Li, Jiang-Qi; Yang, Bin; Chen, Guang-Hua: Systematic studies on transition layers of carbides between CVD diamond films and substrates of strong carbide-forming elements,V1534,207-213(1991)

Jiang, Xuping
; Wu, Guobing: Experimental investigation on the flow behavior of liquid aluminum inside pressure-die-casting dies using high-speed photography,V1358,1237-1244(1991)

Jiang, Yasi See Jiang, Shibin: V1535,143-147(1991)

Jiang, Yingqui
See Zhuang, Song Lin: V1558,28-33(1991)
See Zhuang, Song Lin: V1558,149-153(1991)

Jiang, Z. M. See Lin, Li-Huang: V1346,490-501(1991)

Jiang, Zhi X.
; Whitehurst, Colin; King, Terence A.: Fragmentation methods in laser lithotripsy,V1421,88-99(1991)

Jiang, Zhiying See Li, Yuchuan: V1572,382-385(1991)

Jiao, Jiangzhong See Hsu, Dahsiung: V1238,13-17(1991)

Jiao, Kaili L.
; Zheng, J. P.; Kwok, Hoi-Sing; Anderson, Wayne A.: Optical characterization of InP epitaxial layers on different substrates,V1361,776-783(1991)

Jin, Baohui See Wang, Guomei: V1590,229-236(1991)

Jin, Guofan
; Zhu, Zimin; Yu, Xinglong: Transformation from tristimulus RGB to Munsell notation HVC in a colored computer vision system,V1569,507-512(1991)
See Hung, Yau K.: V1332,738-747(1991)

Jin, Guoliang
; Shen, Ronggui; Ying, Zaisheng: Grating splitter for glass waveguide,V1513,50-55(1991)

Jin, Henghuan See Sun, Dexing: V1572,508-513(1991)

Jin, Huijia J. See Favro, Lawrence D.: V1467,290-294(1991)

Jin, Lei
See Guo, Yong H.: V1519,327-332(1991)
See Li, Futian: V1345,78-88(1991)
See Shao, Jian D.: V1519,298-301(1991)

Jin, Michael S. See Skeldon, Mark D.: V1410,116-124(1991)

Jin, Tianfeng
; Yuan, Youxin: Ophthalmic antireflection coatings with same residual reflective colors on ophthalmic optics with different refractive indices,V1529,132-137(1991)

Jin, Wei
; Uttamchandani, Deepak G.; Culshaw, Brian: Direct readout of dynamic phase changes in a fiber-optic homodyne interferometer,V1504,125-132(1991)

Jin, X. D. See Fang, X. J.: V1572,279-283(1991)

Jin, Xi See Wang, Chihcheng: V1572,406-409(1991)

Jin, Ya-Qiu
: Scattering and thermal emission from spatially inhomogeneous atmospheric rain and cloud,V1487,324-332(1991)

Jin, Yiping See Shen, Zhen-kang: V1481,233-240(1991)

Jin, Yixin See Du, MingZe: V1361,699-705(1991)

Jing, Lian-hua See Wang, Yi-Ming: V1361,325-335(1991)

Jing, Shangzhong See Zhang, Zaixuan: V1572,201-204(1991)

Jing, Xing-Liang
; Zhang, Yong-Tao; Chen, Yi-Xin: Threshold current density of InGaAsP/InP surface-emitting laser diodes with hemispherical resonator,V1418,434-441(1991)

Jipson, Victor B.
: Directions in optical storage,V1499,2-4(1991)

Jitsuno, Takahisa See Nakatsuka, Masahiro: V1411,108-115(1991)

Jocham, D. See Baumgartner, R.: V1525,246-248(1991)

Jochs, Werner W. See Westenberger, Gerhard: V1535,113-120(1991)

Jodlbauer, Heibert See Schad, Hanspeter: V1507,446-449(1991)

Jodoin, Ronald E.
; Loce, Robert P.; Nowak, William J.; Costanza, Daniel W.: Development of a precision high-speed flying spot position detector,V1398,61-70(1991)

Joeckle, Rene C.
; Koeneke, Axel; Sontag, Andre: Quantitative analysis of a CO2 laser beam by PMMA burn patterns,V1397,679-682(1991)
; Rapp, Gerard; Sontag, Andre: Boiling process in PMMA irradiated by CO2, DF and HF laser radiations,V1397,683-687(1991)

Joenathan, Charles
; Khorana, Brij M.: Phase-measuring fiber optic ESPI system: phase-step calibration and error sources,V1554B,56-63(1991)
; Khorana, Brij M.: Phase measuring fiber optic electronic speckle pattern interferometer,V1396,155-163(1991)
See Brewer, Donald R.: V1396,430-434(1991)

Joffre, Pascal See Maze, Sylvie: V1505,20-31(1991)

Johann, Ulrich A.
; Pribil, Klaus; Sontag, Heinz: Novel optical-fiber-based conical scan tracking device,V1522,243-252(1991)
; Seelert, Wolf: 1W cw diode-pumped Nd:YAG laser for coherent space communication systems,V1522,158-168(1991)
See Pribil, Klaus: V1522,36-47(1991)
See Sontag, Heinz: V1417,573-587(1991)
See Sontag, Heinz: V1522,61-69(1991)

Johansson, Bo G. See Johansson, Micael: V1538,180-188(1991)

Johansson, Jonas
See Andersson-Engels, Stefan: V1426,31-43(1991)
See Baert, Luc: V1525,385-390(1991)

Johansson, Micael
; Engman, Per; Johansson, Bo G.: Ericsson digital recce management system,V1538,180-188(1991)

John, Eugine See Shadaram, Mehdi: V1365,108-115(1991)

John, S. See Karuppiah, N.: V1536,215-221(1991)

John, Sarah
: Recovery of aliased signals for high-resolution digital imaging,V1385,39-47(1991)

Johns, Steven T.
See Goel, Kamal K.: V1476,314-319(1991)
See Prucnal, Paul R.: V1389,462-476(1991)

Johnson, Anthony M.
ed.: Eyesafe Lasers: Components, Systems, and Applications,V1419(1991)

Johnson, Barry C. See Jarman, Richard H.: V1561,33-42(1991)

Johnson, Brett W. See Huang, Chin M.: V1447,156-164(1991)

Johnson, C. B.
: Review of ITT/EOPD's special purpose photosensitive devices and technologies,V1396,360-376(1991)
; Abshire, James B.; Zagwodzki, Thomas W.; Hunkler, L. T.; Letzring, Samuel A.; Jaanimagi, Paul A.: Space-qualified streak camera for the Geodynamic Laser Ranging System,V1346,340-370(1991)
; Patton, Stanley B.; Bender, E.: High-resolution microchannel plate image tube development,V1449,2-12(1991)

Johnson, C. R. See Ding, Zhi: V1565,154-165(1991)

Johnson, Carroll See Gonzalez, Orlando: V1381,284-291(1991)

Johnson, Charles L.
; Dietz, Kurtis L.: Effects of the lunar environment on optical telescopes and instruments,V1494,208-218(1991)

Johnson, Craig W. See Campbell, Duncan R.: V1400(1991)

Johnson, Curtis E. See Weimer, Wayne A.: V1534,9-13(1991)

Johnson, D. G. See de Salabert, Arturo: V1521,74-88(1991)

Johnson, Daniel E. See Crannage, R. P.: V1397,261-265(1991)

Johnson, David A.
; Valenzuela, C. F.: Use of fluorescence spectroscopy to elucidate structural features of the nicotinic acetylcholine receptor,V1432,82-90(1991)
See Zidovetzki, Raphael: V1432,76-81(1991)

Johnson, David G. See Traub, Wesley A.: V1491,298-307(1991)

Johnson, Dean R.
; Fredricks, Ronald J.; Vuong, S. C.; Dembinski, David T.; Sabri, Sehbaz H.: Multimode fiber optic rotation sensor with low-cost digital signal processing,V1367,140-154(1991)
See Fredricks, Ronald J.: V1367,127-139(1991)

Johnson, Derwyn C. See Malo, Bernard: V1590,83-93(1991)

Johnson, Donald W. See Hartney, Mark A.: V1466,238-247(1991)

Johnson, Douglas E.
; Levinson, A. K.; Greskovich, Frank J.; Cromeens, Douglas M.; Ro, Jae Y.; Costello, Anthony J.; Wishnow, Kenneth I.: Transurethral laser prostatectomy using a right-angle laser delivery system,V1421,36-41(1991)

Johnson, Edward A. See Murray, Brian W.: V1330,2-24(1991)

Johnson, Edward J.
; Hyer, Paul V.; Culotta, Paul W.; Clark, Ivan O.: Flow-field velocity measurements for nonisothermal systems,V1557,168-179(1991)

Johnson, Eric G.
; Kathman, Alan D.: Rigorous electromagnetic modeling of diffractive optical elements,V1545,209-216(1991)

Johnson, Erik D. See Kao, Chi-Chang: V1548,149-157(1991)

Johnson, Fred M. See Ehrenberg, Benjamin: V1426,244-251(1991)

Johnson, G. A. See Laxer, Cary: V1444,190-195(1991)

Johnson, G. C. See Caudell, Thomas P.: V1469,612-621(1991)

Johnson, Glen W. See Ledermann, Peter G.: V1598,160-163(1991)

Johnson, Jeffrey P. See Kimball, Paulette R.: V1540,687-698(1991)

Johnson, Jerome L.
: Update: high-speed/high-volume radiometric testing of Z-technology focal planes,V1541,210-219(1991)

Johnson, Jerris F.
; Lomheim, Terrence S.: Hybrid infrared focal-plane signal and noise modeling,V1541,110-126(1991)

Johnson, Kevin E. See Jackson, John E.: V1533,75-86(1991)

Johnson, Kristina M.
See Robinson, Michael G.: V1469,240-249(1991)
See Zhang, Lin: V1469,225-229(1991)

Johnson, Leonard M.
; Betts, Gary E.; Roussell, Harold V.: Integrated-optical modulators for bandpass analog links,V1371,2-7(1991)
See Betts, Gary E.: V1371,252-257(1991)
See Betts, Gary E.: V1562,281-302(1991)

Johnson, Michael L. See Lakowicz, Joseph R.: V1435,142-160(1991)

Johnson, Paul A. See Cuellar, Louis: V1542,468-476(1991)

Johnson, Peter F.
: Is it worth it?— statistics of corporate-based IR program results,V1467,47-50(1991)

Johnson, Peter T.
: ; Debney, Brian T.; Carter, Andrew C.: Components and applications for high-speed optical analog links,V1371,87-97(1991)
: ; Hankey, Judith; Debney, Brian T.: Computer-controlled two-segment DFB local oscillator laser tuning for multichannel coherent systems,V1372,188-199(1991)

Johnson, R. See Haigh, N. R.: V1366,259-264(1991)

Johnson, R. B.
: : Knowledge-based environment for optical system design,V1354,346-358(1991)
: ; Ahmad, Anees; Hadaway, James B.; Geiger, Alan L.: Lightweight SXA metal matrix composite collimator,V1535,136-142(1991)
: ; Feng, Chen: History of infrared optics,VCR38,3-18(1991)
: ; Feng, Chen; Ethridge, Edwin C.: Reluctant glass formers and their applications in lens design,V1527,2-18(1991); 1535,231-247(1991)
: ; Feng, Chen: Zoom lenses with a single moving element,V1354,676-683(1991)
: ; Hadaway, James B.; Burleson, Thomas A.; Watts, Bob; Park, Ernest D.: All-reflective four-element zoom telescope: design and analysis,V1354,669-675(1991)
: See Caulfield, H. J.: V1555,154-158(1991)
: See Hoover, Richard B.: V1343,189-202(1991)
: See Jaenisch, Holger M.: V1379,162-167(1991)

Johnson, R. J. See Lowry, Heard S.: V1454,453-464(1991)

Johnson, Richard V. See Skeie, Halvor: V1583,153-164(1991)

Johnson, Rubin See Bennett, Laura F.: V1470,263-274(1991)

Johnson, Stephen G.
: ; Rios, E. L.; Miller, Charles M.; Fearey, Bryan L.: High-efficiency resonance ionization mass spectrometric analysis by external laser cavity enhancement techniques,V1435,292-297(1991)

Johnson, Stewart W.
: ; Chua, Koon M.; Burns, Jack O.; Slane, Frederic A.: Lunar dust: implications for astronomical observatories,V1494,194-207(1991)
: See Chua, Koon M.: V1494,119-134(1991)

Johnson, W. J. See Newnam, Brian E.: V1552,154-174(1991)

Johnson, W. Todd
: ; Lavi, Moshe; Sapir, Eyal: Temperature chamber FLIR and missile test system,V1488,343-354(1991)
: See Kopolovich, Zvi: V1540,565-577(1991)

Johnson, Walter H. See Keenan, W. A.: V1393,354-365(1991)

Johnson-Cole, Helen
: ; Clark, Rodney L.: Validated CCD camera model,V1488,203-211(1991)

Johnston, Alan R. See Barnes, Charles E.: V1366,9-16(1991)

Johnston, Dustin C.
: ; Welsh, Byron M.: Atmospheric turbulence sensing for a multiconjugate adaptive optics system,V1542,76-87(1991)

Johnston, George L.
: ; Chen, Shien C.; Davidson, Ronald C.; Bekefi, George: Models of driven and mutually coupled relativistic magnetrons with nonlinear frequency-shift and growth-saturation effects,V1407,92-99(1991)

Johnston, James S.
: ; Romer, A. E.; Beales, M. S.: Optical encoders using pseudo-random-binary-sequence scales,V1589,126-132(1991)

Johnston, R. E.
: See Beard, David V.: V1446,52-58(1991)
: See Beard, David V.: V1446,289-296(1991)

Johnston, Steve C.
: : Examples of the topographies of the wavefront-variance merit function at different aberration orders,V1354,77-82(1991)
: ; Greynolds, Alan W.: Separation of function in the ASAP software package,V1354,136-141(1991)

Johnston, Timothy J. See Sasnett, Michael W.: V1414,21-32(1991)

Johnstone, Walter
: ; Thursby, G.; Culshaw, Brian; Murray, S.; Gill, M.; McDonach, Alaster; Moodie, D. G.; Fawcett, G. M.; Stewart, George; McCallion, Kevin J.: Multimode approach to optical fiber components and sensors,V1506,145-149(1991)

Johri, Subodh See Yadav, M. S.: V1362,811-819(1991)

Jolleys, Glenn D. See Browne, Mark A.: V1450,338-349(1991)

Jollie, William B.
: : Airborne seeker evaluation and test system,V1482,92-103(1991)

Jolson, Alfred S. See Weeks, Arthur R.: V1567,77-87(1991)

Joly, V. See Barnault, B.: V1397,257-260(1991)

Jonas, Reginald P.
: : Prismatic anamorphic beam expanders for free-space optical communications,V1417,402-411(1991)
: See Gardam, Allan: V1417,381-390(1991)

Jones, Andrew E. See Forbes, Greg W.: V1354,144-153(1991)

Jones, Barry E.
: See Zheng, S. X.: V1572,359-364(1991)
: See Zheng, S. X.: V1572,268-272(1991)

Jones, Coleen T.
: ; Sauer, Ken D.: Bayesian approach to segmentation of temporal dynamics in video data,V1605,522-533(1991)

Jones, Darlena J. See Bandy, Donna K.: V1497,142-152(1991)

Jones, David C.
: ; Ridley, Kevin D.; Cook, Gary; Scott, Andrew M.: Fidelity of Brillouin amplification with Gaussian input beams,V1500,46-52(1991)

Jones, David E. See Krishnan, Ganapathy: V1468,563-572(1991)

Jones, Douglas L.
: ; Baraniuk, Richard G.: Efficient computation of densely sampled wavelet transforms,V1566,202-206(1991)

Jones, E. W. See Lin, TrueLon: V1540,135-139(1991)

Jones, Edwin R.
: ; McLaurin, A. P.; Mason, J. L.: Three-dimensional imaging laparoscope,V1457,318-321(1991)

Jones, Graeme
: ; Boashash, Boualem: Instantaneous quantities and uncertainty concepts for signal-dependent time-frequency distributions,V1566,167-178(1991)

Jones, J. R.
: ; Sharpe, Randall B.: Cost aspects of narrowband and broadband passive optical networks,V1363,106-118(1991)

Jones, Jack C. See Meitzler, Thomas J.: V1486,380-389(1991)

Jones, Judson P. See Sluder, John C.: V1468,642-652(1991)

Jones, Julian D.
: See Atcha, Hashim: V1504,221-232(1991)
: See Farahi, Faramarz: V1584,53-61(1991)
: See Kidd, S. R.: V1504,180-190(1991)
: See Morgan, Russell D.: V1504,118-124(1991)
: See Su, Daoning: V1502,41-51(1991)
: See Valera Robles, Jesus D.: V1508,170-179(1991)
: See Zheng, S. X.: V1572,359-364(1991)

Jones, Katharine J. See Morgan, Robert E.: V1588,189-197(1991)

Jones, M. E. See Thode, Lester E.: V1552,87-106(1991)

Jones, Mark S. See Jones, Mike I.: V1498,110-127(1991)

Jones, Michael G.
: ; Moore, Douglas R.: High-isolation single-taper filters,V1365,43-52(1991)

Jones, Mike I.
: : Tilt-tolerant dual-band interferometer objective lens,V1498,158-162(1991)
: ; Jones, Mark S.: Optical analysis of segmented aircraft windows,V1498,110-127(1991)
: See Osmer, Kurt A.: V1498,138-146(1991)

Jones, O. C.
: ; Glicksman, M. E.; Lin, Jen T.; Kim, G. T.; Singh, N. B.: Development of a laser Doppler system for measurement of velocity fields in PVT crystal growth systems,V1557,202-208(1991)
: See Paradies, C. J.: V1561,2-5(1991)

Jones, Paul W.
: ; Rabbani, Majid: <author>Digital Image Compression Techniques,VTT07(1991)

Jones, Peter A.
: : Alignment and focus control of a telescope using image sharpening,V1542,194-204(1991)

Jones, Philip J.
: ; Tomita, Akira; Wartenberg, Mark: Performance of NCAP projection displays,V1456,6-14(1991)

Jones, R. See Acebal, Robert: V1397,191-196(1991)

Jones, Richard See Dent, Andrew J.: V1550,97-107(1991)

Jones, Richard A.
: See Priebe, Russell: V1482,265-274(1991)
: See Viswanath, Harsha C.: V1482,275-284(1991)

Jones, Robert A.
: ed.: Computer-Controlled Optical Surfacing,VMS40(1991)
: See Zimmerman, Jerrold: VCR38,184-192(1991)

Jones, Robert H.
: Optical laser intelligent verification expert system,V1401,86-93(1991)
; Hadjinicolaou, M. G.; Musgrave, G.: OLAS: optical logic array structures,V1401,138-145(1991)

Jones, Scott A. See Bocker, Richard P.: V1569,298-310(1991)

Jones, Susan K.
; Dudley, Bruce W.; Peters, Charles R.; Kellam, Mark D.; Pavelchek, Edward K.: Characterization of wavelength offset for optimization of deep-UV stepper performance,V1464,546-553(1991)
See Freeman, Peter W.: V1464,377-385(1991)
See Russ, John C.: V1464,10-21(1991)

Jones, Terry L.
: Development of automatic target recognizers for Army applications,VIS07,4-13(1991)

Jones, W. K. See Larkins, Grover L.: V1477,26-33(1991)

Jones, William D. See Theon, John S.: V1492,2-23(1991)

Jones, William L.
See Chow, P. D.: V1475,42-47(1991)
See Liu, Louis: V1475,193-198(1991)
See Minot, Katcha: V1475,309-313(1991)

Jones-King, Yolanda D. See Bacal, Marthe: V1407,605-609(1991)

Jones-Parry, I. H. See de Salabert, Arturo: V1521,74-88(1991)

Jong, Jyh-Ming See Hayden, Leonard A.: V1389,205-214(1991)

Jonker, Stephan A. See Warman, John M.: V1559,159-170(1991)

Jonoska, M. See Janicijevic, Lj.: V1574,167-178(1991)

Joo, Jin W.
; Kwon, Young H.; Park, Seung O.: Quantitative analysis of contact deformation using holographic interferometry and speckle photography,V1554B,113-118(1991)

Jordan, Harry F.
: Digital optical computers at Boulder,V1505,87-98(1991)

Jordan, Jay B.
See Gaughan, Patrick T.: V1469,812-819(1991)
See Watkins, Wendell R.: V1486,17-24(1991)

Jordan, John K. See Muhs, Jeffry D.: V1584,374-386(1991)

Jordan, John R.
; Bovik, Alan C.: Dense stereo correspondence using color,V1382,111-122(1991)

Jordan, Michael
: Neural network for improving terrain elevation measurement from stereo images,V1567,179-187(1991)

Jorgensen, Christian K. See Reisfeld, Renata: V1590,215-228(1991)

Jorgensen, Gary J.
; Govindarajan, Rangaprasad: Ultraviolet reflector materials for solar detoxification of hazardous waste,V1536,194-205(1991)

Jorgensen, Henning
; Olsen, Flemming O.: Process monitoring during CO2 laser cutting,V1412,198-208(1991)

Jorgensen, Steven M.
; Whitlock, S. V.; Thomas, Paul J.; Roessler, R. W.; Ritman, Erik L.: Dynamic spatial reconstructor: a high-speed, stop-action, 3-D, digital radiographic imager of moving internal organs and blood,V1346,180-191(1991)

Jori, Guilio
; Reddi, Elena; Biolo, Roberta; Polo, Laura; Valduga, Giuliana: Phthalocyanines as phototherapeutic agents for tumors,V1525,367-376(1991)

Josek, K. See Werle, Peter W.: V1433,128-135(1991)

Joseph, Alan A. See Udd, Eric: V1418,134-152(1991)

Josephs, Robert See Watowich, Stanley J.: V1396,316-323(1991)

Josepson, J. I. See Aben, Hillar K.: V1554A,298-309(1991)

Joshi, A. B.
; Lo, G. Q.; Shih, Dennis K.; Kwong, Dim-Lee: Performance and reliability of ultrathin reoxidized nitrided oxides fabricated by rapid thermal processing,V1393,122-149(1991)

Joshi, Abhay M. See Olsen, Gregory H.: V1540,596-605(1991)

Joshi, Chan J. See Dodd, James W.: V1407,467-473(1991)

Joss, Brian T.
; Sunak, Harish R.: Modeling direct detection and coherent-detection lightwave communication systems that utilize cascaded erbium-doped fiber amplifiers,V1372,94-117(1991)

Josse, Michel A. See Mainguy, Stephan: V1530,269-282(1991)

Jost, R. G.
ed.: *Medical Imaging V: PACS Design and Evaluation,*V1446(1991)
; Blaine, G. J.; Kocher, Thomas E.; Muka, Edward; Whiting, Bruce R.: High-resolution teleradiology applications within the hospital,V1446,2-9(1991)
See Cox, Jerome R.: V1446,40-51(1991)

Jouan, Jacky See Fratter, C.: V1490,59-73(1991)

Joubert, Olivier P. See Paniez, Patrick J.: V1466,583-591(1991)

Joud, J. C. See Boher, Pierre: V1345,165-179(1991)

Joulain, Franck See Godard, Bruno: V1503,71-77(1991)

Joullie, Andre F.
See Mani, Habib: V1362,38-48(1991)
See Tournie, Eric: V1361,641-656(1991)

Jouny, Ismail
; Garber, Frederick D.; Moses, Randolph L.; Walton, Eric K.: Applications of the bispectrum in radar signature analysis and target identification,V1471,142-153(1991)

Jouve, Philippe See Lequime, Michael: V1511,244-249(1991)

Jouvet, C. See Bacis, Roger: V1397,173-176(1991)

Jovin, Thomas M.
; Arndt-Jovin, Donna J.: Digital imaging microscopy: the marriage of spectroscopy and the solid state CCD camera,V1439,109-120(1991)
See Zidovetzki, Raphael: V1432,76-81(1991)

Joyce, Bruce A.
; Zhang, J.; Foxon, C. T.; Vvedensky, D. D.; Shitara, T.; Myers-Beaghton, A. K.: Growth dynamics of lattice-matched and strained layer III-V compounds in molecular beam epitaxy,V1361,13-22(1991)

Joyner, David J. See Browne, Mark A.: V1450,338-349(1991)

Joyner, M. See Zarnowski, Jeffrey J.: V1447,191-201(1991)

Jozefiak, Thomas H.
; Sailor, Michael J.; Ginsburg, Eric J.; Gorman, Christopher B.; Lewis, Nathan S.; Grubbs, Robert H.: Soluble polyacetylenes derived from the ring-opening metathesis polymerization of substituted cyclooctatetraenes: electrochemical characterization and Schottky barrier devices,V1436,8-19(1991)

Jozwiak, Phillip C. See Taylor, Gordon C.: V1475,103-112(1991)

Ju, M. S. See Chen, Terry Y.: V1554B,92-98(1991)

Ju, Rong-Hauh See Lee, Ching-Long: V1606,728-734(1991)

Juang, F. S.
See Wu, T. S.: V1361,23-33(1991)
See Wu, T. S.: V1519,263-268(1991)

Juang, Jeng-Dang
; Chang, Ming-Wen: Design of double-Gauss objective by using concentric lenses,V1354,273-276(1991)

Juchem, M. See Hitzenberger, Christoph K.: V1423,46-50(1991); 1429,21-25(1991)

Jucks, Kenneth W. See Traub, Wesley A.: V1491,298-307(1991)

Juday, Richard D.
; Florence, James M.: Full-complex modulation with two one-parameter SLMs,V1558,499-504(1991)
See Florence, James M.: V1558,487-498(1991)
See Vijaya Kumar, B. V. K.: V1555,138-145(1991)

Judd, Thomas C. See Silver, Michael D.: V1443,250-260(1991)

Judge, C. P. See Crossland, William A.: V1455,264-273(1991)

Judge, Thomas R.
See Quan, C.: V1507,463-475(1991)
See Towers, Catherine E.: V1358,952-965(1991)

Judy, Daniel C. See Huttlin, George A.: V1407,147-158(1991)

Judy, Millard M. See Sessler, Jonathan L.: V1426,318-329(1991)

Jueptner, Werner P.
ed.: *Industrial Applications of Holographic and Speckle Measuring Techniques,*V1508(1991)
; Bischof, Thomas: Investigation of fiber-reinforced-plastics-based components by means of holographic interferometry,V1400,69-79(1991)
See Bischof, Thomas: V1508,90-95(1991)
See Hoefling, Roland: V1508,135-142(1991)
See Skudayski, Ulf: V1508,68-72(1991)

Jun, Heesung See Keizer, Richard L.: V1406,88-97(1991)

Jung, Christoph
See Dusa, Mircea V.: V1464,447-458(1991)
See Dusa, Mircea V.: V1496,217-223(1991)

Jung, K. H.
; Hsieh, T. Y.; Kwong, Dim-Lee; Spratt, D. B.: Si-based epitaxial growth by rapid thermal processing chemical vapor deposition,V1393,240-251(1991)

Jung, Le-Tien
See Fowler, Burt: V1598,108-117(1991)
See Lian, S.: V1598,98-107(1991)

Jung, Paul See Dusa, Mircea V.: V1464,447-458(1991)

Jung, Peter
; Risken, H.: Use of Fokker-Planck equations for the statistical properties of laser light,V1376,82-93(1991)

Jungbluth, Karl-Heinz See Jahn, Renate: V1424,23-32(1991)

Juodka, B. See Kirveliene, V.: V1403,582-584(1991)

Jurgilewicz, Robert P. See Hillman, Robert L.: V1562,136-142(1991)

Juri, Hugo
; Palma, J. A.; Campana, Vilma; Gavotto, A.; Lapin, R.; Yung, S.; Lillo, J.: Helium-neon effects of laser radiation in rats infected with thromboxane B2,V1422,128-135(1991)

Jurick, Thomas W. See Sarr, Dennis P.: V1469,506-514(1991)

Jursich, Gregory M.
; Von Drusek, William A.; Mulderink, Ken; Olchowka, V.; Reid, John; Brimacombe, Robert K.: Influence of gas composition on XeCl laser performance,V1412,115-122(1991)

Just, Dieter See Ling, Daniel T.: V1452,10-20(1991)

Justus, Brian L.
; Seaver, Mark E.; Ruller, Jacqueline A.; Campillo, Anthony J.: Nonlinear optical susceptibility of CuCl quantum dot glass,V1409,2-8(1991)

Jutamulia, Suganda
; Storti, George M.; Seiderman, William M.; Lindmayer, Joseph: Erasable optical 3D memory using novel electron trapping materials,V1401,113-118(1991)
; Storti, George M.; Seiderman, William M.: Optical neural networks using electron trapping materials,V1558,442-447(1991)

Juvells, Ignacio See Vallmitjana, Santiago: V1564,266-274(1991)

Juza, P. See Hingerl, Kurt: V1361,943-953(1991)

Kaabi, Lotfi
; Loloyan, Mansur; Huang, H. K.: Comparison of morphological and conventional edge detectors in medical imaging applications,V1445,11-23(1991)

Kaaret, Philip E.
; Novick, Robert; Shaw, Ping-Shine; Hanany, Shaul; Liu, Yee; Fleischman, Judith A.; Sunyaev, Rashid; Lapshov, I.; Weisskopf, Martin C.; Elsner, Ronald F.; Ramsey, Brian D.; Silver, Eric H.; Ziock, Klaus P.; Costa, Enrico; Piro, Luigi; Soffitta, Paolo; Manzo, Giuseppe; Giarrusso, Salvatore; Santangelo, Andrea E.; Scarsi, Livio; Fraser, George W.; Pearson, James F.; Lees, John E.; Perola, G. C.; Massaro, Enrico; Matt, Giorgio: Status of the stellar x-ray polarimeter for the Spectrum-X-Gamma mission,V1548,106-117(1991)
See Holley, Jeff: V1343,500-511(1991)
See Shaw, Ping-Shine: V1343,485-499(1991)
See Shaw, Ping-Shine: V1548,118-131(1991)
See Weisskopf, Martin C.: V1343,457-468(1991)

Kaarli, Rein K. See Saari, Peeter M.: V1507,500-508(1991)

Kaasinen, Harri I.
; Kauppi, Ari; Nykanen, Esa: Applications and development of IR techniques for building research in Finland,V1467,90-98(1991)

Kabrisky, Matthew
; Rogers, Steven K.: <author>An Introduction to Biological and Artificial Neural Networks for Pattern Recognition,VTT04(1991)
See Burns, Thomas J.: V1469,208-218(1991)
See Rogers, Steven K.: VIS07,231-243(1991)
See Tarr, Gregory L.: V1469,2-11(1991)
See Veronin, Christopher P.: V1469,281-291(1991)

Kachelmyer, Alan L.
; Nordquist, David P.: Centroid tracking of range-Doppler images,V1416,184-198(1991)
See Eng, Richard S.: V1416,70-85(1991)

Kachurin, O. R. See Antyuhov, V.: V1397,355-365(1991); 1415,48-59(1991)

Kachwalla, Zain
: Frequency response study of traps in III-V compound semiconductors,V1361,784-793(1991)

Kaczelnik, F. See Rotman, Stanley R.: V1442,194-204(1991)

Kadambe, Shubha
; Boudreaux-Bartels, G. F.: Existence of cross terms in the wavelet transform,V1565,423-434(1991)

Kadi, A. M. See Zamorano, Lucia J.: V1428,59-75(1991)

Kadin, Alan M.
; Leung, Michael; Smith, Andrew D.; Murduck, J. M.: Infrared photodetector based on the photofluxonic effect in superconducting thin films,V1477,156-165(1991)
See Mallory, Derek S.: V1477,66-76(1991)

Kadono, Hirofumi
; Toyooka, Satoru: Liquid-crystal phase modulator used in DSPI,V1554A,628-638(1991)
; Toyooka, Satoru: Statistical interferometry based on the statistics of speckle phase,V1554A,718-726(1991)

Kadowaki, Shin-ichi See Kato, Makoto: V1507,36-44(1991)

Kadowaki, Tomoko See Iwasaki, Nobuo: V1490,216-221(1991)

Kadri, Abderrahmane
; Portal, Jean-Claude: Shallow-deep bistability behavior of the DX-centers in n-AlxGa1-xAs and the EL2-defects in n-GaAs,V1362,930-937(1991)

Kaegi, E. M. See Cross, Edward F.: V1540,756-763(1991)

Kaempfer, Niklaus A.
; Bodenmann, P.; Peter, Reto: Ground-based microwave radiometry of ozone,V1491,314-322(1991)
See Gerber, Louis: V1491,211-217(1991)

Kafka, James D.
; Baer, Thomas M.: Ultrafast pulse generation in fiber lasers,V1373,140-149(1991)

Kagann, Robert H. See Russwurm, George M.: V1433,302-314(1991)

Kaganov, S. See Bulichev, A.: V1500,151-162(1991)

Kagel, Joseph H.
: Hardware implementation of a neural network performing multispectral image fusion,V1469,659-664(1991)

Kahan, Lloyd R. See Becerer, Richard J.: V1416,306-313(1991)

Kahlert, Hans-Juergen
; Rebhan, Ulrich; Lokai, Peter; Basting, Dirk: Comparison of 248-nm line narrowing resonator optics for deep-UV lithography lasers,V1463,604-609(1991)

Kahn, Charles E.
See Chen, Chin-Tu: V1444,75-77(1991)
See Chou, Jin-Shin: V1396,45-50(1991)

Kahn, David A. See Hutcheson, Lynn D.: V1363(1991)

Kahn, Steven M.
See Atkinson, Dennis P.: V1343,530-541(1991)
See Bixler, Jay V.: V1549,420-428(1991)

Kahn, Thomas See Bettag, Martin: V1525,409-411(1991)

Kahru, A. See Chekulayev, V.: V1426,367-377(1991)

Kai, Teruhiko
; Takebe, Hiromichi; Morinaga, Kenji: Sputtering of silicate glasses,V1519,99-103(1991)

Kaidalov, S. A. See Bryukhnevitch, G. I.: V1449,109-115(1991)

Kaikati, P. See Boher, Pierre: V1345,165-179(1991)

Kaino, Toshikuni See Shuto, Yoshito: V1560,184-195(1991)

Kaiser, Anne M.
: Optical polysilicon over oxide thickness measurement,V1464,554-565(1991)
: Semiconductor thin-film optical constant determination and thin-film thickness measurement equipment correlation,V1464,386-392(1991)

Kaiser, James F. See Yu, Tian-Hu: V1452,303-309(1991)

Kaiser, Todd J.
; Cardarelli, Donato; Walsh, Joseph: Experimental developments in the RFOG,V1367,121-126(1991)

Kaiser, Winfried
See Freischlad, Klaus R.: V1332,8-17(1991)
See Freischlad, Klaus R.: V1332,18-24(1991)

Kaji, Akihiro See Sakagami, Iwata: V1389,329-339(1991)

Kajii, Makoto See Iwasaki, Nobuo: V1490,192-199(1991)

Kajita, Toru
; Ota, Toshiyuki; Nemoto, Hiroaki; Yumoto, Yoshiji; Miura, Takao: Novel novolak resins using substituted phenols for high-performance positive photoresist,V1466,161-173(1991)

Kajokas, R. See Vaicikauskas, V.: V1440,357-363(1991)

Kajzar, Francois
    See Gadret, G.: V1560,226-237(1991)
    See Morichere, D.: V1560,214-225(1991)

Kakehata, Masayuki
    ; Hashimoto, Etsu; Kannari, Fumihiko; Obara, Minoru: Frequency up-conversion of a discharge-pumped molecular fluorine laser by stimulated Raman scattering in H2,V1397,185-189(1991)

Kakichashvili, Shermazan D.
    : Polarization holographic elements,V1574,101-108(1991)
    ; Kilosanidze, Barbara N.: Three-dimensional holograms in polarization-sensitive media,V1238,74-79(1991)
    ; Wardosanidze, Zurab V.: Spectrally nonselective reflection holograms,V1238,134-137(1991)
    See Imedadze, Theodore S.: V1238,439-441(1991)

Kakihana, Yasuyuki
    ; Tamura, Mamoru: Near-infrared optical monitoring of cardiac oxygen sufficiency through thoracic wall without open-chest surgery,V1431,314-321(1991)

Kakii, Eiji See Uwabu, H.: V1605,928-937(1991)

Kakinoki, Yoshikazu See Koezuka, Tetsuo: V1332,323-331(1991)

Kakizaki, K. See Ishikawa, Ken: V1397,55-58(1991)

Kakizaki, Sunao
    ; Miyazaki, Daisuke; Yoshikawa, Eiji; Tanida, Jun; Ichioka, Yoshiki: Hybrid optical array logic system,V1505,199-205(1991)

Kakuta, Atsushi See Imanishi, Yasuo: V1361,570-580(1991)

Kalata, Kenneth
    ; Phillips, Walter C.; Stanton, Martin J.; Li, Youli: Development of a synchrotron CCD-based area detector for structural biology,V1345,270-280(1991)

Kalhur, Farzeen See Novak, Agneta: V1389,80-86(1991)

Kalin, David A.
    ; Saylor, Danny A.; Street, Troy A.: Simultaneous imaging and interferometric turbule visualization in a high-velocity mixing/shear layer,V1358,780-787(1991)
    See Couch, Lori L.: V1486,417-423(1991)
    See Peters, Bruce R.: V1486,410-416(1991)

Kalinin, Victor P. See Sherstobitov, Vladimir E.: V1415,79-89(1991)

Kalinkin, Vyacheslav V. See Sherstyuk, Valentin P.: V1238,218-223(1991)

Kalinowski, Hypolito J. See Guedes Valente, Luiz C.: V1584,96-102(1991)

Kalisky, Yehoshua Y. See Chuchem, D.: V1397,277-281(1991)

Kallhammer, Jan-Erik O.
    : Digitize your films without losing resolution,V1358,631-636(1991)

Kallikorm, A. P.
    See Chekulayev, V.: V1426,367-377(1991)
    See Nevorotin, Alexey J.: V1427,381-397(1991)

Kallis, N. See Landwehr, Gottfried: V1362,282-290(1991)

Kallivayalil, Jacob A. See Zehnder, Alan T.: V1554A,48-59(1991)

Kallman, Robert R.
    ; Goldstein, Dennis H.: Invariant phase-only filters for phase-encoded inputs,V1564,330-347(1991)

Kalman, Robert F.
    ; Dias, Antonio R.; Chau, Kelvin K.; Goodman, Joseph W.: Optical switches based on semiconductor optical amplifiers,V1563,34-44(1991)
    ; Digonnet, Michel J.; Wysocki, Paul F.: Modeling of three-level laser superfluorescent fiber sources,V1373,209-222(1991)
    See Wysocki, Paul F.: V1373,66-77(1991)

Kalomiris, Vasilios E. See Darden, Bruce V.: V1474,300-308(1991)

Kalson, S. Z.
    : Recent results in adaptive array detection,V1566,406-418(1991)

Kalt, Heinz See Rinker, Michael: V1362,14-23(1991)

Kaltenbacher, Eric See Harding, Kevin G.: V1385,246-255(1991)

Kalvin, Alan D.
    ; Cutting, Court B.; Haddad, Betsy; Noz, Marilyn E.: Constructing topologically connected surfaces for the comprehensive analysis of 3-D medical structures,V1445,247-258(1991)

Kalweit, Edith
    See Bristow, Julian P.: V1389,535-546(1991)
    See Hibbs-Brenner, Mary K.: V1563,10-20(1991)

Kalyashora, L. N.
    ; Michailova, A. G.; Pavlov, A. P.; Rykov, V. S.; Guba, B. S.: High-sensitive layers of dichromated gelatin for hologram recording by continuous wave and pulsed laser radiation,V1238,189-194(1991)

Kalyashov, E. V.
    ; Tyutchev, M. V.: Self-diffraction for active stabilization of interference field for reflection hologram recording,V1238,442-446(1991)

Kalyayev, Anatoli V.
    ; Brukhomitsky, Yuri A.; Galuyev, Gennady A.; Chernukhin, Yu. V.: Preconditions and prospects for the construction of parallel digital neurocomputers with programmable architecture,V1621,299-308(1991)

Kalymnios, Demetrios See Bougas, V.: V1504,298-302(1991)

Kamal, Arvind See Sun, Mei H.: V1420,44-52(1991)

Kamal, Avais
    ; Terhune, Robert W.; Weinberger, Doreen A.: Dynamics of self-organized x(2) gratings in optical fibers,V1516,137-153(1991)

Kamalakar, J. A.
    ; Jain, Yashwant K.; Laxmiprasad, A. S.; Shashikala, M.: Static horizon sensor for remote sensing satellite,V1478,92-100(1991)
    See Alex, T. K.: V1478,106-111(1991)

Kamarainen, Veikko J. See Sipila, Heikki: V1549,246-255(1991)

Kamata, Seiichiro
    See Eason, Richard O.: V1383,566-573(1991)
    See Perez, Arnulfo: V1605,879-884(1991)
    See Perez, Arnulfo: V1567,354-361(1991)
    See Perez, Arnulfo: V1452,430-441(1991)

Kameda, Yoshihiko See Tsuno, Katsuhiko: V1490,222-232(1991)

Kamei, Shuya
    ; Hirata, Masaru: Visualization study on pool boiling heat transfer,V1358,979-983(1991)

Kamemaru, Shun-ichi
    ; Yano, Jun-ichi; Itoh, Haruyasu: Matched spatial filtering by feature-extracted reference patterns using cross-correlated signals,V1564,143-154(1991)

Kamen, Edward W. See Sastry, C. R.: V1481,261-280(1991)

Kamgar-Parsi, Behzad See Rosenblum, Lawrence: V1459,274-282(1991)

Kaminski, Robert L.
    : Integrated C3I optical processor,V1564,156-164(1991)

Kamitsubo, Hiromichi
    : 8-GeV synchrotron radiation facility at Nishi-Harima,V1345,16-25(1991)

Kamiya, Masami See Ogura, Toshihiro: V1443,153-157(1991)

Kamiya, Takeshi See Kodate, Kashiko: V1545,251-260(1991)

Kamizawa, Koh See Koshi, Yutaka: V1605,362-373(1991)

Kamon, Kazuya
    See Miyazaki, Junji: V1464,327-335(1991)
    See Op de Beeck, Maaike: V1463,180-196(1991)

Kamoshida, Minoru
    ; Enomoto, Hajime; Miyamura, Isao: Window-based elaboration language for picture processing and painting,V1606,951-960(1991)

Kamp, B. See Andresen, Klaus: V1554A,93-100(1991)

Kampf, Dirk
    See Graue, Roland: V1494,377-385(1991)
    See Rippel, Harald: V1343,520-529(1991)

Kamrukov, A. S.
    ; Kozlov, N. P.; Opekan, A. G.; Protasov, Yu. S.; Rudoi, I. G.; Soroka, A. M.: Atomic xenon recombination laser excited by thermal ionizing radiation from a magnetoplasma compressor and discharge,V1503,438-452(1991)
    ; Kozlov, N. P.; Protasov, Yu. S.: Visible and UV gas lasers with high-current radiative discharges excitation,V1397,137-144(1991)

Kamshilin, Alexei A. See Vlad, Ionel V.: V1332,236-244(1991)

Kan, Kou-Sou See Lee, Ching-Long: V1606,728-734(1991)

Kanabe, Tadashi See Nakatsuka, Masahiro: V1411,108-115(1991)

Kanai, Toshitaka See Ito, Katsunori: V1499,382-385(1991)

Kanayama, Toshihiko See Tomie, Toshihisa: V1552,254-263(1991)

Kanazawa, Hirotaka
    ; Yamaguchi, Naohito; Nakajima, Takuro; Hamano, Yasunori; Satani, Ryoichi; Taira, Tatsuji: Power stabilization of high-power industrial CO laser using gas exchange,V1397,445-448(1991)
    See Yokozawa, T.: V1397,513-517(1991)

Kandel, Robert S. See Monge, J.L.: V1490,84-93(1991)

Kandelaki, M. D. See Volkov, Alexander G.: V1436,68-79(1991)

Kane, Jonathan S.
    ; Hemmer, Philip R.; Woods, Charles L.; Khoury, Jehad: Binary phase-only filter associative memory,V1564,511-520(1991)

Kane, Thomas J. See Cheng, Emily A.: V1417,300-306(1991)

Kanechika, Hideaki See Nagano, Yasutada: V1605,614-623(1991)

Kaneda, Yushi See Kubota, Shigeo: V1354,572-580(1991)

Kanefsky, Morton See Falk, Joel: V1409,83-86(1991)

Kanehori, Keiichi
; Sugii, Nobuyuki: Growth of oxide superconducting thin films by plasma-enhanced MOCVD,V1394,238-243(1991)

Kaneko, Masahiko
See Aratani, Katsuhisa: V1499,209-215(1991)
See Fukumoto, Atsushi: V1499,216-225(1991)
See Maeda, Fumisada: V1499,62-69(1991)

Kanellopoulos, S. E.
; Guedes Valente, Luiz C.; Handerek, Vincent A.; Rogers, Alan J.: Comparison of photorefractive effects and photogenerated components in polarization-maintaining fibers,V1516,200-210(1991)

Kanemaru, Toyomi
; Nakajima, Takashi; Igarashi, Tadanao; Masuda, Rika; Orita, Nobuyuki: E-beam data compaction method for large-capacity mask ROM production,V1496,118-123(1991)

Kanevsky, M. F.
; Bolshov, L. A.; Chernov, S. Y.; Vorobjev, V. A.: Computer modeling of unsteady gas-dynamical and·optical phenomena in low-temperature laser plasma,V1440,154-165(1991)

Kang, Chih-Chieh
; Fiddy, Michael A.: Nonlinear optical properties of suspensions of microparticles: electrostrictive effect and enhanced backscatter,V1497,372-381(1991)

Kang, Dae I.
; Kwon, Young H.; Ko, Yeong-Uk: Measurement of thermal deformation of square plate by using holographic interferometry,V1554B,139-147(1991)

Kang, Deokman See Hoyle, Charles E.: V1559,202-213(1991)

Kang, Der-Kuan See Honda, Toshio: V1461,156-166(1991)

Kang, Moon G.
; Lay, Kuen-Tsair; Katsaggelos, Aggelos K.: Image restoration algorithms based on the bispectrum,V1606,408-418(1991)

Kang, Songgao
; Lu, Kaichang; Zhu, Yafei: Comments on the Seidel aberration theory,V1527,376-379(1991)
; Lu, Kaichang; Zhu, Yafei: New formulations between spherical aberration and spherical aberration coefficient using the Abbe sine condition,V1527,409-412(1991)
; Lu, Kaichang; Zhu, Yafei: New method for calculating third-, fifth-, and seventh-order spherical aberration coefficients and aberration offenses against sine condition,V1527,400-405(1991)
; Lu, Kaichang; Zhu, Yafei: Revision of the Seidel aberration theory for application in range of small viewing field and large aperture,V1527,406-408(1991)
See Lu, Kaichang: V1527,413-418(1991)

Kang, Yilan
; Jia, Youquan; Du, Ji: Determination of thermal stresses in a bimaterial specimen by moire interferometry,V1554B,514-522(1991)
See Jia, Youquan: V1554B,331-334(1991)
See Lin, Li-Huang: V1410,65-71(1991)
See Lin, Li-Huang: V1346,490-501(1991)

Kangarloo, Hooshang
See Huang, H. K.: V1443,198-202(1991)
See Huang, H. K.: V1444,214-220(1991)
See Razavi, Mahmood: V1446,24-34(1991)

Kanhere, R. See Chua, Soo-Jin: V1401,96-102(1991)

Kanjilal, A. K. See Chitnis, Vijay T.: V1332,613-622(1991)

Kankelborg, Charles C.
See DeForest, Craig E.: V1343,404-414(1991)
See Walker, Arthur B.: V1343,319-333(1991)

Kann, J. L. See Milster, Tom D.: V1499,348-353(1991)

Kannari, Fumihiko
; Obara, Minoru: Characteristics of amplification of ultrashort laser pulses in excimer media,V1397,85-89(1991)
; Sato, F.; Obara, Minoru: Assessment of a discharge-excited supersonic free jet as a laser medium,V1397,493-497(1991)
See Ashidate, Shu-ichi: V1397,283-286(1991)
See Kakehata, Masayuki: V1397,185-189(1991)

Kannewurf, C. R. See Wessels, Bruce W.: V1394,232-237(1991)

Kanofsky, Alvin S.
; Minford, William J.; Watson, James E.: Radiation effects on dynamical behavior of LiNbO3 switching devices,V1374,59-66(1991)
; Spector, Magaly; Remke, Ronald L.; Witmer, Steve B.: Radiation effects on GaAs optical system FET devices,V1374,48-58(1991)

Kanterakis, Emmanuel G. See Zhang, Yan: V1481,23-31(1991)

Kanungo, Tapas
; Jaisimha, Mysore Y.; Haralick, Robert M.; Palmer, John: Experimental methodology for performance characterization of a line detection algorithm,V1385,104-112(1991)

Kao, Chi-Chang
; Johnson, Erik D.; Hastings, Jerome B.; Siddons, D. P.; Vettier, C.: Soft x-ray resonant magnetic scattering study of thin films and multilayers,V1548,149-157(1991)

Kao, Ming-Seng
: Effect of polarization-mode dispersion on coherent optical distribution systems with shared local oscillator,V1579,221-229(1991)
See Lee, Yang-Hang: V1579,155-166(1991)

Kao, Yi-Han See Witanachchi, S.: V1394,161-168(1991)

Kao, Yung C.
See Hwang, Kiu C.: V1371,128-137(1991)
See Saunier, Paul: V1475,86-90(1991)

Kapasi, S.
See Dobbins, B. N.: V1554A,772-780(1991)
See Dobbins, B. N.: V1554B,586-592(1991)

Kapetanakos, Christos A. See Golden, Jeffry: V1407,418-429(1991)

Kaplan, L. See George, John S.: V1443,37-51(1991)

Kaplan, Martin
: Monte Carlo calculation of light distribution in an integrating cavity illuminator,V1448,206-217(1991)

Kaplin, Peter L. See Tregay, George W.: V1589,38-47(1991)

Kapp, Oscar H.
See Ruan, Shengyang: V1396,298-310(1991)
See Ryan, Martin J.: V1396,335-339(1991)
See Zmola, Carl: V1396,51-55(1991)
See Zmola, Carl: V1396,331-334(1991)

Kappers, Lawrence A. See Scripsick, Michael P.: V1561,93-103(1991)

Kapron, Felix P.
: Theory of stressed fiber lifetime calculations,V1366,136-143(1991)
See Yuce, Hakan H.: V1366,144-156(1991)

Kapsenberg, Th. See Nieuwkoop, E.: V1511,255-263(1991)

Karageorgiou, Angelos See Parikh, Jo Ann: V1469,526-538(1991)

Karagozian, Ann See Cadou, Christopher P.: V1434,67-77(1991)

Karahalios, Margarida See Rosenblum, Lawrence: V1459,274-282(1991)

Karaian, A. S. See Arakelian, Sergei M.: V1402,175-191(1991)

Karalyunts, A. See Popov, A.: V1519,457-462(1991)

Karasaki, Hidehiko See Hongu, Hitoshi: V1501,198-204(1991)

Karasev, Viatcheslav B.
See Altshuler, Grigori B.: V1440,116-126(1991)
See Azarenkov, Aleksey N.: V1409,154-164(1991)

Karbassiyoon, Kamran
; Greenwell, Roger A.; Scott, David M.; Spencer, Robert A.: Radiation effects on bend-insensitive fibers at 1300 nm and 1500 nm,V1366,178-183(1991)

Karecki, David R. See Mackay, Gervase I.: V1433,104-119(1991)

Kargin, Igor U. See Altshuler, Grigori B.: V1415,269-280(1991)

Karim, Douglas P.
: Bandwidth measurements of polymer optical fibers,V1592,31-41(1991)

Karim, Mohammad A. See Perez, Osvaldo: V1471,255-264(1991)

Karimi, B.
; Baradaran, T.; Ashenayi, Kaveh; Vogh, James: Comparison of sinusoidal perceptron with multilayer classical perceptron,V1396,226-236(1991)

Karimov, M. G.
: Use of intelligent multichannel analyzer for study of biological objects,V1402,209-215(1991)

Karioja, Pentti
; Tammela, Simo; Hannula, Tapio: Integrated optics bus access module for intramachine communication,V1533,129-137(1991)

Karis, Thomas E.
; Best, Margaret E.; Logan, John A.; Lyerla, James R.; Lynch, Robert T.; McCormack, R. P.: Verification of tracking servo signal simulation from scanning tunneling microscope surface profiles,V1499,366-376(1991)

Karkanis, S.
; Tsoutsou, K.; Vergados, J.; Dimitriadis, Basile D.: Real-time quality inspection system for textile industries,V1500,164-170(1991)

Karlsson, L. M. See Cox, Jerome R.: V1446,40-51(1991)

Karne, Ramesh K.
; Sood, Arun K.: High-level parallel architecture for a rule-based system,V1468,938-949(1991)

Karnis, A. See Stationwala, Mustafa I.: V1358,237-245(1991)

Karol, Mark J.
: Architectures and access protocols for multichannel networks,V1579,2-13(1991)

Karp, Christopher K. See Sellar, R. G.: V1479,140-155(1991)

Karpetscki, V. V. See Alejnikov, Vladislav S.: V1412,173-184(1991)

Karpierz, Miroslaw A. See Domanski, Andrzej W.: V1505,59-66(1991)

Karpov, D. A.
; Nazikov, S. N.: Multicomponent electric-arc source of metallic plasma,V1519,115-121(1991)

Karpov, V. I.
; Korobkin, V. V.; Naboichenko, A. K.; Dolgolenko, D. A.: Multistage XeCl excimer system "Cactus" and some investigations of stimulated scattering in liquids,V1503,492-502(1991)

Karpova, S. G. See Vlasov, N. G.: V1238,332-337(1991)

Karppinen, Arto
; Kopola, Harri K.; Myllyla, Risto A.: Scattering liquid crystal in optical attenuator applications,V1455,179-189(1991)

Karr, Charles L.
: Design of a cart-pole balancing fuzzy logic controller using a genetic algorithm,V1468,26-36(1991)

Karr, Michael A. See Albanese, Andres: V1364,320-326(1991)

Karsch, Karl R.
See Neu, Walter: V1425,28-36(1991)
See Neu, Walter: V1425,37-44(1991)
See Nyga, Ralf: V1525,119-123(1991)

Karssemeijer, Nico
; van Erning, Leon J.: Iso-precision scaling of digitized mammograms to facilitate image analysis,V1445,166-177(1991)

Karstensen, Holger See Zarschizky, Helmut: V1389,484-495(1991)

Karthe, Wolfgang
See Rasch, Andreas: V1511,149-154(1991)
See Rasch, Andreas: V1522,83-92(1991)

Karthik, Sankaran
; Nandhakumar, N.; Aggarwal, Jake K.: Modeling nonhomogeneous 3-D objects for thermal and visual image synthesis,V1468,686-697(1991)

Karube, Norio
: Industrial lasers in Japan,V1517,1-14(1991)

Karuppiah, N.
; John, S.; Natarajan, Sanjay S.; Sivan, V.: Electrodeposited nickel-cobalt thin films for photothermal conversion of solar energy,V1536,215-221(1991)

Karvelis, K. C. See Wang, David C.: V1445,574-575(1991)

Karwoski, Ronald A. See Higgins, William E.: V1445,276-286(1991)

Kasahara, Eiji See Ohue, Hiroshi: V1397,527-530(1991)

Kasahara, Jack S.
; Dusa, Mircea V.; Perera, Thiloma: Evaluation of a photoresist process for 0.75-micron g-line lithography,V1463,492-503(1991)

Kasai, Paul H. See McKean, Dennis R.: V1559,214-221(1991)

Kasama, Kunihiko
See Itoh, Katsuyuki: V1466,485-496(1991)
See Ohfuji, Takeshi: V1463,345-354(1991)

Kasemann, Reiner See Schmidt, Helmut: V1590,36-43(1991)

Kasemset, Dumrong
; Ackerman, Edward I.; Wanuga, Stephen; Herczfeld, Peter R.; Daryoush, Afshin S.: Comparison of alternative modulation techniques for microwave optical links,V1371,104-114(1991)
See Wanuga, Stephen: V1374,97-106(1991)

Kashifian, Payam See Razavi, Mahmood: V1446,24-34(1991)

Kashima, Yasumasa
; Matoba, Akio; Kobayashi, Masao; Takano, Hiroshi: 1.55-um superluminescent diode for a fiber optic gyroscope,V1365,102-107(1991)

Kashimura, Hirotsugu See Kobayashi, Kenichi: V1448,157-163(1991)

Kashin, Peter See Cai, Wen: V1528,118-128(1991)

Kashiwabara, S.
; Watanabe, Kazuhiro; Fujimoto, R.: Highly conductive amorphous-ferrite formed by excimer laser material processing,V1397,803-806(1991)

Kashyap, Raman
; Davey, Steven T.; Williams, Doug L.: Phase-matched second-harmonic generation of infrared wavelengths in optical fibers,V1516,164-174(1991)
See Campbell, Robert J.: V1499,160-164(1991)
See Pantelis, Philip: V1559,2-9(1991)
See Williams, Doug L.: V1513,158-167(1991)
See Williams, Doug L.: V1516,29-37(1991)

Kasiewicz, Allen B. See Annamalai, Kadiresan: V1364(1991)

Kasinski, Jeff
See Koechner, Walter: V1522,169-179(1991)
See Marshall, Larry R.: V1419,141-152(1991)

Kasky, Thomas J. See Gardner, Kenneth R.: V1481,254-260(1991)

Kasle, David B.
; Morgan, Jeffrey S.: High-resolution decoding of multianode microchannel array detectors,V1549,52-58(1991)

Kasody, R. See Liu, Louis: V1475,193-198(1991)

Kasotakis, G. See Hora, Heinrich: V1502,258-269(1991)

Kasparis, Takis
; Tzannes, Nicolaos S.; Bassiouni, Mostafa; Chen, Qing: Fractal-based multifeature texture description,V1521,46-54(1991)

Kasper, Erich
; Presting, Hartmut: Device concepts for SiGe optoelectronics,V1361,302-312(1991)

Kasprzak, Henryk
; Podbielska, Halina; Pennig, Dietmar: Speckle photography for investigation of bones supported by different fixing devices,V1429,55-61(1991)
See Foerster, Werner: V1429,146-151(1991)

Kassam, Samer See Duparre, Angela: V1530,283-286(1991)

Kassi, Hassane
; Hotchandani, Surat; Leblanc, Roger M.; Beaudoin, N.; Dery, M.: Pulsed photoconductivity of chlorophyll a,V1436,58-67(1991)

Kasson, James M.
; Plouffe, Wil: Subsampled device-independent interchange color spaces,V1460,11-19(1991)

Kast, Brian A.
; Dickey, Fred M.: Normalization of correlations,V1564,34-42(1991)
See Connelly, James M.: V1564,572-592(1991)
See Hendrix, Charles D.: V1564,2-13(1991)
See Molley, Perry A.: V1471,224-232(1991)

Kasuya, Koichi
; Horioka, K.; Hikida, N.; Murazi, M.; Nakata, K.; Kawakita, Y.; Miyai, Y.; Kato, S.; Yoshida, S.; Ideno, S.: Measurements of gas flow and gas constituent in a wind-tunnel-type excimer laser under high-repetition-rate operations,V1397,67-70(1991)
See Sato, Heihachi: V1397,331-338(1991)

Katagiri, Souichi See Terasawa, Tsuneo: V1463,197-206(1991)

Katagiri, Yoshitada See Ukita, Hiroo: V1499,248-262(1991)

Kataoka, Kenji See Torao, Akira: V1358,843-850(1991)

Katayama, Norihisa
; Fukui, Masahiko; Ozaki, Yukihiro; Araki, Toshinari; Yokoi, Seiichi; Iriyama, Keiji: Raman and FT-IR characterization of biologically relevant Langmuir-Blodgett films,V1403,147-149(1991)

Katayama, Ryuichi See Yamanaka, Yutaka: V1499,263-268(1991)

Katayama, Takuma See Shimizu, Masaru: V1519,122-127(1991)

Katayama, Yoshifumi See Akita, K.: V1392,576-587(1991)

Katchalov, O. V. See Zaitseva, N. P.: V1402,223-230(1991)

Katehi, Linda P. See van Deventer, T. E.: V1389,285-296(1991)

Kathman, Alan D.
; Pitalo, Stephen K.: Binary optics in lens design,V1354,297-309(1991)
; Temmen, Mark G.; Scott, Miles L.: Microlens array for modification of SLM devices,V1544,58-65(1991)
See Banish, Michele R.: V1557,209-221(1991)
See Brown, Daniel M.: V1555,114-127(1991)
See Johnson, Eric G.: V1545,209-216(1991)

Kato, M. See Yamakoshi, H.: V1397,119-122(1991)

Kato, Makoto
; Kadowaki, Shin-ichi; Komma, Yoshiaki; Hori, Yoshikazu: Recent advances in optical pickup head with holographic optical elements,V1507,36-44(1991)

Kato, Masayuki See Kostuk, Raymond K.: V1389,515-522(1991)

Kato, Naoki
; Sekura, Rieko; Yamanaka, Junko; Ebihara, Teruo; Yamamoto, Shuhei: Characteristics of a ferroelectric liquid crystal spatial light modulator with a dielectric mirror,V1455,190-205(1991)

Kato, S. See Kasuya, Koichi: V1397,67-70(1991)

Kato, Takahito
; Yamada, Mitsuho: Quality factors of handwritten characters based on human visual perception,V1453,43-50(1991)

Katoh, Kikuji See Watabe, Akinori: V1499,226-235(1991)

Katsaggelos, Aggelos K.
; Kleihorst, Richard P.; Efstratiadis, Serafim N.; Lagendijk, Reginald L.: Adaptive image sequence noise filtering methods,V1606,716-727(1991)
See Chan, Cheuk L.: V1450,208-217(1991)
See Efstratiadis, Serafim N.: V1605,16-25(1991)
See Galatsanos, Nikolas P.: V1396,590-600(1991)
See Kang, Moon G.: V1606,408-418(1991)
See Kwak, J. Y.: V1396,32-44(1991)

Katsap, Victor N.
; Koshevoy, Alexander V.; Meerovich, Gennady A.; Ulasjuk, Vladimir N.: Promising applications of scanning electron-beam-pumped laser devices in medicine and biology,V1420,259-265(1991)

Katsouleas, Thomas C. See Dodd, James W.: V1407,467-473(1991)

Katsuma, Hidetoshi See Sato, Koki: V1461,124-131(1991)

Katsuyama, Yoshihiko See Tanii, Jun: V1490,200-206(1991)

Kattawar, George W.
ed.: Selected Papers on Multiple Scattering in Plane Parallel Atmospheres and Oceans: Methods,VMS42(1991)

Katto, Jiro
; Yasuda, Yasuhiko: Performance evaluation of subband coding and optimization of its filter coefficients,V1605,95-106(1991)

Katz, Al See Zhang, Yan: V1481,23-31(1991)

Katz, Avishay
: Rapid thermal processing in the manufacturing technology of contacts to InP-based photonic devices,V1393,67-77(1991)
See Pearton, Stephen J.: V1393,150-160(1991)

Katz, Barton A.
See Chu, Ron: V1465,238-243(1991)
See van den Brink, Martin A.: V1463,709-724(1991)

Katz, David A. See DaPonte, John S.: V1469,441-450(1991)

Katz, Howard E.
; Schilling, M. L.; Ungashe, S.; Putvinski, T. M.; Scheller, G. E.; Chidsey, C. E.; Wilson, William L.: Electrical and nonlinear optical properties of zirconium phosphonate multilayer assemblies,V1560,370-376(1991)

Katz, Philip L. See Haralick, Robert M.: V1472,108-117(1991)

Katzir, Abraham
ed.: Optical Fibers in Medicine VI,V1420(1991)
See Chance, Britton: V1431(1991)
See Drizlikh, S.: V1420,53-62(1991)
See Paiss, Idan: V1420,141-148(1991)

Kaufel, G. See Huelsmann, Axel: V1465,201-208(1991)

Kauffman, Michael T. See Boyd, Joseph T.: V1374,138-143(1991)

Kaufman, Harold R. See Wei, David T.: V1519,47-55(1991)

Kaufman, Howard
See Damour, Kevin: V1452,78-89(1991)
See Steinvorth, Rodrigo: V1387,136-147(1991)

Kaufman, James H.
; Metin, Serhat; Saperstein, David D.: Optical studies of nitrogen-doped amorphous carbon: laboratory and interstellar investigations,V1437,36-41(1991)

Kaufman, Leon See Liao, Wen-gen: V1444,47-55(1991)

Kaufmann, Raimund
See Hennig, Thomas: V1424,99-105(1991)
See Rechmann, Peter: V1424,106-115(1991)
See Schwarzmaier, Hans-Joachim: V1427,128-133(1991)

Kaul, Neeraj
; Biswas, Gautam: Multilevel qualitative reasoning in CMOS circuit analysis,V1468,204-215(1991)

Kaup, Andre
; Aach, Til: Variable-blocksize transform coding of four-color printed images,V1605,420-427(1991)

Kauppi, Ari See Kaasinen, Harri I.: V1467,90-98(1991)

Kauppinen, Timo See Tervo, Matti: V1467,161-168(1991)

Kaura, Sushil K. See Aggarwal, Anil K.: V1238,18-27(1991)

Kautz, Gregory M. See Zoltowski, Michael D.: V1566,452-463(1991)

Kavanagh, Katherine M. See Laxer, Cary: V1444,190-195(1991)

Kavaya, Michael J. See Targ, Russell: V1478,211-227(1991)

Kavechansky, J. See Hianik, T.: V1402,93-96(1991)

Kawabe, Yasumasa See Uenishi, Kazuya: V1466,102-116(1991)

Kawada, Masakuni
; Fujisada, Hiroyuki: Mechanical cooler development program for ASTER,V1490,299-308(1991)

Kawaguchi, Eiji
See Perez, Arnulfo: V1605,879-884(1991)
See Perez, Arnulfo: V1567,354-361(1991)
See Perez, Arnulfo: V1452,430-441(1991)

Kawahara, Hideo
; Sakai, Y.; Goda, Takuji; Hishinuma, Akihiro; Takemura, Kazuo: Preparation of SiO2 film utilizing equilibrium reaction in aqueous solution,V1513,198-203(1991)
See Kitaoka, Masaki: V1519,109-114(1991)

Kawahara, Nobuo See Hatanaka, Hidekazu: V1397,379-382(1991)

Kawai, Akira
; Fujiwara, Keiji; Tsujita, Kouichirou; Nagata, Hitoshi: Characterization of automatic overlay measurement technique for sub-half-micron devices,V1464,267-277(1991)

Kawai, Kenji
; Panarella, Emilio; Mostacci, D.: Spherical pinch x-ray generator prototype for microlithography,V1465,308-314(1991)

Kawai, N. See Matsuoka, Masaru: V1549,2-8(1991)

Kawakami, Hajimu
: Selective edge detection based on harmonic oscillator wave functions,V1468,156-166(1991)

Kawakami, Haruo See Ozawa, Kenji: V1499,180-186(1991)

Kawakami, Shinichiro
See Inokuchi, Ikuo: V1429,39-45(1991)
See Maeta, Manabu: V1429,152-161(1991)

Kawakita, Y. See Kasuya, Koichi: V1397,67-70(1991)

Kawakubo, Yukio See Ogura, Satoshi: V1412,123-128(1991)

Kawamonzen, Yosiaki See Mori, Yasushi: V1560,310-314(1991)

Kawamoto, M. See Misaka, Akio: V1465,174-184(1991)

Kawanishi, Hidenori See Akita, K.: V1392,576-587(1991)

Kawano, Sumio
; Iwamoto, Mutsuo: Advances in R&D in near-infrared spectroscopy in Japan,V1379,2-9(1991)

Kawasaki, Tetsuharu See Miyoshi, Masahiro: V1499,116-119(1991)

Kawase, Liliana R.
See Brooking, Nicholas L.: V1572,88-93(1991)
See Guedes Valente, Luiz C.: V1584,96-102(1991)

Kawashima, Masahisa
; Tominaga, Hideyoshi: Method to convert image resolution using M-band-extended QMF banks,V1605,107-111(1991)

Kawata, Koichi
See Kawauchi, Yoshikazu: V1555,224-227(1991)
See Kawauchi, Yoshikazu: V1507,538-543(1991)

Kawatani, Takahiko See Miyahara, Sueharu: V1384,317-323(1991)

Kawate, K. See Kuroda, Masami: V1458,155-161(1991)

Kawato, Suguru
: Visualization of electron transfer interactions of membrane proteins,V1429,127-131(1991)

Kawauchi, Yoshikazu
; Kawata, Koichi: Vision sensor system for the environment of disaster,V1507,538-543(1991)
; Toyoda, Ryuuichi; Kimura, Minoru; Kawata, Koichi: Holographic disk scanner for active infrared sensors,V1555,224-227(1991)

Kay, Armin V.
: Slow-motion acquisition of laser beam profiles after propagation through gun blast,V1486,8-16(1991)

Kay, David B.
; Chase, Scott B.; Gage, Edward C.; Silverstein, Barry D.: Write noise from optical heads with nonachromatic beam expansion prisms,V1499,281-285(1991)

Kay, Michael G. See Luo, Ren C.: V1383,537-544(1991)

Kayaalp, Ali E.
 ; Moezzi, Saied; Davies, Henry C.: High-speed sensor-based systems for mobile robotics,V1406,98-109(1991)
 See Davies, Henry C.: V1388,165-175(1991)

Kayser, O. See Geurts, Jean: V1361,744-750(1991)

Kazanskiy, Nikolay L.
 See Golub, Mikhail A.: V1500,194-206(1991)
 See Golub, Mikhail A.: V1500,211-221(1991)

Kazansky, Peter G.
 See Dianov, Evgeni M.: V1516,81-98(1991)
 See Dianov, Evgeni M.: V1516,75-80(1991)

Kaziska, Andrew J. See Wittmeyer, Stacey A.: V1435,267-278(1991)

Kazmierski, Christophe See Sermage, Bernard: V1362,617-622(1991)

Kazmirowski, Antoni See Wolinski, Wieslaw: V1391,334-340(1991)

Kazovsky, Leonid G.
 ed.: *Advanced Fiber Communications Technologies*,V1579(1991)
 See Benedetto, Sergio: V1579,112-121(1991)
 See Poggiolini, Pierluigi T.: V1579,14-29(1991)
 See Tonguz, Ozan K.: V1579,179-183(1991)

Ke, Ying See Wong, Kam W.: V1526,17-26(1991)

Keagy, Pamela M. See Schatzki, Thomas F.: V1379,182-188(1991)

Kean, P. N. See Dixon, G. J.: V1561,147-150(1991)

Keane, Christopher J.
 See Amendt, Peter: V1413,59-69(1991)
 See Wilks, S. C.: V1413,131-137(1991)

Keane, William K.
 : Policy issues affecting telephone company opportunities in CATV and deployment of fiber optic cable,V1363,184-185(1991)

Kear, B. H.
 See Chern, Chyi S.: V1394,255-265(1991)
 See Zhao, Jing-Fu: V1362,135-143(1991)

Kearney, Patrick A.
 ; Slaughter, Jon M.; Falco, Charles M.: Selection, growth, and characterization of materials for MBE-produced x-ray optics,V1343,25-31(1991)
 See Slaughter, Jon M.: V1343,73-82(1991)

Kearsley, Andrew J. See Naylor, Graham A.: V1377,107-112(1991)

Keating, Kenneth B. See Booth, Bruce L.: V1377,57-63(1991)

Kebabian, Paul L. See McManus, J. B.: V1433,330-339(1991)

Kechemair, Didier
 See Bataille, F.: V1502,135-139(1991)
 See Bataille, F.: V1397,839-842(1991)
 See Caillibotte, Georges: V1502,117-122(1991)
 See Caillibotte, Georges: V1412,209-211(1991)

Kecik, Tadeusz
 ; Switka-Wieclawska, Iwona; Ciszewska, Joanna; Portacha, Lidia: Use of krypton laser stimulation in the treatment of dry eye syndrome,V1391,341-345(1991)

Kee, Richard C.
 : High-resolution CCD still/video and still-still/video systems,V1448,13-20(1991)

Keeble, Peter J.
 : Optical time-domain reflectometry performance enhancement using erbium-doped fiber amplifiers,V1366,39-44(1991)

Keefer, Christopher W. See Welstead, Stephen T.: V1565,482-491(1991)

Keeler, Elaine K.
 ; Oyen, Ordean J.: Three-dimensional magnetic resonance imaging of the head,V1380,24-32(1991)

Keenan, W. A.
 ; Johnson, Walter H.; Hodul, David T.; Mordo, David: RTP temperature uniformity mapping,V1393,354-365(1991)
 See Raicu, Bruha: V1393,161-171(1991)

Keery, William J. See Postek, Michael T.: V1464,35-47(1991)

Keeter, Howard S.
 ; Gudmundson, Glen A.; Woodall, Milton A.: SIRE (sight-integrated ranging equipment): an eyesafe laser rangefinder for armored vehicle fire control systems,V1419,84-93(1991)

Kegelman, Thomas D.
 : Paraxial electron imaging system,V1454,2-10(1991)

Kehtarnavaz, Nasser
 ; Griswold, Norman C.; Eem, J. K.: Comparison of mono- and stereo-camera systems for autonomous vehicle tracking,V1468,467-478(1991)
 See Griswold, Norman C.: V1388,342-349(1991)

Keijzer, Marleen See Jacques, Steven L.: V1422,2-13(1991)

Keil, Norbert
 ; Strebel, Bernhard N.; Yao, HuiHai; Krauser, Juergen: Applications of optical polymer waveguide devices on future optical communication and signal processing,V1559,278-287(1991)

Keinigs, Rhon K.
 : Tandem betatron accelerator,V1407,456-466(1991)

Keitzer, Scott
 ; Kimbrell, James E.; Greenwald, David: Deterministic errors in pointing and tracking systems I: identification and correction of static errors,V1482,406-414(1991)

Keizer, Richard L.
 ; Jun, Heesung; Dunn, Stanley M.: Structured light: theory and practice and practice and practice...,V1406,88-97(1991)

Kekelidze, George N. See Prokhorov, Kirill A.: V1501,80-84(1991)

Kelderman, E. See van Hulst, Niek F.: V1361,581-588(1991)

Kelkar, Atul G.
 ; Alberts, Thomas E.: Optimal control/structure integrated design of a flexible space platform with articulated appendages,V1489,243-253(1991)

Kell, K.-J. E. See Aben, Hillar K.: V1554A,298-309(1991)

Kellam, Mark D. See Jones, Susan K.: V1464,546-553(1991)

Kelleher, Tracy A. See Spangler, Charles W.: V1560,139-147(1991)

Keller, Catherine E.
 ; Stenger, Anthony J.; Bernstein, Uri: Improved IR image generator for real-time scene simulation,V1486,278-285(1991)
 See Bernstein, Uri: V1486,345-351(1991)

Keller, Gregory S.
 : Rational anatomical treatment of basal cell carcinoma with photodynamic therapy,V1426,266-270(1991)

Keller, James M. See Tahani, Hossein: V1381,379-389(1991)

Keller, Ole See Liu, Ansheng: V1512,226-231(1991)

Keller, Richard A.
 See Fairbank, William M.: V1435,86-89(1991)
 See Soper, Steven A.: V1435,168-178(1991)

Keller, Ulrich
 : Lasers in dentistry,V1525,282-288(1991)
 ; Hibst, Raimund: Tooth pulp reaction following Er:YAG laser application,V1424,127-133(1991)
 See Hibst, Raimund: V1424,120-126(1991)

Kellett, Bruce J.
 See James, Jonathan H.: V1394,45-61(1991)
 See Pavuna, Davor: V1362,102-116(1991)

Kelley, J. D.
 ; Stuff, Michael I.; Hovis, Floyd E.; Linford, Gary J.: Removal of small particles from surfaces by pulsed laser irradiation: observations and a mechanism,V1415,211-219(1991)

Kelley, Robert B.
 ; Tsai, Jodi; Bethel, Jeff; Peiffer, John: Gripper for truss structure assembly,V1387,38-46(1991)

Kellner, Albert L.
 See Bradley, Joe C.: V1476,330-336(1991)
 See Costa, Joannes M.: V1476,74-80(1991)
 See Lam, Benson C.: V1371,36-45(1991)

Kellner, Robert A.
 ; Taga, Karim: New developments in the field of chemical infrared fiber sensors,V1510,232-241(1991)

Kelly, Anne M. See O'Brien, Stephen J.: V1424,62-75(1991)

Kelly, E. See Kozaitis, Samuel P.: V1474,112-115(1991)

Kelly, George See Thomas, Melvin L.: V1456,65-75(1991)

Kelly, Jack R. See Seekola, Desmond: V1455,19-26(1991)

Kelly, John H.
 ; Shoup, Milton J.; Skeldon, Mark D.; Bui, Snow T.: Design and energy characteristics of a multisegment glass-disk amplifier,V1410,40-46(1991)
 See Jaanimagi, Paul A.: V1358,337-343(1991)

Kelly, Michael J. See Syme, Richard T.: V1362,467-476(1991)

Kelly, Thomas J. See Cebula, Richard P.: V1493,91-99(1991)

Kelmar, Cheryl M.
 : Digital image processing for the early localization of cancer,V1426,47-57(1991)

Kelner, M. See Frolov, K.: V1397,461-468(1991)

Kelsall, Robert W. See Powell, Norman J.: V1545,19-30(1991)

Kelsey, J. F. See Jackson, John E.: V1533,75-86(1991)

Kemeny, Sabrina E.
; Eid, El-Sayed I.; Mendis, Sunetra; Fossum, Eric R.: Update on focal-plane image processing research,V1447,243-250(1991)

Kemmer, J. See Braueninger, Heinrich: V1549,330-339(1991)

Kemp, Kevin G.
; King, Charles F.; Wu, Wei; Stager, Charles: "Golden standard" wafer design for optical stepper characterization,V1464,260-266(1991)

Kempe, M. See Damm, Tobias: V1403,686-694(1991)

Kemper, J. See van Gemert, Martin J.: V1427,316-319(1991)

Kemppainen, Seppo See Roning, Juha J.: V1388,350-360(1991)

Kemsley, David See Daniell, Cindy E.: V1471,436-451(1991)

Kenan, Richard P. See Verber, Carl M.: V1374,68-73(1991)

Kendall, G. D. See Kirk, Andrew G.: V1621,320-327(1991)

Kender, John R.
; Kjeldsen, Rick: On seeing spaghetti: a novel self-adjusting seven-parameter Hough space for analyzing flexible extruded objects,V1570,315-321(1991)
; Park, Il-Pyung; Yang, David: Formalization and implementation of topological visual navigation in two dimensions,V1388,476-489(1991)
See Bolle, Ruud M.: V1383,305-318(1991)

Kendziorra, E. See Braueninger, Heinrich: V1549,330-339(1991)

Kenefic, Richard J.
: Local and remote track-file registration using minimum description length,V1481,430-439(1991)

Kennedy, Benjamin J.
: Intensified multispectral imaging measuring in the spatial, frequency, and time domains with a single instrument,V1346,68-74(1991)

Kennedy, Eugene T.
; Costello, John T.; Mosnier, Jean-Paul: Applications of laser plasmas in XUV photoabsorption spectroscopy,V1503,406-415(1991)

Kennedy, Howard V.
: Miscellaneous modulation transfer function effects relating to sample summing,V1488,165-176(1991)
: Recognition criterion for two-dimensional minimum resolvable temperature difference,V1488,196-202(1991)

Kennedy, Michael T.
; Cuellar, Enrique; Roberts, Daniel R.: Effect of color removal on optical fiber reliability,V1366,167-176(1991)
See Roberts, Daniel R.: V1366,129-135(1991)

Kennedy, Randall B. See Kulp, Thomas J.: V1479,352-363(1991)

Kenner, Vern E. See Schimert, Thomas R.: V1484,19-30(1991)

Kennett, Ruth D.
; Frazier, John C.: Air Force program in coherent semiconductor lasers,V1501,57-68(1991)

Kenney, John T.
See Lipscomb, George F.: V1560,388-399(1991)
See Lytel, Richard S.: V1563,122-138(1991)
See Lytel, Richard S.: V1389,547-558(1991)
See Wu, Jeong W.: V1560,196-205(1991)

Kenney, Malcolm E. See Sounik, James R.: V1426,340-349(1991)

Kenney, Steven B.
; Hirleman, Edwin D.: Edge effects in silicon photodiode arrays,V1480,82-93(1991)

Kenny, M. J. See Bell, J. M.: V1536,29-36(1991)

Kent, Richard A.
; McGlone, John T.; Zoltowski, Norbert W.: Comparison of Kodak Professional Digital Camera System images to conventional film, still video, and freeze-frame images,V1448,27-44(1991)

Kenue, Surender K.
: LANELOK: an algorithm for extending the lane sensing operating range to 100-feet,V1388,222-233(1991)
; Wybo, David R.: LaneLok: an improved Hough transform algorithm for lane sensing using strategic search methods,V1468,538-550(1991)

Kenway-Jackson, Damian See Hobbs, Chris P.: V1467,264-277(1991)

Keolian, Robert M. See Brown, David A.: V1584,328-335(1991)

Keranen, Heimo See Malinen, Jouko: V1533,122-128(1991)

Kerbis, Esther
; Morrison, Richard L.; Cloonan, Thomas J.; Downs, Maralene M.: Stability analysis of optomechanical components,V1396,447-452(1991)

Kerekes, John P.
: Parametric analysis of target/decoy performance,V1483,155-166(1991)

Keren, Eliezer
; Kreske, Kathi; Zac, Yaacov; Livnat, Ami: Determination of surface quality using moire methods,V1442,266-274(1991)

Kermani, Ahmad
: Single-wafer integrated processing as a manufacturing tool using rapid thermal chemical vapor deposition technology,V1393,109-119(1991)

Kerns, W. D. See Maclachlan Spicer, J. W.: V1467,311-321(1991)

Kerr, Anthony See Li, Luksun: V1584,170-177(1991)

Kerr, David
; Rodriguez-Vera, Ramon; Mendoza Santoyo, Fernando: Surface contouring using electronic speckle pattern interferometry,V1554A,668-679(1991)

Kerr, Noel C. See Emmony, David C.: V1397,651-659(1991)

Kersey, Alan D.
: Recent progress in interferometric fiber sensor technology,V1367,2-12(1991)
; Berkoff, Timothy A.; Dandridge, Anthony D.: Interferometric optical fiber sensors for absolute measurement of displacement and strain,V1511,40-50(1991)
; Berkoff, Timothy A.: Passive laser phase noise suppression technique for fiber interferometers,V1367,310-318(1991)
See Berkoff, Timothy A.: V1588,169-176(1991)
See Berkoff, Timothy A.: V1367,53-58(1991)

Kerslick, G. See Nation, John A.: V1407,32-43(1991)

Kervick, Gerard N. See Simon, Gabriel: V1423,154-156(1991)

Kervin, Paul W.
See Dierks, Jeffrey S.: V1482,146-158(1991)
See Sydney, Paul F.: V1482,196-208(1991)

Keselbrener, Michel
; Ruschin, Shlomo; Lissak, Boaz; Gover, Avraham: Numerical studies of resonators with on-axis holes in mirrors for FEL applications,V1415,38-47(1991)

Kesik, Jerzy
; Siejca, Antoni; Sokolowski, Maciej: Ion argon laser with metal-ceramic discharge tube: construction, technology, and gas-pumping effect,V1391,34-41(1991)
See Wolinski, Wieslaw: V1391,334-340(1991)

Keski-Kuha, Ritva A.
; Thomas, Roger J.; Neupert, Werner M.; Condor, Charles E.; Gum, Jeffrey S.: EUV performance of a multilayer-coated high-density toroidal grating,V1343,566-575(1991)

Keskinen, Jari See Pessa, Markus: V1361,529-542(1991)

Keskinova, E.
See Angelov, D.: V1403,572-574(1991)
See Angelov, D.: V1403,575-577(1991)
See Angelov, D.: V1403,230-239(1991)

Kessel, David
; Nseyo, Unyime O.; Schulz, Veronique; Sykes, Elizabeth: Pharmacokinetics of Photofrin II distribution in man,V1426,180-187(1991)

Kessler, David See Kurtz, Andrew F.: V1448,191-205(1991)

Kessler, I. See Kretschmer, K.-H.: V1392,246-252(1991)

Kessler, Jochen
; Dill, Roland; Russer, Peter: Characterization of high-Tc coplanar transmission lines and resonators,V1477,45-56(1991)

Kessler, Manfred See Frank, Klaus: V1431,2-11(1991)

Kessler, Manfred A.
: On-line optical determination of water in ethanol,V1510,218-223(1991)

Kessler, William D.
; Abeille, Pierre; Sulzer, Jean-Francois: Pod-mounted MIL-STD-2179B recorder,V1538,104-111(1991)

Kesting, Wolfgang See Bahners, Thomas: V1503,206-214(1991)

Ketsle, G. A.
; Letuta, S. N.: Energy circulation effect between excited singlet- and triplet-state systems in biopigment molecules,V1403,622-630(1991)

Ketterle, Wolfgang See Dibble, R. W.: V1602(1991)

Ketterson, John B. See Rippert, Edward D.: V1549,283-288(1991)

Kevorkian, Antoine P. See Saint-Andre, Francoise: V1583,278-288(1991)

Key, Michael H.
; Bailly-Salins, Rene; Edwards, B.; Harvey, Erol C.; Hirst, Graeme J.; Hooker, Chris J.; Kidd, A. K.; Madraszek, E. M.; Rodgers, P. A.; Ross, Ian N.; Shaw, M. J.; Steyer, M.: High-power KrF lasers,V1397,9-17(1991)
See Tallents, Gregory J.: V1413,70-76(1991)

Key, P. L. See Yuce, Hakan H.: V1366,120-128(1991)

Khaibullin, Ildus B. See Fattakhov, Yakh'ya V.: V1440,16-23(1991)

Khairullina, A. Y. See Korolevich, A. N.: V1403,364-371(1991)

Khaled, Elsayed-Esam M. See Hill, Steven C.: V1497,16-27(1991)

Khalil, Ali E.
: Noise mechanisms of high-temperature superconducting infrared detectors, V1477,148-158(1991)

Khalili, Payman
; Jain, Ramesh C.: Fast algorithm for obtaining dense depth maps for high-speed navigation, V1388,210-221(1991)

Khalsa, Satjeet S.
See Kishore, Sheel: V1446,188-194(1991)
See Kishore, Sheel: V1446,236-242(1991)
See Seshadri, Sridhar B.: V1446,388-395(1991)

Khan, M. A.
; Naumaan, Ahmed; Van Hove, James M.: Integrated optics intensity modulators in the GaAs/AlGaAs system, V1396,753-759(1991)

Khan, M. N. See Bristow, Julian P.: V1389,535-546(1991)

Khan, Mumit
; Anderson, Paul D.; Cerrina, Franco: EXCON: a graphics-based experiment-control manager, V1465,315-323(1991)

Khan, S. See Mang, Thomas S.: V1426,97-110(1991)

Khan Malek, Chantal
; Moreno, T.; Guerin, Philippe; Ladan, F. R.; Rivoira, R.; Barchewitz, Robert J.: Fabrication of multilayer Bragg-Fresnel zone plates for the soft x-ray range, V1343,56-61(1991)
See Berrouane, H.: V1343,428-436(1991)

Khanna, Dinesh N. See Przybilla, Klaus J.: V1466,174-187(1991)

Khanna, Shyam M.
: Cellular vibration measurement with a noninvasive optical system, V1429,9-20(1991)

Kharchenko, Mikhail V. See Glebov, Leonid B.: V1440,24-35(1991)

Kharitonov, Sergey I.
See Golub, Mikhail A.: V1500,194-206(1991)
See Golub, Mikhail A.: V1500,211-221(1991)

Khasho, Fouad See Frazin, Leon J.: V1425,207-207(1991)

Kheddache, Susanne See Angelhed, Jan-Erik: V1444,159-170(1991)

Khesin, G. L.
: Photoelasticity, holography, and moire for strain-stress-state studies of NPS, HPS underground and machinery structures: comparative analysis of physical modeling and numerical methods, V1554A,313-322(1991)
; Sakharov, V. N.; Zhavoronok, I. V.; Rottenkolber, Hans; Schorner, Jurgen: Multipulse polarized-holographic set-up for isodromic fringe registration in the field of running stress waves, V1554B,86-90(1991)

Khloponin, Leonid V. See Altshuler, Grigori B.: V1415,269-280(1991)

Khoka, Mustafa
See VanderMeulen, David L.: V1428,84-90(1991)
See VanderMeulen, David L.: V1428,91-98(1991)

Khokhlov, A. R. See Brazhnik, O. D.: V1402,70-77(1991)

Khomutova, T. G. See Tuchin, Valery V.: V1422,85-96(1991)

Khorana, Brij M.
See Brewer, Donald R.: V1396,430-434(1991)
See Eichelberger, Christopher L.: V1396,678-686(1991)
See Joenathan, Charles: V1554B,56-63(1991)
See Joenathan, Charles: V1396,155-163(1991)

Khoroshilov, E. V.
; Kryukov, I. V.; Kryukov, P. G.; Sharkov, A. V.; Gillbro, T.: Femtosecond processes in allophycocyanin trimers, V1403,431-433(1991)

Khorram, Siamak See Brockhaus, John A.: V1492,200-205(1991)

Khorshev, V. A. See Nadezhdinskii, Alexander I.: V1418,487-495(1991)

Khounsary, Ali M.
; Viccaro, P. J.; Kuzay, Tuncer M.: Filter and window assemblies for high-power insertion device synchrotron radiation sources, V1345,42-54(1991)
See Kuzay, Tuncer M.: V1345,55-70(1991)

Khoury, Jehad See Kane, Jonathan S.: V1564,511-520(1991)

Khramov, Valery See Altshuler, Grigori B.: V1415,269-280(1991)

Khristoforov, O. B. See Borisov, V. M.: V1503,40-47(1991)

Khryapov, V. T.
; Ponomarenko, Vladimir P.; Butkevitch, V. G.; Taubkin, I. I.; Stafeev, V. I.; Popov, S. A.; Osipov, V. V.: ORION semiconductor optical detectors: research and development, V1540,412-423(1991)

Khurgin, Yu. I. See Chikishev, A. Y.: V1403,517-521(1991)

Khuri-Yakub, B. T.
See Bhardwaj, Sanjay: V1392,555-562(1991)
See Lee, Yong J.: V1393,366-371(1991)

Khvilivitzky, A. T.
; Berezin, V. Y.; Lazovsky, L. Y.; Tataurschikov, S. S.; Pisarevsky, A. N.; Vydrevich, M. G.; Kossov, V. G.: Virtual-phase charge-coupled device image sensors for industrial and scientific applications, V1447,184-190(1991)
; Zuev, A. G.; Rybakov, M. I.; Kiryan, G. V.; Berezin, V. Y.: Virtual-phase charge-coupled device imaging system for astronomy, V1447,64-68(1991)

Kiamilev, Fouad E.
; Fan, J.; Catanzaro, Brian E.; Esener, Sadik C.; Lee, Sing H.: Architecture of an integrated computer-aided design system for optoelectronics, V1390,311-329(1991)
See Esener, Sadik C.: V1562,11-20(1991)

Kiang, Jean-Fu
: Analysis of skin effect loss in high-frequency interconnects with finite metallization thickness, V1389,340-351(1991)

Kiang, Ying J.
; Stigliani, Daniel J.: Measurement of mode field diameter and fiber bending loss, V1366,252-258(1991)

Kibeya, Saidi
; Orsal, Bernard; Alabedra, Robert; Lippens, D.: GaAs/GaAlAs superlattice avalanche photodiode at wavelength L = 0.8 um, V1501,97-106(1991)

Kiceniuk, T. See Mason, Peter V.: V1540,88-96(1991)

Kidd, A. K. See Key, Michael H.: V1397,9-17(1991)

Kidd, Keith See Kimbrell, James E.: V1482,415-424(1991)

Kidd, Robert C.
; Zalinski, Charles M.; Nadel, Jerome I.; Klein, Robert D.: Legibility of compressed document images at various spatial resolutions, V1454,414-424(1991)

Kidd, S. R.
; Sinha, P. G.; Barton, James S.; Jones, Julian D.: Fiber optic Fabry-Perot sensors for high-speed heat transfer measurements, V1504,180-190(1991)

Kidger, Michael J.
; Benham, Paul: Optimization of the optical transfer function, V1354,92-96(1991)
; Leamy, Paul T.: Existence of local minima in lens design, V1354,69-76(1991)
See Benham, Paul: V1354,120-125(1991)

Kido, Daniel K. See Ning, Ruola: V1443,236-249(1991)

Kidwell, J. J.
; Berthold, John W.: Metal-embedded optical fiber pressure sensor, V1367,192-196(1991)

Kieffer, J. C. See Chaker, Mohamed: V1465,16-25(1991)

Kielar, Alan See Razdan, Rikki: V1388,361-371(1991)

Kielin, E. J. See Upchurch, Billy T.: V1416,21-29(1991)

Kiely, Philip A. See Taylor, Geoff W.: V1476,2-13(1991)

Kiernan, Howard See Forman, Scott K.: V1424,2-6(1991)

Kihara, Naoko See Hayase, Rumiko H.: V1466,438-445(1991)

Kihara, Tami
; Yokomori, Kiyoshi: Simultaneous measurement of refractive index and thickness of thin film by polarized reflectances, V1332,783-791(1991)

Kiiveri, Pauli See Tammela, Simo: V1373,103-110(1991)

Kijko, Vadim V. See Aksinin, V. I.: V1500,93-104(1991)

Kikuchi, Akira See Darg, David A.: V1538,124-131(1991)

Kikuchi, Hideyuki See Nakazawa, Akira: V1458,115-118(1991)

Kikuchi, K. See Mukasa, Minoru: V1446,253-265(1991)

Kikuchi, Makoto
See Arai, Tsunenori: V1425,191-195(1991)
See Daidoh, Yuichiro: V1421,120-123(1991)
See Sakurada, Masami: V1425,158-164(1991)

Kilcoyne, M. K.
; Whitehead, Mark; Coldren, Larry A.: Asymmetric Fabry-Perot modulators for optical signal processing and optical computing applications, V1389,422-454(1991)

Kilgus, Charles C.
See Peacock, Keith: V1494,147-159(1991)
See Strikwerda, Thomas E.: V1478,13-23(1991)

Kilgus, Donald B. See Svetkoff, Donald J.: V1385,113-122(1991)

Kilian, Arnd
See Malo, Bernard: V1590,83-93(1991)
See Oh, Kyunghwan: V1590,94-100(1991)

Kilibarda, S. See Ng, Hon K.: V1477,15-19(1991)

Kilkenny, Joseph D.
; Bell, Perry M.; Hammel, Bruce A.; Hanks, Roy L.; Landen, Otto L.; McEwan, Thomas E.; Montgomery, David S.; Turner, R. E.; Wiedwald, Douglas J.; Bradley, David K.: Sub-100 psec x-ray gating cameras for ICF imaging applications,V1358,117-133(1991)
See Bell, Perry M.: V1346,456-464(1991)
See Wiedwald, Douglas J.: V1346,449-455(1991)

Kille, Knut
; Ahlers, Rolf-Juergen; Schneider, B.: Experiences with transputer systems for high-speed image processing,V1386,76-83(1991)
; Ahlers, Rolf-Juergen; Hager, K.: Integration of an application accelerator for high-speed inspection,V1386,222-227(1991)

Killian, Kevin M. See Wong, Ka-Kha: V1374,278-286(1991)

Killinger, Dennis K.
; Chan, Kin-Pui: Solid-state lidar measurements at 1 and 2 um,V1416,125-128(1991)
See Chan, Kin-Pui: V1492,111-114(1991)

Killinger, M.
See Cambon, P.: V1562,116-125(1991)
See Guibert, L.: V1505,99-103(1991)

Kilmer, Joyce P.
; DeVito, Anthony; Yuce, Hakan H.; Wieczorek, Casey J.; Varachi, John P.; Anderson, William T.: Optical cable reliability: lessons learned from post-mortem analyses,V1366,85-91(1991)

Kilmer, Neal H.
; Rachele, Henry: Simulation of vertical profiles of extinction and backscatter coefficients in very low stratus clouds and subcloud regions,V1487,109-122(1991)

Kilosanidze, Barbara N. See Kakichashvili, Shermazan D.: V1238,74-79(1991)

Kilpatrick, J. M. See Watson, John: V1461,245-253(1991)

Kilston, Steven
; Kilston, Vera M.; Utsch, Thomas F.: Multimission sensor for the RESERVES small-satellite program,V1495,193-204(1991)

Kilston, Vera M. See Kilston, Steven: V1495,193-204(1991)

Kim, A. H.
; Zeto, Robert J.; Youmans, Robert J.; Kondek, Christine D.; Weiner, Maurice; Lalevic, Bogoliub: High-power/high-pulse repetition frequency (PRF) pulse generation using a laser-diode-activated photoconductive GaAs switch,V1378,173-178(1991)

Kim, Byoung-Yoon
See Wysocki, Paul F.: V1373,66-77(1991)
See Wysocki, Paul F.: V1373,234-245(1991)

Kim, C. See Frear, Darrel R.: V1596,72-82(1991)

Kim, Chai W.
; Ansari, Rashid: Subband decomposition procedure for quincunx sampling grids,V1605,112-123(1991)

Kim, Choong-Ki
See Cho, Byung-Jin: V1393,180-191(1991)
See Lee, Hyouk: V1393,404-410(1991)

Kim, Chung-Won See Antonuk, Larry E.: V1443,108-119(1991)

Kim, Dong-Lok See Yee, David K.: V1444,292-305(1991)

Kim, Dong S.
; Lee, Sang U.: Classified vector quantizer based on minimum-distance partitioning,V1605,190-201(1991)

Kim, G. T.
See Jones, O. C.: V1557,202-208(1991)
See Paradies, C. J.: V1561,2-5(1991)

Kim, Gary
: New CATV fiber-to-the-subscriber architectures,V1363,133-140(1991)

Kim, H. See Lachs, Gerard: V1474,248-259(1991)

Kim, Hyo-Gun See Samarabandu, J. K.: V1450,296-322(1991)

Kim, HyungJun See Lee, Hyung G.: V1361,893-900(1991)

Kim, Jae-Kyoon See Jeon, Joon-hyeon: V1605,954-962(1991)

Kim, Jae H. See Lang, Robert J.: V1563,2-7(1991)

Kim, Jin J.
; Tittel, Frank K.; eds.: *Gas and Metal Vapor Lasers and Applications*,V1412(1991)

Kim, Jong See Lauffenburger, Jim: V1499,104-113(1991)

Kim, Jonglak See Kim, Kyoil: V1605,851-862(1991)

Kim, K. W. See Stroscio, Michael A.: V1362,566-579(1991)

Kim, Kwang-II See Kwon, O'Dae: V1361,802-808(1991)

Kim, Kwang-je
: Polarized nature of synchrotron radiation,V1548,73-79(1991)
: Survey of synchrotron radiation devices producing circular or variable polarization,V1345,116-124(1991)

Kim, Kwang H.
; Shafai, Bahram: Neural data association,V1481,406-417(1991)

Kim, Kyoil
; Kim, Jonglak; Kim, Taejeong: Block arithmetic coding of contour images,V1605,851-862(1991)

Kim, Kyung-Suk See Tsai, Kun-Hsieh: V1554A,529-541(1991)

Kim, Manjin J.
: Unframed via interconnection of nonplanar device structures,V1596,12-22(1991)

Kim, Munsok See Carey, Paul R.: V1403,37-39(1991)

Kim, Myung J. See Han, Joon H.: V1381,122-129(1991)

Kim, Richard C.
; Lin, Freddie S.: Holographic optical backplane hardware implementation for parallel and distributed processors,V1474,27-38(1991)

Kim, Sang S. See Lucovsky, Gerald: V1392,605-616(1991)

Kim, Seong-Han See Gooding, Linda: V1457,259-266(1991)

Kim, Shang H.
: Inverse bremsstrahlung acceleration in an electrostatic wave,V1407,620-634(1991)

Kim, Sung J. See Hayes, Todd R.: V1418,190-202(1991)

Kim, T. W. See Bourne, Lincoln C.: V1477,205-208(1991)

Kim, Taejeong See Kim, Kyoil: V1605,851-862(1991)

Kim, Ug-Sung See Schurr, J. M.: V1403,248-257(1991)

Kim, Ung See Lee, Eun Soo: V1358,262-268(1991)

Kim, Whee-Kuk
; Bevill, Pat; Tesar, Delbert: Implementation and control of a 3 degree-of-freedom, force-reflecting manual controller,V1387,392-406(1991)

Kim, Whoi-Yul
: Range data analysis using cross-stripe structured-light system,V1385,216-228(1991)

Kim, Won S.
See Bejczy, Antal K.: V1387,365-377(1991)
See Schenker, Paul S.: V1383,202-223(1991)

Kim, Y. S. See Mun, In K.: V1444,232-237(1991)

Kim, Yong-Soo
; Mitra, Sunanda: Noise tolerance of adaptive resonance theory neural network for binary pattern recognition,V1565,323-330(1991)

Kim, Yongmin
ed.: *Medical Imaging V: Image Capture, Formatting, and Display*,V1444(1991)
See Gove, Robert J.: V1444,318-333(1991)
See Tian, Qi: V1606,260-268(1991)
See Yee, David K.: V1444,292-305(1991)

Kim, Yun J.
; Jaggard, Dwight L.: Wave interactions with continuous fractal layers,V1558,113-119(1991)

Kimata, Masafumi
; Yutani, Naoki; Yagi, Hirofumi; Nakanishi, Junji; Tsubouchi, Natsuro; Seto, Toshiki: High-fill-factor monolithic infrared image sensor,V1540,238-249(1991)

Kimball, Paulette R.
; Fraser, James C.; Johnson, Jeffrey P.; Siegel, Andrew M.: Dual-band optical system for IR multicolor signal processing,V1540,687-698(1991)

Kimble, H. J. See Polzik, Eugene S.: V1561,143-146(1991)

Kimbrell, James E.
; Greenwald, David; Smith, Robert; Kidd, Keith: Deterministic errors in pointing and tracking systems II: identification and correction of dynamic errors,V1482,415-424(1991)
See Keitzer, Scott: V1482,406-414(1991)
See Liebst, Brad: V1482,425-438(1991)

Kime, Kenjiro See Irie, Mitsuru: V1499,360-365(1991)

Kimel, Sol
; Svaasand, Lars O.; Hammer-Wilson, Marie J.; Gottfried, Varda; Berns, Michael W.: Chick chorioallantoic membrane for the study of synergistic effects of hyperthermia and photodynamic therapy,V1525,341-350(1991)
See Gottfried, Varda: V1442,218-229(1991)

Kimmitt, Maurice F.
: Detection of infrared, free-electron laser radiation,V1501,86-96(1991)
See Ciocci, Franco: V1501,154-162(1991)

Kimoto, Tadahiko
; Yasuda, Yasuhiko: Transmission of the motion of a walker by model-based image coding,V1605,253-262(1991)

Kimpel, B. M.
See Schoenbach, Karl H.: V1362,428-435(1991)
See Schulz, Hans-Joachim: V1361,834-847(1991)

Kimura, Hironobu See Aoki, Nobutada: V1412,2-11(1991)

Kimura, Masakatsu See Ito, Katsunori: V1499,382-385(1991)

Kimura, Minoru See Kawauchi, Yoshikazu: V1555,224-227(1991)

Kimura, Shigeru
; Murakami, Yoshiyuki; Matsuda, Hikaru: Real-time architecture based on the image processing module family,V1483,10-17(1991)

Kimura, Shingo
See Isobe, Hiroshi: V1358,201-208(1991)
See Sato, Eiichi: V1358,146-153(1991)
See Shikoda, Arimitsu: V1358,154-161(1991)

Kimura, Tadamasa See Isshiki, Hideo: V1361,223-227(1991)

Kincaid, Glen T.
See Blackwell, John D.: V1479,324-334(1991)
See Parrish, William J.: V1540,274-284(1991)

King, Charles F. See Kemp, Kevin G.: V1464,260-266(1991)

King, Donald A. See Rinne, Glenn A.: V1389,110-121(1991)

King, F. D.
; Tremblay, Yves: Near-term applications of optical switching in the metropolitan networks,V1364,295-303(1991)

King, H. J. See Shao, C. A.: V1554A,613-618(1991)

King, Jill L.
; Gur, David; Rockette, Howard E.; Curtin, Hugh D.; Obuchowski, Nancy A.; Thaete, F. L.; Britton, Cynthia A.; Metz, Charles E.: Radiologists' confidence in detecting abnormalities on chest images and their subjective judgments of image quality,V1446,268-275(1991)
See Fuhrman, Carl R.: V1446,422-429(1991)
See Gur, David: V1446,284-288(1991)

King, Joseph A.
: Corrosion of ZnSe by alternating high voltage in a saline solution,V1535,216-223(1991)

King, Kevin F. See Crawford, Carl R.: V1443,203-213(1991)

King, Ronold W. See Fikioris, George: V1407,295-305(1991)

King, Steven J.
; Weiman, Carl F.: HelpMate autonomous mobile robot navigation system,V1388,190-198(1991)

King, Terence A.
See Charlton, Andrew: V1427,189-197(1991)
See Jiang, Zhi X.: V1421,88-99(1991)
See Wannop, Neil M.: V1423,163-166(1991)

King, Tsu-Jae See Wood, Samuel C.: V1393,337-348(1991)

Kingsbury, Edward P.
; Francis, Henry A.: Geometric error coupling in instrument ball bearings,V1454,152-158(1991)

Kingsbury, Jeffrey S.
; Margarone, Joseph E.; Satchidanand, S.; Liebow, Charles: Promotional effects of CO2 laser on neoplastic lesions in hamsters,V1427,363-367(1991)

Kingsley, Lawrence E. See Donaldson, William R.: V1378,226-236(1991)

Kink, W. See Braueninger, Heinrich: V1549,330-339(1991)

Kinley, Fred G. See Pacala, Thomas J.: V1411,69-79(1991)

Kinney, Terrance R. See Emo, Stephen M.: V1374,266-276(1991)

Kino, Gordon S.
See Chim, Stanley S.: V1464,138-144(1991)
See Chou, Ching-Hua: V1464,145-154(1991)

Kinoshita, Hiroo See Takenaka, Hisataka: V1345,213-224(1991)

Kinoshita, Hiroyuki See Peyton, David: V1393,295-308(1991)

Kinoshita, Katsuyuki
; Suyama, Motohiro; Inagaki, Yoshinori; Ishihara, Y.; Ito, Masuo: Femtosecond streak tube,V1358,490-496(1991)

Kinoshita, Taizo
; Nakahashi, Tomoko; Takizawa, Masaaki: Variable-bit-rate HDTV coding algorithm for ATM environments for B-ISDN,V1605,604-613(1991)

Kinoshita, Takeshi See Sugihara, Okihiro: V1361,599-605(1991)

Kinoshita, Yoshio See Satoh, Hiroharu: V1499,324-329(1991)

Kinrot, O.
; Friesem, Asher A.: Design of multiple-beam gratings for far-IR applications,V1442,106-108(1991)

Kinser, Jason M.
: Failure of outer-product learning to perform higher-order mapping,V1541,187-198(1991)

Kinsey, J. See Troeltzsch, John R.: V1494,9-15(1991)

Kinsey, Trevor J.
; Bussell, T. J.; Chick, M. C.: Hycam camera study of the features of a deflagrating munition,V1358,914-924(1991)

Kintis, M. See McAdam, Bridget A.: V1475,267-274(1991)

Kinzel, Joseph A.
: Monolithic phased arrays: recent advances,V1475,158-163(1991)

Kioumehr, Farhad See Mankovich, Nicholas J.: V1444,2-8(1991)

Kipp, Lutz
; Manzke, Recardo; Skibowski, Michael: Electronic structure of Ge(001) 2x1 by different angle-resolved photoemission techniques: EDC, CFS and CIS,V1361,794-801(1991)

Kirby, Steve See McWilliams, Gary: V1468,417-428(1991)

Kirchner, G. See Prochazka, Ivan: V1449,116-120(1991)

Kiriencko, Yu. A.
See Galpern, A. D.: V1238,320-323(1991)
See Smaev, V. P.: V1238,311-315(1991)

Kirillov, A. B. See Makarenko, Vladimir I.: V1469,843-845(1991)

Kirillov, Gennadi A. See Dolgopolov, Y. V.: V1412,267-275(1991)

Kirk, Andrew G.
; Imam, H.; Bird, K.; Hall, Trevor J.: Design and fabrication of computer-generated holographic fan-out elements for a matrix/matrix interconnection scheme,V1574,121-132(1991)
; Kendall, G. D.; Imam, H.; Hall, Trevor J.: Optical neural network with reconfigurable holographic interconnection,V1621,320-327(1991)

Kirk, Joseph P.
: Astigmatism and field curvature from pin-bars,V1463,282-291(1991)
; Hibbs, Michael S.: Deep-UV diagnostics using continuous tone photoresist,V1463,575-583(1991)

Kirkham, Anthony J.
: Optical design of dual-combiner head-up displays,V1354,310-315(1991)

Kirkiewicz, Jozef
: Laser application in chosen maritime economy divisions,V1391,351-360(1991)

Kirkovsky, A. N. See Zheltov, Georgi I.: V1403,752-753(1991)

Kirkpatrick, Michael D.
; Mihalakis, George M.: Projection screens for high-definition television,V1456,40-47(1991)

Kirkup, L. See Bell, J. M.: V1536,29-36(1991)

Kirkwood, James J.
; Cartwright, Anne: Behavioral observations in thermal imaging of the big brown bat: Eptesicus fuscus,V1467,369-371(1991)

Kirley, S. D. See Lin, Chi-Wei: V1426,216-227(1991)

Kirlin, Peter S. See Hamaguchi, Norihito: V1394,244-254(1991)

Kirsch, James C.
; Gregory, Don A.; Hudson, Tracy D.; Loudin, Jeffrey A.; Crowe, William M.: Optical correlator field demonstration,V1482,69-78(1991)
; Loudin, Jeffrey A.; Gregory, Don A.: Application of the phase and amplitude modulating properties of LCTVs,V1474,90-100(1991)
; Loudin, Jeffrey A.; Gregory, Don A.: Hybrid modulation properties of the Epson LCTV,V1558,432-441(1991)
See Duffey, Jason N.: V1558,422-431(1991)
See Hudson, Tracy D.: V1474,101-111(1991)
See Hudson, Tracy D.: V1564,54-64(1991)

Kirschner, K. See Kozaitis, Samuel P.: V1474,112-115(1991)

Kirveliene, V.
; Rotomskis, Richardas; Juodka, B.; Piskarskas, Algis S.: Importance of pulsed laser intensity in porphyrin-sensitized NADH photo-oxidation,V1403,582-584(1991)

Kiryan, G. V. See Khvilivitzky, A. T.: V1447,64-68(1991)

Kirykhin, Yu. B. See Borisov, V. M.: V1503,40-47(1991)

Kiryukhin, Yu. B. See Vizir, V. A.: V1411,63-68(1991)

Kisaki, Jyunko See Nakajima, Hajime: V1456,29-39(1991)

Kiselev, A. Y. See Gamaly, Eugene G.: V1413,95-106(1991)

Kiseleva, Margaret S. See Mirzoeva, Larisa A.: V1540,444-449(1991)

Kiselyov, Boris S.
; Kulakov, Nickolay Y.; Mikaelian, Andrei L.; Shkitin, Vladimir A.: Optical auto- and heteroassociative memory based on a high-order neural network,V1621,328-339(1991)
; Mikaelian, Andrei L.; Novoselov, B. A.; Shkitin, Vladimir A.; Arkhontov, L. B.; Evtikhiev, Nickolay N.: LiNbO3-based multichannel electro-optical light modulators,V1621,126-137(1991)
See Gubanov, V. A.: V1621,102-113(1991)

Kishimoto, Satoshi
; Egashira, Mitsuru; Shinya, Norio: Development of electron moire method using a scanning electron microscope,V1554B,174-180(1991)

Kishimoto, Takashi See Oikawa, Masahiro: V1544,226-237(1991)

Kishino, Fumio
See Ishibashi, Satoshi: V1605,242-252(1991)
See Nagashima, Yoshio: V1606,566-573(1991)
See Satoh, Takanori: V1606,1014-1021(1991)

Kishioka, Kiyoshi
: Characterization of ion-exchange waveguide made with diluted KNO3,V1583,19-26(1991)
See Noutsios, Peter C.: V1374,14-22(1991)
See Yip, Gar L.: V1583,14-18(1991)

Kishner, Stanley J.
: High-bandwidth alignment sensing in active optical systems,V1532,215-229(1991)

Kishore, Sheel
; Khalsa, Satjeet S.; Seshadri, Sridhar B.; Arenson, Ronald L.: Digital film library implementation,V1446,188-194(1991)
; Khalsa, Satjeet S.; Seshadri, Sridhar B.; Arenson, Ronald L.: Preliminary results of a PACS implementation,V1446,236-242(1991)
See Seshadri, Sridhar B.: V1446,388-395(1991)

Kislov, V. I. See Apollonov, V. V.: V1502,83-94(1991)

Kiss, Gabor D.
; Pellegrino, Anthony; Leopold, Simon: Analysis of fiber damage and field failures in fiber-grip-type mechanical fiber optic splices,V1366,223-234(1991)

Kisslinger, Annamaria See Andreoni, Alessandra: V1525,351-366(1991)

Kist, Edward H. See Goad, Joseph H.: V1410,107-115(1991)

Kita, Toshiaki See Harada, Tatsuo: V1545,2-10(1991)

Kitabatake, Makoto See Nishimura, Kazuhito: V1534,199-206(1991)

Kitagaki, Kazukuni
; Oto, Takeshi; Demura, Tatsuhiko; Araki, Yoshitsugu; Takada, Tomoji: New address-generation-unit architecture for video signal processing,V1606,891-900(1991)

Kitagawa, Teizo
; Ogura, Takashi: Resonance Raman spectra of transient species of a respiration enzyme detected with an artificial cardiovascular system and Raman/absorption simultaneous measurement system,V1403,563-571(1991)

Kitagawa, Yoichi See Matsumoto, Tetsuya: V1332,530-536(1991)

Kitai, Adrian H.
See Sanders, Brian W.: V1398,81-87(1991)
See Wong, Brian: V1398,261-268(1991)
See Young, Richard: V1398,71-80(1991)

Kitakohji, Toshisuke See Goto, Yasuyuki: V1534,49-58(1991)

Kitamura, Kohei See Hashimoto, Nobuyuki: V1461,291-302(1991)

Kitamura, Shin-ichi See Aoki, Yutaka: V1490,278-284(1991)

Kitaoka, Masaki
; Honda, Hisao; Yoshida, Harunobu; Takigawa, Akio; Kawahara, Hideo: Formation of SiO2 film on plastic substrate by liquid-phase-deposition method,V1519,109-114(1991)

Kitaori, Tomoyuki
See Fukunaga, Seiki: V1466,446-457(1991)
See Koyanagi, Hiroo: V1466,346-361(1991)

Kitazawa, Mototaka
: Recent developments in plastic optical fibers: 135-C heat-resistant fibers and image transfer fibers,V1369,44-47(1991)
; Kreidl, John F.; Steele, Robert E.; eds.: *Plastic Optical Fibers*,V1592(1991)

Kitipichai, Prakob See LaPeruta, Richard: V1497,357-366(1991)

Kiyak, S. G. See Craciun, Valentin: V1392,629-634(1991)

Kjeldsen, Rick
See Bolle, Ruud M.: V1383,305-318(1991)
See Kender, John R.: V1570,315-321(1991)

Kjell, Bradley P.
; Sood, Arun K.; Topkar, V. A.: Scale-space features for object detection,V1468,148-155(1991)
; Wang, Pearl Y.: Noise-tolerant texture classification and image segmentation,V1381,553-560(1991)

Kla, Sylvanus
; Mazenc, Christophe; Merrheim, Xavier; Muller, Jean-Michel M.: New algorithms for on-line computation of elementary functions,V1566,275-285(1991)

Klabunde, C. E. See Lowndes, Douglas H.: V1394,150-160(1991)

Klainer, Stanley M.
; Thomas, Johnny R.; Dandge, Dileep K.; Frank, Chet A.; Butler, Marcus S.; Arman, Helen; Goswami, Kisholoy: In situ monitoring for hydrocarbons using fiber optic chemical sensors,V1434,119-126(1991)

Klarquist, William N.
; Acton, Scott T.; Ghosh, Joydeep: Mean-field stereo correspondence for natural images,V1453,321-332(1991)

Klearman, Debbie
; Schmitz, Kenneth S.: Effect of counterion distribution on the electrostatic component of the persistence length of flexible linear polyions,V1430,236-255(1991)

Klebanow, Edward R. See Panjehpour, Masoud: V1424,179-185(1991)

Klee, H. W. See McDowell, Maurice W.: V1358,227-236(1991)

Kleihorst, Richard P. See Katsaggelos, Aggelos K.: V1606,716-727(1991)

Kleiman, Gary
See Delamere, W. A.: V1479,31-40(1991)
See Delamere, W. A.: V1447,204-213(1991)

Klein, B. See Seka, Wolf D.: V1398,162-169(1991)

Klein, Claude A.
: Diamond windows for the infrared: fact and fallacy,VCR38,218-257(1991)
: Thermal stress modeling for diamond-coated optical windows,V1441,488-509(1991)
; Hartnett, Thomas M.; Robinson, Clifford J.: Critical-point phonons of diamond,V1534,117-138(1991)

Klein, Karl-Friedrich
See Fabian, Heinz: V1513,168-173(1991)
See Hitzler, Hermine: V1503,355-362(1991)

Klein, Lisa C.
: Sol-gel overview: transparent, microporous silica, its synthesis and characterization,V1590,2-7(1991)

Klein, Mary K. See Peavy, George M.: V1424,171-178(1991)

Klein, Robert D. See Kidd, Robert C.: V1454,414-424(1991)

Klein, Stanley A.
; Carney, Thom: "Perfect" displays and "perfect" image compression in space and time,V1453,190-205(1991)

Klein, Yitschak See Wilk, Shalom: V1442,140-148(1991)

Klein, Z. See Marcovitch, Orna: V1442,58-58(1991)

Kleineberg, Ulf See Schmiedeskamp, Bernt: V1343,64-72(1991)

Klement, Ekkehard See Zarschizky, Helmut: V1389,484-495(1991)

Klementyev, V. G.
: Estimation of limit time resolution in image streak camera,V1358,1070-1074(1991)

Klenerman, David
; Spowage, K.; Walpole, B.: Second-harmonic generation as an in situ diagnostic for corrosion,V1437,95-102(1991)

Klepsvik, John O. See Bjarnar, Morten L.: V1537,74-88(1991)

Klevanik, A. V. See Shuvalov, V. A.: V1403,400-406(1991)

Kliem, Karl-Heinz
; Sczepanski, Volker; Michl, Uwe; Hesse, Reiner: Selected performance parameters and functional principles of a new stepper generation,V1463,743-751(1991)

Kliger, David S.
See Goldbeck, Robert A.: V1432,14-27(1991)
See Woodruff, William H.: V1432,205-210(1991)

Klimenko, Vladimir I. See Masychev, Victor I.: V1427,344-356(1991)

Kline-Schoder, Robert J.
; Wright, Michael J.: Design and analysis of a dither mirror control system,V1489,189-200(1991)

Klingenberg, Guenter
; Rockstroh, Helmut: Application of high-speed infrared emission spectroscopy in reacting flows,V1358,851-858(1991)

Klingenberg, Hans H.
; Gekat, Frank: Excimer laser performance under various microwave excitation conditions,V1503,140-145(1991)
; Gekat, Frank: Investigation of microwave-pumped excimer and rare-gas laser transitions,V1412,103-114(1991)
See Gekat, Frank: V1411,47-54(1991)

Klingensmith, H. W. See Neff, Raymond K.: V1364,245-256(1991)

Klingenstein, M. See Rosenzweig, Josef: V1362,168-178(1991)

Klingler, Joseph W. See Andrews, Lee T.: V1459,125-135(1991)

Klingmann, Jeffrey L.
See Atkinson, Dennis P.: V1343,530-541(1991)
See Montesanti, Richard C.: V1343,558-565(1991)

Klingsporn, Paul E. See Setchell, Robert E.: V1441,61-70(1991)

Kliot-Dashinskaya, I. M.
; Mikhailova, V. I.; Paltsev, G. P.; Strigun, V. L.: Display holography with pulsed ruby laser: recording and copying,V1238,465-469(1991)

Klocek, Paul
; Hoggins, James T.; McKenna, T. A.; Trombetta, John M.; Boucher, Maurice W.: Optical properties of GaAs, GaP, and CVD diamond,V1498,147-157(1991)
See Trombetta, John M.: V1534,77-88(1991)

Kloidt, Andreas See Schmiedeskamp, Bernt: V1343,64-72(1991)

Klok, Frits H. See Bulsink, Bennie J.: V1384,215-227(1991)

Klose, George J. See Colavita, Mark M.: V1542,205-212(1991)

Klose, Joerg
; Altena, Oliver: Feature trajectory reduction of integrated autoregressive processes based on a multilayer self-organizing neural network,V1565,504-517(1991)

Klosowicz, Stanislaw See Cesarz, Tadeusz: V1391,244-249(1991)

Klotz, Tamas See Racz, Janos: V1469,778-783(1991)

Klug, Michael A.
See Farmer, William J.: V1461,215-226(1991)
See Halle, Michael W.: V1461,142-155(1991)

Kluge, Karl See Aubert, Didier: V1388,141-151(1991)

Klumpe, Herbert W.
; Lajza-Rooks, Barbara A.; Blum, James D.: Absolute phasing of segmented mirrors using the polarization phase sensor,V1398,95-106(1991); 1532,230-240(1991)

Kluyev, V. G.
; Kushnir, M. A.; Latyshev, A. N.; Voloshina, T. V.: Kinetic properties of adsorbed particles photostimulated migration upon the surface of ionic-covalent-type semiconductors,V1440,303-308(1991)

Klyshko, D. N. See Penin, A. N.: V1562,143-148(1991)

Knabke, Gerhard See Osterfeld, Martin: V1559,49-55(1991)

Knasmillner, R. E. See Rossmanith, H. P.: V1554A,850-860(1991)

Knat'ko, M. V. See Zandberg, E. Y.: V1440,292-302(1991)

Knauer, James P.
; Shenoy, Gopal K.; eds.: *Advanced X-Ray/EUV Radiation Sources and Applications*,V1345(1991)

Knauer, Ulrich See Berek, Stefan: V1463,515-520(1991)

Kneubuhl, F. K. See Wilke, Ingrid: V1442,2-10(1991)

Kniffen, Stacy K.
; Becker, Michael F.; Powers, Edward J.: Bispectral magnitude and phase recovery using a wide bandwidth acousto-optical processor,V1564,617-627(1991)

Knight, Stephen E.
; Humphrey, Dean; Bowley, Reginald R.; Cogley, Robert M.: Half-micrometer linewidth metrology,V1464,119-126(1991)

Knights, Mark G. See Minnigh, Stephen W.: V1528,129-134(1991)

Knoepfle, G. See Sutter, Kurt: V1560,296-301(1991)

Knoesen, Andre See Yankelevich, Diego: V1560,406-415(1991)

Knopf, George K.
See Gupta, Madan M.: V1382,60-73(1991)
See Gupta, Madan M.: V1606,482-495(1991)

Knopp, F. See Zerza, Gerald: V1410,202-208(1991)

Knors, C. J. See MacDonald, Scott A.: V1466,2-12(1991)

Knotts, Micheal E. See O'Donnell, Kevin A.: V1558,362-367(1991)

Knowlden, Robert E.
See Holz, Michael: V1544,75-89(1991)
See Motamedi, M. E.: V1544,22-32(1991)

Knowles, Martyn R.
; Webb, Colin E.; Naylor, Graham A.: Experimental and computer-modeled results of titanium sapphire lasers pumped by copper vapor lasers,V1410,195-201(1991)

Knowles, Simon C.
: Arithmetic processor design for the T9000 transputer,V1566,230-243(1991)

Ko, Kwok See Eppley, Kenneth: V1407,249-259(1991)

Ko, Yeong-Uk See Kang, Dae I.: V1554B,139-147(1991)

Kobakhidze, T. G. See Dmokhovskij, A. V.: V1554A,323-330(1991)

Kobayashi, Akira
; Ishikawa, Masayoshi; Suzuki, Takayoshi; Ikeya, Morihiro; Shimomoto, Yasuharu: New 2/3-inch MF image pick-up tubes for HDTV camera,V1449,148-156(1991)

Kobayashi, Hitoshi
See Isshiki, Hideo: V1361,223-227(1991)
See Nakamura, Hiroyuki: V1456,226-238(1991)

Kobayashi, K. See Cartlidge, Andy G.: V1423,167-174(1991)

Kobayashi, Katsuyoshi See Shirai, Hisatsugu: V1463,256-274(1991)

Kobayashi, Kenichi
; Abe, Tsutomu; Miyake, Hiroyuki; Kashimura, Hirotsugu; Ozawa, Takashi; Hamano, Toshihisa; Fennell, Leonard E.; Turner, William D.; Weisfield, Richard L.: a-Si:H TFT-driven high-gray-scale contact image sensor with a ground-mesh-type multiplex circuit,V1448,157-163(1991)

Kobayashi, Koichi See Yamaguchi, Ichirou: V1554A,240-249(1991)

Kobayashi, Masakazu See Kudlek, Gotthard: V1361,150-158(1991)

Kobayashi, Masao See Kashima, Yasumasa: V1365,102-107(1991)

Kobayashi, Masaru
; Noda, Juichi; Takada, Kazumasa; Taylor, Henry F.: High-spatial-resolution and high-sensitivity interferometric optical-time-domain reflectometer,V1474,278-284(1991)

Kobayashi, Shigeru See Seki, Masafumi: V1583,184-195(1991)

Kobayashi, Soichi
; Sumida, Shin S.; Miyashita, Tadashi M.: Silica optical integrated devices,V1374,300-303(1991)

Kobayashi, Takayoshi T.
; Terauchi, Mamoru; Kouyama, Tsutomu; Yoshizawa, Masayuki; Taiji, Makoto: Femtosecond spectroscopy of acidified and neutral bacteriorhodopsin,V1403,407-416(1991)
See Nomura, Shintaro: V1560,272-277(1991)
See Wada, Tatsuo: V1560,162-171(1991)

Kobayasi, Yukio See Wakabayashi, Osamu: V1463,617-628(1991)

Kobayasi, Syoyu
: Characterization of motion of probes and networks in gels by laser light scattering under sample rotation,V1403,296-305(1991)

Kobhio, M. N.
; Fontaine, Bernard L.; Hueber, Jean-Marc; Delaporte, Philippe C.; Forestier, Bernard M.; Sentis, Marc L.: Kinetics and electrical modeling of a long-pulse high-efficiency XeCl laser with double discharge and fast magnetic switch,V1397,555-558(1991)
; Fontaine, Bernard L.; Hueber, Jean-Marc; Delaporte, Philippe C.; Forestier, Bernard M.; Sentis, Marc L.: Modeling of a long-pulse high-efficiency XeCl laser with double-discharge and fast magnetic switch,V1503,88-97(1991)
See Hueber, Jean-Marc: V1503,62-70(1991)

Kobialka, Torsten See Tschudi, Theo T.: V1500,80-80(1991)

Kobolla, Harald
; Sauer, Frank; Schwider, Johannes; Streibl, Norbert; Voelkel, Reinhard: Holographic space-variant prism arrays for free-space data permutation in digital optical networks,V1507,175-182(1991)

Kobzev, E. F.
; Vorontsov, Michael A.: Optical implementation of winner-take-all models of neural networks,V1402,165-174(1991)

Koch, Christof
: Implementing early vision algorithms in analog hardware: an overview,V1473,2-16(1991)
See Bair, Wyeth: V1473,59-65(1991)
See Mathur, Bimal P.: V1473,153-160(1991)
See Mathur, Bimal P.: V1473(1991)
See Moore, Andrew: V1473,66-75(1991)

Koch, Eric See Koch, Stephen: V1380,163-170(1991)

Koch, J. F.
: Regular doping structures: a Si-based, quantum-well infrared detector,V1362,544-552(1991)

Koch, Karl W.
; Moore, Gerald T.: Two-color interferometry using a detuned frequency-doubling crystal,V1516,67-74(1991)

Koch, Mark W.
; Ramamurthy, Arjun: Extracting features to recognize partially occluded objects,V1567,638-649(1991)

Koch, Rolf See Woods, Andrew J.: V1457,322-326(1991)

Koch, Stephen
; Koch, Eric: Surface digitizing of anatomical subjects with DIGIBOT-4,V1380,163-170(1991)

Kochelap, Viatcheslav A.
See Barmashenko, B. D.: V1397,303-307(1991)
See Mel'nikov, Lev Y.: V1397,603-610(1991)
See Mel'nikov, Lev Y.: V1501,144-149(1991)

Kochemasov, G. G. See Dolgopolov, Y. V.: V1412,267-275(1991)

Kocher, Thomas E.
; Whiting, Bruce R.: Design considerations for a high-resolution film scanner for teleradiology applications,V1446,459-464(1991)
See Jost, R. G.: V1446,2-9(1991)

Kochevar, Irene E.
; Hefetz, Yaron; Dunn, D. A.; Buckley, L. M.; Hillenkamp, Franz: DNA photoproducts formed using high-intensity 532 nm laser radiation,V1403,756-763(1991)

Kociszewski, Longin
; Pysz, Dariusz: Fiber optic image guide rods as ultrathin endoscopy,V1420,212-217(1991)

Kocyba, Krzysztof
: Miniature pulsed TEA CO2 laser with a 20-Hz repetition frequency,V1391,61-64(1991)
See Trojnar, Eugeniusz: V1391,230-237(1991)
See Trojnar, Eugeniusz: V1391,238-243(1991)

Koczera, Stanley See van Gilse, Jan: V1414,45-54(1991)

Koczor, Ronald J.
: NASAs Geostationary Earth Observatory and its optical instruments,V1527,98-109(1991)

Kodama, R. See Tallents, Gregory J.: V1413,70-76(1991)

Kodas, Toivo T.
See Baum, Thomas H.: V1598,122-131(1991)
See Jain, Ajay: V1596,23-33(1991)

Kodate, Kashiko
; Kamiya, Takeshi: Multiple-beam accessor using microzone plate elements for optoelectronic integrated circuits,V1545,251-260(1991)

Koechner, Walter
; Burnham, Ralph L.; Kasinski, Jeff; Bournes, Patrick A.; DiBiase, Don; Le, Khoa; Marshall, Larry R.; Hays, Alan D.: High-power diode-pumped solid state lasers for optical space communications,V1522,169-179(1991)

Koeck, A. See Gornik, Erich: V1362,1-13(1991)

Koehler, Bertrand
; Bourlon, Philippe M.; Dugue, Michel: Active vibration filtering: application to an optical delay line for stellar interferometer,V1489,177-188(1991)

Koehler, Klaus
See Huelsmann, Axel: V1465,201-208(1991)
See Rinker, Michael: V1362,14-23(1991)

Koehn, Otmar See Henschel, Henning: V1399,49-63(1991)

Koelbl, Roy S.
: Reliability assurance of optoelectronic devices in the local loop,V1366,17-25(1991)

Koelink, M. H.
; de Mul, F. F.; Greve, Jan; Graaff, Reindert; Dassel, A. C.; Aarnouds, J. G.: Depth dependent laser Doppler perfusion measurements: theory and instrumentation,V1403,347-349(1991)
; de Mul, F. F.; Greve, Jan; Graaff, Reindert; Dassel, A. C.; Aarnouds, J. G.: Monte Carlo simulations and measurements of signals in laser Doppler flowmetry on human skin,V1431,63-72(1991)
; Slot, M.; de Mul, F. F.; Greve, Jan; Graaff, Reindert; Dassel, A. C.; Aarnouds, J. G.: In-vivo blood flow velocity measurements using the self-mixing effect in a fiber-coupled semiconductor laser,V1511,120-128(1991)

Koellner, Malte See Seidel, Claus: V1432,105-116(1991)

Koeneke, Axel See Joeckle, Rene C.: V1397,679-682(1991)

Koenig, George G.
; Bissonnette, Luc R.: Forward-looking IR and lidar atmospheric propagation in the infrared field program,V1487,240-249(1991)

Koenig, Karsten
; Schneckenburger, Herbert; Rueck, Angelika C.; Auchter, S.: Photoproduct formation of endogeneous protoporphyrin and its photodynamic activity,V1525,412-419(1991)

Koenig, Michel
; Fabre, Edouard; Malka, Victor; Hammerling, Peter; Michard, Alain; Boudenne, Jean-Michel; Fews, Peter: Hydrodynamic efficiency as determined from implosion experiments at wavelength = 0.26 um,V1502,338-342(1991)

Koenraad, Paul See Albrektsen, O.: V1361,338-342(1991)

Koerber, B. See Abel, B.: V1525,110-118(1991)

Koert, Peter See Genuario, Ralph: V1407,553-565(1991)

Koezuka, Tetsuo
; Kakinoki, Yoshikazu; Hashinami, Shinji; Nakashima, Masato: Visual inspection system using multidirectional 3-D imager,V1332,323-331(1991)

Koga, Nobuhiro
; Kudoh, Takeshi; Murakawa, Masao: Application of visioplasticity to an experimental analysis of the shearing phenomenon,V1554A,84-92(1991)

Koga, Toshio
See Tzou, Kou-Hu: V1606(1991)
See Tzou, Kou-Hu: V1605(1991)

Kohanzadeh, Youssef See Hindy, Robert N.: V1414(1991)

Kohl, Charles A.
: Technology transfer in image understanding,V1406,18-29(1991)

Kohl, M.
; Neukammer, Jorg; Sukowski, U.; Rinneberg, Herbert H.; Sinn, H.-J.; Friedrich, E. A.; Graschew, Georgi; Schlag, Peter M.; Woehrle, D.: Imaging of tumors by time-delayed laser-induced fluorescence,V1525,26-34(1991)

Kohler, A.
; Fracasso, B.; Ambs, Pierre; de Bougrenet de la Tocnaye, Jean-Louis M.: Joint transform correlator using nonlinear ferroelectric liquid-crystal spatial light modulator,V1564,236-243(1991)

Kohli, Jagdish C.
: Superhigh-definition image communication: an application perspective,V1605,351-361(1991)

Kohmura, Y. See Ohashi, T.: V1549,9-19(1991)

Kohn, Philip D. See Gruenbaum, F. A.: V1431,232-238(1991)

Kohn, S. See Grunwald, Arthur J.: V1456,132-153(1991)

Kohsiek, W.
: Fast-response water vapor and carbon dioxide sensor,V1511,114-119(1991)

Koht, Lowell
: Modular packaging for FTTC and B-ISDN,V1363,158-162(1991)

Koike, Kouichi See Ogura, Toshihiro: V1443,153-157(1991)

Koike, Masato
; Namioka, Takeshi: Analysis of polarization properties of shallow metallic gratings by an extended Rayleigh-Fano theory,V1545,88-94(1991)

Koike, Tomoyuki See Ohmi, Toshihiko: V1361,606-612(1991)

Koike, Yasuhiro
; Hondo, Yukie; Nihei, Eisuke: Graded-index polymer optical fiber by new random copolymerization technique,V1592,62-72(1991)

Koivunen, Visa
; Pietikainen, Matti: Combined edge- and region-based method for range image segmentation,V1381,501-512(1991)

Kojima, Arata
; Ogawa, Ryokei; Izuchi, N.; Yamamoto, M.; Nishimoto, T.; Matsumoto, Toshiro: Deformation measurement of the bone fixed with external fixator using holographic interferometry,V1429,162-171(1991)

Kojima, Haruhiko
; Akiyama, Teruo: Table recognition for automated document entry system,V1384,285-292(1991)

Kojima, Kotaro See Ito, Katsunori: V1499,382-385(1991)

Kojima, Toshiyuki See Takahashi, Kazuhiro: V1463,696-708(1991)

Kojima, Yoshikatsu See Takeuchi, Ichiro: V1394,96-101(1991)

Kokaj, Jahja O. See Casasent, David P.: V1385,152-164(1991)

Kokin, A. N.
; Libenson, Michail N.; Minaev, Sergei M.: Peculiarities of the formation of periodic structures on silicon under millisecond laser radiation,V1440,338-341(1991)

Kokta, Milan R.
See Pollack, S. A.: V1410,156-164(1991)
See Quarles, Gregory J.: V1410,165-174(1991)

Kokubo, Tadayoshi See Uenishi, Kazuya: V1466,102-116(1991)

Kolb, Charles E.
See McManus, J. B.: V1433,330-339(1991)
See McManus, J. B.: V1433,340-351(1991)

Kolbe, J. See Schink, Harald: V1441,327-338(1991)

Kolenda, Jonas
See Rueckmann, I.: V1513,78-85(1991)
See Woggon, Ulrike: V1362,888-898(1991)

Kolesnikov, S.A. See Mnatsakanyan, Eduard A.: V1414,130-133(1991)

Kolesnikov, Sergey V. See Degtyareva, V. P.: V1358,546-548(1991)

Kolesov, G. V.
See Averin, V. I.: V1358,589-602(1991)
See Averin, V. I.: V1358,603-605(1991)
See Korzhenevich, Irina M.: V1358,1084-1089(1991)
See Korzhenevich, Irina M.: V1358,1090-1095(1991)

Kolitz, Stephan E.
: Passive-sensor data fusion,V1481,329-340(1991)

Kollia, Z.
; Hontzopoulos, Elias I.: Excimer laser patterning of flexible
materials,V1503,215-222(1991)

Kollmer, Charles
See Lane, Gregory J.: V1424,7-11(1991)
See Sazy, John: V1424,50-50(1991)

Kolodziejczyk, Andrzej See Jaroszewicz, Zbigniew: V1555,236-240(1991)

Kolodzy, Paul J. See Rak, Steven J.: V1471,177-184(1991)

Kolomoitsev, D. V.
; Nikitin, S. Y.: New problems of femtosecond time-domain CARS (coherent
antistokes Raman spectroscopy) of large molecules,V1402,31-43(1991)
; Nikitin, S. Y.: Physical effects in time-domain CARS (coherent antistokes
Raman spectroscopy) of molecular gases,V1402,11-30(1991)

Kolonko, J. See Dodd, James W.: V1407,467-473(1991)

Koltracht, Israel See Gohberg, I.: V1566,14-22(1991)

Koma, Atsushi See Wada, Tatsuo: V1560,162-171(1991)

Komarov, Konstantin P.
; Kuch'yanov, Aleksandr S.; Ugozhayev, Vladimir D.: High-performance
picosecond and femtosecond solid-state lasers with feedback-controlled
passive mode-locking,V1501,135-143(1991)

Komarov, Vyacheslav A.
; Melnik, N. E.: Signal and noise characteristics of space-time light
modulators,V1238,275-285(1991)
See Zaichenko, O. V.: V1238,271-274(1991)

Komatsu, Katsuhiko
See Hashimoto, Takashi: V1397,519-522(1991)
See Yokozawa, T.: V1397,513-517(1991)

Komatsu, Ken-ichi See Martinez, Ralph: V1446,100-107(1991)

Komatsu, Takashi
See Aizawa, Kiyoharu: V1605,318-328(1991)
See Saito, Takahiro: V1605,382-393(1991)

Komatsu, Y. See Yamanaka, Yutaka: V1499,263-268(1991)

Komeda, Koji See Yamaguchi, Takao: V1418,363-371(1991)

Kometani, Janet M.
See Nalamasu, Omkaram: V1466,13-25(1991)
See Novembre, Anthony E.: V1466,89-99(1991)

Kometer, R.
See Zerza, Gerald: V1397,107-110(1991)
See Zerza, Gerald: V1410,202-208(1991)

Komine, Yukikuni See Daidoh, Yuichiro: V1421,120-123(1991)

Komissarov, Gennady G. See Gariaev, Peter P.: V1621,280-291(1991)

Komlev, O. U. See Taratorin, B. I.: V1554A,449-456(1991)

Komma, Yoshiaki See Kato, Makoto: V1507,36-44(1991)

Komori, Masaru
; Minato, Kotaro; Takahashi, Takashi; Nakano, Yoshihisa; Sakurai, Tsunetaro:
Possibility of liquid crystal display panels for a space-saving PACS
workstation,V1444,334-337(1991)
See Minato, Kotaro: V1446,195-198(1991)

Komsa, M. L. See Krongelb, Sol: V1389,249-256(1991)

Konagai, Chikara See Aoki, Nobutada: V1412,2-11(1991)

Konak, Cestmir See Sedlak, Marian: V1430,191-202(1991)

Kondek, Christine D. See Kim, A. H.: V1378,173-178(1991)

Kondo, H. See Tomie, Toshihisa: V1552,254-263(1991)

Kondo, M. See Matsuoka, Masaru: V1549,2-8(1991)

Kondo, Motoe
See Kuribayashi, Shizuma: V1397,439-443(1991)
See Noda, Osama: V1397,427-432(1991)
See Sato, Shunichi: V1397,421-425(1991)
See Sato, Shunichi: V1397,433-437(1991)

Kondoh, Hiroshi P. See Umeda, Tokuo: V1446,199-210(1991)

Kondoh, Kayoko See Tsuno, Katsuhiko: V1490,222-232(1991)

Kondratenko, Nina A. See Sherstyuk, Valentin P.: V1238,211-217(1991)

Kondratenko, P. S.
See Bodulinsky, V. K.: V1440,392-396(1991)
See Dolgina, A. N.: V1440,342-353(1991)

Kondratiuk, N. V. See Jankiewicz, Zdzislaw: V1391,101-104(1991)

Konefal, Janusz
: Experimental and theoretical description of DC transverse glow-discharge
excitation of high-power convective CO2 laser and stabilization of its
output power,V1391,135-143(1991)
: Study of characteristics of a multikilowatt CO2 laser operating in flow-
closed cycle: a semiempirical model,Vi397,409-415(1991)

Konforti, Naim
See Mendlovic, David: V1442,182-192(1991)
See Shafir, Ehud: V1442,236-241(1991)

Kong, J. See Shi, W.: V1519,138-141(1991)

Kong, Jin A. See Tsang, Leung: V1558,260-268(1991)

Kong, Keon-Shik
; Bhasin, Kul B.; Itoh, Tatsuo: Design aspects and comparison between high-
Tc superconducting coplanar waveguide and microstrip line,V1477,57-
65(1991)

Kong, Mei-Ying See Chen, Liang-Hui: V1361,60-73(1991)

Kong, Mo-Nga See Chee, Joseph K.: V1413,14-20(1991)

Konishi, Junji See Minato, Kotaro: V1446,195-198(1991)

Konishi, Mashide
; Brooks, Rodney A.: Multiple unfused passive sensors for operating in busy
indoor environments,V1383,448-458(1991)

Kononenko, Vadim L.
; Shimkus, J. K.: Transverse concentrational profiles in dilute erythrocytes
suspension flows in narrow channels measured by integral Doppler
anemometer,V1403,381-383(1991)
See Bek, A. M.: V1403,384-386(1991)

Konopka, Wayne L.
; Soel, Michael A.; Celentano, A.; Calia, V.: Transient radiometric
measurements with a PtSi IR camera,V1488,355-365(1991)

Konoplev, O. A. See Vorobiev, N. S.: V1358,895-901(1991)

Konov, Vitaly I.
; Kulevsky, Lev A.; Lukashev, Alexei V.; Pashinin, Vladimir P.; Silenok,
Alexander S.: Comparative study of gelatin ablation by free-running and
Q-switch modes of Er:YAG laser,V1427,232-242(1991)
; Prokhorov, Alexander M.; Silenok, Alexander S.; Tsarkova, O. G.; Tsvetkov,
V. B.; Shcherbakov, I. A.: Experimental simulation of holmium laser
action on biological tissues,V1427,220-231(1991)
; Prokhorov, Alexander M.; Shcherbakov, I. A.: Perspectives of powerful laser
technique for medicine,V1525,250-252(1991)
See Armeyev, V. Y.: V1621,2-10(1991)
See Artjushenko, Vjacheslav G.: V1420,149-156(1991)
See Artjushenko, Vjacheslav G.: V1420,157-168(1991)
See Artjushenko, Vjacheslav G.: V1420,176-176(1991)
See Artjushenko, Vjacheslav G.: V1590,131-136(1991)
See Bonch-Bruevich, Aleksei M.: V1440(1991)
See Craciun, Valentin: V1392,625-628(1991)
See D'yakonov, G. I.: V1421,156-162(1991)
See D'yakonov, G. I.: V1424,81-86(1991)
See Nikitin, Petr I.: V1584,124-134(1991)
See Tokarev, Vladimir N.: V1503,269-278(1991)

Konstantinov, N. Y. See Artjushenko, Vjacheslav G.: V1590,131-136(1991)

Kontsevoy, B. L. See Apanasevich, P. A.: V1403,475-477(1991)

Konwerska-Hrabowska, Joanna
; Kryszczynski, Tadeusz; Smolka, M.; Olbrysz, P.; Widomski, L.: Moire sensor
as an automatic feeler gauge,V1554B,225-232(1991)
; Kryszczynski, Tadeusz; Tomaszewicz, Tomasz; Lietz, J.; Mazurkiewicz,
Wojciech: Photoelastic sensors for automatic control system of dam
safety,V1554A,388-399(1991)

Koo, Kee P. See Bucholtz, Frank: V1367,226-235(1991)

Koort, Hans J.
: Excimer laser in arthroscopic surgery,V1424,53-59(1991)
: Processing of bioceramic implants with excimer laser,V1427,173-180(1991)
; Frentzen, Matthias: Pulsed lasers in dentistry: sense or nonsense?,V1424,87-98(1991)

Kooyman, Rob P. See Heideman, Rene: V1510,131-137(1991)

Kopchok, George E.
See Tabbara, Marwan R.: V1425,208-216(1991)
See White, Rodney A.: V1422,103-110(1991)

Kopcsay, Gerard V. See Deutsch, Alina: V1389,161-176(1991)

Kopczynski, Krzysztof See Czechowicz, Roman: V1391,52-60(1991)

Kopec, Gary E. See Bagley, Steven C.: V1460,71-79(1991)

Kopeika, Norman S.
See Dror, Itai: V1487,192-202(1991)
See Gottlieb, J.: V1487,184-191(1991)
See Hadar, Ofer: V1482,79-91(1991)
See Hadar, Ofer: V1533,61-74(1991)
See Sadot, Danny: V1442,325-334(1991)
See Sadot, Danny: V1487,40-50(1991)
See Shushan, A.: V1487,300-311(1991)

Kopelman, Raoul
; Tan, Weihong; Lewis, Aaron; Lieberman, Klony: Scanning exciton microscopy and single-molecule resolution and detection,V1435,96-101(1991)

Kopera, Paul M.
ed.: *Components for Fiber Optic Applications V*,V1365(1991)

Koplowitz, Jack
: Scaling of digital shapes with subpixel boundary estimation,V1470,167-174(1991)
; Lee, Xiaobing: Edge detection with subpixel accuracy,V1471,452-463(1991)
See Wong, Ping W.: V1398,39-47(1991)

Kopola, Harri K. See Karppinen, Arto: V1455,179-189(1991)

Kopolovich, Zvi
; Naor, Yoram; Cabib, Dario; Johnson, W. Todd; Sapir, Eyal: State-of-the-art transfer radiometer for testing and calibration of FLIR test equipment,V1540,565-577(1991)

Kopylov, Yuri L. See Gulyaev, Yuri V.: V1621,84-92(1991)

Korb, C. L.
; Gentry, Bruce M.: Edge technique: a new method for atmospheric wind measurements with lidar,V1416,177-182(1991)
See Targ, Russell: V1521,144-157(1991)

Korb, Thomas See Zell, Andreas: V1469,708-718(1991)

Korbelik, Mladen
; Krosl, Gorazd; Lam, Stephen; Chaplin, David J.; Palcic, Branko: Microlocalization of Photofrin in neoplastic lesions,V1426,172-179(1991)

Korchazhkin, S. V.
; Krasnova, L. O.: Interferometric analysis of absorbing objects,V1512,195-197(1991)
; Krasnova, L. O.: Recording of laser radiation power distribution with the use of nonlinear optical effects,V1358,966-972(1991)
See Vlasov, Nikolai G.: V1358,1018-1020(1991)

Kordas, Joseph F. See Lewis, Isabella T.: V1478,2-12(1991)

Koren, Eugen
; Koscec, Mirna; Fugate, Robert D.: Degradation of cholesterol crystals in macrophages: the role of phospholipids,V1428,214-223(1991)

Koren, Yoram See Borenstein, Johann: V1388,74-81(1991)

Korenev, A. S. See Vas'kovsky, Yu. M.: V1440,229-240(1991)

Koreneva, N. A. See Matejec, Vlastimil: V1513,174-179(1991)

Korenowski, G. M. See LaPeruta, Richard: V1497,357-366(1991)

Korjenevitch, I. N. See Feldman, G. G.: V1358,497-502(1991)

Korjov, M. Y. See Apollonova, O. V.: V1501,108-119(1991)

Kormer, S. B. See Zykov, L. I.: V1412,258-266(1991)

Kornack, J.
See Rueckmann, I.: V1513,78-85(1991)
See Woggon, Ulrike: V1362,888-898(1991)

Kornilov, Yu. M. See Manenkov, Alexander A.: V1420,254-258(1991)

Korniski, Ronald J.
: Refractive-index interpolation fit criterion for materials used in optical design,VCR38,193-217(1991)
; Thompson, Kevin P.: Importance of dispersion tolerances in infrared lens design,V1354,402-407(1991)

Korniski, T. J. See Gierczak, Christine A.: V1433,315-328(1991)

Korobkin, V. V. See Karpov, V. I.: V1503,492-502(1991)

Korobowicz, Witold See Wolinski, Wieslaw: V1391,334-340(1991)

Korolevich, A. N.
; Khairullina, A. Y.: Multiple quasi-elastic circular-polarized light scattering by brownian moving aspherical particles,V1403,364-371(1991)

Koronkevich, Voldemar P.
; Palchikova, Irena G.: Kinoforms with increased depth of focus,V1507,110-115(1991)
See Hossfeld, Jens: V1574,159-166(1991)

Korostelin, Y. V. See Nasibov, Alexander S.: V1361,901-908(1991)

Koroteev, Nikolai I.
See Chernyaeva, E. B.: V1402,7-10(1991)
See Ivanov, Anatoliy A.: V1403,174-184(1991)
See Ivanov, Anatoliy A.: V1403,243-245(1991)
See Ivanov, Anatoliy A.: V1429,132-144(1991); 1432,141-153(1991)

Korotkov, V. A. See Apollonova, O. V.: V1501,108-119(1991)

Korovkin, Vladimir P. See Grodnikov, Alexander I.: V1400,186-193(1991)

Korppi-Tommola, Jouko E.
; Hakkarainen, Aulis; Helenius, Vesa F.: Molecular aggregates of quinuclidine and chlorophyll a,V1403,457-465(1991)

Korsch, Dietrich G.
: Optical design considerations for next-generation space and lunar telescopes,V1494,111-118(1991)

Korsten, Maarten J. See Wijbrans, Klaas C.: V1386,197-205(1991)

Korte, Lutz See Gornik, Erich: V1362,1-13(1991)

Kortright, Jeffrey B.
: Mirrors as power filters,V1345,38-41(1991)
; Underwood, James H.: Design considerations for multilayer-coated Schwarzschild objectives for the XUV,V1343,95-103(1991)
See Windt, David L.: V1343,292-308(1991)

Korzhenevich, Irina M.
; Kolesov, G. V.; Lebedev, V. B.; Petrokovich, O. A.; Feldman, G. G.: Examination of EOT operation with spherical elements or slit accelerating diaphragm,V1358,1084-1089(1991)
; Kolesov, G. V.: Time aberrations of combined focusing system of high-speed image converter,V1358,1090-1095(1991)
See Averin, V. I.: V1358,589-602(1991)

Kosa, Nadhir B.
: Key issues in selecting plastic optical fibers used in novel medical sensors,V1592,114-121(1991)

Kosbar, Laura L. See Rao, Veena: V1466,309-323(1991)

Koscec, Mirna See Koren, Eugen: V1428,214-223(1991)

Koschmann, Eric C.
: Uniform energy discharge for pulsed lasers,V1411,118-126(1991)

Koseki, Kenichi See Ishikawa, Toshiharu: V1461,73-78(1991)

Kosel, Frank M. See Snyder, Donald R.: V1346,216-225(1991)

Kosel, Peter B.
; Bozorgebrahimi, Nercy; Iyer, J.: Characteristics of MSM detectors for meander channel CCD imagers on GaAs,V1541,48-59(1991)

Koselja, Michal P. See Zschocke, Wolfgang: V1501,49-56(1991)

Koshevoy, Alexander V. See Katsap, Victor N.: V1420,259-265(1991)

Koshi, Yutaka
; Kunitake, Setsu; Suzuki, Kazuhiro; Kamizawa, Koh; Yamasaki, Toru; Miyake, Hidetaka: High-resolution color image coding scheme for office systems,V1605,362-373(1991)

Koshino, Nagaaki See Naitou, Kazunori: V1499,386-392(1991)

Koshizuka, Kunihiro
; Abe, Takao: Study of thermal dye diffusion,V1458,97-104(1991)

Koshkin, A. A. See Benimetskaya, L. Z.: V1525,210-211(1991)

Koshy, Matthew See Liu, Yi-Hwa: V1428,191-199(1991)

Kosicki, Bernard B. See Huang, Chin M.: V1447,156-164(1991)

Koskinen, Lasse
; Astola, Jaakko; Neuvo, Yrjo A.: Analysis of noise attenuation in morphological image processing,V1451,102-113(1991)
; Astola, Jaakko; Neuvo, Yrjo A.: Soft morphological filters,V1568,262-270(1991)

Kosman, Steven L. See Stevens, Eric G.: V1447,274-282(1991)

Kosobokova, N. L.
; Usanov, Yu. Y.; Shevtsov, M. K.: Properties of volume reflection silver-halide gelatin holograms,V1238,183-188(1991)
See Usanov, Yu. Y.: V1238,178-182(1991)

Kosovsky, L. A.
See Gordienko, Vyacheslav M.: V1416,102-114(1991)
See Pogosov, Gregory A.: V1416,115-124(1991)

Kossentini, Faouzi
; Smith, Mark J.; Barnes, Christopher: Residual VQ (vector quantizaton) with state prediction: a new method for image coding,V1452,383-394(1991)

Kossov, V. G. See Khvilivitzky, A. T.: V1447,184-190(1991)

Kost, Bernd See Beldie, Ion P.: V1457,242-247(1991)

Kostelak, Robert L. See Garofalo, Joseph G.: V1463,151-166(1991)

Koster, Rients See Heideman, Rene: V1510,131-137(1991)

Kostin, Ivan K.
; Dvalishvili, V. V.; Ureneva, E. V.; Freishist, N. A.; Fjodorov, A. V.; Dmitriyenko, O. L.: Application of a dynamic photoelasticity technique for the study of two- and three-dimensional structures under seismic, shock, and explosive loads effect,V1554A,418-425(1991)
See Polyak, Alexander: V1554A,553-556(1991)

Kostishack, Daniel F.
: Small-satellite sensors for multispectral space surveillance,V1495,214-227(1991)

Kostrubiec, Franciszek
: Thermocapillary mechanism of laser pulse alloying of a metal surface layer,V1391,224-227(1991)

Kostrzewa, Stanislaw
See Domanski, Andrzej W.: V1510,72-77(1991)
See Domanski, Andrzej W.: V1420,72-80(1991)

Kostrzewski, Andrew A. See Lin, Freddie S.: V1558,406-413(1991)

Kostuk, Raymond K.
: Packaging issues for free-space interconnects at the board level,V1563,72-80(1991)
: Practical design considerations and performance characteristics of high-numerical-aperture holographic lenses,V1461,24-34(1991)
; Huang, Yang-Tung; Kato, Masayuki: Multiprocessor optical bus,V1389,515-522(1991)
See Haggans, Charles W.: V1499,293-302(1991)
See Huang, Yang-Tung: V1474,39-44(1991)

Kostyk, D. See Feit, Zeev: V1512,164-169(1991)

Kostyljov, Ghennadij D.
: Denisyuk hologram recording with simultaneous use of all spectral components of the white light,V1238,316-319(1991)

Kotani, Kazunori See Gan, Qing: V1605,374-381(1991)

Koteles, Emil S. See Elman, Boris S.: V1361,362-372(1991)

Kothari, Harshad S. See Yadav, M. S.: V1362,811-819(1991)

Kotkov, A. A. See Ionin, A. A.: V1397,453-456(1991)

Kotmel, Robert F. See White, Christopher J.: V1425,130-133(1991)

Kotov, Gennady A.
; Filippov, N.; Shandybina, Galina: Laser-induced optical effects in light-sensitive complexes,V1440,321-324(1991)
See Shestakov, S. D.: V1440,423-435(1991)

Kotowski, Tomasz
; Skubiszak, Wojciech; Soroka, Jacek A.; Soroka, Krystyna B.; Stacewicz, Tadeusz: 9H-Indolo(1,2-f) phenanthridinium hemicyanines: a new group of high-efficiency broadband generating laser dyes,V1391,6-11(1991)

Kotsuka, Hiroshi See Nishimura, Kazuhito: V1534,199-206(1991)

Kottapalli, M. S. See Chu, Chee-Hung H.: V1606,470-481(1991)

Kountouris, Vasilios G.
; Stephanou, Harry E.: Task decomposition, distribution, and localization for intelligent robot coordination,V1387,169-180(1991)

Koutsky, J. A. See Wallenberger, Frederick T.: V1590,72-82(1991)

Kouwenberg, Jef M.
; Ottes, Fenno P.; Bakker, Albert R.: European activities towards a hospital-integrated PACS based on open systems,V1446,357-361(1991)

Kouyama, Tsutomu See Kobayashi, Takayoshi T.: V1403,407-416(1991)

Kovacevic, Milan
; Gerla, Mario: Efficient multiaccess scheme for linear lightwave networks,V1579,74-83(1991)

Kovacic, Stane See Gee, Jim C.: V1445,226-234(1991)

Kovacich, Michael
; Casaletto, Tom; Lutjens, William; McIntyre, David; Ansell, Ralph; VanDyk, Ed: Application of MHT to group-to-object tracking,V1481,357-370(1991)

Kovacs, Emoke
; Marosi, Istvan: Effectiveness of certain features for optical character recognition,V1384,338-343(1991)

Kovacs, Stephen
; Melore, Dan; Cerrina, Franco; Cole, Richard K.: System design considerations for a production-grade, ESR-based x-ray lithography beamline,V1465,88-99(1991)

Koval'chuk, Alexandr V. See Bodnar, Vladimir G.: V1455,61-72(1991)

Koval, G. I. See Artemjev, S. V.: V1238,206-210(1991)

Kovalenko, A. N. See Romanovsky, V. F.: V1358,140-145(1991)

Kovalenko, Sergey N.
See Gorelik, Vladimir S.: V1507,379-382(1991)
See Gorelik, Vladimir S.: V1507,488-490(1991)

Kovalev, A. A. See Dolgina, A. N.: V1440,342-353(1991)

Kovalev, Valeri I.
: Damage to InAs surface from long-pulse 10.6um laser radiation,V1441,536-540(1991)

Kovatchev, Methodi
; Ilieva, R.: Neural networks and computers based on in-phase optics,V1621,259-267(1991)

Kovrigin, Yevgeny
; Potylitsyn, Yevgeny: Preamplifiers of the high-ohmic high-speed photodetector signals,V1362,967-975(1991)

Kowada, Masayoshi See Mukasa, Minoru: V1446,253-265(1991)

Kowaka, Masahiko See Wakabayashi, Osamu: V1463,617-628(1991)

Kowalczyk, M. See Grzegorzewski, Bronislaw: V1391,290-294(1991)

Kowalczyk, T. C. See Cahill, Paul A.: V1560,130-138(1991)

Kowel, Stephen T. See Yankelevich, Diego: V1560,406-415(1991)

Kowsalya, S. See Chitnis, Vijay T.: V1332,613-622(1991)

Koyama, Eiji See Watanabe, Hitoshi: V1499,21-28(1991)

Koyama, Fumio See Tamanuki, Takemasa: V1361,614-617(1991)

Koyamatsu, A. H. See Aviles, Walter A.: V1388,587-597(1991)

Koyanagi, Hiroo
; Umeda, Shin'ichi; Fukunaga, Seiki; Kitaori, Tomoyuki; Nagasawa, Kohtaro: Study of the chemically amplifiable resist materials for electron-beam lithography,V1466,346-361(1991)
See Fukunaga, Seiki: V1466,446-457(1991)

Koyanagi, Mitsumasa
: Optically coupled three-dimensional memory system,V1390,467-476(1991)
See Hirose, Masataka: V1362,316-322(1991)

Koyata, Sakuo See Hattori, Takeshi: V1464,367-376(1991)

Koza, M. A. See Soole, Julian B.: V1474,268-276(1991)

Kozaitis, Samuel P.
; Cofer, Rufus H.: Optical processing of wire-frame models for object recognition,V1471,249-254(1991)
; Kirschner, K.; Kelly, E.; Been, D.; Delgado, J.; Velez, E.; Alkindy, A.; Al-Houra, H.; Ali, F.: Electronic interface for high-frame-rate electrically addressed spatial light modulators,V1474,112-115(1991)
; Petrilak, Robert: Performance of pyramidal phase-only filtering of infrared imagery,V1564,403-413(1991)
; Tepedelenlioglu, N.; Foor, Wesley E.: Analysis and experimental performance of reduced-resolution binary phase-only filters,V1564,373-383(1991)

Kozak, Taras See Cinelli, Joseph L.: V1499,203-208(1991)

Kozar, A. V.
; Krupenko, S. A.: Optical cross-modulation method for diagnostic of powerful microwave radiation,V1476,305-312(1991)

Kozhevnikov, Nikolai M.
: Dynamic holographic microphasometry,V1507,509-516(1991)
; Barmenkov, Yuri O.: Dynamic holography application in fiber optic interferometry,V1584,387-395(1991)
; Barmenkov, Yuri O.; Belyakov, V. A.; Medvedev, A. A.; Razdobarin, G. T.: Fiber optic sensor for plasma current diagnostics in tokamaks,V1584,138-144(1991)
; Barmenkov, Yuri O.; Lipovskaya, Margarita J.: Holographic recording in photorefractive media containing bacteriorhodopsin,V1507,517-524(1991)

Kozich, V. P. See Apanasevich, P. A.: V1403,475-477(1991)

Kozicki, M. N. See Ferry, David K.: V1596,2-11(1991)

Kozionov, Andrew L.
  See Benimetskaya, L. Z.: V1525,242-245(1991)
  See Benimetskaya, L. Z.: V1525,210-211(1991)

Kozlik, Claudius See Bickel, P.: V1503,167-175(1991)

Kozlov, N. P.
  See Kamrukov, A. S.: V1503,438-452(1991)
  See Kamrukov, A. S.: V1397,137-144(1991)

Kozlov, Sergei N. See Bondarenko, Alexander V.: V1376,117-127(1991)

Kozlov, Sergey A.
  See Azarenkov, Aleksey N.: V1409,154-164(1991)
  See Azarenkov, Aleksey N.: V1409,166-177(1991)

Kozlova, T. G.
  ; Lobacheva, M. I.; Pravdin, A. B.; Romakina, M. Y.; Sinichkin, Yury P.; Tuchin, Valery V.: Laser spectroscopy of carotenoids in plant bio-objects,V1403,159-160(1991)

Kozlovsky, William J. See Lenth, Wilfried: V1499,308-313(1991)

Kozlowski, Alan E. See Ostrout, Wayne H.: V1463,54-73(1991)

Kozlowski, Lester J.
  ; Bailey, Robert B.; Cooper, Donald E.; Vural, Kadri; Gertner, E. R.; Tennant, William E.: Large staring IRFPAs of HgCdTe on alternative substrates,V1540,250-261(1991)

Kozlowski, Mark R.
  ; Staggs, Michael C.; Rainer, Frank; Stathis, J. H.: Laser conditioning and electronic defects of HfO2 and SiO2 thin films,V1441,269-282(1991)
  ; Thomas, Ian M.; Edwards, Gary; Stanion, Ken; Fuchs, Baruch A.; Latanich, L.: Influence of diamond turning and surface cleaning processes on the degradation of KDP crystal surfaces,V1561,59-69(1991)
  See Campbell, John H.: V1441,444-456(1991)

Kozousek, V. See Shen, Zhijiang: V1606,632-640(1991)

Kozuka, Takahiro See Umeda, Tokuo: V1446,199-210(1991)

Kpalma, Kidiyo
  ; Bruno, Alain; Haese-Coat, Veronique: Natural texture analysis in a multiscale context using fractal dimension,V1606,55-66(1991)

Kraemer, Udo See Jiang, Jie: V1505,166-174(1991)

Krainak, Michael A. See Davidson, Frederic M.: V1417,75-88(1991)

Krall, Jonathan
  ; Friedman, Moshe; Lau, Yue Y.; Serlin, Victor: Relativistic klystron amplifier IV: simulation studies of a coaxial-geometry RKA,V1407,23-31(1991)
  See Colombant, Denis G.: V1407,13-22(1991)
  See Friedman, Moshe: V1407,474-478(1991)
  See Friedman, Moshe: V1407,2-7(1991)
  See Serlin, Victor: V1407,8-12(1991)

Kramer, Charles J.
  : Holographic deflectors for graphic arts applications: an overview,V1454,68-100(1991)
  ; Szalowski, Rafal; Watkins, Mark: Ultra-high scanning holographic deflector unit,V1454,434-446(1991)

Kramer, David M. See Liao, Wen-gen: V1444,47-55(1991)

Kranenberg, C. F.
  See Reicher, David W.: V1441,106-112(1991)
  See Wilson, Scott R.: V1441,82-86(1991)

Kransteuber, Amy S. See Gilbert, John A.: V1238,412-420(1991)

Kranz, Wolfgang
  : Coherent laser radar for target classification,V1479,270-274(1991)

Krapels, Keith A. See Taylor, James S.: V1478,41-49(1991)

Krasauskas, V.
  See Chirvony, V. S.: V1403,504-506(1991)
  See Chirvony, V. S.: V1403,638-640(1991)

Krasinskaya, A. G. See Ashkinadze, B. M.: V1361,866-873(1991)

Krasner, Brian H. See Lo, Shih-Chung B.: V1444,265-271(1991)

Krasnov, Victor F.
  ; Musher, Semion L.; Prots, V. I.; Rubenchik, Aleksandr M.; Ryabchenko, Vladimir E.; Stupak, Mikhail F.: Nonlinear optic in-situ diagnostics of a crystalline film in molecular-beam-epitaxy devices,V1506,179-187(1991)

Krasnova, L. O.
  See Korchazhkin, S. V.: V1512,195-197(1991)
  See Korchazhkin, S. V.: V1358,966-972(1991)

Krasnovsky, A. A. See Egorov, S. Y.: V1403,611-621(1991)

Krasowski, Michael J. See Beheim, Glenn: V1369,50-59(1991)

Kraus, Eugene J.
  : Domain-variant gray-scale morphology,V1451,171-178(1991)
  See Schott, John R.: V1540,533-549(1991)

Kraus, George A. See Carpenter, Susan: V1426,228-234(1991)

Krause, A. See Bergner, Harald: V1361,723-731(1991)

Krauser, Juergen See Keil, Norbert: V1559,278-287(1991)

Krauss, Matthias G. See Norkus, Volkmar: V1484,98-105(1991)

Krauss, Thomas
  ; Laybourn, Peter J.: Monolithic integration of a semiconductor ring laser and a monitoring photodetector,V1583,150-152(1991)

Krautschik, Christof G. See Dianov, Evgeni M.: V1516,75-80(1991)

Kravchenko, V. A. See Bordo, V. G.: V1440,364-369(1991)

Kravetz, V. G. See Petrov, V. V.: V1621,45-50(1991)

Kravitz, Stanley H.
  See Hietala, Vincent M.: V1476,170-175(1991)
  See Vawter, Gregory A.: V1476,102-106(1991)

Krawczyk, Rodolphe
  ; Chessel, Jean-Pierre; Durpaire, Jean-Pierre; Durieux, Alain; Churoux, Pascal; Briottet, Xavier: On-board calibration device for a wide field-of-view instrument,V1493,2-15(1991)
  See Lange, Robert: V1492,24-37(1991)

Krawiec, Theresa M.
  ; Giammarco, Nicholas J.: Reactive ion etching of deep isolation trenches using sulfur hexafluoride, chlorine, helium, and oxygen,V1392,265-271(1991)

Krchnavek, Robert R. See Lalk, Gail R.: V1389,386-400(1991)

Krebs, Danny J. See Haake, John M.: V1418,298-308(1991)

Krecker, Donald K. See Rosenfeld, J. P.: V1406,147-147(1991) ·

Krehl, Peter
  ; Warken, D.: Flash soft radiography: its adaption to the study of breakup mechanisms of liquid jets into a high-density gas,V1358,162-173(1991)

Kreider, Gregory
  ; Van der Spiegel, Jan; Born, I.; Claeys, Cor L.; Debusschere, Ingrid; Sandini, Giulio; Dario, Paolo: Design and characterization of a space-variant CCD sensor,V1381,242-249(1991)

Kreider, James F.
  ; Preis, Mark K.; Roberts, Peter C.; Owen, Larry D.; Scott, Walter M.; Walmsley, Charles F.; Quin, Alan: Multiplexed mid-wavelength IR long linear photoconductive focal-plane arrays,V1488,376-388(1991)

Kreidl, John F. See Kitazawa, Mototaka: V1592(1991)

Kreidl, Norbert J. See Bray, Phillip: V1590(1991)

Kreis, Thomas M.
  ; Geldmacher, Juergen: Evaluation of interference patterns: a comparison of methods,V1554B,718-724(1991)

Kreisler, Alain J.
  See Baixeras, Joseph M.: V1362,117-126(1991)
  See Hosseini Teherani, Ferechteh: V1362,921-929(1991)
  See Naumaan, Ahmed: V1512(1991)

Kreiss, William T. See Sikes, Terry L.: V1479,199-211(1991)

Kremer, Paul J. See Ruiz, Domingo: V1417,212-222(1991)

Kremser, Christian
  ; Unterrainer, Karl; Gornik, Erich; Strasser, G.; Pidgeon, Carl R.: Tunable cyclotron-resonance laser in germanium,V1501,69-79(1991)

Kreske, Kathi See Keren, Eliezer: V1442,266-274(1991)

Kreskovsky, John P. See Meyyappan, Meyya: V1392,67-76(1991)

Kretlov, Boris S. See Mikaelian, Andrei L.: V1621,194-203(1991)

Kretschmer, K.-H.
  ; Lorenz, Gerhard; Castrischer, G.; Kessler, I.; Baumann, P.: LH electron cyclotron resonance plasma source,V1392,246-252(1991)

Kreutz, Ernst W.
  ; Pirch, Norbert: Melt dynamics and surface deformation in processing with laser radiation,V1502,160-176(1991)

Krieger, Jeffry M. See Ali, Mir A.: V1513,215-223(1991)

Kriegmair, M. See Baumgartner, R.: V1525,246-248(1991)

Kriete, Andres See Masters, Barry R.: V1431,218-223(1991)

Krikelis, A.
  : Multitarget detection and estimation parallel algorithm,V1482,307-316(1991)

Krile, Thomas F. See Ellett, Scott A.: V1564,634-643(1991)

Krinsky, Jeffrey A.
  ; Reddy, Mahesh C.: Ionizing radiation-induced attenuation in optical fibers at multiple wavelengths and temperature extremes,V1366,191-203(1991)
  See Reddy, Mahesh C.: V1369,107-113(1991)

Krishnamoorthy, Ashok V.
  See Esener, Sadik C.: V1562,11-20(1991)
  See Marchand, Philippe J.: V1505,38-49(1991)

Krishnamurthy, Ashok K. See Kumar, Vinod V.: V1469,484-494(1991)

Krishnamurthy, Giridhar See Shynk, John J.: V1565,102-117(1991)

Krishnan, Ganapathy
  ; Jones, David E.: Machine verification of traced signatures,V1468,563-
    572(1991)
  See Randolph, David: V1386,255-264(1991)

Krishnan, S.
  See Fowler, Burt: V1598,108-117(1991)
  See Lian, S.: V1598,98-107(1991)

Krishnapuram, Raghu J.
  ; Chen, Ling-Fan: Iterative neural networks for skeletonization and
    thinning,V1382,271-281(1991)
  See Gupta, Sundeep: V1568,335-346(1991)

Krishnaswamy, Sridhar
  See Ahmadshahi, Mansour A.: V1554A,620-627(1991)
  See Tippur, Hareesh V.: V1554A,176-191(1991)

Krisiloff, Allen J. See Antos, Ronald L.: V1398(1991)

Kriss, Michael A.
  : Image analysis of discrete and continuous systems: film and CCD
    sensors,V1398,4-14(1991)

Kristensen, Anders See Hansen, Ole P.: V1362,192-198(1991)

Kristl, Joseph See Noah, Meg A.: V1479,275-282(1991)

Krivak, T. See Morgan, Alan R.: V1426,350-355(1991)

Krivokhizha, A. M. See Melnik, Ivan S.: V1572,118-122(1991)

Krivonogov, Yu. See Uglov, A.: V1440,310-320(1991)

Krivov, V. V. See Artsimovich, M. V.: V1397,729-733(1991)

Kroes, Roger L. See Springer, John M.: V1557,192-196(1991)

Krohn, David A.
  ; Maklad, Mokhtar S.; Bacon, Fredrick: High-strength optical fiber for medical
    applications,V1420,126-135(1991)

Krol, Denise M.
  ; Atkins, Robert M.; Lemaire, Paul J.: Photoinduced second-harmonic
    generation and luminescence of defects in Ge-doped silica
    fibers,V1516,38-46(1991)

Krol, Mark F.
  See Goel, Kamal K.: V1476,314-319(1991)
  See Prucnal, Paul R.: V1389,462-476(1991)

Kromin, A. K. See Veiko, Vadim P.: V1544,152-163(1991)

Krone-Schmidt, Wilfried See Lippey, Barret: V1530,150-161(1991)

Krongauz, Vadim V.
  ; Schmelzer, E. R.: Peculiarities of anisotropic photopolymerization in
    films,V1559,354-376(1991)

Krongauz, Valeri
  See Berkovic, Garry: V1560,238-242(1991)
  See Berkovic, Garry: V1442,44-52(1991)

Krongelb, Sol
  ; Dukovic, John O.; Komsa, M. L.; Mehdizadeh, S.; Romankiw, Lubomyr T.;
    Andricacos, P. C.; Pfeiffer, A. T.; Wong, K.: Application of
    electrodeposition processes to advanced package fabrication,V1389,249-
    256(1991)

Krook, Lennart See Nixon, Alan J.: V1424,198-208(1991)

Krosl, Gorazd See Korbelik, Mladen: V1426,172-179(1991)

Kross, Juergen
  : Principles of optimization in lens design developed at the Institute of Optics
    in Berlin (West),V1354,165-170(1991)

Krost, Alois See Hingerl, Kurt: V1361,383-393(1991)

Krovvidy, Srinivas
  ; Wee, William G.; Han, Chia Y.: Case-based reasoning approach for
    heuristic search,V1468,216-226(1991)

Krueger, D. J. See Singh, Rajendra: V1394,62-67(1991)

Krueger, G. See Breuer, J.: V1503,223-230(1991)

Kruepper, Peter See Dalsing, Michael C.: V1422,98-102(1991)

Kruer, Melvin R. See Scribner, Dean A.: V1541,100-109(1991)

Kruer, William L. See Wilks, S. C.: V1413,131-137(1991)

Krug, Herbert See Schmidt, Helmut: V1590,36-43(1991)

Kruger, Robert A.
  See Hebden, Jeremy C.: V1431,225-231(1991)
  See Hebden, Jeremy C.: V1443,294-300(1991)
  See Ning, Ruola: V1443,236-249(1991)

Kruglik, S. G.
  See Apanasevich, P. A.: V1403,195-211(1991)
  See Apanasevich, P. A.: V1403,240-242(1991)
  See Gurinovich, G. P.: V1403,139-141(1991)

Krumweide, Gary C.
  ; Brand, Richard A.: Attacking dimensional instability problems in graphite/
    epoxy structures,V1533,252-261(1991)

Krupchitsky, Vladimir P. See Artjushenko, Vjacheslav G.: V1420,157-168(1991)

Krupenko, S. A. See Kozar, A. V.: V1476,305-312(1991)

Kruse, Fred A.
  ; Dietz, John B.: Integration of diverse remote sensing data sets for geologic
    mapping and resource exploration,V1492,326-337(1991)

Krusius, J. P.
  : System interconnection of high-density multichip modules,V1390,261-
    270(1991)

Krutikov, N. I. See Ivanov, K. N.: V1358,732-738(1991)

Krutz, Gary W.
  See Jia, Jiancheng: V1379,246-253(1991)
  See Jia, Jiancheng: V1396,656-663(1991)

Krylov, Vitaly N.
  See Bespalov, V. G.: V1238,457-461(1991)
  See Bryskin, V. Z.: V1238,448-451(1991)

Kryskowski, David See McGlynn, John D.: V1486,141-150(1991)

Kryszczynski, Tadeusz
  See Konwerska-Hrabowska, Joanna: V1554B,225-232(1991)
  See Konwerska-Hrabowska, Joanna: V1554A,388-399(1991)

Kryukov, A. P. See Artjushenko, Vjacheslav G.: V1420,157-168(1991)

Kryukov, I. V. See Khoroshilov, E. V.: V1403,431-433(1991)

Kryukov, P. G. See Khoroshilov, E. V.: V1403,431-433(1991)

Ksendzov, Alexander See Lin, TrueLon: V1540,135-139(1991)

Ku, Ta-Kang
  ; Kuo, C.-C. J.: Minimum-phase LU factorization preconditioner for Toeplitz
    matrices,V1566,59-73(1991)

Ku, Walter H. See Soni, Tarun: V1565,338-344(1991)

Kubecek, Vaclav
  See Agnesi, Antoniangelo: V1415,242-247(1991)
  See Agnesi, Antoniangelo: V1415,104-109(1991)

Kubelka, Jiri See Zschocke, Wolfgang: V1501,49-56(1991)

Kubena, Randall L.
  ; Stratton, F. P.; Mayer, T. M.: Selective metal deposition using low-dose
    focused ion-beam patterning,V1392,595-597(1991)

Kubiak, Glenn D.
  : XUV resist characterization: studies with a laser plasma source,V1343,283-
    291(1991)
  See Windt, David L.: V1343,274-282(1991)

Kubica, Jacek M.
  : Attenuation of leaky waves in GaAs/AlGaAs MQW waveguides formed on a
    GaAs substrate,V1506,134-139(1991)
  ; Domanski, Andrzej W.: Transfer matrix approach to the design of InP/
    InGaAsP ARROW structures,V1391,24-31(1991)

Kubin, R. F.
  See Lindsay, Geoffrey A.: V1497,418-422(1991)
  See Lindsay, Geoffrey A.: V1560,443-453(1991)

Kubinec, James J.
  ; Somerville, James A.; Chown, David P.; Birch, Martin J.: Integrated "Byte-to-
    light" solution for fiber optic data communication,V1364,130-143(1991)

Kubitzek, Ruediger
  ; Froelich, Klaus; Tropartz, Stephan; Stojanoff, Christo G.: Development and
    experimental investigation of a copying procedure for the reproduction of
    large-format transmissive holograms,V1507,365-372(1991)
  See Stojanoff, Christo G.: V1536,206-214(1991)
  See Stojanoff, Christo G.: V1527,48-60(1991)
  See Stojanoff, Christo G.: V1485,274-280(1991)
  See Stojanoff, Christo G.: V1559,321-330(1991)
  See Tropartz, Stephan: V1507,345-353(1991)

Kubo, Uichi
  ; Okada, Kasuyuki; Hashishin, Yuichi: Hollow-tube-guide for UV-power laser
    beams,V1420,102-107(1991)

Kubo, Y. See Yamakoshi, H.: V1397,119-122(1991)

Kubota, F. See Allan, D. S.: V1560,362-369(1991)

Kubota, Hitoshi See Maeda, Shunji: V1567,100-109(1991)

Kubota, Shigeo
; Oka, Michio: Shot-noise-limited high-power green laser for higher density optical disk systems,V1499,314-323(1991)
; Oka, Michio; Kaneda, Yushi; Masuda, Hisashi: Thermal aberration analysis of a laser-diode-pumped Nd:YAG laser,V1354,572-580(1991)

Kubota, Shigeto See Oyama, Yoshiro: V1444,413-423(1991)

Kubota, Yoshiyuki See Ogura, Satoshi: V1412,123-128(1991)

Kubulins, Vilnis E.
See Carome, Edward F.: V1584,110-117(1991)
See Carome, Edward F.: V1589,133-138(1991)

Kuch'yanov, Aleksandr S. See Komarov, Konstantin P.: V1501,135-143(1991)

Kucharska, Alicia I. See Dawson, Philip: V1362,384-390(1991)

Kuchiki, Hirotsuna See Fujii, H.: V1397,213-220(1991)

Kuchinskii, Sergei A.
; Dostenko, Alexander V.: Theoretical models in optics of glass-composite materials,V1513,297-308(1991)

Kuczynski, Andrzej P. See Ashwell, Geoffrey J.: V1361,589-598(1991)

Kuder, James E. See Groh, Werner: V1592,20-30(1991)

Kudlek, Gotthard
; Presser, Nazmir; Gutowski, Juergen; Mathine, David L.; Kobayashi, Masakazu; Gunshor, Robert L.: Optical properties of molecular beam epitaxy grown ZnTe epilayers,V1361,150-158(1991)
See Hingerl, Kurt: V1361,383-393(1991)

Kudoh, Takeshi See Koga, Nobuhiro: V1554A,84-92(1991)

Kudriavtsev, E. M.
: CO2 coupled-mode, CO, and other lasers with supersonic cooling of gas mixture,V1397,475-484(1991)
See Artsimovich, M. V.: V1397,729-733(1991)

Kudriavtsev, N. N. See Doroschenko, V. M.: V1397,503-511(1991)

Kudryavtcev, N. Y. See Dvoryankin, A. N.: V1397,247-250(1991)

Kudryavtsev, I. A. See Ekimov, A. I.: V1513,123-129(1991)

Kudryavtseva, Anna D.
; Tcherniega, Nicolaii V.; Brekhovskikh, Galina L.; Sokolovskaya, Albina I.: Spatial frequency filtering on the basis of the nonlinear optics phenomena,V1385,190-199(1991)

Kudykina, T. A. See Grigor'ev, N. N.: V1440,105-111(1991)

Kuebler, C. W. See Doerband, Bernd: V1332,664-672(1991)

Kuechel, Michael F.
: New Zeiss interferometer,V1332,655-663(1991)
See Freischlad, Klaus R.: V1332,8-17(1991)
See Freischlad, Klaus R.: V1332,18-24(1991)

Kuehne, Mikhael
See Boher, Pierre: V1343,39-55(1991)
See Schmiedeskamp, Bernt: V1343,64-72(1991)

Kueper, Stephan
; Brannon, James H.: KrF laser ablation of polyurethane,V1598,27-35(1991)

Kuepfer, Manfred See Bosshard, Christian: V1560,344-352(1991)

Kuerbitz, Gunther
: Automatic active athermalization of infrared optical systems,V1540,612-621(1991)

Kuhl, Juergen See Rosenzweig, Josef: V1362,168-178(1991)

Kui, Jianming See Black, Linda R.: V1557,54-59(1991)

Kuittinen, Markku
; Jaaskelainen, Timo: Inverse-grating diffraction problems in the coupled-wave analysis,V1507,258-267(1991)
See Jaaskelainen, Timo: V1574,272-281(1991)

Kujawinska, Malgorzata
; Schmidt, Joanna: Spatial-carrier phase-shifting technique of fringe pattern analysis,V1508,61-67(1991)
; Spik, Andrzej: Spatial techniques of fringe pattern analysis in interferometry,V1391,303-312(1991)
; Wojiak, Joanna: Spatial phase-shifting techniques of fringe pattern analysis in photomechanics,V1554B,503-513(1991)
See Salbut, Leszek A.: V1554B,451-460(1991)

Kukiello, P.
; Rabczuk, G.: Output characteristics of a transverse-flow cw CO2 modular laser,V1397,417-419(1991)
; Rabczuk, G.: Transverse-flow cw CO2 modular laser: preliminary investigation,V1391,93-97(1991)

Kuklin, A. E. See Babin, Sergei A.: V1397,589-592(1991)

Kukreja, Lalit M.
; Braun, R.; Hess, P.: Simple analytical model for laser-induced tissue ablation,V1427,243-254(1991)

Kukulies, Joerg See Masters, Barry R.: V1431,218-223(1991)

Kulagin, Yu. A. See Dvoryankin, A. N.: V1397,247-250(1991)

Kulakov, Nickolay Y. See Kiselyov, Boris S.: V1621,328-339(1991)

Kulakovskii, Vladimir D. See Andersson, Thorvald G.: V1361,434-442(1991)

Kulesha, G. A. See Zhao, Jing-Fu: V1362,135-143(1991)

Kuleshov, Nickolay B. See Tarasov, Victor A.: V1559,331-342(1991)

Kulevsky, Lev A. See Konov, Vitaly I.: V1427,232-242(1991)

Kulikov, Stanislav M.
See Dolgopolov, Y. V.: V1412,267-275(1991)
See Zykov, L. I.: V1412,258-266(1991)

Kulikov, Yu. V. See Degtyareva, V. P.: V1358,546-548(1991)

Kulikowski, Casimir A. See Gong, Leiguang: V1450,144-153(1991)

Kulisch, Juergen R. See Frank, Werner F.: V1559,344-353(1991)

Kulkarni, Arun D.
; Byars, P.: Artificial neural network models for image understanding,V1452,512-522(1991)

Kulkarni, N. M.
; Kulkarni, R. N.; Shaligram, Arvind D.: RTP-induced defects in silicon studied by positron annihilation technique,V1393,207-214(1991)

Kulkarni, R. N. See Kulkarni, N. M.: V1393,207-214(1991)

Kull, Mart M. See Nevorotin, Alexey J.: V1427,381-397(1991)

Kulp, John M.
: Critical dimension shift resulting from handling time variation in the track coat process,V1466,630-640(1991)

Kulp, R. L. See Roth, Michael W.: V1469,25-36(1991)

Kulp, Thomas J.
; Garvis, Darrel G.; Kennedy, Randall B.; McRae, Thomas G.: Application of backscatter absorption gas imaging to the detection of chemicals related to drug production,V1479,352-363(1991)

Kuls, Juergen See Guelker, Gerd: V1500,124-134(1991)

Kulya, S. V. See Bogdanov, V. L.: V1403,470-474(1991)

Kumai, Satoshi See Narahara, Tatsuya: V1499,120-128(1991)

Kumar, Chellappa See Cui, Weijia: V1431,180-191(1991)

Kumar, G. N.
; Nandhakumar, N.: Efficient object contour tracing in a quadtree encoded image,V1468,884-895(1991)

Kumar, Harish See Sirohi, Rajpal S.: V1332,50-55(1991)

Kumar, Jayant See Li, Lian: V1560,243-250(1991)

Kumar, Prem
; Huang, Jianming; Aytur, Orhan: Photon-noise reduction experiments with a Q-switched Nd:YAG laser,V1376,192-197(1991)

Kumar, V.
; Chandra, Dinesh: Average energy gap of AIBIIIC2VI optoelectronic materials,V1361,809-811(1991)

Kumar, Vas
See Balasubramanian, Kunjithapa: V1485,245-253(1991)
See Murty, Mantravady V.: V1332,107-114(1991)

Kumar, Vinod V.
; Krishnamurthy, Ashok K.; Ahalt, Stanley C.: Phonetic-to-acoustic and acoustic-to-phonetic mapping using recurrent neural networks,V1469,484-494(1991)

Kume, Hidehiro
; Nakamura, Haruhito; Suzuki, Makoto: Gatable photonic detector and its image processing,V1358,1144-1155(1991)

Kume, Stewart M. See Spears, Kenneth G.: V1429,2-8(1991)

Kumer, John B.
: Review of some non-LTE high-altitude CO2 4.3-micrometer background effects: a ten-year update,V1540,321-330(1991)
See Roche, Aidan E.: V1491,91-103(1991)

Kumpel, Daniel M. See del Castillo, Maria D.: V1468,596-607(1991)

Kun, Zoltan K. See Leksell, David: V1458,133-144(1991)

Kunath, Richard R. See Polifko, David M.: V1476,91-99(1991)

Kundel, Harold L.
; Seshadri, Sridhar B.; Arenson, Ronald L.: Diagnostic value model for the evaluation of PACS: physician ratings of the importance of prompt image access and the utilization of a display station in an intensive care unit,V1446,297-300(1991)
See Nodine, Calvin F.: V1444,56-62(1991)

Kundu, Amlan See Wu, Wen-Rong: V1451,13-23(1991)

Kundu, Sourav K.
; McMath, Linda P.; Zaidan, Jonathan T.; Spears, J. R.: Application of conjugated heparin-albumin microparticles with laser-balloon angioplasty: a potential method for reducing adverse biologic reactivity after angioplasty,V1425,142-148(1991)

Kuneva, M.
See Savatinova, Ivanka T.: V1374,37-46(1991)
See Ziling, C. C.: V1583,90-101(1991)

Kung, Chun C.
; Funk, Eric E.; Chauchard, Eve A.; Rhee, M. J.; Lee, Chi H.; Yan, Li I.: Observation of power gain in an inductive pulsed power system with an optically activated semiconductor closing and opening switch,V1378,250-258(1991)

Kunioka, S. See Sugihara, Okihiro: V1361,599-605(1991)

Kunitake, Setsu See Koshi, Yutaka: V1605,362-373(1991)

Kunt, Murat
See Dufaux, Frederic: V1567,362-379(1991)
See Dufaux, Frederic: V1606,851-864(1991)
See Ebrahimi, Touradj: V1605,2-15(1991)
See Li, Wei: V1605,124-136(1991)
See Popat, Ashok C.: V1567,341-353(1991)
See Popat, Ashok C.: V1605,940-953(1991)

Kunz, Roderick R.
; Bianconi, Patricia A.; Horn, Mark W.; Paladugu, R. R.; Shaver, David C.; Smith, David A.; Freed, Charles A.: Polysilyne resists for 193 nm excimer laser lithography,V1466,218-226(1991)

Kunz, Warren F. See Friedl, Stephan E.: V1425,134-141(1991)

Kunzmann, H. See Tanwar, Lakhan S.: V1332,877-882(1991)

Kuo, C.-C. J. See Ku, Ta-Kang: V1566,59-73(1991)

Kuo, Chai-Pei
: Optical memory using localized photoinduced anisotropy in a synthetic dye-polymer,V1499,148-159(1991)

Kuo, Chien-yu C. See Park, Yong K.: V1372,219-227(1991)

Kuo, Chin-Po
: Deformable-mirror concept for adaptive optics in space,V1542,420-431(1991)
See Helms, Richard G.: V1532,64-80(1991)
See Shih, Choon-Foo: V1532,91-102(1991)

Kuo, Jen-Tsai
; Tzuang, Ching-Kuang C.: Termination for minimal reflection of high-speed pulse propagation along multiple-coupled microstrip lines,V1389,156-160(1991)

Kuo, Jenn-Ming See Levine, Barry F.: V1540,232-237(1991)

Kuo, Pao K. See Favro, Lawrence D.: V1467,290-294(1991)

Kuo, Shyh-Shiaw
; Mammone, Richard J.: Adaptive projection technique for CT images refinement,V1606,641-652(1991)
; Mammone, Richard J.: Image restoration and identification using the EM and adaptive RAP algorithms,V1606,419-430(1991)
; Mammone, Richard J.: Refinement of EM (expectation maximization) restored images,V1452,192-202(1991)
; Mammone, Richard J.: Resolution enhancement of CT images,V1450,18-29(1991)

Kuo, Spencer P.
; Ren, A.; Zhang, Y. S.: Frequency up-conversion for the reflectionless propagation of a high-power microwave pulse in a self-generated plasma,V1407,272-280(1991)
; Zhang, Y. S.: Simulation of intense microwave pulse propagation in air breakdown environment,V1407,260-271(1991)

Kuo, Yue
: New thin-film transistor structure and its processing method for liquid-crystal displays,V1456,288-299(1991)

Kuok, M. H.
; Silva, Donald E.; Tam, Siu-Chung; eds.: *Optical Systems in Adverse Environments*,V1399(1991)

Kuper, Thomas G. See Rimmer, Matthew P.: V1354,83-91(1991)

Kupfer, Roland
; Bergmann, Hans W.; Lingenauer, Marion: Material influence on cutting and drilling of metals using copper vapor lasers,V1598,46-60(1991)
See Hartmann, Martin: V1598,175-185(1991)

Kuppenheimer, John D.
; Lawson, Robert I.: Two-dimensional control of radiant intensity by use of nonimaging optics,V1528,93-103(1991)

Kuprenyuk, Victor I. See Sherstobitov, Vladimir E.: V1415,79-89(1991)

Kurahashi, Akira See Ohara, Shunji: V1499,187-194(1991)

Kuramitsu, Yoko See Yamagishi, Yasuo: V1461,68-72(1991)

Kurata, Yukio See Yoshida, Yoshio: V1401,58-65(1991)

Kurban, Amal K.
; Morrison, Paul R.; Trainor, S. W.; Cheong, Wai-Fung; Yasuda, Yukio; Tan, Oon T.: Importance of pulse duration in laser-tissue interactions: a histological study,V1422,43-49(1991)
See Cheong, Wai-Fung: V1422,19-26(1991)
See Yasuda, Yukio: V1422,50-55(1991)

Kurgansky, A. D. See Anufrik, S. S.: V1391,87-92(1991)

Kuribayashi, Kazuhiko See Inatomi, Yuko: V1557,132-139(1991)

Kuribayashi, Shizuma
; Noda, Osama; Ogino, M.; Imatake, Shigenori; Kondo, Motoe; Sato, Shunichi; Takahashi, Kunimitsu: Development of the aerodynamic window for high-power CO laser,V1397,439-443(1991)
See Noda, Osama: V1397,427-432(1991)
See Sato, Shunichi: V1397,421-425(1991)
See Sato, Shunichi: V1397,433-437(1991)

Kurihara, Kazuaki See Goto, Yasuyuki: V1534,49-58(1991)

Kurihara, Kenji See Takenaka, Hisataka: V1345,213-224(1991)

Kurin, V. See Andronov, Alexander A.: V1362,684-695(1991)

Kurita, Akira
See Arai, Tsunenori: V1425,191-195(1991)
See Sakurada, Masami: V1425,158-164(1991)

Kuriyama, Masao See Steiner, Bruce W.: V1557,156-167(1991)

Kurizki, Gershon
; Goldner, E.: Nonlinear optical amplification and oscillation in spherical microdroplets,V1497,48-62(1991)

Kurki, Jouko See Sipila, Heikki: V1549,246-255(1991)

Kurochkin, Nikolai N.
See Gordienko, Vyacheslav M.: V1416,102-114(1991)
See Pogosov, Gregory A.: V1416,115-124(1991)

Kurochkin, Vadim Y.
; Petrovsky, Victor N.; Protsenko, Evgeniy D.; Rurukin, Alexander N.; Golovchenko, Anotoly M.: Double-mode CO2 laser with complex cavity for ultrasensitive sub-Doppler spectroscopy,V1435,322-330(1991)

Kuroda, Masami
; Kawate, K.; Nabeta, Osamu; Furusho, N.: Electrophotographic properties of thiophene derivatives as charge transport material,V1458,155-161(1991)

Kurokawa, Akihiro
; Hayakawa, Yoshiaki: Fiber optic gyroscopes in Japan,V1504,156-164(1991)
See Hayakawa, Yoshiaki: V1572,353-358(1991)

Kurokawa, Haruhisa
; Ichikawa, Naoki; Yajima, Nobuyuki: Fringe-scanning moire system using a servo-controlled grating,V1332,643-654(1991)

Kurosawa, Kiyoshi
; Sawa, Takeshi; Sawada, Hisashi; Tanaka, Akira; Wakatsuki, Noboru: Diagnostic technique for electrical power equipment using fluorescent fiber,V1368,150-156(1991)

Kurtz, Andrew F.
; Kessler, David: Optical scanning system for a CCD telecine for HDTV,V1448,191-205(1991)

Kurylo, Michael J.
: Network for the detection of stratospheric change,V1491,168-174(1991)

Kusakin, Vladimir I. See Alimpiev, Sergei S.: V1503,154-158(1991)

Kusba, Jozef See Lakowicz, Joseph R.: V1435,142-160(1991)

Kushibiki, J. See Jen, Cheng K.: V1590,107-119(1991)

Kushlevsky, S. V.
; Patrin, V. V.; Provorov, Alexander S.; Salmin, V. V.: UV waveguide gas laser for biological and medical diagnostic methods,V1403,799-800(1991)

Kushnarev, Igor See Bokov, Lev: V1505,186-198(1991)

Kushnir, M. A. See Kluyev, V. G.: V1440,303-308(1991)

Kusinski, Jan P.
: Influence of laser melting on microstructure and properties of M2 high-speed tool steel,V1391,387-392(1991)

Kuszelewicz, Robert See Oudar, Jean-Louis: V1361,490-498(1991)

Kutal, Charles
; Weit, Scott K.; Allen, Robert D.; MacDonald, Scott A.; Willson, C. G.: Novel base-generating photoinitiators for deep-UV lithography,V1466,362-367(1991)

Kutami, Michinori See Nakazawa, Akira: V1458,115-118(1991)

Kutanov, Askar A.
; Akaev, Askar A.; Abdrisaev, Baktybek D.; Snimshikov, Igor A.: Local microholograms recording on the moving photothermoplastic disk,V1507,94-98(1991)
See Isaev, Urkaly T.: V1507,198-201(1991)

Kutt, R. N. See Baranov, Igor M.: V1361,1110-1115(1991)

Kuwabara, Y. See Ishikawa, Toshiharu: V1461,73-78(1991)

Kuwahara, Hideo See Naito, Takao: V1372,200-207(1991)

Kuwahara, Kazuyuki See Ohtsuka, Hiroshi: V1463,112-123(1991)

Kuwamoto, Hide See Jiang, Ching-Long: V1418,261-271(1991)

Kuwano, Yukinori
; Nakano, Shoichi; Tsuda, Shinya: Recent progress in Si thin film technology for solar cells,V1519,200-209(1991)

Kuyel, Birol
; Barrick, Mark W.; Hong, Alexander; Vigil, Joseph: 0.5-micron deep-UV lithography using a Micrascan-90 step-and-scan exposure tool,V1463,646-665(1991)
See Doi, Takeshi: V1464,336-345(1991)

Kuz'min, G. A.
See Bryukhnevitch, G. I.: V1358,739-749(1991)
See Bryukhnevitch, G. I.: V1449,109-115(1991)
See Dalinenko, I. N.: V1449,167-172(1991)

Kuzay, Tuncer M.
; Collins, Jeffrey T.; Khounsary, Ali M.; Viccaro, P. J.: Experimental and analytical studies on fixed mask assembly for APS with enhanced cooling,V1345,55-70(1991)
See Khounsary, Ali M.: V1345,42-54(1991)

Kuzma, Marian
See Adamiak, S.: V1391,382-386(1991)
See Szeregij, E. M.: V1391,199-203(1991)

Kuzmak, Peter See Dayhoff, Ruth E.: V1446,323-329(1991)

Kuzmichov, A. V.
See Ageev, Vladimir P.: V1503,453-462(1991)
See Craciun, Valentin: V1392,625-628(1991)

Kuzmin, I. V. See Zaichenko, O. V.: V1238,271-274(1991)

Kuzmin, Vladimir S. See Bulichev, A.: V1500,151-162(1991)

Kuznetcov, R. See Artjushenko, Vjacheslav G.: V1420,157-168(1991)

Kuznetsov, L. I.
: Screening properties of the erosion torch and pressure oscillations at a laser-irradiated target,V1440,222-228(1991)

Kuznetsov, S. G. See Borisov, V. M.: V1503,40-47(1991)

Kuznetsov, V. I. See Agranat, M. B.: V1440,397-400(1991)

Kuznetzov, Alexsey V. See Fediay, Sergey G.: V1420,41-43(1991)

Kuznetzov, Andrian I. See Nadezhdinskii, Alexander I.: V1418,487-495(1991)

Kuzyk, Mark G.
: Progress in organic third-order nonlinear optical materials,V1436,160-168(1991)
; Andrews, Mark P.; Paek, Un-Chul; Dirk, Carl W.: Guest-host polymer fibers and fractal clusters for nonlinear optics,V1560,44-53(1991)

Kvach, V. V.
See Apanasevich, P. A.: V1403,195-211(1991)
See Apanasevich, P. A.: V1403,240-242(1991)
See Gurinovich, G. P.: V1403,139-141(1991)

Kvapil, Jiri See Zschocke, Wolfgang: V1501,49-56(1991)

Kvasnichka, P.
See Chorvat, Dusan: V1402,89-92(1991)
See Shvec, Peter: V1402,78-81(1991)

Kvasnik, Frank
See Christie, Simon: V1507,202-209(1991)
See Ediriweera, Sanath R.: V1399,64-75(1991)
See Ediriweera, Sanath R.: V1504,110-117(1991)

Kwak, J. Y.
; Efstratiadis, Serafim N.; Katsaggelos, Aggelos K.; Sahakian, Alan V.; Sullivan, Barry J.; Swiryn, Steven; Hueter, David C.; Frohlich, Thomas: Motion estimation in digital angiographic images using skeletons,V1396,32-44(1991)

Kwiat, Doron
; Einav, Shmuel: Decoupled coil detector array in magnetic resonance imaging,V1443,2-28(1991)

Kwo, Deborah P.
; Damas, George; Zmek, William P.; Haller, Mitch: Hartmann-Shack wavefront sensor using a binary optic lenslet array,V1544,66-74(1991)

Kwok, Hoi-Sing
; Shi, Lei; Zheng, J. P.; Dong, S. Y.; Pang, Y.; Prasad, Paras N.: Speed of optically-controlled superconducting devices,V1394,196-200(1991)
See Jiao, Kaili L.: V1361,776-783(1991)

Kwok, John C.
; Yu, Peter K.; Cheng, Andrew Y.; Ho, Wai-Chin: Microcomputer-based image processing system for CT/MRI scans: II. Expert system,V1445,446-455(1991)
See Cheng, Andrew Y.: V1444,400-406(1991)

Kwok, Johnny H.
; Osmolovsky, Michael G.: Mission design for the Space Infrared Telescope Facility,V1540,27-37(1991)

Kwok, Kam-cheung
; Chan, Ming-kam: Parallel DC notch filter,V1567,709-719(1991)

Kwok, Thomas Y.
: Electromigration in VLSI metallization,V1596,60-71(1991)

Kwon, O'Dae
; Lee, Seung-Won; Choi, Woong-Lim; Kim, Kwang-Il; Jeong, Yoon-Ha: Excitonic photoabsorption study of AlGaAs/GaAs multiple-quantum-well grown by low-pressure MOCVD,V1361,802-808(1991)

Kwon, Taek M.
; Zervakis, Michael E.: Robust regularized image restoration,V1569,317-328(1991)

Kwon, Young H.
; Park, Seung O.; Park, B. C.: Simple technique for 3-D displacements measurement using synthesis of holographic interferometry and speckle photography,V1554A,639-644(1991)
See Joo, Jin W.: V1554B,113-118(1991)
See Kang, Dae I.: V1554B,139-147(1991)

Kwong, Dim-Lee
See Joshi, A. B.: V1393,122-149(1991)
See Jung, K. H.: V1393,240-251(1991)
See Peyton, David: V1393,295-308(1991)

Kwong, Henry Y. See Fielden, John: V1445,145-154(1991)

Kyrazis, Demos T.
; Weiland, Timothy L.: Determination of SBS-induced damage limits in large fused silica optics for intense, time-varying laser pulses,V1441,469-477(1991)
See Pitts, John H.: V1441,71-81(1991)

Kyriakopoulos, Konstantinos J.
; Saridis, George N.: Minimum jerk trajectory planning for robotic manipulators,V1387,159-164(1991)

Kyser, David F. See Dunbrack, Steven K.: V1464,314-326(1991)

Kyser, Jeff See Wesley, Michael: V1481,209-220(1991)

Kzuzhilin, Yu E.
; Mel'nichenko, Yu. B.; Shilov, V. V.: Kinetics of formation of holographic structure of a hologram mirror in dichromated gelatin,V1238,200-205(1991)

Laamann, K. See Weiss, Howard: V1493,80-90(1991)

Labanti, Claudio See Di Cocco, Guido: V1549,102-112(1991)

Labaune, C.
; Rozmus, Wojtek; Baldis, Hector A.; Mounaix, P.; Pesme, Denis; Baton, S.; LaFontaine, Bruno; Villeneuve, D. M.; Enright, G. D.: Reflectivity of stimulated Brillouin scattering in picosecond time scales,V1413,138-143(1991)

LaBelle, Gary L.
; McDonald, Mark D.: Applications of high speed silicon bipolar ICs in fiber optic systems,V1365,116-121(1991)

LaBelle, R. D. See Fairbank, William M.: V1435,86-89(1991)

Labit, Claude
; Nicolas, Henri: Compact motion representation based on global features for semantic image sequence coding,V1605,697-708(1991)
See Tamtaoui, Ahmed: V1605,720-731(1991)

Labo, Jack A.
; Mayo, Michael W.: Testing laser eye protection,V1419,32-39(1991)

Laborde, Pascale See Nissim, Carlos: V1365,13-19(1991)

Labourdette, Jean-Francois P. See Acampora, Anthony S.: V1579,30-39(1991)

Labranche, Bruno See Snell, Kevin J.: V1410,99-106(1991)

Lach, Zbigniew
; Zientkiewicz, Jacek K.: Fiber optic network for mining seismology,V1364,209-220(1991)
See Zientkiewicz, Jacek K.: V1366,45-56(1991)

Lachin, Yu Y. See Dodd, James W.: V1407,467-473(1991)

Lachmann, F. See Barillot, Christian: V1445,54-65(1991)

Lachs, Gerard
; Zaidi, Syed M.; Singh, Amit K.; Henning, Rudolf E.; Trascritti, D.; Kim, H.; Bhattacharya, Pallab K.; Pamulapati, J.; McCleer, P. J.; Haddad, George I.; Peng, S.: Integration of a coherent optical receiver with adaptive image rejection capability,V1474,248-259(1991)

Lackner, Anna M. See Margerum, J. D.: V1455,27-38(1991)

Lacombe, R. See Uwabu, H.: V1605,928-937(1991)

Lacovara, Phil
; Gleckman, Philip; Holman, Robert L.; Winston, Roland: Nonimaging concentrators for diode-pumped slab lasers,V1528,135-141(1991)

Laczko, Gabor See Lakowicz, Joseph R.: V1435,142-160(1991)

Ladan, F. R.
See Berrouane, H.: V1343,428-436(1991)
See Khan Malek, Chantal: V1343,56-61(1991)

Ladkin, Peter B. See Anger, Frank D.: V1468,122-136(1991)

Ladokhin, Alexey S. See Chikishev, A. Y.: V1403,448-456(1991)

Laegsgaard, Erik See Sipila, Heikki: V1549,246-255(1991)

Laermann, Karl-Hans
: Analysis of the displacements of cylindrical shells by moire techniques,V1554B,248-256(1991)
: Evaluation of interference fringe pattern on spatially curved objects,V1554A,522-528(1991)
: Hybrid method to analyze the stress state in piecewise homogeneous two-dimensional objects,V1554A,143-150(1991)

LaFollette, Robert
; Horger, John D.: Thermal signature training for military observers,V1488,289-299(1991)

LaFontaine, Bruno See Labaune, C.: V1413,138-143(1991)

Lafontaine, H. See Chaker, Mohamed: V1465,16-25(1991)

Lagage, Pierre O. See Mottier, Patrick: V1494,419-426(1991); 1512,60-67(1991)

Lagarde, Alexis
See Bremand, Fabrice J.: V1554B,650-660(1991)
See Dupre, Jean-Christophe: V1554A,766-771(1991)
See Hammami, Slimane: V1554A,136-142(1991)
See Mauvoisin, Gerard: V1554B,181-187(1991)

LaGasse, M. J. See Freeman, J. L.: V1476,320-325(1991)

Lagendijk, Reginald L.
See Belfor, Ricardo A.: V1605,274-284(1991)
See Bosveld, Frank: V1605,769-780(1991)
See Katsaggelos, Aggelos K.: V1606,716-727(1991)

Lago, Paolo See Cantu, Laura: V1430,144-159(1991)

Laguesse, Michel F.
; Bourdinaud, Michel J.: Characterization of fluorescent plastic optical fibers for x-ray beam detection,V1592,96-107(1991)

LaHaie, Ivan J. See Stewart, Stephen R.: VIS07,57-97(1991)

Lai, Barry P.
; Chrzas, John J.; Yun, Wen-Bing; Legnini, Dan; Viccaro, P. J.; Bionta, Richard M.; Skulina, Kenneth M.: Experimental characterization of Fresnel zone plate for hard x-ray applications,V1550,46-49(1991)

Lai, Ming See Diels, Jean-Claude: V1376,198-205(1991)

Lai, Shurong See Liao, Yan-Biao: V1584,400-404(1991)

Lai, Tianshu
; Tan, Yushan: Double-exposure phase-shifting holography applied to particle velocimetry,V1554B,580-585(1991)
; Tan, Yushan: Improvement of recordable depth of field in far-field holography for analysis of particle size,V1461,286-290(1991)

Lai, Winnie C. See Jaeger, Nicolas A.: V1583,202-209(1991)

Laidler, Ian See Baker, Howard J.: V1397,545-548(1991)

LaiHing, Kenneth See McDonald, Joseph K.: V1497,367-370(1991)

Laird, Alan
; Miller, James: Hierarchial symmetry segmentation,V1381,536-544(1991)

Laird, Daniel L.
; Engelstad, Roxann L.; Palmer, Shane R.: Optimization of an x-ray mask design for use with horizontal and vertical kinematic mounts,V1465,134-144(1991)

Laird, Robin T.
See Aviles, Walter A.: V1388,587-597(1991)
See Smurlo, Richard: V1388,566-577(1991)

Lajza-Rooks, Barbara A. See Klumpe, Herbert W.: V1398,95-106(1991); 1532,230-240(1991)

Lajzerowicz, Jean
; Tedesco, Serge V.; Pierrat, Christophe; Muyard, D.; Taccussel, M. C.; Laporte, Philippe: Polysilicon etching for nanometer-scale features,V1392,222-231(1991)

Lakdawala, Vishnu K.
; Schoenbach, Karl H.; Roush, Randy A.; Barevadia, Gordon R.; Mazzola, Michael S.: Photoquenching and characterization studies in a bulk optically controlled GaAs semiconductor switch,V1378,259-270(1991)
See Brinkmann, Ralf P.: V1378,203-208(1991)
See Schoenbach, Karl H.: V1362,428-435(1991)

Lake, David S. See Roos, Kenneth P.: V1428,159-168(1991)

Lake, Stephen P.
; Pritchard, Alan P.; Sturland, Ian M.; Murray, Anthony R.; Prescott, Anthony J.; Gough, David W.: Description and performance of a 256 x 256 electrically heated pixel IR scene generator,V1486,286-293(1991)

Lakew, Brook See Brasunas, John C.: V1477,166-173(1991)

Lakhno, V. D. See Balabaev, N. K.: V1403,478-486(1991)

Lakhtakia, Akhlesh
: Strategies for tunable frequency-selective surfaces,V1489,108-111(1991)
; Varadan, Vijay K.; Varadan, Vasundara V.: Normally incident plane waves on a chiral slab with linear property variations,V1558,120-126(1991)
; Varadan, Vijay K.; Varadan, Vasundara V.: Sparse random distribution of noninteracting small chiral spheres in a chiral host medium,V1558,22-27(1991)

Lakowicz, Joseph R.
; Gryczynski, Ignacy; Malak, Henryk; Johnson, Michael L.; Laczko, Gabor; Wiczk, Wieslaw M.; Szmacinski, Henryk; Kusba, Jozef: Frequency-domain fluorescence spectroscopy: instrumentation and applications to the biosciences,V1435,142-160(1991)
See Berndt, Klaus W.: V1431,149-160(1991)

Lakshmanan, A. S. See Murali, K. R.: V1536,289-295(1991)

Lakshminarayanan, Vasudevan See Mondal, Pronab K.: V1429,108-116(1991)

Lal, Ravindra B.
See Rao, S. M.: V1557,283-292(1991)
See Steiner, Bruce W.: V1557,156-167(1991)
See Trolinger, James D.: V1557,250-258(1991)
See Trolinger, James D.: V1557(1991)
See Trolinger, James D.: V1332,151-165(1991)
See Trolinger, James D.: V1557,98-109(1991)

Lalanne, Frederic P. See Weill, Andre P.: V1465,264-270(1991)

Lalanne, Philippe
; Chavel, Pierre: Optoelectronic hardware issues for implementation of simulated annealing or Boltzmann machines,V1621,388-401(1991)

Lalevic, Bogoliub See Kim, A. H.: V1378,173-178(1991)

Lalk, Gail R.
; Habiby, Sarry F.; Hartman, Davis H.; Krchnavek, Robert R.; Wilson, Donald K.; Young, Kenneth C.: Potential roles of optical interconnections within broadband switching modules,V1389,386-400(1991)
See Hartman, Davis H.: V1390,368-376(1991)

Lallauret, Fabrice
; Barba, Dominique: Motion compensation by block matching and vector postprocessing in subband coding of TV signals at 15 Mbit/s,V1605,26-36(1991)

Lallier, Eric
; Pocholle, Jean-Paul; Papuchon, Michel R.; De Micheli, Marc; Li, M. J.; He, Q.; Ostrowsky, Daniel B.; Grezes-Besset, C.; Pelletier, Emile P.: LiNbO3 with rare earths: lasers and amplifiers,V1506,71-79(1991)

Lalor, Michael J.
See Harvey, David M.: V1400,86-93(1991)
See Wood, Christopher M.: V1332,301-313(1991)

Lam, Benson C.
; Kellner, Albert L.; Campion, David C.; Costa, Joannes M.; Yu, Paul K.: Analysis and measurement of the external modulation of modelocked laser diodes (relative noise performance),V1371,36-45(1991)
See Costa, Joannes M.: V1476,74-80(1991)

Lam, David T.
; Carroll, John E.: Optical higher order double-layer associative memory,V1505,104-114(1991)

Lam, Stephen
See Korbelik, Mladen: V1426,172-179(1991)
See Palcic, Branko: V1448,113-117(1991)
See Profio, A. E.: V1426,44-46(1991)

Lam, Wayne W. See Liu, Louis: V1475,193-198(1991)

LaMarche, R. E. See Dickinson, Alex G.: V1389,503-514(1991)

Lamarre, Daniel
; Giroux, Jean: Radiometric calibration of space-borne Fourier transform infrared-emission spectrometers: proposed scenario for European Space Agency's Michelson interferometer for passive atmospheric sounding,V1493,28-36(1991)

Lamarre, J. See Bacis, Roger: V1397,173-176(1991)

Lamb, Bryan K. See Park, Eric D.: V1366,294-303(1991)

Lamb, D. W. See Woolsey, G. A.: V1584,243-253(1991)

Lamb, James E. See Sethi, Satyendra A.: V1463,30-40(1991)

Lamb, Lowell See Sinha, Kislay: V1437,32-35(1991)

Lamb, Willis E.
See Fearn, Heidi: V1497,245-254(1991)
See Fearn, Heidi: V1497,283-290(1991)

Lambeck, Paul V.
: Chemo-optical microsensing systems,V1511,100-113(1991)

Lambert, Gary M. See Zemon, Stanley A.: V1373,21-32(1991)

Lambert, James L. See Bergman, Larry A.: V1364,14-21(1991)

Lambert, Robin A.
; Batchelor, Bruce G.: Method of preprocessing color images using a Peano curve on a Transputer array,V1381,582-588(1991)

Lamberti, Antonio See Furgiuele, Franco M.: V1554A,275-284(1991)

Lambeth, David N. See Ross, William E.: V1562,93-102(1991)

Lambros, John M.
; Mason, James J.; Rosakis, Ares J.: Experimental investigation of dynamic mixed-mode fracture initiation,V1554A,70-83(1991)

Lambry, J.-C. See Woodruff, William H.: V1432,205-210(1991)

Lambsdorff, M. See Rosenzweig, Josef: V1362,168-178(1991)

Laming, Richard I. See Morkel, Paul R.: V1373,224-233(1991)

LaMont, Douglas V.
; Mar, Lim O.; Rodden, Jack J.: Development and test of the Starlab control system,V1482,2-12(1991)
See Eller, E. D.: V1415,2-12(1991)

Lamont, Gary B. See Leahy, Michael B.: V1387,148-158(1991)

Lampert, Carl M.
; Granqvist, Claes G.; eds.: *Optical Materials Technology for Energy Efficiency and Solar Energy Conversion X*,V1536(1991)
See Ma, Y. P.: V1536,93-103(1991)

Lampinen, Jouko
: Feature extractor giving distortion invariant hierarchical feature space,V1469,832-842(1991)

LaMuraglia, Glenn M. See Obremski, Susan M.: V1427,327-334(1991)

Lan, Guey-Liu
; Pao, Cheng K.; Wu, Chan-Shin; Mandolia, G.; Hu, M.; Yuan, Steve; Leonard, Regis F.: Millimeter wave pseudomorphic HEMT MMIC phased-array components for space communications,V1475,184-192(1991)

Lan, Zu-yun See Zhang, Renxiang: V1380,116-121(1991)

Lancini, R. See Dal Degan, Nevaino: V1606,934-940(1991)

Land, Donald P. See McIver, Robert T.: V1437,124-128(1991)

Landen, Otto L.
; Vu, Brian-Tinh; Stearns, Daniel G.; Alley, W. E.: Hydrodynamic evolution of picosecond laser plasmas,V1413,120-130(1991)
See Bell, Perry M.: V1346,456-464(1991)
See Kilkenny, Joseph D.: V1358,117-133(1991)

Lander, Mike L. See Reilly, James P.: V1397,339-354(1991)

Landini, G. See Frontera, Filippo: V1549,113-119(1991)

Lando, Mordechai
; Belker, D.; Lerrer, A.; Lotem, Haim; Dikman, A.; Bialolanker, Gabriel; Lavi, S.; Gabay, Shimon: Modified off-axis unstable resonator for copper vapor laser,V1412,19-26(1991)
; Belker, D.; Lerrer, A.; Lotem, Haim; Dikman, A.; Bialolanker, Gabriel; Lavi, S.; Gabay, Shimon: Pointing stability of copper vapor laser with novel off-axis unstable resonator,V1442,172-180(1991)

Landraud, Anne M.
: Vision-based model of artificial texture perception,V1453,314-320(1991)

Landreth, Bruce See Perlmutter, S. H.: V1562,74-84(1991)

Landwehr, Gottfried
; Waag, Andreas; Hofmann, K.; Kallis, N.; Bicknell-Tassius, Robert N.: Recent progress in device-oriented II-VI research at the University of Wuerzburg,V1362,282-290(1991)

Landy, Michael S.
; Maloney, Laurence T.; Young, Mark J.: Psychophysical estimation of the human depth combination rule,V1383,247-254(1991)

Lane, Alan See Swairjo, Manal: V1437,60-65(1991)

Lane, Gregory J.
; Sherk, Henry H.; Kollmer, Charles; Uppal, Gurvinder S.; Rhodes, Anthony; Sazy, John; Black, Johnathan D.; Lee, Steven: Stimulatory effects of Nd:YAG lasers on canine articular cartilage,V1424,7-11(1991)
See Black, Johnathan D.: V1424,20-22(1991)
See Black, Johnathan D.: V1424,12-15(1991)
See Meller, Menachem M.: V1424,60-61(1991)
See Sazy, John: V1424,50-50(1991)
See Uppal, Gurvinder S.: V1424,51-52(1991)

Lane, Stephen M. See Yorkey, Thomas J.: V1345,255-259(1991)

Lang, Graham K.
See Gale, Michael T.: V1506,65-70(1991)
See Seitz, Peter: V1606,252-259(1991)

Lang, Harold R.
: Spectral stratigraphy,V1492,351-357(1991)

Lang, Hengyuan See Zhang, Jingfang: V1238,306-310(1991)

Lang, Peter T. See Beregulin, Eugene V.: V1362,853-862(1991)

Lang, Robert J.
; Kim, Jae H.; Larsson, Anders G.; Nouhi, Akbar; Cody, Jeffrey; Lin, Steven H.; Psaltis, Demetri; Tiberio, Richard C.; Porkolab, Gyorgy A.; Wolf, Edward D.: Optoelectronic master chip for optical computing,V1563,2-7(1991)
See Larsson, Anders G.: V1418,292-297(1991)

Lang, Zhengping
; Zhang, Zhen; Scarberry, Randell E.; Shao, Weimin; Sun, Xu-Mei: Knowledge-based direct 3-D texture segmentation system for confocal microscopic images,V1468,826-833(1991)

Langdon, A. B. See Wilks, S. C.: V1413,131-137(1991)

Langdon, Glen G.
: Sunset: a hardware-oriented algorithm for lossless compression of gray-scale images,V1444,272-282(1991)

Lange, John A.
; Allen, Chris: Application and integration of a focused ion beam circuit repair system,V1465,50-56(1991)

Lange, M. D. See Sporken, R.: V1512,155-163(1991)

Lange, M. J. See Olsen, Gregory H.: V1419,24-31(1991)

Lange, Robert
; Endemann, Martin J.; Reiland, Werner; Krawczyk, Rodolphe; Hofer, Bruno: ATLID: the first preoperational ATmospheric LIDar for the European polar platform,V1492,24-37(1991)

Langen, K. J. See Ulrich, Frank: V1428,135-135(1991)

Langer, Dietrich W. See Lee, Hyung G.: V1361,893-900(1991)

Langford, A. O. See Proffitt, Michael H.: V1491,2-6(1991)

Langford, C. H. See Huang, Zhiqing: V1436,103-113(1991)

Langford, S. C. See Dickinson, J. T.: V1598,72-83(1991)

Langhoff, Charles A. See Thompson, Mark E.: V1497,423-429(1991)

Langinmaa, Anu
: Automatic analysis of heliostat strips,V1468,573-580(1991)

Langlois, Richard G. See Jensen, Ronald H.: V1403,372-380(1991)

Langry, Kevin See Angel, S. M.: V1368,98-104(1991)

Lanig, E. M. See Eidmann, K.: V1502,320-330(1991)

Lankard, John R.
: What industry needs in a high-power excimer laser,V1377,2-5(1991)

Lanni, Thomas R.
; Demerjian, Kenneth L.: Developing a long-path diode array spectrometer for tropospheric chemistry studies,V1433,21-24(1991)

Lanteigne, David J.
: Development of an IR projector using a deformable mirror device,V1486,376-379(1991)

Lanza, George M. See Frazin, Leon J.: V1425,207-207(1991)

Lanzafame, Raymond J.
: Applications of lasers in laparoscopic cholecystectomy: technical considerations and future directions,V1421,189-196(1991)
See Seka, Wolf D.: V1398,162-169(1991)

Lanzl, Franz See Ackermann, F.: V1490,94-101(1991)

Laor, Herzel See Wolfe, Ronald: V1363,125-132(1991)

LaPeruta, Richard
; Van Wagenen, E. A.; Roche, J. J.; Kitipichai, Prakob; Wnek, Gary E.; Korenowski, G. M.: Preparation and characterization of silver-colloid/polymer-composite nonlinear optical materials,V1497,357-366(1991)

Lapeyre, G. J. See Allinger, Thomas: V1361,935-942(1991)

Lapierre, A. See Varshneya, Deepak: V1584,188-201(1991)

Lapin, R. See Juri, Hugo: V1422,128-135(1991)

Laplante, Phillip A.
; Giardina, Charles R.: Fast dilation and erosion of time-varying grey-valued images with uncertainty,V1568,295-302(1991)

LaPointe, Michael R.
: Review of nondiffracting Bessel beams,V1527,258-276(1991)

La Porta, A. See Shaw, Earl D.: V1552,14-23(1991)

Laporte, Philippe
; Van den hove, Luc; Melaku, Yosias: Dry etching for silylated resist development,V1392,196-207(1991)
See Heitzmann, Michel: V1392,272-279(1991)
See Lajzerowicz, Jean: V1392,222-231(1991)

Lappe, Dirk See Stiller, Christoph: V1605,47-57(1991)

Lappin, Joseph S. See Mowafy, Lyn: V1453,177-187(1991)

Lapshov, I. See Kaaret, Philip E.: V1548,106-117(1991)

Larchenko, Yu. V. See Erokhovets, Valerii K.: V1621,227-236(1991)

Larciprete, Rosanna See Borsella, E.: V1503,312-320(1991)

Lareau, Andy G.
; Collins, Ross: Tactical reconnaissance mission survivability requirements,V1538,81-98(1991)

Lareau, Richard See Soref, Richard A.: V1389,408-421(1991)

Largent, Craig C.
; Gallant, David J.; Yang, Jane; Allen, Michael S.; Jansen, Michael: Fabrication of unstable resonator diode lasers,V1418,40-45(1991)

Larichev, A. V.
See Ivanov, Vladimir Y.: V1402,145-153(1991)
See Vorontsov, Michael A.: V1409,260-266(1991)
See Vorontsov, Michael A.: V1402,154-164(1991)

Larimer, James O.
; Prevost, Michael P.; Arditi, Aries R.; Azueta, Steven; Bergen, James R.; Lubin, Jeffrey: Human visual performance model for crewstation design,V1456,196-210(1991)

Larkin, Alexander I.
; Matveev, Alexander; Mironov, Yury: Time-integrating optical raster spectrum analyzer using semiconductor laser as input modulator,V1621,414-423(1991)

Larkin, Eric W.
: Quadric surfaces: some derivations and applications,V1354,222-231(1991)

Larkins, Grover L.
; Jones, W. K.; Lu, Q.; Levay, C.; Albaijes, D.: Aging and surface instability in high-Tc superconductors,V1477,26-33(1991)

LaRochelle, Sophie
: Fiber defects in Ge-doped fibers: towards a coherent picture,V1516,55-57(1991)
See Ouellette, Francois: V1516,2-13(1991)

Laroussi, M. See Rouhani, Mehdi D.: V1361,954-962(1991)

Larrabee, Robert D. See Postek, Michael T.: V1464,35-47(1991)

Larsen, Robert G. See Schaeffer, A. R.: V1447,165-176(1991)

Larson, Allen L. See Grimes, Gary J.: V1592,139-149(1991)

Larson, Bennett C.
; Tischler, J. Z.: Time-resolved techniques: an overview,V1345,90-100(1991)

Larson, David J. See Gillies, Donald C.: V1484,2-10(1991)

Larson, Donald C.
; Bibby, Y. W.; Tyagi, S.: Metallic-glass-coated optical fibers as magnetic-field sensors,V1572,517-522(1991)

Larson, Larry E. See Chou, Chia-Shing: V1475,151-156(1991)

Larsson, Anders G.
; Forouhar, Siamak; Cody, Jeffrey; Lang, Robert J.; Andrekson, Peter A.: High-power single-element pseudomorphic InGaAs/GaAs/AlGaAs single-quantum-well lasers for pumping Er-doped fiber amplifiers,V1418,292-297(1991)
See Lang, Robert J.: V1563,2-7(1991)
See Maserjian, Joseph L.: V1562,85-92(1991)

Larsson, Michael
; Beckman, Claes; Nystrom, Alf; Hard, Sverker; Sjostrand, Johan: Optical properties of diffractive, bifocal intraocular lenses,V1529,63-70(1991)
See Ekberg, Mats: V1527,236-239(1991)

Larsson, T.
; Perev, K.; Valavanis, Kimon P.; Gardner, S.: Stability evaluation of the PUMA-560 robot arm under model mismatch,V1387,165-168(1991)

LaSala, Paul V.
; McLaughlin, Chris: Beam-tracker and point-ahead system for optical communications II: servo performance,V1482,121-137(1991)

Lasher, Mark E. See Phillips, Thomas E.: V1454,290-298(1991)

Lassahn, Gordon D.
: Automatic, high-resolution analysis of low-noise fringes,V1332,690-695(1991)

Lassiter, Terrell N. See Chen, ChuXin: V1468,354-366(1991)

Latanich, L. See Kozlowski, Mark R.: V1561,59-69(1991)

Latham, Geoff A. See Gruenbaum, F. A.: V1431,232-238(1991)

Latham, Steve C. See Underwood, Ian: V1562,107-115(1991)

Latham, William P. See Bohn, Willy L.: V1397,235-238(1991)

Latysh, E. G.
See Askadskij, A. A.: V1554A,426-431(1991)
See Shvej, E. M.: V1554A,488-495(1991)

Latyshev, A. N. See Kluyev, V. G.: V1440,303-308(1991)

Lau, A.
; Pfeiffer, M.; Werncke, W.: Adaptation of coherent Raman methods for the investigation of biological samples in vivo,V1403,212-220(1991)

Lau, Christina See Beeson, Karl W.: V1559,258-266(1991)

Lau, Christina C. See Chan, Kelby K.: V1444,250-255(1991)

Lau, Hon W.
; Davies, Neil A.; McCormick, Malcolm: Microlens array fabricated in surface relief with high numerical aperture,V1544,178-188(1991)
See Davies, Neil A.: V1544,189-198(1991)

Lau, Kam-Yin
See Gidron, Rafael: V1579,40-48(1991)
See Pepeljugoski, Petar K.: V1371,233-243(1991)

Lau, Kenneth See Redding, David C.: V1489,201-215(1991)

Lau, Manhot
; Okagaki, Takashi: Divergence as a measure of visual recognition: bias and errors caused by small samples,V1606,705-713(1991)

Lau, Yue Y.
; Chernin, David P.; Colombant, Denis G.; Ho, Ping-Tong: Quantum extension of Child-Langmuir law,V1407,546-552(1991)
; Colombant, Denis G.; Pilloff, Mark D.: Beam divergence from sharp emitters in a general longitudinal magnetic field,V1407,635-646(1991); 1552,182-184(1991)
; Colombant, Denis G.: Some issues on beam breakup in linear accelerators,V1407,479-483(1991)
See Colombant, Denis G.: V1407,484-495(1991)
See Colombant, Denis G.: V1407,13-22(1991)
See Friedman, Moshe: V1407,474-478(1991)
See Friedman, Moshe: V1407,2-7(1991)
See Krall, Jonathan: V1407,23-31(1991)
See Serlin, Victor: V1407,8-12(1991)

Lauchlan, Laurie J. See Bennett-Lilley, Marylyn H.: V1464,127-136(1991)

Lauck, Teresa L.
; Nomura, Masafumi; Omori, Tsutae; Yoshioka, Kajutoshi: Effects of wafer cooling characteristics after post-exposure bake on critical dimensions,V1464,527-538(1991)

Laucournet, A. See Froehly, Claude: V1358,532-540(1991)

Laude, Lucien D.
See Letardi, Tommaso: V1503(1991)
See Shafeev, George A.: V1503,321-329(1991)

Laudenslager, James B. See Pacala, Thomas J.: V1411,69-79(1991)

Laudwein, Norbert See Brinkop, Axel: V1468,227-234(1991)

Lauer, Robert B. See Eichen, Elliot G.: V1474,260-267(1991)

Lauffenburger, Jim
; Arachtingi, John W.; Robbins, Jamey L.; Kim, Jong: Margin measurements of pulse amplitude modulation channels,V1499,104-113(1991)

Laulicht, Israel
See Friedmann, Harry: V1427,357-362(1991)
See Lubart, Rachel: V1422,140-146(1991)

Laumann, Curt W.
; Caird, John A.; Smith, James R.; Horton, R. L.; Nielsen, Norman D.: Development of third-harmonic output beam diagnostics on NOVA,V1414,151-160(1991)
See Henesian, Mark A.: V1415,90-103(1991)

Launay, Jean C. See Mathey, P.: V1500,26-33(1991)

Laurance, R. J. See Greenfield, Perry E.: V1494,16-39(1991)

Laurendeau, Denis See Houde, Regis: V1332,343-354(1991)

Laurent, Bernard
; Duchmann, Olivier: SILEX project: the first European optical intersatellite link experiment,V1417,2-12(1991)

Laurent, Jean See Michau, Vincent: V1487,64-71(1991)

Lavagetto, Fabio
; Leonardi, Riccardo: Block-adaptive quantization of multiple-frame motion field,V1605,534-545(1991)

Laval, Suzanne
: Semiconductor waveguides for optical switching,V1362,82-92(1991)

LaValle, Steven M.
; Hutchinson, Seth A.: Considering multiple-surface hypotheses in a Bayesian hierarchy,V1569,2-15(1991)

Lavallee, Daniel
; Lovejoy, Shaun; Schertzer, Daniel: Universal multifractal theory and observations of land and ocean surfaces, and of clouds,V1558,60-75(1991)

Lavallee, Stephane
; Szeliski, Richard; Brunie, Lionel: Matching 3-D smooth surfaces with their 2-D projections using 3-D distance maps,V1570,322-336(1991)

Lavelle, Stephen P.
; Jackman, Louise A.; Nolan, P. F.: High-speed microcinematography of aerosols,V1358,821-830(1991)
See Jackman, Louise A.: V1358,831-842(1991)

Laverick, Timothy See Bayruns, Robert J.: V1541,83-90(1991)

Lavi, Adina See Ehrenberg, Benjamin: V1426,244-251(1991)

Lavi, Moshe See Johnson, W. Todd: V1488,343-354(1991)

Lavi, S.
See Lando, Mordechai: V1412,19-26(1991)
See Lando, Mordechai: V1442,172-180(1991)

Lavin, Mark A. See Flickner, Myron D.: V1568,113-124(1991)

Lavine, T. L. See Farkas, Zoltan D.: V1407,502-511(1991)

LaViolette, Kerry D.
; Bossler, Franklin B.: Interferometric fiber optic gyroscopes for today's market,V1398,213-218(1991)

Lavoie, Matt J. See Hackett, Jay K.: V1383,611-622(1991)

Lavoine, J. P. See Hoerner, Claudine: V1362,863-869(1991)

Lavrentovich, Oleg D. See Bodnar, Vladimir G.: V1455,61-72(1991)

Lavrov, Nikolaj A. See Glebova, Svetlana N.: V1500,275-280(1991)

Law, K. K. See Coldren, Larry A.: V1362,24-37(1991)

Lawandy, Nabil M.
: What we can learn about second-harmonic generation in germanosilicate glass from the analogous effect in semiconductor-doped glasses,V1516,99-112(1991)

Lawrence, George N.
ed.: 1990 Intl Lens Design Conf,V1354(1991)
: Optical design with physical optics using GLAD,V1354,126-135(1991)
; Moore, Kenneth E.: Integration of geometrical and physical optics,V1415,322-329(1991)
; Moore, Kenneth E.: Optical design and optimization with physical optics,V1354,15-22(1991)
See Moore, Kenneth E.: V1354,581-587(1991)

Lawrence, Marcus F.
See Frank, Arthur J.: V1436(1991)
See Huang, Zhiqing: V1436,103-113(1991)

Lawrence, Robert C. See Dallabetta, Kyle A.: V1371,116-127(1991)

Lawrie, David J. See Chambers, Robert J.: V1488,312-326(1991)

Lawson, Christopher M.
; Michael, Robert R.; Dressel, Earl M.; Harmony, David W.: Nd:YAG-laser-based time-domain reflectometry measurements of the intrinsic reflection signature from PMMA fiber splices,V1592,73-83(1991)

Lawson, David G. See Brodsky, Aaron: V1482,159-169(1991)

Lawson, Janice K. See Henesian, Mark A.: V1415,90-103(1991)

Lawson, Michael A.
; Wika, Kevin G.; Gillies, George T.; Ritter, Rogers C.: Near-real-time biplanar fluoroscopic tracking system for the video tumor fighter,V1445,265-275(1991)
See Ramos, P. A.: V1443,160-170(1991)

Lawson, Robert I. See Kuppenheimer, John D.: V1528,93-103(1991)

Lawson, Shelah M.
; Clark, Rodney L.; Banish, Michele R.; Crouse, Randy F.: Wave-optic model to determine image quality through supersonic boundary and mixing layers,V1488,268-278(1991)
See Banish, Michele R.: V1557,209-221(1991)

Lax, Melvin
: Theory of laser noise,V1376,2-20(1991)

Laxer, Cary
; Johnson, G. A.; Kavanagh, Katherine M.; Simpson, Edward V.; Ideker, Raymond E.; Smith, William M.: Interactive graphics system for locating plunge electrodes in cardiac MRI images,V1444,190-195(1991)

Laxmiprasad, A. S. See Kamalakar, J. A.: V1478,92-100(1991)

Lay, Kuen-Tsair See Kang, Moon G.: V1606,408-418(1991)

Laybourn, Peter J.
See Barbarossa, Giovanni: V1583,122-128(1991)
See Barbarossa, Giovanni: V1513,37-43(1991)
See Jiang, Pisu: V1362,899-906(1991)
See Krauss, Thomas: V1583,150-152(1991)

Lazarchuk, V. P. See Egorov, V. V.: V1358,984-991(1991)

Lazard, Carl J.
; Pereira, Nino R.; Huttlin, George A.; Litz, Marc S.: Silicon calorimeter for high-power microwave measurements,V1407,167-171(1991)
See Huttlin, George A.: V1407,147-158(1991)
See Litz, Marc S.: V1407,159-166(1991)

Lazarov, M.
; Roehle, B.; Eisenhammer, T.; Sizmann, R.: TiNxOy-Cu coatings for low-emissive solar-selective absorbers,V1536,183-191(1991)

Lazarovici, Doina N. See Halmagean, Eugenia T.: V1484,106-114(1991)

Lazaruk, A. M.
; Rubanov, Alexander S.; Serebryakova, L. M.: Possibility of associative reconstruction of reflection hologram in phase-conjugate interferometer,V1621,114-124(1991)

Lazewatsky, J. See Shepp, Larry A.: V1452,216-224(1991)

Lazovsky, L. Y.
See Khvilivitzky, A. T.: V1447,184-190(1991)
See Vishnevsky, G. I.: V1447,34-43(1991)

Lazzari, J. L. See Tournie, Eric: V1361,641-656(1991)

Le, Khoa See Koechner, Walter: V1522,169-179(1991)

Le, Shixiao
See Xiao, Guohua: V1409,106-113(1991)
See Yang, Darang: V1417,440-450(1991)

Le, Tam V. See Balasubramanian, Kunjithapa: V1485,245-253(1991)

Lea, M. J. See Zammit, C. C.: V1549,274-282(1991)

Leadbeater, M. L. See Eaves, Lawrence: V1362,520-533(1991)

Leahy, Michael B.
; Whalen, P. V.; Lamont, Gary B.: Adaptive gross motion control: a case study,V1387,148-158(1991)

Leahy, Richard M.
See Wu, Zhenyu: V1450,120-132(1991)
See Yan, Xiao-Hong: V1452,158-169(1991)

Leamy, Paul T. See Kidger, Michael J.: V1354,69-76(1991)

Lean, Meng H. See Andrews, John R.: V1461,110-123(1991)

Leang, Sovarong See Ling, Zhi-Min: V1392,660-669(1991)

Leard, Francis L. See Hillman, Robert L.: V1562,136-142(1991)

Learner, R. C. See Al-Baho, Tareq I.: V1354,417-428(1991)

Leary, James F. See Smith, Warren E.: V1398,142-150(1991)

Leatham, James G.
: Optical power budget and device time-constant considerations in undersea laser-based sensor design,V1537,194-202(1991)

LeBeau, Christopher J.
: Machine vision platform requirements for successful implementation and support in the semiconductor assembly manufacturing environment,V1386,228-231(1991)

LeBeau, M. M. See Hallegot, Philippe: V1396,311-315(1991)

Lebedev, A. V. See Benimetskaya, L. Z.: V1525,210-211(1991)

Lebedev, Fedor V.
See Antyuhov, V.: V1397,355-365(1991); 1415,48-59(1991)
See Bondarenko, Alexander V.: V1376,117-127(1991)

Lebedev, V. B.
; Saulevich, S. V.: Thomson parabolic spectrograph with microchannel plate framing camera as register of ionic parabolae,V1358,874-877(1991)
See Averin, V. I.: V1358,589-602(1991)
See Averin, V. I.: V1358,603-605(1991)
See Bass, V. I.: V1358,1075-1083(1991)
See Korzhenevich, Irina M.: V1358,1084-1089(1991)

Le Besnerais, Guy
; Navaza, Jorge; Demoment, Guy: Aperture synthesis in astronomical radio-interferometry using maximum entropy on the mean,V1569,386-395(1991)

Lebich, Lan See Rayces, Juan L.: V1354,752-759(1991)

Lebiush, E. See Aharon, O.: V1397,251-255(1991)

LeBlanc, Herve P. See Soole, Julian B.: V1474,268-276(1991)

Leblanc, Jean-M.
; Rozelot, Jean-Pierre: Large active mirror in aluminium,V1535,122-129(1991)
See Rozelot, Jean-Pierre: V1494,481-490(1991)

LeBlanc, Michel
See Measures, Raymond M.: V1489,86-96(1991)
See Measures, Raymond M.: V1332,431-443(1991)

Leblanc, Roger M.
; Blanchet, P.-F.; Cote, D.; Gugeshashvili, M. I.; Munger, G.; Volkov, Alexander G.: Light energy conversion with pheophytin a and chlorophyll a monolayers at the optical transparent electrode,V1436,92-102(1991)
See Kassi, Hassane: V1436,58-67(1991)
See Volkov, Alexander G.: V1436,68-79(1991)

Lebland, F. See Nissim, Yves I.: V1393,216-228(1991)

Lebow, Paul S.
; Ackerman, John R.: Spectral characteristics of Brillouin-enhanced four-wave mixing for pulsed and CW inputs,V1409,60-66(1991)

Lebreton, Guy J.
ed.: *Optics for Computers: Architectures and Technologies,*V1505(1991)

LeCarpentier, Gerald L.
; Motamedi, Massoud E.; Welch, Ashley J.: Effects of pressure rise on cw laser ablation of tissue,V1427,273-278(1991)

Lechner, P. See Braeuninger, Heinrich: V1549,330-339(1991)

Leclerc, Luc
; Sheng, Yunlong; Arsenault, Henri H.: Bipolar correlations in composite circular harmonic filters,V1564,78-85(1991)
See Sheng, Yunlong: V1564,320-329(1991)

Leclerc, Norbert See Hitzler, Hermine: V1503,355-362(1991)

Leclere, Philippe
; Renotte, Yvon L.; Lion, Yves F.: Characterization of DC-PVA films for holographic recording materials,V1507,339-344(1991)
; Renotte, Yvon L.; Lion, Yves F.: New technique for characterizing holographic recording materials,V1559,298-307(1991)

Lecluse, Yves See Boyer, Corinne: V1483,77-91(1991)

LeCompte, J. P. See Meyrueix, Remi: V1560,454-466(1991)

Lecot, C. See Lequime, Michael: V1511,244-249(1991)

Lecoy, Pierre See Bocquet, Jean-Claud: V1588,210-217(1991)

Lecuellet, Jerome See Bonniau, Philippe: V1588,52-63(1991)

Ledanois, Guy See Gaffard, Jean-Paul: V1542,34-45(1991)

Ledebuhr, Arno G.
See Lewis, Isabella T.: V1530,22-34(1991)
See Lewis, Isabella T.: V1530,306-324(1991)
See Lewis, Isabella T.: V1478,2-12(1991)

Ledermann, Peter G.
; Johnson, Glen W.; Ritter, Mark B.: Laser-formed structures to facilitate TAB bonding,V1598,160-163(1991)

Ledet, Earl G. See Gregory, Kenton W.: V1425,217-225(1991)

Ledger, N. R. See Waters, Ruth A.: V1427,336-343(1991)

Ledig, Mario See Zschocke, Wolfgang: V1501,49-56(1991)

Lednum, Eugene E.
; McDermott, David B.; Lin, Anthony T.; Luhmann, Neville C.: Negative energy cyclotron resonance maser,V1407,202-208(1991)

Ledoyen, Fernand
; Lewandowski, Jacques; Cormier, Maurice: Behavior of a thin liquid film under thermal stimulation: theory and applications to infrared interferometry,V1507,328-338(1991)

Lee, Andrew J.
; Casasent, David P.: Pose determination of spinning satellites using tracks of novel regions,V1383,72-83(1991)

Lee, Bonita G.
; Tom, Victor T.: Pattern spectrum morphology for texture discrimination and object recognition,V1381,80-91(1991)

Lee, Byung-Uk
; Adler, John R.; Binford, Thomas O.: Error analysis on target localization from two projection images,V1380,96-107(1991)

Lee, C. Y. See Chen, Terry Y.: V1554B,92-98(1991)

Lee, Changhun H. See Imen, Kamran: V1365,60-64(1991)

Lee, Chi H.
: Generation and sampling of high-repetition-rate/high-frequency electrical waveforms in microstrip circuits by picosecond optoelectronic technique,V1390,377-387(1991)
See Hung, Hing-Loi A.: V1476,276-281(1991)
See Kung, Chun C.: V1378,250-258(1991)
See Ling, Junda D.: V1454,353-362(1991)

Lee, Ching-Long
; Jeng, Bor S.; Ju, Rong-Hauh; Huang, Huang-Cheng; Kan, Kou-Sou; Huang, Jei-Shyong; Liu, Tsann-Shyong: Postprocessing of video sequence using motion-dependent median filters,V1606,728-734(1991)

Lee, Chongmok See Bartels, Keith A.: V1450,30-39(1991)

Lee, Choo-Hie
; Choi, Boo-Yeon: Power gain characteristics of discharge-excited KrF laser amplifier system,V1397,91-95(1991)

Lee, Chung E.
; Alcoz, J. J.; Gibler, William N.; Atkins, Robert A.; Taylor, Henry F.: Method for embedding optical fibers and optical fiber sensors in metal parts and structures,V1588,110-116(1991)
; Gibler, William N.; Atkins, Robert A.; Taylor, Henry F.: In-line fiber Fabry-Perot interferometer with high-reflectance internal mirrors,V1584,396-399(1991)
See Taylor, Henry F.: V1562,264-275(1991)
See Yeh, Yunhae: V1584,72-78(1991)

Lee, David T.
; Shiau, Jyh-Jen H.: Surface reconstruction with discontinuities,V1383,297-304(1991)

Lee, El-Hang
: Lattice-mismatched elemental and compound semiconductor heterostructures for 2-D and 3-D applications,V1362,499-509(1991)

Lee, Eun Ho See Dickerson, Stephen L.: V1381,201-209(1991)

Lee, Eun Soo
; Song, So-Young; Jhung, Kyu Soo; Kim, Ung; Lee, Sang-Soo: Wire explosion at reduced pressures,V1358,262-268(1991)

Lee, Hae-Seung See Hakkarainen, J. M.: V1473,173-184(1991)

Lee, Heun J.
; Haus, Joseph W.: Vector squeezed states,V1497,228-239(1991)

Lee, Hua See Goldgof, Dmitry B.: V1383,109-121(1991)

Lee, Huai-Chuan
; Meissner, Helmuth E.: Ion-exchange strengthening of high-average-power phosphate laser glass,V1441,87-103(1991)

Lee, Hyo S. See Schwemmer, Geary K.: V1492,52-62(1991)

Lee, Hyouk
; Yoo, Young-Don; Shin, Hyun-Dong; Earmme, Youn-Young; Kim, Choong-Ki: Analysis of temperature distribution and slip in rapid thermal processing,V1393,404-410(1991)

Lee, Hyung-Jae See Son, Yung-Sung: V1374,23-29(1991)

Lee, Hyung G.
; Kim, HyungJun; Park, S. H.; Langer, Dietrich W.: Study of GaAs/AlGaAs quantum-well structures grown by MOVPE using tertiarybutylarsine,V1361,893-900(1991)

Lee, J. H. See Yeh, Yunhae: V1584,72-78(1991)

Lee, J. Y. See Li, Lian: V1560,243-250(1991)

Lee, Jar J. See Ng, Willie W.: V1371,205-211(1991)

Lee, Jin-Fa See Schutt-Aine, Jose E.: V1389,138-143(1991)

Lee, Jueen See Wang, David T.: V1386,206-219(1991)

Lee, Julienne Y. See Pease, R. F.: V1496,234-238(1991)

Lee, Jun Wung See Ahn, Jae W.: V1358,269-277(1991)

Lee, K. D. See Bell, J. M.: V1536,29-36(1991)

Lee, Kok-Meng See Dickerson, Stephen L.: V1381,201-209(1991)

Lee, Meehye See Heikes, Brian G.: V1433,253-262(1991)

Lee, Michael J. See Crossland, William A.: V1455,264-273(1991)

Lee, Michael S.
: Lidar for expendable launch vehicles,V1480,23-34(1991)
See Adler, James M.: V1480,11-22(1991)

Lee, Mowchen C. See Song, Q. W.: V1558,143-148(1991)

Lee, Nahm S.
; Park, Kyung S.: Anthropomorphic classification using three-dimensional Fourier descriptor,V1450,133-143(1991)

Lee, Nicholas A.
: Fast, epoxiless bonding system for fiber optic connectors,V1365,139-143(1991)

Lee, Paul A. See Armstrong, Neal R.: V1559,18-26(1991)

Lee, Robert A.
: Pixelgram: an application of electron-beam lithography for the security printing industry,V1509,48-54(1991)

Lee, Robert B.
: Flight solar calibrations using the mirror attenuator mosaic: low-scattering mirror,V1493,267-280(1991)

Lee, Robert R. See Passner, Jeffrey E.: V1468,2-10(1991)

Lee, S. Y. See Witanachchi, S.: V1394,161-168(1991)

Lee, Sae H. See Davis, Richard L.: V1474,20-26(1991)

Lee, Sang-Soo See Lee, Eun Soo: V1358,262-268(1991)

Lee, Sang U. See Kim, Dong S.: V1605,190-201(1991)

Lee, Seung-Won See Kwon, O'Dae: V1361,802-808(1991)

Lee, Shyan J.
; Imen, Kamran; Allen, Susan D.: Threshold measurements in laser-assisted particle removal,V1598,2-12(1991)

Lee, Si-Chen See Sun, Tai-Ping: V1361,1033-1037(1991)

Lee, Sing H.
See Cindrich, Ivan: V1555(1991)
See Kiamilev, Fouad E.: V1390,311-329(1991)
See Marchand, Philippe J.: V1505,38-49(1991)
See Ozguz, Volkan H.: V1390,477-488(1991)
See Urquhart, Kristopher S.: V1555,214-223(1991)
See Urquhart, Kristopher S.: V1555,13-22(1991)

Lee, Steven See Lane, Gregory J.: V1424,7-11(1991)

Lee, Sukhan
; Schenker, Paul S.; Park, Jun S.: Sensor-knowledge-command fusion paradigm for man-machine systems,V1383,391-402(1991)

Lee, T. S.
: Impact of fiber-in-the-loop architecture on predicted system reliability,V1366,28-38(1991)

Lee, Terence B. See Sydney, Paul F.: V1482,196-208(1991)

Lee, Terence J.
: Acquisition and processing of digital images and spectra in astronomy,V1439,152-158(1991)

Lee, Tong J. See Bojanczyk, Adam W.: V1566,74-83(1991)

Lee, Tzuo-chang
; Chen, Di: Performance comparison of various data codes in Z-CAV optical recording,V1499,87-103(1991)

Lee, W. D.
; Campbell, Joe C.: Frequency stabilization of AlxGa1-xAs/GaAs lasers using magnetically induced birefringence in an atomic vapor,V1365,96-101(1991)

Lee, W. Y. See Hay, Stephen O.: V1435,352-358(1991)

Lee, Wai-Hon See Tang, Yaw-Tzong: V1555,194-199(1991)

Lee, William See Cartlidge, Andy G.: V1423,167-174(1991)

Lee, Woobin
See Gove, Robert J.: V1444,318-333(1991)
See Yee, David K.: V1444,292-305(1991)

Lee, Xiaobing See Koplowitz, Jack: V1471,452-463(1991)

Lee, Y.-H. See Bell, John A.: V1476,326-329(1991)

Lee, Y. S.
See Liu, W. C.: V1519,670-674(1991)
See Liu, W. C.: V1519,640-644(1991)

Lee, Yang-Hang
; Wu, Jingshown; Tsao, Hen-Wai: Analysis of the multichannel coherent FSK subcarrier multiplexing system with pilot carrier and phase noise cancelling scheme,V1372,140-149(1991)
; Wu, Jingshown; Kao, Ming-Seng; Tsao, Hen-Wai: Wavelength division and subcarrier system based on Brillouin amplification,V1579,155-166(1991)

Lee, Yimkul See Masters, Barry R.: V1429,82-90(1991)

Lee, Yong H.
See Jewell, Jack L.: V1389,401-407(1991)
See Morgan, Robert A.: V1562,149-159(1991)

Lee, Yong J.
; Chou, Ching-Hua; Khuri-Yakub, B. T.; Saraswat, Krishna C.: Noncontacting acoustics-based temperature measurement techniques in rapid thermal processing,V1393,366-371(1991)

Lee, Yung-Cheng See Chandra, S.: V1390,548-559(1991)

Leeb, Walter R.
See Hueber, Martin F.: V1417,233-239(1991)
See Hueber, Martin F.: V1522,259-267(1991)
See Neubert, Wolfgang M.: V1522,93-102(1991)
See Neubert, Wolfgang M.: V1417,122-130(1991)

Leebrick, David H.
: Focus considerations with high-numerical-aperture widefield lenses,V1463,275-280(1991)

Leemann, Thomas See Muser, Markus H.: V1448,106-112(1991)

Leen, Todd K.
: Learning in linear feature-discovery networks,V1565,472-481(1991)

Leep, David A.
See Bell, John A.: V1476,326-329(1991)
See Bell, John A.: V1562,276-280(1991)

Leepa, Douglas C. See Turk, Harris: V1454,344-352(1991)

Lees, John E.
See Fraser, George W.: V1343,438-456(1991)
See Fraser, George W.: V1548,132-148(1991)
See Kaaret, Philip E.: V1548,106-117(1991)

Lefaucheux, Francoise See Mary, Joel: V1557,147-155(1991)

Lefebvre, P.
; Vahid-Shahidi, A.; Albert, Jacques; Najafi, S. I.: Integrated optical Mach-Zehnder interferometers in glass,V1583,221-225(1991)

LeFevre, B. G. See Darden, Bruce V.: V1474,300-308(1991)

Lefevre, Herve C.
; Martin, Philippe; Morisse, J.; Simonpietri, Pascal; Vivenot, P.; Arditty, Herve J.: High-dynamic-range fiber gyro with all-digital signal processing,V1367,72-80(1991)
; Simonpietri, Pascal; Martin, Philippe: Applications and characteristics of polished polarization-splitting couplers,V1365,65-73(1991)

Leffler, Richard G.
See Birth, Gerald S.: V1379,10-15(1991)
See Jaenisch, Holger M.: V1379,162-167(1991)
See Windham, William R.: V1379,39-44(1991)

Le Flem, Gille See Boutinaud, P.: V1590,168-178(1991)

Le Gall, Didier J.
: MPEG (moving pictures expert group) video compression algorithm: a review,V1452,444-457(1991)

Le Gall, G. See Fresne, Francoise: V1444,26-36(1991)

le Gall, Perrine M. See van Dongen, A. M.: V1513,330-339(1991)

Legan, R. G. See Gibbons, Robert C.: V1498,64-69(1991)

Leggieri, Gilberto See D'Anna, Emilia: V1503,256-268(1991)

Legnini, Dan See Lai, Barry P.: V1550,46-49(1991)

LeGrice, Yvonne M. See Nemanich, Robert J.: V1437,2-12(1991)

Legros, J. C. See Regel, L. L.: V1557,182-191(1991)

Lehar, Steve M.
; Worth, Andrew J.: Multiple resonant boundary contour system,V1469,50-62(1991)

Lehner, David L. See Gilbert, John A.: V1554B,202-209(1991)

Lehoczky, S. L. See Gillies, Donald C.: V1484,2-10(1991)

Lei, Gang See Xu, Xu R.: V1519,525-528(1991)

Lei, Jiaheng See Wang, Guomei: V1590,229-236(1991)

Lei, Shaw-Min
: Construction of efficient variable-length codes with clear synchronizing codewords for digital video applications,V1605,863-873(1991)

Lei, Zhiyuan
See Chang, Zenghu: V1358,614-618(1991)
See Hou, Xun: V1358,868-873(1991)

Leib, Kenneth G.
; Brandstetter, Robert W.; Drake, Marvin D.; Franks, Glen B.; Siewert, Ronald O.: Experimental comparison of optical binary phase-only filter and high-pass matched filter correlation,V1483,140-154(1991)
; Pernick, Benjamin J.: Holographic optical elements as laser irradiation sensor components,V1532,261-270(1991)

Leibenguth, R. E.
See Levine, Barry F.: V1540,232-237(1991)
See Morgan, Robert A.: V1562,149-159(1991)

Leiderer, Paul See Boneberg, J.: V1598,84-90(1991)

Leier, H. See Rosenzweig, Josef: V1362,168-178(1991)

Leigh, John S. See Haselgrove, John C.: V1431,30-41(1991)

Leite, Antonio P. See Santos, Jose L.: V1511,179-189(1991)

Leith, Emmett N.
; Chen, Chiaohsiang; Cunha, Andre: Image formation through inhomogeneities,V1396,80-84(1991)

Le Joie, Elaine N. See Williams, Gareth T.: V1332,802-804(1991)

Lekhtsier, Eugeny N. See Furzikov, Nickolay P.: V1427,288-297(1991)

Leksell, David
; Kun, Zoltan K.; Asars, Juris A.; Phillips, Norman J.; Brandt, Gerald B.; Stringer, J. T.; Matty, T. C.; Gigante, Joseph R.: Gray scale and resolution enhancement capabilities of edge emitter imaging stations,V1458,133-144(1991)

Lelievre, Sylviane
; Pawluczyk, Romuald: Wavelength shift in DCG holograms,V1559,288-297(1991)

Lemaire, Paul J. See Krol, Denise M.: V1516,38-46(1991)

Lemaitre, Gerard R. See Huber, Martin C.: V1494,472-480(1991)

Lemak, A. S. See Balabaev, N. K.: V1402,53-69(1991)

Le Masson, Patrick See Grangeat, Pierre: V1445,320-331(1991)

Lemberg, Howard L. See Menendez, Ronald C.: V1363,97-105(1991)

Lemoine, D. See Fresne, Francoise: V1444,26-36(1991)

Lemoine, F. See Leporcq, B.: V1397,153-156(1991)

Lemoine, Patrick
; Magan, John D.; Blau, Werner: Photoablative etching of materials for optoelectronic integrated devices,V1377,45-56(1991)

Lemole, G. M.
; Anderson, R. R.; DeCoste, Sue: Preliminary evaluation of collagen as a component in the thermally induced 'weld',V1422,116-122(1991)

Lemole, Gerald M. See Ashton, Robert C.: V1422,151-155(1991)

Len, Lek K. See Golden, Jeffry: V1407,418-429(1991)

Lena, Pierre J.
See Gendron, Eric: V1542,297-307(1991)
See Merkle, Fritz: V1542,308-318(1991)

Lenkova, Galina A.
; Churin, Evgeny G.: High-resolution micro-objective lens with kinoform corrector,V1574,235-242(1991)

Lennmarken, Claes See Vegfors, Magnus: V1426,79-83(1991)

Lenski, Harald
; Braun, Michael: Laser beam deflection: a method to investigate convection in vapor growth experiments,V1557,124-131(1991)

Lenstra, Daan
; Cohen, Julius S.: Feedback noise in single-mode semiconductor lasers,V1376,245-258(1991)

Lenth, Wilfried
; Kozlovsky, William J.; Risk, William P.: Blue laser devices for optical data storage,V1499,308-313(1991)

Lentine, Anthony L.
See Brubaker, John L.: V1533,88-96(1991)
See McCormick, Frederick B.: V1396,508-521(1991)
See McCormick, Frederick B.: V1533,97-114(1991)

Lentz, William J.
See Cooper, Alfred W.: V1486,47-57(1991)
See Cooper, Alfred W.: V1486,37-46(1991)

Lenz, P.
: Determination of the optical penetration depth in tumors from biopsy samples,V1525,183-191(1991)
: Light distributor for endoscopic photochemotherapy,V1525,192-195(1991)

Lenz, Reimar K.
; Lenz, Udo: Geometrical and radiometrical signal transfer characteristics of a color CCD camera with 21-million pixels,V1526,123-132(1991)

Lenz, Reiner
; Osterberg, Mats: Learning filter systems with maximum correlation and maximum separation properties,V1469,784-795(1991)

Lenz, Udo See Lenz, Reimar K.: V1526,123-132(1991)

Lenzing, Erik See Rachlin, Adam: V1477,101-114(1991)

Lenzner, Matthias See Brueckner, Volkmar: V1362,510-517(1991)

Leon, Francisco A. See Wojcik, Gregory L.: V1463,292-303(1991)

Leonard, Regis F.
; Bhasin, Kul B.; eds.: *Monolithic Microwave Integrated Circuits for Sensors, Radar, and Communications Systems*,V1475(1991)
; Connolly, Denis J.; Bhasin, Kul B.; Warner, Joseph D.; Alterovitz, Samuel A.: High-temperature superconductive microwave technology for space applications,V1394,114-125(1991)
See Lan, Guey-Liu: V1475,184-192(1991)

Leonardi, Riccardo
See Lavagetto, Fabio: V1605,534-545(1991)
See Radha, Hayder: V1605,475-486(1991)

Leonelli, Richard See Li, Ming-Jun: V1513,410-417(1991)

Leong, D. Y. See Carduner, Keith R.: V1433,190-201(1991)

Leonov, A. M. See Erokhovets, Valerii K.: V1621,227-236(1991)

Leonov, A. S. See Petrov, Dmitry V.: V1374,152-159(1991)

Leopold, Louis See Arndt, George D.: V1475,231-242(1991)

Leopold, Simon See Kiss, Gabor D.: V1366,223-234(1991)

Leow, S. K. See Chua, Soo-Jin: V1401,96-102(1991)

Le Pannerer, Yves-Marie
; Tourtier, Philippe: Digital picture processing for the transmission of HDTV: the progress in Europe,V1567,556-565(1991)

Le Person, H. See Minot, Christophe: V1362,301-308(1991)

Le Pesant, Jean-Pierre See Bonniau, Philippe: V1588,52-63(1991)

Lepesheva, Galina I. See Turko, Illarion V.: V1572,419-423(1991)

Lepetre, Yves See Berrouane, H.: V1343,428-436(1991)

Leporcq, B.
; Verdier, C.; Arbus, R.; Lemoine, F.: Potential IF chemical laser,V1397,153-156(1991)

Leppelmeier, Gilbert W. See Sipila, Heikki: V1549,246-255(1991)

Lequime, Michael
; Lecot, C.; Jouve, Philippe; Pouleau, J.: Fiber optic pressure and temperature sensor for down-hole applications,V1511,244-249(1991)
; Meunier, Carole; Giovannini, Hugues: Fiber optic magnetic field sensor using spectral modulation encoding,V1367,236-242(1991)
See Sansonetti, Pierre: V1588,198-209(1991)
See Sansonetti, Pierre: V1588,143-149(1991)

Lerbet, F. See Plichon, V.: V1536,37-47(1991)

Lerche, Richard A.
: Timing between streak cameras with a precision of 10 ps,V1346,376-383(1991)

Lerman, A. A. See Artjushenko, Vjacheslav G.: V1420,157-168(1991)

Lerminiaux, Christian See Nissim, Carlos: V1365,13-19(1991)

Lerner, Jeremy M.
; McKinney, Wayne R.; eds.: *Intl Conf on the Application and Theory of Periodic Structures*,V1545(1991)

Lerner, Mark
; Boult, Terrance E.: Energy-based segmentation of very sparse range surfaces,V1383,277-284(1991)

Leroy, Renan See Bacal, Marthe: V1407,605-609(1991)

Lerrer, A.
See Lando, Mordechai: V1412,19-26(1991)
See Lando, Mordechai: V1442,172-180(1991)

Lesche, Bernhard
: Physical models of second-harmonic generation in optical fibers,V1516,125-136(1991)
See Margulis, Walter: V1516,60-66(1991)

Lesco, Daniel J.
: Research and development of optical measurement techniques for aerospace propulsion research: a NASA/Lewis Research Center perspective,V1554B,530-539(1991)

Leseberg, Detlef
: Position detection using computer-generated holograms,V1500,171-171(1991)

Le Sergent, Christian See Briancon, Anne-Marie: V1373,9-20(1991)

Lesh, James R.
; Deutsch, L. J.; Weber, W. J.: Plan for the development and demonstration of optical communications for deep space,V1522,27-35(1991)
See Chen, Chien-Chung: V1417,170-181(1991)
See Chen, Chien-Chung: V1417,240-250(1991)
See Hemmati, Hamid: V1417,476-483(1991)
See Wilson, Keith E.: V1417,22-28(1991)

Lesko, Camille
; Hill, William A.; Dietrich, Fred; Nelson, William: Compact time-delay shifters that are process insensitive,V1475,330-339(1991)

Lesko, John J. See Fogg, Brian R.: V1588,14-25(1991)

L'Esperance, Drew M. See Chronister, Eric L.: V1559,56-64(1991)

Lessard, Michel See Jiang, Jing-Wen: V1363,191-195(1991)

Lessard, Roger A.
ed.: *Photopolymer Device Physics, Chemistry, and Applications II*,V1559(1991)
; Changkakoti, Rupak; Manivannan, Gurusamy: Metal-ion-doped polymer systems for real-time holographic recording,V1559,438-448(1991)
See Cappolla, Nadia: V1389,612-620(1991)
See Frank, Werner F.: V1559,344-353(1991)
See Malouin, Christian: V1559,385-392(1991)
See Sheng, Yunlong: V1559,222-228(1991)

Lesser, Michael P.
: Array detectors in astronomy,V1439,144-151(1991)
: Back-illuminated large-format Loral CCDs,V1447,177-182(1991)

Lester, J. See Chow, P. D.: V1475,42-47(1991)

Lesterlin, L. See Berrouane, H.: V1343,428-436(1991)

Letalick, Dietmar See Renhorn, Ingmar G.: V1480,35-45(1991)

Letardi, Tommaso
; Laude, Lucien D.; eds.: *Excimer Lasers and Applications III*,V1503(1991)
See Bollanti, Sarah: V1503,80-87(1991)
See Bollanti, Sarah: V1397,97-102(1991)

Letartre, Xavier
; Stievenard, Didier; Barbier, E.: Behaviour of a single quantum-well under deep level transient spectroscopy (DLTS) measurement: a new theoretical model,V1362,778-789(1991)
; Stievenard, Didier: Observation of tunneling emission from a single-quantum-well using deep-level transient spectroscopy,V1361,1144-1155(1991)

Letellier, H. See Orsal, Bernard: V1512,112-123(1991)

Letellier, J. P. See Alter, James J.: V1566,296-301(1991)

Letokhov, Vladilen S.
: Laser femtosecond resonant ionization and mapping of biomolecules,V1403,4-4(1991)

Letterer, Rudolf
; Wallmeroth, Klaus: Single-frequency Nd:YAG laser development for space communication,V1522,154-157(1991)

Lettieri, Thomas R.
See Marx, Egon: V1332,826-834(1991)
See Marx, Egon: V1530,15-21(1991)

Lettington, Alan H. See Gleed, David G.: V1567,65-72(1991)

Letuta, S. N. See Ketsle, G. A.: V1403,622-630(1991)

Letzring, Samuel A.
See Brown, David L.: V1410,209-214(1991)
See Johnson, C. B.: V1346,340-370(1991)

Leu, Fang Y.
; Chang, Shi-Kuo: Internal protocol assistant for distributed systems,V1468,620-631(1991)

Leu, R. L. See Huang, C. J.: V1519,70-73(1991)

Leung, Billy C. See Smith, Warren E.: V1398,142-150(1991)

Leung, Chung-yee
; Wu, Jiunn-Shyong; Ho, M. Y.; Chen, Kuang-yi: Optimizing the performance of a frequency-division distributed-optical-fiber sensing system,V1572,566-571(1991)

Leung, Dominic S.
; Williams, D. S.: Multiple-hypothesis-based multiple-sensor spatial data fusion algorithm,V1471,314-325(1991)

Leung, Doris See Ermer, Susan P.: V1560,120-129(1991)

Leung, Henry
; Haykin, Simon: Neural network modeling of radar backscatter from an ocean surface using chaos theory,V1565,279-286(1991)

Leung, Michael See Kadin, Alan M.: V1477,156-165(1991)

Leung, Pak K. See Chaker, Mohamed: V1465,16-25(1991)

Leuschner, Rainer See Sebald, Michael: V1466,227-237(1991)

Leuw, David H.
: Fabrication of special-purpose optical components,V1442,31-41(1991)

LeVan, M. See Wiepking, Mark: V1392,139-150(1991)

Levay, C. See Larkins, Grover L.: V1477,26-33(1991)

Levenson, Marc D.
: Beam position noise and other fundamental noise processes that affect optical storage,V1376,259-271(1991)
: What IS a phase-shifting mask?,V1496,20-26(1991)
See Brock, Phillip J.: V1463,87-100(1991)

Levenson, R.
; Liang, J.; Toussaere, E.; Carenco, Alain; Zyss, Joseph: Quasirelaxation-free guest-host poled-polymer waveguide modulator: material, technology, and characterization,V1560,251-261(1991)

Levenstein, Harold
: Laser radar based on diode lasers,V1416,30-43(1991)

Levesque, Martin P.
: Dynamic sea image generation,V1486,294-300(1991)
: Generation of IR sky background images,V1486,200-209(1991)

Levi, Zelma See Savatinova, Ivanka T.: V1374,37-46(1991)

Levi-Setti, Riccardo See Hallegot, Philippe: V1396,311-315(1991)

Levin, Paul A. See Cheng, Xin: V1363,177-181(1991)

Levina, Olga V. See Belov, Sergei N.: V1503,503-509(1991)

Levine, Alfred M.
; Ozizmir, Ercument; Zaibel, Reuben; Prior, Yehiam: General jump model for laser noise: non-Markovian phase and frequency jumps,V1376,47-53(1991)

Levine, Barry F.
: Long-wavelength GaAs quantum-well infrared photodetectors,V1362,163-167(1991)
; Bethea, Clyde G.; Stayt, J. W.; Glogovsky, K. G.; Leibenguth, R. E.; Gunapala, S. D.; Pei, S. S.; Kuo, Jenn-Ming: Long-wavelength GaAs/AlxGa1-xAs quantum-well infrared photodetectors,V1540,232-237(1991)
See Gunapala, S. D.: V1541,11-23(1991)

Levine, Betty A.
See Horii, Steven C.: V1446,10-15(1991)
See Horii, Steven C.: V1446,475-480(1991)

Levine, Bruce M.
See Colavita, Mark M.: V1494,168-181(1991)
See Redding, David C.: V1489,201-215(1991)

Levine, Howard See Rava, Richard P.: V1426,68-78(1991)

Levine, Jerrold S.
; Benford, James N.; Courtney, R.; Harteneck, Bruce D.: Operational characteristics of a phase-locked module of relativistic magnetrons,V1407,74-82(1991)
See Smith, Richard R.: V1407,83-91(1991)

Levine, Peter A. See Sauer, Donald J.: V1540,285-296(1991)

Levinsen, Mogens T. See Barholm-Hansen, Claus: V1512,218-225(1991)

Levinson, A. K. See Johnson, Douglas E.: V1421,36-41(1991)

Levinson, R. M. See Liu, Yung S.: V1377,126-133(1991)

Levit, Creon C. See Bryson, Steve T.: V1457,161-168(1991)

Levitan, Steven P.
See Chiarulli, Donald M.: V1390,403-414(1991)
See Guo, Zicheng: V1390,415-426(1991)

Levitt, Morris R.
: Laser market in the 1990s,V1520,1-36(1991)
See Belforte, David A.: V1517(1991)
See Forrest, Gary T.: V1520(1991)

Levshin, L. V.
See Berezin, M. V.: V1403,335-337(1991)
See Bisenbaev, A. K.: V1403,606-610(1991)

Levush, Baruch See Gregoire, Daniel J.: V1552,118-126(1991)

Levy, Dorron
; Harris, Karl L.: Charging phenomena in e-beam metrology,V1464,413-423(1991)

Levy, I. See Shushan, A.: V1487,300-311(1991)

Levy, Joseph L. See Haake, John M.: V1418,298-308(1991)

Levy, Julia G. See Allison, Beth: V1426,200-207(1991)

Levy, Ram L. See Udd, Eric: V1418,134-152(1991)

Levy, Saul Y. See Murdocca, Miles J.: V1563,255-266(1991)

Levy, Yves
See Dumont, Michel L.: V1559,127-138(1991)
See Morichere, D.: V1560,214-225(1991)

Lew, Michael See Wong, Kam W.: V1526,17-26(1991)

Lewandowski, Jacques See Ledoyen, Fernand: V1507,328-338(1991)

Lewis, Aaron
: ; Palanker, Daniel; Hemo, Itzhak; Pe'er, Jacob; Zauberman, Hanan: Cold-laser microsurgery of the retina with a syringe-guided 193 nm excimer laser,V1423,98-102(1991)

: See Kopelman, Raoul: V1435,96-101(1991)

Lewis, Catherine C. See Hartz, William G.: V1398,52-60(1991)

Lewis, Dorothy E. See Smith, Louis C.: V1428,224-232(1991)

Lewis, Gary See McCormick, Patrick W.: V1431,294-302(1991)

Lewis, George R.
: Direct-drive film magazines,V1538,167-179(1991)

Lewis, H. G. See Forsyth, William B.: V1472,18-25(1991)

Lewis, Isabella T.
: ; Ledebuhr, Arno G.; Bernt, Marvin L.: Stray-light implications of scratch/dig specifications,V1530,22-34(1991)

: ; Ledebuhr, Arno G.; Axelrod, Timothy S.; Ruddell, Scott A.: Stray-light reduction in a WFOV star tracker lens,V1530,306-324(1991)

: ; Ledebuhr, Arno G.; Axelrod, Timothy S.; Kordas, Joseph F.; Hills, Robert F.: Wide-field-of-view star tracker camera,V1478,2-12(1991)

Lewis, Keith L. See Wood, Roger M.: V1441,316-326(1991)

Lewis, Nathan S. See Jozefiak, Thomas H.: V1436,8-19(1991)

Lewis, Norris E.
: ; Moore, Emery L.; eds.: *Fiber Optic Systems for Mobile Platforms IV,*V1369(1991)

: ; Moore, Emery L.; eds.: *Specialty Fiber Optic Systems for Mobile Platforms,*V1589(1991)

Lewis, Paul S. See George, John S.: V1443,37-51(1991)

Lewis, Tom R.
: ; Mitra, Sunanda: Application of a blind-deconvolution restoration technique to space imagery,V1565,221-226(1991)

Lewis, Warren H.
: ; Miller, Michael B.: Single-mode fiber optic rotary joint for aircraft applications,V1369,79-86(1991)

Lewis, Wilfred See Tobin, Kenneth W.: V1584,23-31(1991)

Lewotsky, Kristin L. See Harvey, James E.: V1530,35-41(1991)

Leys, C.
: ; Sona, P.; Muys, Peter F.: Probe measurements in a CO2 laser plasma,V1397,399-402(1991)

: See Sona, P.: V1397,373-377(1991)

Li, Aizhen See Zhu, Wen H.: V1519,423-427(1991)

Li, Aizhong
: Optical fibre image sensor,V1572,548-551(1991)

Li, Andrew See Liao, Wen-gen: V1444,47-55(1991)

Li, Bing Z. See Gu, Zhi G.: V1519,247-251(1991)

Li, Bo See Dougherty, Edward R.: V1606,141-152(1991)

Li, C.
: See Fowler, Burt: V1598,108-117(1991)

: See Lian, S.: V1598,98-107(1991)

Li, Chang J.
: ; Liu, Ji M.; Yao, Qi: Bi(Pb)-Sr-Ca-Cu-O superconducting films prepared by chemical spray deposition,V1519,779-787(1991)

Li, Cheng F.
: See Li, Zhong Y.: V1519,374-379(1991)

: See Sun, Yang: V1519,554-558(1991)

Li, Chunfei See Liu, Yudong: V1362,436-447(1991)

Li, Da-ren See Dickerson, Stephen L.: V1381,201-209(1991)

Li, Dapeng
: See Niu, Aiqun: V1469,495-505(1991)

: See Wan, Liqun: V1469,432-440(1991)

Li, Da Y. See Lu, Yu C.: V1519,463-466(1991)

Li, Dian-en
: See Zhang, Yue-qing: V1418,444-447(1991)

: See Zhang, Yue-qing: V1400,137-143(1991)

Li, Dong
: Max-polynomials and template decomposition,V1568,149-156(1991)

Li, Faxian See Jiang, Desheng: V1572,313-317(1991)

Li, Futian
: ; Cao, Jianlin; Chen, Po; Qian, Limin; Jin, Lei; Chen, Xingdan: VUV wall stabilized argon arc discharge source,V1345,78-88(1991)

: See Cao, Jianlin: V1345,71-77(1991)

: See Cao, Jianlin: V1345,225-232(1991)

Li, Gabriel
: Solving clock distribution problems in FDDI concentrators,V1364,72-83(1991)

Li, Gongde See Yang, Daren: V1527,456-461(1991)

Li, Guoguang See Bridges, Frank: V1550,76-84(1991)

Li, Guo Z. See Huang, Zong T.: V1519,788-789(1991)

Li, Guozhu
: ; Wu, Wenming; Su, Degong: Laser high-speed photography in cylindrical charge shell burst process,V1358,1120-1122(1991)

Li, Hanjie See Yan, Hongtao: V1572,396-398(1991)

Li, Hong T.
: See Fan, Xiang J.: V1519,808-812(1991)

: See Fan, Xiang J.: V1519,805-807(1991)

Li, Hua
: ; Park, J. D.; Seo, Dongsun; Marin, L. D.; McInerney, John G.; Telle, H. R.: Reduction of spectral linewidth and FM noise in semiconductor lasers by application of optical feedback,V1376,172-179(1991)

Li, J. P. See Sumner, Timothy J.: V1549,256-264(1991)

Li, J. W. See Hsu, C. T.: V1519,391-395(1991)

Li, Jiang-Qi See Jiang, Xiang-Liu: V1534,207-213(1991)

Li, Jianshu See Zhong, Xian-Xin: V1572,84-87(1991)

Li, Jie
: ; He, Li; Tang, Wen G.; Shan, W.; Yuan, Shi X.: High-quality heavily strained II-VI quantum well,V1519,660-664(1991)

Li, Jie-gu
: ; Yuan, Qiang: Application of perceptron to the detecting of particle motion,V1469,178-187(1991)

Li, Jinghua H. See Lin, Cheng L.: V1519,104-108(1991)

Li, Jinian
: Analysis of the accuracy of aerostatic slideway made from granite,V1533,150-154(1991)

Li, Kam See Lingua, Robert W.: V1423,58-61(1991)

Li, L. See Chen, Shang-Liang: V1502,123-134(1991)

Li, Le See Yuan, Haiji J.: V1455,73-83(1991)

Li, Li See Zhuang, Qi: V1397,157-160(1991)

Li, Li-Ping See Zhang, Yu: V1384,60-65(1991)

Li, Lian
: ; Jeng, Ru J.; Lee, J. Y.; Kumar, Jayant; Tripathy, Sukant K.: Photoconducting nonlinear optical polymers,V1560,243-250(1991)

Li, Luksun
: ; Kerr, Anthony; Giles, Ian P.: Polarimetric optical fiber pressure sensor,V1584,170-177(1991)

Li, M. J. See Lallier, Eric: V1506,71-79(1991)

Li, Min See Zhengmin, Li: V1401,66-73(1991)

Li, Ming-Guang See Hung, Hing-Loi A.: V1476,276-281(1991)

Li, Ming-Jun
: ; Honkanen, Seppo; Wang, Wei-Jian; Najafi, S. I.; Tervonen, Ari; Poyhonen, Pekka: Buried-glass waveguides by ion exchange through ionic barrier,V1506,52-57(1991)

: ; Honkanen, Seppo; Wang, Wei-Jian; Leonelli, Richard; Najafi, S. I.: Dual-core ion-exchanged glass waveguides,V1513,410-417(1991)

: See Najafi, S. I.: V1583,32-36(1991)

: See Wang, Wei-Jian: V1513,434-440(1991)

: See Zhang, H.: V1583,83-89(1991)

Li, N. Y.
: See Wu, T. S.: V1361,23-33(1991)

: See Wu, T. S.: V1519,263-268(1991)

Li, Naiji
: ; Wang, Huiwen; Yang, Yang; Gu, Shenghua: Study on quasi-instantaneous converse piezoelectric effect of piezoelectric ceramics with the sinusoidal phase-modulating interferometer using optical fibers,V1572,47-51(1991)

: See Yang, Yang: V1572,205-210(1991)

Li, Penghui See Chu, Kunliang: V1554A,192-195(1991)

Li, Q. See Xi, Xiaoxing: V1477,20-25(1991)

Li, Qi See Nie, Chao-Jiang: V1579,264-267(1991)

Li, Qiang See Piestrup, Melvin A.: V1552,214-239(1991)

Li, Qin See Yu, Tong: V1572,469-471(1991); 1584,135-137(1991)

Li, Qun Z.
: See He, Anzhi: V1567,698-702(1991)

: See He, Anzhi: V1545,278-281(1991)

Li, S. T. See Pappert, Stephen A.: V1476,282-293(1991)

Li, Shaohui
; Wen, Shengping; Zhang, Zhipeng; Lu, Feizhen: Optical fibre system with CCLID for the transmission of two-dimensional images,V1572,543-547(1991)

Li, Shen See Chen, Duanjun: V1554A,922-931(1991)

Li, Shenghong See Sheng, Lie-yi: V1572,273-278(1991)

Li, Sheng S. See Hwang, Kiu C.: V1371,128-137(1991)

Li, Shiqing See Wu, Lin: V1388,598-601(1991)

Li, Shiying
; Chen, Shisheng; Lin, Li-Huang; Wang, Shijie; Bin, Ouyang; Xu, Zhizhan: Active-passive colliding pulse mode-locked Nd:YAG laser,V1410,215-220(1991)
See Lin, Li-Huang: V1410,65-71(1991)

Li, Shi Z. See Xu, Kewei: V1519,765-770(1991)

Li, Shuqiu See Qian, Shen-en: V1483,196-206(1991)

Li, Wei
; Basso, Andrea; Popat, Ashok C.; Nicoulin, Andre; Kunt, Murat: Design of parallel multiresolution filter banks by simulated annealing,V1605,124-136(1991)
See Popat, Ashok C.: V1567,341-353(1991)
See Popat, Ashok C.: V1605,940-953(1991)

Li, Wei-min See Liu, Rui-Fu: V1572,180-184(1991)

Li, Wei Z.
See Pan, J. J.: V1476,133-142(1991)
See Pan, J. J.: V1476,122-132(1991)
See Pan, J. J.: V1476,32-43(1991)

Li, Wen
See Wamser, Carl C.: V1436,114-124(1991)
See Wu, Gengsheng: V1572,497-502(1991)

Li, Wenchong
; Ma, Chunhua: Imaging spectroscope with an optical recombination system,V1428,242-248(1991)

Li, Wenguang See Gussenhoven, Elma J.: V1425,203-206(1991)

Li, Wen L.
; Zhong, An: Production of gratings with desired pitch using the Talbot effect,V1554B,275-280(1991)

Li, X. M.
; Chiang, Fu-Pen: Void-crack interaction in aluminum single crystal,V1554A,285-296(1991)

Li, Xian See Hu, Yu: V1409,230-239(1991)

Li, Xiang Ying
; Shi, Zhenbang; Sun, Ning: Double-pulsed TEA CO2 laser,V1412,246-251(1991)

Li, Xiao-Rong
; Chang, Shi-Kuo: Visualization of image from 2-D strings using visual reasoning,V1468,720-731(1991)

Li, Xiao L. See Zhang, Si J.: V1519,744-751(1991)

Li, Xiao Q.
; Lin, Cheng L.; Zou, Shi C.; Weng, Hei M.; Han, Xue D.: SPAT studies of near-surface defects in silicon induced by BF2+ and F++B+ implantation,V1519,14-17(1991)

Li, Xide
; Tan, Yushan: Theory of electronic projection correlation and its application in measurement of rigid body displacement and rotation,V1554B,661-668(1991)

Li, Xingkang See Cadzow, James A.: V1486,352-363(1991)

Li, Y. See McIver, Robert T.: V1437,124-128(1991)

Li, Y. Q.
See Chern, Chyi S.: V1394,255-265(1991)
See Zhao, Jing-Fu: V1362,135-143(1991)

Li, Yao
; Zhang, Yan: Optical A/D conversion based on acousto-optic theta modulation,V1474,167-173(1991)
See Zhang, Yan: V1474,188-198(1991)
See Zhang, Yan: V1481,23-31(1991)

Li, Yao-lin See Chen, Shisheng: V1552,288-295(1991)

Li, Yao W. See Zhang, Xi: V1554A,444-448(1991)

Li, YaTing See Cui, DaFu: V1572,386-391(1991)

Li, Yi J.
; Ren, Cong X.; Chen, Guo L.; Chen, Jian M.; Zou, Shi C.: Superconductivity and structural changes of Ar ion-implanted YBa2Cu3O7-x thin films,V1519,800-804(1991)

Li, Ying See Nie, Chao-Jiang: V1579,264-267(1991)

Li, Ying-Sing See Vo-Dinh, Tuan: V1368,203-209(1991)

Li, Ying X. See Lin, Hong Y.: V1519,210-213(1991)

Li, Yi Q.
See Pan, J. J.: V1476,22-31(1991)
See Pan, J. J.: V1476,32-43(1991)

Li, Yongchun L. See Lin, Li-Huang: V1346,490-501(1991)

Li, Yonghong See Yan, Hongtao: V1572,396-398(1991)

Li, Yong N. See He, Anzhi: V1423,117-120(1991)

Li, Yongqing See Wang, Yuzhu: V1501,40-48(1991)

Li, Youli See Kalata, Kenneth: V1345,270-280(1991)

Li, Yu See Yu, Bing Kun: V1530,363-369(1991)

Li, Yuan-Yuan See Weng, Zhicheng: V1527,349-356(1991)

Li, Yuan J. See Zhao, Shu L.: V1519,275-280(1991)

Li, Yuchuan
; Xiong, Guiguang; Yu, Guoping; Jiang, Zhiying; Wang, Fang: Investigation of optical fibre sensor for remote measuring optical activity,V1572,382-385(1991)

Li, Yulin
; Zhao, Mingjun; Zhao, Li: Hybrid pattern recognition system for the robotic vision,V1385,200-205(1991)
See Zhao, Mingjun: V1558,529-534(1991)

Li, Yu M. See Liu, Wei Y.: V1519,172-174(1991)

Li, Yu Z. See Zhuang, Weihua: V1361,980-986(1991)

Li, Zai Q.
See An, Cheng W.: V1519,818-821(1991)
See Fan, Yong C.: V1519,775-778(1991)
See Fan, Yong C.: V1519,813-817(1991)
See Pan, Yingtain: V1572,477-482(1991)

Li, Ze-Nian See Tate, Kevin J.: V1383,491-502(1991)

Li, Zhao-zhang
; Van De Merwe, Willem P.; Reinisch, Lou: Comparison of the excimer laser with the erbium yttrium aluminum garnet laser for applications in osteotomy,V1427,152-161(1991)

Li, Zheng Q. See Zhao, Shu L.: V1519,275-280(1991)

Li, Zhi-yong
; Sun, Zhong-kang; Shen, Zhen-kang: Recognition of a moving planar shape in space from two perspective images,V1472,97-105(1991)

Li, Zhi F. See Chen, Kun J.: V1519,632-639(1991)

Li, Zhigang
See Boher, Pierre: V1345,165-179(1991)
See Boher, Pierre: V1343,39-55(1991)

Li, Zhiquan
; Shi, Jinshan; Wang, Yutian: Study on pyrometer with double Y-type optical fibers,V1572,185-188(1991)
See Wang, Yutian: V1572,192-196(1991)

Li, Zhixin See Cheever, Charles J.: V1454,139-151(1991)

Li, Zhong-Zhe See Hong, Xiao-Yin: V1466,546-557(1991)

Li, Zhong Y.
; Li, Cheng F.; Guo, Ju P.: Multiple-pulse laser damage to thin film optical coating,V1519,374-379(1991)

Li, Zong Yan See Yin, Zhi Xiang: V1559,487-496(1991)

Liakhou, G.
See Andriesh, A.: V1513,137-147(1991)
See Montenero, Angelo: V1513,234-242(1991)

Lian, Jing See Rowberg, Alan H.: V1444,125-127(1991)

Lian, Jiunn-Long See Shie, Jin-Shown: V1362,655-663(1991)

Lian, S.
; Fowler, Burt; Krishnan, S.; Jung, Le-Tien; Li, C.; Banerjee, Sanjay: Characterization of microstructure of Si films grown by laser-enhanced photo-CVD using Si2H6,V1598,98-107(1991)
See Fowler, Burt: V1598,108-117(1991)

Lian, Tongshu See Yao, Wu: V1527,448-455(1991)

Liang, Chen See Liu, Rui-Fu: V1572,399-402(1991)

Liang, Cheng-Chung
; Chen, Chin-Tu; Lin, Wei-Chung: Intensity interpolation for branching in reconstructing three-dimensional objects from serial cross-sections,V1445,456-467(1991)

Liang, Dawei
; Fraser Monteiro, L.; Fraser Monteiro, M. L.: Novel device for increasing the laser pulse intensity in multiphoton ionization mass spectrometry,V1501,129-134(1991)
; Fraser Monteiro, L.; Fraser Monteiro, M. L.; Lu, Boyin: Three-dimensional interferometric and fiber-optic displacement measuring probe,V1511,90-97(1991)

Liang, J. See Levenson, R.: V1560,251-261(1991)

Liang, Jinwen See Zheng, Gang: V1572,299-303(1991)

Liang, Junzhong
; Grimm, B.; Goelz, Stefan; Bille, Josef F.: Hartmann-Shack sensor as a component in active optical system to improve the depth resolution of the laser tomographic scanner,V1542,543-554(1991)
See Goelz, Stefan: V1542,502-510(1991)

Liang, Ping See Taubes, C. H.: V1570,96-102(1991)

Liang, R. M. See Arun, K. S.: V1566,157-166(1991)

Liang, Tajen See Shemlon, Stephen: V1381,470-481(1991)

Liang, Victor See Vijaya Kumar, B. V. K.: V1555,138-145(1991)

Liang, W. Y.
: Metal-superconductor-insulator transitions in oxide materials,V1362,127-134(1991)

Liang, Yue-kun See Chen, Yu-Feng: V1572,113-117(1991)

Liang, Zhujian See Dahe, Liu: V1507,310-315(1991)

Liao, Chang G.
; Zheng, Zhi H.; Wang, Yong Q.; Yang, Sheng S.: Boron depth profiles in a-Si1-xCx:H(B) films after thermal annealing,V1519,152-155(1991)
See Xiang, Jin Z.: V1519,683-687(1991)

Liao, Wen-gen
; Simovsky, Ilya; Li, Andrew; Kramer, David M.; Kaufman, Leon; Rhodes, Michael L.: Surface definition technique for clinical imaging,V1444,47-55(1991)

Liao, Yan-Biao
; Lai, Shurong; Zhao, Huafeng; Wu, Gengsheng: Study of OFS for gas-liquid two-phase flow,V1584,400-404(1991)
See Culshaw, Brian: V1572(1991)
See Fang, Cui-Chang: V1554A,915-921(1991)
See Wu, Gengsheng: V1572,497-502(1991)
See Zhang, Peng-Gang: V1572,528-533(1991)
See Zhao, Huafeng: V1572,503-507(1991)

Liau, Zong-Long See Diadiuk, Vicky: V1354,496-500(1991)

Libby, Vibeke
ed.: *Data Structures and Target Classification*,V1470(1991)
; Bardin, R. K.: Conversion of sensor data for real-time scene generation,V1470,59-64(1991)
See Bardin, R. K.: V1566,394-404(1991)

Libeau, James See Shepp, Larry A.: V1452,216-224(1991)

Libenson, B. N. See Bryukhnevitch, G. I.: V1358,739-749(1991)

Libenson, Michail N.
; Makin, Vladimir S.; Trubaev, Vladimir V.: Measurement of SEW phase velocity by optical heterodyning method,V1440,354-356(1991)
See Bazhenov, V. V.: V1440,332-337(1991)
See Bonch-Bruevich, Aleksei M.: V1440(1991)
See Kokin, A. N.: V1440,338-341(1991)
See Minaeva, E. M.: V1440,63-70(1991)
See Vorobyov, A. Y.: V1440,197-205(1991)

Libert, John M. See Tsao, Tien-Ren J.: V1470,37-47(1991)

Libutti, Steven K.
; Williams, Mathew R.; Oz, Mehmet C.; Forde, Kenneth A.; Bass, Lawrence S.; Weinstein, Samuel; Auteri, Joseph S.; Treat, Michael R.; Nowygrod, Roman: Preliminary results with sutured colonic anastomoses reinforced with dye-enhanced fibrinogen and a diode laser,V1421,169-172(1991)
See Bass, Lawrence S.: V1421,164-168(1991)
See Eaton, Alexander M.: V1423,52-57(1991)
See Oz, Mehmet C.: V1422,147-150(1991)
See Wider, Todd M.: V1422,56-61(1991)

Lichtenauer, Gert See Elstner, Christian: V1478,150-159(1991)

Licoppe, C. See Nissim, Yves I.: V1393,216-228(1991)

Lidestri, Joseph P. See Bailey, Vernon L.: V1407,400-406(1991)

Lidke, Steve L.
; Haskett, Michael C.: Lessons learned from a commercial module approach to real-time ATR development,V1406,203-203(1991)
See Haskett, Michael C.: VIS07,14-23(1991)

Lieberman, Klony See Kopelman, Raoul: V1435,96-101(1991)

Lieberman, Robert A.
: Recent progress in intrinsic fiber optic chemical sensing,V1368,15-24(1991)
; Wlodarczyk, Marek T.; eds.: *Chemical, Biochemical, and Environmental Fiber Sensors II*,V1368(1991)

Lieberman, Stephen H. See Lurk, Paul W.: V1434,114-118(1991)

Liebow, Charles
; Maloney, M. J.: Histochemical identification of malignant and premalignant lesions,V1426,22-30(1991)
See Braun, Robert E.: V1424,138-144(1991)
See Kingsbury, Jeffrey S.: V1427,363-367(1991)
See Mang, Thomas S.: V1426,97-110(1991)

Liebst, Brad
; Verbanets, William R.; Kimbrell, James E.: Compact high-accuracy Inductosyn-based gimbal control system,V1482,425-438(1991)

Liessi, Guido See Visona, Adriana: V1425,75-83(1991)

Lietz, J. See Konwerska-Hrabowska, Joanna: V1554A,388-399(1991)

Liew, So K. See Evans, Gary A.: V1418,406-413(1991)

Liew, Soung C.
: Multiwavelength optical switch based on the acousto-optic tunable filter,V1363,57-61(1991)

Lifshitz, N.
; Pinto, Mark R.: Spin-on glasses in the silicon IC: plague or panacea?,V1596,96-105(1991)

Ligachev, Valery A.
; Filikov, V. A.; Gordeev, V. N.: Condensation mechanisms and properties of rf-sputtered a-Si:H,V1519,214-219(1991)

Lightner, David A.
; Person, Richard; Peterson, Blake; Puzicha, Gisbert; Pu, Yu-Ming; Bojadziev, Stefan: Conformational analysis and circular dichroism of bilirubin, the yellow pigment of jaundice,V1432,2-13(1991)

Lightsey, Paul A.
; Anspach, Joel E.; Sydney, Paul F.: Observations of uplink and retroreflected scintillation in the Relay Mirror Experiment,V1482,209-222(1991)

Ligier, Yves See Ratib, Osman M.: V1446,330-340(1991)

Lignon, Dominique See Reynes, Anne M.: V1525,177-182(1991)

Ligomenides, Panos A.
: On-line visual prosthesis for a decision maker,V1382,145-156(1991)
: Structural identity in visual-perceptual recognition,V1382,14-25(1991)
See Liu, Lurng-Kuo: V1606,496-507(1991)
See Teng, Chungte: V1382,74-86(1991)

Li Kam Wa, Patrick
; Park, Choong-Bum; Miller, Alan: All-optical switching of picosecond pulses in GaAs MQW waveguides,V1474,2-7(1991)
See Park, Choong-Bum: V1474,8-17(1991)

Likhanskii, Vladimir V.
See Antyuhov, V.: V1397,355-365(1991); 1415,48-59(1991)
See Apollonova, O. V.: V1501,108-119(1991)
See Bondarenko, Alexander V.: V1376,117-127(1991)

Likhansky, Sergey V. See Alimpiev, Sergei S.: V1503,154-158(1991)

Lilja, J. See Hingerl, Kurt: V1361,943-953(1991)

Liljeborg, Anders See Aslund, Nils R.: V1450,329-337(1991)

Liljegren, Douglas R.
: Defect reduction strategies for process control and yield improvement,V1392,681-687(1991)

Lilley, F. See Hartley, David A.: V1567,277-282(1991)

Lillo, J. See Juri, Hugo: V1422,128-135(1991)

Lim, Art J. See Huang, H. K.: V1443,198-202(1991)

Lim, Eric J.
; Matsumoto, S.; Bortz, M. L.; Fejer, Martin M.: Quasi-phase-matched frequency conversion in lithium niobate and lithium tantalate waveguides,V1561,135-142(1991)

Lim, G.
; Park, Youngsoh; Zmudzinski, C. A.; Zory, Peter S.; Miller, L. M.; Cockerill, T. M.; Coleman, James J.; Hong, C. S.; Figueroa, Luis: Predicting diode laser performance,V1418,123-131(1991)

Lim, Hong-Seh
: Range data from stereo images of edge points,V1382,434-442(1991)

Lim, Siak-Piang See Yak, A. S.: V1401,74-81(1991)

Lim, William Y.
; Eilbert, James L.: Plan-behavior interaction in autonomous navigation,V1388,464-475(1991)

Limmongkol, Siriluk See Forbes, Fred F.: V1532,146-160(1991)

Lin, Anthony T.
: Magnetic field effects on plasma-filled backward-wave oscillators,V1407,234-241(1991)
See Lednum, Eugene E.: V1407,202-208(1991)

Lin, Bochien
; Matthewson, M. J.; Nelson, G. J.: Indentation experiments on silica optical fibers,V1366,157-166(1991)

Lin, Burn J.
: Optimum numerical aperture for optical projection microlithography,V1463,42-53(1991)
: Phase-shifting and other challenges in optical mask technology,V1496,54-79(1991)

Lin, Cheng L.
; Li, Jinghua H.; Zou, Shi C.: Thin film SIMNI material formed by low energy nitrogen implantation and epitaxial growth,V1519,104-108(1991)
See Li, Xiao Q.: V1519,14-17(1991)
See Tan, Hui: V1519,752-756(1991)
See Zhu, Wen H.: V1519,423-427(1991)

Lin, Chi-Wei
; Shulok, Janine R.; Kirley, S. D.; Cincotta, Louis; Foley, James W.: Nile Blue derivatives as lysosomotropic photosensitizers,V1426,216-227(1991)

Lin, Dan See Zhang, Zaixuan: V1572,201-204(1991)

Lin, Dhei-Jhai See Tzeng, Chao H.: V1466,469-476(1991)

Lin, Eleanor
; Lu, Cheng-Chang: Study on objective fidelity for progressive image transmission systems,V1453,206-209(1991)

Lin, Freddie S.
; Chen, Jenkins C.; Nguyen, Cong T.; Liu, William Y.: Engineering fabrication technology of multiplexed hologram in single-mode optical waveguide,V1461,39-50(1991)
; Lu, Taiwei; Kostrzewski, Andrew A.; Chou, Hung: Optical interconnect and neural network applications based on a simplified holographic N4 recording technique,V1558,406-413(1991)
; Strzelecki, Eva M.; Liu, William Y.: Integrated compact matrix-vector multipliers for optical interconnect applications,V1389,642-647(1991)
; Zaleta, David E.; Jannson, Tomasz: Wavelength-sensitive holographic optical interconnects,V1474,45-56(1991)
See Jannson, Tomasz: V1555,159-176(1991)
See Kim, Richard C.: V1474,27-38(1991)
See Liu, William Y.: V1365,20-24(1991)

Lin, Fucheng See Guo, Ting: V1409,24-27(1991)

Lin, H.-B. See Campillo, Anthony J.: V1497,78-89(1991)

Lin, Hao-Hsiung See Wu, Chao-Wen: V1362,768-777(1991)

Lin, Hong Y.
; Yang, Dao M.; Li, Ying X.: Crystallization of hydrogenated amorphous silicon film and its fractal structure,V1519,210-213(1991)

Lin, Hwang-Kuen
See Chu, Ron: V1465,238-243(1991)
See Tzeng, Chao H.: V1466,469-476(1991)

Lin, J. T. See Ren, Qiushi: V1423,129-139(1991)

Lin, J. X. See Fang, X. J.: V1572,279-283(1991)

Lin, Jen T.
See Jones, O. C.: V1557,202-208(1991)
See Paradies, C. J.: V1561,2-5(1991)

Lin, Jin T.
; Gambling, William A.: Polarization effects in fiber lasers: phenomena, theory, and applications,V1373,42-53(1991)

Lin, Jung-Hui See Cooper, Kent: V1392,253-264(1991)

Lin, Li-Huang
; Chen, Shisheng; Jiang, Z. M.; Ge, Wen; Qian, Aidi D.; Bin, Ouyang; Li, Yongchun L.; Kang, Yilan; Xu, Zhizhan: Optical diagnostics of line-focused laser-produced plasmas,V1346,490-501(1991)
; Ge, Wen; Kang, Yilan; Chen, Shisheng; Bin, Ouyang; Li, Shiying; Wang, Shijie: Nd-doped lasers with widely variable pulsewidths,V1410,65-71(1991)
See Li, Shiying: V1410,215-220(1991)

Lin, S. C. See Pappert, Stephen A.: V1476,282-293(1991)

Lin, Shao-An See Chang, Wen-Thong: V1481,242-253(1991)

Lin, Song S. See Tzeng, Chao H.: V1466,469-476(1991)

Lin, Steven H.
See Lang, Robert J.: V1563,2-7(1991)
See Psaltis, Demetri: V1562,204-212(1991)

Lin, T. H.
See Samarabandu, J. K.: V1450,296-322(1991)
See Wang, Ge: V1398,180-190(1991)

Lin, TrueLon
; Jones, E. W.; George, T.; Ksendzov, Alexander; Huberman, M. L.: Advanced Si IR detectors using molecular beam epitaxy,V1540,135-139(1991)

Lin, W. P. See Allan, D. S.: V1560,362-369(1991)

Lin, Wei-Chung
; Tsao, Chen-Kuo; Chen, Chin-Tu; Feng, Yu-Jen: Neural networks for medical image segmentation,V1445,376-385(1991)
See Chen, Chin-Tu: V1444,75-77(1991)
See Chen, Shiuh-Yung: V1445,386-397(1991)
See Chou, Jin-Shin: V1396,45-50(1991)
See Liang, Cheng-Chung: V1445,456-467(1991)
See Neiw, Han-Min: V1445,259-264(1991)

Lin, Wenzheng
; Jiang, Aibao; Zhou, Menzhen: Ratio-telecontrolled strobolume for high-speed photography of depressurized towing tanks,V1358,29-36(1991)

Lin, Xi-Zhang See Sun, Yung-Nien: V1606,653-664(1991)

Lin, Zhi H. See Gao, Jian C.: V1519,159-163(1991)

Linares, Jesus See Gomez-Reino, Carlos: V1332,468-473(1991)

Linchuck, Barry A.
See Dreyer, Donald R.: V1366,372-379(1991)
See Vethanayagam, Thirukumar K.: V1366,343-350(1991)

Lincoln, John See Tropper, Anne C.: V1373,152-157(1991)

Lind, John A. See Cantrell, Chris A.: V1433,263-268(1991)

Lind, M. D.
; Nielsen, K. F.: Crystal growth by solute diffusion in Earth orbit,V1557,259-270(1991)

Lindacher, Joseph M. See Wike, Charles K.: V1398,119-126(1991)

Lindberg, Lars-Goran See Vegfors, Magnus: V1426,79-83(1991)

Lindberg, Perry C.
; Gregory, Don A.: Building an optical pattern recognizer,V1470,220-225(1991)

Lindblom, Joakim F.
; O'Neal, Ray H.; Walker, Arthur B.; Powell, Forbes R.; Barbee, Troy W.; Hoover, Richard B.; Powell, Stephen F.: Performance of the Multi-Spectral Solar Telescope Array IV: the soft x-ray and extreme ultraviolet filters,V1343,544-557(1991)
See DeForest, Craig E.: V1343,404-414(1991)
See Hoover, Richard B.: V1343,189-202(1991)
See Hoover, Richard B.: V1343,175-188(1991)
See Walker, Arthur B.: V1494,320-333(1991)
See Walker, Arthur B.: V1343,319-333(1991)

Lindenbaum, Ella S. See Gottfried, Varda: V1442,218-229(1991)

Linder, Stefan See Patrick, Roger: V1392,506-513(1991)

Lindler, Don See Troeltzsch, John R.: V1494,9-15(1991)

Lindmayer, Joseph
; Goldsmith, Paul; Gross, Kirk: Electron trapping for mass-data-storage memory,V1401,103-112(1991)
See Jutamulia, Suganda: V1401,113-118(1991)

Lindner, Douglas K.
See Reichard, Karl M.: V1480,115-125(1991)
See Reichard, Karl M.: V1489,218-229(1991)

Lindsay, Geoffrey A.
; Nee, Soe-Mie F.; Hoover, James M.; Stenger-Smith, John D.; Henry, Ronald A.; Kubin, R. F.; Seltzer, Michael D.: New syndioregic main-chain, nonlinear optical polymers, and their ellipsometric characterization,V1560,443-453(1991)
; Stenger-Smith, John D.; Henry, Ronald A.; Hoover, James M.; Kubin, R. F.: New class of mainchain chromophoric nonlinear optical polymers,V1497,418-422(1991)

Lindsay, Tracy K.
; Orvek, Kevin J.; Mumaw, Richard T.: 0.50 um contact measurement and characterization,V1464,104-118(1991)

Lindsey, N. E. See Clarke, Timothy A.: V1332,474-486(1991)

Lindstrom, Kjell M.
; Wallin, Bo: Advanced real-time scanning concept for full dynamics recording, high image quality, and superior measurement accuracy,V1488,389-398(1991)

Linford, Gary J.
: Simulations of intracavity laser heating of particles,V1415,196-210(1991)
See Kelley, J. D.: V1415,211-219(1991)

Ling, Daniel T.
; Just, Dieter: Neural networks for halftoning of color images,V1452,10-20(1991)

Ling, Fuyun
; Proakis, John G.; Zhao, Ke.: Systematic treatment of order-recursive least-squares algorithms,V1565,296-306(1991)

Ling, Junda D.
; Yan, Li I.; Liu, YuanQun; Lee, Chi H.; Soos, Jolanta I.: 0.5-GHz cw mode-locked Nd:glass laser,V1454,353-362(1991)

Ling, Ning See Jiang, Wen-Han: V1542,130-137(1991)

Ling, Zhi-Min
; Leang, Sovarong; Spanos, Costas J.: In-line supervisory control in a photolithographic workcell,V1392,660-669(1991)

Lingenauer, Marion See Kupfer, Roland: V1598,46-60(1991)

Lingle, Robert See Hopkins, John B.: V1432,221-226(1991)

Lingua, Robert W.
; Parel, Jean-Marie; Simon, Gabriel; Li, Kam: Photodynamic treatment of lens epithelial cells for cataract surgery,V1423,58-61(1991)

Linholm, Loren See Troccolo, Patrick M.: V1464,90-103(1991)

Link, Kerry M. See Logenthiran, Ambalavaner: V1452,225-243(1991)

Linn, Robert J.
; Hall, David L.; Llinas, James: Survey of multisensor data fusion systems,V1470,13-29(1991)

Linnainmaa, Seppo See Vepsalainen, Ari M.: V1568,2-13(1991)

Linney, Alfred D.
; Moss, James P.; Richards, Robin; Mosse, C. A.; Grindrod, S. R.; Coombes, Anne M.: Use of a 3-D visualization system in the planning and evaluation of facial surgery,V1380,190-199(1991)
See Coombes, Anne M.: V1380,180-189(1991)

Linsky, J. B. See Rosenblatt, Edward I.: V1449,72-86(1991)

Linton, R. S. See Haigh, N. R.: V1366,259-264(1991)

Linzer, Elliot See Feig, Ephraim: V1452,458-467(1991)

Lion, Yves F.
See Defosse, Yves: V1507,277-287(1991)
See Leclere, Philippe: V1507,339-344(1991)
See Leclere, Philippe: V1559,298-307(1991)

Liou, James J. See Rao, Kashi: V1482,293-306(1991)

Liou, Shih-Ping
; Jain, Ramesh C.: Parallel algorithm for volumetric segmentation,V1381,447-458(1991)

Lipari, Charles A.
: Geometric modeling of noisy image objects,V1468,905-917(1991)

Lipovskaya, Margarita J. See Kozhevnikov, Nikolai M.: V1507,517-524(1991)

Lipp, John I. See Namazi, Nader M.: V1567,659-669(1991)

Lippens, C. See De Maziere, Martine M.: V1491,288-297(1991)

Lippens, D. See Kibeya, Saidi: V1501,97-106(1991)

Lippert, Thomas M. See Assenheim, Harry M.: V1456(1991)

Lippey, Barret
; Krone-Schmidt, Wilfried: Cryoscatter measurements of beryllium,V1530,150-161(1991)

Lipscomb, George F.
; Lytel, Richard S.; Ticknor, Anthony J.; Van Eck, Timothy E.; Girton, Dexter G.; Ermer, Susan P.; Valley, John F.; Kenney, John T.; Binkley, E. S.: Organic electro-optic devices for optical interconnnection,V1560,388-399(1991)
See Lytel, Richard S.: V1563,122-138(1991)
See Lytel, Richard S.: V1389,547-558(1991)
See Wu, Jeong W.: V1560,196-205(1991)

Lipsky, S. See Even-Or, Baruch: V1371,161-169(1991)

Lipson, David
; McLeaster, Kevin D.; Cohn, Brian; Fischer, Robert E.: Drilled optical fiber sensors: a novel single-fiber sensor,V1368,36-43(1991)

Lipson, Steven G. See Devir, Adam D.: V1442,347-359(1991)

Lipton, Lenny
: Selection devices for field-sequential stereoscopic displays: a brief history,V1457,274-282(1991)

Lischka, Klaus See Gresslehner, Karl-Heinz: V1361,1087-1093(1991)

Liseth, Ole J.
; Bezdek, James C.; Ford, Kenneth M.; Adams-Webber, Jack R.: New techniques for repertory grid analysis,V1468,256-267(1991)

Lisi, Nicola
See Bollanti, Sarah: V1503,80-87(1991)
See Bollanti, Sarah: V1397,97-102(1991)

Lissak, Boaz See Keselbrener, Michel: V1415,38-47(1991)

Lit, John W. See Zhang, Feng: V1588,100-109(1991)

Little, Gordon R. See Gustafson, Steven C.: V1469,268-274(1991)

Little, James J. See Hakkarainen, J. M.: V1473,173-184(1991)

Little, William R.
: Aerospace resource document: fiber optic interconnection hardware,V1589,20-23(1991)

Littlejohn, Michael A.
See Grider, Douglas T.: V1393,229-239(1991)
See Ozturk, Mehmet C.: V1393,260-269(1991)
See Stroscio, Michael A.: V1362,566-579(1991)

Littlewood, Paul A. See Cubbage, Robert W.: V1363,163-171(1991)

Littwitz, Brigitte See Wand, Michael D.: V1455,97-104(1991)

Litvak, Alexandr M. See Charykov, N. A.: V1512,198-203(1991)

Litvak, Marvin M.
; Wagner, Richard J.: Stimulated Brillouin scatter SIX-code,V1415,62-71(1991)

Litvin, K. See Chou, Chia-Shing: V1475,151-156(1991)

Litvin, S. A. See Zhavoronok, I. V.: V1554A,371-379(1991)

Litvinenko, E. G. See Artjushenko, Vjacheslav G.: V1420,157-168(1991)

Litvinenko, Vladimir N.
; Madey, John M.; Benson, Stephen V.; Burnham, B.; Wu, Y.: Duke storage ring FEL program,V1552,2-6(1991)

Litwin, Todd See Bon, Bruce: V1387,337-342(1991)

Litz, Marc S.
; Huttlin, George A.; Lazard, Carl J.; Golden, Jeffry; Pereira, Nino R.; Hahn, Terry D.: Comparison of negative and positive polarity reflex diode microwave source,V1407,159-166(1991)
See Huttlin, George A.: V1407,147-158(1991)
See Lazard, Carl J.: V1407,167-171(1991)

Liu, Ansheng
; Keller, Ole: Ellipsometric studies of the optical anisotropy of GdBa2Cu3O7-x epitaxial films,V1512,226-231(1991)

Liu, B. See Wengler, Michael J.: V1477,209-220(1991)

Liu, Bai Y.
See Wong, H.: V1519,494-498(1991)
See Yang, Bing L.: V1519,241-246(1991)
See Yang, Bing L.: V1519,269-274(1991)

Liu, Bo
; Deng, Xingzhong; Cao, Wei; Cheng, Xianping; Xie, Tuqiang; Zhong, Zugen: Fiber optic temperature probe system for inner body,V1572,211-215(1991)
See Cheng, Xianping: V1572,216-219(1991)
See Deng, Xingzhong: V1572,220-223(1991)

Liu, C. H. See Pradhan, Asima: V1425,2-5(1991)

Liu, C. S. See Hsu, C. T.: V1519,391-395(1991)

Liu, C. Y. See Chan, W. K.: V1399,82-89(1991)

Liu, Chang
See Fan, Xiang J.: V1519,808-812(1991)
See Fan, Xiang J.: V1519,805-807(1991)

Liu, Changjun See Liu, Dingyu: V1443,191-196(1991)

Liu, Cheng
; Rosakis, Ares J.: Interpretation of optical caustic patterns obtained during unsteady crack growth: an analysis based on a higher-order transient expansion,V1554A,814-825(1991)

Liu, Chen Z. See Jiang, Jie: V1519,717-724(1991)

Liu, D. See Wang, Xianhua: V1358,775-779(1991)

Liu, Dan See Hong, Xiao-Yin: V1466,546-557(1991)

Liu, Dayou
; Zheng, Fangqing; Ma, Zhifang; Shi, Qiaotin: Conflict resolution in multi-ES cooperation systems*,V1468,37-49(1991)

Liu, Dingquan See Wang, Tianmin: V1519,890-900(1991)

Liu, Dingyu
; Yang, Xiaobo; Liu, Changjun; Zhang, Hongguo: Color analysis of nonlinear-phase-modulation method for density pseudocolor encoding technique in medical application,V1443,191-196(1991)

Liu, Duncan T.
; Cheng, Li-Jen; Luke, Keung L.: Optical processing using photorefractive GaAs and InP,V1409,116-126(1991)

Liu, Guang T. See Wen, Mei-Yuan: V1332,673-675(1991)

Liu, Guidng See Wang, Tianmin: V1519,890-900(1991)

Liu, Hua-Kuang See Casasent, David P.: V1555,23-33(1991)

Liu, Iching
; Sun, Ying: Rule-based system to reconstruct 3-D tree structure from two views,V1606,67-77(1991)

Liu, Jia-ming See Chee, Joseph K.: V1413,14-20(1991)

Liu, Jian
; Bao, Xue-Cheng; Zhang, Cai-Gen; Zhang, You-Wen: Universal equation for IR thermometer and its applications,V1540,744-755(1991)

Liu, Jian-Hua See Wang, Jianshe: V1533,175-184(1991)

Liu, Jinsheng See Qu, Zhi-Min: V1238,406-411(1991)

Liu, K. See Wysocki, Paul F.: V1373,234-245(1991)

Liu, Ke
; Quan, Jun; Yang, Jing-Yu; Cheng, Yong-Qing: Identification and restoration of images with out-of-focus blurs,V1567,720-728(1991)

Liu, Kei-Lee See Morris, Michael D.: V1439,95-101(1991)

Liu, Kexing
: Design of optically activated conventional sensors and actuators,V1398,269-275(1991)
; Ferguson, Suzanne M.; Davis, Andrew; McEwen, Keith; Measures, Raymond M.: Ultrasonic NDE (nondestructive evaluation) for composite materials using embedded fiber optic interferometric sensors,V1398,206-212(1991)
; Measures, Raymond M.: Detection of high-frequency elastic waves with embedded ordinary single-mode fibers,V1584,226-234(1991)
See Davis, Andrew: V1489,33-43(1991)
See Davis, Andrew: V1588,264-274(1991)
See Measures, Raymond M.: V1489,86-96(1991)
See Measures, Raymond M.: V1332,421-430(1991)
See Melle, Serge M.: V1588,255-263(1991)
See Zuliani, Gary: V1588,308-313(1991)

Liu, Kou-Chen See Sun, Tai-Ping: V1361,1033-1037(1991)

Liu, Li M. See Li, Chang J.: V1519,779-787(1991)

Liu, Li Y. See Zheng, Tian S.: V1519,339-346(1991)

Liu, Louis
; Jones, William L.; Carandang, R.; Lam, Wayne W.; Yonaki, J.; Streit, Dwight C.; Kasody, R.: Phased-array receiver development using high-performance HEMT MMICs,V1475,193-198(1991)

Liu, Lurng-Kuo
; Ligomenides, Panos A.: Less interclass disturbance learning for unsupervised neural computing,V1606,496-507(1991)

Liu, Maili See He, Da-Ren: V1362,696-701(1991)

Liu, Pao-Lo
: Photonic computer-aided design tools for high-speed optical modulators,V1371,46-55(1991)

Liu, Po-Hsin P.
See Chow, P. D.: V1475,48-54(1991)
See Minot, Katcha: V1475,309-313(1991)

Liu, Pu L.
See Liu, Wei J.: V1519,415-418(1991)
See Liu, Wei J.: V1519,481-488(1991)

Liu, Qihai See He, Da-Ren: V1362,696-701(1991)

Liu, Rui-Fu
; Jiang, G.; Jiang, H. Y.; Xi, Xiao-chun; Song, W. S.: Reflective fiber temperature sensor using a bimetallic transducer,V1572,189-191(1991)
; Xi, Xiao-chun; Li, Wei-min; Tian, Da-chao: Fiber optic temperature sensor based on bend losses,V1572,180-184(1991)
; Xi, Xiao-chun; Liang, Chen: Optical fiber densimeter for water in oil,V1572,399-402(1991)

Liu, Shih-Chii
; Harris, John G.: Negative fuse network,V1473,185-193(1991)
See Mathur, Bimal P.: V1473,153-160(1991)

Liu, Shing G. See Taylor, Gordon C.: V1475,103-112(1991)

Liu, Shu Q. See Zhao, Shu L.: V1519,275-280(1991)

Liu, Shutian
See Liu, Yudong: V1362,436-447(1991)
See Mao, Xianjun: V1563,58-63(1991)

Liu, Tsann-Shyong See Lee, Ching-Long: V1606,728-734(1991)

Liu, Tsung-Ching
; Van Veen, Barry D.: Modular implementations of linearly constrained beamformers,V1566,419-426(1991)

Liu, W. C.
; Lour, W. S.; Sun, C. Y.; Lee, Y. S.; Guo, D. F.: Application of AlGaAs/GaAs superlattice for negative-differential-resistance transistor,V1519,670-674(1991)
; Sun, C. Y.; Lour, W. S.; Guo, D. F.; Lee, Y. S.: Applications of GaAs grade-period doping superlattice for negative-differential-resistance device,V1519,640-644(1991)

Liu, Wei G.
: Specific properties of ferroelectric thin films,V1519,704-706(1991)

Liu, Wei J.
; Liu, Pu L.; Shi, Guo L.; Zhu, Jing-Bing; Yuan, Shi X.; Xie, Qin X.; He, Li: Magneto-optical studies of n-type Hg0.622Cd0.378Te grown by molecular beam epitaxy,V1519,415-418(1991)
; Liu, Pu L.; Shi, Guo L.; Zhu, Jing-Bing; He, Li; Xie, Qin X.; Yuan, Shi X.: Optical characterization of Hg1-xCdxTe/CdTe/GaAs multilayers grown by molecular beam epitaxy,V1519,481-488(1991)

Liu, Wei Y.
; Li, Yu M.; Cui, Zhan J.; He, Tian X.: Novel cathodic arc plasma PVD system with column target for the deposition of TiN film and other metallic films,V1519,172-174(1991)

Liu, Wen-Qing
; Jiang, Rong-Xi; Wang, Ya-Ping; Xia, Yu-Xing: Experimental study of the laser retroreflection of various surfaces,V1530,240-243(1991)

Liu, Wi See Zhong, Di S.: V1519,350-358(1991)

Liu, William Y.
; Strzelecki, Eva M.; Lin, Freddie S.; Jannson, Tomasz: High-resolution integrated optic holographic wavelength division multiplexer,V1365,20-24(1991)
See Lin, Freddie S.: V1461,39-50(1991)
See Lin, Freddie S.: V1389,642-647(1991)

Liu, Xiande
See Pan, Yingtain: V1572,477-482(1991)
See Xiong, Zhengjun: V1467,410-415(1991)

Liu, Xiang H.
See Feng, Yi P.: V1519,440-443(1991)
See Wang, Xi: V1519,740-743(1991)

Liu, Xiang N. See Yuan, Ren K.: V1519,831-834(1991)

Liu, Xianjun
; Ansari, Nirwan: Landmark-based partial shape recognition by a BAM neural network,V1606,1069-1079(1991)

Liu, Xiouqin
See Chang, Zenghu: V1358,614-618(1991)
See Chang, Zenghu: V1358,541-545(1991)
See Hou, Xun: V1358,868-873(1991)

Liu, Xu
; Tang, Jinfa; Pelletier, Emile P.: Analysis of thin-film losses from guided wave attenuations with photothermal deflection technique,V1554A,558-569(1991)
See Mei, Ting: V1554A,570-578(1991)

Liu, Y.
See Olsen, Gregory H.: V1419,24-31(1991)
See Wang, LeMing: V1431,97-101(1991)

Liu, Yanbing
; Zhang, Jinru: Phase compensation of PZT in an optical fibre Mach-Zehnder interferometer,V1572,61-64(1991)

Liu, Yashu See Wang, Yuzhu: V1501,40-48(1991)

Liu, Yee
See Kaaret, Philip E.: V1548,106-117(1991)
See Shaw, Ping-Shine: V1343,485-499(1991)
See Shaw, Ping-Shine: V1548,118-131(1991)

Liu, Yi-Hwa
; Holmes, Timothy J.; Koshy, Matthew: Preliminary tests of maximum likelihood image reconstruction method on 3-D real data and some practical considerations for the data corrections,V1428,191-199(1991)

Liu, Ying
; Ma, Hede: Pattern recognition using w-orbit finite automata,V1606,226-240(1991)
; Ma, Shining; Du, Fuli; Peng, Xiang; Ye, Shenhua: Study of measuring skin blood flow using speckle counting,V1554A,610-612(1991)
See Ma, Shining: V1554A,645-648(1991)

Liu, Yipeng See Caglar, Hakan: V1605,86-94(1991)

Liu, Yong
; Zakhor, Avideh: Binary and phase-shifting image design for optical
lithography,V1463,382-399(1991)

Liu, YuanQun See Ling, Junda D.: V1454,353-362(1991)

Liu, Yudong
; Liu, Shutian; Li, Chunfei; Shen, Dezen; Fan, Xi W.; Fan, Guang H.; Chen,
Lian C.: Optical nonlinearities in ZnSe multiple-quantum-
wells,V1362,436-447(1991)

Liu, Yue See Chou, Stephen Y.: V1474,243-247(1991)

Liu, Yueping
; Sibbett, Wilson: Profile-related time resolution for a femtosecond streak
tube,V1358,503-510(1991)
; Sibbett, Wilson; Walker, David R.: Theoretical and experimental
performance evaluations of Picoframe framing cameras,V1358,300-
310(1991)

Liu, Yung S.
; Levinson, R. M.; Rose, J. W.: Novel excimer laser beam delivery technique
using binary masks,V1377,126-133(1991)

Liu, Yupin See Gao, Yiqing: V1554B,193-199(1991)

Liu, Yu X. See Yuan, Ren K.: V1519,396-399(1991)

Liu, Zhi-Shen
; Ma, Jun; Zhang, Jin-Long; Chen, Wen-Zhong: New, simple method of
extracting temperature of liquid water from Raman scattering,V1558,306-
316(1991)
; Zhang, Jin-Long; Chen, Wen-Zhong; Huang, Xiao-Sheng; Ma, Jun: Raman
lidar for measuring backscattering in the China Sea,V1558,379-
383(1991)

Liu, Zhili
: Multitarget adaptive gate tracker with linear prediction,V1482,285-
292(1991)

Liu, Zhiyuan See He, Da-Ren: V1362,696-701(1991)

Livezey, Darrell L. See Hong, C. S.: V1418,177-187(1991)

Livi, S.
; Magnani, M.; Pieri, Silvano; Romoli, Andrea: Optical system for automatic
inspection of curved surfaces,V1500,144-150(1991)

Livingston, John M. See Uthe, Edward E.: V1487,228-239(1991)

Livingston, William C.
: Proposed conversion of the McMath Telescope to 4.0-meter aperture for
solar observations in the IR,V1494,498-502(1991)
; Wallace, Lloyd V.: Spectroscopic observations of CO2, CH4, N2O, and CO
from Kitt Peak, 1979-1990,V1491,43-47(1991)

Livnat, Ami See Keren, Eliezer: V1442,266-274(1991)

Ljungberg, Sven-Ake See Lyberg, Mats D.: V1467,104-115(1991)

Llacer, Jorge See Coakley, Kevin J.: V1443,226-233(1991)

Llario, Vicenc See Frau, Juan: V1388,329-340(1991)

Llewellyn, Steven A.
: Nd:YAG laser market,V1517,85-99(1991)

Llinas, James See Linn, Robert J.: V1470,13-29(1991)

Lloyd, Ashley D. See Wang, Chang H.: V1505,130-140(1991)

Lloyd, James R.
: Electromigration physical modeling of failure in thin film
structures,V1596,106-117(1991)

Lloyd-Hart, Michael See Wizinowich, Peter L.: V1542,148-158(1991)

Lo, Ben See Horii, Steven C.: V1446,475-480(1991)

Lo, G. Q.
See Joshi, A. B.: V1393,122-149(1991)
See Peyton, David: V1393,295-308(1991)

Lo, M. P. See Chen, Terry Y.: V1554B,92-98(1991)

Lo, Shih-Chung B.
; Krasner, Brian H.; Mun, Seong Ki; Horii, Steven C.: Full-frame entropy
coding for radiological image compression,V1444,265-271(1991)

Lo, Yee-Man V.
: Space-time system architecture for the neural optical
computing,V1382,199-208(1991)

Lobacheva, M. I. See Kozlova, T. G.: V1403,159-160(1991)

Lobanov, O. V. See Chernyaeva, E. B.: V1402,7-10(1991)

Lobb, Daniel R.
See Baudin, Gilles: V1493,16-27(1991)
See Baudin, Gilles: V1490,102-113(1991)

Lober, K. See Bauer, Dietrich: V1533,277-285(1991)

Locascio-Brown, Laurie See Choquette, Steven J.: V1368,258-263(1991)

Loce, Robert P.
; Dougherty, Edward R.: Using structuring-element libraries to design
suboptimal morphological filters,V1568,233-246(1991)
See Jodoin, Ronald E.: V1398,61-70(1991)

Locker, M. H. See Blatherwick, Ronald D.: V1491,203-210(1991)

Lockwood, Harry F.
: Hybrid wafer scale optoelectronic integration,V1389,55-67(1991); 1390,55-
67(1991)

Lodder, Herman
; van Poppel, Bas M.; Bakker, Albert R.: Prefetching: PACS image
management optimization using HIS/RIS information,V1446,227-
233(1991)

Loe, K. F. See Hsu, L. S.: V1469,197-207(1991)

Loeffler, Mary F. See Salter, Jeffery L.: V1555,268-278(1991)

Loeffler, Roland See Menke, Bodo: V1417,316-327(1991)

Loew, Murray H.
ed.: Medical Imaging V: Image Processing,V1445(1991)
See Anderson, Mary P.: V1444,407-412(1991)
See Zhang, Ya-Qin: V1445,358-366(1991)
See Zhang, Ya-Qin: V1445,367-373(1991)

Loewen, Erwin G. See Hoose, John F.: V1545,189-199(1991)

Logak, L. G.
; Fassakhova, H. H.; Antonova, N. E.; Minina, L. A.; Gainutdinov, R. K.:
Ultra-fine-grain silver-halide photographic materials for holography on
the flexible film base,V1238,171-176(1991)

Logan, John A. See Karis, Thomas E.: V1499,366-376(1991)

Logan, Pamela See Cadou, Christopher P.: V1434,67-77(1991)

Logan, Ralph A. See Temkin, Henryk: V1418,88-98(1991)

Logenthiran, Ambalavaner
; Snyder, Wesley E.; Santago, Peter; Link, Kerry M.: MAP segmentation of
magnetic resonance images using mean field annealing,V1452,225-
243(1991)

LoGerfo, Paul See Bass, Lawrence S.: V1422,123-127(1991)

Logozinskii, Valerii N. See Semenov, Alexandr T.: V1584,348-352(1991)

Lohrenz, Maura C. See Trenchard, Michael E.: V1456,318-326(1991)

Loi, K. K.
See Pun, Edwin Y.: V1583,102-108(1991)
See Pun, Edwin Y.: V1583,64-70(1991)

Lok, Y. F. See Young, Donald S.: V1416,221-228(1991)

Lokai, Peter See Kahlert, Hans-Juergen: V1463,604-609(1991)

Lokberg, Ole J.
: Biomedical applications of video-speckle techniques,V1525,9-16(1991)
; Ellingsrud, Svein; Vikhagen, Eiolf: TV holography and image processing in
practical use,V1332,142-150(1991)
; Rosvold, Geir O.; Malmo, Jan T.; Ellingsrud, Svein: Computerized vibration
analysis of hot objects,V1508,153-160(1991)
See Ellingsrud, Svein: V1399,30-41(1991)

Lokerson, Donald See Hayden, William L.: V1417,182-199(1991)

Loloyan, Mansur
See Hayrapetian, Alek: V1446,243-247(1991)
See Huang, H. K.: V1443,198-202(1991)
See Kaabi, Lotfi: V1445,11-23(1991)
See Lou, Shyh-Liang: V1446,302-311(1991)
See Weinberg, Wolfram S.: V1446,430-435(1991)

Lombard, Calvin M. See Wells, James A.: V1485,2-12(1991)

Lombardi, Loreto See Guadagni, Stefano: V1420,178-182(1991)

Lomheim, Terrence S.
See Chambers, Robert J.: V1488,312-326(1991)
See Johnson, Jerris F.: V1541,110-126(1991)

Lompre, Louis A. See Andre, Michel L.: V1502,286-298(1991)

Londo, T. R. See Barisas, B. G.: V1432,52-63(1991)

Londono, Carmina
; Clark, Peter P.: Design of achromatized hybrid diffractive lens
systems,V1354,30-37(1991)
See Clark, Peter P.: V1354,555-569(1991)

Loney, Gregory C.
: Design and performance of a small two-axis high-bandwidth steering
mirror,V1454,198-206(1991)

Long, Allan C. See Pausch, Randy: V1457,198-208(1991)

Long, Greg See Tanaka, Satoru C.: V1448,21-26(1991)

Long, J. A.
; Manousos, S. N.: Use of a knowledge-based system for the valuation of unlisted shares,V1468,446-454(1991)

Long, James See Zucker, Oved S.: V1378,22-33(1991)

Long, Jerry E.
; Hills, Louis S.; Gebhardt, Frederick G.: Realistic wind effects on turbulence and thermal blooming compensation,V1408,58-71(1991)
See Hills, Louis S.: V1408,41-57(1991)

Long, Kah W.
; Hung, Yau Y.; Der Hovanesian, Joseph: Measurement of residual stresses in plastic materials by electronic shearography,V1554A,116-123(1991)

Long, Margaret See Wong, Ka-Kha: V1374,278-286(1991)

Long, Pin
; Hsu, Dahsiung: Quantization and sampling considerations of computer-generated hologram for optical interconnection,V1461,270-277(1991)
See Hsu, Dahsiung: V1238,13-17(1991)

Long, Theresa W.
See Salmon, J. T.: V1542,459-467(1991)
See Toeppen, John S.: V1544,218-225(1991)

Long, Treva
; Rodriguez, Ferdinand: Dissolution of poly(p-hydroxystyrene): molecular weight effects,V1466,188-198(1991)

Longbotham, Harold G.
; Barsalou, Norman: Class of GOS (generalized order-statistic) filters, similar to the median, that provide edge enhancement,V1451,36-47(1991)

Longo, M. See Antonuk, Larry E.: V1443,108-119(1991)

Longo, Savino See Capitelli, Mario: V1503,126-131(1991)

Longoni, Antonio See Braueninger, Heinrich: V1549,330-339(1991)

Longshore, Randolph E.
; Baars, Jan W.; eds.: *Growth and Characterization of Materials for Infrared Detectors and Nonlinear Optical Switches*,V1484(1991)

Longworth, James W. See Haddad, Waleed S.: V1448,81-88(1991)

Lontz, John F.
See Ashton, Robert C.: V1422,151-155(1991)
See Bass, Lawrence S.: V1421,164-168(1991)

Looney, Larry D.
; Lyons, Peter B.: Measurement of radiation-induced attenuation in optical fibers by optical-time-domain reflectometry,V1474,132-137(1991)

Loosen, Peter
See Du, Keming: V1397,639-643(1991)
See Franek, Joachim: V1397,791-795(1991)
See Habich, Uwe: V1397,383-386(1991)

Lopasov, V. P. See Burakov, V. S.: V1440,270-276(1991)

Lopez, Adolphe
; Baniel, P.; Gall, Pascal; Granier, Jean E.: Scattering properties of ZrF4-based glasses prepared by the gas film levitation technique,V1590,191-202(1991)

Lopez, C.
; Meseguer, Francisco; Sanchez-Dehesa, Jose; Ruehle, Wolfgang W.; Ploog, Klaus H.: Time-resolved luminescence experiments on modulation n-doped GaAs quantum wells,V1361,89-95(1991)
See Blanco, J.: V1508,180-190(1991)

Lopez-Amo, Manuel
; Menendez-Valdes, Pedro; Sanz, Inmaculada; Muriel, Miguel A.: Design of novel integrated optic devices utilizing depressed index waveguides,V1374,74-85(1991)
See Lopez-Higuera, Jose M.: V1374,144-151(1991)

Lopez-Higuera, Jose M.
; Lopez-Amo, Manuel; Muriel, Miguel A.: Novel technique for analysis and design of diffused Ti:LiNbO3 optical planar waveguides,V1374,144-151(1991)

Lopez Famozo, C. See Tentori, Diana: V1535,209-215(1991)

Lord, Jeffrey R. See Spillman, William B.: V1367,197-203(1991)

Loree, Thomas R. See Bigio, Irving J.: V1403,776-780(1991)

Lorenz, Gerhard See Kretschmer, K.-H.: V1392,246-252(1991)

Lorenz, M. See Becker, J. S.: V1598,227-238(1991)

Lorenzen, Manfred See Campbell, Duncan R.: V1400(1991)

Lorenzo, Maximo
; Der, Sandor Z.; Moulton, Joseph R.: Information-theoretic approach to optimal quantization,V1483,118-137(1991)
See Walters, Clarence P.: VIS07,181-201(1991)

Loria, A. See Barresi, Giangrande: V1495,246-258(1991)

Lorsignol, Jean
; Hollier, Pierre A.; Deshayes, Jean-Pierre: Polarization and directionality of the Earth's reflectances: the POLDER instrument,V1490,155-163(1991)

Losacco, Aurora M.
See Antuofermo, Pasquale: V1583,143-149(1991)
See Battaglin, Giancarlo: V1513,441-450(1991)

Loschialpo, Peter See Golden, Jeffry: V1407,418-429(1991)

Loshchenov, V. B.
; Baryshev, M. V.; Svystushkin, V. M.; Ovchinnikov, U. M.; Babyn, A. V.; Schaldach, Brita J.; Mueller, Gerhard J.: Spectral-luminescence analysis as method of tissue states evaluation and laser influence on tissues before and after transplantation,V1420,271-281(1991)

Lotem, Haim
See Lando, Mordechai: V1412,19-26(1991)
See Lando, Mordechai: V1442,172-180(1991)

Lotkova, E. N. See Artsimovich, M. V.: V1397,729-733(1991)

Lou, Da-li See Wang, He-Chen: V1386,273-276(1991)

Lou, Michael C.
See MacNeal, Paul D.: V1540,68-85(1991)
See MacNeal, Paul D.: V1494,236-254(1991)
See Satter, Celeste M.: V1494,279-300(1991)
See Shih, Choon-Foo: V1532,81-90(1991)

Lou, Pei-De See Zou, Zi-Li: V1572,18-26(1991)

Lou, Qihong
: Gas flow effects in pulse avalanche discharge XeCl excimer laser,V1397,103-106(1991)
; Guo, Hongping: Cumulative effect and cutting quality improvement of XeCl laser ablation of PMMA,V1598,221-226(1991)

Lou, Shyh-Liang
; Loloyan, Mansur; Weinberg, Wolfram S.; Valentino, Daniel J.; Lufkin, Robert B.; Hanafee, William; Bentson, John R.; Jabour, Bradly; Huang, H. K.: Image delivery performance of a CT/MR PACS module applied in neuroradiology,V1446,302-311(1991)
See Stewart, Brent K.: V1446,141-153(1991)
See Wong, Albert W.: V1446,73-80(1991)

Lou, Wei G. See Zhou, Shi P.: V1519,793-799(1991)

Loubriel, Guillermo M.
; Buttram, Malcolm T.; Helgeson, Wes D.; McLaughlin, Dan L.; O'Malley, Marty W.; Zutavern, Fred J.; Rosen, Arye; Stabile, Paul J.: Triggering GaAs lock-on switches with laser diode arrays,V1378,179-186(1991)
See Rosen, Arye: V1378,187-194(1991)
See Zutavern, Fred J.: V1378,271-279(1991)

Loudin, Jeffrey A.
See Hudson, Tracy D.: V1564,54-64(1991)
See Kirsch, James C.: V1474,90-100(1991)
See Kirsch, James C.: V1558,432-441(1991)
See Kirsch, James C.: V1482,69-78(1991)

Loughlin, Patrick See Atlas, Les E.: V1566,134-143(1991)

Louis, Eric See Turcu, I. C.: V1503,391-405(1991)

Lour, W. S.
See Liu, W. C.: V1519,670-674(1991)
See Liu, W. C.: V1519,640-644(1991)

Lourtie, Pedro M. See Tavora, Jose: V1468,524-535(1991)

Louvet, Y.
See Bacis, Roger: V1397,173-176(1991)
See Barnault, B.: V1397,231-234(1991)

Love, G. D. See Powell, Norman J.: V1545,19-30(1991)

Love, John C. See Norbury, David H.: V1463,558-573(1991)

Love, Walter F. See Walczak, Irene M.: V1420,2-12(1991)

Lovejoy, Shaun
See Davis, Anthony B.: V1558,37-59(1991)
See Lavallee, Daniel: V1558,60-75(1991)

Lovejoy, Steven M. See Ermer, Susan P.: V1560,120-129(1991)

Lovell, Robb E. See Uttal, William: V1453,258-269(1991)

Low, J. P. See Subramanian, K. R.: V1364,190-201(1991)

Low, Toh-Siew See Yak, A. S.: V1401,74-81(1991)

Lowdermilk, W. H. See Murray, John R.: V1410,28-39(1991)

Lowe, David
: What have neural networks to offer statistical pattern processing?,V1565,460-471(1991)

Lowe, Walter P.
    See Clarke, Roy: V1361,2-12(1991)
    See Clarke, Roy: V1345,101-114(1991)
    See Rodricks, Brian G.: V1550,18-26(1991)

Lowery, Joseph A. See Cartlidge, Andy G.: V1423,167-174(1991)

Lowndes, Douglas H.
    ; Norton, D. P.; Budai, J. D.; Christen, D. K.; Klabunde, C. E.; Warmack, R. J.;
        Pennycook, Stephen J.: Growth and transport properties of Y-Ba-Cu-O/Pr-
        Ba-Cu-O superlattices,V1394,150-160(1991)

Lowry, Heard S.
    ; Elrod, Parker D.; Johnson, R. J.: AEDC direct-write scene generation test
        capabilities,V1454,453-464(1991)

Lowry, Jay H.
    ; Iffrig, C. D.: Radiation effects on various optical components for the Mars
        Observer Spacecraft,V1330,132-141(1991)
    ; Mendlowitz, Joseph S.; Subramanian, N. S.: Optical characteristics of the
        Teflon AF fluoro-plastic materials,V1330,142-151(1991)

Lowry, Steven R. See Herget, William F.: V1433,275-289(1991)

Lozes, Richard See Dunbrack, Steven K.: V1464,314-326(1991)

Lozovoi, V. I.
    See Briscoe, Dennis: V1358,329-336(1991)
    See Prochazka, Ivan: V1358,574-577(1991)

Lu, Baolong See Wang, Yuzhu: V1501,40-48(1991)

Lu, Bo-kao See Wang, Rui-zhong: V1421,203-207(1991)

Lu, Boyin See Liang, Dawei: V1511,90-97(1991)

Lu, Cheng-Chang See Lin, Eleanor: V1453,206-209(1991)

Lu, Dong See Yang, Yong: V1519,444-448(1991)

Lu, Dong S.
    See An, Cheng W.: V1519,818-821(1991)
    See Fan, Yong C.: V1519,775-778(1991)
    See Fan, Yong C.: V1519,813-817(1991)

Lu, Dunwu See Zhang, Yudong: V1463,456-463(1991)

Lu, Eh
    ; Yang, Hongbo; Meng, Qinglai; Han, Rong-jiu: Stability analysis on 3-axes
        servo revolution pedestal system,V1533,155-162(1991)

Lu, Feizhen See Li, Shaohui: V1572,543-547(1991)

Lu, Grant See Gray, Kevin J.: V1534,60-66(1991)

Lu, Hong-Qian
    : Quantitative evaluation of image enhancement algorithms,V1453,223-
        234(1991)

Lu, Huanzhang
    ; Sun, Zhong-kang: Precision tracking of small target in IR
        systems,V1481,398-405(1991)

Lu, Huey T.
    See Chen, Ray T.: V1559,110-117(1991)
    See Chen, Ray T.: V1583,362-374(1991)
    See Chen, Ray T.: V1583,135-142(1991)

Lu, Jian-Feng
    ; Ni, Xiao W.; Wang, Hong Y.; He, Anzhi: Interferometric measurement for
        the plasma produced by Q-switched laser in air near the surface of an Al
        Target,V1554B,593-597(1991)
    ; Wang, Chang Xing; Miao, Peng-Cheng; Ni, Xiao W.; He, Anzhi:
        Interferential diagnosis of self-focusing of Q-switched YAG laser in
        liquid,V1415,220-224(1991)
    See He, Anzhi: V1554A,547-552(1991)
    See He, Anzhi: V1423,117-120(1991)
    See Ni, Xiao W.: V1554B,632-635(1991)
    See Ni, Xiao W.: V1527,437-441(1991)
    See Ni, Xiao W.: V1519,365-369(1991)

Lu, Jiuhuai
    ; Hansen, Eric W.; Gazzaniga, Michael S.: Brain surface maps from 3-D
        medical images,V1459,117-124(1991)

Lu, Kaichang
    ; Zhu, Yafei; Kang, Songgao: New type of large-angle binocular
        microtelescopes,V1527,413-418(1991)
    See Kang, Songgao: V1527,376-379(1991)
    See Kang, Songgao: V1527,409-412(1991)
    See Kang, Songgao: V1527,400-405(1991)
    See Kang, Songgao: V1527,406-408(1991)

Lu, Kevin W. See Menendez, Ronald C.: V1363,97-105(1991)

Lu, Kuang-sheng See Wang, Rui-zhong: V1421,203-207(1991)

Lu, Liang-ju See Roberts, Harold A.: V1363,62-69(1991)

Lu, M. See Zhai, H. R.: V1519,575-579(1991)

Lu, P. See Zhao, Jing-Fu: V1362,135-143(1991)

Lu, Q. See Larkins, Grover L.: V1477,26-33(1991)

Lu, Simon W.
    : Frequency-based pattern recognition using neural networks,V1569,452-
        462(1991)
    : Multiple target tracking system,V1388,299-305(1991)

Lu, Song W. See Hou, Li S.: V1519,580-588(1991)

Lu, Taiwei
    See Lin, Freddie S.: V1558,406-413(1991)
    See Yu, Francis T.: V1507,210-221(1991)

Lu, Toh-Ming See McDonald, John F.: V1390,286-301(1991)

Lu, Tongxin
    ; Udpa, Satish S.; Udpa, L.: Optoelectronic implementation of filtered-back-
        projection tomography algorithm,V1564,704-713(1991)

Lu, Wei-Wei
    ; Gough, M. P.; Davies, Peter N.: Scientific data compression for space: a
        modified block truncation coding algorithm,V1470,197-205(1991)

Lu, Weiping See Harrison, Robert G.: V1376,94-102(1991)

Lu, X. J. See Zhang, Yan: V1481,23-31(1991)

Lu, Xiaoming
    ; Ren, Xin; Wang, Peizheng; Chi, Rongsheng: Reflective optical fibre
        displacement sensor,V1572,248-251(1991)
    ; Ren, Xin; Chen, Yu-bao; Chi, Rongsheng: Research of distributed-fiber-optic
        pressure sensor,V1572,304-307(1991)

Lu, Xun See Bhattacharya, Prabir: V1468,734-739(1991)

Lu, Y. M. See Guan, Z. P.: V1519,26-32(1991)

Lu, Yi
    ; Vogt, Robert C.: Multiscale analysis based on mathematical
        morphology,V1568,14-25(1991)

Lu, Yin-Cheng See Rinker, Michael: V1362,14-23(1991)

Lu, Yu C.
    ; Li, Da Y.; Chen, Jian G.; Luo, Bin: Establishment of new criterion aiding the
        control of antireflection coating semiconductor diodes,V1519,463-
        466(1991)

Lu, Yue-guang
    ; Jiang, Lingzhen; Zou, Lixun; Zhao, Xia; Sun, Junyong: Nondestructive
        testing of printed circuit board by phase-shifting
        interferometry,V1332,287-291(1991)

Lu, Yue M. See Fan, Ru Y.: V1519,134-137(1991)

Lu, Yutian See He, Huijuan: V1409,18-23(1991)

Lu, Zhen-Wu
    : Moire pattern and interferogram transform,V1554B,383-388(1991)

Lu, Zhiyi See Yuan, Libo: V1572,258-263(1991)

Lu, Zhuo J.
    ; Blaha, Franz A.: Application issues of fiber-optic sensors in aircraft
        structures,V1588,276-281(1991)

Lubart, Rachel
    ; Wollman, Yoram; Friedmann, Harry; Rochkind, Simeone; Laulicht, Israel:
        Role of various wavelengths in phototherapy,V1422,140-146(1991)
    See Friedmann, Harry: V1427,357-362(1991)
    See Rochkind, Simeone: V1428,52-58(1991)

Lubell, Bradford A. See Roos, Kenneth P.: V1428,159-168(1991)

Lubezky, I. See Marcovitch, Orna: V1442,58-58(1991)

Lubicz, Stephanie See Das, Bidyut B.: V1427,368-373(1991)

Lubin, Jeffrey See Larimer, James O.: V1456,196-210(1991)

Lubowsky, Jack See Barbour, Randall L.: V1431,52-62(1991); 1431,192-
        203(1991)

Lucas, Dean
    ; Gibson, Laurie: Template decomposition and inversion over hexagonally
        sampled images,V1568,157-163(1991)

Lucas, Jacques
    See Adam, Jean-Luc: V1513,150-157(1991)
    See Ma, Hong-Li: V1590,146-151(1991)
    See Zhang, Xhang H.: V1513,209-214(1991)

Lucassen, G. W.
    ; Scholten, T. A.; de Boey, W. P.; de Mul, F. F.; Greve, Jan: Nonresonant
        background suppression in coherent anti-Stokes Raman scattering spectra
        of dissolved molecules,V1403,185-194(1991)

Lucente, Mark See St.-Hilaire, Pierre: V1461,254-261(1991)

Luches, Armando
    See Craciun, Valentin: V1392,629-634(1991)
    See D'Anna, Emilia: V1503,256-268(1991)

Lucier, Ronald D.
: So now what?— things to do if your IR program stops producing results,V1467,59-62(1991)

Luck, Randall L.
See Rangachar, Ramesh M.: V1382,376-385(1991)
See Rangachar, Ramesh M.: V1382,331-340(1991)

Lucoro, Jacobo P. See Fry, David A.: V1346,270-275(1991)

Lucovsky, Gerald
; Kim, Sang S.; Fitch, J. T.; Wang, Cheng: Formation of heterostructure devices in a multichamber processing environment with in-vacuo surface analysis diagnostics and in-situ process monitoring,V1392,605-616(1991)

Ludikov, V. V.
; Prokhorov, Alexander M.; Chevokin, Victor K.: Development of subnanosecond framing cameras in IOFAN,V1346,418-436(1991)

Ludwig, David E.
; Skow, Michael: Compensated digital readout family,V1541,73-82(1991)
See Shanken, Stuart N.: V1346,210-215(1991)

Ludwig, Reinhold See Schunk, Nikolaus: V1362,391-397(1991)

Ludwiszewski, Alan P.
: Standards for oscillatory scanners,V1454,174-185(1991)

Luebbers, Hubertus
See Stojanoff, Christo G.: V1507,426-434(1991)
See Tholl, Hans D.: V1456,262-273(1991)

Luebbers, Paul G.
; Pandya, Abhijit S.: Video-image-based neural network guidance system with adaptive view-angles for autonomous vehicles,V1469,756-765(1991)
See Pandya, Abhijit S.: V1468,802-811(1991)

Luehrmann, Paul F.
; de Mol, Chris G.; van Hout, Frits J.; George, Richard A.; van der Putten, Harrie B.: Improvements in 0.5-micron production wafer steppers,V1463,434-445(1991)
See Samarakone, Nandasiri: V1463,16-29(1991)

Luepke, G. See Marowsky, Gerd: V1560,328-334(1991)

Lueth, Hans See Rizzi, Angela: V1361,827-833(1991)

Lufkin, Robert B. See Lou, Shyh-Liang: V1446,302-311(1991)

Lugara, M. See Ferrara, M.: V1513,130-136(1991)

Lugiez, F. See Scavennec, Andre: V1362,331-337(1991)

Luhmann, Neville C.
See Chong, Chae K.: V1407,226-233(1991)
See Dodd, James W.: V1407,467-473(1991)
See Lednum, Eugene E.: V1407,202-208(1991)
See Wang, Qinsong: V1407,209-216(1991)

Lui, M. See Chou, Chia-Shing: V1475,151-156(1991)

Lui, Po-Chang See Fan, Kuo-Chin: V1468,1010-1021(1991)

Luk, Franklin T.
ed.: *Advanced Signal Processing Algorithms, Architectures, and Implementations II,*V1566(1991)
See Bojanczyk, Adam W.: V1566,74-83(1991)
See Chun, Joohwan: V1566,483-492(1991)

Lukac, Matjaz
: Energy storage efficiency and population dynamics in flashlamp-pumped sensitized erbium glass laser,V1419,55-62(1991)

Lukashev, Alexei V. See Konov, Vitaly I.: V1427,232-242(1991)

Lukasiewicz, Stan
: Lasers applied to photoelastic stress measurements,V1554A,349-358(1991)

Luke, Keung L. See Liu, Duncan T.: V1409,116-126(1991)

Lukin, Vladimir P.
: Use of adaptive optics for minimizing atmospheric distortion of optical waves,V1408,86-95(1991)

Lukishova, Svetlana G.
; Mendez, Nestor R.; Ter-Mikirtychev, Valery V.; Tulajkova, Tamara V.: Brightness enhancement of solid state laser oscillators in single-mode lasing using novel inside-resonator optical elements with radially variable transmission,V1527,380-391(1991)
; Obidin, Alexey Z.; Vartapetov, Serge K.; Veselovsky, Igor A.; Osiko, Anatoly V.; Tulajkova, Tamara V.; Ter-Mikirtychev, Valery V.; Mendez, Nestor R.: Photochemical changes of rare-earth valent state in gamma-irradiated CaF2:Pr crystals by the excimer laser radiation: investigation and application,V1503,338-345(1991)

Lumia, John J.
: Mobile system for measuring retroreflectance of traffic signs,V1385,15-26(1991)

Lumley, John M. See Rando, Nicola: V1549,340-356(1991)

Lumpkin, Alex H.
: Applications of synchroscan and dual-sweep streak camera techniques to free-electron laser experiments,V1552,42-49(1991)

Lunchev, V. A. See Simonov, Alexander P.: V1503,330-337(1991)

Lund, Glenn I.
; Gaignebet, Jean: Potential of tunable lasers for optimized dual-color laser ranging,V1492,166-175(1991)
; Renault, Herve: Toward the optimization of passive ground targets in spaceborne laser ranging,V1492,153-165(1991)

Lundahl, Scott L. See Nseyo, Unyime O.: V1426,287-292(1991)

Lundgren, Mark A.
; Wolfe, William L.: Simultaneous alignment and multiple surface figure testing of optical system components via wavefront aberration measurement and reverse optimization,V1354,533-539(1991)

Lunn, George H.
: Appreciation of the Society of Motion Picture and Television Engineers: the originators of the Congress,V1358,1161-1163(1991)

Lunter, Sergei G.
; Dymnikov, Alexander A.; Przhevuskii, Alexander K.; Fedorov, Yurii K.: Laser glasses,V1513,349-359(1991)

Luo, Bikai
; Ni, Xiao W.; Zhang, Qi; He, Anzhi: Interference measurement of thermal effect induced by optical pumping on YAG crystal rod,V1554A,542-546(1991)

Luo, Bin See Lu, Yu C.: V1519,463-466(1991)

Luo, Chanzou See Luo, Zhishan: V1554B,523-528(1991)

Luo, Chuntai See Wang, Tianmin: V1519,890-900(1991)

Luo, Da L. See Wang, He-Chen: V1384,133-136(1991)

Luo, Fei
; Yan, Muolin; Huang, Shang-Lian: Distributed fiber optic pressure sensor,V1367,221-224(1991)

Luo, Gei-peng
; Xu, Sen-lu: Research on the characteristics of temperature drift in fiber-optic gyroscope,V1572,337-341(1991)
See Xu, Sen-lu: V1367,59-69(1991)

Luo, Jhy-Ming
: Coherent coupling of lasers using a photorefractive ring oscillator,V1409,100-105(1991)

Luo, Jin
See Mathur, Bimal P.: V1473,153-160(1991)
See Tanner, John E.: V1473,76-87(1991)

Luo, Nan
; Wang, Changgui; Xiao, Wen; Zhao, Yanhan; Chen, Yaosheng: Wavelength distribution optical fiber sensor,V1572,2-4(1991)

Luo, Ren C.
; Kay, Michael G.: Hierarchical target representation for autonomous recognition using distributed sensors,V1383,537-544(1991)

Luo, Rolland See Aragon, Sergio R.: V1430,65-84(1991)

Luo, Tao
; Zhang, Wei Q.; Gan, Fuxi: Langmuir-Blodgett films of tetra-tert-butyl-phenoxy phthalocyanine iron [II],V1519,826-830(1991)

Luo, Wei A. See Zhang, Z. J.: V1519,790-792(1991)

Luo, Wei G. See Zhang, Rui T.: V1519,757-760(1991)

Luo, Xiangyang See Fang, Qiang: V1332,216-222(1991)

Luo, Y. Q. See Chen, H. Y.: V1519,761-764(1991)

Luo, Yi See Nakano, Yoshiaki: V1418,250-260(1991)

Luo, Zhishan
; Luo, Chanzou; Hu, Qinghua; Mu, Zongxue: Moire interferometry for subdynamic tests in normal light environment,V1554B,523-528(1991)
; Zhang, Souyi; Tuo, Sueshang: Moire interferometry of sticking film using white light,V1554B,339-342(1991)

Luo, Zhongkuan
: Study on the special properties of electrochromic film of a-WO3,V1489,124-134(1991)

Luppino, Gerard A. See Geary, John C.: V1447,264-273(1991)

Luque, Antonio See Minano, Juan C.: V1528,58-69(1991)

Luquet, Philippe See Duran, Michel: V1494,357-376(1991)

Lure, Yuan-Ming F.
: On the determination of optimal window for registration of nonlinear distributed images,V1452,292-302(1991)

Luridiana, S.
  See Andreani, F.: V1519,63-69(1991)
  See Andreani, F.: V1519,18-22(1991)

Lurk, Paul W.
  ; Cooper, Stafford S.; Malone, Philip G.; Olsen, R. S.; Lieberman, Stephen H.:
  Laser-induced fluorescence in contaminated soils,V1434,114-118(1991)

Lusiani, Luigi See Visona, Adriana: V1425,75-83(1991)

Lusis, Andrejs R.
  : Solid state ionics and optical materials technology for energy efficiency,
  solar energy conversion, and environment control,V1536,116-124(1991)

Luther, Klaus See Ihlemann, Juergen: V1361,1011-1019(1991)

Lutjens, William See Kovacich, Michael: V1481,357-370(1991)

Lutschin, V. I. See Brailovsky, A. B.: V1440,84-89(1991)

Luttrell, Stephen P.
  : Hierarchical network for clutter and texture modeling,V1565,518-
  528(1991)

Lutz, G. See Braeuninger, Heinrich: V1549,330-339(1991)

Lutz, Marc
  ; Mattioli, Tony; Moenne-Loccoz, Pierre; Zhou, Qing; Robert, Bruno:
  Resonance Raman studies of photosynthetic membrane
  proteins,V1403,59-65(1991)

Lutz, Michael A. See Wojcik, Gregory L.: V1463,292-303(1991)

Lutz, Norbert See Geiger, Manfred: V1503,238-248(1991)

Lutz, Stephen S. See Di Benedetto, John A.: V1492,115-125(1991)

Lutze, Frederick H. See Hartman, Richard D.: V1478,64-75(1991)

Luu, K. T. See Chambers, Robert J.: V1488,312-326(1991)

Luy, Johann-Freidrich
  ; Strohm, Karl M.; Buechler, J.: Coplanar SIMMWIC circuits,V1475,129-
  139(1991)

Lybanon, Matthew
  ; Molinelli, Eugene J.; Muncill, G.; Pepe, Kevin: Neural network labeling of
  the Gulf Stream,V1469,637-647(1991)
  ; Peckinpaugh, Sarah; Holyer, Ronald J.; Cambridge, Vivian: Integrated
  oceanographic image understanding system,V1406,180-189(1991)

Lyberg, Mats D.
  ; Ljungberg, Sven-Ake: Thermography and complementary method: a tool for
  cost-effective measures in retrofitting buildings,V1467,104-115(1991)

Lyerla, James R.
  See Brock, Phillip J.: V1463,87-100(1991)
  See Karis, Thomas E.: V1499,366-376(1991)

Lyman, Dwight D.
  See Hiddleston, H. R.: V1542,20-33(1991)
  See Schafer, Eric L.: V1415,310-316(1991)

Lymar, V. I. See Miloslavsky, V. K.: V1440,90-96(1991)

Lynch, Robert T. See Karis, Thomas E.: V1499,366-376(1991)

Lynch, William E.
  ; Reibman, Amy R.: Lapped orthogonal transform for motion-compensated
  video compression,V1605,285-296(1991)

Lynn, J. G. See Stoneman, Robert C.: V1410,148-155(1991)

Lynn, Mark A.
  See Bulusu, Dutt V.: V1364,49-60(1991)
  See Cirillo, James R.: V1592,53-59(1991)
  See Cirillo, James R.: V1592,42-52(1991)

Lynne, Kenton J.
  : Distributing the server function in a multiring PAC system,V1446,177-
  187(1991)

Lyon, Douglas
  : Ad-hoc and derived parking curves,V1388,39-49(1991)

Lyon, Richard G.
  ; Miller, Peter E.; Grusczak, Anthony: HST phase retrieval: a parameter
  estimation,V1567,317-326(1991)

Lyon, Ronald J.
  ; Honey, Frank R.: Thermal infrared imagery from the Geoscan Mk II scanner
  and its calibration: two case histories from Nevada—Ludwig Skarn
  (Yerington District) & Virginia City,V1492,339-350(1991)

Lyons, Damian M.
  ; Allton, Judith H.: Achieving a balance between autonomy and tele-
  operation in specifying plans for a planetary rover,V1387,124-133(1991)

Lyons, E. R. See Buckley, Robert H.: V1371,212-222(1991)

Lyons, Jayne T. See Stewart, Stephen R.: VIS07,57-97(1991)

Lyons, Peter B. See Looney, Larry D.: V1474,132-137(1991)

Lyons, R. W. See Rosenblatt, Edward I.: V1449,72-86(1991)

Lypsky, Volodymyr V.
  : Design characters and radiation parameters of a gas laser as a holographic
  tool,V1238,390-395(1991)

Lysogorski, Charles See Turney, Jerry L.: V1380,53-63(1991)

Lytel, Richard S.
  ; Binkley, E. S.; Girton, Dexter G.; Kenney, John T.; Lipscomb, George F.;
  Ticknor, Anthony J.; Van Eck, Timothy E.: Electro-optic polymer devices
  for optical interconnects,V1389,547-558(1991)
  ; Lipscomb, George F.; Kenney, John T.; Ticknor, Anthony J.: Applications of
  electro-optic polymers to optical interconnects,V1563,122-138(1991)
  See Lipscomb, George F.: V1560,388-399(1991)
  See Wu, Jeong W.: V1560,196-205(1991)

Lytle, John D.
  : Status and future of polymeric materials in imaging systems,V1354,388-
  394(1991)

Lyubomirsky, I.
  ; Durfee, C.; Milchberg, Howard M.: Hydrodynamic evolution and radiation
  emission from an impulse-heated solid-density plasma,V1413,108-
  111(1991)

Ma, Chun-Chi See Becker, Michael F.: V1441,174-187(1991)

Ma, Chunhua See Li, Wenchong: V1428,242-248(1991)

Ma, Chunrong See Zhang, Jingfang: V1238,306-310(1991)

Ma, De-Yuan See Fang, Cui-Chang: V1554A,915-921(1991)

Ma, Hede See Liu, Ying: V1606,226-240(1991)

Ma, Hong-Li
  ; Zhang, Xhang H.; Lucas, Jacques: Thermal relaxation of tellurium-halide-
  based glasses,V1590,146-151(1991)
  See Zhang, Xhang H.: V1513,209-214(1991)

Ma, Jiancheng See Gu, Lizhong: V1572,450-452(1991)

Ma, Jiankang See Zhou, Renkui: V1358,1245-1251(1991)

Ma, Jing S. See Nishimura, Kazuhito: V1534,199-206(1991)

Ma, Jun
  See Liu, Zhi-Shen: V1558,306-316(1991)
  See Liu, Zhi-Shen: V1558,379-383(1991)
  See Song, Zhengfang: V1487,382-386(1991)

Ma, Junxian
  ; Yang, Shuwen: Optical fiber refractometer and its application in the sugar
  industry,V1572,377-381(1991)

Ma, Ke J.
  ; Yu, Zhen Z.; Yanh, Jian R.; Shen, Shou Z.; He, Jin; Chen, Wei M.; Song,
  Xiang Y.: Study of HgCdTe/CdTe interface structure grown by metal-
  organic chemical vapor deposition,V1519,489-493(1991)
  See Chen, Wei M.: V1519,521-524(1991)
  See He, Jin: V1519,499-507(1991)

Ma, Li See Wang, Ming: V1554B,242-246(1991)

Ma, Marco See Ho, Bruce K.: V1445,95-100(1991)

Ma, Min W. See Zhang, Xiao P.: V1519,514-520(1991)

Ma, Pin-Zhong
  : Rigidity test of large and high-precision instruments,V1532,177-186(1991)
  : Scheme of optical synthetic-aperture telescope,V1533,163-174(1991)

Ma, Shining
  ; Liu, Ying; Du, Fuli: Space-time correlation properties and their applications
  of dynamic speckles after propagation through an imaging system and
  double-random modulation,V1554A,645-648(1991)
  See Liu, Ying: V1554A,610-612(1991)

Ma, W. G. See Shu, Q. Q.: V1519,675-679(1991)

Ma, Y. P.
  ; Yu, Phillip C.; Lampert, Carl M.: Development of laminated nickel/
  manganese oxide and nickel/niobium oxide electrochromic
  devices,V1536,93-103(1991)

Ma, Y. Z. See Pan, Chuan K.: V1519,645-651(1991)

Ma, Yueying
  See Cao, Jianlin: V1345,71-77(1991)
  See Cao, Jianlin: V1345,225-232(1991)

Ma, Yushieh
  ; Varadan, Vijay K.; Varadan, Vasundara V.: Effective refractive indices of
  three-phase optical coatings,V1558,138-142(1991)
  ; Varadan, Vijay K.; Varadan, Vasundara V.: Frequency-selective devices
  using a composite multilayer design,V1558,132-137(1991)
  ; Varadan, Vijay K.; Varadan, Vasundara V.: Multiple-scattering effects on
  pulse propagation through burning particles,V1487,220-225(1991)

Ma, Zhifang See Liu, Dayou: V1468,37-49(1991)

Ma, Zi
See Ni, Xiao W.: V1554B,632-635(1991)
See Ni, Xiao W.: V1527,437-441(1991)
See Ni, Xiao W.: V1519,365-369(1991)

Maag, Carl R.
; Hansen, Patricia A.: Flight experiment to investigate microgravity effects on solidification phenomena of selected materials,V1557,24-30(1991)

Maas, Hans-Gerd
: Automated photogrammetric surface reconstruction with structured light,V1526,70-77(1991)

Maassen, M. See Geurts, Jean: V1361,744-750(1991)

Mabru, Michel See Brunet, Henri: V1397,273-276(1991)

MacAndrew, J. A.
; Humphries, Mark R.; Welford, W. T.; Golby, John A.; Dickinson, P. H.; Wheeler, J. R.: Beam shaping system,V1500,172-176(1991)

MacArthur, Thomas D.
: Comparing laser printing and barcode scanning designs,V1384,176-184(1991)

MacAulay, Calum See Palcic, Branko: V1448,113-117(1991)

Macchetto, F. See Greenfield, Perry E.: V1494,16-39(1991)

MacCraith, Brian D.
; Maxwell, J.: Fiber optic sensor for nitrates in water,V1510,195-203(1991)
; Ruddy, Vincent; Potter, C.; McGilp, J. F.; O'Kelley, B.: Fiber optic fluorescence sensors based on sol-gel entrapped dyes,V1510,104-109(1991)

MacDavid, Jacob H.
: Buried-steam-line temperature and heat loss calculation,V1467,11-17(1991)

Macdonald, Burns
; Helliwell, William S.; Sanborn, James; Voss, Kenneth J.: Detector perturbation of ocean radiance measurements,V1537,104-114(1991)

MacDonald, Daniel R.
See Hink, Paul L.: V1549,193-202(1991)
See Pelling, Michael R.: V1549,134-146(1991)

MacDonald, K. A. See Bell, J. M.: V1536,29-36(1991)

MacDonald, Scott A.
; Clecak, Nicholas J.; Wendt, H. R.; Willson, C. G.; Snyder, C. D.; Knors, C. J.; Deyoe, N. B.; Maltabes, John G.; Morrow, James R.; McGuire, Anne E.; Holmes, Steven J.: Airborne chemical contamination of a chemically amplified resist,V1466,2-12(1991)
See Kutal, Charles: V1466,362-367(1991)
See McKean, Dennis R.: V1559,214-221(1991)

MacDowell, Alastair A.
See Raab, Eric L.: V1343,104-109(1991)
See White, Donald L.: V1343,204-213(1991)

Macedo, Marcelo A. See Valla, Bruno: V1536,48-62(1991)

Macenka, Steve A. See Chrisp, Michael P.: V1540,213-218(1991)

Macfadyen, Allan J.
; Jennings, B. R.: Photon correlation spectroscopy and electrophoretic light scattering using optical fibers,V1367,319-328(1991)

MacFarlane, Duncan L.
; Tatum, Jim A.: Optimization of gain-switched diode lasers for high-speed fiber optics,V1365,88-95(1991)
See Tatum, Jim A.: V1497,320-325(1991)

MacGregor, Andrew D.
; Dion, Bruno; Noeldeke, Christoph; Duchmann, Olivier: A 39-photon/bit direct-detection receiver at 810 nm, BER = 1x10-6, 60-Mbit/s QPPM,V1417,374-380(1991)

Mach, J.-P. See Wagnieres, G.: V1525,219-236(1991)

Mache, Neils See Zell, Andreas: V1469,708-718(1991)

Machida, Tsuneo See Tanii, Jun: V1490,200-206(1991)

Machintsev, G. A. See Nasibov, Alexander S.: V1361,901-908(1991)

Machol, Steven
: Sola ASL in Spectralite and polycarbonate aspheric lens designs,V1529,45-56(1991)
; Modglin, Luan: Sola ASL in Spectralite strikes the perfect balance between cosmetics and optics,V1529,38-44(1991)

Maciunas, Robert J.
See Edwards, Charles A.: V1444,38-46(1991)
See Galloway, Robert L.: V1444,9-18(1991)
See Thomas, Judith G.: V1444,379-388(1991)

Mack, Chris A.
See Trefonas, Peter: V1466,117-131(1991)
See Ziger, David H.: V1466,270-282(1991)

Mack, Russell T. See Baird, George S.: V1467,63-63(1991)

Mackara, Steven R. See Gambogi, William J.: V1555,256-267(1991)

Mackay, Gervase I.
; Karecki, David R.; Schiff, Harold I.: Tunable diode laser systems for trace gas monitoring,V1433,104-119(1991)
See Carduner, Keith R.: V1433,190-201(1991)
See Drummond, John W.: V1433,242-252(1991)
See Drummond, John W.: V1433,224-231(1991)

Mackenzie, H. S. See Payne, Frank P.: V1504,165-175(1991)

Mackinnon, G. See Craig, Robert G.: V1505,76-86(1991)

Maclachlan Spicer, J. W.
; Kerns, W. D.; Aamodt, Leonard C.; Murphy, John C.: Time-resolved infrared radiometry of multilayer organic coatings using surface and subsurface heating,V1467,311-321(1991)

MacLaren, J. M. See Victora, Randall H.: V1499,378-381(1991)

MacLean, Steve G.
; Pinkney, H. F.: Calibration procedures for the space vision system experiment,V1526,113-122(1991)
See Blais, Francois: V1482,473-479(1991)

Macleod, H. A.
: Optics in adverse environments,V1399,2-6(1991)
See Bovard, Bertrand G.: V1534,216-222(1991)
See Milanovic, Zoran: V1535,160-170(1991)

MacLeod, Roderick W.
; Sotomayor Torres, Clivia M.; Razeghi, Manijeh; Stanley, C. R.; Wilkinson, Chris D.: Reactive ion etching of InP and its optical assessment,V1361,562-567(1991)

MacLeod, Wade See Sinnott, Heather: V1363,196-200(1991)

MacMahon, Heber
See Giger, Maryellen L.: V1445,101-103(1991)
See Nishikawa, Robert M.: V1444,180-189(1991)
See Yoshimura, Hitoshi: V1445,47-51(1991)

MacMullan, Jay See Greene, Herbert G.: V1470,98-102(1991)

MacNeal, Paul D.
; Lou, Michael C.; Chen, Gun-Shing: Space Infrared Telescope Facility structural design requirements,V1540,68-85(1991)
; Lou, Michael C.: Structural design considerations for the Space Infrared Telescope Facility,V1494,236-254(1991)

MacNeil, Ronald L.
: Capturing multimedia design knowledge using TYRO, the constraint-based designer's apprentice,V1460,94-102(1991)

Macomber, Steven H. See Akkapeddi, Prasad R.: V1416,44-49(1991)

Macovski, Albert
See Noll, Douglas C.: V1443,29-36(1991)
See Plevritis, Sylvia: V1445,118-127(1991)

Macq, Benoit M.
; Ronse, Christian; Van Dongen, V.: Multiscale morphological region coding,V1606,165-173(1991)

Macrander, Albert T.
; Schwartz, Gary P.; Gualteri, Gregory J.; Gilmer, George: X-ray diffraction study of GaSb/AlSb strained-layer-superlattices grown on miscut (100) substrates,V1550,122-133(1991)

MacRobert, Alexander J.
; Phillips, David: Fluorescence imaging in photodynamic therapy,V1439,79-87(1991)

Mactaggart, R. See Swirhun, S.: V1475,223-230(1991)

Macukow, Bohdan See Berus, Tomasz: V1564,562-570(1991)

Madarazo, R.
; Pedrine, A. G.; Sol, A. A.; Baranauskas, Vitor: Epitaxial regrowth of silicon on sapphire by rapid isothermal processing,V1393,270-277(1991)

Maddalena, P.
See Barone, Fabrizio: V1584,304-307(1991)
See Carbonara, Giuseppe: V1361,688-691(1991)

Maddox, Christine E. See Rieger, Paul L.: V1433,290-301(1991)

Mader, Leonhard See Noelscher, Christoph: V1463,135-150(1991)

Mader, Tom See Bundy, Scott: V1475,319-329(1991)

Madey, John M. See Litvinenko, Vladimir N.: V1552,2-6(1991)

Madjar, Asher See Paolella, Arthur: V1378,195-202(1991)

Madore, Michel See Eckes, Charlotte: V1466,394-407(1991)

Madraszek, E. M. See Key, Michael H.: V1397,9-17(1991)

Madrid, Angel
: Quantitative evaluation of cavities and inclusions in solids using IR thermography,V1467,322-336(1991)

Madsen, M. M. See Budtz-Jorgensen, Carl: V1549,429-437(1991)

Madsen, Steen J.
; Patterson, Michael S.; Wilson, Brian C.; Park, Young D.; Moulton, J. D.; Jacques, Steven L.; Hefetz, Yaron: Time-resolved diffuse reflectance and transmittance studies in tissue simulating phantoms: a comparison between theory and experiment,V1431,42-51(1991)

Maeda, Akira See Nagano, Yasutada: V1605,614-623(1991)

Maeda, Fumisada
; Arai, Masayuki; Owa, Hideo; Takahashi, Hiroo; Kaneko, Masahiko: Direct overwriting system by light-intensity modulation using triple-layer disks,V1499,62-69(1991)

Maeda, Koichi
; Ishizuka, S.; Tsujino, T.; Yamamoto, H.; Takigawa, Akio: Optical performance of angle-dependent light-control glass,V1536,138-148(1991)

Maeda, Mari W. See Sessa, William B.: V1372,208-218(1991)

Maeda, Mitsuo See Yamakoshi, H.: V1397,119-122(1991)

Maeda, Miyozo See Naitou, Kazunori: V1499,386-392(1991)

Maeda, Shunji
; Hiroi, Takashi; Makihira, Hiroshi; Kubota, Hitoshi: Automated visual inspection for LSI wafer patterns using a derivative-polarity comparison algorithm,V1567,100-109(1991)

Maeda, T.
; Doi, Masao: Dynamical structure factor of a solution of charged rod-like polymers in the isotropic phase,V1403,268-277(1991)

Maeda, Takeshi
; Saito, Atsushi; Sugiyama, Hisataka; Ojima, Masahiro; Arai, Shinichi; Shigematu, Kazuo: High-speed large-capacity optical disk using pit-edge recording and MCAV method,V1499,414-418(1991)

Maeda, Tetsuo See Ogata, Shiro: V1544,92-100(1991)

Maehara, Hideaki See Nagano, Yasutada: V1605,614-623(1991)

Maejima, Tsukasa See Chandran, Sharat: V1570,391-402(1991)

Maestre, Marcos F.
; Bustamante, Carlos; Snyder, Patricia A.; Rowe, Ednor M.; Hansen, Roger W.: Circular intensity differential scattering measurements in the soft x-ray region of the spectrum (~16 EV to 500 EV),V1548,179-187(1991)

Maeta, Manabu
; Kawakami, Shinichiro; Ogawara, Toshiaki; Masuda, Yu: Vibration analysis of the tympanic membrane with a ventilation tube and a perforation by holography,V1429,152-161(1991)
See Inokuchi, Ikuo: V1429,39-45(1991)

Maex, Karen
See Vandenabeele, Peter: V1393,316-336(1991)
See Vandenabeele, Peter: V1393,372-394(1991)

Maffione, Robert A.
; Dana, David R.; Honey, Richard C.: Instrument for underwater measurement of optical backscatter,V1537,173-184(1991)
; Honey, Richard C.; Brown, Robert A.: Experiment for testing the closure property in ocean optics,V1537,115-126(1991)
See Brown, Robert A.: V1537,147-150(1991)

Magan, John D. See Lemoine, Patrick: V1377,45-56(1991)

Magee, Charles W. See Sturm, James C.: V1393,252-259(1991)

Magee, Michael
See Sklair, Cheryl W.: V1387,380-391(1991)
See Wolfe, William J.: V1383,20-25(1991)

Maggi, C. See Serri, Laura: V1397,469-472(1991)

Magnani, M. See Livi, S.: V1500,144-150(1991)

Magner, Thomas J.
: Moderate-resolution imaging spectrometer-tilt baseline concept,V1492,272-285(1991)

Magni, Vittorio C. See Serri, Laura: V1397,469-472(1991)

Magnin, Isabelle E.
See Baskurt, Atilla: V1444,240-249(1991)
See El Alaoui, Mohsine: V1569,80-87(1991)
See Friboulet, Denis: V1445,106-117(1991)

Magome, Nobutaka See Ota, Kazuya: V1463,304-314(1991)

Magrini, Sandro See Cifarelli, Salvatore: V1482,480-490(1991)

Magun, I. I. See Angelsky, Oleg V.: V1402,231-235(1991)

Magyar, James T.
: History of and potential for optical bonding agents in the visible,V1535,55-58(1991)

Mahadevan, K. See Umadevi, P.: V1484,125-135(1991); 1485,195-205(1991)

Mahajan, Virendra N.
: <author>Aberration Theory Made Simple,VTT06(1991)

Mahalanobis, Abhijit
; Nadar, Mariappan S.: Application of neural networks for the synthesis of binary correlation filters for optical pattern recognition,V1469,292-302(1991)
; Song, Sewoong: Purely real correlation filters,V1564,14-21(1991)

Mahapatra, Amaresh
; Cooper, Ronald F.: Design and fabrication of amplitude modulators for CATV,V1374,296-299(1991)

Mahbobzadeh, Mohammad
; Osinski, Marek A.: Novel distributed feedback structure for surface-emitting semiconductor lasers,V1418,25-31(1991)

Mahin, K. W. See Fuchs, Elizabeth A.: V1467,136-149(1991)

Mahle, Christoph E. See Hung, Hing-Loi A.: V1475,212-222(1991)

Mahmood Reza, Syed
; Willett, Peter K.: Ll-filters in CFAR (constant false-alarm rate) detection,V1451,298-308(1991)

Mahon, R. See Duncan, Michael D.: V1409,127-134(1991)

Mahoney, Michael J.
; Marsh, Kenneth A.: SALSA: a synthesis array for lunar submillimeter astronomy,V1494,182-193(1991)
See Gershman, Robert: V1494,160-167(1991)

Mahowald, Misha A.
: Silicon retina with adaptive photoreceptors,V1473,52-58(1991)

Mahrane, Achour See Simonne, John J.: V1589,139-147(1991)

Maida, Anthony See Dunn, Dennis F.: V1606,541-552(1991)

Maida, John L. See Varshneya, Deepak: V1367,181-191(1991)

Maienschein, Jon L.
; Barry, Patrick E.; McMurphy, Frederick E.; Bowers, John S.: New low-density, high-porosity lithium hydride-beryllium hydride foam: properties and applications to x-ray astronomy,V1343,477-484(1991)

Maile, Michael
; Weidel, Edgar: Multimode stripe waveguides for optical interconnections,V1563,188-196(1991)

Mainfray, Gerard L. See Andre, Michel L.: V1502,286-298(1991)

Mainguy, Stephan
; Olivier, Michel; Josse, Michel A.; Guidon, Michel: Description and calibration of a fully automated infrared scatterometer,V1530,269-282(1991)

Maissen, Clau
See Zogg, Hans: V1361,406-413(1991)
See Zogg, Hans: V1361,1079-1086(1991)

Mait, Joseph N.
: Designs for two-dimensional nonseparable array generators,V1555,43-52(1991)
: Upper bound on the diffraction efficiency of phase-only array generators,V1555,53-62(1991)

Maitan, Jacek
: Flow-control mechanism for distributed systems,V1470,88-97(1991)

Maitland, Arthur See Dawber, William N.: V1476,81-90(1991)

Maitz, Glenn S. See Rosenthal, Marc S.: V1443,132-142(1991)

Maiya, Bhaskar G. See Sessler, Jonathan L.: V1426,318-329(1991)

Majani, Eric See Eberlein, Susan: V1388,578-586(1991)

Majercak, David
; Sernas, Valentinas; Polymeropoulos, Constantine E.; Sigel, George H.: Microbend pressure sensor for high-temperature environments,V1584,162-169(1991)

Majid, W. H. See Muhamad, M. R.: V1519,872-877(1991)

Majima, T. See Tomie, Toshihisa: V1552,254-263(1991)

Majni, G. See D'Anna, Emilia: V1503,256-268(1991)

Major, J. V. See Doel, A. P.: V1542,319-326(1991)

Majumder, Fariduddin A. See Gregory, Andrew: V1362,268-274(1991)

Mak, Arthur A.
; Orlov, Oleg A.; Ustyugov, Vladimir I.; Vitrishchak, Il'ya B.: Single-frequency solid state lasers and amplifiers,V1410,233-243(1991)
; Soms, Leonid N.: Optical methods for laser beams control,V1415,110-119(1991)

Mak, Paul S. See Feit, Zeev: V1512,164-169(1991)

Makarenko, Vladimir I.
; Kirillov, A. B.: Self-organized criticality in neural networks,V1469,843-845(1991)

Makarov, V. N. See Dvoryankin, A. N.: V1397,177-180(1991)

Makarov, V. V. See Anisimov, N. R.: V1440,206-210(1991)

Makashev, N. K. See Anisimov, N. R.: V1440,206-210(1991)

Makeev, B. H. See Artsimovich, M. V.: V1397,729-733(1991)

Makeev, V. Y.
; Poponin, Vladimir P.; Schzeglov, V. A.: Dynamics of nonlinear excitations in soft-quasi-one-dimensional molecular chains,V1403,522-527(1991)

Makekau, Charles K. See Fielding, Kenneth H.: V1564,224-230(1991)

Maki, August H.
; Tsao, Desiree H.: Microwave-optical study of an As(III) derivative of Eco RI methylase,V1432,119-128(1991)

Makihira, Hiroshi See Maeda, Shunji: V1567,100-109(1991)

Makin, Vladimir S.
See Bazhenov, V. V.: V1440,332-337(1991)
See Libenson, Michail N.: V1440,354-356(1991)

Makishima, K. See Ohashi, T.: V1549,9-19(1991)

Maklad, Mokhtar S. See Krohn, David A.: V1420,126-135(1991)

Makowiecki, Daniel M. See Jankowski, Alan F.: V1343,32-38(1991)

Maksimova, G. V. See Denker, Boris I.: V1419,50-54(1991)

Maksimova, I. L.
; Shubochkin, L. P.: Light scattering matrices of a densely packed binary system of hard spheres,V1403,749-751(1991)

Maksimovskiy, Y. V. See D'yakonov, G. I.: V1424,81-86(1991)

Maksimyak, Peter P.
See Angelsky, Oleg V.: V1403,667-673(1991)
See Angelsky, Oleg V.: V1402,231-235(1991)

Makuchowski, Jozef
; Pokora, Ludwik J.; Ujda, Zbigniew; Wawer, Janusz: Laser set for the investigations of the NO2 contents in atmosphere,V1391,348-350(1991)
; Pokora, Ludwik J.; Wawer, Janusz: Low-pressure nitrogen laser pulse repetition frequency to 100 Hz,V1391,79-86(1991)

Malacara Hernandez, Daniel
: Diffraction performance calculations in lens design,V1354,2-14(1991)
: Review of interferogram analysis methods,V1332,678-689(1991)
; Flores, Ricardo: Simple test for the 90 degree angle in prisms,V1332,36-40(1991)

Malaguti, Giuseppe
See Di Cocco, Guido: V1549,102-112(1991)
See Swinyard, Bruce M.: V1548,94-105(1991)

Malak, Henryk See Lakowicz, Joseph R.: V1435,142-160(1991)

Malakian, Kourken
; Vidmar, Anthony: New track-to-track association logic for almost identical multiple sensors,V1481,315-328(1991)

Malatesta, Vincenzo See Andreoni, Alessandra: V1525,351-366(1991)

Malchow, Douglas S.
: Characterization of plasma processes with optical emission spectroscopy,V1392,498-505(1991)

Maldague, Xavier
; Fortin, Louis; Picard, J.: Applications of tridimensional heat calibration to a thermographic nondestructive evaluation station,V1467,239-251(1991)
See Vavilov, Vladimir P.: V1540,488-495(1991)

Maldonado, Juan R.
: X-ray lithography system development at IBM: overview and status,V1465,2-15(1991)
See Oertel, Heinrich K.: V1465,244-253(1991)

Maldutis, Evaldas K. See Vaicikauskas, V.: V1440,357-363(1991)

Malhotra, H. S. See Jain, Subhash C.: V1488,410-413(1991)

Malhotra, Sandeep See Spragg, Peggy M.: V1466,283-296(1991)

Malicka, Marianna
; Parma, Ludvik: Dynamics of laser light transmission losses in aerosols,V1391,181-189(1991)
See Parma, Ludvik: V1391,190-198(1991)

Malik, Raashid
; Polkowski, Edward T.: Robot location densities,V1388,280-290(1991)
; Polkowski, Edward T.: Robot self-location based on corner detection,V1388,306-316(1991)

Malik, Zvi
See Ehrenberg, Benjamin: V1426,244-251(1991)
See Sessler, Jonathan L.: V1426,318-329(1991)

Malina, Roger F. See Welsh, Barry Y.: V1343,166-174(1991)

Malinconico, Mario
See Carbonara, Giuseppe: V1361,1038-1040(1991)
See Carbonara, Giuseppe: V1361,688-691(1991)

Malinen, Jouko
; Hyvarinen, Timo S.: Advanced portable four-wavelength NIR analyzer for rapid chemical composition analysis,V1510,204-209(1991)
; Keranen, Heimo; Hannula, Tapio; Hyvarinen, Timo S.: Thirty-two-channel LED array spectrometer module with compact optomechanical construction,V1533,122-128(1991)

Malinin, Boris G. See Bouchenkov, Vyatcheslav A.: V1410,185-188(1991); 1427,409-412(1991)

Malins, Robert J. See Acebal, Robert: V1397,191-196(1991)

Malka, Victor See Koenig, Michel: V1502,338-342(1991)

Mallek, Janusz See Grzegorzewski, Bronislaw: V1391,290-294(1991)

Malley, Michael M.
See Simpson, Thomas B.: V1410,133-140(1991)
See Simpson, Thomas B.: V1416,2-9(1991)

Mallory, Derek S.
; Kadin, Alan M.; Ballentine, Paul H.: Performance of stripline resonators using sputtered YBCO films,V1477,66-76(1991)

Mallory, William R.
: Atmospheric laser-transmission tables simply generated,V1540,359-364(1991)
: Influence of atmospheric conditions on two-color temperature discrimination,V1540,365-369(1991)

Mallot, Hanspeter A.
; Zielke, Thomas; Storjohann, Kai; von Seelen, Werner: Topographic mapping for stereo and motion processing,V1382,397-408(1991)

Malloy, Kevin J. See Zou, Lian C.: V1519,707-711(1991)

Malloy, Terrence P.
: Bladder outlet obstruction treated with transurethral ultrasonic aspiration,V1421,46-46(1991)
: KTP-532 laser ablation of urethral strictures,V1421,72-72(1991)
: KTP-532 laser utilization in endoscopic pelvic lymphadenectomy,V1421,78-78(1991)

Malloy, W. J. See Musikant, Solomon: V1330,119-130(1991)

Malmo, Jan T. See Lokberg, Ole J.: V1508,153-160(1991)

Malo, Bernard
; Bilodeau, Francois; Johnson, Derwyn C.; Skinner, Iain M.; Hill, Kenneth O.; Morse, Ted F.; Kilian, Arnd; Reinhart, Larry J.; Oh, Kyunghwan: Photosensitivity in optical fibers: detection, characterization, and application to the fabrication of in-core fiber index gratings,V1590,83-93(1991)

Maloletov, Sergei M.
See Sherstyuk, Valentin P.: V1238,211-217(1991)
See Sherstyuk, Valentin P.: V1238,218-223(1991)

Malone, Philip G. See Lurk, Paul W.: V1434,114-118(1991)

Malone, Steven A. See Polifko, David M.: V1476,91-99(1991)

Maloney, Daniel L. See Dayhoff, Ruth E.: V1446,323-329(1991)

Maloney, Laurence T. See Landy, Michael S.: V1383,247-254(1991)

Maloney, M. J. See Liebow, Charles: V1426,22-30(1991)

Malotaux, Eric
; Alimenti, Rodolphe; Bogaert, Marc; Gaspart, Pierre: Intelligent piloting tools for control of an autonomous mobile robot,V1388,372-383(1991)

Malouin, Christian
; Song, Li; Thibault, S.; Denariez-Roberge, Marguerite M.; Lessard, Roger A.: Degenerate four-wave mixing using wave pump beams near the critical angle: two distinct behaviors,V1559,385-392(1991)

Malov, Alexander N. See Sherstyuk, Valentin P.: V1238,218-223(1991)

Malstrom, Eric M. See Chen, Robert S.: V1381,189-200(1991)

Maltabes, John G.
; Norris, Katherine C.; Writer, Dean: Computer simulation of 0.5-micrometer lithography for a 16-megabit DRAM,V1463,326-335(1991)
See MacDonald, Scott A.: V1466,2-12(1991)

Maluf, Nadim I. See Pease, R. F.: V1496,234-238(1991)

Malyarov, A. V.
See Bryukhnevitch, G. I.: V1358,739-749(1991)
See Dalinenko, I. N.: V1449,167-172(1991)

Malzahn, Eric
; Heuken, Michael; Gruetzmacher, Dettev; Stollenwerk, M.; Heime, Klaus: Advanced InGaAs/InP p-type pseudomorphic MODFET,V1362,199-204(1991)

Maman, K. H.
; Haines, Robert; Chatterjee, Samir: Research and development of a NYNEX switched multi-megabit data service prototype system,V1364,304-311(1991)

Mamedov, S. B. See Yusupov, I. Y.: V1238,240-247(1991)

Mammone, Richard J.
See Kuo, Shyh-Shiaw: V1606,641-652(1991)
See Kuo, Shyh-Shiaw: V1606,419-430(1991)
See Kuo, Shyh-Shiaw: V1452,192-202(1991)
See Kuo, Shyh-Shiaw: V1450,18-29(1991)
See Prasad, K. V.: V1385,280-291(1991)
See Sankar, Ananth: V1469,374-383(1991)

Man, Hong-Tai See Haas, David R.: V1371,56-67(1991)

Manalo, Natividad D.
; Smith, G. L.: Spatial sampling errors for a satellite-borne scanning radiometer,V1493,281-291(1991)
; Smith, G. L.; Barkstrom, Bruce R.: Transfer function considerations for the CERES (cloud's and earth's radiant energy system) scanning radiometer,V1521,106-116(1991)

Manasses, J. See Guldner, Yves: V1361,657-668(1991)

Manasson, Vladimir A.
; Sadovnik, Lev S.; Chen, Ray T.: Heterostructure photosensitive memory,V1559,194-201(1991)

Mancinelli Degli Esposti, Luca See Guglielmi, Massimo: V1513,44-49(1991)

Mancini, G. B. See Tehrani, Saeid: V1445,434-445(1991)

Mancini, Richard See Bergman, Larry A.: V1364,14-21(1991)

Mandal, Anup K.
See Das, Alok K.: V1572,572-580(1991)
See Das, Alok K.: V1365,144-155(1991)

Mandeville, Jon R.
See Rodriguez, Arturo A.: V1384,2-14(1991)
See Rodriguez, Arturo A.: V1332,25-35(1991)

Mandolia, G. See Lan, Guey-Liu: V1475,184-192(1991)

Manenkov, Alexander A.
; Denisov, N. N.; Bagdasarov, V. H.; Starkovsky, A. N.; Yurchenko, S. V.; Kornilov, Yu. M.; Mikaberidze, V. M.; Sarkisov, S. E.: Potentials for pulsed YAG:Nd laser application to endoscopic surgery,V1420,254-258(1991)
; Nechitailo, V. S.: Physics of multishot laser damage to optical materials,V1441,392-405(1991)

Manens, J. P. See Fresne, Francoise: V1444,26-36(1991)

Maneval, Jean-Paul See Gilabert, Alain: V1477,140-147(1991)

Manfait, Michel
; Morjani, Hamid; Millot, Jean-Marc; Debal, Vincent; Angiboust, Jean-Francois; Nabiev, I. R.: Drug-target interactions on a single living cell: an approach by optical microspectroscopy,V1403,695-707(1991)

Mang, Thomas S.
; McGinnis, Carolyn; Crean, David H.; Khan, S.; Liebow, Charles: Fluorescence detection of tumors: studies on the early diagnosis of microscopic lesions in preclinical and clinical studies,V1426,97-110(1991)
; Wieman, T. J.; Crean, David H.: Studies on the absence of photodynamic mechanism in the normal pancreas,V1426,188-199(1991)

Mangili, Fulvia See Bruzzone, Elisabetta: V1526,96-112(1991)

Manhart, Paul K. See Schember, Helene R.: V1540,51-62(1991)

Manhart, Sigmund See Birkl, Reinhard: V1522,252-258(1991)

Mani, Habib
; Joullie, Andre F.: Some characteristics of 3.2 um injection lasers based on InAsSb/InAsSbP system,V1362,38-48(1991)
See Tournie, Eric: V1361,641-656(1991)

Mani, Siva A. See Dansereau, Jeffrey P.: V1409,67-82(1991)

Manikalo, V. V. See Vlasov, Nikolai G.: V1358,1018-1020(1991)

Manikopoulos, Constantine N. See Sun, Huifang: V1605,214-220(1991)

Manivannan, Gurusamy See Lessard, Roger A.: V1559,438-448(1991)

Mankovich, Nicholas J.
; Yue, Alvin; Ammirati, Mario; Kioumehr, Farhad; Turner, Scott: Solid models for CT/MR image display: accuracy and utility in surgical planning,V1444,2-8(1991)

Mann, Allen
: Infrared zoom lenses in the eighties and beyond,V1540,338-349(1991)

Mann, David M. See Caveny, Leonard H.: V1479,102-110(1991)

Mann, James
: Implementing early visual processing in analog VLSI: light adaptation,V1473,128-136(1991)

Mann, Klaus R.
; Gerhardt, Harald: Damage testing of optical components for high-power excimer lasers,V1503,176-184(1991)

Mann, L. G. See Bellman, Robert H.: V1544,209-217(1991)

Mann, Steven
; Haykin, Simon: Adaptive chirplet: an adaptive generalized wavelet-like transform,V1565,402-413(1991)

Mannik, Len See Brown, Stewart K.: V1584,15-22(1991)

Manning, Raymond A. See Betros, Robert S.: V1542,406-419(1991)

Manno, D.
See Ferrara, M.: V1513,130-136(1991)
See Righini, Giancarlo C.: V1513,105-111(1991)

Manousos, S. N. See Long, J. A.: V1468,446-454(1991)

Manseur, Zohra Z.
; Wilson, David C.: Decomposition and inversion of von Neumann-like convolution operators,V1568,164-173(1991)

Mansfield, William M.
See Taylor, Gary N.: V1343,258-273(1991)
See White, Donald L.: V1343,204-213(1991)

Manske, Loni M.
; Graves, David B.: Origins of asymmetry in spin-cast films over topography,V1463,414-422(1991)

Mansson, Lars G. See Angelhed, Jan-Erik: V1444,159-170(1991)

Mansuripur, Masud
: Enumerative modulation coding with arbitrary constraints and postmodulation error correction coding for data storage systems,V1499,72-86(1991)

Mantalas, Lynda C.
See Bennett-Lilley, Marylyn H.: V1464,127-136(1991)
See Troccolo, Patrick M.: V1464,90-103(1991)

Mantei, Thomas D.
: Electrical probe diagnostics for processing discharges,V1392,466-473(1991)

Mantini, A. R. See Smulevich, G.: V1403,125-127(1991)

Mantulin, William W. See Fishkin, Joshua B.: V1431,122-135(1991)

Mantz, Joseph L. See Hong, C. S.: V1418,177-187(1991)

Manus, Claude See Andre, Michel L.: V1502,286-298(1991)

Manykin, Edward A.
; Chernishev, Nicholas A.: Photon echo as a tool for optical digital processing,V1505,141-146(1991)
See Belov, M. N.: V1621,268-279(1991)

Manylov, V. I. See Vizir, V. A.: V1411,63-68(1991)

Manzke, Recardo See Kipp, Lutz: V1361,794-801(1991)

Manzo, Giuseppe
See Costa, Enrico: V1343,469-476(1991)
See Kaaret, Philip E.: V1548,106-117(1991)
See Shaw, Ping-Shine: V1343,485-499(1991)

Mao, Chi-Wu See Sun, Yung-Nien: V1606,653-664(1991)

Mao, Shu Z.
: Optical thin film devices,V1519,288-297(1991)
See Andreani, F.: V1519,63-69(1991)
See Andreani, F.: V1519,18-22(1991)

Mao, Xianjun
; Liu, Shutian; Wang, Ruibo: Optoelectronic compare-and-exchange switches based on BILED circuits,V1563,58-63(1991)

Mao, Xuguang See Shen, Zhen-kang: V1481,233-240(1991)

Maoz, O. See Rotman, Stanley R.: V1442,194-204(1991)

Mar, Lim O. See LaMont, Douglas V.: V1482,2-12(1991)

Marable, William P.
; Tang, Cha-Mei; Esarey, Eric: Simulation of free-electron lasers in the presence of measured magnetic field errors of the undulator,V1552,80-86(1991)

Maradudin, Alexei A.
; Michel, T.: Role of the surface height correlation function in the enhanced backscattering of light from random metallic surfaces,V1558,233-250(1991)
See Gu, Zu-Han: V1530,60-70(1991)
See Ruiz-Cortes, Victor: V1558,222-232(1991)

Maragos, Petros See Yang, Ping-Fai: V1606,294-308(1991)

Marais, T. See Michau, Vincent: V1487,64-71(1991)

Maranichenko, N. I. See Averin, V. I.: V1358,589-602(1991)

Marcal, Joao P. See Freitas, Jose C.: V1399,16-23(1991)

Marcelja, Frane See Rainer, Frank: V1441,247-254(1991)

Marcerou, Jean-Francois
  ; Fevrier, Herve A.; Ramos, Josiane; Auge, Jacques C.; Bousselet, Philippe: General theoretical approach describing the complete behavior of the erbium-doped fiber amplifier,V1373,168-186(1991)
  See Briancon, Anne-Marie: V1373,9-20(1991)

Marchand, Jean J. See Ababou, Soraya: V1361,706-711(1991)

Marchand, Philippe J.
  ; Krishnamoorthy, Ashok V.; Ambs, Pierre; Gresser, Julien; Esener, Sadik C.; Lee, Sing H.: Design of a motionless head for parallel readout optical disk,V1505,38-49(1991)
  See Esener, Sadik C.: V1562,11-20(1991)
  See Urquhart, Kristopher S.: V1555,214-223(1991)

Marchant, John A.
  See Schofield, C. P.: V1379,209-219(1991)
  See Tillett, Robin D.: V1379,201-208(1991)

Marchant, Roy See Cadou, Christopher P.: V1434,67-77(1991)

Marchegiano, Joseph E.
  See Booth, Bruce L.: V1377,57-63(1991)
  See Hartman, Davis H.: V1390,368-376(1991)

Marchenko, S. N.
  ; Smirnova, S. N.: Registration of 3-D holograms of diamond crystals,V1238,371-371(1991)
  See Smirnova, S. N.: V1238,370-370(1991)

Marchiori, Livio D. See Nava, Enzo: V1417,307-315(1991)

Marchon, Bruno See Ager, Joel W.: V1437,24-31(1991)

Marcin, John See Udd, Eric: V1418,134-152(1991)

Marcovitch, Orna
  ; Zipin, Hedva; Klein, Z.; Lubezky, I.: Infrared transparent conductive coatings deposited by activated reactive evaporation,V1442,58-58(1991)

Marcozzi, Elena See Brofferio, Sergio C.: V1605,894-905(1991)

Marcus, Beth A. See Patrick, Nicholas J.: V1387,215-222(1991)

Marcus, Michael A.
  : Fiber optic sensors for process monitoring and control,V1398,194-205(1991)
  : Process monitoring and control with fiber optics,V1368,191-202(1991)

Marcus, Stuart L.
  : Current status of photodynamic therapy for human cancer,V1426,301-310(1991)

Marczak, Jan
  ; Rycyk, Antoni; Sarzynski, Antoni: Positive-branch unstable resonator for Nd:YAG laser with Q-switching,V1391,48-51(1991)
  See Firak, Jozef: V1391,42-47(1991)

Marczuk, Krystyna
  ; Prokopovich, Ludvig P.; Roizin, Yacov O.; Rysiakiewicz-Pasek, Ewa; Sviridov, Victor N.: Porous glass structure as revealed by capacitance and conductance measurements,V1513,291-296(1991)
  See Rysiakiewicz-Pasek, Ewa: V1513,283-290(1991)

Marder, James M. See Gildner, Donald: V1485,46-53(1991)

Marder, Seth R.
  ; Cheng, Lap Tak A.; Tiemann, Bruce G.; Beratan, David N.: Structure/property relationships for molecular second-order nonlinear optics,V1560,86-97(1991)
  See Perry, Joseph W.: V1560,302-309(1991)

Maresca, Massimo
  : Packet-switching algorithm for SIMD computers and its application to parallel computer vision,V1384,206-214(1991)

Maresi, Luca See Bonsignori, Roberto: V1478,76-91(1991)

Margaritondo, Giorgio See Tolk, Norman H.: V1552,7-13(1991)

Margarone, Joseph E. See Kingsbury, Jeffrey S.: V1427,363-367(1991)

Margerum, J. D.
  ; Lackner, Anna M.; Erdmann, John H.; Sherman, E.: Addressing factors for polymer-dispersed liquid-crystal displays,V1455,27-38(1991)

Margitan, James See McGee, Thomas J.: V1491,182-187(1991)

Marglin, Stephen I.
  : Clinical aspects of quality assurance in PACS,V1444,83-86(1991)

Margulis, Walter
  ; Carvalho, Isabel C.; Lesche, Bernhard: Frequency doubling in optical fibers: a complex puzzle,V1516,60-66(1991)

Marhic, Michel E.
  See Pan, Anpei: V1396,99-106(1991)
  See Pan, Anpei: V1461,8-16(1991)

Maricot, S. See Aboulhouda, S.: V1362,494-498(1991)

Mariella, Raymond P. See Morse, Jeffrey D.: V1378,55-59(1991)

Marin, L. D. See Li, Hua: V1376,172-179(1991)

Marine, W. See Gerri, Mireille: V1503,280-291(1991)

Marineau, Peter See Bhagvati, Chakravarthy: V1606,112-119(1991)

Marinelli, M. See Pizzoferrato, R.: V1409,192-201(1991)

Marini, Andrea E.
  ; Della Torre, Antonio: Definition of interorbit link optical terminals with diode-pumped Nd:host laser technology,V1522,222-233(1991)
  ; Della Torre, Antonio; Popescu, Alexandru F.: Nd:host laser-based optical terminal development study for intersatellite links,V1417,200-211(1991)
  See Nava, Enzo: V1417,307-315(1991)

Marini, Michele
  ; Albanese, Andres: Management of an adaptable-bit-rate video service in a MAN environment,V1364,289-294(1991)

Marino, S. A. See Tsaur, Bor-Yeu: V1540,580-595(1991)

Maris, Michael B.
  ; Mayevsky, Avraham; Sevick, Eva M.; Chance, Britton: Frequency-domain measurements of changes of optical pathlength during spreading depression in a rodent brain model,V1431,136-148(1991)
  See Chance, Britton: V1525,68-82(1991)
  See Haselgrove, John C.: V1431,30-41(1991)
  See Sevick, Eva M.: V1431,264-275(1991)

Markandey, Vishal See Russo, Paul: V1383,60-71(1991)

Marker, Alexander J.
  : Optical properties: a trip through the glass map,V1535,60-77(1991)
  ; Hayden, Joseph S.; Speit, Burkhard: Radiation-resistant optical glasses,V1485,160-172(1991)

Markert, Thomas H.
  ; Bauer, J.; Canizares, Claude R.; Isobe, T.; Nenonen, Seppo A.; O'Connor, J.; Schattenburg, Mark L.; Flanagan, Kathryn A.; Zombeck, Martin V.: Proportional counter windows for the Bragg crystal spectrometer on AXAF,V1549,408-419(1991)
  See Flanagan, Kathryn A.: V1549,395-407(1991)

Markham, Brian L.
  ; Ahmad, Suraiya P.; Jackson, Ray D.; Moran, M. S.; Biggar, Stuart F.; Gellman, David I.; Slater, Philip N.: Radiometric calibration of an airborne multispectral scanner,V1493,207-214(1991)

Markhvida, Igor V.
  ; Tanin, Leonid V.; Drobot, Igor L.: Experimental investigation of speckle-size distribution,V1508,128-134(1991)
  See Dick, Sergei C.: V1454,447-452(1991)

Markin, V. S. See Volkov, Alexander G.: V1436,68-79(1991)

Markland, Chris A.
  : METEOSAT second-generation program,V1490,39-50(1991)

Markmiller, U. See Waidhauser, Erich: V1428,75-83(1991)

Markov, L. S. See Nasibov, Alexander S.: V1361,901-908(1991)

Markov, Vladimir B.
  : Display and applied holography in culture development,VIS08,268-304(1991)
  ; Mironyuk, G. I.: Holography in museums of the Ukraine,V1238,340-347(1991)
  ; Shishkov, Vladimir F.: Backward diffraction of the light by the phase-transmission holographic grating,V1238,41-43(1991)
  ; Shishkov, Vladimir F.: Bragg diffraction with multiple internal reflections,V1238,30-40(1991)
  ; Shishkov, Vladimir F.; Voroshnin, A. B.: Comparison of images on the basis of structural features,V1238,118-122(1991)
  See Darsky, Alexei M.: V1238,54-61(1991)
  See Darsky, Alexei M.: V1509,36-46(1991)
  See Jeong, Tung H.: V1238(1991)

Markowitz, Raymond See Even-Or, Baruch: V1371,161-169(1991)

Marks, Tobin J.
  See Allan, D. S.: V1560,362-369(1991)
  See Wessels, Bruce W.: V1394,232-237(1991)

Marmarelis, Vasilis Z. See Courellis, Spiridon H.: V1606,336-349(1991)

Marmur, I. Y. See Amosova, L. P.: V1440,406-413(1991)

Marom, Emanuel
    See Mendlovic, David: V1442,182-192(1991)
    See Sneh, Anat: V1442,252-257(1991)

Marone, Giuseppe
    ; Calvani, Riccardo A.; Caponi, Renato: Polarization modulation with
    frequency shift redundancy and frequency modulation with polarization
    redundancy for POLSK and FSK systems,V1579,122-132(1991)

Marosi, Istvan See Kovacs, Emoke: V1384,338-343(1991)

Marowsky, Gerd
    ; Luepke, G.: Second-order and third-order processes as diagnostic tools for
    the analysis of organic polymers: an overview,V1560,328-334(1991)

Marquis, Jacques See Jaggi, Bruno: V1448,89-97(1991)

Marquis-Weible, Fabienne D. See Chambettaz, Francois: V1427,134-140(1991)

Marrett, Sean See Evans, Alan C.: V1445,236-246(1991)

Marron, Joseph C. See Cederquist, Jack N.: V1416,266-277(1991)

Marsh, Harry H.
    ; Hayes, Raymond; Blouke, Morley M.; Yang, Fanling H.: Back-illuminated
    1024 x 1024 quadrant readout imager: operation and screening test
    results,V1447,298-309(1991)

Marsh, John H. See Hansen, Stein I.: V1362,361-369(1991)

Marsh, Kenneth A. See Mahoney, Michael J.: V1494,182-193(1991)

Marsh, P. F. See Shewchun, John: V1477,115-138(1991)

Marsh, S. J. See Golden, Jeffry: V1407,418-429(1991)

Marshalek, Robert G.
    : Optical approach to proximity-operations communications for Space Station
    Freedom,V1417,53-62(1991)
    See Begley, David L.: V1522,234-242(1991)

Marshalkovich, A. S. See Askadskij, A. A.: V1554A,426-431(1991)

Marshall, Andrew D.
    : Automatically inspecting gross features of machined objects using three-
    dimensional depth data,V1386,243-254(1991)

Marshall, Daniel R.
    : Focus sensing method with improved pattern noise rejection,V1499,332-
    339(1991)

Marshall, David B. See Dadkhah, Mahyar S.: V1554A,164-175(1991)

Marshall, Gerald F.
    ; Vettese, Thomas; Carosella, John H.: Butterfly line scanner: rotary twin
    reflective deflector that desensitizes scan-line jitter to wobble of the
    rotational axis,V1454,37-45(1991)
    See Beiser, Leo: V1454(1991)

Marshall, Ian W.
    See Barnsley, Peter E.: V1378,116-126(1991)
    See Spirit, David M.: V1373,197-208(1991)

Marshall, Jonathan A.
    : Adaptive neural methods for multiplexing oriented edges,V1382,282-
    291(1991)
    : Challenges of vision theory: self-organization of neural mechanisms for
    stable steering of object-grouping data in visual motion
    perception,V1569,200-215(1991)

Marshall, Larry R.
    ; Hays, Alan D.; Kasinski, Jeff; Burnham, Ralph L.: Highly efficient optical
    parametric oscillators,V1419,141-152(1991)
    See Koechner, Walter: V1522,169-179(1991)

Marshall, Martin S.
    ; Benner, Robert E.: Spherical and nonspherical aerosol and particulate
    characterization using optical pattern recognition techniques,V1564,121-
    134(1991)

Marshall, Paul W. See Dale, Cheryl J.: V1447,70-86(1991)

Marshall, William C.
    : Evaluation of image tracker algorithms,V1483,207-218(1991)

Marshall, William K. See Chen, Chien-Chung: V1417,170-181(1991)

Marston, Robert K. See Casey, William L.: V1417,89-98(1991)

Marszalec, Janusz A.
    ; Heikkila, Tapio A.; Jarviluoma, Markku: Dynamic range data acquisition
    and pose estimation for 3-D regular objects,V1382,443-452(1991)

Martens, Jon S.
    ; Ginley, David S.; Zipperian, Thomas E.; Hietala, Vincent M.; Tigges, Chris
    P.: Thin film processing and device fabrication in the Tl-Ca-Ba-Cu-O
    system,V1394,140-149(1991)

Martin, Andrew See Mueller, Walter: V1472,66-75(1991)

Martin, Brian
    ; Snowden, Ian M.; Mortimer, Simon H.: Two-layer 1.2-micron pitch
    multilevel metal demonstrator using resist patterning by surface imaging
    and dry development,V1463,584-594(1991)

Martin, Charles J. See Fischer, Robert C.: V1494,414-418(1991)

Martin, Christopher
    : X-ray interferometric observatory,V1549,203-220(1991)

Martin, Claude E. See Wells, James A.: V1485,2-12(1991)

Martin, E.
    ; Pardo, A.; Poyato, J. M.; Weigand, R.; Guerra, J. M.: N-substituted 1,8-
    naphthalimide derivatives as high-efficiency laser-dye: dependence of
    dye laser emission of protonated solvent,V1397,835-838(1991)

Martin, Gerard M.
    ; Frijlink, P. M.: New materials for high-performance III-V ICs and OEICs: an
    industrial approach,V1362,67-74(1991)

Martin, James P.
    : Unique symbol for marking and tracking very small semiconductor
    products,V1598,206-220(1991)

Martin, Jean-Louis See Woodruff, William H.: V1432,205-210(1991)

Martin, John C. See Soper, Steven A.: V1435,168-178(1991)

Martin, L. R.
    : Direct measurement of H-atom sticking coefficient during diamond film
    growth,V1534,175-182(1991)

Martin, P. See Pierrat, Christophe: V1466,248-256(1991)

Martin, Peter M. See Wilson, Scott R.: V1441,82-86(1991)

Martin, Philip J.
    See Ashwell, Geoffrey J.: V1361,589-598(1991)
    See Smith, Geoffrey B.: V1536,126-137(1991)

Martin, Philippe
    See Lefevre, Herve C.: V1365,65-73(1991)
    See Lefevre, Herve C.: V1367,72-80(1991)

Martin, R. See Paek, Eung Gi: V1621,340-350(1991)

Martin, Stephen W. See Aviles, Walter A.: V1388,587-597(1991)

Martin, Tom A.
    : SAW real-time Doppler analysis,V1416,52-58(1991)

Martin, William D.
    : Electronic materials basic research program managed by the Advanced
    Technology Directorate of the U.S. Army Strategic Defense
    Command,V1559,10-17(1991)

Martinelli, Mario
    See De Maria, Letizia: V1366,304-312(1991)
    See Sansonetti, Pierre: V1588,198-209(1991)

Martinelli, Michael A. See Gregory, Kenton W.: V1425,217-225(1991)

Martinez, A. S. See Hemmati, Hamid: V1417,476-483(1991)

Martinez, Andrew B. See Zhang, Peng: V1569,398-409(1991)

Martinez, James C. See Bergman, Larry A.: V1364,14-21(1991)

Martinez, Ralph
    ; Nam, Jiseung; Dallas, William J.; Osada, Masakazu; McNeill, Kevin M.;
    Ozeki, Takeshi; Komatsu, Ken-ichi: Picture archiving and
    communications systems protocol based on ISO-OSI
    standard,V1446,100-107(1991)
    See Alsafadi, Yasser H.: V1446,129-140(1991)

Martinez-Herrero, R.
    ; Mejias, P. M.: Beam quality in laser amplifiers,V1397,623-626(1991)
    ; Mejias, P. M.: Laser beam propagation through inhomogeneous amplifying
    media,V1397,619-622(1991)
    ; Mejias, P. M.; Hernandez Neira, Jose Luis; Sanchez, Miguel: Propagation
    invariance of laser beam parameters through optical systems,V1397,627-
    630(1991)
    See Serna, Julio: V1397,631-634(1991)

Martinez-Uriegas, Eugenio
    : Simulation of parvocellular demultiplexing,V1453,300-313(1991)

Martinez-Val, J. M. See Velarde, G.: V1502,242-257(1991)

Martini, Giuseppe See Donati, Silvano: V1389,665-671(1991)

Martino, Anthony J. See Short, Ralph C.: V1417,131-141(1991)

Martino, M. See D'Anna, Emilia: V1503,256-268(1991)

Martino, Ronald M. See Conley, Willard E.: V1466,53-66(1991)

Martone, Joseph P. See Croskey, Charles L.: V1491,323-334(1991)

Martucci, Stephen A.
    : Signal extension and noncausal filtering for subband coding of
    images,V1605,137-148(1991)

Martz, A. See Cunin, B.: V1358,606-613(1991)

Maruta, Akihiro
; Matsuhara, Masanori: Analysis of lightwave propagation in the bent waveguide by the Galerkin Method,V1583,307-313(1991)

Maruyama, Masanori See Uwabu, H.: V1605,928-937(1991)

Maruyama, Mitsuru
; Sakamoto, Hideki; Ishibashi, Yutaka; Nishimura, Kazutoshi: High-speed hardware architecture for high-definition videotex system,V1605,916-927(1991)

Maruyama, Takesuke See Yatsu, Masahiko: V1354,663-668(1991)

Maruyama, Xavier K.
See Piestrup, Melvin A.: V1552,214-239(1991)
See Rule, Donald W.: V1552,240-251(1991)
See Uberall, Herbert: V1552,198-213(1991)

Marx, Egon
; Lettieri, Thomas R.; Vorburger, Theodore V.; McIntosh, Malcolm B.: Sinusoidal surfaces as standards for BRDF instruments,V1530,15-21(1991)
; Song, J. F.; Vorburger, Theodore V.; Lettieri, Thomas R.: Light scattered by coated paper,V1332,826-834(1991)
See Postek, Michael T.: V1464,35-47(1991)

Marx, Jay N.
: Status of the Advanced Light Source,V1345,2-10(1991)

Marx, Thomas A.
: Advanced processes for the shaping of the Zerodur glass ceramic,V1398,94-94(1991)
: Advanced processing of the Zerodur R glass ceramic,V1535,130-135(1991)

Mary, Joel
; Bernard, Yves; Lefaucheux, Francoise; Gonzalez, Francois: HOLIDDO: an interferometer for space experiments,V1557,147-155(1991)

Marzin, Jean Y. See Sermage, Bernard: V1361,131-135(1991)

Marzocchi, M. P. See Smulevich, G.: V1403,125-127(1991)

Masajada, Jan
: Aberrations of holographic lens recorded on surface of revolution with shifted pupil,V1574,188-196(1991)

Masaki, Atsushi
See Okada, Shuji: V1560,25-34(1991)
See Suda, Yasumasa: V1560,75-83(1991)

Masaki, Ichiro
: Industrial vision systems based on a digital image correlation chip,V1473,90-94(1991)

Masarikova, D. See Hianik, T.: V1402,85-88(1991)

Masclet, Philippe See Herve, Philippe: V1487,387-395(1991)

Masek, Jiri
See Zogg, Hans: V1361,406-413(1991)
See Zogg, Hans: V1361,1079-1086(1991)

Masek, Karel See Rohlena, Karel: V1415,259-268(1991)

Maserjian, Joseph L.
: Long-wave infrared detectors based on III-V materials,V1540,127-134(1991)
; Larsson, Anders G.: Low-power optically addressed spatial light modulators using MBE-grown III-V structures,V1562,85-92(1991)

Mash, Eugene A.
See Ito, Hiroshi: V1466,408-418(1991)
See Sooriyakumaran, Ratna: V1466,419-428(1991)

Maslenitsyn, S. F.
; Svetovoy, V. B.: Thermo-activated photoetching of PMMA in direct writing by laser beam,V1440,436-441(1991)

Maslennikov, V. L. See Bazakutsa, P. V.: V1440,370-376(1991)

Masliah, Denis A.
; Cole, Brad; Platzker, Aryeh; Schindler, Manfred: High-efficiency dual-band power amplifier for radar applications,V1475,113-120(1991)

Mason, D. See de Salabert, Arturo: V1521,74-88(1991)

Mason, J. L. See Jones, Edwin R.: V1457,318-321(1991)

Mason, James J. See Lambros, John M.: V1554A,70-83(1991)

Mason, Peter V.
; Kiceniuk, T.; Plamondon, Joseph A.; Petrick, Walt: Space Infrared Telescope Facility cryogenic and optical technology,V1540,88-96(1991)

Masory, Oren
: Detection of tool wear using multisensor readings defused by artificial neural network,V1469,515-525(1991)

Massarella, Alistair J. See Cryan, Robert A.: V1372,64-71(1991)

Massaro, Enrico
See Costa, Enrico: V1343,469-476(1991)
See Kaaret, Philip E.: V1548,106-117(1991)

Massicott, Jennifer F.
; Wyatt, Richard; Ainslie, B. J.; Craig-Ryan, S. P.: Efficient, high-power, high-gain Er3+-doped silica fiber amplifier,V1373,93-102(1991)

Massies, J. See Chen, Yong: V1361,860-865(1991)

Massimino, Michael J. See Patrick, Nicholas J.: V1387,215-222(1991)

Masso, Jon D. See Ning, Xiaohui: V1545,125-129(1991)

Massoni, Jean A. See Singer, Christian: V1494,255-264(1991)

Masten, Michael K.
; Stockum, Larry A.; eds.: *Acquisition, Tracking, and Pointing V*,V1482(1991)

Masters, Barry R.
: Confocal light microscopy of the living in-situ ocular lens: two- and three-dimensional imaging,V1443,288-293(1991)
: Design considerations of a real-time clinical confocal microscope,V1423,8-14(1991)
: Fractal patterns in the human retina and their physiological correlates,V1380,218-226(1991)
: New developments in CCD imaging devices for low-level confocal light imaging,V1428,169-176(1991)
: Performance of a thinned back-illuminated CCD coupled to a confocal microscope for low-light-level fluorescence imaging,V1447,56-63(1991)
: Three-dimensional confocal microscopy of the living cornea and ocular lens,V1450,286-294(1991)
: Use of a cooled CCD camera for confocal light microscopy,V1448,98-105(1991)
; Kriete, Andres; Kukulies, Joerg: Confocal redox fluorescence microscopy for the evaluation of corneal hypoxia,V1431,218-223(1991)
; Lee, Yimkul; Rhodes, William T.: Fourier transform method to determine human corneal endothelial morphology,V1429,82-90(1991)

Masthay, M. B. See Birge, Robert R.: V1432,129-140(1991)

Mastin, Gary A.
; Plimpton, Steven J.; Ghiglia, Dennis C.: Massively parallel synthetic-aperture radar autofocus,V1566,341-352(1991)

Mastrocinque, Michele See Andreoni, Alessandra: V1525,351-366(1991)

Mastroianni, Marta See Morgan, Alan R.: V1426,350-355(1991)

Masuda, Hisashi See Kubota, Shigeo: V1354,572-580(1991)

Masuda, Isao See Yanagita, Hiroaki: V1513,386-395(1991)

Masuda, Rika See Kanemaru, Toyomi: V1496,118-123(1991)

Masuda, Shoichi See Watanabe, Hiroshi: V1490,317-323(1991)

Masuda, Shun-ichi See Iwahashi, Masahiro: V1605,844-850(1991)

Masuda, Wataru
: Visual observation and numerical analysis on the reaction zone structure of a supersonic-flow CO chemical laser,V1397,531-534(1991)

Masuda, Yu
See Inokuchi, Ikuo: V1429,39-45(1991)
See Maeta, Manabu: V1429,152-161(1991)

Masuda, Yukihisa See Ogura, Toshihiro: V1443,153-157(1991)

Masuhara, Hiroshi See Toriumi, Minoru: V1466,458-468(1991)

Masui, Nobuhiko See Akamatsu, Shigeru: V1606,204-216(1991)

Masulli, Francesco See Riani, Massimo: V1469,166-177(1991)

Masychev, Victor I.
; Alejnikov, Vladislav S.: Alternated or simultaneous sealed-off room temperature CO/CO2 laser tuning by chemical reactions,V1412,227-235(1991)
; Alejnikov, Vladislav S.; Klimenko, Vladimir I.: Coagulation and precise ablation of biotissues by pulsed sealed-off carbon monoxide laser,V1427,344-356(1991)
See Alejnikov, Vladislav S.: V1412,173-184(1991)

Maszczak, M. W. See Albanese, Andres: V1364,320-326(1991)

Matake, S. See Hayase, Rumiko H.: V1466,438-445(1991)

Mataloni, Raymond J.
: Application of lighter-than-air platforms to law enforcement,V1479,306-315(1991)
See Gowrinathan, Sankaran: V1479(1991)

Matejec, Vlastimil
; Sasek, Ladislav; Gotz, J.; Ivanov, G. A.; Koreneva, N. A.; Grigor'yants, Vil V.: New gas-phase etching method for preparation of polarization-maintaining fibers,V1513,174-179(1991)

Matejka, Frantisek See Urban, Frantisek: V1574,58-65(1991)

Mathae, J. C. See Mens, Alain: V1358,719-731(1991)

Mathew, A. See Dougherty, Edward R.: V1451,137-147(1991)

Mathew, Joseph See Golden, Jeffry: V1407,418-429(1991)

Mathews, Allen R.
; Warnes, Richard H.; Hemsing, Willard F.; Whittemore, Gerald R.: Line-imaging Fabry-Perot interferometer,V1346,122-132(1991)
See Hemsing, Willard F.: V1346,133-140(1991)

Mathews, Eric D. See Friedl, Stephan E.: V1425,134-141(1991)

Mathews, V. J. See Syed, Mushtaq A.: V1565,25-34(1991)

Mathews, V. K. See Yu, Chang: V1598,186-197(1991)

Mathey, P.
; Launay, Jean C.; Pauliat, G.; Roosen, Gerald: Novel technique for efficient wave mixing in photorefractive materials,V1500,26-33(1991)

Mathey, Yves
; Pailharey, Daniel; Gerri, Mireille; Bonnot, Anne-Marie; Sorbier, J. P.: Anisotropic electrical and optical behavior and preferential orientation in early transition metal tellurides thin films,V1361,909-916(1991)
See Gerri, Mireille: V1503,280-291(1991)

Mathies, Richard A.
See Doig, Stephen J.: V1432,184-196(1991)
See Reid, Philip J.: V1432,172-183(1991)

Mathieu, Christophe See Friboulet, Denis: V1445,106-117(1991)

Mathine, David L. See Kudlek, Gotthard: V1361,150-158(1991)

Mathis, Donald W. See Wolfe, William J.: V1382,240-254(1991)

Mathorne, Jan See Jensen, Jorgen A.: V1444,221-231(1991)

Mathur, Anoop See Samad, Tariq: V1469,766-777(1991)

Mathur, B. P. See Mehrotra, R.: V1463,487-491(1991)

Mathur, Bimal P.
; Koch, Christof; eds.: *Visual Information Processing: From Neurons to Chips,*V1473(1991)
; Wang, H. T.; Liu, Shih-Chii; Koch, Christof; Luo, Jin: Pixel level data fusion: from algorithm to chip,V1473,153-160(1991)

Mathur, Rajive K.
; Sanderson, Arthur C.: Hierarchical planner for space truss assembly,V1387,47-57(1991)

Matic, P. See Merzbacher, Celia I.: V1533,222-228(1991)

Matoba, Akio See Kashima, Yasumasa: V1365,102-107(1991)

Matson, Charles L.
See Carmona, Edward A.: V1566,329-340(1991)
See Roggemann, Michael C.: V1542,477-487(1991)

Matsuda, Hikaru See Kimura, Shigeru: V1483,10-17(1991)

Matsuda, Hiro
See Nomura, Shintaro: V1560,272-277(1991)
See Okada, Shuji: V1560,25-34(1991)
See Suda, Yasumasa: V1560,75-83(1991)

Matsuda, Kenichi
; Shibata, Jun: Optoelectronic approach to optical parallel processing based on the photonic parallel memory,V1562,21-29(1991)

Matsuda, Kiyofumi
; Tenjimbayashi, Koji: Holographic measurement of the angular error of a table moving along a slideway,V1332,230-235(1991)
See Muramatsu, Mikiya: V1332,792-797(1991)

Matsuda, Koyo See Oyama, Yoshiro: V1444,413-423(1991)

Matsuda, Minoru See Gozdz, Antoni S.: V1466,200-205(1991)

Matsuda, Shuichi See Miyazaki, Junji: V1464,327-335(1991)

Matsuda, Yasuhiko See Tanazawa, Takeshi: V1435,310-321(1991)

Matsuhara, Masanori See Maruta, Akihiro: V1583,307-313(1991)

Matsuhata, H. See Hugsted, B.: V1361,751-757(1991)

Matsui, Kenichi
See Adachi, Yuzi: V1527,225-233(1991)
See Honda, Tatsuro: V1540,709-716(1991)

Matsui, Kineo See Tanaka, Kiyoshi: V1605,646-649(1991)

Matsui, Shinji See Takeuchi, Ichiro: V1394,96-101(1991)

Matsui, Toshikazu
; Hirahara, Shuzo: New human vision system model for spatio-temporal image signals,V1453,282-289(1991)

Matsukura, Noritsuga See Takeuchi, Ichiro: V1394,96-101(1991)

Matsumoto, Hiroshi See Yamazaki, Satomi: V1466,538-545(1991)

Matsumoto, Kiyoshi See Seya, Eiichi: V1499,269-273(1991)

Matsumoto, S. See Lim, Eric J.: V1561,135-142(1991)

Matsumoto, Shuichi
; Hamada, Takahiro; Saito, Masahiro; Murakami, Hitomi: 45-Mbps multichannel TV coding system,V1605,37-46(1991)

Matsumoto, Tetsuya
; Kitagawa, Yoichi; Adachi, Masaaki; Hayashi, Akihiro: Laser moire topography for 3-D contour measurement,V1332,530-536(1991)

Matsumoto, Toshiro See Kojima, Arata: V1429,162-171(1991)

Matsumoto, Tsuneo See Yoshimura, Hitoshi: V1445,47-51(1991)

Matsumoto, Y. See Izawa, Takao: V1441,339-344(1991)

Matsunaga, Tadayo See Suganuma, Heiroku: V1592,12-17(1991)

Matsunawa, Akira
: Present and future trends of laser materials processing in Japan,V1502,60-71(1991)

Matsuo, Seitaro See Nishioka, Teiichi: V1554A,802-813(1991)

Matsuo, Takahiro See Misaka, Akio: V1465,174-184(1991)

Matsuoka, Masaru
; Yamauchi, Masamitsu; Nakamura, Haruo; Kondo, M.; Kawai, N.; Yoshida, A.; Imai, Tohru: One-dimensional proportional counters for x-ray all-sky monitors,V1549,2-8(1991)

Matsuoka, Yasushi See Aibara, Tsunehiro: V1606,198-203(1991)

Matsushiro, Nobuhito
; Asada, Osamu; Tsuji, Kenzo: Arithmetic coding model for color images processed by error diffusion,V1452,21-26(1991)

Matsushita, Tadashi
; Suzuki, Hiroshi; Wakabayashi, Satoshi; Tajime, Toru: High-resolution thermal imager with a field-of-view of 112 degrees,V1488,368-375(1991)

Matsuzaka, Fumio
; Nigawara, Kazushige; Terasawa, Ken; Uchiyama, Taro: Second harmonic generation of chemical oxygen-iodine laser,V1397,239-242(1991)

Matsuzaki, Akiyoshi See Sasano, Yasuhiro: V1490,233-242(1991)

Matt, Giorgio
See Costa, Enrico: V1343,469-476(1991)
See Kaaret, Philip E.: V1548,106-117(1991)

Mattaini, E. See Citterio, Oberto: V1343,145-154(1991)

Matteo, Annamaria
See Armenise, Mario N.: V1583,256-267(1991)
See Armenise, Mario N.: V1583,289-297(1991)
See Conese, Tiziana: V1583,249-255(1991)

Mattheus, Rudy A.
; Temmerman, Yvan; Verhellen, P.; Osteaux, Michel: Management system for a PACS network in a hospital environment,V1446,341-351(1991)

Matthews, James E.
; Cucalon, Antoine: Reliability of planar optical couplers,V1366,206-214(1991)

Matthews, James L. See Sessler, Jonathan L.: V1426,318-329(1991)

Matthewson, M. J.
See Lin, Bochien: V1366,157-166(1991)
See Rondinella, Vincenzo V.: V1366,77-84(1991)

Matthias, Eckart
See Hunger, Hans E.: V1441,283-286(1991)
See Hunger, Hans E.: V1598,19-26(1991)

Matthies, Larry H.
: Stereo vision for planetary rovers: stochastic modeling to near-real-time implementation,V1570,187-200(1991)

Matthiesen, David See Steiner, Bruce W.: V1557,156-167(1991)

Matthys, Donald R.
; Gilbert, John A.; Puliparambil, Joseph T.: Endoscopic inspection using a panoramic annular lens,V1554B,736-742(1991)
; Gilbert, John A.; Puliparambil, Joseph T.: Measurement of fluid velocity fields using digital correlation techniques,V1332,850-861(1991)
See Gilbert, John A.: V1554B,128-134(1991)
See Gilbert, John A.: V1554B,202-209(1991)

Mattingly, G. E.
: Fluid-flow-rate metrology: laboratory uncertainties and traceabilities,V1392,386-401(1991)

Mattioli, Juliette See Schmitt, Michel: V1469,392-403(1991)

Mattioli, Stefano
; Cremona, M.; Benaim, George; Ferrario, Angelo: Alexandrite laser and blind lithotripsy: initial experience—first clinical results,V1421,114-119(1991)

Mattioli, Tony See Lutz, Marc: V1403,59-65(1991)

Mattson, J. E. See Gluskin, Efim S.: V1548,56-68(1991)

Matty, T. C. See Leksell, David: V1458,133-144(1991)

Matull, Ralph See Ibel, Konrad: V1559,393-402(1991)

Matveev, Alexander See Larkin, Alexander I.: V1621,414-423(1991)

Mauche, C. W. See Bixler, Jay V.: V1549,420-428(1991)

Mauger, Philip E. See Thomas, Stan W.: V1358,91-99(1991)

Mauldin, Lemuel E. See McCormick, M. P.: V1491,125-141(1991)

Maurin, T.
; Devos, F.: Optical approaches to overcome present limitations for interconnection and control in parallel electronic architectures,V1505,158-165(1991)

Maurincomme, Eric See Algazi, V. R.: V1452,364-370(1991)

Mauritz, F. L. See Cross, Edward F.: V1540,756-763(1991)

Mauro, Billie R.
See Boyer, James D.: V1377,92-98(1991)
See Boyer, James D.: V1441,255-261(1991)

Mauro, M. See Beard, David V.: V1446,52-58(1991)

Mauroschat, Andreas
: Reliability analysis of a multiple-laser-diode beacon for intersatellite links,V1417,513-524(1991)

Maus, Steven See Galic, George J.: V1529,13-21(1991)

Mauvoisin, Gerard
; Bremand, Fabrice J.; Lagarde, Alexis: Improvement on phase-shifting method precision and application on shadow moire method,V1554B,181-187(1991)

Mavroidis, Theo See Dainty, J. C.: V1487,2-9(1991)

Mawst, Luke J.
; Botez, Dan; Anderson, Eric R.; Jansen, Michael; Ou, Szutsun S.; Sergant, Moshe; Peterson, Gary L.; Roth, Thomas J.; Rozenbergs, John: 0.36-W cw diffraction-limited-beam operation from phase-locked arrays of antiguides,V1418,353-357(1991)
See Anderson, Eric R.: V1417,543-549(1991)
See Botez, Dan: V1474,64-74(1991)
See Jansen, Michael: V1418,32-39(1991)

Maxey, L. C. See Thomas, Brigham B.: V1485,20-30(1991)

Maximov, G. A. See Nadezhdinskii, Alexander I.: V1418,487-495(1991)

Maximova, I. L. See Guseva, N. P.: V1403,332-334(1991)

Maxwell, Graeme D. See McGoldrick, Elizabeth: V1374,118-125(1991)

Maxwell, J. See MacCraith, Brian D.: V1510,195-203(1991)

Maxwell, Jonathan
: Tertiary spectrum manipulation in apochromats,V1354,408-416(1991)
; Hull, Chris S.: Simple multidimensional quadratic extrapolation method for the correction of specific aberrations in lens systems,V1354,277-285(1991)
See Al-Baho, Tareq I.: V1354,417-428(1991)

Maxwell, K. See Reilly, James P.: V1397,339-354(1991)

May, B. S. See Peacock, Keith: V1494,147-159(1991)

May, G. B. See Perry, Kenneth E.: V1554A,209-227(1991)

May, R. D. See Webster, Chris R.: V1540,187-194(1991)

Mayayo, E. See Trelles, Mario A.: V1403,781-798(1991)

Maydell, Ursula M.
; Hassanein, Hossam S.; Deng, Shuang: On enhancing the performance of the ACR-NEMA protocol,V1446,81-92(1991)

Mayer, Arnold
: European industrial laser market,V1517,15-27(1991)

Mayer, Max See Freischlad, Klaus R.: V1332,8-17(1991)

Mayer, Richard C.
; Dreisewerd, Douglas W.: End-to-end model of a diode-pumped Nd:YAG pulsed laser,V1415,248-258(1991)

Mayer, T. M. See Kubena, Randall L.: V1392,595-597(1991)

Mayer-Costa, Carlos M. See Feldblum, Avi Y.: V1544,200-208(1991)

Mayerhofer, Wilhelm See Beth, Mark-Udo: V1397,577-580(1991)

Mayet, Louis See Ababou, Soraya: V1361,706-711(1991)

Mayevsky, Avraham
; Flamm, E. S.; Pennie, W.; Chance, Britton: Fiber optic based multiprobe system for intraoperative monitoring of brain functions,V1431,303-313(1991)
See Maris, Michael B.: V1431,136-148(1991)

Mayhall, David J.
; Yee, Jick H.; Villa, Francesco: Computational and experimental progress on laser-activated gas avalanche switches for broadband, high-power electromagnetic pulse generation,V1378,101-114(1991)

Maymon, Peter W.
: Optical design of the moderate-resolution imaging spectrometer-tilt for the Earth Observing System,V1492,286-297(1991)

Maynard, C. D. See Boehme, Johannes M.: V1446,312-317(1991)

Maynard, William L. See Fox, Neil D.: V1417,452-463(1991)

Maynard de Lavalette, Guy M. See Tomasini, Bernard: V1468,60-71(1991)

Mayo, Michael W. See Labo, Jack A.: V1419,32-39(1991)

Mayo, Phyllis See Wiepking, Mark: V1392,139-150(1991)

Mayor, J. M. See Buczek, Harthmuth: V1574,48-57(1991)

Mazar, Reuven
; Bronshtein, Alexander: Finite aperture effects on intensity fluctuations of laser radiation in a turbulent atmosphere,V1487,361-371(1991)
; Rozental, Mark: Two-frequency correlation in a turbulent atmosphere,V1487,31-39(1991)

Mazataud, D. See Mens, Alain: V1358,719-731(1991)

Maze, Sylvie
; Joffre, Pascal; Refregier, Philippe: Influence of input information coding for correlation operations,V1505,20-31(1991)

Mazenc, Christophe See Kla, Sylvanus: V1566,275-285(1991)

Mazumdar, Sumit
; Guo, Dandan; Dixit, Sham N.: Microscopic mechanism of optical nonlinearity in conjugated polymers and other quasi-one-dimensional systems,V1436,136-149(1991)

Mazumder, Mohiuddin See Hill, Steven C.: V1497,16-27(1991)

Mazurenko, Yuri T.
: Three-dimensional holography of nonstationary waves,V1238,85-96(1991)
; Smirnov, V. V.: Theory of non-Condon femtosecond quantum beats in electronic transitions of dye molecules,V1403,466-469(1991)

Mazurkiewicz, Wojciech See Konwerska-Hrabowska, Joanna: V1554A,388-399(1991)

Mazzi, Giulio See Gonella, Francesco: V1513,425-433(1991)

Mazzinghi, Piero
: LEAF: a fiber-optic fluorometer for field measurement of chlorophyll fluorescence,V1510,187-194(1991)

Mazzola, M.
; Montrosset, Ivo; Fincato, Antonello: Design and optimization of demultiplexer in ion-exchanged glass waveguides,V1365,2-12(1991)

Mazzola, Michael S.
See Lakdawala, Vishnu K.: V1378,259-270(1991)
See Stoudt, David C.: V1378,280-285(1991)

Mazzoldi, Andrea See Grinzato, Ermanno G.: V1467,75-82(1991)

Mazzoldi, Paolo
; Carnera, Alberto; Caccavale, F.; Granozzi, G.; Bertoncello, R.; Battaglin, Giancarlo; Boscolo-Boscoletto, A.; Polato, P.: Ion beam modification of glasses,V1513,182-197(1991)
See Battaglin, Giancarlo: V1513,441-450(1991)

Mazzoleni, S. See Milani, Marziale: V1474,83-97(1991)

Mazzoni, Alessandro See Melozzi, Mauro: V1512,178-188(1991)

McAdam, Bridget A.
; Sharma, Arvind K.; Allen, B.; Kintis, M.: Insertion of emerging GaAs HBT technology in military and communication system applications,V1475,267-274(1991)

McAdam, Wylie See Daher, Reinhard: V1526,90-93(1991)

McAllister, David F.
See Barham, Paul T.: V1457,18-26(1991)
See Carver, Donald E.: V1457,54-65(1991)
See Devarajan, Ravinder: V1457,37-48(1991)
See Hebbar, Prashant D.: V1457,233-241(1991)

McArdle, N. See Craig, Robert G.: V1505,76-86(1991)

McArthur, S. P. See Aviles, Walter A.: V1388,587-597(1991)

McAulay, Alastair D.
; Wang, Junqing; Xu, Xin; Zeng, Ming: Optical parallel set of half adders using spatial light rebroadcasters,V1564,685-690(1991)
See Chen, C. L. P.: V1468,394-405(1991)

McBride, Dennis J.
: CAD in new areas of the package and interconnect design space,V1390,330-335(1991)

McBride, R. See Zheng, S. X.: V1572,359-364(1991)

McCafferty, Robert H.
: Real-time automation of a dry etching system,V1392,331-339(1991)

McCall, Samuel L. See Jewell, Jack L.: V1389,401-407(1991)

McCallion, Kevin J. See Johnstone, Walter: V1506,145-149(1991)

McCallum, John C. See Chua, Soo-Jin: V1401(1991)

McCammon, Kent G.
; Hagans, Karla G.; Hankla, A.: Noise performance of microchannel plate imaging systems,V1346,398-403(1991)

McCann, Brian P.
: Comparison of silica-core optical fibers,V1420,116-125(1991)
: Power transmission for silica fiber laser delivery systems,V1398,230-237(1991)

McCann, James T.
: Applications of diamond-turned null reflectors for generalized aspheric metrology,V1332,843-849(1991)
See Riedl, Max J.: VCR38,153-163(1991)

McCann, John J.
; Savoy, Robert L.: Measurements of lightness: dependence on the position of a white in the field of view,V1453,402-411(1991)

McCanny, J. V. See McQuillan, S. E.: V1566,220-229(1991)

McCargar, James W.
; Doane, J. W.; West, John L.; Anderson, Thomas W.: Polymer-dispersed liquid-crystal shutters for IR imaging,V1455,54-60(1991)

McCarthy, Donald W. See Wizinowich, Peter L.: V1542,148-158(1991)

McCarthy, G. See Schamiloglu, Edl: V1407,242-248(1991)

McCarthy, Kieran J. See Wells, Alan A.: V1549,357-373(1991)

McCartney, David J. See Sheat, Dennis E.: V1461,35-38(1991)

McCaughan, James S.
; Barabash, Rostislav D.; Hawley, Philip: Stage III endobronchial squamous cell cancer: survival after Nd:YAG laser combined with photodynamic therapy versus Nd:YAG laser or photodynamic therapy alone,V1426,279-287(1991)

McCauley, Howard M.
: Target cuing: a heterogeneous neural network approach,V1469,69-76(1991)
; Auborn, John E.: Image enhancement of infrared uncooled focal plane array imagery,V1479,416-422(1991)
See Auborn, John E.: V1482,246-252(1991)

McCleer, P. J. See Lachs, Gerard: V1474,248-259(1991)

McCleese, Daniel J. See Rider, David M.: V1540,142-147(1991)

McClenny, William A. See Russwurm, George M.: V1433,302-314(1991)

McClure, W. F.
: Fourier analysis of near-infrared spectra,V1379,45-51(1991)

McColl, P. See Al-Jumaily, Ghanim A.: V1441,360-365(1991)

McConville, C. F. See Ashley, Timothy: V1361,238-244(1991)

McCorkle, R. See Spiller, Eberhard: V1343,134-144(1991)

McCormack, R. P. See Karis, Thomas E.: V1499,366-376(1991)

McCormack, S. M. See Featherstone, John D.: V1424,145-149(1991)

McCormick, Frederick B.
; Tooley, Frank A.; Cloonan, Thomas J.; Brubaker, John L.; Lentine, Anthony L.; Morrison, Richard L.; Hinterlong, Steve J.; Herron, Michael J.; Walker, Sonya L.: Free space cascaded optical logic demonstration,V1396,508-521(1991)
; Tooley, Frank A.; Brubaker, John L.; Sasian, Jose M.; Cloonan, Thomas J.; Lentine, Anthony L.; Morrison, Richard L.; Crisci, R. J.; Walker, Sonya L.; Hinterlong, Steve J.; Herron, Michael J.: Optomechanics of a free-space photonic switching fabric: the system,V1533,97-114(1991)
See Brubaker, John L.: V1533,88-96(1991)
See Cloonan, Thomas J.: V1396,488-500(1991)

McCormick, M. P.
; Chu, William P.; Zawodny, J. M.; Mauldin, Lemuel E.; McMaster, Leonard R.: Stratospheric aerosol and gas experiment III: aerosol and trace gas measurements for the Earth Observing System,V1491,125-141(1991)
See Chu, William P.: V1491,243-250(1991)
See Delamere, W. A.: V1479,31-40(1991)
See Delamere, W. A.: V1447,204-213(1991)
See Theon, John S.: V1492,2-23(1991)

McCormick, Malcolm
See Davies, Neil A.: V1544,189-198(1991)
See Lau, Hon W.: V1544,178-188(1991)

McCormick, Patrick W.
; Stewart, Melville; Lewis, Gary: Noninvasive measurement of regional cerebrovascular oxygen saturation in humans using optical spectroscopy,V1431,294-302(1991)

McCown, Andrew W. See Bigio, Irving J.: V1397,47-53(1991)

McCracken, Thomas O.
; Roper, Stephen D.; Spurgeon, Thomas L.: Vesalius project: interactive computers in anatomical instruction,V1380,6-10(1991)
See Miranda, Rick: V1380,210-217(1991)

McCray, Allan G. See Yeung, Keith K.: V1569,133-146(1991)

McCready, David E. See Wilson, Scott R.: V1441,82-86(1991)

McCredie, B. D. See Merkelo, Henri: V1390,91-163(1991)

McCreery, Richard L.
: NIR/CCD Raman spectroscopy: second battle of a revolution,V1439,25-36(1991)

McCurnin, Thomas W.
; Schooley, Larry C.; Sims, Gary R.: Signal processing for low-light-level, high-precision CCD imaging,V1448,225-236(1991)

McCutchen, William C.
: Photomask fabrication utilizing a Philips/Cambridge vector scan e-beam system,V1496,97-106(1991)

McDaniel, Olin K.
: Slapper detonator flyer microphotography with a multiframe Kerr cell and Cranz-Schardin camera,V1358,1164-1179(1991)

McDermid, I. S.
; Schmoe, Martha S.; Walsh, T. D.: Ground-based lidar for long-term and network measurements of ozone,V1491,175-181(1991)

McDermott, David B.
See Chong, Chae K.: V1407,226-233(1991)
See Dodd, James W.: V1407,467-473(1991)
See Lednum, Eugene E.: V1407,202-208(1991)
See Wang, Qinsong: V1407,209-216(1991)

McDonach, Alaster See Johnstone, Walter: V1506,145-149(1991)

McDonald, Carlos
: Measurement of modulation transfer function of desert atmospheres,V1487,203-219(1991)

McDonald, Hillary L. See Abrams, Susan B.: V1420,13-21(1991)

McDonald, John F.
; Vlannes, Nickolas P.; Lu, Toh-Ming; Wnek, Gary E.; Nason, Theodore C.; You, Lu: Photonic Multichip Packaging (PMP) using electro-optic organic materials and devices,V1390,286-301(1991)

McDonald, Joseph K.
; LaiHing, Kenneth: Fabrication and nonlinear optical properties of mixed and layered colloidal particles,V1497,367-370(1991)

McDonald, Marguerite B. See Shimmick, John K.: V1423,140-153(1991)

McDonald, Mark D. See LaBelle, Gary L.: V1365,116-121(1991)

McDonald, Thomas E. See Rose, Evan A.: V1411,15-27(1991)

McDonald, Timothy P.
; Chen, Yud-Ren: Comparison of methods to treat nonuniform illumination in images,V1379,89-98(1991)

McDonnell, Peter J. See Clapham, Terrance N.: V1423,2-7(1991)

McDowell, David Q.
: Color standards for electronic imaging,VCR37,40-53(1991)
: Summary of color definition activity in the graphic arts,V1460,29-35(1991)

McDowell, Maurice W.
; Klee, H. W.; Griffith, Derek J.: New approach to synchroballistic photography,V1358,227-236(1991)

McDuff, G. G.
ed.: *Pulse Power for Lasers III*,V1411(1991)
See Harris, Kevin: V1411,87-99(1991)

McEachern, Robert
; Eckhardt, Andrew J.; Nauda, Alexander: Efficient use of data structures for digital monopulse feature extraction,V1470,226-232(1991)

McElroy, James L.
; McNeal, Robert J.; eds.: *Remote Sensing of Atmospheric Chemistry*,V1491(1991)
See Bristow, Michael P.: V1491,68-74(1991)

McEwan, Thomas E.
; Hanks, Roy L.: High-voltage picosecond pulse generation using avalanche diodes,V1346,465-470(1991)
See Kilkenny, Joseph D.: V1358,117-133(1991)

McEwen, Keith
See Liu, Kexing: V1398,206-212(1991)
See Measures, Raymond M.: V1489,86-96(1991)
See Measures, Raymond M.: V1332,431-443(1991)

McFall, John D. See Mintzer, Fred: V1460,38-49(1991)

McFarland, Michael J.
; Beeson, Karl W.; Horn, Keith A.; Nahata, Ajay; Wu, Chengjiu; Yardley, James T.: Polymeric optical waveguides for device applications,V1583,344-354(1991)
See Beeson, Karl W.: V1559,258-266(1991)
See Beeson, Karl W.: V1374,176-185(1991)

McFarland, Robert D. See McVey, Brian D.: V1441,457-468(1991)

McFee, John E.
; Russell, Kevin L.; Ito, Mabo R.: Detection of surface-laid minefields using a hierarchical image processing algorithm,V1567,42-52(1991)

McGahan, William A. See He, Ping: V1499,401-411(1991)

McGarvey, John A. See Soreide, David C.: V1416,280-285(1991)

McGary, Douglas E. See Rudberg, Donald A.: V1530,232-239(1991)

McGavran, Loris See Jericevic, Zeljko: V1428,200-213(1991)

McGee, Thomas J.
; Ferrare, Richard; Butler, James J.; Frost, Robert L.; Gross, Michael; Margitan, James: Recent lidar measurements of stratospheric ozone and temperature within the Network for the Detection of Stratospheric Change,V1491,182-187(1991)

McGeehin, Peter
: Are optical fiber sensors intrinsically, inherently, or relatively safe?,V1504,75-79(1991)

McGilp, J. F. See MacCraith, Brian D.: V1510,104-109(1991)

McGinnis, Carolyn See Mang, Thomas S.: V1426,97-110(1991)

McGlone, John T. See Kent, Richard A.: V1448,27-44(1991)

McGlynn, John D.
; Ellis, Kenneth K.; Kryskowski, David: High-resolution spectral analysis and modeling of infrared ocean surface radiometric clutter,V1486,141-150(1991)

McGoldrick, Elizabeth
; Hubbard, Steven D.; Maxwell, Graeme D.; Thomas, N.: Passive optical silica-on-silicon waveguide components,V1374,118-125(1991)

McGouey, R. P. See Deutsch, Alina: V1389,161-176(1991)

McGowan, Steve See Duffy, Neil D.: V1380,46-52(1991)

McGrann, John V. See Neev, Joseph: V1427,162-172(1991)

McGraw, Robert L.
: Stochastic simulations of light-scattering noise in two-wave mixing: application to artificial Kerr media,V1409,135-147(1991)

McGreevy, Michael W.
; Stoker, Carol R.: Telepresence for planetary exploration,V1387,110-123(1991)

McGuire, Anne E. See MacDonald, Scott A.: V1466,2-12(1991)

McHugh, Stephen W.
: High-performance FLIR testing using reflective-target technology,V1540,775-780(1991)

McInerney, John G.
: Low-frequency intensity fluctuations in external cavity semiconductor lasers,V1376,236-244(1991)
See Li, Hua: V1376,172-179(1991)

McIntire, Harold D. See Schneeberger, Timothy J.: V1567,2-8(1991)

McIntosh, Bruce C. See Howe, Gordon S.: V1434,90-103(1991)

McIntosh, D. See Trolinger, James D.: V1557,98-109(1991)

McIntosh, Malcolm B.
; McNeely, Joseph R.: Bidirectional reflectance distribution function raster scan technique for curved samples,V1530,263-268(1991)
See Marx, Egon: V1530,15-21(1991)
See McNeely, Joseph R.: V1530,288-298(1991)

McIntosh, Thomas F.
: Building and campus networks for fiber distributed data interface,V1364,84-93(1991)

McIntyre, David See Kovacich, Michael: V1481,357-370(1991)

McIntyre, Iain A.
; Giorgi, David M.; Hargis, David E.; Zucker, Oved S.: Optical requirements for light-activated switches,V1378,162-172(1991)
See Zucker, Oved S.: V1378,22-33(1991)

McIver, John K. See Becker, Wilhelm: V1441,541-552(1991)

McIver, Robert T.
; Hemminger, John C.; Parker, D.; Li, Y.; Land, Donald P.; Pettiette-Hall, C. L.: Analysis of molecular adsorbates by laser-induced thermal desorption,V1437,124-128(1991)

McKay, James See Powell, Richard K.: V1419,126-140(1991)

McKean, Dennis R.
; Allen, Robert D.; Kasai, Paul H.; MacDonald, Scott A.: Polymer effects on the photochemistry of triarylsulfonium salts,V1559,214-221(1991)

McKechnie, Thomas S.
: Focusing infrared laser beams on targets in space without using adaptive optics,V1408,119-135(1991)

McKee, Paul
; Wood, David; Dames, Mark P.; Dix, C.: Fabrication of multiphase optical elements for weighted array spot generation,V1461,17-23(1991)
See Barnes, Nigel: V1389,477-483(1991)

McKendall, Raymond
; Mintz, Max: Using robust statistics for sensor fusion,V1383,547-565(1991)

McKenna, Daniel
See Graves, J. E.: V1542,262-272(1991)
See Roddier, Francois J.: V1542,248-253(1991)

McKenna, T. A.
See Klocek, Paul: V1498,147-157(1991)
See Trombetta, John M.: V1534,77-88(1991)

McKernan, M. A. See Jankowski, Alan F.: V1343,32-38(1991)

McKie, Andrew D. See Wagner, James W.: V1332,491-501(1991)

McKiernan, M. L. See Selser, James C.: V1430,85-88(1991)

McKinley, J. T. See Tolk, Norman H.: V1552,7-13(1991)

McKinney, Wayne R. See Lerner, Jeremy M.: V1545(1991)

McKnight, D. J. See Craig, Robert G.: V1505,76-86(1991)

McKnight, William H. See Sun, C. K.: V1476,294-300(1991)

McLandrich, Matthew N. See Pappert, Stephen A.: V1476,282-293(1991)

McLaughlin, Chris See LaSala, Paul V.: V1482,121-137(1991)

McLaughlin, Dan L.
See Loubriel, Guillermo M.: V1378,179-186(1991)
See Zutavern, Fred J.: V1378,271-279(1991)

McLaughlin, Edward R. See Flanagan, Kathryn A.: V1549,395-407(1991)

McLaughlin, Kevin J. See Butler, Stephanie W.: V1392,361-372(1991)

McLaughlin, S. See Nisbet, K. C.: V1565,244-254(1991)

McLaurin, A. P. See Jones, Edwin R.: V1457,318-321(1991)

McLean, Craig See Springer, Steven C.: V1479,372-379(1991)

McLean, James T. See Walker, James H.: V1493,224-230(1991)

McLeaster, Kevin D. See Lipson, David: V1368,36-43(1991)

McLeod, Brian A. See Wizinowich, Peter L.: V1542,148-158(1991)

McMahon, A. W. See Arlinghaus, Heinrich F.: V1435,26-35(1991)

McMahon, David M. See Patty, Charles E.: V1535,13-26(1991)

McManus, J. B.
; Kebabian, Paul L.; Kolb, Charles E.: Aerodyne research mobile infrared methane monitor,V1433,330-339(1991)
; Kolb, Charles E.: Long-path intracavity laser for the measurement of atmospheric trace gases,V1433,340-351(1991)

McMaster, Leonard R.
See Chu, William P.: V1491,243-250(1991)
See McCormick, M. P.: V1491,125-141(1991)

McMath, Linda P.
; Schuger, Claudio D.; Crilly, Richard J.; Spears, J. R.: Percutaneous laser balloon coagulation of accessory pathways,V1425,165-171(1991)
See Kundu, Sourav K.: V1425,142-148(1991)

McMillan, Charles F.
; Steinmetz, Lloyd L.: Striped Fabry-Perots: improved efficiency for velocimetry,V1346,113-121(1991)

McMillen, Donald V.
: Model "T" satellite series: small satellites designed for scientific and commercial use,V1495,95-102(1991)

McMillin, Douglas J. See Dewey, John M.: V1358,246-253(1991)

McMillin, Pat See Neal, Daniel R.: V1542,449-458(1991)

McMullan, Phillip C.
: Field documentation and client presentation of IR inspections on new masonry structures,V1467,66-74(1991)

McMullin, James N.
: Propagating beam method simulation of planar multimode optical waveguides,V1389,621-629(1991)

McMurphy, Frederick E. See Maienschein, Jon L.: V1343,477-484(1991)

McMurray, Tom J. See Pearce, John A.: V1422,27-33(1991)

McNeal, Robert J.
: Upper-atmosphere research satellite: an overview,V1491,84-90(1991)
See McElroy, James L.: V1491(1991)

McNeal, Terry See Couch, Lori L.: V1486,417-423(1991)

McNeely, Joseph R.
; McIntosh, Malcolm B.; Akerman, M. A.: Scatter and contamination of a low-scatter mirror,V1530,288-298(1991)
See McIntosh, Malcolm B.: V1530,263-268(1991)

McNeff, W. See Gilbert, Barry K.: V1390,235-248(1991)

McNeil, John R.
See Bishop, Kenneth P.: V1545,64-73(1991)
See Hickman, Kirt C.: V1464,245-257(1991)
See Reicher, David W.: V1441,106-112(1991)
See Wilson, Scott R.: V1441,82-86(1991)

McNeill, D. See Ruddell, F. H.: V1361,159-170(1991)

McNeill, John A. See Raanes, Chris A.: V1358,637-643(1991)

McNeill, Kevin M. See Martinez, Ralph: V1446,100-107(1991)

McNicholas, Thomas A.
; O'Donoghue, Neil: Endoscopic YAG laser coagulation for early prostate cancer,V1421,56-67(1991)
; Steger, Adrian C.; Bown, Stephen G.; O'Donoghue, Neil: Interstitial laser coagulation of the prostate: experimental studies,V1421,30-35(1991)

McNitt-Gray, Michael F.
; Taira, Ricky K.; Eldredge, Sandra L.; Razavi, Mahmood: Brightness and contrast adjustments for different tissue densities in digital chest radiographs,V1445,468-478(1991)

McOwan, Peter W.
; Powell, A. K.; Burge, Ronald E.: Holographic labeling for automated identification,V1384,75-82(1991)
See Bostel, Ashley J.: V1385,165-172(1991)

McPherson, Armon See Haddad, Waleed S.: V1448,81-88(1991)

McPherson, David D. See Frazin, Leon J.: V1425,207-207(1991)

McQueen, A. E. See Mullin, John B.: V1361,1116-1127(1991)

McQuillan, S. E.
; McCanny, J. V.: VLSI processor for high-performance arithmetic computations,V1566,220-229(1991)

McRae, Thomas G. See Kulp, Thomas J.: V1479,352-363(1991)

McSheery, Tracy D.
: Diamond multichip modules,V1563,21-26(1991)

McStay, Daniel
; Chu, W. W.; Rogers, Alan J.: Temperature compensation of a highly birefringent optical fiber current sensor,V1584,118-123(1991)
See Chu, W. W.: V1572,523-527(1991)
See Yao, Jialing: V1572,428-433(1991)

McSwain, G. G.
; Fernandes, Stan T.; Doane, Kent B.: Autonomous guidance, navigation, and control bridging program plan,V1478,228-238(1991)

McVey, Brian D.
; Goldstein, John C.; McFarland, Robert D.; Newnam, Brian E.: Thermal analysis of multifacet-mirror ring resonator for XUV free-electron lasers,V1441,457-468(1991)
See Newnam, Brian E.: V1552,154-174(1991)
See Thode, Lester E.: V1552,87-106(1991)

McVey, Eugene S.
See Minnix, Jay I.: V1606,241-251(1991)
See Xu, Qing: V1471,378-389(1991)

McVittie, James P.
; Rey, J. C.; Bariya, A. J.; IslamRaja, M. M.; Cheng, L. Y.; Ravi, S.; Saraswat, Krishna C.: SPEEDIE: a profile simulator for etching and deposition,V1392,126-138(1991)

McWhorter, Shane W.
; Hodges, Larry F.; Rodriguez, Walter E.: Comparison of 3-D display formats for CAD applications,V1457,85-90(1991)

McWilliams, Gary
; Kirby, Steve; Eskridge, Thomas; Newberry, Jeff: Expert system for fusing weather and doctrinal information used in the intelligence preparation of the battlefield,V1468,417-428(1991)

Mead, Carver A. See Delbruck, Tobi: V1541,92-99(1991)

Meador, Jack L. See Barga, Roger S.: V1469,602-611(1991)

Meadows, Michael R. See Zhuang, Zhiming: V1455,105-109(1991)

Means, Robert W.
: Systolic array architecture of a new VLSI vision chip,V1566,388-393(1991)
See Qing, Kent P.: V1569,111-120(1991)
See Qing, Kent P.: V1567,390-396(1991)

Mears, Carl A.
; Hu, Qing; Richards, Paul L.; Worsham, A.; Prober, Daniel E.; Raisanen, Antti: Quantum-limited quasiparticle mixers at 100 GHz,V1477,221-233(1991)

Mears, Richard B. See Heavin, Scott D.: V1455,12-18(1991)

Mears, Robert J. See O'Brien, Dominic C.: V1505,32-37(1991)

Measures, Raymond M.
: Fiber optic sensor considerations and developments for smart structures,V1588,282-299(1991)
: Fiber optic smart structures: structures that see the light,V1332,377-398(1991)
; LeBlanc, Michel; Hogg, W. D.; McEwen, Keith; Park, B. K.: Fiber optic damage detection for an aircraft leading edge,V1332,431-443(1991)
; Liu, Kexing; LeBlanc, Michel; McEwen, Keith; Shankar, K.; Tennyson, R. C.; Ferguson, Suzanne M.: Damage assessment in composites with embedded optical fiber sensors,V1489,86-96(1991)
; Valis, Tomas; Liu, Kexing; Hogg, W. D.; Ferguson, Suzanne M.; Tapanes, Edward: Interferometric fiber optic sensors for use with composite materials,V1332,421-430(1991)
See Davis, Andrew: V1489,33-43(1991)
See Davis, Andrew: V1588,264-274(1991)
See Hogg, W. D.: V1588,300-307(1991)
See Liu, Kexing: V1584,226-234(1991)
See Liu, Kexing: V1398,206-212(1991)
See Melle, Serge M.: V1588,255-263(1991)
See Zuliani, Gary: V1588,308-313(1991)

Mecherikunnel, Ann
: Calibration of EOS multispectral imaging sensors and solar irradiance variability,V1493,292-302(1991)

Mecherle, G. S.
; Henderson, Robert J.: Comparison of simulation and experimental measurements of avalanche photodiode receiver performance,V1417,537-542(1991)
; Henderson, Robert J.: Homodyne PSK receivers with laser diode sources,V1417,99-107(1991)

Mechtenberg, Monica L.
; Watson, Larry J.: Performance appraisal of the ATEQ CORE-2500 in production,V1496,124-155(1991)

Meckley, John R.
: Application of image processing technology to 2-D signal processing,V1406,72-73(1991)
: Linear feature SNR enhancement in radon transform space,V1569,375-385(1991)

Medbery, James D.
; Germann, Lawrence M.: Six degree-of-freedom magnetically suspended fine-steering mirror,V1482,397-405(1991)
; Germann, Lawrence M.: Specification of precision optical pointing systems,V1489,163-176(1991)

Meddeb, Jaafar See Bremond, Georges E.: V1361,732-743(1991)

Medeiros, Shirley See Holz, Michael: V1544,75-89(1991)

Medici, Gastone See Costa, Enrico: V1343,469-476(1991)

Medioni, Gerard See Rom, Hillel: V1570,262-273(1991)

Medvedev, A. A. See Kozhevnikov, Nikolai M.: V1584,138-144(1991)

Medvedev, Boris A.
; Tuchin, Valery V.; Yaroslavsky, Ilya V.: Mathematical modeling for laser PUVA treatment of psoriasis,V1422,73-84(1991)
; Tuchin, Valery V.; Yaroslavsky, Ilya V.: Mathematical model of laser PUVA psoriasis treatment,V1403,682-685(1991)

Meeks, Kent D. See Setchell, Robert E.: V1441,61-70(1991)

Meer, Peter
; Mintz, Doron: Robust regression in computer vision,V1381,424-435(1991)

Meerovich, Gennady A. See Katsap, Victor N.: V1420,259-265(1991)

Mees, W. See Acheroy, Marc P.: V1569,121-132(1991)

Meeusen, G. J.
; Qing, Z.; Wilbers, A. T.; Schram, D. C.: Deposition of a-Si:H using a supersonically expanding argon plasma,V1519,252-257(1991)

Meggitt, Beverley T.
See Boyle, William J.: V1511,51-56(1991)
See Chen, Shiping: V1504,191-201(1991)
See Chen, Shiping: V1511,67-77(1991)
See Ning, Yanong N.: V1367,347-356(1991)
See Weir, Kenneth: V1584,220-225(1991)

Megill, L. R. See Claybaugh, William R.: V1495,81-94(1991)

Megorskaja, K. D. See Shevchenko, N. P.: V1574,66-71(1991)

Meguro, Shin-Ichi
; Nunotani, Masakatu; Tanimizu, Katsuyuki; Sano, Mutsuo; Ishii, Akira: Automatic inspection system for full-color printed matter,V1384,27-37(1991)
See Sano, Mutsuo: V1381,101-110(1991)

Mehanian, Couroush
; Aull, Brian F.; Nichols, Kirby B.: Optoelectronically implemented neural network for early visual processing,V1469,275-280(1991)
; Rak, Steven J.: Bidirectional log-polar mapping for invariant object recognition,V1471,200-209(1991)

Mehdizadeh, E. See Bahram-pour, A. R.: V1397,539-542(1991)

Mehdizadeh, S. See Krongelb, Sol: V1389,249-256(1991)

Mehliss, Georg G. See Bauch, Lothar: V1466,510-519(1991)

Mehri, F. See Carru, J. C.: V1512,232-239(1991)

Mehrotra, R.
; Mathur, B. P.; Sharan, Sunil: Reduction of the standing wave effect in positive photoresist using an antireflection coating,V1463,487-491(1991)

Mehta, Naresh C.
: GRAND: a 4-D wave optics code for atmospheric laser propagation,V1487,398-409(1991)
: Remote alignment of adaptive optical systems with far-field optimization,V1408,96-111(1991)
See Digumarthi, Ramji V.: V1408,136-147(1991)

Mehuys, David G.
; Welch, David F.; Parke, Ross; Waarts, Robert G.; Hardy, Amos; Scifres, Donald R.: High-power coherent operation of 2-D monolithically integrated master-oscillator power amplifiers,V1418,57-63(1991)
See Zucker, Erik P.: V1563,223-228(1991)

Mei, Ting
; Liu, Xu; Tang, Jinfa; Gu, Peifu: Study of TEM micrographs of thin-film cross-section replica using spectral analysis,V1554A,570-578(1991)

Mei, Yuan S.
; Shang, Shi X.; Shan, Jin A.; Sun, Jian G.: New approach to obtain uniform thickness ZnS thin film interference filters,V1519,370-373(1991)

Meidan, Moshe
; Schwartz, Roni; Sher, Assaf; Zhaiek, Sasson; Gal, Eli; Neugarten, Michael L.; Afik, Zvi; Baer, C.: PtSi camera: performance model validation,V1540,729-737(1991)

Meidinger, Norbert See Braueninger, Heinrich: V1549,330-339(1991)

Meier, G. E. See Stasicki, Boleslaw: V1358,1222-1227(1991)

Meier, Hans
; Albrecht, Wolfgang; Hanack, Michael: Photoconductivity of bridged polymeric phthalocyanines,V1559,89-100(1991)

Meier, Karl L.
See Chan, Kwok-Chi D.: V1552,69-78(1991)
See Newnam, Brian E.: V1552,154-174(1991)

Meier, Thomas H. See Steiner, Rudolf W.: V1421,124-128(1991)

Meier, Winfried See Eckes, Charlotte: V1466,394-407(1991)

Meier, Wolfgang See Hirschmann, Harald: V1559,27-38(1991)

Meigs, G. See Tjeng, L. H.: V1548,160-167(1991)

Meijer, Gerard See de Vries, Mattanjah S.: V1437,129-137(1991)

Meinel, Aden B.
; Meinel, Marjorie P.: Infrared and the search for extrasolar planets,V1540,196-201(1991)

Meinel, Marjorie P. See Meinel, Aden B.: V1540,196-201(1991)

Meinke, Gary E. See Grieser, James L.: V1330,102-110(1991)

Meiri, A. Z. See Francos, Joseph M.: V1606,553-565(1991)

Meisenzahl, Eric J. See Wong, Kwok Y.: V1447,283-287(1991)

Meisse, Pascal See Pyee, Maurice: V1512,240-248(1991)

Meissner, D. See Ackermann, F.: V1490,94-101(1991)

Meissner, Helmuth E. See Lee, Huai-Chuan: V1441,87-103(1991)

Meissner, P.
: Coherent optical fiber communications,V1522,182-193(1991)

Meitzler, Thomas J.
; Gonda, Teresa G.; Jones, Jack C.; Reynolds, William R.: Variable emissivity plates under a three-dimensional sky background,V1486,380-389(1991)

Mejias, P. M.
See Martinez-Herrero, R.: V1397,623-626(1991)
See Martinez-Herrero, R.: V1397,619-622(1991)
See Martinez-Herrero, R.: V1397,627-630(1991)
See Serna, Julio: V1397,631-634(1991)

Mekryukov, A. M.
See Bosyi, O. N.: V1440,57-62(1991)
See Glebov, Leonid B.: V1513,274-282(1991)

Mel'nichenko, Yu. B. See Kzuzhilin, Yu E.: V1238,200-205(1991)

Mel'nikov, Lev Y.
; Kochelap, Viatcheslav A.; Izmailov, I. A.: Cavityless dielectric-waveguide-mode generation in a weakly amplifying gaseous medium,V1397,603-610(1991)
; Kochelap, Viatcheslav A.; Izmailov, I. A.: Theory of solar-powered, cavityless waveguide laser,V1501,144-149(1991)

Melaku, Yosias See Laporte, Philippe: V1392,196-207(1991)

Melamed, Nathan T.
; Gottlieb, Milton S.: Investigation of higher-order diffraction in a one-crystal 2-D scanner,V1454,306-312(1991)

Meland, Edmund
See Eichen, Elliot G.: V1474,260-267(1991)
See Holmstrom, Roger P.: V1418,223-230(1991)

Melchior, Hans M. See Wieland, Joerg B.: V1389,659-664(1991)

Meldrum, Gerald L. See Ali, Mir A.: V1513,215-223(1991)

Melendes, M. See Chou, Chia-Shing: V1475,151-156(1991)

Melendez, J. See Recio, Miguel: V1361,469-478(1991)

Melennec, Pierre See Grangeat, Pierre: V1445,320-331(1991)

Melfi, Samuel H. See Theon, John S.: V1492,2-23(1991)

Melhem, Rami G.
See Chiarulli, Donald M.: V1390,403-414(1991)
See Guo, Zicheng: V1390,415-426(1991)

Meliga, Marina See Caldera, Claudio: V1372,82-87(1991)

Mell, Richard J. See Harada, Yoshiro: V1330,90-101(1991)

Melle, Serge M.
; Liu, Kexing; Measures, Raymond M.: Strain sensing using a fiber-optic Bragg grating,V1588,255-263(1991)

Meller, Menachem M.
; Sherk, Henry H.; Rhodes, Anthony; Sazy, John; Uppal, Gurvinder S.; Lane, Gregory J.: Technique of CO2 laser arthroscopic surgery,V1424,60-61(1991)
See Black, Johnathan D.: V1424,20-22(1991)
See Black, Johnathan D.: V1424,12-15(1991)

Mellichamp, Duncan A. See Crisalle, Oscar D.: V1464,508-526(1991)

Mellor, J. R. See Moehlmann, Gustaaf R.: V1560,426-433(1991)

Melman, Paul See Elman, Boris S.: V1361,362-372(1991)

Melngailis, John
: Focused ion beam induced deposition: a review,V1465,36-49(1991)

Melnik, George A. See Hillman, Robert L.: V1562,136-142(1991)

Melnik, Ivan S.
; Krivokhizha, A. M.; Ptashnik, O. V.: Multipurpose fiber optic sensor with sloped tip,V1572,118-122(1991)

Melnik, N. E. See Komarov, Vyacheslav A.: V1238,275-285(1991)

Melodia, Tommaso See Gaboardi, Franco: V1421,79-85(1991)

Melore, Dan See Kovacs, Stephen: V1465,88-99(1991)

Melozzi, Mauro
; Mazzoni, Alessandro; Curti, G.: Bidirectional transmittance distribution function measurements on ZnSe and on ZnS Cleartran,V1512,178-188(1991)

Melroy, Owen R. See Toney, Michael F.: V1550,140-150(1991)

Meltz, Gerald
; Morey, William W.: Bragg grating formation and germanosilicate fiber photosensitivity,V1516,185-199(1991)
See Giesler, Leslie E.: V1480,138-142(1991)

Melvin, Leland D. See Baldini, S. E.: V1588,125-131(1991)

Melzer, James E.
; Moffitt, Kirk W.: Ecological approach to partial binocular overlap,V1456,124-131(1991)
See Casey, Curtis J.: V1456,175-178(1991)

Memola Capece Minutolo, F. See Daurelio, Giuseppe: V1397,783-786(1991)

Mena, Manolo G.
See Baltazar, Inmaculada C.: V1392,670-680(1991)
See Gonzalez, Marcelo S.: V1390,568-579(1991)

Menache, Ram See Eisenberg, Shai: V1442,133-138(1991)

Mencaglia, Andrea
; Brenci, Massimo; Falciai, Riccardo; Guzzi, D.; Pascucci, Tania: Heat monitoring by fiber-optic microswitches,V1506,140-144(1991)
See Brenci, Massimo: V1572,318-324(1991)
See Brenci, Massimo: V1504,212-220(1991)

Menczigar, Ulrich
; Dahmen, M.; Zachai, Reinhard; Eberl, K.; Abstreiter, Gerhard: Optical properties of short-period Si/Ge superlattices grown on (001) Ge studied with photoreflectance,V1361,282-292(1991)

Mendez, Antonio J.
; Gagliardi, Robert M.: Code Division Multiple Access system candidate for integrated modular avionics,V1369,67-71(1991)
; Gagliardi, Robert M.: Performance of pseudo-orthogonal codes in temporal, spatial, and spectral code division multiple access systems,V1364,163-169(1991)
See Sherman, Bradley D.: V1369,60-66(1991)

Mendez, Eugenio R.
See Gu, Zu-Han: V1530,60-70(1991)
See Ruiz-Cortes, Victor: V1558,222-232(1991)

Mendez, Nestor R.
See Lukishova, Svetlana G.: V1527,380-391(1991)
See Lukishova, Svetlana G.: V1503,338-345(1991)

Mendis, Sunetra See Kemeny, Sabrina E.: V1447,243-250(1991)

Mendlovic, David
; Konforti, Naim; Deutsch, Meir; Marom, Emanuel: Composite image joint transform correlator,V1442,182-192(1991)

Mendlowitz, Joseph S. See Lowry, Jay H.: V1330,142-151(1991)

Mendoza, Edgar A.
; Gafney, Harry D.; Morse, David L.: Photochemical generation of gradient indices in glass,V1378,139-144(1991)
; Gafney, Harry D.; Morse, David L.: Photolithographic processing of integrated optic devices in glasses,V1583,43-51(1991)

Mendoza Santoyo, Fernando
; Shellabear, Michael C.; Tyrer, John R.: Four cases of engineering vibration studies using pulsed ESPI,V1508,143-152(1991)
See Kerr, David: V1554A,668-679(1991)

Menegat, A. See Farkas, Zoltan D.: V1407,502-511(1991)

Menegay, Harry J. See Oleinick, Nancy L.: V1426,235-243(1991)

Menendez, Jose See Sinha, Kislay: V1437,32-35(1991)

Menendez, Ronald C.
; Lu, Kevin W.; Rizzo, Annmarie; Lemberg, Howard L.: Evolution of fiber-to-the-curb networks toward broadband capabilities,V1363,97-105(1991)
See Wang, Lon A.: V1363,85-91(1991)

Menendez-Valdes, Pedro See Lopez-Amo, Manuel: V1374,74-85(1991)

Menenkov, V. D. See Gaiduk, Mark I.: V1403,674-675(1991)

Meng, A. See Shepard, Steven M.: V1467,234-238(1991)

Meng, Qinglai See Lu, Eh: V1533,155-162(1991)

Meng, Y. See Allinger, Thomas: V1361,935-942(1991)

Menge, Peter R.
; Bosch, Robert A.; Gilgenbach, Ronald M.; Choi, J. J.; Ching, Hong; Spencer, Thomas A.: Experiments on the beam breakup instability in long-pulse electron beam transport through cavity systems,V1407,578-588(1991)

Mengucci, P. See D'Anna, Emilia: V1503,256-268(1991)

Meninberg, Y. See Shushan, A.: V1487,300-311(1991)

Menke, Bodo
; Loeffler, Roland: Comparative life test of 0.8-um laser diodes for SILEX under NRZ and QPPM modulation,V1417,316-327(1991)

Mennig, Martin
; Schmidt, Helmut; Fink, Claudia: Synthesis and properties of sol-gel-derived AgClxBr1-x colloid containing sodium alumo borosilicate glasses,V1590,152-159(1991)

Menninger, W. L. See Danly, Bruce G.: V1407,192-201(1991)

Menon, Murali M.
: Massively parallel image restoration,V1471,185-190(1991)
; Van Allen, Eric J.: Automatic design of signal processors using neural networks,V1469,322-328(1991)
See Green, Thomas J.: V1471,328-341(1991)

Menon, P. K.
; Chatterji, Gano B.; Sridhar, Banavar: Fast algorithm for image-based ranging,V1478,190-200(1991)

Mens, Alain
; Dalmasso, J. M.; Sauneuf, Richard; Verrecchia, R.; Roth, J. M.; Tomasini, F.; Miehe, Joseph A.; Rebuffie, Jean-Claude: C 850X picosecond high-resolution streak camera,V1358,315-328(1991)
; Ducrocq, N.; Mazataud, D.; Mugnier, A.; Eouzan, J. Y.; Heurtaux, J. C.; Tomasini, F.; Mathae, J. C.: High-dynamic-range image readout system,V1358,719-731(1991)
; Gontier, D.; Giraud, P.; Thebault, J. P.: Temporal fiducial for picosecond streak cameras in laser fusion experiments,V1358,878-887(1991)
See Rebuffie, Jean-Claude: V1358,511-523(1991)

Mentle, Robert E.
; Shapiro, Jeffrey H.: Track-while-image in the presence of background,V1471,342-353(1991)

Menzel, E. R. See Ordonez, Ishmael D.: V1437,184-193(1991)

Menzies, Robert T.
: Infrared lidars for atmospheric remote sensing,V1540,160-163(1991)
; Post, Madison J.: GLOBE backscatter: climatologies and mission results,V1416,139-146(1991)
See Theon, John S.: V1492,2-23(1991)

Merat, Frederic See Voignier, Francois: V1397,297-301(1991)

Meray, G. M. See Sauer, Donald J.: V1540,285-296(1991)

Mercado, Romeo I.
: Correction of secondary and higher-order spectrum using special materials,V1535,184-198(1991)
: Designs of two-glass apochromats and superachromats,V1354,262-272(1991)

Mercier, Jean-Louis M. See Bourdoncle, Bernard: V1354,194-199(1991)

Mercier, P. P. See Chaker, Mohamed: V1465,16-25(1991)

Meredith, P. See Craig, Robert G.: V1505,76-86(1991)

Merhav, S. J. See Grunwald, Arthur J.: V1456,132-153(1991)

Meringdal, Frode
; Slinde, Harald: Ablation of ITO and TO films from glass substrates,V1503,292-298(1991)

Merkelo, Henri
; McCredie, B. D.; Veatch, M. S.; Quinn, D. L.; Dorneich, M.; Doi, Yutaka: Methods for comparative analysis of waveform degradation in electrical and optical high-performance interconnections,V1390,91-163(1991)

Merker, M. See Read, Harold E.: V1441,345-359(1991)

Merkle, Fritz
; Gehring, G.; Rigaut, Francois; Lena, Pierre J.; Rousset, Gerard; Fontanella, Jean-Claude; Gaffard, Jean-Paul: Adaptive optics system tests at the ESO 3.6-m telescope,V1542,308-318(1991)
; Hubin, Norbert: Adaptive optics for the European very large telescope,V1542,283-292(1991)
See Gendron, Eric: V1542,297-307(1991)
See Noethe, L.: V1542,293-296(1991)

Merlet, Nicolas See Preteux, Francoise: V1568,66-77(1991)

Merrheim, Xavier See Kla, Sylvanus: V1566,275-285(1991)

Merrill, Daniel C. See Nseyo, Unyime O.: V1426,287-292(1991)

Merritt, C. D. See Campillo, Anthony J.: V1497,78-89(1991)

Merritt, Edward See Bietry, Joseph R.: V1332,537-543(1991)

Merritt, John O.
; CuQlock-Knopp, V. G.: Perceptual training with cues for hazard detection in off-road driving,V1457,133-138(1991)
; Fisher, Scott S.; eds.: *Stereoscopic Displays and Applications II,*V1457(1991)
See Cole, Robert E.: V1457,111-119(1991)

Merrow, Scott F. See Cambria, Terrell D.: V1466,670-675(1991)

Mersereau, Russell M. See Reeves, Stanley J.: V1452,127-138(1991)

Mertelmeier, Thomas See Barden, Raimund: V1449,136-147(1991)

Mertens, Hans W. See Moehlmann, Gustaaf R.: V1560,426-433(1991)

Mertz, Lawrence N.
: Concentric systems for adaptation as spectrographs,V1354,457-459(1991)

Merwin, David H.
; Wickens, Christopher D.: Comparison of 2-D planar and 3-D perspective display formats in multidimensional data visualization,V1456,211-218(1991)

Merz, James L.
; Holtz, Per O.: Radiative processes in quantum-confined structures,V1361,76-88(1991)
See Coldren, Larry A.: V1362,24-37(1991)

Merzbacher, Celia I.
; Friebele, E. J.; Ruller, Jacqueline A.; Matic, P.: Finite element analysis of deformation in large optics due to space environment radiation,V1533,222-228(1991)

Mesa, Juan E.
See Ramee, Stephen R.: V1420,199-202(1991)
See White, Christopher J.: V1425,130-133(1991)

Meseguer, Francisco See Lopez, C.: V1361,89-95(1991)

Meshalkin, E. A.
: Evolution of a space-charge layer, its instability, and ignition of arc gas discharge under photoemission from a target into a gas,V1440,211-221(1991)

Meshalkina, M. N.
; Smirnov, V. V.: Study of the structure of reflection holograms on a dichromate gelatin layer,V1238,195-199(1991)

Messenger, Heather
: Scientific laser trends,V1520,58-92(1991)

Messier, Jean See Charra, Fabrice: V1516,211-219(1991)

Messier, Russell F. See Zhu, Wei: V1534,230-242(1991)

Messina, G. See Ciocci, Franco: V1501,154-162(1991)

Messner, Richard A.
; Whitney, Erich C.: Development of a low-cost VME-based Nth order 2-D warper,V1381,261-271(1991)
See Abbott, Mark A.: V1381,272-282(1991)

Messuri, Dominic A.
See Cirillo, James R.: V1592,53-59(1991)
See Cirillo, James R.: V1592,42-52(1991)

Mestre, J. P. See Ravat, Christian J.: V1478,239-246(1991)

Mesyats, Gennady A.
: High-power particle beams for gas lasers,V1411,2-14(1991)

Meszaros, Peter
: Theoretical models for stellar x-ray polarization in compact objects,V1548,13-22(1991)

Metaxas, Dimitri
; Terzopoulos, Demetri: Shape representation and nonrigid motion tracking using deformable superquadrics,V1570,12-20(1991)

Metev, S. See Breuer, J.: V1503,223-230(1991)

Meth, Jeffrey S.
; Vanherzeele, Herman A.; Jenekhe, Samson A.; Roberts, Michael F.; Agrawal, A. K.; Yang, Chen-Jen: Dispersion of x(3) in fused aromatic ladder polymers and their precursors probed by third-harmonic generation,V1560,13-24(1991)

Metin, Serhat See Kaufman, James H.: V1437,36-41(1991)

Metz, Charles E.
See Chen, Chin-Tu: V1445,222-225(1991)
See Giger, Maryellen L.: V1445,101-103(1991)
See King, Jill L.: V1446,268-275(1991)
See Yin, Fang-Fang: V1396,2-4(1991)

Metz, Jurgen See Perger, Andreas: V1419,75-83(1991)

Metz, Thomas E. See Davies, John T.: V1392,551-554(1991)

Metzger, Don W. See Norgard, John D.: V1540,699-708(1991)

Metzner, G. See Braueninger, Heinrich: V1549,330-339(1991)

Meulenberg, A.
; Hung, Hing-Loi A.; Singer, J. L.; Anderson, Wallace T.: Effects of radiation on millimeter wave monolithic integrated circuits,V1475,280-285(1991)

Meunier, Carole See Lequime, Michael: V1367,236-242(1991)

Meyer, E. H. See Crostack, Horst-Artur: V1508,101-109(1991)

Meyer, George E.
; Troyer, W. W.; Fitzgerald, Jay B.: Vision methods for inspection of greenhouse poinsettia plants,V1379,99-105(1991)
See DeShazer, James A.: V1379(1991)

Meyer, Rudolph See Huenermann, Lucia: V1503,134-139(1991)

Meyer, Tim J. See Nagy, Peter A.: V1365,33-37(1991)

Meyer, Trevor E.
; Davidson, Jennifer L.: Image algebra preprocessor for the MasPar parallel computer,V1568,125-136(1991)

Meyer, W. J.
See Hietala, Vincent M.: V1476,170-175(1991)
See Vawter, Gregory A.: V1476,102-106(1991)

Meyer-Ilse, Werner See Germer, Rudolf K.: V1358,346-350(1991)

Meyerhofer, David D.
See Chen, Hong: V1413,112-119(1991)
See Chuang, Yung-Ho: V1413,32-40(1991)

Meyers, F. See Bredas, Jean-Luc: V1560,98-110(1991)

Meyrueis, Patrick
See Twardowski, Patrice J.: V1456,164-174(1991)
See Twardowski, Patrice J.: V1507,55-65(1991)

Meyrueix, Remi
; LeCompte, J. P.; Tapolsky, Gilles: Decay of the nonlinear susceptibility components in main-chain functionalized poled polymers,V1560,454-466(1991)

Meyyappan, Meyya
; Govindan, T. R.; Kreskovsky, John P.: Plasma modeling in microelectronic processing,V1392,67-76(1991)

Mezhoudi, Mohcene See Herskowitz, Gerald J.: V1368,55-60(1991)

Mezrich, Reuben S. See Gong, Leiguang: V1450,144-153(1991)

Miao, Cheng-Hsi
: Design of array systems using shared symmetry,V1354,447-456(1991)

Miao, Peng-Cheng
; Wang, Hai-Lin; Yan, Da-Peng; He, Anzhi: Aerodynamic interferograms of explosive field,V1554B,641-644(1991)
See He, Anzhi: V1527,334-337(1991)
See He, Anzhi: V1527,423-426(1991)
See He, Anzhi: V1567,698-702(1991)
See He, Anzhi: V1545,278-281(1991)
See He, Anzhi: V1554B,429-434(1991)
See Lu, Jian-Feng: V1415,220-224(1991)
See Wang, Hai-Lin: V1527,419-422(1991)
See Wang, Hai-Lin: V1545,268-273(1991)
See Wang, Hai-Lin: V1545,274-277(1991)
See Yan, Da-Peng: V1527,442-447(1991)

Miao, Tongqun See Cao, Jianlin: V1345,225-232(1991)

Miao, Y. Z. See Zhai, H. R.: V1519,575-579(1991)

Miaoulis, Ioannis See Papamichael, Haris: V1590,122-130(1991)

Michael, David J.
: Simple iterative method for finding the foe using depth-is-positive constraint,V1388,234-245(1991)

Michael, J. R. See Frear, Darrel R.: V1596,72-82(1991)

Michael, Jason See VanderMeulen, David L.: V1428,91-98(1991)

Michael, Robert R. See Lawson, Christopher M.: V1592,73-83(1991)

Michaelis, Alexander
; Uhlenbusch, Juergen; Vioel, Wolfgang: Electron density measurements of laser-induced surface plasma by means of a beam deflection technique,V1397,709-712(1991)

Michaels, Charles E. See Dew, Douglas K.: V1422,111-115(1991)

Michailov, Vladimir I.
; Sigaryov, Sergei E.; Terziev, Vladimir G.: Possibility of a "lithium glass" state appearance in the Li3Sc2-xFex(PO4)3 superionic solid solutions,V1590,203-214(1991)

Michailova, A. G. See Kalyashora, L. N.: V1238,189-194(1991)

Michalewicz, Zbigniew
: Step towards optimal topology of communication networks,V1470,112-122(1991)

Michalka, Timothy See Rajasekharan, K.: V1389,568-579(1991)

Michalkin, Igor See Zalessky, Viacheslav N.: V1426,162-169(1991)

Michard, Alain See Koenig, Michel: V1502,338-342(1991)

Michau, Vincent
; Marais, T.; Laurent, Jean; Primot, Jerome; Fontanella, Jean-Claude; Tallon, M.; Fuensalida, Jesus J.: High-resolution astronomical observations using deconvolution from wavefront sensing,V1487,64-71(1991)

Michel, Bernd See Vogel, Dietmar: V1554A,262-274(1991)

Michel, T. See Maradudin, Alexei A.: V1558,233-250(1991)

Michelman, Paul See Allen, Peter K.: V1383,176-188(1991)

Michette, Alan G.
: Diffractive optics for x rays: the state of the art,V1574,8-21(1991)
See Turcu, I. C.: V1503,391-405(1991)

Michie, Robert B. See Neal, Daniel R.: V1542,449-458(1991)

Michie, W. C.
; Culshaw, Brian; Roberts, Scott S.; Davidson, Roger: Fiber optic technique for simultaneous measurement of strain and temperature variations in composite materials,V1588,342-355(1991)
See Sansonetti, Pierre: V1588,198-209(1991)
See Vengsarkar, Ashish M.: V1367,249-260(1991)

Michioka, Hidekazu See Iwasaki, Nobuo: V1490,216-221(1991)

Michl, J. See Burland, Donald M.: V1560,111-119(1991)

Michl, Uwe See Kliem, Karl-Heinz: V1463,743-751(1991)

Michon, G. J. See Tomlinson, Harold W.: V1541,178-186(1991)

Mickelson, Alan R.
See Bundy, Scott: V1475,319-329(1991)
See Feuerstein, Robert J.: V1583,196-201(1991)

Midday, S. See Sen, Susanta: V1362,750-759(1991)

Middleton, John See Chronister, Eric L.: V1559,56-64(1991)

Midorikawa, Katsumi See Hatanaka, Hidekazu: V1397,379-382(1991)

Miehe, Joseph A.
    See Cunin, B.: V1358,606-613(1991)
    See Froehly, Claude: V1358,532-540(1991)
    See Mens, Alain: V1358,315-328(1991)

Mierswa, B. See Boneberg, J.: V1598,84-90(1991)

Mieth, Ulrike See Andrae, P.: V1508,50-60(1991)

Migitko, I. A. See Shestakov, S. D.: V1440,423-435(1991)

Migliorati, Pierangelo See Dal Degan, Nevaino: V1606,934-940(1991)

Mignani, Anna G.
    See Brenci, Massimo: V1572,318-324(1991)
    See Brenci, Massimo: V1504,212-220(1991)
    See Scheggi, Annamaria V.: V1510,40-45(1991)

Migus, Arnold See Andre, Michel L.: V1502,286-298(1991)

Mihailescu, Ion N.
    See Craciun, Valentin: V1392,629-634(1991)
    See Craciun, Valentin: V1392,625-628(1991)
    See D'Anna, Emilia: V1503,256-268(1991)

Mihalakis, George M. See Kirkpatrick, Michael D.: V1456,40-47(1991)

Mihara, Teruyoshi See Ohashi, T.: V1549,9-19(1991)

Mihkelsoo, V. T.
    See Chekulayev, V.: V1426,367-377(1991)
    See Nevorotin, Alexey J.: V1427,381-397(1991)

Mikaberidze, V. M. See Manenkov, Alexander A.: V1420,254-258(1991)

Mikaelian, Andrei L.
    ed.: Optical Memory and Neural Networks,V1621(1991)
    ; Barachevsky, Valery A.: Photopolymers for optical devices in the
        USSR,V1559,246-257(1991)
    ; Gulanian, Emin H.; Kretlov, Boris S.; Molchanova, L. V.; Semichev, V. A.:
        One-dimensional hologram recording in thick photolayer
        disk,V1621,194-203(1991)
    ; Salakhutdinov, Viktor K.; Vsevolodov, N. N.; Dyukova, T. V.: High-capacity
        optical spatial switch based on reversible holograms,V1621,148-
        157(1991)
    See Kiselyov, Boris S.: V1621,126-137(1991)
    See Kiselyov, Boris S.: V1621,328-339(1991)

Mikamura, Yasuki See Suzuki, Tomihiro: V1389,455-461(1991)

Miketic, Linda M. See Gur, David: V1446,284-288(1991)

Mikhailov, A. V. See Garbuzov, Dmitriy Z.: V1418,386-393(1991)

Mikhailov, I. A.
    : Monochromatic aberrations of an off-axis hologram with carry out
        pupil,V1238,123-128(1991)
    : Parameters of recording media for 3-D holograms,V1238,140-143(1991)

Mikhailov, M. D. See Yusupov, I. Y.: V1238,240-247(1991)

Mikhailov, V. A.
    See D'yakonov, G. I.: V1421,156-162(1991)
    See D'yakonov, G. I.: V1424,81-86(1991)
    See D'yakonov, G. I.: V1421,153-155(1991)

Mikhailov, V. N.
    ; Grinevitskaya, O. V.; Zagorskaya, Z. A.; Mikhailova, V. I.: Study of
        electronic stage of the Herschel effect in holographic emulsions with
        different types of chemical sensitization,V1238,144-152(1991)

Mikhailova, G. N. See Craciun, Valentin: V1392,629-634(1991)

Mikhailova, Maya P.
    ; Baranov, Alexej N.; Imenkov, Albert N.; Yakovlev, Yurii P.: Type-II
        heterojunctions in GaSb-InAs solid solutions: physics and
        applications,V1361,674-685(1991)
    See Yakovlev, Yurii P.: V1510,170-177(1991)

Mikhailova, V. I.
    See Kliot-Dashinskaya, I. M.: V1238,465-469(1991)
    See Mikhailov, V. N.: V1238,144-152(1991)

Mikhalchenko, O. E. See Shvej, E. M.: V1554A,488-495(1991)

Mikhalev, N. See Popov, A.: V1519,457-462(1991)

Mikhaylenko, B. M. See Bass, V. I.: V1358,1075-1083(1991)

Mikhaylov, A. V. See Vishnevsky, G. I.: V1448,69-72(1991)

Miki, Atsushi See Ogura, Satoshi: V1412,123-128(1991)

Miklovicova, J. See Shvec, Peter: V1403,635-637(1991)

Milam, David See Elder, Melanie L.: V1441,237-246(1991)

Milanesi, Francesco See Baldini, Francesco: V1510,58-62(1991)

Milani, Marziale
    ; Mazzoleni, S.; Brivio, Franca: Semiconductor laser diode with weak optical
        feedback: self-coupling effects on P-I characteristics,V1474,83-97(1991)

Milano, L. See Barone, Fabrizio: V1584,304-307(1991)

Milanovic, Zoran
    ; Jacobson, Michael R.; Macleod, H. A.: Solid matrix Christiansen
        filters,V1535,160-170(1991)

Milanovich, Fred P.
    See Angel, S. M.: V1368,98-104(1991)
    See Campbell, John H.: V1441,444-456(1991)

Milch, James R. See Chang, Win-Chyi: V1448(1991)

Milchberg, Howard M.
    See Durfee, C.: V1413,78-80(1991)
    See Lyubomirsky, I.: V1413,108-111(1991)

Milder, D. M.
    : Improved formalism for rough-surface scattering of acoustic and
        electromagnetic waves,V1558,213-221(1991)

Miler, Miroslav
    ; Aubrecht, I.: Holographic crossed gratings: their nature and
        applications,V1574,22-33(1991)

Miley, George H.
    ; Barnouin, O.; Procoli, Alfredo: Effect of ionizing radiations and thermal
        treatment on the infrared transmittance of polycrystal CsI,V1441,16-
        26(1991)

Milgazo, H. See Bharj, Sarjit: V1475,340-349(1991)

Milgram, Paul See Drascic, David: V1457,302-313(1991)

Milios, Evangelos E.
    See Prassler, Erwin E.: V1388,2-13(1991)
    See Wilkes, David R.: V1388,536-542(1991)

Millan, Maria S.
    ; Yzuel, Maria J.; Campos, Juan; Ferreira, Carlos: Strategies for the color
        character recognition by optical multichannel correlation,V1507,183-
        197(1991)
    See Campos, Juan: V1564,189-198(1991)

Miller, Alan
    See Li Kam Wa, Patrick: V1474,2-7(1991)
    See Park, Choong-Bum: V1474,8-17(1991)

Miller, Andrew S.
    ; Sawchuk, Alexander A.: Capabilities of simple lenses in a free-space perfect
        shuffle,V1563,81-92(1991)

Miller, Arthur C. See Thomas, Brigham B.: V1485,20-30(1991)

Miller, Charles E. See Stationwala, Mustafa I.: V1358,237-245(1991)

Miller, Charles M.
    See Fairbank, William M.: V1435,86-89(1991)
    See Johnson, Stephen G.: V1435,292-297(1991)

Miller, David A.
    : Quantum-well devices for optics in digital systems,V1389,496-502(1991)

Miller, David P.
    ; Gat, Erann: Exploiting known topologies to navigate with low-computation
        sensing,V1383,425-435(1991)

Miller, David W.
    See Blackwood, Gary H.: V1542,371-391(1991)
    See How, Jonathan P.: V1489,148-162(1991)

Miller, Drew V. See O'Brien, Stephen J.: V1424,62-75(1991)

Miller, Edward A. See Guzman, Carmen T.: V1493,120-131(1991)

Miller, Eileen O. See Becker, Richard A.: V1459,150-154(1991)

Miller, Glen E.
    : Application of analog fiber optic position sensors to flight control
        systems,V1367,165-173(1991)
    : A very unique pre-1970 fiber optic instrumentation system,V1589,2-
        10(1991)
    See Figueroa, Luis: V1418,153-176(1991)

Miller, Gordon H. See Vo-Dinh, Tuan: V1368,203-209(1991)

Miller, Gregory D.
    See Bulusu, Dutt V.: V1364,49-60(1991)
    See Jennings, Kurt L.: V1369,36-42(1991)

Miller, J. M.
    See Craig, Robert G.: V1505,76-86(1991)
    See Vari, Sandor G.: V1424,33-42(1991)
    See Vasara, Antti H.: V1507,224-238(1991)

Miller, James See Laird, Alan: V1381,536-544(1991)

Miller, Kurt
  ; Reich, Ella; Grau, T.: Intracellular uptake and ultrastructural phototoxic effects of sulfonated chlor-aluminum phthalocyanine on bladder tumor cells in vitro,V1426,378-383(1991)
  ; Weber, Hans M.; Rueschoff, Josef; Hautmann, Richard E.: Experimental and first clinical results with the alexandrite laser lithotripter,V1421,108-113(1991)

Miller, L. M. See Lim, G.: V1418,123-131(1991)

Miller, Larry S. See Boiarski, Anthony A.: V1368,264-272(1991)

Miller, Michael B. See Lewis, Warren H.: V1369,79-86(1991)

Miller, Michael E. See Gooding, Linda: V1457,259-266(1991)

Miller, Peter E. See Lyon, Richard G.: V1567,317-326(1991)

Miller, Peter J.
  : Methods and applications for intensity stabilization of pulsed and cw lasers from 257 nm to 10.6 microns,V1376,180-191(1991)

Miller, R. L. See Deason, Vance A.: V1332,868-876(1991)

Miller, Richard B.
  : Ground systems and operations concepts for the Space Infrared Telescope Facility,V1540,38-46(1991)

Miller, Richard M.
  ; Birmingham, John J.; Cummins, Philip G.; Singleton, Scott: Industrial applications of spectroscopic imaging,V1439,66-78(1991)

Miller, Robert D.
  See Neher, Dieter: V1560,335-343(1991)
  See Wallraff, Gregory M.: V1466,211-217(1991)

Miller, Robert H.
  ; Beaton, Robert J.: Some effects on depth-position and course-prediction judgments in 2-D and 3-D displays,V1457,248-258(1991)

Miller, Robert J. See Dougherty, Edward R.: V1567,88-99(1991)

Miller, Roger H. See Farkas, Zoltan D.: V1407,502-511(1991)

Miller, Ronald E. See Wengler, Michael J.: V1477,209-220(1991)

Miller, V. A. See Averin, V. I.: V1358,603-605(1991)

Miller, Walter B.
  See Bissonnette, Luc R.: V1487(1991)
  See Tunick, Arnold: V1487,51-62(1991)

Miller, Walter E.
  : Sensor fusion approach to optimization for human perception: an observer-optimized tricolor IR target locating sensor,V1482,224-233(1991)

Miller, Wesley A. See Stevens, Eric G.: V1447,274-282(1991)

Miller, William G. See Hadfield, Michael J.: V1478,126-144(1991)

Miller, William L. See Heikes, Brian J.: V1433,253-262(1991)

Miller, William V.
  See Fogg, Brian R.: V1588,14-25(1991)
  See Tran, Tuan A.: V1584,178-186(1991)

Millet, P. See Saissac, M.: V1397,739-742(1991)

Millnert, Mille See Renhorn, Ingmar G.: V1480,35-45(1991)

Millot, Jean-Marc See Manfait, Michel: V1403,695-707(1991)

Mills, David R.
  ; Monger, A. G.; Morrison, G. L.: Comparison of asymmetrical and symmetrical nonimaging reflectors for east-west circular cylindrical solar receivers,V1528,44-55(1991)

Mills, Dennis M.
  : Techniques of production and analysis of polarized synchrotron radiation,V1345,125-136(1991)
  ed.: *X Rays in Materials Analysis II: Novel Applications and Recent Developments,*V1550(1991)

Mills, F. See Watanabe, Hiroshi: V1490,317-323(1991)

Mills, Robert E. See Gates, James L.: V1540,262-273(1991)

Milne, Edmund A.
  ; Cooper, Alfred W.; Reategui, Rodolfo; Walker, Philip L.: Sensitivity analysis of Navy tactical decision-aid FLIR performance codes,V1486,151-161(1991)
  See Cooper, Alfred W.: V1486,37-46(1991)
  See Crittenden, Eugene C.: V1492,187-197(1991)

Milne, James M. See Hobbs, Chris P.: V1467,264-277(1991)

Milne, Peter J.
  ; Zika, Rod G.; Parel, Jean-Marie; Denham, David B.; Penney, Carl M.: Measurement of fluorescence spectra and quantum yields of 193 nm ArF laser photoablation of the cornea and synthetic lenticules,V1423,122-129(1991)

Milner, Kathy S.
  ; Chipman, Paul S.: 5X reticle fabrication using MEBES multiphase virtual address and AZ5206 resist,V1496,180-196(1991)

Milner, Lisa-Michelle See Bishop, Kenneth P.: V1545,64-73(1991)

Milnes, Peter See Smith, Roy E.: V1425,116-121(1991)

Milonni, Peter W.
  : QED theory of excess spontaneous emission noise,V1376,164-169(1991)

Miloslavsky, V. K.
  ; Ageev, L. A.; Lymar, V. I.: Self-organization of spontaneous structures in photosensitive layers,V1440,90-96(1991)

Milsted, Carl S.
  ; Cantrell, Cyrus D.: Vector beam propagation using Hertz vectors,V1497,202-215(1991)

Milster, Tom D.
  ; Treptau, Jeffrey P.: Measurement of laser spot quality,V1414,91-96(1991)
  ; Trusty, Robert M.; Wang, Mark S.; Froehlich, Fred F.; Erwin, J. K.: Micro-optic lens for data storage,V1499,286-292(1991)
  ; Wang, Mark S.; Froehlich, Fred F.; Kann, J. L.; Treptau, Jeffrey P.; Erwin, K. E.: Differential spot-size focus servo,V1499,348-353(1991)
  See Treptau, Jeffrey P.: V1415,317-321(1991)

Milton, James E. See Anderson, Roland C.: V1358,992-1002(1991)

Miltonberger, Thomas W. See Bullock, Michael E.: V1471,291-302(1991)

Min'ko, L. Y.
  ; Chumakov, A. N.: Dynamics of near-surface plasma formation and laser absorption waves under the action of microsecond laser radiation with different wavelengths on absorbing condensed media,V1440,166-178(1991)

Min, Hisook L. See Min, Kwang-Shik: V1469,129-136(1991)

Min, Kwang-Shik
  ; Min, Hisook L.: Fast identification of images using neural networks,V1469,129-136(1991)

Min, Szuk W. See Yuan, Xiang L.: V1519,167-171(1991)

Min, Zhu-Yue See Baskurt, Atilla: V1445,485-495(1991)

Minaev, Sergei M. See Kokin, A. N.: V1440,338-341(1991)

Minaeva, E. M.
  ; Libenson, Michail N.: Heating and damage of thin metal films under the conditions of disturbed equilibrium,V1440,63-70(1991)

Minakuchi, Kimihide See Yamaguchi, Takao: V1418,363-371(1991)

Minami, Toshi
  See Numagami, Yasuhiko: V1606,970-979(1991)
  See So, Ikken: V1605,263-272(1991)

Minano, Juan C.
  ; Gonzalez, Juan C.: Design of nonimaging lenses and lens-mirror combinations,V1528,104-115(1991)
  ; Luque, Antonio: Angular confining cavities for photovoltaics,V1528,58-69(1991)

Minato, Atsushi See Sasano, Yasuhiro: V1490,233-242(1991)

Minato, Kotaro
  ; Komori, Masaru; Nakano, Yoshihisa; Yonekura, Yoshiharu; Sasayama, Satoshi; Takahashi, Takashi; Konishi, Junji; Abe, Mituyuki; Sato, Kazuhiro; Hosoba, Minoru: Off-line image exchange between two PACS modules using the "ISAC" magneto-optical disk,V1446,195-198(1991)
  See Komori, Masaru: V1444,334-337(1991)

Mindak, Marek K. See Jabczynski, Jan K.: V1391,109-116(1991)

Mine, Hironori See Miyoshi, Masahiro: V1499,116-119(1991)

Mine, Katsutoshi See Yamazaki, Satomi: V1466,538-545(1991)

Minelli, R. J. See Jackson, John E.: V1533,75-86(1991)

Minemoto, Takumi See Nakagawa, Kiyoshi: V1508,191-200(1991)

Minemura, Tetsuroh
  ; Yazawa, Y.; Asano, J.; Unno, T.: Molecular beam epitaxy/liquid phase epitaxy hybrid growth for GaAs-LED on Si,V1361,344-353(1991)

Miner, Carla J. See Puetz, Norbert: V1361,692-698(1991)

Minford, William J.
  ; DePaula, Ramon P.: Integrated optics for fiber optic sensors,V1367,46-52(1991)
  See Kanofsky, Alvin S.: V1374,59-66(1991)

Ming, Hai
  ; Sun, Yuesheng; Ren, Baorui; Xie, Jianping; Nakajima, Toshinori: Fizeau-type of gradient-index rod lens interferometer by using semiconductor laser,V1572,27-31(1991)

Minguez, E. See Velarde, G.: V1502,242-257(1991)

Mini, Susan M.
  ; Alp, Esen E.; Ramanathan, Mohan; Bommannavar, A.; Hyun, O. B.: Polarization-dependent EXAFS studies in layered copper oxide superconductors,V1345,260-269(1991)
  See Alp, Esen E.: V1345,137-145(1991)
  See Ramanathan, Mohan: V1548,168-178(1991)

Minihan, Charles E.
: Competitiveness,V1496,2-18(1991)

Minina, L. A. See Logak, L. G.: V1238,171-176(1991)

Mininni, Robert M. See Banash, Mark A.: V1590,8-13(1991)

Miniscalco, William J.
See Dakss, Mark L.: V1373,111-124(1991)
See Zemon, Stanley A.: V1373,21-32(1991)

Minkovsky, M. G. See Ionin, A. A.: V1397,453-456(1991)

Minnigh, Stephen W.
; Knights, Mark G.; Avidor, Joel M.; Chicklis, Evan P.: Efficient, uniform transverse coupling of diode arrays to laser rods using nonimaging optics,V1528,129-134(1991)

Minnix, Jay I.
; McVey, Eugene S.; Inigo, Rafael M.: Rotation and scale invariant pattern recognition using a multistaged neural network,V1606,241-251(1991)

Minor, John L.
; Jenquin, Michael J.: Developmental test and evaluation plans for the advanced tactical air reconnaissance system,V1538,18-39(1991)

Minot, Christophe
; Le Person, H.; Mollot, Francis; Palmier, Jean F.: Picosecond photocurrent measurements of negative differential velocity in GaAs/AlAs superlattices,V1362,301-308(1991)

Minot, Katcha
; Cochrane, Mike; Nelson, Bradford; Jones, William L.; Streit, Dwight C.; Liu, Po-Hsin P.: Low-noise high-yield octave-band feedback amplifiers to 20 GHz,V1475,309-313(1991)

Minott, Peter O. See Hayden, William L.: V1417,182-199(1991)

Minoura, Kazuo See Suzuki, Masayuki: V1454,370-381(1991)

Minti, Harry See Reisfeld, Renata: V1513,360-367(1991)

Mintz, Doron See Meer, Peter: V1381,424-435(1991)

Mintz, Max See McKendall, Raymond: V1383,547-565(1991)

Mintzer, Fred
; McFall, John D.: Organization of a system for managing the text and images that describe an art collection,V1460,38-49(1991)

Mintzer, Robert See Pan, Fu-shih: V1396,5-8(1991)

Minvielle, Anna M. See Dunbrack, Steven K.: V1464,314-326(1991)

Minyeev, V. N. See Miroshnikov, Mikhail M.: V1540,496-505(1991)

Miotello, Antonio See Battaglin, Giancarlo: V1513,441-450(1991)

Miracky, Robert F. See Mueller, Heinrich G.: V1598,132-140(1991)

Miranda, Rick
; McCracken, Thomas O.; Fedde, Chris: Triangulating between parallel splitting contours using a simplicial algorithm,V1380,210-217(1991)

Mircea, Andrei See Sermage, Bernard: V1362,617-622(1991)

Miridonov, S. V. See Petrov, Mikhail P.: V1621,402-413(1991)

Miron, G. See Bar, I.: V1397,169-172(1991)

Miron, Nicolae
; Sporea, Dan G.: Graduating rules checking up by laser interferometry,V1500,334-338(1991)

Mironchenko, Vladislav Y. See Boiko, Yuri B.: V1238,258-265(1991)

Mironov, S. F. See Ivanov, Anatoliy A.: V1403,243-245(1991)

Mironov, Yury See Larkin, Alexander I.: V1621,414-423(1991)

Mironova, Yu. V. See Balabaev, N. K.: V1402,53-69(1991)

Mironyuk, G. I. See Markov, Vladimir B.: V1238,340-347(1991)

Miroshnikov, Mikhail M.
: Infrared in the USSR: brief historical survey of infrared development in the Soviet Union,V1540,372-400(1991)
: Mirrors for optical telescopes,V1533,286-298(1991)
; Minyeev, V. N.; Povarkov, V. I.; Samkov, V. M.; Solovyev, V. I.: Infrared instrumental complex for remote measurement of ocean surface temperature distribution,V1540,496-505(1991)
; Nesteruk, Vsevolod F.: B-transformation and Fibonacci-transformations of optical images for the information conformity of their perception,V1500,322-333(1991)
; Nesteruk, Vsevolod F.: Optical image transformation including IR region for information conformity of their formation and perception processes based on Fibonacci polynomials and series,V1540,477-487(1991)

Mirov, Sergey B. See Basiev, Tasoltan T.: V1500,65-71(1991)

Mirowska, N. See Misiewicz, Jan: V1561,6-18(1991)

Mirzoeva, Larisa A.
; Kiseleva, Margaret S.; Reshetnikova, Irina N.: Balloon-born investigations of total and aerosol attenuation continuous spectra in the stratosphere,V1540,444-449(1991)

Misaka, Akio
; Hashimoto, Kazuhiko; Kawamoto, M.; Yamashita, H.; Matsuo, Takahiro; Sakashita, Toshihiko; Harafuji, Kenji; Nomura, Noboru: Hierarchical proximity effect correction for e-beam direct writing of 64-Mbit DRAM,V1465,174-184(1991)

Misemer, David K. See Smith, Terrance L.: V1512,92-100(1991)

Miserocchi, Luigi See Visona, Adriana: V1425,75-83(1991)

Mishra, R. K.
; Pillai, A. M.; Sheshadri, M. R.; Sarma, C. G.: Airborne electro-optical sensor: performance predictions and design considerations,V1482,138-145(1991)

Mishra, Sanjoy K.
; Goldgof, Dmitry B.: Extracting local stretching from left ventricle angiography data,V1450,218-230(1991)

Mishra, Umesh K. See Chou, Chia-Shing: V1475,151-156(1991)

Mishurda, Helen L.
; Hershkowitz, Noah: Instantaneous etch rate measurement of thin transparent films by interferometry for use in an algorithm to control a plasma etcher,V1392,563-569(1991)

Misiewicz, Jan
; Szatkowski, Jan; Mirowska, N.; Gumienny, Zbigniew; Placzek-Popko, E.: Zn3P2: new material for optoelectronic devices,V1561,6-18(1991)
See Gumienny, Zbigniew: V1527,462-465(1991)

Misra, Prabhakar See VanderMeulen, David L.: V1428,91-98(1991)

Misra, Ram B. See Beard, David V.: V1446,52-58(1991)

Mita, Seiichi
See Seya, Eiichi: V1499,269-273(1991)
See Sukeda, Hirofumi: V1499,419-425(1991)

Mitchell, A. C.
See Morgan, Chris G.: V1525,83-90(1991)
See Morgan, Chris G.: V1525,391-396(1991)

Mitchell, Brian T.
ed.: *Image Understanding in the '90s: Building Systems that Work*,V1406(1991)

Mitchell, David C.
; Smith, Roy E.; Walters, Tena K.; Murray, Alan; Wood, Richard F.: Tissue interactions of ball-tipped and multifiber catheters,V1427,181-188(1991)
See Smith, Roy E.: V1425,116-121(1991)

Mitchell, George W. See Swift, Kerry M.: V1431,171-178(1991)

Mitchell, Gordon L.
; Saaski, Elric W.; Pace, John W.: GigaHertz RMS current sensors for electromagnetic compatibility testing,V1367,266-272(1991)

Mitchell, Harvey B. See Hauser, Neal A.: V1605,428-433(1991)

Mitchell, J. R. See Picot, Paul A.: V1444,206-213(1991)

Mitchell, Joan L.
; Pennebaker, William B.: Evolving JPEG color data compression standard,VCR37,68-97(1991)

Mitchell, John H.
; Cohen, Harold L.; Hanley, Stephen T.: High-fidelity diagnostic beam sampling of a tunable high-energy laser,V1414,141-150(1991)

Mitchell, Joseph S.
: Algorithmic approaches to optimal route planning,V1388,248-259(1991)

Mitchell, O. R. See Rodriguez, Arturo A.: V1569,182-199(1991)

Mitchell, Robert J. See Tucker, Christopher J.: V1498,92-98(1991)

Miteva, Margarita G. See Tontchev, Dimitar A.: V1429,76-80(1991)

Mito, Keiichi See Ohue, Hiroshi: V1397,527-530(1991)

Mitra, Bhaswati
See Ghatak, Kamakhya P.: V1415,281-297(1991)
See Ghatak, Kamakhya P.: V1411,137-148(1991)

Mitra, Sanjit K.
; Yu, Tian-Hu: Enhancement of dental x-ray images by two-channel image processing,V1445,156-165(1991)
See Civanlar, Mehmet R.: V1452(1991)
See Yu, Tian-Hu: V1452,420-429(1991)
See Yu, Tian-Hu: V1452,303-309(1991)

Mitra, Shashanka S. See Singh, Anjali: V1366,184-190(1991)

Mitra, Sunanda
See Kim, Yong-Soo: V1565,323-330(1991)
See Lewis, Tom R.: V1565,221-226(1991)
See Newton, Scott C.: V1565,331-337(1991)
See Newton, Scott C.: V1382,209-218(1991)
See Ramirez, Manuel: V1567,632-637(1991)

Mitrofanova, Natalya N. See Vysogorets, Mikhail V.: V1358,1066-1069(1991)

Mitsa, Theophano
; Parker, Kevin J.: Digital halftoning using a blue-noise mask,V1452,47-56(1991)
See Riek, Jonathan K.: V1398,130-141(1991)
See Riek, Jonathan K.: V1445,198-206(1991)

Mitsuoka, Yasuyuki
; Iwaki, Tadao; Yamamoto, Shuhei: Rotation- and scale-invariant joint transform correlator using FLC-SLMs,V1564,244-252(1991)

Mittler-Neher, S. See Neher, Dieter: V1560,335-343(1991)

Miura, Masayuki See Horikawa, Hiroshi: V1454,20-32(1991)

Miura, Michio See Watanabe, Hitoshi: V1499,21-28(1991)

Miura, Sadahiko See Takeuchi, Ichiro: V1394,96-101(1991)

Miura, Takao See Kajita, Toru: V1466,161-173(1991)

Miwa, Mitsuharu See Sawaki, Akihiro: V1366,324-331(1991)

Mixon, D. A. See Nalamasu, Omkaram: V1466,13-25(1991)

Mixon, Harold D. See Sabol, Bruce M.: V1486,258-270(1991)

Miyabo, Katsuaki See Oyaizu, Ikuro: V1606,990-1001(1991)

Miyachi, Yuji
See Iwasaki, Nobuo: V1490,216-221(1991)
See Iwasaki, Nobuo: V1490,192-199(1991)
See Nakayama, Masao: V1490,207-215(1991)
See Tanii, Jun: V1490,200-206(1991)

Miyagi, Yoshihiro See Ogura, Toshihiro: V1443,153-157(1991)

Miyahara, Makoto
See Gan, Qing: V1605,374-381(1991)
See Horita, Yuukou: V1606,607-620(1991)

Miyahara, Sueharu
; Suzuki, Akira; Tada, Shunkichi; Kawatani, Takahiko: Japanese document recognition and retrieval system using programmable SIMD processor,V1384,317-323(1991)

Miyai, Y. See Kasuya, Koichi: V1397,67-70(1991)

Miyake, Akira See Niibe, Masahito: V1343,2-13(1991)

Miyake, Hidetaka See Koshi, Yutaka: V1605,362-373(1991)

Miyake, Hiroyuki
; Sakai, Kazuhiro; Abe, Tsutomu; Sakai, Yoshihiko; Hotta, Hiroyuki; Sugino, Hajime; Ito, Hisao; Ozawa, Takashi: High-performance a-Si:H TFT-driven linear image sensor and its application to a compact scanner,V1448,150-156(1991)
See Kobayashi, Kenichi: V1448,157-163(1991)

Miyake, Takahiro See Yoshida, Yoshio: V1401,58-65(1991)

Miyake, Yoichi See Yamaba, Kazuo: V1453,290-299(1991)

Miyamae, Hiroshi See Tanaka, Yuki: V1354,395-401(1991)

Miyamoto, Akira
See Arai, Tsunenori: V1425,191-195(1991)
See Sakurada, Masami: V1425,158-164(1991)

Miyamoto, Shigeaki See Fujime, Satoru: V1403,306-315(1991)

Miyamura, Isao See Kamoshida, Minoru: V1606,951-960(1991)

Miyanaga, Noriaki See Nakatsuka, Masahiro: V1411,108-115(1991)

Miyasaka, Nobuji See Takemori, Toshikazu: V1469,157-165(1991)

Miyashita, Atsumi See Yoda, Osamu: V1503,463-466(1991)

Miyashita, Tadashi M.
See Kobayashi, Soichi: V1374,300-303(1991)
See Schmidt, Kevin M.: V1396,744-752(1991)

Miyauchi, Yasushi See Okamine, Shigenori: V1499,166-170(1991)

Miyazaki, Daisuke See Kakizaki, Sunao: V1505,199-205(1991)

Miyazaki, Eiichi See Yoshimura, Takeaki: V1332,835-842(1991)

Miyazaki, Junji
; Kamon, Kazuya; Yoshioka, Nobuyuki; Matsuda, Shuichi; Fujinaga, Masato; Watakabe, Yaichiro; Nagata, Hitoshi: New phase-shifting mask structure for positive resist process,V1464,327-335(1991)
See Wiley, James N.: V1464,346-355(1991)

Miyazaki, Sanae See Chandran, Sharat: V1570,391-402(1991)

Miyazawa, Tatsuo
: High-speed integrated rendering algorithm for interpreting multiple-variable 3-D data,V1459,36-47(1991)
See Aoki, Nobutada: V1412,2-11(1991)

Miyoshi, Masahiro
; Kawasaki, Tetsuharu; Takahashi, Satoru; Mine, Hironori: 90-mm magneto-optical disk drive with digital servo,V1499,116-119(1991)

Miyoshi, Tamihiro See Tachihara, Satoru: V1527,305-314(1991)

Mizell, Gregory J. See Gilbreath, G. C.: V1409,87-99(1991)

Mizerka, Larry J. See Al-Jumaily, Ghanim A.: V1441,360-365(1991)

Mizes, Howard See Dougherty, Edward R.: V1567,88-99(1991)

Mizrahi, Victor See Anderson, Dana Z.: V1516,154-161(1991)

Mizuno, Aritake See Ozaki, Yukihiro: V1403,710-719(1991)

Mizuno, Koichi
; Higashino, Hidetaka; Setsune, Kentaro; Wasa, Kiyotaka: Characteristics of thin-film-type Josephson junctions using Bi2Sr2CaCu2Ox/Bi2Sr2CuOy/Bi2Sr2CaCu2Oz structure,V1477,197-204(1991)
See Setsune, Kentaro: V1394,79-88(1991)

Mizuno, Kyoichi
See Arai, Tsunenori: V1425,191-195(1991)
See Sakurada, Masami: V1425,158-164(1991)

Mizushima, Akiko See Nate, Kazuo: V1466,206-210(1991)

Mnatsakanyan, Eduard A.
; Andreev, V. G.; Bestalanny, S. I.; Velikoselsky, V.V.; Grinyakin, A.P.; Grinyakin, V.P.; Kolesnikov, S.A.; Rozhnov, U.V.: Measuring device for space-temporal characteristics of technological lasers radiation,V1414,130-133(1991)

Moacanin, Jovan See Dolgin, Benjamin P.: V1533,229-239(1991)

Moazami, Nader See Bass, Lawrence S.: V1422,123-127(1991)

Mobasser, Sohrab
; Horwitz, Larry S.; Griffith, O'Dale: Development of a low-pass far-infrared filter for lunar observer horizon sensor application,V1540,764-774(1991)

Mobasseri, Bijan G.
: Impact of uncertain terrain models on the weighted region problem,V1388,270-277(1991)

Mobley, Scottie B.
: Dynamic infrared scene projection technology,V1486,325-332(1991)

Moccagatta, Iole See Ebrahimi, Touradj: V1605,2-15(1991)

Mocenter, Michael M.
: Laptop image transmission equipment,V1538,132-139(1991)

Mochizuki, Takashi
; Yano, Mitsuharu; Nishitani, Takao: Image coding using adaptive-blocksize Princen-Bradley transform,V1605,434-444(1991)

Mockler, Daniel J. See Feldman, Mark: V1542,490-501(1991)

Moddel, Garrett See Perlmutter, S. H.: V1562,74-84(1991)

Modern, P. J. See Chen, Shang-Liang: V1502,123-134(1991)

Modglin, Luan See Machol, Steven: V1529,38-44(1991)

Modi, M. B. See Razdan, Anil K.: V1558,384-388(1991)

Moehlmann, Gustaaf R.
; Horsthuis, Winfried H.; Hams, Benno H.: Nonlinear polymers and devices,V1512,34-39(1991)
; Horsthuis, Winfried H.; Mertens, Hans W.; Diemeer, Mart B.; Suyten, F. M.; Hendriksen, B.; Duchet, Christian; Fabre, P.; Brot, Christian; Copeland, J. M.; Mellor, J. R.; van Tomme, E.; van Daele, Peter; Baets, Roel G.: Optically nonlinear polymeric devices,V1560,426-433(1991)

Moehrle, M. See Rosenzweig, M.: V1362,876-887(1991)

Moeller, Robert P.
See Berkoff, Timothy A.: V1367,53-58(1991)
See Burns, William K.: V1367,87-92(1991)

Moeller, Werner
; Heitmann, Walter; Reich, M.: OTDR calibration for attenuation measurement,V1504,47-54(1991)

Moenne-Loccoz, Pierre See Lutz, Marc: V1403,59-65(1991)

Moerman, Ingrid See Demeester, Piet M.: V1361,1132-1143(1991)

Moerner, William E.
; Ambrose, William P.: Finding a single molecule in a haystack: laser spectroscopy of solids from (square root of)N to N=1,V1435,244-251(1991)
; Walsh, C. P.; Scott, J. C.; Ducharme, Stephen P.; Burland, Donald M.; Bjorklund, Gary C.; Twieg, Robert J.: Photorefractivity in doped nonlinear organic polymers,V1560,278-289(1991)

Moezzi, Saied
See Davies, Henry C.: V1388,165-175(1991)
See Kayaalp, Ali E.: V1406,98-109(1991)
See Sinha, Saravajit S.: V1385,259-267(1991)

Moffitt, Kirk W. See Melzer, James E.: V1456,124-131(1991)

Moghaddam, Baback
; Hintz, Kenneth J.; Stewart, Clayton V.: Dimension and lacunarity measurement of IR images using Hilbert scanning,V1486,115-126(1991)
; Hintz, Kenneth J.; Stewart, Clayton V.: Fractal image compression and texture analysis,V1406,42-57(1991)
; Hintz, Kenneth J.; Stewart, Clayton V.: Space-filling curves for image compression,V1471,414-421(1991)

Mogilnik, I. F. See Artsimovich, M. V.: V1397,729-733(1991)

Moglestue, C. See Rosenzweig, Josef: V1362,168-178(1991)

Moh, Jenlong
See Shih, Frank Y.: V1382,99-110(1991)
See Shih, Frank Y.: V1382,26-36(1991)

Mohan, Rakesh See Bolle, Ruud M.: V1383,305-318(1991)

Moharam, M. G.
: Analysis and design of binary gratings for broadband, infrared, low-reflectivity surfaces,V1485,254-259(1991)

Moheimany, O. R. See Rahnavard, Mohammad H.: V1367,374-381(1991)

Mohr, Joachim See Germer, Rudolf K.: V1358,925-931(1991)

Mohr, Jurgen See Goettert, Jost: V1506,170-178(1991)

Mohr, Roger See Wu, Chengke: V1570,362-370(1991)

Mohr, S. See Buchal, Christoph: V1361,881-892(1991)

Mohsenian, Nader
; Nasrabadi, Nasser M.: Edge-based subband image coding technique for encoding the upper-frequency bands,V1605,781-792(1991)

Mohsin, Mohammad See Dhali, Shirshak K.: V1396,353-359(1991)

Moin, M.
: Influence of defects on dynamics of semiconductors (Ge, Si, GaAs) heating by laser radiation,V1440,2-7(1991)

Moine, Bernard See Boutinaud, P.: V1590,168-178(1991)

Moisan, Jean-Yves See Gravey, Philippe: V1507,239-246(1991)

Moiseev, K. D. See Charykov, N. A.: V1512,198-203(1991)

Moison, Jean M. See Nissim, Yves I.: V1393,216-228(1991)

Moissl, M.
; Paul, R.; Breining, K.; Giesen, Adolf; Huegel, Helmut: Thermal lensing effect in fast-axial flow CO2 lasers,V1397,395-398(1991)

Moix, Dominique See Wilke, Ingrid: V1442,2-10(1991)

Moiz, Saifuddin See Celenk, Mehmet: V1468,764-775(1991)

Mojzes, Peter
: Differential Raman spectroscopic study of the interaction of nickel (II) cation with adenine nucleotides,V1403,167-171(1991)

Molchanova, L. V. See Mikaelian, Andrei L.: V1621,194-203(1991)

Molinelli, Eugene J. See Lybanon, Matthew: V1469,637-647(1991)

Molley, Perry A.
; Kast, Brian A.: Automatic target recognition using acousto-optic image correlator,V1471,224-232(1991)
; Sweatt, William C.; Strong, David S.: Miniature acousto-optic image correlator,V1564,610-616(1991)
See Connelly, James M.: V1564,572-592(1991)

Mollot, Francis
See Ababou, Soraya: V1361,706-711(1991)
See Minot, Christophe: V1362,301-308(1991)

Molloy, J. A. See Ramos, P. A.: V1443,160-170(1991)

Molteni, William J.
: Shear-lens photography for holographic stereograms,V1461,132-141(1991)

Momchilova, Maia M. See Baleva, Mitra I.: V1361,712-722(1991)

Momin, S. A.
; Narayanaswamy, R.: Chlorine sensing by optical techniques,V1510,180-186(1991)

Mommsen, Jens
; Stuermer, Martin: Use of excimer lasers in medicine: applications, problems, and dangers,V1503,348-354(1991)

Monahan, Kevin M.
; Benschop, Jozef P.; Harris, Tom A.: Charging effects in low-voltage SEM metrology,V1464,2-9(1991)
See Benschop, Jozef P.: V1464,62-70(1991)
See Davidson, Mark P.: V1464,155-176(1991)
See Rodgers, Mark R.: V1464,358-366(1991)

Monastyrski, M. A. See Degtyareva, V. P.: V1358,546-548(1991)

Moncton, David E.
: Status report on the Advanced Photon Source, summer 1990,V1345,11-15(1991)

Mondal, J. P.
; Contolatis, T.; Geddes, John J.; Swirhun, S.; Sokolov, Vladimir: Low-noise MMIC performance in Kaband using ion implantation technology,V1475,314-318(1991)
See Swirhun, S.: V1475,223-230(1991)

Mondal, Manabendra See Ghatak, Kamakhya P.: V1346,471-489(1991)

Mondal, Pronab K.
; Calvo, Maria L.; Chevalier, Margarita L.; Lakshminarayanan, Vasudevan: Theoretical approach to hyperacuity tests based on resolution criteria for two-line images,V1429,108-116(1991)

Mondot, Michel
; Remiot, Christian: Free-surface temperature measurement of shock-loaded tin using ultrafast infrared pyrometry,V1558,351-361(1991)

Monga, Olivier
; Ayache, Nicholas; Sander, Peter T.: From voxel to curvature,V1570,382-390(1991)

Monge, J.L.
; Kandel, Robert S.; Pakhomov, L. A.; Bauche, B.: ScaRaB Earth radiation budget scanning radiometer,V1490,84-93(1991)

Monger, A. G. See Mills, David R.: V1528,44-55(1991)

Moniwa, Akemi
; Okazaki, Shinji: Investigation on the effect of electron-beam acceleration voltage and electron-beam sharpness on 0.2-um patterns,V1465,154-163(1991)

Monjoux, E.
; Brunet, Gerard; Rudant, J. P.: Texture segmentation in aerial images,V1452,310-318(1991)

Monnier, Philippe
See Bays, Roland: V1525,397-408(1991)
See Braichotte, D.: V1525,212-218(1991)
See Wagnieres, G.: V1525,219-236(1991)

Monno, Giuseppe See Pappalettere, Carmine: V1554B,99-105(1991)

Monsees, Thomas L. See Cox, Jerome R.: V1446,40-51(1991)

Montagu, Jean I.
: Update of scanner selection, performances, and multiaxis configurations,V1454,160-173(1991)

Monteiro, Ricardo M. See Trunk, Jonah: V1365,124-130(1991)

Montel, J. See Ayral, Jean-Luc: V1500,81-92(1991)

Montenero, Angelo
; Gnappi, G.; Bertolotti, Mario; Sibilia, C.; Fazio, E.; Liakhou, G.: Crystallization and optothermal characteristics of germanate glasses,V1513,234-242(1991)

Montesanti, Richard C.
; Atkinson, Dennis P.; Edwards, David F.; Klingmann, Jeffrey L.: XMM space telescope: development plan for the lightweight replicated x-ray gratings,V1343,558-565(1991)
See Atkinson, Dennis P.: V1343,530-541(1991)

Monteverde, Robert J.
See Davidson, Mark P.: V1464,155-176(1991)
See Wojcik, Gregory L.: V1464,187-203(1991)

Montgomery, David S.
See Hatch, Jeffrey A.: V1346,371-375(1991)
See Kilkenny, Joseph D.: V1358,117-133(1991)
See Wiedwald, Douglas J.: V1346,449-455(1991)

Montgomery, G. P.
; West, John L.; Tamura-Lis, Winifred: Droplet-size effects in light scattering from polymer-dispersed liquid-crystal films,V1455,45-53(1991)
See Vaz, Nuno A.: V1455,110-122(1991)

Montgomery, Paul C.
; Fillard, Jean-Pierre B.; Tchandjou, N.; Ardisasmita, Moh S.: Three-dimensional nanoprofiling of semiconductor surfaces,V1332,515-524(1991)
; Gall, Pascal; Ardisasmita, Moh S.; Castagne, Michel; Bonnafe, Jacques; Fillard, Jean-Pierre B.: Laser-scanning tomography and related dark-field nanoscopy method,V1332,563-570(1991)

Montgomery, Raymond C.
; Ghosh, David: Studies on distributed sensing and processing for the control of large flexible spacecraft,V1480,126-137(1991)

Montgomery, Warren See Brunsvold, William R.: V1466,368-376(1991)

Monti, P. See Bertoluzza, Alessandro: V1403,743-745(1991)

Montner, Steven M. See Yoshimura, Hitoshi: V1445,47-51(1991)

Montoya, Nelson I.
See Paisley, Dennis L.: V1358,760-765(1991)
See Paisley, Dennis L.: V1346,172-178(1991)

Montoya, Wayne See Reamey, Robert H.: V1455,39-44(1991)

Montrosset, Ivo
See Mazzola, M.: V1365,2-12(1991)
See Rocca, Corrado: V1372,39-47(1991)

Moodie, D. G. See Johnstone, Walter: V1506,145-149(1991)

Mookherji, Tripty K. See Rao, S. M.: V1557,283-292(1991)

Mooney, Jonathan M.
: Staring sensor MRT measurement and modeling,V1540,550-564(1991)

Moore, Andrew
; Koch, Christof: Multiplication-based analog motion detection
chip,V1473,66-75(1991)

Moore, Chris J. See Puetz, Norbert: V1361,692-698(1991)

Moore, Douglas R.
: Long-term reliability tests on fused biconical taper couplers,V1366,241-
250(1991)
See Daniel, Hani S.: V1365,53-59(1991)
See Jones, Michael G.: V1365,43-52(1991)

Moore, Duncan T.
See Pollicove, Harvey M.: V1398,90-93(1991)
See Pollicove, Harvey M.: V1354,482-486(1991)
See Wang, David Y.: V1354,599-605(1991)

Moore, E. See Cooper, Alfred W.: V1486,37-46(1991)

Moore, Emery L.
See Lewis, Norris E.: V1369(1991)
See Lewis, Norris E.: V1589(1991)

Moore, Gerald T.
; Bochove, Erik J.; Scully, Marlan O.: Laser linac in vacuum: assessment of a
high-energy particle accelerator,V1497,328-337(1991)
See Bochove, Erik J.: V1497,338-347(1991)
See Koch, Karl W.: V1516,67-74(1991)

Moore, Jana See Gooding, Linda: V1457,259-266(1991)

Moore, Kenneth E.
; Lawrence, George N.: Focus grating coupler construction optics: theory,
design, and tolerancing,V1354,581-587(1991)
See Lawrence, George N.: V1415,322-329(1991)
See Lawrence, George N.: V1354,15-22(1991)

Moore, Larry J. See Xiong, Xiaoxiong: V1435,188-196(1991)

Moore, Leslie K. See Sulya, Andrew W.: V1434,147-158(1991)

Moore, Russell L. See Yelamarty, Rao V.: V1398,170-179(1991)

Moore, Stephen M. See Cox, Jerome R.: V1446,40-51(1991)

Moore, Wally W. See Wells, James A.: V1485,2-12(1991)

Moore, Wesley S. See Eton, Darwin: V1425,182-187(1991)

Moorhead, Robert J.
See Civanlar, Mehmet R.: V1452(1991)
See Welch, Eric B.: V1453,235-243(1991)

Mopsik, Fred See Schen, Michael A.: V1560,315-325(1991)

Moraes, E. S. See Gomes, Anderson S.: V1579,257-263(1991)

Morais, F. V. See Soares, Oliverio D.: V1332,166-184(1991)

Morales, Aldo W.
; Acharya, Raj S.: Morphological pyramid with alternating sequential
filters,V1452,258-269(1991)

Morales-Romero, Arquimedes A. See Rodriguez-Vera, Ramon: V1507,416-
424(1991)

Moran, Kelly B. See Artjushenko, Vjacheslav G.: V1420,157-168(1991)

Moran, M. J. See Piestrup, Melvin A.: V1552,214-239(1991)

Moran, M. S.
See Biggar, Stuart F.: V1493,155-162(1991)
See Markham, Brian L.: V1493,207-214(1991)
See Nianzeng, Che: V1493,182-194(1991)

Morasca, Salvatore
See Caldera, Claudio: V1372,82-87(1991)
See Cognolato, Livio: V1513,368-377(1991)

Moravec, Thomas J.
: Challenges of using advanced multichip packaging for next generation
spaceborne computers,V1390,195-201(1991)

Mordaunt, David W.
; Nieuwsma, Daniel E.: Development and implementation of MIS-36477
laser damage certification of designator optical components,V1441,27-
30(1991)

Mordo, David
See Keenan, W. A.: V1393,354-365(1991)
See Raicu, Bruha: V1393,161-171(1991)

Moreau, Louis See Sheng, Yunlong: V1559,222-228(1991)

Moreen, Harry A.
: Control of thermally induced porosity for the fabrication of beryllium
optics,V1485,54-61(1991)

Moreira, Roberto P. See Trunk, Jonah: V1365,124-130(1991)

Morelli, M. A.
See Bertoluzza, Alessandro: V1403,150-152(1991)
See Bertoluzza, Alessandro: V1403,153-155(1991)

Moreno, T. See Khan Malek, Chantal: V1343,56-61(1991)

Moreno-Bondi, Maria C.
; Orellana, Guillermo; Camara, Carmen; Wolfbeis, Otto S.: New luminescent
metal complex for pH transduction in optical fiber sensing: application to
a CO_2-sensitive device,V1368,157-164(1991)

Moresco, Larry L.
: System interconnect issues for subnanosecond signal
transmission,V1390,202-213(1991)

Moretti, Michael
: Medical marketplace,V1520,118-131(1991)

Morey, William W.
See Giesler, Leslie E.: V1480,138-142(1991)
See Meltz, Gerald: V1516,185-199(1991)

Morf, Werner E. See Spichiger, Ursula E.: V1510,118-130(1991)

Morgan, A. J. See Rainer, Frank: V1441,247-254(1991)

Morgan, Alan R.
; Garbo, Greta M.; Krivak, T.; Mastroianni, Marta; Petousis, Nikolaos H.; St
Clair, T.; Weisenberger, M.; van Lier, Johan E.: New sensitizers for
PDT,V1426,350-355(1991)

Morgan, Chris G.
; Mitchell, A. C.; Chowdhary, R. K.: Photosensitive liposomes as potential
drug delivery vehicles for photodynamic therapy,V1525,391-396(1991)
; Murray, J. G.; Mitchell, A. C.: Frequency-domain imaging using array
detectors: present status and prospects for picosecond
resolution,V1525,83-90(1991)

Morgan, David A.
: 3M's Dry Silver technology: an ideal media for electronic
imaging,V1458,62-67(1991)

Morgan, J. H. See Peterson, Heidi A.: V1453,210-222(1991)

Morgan, Jeffrey S.
See Huber, Martin C.: V1494,472-480(1991)
See Kasle, David B.: V1549,52-58(1991)
See Timothy, J. G.: V1343,350-358(1991)

Morgan, Robert A.
: Improvements in self electro-optic effect devices: toward system
implementation,V1562,213-227(1991)
; Chirovsky, Leo M.; Focht, Marlin W.; Guth, Gregory O.; Asom, Moses T.;
Leibenguth, R. E.; Robinson, K. C.; Lee, Yong H.; Jewell, Jack L.: Progress
in planarized vertical-cavity surface-emitting laser devices and
arrays,V1562,149-159(1991)

Morgan, Robert E.
; Ehlers, Sandy L.; Jones, Katharine J.: Composite-embedded fiber-optic data
links and related material/connector issues,V1588,189-197(1991)

Morgan, Russell D.
; Jones, Julian D.; Harper, Philip G.; Barton, James S.: Wavelength and
temperature dependence of bending loss in monomode optical
fibers,V1504,118-124(1991)

Morgan, S. H. See Burger, Arnold: V1557,245-249(1991)

Morgavi, Giovanna See Basti, Gianfranco: V1469,719-736(1991)

Morgenthaler, Matthew K.
; Bruno, Guy; Spofford, John R.; Greunke, Roy G.; Gatrell, Lance B.: Testbed
for tele-autonomous operation of multiarmed robotic servicers in
space,V1387,82-95(1991)
See Bruno, Guy: V1387,352-358(1991)

Morguet, A. See Abel, B.: V1525,110-118(1991)

Morgunov, A. N. See Shvej, E. M.: V1554A,488-495(1991)

Mori, Kazushi See Yamaguchi, Takao: V1418,363-371(1991)

Mori, Luigi See Brofferio, Sergio C.: V1605,894-905(1991)

Mori, Masahiko See Ishihara, Satoshi: V1621,362-372(1991)

Mori, Masahiro See Nakano, Yasuhiko: V1605,874-878(1991)

Mori, Masaki
See Inaoka, Noriko: V1450,2-12(1991)
See Suzuki, Hideo: V1450,99-107(1991)

Mori, Yasushi
; Sano, Kenji; Todori, Kenji; Kawamonzen, Yosiaki: New organic crystal material for SHG, 2-cyano-3-(3,4-methylene dioxy phenyl)-2-propionic acid ethyl ester,V1560,310-314(1991)

Mori, Yusuke See Nishimura, Kazuhito: V1534,199-206(1991)

Morian, Hans F.
: New glasses for optics and optoelectronics,V1400,146-157(1991)

Morichere, D.
; Dumont, Michel L.; Levy, Yves; Gadret, G.; Kajzar, Francois: Nonlinear properties of poled-polymer films: SHG and electro-optic measurements,V1560,214-225(1991)
See Dumont, Michel L.: V1559,127-138(1991)

Morii, Fujiki
: Generalization of Lloyd's algorithm for image segmentation,V1381,545-552(1991)

Morikawa, Yoshitaka
; Yamane, Nobumoto; Hamada, Hiroshi: Overlapping block transform for offset-sampled image compression,V1605,445-455(1991)
See Yamane, Nobumoto: V1605,679-686(1991)

Morimoto, Yoshiharu
; Post, Daniel; Gascoigne, Harold E.: Carrier pattern analysis of moire interferometry using Fourier transform,V1554B,493-502(1991)
See Sogabe, Yasushi: V1554B,289-297(1991)

Morin, Richard L.
; Forbes, Glenn S.; Gehring, Dale G.; Salutz, James R.; Pavlicek, William: Present status and future directions of the Mayo/IBM PACS project,V1446,436-441(1991)
See Forbes, Glenn S.: V1446,318-322(1991)
See Gehring, Dale G.: V1446,248-252(1991)

Morinaga, Kenji See Kai, Teruhiko: V1519,99-103(1991)

Moring, Ilkka See Ailisto, Heikki J.: V1384,50-59(1991)

Morioka, Craig A.
; Chan, Kelby K.; Huang, H. K.: Parallel constructs for three-dimensional registration on a SIMD (single-instruction stream/multiple-data stream) processor,V1445,534-538(1991)
See Chan, Kelby K.: V1444,250-255(1991)

Morisse, J. See Lefevre, Herve C.: V1367,72-80(1991)

Morita, Shinzo See Sugiura, Norio: V1358,442-446(1991)

Morita, Yuji See Tchoryk, Peter: V1479,164-182(1991)

Morito, K. See Tamanuki, Takemasa: V1361,614-617(1991)

Moritsugu, Masaharu See Ozawa, Yasuyuki: V1499,136-142(1991)

Moritz, Heiko See Rizzi, Angela: V1361,827-833(1991)

Moriya, Mitsuro See Ishibashi, Hiromichi: V1499,340-347(1991)

Moriya, Susumu See Torao, Akira: V1358,843-850(1991)

Moriyama, Shigeo See Seya, Eiichi: V1499,269-273(1991)

Moriyama, Takashi
; Nakayama, Kimihiko; Homma, M.; Haruyama, Yukio: Mission study overview of Japanese polar orbiting platform program,V1490,310-316(1991)

Morjani, Hamid See Manfait, Michel: V1403,695-707(1991)

Morkel, Paul R.
; Laming, Richard I.; Cowle, Gregory J.; Payne, David N.: Noise characteristics of rare-earth-doped fiber sources and amplifiers,V1373,224-233(1991)
See Cowle, Gregory J.: V1373,54-65(1991)

Morley, Stefan
; Eickhoff, Thomas; Zahn, Dietrich R.; Richter, W.; Woolf, D.; Westwood, D. I.; Williams, R. H.: Far-IR studies of moderately doped molecular beam epitaxy grown GaAs on Si(100),V1361,213-222(1991)

Mormile, Pasquale
See Carbonara, Giuseppe: V1361,1038-1040(1991)
See Carbonara, Giuseppe: V1361,688-691(1991)

Moro, C. See Ferrara, M.: V1513,130-136(1991)

Morokawa, Shigeru See Hashimoto, Nobuyuki: V1461,291-302(1991)

Moronaga, Kenji See Watanabe, Mikio: V1452,27-36(1991)

Morookian, John-Michael See Sherman, Bradley D.: V1369,60-66(1991)

Morosova, D. V. See Savostjanov, V. N.: V1554A,579-585(1991)

Morozov, N. V.
; Sagitov, S. I.; Sergeyev, P. B.: Excimer laser optics,V1441,557-564(1991)

Morozov, Valentin N. See Guilfoyle, Peter S.: V1563,279-283(1991)

Morreale, Jeanne See Weisenbach, Lori: V1590,50-58(1991)

Morrel, William G.
: Passive components for the subscriber loop,V1363,40-47(1991)

Morris, David E. See Berg, John M.: V1435,331-337(1991)

Morris, David R.
; Watson, A. J.: Dynamic photoelasticity with a split Hopkinson pressure bar,V1358,254-261(1991)

Morris, G. M.
ed.: *Holographic Optics III: Principles and Applications,*V1507(1991)
See Buralli, Dale A.: V1354,292-296(1991)
See Spaulding, Kevin E.: V1507,45-54(1991)

Morris, J. W. See Frear, Darrel R.: V1596,72-82(1991)

Morris, Jacqueline H.
; Belopolsky, Yakov: Characterization, modeling, and design of dielectric resonators based on thin ceramic tape,V1389,236-248(1991)

Morris, Michael D.
; Govil, Anurag; Liu, Kei-Lee; Sheng, Rong-Sheng: Hadamard transform Raman imaging,V1439,95-101(1991)

Morris, Patricia A.
; Crawford, Michael K.; Roelofs, Mark G.; Bierlein, John D.; Baer, Thomas M.: Ionic conductivity and damage mechanisms in KTiOPO4 crystals,V1561,104-111(1991)

Morris, Stephen J.
; Beatty, Diane M.; Welling, Larry W.; Wiegmann, Thomas B.: Instrumentation for simultaneous kinetic imaging of multiple fluorophores in single living cells,V1428,148-158(1991)

Morris, V. A. See Quesenberry, Laura A.: V1428,177-190(1991)

Morris, Winfred L. See Dadkhah, Mahyar S.: V1554A,164-175(1991)

Morrish, Arthur A. See Hickey, Carolyn F.: V1534,67-76(1991)

Morrison, G. L. See Mills, David R.: V1528,44-55(1991)

Morrison, Jasper J. See Poulsen, Peter: V1407,172-182(1991)

Morrison, Mons D. See Acton, Loren W.: V1447,123-139(1991)

Morrison, Paul R.
See Cheong, Wai-Fung: V1422,19-26(1991)
See Kurban, Amal K.: V1422,43-49(1991)

Morrison, Philip W. See Bates, Stephen C.: V1434,28-38(1991)

Morrison, Richard L.
: Image analysis for diagnostics in photonic switching,V1396,568-574(1991)
See Kerbis, Esther: V1396,447-452(1991)
See McCormick, Frederick B.: V1396,508-521(1991)
See McCormick, Frederick B.: V1533,97-114(1991)

Morrison, Steve A. See Dyott, Richard B.: V1482,439-443(1991)

Morrone, Barbara L. See Soper, Steven A.: V1435,168-178(1991)

Morrow, James R. See MacDonald, Scott A.: V1466,2-12(1991)

Morrow, Joel See Coden, Michael H.: V1364,22-39(1991)

Morse, David L.
See Mendoza, Edgar A.: V1378,139-144(1991)
See Mendoza, Edgar A.: V1583,43-51(1991)

Morse, Jeffrey D.
; Mariella, Raymond P.; Dutton, Robert W.: Graded AlxGa1-xAs photoconductive devices for high-efficiency picosecond optoelectronic switching,V1378,55-59(1991)

Morse, Ted F.
See Malo, Bernard: V1590,83-93(1991)
See Nath, Dilip K.: V1489,17-32(1991)
See Oh, Kyunghwan: V1590,94-100(1991)
See Risen, William M.: V1590,44-49(1991)

Mortazavi, Mansour See Singh, Surendra: V1376,143-152(1991)

Mortensen, David A. See Shropshire, Geoffrey J.: V1379,222-235(1991)

Mortimer, Simon H. See Martin, Brian: V1463,584-594(1991)

Morton, E. J. See Antonuk, Larry E.: V1443,108-119(1991)

Morvan, D.
; Cipriani, Francois D.; Dufresne, Daniel; Garino, A.: Thermocapillary effects in a melted pool during laser surface treatment,V1397,689-692(1991)

Mosbrooker, Michael L.
: Eyesafe laser application in military and law enforcement training,V1419,107-114(1991)

Moseley, Andrew J. See Goodfellow, Robert C.: V1389,594-599(1991)

Moser, A. See Webb, David J.: V1418,231-239(1991)

Moser, Frank
See Drizlikh, S.: V1420,53-62(1991)
See Paiss, Idan: V1420,141-148(1991)

Moses, Harry E.
; Prosser, Reese T.: Time development of acoustic bullets, wave-zone form of focus wave modes, and other solutions of the acoustic equations,V1407,354-374(1991)

Moses, Randolph L. See Jouny, Ismail: V1471,142-153(1991)

Moshin, J. N. See Fakhrutdinov, I. H.: V1399,98-106(1991)

Mosier, Marty R.
; Harris, Gary N.; Whitmeyer, Charlie: Pegasus air-launched space booster payload interfaces and processing procedures for small optical payloads,V1495,177-192(1991)

Moskalenko, Konstantin L.
; Nadezhdinskii, Alexander I.; Stepanov, Eugene V.: High-sensitivity sensor of gases based on IR tunable diode lasers for human exhalation monitoring,V1426,121-132(1991)

Moskalev, B. B.
See Bryukhnevitch, G. I.: V1358,739-749(1991)
See Bryukhnevitch, G. I.: V1449,109-115(1991)
See Degtyareva, V. P.: V1358,524-531(1991)

Moslehi, Behzad M. See Jannson, Tomasz: V1555,159-176(1991)

Moslehi, Mehrdad M.
: Low-temperature in-situ dry cleaning process for epitaxial layer multiprocessing,V1393,90-108(1991)
: Noninvasive sensors for in-situ process monitoring and control in advanced microelectronics manufacturing,V1393,280-294(1991)
; Singh, Rajendra; eds.: *Rapid Thermal and Related Processing Techniques*,V1393(1991)
See Wood, Samuel C.: V1393,337-348(1991)

Mosnier, Jean-Paul See Kennedy, Eugene T.: V1503,406-415(1991)

Moss, James P.
See Coombes, Anne M.: V1380,180-189(1991)
See Linney, Alfred D.: V1380,190-199(1991)

Moss, Rodney H.
: MOVPE technology in device applications for telecommunication,V1361,1170-1181(1991)

Mossbacher, Bernard See Singer, Christian: V1494,255-264(1991)

Mosse, C. A. See Linney, Alfred D.: V1380,190-199(1991)

Mostacci, D. See Kawai, Kenji: V1465,308-314(1991)

Mostafavi, Hassan
: Wind tunnel model aircraft attitude and motion analysis,V1483,104-111(1991)

Mostafavi, M. T.
: Linear interconnection architecture in parallel implementation of neural network models,V1396,193-201(1991)

Motamed, Cina
; Schmitt, Alain: Visual surveillance system based on spatio-temporal model of moving objects in industrial workroom environments,V1606,961-969(1991)

Motamedi, M. E.
; Griswold, Marsden P.; Knowlden, Robert E.: Silicon microlenses for enhanced optical coupling to silicon focal planes,V1544,22-32(1991)
; Southwell, William H.; Anderson, Robert J.; Hale, Leonard G.; Gunning, William J.; Holz, Michael: High-speed binary optic microlens array in GaAs,V1544,33-44(1991)

Motamedi, Massoud E.
See Agah, Ramtin: V1425,172-179(1991)
See LeCarpentier, Gerald L.: V1427,273-278(1991)

Motenko, Boris
; Ermakov, Boris A.; Berezin, Boris: Solid state lasers for field application,V1399,78-81(1991)

Mott, Leonard P. See Rainer, Frank: V1441,247-254(1991)

Mottier, Patrick
; Agnese, Patrick; Lagage, Pierre O.: Si:Ga focal-plane arrays for satellite and ground-based telescopes,V1494,419-426(1991); 1512,60-67(1991)

Mou, Hai C. See Chen, You S.: V1519,56-62(1991)

Mouanda, B. See Vinet, Francoise: V1466,558-569(1991)

Moubarak, Michel See Preteux, Francoise: V1450,72-83(1991)

Mould, John
See Wojcik, Gregory L.: V1463,292-303(1991)
See Wojcik, Gregory L.: V1464,187-203(1991)

Moulton, J. D. See Madsen, Steen J.: V1431,42-51(1991)

Moulton, Joseph R. See Lorenzo, Maximo: V1483,118-137(1991)

Mounaix, P. See Labaune, C.: V1413,138-143(1991)

Mount, George H.
: Measurement of the tropospheric hydroxyl radical by long-path absorption,V1491,26-32(1991)
; Jakoubek, Roger O.; Sanders, Ryan W.; Harder, Jerald W.; Solomon, Susan; Winkler, Richard; Thompson, Thomas; Harrop, Walter: New spectroscopic instrumentation for measurement of stratospheric trace species by remote sensing of scattered skylight,V1491,188-193(1991)
See Harder, Jerald W.: V1491,33-42(1991)

Moura, Jose M. See Balram, Nikhil: V1606,374-385(1991)

Mourad, Pierre D. See Porter, Robert P.: V1558,91-102(1991)

Mourrain, Jean-Loic See Panayotatos, Paul: V1361,1100-1109(1991)

Mousavi, Madjid S.
; Schalkoff, Robert J.: Stereo vision: a neural network application to constraint satisfaction problem,V1382,228-239(1991)

Moustaizis, Stavros D.
; Fotakis, Costas; Girardeau-Montaut, Jean-Pierre: Laser photocathode development for high-current electron source,V1552,50-56(1991)

Mouthon, Alain
See Barnault, B.: V1397,231-234(1991)
See Georges, Eric: V1397,243-246(1991)

Mowafy, Lyn
; Lappin, Joseph S.: Perceiving the coherent movements of spatially separated features,V1453,177-187(1991)

Moy, Richard Q.
; Vargas, Raymund; Eubanks, Charles: Predicting electronic component lifetime using thermography,V1467,154-160(1991)

Moynagh, P. See Rosser, Paul J.: V1393,49-66(1991)

Mozina, Janez See Grad, Ladislav: V1525,206-209(1991)

Mozumdar, Subir
; Bond, Robert L.: Improvement in spatial resolution of a forward-scatter laser Doppler velocimeter,V1584,254-261(1991)

Mozzi, Massimo See Angrilli, Francesco: V1482,26-39(1991)

Mu, Guoguang See Wang, Xu-Ming: V1558,518-528(1991)

Mu, Lemin See Huang, Yue-Huai: V1572,539-542(1991)

Mu, Zongxue See Luo, Zhishan: V1554B,523-528(1991)

Mudge, Trevor N. See Gottschalk, Paul G.: V1383,84-96(1991)

Muecke, Robert J.
; Werle, Peter W.; Slemr, Franz; Prettl, William: Comparison of time and frequency multiplexing techniques in multicomponent FM spectroscopy,V1433,136-144(1991)

Muecksch, Michaela C.
: Analysis of vegetation stress and damage from images of the high-resolution airborne pushbroom image spectrograph compact airborne spectrographic imager,V1399,157-161(1991)

Mueller, Adrian
; Zell, Andreas: Connectionist natural language parsing with BrainC,V1469,188-196(1991)
; Zell, Andreas: Natural language parsing in a hybrid connectionist-symbolic architecture,V1468,875-881(1991)

Mueller, Edward P. See Fitzpatrick, Colleen M.: V1332,185-192(1991)

Mueller, Gerhard J.
See Artjushenko, Vjacheslav G.: V1420,149-156(1991)
See Artjushenko, Vjacheslav G.: V1420,157-168(1991)
See Artjushenko, Vjacheslav G.: V1420,176-176(1991)
See Artjushenko, Vjacheslav G.: V1590,131-136(1991)
See Beuthan, Jurgen: V1420,234-241(1991)
See Beuthan, Jurgen: V1420,225-233(1991)
See Doerschel, Klaus: V1525,253-279(1991)
See Loshchenov, V. B.: V1420,271-281(1991)

Mueller, H. See Schmiedeskamp, Bernt: V1343,64-72(1991)

Mueller, Heinrich G.
; Galanakis, Claire T.; Sommerfeldt, Scott C.; Hirsch, Tom J.; Miracky, Robert F.: Laser process for personalization and repair of multichip modules,V1598,132-140(1991)

Mueller, Joerg See Peters, Dethard: V1362,338-349(1991)

Mueller, L. See Bowers, C. R.: V1435,36-50(1991)

Mueller, Matthias
See Rueckmann, I.: V1513,78-85(1991)
See Woggon, Ulrike: V1362,888-898(1991)

Mueller, Peter
See Boher, Pierre: V1343,39-55(1991)
See Schmiedeskamp, Bernt: V1343,64-72(1991)

Mueller, Walter
; Olson, James; Martin, Andrew; Hinchman, John H.: Focusing on targets through exclusion,V1472,66-75(1991)

Mueller-Horsche, Elmar
; Oesterlin, Peter; Basting, Dirk: Recent progress towards multikilowatt output,V1503,28-39(1991)

Muenchausen, Ross E.
; Foltyn, Stephen R.; Wu, Xin D.; Dye, Robert C.; Nogar, Nicholas S.; Carim, A. H.; Heidelbach, F.; Cooke, D. W.; Taber, R. C.; Quinn, Rod K.: Pulsed-laser deposition of YBa2Cu3O7-x thin films: processing, properties, and performance,V1394,221-229(1991)
See Wu, Xin D.: V1477,8-14(1991)

Mugnier, A. See Mens, Alain: V1358,719-731(1991)

Muhamad, M. R.
; Majid, W. H.; Ariffin, Z.: Optical transition characteristic energies of amorphous and polycrystalline tin oxide films,V1519,872-877(1991)

Muhonen, John See Casey, William L.: V1417,89-98(1991)

Muhs, Jeffry D.
; Jordan, John K.; Scudiere, M. B.; Tobin, Kenneth W.: Results of a portable fiber optic weigh-in-motion system,V1584,374-386(1991)
See Tobin, Kenneth W.: V1589,102-109(1991)

Muhuri, K. See Das, Alok K.: V1364,61-69(1991)

Muka, Edward See Jost, R. G.: V1446,2-9(1991)

Mukasa, Minoru
; Aoki, Makoto; Satoh, Minoru; Kowada, Masayoshi; Kikuchi, K.: Development of computer-aided functions in clinical neurosurgery with PACS,V1446,253-265(1991)

Mukhamedgalieva, Anel F.
: Variation of magnetic properties of iron-contained quartzite irradiated by CO2 laser,V1502,223-225(1991)

Mukhedkar, Dinkar See Rayapati, Venkatapathi N.: V1389,98-109(1991)

Mukherjee, Sayan D.
See Bristow, Julian P.: V1389,535-546(1991)
See Hibbs-Brenner, Mary K.: V1563,10-20(1991)
See Swirhun, S.: V1475,223-230(1991)

Mukoh, Akio See Imanishi, Yasuo: V1361,570-580(1991)

Mulak, Grazyna
: Boundary diffraction wave in imaging by small holograms,V1574,266-271(1991)

Mulderink, Ken See Jursich, Gregory M.: V1412,115-122(1991)

Mulero, Manuel A.
; Barreiros, Manuel A.: Experimental simulation of IR signatures,V1540,519-526(1991)

Mulet, Miguel See Friedlander, Miles H.: V1423,62-69(1991); 1429,229-236(1991)

Mulgaonkar, Prasanna G.
; Decurtins, Jeff; Cowan, Cregg K.: Scene description: an iterative approach,V1382,320-330(1991)

Mulgrew, B. See Nisbet, K. C.: V1565,244-254(1991)

Muller, Gerard See Buvat, Daniel: V1417,251-261(1991)

Muller, Jean-Claude See Hartiti, Bouchaib: V1393,200-206(1991)

Muller, Jean-Michel M. See Kla, Sylvanus: V1566,275-285(1991)

Muller, K. See Beard, David V.: V1446,52-58(1991)

Muller, Karl P. See Pilz, Wolfgang: V1392,84-94(1991)

Muller, Paul J.
; Wilson, Brian C.: Photodynamic therapy of malignant brain tumors: supplementary postoperative light delivery by implanted optical fibers: field fractionation,V1426,254-265(1991)

Mulligan, Jeffrey B. See Ahumada, Albert J.: V1453,134-146(1991)

Mullin, John B.
: Advanced infrared detector materials,V1512,144-154(1991)
; Cole-Hamilton, D. J.; McQueen, A. E.; Hails, J. E.: Developments in precursors for II-VI semiconductors,V1361,1116-1127(1991)

Mumaw, Richard T. See Lindsay, Tracy K.: V1464,104-118(1991)

Mumford, David
: Mathematical theories of shape: do they model perception?,V1570,2-10(1991)

Mun, In K.
; Kim, Y. S.; Mun, Seong Ki: Direct digital image transfer gateway,V1444,232-237(1991)

Mun, Seong Ki
See Horii, Steven C.: V1446,10-15(1991)
See Horii, Steven C.: V1446,475-480(1991)
See Lo, Shih-Chung B.: V1444,265-271(1991)
See Mun, In K.: V1444,232-237(1991)

Munch, Jesper See Wickham, Michael: V1414,80-90(1991)

Muncill, G. See Lybanon, Matthew: V1469,637-647(1991)

Mundy, Joseph L.
: Model-based vision: an operational reality?,V1567,124-141(1991)

Munger, G.
See Leblanc, Roger M.: V1436,92-102(1991)
See Volkov, Alexander G.: V1436,68-79(1991)

Munger, Robert See Snyder, Patricia A.: V1548,188-196(1991)

Munier, Bernard
; Prieur-Drevon, P.; Chabbal, Jean: Digital radiology with solid state linear x-ray detectors,V1447,44-55(1991)

Munnerlyn, Charles R.
See Clapham, Terrance N.: V1423,2-7(1991)
See Shimmick, John K.: V1423,140-153(1991)

Munoz-Yague, Antonio See Fontaine, Chantal: V1362,59-66(1991)

Murahata, M. See Yamaguchi, H.: V1499,29-38(1991)

Murai, Fumio See Saitou, Norio: V1465,185-191(1991)

Murai, Masaru See Daidoh, Yuichiro: V1421,120-123(1991)

Murakami, Hitomi
: Present state of HDTV coding in Japan and future prospect,V1567,544-555(1991)
See Matsumoto, Shuichi: V1605,37-46(1991)
See Takishima, Yasuhiro: V1605,635-645(1991)

Murakami, Kouichi See Yoda, Osamu: V1503,463-466(1991)

Murakami, R. See Nishioka, Teiichi: V1554A,802-813(1991)

Murakami, Yasunori See Suzuki, Tomihiro: V1389,455-461(1991)

Murakami, Yoshiyuki See Kimura, Shigeru: V1483,10-17(1991)

Murakawa, Masao See Koga, Nobuhiro: V1554A,84-92(1991)

Murali, K. R.
; Subramanian, V.; Rangarajan, N.; Lakshmanan, A. S.; Rangarajan, S. K.: Photoelectrochemical characteristics of slurry-coated CdSeTe films,V1536,289-295(1991)

Muramatsu, Mikiya
; Guedes, G. H.; Matsuda, Kiyofumi; Barnes, Thomas H.: Study of oxidization process in real time using speckle correlation,V1332,792-797(1991)

Muramoto, Naohiro See Yamazaki, Satomi: V1466,538-545(1991)

Murano, Hiroshi
; Watari, Toshihiko: Packaging technology for the NEC SX-3 supercomputers,V1390,78-90(1991)

Muraoka, K. See Yamakoshi, H.: V1397,119-122(1991)

Murata, Kazuo See Suzuki, Tomihiro: V1389,455-461(1991)

Murata, Shigeki See Sogabe, Yasushi: V1554B,289-297(1991)

Murata-Seawalt, Debbie See Delamere, W. A.: V1447,288-297(1991); 1479,21-30(1991)

Muratori, Mario See Barbero, Marzio: V1567,566-577(1991)

Muraviev, S. V. See Aksinin, V. I.: V1500,93-104(1991)

Muray, Andrew
; Dean, Robert L.: Proximity effect correction on MEBES for 1x mask fabrication: lithography issues and tradeoffs at 0.25 micron,V1496,171-179(1991)
See Dunbrack, Steven K.: V1464,314-326(1991)

Murazi, M. See Kasuya, Koichi: V1397,67-70(1991)

Murcray, David G.
See Blatherwick, Ronald D.: V1491,203-210(1991)
See Goldman, Aaron: V1491,194-202(1991)
See Murcray, Frank J.: V1491,282-287(1991)

Murcray, F. H.
See Goldman, Aaron: V1491,194-202(1991)
See Murcray, Frank J.: V1491,282-287(1991)

Murcray, Frank J.
; Murcray, F. H.; Goldman, Aaron; Blatherwick, Ronald D.; Murcray, David G.: High-resolution studies of atmospheric IR emission spectra,V1491,282-287(1991)
See Blatherwick, Ronald D.: V1491,203-210(1991)
See Goldman, Aaron: V1491,194-202(1991)

Murdaugh, William O. See Troyer, David E.: V1486,396-409(1991)

Murdocca, Miles J.
: High-level design of digital computers using optical logic arrays,V1474,176-187(1991)
; Levy, Saul Y.: Design of a Gaussian elimination architecture for the DOC II processor,V1563,255-266(1991)

Murdock, R. H. See Ordonez, Ishmael D.: V1437,184-193(1991)

Murdock, Tom L. See Chalupa, John: V1530,343-351(1991)

Murduck, J. M. See Kadin, Alan M.: V1477,156-165(1991)

Murer, Pierre See Godard, Bruno: V1503,71-77(1991)

Muriel, Miguel A.
See Lopez-Amo, Manuel: V1374,74-85(1991)
See Lopez-Higuera, Jose M.: V1374,144-151(1991)

Muro, Mikio See Fujii, H.: V1397,213-220(1991)

Murota, Tomoya
See Iyoda, Mitsuhiro: V1415,342-349(1991)
See Iyoda, Mitsuhiro: V1397,457-460(1991)

Murphy, Arthur T. See Rahal Arabi, Tawfik R.: V1389,302-313(1991)

Murphy, C. J. See Herman, Warren N.: V1560,206-213(1991)

Murphy, Eamonn See Arshak, Khalil I.: V1463,521-533(1991)

Murphy, Elizabeth See Conley, Willard E.: V1466,53-66(1991)

Murphy, John C. See Maclachlan Spicer, J. W.: V1467,311-321(1991)

Murphy, Kent A.
; Fogg, Brian R.; Wang, George Z.; Vengsarkar, Ashish M.; Claus, Richard O.: Sapphire fiber interferometer for microdisplacement measurements at high temperatures,V1588,117-124(1991)
; Gunther, Michael F.; Vengsarkar, Ashish M.; Claus, Richard O.: Fabry-Perot fiber-optic sensors in full-scale fatigue testing on an F-15 aircraft,V1588,134-142(1991)
See Claus, Richard O.: V1588,159-168(1991)
See Tran, Tuan A.: V1584,178-186(1991)
See Vengsarkar, Ashish M.: V1588,2-13(1991)

Murphy, Kevin S.
; Bradley, D. J.; Dennis, Peter N.: Relative performance studies for focal-plane arrays,V1488,178-185(1991)
See Bourne, Robert W.: V1488,73-79(1991)
See Bradley, D. J.: V1488,186-195(1991)

Murphy, Robert A.
; Christodoulou, Christos G.; Phillips, Ronald L.: Electromagnetic scattering from a finite cylinder with complex permittivity,V1558,295-305(1991)

Murphy, Robin R.
: Application of Dempster-Shafer theory to a novel control scheme for sensor fusion,V1569,55-68(1991)
: Control scheme for sensor fusion for navigation of autonomous mobile robots,V1383,436-447(1991)

Murphy, S. P. See Scheiwiller, Peter M.: V1457,2-8(1991)

Murphy, Steve H.
; Wen, John T.; Saridis, George N.: Analysis of cooperative robot manipulators on a mobile platform,V1387,14-25(1991)

Murphy-Chutorian, Douglas
: Lesion-specific laser catheters for angioplasty,V1420,244-248(1991)

Murray, Alan See Mitchell, David C.: V1427,181-188(1991)

Murray, Annie L. See Dubowsky, Steven: V1386,10-20(1991)

Murray, Anthony R. See Lake, Stephen P.: V1486,286-293(1991)

Murray, Brian W.
; Floyd, Dennis R.; Ulph, Eric: Light-absorbing, lightweight beryllium baffle materials,V1485,88-95(1991)
; Johnson, Edward A.: Pulsed electron-beam testing of optical surfaces,V1330,2-24(1991)
; Ulph, Eric; Richard, Peter N.: Thick, fine-grained beryllium optical coatings,V1485,106-115(1991)

Murray, J. G. See Morgan, Chris G.: V1525,83-90(1991)

Murray, John R.
; Trenholme, J. B.; Hunt, John T.; Frank, D. N.; Lowdermilk, W. H.; Storm, E.: Upgrade of the LLNL Nova laser for inertial confinement fusion,V1410,28-39(1991)

Murray, S. See Johnstone, Walter: V1506,145-149(1991)

Murthy, S. N. See Soh, Young-Sung: V1381,561-573(1991)

Murty, M. V.
; Shukla, Ram P.; Apparao, K. V.: Set of two 45 - 90 - 45 prisms equivalent to the Fresnel rhomb,V1332,41-49(1991)

Murty, Mantravady V.
; Kumar, Vas; von Handorf, Robert J.: Aspheric testing using null mirrors,V1332,107-114(1991)

Murugov, V. M.
See Berkovsky, A. G.: V1358,750-755(1991)
See Dolgopolov, Y. V.: V1412,267-275(1991)
See Egorov, V. V.: V1358,984-991(1991)

Musatov, L. G. See Vardanjan, G. S.: V1554A,496-502(1991)

Muser, Markus H.
; Leemann, Thomas; Anliker, M.: High-resolution image digitizing camera for use in quantitative coronary arteriography,V1448,106-112(1991)

Musgrave, G. See Jones, Robert H.: V1401,138-145(1991)

Musher, Semion L. See Krasnov, Victor F.: V1506,179-187(1991)

Musial, C. E. See Samarabandu, J. K.: V1450,296-322(1991)

Music, Wayne See Atlas, Les E.: V1566,134-143(1991)

Musikant, Solomon
ed.: *Optical Surfaces Resistant to Severe Environments*,V1330(1991)
; Malloy, W. J.: Environments stressful to optical materials in low earth orbit,V1330,119-130(1991)

Musolino, Mario See Cantu, Laura: V1430,144-159(1991)

Mussche, Paul L.
; Siegman, Anthony E.: Enhanced Schawlow-Townes linewidth in lasers with nonorthogonal transverse eigenmodes,V1376,153-163(1991)

Musselman, Martin L.
: Evaluation of commercial fiber optic sensors in a marine boiler room,V1589,56-57(1991)

Mustafa, Mazidah B.
; Russo, Paul S.: Diffusion of spherical probes in aqueous systems containing the semiflexible polymer hydroxypropylcellulose,V1430,132-141(1991)

Mustre de Leon, Jose
; Conradson, Steven D.; Batistic, I.; Bishop, A. R.; Raistrick, Ian D.; Jackson, W. E.; Brown, George S.: X-ray absorption fine structure of systems in the anharmonic limit,V1550,85-96(1991)

Mutikainen, Risto See Viitanen, Veli-Pekka: V1549,28-34(1991)

Muuss, Michael J. See Deitz, Paul: VIS07,40-56(1991)

Muyard, D. See Lajzerowicz, Jean: V1392,222-231(1991)

Muys, Peter F.
See Leys, C.: V1397,399-402(1991)
See Sona, P.: V1397,373-377(1991)

Myakinin, Vladimir A. See Gurvich, Alexander S.: V1408,10-18(1991)

Myatezh, O. V. See Ryabova, R. V.: V1238,166-170(1991)

Myers, John D. See Hamlin, Scott J.: V1419,100-106(1991)

Myers, John M. See Shen, Hao-Ming: V1407,286-294(1991)

Myers, Kyle J.
; Hanson, Kenneth M.: Task performance based on the posterior probability of maximum-entropy reconstructions obtained with MEMSYS 3,V1443,172-182(1991)

Myers, Lori K. See Thompson, Mark E.: V1497,423-429(1991)

Myers, M. B.
; Bickley, Harmon: Holography of human pathologic specimens with continuous-beam lasers through plastination,V1461,242-244(1991)

Myers, Michael J. See Hamlin, Scott J.: V1419,100-106(1991)

Myers, Richard F.
: Laser disk: a practicum on effective image management,V1460,50-58(1991)

Myers, Richard M. See Doel, A. P.: V1542,319-326(1991)

Myers-Beaghton, A. K. See Joyce, Bruce A.: V1361,13-22(1991)

Myl'nikov, M. Y. See Aleksandrov, A. P.: V1440,442-453(1991)

Myler, Harley R.
; Weeks, Arthur R.; Van Dyke-Lewis, Michelle: Decision-directed entropy-based adaptive filtering,V1565,57-68(1991)
See Weeks, Arthur R.: V1567,77-87(1991)
See Wenaas, Holly: V1569,410-421(1991)

Myllyla, Risto A. See Karppinen, Arto: V1455,179-189(1991)

Myrick, Michael L. See Angel, S. M.: V1435,72-81(1991)

Mysyrowicz, Andre See Geindre, Jean-Paul: V1502,311-318(1991)

Naarmann, Herbert See Gierulski, Alfred: V1560,172-182(1991)

Nabatov, A. O. See Artjushenko, Vjacheslav G.: V1420,157-168(1991)

Nabet, Bahram See Bouzerdoum, Abdesselam: V1473,29-40(1991)

Nabeta, Osamu See Kuroda, Masami: V1458,155-161(1991)

Nabiev, I. R.
; Sokolov, K. V.; Efremov, R. G.; Chumanov, G. D.: Surface-enhanced Raman spectroscopy in the structural studies of biomolecules: the state of the art,V1403,85-92(1991)

See Baranov, A. V.: V1403,128-131(1991)
See Efremov, R. G.: V1403,161-163(1991)
See Manfait, Michel: V1403,695-707(1991)
See Oleynikov, V. A.: V1403,164-166(1991)

Naboichenko, A. K. See Karpov, V. I.: V1503,492-502(1991)

Nace, Dave See Hayden, William L.: V1417,182-199(1991)

Nadar, Mariappan S. See Mahalanobis, Abhijit: V1469,292-302(1991)

Naday, Istvan See Strauss, Michael G.: V1447,12-27(1991)

Nadeau, Pierre
; Pottier, R.; Szabo, Arthur G.; Brault, Daniel; Vever-Bizet, C.:
    Characterization of the fluorescence lifetimes of the ionic species found
    in aqueous solutions of hematoporphyrin IX as a function of
    pH,V1398,151-161(1991)

Nadeau, Theodore R. See Guadagni, Stefano: V1420,178-182(1991)

Nadel, Jerome I. See Kidd, Robert C.: V1454,414-424(1991)

Nader-Rezvani, Navid
; Claus, Richard O.; Sarrafzadeh, A. K.: Low-frequency fiber optic magnetic
    field sensors,V1584,405-414(1991)

Nadezhdinskii, Alexander I.
; Stepanov, Eugene V.; Kuznetzov, Andrian I.; Devyatykh, Grigory G.;
    Maximov, G. A.; Khorshev, V. A.; Shapin, S. M.: Application of tunable
    diode lasers in control of high-pure-material technologies,V1418,487-
    495(1991)
; Stepanov, Eugene V.: Diode laser spectroscopy of atmospheric
    pollutants,V1433,202-210(1991)
See Moskalenko, Konstantin L.: V1426,121-132(1991)

Nadler, Ch. K. See Risk, William P.: V1561,130-134(1991)

Naether, Dirk See Roeder, Beate: V1525,377-384(1991)

Nafie, Laurence A.
; Che, Diping; Yu, Gu-Sheng; Freedman, Teresa B.: New experimental
    methods and theory of Raman optical activity,V1432,37-49(1991)
See Birge, Robert R.: V1432(1991)

Nag, B. R. See Sen, Susanta: V1362,750-759(1991)

Nagae, Takanori See Agui, Takeshi: V1606,912-916(1991)

Nagahashi, Hiroshi See Agui, Takeshi: V1606,188-197(1991)

Nagai, Haruhiko
; Haruta, K.; Sato, Y.; Inoue, M.; Suzuki, A.: High-average power XeCl laser
    with surface corona-discharge preionization,V1397,31-36(1991)

Nagai, Katumi See Oda, Ichiro: V1431,284-293(1991)

Nagai, Toshiaki See Wakana, Shin-ichi: V1479,283-290(1991)

Nagano, Yasutada
; Kanechika, Hideaki; Tanaka, Satoshi; Maehara, Hideaki; Maeda, Akira:
    Experimental system using an interactive drawing input
    method,V1605,614-623(1991)

Nagarajan, Rajamani
See Hoyle, Charles E.: V1559,101-109(1991)
See Hoyle, Charles E.: V1559,202-213(1991)

Nagasawa, Kohtaro
See Fukunaga, Seiki: V1466,446-457(1991)
See Koyanagi, Hiroo: V1466,346-361(1991)

Nagashima, Yoshio
; Agawa, Hiroshi; Kishino, Fumio: Three-dimensional face model
    reproduction method using multiview images,V1606,566-573(1991)

Nagata, Hitoshi
See Kawai, Akira: V1464,267-277(1991)
See Miyazaki, Junji: V1464,327-335(1991)

Nagatani, Hiroyuki See Satoh, Hiroharu: V1499,324-329(1991)

Nagato, Hitoshi
; Ohno, Tadayoshi: Bidirectional printing method for a thermal ink transfer
    printer,V1458,84-91(1991)

Nagatomo, Takao
; Ochiai, Ichiro; Ookoshi, Shigeo; Omoto, Osamu: GaN single-crystal films
    on silicon substrates grown by MOVPE,V1519,90-95(1991)

Nagayama, Akira See Sakai, Masao: V1385,8-14(1991)

Nagendra, C. L.
See Annapurna, M. N.: V1485,260-271(1991)
See Umadevi, P.: V1484,125-135(1991); 1485,195-205(1991)

Nagle, Daniel T.
; Saniie, Jafar: Robust CFAR detection using order statistic processors for
    Weibull-distributed clutter,V1481,49-63(1991)

Nagura, Riichi
See Hino, Hideo: V1490,166-176(1991)
See Suzuki, Naoshi: V1493,48-57(1991)
See Takahashi, Fumiho: V1490,255-268(1991)

Nagy, Andrew
: Vertical oxide etching without inducing change in critical
    dimensions,V1392,165-179(1991)

Nagy, James G.
; Plemmons, Robert J.: Some fast Toeplitz least-squares algorithms,V1566,35-
    46(1991)

Nagy, Peter A.
; Meyer, Tim J.; Tekippe, Vincent J.: Wavelength division multiplexers for
    optical fiber amplifiers,V1365,33-37(1991)

Nahata, Ajay See McFarland, Michael J.: V1583,344-354(1991)

Naik, H. M. See Zhai, H. R.: V1519,575-579(1991)

Naimark, Michael
: Elements of real-space imaging: a proposed taxonomy,V1457,169-
    179(1991)

Nair, M. T. See Nair, Padmanabhan K.: V1485,228-239(1991)

Nair, Padmanabhan K.
; Nair, M. T.; Fernandez, A. M.; Garcia, V. M.; Hernandez, A. B.: Metal
    sulfide thin films on glass as solar control, solar absorber, decorative, and
    photographic coatings,V1485,228-239(1991)

Naito, Kenta See Yamanaka, Masanobu: V1501,30-39(1991)

Naito, Takao
; Chikama, Terumi; Kuwahara, Hideo: New modulation scheme for coherent
    lightwave communication: Direct-Modulation PSK,V1372,200-207(1991)

Naito, Yoichi See Tanimoto, Masayuki: V1605,394-405(1991)

Naitou, Kazunori
; Numata, Takehiko; Nakashima, Kazuo; Maeda, Miyozo; Koshino, Nagaaki:
    DyFeCo magneto-optical disks with a Ce-SiO2 protective
    film,V1499,386-392(1991)

Najafi, S. I.
; Wang, Wei-Jian; Orcel, Gerard F.; Albert, Jacques; Honkanen, Seppo;
    Poyhonen, Pekka; Li, Ming-Jun: Nd- and Er-doped glass integrated optical
    amplifiers and lasers,V1583,32-36(1991)
See Albert, Jacques: V1583,27-31(1991)
See Lefebvre, P.: V1583,221-225(1991)
See Li, Ming-Jun: V1506,52-57(1991)
See Li, Ming-Jun: V1513,410-417(1991)
See Wang, Wei-Jian: V1513,434-440(1991)
See Zhang, H.: V1583,83-89(1991)

Nakada, Hiroshi See Ukita, Hiroo: V1499,248-262(1991)

Nakagawa, Hachiro See Oda, Ichiro: V1431,284-293(1991)

Nakagawa, Kazuo
; Egami, Chikara; Suzuki, Takayoshi; Fujiwara, Hirofumi: Phase-conjugate
    interferometry by using dye-doped polymer films,V1332,267-273(1991)

Nakagawa, Kenji
See Shirai, Hisatsugu: V1463,256-274(1991)
See Yanagishita, Yuichiro: V1463,207-217(1991)

Nakagawa, Kiyoshi
; Minemoto, Takumi: Real-time speckle photography using photorefractive
    Bi12SiO20 crystal,V1508,191-200(1991)

Nakahara, Ktsuhiko See Tanazawa, Takeshi: V1435,310-321(1991)

Nakahara, Motohiro See Ikegami, Tetsuhiko: V1362,350-360(1991)

Nakahara, Sumio
; Hisada, Shigeyoshi; Fujita, Takeyoshi; Sugihara, Kiyoshi: Low-temperature
    deformation measurement of materials by the use of laser speckle
    photography method,V1554A,602-609(1991)

Nakahashi, Tomoko See Kinoshita, Taizo: V1605,604-613(1991)

Nakai, Akira See Ohmori, Yasuhiro: V1490,177-183(1991)

Nakai, Sadao S.
See Nakatsuka, Masahiro: V1411,108-115(1991)
See Yamanaka, Masanobu: V1501,30-39(1991)
See Yoshida, Kunio: V1441,9-9(1991)

Nakajima, H. See Yokozawa, T.: V1397,513-517(1991)

Nakajima, Hajime
; Kisaki, Jyunko; Tahata, Shin; Horikawa, Tsuyoshi; Nishi, Kazuro: High-
    resolution display using a laser-addressed ferroelectric liquid-crystal light
    valve,V1456,29-39(1991)

Nakajima, Masayuki
    See Agui, Takeshi: V1606,912-916(1991)
    See Agui, Takeshi: V1606,188-197(1991)

Nakajima, Shigeyoshi
    ; Zhou, Mingyong; Hama, Hiromitsu; Yamashita, Kazumi: Three-dimensional
      motion analysis and structure recovering by multistage Hough
      transform,V1605,709-719(1991)

Nakajima, Takashi See Kanemaru, Toyomi: V1496,118-123(1991)

Nakajima, Takuro See Kanazawa, Hirotaka: V1397,445-448(1991)

Nakajima, Toshinori See Ming, Hai: V1572,27-31(1991)

Nakajima, Yoshiki See Shimamoto, Masayoshi: V1499,393-400(1991)

Nakaki, Y. See Tsutsumi, K.: V1499,55-61(1991)

Nakama, Kenichi See Seki, Masafumi: V1583,184-195(1991)

Nakamura, Haruhito See Kume, Hidehiro: V1358,1144-1155(1991)

Nakamura, Haruo
    See Arai, Tsunenori: V1425,191-195(1991)
    See Matsuoka, Masaru: V1549,2-8(1991)
    See Sakurada, Masami: V1425,158-164(1991)

Nakamura, Hiroshi See Daidoh, Yuichiro: V1421,120-123(1991)

Nakamura, Hiroyuki
    ; Kobayashi, Hitoshi; Taya, Katsuo; Ishigami, Shigenobu: Design of eye
      movement monitoring system for practical environment,V1456,226-
      238(1991)

Nakamura, Osamu
    See Numagami, Yasuhiko: V1606,970-979(1991)
    See So, Ikken: V1605,263-272(1991)

Nakamura, Shigeru
    See Seya, Eiichi: V1499,269-273(1991)
    See Sukeda, Hirofumi: V1499,419-425(1991)

Nakamura, Taichi See Uchiyama, Toshio: V1451,192-203(1991)

Nakamura, Yasuhiro See Tanaka, Kiyoshi: V1605,646-649(1991)

Nakanishi, Hachiro
    See Nomura, Shintaro: V1560,272-277(1991)
    See Okada, Shuji: V1560,25-34(1991)
    See Suda, Yasumasa: V1560,75-83(1991)

Nakanishi, Junji See Kimata, Masafumi: V1540,238-249(1991)

Nakano, Shoichi See Kuwano, Yukinori: V1519,200-209(1991)

Nakano, Susumu See Hashimoto, Takashi: V1397,519-522(1991)

Nakano, Yasuhiko
    ; Chiba, Hirotaka; Okada, Yoshiyuki; Yoshida, Shigeru; Mori, Masahiro:
      Study of binary image compression using universal coding,V1605,874-
      878(1991)

Nakano, Yoshiaki
    ; Luo, Yi; Tada, Kunio: Laser diodes with gain-coupled distributed optical
      feedback,V1418,250-260(1991)

Nakano, Yoshihisa
    See Komori, Masaru: V1444,334-337(1991)
    See Minato, Kotaro: V1446,195-198(1991)

Nakao, Takashi See Narahara, Tatsuya: V1499,120-128(1991)

Nakao, Takeshi
    ; Arimoto, Akira A.; Takahashi, Masahiko: Magneto-optical signal detection
      with elliptically polarized light,V1499,433-437(1991)

Nakashima, Kazuo See Naitou, Kazunori: V1499,386-392(1991)

Nakashima, Masato
    See Koezuka, Tetsuo: V1332,323-331(1991)
    See Tsukahara, Hiroyuki: V1384,15-26(1991)

Nakata, K. See Kasuya, Koichi: V1397,67-70(1991)

Nakatsuka, Masahiro
    ; Jitsuno, Takahisa; Kanabe, Tadashi; Urushihara, Shinji; Miyanaga, Noriaki;
      Nakai, Sadao S.: Beam control and power conditioning of GEKKO glass
      laser system,V1411,108-115(1991)
    See Yamanaka, Masanobu: V1501,30-39(1991)

Nakaya, Yuichiro
    ; Harashima, Hiroshi: Iterative motion estimation method using triangular
      patches for motion compensation,V1605,546-557(1991)

Nakayama, Kimihiko See Moriyama, Takashi: V1490,310-316(1991)

Nakayama, Masao
    ; Izawa, Toshiyuki; Fujisada, Hiroyuki; Tange, Yoshio; Miyachi, Yuji; Sato,
      Ryota; Ishida, Juro; Tanii, Jun: Cooled focal-plane assembly for ocean
      color and temperature scanner,V1490,207-215(1991)

Nakazawa, Akira
    ; Kutami, Michinori; Ozaki, Mitsuo; Suzuki, Shigeharu; Kikuchi, Hideyuki:
      Electrostatic screen-through ink jet printing technique,V1458,115-
      118(1991)

Nalamasu, Omkaram
    ; Reichmanis, Elsa; Cheng, May; Pol, Victor; Kometani, Janet M.; Houlihan,
      Frank M.; Neenan, Thomas X.; Bohrer, M. P.; Mixon, D. A.; Thompson,
      Larry F.; Takemoto, Cliff H.: Preliminary lithographic characteristics of an
      all-organic chemically amplified resist formulation for single-layer deep-
      UV lithography,V1466,13-25(1991)
    See Novembre, Anthony E.: V1466,89-99(1991)

Naletto, Giampiero
    See Huber, Martin C.: V1494,472-480(1991)
    See Timothy, J. G.: V1343,350-358(1991)

Nam, Derek W. See Zucker, Erik P.: V1563,223-228(1991)

Nam, J. Y.
    ; Rao, K. R.: Classified transform coding of images using two-channel
      conjugate vector quantization,V1605,202-213(1991)
    ; Rao, K. R.: Image coding based on two-channel conjugate vector
      quantization,V1452,485-496(1991)

Nam, Jiseung See Martinez, Ralph: V1446,100-107(1991)

Namavar, Fereydoon See Soref, Richard A.: V1389,408-421(1991)

Namazi, Nader M.
    ; Lipp, John I.: Nonuniform image motion estimation in transformed-
      domain,V1567,659-669(1991)
    See Shaltaf, Samir J.: V1567,609-620(1991)

Namikawa, Iwao See Sasaki, Nobuyuki: V1606,1002-1013(1991)

Namioka, Takeshi
    See Koike, Masato: V1545,88-94(1991)
    See Sudoh, Masaaki: V1343,14-24(1991)

Nan, Zhilin See Zhang, Zhongxian: V1572,5-10(1991)

Nandhakumar, N.
    See Karthik, Sankaran: V1468,686-697(1991)
    See Kumar, G. N.: V1468,884-895(1991)

Nantista, C. See Farkas, Zoltan D.: V1407,502-511(1991)

Naor, Yoram See Kopolovich, Zvi: V1540,565-577(1991)

Napartovich, Anatoly P.
    See Antyuhov, V.: V1397,355-365(1991); 1415,48-59(1991)
    See Apollonova, O. V.: V1501,108-119(1991)
    See Bondarenko, Alexander V.: V1376,117-127(1991)

Nappi, Bruce See Swairjo, Manal: V1437,60-65(1991)

Naqvi, H. S.
    See Bishop, Kenneth P.: V1545,64-73(1991)
    See Hickman, Kirt C.: V1464,245-257(1991)
    See Zaidi, Saleem H.: V1343,245-255(1991)

Narahara, Tatsuya
    ; Kumai, Satoshi; Nakao, Takashi; Ozue, Tadashi: Tracking method of optical
      tape recorder using acousto-optic scanning,V1499,120-128(1991)

Narain, Ram See Chitnis, Vijay T.: V1332,613-622(1991)

Narasimhan, Anand See Seetharaman, Guna S.: V1569,269-273(1991)

Narathong, Chiewcharn
    ; Inigo, Rafael M.: Design of an analog VLSI chip for a neural network target
      tracker,V1452,523-531(1991)
    See Himes, Glenn S.: V1469,671-682(1991)

Narayan, Jagdish
    ; Shaw, David T.; Singh, Rajendra; eds.: *Progress In High-Temperature
      Superconducting Transistors and Other Devices,*V1394(1991)
    See Singh, Rajiv K.: V1394,203-213(1991)
    See Sinha, Sanjai: V1394,266-276(1991)

Narayanan, Authi A. See Yap, Daniel: V1418,471-476(1991)

Narayanan, Ram M.
    ; Green, Steven E.; Alexander, Dennis R.: Midinfrared backscatter spectra of
      selected agricultural crops,V1379,116-122(1991)

Narayanaswamy, R. See Momin, S. A.: V1510,180-186(1991)

Narayanswamy, Ramkumar
    ; Alter-Gartenberg, Rachel; Huck, Friedrich O.: Robust image coding with a
      model of adaptive retinal processing,V1385,93-103(1991)

Nardo, Laurent See Pugnet, Michel: V1361,96-108(1991)

Narducci, Lorenzo M. See Scully, Marlan O.: V1497,264-276(1991)

Narimatsu, Yoshito
    See Hino, Hideo: V1490,166-176(1991)
    See Suzuki, Naoshi: V1493,48-57(1991)
    See Takahashi, Fumiho: V1490,255-268(1991)

Narumi, E. See Shaw, David T.: V1394,214-220(1991)

Nasburg, Robert E.
; Stillman, Steve M.; Nguyen, M. T.: Integrated processor architecture for multisensor signal processing,V1481,84-95(1991)

Nasedkhina, N. V. See Anikin, V. I.: V1238,286-296(1991)

Nash, Graeme
: Effect of resampling on the MUSIC algorithm when applied to inverse-synthetic-aperture radar,V1566,476-482(1991)

Nashimoto, Ichiro See Araki, Ryuichiro: V1431,321-332(1991)

Nashman, Marilyn
; Chaconas, Karen: Three-dimensional position determination from motion,V1383,166-175(1991)

Nasibov, Alexander S.
; Markov, L. S.; Fedorov, D. L.; Shapkin, P. V.; Korostelin, Y. V.; Machintsev, G. A.: Exciton spectroscopy of semiconductor materials used in laser elements,V1361,901-908(1991)

Nason, Donald
See Burger, Arnold: V1557,245-249(1991)
See Sawyer, Curry R.: V1567,254-263(1991)

Nason, Theodore C. See McDonald, John F.: V1390,286-301(1991)

Nasr, Hatem N.
: ATR performance modeling for building multiscenario adaptive systems,V1483,112-117(1991)
: Automated instrumentation, evaluation, and diagnostics of automatic target recognition systems,VIS07,202-213(1991)
ed.: Automatic Object Recognition,VIS07(1991)
: Contextual image understanding of airport photographs,V1521,24-33(1991)
ed.: Image Understanding for Aerospace Applications,V1521(1991)
ed.: Selected Papers on Automatic Object Recognition,VMS41(1991)
; Amehdi, Hossien: Model-based automatic target recognition development tools,V1471,283-290(1991)
; Bazakos, Michael E.; Sadjadi, Firooz A.; Amehdi, Hossien: Knowledge- and model-based ATR (automatic target recognition) algorithms adaptation,VIS07,122-129(1991)
; Bazakos, Michael E.; eds.: Signal and Image Processing Systems Performance Evaluation, Simulation, and Modeling,V1483(1991)
See Amehdi, Hossien: V1483,177-184(1991)

Nasrabadi, Nasser M.
See Mohsenian, Nader: V1605,781-792(1991)
See Smith, John M.: V1468,493-501(1991)

Nassisi, Vincenzo
: Experimental studies of an XeCl laser with UV preionization perpendicular and parallel to the electrode surfaces,V1503,115-125(1991)
: Impedance of a UV preionized excimer laser,V1527,291-304(1991)

Natarajan, Sanjay S.
; Casasent, David P.; Smokelin, John-Scott: Multifunctional hybrid neural network: real-time laboratory results,V1564,474-488(1991)
See Casasent, David P.: V1555,23-33(1991)
See Karuppiah, N.: V1536,215-221(1991)
See Sasaki, Kenji: V1384,228-233(1991)

Nate, Kazuo
; Mizushima, Akiko; Sugiyama, Hisashi: Bilayer resist system utilizing alkali-developable organosilicon positive photoresist,V1466,206-210(1991)

Nath, Dilip K.
; Nelson, Gary W.; Griffin, Stephen E.; Harrington, C. T.; He, Yi-Fei; Reinhart, Larry J.; Paine, D. C.; Morse, Ted F.: Polyimide-coated embedded optical fiber sensors,V1489,17-32(1991)

Nathan, Vaidya
: Two-photon absorption calculations in HgCdTe,V1441,553-556(1991)

Nathanson, F. E. See Bruder, Joseph A.: V1479,316-321(1991)

Nathel, Howard See Gold, David: V1413,41-52(1991)

Nation, John A.
; Ivers, J. D.; Kerslick, G.; Shiffler, Donald; Schachter, Levi: High-gain high-efficiency TWT (traveling wave tube) amplifiers,V1407,32-43(1991)
See Schachter, Levi: V1407,44-56(1991)

Natori, Hiroshi See Suzuki, Hideo: V1450,99-107(1991)

Nauda, Alexander See McEachern, Robert: V1470,226-232(1991)

Naudy, Michel See Andre, Michel L.: V1502,286-298(1991)

Naumaan, Ahmed
; Corsi, Carlo; Baixeras, Joseph M.; Kreisler, Alain J.; eds.: Infrared and Optoelectronic Materials and Devices,V1512(1991)
See Khan, M. A.: V1396,753-759(1991)

Naumenkov, P. A. See Burakov, V. S.: V1440,270-276(1991)

Naumov, B. L.
: Didactic problems of using holographic means of education,V1238,365-367(1991)

Naumov, S. K.
See Borodin, A. M.: V1358,756-758(1991)
See Bryukhnevitch, G. I.: V1449,109-115(1991)
See Ivanov, K. N.: V1358,732-738(1991)

Naundorf, M. See Schramm, Werner: V1525,237-241(1991)

Nava, Enzo
; Re Garbagnati, Giuseppe; Garbi, Maurizio; Marchiori, Livio D.; Marini, Andrea E.: Diode-pumped Nd:YAG laser transmitter for free-space optical communications,V1417,307-315(1991)

Navarro, Maria T.
; Egozcue, Juan J.; Fimia-Gil, Antonio: Computer-generated holograms of linear segments,V1507,142-148(1991)
See Fimia-Gil, Antonio: V1507,153-157(1991)

Navaza, Jorge See Le Besnerais, Guy: V1569,386-395(1991)

Naylor, David C.
; Daemi, M. F.: Modified Laplacian enhancement of low-resolution digital images,V1567,522-532(1991)

Naylor, David L.
See Amleshi, Peerouz M.: V1396,396-403(1991)
See Goel, Sanjay: V1396,404-410(1991)

Naylor, Graham A.
; Kearsley, Andrew J.: Progress in gas purifiers for industrial excimer lasers,V1377,107-112(1991)
See Knowles, Martyn R.: V1410,195-201(1991)

Nazarian, David See Garino, Jonathan P.: V1424,43-47(1991)

Nazarian, Soheil See Shadaram, Mehdi: V1332,487-490(1991)

Nazikov, S. N. See Karpov, D. A.: V1519,115-121(1991)

Neal, Daniel R.
; McMillin, Pat; Michie, Robert B.: Astigmatic unstable resonator with an intracavity deformable mirror,V1542,449-458(1991)

Neapolitan, Richard E.
: Propagation of variances in belief networks,V1468,333-344(1991)

Neat, Gregory W. See Steinvorth, Rodrigo: V1387,136-147(1991)

Nebeker, Sidney J. See Briscoe, Dennis: V1358,329-336(1991)

Nebesny, Ken W. See Armstrong, Neal R.: V1559,18-26(1991)

Nechitailo, V. S. See Manenkov, Alexander A.: V1441,392-405(1991)

Nechval, Nicholas A.
: Algorithm for statistical classification of radar clutter into one of several categories,V1470,282-293(1991)

Nederlof, Michel A.
; Witkin, Andrew; Taylor, D. L.: Knowledge-driven image analysis of cell structures,V1428,233-241(1991)

Nedkova, Rumiana
: Possibilities of the fringe pattern learning system VARNA,V1429,214-216(1991)
; Delchev, Georgy: Automated cyst recognition from x-ray photographs,V1429,105-107(1991)

Nee, Soe-Mie F.
; Bennett, Harold E.: Simple high-precision extinction method for measuring refractive index of transparent materials,V1441,31-37(1991)
See Lindsay, Geoffrey A.: V1560,443-453(1991)

Neelin, Peter See Evans, Alan C.: V1445,236-246(1991)

Neely, Jerry See Caimi, Frank M.: V1589,90-99(1991)

Neenan, Thomas X. See Nalamasu, Omkaram: V1466,13-25(1991)

Neering, Ian R. See Quesenberry, Laura A.: V1428,177-190(1991)

Neev, Joseph
; Raney, Daniel; Whalen, William E.; Fujishige, Jack T.; Ho, Peter D.; McGrann, John V.; Berns, Michael W.: Ablation of hard dental tissues with an ArF-pulsed excimer laser,V1427,162-172(1991)

Nefedov, I. See Andronov, Alexander A.: V1362,684-695(1991)

Neff, John A.
: Massive optical interconnections for computer applications,V1389,27-38(1991); 1390,27-38(1991)
ed.: Optical Enhancements to Computing Technology,V1563(1991)

Neff, Raymond K.
; Klingensmith, H. W.; Gumpf, Jeffrey A.; Haigh, Peter J.: CWRUnet: case history of a campus-wide fiber-to-the-desktop network,V1364,245-256(1991)

Neger, Theo
; Jaeger, Helmut; Philipp, Harald; Pretzler, Georg; Widmann, Klaus; Woisetschlaeger, Jakob: Application of spatially resolving holographic interferometry to plasma diagnostics,V1507,476-487(1991)
See Vukicevic, Dalibor: VIS08,160-193(1991)

Negrete, George V.
; Hammond, Robert B.: Epitaxial Tl2Ba2CaCu2O8 thin films on LaAlO3 and their microwave device properties,V1477,36-44(1991)

Negri, A. J. See Elman, Boris S.: V1362,610-616(1991)

Neher, Dieter
; Mittler-Neher, S.; Cha, M.; Stegeman, George I.; Embs, F. W.; Wegner, Gerhard; Miller, Robert D.; Willson, C. G.: Determination of the orientational order parameters <P*2>, <P*4> in a polysilane LB film via polarization-dependent THG,V1560,335-343(1991)

Neifeld, Mark A.
; Rakshit, S.; Psaltis, Demetri: Handwritten zip code recognition using an optical radial basis function classifier,V1469,250-255(1991)

Neil, Iain A.
: Ultra-high-performance zoom lens for the visible waveband,V1354,684-694(1991)

Nein, Max E.
; Davis, Billy: System concepts for a large UV/optical/IR telescope on the moon,V1494,98-110(1991)

Neiw, Han-Min
; Chen, Chin-Tu; Lin, Wei-Chung; Pelizzari, Charles A.: Automated three-dimensional registration of medical images,V1445,259-264(1991)

Nejedly, Pavel See Hudec, Rene: V1343,162-163(1991)

Nekrasov, Victor V.
: Fast holographic correlator for machine vision systems,V1507,170-174(1991)

Nelson, Bradford See Minot, Katcha: V1475,309-313(1991)

Nelson, Edward T. See Clark, David L.: V1540,303-311(1991)

Nelson, G. J. See Lin, Bochien: V1366,157-166(1991)

Nelson, Gary W.
See Nath, Dilip K.: V1489,17-32(1991)
See York, Jim F.: V1584,308-319(1991)

Nelson, J. S. See Fong, C. Y.: V1361,479-488(1991)

Nelson, Jeffrey G.
; Neurgaonkar, Ratnakar R.; Crooks, R.; Rhodes, C. G.: Grain-oriented high-Tc superconductors and their applications,V1394,191-195(1991)

Nelson, John C. See Nie, Qiuhua: V1332,409-420(1991)

Nelson, Matthew See Richards, Evan: V1494,40-48(1991)

Nelson, William See Lesko, Camille: V1475,330-339(1991)

Nemanich, Robert J.
; Bergman, Larry; LeGrice, Yvonne M.; Turner, K. F.; Humphreys, T. P.: Microstructures and domain size effects in diamond films characterized by Raman spectroscopy,V1437,2-12(1991)

Nemat-Nasser, Siavouche See Ahmadshahi, Mansour A.: V1554A,620-627(1991)

Nemati, Babak See Sardar, Dhiraj K.: V1427,374-380(1991)

Nemchinov, V. V. See Zhavoronok, I. V.: V1554A,371-379(1991)

Nemirovsky, Yael See Bloom, I.: V1442,286-297(1991)

Nemkovich, N. A.
; Rubinov, A. N.; Savvidi, M. G.; Tomin, V. I.; Shcherbatska, Nina V.: Distribution of fluorescent probe molecules throughout the phospholipid membrane depth,V1403,578-581(1991)

Nemoto, Hiroaki See Kajita, Toru: V1466,161-173(1991)

Nemoto, Hiroyuki See Oikawa, Masahiro: V1544,226-237(1991)

Nemoto, Koshichi See Fujii, Takashi: V1412,50-57(1991)

Nemoto, Yoshiaki See Ono, Makoto: V1490,184-190(1991)

Nenonen, Seppo A.
See Markert, Thomas H.: V1549,408-419(1991)
See Viitanen, Veli-Pekka: V1549,28-34(1991)

Neporent, B. S.
: Laser investigation of molecular dynamic processes of the relaxational spectra formation,V1403,600-605(1991)

Nerad, Bruce A. See Shlichta, Paul J.: V1557,10-23(1991)

Nerheim, Noble M.
: Fiber coupled heterodyne interferometric displacement sensor,V1542,523-533(1991)

Neri, Alessandro
; Scarano, Gaetano; Jacovitti, Giovanni: Bayesian iterative method for blind deconvolution,V1565,196-208(1991)

Neron, C. See Jen, Cheng K.: V1590,107-119(1991)

Nesterova, Z. V. See Aleksandrov, I. V.: V1513,309-312(1991)

Nesteruk, Vsevolod F.
: General principles of constructing the nuclei of nonlinear IR-image transformations,V1540,468-476(1991)
See Miroshnikov, Mikhail M.: V1500,322-333(1991)
See Miroshnikov, Mikhail M.: V1540,477-487(1991)

Netrebko, N. V.
; Romanovsky, Yuri M.; Shidlovskaya, E. G.; Tereshko, V. M.: Damping in the models for molecular dynamics,V1403,512-516(1991)

Netterfield, Roger P. See Smith, Geoffrey B.: V1536,126-137(1991)

Netz, Yoel
; Hoffman, Arnold: Normal-contrast lith,V1458,61-61(1991)

Neu, G. See Chen, Yong: V1361,860-865(1991)

Neu, John T.
; Bressler, Martin: Design considerations for multipurpose bidirectional reflectometers,V1530,244-254(1991)

Neu, Walter
: New method for tissue indentification: resonance fluorescence spectroscopy,V1525,124-131(1991)
; Haase, Karl K.; Tischler, Christian; Nyga, Ralf; Karsch, Karl R.: Diagnosis of atherosclerotic tissue by resonance fluorescence spectroscopy,V1425,28-36(1991)
; Nyga, Ralf; Tischler, Christian; Haase, Karl K.; Karsch, Karl R.: Ultrafast imaging of vascular tissue ablation by an XeCl excimer laser,V1425,37-44(1991)
See Abel, B.: V1525,110-118(1991)
See Jahn, Renate: V1424,23-32(1991)
See Nyga, Ralf: V1525,119-123(1991)

Neuberger, Wolfgang
See Artjushenko, Vjacheslav G.: V1420,149-156(1991)
See Artjushenko, Vjacheslav G.: V1420,157-168(1991)
See Artjushenko, Vjacheslav G.: V1590,131-136(1991)

Neubert, Wolfgang M.
; Leeb, Walter R.; Scholtz, Arpad L.: Experimental implementation of an optical multiple-aperture antenna for space communications,V1522,93-102(1991)
; Leeb, Walter R.; Scholtz, Arpad L.: Nonmechanical steering of laser beams by multiple aperture antennas: tolerance analysis,V1417,122-130(1991)

Neugarten, Michael L.
See Goldmunz, Menachem: V1442,149-153(1991)
See Meidan, Moshe: V1540,729-737(1991)

Neuhoff, David L. See Pappas, Thrasyvoulos N.: V1453,244-255(1991)

Neukammer, Jorg See Kohl, M.: V1525,26-34(1991)

Neuman, William A.
; Fincke, James R.: Characteristics of pulsed nuclear-reactor-pumped flowing gas lasers,V1411,28-40(1991)

Neumann, M. See Devir, Adam D.: V1442,347-359(1991)

Neumann, Norbert See Norkus, Volkmar: V1484,98-105(1991)

Neupert, Werner M. See Keski-Kuha, Ritva A.: V1343,566-575(1991)

Neureuther, Andrew R.
: Modeling phase-shifting masks,V1496,80-88(1991)
See Chiu, Anita S.: V1466,641-652(1991)
See Doi, Takeshi: V1464,336-345(1991)
See Pfau, Anton K.: V1463,124-134(1991)
See Tadros, Karim H.: V1464,177-186(1991)
See Toh, Kenny K.: V1463,402-413(1991)
See Toh, Kenny K.: V1463,356-367(1991)
See Wong, Alfred K.: V1463,315-323(1991)

Neurgaonkar, Ratnakar R. See Nelson, Jeffrey G.: V1394,191-195(1991)

Neuroth, Norbert N. See Speit, Burkhard: V1361,1128-1131(1991)

Neuvo, Yrjo
See Koskinen, Lasse: V1451,102-113(1991)
See Koskinen, Lasse: V1568,262-270(1991)
See Oistamo, Kai: V1606,735-742(1991)
See Wang, Qiaofei: V1450,47-58(1991)
See Yin, Lin: V1451,216-227(1991)
See Yin, Lin: V1606,431-442(1991)
See Zeng, Bing: V1606,443-454(1991)
See Zhou, Hongbing: V1451,2-12(1991)

Neve, Peter
: Distributed industrial vision systems,VCR36,78-92(1991)

Neviere, Michel
: Blazing of transmission gratings for astronomical use,V1545,11-18(1991)
; den Boggende, Antonius J.; Padmore, Howard A.; Hollis, K.: Grating efficiency theory versus experimental data in extreme situations,V1545,76-87(1991)
; den Boggende, Antonius J.: X-ray multilayer-coated reflection gratings: theory and applications,V1545,116-124(1991)
See Akhouayri, Hassan: V1545,140-144(1991)
See Goudonnet, Jean-Pierre: V1545,130-139(1991)
See Vitrant, Guy: V1545,225-231(1991)

Nevorotin, Alexey J.
; Kallikorm, A. P.; Zeltzer, G. L.; Kull, Mart M.; Mihkelsoo, V. T.: Mechanism of injurious effect of excimer (308 nm) laser on the cell,V1427,381-397(1991)

Newberg, Irwin L. See Ng, Willie W.: V1371,205-211(1991)

Newberger, Ed See Srikantan, Geetha: V1459,258-267(1991)

Newberry, Jeff See McWilliams, Gary: V1468,417-428(1991)

Newberry, S. P. See Wang, Ge: V1398,180-190(1991)

Newhouse, Mark A. See Thevenot, Clarel: V1398,250-260(1991)

Newman, Brian See Bennett, Harold E.: V1441(1991)

Newman, H. C. See Peavy, George M.: V1424,171-178(1991)

Newman, Jay E.
; San Biagio, Pier L.; Schick, Kenneth L.: Photon correlation spectroscopic studies of filamentous actin networks,V1430,89-108(1991)

Newman, N. See Fork, David K.: V1394,202-202(1991)

Newman, Timothy S.
; Flynn, Patrick J.; Jain, Anil K.: Model-based surface classification,V1570,250-261(1991)

Newmiller, Chris
: Thermal dye transfer color hard-copy image stability,V1458,92-96(1991)

Newnam, Brian E.
: XUV free-electron laser-based projection lithography systems,V1343,214-228(1991)
; Warren, Roger W.; Conradson, Steven D.; Goldstein, John C.; McVey, Brian D.; Schmitt, Mark J.; Elliott, C. J.; Burns, M. J.; Carlsten, Bruce E.; Chan, Kwok-Chi D.; Johnson, W. J.; Wang, Tai-San; Sheffield, Richard L.; Meier, Karl L.; Olsher, R. H.; Scott, Marion L.; Griggs, J. E.: Development of an XUV-IR free-electron laser user facility for scientific research and industrial applications,V1552,154-174(1991)
See McVey, Brian D.: V1441,457-468(1991)

Newson, Trevor P.
See Chu, Beatrice C.: V1504,251-257(1991)
See Gerges, Awad S.: V1504,176-179(1991)
See Santos, Jose L.: V1511,179-189(1991)

Newstead, Emma
See Trigano, Philippe: V1468,866-874(1991)
See Trigano, Philippe: V1468,408-416(1991)

Newt, J. E.
See Futterman, Walter I.: V1486,127-140(1991)
See Strugala, Lisa A.: V1486,176-187(1991)

Newton, Scott C.
; Mitra, Sunanda: Self-organizing leader clustering in a neural network using a fuzzy learning rule,V1565,331-337(1991)
; Nutter, Brian S.; Mitra, Sunanda: Statistical and neural network classifiers in model-based 3-D object recognition,V1382,209-218(1991)

Neyer, Barry T. See Jaanimagi, Paul A.: V1346(1991)

Ng, Hon K.
; Kilibarda, S.: Possible enhancement in bolometric response using free-standing film of YBa2Cu3Ox,V1477,15-19(1991)

Ng, J. T. See Sinha, Sanjai: V1394,266-276(1991)

Ng, M. W. See Smith, Geoffrey B.: V1536,126-137(1991)

Ng, Peter A. See Wang, David T.: V1386,206-219(1991)

Ng, T. B. See Chua, Soo-Jin: V1401,96-102(1991)

Ng, Terence K. See Zakaria, Marwan F.: V1386,121-127(1991)

Ng, Willie W.
; Tangonan, Gregory L.; Walston, Andrew; Newberg, Irwin L.; Lee, Jar J.; Bernstein, Norman P.: True-time-delay steering of dual-band phased-array antenna using laser-switched optical beam forming networks,V1371,205-211(1991)

Ngo, Catherine M. See Chou, Chia-Shing: V1475,151-156(1991)

Ngo, Hoc D. See Rino, Charles L.: V1558,339-350(1991)

Nguyen, Bich-Yen See Cooper, Kent: V1392,253-264(1991)

Nguyen, Cattien See Twieg, Robert J.: V1455,86-96(1991)

Nguyen, Cong T. See Lin, Freddie S.: V1461,39-50(1991)

Nguyen, Dinh C. See Chan, Kwok-Chi D.: V1552,69-78(1991)

Nguyen, H. T. See Twieg, Robert J.: V1455,86-96(1991)

Nguyen, Hat See Cheng, Xin: V1363,177-181(1991)

Nguyen, L. See Scavennec, Andre: V1362,331-337(1991)

Nguyen, Loi D.
See Chou, Chia-Shing: V1475,151-156(1991)
See Greiling, Paul T.: V1475,34-41(1991)

Nguyen, Luong A.
; Walker, Ian D.; De Figueiredo, Rui; deFigueiredo, Rui J.: Control of flexible, kinematically redundant robot manipulators,V1387,296-312(1991)

Nguyen, M. T. See Nasburg, Robert E.: V1481,84-95(1991)

Nguyen, Tan T.
; Veuillen, J. Y.: Formation and electronic properties of epitaxial erbium silicide,V1512,289-298(1991)

Ni, Xiao W.
; Lu, Jian-Feng; He, Anzhi; Ma, Zi; Zhou, Jiu L.: Interference detection of plasma of laser field interaction with optical thin films,V1554B,632-635(1991)
; Lu, Jian-Feng; He, Anzhi; Ma, Zi; Zhou, Jiu L.: New definition of laser damage threshold of thin film,V1527,437-441(1991)
; Lu, Jian-Feng; He, Anzhi; Ma, Zi; Zhou, Jiu L.: New definition of laser damage threshold of thin film: optical breakdown threshold,V1519,365-369(1991)
See He, Anzhi: V1554A,747-749(1991)
See He, Anzhi: V1554A,547-552(1991)
See He, Anzhi: V1423,117-120(1991)
See Lu, Jian-Feng: V1415,220-224(1991)
See Lu, Jian-Feng: V1554B,593-597(1991)
See Luo, Bikai: V1554A,542-546(1991)
See Wang, Hai-Lin: V1545,274-277(1991)

Nianzeng, Che
; Grant, Barbara G.; Flittner, David E.; Slater, Philip N.; Biggar, Stuart F.; Jackson, Ray D.; Moran, M. S.: Results of calibrations of the NOAA-11 AVHRR made by reference to calibrated SPOT imagery at White Sands, N.M.,V1493,182-194(1991)

Niazy, Nagla N. See Schwartzkopf, George: V1466,26-38(1991)

Niblack, Curtiss A.
; Blessinger, Michael A.; Forsthoefel, John J.; Staller, Craig O.; Sobel, Harold R.: InSb linear multiplexed FPAs for the CRAF/Cassini visible and infrared mapping spectrometer,V1494,403-413(1991)
See Fischer, Robert C.: V1494,414-418(1991)
See Staller, Craig O.: V1540,219-230(1991)

Nicastri, Edward See Bonometti, Robert J.: V1495,166-176(1991)

Nicholas, Mark See Havrilla, George J.: V1435,12-18(1991)

Nichols, B. A. See Wood, Roger M.: V1441,316-326(1991)

Nichols, Joseph D.
: Firefly system concept,V1540,202-206(1991)

Nichols, Kirby B. See Mehanian, Courosh: V1469,275-280(1991)

Nickel, Eric G.
; Spangler, Charles W.: Extended pi-electron systems incorporating stabilized quinoidal bipolarons: anthracenyl polyene-polycarbonate composites,V1560,35-43(1991)
See Spangler, Charles W.: V1497,408-417(1991)

Nickel, Michael
; Husoy, John H.: Hybrid coder for image sequences using detailed motion estimates,V1605,963-971(1991)

Nicolaev, S. D. See Starobogatov, Igor O.: V1238,153-157(1991)

Nicolas, Henri See Labit, Claude: V1605,697-708(1991)

Nicolau, Dan V.
; Dusa, Mircea V.; Fulga, Florin: Simulation of connected image reversal and DESIRE techniques for submicron lithography,V1466,663-669(1991)
; Fulga, Florin; Dusa, Mircea V.: DESIRE technology with electron-beam resists: fundamentals, experiments, and simulation,V1465,282-288(1991)
; Fulga, Florin; Dusa, Mircea V.: Expert system for diagnosis/optimization of microlithography process*,V1468,345-351(1991)

Nicoli, David F.
: High-resolution submicron particle sizing by dynamic light scattering,V1430,19-36(1991)

Nicollian, Edward H. See Tsu, Raphael: V1361,232-235(1991)

Nicoulin, Andre
See Li, Wei: V1605,124-136(1991)
See Popat, Ashok C.: V1605,940-953(1991)

Nie, Chao-Jiang
; Li, Ying; Wu, Fang D.; Chen, Ying-Li; Li, Qi; Hua, Yi-Min: Generation of 1-GHz ps optical pulses by direct modulation of a 1.532-um DFB-LD,V1579,264-267(1991)
; Yu, Jing-ming; Wu, Fang D.; Gao, Yu-ping: Basic principle and system research of self-emission fiber optic temperature sensor,V1584,87-93(1991)

Nie, Qiuhua
; Nelson, John C.; Fleming, Simon C.: Geometric measurement of optical fibers with pulse-counting method,V1332,409-420(1991)

Nie, Shuming See Yu, Nai-Teng: V1403,112-124(1991)

Niechoda, Zygmunt J. See Nowicki, Marian: V1391,370-377(1991)

Niederer, Peter See Rol, Pascal O.: V1423,84-88(1991)

Niedrauer, Terren M.
: Coastal survey with a multispectral video system,V1492,240-251(1991)

Niedzwiecki, Abraham J. See Jaenisch, Holger M.: V1379,162-167(1991)

Nielsen, Bruce J. See Arendale, William F.: V1434,159-170(1991)

Nielsen, K. F. See Lind, M. D.: V1557,259-270(1991)

Nielsen, Norman D. See Laumann, Curt W.: V1414,151-160(1991)

Nielsen, Peter L.
: Investigation of optical fibers as sensors for condition monitoring of composite materials,V1588,229-240(1991)

Nielson, Gregory M.
; Dierks, Tim: Modeling and visualization of scattered volumetric data,V1459,22-33(1991)

Niemela, Arto See Sipila, Heikki: V1549,246-255(1991)

Nien, Chia-Fu See Plane, John M.: V1433,8-20(1991)

Nier, Michael
; Courtot, Marilyn E.; eds.: *Standards for Electronic Imaging Systems,*VCR37(1991)
See Courtot, Marilyn E.: VCR37,221-246(1991)

Nierat, G. See David, Jean: V1397,697-700(1991)

Niesterowicz, Andrzej See Gajda, Jerzy K.: V1391,329-331(1991)

Nieu, Jian-guo See Xue, Qing-yu: V1491,75-82(1991)

Nieuwkoop, E.
; Kapsenberg, Th.; Steenvoorden, G. K.; Bruinsma, Anastasius J.: Optically powered sensor system using conventional electrical sensors,V1511,255-263(1991)

Nieuwsma, Daniel E.
See Mordaunt, David W.: V1441,27-30(1991)
See Stultz, Robert D.: V1419,64-74(1991)

Nigawara, Kazushige See Matsuzaka, Fumio: V1397,239-242(1991)

Nihei, Eisuke See Koike, Yasuhiro: V1592,62-72(1991)

Nihei, Hideki See Seya, Eiichi: V1499,269-273(1991)

Niibe, Masahito
; Hayashida, Masami; Iizuka, Takashi; Miyake, Akira; Watanabe, Yutaka; Takahashi, Rie; Fukuda, Yasuaki: Suppression of columnar-structure formation in Mo-Si layered synthetic microstructures,V1343,2-13(1991)

Niihara, Toshio See Sukeda, Hirofumi: V1499,419-425(1991)

Niizeki, H. See Sakurai, Hirohisa: V1343,512-518(1991)

Nijander, Casimir R. See Feldblum, Avi Y.: V1544,200-208(1991)

Nijdam, D. C. See van Gemert, Martin J.: V1427,316-319(1991)

Nikias, Chrysostomos L.
See Bessios, Anthony G.: V1565,166-177(1991)
See Proakis, John G.: V1565,76-87(1991)

Nikirui, Ernest Y. See Danilov, Alexander A.: V1362,916-920(1991)

Nikitin, Petr I.
; Grigorenko, A. N.; Konov, Vitaly I.; Savchuk, A. I.: Fiber optic magnetic field sensors based on Faraday effect in new materials,V1584,124-134(1991)

Nikitin, S. Y.
See Kolomoitsev, D. V.: V1402,31-43(1991)
See Kolomoitsev, D. V.: V1402,11-30(1991)

Nikitina, Irina P. See Baranov, Igor M.: V1361,1110-1115(1991)

Nikitov, S. A.
See Boardman, A. D.: V1519,609-615(1991)
See Boardman, A. D.: V1519,597-604(1991)

Niklason, Loren T. See Cheng, Shirley N.: V1450,90-98(1991)

Niklasson, Gunnar A. See Nilsson, Torbjorn M.: V1536,169-182(1991)

Nikles, David E.
; Forbes, Charles E.: Accelerated aging studies for polycarbonate optical disk substrates,V1499,39-41(1991)

Nikolaev, D. A. See D'yakonov, G. I.: V1421,156-162(1991)

Nikolaev, V. A. See Ignatyev, Alexander V.: V1584,336-345(1991)

Nikolopoulos, John
; Yip, Gar L.: Accurate modeling of the index profile in annealed proton-exchanged LiNbO3 waveguides,V1583,71-82(1991)
; Yip, Gar L.: Characterization of proton-exchanged and annealed proton-exchanged optical waveguides in z-cut LiNbO3,V1374,30-36(1991)

Nikonchuk, Michael O.
: Copper vapor laser precision processing,V1412,38-49(1991)
; Polyakov, Igor V.: Focal plane intensity distribution of copper vapor laser with different unstable resonators,V1412,72-78(1991)

Nikonorov, Nikolai V.
See Glebov, Leonid B.: V1440,24-35(1991)
See Glebov, Leonid B.: V1621,21-32(1991)
See Glebov, Leonid B.: V1513,56-70(1991)

Nikzad, Arman See Subbarao, Murali: V1385,70-84(1991)

Nilsen, Jan E. See Bjarnar, Morten L.: V1537,74-88(1991)

Nilsson, Torbjorn M.
; Niklasson, Gunnar A.: Optimization of optical properties of pigmented foils for radiative cooling applications: model calculations,V1536,169-182(1991)

Ning, Gang See Nishimura, Kazuhito: V1534,199-206(1991)

Ning, Paul C.
; Hesselink, Lambertus B.: Adaptive isosurface generation in a distortion-rate framework,V1459,11-21(1991)

Ning, Ruola
; Barsotti, John B.; Kido, Daniel K.; Kruger, Robert A.: Clinical image-intensifier-based volume CT imager for angiography,V1443,236-249(1991)

Ning, Xiaohui
: Application of nonimaging optical concentrators to infrared energy detection,V1528,88-92(1991)
; Masso, Jon D.: Multiline holographic notch filters,V1545,125-129(1991)

Ning, Yanong N.
; Grattan, Kenneth T.; Palmer, Andrew W.; Meggitt, Beverley T.: Characteristics of a multimode laser diode source in several types of dual-interferometer configuration,V1367,347-356(1991)

Nisbet, K. C.
; McLaughlin, S.; Mulgrew, B.: Nonlinear prediction and the Wiener process,V1565,244-254(1991)

Nisenoff, Martin
; Gubser, Don U.; Wolf, Stuart A.; Ritter, J. C.; Price, George E.: High-temperature superconductivity space experiment: passive millimeter wave devices,V1394,104-113(1991)

Nisenson, Peter See Gonsalves, Robert A.: V1567,294-307(1991)

Nishi, Kazuro See Nakajima, Hajime: V1456,29-39(1991)

Nishi, Kenji See Ota, Kazuya: V1463,304-314(1991)

Nishi, Seiki See Watanabe, Mikio: V1452,27-36(1991)

Nishi, Y. See Ohtsuka, Yoshihiro: V1572,347-352(1991)

Nishida, T. See Watanabe, Hitoshi: V1499,21-28(1991)

Nishihara, H. K. See Bonasso, R. P.: V1387,237-244(1991)

Nishiie, Edward S. See Pelling, Michael R.: V1549,134-146(1991)

Nishikawa, Robert M.
; MacMahon, Heber; Doi, Kunio; Bosworth, Eric: Potential usefulness of a video printer for producing secondary images from digitized chest radiographs,V1444,180-189(1991)
See Giger, Maryellen L.: V1445,101-103(1991)

Nishiki, Masayuki See Silver, Michael D.: V1443,250-260(1991)

Nishimoto, T. See Kojima, Arata: V1429,162-171(1991)

Nishimura, Dwight G. See Noll, Douglas C.: V1443,29-36(1991)

Nishimura, Kazuhito
; Ma, Jing S.; Yokota, Yoshihiro; Mori, Yusuke; Kotsuka, Hiroshi; Hirao, Takashi; Kitabatake, Makoto; Deguchi, Masahiro; Ogawa, Kazuo; Ning, Gang; Tomimori, Hiroshi; Hiraki, Akio: Study of impurities in CVD diamond using cathodoluminescence,V1534,199-206(1991)

Nishimura, Kazutoshi See Maruyama, Mitsuru: V1605,916-927(1991)

Nishimura, Tetsuya
; Yagyu, Eiji; Yoshimura, Motomu; Tsukada, Noriaki; Takeyama, Tetsu: Electric field effect on the persistent hole burning of quinone derivatives,V1436,31-37(1991)

Nishino, Taneo
: Interface stress at thin film semiconductor heterostructures,V1519,382-390(1991)

Nishino, Toshikazu
; Hatano, Mutsuko; Hasegawa, Haruhiro: Superconductor/semiconductor structure and its application to superconducting devices,V1394,36-44(1991)

Nishio, Izumi See Tanaka, Toyoichi: V1403,280-287(1991)

Nishioka, Teiichi
; Murakami, R.; Matsuo, Seitaro; Ohishi, Y.: Higher-order theory of caustics for fast running cracks under general loadings,V1554A,802-813(1991)

Nishitani, Takao
See Mochizuki, Takashi: V1605,434-444(1991)
See Ohta, Mutsumi: V1605,456-466(1991)

Nishiuchi, Kenichi See Ohno, Eiji: V1499,171-179(1991)

Nissim, Carlos
; Beguin, Alain; Laborde, Pascale; Lerminiaux, Christian: Optical performance of wavelength division multiplexers made by ion-exchange in glass,V1365,13-19(1991)

Nissim, Yves I.
; Moison, Jean M.; Houzay, Francoise; Lebland, F.; Licoppe, C.; Bensoussan, M.: Multiple photo-assisted CVD of thin-film materials for III-V device technology,V1393,216-228(1991)

Nistler, John L.
See Dunbrack, Steven K.: V1464,314-326(1991)
See Phan, Khoi: V1464,424-437(1991)

Nittoh, Koichi See Tanazawa, Takeshi: V1435,310-321(1991)

Nitzan, Yeshayahu See Sessler, Jonathan L.: V1426,318-329(1991)

Niu, Aiqun
; Li, Dapeng; Wan, Liqun; Wee, William G.; Carrier, Charles W.: Comparative study of model fitting by using neural network and regression,V1469,495-505(1991)

Niu, Hanben See Degtyareva, V. P.: V1358,546-548(1991)

Niu, Lihong
See Chang, Zenghu: V1358,614-618(1991)
See Chang, Zenghu: V1358,541-545(1991)

Nixon, Alan J.
; Roth, Jerry E.; Krook, Lennart: Pulsed CO2 laser for intra-articular cartilage vaporization and subchondral bone perforation in horses,V1424,198-208(1991)

Noah, Meg A.
; Kristl, Joseph; Schroeder, John W.; Sandford, B. P.: NIRATAM-NATO infrared air target model,V1479,275-282(1991)

Noble, J. P. See Ruiz, D.: V1358,1134-1143(1991)

Noda, Etsuo See Fujii, Takashi: V1412,50-57(1991)

Noda, Juichi See Kobayashi, Masaru: V1474,278-284(1991)

Noda, Osama
; Kuribayashi, Shizuma; Imatake, Shigenori; Kondo, Motoe; Sato, Shunichi; Takahashi, Kunimitsu: Long-time operation of a 5kW cw CO laser,V1397,427-432(1991)
See Kuribayashi, Shizuma: V1397,439-443(1991)
See Sato, Shunichi: V1397,421-425(1991)
See Sato, Shunichi: V1397,433-437(1991)

Noda, T. See Yoshida, Kunio: V1441,9-9(1991)

Nodine, Calvin F.
; Brikman, Inna; Kundel, Harold L.; Douglas, A.; Seshadri, Sridhar B.; Arenson, Ronald L.: Image-display optimization using clinical history,V1444,56-62(1991)

Noel, Bruce W. See Tobin, Kenneth W.: V1584,23-31(1991)

Noel, Eric See Dhadwal, Harbans S.: V1564,664-673(1991)

Noeldeke, Christoph See MacGregor, Andrew D.: V1417,374-380(1991)

Noelscher, Christoph
; Mader, Leonhard: Investigation of self-aligned phase-shifting reticles by simulation techniques,V1463,135-150(1991)
See Sebald, Michael: V1466,227-237(1991)

Noethe, L.
; Andreoni, G.; Franza, F.; Giordano, P.; Merkle, Fritz; Wilson, Raymond N.: Latest developments of active optics of the ESO NTT and the implications for the ESO VLT,V1542,293-296(1991)

Nogar, Nicholas S.
See Muenchausen, Ross E.: V1394,221-229(1991)
See Wu, Xin D.: V1477,8-14(1991)

Noguchi, Miyoko See Takahashi, Kazuhiro: V1463,696-708(1991)

Noguchi, Tsutomu
; Tomita, Hidemi: Novolak resin design concept for high-resolution positive resists,V1466,149-160(1991)

Noh, T. W. See Xue, L. A.: V1534,183-196(1991)

Nohara, Mitsuo
; Harada, Takashi; Fujise, Masayuki: Optical ISL transmitter design that uses a high-power LD amplifier,V1417,338-345(1991)
See Inagaki, Keizo: V1417,160-169(1991)

Noirot, Laurence See Balageas, Daniel L.: V1467,278-289(1991)

Nolan, Adam R.
; Hu, Yong-Lin; Wee, William G.: X-ray inspection utilizing knowledge-based feature isolation with a neural network classifier,V1472,157-164(1991)

Nolan, P. F.
See Jackman, Louise A.: V1358,831-842(1991)
See Lavelle, Stephen P.: V1358,821-830(1991)

Nolhier, N.[$BFE] See Dilhac, Jean-Marie R.: V1393,349-353(1991)

Noll, Douglas C.
; Pauly, John M.; Nishimura, Dwight G.; Macovski, Albert: Magnetic resonance reconstruction from projections using half the data,V1443,29-36(1991)

Nolte, Loren W. See Rajala, Sarah A.: V1606,360-371(1991)

Nolting, Eugene E. See Ho, Ping-Tong: V1378,210-216(1991)

Nolting, Kerstin See Schmiedeskamp, Bernt: V1343,64-72(1991)

Noma, M. See Sakurai, Hirohisa: V1343,512-518(1991)

Nomura, Masafumi See Lauck, Teresa L.: V1464,527-538(1991)

Nomura, Noboru See Misaka, Akio: V1465,174-184(1991)

Nomura, Shintaro
; Kobayashi, Takayoshi T.; Nakanishi, Hachiro; Matsuda, Hiro; Okada, Shuji; Tomiyama, Hiromitsu: Nonlinear susceptibilities investigated by the electroabsorption of polymer ion-hemicyanine dye complexes,V1560,272-277(1991)

Nomura, Yasutomo
; Tamura, Mamoru: Quantitative analysis of hemoglobin oxygenation state of rat head by time-resolved photometry using picosecond laser pulse at 1064 nm,V1431,102-109(1991)

Nonaka, Y. See Sugihara, Okihiro: V1361,599-605(1991)

Noordam, Jan E.
: Toward an ESA strategy for optical interferometry outside the atmosphere: recommendations by the ESA space interferometry study team,V1494,344-346(1991)

Noponen, Eero
See Saarinen, Jyrki V.: V1555,128-137(1991)
See Vasara, Antti H.: V1507,224-238(1991)

Norbury, David H.
; Love, John C.: Photoresist bake conditions and their effects on lithography process control,V1463,558-573(1991)
See Flores, Gary E.: V1464,610-627(1991)

Nordin, C. W. See Egert, Charles M.: V1485,64-77(1991)

Nordquist, David P. See Kachelmyer, Alan L.: V1416,184-198(1991)

Noreils, Fabrice R.
: Towards a versatile control system for mobile robots,V1388,384-396(1991)

Norgard, John D.
; Metzger, Don W.; Cleary, John C.; Seifert, Michael: Infrared/microwave correlation measurements,V1540,699-708(1991)

Norkus, Volkmar
; Neumann, Norbert; Walther, Ludwig; Hofmann, Guenter; Schieferdecker, Jorg; Krauss, Matthias G.; Budzier, Helmut; Hess, Norbert: Pyroelectric linear array IR detectors with CCD multiplexer,V1484,98-105(1991)

Normand, Didier See Andre, Michel L.: V1502,286-298(1991)

Normann, Richard A.
: Recent progress in artificial vision,V1423,40-45(1991)

Normile, James
; Wright, Dan; Chu, Ke-Chiang; Yeh, Chia L.: Parallel processing approach to transform-based image coding,V1452,480-484(1991)

Norreys, P. See Tallents, Gregory J.: V1413,70-76(1991)

Norris, Katherine C. See Maltabes, John G.: V1463,326-335(1991)

Norris, Peter E.
See Chern, Chyi S.: V1394,255-265(1991)
See Zhao, Jing-Fu: V1362,135-143(1991)

Northcott, Malcolm J.
: University of Hawaii adaptive optics system: II. Computer simulation,V1542,254-261(1991)
See Roddier, Francois J.: V1542,248-253(1991)

Norton, D. P. See Lowndes, Douglas H.: V1394,150-160(1991)

Norton, Douglas A. See Thiem, Clare D.: V1532,39-47(1991)

Norton, Timothy J. See Fordham, John L.: V1449,87-98(1991)

Norwood, Robert A.
; Sounik, James R.; Popolo, J.; Holcomb, D. R.: Third-order nonlinear optical characterization of side-chain copolymers,V1560,54-65(1991)
See Spangler, Charles W.: V1497,408-417(1991)

Nory, Pierre
: Key technologies for IR zoom lenses: aspherics and athermalization,VCR38,142-152(1991)

Nosachev, L. V. See Anisimov, N. R.: V1440,206-210(1991)

Nose, I. See Cartlidge, Andy G.: V1423,167-174(1991)

Nosenko, S. P. See Vizir, V. A.: V1411,63-68(1991)

Noseworthy, J. R.
; Gerhardt, Lester A.: Three-dimensional vision: requirements and applications in a space environment,V1387,26-37(1991)

Nossal, Ralph J.
; Bonner, Robert F.: Differential time-resolved detection of absorbance changes in composite structures,V1431,21-28(1991)
; Schmitt, Joseph M.: Measuring photon pathlengths by quasielastic light scattering in a multiply scattering medium,V1430,37-47(1991)

Nota, Anatonella See Greenfield, Perry E.: V1494,16-39(1991)

Nota, M. See Bacis, Roger: V1397,173-176(1991)

Nouhi, Akbar See Lang, Robert J.: V1563,2-7(1991)

Nourshargh, Noorallah See Wood, Roger M.: V1441,316-326(1991)

Noutsios, Peter C.
; Yip, Gar L.; Kishioka, Kiyoshi: Propagation and diffusion characteristics of optical waveguides made by electric field-assisted k+-ion exchange in glass,V1374,14-22(1991)

Novak, Agneta
; Glaes, Anders; Blom, Claes; Hentzell, Hans; Hodges, Charles; Kalhur, Farzeen: Silicon-on-silicon microsystem in plastic packages,V1389,80-86(1991)

Novak, Carol L.
; Shafer, Steven A.: Supervised color constancy for machine vision,V1453,353-368(1991)

Novak, Leslie M. See Burl, Michael C.: V1471,92-115(1991)

Novaro, Marc
; Andre, Michel L.: Advanced technology development for high-fluence laser applications,V1501,183-193(1991)
See Floch, Herve G.: V1441,304-315(1991)

Novembre, Anthony E.
; Tai, Woon W.; Kometani, Janet M.; Hanson, James E.; Nalamasu, Omkaram; Taylor, Gary N.; Reichmanis, Elsa; Thompson, Larry F.: Single-component chemically amplified resist materials for electron-beam and x-ray lithography,V1466,89-99(1991)
See Chen, Wen-Chih: V1496,266-283(1991)

Novick, Robert
See Holley, Jeff: V1343,500-511(1991)
See Kaaret, Philip E.: V1548,106-117(1991)
See Shaw, Ping-Shine: V1343,485-499(1991)
See Shaw, Ping-Shine: V1548,118-131(1991)
See Weisskopf, Martin C.: V1343,457-468(1991)

Novini, Amir R.
: Fundamentals of on-line gauging for machine vision,V1526,2-16(1991)
: Lighting and optics expert system for machine vision,V1386,2-9(1991)

Novoselets, Mikhail K. See Tarasov, Victor A.: V1559,331-342(1991)

Novoselov, B. A. See Kiselyov, Boris S.: V1621,126-137(1991)

Novoselskiy, Vadim V. See Gan, Michael A.: V1574,254-260(1991)

Novozhilov, S. Y.
See Benimetskaya, L. Z.: V1525,242-245(1991)
See Benimetskaya, L. Z.: V1525,210-211(1991)

Nowak, Jerzy
; Zajac, Marek; eds.: *Intl Colloquium on Diffractive Optical Elements,*V1574(1991)
See Zajac, Marek: V1507,73-80(1991)
See Zajac, Marek: V1574,197-204(1991)

Nowak, William J. See Jodoin, Ronald E.: V1398,61-70(1991)

Nowakowski, Edward See Baldini, S. E.: V1588,125-131(1991)

Nowakowski, Jerzy
; Wlodawski, Mitchell: Optimal algorithms for holographic image reconstruction of an object with a glint using nonuniform illumination,V1416,229-240(1991)
See Wlodawski, Mitchell: V1416,241-249(1991)

Nowicki, Marian
; Niechoda, Zygmunt J.: Laser systems for precision micromachining,V1391,370-377(1991)

Nowygrod, Roman
See Bass, Lawrence S.: V1421,164-168(1991)
See Libutti, Steven K.: V1421,169-172(1991)
See Oz, Mehmet C.: V1422,147-150(1991)

Noz, Arthur See Firstein, Leon A.: V1464,81-88(1991)

Noz, Marilyn E. See Kalvin, Alan D.: V1445,247-258(1991)

Nozaki, Youichi See Sakurada, Masami: V1425,158-164(1991)

Nozue, Hiroshi
See Itoh, Katsuyuki: V1466,485-496(1991)
See Ohfuji, Takeshi: V1463,345-354(1991)

Nseyo, Unyime O.
; Lundahl, Scott L.; Merrill, Daniel C.: Photodynamic therapy in the prophylactic management of bladder cancer,V1426,287-292(1991)
See Kessel, David: V1426,180-187(1991)

Nulty, James E. See Bogle-Rohwer, Elizabeth: V1392,280-290(1991)

Numagami, Yasuhiko
; Hattori, Yasuyuki; Nakamura, Osamu; Minami, Toshi: Algorithm for quality inspection of characters printed on chip resistors,V1606,970-979(1991)

Numajiri, Yasuyuki See Wright, William H.: V1427,279-287(1991)

Numata, Takehiko See Naitou, Kazunori: V1499,386-392(1991)

Nunes, S. L. See Deutsch, Alina: V1389,161-176(1991)

Nunn, John W.
; Turner, Nicholas P.: Linewidth measurement comparison between a photometric optical microscope and a scanning electron microscope backed with Monte Carlo trajectory computations,V1464,50-61(1991)

Nunnally, William C.
: Design of frozen-wave and injected-wave microwave generators using optically controlled switches,V1378,10-21(1991)
See Sardesai, Harshad P.: V1378,237-248(1991)

Nunokawa, Mitsuji See Hanyu, Isamu: V1463,595-601(1991)

Nunomura, T. See Watanabe, Hitoshi: V1499,21-28(1991)

Nunotani, Masakatu See Meguro, Shin-Ichi: V1384,27-37(1991)

Nunzi, Jean-Michel See Charra, Fabrice: V1516,211-219(1991)

Nurre, Joseph H.
: Active stereo inspection using computer solids models,V1381,171-176(1991)

Nuttall, Albert H. See Griffin, Chintana: V1566,439-451(1991)

Nutter, Brian S. See Newton, Scott C.: V1382,209-218(1991)

Nutter, Harvey L. See Soper, Steven A.: V1435,168-178(1991)

Nuwaysir, Lydia M.
; Wilkins, Charles L.: Matrix-assisted laser desorption by Fourier transform mass spectrometry,V1437,112-123(1991)

Nwagboso, Christopher O.
: Diagnostic image processing of remote operating seals for aerospace application,V1521,55-63(1991)
: Digital imaging of aircraft dynamic seals: a fiber optics solution,V1500,234-245(1991)
: Proposals for a computer-controlled orbital scanning camera for remote image aquisition,V1454,111-122(1991)
; Whomes, Terence L.; Davies, P. B.: Remote visual monitoring of seal performance in aircraft jacks using fiber optics,V1386,30-41(1991)

Nyga, Ralf
; Neu, Walter; Preisack, M.; Haase, Karl K.; Karsch, Karl R.: Imaging of excimer laser vascular tissue ablation by ultrafast photography,V1525,119-123(1991)
See Neu, Walter: V1425,28-36(1991)
See Neu, Walter: V1425,37-44(1991)

Nykanen, Esa See Kaasinen, Harri I.: V1467,90-98(1991)

Nystrom, Alf See Larsson, Michael: V1529,63-70(1991)

Nystrom, George U. See Spiller, Eberhard: V1343,134-144(1991)

O'Bannon, Jim See Druce, Robert L.: V1378,43-54(1991)

O'Brien, Dominic C.
; Mears, Robert J.: Computer-generated holograms optimized for illumination with partially coherent light using silicon backplane spatial light modulators as the recording device,V1505,32-37(1991)

O'Brien, Kenneth M.
See Pradhan, Asima: V1425,2-5(1991)
See Stetz, Mark L.:V1425,55-62(1991)

O'Brien, Stephen J.
; Miller, Drew V.; Fealy, Stephen V.; Gibney, Mary A.; Kelly, Anne M.: Arthroscopic contact Nd:YAG laser meniscectomy: surgical technique and clinical follow-up,V1424,62-75(1991)
See Dederich, Douglas N.: V1424(1991)

O'Byrne, Vincent A.
; Tatlock, Timothy; Stone, Samuel M.: Optical decoding and coherent detection of a four-level FSK signal,V1372,88-93(1991)
See Childs, Richard B.: V1371,223-232(1991)

O'Connell, Robert M.
: Closed-form onset threshold analysis of defect-driven surface and bulk laser damage,V1441,406-419(1991)

O'Connor, Carol See Albert, Thomas: V1450,84-89(1991)

O'Connor, J. See Markert, Thomas H.: V1549,408-419(1991)

O'Connor, R. P.
; Bohling, Edward H.: User interface development for semiautomated imagery exploitation,V1472,26-37(1991)
See Bohling, Edward H.: V1406,164-168(1991)

O'Donnell, Kevin A.
; Knotts, Micheal E.: Polarization dependence of light scattered from rough surfaces with steep slopes,V1558,362-367(1991)

O'Donnell, P. J. See Cantella, Michael J.: V1540,634-652(1991)

O'Donnell, Timothy P.
See Coulter, Daniel R.: V1540,119-126(1991)
See Dolgin, Benjamin P.: V1533,229-239(1991)
See Hsieh, Cheng: V1533,240-251(1991)

O'Donoghue, Neil
See McNicholas, Thomas A.: V1421,56-67(1991)
See McNicholas, Thomas A.: V1421,30-35(1991)

O'Haver, Thomas C. See Xiong, Xiaoxiong: V1435,188-196(1991)

O'Kelley, B. See MacCraith, Brian D.: V1510,104-109(1991)

O'Kelly, John See Weisenbach, Lori: V1590,50-58(1991)

O'Key, Michael A.
; Osborne, Michael R.: High-fidelity fast-response phase conjugators for visible and ultraviolet applications,V1500,53-64(1991)
See Fieret, Jim: V1503,53-61(1991)

O'Loughlin, Michael J. See Bloss, Walter L.: V1541,2-10(1991)

O'Malley, Marty W.
See Loubriel, Guillermo M.: V1378,179-186(1991)
See Zutavern, Fred J.: V1378,271-279(1991)

O'Mara, Daniel M.
; Schiller, Craig M.; Warde, Cardinal: Membrane light modulators: engineering design considerations,V1527,110-117(1991)
See Horsky, Thomas N.: V1540,527-532(1991)

O'Neal, Michael C.
; Spanos, John T.: Optical pathlength control in the nanometer regime on the JPL phase-B interferometer testbed,V1542,359-370(1991)

O'Neal, Ray H.
See DeForest, Craig E.: V1343,404-414(1991)
See Hoover, Richard B.: V1343,189-202(1991)
See Lindblom, Joakim F.: V1343,544-557(1991)
See Walker, Arthur B.: V1343,319-333(1991)

O'Neil, Jason
; Goralczyk, Steven M.: Eagle-class small satellite for LEO applications,V1495,72-80(1991)

O'Neill, Alan See Barnes, Nigel: V1389,477-483(1991)

O'Neill, James A.
; Holber, William M.; Caughman, John: Radial ion energy measurements in an electron cyclotron resonance reactor,V1392,516-528(1991)

O'Neill, Mark B. See Sultana, John A.: V1532,27-38(1991)

O'Rourke, Michael J. See Sultan, Michel F.: V1584,212-219(1991)

O'Shea, Donald C.
ed.: *Analysis of Optical Structures*,V1532(1991)
: Monochromatic quartet: a search for the global optimum,V1354,548-554(1991)
See Sturlesi, Doron: V1354,54-68(1991)

O'Shea, Patrick G.
: Los Alamos photoinjector-driven free-electron laser,V1552,36-41(1991)

O'Shea, Peter See Boashash, Boualem: V1566,98-108(1991)

O'Sullivan, Gerard D. See Carroll, P. K.: V1503,416-427(1991)

Oakley, Philip J. See Armstrong, George R.: VCR38,120-141(1991)

Oba, Koichiro See Yokoi, Takane: V1449,30-39(1991)

Oba, Masayuki See Hayase, Rumiko H.: V1466,438-445(1991)

Obaidat, Mohammed T. See Wong, Kam W.: V1526,17-26(1991)

Obara, Minoru
See Ashidate, Shu-ichi: V1397,283-286(1991)
See Hatanaka, Hidekazu: V1397,379-382(1991)
See Kakehata, Masayuki: V1397,185-189(1991)
See Kannari, Fumihiko: V1397,493-497(1991)
See Kannari, Fumihiko: V1397,85-89(1991)

Obara, Tetsuro See Takayama, Kazuyoshi: V1358,1180-1190(1991)

Oberg, Per A.
See Pettersson, Hans: V1424,116-119(1991)
See Stenow, Eric: V1420,29-33(1991)
See Vegfors, Magnus: V1426,79-83(1991)

Oberhuber, Josef M. See Farrell, Edward J.: V1459,136-147(1991)

Oberlin, Jean C. See Paniez, Patrick J.: V1466,583-591(1991)

Obidin, Alexey Z.
See Alimpiev, Sergei S.: V1503,154-158(1991)
See Lukishova, Svetlana G.: V1503,338-345(1991)

Obremski, Susan M.
; LaMuraglia, Glenn M.; Bruggemann, Ulrich H.; Anderson, R. R.: Comparison of thermal and optical techniques for describing light interaction with vascular grafts, sutures, and thrombus,V1427,327-334(1991)

Obuchowski, Nancy A.
See Fuhrman, Carl R.: V1446,422-429(1991)
See King, Jill L.: V1446,268-275(1991)
See Rockette, Howard E.: V1446,276-283(1991)

Obyknovennaja, I. E. See Artemjev, S. V.: V1238,206-210(1991)

Ocana, Jose L.
: Numerical modeling of laser-matter interaction in high-intensity laser applications,V1397,813-820(1991)
: Numerical simulation of energy transport mechanisms in high-intensity laser-matter interaction experiments,V1502,299-310(1991)

Ocenasek, Josef See Das, Santanu K.: V1363,172-176(1991)

Ochiai, Ichiro See Nagatomo, Takao: V1519,90-95(1991)

Oda, Ichiro
; Ito, Yasunobu; Eda, Hideo; Tamura, Tomomi; Takada, Michinosuke; Abumi, Rentaro; Nagai, Katumi; Nakagawa, Hachiro; Tamura, Masahide: Noninvasive hemoglobin oxygenation monitor and computed tomography by NIR spectrophotometry,V1431,284-293(1991)

Oda, Kazuyuki See Ogura, Toshihiro: V1443,153-157(1991)

Oddy, C. J. See de Salabert, Arturo: V1521,74-88(1991)

Odinokov, S. B.
; Poddubnaya, T. E.; Rozhkov, Oleg V.; Yakimovich, A. P.: Gray level distortions induced by 3-D reflective hologram,V1238,109-117(1991)
; Poddubnaya, T. E.; Rozhkov, Oleg V.: Transer function of a 3-D reflective hologram,V1238,103-108(1991)

Oehler, Karen See Cosman, Pamela C.: V1605,162-171(1991)

Oehrlein, Gottlieb S. See Angell, David: V1392,543-550(1991)

Oertel, E. See Rahe, Manfred: V1441,113-126(1991)

Oertel, Heinrich K.
; Weiss, M.; Huber, Hans L.; Vladimirsky, Yuli; Maldonado, Juan R.: Modeling of illumination effects on resist profiles in x-ray lithography,V1465,244-253(1991)

Oesterberg, Ulf L.
: Measurement of nonlinear constants in photosensitive glass optical fibers,V1504,107-109(1991)

Oesterlin, Peter
See Basting, Dirk: V1412,80-83(1991)
See Mueller-Horsche, Elmar: V1503,28-39(1991)

Ogata, Hakaru See Ohhashi, Akinami: V1444,63-74(1991)

Ogata, Shiro
; Yoneda, Masahiro; Maeda, Tetsuo; Imanaka, Koichi: Low-cost and compact fiber-to-laser coupling with micro-Fresnel lens,V1544,92-100(1991)

Ogawa, Hisahito See Shiono, Teruhiro: V1545,232-240(1991)

Ogawa, Kazuo See Nishimura, Kazuhito: V1534,199-206(1991)

Ogawa, Kinichiro K. See Park, Yong K.: V1372,219-227(1991)

Ogawa, Ryokei See Kojima, Arata: V1429,162-171(1991)

Ogawara, Toshiaki See Maeta, Manabu: V1429,152-161(1991)

Ogikubo, Kazuhiro See Hino, Hideo: V1490,166-176(1991)

Ogino, M. See Kuribayashi, Shizuma: V1397,439-443(1991)

Ogino, Shinji See Ozawa, Kenji: V1499,180-186(1991)

Ogiwara, Akifumi
; Ohtsubo, Junji: Real-time nonlinear optical correlator in speckle metrology,V1564,294-305(1991)

Ogmen, Haluk
: Neural network model of dynamic form perception: implications of retinal persistence and extraretinal sharpening for the perception of moving boundaries,V1606,350-359(1991)

Ogura, Mutsuo
: Progress of surface-emitting lasers in Japan,V1418,396-405(1991)

Ogura, Satoshi
; Kawakubo, Yukio; Sasaki, Kouji; Kubota, Yoshiyuki; Miki, Atsushi: Output power stabilization of a XeCl excimer laser by HCl gas injection,V1412,123-128(1991)

Ogura, Takashi See Kitagawa, Teizo: V1403,563-571(1991)

Ogura, Toshihiro
; Masuda, Yukihisa; Fujita, Hiroshi; Inoue, Nobuo; Yonekura, Fukuo; Miyagi, Yoshihiro; Takatsu, Kazuaki; Akahira, Katsuyoshi; Tsuruta, Shigehiko; Kamiya, Masami; Takahashi, Fumitaka; Oda, Kazuyuki; Ikeda, Shigeyuki; Koike, Kouichi: Technical and clinical evaluations of a 2048 x 2048-matrix digital radiography system for gastrointestinal examinations,V1443,153-157(1991)

Oh, Kyunghwan
; Morse, Ted F.; Reinhart, Larry J.; Kilian, Arnd: Spectroscopic characteristics of Eu-doped aluminosilicate optical fiber preform,V1590,94-100(1991)
See Malo, Bernard: V1590,83-93(1991)

Ohara, M. See Yokozawa, T.: V1397,513-517(1991)

Ohara, Shunji
; Ishida, Takashi; Inokuchi, Chikashi; Furutani, Tadashige; Ishibashi, Kenzo; Kurahashi, Akira; Yoshida, Tomio: Compatability test for phase-change erasable and WORM media in a multifunction drive,V1499,187-194(1991)

Ohashi, T.
; Makishima, K.; Ishida, M.; Tsuru, T.; Tashiro, M.; Mihara, Teruyoshi; Kohmura, Y.; Inoue, Hiroyuki: Imaging gas scintillation proportional counters for ASTRO-D,V1549,9-19(1991)

Ohba, Yuji See Otsuji, Kiyotaka: V1606,980-989(1991)

Ohbayashi, Keiji See Suda, Yoshihiko: V1458,76-78(1991)

Ohfuji, Takeshi
; Soenosawa, Masanobu; Nozue, Hiroshi; Kasama, Kunihiko: Simulation analysis of deep-UV chemically amplified resist,V1463,345-354(1991)

Ohhashi, Akinami
; Yamada, Shinichi; Haruki, Kazuhito; Hatano, Hisaaki; Fujii, Yumi; Yamaguchi, Koujiro; Ogata, Hakaru: Automatic adjustment of display window (gray level) for MR images using a neural network,V1444,63-74(1991)

Ohishi, Satoru
; Yamaguchi, Masahiro; Ohyama, Nagaaki; Honda, Toshio: Three-dimensional reconstruction from cone beam projections,V1443,280-285(1991)

Ohishi, Y. See Nishioka, Teiichi: V1554A,802-813(1991)

Ohki, Y. See Akita, K.: V1392,576-587(1991)

Ohkuma, Tohru See Hagi, Toshio: V1464,215-221(1991)

Ohlhaber, Jack See Bagg, John C.: V1369,126-137(1991)

Ohmae, Hirokazu See Aoki, Yutaka: V1490,278-284(1991)

Ohmi, Toshihiko
; Yoshikawa, Nobuo; Sakai, Koji; Koike, Tomoyuki; Umegaki, Shinsuke: Nonlinear optical properties of a fiber with an organic core crystal grown from solution,V1361,606-612(1991)

Ohmori, Yasuhiro
; Arakawa, Isao; Nakai, Akira; Tsubosaka, Kazuyoshi: Cooling system for short-wave infrared radiometer of JERS-1 optical sensor,V1490,177-183(1991)

Ohn, Myo M.
See Davis, Andrew: V1489,33-43(1991)
See Davis, Andrew: V1588,264-274(1991)

Ohno, Eiji
; Nishiuchi, Kenichi; Yamada, Noboru; Akahira, Nobuo: Erasable compact disk utilizing phase-change material and multipulse recording method,V1499,171-179(1991)

Ohno, Tadayoshi See Nagato, Hitoshi: V1458,84-91(1991)

Ohshima, Iwao See Yasuoka, Koichi: V1412,32-37(1991)

Ohta, Masakatsu See Takahashi, Kazuhiro: V1463,696-708(1991)

Ohta, Masumi
See Aratani, Katsuhisa: V1499,209-215(1991)
See Fukumoto, Atsushi: V1499,216-225(1991)

Ohta, Mutsumi
; Yano, Mitsuharu; Nishitani, Takao: Entropy coding for wavelet transform of image and its application for motion picture coding,V1605,456-466(1991)

Ohta, Naohisa See Fujii, Tetsurou: V1605,339-350(1991)

Ohta, Norio
See Sukeda, Hirofumi: V1499,419-425(1991)
See Watanabe, Hitoshi: V1499,21-28(1991)

Ohta, Shigekata See Abe, Makoto: V1332,366-376(1991)

Ohtsubo, Junji See Ogiwara, Akifumi: V1564,294-305(1991)

Ohtsuka, Hiroshi
; Abe, Kazutoshi; Onodera, Toshio; Kuwahara, Kazuyuki; Taguchi, Takashi: Conjugate twin-shifter for the new phase-shift method to high-resolution lithography,V1463,112-123(1991)

Ohtsuka, Yoshihiro
; Nishi, Y.; Sawae, S.; Tanaka, Satoshi: Optical heterodyne fiber-coil deformation sensor operating in a wide dynamic range,V1572,347-352(1991)
See Imai, Masaaki M.: V1371,13-20(1991)

Ohue, Hiroshi
; Kasahara, Eiji; Uemura, Kamon; Mito, Keiichi: Estimation of CO2/N2 mixing region by the small-signal gain measurement,V1397,527-530(1991)

Ohue, Kenji See Aibara, Tsunehiro: V1606,198-203(1991)

Ohyama, Nagaaki
See Honda, Toshio: V1461,156-166(1991)
See Ohishi, Satoru: V1443,280-285(1991)

Oi, F. See Hanabata, Makoto: V1466,132-140(1991)

Oicles, Jeffrey A.
; Helava, Heikki I.; Grant, Jon R.; Ragle, Larry O.; Wessman, Susan C.: High-power waveform generation using photoconductive switches,V1378,60-69(1991)

Oikawa, Masahiro
; Nemoto, Hiroyuki; Hamanaka, Kenjiro; Imanishi, Hideki; Kishimoto, Takashi: Light coupling characteristics of planar microlens,V1544,226-237(1991)

Oinen, Donald E.
; Billow, Nick W.: New approach to the simulation of optical manufacturing processes,V1354,487-493(1991)

Oistamo, Kai
; Neuvo, Yrjo A.: Reconstruction of quincunx-coded image sequences using vector median,V1606,735-742(1991)

Ojeda-Castaneda, Jorge
; Ramirez, G.; Ibarra, J.: Zero axial irradiance with Rademacher's zone plates,V1500,252-255(1991)
; Rodriguez-Montero, Ponciano: Axial apodization using polar curves,V1500,256-261(1991)
; Szwaykowski, P.: Novel modes in a-power GRIN,V1500,246-251(1991)

Ojima, Masahiro See Maeda, Takeshi: V1499,414-418(1991)

Oka, Michio
See Kubota, Shigeo: V1499,314-323(1991)
See Kubota, Shigeo: V1354,572-580(1991)

Okabe, M. See Sugihara, Okihiro: V1361,599-605(1991)

Okada, Kasuyuki See Kubo, Uichi: V1420,102-107(1991)

Okada, Katsuyuki
; Tsujiuchi, Jumpei: Optical testing with wavelength scanning interferometer,V1400,33-37(1991)

Okada, Shuji
; Matsuda, Hiro; Masaki, Atsushi; Nakanishi, Hachiro; Hayamizu, Kikuko: Novel linear and ladder polymers from tetraynes for nonlinear optics,V1560,25-34(1991)
See Nomura, Shintaro: V1560,272-277(1991)
See Suda, Yasumasa: V1560,75-83(1991)

Okada, Yoshiyuki See Nakano, Yasuhiko: V1605,874-878(1991)

Okagaki, Takashi See Lau, Manhot: V1606,705-713(1991)

Okai, Makoto O.
; Harada, Tatsuo: Novel method to fabricate corrugation for distributed-feedback lasers using a grating photomask,V1545,218-224(1991)

Okajima, Masaki See Ishikawa, Masayuki M.: V1418,344-352(1991)

Okamine, Shigenori
; Terao, Motoyasu; Andoo, Keikichi; Miyauchi, Yasushi: Reversible phase-change optical recording by using microcellular GeSbTeCo recording film,V1499,166-170(1991)

Okamoto, N. See Ishikawa, Ken: V1397,55-58(1991)

Okamoto, Satoru See Watanabe, Mikio: V1452,27-36(1991)

Okamoto, Takashi
; Asakura, Toshimitsu: Detection of the object velocity using the time-varying scattered speckles,V1399,192-199(1991)

Okao, Keiichi See Satoh, Hiroharu: V1499,324-329(1991)

Okazaki, A. See Tsumanuma, Takashi: V1420,193-198(1991)

Okazaki, M. See Tsumanuma, Takashi: V1420,193-198(1991)

Okazaki, Shinji See Moniwa, Akemi: V1465,154-163(1991)

Okishev, Andrey V. See Altshuler, Grigori B.: V1440,116-126(1991)

Okomelkov, A. See Andronov, Alexander A.: V1362,684-695(1991)

Oksman, Ya. A.
See Amosova, L. P.: V1440,406-413(1991)
See Verechshagin, I. I.: V1440,401-405(1991)

Oktu, Ozcau
; Usala, Sandro; Adriaenssens, Guy J.; Tolunay, H.; Eray, A.: Density of localized states in glow-discharge a-Si1-x Cx:H,V1361,812-818(1991)

Okuda, Yoshimitsu See Hagi, Toshio: V1464,215-221(1991)

Okulov, Alexey Y.
: Two-dimensional periodic structures in solid state laser resonator,V1410,221-232(1991)

Okumura, Yoshiyuki See Isuzugawa, Kohji: V1358,1003-1010(1991)

Olbrysz, P. See Konwerska-Hrabowska, Joanna: V1554B,225-232(1991)

Olchowka, V. See Jursich, Gregory M.: V1412,115-122(1991)

Olczak, Eugene G. See Gustafson, Steven C.: V1469,268-274(1991)

Oldekop, Erik
; Siahmakoun, Azad: Real-time one-pass distortion correction,V1396,174-177(1991)

Oldham, William G. See Pfau, Anton K.: V1463,124-134(1991)

Oldmixon, E. H. See Aslund, Nils R.: V1450,329-337(1991)

Oleinick, Nancy L.
; Agarwal, Munna L.; Antunez, Antonio R.; Clay, Marian E.; Evans, Helen H.; Harvey, Ella Jo; Rerko, Ronald M.; Xue, Liang-yan: Effects of photodynamic treatment on DNA,V1427,90-100(1991)
; Varnes, Marie E.; Clay, Marian E.; Menegay, Harry J.: Interaction of phthalocyanine photodynamic treatment with ionophores and lysosomotrophic agents,V1426,235-243(1991)

Olesen, C. See Budtz-Jorgensen, Carl: V1549,429-437(1991)

Oleski, Paul J. See Short, Ralph C.: V1417,464-475(1991)

Oleynikov, V. A.
; Sokolov, K. V.; Hodorchenko, P. V.; Nabiev, I. R.: Polynuclear membranes as a substrate for obtaining surface-enhanced Raman scattering films,V1403,164-166(1991)

Oliensis, John
; Dupuis, Paul: Direct method for reconstructing shape from shading,V1570,116-128(1991)

Olivares-Perez, A. See Rodriguez-Vera, Ramon: V1507,416-424(1991)

Oliveira, Eugenio C. See Gouveia, Feliz A.: V1468,516-523(1991)

Oliveira, Joao See Rodrigues, F. C.: V1399,90-97(1991)

Oliver, Gabriel See Frau, Juan: V1388,329-340(1991)

Oliver, Jeffrey W. See Suter, Kevin J.: V1414,33-65(1991)

Olivero, John J. See Croskey, Charles L.: V1491,323-334(1991)

Olivier, A. See Henri, Christopher J.: V1444,306-317(1991)

Olivier, Lutz See Gieschen, Nikolaus: V1579,237-248(1991)

Olivier, Michel See Mainguy, Stephan: V1530,269-282(1991)

Ollikainen, Olavi
; Rebane, A.: Optical implementation of a Hopfield-type neural network by the use of persistent spectral hole-burning media,V1621,351-361(1991)

Olsen, Flemming O. See Jorgensen, Henning: V1412,198-208(1991)

Olsen, Gregory H.
; Ackley, Donald A.; Hladky, J.; Spadafora, J.; Woodruff, K. M.; Lange, M. J.; Van Orsdel, Brian T.; Forrest, Stephen R.; Liu, Y.: High-performance InGaAs PIN and APD (avalanche photdiode) detectors for 1.54 um eyesafe rangefinding,V1419,24-31(1991)
; Joshi, Abhay M.; Ban, Vladimir S.: Current status of InGaAs detector arrays for 1-3 um,V1540,596-605(1991)

Olsen, R. S. See Lurk, Paul W.: V1434,114-118(1991)

Olshansky, Robert See Hill, Paul M.: V1579,210-220(1991)

Olshausen, B. See Van Essen, David: V1473,17-28(1991)

Olsher, R. H. See Newnam, Brian E.: V1552,154-174(1991)

Olson, James See Mueller, Walter: V1472,66-75(1991)

Olson, Jeffrey See Hoose, John F.: V1545,160-166(1991)

Olsson, N. A. See Jewell, Jack L.: V1389,401-407(1991)

Olstad, Bjorn
: Automatic wall motion detection in the left ventricle using ultrasonic images,V1450,243-254(1991)
: Noise reduction in ultrasound images using multiple linear regression in a temporal context,V1451,269-281(1991)

Olthoff, James K. See Roberts, James R.: V1392,428-436(1991)

Oman, James See Hutt, Daniel L.: V1487,312-323(1991)

Omar, Basil A. See Emmony, David C.: V1397,651-659(1991)

Omar, Husam A.
: Construction of the 16-meter large lunar telescope,V1494,135-146(1991)

Omidvar, Omid M.
; Wilson, Charles L.: Massively parallel implementation of neural network architectures,V1452,532-543(1991)

Omori, Tsutae See Lauck, Teresa L.: V1464,527-538(1991)

Omoto, Osamu See Nagatomo, Takao: V1519,90-95(1991)

Omura, Etsuji E. See Aiga, Masao: V1418,217-222(1991)

Onda, Hajime See Yatagai, Toyohiko: V1555,8-12(1991)

Ondris-Crawford, Renate See Crawford, Gregory P.: V1455,2-11(1991)

Ong, Sim-Heng
; Giam, S. T.; Jayasooriah, Mr.; Sinniah, R.: Semiautomated detection and measurement of glomerular basement membrane from electron micrographs,V1445,564-573(1991)

Onischenko, Yuri I. See Dovgalenko, George Y.: V1559,479-486(1991)

Onishi, Randall M.
: SCORPIUS: final report,V1472,56-65(1991)
; Bogdanowicz, Julius F.; Watanabe, Miki: SCORPIUS: lessons learned in managing an image understanding system,V1406,171-178(1991)

Onn, Ruth
; Steinhardt, Allan O.: Multiwindow method for spectrum estimation and sinusoid detection in an array environment,V1566,427-438(1991)

Ono, Akira
; Sakuma, Fumihiro: ASTER calibration concept,V1490,285-298(1991)
See Fujisada, Hiroyuki: V1490,244-254(1991)
See Sakuma, Fumihiro: V1493,37-47(1991)
See Sasaki, Kenji: V1332,97-106(1991)
See Suzuki, Naoshi: V1493,48-57(1991)

Ono, Hiromi See Hino, Hideo: V1490,166-176(1991)

Ono, Hiroshi See Gozdz, Antoni S.: V1466,200-205(1991)

Ono, Makoto
; Nemoto, Yoshiaki: Synthetic aperture radar of JERS-1,V1490,184-190(1991)
See Akasaka, Akira: V1490,269-277(1991)

Ono, Sadayasu See Fujii, Tetsurou: V1605,339-350(1991)

Ono, Yuzo
: Polarizing holographic optical elements for optical data storage,V1555,177-181(1991)
See Yamanaka, Yutaka: V1499,263-268(1991)

Onodera, Kaoru See Suda, Yoshihiko: V1458,76-78(1991)

Onodera, Osamu See Takayama, Kazuyoshi: V1358,1180-1190(1991)

Onodera, Toshio See Ohtsuka, Hiroshi: V1463,112-123(1991)

Onuma, Kazuo See Tsukamoto, Katsuo: V1557,112-123(1991)

Ooi, B. C.
: Fail-safe WORM file system,V1401,27-34(1991)

Ooi, James
; Rao, Kashi: New insights into correlation-based template matching,V1468,740-751(1991)

Ooki, Hiroshi
: Rank-down method for automatic lens design,V1354,171-176(1991)

Ookoshi, Shigeo See Nagatomo, Takao: V1519,90-95(1991)

Oomen, Emmanuel W. See van Dongen, A. M.: V1513,330-339(1991)

Oona, H. See Fairbank, William M.: V1435,86-89(1991)

Oosterlinck, Andre J. See Xie, Kan: V1567,380-389(1991)

Opalinska, M. M. See Gomes, Anderson S.: V1579,257-263(1991)

Op de Beeck, Maaike
; Tokui, Akira; Fujinaga, Masato; Yoshioka, Nobuyuki; Kamon, Kazuya; Hanawa, Tetsuro; Tsukamoto, Katsuhiro: Improvement of focus and exposure latitude by the use of phase-shifting masks for DUV applications,V1463,180-196(1991)

Opekan, A. G. See Kamrukov, A. S.: V1503,438-452(1991)

Opitz, B. See Geurts, Jean: V1361,744-750(1991)

Oppenhaeuser, Gotthard
; Wittig, Manfred E.; Popescu, Alexandru F.: European SILEX project and other advanced concepts for optical space communications,V1522,2-13(1991)

Oppenheim, Uri P. See Devir, Adam D.: V1442,347-359(1991)

Oppenlaender, A. See Hochstrasser, Robin M.: V1403,221-229(1991)

Oprysko, Modest M.
See Chiu, George: V1389,364-374(1991)
See Trewhella, Jeannine M.: V1377,64-72(1991)

Opsal, Jon L.
: Application of thermal wave technology to thickness and grain size monitoring of aluminum films,V1596,120-131(1991)

Opsasnick, Michael N.
; Stancil, Daniel D.; White, Sean T.; Tsai, Ming-Horn: Optical fibers for magneto-optical recording,V1499,276-280(1991)

Orazi, Richard J. See Pappert, Stephen A.: V1476,282-293(1991)

Orcel, Gerard F. See Najafi, S. I.: V1583,32-36(1991)

Ordonez, Caesar
See Chen, Chin-Tu: V1445,222-225(1991)
See Yu, Xiaolin: V1396,56-58(1991)

Ordonez, I. See Huang, Zhiqing: V1436,103-113(1991)

Ordonez, Ishmael D.
; Crafton, J.; Murdock, R. H.; Hatfield, Lynn L.; Menzel, E. R.: Laser-excited fluorescence and fluorescence probes for diagnosing bulk damage in cable insulation,V1437,184-193(1991)

Orellana, Guillermo See Moreno-Bondi, Maria C.: V1368,157-164(1991)

Oren, Ehud See Wilk, Shalom: V1442,140-148(1991)

Oren, Moshe
: Computer-generated holograms fabricated on photopolymer for optical interconnect applications,V1389,527-534(1991)
See Shvartsman, Felix P.: V1555,71-79(1991)

Orgeret, Marc
: BEST: a new satellite for a climatology study in the tropics,V1490,14-22(1991)

Orham, Edward L. See Salmon, J. T.: V1542,459-467(1991)

Orhanovic, Neven
; Hayden, Leonard A.; Tripathi, Vijai K.: Analysis and modeling of uniformly- and nonuniformly-coupled lossy lines for interconnections and packaging in hybrid and monolithic circuits,V1389,273-284(1991)

Orita, Nobuyuki See Kanemaru, Toyomi: V1496,118-123(1991)

Orlov, M. Y. See Furzikov, Nickolay P.: V1427,288-297(1991)

Orlov, Oleg A. See Mak, Arthur A.: V1410,233-243(1991)

Orlova, Natalia G. See Golub, Mikhail A.: V1500,194-206(1991)

Orlovich, V. A.
See Apanasevich, P. A.: V1403,195-211(1991)
See Apanasevich, P. A.: V1403,240-242(1991)

Ormond, Karen A. See Peacock, Keith: V1494,147-159(1991)

Oron, Moshe
; Shladov, Itzhak; eds.: *7th Mtg in Israel on Optical Engineering*,V1442(1991)
See Bernstein, Hana: V1442,81-88(1991)

Orr, Richard S.
: Dimensionality of signal sets,V1565,435-446(1991)
See An, Myoung H.: V1565,383-401(1991)

Orsal, Bernard
; Alabedra, Robert; Signoret, P.; Letellier, H.: InGaAsP/InP distributed-feedback lasers for long-wavelength optical communication systems, lambda=1.55 um: electrical and optical noises study,V1512,112-123(1991)
See Kibeya, Saidi: V1501,97-106(1991)

Orsula, George W. See Thackeray, James W.: V1466,39-52(1991)

Orszag, Steven A.
See Barouch, Eytan: V1465,254-262(1991)
See Barouch, Eytan: V1463,336-344(1991)
See Barouch, Eytan: V1463,464-474(1991)

Ortega, A. R. See Fuchs, Elizabeth A.: V1467,136-149(1991)

Orvek, Kevin J. See Lindsay, Tracy K.: V1464,104-118(1991)

Ory, M. See Barnault, B.: V1397,231-234(1991)

Orza, Jose M.
See de Frutos, Angel M.: V1397,717-720(1991)
See Domingo, Concepcion: V1397(1991)
See Poueyo, Anne: V1502,140-147(1991)
See Quenzer, A.: V1397,753-759(1991)

Orzeszko, Gabriel R. See Hill, Peter C.: V1469,329-340(1991)

Osada, Masakazu See Martinez, Ralph: V1446,100-107(1991)

Osaka, Yukio See Tsunetomo, Keiji: V1513,93-104(1991)

Osawa, Atsuo
; Fukuda, Kyohei; Hirata, Kouji: Optical design of high-aperture aspherical projection lens,V1354,337-343(1991)

Osborne, Michael R.
See Fieret, Jim: V1503,53-61(1991)
See O'Key, Michael A.: V1500,53-64(1991)

Osbourn, S. J. See Fieret, Jim: V1503,53-61(1991)

Osche, Gregory R. See Young, Donald S.: V1416,221-228(1991)

Osher, John E. See Chau, Henry H.: V1346,103-112(1991)

Osher, Stanley
; Rudin, Leonid I.: Shocks and other nonlinear filtering applied to image processing,V1567,414-431(1991)

Oshida, Yoshitada See Yoshitake, Yasuhiro: V1463,678-687(1991)

Osiko, Anatoly V. See Lukishova, Svetlana G.: V1503,338-345(1991)

Osiko, Vyacheslav V.
See Danilov, Alexander A.: V1362,916-920(1991)
See Denker, Boris I.: V1419,50-54(1991)

Osinski, Marek A.
: Vertical-cavity surface-emitting semiconductor lasers: present status and future prospects,V1418,2-24(1991)
See Cherng, Chung-Pin: V1418,372-385(1991)
See Mahbobzadeh, Mohammad: V1418,25-31(1991)

Osipov, V. V. See Khryapov, V. T.: V1540,412-423(1991)

Oskard, David N. See Grupen, Roderic A.: V1383,189-201(1991)

Oskotsky, Mark L.
: Theory of two-component zoom systems,V1527,37-47(1991)

Osmer, Kurt A.
; Jones, Mike I.: Optical characterization of photolithographic metal grids,V1498,138-146(1991)

Osmolovsky, Michael G. See Kwok, Johnny H.: V1540,27-37(1991)

Ossin, A. B. See Dubrovsky, V.: V1403,344-346(1991)

Ostaszewski, Miroslaw A.
; Guy, Larry J.: Angular positioning mechanism for the ultraviolet coronagraph spectrometer,V1482,13-25(1991)

Osteaux, Michel See Mattheus, Rudy A.: V1446,341-351(1991)

Osten, Wolfgang See Andrae, P.: V1508,50-60(1991)

Osterberg, Mats See Lenz, Reiner: V1469,784-795(1991)

Osterfeld, Martin
; Knabke, Gerhard; Franke, Hilmar: Dye orientation in organic guest host systems on ferroelectric polymers,V1559,49-55(1991)

Osterheld, Robert See Wesley, Michael: V1481,209-220(1991)

Ostrout, Wayne H.
; Hiatt, William M.; Kozlowski, Alan E.: Process enhancement for a new generation g-line photolithographic system,V1463,54-73(1991)
; Hunkler, Sean; Ward, Steven D.: Enhanced process control of submicron contact definition,V1392,151-164(1991)

Ostrowski, Tomasz
: Custom-made filters in digital image analysis system,V1391,264-266(1991)

Ostrowsky, Daniel B. See Lallier, Eric: V1506,71-79(1991)

Oswald, Josef
; Pippan, Manfred; Tranta, Beate; Bauer, Guenther E.: Novel doping superlattice-based PbTe-IR detector device,V1362,534-543(1991)

Ota, Kazuya
; Magome, Nobutaka; Nishi, Kenji: New alignment sensors for wafer stepper,V1463,304-314(1991)

Ota, Osamu See Tsuchiya, Yoichi: V1499,450-456(1991)

Ota, Toshiyuki See Kajita, Toru: V1466,161-173(1991)

Ota, Yusuke
; Swartz, Robert G.: Multichannel optical data link,V1364,146-152(1991)

Oto, Takeshi See Kitagaki, Kazukuni: V1606,891-900(1991)

Otsuji, Kiyotaka
; Tonomura, Yoshinobu; Ohba, Yuji: Video browsing using brightness data,V1606,980-989(1991)

Otsuka, Kenju
: Nonlinear phenomena in semiconductor lasers,V1497,432-443(1991)
: Transition from homoclinic to heteroclinic chaos in coupled laser arrays,V1497,300-312(1991)

Otsuki, Taisuke
; Yoshimoto, Takashi: New approach for endoscopic stereotactic brain surgery using high-power laser,V1420,220-224(1991)

Ott, Carl See Smolka, Greg L.: V1498,70-80(1991)

Ottes, Fenno P.
See Kouwenberg, Jef M.: V1446,357-361(1991)
See van Gennip, Elisabeth M.: V1446,442-450(1991)

Otto, C. See Puppels, G. J.: V1403,146-146(1991)

Otto, Robert G. See Targ, Russell: V1478,211-227(1991)

Ou, Szutsun S.
See Botez, Dan: V1474,64-74(1991)
See Jansen, Michael: V1418,32-39(1991)
See Mawst, Luke J.: V1418,353-357(1991)

Ouaknine, G. E. See Rochkind, Simeone: V1428,52-58(1991)

Oudar, Jean-Louis
; Sfez, B. G.; Kuszelewicz, Robert: Bistable optical switching in GaAs multiple-quantum-well epitaxial etalons,V1361,490-498(1991)

Ouellet, Rene See Paquette, Benoit: V1426,362-366(1991)

Ouellette, Francois
ed.: *Intl Workshop on Photoinduced Self-Organization Effects in Optical Fiber*,V1516(1991)
: SHG in fiber: is a high-conversion efficiency possible?,V1516,113-114(1991)
; Duguay, Michel A.: Wavelength division multiplexing based on mode-selective coupling,V1365,25-32(1991)
; Gagnon, Daniel; LaRochelle, Sophie; Poirier, Michel: Photoinduced self-organization in optical fiber: some answered and unanswered questions,V1516,2-13(1991)

Ougazzaden, A. See Sermage, Bernard: V1362,617-622(1991)

Ouhayoun, Michel O. See Vigroux, Luc M.: V1500,74-79(1991)

Ouyang, Xiaolong See Chen, Chin-Tu: V1445,222-225(1991)

Ovchinnikov, Alexander V. See Garbuzov, Dmitriy Z.: V1418,386-393(1991)

Ovchinnikov, U. M. See Loshchenov, V. B.: V1420,271-281(1991)

Overfelt, Tony See Sahai, Vivek: V1557,60-70(1991)

Overholt, Bergein F.
See Panjehpour, Masoud: V1424,179-185(1991)
See Panjehpour, Masoud: V1427,307-315(1991)

Overstreet, Mark A. See Varshneya, Deepak: V1367,181-191(1991)

Ovryn, Benjie
; Haacke, E. M.: Measurement of the piezoelectric effect in bone using quasiheterodyne holographic interferometry,V1429,172-182(1991)

Ovsyuk, Victor N.
; Svitashev, Konstantin K.: Selective photodetectors: a view from the USSR,V1540,424-431(1991)

Owa, Hideo See Maeda, Fumisada: V1499,62-69(1991)

Owadano, Yoshirou See Izawa, Takao: V1441,339-344(1991)

Owechko, Yuri
; Soffer, Bernard H.: Optical neural networks based on liquid-crystal light valves and photorefractive crystals,V1455,136-144(1991)

Owen, A. E. See Ewen, Peter J.: V1512,101-111(1991)

Owen, B. See Park, Yong K.: V1372,219-227(1991)

Owen, Geraint See Pease, R. F.: V1496,234-238(1991)

Owen, Harry
: Holographic notch filter,V1555,228-235(1991)

Owen, John
; Ulph, Eric: Beryllium and titanium cost-adjustment report,V1485,128-137(1991)

Owen, Larry D. See Kreider, James F.: V1488,376-388(1991)

Owen, Philip R.
; Dawson, James A.; Borg, Eric J.: Solutions to modeling of imaging IR systems for missile applications: MICOM imaging IR system performance model-90,V1488,122-132(1991)

Owen, R. C.
: Easily fabricated wide-angle telescope,V1354,430-433(1991)

Owens, John K. See Welch, Eric B.: V1453,235-243(1991)

Owens, Robert M.
; Irwin, Mary J.: Parallel uses for serial arithmetic in signal processors,V1566,252-262(1991)

Owens, T. G. See Shreve, A. P.: V1403,394-399(1991)

Owner-Petersen, Mette
: Phase-map unwrapping: a comparison of some traditional methods and a presentation of a new approach,V1508,73-82(1991)

Oxley, Mark E.
See Rogers, Steven K.: VIS07,231-243(1991)
See Tarr, Gregory L.: V1469,2-11(1991)

Oyaizu, Ikuro
; Tanaka, Kiyoto; Yamaguchi, Toshikazu; Miyabo, Katsuaki; Takahashi, Mamoru: ISDN audio color-graphics teleconferencing system,V1606,990-1001(1991)

Oyama, Yoshiro
; Tani, Yuichiro; Shigemura, Naoshi; Abe, Toshio; Matsuda, Koyo; Kubota, Shigeto; Inami, Takashi: Removing vertical lines generated when x-ray images are digitized,V1444,413-423(1991)

Oyasato, Naohiko See Hayase, Rumiko H.: V1466,438-445(1991)

Oyen, Ordean J. See Keeler, Elaine K.: V1380,24-32(1991)

Oz, Mehmet C.
; Bass, Lawrence S.; Williams, Mathew R.; Benvenisty, Alan I.; Hardy, Mark A.; Libutti, Steven K.; Eaton, Alexander M.; Treat, Michael R.; Nowygrod, Roman: Preliminary experience with laser reinforcement of vascular anastomoses,V1422,147-150(1991)
See Ashton, Robert C.: V1422,151-155(1991)
See Auteri, Joseph S.: V1421,182-184(1991)
See Bass, Lawrence S.: V1422,123-127(1991)
See Bass, Lawrence S.: V1421,164-168(1991)
See Forman, Scott K.: V1424,2-6(1991)
See Libutti, Steven K.: V1421,169-172(1991)
See Wider, Todd M.: V1422,56-61(1991)

Ozaki, Mitsuo See Nakazawa, Akira: V1458,115-118(1991)

Ozaki, Yukihiro
; Mizuno, Aritake: Raman and FT-IR studies of ocular tissues,V1403,710-719(1991)
See Katayama, Norihisa: V1403,147-149(1991)

Ozawa, Kenji
; Ogino, Shinji; Satoh, Yoshikazu; Urushidani, Tatuo; Ueda, Atushi; Deno, Hiroshi; Kawakami, Haruo: Crystalization kinetics in Ge2Sb2Te5 phase-change recording films,V1499,180-186(1991)

Ozawa, Takashi
See Kobayashi, Kenichi: V1448,157-163(1991)
See Miyake, Hiroyuki: V1448,150-156(1991)

Ozawa, Yasuyuki
; Hamaguchi, S.; Moritsugu, Masaharu; Futamata, Akio: Clock generation circuit of a sampled servo optical disk drive,V1499,136-142(1991)

Ozeki, Takeshi See Martinez, Ralph: V1446,100-107(1991)

Ozguz, Volkan H.
; Esener, Sadik C.; Lee, Sing H.: Packaging issues for free-space optically interconnected multiprocessors,V1390,477-488(1991)

Ozizmir, Ercument See Levine, Alfred M.: V1376,47-53(1991)

Ozkan, Mehmet M.
; Sezan, M. I.; Erdem, A. T.; Tekalp, Ahmet M.: LMMSE restoration of blurred and noisy image sequences,V1606,743-754(1991)
; Tekalp, Ahmet M.; Sezan, M. I.: Identification of a class of space-variant image blurs,V1452,146-156(1991)

Ozkul, Cafer See Abdelghani-Idrissi, Ahmed M.: V1362,417-427(1991)

Ozolinsh, H. See Purinsh, Juris: V1525,289-308(1991)

Ozturk, Mehmet C.
; Zhong, Y.; Grider, Douglas T.; Sanganeria, M.; Wortman, Jim J.; Littlejohn, Michael A.: Selective deposition of polycrystalline SixGe1-x by rapid thermal processing,V1393,260-269(1991)
See Grider, Douglas T.: V1393,229-239(1991)

Ozue, Tadashi See Narahara, Tatsuya: V1499,120-128(1991)

Paakkari, Jussi See Ailisto, Heikki J.: V1384,50-59(1991)

Pacala, Thomas J.
; Tranis, Art; Laudenslager, James B.; Kinley, Fred G.: Thyratron-switched, L-C inverter, prepulse-sustainer, laser discharge circuit,V1411,69-79(1991)

Pace, John W. See Mitchell, Gordon L.: V1367,266-272(1991)

Pace, M. See Zarnowski, Jeffrey J.: V1447,191-201(1991)

Pacelli, Jean L.
; Geyer, Steve L.; Ramsey, Timothy S.: Parallel rule inferencing for automatic target recognition,V1472,76-84(1991)

Pacheco, Monica
; Barticevic, Zdenka; Claro, Francisco: Optical resonances of a semiconductor superlattice in parallel magnetic and electric fields,V1361,819-826(1991)

Padden, Richard J.
; Taylor, Edward W.; Sanchez, Anthony D.; Berry, J. N.; Chapman, S. P.; DeWalt, Steve A.; Wong, Ka-Kha: LiTaO3 and LiNbO3:Ti responses to ionizing radiation,V1474,148-159(1991)
See Taylor, Edward W.: V1474,126-131(1991)

Paddock, Stephen W. See Smith, Louis C.: V1428,224-232(1991)

Paddon-Row, Michael N. See Warman, John M.: V1559,159-170(1991)

Padgett, Joseph
; Pritchett, Thomas H.: Applicability of open-path monitors at Superfund sites,V1433,352-364(1991)

Padgett, Mary Lou
ed.: *2nd Workshop on Neural Networks: Academic/Industrial/NASA/Defense,*V1515(1991)

Padmore, Howard A.
See den Boggende, Antonius J.: V1345,189-197(1991)
See Neviere, Michel: V1545,76-87(1991)

Paek, Eung Gi
; Wullert, John R.; Von Lehman, A.; Patel, J. S.; Martin, R.: Compact optical neuro-processors,V1621,340-350(1991)

Paek, Un-Chul See Kuzyk, Mark G.: V1560,44-53(1991)

Paerels, F. B.
See Atkinson, Dennis P.: V1343,530-541(1991)
See Bixler, Jay V.: V1549,420-428(1991)

Paesler, Michael A. See Singh, Rajiv K.: V1394,203-213(1991)

Paez-Leon, Cristobal J.
; Patino, Antonio R.; Aguillon, Luis: Answer to the dynamic (fretting effect) and static (oxide) behavior of electric contact surfaces: based on a five-year infrared thermographic study,V1467,188-194(1991)

Paffett, M. F. See Jain, Ajay: V1596,23-33(1991)

Pagano, Robert J.
; Hatch, Marcus R.: Multiaperture spectrometer design for the Atmospheric Infrared Sounder,V1354,460-471(1991)

Page, J. B. See Sinha, Kislay: V1437,32-35(1991)

Page, Jerry L.
: Low-loss Y-couplers for fiber optic gyro applications,V1374,287-293(1991)
: Multiplexed approach for the fiber optic gyro inertial measurement unit,V1367,93-102(1991)

Page, Kevin J. See Zucker, Oved S.: V1378,22-33(1991)

Page, Lance A.
; Shen, Chi N.: Analysis of terrain-map-matching using multisensing techniques for applications to autonomous vehicle navigation,V1383,471-482(1991)

Paghdiwala, A. F.
: Root resection of endodontically treated teeth by Erbium:YAG laser radiation,V1424,150-159(1991)

Paglieroni, David W. See Ford, Gary E.: V1452,244-255(1991)

Pagnan, Antonio See Visona, Adriana: V1425,75-83(1991)

Pagnotta, Leonard See Furgiuele, Franco M.: V1554A,275-284(1991)

Pagnoux, Dominique
; Blondy, Jean M.; Facq, Paul: Wavelength and mode-adjustable source for modal characterization of optical fiber components: application to a new alignment method,V1504,98-106(1991)

Pahlajrai, Prem See Hughes, Joseph L.: V1389,87-97(1991)

Paige, Jeffrey L.
; Howard, Edward A.: SAFENET II: The Navy's FDDI-based computer network standard,V1364,7-13(1991)

Pailharey, Daniel See Mathey, Yves: V1361,909-916(1991)

Pain, M. D. See Fraser, George W.: V1548,132-148(1991)

Paindavoine, M. See Fauvet, E.: V1358,620-630(1991)

Paine, D. C. See Nath, Dilip K.: V1489,17-32(1991)

Pais, Cassiano P.
; Carvalho, Fernando D.; Silvestre, Victor M.: Architecture for surveillance in real time using nonlinear image processing hardware,V1451,282-288(1991)

Paisley, Dennis L.
; Montoya, Nelson I.; Stahl, David B.; Garcia, Ismel A.: Interferometry, streak photography, and stereo photography of laser-driven miniature flying plates,V1358,760-765(1991)
; Montoya, Nelson I.; Stahl, David B.; Garcia, Ismel A.; Hemsing, Willard F.: Velocity interferometry of miniature flyer plates with subnanosecond time resolution,V1346,172-178(1991)

Paiss, Idan
; Moser, Frank; Katzir, Abraham: Core-clad silver halide fibers for CO2 laser power transmission,V1420,141-148(1991)

Pak, S. K.
See D'yakonov, G. I.: V1421,156-162(1991)
See D'yakonov, G. I.: V1424,81-86(1991)
See D'yakonov, G. I.: V1421,153-155(1991)

Pakhomov, L. A. See Monge, J.L.: V1490,84-93(1991)

Pakhomova, Tat'yana A. See Tamkivi, Raivo: V1503,375-378(1991)

Pal, Suranjan See Basiev, Tasoltan T.: V1500,65-71(1991)

Palacios, Fernando R.
See Crow, Samuel B.: V1486,333-344(1991)
See Watkins, Wendell R.: V1486,17-24(1991)

Paladugu, R. R. See Kunz, Roderick R.: V1466,218-226(1991)

Palais, Joseph C. See Yuk, Tung Ip: V1401,130-137(1991)

Palakal, Mathew J.
; Chien, Yung-Ping; Chittajallu, Siva K.; Xue, Qing L.: Mobile robot system for the handicapped,V1468,456-466(1991)

Palaniappan, K.
; Huang, Thomas S.: Image analysis for DNA sequencing,V1450,186-197(1991)

Palanker, Daniel See Lewis, Aaron: V1423,98-102(1991)

Palchikova, Irena G. See Koronkevich, Voldemar P.: V1507,110-115(1991)

Palcic, Branko
; Lam, Stephen; MacAulay, Calum; Hung, Jacklyn; Jaggi, Bruno; Radjinia, Massud; Pon, Alfred; Profio, A. E.: Lung imaging fluorescence endoscope: development and experimental prototype,V1448,113-117(1991)
See Jaggi, Bruno: V1448,89-97(1991)
See Korbelik, Mladen: V1426,172-179(1991)

Paleev, V. I. See Zandberg, E. Y.: V1440,292-302(1991)

Palena, Patricia
: Variable-gain MMIC module for space application,V1475,91-102(1991)

Paley, M. S. See Frazier, Donald O.: V1557,86-97(1991)

Palffy-Muhoray, Peter See Yuan, Haiji J.: V1455,73-83(1991)

Palis, Michael A. See Alameldin, Tarek K.: V1381,217-225(1991)

Palladino, Libero See Batani, D.: V1503,479-491(1991)

Palma, J. A. See Juri, Hugo: V1422,128-135(1991)

Palmer, Andrew W.
See Boyle, William J.: V1511,51-56(1991)
See Ning, Yanong N.: V1367,347-356(1991)
See Weir, Kenneth: V1584,220-225(1991)
See Zhang, Zhiyi: V1511,264-274(1991)

Palmer, Graham See Woodruff, William H.: V1432,205-210(1991)

Palmer, James M.
: Stray light effects on calibrations using a solar diffuser,V1493,143-154(1991)
; Slater, Philip N.: Ratioing radiometer for use with a solar diffuser,V1493,106-117(1991)
See Guzman, Carmen T.: V1493,120-131(1991)
See Slater, Philip N.: V1493,100-105(1991)

Palmer, John See Kanungo, Tapas: V1385,104-112(1991)

Palmer, Scott E.
: Transendoscopic and freehand use of flexible hollow fibers for carbon dioxide laser surgery in the upper airway of the horse: a preliminary report,V1424,218-220(1991)

Palmer, Shane R.
See Babcock, Carl P.: V1466,653-662(1991)
See Laird, Daniel L.: V1465,134-144(1991)
See Wells, Gregory M.: V1465,124-133(1991)

Palmier, Jean F. See Minot, Christophe: V1362,301-308(1991)

Palmieri, Francesco
; Croteau, R. E.: Image restoration based on perception-related cost functions,V1451,24-35(1991)

Palowitch, Andrew W.
; Jaffe, Jules S.: Computer model for predicting underwater color images,V1537,128-139(1991)

Paltauf, Guenther See Reichel, Erich: V1421,129-133(1991)

Paltsev, G. P. See Kliot-Dashinskaya, I. M.: V1238,465-469(1991)

Palucci, Antonio See Barbini, Roberto: V1503,363-374(1991)

Palumbo, L. J. See Walter, Robert F.: V1397,71-76(1991)

Palusinski, Olgierd A. See Prince, John L.: V1390,271-285(1991)

Pamulapati, J. See Lachs, Gerard: V1474,248-259(1991)

Pan, Anpei
; Marhic, Michel E.; Epstein, Max: High-fidelity phase conjugation generated by holograms: application to imaging through multimode fibers,V1396,99-106(1991)
; Marhic, Michel E.: Holographic precompensation for one-way transmission of diffraction-limited beams through diffusing media,V1461,8-16(1991)

Pan, Bailiang
See Wang, Yongjiang: V1412,67-71(1991)
See Wang, Yongjiang: V1412,60-66(1991)

Pan, Chao See Sun, Dexing: V1572,508-513(1991)

Pan, Chuan K.
; Jiang, F. Y.; Fan, Guang H.; Ma, Y. Z.; Fan, Xi W.: Growth of ZnSe-ZnTe strained-layer supperlattices by atmospheric pressure MOCVD on transparent substrate CaF2 (111),V1519,645-651(1991)
See Wang, Ming: V1554B,242-246(1991)

Pan, Fu-shih
; Chen, Stephen; Mintzer, Robert; Chen, Chin-Tu; Schumacker, Paul: Studies of yeast cell oxygenation and energetics by laser fluorometry of reduced nicotinamide adenine dinucleotide,V1396,5-8(1991)

Pan, George W. See Gilbert, Barry K.: V1390,235-248(1991)

Pan, Horng-Lon See Fox, Marye A.: V1436,2-7(1991)

Pan, J. J.
: Microwave fiber optic RF/IF link,V1371,195-204(1991)
: Semiconductor laser transmitters for millimeter-wave fiber-optic links,V1476,63-73(1991)
; Chau, Seung L.; Li, Wei Z.; Grove, Charles H.: Cost-effective optical switch matrix for microwave phased array,V1476,133-142(1991)
; Cordeiro, D.: Adaptive semiconductor laser phased arrays for real-time multiple-access communications,V1476,157-169(1991)
; Garafalo, David A.: Microwave high-dynamic-range EO modulators,V1371,21-35(1991)
; Li, Wei Z.: High on-off ratio, ultrafast optical switch for optically controlled phased array,V1476,122-132(1991)
; Li, Wei Z.; Li, Yi Q.: Ultralinear electro-optic modulators for microwave fiber-optic communications,V1476,32-43(1991)
; Li, Yi Q.: Broadband microwave and millimeter-wave EOMs with ultraflat frequency response,V1476,22-31(1991)

Pan, Jingming
; Yin, Zongming: Intelligent optical fiber sensor system for measurement of gas concentration,V1572,403-405(1991)

Pan, X. J. See Fairbank, William M.: V1435,86-89(1991)

Pan, Xiao R.
; Gu, Zhi X.: Thin films of solid electrolytes and studies of their surface,V1519,85-89(1991)

Pan, Yao L. See Wang, Yun Z.: V1519,860-863(1991)

Pan, Yingtain
; Liu, Xiande; Du, Chongwu; Li, Zai Q.: Fiber optic magnetic field and current sensor using magneto-birefringence of dense ferrofluid thin films,V1572,477-482(1991)

Panabiere, Jean-Pierre See Weill, Andre P.: V1465,264-270(1991)

Panarella, Emilio See Kawai, Kenji: V1465,308-314(1991)

Panasenko, N. A. See Skopinov, S. A.: V1403,676-679(1991)

Panasiuk, L. M. See Anikin, V. I.: V1238,286-296(1991)

Panasyuk, L. M.
; Forsh, A. A.: Optical information recording on vitreous semiconductors with a thermoplastic method of visualization,V1621,74-82(1991)

Panayotatos, Paul
; Georgakilas, Alexandros; Mourrain, Jean-Loic; Christou, Aristos: Molecular beam epitaxy GaAs on Si: material and devices for optical interconnects,V1361,1100-1109(1991)

Pance, A. See Wengler, Michael J.: V1477,209-220(1991)

Pance, Gordana See Wengler, Michael J.: V1477,209-220(1991)

Pandey, Ravindra K.
; Vicente, M. G.; Shiau, Fuu-Yau; Dougherty, Thomas J.; Smith, Kevin M.: Syntheses of porphyrin and chlorin dimers for photodynamic therapy,V1426,356-361(1991)
See Shiau, Fuu-Yau: V1426,330-339(1991)

Pandian, Natesa G.
: Intravascular ultrasound imaging and intracardiac echocardiography: recent developments and future directions,V1425,198-202(1991)
See Sehgal, Chandra M.: V1425,226-233(1991)

Pandit, Kirit See Pearce, John A.: V1422,27-33(1991)

Pandit, Malay K.
See Das, Alok K.: V1365,74-85(1991)
See Das, Alok K.: V1572,572-580(1991)

Pandya, Abhijit S.
; Luebbers, Paul G.: Neural networks for robot navigation,V1468,802-811(1991)
See Luebbers, Paul G.: V1469,756-765(1991)
See Szabo, Raisa R.: V1468,794-801(1991)
See Venugopal, Kootala P.: V1471,44-53(1991)

Panero, Stefania See Peres, Rosa C.: V1559,151-158(1991)

Panesh, Anatoly M.
See Simonov, Alexander P.: V1503,330-337(1991)
See Simonov, Alexander P.: V1436,20-30(1991)

Pang, Da-wen See Wang, Yi-Ming: V1361,325-335(1991)

Pang, Khee K. See Sikora, Thomas: V1605,624-634(1991)

Pang, Shi J. See Zhao, Yun F.: V1519,411-414(1991)

Pang, Stella W. See Demos, Alexandros T.: V1392,291-297(1991)

Pang, Y. See Kwok, Hoi-Sing: V1394,196-200(1991)

Pang, Yen-Ming See Sun, Tai-Ping: V1361,1033-1037(1991)

Paniez, Patrick J.
; Joubert, Olivier P.; Pons, Michel J.; Oberlin, Jean C.; Weill, Andre P.: Dry development and plasma durability of resists: melt viscosity and self-diffusion effects,V1466,583-591(1991)
; Weill, Andre P.; Cohendoz, Stephane D.: Physical aging of resists: the continual evolution of lithographic material,V1466,336-344(1991)
See Pierrat, Christophe: V1466,248-256(1991)

Panish, Morton B.
See Gunapala, S. D.: V1541,11-23(1991)
See Harriott, Lloyd R.: V1465,57-63(1991)

Panjehpour, Masoud
; Overholt, Bergein F.; Frazier, Donita L.; Klebanow, Edward R.: Hyperthermia treatment of spontaneously occurring oral cavity tumors using a computer-controlled Nd:YAG laser system,V1424,179-185(1991)
; Wilke, August; Frazier, Donita L.; Overholt, Bergein F.: Hyperthermia treatment using a computer-controlled Nd:YAG laser system in combination with surface cooling,V1427,307-315(1991)

Pankove, Jacques I.
; Radehaus, C.; Borghs, Gustaaf: Optoelectronic neuron,V1361,620-627(1991)

Pankratov, Nickolai A.
: Nonselective thermal detectors of radiation,V1540,432-443(1991)

Pannell, Christopher N.
See Berwick, Michael: V1584,364-373(1991)
See Berwick, Michael: V1572,157-162(1991)

Pansini, Evangelista See Armenise, Mario N.: V1562,160-171(1991)

Pantelis, Philip
; Hill, Julian R.; Kashyap, Raman: Polymer waveguide systems for nonlinear and electro-optic applications,V1559,2-9(1991)

Panysheva, Elena I. See Glebov, Leonid B.: V1621,21-32(1991)

Pao, Cheng K. See Lan, Guey-Liu: V1475,184-192(1991)

Paolacci, Sylvie
; Hugenschmidt, Manfred; Bournot, Philippe: Mechanical effects induced by high-power HF laser pulses on different materials under normal atmospheric conditions,V1397,705-708(1991)

Paolella, Arthur
; Madjar, Asher; Herczfeld, Peter R.; Sturzebecher, Dana: Optically activated GaAs MMIC switch for microwave and millimeter wave applications,V1378,195-202(1991)

Paoletti, Domenica
; Schirripa Spagnolo, Giuseppe: Contouring by DSPI for surface inspection,V1554A,660-667(1991)

Paone, Nicola
; Rossi, G.: Fiber optic ice sensors for refrigerators,V1511,129-139(1991)

Papadopoulos, George A.
: Influence of the orthotropy of the ductile materials and the stress-assisted diffusion on the caustics,V1554A,826-834(1991)

Papaioannou, Thanassis
; Papazoglou, Theodore G.; Daykhovsky, Leon; Gershman, Alex; Segalowitz, Jacob; Reznik, G.; Beeder, Clain; Chandra, Mudjianto; Grundfest, Warren S.: Practical considerations for effective microendoscopy,V1420,203-211(1991)
See Buchelt, Martin: V1420,249-253(1991)
See Chandra, Mudjianto: V1421,68-71(1991)
See Vari, Sandor G.: V1424,33-42(1991)
See Vari, Sandor G.: V1426,58-65(1991)
See Vari, Sandor G.: V1426,111-120(1991)

Papamichael, Haris
; Miaoulis, Ioannis: Mixed-convection effects during the drawing of optical fibers,V1590,122-130(1991)

Paparao, Palacharla See Ghosh, Anjan K.: V1371,170-181(1991)

Papavaritis, P. See Danly, Bruce G.: V1407,192-201(1991)

Papazoglou, Theodore G.
See Chandra, Mudjianto: V1421,68-71(1991)
See Papaioannou, Thanassis: V1420,203-211(1991)
See Vari, Sandor G.: V1426,58-65(1991)
See Vari, Sandor G.: V1426,111-120(1991)

Pape, Dennis R.
; Vlannes, Nickolas P.; Patel, Dharmesh P.; Phuvan, Sonlinh: Compact low-power acousto-optic range-Doppler-angle processor for a pulsed-Doppler radar,V1476,201-213(1991)

Papineau, N. See De Maziere, Martine M.: V1491,288-297(1991)

Pappalettere, Carmine
; Trentadue, Bartolo; Monno, Giuseppe: Moire-holographic evaluation of the strain fields at the weld toes of welded structures,V1554B,99-105(1991)

Pappas, Thrasyvoulos N.
; Neuhoff, David L.: Model-based halftoning,V1453,244-255(1991)

Pappert, Stephen A.
; Lin, S. C.; Orazi, Richard J.; McLandrich, Matthew N.; Yu, Paul K.; Li, S. T.: Broadband electromagnetic environment monitoring using semiconductor electroabsorption modulators,V1476,282-293(1991)

Paprocki, K. See Brylowska, Irena: V1391,164-169(1991)

Papuchon, Michel R. See Lallier, Eric: V1506,71-79(1991)

Paquette, Benoit
; Rousseau, Jacques; Ouellet, Rene; van Lier, Johan E.: Radiolabeled red blood cells for the direct measurement of the blood flow kinetics in experimental tumors after photodynamic therapy,V1426,362-366(1991)

Paquin, Roger A.
; Gardopee, George J.: Fabrication of a fast, aspheric beryllium mirror,V1485,39-45(1991)
; Vukobratovich, Daniel; eds.: *Optomechanics and Dimensional Stability*,V1533(1991)

Paradies, C. J.
; Glicksman, M. E.; Jones, O. C.; Kim, G. T.; Lin, Jen T.; Gottlieb, Milton S.; Singh, N. B.: Latest developments in crystal growth and characterization of efficient acousto-optic materials,V1561,2-5(1991)

Paramonov, A. A. See Galpern, A. D.: V1238,320-323(1991)

Paraszczak, Jurij R. See Deutsch, Alina: V1389,161-176(1991)

Parcell, E. W. See Gorton, E. K.: V1397,291-295(1991)

Pardo, A. See Martin, E.: V1397,835-838(1991)

Pardo, C. See Bergman, Larry A.: V1364,14-21(1991)

Parel, Jean-Marie
See Cartlidge, Andy G.: V1423,167-174(1991)
See Lingua, Robert W.: V1423,58-61(1991)
See Milne, Peter J.: V1423,122-129(1991)
See Rol, Pascal O.: V1423,89-93(1991)
See Simon, Gabriel: V1423,154-156(1991)

Parent, Andre See Snell, Kevin J.: V1410,99-106(1991)

Parent, C. See Boutinaud, P.: V1590,168-178(1991)

Paresce, Francesco See Greenfield, Perry E.: V1494,16-39(1991)

Parikh, Jo Ann
; DaPonte, John S.; Damodaran, Meledath; Karageorgiou, Angelos; Podaras, Petros: Comparison of backpropagation neural networks and statistical techniques for analysis of geological features in Landsat imagery,V1469,526-538(1991)
See DaPonte, John S.: V1469,441-450(1991)

Paris, E. See Etienne, Bernard: V1362,256-267(1991)

Paris, Elizabeth S.
See DeForest, Craig E.: V1343,404-414(1991)
See Walker, Arthur B.: V1343,319-333(1991)

Paris, Robert D. See English, R. Edward: V1527,174-179(1991)

Parizzi, Francesco See Vernazza, Gianni L.: V1492,206-212(1991)

Park, B. C. See Kwon, Young H.: V1554A,639-644(1991)

Park, B. K. See Measures, Raymond M.: V1332,431-443(1991)

Park, C. S. See Gopalakrishnan, G. K.: V1476,270-275(1991)

Park, Choong-Bum
; Li Kam Wa, Patrick; Miller, Alan: Ultrafast measurements of carrier transport optical nonlinearities in a GaAs/AlGaAs MQW SEED device,V1474,8-17(1991)
See Li Kam Wa, Patrick: V1474,2-7(1991)

Park, Eric D.
; Swafford, William J.; Lamb, Bryan K.: Reliability of fiber optic position sensors,V1366,294-303(1991)

Park, Ernest D. See Johnson, R. B.: V1354,669-675(1991)

Park, Gun-Sik See Hirshfield, Jay L.: V1552,138-146(1991)

Park, Hong-Woo
See Cho, Young-Chung: V1346,160-171(1991)
See Short, Ralph C.: V1417,131-141(1991)

Park, Il-Pyung See Kender, John R.: V1388,476-489(1991)

Park, J. D. See Li, Hua: V1376,172-179(1991)

Park, Joon B.
; Benatar, Avraham: Measurement of residual stress during implant resistance welding of plastics,V1554B,357-370(1991)

Park, Jun S. See Lee, Sukhan: V1383,391-402(1991)

Park, Kyung S. See Lee, Nahm S.: V1450,133-143(1991)

Park, R. E. See Garside, Brian K.: V1588,150-158(1991)

Park, S. H.
; Casey, Michael P.; Falk, Joel: Degenerate four-wave mixing and optical nonlinearities in quantum-confined CdSe microcrystallites,V1409,9-13(1991)
See Lee, Hyung G.: V1361,893-900(1991)

Park, Sanghyun See Dodd, James W.: V1407,467-473(1991)

Park, Seung O.
See Joo, Jin W.: V1554B,113-118(1991)
See Kwon, Young H.: V1554A,639-644(1991)

Park, Stephen K. See Reichenbach, Stephen E.: V1569,422-433(1991)

Park, Yong K.
; Delavaux, Jean-Marc P.; Tench, Robert E.; Cline, Terry W.; Tzeng, Liang D.; Kuo, Chien-yu C.; Wagner, Earl J.; Flores, Carlos F.; Van Eijk, Peter; Pleiss, T. C.; Barski, S.; Owen, B.; Twu, Yih-Jye; Dutta, Niloy K.; Riggs, R. S.; Ogawa, Kinichiro K.: Coherent communication systems research and development at AT&T Bell Laboratories, Solid State Technology Center,V1372,219-227(1991)

Park, Young D. See Madsen, Steen J.: V1431,42-51(1991)

Park, Youngsoh See Lim, G.: V1418,123-131(1991)

Parke, Ross See Mehuys, David G.: V1418,57-63(1991)

Parker, D. See McIver, Robert T.: V1437,124-128(1991)

Parker, James D.
: Psychometrics for quantitative print quality studies,V1398,15-23(1991)

Parker, Kevin J.
See Mitsa, Theophano: V1452,47-56(1991)
See Riek, Jonathan K.: V1445,190-197(1991)
See Riek, Jonathan K.: V1398,130-141(1991)
See Riek, Jonathan K.: V1445,198-206(1991)

Parker, Richard J.
See Brownell, John B.: V1554B,481-492(1991)
See Cullis, I. C.: V1358,52-64(1991)

Parker, Sandra C. See Chen, Robert S.: V1381,189-200(1991)

Parkinson, Bruce See Armstrong, Neal R.: V1559,18-26(1991)

Parkkinen, Jussi P. See Honkonen, Veijo: V1564,43-51(1991)

Parks, Robert E.
: Status and importance of optical standards,V1346,76-78(1991)

Parkyn, William A.
; Pelka, David G.: Compact nonimaging lens with totally internally reflecting facets, V1528,70-81(1991)

Parma, Ludvik
; Malicka, Marianna: Multiple scattering of laser light in dense aerosol, V1391,190-198(1991)
See Malicka, Marianna: V1391,181-189(1991)

Parmar, A. N. See Bavdaz, Marcos: V1549,35-44(1991)

Parra-Loera, Ramon
; Thompson, Wiley E.; Salvi, Ajit P.: Adaptive selection of sensors based on individual performances in a multisensor environment, V1470,30-36(1991)
See Thompson, Wiley E.: V1470,48-58(1991)

Parriaux, Olivier M.
: Passive integrated optics in optical sensor systems, V1506,111-119(1991)
; Sychugov, V. A.: Integrated optic flat antennae: early applications and design tools, V1583,376-382(1991)
See Poscio, Patrick: V1510,112-117(1991)

Parrish, Alan See Connor, Brian J.: V1491,218-230(1991)

Parrish, Russell V. See Busquets, Anthony M.: V1457,91-102(1991)

Parrish, William J.
; Blackwell, John D.; Paulson, Robert C.; Arnold, Harold: 128 x 128 MWIR InSb focal plane and camera system, V1512,68-77(1991)
; Blackwell, John D.; Kincaid, Glen T.; Paulson, Robert C.: Low-cost high-performance InSb 256 x 256 infrared camera, V1540,274-284(1991)
See Blackwell, John D.: V1479,324-334(1991)

Parry, David J.
; Stewardson, H. R.; Ahmad, S. H.; Al-Maliky, Noori S.: Application of cylindrical blast waves to impact studies of materials, V1358,1057-1064(1991)

Parshall, Robert F.
; Sadler, Lewis L.: Computer-aided design and drafting visualization of anatomical structure of the human eye and orbit, V1380,200-207(1991)

Parson, Dale E.
: Parallel reduced-instruction-set-computer architecture for real-time symbolic pattern matching, V1468,960-971(1991)

Parsonage, Thomas B. See Riedl, Max J.: V1485(1991)

Parsons, C. See Ho, Ping-Tong: V1378,210-216(1991)

Parthenios, J. See Persephonis, Peter: V1503,185-196(1991)

Partovi, Afshin
; Glass, Alastair M.; Feldman, Robert D.: High-sensitivity photorefractivity in bulk and multiple-quantum-well semiconductors, V1561,20-32(1991)

Parulski, Kenneth A.
; Benamati, Brian L.; D'Luna, Lionel J.; Shelley, Paul R.: High-performance digital color video camera, V1448,45-58(1991)

Parvaneh, Farhad See Rogers, Alan J.: V1511,190-200(1991)

Pascual, I.
; Belendez, A.; Fimia-Gil, Antonio: Holographic optical system to copy holographic optical elements, V1507,373-378(1991)
; Belendez, A.; Fimia-Gil, Antonio: Reflection holographic optical elements in silver-halide-sensitized gelatin, V1574,72-76(1991)
See Belendez, A.: V1507,268-276(1991)
See Belendez, A.: V1574,77-83(1991)

Pascucci, Tania
See Falciai, Riccardo: V1506,120-125(1991)
See Mencaglia, Andrea: V1506,140-144(1991)

Pashev, I. See Angelov, D.: V1403,230-239(1991)

Pashinin, Vladimir P.
See Artjushenko, Vjacheslav G.: V1420,149-156(1991)
See Artjushenko, Vjacheslav G.: V1420,176-176(1991)
See Artjushenko, Vjacheslav G.: V1590,131-136(1991)
See Konov, Vitaly I.: V1427,232-242(1991)

Pasik, Michael F.
; Cangellaris, Andreas C.; Prince, John L.: Dispersion of picosecond pulses propagating on microstrip interconnections on semiconductor integrated-circuit substrates, V1389,297-301(1991)

Pasqualini, G. See Frontera, Filippo: V1549,113-119(1991)

Passaro, Vittorio M.
See Armenise, Mario N.: V1583,256-267(1991)
See Armenise, Mario N.: V1374,86-96(1991)
See Armenise, Mario N.: V1562,160-171(1991)
See Armenise, Mario N.: V1583,289-297(1991)

Passerini, S. See Scarminio, J.: V1536,70-80(1991)

Passner, Jeffrey E.
; Lee, Robert R.: Use of an expert system to predict thunderstorms and severe weather, V1468,2-10(1991)

Pastor, Rickey G.
: Understanding metal-dielectric-metal absorption interference filters using lumped circuit theory and transmission line theory, V1396,501-507(1991)

Pastukhov, A. V. See Askadskij, A. A.: V1554A,426-431(1991)

Pastukhov, S. A. See Zaitseva, N. P.: V1402,223-230(1991)

Patel, Devesh
; Stonham, T. J.: Low-level image segmentation via texture recognition, V1606,621-629(1991)

Patel, Dharmesh P. See Pape, Dennis R.: V1476,201-213(1991)

Patel, J. S. See Paek, Eung Gi: V1621,340-350(1991)

Patel, Jamshed N. See Cadzow, James A.: V1486,352-363(1991)

Patel, Kalpesh J. See Dave, Rajesh N.: V1381,320-333(1991)

Patel, Sushil
See Shaw, David T.: V1394,214-220(1991)
See Xie, Gong-Wie: V1385,132-141(1991)

Patino, Antonio R. See Paez-Leon, Cristobal J.: V1467,188-194(1991)

Paton, Andrew T.
; Scott, Chris J.: UV optical fiber distributed temperature sensor, V1367,274-281(1991)

Paton, Barry E.
: Low- cost fiber optic sensing systems using spatial division multiplexing, V1332,446-455(1991)

Patonay, Gabor
; Antoine, Miquel D.; Boyer, A. E.: Semiconductor lasers in analytical chemistry, V1435,52-63(1991)

Patorski, Krzysztof See Salbut, Leszek A.: V1554B,451-460(1991)

Patrick, Duane L.
: Laser Tracker II: Sandia National Laboratories second-generation laser tracking system, V1482,61-68(1991)

Patrick, Nicholas J.
; Sheridan, Thomas B.; Massimino, Michael J.; Marcus, Beth A.: Design and testing of a nonreactive, fingertip, tactile display for interaction with remote environments, V1387,215-222(1991)

Patrick, Roger
; Schoenborn, Philippe; Linder, Stefan; Baltes, Henry P.: Fast-injection Langmuir probe for process diagnostic and control, V1392,506-513(1991)

Patrin, V. V. See Kushlevsky, S. V.: V1403,799-800(1991)

Patriquin, Douglas R. See Spillman, William B.: V1367,197-203(1991)

Patron, Zbigniew
: Shaping of the second-harmonic pulse temporal profile, V1391,259-263(1991)

Patterson, A. See Bell, J. M.: V1536,29-36(1991)

Patterson, Angela C. See Slama, M. M.: V1464,602-609(1991)

Patterson, Dan W. See Cooke, Daniel E.: V1381,299-305(1991)

Patterson, Michael S.
See Chen, Qun: V1426,156-161(1991)
See Farrell, Thomas J.: V1426,146-155(1991)
See Madsen, Steen J.: V1431,42-51(1991)

Patterson, R. G. See Jankowski, Alan F.: V1343,32-38(1991)

Patterson, S. See Acebal, Robert: V1397,191-196(1991)

Patterson, Tim J.
: Three-dimensional morphology for target detection, V1471,358-368(1991)

Patton, Ronald See Akerman, Alexander: V1406,30-39(1991)

Patton, Stanley B. See Johnson, C. B.: V1449,2-12(1991)

Patty, Charles E.
; McMahon, David M.: Optical materials: evaluation methodology and data base utility, V1535,13-26(1991)

Pauer, Eric K. See Rorabaugh, Terry L.: V1566,312-322(1991)

Paul, Dilip K. See Greenwell, Roger A.: V1366(1991)

Paul, R. See Moissl, M.: V1397,395-398(1991)

Pauliat, G. See Mathey, P.: V1500,26-33(1991)

Paulicka, Ivan See Sasek, Ladislav: V1572,151-156(1991)

Pauls, Markus See Freisinger, Bernhard: V1397,311-318(1991)

Pauls, Thomas A. See Bevilacqua, Richard M.: V1491,231-242(1991)

Paulsen, Otto See Perch-Nielsen, Thomas: V1467,169-179(1991)

Paulson, Robert C.
See Parrish, William J.: V1512,68-77(1991)
See Parrish, William J.: V1540,274-284(1991)
Pauly, John M. See Noll, Douglas C.: V1443,29-36(1991)
Pausch, Randy
; Dwivedi, Pramod; Long, Allan C.: Practical low-cost stereo head-mounted display,V1457,198-208(1991)
Pavelchek, Edward K.
See Calabrese, Gary C.: V1466,528-537(1991)
See Freeman, Peter W.: V1464,377-385(1991)
See Jones, Susan K.: V1464,546-553(1991)
See Thackeray, James W.: V1466,39-52(1991)
Pavicic, Mark J.
; Ding, Yingjai: Multicomputer performance evaluation tool and its application to the Mayo/IBM image archival system,V1446,370-378(1991)
See Persons, Kenneth R.: V1446,60-72(1991)
Pavlicek, William
See Forbes, Glenn S.: V1446,318-322(1991)
See Morin, Richard L.: V1446,436-441(1991)
Pavlidis, Dimitris
: Millimeter-wave and optoelectronic applications of heterostructure integrated circuits,V1362,450-466(1991)
See Clarke, Roy: V1361,2-12(1991)
Pavlidis, Theo
See Wang, Ynjiun P.: V1384,145-160(1991)
See Wang, Ynjiun P.: V1384,169-175(1991)
Pavlopoulos, Theodore G.
: Measurement of triplet optical densities of organic compounds by means of CW laser excitation,V1437,168-183(1991)
Pavlov, A. P. See Kalyashora, L. N.: V1238,189-194(1991)
Pavlov, V. V. See Zhavoronok, I. V.: V1554A,371-379(1991)
Pavlovich, V. N. See Artsimovich, M. V.: V1397,729-733(1991)
Pavuna, Davor
; Dwir, Benjamin; Gauzzi, Andrea; James, Jonathan H.; Kellett, Bruce J.: Superconducting YBa2Cu307 films for novel optoelectronic device structures,V1362,102-116(1991)
See James, Jonathan H.: V1394,45-61(1991)
Pavy, Henry G.
; Smith, Stephen W.; von Ramm, Olaf T.: Improved real-time volumetric ultrasonic imaging system,V1443,54-61(1991)
Pawlak, Ryszard
; Raczynski, Tomasz: Surface layers of metals alloyed with a pulsed laser,V1391,170-173(1991)
Pawlovski, C. See Bataille, F.: V1397,839-842(1991)
Pawlowski, Georg
See Eckes, Charlotte: V1466,394-407(1991)
See Przybilla, Klaus J.: V1466,174-187(1991)
Pawluczyk, Romuald See Lelievre, Sylviane: V1559,288-297(1991)
Paxton, Jack G.
See Brown, Gair D.: V1366,351-360(1991)
See Brown, Gair D.: V1589,58-68(1991)
Payne, David N.
See Cowle, Gregory J.: V1373,54-65(1991)
See Morkel, Paul R.: V1373,224-233(1991)
Payne, Frank P.
; Mackenzie, H. S.: Novel applications of monomode fiber tapers,V1504,165-175(1991)
Payton, Paul M.
; Haines, Barry K.; Smedley, Kirk G.; Barrett, Eamon B.: Machine vision applications of image invariants: real-time processing experiments,V1406,58-71(1991)
See Barrett, Eamon B.: V1567,142-169(1991)
See Digumarthi, Ramji V.: V1472,128-138(1991)
Pazaris, James See Arjavalingam, Gnanalingam: V1389(1991)
Pe'er, Jacob See Lewis, Aaron: V1423,98-102(1991)
Peacock, Anthony J. See Rando, Nicola: V1549,340-356(1991)
Peacock, Keith
; Giannini, Judith A.; Kilgus, Charles C.; Bely, Pierre Y.; May, B. S.; Cooper, Shannon A.; Schlimm, Gerard H.; Sounder, Charles; Ormond, Karen A.; Cheek, Eric A.: Four-meter lunar engineering telescope,V1494,147-159(1991)

Pearce, John A.
; Cheong, Wai-Fung; Pandit, Kirit; McMurray, Tom J.; Thomsen, Sharon L.: Kinetic models for coagulation processes: determination of rate coefficients in vivo,V1422,27-33(1991)
See Schulze, Mark A.: V1451,48-57(1991)
See Thomsen, Sharon L.: V1422,34-42(1991)
See Thomsen, Sharon L.: V1422,14-18(1991)
Pearlman, William A. See de Garrido, Diego P.: V1605,221-232(1991)
Pearson, Guy N.
: Design and performance characteristics of a compact CO2 Doppler lidar transmitter,V1416,147-150(1991)
Pearson, James E. See Dubiel, Mary K.: V1396(1991)
Pearson, James F.
See Fraser, George W.: V1343,438-456(1991)
See Fraser, George W.: V1548,132-148(1991)
See Kaaret, Philip E.: V1548,106-117(1991)
Pearson, James J. See Barrett, Eamon B.: V1472(1991)
Pearson, John C. See Spence, Clay D.: V1469,665-670(1991)
Pearton, Stephen J.
; Katz, Avishay; Geva, M.: High-temperature degradation-free rapid thermal annealing of GaAs and InP,V1393,150-160(1991)
Pease, R. F.
; Owen, Geraint; Browning, Raymond; Hsieh, Robert L.; Lee, Julienne Y.; Maluf, Nadim I.; Berglund, C. N.: Sophisticated masks,V1496,234-238(1991)
See Blennemann, Heinrich C.: V1389,215-235(1991)
See Rao, Veena: V1466,309-323(1991)
Peatross, J. See Chuang, Yung-Ho: V1413,32-40(1991)
Peavy, George M.
; Klein, Mary K.; Newman, H. C.; Roberts, Walter G.; Berns, Michael W.: Use of chloro-aluminum sulfonated phthalocyanine as a photosensitizer in the treatment of malignant tumors in dogs and cats,V1424,171-178(1991)
Peck, Alex N.
; Breul, Harry T.: Vehicle path planning via dual-world representations,V1388,30-38(1991)
Peck, C. See Troeltzsch, John R.: V1494,9-15(1991)
Peckerar, Martin
ed.: *Electron-Beam, X-Ray, and Ion-Beam Submicrometer Lithographies for Manufacturing*,V1465(1991)
See Barouch, Eytan: V1465,254-262(1991)
Peckinpaugh, Sarah See Lybanon, Matthew: V1406,180-189(1991)
Pedersen, Frands B. See Vendeltorp-Pommer, Helle: V1373,254-265(1991)
Pedersen, Hans M. See Ellingsrud, Svein: V1399,30-41(1991)
Pedersen, Rune J. See Schiellerup, Gert: V1372,27-38(1991)
Pedrine, A. G. See Madarazo, R.: V1393,270-277(1991)
Pedrini, Christian See Boutinaud, P.: V1590,168-178(1991)
Pedrini, G. See Collings, Neil: V1505,12-19(1991)
Peet, Viktor E.
; Treshchalov, Alexei B.; Slivinskij, E. V.: Diagnostics of a compact UV-preionized XeCl laser with BCl3 halogen donor,V1412,138-148(1991)
Peetermans, Joyce See Tanaka, Toyoichi: V1403,280-287(1991)
Pegneaux, Jean-Claude See Desse, Jean-Michel: V1358,766-774(1991)
Pei, S. S. See Levine, Barry F.: V1540,232-237(1991)
Peiffer, John See Kelley, Robert B.: V1387,38-46(1991)
Pekala, Rick See Cotts, Patricia M.: V1430,181-190(1991)
Pekruhn, Wolfgang See Drees, Friedrich-Wilhelm: V1458,80-83(1991)
Pelakauskas, A.
See Chirvony, V. S.: V1403,504-506(1991)
See Chirvony, V. S.: V1403,638-640(1991)
Pelegrin, A. See Wagnieres, G.: V1525,219-236(1991)
Pelekhaty, V. M. See Baranov, D. V.: V1583,389-394(1991)
Peli, Eli
: Display nonlinearity in digital image processing for visual communications,V1606,508-519(1991)
: New perspectives on image enhancement for the visually impaired,V1382,49-59(1991)
Pelizzari, Charles A.
See Neiw, Han-Min: V1445,259-264(1991)
See Yu, Xiaolin: V1396,56-58(1991)
Pelka, David G. See Parkyn, William A.: V1528,70-81(1991)

Pelka, Joachim
: Simulation of ion-enhanced dry-etch processes,V1392,55-66(1991)
See Pilz, Wolfgang: V1392,84-94(1991)

Pelle, Gabriel See Dong, LiXin: V1445,178-187(1991)

Pellegrini, Claudio See Dodd, James W.: V1407,467-473(1991)

Pellegrino, Anthony
See Kiss, Gabor D.: V1366,223-234(1991)
See Wei, Ta-Sheng: V1366,235-240(1991)

Pellegrino, John M. See Smith, Dale J.: V1562,242-250(1991)

Pellerin, Sharon L.
: Network powering architecture for fiber-to-the-subscriber systems,V1363,186-190(1991)

Pelletier, Emile P.
See Lallier, Eric: V1506,71-79(1991)
See Liu, Xu: V1554A,558-569(1991)

Pelleymounter, Douglas R. See Grieser, James L.: V1330,111-118(1991)

Pelli, Stefano
See Guglielmi, Massimo: V1513,44-49(1991)
See Righini, Giancarlo C.: V1513,418-424(1991)
See Righini, Giancarlo C.: V1513,105-111(1991)

Pellicori, Samuel F.
: Optical bonding agents for IR and UV refracting elements,V1535,48-54(1991)

Pelling, Michael R.
; Rothschild, Richard E.; MacDonald, Daniel R.; Hertel, Robert H.; Nishiie, Edward S.: Design and performance of an automatic gain control system for the high-energy x-ray timing experiment,V1549,134-146(1991)
See Hink, Paul L.: V1549,193-202(1991)
See Rothschild, Richard E.: V1549,120-133(1991)

Pelo, John See Chronister, Eric L.: V1559,56-64(1991)

Pelous, G.
; Guern, Yves; Gobleid, D.; David, J.; Chion, A.; Tonneau, Didier: IC rewiring by laser microchemistry,V1598,149-158(1991)

Pen, Shengyang See Hu, Yu: V1409,230-239(1991)

Pen, You G.
See Fan, Xiang J.: V1519,808-812(1991)
See Fan, Xiang J.: V1519,805-807(1991)

Pen, Zhi L.
See Fan, Xiang J.: V1519,808-812(1991)
See Fan, Xiang J.: V1519,805-807(1991)

Penasa, Mauro See Del Bello, Umberto: V1502,104-116(1991)

Pendergast, K. D. See Danly, Bruce G.: V1407,192-201(1991)

Pendleton, Thomas W. See deFigueiredo, Rui J.: V1387,330-336(1991)

Peng, Chuan C. See Huang, Guang L.: V1519,179-182(1991)

Peng, F. See Ho, Ping-Tong: V1378,210-216(1991)

Peng, Hui See Peterson, Heidi A.: V1453,210-222(1991)

Peng, Huihong See Qu, Zhihao: V1554A,503-510(1991)

Peng, Ming-Chien See Wang, David T.: V1386,206-219(1991)

Peng, S. See Lachs, Gerard: V1474,248-259(1991)

Peng, Song-Tsuen See Gupta, Mool C.: V1499,303-306(1991)

Peng, Xiang See Liu, Ying: V1554A,610-612(1991)

Penin, A. N.
; Klyshko, D. N.: Absolute measurements of radiation sources spectral brightness and detectors quantum efficiency,V1562,143-148(1991)

Penn, Benjamin G. See Frazier, Donald O.: V1557,86-97(1991)

Pennebaker, William B.
See Mitchell, Joan L.: VCR37,68-97(1991)
See Peterson, Heidi A.: V1453,210-222(1991)

Penner, Thomas L.
; Armstrong, Nancy J.; Willand, Craig S.; Schildkraut, Jay S.; Robello, Douglas R.: Langmuir-Blodgett films for second-order nonlinear optics,V1560,377-386(1991)
; Willand, Craig S.; Robello, Douglas R.; Schildkraut, Jay S.; Ulman, Abraham: Linear and nonlinear optical properties of polymeric Langmuir-Blodgett films,V1436,169-178(1991)

Penney, Carl M.
See Milne, Peter J.: V1423,122-129(1991)
See Ren, Qiushi: V1423,129-139(1991)

Pennie, W. See Mayevsky, Avraham: V1431,303-313(1991)

Pennig, Dietmar See Kasprzak, Henryk: V1429,55-61(1991)

Pennycook, Stephen J. See Lowndes, Douglas H.: V1394,150-160(1991)

Pensabene, Joseph See Eton, Darwin: V1425,182-187(1991)

Pentland, Alexander P.
: Spatial and temporal surface interpolation using wavelet bases,V1570,43-62(1991)
; Horowitz, Bradley: Practical approach to fractal-based image compression,V1605,467-474(1991)
See Turk, Matthew A.: V1381,43-54(1991)

Pepe, Kevin See Lybanon, Matthew: V1469,637-647(1991)

Pepeljugoski, Petar K.
; Lau, Kam-Yin: Modal noise reduction in analog fiber optic links by superposition of high-frequency modulation,V1371,233-243(1991)

Pepi, John W.
; Golini, Donald: Delayed elasticity in Zerodur at room temperature,V1533,212-221(1991)

Pepin, Henri See Chaker, Mohamed: V1465,16-25(1991)

Perbellini, Antonio See Barbieri, Enrico: V1425,122-127(1991)

Perch-Nielsen, Thomas
; Paulsen, Otto; Drivsholm, Christian: Surveying and damping heat loss from machines with high surface temperatures: thermography as a tool,V1467,169-179(1991)

Percival, Jeff See Richards, Evan: V1494,40-48(1991)

Perea, Ernesto H.
: Current technologies for very high performance VLSI ICs,V1362,477-483(1991)

Pereira, Carl See Rando, Nicola: V1549,340-356(1991)

Pereira, Nino R.
See Huttlin, George A.: V1407,147-158(1991)
See Lazard, Carl J.: V1407,167-171(1991)
See Litz, Marc S.: V1407,159-166(1991)

Perera, Thiloma See Kasahara, Jack S.: V1463,492-503(1991)

Peres, Rosa C.
; De Paoli, Marco-Aurelio; Panero, Stefania; Scrosati, Bruno: Electrochromic properties of poly(pyrrole)/dodecylbenzenesulfonate,V1559,151-158(1991)

Perev, K. See Larsson, T.: V1387,165-168(1991)

Perez, Arnulfo
; Kamata, Seiichiro; Kawaguchi, Eiji: Arithmetic coding model for compression of LANDSAT images,V1605,879-884(1991)
; Kamata, Seiichiro; Kawaguchi, Eiji: Hilbert scanning arithmetic coding for multispectral image compression,V1567,354-361(1991)
; Kamata, Seiichiro; Kawaguchi, Eiji: N-dimensional Hilbert scanning and its application to data compression,V1452,430-441(1991)

Perez, Eric
See Bailly, Michel: V1417,142-157(1991)
See Cossec, Francois R.: V1417,262-276(1991)

Perez, Joseph D. See Fisher, Thornton R.: V1479,212-225(1991)

Perez, Osvaldo
; Karim, Mohammad A.: Optical enhancements of joint-Fourier-transform correlator by image subtraction,V1471,255-264(1991)

Perez-Amor, Mariano See Blanco, J.: V1508,180-190(1991)

Pergamenshchik, V. M. See Bodnar, Vladimir G.: V1455,61-72(1991)

Perger, Andreas
; Metz, Jurgen; Tiedeke, J.; Rille, Eduard P.: Eyesafe diode laser rangefinder technology,V1419,75-83(1991)

Perilloux, Bruce E.
: Helium neon laser optics: scattered light measurements and process control,V1530,255-262(1991)

Perina, Jan
ed.: *Photon Statistics and Coherence in Nonlinear Optics,*VMS39(1991)
: Stimulated Raman scattering with initially nonclassical light,V1402,192-192(1991)

Perkins, Frederick W.
: Low-cost space flight for attached payloads,V1495,157-163(1991)

Perkins, Walton A.
: Representing sentence information,V1468,854-865(1991)

Perlado, J. M. See Velarde, G.: V1502,242-257(1991)

Perlinski, Leszek See Badziak, Jan: V1397,81-84(1991)

Perlmutter, S. H.
; Doroski, D.; Landreth, Bruce; Gabor, A. M.; Barbier, Pierre R.; Moddel, Garrett: Tradeoffs in the design and operation of optically addressed spatial light modulators,V1562,74-84(1991)

Perlovsky, Leonid I.
: Neural networks and model-based approaches to object identification,V1606,1080-1085(1991)

Pernick, Benjamin J. See Leib, Kenneth G.: V1532,261-270(1991)

Perola, G. C. See Kaaret, Philip E.: V1548,106-117(1991)

Perov, Polievkt I. See Gulyaev, Yuri V.: V1621,84-92(1991)

Perov, S. N. See Skopinov, S. A.: V1403,676-679(1991)

Perrier, Gerard
: Close-spaced vapor transport of II-VI semiconductors,V1536,258-267(1991)

Perrone, A. See D'Anna, Emilia: V1503,256-268(1991)

Perrone, Antonio See Basti, Gianfranco: V1469,719-736(1991)

Perrott, Colin M.
: Color vision enhancement with spectacles,V1529,31-36(1991)
ed.: Opthalmic Lens Design and Fabrication,V1529(1991)

Perry, Ian R. See Tropper, Anne C.: V1373,152-157(1991)

Perry, John R. See Beard, David V.: V1446,52-58(1991)

Perry, Joseph W.
; Marder, Seth R.; Perry, Kelly J.; Sleva, E. T.; Yakymyshyn, Christopher P.; Stewart, Kevin R.; Boden, Eugene P.: Organic salts with large electro-optic coefficients,V1560,302-309(1991)

Perry, Kelly J. See Perry, Joseph W.: V1560,302-309(1991)

Perry, Kenneth E.
; Epstein, Jonathan S.; May, G. B.; Shull, J. E.: Correspondence in damage phenomena and R-curve behavior in ceramics and geomaterials using moire interferometry,V1554A,209-227(1991)

Perry, Meg W.
; Rumbaugh, Richard C.; Frost, David P.: Computer animation method for simulating polymer flow for injection-molded parts,V1459,155-156(1991)

Perry Beng, S. J. See Chan, John P.: V1386,171-179(1991)

Persch, V.
See Allinger, Thomas: V1361,935-942(1991)
See Schaefer, Juergen A.: V1361,1026-1032(1991)

Persephonis, Peter
; Giannetas, B.; Parthenios, J.; Georgiades, C.: Electrical discharges investigation in gas-pulsed laser,V1503,185-196(1991)

Persiano, G. See Buyukkoc, Cagatay: V1452,170-179(1991)

Persoff, Jeffrey J. See Goelz, Stefan: V1542,502-510(1991)

Person, Richard See Lightner, David A.: V1432,2-13(1991)

Persons, Kenneth R.
; Gehring, Dale G.; Pavicic, Mark J.; Ding, Yingjai: Performance characteristics of the Mayo/IBM PACS,V1446,60-72(1991)
See Gehring, Dale G.: V1446,248-252(1991)

Persoons, Andre P.
; Van Wonterghem, Bruno M.; Tackx, Peter: Measurements of second hyperpolarisabilities of diphenylpolyenes by means of phase-conjugate interferometry,V1409,220-229(1991)
See Verbiest, Thierry: V1560,353-361(1991)

Pertuz-Campo, Jairo See Rodriguez, Domingo: V1565,492-503(1991)

Perun, T. O. See Angelsky, Oleg V.: V1402,231-235(1991)

Perveev, Anatoly F. See Gan, Michael A.: V1574,243-249(1991)

Pervez, Anjum
: Multibit optical sensor networking,V1511,220-231(1991)

Pesavento, Gerry A.
: Optomechanical M x N fiber-optic matrix switch,V1474,57-61(1991)

Pesce, Anastasia See Armenise, Mario N.: V1583,210-220(1991)

Pesek, Andreas See Hingerl, Kurt: V1361,383-393(1991)

Peskin, Richard L.
See Gertner, Izidor: V1565,414-422(1991)
See Walther, Sandra S.: V1459,232-243(1991)

Pesl, Ales A.
: Electro-optic illuminating module,V1454,299-305(1991)
: New design of the illuminating system for transmission film copy,V1448,218-224(1991)

Pesme, Denis See Labaune, C.: V1413,138-143(1991)

Pessa, Markus
; Hakkarainen, T.; Keskinen, Jari; Rakennus, K.; Salokatve, A.; Zhang, G.; Asonen, Harry M.: Current state of gas-source molecular beam epitaxy for growth of optoelectronic materials,V1361,529-542(1991)
See Hingerl, Kurt: V1361,943-953(1991)

Pessino, P. See Brenci, Massimo: V1572,318-324(1991)

Pestak, Christopher J. See Hartz, William G.: V1398,52-60(1991)

Peter, Reto See Kaempfer, Niklaus A.: V1491,314-322(1991)

Peterkin, Frank E. See Williams, Paul F.: V1378,217-225(1991)

Petermann, Klaus See Splett, Armin O.: V1362,827-833(1991)

Peters, Bruce R.
; Kalin, David A.: Interferometric measurements of a high-velocity mixing/shear layer,V1486,410-416(1991)

Peters, Charles R. See Jones, Susan K.: V1464,546-553(1991)

Peters, Dethard
; Mueller, Joerg: Integrated optical devices with silicon oxynitride prepared by plasma-enhanced chemical vapor deposition (PECVD) on Si and GaAs substrates,V1362,338-349(1991)

Peters, Peter J.
See Bastiaens, H. M.: V1397,77-80(1991)
See Botma, H.: V1397,573-576(1991)

Peters, Rainer See Schaetzel, Klaus: V1430,109-115(1991)

Peters, Richard A. See Cadzow, James A.: V1486,352-363(1991)

Peters, Terence M. See Henri, Christopher J.: V1444,306-317(1991)

Peters, Werner See Buchelt, Martin: V1420,249-253(1991)

Petersen, James V.
; Dessy, Raymond E.: Direct exchange of metal ions onto silica waveguides,V1368,61-72(1991)

Petersen, Paul M.
; Buchhave, Preben: Amplification of amplitude modulated signals in a self-pumped photorefractive phase conjugator,V1362,582-585(1991)

Peterson, Blake See Lightner, David A.: V1432,2-13(1991)

Peterson, Carsten See Yuille, Alan L.: V1570,166-174(1991)

Peterson, Cynthia See Hoover, Richard B.: V1343,189-202(1991)

Peterson, Dean B. See Pickett, Herbert M.: V1491,308-313(1991)

Peterson, Ed See Flynn, Michael J.: V1444,172-179(1991)

Peterson, Gary L.
See Botez, Dan: V1474,64-74(1991)
See Mawst, Luke J.: V1418,353-357(1991)

Peterson, Heidi A.
; Peng, Hui; Morgan, J. H.; Pennebaker, William B.: Quantization of color image components in the DCT domain,V1453,210-222(1991)

Peterson, James R. See Ivey, Jim: V1567,170-178(1991)

Peterson, Kristen A.
See Dyer, R. B.: V1432,197-204(1991)
See Woodruff, William H.: V1432,205-210(1991)

Peterson, William A.
See Eaton, Frank D.: V1487,84-90(1991)
See Waldie, Arthur H.: V1487,103-108(1991)

Peticolas, Warner L.
; Dai, Z.; Thomas, G. A.: Sequence dependence of the length of the B to Z junctions in DNA,V1403,22-26(1991)

Petit, Roger
; Bouchitte, G.; Tayeb, Gerard; Zolla, F.: Diffraction by one-dimensional or two-dimensional periodic arrays of conducting plates,V1545,31-41(1991)
See Tayeb, Gerard: V1545,95-105(1991)

Petousis, Nikolaos H. See Morgan, Alan R.: V1426,350-355(1991)

Petras, Robert See Rava, Richard P.: V1426,68-78(1991)

Petrauskas, Mendogas
See Rueckmann, I.: V1513,78-85(1991)
See Woggon, Ulrike: V1362,888-898(1991)

Petrick, Stanley W. See Bhandari, Pradeep: V1540,97-108(1991)

Petrick, Walt See Mason, Peter V.: V1540,88-96(1991)

Petrilak, Robert See Kozaitis, Samuel P.: V1564,403-413(1991)

Petro, David See Henson, Michael A.: V1396,582-589(1991)

Petrokovich, O. A.
See Korzhenevich, Irina M.: V1358,1084-1089(1991)
See Silkis, E. G.: V1358,46-49(1991)

Petros, Mulugeta See Chang, John H.: V1492,43-46(1991)

Petrosky, Anne See Conley, Willard E.: V1466,53-66(1991)

Petrov, A. P.
: Color and Grassmann-Cayley coordinates of shape,V1453,342-352(1991)

Petrov, Dmitry V.
; Belostotsky, A. L.; Dolgopolov, V. G.; Leonov, A. S.; Fedjukhin, L. A.: Acousto-optics in integrated-optic devices for optical recording,V1374,152-159(1991)

Petrov, E. P. See Gakamsky, D. M.: V1403,641-643(1991)

Petrov, M. P. See Vlad, Ionel V.: V1332,236-244(1991)

Petrov, Mikhail P.
; Miridonov, S. V.: Discrete analog and digital devices using fiber-optic logic elements,V1621,402-413(1991)

Petrov, Mikhail Y. See Vysogorets, Mikhail V.: V1358,1066-1069(1991)

Petrov, S. I. See Berkovsky, A. G.: V1358,750-755(1991)

Petrov, V. V.
; Zymenko, V. I.; Kravetz, V. G.; Polishchuk, E. Y.; Sushko, A. M.: Study of optical data recording based on photoluminescence effect,V1621,45-50(1991)

Petrov, Yu. N. See Bordo, V. G.: V1440,364-369(1991)

Petrova, G. P.
; Petrusevich, Yu. M.; Borisov, B. A.: Anisotropic polarizability and diffusion of protein in water solutions studied by laser light scattering,V1403,387-389(1991)

Petrova, Maria V. See Tkachuk, A. M.: V1403,801-804(1991)

Petrovich, Zbigniew
; Astrahan, Melvin; Baert, Luc: Transurethral microwave heating without urethral cooling: theory and experimental results,V1421,14-17(1991)
See Baert, Luc: V1421,18-29(1991)

Petrovskii, Gurii T.
: Optical materials for infrared range of spectrum,V1540,401-411(1991)
See Aleksandrov, I. V.: V1513,309-312(1991)
See Glebov, Leonid B.: V1440,24-35(1991)
See Glebov, Leonid B.: V1621,21-32(1991)
See Glebov, Leonid B.: V1399,200-206(1991)
See Glebov, Leonid B.: V1513,224-231(1991)
See Glebov, Leonid B.: V1513,56-70(1991)

Petrovsky, Victor N. See Kurochkin, Vadim Y.: V1435,322-330(1991)

Petrusevich, Yu. M. See Petrova, G. P.: V1403,387-389(1991)

Petruziello, David
: Passive optic solution for an urban rehabilitation topology,V1363,30-37(1991)

Petruzzelli, Vincenzo
See d'Alessandro, Antonio: V1378,127-138(1991)
See d'Alessandro, Antonio: V1332,554-562(1991)
See d'Alessandro, Antonio: V1366,313-323(1991)
See d'Alessandro, Antonio: V1399,184-191(1991)

Petsch, Thomas
: Electro-optic modulator for high-speed Nd:YAG laser communication,V1522,72-82(1991)

Petschik, Benno
: Color hard copy: a self-tuning color correction algorithm based on a colorimetric model,V1458,108-114(1991)

Pettersson, Hans
; Oberg, Per A.: Pulp blood flow assessment in human teeth by laser Doppler flowmetry,V1424,116-119(1991)

Pettiette-Hall, C. L. See McIver, Robert T.: V1437,124-128(1991)

Pettiford, Alvin A. See Alukaidey, Talib A.: V1390,513-522(1991)

Pettit, Christopher J.
: Line-of-sight stabilization: sensor blending,V1489,278-287(1991)

Petushkov, V. N. See Benimetskaya, L. Z.: V1525,242-245(1991)

Petzold, Theodore J. See Austin, Roswell W.: V1537,57-73(1991)

Petzoldt, Stefan
See Hunger, Hans E.: V1441,283-286(1991)
See Hunger, Hans E.: V1598,19-26(1991)

Pevny, S. N. See Dolgopolov, Y. V.: V1412,267-275(1991)

Pevzner, Eliyahu See Ehrenberg, Benjamin: V1432,154-163(1991)

Peyrin, Francoise
; Goutte, Robert; Amiel, Michel: Three-dimensional reconstruction from cone beam projection by a block iterative technique,V1443,268-279(1991)
See Baskurt, Atilla: V1445,485-495(1991)

Peyrot, Patrick See Buvat, Daniel: V1417,251-261(1991)

Peyton, David
; Kinoshita, Hiroyuki; Lo, G. Q.; Kwong, Dim-Lee: Systems-oriented survey of noncontact temperature measurement techniques for rapid thermal processing,V1393,295-308(1991)

Peyton, James A. See Bieman, Leonard H.: VCR36,3-19(1991)

Pezeshki, Bardia
; Thomas, Dominique; Harris, James S.: Optimization of reflection electro-absorption modulators,V1362,559-565(1991)

Pfau, Anton K.
; Oldham, William G.; Neureuther, Andrew R.: Exploration of fabrication techniques for phase-shifting masks,V1463,124-134(1991)

Pfeiffer, A. T. See Krongelb, Sol: V1389,249-256(1991)

Pfeiffer, Douglas E. See Chakraborty, Dev P.: V1443,183-190(1991)

Pfeiffer, M. See Lau, A.: V1403,212-220(1991)

Pfeiffer, Wolfgang
; Stolz, Dieter; Zipfl, P.: Subnanosecond high-speed framing of prebreakdown phenomena,V1358,1191-1201(1991)

Pfleiderer, Christoph See Hitzler, Hermine: V1503,355-362(1991)

Pflibsen, Kent P. See Gleckler, Anthony D.: V1494,454-471(1991)

Pflueger, Silke See Franek, Joachim: V1397,791-795(1991)

Pflug, Leopold See Rastogi, Pramod K.: V1554B,48-55(1991)

Pflugrath, J. W. See Strauss, Michael G.: V1447,12-27(1991)

Pham, Vui V.
See Simonne, John J.: V1374,107-115(1991)
See Simonne, John J.: V1589,139-147(1991)

Phan, Khoi
; Nistler, John L.; Singh, Bhanwar: Metrology issues associated with submicron linewidths,V1464,424-437(1991)

Phat, Darith
; Vuong, Phat N.; Plaza, Pascal; Cheilan, Francis; Dao, Nguyen Q.: Remote Raman spectroscopic imaging of human artery wall,V1525,196-205(1991)
See Hong, Nguyen T.: V1525,132-142(1991)

Phatak, Anil See Sridhar, Banavar: V1478,178-189(1991)

Phelan, Paul J.
; Hegarty, John; Elsasser, Wolfgang E.: Novel device for short-pulse generation using optoelectronic feedback,V1362,623-630(1991)

Philbrick, Charles R.
: Lidar profiles of atmospheric structure properties,V1492,76-84(1991)

Philbrick, Daniel A. See Watt, David W.: V1467,357-368(1991)

Philip, Gary P. See Herd, James T.: V1387,194-201(1991)

Philipp, Harald
See Neger, Theo: V1507,476-487(1991)
See Vukicevic, Dalibor: VIS08,160-193(1991)

Philippov, S. A. See Borodin, A. M.: V1358,756-758(1991)

Phillies, George D.
: Probe diffusion in polymer solutions,V1430,118-131(1991)

Phillips, David See MacRobert, Alexander J.: V1439,79-87(1991)

Phillips, Julia M.
; Siegal, Michael P.: Optimizing the structural and electrical properties of Ba2YCu3O7-delta,V1394,186-190(1991)
See Fork, David K.: V1394,202-202(1991)

Phillips, K. J. See Singh, Surendra: V1376,143-152(1991)

Phillips, Mark L.
; Harrison, William T.; Stucky, Galen D.: Nonlinear optical properties of new KTiOPO4 isostructures,V1561,84-92(1991)

Phillips, Nicholas J.
: Bridging the gap between Soviet and Western holography,VIS08,206-214(1991)
; Barnett, Christopher A.: Micro-optic studies using photopolymers,V1544,10-21(1991)
See Bjelkhagen, Hans I.: V1461,321-328(1991)

Phillips, Norman J. See Leksell, David: V1458,133-144(1991)

Phillips, Richard T. See Gregory, Andrew: V1362,268-274(1991)

Phillips, Richard W. See Jensen, Stephen C.: V1369,87-95(1991)

Phillips, Ronald L. See Murphy, Robert A.: V1558,295-305(1991)

Phillips, Thomas E.
; Trias, John A.; Lasher, Mark E.; Poirier, Peter M.; Dahlke, Weldon J.; Robinson, Waldo R.: Laser-based display technology development at the Naval Ocean Systems Center,V1454,290-298(1991)

Phillips, Walter C. See Kalata, Kenneth: V1345,270-280(1991)

Phillips, Wayne S. See Coles, Christopher L.: V1488,327-333(1991)

Phillips, William E. See Flannery, David L.: V1564,65-77(1991)

Phipps, James A. See Grieser, James L.: V1330,111-118(1991)

Phuvan, Sonlinh See Pape, Dennis R.: V1476,201-213(1991)

Piao, Yue-zhi
See Zhang, Yue-qing: V1418,444-447(1991)
See Zhang, Yue-qing: V1400,137-143(1991)

Piatkowski, Tadeusz See Schmit, Joanna: V1391,313-317(1991)

Picard, A. See Dilhac, Jean-Marie R.: V1393,395-403(1991)

Picard, J. See Maldague, Xavier: V1467,239-251(1991)

Picard, Rosalind W. See Elfadel, Ibrahim M.: V1381,524-535(1991)

Piccolo, R. See Valentini, Antonio: V1400,164-170(1991)

Pichugin, Sergei See Igoshin, Valery I.: V1501,150-152(1991)

Pickering, Michael A. See Goela, Jitendra S.: V1330,25-38(1991)

Pickett, Herbert M.
; Peterson, Dean B.: Far-IR Fabry-Perot spectrometer for OH measurements,V1491,308-313(1991)

Pickett, Ronald M. See Smith, Stuart: V1459,192-206(1991)

Pickholtz, Raymond L.
See Zhang, Ya-Qin: V1445,358-366(1991)
See Zhang, Ya-Qin: V1445,367-373(1991)

Pickles, A. J. See Cavedoni, Charles P.: V1542,273-282(1991)

Picoli, Gilbert See Abdelghani-Idrissi, Ahmed M.: V1362,417-427(1991)

Picot, Paul A.
; Rickey, Daniel W.; Mitchell, J. R.; Rankin, Richard N.; Fenster, Aaron: Three-dimensional color Doppler imaging of the carotid artery,V1444,206-213(1991)

Pidgeon, Carl R.
: Applications of the FEL to NLO spectroscopy of semiconductors,V1501,178-182(1991)
See Kremser, Christian: V1501,69-79(1991)

Pien, P. See Grebel, Haim: V1583,331-337(1991)

Pieper, Ronald J.
; Poon, Ting-Chung: Frequency-dependent optical beam distortion generated by acousto-optic Bragg cells,V1454,324-335(1991)

Pieper, Steve
; Delp, Scott; Rosen, Joseph; Fisher, Scott S.: Virtual environment system for simulation of leg surgery,V1457,188-197(1991)

Pierce, Bill
: Application results for an augmented video tracker,V1482,182-195(1991)

Pierce, Brian M.
: Theoretical analysis of the third-order nonlinear optical properties of linear cyanines and polyenes,V1560,148-161(1991)

Pieri, David C.
: Role of orbital observations in detecting and monitoring geological hazards: prospects for the future,V1492,410-417(1991)

Pieri, Silvano See Livi, S.: V1500,144-150(1991)

Pierrat, Christophe
; Paniez, Patrick J.; Martin, P.: In-situ monitoring of silylation mechanisms by laser interferometry,V1466,248-256(1991)
See Lajzerowicz, Jean: V1392,222-231(1991)
See Vinet, Francoise: V1466,558-569(1991)

Pierre, Guillaume
; Jarret, Bertrand; Brun, Eric: Behavior of WDM system for intensity modulation,V1511,201-211(1991)

Pierre, M. See Hochstrasser, Robin M.: V1403,221-229(1991)

Piestrup, Melvin A.
; Boyers, D. G.; Pincus, Cary I.; Li, Qiang; Harris, J. L.; Bergstrom, J. C.; Caplan, H. S.; Silzer, R. M.; Skopik, D. M.; Moran, M. J.; Maruyama, Xavier K.: High-power x-ray generation using transition radiation,V1552,214-239(1991)
See Rule, Donald W.: V1552,240-251(1991)

Pieszynski, Krzysztof See Drobnik, Antoni: V1391,378-381(1991)

Pietermann, Herman See Gussenhoven, Elma J.: V1425,203-206(1991)

Pietikainen, Matti See Koivunen, Visa: V1381,501-512(1991)

Pietrafitta, Joseph J. See Watson, Graham M.: V1421(1991)

Pietsch, H.
See Hunger, Hans E.: V1441,283-286(1991)
See Hunger, Hans E.: V1598,19-26(1991)

Pigache, Daniel R.
See Bacis, Roger: V1397,173-176(1991)
See Barnault, B.: V1397,231-234(1991)
See Barnault, B.: V1397,257-260(1991)
See Godard, Bruno: V1503,71-77(1991)
See Godard, Bruno: V1397,59-62(1991)

Piglia, Roberto See Calvani, Riccardo A.: V1504,258-263(1991)

Pike, G. B. See Henri, Christopher J.: V1444,306-317(1991)

Pike, T. K. See de Salabert, Arturo: V1521,74-88(1991)

Pikhtin, N. A. See Garbuzov, Dmitriy Z.: V1418,386-393(1991)

Piknova, Barbora See Hianik, T.: V1402,93-96(1991)

Pikuleva, Irene A. See Turko, Illarion V.: V1510,53-56(1991)

Pileggi, R. See Scarminio, J.: V1536,70-80(1991)

Pilgrim, C. See Hennig, Thomas: V1424,99-105(1991)

Pilinskaya, Maria A. See Jensen, Ronald H.: V1403,372-380(1991)

Pillai, A. M. See Mishra, R. K.: V1482,138-145(1991)

Pillai, P.K. C. See Vishnoi, Gargi: V1572,94-100(1991)

Pilloff, Mark D. See Lau, Yue Y.: V1407,635-646(1991); 1552,182-184(1991)

Pillsbury, Allen D.
; Taylor, John A.: Qualification testing of a diode-laser transmitter for free-space coherent communications,V1417,292-299(1991)

Pilon, Michael J. See Carbone, Joseph: V1447,229-242(1991)

Pilz, Wolfgang
; Janes, Joachim; Muller, Karl P.; Pelka, Joachim: Oxygen reactive ion etching of polymers: profile evolution and process mechanisms,V1392,84-94(1991)

Pincosy, Phillip A. See Poulsen, Peter: V1407,172-182(1991)

Pincus, Cary I. See Piestrup, Melvin A.: V1552,214-239(1991)

Pindera, Jerzy T.
: Scattered-light optical isodynes: basis for 3-D isodyne stress analysis,V1554A,458-471(1991)
; Wen, Baicheng: Optical isodyne measurements in fracture mechanics,V1554A,196-205(1991)
See Hecker, Friedrich W.: V1554A,151-162(1991)

Pine, Nicholson L. See Wesson, Laurence N.: V1367,204-213(1991)

Pines, Sheldon See Dickinson, Robert R.: V1459,178-189(1991)

Pinfold, Wilfred
: Use of heterogeneous distributed memory parallel systems in image processing,V1406,132-137(1991)

Pini, Roberto See Viligiardi, Riccardo: V1425,72-74(1991)

Pinkevich, Igor P.
; Reshetnyak, Victor Y.; Reznikov, Yuriy: Influence of phototransformed molecules on optical properties of finite cholesteric liquid-crystal cell,V1455,122-133(1991)

Pinkney, H. F. See MacLean, Steve G.: V1526,113-122(1991)

Pino, F. See Blanco, J.: V1508,180-190(1991)

Pinsky, Howard J.
: Determination of FLIR LOS stabilization errors,V1488,334-342(1991)

Pinter, Jeff H. See Bredthauer, Richard A.: V1447,310-315(1991)

Pinter, Robert B. See Bouzerdoum, Abdesselam: V1473,29-40(1991)

Pinto, Joao L. See Ferreira, Mario F.: V1372,14-26(1991)

Pinto, Joseph F.
; Esterowitz, Leon; Bonner, Robert F.: Reduction of acoustic transients in tissue with a 2 um thulium laser,V1420,242-243(1991)
; Esterowitz, Leon: Suppression of relaxation oscillations in flash-pumped 2-um lasers,V1410,175-178(1991)
See Quarles, Gregory J.: V1410,165-174(1991)

Pinto, M. I. See Soares, Oliverio D.: V1332,166-184(1991)

Pinto, Mark R. See Lifshitz, N.: V1596,96-105(1991)

Pinz, Ingo See Budzinski, Christel: V1500,264-274(1991)

Piotrowski, Jan
: New configuration of a generator and regenerative amplifier built on three mirrors,V1391,272-278(1991)
See Badziak, Jan: V1391,117-126(1991)
See Brzezinski, Ryszard: V1391,127-134(1991)

Piotrowski, Jozef
; Djuric, Zoran G.: Magnetoconcentration nonequilibrium IR photodetectors,V1512,84-90(1991)
See Djuric, Zoran G.: V1540,622-632(1991)

Pippan, Manfred See Oswald, Josef: V1362,534-543(1991)

Pique, A. See Wu, Xin D.: V1477,8-14(1991)

Pirch, Norbert See Kreutz, Ernst W.: V1502,160-176(1991)

Pires, Joao J.
; Rocha, Jose R.: Carrier recovery and filtering in optical BPSK systems using external cavity semiconductor lasers,V1372,118-127(1991)
; Rocha, Jose R.: Performance analysis of direct-detection optical DPSK systems using a dual-detector optical receiver,V1579,144-154(1991)

Pirlo, Giuseppe See Impedovo, Sebastiano: V1384,280-284(1991)

Piro, Luigi
See Costa, Enrico: V1343,469-476(1991)
See Kaaret, Philip E.: V1548,106-117(1991)

Pirri, C. F. See Demichelis, Francesca: V1534,140-147(1991)

Pirsch, Peter
; Jeschke, Hartwig: MIMD (multiple instruction multiple data) multiprocessor system for real-time image processing,V1452,544-555(1991)

Pis'menny, Vladislav D.
See Antyuhov, V.: V1397,355-365(1991); 1415,48-59(1991)
See Bondarenko, Alexander V.: V1376,117-127(1991)

Pisano, E. See Beard, David V.: V1446,289-296(1991)

Pisarevsky, A. N. See Khvilivitzky, A. T.: V1447,184-190(1991)

Pischelin, E. V.
See Bryukhnevitch, G. I.: V1449,109-115(1991)
See Ivanov, K. N.: V1358,732-738(1991)

Piskarskas, Algis S.
See Danelius, R.: V1402,198-208(1991)
See Kirveliene, V.: V1403,582-584(1991)

Pistoia, Francesco See Guadagni, Stefano: V1420,178-182(1991)

Pistoia, Maria A. See Guadagni, Stefano: V1420,178-182(1991)

Pistoni, Natale C. See De Maria, Letizia: V1366,304-312(1991)

Pitalo, Stephen K.
See Brown, Daniel M.: V1527,73-84(1991)
See Kathman, Alan D.: V1354,297-309(1991)

Pitard, F. See Tournie, Eric: V1361,641-656(1991)

Pitas, Alan A. See Gluskin, Efim S.: V1548,56-68(1991)

Pitaval, M. See Bremond, Georges E.: V1361,732-743(1991)

Pitts, John H.
; Kyrazis, Demos T.; Seppala, Lynn G.; Bumpas, Stanley E.: Use of dome (meniscus) lenses to eliminate birefringence and tensile stresses in spatial filters for the Nova laser,V1441,71-81(1991)

Piwczyk, Bernhard P.
ed.: Excimer Laser Materials Processing and Beam Delivery Systems,V1377(1991)

Pizarro, P. J. See Bowers, C. R.: V1435,36-50(1991)

Pizer, Stephen M. See Beard, David V.: V1446,289-296(1991)

Pizey, Gordon See Daher, Reinhard: V1526,90-93(1991)

Pizzi, Nicolino J.
; El-Hakim, Sabry F.: Object-oriented strategies for a vision dimensional metrology system,V1468,296-304(1991)

Pizzoferrato, R.
; De Spirito, M.; Zammit, Ugo; Marinelli, M.; Rogovin, Dan N.; Scholl, James F.: Nondegenerate two-wave mixing in shaped microparticle suspensions,V1409,192-201(1991)

Placzek-Popko, E. See Misiewicz, Jan: V1561,6-18(1991)

Plamondon, Joseph A. See Mason, Peter V.: V1540,88-96(1991)

Planche, G. See Duchmann, Olivier: V1417,30-41(1991)

Plane, John M.
; Nien, Chia-Fu: Study of nighttime NO3 chemistry by differential optical absorption spectroscopy,V1433,8-20(1991)

Planel, Richard See Ababou, Soraya: V1361,706-711(1991)

Plano, Linda S.
: Optical emission spectroscopy of diamond-producing plasmas,V1437,13-23(1991)

Plant, David V. See Chen, Ray T.: V1559,110-117(1991)

Plante, Angela J. See Vengsarkar, Ashish M.: V1588,2-13(1991)

Plantey, J. See Sansonetti, Pierre: V1588,198-209(1991)

Plassard, T. See de Salabert, Arturo: V1521,74-88(1991)

Platen, Winfried See Guelker, Gerd: V1500,124-134(1991)

Platonov, N. S. See Ryabushkin, Oleg A.: V1362,75-79(1991)

Platonov, Valeri N. See Vysogorets, Mikhail V.: V1358,1066-1069(1991)

Platonov, Yu. Y.
; Salashchenko, N. N.; Shmaenok, L. A.: Spectrum investigation and imaging of laser-produced plasma by multilayer x-ray optics,V1440,188-196(1991)

Platzker, Aryeh See Masliah, Denis A.: V1475,113-120(1991)

Platzman, P. M.
: What can we learn that's new and interesting about condensed matter systems using polarized x rays?,V1548,34-34(1991)

Plaza, Pascal
See Hong, Nguyen T.: V1525,132-142(1991)
See Phat, Darith: V1525,196-205(1991)

Pleiss, T. C. See Park, Yong K.: V1372,219-227(1991)

Plemmons, Robert J. See Nagy, James G.: V1566,35-46(1991)

Pleshanov, Pavel G. See Jensen, Ronald H.: V1403,372-380(1991)

Plessis, Brigitte
; Goldberg, Morris; Belanger, Garry; Hickey, Nancy M.: Subjective evaluation of image enhancements in improving the visibility of pathology in chest radiographs,V1445,539-554(1991)

Plevritis, Sylvia
; Macovski, Albert: Resolution improvement for in-vivo magnetic resonance spectroscopic images,V1445,118-127(1991)

Plichon, V.
; Giron, J. C.; Delboulbe, J. P.; Lerbet, F.: Detection by mirage effect of the counter-ion flux between an electrochrome and a liquid electrolyte: application to WO3, Prussian blue, and lutetium diphthalocyanine films,V1536,37-47(1991)

Plimpton, Steven J. See Mastin, Gary A.: V1566,341-352(1991)

Pline, Alexander D.
; Wernet, Mark P.; Hsieh, Kwang-Chung: Ground-based PIV and numerical flow visualization results from the surface-tension-driven convection experiment,V1557,222-234(1991)

Ploetz, F.
; Schelp, C.; Anders, K.; Eberhardt, F.; Scheper, Thomas-Helmut; Bueckmann, F.: Optical sensors for process monitoring in biotechnology,V1510,224-230(1991)

Plogstedt, Allen E.
See Anderson, Barry K.: V1488,416-425(1991)
See Barnard, Kenneth J.: V1488,426-431(1991)

Ploog, Klaus H.
See Lopez, C.: V1361,89-95(1991)
See Richards, David R.: V1361,246-254(1991)

Plotnichenko, V. G. See Artjushenko, Vjacheslav G.: V1420,157-168(1991)

Plotnik, Irving See Stewart, Diane K.: V1465,64-77(1991)

Plouffe, Wil
See Bender, Walter J.: V1460(1991)
See Kasson, James M.: V1460,11-19(1991)

Plum, Heinz-Dieter See Habich, Uwe: V1397,383-386(1991)

Plumb, Dean
See Babcock, Carl P.: V1466,653-662(1991)
See Berry, Amanda K.: V1465,210-220(1991)

Plummer, A. P.
: Inspecting colored objects using gray-level vision systems,VCR36,64-77(1991)

Plummer, Grant M.
: Field and laboratory studies of Fourier transform infrared spectroscopy in continuous emissions monitoring applications,V1434,78-89(1991)

Pluta, Mieczyskaw See Talatinian, A.: V1574,205-217(1991)

Po, Hong See Zenteno, Luis A.: V1373,246-253(1991)

Pocha, Michael D.
; Hofer, Wayne W.: Photoconductive switching for high-power microwave generation,V1378,2-9(1991)
See Druce, Robert L.: V1378,43-54(1991)

Pocholle, Jean-Paul See Lallier, Eric: V1506,71-79(1991)

Pocsidio, Joanne O. See Bass, Lawrence S.: V1422,123-127(1991)

Poczulp, Gary A. See Cho, Myung K.: V1532,137-145(1991)

Podaras, Petros See Parikh, Jo Ann: V1469,526-538(1991)

Podbielska, Halina
ed.: Holography, Interferometry, and Optical Pattern Recognition in Biomedicine,V1429(1991)
: Trends in holographic endoscopy,V1429,207-213(1991)
See Kasprzak, Henryk: V1429,55-61(1991)

Poddubnaya, T. E.
See Odinokov, S. B.: V1238,109-117(1991)
See Odinokov, S. B.: V1238,103-108(1991)

Podenas, D. See Danelius, R.: V1402,198-208(1991)

Podgorski, William A. See Flanagan, Kathryn A.: V1549,395-407(1991)

Podladchikova, Lubov N. See Rybak, Ilya A.: V1469,737-748(1991)

Podlogar, Carl V. See Brock, Phillip J.: V1463,87-100(1991)

Podol'tsev, A. S. See Zheltov, Georgi I.: V1403,752-753(1991)

Podoleanu, Adrian G.
; Sterian, Paul E.; Popescu, Ion M.; Puscas, Niculae Tiberiu N.: Equivalent capacitance photodiode determination by an optical method,V1362,721-726(1991)
See Popescu, Ion M.: V1361,1041-1047(1991)

Podvyaznikov, V. A. See Dashevsky, Boris E.: V1346,437-442(1991)

Poelzleitner, Wolfgang
: Geometric property measurement of convex objects using fuzzy sets,V1381,411-422(1991)
; Schwingshakl, Gert: Real-time classification of wooden boards,V1384,38-49(1991)

Poggiolini, Pierluigi T.
; Kazovsky, Leonid G.: STARNET: an integrated services broadband optical network with physical star topology,V1579,14-29(1991)
See Benedetto, Sergio: V1579,112-121(1991)

Pogorelsky, Igor V.
; Fisher, A. S.; Veligdan, James T.; Russell, P.: Spatial dynamics of picosecond CO2 laser pulses produced by optical switching in Ge,V1413,21-31(1991)

Pogosov, Gregory A.
; Akhmanov, Sergei A.; Gordienko, Vyacheslav M.; Kosovsky, L. A.; Kurochkin, Nikolai N.; Priezzhev, Alexander V.: Detection and parameter estimation of atmospheric turbulence by ground-based and airborne CO2 Doppler lidars,V1416,115-124(1991)
See Gordienko, Vyacheslav M.: V1416,102-114(1991)

Pohjonen, Harri See Tammela, Simo: V1583,37-42(1991)

Pohl, Klaus-Juergen See Crostack, Horst-Artur: V1508,101-109(1991)

Pohlig, Kenneth See Chudyk, Wayne: V1368,105-114(1991)

Pointu, Bernard See Daugy, Eric: V1502,203-212(1991)

Poirier, Michel See Ouellette, Francois: V1516,2-13(1991)

Poirier, Peter M. See Phillips, Thomas E.: V1454,290-298(1991)

Pokasov, Vladimir See Gurvich, Alexander S.: V1408,10-18(1991)

Pokora, Ludwik J.
: Excimer and nitrogen lasers with low-average power for technology applications,V1397,111-114(1991)
; Puzewicz, Zbigniew: Excimer-dye laser system for diagnosis and therapy of cancer,V1503,467-478(1991)
See Iwanejko, Leszek: V1391,105-108(1991)
See Iwanejko, Leszek: V1391,98-100(1991)
See Makuchowski, Jozef: V1391,348-350(1991)
See Makuchowski, Jozef: V1391,79-86(1991)

Pokorski, Joseph D.
: Loss tolerant, self-monitoring fiber optic discrete position sensor,V1398,219-229(1991)

Pokrovskii, L. See Ziling, C. C.: V1583,90-101(1991)

Pol, Victor
ed.: Optical/Laser Microlithography IV,V1463(1991)
See Nalamasu, Omkaram: V1466,13-25(1991)

Poland, Arthur I.
; Domingo, Vicente: SOHO space satellite: UV instrumentation,V1343,310-318(1991)

Polato, P. See Mazzoldi, Paolo: V1513,182-197(1991)

Poleshchuk, Alexander G.
: Fabrication of phase structures with continuous and multilevel profile for diffraction optics,V1574,89-100(1991)

Polifko, David M.
; Malone, Steven A.; Daryoush, Afshin S.; Kunath, Richard R.: Fiber optic link for millimeter-wave communication satellites,V1476,91-99(1991)

Polis, David W. See Spangler, Charles W.: V1497,408-417(1991)

Polishchuk, E. Y. See Petrov, V. V.: V1621,45-50(1991)

Polkowski, Edward T.
See Malik, Raashid: V1388,280-290(1991)
See Malik, Raashid: V1388,306-316(1991)

Pollack, Ashok K.
; Chang, David B.; Birnbaum, Milton; Kokta, Milan R.: Upconversion-pumped IR (2.8-2.9 microns) lasing of Er3+ in garnets,V1410,156-164(1991)

Pollak, Fred H.
: Photoreflectance for the in-situ monitoring of semiconductor growth and processing,V1361,109-130(1991)

Pollentier, Ivan K.
; Ackaert, A.; De Dobbelaere, P.; Buydens, Luc; van Daele, Peter; Demeester, Piet M.: Fabrication of high-radiance LEDs by epitaxial lift-off,V1361,1056-1062(1991)
See Demeester, Piet M.: V1361,987-998(1991)

Pollicove, Harvey M.
; Moore, Duncan T.: Automation in optics manufacturing,V1354,482-486(1991)
; Moore, Duncan T.: Center for Optics Manufacturing,V1398,90-93(1991)

Pollitt, Stuart
: Fiber optic measurement standards,V1504,80-87(1991)

Pollock, Michael A. See Chang, Kai: V1475,257-266(1991)

Polo, Laura See Jori, Guilio: V1525,367-376(1991)

Polo Da Fonseca, C. N. See Gorenstein, Annette: V1536,104-115(1991)

Poltavsky, V. V. See Fakhrutdinov, I. H.: V1399,98-106(1991)

Polushkin, Valery G. See Danilov, Alexander A.: V1362,916-920(1991)

Polvani, Robert S.
; Evans, Chris J.: Characterization of subsurface damage in glass and metal optics,V1441,173-173(1991)

Polyak, Alexander
; Kostin, Ivan K.: Optical research of elastic anisotropy of monocrystals with a cube structure,V1554A,553-556(1991)

Polyakov, Igor V. See Nikonchuk, Michael O.: V1412,72-78(1991)

Polyakova, M. S. See Fedoseeva, E. V.: V1403,355-358(1991)

Polyansky, S. V. See Drozhbin, Yu. A.: V1358,1029-1034(1991)

Polymeropoulos, Constantine E. See Majercak, David: V1584,162-169(1991)

Polzik, Eugene S.
; Kimble, H. J.: Frequency doubling and optical parametric oscillation with potassium niobate,V1561,143-146(1991)

Pomalaza-Raez, Carlos A.
See Fong, David Y.: V1569,156-161(1991)
See Sayegh, Samir I.: V1569,100-110(1991)

Pomerleau, Dean A.
: Neural-network-based vision processing for autonomous robot guidance,V1469,121-128(1991)

Pomeroy, Robert S.
; Baker, Mark E.; Radspinner, David A.; Denton, M. B.: Fluorescence imaging of latent fingerprints with a cooled charge-coupled-device detector,V1439,60-65(1991)

Pommeret, Stanislas See Gauduel, Yann: V1403,417-426(1991)

Pomplun, Alan R. See Fuchs, Elizabeth A.: V1467,136-149(1991)

Pon, Alfred See Palcic, Branko: V1448,113-117(1991)

Ponomarenko, Vladimir P. See Khryapov, V. T.: V1540,412-423(1991)

Ponomaryov, A. M. See Ushakov, Leonid S.: V1358,447-450(1991)

Pons, Alicia A.
See Campos, Joaquin: V1504,66-74(1991)
See Corredera, Pedro: V1504,281-286(1991)

Pons, Michel J. See Paniez, Patrick J.: V1466,583-591(1991)

Ponsard, Benoit See Gagnaire, A.: V1364,277-288(1991)

Ponsardin, Patrick
; Higdon, Noah S.; Grossmann, Benoist E.; Browell, Edward V.: Alexandrite laser characterization and airborne lidar developments for water vapor DIAL measurements,V1492,47-51(1991)

Pont, W. F.
; Gader, Paul D.: Gradient descent techniques for feature detection template generation,V1568,247-260(1991)
See Reiley, Michael F.: V1416,295-303(1991)

Pontifex, Brian See Jaggi, Bruno: V1448,89-97(1991)

Ponyavina, A. N. See Borisevich, Nikolai A.: V1500,222-231(1991)

Poon, Steven S. See Jaggi, Bruno: V1448,89-97(1991)

Poon, Ting-Chung See Pieper, Ronald J.: V1454,324-335(1991)

Poore, Aubrey B.
; Rijavec, Nenad: Multitarget tracking and multidimensional assignment problems,V1481,345-356(1991)

Popa, Dragos See Vlad, Ionel V.: V1332,236-244(1991)

Popat, Ashok C.
; Li, Wei; Kunt, Murat: Numerical design of parallel multiresolution filter banks for image coding applications,V1567,341-353(1991)
; Nicoulin, Andre; Basso, Andrea; Li, Wei; Kunt, Murat: Subband coding of video using energy-adaptive arithmetic coding and statistical feedback-free rate control,V1605,940-953(1991)
See Li, Wei: V1605,124-136(1991)

Pope, Edward J. See Banash, Mark A.: V1590,8-13(1991)

Popenko, N. A. See Vertiy, Alexey A.: V1361,1070-1078(1991)

Popescu, Alexandru F.
See Garcia-Prieto, Rafael: V1522,267-276(1991)
See Marini, Andrea E.: V1417,200-211(1991)
See Oppenhaeuser, Gotthard: V1522,2-13(1991)

Popescu, Ion M.
; Puscas, Niculae Tiberiu N.; Sterian, Paul E.; Irimescu, Dorin I.; Podoleanu, Adrian G.: Theoretical study of the fifth harmonic generation in organic liquids, V1361, 1041-1047(1991)
See Podoleanu, Adrian G.: V1362, 721-726(1991)

Popolo, J. See Norwood, Robert A.: V1560, 54-65(1991)

Poponin, Vladimir P. See Makeev, V. Y.: V1403, 522-527(1991)

Popov, A.
; Mikhalev, N.; Karalyunts, A.; Smirnov, O.; Vasilyeva, N.: Photosensitivity of selenium-bismuth films with varigap structure, V1519, 457-462(1991)
; Vasiljeva, Natalja V.: Studies of correlation of molecular structure under preparation conditions for noncrystalline selenium thin films with aid of computer simulation, V1519, 37-42(1991)

Popov, A. G. See Budnik, A. P.: V1440, 135-145(1991)

Popov, Alexander P. See Goncharov, V. F.: V1238, 97-102(1991)

Popov, S. A. See Khryapov, V. T.: V1540, 412-423(1991)

Popov, Yury V.
: Integrated optics in optical engineering, V1399, 207-213(1991)
See Berezhnoy, Anatoly A.: V1401, 44-49(1991)

Popovic, Zoya See Bundy, Scott: V1475, 319-329(1991)

Poppel, Gary L. See Dasgupta, Samhita: V1374, 211-222(1991)

Porat, Boaz See Francos, Joseph M.: V1606, 553-565(1991)

Porden, Mark
: Antireflection coating standards of ophthalmic resin lens materials, V1529, 115-123(1991)

Porembka, D. T. See Wan, Liqun: V1469, 432-440(1991)

Porindla, Sridhar N.
; Rylander, Henry G.; Welch, Ashley J.: Acoustic effects of Q-switched Nd:YAG on crystalline lens, V1427, 267-272(1991)

Porjadin, V. See Bulichev, A.: V1500, 151-162(1991)

Porkolab, Gyorgy A. See Lang, Robert J.: V1563, 2-7(1991)

Poroshina, Marina Y.
See Akhmanov, Sergei A.: V1403(1991)
See Hianik, T.: V1402, 85-88(1991)

Porras, Miguel A.
; Alda, Javier; Bernabeu, Eusebio: Multimode laser beams behaviour through variable reflectivity mirrors, V1397, 645-648(1991)
See Alda, Javier: V1527, 240-251(1991)

Porta, Paola M.
: Evaluation of the IR signature of dynamic air targets, V1540, 508-518(1991)

Portacha, Lidia See Kecik, Tadeusz: V1391, 341-345(1991)

Portal, Jean-Claude See Kadri, Abderrahmane: V1362, 930-937(1991)

Porter, Charles R. See Figueroa, Luis: V1418, 153-176(1991)

Porter, Christopher C. See Helms, Richard G.: V1532, 64-80(1991)

Porter, Robert F.
; Mourad, Pierre D.; Al-Kurd, Azmi: Wavefront reconstruction of acoustic waves in a variable ocean, V1558, 91-102(1991)

Porter, Wallace M. See Eastwood, Michael L.: V1540, 164-175(1991)

Poryvkina, Larisa See Babichenko, Sergey M.: V1492, 319-323(1991)

Poscio, Patrick
; Depeursinge, Ch.; Emery, Y.; Parriaux, Olivier M.; Voirin, G.: Biochemical measurement of bilirubin with an evanescent wave optical sensor, V1510, 112-117(1991)

Pospisilova, Marie
; Schneiderova, Martina: Comparison of Gauss' and Petermann's formulas for real single-mode fibers by far-field pattern technique, V1504, 287-291(1991)

Posselt, Winfried
: Michelson interferometer for passive atmosphere sounding, V1490, 114-125(1991)

Possin, G. E. See Tomlinson, Harold W.: V1541, 178-186(1991)

Possner, Torsten
; Ehrt, Doris; Sargsjan, Geworg; Unger, Clemens: Stripe waveguides by Cs+ and K+-exchange in neodymium-doped soda silicate glasses for laser application, V1513, 378-385(1991)
See Guntau, Matthias: V1513, 112-122(1991)

Post, Daniel See Morimoto, Yoshiharu: V1554B, 493-502(1991)

Post, Madison J.
See Menzies, Robert T.: V1416, 139-146(1991)
See Zhao, Yanzeng: V1492, 85-90(1991)

Post, Mark J. See Smits, Pieter C.: V1425, 188-190(1991)

Postek, Michael T.
; Larrabee, Robert D.; Keery, William J.; Marx, Egon: Application of transmission electron detection to x-ray mask calibrations and inspection, V1464, 35-47(1991)

Postovalov, V. E.
See Borodin, A. M.: V1358, 756-758(1991)
See Briscoe, Dennis: V1358, 329-336(1991)
See Bryukhnevitch, G. I.: V1358, 739-749(1991)
See Bryukhnevitch, G. I.: V1449, 109-115(1991)
See Degtyareva, V. P.: V1358, 524-531(1991)
See de Souza, Eunezio A.: V1358, 556-560(1991)
See Ivanov, K. N.: V1358, 732-738(1991)
See Prochazka, Ivan: V1449, 116-120(1991)
See Prochazka, Ivan: V1358, 574-577(1991)

Poszinskaja, Irina I. See Gan, Michael A.: V1507, 116-125(1991)

Potasek, M. J.
; Tabor, Mark: Multigigabit solitary-wave propagation in both the normal and anomalous dispersion regions of optical fibers, V1579, 232-236(1991)

Pothier, Steven
: Image annotation under X Windows, V1472, 46-53(1991)

Potter, B. L.
; Walker, D. S.; Greer, L.; Saavedra, Steven S.; Reichert, William M.: Multiple mode and multiple source coupling into polymer thin-film waveguides, V1368, 251-257(1991)

Potter, C. See MacCraith, Brian D.: V1510, 104-109(1991)

Potter, Duncan J.
; Ranshaw, M. J.; Al-Chalabi, Adil O.; Fancey, Norman E.; Sillitto, Richard M.; Vass, David G.: Optical correlation using a phase-only liquid-crystal-over-silicon spatial light modulator, V1564, 363-372(1991)

Pottier, R. See Nadeau, Pierre: V1398, 151-161(1991)

Potts, Michael A.
; Broomhead, David S.: Time series prediction with a radial basis function neural network, V1565, 255-266(1991)

Potukuchi, J. R. See Sorbello, Robert M.: V1475, 175-183(1991)

Potyemin, Igor S.
See Gan, Michael A.: V1507, 116-125(1991)
See Gan, Michael A.: V1574, 254-260(1991)
See Gan, Michael A.: V1574, 243-249(1991)
See Gan, Michael A.: V1507, 549-560(1991)

Potylitsyn, Yevgeny See Kovrigin, Yevgeny: V1362, 967-975(1991)

Potyondy, David O. See Andrews, John R.: V1461, 110-123(1991)

Potyrailo, Radislav A.
; Golubkov, Sergei P.; Borsuk, Pavel S.: Fiber optic sensor for ammonia vapors of variable temperature, V1572, 434-438(1991)

Poublan, Helene
; Castanie, Francis: Error probabilities of minimum-distance classifiers, V1569, 329-340(1991)

Poueyo, Anne
; Sabatier, Lilian; Deshors, G.; Fabbro, Remy; de Frutos, Angel M.; Bermejo, Dionisio; Orza, Jose M.: Experimental study of the parameters of the laser-induced plasma observed in welding of iron targets with continuous high-power CO2 lasers, V1502, 140-147(1991)
See de Frutos, Angel M.: V1397, 717-720(1991)

Poulain, D. E. See Alexander, Dennis R.: V1497, 90-97(1991)

Pouleau, J. See Lequime, Michael: V1511, 244-249(1991)

Poulsen, Peter
; Pincosy, Phillip A.; Morrison, Jasper J.: Progress toward steady-state high-efficiency vircators, V1407, 172-182(1991)

Pour, I. K. See Bahuguna, Ramendra D.: V1332, 805-807(1991)

Pourboghrat, Farzad
: Neural networks application in autonomous path generation for mobile robots, V1396, 243-251(1991)

Pouskouleli, G.
; Wheat, T. A.; Ahmad, A.; Varma, Sudhanshu; Prasad, S. E.: Application of statistical design in materials development and production, V1590, 179-190(1991)

Poussart, Denis
See Houde, Regis: V1332, 343-354(1991)
See Wang, Shengrui: V1382, 37-48(1991)

Povarkov, V. I. See Miroshnikov, Mikhail M.: V1540, 496-505(1991)

Powazinik, William
See Eichen, Elliot G.: V1474, 260-267(1991)
See Holmstrom, Roger P.: V1418, 223-230(1991)

Powell, A. K. See McOwan, Peter W.: V1384,75-82(1991)

Powell, Forbes R.
See Lindblom, Joakim F.: V1343,544-557(1991)
See Walker, Arthur B.: V1494,320-333(1991)

Powell, Howard T. See Woods, Bruce W.: V1410,47-54(1991)

Powell, J. H. See Dowell, Floyd E.: V1379,136-140(1991)

Powell, Norman J.
; Kelsall, Robert W.; Love, G. D.; Purvis, Alan: Investigation of fringing fields in liquid-crystal devices,V1545,19-30(1991)
See Purvis, Alan: V1455,145-149(1991)

Powell, Richard C.
ed.: Selected Papers on Solid State Lasers,VMS31(1991)

Powell, Richard K.
; Berdanier, Barry N.; McKay, James: Carbon dioxide eyesafe laser rangefinders,V1419,126-140(1991)

Powell, Stephen F. See Lindblom, Joakim F.: V1343,544-557(1991)

Powell, William R.
: Analysis of thermal stability of fused optical structure,V1532,126-136(1991)

Powers, Edward J. See Kniffen, Stacy K.: V1564,617-627(1991)

Powers, J. See Elman, Boris S.: V1362,610-616(1991)

Poyato, J. M. See Martin, E.: V1397,835-838(1991)

Poyet, Patrice See Tomasini, Bernard: V1468,60-71(1991)

Poyhonen, Pekka
See Li, Ming-Jun: V1506,52-57(1991)
See Najafi, S. I.: V1583,32-36(1991)
See Tervonen, Ari: V1513,71-75(1991)

Poyntz-Wright, L. J.
See Russell, Philip S.: V1373,126-139(1991)
See Russell, Philip S.: V1516,47-54(1991)

Pozzi, S. See Dal Degan, Nevaino: V1606,934-940(1991)

Pradhan, Asima
; Das, Bidyut B.; Liu, C. H.; Alfano, Robert R.; O'Brien, Kenneth M.; Stetz, Mark L.; Scott, John J.; Deckelbaum, Lawrence I.: Time-resolved fluorescence of normal and atherosclerotic arteries,V1425,2-5(1991)

Pramanik, Dipankar
; Jain, Vivek: Observation of stress voids and grain structure in laser-annealed aluminum using focused ion-beam microscopy,V1596,132-138(1991)

Prasad, Coorg R. See Schwemmer, Geary K.: V1492,52-62(1991)

Prasad, K. V.
; Mammone, Richard J.: Monocular passive range sensing using defocus information,V1385,280-291(1991)

Prasad, Paras N.
See Kwok, Hoi-Sing: V1394,196-200(1991)
See Zhang, Yue: V1560,264-271(1991)

Prasad, S. E. See Pouskouleli, G.: V1590,179-190(1991)

Praseuth, Jean-Pierre See Scavennec, Andre: V1362,331-337(1991)

Prassler, Erwin E.
; Milios, Evangelos E.: Parallel path planning in unknown terrains,V1388,2-13(1991)

Prat, Ch.
; Autric, Michel L.; Inglesakis, Georges; Astic, Dominique: Pulsed CO2 laser-material interaction: mechanical coupling and reflected and scattered radiation,V1397,701-704(1991)
See Astic, Dominique: V1397,713-716(1991)

Pratt, Jonathan P.
; Heuring, Vincent P.: Designing digital optical computing systems: power distribution and crosstalk,V1505,124-129(1991)

Pratt, Stephen R.
; Tucker, Robert: Reconnaissance and imaging sensor test facilities at Eglin Air Force Base,V1538,40-45(1991)

Praus, Petr
; Stepanek, Josef: Statistical signal processing in Raman spectroscopy of biological samples,V1403,76-84(1991)

Pravdin, A. B. See Kozlova, T. G.: V1403,159-160(1991)

Pravilov, A. M. See Ivanov, U. V.: V1397,181-184(1991)

Preater, Richard W.
: In-plane displacement measurement on rotating components using pulsed laser ESPI (electronic speckle pattern interferometry),V1399,164-171(1991)
; Swain, Robin C.: In-plane ESPI measurements in hostile environments,V1554A,727-738(1991)

Predehl, Peter See Braeuninger, Heinrich: V1549,330-339(1991)

Predtechensky, M.
; Smal, A.; Varlamov, Yu.: Epitaxial films YBa2Cu3O7-delta(jc(78K)>106A/cm2) on sapphire and SrTiO3: peculiarities and differences in conditions of film growth and properties,V1477,234-241(1991)

Preece, Alun D.
; Shinghal, Rajjan: Practical approach to knowledge base verification,V1468,608-619(1991)

Preil, Moshe E. See Dunbrack, Steven K.: V1464,314-326(1991)

Preis, Mark K. See Kreider, James F.: V1488,376-388(1991)

Preisack, M. See Nyga, Ralf: V1525,119-123(1991)

Prentice, Jeffrey A. See Gilbert, Barry K.: V1390,235-248(1991)

Prescott, Anthony J. See Lake, Stephen P.: V1486,286-293(1991)

Pressel, Philip
: Generic telescope truss,V1532,50-56(1991)

Presser, Nazmir See Kudlek, Gotthard: V1361,150-158(1991)

Presta, Robert W. See Salmon, J. T.: V1542,459-467(1991)

Presting, Hartmut
; Jaros, Milan; Abstreiter, Gerhard: Growth and characterization of ultrathin SimGen strained-layer superlattices,V1512,250-277(1991)
See Kasper, Erich: V1361,302-312(1991)

Preston, Kendall
: Residue-producing E-filters and their applications in medical image analysis,V1450,59-70(1991)
: Trade-offs between pseudohexagonal and pseudocubic filters,V1451,75-90(1991)

Preston, Ralph G.
; Colquhoun, Allan B.; Shepherd, Joseph: Bearings for rotary scanners: an overview,V1454,124-131(1991)
; Robinson, D. H.; Shepherd, Joseph: Self-acting gas bearings for high-speed scanners,V1454,132-138(1991)

Presuhn, Gary G.
; Zeis, Joseph E.: Flight test integration and evaluation of the LANTIRN system on the F-15E,V1479,249-258(1991)

Preteux, Francoise
; Merlet, Nicolas: New concepts in mathematical morphology: the topographical and differential distance functions,V1568,66-77(1991)
; Moubarak, Michel; Grenier, Philippe: Pattern recognition in pulmonary computerized tomography images using Markovian modeling,V1450,72-83(1991)
See Rougon, Nicolas F.: V1568,78-89(1991)

Pretterebner, Julius See Chong, Chae K.: V1407,226-233(1991)

Prettl, William See Muecke, Robert J.: V1433,136-144(1991)

Pretzler, Georg See Neger, Theo: V1507,476-487(1991)

Prevost, Michael P.
; Banda, Carolyn P.: Visualization tool for human-machine interface designers,V1459,58-66(1991)
See Larimer, James O.: V1456,196-210(1991)

Pribil, Klaus
; Johann, Ulrich A.; Sontag, Heinz: SOLACOS: a diode-pumped Nd:YAG laser breadboard for coherent space communication system verification,V1522,36-47(1991)
See Johann, Ulrich A.: V1522,243-252(1991)
See Sontag, Heinz: V1417,573-587(1991)
See Sontag, Heinz: V1522,61-69(1991)

Price, Edwin P. See Xu, Yian-sun: V1407,648-652(1991)

Price, Frederick W. See Gates, James L.: V1540,262-273(1991)

Price, Garth L.
; Usher, Brian F.: Application of epitaxial lift-off to optoelectronic material studies,V1361,543-550(1991)

Price, George E. See Nisenoff, Martin: V1394,104-113(1991)

Price, R. S.
; Chun, Wendell H.; Hammond, Mark W.; Hubert, Alexis: Wheeled planetary rover testbed,V1388,550-559(1991)

Priddy, Kevin L.
See Tarr, Gregory L.: V1469,2-11(1991)
See Veronin, Christopher P.: V1469,281-291(1991)

Priebe, Russell
; Jones, Richard A.: Fuzzy logic approach to multitarget tracking in clutter,V1482,265-274(1991)

Priest, J. A. See Zediker, Mark S.: V1418,309-315(1991)

Prieur-Drevon, P. See Munier, Bernard: V1447,44-55(1991)

Priezzhev, Alexander V.
; Proskurin, S. G.; Romanovsky, Yuri M.: Laser Doppler measurements of ameboid cytoplasmic streaming and problems of mathematical modeling of intracellular hydrodynamics,V1402,107-113(1991)
See Firsov, N. N.: V1403,350-354(1991)
See Gordienko, Vyacheslav M.: V1416,102-114(1991)
See Pogosov, Gregory A.: V1416,115-124(1991)

Prigent, Pascale
; Brunet, Henri: Study of the production of S2(B) from the recombination of sulfur atoms,V1397,197-201(1991)

Primot, Jerome See Michau, Vincent: V1487,64-71(1991)

Prince, John L.
; Cangellaris, Andreas C.; Palusinski, Olgierd A.: Modeling progress and trends in electrical interconnects,V1390,271-285(1991)
See Pasik, Michael F.: V1389,297-301(1991)

Prinz, Erwin J. See Sturm, James C.: V1393,252-259(1991)

Priolo, Pierluigi See Aymerich, Francesco: V1554B,304-314(1991)

Prior, John A.
See Hatch, Jeffrey A.: V1346,371-375(1991)
See Richards, James B.: V1346,384-389(1991)

Prior, Yehiam See Levine, Alfred M.: V1376,47-53(1991)

Prise, Michael E. See Dickinson, Alex G.: V1389,503-514(1991)

Pritchard, Alan P. See Lake, Stephen P.: V1486,286-293(1991)

Pritchett, Thomas H. See Padgett, Joseph: V1433,352-364(1991)

Proakis, John G.
; Nikias, Chrysostomos L.: Blind equalization,V1565,76-87(1991)
See Ling, Fuyun: V1565,296-306(1991)
See Zervas, E.: V1565,178-187(1991)

Prober, Daniel E. See Mears, Carl A.: V1477,221-233(1991)

Probst, David K.
; Rice, Robert R.: Performance of a phased array semiconductor laser source for coherent laser communications,V1417,346-357(1991)

Probst, O. See Etienne, Bernard: V1362,256-267(1991)

Prochazka, Ivan
; Hamal, Karel; Kirchner, G.; Schelev, Mikhail Y.; Postovalov, V. E.: Circular streak camera application for satellite laser ranging,V1449,116-120(1991)
; Hamal, Karel; Schelev, Mikhail Y.; Lozovoi, V. I.; Postovalov, V. E.: Modular streak camera for laser ranging,V1358,574-577(1991)

Prochazka, Jaroslav J. See Wojcik, Gregory L.: V1464,187-203(1991)

Procoli, Alfredo See Miley, George H.: V1441,16-26(1991)

Proesmans, Marc
; Verlinde, Patrick S.: Numerical evaluation of the efficiency of camouflage systems in the thermal infrared,V1486,102-114(1991)
See Verlinde, Patrick S.: V1486,58-65(1991)

Proffitt, Michael H.
; Langford, A. O.: Measuring tropospheric ozone using differential absorption lidar technique,V1491,2-6(1991)

Profio, A. E.
; Balchum, Oscar J.; Lam, Stephen: Endoscopic fluorescence detection of early lung cancer,V1426,44-46(1991)
See Palcic, Branko: V1448,113-117(1991)

Prohaska, Robert M.
; Fisher, Amnon; Rostoker, Norman: Negative ions from magnetically insulated diodes,V1407,598-604(1991)
See Song, Yuanxu Y.: V1407,430-441(1991)

Prokhorov, Alexander M.
; Schelev, Mikhail Y.: Recent research and development in electron image tubes/cameras/systems,V1358,280-289(1991)
; Shcherbakov, I. A.: Soviet developments in solid state lasers,V1410,70-88(1991)
See Artjushenko, Vjacheslav G.: V1420,157-168(1991)
See Atjezhev, Vladimir V.: V1503,197-199(1991)
See Babushkin, A. V.: V1346,410-417(1991)
See Borodin, A. M.: V1358,756-758(1991)
See Bowley, David J.: V1358,550-555(1991)
See Briscoe, Dennis: V1358,329-336(1991)
See Bryukhnevitch, G. I.: V1358,739-749(1991)
See Bryukhnevitch, G. I.: V1449,109-115(1991)
See Dalinenko, I. N.: V1449,167-172(1991)
See Dashevsky, Boris E.: V1346,437-442(1991)
See Degtyareva, V. P.: V1358,524-531(1991)

See Denker, Boris I.: V1419,50-54(1991)
See de Souza, Eunezio A.: V1358,556-560(1991)
See Dubovoi, I. A.: V1358,134-138(1991)
See Ivanov, K. N.: V1358,732-738(1991)
See Konov, Vitaly I.: V1427,220-231(1991)
See Konov, Vitaly I.: V1525,250-252(1991)
See Ludikov, V. V.: V1346,418-436(1991)
See Prokhorov, Kirill A.: V1501,80-84(1991)

Prokhorov, Kirill A.
; Prokhorov, Alexander M.; Djibladze, Merab I.; Kekelidze, George N.; Gorelik, Vladimir S.: Multiphoton excited emission in zinc selenide and other crystals,V1501,80-84(1991)

Prokopovich, Ludvig P. See Marczuk, Krystyna: V1513,291-296(1991)

Prongue, Damien See Herzig, Hans-Peter: V1507,247-255(1991)

Pronios, Nikolaos-John B.
; Yovanof, Gregory S.: Effects of transmission errors on medical images,V1446,108-128(1991)

Proskurin, S. G. See Priezzhev, Alexander V.: V1402,107-113(1991)

Prosser, Reese T. See Moses, Harry E.: V1407,354-374(1991)

Prostev, A. See Bryskin, V. Z.: V1238,368-369(1991)

Protasov, Yu. S.
See Kamrukov, A. S.: V1503,438-452(1991)
See Kamrukov, A. S.: V1397,137-144(1991)

Prots, V. I. See Krasnov, Victor F.: V1506,179-187(1991)

Protsenko, Evgeniy D. See Kurochkin, Vadim Y.: V1435,322-330(1991)

Provorov, Alexander S. See Kushlevsky, S. V.: V1403,799-800(1991)

Prucnal, Paul R.
; Johns, Steven T.; Krol, Mark F.; Stacy, John L.: Time-division optical microarea networks,V1389,462-476(1991)
See Goel, Kamal K.: V1476,314-319(1991)

Prudente, R. See Das, Bidyut B.: V1427,368-373(1991)

Pruett, J. G. See Havrilla, George J.: V1435,12-18(1991)

Pruszynski, Charles J.
: Overpressure proof testing of large infrared windows for aircraft applications,V1498,163-170(1991)

Prutzer, Steven
; Biron, David G.; Quist, Theodore M.: High-resolution airborne multisensor system,V1480,46-61(1991)

Pryanichnikov, Victor See Gurvich, Alexander S.: V1408,10-18(1991)

Pryanishnikov, Ivan G. See Berkovsky, A. G.: V1358,750-755(1991)

Pryputniewicz, Ryszard J.
: Automated systems for quantitative analysis of holograms,VIS08,215-246(1991)
: Static and dynamic measurements using electro-optic holography,V1554B,790-798(1991)

Przhevuskii, Alexander K. See Lunter, Sergei G.: V1513,349-359(1991)

Przybilla, Gudrun See Westenberger, Gerhard: V1535,113-120(1991)

Przybilla, Klaus J.
; Roeschert, Heinz; Spiess, Walter; Eckes, Charlotte; Chatterjee, Subhankar; Khanna, Dinesh N.; Pawlowski, Georg; Dammel, Ralph: Progress in DUV resins,V1466,174-187(1991)
See Eckes, Charlotte: V1466,394-407(1991)

Psaltis, Demetri
; Lin, Steven H.: Optoelectronic neuron arrays,V1562,204-212(1991)
; Qiao, Yong: Optical multilayer neural networks,V1564,489-494(1991)
See Lang, Robert J.: V1563,2-7(1991)
See Neifeld, Mark A.: V1469,250-255(1991)
See Stirk, Charles W.: V1389,580-593(1991)

Ptashnik, O. V. See Melnik, Ivan S.: V1572,118-122(1991)

Pu, Yu-Ming See Lightner, David A.: V1432,2-13(1991)

Puetz, Norbert
; Miner, Carla J.; Hingston, G.; Moore, Chris J.; Watt, B.; Hillier, Glen: Investigation of uniform deposition of GaInAsP quantum wells by MOCVD,V1361,692-698(1991)

Pugnet, Michel
; Collet, Jacques; Nardo, Laurent: Photocurrent response to picosecond pulses in semiconductors: application to EL2 in gallium arsenide,V1361,96-108(1991)

Puhov, Anatoly M. See Belov, Sergei N.: V1503,503-509(1991)

Puig, Thierry T. See Bourda, C.: V1502,148-159(1991)

Puik, Eric J. See den Boggende, Antonius J.: V1345,189-197(1991)

Puliafito, Carmen A.
ed.: Ophthalmic Technologies,V1423(1991)

Puliparambil, Joseph T.
See Matthys, Donald R.: V1554B,736-742(1991)
See Matthys, Donald R.: V1332,850-861(1991)

Puliti, Paolo See Tascini, Guido: V1450,178-185(1991)

Puls, J. See Spiegelberg, Christine: V1362,951-958(1991)

Puls, Juergen
: German ATMOS program,V1490,2-13(1991)

Pun, Edwin Y.
: Recent development of proton-exchanged waveguides and devices in lithium niobate using phosphoric acid,V1374,2-13(1991)
; Loi, K. K.; Zhao, S.; Chung, P. S.: Annealing properties of proton-exchanged waveguides in LiNbO3 fabricated using stearic acid,V1583,102-108(1991)
; Loi, K. K.; Chung, P. S.: Experimental studies of proton-exchanged waveguides in lithium niobate using toluic acid,V1583,64-70(1991)
See Chan, Hau P.: V1583,129-134(1991)

Puppels, G. J.
; Otto, C.; de Mul, F. F.; Greve, Jan: Investigation of single biological cells and chromosomes by confocal Raman microspectroscopy,V1403,146-146(1991)

Puri, Atul
; Aravind, R.: Adaptive perceptual quantization for video compression,V1605,297-300(1991)

Purinsh, Juris
; Elksninsh, N.; Dzenis, J.; Tomass, V.; Teivans, A.; Freimanis, R.; Garsha, I.; Ozolinsh, H.: Biomechanical and structural studies of rabbits carotid arteries after endovascular laser exposition,V1525,289-308(1991)

Puro, A. E. See Aben, Hillar K.: V1554A,298-309(1991)

Purvis, Alan
; Williams, Geoffrey; Powell, Norman J.; Clark, Michael G.; Wiltshire, Michael C.: Liquid-crystal phase modulators for active micro-optic devices,V1455,145-149(1991)
See Powell, Norman J.: V1545,19-30(1991)

Puscas, Niculae Tiberiu N.
See Podoleanu, Adrian G.: V1362,721-726(1991)
See Popescu, Ion M.: V1361,1041-1047(1991)

Putilin, Andrei N.
: Optical interconnections based on waveguide holograms,V1621,93-101(1991)

Putivskii, Yu. Y. See Gordienko, Vyacheslav M.: V1416,102-114(1991)

Putnam, Martin A. See Baldini, S. E.: V1588,125-131(1991)

Putnam, Sidney D. See Bailey, Vernon L.: V1407,400-406(1991)

Putvinski, T. M. See Katz, Howard E.: V1560,370-376(1991)

Puzewicz, Zbigniew See Pokora, Ludwik J.: V1503,467-478(1991)

Puzicha, Gisbert See Lightner, David A.: V1432,2-13(1991)

Pye, L. D. See Capozzi, Carol A.: V1513,320-329(1991)

Pyee, Maurice
; Meisse, Pascal; Baudrand, Henry; Chaubet, Michel: Characterization of a high-critical-temperature superconducting thin film by the ring resonator method,V1512,240-248(1991)

Pylnov, I. L. See Artjushenko, Vjacheslav G.: V1420,157-168(1991)

Pypkin, B. N. See Ageev, Vladimir P.: V1503,453-462(1991)

Pysz, Dariusz See Kociszewski, Longin: V1420,212-217(1991)

Qi, GuangXue
; Deng, Zesheng: Width measurement of cold-rolling strip,V1572,536-538(1991)

Qi, Lei See Yuan, Xiang L.: V1519,167-171(1991)

Qi, Zhen Z.
See Hou, Bi H.: V1519,822-823(1991)
See Shi, W.: V1519,138-141(1991)

Qian, Aidi D.
See Chen, Shisheng: V1552,288-295(1991)
See Lin, Li-Huang: V1346,490-501(1991)

Qian, Anping
; Dong, Linjun: Studies on antireflection technology of end surfaces for coherent fibre bundles,V1572,144-147(1991)

Qian, Jingren See Dong, Xiaopeng: V1572,56-60(1991)

Qian, Kai
See Bhattacharya, Prabir: V1468,918-922(1991)
See Bhattacharya, Prabir: V1383,530-536(1991)

Qian, Limin
See Cao, Jianlin: V1345,71-77(1991)
See Cao, Jianlin: V1345,225-232(1991)
See Li, Futian: V1345,78-88(1991)

Qian, Longsheng See Cao, Jianlin: V1345,225-232(1991)

Qian, Qi-de
See Hansen, Steven G.: V1463,230-244(1991)
See Wojcik, Gregory L.: V1463,292-303(1991)

Qian, Shen-en
; Wang, Ruqin; Li, Shuqiu; Dai, Yisong: Computer-aided performance evaluation system for the on-board data compression system in HIRIS,V1483,196-206(1991)

Qian, Yujun
See Wang, Yongjiang: V1412,67-71(1991)
See Wang, Yongjiang: V1412,60-66(1991)

Qian, Zhaogang See Cheng, Jie: V1392,373-384(1991)

Qian, Zheng X. See Wu, Guang M.: V1519,315-320(1991)

Qiang, Xifu See Tan, Jiubin: V1572,552-557(1991)

Qiang, Xue-li
; Yang, Qing: Automatic searching center measurement of profile of a line,V1567,670-679(1991)

Qiao, Sanzheng
: Fast recursive-least-squares algorithms for Toeplitz matrices,V1566,47-58(1991)

Qiao, Yong See Psaltis, Demetri: V1564,489-494(1991)

Qin, Jing See Shi, Dahuan: V1554A,680-689(1991)

Qin, Pingling See Fang, Yin: V1572,197-200(1991)

Qin, Yuwen
; Wang, Jinqi; Ji, Xinhua: Study of electronic shearing speckle technique,V1554A,739-746(1991)
See Zhao, Mingjun: V1558,529-534(1991)

Qing, Kent P.
; Means, Robert W.: Fast algorithm for a neocognitron neural network with back-propagation,V1569,111-120(1991)
; Means, Robert W.: New method for chain coding based on convolution,V1567,390-396(1991)

Qing, Z. See Meeusen, G. J.: V1519,252-257(1991)

Qiu, Anping See Yuan, Libo: V1572,258-263(1991)

Qiu, Guanming
; DuanMu, Qingduo: Study of fluorescent glass-ceramics,V1513,396-407(1991)

Qiu, Peixia See Guo, Ting: V1409,24-27(1991)

Qiu, Ying See Han, Xin Z.: V1479,156-161(1991)

Qiu, Yinggang
See Zhuang, Song Lin: V1558,28-33(1991)
See Zhuang, Song Lin: V1558,149-153(1991)

Qu, Dong-Ning
; Burge, Ronald E.; Yuan, X.: Diffractive properties of surface-relief microstructures,V1506,152-159(1991)

Qu, Gen See Hu, Shichuang: V1572,492-496(1991)

Qu, Guangjian See Zhu, Zhenhai: V1554A,472-481(1991)

Qu, Wen-ji See Zhang, Renxiang: V1380,116-121(1991)

Qu, Zhi-Min
; Liu, Jinsheng; Xu, Liangying: Flat holographic stereograms synthesized from computer-generated images by using LiNbO3 crystal,V1238,406-411(1991)
See Jeong, Tung H.: V1396,60-70(1991)

Qu, Zhihao
; Jiang, Weixing; Peng, Huihong: Photoelastical mixed-solution method of contact problems of the roll-shape member of limited length,V1554A,503-510(1991)

Qu, Zhijin See Zhuang, Song Lin: V1354,177-179(1991)

Quach, Viet See Sawyer, Curry R.: V1567,254-263(1991)

Quadflieg, P. See Hillen, Walter: V1443,120-131(1991)

Quadrini, E. See Di Cocco, Guido: V1549,102-112(1991)

Quaglia, Enrico See Ferrari, Marco: V1431,276-283(1991)

Qualtrough, John A.
See Eaton, Frank D.: V1487,84-90(1991)
See Waldie, Arthur H.: V1487,103-108(1991)

Quan, C.
; Judge, Thomas R.; Bryanston-Cross, Peter J.: Holographic measurement of deformation using carrier fringe and FFT techniques,V1507,463-475(1991)

Quan, Jun See Liu, Ke: V1567,720-728(1991)

Quaranta, Alberto
  See Battaglin, Giancarlo: V1513,441-450(1991)
  See Gonella, Francesco: V1513,425-433(1991)

Quaranta, Fabio See Valentini, Antonio: V1400,164-170(1991)

Quarles, Gregory J.
  ; Pinto, Joseph F.; Esterowitz, Leon; Kokta, Milan R.: Comparison of flash-
  pumped Cr;Tm 2-um laser action in garnet hosts,V1410,165-174(1991)

Quate, E. G. See Ramos, P. A.: V1443,160-170(1991)

Queeckers, P. See Regel, L. L.: V1557,182-191(1991)

Quelle, Fred W.
  : Superconducting IR focal plane arrays,V1449,157-166(1991)

Quen, R. W. See Yap, Daniel: V1418,471-476(1991)

Quenzer, A.
  ; Beyer, Eckhard; Orza, Jose M.; Ricciardi, G.; Russell, J. D.; Sanz Justes,
  Pedro; Serafetinides, A.; Schuoecker, Dieter; Thorstensen, B.: The EU 194
  project: industrial applications of high-power CO2 cw lasers,V1397,753-
  759(1991)

Quesada, Jose L. See Casals, Alicia: V1482,317-324(1991)

Quesenberry, Laura A.
  ; Morris, V. A.; Neering, Ian R.; Taylor, Stuart R.: Reduced defocus
  degradation in a system for high-speed three-dimensional digital
  microscopy,V1428,177-190(1991)

Quillec, Maurice
  : Material for future InP-based optoelectronics: InGaAsP versus
  InGaAlAs,V1361,34-46(1991)

Quin, Alan See Kreider, James F.: V1488,376-388(1991)

Quinby-Hunt, Mary S.
  See Hull, Patricia G.: V1537,21-29(1991)
  See Shapiro, Daniel B.: V1537,30-41(1991)

Quinn, Anna M.
  ; Eastman, Jay M.: Optical properties of barcode symbols for laser
  scanning,V1384,138-144(1991)
  See Eastman, Jay M.: V1384,185-194(1991)

Quinn, D. L. See Merkelo, Henri: V1390,91-163(1991)

Quinn, James
  : DSP-based stabilization systems for LOROP cameras,V1538,150-166(1991)

Quinn, Peter J. See Yao, Jialing: V1572,428-433(1991)

Quinn, Rod K. See Muenchausen, Ross E.: V1394,221-229(1991)

Quintal, J. M. See Bellman, Robert H.: V1544,209-217(1991)

Quiroga, J. See Yonte, T.: V1554B,233-241(1991)

Quist, Theodore M. See Prutzer, Steven: V1480,46-61(1991)

Raab, Eric L.
  ; Tennant, Donald M.; Waskiewicz, Warren K.; MacDowell, Alastair A.;
  Freeman, Richard R.: Performance of multilayer-coated figured optics for
  soft x-rays near the diffraction limit,V1343,104-109(1991)
  See White, Donald L.: V1343,204-213(1991)

Raanes, Chris A.
  ; McNeill, John A.; Cunningham, Andrew P.: Multiple-channel correlated
  double sampling amplifier hybrid to support a 64 parallel output CCD
  array,V1358,637-643(1991)

Rabbani, Majid See Jones, Paul W.: VTT07(1991)

Rabczuk, G.
  : Mode analysis of an unstable resonator with an internal
  aperture,V1391,267-271(1991)
  : Multipass unstable negative branch resonator with a spatial filter for a
  transverse-flow CO2 laser,V1397,387-389(1991)
  See Kukiello, P.: V1397,417-419(1991)
  See Kukiello, P.: V1391,93-97(1991)

Rabin, Harold See Doyle, Michael D.: V1380,227-235(1991)

Rabinovich, Emmanuil M. See Tuchin, Valery V.: V1420,81-92(1991); 1429,62-
73(1991)

Rabinovitch, Kopel See Gilo, Mordechai: V1442,90-104(1991)

Rabl, Ari
  See Cai, Wen: V1528,118-128(1991)
  See Gordon, Jeff M.: V1528,152-162(1991)

Racca, Roberto G.
  ; Dewey, John M.: High-speed time-resolved holographic interferometer
  using solid state shutters,V1358,932-939(1991)

Racette, James See Anderson, Stephen J.: V1364,94-100(1991)

Rachele, Henry
  See Kilmer, Neal H.: V1487,109-122(1991)
  See Tunick, Arnold: V1487,51-62(1991)

Rachkovsky, Leonid I.
  ; Drobot, Igor L.; Tanin, Leonid V.: New multisource contouring
  method,V1507,450-457(1991)
  ; Tanin, Leonid V.; Rubanov, Alexander S.: About investigation of absolute
  surface relief by holographic method,V1461,232-240(1991)

Rachlin, Adam
  ; Babbitt, Richard; Lenzing, Erik; Cadotte, Roland: Superconducting devices
  and system insertion,V1477,101-114(1991)

Racz, Janos
  ; Klotz, Tamas: Knowledge representation by dynamic competitive learning
  techniques,V1469,778-783(1991)

Raczynski, Tomasz See Pawlak, Ryszard: V1391,170-173(1991)

Radcliffe, Jerry K.
  : Fiber optic link performance in the presence of reflection-induced intensity
  noise,V1366,361-371(1991)

Radehaus, C. See Pankove, Jacques I.: V1361,620-627(1991)

Radeka, V. See Braueninger, Heinrich: V1549,330-339(1991)

Radens, Carl J. See Boyd, Joseph T.: V1374,138-143(1991)

Radford, John C. See Varga, Margaret J.: V1381,92-100(1991)

Radha, Hayder
  ; Vetterli, Martin; Leonardi, Riccardo: Fast piecewise-constant approximation
  of images,V1605,475-486(1991)

Radjinia, Massud See Palcic, Branko: V1448,113-117(1991)

Radl, Bruce M.
  : Simplified pupil enlargement technique,V1457,314-317(1991)

Radspinner, David A.
  ; Fields, Robert E.; Earle, Colin W.; Denton, M. B.: Revolutionary impact of
  today's array detector technology on chemical analysis,V1439,2-
  14(1991)
  See Pomeroy, Robert S.: V1439,60-65(1991)

Ragazzoni, Roberto
  ; Bortoletto, Favio: Moving M2 mirror without pointing offset,V1542,236-
  246(1991)
  See Bortoletto, Favio: V1542,225-235(1991)

Raghavan, Raghu
  : Image recognition, learning, and control in a cellular automata
  network,V1469,89-101(1991)
  : Linear programming for learning in neural networks,V1472,139-148(1991)

Ragle, Larry O. See Oicles, Jeffrey A.: V1378,60-69(1991)

Ragu, A. See Sayeh, Mohammad R.: V1396,276-284(1991)

Rahal Arabi, Tawfik R.
  ; Murphy, Arthur T.; Sarkar, Tapan K.: Efficient and accurate dynamic
  analysis of microstrip integrated circuits,V1389,302-313(1991)

Rahe, Manfred
  ; Oertel, E.; Reinhardt, L.; Ristau, Detleu; Welling, Herbert: Absorption
  calorimetry and laser-induced damage threshold measurements of
  antireflective-coated ZnSe and metal mirrors at 10.6 um,V1441,113-
  126(1991)

Rahman, Hafiz-ur
  ; Yur, Gung; White, R. S.; Wessel, Frank J.; Song, Joshua J.; Rostoker,
  Norman: Propagation of plasma beams across the magnetic
  field,V1407,589-597(1991)

Rahman, N. A. See Barisas, B. G.: V1432,52-63(1991)

Rahman, Zia-ur
  ; Alter-Gartenberg, Rachel; Fales, Carl L.; Huck, Friedrich O.: Wiener-matrix
  image restoration beyond the sampling passband,V1488,237-248(1991)
  See Alter-Gartenberg, Rachel: V1605,745-756(1991)
  See Reichenbach, Stephen E.: V1569,422-433(1991)

Rahmati, Mohammad
  ; Hassebrook, Laurence G.; Bhushan, M.: Distortion- and intensity-invariant
  optical correlation filter system,V1567,480-489(1991)

Rahnavard, Mohammad H.
  ; Bakhtazad, A.: Excess carrier profile in a moving Gaussian spot illumination
  semiconductor panel and its use as a laser sensor,V1385,123-130(1991)
  ; Bakhtazad, A.; Abiri-Jahromi, H.: Tapered DBR for improving characteristics
  of DFB and DBR lasers,V1562,327-337(1991)
  ; Moheimany, O. R.; Abiri-Jahromi, H.: Tunability of cascaded grating that is
  used in distributed feedback laser,V1367,374-381(1991)

Raicu, Bruha
  ; Keenan, W. A.; Current, Michael I.; Mordo, David; Brennan, Roger: High-
  dose boron implantation and RTP anneal of polysilicon films for shallow
  junction diffusion sources and interconnects,V1393,161-171(1991)

Raikov, S. N. See Burakov, V. S.: V1440,270-276(1991)

Raimond, P. See Gadret, G.: V1560,226-237(1991)

Rainen, R. See Coulter, Daniel R.: V1540,119-126(1991)

Rainer, Frank
; De Marco, Frank P.; Hunt, John T.; Morgan, A. J.; Mott, Leonard P.; Marcelja, Frane; Greenberg, Michael R.: High-threshold highly reflective coatings at 1064 nm,V1441,247-254(1991)
See Campbell, John H.: V1441,444-456(1991)
See Kozlowski, Mark R.: V1441,269-282(1991)

Rainsdon, Michael D. See Andrews, John R.: V1461,110-123(1991)

Raisanen, Antti See Mears, Carl A.: V1477,221-233(1991)

Raistrick, Ian D. See Mustre de Leon, Jose: V1550,85-96(1991)

Raizen, M. G. See Bergquist, James C.: V1435,82-85(1991)

Raizer, Yu. P. See Budnik, A. P.: V1397,721-724(1991)

Raj, K. See Cheever, Charles J.: V1454,139-151(1991)

Raj, Rishi See Xue, L. A.: V1534,183-196(1991)

Raja, M. Y. See Reicher, David W.: V1441,106-112(1991)

Rajala, Sarah A.
; Nolte, Loren W.; Aanstoos, James V.: Detection of unresolved target tracks in infrared imagery,V1606,360-371(1991)

Rajan, Sumati See Wu, Shudong: V1584,415-424(1991)

Rajan, Sunder See Carvlin, Mark J.: V1426,13-21(1991)

Rajaratnam, Martha M. See Thackeray, James W.: V1466,39-52(1991)

Rajasekaran, Sanguthevar See Alameldin, Tarek K.: V1381,217-225(1991)

Rajasekaran, Suresh See Ward, Matthew O.: V1381,490-500(1991)

Rajasekharan, K.
; Michalka, Timothy: Proposed electro-optic package with bidirectional lensed coupling,V1389,568-579(1991)

Rajavel, D. See Summers, Christopher J.: V1512,170-176(1991)

Rajbenbach, Henri J. See Figue, J.: V1564,550-561(1991)

Raju, G. V.
; Rudraraju, Prasad: Camera calibration using distance invariance principles,V1385,50-56(1991)
; Rudraraju, Prasad: Rigid body motion estimation using a sequence of images from a static camera,V1388,415-424(1991)

Rak, Steven J.
; Kolodzy, Paul J.: Performance of a neural-network-based 3-D object recognition system,V1471,177-184(1991)
See Baum, Jerrold E.: V1416,209-220(1991)
See Mehanian, Courosh: V1471,200-209(1991)

Rakennus, K. See Pessa, Markus: V1361,529-542(1991)

Rakshit, S. See Neifeld, Mark A.: V1469,250-255(1991)

Ralchenko, V. G. See Armeyev, V. Y.: V1621,2-10(1991)

Rallison, Richard D.
; Schicker, Scott R.: Combat vehicle stereo HMD,V1456,179-190(1991)

Ramabadran, Tenkasi V. See Chen, Keshi: V1567,397-401(1991)

Ramachandran, Umakishore See Davis, Martin H.: V1563,176-187(1991)

Ramamurthy, Arjun See Koch, Mark W.: V1567,638-649(1991)

Raman, R. See Bahuguna, Ramendra D.: V1332,805-807(1991)

Ramanathan, Mohan
; Alp, Esen E.; Mini, Susan M.; Salem-Sugui, S.; Bommannavar, A.: Polarized x-ray absorption spectroscopy for the study of superconductors and magnetic materials,V1548,168-178(1991)
See Alp, Esen E.: V1345,137-145(1991)
See Mini, Susan M.: V1345,260-269(1991)

Ramasesha, S.
See Albert, I. D.: V1436,179-189(1991)
See Frank, Arthur J.: V1436(1991)

Ramaswamy, Raju
: Performance analysis of lightwave packet communication networks,V1364,153-162(1991)

Ramaswamy, Ramu V.
See Cheng, Hsing C.: V1474,291-297(1991)
See Srivastava, Ramakant: V1583,2-13(1991)

Rambaud, C. See Hochstrasser, Robin M.: V1403,221-229(1991)

Ramee, Stephen R.
; White, Christopher J.; Mesa, Juan E.; Jain, Ashit; Collins, Tyrone J.: Percutaneous coronary angioscopy during coronary angioplasty: clinical findings and implications,V1420,199-202(1991)
See White, Christopher J.: V1425,130-133(1991)

Ramesh, Nagarajan
; Sethi, Ishwar K.; Cheung, Huey: Automatic target detection for surveillance,V1468,72-80(1991)

Ramesh, S. K.
; Smith, Thomas D.: Design and evaluation of optical switching architectures,V1474,208-211(1991)

Ramesh, Visvanathan See Haralick, Robert M.: V1406,2-16(1991)

Ramian, Gerald
: Properties of the new UCSB free-electron lasers,V1552,57-68(1991)

Ramirez, G. See Ojeda-Castaneda, Jorge: V1500,252-255(1991)

Ramirez, Manuel
; Mitra, Sunanda: Three-dimensional target recognition from fusion of dense range and intensity images,V1567,632-637(1991)

Ramirez, W. F. See Carroll, Thomas A.: V1464,222-231(1991)

Ramola, Ramesh C. See Yadav, M. S.: V1362,811-819(1991)

Ramos, Josiane See Marcerou, Jean-Francois: V1373,168-186(1991)

Ramos, P. A.
; Lawson, Michael A.; Wika, Kevin G.; Allison, Stephen W.; Quate, E. G.; Molloy, J. A.; Ritter, Rogers C.; Gillies, George T.: Low-dose magnetic-field-immune biplanar fluoroscopy for neurosurgery,V1443,160-170(1991)

Ramsdale, Christine A. See Dent, Andrew J.: V1550,97-107(1991)

Ramsey, Brian See Coden, Michael H.: V1364,22-39(1991)

Ramsey, Brian D.
See Kaaret, Philip E.: V1548,106-117(1991)
See Sakurai, Hirohisa: V1549,20-27(1991)

Ramsey, J. M. See Shaw, Robert W.: V1534,170-174(1991)

Ramsey, Timothy S. See Pacelli, Jean L.: V1472,76-84(1991)

Ramstad, Tor A. See Aase, Sven O.: V1605,734-744(1991)

Ran, Jun G. See Yang, Bang C.: V1519,864-865(1991)

Ranat, Bhadrayu M. See Armstrong, George R.: VCR38,120-141(1991)

Randall, David See Hoffman, Alan: V1540,297-302(1991)

Rando, Nicola
; Peacock, Anthony J.; Foden, Clare; van Dordrecht, Axel; Engelhardt, Ralph; Lumley, John M.; Pereira, Carl: Nb tunnel junctions as x-ray spectrometers,V1549,340-356(1991)

Randolph, D.
See Aridgides, Athanasios: V1481,110-116(1991)
See Fernandez, Manuel F.: V1481,172-179(1991)

Randolph, David
; Krishnan, Ganapathy: Off-line machine recognition of forgeries,V1386,255-264(1991)

Randorf, Jeffrey A.
: Thermal analysis of a small expendable tether satellite package,V1495,259-267(1991)

Randoshkin, Vladimir V.
: Ways of the high-speed increasing of magneto-optical spatial light modulators,V1469,796-803(1991)

Raney, Daniel See Neev, Joseph: V1427,162-172(1991)

Raney, R. K.
; Specter, Christine N.: TREIS: a concept for a user-affordable, user-friendly radar satellite system for tropical forest monitoring,V1492,298-306(1991)

Rangachar, Ramesh M.
; Hong, Tsai-Hong; Herman, Martin; Luck, Randall L.: Analysis of optical flow estimation using epipolar plane images,V1382,376-385(1991)
; Hong, Tsai-Hong; Herman, Martin; Luck, Randall L.: Three-dimensional reconstruction from optical flow using temporal integration,V1382,331-340(1991)

Rangarajan, G. See Sharp, William M.: V1407,535-545(1991)

Rangarajan, N. See Murali, K. R.: V1536,289-295(1991)

Rangarajan, S. K. See Murali, K. R.: V1536,289-295(1991)

Rangelow, Iwilo W.
: High-resolution tri-level process by downstream-microwave rf-biased etching,V1392,180-184(1991)
; Fichelscher, Andreas: Chlorine or bromine chemistry in reactive ion etching Si-trench etching,V1392,240-245(1991)
See Fichelscher, Andreas: V1392,77-83(1991)

Ranieri, V. A. See Deutsch, Alina: V1389,161-176(1991)

Ranka, Sanjay See Choudhary, Alok: V1406,148-161(1991)

Ranken, D. M. See George, John S.: V1443,37-51(1991)

Rankin, Richard N. See Picot, Paul A.: V1444,206-213(1991)

Ranshaw, M. J. See Potter, Duncan J.: V1564,363-372(1991)

Rantala, Aarne E. See Vepsalainen, Ari M.: V1452,64-75(1991)

Rao, Bhaskar D. See Soni, Tarun: V1565,338-344(1991)

Rao, K. R.
See Ben-Arie, Jezekiel: V1569,367-374(1991)
See Nam, J. Y.: V1605,202-213(1991)
See Nam, J. Y.: V1452,485-496(1991)

Rao, Kashi
; Liou, James J.: Salient contour extraction for target recognition,V1482,293-306(1991)
See Ooi, James: V1468,740-751(1991)

Rao, M. K.
; Tam, Siu-Chung: Microprocessor-based laser range finder,V1399,116-121(1991)

Rao, Navalgund A.
; Zhu, Hui: Modeling ultrasound speckle formation and its dependence on imaging system's response,V1443,81-95(1991)

Rao, P. N. See Chandrasekhar, N. S.: V1390,523-536(1991)

Rao, S. M.
; Cao, C.; Batra, Ashok K.; Lal, Ravindra B.; Mookherji, Tripty K.: Ground-based experiments on the growth and characterization of L-arginine phosphate crystals,V1557,283-292(1991)

Rao, V. K. See Alex, T. K.: V1478,101-105(1991)

Rao, Veena
; Kosbar, Laura L.; Frank, Curtis W.; Pease, R. F.: Effect of sensitizer spatial distribution on dissolution inhibition in novolak/diazonaphthoquinone resists,V1466,309-323(1991)

Rao, Xuejun See Jiang, Wen-Han: V1542,130-137(1991)

Rao, Yun-Jiang
; Culshaw, Brian: Analysis of the self-oscillation phenomenon of fiber optically addressed silicon microresonators,V1506,126-133(1991)
; Uttamchandani, Deepak G.; Culshaw, Brian: Compact probe for all-fiber optically addressed silicon cantilever microresonators,V1572,287-292(1991)

Rapoport, Eliezer
; Bar, Doron; Shiloh, Klara: Rapid procedure for obtaining time-average interferograms of vibrating bodies,V1442,383-391(1991)

Rapp, Donald
: Need for active structures in future large IR and sub-mm telescopes,V1542,328-358(1991)

Rapp, Gerard See Joeckle, Rene C.: V1397,683-687(1991)

Raptakis, E. See Hontzopoulos, Elias I.: V1397,761-768(1991)

Raqueno, Rolando See Schott, John R.: V1540,533-549(1991)

Raquet, Charles A. See Shalkhauser, Kurt A.: V1475,204-209(1991)

Rarenko, I. M. See Vertiy, Alexey A.: V1361,1070-1078(1991)

Ras, Zbigniew W.
: Routing in distributed information systems,V1470,76-87(1991)

Rasanen, Jari See Silvennoinen, Raimo: V1574,84-88(1991)

Rasch, Andreas
; Buss, Wolfgang; Goering, Rolf; Steinberg, Steffen; Karthe, Wolfgang: Optical carrier modulation by integrated optical devices in lithium niobate,V1522,83-92(1991)
; Goering, Rolf; Karthe, Wolfgang; Schroeter, Siegmund; Ecke, Wolfgang; Schwotzer, Guenter; Willsch, Reinhardt: Integrated optical device in a fiber gyroscope,V1511,149-154(1991)

Rascoe, Daniel L.
See Chew, Wilbert: V1477,95-100(1991)
See Cooley, Thomas W.: V1475,243-247(1991)

Rashmi, Dr. See Chitnis, Vijay T.: V1332,613-622(1991)

Rasmussen, Mary See Evenhouse, Raymond J.: V1380,147-156(1991)

Rast, Michael
; Readings, C. J.: ESA Earth observation polar platform program,V1490,51-58(1991)

Rastegar, Sohi
; Glenn, T.: Analysis of kinetic rate modeling of thermal damage in laser-irradiated tissue,V1427,300-306(1991)
See Zoghi, Behbood: V1420,63-71(1991)

Rastogi, Pramod K.
; Pflug, Leopold: Broad-range holographic contouring of diffuse surfaces by dual-beam illumination: study of two related techniques,V1554B,48-55(1991)

Rastopov, S. F.
; Sukhodolsky, A. T.: Sound generation by thermocavitation-induced cw laser in solutions,V1440,127-134(1991)

Ratib, Osman M.
; Ligier, Yves; Hochstrasser, Denis; Scherrer, Jean-Raoul: Hospital-integrated PACS at the University Hospital of Geneva,V1446,330-340(1991)
See Stut, W. J.: V1446,396-404(1991)

Ratier, A. See Durpaire, Jean-Pierre: V1490,23-38(1991)

Ratner, A. See Detwiler, Paul W.: V1425,149-155(1991)

Ratzlaff, Eugene H.
; Grinvald, Amiran: Optical imaging of cortical activity in the living brain,V1439,88-94(1991)

Rauchmiller, Robert F.
; Vanderbok, Raymond S.: Spectral signature analysis for industrial inspection,V1384,100-114(1991)

Rauh, R. D. See Carrabba, Michael M.: V1434,127-134(1991)

Raupp, Gregory B. See Ferry, David K.: V1596,2-11(1991)

Rava, Richard P.
; Richards-Kortum, Rebecca; Fitzmaurice, Maryann; Cothren, Robert M.; Petras, Robert; Sivak, Michael V.; Levine, Howard: Early detection of dysplasia in colon and bladder tissue using laser-induced fluorescence,V1426,68-78(1991)

Ravat, Christian J.
; Mestre, J. P.; Rose, C.; Schorter, M.: Radar-optronic tracking experiment for short- and medium-range aerial combat,V1478,239-246(1991)

Raveglia, Dalmazio See Brofferio, Sergio C.: V1605,894-905(1991)

Ravel, Mihir K.
; Reinheimer, Alice L.: Backside-thinned CCDs for keV electron detection,V1447,109-122(1991)

Ravi, S. See McVittie, James P.: V1392,126-138(1991)

Ravi-Chandar, K.
: Use of scattered-light photoelasticity at crack tips,V1554A,228-238(1991)

Ravichandran, Gopalan
; Casasent, David P.: Advanced in-plane rotation-invariant filter results,V1567,466-479(1991)
; Casasent, David P.: Noise and discrimination performance of the MINACE optical correlation filter,V1471,233-248(1991)
See Casasent, David P.: V1384,324-337(1991)
See Casasent, David P.: V1384,344-354(1991)

Raviv, Daniel
: Reconstruction during camera fixation,V1382,312-319(1991)

Ray, Frank B.
: Active optics technology: an overview,V1532,188-206(1991)

Ray, Knute A. See Crocker, James H.: V1494,434-439(1991)

Ray, Sankar See Freeman, J. L.: V1476,320-325(1991)

Ray, Wayne See Cooper, Kent: V1392,253-264(1991)

Rayapati, Venkatapathi N.
; Mukhedkar, Dinkar: Interconnection problems in VLSI random access memory chip,V1389,98-109(1991)

Rayces, Juan L.
; Lebich, Lan: Thermal compensation of infrared achromatic objectives with three optical materials,V1354,752-759(1991)

Rayfield, George W.
: Nonlinear absorbance effects in bacteriorhodopsin,V1436,150-159(1991)

Rayman, M. D. See Gershman, Robert: V1494,160-167(1991)

Raymer, M. G.
; Gadomski, Wojciech; Hodges, Steven E.; Adkison, D.: Deterministic and quantum noise in dye lasers,V1376,128-131(1991)

Raymond, Andre
; Chaubet, C.; Razeghi, Manijeh: Far-IR magneto-emission study of the quantum-hall state and breakdown of the quantum-hall effect,V1362,275-281(1991)

Raynes, Alan S. See Cranmer, David C.: V1330,152-163(1991)

Raynor, Brian See Huelsmann, Axel: V1465,201-208(1991)

Raynor, Jeffrey M. See Gale, Michael T.: V1506,65-70(1991)

Razavi, Mahmood
; Sayre, James W.; Simons, Margaret A.; Hamedaninia, Azar; Boechat, Maria I.; Hall, Theodore R.; Kangarloo, Hooshang; Taira, Ricky K.; Chuang, Keh-Shih; Kashifian, Payam: Clinical evaluation of a 2K x 2K workstation for primary diagnosis in pediatric radiology,V1446,24-34(1991)
See McNitt-Gray, Michael F.: V1445,468-478(1991)

Razdan, Anil K.
; Singh, Brahm P.; Chopra, S.; Modi, M. B.: Characterization of simulated and open-air atmospheric turbulence,V1558,384-388(1991)

Razdan, Rikki
; Kielar, Alan: Eye-slaved pointing system for tele-operator control,V1388,361-371(1991)

Razdobarin, G. T. See Kozhevnikov, Nikolai M.: V1584,138-144(1991)

Razeghi, M. M. See Chong, Chae K.: V1407,226-233(1991)

Razeghi, Manijeh
: <author>Optoelectronic Materials and Device Concepts,VPM05(1991)
ed.: Physical Concepts of Materials for Novel Optoelectronic Device Applications I: Materials Growth and Characterization,V1361(1991)
ed.: Physical Concepts of Materials for Novel Optoelectronic Device Applications II: Device Physics and Applications,V1362(1991)
See Aboulhouda, S.: V1362,494-498(1991)
See Acher, O.: V1361,1156-1163(1991)
See Drevillon, Bernard: V1361,200-212(1991)
See Hansen, Ole P.: V1362,192-198(1991)
See MacLeod, Roderick W.: V1361,562-567(1991)
See Raymond, Andre: V1362,275-281(1991)

Razon, N. See Rochkind, Simeone: V1428,52-58(1991)

Razzaghi-Masoud, Mahmoud
: High-resolution high-speed laser recording and printing using low-speed polygons,V1458,145-153(1991)

Re, Stefano See Costa, Enrico: V1343,469-476(1991)

Read, C. See Brecher, Virginia H.: V1473,39-51(1991)

Read, Harold E.
; Merker, M.; Gurtman, G. A.; Wilson, Russell S.: Laser-induced damage of diamond films,V1441,345-359(1991)

Readings, C. J. See Rast, Michael: V1490,51-58(1991)

Reali, Giancarlo C.
See Agnesi, Antoniangelo: V1415,242-247(1991)
See Agnesi, Antoniangelo: V1415,104-109(1991)

Ream, Stanley L.
: Market for multiaxis laser machine tools,V1517,136-149(1991)
; Dausinger, Friedrich; Fujioka, Tomoo; eds.: <author>ICALEO '90 Laser Materials Processing,V1601(1991)

Reamey, Robert H.
; Montoya, Wayne; Wartenberg, Mark: Switching speeds in NCAP displays: dependence on collection angle and wavelength,V1455,39-44(1991)

Reardon, A. C. See Headley, Clifford: V1497,197-201(1991)

Reardon, Frank J.
; Salutz, James R.: Image acquisition unit for the Mayo/IBM PACS project,V1446,481-491(1991)

Reardon, Janine See Caulfield, H. J.: V1562,103-106(1991)

Reardon, Patrick J.
; Chipman, Russell A.: Lens with maximum power occupying a given cylindrical volume,V1354,234-239(1991)

Reason, C. J. See Turcu, I. C.: V1503,391-405(1991)

Reategui, Rodolfo See Milne, Edmund A.: V1486,151-161(1991)

Reaves, Casper M. See Sturm, James C.: V1393,309-315(1991)

Rebane, A. See Ollikainen, Olavi: V1621,351-361(1991)

Rebeiz, Gabriel M.
; Ulaby, Fawwaz T.: Millimeter wave monolithic antenna and receiver arrays for space-based applications,V1475,199-203(1991)

Rebhan, Ulrich See Kahlert, Hans-Juergen: V1463,604-609(1991)

Rebuffie, Jean-Claude
; Mens, Alain: New streak tubes of the P500 series: features and experimental results,V1358,511-523(1991)
See Froehly, Claude: V1358,532-540(1991)
See Mens, Alain: V1358,315-328(1991)

Rechmann, Peter
; Tourmann, J. L.; Kaufmann, Raimund: LAser Microprobe Mass Spectrometry (LAMMS) in dental science: basic principles, instrumentation, and applications,V1424,106-115(1991)
See Hennig, Thomas: V1424,99-105(1991)

Recio, Miguel
; Ruiz, Ana; Melendez, J.; Rodriguez, Jose M.; Armelles, Gaspar; Dotor, M. L.; Briones, Fernando: Novel GaP/InP strained heterostructures: growth, characterization,and technological perspectives,V1361,469-478(1991)

Reddi, Elena See Jori, Guilio: V1525,367-376(1991)

Redding, David C.
; Breckenridge, William G.: Linearized ray-trace analysis,V1354,216-221(1991)
; Breckenridge, William G.; Lau, Kenneth; Sevaston, George E.; Levine, Bruce M.; Shaklan, Stuart B.: Segmented mirror figure control for a space-based far-IR astronomical telescope,V1489,201-215(1991)

Reddy, Mahesh C.
; Krinsky, Jeffrey A.: Effects of ionizing radiation on fiber optic systems and components for use in mobile platforms,V1369,107-113(1991)
See Krinsky, Jeffrey A.: V1366,191-203(1991)

Reddy, Subhash C. See Huang, Thomas S.: V1605,234-241(1991)

Redfern, R. M. See Glynn, Thomas J.: V1598,200-205(1991)

Redko, S. V. See Guo, S. P.: V1519,400-404(1991)

Redmond, Ian R. See Craig, Robert G.: V1505,76-86(1991)

Redner, Alex S.
; Adamovsky, Grigory; Wesson, Laurence N.: Photoelastic transducer for high-temperature applications,V1332,775-782(1991)

Redus, Wesley D.
: Airborne imaging system performance model,V1498,2-23(1991)

Reed, Irving S. See Sperling, I.: V1481,2-11(1991)

Reed, Murray K.
; Tyminski, Jacek K.; Bischel, William K.: Second-harmonic generation of mode-locked Nd:YAG and Nd:YLF lasers using LiB3O5,V1410,179-184(1991)

Reed, Richard See Delamere, W. A.: V1447,288-297(1991); 1479,21-30(1991)

Reed, Todd R.
See Algazi, V. R.: V1567,589-598(1991)
See Algazi, V. R.: V1605,329-338(1991)

Reedy, Herman E. See Berger, R. M.: V1397,611-618(1991)

Reekie, Laurence See Cowle, Gregory J.: V1373,54-65(1991)

Reel, Richard L. See Flannery, David L.: V1564,65-77(1991)

Reeves, Adam J. See Yang, Jian: V1606,520-530(1991)

Reeves, Geoffrey K. See Harrison, H. B.: V1596,52-59(1991)

Reeves, James B.
; Blosser, Timothy H.; Colenbrander, V. F.: Analysis of silage composition by near-infrared reflectance spectroscopy,V1379,28-38(1991)

Reeves, Richard C.
; Schaibly, John H.: Microcomputer-based workstation for simulation and analysis of background and target IR signatures,V1486,85-101(1991)

Reeves, Stanley J.
; Mersereau, Russell M.: Optimal regularization parameter estimation for image restoration,V1452,127-138(1991)

Refenes, A. N.
; Jain, N.; Alsulaiman, Mansour M.: Integrated neural network system for histological image understanding,V1386,62-74(1991)

Reffner, John A.
: Infrared microanalysis of contaminants at grazing incidence,V1437,89-94(1991)

Refregier, Philippe
See Figue, J.: V1564,550-561(1991)
See Maze, Sylvie: V1505,20-31(1991)

Regal, Anne-Marie
; Takita, Hiroshi: Endobronchial occlusive disease: Nd:YAG or PDT?,V1426,271-278(1991)

Regalia, Phillip A.
: Numerical stability issues in fast least-squares adaptive algorithms,V1565,2-13(1991)

Re Garbagnati, Giuseppe See Nava, Enzo: V1417,307-315(1991)

Regel, L. L.
; Vedernikov, A. A.; Queeckers, P.; Legros, J. C.: Crystal separation from mother solution and conservation under microgravity conditions using inert liquid,V1557,182-191(1991)

Regensburger, Uwe
; Graefe, Volker: Object classification for obstacle avoidance,V1388,112-119(1991)

Regnault, P. See Buczek, Harthmuth: V1574,48-57(1991)

Rehak, P. See Braeuninger, Heinrich: V1549,330-339(1991)

Rehberg, Raymond A. See Clark, David L.: V1540,303-311(1991)

Rehkugler, Gerald E. See Throop, James A.: V1379,124-135(1991)

Rehm, Kelly
; Seeley, George W.; Dallas, William J.: Image and modality control issues in the objective evaluation of manipulation techniques for digital chest images,V1445,24-35(1991)

Reiber, J. H. See Dumay, Adrie C.: V1445,38-46(1991)

Reibman, Amy R. See Lynch, William E.: V1605,285-296(1991)

Reich, Ella See Miller, Kurt: V1426,378-383(1991)

Reich, M. See Moeller, Werner: V1504,47-54(1991)

Reichard, Karl M.
; Lindner, Douglas K.: Distributed-effect optical fiber sensors for trusses and plates,V1480,115-125(1991)
; Lindner, Douglas K.; Claus, Richard O.: Vibration sensing in flexible structures using a distributed-effect modal domain optical fiber sensor,V1489,218-229(1991)

Reichel, Erich
; Schmidt-Kloiber, Heinz; Paltauf, Guenther; Groke, Karl: Bifunctional irrigation liquid as an ideal energy converter for laser lithotripsy with nanosecond laser pulses,V1421,129-133(1991)
See Gratzl, Thomas: V1427,55-62(1991)

Reichenbach, Stephen E.
; Park, Stephen K.; Alter-Gartenberg, Rachel; Rahman, Zia-ur: Artificial scenes and simulated imaging,V1569,422-433(1991)
See Alter-Gartenberg, Rachel: V1605,745-756(1991)

Reicher, David W.
; Wilson, Scott R.; Kranenberg, C. F.; Raja, M. Y.; McNeil, John R.; Brueck, Steven R.: Photoacoustic characterization of surface absorption,V1441,106-112(1991)
See Wilson, Scott R.: V1441,82-86(1991)

Reichert, Abraham See Cohen, Jonathan: V1456,250-261(1991)

Reichert, Patrick See Snyder, James J.: V1544,146-151(1991)

Reichert, William M. See Potter, B. L.: V1368,251-257(1991)

Reichle, Henry G.
: Measurement of tropospheric carbon monoxide using gas filter radiometers,V1491,15-25(1991)

Reichling, M.
See Wu, Zhouling: V1441,214-227(1991)
See Wu, Zhouling: V1441,200-213(1991)

Reichman, Joseph
: Photoconductivity decay method for determining minority carrier lifetime of p-type HgCdTe,V1484,31-38(1991)

Reichmanis, Elsa
See Nalamasu, Omkaram: V1466,13-25(1991)
See Novembre, Anthony E.: V1466,89-99(1991)

Reid, John See Jursich, Gregory M.: V1412,115-122(1991)

Reid, Max B.
; Hine, Butler P.: Planetary lander guidance using binary phase-only filters,V1564,384-394(1991)
See Downie, John D.: V1564,308-319(1991)
See Hine, Butler P.: V1564,416-426(1991)

Reid, Philip J.
; Doig, Stephen J.; Mathies, Richard A.: Pericyclic photochemical ring-opening reactions are complete in picoseconds: a time-resolved UV resonance Raman study,V1432,172-183(1991)
See Doig, Stephen J.: V1432,184-196(1991)

Reif, J. See Hunger, Hans E.: V1441,283-286(1991)

Reif, Philip G. See Sayyah, Keyvan: V1455,249-254(1991)

Reifsnider, Kenneth L. See Fogg, Brian R.: V1588,14-25(1991)

Reihani, Kamran
; Thompson, Wiley E.: Clue derivation and selection activities in a robot vision system,V1468,305-312(1991)
; Thompson, Wiley E.: Geometrical and morphological image processing algorithm,V1452,319-329(1991)
; Thompson, Wiley E.: Real-time region hierarchy and identification algorithm,V1449,99-108(1991)

Reiland, Werner See Lange, Robert: V1492,24-37(1991)

Reiley, Michael F.
; Carmer, Dwayne C.; Pont, W. F.: Three-dimensional laser radar simulation for autonomous spacecraft landing,V1416,295-303(1991)

Reilly, James F. See Behlau, Jerry L.: V1530,218-230(1991)

Reilly, James P.
: Debris plume phenomenology for laser-material interaction in high-speed flowfields,V1397,661-674(1991)
; Lander, Mike L.; Maxwell, K.; Hull, R. J.: Design, construction, and operation of 65 kilowatt carbon dioxide electric discharge coaxial laser device,V1397,339-354(1991)
See Chen, Wei G.: V1566,464-475(1991)

Reinen, Tor A.
: Noise reduction in heart movies by motion-compensated filtering,V1606,755-763(1991)

Reinhardt, L. See Rahe, Manfred: V1441,113-126(1991)

Reinhart, Larry J.
See Malo, Bernard: V1590,83-93(1991)
See Nath, Dilip K.: V1489,17-32(1991)
See Oh, Kyunghwan: V1590,94-100(1991)

Reinhart, Werner
: Advanced hot stamping foil-based OVD technology: an overview about security applications,V1509,67-72(1991)

Reinhart, William F.
: Depth cueing for visual search and cursor positioning,V1457,221-232(1991)

Reinheimer, Alice L. See Ravel, Mihir K.: V1447,109-122(1991)

Reinhold, Ralph R.
: Direct view thermal imager,V1447,251-262(1991)

Reinholdtsen, Paul A. See Bullock, Michael E.: V1471,291-302(1991)

Reinhoudt, D. N. See van Hulst, Niek F.: V1361,581-588(1991)

Reining, Gale See Thomas, Melvin L.: V1456,65-75(1991)

Reinisch, Lou See Li, Zhao-zhang: V1427,152-161(1991)

Reinisch, Raymond See Vitrant, Guy: V1545,225-231(1991)

Reintjes, John F.
See Duncan, Michael D.: V1409,127-134(1991)
See Fisher, Robert A.: V1409(1991)

Reiser, Martin P. See Guharay, Samar K.: V1407,610-619(1991)

Reisert, Norbert
: Application and machining of Zerodur for optical purposes,V1400,171-177(1991)

Reisfeld, Renata
; Eyal, Marek; Chernyak, Valery; Jorgensen, Christian K.: Glasses including quantum dots of cadmium sulfide, silver, and laser dyes,V1590,215-228(1991)
; Minti, Harry; Eyal, Marek: Active glasses prepared by the sol-gel method including islands of CdS or silver,V1513,360-367(1991)
See Rotman, Stanley R.: V1442,194-204(1991)

Reiss, Don See Springer, John M.: V1557,192-196(1991)

Reister, K. R. See Twiggs, Robert J.: V1495,12-18(1991)

Reith, Leslie A. See Hartman, Davis H.: V1390,368-376(1991)

Reithmaier, Karl D. See Chang, John H.: V1492,38-42(1991)

Reivich, Martin See Gee, Jim C.: V1445,226-234(1991)

Rejman-Greene, Marek A. See Barnes, Nigel: V1389,477-483(1991)

Remiens, D.
; Hornung, V.; Rose, B.; Robein, D.: Buried-ridge-stripe lasers monolithically integrated with butt-coupled passive waveguides for OEIC,V1362,323-330(1991)

Remiot, Christian See Mondot, Michel: V1558,351-361(1991)

Remitz, K. E. See Speit, Burkhard: V1361,1128-1131(1991)

Remke, Ronald L. See Kanofsky, Alvin S.: V1374,48-58(1991)

Remo, John L. See Turner, Robert E.: V1377,99-106(1991)

Rempel, U.
; von Maltzan, B.; von Borczykowski, C.: Fluorescence detected energy transport in selectively excited porphyrin dimers,V1403,631-634(1991)

Ren, A. See Kuo, Spencer P.: V1407,272-280(1991)

Ren, Baorui See Ming, Hai: V1572,27-31(1991)

Ren, Biao See Gennert, Michael A.: V1385,268-279(1991)

Ren, Cong X.
See Li, Yi J.: V1519,800-804(1991)
See Wu, Zi L.: V1519,618-624(1991)

Ren, Qiushi
; Gailitis, Raymond P.; Thompson, Keith P.; Penney, Carl M.; Lin, J. T.; Waring, George O.: Corneal refractive surgery using an ultraviolet (213 nm) solid state laser,V1423,129-139(1991)

Ren, Tao See Weng, Zhicheng: V1527,349-356(1991)

Ren, Victor
; Hatfield, Donald J.; Deacutis, Martin: Preliminary review of imaging standards,VCR37,54-67(1991)

Ren, Xin
See Lu, Xiaoming: V1572,248-251(1991)
See Lu, Xiaoming: V1572,304-307(1991)

Renard, Stephane
   ; Valette, Serge: Magneto-optical reading and writing integrated heads: a way to a multigigabyte multi-rigid-disk drive,V1499,238-247(1991)

Renault, Herve See Lund, Glenn I.: V1492,153-165(1991)

Rendell, John T.
   : New rippleflash system for large-area high-intensity lighting in harsh environments,V1358,806-810(1991)

Rendina, Ivo
   ; Cocorullo, Giuseppe: All-silicon Fabry-Perot modulator based on thermo-optic effect,V1583,338-343(1991)
   See Cocorullo, Giuseppe: V1374,132-137(1991)

Renhorn, Ingmar G.
   ; Letalick, Dietmar; Millnert, Mille: Terrain and target segmentation using coherent laser radar,V1480,35-45(1991)

Renieri, Alberto See Ciocci, Franco: V1501,154-162(1991)

Renk, Karl F. See Beregulin, Eugene V.: V1362,853-862(1991)

Renn, Curtiss N. See Synovec, Robert E.: V1435,128-139(1991)

Renner, Daniel
   ed.: Laser Diode Technology and Applications III,V1418(1991)
   See Cheng, Wood-Hi: V1418,279-283(1991)
   See Jiang, Ching-Long: V1418,261-271(1991)

Rennie, A. R. See Banfi, Giampiero P.: V1361,874-880(1991)

Renotte, Yvon L.
   See Defosse, Yves: V1507,277-287(1991)
   See Leclere, Philippe: V1507,339-344(1991)
   See Leclere, Philippe: V1559,298-307(1991)

Rentzepis, Peter M.
   ; Dvornikov, Alexander S.: Volume optical memory by two-photon interaction,V1563,198-207(1991)
   See Esener, Sadik C.: V1499,144-147(1991)

Repeyev, Yu A. See Terenetskaya, I. P.: V1403,500-503(1991)

Reppin, C. See Braueninger, Heinrich: V1549,330-339(1991)

Requena, Philippe See Fontaine, Chantal: V1362,59-66(1991)

Rerko, Ronald M. See Oleinick, Nancy L.: V1427,90-100(1991)

Resa, A. M. See Trelles, Mario A.: V1403,781-798(1991)

Rescia, S. See Braueninger, Heinrich: V1549,330-339(1991)

Reshetnikova, Irina N.
   See Mirzoeva, Larisa A.: V1540,444-449(1991)
   See Shevchenko, N. P.: V1574,66-71(1991)

Reshetnyak, Victor Y. See Pinkevich, Igor P.: V1455,122-133(1991)

Resnikoff, Howard L.
   : Wavelets and adaptive signal processing,V1565,370-382(1991)

Restall, Edward J. See Craig, Robert G.: V1505,76-86(1991)

Restrepo, Alfredo
   : Nonlinear regression for signal processing,V1451,258-268(1991)

Rettig, John B. See Hayden, Leonard A.: V1389,205-214(1991)

Reulet, Jean-Francois See Fratter, C.: V1490,59-73(1991)

Revankar, Shriram
   ; Sher, David B.; Rosenthal, Steven: Collaborative processing to extract myocardium from a sequence of two-dimensional echocardiograms,V1459,268-273(1991)

Revel, D. See Friboulet, Denis: V1445,106-117(1991)

Rey, J. C. See McVittie, James P.: V1392,126-138(1991)

Reyes, Roy M. See Spence, Scott E.: V1474,199-207(1991)

Reyes-Mena, Arturo See Williams, Memorie: V1549,147-154(1991)

Reynaud, Roger
   ; Chebira, Abdennasser; Demoment, Guy: Parallel implementation of some fast adaptive algorithms on a digital signal processor network,V1566,302-311(1991)
   See Demoment, Guy: V1565,357-367(1991)

Reynes, Anne M.
   ; Diebold, Simon; Lignon, Dominique; Granjon, Yves; Guillemin, Francois: Light dosimetry in vivo in interstitial photodynamic therapy of human tumors,V1525,177-182(1991)

Reynolds, Larry D.
   ; Shaw, C. B.: Experimental method for gas kinetic temperature measurements in a thermal plasma,V1554B,622-631(1991)

Reynolds, Rodger L.
   ; Hageniers, Omer L.: Rigorous optical theory of the D Sight phenomenon,V1332,85-96(1991)

Reynolds, William R. See Meitzler, Thomas J.: V1486,380-389(1991)

Reznik, G. See Papaioannou, Thanassis: V1420,203-211(1991)

Reznikov, Yuriy See Pinkevich, Igor P.: V1455,122-133(1991)

Rhee, M. J. See Kung, Chun C.: V1378,250-258(1991)

Rhodes, Anthony
   See Black, Johnathan D.: V1424,20-22(1991)
   See Black, Johnathan D.: V1424,12-15(1991)
   See Lane, Gregory J.: V1424,7-11(1991)
   See Meller, Menachem M.: V1424,60-61(1991)
   See Uppal, Gurvinder S.: V1424,51-52(1991)

Rhodes, C. G. See Nelson, Jeffrey G.: V1394,191-195(1991)

Rhodes, Charles K. See Haddad, Waleed S.: V1448,81-88(1991)

Rhodes, Marvin D. See Will, Ralph W.: V1387,60-71(1991)

Rhodes, Michael L. See Liao, Wen-gen: V1444,47-55(1991)

Rhodes, William T. See Masters, Barry R.: V1429,82-90(1991)

Riani, Massimo
   ; Masulli, Francesco; Simonotto, Enrico: Perceptual alternation of ambiguous patterns: a model based on an artificial neural network,V1469,166-177(1991)

Ribak, Erez N. See Ben-Shalom, Ami: V1486,238-257(1991)

Ribezzo, Sergio See Barbini, Roberto: V1503,363-374(1991)

Riblett, Susan E. See Bugner, Douglas E.: V1458,162-178(1991)

Ricciardi, G. See Quenzer, A.: V1397,753-759(1991)

Rice, J. E. See Burland, Donald M.: V1560,111-119(1991)

Rice, Patrick D.
   ; Thorne, John B.; Bobbitt, Donald R.: New approaches to signal-to-noise ratio optimization in background-limited photothermal measurements,V1435,104-113(1991)

Rice, Robert R. See Probst, David K.: V1417,346-357(1991)

Rice, Theodore R. See Blair, William D.: V1482,234-245(1991)

Rich, Brian W.
   : Infrared simulation of missile dome heating,V1540,781-786(1991)

Rich, Chris C.
   ; Cook, David M.: Lippmann volume holographic filters for Rayleigh line rejection in Raman spectroscopy,V1461,2-7(1991)
   See Jannson, Tomasz: V1545,42-63(1991)
   See Sadovnik, Lev S.: V1559,424-428(1991)

Richard, Jean-Claude See Gendron, Eric: V1542,297-307(1991)

Richard, Peter N. See Murray, Brian W.: V1485,106-115(1991)

Richard, Ralph M.
   ; Valente, Tina M.: Interference fit equations for lens cell design,V1533,12-20(1991)
   See Valente, Tina M.: V1533,21-26(1991)
   See Vukobratovich, Daniel: V1396,522-534(1991)

Richards, David R.
   ; Fasol, Gerhard; Ploog, Klaus H.: Raman scattering determination of nonpersistent optical control of electron density in a heterojunction,V1361,246-254(1991)

Richards, Evan
   ; Percival, Jeff; Nelson, Matthew; Hatter, Edward; Fitch, John E.; White, Richard L.: Initial performance of the high-speed photometer,V1494,40-48(1991)

Richards, James B.
   ; Weiland, Timothy L.; Prior, John A.: Evaluation of dynamic range for LLNL streak cameras using high-contrast pulses and "pulse podiatry" on the NOVA laser system,V1346,384-389(1991)

Richards, Kent F.
   ; Christensen, Brent J.: Laparoscopic appendectomy,V1421,198-202(1991)

Richards, Paul L.
   : Superconducting bolometers: high-Tc and low-Tc,V1477,2-6(1991)
   See Mears, Carl A.: V1477,221-233(1991)

Richards, Robin
   See Coombes, Anne M.: V1380,180-189(1991)
   See Linney, Alfred D.: V1380,190-199(1991)

Richards-Kortum, Rebecca See Rava, Richard P.: V1426,68-78(1991)

Richardson, A. D. See Boardman, A. D.: V1503,160-166(1991)

Richardson, Christina E.
   ; Andrews, Larry C.: Expressions for the spherical-wave-structure function based on a bump spectrum model for the index of refraction,V1487,19-30(1991)

Richardson, David
   : Business, manufacturing, and system integration issues in cluster tool process control,V1392,302-314(1991)

Richardson, E. H.
: Optical design of an off-axis low-distortion UV telescope,V1494,314-319(1991)

Richardson, M. B. See Boardman, A. D.: V1503,160-166(1991)

Richardson, Martin C.
: New opportunities with intense ultra-short-pulse lasers,V1410,15-25(1991)

Richardson, Philip I.
: Radiometric versus thermometric calibration of IR test systems: which is best?,V1488,80-88(1991)

Riches, Mark J. See Garfield, Brian R.: V1358,290-299(1991)

Richeson, D. S. See Wessels, Bruce W.: V1394,232-237(1991)

Richier, R.
; Amra, Claude: High-selectivity spectral multiplexers-demultiplexers usable in optical telecommunications obtained from multidielectric coatings at the end of optical fibers,V1504,202-210(1991)

Richmond, Jeff See Balasubramanian, Kunjithapa: V1529,106-114(1991)

Richter, Franz See Huenermann, Lucia: V1503,134-139(1991)

Richter, G.
: High-speed oscillation free lapping and polishing process for optical lenses,V1400,158-163(1991)
: New methods for economic production of prisms and lenses,V1400,11-23(1991)

Richter, W.
See Hingerl, Kurt: V1361,383-393(1991)
See Morley, Stefan: V1361,213-222(1991)

Richtsmeier, Joan T.
: Quantitative analysis of three-dimensional landmark coordinate data,V1380,12-23(1991)

Rickett, Ph. See Bowley, David J.: V1358,550-555(1991)

Rickey, Daniel W. See Picot, Paul A.: V1444,206-213(1991)

Ricks, Douglas W.
; Ajmera, Ramesh: Light scattering from binary optics,V1555,89-100(1991)

Ride, Sally K.
; Golightly, W.: Compact far-infrared free-electron laser,V1552,128-137(1991)

Rideout, William C.
: Experimental determination of recombination mechanisms in strained and unstrained quantum-well lasers,V1418,99-107(1991)
See Eichen, Elliot G.: V1474,260-267(1991)

Rider, David M.
; McCleese, Daniel J.: Stratospheric wind infrared limb sounder,V1540,142-147(1991)

Ridgway, Richard W. See Boiarski, Anthony A.: V1368,264-272(1991)

Riding, Thomas J. See Downey, George A.: V1482,40-47(1991)

Ridley, Brian K.
See Balkan, N.: V1361,927-934(1991)
See Gupta, Rita P.: V1362,798-803(1991)
See Gupta, Rita P.: V1362,790-797(1991)
See Walega, James G.: V1433,232-241(1991)

Ridley, Kevin D. See Jones, David C.: V1500,46-52(1991)

Ridley, W. C.
: Integrating thermography into the Palisades Nuclear Plant's electrical predictive maintenance program,V1467,51-58(1991)

Ridolfi, Tim See Williams, Paul F.: V1378,217-225(1991)

Riedel, Wolfgang J.
: Optics for tunable diode laser spectrometers,V1433,179-189(1991)

Riedl, Max J.
; Hale, Robert R.; Parsonage, Thomas B.; eds.: <author>Reflective and Refractive Optical Materials for Earth and Space Applications,V1485(1991)
; McCann, James T.: Analysis and performance limits of diamond-turned diffractive lenses for the 3-5 and 8-12 micrometer regions,VCR38,153-163(1991)

Rieger, Michael L. See Buck, Peter D.: V1463,218-228(1991)

Rieger, Paul L.
; Maddox, Christine E.: Analysis of exhaust from clean-fuel vehicles using FTIR spectroscopy,V1433,290-301(1991)

Riek, Jonathan K.
; Mitsa, Theophano; Parker, Kevin J.; Smith, Warren E.; Tekalp, Ahmet M.; Szumowski, J.: Modeling and suppression of amplitude artifacts due to z motion in MR imaging,V1398,130-141(1991)

; Smith, Warren E.; Tekalp, Ahmet M.; Parker, Kevin J.; Mitsa, Theophano; Szumowski, J.: Modeling data acquisition and the effects of patient motion in magnetic resonance imaging,V1445,198-206(1991)
; Tekalp, Ahmet M.; Smith, Warren E.; Parker, Kevin J.: Correction of image-phase aberrations in MRI with applications,V1445,190-197(1991)

Riekki, Jukka P. See Roning, Juha J.: V1388,350-360(1991)

Riendeau, J. R. See Sullivan, Timothy D.: V1596,83-95(1991)

Rienks, Rienk See Smits, Pieter C.: V1425,188-190(1991)

Ries, Harald
; Smestad, Greg P.; Winston, Roland: Thermodynamics of light concentrators,V1528,7-14(1991)
See Smestad, Greg P.: V1536,234-245(1991)

Rigau, J. See Trelles, Mario A.: V1403,781-798(1991)

Rigaut, Francois
See Gendron, Eric: V1542,297-307(1991)
See Merkle, Fritz: V1542,308-318(1991)

Riggs, Bud See Andrews, Lee T.: V1459,125-135(1991)

Riggs, R. S. See Park, Yong K.: V1372,219-227(1991)

Righini, Giancarlo C.
ed.: Glasses for Optoelectronics II,V1513(1991)
; Pelli, Stefano; Saracini, R.; Battaglin, Giancarlo; Scaglione, Antonio: Analysis of the refractive-index profile in ion-exchanged waveguides,V1513,418-424(1991)
; Pelli, Stefano; De Blasi, C.; Fagherazzi, Giuliano; Manno, D.: Ion-exchanged waveguides in semiconductor-doped glasses,V1513,105-111(1991)
See Banfi, Giampiero P.: V1361,874-880(1991)
See Ferrara, M.: V1513,130-136(1991)
See Guglielmi, Massimo: V1513,44-49(1991)
See Jiang, Pisu: V1362,899-906(1991)

Rigo, Cesare F. See Caldera, Claudio: V1372,82-87(1991)

Rigsbee, Daniel R.
; Dubin, Paul L.: Quasielastic and electrophoretic light scattering studies of polyelectrolyte-micelle complexes,V1430,203-215(1991)

Rihter, Boris D. See Sounik, James R.: V1426,340-349(1991)

Rijavec, Nenad See Poore, Aubrey B.: V1481,345-356(1991)

Riley, A. L.
See Chew, Wilbert: V1477,95-100(1991)
See Cooley, Thomas W.: V1475,243-247(1991)

Riley, Susan
; Schick, Larry A.: Laser drilling vias in GaAs wafers,V1598,118-120(1991)

Rille, Eduard P. See Perger, Andreas: V1419,75-83(1991)

Rimmele, Thomas
; von der Luhe, Oskar; Wiborg, P. H.; Widener, A. L.; Dunn, Richard B.; Spence, G.: Solar feature correlation tracker,V1542,186-193(1991)

Rimmer, Matthew P.
; Bruegge, Thomas J.; Kuper, Thomas G.: MTF optimization in lens design,V1354,83-91(1991)

Rinaldo, R.
; Bernardini, Riccardo; Cortelazzo, Guido M.: Extension of Rader's algorithm for high-speed multidimensional autocorrelation,V1606,773-787(1991)

Rinker, Michael
; Kalt, Heinz; Lu, Yin-Cheng; Bauser, Elizabeth; Koehler, Klaus: Indirect stimulated emission at room temperature in the visible range,V1362,14-23(1991)

Rinne, Glenn A.
; Hwang, Lih-Tyng; Adema, G. M.; King, Donald A.; Turlik, Iwona: Design and process impact on thin-film interconnection performance,V1389,110-121(1991)

Rinneberg, Herbert H. See Kohl, M.: V1525,26-34(1991)

Rino, Charles L.
; Ngo, Hoc D.: Scattering from objects near a rough surface,V1558,339-350(1991)

Rinsland, C. P. See Goldman, Aaron: V1491,194-202(1991)

Rinsland, Pamela L. See Goad, Joseph H.: V1410,107-115(1991)

Riordan, Richard J. See Brown, James C.: V1488,300-311(1991)

Rios, E. L. See Johnson, Stephen G.: V1435,292-297(1991)

Rioux, Marc
See Blais, Francois: V1482,473-479(1991)
See Blais, Francois: V1480,94-101(1991)

Rippel, Harald
; Kampf, Dirk; Graue, Roland: ORFEUS: orbiting and retrievable far and extreme ultraviolet spectrometer,V1343,520-529(1991)
See Graue, Roland: V1494,377-385(1991)

Rippert, Edward D.
; Song, S. N.; Ketterson, John B.; Ulmer, Melville P.: Multilayered superconducting tunnel junctions for use as high-energy-resolution x-ray detectors,V1549,283-288(1991)

Riris, Haris See Cooper, David E.: V1433,120-127(1991)

Risen, William M.
; Morse, Ted F.; Tsagaropoulos, George: Low-temperature ion exchange of dried gels for potential waveguide fabrication in glasses,V1590,44-49(1991)

Risk, William P.
; Nadler, Ch. K.: KTP waveguides for frequency upconversion of strained-layer InGaAs laser diodes,V1561,130-134(1991)
See Lenth, Wilfried: V1499,308-313(1991)

Risken, H. See Jung, Peter: V1376,82-93(1991)

Rissberger, Ed See Brown, David A.: V1369,2-8(1991)

Risser, Steven M.
; Ferris, Kim F.: Defect enhancement of local electric fields in dielectric films,V1441,262-268(1991)
See Ferris, Kim F.: V1441,510-520(1991)

Ristau, Detleu
See Rahe, Manfred: V1441,113-126(1991)
See Schink, Harald: V1441,327-338(1991)

Ritchie, Douglas R. See Samarakone, Nandasiri: V1463,16-29(1991)

Ritman, Erik L.
See Higgins, William E.: V1445,276-286(1991)
See Jorgensen, Steven M.: V1346,180-191(1991)

Ritsch, Helmut
; Zoller, Peter: Quantum noise reduction in lasers with nonlinear absorbers,V1376,206-215(1991)

Ritsko, J. J. See Deutsch, Alina: V1389,161-176(1991)

Ritter, D. See Gunapala, S. D.: V1541,11-23(1991)

Ritter, Gerhard X.
: Heterogeneous matrix products,V1568,92-100(1991)
; Davidson, Jennifer L.: Recursion and feedback in image algebra,V1406,74-86(1991)

Ritter, J. A. See Sachse, Glen W.: V1433,157-166(1991)

Ritter, J. C. See Nisenoff, Martin: V1394,104-113(1991)

Ritter, Mark B. See Ledermann, Peter G.: V1598,160-163(1991)

Ritter, Reinold
See Andresen, Klaus: V1554A,93-100(1991)
See Hilbig, Jens: V1554A,588-592(1991)

Ritter, Rogers C.
See Lawson, Michael A.: V1445,265-275(1991)
See Ramos, P. A.: V1443,160-170(1991)

Rittinger, K. See Seidel, Claus: V1432,105-116(1991)

Rivela, Cristina See Del Bello, Umberto: V1502,104-116(1991)

Rivera, Christopher See Cookson, John P.: V1444,374-378(1991)

Rivera, Giovanni See Canestrari, Paolo: V1463,446-455(1991)

Rivest, Jean-Francois
; Beucher, Serge; Delhomme, J.: Marker-controlled picture segmentation applied to electrical logging images,V1451,179-190(1991)

Rivkind, V. L. See Vishnevsky, G. I.: V1447,34-43(1991)

Rivoira, R.
See Berrouane, H.: V1343,428-436(1991)
See Khan Malek, Chantal: V1343,56-61(1991)

Riza, Nabeel A.
: Acousto-optic architectures for multidimensional phased-array antenna processing,V1476,144-156(1991)

Rizikov, Igor V.
; Svechnikov, Georgy S.; Bulyarsky, Sergey V.; Ambrozevich, Alexander S.: Comparative analysis of external factors' influences on the GaP light-emitting p-n-junctions,V1362,664-673(1991)
See Svechnikov, Georgy S.: V1362,674-683(1991)

Rizki, Mateen M.
; Tamburino, Louis A.; Zmuda, Michael A.: Applications of learning strategies to pattern recognition,V1469,384-391(1991)
See Zmuda, Michael A.: V1470,183-189(1991)

Rizzi, Angela
; Moritz, Heiko; Lueth, Hans: Growth and characterization of semiconducting Fe-Si2 thin layers on Si(111),V1361,827-833(1991)

Rizzi, Maria L. See Armenise, Mario N.: V1583,210-220(1991)

Rizzo, Annmarie See Menendez, Ronald C.: V1363,97-105(1991)

Rjabinkina, V. A. See Babaeva, N. A.: V1440,260-269(1991)

Rjabova, R. V.
See Vorzobova, N. D.: V1238,476-477(1991)
See Vorzobova, N. D.: V1238,462-464(1991)

Ro, Jae Y. See Johnson, Douglas E.: V1421,36-41(1991)

Ro, Ru-Yen
; Varadan, Vasundara V.; Varadan, Vijay K.: Experimental study of chiral composites,V1558,269-287(1991)
; Varadan, Vasundara V.; Varadan, Vijay K.: Low-frequency chiral coatings,V1489,46-55(1991)

Robbins, D. J. See Abram, Richard A.: V1361,424-433(1991)

Robbins, Jamey L. See Lauffenburger, Jim: V1499,104-113(1991)

Robbins, JoAnne See Dengel, Gail A.: V1445,88-94(1991)

Robein, D. See Remiens, D.: V1362,323-330(1991)

Robello, Douglas R.
See Penner, Thomas L.: V1560,377-386(1991)
See Penner, Thomas L.: V1436,169-178(1991)

Robert, Bruno See Lutz, Marc: V1403,59-65(1991)

Robert, Philippe
: Temporal projection for motion estimation and motion compensating interpolation,V1605,558-569(1991)

Roberti, Giuseppe See Andreoni, Alessandra: V1525,351-366(1991)

Roberts, A. S. See Gartenberg, Ehud: V1467,338-356(1991)

Roberts, Barry
; Bazakos, Michael E.: INS integrated motion analysis for autonomous vehicle navigation,V1521,2-13(1991)

Roberts, Charles C.
: Infrared thermographic analysis of snow ski tracks,V1467,207-218(1991)

Roberts, Daniel R.
; Cuellar, Enrique; Kennedy, Michael T.; Tomita, Akira: Fiber construction for improved mechanical reliabiltiy,V1366,129-135(1991)
See Kennedy, Michael T.: V1366,167-176(1991)

Roberts, Harold A.
; Emmons, David J.; Beard, Michael S.; Lu, Liang-ju: Low-back-reflection, low-loss fiber switch,V1363,62-69(1991)

Roberts, Harry H. See Huh, Oscar K.: V1492,378-386(1991)

Roberts, James R.
; Olthoff, James K.; Van Brunt, R. J.; Whetstone, James R.: Measurements on the NIST GEC reference cell,V1392,428-436(1991)

Roberts, John S. See Hansen, Stein I.: V1362,361-369(1991)

Roberts, Michael F. See Meth, Jeffrey S.: V1560,13-24(1991)

Roberts, Peter C. See Kreider, James F.: V1488,376-388(1991)

Roberts, Scott S.
; Davidson, Roger: Mechanical properties of composite materials containing embedded fiber-optic sensors,V1588,326-341(1991)
See Michie, W. C.: V1588,342-355(1991)
See Sansonetti, Pierre: V1588,198-209(1991)

Roberts, Walter G. See Peavy, George M.: V1424,171-178(1991)

Robertson, James A. See Worrell, Steven W.: V1382,219-227(1991)

Robertson, Michael J. See Barnsley, Peter E.: V1378,116-126(1991)

Robertson, Stewart A.
; Stevenson, J. T.; Holwill, Robert J.; Thirsk, Mark; Daraktchiev, Ivan S.; Hansen, Steven G.: Photoresist dissolution rates: a comparison of puddle, spray, and immersion processes,V1464,232-244(1991)

Robin, Laura
: Temporal adaptation of multimedia scripts,V1460,103-113(1991)

Robinett, Warren
; Rolland, Jannick P.: Computational model for the stereoscopic optics of a head-mounted display,V1457,140-160(1991)

Robins, Lawrence H.
; Farabaugh, Edward N.; Feldman, Albert: Determination of the optical constants of thin chemical-vapor-deposited diamond windows from 0.5 to 6.5 eV,V1534,105-116(1991)
See Shechtman, Dan: V1534,26-43(1991)

Robinson, Clifford J. See Klein, Claude A.: V1534,117-138(1991)

Robinson, D. H. See Preston, Ralph G.: V1454,132-138(1991)

Robinson, Daniel
  See Chen, Ray T.: V1559,110-117(1991)
  See Chen, Ray T.: V1583,362-374(1991)
  See Chen, Ray T.: V1583,135-142(1991)

Robinson, Deborah L.
  : Environmental testing of a Q-switched Nd:YLF laser and a Nd:YAG ring laser,V1417,562-572(1991)
  ; Chen, Chien-Chung; Hemmati, Hamid: Electro-optic resonant modulator for coherent optical communication,V1417,421-430(1991)

Robinson, Gary Y. See Crawford, Deborah L.: V1371,138-141(1991)

Robinson, K. C. See Morgan, Robert A.: V1562,149-159(1991)

Robinson, Lloyd B.
  ; Brown, William E.; Gilmore, Kirk; Stover, Richard J.; Wei, Mingzhi; Geary, John C.: Performance tests of large CCDs,V1447,214-228(1991)
  See Bredthauer, Richard A.: V1447,310-315(1991)
  See Geary, John C.: V1447,264-273(1991)

Robinson, Max See Godber, Simon X.: V1526,170-189(1991)

Robinson, Michael G.
  ; Zhang, Lin; Johnson, Kristina M.: Optical neurocomputer architectures using spatial light modulators,V1469,240-249(1991)
  See Zhang, Lin: V1469,225-229(1991)

Robinson, Waldo R. See Phillips, Thomas E.: V1454,290-298(1991)

Robur, I. See Indutnyi, I. Z.: V1555,243-253(1991)

Rocca, Corrado
  ; Montrosset, Ivo; Gollinucci, Stefano; Ghione, Giovanni: Modeling of traveling wave Ti:LiNbO3 waveguide modulators,V1372,39-47(1991)

Rocca Serra, J. See Brunet, Henri: V1397,273-276(1991)

Rocha, Jose R.
  See Ferreira, Mario F.: V1372,14-26(1991)
  See Pires, Joao J.: V1372,118-127(1991)
  See Pires, Joao J.: V1579,144-154(1991)

Roche, Aidan E.
  ; Kumer, John B.: Cryogenic limb array etalon spectrometer: experiment description,V1491,91-103(1991)

Roche, C. T. See Delayen, Jean R.: V1407,524-534(1991)

Roche, J. J. See LaPeruta, Richard: V1497,357-366(1991)

Rochester, G. K. See Sumner, Timothy J.: V1549,265-273(1991)

Rochkind, Simeone
  ; Alon, M.; Ouaknine, G. E.; Weiss, S.; Avram, J.; Razon, N.; Lubart, Rachel; Friedmann, Harry: Intraoperative clinical use of low-power laser irradiation following surgical treatment of the tethered spinal cord,V1428,52-58(1991)
  See Friedmann, Harry: V1427,357-362(1991)
  See Lubart, Rachel: V1422,140-146(1991)

Rochon, Paul L.
  ; Bissonnette, Daniel: Backscattering image resolution as a function of particle density,V1530,50-57(1991)

Rockafellow, David R.
  ; Teppo, Edward A.; Jacco, John C.: Bulk darkening of flux-grown KTiOPO4,V1561,112-118(1991)

Rockette, Howard E.
  ; Obuchowski, Nancy A.; Gur, David; Good, Walter F.: Effect of experimental design on sample size,V1446,276-283(1991)
  See Fuhrman, Carl R.: V1446,422-429(1991)
  See Gur, David: V1446,284-288(1991)
  See King, Jill L.: V1446,268-275(1991)

Rocks, Manfred See Gieschen, Nikolaus: V1579,237-248(1991)

Rockstroh, Helmut See Klingenberg, Guenter: V1358,851-858(1991)

Rodden, Jack J.
  See Eller, E. D.: V1415,2-12(1991)
  See LaMont, Douglas V.: V1482,2-12(1991)

Roddier, Claude A.
  ; Roddier, Francois J.: Reconstruction of the Hubble Space Telescope mirror figure from out-of-focus stellar images,V1494,78-84(1991)

Roddier, Francois J.
  : Wavefront curvature sensing and compensation methods in adaptive optics,V1487,123-128(1991)
  ; Graves, J. E.; McKenna, Daniel; Northcott, Malcolm J.: University of Hawaii adaptive optics system: I. General approach,V1542,248-253(1991)
  See Roddier, Claude A.: V1494,78-84(1991)

Roddier, Nicolas
  : Algorithms for wavefront reconstruction out of curvature sensing data,V1542,120-129(1991)
  See Forbes, Fred F.: V1542,140-147(1991)
  See Stepp, Larry M.: V1542,175-185(1991)

Rodeback, G. W. See Crittenden, Eugene C.: V1492,187-197(1991)

Rodgers, John M.
  ; Jewell, Tanya E.: Design of reflective relay for soft x-ray lithography,V1354,330-336(1991)
  See Jewell, Tanya E.: V1527,163-173(1991)

Rodgers, Mark R.
  ; Monahan, Kevin M.: Using the Atomic Force Microscope to measure submicron dimensions of integrated circuit devices and processes,V1464,358-366(1991)

Rodgers, Michael A. See Sounik, James R.: V1426,340-349(1991)

Rodgers, P. A. See Key, Michael H.: V1397,9-17(1991)

Rodionova, N. S. See Benimetskaya, L. Z.: V1525,242-245(1991)

Rodricks, Brian G.
  ; Brizard, Christine M.; Clarke, Roy; Lowe, Walter P.: X-ray detector for time-resolved studies,V1550,18-26(1991)
  See Campuzano, Juan C.: V1345,245-254(1991)
  See Clarke, Roy: V1361,2-12(1991)
  See Clarke, Roy: V1345,101-114(1991)

Rodrigues, F. C.
  ; Simao, Jose V.; Oliveira, Joao; Freitas, Jose C.; Carvalho, Fernando D.: Production of laser simulation systems for adverse environments,V1399,90-97(1991)
  See Carvalho, Fernando D.: V1399,130-136(1991)
  See Correia, Bento A.: V1567,15-24(1991)
  See Davies, Roger: V1459,283-291(1991)
  See de Lourdes Quinta, Maria: V1399,24-29(1991)
  See Freitas, Jose C.: V1399,16-23(1991)
  See Freitas, Jose C.: V1399,42-48(1991)

Rodriguez, Arturo A.
  ; Mandeville, Jon R.: Image registration for automated inspection of 2-D electronic circuit patterns,V1384,2-14(1991)
  ; Mandeville, Jon R.; Wu, Frederick Y.: System calibration and part alignment for inspection of 2-D electronic circuit patterns,V1332,25-35(1991)
  ; Mitchell, O. R.: Robust statistical method for background extraction in image segmentation,V1569,182-199(1991)

Rodriguez, Domingo
  ; Pertuz-Campo, Jairo: Artificial neural networks and Abelian harmonic analysis,V1565,492-503(1991)

Rodriguez, Ferdinand See Long, Treva: V1466,188-198(1991)

Rodriguez, J. R. See Bommarito, G. M.: V1550,156-170(1991)

Rodriguez, Jose M. See Recio, Miguel: V1361,469-478(1991)

Rodriguez, Leonard J. See Rushmeier, Holly E.: V1486,210-216(1991)

Rodriguez, Rita V. See Anger, Frank D.: V1468,122-136(1991)

Rodriguez, Walter E. See McWhorter, Shane W.: V1457,85-90(1991)

Rodriguez-Montero, Ponciano See Ojeda-Castaneda, Jorge: V1500,256-261(1991)

Rodriguez-Vera, Ramon
  ; Olivares-Perez, A.; Morales-Romero, Arquimedes A.: Holographic pseudocoloring of schlieren images,V1507,416-424(1991)
  See Kerr, David: V1554A,668-679(1991)

Roe, Jeffrey N. See Angel, S. M.: V1368,98-104(1991)

Roe, S. See Sumner, Timothy J.: V1549,265-273(1991)

Roeder, Beate
  ; Naether, Dirk: Characterization of photobiophysical properties of sensitizers used in photodynamic therapy,V1525,377-384(1991)

Roehle, B. See Lazarov, M.: V1536,183-191(1991)

Roehrenbeck, Paul W. See Sawaki, Akihiro: V1366,324-331(1991)

Roehring, Hans See Ji, Tinglan: V1444,136-150(1991)

Roelofs, Mark G. See Morris, Patricia A.: V1561,104-111(1991)

Roeschert, Heinz See Przybilla, Klaus J.: V1466,174-187(1991)

Roeser, Hans-Peter
  ; van der Wal, Peter: FIR lasers as local oscillators in submillimeter astronomy,V1501,194-197(1991)

Roess, Debrah A. See Barisas, B. G.: V1432,52-63(1991)

Roessler, R. W. See Jorgensen, Steven M.: V1346,180-191(1991)

Roeth, Klaus-Dieter
See Dusa, Mircea V.: V1464,447-458(1991)
See Dusa, Mircea V.: V1496,217-223(1991)

Roetling, Paul G.
: Systems considerations in color printing,V1458,17-24(1991)

Rogal, Fannie A. See Tsai, Ming-Jer: V1481,281-291(1991)

Rogalski, Antoni
: Intrinsic carrier concentration and effective masses in Hg1-
xMnxTe,V1512,189-194(1991)

Rogers, Alan J.
: Distributed optical fiber sensing,V1504,2-24(1991); 1506,2-24(1991);
1507,2-24(1991)
; Handerek, Vincent A.; Parvaneh, Farhad: Frequency-derived distributed
optical fiber sensing: backscatter analysis,V1511,190-200(1991)
See Chen, Shiping: V1504,191-201(1991)
See Chen, Shiping: V1511,67-77(1991)
See Chu, W. W.: V1572,523-527(1991)
See Kanellopoulos, S. E.: V1516,200-210(1991)
See McStay, Daniel: V1584,118-123(1991)
See Yao, Jialing: V1572,428-433(1991)

Rogers, David W. See Seka, Wolf D.: V1398,162-169(1991)

Rogers, Harvey N.
: Cyclic fatigue behavior of silica fiber,V1366,112-117(1991)

Rogers, Mark E.
; Rought, Nathan W.: Numerical study of the onset of chaos in coupled
resonators,V1415,24-37(1991)

Rogers, Philip J.
: Athermalization of IR optical systems,VCR38,69-94(1991)
: Athermalized FLIR optics,V1354,742-751(1991)

Rogers, Stephen B. See Troyer, David E.: V1486,396-409(1991)

Rogers, Steven K.
ed.: **Applications of Artificial Neural Networks II,V1469(1991)**
; Ruck, Dennis W.; Kabrisky, Matthew; Tarr, Gregory L.; Oxley, Mark E.:
Artificial neural networks for automatic object recognition,VIS07,231-
243(1991)
See Burns, Thomas J.: V1469,208-218(1991)
See Kabrisky, Matthew: VTT04(1991)
See Tarr, Gregory L.: V1469,2-11(1991)
See Veronin, Christopher P.: V1469,281-291(1991)

Roggemann, Michael C.
; Matson, Charles L.: Partially compensated speckle imaging: Fourier phase
spectrum estimation,V1542,477-487(1991)

Rogne, Timothy J. See Cook, Thomas H.: V1488,214-225(1991)

Rogner, Arnd
; Ehrfeld, Wolfgang: Fabrication of light-guiding devices and fiber-coupling
structures by the LIGA process,V1506,80-91(1991)

Rogovin, Dan N. See Pizzoferrato, R.: V1409,192-201(1991)

Rogowitz, Bernice E.
; Brill, Michael H.; Allebach, Jan P.; eds.: <author>Human Vision, Visual
Processing, and Digital Display II,V1453(1991)

Rogowski, Robert S.
See Baldini, S. E.: V1588,125-131(1991)
See Egalon, Claudio O.: V1489,9-16(1991)
See Egalon, Claudio O.: V1588,241-254(1991)
See Egalon, Claudio O.: V1368,134-149(1991)

Rohde, Axel
; Saffert, Ralf; Fitch, John: Advanced confocal technique for submicron CD
measurements,V1464,438-446(1991)

Rohlena, Karel
; Beranek, J.; Masek, Karel: Modeling of pumping kinetics of an iodine
photodissociation laser with long pumping pulse,V1415,259-268(1991)

Rohman, H. See Stenow, Eric: V1420,29-33(1991)

Roizin, Yacov O. See Marczuk, Krystyna: V1513,291-296(1991)

Rojer, Alan S.
; Schwartz, Eric L.: Biological basis for space-variant sensor design I:
parameters of monkey and human spatial vision,V1382,132-144(1991)
; Schwartz, Eric L.: Biological basis for space-variant sensor design II:
implications for VLSI sensor design,V1386,44-52(1991)
; Schwartz, Eric L.: Fusion of multiple fixations with a space-variant sensor:
conditional optimality of maximum-resolution blending,V1381,250-
260(1991)

Rol, Pascal O.
: Design characteristics and point-spread function evaluation of bifocal
intraocular lenses,V1423,15-19(1991)
; Beck, Dominik; Niederer, Peter: Endocular ophthalmoscope:
miniaturization and optical imaging quality,V1423,84-88(1991)
; Parel, Jean-Marie; Hanna, Khalil: Computed estimation of visual acuity after
laser refractive keratectomy,V1423,89-93(1991)
See Simon, Gabriel: V1423,154-156(1991)

Rolain, Yves See Schrever, Koen: V1572,107-112(1991)

Roland, David A. See Wu, Chwan-Hwa: V1471,210-221(1991)

Roli, Fabio See Vernazza, Gianni L.: V1492,206-212(1991)

Rolland, Jannick P. See Robinett, Warren: V1457,140-160(1991)

Rollett, A. D. See Wu, Xin D.: V1477,8-14(1991)

Roltsch, Tom J.
: Cost-effective x-ray lithography,V1465,289-307(1991)

Rom, Hillel
; Medioni, Gerard: Hierarchical decomposition and axial representation of
shape,V1570,262-273(1991)

Romakina, M. Y. See Kozlova, T. G.: V1403,159-160(1991)

Roman, Bernard J. See Cooper, Kent: V1392,253-264(1991)

Roman, Ward C. See Hay, Stephen O.: V1435,352-358(1991)

Romanenko, Peter F. See Indutnyi, I. Z.: V1555,243-253(1991)

Romaniuk, Ryszard S.
; Stepien, Ryszard: Glass-ceramic fiber optic sensors,V1368,73-84(1991)
See Wolinski, Wieslaw: V1391(1991)

Romankiw, Lubomyr T. See Krongelb, Sol: V1389,249-256(1991)

Romano, Valerio
; Greber, Charlotte M.; Frenz, Martin; Forrer, Martin; Weber, Heinz P.:
Measurement of temperature distributions after pulsed IR radiation
impact in biological tissue models with fluorescent thin films,V1427,16-
26(1991)
See Frenz, Martin: V1427,9-15(1991)

Romanofsky, Robert R. See Tabib-Azar, Massood: V1477,85-94(1991)

Romanov, Nikolay A. See Sherstobitov, Vladimir E.: V1415,79-89(1991)

Romanov, S. V. See Guseva, N. P.: V1403,332-334(1991)

Romanovsky, V. F.
; Kovalenko, A. N.; Sapozhnikova, T. I.; Tushev, N. R.; Abgaryan, A. A.:
Generation of hard x-ray pulse trains with the help of high-frequency
oscillating systems for fast processes recording,V1358,140-145(1991)

Romanovsky, Yuri M.
; Stepanian, A. S.; Shogenov, Yu. H.: Measurement of bleeding sap flow
velocity in xylem bundle of herbs by laser probing,V1403,359-362(1991)
See Chikishev, A. Y.: V1403,517-521(1991)
See Netrebko, N. V.: V1403,512-516(1991)
See Priezzhev, Alexander V.: V1402,107-113(1991)

Romanuk, B. N. See Artsimovich, M. V.: V1397,729-733(1991)

Romash, E. V. See Ryabova, R. V.: V1238,166-170(1991)

Romer, A. E. See Johnston, James S.: V1589,126-132(1991)

Romig, A. D. See Frear, Darrel R.: V1596,72-82(1991)

Rominger, James P.
: Application of a reduced area electrical test pattern to precise pattern
registration measurements,V1496,224-231(1991)

Romoli, Andrea See Livi, S.: V1500,144-150(1991)

Roncone, Ronald L.
; Burke, James J.; Weisenbach, Lori; Zelinski, Brian J.: Experimental and
theoretical investigation of surface- and bulk-induced attenuation in
solution-deposited waveguides,V1590,14-25(1991)
See Weisenbach, Lori: V1590,50-58(1991)

Rondinella, Vincenzo V.
; Matthewson, M. J.: Ionic effects on silica optical fiber strength and models
for fatigue,V1366,77-84(1991)

Rondinelli, Giuseppe
; Di Girolamo, Sergio; Barresi, Giangrande: Small satellites for water cycle
experiments,V1495,19-31(1991)
See Barresi, Giangrande: V1495,246-258(1991)

Rong, Jian See Fang, Yin: V1572,197-200(1991)

Roning, Juha J.
; Riekki, Jukka P.; Kemppainen, Seppo: Simulator for developing mobile
robot control systems,V1388,350-360(1991)

Ronning, E.
; Fjarlie, E. J.: Passive range and azimuth measuring system,V1399,178-
183(1991)

Ronse, Christian See Macq, Benoit M.: V1606,165-173(1991)

Roos, Arne
: Optical characterization of solar-selective transmitting coatings,V1536,158-168(1991)
See Bergkvist, Mikael: V1530,352-362(1991)

Roos, Kenneth P.
; Lake, David S.; Lubell, Bradford A.: Microscopic feature extraction from optical sections of contracting cardiac muscle cells recorded at high speed,V1428,159-168(1991)

Roos, Maurits See Anderson, John E.: V1492,252-262(1991)

Roos, Paul
; Viergever, Max A.: Reversible image data compression based on HINT (hierarchical interpolation) decorrelation and arithmetic coding,V1444,283-290(1991)

Roosen, Gerald
: Nonlinear energy transfer between nanosecond pulses in iron-doped InP crystals,V1362,398-416(1991)
See Mathey, P.: V1500,26-33(1991)

Roper, Stephen D. See McCracken, Thomas O.: V1380,6-10(1991)

Rorabaugh, Terry L.
; Vaccaro, John J.; Grace, Kevin H.; Pauer, Eric K.: DSP array for real-time adaptive sidelobe cancellation,V1566,312-322(1991)

Rosa, Louis See Carvlin, Mark J.: V1426,13-21(1991)

Rosakis, Ares J.
See Lambros, John M.: V1554A,70-83(1991)
See Liu, Cheng: V1554A,814-825(1991)
See Tippur, Hareesh V.: V1554A,176-191(1991)

Rosche, Henry See Trenchard, Michael E.: V1456,318-326(1991)

Rose, Allen H. See Deeter, Merritt N.: V1367,243-248(1991)

Rose, B. See Remiens, D.: V1362,323-330(1991)

Rose, C. See Ravat, Christian J.: V1478,239-246(1991)

Rose, Eric A. See Detwiler, Paul W.: V1425,149-155(1991)

Rose, Evan A.
; McDonald, Thomas E.; Rosocha, Louis A.; Harris, David B.; Sullivan, J. A.; Smith, I. D.: Pulsed-power considerations for electron-beam-pumped krypton-fluoride lasers for inertial confinement fusion applications,V1411,15-27(1991)

Rose, G. E. See Jackson, John E.: V1533,75-86(1991)

Rose, J. W. See Liu, Yung S.: V1377,126-133(1991)

Rose, Todd S.
; Fincher, Curtis L.; Fields, Renny A.: Performance and spectroscopic characterization of irradiated Nd:YAG,V1561,43-49(1991)

Rose, V. See Flanagan, Kathryn A.: V1549,395-407(1991)

Rose, William I.
; Schneider, David J.: Remote sensing of volcanic ash hazards to aircraft,V1492,387-390(1991)

Rosen, Arye
; Stabile, Paul J.; Zutavern, Fred J.; Loubriel, Guillermo M.: Generic applications for Si and GaAs optical switching devices utilizing semiconductor lasers as an optical source,V1378,187-194(1991)
See Ayekavadi, Raj: V1418,74-85(1991)
See Evans, Gary A.: V1378,146-161(1991)
See Loubriel, Guillermo M.: V1378,179-186(1991)

Rosen, Hal J. See Tang, Wade C.: V1418,338-342(1991)

Rosen, Jeffrey See Bass, Lawrence S.: V1421,164-168(1991)

Rosen, Joseph See Pieper, Steve: V1457,188-197(1991)

Rosen, Robert S. See Stearns, Daniel G.: V1465,80-87(1991)

Rosen, Warren A. See Herman, Warren N.: V1560,206-213(1991)

Rosenbaum, Steven E.
See Chou, Chia-Shing: V1475,151-156(1991)
See Yap, Daniel: V1418,471-476(1991)

Rosenbek, John C. See Dengel, Gail A.: V1445,88-94(1991)

Rosenberg, Charles See Bender, Walter J.: V1460,59-70(1991)

Rosenblatt, Edward I.
; Beaver, Edward A.; Cohen, R. D.; Linsky, J. B.; Lyons, R. W.: Cerenkov background radiation in imaging detectors,V1449,72-86(1991)

Rosenblatt, Gern M. See Ager, Joel W.: V1437,24-31(1991)

Rosenblum, Karl W. See Zhou, Quan: V1592,108-113(1991)

Rosenblum, Lawrence
; Kamgar-Parsi, Behzad; Karahalios, Margarida; Heitmeyer, Richard: Visualizing underwater acoustic matched-field processing,V1459,274-282(1991)

Rosenbluth, Mary See Bloss, Walter L.: V1541,2-10(1991)

Rosenbruch, Klaus J. See Yang, Zishao: V1354,323-327(1991)

Rosenfeld, J. P.
; Krecker, Donald K.; Hord, R. M.: Applications of a minimum sum path algorithm implemented on the connection machine,V1406,147-147(1991)

Rosenholtz, Ruth E.
; Zakhor, Avideh: Iterative procedures for reduction of blocking effects in transform image coding,V1452,116-126(1991)

Rosenschein, Uri See Detwiler, Paul W.: V1425,149-155(1991)

Rosenthal, Marc S.
; Sashin, Donald; Herron, John M.; Maitz, Glenn S.; Boyer, Joseph W.; Gur, David: Evaluation of a moving slit technique for mammography,V1443,132-142(1991)

Rosenthal, Steven
; Stahlberg, Larry: Integrated approach to machine vision application development,V1386,158-162(1991)
See Revankar, Shriram: V1459,268-273(1991)

Rosenwaks, Salman
See Aharon, O.: V1397,251-255(1991)
See Bar, I.: V1397,169-172(1991)

Rosenzweig, Josef
; Moglestue, C.; Axmann, A.; Schneider, Joachim J.; Huelsmann, Axel; Lambsdorff, M.; Kuhl, Juergen; Klingenstein, M.; Leier, H.; Forchel, Alfred W.: Characterization of picosecond GaAs metal-semiconductor-metal photodetectors,V1362,168-178(1991)

Rosenzweig, M.
; Moehrle, M.; Dueser, H.; Tischel, M.; Heitz, R.; Hoffmann, Axel: Nonlinear optical gain in InGaAs/InGaAsP quantum-wells,V1362,876-887(1991)
See Tischel, M.: V1361,917-926(1991)

Rosilio, A. See Vinet, Francoise: V1466,558-569(1991)

Rosilio, Charles See Vinet, Francoise: V1466,558-569(1991)

Rosiwal, S. See Schubert, Emil: V1503,299-309(1991)

Rosocha, Louis A. See Rose, Evan A.: V1411,15-27(1991)

Rospendowski, Bernard See Cotton, Therese M.: V1403,93-101(1991)

Ross, A. J. See Bacis, Roger: V1397,173-176(1991)

Ross, Ian M.
: Electronics/photonics technology: vision and reality,V1389,2-26(1991); 1390,2-26(1991)

Ross, Ian N. See Key, Michael H.: V1397,9-17(1991)

Ross, Susan E. See Dierks, Jeffrey S.: V1482,146-158(1991)

Ross, William E.
; Lambeth, David N.: Advanced magneto-optic spatial light modulator device development,V1562,93-102(1991)

Rosser, Paul J.
; Moynagh, P.; Affolter, K. B.: Multichamber rapid thermal processing,V1393,49-66(1991)

Rosser, Roy J. See Homer, Michael: V1527,134-144(1991)

Rossi, David V. See Fossum, Eric R.: V1447,202-203(1991)

Rossi, Elio See Di Cocco, Guido: V1549,102-112(1991)

Rossi, G. See Paone, Nicola: V1511,129-139(1991)

Rossi, M. See Di Cocco, Guido: V1549,102-112(1991)

Rossin, Victor V. See Zhilyaev, Yuri V.: V1361,848-859(1991)

Rossina, Tatiana V. See Zhilyaev, Yuri V.: V1361,848-859(1991)

Rossing, Thomas D.
; Hampton, D. S.: Modal analysis of musical instruments with holographic interferometry,V1396,108-121(1991)

Rossmanith, H. P.
: Interaction of cracks, waves, and contacts: a photomechanics study,V1554A,2-28(1991); 1554B,2-28(1991)
: Method of caustics for anisotropic materials,V1554A,835-849(1991)
; Beer, Rudolf J.; Knasmillner, R. E.: Interface cracks close to free surfaces: a caustic study,V1554A,850-860(1991)

Rostaing, M. See David, Jean: V1397,697-700(1991)

Rosteck, Helmut See Schmitt, Dirk-Roger: V1530,104-110(1991)

Rosten, David P.
; Yuen, P.; Hunt, Bobby R.: Hierarchial neural net with pyramid data structures for region labeling of images,V1472,118-127(1991)

Rostoker, Norman
See Ishizuka, Hiroshi: V1407,442-455(1991)
See Prohaska, Robert M.: V1407,598-604(1991)
See Rahman, Hafiz-ur: V1407,589-597(1991)
See Song, Yuanxu Y.: V1407,430-441(1991)

Rosvold, Geir O. See Lokberg, Ole J.: V1508,153-160(1991)

Roszko, Marcin
  See Domanski, Andrzej W.: V1362,844-852(1991)
  See Domanski, Andrzej W.: V1362,907-915(1991)

Rot, A. S. See Barachevsky, Valery A.: V1621,33-44(1991)

Roth, Hans-Klaus See Baumann, Reinhard: V1463,638-643(1991)

Roth, J. M.
  See Froehly, Claude: V1358,532-540(1991)
  See Mens, Alain: V1358,315-328(1991)

Roth, Jerry E. See Nixon, Alan J.: V1424,198-208(1991)

Roth, Michael See Azoulay, Moshe: V1535,35-45(1991)

Roth, Michael W.
  ; Thompson, K. E.; Kulp, R. L.; Alvarez, E. B.: High-density CCD
    neurocomputer chip for accurate real-time segmentation of noisy
    images,V1469,25-36(1991)

Roth, Richard S. See Wu, Kuang-Yi: V1563,168-175(1991)

Roth, Robert A. See Babayan, Richard K.: V1421,42-44(1991)

Roth, Sylvia See Stenzel, Olaf: V1534,148-157(1991)

Roth, Thomas J.
  See Anderson, Eric R.: V1417,543-549(1991)
  See Botez, Dan: V1474,64-74(1991)
  See Mawst, Luke J.: V1418,353-357(1991)

Roth-Tabak, Yuval
  ; Weymouth, Terry E.: Environment model for mobile robots indoor
    navigation,V1388,453-463(1991)
  See Jain, Ramesh C.: V1468,188-201(1991)

Rothemund, W. See Hingerl, Kurt: V1361,943-953(1991)

Rothenberg, E. See Brown, David A.: V1369,2-8(1991)

Rothman, Jay See Wolfe, William J.: V1382,240-254(1991)

Rothman, Mark A.
  ; Thompson, John A.; Armiento, Craig A.: Multichamber reactive ion etching
    processing for III-V optoelectronic devices,V1392,598-604(1991)
  See Elman, Boris S.: V1361,362-372(1991)

Rothman, Melvyn L.
  See Brooks, Everett G.: V1446,217-226(1991)
  See Gehring, Dale G.: V1446,248-252(1991)

Rothman, Peter L.
  ; Denton, Richard V.: Fusion or confusion: knowledge or nonsense?,V1470,2-
    12(1991)

Rothschild, Kenneth J. See Swairjo, Manal: V1437,60-65(1991)

Rothschild, Mordechai See Shaver, David C.: V1596,46-50(1991)

Rothschild, Richard E.
  ; Pelling, Michael R.; Hink, Paul L.: UCSD high-energy x-ray timing
    experiment magnetic shield design and test results,V1549,120-133(1991)
  See Hink, Paul L.: V1549,193-202(1991)
  See Pelling, Michael R.: V1549,134-146(1991)
  See Siegmund, Oswald H.: V1549(1991)

Rotman, Stanley R.
  ; Aizenberg, G. E.; Gordon, E. S.; Tuller, H. L.; Warde, Cardinal: Mechanisms
    for the cathodoluminescence of cerium-doped yttrium aluminum garnet
    phosphors,V1442,205-215(1991)
  ; Gordon, E. S.: Target acquisition model appropriate for dynamically
    changing scenarios,V1442,335-346(1991)
  ; Maoz, O.; Arnon, S.; Kaczelnik, F.; Felus, Y.; Weiss, Aryeh M.; Reisfeld,
    Renata; Eyal, Marek; Hartmann, F. X.: Unusually fast energy transfer in
    solid state crystals and glasses,V1442,194-204(1991)
  See Tidhar, Gil: V1442,310-324(1991)
  See Tidhar, Gil: V1486,188-199(1991)

Rotomskis, Richardas See Kirveliene, V.: V1403,582-584(1991)

Rottenkolber, Hans See Khesin, G. L.: V1554B,86-90(1991)

Rottmann, Hans See Bennett-Lilley, Marylyn H.: V1464,127-136(1991)

Rought, Nathan W. See Rogers, Mark E.: V1415,24-37(1991)

Rougon, Nicolas F.
  ; Preteux, Francoise: Deformable markers: mathematical morphology for
    active contour models control,V1568,78-89(1991)

Rouhani, Mehdi D.
  ; Laroussi, M.; Gue, A. M.; Esteve, Daniel: Atomic scale simulation of the
    growth of CdTe layers on GaAs,V1361,954-962(1991)

Rouse, Gordon F.
  ; Stauffer, Donald R.; French, Howard B.: Solid state magnetic azimuth sensor
    for small satellites,V1495,240-245(1991)

Rouse, Lawrence J. See Huh, Oscar K.: V1492,378-386(1991)

Roush, Randy A.
  See Brinkmann, Ralf P.: V1378,203-208(1991)
  See Lakdawala, Vishnu K.: V1378,259-270(1991)

Rousseau, Jacques See Paquette, Benoit: V1426,362-366(1991)

Roussell, Harold V. See Johnson, Leonard M.: V1371,2-7(1991)

Rousset, Gerard
  See Gendron, Eric: V1542,297-307(1991)
  See Merkle, Fritz: V1542,308-318(1991)

Roustan, Jean-Claude See Gilabert, Alain: V1477,140-147(1991)

Roux, Agnes
  ; Cipriani, Francois D.: Laser surface treatment: numerical simulation of
    thermocapillary flows,V1397,693-696(1991)

Rouyer, C. See Fleurot, Noel A.: V1502,230-241(1991)

Rovinsky, R. E.
  See Babaeva, N. A.: V1440,260-269(1991)
  See Vas'kovsky, Yu. M.: V1440,241-249(1991)
  See Vas'kovsky, Yu. M.: V1440,229-240(1991)

Rowan, Paul
  : Fiber hub in a second-generation ethernet system at Taylor
    University,V1364,262-266(1991)

Rowberg, Alan H.
  : Need for quality assurance related to PACS,V1444,80-82(1991)
  ; Lian, Jing: Quality assessment of video image capture systems,V1444,125-
    127(1991)
  See Yee, David K.: V1444,292-305(1991)

Rowe, Ednor M.
  See Maestre, Marcos F.: V1548,179-187(1991)
  See Snyder, Patricia A.: V1548,188-196(1991)

Rowe, W. J. See Snyder, Donald R.: V1346,226-236(1991)

Rowlands, David O. See Gilbert, Barry K.: V1390,235-248(1991)

Rowlands, Robert E. See Hyzer, Jim B.: V1554B,371-382(1991)

Roy, Adrian G. See Flanagan, Kathryn A.: V1549,395-407(1991)

Roy, Rajarshi
  ed.: Laser Noise,V1376(1991)
  See Bracikowski, Christopher: V1376,103-116(1991)

Roy, Sumit
  : L-filter design using the gradient search algorithm,V1451,254-256(1991)

Roychoudhuri, Chandrasekhar
  ; Veldkamp, Wilfrid B.; eds.: <author>Miniature and Micro-Optics:
    Fabrication and System Applications,V1544(1991)

Rozelot, Jean-Pierre
  ; Leblanc, Jean-M.: Metallic alternative to glass mirrors (active mirrors in
    aluminum): a review,V1494,481-490(1991)
  See Leblanc, Jean-M.: V1535,122-129(1991)

Rozenbergs, John
  See Botez, Dan: V1474,64-74(1991)
  See Jansen, Michael: V1418,32-39(1991)
  See Mawst, Luke J.: V1418,353-357(1991)

Rozental, Mark See Mazar, Reuven: V1487,31-39(1991)

Rozhkov, Oleg V.
  See Odinokov, S. B.: V1238,109-117(1991)
  See Odinokov, S. B.: V1238,103-108(1991)

Rozhnov, U.V. See Mnatsakanyan, Eduard A.: V1414,130-133(1991)

Rozhyki, Alicia See Yaeli, Joseph: V1442,378-382(1991)

Rozmus, Wojtek See Labaune, C.: V1413,138-143(1991)

Rozniakowski, Kazimierz
  : Characteristic features of melting surface of tin produced with laser
    pulse,V1391,215-223(1991)
  See Drobnik, Antoni: V1391,361-369(1991)

Ruan, Ke F. See Yan, Zu Q.: V1519,183-191(1991)

Ruan, S. See Wang, Xianhua: V1358,775-779(1991)

Ruan, Shengyang
  ; Kapp, Oscar H.: Progress on the subangstrom field emission scanning
    transmission electron microscope,V1396,298-310(1991)

Rubanov, Alexander S.
  See Lazaruk, A. M.: V1621,114-124(1991)
  See Rachkovsky, Leonid I.: V1461,232-240(1991)

Rubenchik, Aleksandr M. See Krasnov, Victor F.: V1506,179-187(1991)

Rubens, Deborah See Hornak, Joseph P.: V1445,523-533(1991)

Rubin, Barry
: Electrical characterization of the interconnects inside a computer,V1389,314-328(1991)

Rubin, Lawrence M.
: CAE tools for verifying high-performance digital systems,V1390,336-358(1991)

Rubini, Alda See Costa, Enrico: V1343,469-476(1991)

Rubinov, A. N.
See Gakamsky, D. M.: V1403,641-643(1991)
See Nemkovich, N. A.: V1403,578-581(1991)

Ruch, Eric
: Manufacture of ISO mirrors,V1494,265-278(1991)

Rucinski, Andrzej
; Drexel, Peter G.; Dziurla, Barbara: Chaotic nature of mesh networks with distributed routing,V1390,388-398(1991)

Ruck, Dennis W. See Rogers, Steven K.: VIS07,231-243(1991)

Rudant, J. P. See Monjoux, E.: V1452,310-318(1991)

Rudberg, Donald A.
; Stover, John C.; McGary, Douglas E.: Mapping of imbedded contaminants in transparent material by optical scatter,V1530,232-239(1991)

Rudd, Robert E. See Spillman, William B.: V1367,197-203(1991)

Ruddell, F. H.
; McNeill, D.; Armstrong, Brian M.; Gamble, Harold S.: Silicon carbide layers produced by rapid thermal chemical vapor deposition,V1361,159-170(1991)

Ruddell, Scott A. See Lewis, Isabella T.: V1530,306-324(1991)

Ruddy, Vincent See MacCraith, Brian D.: V1510,104-109(1991)

Rudenko, O. V. See D'yakonov, G. I.: V1421,153-155(1991)

Rudin, Leonid I. See Osher, Stanley: V1567,414-431(1991)

Rudoi, I. G. See Kamrukov, A. S.: V1503,438-452(1991)

Rudolf, P. See Tjeng, L. H.: V1548,160-167(1991)

Rudolph, John D.
: Developments in projection lenses for HDTV,V1456,15-28(1991)

Rudolph, Larry
: Role for optics in future parallel processing,V1505,175-185(1991)

Rudraraju, Prasad
See Raju, G. V.: V1385,50-56(1991)
See Raju, G. V.: V1388,415-424(1991)

Rueck, Angelika C.
See Koenig, Karsten: V1525,412-419(1991)
See Schneckenburger, Herbert: V1403,646-652(1991)
See Schneckenburger, Herbert: V1525,91-98(1991)
See Sessler, Jonathan L.: V1426,318-329(1991)

Rueck, Dorothee M. See Frank, Werner F.: V1559,344-353(1991)

Rueckmann, I.
; Woggon, Ulrike; Kornack, J.; Mueller, Matthias; Cesnulevicius, J.; Kolenda, Jonas; Petrauskas, Mendogas: Influence of surface passivation on the optical bleaching of CdSe microcrystallites embedded in glass,V1513,78-85(1991)
See Woggon, Ulrike: V1362,888-898(1991)

Rueff, Manfred
; Frankhauser, P.; Dettki, Frank: Joint space/spatial-frequency representations as preprocessing steps for neural nets; joint recognition of separately learned patterns; results and limitations,V1382,255-270(1991)

Ruehle, Wolfgang W. See Lopez, C.: V1361,89-95(1991)

Rueschoff, Josef See Miller, Kurt: V1421,108-113(1991)

Ruffin, Paul B.
; Sung, C. C.: Sensitivity of polarization-maintaining fibers to temperature variations,V1478,160-167(1991)

Ruiz, Ana See Recio, Miguel: V1361,469-478(1991)

Ruiz, D.
; Harding, John; Noble, J. P.; Hillsdon, G. K.: Application of high-speed photography to the study of high-strain-rate materials testing,V1358,1134-1143(1991)

Ruiz, Domingo
; Kremer, Paul J.: Equipment development for an atmospheric-transmission measurement campaign,V1417,212-222(1991)

Ruiz-Cortes, Victor
; Mendez, Eugenio R.; Gu, Zu-Han; Maradudin, Alexei A.: Light scattering from gold-coated ground glass and chemically etched surfaces,V1558,222-232(1991)

Ruizhong, Rao
; Song, Zhengfang: Natural terrain infrared radiance statistics in a wind field,V1486,390-395(1991)

Rukhman, N. V. See Aleksandrov, A. P.: V1440,442-453(1991)

Rule, Donald W.
; Fiorito, Ralph B.; Piestrup, Melvin A.; Gary, Charles K.; Maruyama, Xavier K.: Production of x-rays by the interaction of charged particle beams with periodic structures and crystalline materials,V1552,240-251(1991)

Ruller, Jacqueline A.
See Justus, Brian L.: V1409,2-8(1991)
See Merzbacher, Celia I.: V1533,222-228(1991)

Rumbaugh, Richard C. See Perry, Meg W.: V1459,155-156(1991)

Rummel, Paul
: Monitoring and control of rf electrical parameters near plasma loads,V1392,411-420(1991)

Rumsby, Phil T.
; Gower, Malcolm C.: Excimer laser projector for microelectronics applications,V1598,36-45(1991)

Rumyantsev, B. L.
See Sreseli, Olga M.: V1440,326-331(1991)
See Sreseli, Olga M.: V1545,149-158(1991)

Rumyantsev, V. A.
: Problems of production of display holograms,V1238,381-381(1991)

Rundle, Debra A.
; Evers, K.; Seshadri, Sridhar B.; Arenson, Ronald L.: DRILL: a standardized radiology-teaching knowledge base,V1446,405-413(1991)
; Watson, Carolyn K.; Seshadri, Sridhar B.; Wehrli, Felix W.: Signa Tutor: results and future directions,V1446,379-387(1991)

Rupp, Romano A. See Ibel, Konrad: V1559,393-402(1991)

Rupp, Wiktor J. See Zimmerman, Jerrold: VCR38,184-192(1991)

Rurukin, Alexander N. See Kurochkin, Vadim Y.: V1435,322-330(1991)

Rusche, Gerald A. See Der, Sandor Z.: V1483,167-176(1991)

Ruschin, Shlomo
See Keselbrener, Michel: V1415,38-47(1991)
See Sneh, Anat: V1442,252-257(1991)

Rushmeier, Holly E.
; Rodriguez, Leonard J.: Generic models for rapid calculation of target signatures,V1486,210-216(1991)

Rusinek, Henry See Chandra, Ramesh: V1445,133-144(1991)

Russ, John C.
; Dudley, Bruce W.; Jones, Susan K.: Monte Carlo modeling of secondary electron signals from heterogeneous specimens with nonplanar surfaces,V1464,10-21(1991)

Russanova, V. See Angelov, D.: V1403,230-239(1991)

Russell, Derrek
; Taborek, Peter: Diagnostics of a DC plasma torch,V1534,14-23(1991)

Russell, J. D. See Quenzer, A.: V1397,753-759(1991)

Russell, James M.
: Overview of the halogen occultation experiment,V1491,110-116(1991)
: Overview of the spectroscopy of the atmosphere using far-infrared emission experiment,V1491,142-150(1991)

Russell, Kevin L. See McFee, John E.: V1567,42-52(1991)

Russell, P. See Pogorelsky, Igor V.: V1413,21-31(1991)

Russell, Philip S.
; Hand, Duncan P.; Chow, Y. T.; Poyntz-Wright, L. J.: Optically induced creation, transformation, and organization of defects and color centers in optical fibers,V1516,47-54(1991)
; Poyntz-Wright, L. J.; Hand, Duncan P.: Frequency doubling, absorption, and grating formation in glass fibers: effective defects or defective effects?,V1373,126-139(1991)
See Berwick, Michael: V1584,364-373(1991)
See Berwick, Michael: V1572,157-162(1991)

Russer, Peter See Kessler, Jochen: V1477,45-56(1991)

Russo, G. See Barone, Fabrizio: V1584,304-307(1991)

Russo, Paolo
See Barone, Fabrizio: V1584,304-307(1991)
See Bernini, Umberto: V1427,398-404(1991)
See Carbonara, Giuseppe: V1361,1038-1040(1991)

Russo, Paul
; Markandey, Vishal; Bui, Trung H.; Shrode, David: Optical flow techniques for moving target detection,V1383,60-71(1991)

Russo, Paul S. See Mustafa, Mazidah B.: V1430,132-141(1991)

Russo, Rick See Berdahl, Paul H.: V1394,180-185(1991)

Russwurm, George M.
; Kagann, Robert H.; Simpson, Orman A.; McClenny, William A.: Use of a
Fourier transform spectrometer as a remote sensor at Superfund
sites,V1433,302-314(1991)

Rust, Kenneth R.
: Modifications to achieve subnanosecond jitter from standard commercial
excimer lasers,V1411,80-86(1991)

Ruth, Bernhard
: Reflectance at the red edge as a sensitive indicator of the damage of trees
and its correlation to the state of the photosynthetic system,V1521,131-
142(1991)

Rutledge, Sharon K. See Banks, Bruce A.: V1330,66-77(1991)

Rutt, James
: Industrial lasers in the United States,V1517,28-43(1991)

Ryabchenko, Vladimir E. See Krasnov, Victor F.: V1506,179-187(1991)

Ryabova, R. V.
; Barinova, E. S.; Myatezh, O. V.; Zaborov, A. N.; Romash, E. V.; Chernyi, D.
I.: High-resolution photographic material of Institute of Atomic Energy for
holography in intersecting beams,V1238,166-170(1991)

Ryabukho, Vladimir P. See Tuchin, Valery V.: V1420,81-92(1991); 1429,62-
73(1991)

Ryabushkin, Oleg A.
; Bader, Vladimir A.: Photocryosar: bistable element for optoelectronic
computing,V1505,67-74(1991)
; Platonov, N. S.; Sablikov, V. A.; Sergeyev, V. I.; Bader, Vladimir A.:
Optoelectronic and optical bistabilities of photocurrent and
photoluminescence at low-temperature avalanche breakdown in GaAs
epitaxial films,V1362,75-79(1991)

Ryan, James G. See Sullivan, Timothy D.: V1596,83-95(1991)

Ryan, Martin J.
; Kapp, Oscar H.: Development of an image processing system on a second-
generation RISC workstation,V1396,335-339(1991)

Ryan, Robert E. See Bergeman, Thomas H.: V1376,54-67(1991)

Ryan, Samantha See Exarhos, Gregory J.: V1441,190-199(1991)

Ryan, Thomas W.
See Hunt, Bobby R.: V1472,208-218(1991)
See Hunt, Bobby R.: V1472,190-200(1991)

Ryan, Vincent
; Fielding, Kenneth H.: Position-, scale-, and rotation-invariant
photorefractive correlator,V1564,86-97(1991)

Ryan, Vivian See Siddiqui, Humayun R.: V1596,139-157(1991)

Ryba-Romanowski, Witold
; Golab, Stanislaw; Berkowski, Marek: Crystal growth and characterization of
rare-earth-doped gallates of alkaline earth and lanthanum,V1391,2-
5(1991)

Rybak, Ilya A.
; Golovan, Alexander V.; Gusakova, Valentina I.: Approach to invariant
object recognition on grey-level images by exploiting neural network
models,V1469,472-482(1991)
; Shevtsova, Natalia A.; Podladchikova, Lubov N.: Modeling of local neural
networks of the visual cortex and applications to image
processing,V1469,737-748(1991)

Rybakov, M. I.
See Khvilivitzky, A. T.: V1447,64-68(1991)
See Vishnevsky, G. I.: V1448,69-72(1991)

Rybka, Theodore W. See Bradley, Eric M.: V1418,272-278(1991)

Rycyk, Antoni See Marczak, Jan: V1391,48-51(1991)

Rye, A. J. See de Salabert, Arturo: V1521,74-88(1991)

Rykov, V. S. See Kalyashora, L. N.: V1238,189-194(1991)

Rylander, Henry G. See Porindla, Sridhar N.: V1427,267-272(1991)

Rymer, William Z. See Arber, Simon: V1428,23-29(1991)

Ryne, Robert D. See Westenskow, Glen A.: V1407,496-501(1991)

Rynkowski, Gerald
: Overview of a high-performance polygonal scanner subsystem,V1454,102-
110(1991)

Rysiakiewicz-Pasek, Ewa
; Marczuk, Krystyna: Electrical and optical properties of porous
glass,V1513,283-290(1991)
See Marczuk, Krystyna: V1513,291-296(1991)

Ryu, Zee Man
: Measurement of point spread function of thermal imager,V1467,469-
474(1991)

Saad, Ricardo E.
; Souza, Rui F.: Method to find the transimpedance gain of optical receivers
using measured S-parameters,V1371,142-148(1991)

Saari, Peeter M.
; Kaarli, Rein K.; Sonajalg, Heiki: Four-dimensional photosensitive materials:
applications for shaping, time-domain holographic dissection, and
scanning of picosecond pulses,V1507,500-508(1991)

Saarinen, Jyrki V.
; Huttunen, Juhani; Vasara, Antti H.; Noponen, Eero; Turunen, Jari P.; Salin,
Arto U.: Synthetic holographic beamsplitters for integrated
optics,V1555,128-137(1991)

Saaski, Elric W. See Mitchell, Gordon L.: V1367,266-272(1991)

Saatdjian, E.
; Caressa, J. P.; Andre, Jean-Claude: Numerical simulation of the combustion
chamber of a chemical laser,V1397,535-538(1991)

Saavedra, Steven S. See Potter, B. L.: V1368,251-257(1991)

Sabatier, Lilian
See Caillibotte, Georges: V1502,117-122(1991)
See Caillibotte, Georges: V1412,209-211(1991)
See de Frutos, Angel M.: V1397,717-720(1991)
See Poueyo, Anne: V1502,140-147(1991)

Sabia, E. See Ciocci, Franco: V1501,154-162(1991)

Sablikov, V. A. See Ryabushkin, Oleg A.: V1362,75-79(1991)

Sabol, Bruce M.
; Mixon, Harold D.: Simulation study to characterize thermal infrared sensor
false alarms,V1486,258-270(1991)

Sabri, Sehbaz H.
See Fredricks, Ronald J.: V1367,127-139(1991)
See Johnson, Dean R.: V1367,140-154(1991)

Sacco, Albert
: Large zeolites: why and how to grow in space,V1557,6-9(1991)

Sacco, B. See Citterio, Oberto: V1343,145-154(1991)

Saccomanno, Geno
; Bechtel, Joel J.: Early diagnosis of lung cancer,V1426,2-12(1991)

Sachdev, Krishna See Aviram, Ari: V1458,105-107(1991)

Sachse, Glen W.
; Collins, Jim E.; Hill, G. F.; Wade, L. O.; Burney, Lewis G.; Ritter, J. A.:
Airborne tunable diode laser sensor for high-precision concentration and
flux measurements of carbon monoxide and methane,V1433,157-
166(1991)

Sadjadi, Firooz A.
ed.: *Automatic Object Recognition*,V1471(1991)
: Automatic object recognition: critical issues and current
approaches,V1471,303-313(1991)
: Landmark recognition using motion-derived scene structures,V1521,98-
105(1991)
: Multispectral and multisensor adaptive automatic object
recognition,VIS07,218-230(1991)
: Object recognition using coding schemes,V1471,428-434(1991)
: Performance evaluation of a texture-based segmentation
algorithm,V1483,185-195(1991)
See Nasr, Hatem N.: VIS07,122-129(1991)

Sadler, Brian M.
; Giannakis, Georgios B.; Smith, Dale J.: Acousto-optic estimation of
autocorrelation and spectra using triple correlations and
bispectra,V1476,246-256(1991)
See Garvin, Charles G.: V1562,303-318(1991)

Sadler, Lewis L.
; Chen, Xiaoming; Figueroa, Alvaro A.; Aduss, Howard: Medical applications
of three-dimensional and four-dimensional laser scanning of facial
morphology,V1380,158-162(1991)
; Chen, Xiaoming; Fyler, Ann: Medical prosthetic applications of growth
simulations in four-dimensional facial morphology,V1380,137-146(1991)
See Evenhouse, Raymond J.: V1380,147-156(1991)
See Parshall, Robert F.: V1380,200-207(1991)

Sadot, Danny
; Kopeika, Norman S.: Forecasting optical turbulence strength: effects of
macroscale meteorology and aerosols,V1442,325-334(1991)
; Kopeika, Norman S.: Prediction of Cn2 on the basis of macroscale
meteorology including aerosols,V1487,40-50(1991)

Sadovnik, Lev S.
; Demichovskaya, Olga; Chen, Ray T.: Nonlinear optical processing using
phase grating,V1545,200-208(1991)

; Rich, Chris C.: Instant measurement of phase-characteristic curve,V1559,424-428(1991)
See Chen, Ray T.: V1559,449-460(1991)
See Chen, Ray T.: V1583,362-374(1991)
See Jannson, Tomasz: V1545,42-63(1991)
See Manasson, Vladimir A.: V1559,194-201(1991)

Sadovskaya, O. See Zhvavyi, S.: V1440,8-15(1991)

Sadowski, Thomas J.
: Laser beam diagnostics: a conference overview,V1414,136-140(1991)

Sadyigov, Z. Y.
; Gasanov, A. G.; Yusipov, N. Y.; Golovin, V. M.; Gulanian, Emin H.; Vinokurov, Y. S.; Simonov, A. V.: Investigation of the possibility of creating a multichannel photodetector based on the avalanche MRS-structure,V1621,158-168(1991)

Saedi, Reza See Even-Or, Baruch: V1371,161-169(1991)

Safabakhsh, Reza
: Determination of flint wheel orientation for the automated assembly of lighters,V1472,185-189(1991)

Saffert, Ralf See Rohde, Axel: V1464,438-446(1991)

Sage, Maurice G.
; Hartley, Neil: System issues for multichip packaging,V1390,302-310(1991)

Saggese, Steven J.
; Harrington, James A.; Sigel, George H.: Hollow and dielectric waveguides for infrared spectroscopic applications,V1437,44-53(1991)
; Harrington, James A.; Sigel, George H.: Hollow waveguides for sensor applications,V1368,2-14(1991)

Sagie, Daniel See Jacobson, Benjamin A.: V1528,82-85(1991)

Sagitov, S. I. See Morozov, N. V.: V1441,557-564(1991)

Sahai, Viveik
; Williamson, John W.; Overfelt, Tony: Steady-state modeling of large-diameter crystal growth using baffles,V1557,60-70(1991)

Sahakian, Alan V.
See Chan, Cheuk L.: V1450,208-217(1991)
See Kwak, J. Y.: V1396,32-44(1991)

Sahinoglou, Haralambos
; Cabrera, Sergio D.: High-speed two-dimensional pyramid image coding method and its implementation,V1605,793-804(1991)

Sahni, Omesh
: Color printing technologies,V1458,4-16(1991)

Sahyun, Melville R.
; Crooks, G. P.; Sharma, D. K.: Photophysics of 1,3,5-triaryl-2-pyrazolines,V1436,125-133(1991)

Said, Ali A. See Soileau, M. J.: V1441,10-15(1991)

Said, Hicham See Bremond, Georges E.: V1361,732-743(1991)

Saif, Babak N. See Cornwell, Donald M.: V1417,431-439(1991)

Saifi, Mansoor A.
: Long-term fiber reliability,V1366,58-70(1991)

Saijo, Y. See Saito, Juichi: V1496,284-301(1991)

Saikkonen, Stuart L. See Dreyer, Donald R.: V1366,372-379(1991)

Sailer, Donald R.
: Planning for fiber optic use at the University of Massachusetts,V1364,257-261(1991)

Sailor, Michael J. See Jozefiak, Thomas H.: V1436,8-19(1991)

Saindon, Michelle L. See Spangler, Charles W.: V1497,408-417(1991)

Sainov, Ventseslav C.
; Simova, Eli S.: Holographic in-plane displacement measurement in cracked specimens in plane stress state,V1461,175-183(1991)

Saint-Andre, Francoise
; Benech, Pierre; Kevorkian, Antoine P.: Modelization of a semi-leaky waveguide: application to a polarizer,V1583,278-288(1991)

Saint-Jacques, Robert G. See Dodelet, Jean-Pol: V1436,38-49(1991)

Saissac, M.
; Berejny, P.; Millet, P.; Yousfi, M.; Salamero, Y.: Luminescence and ionization of krypton by multiphotonic excitation near the 3P1 resonant state,V1397,739-742(1991)

Saita, Masahiro See Hattori, Shuzo: V1463,539-550(1991)

Saito, Atsushi See Maeda, Takeshi: V1499,414-418(1991)

Saito, Juichi
; Saijo, Y.: Improvements in sensitivity and discrimination capability of the PD reticle/mask inspection system,V1496,284-301(1991)

Saito, Jun
; Akasaka, Hideki: Direct overwritable magneto-optical exchange-coupled multilayered disk by laser power modulation recording,V1499,44-54(1991)

Saito, M. See Yamaguchi, H.: V1499,29-38(1991)

Saito, Masahiro See Matsumoto, Shuichi: V1605,37-46(1991)

Saito, Osamu See Watanabe, Mikio: V1452,27-36(1991)

Saito, Riichiro See Isshiki, Hideo: V1361,223-227(1991)

Saito, Takahiro
; Higuchi, Hirofumi; Komatsu, Takashi: High-fidelity subband coding for superhigh-resolution images,V1605,382-393(1991)
See Aizawa, Kiyoharu: V1605,318-328(1991)

Saito, Tomohide See Uchiyama, Toshio: V1451,192-203(1991)

Saitou, Norio
; Iwasaki, Teruo; Murai, Fumio: Multiple scattered electron-beam effect in electron-beam lithography,V1465,185-191(1991)

Sakagami, Iwata
; Kaji, Akihiro: Transmission of high-speed voltage waves at the junction of three transmission lines,V1389,329-339(1991)

Sakaguchi, Shinji See Honda, Kenji: V1466,141-148(1991)

Sakai, Katsuo
: Multilayer OPC for one-shot two-color printer,V1458,179-191(1991)

Sakai, Kazuhiro See Miyake, Hiroyuki: V1448,150-156(1991)

Sakai, Koji See Ohmi, Toshihiko: V1361,606-612(1991)

Sakai, Masao
; Nagayama, Akira; Sasakura, Kunihiko: Positioning method using polarization-detecting optical sensor for precision robot systems,V1385,8-14(1991)

Sakai, Mitsuhiro See Uchiyama, Toshio: V1451,192-203(1991)

Sakai, Y. See Kawahara, Hideo: V1513,198-203(1991)

Sakai, Yoshihiko See Miyake, Hiroyuki: V1448,150-156(1991)

Sakamoto, Hideki See Maruyama, Mitsuru: V1605,916-927(1991)

Sakashita, Toshihiko See Misaka, Akio: V1465,174-184(1991)

Sakharov, V. N.
See Khesin, G. L.: V1554B,86-90(1991)
See Savostjanov, V. N.: V1554A,380-386(1991)
See Taratorin, B. I.: V1554A,449-456(1991)

Sakuma, Fumihiro
; Ono, Akira: Prelaunch calibration system for optical sensors of Japanese Earth Resources Satellite,V1493,37-47(1991)
See Ono, Akira: V1490,285-298(1991)
See Suzuki, Naoshi: V1493,48-57(1991)

Sakurada, Masami
; Miyamoto, Akira; Mizuno, Kyoichi; Nozaki, Youichi; Tabata, Hirotsugu; Etsuda, Hirokuni; Kurita, Akira; Nakamura, Haruo; Arai, Tsunenori; Suda, Akira; Kikuchi, Makoto; Watanabe, Tamishige; Utsumi, Atsushi; Akai, Yoshiro; Takeuchi, Kiyoshi: Efficacy of argon-laser-mediated hot-balloon angioplasty,V1425,158-164(1991)
See Arai, Tsunenori: V1425,191-195(1991)

Sakurai, Hirohisa
; Noma, M.; Niizeki, H.: Hard x-ray polarimeter utilizing Compton scattering,V1343,512-518(1991)
; Ramsey, Brian D.: Characteristics of a high-pressure gas proportional counter filled with xenon,V1549,20-27(1991)

Sakurai, Tsunetaro See Komori, Masaru: V1444,334-337(1991)

Sakurai, Yasushi See Akasaka, Akira: V1490,269-277(1991)

Salakhutdinov, Viktor K. See Mikaelian, Andrei L.: V1621,148-157(1991)

Salamero, Y. See Saissac, M.: V1397,739-742(1991)

Salari, Ezzatollah
; Balaji, Sridhar: Recognition of partially occluded objects using B-spline representation,V1384,115-123(1991)

Salari, Valiollah
: Sperm motion analysis,V1450,255-263(1991)

Salashchenko, N. N. See Platonov, Yu. Y.: V1440,188-196(1991)

Salathe, R. P. See Chambettaz, Francois: V1427,134-140(1991)

Salazar, Robert A. See Carson, Susan D.: V1592,134-138(1991)

Salbut, Leszek A.
; Patorski, Krzysztof; Kujawinska, Malgorzata: Polarization approach to high-sensitivity moire interferometry,V1554B,451-460(1991)

Salem, H. K. See Gussenhoven, Elma J.: V1425,203-206(1991)

Salem-Sugui, S. See Ramanathan, Mohan: V1548,168-178(1991)

Salembier, Philippe
; Jaquenoud, Laurent: Adaptive morphological multiresolution decomposition,V1568,26-37(1991)

Salemink, Huub W. See Albrektsen, O.: V1361,338-342(1991)

Saletsky, A. M.
See Berezin, M. V.: V1403,335-337(1991)
See Bisenbaev, A. K.: V1403,606-610(1991)

Salimbeni, Renzo See Viligiardi, Riccardo: V1425,72-74(1991)

Salin, Arto U. See Saarinen, Jyrki V.: V1555,128-137(1991)

Salk, Ants A. See Vill, Arnold A.: V1503,110-114(1991)

Salmin, V. V. See Kushlevsky, S. V.: V1403,799-800(1991)

Salmon, J. T.
; Bliss, Erlan S.; Long, Theresa W.; Orham, Edward L.; Presta, Robert W.; Swift, Charles D.; Ward, Richard S.: Real-time wavefront correction system using a zonal deformable mirror and a Hartmann sensor,V1542,459-467(1991)
See Feldman, Mark: V1542,490-501(1991)
See Toeppen, John S.: V1544,218-225(1991)

Salokatve, A. See Pessa, Markus: V1361,529-542(1991)

Saloman, Edward B. See Slaughter, Jon M.: V1343,73-82(1991)

Salomon, L.
See Goudonnet, Jean-Pierre: V1545,130-139(1991)
See Goudonnet, Jean-Pierre: V1400,116-123(1991)

Salt, Kimberly L. See Iannone, Mark A.: V1559,172-183(1991)

Salter, Jeffery L.
; Loeffler, Mary F.: Comparison of dichromated gelatin and Du Pont HRF-700 photopolymer as media for holographic notch filters,V1555,268-278(1991)

Salutz, James R.
See Gehring, Dale G.: V1446,248-252(1991)
See Morin, Richard L.: V1446,436-441(1991)
See Reardon, Frank J.: V1446,481-491(1991)

Salvadori, L. See Bortoletto, Favio: V1542,225-235(1991)

Salvaggio, Carl See Schott, John R.: V1540,533-549(1991)

Salvi, Ajit P. See Parra-Loera, Ramon: V1470,30-36(1991)

Salvi, Theodore C.
; Shakir, Sami A.: Modeling of coupled grating surface-emitting diodes,V1418,64-73(1991)

Samad, Tariq
; Mathur, Anoop: Parameter estimation for process control with neural networks,V1469,766-777(1991)

Samadani, Ramin
: Adaptive snakes: control of damping and material parameters,V1570,202-213(1991)

Samal, Ashok
: Minimum resolution for human face detection and identification,V1453,81-89(1991)

Samarabandu, J. K.
; Acharya, Raj S.; Edirisinghe, Chandima D.; Cheng, Ping-Chin; Kim, Hyo-Gun; Lin, T. H.; Summers, R. G.; Musial, C. E.: Analysis of multidimensional confocal images,V1450,296-322(1991)
See Acharya, Raj S.: V1450,170-177(1991)

Samarakone, Nandasiri
; Van Driessche, Veerle; Jaenen, Patrick; Van den hove, Luc; Ritchie, Douglas R.; Luehrmann, Paul F.: Improving the performance and usability of a wet-developable DUV resist for sub-500nm lithography,V1463,16-29(1991)

Sambles, J. R. See Bryan-Brown, G. P.: V1545,167-178(1991)

Samkov, V. M. See Miroshnikov, Mikhail M.: V1540,496-505(1991)

Samoriski, Brian
; Wiedeger, S.; Villarica, M.; Chaiken, Joseph: Metal vapor gain media based on multiphoton dissociation of organometallics,V1412,12-18(1991)

Sampietro, Marco See Braeuninger, Heinrich: V1549,330-339(1991)

Sampson, Robert E. See Gleichman, Kurt W.: V1416,286-294(1991)

Samson, Marc
; Dufour, Marc L.: New stereo laser triangulation device for specular surface inspection,V1332,314-322(1991)

Samson, Peter J. See Hills, Peter C.: V1589,110-119(1991)

Samyn, Celest See Verbiest, Thierry: V1560,353-361(1991)

Sanada, Kazuo See Tsumanuma, Takashi: V1420,193-198(1991)

San Biagio, Pier L. See Newman, Jay E.: V1430,89-108(1991)

Sanborn, James See Macdonald, Burns: V1537,104-114(1991)

Sanborn, Timothy A. See Golobic, Robert A.: V1425,84-92(1991)

Sanches-Cortes, S.
; Garcia-Ramos, J. V.: Surface-enhanced Raman spectroscopy of adenosine and 5'AMP: evolution in time,V1403,142-145(1991)

Sanchez, Anthony D.
See Padden, Richard J.: V1474,148-159(1991)
See Taylor, Edward W.: V1474,126-131(1991)

Sanchez, Elie
: Fuzzy logic and neural networks in artificial intelligence and pattern recognition,V1569,474-483(1991)

Sanchez, Juan A. See Auteri, Joseph S.: V1421,182-184(1991)

Sanchez, Luis See Casals, Alicia: V1482,317-324(1991)

Sanchez, Miguel
; Hernandez Neira, Jose Luis; Delgado, J.; Calvo, G.: Free propagation of high-order moments of laser beam intensity distribution,V1397,635-638(1991)
See Martinez-Herrero, R.: V1397,627-630(1991)

Sanchez-Dehesa, Jose See Lopez, C.: V1361,89-95(1991)

Sand, Francis M. See Dougherty, Edward R.: V1451,114-124(1991)

Sand, Rolf
: Investigation of unobscured mirror telescope for telecom purposes,V1522,103-110(1991)
See Hildebrand, Ulrich: V1522,50-60(1991)

Sand, Ruediger
: 3-DTV research and development in Europe,V1457,76-84(1991)

Sandborn, Peter A.
; Hashemi, Seyed H.; Weigler, William: Switching noise in a medium-film copper/polyimide multichip module,V1389,177-186(1991)

Sandel, Bill R. See Schulze, Dean W.: V1549,319-328(1991)

Sander, Peter T.
See Fua, Pascal: V1570,286-296(1991)
See Monga, Olivier: V1570,382-390(1991)

Sanders, Brian W.
; Kitai, Adrian H.: Atomic layer growth of zinc oxide and zinc sulphide,V1398,81-87(1991)

Sanders, Jeffrey S.
; Currin, Michael S.: Human recognition of infrared images II,V1488,144-155(1991)
; Halford, Carl E.: Multispectral imaging with frequency-modulated reticles,V1478,52-63(1991)

Sanders, Ryan W. See Mount, George H.: V1491,188-193(1991)

Sanders, Virgil E.
See Boyer, James D.: V1377,92-98(1991)
See Boyer, James D.: V1441,255-261(1991)

Sanders, William H. See Alsafadi, Yasser H.: V1446,129-140(1991)

Sanderson, Arthur C. See Mathur, Rajive K.: V1387,47-57(1991)

Sandford, B. P. See Noah, Meg A.: V1479,275-282(1991)

Sandholm, Scott T.
; Bradshaw, John D.: VUV/photofragmentation laser-induced fluorescence sensor for the measurement of atmospheric ammonia,V1433,69-80(1991)

Sandhu, Gurtej S. See Yu, Chang: V1598,186-197(1991)

Sandini, Giulio
See Grosso, Enrico: V1570,371-381(1991)
See Kreider, Gregory: V1381,242-249(1991)

Sandler, David G.
See Cuellar, Louis: V1542,468-476(1991)
See Wizinowich, Peter L.: V1542,148-158(1991)

Sandoz, Patrick See Tribillon, Gilbert M.: V1332,632-642(1991)

Sandrik, John M.
: Standardization of image quality measurements of medical x-ray image intensifier systems,VCR37,180-206(1991)

Sandroff, Claude J. See Jewell, Jack L.: V1389,401-407(1991)

Sandstrom, Richard L.
: Argon fluoride excimer laser source for sub-0.25 mm optical lithography,V1463,610-616(1991)

Sandstrom, Torbjorn
; Tison, James K.: Highly accurate pattern generation using acousto-optical deflection,V1463,629-637(1991)

Sanford, N. A. See Hickernell, Robert K.: V1474,138-147(1991)

Sang, Fengting
; Zhuang, Qi; Wang, Chengdong; Feng, Hao; Zhang, Cunhao: Study of visible chemical lasers: modeling of an IF chemical laser within the F-NH3-IF system,V1412,252-257(1991)
See Zhuang, Qi: V1397,157-160(1991)

Sanganeria, M. See Ozturk, Mehmet C.: V1393,260-269(1991)

Saniie, Jafar
See Algrain, Marcelo C.: V1478,201-210(1991)
See Nagle, Daniel T.: V1481,49-63(1991)

Sankar, Ananth
; Mammone, Richard J.: Combining neural networks and decision trees,V1469,374-383(1991)

Sankey, Otto F. See Sinha, Kislay: V1437,32-35(1991)

San Miguel, Maximino
: Statistics of laser switch-on,V1376,272-283(1991)

Sano, Kenji See Mori, Yasushi: V1560,310-314(1991)

Sano, Masaharu See Watanabe, Hiroshi: V1490,317-323(1991)

Sano, Mutsuo
; Ishii, Akira; Meguro, Shin-Ichi: Flexible gray-level vision system based on multiple cell-feature description and generalized Hough transform,V1381,101-110(1991)
See Meguro, Shin-Ichi: V1384,27-37(1991)

Sanson, Henri
: Motion affine models identification and application to television image coding,V1605,570-581(1991)

Sansonetti, Pierre
; Guerin, J. J.; Lequime, Michael; Debrie, J.: Parallel coherence receiver for quasidistributed optical sensor,V1588,143-149(1991)
; Lequime, Michael; Engrand, D.; Guerin, J. J.; Davidson, Roger; Roberts, Scott S.; Fornari, B.; Martinelli, Mario; Escobar Rojo, Priscilla; Gusmeroli, Valeria; Ferdinand, Pierre; Plantey, J.; Crowther, Margaret F.; Culshaw, Brian; Michie, W. C.: Intelligent composites containing measuring fiber-optic networks for continuous self-diagnosis,V1588,198-209(1991)

Santago, Peter
; Slade, James N.: Using MRI to calculate cardiac velocity fields,V1445,555-563(1991)
See Logenthiran, Ambalavaner: V1452,225-243(1991)

Santambrogio, E. See Citterio, Oberto: V1343,145-154(1991)

Santangelo, A. See Shaw, Ping-Shine: V1548,118-131(1991)

Santangelo, Andrea E. See Kaaret, Philip E.: V1548,106-117(1991)

Santoleri, Joseph J.
ed.: *Environmental Sensing and Combustion Diagnostics,*V1434(1991)
: Incinerator technology overview,V1434,2-13(1991)

Santos, Jose L.
; Farahi, Faramarz; Newson, Trevor P.; Leite, Antonio P.; Jackson, David A.: Multiplexing of remote all-fiber Michelson interferometers with lead insensitivity,V1511,179-189(1991)
; Jackson, David A.: Time-addressing of coherence-tuned optical fibre sensors based on a multimode laser diode,V1572,325-330(1991)

Sanz, Inmaculada See Lopez-Amo, Manuel: V1374,74-85(1991)

Sanz Justes, Pedro See Quenzer, A.: V1397,753-759(1991)

Saperstein, David D.
ed.: *Applied Spectroscopy in Material Science,*V1437(1991)
See Kaufman, James H.: V1437,36-41(1991)

Sapir, Eyal
See Johnson, W. Todd: V1488,343-354(1991)
See Kopolovich, Zvi: V1540,565-577(1991)

Sapochak, Linda S. See Spangler, Charles W.: V1497,408-417(1991)

Sapozhnikov, A. V. See D'yakonov, G. I.: V1421,153-155(1991)

Sapozhnikova, T. I. See Romanovsky, V. F.: V1358,140-145(1991)

Sapritsky, Victor I.
: Radiometric standards in the USSR,V1493,58-69(1991)

Saptzin, V. M. See Vavilov, Vladimir P.: V1540,488-495(1991)

Saracini, R. See Righini, Giancarlo C.: V1513,418-424(1991)

Saraswat, Krishna C.
See Lee, Yong J.: V1393,366-371(1991)
See McVittie, James P.: V1392,126-138(1991)
See Wood, Samuel C.: V1393,36-48(1991)
See Wood, Samuel C.: V1393,337-348(1991)

Saravanos, C. See Jen, Cheng K.: V1590,107-119(1991)

Sardar, Dhiraj K.
; Nemati, Babak; Barrera, Frederick J.: Use of polarization to separate on-axis scattered and unscattered light in red blood cells,V1427,374-380(1991)

Sardesai, Harshad P.
; Nunnally, William C.; Williams, Paul F.: Surface field measurement of photoconductive power switches using the electro-optic Kerr effect,V1378,237-248(1991)

Sarebahi, K. N. See Jain, Subhash C.: V1488,410-413(1991)

Sargis, Paul G. See Hagans, Karla G.: V1346,404-408(1991)

Sargood, Stephen K. See Taylor, Geoff W.: V1476,2-13(1991)

Sargsjan, Geworg See Possner, Torsten: V1513,378-385(1991)

Sari-Sarraf, Hamed See Beck, Hal E.: V1381,600-609(1991)

Saridis, George N.
See Kyriakopoulos, Konstantinos J.: V1387,159-164(1991)
See Murphy, Steve H.: V1387,14-25(1991)

Sarkady, Kenneth A. See Scribner, Dean A.: V1541,100-109(1991)

Sarkar, N. R.
: Images: from a printer's perspective,V1458,42-50(1991)

Sarkar, Tapan K.
; Banerjee, Partha P.: Generation of stable low-jitter kilovolt amplitude picosecond pulses,V1378,34-42(1991)
See Rahal Arabi, Tawfik R.: V1389,302-313(1991)

Sarkisov, S. E. See Manenkov, Alexander A.: V1420,254-258(1991)

Sarkisov, Sergey S. See Tarasov, Victor A.: V1559,331-342(1991)

Sarkissian, K. A.
See Arutunian, A. H.: V1402,102-106(1991)
See Arutunian, A. H.: V1402,2-6(1991)
See Arutunian, A. H.: V1403,585-587(1991)

Sarma, C. G. See Mishra, R. K.: V1482,138-145(1991)

Sarnakov, S. M.
: Low-energy hydrodynamic mechanism of laser destruction of thin films,V1440,112-114(1991)

Sarr, Dennis P.
: Aircraft exterior scratch measurement system using machine vision,V1472,177-184(1991)
; Jurick, Thomas W.: Faying surface-gap measurement of aircraft structures for shim fabrication and installation,V1469,506-514(1991)

Sarrabay, Laurence See Duchet, Christian: V1372,72-81(1991)

Sarrafzadeh, A. K. See Nader-Rezvani, Navid: V1584,405-414(1991)

Sarture, Charles M. See Eastwood, Michael L.: V1540,164-175(1991)

Sarubbi, Thomas R. See Toukhy, Medhat A.: V1466,497-507(1991)

Sarzynski, Antoni
See Firak, Jozef: V1391,42-47(1991)
See Marczak, Jan: V1391,48-51(1991)

Sasabe, Hiroyuki See Wada, Tatsuo: V1560,162-171(1991)

Sasakawa, Koichi
; Isogai, Fumihiko; Ikebata, Sigeki: Personal verification system with high tolerance of poor-quality fingerprints,V1386,265-272(1991)

Sasaki, Keisuke See Sugihara, Okihiro: V1361,599-605(1991)

Sasaki, Kenji
; Casasent, David P.; Natarajan, Sanjay S.: Neural net selection of features for defect inspection,V1384,228-233(1991)
; Ono, Akira: Optical aspheric surface profiler using phase shift interferometry,V1332,97-106(1991)

Sasaki, Kouji See Ogura, Satoshi: V1412,123-128(1991)

Sasaki, Nobuyuki
; Namikawa, Iwao: Measuring and display system of a marathon runner by real-time digital image processing,V1606,1002-1013(1991)

Sasaki, Tsutomu See Akamatsu, Shigeru: V1606,204-216(1991)

Sasakura, Kunihiko See Sakai, Masao: V1385,8-14(1991)

Sasano, Yasuhiro
; Asada, Kazuya; Sugimoto, Nobuo; Yokota, Tatsuya; Suzuki, Makoto; Minato, Atsushi; Matsuzaki, Akiyoshi; Akimoto, Hajime: Improved limb atmospheric spectrometer and retroreflector in-space for ADEOS,V1490,233-242(1991)

Sasayama, Satoshi See Minato, Kotaro: V1446,195-198(1991)

Sasek, Ladislav
; Vohryzek, Jachym: Polarization-maintaining single-mode fibers: measurement and prediction of fundamental characteristics,V1504,147-154(1991)
; Vohryzek, Jachym; Sochor, Vaclav; Paulicka, Ivan; van Nhac, Nguyen; Franek, Alexandr: Polarization-maintaining single-mode fibers with layered core,V1572,151-156(1991)
See Matejec, Vlastimil: V1513,174-179(1991)

Sashin, Donald
See Fuhrman, Carl R.: V1446,422-429(1991)
See Rosenthal, Marc S.: V1443,132-142(1991)

Sasian, Jose M.
: Image plane tilt in optical systems,V1527,85-95(1991)
: Review of methods for the design of unsymmetrical optical systems,V1396,453-466(1991)
See Brubaker, John L.: V1533,88-96(1991)
See McCormick, Frederick B.: V1533,97-114(1991)

Sasnett, Michael W.
; Johnston, Timothy J.: Beam characterization and measurement of propagation attributes,V1414,21-32(1991)

Sass, David T. See Shepard, Steven M.: V1467,234-238(1991)

Sasso, G. See Webb, David J.: V1418,231-239(1991)

Sastry, C. R.
; Kamen, Edward W.: Multiple-target tracking using the SME filter with polar coordinate measurements,V1481,261-280(1991)

Satani, Ryoichi See Kanazawa, Hirotaka: V1397,445-448(1991)

Satchidanand, S. See Kingsbury, Jeffrey S.: V1427,363-367(1991)

Sathe, Anand See Seetharaman, Guna S.: V1569,269-273(1991)

Sato, A. See Yamaguchi, H.: V1499,29-38(1991)

Sato, Eiichi
; Isobe, Hiroshi; Takahashi, Kei; Tamakawa, Yoshiharu; Yanagisawa, Toru: Fundamental studies for the high-intensity long-duration flash x-ray generator for biomedical radiography,V1358,193-200(1991)
; Isobe, Hiroshi; Takahashi, Kei; Tamakawa, Yoshiharu; Yanagisawa, Toru: Repetitive flash x-ray generator operated at low-dose rates for a medical x-ray television system,V1358,462-470(1991)
; Kimura, Shingo; Isobe, Hiroshi; Takahashi, Kei; Tamakawa, Yoshiharu; Yanagisawa, Toru: Disk-cathode flash x-ray tube driven by a repetitive type of Blumlein pulser,V1358,146-153(1991)
; Shikoda, Arimitsu; Isobe, Hiroshi; Tamakawa, Yoshiharu; Yanagisawa, Toru; Honda, Keiji; Yokota, Yoshiharu: Kilohertz range pulsed x-ray generator having a hot cathode triode,V1358,479-487(1991)
See Isobe, Hiroshi: V1358,471-478(1991)
See Isobe, Hiroshi: V1358,201-208(1991)
See Shikoda, Arimitsu: V1358,154-161(1991)

Sato, F. See Kannari, Fumihiko: V1397,493-497(1991)

Sato, Heihachi
; Tsuchida, Eiichi; Kasuya, Koichi: Transient gain phenomena and gain enhancement in a fast-axial flow CO2 laser amplifier,V1397,331-338(1991)

Sato, Isao See Yamaguchi, Yasushi: V1490,324-334(1991)

Sato, Kazuhiro See Minato, Kotaro: V1446,195-198(1991)

Sato, Koki
; Akiyama, Iwaki; Shoji, Hideo; Sumiya, Daigaku; Wada, Tuneyo; Katsuma, Hidetoshi; Itoh, Koichi: Three-dimensional display of inside of human body by holographic stereogram,V1461,124-131(1991)
See Hoshino, Hideshi: V1461,227-231(1991)

Sato, Mitsuyoshi See Tamura, Akira: V1465,271-281(1991)

Sato, Ryota
See Iwasaki, Nobuo: V1490,192-199(1991)
See Nakayama, Masao: V1490,207-215(1991)
See Tanii, Jun: V1490,200-206(1991)

Sato, S. See Ishikawa, Ken: V1397,55-58(1991)

Sato, Shiro See Seki, Masafumi: V1583,184-195(1991)

Sato, Shunichi
; Takahashi, Kunimitsu; Tanaka, Ikuzo; Noda, Osama; Kuribayashi, Shizuma; Imatake, Shigenori; Kondo, Motoe: High-power unstable resonator CO laser,V1397,421-425(1991)
; Takahashi, Kunimitsu; Noda, Osama; Kuribayashi, Shizuma; Imatake, Shigenori; Kondo, Motoe: Scaling of self-sustained discharge-excited cw CO laser,V1397,433-437(1991)
See Iyoda, Mitsuhiro: V1415,342-349(1991)
See Iyoda, Mitsuhiro: V1397,457-460(1991)
See Kuribayashi, Shizuma: V1397,439-443(1991)
See Noda, Osama: V1397,427-432(1991)

Sato, Tadamitu See Shimada, Wataru: V1332,525-529(1991)

Sato, Y. See Nagai, Haruhiko: V1397,31-36(1991)

Satoh, Hiroharu
; Kinoshita, Yoshio; Okao, Keiichi; Tanaka, Masahiko; Nagatani, Hiroyuki; Honguh, Yoshinori: Fast-laser-power control for high-density optical disk systems,V1499,324-329(1991)

Satoh, Minoru See Mukasa, Minoru: V1446,253-265(1991)

Satoh, Takanori
; Tomono, Akira; Kishino, Fumio: Allowable delay time of images with motion parallax and high-speed image generation,V1606,1014-1021(1991)

Satoh, Yoshikazu See Ozawa, Kenji: V1499,180-186(1991)

Satter, Celeste M.
; Lou, Michael C.: Structural design of the large deployable reflector,V1494,279-300(1991)

Sauer, Charles See Dunbrack, Steven K.: V1464,314-326(1991)

Sauer, Donald J.
; Shallcross, Frank V.; Hsueh, Fu-Lung; Meray, G. M.; Levine, Peter A.; Gilmartin, Harvey R.; Villani, Thomas S.; Esposito, Benjamin J.; Tower, John R.: 640 x 480 MOS PtSi IR sensor,V1540,285-296(1991)

Sauer, Frank See Kobolla, Harald: V1507,175-182(1991)

Sauer, Jon R.
: Multi-Gb/s optical computer interconnect,V1579,49-61(1991)
; Islam, Mohammed N.; Dijaili, Sol P.: Soliton ring network,V1579,84-97(1991)
See Feuerstein, Robert J.: V1583,196-201(1991)

Sauer, Ken D. See Jones, Coleen T.: V1605,522-533(1991)

Sauerbrey, Roland A. See Tittel, Frank K.: V1397,21-29(1991)

Saugen, John D. See Adler, James M.: V1480,11-22(1991)

Saulevich, S. V.
See Averin, V. I.: V1358,603-605(1991)
See Bass, V. I.: V1358,1075-1083(1991)
See Lebedev, V. B.: V1358,874-877(1991)

Sauneuf, Richard See Mens, Alain: V1358,315-328(1991)

Saunier, Paul
; Tserng, Hua Q.; Kao, Yung C.: High-efficiency Kaband monolithic pseudomorphic HEMT amplifier,V1475,86-90(1991)
See Tserng, Hua Q.: V1475,74-85(1991)

Saupper, N. See Sutter, Kurt: V1560,296-301(1991)

Sauteret, Christian See Andre, Michel L.: V1502,286-298(1991)

Savage, Richard N. See Davies, John T.: V1392,551-554(1991)

Savant, Gajendra D.
; Jannson, Joanna: Optical recording materials,V1461,79-90(1991)
See Yacoubian, Araz: V1559,403-409(1991)

Savary, M.
See Bays, Roland: V1525,397-408(1991)
See Braichotte, D.: V1525,212-218(1991)
See Wagnieres, G.: V1525,219-236(1991)

Savatinova, Ivanka T.
; Kuneva, M.; Levi, Zelma; Atuchin, V.; Ziling, K.; Armenise, Mario N.: Proton exchange in LiTaO3 with different stoichiometric composition,V1374,37-46(1991)
See Ziling, C. C.: V1583,90-101(1991)

Savchuk, A. I. See Nikitin, Petr I.: V1584,124-134(1991)

Savinykh, Viktor P.
; Glushko, A. A.: Functional row of pyroelectric sensors for infrared devices used in ecological monitoring,V1540,450-454(1991)

Savory, Eric
See Disimile, Peter J.: V1489,66-74(1991)
See Shoe, Bridget: V1521,34-45(1991)
See Toy, Norman: V1489,112-123(1991)

Savostjanov, V. N.
; Dvalishvili, V. V.; Sakharov, V. N.; Isajkin, A. S.; Frishter, L.; Starchevsky, A. V.: Photoelastic stress investigation in underground large hole in permafrost soil (statics, thermoelasticity, dynamics, photoelastic strain-gauges),V1554A,380-386(1991)
; Zavalishin, S. I.; Smirnov, S. B.; Morosova, D. V.: Stress-static and strength research of building of NPS and nuclear reactor under power and thermal loads including creep and relaxation influence,V1554A,579-585(1991)

Savoy, Robert L. See McCann, John J.: V1453,402-411(1991)

Savrukov, N. T.
: Holographic technique inculcation economic assessment,V1238,382-389(1991)

Savvidi, M. G. See Nemkovich, N. A.: V1403,578-581(1991)

Savvin, Vladimir V. See Glebov, Leonid B.: V1621,21-32(1991)

Sawa, Takeshi See Kurosawa, Kiyoshi: V1368,150-156(1991)

Sawabe, Masaji See Abe, Makoto: V1332,366-376(1991)

Sawabe, Tomoko See Fujii, Tetsurou: V1605,339-350(1991)

Sawada, Hisashi See Kurosawa, Kiyoshi: V1368,150-156(1991)

Sawae, S. See Ohtsuka, Yoshihiro: V1572,347-352(1991)

Sawaki, Akihiro
; Miwa, Mitsuharu; Roehrenbeck, Paul W.: Ultrahigh resolution OTDR using streak camera technology,V1366,324-331(1991)

Sawamoto, Yumiko See Goto, Yasuyuki: V1534,49-58(1991)

Sawchuk, Alexander A. See Miller, Andrew S.: V1563,81-92(1991)

Sawyer, Curry R.
; Quach, Viet; Nason, Donald; van den Berg, Lodewijk: Crystal surface analysis using matrix textural features classified by a probabilistic neural network,V1567,254-263(1991)

Sawyer, F. G. See de Salabert, Arturo: V1521,74-88(1991)

Saxena, Ajay K. See Timothy, J. G.: V1343,350-358(1991)

Saxena, Indu
: High-speed digital ellipsometer for the study of fiber optic sensor systems,V1367,367-373(1991)

Sayegh, Samir I.
; Pomalaza-Raez, Carlos A.; Tepper, E.; Beer, B. A.: Timbre discrimination of signals with identical pitch using neural networks,V1569,100-110(1991)

Sayeh, Mohammad R.
: Optical engineering for neural networks: an emerging technology,V1396,734-743(1991)
; Ragu, A.: Generating good design from bad design: dynamical network approach,V1396,276-284(1991)
; Viswanathan, R.; Dhali, Shirshak K.: Neural networks for smart structures with fiber optic sensors,V1396,417-429(1991)
See Abbasi, M.: V1396,237-242(1991)
See Gupta, Lalit: V1396,266-269(1991)
See Yuan, Xiao L.: V1396,178-187(1991)
See Zargham, Mehdi R.: V1396,202-208(1991)

Saylor, Danny A. See Kalin, David A.: V1358,780-787(1991)

Sayre, James W. See Razavi, Mahmood: V1446,24-34(1991)

Sayyah, Keyvan
; Efron, Uzi; Forber, Richard A.; Goodwin, Norman W.; Reif, Philip G.: Schottky diode silicon liquid-crystal light valve,V1455,249-254(1991)
See Efron, Uzi: V1455,237-247(1991)

Sazy, John
; Kollmer, Charles; Uppal, Gurvinder S.; Lane, Gregory J.; Sherk, Henry H.: Endoscopic removal of PMMA in hip revision surgery with a CO2 laser,V1424,50-50(1991)
See Black, Johnathan D.: V1424,20-22(1991)
See Black, Johnathan D.: V1424,12-15(1991)
See Lane, Gregory J.: V1424,7-11(1991)
See Meller, Menachem M.: V1424,60-61(1991)
See Uppal, Gurvinder S.: V1424,51-52(1991)

Scaglione, Antonio See Righini, Giancarlo C.: V1513,418-424(1991)

Scaletti, Carla
; Craig, Alan B.: Using sound to extract meaning from complex data,V1459,207-219(1991)

Scalora, Michael
; Haus, Joseph W.: Diffraction effects in stimulated Raman scattering,V1497,153-164(1991)

Scalzi, Gary J.
; Turtle, John P.; Carr, Paul H.: MMICs for airborne phased arrays,V1475,2-9(1991)

Scammon, R. J. See Dingus, Ronald S.: V1427,45-54(1991)

Scarabin, Jean-Marie
See Aubry, Florent: V1446,168-176(1991)
See Barillot, Christian: V1445,54-65(1991)
See Bizais, Yves: V1446,156-167(1991)
See Fresne, Francoise: V1444,26-36(1991)

Scarano, Domenica See Cognolato, Livio: V1513,368-377(1991)

Scarano, Gaetano See Neri, Alessandro: V1565,196-208(1991)

Scarberry, Randell E. See Lang, Zhengping: V1468,826-833(1991)

Scarminio, J.
; Gorenstein, Annette; Decker, Franco; Passerini, S.; Pileggi, R.; Scrosati, Bruno: Cation intercalation in electrochromic NiOx films,V1536,70-80(1991)

Scarparo, Marco A. See de Souza, Eunezio A.: V1358,556-560(1991)

Scarsi, Livio See Kaaret, Philip E.: V1548,106-117(1991)

Scavennec, Andre
; Billard, M.; Blanconnier, P.; Caquot, E.; Carer, P.; Giraudet, Louis; Nguyen, L.; Lugiez, F.; Praseuth, Jean-Pierre: InGaAs/InP monolithic photoreceivers for 1.3-1.5 um optical fiber transmission,V1362,331-337(1991)

Scerbak, David G. See Schmitt, Randal L.: V1410,55-64(1991)

Schaarschmidt, Guenther See Stenzel, Olaf: V1534,148-157(1991)

Schachter, Levi
; Nation, John A.: Efficiency increase in a traveling wave tube by tapering the phase velocity of the wave,V1407,44-56(1991)
See Nation, John A.: V1407,32-43(1991)

Schad, Hanspeter
; Schweizer, Edwin; Jodlbauer, Heibert: PC-based determination of 3-D deformation using holographic interferometry,V1507,446-449(1991)

Schadt, Martin
: Novel supertwisted nematic liquid-crystal-display operating modes and electro-optical performance of generally twisted nematic configurations,V1455,214-224(1991)

Schaefer, Dieter See Budzinski, Christel: V1500,264-274(1991)

Schaefer, Johannes H. See Freisinger, Bernhard: V1397,311-318(1991)

Schaefer, Juergen A.
; Persch, V.; Stock, S.; Allinger, Thomas; Goldmann, A.: Oxide removal from GaAs(100) by atomic hydrogen,V1361,1026-1032(1991)
See Allinger, Thomas: V1361,935-942(1991)

Schaefer, Martin See Corlatan, Dorina: V1507,354-364(1991)

Schaefer, Roland H.
See Casasent, David P.: V1567,683-690(1991)
See Casasent, David P.: V1385,152-164(1991)
See Casasent, David P.: V1568,313-326(1991)

Schaeffer, A. R.
; Varian, Richard H.; Cover, John R.; Larsen, Robert G.: Megapixel CCD thinning/backside progress at SAIC,V1447,165-176(1991)

Schaetzel, Klaus
; Peters, Rainer: Noise on multiple-tau photon correlation data,V1430,109-115(1991)

Schafer, Eric L.
; Lyman, Dwight D.: Effects of deformable mirror/wavefront sensor separation in laser beam trains,V1415,310-316(1991)
See Hiddleston, H. R.: V1542,20-33(1991)

Schaffner, James H. See Bridges, William B.: V1371,68-77(1991)

Schaibly, John H. See Reeves, Richard C.: V1486,85-101(1991)

Schaldach, Brita J.
See Artjushenko, Vjacheslav G.: V1420,149-156(1991)
See Artjushenko, Vjacheslav G.: V1420,157-168(1991)
See Artjushenko, Vjacheslav G.: V1420,176-176(1991)
See Artjushenko, Vjacheslav G.: V1590,131-136(1991)
See Beuthan, Jurgen: V1420,234-241(1991)
See Beuthan, Jurgen: V1420,225-233(1991)
See Loshchenov, V. B.: V1420,271-281(1991)

Schalkoff, Robert J. See Mousavi, Madjid S.: V1382,228-239(1991)

Schaller, J. K. See Selser, James C.: V1430,85-88(1991)

Schamiloglu, Edl
; Gahl, John M.; McCarthy, G.: Ku-band radiation in the UNM backward-wave oscillator experiment,V1407,242-248(1991)

Schamschula, Marius P. See Caulfield, H. J.: V1563,103-111(1991)

Scharf, Bruce R. See Hong, C. S.: V1418,177-187(1991)

Scharff, Wolfram See Stenzel, Olaf: V1534,148-157(1991)

Schattenburg, Mark L.
See Flanagan, Kathryn A.: V1549,395-407(1991)
See Markert, Thomas H.: V1549,408-419(1991)

Schatz, Wolfgan See Beregulin, Eugene V.: V1362,853-862(1991)

Schatzki, Thomas F.
; Keagy, Pamela M.: Effect of image size and contrast on the recognition of insects in radiograms,V1379,182-188(1991)

Schaub, Scott A. See Alexander, Dennis R.: V1497,90-97(1991)

Schaus, Christian F. See Shiau, T. H.: V1418,116-122(1991)

Scheggi, Annamaria V.
ed.: *Micro-Optics II*,V1506(1991)
; Mignani, Anna G.: Fiber optic biosensors: the situation of the European market,V1510,40-45(1991)
See Falciai, Riccardo: V1365,38-42(1991)

Scheinberg, Norman See Bayruns, Robert J.: V1541,83-90(1991)

Scheiwiller, Peter M.
; Murphy, S. P.; Dumbreck, Andrew A.: Compact zoom lens for stereoscopic television,V1457,2-8(1991)

Scheler, Ulrich See Sebald, Michael: V1466,227-237(1991)

Schelev, Mikhail Y.
; Serdyuchenko, Yuri N.; Vaschenko, G. O.: 328 MHz synchroscan streak camera,V1358,569-573(1991)
See Babushkin, A. V.: V1346,410-417(1991)
See Borodin, A. M.: V1358,756-758(1991)
See Bowley, David J.: V1358,550-555(1991)
See Briscoe, Dennis: V1358,329-336(1991)
See Bryukhnevitch, G. I.: V1358,739-749(1991)
See Bryukhnevitch, G. I.: V1449,109-115(1991)
See Dalinenko, I. N.: V1449,167-172(1991)
See Degtyareva, V. P.: V1358,524-531(1991)
See Degtyareva, V. P.: V1358,546-548(1991)
See de Souza, Eunezio A.: V1358,556-560(1991)
See Dubovoi, I. A.: V1358,134-138(1991)
See Ivanov, K. N.: V1358,732-738(1991)
See Prochazka, Ivan: V1449,116-120(1991)
See Prochazka, Ivan: V1358,574-577(1991)
See Prokhorov, Alexander M.: V1358,280-289(1991)

Schell, David L. See Grove, Steven L.: V1499,354-359(1991)

Scheller, G. E. See Katz, Howard E.: V1560,370-376(1991)

Scheller, Richard H. See Sweedler, Jonathan V.: V1439,37-46(1991)

Schelp, C. See Ploetz, F.: V1510,224-230(1991)

Schember, Helene R.
; Manhart, Paul K.; Guiar, Cecilia N.; Stevens, James H.: Space Infrared Telescope Facility telescope overview,V1540,51-62(1991)
See Bhandari, Pradeep: V1540,97-108(1991)

Schempp, Walter
: Quantum holography and neurocomputer architectures,VIS08,62-144(1991)

Schempp, William V.
: CCD camera for an autoguider,V1448,129-133(1991)

Schen, Michael A.
; Mopsik, Fred: Dielectric relaxation studies of x2 dye containing polystyrene films,V1560,315-325(1991)

Schenk, Anton F.
; Toth, Charles: Reconstructing visible surfaces,V1526,78-89(1991)
; Toth, Charles: Recovering epipolar geometry in 3-D vision systems,V1457,66-73(1991)

Schenker, Paul S.
ed.: Sensor Fusion III: 3-D Perception and Recognition,V1383(1991)
; Kim, Won S.; Venema, Steven; Bejczy, Antal K.: Fusing human and machine skills for remote robotic operations,V1383,202-223(1991)
See Lee, Sukhan: V1383,391-402(1991)

Scheper, Thomas-Helmut See Ploetz, F.: V1510,224-230(1991)

Scheps, Richard
: Diode pumping of tunable Cr-doped lasers,V1410,190-194(1991)

Scherer, Axel
See Jewell, Jack L.: V1389,401-407(1991)
See Soole, Julian B.: V1474,268-276(1991)

Scherf, Alan V.
; Scott, Peter A.: Target detection using multilayer feedforward neural networks,V1469,63-68(1991)

Schermer, Mack J.
: Stationary platen 2-axis scanner,V1454,257-264(1991)

Scherrer, Jean-Raoul See Ratib, Osman M.: V1446,330-340(1991)

Schertzer, Daniel
See Davis, Anthony B.: V1558,37-59(1991)
See Lavallee, Daniel: V1558,60-75(1991)

Scheu, M.
; Engelhardt, R.: Frequency-doubled alexandrite laser for tissue differentiation in angioplasty,V1425,63-69(1991)
; Flemming, G.; Engelhardt, R.: Stone/ureter identification during alexandrite laser lithotripsy,V1421,100-107(1991)

Schiavone, Filomena See Di Cocco, Guido: V1549,102-112(1991)

Schick, Kenneth L. See Newman, Jay E.: V1430,89-108(1991)

Schick, Larry A. See Riley, Susan: V1598,118-120(1991)

Schicker, Scott R. See Rallison, Richard D.: V1456,179-190(1991)

Schiebler, M. See Beard, David V.: V1446,52-58(1991)

Schieferdecker, Jorg See Norkus, Volkmar: V1484,98-105(1991)

Schiellerup, Gert
; Pedersen, Rune J.: ASK transmitter for high-bit-rate systems,V1372,27-38(1991)

Schietinger, Chuck W.
; Adams, Bruce: Reliability improvement methods for sapphire fiber temperature sensors,V1366,284-293(1991)

Schiff, Harold I.
ed.: Measurement of Atmospheric Gases,V1433(1991)
See Drummond, John W.: V1433,242-252(1991)
See Drummond, John W.: V1433,224-231(1991)
See Mackay, Gervase I.: V1433,104-119(1991)

Schiffler, Richard A.
: Proposed one- and two-fiber-to-the-pedestal architectural evolution,V1363,13-18(1991)

Schildbach, M. A.
; Chase, Lloyd L.; Hamza, Alex V.: Investigation of neutral atom and ion emission during laser conditioning of multilayer HfO2-SiO2 coatings,V1441,287-293(1991)
; Hamza, Alex V.: Interaction of 1064 nm photons with the Al2O3(1120) surface,V1441,139-145(1991)

Schildknegt, Klaas See Ito, Hiroshi: V1466,408-418(1991)

Schildkraut, Jay S.
See Penner, Thomas L.: V1560,377-386(1991)
See Penner, Thomas L.: V1436,169-178(1991)

Schildwachter, Eric F.
; Boreman, Glenn D.: MTF characteristics of a Scophony scene projector,V1488,48-57(1991)

Schiller, Craig M.
See Horsky, Thomas N.: V1540,527-532(1991)
See O'Mara, Daniel M.: V1527,110-117(1991)

Schilling, M. L. See Katz, Howard E.: V1560,370-376(1991)

Schimert, Thomas R.
; Tyan, John; Barnes, Scott L.; Kenner, Vern E.; Brouns, Austin J.; Wilson, H. L.: Noncontact lifetime characterization technique for LWIR HgCdTe using transient millimeter-wave reflectance,V1484,19-30(1991)

Schina, Giovanni
See Bollanti, Sarah: V1503,80-87(1991)
See Bollanti, Sarah: V1397,97-102(1991)

Schindler, Jeffery A. See Andrews, Lee T.: V1459,125-135(1991)

Schindler, Manfred See Masliah, Denis A.: V1475,113-120(1991)

Schink, Harald
; Kolbe, J.; Zimmermann, F.; Ristau, Detleu; Welling, Herbert: Reactive ion-beam-sputtering of fluoride coatings for the UV/VUV range,V1441,327-338(1991)

Schippnick, Paul F.
: Extension of model of beam spreading in seawater to include dependence on the scattering phase function,V1537,185-193(1991)

Schirripa Spagnolo, Giuseppe See Paoletti, Domenica: V1554A,660-667(1991)

Schirz, W. See Gresslehner, Karl-Heinz: V1361,1087-1093(1991)

Schlafer, John See Eichen, Elliot G.: V1474,260-267(1991)

Schlag, Peter M. See Kohl, M.: V1525,26-34(1991)

Schlechtweg, M. See Huelsmann, Axel: V1465,201-208(1991)

Schlegel, J. D. See Chambers, Robert J.: V1488,312-326(1991)

Schlegel, Leo See Hayashi, Nobuaki: V1466,377-383(1991)

Schlegel, Vicki See Cotton, Therese M.: V1403,93-101(1991)

Schlimm, Gerard H. See Peacock, Keith: V1494,147-159(1991)

Schlosser, Steve See Trenkle, John M.: V1568,212-223(1991)

Schmalz, Mark S.
: Errors inherent in the restoration of imagery acquired through remotely sensed refractive interfaces and scattering media,V1479,183-198(1991)
; Wilson, Joseph N.: Relationship of image algebra to the optical processing of signals and imagery,V1474,212-234(1991)

Schmelzer, E. R. See Krongauz, Vadim V.: V1559,354-376(1991)

Schmidt, Guenther See Stenzel, Olaf: V1534,148-157(1991)

Schmidt, Hans U. See Henschel, Henning: V1399,49-63(1991)

Schmidt, Helmut
; Krug, Herbert; Kasemann, Reiner; Tiefensee, Frank: Development of optical waveguides by sol-gel techniques for laser patterning,V1590,36-43(1991)
See Mennig, Martin: V1590,152-159(1991)

Schmidt, Joanna See Kujawinska, Malgorzata: V1508,61-67(1991)

Schmidt, Kevin M.
; Sumida, Shin S.; Miyashita, Tadashi M.: Recent advances in single-mode 1 x n splitters using high-silica optical waveguide circuit technology,V1396,744-752(1991)

Schmidt, Robert A.
See Giger, Maryellen L.: V1445,101-103(1991)
See Yin, Fang-Fang: V1396,2-4(1991)

Schmidt, Walter See Fuerstenau, Norbert: V1367,357-366(1991)

Schmidt, William A. See Davis, Jon P.: V1469,804-811(1991)

Schmidt-Kloiber, Heinz
See Gratzl, Thomas: V1427,55-62(1991)
See Reichel, Erich: V1421,129-133(1991)

Schmidtchen, Joachim See Splett, Armin O.: V1362,827-833(1991)

Schmiedeskamp, Bernt
; Heidemann, B.; Kleineberg, Ulf; Kloidt, Andreas; Kuehne, Mikhael; Mueller, H.; Mueller, Peter; Nolting, Kerstin; Heinzmann, Ulrich: Fabrication and characterization of Si-based soft x-ray mirrors,V1343,64-72(1991)

Schmit, Joanna
; Piatkowski, Tadeusz: Testing of PZT shifters for interferometric measurements,V1391,313-317(1991)

Schmitt, Alain See Motamed, Cina: V1606,961-969(1991)

Schmitt, Dirk-Roger
; Swoboda, Helmut; Rosteck, Helmut: Scatter and roughness measurements on optical surfaces exposed to space,V1530,104-110(1991)

Schmitt, Joseph M. See Nossal, Ralph J.: V1430,37-47(1991)

Schmitt, Mark J.
: Spontaneous emission from free-electron lasers,V1552,107-117(1991)
See Newnam, Brian E.: V1552,154-174(1991)
See Thode, Lester E.: V1552,87-106(1991)

Schmitt, Michel
: Two inverse problems in mathematical morphology,V1568,283-294(1991)
; Mattioli, Juliette: Shape recognition combining mathematical morphology and neural networks,V1469,392-403(1991)

Schmitt, Randal L.
; Spence, Paul A.; Scerbak, David G.: Thermal effects in diode-laser-pumped monolithic Nd:glass lasers,V1410,55-64(1991)

Schmitz, Kenneth S.
ed.: *Photon Correlation Spectroscopy: Multicomponent Systems,* V1430(1991)
: Role of counterion size and distribution on the electrostatic component to the persistence length of wormlike polyions,V1430,216-235(1991)
See Klearman, Debbie: V1430,236-255(1991)

Schmoe, Martha S. See McDermid, I. S.: V1491,175-181(1991)

Schmoyer, D. D. See Egert, Charles M.: V1485,64-77(1991)

Schnabel, Ronald See Schunk, Nikolaus: V1362,391-397(1991)

Schnase, Alexander See Huenermann, Lucia: V1503,134-139(1991)

Schneckenburger, Herbert
; Seidlitz, Harold K.; Wessels, Jurina; Rueck, Angelika C.: Intracellular location, picosecond kinetics, and light-induced reactions of photosensitizing porphyrins,V1403,646-652(1991)
; Seidlitz, Harold K.; Wessels, Jurina; Strauss, Wolfgang; Rueck, Angelika C.: Microscopic fluorescence spectroscopy and diagnosis,V1525,91-98(1991)
See Koenig, Karsten: V1525,412-419(1991)

Schneeberger, Timothy J.
; McIntire, Harold D.: Integrated image processing and tracker performance prediction workstation,V1567,2-8(1991)
See Erteza, Ahmed: V1527,182-187(1991)

Schneider, B. See Kille, Knut: V1386,76-83(1991)

Schneider, David J. See Rose, William I.: V1492,387-390(1991)

Schneider, Hartmut W. See Zarschizky, Helmut: V1389,484-495(1991)

Schneider, Joachim J. See Rosenzweig, Josef: V1362,168-178(1991)

Schneider, Roger H.
ed.: *Medical Imaging V: Image Physics,* V1443(1991)

Schneider, Wolfgang See Chance, Kelly V.: V1491,151-165(1991)

Schneiderova, Martina See Pospisilova, Marie: V1504,287-291(1991)

Schnopper, Herbert W. See Budtz-Jorgensen, Carl: V1549,429-437(1991)

Schnurr, Alvin D.
ed.: *Modeling and Simulation of Laser Systems II,* V1415(1991)

Schock, Wolfram See Wildermuth, Eberhard: V1397,367-371(1991)

Schoen, Christian
; Sharma, Shiv K.; Seki, Arthur; Angel, S. M.: Near-infrared fiber optic temperature sensor,V1584,79-86(1991)

Schoenbach, Karl H.
; Schulz, Hans-Joachim; Lakdawala, Vishnu K.; Kimpel, B. M.; Brinkmann, Ralf P.; Germer, Rudolf K.; Barevadia, Gordon R.: Deep-level configuration of GaAs:Si:Cu: a material for a new type of optoelectronic switch,V1362,428-435(1991)
See Brinkmann, Ralf P.: V1361,274-280(1991)
See Brinkmann, Ralf P.: V1378,203-208(1991)
See Germer, Rudolf K.: V1358,925-931(1991)
See Lakdawala, Vishnu K.: V1378,259-270(1991)
See Schulz, Hans-Joachim: V1361,834-847(1991)

Schoenborn, Philippe See Patrick, Roger: V1392,506-513(1991)

Schoennagel, Horst See Budzinski, Christel: V1500,264-274(1991)

Schoenwald, Jeffrey S.
: Amplitude-modulated laser-driven fiber-optic RF interferometric strain sensor,V1418,450-458(1991)

Schofield, C. P.
; Marchant, John A.: Image analysis for estimating the weight of live animals,V1379,209-219(1991)

Scholen, Douglas E.
; Clerke, William H.; Burns, Gregory S.: Remote spectral identification of surface aggregates by thermal imaging techniques: progress report,V1492,358-369(1991)

Scholes, R. See Wareberg, P. G.: V1538,112-123(1991)

Scholl, Frederick W.
See Anderson, Stephen J.: V1364,94-100(1991)
See Bulusu, Dutt V.: V1364,49-60(1991)

Scholl, James F. See Pizzoferrato, R.: V1409,192-201(1991)

Scholl, James W. See Scholl, Marija S.: V1540,109-118(1991)

Scholl, Laura See Colby, Grace: V1460,114-125(1991)

Scholl, Marija S.
; Scholl, James W.: Stray light issues for background-limited infrared telescope operation,V1540,109-118(1991)
; Shumate, Michael S.; Udomkesmalee, Suraphol: Object-enhanced optical correlation,V1564,165-176(1991)
; Udomkesmalee, Suraphol: Background characterization using a second-order moment function,V1468,92-98(1991)
See Andresen, Bjorn F.: V1540(1991)
See Udomkesmalee, Suraphol: V1468,81-91(1991)

Schollmeyer, Eckhard See Bahners, Thomas: V1503,206-214(1991)

Scholten, T. A. See Lucassen, G. W.: V1403,185-194(1991)

Scholtz, Arpad L.
See Hueber, Martin F.: V1417,233-239(1991)
See Hueber, Martin F.: V1522,259-267(1991)
See Neubert, Wolfgang M.: V1522,93-102(1991)
See Neubert, Wolfgang M.: V1417,122-130(1991)

Scholtz, Robert A. See Win, Moe Z.: V1417,42-52(1991)

Schooley, Larry C. See McCurnin, Thomas W.: V1448,225-236(1991)

Schoonover, J. R. See Woodruff, William H.: V1432,205-210(1991)

Schor, Matthew J.
: Space experiments using small satellites,V1495,146-148(1991)

Schorner, Jurgen See Khesin, G. L.: V1554B,86-90(1991)

Schorter, M. See Ravat, Christian J.: V1478,239-246(1991)

Schott, John R.
; Raqueno, Rolando; Salvaggio, Carl; Kraus, Eugene J.: Incorporation of time-dependent thermodynamic models and radiation propagation models into IR 3-D synthetic image generation models,V1540,533-549(1991)

Schotter, Leo L. See Tamkivi, Raivo: V1503,375-378(1991)

Schowengerdt, Robert A. See Wu, Hsien-Huang: V1569,147-154(1991)

Schram, D. C. See Meeusen, G. J.: V1519,252-257(1991)

Schramm, Werner
; Naundorf, M.: NADH-fluorescence in medical diagnostics: first experimental results,V1525,237-241(1991)

Schranner, R. See Boehm, Hans-Dieter V.: V1456,95-123(1991)

Schreiber, H. See Acebal, Robert: V1397,191-196(1991)

Schreiner, Steven
See Edwards, Charles A.: V1444,38-46(1991)
See Galloway, Robert L.: V1444,9-18(1991)

Schreppel, Ulrich See Vogel, Dietmar: V1554A,262-274(1991)

Schrever, Koen
; De Vilder, Jan; Rolain, Yves; Voet, Marc R.; Barel, Alain R.: Designing enhanced maintainability fiber-optic networks,V1572,107-112(1991)

Schreyer, H. See Boehm, Hans-Dieter V.: V1456,95-123(1991)

Schroeder, John W. See Noah, Meg A.: V1479,275-282(1991)

Schroeter, Siegmund
See Ecke, Wolfgang: V1511,57-66(1991)
See Rasch, Andreas: V1511,149-154(1991)

Schrof, Wolfgang See Gierulski, Alfred: V1560,172-182(1991)

Schryer, David R. See Upchurch, Billy T.: V1416,21-29(1991)

Schubert, Emil
; Rosiwal, S.; Bergmann, Hans W.: Investigations on excimer-laser-treated Cu/Cr contact materials,V1503,299-309(1991)

Schubert, Herman D. See Eaton, Alexander M.: V1423,52-57(1991)

Schueppert, B. See Splett, Armin O.: V1362,827-833(1991)

Schuette, H. See Stojanoff, Christo G.: V1559,321-330(1991)

Schuetz, Helmut See Gale, Michael T.: V1506,65-70(1991)

Schuetzle, Dennis See Carduner, Keith R.: V1433,190-201(1991)

Schuger, Claudio D. See McMath, Linda P.: V1425,165-171(1991)

Schultz, Kenneth I.
: Analytic approach to centroid performance analysis,V1416,199-208(1991)

Schulz, Hans-Joachim
; Schoenbach, Karl H.; Kimpel, B. M.; Brinkmann, Ralf P.: Energy-level structure and electron transitions of GaAs:Cr optoelectronic materials,V1361,834-847(1991)
See Brinkmann, Ralf P.: V1361,274-280(1991)
See Schoenbach, Karl H.: V1362,428-435(1991)

Schulz, M. See Turcu, I. C.: V1503,391-405(1991)

Schulz, Max J. See Cabanski, Wolfgang A.: V1484,81-97(1991)

Schulz, Peter A. See Fujimoto, James G.: V1413,2-13(1991)

Schulz, Timothy J. See Cederquist, Jack N.: V1416,266-277(1991)

Schulz, Veronique See Kessel, David: V1426,180-187(1991)

Schulze, Dean W.
; Sandel, Bill R.; Broadfoot, A. L.: Multilayer mirrors and filters for imaging the earth's inner magnetosphere,V1549,319-328(1991)
See Slaughter, Jon M.: V1343,73-82(1991)

Schulze, Mark A.
; Pearce, John A.: Some properties of the two-dimensional pseudomedian filter,V1451,48-57(1991)

Schumacker, Paul See Pan, Fu-shih: V1396,5-8(1991)

Schumann, Walter
: Approach for applying holographic interferometry to large deformations and modifications,V1507,526-537(1991)

Schunck, Brian G.
See Jiang, Fan: V1569,162-173(1991)
See Sinha, Saravajit S.: V1385,259-267(1991)
See Tirumalai, Arun P.: V1569,28-42(1991)
See Tirumalai, Arun P.: V1383,345-358(1991)

Schunk, Nikolaus
; Grosskopf, Gerd; Ludwig, Reinhold; Schnabel, Ronald; Weber, Hans-Georg: Semiconductor laser amplifiers as all-optical frequency converters,V1362,391-397(1991)

Schuoecker, Dieter
: Modeling of high-power laser welding,V1397,745-751(1991)
See Quenzer, A.: V1397,753-759(1991)

Schurr, J. M.
; Song, Lu; Kim, Ug-Sung: Dynamic vs static bending rigidities for DNA and M13 virus,V1403,248-257(1991)

Schuster, Karl-Heinz See Freischlad, Klaus R.: V1332,18-24(1991)

Schutt-Aine, Jose E.
; Lee, Jin-Fa: V-line: a new interconnect for packaging and microwave applications,V1389,138-143(1991)

Schuurman, Dirk
: Fraud involving OVDs and possible security measures,V1509,126-130(1991)

Schvarzvald, A. I. See Vorzobova, N. D.: V1238,476-477(1991)

Schwab, Scott D. See Udd, Eric: V1418,134-152(1991)

Schwanda, W. See Eidmann, K.: V1502,320-330(1991)

Schwartz, Bernard See Takamoto, Takenori: V1380,64-74(1991)

Schwartz, Eric L.
See Rojer, Alan S.: V1382,132-144(1991)
See Rojer, Alan S.: V1386,44-52(1991)
See Rojer, Alan S.: V1381,250-260(1991)

Schwartz, Gary P.
See Gildenblat, Gennady S.: V1596(1991)
See Macrander, Albert T.: V1550,122-133(1991)

Schwartz, Jon A. See Wilson, Keith E.: V1417,22-28(1991)

Schwartz, P. V.
See Sturm, James C.: V1393,252-259(1991)
See Xiao, Xiaodong: V1476,301-304(1991)

Schwartz, Philip R. See Bevilacqua, Richard M.: V1491,231-242(1991)

Schwartz, Roni
See Afik, Zvi: V1442,392-398(1991)
See Meidan, Moshe: V1540,729-737(1991)

Schwartz, Stanley J. See Gowrinathan, Sankaran: V1479(1991)

Schwartzkopf, George
; Niazy, Nagla N.; Das, Siddhartha; Surendran, Geetha; Covington, John B.: Onium salt structure/property relationships in poly(4-tert-butyloxycarbonyloxystyrene) deep-UV resists,V1466,26-38(1991)

Schwarzmaier, Hans-Joachim
; Heintzen, Matthias P.; Zumdick, Mathias; Kaufmann, Raimund; Wolbarsht, Myron L.: Changes in optical density of normal vessel wall and lipid atheromatous plaque after Nd:YAG laser irradiation,V1427,128-133(1991)
See Hennig, Thomas: V1424,99-105(1991)

Schwarzmann, Peter See Wu, Xiangchen: V1450,278-285(1991)

Schweiter, G. A. See Taff, Laurence G.: V1481,440-448(1991)

Schweitzer, Eric L.
See Futterman, Walter I.: V1486,127-140(1991)
See Strugala, Lisa A.: V1486,176-187(1991)

Schweizer, Edwin See Schad, Hanspeter: V1507,446-449(1991)

Schweizer, T. See Huelsmann, Axel: V1465,201-208(1991)

Schwelb, Otto See Zhang, H.: V1583,83-89(1991)

Schwemmer, Geary K.
; Lee, Hyo S.; Prasad, Coorg R.: Narrowband alexandrite laser injection seeded with frequency-dithered diode laser,V1492,52-62(1991)
See Theon, John S.: V1492,2-23(1991)

Schwentner, N.
See Zerza, Gerald: V1397,107-110(1991)
See Zerza, Gerald: V1410,202-208(1991)

Schwering, Piet B.
: Infrared clutter measurements of marine backgrounds,V1486,25-36(1991)

Schwettman, H. A. See Swent, Richard L.: V1552,24-35(1991)

Schwider, Johannes See Kobolla, Harald: V1507,175-182(1991)

Schwind, David See Beeson, Karl W.: V1559,258-266(1991)

Schwingshakl, Gert See Poelzleitner, Wolfgang: V1384,38-49(1991)

Schwotzer, Guenter
See Ecke, Wolfgang: V1511,57-66(1991)
See Rasch, Andreas: V1511,149-154(1991)

Schzeglov, V. A. See Makeev, V. Y.: V1403,522-527(1991)

Sciammarella, Cesar A.
; Ahmadshahi, Mansour A.: Nondestructive evaluation of turbine blades vibrating in resonant modes,V1554B,743-753(1991)
; Bhat, Gopalakrishna K.; Albertazzi, Armando A.: Electro-optical system for the nondestructive evaluation of bioengineering materials,V1429,183-194(1991)
; Bhat, Gopalakrishna K.: High-resolution computer-aided moire,V1554B,162-173(1991)
; Bhat, Gopalakrishna K.; Albertazzi, Armando A.: Measurement of strains by means of electro-optics holography,V1396,143-154(1991)

Scifres, Donald R.
See Mehuys, David G.: V1418,57-63(1991)
See Yaeli, Joseph: V1442,378-382(1991)
See Zucker, Erik P.: V1563,223-228(1991)

Sckapa, A. F. See Zykov, L. I.: V1412,258-266(1991)

Scott, Andrew M. See Jones, David C.: V1500,46-52(1991)

Scott, Chris J.
: Laser sensing in the iron-making blast furnace,V1399,137-144(1991)
See Paton, Andrew T.: V1367,274-281(1991)

Scott, David M. See Karbassiyoon, Kamran: V1366,178-183(1991)

Scott, David R.
; Flachs, Gerald M.; Gaughan, Patrick T.: Sensor fusion using K-nearest neighbor concepts,V1383,367-378(1991)

Scott, Gary W. See Iannone, Mark A.: V1559,172-183(1991)

Scott, Geoff See Barnes, Nigel: V1389,477-483(1991)

Scott, J. C. See Moerner, William E.: V1560,278-289(1991)

Scott, Jeffrey W. See Coldren, Larry A.: V1362,24-37(1991)

Scott, John J.
See Pradhan, Asima: V1425,2-5(1991)
See Stetz, Mark L.: V1425,55-62(1991)

Scott, Luke B.
; Tomkinson, David M.: Update on the C2NVEO FLIR90 and ACQUIRE sensor performance models,V1488,99-109(1991)

Scott, Marion L. See Newnam, Brian E.: V1552,154-174(1991)

Scott, Miles L. See Kathman, Alan D.: V1544,58-65(1991)

Scott, Peter A. See Scherf, Alan V.: V1469,63-68(1991)

Scott, Peter D.
See Hua, Lifan: V1385,142-151(1991)
See Xie, Gong-Wie: V1500,310-321(1991)
See Xie, Gong-Wie: V1385,132-141(1991)

Scott, Walter M. See Kreider, James F.: V1488,376-388(1991)

Scott, Waymond R. See Fulton, Robert E.: V1389,144-155(1991)

Scribner, Dean A.
; Sarkady, Kenneth A.; Kruer, Melvin R.; Caulfield, John T.; Hunt, J. D.; Herman, Charles: Adaptive nonuniformity correction for IR focal-plane arrays using neural networks,V1541,100-109(1991)

Scripsick, Michael P.
; Edwards, Gary J.; Halliburton, Larry E.; Belt, Roger F.; Kappers, Lawrence A.: Point defects in KTP and their possible role in laser damage,V1561,93-103(1991)

Scrosati, Bruno
See Peres, Rosa C.: V1559,151-158(1991)
See Scarminio, J.: V1536,70-80(1991)

Scudiere, M. B. See Muhs, Jeffry D.: V1584,374-386(1991)

Scully, Marlan O.
; Zhu, Shi-Yao; Narducci, Lorenzo M.; Fearn, Heidi: Gain and threshold in noninversion lasers,V1497,264-276(1991)
See Bochove, Erik J.: V1497,338-347(1991)
See Fearn, Heidi: V1497,283-290(1991)
See Moore, Gerald T.: V1497,328-337(1991)
See Zhu, Shi-Yao: V1497,255-262(1991)
See Zhu, Shi-Yao: V1497,277-282(1991)

Sczepanski, Volker See Kliem, Karl-Heinz: V1463,743-751(1991)

Se, H. See Fang, X. J.: V1572,279-283(1991)

Seales, W. B.
; Dyer, Charles R.: Representing the dynamics of the occluding contour,V1383,47-58(1991)

Seals, Roland D.
: Advanced broadband baffle materials,V1485,78-87(1991)
; Egert, Charles M.; Allred, David D.: Advanced infrared optically black baffle materials,V1330,164-177(1991)

Searcy, Stephen W. See Howarth, M. S.: V1379,141-150(1991)

Searfus, Robert M.
; Eeckman, Frank H.; Colvin, Michael E.; Axelrod, Timothy S.: Analog retina model for detecting moving objects against a moving background,V1473,95-101(1991)

Sears, Robert D. See Strugala, Lisa A.: V1486,176-187(1991)

Seastone, A. J.
: New concepts in remote sensing and geolocation,V1495,228-239(1991)

Seaver, Mark E. See Justus, Brian L.: V1409,2-8(1991)

Sebald, Michael
; Berthold, Joerg; Beyer, Michael; Leuschner, Rainer; Noelscher, Christoph; Scheler, Ulrich; Sezi, Recai; Ahne, Hellmut; Birkle, Siegfried: Application aspects of the Si-CARL bilayer process,V1466,227-237(1991)

Sebille, Bernard See Boisde, Gilbert: V1510,80-94(1991)

Sebillotte-Caron, Claudine
: Holographic optical backplane for boards interconnection,V1389,600-611(1991)

Seborg, Dale E. See Crisalle, Oscar D.: V1464,508-526(1991)

Sebring, Page E. See Bustamante, Carlos: V1435,179-187(1991)

Sedayao, J. See Cox, J. N.: V1392,650-659(1991)

Seddon, Angela B.
; Cardoso, A. V.: Low-temperature viscosity measurements of infrared transmitting halide glasses,V1513,255-263(1991)

Sedlacek, J. H. See Shaver, David C.: V1596,46-50(1991)

Sedlacek, Tomas See Gregory, Kenton W.: V1425,217-225(1991)

Sedlak, Marian
; Amis, Eric J.; Konak, Cestmir; Stepanek, Petr: Dynamic light scattering from strongly interacting multicomponent systems: salt-free polyelectrolyte solutions,V1430,191-202(1991)

Seeds, A. J. See Fernando, P. N.: V1372,152-163(1991)

Seekola, Desmond
; Kelly, Jack R.: Comparative study of the dielectric and optical response of PDLC films,V1455,19-26(1991)

Seelert, Wolf See Johann, Ulrich A.: V1522,158-168(1991)

Seeley, George W.
See Ji, Tinglan: V1444,136-150(1991)
See Rehm, Kelly: V1445,24-35(1991)

Seeley, John S. See Jaenisch, Holger M.: V1379,162-167(1991)

Seeliger, Reinhard
See Bauer, Dietrich: V1533,277-285(1991)
See Hildebrand, Ulrich: V1522,50-60(1991)

Seery, Bernard D.
See Begley, David L.: V1417(1991)
See Short, Ralph C.: V1417,131-141(1991)

Seetharaman, Guna S.
; Chu, Chee-Hung H.: Hierarchical fusion of geometric constraints for image segmentation,V1383,582-588(1991)
; Narasimhan, Anand; Sathe, Anand; Storc, Lisa: Image segmentation with genetic algorithms: a formulation and implementation,V1569,269-273(1991)

Segal, Andrew C.
: Heterogeneous parallel processor for a model-based vision system,V1396,601-614(1991)

Segall, Ilana
; Zeevi, Yehoshua Y.: Application of wavelet-type functions in image processing,V1606,1048-1058(1991)

Segalowitz, Jacob See Papaioannou, Thanassis: V1420,203-211(1991)

Segman, Joseph
: Fourier cross-correlation and invariance transformation for affine groups,V1606,788-802(1991)
; Zeevi, Yehoshua Y.: Image representation by group theoretic approach,V1606,97-109(1991)

Sehgal, Chandra M.
; Chandrasekaran, K.; Pandian, Natesa G.: Volumetric-intravascular imaging by high-frequency ultrasound,V1425,226-233(1991)

Seiber, James N.
See Biermann, Heinz W.: V1433,2-7(1991)
See Green, Martina: V1433,270-274(1991)

Seiberling, Walter E.
; Traxler-Lee, Laura A.; Collins, Sean K.: Small-satellite constellation for many uses,V1495,32-41(1991)

Seibert, Holger
; Sohler, Wolfgang: Ferroelectric microdomain reversal on Y-cut LiNbO3 surfaces,V1362,370-376(1991)

Seibert, Michael
; Waxman, Allen M.: Aspect networks: using multiple views to learn and recognize 3-D objects,V1383,10-19(1991)

Seidel, Claus
: Nucleic acid base specific quenching of coumarin-120-derivative in nucleotid-conjugates-photoinduced electron transfer,V1432,91-104(1991)
; Rittinger, K.; Cortes, J.; Goody, R. S.; Koellner, Malte; Wolfrum, Juergen M.; Greulich, Karl O.: Characterization of fluorescence-labeled DNA by time-resolved fluorescence spectroscopy,V1432,105-116(1991)

Seidel, David See Hadfield, Michael J.: V1478,126-144(1991)

Seiderman, William M.
See Jutamulia, Suganda: V1401,113-118(1991)
See Jutamulia, Suganda: V1558,442-447(1991)

Seidlitz, Harold K.
See Schneckenburger, Herbert: V1403,646-652(1991)
See Schneckenburger, Herbert: V1525,91-98(1991)

Seifert, Michael See Norgard, John D.: V1540,699-708(1991)

Seige, Peter See Ackermann, F.: V1490,94-101(1991)

Seil, Konrad
: Progress in binocular design,V1533,48-60(1991)

Seiler, Kurt See Spichiger, Ursula E.: V1510,118-130(1991)

Seiler, Theo
: Keratorefractive procedures,V1525,280-281(1991)

Seitz, Peter
; Lang, Graham K.: Using local orientation and hierarchical spatial feature matching for the robust recognition of objects,V1606,252-259(1991)

Seitz, R. See Bettag, Martin: V1525,409-411(1991)

Seitzler, Thomas M.
: Binary object analysis hardware area parameter acceleration,V1567,25-30(1991)
: Vision-based strip inspection hardware for metal production,V1567,73-76(1991)

Seka, Wolf D.
; Golding, Douglas J.; Klein, B.; Lanzafame, Raymond J.; Rogers, David W.: Laser energy repartition inside metal, sapphire, and quartz surgical laser tips,V1398,162-169(1991)
See Brown, David L.: V1410,209-214(1991)
See Jaanimagi, Paul A.: V1358,337-343(1991)

Sekhar, Prayaga C. See Venkateswarlu, Putcha: V1332,245-266(1991)

Seki, Arthur See Schoen, Christian: V1584,79-86(1991)

Seki, Masafumi
; Sato, Shiro; Nakama, Kenichi; Wada, Hiroshi; Hashizume, Hideki; Kobayashi, Shigeru: New approaches to practical guided-wave passive devices based on ion-exchange technologies in glass,V1583,184-195(1991)

Sekiguchi, Takeshi See Suzuki, Tomihiro: V1389,455-461(1991)

Sekkat, Z. See Dumont, Michel L.: V1559,127-138(1991)

Sekura, Rieko See Kato, Naoki: V1455,190-205(1991)

Selberg, Lars A.
: Interferometer accuracy and precision,V1400,24-32(1991)

Selby, Vaughn H. See Fallon, James J.: V1495,268-279(1991)

Seldin, J. H. See Cederquist, Jack N.: V1416,266-277(1991)

Seleznev, V. P. See Bass, V. I.: V1358,1075-1083(1991)

Seliger, Mark See Bennett-Lilley, Marylyn H.: V1464,127-136(1991)

Sellar, R. G.
; Gault, William A.; Karp, Christopher K.: Optical filters for the wind imaging interferometer,V1479,140-155(1991)

Sellen, D. B. See Burne, P. M.: V1403,288-295(1991)

Selman, Steven H. See Hampton, James A.: V1426,134-145(1991)

Selmi, Fathi
; Ghodgaonkar, Deepak K.; Hughes, Raymond; Varadan, Vasundara V.; Varadan, Vijay K.: Ceramic phase-shifters for electronically steerable antenna systems,V1489,97-107(1991)

Selser, James C.
; Ellis, Albert R.; Schaller, J. K.; McKiernan, M. L.; Devanand, Krisha: Using photon correlation spectroscopy to study polymer coil internal dynamic behavior,V1430,85-88(1991)

Seltzer, Michael D. See Lindsay, Geoffrey A.: V1560,443-453(1991)

Semanovich, Sharon A. See Allen, Lee R.: V1467,99-103(1991)

Semenov, Alexandr T.
; Elenkrig, Boris B.; Logozinskii, Valerii N.: Universal light source for optical fiber sensors,V1584,348-352(1991)

Semenov, L. P. See Almaev, R. K.: V1397,831-834(1991)

Semenov, V. A. See Ivanov, K. N.: V1358,732-738(1991)

Semenov, V. E. See Brailovsky, A. B.: V1440,84-89(1991)

Sementilli, P. See Hunt, Bobby R.: V1472,208-218(1991)

Semichev, V. A. See Mikaelian, Andrei L.: V1621,194-203(1991)

Semyenov, Alexander D. See Furzikov, Nickolay P.: V1427,288-297(1991)

Sen, Amit See Crocker, James H.: V1494,434-439(1991)

Sen, Susanta
; Nag, B. R.; Midday, S.: Current-voltage characteristics of resonant tunneling diodes,V1362,750-759(1991)

Sengulye, A. A. See Vavilov, Vladimir P.: V1467,230-233(1991)

Sengupta, Sonnath See Chang, Robert S.: V1410,125-132(1991)

Sengupta, Uday K. See Elliott, David J.: V1377,6-17(1991)

Senik, A. V. See Berkovsky, A. G.: V1358,750-755(1991)

Sentenac, Anne See Greffet, Jean-Jacques: V1558,288-294(1991)

Senthilathipan, Velu See Wamser, Carl C.: V1436,114-124(1991)

Sentis, Marc L.
See Delaporte, Philippe C.: V1397,485-492(1991)
See Gerri, Mireille: V1503,280-291(1991)
See Hueber, Jean-Marc: V1503,62-70(1991)
See Kobhio, M. N.: V1397,555-558(1991)
See Kobhio, M. N.: V1503,88-97(1991)
See Zeitoun, David: V1397,585-588(1991)

Seo, Dongsun See Li, Hua: V1376,172-179(1991)

Sepold, Gerd See Breuer, J.: V1503,223-230(1991)

Seppala, Lynn G. See Pitts, John H.: V1441,71-81(1991)

Serafetinides, A. See Quenzer, A.: V1397,753-759(1991)

Serafimovich, O. A. See Vorobiev, N. S.: V1358,698-702(1991)

Serafino, Anthony J. See Coyle, Richard J.: V1412,129-137(1991)

Seraphim, Donald
; Barr, Donald E.: Interconnect and packaging technology in the '90s,V1389,39-54(1991); 1390,39-54(1991)

Serdyuchenko, Yuri N.
See Briscoe, Dennis: V1358,329-336(1991)
See Frontov, H. N.: V1358,311-314(1991)
See Schelev, Mikhail Y.: V1358,569-573(1991)

Serdyuk, I. N. See Timchenko, A. A.: V1403,340-343(1991)

Serebrennikov, Leonid J. See Bokov, Lev: V1505,186-198(1991)

Serebryakov, Victor A.
See Bouchenkov, Vyatcheslav A.: V1410,185-188(1991); 1427,409-412(1991)
See Bouchenkov, Vyatcheslav A.: V1427,405-408(1991)
See Bouchenkov, Vyatcheslav A.: V1410,244-247(1991)

Serebryakova, L. M. See Lazaruk, A. M.: V1621,114-124(1991)

Seredkin, V. A. See Aleksandrov, K. S.: V1621,51-61(1991)

Serenko, M. Y. See Spornik, Nikolai M.: V1508,162-169(1991)

Serfas, D. A. See Bussard, Anne B.: V1366,380-386(1991)

Sergan, V. V. See Bodnar, Vladimir G.: V1455,61-72(1991)

Sergant, Moshe
See Botez, Dan: V1474,64-74(1991)
See Jansen, Michael: V1418,32-39(1991)
See Mawst, Luke J.: V1418,353-357(1991)

Sergeyev, P. B. See Morozov, N. V.: V1441,557-564(1991)

Sergeyev, V. I. See Ryabushkin, Oleg A.: V1362,75-79(1991)

Sergi, Vincenzo
: Quality control of laser cutting process by surface morphology,V1397,775-781(1991)
See De Iorio, I.: V1397,787-790(1991)

Serino, R. J. See Deutsch, Alina: V1389,161-176(1991)

Serlin, Victor
; Friedman, Moshe; Lau, Yue Y.; Krall, Jonathan: Relativistic klystron amplifier II: high-frequency operation,V1407,8-12(1991)
See Colombant, Denis G.: V1407,13-22(1991)
See Friedman, Moshe: V1407,474-478(1991)
See Friedman, Moshe: V1407,2-7(1991)
See Krall, Jonathan: V1407,23-31(1991)

Sermage, Bernard
; Blez, M.; Kazmierski, Christophe; Ougazzaden, A.; Mircea, Andrei; Bouley, J. C.: Influence of the p-type doping of the InP cladding layer on the threshold current density in 1.5 um QW lasers,V1362,617-622(1991)
; Gerard, Jean M.; Bergomi, Lorenzo; Marzin, Jean Y.: Differentiation of the nonradiative recombination properties of the two interfaces of molecular beam epitaxy grown GaAs-GaAlAs quantum wells,V1361,131-135(1991)

Serna, Julio
; Martinez-Herrero, R.; Mejias, P. M.: Spatial characterization of high-power multimode laser beams,V1397,631-634(1991)

Sernas, Valentinas See Majercak, David: V1584,162-169(1991)

Serocki, John See Garrett, Steven L.: V1367,13-29(1991)

Serpico, Sebastiano B.
See de Salabert, Arturo: V1521,74-88(1991)
See Vernazza, Gianni L.: V1492,206-212(1991)

Serra, Jean C.
: Lipschitz lattices and numerical morphology,V1568,54-65(1991)

Serreze, Harvey B. See Baumann, John A.: V1418,328-337(1991)

Serri, Laura
; Maggi, C.; Garifo, Luciano; De Silvestri, S.; Magni, Vittorio C.; Svelto, Orazio: Multikilowatt transverse-flow CO2 laser with variable reflectivity mirrors,V1397,469-472(1991)

Servadio, Ciro
: Local hyperthermia for the treatment of diseases of the prostate,V1421,6-11(1991)

Seshadri, Sridhar B.
; Kishore, Sheel; Khalsa, Satjeet S.; Feingold, Eric; Arenson, Ronald L.: PACS modeling and development: requirements versus reality,V1446,388-395(1991)
See Carey, Bruce: V1446,414-419(1991)
See Feingold, Eric: V1446,211-216(1991)
See Kishore, Sheel: V1446,188-194(1991)
See Kishore, Sheel: V1446,236-242(1991)
See Kundel, Harold L.: V1446,297-300(1991)
See Nodine, Calvin F.: V1444,56-62(1991)
See Rundle, Debra A.: V1446,405-413(1991)
See Rundle, Debra A.: V1446,379-387(1991)

Seshamani, Ramani
: Diode-laser-based range sensor,V1471,354-356(1991)
; Alex, T. K.: Use of 3 X 3 integrated optic polarizer/splitters for a smart aerospace plane structure,V1489,56-64(1991)

Sessa, William B.
; Welter, Rudy; Wagner, Richard E.; Maeda, Mari W.: Recent progress of coherent lightwave systems at Bellcore,V1372,208-218(1991)

Sessler, Andrew M. See Sharp, William M.: V1407,535-545(1991)

Sessler, Jonathan L.
; Hemmi, Gregory; Maiya, Bhaskar G.; Harriman, Anthony; Judy, Millard M.; Boriak, Richard; Matthews, James L.; Ehrenberg, Benjamin; Malik, Zvi; Nitzan, Yeshayahu; Rueck, Angelika C.: Tripyrroledimethine-derived ("texaphyrin"-type) macrocycles: potential photosensitizers which absorb in the far-red spectral region,V1426,318-329(1991)

Setchell, Robert E.
; Meeks, Kent D.; Trott, Wayne M.; Klingsporn, Paul E.; Berry, Dante M.: High-power transmission through step-index multimode fibers,V1441,61-70(1991)

Sethi, Ishwar K. See Ramesh, Nagarajan: V1468,72-80(1991)

Sethi, Satyendra A.
; Distasio, Romelia; Ziger, David H.; Lamb, James E.; Flaim, Tony: Use of antireflective coatings in deep-UV lithography,V1463,30-40(1991)

Seto, K. See Tsumanuma, Takashi: V1420,193-198(1991)

Seto, Lloyd See Golden, Jeffry: V1407,418-429(1991)

Seto, Toshiki See Kimata, Masafumi: V1540,238-249(1991)

Setoguchi, Toshihiko
; Eguchi, Koichi; Arai, Hiromichi: Thin film fabrication of stabilized zirconia for solid oxide fuel cells,V1519,74-79(1991)

Setra, M. See Bacis, Roger: V1397,173-176(1991)

Setsune, Kentaro
; Mizuno, Koichi; Higashino, Hidetaka; Wasa, Kiyotaka: Planar SNS Josephson junctions using multilayer Bi system,V1394,79-88(1991)
See Mizuno, Koichi: V1477,197-204(1991)

Sette, Francesco See Tjeng, L. H.: V1548,160-167(1991)

Sevaston, George E. See Redding, David C.: V1489,201-215(1991)

Severinsky, J. Y. See Detwiler, Paul W.: V1425,149-155(1991)

Severon, Burkhard
: New technologies in lighting systems for high-speed film and photography regarding high-intensity and heat problems,V1358,1202-1208(1991)

Severson, William E.
; Douglass, Robert J.; Hennessy, Stephen J.; Boyd, Robert; Anhalt, David J.: Surface property determination for planetary rovers,V1388,490-501(1991)

Sevick, Eva M.
; Chance, Britton: Photon migration in a model of the head measured using time- and frequency-domain techniques: potentials of spectroscopy and imaging,V1431,84-96(1991)
; Weng, Jian; Maris, Michael B.; Chance, Britton: Analysis of absorption, scattering, and hemoglobin saturation using phase-modulation spectroscopy,V1431,264-275(1991)
See Chance, Britton: V1525,68-82(1991)
See Maris, Michael B.: V1431,136-148(1991)

Sewell, Derek
: Low-cost high-quality range camera system,V1358,1209-1214(1991)
See Cullis, I. C.: V1358,52-64(1991)

Sewell, Harry
: Impact of phase masks on deep-UV lithography,V1463,168-179(1991)

Seya, Eiichi
; Matsumoto, Kiyoshi; Nihei, Hideki; Ichikawa, Atsushi; Moriyama, Shigeo; Nakamura, Shigeru; Mita, Seiichi: 10-mm-thick head mechanism for a stacked optical disk system,V1499,269-273(1991)

Sezan, M. I.
See Ozkan, Mehmet K.: V1452,146-156(1991)
See Ozkan, Mehmet K.: V1606,743-754(1991)

Sezi, Recai See Sebald, Michael: V1466,227-237(1991)

Sfez, B. G. See Oudar, Jean-Louis: V1361,490-498(1991)

Shaarawi, Amr M. See Ziolkowski, Richard W.: V1407,387-397(1991)

Shacham-Diamand, Yosef Y.
: Narrow (0.1 um to 0.5 um) copper lines for ultra-large-scale integration technology,V1442,11-19(1991)

Shadaram, Mehdi
; John, Eugine: Effect of input pulse shape on FSK optical coherent communication system,V1365,108-115(1991)
; Solehjou, Amin; Nazarian, Soheil: Application of fiber optic sensors in pavement maintenance,V1332,487-490(1991)

Shafai, Bahram See Kim, Kwang H.: V1481,406-417(1991)

Shafeev, George A.
; Laude, Lucien D.: Laser-assisted deposition of thin films onto transparent substrates from liquid-phase organometallic precursor: iron acetylacetonate,V1503,321-329(1991)

Shafer, David R.
: Design challenges for the 1990s,V1354,608-616(1991)

Shafer, Steven A. See Novak, Carol L.: V1453,353-368(1991)

Shafir, Ehud
; Ben-Kish, A.; Konforti, Naim; Tur, Moshe: Portable fiber optic current sensor,V1442,236-241(1991)

Shah, Mubarak A. See Hackett, Jay K.: V1383,611-622(1991)

Shah, S. P. See Castro-Montero, Alberto: V1396,122-130(1991)

Shahaf, Nachum See Wilk, Shalom: V1442,140-148(1991)

Shainoga, I. S.
; Shentsev, N. I.; Zakharov, N. S.: Magnetic fields generated in laser-produced plasmas,V1440,277-290(1991)

Shakhbazyan, V. Y. See Vsevolodov, N. N.: V1621,11-20(1991)

Shakir, Sami A. See Salvi, Theodore C.: V1418,64-73(1991)

Shaklan, Stuart B. See Redding, David C.: V1489,201-215(1991)

Shakola, A. T. See Veiko, Vadim P.: V1544,152-163(1991)

Shaligram, Arvind D. See Kulkarni, N. M.: V1393,207-214(1991)

Shalkhauser, Kurt A.
; Raquet, Charles A.: System-level integrated circuit development for phased-array antenna applications,V1475,204-209(1991)

Shallcross, Frank V. See Sauer, Donald J.: V1540,285-296(1991)

Shaltaf, Samir J.
; Namazi, Nader M.: Maximum likelihood estimation of affine-modeled image motion,V1567,609-620(1991)

Shalvi, Ofir
; Weinstein, Ehud: Super-exponential methods for blind deconvolution,V1565,143-152(1991)

Shammas, A. See Afik, Zvi: V1442,392-398(1991)

Shamoto, N. See Tsumanuma, Takashi: V1420,193-198(1991)

Shamouilian, Shamouil See Deshpande, Ujwal A.: V1390,489-501(1991)

Shamray-Bertaud, Patricia See Carome, Edward F.: V1584,110-117(1991)

Shams, Iden
; Butler, Clive: Performance of an optoelectronic probe used with coordinate measuring machines,V1589,120-125(1991)

Shan, Bin See Hou, Xun: V1358,868-873(1991)

Shan, Jin A. See Mei, Yuan S.: V1519,370-373(1991)

Shan, W. See Li, Jie: V1519,660-664(1991)

Shan, Xuekang See Cheng, Yuqi: V1572,392-395(1991)

Shandybina, Galina
: Dislocation mechanism of periodical relief formation in laser-irradiated silicon,V1440,384-389(1991)
See Kotov, Gennady A.: V1440,321-324(1991)

Shang, Alain See Jen, Cheng K.: V1590,107-119(1991)

Shang, H. M.
; Chau, Fook S.; Tay, C. J.; Toh, Siew-Lok: Interpretation of holographic and shearographic fringes for estimating the size and depth of debonds in laminated plates,V1554B,680-691(1991)

Shang, Jian See Zhu, Zhenhai: V1554A,472-481(1991)

Shang, Shi X. See Mei, Yuan S.: V1519,370-373(1991)

Shankar, K. See Measures, Raymond M.: V1489,86-96(1991)

Shanken, Stuart N.
; Ludwig, David E.: Applications of Z-Plane memory technology to high-frame rate imaging systems,V1346,210-215(1991)

Shao, C. A.
; King, H. J.; Wang, Yeong-Kang; Chiang, Fu-Pen: Accumulating displacement fields from different steps in laser or white-light speckle methods,V1554A,613-618(1991)

Shao, Jian D.
; Fan, Zheng X.; Guo, Yong H.; Jin, Lei: Soft X-UV silver silicon multilayer mirrors,V1519,298-301(1991)
See Guo, Yong H.: V1519,327-332(1991)

Shao, Michael
: Hubble extra-solar planet interferometer,V1494,347-356(1991)
See Colavita, Mark M.: V1494,168-181(1991)
See Colavita, Mark M.: V1542,205-212(1991)
See Gershman, Robert: V1494,160-167(1991)

Shao, Weimin See Lang, Zhengping: V1468,826-833(1991)

Shaout, Adnan
: Image analysis using attributed fuzzy tournament matching algorithm,V1381,357-367(1991)

Shapar, Vladimir N. See Svechnikov, Georgy S.: V1589,24-31(1991)

Shapin, S. M. See Nadezhdinskii, Alexander I.: V1418,487-495(1991)

Shapiro, Alexander G.
; Yaminsky, Igor V.: Nondestructive investigations of multilayer dielectrical coatings,V1362,834-843(1991)

Shapiro, Daniel B.
; Hunt, Arlon J.; Quinby-Hunt, Mary S.; Hull, Patricia G.: Circular polarization effects in the light scattering from single and suspensions of dinoflagellates,V1537,30-41(1991)
See Hull, Patricia G.: V1537,21-29(1991)

Shapiro, Jeffrey H.
See Green, Thomas J.: V1471,328-341(1991)
See Mentle, Robert E.: V1471,342-353(1991)

Shapkin, P. V. See Nasibov, Alexander S.: V1361,901-908(1991)

Sharan, Sunil See Mehrotra, R.: V1463,487-491(1991)

Sharangovich, Sergey See Bokov, Lev: V1505,186-198(1991)

Shariat, Hormoz
: Knowledge-based system for analysis of aerial images,V1381,306-317(1991)
: Learning object shapes from examples,V1567,194-203(1991)

Shariati, M. See Featherstone, John D.: V1424,145-149(1991)

Shariv, I.
; Friesem, Asher A.: Optical thresholding with a liquid crystal light valve,V1442,258-263(1991)

Sharkov, A. V.
; Gillbro, T.: Nonlinear properties of oriented purple membrane films derived from second-harmonic generation under picosecond excitation: prospect of electro-optical measurements of ultrafast photoelectric respon,V1403,434-438(1991)
See Khoroshilov, E. V.: V1403,431-433(1991)

Sharma, Arvind K. See McAdam, Bridget A.: V1475,267-274(1991)

Sharma, D. K. See Sahyun, Melville R.: V1436,125-133(1991)

Sharma, Minoti
; Freund, Harold G.: Development of laser-induced fluorescence detection to assay DNA damage,V1435,280-291(1991)

Sharma, Ramesh D.
; Healey, Rebecca J.: Earthlimb emission analysis of spectral infrared rocket experiment data at 2.7 micrometers: a ten-year update,V1540,314-320(1991)

Sharma, Shiv K. See Schoen, Christian: V1584,79-86(1991)

Sharp, Tracy E.
See Hofmann, Thomas: V1412,84-90(1991)
See Tittel, Frank K.: V1397,21-29(1991)

Sharp, William M.
; Rangarajan, G.; Sessler, Andrew M.; Wurtele, Jonathan S.: Phase stability of a standing-wave free-electron laser,V1407,535-545(1991)

Sharpe, Randall B. See Jones, J. R.: V1363,106-118(1991)

Sharples, R. M. See Doel, A. P.: V1542,319-326(1991)

Shashikala, M. See Kamalakar, J. A.: V1478,92-100(1991)

Shashkin, Viktor S. See Glebov, Leonid B.: V1513,224-231(1991)

Shatalin, Igor D.
: High-efficiency holograms on the basis of polarization recording by means of beams from opposite directions,V1238,68-73(1991)

Shaver, David C.
; Doran, S. P.; Rothschild, Mordechai; Sedlacek, J. H.: Laser-induced metal deposition and laser cutting techniques for fixing IC design errors,V1596,46-50(1991)
See Kunz, Roderick R.: V1466,218-226(1991)

Shaw, C. B. See Reynolds, Larry D.: V1554B,622-631(1991)

Shaw, David T.
; Narumi, E.; Yang, F.; Patel, Sushil: Role of buffer layers in the laser-ablated films on metallic substrates,V1394,214-220(1991)
See Hua, Lifan: V1385,142-151(1991)
See Narayan, Jagdish: V1394(1991)
See Witanachchi, S.: V1394,161-168(1991)
See Xie, Gong-Wie: V1500,310-321(1991)
See Xie, Gong-Wie: V1385,132-141(1991)

Shaw, Earl D.
; Chichester, Robert J.; La Porta, A.: FIR optical cavity oscillation is observed with the AT&T Bell Laboratories free-electron laser,V1552,14-23(1991)

Shaw, J. K.
; Vengsarkar, Ashish M.; Claus, Richard O.: Theoretical analysis of two-mode, elliptical-core optical fiber sensors,V1367,337-346(1991)

Shaw, Joseph A.
; Churnside, James H.; Westwater, Edward R.: Infrared spectrometer for ground-based profiling of atmospheric temperature and humidity,V1540,681-686(1991)

Shaw, Larry L. See Jaanimagi, Paul A.: V1346(1991)

Shaw, Leslie B. See Chang, Robert S.: V1410,125-132(1991)

Shaw, M. J. See Key, Michael H.: V1397,9-17(1991)

Shaw, Michael M.
See Harvey, David M.: V1400,86-93(1991)
See Wood, Christopher M.: V1332,301-313(1991)

Shaw, Ping-Shine
; Church, Eric D.; Hanany, Shaul; Liu, Yee; Fleischman, Judith R.; Kaaret, Philip E.; Novick, Robert; Manzo, Giuseppe: Photoelectric effect from CsI by polarized soft x-rays,V1343,485-499(1991)
; Hanany, Shaul; Liu, Yee; Church, Eric D.; Fleischman, Judith R.; Kaaret, Philip E.; Novick, Robert; Santangelo, A.: Vectorial photoelectric effect at 2.69 keV,V1548,118-131(1991)
See Fraser, George W.: V1548,132-148(1991)
See Kaaret, Philip E.: V1548,106-117(1991)

Shaw, Robert W.
; Whitten, W. B.; Ramsey, J. M.; Heatherly, Lee: Fundamental studies of chemical-vapor-deposition diamond growth processes,V1534,170-174(1991)

Shaw, Roland See Arndt, George D.: V1475,231-242(1991)

Shayegan, Mansour
: Quasi-three-dimensional electron systems and superlattices in wide parabolic wells: fabrication and physics,V1362,228-239(1991)

Shazeer, Dov J.
; Bello, M.: Detection and classification of undersea objects using multilayer perceptrons,V1469,622-636(1991)

Shcherbakov, A. S.
: Some problems of implementation of the memory system based on optical solitons in fibers,V1621,204-214(1991)

Shcherbakov, I. A.
See D'yakonov, G. I.: V1421,156-162(1991)
See D'yakonov, G. I.: V1424,81-86(1991)
See D'yakonov, G. I.: V1421,153-155(1991)
See Konov, Vitaly I.: V1427,220-231(1991)
See Konov, Vitaly I.: V1525,250-252(1991)
See Prokhorov, Alexander M.: V1410,70-88(1991)

Shcherbatska, Nina V. See Nemkovich, N. A.: V1403,578-581(1991)

Shchuka, Maria I. See Wittmeyer, Stacey A.: V1435,267-278(1991)

She, K.
; Created, C. A.; Easson, William J.; Skyner, D.; Xu, M. C.: Development of a fibre optics flowmeter,V1572,581-587(1991)

Shealy, David L.
; Jiang, Wu; Hoover, Richard B.: Design and analysis of aspherical multilayer imaging x-ray microscope,V1343,122-132(1991)
; Viswanathan, Vriddhachalam K.: Design survey of x-ray/XUV projection lithography systems,V1343,229-240(1991)
See Hoover, Richard B.: V1426,84-96(1991)
See Hoover, Richard B.: V1435,338-351(1991)

Shear, Jason B. See Sweedler, Jonathan V.: V1439,37-46(1991)

Sheat, Dennis E.
; Chamberlin, Giles R.; McCartney, David J.: Double dispersion from dichromated gelatin volume transmission gratings,V1461,35-38(1991)

Shechterman, Mark
: High-performance, wide-magnification-range IR zoom telescope with automatic compensation for temperature effects,V1442,276-285(1991)

Shechtman, Dan
; Farabaugh, Edward N.; Robins, Lawrence H.; Feldman, Albert; Hutchison, Jerry L.: High-resolution electron microscopy of diamond film growth defects and their interactions,V1534,26-43(1991)

Sheehy, Finbar T. See Bridges, William B.: V1371,68-77(1991)

Sheffer, Dan
; Ben-Shalom, Ami; Devir, Adam D.: Remote colorimetry and its applications,V1493,232-243(1991)

Sheffield, Richard L.
See Chan, Kwok-Chi D.: V1552,69-78(1991)
See Newnam, Brian E.: V1552,154-174(1991)

Sheik-Bahae, Mansoor
; Hutchings, David C.; Hagan, David J.; Soileau, M. J.; Van Stryland, Eric W.: Dispersion of n2 in solids,V1441,430-443(1991)

Shelburne, John A.
See Gozdz, Antoni S.: V1466,200-205(1991)
See Gozdz, Antoni S.: V1466,520-527(1991)

Shell, Forney L.
: Mother's recipe is not a standard,V1346,86-89(1991)

Shell, Richard See Hall, Ernest L.: V1381,162-170(1991)

Shellabear, Michael C. See Mendoza Santoyo, Fernando: V1508,143-152(1991)

Shellan, Jeffrey B.
; Yeh, Pochi A.: Reflection spectrum of multiple chirped gratings,V1545,179-188(1991)

Shelley, Paul R. See Parulski, Kenneth A.: V1448,45-58(1991)

Shelton, Douglas S.
: Reliability considerations for fiber optic systems in telecommunications,V1366,2-8(1991)

Shelyakov, Alexander V. See Antonov, Victor A.: V1474,116-123(1991)

Shemlon, Stephen
; Liang, Tajen; Cho, Kyugon; Dunn, Stanley M.: Segmentation using models of expected structure,V1381,470-481(1991)

Shen, Ai D.
; Cui, Jie; Wang, Hai L.; Wang, Zhi-Jiang: Molecular beam epitaxial growth of ZnSe-ZnS strained-layer superlattices,V1519,656-659(1991)

Shen, Chi N. See Page, Lance A.: V1383,471-482(1991)

Shen, Dezen See Liu, Yudong: V1362,436-447(1991)

Shen, F. Z. See Cardenas-Garcia, Jaime F.: V1554B,210-224(1991)

Shen, H. See Shi, W.: V1519,138-141(1991)

Shen, Hao-Ming
: Experimental study of the resonance of a circular array,V1407,306-315(1991)
; Wu, Tai T.; Myers, John M.: Experimental study of electromagnetic missiles from a hyperboloidal lens,V1407,286-294(1991)
See Fikioris, George: V1407,295-305(1991)

Shen, Helen C.
; Basir, O. A.: Scheme for sensory data integration,V1383,403-408(1991)

Shen, Janice See Boardman, A. D.: V1519,609-615(1991)

Shen, Ji-Yao See Huang, Jen-Kuang: V1489,266-277(1991)

Shen, Jinyuan See Zhang, Yanxin: V1469,303-307(1991)

Shen, Junquan See Du, Keming: V1397,639-643(1991)

Shen, KangKang See Gole, James L.: V1397,125-135(1991)

Shen, Meng Y. See Xu, Xu R.: V1519,525-528(1991)

Shen, Qun
: Polarization-state mixing in multiple-beam diffraction and its application to solving the phase problem,V1550,27-33(1991)

Shen, Ronggui See Jin, Guoliang: V1513,50-55(1991)

Shen, Shou Z.
See He, Jin: V1519,499-507(1991)
See Ma, Ke J.: V1519,489-493(1991)

Shen, Sylvia S.
: Three-dimensional scene interpretation through information fusion,V1382,427-433(1991)
; Trang, Bonnie Y.: Feature selection technique for classification of hyperspectral AVIRIS data,V1567,188-193(1991)

Shen, Weisheng
; Zhang, Shen; Tao, Chunkan: Inverse filtering technique for the synthesis of distortion-invariant optical correlation filters,V1567,691-697(1991)

Shen, X. L. See Zhang, Z. J.: V1519,790-792(1991)

Shen, Xuanguo See Siahmakoun, Azad: V1396,535-538(1991)

Shen, Xue C. See Yuan, Ren K.: V1519,831-834(1991)

Shen, Yu-qing See Fang, Zhen-he: V1572,342-346(1991)

Shen, Yuanhua See Zheng, Tian S.: V1519,339-346(1991)

Shen, Zhen-kang
; Han, Zhenduo; Zhao, Mingsheng: High-accuracy tracking algorithm based on iterating tracking window center towards the target's center,V1482,337-347(1991)
; Mao, Xuguang; Jin, Yiping: Moving-point-target tracking in low SNR,V1481,233-240(1991)
; Zhao, Mingsheng: High-accuracy target tracking algorithm based on deviation vector of the local window grey center,V1482,325-336(1991)
See Li, Zhi-yong: V1472,97-105(1991)
See Zhao, Mingsheng: V1471,464-473(1991)

Shen, Zhijiang
; Gregson, Peter H.; Cheng, Heng-Da; Kozousek, V.: Automated grading of venous beading: an algorithm and parallel implementation,V1606,632-640(1991)

Sheng, D. See Hoyle, Charles E.: V1559,202-213(1991)

Sheng, Kemin
See Fang, Yin: V1572,197-200(1991)
See Fang, Yin: V1572,453-456(1991)

Sheng, Lie-yi
; Li, Shenghong; Xu, Sen-lu; Zhu, Lie-wei: Studies of displacement sensing based on the deformation loss of an optical fiber ring,V1572,273-278(1991)
See Xu, Sen-lu: V1367,303-308(1991)

Sheng, Rong-Sheng See Morris, Michael D.: V1439,95-101(1991)

Sheng, Yunlong
; Delisle, Claude; Moreau, Louis; Song, Li; Lessard, Roger A.; Arsenault, Henri H.: Microreflective elements for integrated planar optics interconnects,V1559,222-228(1991)
; Leclerc, Luc; Arsenault, Henri H.: Multiplexed binary phase-only circular harmonic filters,V1564,320-329(1991)
See Leclerc, Luc: V1564,78-85(1991)

Shenker, Martin
: Power of one power,V1354,647-653(1991)

Shenoy, Gopal K. See Knauer, James P.: V1345(1991)

Shentsev, N. I. See Shainoga, I. S.: V1440,277-290(1991)

Shepard, Allan H. See Baumann, John A.: V1418,328-337(1991)

Shepard, Barclay M. See Dayhoff, Ruth E.: V1446,323-329(1991)

Shepard, Steven M.
; Sass, David T.; Imirowicz, Thomas P.; Meng, A.: Time-resolved videothermography at above-frame-rate frequencies,V1467,234-238(1991)

Shepherd, David W. See Skipper, Richard S.: V1529,22-30(1991)

Shepherd, Joseph
See Colquhoun, Allan B.: V1454,12-19(1991)
See Preston, Ralph G.: V1454,124-131(1991)
See Preston, Ralph G.: V1454,132-138(1991)

Shepherd, Joseph E. See Tao, William C.: V1346,300-310(1991)

Shepherd, Thomas See Uttal, William: V1453,258-269(1991)

Shepherd, W. J. See Jackson, John E.: V1533,75-86(1991)

Shepp, Larry A.
; Vardi, Y.; Lazewatsky, J.; Libeau, James; Stein, Jay A.: Automatic recognition of bone for x-ray bone densitometry,V1452,216-224(1991)

Sheppard, Colin J. See Cogswell, Carol J.: V1450,323-328(1991)

Sher, Assaf See Meidan, Moshe: V1540,729-737(1991)

Sher, David B.
See Revankar, Shriram: V1459,268-273(1991)
See Srikantan, Geetha: V1459,258-267(1991)

Shera, E. B. See Soper, Steven A.: V1435,168-178(1991)

Sheremetyev, Yu. N. See Egorov, V. V.: V1358,984-991(1991)

Shereshevsky, I. See Andronov, Alexander A.: V1362,684-695(1991)

Shergill, Gurmeet S. See Cox, J. N.: V1392,650-659(1991)

Sheridan, Thomas B. See Patrick, Nicholas J.: V1387,215-222(1991)

Sherk, Henry H.
See Black, Johnathan D.: V1424,20-22(1991)
See Black, Johnathan D.: V1424,12-15(1991)
See Lane, Gregory J.: V1424,7-11(1991)
See Meller, Menachem M.: V1424,60-61(1991)
See Sazy, John: V1424,50-50(1991)
See Uppal, Gurvinder S.: V1424,51-52(1991)

Sherman, Bradley D.
; Mendez, Antonio J.; Morookian, John-Michael: Efficiency of a 5V/5-mW power by light power supply for avionics applications,V1369,60-66(1991)

Sherman, E. See Margerum, J. D.: V1455,27-38(1991)

Sherman, Glenn H. See Sona, P.: V1397,373-377(1991)

Sherman, James W.
: Development and use of confidence intervals for automatic target recognition evaluation,VIS07,144-169(1991)
See Gleason, Jim M.: V1406,169-170(1991)

Sherstobitov, Vladimir E.
; Kalinin, Victor P.; Goryachkin, Dmitriy A.; Romanov, Nikolay A.; Dimakov, Sergey A.; Kuprenyuk, Victor I.: CO2 lasers and phase conjugation,V1415,79-89(1991)

Sherstyuk, Valentin P.
; Maloletov, Sergei M.; Kondratenko, Nina A.: Physicochemistry of holographic recording and processing in dichromated gelatin,V1238,211-217(1991)
; Malov, Alexander N.; Maloletov, Sergei M.; Kalinkin, Vyacheslav V.: Some principles for formation of self-developing dichromate media,V1238,218-223(1991)

Sherwood, Robert J. See Del Grande, Nancy K.: V1479,335-351(1991)

Sheshadri, M. R. See Mishra, R. K.: V1482,138-145(1991)

Shestakov, S. D.
; Kotov, Gennady A.; Hekalo, A. V.; Migitko, I. A.; Bajkov, A. V.; Yurkevith, B. M.: Laser-induced mass transfer simulation and experiment,V1440,423-435(1991)

Sheth, Devang See Agah, Ramtin: V1425,172-179(1991)

Shetter, Richard E. See Cantrell, Chris A.: V1433,263-268(1991)

Shevchenko, N. P.
; Megorskaja, K. D.; Reshetnikova, Irina N.: Ion treatment in technology of diffraction optical elements,V1574,66-71(1991)

Shevchuk, Igor See Chekulayev, V.: V1426,367-377(1991)

Shevelko, A. P. See Bijkerk, Fred: V1503,380-390(1991)

Shevtsov, M. K.
See Artemjev, S. V.: V1238,206-210(1991)
See Kosobokova, N. L.: V1238,183-188(1991)
See Usanov, Yu. Y.: V1238,178-182(1991)

Shevtsova, Natalia A. See Rybak, Ilya A.: V1469,737-748(1991)

Shewchun, John
; Marsh, P. F.: Performance characteristics of Y-Ba-Cu-O microwave superconducting detectors,V1477,115-138(1991)

Shi, Bao A. See Zhou, Bing: V1519,454-456(1991)

Shi, Dahuan
; Qin, Jing; Hung, Yau Y.: Automated measurement of 3-D shapes by a dual-beam digital speckle interferometric technique,V1554A,680-689(1991)
See Hung, Yau Y.: V1332,332-342(1991)

Shi, Dexiu
; Xing, Xiaozheng; Wolbarsht, Myron L.: Holographic filter for coherent radiation,V1419,40-49(1991)

Shi, Fan See Jiang, Wen-Han: V1542,130-137(1991)

Shi, Gao-yi See Wei, Yan-nian: V1358,457-460(1991)

Shi, Guo L.
See Liu, Wei J.: V1519,415-418(1991)
See Liu, Wei J.: V1519,481-488(1991)

Shi, Jinshan
; Wang, Yutian: Optically powered thermistor with optical fiber link,V1572,175-179(1991)
See Li, Zhiquan: V1572,185-188(1991)
See Wang, Yutian: V1572,192-196(1991)

Shi, Lei See Kwok, Hoi-Sing: V1394,196-200(1991)

Shi, Qiaotin See Liu, Dayou: V1468,37-49(1991)

Shi, Qinyun See Gong, Wei: V1606,153-164(1991)

Shi, Shuyi
See Wang, Yongjiang: V1412,67-71(1991)
See Wang, Yongjiang: V1412,60-66(1991)

Shi, T. Y. See Zhang, L. B.: V1572,240-242(1991)

Shi, W.
; Kong, J.; Shen, H.; Du, G.; Yao, W.; Qi, Zhen Z.: Investigation on the anomalous structure of the nanocrystal Ti and Zr films,V1519,138-141(1991)

Shi, Wei-Qiang
See Vari, Sandor G.: V1424,33-42(1991)
See Vari, Sandor G.: V1426,58-65(1991)
See Vari, Sandor G.: V1426,111-120(1991)

Shi, Weimin
; Cui, Ting; Sigel, George H.: Novel method for preventing solar ultraviolet-radiation-induced skin cancer,V1422,62-72(1991)

Shi, Xiao D.
: Ionization potentials effects on CGL/CTL photoconductors,V1519,884-889(1991)

Shi, Yi-Wei
; Wang, Yao-Cai; Jiang, Hong-Tao; Yao, Cheng-Shan; Wu, Zhen-Chun: Distributed-fibre-optic methane gas concentration detection,V1572,308-312(1991)
See Wang, Yao-Cai: V1572,32-37(1991)

Shi, Youngqiang
; Steier, William H.; Yu, Luping; Chen, Mai; Dalton, Larry R.: Photo-induced refractive-index changes and birefringence in optically nonlinear polyester,V1559,118-126(1991)
See Chen, Mai: V1409,202-213(1991)

Shi, Zeng R. See Fan, Zheng X.: V1519,359-364(1991)

Shi, Zhenbang See Li, Xiang Ying: V1412,246-251(1991)

Shiau, Fuu-Yau
; Pandey, Ravindra K.; Dougherty, Thomas J.; Smith, Kevin M.: Synthetic approaches to long-wavelength photosensitizers for photodynamic therapy,V1426,330-339(1991)
See Pandey, Ravindra K.: V1426,356-361(1991)

Shiau, Jyh-Jen H. See Lee, David T.: V1383,297-304(1991)

Shiau, T. H.
; Sun, Shang-zhu; Schaus, Christian F.; Zheng, Kang; Hadley, G. R.; Hohimer, John P.: Strained quantum-well leaky-mode diode laser arrays,V1418,116-122(1991)

Shibata, Jun See Matsuda, Kenichi: V1562,21-29(1991)

Shibata, Yasumasa See Ito, Katsunori: V1499,382-385(1991)

Shibuya, Masato
: Exact sine condition in the presence of spherical aberration,V1354,240-245(1991)

Shidlovskaya, E. G.
See Chikishev, A. Y.: V1403,517-521(1991)
See Netrebko, N. V.: V1403,512-516(1991)

Shie, Jin-Shown
; Lian, Jiunn-Long: Temperature instability in silicon-based microheating device,V1362,655-663(1991)

Shields, Frank J. See Swan, John: V1483,219-230(1991)

Shields, Steven E.
; Bleha, William P.: Liquid-crystal light valves for projection displays,V1455,225-236(1991)

Shiffler, Donald See Nation, John A.: V1407,32-43(1991)

Shiga, Nobuo See Suzuki, Tomihiro: V1389,455-461(1991)

Shigehara, Kiyotaka See Suda, Yasumasa: V1560,75-83(1991)

Shigematsu, Kazumasa See Yanagishita, Yuichiro: V1463,207-217(1991)

Shigematu, Kazuo See Maeda, Takeshi: V1499,414-418(1991)

Shigemura, Naoshi See Oyama, Yoshiro: V1444,413-423(1991)

Shih, Choon-Foo
; Kuo, Chin-Po: Dynamic characteristics of joint-dominated space trusses,V1532,91-102(1991)
; Lou, Michael C.: Mechanical and thermal disturbances of the PSR moderate focus-mission structure,V1532,81-90(1991)

Shih, Chun-Ching
: Wave-optics analysis of fast-beam focusing,V1415,150-153(1991)

Shih, Chun-Hsi See Chien, Bing-Shan: V1606,588-598(1991)

Shih, D. Y. See Deutsch, Alina: V1389,161-176(1991)

Shih, Dennis K. See Joshi, A. B.: V1393,122-149(1991)

Shih, Frank Y.
; Moh, Jenlong: Implementing neural-morphological operations using programmable logic,V1382,99-110(1991)
; Moh, Jenlong: Improved adaptive resonance theory,V1382,26-36(1991)

Shih, J. See Hyzer, Jim B.: V1554B,371-382(1991)

Shih, Kwang K. See Aviram, Ari: V1458,105-107(1991)

Shikoda, Arimitsu
; Sato, Eiichi; Kimura, Shingo; Isobe, Hiroshi; Takahashi, Kei; Tamakawa, Yoshiharu; Yanagisawa, Toru: Repetitive flash x-ray generator as an energy transfer source utilizing a compact-glass body diode,V1358,154-161(1991)
See Isobe, Hiroshi: V1358,471-478(1991)
See Sato, Eiichi: V1358,479-487(1991)

Shiloh, Klara See Rapoport, Eliezer: V1442,383-391(1991)

Shilov, I. P. See Artjushenko, Vjacheslav G.: V1420,149-156(1991)

Shilov, V. V. See Kzuzhilin, Yu E.: V1238,200-205(1991)

Shima, Ralph M. See Chambers, Robert J.: V1488,312-326(1991)

Shimada, Wataru
; Sato, Tadamitsu; Yatagai, Toyohiko: Optical surface microtopography using phase-shifting Nomarski microscope,V1332,525-529(1991)

Shimamoto, Masayoshi
; Yamada, Koichi; Watanabe, Isao; Nakajima, Yoshiki; Ito, Osamu; Tanaka, Kunimaro: Interchangeability of optical disks,V1499,393-400(1991)

Shimazu, Hideto See Toba, Eiji: V1584,353-363(1991)

Shimer, Julie A. See Siddiqui, Humayun R.: V1596,139-157(1991)

Shimizu, Hazime See Tomie, Toshihisa: V1552,254-263(1991)

Shimizu, K. See Yoshida, S.: V1397,205-212(1991)

Shimizu, Keijiro See Ito, Katsunori: V1499,382-385(1991)

Shimizu, Masaru
; Katayama, Takuma; Fujimoto, Masashi; Shiosaki, Tadashi: Growth of PbTiO3 films by photo-MOCVD,V1519,122-127(1991)

Shimizu, Ryuichiro See Tsunetomo, Keiji: V1513,93-104(1991)

Shimizu, Yasuhiro
See Egashira, Makoto: V1519,467-476(1991)
See Feng, Chang D.: V1519,8-13(1991)

Shimkunas, Alex R. See Thomas, Stan W.: V1358,91-99(1991)

Shimkus, J. K. See Kononenko, Vadim L.: V1403,381-383(1991)

Shimmick, John K.
; Munnerlyn, Charles R.; Clapham, Terrance N.; McDonald, Marguerite B.: Axial and transverse displacement tolerances during excimer laser surgery for myopia,V1423,140-153(1991)

Shimomoto, Yasuharu See Kobayashi, Akira: V1449,148-156(1991)

Shimosaka, Tetsuya See Toba, Eiji: V1584,353-363(1991)

Shimura, Kei See Honda, Toshio: V1461,156-166(1991)

Shin, H. K. See Jain, Ajay: V1596,23-33(1991)

Shin, Hyun-Dong See Lee, Hyouk: V1393,404-410(1991)

Shin, Sang-Yung See Son, Yung-Sung: V1374,23-29(1991)

Shines, Janet E. See Balick, Lee K.: V1493,215-223(1991)

Shinghal, Rajjan See Preece, Alun D.: V1468,608-619(1991)

Shinozaki, D. M. See Wang, Ge: V1398,180-190(1991)

Shinya, Norio See Kishimoto, Satoshi: V1554B,174-180(1991)

Shiono, Teruhiro
; Ogawa, Hisahito: Electron-beam-written reflection diffractive microlenses for oblique incidence,V1545,232-240(1991)

Shiosaki, Tadashi
: Properties and applications of ferroelectric and piezoelectric thin films,V1519,694-703(1991)
See Shimizu, Masaru: V1519,122-127(1991)

Shiota, Alan T.
; Inada, Koichi: Measurement of physical parameters with special fibers,V1504,90-97(1991)
See Suzaki, Shinzoh: V1374,126-131(1991)

Shipocz, Tibor
See Chorvat, Dusan: V1402,89-92(1991)
See Shvec, Peter: V1402,78-81(1991)

Shirai, Hisatsugu
; Kobayashi, Katsuyoshi; Nakagawa, Kenji: 64-Mbit DRAM production with i-line stepper,V1463,256-274(1991)

Shiraishi, Hiroshi See Hayashi, Nobuaki: V1466,377-383(1991)

Shiraiwa, Yoshinobu See Suzuki, Masayuki: V1454,370-381(1991)

Shirayama, Shimpey See Aoki, Nobutada: V1412,2-11(1991)

Shirazi, Behrooz See Wang, Chung Ching: V1396,209-214(1991)

Shirk, Kevin W. See Jannson, Tomasz: V1555,159-176(1991)

Shirmulis, E. See Amosova, L. P.: V1440,406-413(1991)

Shirokova, I. P. See Vas'kovsky, Yu. M.: V1440,229-240(1991)

Shirvaikar, Mukul V.
; Trivedi, Mohan M.: Studies in robust approaches to object detection in high-clutter background,V1468,52-59(1991)

Shishkov, Vladimir F.
See Markov, Vladimir B.: V1238,41-43(1991)
See Markov, Vladimir B.: V1238,30-40(1991)
See Markov, Vladimir B.: V1238,118-122(1991)

Shitara, T. See Joyce, Bruce A.: V1361,13-22(1991)

Shiu, Yiu C.
: Experiments with perceptual grouping,V1381,130-141(1991)

Shkitin, Vladimir A.
See Kiselyov, Boris S.: V1621,126-137(1991)
See Kiselyov, Boris S.: V1621,328-339(1991)

Shkurinov, A. P. See Chikishev, A. Y.: V1403,448-456(1991)

Shkuropatov, A. Y. See Shuvalov, V. A.: V1403,400-406(1991)

Shladov, Itzhak See Oron, Moshe: V1442(1991)

Shlichta, Paul J.
; Nerad, Bruce A.: Advantages of drawing crystal-core fibers in microgravity,V1557,10-23(1991)

Shmaenok, L. A. See Platonov, Yu. Y.: V1440,188-196(1991)

Shoe, Bridget
; Disimile, Peter J.; Savory, Eric; Toy, Norman; Tahouri, Bahman: Image analysis of two impinging jets using laser-induced fluorescence and smoke flow visualization,V1521,34-45(1991)
See Disimile, Peter J.: V1489,66-74(1991)
See Disimile, Peter J.: V1483,39-48(1991)

Shogenov, Yu. H. See Romanovsky, Yuri M.: V1403,359-362(1991)

Shojaey, M. See Bahram-pour, A. R.: V1397,539-542(1991)

Shoji, Hideo See Sato, Koki: V1461,124-131(1991)

Shoji, Kenji
: Quadtree decomposition of binary structuring elements,V1451,148-157(1991)

Shoman, Mineo See Fujimoto, Tsuyoshi: V1606,599-606(1991)

Shopova, Maria See Angelov, D.: V1403,572-574(1991)

Shori, Raj K. See Hendrix, Charles D.: V1564,2-13(1991)

Short, Michael
: Use of absorption spectroscopy for refined petroleum product discrimination,V1480,72-79(1991)

Short, Ralph C.
; Cosgrove, Michael A.; Clark, David L.; Martino, Anthony J.; Park, Hong-Woo; Seery, Bernard D.: Acquisition and tracking performance measurements for a high-speed area array detector system,V1417,131-141(1991)
; Cosgrove, Michael A.; Clark, David L.; Oleski, Paul J.: Performance of a demonstration system for simultaneous laser beacon tracking and low-data-rate optical communications with multiple platforms,V1417,464-475(1991)

Shotts, K. W. See Tesar, Aleta A.: V1441,228-236(1991)

Shou, Reilan See Yuan, Libo: V1572,258-263(1991)

Shoup, Milton J. See Kelly, John H.: V1410,40-46(1991)

Shreve, A. P.
; Trautman, Jay K.; Owens, T. G.; Frank, Harry A.; Albricht, Andreas C.: In vivo energy transfer studies in photosynthetic systems by subpicosecond timing,V1403,394-399(1991)

Shrivastava, Chinmaya A. See Briscoe, Dennis: V1358,329-336(1991)

Shrivastava, Udy A.
: Design, simulation model, and measurements for high-density interconnections,V1389,122-137(1991)

Shrode, David See Russo, Paul: V1383,60-71(1991)

Shropshire, Geoffrey J.
; Von Bargen, Kenneth; Mortensen, David A.: Optical reflectance sensor for detecting plants,V1379,222-235(1991)

Shtepanek, J. See Zachova, J.: V1402,216-222(1991)

Shu, Ju P.
; Zhou, Jian P.; Xu, Shi Z.: Organo-metallic thin film for erasable optical recording medium,V1519,565-569(1991)
See Zhou, Jian P.: V1519,559-564(1991)

Shu, Peter K. See Ewin, Audrey J.: V1479,12-20(1991)

Shu, Q. Q.
; Tian, X. M.; Chen, X. Y.; Cai, C. Z.; Zheng, K. Q.; Ma, W. G.: Optical and electrical properties of Al-Al2O3-Cu tunnel junctions,V1519,675-679(1991)

Shubkin, N. G. See Vizir, V. A.: V1411,63-68(1991)

Shubochkin, L. P.
: Tuchin, Valery V.: New results in human eye laser diagnostics,V1403,720-731(1991)
See Guseva, N. P.: V1403,332-334(1991)
See Maksimova, I. L.: V1403,749-751(1991)

Shukla, Oodaye B. See Boone, Bradley G.: V1471,390-403(1991)

Shukla, Ram P.
: Dokhanian, Mostafa; Venkateswarlu, Putcha; George, M. C.: Phase-conjugate Twyman-Green interferometer for testing conicoidal surfaces,V1332,274-286(1991)
See Murty, M. V.: V1332,41-49(1991)

Shulika, A. N. See Averin, V. I.: V1358,603-605(1991)

Shull, J. E. See Perry, Kenneth E.: V1554A,209-227(1991)

Shull, Thomas A. See Burke, James J.: V1499(1991)

Shulok, Janine R. See Lin, Chi-Wei: V1426,216-227(1991)

Shumaker, Bryan P.
: Clinical applications of PDT in urology: present and near future,V1426,293-300(1991)

Shumate, Michael S.
See Scholl, Marija S.: V1564,165-176(1991)
See Udomkesmalee, Suraphol: V1468,81-91(1991)

Shumilin, V. V.
See Tkachuk, A. M.: V1403,801-804(1991)
See Tkachuk, T. M.: V1403,805-808(1991)

Shumilov, S. K. See Ekimov, A. I.: V1513,123-129(1991)

Shushan, A.
: Meninberg, Y.; Levy, I.; Kopeika, Norman S.: Prediction of thermal-image quality as a function of weather forecast,V1487,300-311(1991)

Shuto, Yoshito
: Amano, Michiyuki; Kaino, Toshikuni: Electro-optical light modulation in novel azo-dye-substituted poled polymers,V1560,184-195(1991)

Shuvalov, V. A.
: Ganago, A. O.; Shkuropatov, A. Y.; Klevanik, A. V.: Subpicosecond electron transfer in reaction centers of photosynthetic bacteria,V1403,400-406(1991)

Shvartsburg, Alexandre B.
: Physical foundations of high-speed thermomagnetic tuning of spatial-temporal structure of far-IR and submillimeter beams,V1488,28-35(1991)

Shvartsman, Felix P.
: Dry photopolymer embossing: novel photoreplication technology for surface relief holographic optical elements,V1507,383-391(1991)
: Holographic optical elements by dry photopolymer embossing,V1461,313-320(1991)
: Oren, Moshe: Photolithographic imaging of computer-generated holographic optical elements,V1555,71-79(1991)

Shvec, Peter
: Kvasnichka, P.; Shipocz, Tibor; Chorvat, Dusan: Method of laser fluorescence microphotolysis,V1402,78-81(1991)
: Miklovicova, J.; Hunakova, L.; Chorvath, B.: Translational dynamics of immune system components by fluorescence recovery after photobleaching,V1403,635-637(1991)
See Chorvat, Dusan: V1403,659-666(1991)
See Chorvat, Dusan: V1402,89-92(1991)

Shvej, E. M.
: Latysh, E. G.; Morgunov, A. N.; Mikhalchenko, O. E.; Stchetinin, A. L.; Volokh, K. Y.: Solution of optimization problems of perforated and box-shaped structures by photoelasticity and numerical methods,V1554A,488-495(1991)

Shykind, David See Bowers, C. R.: V1435,36-50(1991)

Shynk, John J.
: Gooch, Richard P.; Krishnamurthy, Giridhar; Chan, Christina K.: Comparative performance study of several blind equalization algorithms,V1565,102-117(1991)

Shyu, Keh-Haw See Chien, Bing-Shan: V1606,588-598(1991)

Shyu, William M. See Becker, Richard A.: V1459,48-56(1991)

Siahmakoun, Azad
: Blair, Steven M.; Weiss, Markus R.: Optical neural networks: an implementation,V1396,190-192(1991)
: Shen, Xuanguo: Detection of phase objects in transparent liquids using nonlinear coupling in BaTiO3 crystal,V1396,535-538(1991)
See Oldekop, Erik: V1396,174-177(1991)
See Yang, Guanglu: V1396,552-556(1991)

Sibbett, Wilson
See Liu, Yueping: V1358,503-510(1991)
See Liu, Yueping: V1358,300-310(1991)
See Walker, David R.: V1358,860-867(1991)

Sibilia, C.
See Andriesh, A.: V1513,137-147(1991)
See Montenero, Angelo: V1513,234-242(1991)

Sibley, Martin J. See Cryan, Robert A.: V1372,64-71(1991)

Siddiqui, Humayun R.
: Ryan, Vivian; Shimer, Julie A.: Interfacial adhesive strength measurement in a multilayered two-level metal device structure,V1596,139-157(1991)

Siddons, D. P.
ed.: *Production and Analysis of Polarized X Rays*,V1548(1991)
See Kao, Chi-Chang: V1548,149-157(1991)

Sidiropoulos, N. D.
: Baras, John S.; Berenstein, C. A.: Two-dimensional signal deconvolution: design issues related to a novel multisensor-based approach,V1569,356-366(1991)

Sidorova, G. I. See Askadskij, A. A.: V1554A,426-431(1991)

Siebinga, I.
: de Mul, F. F.; Vrensen, G. F.; Greve, Jan: Spatially resolved water concentration determination in human eye lenses using Raman microspectroscopy,V1403,746-748(1991)

Siegal, Michael P. See Phillips, Julia M.: V1394,186-190(1991)

Siegel, Andrew M. See Kimball, Paulette R.: V1540,687-698(1991)

Siegel, David M.
: Finding a grasped object's pose using joint angle and torque constraints,V1383,151-165(1991)

Siegman, Anthony E. See Mussche, Paul L.: V1376,153-163(1991)

Siegmund, Oswald H.
: Rothschild, Richard E.; eds.: *EUV, X-Ray, and Gamma-Ray Instrumentation for Astronomy II*,V1549(1991)
: Stock, Joseph: Performance of low-resistance microchannel-plate stacks,V1549,81-89(1991)
See Bush, Brett C.: V1549,290-301(1991)

Siegmund, Walter A. See Forbes, Fred F.: V1532,146-160(1991)

Siejca, Antoni See Kesik, Jerzy: V1391,34-41(1991)

Siekhaus, Wigbert J. See Tesar, Aleta A.: V1441,228-236(1991)

Sierak, Paul
: Cascaded optical modulators: any advantages?,V1398,238-249(1991)
ed.: *High-Frequency Analog Fiber Optic Systems*,V1371(1991)

Sierakowski, Marek W.
See Domanski, Andrzej W.: V1362,844-852(1991)
See Domanski, Andrzej W.: V1362,907-915(1991)

Sievers, A. J. See Xue, L. A.: V1534,183-196(1991)

Siewert, Ronald O. See Leib, Kenneth G.: V1483,140-154(1991)

Siffert, Paul See Hartiti, Bouchaib: V1393,200-206(1991)

Siffredi, Brigitte See Tomasini, Bernard: V1468,60-71(1991)

Sigaryov, Sergei E. See Michailov, Vladimir I.: V1590,203-214(1991)

Sigel, George H.
See Majercak, David: V1584,162-169(1991)
See Saggese, Steven J.: V1437,44-53(1991)
See Saggese, Steven J.: V1368,2-14(1991)
See Shi, Weimin: V1422,62-72(1991)

Sigel, Richard See Eidmann, K.: V1502,320-330(1991)

Sigg, Hans See Diessel, Edgar: V1361,1094-1099(1991)

Sigler, Robert D.
: Designing apochromatic telescope objectives with liquid lenses,V1535,89-112(1991)

Signorazzi, Mario See Varasi, Mauro: V1583,165-169(1991)

Signoret, P. See Orsal, Bernard: V1512,112-123(1991)

Sikes, Terry L.
: Kreiss, William T.: Graphical interface for multispectral simulation, scene generation, and analysis,V1479,199-211(1991)

Sikora, Thomas
: Tan, T. K.; Pang, Khee K.: Two-layer pyramid image coding scheme for interworking of video services in ATM,V1605,624-634(1991)

Silberberg, Teresa M.
: Automatic inspection of optical fibers,V1472,150-156(1991)
See Daily, Michael J.: V1472,85-96(1991)

Silberman, Donn M.
: Image quality evaluation of multifocal intraocular lenses,V1423,20-28(1991)

Silberman, Enrique
See Burger, Arnold: V1557,245-249(1991)
See Springer, John M.: V1557,192-196(1991)

Silbey, R. See Hochstrasser, Robin M.: V1403,221-229(1991)

Silenok, Alexander S.
See Artjushenko, Vjacheslav G.: V1420,149-156(1991)
See Artjushenko, Vjacheslav G.: V1420,176-176(1991)
See Artjushenko, Vjacheslav G.: V1590,131-136(1991)
See Konov, Vitaly I.: V1427,232-242(1991)
See Konov, Vitaly I.: V1427,220-231(1991)

Silfvast, William T.
: Tutorial on x-ray lasers,V1397,3-8(1991)
See White, Donald L.: V1343,204-213(1991)

Silin, Andrejs R.
: Multiphoton-absorption-induced structural changes in fused silica,V1513,270-273(1991)

Silkis, E. G.
; Titov, V. D.; Feldman, G. G.; Zhilkina, V. M.; Petrokovich, O. A.; Syrtzev, V. N.: One-frame subnanosecond spectroscopy camera,V1358,46-49(1991)

Silletti, Andrew See Elman, Boris S.: V1362,610-616(1991)

Sillitto, Richard M.
See Potter, Duncan J.: V1564,363-372(1991)
See Underwood, Ian: V1562,107-115(1991)

Sills, Robert M. See Weiss, Rudolf M.: V1465,192-200(1991)

Silsbee, Peter L.
; Bovik, Alan C.: Nonuniform visual pattern image sequence coding,V1452,409-419(1991)

Silva, Donald E.
See Kuok, M. H.: V1399(1991)
See Tam, Siu-Chung: V1400,38-48(1991)
See Wihardjo, Erning: V1529,57-62(1991)

Silva, Jorge A. See de Lourdes Quinta, Maria: V1399,24-29(1991)

Silvennoinen, Raimo
; Hamalainen, Rauno M.; Rasanen, Jari: Computer-generated diffraction gratings in optical region,V1574,84-88(1991)
; Hamalainen, Rauno M.: Matching of Bragg condition of holographic phase gratings in 1.3-1.5um region,V1574,261-265(1991)

Silver, Deborah E.
; Zabusky, Norman J.: Three-dimensional visualization and quantification of evolving amorphous objects,V1459,97-108(1991)

Silver, Eric H.
See Holley, Jeff: V1343,500-511(1991)
See Kaaret, Philip E.: V1548,106-117(1991)
See Weisskopf, Martin C.: V1343,457-468(1991)

Silver, Joel A.
; Bomse, David S.; Stanton, Alan C.: Frequency-modulation absorption spectroscopy for trace species detection: theoretical and experimental comparison among methods,V1435,64-71(1991)

Silver, Michael D.
; Nishiki, Masayuki; Tochimura, Katsumi; Arita, Masataka; Drawert, Bruce M.; Judd, Thomas C.: CT imaging with an image intensifier: using a radiation therapy simulator as a CT scanner,V1443,250-260(1991)

Silverstein, Barry D. See Kay, David B.: V1499,281-285(1991)

Silverstein, Seth D. See Zoltowski, Michael D.: V1566,452-463(1991)

Silvestre, Victor M. See Pais, Cassiano P.: V1451,282-288(1991)

Silvey, Tom See Tanaka, Satoru C.: V1448,21-26(1991)

Silzer, R. M. See Piestrup, Melvin A.: V1552,214-239(1991)

Simao, Jose V. See Rodrigues, F. C.: V1399,90-97(1991)

Simchick, Richard T.
See Fripp, Archibald L.: V1557,236-244(1991)
See Steiner, Bruce W.: V1557,156-167(1991)

Simeoni, Denis
: New concept for a highly compact imaging Fourier transform spectrometer,V1479,127-138(1991)

Simes, Robert J. See Coldren, Larry A.: V1362,24-37(1991)

Simko, Joseph See Dreyer, Keith J.: V1445,398-408(1991)

Simmonds, P. E. See Eaves, Lawrence: V1362,520-533(1991)

Simmons, Horace O. See Davies, John T.: V1392,551-554(1991)

Simon, Deborah R. See Smith, Dale J.: V1562,242-250(1991)

Simon, Gabriel
; Parel, Jean-Marie; Kervick, Gerard N.; Rol, Pascal O.; Hanna, Khalil; Thompson, Keith P.: Corneal epithelium, visual acuity, and laser refractive keratectomy,V1423,154-156(1991)
See Cartlidge, Andy G.: V1423,167-174(1991)
See Lingua, Robert W.: V1423,58-61(1991)

Simon, John D.
; Xie, Xiaoliang; Dunn, Robert C.: Picosecond absorption and circular dichroism studies of proteins,V1432,211-220(1991)

Simon, Randy W.
: High-temperature superconducting Josephson junction devices,V1394,70-78(1991)
: High-temperature superconductor junction technology,V1477,184-191(1991)

Simon, Wayne E.
; Carter, Jeffrey R.: Generalizing from a small set of training exemplars for handwritten digit recognition,V1469,592-601(1991)

Simon, Wilhelm See Spichiger, Ursula E.: V1510,118-130(1991)

Simoni, R. See Bertoluzza, Alessandro: V1403,743-745(1991)

Simonis, Roland
: Scanner analyzer target,V1454,364-369(1991)

Simonne, John J.
; Pham, Vui V.; Esteve, Daniel; Alaoui-Amine, Mohammed; Bousbiat, Essaid: Silicon/PVDF integrated double detector: application to obstacle detection in automotive,V1374,107-115(1991)
; Pham, Vui V.; Esteve, Daniel; Clot, Jean; Mahrane, Achour; Beconne, Jean P.: Visible/infrared integrated double detector: application to obstacle detection in automotive (Phase 2),V1589,139-147(1991)

Simonotto, Enrico See Riani, Massimo: V1469,166-177(1991)

Simonov, A. V. See Sadyigov, Z. Y.: V1621,158-168(1991)

Simonov, Alexander P.
; Varakin, Vladimir N.; Panesh, Anatoly M.; Lunchev, V. A.: Laser chemistry of dimethylcadmium adsorbed on silicon: 308- versus 222-nm laser excitation,V1503,330-337(1991)
; Varakin, Vladimir N.; Panesh, Anatoly M.: UV laser-induced photofragmentation and photoionization of dimethylcadmium chemisorbed on silicon,V1436,20-30(1991)

Simonov, V. P. See Ignatosyan, S. S.: V1358,100-108(1991)

Simonpietri, Pascal
See Lefevre, Herve C.: V1365,65-73(1991)
See Lefevre, Herve C.: V1367,72-80(1991)

Simons, K. See Weng, Jian: V1431,161-170(1991)

Simons, Margaret A. See Razavi, Mahmood: V1446,24-34(1991)

Simova, Eli S. See Sainov, Ventseslav C.: V1461,175-183(1991)

Simovsky, Ilya See Liao, Wen-gen: V1444,47-55(1991)

Simpson, Daniel J. See Soper, Steven A.: V1435,168-178(1991)

Simpson, Edward V. See Laxer, Cary: V1444,190-195(1991)

Simpson, Orman A. See Russwurm, George M.: V1433,302-314(1991)

Simpson, Thomas B.
; Doft, F.; Malley, Michael M.; Sutton, George W.; Day, Timothy: Frequency stability of a solid state mode-locked laser system,V1410,133-140(1991)
; Malley, Michael M.; Sutton, George W.; Doft, F.: Mode-locked Nd:YLF laser for precision range-Doppler imaging,V1416,2-9(1991)

Sims, Gary R.
See Doherty, Peter E.: V1448,118-128(1991)
See McCurnin, Thomas W.: V1448,225-236(1991)

Sinclair, Douglas C.
: Optimization using the OSLO and Super-OSLO programs,V1354,116-119(1991)

Singer, Christian
; Massoni, Jean A.; Mossbacher, Bernard; Cinotti, Ciro: Infrared Space Observatory optical subsystem,V1494,255-264(1991)

Singer, J. L. See Meulenberg, A.: V1475,280-285(1991)

Singer, Jay R. See Gruenbaum, F. A.: V1431,232-238(1991)

Singer, Jon See Horii, Steven C.: V1446,475-480(1991)

Singer, Kenneth D.
ed.: *Nonlinear Optical Properties of Organic Materials IV*,V1560(1991)
See Cahill, Paul A.: V1560,130-138(1991)

Singer, Sidney
ed.: *Advanced Laser Concepts and Applications*,V1501(1991)

Singh, Ajit
: Computing image flow and scene depth: an estimation-theoretic fusion-based framework,V1383,122-140(1991)

Singh, Amit K. See Lachs, Gerard: V1474,248-259(1991)

Singh, Anjali
; Banerjee, Pranab K.; Mitra, Shashanka S.: Neutron irradiation effects on fibers operating at 1.3 um and 1.55 um,V1366,184-190(1991)

Singh, Babu R. See Yadav, M. S.: V1362,811-819(1991)

Singh, Bhanwar
See Bennett-Lilley, Marylyn H.: V1464,127-136(1991)
See Dunbrack, Steven K.: V1464,314-326(1991)
See Phan, Khoi: V1464,424-437(1991)

Singh, Brahm P.
; Chitnis, Vijay T.: Angle measurement by moire interference technique,V1554B,335-338(1991)
; Chopra, S.: Experimental investigations of the autocorrelation function for inhomogeneous scatterers,V1558,317-321(1991)
See Razdan, Anil K.: V1558,384-388(1991)

Singh, Harcharan See Sung, Eric: V1385,57-69(1991)

Singh, Hemant
; Sirkis, James S.; Dasgupta, Abhijit: Microinteraction of optical fibers embedded in laminated composites,V1588,76-85(1991)

Singh, Jasprit
; Bhattacharya, Pallab K.: Quantum-well excitonic devices for optical computing,V1362,586-597(1991)
See Bhattacharya, Pallab K.: V1361,394-405(1991)

Singh, Jyothi
: In-situ measurements of radicals and particles in a selective silicon oxide etching plasma,V1392,474-486(1991)

Singh, Mahendra P. See Zemon, Stanley A.: V1373,21-32(1991)

Singh, N. B.
See Brandt, Gerald B.: V1454,336-343(1991)
See Jones, O. C.: V1557,202-208(1991)
See Paradies, C. J.: V1561,2-5(1991)

Singh, Rajeev
See Toh, Kenny K.: V1496,27-53(1991)
See Toh, Kenny K.: V1463,74-86(1991)

Singh, Rajendra
: Rapid isothermal process technology for optoelectronic applications,V1418,203-216(1991)
; Sinha, Sanjai; Thakur, Randhir P.; Hsu, N. J.: Integrated rapid isothermal processing,V1393,78-89(1991)
; Sinha, Sanjai; Krueger, D. J.: Prospects for hybrid electronics,V1394,62-67(1991)
See Moslehi, Mehrdad M.: V1393(1991)
See Narayan, Jagdish: V1394(1991)
See Sinha, Sanjai: V1394,266-276(1991)

Singh, Rajiv K.
; Bhattacharya, Deepika; Narayan, Jagdish; Jahncke, Catherine; Paesler, Michael A.: Effect of laser irradiation on superconducting properties of laser-deposited YBa2Cu3O7 thin films,V1394,203-213(1991)

Singh, Surendra
; Mortazavi, Mansour; Phillips, K. J.; Young, M. R.: Noise in He:Ne lasers near threshold,V1376,143-152(1991)

Singhal, Sharad See Watanabe, Hiroshi: V1605,582-589(1991)

Single, Thomas See Dickerson, Stephen L.: V1381,201-209(1991)

Singleton, Scott See Miller, Richard M.: V1439,66-78(1991)

Singman, Leif V.
; Davis, Brent A.; Holmberg, D. L.; Blake, Richard L.; Hockaday, Robert G.: Measurement of synchrotron beam polarization,V1548,80-92(1991)

Sinha, Kislay
; Menendez, Jose; Adams, G. B.; Page, J. B.; Sankey, Otto F.; Lamb, Lowell; Huffman, Donald R.: Raman study of icosahedral C60,V1437,32-35(1991)

Sinha, Mahadeva P.
: Laser-induced volatilization and ionization of aerosol particles for their mass spectral analysis in real time,V1437,150-156(1991)

Sinha, P. G. See Kidd, S. R.: V1504,180-190(1991)

Sinha, Sanjai
; Singh, Rajendra; Hsu, N. J.; Ng, J. T.; Chou, P.; Narayan, Jagdish: Reduced thermal budget processing of high-Tc superconducting thin films and related materials by MOCVD,V1394,266-276(1991)
See Singh, Rajendra: V1393,78-89(1991)
See Singh, Rajendra: V1394,62-67(1991)

Sinha, Saravajit S.
: Automatic reconstruction of buildings from aerial imagery,V1468,698-709(1991)
: Differential properties from adaptive thin-plate splines,V1570,64-74(1991)
; Moezzi, Saied; Schunck, Brian G.: Robust stereo vision,V1385,259-267(1991)

Siniard, Sheri M. See Greenleaf, William G.: V1486,364-375(1991)

Sinichkin, Yury P. See Kozlova, T. G.: V1403,159-160(1991)

Sinitsin, D. V.
See Frolov, K.: V1397,461-468(1991)
See Ionin, A. A.: V1397,453-456(1991)

Sinn, H.-J. See Kohl, M.: V1525,26-34(1991)

Sinniah, R. See Ong, Sim-Heng: V1445,564-573(1991)

Sinnott, Heather
; MacLeod, Wade: Fiber access maintenance leverages,V1363,196-200(1991)

Sinta, Roger F. See Thackeray, James W.: V1466,39-52(1991)

Sintov, Yoav See Yatsiv, Shaul: V1397,319-329(1991)

Sinyakov, G. N. See Chirvony, V. S.: V1403,504-506(1991)

Sinyakova, Elena F. See Yelisseyev, Alexander P.: V1512,204-212(1991)

Siomos, Konstadinos
: Laser/light tissue interaction: on the mechanism of optical breakdown,V1525,154-161(1991)

Sipe, John E.
See An, Sunghyuck: V1516,175-184(1991)
See Anderson, Dana Z.: V1516,154-161(1991)

Sipila, Heikki
; Huttunen, Pekka; Kamarainen, Veikko J.; Vilhu, Osmi; Kurki, Jouko; Leppelmeier, Gilbert W.; Taylor, Ivor; Niemela, Arto; Laegsgaard, Erik; Sunyaev, Rashid: Silicon x-ray array detector on spectrum-x-gamma satellite,V1549,246-255(1991)
See Viitanen, Veli-Pekka: V1549,28-34(1991)

Sire, Pascal See Grangeat, Pierre: V1445,320-331(1991)

Sirkis, James S.
: Phase-strain-temperature model for structurally embedded interferometric optical fiber strain sensors with applications,V1588,26-43(1991)
; Dasgupta, Abhijit: Thermal plastic metal coatings on optical fiber sensors,V1588,88-99(1991)
See Singh, Hemant: V1588,76-85(1991)

Sirmulis, Edmundas See Asmontas, Steponas: V1512,131-134(1991)

Sirohi, Rajpal S.
ed.: *Selected Papers on Speckle Metrology*,VMS35(1991)
; Kumar, Harish; Jain, Narinder K.: Focal length measurement using diffraction at a grating,V1332,50-55(1991)

Sirois, J. See Teillet, Philippe M.: V1492,213-223(1991)

Sirota, J. M. See Christiansen, Walter H.: V1397,821-825(1991)

Sirutkaitis, V. See Danelius, R.: V1402,198-208(1991)

Sisakian, Iosif N.
See Boiko, Yuri B.: V1238,253-257(1991)
See Golub, Mikhail A.: V1500,194-206(1991)
See Golub, Mikhail A.: V1365,156-165(1991)
See Golub, Mikhail A.: V1500,211-221(1991)
See Golub, Mikhail A.: V1366,273-282(1991)

Sisakyan, Iosiph N. See Golub, Mikhail A.: V1572,101-106(1991)

Sisterman, Elizabeth A.
; Andrews, Larry C.: New look at bump spectra models for temperature and refractive-index fluctuations,V1487,345-355(1991)

Sitter, Helmut
See Gresslehner, Karl-Heinz: V1361,1087-1093(1991)
See Hingerl, Kurt: V1361,943-953(1991)
See Hingerl, Kurt: V1361,383-393(1991)

Sivak, Michael V. See Rava, Richard P.: V1426,68-78(1991)

Sivan, V. See Karuppiah, N.: V1536,215-221(1991)

Sizmann, R. See Lazarov, M.: V1536,183-191(1991)

Sizov, V. N.
See Bespalov, V. G.: V1238,457-461(1991)
See Vorzobova, N. D.: V1238,462-464(1991)

Sjollema, J. I.
; den Exter, Ir. T.; Zijp, Jaap R.; Ten Bosch, Jaap J.: Measurement of color and scattering phenomena of translucent materials,V1500,177-188(1991)

Sjostrand, Johan See Larsson, Michael: V1529,63-70(1991)

Skanavi, A. M. See Zhavoronok, I. V.: V1554A,371-379(1991)

Skarus, Waldemar See Elstner, Christian: V1478,150-159(1991)

Skeie, Halvor
; Johnson, Richard V.: Linearization of electro-optic modulators by a cascade coupling of phase-modulating electrodes,V1583,153-164(1991)

Skeldon, Mark D.
; Jin, Michael S.; Smith, Douglas H.; Bui, Snow T.: Performance of longitudinal-mode KD*P Pockels cells with transparent conductive coatings,V1410,116-124(1991)
See Kelly, John H.: V1410,40-46(1991)

Skibowski, Michael See Kipp, Lutz: V1361,794-801(1991)

Skinner, Iain M. See Malo, Bernard: V1590,83-93(1991)

Skinner, J. L. See Hochstrasser, Robin M.: V1403,221-229(1991)

Skipper, M. C. See Clark, Wally T.: V1371,258-265(1991)

Skipper, Richard S.
; Shepherd, David W.: Moulding process for contact lens,V1529,22-30(1991)

Sklair, Cheryl W.
; Gatrell, Lance B.; Hoff, William A.; Magee, Michael: Optical target location using machine vision in space robotics tasks,V1387,380-391(1991)

Sklansky, Jack See Gutfinger, Dan: V1445,288-296(1991)

Sklar, Michael E.
See Hudgens, Jeffrey C.: V1387,271-282(1991)
See Vargo, Rick C.: V1387,72-81(1991)

Sklodowska, Malgorzata
See Szczepanski, Pawel: V1391,72-78(1991)
See Szczepanski, Pawel: V1391,65-71(1991)

Skofteland, G. See Vemuri, Baba C.: V1383,97-108(1991)

Skogen, J. See Hibbs-Brenner, Mary K.: V1563,10-20(1991)

Skolnick, Maurice S. See Eaves, Lawrence: V1362,520-533(1991)

Skolnick, Michael M. See Bhagvati, Chakravarthy: V1606,112-119(1991)

Skoneczny, Slawomir See Swiniarski, Roman W.: V1451,234-241(1991)

Skopik, D. M. See Piestrup, Melvin A.: V1552,214-239(1991)

Skopinov, S. A.
; Yakovleva, S. V.: Crystallo-optic diagnostics method of the soft laser-induced effects in biological fluids,V1403,680-681(1991)
; Zakharov, S. D.; Volf, E. B.; Perov, S. N.; Panasenko, N. A.: Cellular and extracellular effects of soft laser irradiation on an erythrocytes suspension,V1403,676-679(1991)

Skorczakowski, M. See Jankiewicz, Zdzislaw: V1391,101-104(1991)

Skow, Michael See Ludwig, David E.: V1541,73-82(1991)

Skrehot, Michael K. See Chang, Kai: V1475,257-266(1991)

Skripko, G. A. See Jankiewicz, Zdzislaw: V1391,101-104(1991)

Skropkin, S. A. See Dmokhovskij, A. V.: V1554A,323-330(1991)

Skrzypczak, J. A. See Troyer, David E.: V1486,396-409(1991)

Skubiszak, Wojciech See Kotowski, Tomasz: V1391,6-11(1991)

Skudayski, Ulf
; Jueptner, Werner P.: Synthetic wavelength interferometry for the extension of the dynamic range,V1508,68-72(1991)

Skulina, Kenneth M. See Lai, Barry P.: V1550,46-49(1991)

Skumanich, Andrew
: Highly sensitive absorption measurements in organic thin films and optical media,V1559,267-277(1991)

Skyner, D. See She, K.: V1572,581-587(1991)

Slack, Marc G.
: Coordinating sensing and local navigation,V1383,459-470(1991)

Slade, James N. See Santago, Peter: V1445,555-563(1991)

Slama, M. M.
; Patterson, Angela C.: Automated approach to the correlation of defect locations to electrical test results to determine yield reducing defects,V1464,602-609(1991)

Slane, Frederic A. See Johnson, Stewart W.: V1494,194-207(1991)

Slasky, B. S. See Gur, David: V1446,284-288(1991)

Slater, David C.
; Timothy, J. G.: High-dynamic-range MCP structures,V1549,68-80(1991)

Slater, Philip N.
ed.: *Future European and Japanese Remote-Sensing Sensors and Programs*,V1490(1991)
; Palmer, James M.: Solar-diffuser panel and ratioing radiometer approach to satellite sensor on-board calibration,V1493,100-105(1991)
See Balick, Lee K.: V1493,215-223(1991)

See Biggar, Stuart F.: V1493,155-162(1991)
See Gellman, David I.: V1493,175-180(1991)
See Guzman, Carmen T.: V1493,120-131(1991)
See Markham, Brian L.: V1493,207-214(1991)
See Nianzeng, Che: V1493,182-194(1991)
See Palmer, James M.: V1493,106-117(1991)

Slater, Sydney G. See Uenishi, Kazuya: V1466,102-116(1991)

Slaughter, Jon M.
; Kearney, Patrick A.; Schulze, Dean W.; Falco, Charles M.; Hills, C. R.; Saloman, Edward B.; Watts, Richard N.: Interfaces in Mo/Si multilayers,V1343,73-82(1991)
See Kearney, Patrick A.: V1343,25-31(1991)

Sleat, William E. See Walker, David R.: V1358,860-867(1991)

Slegtenhorst, Ronald P. See Gielen, Robert M.: V1419,153-159(1991)

Sleigh, William J.
: Improvement in detection of small wildfires,V1540,207-212(1991)

Slemr, Franz
See Muecke, Robert J.: V1433,136-144(1991)
See Werle, Peter W.: V1433,128-135(1991)

Slepicka, James S.
; Cha, Soyoung S.: Stabilization of ill-posed nonlinear regression model and its application to interferogram reduction,V1554B,574-579(1991)

Slesarev, A. G. See Almaev, R. K.: V1397,831-834(1991)

Sleva, E. T. See Perry, Joseph W.: V1560,302-309(1991)

Slinde, Harald See Meringdal, Frode: V1503,292-298(1991)

Sliney, David H.
; Clapham, Terrance N.: Safety of medical excimer lasers with an emphasis on compressed gases,V1423,157-162(1991)

Slinger, Christopher W. See Ewen, Peter J.: V1512,101-111(1991)

Slivinskij, E. V. See Peet, Viktor E.: V1412,138-148(1991)

Sliwinski, G.
See Zerza, Gerald: V1397,107-110(1991)
See Zerza, Gerald: V1410,202-208(1991)

Sloan, George B. See Wells, James A.: V1485,2-12(1991)

Sloan, Kenneth R.
; Curcio, Christine A.: Packing geometry of human cone photoreceptors: variations with eccentricity and evidence for local anisotropy,V1453,124-133(1991)

Slock, Dirk T.
: Backward consistency concept and a new decomposition of the error propagation dynamics in RLS algorithms,V1565,14-24(1991)

Slocum, Robert E.
See Drawert, Bruce M.: V1396,566-567(1991)
See Hoeft, Gregory L.: V1396,638-645(1991)
See Hughes, Simon H.: V1396,575-581(1991)

Slot, M. See Koelink, M. H.: V1511,120-128(1991)

Sloth, Poul H. See Dougherty, Edward R.: V1606,141-152(1991)

Slovacek, Rudolf E. See Walczak, Irene M.: V1420,2-12(1991)

Sluder, John C.
; Bidlack, Clint R.; Abidi, Mongi A.; Trivedi, Mohan M.; Jones, Judson P.; Sweeney, Frank J.: Range image-based object detection and localization for HERMIES III mobile robot,V1468,642-652(1991)

Sluss, James J. See Heavin, Scott D.: V1455,12-18(1991)

Slutzky, Gale D. See Hall, Ernest L.: V1381,162-170(1991)

Smaev, V. P.
; Galpern, A. D.; Kiriencko, Yu. A.: Three-layer material for the registration of colored holograms,V1238,311-315(1991)
See Galpern, A. D.: V1238,320-323(1991)

Smal, A. See Predtechensky, M.: V1477,234-241(1991)

Smalanskas, Joseph P.
; Valentine, Gary W.; Wolfson, Ronald I.: Wideband embedded/conformal antenna subsystem concept,V1489,2-8(1991)

Small, David
: Simulating watercolor by modeling diffusion, pigment, and paper fibers,V1460,140-146(1991)

Small, Gerald J.
See Hayes, John M.: V1435,258-266(1991)
See Jankowiak, Ryszard: V1435,203-213(1991)

Small, Susan M. See Flock, Stephen T.: V1427,77-89(1991)

Smart, Anthony E.
: Optical velocity sensor for air data applications,V1480,62-71(1991)

Smart, Richard G.
; Carter, John N.; Tropper, Anne C.; Hanna, David C.: Fluoride fiber lasers,V1373,158-165(1991)
See Tropper, Anne C.: V1373,152-157(1991)

Smedley, Kirk G. See Payton, Paul M.: V1406,58-71(1991)

Smeins, Larry G.
; Stechman, John M.; Cole, Edward H.: Application-specific integrated-circuit-based multinode microchannel array readout system,V1549,59-65(1991)
See Cole, Edward H.: V1549,46-51(1991)

Smektala, Frederic See Adam, Jean-Luc: V1513,150-157(1991)

Smestad, Greg P.
; Ries, Harald: Luminescence and chemical potential of solar cells,V1536,234-245(1991)
See Ries, Harald: V1528,7-14(1991)

Smetana, Daryl L.
; Carson, John C.: On-focal-plane superconducting signal processing for low- and intermediate-temperature operation,V1541,220-226(1991)

Smigielski, Paul
: Holography as a tool for widespread industrial applications: analysis and comments,V1508,38-49(1991)
See Albe, Felix: V1358,1098-1102(1991)

Smilanski, Israel See Chuchem, D.: V1397,277-281(1991)

Smirnov, A. V. See Vorobiev, N. S.: V1358,698-702(1991)

Smirnov, O. See Popov, A.: V1519,457-462(1991)

Smirnov, S. B. See Savostjanov, V. N.: V1554A,579-585(1991)

Smirnov, V. V.
See Mazurenko, Yuri T.: V1403,466-469(1991)
See Meshalkina, M. N.: V1238,195-199(1991)

Smirnova, L. G. See Ivanov, U. V.: V1397,181-184(1991)

Smirnova, S. N.
; Marchenko, S. N.: Display holograms of crystals recording,V1238,370-370(1991)
See Marchenko, S. N.: V1238,371-371(1991)

Smirnova, Tatiana N. See Gulnazarov, Eduard S.: V1238,235-239(1991)

Smit, W. See Frietman, Edward E.: V1390,434-453(1991)

Smith, A.
; Dagli, Cihan H.: Relating binary and continuous problem entropy to backpropagation network architecture,V1469,551-562(1991)

Smith, Alan See Bavdaz, Marcos: V1549,35-44(1991)

Smith, Alan L. See Chaudhri, Mohammad M.: V1358,683-689(1991)

Smith, Andrew D. See Kadin, Alan M.: V1477,156-165(1991)

Smith, Barton A.
; Herminghaus, Stephan; Swalen, Jerome D.: Electro-optic coefficients in electric-field poled-polymer waveguides,V1560,400-405(1991)

Smith, Bruce D.
: Cone beam for medical imaging and NDE,V1450,13-17(1991)

Smith, C. W.
; Constantinescu, D. M.: Intersection of crack borders with free surfaces: an engineering interpretation of optical experimental results,V1554A,102-115(1991)

Smith, Chadwick F. See Vangsness, C. T.: V1424,16-19(1991)

Smith, Craig R. See Auteri, Joseph S.: V1421,182-184(1991)

Smith, Dale J.
; Harrison, Lorna J.; Pellegrino, John M.; Simon, Deborah R.: Performance of high-dynamic-range CCD arrays with various epilayer structures,V1562,242-250(1991)
See Sadler, Brian M.: V1476,246-256(1991)

Smith, David A. See Kunz, Roderick R.: V1466,218-226(1991)

Smith, David C.
; Townsend, Sallie S.: Thermal blooming critical power and adaptive optics correction for the ground-based laser,V1408,112-118(1991)

Smith, David J.
See Boher, Pierre: V1345,165-179(1991)
See Boher, Pierre: V1343,39-55(1991)

Smith, David K. See Figueroa, Luis: V1418,153-176(1991)

Smith, David W. See Barnes, Nigel: V1389,477-483(1991)

Smith, Donald V.
; Smith, Suzy; Cawthon, Michael A.: MDIS (medical diagnostic imaging support) workstation issues: clinical perspective,V1444,357-362(1991)

Smith, Douglas H. See Skeldon, Mark D.: V1410,116-124(1991)

Smith, Duane D. See Gleckler, Anthony D.: V1494,454-471(1991)

Smith, Frederick G. See Cook, Thomas H.: V1488,214-225(1991)

Smith, G. L.
See Chu, Shirley S.: V1361,1020-1025(1991)
See Manalo, Natividad D.: V1493,281-291(1991)
See Manalo, Natividad D.: V1521,106-116(1991)

Smith, Geoffrey B.
; Ng, M. W.; Ditchburn, Robert J.; Martin, Philip J.; Netterfield, Roger P.: Angular-selective cermet films produced from a magnetically filtered cathodic arc,V1536,126-137(1991)
See Bell, J. M.: V1536,29-36(1991)

Smith, George W. See Vaz, Nuno A.: V1455,110-122(1991)

Smith, Graham W. See Ball, K.: V1358,409-420(1991)

Smith, Herb G.
See Baldini, S. E.: V1588,125-131(1991)
See Udd, Eric: V1418,134-152(1991)

Smith, I. D. See Rose, Evan A.: V1411,15-27(1991)

Smith, James A. See Curran, Robert J.: V1492(1991)

Smith, James R. See Laumann, Curt W.: V1414,151-160(1991)

Smith, Jay L.
: WEBERSAT: data analysis and dynamic behavior,V1495,59-70(1991)

Smith, John M.
; Choo, Chang Y.; Nasrabadi, Nasser M.: Terrain acquisition algorithm for an autonomous mobile robot with finite-range sensors,V1468,493-501(1991)

Smith, K. See Sumner, Timothy J.: V1549,256-264(1991)

Smith, Kevin See Golden, Jeffry: V1407,418-429(1991)

Smith, Kevin M.
See Pandey, Ravindra K.: V1426,356-361(1991)
See Shiau, Fuu-Yau: V1426,330-339(1991)

Smith, Louis C.
; Jericevic, Zeljko; Cuellar, Roland; Paddock, Stephen W.; Lewis, Dorothy E.: HIV detection by in-situ hybridization based on confocal reflected light microscopy,V1428,224-232(1991)
See Boggan, James E.: V1428(1991)
See Jericevic, Zeljko: V1428,200-213(1991)

Smith, Mark J.
See Bamberger, Roberto H.: V1605,757-768(1991)
See Kossentini, Faouzi: V1452,383-394(1991)

Smith, Owen I. See Cadou, Christopher P.: V1434,67-77(1991)

Smith, Patrick See Calabrese, Gary C.: V1466,528-537(1991)

Smith, Paul W. See Fairlie, S. A.: V1411,56-62(1991)

Smith, R. C. See Brown, David A.: V1369,2-8(1991)

Smith, Raymond C. See Austin, Roswell W.: V1537,57-73(1991)

Smith, Richard C. See Tiszauer, Detlev H.: V1408,72-83(1991)

Smith, Richard R.
; Benford, James N.; Cooksey, N. J.; Aiello, Norm; Levine, Jerrold S.; Harteneck, Bruce D.: Operation of an L-band relativistic magnetron at 100 hz,V1407,83-91(1991)

Smith, Robert See Kimbrell, James E.: V1482,415-424(1991)

Smith, Robin W. See West, Andrew A.: V1507,158-167(1991)

Smith, Ron See Wheeler, James R.: V1495,280-285(1991)

Smith, Ronald H.
: Reduced cost coil windings for interferometric fiber-optic gyro sensors,V1478,145-149(1991)

Smith, Ronald T.
: Holographic center high-mounted stoplight,V1461,186-198(1991)

Smith, Ross
: Survey of parallel architectures used for three image processing algorithms,V1396,615-623(1991)

Smith, Roy E.
; Milnes, Peter; Edwards, David H.; Mitchell, David C.; Wood, Richard F.: In-line power meter for use during laser angioplasty,V1425,116-121(1991)
See Mitchell, David C.: V1427,181-188(1991)

Smith, S. D. See Craig, Robert G.: V1505,76-86(1991)

Smith, S. P.
; Zarinetchi, F.; Ezekiel, Shaoul: Recent developments in fiber optic ring laser gyros,V1367,103-106(1991)

Smith, Scott D. See Caudell, Thomas P.: V1469,612-621(1991)

Smith, Stephen W. See Pavy, Henry G.: V1443,54-61(1991)

Smith, Steven B. See Bustamante, Carlos: V1435,179-187(1991)

Smith, Stuart
; Grinstein, Georges G.; Pickett, Ronald M.: Global geometric, sound, and color controls for iconographic displays of scientific data,V1459,192-206(1991)

Smith, Suzy See Smith, Donald V.: V1444,357-362(1991)

Smith, T. I. See Swent, Richard L.: V1552,24-35(1991)

Smith, Tab J. See Golden, Jeffry: V1407,418-429(1991)

Smith, Terrance L.
; Misemer, David K.; Attanasio, Daniel V.; Crow, Gretchen L.; Smith, Wiley K.; Swierczek, Mary J.; Watson, James E.: Symmetry properties of reverse-delta-beta directional couplers,V1512,92-100(1991)

Smith, Thomas D. See Ramesh, S. K.: V1474,208-211(1991)

Smith, W. See Acebal, Robert: V1397,191-196(1991)

Smith, W. L. See Yarling, Charles B.: V1393,192-199(1991)

Smith, Walter S.
; Wong, Tony; Williams, Gary L.; Brintle, David G.: Developing a trial burn plan,V1434,14-25(1991)

Smith, Warren E.
; Leung, Billy C.; Leary, James F.: Automatic counting of chromosome fragments for the determination of radiation dose,V1398,142-150(1991)
See Riek, Jonathan K.: V1445,190-197(1991)
See Riek, Jonathan K.: V1398,130-141(1991)
See Riek, Jonathan K.: V1445,198-206(1991)

Smith, Warren J.
: Fundamentals of the optical tolerance budget,V1354,474-481(1991)
See Fischer, Robert E.: V1527(1991)
See Hartmann, Rudolf: VCR38(1991)

Smith, Wiley K. See Smith, Terrance L.: V1512,92-100(1991)

Smith, William M. See Laxer, Cary: V1444,190-195(1991)

Smithson, Robert C. See Acton, D. S.: V1542,159-164(1991)

Smits, Pieter C.
; Post, Mark J.; Velema, Evelyn; Rienks, Rienk; Borst, Cornelius: Coronary and peripheral angioscopy with carbon dioxide gas and saline in animals,V1425,188-190(1991)

Smokelin, John-Scott See Natarajan, Sanjay S.: V1564,474-488(1991)

Smolarz, Andrzej See Wojcik, Waldemar: V1504,292-297(1991)

Smolka, Greg L.
; Strother, George T.; Ott, Carl: Laser applications for multisensor systems,V1498,70-80(1991)

Smolka, M. See Konwerska-Hrabowska, Joanna: V1554B,225-232(1991)

Smotroff, Ira G.
; Friedman, David H.; Connolly, Dennis: Large-scale networks via self-organizing hierarchical networks,V1469,544-550(1991)

Smulevich, G.
; English, A. M.; Spiro, T. G.: Structure and dynamics of the active site of peroxidases as revealed by resonance Raman spectroscopy,V1403,440-447(1991)
; Mantini, A. R.; Casu, M.; Marzocchi, M. P.: Intercalation between antitumor anthracyclines and DNA as probed by resonance and surface-enhanced Raman spectroscopy,V1403,125-127(1991)

Smurlo, Richard
; Laird, Robin T.: Modular robotic architecture,V1388,566-577(1991)

Smutny, Berry See Hildebrand, Ulrich: V1522,50-60(1991)

Snail, Keith A.
: CVD diamond as an optical material for adverse environments,V1330,46-64(1991)
See Hanssen, Leonard M.: V1528,142-150(1991)
See Hickey, Carolyn F.: V1534,67-76(1991)

Sneh, Anat
; Ruschin, Shlomo; Marom, Emanuel: Single-mode fibers to single-mode waveguides coupling with minimum Fresnel back-reflection,V1442,252-257(1991)

Sneiderman, Charles See Cookson, John P.: V1444,374-378(1991)

Snell, Hilary E.
; Hays, Paul B.: Multiplex Fabry-Perot interferometer,V1492,403-407(1991)

Snell, John R.
; Spring, Robert W.: Surveying the elements of successful infrared predictive maintenance programs,V1467,2-10(1991)

Snell, Kevin J.
; Duplain, Gaetan; Parent, Andre; Labranche, Bruno; Galarneau, Pierre: Diffraction-limited Nd:glass and alexandrite lasers using graded reflectivity mirror unstable resonators,V1410,99-106(1991)

Snimshikov, Igor A. See Kutanov, Askar A.: V1507,94-98(1991)

Snorrason, Ogmundur
; Garber, Frederick D.: Formulation and performance evaluation of adaptive, sequential radar-target-recognition algorithms,V1471,116-127(1991)

Snowden, Ian M. See Martin, Brian: V1463,584-594(1991)

Snowdon, John F. See Craig, Robert G.: V1505,76-86(1991)

Snyder, C. D. See MacDonald, Scott A.: V1466,2-12(1991)

Snyder, Donald R.
; Kosel, Frank M.: Applications of high-resolution still video cameras to ballistic imaging,V1346,216-225(1991)
; Rowe, W. J.: Systems analysis and design for next generation high-speed video systems,V1346,226-236(1991)

Snyder, Gerald C. See Gershman, Robert: V1494,160-167(1991)

Snyder, James J.
; Reichert, Patrick: Fast, inexpensive, diffraction-limited cylindrical microlenses,V1544,146-151(1991)

Snyder, John L.
: Signal and background models in nonstandard IR systems,V1498,52-63(1991)

Snyder, Patricia A.
; Munger, Robert; Hansen, Roger W.; Rowe, Ednor M.: Magnetic circular dichroism measurements of benzene in the 9-eV region with synchrotron radiation,V1548,188-196(1991)
See Maestre, Marcos F.: V1548,179-187(1991)

Snyder, Robert F. See Turk, Harris: V1454,344-352(1991)

Snyder, Wesley E. See Logenthiran, Ambalavaner: V1452,225-243(1991)

So, Ikken
; Nakamura, Osamu; Minami, Toshi: Model-based coding of facial images based on facial muscle motion through isodensity maps,V1605,263-272(1991)

So, Vincent See Jiang, Jing-Wen: V1363,191-195(1991)

Soares, Edmundo A.
; Dantas, Tarcisio M.: Bare fiber temperature sensor,V1367,261-265(1991)

Soares, Oliverio D.
ed.: Fiber Optic Metrology and Standards,V1504(1991)
; Bernardo, Luis M.; Pinto, M. I.; Morais, F. V.: Holotag: a novel holographic label,V1332,166-184(1991)

Sobel, Harold R. See Niblack, Curtiss A.: V1494,403-413(1991)

Sobey, Peter J.
: Automated optical grading of timber,V1379,168-179(1991)
: Determining range information from self-motion: the template model,V1382,123-131(1991)

Sobh, Tarek M.
; Alameldin, Tarek K.: Efficient system for 3-D object recognition,V1383,359-366(1991)
; Alameldin, Tarek K.: Operator/system communication: an optimizing decision tool,V1388,524-535(1991)
; Alameldin, Tarek K.: Structure and motion of entire polyhedra,V1388,425-431(1991)
See Alameldin, Tarek K.: V1386,112-120(1991)

Sobhan, M. A.
; Islam, M. N.: Two-dimensional model for high-efficiency microgroove silicon solar cells,V1536,246-257(1991)

Sobolev, A. A. See Averin, V. I.: V1358,589-602(1991)

Sobolev, V. B. See Zaitseva, N. P.: V1402,223-230(1991)

Sobolewski, J. M. See Harris, Neville W.: V1416,59-69(1991)

Sobolewski, Roman
: Applications of high-Tc superconductors in optoelectronics,V1501,14-27(1991); 1502,14-27(1991); 1506,25-38(1991)

Sochor, Vaclav See Sasek, Ladislav: V1572,151-156(1991)

Soddu, Claudio See Barresi, Giangrande: V1495,246-258(1991)

Soel, Michael A. See Konopka, Wayne L.: V1488,355-365(1991)

Soellner, Joerg
; Heuken, Michael; Heime, Klaus: Properties of ZnSe/ZnS grown by MOVPE on a rotating substrate,V1361,963-971(1991)

Soenosawa, Masanobu See Ohfuji, Takeshi: V1463,345-354(1991)

Soffer, Bernard H. See Owechko, Yuri: V1455,136-144(1991)

Soffitta, Paolo
See Costa, Enrico: V1343,469-476(1991)
See Kaaret, Philip E.: V1548,106-117(1991)

Soga, Naohiro See Tanabe, Setsuhisa: V1513,340-348(1991)

Sogabe, Yasushi
; Morimoto, Yoshiharu; Murata, Shigeki: Displacement measurement of track on magnetic tape by Fourier transform grid method,V1554B,289-297(1991)

Soh, W. K.
: High-speed photographic study of a cavitation bubble,V1358,1011-1015(1991)

Soh, Young-Sung
; Murthy, S. N.; Huntsberger, Terrance L.: Development of criteria to compare model-based texture analysis methods,V1381,561-573(1991)

Sohler, Wolfgang
: Rare-earth-doped LiNbO3 waveguide amplifiers and lasers,V1583,110-121(1991)
See Brinkmann, R.: V1362,377-382(1991)
See Seibert, Holger: V1362,370-376(1991)
See Volk, Raimund: V1362,820-826(1991)

Sohlstrom, Hans B.
; Svantesson, Kjell G.: Waveguide-based fiber optic magnetic field sensor with directional sensitivity,V1511,142-148(1991)

Soifer, Victor A.
See Boiko, Yuri B.: V1238,253-257(1991)
See Golub, Mikhail A.: V1500,194-206(1991)
See Golub, Mikhail A.: V1365,156-165(1991)
See Golub, Mikhail A.: V1500,211-221(1991)
See Golub, Mikhail A.: V1572,101-106(1991)
See Golub, Mikhail A.: V1366,273-282(1991)

Soileau, M. J.
; Wei, Tai-Huei; Said, Ali A.; Chapliev, N. I.; Garnov, Sergei V.; Epifanov, Alexandre S.: Comparison of laser-induced damage of optical crystals from the USA and USSR,V1441,10-15(1991)
See Bennett, Harold E.: V1441(1991)
See Sheik-Bahae, Mansoor: V1441,430-443(1991)

Sokolov, K. V.
See Nabiev, I. R.: V1403,85-92(1991)
See Oleynikov, V. A.: V1403,164-166(1991)

Sokolov, Vladimir
: Phase-shifter technology assessment: prospects and applications,V1475,288-302(1991)
See Baldwin, David L.: V1476,46-55(1991)
See Mondal, J. P.: V1475,314-318(1991)
See Swirhun, S.: V1475,303-308(1991)
See Swirhun, S.: V1475,223-230(1991)

Sokolovskaya, Albina I. See Kudryavtseva, Anna D.: V1385,190-199(1991)

Sokolowski, Maciej See Kesik, Jerzy: V1391,34-41(1991)

Sokolowski, Robert S.
: NASA microgravity materials science program,V1557,2-5(1991)

Sol, A. A. See Madarazo, R.: V1393,270-277(1991)

Soldatkin, Alexey E. See Atjezhev, Vladimir V.: V1503,197-199(1991)

Soldatov, N. F. See Degtyareva, V. P.: V1358,524-531(1991)

Solder, Ulrich
; Graefe, Volker: Object detection in real-time,V1388,104-111(1991)

Solehjou, Amin See Shadaram, Mehdi: V1332,487-490(1991)

Solem, Johndale C. See Haddad, Waleed S.: V1448,81-88(1991)

Soli, George See Janesick, James R.: V1447,87-108(1991)

Solimeno, Salvatore See Bruzzese, Riccardo: V1397,735-738(1991)

Solinov, Vladimir F. See Belostotsky, Vladimir I.: V1513,313-318(1991)

Solis, Javier See Vega, Fidel: V1397,807-811(1991)

Solomatin, A. M. See Artjushenko, Vjacheslav G.: V1420,149-156(1991)

Solomon, Christopher J. See Dainty, J. C.: V1487,2-9(1991)

Solomon, Peter R. See Bates, Stephen C.: V1434,28-38(1991)

Solomon, Susan See Mount, George H.: V1491,188-193(1991)

Solone, Paul J. See Zucker, Oved S.: V1378,22-33(1991)

Solonko, Alexander G.
: Investigation of hot electron emission in MOS structure under avalanche conditions,V1435,360-365(1991)

Solorzano, M. See Aviles, Walter A.: V1388,587-597(1991)

Solovjev, Vladimir S. See Boiko, Yuri B.: V1238,253-257(1991)

Solovyev, V. I. See Miroshnikov, Mikhail M.: V1540,496-505(1991)

Solovyov, K. N.
; Gladkov, L. L.: Interpretation of the resonance Raman spectra of hemoproteins and model compounds,V1403,132-133(1991)

Soltan, Hesham See El-Konyaly, Sayed H.: V1388,317-328(1991)

Soltz, David
; Cescato, Lucila H.; Decker, Franco: Photoelectrochemical etching of n-InP producing antireflecting structures for solar cells,V1536,268-276(1991)

Som, Zainal A. See Hassan, Azmi: V1526,142-156(1991)

Somerville, James A. See Kubinec, James J.: V1364,130-143(1991)

Sommar, K. See Taff, Laurence G.: V1481,440-448(1991)

Sommer, Tilman See Zell, Andreas: V1469,708-718(1991)

Sommerfeldt, Scott C. See Mueller, Heinrich G.: V1598,132-140(1991)

Soms, Leonid N. See Mak, Arthur A.: V1415,110-119(1991)

Son, Yung-Sung
; Lee, Hyung-Jae; Yi, Sang-Yoon; Shin, Sang-Yung: Proton-diffused channel waveguides on Y-cut LiNbO3 using a self-aligned SiO2-cap diffusion method,V1374,23-29(1991)

Sona, P.
; Muys, Peter F.; Leys, C.; Sherman, Glenn H.: Variable reflectivity output coupler for improvement of the beam quality of a fast-axial flow CO2 laser,V1397,373-377(1991)
See Leys, C.: V1397,399-402(1991)

Sonajalg, Heiki See Saari, Peeter M.: V1507,500-508(1991)

Sonek, Gregory J. See Wright, William H.: V1427,279-287(1991)

Song, Dehui See Wang, Erqi: V1400,124-128(1991)

Song, Hongjun
; Zhou, Lijia: Research and improvement of Vogl's acceleration algorithm,V1469,581-584(1991)
See Zhou, Lijia: V1469,404-411(1991)

Song, J. F. See Marx, Egon: V1332,826-834(1991)

Song, Jong I. See Fossum, Eric R.: V1447,202-203(1991)

Song, Joshua J. See Rahman, Hafiz-ur: V1407,589-597(1991)

Song, L. W. See Witanachchi, S.: V1394,161-168(1991)

Song, Li
See Malouin, Christian: V1559,385-392(1991)
See Sheng, Yunlong: V1559,222-228(1991)

Song, Lu See Schurr, J. M.: V1403,248-257(1991)

Song, Pill-Soon
: Picosecond time-resolved fluorescence spectroscopy of phytochrome and stentorin,V1403,590-599(1991)

Song, Q. W.
; Lee, Mowchen C.; Talbot, Peter J.: Holographic optical switching with photorefractive crystals,V1558,143-148(1991)

Song, Ruhua H.
See Xiao, Guohua: V1409,106-113(1991)
See Yang, Darang: V1417,440-450(1991)

Song, S. N. See Rippert, Edward D.: V1549,283-288(1991)

Song, Sewoong See Mahalanobis, Abhijit: V1564,14-21(1991)

Song, So-Young
See Ahn, Jae W.: V1358,269-277(1991)
See Lee, Eun Soo: V1358,262-268(1991)

Song, W. S. See Liu, Rui-Fu: V1572,189-191(1991)

Song, Xiang
See Wang, He-Chen: V1386,273-276(1991)
See Wang, He-Chen: V1384,133-136(1991)

Song, Xiang Y.
See Fan, Ru Y.: V1519,134-137(1991)
See Ma, Ke J.: V1519,489-493(1991)

Song, Yi Z. See Zhang, Wei P.: V1519,23-25(1991)

Song, Yuanxu Y.
; Fisher, Amnon; Prohaska, Robert M.; Rostoker, Norman: Electron trapping and acceleration in a modified elongated betatron,V1407,430-441(1991)

Song, Zhengfang
; Ma, Jun: Correlation of bifrequency beam propagating in a folded turbulent path,V1487,382-386(1991)
See Feng, Yue-Zhong: V1417,370-372(1991)
See Feng, Yue-Zhong: V1487,356-360(1991)
See Ruizhong, Rao: V1486,390-395(1991)

Soni, Tarun
; Zeidler, James R.; Rao, Bhaskar D.; Ku, Walter H.: Signal enhancement in noise- and clutter-corrupted images using adaptive predictive filtering techniques,V1565,338-344(1991)

Sontag, Andre
; Baronnet, Jean M.: Arc heater for thermal driven HF/DF chemical laser, V1397,287-290(1991)
See Joeckle, Rene C.: V1397,683-687(1991)
See Joeckle, Rene C.: V1397,679-682(1991)

Sontag, Heinz
; Johann, Ulrich A.; Pribil, Klaus: Design of a diode-pumped Nd:YAG laser communication system, V1417,573-587(1991)
; Johann, Ulrich A.; Pribil, Klaus: Second-generation high-data-rate interorbit link based on diode-pumped Nd:YAG laser technology, V1522,61-69(1991)
See Johann, Ulrich A.: V1522,243-252(1991)
See Pribil, Klaus: V1522,36-47(1991)
See Walker, Karl-Heinz: V1510,212-217(1991)

Sood, Arun K.
See Al-Hujazi, Ezzet H.: V1381,589-599(1991)
See Karne, Ramesh K.: V1468,938-949(1991)
See Kjell, Bradley P.: V1468,148-155(1991)

Soole, Julian B.
; Scherer, Axel; LeBlanc, Herve P.; Andreadakis, Nicholas C.; Bhat, Rajaram; Koza, M. A.: Spectrometer on a chip: InP-based integrated grating spectrograph for wavelength-multiplexed optical processing, V1474,268-276(1991)

Sooriyakumaran, Ratna
; Ito, Hiroshi; Mash, Eugene A.: Acid-catalyzed pinacol rearrangement: chemically amplified reverse polarity change, V1466,419-428(1991)

Soos, Jolanta I.
See Ling, Junda D.: V1454,353-362(1991)
See Turk, Harris: V1454,344-352(1991)

Soper, Robert A. See Crisalle, Oscar D.: V1464,508-526(1991)

Soper, Steven A.
; Davis, Lloyd M.; Fairfield, Frederick R.; Hammond, Mark L.; Harger, Carol A.; Jett, James H.; Keller, Richard A.; Morrone, Barbara L.; Martin, John C.; Nutter, Harvey L.; Shera, E. B.; Simpson, Daniel J.: Rapid sequencing of DNA based on single-molecule detection, V1435,168-178(1991)

Sopko, Victor
: Discussion of the standard practice for the location of wet insulation in roofing systems using infrared imaging (ASTM C1153-90), V1467,83-89(1991)

Sorbello, Robert M.
; Zaghloul, A. I.; Gupta, R. K.; Geller, B. D.; Assal, F. T.; Potukuchi, J. R.: MMIC: a key technology for future communications satellite antennas, V1475,175-183(1991)

Sorbier, J. P. See Mathey, Yves: V1361,909-916(1991)

Sordo, Bruno See Cognolato, Livio: V1579,249-256(1991)

Soref, Richard A.
; Namavar, Fereydoon; Cortesi, Elisabetta; Friedman, Lionel R.; Lareau, Richard: Vertical 3-D integration of silicon waveguides in a Si-SiO2-Si-SiO2-Si structure, V1389,408-421(1991)

Soreide, David C.
; McGarvey, John A.: Laser velocimetry applications, V1416,280-285(1991)

Sorlie, Paul H. See Clark, David L.: V1540,303-311(1991)

Soroka, A. M. See Kamrukov, A. S.: V1503,438-452(1991)

Soroka, Jacek A. See Kotowski, Tomasz: V1391,6-11(1991)

Soroka, Krystyna B. See Kotowski, Tomasz: V1391,6-11(1991)

Soroka, V. I. See Artsimovich, M. V.: V1397,729-733(1991)

Sorokach, S. K. See Fripp, Archibald L.: V1557,236-244(1991)

Sorokin, Svjatoslav N.
See Danilov, Alexander A.: V1362,916-920(1991)
See Danilov, Alexander A.: V1362,647-654(1991)

Sorokina, Svetlana
; Dikov, Juriy: Structural investigations of the (Si1-x,Gex)O2 single-crystal thin films by x-ray photoelectron spectroscopy, V1519,128-133(1991)

Soskey, Paul R. See Banash, Mark A.: V1590,8-13(1991)

Soskin, M. S. See Bazhenov, V. Y.: V1574,148-153(1991)

Sotomayor, Jorge L.
See Beheim, Glenn: V1584,64-71(1991)
See Beheim, Glenn: V1369,50-59(1991)

Sotomayor Torres, Clivia M. See MacLeod, Roderick W.: V1361,562-567(1991)

Soucy, L. See Brule, Michel J.: V1388,432-441(1991)

Souilhac, Dominique J.
; Billerey, Dominique: Comparison of the angular resolution limit and SNR of the Hubble Space Telescope and the large ground-based telescopes, V1494,503-526(1991)
; Billerey, Dominique: Infrared tellurium two-dimensional acousto-optic processor for synthetic aperture radar, V1521,158-174(1991)
See Targ, Russell: V1521,144-157(1991)

Soules, David B.
See Eaton, Frank D.: V1487,84-90(1991)
See Waldie, Arthur H.: V1487,103-108(1991)

Soumekh, Mehrdad
: Scan-free echo imaging of dynamic objects, V1443,96-106(1991)
; Yang, H.: Complex phase error and motion estimation in synthetic aperture radar imaging, V1452,104-113(1991)

Sounder, Charles See Peacock, Keith: V1494,147-159(1991)

Sounik, James R.
; Rihter, Boris D.; Ford, William E.; Rodgers, Michael A.; Kenney, Malcolm E.: Naphthalocyanines relevant to the search for second-generation PDT sensitizers, V1426,340-349(1991)
See Norwood, Robert A.: V1560,54-65(1991)

Soures, John M.
ed.: Selected Papers on High Power Lasers, VMS43(1991)

Sousa, Michael J. See Games, Richard A.: V1566,323-328(1991)

Southwell, William H.
: Binary optics from a ray-tracing point of view, V1354,38-42(1991)
See Motamedi, M. E.: V1544,33-44(1991)

Souza, Rui F. See Saad, Ricardo E.: V1371,142-148(1991)

Sowada, Ulrich See Basting, Dirk: V1412,80-83(1991)

Spaar, M. T. See Arlinghaus, Heinrich F.: V1435,26-35(1991)

Spadafora, J. See Olsen, Gregory H.: V1419,24-31(1991)

Spain, Edward H.
; Holzhausen, Klause-Peter: Stereoscopic versus orthogonal view displays for performance of a remote manipulation task, V1457,103-110(1991)

Spal, Richard See Steiner, Bruce W.: V1557,156-167(1991)

Spalek, G. See Farkas, Zoltan D.: V1407,502-511(1991)

Spammer, Stephanus J. See Booysen, Andre: V1584,273-279(1991)

Spangler, Charles W.
; Havelka, Kathleen O.; Becker, Mark W.; Kelleher, Tracy A.; Cheng, Lap Tak A.: Relationship between conjugation length and third-order nonlinearity in bis-donor substituted diphenyl polyenes, V1560,139-147(1991)
; Saindon, Michelle L.; Nickel, Eric G.; Sapochak, Linda S.; Polis, David W.; Dalton, Larry R.; Norwood, Robert A.: Synthesis and incorporation of ladder polymer subunits in copolyamides, pendant polymers, and composites for enhanced nonlinear optical response, V1497,408-417(1991)
See Havelka, Kathleen O.: V1560,66-74(1991)
See Nickel, Eric G.: V1560,35-43(1991)

Spannenburg, S.
: Frequency modulation of printed gratings as a protection against copying, V1509,88-104(1991)

Spanos, Costas J. See Ling, Zhi-Min: V1392,660-669(1991)

Spanos, John T. See O'Neal, Michael C.: V1542,359-370(1991)

Sparks, Adrian P.
See Crossland, William A.: V1455,264-273(1991)
See Underwood, Ian: V1562,107-115(1991)

Sparks, W. B. See Greenfield, Perry E.: V1494,16-39(1991)

Spassova, E. See Guttmann, Peter: V1361,999-1010(1991)

Spatscheck, Thomas
See Bopp, Matthias: V1522,199-209(1991)
See Greulich, Peter: V1417,358-369(1991)
See Greulich, Peter: V1522,144-153(1991)

Spaulding, Kevin E.
; Morris, G. M.: Achromatization of optical waveguide components, V1507,45-54(1991)

Spears, J. R.
See Kundu, Sourav K.: V1425,142-148(1991)
See McMath, Linda P.: V1425,165-171(1991)

Spears, Kenneth G.
; Kume, Stewart M.; Winakur, Eric: Imaging inside scattering media: chronocoherent imaging, V1429,2-8(1991)
See VanderMeulen, David L.: V1428,84-90(1991)
See VanderMeulen, David L.: V1428,91-98(1991)
See Yang, Xueyu: V1367,382-386(1991)

Speck, David R. See Henesian, Mark A.: V1415,90-103(1991)

Specker, Harald See Bopp, Matthias: V1522,199-209(1991)

Specter, Christine N. See Raney, R. K.: V1492,298-306(1991)

Spector, Magaly See Kanofsky, Alvin S.: V1374,48-58(1991)

Speer, R. J. See Homer, Michael: V1527,134-144(1991)

Spehalski, Richard J.
; Werner, Michael J.: Objectives for the Space Infrared Telescope Facility,V1540,2-14(1991)

Speiser, Shammai
: Model systems for optoelectronic devices based on nonlinear molecular absorption,V1560,434-442(1991)
: Nonlinear optical properties of phenosafranin-doped substrates,V1559,238-244(1991)

Speit, Burkhard
; Remitz, K. E.; Neuroth, Norbert N.: Semiconductor-doped glass as a nonlinear material,V1361,1128-1131(1991)
See Marker, Alexander J.: V1485,160-172(1991)

Spence, A. J. See Boardman, A. D.: V1503,160-166(1991)

Spence, Christopher A.
; Ferguson, Richard A.: Some experimental techniques for characterizing photoresists,V1466,324-335(1991)
See Dao, T. T.: V1466,257-268(1991)

Spence, Clay D.
; Pearson, John C.; Sverdlove, Ronald: Artificial neural networks as TV signal processors,V1469,665-670(1991)

Spence, G. See Rimmele, Thomas: V1542,186-193(1991)

Spence, Paul A. See Schmitt, Randal L.: V1410,55-64(1991)

Spence, Rodney L.
: Comparison of optical technologies for a high-data-rate Mars link,V1417,550-561(1991)

Spence, Scott E.
; Reyes, Roy M.: Algorithms and architectures for performing Boolean equations using self electro-optic effect devices,V1474,199-207(1991)
See Brown, Gair D.: V1366,351-360(1991)

Spencer, Allen C. See Hartney, Mark A.: V1466,238-247(1991)

Spencer, Kevin M. See Carrabba, Michael M.: V1434,127-134(1991)

Spencer, Paul E.
; Zaharakis, Steven C.; Denton, Richard T.: Fault management of a fiber optic LAN,V1364,228-234(1991)

Spencer, Robert A. See Karbassiyoon, Kamran: V1366,178-183(1991)

Spencer, Thomas A. See Menge, Peter R.: V1407,578-588(1991)

Sperling, I.
; Drummond, Oliver E.; Reed, Irving S.: Hexagonal sampling and filtering for target detection with a scanning E-O sensor,V1481,2-11(1991)

Sperling, L. H. See Herman, Warren N.: V1560,206-213(1991)

Speser, Philip L.
: U.S. government market,V1520,132-150(1991)

Speyer, Brian A.
: High-speed still video photography for ballistic range applications,V1358,1215-1221(1991)

Spiccia, L. See Bell, J. M.: V1536,29-36(1991)

Spicer, James B. See Wagner, James W.: V1332,491-501(1991)

Spichiger, Ursula E.
; Seiler, Kurt; Wang, Kemin; Suter, Gaby; Morf, Werner E.; Simon, Wilhelm: Optical quantification of sodium, potassium, and calcium ions in diluted human plasma based on ion-selective liquid membranes,V1510,118-130(1991)

Spiegelberg, Christine
; Henneberger, Fritz; Puls, J.: Spectral hole burning of strongly confined CdSe quantum dots,V1362,951-958(1991)

Spiero, Francois See Garcia-Prieto, Rafael: V1522,267-276(1991)

Spiers, Gary D. See Jaenisch, Holger M.: V1411,127-136(1991)

Spiess, Walter See Przybilla, Klaus J.: V1466,174-187(1991)

Spik, Andrzej See Kujawinska, Malgorzata: V1391,303-312(1991)

Spiller, Eberhard
; McCorkle, R.; Wilczynski, Janusz S.; Golub, Leon; Nystrom, George U.; Takacs, Peter Z.; Welch, Charles W.: Imaging performance and tests of soft x-ray telescopes,V1343,134-144(1991)

Spillman, William B.
; Fuhr, Peter L.: Noncontact technique for the measurement of linear displacement using chirped diffraction gratings,V1332,591-601(1991)
; Rudd, Robert E.; Hoff, Frederick G.; Patriquin, Douglas R.; Lord, Jeffrey R.: Wavelength-encoded fiber optic angular displacement sensor,V1367,197-203(1991)

Spinar, Karen K. See Brand, Richard A.: V1532,57-63(1991)

Spinelli, N. See Bruzzese, Riccardo: V1397,735-738(1991)

Spinhirne, James D. See Theon, John S.: V1492,2-23(1991)

Spinillo, Gary See Conley, Willard E.: V1466,53-66(1991)

Spinrad, Richard W.
ed.: *Underwater Imaging, Photography, and Visibility,*V1537(1991)

Spirit, David M.
; Brown, Graeme N.; Marshall, Ian W.; Blank, Lutz C.: Applications of fiber amplifiers to high-data-rate nonlinear transmission,V1373,197-208(1991)

Spiro, A. G. See Bogdanov, V. L.: V1403,470-474(1991)

Spiro, Irving J. See Andresen, Bjorn F.: V1540(1991)

Spiro, T. G. See Smulevich, G.: V1403,440-447(1991)

Spitzer, Ronnie
See Ermer, Susan P.: V1560,120-129(1991)
See Swanson, David A.: V1560,416-425(1991)

Spizarny, David
See Chen, Ji: V1444,256-264(1991)
See Flynn, Michael J.: V1444,172-179(1991)

Spizzichino, A. See Di Cocco, Guido: V1549,102-112(1991)

Splett, Armin O.
; Schmidtchen, Joachim; Schueppert, B.; Petermann, Klaus: Integrated optical channel waveguides in silicon using SiGe alloys,V1362,827-833(1991)

Spofford, John R. See Morgenthaler, Matthew K.: V1387,82-95(1991)

Spooner, N. J. See Sumner, Timothy J.: V1549,256-264(1991)

Spooner, Susan P.
; Whitten, David G.: Photochemistry and photophysics of stilbene and diphenylpolyene surfactants in supported multilayer films,V1436,82-91(1991)

Spooren, Rudie
: Fringe quality in pulsed TV-holography,V1508,118-127(1991)

Sporea, Dan G. See Miron, Nicolae: V1500,334-338(1991)

Sporken, R.
; Lange, M. D.; Faurie, Jean-Pierre: Molecular beam epitaxy of CdTe and HgCdTe on large-area Si(100),V1512,155-163(1991)

Spornik, Nikolai M.
; Serenko, M. Y.: Shadow method of scale with a-posteriori increase of measurements sensitivity,V1508,162-169(1991)
; Tujev, A. F.: Universal interferometer with synthesized reference wave,V1508,83-88(1991)

Spowage, K. See Klenerman, David: V1437,95-102(1991)

Spragg, J. E. See Wells, Alan A.: V1549,357-373(1991)

Spragg, Peggy M.
; Hurditch, Rodney J.; Toukhy, Medhat A.; Helbert, John N.; Malhotra, Sandeep: Reliability of contrast and dissolution-rate-derived parameters as predictors of photoresist performance,V1466,283-296(1991)
See Hansen, Steven G.: V1463,230-244(1991)

Spraggins, Thomas See Henning, Michael R.: V1420,34-40(1991)

Sprangers, Rene L. See Gijsbers, Geert H.: V1425,94-101(1991)

Sprangle, Phillip
ed.: *Short-Wavelength Radiation Sources,*V1552(1991)
; Esarey, Eric: Stimulated relativistic harmonic generation from intense laser interactions with beams and plasmas,V1552,147-153(1991)
See Esarey, Eric: V1407,407-417(1991)
See Hafizi, Bahman: V1407,316-321(1991)

Spratt, D. B. See Jung, K. H.: V1393,240-251(1991)

Spreeuwers, Luuk J.
: Neural network edge detector,V1451,204-215(1991)

Spring, Robert W. See Snell, John R.: V1467,2-10(1991)

Springer, John M.
; Silberman, Enrique; Kroes, Roger L.; Reiss, Don: Mapping crystal defects with a digital scanning ultramicroscope,V1557,192-196(1991)

Springer, Steven C.
; McLean, Craig: Overview of U.S. Fishery requirements,V1479,372-379(1991)

Springsteen, Arthur W. See Bruegge, Carol J.: V1493,132-142(1991)

Spurgeon, Thomas L. See McCracken, Thomas O.: V1380,6-10(1991)

Spyra, Wolfgang J. See Higgins, William E.: V1445,276-286(1991)

Spytkowski, Wojciech See Wolinski, Wieslaw: V1391,334-340(1991)

Sreseli, Olga M.
; Belyakov, Ludvig V.; Goryachev, D. N.; Rumyantsev, B. L.; Yaroshetskii, Ilya D.: Double-resonant tunneling via surface plasmons in a metal-semiconductor system,V1440,326-331(1991)
; Belyakov, Ludvig V.; Goryachev, D. N.; Rumyantsev, B. L.; Yaroshetskii, Ilya D.: Double-resonant tunneling via surface plasmons in layered gratings,V1545,149-158(1991)

Sricharoenchaikit, Prasit See Calabrese, Gary C.: V1466,528-537(1991)

Sridhar, Banavar
; Phatak, Anil; Chatterji, Gano B.: Passive range sensor refinement using texture and segmentation,V1478,178-189(1991)
See Menon, P. K.: V1478,190-200(1991)
See Suorsa, Raymond E.: V1388,90-103(1991)

Srikantan, Geetha
; Sher, David B.; Newberger, Ed: Efficient extraction of local myocardial motion with optical flow and a resolution hierarchy,V1459,258-267(1991)

Srikanth, Usha
; Sundararajan, Srikanth: Photolith analysis and control system,V1468,429-433(1991)

Srivastava, Ramakant
; Ramaswamy, Ramu V.: Ion-exchanged waveguides: current status,V1583,2-13(1991)

Sroka, Ronald See Gottschalk, Wolfgang: V1427,320-326(1991)

St.-Hilaire, Pierre
; Benton, Stephen A.; Lucente, Mark; Underkoffler, John S.; Yoshikawa, Hiroshi: Real-time holographic display: improvements using a multichannel acousto-optic modulator and holographic optical elements,V1461,254-261(1991)

St. Clair Dinger, Ann See Elliott, David G.: V1540,63-67(1991)

St. George, Dennis R.
; Feddes, John J.: Fiber optic lighting system for plant production,V1379,69-80(1991)

Staab, Edward V.
; Honeyman, Janice C.; Frost, Meryll M.; Bidgood, W. D.: Teleradiology in the local environment,V1446,16-22(1991)
See Honeyman, Janice C.: V1446,362-368(1991)

Stabile, Paul J.
See Ayekavadi, Raj: V1418,74-85(1991)
See Evans, Gary A.: V1378,146-161(1991)
See Loubriel, Guillermo M.: V1378,179-186(1991)
See Rosen, Arye: V1378,187-194(1991)

Stacewicz, Tadeusz See Kotowski, Tomasz: V1391,6-11(1991)

Stacy, John L.
See Goel, Kamal K.: V1476,314-319(1991)
See Prucnal, Paul R.: V1389,462-476(1991)

Stafeev, V. I. See Khryapov, V. T.: V1540,412-423(1991)

Stager, Charles See Kemp, Kevin G.: V1464,260-266(1991)

Staggs, Michael C. See Kozlowski, Mark R.: V1441,269-282(1991)

Stahl, Charles G. See Ivey, Jim: V1567,170-178(1991)

Stahl, David B.
See Paisley, Dennis L.: V1358,760-765(1991)
See Paisley, Dennis L.: V1346,172-178(1991)

Stahl, H. P.
: Aspheric surface testing techniques,V1332,66-76(1991)
: Review of phase-measuring interferometry,V1332,704-719(1991)
: White-light moire phase-measuring interferometry,V1332,720-730(1991)

Stahlberg, Larry See Rosenthal, Steven: V1386,158-162(1991)

Stalker, K. T.
See Connelly, James M.: V1564,572-592(1991)
See Hendrix, Charles D.: V1564,2-13(1991)

Stallard, Brian R. See Hickman, Kirt C.: V1464,245-257(1991)

Staller, Craig O.
; Niblack, Curtiss A.; Evans, Thomas G.; Blessinger, Michael A.; Westrick, Anthony: Infrared focal-plane design for the comet rendezvous/asteroid flyby and Cassini visible and infrared mapping spectrometers,V1540,219-230(1991)
See Niblack, Curtiss A.: V1494,403-413(1991)

Stallings, Casson See Brockhaus, John A.: V1492,200-205(1991)

Stallmann, Dirk See Gruen, Armin: V1526,42-55(1991)

Stamm, A. See Fichelscher, Andreas: V1392,77-83(1991)

Stamm, Uwe
See Bergner, Harald: V1362,484-493(1991)
See Bergner, Harald: V1361,723-731(1991)
See Damm, Tobias: V1403,686-694(1991)
See Zschocke, Wolfgang: V1501,49-56(1991)

Stancil, Daniel D. See Opsasnick, Michael N.: V1499,276-280(1991)

Standley, David L.
; Horn, Berthold K.: Analog CMOS IC for object position and orientation,V1473,194-201(1991)

Stanion, Ken See Kozlowski, Mark R.: V1561,59-69(1991)

Staniszewski, A. See Gilbert, Barry K.: V1390,235-248(1991)

Stankiewicz, Maria
; Zachorowski, Jan: Simulation of the process of laser beam machining based on the model of surface vaporization,V1391,174-180(1991)

Stankovich, Marion T. See Cotton, Therese M.: V1403,93-101(1991)

Stanley, C. R. See MacLeod, Roderick W.: V1361,562-567(1991)

Stanley, Pamela S. See Hough, Stewart E.: V1456,240-249(1991)

Stano, Alessandro See Caldera, Claudio: V1372,82-87(1991)

Stanton, Alan C. See Silver, Joel A.: V1435,64-71(1991)

Stanton, Martin J.
See Kalata, Kenneth: V1345,270-280(1991)
See Strauss, Michael G.: V1447,12-27(1991)

Stanton, Philip L. See Sweatt, William C.: V1346,151-159(1991)

Starchevsky, A. V.
See Savostjanov, V. N.: V1554A,380-386(1991)
See Taratorin, B. I.: V1554A,449-456(1991)

Stark, Louise
; Hall, Lawrence O.; Bowyer, Kevin W.: Investigation of methods of combining functional evidence for 3-D object recognition,V1381,334-345(1991)

Stark, Margaret A. See Dosser, Larry R.: V1346,293-299(1991)

Stark, Robert A.
; Chen, H. C.; Uhm, Han S.: Simulation studies of the relativistic magnetron,V1407,128-138(1991)
See Chen, H. C.: V1407,139-146(1991)
See Uhm, Han S.: V1407,113-127(1991)

Starkovsky, A. N. See Manenkov, Alexander A.: V1420,254-258(1991)

Starks, Michael R.
: Stereoscopic video and the quest for virtual reality: an annotated bibliography of selected topics,V1457,327-342(1991)

Starks, Scott A.
See Cooke, Daniel E.: V1381,299-305(1991)
See Gonzalez, Orlando: V1381,284-291(1991)

Starkweather, Gary K.
: Technology trends in electrophotographic printers,V1458,120-127(1991)

Starobogatov, Igor O.
; Nicolaev, S. D.: Research of fast stages of latent image formation in holographic photoemulsions influenced by ultrashort radiation pulses,V1238,153-157(1991)

Staselko, D. I. See Bryskin, V. Z.: V1238,448-451(1991)

Stasenko, A. L. See Guryashkin, L. P.: V1358,974-978(1991)

Stasicki, Boleslaw
; Meier, G. E.: Miniaturized semiconductor light source system for Cranz-Schardin applications,V1358,1222-1227(1991)

Stassinopoulos, E. See Barnes, Charles E.: V1366,9-16(1991)

Stathis, J. H. See Kozlowski, Mark R.: V1441,269-282(1991)

Stationwala, Mustafa I.
; Miller, Charles E.; Atack, Douglas; Karnis, A.: Application of high-speed photography to chip refining,V1358,237-245(1991)

Staubert, R. See Braueninger, Heinrich: V1549,330-339(1991)

Stauffer, Donald R. See Rouse, Gordon F.: V1495,240-245(1991)

Stavroudis, Orestes N.
: Ellipsoidal mirrors as modular elements in the design of off-axis optical systems,V1354,627-646(1991)

Stayt, J. W. See Levine, Barry F.: V1540,232-237(1991)

Stcherbakov, V. N. See Taratorin, B. I.: V1554A,449-456(1991)

Stchetinin, A. L. See Shvej, E. M.: V1554A,488-495(1991)

St Clair, T. See Morgan, Alan R.: V1426,350-355(1991)

Stearns, Daniel G.
; Ceglio, Natale M.; Hawryluk, Andrew M.; Rosen, Robert S.; Vernon, Stephen P.: Multilayer optics for soft x-ray projection lithography: problems and prospects,V1465,80-87(1991)
See Landen, Otto L.: V1413,120-130(1991)

Stechman, John M. See Smeins, Larry G.: V1549,59-65(1991)

Steckenrider, John S.
; Ehrlich, Michael J.; Wagner, James W.: Pulsed holographic recording of very high speed transient events,V1554B,106-112(1991)

Steel, William H. See James, William E.: V1400,129-136(1991)

Steele, Ann F.
: Application of high-speed photography in the design of new initiating systems,V1358,1123-1133(1991)

Steele, James L. See Zayas, Inna Y.: V1379,151-161(1991)

Steele, Robert E.
See Bulusu, Dutt V.: V1364,49-60(1991)
See Cirillo, James R.: V1592,53-59(1991)
See Cirillo, James R.: V1592,42-52(1991)
See Kitazawa, Mototaka: V1592(1991)

Steele, Roger C.
: Overview of coherent lightwave communications,V1372,2-3(1991)
; Sunak, Harish R.; eds.: *Coherent Lightwave Communications: Fifth in a Series*,V1372(1991)
; Walker, Nigel G.: Coherent optical transmission systems with optical amplifiers,V1372,173-187(1991)

Steen, William M.
See Chen, Shang-Liang: V1502,123-134(1991)
See Fellowes, Fiona C.: V1502,213-222(1991)

Steenvoorden, G. K. See Nieuwkoop, E.: V1511,255-263(1991)

Stefanovsky, Yu. See Angelov, D.: V1403,230-239(1991)

Stegeman, George I. See Neher, Dieter: V1560,335-343(1991)

Steger, Adrian C. See McNicholas, Thomas A.: V1421,30-35(1991)

Stehle, Marc X.
See Godard, Bruno: V1503,71-77(1991)
See Godard, Bruno: V1397,59-62(1991)

Steier, William H.
See Chen, Mai: V1409,202-213(1991)
See Shi, Youngqiang: V1559,118-126(1991)

Steiger, Erwin
: New concept of a compact multiwavelength solid state laser for laser-induced shock wave lithotripsy,V1421,140-146(1991)

Stein, Jay A. See Shepp, Larry A.: V1452,216-224(1991)

Steinberg, Steffen See Rasch, Andreas: V1522,83-92(1991)

Steinbichler, Hans
: New options of holographic metrology,V1507,435-445(1991)

Steiner, Bruce W.
; Dobbyn, Ronald C.; Black, David; Burdette, Harold; Kuriyama, Masao; Spal, Richard; van den Berg, Lodewijk; Fripp, Archibald L.; Simchick, Richard T.; Lal, Ravindra B.; Batra, Ashok K.; Matthiesen, David; Ditchek, Brian M.: High-resolution synchrotron x-radiation diffraction imaging of crystals grown in microgravity and closely related terrestrial crystals,V1557,156-167(1991)

Steiner, Rudolf W.
; Meier, Thomas H.: Dual wavelengths (750/375 nm) laser lithotripsy,V1421,124-128(1991)
See Watson, Graham M.: V1421(1991)

Steinhardt, Allan O. See Onn, Ruth: V1566,427-438(1991)

Steinmetz, Lloyd L. See McMillan, Charles F.: V1346,113-121(1991)

Steinvorth, Rodrigo
; Kaufman, Howard; Neat, Gregory W.: Model reference adaptive control of flexible robots in the presence of sudden load changes,V1387,136-147(1991)

Steliou, Kypros See Ishaq, Naseem: V1381,153-159(1991)

Stella, Ettore
See Attolico, Giovanni: V1383,34-46(1991)
See Distante, Arcangelo: V1381,513-523(1991)

Stelmach, Lew B.
; Tam, Wa J.; Hearty, Paul J.: Static and dynamic spatial resolution in image coding: an investigation of eye movements,V1453,147-152(1991)

Stenger, Anthony J. See Keller, Catherine E.: V1486,278-285(1991)

Stenger-Smith, John D.
See Lindsay, Geoffrey A.: V1497,418-422(1991)
See Lindsay, Geoffrey A.: V1560,443-453(1991)

Stening, R. L. See Hora, Heinrich: V1502,258-269(1991)

Stenow, Eric
; Rohman, H.; Eriksson, L.-E.; Oberg, Per A.: Venous occlusion plethysmography based on fiber-optic sensor using the microbending principle,V1420,29-33(1991)

Stenram, U. See Andersson-Engels, Stefan: V1426,31-43(1991)

Stenstrom, J. R.
; Connolly, C. I.: Automatic and operator-assisted solid modeling of objects for automatic recognition,V1470,275-281(1991)

Stentz, Anthony See Gowdy, Jay W.: V1388,131-140(1991)

Stenzel, Olaf
; Schaarschmidt, Guenther; Roth, Sylvia; Schmidt, Guenther; Scharff, Wolfram: Optical properties of amorphous hydrogenated carbon layers,V1534,148-157(1991)

Stepanek, Josef See Praus, Petr: V1403,76-84(1991)

Stepanek, Petr See Sedlak, Marian: V1430,191-202(1991)

Stepanian, A. S.
See Firsov, N. N.: V1403,350-354(1991)
See Romanovsky, Yuri M.: V1403,359-362(1991)

Stepanov, Boris M. See Ignatosyan, S. S.: V1358,100-108(1991)

Stepanov, D. Y.
See Dianov, Evgeni M.: V1516,81-98(1991)
See Dianov, Evgeni M.: V1516,75-80(1991)

Stepanov, Eugene V.
See Moskalenko, Konstantin L.: V1426,121-132(1991)
See Nadezhdinskii, Alexander I.: V1418,487-495(1991)
See Nadezhdinskii, Alexander I.: V1433,202-210(1991)

Stepanov, M. S.
See Borodin, A. M.: V1358,756-758(1991)
See Ivanov, K. N.: V1358,732-738(1991)

Stepanov, Yu. Y. See Borisov, V. M.: V1503,40-47(1991)

Stepanov, Yuri D. See Atjezhev, Vladimir V.: V1503,197-199(1991)

Stephanou, Harry E. See Kountouris, Vasilios G.: V1387,169-180(1991)

Stephens, Graeme L. See Gabriel, Philip: V1558,76-90(1991)

Stephens, Peter See Eng, Peter J.: V1550,110-121(1991)

Stepien, Ryszard See Romaniuk, Ryszard S.: V1368,73-84(1991)

Stepp, Herbert See Baumgartner, R.: V1525,246-248(1991)

Stepp, Larry M.
; Roddier, Nicolas; Dryden, David M.; Cho, Myung K.: Active optics system for a 3.5-meter structured mirror,V1542,175-185(1991)

Sterenborg, H. J.
: Theoretical analysis of stone fragmentation rates,V1525,170-176(1991)
; van Swol, Christiaan F.; Biganic, Dane D.; van Gemert, Martin J.: Laser lithotripsy with a pulsed dye laser: correlation between threshold energy and optical properties,V1427,256-266(1991)

Sterian, Livia See Dumitru, Mihaela A.: V1500,339-348(1991)

Sterian, Paul E.
See Podoleanu, Adrian G.: V1362,721-726(1991)
See Popescu, Ion M.: V1361,1041-1047(1991)

Sterman, Baruch See Yatsiv, Shaul: V1397,319-329(1991)

Stern, Adin See Biderman, Shlomo: V1535,27-34(1991)

Stern, Jill A.
: Small satellites: current legal issues in remote sensing,V1495,42-51(1991)

Stern, Margaret B. See Holz, Michael: V1544,75-89(1991)

Stern, Raul A. See Bacal, Marthe: V1407,605-609(1991)

Stern, Scott J. See Flock, Stephen T.: V1427,77-89(1991)

Stern, Thomas E. See Bala, Krishna: V1579,62-73(1991)

Sternad, Mikael See Ahlen, Anders: V1565,130-142(1991)

Sternberg, Stanley See Bhagvati, Chakravarthy: V1606,112-119(1991)

Stetz, Mark L.
; O'Brien, Kenneth M.; Scott, John J.; Baker, Glenn S.; Deckelbaum, Lawrence I.: Optical and mechanical parameter detection of calcified plaque for laser angioplasty,V1425,55-62(1991)
See Pradhan, Asima: V1425,2-5(1991)

Stevens, Eric G.
; Kosman, Steven L.; Cassidy, John C.; Chang, Win-Chyi; Miller, Wesley A.: Large-format 1280 x 1024 full-frame CCD image sensor with a lateral-overflow drain and transparent gate electrode,V1447,274-282(1991)

Stevens, J. L. See Cheah, Chun-Wah: V1392,487-497(1991)

Stevens, James H. See Schember, Helene R.: V1540,51-62(1991)

Stevens, R. S.
: FDDI network cabling,V1364,101-114(1991)

Stevens, Richard F.
: Improved image quality from retroreflective screens by spectral smearing,V1358,1265-1267(1991)

Stevens, Robert K.
; Conner, Teri L.: Air quality monitoring with the differential optical absorption spectrometer,V1491,56-67(1991)
; Vossler, T. L.: DOAS (differential optical absorption spectroscopy) urban pollution measurements,V1433,25-35(1991)

Stevenson, J. T. See Robertson, Stewart A.: V1464,232-244(1991)

Stevenson, Robert L.
; Delp, Edward J.: Invariant reconstruction of 3-D curves and surfaces,V1382,364-375(1991)

Stevenson, William A.
See Druy, Mark A.: V1584,48-52(1991)
See Druy, Mark A.: V1437,66-74(1991)

Steward, Robert G. See Costello, David K.: V1537,161-172(1991)

Stewardson, H. R. See Parry, David J.: V1358,1057-1064(1991)

Stewart, Brent K.
; Lou, Shyh-Liang; Wong, Albert W.; Chan, Kelby K.; Huang, H. K.: Performance characteristics of an ultrafast network for PACS,V1446,141-153(1991)
See Taira, Ricky K.: V1446,451-458(1991)
See Wong, Albert W.: V1446,73-80(1991)

Stewart, Charles V.
: Stereo matching, error detection, and surface reconstruction,V1383,285-296(1991)

Stewart, Clayton V.
See Moghaddam, Baback: V1486,115-126(1991)
See Moghaddam, Baback: V1406,42-57(1991)
See Moghaddam, Baback: V1471,414-421(1991)

Stewart, Diane K.
; Doherty, John A.: Issues in the repair of x-ray masks,V1496,247-265(1991)
; Fuchs, Jacob; Grant, Robert A.; Plotnik, Irving: Defect repair for gold absorber/silicon membrane x-ray masks,V1465,64-77(1991)

Stewart, George
; Culshaw, Brian; Clark, Douglas F.; Andonovic, Ivan: Improvement in the performance of evanescent wave chemical sensors by special waveguide structures,V1368,230-238(1991)
See Johnstone, Walter: V1506,145-149(1991)

Stewart, Kevin See Conley, Willard E.: V1466,53-66(1991)

Stewart, Kevin R. See Perry, Joseph W.: V1560,302-309(1991)

Stewart, Melville See McCormick, Patrick W.: V1431,294-302(1991)

Stewart, P. L. See Young, Raymond P.: V1530,335-342(1991)

Stewart, Stephen R.
; LaHaie, Ivan J.; Lyons, Jayne T.: Unified approach to multisensor simulation of target signatures,VIS07,57-97(1991)

Steyer, M. See Key, Michael H.: V1397,9-17(1991)

Stiegman, Albert E. See Bruegge, Carol J.: V1493,132-142(1991)

Stievenard, Didier
See Letartre, Xavier: V1362,778-789(1991)
See Letartre, Xavier: V1361,1144-1155(1991)

Stigliani, Daniel J. See Kiang, Ying J.: V1366,252-258(1991)

Stiles, Tom See Hadfield, Michael J.: V1478,126-144(1991)

Stiller, Christoph
; Lappe, Dirk: Laplacian pyramid coding of prediction error images,V1605,47-57(1991)

Stiller, Marc A.
: Excimer laser fabrication of waveguide devices,V1377,73-78(1991)
See Wu, Jeong W.: V1560,196-205(1991)

Stillman, Steve M. See Nasburg, Robert E.: V1481,84-95(1991)

Stinehour, Judith E. See Andrews, John R.: V1461,110-123(1991)

Stinson, Michael C.
: Pyramid nets for computer vision,V1386,53-61(1991)

Stirk, Charles W.
; Psaltis, Demetri: Comparison of optical and electronic 3-dimensional circuits,V1389,580-593(1991)

Stjerna, B. A. See Yin, Zhiqiang: V1536,149-157(1991)

Stock, Joseph See Siegmund, Oswald H.: V1549,81-89(1991)

Stock, S. See Schaefer, Juergen A.: V1361,1026-1032(1991)

Stocker, Alan D.
; Jensen, Preben D.: Algorithms and architectures for implementing large-velocity filter banks,V1481,140-155(1991)
; Yu, Xiaoli; Winter, Edwin M.; Hoff, Lawrence E.: Adaptive detection of subpixel targets using multiband frame sequences,V1481,156-169(1991)

Stockman, M. I.
See Benimetskaya, L. Z.: V1525,242-245(1991)
See Benimetskaya, L. Z.: V1525,210-211(1991)

Stockum, Larry A.
See Gorenflo, Ronald L.: V1346,42-53(1991)
See Masten, Michael K.: V1482(1991)
See Williams, Elmer F.: V1482,104-111(1991)

Stoehr, M. See Zeyfang, E.: V1397,449-452(1991)

Stofman, Daniel See Bayruns, Robert J.: V1541,83-90(1991)

Stojanoff, Christo G.
; Kubitzek, Ruediger; Tropartz, Stephan; Froehlich, K.; Brasseur, Olivier: Design, fabrication, and integration of holographic dispersive solar concentrator for terrestrial applications,V1536,206-214(1991)
; Kubitzek, Ruediger; Tropartz, Stephan; Brasseur, Olivier; Froehlich, K.: Design and fabrication of large-format holographic lenses,V1527,48-60(1991)
; Schuette, H.; Brasseur, Olivier; Kubitzek, Ruediger; Tropartz, Stephan: Photochemical and thermal treatment of dichromated gelatin film for the manufacturing of holographic optical elements for operation in the IR,V1559,321-330(1991)
; Tholl, Hans D.; Luebbers, Hubertus; Windeln, Wilbert: Development, manufacturing, and integration of holographic optical elements for laser Doppler velocimetry applications,V1507,426-434(1991)
; Tropartz, Stephan; Brasseur, Olivier; Kubitzek, Ruediger: Manufacturing and reproduction of holographic optical elements in dichromated gelatin films for operation in the infrared,V1485,274-280(1991)
See Kubitzek, Ruediger: V1507,365-372(1991)
See Tholl, Hans D.: V1456,262-273(1991)
See Tropartz, Stephan: V1507,345-353(1991)

Stoker, Carol R. See McGreevy, Michael W.: V1387,110-123(1991)

Stokes, Brian P.
: High-accuracy capacitive position sensing for low-inertia actuators,V1454,223-229(1991)

Stokes, D. L. See Vo-Dinh, Tuan: V1368,203-209(1991)

Stokowski, Stanley E. See Engstrom, Herbert L.: V1464,566-573(1991)

Stolberg, K. P. See Damm, Tobias: V1403,686-694(1991)

Stollenwerk, M. See Malzahn, Eric: V1362,199-204(1991)

Stoller, Meryl D. See Harrigan, Jeanne E.: V1464,596-601(1991)

Stoltz, John R.
; Cole, Donald C.: Design and testing of data fusion systems for the U.S. Customs Service drug interdiction program,V1479,423-434(1991)
See Bubb, Daniel: V1483,18-28(1991)

Stolz, Christopher J. See Tesar, Aleta A.: V1441,154-172(1991)

Stolz, Dieter See Pfeiffer, Wolfgang: V1358,1191-1201(1991)

Stone, Maureen C. See Wallace, William E.: V1460,20-28(1991)

Stone, Richard E.
See Ermer, Susan P.: V1560,120-129(1991)
See Swanson, David A.: V1560,416-425(1991)

Stone, Richard V.
; Zeise, Frederick F.; Guilfoyle, Peter S.: DOC II 32-bit digital optical computer: optoelectronic hardware and software,V1563,267-278(1991)
See Guilfoyle, Peter S.: V1563,214-222(1991)

Stone, Samuel M. See O'Byrne, Vincent A.: V1372,88-93(1991)

Stone, Thomas W.
; Thompson, Brian J.; eds.: *Selected Papers on Holographic and Diffractive Lenses and Mirrors,*VMS34(1991)

Stoneman, Robert C.
; Esterowitz, Leon; Lynn, J. G.: 2.8-um Er3+:YLiF4 laser resonantly pumped at 970 nm,V1410,148-155(1991)

Stoner, William W. See Chao, Tien-Hsin: V1558,505-517(1991)

Stoney, William E.
; deFigueiredo, Rui J.; eds.: *Cooperative Intelligent Robotics in Space,*V1387(1991)
See Castellano, Anthony R.: V1387,343-350(1991)

Stonham, T. J. See Patel, Devesh: V1606,621-629(1991)

Stoop, Karel W. See van Geest, Lambertus K.: V1449,121-134(1991)

Storc, Lisa See Seetharaman, Guna S.: V1569,269-273(1991)

Storjohann, Kai See Mallot, Hanspeter A.: V1382,397-408(1991)

Storm, E. See Murray, John R.: V1410,28-39(1991)

Storm, Mark E. See Allario, Frank: V1492,92-110(1991)

Storm, William D. See Adelstein, Peter Z.: VCR37,159-179(1991)

Storti, George M.
See Jutamulia, Suganda: V1401,113-118(1991)
See Jutamulia, Suganda: V1558,442-447(1991)
See Yu, Dong X.: V1584,236-242(1991)

Stoudt, David C.
; Mazzola, Michael S.; Griffiths, Scott F.: Characterization and switching study of an optically controlled GaAs switch,V1378,280-285(1991)
See Brinkmann, Ralf P.: V1378,203-208(1991)

Stournaras, C. See Hourdakis, G.: V1503,249-255(1991)

Stoutland, Page O.
See Dyer, R. B.: V1432,197-204(1991)
See Woodruff, William H.: V1432,205-210(1991)

Stover, John C.
: Optical scatter: an overview,V1530,2-6(1991)
ed.: *Optical Scatter: Applications, Measurement, and Theory,*V1530(1991)
; Bernt, Marvin L.; Henning, Timothy D.: Study of anomalous scatter characteristics,V1530,185-195(1991)
See Bernt, Marvin L.: V1530,42-49(1991)
See Rudberg, Donald A.: V1530,232-239(1991)

Stover, Richard J. See Robinson, Lloyd B.: V1447,214-228(1991)

Straboni, Alain See Baixeras, Joseph M.: V1362,117-126(1991)

Stradling, R. A.
: Novel narrow-gap semiconductor systems,V1361,630-640(1991)

Straforini, Marco
See Campani, Marco: V1388,409-414(1991)
See Coelho, Christopher: V1388,398-408(1991)

Strand, Michael P.
: Imaging model for underwater range-gated imaging systems,V1537,151-160(1991)

Strang, Steven E.
: Adaptive imager: a real-time locally adaptive edge enhancement system,V1384,246-256(1991)

Strasser, G. See Kremser, Christian: V1501,69-79(1991)

Stratton, F. P. See Kubena, Randall L.: V1392,595-597(1991)

Stratton, Norman A.
; Vaina, Lucia M.: Generalization of the problem of correspondence in long-range motion and the proposal for a solution,V1468,176-185(1991)

Straub, William H. See Gur, David: V1446,284-288(1991)

Strauss, Michael G.
; Westbrook, Edwin M.; Naday, Istvan; Coleman, T. A.; Westbrook, M. L.; Travis, D. J.; Sweet, Robert M.; Pflugrath, J. W.; Stanton, Martin J.: Large-aperture CCD x-ray detector for protein crystallography using a fiber-optic taper,V1447,12-27(1991)

Strauss, Wolfgang See Schneckenburger, Herbert: V1525,91-98(1991)

Strebel, Bernhard N. See Keil, Norbert: V1559,278-287(1991)

Street, Robert A. See Antonuk, Larry E.: V1443,108-119(1991)

Street, Troy A. See Kalin, David A.: V1358,780-787(1991)

Streibl, Norbert
: Diffractive optical elements for optoelectronic interconnections,V1574,34-47(1991)
See Kobolla, Harald: V1507,175-182(1991)

Streifer, William S. See Yaeli, Joseph: V1442,378-382(1991)

Streit, Dwight C.
See Chow, P. D.: V1475,48-54(1991)
See Liu, Louis: V1475,193-198(1991)
See Minot, Katcha: V1475,309-313(1991)

Strelnitsky, V. E. See Armeyev, V. Y.: V1621,2-10(1991)

Strickland, Donna
; Corkum, Paul B.: Short pulse self-focusing,V1413,54-58(1991)

Strickler, James H.
; Webb, Watt W.: Two-photon excitation in laser scanning fluorescence microscopy,V1398,107-118(1991)

Strifors, Hans C. See Abrahamsson, S.: V1471,130-141(1991)

Strigalev, V. E. See Ignatyev, Alexander V.: V1584,336-345(1991)

Strigun, V. L. See Kliot-Dashinskaya, I. M.: V1238,465-469(1991)

Strikwerda, Thomas E.
; Fisher, H. L.; Frank, L. J.; Kilgus, Charles C.; Gray, Connie B.; Barnes, Donald L.: Evaluation of a CCD star camera at Table Mountain Observatory,V1478,13-23(1991)

Stringer, J. T. See Leksell, David: V1458,133-144(1991)

Strobach, Peter
: Schur RLS adaptive filtering using systolic arrays,V1565,307-322(1991)

Strobel, Reiner See Brueckner, Volkmar: V1362,510-517(1991)

Strohm, Karl M. See Luy, Johann-Freidrich: V1475,129-139(1991)

Strojewski, Dariusz See Domanski, Andrzej W.: V1505,59-66(1991)

Strong, David S. See Molley, Perry A.: V1564,610-616(1991)

Strong, James P. See Fischer, James R.: V1492,229-238(1991)

Stronski, Alexander V. See Indutnyi, I. Z.: V1555,243-253(1991)

Stroscio, Michael A.
; Kim, K. W.; Littlejohn, Michael A.: Theory of optical-phonon interactions in a rectangular quantum wire,V1362,566-579(1991)

Strother, George T. See Smolka, Greg L.: V1498,70-80(1991)

Stroup, Eric See Wang, Jinyu: V1519,835-841(1991)

Stroyer-Hansen, T. See Svendsen, Erik N.: V1361,1048-1055(1991)

Strueder, Lothar See Braueninger, Heinrich: V1549,330-339(1991)

Strugala, Lisa A.
; Newt, J. E.; Futterman, Walter I.; Schweitzer, Eric L.; Herman, Bruce J.; Sears, Robert D.: Development of high resolution statistically nonstationary infrared earthlimb radiance scenes,V1486,176-187(1991)

Strunge, Ch. See Flemming, G.: V1421,146-152(1991)

Strzelecki, Eva M.
See Lin, Freddie S.: V1389,642-647(1991)
See Liu, William Y.: V1365,20-24(1991)

Stuart, Jeffrey A. See Birge, Robert R.: V1432,129-140(1991)

Stubbs, David M.
; Hsu, Ike C.: Rapid-cooled lens cell,V1533,36-47(1991)

Stucky, Galen D. See Phillips, Mark L.: V1561,84-92(1991)

Stuermer, Martin See Mommsen, Jens: V1503,348-354(1991)

Stuff, Michael I. See Kelley, J. D.: V1415,211-219(1991)

Stuhlinger, Tilman W.
: All-reflective phased array imaging telescopes,V1354,438-446(1991)

Stultz, Robert D.
; Nieuwsma, Daniel E.; Gregor, Eduard: Eyesafe high-pulse-rate laser progress at Hughes,V1419,64-74(1991)

Stupak, Mikhail F. See Krasnov, Victor F.: V1506,179-187(1991)

Sturgill, Robert See Casasent, David P.: V1567,683-690(1991)

Sturland, Ian M. See Lake, Stephen P.: V1486,286-293(1991)

Sturlesi, Doron
; O'Shea, Donald C.: Future of global optimization in optical design,V1354,54-68(1991)

Sturm, James C.
; Reaves, Casper M.: Fundamental mechanisms and doping effects in silicon infrared absorption for temperature measurement by infrared transmission,V1393,309-315(1991)
; Schwartz, P. V.; Prinz, Erwin J.; Magee, Charles W.: Control of oxygen incorporation and lifetime measurement in Si1-xGex epitaxial films grown by rapid thermal chemical vapor deposition,V1393,252-259(1991)
See Xiao, Xiaodong: V1476,301-304(1991)

Sturz, Richard A.
: Electronically gated airborne video camera,V1538,77-80(1991)
: Practical videography,V1346,64-67(1991)

Sturzebecher, Dana See Paolella, Arthur: V1378,195-202(1991)

Stut, W. J.
; van Steen, M. R.; Groenewegen, L. P.; Ratib, Osman M.; Bakker, Albert R.: Simulation-based PACS development,V1446,396-404(1991)

Styczynski, K. See Campos, Juan: V1574,141-147(1991)

Styer, Stephen B. See Christie, Phillip: V1390,359-367(1991)

Su, Chang See Zhu, Shi-Yao: V1497,255-262(1991)

Su, Chin B. See Jiang, Ching-Long: V1418,261-271(1991)

Su, Ching-Hua See Gillies, Donald C.: V1484,2-10(1991)

Su, Daoning
; Boechat, Alvaro A.; Jones, Julian D.: Dependence of output beam profile on launching conditions in fiber-optic beam delivery systems for Nd:YAG lasers,V1502,41-51(1991)

Su, Degong See Li, Guozhu: V1358,1120-1122(1991)

Su, Kai L. See Wu, Fang F.: V1519,347-349(1991)

Su, Pin
: Temperature stress testing of laser modules for the uncontrolled environment,V1366,94-106(1991)

Su, Renjeng See Chapel, Jim D.: V1387,284-295(1991)

Su, Wan-Yong
; Su, Xianyu: Optical 3-D sensing for measurement of bottomhole pattern,V1567,680-682(1991)
; Su, Xianyu: Optical three-dimensional sensing for measurement of bottomhole pattern,V1332,820-823(1991)

Su, Xianyu
; Zhou, Wen-Sheng: Algorithm for the generation of look-up range table in 3-D sensing,V1332,355-357(1991)
See Su, Wan-Yong: V1567,680-682(1991)
See Su, Wan-Yong: V1332,820-823(1991)

Su, Xing
; Fan, Zheng X.: Photothermal displacement spectroscopy of optical coatings,V1519,80-84(1991)

Su, Yan K.
See Hsu, C. T.: V1519,391-395(1991)
See Huang, C. J.: V1519,70-73(1991)
See Wu, T. S.: V1361,23-33(1991)
See Wu, T. S.: V1519,263-268(1991)

Subari, Mustofa See Hassan, Azmi: V1526,142-156(1991)

Subbarao, Murali
; Nikzad, Arman: Model for image sensing and digitization in machine vision,V1385,70-84(1991)

Subramanian, K. R.
; Dubey, V. K.; Low, J. P.; Tan, L. S.: Line coding for high-speed fiber optic transmission systems,V1364,190-201(1991)

Subramanian, N. S. See Lowry, Jay H.: V1330,142-151(1991)

Subramanian, P.
; Creed, David; Hoyle, Charles E.; Venkataram, Krishnan: Triplet-sensitized reactions of some main chain liquid-crystalline polyaryl cinnamates,V1559,461-469(1991)
See Hoyle, Charles E.: V1559,101-109(1991)

Subramanian, S.
; Chakravarty, S.; Anand, S.; Arora, B. M.: Use of admittance spectroscopy to probe the DX-centers in AlGaAs,V1362,205-216(1991)

Subramanian, V. See Murali, K. R.: V1536,289-295(1991)

Suche, Hubertus See Brinkmann, R.: V1362,377-382(1991)

Suchkov, A. F. See Frolov, K.: V1397,461-468(1991)

Sud'enkov, Yu. V. See Baloshin, Yu. A.: V1440,71-77(1991)

Suda, Akira
See Arai, Tsunenori: V1425,191-195(1991)
See Daidoh, Yuichiro: V1421,120-123(1991)
See Sakurada, Masami: V1425,158-164(1991)

Suda, Yasumasa
; Shigehara, Kiyotaka; Yamada, Akira; Matsuda, Hiro; Okada, Shuji; Masaki, Atsushi; Nakanishi, Hachiro: Reversible phase transition and third-order nonlinearity of phthalocyanine derivatives,V1560,75-83(1991)

Suda, Yoshihiko
; Ohbayashi, Keiji; Onodera, Kaoru: Chemistry of the Konica Dry Color System,V1458,76-78(1991)

Sudhakar, Raghavan See Venugopal, Kootala P.: V1471,44-53(1991)

Sudoh, Masaaki
; Yokoyama, Ryouhei; Sumiya, Mitsuo; Yamamoto, Masaki; Yanagihara, Mihiro; Namioka, Takeshi: Soft x-ray multilayers fabricated by electron-beam deposition,V1343,14-24(1991)

Sudol, R. See Bonometti, Robert J.: V1495,166-176(1991)

Suen, Eric S. See Guralnick, Sidney A.: V1396,664-677(1991)

Suenaga, Yasuhito
See Akamatsu, Shigeru: V1606,204-216(1991)
See Wallace, Richard S.: V1381,436-446(1991)

Suga, Hisaaki See Yarling, Charles B.: V1393,192-199(1991)

Suganuma, Heiroku
; Matsunaga, Tadayo: Plastic optical fibers for automotive applications,V1592,12-17(1991)

Sugawara, Takehisa See Tsukahara, Hiroyuki: V1384,15-26(1991)

Sugihara, Kiyoshi See Nakahara, Sumio: V1554A,602-609(1991)

Sugihara, Okihiro
; Kinoshita, Takeshi; Okabe, M.; Kunioka, S.; Nonaka, Y.; Sasaki, Keisuke: Phase-matched second harmonic generations in poled dye-polymer waveguides,V1361,599-605(1991)

Sugii, Nobuyuki See Kanehori, Keiichi: V1394,238-243(1991)

Sugimoto, Dai See Hattori, Shuzo: V1463,539-550(1991)

Sugimoto, Nobuo See Sasano, Yasuhiro: V1490,233-242(1991)

Sugimoto, Tetsuo See Yuuki, Hayato: V1592,2-11(1991)

Sugimoto, Yoshimasa See Akita, K.: V1392,576-587(1991)

Sugino, Hajime See Miyake, Hiroyuki: V1448,150-156(1991)

Sugisaki, Iwao See Horikawa, Hiroshi: V1454,46-59(1991)

Sugiura, Norio
; Morita, Shinzo: Auto-focus video camera system with bag-type lens,V1358,442-446(1991)

Sugiyama, Hisashi See Nate, Kazuo: V1466,206-210(1991)

Sugiyama, Hisataka See Maeda, Takeshi: V1499,414-418(1991)

Suh, Doowon
See Babcock, Carl P.: V1466,653-662(1991)
See Berry, Amanda K.: V1465,210-220(1991)

Suk, Minsoo See Bhandarkar, Suchendra M.: V1384,234-245(1991)

Sukamto, Lin See Cooley, Thomas W.: V1475,243-247(1991)

Sukeda, Hirofumi
; Tsuchinaga, Hiroyuki; Tanaka, Satoshi; Niihara, Toshio; Nakamura, Shigeru; Mita, Seiichi; Yamada, Yukinori; Ohta, Norio; Fukushima, Mitsugi: High-speed recording technologies for a magneto-optical disk system,V1499,419-425(1991)

Sukhanov, Vitaly I.
: Heterogeneous recording media,V1238,226-230(1991)

Sukharev, S. A.
See Dolgopolov, Y. V.: V1412,267-275(1991)
See Zykov, L. I.: V1412,258-266(1991)

Sukhodolsky, A. T. See Rastopov, S. F.: V1440,127-134(1991)

Sukhov, A. M. See Doroschenko, V. M.: V1397,503-511(1991)

Sukowski, U. See Kohl, M.: V1525,26-34(1991)

Sullivan, Barry J.
See Brailean, James C.: V1450,40-46(1991)
See Chan, Cheuk L.: V1450,208-217(1991)
See Chiang, John Y.: V1396,15-26(1991)
See Hayes, Jeffrey A.: V1396,324-330(1991)
See Kwak, J. Y.: V1396,32-44(1991)

Sullivan, Brian J. See Swanson, David A.: V1560,416-425(1991)

Sullivan, Charles T.
See Bristow, Julian P.: V1389,535-546(1991)
See Guha, Aloke: V1389,375-385(1991)
See Swirhun, S.: V1475,223-230(1991)

Sullivan, Daniel P.
; Weber, Charles L.: Analog optical processing of radio frequency signals,V1476,234-245(1991)

Sullivan, J. A. See Rose, Evan A.: V1411,15-27(1991)

Sullivan, Monroe See Babcock, Carl P.: V1466,653-662(1991)

Sullivan, Timothy D.
; Ryan, James G.; Riendeau, J. R.; Bouldin, Dennis P.: Stress-induced voiding in aluminum alloy metallizations,V1596,83-95(1991)

Sultan, Michel F.
; O'Rourke, Michael J.: Linear position sensing by light exchange between two lossy waveguides,V1584,212-219(1991)

Sultana, John A.
; O'Neill, Mark B.: Design, analysis, and testing of a CCD array mounting structure,V1532,27-38(1991)

Sulya, Andrew W.
; Moore, Leslie K.; Synovec, Robert E.: Gradient microbore liquid chromatography with dual-wavelength absorbance detection: tunable analyzers for remote chemical monitoring,V1434,147-158(1991)

Sulzer, Jean-Francois See Kessler, William D.: V1538,104-111(1991)

Sumen, Juha See Hyvarinen, Timo S.: V1530,325-334(1991)

Sumida, Shin S.
See Kobayashi, Soichi: V1374,300-303(1991)
See Schmidt, Kevin M.: V1396,744-752(1991)

Sumiya, Daigaku See Sato, Koki: V1461,124-131(1991)

Sumiya, Mitsuo See Sudoh, Masaaki: V1343,14-24(1991)

Sumkin, Jules H. See Gur, David: V1446,284-288(1991)

Summers, Charles L. See Harris, Neville W.: V1416,59-69(1991)

Summers, Christopher J.
; Wagner, Brent K.; Benz, Rudolph G.; Rajavel, D.: Metal-organic molecular beam epitaxy of II-VI materials,V1512,170-176(1991)

Summers, R. G. See Samarabandu, J. K.: V1450,296-322(1991)

Summers, Robert A. See Hansen, Verne W.: V1495,115-122(1991)

Sumner, Timothy J.
; Grant, S. M.; Bewick, A.; Li, J. P.; Spooner, N. J.; Smith, K.; Beaumont, Steven P.: Recent developments using GaAs as an x-ray detector,V1549,256-264(1991)
; Roe, S.; Rochester, G. K.; Hall, G.; Evensen, Per; Avset, B. S.: Low-temperature operation of silicon drift detectors,V1549,265-273(1991)
See Zammit, C. C.: V1549,274-282(1991)

Sun, Bingrong
; Xu, Jiafang: Image processing system for brain and neural tissue,V1606,1022-1026(1991)

Sun, C. K.
; Wu, Chao-Chia C.; Chang, Ching T.; Yu, Paul K.; McKnight, William H.: Bridge-type optoelectronic sample and hold circuit,V1476,294-300(1991)

Sun, C. Y.
See Liu, W. C.: V1519,670-674(1991)
See Liu, W. C.: V1519,640-644(1991)

Sun, Ching-Cherng
; Chang, Ming-Wen; Yeh, Smile; Cheng, Nai-Jen: Computer-aided photorefractive pattern recognition,V1564,199-210(1991)

Sun, Da M.
: Study of crystal structure of vacuum-evaporated Ag-Cu thin film,V1519,688-691(1991)

Sun, David See Claus, Richard O.: V1588,159-168(1991)

Sun, Deming See Zhou, Changxin: V1345,281-287(1991)

Sun, Dexing
; Chen, Shouliu; Pan, Chao; Jin, Henghuan: High-accurate optical fiber liquid-level sensor,V1572,508-513(1991)

Sun, Fang-kui
; Yang, Mao-hua; Gao, Shao-hong; Zhao, Shi-jie: New type of IR to visible real-time image converter: design and fabrication,V1488,2-5(1991)

Sun, Fang-Kuo See Trainer, Thomas J.: V1567,650-658(1991)

Sun, Geng-tian See Chang, Hsiao T.: V1388,442-446(1991)

Sun, Han D.
; Zhou, Fang Q.; Zhao, Xing R.; Wang, Lingjie; Yi, Xin J.: Infrared optical response of superconducting YBaCuO thin films,V1477,174-177(1991)
See Wang, Lingjie: V1540,738-741(1991)
See Zhao, Xing R.: V1519,772-774(1991)
See Zhou, Fang Q.: V1477,178-181(1991)

Sun, Huifang
; Manikopoulos, Constantine N.; Hsu, Hwei P.: Subsampled vector quantization with nonlinear estimation using neural network approach,V1605,214-220(1991)

Sun, Jian G. See Mei, Yuan S.: V1519,370-373(1991)

Sun, Junyong
See Lu, Yue-guang: V1332,287-291(1991)
See Zhao, Feng: V1461,262-264(1991)

Sun, K. See Davidson, Jennifer L.: V1568,176-187(1991)

Sun, Mei H.
; Kamal, Arvind: Small single-sensor for temperature, flow, and pressure measurement,V1420,44-52(1991)
See Ingold, Joseph P.: V1589,83-89(1991)

Sun, Ning See Li, Xiang Ying: V1412,246-251(1991)

Sun, Q. See Field, John E.: V1358,703-712(1991)

Sun, San-Wei See Chien, Bing-Shan: V1606,588-598(1991)

Sun, Shang-zhu See Shiau, T. H.: V1418,116-122(1991)

Sun, Shu Z. See Yin, Dong: V1519,164-166(1991)

Sun, Tai-Ping
; Lee, Si-Chen; Liu, Kou-Chen; Pang, Yen-Ming; Yang, Sheng-Jenn: High-performance metal/SiO2/InSb capacitor fabricated by photoenhanced chemical vapor deposition,V1361,1033-1037(1991)

Sun, Xiaohan
; Zhang, Mingde; Wan, Suiren; Wang, Shuhua: Fiber optic differential pressure sensor for leak-detection system,V1572,243-247(1991)

Sun, Xiaoli See Davidson, Frederic M.: V1417,75-88(1991)

Sun, Xu-Mei See Lang, Zhengping: V1468,826-833(1991)

Sun, Yang
; Li, Cheng F.; Deng, He; Gan, Fuxi: Laser-induced phase transition in crystal InSb films as used in optical storage,V1519,554-558(1991)

Sun, Yi L. See Chen, You S.: V1519,56-62(1991)

Sun, Ying See Liu, Iching: V1606,67-77(1991)

Sun, Yongda See Jia, Youquan: V1554B,135-138(1991)

Sun, Yuesheng See Ming, Hai: V1572,27-31(1991)

Sun, Yue Z.
; Chen, Chun X.; Xie, Qi Y.; Yin, Chen Z.; He, Yu L.: Thermal stabilities of a-Si:H films and its application to thyristor elements,V1519,234-240(1991)

Sun, Yung-Nien
; Chen, Jiann-Jone; Lin, Xi-Zhang; Mao, Chi-Wu: New method for constructing 3-D liver from CT images,V1606,653-664(1991)

Sun, Yunlong
: Interference effect on laser trimming and layer thickness optimization,V1598,91-97(1991)

Sun, Zhong-kang
See Li, Zhi-yong: V1472,97-105(1991)
See Lu, Huanzhang: V1481,398-405(1991)

Sun, Zhu Z. See Zhao, Yun F.: V1519,411-414(1991)

Sunak, Harish R.
See Joss, Brian T.: V1372,94-117(1991)
See Steele, Roger C.: V1372(1991)

Sundararajan, Srikanth See Srikanth, Usha: V1468,429-433(1991)

Sundari, V. See Chandrasekhar, N. S.: V1390,523-536(1991)

Sundell, Per-Erik See Hoyle, Charles E.: V1559,202-213(1991)

Sung, C. C.
: Nonlinear optical properties of quantum-well structures and some of their applications,V1497,456-466(1991)
See Ruffin, Paul B.: V1478,160-167(1991)

Sung, Eric
: Stereovision and color segmentation for autonomous navigation,V1388,176-187(1991)
; Singh, Harcharan; Tan, Daniel H.: Semiautomatic calibration of the general camera model for stereovision,V1385,57-69(1991)

Sunyaev, Rashid
See Kaaret, Philip E.: V1548,106-117(1991)
See Sipila, Heikki: V1549,246-255(1991)

Suorsa, Raymond E.
; Sridhar, Banavar: Validation of vision-based obstacle detection algorithms for low-altitude helicopter flight,V1388,90-103(1991)

Super, Boaz J.
; Bovik, Alan C.: Three-dimensional orientation from texture using Gabor wavelets,V1606,574-586(1991)

Surendran, Geetha See Schwartzkopf, George: V1466,26-38(1991)

Surette, Mark R. See Bundy, Scott: V1475,319-329(1991)

Surget, Jean
; Dunet, G.: Multipass holographic interferometry for low-density gas flow analysis,V1358,65-72(1991)

Surhigh, James W. See Gammarino, Rudolph R.: V1419,115-125(1991)

Suri, Jasjit S. See Doyle, Michael D.: V1380,227-235(1991)

Surjhikov, S. T. See Budnik, A. P.: V1397,721-724(1991)

Sushchikh, M. M. See Zandberg, E. Y.: V1440,292-302(1991)

Sushko, A. M. See Petrov, V. V.: V1621,45-50(1991)

Suter, David
: Coupled depth-slope model based on augmented Lagrangian techniques,V1570,129-139(1991)

Suter, Gaby See Spichiger, Ursula E.: V1510,118-130(1991)

Suter, Kevin J.
; Oliver, Jeffrey W.; Cunningham, Philip R.: Diagnostic instrumentation suite for the characterization of two coupled lasers,V1414,33-65(1991)

Sutter, Kurt
; Hulliger, J.; Knoepfle, G.; Saupper, N.; Guenter, Peter: Nonlinear optical properties of N-(4-nitro-2-pyridinyl)-phenylalaninol single crystals,V1560,296-301(1991)
; Hulliger, J.; Guenter, Peter: Photorefractive gratings in the organic crystal 2-cyclooctylamino-5-nitropyridine doped with 7,7,8,8-tetracyanoquinodimethane,V1560,290-295(1991)

Sutton, George W.
See Simpson, Thomas B.: V1410,133-140(1991)
See Simpson, Thomas B.: V1416,2-9(1991)

Sutton, Phillip See Dawber, William N.: V1476,81-90(1991)

Suyama, Motohiro See Kinoshita, Katsuyuki: V1358,490-496(1991)

Suyten, F. M. See Moehlmann, Gustaaf R.: V1560,426-433(1991)

Suzaki, Shinzoh
; Ishii, Masanori; Watanabe, Tsutomu; Shiota, Alan T.: Temperature and
modulation dependence of spectral linewidth in distributed Bragg
reflector laser diodes,V1374,126-131(1991)

Suzuki, A. See Nagai, Haruhiko: V1397,31-36(1991)

Suzuki, Akira
See Inaoka, Noriko: V1450,2-12(1991)
See Miyahara, Sueharu: V1384,317-323(1991)
See Suzuki, Hideo: V1450,99-107(1991)

Suzuki, Fumio
: Novel plastic image-transmitting fiber,V1592,150-157(1991)

Suzuki, Hideo
; Inaoka, Noriko; Takabatake, Hirotsugu; Mori, Masaki; Natori, Hiroshi;
Suzuki, Akira: Experimental system for detecting lung nodules by chest x-
ray image processing,V1450,99-107(1991)
See Inaoka, Noriko: V1450,2-12(1991)

Suzuki, Hiroshi See Matsushita, Tadashi: V1488,368-375(1991)

Suzuki, Katsumi See Hasegawa, Shinya: V1465,145-151(1991)

Suzuki, Kazuhiro See Koshi, Yutaka: V1605,362-373(1991)

Suzuki, Kenji See Yokoi, Takane: V1449,30-39(1991)

Suzuki, Makoto
See Kume, Hidehiro: V1358,1144-1155(1991)
See Sasano, Yasuhiro: V1490,233-242(1991)

Suzuki, Masayuki
; Shiraiwa, Yoshinobu; Minoura, Kazuo: High-quality image recorder and
evaluation,V1454,370-381(1991)

Suzuki, Naoshi
; Narimatsu, Yoshito; Nagura, Riichi; Sakuma, Fumihiro; Ono, Akira: Large
integrating sphere of prelaunch calibration system for Japanese Earth
Resources Satellite optical sensors,V1493,48-57(1991)

Suzuki, Shigeharu See Nakazawa, Akira: V1458,115-118(1991)

Suzuki, Takayoshi
See Kobayashi, Akira: V1449,148-156(1991)
See Nakagawa, Kazuo: V1332,267-273(1991)

Suzuki, Tomihiro
; Mikamura, Yasuki; Murata, Kazuo; Sekiguchi, Takeshi; Shiga, Nobuo;
Murakami, Yasunori: Compact PIN-amplifier module for giga bit rates
optical interconnection,V1389,455-461(1991)

Svaasand, Lars O.
ed.: *Future Trends in Biomedical Applications of Lasers*,V1525(1991)
; Tromberg, Bruce J.: Properties of optical waves in turbid media,V1525,41-
51(1991)
See Bays, Roland: V1525,397-408(1991)
See Kimel, Sol: V1525,341-350(1991)
See Tromberg, Bruce J.: V1525,52-58(1991)

Svanberg, Katarina
See Andersson-Engels, Stefan: V1426,31-43(1991)
See Baert, Luc: V1525,385-390(1991)

Svanberg, Sune
See Andersson-Engels, Stefan: V1426,31-43(1991)
See Baert, Luc: V1525,385-390(1991)
See Berg, Roger: V1431,110-119(1991)
See Berg, Roger: V1525,59-67(1991)

Svantesson, Kjell G. See Sohlstrom, Hans B.: V1511,142-148(1991)

Svatek, Lubomir See Hudec, Rene: V1343,162-163(1991)

Svechnikov, Georgy S.
; Rizikov, Igor V.: External factors' influences on AIIIBV light-emitting
structures,V1362,674-683(1991)
; Shapar, Vladimir N.: Optical rotary connector for transfer of data signals
from fiber optic sensors plasing on rotary objects,V1589,24-31(1991)
See Rizikov, Igor V.: V1362,664-673(1991)

Svelto, Orazio See Serri, Laura: V1397,469-472(1991)

Svendsen, Erik N.
; Stroyer-Hansen, T.: Semiempirical method for calculation of dynamic
susceptibilities,V1361,1048-1055(1991)

Sverchkov, Sergey E. See Denker, Boris I.: V1419,50-54(1991)

Sverchkov, Yuri E. See Denker, Boris I.: V1419,50-54(1991)

Sverdlove, Ronald See Spence, Clay D.: V1469,665-670(1991)

Svetkoff, Donald J.
; Xydis, Thomas G.; Kilgus, Donald B.: Noise statistics of ratiometric signal
processing systems,V1385,113-122(1991)
See Harding, Kevin G.: V1385(1991)

Svetovoy, V. B. See Maslenitsyn, S. F.: V1440,436-441(1991)

Sviridov, Victor N. See Marczuk, Krystyna: V1513,291-296(1991)

Svitashev, Konstantin K. See Ovsyuk, Victor N.: V1540,424-431(1991)

Svystushkin, V. M. See Loshchenov, V. B.: V1420,271-281(1991)

Swafford, William J. See Park, Eric D.: V1366,294-303(1991)

Swain, Robin C. See Preater, Richard W.: V1554A,727-738(1991)

Swairjo, Manal
; Rothschild, Kenneth J.; Nappi, Bruce; Lane, Alan; Gold, Harris: Infrared
fiber optic sensors: new applications in biology and medicine,V1437,60-
65(1991)

Swalen, Jerome D. See Smith, Barton A.: V1560,400-405(1991)

Swan, John
; Shields, Frank J.: Multisensor fusion methodologies compared,V1483,219-
230(1991)

Swanberg, Norman
; Urbach, Michael K.; Ginaven, Robert O.: Gating techniques for imaging
digicon tubes,V1346,249-267(1991)

Swanson, Curtis J.
; Wingard, Christopher J.: Quantitative thermal gradient imaging of biological
surfaces,V1467,372-383(1991)

Swanson, David A.
; Altman, Joe C.; Sullivan, Brian J.; Spitzer, Ronnie; Hansen, Glenn A.; Stone,
Richard E.: Multilayered optically activated devices based on organic
third-order nonlinear optical materials,V1560,416-425(1991)

Swanson, Gary J.
See Hector, Scott D.: V1555,200-213(1991)
See Wong, Vincent V.: V1544,123-133(1991)

Swanson, Robert A. See Bowmaster, Thomas A.: V1363,119-124(1991)

Swanson, S. See Werner, Thomas R.: V1544,46-57(1991)

Swarnakar, Vivek See Dougherty, Edward R.: V1451,137-147(1991)

Swart, Pieter L.
; von Bergmann, Hubertus M.: Thyristor driven pulser for multikilowatt
lasers,V1397,559-562(1991)
See Aharoni, Herzl: V1442,118-125(1991)
See Booysen, Andre: V1584,273-279(1991)
See von Bergmann, Hubertus M.: V1397,63-66(1991)

Swartz, Barry A.
; Cummings, James D.: Laser range-gated underwater imaging including
polarization discrimination,V1537,42-56(1991)

Swartz, Jerome
See Wang, Ynjiun P.: V1384,145-160(1991)
See Wang, Ynjiun P.: V1384,169-175(1991)

Swartz, Robert G. See Ota, Yusuke: V1364,146-152(1991)

Sweatt, William C.
; Stanton, Philip L.; Crump, O. B.: Simplified VISAR system,V1346,151-
159(1991)
See Molley, Perry A.: V1564,610-616(1991)

Sweedler, Jonathan V.
; Shear, Jason B.; Fishman, Harvey A.; Zare, Richard N.; Scheller, Richard H.:
Analysis of neuropeptides using capillary zone electrophoresis with
multichannel fluorescence detection,V1439,37-46(1991)

Sweeney, Frank J. See Sluder, John C.: V1468,642-652(1991)

Sweeney, Michael N.
: Manufacture of fast, aspheric, bare beryllium optics for radiation hard,
spaceborne sensor systems,V1485,116-127(1991)

Sweet, Robert M. See Strauss, Michael G.: V1447,12-27(1991)

Swent, Richard L.
; Schwettman, H. A.; Smith, T. I.: Stanford picosecond FEL center,V1552,24-
35(1991)

Swerdlow, H. See Chen, Da Yong: V1435,161-167(1991)

Swierczek, Mary J. See Smith, Terrance L.: V1512,92-100(1991)

Swift, Charles D. See Salmon, J. T.: V1542,459-467(1991)

Swift, Kerry M.
; Mitchell, George W.: Background correction in multiharmonic Fourier
transform fluorescence lifetime spectroscopy,V1431,171-178(1991)

Swimm, Randall T.
: Thermal transport properties of optical thin films,V1441,45-55(1991)

Swiniarski, Roman W.
; Dzielinski, Andrzej; Skoneczny, Slawomir; Butler, Michael P.: Recurrent neural network application to image filtering: 2-D Kalman filtering approach,V1451,234-241(1991)

Swinyard, Bruce M.
; Malaguti, Giuseppe; Caroli, Ezio; Dean, Anthony J.; Di Cocco, Guido: Spectroscopy and polarimetry capabilities of the INTEGRAL imager: Monte Carlo simulation results,V1548,94-105(1991)
See Di Cocco, Guido: V1549,102-112(1991)

Swirhun, S.
; Bendett, Mark P.; Sokolov, Vladimir; Bauhahn, Paul E.; Sullivan, Charles T.; Mactaggart, R.; Mukherjee, Sayan D.; Hibbs-Brenner, Mary K.; Mondal, J. P.: Mixed application MMIC technologies: progress in combining RF, digital, and photonic circuits,V1475,223-230(1991)
; Geddes, John J.; Sokolov, Vladimir; Bosch, D.; Gawronski, M. J.; Anholt, R.: Comparison of MESFET and HEMT MMIC technologies using a compact Kaband voltage-controlled oscillator,V1475,303-308(1991)
See Mondal, J. P.: V1475,314-318(1991)

Swiryn, Steven See Kwak, J. Y.: V1396,32-44(1991)

Swisher, Richard L. See Grieser, James L.: V1330,111-118(1991)

Switka-Wieclawska, Iwona See Kecik, Tadeusz: V1391,341-345(1991)

Swoboda, Helmut See Schmitt, Dirk-Roger: V1530,104-110(1991)

Sychev, S. P. See Vizir, V. A.: V1411,63-68(1991)

Sychugov, G. V. See Drozhbin, Yu. A.: V1358,1029-1034(1991)

Sychugov, V. A.
See Bazakutsa, P. V.: V1440,370-376(1991)
See Parriaux, Olivier M.: V1583,376-382(1991)

Sydney, Paul F.
; Dillow, Michael A.; Anspach, Joel E.; Kervin, Paul W.; Lee, Terence B.: Relay Mirror Experiment scoring analysis and the effects of atmospheric turbulence,V1482,196-208(1991)
See Anspach, Joel E.: V1482,170-181(1991)
See Lightsey, Paul A.: V1482,209-222(1991)

Syed, Mushtaq A.
; Mathews, V. J.: Finite-precision error analysis of a QR-decomposition-based lattice predictor,V1565,25-34(1991)

Syeda, Tanveer F.
: Finding distinctive colored regions in images,V1381,574-581(1991)

Sykes, Elizabeth See Kessel, David: V1426,180-187(1991)

Syme, Richard T.
; Kelly, Michael J.; Condie, Angus; Dale, Ian: Asymmetric superlattices for microwave detection,V1362,467-476(1991)

Synnott, Stephen P.
: Filled-arm Fizeau telescope,V1494,334-343(1991)

Synovec, Robert E.
; Renn, Curtiss N.: Novel approach for the refractive index gradient measurement in microliter volumes using fiber-optic technology,V1435,128-139(1991)
See Sulya, Andrew W.: V1434,147-158(1991)

Syrtzev, V. N.
See Feldman, G. G.: V1358,497-502(1991)
See Silkis, E. G.: V1358,46-49(1991)

Szabo, Arthur G. See Nadeau, Pierre: V1398,151-161(1991)

Szabo, Bela See Szabo, Raisa R.: V1468,794-801(1991)

Szabo, Gabor
See Hofmann, Thomas: V1412,84-90(1991)
See Tittel, Frank K.: V1397,21-29(1991)

Szabo, Raisa R.
; Pandya, Abhijit S.; Szabo, Bela: Knowledge-based system using a neural network,V1468,794-801(1991)

Szalowski, Rafal See Kramer, Charles J.: V1454,434-446(1991)

Szatkowski, Jan See Misiewicz, Jan: V1561,6-18(1991)

Szczepanski, Pawel
; Sklodowska, Malgorzata; Wolinski, Wieslaw: Mode frequency behaviour in distributed-feedback lasers,V1391,72-78(1991)
; Sklodowska, Malgorzata; Wolinski, Wieslaw: Nonlinear operation of fiber distributed-feedback lasers,V1391,65-71(1991)

Sze, Robert C.
: Fast-iterative technique for the calculation of frequency-dependent gain in excimer laser amplifiers,V1412,164-172(1991)

Sze, Wah Wai See Holm-Kennedy, James W.: V1527,322-331(1991)

Szeliski, Richard
: Probabilistic modeling of surfaces,V1570,154-165(1991)
; Terzopoulos, Demetri: Physically based and probabilistic models for computer vision,V1570,140-152(1991)
See Lavallee, Stephane: V1570,322-336(1991)

Szeregij, E. M.
; Ugrin, J. O.; Virt, I. S.; Abeynayake, C.; Kuzma, Marian: Influence of a high-power laser beam on electrophysical and photoelectrical properties of epitaxial films of CdxHg1-xTe (x is approximately equal to 0.2),V1391,199-203(1991)

Szeto, L. H. See White, Donald L.: V1343,204-213(1991)

Szeto, Roque K.
See Fried, David L.: V1416,163-176(1991)
See Fried, David L.: V1408,150-166(1991)

Szmacinski, Henryk See Lakowicz, Joseph R.: V1435,142-160(1991)

Szmanda, Charles R. See Barouch, Eytan: V1463,464-474(1991)

Szofran, Frank R. See Gillies, Donald C.: V1484,2-10(1991)

Szoplik, Tomasz See Garcia, Javier: V1574,133-140(1991)

Sztuka, Edward See Coden, Michael H.: V1364,22-39(1991)

Szumowski, J.
See Riek, Jonathan K.: V1398,130-141(1991)
See Riek, Jonathan K.: V1445,198-206(1991)

Szwaykowski, P. See Ojeda-Castaneda, Jorge: V1500,246-251(1991)

Szydlak, J. See Jankiewicz, Zdzislaw: V1391,101-104(1991)

Taalebinezhaad, M. A.
: Using fixation for direct recovery of motion and shape in the general case,V1388,199-209(1991)

Tabacco, Mary B. See Zhou, Quan: V1592,108-113(1991)

Tabata, A. See Banvillet, Henri: V1361,972-979(1991)

Tabata, Hirotsugu See Sakurada, Masami: V1425,158-164(1991)

Tabata, Yasuko See Yanagishita, Yuichiro: V1463,207-217(1991)

Tabbara, Marwan R.
; Cavaye, Douglas; Kopchok, George E.; White, Rodney A.: In-vivo intravascular ultrasound in human ileo-femoral vessels,V1425,208-216(1991)

Taber, R. C. See Muenchausen, Ross E.: V1394,221-229(1991)

Tabib-Azar, Massood
; Bhasin, Kul B.; Romanofsky, Robert R.: Comparative study for bolometric and nonbolometric switching elements for microwave phase shifters,V1477,85-94(1991)

Taboada, John See Griffin, Paul M.: V1381,292-298(1991)

Tabor, Mark See Potasek, M. J.: V1579,232-236(1991)

Taborek, Peter See Russell, Derrek: V1534,14-23(1991)

Tabouret, Evelyne See Heitzmann, Michel: V1392,272-279(1991)

Taboury, Jean
; Chavel, Pierre: Integration of an edge extraction cellular automaton,V1505,115-123(1991)

Tabuchi, Norio See Yamaguchi, Takao: V1418,363-371(1991)

Taccussel, M. C. See Lajzerowicz, Jean: V1392,222-231(1991)

Tachihara, Satoru
; Miyoshi, Tamihiro: Development of a high-speed high-precision laser plotter,V1527,305-314(1991)

Tackx, Peter See Persoons, Andre P.: V1409,220-229(1991)

Taczak, William J. See Troyer, David E.: V1486,396-409(1991)

Tada, Hiroaki See Wada, Tatsuo: V1560,162-171(1991)

Tada, Kunio See Nakano, Yoshiaki: V1418,250-260(1991)

Tada, Shunkichi See Miyahara, Sueharu: V1384,317-323(1991)

Taddei, Giuseppe See Barbieri, Enrico: V1425,122-127(1991)

Tadokoro, T. See Tamanuki, Takemasa: V1361,614-617(1991)

Tadros, Karim H.
; Neureuther, Andrew R.; Guerrieri, Roberto: Understanding metrology of polysilicon gates through reflectance measurements and simulation,V1464,177-186(1991)
See Doi, Takeshi: V1464,336-345(1991)

Tadros, Sobhy P. See Honda, Kenji: V1466,141-148(1991)

Taff, Laurence G.
: Scientific results from the Hubble Space Telescope fine-guidance sensors,V1494,66-77(1991)
; Belkin, Barry; Schweiter, G. A.; Sommar, K.: Statistical initial orbit determination,V1481,440-448(1991)

Tafto, J. See Hugsted, B.: V1361,751-757(1991)

Taga, Karim See Kellner, Robert A.: V1510,232-241(1991)

Taghizadeh, Mohammad R.
See Craig, Robert G.: V1505,76-86(1991)
See Vasara, Antti H.: V1507,224-238(1991)

Tagliaferri, V.
See Covelli, L.: V1397,797-802(1991)
See De Iorio, I.: V1397,787-790(1991)

Tagliaferro, Alberto See Demichelis, Francesca: V1534,140-147(1991)

Taguchi, Takashi See Ohtsuka, Hiroshi: V1463,112-123(1991)

Tahani, Hossein
; Keller, James M.: Automated calculation of nonadditive measures for object recognition,V1381,379-389(1991)

Tahata, Shin See Nakajima, Hajime: V1456,29-39(1991)

Tahkokorpi, Markku T. See Tervonen, Ari: V1513,71-75(1991)

Tahmoush, Donald J.
See Huguenin, Robert L.: V1479,403-411(1991)
See Huguenin, Robert L.: V1481,64-72(1991)

Tahouri, Bahman
See Disimile, Peter J.: V1489,66-74(1991)
See Shoe, Bridget: V1521,34-45(1991)

Tai, Alan C. See Egalon, Claudio O.: V1489,9-16(1991)

Tai, Woon W. See Novembre, Anthony E.: V1466,89-99(1991)

Taii, T. See Watanabe, Hitoshi: V1499,21-28(1991)

Taiji, Makoto See Kobayashi, Takayoshi T.: V1403,407-416(1991)

Taira, Kazuo
; Takahashi, Junichi; Yanagihara, Kenji: Effect of silylation condition on the silylated image in the DESIRE process,V1466,570-582(1991)

Taira, Ricky K.
; Chan, Kelby K.; Stewart, Brent K.; Weinberg, Wolfram S.: Reliability issues in PACS,V1446,451-458(1991)
See Hayrapetian, Alek: V1446,243-247(1991)
See McNitt-Gray, Michael F.: V1445,468-478(1991)
See Razavi, Mahmood: V1446,24-34(1991)
See Weinberg, Wolfram S.: V1446,35-39(1991)

Taira, Tatsuji See Kanazawa, Hirotaka: V1397,445-448(1991)

Tait, C. D. See Berg, John M.: V1435,331-337(1991)

Tait, Mary B. See Greenleaf, William G.: V1486,364-375(1991)

Tajbakhsh, Shahram
; Boyce, James F.: Interframe registration and preprocessing of image sequences,V1521,14-22(1991)

Tajime, Toru See Matsushita, Tadashi: V1488,368-375(1991)

Tajiri, Atsushi See Yamaguchi, Takao: V1418,363-371(1991)

Takabatake, Hirotsugu
See Inaoka, Noriko: V1450,2-12(1991)
See Suzuki, Hideo: V1450,99-107(1991)

Takacs, Peter Z.
See Church, Eugene L.: V1332,504-514(1991)
See Church, Eugene L.: V1530,71-85(1991)
See Spiller, Eberhard: V1343,134-144(1991)

Takada, Hisashi See Fukuoka, Takashi: V1364,40-48(1991)

Takada, Kazumasa See Kobayashi, Masaru: V1474,278-284(1991)

Takada, Michinosuke See Oda, Ichiro: V1431,284-293(1991)

Takada, Tomoji See Kitagaki, Kazukuni: V1606,891-900(1991)

Takada, Yutaka
; Uchida, Yoshiyuki; Akao, Yasuo; Yamada, Jun; Hattori, Shuzo: Super-accurate positioning technique using diffracted moire signals,V1332,571-576(1991)

Takagi, Naofumi
: Arithmetic unit based on a high-speed multiplier with a redundant-binary addition tree,V1566,244-251(1991)

Takagi, S. See Ishikawa, Ken: V1397,55-58(1991)

Takagi, Yuuji
; Hata, Seiji; Hibi, Susumu; Beutel, Wilhelm: Visual inspection machine for solder joints using tiered illumination,V1386,21-29(1991)

Takahashi, A. See Yamakoshi, H.: V1397,119-122(1991)

Takahashi, Eietsu
; Tanji, Shigeo; Tanaka, Akira; Hayashi, Yuji: Plastic lens array with the function of forming unit magnification erect image using roof prisms,V1527,145-154(1991)

Takahashi, Fumiho
; Hiramatsu, Masaru; Watanabe, Fumito; Narimatsu, Yoshito; Nagura, Riichi: Visible and near-infrared subsystem and common signal processor design status of ASTER,V1490,255-268(1991)

Takahashi, Fumitaka See Ogura, Toshihiro: V1443,153-157(1991)

Takahashi, Hidenori See Fukuoka, Takashi: V1364,40-48(1991)

Takahashi, Hiroki See Agui, Takeshi: V1606,188-197(1991)

Takahashi, Hiroo See Maeda, Fumisada: V1499,62-69(1991)

Takahashi, Hiroyasu See Amano, Tomio: V1452,330-339(1991)

Takahashi, Junichi See Taira, Kazuo: V1466,570-582(1991)

Takahashi, Kazuhiro
; Ohta, Masakatsu; Kojima, Toshiyuki; Noguchi, Miyoko: New i-line lens for half-micron lithography,V1463,696-708(1991)

Takahashi, Kei
See Isobe, Hiroshi: V1358,471-478(1991)
See Sato, Eiichi: V1358,146-153(1991)
See Sato, Eiichi: V1358,193-200(1991)
See Sato, Eiichi: V1358,462-470(1991)
See Shikoda, Arimitsu: V1358,154-161(1991)

Takahashi, Kunimitsu
See Kuribayashi, Shizuma: V1397,439-443(1991)
See Noda, Osama: V1397,427-432(1991)
See Sato, Shunichi: V1397,421-425(1991)
See Sato, Shunichi: V1397,433-437(1991)

Takahashi, Mamoru See Oyaizu, Ikuro: V1606,990-1001(1991)

Takahashi, Masahiko See Nakao, Takeshi: V1499,433-437(1991)

Takahashi, Rie See Niibe, Masahito: V1343,2-13(1991)

Takahashi, Satoru See Miyoshi, Masahiro: V1499,116-119(1991)

Takahashi, Susumu
; Toda, Toshiki; Iwata, Fujio: Three-dimensional grating images,V1461,199-205(1991)

Takahashi, Takashi
See Komori, Masaru: V1444,334-337(1991)
See Minato, Kotaro: V1446,195-198(1991)

Takahashi, Tomoyuki See Aoki, Nobutada: V1412,2-11(1991)

Takamoto, Takenori
; Schwartz, Bernard: Photogrammetric measurements of retinal nerve fiber layer thickness along the disc margin and peripapillary region,V1380,64-74(1991)

Takano, Hiroshi See Kashima, Yasumasa: V1365,102-107(1991)

Takashima, Toshiaki See Ashidate, Shu-ichi: V1397,283-286(1991)

Takata, H. See Hirose, Masataka: V1362,316-322(1991)

Takatsu, Kazuaki See Ogura, Toshihiro: V1443,153-157(1991)

Takayama, Kazuyoshi
; Obara, Tetsuro; Onodera, Osamu: Holographic interferometric observation of shock wave focusing to extracorporeal shock wave lithotripsy,V1358,1180-1190(1991)

Takebe, Hiromichi See Kai, Teruhiko: V1519,99-103(1991)

Takeda, H. See Thode, Lester E.: V1552,87-106(1991)

Takeda, Haruo
: Extracting characters from illustration document by relaxation,V1386,128-134(1991)

Takei, Mitsuru See Hino, Hideo: V1490,166-176(1991)

Takemori, Toshikazu
; Miyasaka, Nobuji; Hirose, Shozo: Neural network air-conditioning system for individual comfort,V1469,157-165(1991)

Takemoto, Cliff H.
; Ziger, David H.; Connor, William; Distasio, Romelia: Resist tracking: a lithographic diagnostic tool,V1464,206-214(1991)
See Nalamasu, Omkaram: V1466,13-25(1991)

Takemura, Kazuo See Kawahara, Hideo: V1513,198-203(1991)

Takenaka, Hisataka
; Ishii, Yoshikazu: Lateral-periodicity evaluation of multilayer Bragg reflector surface roughness using x-ray diffraction,V1345,180-188(1991)
; Ishii, Yoshikazu; Kinoshita, Hiroo; Kurihara, Kenji: Soft and hard x-ray reflectivities of multilayers fabricated by alternating-material sputter deposition,V1345,213-224(1991)

Takeo, Takashi
; Hattori, Hajime: U-shaped fiber-optic refractive-index sensor and its applications,V1544,282-286(1991)

Takeshita, Nobuo See Irie, Mitsuru: V1499,360-365(1991)

Takeuchi, Ichiro
; Tsai, Jaw S.; Tsuge, Hisanao; Matsukura, Noritsuga; Miura, Sadahiko; Yoshitake, T.; Kojima, Yoshikatsu; Matsui, Shinji: Cleaved surfaces of high Tc films for making SNS structures,V1394,96-101(1991)

Takeuchi, Kiyoshi
See Arai, Tsunenori: V1425,191-195(1991)
See Sakurada, Masami: V1425,158-164(1991)

Takeuchi, Susumu See Wiley, James N.: V1464,346-355(1991)

Takeyama, Tetsu See Nishimura, Tetsuya: V1436,31-37(1991)

Takigawa, Akio
See Kitaoka, Masaki: V1519,109-114(1991)
See Maeda, Koichi: V1536,138-148(1991)

Takishima, Yasuhiro
; Wada, M.; Murakami, Hitomi: Analysis of optimum-frame-rate in low-bit-rate video coding,V1605,635-645(1991)

Takita, Hiroshi See Regal, Anne-Marie: V1426,271-278(1991)

Takizawa, Masaaki See Kinoshita, Taizo: V1605,604-613(1991)

Talano, James V. See Frazin, Leon J.: V1425,207-207(1991)

Talatinian, A.
; Pluta, Mieczyskaw: Analytical design of curved holographic optical elements for Fourier transform,V1574,205-217(1991)

Talbot, Peter J. See Song, Q. W.: V1558,143-148(1991)

Talisa, Salvador H.
: Status of high-temperature superconducting analog devices,V1477,78-83(1991)

Tallant, David R. See Cahill, Paul A.: V1560,130-138(1991)

Tallent, Jack R. See Birge, Robert R.: V1432,129-140(1991)

Tallents, Gregory J.
; Key, Michael H.; Norreys, P.; Jacoby, J.; Kodama, R.; Tragin, N.; Baldis, Hector A.; Dunn, James; Brown, D.: Experiments with SPRITE 12 ps facility,V1413,70-76(1991)
See Batani, D.: V1503,479-491(1991)
See Turcu, I. C.: V1503,391-405(1991)

Tallon, M. See Michau, Vincent: V1487,64-71(1991)

Talmore, Eli T.
: CiNeRaMa model: a useful tool for detection range estimate,V1442,362-371(1991)
See Ben-Shalom, Ami: V1486,238-257(1991)

Talon, Benedicte See Trigano, Philippe: V1468,866-874(1991)

Tam, Andrew C.
: Photothermal spectroscopy as a sensitive spectroscopic tool,V1435,114-127(1991)
; Zapka, Werner; Ziemlich, Winfrid: Efficient laser cleaning of small particulates using pulsed laser irradiation synchronized with liquid-film deposition,V1598,13-18(1991)

Tam, Nelson See Chiu, Anita S.: V1466,641-652(1991)

Tam, Siu-Chung
; Silva, Donald E.; Wong, H. L.: Digital Talbot interferometer,V1400,38-48(1991)
See Kuok, M. H.: V1399(1991)
See Rao, M. K.: V1399,116-121(1991)

Tam, Wa J. See Stelmach, Lew B.: V1453,147-152(1991)

Tamagawa, Tohru See Yasuoka, Koichi: V1412,32-37(1991)

Tamakawa, Yoshiharu
See Isobe, Hiroshi: V1358,471-478(1991)
See Isobe, Hiroshi: V1358,201-208(1991)
See Sato, Eiichi: V1358,146-153(1991)
See Sato, Eiichi: V1358,193-200(1991)
See Sato, Eiichi: V1358,479-487(1991)
See Sato, Eiichi: V1358,462-470(1991)
See Shikoda, Arimitsu: V1358,154-161(1991)

Tamanuki, Takemasa
; Tadokoro, T.; Morito, K.; Koyama, Fumio; Iga, Kenichi: Microfabrication techniques for semiconductor lasers,V1361,614-617(1991)

Tamburino, Louis A.
See Rizki, Mateen M.: V1469,384-391(1991)
See Zmuda, Michael A.: V1470,183-189(1991)

Tamkivi, Raivo
; Schotter, Leo L.; Pakhomova, Tat'yana A.: Excimer laser cutting of corneal transplants,V1503,375-378(1991)

Tammela, Simo
; Pohjonen, Harri; Honkanen, Seppo; Tervonen, Ari: Fabrication of large multimode glass waveguides by dry silver ion exchange in vacuum,V1583,37-42(1991)
; Zhan, Xiaowei; Kiiveri, Pauli: Comparison of gain dependence of different Er-doped fiber structures,V1373,103-110(1991)
See Karioja, Pentti: V1533,129-137(1991)

Tamtaoui, Ahmed
; Labit, Claude: 3-D TV: joined identification of global motion parameters for stereoscopic sequence coding,V1605,720-731(1991)

Tamura, Akira
; Yonezawa, Masaji; Sato, Mitsuyoshi; Fujimoto, Yoshiaki: High-sensitivity and high-dry-etching durability positive-type electron-beam resist,V1465,271-281(1991)

Tamura, Mamoru
See Kakihana, Yasuyuki: V1431,314-321(1991)
See Nomura, Yasutomo: V1431,102-109(1991)

Tamura, Masahide See Oda, Ichiro: V1431,284-293(1991)

Tamura, Tomomi See Oda, Ichiro: V1431,284-293(1991)

Tamura-Lis, Winifred See Montgomery, G. P.: V1455,45-53(1991)

Tan, Chin
; Carlson, Robert T.: Liquid crystals for lasercom applications,V1417,391-401(1991)

Tan, Daniel H. See Sung, Eric: V1385,57-69(1991)

Tan, Hui
; Tao, Ming D.; Gin, Dong; Han, Ying; Lin, Cheng L.: Small polaron conduction in amorphous CoMnNiO thin film,V1519,752-756(1991)

Tan, Jiubin
; Qiang, Xifu; Ding, Xuemei: Clinical measuring system for the form and position errors of circular workpieces using optical fiber sensors,V1572,552-557(1991)

Tan, K. See Chow, P. D.: V1475,48-54(1991)

Tan, L. S. See Subramanian, K. R.: V1364,190-201(1991)

Tan, Michael T. See Wang, S. Y.: V1371,98-103(1991)

Tan, Oon T.
; White, Rodney A.; White, John V.; eds.: *Lasers in Dermatology and Tissue Welding*,V1422(1991)
See Cheong, Wai-Fung: V1422,19-26(1991)
See Kurban, Amal K.: V1422,43-49(1991)
See Yasuda, Yukio: V1422,50-55(1991)

Tan, Ronson K. See Verber, Carl M.: V1374,68-73(1991)

Tan, Shiro See Honda, Kenji: V1466,141-148(1991)

Tan, Shun See Zhang, Shu Y.: V1519,43-46(1991)

Tan, Swie-In
; Colavito, D. B.: Contamination and damage of silicon surfaces during magnetron-enhanced reactive ion etching in a single-wafer system,V1392,106-118(1991)

Tan, T. K. See Sikora, Thomas: V1605,624-634(1991)

Tan, Weihong See Kopelman, Raoul: V1435,96-101(1991)

Tan, Yushan
See Chen, Duanjun: V1554A,706-717(1991)
See Chen, Wenyi: V1554A,879-885(1991)
See Fang, Qiang: V1332,216-222(1991)
See Fang, Qiang: V1554A,696-704(1991)
See Lai, Tianshu: V1554B,580-585(1991)
See Lai, Tianshu: V1461,286-290(1991)
See Li, Xide: V1554B,661-668(1991)
See Wang, Xizhou: V1554A,907-914(1991)
See Wen, Zheng: V1461,278-285(1991)
See Yang, Yu X.: V1554B,768-773(1991)

Tanabe, Setsuhisa
; Hirao, Kazuyuki; Soga, Naohiro; Hanada, Teiichi: Upconversion intensity and multiphonon relaxation of Er3+-doped glasses,V1513,340-348(1991)

Tanaka, Akira
See Kurosawa, Kiyoshi: V1368,150-156(1991)
See Takahashi, Eietsu: V1527,145-154(1991)

Tanaka, Hirokazu See Iwasaki, Nobuo: V1490,216-221(1991)

Tanaka, Ikuzo See Sato, Shunichi: V1397,421-425(1991)

Tanaka, Kiyoshi
; Nakamura, Yasuhiro; Matsui, Kineo: Secret transmission method of character data in motion picture communication,V1605,646-649(1991)

Tanaka, Kiyoto See Oyaizu, Ikuro: V1606,990-1001(1991)

Tanaka, Kunimaro See Shimamoto, Masayoshi: V1499,393-400(1991)

Tanaka, Masahiko See Satoh, Hiroharu: V1499,324-329(1991)

Tanaka, Minoru See Yoshitake, Yasuhiro: V1463,678-687(1991)

Tanaka, Satoru C.
; Silvey, Tom; Long, Greg; Braze, Bill: High-resolution, low-light, image-intensified CCD camera,V1448,21-26(1991)

Tanaka, Satoshi
See Nagano, Yasutada: V1605,614-623(1991)
See Ohtsuka, Yoshihiro: V1572,347-352(1991)
See Sukeda, Hirofumi: V1499,419-425(1991)

Tanaka, Shin-ichi See Ishibashi, Hiromichi: V1499,340-347(1991)

Tanaka, T. See Tsumanuma, Takashi: V1420,193-198(1991)

Tanaka, Takashi See Wiley, James N.: V1464,346-355(1991)

Tanaka, Toshihiko See Terasawa, Tsuneo: V1463,197-206(1991)

Tanaka, Toyoichi
; Nishio, Izumi; Peetermans, Joyce; Gorti, Sridhar: Quasi-elastic light scattering spectroscopy of single biological cells under a microscope,V1403,280-287(1991)

Tanaka, Yuki
; Miyamae, Hiroshi: Analysis on image performance of a moisture-absorbed plastic singlet for an optical disk,V1354,395-401(1991)

Tanazawa, Takeshi
; Adachi, Hajime A.; Nakahara, Ktsuhiko; Nittoh, Koichi; Yoshida, Toshifumi; Yoshida, Tadashi; Matsuda, Yasuhiko: Photo resonance excitation and ionization characteristics of atoms by pulsed laser irradiation,V1435,310-321(1991)

Tanbun-Ek, Tawaee See Temkin, Henryk: V1418,88-98(1991)

Taneya, M. See Akita, K.: V1392,576-587(1991)

Tang, Cha-Mei See Marable, William P.: V1552,80-86(1991)

Tang, Chong Z.
: Amorphous microcellular polytetrafluoroethylene foam film,V1519,842-846(1991)

Tang, D. See Hayes, John M.: V1435,258-266(1991)

Tang, Di-zhu See Xia, Sheng-jie: V1358,43-45(1991)

Tang, Jia T. See Yu, Ju X.: V1519,308-314(1991)

Tang, Jinfa
See Liu, Xu: V1554A,558-569(1991)
See Mei, Ting: V1554A,570-578(1991)
See Zhengmin, Li: V1401,66-73(1991)

Tang, Kai See Haglund, Richard F.: V1441,127-138(1991)

Tang, Minguang See Zou, Zi-Li: V1572,18-26(1991)

Tang, Qing See Javidi, Bahram: V1564,212-223(1991)

Tang, Ru-Shyah See Yeh, Long-Ching: V1527,361-367(1991)

Tang, Shou-Hong
; Hung, Yau Y.; Zhu, Qiuming: New phase measurement for nonmonotonical fringe patterns,V1332,731-737(1991)
See Hung, Yau Y.: V1332,332-342(1991)
See Hung, Yau Y.: V1332,738-747(1991)
See Hung, Yau Y.: V1332,696-703(1991)

Tang, Shunqing See Zhang, Jingfang: V1238,401-405(1991)

Tang, Wade C.
; Rosen, Hal J.; Vettiger, Peter; Webb, David J.: Time development of AlGaAs single-quantum-well laser facet temperature on route to catastrophical breakdown,V1418,338-342(1991)

Tang, Weiguo See Dahe, Liu: V1507,310-315(1991)

Tang, Weizhong See Chen, Xiaoguang: V1572,294-298(1991)

Tang, Wen G. See Li, Jie: V1519,660-664(1991)

Tang, Wu-bin See Chen, Su-Shing: V1569,446-450(1991)

Tang, Xia See Yang, Bang C.: V1519,864-865(1991)

Tang, Xue F.
; Fan, Zheng X.; Wang, Zhi-Jiang: Optical properties of oxide films prepared by ion-beam-sputter deposition,V1519,96-98(1991)

Tang, Yaw-Tzong
; Lee, Wai-Hon; Chang, Ming-Wen: Design of CGH for special image formation,V1555,194-199(1991)

Tang, Yonghong
; Wee, William G.; Han, Chia Y.: Application of a multilayer network in image object classification,V1469,113-120(1991)

Tang, Zhi J. See Wang, Yue: V1519,428-433(1991)

Tanganelli, Pietro See Barbieri, Enrico: V1425,122-127(1991)

Tange, Yoshio
See Iwasaki, Nobuo: V1490,216-221(1991)
See Iwasaki, Nobuo: V1490,192-199(1991)
See Nakayama, Masao: V1490,207-215(1991)
See Tanii, Jun: V1490,200-206(1991)

Tangonan, Gregory L. See Ng, Willie W.: V1371,205-211(1991)

Tanguy, Mireille
; Bonhommet, Herve; Autric, Michel L.; Vigliano, Patrick: Correlation between the aerosol profiles measurements, the meteorological conditions, and the atmospheric IR transmission in a mediterranean marine atmosphere,V1487,172-183(1991)

Tani, Yuichiro See Oyama, Yoshiro: V1444,413-423(1991)

Tanida, Jun See Kakizaki, Sunao: V1505,199-205(1991)

Tanii, Jun
; Machida, Tsuneo; Ayada, Haruki; Katsuyama, Yoshihiko; Ishida, Juro; Iwasaki, Nobuo; Tange, Yoshio; Miyachi, Yuji; Sato, Ryota: Ocean color and temperature scanner for ADEOS,V1490,200-206(1991)
See Nakayama, Masao: V1490,207-215(1991)

Tanimizu, Katsuyuki See Meguro, Shin-Ichi: V1384,27-37(1991)

Tanimoto, Masayuki
; Yamada, Akio; Naito, Yoichi: New subband scheme for super-HDTV coding,V1605,394-405(1991)

Tanimoto, Tetsuzou See Yoshitake, Yasuhiro: V1463,678-687(1991)

Tanin, Leonid V.
See Dick, Sergei C.: V1454,447-452(1991)
See Markhvida, Igor V.: V1508,128-134(1991)
See Rachkovsky, Leonid I.: V1461,232-240(1991)
See Rachkovsky, Leonid I.: V1507,450-457(1991)

Tanji, Shigeo See Takahashi, Eietsu: V1527,145-154(1991)

Tankersley, Lawrence L. See Duncan, Michael D.: V1409,127-134(1991)

Tannen, Peter D. See Walter, Robert F.: V1397,71-76(1991)

Tanner, John E.
; Luo, Jin: Single-chip imager and feature extractor,V1473,76-87(1991)

Tantawy, Ahmed N.
: Bridging issues in DQDB subnetworks,V1364,268-276(1991)

Tanwar, Lakhan S.
; Jain, P. C.; Kunzmann, H.: Design parameters of an EO sensor,V1332,877-882(1991)

Tao, Chin-Wang See Thompson, Wiley E.: V1470,48-58(1991)

Tao, Chunkan See Shen, Weisheng: V1567,691-697(1991)

Tao, Huiying
See Hsu, Dahsiung: V1238,13-17(1991)
See Zou, Yunlu: V1238,452-456(1991)

Tao, Ming D. See Tan, Hui: V1519,752-756(1991)

Tao, William C.
; Frank, Alan M.; Clements, Rochelle E.; Shepherd, Joseph E.: Aluminum metal combustion in water revealed by high-speed microphotography,V1346,300-310(1991)

Tapanes, Edward
: Real-time structural integrity monitoring using a passive quadrature demodulated, localized Michelson optical fiber interferometer capable of simultaneous strain and acoustic emission sensing,V1588,356-367(1991)
See Measures, Raymond M.: V1332,421-430(1991)

Tapolsky, Gilles See Meyrueix, Remi: V1560,454-466(1991)

Tarabelli, D.
; Zeitoun, David; Imbert, Michel P.: Flowfield in a CO2-N2 gas-dynamics laser with staggered nozzles,V1397,523-526(1991)
See Zeitoun, David: V1397,585-588(1991)

Taranenko, Victor B.
; Vasnetsov, M. V.: Optical bistability and signal competition in active cavity with photochromic nonlinearity of bacteriorhodopsin,V1621,169-179(1991)
See Bazhenov, V. Y.: V1574,148-153(1991)

Tarapov, Sergey I. See Vertiy, Alexey A.: V1361,1070-1078(1991)

Tarasenko, Victor F.
: Laser action of Xe and Ne pumped by electron beam,V1412,185-196(1991)

Tarasov, I. S. See Garbuzov, Dmitriy Z.: V1418,386-393(1991)

Tarasov, Victor A.
; Kuleshov, Nickolay B.; Novoselets, Mikhail K.; Sarkisov, Sergey S.: Optimal characteristics of rheology and electric field in deformable polymer films of optoelectronic image formation devices,V1559,331-342(1991)

Taratorin, B. I.
; Sakharov, V. N.; Komlev, O. U.; Stcherbakov, V. N.; Starchevsky, A. V.: Photoelastic investigation of statics, kinetics, and dynamics of crack formation in transparent models and natural structural elements,V1554A,449-456(1991)

Targ, Russell
; Bowles, Roland L.; Korb, C. L.; Gentry, Bruce M.; Souilhac, Dominique J.: Infrared lidar windshear detection for commercial aircraft and the edge technique: a new method for atmospheric wind measurement,V1521,144-157(1991)
; Bowles, Roland L.: Lidar wind shear detection for commercial aircraft,V1416,130-138(1991)
; Hawley, James G.; Otto, Robert G.; Kavaya, Michael J.: Coherent launch-site atmospheric wind sounder,V1478,211-227(1991)

Targowski, Piotr
; Zietek, Bernard: Energy of the lowest triplet states of rhodamines in ethanolic solutions,V1391,12-17(1991)
See Zietek, Bernard: V1391,18-23(1991)

Taroni, Paola See Cubeddu, Rinaldo: V1525,17-25(1991)

Tarr, Gregory L.
; Rogers, Steven K.; Kabrisky, Matthew; Oxley, Mark E.; Priddy, Kevin L.: Generalized neural networks for tactical target image segmentation,V1469,2-11(1991)
See Rogers, Steven K.: VIS07,231-243(1991)

Tarr, Tomas See Holland, Orgal T.: V1469,102-112(1991)

Tarrant, Peter J.
; Truman, Alan K.: Implementation of FDDI in the intelligent wiring hub,V1364,2-6(1991)

Tarry, H. A. See Ashley, Timothy: V1361,238-244(1991)

Tarver, Craig M. See Breithaupt, R. D.: V1346,96-102(1991)

Tas, Jeroen
: Commercial applications for optical data storage,V1401,10-18(1991)

Tascini, Guido
; Puliti, Paolo; Zingaretti, Primo: Decision support system for capillaroscopic images,V1450,178-185(1991)

Tashiro, Hideo See Hatanaka, Hidekazu: V1397,379-382(1991)

Tashiro, M. See Ohashi, T.: V1549,9-19(1991)

Tashiro, Masaru See Horikawa, Hiroshi: V1454,46-59(1991)

Tasker, Paul J. See Huelsmann, Axel: V1465,201-208(1991)

Tatam, Ralph P.
See Atcha, Hashim: V1504,221-232(1991)
See Atcha, Hashim: V1584,425-434(1991)
See Duffy, Christopher J.: V1511,155-165(1991)
See Hamid, Sohail: V1511,78-89(1991)

Tatarintsev, S. V. See Guseva, N. P.: V1403,332-334(1991)

Tatarkiewicz, Jakub J. See Iwanowski, Ryszard J.: V1361,765-775(1991)

Tatarskii, V. I.
: Turbulence at the inner scale,V1408,2-9(1991)

Tataurschikov, S. S. See Khvilivitzky, A. T.: V1447,184-190(1991)

Tate, Kevin J.
; Li, Ze-Nian: Multiresolution range-guided stereo matching,V1383,491-502(1991)

Tate, Lloyd P.
; Glasser, Mardi: Six years of transendoscopic Nd:YAG application in large animals,V1424,209-217(1991)

Tatian, Berge
: Nonlinearity and lens design,V1354,154-164(1991)

Tatlock, Timothy See O'Byrne, Vincent A.: V1372,88-93(1991)

Tatum, Jim A.
; MacFarlane, Duncan L.: Ultrashort pulse propagation in visible semiconductor diode laser amplifiers,V1497,320-325(1991)
See MacFarlane, Duncan L.: V1365,88-95(1991)

Tatum, William F.
: Model-based ATR (automatic target recognition) systems for the military,VIS07,130-139(1991)

Taubes, C. H.
; Liang, Ping: Orientation-based differential geometric representations for computer vision applications,V1570,96-102(1991)

Taubin, Gabriel
; Cooper, David B.: Recognition and positioning of rigid objects using algebraic moment invariants,V1570,175-186(1991)

Taubkin, I. I. See Khryapov, V. T.: V1540,412-423(1991)

Taubman, Matthew See Greene, Ben: V1492,126-139(1991)

Tauschek, D. See Frank, Klaus: V1431,2-11(1991)

Tavakoli, Nassrin
: Information preserving image data compression,V1452,371-382(1991)

Tavassoly, M. T.
: Interference grating with an arbitrary transparent-to-opaque groove ratio,V1527,392-399(1991)

Tavlykaev, R. F. See Baranov, D. V.: V1583,389-394(1991)

Tavora, Jose
; Lourtie, Pedro M.: Parallel message-passing architecture for path planning,V1468,524-535(1991)

Tavrov, Alexander V. See Tychinsky, Vladimir P.: V1500,207-210(1991)

Tay, C. J. See Shang, H. M.: V1554B,680-691(1991)

Taya, Katsuo See Nakamura, Hiroyuki: V1456,226-238(1991)

Taya, Makoto See Gemma, Takashi: V1332,77-84(1991)

Tayeb, Gerard
; Petit, Roger; Cadilhac, M.: Synthesis method applied to the problem of diffraction by gratings: the method of fictitious sources,V1545,95-105(1991)
See Petit, Roger: V1545,31-41(1991)

Taylor, Andrew T.
; Wang, Paul K.: Machine vision feedback for on-line correction of manufacturing processes: a control formulation,V1567,220-231(1991)

Taylor, Antoinette J.
See Bigio, Irving J.: V1397,47-53(1991)
See Gosnell, Timothy R.: VMS44(1991)

Taylor, Arthur F. See Fieret, Jim: V1503,53-61(1991)

Taylor, Christopher J. See Davis, Darryl N.: V1445,421-432(1991)

Taylor, D. L. See Nederlof, Michel A.: V1428,233-241(1991)

Taylor, Donna L. See Vethanayagam, Thirukumar K.: V1366,343-350(1991)

Taylor, Edward W.
; Padden, Richard J.; Sanchez, Anthony D.; Chapman, S. P.; Berry, J. N.; DeWalt, Steve A.: Radiation-induced crosstalk in guided-wave devices,V1474,126-131(1991)
See Padden, Richard J.: V1474,148-159(1991)

Taylor, Gary N.
; Hutton, Richard S.; Windt, David L.; Mansfield, William M.: Resist schemes for soft x-ray lithography,V1343,258-273(1991)
See Novembre, Anthony E.: V1466,89-99(1991)

Taylor, Geoff L.
; Derksen, Grant: Improved precision/resolution by camera movement,V1526,27-34(1991)

Taylor, Geoff W.
; Cooke, Paul W.; Kiely, Philip A.; Claisse, Paul R.; Sargood, Stephen K.; Doctor, D. P.; Vang, T.; Evaldsson, P.; Daryanani, Sonu L.: Electronic/photonic inversion channel technology for optoelectronic ICs and photonic switching,V1476,2-13(1991)

Taylor, Gordon C.
; Bechtle, Daniel W.; Jozwiak, Phillip C.; Liu, Shing G.; Camisa, Raymond L.: Novel selective-plated heatsink, key to compact 2-watt MMIC amplifier,V1475,103-112(1991)
See Bharj, Sarjit: V1475,340-349(1991)

Taylor, Henry F.
: Application of fiber optic delay lines and semiconductor optoelectronics to microwave signal processing,V1371,150-160(1991)
; Gweon, S.; Fang, S. P.; Lee, Chung E.: Fiber optic delay lines for wideband signal processing,V1562,264-275(1991)
See Bell, John A.: V1476,326-329(1991)
See Gong, Xue-Mei: V1418,422-433(1991)
See Gopalakrishnan, G. K.: V1476,270-275(1991)
See Kobayashi, Masaru: V1474,278-284(1991)
See Lee, Chung E.: V1584,396-399(1991)
See Lee, Chung E.: V1588,110-116(1991)
See Yeh, Yunhae: V1584,72-78(1991)
See Yuan, Ruixi: V1497,313-319(1991)

Taylor, Ivor See Sipila, Heikki: V1549,246-255(1991)

Taylor, James S.
; Krapels, Keith A.: Frequency-modulated reticle trackers in the pupil plane,V1478,41-49(1991)

Taylor, James W.
See Babcock, Carl P.: V1466,653-662(1991)
See Berry, Amanda K.: V1465,210-220(1991)

Taylor, Joann M.
: Integrated color management: the key to color control in electronic imaging and graphic systems,V1460,2-10(1991)

Taylor, John A. See Pillsbury, Allen D.: V1417,292-299(1991)

Taylor, John R. See Tesar, Aleta A.: V1441,154-172(1991)

Taylor, Kenneth See Ishaq, Naseem: V1381,153-159(1991)

Taylor, Lawrence W. See Huang, Jen-Kuang: V1489,266-277(1991)

Taylor, Michael S.
: High-level PC-based laser system modeling,V1415,300-309(1991)

Taylor, R. See Wareberg, P. G.: V1538,112-123(1991)

Taylor, Raymond L. See Goela, Jitendra S.: V1330,25-38(1991)

Taylor, Robert A.
; Godbey, Luther C.: Using digital images to measure and discriminate small particles in cotton,V1379,16-27(1991)

Taylor, Roderick S.
; Gladysz, D.; Brown, D.; Higginson, Lyall A.: Laser-induced fluorescence imaging of coronary arteries for open-heart surgery applications,V1420,183-192(1991)

Taylor, S. See Acebal, Robert: V1397,191-196(1991)

Taylor, Stuart R. See Quesenberry, Laura A.: V1428,177-190(1991)

Taylor, William W.
: Quantifying predictability for applications in signal separation,V1565,267-278(1991)

Tchandjou, N. See Montgomery, Paul C.: V1332,515-524(1991)

Tcherniega, Nicolaii V. See Kudryavtseva, Anna D.: V1385,190-199(1991)

Tchoryk, Peter
; Gleichman, Kurt W.; Carmer, Dwayne C.; Morita, Yuji; Trichel, Milton; Gilbert, R. K.: Passive and active sensors for autonomous space applications,V1479,164-182(1991)
See Gleichman, Kurt W.: V1416,286-294(1991)

Teaford, Mark F.
: Measurements of teeth using the Reflex Microscope,V1380,33-44(1991)

Teague, Peter F. See Bixler, Jay V.: V1549,420-428(1991)

Teat, William See Yeh, Chao-Pin: V1389,187-198(1991)

Tecotzky, Raymond H. See Huang, H. K.: V1444,214-220(1991)

Tedesco, Serge V.
See Goethals, Anne-Marie: V1466,604-615(1991)
See Lajzerowicz, Jean: V1392,222-231(1991)

Tedjojuwono, Ken K.
: Microdisplacement fiber sensor using two-frequency interferometry,V1584,146-151(1991)

Teeters, Jeffrey L.
; Werblin, Frank S.: Real-time simulation of the retina allowing visualization of each processing stage,V1472,6-17(1991); 1473,102-113(1991)

Teh, H. H. See Hsu, L. S.: V1469,197-207(1991)

Tehrani, Saeid
; Weymouth, Terry E.; Mancini, G. B.: Model generation and partial matching of left ventricular boundaries,V1445,434-445(1991)

Teillet, Philippe M.
; Fedosejevs, Gunar; Ahern, Francis J.; Gauthier, Robert P.; Sirois, J.: Atmospheric code sensitivity to uncertainties in aerosol optical depth characteristics,V1492,213-223(1991)

Teivans, A. See Purinsh, Juris: V1525,289-308(1991)

Tejika, Yasuhiro See Fukuoka, Takashi: V1364,40-48(1991)

Tekalp, Ahmet M.
See Ozkan, Mehmet K.: V1452,146-156(1991)
See Ozkan, Mehmet K.: V1606,743-754(1991)
See Riek, Jonathan K.: V1445,190-197(1991)
See Riek, Jonathan K.: V1398,130-141(1991)
See Riek, Jonathan K.: V1445,198-206(1991)

Tekippe, Vincent J. See Nagy, Peter A.: V1365,33-37(1991)

Telfer, Brian A. See Casasent, David P.: V1382,158-166(1991)

Telle, H. R. See Li, Hua: V1376,172-179(1991)

Temkin, Henryk
; Logan, Ralph A.; Tanbun-Ek, Tawaee: InGaAs/InP distributed feedback quantum-well lasers,V1418,88-98(1991)
See Harriott, Lloyd R.: V1465,57-63(1991)
See Wang, Yuh-Lin: V1392,588-594(1991)

Temkin, Richard J.
See Chen, Shien C.: V1407,67-73(1991)
See Danly, Bruce G.: V1407,192-201(1991)

Temmen, Mark G. See Kathman, Alan D.: V1544,58-65(1991)

Temmerman, Yvan See Mattheus, Rudy A.: V1446,341-351(1991)

Templeton, Michael K. See Dunbrack, Steven K.: V1464,314-326(1991)

Ten Bosch, Jaap J. See Sjollema, J. I.: V1500,177-188(1991)

Tench, Robert E. See Park, Yong K.: V1372,219-227(1991)

Tenek, Lazarus H.
; Henneke, Edmund G.: Flaw dynamics and vibrothermographic-thermoelastic nondestructive evaluation of advanced composite materials,V1467,252-263(1991)

Teneketges, Nicholas J.
See Jacobi, William J.: V1479,111-119(1991)
See Jacobi, William J.: V1495,205-213(1991)

Teng, Chia-Chi
See Findakly, Talal K.: V1476,14-21(1991)
See Haas, David R.: V1371,56-67(1991)

Teng, Chungte
; Ligomenides, Panos A.: ANN-implemented robust vision model,V1382,74-86(1991)

Teng, Da See Zhuang, Weihua: V1361,980-986(1991)

Tenjimbayashi, Koji See Matsuda, Kiyofumi: V1332,230-235(1991)

Tennant, Donald M.
See Raab, Eric L.: V1343,104-109(1991)
See White, Donald L.: V1343,204-213(1991)

Tennant, William E. See Kozlowski, Lester J.: V1540,250-261(1991)

Tennyson, R. C. See Measures, Raymond M.: V1489,86-96(1991)

Tentori, Diana
: Critical-angle refractometry: accuracy analysis,V1527,216-224(1991)
; Lopez Famozo, C.: Refractive-index measurement using moire deflectometry: working conditions,V1535,209-215(1991)

Tepedelenlioglu, N. See Kozaitis, Samuel P.: V1564,373-383(1991)

Tepper, E. See Sayegh, Samir I.: V1569,100-110(1991)

Teppo, Edward A. See Rockafellow, David R.: V1561,112-118(1991)

Ter-Mikirtychev, Valery V.
See Lukishova, Svetlana G.: V1527,380-391(1991)
See Lukishova, Svetlana G.: V1503,338-345(1991)

Teral, Stephane See Bourhis, Jean-Francois: V1583,383-388(1991)

Terao, Motoyasu See Okamine, Shigenori: V1499,166-170(1991)

Terasaki, A. See Wada, Tatsuo: V1560,162-171(1991)

Terasaki, Hitoshi See Tsuchiya, Yoichi: V1499,450-456(1991)

Terasawa, Ken See Matsuzaka, Fumio: V1397,239-242(1991)

Terasawa, Tsuneo
; Hasegawa, Norio; Imai, Akira; Tanaka, Toshihiko; Katagiri, Souichi: Variable phase-shift mask for deep-submicron optical lithography,V1463,197-206(1991)

Terauchi, Mamoru See Kobayashi, Takayoshi T.: V1403,407-416(1991)

Terekhov, S. N. See Gurinovich, G. P.: V1403,139-141(1991)

Terenetskaya, I. P.
; Repeyev, Yu A.: Provitamin D photoisomerization kinetics upon picosecond laser irradiation: role of previtamin conformational nonequilibrium,V1403,500-503(1991)

Tereshko, V. M. See Netrebko, N. V.: V1403,512-516(1991)

Terhune, Robert W. See Kamal, Avais: V1516,137-153(1991)

Terpstra, Jacob See Fidder, Henk: V1403,530-544(1991)

Terpugov, N. V. See Ziling, C. C.: V1583,90-101(1991)

Terrien, J. C. See Dodd, James W.: V1407,467-473(1991)

Terry, David H.
See Boone, Bradley G.: V1471,390-403(1991)
See Christens-Barry, William A.: V1564,177-188(1991)

Terry, Fred L. See Grimard, Dennis S.: V1392,535-542(1991)

Teruya, Alan T. See Thomas, Stan W.: V1358,578-588(1991)

Tervo, Matti
; Kauppinen, Timo: Supervision of self-heating in peat stockpiles by aerial thermography,V1467,161-168(1991)

Tervonen, Ari
; Poyhonen, Pekka; Honkanen, Seppo; Tahkokorpi, Markku T.: Channel waveguide Mach-Zehnder interferometer for wavelength splitting and combining,V1513,71-75(1991)
See Li, Ming-Jun: V1506,52-57(1991)
See Tammela, Simo: V1583,37-42(1991)
See Wang, Wei-Jian: V1513,434-440(1991)

Terziev, Vladimir G. See Michailov, Vladimir I.: V1590,203-214(1991)

Terzopoulos, Demetri
; Vasilescu, Manuela: Adaptive surface reconstruction,V1383,257-264(1991)
See Metaxas, Dimitri: V1570,12-20(1991)
See Szeliski, Richard: V1570,140-152(1991)

Tesar, Aleta A.
; Balooch, M.; Shotts, K. W.; Siekhaus, Wigbert J.: Morphology and laser damage studies by atomic force microscopy of e-beam evaporation deposited antireflection and high-reflection coatings,V1441,228-236(1991)
; Brown, Norman J.; Taylor, John R.; Stolz, Christopher J.: Subsurface polishing damage of fused silica: nature and effect on laser damage of coated surfaces,V1441,154-172(1991)

Tesar, Delbert
See Hudgens, Jeffrey C.: V1387,271-282(1991)
See Kim, Whee-Kuk: V1387,392-406(1991)

Tescher, Andrew G.
ed.: *Applications of Digital Image Processing XIV*,V1567(1991)

Tessendorf, Jerry A.
: Underwater solar light field: analytical model from a WKB evaluation,V1537,10-20(1991)

Testan, Peter
: Trends in color hard copy,V1458,25-28(1991)

Testud, P. See Alloncle, Anne P.: V1397,675-678(1991)

Tewksbury, Stuart K. See Carruthers, John R.: V1390(1991)

Thacker, Dorsey L. See Bevilacqua, Richard M.: V1491,231-242(1991)

Thackeray, James W.
; Orsula, George W.; Rajaratnam, Martha M.; Sinta, Roger F.; Herr, Daniel J.; Pavelchek, Edward K.: Dissolution inhibition mechanism of ANR photoresists: crosslinking vs. -OH site consumption,V1466,39-52(1991)
See Barouch, Eytan: V1463,464-474(1991)

Thaete, F. L.
See Gur, David: V1446,284-288(1991)
See King, Jill L.: V1446,268-275(1991)

Thai, Tan Q.
: Fast one-pass algorithm to label objects and compute their features,V1567,533-541(1991)

Thakur, Randhir P. See Singh, Rajendra: V1393,78-89(1991)

Thanailakis, A. See Tzionas, Panagiotis: V1606,269-280(1991)

Thanh, V. L. See Chevrier, Joel S.: V1512,278-288(1991)

Thaniyavarn, Suwat
; Abbas, Gregory L.; Dougherty, William A.: Millimeter-wave signal generation and control using optical heterodyne techniques and electro-optic devices,V1371,250-251(1991)

Thanner, C. See Gornik, Erich: V1362,1-13(1991)

Thau, Robert S.
: Application of generalized radial basis functions to the problem of object recognition,V1469,37-47(1991)

Thebault, J. P.
See Fleurot, Noel A.: V1502,230-241(1991)
See Mens, Alain: V1358,878-887(1991)

Theis, Juergen See Groh, Werner: V1592,20-30(1991)

Theiss, W. See Clasen, Rolf: V1513,243-254(1991)

Theon, John S.
; Vaughan, William W.; Browell, Edward V.; Jones, William D.; McCormick, M. P.; Melfi, Samuel H.; Menzies, Robert T.; Schwemmer, Geary K.; Spinhirne, James D.: NASA's program in lidar remote sensing,V1492,2-23(1991)

Theriault, Joseph R. See Huang, Chin M.: V1447,156-164(1991)

Thevenot, Clarel
; Newhouse, Mark A.; Annunziata, Frank A.: Applications and testing of an opto-mechanical switch,V1398,250-260(1991)

Thibaudeau, Laurent See Ingers, Joakim P.: V1400,178-185(1991)

Thibault, S. See Malouin, Christian: V1559,385-392(1991)

Thiede, David A.
: Optical response in high-temperature superconducting thin films,V1484,72-80(1991)

Thiede, Edwin C.
: IR/MMW fusion ATR (automatic target recognition),VIS07,24-35(1991)

Thiell, Gaston See Fleurot, Noel A.: V1502,230-241(1991)

Thiem, Clare D.
; Norton, Douglas A.: Finite element analysis enhancement of cryogenic testing,V1532,39-47(1991)

Thierry-Mieg, V. See Etienne, Bernard: V1362,256-267(1991)

Thirsk, Mark See Robertson, Stewart A.: V1464,232-244(1991)

Thode, Lester E.
; Carlsten, Bruce E.; Chan, Kwok-Chi D.; Cooper, Richard K.; Elliott, C. J.; Gitomer, Steven J.; Goldstein, John C.; Jones, M. E.; McVey, Brian D.; Schmitt, Mark J.; Takeda, H.; Tokar, Robert L.; Wang, Tai-San; Young, Lloyd M.: Comparison of integrated numerical experiments with accelerator and FEL experiments,V1552,87-106(1991)

Thoe, Robert S. See Bixler, Jay V.: V1549,420-428(1991)

Tholl, Hans D.
: Two-wave coupled-wave theory of the polarizing properties of volume phase gratings,V1555,101-111(1991)
; Luebbers, Hubertus; Stojanoff, Christo G.: Analysis of the chromatic aberrations of imaging holographic optical elements,V1456,262-273(1991)
See Stojanoff, Christo G.: V1507,426-434(1991)

Thomas, Brigham B.
; Maxey, L. C.; Miller, Arthur C.: Design and optical performance of beryllium assessment mirrors,V1485,20-30(1991)

Thomas, David C. See Jensen, Stephen C.: V1369,87-95(1991)

Thomas, Dominique See Pezeshki, Bardia: V1362,559-565(1991)

Thomas, Don L.
: High-speed electronic memory video recording,V1448,140-147(1991)

Thomas, G. See Zediker, Mark S.: V1418,309-315(1991)

Thomas, G. A. See Peticolas, Warner L.: V1403,22-26(1991)

Thomas, George J. See Towse, Stacy A.: V1403,6-14(1991)

Thomas, Ian M.
: Optical and environmentally protective coatings for potassium dihydrogen phosphate harmonic converter crystals,V1561,70-82(1991)
; Campbell, John H.: Novel perfluorinated antireflective and protective coating for KDP and other optical materials,V1441,294-303(1991)
See Campbell, John H.: V1441,444-456(1991)
See Floch, Herve G.: V1441,304-315(1991)
See Kozlowski, Mark R.: V1561,59-69(1991)
See Woods, Bruce W.: V1410,47-54(1991)

Thomas, J. M. See Waters, Ruth A.: V1427,336-343(1991)

Thomas, John
: MITAS: multisensor imaging technology for airborne surveillance,V1470,65-74(1991)

Thomas, John C.
: Photon correlation spectroscopy: technique and instrumentation,V1430,2-18(1991)

Thomas, John M. See Dent, Andrew J.: V1550,97-107(1991)

Thomas, Johnny R. See Klainer, Stanley M.: V1434,119-126(1991)

Thomas, Judith G.
; Galloway, Robert L.; Edwards, Charles A.; Haden, Gerald L.; Maciunas, Robert J.: Comparison of three-dimensional surface rendering techniques,V1444,379-388(1991)
See Edwards, Charles A.: V1444,38-46(1991)
See Galloway, Robert L.: V1444,9-18(1991)

Thomas, Melvin L.
; Reining, Gale; Kelly, George: Display for advanced research and training: an inexpensive answer to tactical simulation,V1456,65-75(1991)

Thomas, Michael E. See Delaye, Corinne T.: V1487,291-298(1991)

Thomas, Michael J. See Fischer, Robert E.: V1535,78-88(1991)

Thomas, N. See McGoldrick, Elizabeth: V1374,118-125(1991)

Thomas, Paul J.
; Hollinger, Allan B.; Chu, Kan M.; Harron, John W.: ISTS array detector test facility,V1488,36-47(1991)
See Jorgensen, Steven M.: V1346,180-191(1991)

Thomas, Robert L. See Favro, Lawrence D.: V1467,290-294(1991)

Thomas, Roger J. See Keski-Kuha, Ritva A.: V1343,566-575(1991)

Thomas, Stan W.
; Griffith, Roger L.; Teruya, Alan T.: Avalanche transistor selection for long-term stability in streak camera sweep and pulser applications,V1358,578-588(1991)
; Shimkunas, Alex R.; Mauger, Philip E.: Subnanosecond intensifier gating using heavy and mesh cathode underlays,V1358,91-99(1991)
; Trevino, Jimmy: Picosecond intensifier gating with a plated webbing cathode underlay,V1358,84-90(1991)

Thomas, Stephen R. See Arata, Louis K.: V1446,465-474(1991)

Thomopoulos, Stelios C.
: Theories in distributed decision fusion: comparison and generalization, V1383, 623-634(1991)
; Bougoulias, Dimitrios K.: DIGNET: a self-organizing neural network for automatic pattern recognition and classification, V1470, 253-262(1991)

Thompson, Alan G. See Freeman, J. L.: V1476, 320-325(1991)

Thompson, Brian J. See Stone, Thomas W.: VMS34(1991)

Thompson, George A. See Bernstein, Hana: V1442, 81-88(1991)

Thompson, Jill C.
: Fluorescent imaging, V1482, 253-257(1991)
: Ultraviolet-laser-induced fluorescence imaging sensor, V1479, 412-414(1991)

Thompson, John A. See Rothman, Mark A.: V1392, 598-604(1991)

Thompson, K. E. See Roth, Michael W.: V1469, 25-36(1991)

Thompson, Keith P.
See Ren, Qiushi: V1423, 129-139(1991)
See Simon, Gabriel: V1423, 154-156(1991)

Thompson, Kevin P.
See Jewell, Tanya E.: V1527, 163-173(1991)
See Korniski, Ronald J.: V1354, 402-407(1991)

Thompson, Laird A.
; Castle, Richard; Carroll, David L.: Laser guide stars for adaptive optics systems: Rayleigh scattering experiments, V1542, 110-119(1991)

Thompson, Larry F.
See Nalamasu, Omkaram: V1466, 13-25(1991)
See Novembre, Anthony E.: V1466, 89-99(1991)

Thompson, Louise A. See Thompson, Robert A.: V1469, 451-462(1991)

Thompson, M. See Chou, Chia-Shing: V1475, 151-156(1991)

Thompson, Mark E.
; Chiang, William; Myers, Lori K.; Langhoff, Charles A.: Nonlinear optical properties of inorganic coordination polymers and organometallic complexes, V1497, 423-429(1991)

Thompson, Paul A. See Calloway, Terry M.: V1566, 353-364(1991)

Thompson, R. See Gilbert, Barry K.: V1390, 235-248(1991)

Thompson, Robert A.
; Thompson, Louise A.: EGOLOGY: psychological spatial breakthrough for social redirection—multidisciplinary spatial focus for individuals/humankind, V1469, 451-462(1991)

Thompson, Stephen D.
See Berry, Amanda K.: V1465, 210-220(1991)
See Graziano, Karen A.: V1466, 75-88(1991)

Thompson, Thomas See Mount, George H.: V1491, 188-193(1991)

Thompson, Wiley E.
; Parra-Loera, Ramon; Tao, Chin-Wang: Pseudo K-means approach to the multisensor multitarget tracking problem, V1470, 48-58(1991)
See Parra-Loera, Ramon: V1470, 30-36(1991)
See Reihani, Kamran: V1468, 305-312(1991)
See Reihani, Kamran: V1452, 319-329(1991)
See Reihani, Kamran: V1449, 99-108(1991)

Thomsen, Sharon L.
: Practical and theoretical considerations on the use of quantitative histology to measure thermal damage of collagen in cardiovascular tissues, V1425, 110-113(1991)
; Cheong, Wai-Fung; Pearce, John A.: Changes in collagen birefringence: a quantitative histologic marker of thermal damage in skin, V1422, 34-42(1991)
; Cheong, Wai-Fung; Pearce, John A.: Histopathologic assessment of water-dominated photothermal effects produced with laser irradiation, V1422, 14-18(1991)
See Pearce, John A.: V1422, 27-33(1991)

Thomson, J. See Demeester, Piet M.: V1361, 1132-1143(1991)

Thonnard, N. See Arlinghaus, Heinrich F.: V1435, 26-35(1991)

Thorne, John B. See Rice, Patrick D.: V1435, 104-113(1991)

Thornton, William A.
: Colorimetry, normal human vision, and visual display, V1456, 219-225(1991)
: Visual processing, transformability of primaries, and visual efficiency of display devices, V1453, 390-401(1991)

Thorpe, Charles E.
See Amidi, Omead: V1388, 504-523(1991)
See Aubert, Didier: V1388, 141-151(1991)
See Crisman, Jill D.: V1388, 152-164(1991)

Thorpe, Thomas P. See Hickey, Carolyn F.: V1534, 67-76(1991)

Thorstensen, B. See Quenzer, A.: V1397, 753-759(1991)

Throop, James A.
; Rehkugler, Gerald E.; Upchurch, Bruce L.: Using computer vision for detecting watercore in apples, V1379, 124-135(1991)

Thumm, M. See Chong, Chae K.: V1407, 226-233(1991)

Thunen, John G. See Elerding, George T.: V1479, 380-392(1991)

Thurmes, William See Wand, Michael D.: V1455, 97-104(1991)

Thursby, G. See Johnstone, Walter: V1506, 145-149(1991)

Thursby, Michael H.
; Yoo, Kisuck; Grossman, Barry G.: Neural control of smart electromagnetic structures, V1588, 218-228(1991)
See Grossman, Barry G.: V1588, 64-75(1991)

Thursby, William R. See Hough, Gary R.: V1346, 194-199(1991)

Thutupalli, G. K.
See Annapurna, M. N.: V1485, 260-271(1991)
See Umadevi, P.: V1484, 125-135(1991); 1485, 195-205(1991)

Tian, Da-chao See Liu, Rui-Fu: V1572, 180-184(1991)

Tian, Qi
; Zhang, Peng; Alexander, Thomas; Kim, Yongmin: Survey: omnifont-printed character recognition, V1606, 260-268(1991)

Tian, Qian See Zheng, Gang: V1572, 299-303(1991)

Tian, Ronglong
See Yatagai, Toyohiko: V1555, 8-12(1991)
See Yatagai, Toyohiko: V1564, 691-696(1991)

Tian, X. M. See Shu, Q. Q.: V1519, 675-679(1991)

Tian, Xingkang See Yatagai, Toyohiko: V1555, 8-12(1991)

Tiberio, Richard C. See Lang, Robert J.: V1563, 2-7(1991)

Ticknor, Anthony J.
See Armitage, David: V1455, 206-212(1991)
See Lipscomb, George F.: V1560, 388-399(1991)
See Lytel, Richard S.: V1563, 122-138(1991)
See Lytel, Richard S.: V1389, 547-558(1991)

Ticktin, Anton See Gierulski, Alfred: V1560, 172-182(1991)

Tidhar, Gil
; Rotman, Stanley R.: Clutter metrics in infrared target acquisition, V1442, 310-324(1991)
; Rotman, Stanley R.: New method of target acquisition in the presence of clutter, V1486, 188-199(1991)

Tiedeke, J. See Perger, Andreas: V1419, 75-83(1991)

Tiefenback, Michael G. See Bailey, Vernon L.: V1407, 400-406(1991)

Tiefensee, Frank See Schmidt, Helmut: V1590, 36-43(1991)

Tiemann, Bruce G. See Marder, Seth R.: V1560, 86-97(1991)

Tigges, Chris P. See Martens, Jon S.: V1394, 140-149(1991)

Tikhomirov, S. A. See Bondarev, S. L.: V1403, 497-499(1991)

Tikhonchuk, V. T. See Gamaly, Eugene G.: V1413, 95-106(1991)

Tikhonov, Evgenij A. See Gulnazarov, Eduard S.: V1238, 235-239(1991)

Tillett, Robin D.
; Marchant, John A.: Model-based image processing for characterizing pigs in scenes, V1379, 201-208(1991)

Tilstra, Shelle D.
: Fluorescence-based fiber optic temperature sensor for aerospace applications, V1589, 32-37(1991)
See Jensen, Stephen C.: V1369, 87-95(1991)

Tilton, James C. See Fischer, James R.: V1492, 229-238(1991)

Timcenko, Alexander See Allen, Peter K.: V1383, 176-188(1991)

Timchenko, A. A.
; Griko, N. B.; Serdyuk, I. N.: Possibility of resolution of internal macromolecular relaxation by dynamic light scattering, V1403, 340-343(1991)

Timlin, Harold A. See Fischer, Robert C.: V1494, 414-418(1991)

Timmermann, D. See Boehme, Johann F.: V1566, 208-219(1991)

Timoshechkin, M. I. See Danilov, Alexander A.: V1362, 916-920(1991)

Timothy, J. G.
: Imaging pulse-counting detector systems for space ultraviolet astrophysics missions, V1494, 394-402(1991)
: Performance characteristics of the imaging MAMA detector systems for SOHO, STIS, and FUSE/Lyman, V1549, 221-233(1991)
; Berger, Thomas E.; Morgan, Jeffrey S.; Walker, Arthur B.; Bhattacharyya, Jagadish C.; Jain, Surendra K.; Saxena, Ajay K.; Huber, Martin C.; Tondello, Giuseppe; Naletto, Giampiero: High-resolution stigmatic EUV spectroheliometer for studies of the fine scale structure of the solar chromosphere, transition region, and corona, V1343, 350-358(1991)

See Huber, Martin C.: V1494,472-480(1991)
See Slater, David C.: V1549,68-80(1991)
See Walker, Arthur B.: V1343,334-347(1991)
See Walker, Arthur B.: V1494,320-333(1991)
See Walker, Arthur B.: V1343,319-333(1991)

Tindal, Nan E. See Bradley, Lester M.: V1482,48-60(1991)

Ting, Antonio C. See Esarey, Eric: V1407,407-417(1991)

Tingstad, James S.
: Alignment of an aspheric mirror subsystem for an advanced infrared catadioptric system,V1527,194-198(1991)

Tinti, A. See Bottura, Giorgio: V1403,156-158(1991)

Tippur, Hareesh V.
; Krishnaswamy, Sridhar; Rosakis, Ares J.: Crack-tip deformation field measurements using coherent gradient sensing,V1554A,176-191(1991)

Tipton, Gary D. See Hickman, Kirt C.: V1464,245-257(1991)

Tirumalai, Arun P.
; Schunck, Brian G.; Jain, Ramesh C.: Recursive computation of a wire-frame representation of a scene from dynamic stereo using belief functions,V1569,28-42(1991)
; Schunck, Brian G.; Jain, Ramesh C.: Robust self-calibration and evidential reasoning for building environment maps,V1383,345-358(1991)

Tischel, M.
; Rosenzweig, M.; Hoffmann, Axel; Venghaus, Herbert; Fidorra, F.: Threshold current and carrier lifetime in MOVPE regrown 1.5 um GaInAsP buried ridge structure lasers,V1361,917-926(1991)
See Rosenzweig, M.: V1362,876-887(1991)

Tischler, Christian
See Neu, Walter: V1425,28-36(1991)
See Neu, Walter: V1425,37-44(1991)

Tischler, J. Z. See Larson, Bennett C.: V1345,90-100(1991)

Tison, James K. See Sandstrom, Torbjorn: V1463,629-637(1991)

Tistarelli, Massimo See Grosso, Enrico: V1570,371-381(1991)

Tiszauer, Detlev H.
; Smith, Richard C.: Wide bandwidth spectral measurements of atmospheric tilt turbulence,V1408,72-83(1991)

Titkov, A. N.
; Yakovlev, Yurii P.; Baranov, Alexej N.; Cheban, V. N.: Tunneling recombination of carriers at type-II interface in GaInAsSb-GaSb heterostructures,V1361,669-673(1991)

Titov, V. D. See Silkis, E. G.: V1358,46-49(1991)

Tittel, Frank K.
; Canarelli, P.; Dane, C. B.; Hofmann, Thomas; Sauerbrey, Roland A.; Sharp, Tracy E.; Szabo, Gabor; Wilson, William L.; Wisoff, P. J.; Yamaguchi, Shigeru: Advanced concepts of electron-beam-pumped excimer lasers,V1397,21-29(1991)
See Hofmann, Thomas: V1412,84-90(1991)
See Kim, Jin J.: V1412(1991)

Tittle, James S.
; Braunstein, Myron L.: Shape perception from binocular disparity and structure-from-motion,V1383,225-234(1991)

Tiwari, A. N.
See Zogg, Hans: V1361,406-413(1991)
See Zogg, Hans: V1361,1079-1086(1991)

Tjeng, L. H.
; Rudolf, P.; Meigs, G.; Sette, Francesco; Chen, Chien-Te; Idzerda, Y. U.: Magnetic circular dichroism studies with soft x-rays,V1548,160-167(1991)

Tkachenko, Vadim V. See Erokhovets, Valerii K.: V1621,227-236(1991)

Tkachuk, A. M.
; Petrova, Maria V.; Shumilin, V. V.: Pulsed YLF-Ho laser emission at 0.75 to 2.9 um,V1403,801-804(1991)

Tkachuk, T. M.
; Shumilin, V. V.: Operation of a YLF-Er laser at near infrared (0.85 to 2.9 um),V1403,805-808(1991)

Toba, Eiji
; Shimosaka, Tetsuya; Shimazu, Hideto: Noncontact measurement of microscopic displacement and vibration by means of fiber optics bundle,V1584,353-363(1991)

Tobin, Kenneth W.
; Beshears, David L.; Turley, W. D.; Lewis, Wilfred; Noel, Bruce W.: Fiber sensor design for turbine engines,V1584,23-31(1991)
; Muhs, Jeffry D.: Algorithm for a novel fiber optic weigh-in-motion sensor system,V1589,102-109(1991)
See Muhs, Jeffry D.: V1584,374-386(1991)

Tobin, Phil See Cooper, Kent: V1392,253-264(1991)

Toborg, Scott T.
: Neural network vision integration with learning,V1469,77-88(1991)

Tochimura, Katsumi See Silver, Michael D.: V1443,250-260(1991)

Toda, Toshiki See Takahashi, Susumu: V1461,199-205(1991)

Todd, John R.
: Power-by-light flight control: an EMI-immune backup,V1589,48-53(1991)
; Yount, Larry J.: Development of fly-by-light systems for commercial aircraft,V1369,72-78(1991)
See Udd, Eric: V1418,134-152(1991)

Todokoro, Yoshihiro See Watanabe, Hisashi: V1463,101-110(1991)

Todori, Kenji See Mori, Yasushi: V1560,310-314(1991)

Toennies, Klaus D.
; Tronnier, Uwe: Application of a discrete-space representation to three-dimensional medical imaging,V1444,19-25(1991)

Toenshoff, Hans K.
; Gedrat, Olaf: Absorption behavior of ceramic materials irradiated with excimer lasers,V1377,38-44(1991)

Toeppen, John S.
; Bliss, Erlan S.; Long, Theresa W.; Salmon, J. T.: Video Hartmann wavefront diagnostic that incorporates a monolithic microlens array,V1544,218-225(1991)

Tofani, Alessandro
See Barani, Gianni: V1467,458-468(1991)
See Uda, Gianni: V1488,257-266(1991)

Tofsted, David H.
: Empirical modeling of laser propagation effects in intermediate turbulence,V1487,372-381(1991)

Toh, Kenny K.
; Dao, Giang T.; Singh, Rajeev; Gaw, Henry T.: Chromeless phase-shifted masks: a new approach to phase-shifting masks,V1496,27-53(1991)
; Dao, Giang T.; Gaw, Henry T.; Neureuther, Andrew R.; Fredrickson, Larry R.: Design methodology for dark-field phase-shifted masks,V1463,402-413(1991)
; Dao, Giang T.; Singh, Rajeev; Gaw, Henry T.: Optical lithography with chromeless phase-shifted masks,V1463,74-86(1991)
; Neureuther, Andrew R.: Three-dimensional simulation of optical lithography,V1463,356-367(1991)

Toh, Siew-Lok See Shang, H. M.: V1554B,680-691(1991)

Toida, Masahiro See Inaba, Humio: V1399,108-115(1991)

Toivanen, Pekka J. See Vepsalainen, Ari M.: V1606,282-293(1991)

Toivonen, M. See Hingerl, Kurt: V1361,943-953(1991)

Tokar, Robert L. See Thode, Lester E.: V1552,87-106(1991)

Tokarev, Vladimir N.
; Konov, Vitaly I.: Light-induced polishing of evaporating surface,V1503,269-278(1991)

Tokarevski, V. V. See Artsimovich, M. V.: V1397,729-733(1991)

Toker, Gregory See Velger, Mordekhai: V1478,168-176(1991)

Tokui, Akira See Op de Beeck, Maaike: V1463,180-196(1991)

Tokunaga, T. See Tsutsumi, K.: V1499,55-61(1991)

Toler, J. R. See Vethanayagam, Thirukumar K.: V1366,343-350(1991)

Tolimieri, Richard
; Conner, Michael: Zak transform as an adaptive tool,V1565,345-356(1991)
See An, Myoung H.: V1565,383-401(1991)
See Zhang, Yan: V1481,23-31(1991)

Tolk, Norman H.
; Brau, Charles A.; Edwards, Glenn S.; Margaritondo, Giorgio; McKinley, J. T.: Vanderbilt Free-Electron Laser Center for Biomedical and Materials Research,V1552,7-13(1991)

Tolunay, H. See Oktu, Ozcau: V1361,812-818(1991)

Tolxdorff, T. See Gutfinger, Dan: V1445,288-296(1991)

Tom, James L.
; Cotton, Daniel M.; Bush, Brett C.; Chung, Ray; Chakrabarti, Supriya: Modular removable precision mechanism for alignment of an FUV spatial heterodyne interferometer,V1549,308-312(1991)
; Cotton, Daniel M.; Bush, Brett C.; Chung, Ray; Chakrabarti, Supriya: Rigid lightweight optical bench for a spaceborne FUV spatial heterodyne interferometer,V1549,302-307(1991)

Tom, Victor T. See Lee, Bonita G.: V1381,80-91(1991)

Tomasini, Bernard
; Cassassolles, Emmanuel; Poyet, Patrice; Maynard de Lavalette, Guy M.; Siffredi, Brigitte: Multiple-target tracking in a cluttered environment and intelligent track record,V1468,60-71(1991)

Tomasini, F.
    See Froehly, Claude: V1358,532-540(1991)
    See Mens, Alain: V1358,315-328(1991)
    See Mens, Alain: V1358,719-731(1991)

Tomass, V. See Purinsh, Juris: V1525,289-308(1991)

Tomaszewicz, Tomasz See Konwerska-Hrabowska, Joanna: V1554A,388-399(1991)

Tomchuk, P. M. See Grigor'ev, N. N.: V1440,105-111(1991)

Tomie, Toshihisa
    ; Shimizu, Hazime; Majima, T.; Yamada, Mitsuo; Kanayama, Toshihiko; Yano, M.; Kondo, H.: Observation of living cells by x-ray microscopy with a laser-plasma x-ray source,V1552,254-263(1991)

Tomimori, Hiroshi See Nishimura, Kazuhito: V1534,199-206(1991)

Tomimuro, Hisashi
    : Packaging technology for GaAs MMIC (monolithic microwave integrated circuits) modules,V1390,214-222(1991)

Tomin, V. I. See Nemkovich, N. A.: V1403,578-581(1991)

Tominaga, Hideyoshi See Kawashima, Masahisa: V1605,107-111(1991)

Tominaga, Koji See Yamaguchi, Takao: V1418,363-371(1991)

Tomisawa, Teiyu See Yoshizawa, Toru: V1554B,441-450(1991)

Tomita, Akira
    See Jones, Philip J.: V1456,6-14(1991)
    See Roberts, Daniel R.: V1366,129-135(1991)

Tomita, Hidemi See Noguchi, Tsutomu: V1466,149-160(1991)

Tomiyama, Hiromitsu See Nomura, Shintaro: V1560,272-277(1991)

Tomkinson, David M. See Scott, Luke B.: V1488,99-109(1991)

Tomlinson, Harold W.
    ; Weir, Michael P.; Michon, G. J.; Possin, G. E.; Chovan, J.: Focal-plane processing algorithm and architecture for laser speckle interferometry,V1541,178-186(1991)

Tommasi, Tullio
    : Structure of a scene from two and three projections,V1383,26-33(1991)
    ; Bianco, Bruno: Spatial frequency analysis for the computer-generated holography of 3-D objects,V1507,136-141(1991)

Tomono, Akira See Satoh, Takanori: V1606,1014-1021(1991)

Tomozawa, Minoru
    ; Han, Won-Taek; Davis, Kenneth M.: Fatigue-resistant coating of SiO2 glass,V1590,160-167(1991)

Tompkins, Jim M. See Birk, Ronald J.: V1495,2-11(1991)

Tompkins, Neal
    : Advances in color laser printing,V1458,154-154(1991)

Toms, Dennis J. See Baettig, Rainer K.: V1563,93-102(1991)

Tonazzi, Juan C. See Valla, Bruno: V1536,48-62(1991)

Tondello, Giuseppe
    See Huber, Martin C.: V1494,472-480(1991)
    See Timothy, J. G.: V1343,350-358(1991)

Toney, Michael F.
    ; Gordon, Joseph G.; Melroy, Owen R.: In-situ surface x-ray scattering of metal monolayers adsorbed at solid-liquid interfaces,V1550,140-150(1991)

Tong, William G.
    : Subattomole detection in the condensed phase by nonlinear laser spectroscopy based on degenerate four-wave mixing,V1435,90-94(1991)

Tonge, Peter J. See Carey, Paul R.: V1403,37-39(1991)

Tonguz, Ozan K.
    ; Kazovsky, Leonid G.: Sensitivity of direct-detection lightwave receivers using optical preamplifiers,V1579,179-183(1991)

Tonneau, Didier See Pelous, G.: V1598,149-158(1991)

Tonomura, Yoshinobu See Otsuji, Kiyotaka: V1606,980-989(1991)

Tontchev, Dimitar A.
    ; Zhivkova, Svetla; Miteva, Margarita G.; Grigoriev, Ivo D.; Ivanov, I.: Investigation on phase biological micro-objects with a holographic interferometric microscope on the basis of the photorefractive Bi12TiO20 crystal,V1429,76-80(1991)

Tooley, Frank A.
    See Brubaker, John L.: V1533,88-96(1991)
    See McCormick, Frederick B.: V1396,508-521(1991)
    See McCormick, Frederick B.: V1533,97-114(1991)

Toomarian, Nikzad See Zak, Michail: V1481,386-397(1991)

Toossi, Reza
    : Thermal sensing of fireball plumes,V1467,384-393(1991)

Topham, L. A. See Drummond, John W.: V1433,224-231(1991)

Topkar, V. A. See Kjell, Bradley P.: V1468,148-155(1991)

Topp, Michael R. See Wittmeyer, Stacey A.: V1435,267-278(1991)

Torao, Akira
    ; Uchida, Hiroyuki; Moriya, Susumu; Ichikawa, Fumihiko; Kataoka, Kenji; Wakui, Tsuneyoshi: Inspection system using still vision for a rotating laser-textured dull roll,V1358,843-850(1991)

Toratani, Hisayoshi See Yanagita, Hiroaki: V1513,386-395(1991)

Torbey, Habib H.
    ; Zhang, Zhensheng: Model for packet image communication in a centralized distribution system,V1605,650-666(1991)

Toriumi, Minoru
    ; Yanagimachi, Masatoshi; Masuhara, Hiroshi: Structure of poly(p-hydroxystyrene) film,V1466,458-468(1991)

Toriya, T. See Tsumanuma, Takashi: V1420,193-198(1991)

Torok, Georgia R. See Wong, Kwok Y.: V1447,283-287(1991)

Torr, Douglas G.
    See Zukic, Muamer: V1485,216-227(1991)
    See Zukic, Muamer: V1549,234-244(1991)

Torr, Marsha R. See Zukic, Muamer: V1549,234-244(1991)

Torresi, R. M. See Gorenstein, Annette: V1536,104-115(1991)

Torzicky, P. See Wang, H. F.: V1519,405-410(1991)

Tosi, M. R. See Bertoluzza, Alessandro: V1403,150-152(1991)

Tossberg, John See Carpenter, Susan: V1426,228-234(1991)

Toth, Charles
    See Schenk, Anton F.: V1526,78-89(1991)
    See Schenk, Anton F.: V1457,66-73(1991)

Totterman, Saara See Hornak, Joseph P.: V1445,523-533(1991)

Totzauer, Werner F.
    See Hoefling, Roland: V1508,135-142(1991)
    See Vogel, Dietmar: V1554A,262-274(1991)

Tou, Julius T.
    See Ahn, Hong-Young: V1468,896-904(1991)
    See Chung, Kwangsue: V1468,323-332(1991)

Toughlian, Edward N.
    ; Zmuda, Henry: Variable time delay for RF/microwave signal processing,V1476,107-121(1991)

Toukhy, Medhat A.
    ; Sarubbi, Thomas R.; Brzozowy, David J.: Technology and chemistry of high-temperature positive resist,V1466,497-507(1991)
    See Spragg, Peggy M.: V1466,283-296(1991)

Toumodge, Shawn S.
    : Detection and tracking of small targets in persistence,V1481,221-232(1991)

Toumoulin, Christine See Fresne, Francoise: V1444,26-36(1991)

Tourillon, Gerard See Dodelet, Jean-Pol: V1436,38-49(1991)

Tourmann, J. L. See Rechmann, Peter: V1424,106-115(1991)

Tournie, Eric
    ; Lazzari, J. L.; Mani, Habib; Pitard, F.; Alibert, Claude L.; Joullie, Andre F.: Growth by liquid phase epitaxy and characterization of GaInAsSb and InAsSbP alloys for mid-infrared applications (2-3 um),V1361,641-656(1991)

Tourtier, Philippe See Le Pannerer, Yves-Marie: V1567,556-565(1991)

Toussaere, E. See Levenson, R.: V1560,251-261(1991)

Tower, John R. See Sauer, Donald J.: V1540,285-296(1991)

Towers, Catherine E.
    ; Towers, David P.; Judge, Thomas R.; Bryanston-Cross, Peter J.: Laser light sheet investigation into transonic external aerodynamics,V1358,952-965(1991)

Towers, David P. See Towers, Catherine E.: V1358,952-965(1991)

Towers, Nigel M. See Greenfield, Perry E.: V1494,16-39(1991)

Townsend, Sallie S.
    ; Cunningham, Philip R.: Gaussian scaling laws for diffraction: top-hat irradiance and Gaussian beam propagation through a paraxial optical train,V1415,154-194(1991)
    See Smith, David C.: V1408,112-118(1991)

Townsend, Steven W. See Walter, Robert F.: V1397,71-76(1991)

Townsend, Wesley P. See Feldblum, Avi Y.: V1544,200-208(1991)

Towse, Stacy A.
    ; Benevides, James M.; Thomas, George J.: Advances in structural studies of viruses by Raman spectroscopy,V1403,6-14(1991)

Toy, Norman
; Savory, Eric; Disimile, Peter J.: Surface temperature and shear stress measurement using liquid crystals,V1489,112-123(1991)
See Disimile, Peter J.: V1489,66-74(1991)
See Disimile, Peter J.: V1483,39-48(1991)
See Disimile, Peter J.: V1492,64-75(1991)
See Shoe, Bridget: V1521,34-45(1991)

Toyoda, Ryuuichi See Kawauchi, Yoshikazu: V1555,224-227(1991)

Toyooka, Satoru
See Kadono, Hirofumi: V1554A,628-638(1991)
See Kadono, Hirofumi: V1554A,718-726(1991)

Tozzi, Clesio L.
; Castanho, Jose Eduardo C.; Gutierrez da Costa, Henrique S.: Image-processing system based on algorithmically dedicated functional units,V1384,124-132(1991)

Trachtenberg, Isaac See Butler, Stephanie W.: V1392,361-372(1991)

Traci, A. See Di Cocco, Guido: V1549,102-112(1991)

Tragin, N. See Tallents, Gregory J.: V1413,70-76(1991)

Trahan, Daniel J. See FitzSimons, Philip M.: V1489,230-242(1991)

Trainer, Thomas J.
; Sun, Fang-Kuo: Image resampling in remote sensing and image visualization applications,V1567,650-658(1991)

Trainor, S. W.
See Cheong, Wai-Fung: V1422,19-26(1991)
See Kurban, Amal K.: V1422,43-49(1991)

Tran, Minh Q.
ed.: *16th Intl Conf on Infrared and Millimeter Waves*,V1576(1991)

Tran, Tuan A.
; Miller, William V.; Murphy, Kent A.; Vengsarkar, Ashish M.; Claus, Richard O.: Stabilized extrinsic fiber optic Fabry-Perot sensor for surface acoustic wave detection,V1584,178-186(1991)
See Baldini, S. E.: V1588,125-131(1991)

Trang, Bonnie Y. See Shen, Sylvia S.: V1567,188-193(1991)

Tranis, Art See Pacala, Thomas J.: V1411,69-79(1991)

Tranquada, John M. See Heald, Steven M.: V1550,67-75(1991)

Tranta, Beate See Oswald, Josef: V1362,534-543(1991)

Trapp, Martin A. See Hoyle, Charles E.: V1559,202-213(1991)

Trascritti, D. See Lachs, Gerard: V1474,248-259(1991)

Trask, Philip A.
: High-density multichip interconnect: military packaging for the 1990s,V1390,223-234(1991)

Traub, Wesley A.
; Chance, Kelly V.; Johnson, David G.; Jucks, Kenneth W.: Stratospheric spectroscopy with the far-infrared spectrometer: overview and recent results,V1491,298-307(1991)

Traupe, U. See Barden, Raimund: V1449,136-147(1991)

Trautman, Jay K. See Shreve, A. P.: V1403,394-399(1991)

Trautwein, Charles M.
; Dwyer, John L.: Techniques and strategies for data integration in mineral resource assessment,V1492,338-338(1991)

Travis, D. J. See Strauss, Michael G.: V1447,12-27(1991)

Travnikov, V. V. See Zhilyaev, Yuri V.: V1361,848-859(1991)

Traxler-Lee, Laura A. See Seiberling, Walter E.: V1495,32-41(1991)

Treat, Michael R.
See Auteri, Joseph S.: V1421,182-184(1991)
See Bass, Lawrence S.: V1422,123-127(1991)
See Bass, Lawrence S.: V1421,164-168(1991)
See Eaton, Alexander M.: V1423,52-57(1991)
See Forman, Scott K.: V1424,2-6(1991)
See Garino, Jonathan P.: V1424,43-47(1991)
See Libutti, Steven K.: V1421,169-172(1991)
See Oz, Mehmet C.: V1422,147-150(1991)
See Wider, Todd M.: V1422,56-61(1991)

Trebes, James E. See Yorkey, Thomas J.: V1345,255-259(1991)

Trefonas, Peter
; Mack, Chris A.: Exposure dose optimization for a positive resist containing polyfunctional photoactive compound,V1466,117-131(1991)

Tregay, George W.
; Calabrese, Paul R.; Kaplin, Peter L.; Finney, Mark J.: Optical fiber sensor for temperature measurement from 600 to 1900 C in gas turbine engines,V1589,38-47(1991)

Trelles, Mario A.
; Mayayo, E.; Resa, A. M.; Rigau, J.; Calvo, G.: Laboratory and clinical data on wound healing by low-power laser from the Medical Institute of Vilafortuny, Spain,V1403,781-798(1991)

Tremblay, Gary A.
; Derby, Eddy A.: Design and manufacture of an ultralightweight solid deployable reflector,V1532,114-123(1991)

Tremblay, Yves See King, F. D.: V1364,295-303(1991)

Trenchard, Michael E.
; Lohrenz, Maura C.; Rosche, Henry; Wischow, Perry B.: Digital map databases in support of avionic display systems,V1456,318-326(1991)

Trenholme, J. B. See Murray, John R.: V1410,28-39(1991)

Trenkle, John M.
; Schlosser, Steve; Vogt, Robert C.: Morphological feature-set optimization using the genetic algorithm,V1568,212-223(1991)

Trent, Ava M. See Dederich, Douglas N.: V1424(1991)

Trentadue, Bartolo See Pappalettere, Carmine: V1554B,99-105(1991)

Trentelman, M.
; Ekelmans, G. B.; van Goor, Frederik A.; Witteman, Wilhelmus J.: Discharge studies with a high-efficiency XeCl excimer laser,V1397,115-118(1991)
See Ekelmans, G. B.: V1397,569-572(1991)
See Witteman, Wilhelmus J.: V1397,37-45(1991)

Trepagnier, Pierre C. See Anderson, James C.: V1479,78-92(1991)

Treptau, Jeffrey P.
; Milster, Tom D.; Flagello, Donis G.: Laser beam modeling in optical storage systems,V1415,317-321(1991)
See Milster, Tom D.: V1499,348-353(1991)
See Milster, Tom D.: V1414,91-96(1991)

Treshchalov, Alexei B. See Peet, Viktor E.: V1412,138-148(1991)

Trevino, Jimmy See Thomas, Stan W.: V1358,84-90(1991)

Trewhella, Jeannine M.
; Oprysko, Modest M.: Total internal reflection mirrors fabricated in polymeric optical waveguides via excimer laser ablation,V1377,64-72(1991)

Trias, John A. See Phillips, Thomas E.: V1454,290-298(1991)

Tribillon, Gilbert M.
; Calatroni, Jose; Sandoz, Patrick: Multichannel chromatic interferometry: metrology applications,V1332,632-642(1991)

Trichel, Milton See Tchoryk, Peter: V1479,164-182(1991)

Trigano, Philippe
; Boufflet, Jean-Paul; Newstead, Emma: Ten years of failure in automatic time tables scheduling at the UTC,V1468,408-416(1991)
; Talon, Benedicte; Baltazart, Didier; Demko, Christophe; Newstead, Emma: LCS: a natural language comprehension system,V1468,866-874(1991)

Tripathi, Vijai K.
See Hayden, Leonard A.: V1389,205-214(1991)
See Orhanovic, Neven: V1389,273-284(1991)

Tripathy, Sukant K. See Li, Lian: V1560,243-250(1991)

Trisno, Yudhi S.
; Chen, Lian K.; Huber, David R.: Linearized external modulator for analog applications,V1371,8-12(1991)

Trivedi, Mohan M.
ed.: *Applications of Artificial Intelligence IX*,V1468(1991)
; Vidyasagar, M.; eds.: *Intelligent Robotics*,V1571(1991)
See Bidlack, Clint R.: V1468,270-280(1991)
See Chen, ChuXin: V1468,354-366(1991)
See Gadagkar, Hrishikesh P.: V1383,142-150(1991)
See Shirvaikar, Mukul V.: V1468,52-59(1991)
See Sluder, John C.: V1468,642-652(1991)

Troccolo, Patrick M.
; Mantalas, Lynda C.; Allen, Richard A.; Linholm, Loren: Extending electrical measurements to the 0.5 um regime,V1464,90-103(1991)

Troe, Juergen See Ihlemann, Juergen: V1361,1011-1019(1991)

Troeltzsch, John R.
; Ebbets, Dennis C.; Garner, Harry W.; Tuffli, A.; Breyer, R.; Kinsey, J.; Peck, C.; Lindler, Don; Feggans, J.: In-flight performance of the Goddard high-resolution spectrograph of the Hubble Space Telescope,V1494,9-15(1991)

Trofimenko, Vladimir V.
See Drozhbin, Yu. A.: V1358,1029-1034(1991)
See Drozhbin, Yu. A.: V1358,454-456(1991)
See Drozhbin, Yu. A.: V1358,451-453(1991)
See Ushakov, Leonid S.: V1358,447-450(1991)

Trojanowski, Stanislaw
See Trojnar, Eugeniusz: V1391,230-237(1991)
See Trojnar, Eugeniusz: V1391,238-243(1991)

Trojnar, Eugeniusz
; Trojanowski, Stanislaw; Czechowicz, Roman; Derwiszynski, Mariusz; Kocyba, Krzysztof: Liquid-nitrogen-cooled low-noise radiation pulse detector amplifier,V1391,230-237(1991)
; Trojanowski, Stanislaw; Czechowicz, Roman; Derwiszynski, Mariusz; Kocyba, Krzysztof: Properties of liquid-nitrogen-cooled electronic elements,V1391,238-243(1991)

Trokel, Stephen L.
: Unresolved issues in excimer laser corneal surgery,V1423,94-97(1991)

Trolinger, James D.
; Lal, Ravindra B.; Vikram, Chandra S.; Witherow, William K.: Compact spaceflight solution crystal-growth system,V1557,250-258(1991)
; Lal, Ravindra B.; eds.: *Crystal Growth in Space and Related Optical Diagnostics*,V1557(1991)
; Lal, Ravindra B.; Batra, Ashok K.: Holographic instrumentation for monitoring crystal growth in space,V1332,151-165(1991)
; Lal, Ravindra B.; Batra, Ashok K.; McIntosh, D.: Particle image velocimetry experiments for the IML-I spaceflight,V1557,98-109(1991)
See Vikram, Chandra S.: V1557,197-201(1991)
See Wood, Craig P.: V1332,122-131(1991)

Tromberg, Bruce J.
; Dvornikov, Tatiana; Berns, Michael W.: Indirect spectroscopic detection of singlet oxygen during photodynamic therapy,V1427,101-108(1991)
; Svaasand, Lars O.; Tsay, Tsong-Tseh; Haskell, Richard C.; Berns, Michael W.: Optical property measurements in turbid media using frequency-domain photon migration,V1525,52-58(1991)
See Svaasand, Lars O.: V1525,41-51(1991)

Trombetta, John M.
; Hoggins, James T.; Klocek, Paul; McKenna, T. A.: Optical properties of DC arc-discharge plasma CVD diamond,V1534,77-88(1991)
See Klocek, Paul: V1498,147-157(1991)

Trombetta, Steven P. See Erdekian, Vahram V.: V1444,151-158(1991)

Trommsdorff, H. P. See Hochstrasser, Robin M.: V1403,221-229(1991)

Tronnier, Uwe See Toennies, Klaus D.: V1444,19-25(1991)

Tropartz, Stephan
; Brasseur, Olivier; Kubitzek, Ruediger; Stojanoff, Christo G.: Development and investigation of dichromated gelatin film for the fabrication of large-format holograms operating at 400-900 nm,V1507,345-353(1991)
See Kubitzek, Ruediger: V1507,365-372(1991)
See Stojanoff, Christo G.: V1536,206-214(1991)
See Stojanoff, Christo G.: V1527,48-60(1991)
See Stojanoff, Christo G.: V1485,274-280(1991)
See Stojanoff, Christo G.: V1559,321-330(1991)

Tropper, Anne C.
; Smart, Richard G.; Perry, Ian R.; Hanna, David C.; Lincoln, John; Brocklesby, Bill: Thulium-doped silica fiber lasers,V1373,152-157(1991)
See Smart, Richard G.: V1373,158-165(1991)

Troshchiev, V. E. See Apollonova, O. V.: V1501,108-119(1991)

Trott, G.E. See Cunningham, Philip R.: V1414,97-129(1991)

Trott, Wayne M. See Setchell, Robert E.: V1441,61-70(1991)

Trowbridge, Christian A. See Turner, Vernon: V1487,262-271(1991)

Troyer, David E.
; Fouse, Timothy; Murdaugh, William O.; Zammit, Michael G.; Rogers, Stephen B.; Skrzypczak, J. A.; Colley, Charles B.; Taczak, William J.: Infrared background measurements at White Sands Missile Range, NM,V1486,396-409(1991)

Troyer, W. W. See Meyer, George E.: V1379,99-105(1991)

Truax, Bruce E.
: Absolute interferometric testing of spherical surfaces,V1400,61-68(1991)

Trubaev, Vladimir V.
See Bazhenov, V. V.: V1440,332-337(1991)
See Libenson, Michail N.: V1440,354-356(1991)

Truckenbrodt, Horst See Gebhardt, Michael: V1500,135-143(1991)

Truemper, Joachim See Braueninger, Heinrich: V1549,330-339(1991)

Truesdell, Keith See Bohn, Willy L.: V1397,235-238(1991)

Trukhanov, V. A. See Vanin, V. A.: V1238,372-380(1991)

Truman, Alan K. See Tarrant, Peter J.: V1364,2-6(1991)

Trunk, Jonah
; Moreira, Roberto P.; Monteiro, Ricardo M.: Protection with heat-shrinkable sleeves for optical fiber arc fusion splicing,V1365,124-130(1991)

Trunov, Vitaliy See Zalessky, Viacheslav N.: V1426,162-169(1991)

Truong, J. P. See Zeitoun, David: V1397,585-588(1991)

Truong, Vo-Van See Ashrit, Pandurang V.: V1401,119-129(1991)

Trussell, Henry J.
; Fogel, Sergei: Restoration of spatially variant motion blurs in sequential imagery,V1452,139-145(1991)
; Vrhel, Michael J.: Color correction using principle components,V1452,2-9(1991)

Trusty, Robert M. See Milster, Tom D.: V1499,286-292(1991)

Tsagaropoulos, George See Risen, William M.: V1590,44-49(1991)

Tsai, Chen S.
: Guided-wave magneto-optic and acousto-optic Bragg cells for RF signal processing,V1562,55-65(1991)

Tsai, Jaw S. See Takeuchi, Ichiro: V1394,96-101(1991)

Tsai, Jodi See Kelley, Robert B.: V1387,38-46(1991)

Tsai, Kun-Hsieh
; Kim, Kyung-Suk: Study of stick slip behavior in interface friction using optical fiber pull-out experiment,V1554A,529-541(1991)

Tsai, Ming-Horn See Opsasnick, Michael N.: V1499,276-280(1991)

Tsai, Ming-Jer
; Rogal, Fannie A.: Angle-only tracking and prediction of boost vehicle position,V1481,281-291(1991)

Tsai, Mong-Jean See Chang, Long-Wen: V1606,120-131(1991)

Tsai, T. E.
; Griscom, David L.: Defect centers and photoinduced self-organization in Ge-doped silica core fiber,V1516,14-28(1991)

Tsai, Yong-Song See Yeh, Long-Ching: V1527,361-367(1991)

Tsakiris, George D.
: Applications of a soft x-ray streak camera in laser-plasma interaction studies,V1358,174-192(1991)
See Eidmann, K.: V1502,320-330(1991)

Tsakiris, Todd N. See Hillman, Robert L.: V1562,136-142(1991)

Tsalides, Ph. See Tzionas, Panagiotis: V1606,269-280(1991)

Tsang, Dean Z.
: Free-space board-to-board optical interconnections,V1563,66-71(1991)

Tsang, Koon Wing See Holm-Kennedy, James W.: V1527,322-331(1991)

Tsang, Leung
; Ding, Kung-Hau; Kong, Jin A.; Winebrenner, Dale P.: Scattering of waves from dense discrete random media: theory and applications in remote sensing,V1558,260-268(1991)

Tsao, Chen-Kuo See Lin, Wei-Chung: V1445,376-385(1991)

Tsao, Desiree H. See Maki, August H.: V1432,119-128(1991)

Tsao, Hen-Wai
See Lee, Yang-Hang: V1372,140-149(1991)
See Lee, Yang-Hang: V1579,155-166(1991)
See Yang, Shien-Chi: V1372,128-139(1991)

Tsao, Tien-Ren J.
; Libert, John M.: Fusion of multiple-sensor imagery based on target motion characteristics,V1470,37-47(1991)

Tsarkova, O. G. See Konov, Vitaly I.: V1427,220-231(1991)

Tsaur, Bor-Yeu
; Chen, C. K.; Marino, S. A.: Long-wavelength GexSi1-x/Si heterojunction infrared detectors and focal-plane arrays,V1540,580-595(1991)
See Cantella, Michael J.: V1540,634-652(1991)

Tsay, Si-Chee See Gabriel, Philip: V1558,76-90(1991)

Tsay, Tsong-Tseh See Tromberg, Bruce J.: V1525,52-58(1991)

Tschudi, Theo T.
; Denz, Cornelia; Kobialka, Torsten: Phase-conjugating elements in optical information processing networks,V1500,80-80(1991)
See Ahlers, Rolf-Juergen: V1500(1991)
See Hossfeld, Jens: V1574,159-166(1991)

Tse, Teddy See Eng, Peter J.: V1550,110-121(1991)

Tse, Yi-tong
; Baker, Richard L.: Camera zoom/pan estimation and compensation for video compression,V1452,468-479(1991)

Tsekhomskii, Victor A. See Glebov, Leonid B.: V1621,21-32(1991)

Tseng, Chien-Chao
; Hwang, Shu-Yuen: Data-driven parallel architecture for syntactic pattern recognition,V1384,257-268(1991)

Tserng, Hua Q.
; Saunier, Paul: Advances in power MMIC amplifier technology in space communications,V1475,74-85(1991)
See Saunier, Paul: V1475,86-90(1991)

Tsetsekou, A. See Hourdakis, G.: V1503,249-255(1991)

Tshavelev, Oleg S. See Glebov, Leonid B.: V1399,200-206(1991)

Tsibulya, Andrew B. See Artjushenko, Vjacheslav G.: V1420,157-168(1991)

Tsou, Jung-Jung See Connor, Brian J.: V1491,218-230(1991)

Tsoutsou, K. See Karkanis, S.: V1500,164-170(1991)

Tsu, Hiroji See Yamaguchi, Yasushi: V1490,324-334(1991)

Tsu, Raphael
: Some new insights in the physics of quantum-well devices,V1361,313-324(1991)
; Ye, Qui-Yi; Nicollian, Edward H.: Resonant tunneling in microcrystalline silicon quantum box diode,V1361,232-235(1991)

Tsubosaka, Kazuyoshi See Ohmori, Yasuhiro: V1490,177-183(1991)

Tsubouchi, Natsuro See Kimata, Masafumi: V1540,238-249(1991)

Tsuchida, Eiichi See Sato, Heihachi: V1397,331-338(1991)

Tsuchida, Hirofumi
; Aoki, Norihiko; Hyakumura, Kazushi; Yamamoto, Kimiaki: Zoom lens design using GRIN materials,V1354,246-251(1991)

Tsuchinaga, Hiroyuki See Sukeda, Hirofumi: V1499,419-425(1991)

Tsuchiya, Toshiyuki See Wataya, Hideo: V1363,72-84(1991)

Tsuchiya, Yoichi
; Terasaki, Hitoshi; Ota, Osamu: High-density optical MUSE disk,V1499,450-456(1991)

Tsuda, Shinya See Kuwano, Yukinori: V1519,200-209(1991)

Tsuge, Hisanao See Takeuchi, Ichiro: V1394,96-101(1991)

Tsui, Ernest S.
; Vickers, Anthony J.: Study of novel oscillations in degenerate GaAs/A1GaAs quantum-wells using electro-optic voltage probing,V1362,804-810(1991)

Tsui, J. B. See Gill, E. T.: V1476,190-200(1991)

Tsui, King H.
: Superradiant Raman free-electron lasers,V1407,281-284(1991)

Tsuji, Kenzo See Matsushiro, Nobuhito: V1452,21-26(1991)

Tsujimoto, Eric M. See Ford, Gary E.: V1452,244-255(1991)

Tsujino, T. See Maeda, Koichi: V1536,138-148(1991)

Tsujita, Kouichirou See Kawai, Akira: V1464,267-277(1991)

Tsujiuchi, Jumpei
: Medical applications of holographic stereograms,V1238,398-400(1991)
: Synthesized holograms in medicine and industry,VIS08,326-334(1991)
See Okada, Katsuyuki: V1400,33-37(1991)

Tsukada, Noriaki See Nishimura, Tetsuya: V1436,31-37(1991)

Tsukada, Sadao See Yasuda, Yukio: V1422,50-55(1991)

Tsukahara, Hiroyuki
; Nakashima, Masato; Sugawara, Takehisa: Automated visual inspection system for IC bonding wires using morphological processing,V1384,15-26(1991)

Tsukahara, Makoto See Ito, Katsunori: V1499,382-385(1991)

Tsukamoto, Katsuhiro See Op de Beeck, Maaike: V1463,180-196(1991)

Tsukamoto, Katsuo
; Onuma, Kazuo: In-situ observation of crystal growth in microgravity by high-resolution microscopies,V1557,112-123(1991)

Tsukamoto, Y. See Yamaguchi, H.: V1499,29-38(1991)

Tsumanuma, Takashi
; Toriya, T.; Tanaka, T.; Shamoto, N.; Seto, K.; Sanada, Kazuo; Okazaki, A.; Okazaki, M.: New image diagnosis system with ultrathin endoscope and clinical results,V1420,193-198(1991)

Tsunetomo, Keiji
; Shimizu, Ryuichiro; Yamamoto, Masaki; Osaka, Yukio: Structural and optical properties of semiconducting microcrystallite-doped SiO2 glass films prepared by rf-sputtering,V1513,93-104(1991)

Tsuno, Katsuhiko
; Kameda, Yoshihiko; Kondoh, Kayoko; Hirai, Shoichi: Interferometric monitor for greenhouse gasses for ADEOS,V1490,222-232(1991)

Tsuno, Toshio See Yamaguchi, H.: V1499,29-38(1991)

Tsuru, T. See Ohashi, T.: V1549,9-19(1991)

Tsuruta, Shigehiko See Ogura, Toshihiro: V1443,153-157(1991)

Tsutsui, Kumiko See Aisaka, Kazuo: V1445,312-317(1991)

Tsutsumi, K.
; Nakaki, Y.; Fukami, T.; Tokunaga, T.: Directly overwritable magneto-optical disk with light-power-modulation method using no initializing magnet,V1499,55-61(1991)

Tsuyuki, Glenn T. See Helms, Richard G.: V1532,64-80(1991)

Tsvetkov, V. B. See Konov, Vitaly I.: V1427,220-231(1991)

Tu, James Z. See Zhang, Yu: V1384,60-65(1991)

Tu, Meirong
; Gielisse, Peter J.; Xu, Wei: Whole field displacement and strain rosettes by grating objective speckle method,V1554A,593-601(1991)

Tu, Yijun See Zhuang, Song Lin: V1558,149-153(1991)

Tubaro, Stefano See Brofferio, Sergio C.: V1605,500-510(1991)

Tuchin, Valery V.
; Ampilogov, Andrey V.; Bogoroditsky, Alexander G.; Rabinovich, Emmanuil M.; Ryabukho, Vladimir P.; Ul'yanov, Sergey S.; V'yushkin, Maksim E.: Laser speckle and optical fiber sensors for micromovements monitoring in biotissues,V1420,81-92(1991); 1429,62-73(1991)
; Utz, Sergey R.; Barabanov, Alexander J.; Dovzansky, S. I.; Ulyanov, A. N.; Aravin, Vladislav A.; Khomutova, T. G.: Laser photochemotherapy of psoriasis,V1422,85-96(1991)
See Kozlova, T. G.: V1403,159-160(1991)
See Medvedev, Boris A.: V1422,73-84(1991)
See Medvedev, Boris A.: V1403,682-685(1991)
See Shubochkin, L. P.: V1403,720-731(1991)

Tuchman, Avi See Wade, Jack: V1478(1991)

Tuck, Adrian F.
: Summary of atmospheric chemistry observations from the Antarctic and Arctic aircraft campaigns,V1491,252-272(1991)

Tucker, Christopher J.
; Mitchell, Robert J.: High-spatial-resolution FLIR,V1498,92-98(1991)

Tucker, Michael R.
; Christenson, Eric: Absolute interferometer for manufacturing applications,V1367,289-299(1991)

Tucker, Robert See Pratt, Stephen R.: V1538,40-45(1991)

Tucker, Rodney S. See Jewell, Jack L.: V1389,401-407(1991)

Tucker, Russ See Hart, Patrick W.: V1366,334-342(1991)

Tuffli, A. See Troeltzsch, John R.: V1494,9-15(1991)

Tugnait, Jitendra K.
: Adaptive filters and blind equalizers for mixed-phase channels,V1565,209-220(1991)
: Estimation of linear parametric models of non-Gaussian discrete random fields,V1452,204-215(1991)

Tugnoli, V. See Bertoluzza, Alessandro: V1403,150-152(1991)

Tujev, A. F. See Spornik, Nikolai M.: V1508,83-88(1991)

Tukhvatulin, R. T.
; Vaulin, P. P.: Diagnostics of functional state of blood by registration of light scattering intensity variations due to the reversible aggregation of red blood cells,V1403,390-391(1991)

Tukkiniemi, Kari See Heikkinen, Veli: V1533,115-121(1991)

Tulajkova, Tamara V.
See Lukishova, Svetlana G.: V1527,380-391(1991)
See Lukishova, Svetlana G.: V1503,338-345(1991)

Tulip, John See Dederich, Douglas N.: V1424,134-137(1991)

Tuller, H. L. See Rotman, Stanley R.: V1442,205-215(1991)

Tulpule, Sharayu See Griffin, Michael F.: V1566,179-189(1991)

Tumolillo, Thomas A.
; Ashley, Paul R.: Fabrication techniques of photopolymer-clad waveguides for nonlinear polymeric modulators,V1559,65-73(1991)
See Ashley, Paul R.: V1583,316-326(1991)

Tuncer, Temel E. See Vaina, Lucia M.: V1468,990-999(1991)

Tunick, Arnold
; Rachele, Henry; Miller, Walter B.: Cn2 estimates in the boundary layer for damp unstable conditions,V1487,51-62(1991)

Tunimanova, Irina V. See Glebov, Leonid B.: V1621,21-32(1991)

Tuo, Sueshang See Luo, Zhishan: V1554B,339-342(1991)

Tuovinen, Jussi See Vasara, Antti H.: V1507,224-238(1991)

Tuppen, C. G. See White, Julian D.: V1361,293-301(1991)

Tur, Moshe See Shafir, Ehud: V1442,236-241(1991)

Turcu, I. C.
; Gower, Malcolm C.; Reason, C. J.; Huntington, P.; Schulz, M.; Michette, Alan G.; Bijkerk, Fred; Louis, Eric; Tallents, Gregory J.; Al-Hadithi, Yas; Batani, D.: 100-Hz KrF laser plasma x-ray source,V1503,391-405(1991)
See Batani, D.: V1503,479-491(1991)

Turk, F. J. See Vivekanandan, J.: V1558,324-338(1991)

Turk, Harris
; Leepa, Douglas C.; Snyder, Robert F.; Soos, Jolanta I.; Bronstein, Sam: Low-power optical correlator for sonar range finding,V1454,344-352(1991)

Turk, Matthew A.
; Pentland, Alexander P.: Recognition in face space,V1381,43-54(1991)

Turko, Illarion V.
; Lepesheva, Galina I.; Chashchin, Vadim L.: Direct fluoroimmunoassay in Langmuir-Blodgett films of immunoglobulin G,V1572,419-423(1991)
; Pikuleva, Irene A.; Yurkevich, Igor S.; Chashchin, Vadim L.: Langmuir-Blodgett films of immunoglobulin G and direct immunochemical sensing,V1510,53-56(1991)

Turley, W. D. See Tobin, Kenneth W.: V1584,23-31(1991)

Turlik, Iwona
See Hwang, Lih-Tyng: V1390,249-260(1991)
See Rinne, Glenn A.: V1389,110-121(1991)

Turner, Elbert L.
; Hill, William A.: Microwave monolithic integrated circuits for high-data-rate satellite communications,V1475,248-256(1991)

Turner, K. F. See Nemanich, Robert J.: V1437,2-12(1991)

Turner, Nicholas P. See Nunn, John W.: V1464,50-61(1991)

Turner, R. E. See Kilkenny, Joseph D.: V1358,117-133(1991)

Turner, Robert E.
; Remo, John L.; Bradford, Elaine; Dietz, Alvin: Gas handling technology for excimer lasers,V1377,99-106(1991)

Turner, Ruldopha M. See Bristow, Michael P.: V1491,68-74(1991)

Turner, Scott See Mankovich, Nicholas J.: V1444,2-8(1991)

Turner, Terry R. See Bondur, James: V1392(1991)

Turner, Vernon
; Trowbridge, Christian A.: Wide-spectral-range transmissometer used for fog measurements,V1487,262-271(1991)

Turner, William D. See Kobayashi, Kenichi: V1448,157-163(1991)

Turner-Smith, Alan R.
; White, Steven P.; Bulstrode, Christopher: X-ray photogrammetry of the hip revisited,V1380,75-84(1991)

Turney, Jerry L.
; Lysogorski, Charles; Gottschalk, Paul G.; Chiu, Arnold H.: Phase-shift moire camera for real-time measurements of three-dimensional shape information,V1380,53-63(1991)
See Gottschalk, Paul G.: V1383,84-96(1991)

Turon, F.
See Campos, Juan: V1574,141-147(1991)
See Chalasinska-Macukow, Katarzyna: V1564,285-293(1991)

Turpin, Marc See Bonniau, Philippe: V1588,52-63(1991)

Turpin, Terry M.
: Impact of device characteristics on optical processor design,V1562,2-10(1991)

Turtle, John P. See Scalzi, Gary J.: V1475,2-9(1991)

Turton, Richard M. See Jaros, Milan: V1362,242-253(1991)

Turukhano, Boris G.
: Synthesis of large-aperture interference fields,V1500,305-308(1991)
; Gorelik, Vladimir S.; Turukhano, Nikulina: Phase synthesis of elongated holographic diffraction gratings,V1500,290-292(1991)
See Gorelik, Vladimir S.: V1507,379-382(1991)
See Gorelik, Vladimir S.: V1507,488-490(1991)

Turukhano, Nikulina See Turukhano, Boris G.: V1500,290-292(1991)

Turunen, Jari P.
See Saarinen, Jyrki V.: V1555,128-137(1991)
See Vasara, Antti H.: V1507,224-238(1991)

Tushev, N. R. See Romanovsky, V. F.: V1358,140-145(1991)

Tuttle, Jerry W.
ed.: *Tactical Infrared Systems,*V1498(1991)

Twardowski, Patrice J.
; Meyrueis, Patrick: Design of an optimal single-reflective holographic helmet display element,V1456,164-174(1991)
; Meyrueis, Patrick: Design of some achromatic imaging hybrid diffractive-refractive lenses,V1507,55-65(1991)

Twieg, Robert J.
; Betterton, K.; DiPietro, Richard; Gravert, D.; Nguyen, Cattien; Nguyen, H. T.; Babeau, A.; Destrade, C.: Smectic liquid crystals modified by tail and core fluorination,V1455,86-96(1991)
See Moerner, William E.: V1560,278-289(1991)

Twiggs, Robert J.
; Reister, K. R.: WEBERSAT: a low-cost imaging satellite,V1495,12-18(1991)

Twu, Yih-Jye See Park, Yong K.: V1372,219-227(1991)

Tyagi, S. See Larson, Donald C.: V1572,517-522(1991)

Tyan, John See Schimert, Thomas R.: V1484,19-30(1991)

Tychinsky, Vladimir P.
: Phase object imaging inside the airy disc,V1392,570-572(1991)
; Tavrov, Alexander V.: Wavefront dislocations and phase object registering inside the airy disk,V1500,207-210(1991)

Tyminski, Jacek K. See Reed, Murray K.: V1410,179-184(1991)

Tyrer, John R. See Mendoza Santoyo, Fernando: V1508,143-152(1991)

Tyson, Robert K.
: Measuring phase errors of an array or segmented mirror with a single far-field intensity distribution,V1542,62-75(1991)

Tyurin, Vladimir S. See Furzikov, Nickolay P.: V1427,288-297(1991)

Tyutchev, M. V. See Kalyashov, E. V.: V1238,442-446(1991)

Tzannes, Alexis P.
; Tzannes, Michael A.; Tzannes, Marcos C.; Tzannes, Nicolaos S.: Cascade coding with error-constrained relative entropy decoding,V1452,497-500(1991)

Tzannes, Marcos C. See Tzannes, Alexis P.: V1452,497-500(1991)

Tzannes, Michael A. See Tzannes, Alexis P.: V1452,497-500(1991)

Tzannes, Nicolaos S.
See Kasparis, Takis: V1521,46-54(1991)
See Tzannes, Alexis P.: V1452,497-500(1991)

Tzeng, Chao H.
; Lin, Dhei-Jhai; Lin, Song S.; Huang, Dong-Tsair; Lin, Hwang-Kuen: DQN photoresist with tetrahydroxydiphenylmethane as ballasting group in PAC,V1466,469-476(1991)

Tzeng, Liang D. See Park, Yong K.: V1372,219-227(1991)

Tzionas, Panagiotis
; Tsalides, Ph.; Thanailakis, A.: Cellular-automata-based learning network for pattern recognition,V1606,269-280(1991)

Tzou, Kou-Hu
; Koga, Toshio; eds.: *Visual Communications and Image Processing '91: Image Processing,*V1606(1991)
; Koga, Toshio; eds.: *Visual Communications and Image Processing '91: Visual Communication,*V1605(1991)

Tzuang, Ching-Kuang C. See Kuo, Jen-Tsai: V1389,156-160(1991)

Tzuk, Y. See Bar, I.: V1397,169-172(1991)

Uber, Gordon T.
: Flexible object-centered illuminator,V1385,2-7(1991)
; Harding, Kevin G.: Illumination and viewing methods for machine vision,VCR36,20-33(1991)
See Harding, Kevin G.: V1385(1991)

Uberall, Herbert
; Faraday, Bruce J.; Maruyama, Xavier K.; Berman, Barry L.: Channeling radiation as an x-ray source for angiography, x-ray lithography, molecular structure determination, and elemental analysis,V1552,198-213(1991)

Uchida, Hiroyuki See Torao, Akira: V1358,843-850(1991)

Uchida, S. See Chen, Hong: V1413,112-119(1991)

Uchida, Toshiya See Horikawa, Hiroshi: V1454,20-32(1991)

Uchida, Yoshiyuki See Takada, Yutaka: V1332,571-576(1991)

Uchimura, R. See Izawa, Takao: V1441,339-344(1991)

Uchino, K. See Yamakoshi, H.: V1397,119-122(1991)

Uchino, Shou-ichi
; Iwayanagi, Takao; Ueno, Takumi; Hayashi, Nobuaki: Negative resist systems using acid-catalyzed pinacol rearrangement reaction in a phenolic resin matrix,V1466,429-435(1991)

Uchiyama, Taro
See Endo, Masamori: V1397,267-270(1991)
See Matsuzaka, Fumio: V1397,239-242(1991)

Uchiyama, Toshio
; Sakai, Mitsuhiro; Saito, Tomohide; Nakamura, Taichi: Optimum structure learning algorithms for competitive learning neural network,V1451,192-203(1991)

Uda, Gianni
; Tofani, Alessandro: Computer analysis of signal-to-noise ratio and detection probability for scanning IRCCD arrays,V1488,257-266(1991)

Udagawa, Toshiki See Fukumoto, Atsushi: V1499,216-225(1991)

Udd, Eric
: Embedded fiber-optic sensors in large structures,V1588,178-181(1991)
; Clark, Timothy E.; Joseph, Alan A.; Levy, Ram L.; Schwab, Scott D.; Smith, Herb G.; Balestra, Chet L.; Todd, John R.; Marcin, John: Fiber optics development at McDonnell Douglas,V1418,134-152(1991)
See Claus, Richard O.: V1588(1991)
See DePaula, Ramon P.: V1584(1991)
See DePaula, Ramon P.: V1367(1991)

Udomkesmalee, Suraphol
: Shape recognition in the Fourier domain,V1564,464-472(1991)
; Scholl, Marija S.; Shumate, Michael S.: Hybrid solution for high-speed target acquisition and identification systems,V1468,81-91(1991)
See Scholl, Marija S.: V1468,92-98(1991)
See Scholl, Marija S.: V1564,165-176(1991)

Udpa, L.
See Boddu, Jayabharat: V1562,251-262(1991)
See Lu, Tongxin: V1564,704-713(1991)

Udpa, Satish S.
See Al-Hudaithi, Aziz: V1567,490-501(1991)
See Boddu, Jayabharat: V1562,251-262(1991)
See Lu, Tongxin: V1564,704-713(1991)

Udrea-Spenea, Marian N. See Halmagean, Eugenia T.: V1484,106-114(1991)

Ueda, Atushi See Ozawa, Kenji: V1499,180-186(1991)

Uehira, Kazutake
: High-speed, high-resolution image reading technique using multi-area sensors,V1448,182-190(1991)

Uemura, Kamon See Ohue, Hiroshi: V1397,527-530(1991)

Uenishi, Kazuya
; Kawabe, Yasumasa; Kokubo, Tadayoshi; Slater, Sydney G.; Blakeney, Andrew J.: Structural effects of DNQ-PAC backbone on resist lithographic properties,V1466,102-116(1991)

Ueno, Takumi
See Hayashi, Nobuaki: V1466,377-383(1991)
See Uchino, Shou-ichi: V1466,429-435(1991)

Uglov, A.
; Krivonogov, Yu.: Influence of laser irradiation on the kinetics of oxide layer growth and their optical characteristics,V1440,310-320(1991)

Uglov, S. A. See Craciun, Valentin: V1392,625-628(1991)

Ugozhayev, Vladimir D. See Komarov, Konstantin P.: V1501,135-143(1991)

Ugrin, J. O. See Szeregij, E. M.: V1391,199-203(1991)

Uhl, Bernd
: Day/night aerial surveillance system for fishery patrol,V1538,140-147(1991)

Uhlenbusch, Juergen
See Freisinger, Bernhard: V1397,311-318(1991)
See Michaelis, Alexander: V1397,709-712(1991)

Uhm, Han S.
; Chen, H. C.; Stark, Robert A.: Kinetic stability analysis of the extraordinary mode perturbations in a cylindrical magnetron,V1407,113-127(1991)
See Chen, H. C.: V1407,139-146(1991)
See Stark, Robert A.: V1407,128-138(1991)

Ujda, Zbigniew See Makuchowski, Jozef: V1391,348-350(1991)

Ukita, Hiroo
; Katagiri, Yoshitada; Nakada, Hiroshi: Flying head read/write characteristics using a monolithically integrated laser diode/photodiode at a wavelength of 1.3 um,V1499,248-262(1991)

Ukrainec, Andrew
; Haykin, Simon: Signal processing with radial basis function networks using expectation-maximization algorithm clustering,V1565,529-539(1991)

Ul'yanov, Sergey S. See Tuchin, Valery V.: V1420,81-92(1991); 1429,62-73(1991)

Ulaby, Fawwaz T. See Rebeiz, Gabriel M.: V1475,199-203(1991)

Ulasjuk, Vladimir N. See Katsap, Victor N.: V1420,259-265(1991)

Ulbers, Gerd
: Integrated optics sensor on silicon for the measurement of displacement, force, and refractive index,V1506,99-110(1991)

Ulfsparre, M. See Aslund, Nils R.: V1450,329-337(1991)

Ulich, Bobby L. See Gleckler, Anthony D.: V1494,454-471(1991)

Ulman, Abraham See Penner, Thomas L.: V1436,169-178(1991)

Ulmer, Melville P. See Rippert, Edward D.: V1549,283-288(1991)

Ulph, Eric
See Murray, Brian W.: V1485,88-95(1991)
See Murray, Brian W.: V1485,106-115(1991)
See Owen, John: V1485,128-137(1991)

Ulrich, Frank
; Bettag, Martin; Langen, K. J.: Positron emission tomography of laser-induced interstitial hyperthermia in cerebral gliomas,V1428,135-135(1991)
See Bettag, Martin: V1525,409-411(1991)
See Waidhauser, Erich: V1428,75-83(1991)

Ulrich, Peter B.
: Stability analysis of semidiscrete schemes for thermal blooming computation,V1408,192-202(1991)
; Wilson, LeRoy E.; eds.: *Propagation of High-Energy Laser Beams Through the Earth's Atmosphere II*,V1408(1991)

Ulrich, R.
See Artjushenko, Vjacheslav G.: V1420,149-156(1991)
See Artjushenko, Vjacheslav G.: V1420,176-176(1991)
See Artjushenko, Vjacheslav G.: V1590,131-136(1991)

Ulyanov, A. N. See Tuchin, Valery V.: V1422,85-96(1991)

Umadevi, P.
; Nagendra, C. L.; Thutupalli, G. K.; Mahadevan, K.; Yadgiri, G.: New thermistor material for thermistor bolometer: material preparation and characterization,V1484,125-135(1991); 1485,195-205(1991)

Umeagukwu, Charles
See Fulton, Robert E.: V1389,144-155(1991)
See Yeh, Chao-Pin: V1389,187-198(1991)

Umeda, A. Y. See Aviles, Walter A.: V1388,587-597(1991)

Umeda, Shin'ichi
See Fukunaga, Seiki: V1466,446-457(1991)
See Koyanagi, Hiroo: V1466,346-361(1991)

Umeda, Tokuo
; Inamura, Kiyonari; Inamoto, Kazuo; Kondoh, Hiroshi P.; Kozuka, Takahiro: Multi-media PACS integrated with HIS/RIS employing magneto-optical disks,V1446,199-210(1991)

Umegaki, Shinsuke See Ohmi, Toshihiko: V1361,606-612(1991)

Underkoffler, John S.
See Halle, Michael W.: V1461,142-155(1991)
See St.-Hilaire, Pierre: V1461,254-261(1991)

Underwood, Ian
; Vass, David G.; Sillitto, Richard M.; Bradford, George; Fancey, Norman E.; Al-Chalabi, Adil O.; Birch, Martin J.; Crossland, William A.; Sparks, Adrian P.; Latham, Steve C.: High-performance spatial light modulator,V1562,107-115(1991)
See Hossack, William J.: V1564,697-702(1991)

Underwood, James H. See Kortright, Jeffrey B.: V1343,95-103(1991)

Uneus, Leif See Hallstadius, Hans: V1433,36-43(1991)

Ungashe, S. See Katz, Howard E.: V1560,370-376(1991)

Unger, Clemens See Possner, Torsten: V1513,378-385(1991)

Unger, Robert
; DiSessa, Peter A.: New i-line and deep-UV optical wafer stepper,V1463,725-742(1991)

Unno, T. See Minemura, Tetsuroh: V1361,344-353(1991)

Unser, Michael A. See Dong, LiXin: V1445,178-187(1991)

Unsoeld, Eberhard
See Baumgartner, R.: V1525,246-248(1991)
See Gottschalk, Wolfgang: V1427,320-326(1991)

Unterleitner, Fred C.
: Wavelength division multiplexing of services in a fiber-to-the-home system,V1363,92-96(1991)

Unterrainer, Karl See Kremser, Christian: V1501,69-79(1991)

Unwin, Rodney T.
See Cryan, Robert A.: V1372,64-71(1991)
See Cryan, Robert A.: V1579,133-143(1991)

Uozumi, Jun See Widjaja, Joewono: V1400,94-100(1991)

Upadhye, Anand M. See Gupta, Lalit: V1396,266-269(1991)

Upchurch, Billy T.
; Schryer, David R.; Brown, K. G.; Kielin, E. J.; Hoflund, Gar B.; Gardner, Steven D.: Recent advances in CO2 laser catalysts,V1416,21-29(1991)

Upchurch, Bruce L. See Throop, James A.: V1379,124-135(1991)

Uphaus, Robert A. See Cotton, Therese M.: V1403,93-101(1991)

Uppal, Gurvinder S.
; Sherk, Henry H.; Black, Johnathan D.; Rhodes, Anthony; Sazy, John; Lane, Gregory J.: CO2 partial matricectomy in the treatment of ingrown toenails,V1424,51-52(1991)
See Black, Johnathan D.: V1424,20-22(1991)
See Black, Johnathan D.: V1424,12-15(1991)
See Lane, Gregory J.: V1424,7-11(1991)
See Meller, Menachem M.: V1424,60-61(1991)
See Sazy, John: V1424,50-50(1991)

Uppal, J. S. See Harrison, Robert G.: V1376,94-102(1991)

Urakawa, Yoshinori See Fujita, Goro: V1499,426-432(1991)

Uray, Georg
See He, Huarui: V1368,175-180(1991)
See He, Huarui: V1510,95-103(1991)

Urbach, Michael K. See Swanberg, Norman: V1346,249-267(1991)

Urban, Frantisek
; Matejka, Frantisek: Planar diffractive optical elements prepared by electron-beam lithography,V1574,58-65(1991)

Urban, Stephen J.
: Standards for electronic imaging for facsimile systems,VCR37,113-145(1991)

Urciuoli, Rosa See Fasano, Victor A.: V1428,2-12(1991)

Ureneva, E. V. See Kostin, Ivan K.: V1554A,418-425(1991)

Urmson, John
: New apparatus and method for fluid composition monitoring and control,V1392,421-427(1991)

Urquhart, Kristopher S.
; Lee, Sing H.: Phase-only encoding method for complex wavefronts,V1555,13-22(1991)
; Marchand, Philippe J.; Lee, Sing H.; Esener, Sadik C.: Orthogonal cylindrical diffractive lens for parallel readout optical disk system,V1555,214-223(1991)

Urushidani, Tatuo See Ozawa, Kenji: V1499,180-186(1991)

Urushihara, Shinji See Nakatsuka, Masahiro: V1411,108-115(1991)

Usala, Sandro See Oktu, Ozcau: V1361,812-818(1991)

Usanov, Yu. Y.
; Vavilova, Ye. A.; Kosobokova, N. L.; Shevtsov, M. K.: Reflection silver-halide gelatin holograms,V1238,178-182(1991)
See Kosobokova, N. L.: V1238,183-188(1991)

Ushakov, Leonid S.
; Ponomaryov, A. M.; Trofimenko, Vladimir V.; Drozhbin, Yu. A.: Multichannel mirror systems for high-speed framing recording,V1358,447-450(1991)

Ushakov, V. K. See Dubovoi, I. A.: V1358,134-138(1991)

Usher, Brian F. See Price, Garth L.: V1361,543-550(1991)

Ustinov, Sergey I. See Gan, Michael A.: V1574,254-260(1991)

Ustyugov, Vladimir I. See Mak, Arthur A.: V1410,233-243(1991)

Utenkov, Boris I.
See Bouchenkov, Vyatcheslav A.: V1410,185-188(1991); 1427,409-412(1991)
See Bouchenkov, Vyatcheslav A.: V1427,405-408(1991)
See Bouchenkov, Vyatcheslav A.: V1410,244-247(1991)

Uthe, Edward E.
: Airborne lidar elastic scattering, fluorescent scattering, and differential absorption observations,V1479,393-402(1991)
; Livingston, John M.: Lidar evaluation of the propagation environment,V1487,228-239(1991)

Utsch, Thomas F. See Kilston, Steven: V1495,193-204(1991)

Utsumi, Atsushi
See Arai, Tsunenori: V1425,191-195(1991)
See Sakurada, Masami: V1425,158-164(1991)

Uttal, William
; Shepherd, Thomas; Lovell, Robb E.; Dayanand, Sriram: Computational model of an integrated vision system,V1453,258-269(1991)

Uttamchandani, Deepak G.
; Al-Raweshidy, H. S.: Spread spectrum technique for passive multiplexing of interferometric optical fiber sensors,V1511,212-219(1991)
See Al-Raweshidy, H. S.: V1367,329-336(1991)
See Jin, Wei: V1504,125-132(1991)
See Rao, Yun-Jiang: V1572,287-292(1991)

Utz, Sergey R. See Tuchin, Valery V.: V1422,85-96(1991)

Uvarov, G. V.
See Golub, Mikhail A.: V1572,101-106(1991)
See Golub, Mikhail A.: V1366,273-282(1991)

Uwabu, H.
; Kakii, Eiji; Lacombe, R.; Maruyama, Masanori; Fujiwara, Hiroshi: VLSI implementation of a buffer, universal quantizer, and frame-rate-control processor,V1605,928-937(1991)

Uyemura, T. See Yang, Jie: V1358,672-676(1991)

Uyemura, Tsuneyoshi See Yokoyama, Naoki: V1358,351-357(1991)

V'yushkin, Maksim E. See Tuchin, Valery V.: V1420,81-92(1991); 1429,62-73(1991)

Vaccaro, John J. See Rorabaugh, Terry L.: V1566,312-322(1991)

Vacha, M. See Adamec, F.: V1402,82-84(1991)

Vachev, V. D.
; Zadkov, Victor N.: Molecular dynamics of stilbene molecule under laser excitation,V1403,487-496(1991)
See Grishanin, B. A.: V1402,44-52(1991)

Vachon, Bertrand
; Berge-Cherfaoui, Veronique: Multiagent collaboration for experimental calibration of an autonomous mobile robot,V1468,483-492(1991)

Vaezi, Matt M.
; Bavarian, Behnam; Healey, Glenn: Image reconstruction of IDS filter response,V1606,803-809(1991)
; Bavarian, Behnam; Healey, Glenn: Optimum intensity-dependent spread filters in image processing,V1452,57-63(1991)

Vahakangas, Jouko See Heikkinen, Veli: V1533,115-121(1991)

Vahid-Shahidi, A. See Lefebvre, P.: V1583,221-225(1991)

Vaicikauskas, V.
; Maldutis, Evaldas K.; Kajokas, R.: Study of the optical features of YBa2Cu3O7-x films by SEW spectroscopy,V1440,357-363(1991)

Vaidyanathan, C. S. See Wood, Hugh C.: V1387,245-254(1991)

Vaina, Lucia M.
; Tuncer, Temel E.: Neural networks for the recognition of skilled arm and hand movements,V1468,990-999(1991)
; Zlateva, Stoyanka D.: Convexity-based method for extracting object parts from 3-D surfaces,V1468,710-719(1991)
See Citkusev, Ljubomir: V1468,786-793(1991)
See Stratton, Norman A.: V1468,176-185(1991)
See Zlateva, Stoyanka D.: V1468,379-393(1991)

Valavanis, Kimon P.
See Larsson, T.: V1387,165-168(1991)
See Zheng, Joe: V1606,1037-1047(1991)

Valderrama, Giuseppe L.
; Fredin, Leif G.; Berry, Michael J.; Dempsey, B. P.; Harpole, George M.: Temperature distributions in laser-irradiated tissues,V1427,200-213(1991)

Valduga, Giuliana See Jori, Guilio: V1525,367-376(1991)

Valente, Tina M.
; Richard, Ralph M.: Analysis of elastomer lens mountings,V1533,21-26(1991)
See Richard, Ralph M.: V1533,12-20(1991)

Valentine, Gary W. See Smalanskas, Joseph P.: V1489,2-8(1991)

Valentini, Antonio
; Quaranta, Fabio; Vasanelli, L.; Piccolo, R.: Optical properties of Li-doped ZnO films,V1400,164-170(1991)
See Battaglin, Giancarlo: V1513,441-450(1991)

Valentini, G. See Cubeddu, Rinaldo: V1525,17-25(1991)

Valentino, Daniel J.
See Chuang, Keh-Shih: V1445,341-347(1991)
See Lou, Shyh-Liang: V1446,302-311(1991)
See Weinberg, Wolfram S.: V1446,35-39(1991)

Valenzuela, C. F. See Johnson, David A.: V1432,82-90(1991)

Valera Robles, Jesus D.
; Harvey, David; Jones, Julian D.: Automatic heterodyning of fiber-optic speckle pattern interferometry,V1508,170-179(1991)

Valeton, J. M.
; van Meeteren, Aart: Target acquisition modeling based on human visual system performance,V1486,68-84(1991)

Valette, Serge See Renard, Stephane: V1499,238-247(1991)

Valis, Tomas
See Hogg, W. D.: V1588,300-307(1991)
See Measures, Raymond M.: V1332,421-430(1991)

Valla, Bruno
; Tonazzi, Juan C.; Macedo, Marcelo A.; Dall'Antonia, L. H.; Aegerter, Michel A.; Gomes, M. A.; Bulhoes, Luis O.: Transparent storage layers for H+ and Li+ ions prepared by sol-gel technique, V1536,48-62(1991)

Vallerga, John V. See Welsh, Barry Y.: V1343,166-174(1991)

Valley, John F.
See Lipscomb, George F.: V1560,388-399(1991)
See Wu, Jeong W.: V1560,196-205(1991)

Vallmitjana, Santiago
; Juvells, Ignacio; Carnicer, Arturo; Campos, Juan: Pattern detection using a modified composite filter with nonlinear joint transform correlator, V1564,266-274(1991)

Valnicek, Boris See Hudec, Rene: V1343,162-163(1991)

Van Allen, Eric J. See Menon, Murali M.: V1469,322-328(1991)

van Amersfoort, P. W.
: Free-electron lasers as light sources for basic research, V1500,2-13(1991); 1501,2-13(1991); 1502,2-13(1991)

van Bree, J. L.
; van Riet, E. J.: Use of an image converter camera for analysis of ballistic resistance of lightweight armor materials, V1358,692-697(1991)

Van Brunt, R. J. See Roberts, James R.: V1392,428-436(1991)

van Daele, Peter
See Buydens, Luc: V1362,50-58(1991)
See Demeester, Piet M.: V1361,987-998(1991)
See Demeester, Piet M.: V1361,1132-1143(1991)
See Frietman, Edward E.: V1390,434-453(1991)
See Moehlmann, Gustaaf R.: V1560,426-433(1991)
See Pollentier, Ivan K.: V1361,1056-1062(1991)

van Damme, B. See Baert, Luc: V1525,385-390(1991)

Van De Merwe, Willem P. See Li, Zhao-zhang: V1427,152-161(1991)

Vandenabeele, Peter
; Maex, Karen: Emissivity of silicon wafers during rapid thermal processing, V1393,316-336(1991)
; Maex, Karen: Round-robin comparison of temperature nonuniformity during RTP due to patterned layers, V1393,372-394(1991)

Vandenberg, Donald E.
; Humbel, William D.; Wertheimer, Alan: Quantitative evaluation of optical surfaces using an improved Foucault test approach, V1542,534-542(1991)

van den Berg, Lodewijk
See Sawyer, Curry R.: V1567,254-263(1991)
See Steiner, Bruce W.: V1557,156-167(1991)

van den Bergh, Hubert
See Bays, Roland: V1525,397-408(1991)
See Braichotte, D.: V1525,212-218(1991)
See Wagnieres, G.: V1525,219-236(1991)

van den Brink, Martin A.
; Katz, Barton A.; Wittekoek, Stefan: New 0.54 aperture i-line wafer stepper with field-by-field leveling combined with global alignment, V1463,709-724(1991)

van den Broecke, Duco G. See Gijsbers, Geert H.: V1425,94-101(1991)

Van den hove, Luc
See Goethals, Anne-Marie: V1466,604-615(1991)
See Laporte, Philippe: V1392,196-207(1991)
See Samarakone, Nandasiri: V1463,16-29(1991)

Vanderbok, Raymond S. See Rauchmiller, Robert F.: V1384,100-114(1991)

Vandergriff, Linda J. See Wesley, Michael: V1481,209-220(1991)

van der Laan, Jan E. See Cooper, David E.: V1433,120-127(1991)

VanderMeulen, David L.
; Khoka, Mustafa; Spears, Kenneth G.: Laser neural tissue interactions using bilayer membrane models, V1428,84-90(1991)
; Misra, Prabhakar; Michael, Jason; Spears, Kenneth G.; Khoka, Mustafa: Quantitative analysis at the molecular level of laser/neural tissue interactions using a liposome model system, V1428,91-98(1991)

van der Meulen, F. W. See van Gemert, Martin J.: V1427,316-319(1991)

van der Putten, Harrie B. See Luehrmann, Paul F.: V1463,434-445(1991)

Van Der Sluys, William G. See Berg, John M.: V1435,331-337(1991)

Van der Spiegel, Jan See Kreider, Gregory: V1381,242-249(1991)

van der Steen, Hendricus J. See Barbini, Roberto: V1503,363-374(1991)

Van der Stuyft, Emmanuel
; Goedseels, Vic; Geers, Rony: Digital restoration of distorted geometric features of pigs, V1379,189-200(1991)

van der Veen, Maurits J.
See van Leeuwen, Ton G.: V1427,214-219(1991)
See Vari, Sandor G.: V1424,33-42(1991)
See Vari, Sandor G.: V1426,58-65(1991)
See Vari, Sandor G.: V1426,111-120(1991)

van der Voort, H. T. See Brakenhoff, G. J.: V1439,121-127(1991)

van der Wal, Peter See Roeser, Hans-Peter: V1501,194-197(1991)

Vanderweit, Don See Huang, H. K.: V1444,214-220(1991)

van der Zee, Pieter
See Arridge, Simon R.: V1431,204-215(1991)
See Cope, Mark: V1431,251-262(1991)

van de Vaart, Herman See Wong, Ka-Kha: V1374,278-286(1991)

vandeVen, Martin J. See Fishkin, Joshua B.: V1431,122-135(1991)

van Deventer, T. E.
; Katehi, Linda P.; Cangellaris, Andreas C.: Effects of conductor losses on cross-talk in multilevel-coupled VLSI interconnections, V1389,285-296(1991)

van Dijk, Cornelius A. See Bradshaw, John D.: V1433,81-91(1991)

van Dongen, A. M.
; Oomen, Emmanuel W.; le Gall, Perrine M.: Upconversion in rare-earth-doped fluoride glasses, V1513,330-339(1991)

Van Dongen, V. See Macq, Benoit M.: V1606,165-173(1991)

van Dordrecht, Axel See Rando, Nicola: V1549,340-356(1991)

Van Driessche, Veerle See Samarakone, Nandasiri: V1463,16-29(1991)

VanDyk, Ed See Kovacich, Michael: V1481,357-370(1991)

Van Dyke-Lewis, Michelle See Myler, Harley R.: V1565,57-68(1991)

Van Eck, Timothy E.
See Lipscomb, George F.: V1560,388-399(1991)
See Lytel, Richard S.: V1389,547-558(1991)

Van Eijk, Peter See Park, Yong K.: V1372,219-227(1991)

van Erning, Leon J. See Karssemeijer, Nico: V1445,166-177(1991)

Van Essen, David
; Olshausen, B.; Anderson, Clifford H.; Gallant, J. T. L.: Pattern recognition, attention, and information bottlenecks in the primate visual system, V1473,17-28(1991)

Van Eycken, Luc See Xie, Kan: V1567,380-389(1991)

Vang, T. See Taylor, Geoff W.: V1476,2-13(1991)

van Geest, Lambertus K.
; Stoop, Karel W.: Hybrid phototube with Si target, V1449,121-134(1991)

van Gemert, Martin J.
: Principles of optical dosimetry: fluorescence diagnostics, V1525,100-109(1991)
; Brugmans, Marco J.; Gijsbers, Geert H.; Kemper, J.; van der Meulen, F. W.; Nijdam, D. C.: Temperature response of biological tissues to nonablative pulsed CO2 laser irradiation, V1427,316-319(1991)
See Gijsbers, Geert H.: V1425,94-101(1991)
See Sterenborg, H. J.: V1427,256-266(1991)

van Gennip, Elisabeth M.
; van Poppel, Bas M.; Bakker, Albert R.; Ottes, Fenno P.: Comparison of worldwide opinions on the costs and benefits of PACS, V1446,442-450(1991)

van Gilse, Jan
; Koczera, Stanley; Greby, Daniel F.: Direct laser beam diagnostics, V1414,45-54(1991)

Vangonen, Albert I. See Belov, Sergei N.: V1503,503-509(1991)

van Goor, Frederik A.
: High-repetition-rate x-ray preionization source, V1397,563-568(1991)
; Witteman, Wilhelmus J.: X-ray preionization for high-repetition-rate discharge excimer lasers, V1412,91-102(1991)
See Ekelmans, G. B.: V1397,569-572(1991)
See Trentelman, M.: V1397,115-118(1991)
See Witteman, Wilhelmus J.: V1397,37-45(1991)

VanGorden, Steve See Carbone, Joseph: V1447,229-242(1991)

Vangsness, C. T.
; Huang, Jay; Smith, Chadwick F.: Light absorption characteristics of the human meniscus: applications for laser ablation, V1424,16-19(1991)

Vanhauwermeiren, Luc See Delbare, William: V1389,257-272(1991)

Vanherzeele, Herman A. See Meth, Jeffrey S.: V1560,13-24(1991)

Van Hoof, Chris A.
; Goovaerts, Etienne; Borghs, Gustaaf: Bias dependence of the hole tunneling time in AlAs/GaAs resonant tunneling structures, V1362,291-300(1991)

Van Hook, H. J. See Greer, James A.: V1377,79-90(1991)

van Hout, Frits J. See Luehrmann, Paul F.: V1463,434-445(1991)

Van Hove, James M. See Khan, M. A.: V1396,753-759(1991)

van Hulst, Niek F.
; Heesink, Gerard J.; Bolger, Bouwe; Kelderman, E.; Verboom, W.; Reinhoudt, D. N.: Linear and nonlinear optical properties of substituted pyrrolo[1,2-a]quinolines,V1361,581-588(1991)
; Heesink, Gerard J.; de Leeuw, H.; Bolger, Bouwe: Second harmonic generation of diode laser radiation in KNbO3,V1362,631-646(1991)

Van Ijzendoorn, L. J. See Boher, Pierre: V1345,165-179(1991)

Vanin, V. A.
; Trukhanov, V. A.: Display holography commercial prospects,V1238,372-380(1991)
; Vorobjov, S. P.: Pseudocolor reflection hologram properties recorded using monochrome photographic materials,V1238,324-331(1991)

Vanini, P. See Bortoletto, Favio: V1542,225-235(1991)

Van Isselt, J. W. See Anema, P. C.: V1446,352-356(1991)

Van Laethem, Marc
; Xia, Jia-zhi; Clauwaert, Julius: Analysis of photon correlation functions obtained from eye lenses in vivo,V1403,732-742(1991)

van Leeuwen, Ton G.
; van der Veen, Maurits J.; Verdaasdonk, Rudolf M.; Borst, Cornelius: Tissue ablation by holmium:YSGG laser pulses through saline and blood,V1427,214-219(1991)

van Lier, Johan E.
See Morgan, Alan R.: V1426,350-355(1991)
See Paquette, Benoit: V1426,362-366(1991)

van Meeteren, Aart See Valeton, J. M.: V1486,68-84(1991)

Van Metre, Richard
; Curran, Mark E.; Cole, George R.; Williams, Ken: Fiber optic backbone for a submarine combat system,V1369,9-18(1991)

van Nhac, Nguyen See Sasek, Ladislav: V1572,151-156(1991)

Vanni, Paolo
; Isoldi, Felice: Visual characteristics of LED display pushbuttons for avionic applications,V1456,300-309(1991)

Vannier, C. See Brunet, Henri: V1397,273-276(1991)

van Nifterick, W.
See Frietman, Edward E.: V1390,434-453(1991)
See Frietman, Edward E.: V1401,19-26(1991)

Vannucci, Antonello See Varasi, Mauro: V1583,165-169(1991)

Van Orsdel, Brian T. See Olsen, Gregory H.: V1419,24-31(1991)

van Overbeek, Thomas T.
: Document viewing: display requirements in image management,V1454,406-413(1991)

van Poppel, Bas M.
See Lodder, Herman: V1446,227-233(1991)
See van Gennip, Elisabeth M.: V1446,442-450(1991)

van Renesse, Rudolf L.
: Nondiffractive optically variable security devices,V1509,113-125(1991)

van Riet, E. J. See van Bree, J. L.: V1358,692-697(1991)

van Rijk, P. P. See Anema, P. C.: V1446,352-356(1991)

Van Rossum, Marc
: Fabrication technology of strained layer heterostructure devices,V1361,373-382(1991)

van Steen, M. R. See Stut, W. J.: V1446,396-404(1991)

VanSteenkiste, T. H. See Vaz, Nuno A.: V1455,110-122(1991)

Van Stryland, Eric W. See Sheik-Bahae, Mansoor: V1441,430-443(1991)

van Suylen, Robert J. See Gussenhoven, Elma J.: V1425,203-206(1991)

van Swaaij, Michael F.
; Catthoor, Francky V.; De Man, Hugo J.: Novel regular-array ASIC architecture for 2-D ROS sorting,V1606,901-910(1991)

van Swol, Christiaan F. See Sterenborg, H. J.: V1427,256-266(1991)

van Tomme, E. See Moehlmann, Gustaaf R.: V1560,426-433(1991)

van Urk, Hero See Gussenhoven, Elma J.: V1425,203-206(1991)

Van Veen, Barry D. See Liu, Tsung-Ching: V1566,419-426(1991)

Van Wagenen, E. A. See LaPeruta, Richard: V1497,357-366(1991)

van Wieringen, Niek See Gijsbers, Geert H.: V1425,94-101(1991)

Van Woerkom, Linn D.
See Fittinghoff, David N.: V1413,81-88(1991)
See Gold, David: V1413,41-52(1991)

Van Wonterghem, Bruno M. See Persoons, Andre P.: V1409,220-229(1991)

Varachi, John P.
See Hart, Patrick W.: V1366,334-342(1991)
See Kilmer, Joyce P.: V1366,85-91(1991)
See Wei, Ta-Sheng: V1366,235-240(1991)

Varadan, Vasundara V.
; Varadan, Vijay K.: Chirality and its applications to engineered materials,V1558,156-181(1991)
See Apparao, R. T.: V1558,2-13(1991)
See Bayer, Janice I.: V1480,102-114(1991)
See Hong, Suk-Yoon: V1489,75-83(1991)
See Lakhtakia, Akhlesh: V1558,120-126(1991)
See Lakhtakia, Akhlesh: V1558,22-27(1991)
See Ma, Yushieh: V1558,138-142(1991)
See Ma, Yushieh: V1558,132-137(1991)
See Ma, Yushieh: V1487,220-225(1991)
See Ro, Ru-Yen: V1558,269-287(1991)
See Ro, Ru-Yen: V1489,46-55(1991)
See Selmi, Fathi: V1489,97-107(1991)

Varadan, Vijay K.
ed.: Wave Propagation and Scattering in Varied Media II,V1558(1991)
See Apparao, R. T.: V1558,2-13(1991)
See Bayer, Janice I.: V1480,102-114(1991)
See Breakwell, John: V1489(1991)
See Hong, Suk-Yoon: V1489,75-83(1991)
See Lakhtakia, Akhlesh: V1558,120-126(1991)
See Lakhtakia, Akhlesh: V1558,22-27(1991)
See Ma, Yushieh: V1558,138-142(1991)
See Ma, Yushieh: V1558,132-137(1991)
See Ma, Yushieh: V1487,220-225(1991)
See Ro, Ru-Yen: V1558,269-287(1991)
See Ro, Ru-Yen: V1489,46-55(1991)
See Selmi, Fathi: V1489,97-107(1991)
See Varadan, Vasundara V.: V1558,156-181(1991)

Varakin, Vladimir N.
See Simonov, Alexander P.: V1503,330-337(1991)
See Simonov, Alexander P.: V1436,20-30(1991)

Varamit, Srisuda P. See Almeida, Silverio P.: V1500,34-45(1991)

Varasi, Mauro
; Vannucci, Antonello; Signorazzi, Mario: Lithium niobate proton-exchange technology for phase-amplitude modulators,V1583,165-169(1991)

Vardanjan, G. S.
; Musatov, L. G.: Photoelastic modeling of linear and nonlinear creep using two- and three-dimensional models,V1554A,496-502(1991)

Vardanyan, A. G. See Chernyaeva, E. B.: V1402,7-10(1991)

Vardi, Y. See Shepp, Larry A.: V1452,216-224(1991)

Varfolomeev, A. A. See Dodd, James W.: V1407,467-473(1991)

Varga, Margaret J.
; Radford, John C.: Automatic vehicle model identification,V1381,92-100(1991)

Vargas, Raymund See Moy, Richard Q.: V1467,154-160(1991)

Vargo, Rick C.
; Sklar, Michael E.; Wegerif, Daniel G.: Automation of vehicle processing at Space Station Freedom,V1387,72-81(1991)

Vari, Sandor G.
; Papazoglou, Theodore G.; van der Veen, Maurits J.; Papaioannou, Thanassis; Fishbein, Michael C.; Chandra, Mudjianto; Beeder, Clain; Shi, Wei-Qiang; Grundfest, Warren S.: Detection of atheroma using Photofrin II and laser-induced fluorescence spectroscopy,V1426,58-65(1991)
; Papazoglou, Theodore G.; van der Veen, Maurits J.; Fishbein, Michael C.; Young, J. D.; Chandra, Mudjianto; Papaioannou, Thanassis; Beeder, Clain; Shi, Wei-Qiang; Grundfest, Warren S.: Intraoperative metastases detection by laser-induced fluorescence spectroscopy,V1426,111-120(1991)
; Shi, Wei-Qiang; van der Veen, Maurits J.; Fishbein, Michael C.; Miller, J. M.; Papaioannou, Thanassis; Grundfest, Warren S.: Comparative study of excimer and erbium:YAG lasers for ablation of structural components of the knee,V1424,33-42(1991)
See Chandra, Mudjianto: V1421,68-71(1991)

Varian, Richard H. See Schaeffer, A. R.: V1447,165-176(1991)

Varlamov, Yu. See Predtechensky, M.: V1477,234-241(1991)

Varma, Sudhanshu See Pouskouleli, G.: V1590,179-190(1991)

Varner, Thomas L. See Worrell, Steven W.: V1382,219-227(1991)

Varnes, Marie E. See Oleinick, Nancy L.: V1426,235-243(1991)

Varshneya, Deepak
; Lapierre, A.: Characterization of time-division-multiplexed digital optical position transducer,V1584,188-201(1991)
; Maida, John L.; Overstreet, Mark A.: Fiber optic speed sensor for advanced gas turbine engine control,V1367,181-191(1991)
See York, Jim F.: V1584,308-319(1991)

Vartapetov, Serge K.
See Alimpiev, Sergei S.: V1503,154-158(1991)
See Atjezhev, Vladimir V.: V1503,197-199(1991)
See Lukishova, Svetlana G.: V1503,338-345(1991)

Vas'kovsky, Yu. M.
; Gordeeva, I. A.; Korenev, A. S.; Rovinsky, R. E.; Cenina, I. S.; Shirokova, I. P.: Plasma parameter determination formed under the influence of CO2 laser radiation on the obstacle in the air using optical methods,V1440,229-240(1991)
; Rovinsky, R. E.; Fedjushin, B. T.; Filatova, S. A.: Near-surface plasma initiation model for short laser pulses,V1440,241-249(1991)
See Babaeva, N. A.: V1440,260-269(1991)

Vasanelli, L. See Valentini, Antonio: V1400,164-170(1991)

Vasara, Antti H.
; Noponen, Eero; Turunen, Jari P.; Miller, J. M.; Taghizadeh, Mohammad R.; Tuovinen, Jussi: Rigorous diffraction theory of binary optical interconnects,V1507,224-238(1991)
See Saarinen, Jyrki V.: V1555,128-137(1991)

Vaschenko, G. O.
See Briscoe, Dennis: V1358,329-336(1991)
See Schelev, Mikhail Y.: V1358,569-573(1991)

Vasilescu, Manuela See Terzopoulos, Demetri: V1383,257-264(1991)

Vasiliev, Antoly A. See Gariaev, Peter P.: V1621,280-291(1991)

Vasiljeva, Natalja V. See Popov, A.: V1519,37-42(1991)

Vasilyeva, N. See Popov, A.: V1519,457-462(1991)

Vasnetsov, M. V.
See Bazhenov, V. Y.: V1574,148-153(1991)
See Taranenko, Victor B.: V1621,169-179(1991)

Vasquez, R. P. See Hunt, Brian D.: V1394,89-95(1991)

Vass, David G.
See Hossack, William J.: V1564,697-702(1991)
See Potter, Duncan J.: V1564,363-372(1991)
See Underwood, Ian: V1562,107-115(1991)

Vasson, A. M. See Banvillet, Henri: V1361,972-979(1991)

Vastra, I.
; Boudot, C.: Influence of the shielding gas internal bore welding,V1502,190-202(1991)
See Boudot, C.: V1502,177-189(1991)
See Boudot, C.: V1502,72-82(1991)

Vaughan, Bobbi See Andrews, Lee T.: V1459,125-135(1991)

Vaughan, David K. See Wojcik, Gregory L.: V1463,292-303(1991)

Vaughan, William W. See Theon, John S.: V1492,2-23(1991)

Vaughn, David L.
; Arkin, Ronald C.: Workstation recognition using a constrained edge-based Hough transform for mobile robot navigation,V1383,503-514(1991)

Vaulin, P. P. See Tukhvatulin, R. T.: V1403,390-391(1991)

Vavilov, Vladimir P.
: Soviet IR imagers and their applications: short state of the art,V1540,460-465(1991)
; Ivanov, A. I.; Sengulye, A. A.: Qualitative and quantitative evaluation of moisture in thermal insulation by using thermography,V1467,230-233(1991)
; Maldague, Xavier; Saptzin, V. M.: Postprocessing of thermograms in infrared nondestructive testing,V1540,488-495(1991)

Vavilova, Ye. A. See Usanov, Yu. Y.: V1238,178-182(1991)

Vawter, Gregory A.
; Hietala, Vincent M.; Kravitz, Stanley H.; Meyer, W. J.: Direct optical phase shifter for phased-array systems,V1476,102-106(1991)
See Hietala, Vincent M.: V1476,170-175(1991)

Vaz, Nuno A.
; Smith, George W.; VanSteenkiste, T. H.; Montgomery, G. P.: Droplet-size polydispersity in polymer-dispersed liquid-crystal films,V1455,110-122(1991)

Vaz, Richard F. See Wheeler, Frederick W.: V1606,78-85(1991)

Vdovin, Gleb V.
See Aksinin, V. I.: V1500,93-104(1991)
See Apollonov, V. V.: V1502,83-94(1991)

Veatch, M. S. See Merkelo, Henri: V1390,91-163(1991)

Veaux, Jacqueline
; Cavailler, Claude; Gex, Jean-Pierre; Hauducoeur, Alain; Hivernage, M.: MCP image intensifier in the 100-KeV to 1-MeV x-ray range,V1449,13-24(1991)

Vecchi, G. See Felsen, Leopold B.: V1471,154-162(1991)

Vecher, Jaroslav
: Luminescence probes in aqueous micellar solutions,V1402,97-101(1991)

Vedder, Peter W. See Welsh, Barry Y.: V1343,166-174(1991)

Vedernikov, A. A. See Regel, L. L.: V1557,182-191(1991)

Vega, Fidel
; Solis, Javier; Afonso, Carmen N.: Time-space resolved optical study of the plasma produced by laser ablation of Ge: the role of oxygen pressure,V1397,807-811(1991)

Vegfors, Magnus
; Lindberg, Lars-Goran; Lennmarken, Claes; Oberg, Per A.: In-vitro model for evaluation of pulse oximetry,V1426,79-83(1991)

Veiga, Robert See Fishman, Jack: V1491,348-359(1991)

Veiko, Vadim P.
; Yakovlev, Evgeni B.; Frolov, V. V.; Chujko, V. A.; Kromin, A. K.; Abbakumov, M. O.; Shakola, A. T.; Fomichov, P. A.: Laser heating and evaporation of glass and glass-borning materials and its application for creating MOC,V1544,152-163(1991)

Veilleux, George See Dodelet, Jean-Pol: V1436,38-49(1991)

Veirs, D. K. See Ager, Joel W.: V1437,24-31(1991)

Velarde, G.
; Perlado, J. M.; Aragones, J. M.; Honrubia, J. J.; Martinez-Val, J. M.; Minguez, E.: High-power lasers and the production of energy by inertial fusion,V1502,242-257(1991)

Veldkamp, Wilfrid B.
: Overview of micro-optics: past, present, and future,V1544,287-299(1991)
See Roychoudhuri, Chandrasekhar: V1544(1991)

Velema, Evelyn See Smits, Pieter C.: V1425,188-190(1991)

Velez, E. See Kozaitis, Samuel P.: V1474,112-115(1991)

Velger, Mordekhai
; Toker, Gregory: Study of active 3-D terrain mapping for helicopter landings,V1478,168-176(1991)

Veligdan, James T. See Pogorelsky, Igor V.: V1413,21-31(1991)

Velikoselsky, V.V. See Mnatsakanyan, Eduard A.: V1414,130-133(1991)

Vella, Paul J. See Jiang, Jing-Wen: V1363,191-195(1991)

Velten, Vincent J.
: Signature prediction models for flir target recognition,VIS07,98-107(1991)

Veltri, R. D. See Hay, Stephen O.: V1435,352-358(1991)

Vemuri, Baba C.
ed.: *Geometric Methods in Computer Vision*,V1570(1991)
; Skofteland, G.: Computing motion parameters from sparse multisensor range data for telerobotics,V1383,97-108(1991)

Vemuri, Gautam
: Fluctuations in atomic fluorescence induced by laser noise,V1376,34-46(1991)

Vemuri, V.
; Jang, Gyu-Sang: Neural networks for Fredholm-type integral equations,V1469,563-574(1991)

Vendeltorp-Pommer, Helle
; Pedersen, Frands B.; Bjarklev, Anders; Hedegaard Povlsen, Joern: Noise and gain performance for an Er3+-doped fiber amplifier pumped at 980 nm or 1480 nm,V1373,254-265(1991)

Venema, Steven
See Bejczy, Antal K.: V1387,365-377(1991)
See Schenker, Paul S.: V1383,202-223(1991)

Venetsanopoulos, Anastasios N.
See Cheng, Fulin: V1483,49-59(1991)
See Zervakis, Michael E.: V1452,90-103(1991)
See Zhou, Ziheng: V1606,309-319(1991)

Vengal, Jacob V. See Chandrasekhar, N. S.: V1390,523-536(1991)

Venghaus, Herbert See Tischel, M.: V1361,917-926(1991)

Vengsarkar, Ashish M.
; Michie, W. C.; Jankovic, Lilja; Culshaw, Brian; Claus, Richard O.: Fiber optic sensor for simultaneous measurement of strain and temperature,V1367,249-260(1991)
; Murphy, Kent A.; Gunther, Michael F.; Plante, Angela J.; Claus, Richard O.: Low-profile fibers for embedded smart structure applications,V1588,2-13(1991)

See Claus, Richard O.: V1588,159-168(1991)
See Fogg, Brian R.: V1588,14-25(1991)
See Murphy, Kent A.: V1588,134-142(1991)
See Murphy, Kent A.: V1588,117-124(1991)
See Shaw, J. K.: V1367,337-346(1991)
See Tran, Tuan A.: V1584,178-186(1991)
See Wang, Anbo: V1584,294-303(1991)

Veniaminov, Andrei V.
: Dark stability of holograms and holographic investigation of slow diffusion in polymers,V1238,266-270(1991)
See Goncharov, V. F.: V1238,97-102(1991)

Venkataram, Krishnan See Subramanian, P.: V1559,461-469(1991)

Venkatesan, T. See Xi, Xiaoxing: V1477,20-25(1991)

Venkateswara Rao, B.
: Studies on defocus in thermal imaging systems,V1399,145-156(1991)

Venkateswarlu, Putcha
; Dokhanian, Mostafa; Sekhar, Prayaga C.; George, M. C.; Jagannath, H.: Optical phase-conjugate resonators, bistabilities, and applications,V1332,245-266(1991)
See Shukla, Ram P.: V1332,274-286(1991)

Venturi, Giovanni
; Dellepiane, Silvana G.; Vernazza, Gianni L.: Biomedical structure recognition by successive approximations,V1606,217-225(1991)

Venugopal, Kootala P.
; Pandya, Abhijit S.; Sudhakar, Raghavan: Continuous recognition of sonar targets using neural networks,V1471,44-53(1991)

Vepsalainen, Ari M.
; Linnainmaa, Seppo; Yli-Harja, Olli P.: Hierarchical image decomposition based on modeling of convex hulls corresponding to a set of order statistic filters,V1568,2-13(1991)
; Rantala, Aarne E.: Efficient transformation algorithm for 3-D images,V1452,64-75(1991)
; Toivanen, Pekka J.: Two new image compression methods utilizing mathematical morphology,V1606,282-293(1991)

Verbanets, William R.
; Greenwald, David: New generation control system for ultra-low-jitter satellite tracking,V1482,112-120(1991)
See Liebst, Brad: V1482,425-438(1991)

Verber, Carl M.
; Kenan, Richard P.; Tan, Ronson K.; Bao, Y.: Integrated optical devices for high-data-rate serial-to-parallel conversion,V1374,68-73(1991)
See Caulfield, H. J.: V1563,103-111(1991)

Verbiest, Thierry
; Persoons, Andre P.; Samyn, Celest: Synthesis and nonlinear optical properties of preformed polymers forming Langmuir-Blodgett films,V1560,353-361(1991)

Verboom, W. See van Hulst, Niek F.: V1361,581-588(1991)

Verbruggen, Patricia E. See Dumay, Adrie C.: V1445,38-46(1991)

Verdaasdonk, Rudolf M.
; Borst, Cornelius: Optical characteristics of sapphire laser scalpels analysed by ray-tracing,V1420,136-140(1991)
; Borst, Cornelius: Ray-tracing of optically modified fiber tips for laser angioplasty,V1425,102-109(1991)
See van Leeuwen, Ton G.: V1427,214-219(1991)

Verdier, C. See Leporcq, B.: V1397,153-156(1991)

Verechshagin, I. I.
; Oksman, Ya. A.: Thermoelectric voltage in slant-angle-deposited metallic films,V1440,401-405(1991)

Vergados, J. See Karkanis, S.: V1500,164-170(1991)

Vergnolle, Claude
; Houssay, Bruno: Interconnection requirements in avionic systems,V1389,648-658(1991)

Verhellen, P. See Mattheus, Rudy A.: V1446,341-351(1991)

Verhoeven, Jan See den Boggende, Antonius J.: V1345,189-197(1991)

Verhoeven, Jan W. See Warman, John M.: V1559,159-170(1991)

Verlinde, Patrick S.
; Proesmans, Marc: Global approach toward the evaluation of thermal infrared countermeasures,V1486,58-65(1991)
See Proesmans, Marc: V1486,102-114(1991)

Verly, Jacques G.
; Delanoy, Richard L.; Dudgeon, Dan E.: Model-based system for automatic target recognition,V1471,266-282(1991)

Vermillion, Charles H.
; Chan, Paul H.: Design considerations for EOS direct broadcast,V1492,224-228(1991)

Vernazza, Gianni L.
; Dambra, Carlo; Parizzi, Francesco; Roli, Fabio; Serpico, Sebastiano B.: Territorial analysis by fusion of LANDSAT and SAR data,V1492,206-212(1991)
See Venturi, Giovanni: V1606,217-225(1991)

Vernold, Cynthia L.
: Solution for anomalous scattering of bare HIP Be and CVD SiC mirrors,V1530,130-143(1991)
; Harvey, James E.: Effective surface PSD for bare hot-isostatic-pressed beryllium mirrors,V1530,144-149(1991)

Vernon, Stephen P. See Stearns, Daniel G.: V1465,80-87(1991)

Veron, Didier See Fleurot, Noel A.: V1502,230-241(1991)

Veronin, Christopher P.
; Rogers, Steven K.; Kabrisky, Matthew; Welsh, Byron M.; Priddy, Kevin L.; Ayer, Kevin W.: Optical image segmentor using wavelet filtering techniques as the front-end of a neural network classifier,V1469,281-291(1991)

Verrecchia, R. See Mens, Alain: V1358,315-328(1991)

Verri, Alessandro See Campani, Marco: V1388,409-414(1991)

Vertiy, Alexey A.
; Beletskii, N. N.; Gorbatyuk, I. N.; Ivanchenko, I. V.; Popenko, N. A.; Rarenko, I. M.; Tarapov, Sergey I.: Spectrum of surface electromagnetic waves in CdxHg1-xTe crystals at 0.3 K < T < 77 K,V1361,1070-1078(1991)
; Gavrilov, Sergey P.: Nonlinear properties of quasi-optic open resonator with a layer of Crv solution in heavy alcohol under the conditions of magnetic resonance,V1362,702-709(1991)

Veselovsky, Igor A.
See Alimpiev, Sergei S.: V1503,154-158(1991)
See Lukishova, Svetlana G.: V1503,338-345(1991)

Vess, Thomas M. See Angel, S. M.: V1435,72-81(1991)

Vethanayagam, Thirukumar K.
; Linchuck, Barry A.; Dreyer, Donald R.; Toler, J. R.; Amos, Lynn G.; Taylor, Donna L.: Mechanical performance and reliability of Corning Titan SMF CPC5 fiber after exposure to a variety of environments,V1366,343-350(1991)

Vetterli, Martin See Radha, Hayder: V1605,475-486(1991)

Vettese, Thomas See Marshall, Gerald F.: V1454,37-45(1991)

Vettier, C. See Kao, Chi-Chang: V1548,149-157(1991)

Vettiger, Peter
See Tang, Wade C.: V1418,338-342(1991)
See Webb, David J.: V1418,231-239(1991)

Veuillen, J. Y. See Nguyen, Tan T.: V1512,289-298(1991)

Vever-Bizet, C. See Nadeau, Pierre: V1398,151-161(1991)

Vial, J.-C. See Hochstrasser, Robin M.: V1403,221-229(1991)

Viardi, Marzia See Beretta, Stefano: V1544,2-9(1991)

Vibert, Patrick
See Burq, Catherine: V1346,276-285(1991)
See Gerstenmayer, Jean-Louis: V1346,286-292(1991)

Viccaro, P. J.
: Power distribution from insertion device x-ray sources,V1345,28-37(1991)
See Gluskin, Efim S.: V1548,56-68(1991)
See Khounsary, Ali M.: V1345,42-54(1991)
See Kuzay, Tuncer M.: V1345,55-70(1991)
See Lai, Barry P.: V1550,46-49(1991)
See Yun, Wen-Bing: V1345,146-164(1991)

Vicente, M. G. See Pandey, Ravindra K.: V1426,356-361(1991)

Vickers, Anthony J. See Tsui, Ernest S.: V1362,804-810(1991)

Vickers, James S. See Bush, Brett C.: V1549,376-384(1991)

Victora, Randall H.
; MacLaren, J. M.: Magnetic and electronic properties of Co/Pd superlattices,V1499,378-381(1991)

Videen, Gorden W. See Bickel, William S.: V1530,7-14(1991)

Vidmar, Anthony See Malakian, Kourken: V1481,315-328(1991)

Vidyasagar, M. See Trivedi, Mohan M.: V1571(1991)

Viergever, Max A.
See Anema, P. C.: V1446,352-356(1991)
See Roos, Paul: V1444,283-290(1991)

Viernes, Jacques See Alloncle, Anne P.: V1397,675-678(1991)

Vieth, John O. See Chan, Keith C.: V1469,359-372(1991)

Viggh, Herbert E.
: Building and maintaining a local on-orbit reference frame
  model,V1387,224-236(1991)
: Subsumption architecture control system for space proximity
  maneuvering,V1387,202-214(1991)

Vigil, Joseph See Kuyel, Birol: V1463,646-665(1991)

Vigliano, Patrick
See Astic, Dominique: V1397,713-716(1991)
See Tanguy, Mireille: V1487,172-183(1991)

Vigny, Paul See Angelov, D.: V1403,575-577(1991)

Vigroux, Luc M.
; Bourdet, Gilbert L.; Cassard, Philippe; Ouhayoun, Michel O.: TEA CO2
  laser mirror by degenerate four-wave mixing,V1500,74-79(1991)

Viitanen, Veli-Pekka
; Nenonen, Seppo A.; Sipila, Heikki; Mutikainen, Risto: Soft x-ray windows
  for position-sensitive proportional counters,V1549,28-34(1991)

Vijaya Kumar, B. V. K.
; Carlson, Daniel W.: Optimal correlation filters for implementation on
  deformable mirror devices,V1558,476-486(1991)
; Liang, Victor; Juday, Richard D.: Optimal phase-only correlation filters in
  colored scene noise,V1555,138-145(1991)
See Connelly, James M.: V1564,572-592(1991)
See Hendrix, Charles D.: V1564,2-13(1991)

Vikhagen, Eiolf See Lokberg, Ole J.: V1332,142-150(1991)

Vikram, Chandra S.
; Witherow, William K.; Trolinger, James D.: Refractive properties of TGS
  aqueous solution for two-color interferometry,V1557,197-201(1991)
See Trolinger, James D.: V1557,250-258(1991)

Vilcot, Jean-Pierre
See Aboulhouda, S.: V1362,494-498(1991)
See Constant, Monique T.: V1362,156-162(1991)
See Decoster, Didier: V1362,959-966(1991)

Vilesov, A. F. See Ivanov, U. V.: V1397,181-184(1991)

Vilhu, Osmi See Sipila, Heikki: V1549,246-255(1991)

Viligiardi, Riccardo
; Pini, Roberto; Salimbeni, Renzo; Galiberti, Sandra: Advances in clinical
  percutaneous excimer laser angioplasty,V1425,72-74(1991)

Vill, Arnold A.
; Salk, Ants A.; Berik, Irina K.: Super small excimer laser,V1503,110-
  114(1991)

Villa, Francesco See Mayhall, David J.: V1378,101-114(1991)

Villa, G. E. See Di Cocco, Guido: V1549,102-112(1991)

Villaeys, A. A. See Hoerner, Claudine: V1362,863-869(1991)

Villani, Thomas S. See Sauer, Donald J.: V1540,285-296(1991)

Villarica, M. See Samoriski, Brian: V1412,12-18(1991)

Villasol, R. See Cox, J. N.: V1392,650-659(1991)

Villate, Denis
: Optical method of detection for the magnetic alignment of an electron
  accelerator,V1533,193-196(1991)

Villela, Gerard See Daugy, Eric: V1502,203-212(1991)

Villeneuve, D. M. See Labaune, C.: V1413,138-143(1991)

Vinas, Federico See Chavantes, Maria C.: V1428,13-22(1991)

Vincent, Bernard See Daugy, Eric: V1502,203-212(1991)

Vincent, John D. See Coles, Christopher L.: V1488,327-333(1991)

Vincent, Luc M.
: Efficient computation of various types of skeletons,V1445,297-311(1991)
: New trends in morphological algorithms,V1451,158-170(1991)
See Yuille, Alan L.: V1568,271-282(1991)

Vincent, Patrick
See Akhouayri, Hassan: V1545,140-144(1991)
See Vitrant, Guy: V1545,225-231(1991)

Vinet, Francoise
; Chevallier, M.; Pierrat, Christophe; Guibert, Jean C.; Rosilio, Charles;
  Mouanda, B.; Rosilio, A.: Resist design for dry-developed positive
  working systems in deep-UV and e-beam lithography,V1466,558-
  569(1991)

Vinokhodov, A. Y. See Borisov, V. M.: V1503,40-47(1991)

Vinokurov, Y. S. See Sadyigov, Z. Y.: V1621,158-168(1991)

Vinokurova, Elena G. See Belov, Sergei N.: V1552,264-267(1991)

Vinuela, Fernando See Close, Robert A.: V1444,196-203(1991)

Vioel, Wolfgang See Michaelis, Alexander: V1397,709-712(1991)

Virt, I. S. See Szeregij, E. M.: V1391,199-203(1991)

Visa, Ari J.
: Texture classification and neural network methods,V1469,820-831(1991)

Viscito, Eric
; Gonzales, Cesar A.: Video compression algorithm with adaptive bit
  allocation and quantization,V1605,58-72(1991)

Vishnevsky, G. I.
; Berezin, V. Y.; Lazovsky, L. Y.; Vydrevich, M. G.; Rivkind, V. L.; Zhemerov,
  B. N.: Digital charge-coupled device color TV system for
  endoscopy,V1447,34-43(1991)
; Ioffe, S. A.; Berezin, V. Y.; Rybakov, M. I.; Mikhaylov, A. V.; Belyaev, L. V.:
  High-precision digital charge-coupled device TV system,V1448,69-
  72(1991)

Vishnoi, Gargi
; Pillai, P.K. C.; Goel, T. C.: Fiber optic sensor for the study of temperature
  and structural integrity of PZT: epoxy composite materials,V1572,94-
  100(1991)

Visona, Adriana
; Liessi, Guido; Bonanome, Andrea; Lusiani, Luigi; Miserocchi, Luigi; Pagnan,
  Antonio: Percutaneous peripheral excimer laser angioplasty: immediate
  success rate and short-term outcome,V1425,75-83(1991)

Visscher, Koen See Brakenhoff, G. J.: V1439,121-127(1991)

Viswanath, Harsha C.
; Jones, Richard A.: Kalman-based computation of optical flow
  fields,V1482,275-284(1991)

Viswanathan, R.
See Amirmehrabi, Hamid: V1396,252-265(1991)
See Sayeh, Mohammad R.: V1396,417-429(1991)

Viswanathan, Vriddhachalam K. See Shealy, David L.: V1343,229-240(1991)

Vitrant, Guy
; Vincent, Patrick; Reinisch, Raymond; Neviere, Michel: Modal analysis of
  grating-induced optical bistability,V1545,225-231(1991)

Vitrishchak, Il'ya B. See Mak, Arthur A.: V1410,233-243(1991)

Vitsnudel, Ilia
; Ginosar, Ran; Zeevi, Yehoshua Y.: Neural-network-aided design for image
  processing,V1606,1086-1091(1991)

Vittot, M. See Gendron, Eric: V1542,297-307(1991)

Vityaz, Irena V. See Altshuler, Grigori B.: V1429,95-104(1991)

Vivekanandan, J.
; Turk, F. J.; Bringi, Viswanathan N.: Remote sensing of precipitation
  structures using combined microwave radar and radiometric
  techniques,V1558,324-338(1991)

Vivenot, P. See Lefevre, Herve C.: V1367,72-80(1991)

Vizir, V. A.
; Kiryukhin, Yu. B.; Manylov, V. I.; Nosenko, S. P.; Sychev, S. P.; Chervyakov,
  V. V.; Shubkin, N. G.: Magnetic pulse compressor for high-power
  excimer discharge laser pumping,V1411,63-68(1991)

Vizvary, Gerald See Calabrese, Gary C.: V1466,528-537(1991)

Vlad, Ionel V.
; Popa, Dragos; Petrov, M. P.; Kamshilin, Alexei A.: Optical testing by
  dynamic holographic interferometry with photorefractive crystals and
  computer image processing,V1332,236-244(1991)

Vladimirsky, Yuli See Oertel, Heinrich K.: V1465,244-253(1991)

Vlannes, Nickolas P.
See McDonald, John F.: V1390,286-301(1991)
See Pape, Dennis R.: V1476,201-213(1991)

Vlasbloem, Hugo See Geluk, Ronald J.: V1443,143-152(1991)

Vlasov, N. G.
; Vorobjov, S. P.; Karpova, S. G.: Hologram as means of color
  reproduction,V1238,332-337(1991)

Vlasov, Nikolai G.
; Korchazhkin, S. V.; Manikalo, V. V.: Interference visualization of infrared
  images,V1358,1018-1020(1991)

Vo-Dinh, Tuan
: Advanced approaches in luminescence and Raman
  spectroscopy,V1435,197-202(1991)
; Stokes, D. L.; Li, Ying-Sing; Miller, Gordon H.: Fiber optic sensor probe for
  in-situ surface-enhanced Raman monitoring,V1368,203-209(1991)

Vodchitz, A. I. See Apanasevich, P. A.: V1403,475-477(1991)

Voegeli, O. See Webb, David J.: V1418,231-239(1991)

Voelkel, Reinhard See Kobolla, Harald: V1507,175-182(1991)

Voelz, David G.
; Idell, Paul S.; Bush, Keith A.: Illumination coherence effects in laser-speckle imaging,V1416,260-265(1991)

Voet, Marc R. See Schrever, Koen: V1572,107-112(1991)

Vogel, Dietmar
; Michel, Bernd; Totzauer, Werner F.; Schreppel, Ulrich; Clos, Rainer: Laser metrological measurement of transient strain fields in Hopkinson-bar experiments,V1554A,262-274(1991)

Vogel, George A. See Burns, Thomas J.: V1469,208-218(1991)

Vogh, James
See Ashenayi, Kaveh: V1396,215-225(1991)
See Karimi, B.: V1396,226-236(1991)

Vogt, Robert C.
: Set discrimination analysis tools for grey-level morphological operators,V1568,200-211(1991)
See Lu, Yi: V1568,14-25(1991)
See Trenkle, John M.: V1568,212-223(1991)

Vohra, Rohini See Wand, Michael D.: V1455,97-104(1991)

Vohra, Sandeep T. See Bucholtz, Frank: V1367,226-235(1991)

Vohryzek, Jachym
See Sasek, Ladislav: V1504,147-154(1991)
See Sasek, Ladislav: V1572,151-156(1991)

Voignier, Francois
; Merat, Frederic; Brunet, Henri: Mixing diagnostic in a cw DF chemical laser operating at high-cavity pressure,V1397,297-301(1991)

Voirin, G. See Poscio, Patrick: V1510,112-117(1991)

Voisin, Paul
See Bigan, Erwan: V1362,553-558(1991)
See Voos, Michel: V1361,416-423(1991)

Voitek, Mark See Hong, C. S.: V1418,177-187(1991)

Vojtsekhovsky, V. V. See Artjushenko, Vjacheslav G.: V1420,157-168(1991)

Vold, C. See Hickey, Carolyn F.: V1534,67-76(1991)

Volf, E. B. See Skopinov, S. A.: V1403,676-679(1991)

Volk, Raimund
; Sohler, Wolfgang: Characterization of the photorefractive effect in Ti:LiNbO3 stripe waveguides,V1362,820-826(1991)

Volkov, Alexander G.
; Gugeshashvili, M. I.; Kandelaki, M. D.; Markin, V. S.; Zelent, B.; Munger, G.; Leblanc, Roger M.: Artificial photosynthesis at octane/water interface in the presence of hydrated chlorophyll a oligomer thin film,V1436,68-79(1991)
See Leblanc, Roger M.: V1436,92-102(1991)

Volkov, Gennady S.
; Zaroslov, D. Y.: Low-loss line-narrowed excimer oscillator for projection photolithography: experiments and simulation,V1503,146-153(1991)

Volkovitsky, O. A. See Almaev, R. K.: V1397,831-834(1991)

Volodko, V. V. See Artjushenko, Vjacheslav G.: V1420,149-156(1991)

Volokh, K. Y. See Shvej, E. M.: V1554A,488-495(1991)

Voloshin, Arkady S.
; Bastawros, Adel F.: High-sensitivity interferometric technique for strain measurements,V1400,50-60(1991)

Voloshina, T. V. See Kluyev, V. G.: V1440,303-308(1991)

Volosov, Vladimir D.
: Nonlinear optic frequency converters with lowered sensibility to spectral width of laser radiation,V1500,105-110(1991)

Volpe, M. G. See Carbonara, Giuseppe: V1361,1038-1040(1991)

von Bally, Gert
: Holography in endoscopy: illuminating dark holes with Gabor's principle,VIS08,335-346(1991)
: Medical applications of holography,V1525,2-8(1991)
: Dirksen, D.; Zou, Y.: Micro-optical elements in holography,V1507,66-72(1991)
See Foerster, Werner: V1429,146-151(1991)

von Bardeleben, H. J. See Bourgoin, J. C.: V1361,184-194(1991)

Von Bargen, Kenneth See Shropshire, Geoffrey J.: V1379,222-235(1991)

von Bergmann, Hubertus M.
; Swart, Pieter L.: Industrial excimer and CO2 TEA lasers with kilowatt average output power,V1397,63-66(1991)
See Swart, Pieter L.: V1397,559-562(1991)

von Borczykowski, C. See Rempel, U.: V1403,631-634(1991)

von Buelow, H.
; Zeyfang, E.: Gas-dynamically cooled CO laser with rf-excitation: design and performance,V1397,499-502(1991)
See Zeyfang, E.: V1397,449-452(1991)

Vonder Haar, Thomas H. See Behunek, Jan L.: V1479,93-100(1991)

von der Luhe, Oskar See Rimmele, Thomas: V1542,186-193(1991)

Von Drusek, William A. See Jursich, Gregory M.: V1412,115-122(1991)

Vonesh, Michael J. See Frazin, Leon J.: V1425,207-207(1991)

von Handorf, Robert J. See Murty, Mantravady V.: V1332,107-114(1991)

von Klitzing, Klaus See Diessel, Edgar: V1361,1094-1099(1991)

Von Lehman, A. See Paek, Eung Gi: V1621,340-350(1991)

von Maltzan, B. See Rempel, U.: V1403,631-634(1991)

von Ramm, Olaf T. See Pavy, Henry G.: V1443,54-61(1991)

von Seelen, Werner See Mallot, Hanspeter A.: V1382,397-408(1991)

von Zanthier, C. See Braueninger, Heinrich: V1549,330-339(1991)

Voos, Michel
; Voisin, Paul: Electro-optical effects in semiconductor superlattices,V1361,416-423(1991)

Vorburger, Theodore V.
See Marx, Egon: V1332,826-834(1991)
See Marx, Egon: V1530,15-21(1991)

Vorob'ev, Valerii V. See Gurvich, Alexander S.: V1408,10-18(1991)

Vorobiev, N. S.
; Konoplev, O. A.: Two-frequency picosecond laser with electro-optical feedback,V1358,895-901(1991)
; Serafimovich, O. A.; Smirnov, A. V.: Picosecond techniques application for definition of nonuniform ingradients inside turbid medium,V1358,698-702(1991)
See Babushkin, A. V.: V1346,410-417(1991)
See Bowley, David J.: V1358,550-555(1991)
See de Souza, Eunezio A.: V1358,556-560(1991)

Vorobjev, V. A. See Kanevsky, M. F.: V1440,154-165(1991)

Vorobjov, S. P.
: Observation of reflection holograms,V1238,62-67(1991)
See Vanin, V. A.: V1238,324-331(1991)
See Vlasov, N. G.: V1238,332-337(1991)

Vorobyov, A. Y.
; Libenson, Michail N.: Energy release in interactions of laser pulse with solid fuels and metals,V1440,197-205(1991)

Voronin, G. See Bulichev, A.: V1500,151-162(1991)

Vorontsov, Michael A.
: Problems of large neurodynamics system modeling: optical synergetics and neural networks,V1402,116-144(1991)
; Iroshnikov, N. G.: Nonlinear dynamics of neuromorphic optical system with spatio-temporal interactions,V1621,292-298(1991)
; Larichev, A. V.: Intelligent laser systems: adaptive compensation of phase distortions in nonlinear system with two-dimensional feedback,V1409,260-266(1991)
; Zheleznykh, N. I.; Larichev, A. V.: Two-dimensional dynamic neural network optical system with simplest types of large-scale interactions,V1402,154-164(1991)
See Ivanov, Vladimir Y.: V1402,145-153(1991)
See Kobzev, E. F.: V1402,165-174(1991)

Voroshnin, A. B. See Markov, Vladimir B.: V1238,118-122(1991)

Vorzobova, N. D.
; Rjabova, R. V.; Schvarzvald, A. I.: Holographic characteristics of IAE and PFG-01 photoplates for colored pulsed holography,V1238,476-477(1991)
; Sizov, V. N.; Rjabova, R. V.: Monochromatic and two-color recording of holographic portraits with the use of pulsed lasers,V1238,462-464(1991)

Voss, F. See Basting, Dirk: V1412,80-83(1991)

Voss, Kenneth J.
: Variation of the point spread function in the Sargasso Sea,V1537,97-103(1991)
See Giles, John W.: V1537,140-146(1991)
See Macdonald, Burns: V1537,104-114(1991)

Vossler, T. L. See Stevens, Robert K.: V1433,25-35(1991)

Vrensen, G. F. See Siebinga, I.: V1403,746-748(1991)

Vrhel, Michael J. See Trussell, Henry J.: V1452,2-9(1991)

Vsevolodov, N. N.
; Dyukova, T. V.; Druzhko, A. B.; Shakhbazyan, V. Y.: Optical recording material based on bacteriorhodopsin modified with hydroxylamine,V1621,11-20(1991)
See Mikaelian, Andrei L.: V1621,148-157(1991)

Vu, Brian-Tinh See Landen, Otto L.: V1413,120-130(1991)

Vu, Louis P. See Detwiler, Paul W.: V1425,149-155(1991)

Vu, Thuong-Quat See Hartiti, Bouchaib: V1393,200-206(1991)

Vukicevic, Dalibor
; Neger, Theo; Jaeger, Helmut; Woisetschlaeger, Jakob; Philipp, Harald: Optical tomography by heterodyne holographic interferometry,VIS08,160-193(1991)

Vukobratovich, Daniel
: Advances in optomechanics,V1396,436-446(1991)
; Richard, Ralph M.: Roller chain supports for large optics,V1396,522-534(1991)
See Paquin, Roger A.: V1533(1991)

Vuong, Phat N. See Phat, Darith: V1525,196-205(1991)

Vuong, S. C. See Johnson, Dean R.: V1367,140-154(1991)

Vural, Kadri See Kozlowski, Lester J.: V1540,250-261(1991)

Vvedensky, D. D. See Joyce, Bruce A.: V1361,13-22(1991)

Vyborny, Carl J.
See Giger, Maryellen L.: V1445,101-103(1991)
See Yin, Fang-Fang: V1396,2-4(1991)

Vydrevich, M. G.
See Khvilivitzky, A. T.: V1447,184-190(1991)
See Vishnevsky, G. I.: V1447,34-43(1991)

Vysogorets, Mikhail V.
; Mitrofanova, Natalya N.; Petrov, Mikhail Y.; Platonov, Valeri N.; Chulkin, Alexey D.: Peculiarities of frame memory design for slow-scan readout system for scientific application,V1358,1066-1069(1991)

Waag, Andreas See Landwehr, Gottfried: V1362,282-290(1991)

Waarts, Robert G. See Mehuys, David G.: V1418,57-63(1991)

Wabnitz, H. See Damm, Tobias: V1403,686-694(1991)

Wada, Hiroshi See Seki, Masafumi: V1583,184-195(1991)

Wada, M. See Takishima, Yasuhiro: V1605,635-645(1991)

Wada, Osamu
: III-V semiconductor integrated optoelectronics for optical computing,V1362,598-607(1991)

Wada, Tatsuo
; Hosoda, Masahiro; Garito, Anthony F.; Sasabe, Hiroyuki; Terasaki, A.; Kobayashi, Takayoshi T.; Tada, Hiroaki; Koma, Atsushi: Third-order optical nonlinearities and femtosecond responses in metallophthalocyanine thin films made by vacuum deposition, molecular beam epitaxy, and spin coating,V1560,162-171(1991)

Wada, Tuneyo See Sato, Koki: V1461,124-131(1991)

Wade, H. H. See Harriott, Lloyd R.: V1465,57-63(1991)

Wade, Jack
; Tuchman, Avi; eds.: *Sensors and Sensor Systems for Guidance and Navigation*,V1478(1991)

Wade, L. O. See Sachse, Glen W.: V1433,157-166(1991)

Wadsworth, Leo A.
See Jacobi, William J.: V1479,111-119(1991)
See Jacobi, William J.: V1495,205-213(1991)

Wadsworth, Mark V.
See Carbone, Joseph: V1447,229-242(1991)
See Zarnowski, Jeffrey J.: V1447,191-201(1991)

Waeber, Bruce See Donn, Matthew: V1538,189-200(1991)

Wagendristel, A. See Wang, H. F.: V1519,405-410(1991)

Wagner, Brent K. See Summers, Christopher J.: V1512,170-176(1991)

Wagner, Earl J. See Park, Yong K.: V1372,219-227(1991)

Wagner, James W.
; Deaton, John B.; McKie, Andrew D.; Spicer, James B.: Laser ultrasonics: generation and detection considerations for improved signal-to-noise ratio,V1332,491-501(1991)
See Steckenrider, John S.: V1554B,106-112(1991)

Wagner, Jerome F. See Hsieh, Hung-yu: V1396,467-472(1991)

Wagner, Kelvin H.
See Huang, Tizhi: V1559,377-384(1991)
See Huang, Tizhi: V1562,44-54(1991)
See Weverka, Robert T.: V1564,676-684(1991)
See Weverka, Robert T.: V1562,66-72(1991)
See Wu, Kuang-Yi: V1563,168-175(1991)

Wagner, Richard E. See Sessa, William B.: V1372,208-218(1991)

Wagner, Richard J. See Litvak, Marvin M.: V1415,62-71(1991)

Wagnieres, G.
; Braichotte, D.; Chatelain, Andre; Depeursinge, Ch.; Monnier, Philippe; Savary, M.; Fontolliet, Ch.; Calmes, J.-M.; Givel, J.-C.; Chapuis, G.; Folli, S.; Pelegrin, A.; Buchegger, F.; Mach, J.-P.; van den Bergh, Hubert: Photodetection of early cancer in the upper aerodigestive tract and the bronchi using photofrin II and colorectal adenocarcinoma with fluoresceinated monoclonal antibodies,V1525,219-236(1991)
See Bays, Roland: V1525,397-408(1991)
See Braichotte, D.: V1525,212-218(1991)

Wahl, Daniel E.
; Jakowatz, Charles V.; Ghiglia, Dennis C.; Eichel, Paul H.: Relationships between autofocus methods for SAR and self-survey techniques for SONAR,V1567,32-40(1991)

Wahl, Roger L. See Bender, John W.: V1533,264-276(1991)

Waidhauser, Erich
; Markmiller, U.; Enders, S.; Hessel, Stefan F.; Ulrich, Frank; Beck, Oskar J.; Feld, Michael S.: Photoablation using the Holmium:YAG laser: a laboratory and clinical study,V1428,75-83(1991)

Wakabayashi, Osamu
; Kowaka, Masahiko; Kobayashi, Yukio: High-average-power narrow-band KrF excimer laser,V1463,617-628(1991)

Wakabayashi, Satoshi See Matsushita, Tadashi: V1488,368-375(1991)

Wakana, Shin-ichi
; Nagai, Toshiaki; Hama, Soichi; Goto, Yoshiro: Optical delay tester,V1479,283-290(1991)

Wakatsuki, Noboru See Kurosawa, Kiyoshi: V1368,150-156(1991)

Wakeley, Joseph See Dunn, Dennis F.: V1606,541-552(1991)

Wakelin, S. See Craig, Robert G.: V1505,76-86(1991)

Wakui, Tsuneyoshi See Torao, Akira: V1358,843-850(1991)

Walba, David M. See Wand, Michael D.: V1455,97-104(1991)

Walczak, Irene M.
; Love, Walter F.; Slovacek, Rudolf E.: Sensitive fiber-optic immunoassay,V1420,2-12(1991)

Walden, Robert H. See Yap, Daniel: V1418,471-476(1991)

Walder, Brian T.
: New generation of copper vapor lasers for high-speed photography,V1358,811-820(1991)

Waldie, Arthur H.
; Drexler, James J.; Qualtrough, John A.; Soules, David B.; Eaton, Frank D.; Peterson, William A.; Hines, John R.: Characterization and effect of system noise in a differential angle of arrival measurement device,V1487,103-108(1991)
See Eaton, Frank D.: V1487,84-90(1991)

Walega, James G.
; Dye, James E.; Grahek, Frank E.; Ridley, Brian K.: Compact measurement system for the simultaneous determination of NO, NO2, NOy, and O3 using a small aircraft,V1433,232-241(1991)

Wali, Rahman
; Colef, Michael; Barba, Joseph: Best fit ellipse for cell shape analysis,V1606,665-674(1991)

Walker, Andrew C. See Craig, Robert G.: V1505,76-86(1991)

Walker, Arthur B.
; Allen, Maxwell J.; Barbee, Troy W.; Hoover, Richard B.: Design of narrow band XUV and EUV coronagraphs using multilayer optics,V1343,415-427(1991)
; Lindblom, Joakim F.; Timothy, J. G.; Hoover, Richard B.; Barbee, Troy W.; Baker, Phillip C.; Powell, Forbes R.: High-resolution imaging with multilayer soft x-ray, EUV, and FUV telescopes of modest aperture and cost,V1494,320-333(1991)
; Lindblom, Joakim F.; Timothy, J. G.; Allen, Maxwell J.; DeForest, Craig E.; Kankelborg, Charles C.; O'Neal, Ray H.; Paris, Elizabeth S.; Willis, Thomas D.; Barbee, Troy W.; Hoover, Richard B.: Ultra-High-Resolution XUV Spectroheliograph II: predicted performance,V1343,319-333(1991)
; Timothy, J. G.; Barbee, Troy W.; Hoover, Richard B.: Active sun telescope array,V1343,334-347(1991)
See DeForest, Craig E.: V1343,404-414(1991)
See Fineschi, Silvano: V1343,376-388(1991)
See Hoover, Richard B.: V1426,84-96(1991)
See Hoover, Richard B.: V1435,338-351(1991)
See Hoover, Richard B.: V1343,189-202(1991)
See Hoover, Richard B.: V1343,175-188(1991)
See Hoover, Richard B.: V1343,389-403(1991)

See Hoover, Richard B.: V1343(1991)
See Lindblom, Joakim F.: V1343,544-557(1991)
See Timothy, J. G.: V1343,350-358(1991)

Walker, D. S. See Potter, B. L.: V1368,251-257(1991)

Walker, David R.
; Sleat, William E.; Evans, J.; Sibbett, Wilson: Ultrafast streak camera evaluations of phase noise from an actively stabilized colliding-pulse-mode-locked ring dye laser,V1358,860-867(1991)
See Liu, Yueping: V1358,300-310(1991)

Walker, Ellen L.
: Exploiting geometric relationships for object modeling and recognition,V1382,353-363(1991)

Walker, Ian D.
; Cheatham, John B.; Chen, Yu-Che: Efficient method for computing the force distribution of a three-fingered grasp,V1387,256-270(1991)
; Cheatham, John B.; Chen, Yu-Che: Intelligent grasp planning strategy for robotic hands,V1468,974-989(1991)
See Nguyen, Luong A.: V1387,296-312(1991)

Walker, James H.
; Cromer, Chris L.; McLean, James T.: Technique for improving the calibration of large-area sphere sources,V1493,224-230(1991)

Walker, Karl-Heinz
; Sontag, Heinz: Fiber optic based chemical sensor system for in-situ process measurements using the photothermal effect,V1510,212-217(1991)

Walker, Nigel G. See Steele, Roger C.: V1372,173-187(1991)

Walker, Philip L.
See Cooper, Alfred W.: V1486,37-46(1991)
See Milne, Edmund A.: V1486,151-161(1991)

Walker, R. See Wolfe, William J.: V1382,240-254(1991)

Walker, Sonya L.
See McCormick, Frederick B.: V1396,508-521(1991)
See McCormick, Frederick B.: V1533,97-114(1991)

Walker, Stuart D. See Yaseen, Mohammed: V1562,319-326(1991)

Walkup, John F. See Ellett, Scott A.: V1564,634-643(1991)

Wall, David L.
: Advances in tunable diode laser technology for atmospheric monitoring applications,V1433,94-103(1991)

Wallace, James
: Modeling turbulent transport in laser beam propagation,V1408,19-27(1991)

Wallace, Lloyd V. See Livingston, William C.: V1491,43-47(1991)

Wallace, Richard S.
; Suenaga, Yasuhito: Color segmentation using MDL clustering,V1381,436-446(1991)

Wallace, Richard W. See Cheng, Emily A.: V1417,300-306(1991)

Wallace, Rick P. See Dickerson, Gary: V1464,584-595(1991)

Wallace, William E.
; Stone, Maureen C.: Gamut mapping computer-generated imagery,V1460,20-28(1991)

Wallenberger, Frederick T.
; Koutsky, J. A.; Brown, Sherman D.: Melt processing of calcium aluminate fibers with sapphirelike infrared transmission,V1590,72-82(1991)
; Weston, Norman E.; Brown, Sherman D.: Melt-processed calcium aluminate fibers: structural and optical properties,V1484,116-124(1991)

Wallin, Bo
: Real-time temperature measurement on PCB:s, hybrids, and microchips,V1467,180-187(1991)
See Lindstrom, Kjell M.: V1488,389-398(1991)

Wallin, Suante See Hallstadius, Hans: V1433,36-43(1991)

Wallmeroth, Klaus See Letterer, Rudolf: V1522,154-157(1991)

Wallraff, Gregory M.
; Miller, Robert D.; Clecak, Nicholas J.; Baier, M.: Polysilanes for microlithography,V1466,211-217(1991)

Walmsley, Charles F. See Kreider, James F.: V1488,376-388(1991)

Walocha, Jerzy
: Fabry-Perot etalon as a CO2 laser Q-switch modulator,V1391,286-289(1991)

Walpita, Lak M. See Findakly, Talal K.: V1476,14-21(1991)

Walpole, B. See Klenerman, David: V1437,95-102(1991)

Walpole, James N.
See Diadiuk, Vicky: V1354,496-500(1991)
See Evans, Gary A.: V1418,406-413(1991)

Walser, Rodger M.
See Becker, Michael F.: V1441,174-187(1991)
See Huffaker, Diana L.: V1441,365-380(1991)

Walsh, C. P. See Moerner, William E.: V1560,278-289(1991)

Walsh, John E.
See Ciocci, Franco: V1501,154-162(1991)
See Xu, Yian-sun: V1407,648-652(1991)

Walsh, Joseph See Kaiser, Todd J.: V1367,121-126(1991)

Walsh, Joseph T.
; Hill, D. A.: Erbium laser ablation of bone: effect of water content,V1427,27-33(1991)
See Heiferman, Kenneth S.: V1428,128-134(1991)

Walsh, Peter M.
: Rapid system integration with symbolic programming,V1386,84-89(1991)

Walsh, T. D. See McDermid, I. S.: V1491,175-181(1991)

Walston, Andrew See Ng, Willie W.: V1371,205-211(1991)

Walt, David R. See Barnard, Steven M.: V1368,86-92(1991)

Walter, Robert F.
; Palumbo, L. J.; Townsend, Steven W.; Tannen, Peter D.: Output characteristics of a multikilowatt repetitively pulsed XeF laser,V1397,71-76(1991)
See Dente, Gregory C.: V1397,403-408(1991)

Walterman, Michael T.
; Weinhaus, Frederick M.: Antialiasing-warped imagery using lookup-table-based methods for adaptive resampling,V1567,204-214(1991)

Walternberg, P. T. See Hilton, Peter J.: V1385,27-31(1991)

Walters, Clarence P.
; Lorenzo, Maximo: Development of an electronic terrain board as a processor test and evaluation tool,VIS07,181-201(1991)

Walters, Tena K. See Mitchell, David C.: V1427,181-188(1991)

Walters, W. See Gilbert, Barry K.: V1390,235-248(1991)

Walther, Ludwig See Norkus, Volkmar: V1484,98-105(1991)

Walther, Sandra S.
; Peskin, Richard L.: Object-oriented data management for interactive visual analysis of three-dimensional fluid-flow models,V1459,232-243(1991)
See Gertner, Izidor: V1565,414-422(1991)

Waltman, William B. See Bevilacqua, Richard M.: V1491,231-242(1991)

Walton, Eric K. See Jouny, Ismail: V1471,142-153(1991)

Waltz, Frederick M.
: User interfaces for automated visual inspection systems,VCR36,105-137(1991)
See Batchelor, Bruce G.: VCR36(1991)
See Batchelor, Bruce G.: V1386(1991)
See Floeder, Steven P.: V1386,90-101(1991)

Walz, B. See Wildermuth, Eberhard: V1397,367-371(1991)

Wamser, Carl C.
; Senthilathipan, Velu; Li, Wen: Asymmetric photopotentials from thin polymeric porphyrin films,V1436,114-124(1991)
See Frank, Arthur J.: V1436(1991)

Wan, Liqun
; Li, Dapeng; Wee, William G.; Han, Chia Y.; Porembka, D. T.: ANN approach for 2-D echocardiographic image processing: application of neocognitron model to LV boundary formation,V1469,432-440(1991)
See Niu, Aiqun: V1469,495-505(1991)

Wan, Suiren See Sun, Xiaohan: V1572,243-247(1991)

Wan, William See Ballik, Edward A.: V1488,249-256(1991)

Wand, Michael D.
; Vohra, Rohini; Thurmes, William; Walba, David M.; Geelhaar, Thomas; Littwitz, Brigitte: Use of the Boulder model for the design of high-polarization fluorinated ferroelectric liquid crystals,V1455,97-104(1991)

Wandel, K. See Gruska, Bernd: V1361,758-764(1991)

Wander, Joseph M. See Dubowsky, Steven: V1386,10-20(1991)

Wandernoth, Bernhard
; Franz, Juergen: Realization of a coherent optical DPSK (differential phase-shift keying) heterodyne transmission system with 565 MBit/s at 1.064 um,V1522,194-198(1991)

Wang, A. B. See Fang, X. J.: V1572,279-283(1991)

Wang, An
; Xie, Haiming: Coupled fiber ring interferometer array: theory,V1572,365-369(1991)
; Xie, Haiming: Multimode fiber-optic Mach-Zehnder interferometric strain sensor,V1572,444-449(1991)
; Xie, Haiming: Single-mode fiber Mach-Zehnder interferometer as an earth strain sensor,V1572,440-443(1991)

Wang, Anbo
; Wang, Zhiguang; Vengsarkar, Ashish M.; Claus, Richard O.: Two-mode elliptical-core fiber sensors for measurement of strain and temperature,V1584,294-303(1991)

Wang, Baishi See Zhang, Xi: V1554A,444-448(1991)

Wang, Ben See Zou, Yunlu: V1238,452-456(1991)

Wang, Biao See Jiang, Shibin: V1535,143-147(1991)

Wang, C. W. See Xiong, F. K.: V1519,665-669(1991)

Wang, Changgui See Luo, Nan: V1572,2-4(1991)

Wang, Chang H.
; Lloyd, Ashley D.; Wherrett, Brian S.: Microwatt all-optical switches, array memories, and flip-flops,V1505,130-140(1991)

Wang, Chang S.
See Cui, Jing B.: V1519,419-422(1991)
See Zhang, Shu Y.: V1519,43-46(1991)
See Zhang, Wei P.: V1519,23-25(1991)

Wang, Chang Xing
See He, Anzhi: V1554A,547-552(1991)
See Lu, Jian-Feng: V1415,220-224(1991)

Wang, Cheng See Lucovsky, Gerald: V1392,605-616(1991)

Wang, Chengdong See Sang, Fengting: V1412,252-257(1991)

Wang, Chihcheng
; Jin, Xi: Studies of blood gas determination and intelligent image,V1572,406-409(1991)

Wang, Chin H.
; Guan, H. W.; Zhang, J. F.: Electro-optic measurements of dye/polymer systems,V1559,39-48(1991)

Wang, Christine A. See Evans, Gary A.: V1418,406-413(1991)

Wang, Chung Ching
; Shirazi, Behrooz: Neural networks implementation on a parallel machine,V1396,209-214(1991)

Wang, Chunhua
See Huang, Zhaoming: V1572,140-143(1991)
See Zhang, Jinghua: V1572,69-73(1991)

Wang, Danli L. See Cotton, Therese M.: V1403,93-101(1991)

Wang, David C.
; Karvelis, K. C.: Computer interpretation of thallium SPECT studies based on neural network analysis,V1445,574-575(1991)

Wang, David T.
; Peng, Ming-Chien; Lee, Jueen; Chen, Jyh-Woei; Ng, Peter A.: Use of syntactic recognition with sampled boundary distances,V1386,206-219(1991)

Wang, David Y.
; Moore, Duncan T.: Systematic approach to axial gradient lens design,V1354,599-605(1991)

Wang, Defang
; Xu, Yan; Zhu, Minjun: Intelligent magnetometer with photoelectric sampler,V1572,514-516(1991)

Wang, Dexter
; Gardner, Leo R.; Wong, Wallace K.; Hadfield, Peter: Space-based visible all-reflective stray light telescope,V1479,57-70(1991)
See Wong, Wallace K.: V1530,86-103(1991)

Wang, Duo See Chen, Zengtao: V1554A,206-208(1991)

Wang, Erqi
; Song, Dehui: Combination-matching problems in the layout design of minilaser rangefinder,V1400,124-128(1991)

Wang, Fang See Li, Yuchuan: V1572,382-385(1991)

Wang, Feng
; Dai, Fu-long: Photoelastic experimental research on the Wan-An lock by oblique incidence method,V1554A,359-370(1991)

Wang, Gang
: Integrating acoustical and optical sensory data for mobile robots,V1468,479-482(1991)
; Dubant, Olivier: BUOSHI: a tool for developing active expert systems,V1468,11-15(1991)

Wang, Ge
; Lin, T. H.; Cheng, Ping-Chin; Shinozaki, D. M.; Newberry, S. P.: X-ray projection microscopy and cone-beam microtomography,V1398,180-190(1991)

Wang, George Z. See Murphy, Kent A.: V1588,117-124(1991)

Wang, Gunnar See Ballangrud, Ase: V1521,89-96(1991)

Wang, Guomei
; Lei, Jiaheng; Yun, Huaishun; Guo, Liping; Jin, Baohui: XPS, IR, and Mossbauer studies of lithium phosphate glasses containing iron oxides,V1590,229-236(1991)

Wang, Guozhi
; Feng, San; Wang, Zhengrong; Wang, Shuyan: Dynamic particle holographic instrument,V1358,73-81(1991)
; Zhu, Guangkuan: Applied research of optical fiber pulsed-laser holographic interferometry,V1554B,119-125(1991)

Wang, H. F.
; Wagendristel, A.; Yang, X.; Torzicky, P.; Bangert, H.: Finite element study on indentations into TiN- and multiple TiN/Ti- layers on steel,V1519,405-410(1991)

Wang, H. T. See Mathur, Bimal P.: V1473,153-160(1991)

Wang, Hai-Lin
; Miao, Peng-Cheng; He, Anzhi: Analysis of measurement principle of moire interferometer using Fourier method,V1527,419-422(1991)
; Miao, Peng-Cheng; Yan, Da-Peng; He, Anzhi: Analysis of moire deflectometry by wave optics,V1545,268-273(1991)
; Miao, Peng-Cheng; Yan, Da-Peng; Ni, Xiao W.; He, Anzhi: Generalized Talbot effect,V1545,274-277(1991)
See He, Anzhi: V1527,334-337(1991)
See He, Anzhi: V1527,423-426(1991)
See He, Anzhi: V1554B,429-434(1991)
See Miao, Peng-Cheng: V1554B,641-644(1991)
See Yan, Da-Peng: V1527,442-447(1991)

Wang, Hai L.
See Cui, Jie: V1519,652-655(1991)
See Shen, Ai D.: V1519,656-659(1991)

Wang, He-Chen
; Lou, Da-li; Xian, Wu; Song, Xiang; Yong, Jiang: AI application in shoe industry CAD/CAM,V1386,273-276(1991)
; Xian, Wu; Luo, Da L.; Song, Xiang: New search method based on hash table and heuristic search method,V1384,133-136(1991)

Wang, Hong Y.
See He, Anzhi: V1554A,547-552(1991)
See Lu, Jian-Feng: V1554B,593-597(1991)

Wang, Houshu See Yang, Yibing: V1554A,781-788(1991)

Wang, Huirong See Jiang, Shibin: V1535,143-147(1991)

Wang, Huiwen
See Li, Naiji: V1572,47-51(1991)
See Yang, Yang: V1572,205-210(1991)

Wang, Ji-Zhong
; Wang, Yun-Shan; Yin, Yuan-Cheng; Wu, Rui-Lan: Geometrical moire method for displaying directly the stress field of a circular disk,V1554B,188-192(1991)

Wang, Jianhua
: Fiber optic displacement sensor for measurement of thin film thickness,V1572,264-267(1991)

Wang, Jianping
See Javidi, Bahram: V1564,212-223(1991)
See Javidi, Bahram: V1564,254-265(1991)

Wang, Jianshe
; Liu, Jian-Hua: Dynamic behavior and structure optimum of high-speed gear mechanism of high-speed photography apparatus,V1533,175-184(1991)

Wang, Jiazhen See Gu, Lizhong: V1572,450-452(1991)

Wang, Jih-Fang See Wang, Yuan-Fang: V1383,265-276(1991)

Wang, Jin-Xue
; Hays, Paul B.; Drayson, S. R.: Multiorder etalon sounder for vertical temperature profiling: technique and performance analysis,V1492,391-402(1991)

Wang, Jinqi See Qin, Yuwen: V1554A,739-746(1991)

Wang, Jinyu
; Stroup, Eric; Wang, Xing F.; Andrade, Joseph D.: Study of PEO on LTI carbon surfaces by ellipsometry and tribometry,V1519,835-841(1991)

Wang, Ji S.
; Xu, Chun F.: New results on proton nuclear magnetic resonance of a-SiN:H films,V1519,857-859(1991)

Wang, Joe See Hilkert, James M.: V1498,24-38(1991)

Wang, Ju See Dong, Benhan: V1554A,400-406(1991)

Wang, Junqing See McAulay, Alastair D.: V1564,685-690(1991)

Wang, Ju Y. See Yang, Bang C.: V1519,725-728(1991)

Wang, Kemin See Spichiger, Ursula E.: V1510,118-130(1991)

Wang, Kuoping
　　See Guo, Lu Rong: V1461,91-92(1991)
　　See Guo, Lu Rong: V1555,291-292(1991)

Wang, LeMing
　　; Liu, Y.; Ho, Ping-Pei; Alfano, Robert R.: Ballistic imaging of biomedical
　　　samples using picosecond optical Kerr gate,V1431,97-101(1991)

Wang, LiHong See Yu, He: V1469,412-417(1991)

Wang, Liming
　　; Hou, Xun; Cheng, Zhao: Observation of multiphoton photoemission from a
　　　NEA GaAs photocathode,V1358,1156-1160(1991)
　　; Hou, Xun; Cheng, Zhao: Photoemission under three-photon excitation in a
　　　NEA GaAs photocathode,V1415,120-126(1991)

Wang, Ling-jun See Haglund, Richard F.: V1441,127-138(1991)

Wang, Lingjie
　　; Zhou, Fang Q.; Zhao, Xing R.; Sun, Han D.; Yi, Xin J.: Optical detector
　　　prepared by high-Tc superconducting thin film,V1540,738-741(1991)
　　See Sun, Han D.: V1477,174-177(1991)
　　See Zhao, Xing R.: V1519,772-774(1991)
　　See Zhou, Fang Q.: V1477,178-181(1991)

Wang, Lingli
　　; Yun, Dazhen: Curvature of bending shells by moire
　　　interferometry,V1554B,436-440(1991)

Wang, Litszin See Golubeva, N. G.: V1403,134-138(1991)

Wang, Liu Sheng
　　See Dobbins, B. N.: V1554A,772-780(1991)
　　See Dobbins, B. N.: V1554B,586-592(1991)

Wang, Lon A.
　　; Chapuran, Thomas E.; Menendez, Ronald C.: 4-channel, 662-Mb/s
　　　medium-density WDM system with Fabry-Perot laser diodes for
　　　subscriber loop applications,V1363,85-91(1991)

Wang, Mark S.
　　See Milster, Tom D.: V1499,348-353(1991)
　　See Milster, Tom D.: V1499,286-292(1991)

Wang, Michael R.
　　See Chen, Ray T.: V1374,223-236(1991)
　　See Chen, Ray T.: V1583,135-142(1991)

Wang, Ming
　　; Ma, Li; Zeng, Jing-gen; Cheng, Qi-Xian; Pan, Chuan K.: Projection moire
　　　deflectometry for mapping phase objects,V1554B,242-246(1991)

Wang, Nai-Guang
　　See Chance, Britton: V1525,68-82(1991)
　　See Haselgrove, John C.: V1431,30-41(1991)

Wang, Patrick S.
　　: Three-dimensional object representation by array grammars,V1381,210-
　　　216(1991)
　　; Zhang, Y. Y.: Parsing algorithm for line-drawing pattern
　　　recognition,V1384,68-74(1991)

Wang, Paul K. See Taylor, Andrew T.: V1567,220-231(1991)

Wang, Pearl Y. See Kjell, Bradley P.: V1381,553-560(1991)

Wang, Peizheng See Lu, Xiaoming: V1572,248-251(1991)

Wang, Q. See Clarke, Richard H.: V1437,198-204(1991)

Wang, Qi
　　; Bao, Jia S.; Boardman, A. D.: Magnetic steering of energy flow of linear and
　　　nonlinear magnetostatic waves in ferrimagnetic films,V1519,589-
　　　596(1991)
　　See Boardman, A. D.: V1519,609-615(1991)
　　See Boardman, A. D.: V1519,597-604(1991)

Wang, Qiaofei
　　; Neuvo, Yrjo A.: Application of median-type filtering to image segmentation
　　　in electrophoresis,V1450,47-58(1991)

Wang, Qimin See Zhou, Shaoxiang: V1358,788-792(1991)

Wang, Qinsong
　　; McDermott, David B.; Luhmann, Neville C.: Magnetically tapered CARM
　　　(cyclotron autoresonance maser) for high power,V1407,209-216(1991)

Wang, Ren
　　; Yang, Guang Y.; Wu, Bei X.; Fu, Bao W.; Zhan, Yun C.; Zhang, Yun H.:
　　　Investigation of Ti-Al-Mo-V alloy nitride coatings by ARC
　　　technique,V1519,146-151(1991)
　　See Zhang, Yun H.: V1519,729-734(1991)

Wang, Rui-zhong
　　; Wang, Zhen-he; Lu, Kuang-sheng; Yang, Xiao-zhi; Lu, Bo-kao: Clinical
　　　experience in applying endoscopic Nd:YAG laser to treat 451
　　　esophagostenotic cases,V1421,203-207(1991)

Wang, Ruibo See Mao, Xianjun: V1563,58-63(1991)

Wang, Ruli See Chen, Wei: V1450,198-205(1991)

Wang, Ruo Bao See He, Da-Ren: V1362,696-701(1991)

Wang, Rupeng See Bo, Weiyun: V1554A,750-754(1991)

Wang, Ruqin See Qian, Shen-en: V1483,196-206(1991)

Wang, S. C.
　　; Dziura, T. G.; Yang, Y. J.: Vertical-cavity surface-emitting lasers and arrays
　　　for optical interconnect and optical computing applications,V1563,27-
　　　33(1991)
　　See Yang, Y. J.: V1418,414-421(1991)

Wang, S. M. See Andersson, Thorvald G.: V1361,434-442(1991)

Wang, S. Y.
　　; Tan, Michael T.; Houng, Y. M.: Velocity-matched III-V travelling wave
　　　electro-optic modulator,V1371,98-103(1991)

Wang, Shaomin
　　; Bernabeu, Eusebio; Alda, Javier: Advanced matrix optics and its incidence
　　　in laser optics,V1397,595-602(1991)

Wang, Shengrui
　　; Poussart, Denis; Gagne, Simon: Neural model for feature matching in stereo
　　　vision,V1382,37-48(1991)

Wang, Shijie
　　See Li, Shiying: V1410,215-220(1991)
　　See Lin, Li-Huang: V1410,65-71(1991)

Wang, Shizhuo See Yang, Quanzu: V1513,264-269(1991)

Wang, Shu H. See Zhou, Shi P.: V1519,793-799(1991)

Wang, Shuhua See Sun, Xiaohan: V1572,243-247(1991)

Wang, Shuyan See Wang, Guozhi: V1358,73-81(1991)

Wang, Shyh See Yang, Y. J.: V1418,414-421(1991)

Wang, Tai-San
　　See Chan, Kwok-Chi D.: V1552,69-78(1991)
　　See Newnam, Brian E.: V1552,154-174(1991)
　　See Thode, Lester E.: V1552,87-106(1991)

Wang, Tao See Zhuang, Xinhua: V1569,434-445(1991)

Wang, Tianmin
　　; Wang, Weijie; Liu, Guidng; Huang, Liangpu; Luo, Chuntai; Liu, Dingquan;
　　　Xu, Ming; Yang, Yimin: Investigation on the DLC films prepared by dual-
　　　ion-beam-sputtering deposition,V1519,890-900(1991)

Wang, Wangdi See He, Da-Ren: V1362,696-701(1991)

Wang, Wei-Chung
　　; Chen, Jin-Tzaih: Effects of material combinations on the bimaterial fracture
　　　behavior,V1554A,60-69(1991)
　　; Chen, Tsai-Lin; Hwang, Chi-Hung: Stress intensity factors of edge-cracked
　　　semi-infinite plates under transient thermal loading,V1554A,124-
　　　135(1991)

Wang, Wei-Jian
　　; Li, Ming-Jun; Honkanen, Seppo; Najafi, S. I.; Tervonen, Ari: Glass
　　　waveguides by ion exchange with ionic masking,V1513,434-440(1991)
　　See Albert, Jacques: V1583,27-31(1991)
　　See Li, Ming-Jun: V1506,52-57(1991)
　　See Li, Ming-Jun: V1513,410-417(1991)
　　See Najafi, S. I.: V1583,32-36(1991)

Wang, Wei-Yen See Hao, Tianyou: V1584,32-38(1991)

Wang, Weijie See Wang, Tianmin: V1519,890-900(1991)

Wang, Wen-Gui See Chen, Yi-Sheng: V1530,111-117(1991)

Wang, Wen C. See Zheng, Tian S.: V1519,339-346(1991)

Wang, Weng-Lyang See Yeh, Long-Ching: V1527,361-367(1991)

Wang, X. See Favro, Lawrence D.: V1467,290-294(1991)

Wang, Xi
　　; Yang, Gen Q.; Liu, Xiang H.; Zheng, Zhi H.; Huang, Wei; Zhou, Zu Y.; Zou,
　　　Shi C.: Formation of titanium nitride films by Xe+ ion-beam-enhanced
　　　deposition in a N2 gas environment,V1519,740-743(1991)

Wang, Xiang D. See Deng, Hong: V1519,735-739(1991)

Wang, Xianhua
; Chen, Guofu; Liu, D.; Xu, L.; Ruan, S.: Active mode-locking of external cavity semiconductor laser with 1 GHz repetition rate, V1358,775-779(1991)

Wang, Xiao-Ru See Zhang, Wei: V1572,15-17(1991)

Wang, Xiaofang See Chen, Shisheng: V1552,288-295(1991)

Wang, Xiaolin See Yang, Hou-Min: V1533,185-192(1991)

Wang, Xing F. See Wang, Jinyu: V1519,835-841(1991)

Wang, Xizhou
; Tan, Yushan: Vibration modal analysis using stroboscopic digital speckle pattern interferometry, V1554A,907-914(1991)

Wang, Xu-Ming
; Mu, Guoguang: Holographic associative memory with bipolar features, V1558,518-528(1991)

Wang, Y. See Izumitani, Tetsuro: V1535,150-159(1991)

Wang, Y. Q. See Bridwell, Lynn B.: V1519,878-883(1991)

Wang, Y. Y.
; Chiang, Fu-Pen; Barsoum, Roshdy S.; Chou, S. T.: Study of displacement and residual displacement field of an interface crack by moire interferometry, V1554B,344-356(1991)

Wang, Ya-Ping See Liu, Wen-Qing: V1530,240-243(1991)

Wang, Yanqun
: More realistic and efficient algorithm for the drawing of 3-D space-filling molecular models, V1606,1027-1036(1991)

Wang, Yao
; Zhu, Qin-Fan: Signal loss recovery in DCT-based image and video codecs, V1605,667-678(1991)

Wang, Yao-Cai
; Shi, Yi-Wei; Jiang, Hong-Tao: Passive optical fibre sensor based on Cerenkov effect, V1572,32-37(1991)
See Shi, Yi-Wei: V1572,308-312(1991)

Wang, Yeong-Kang See Shao, C. A.: V1554A,613-618(1991)

Wang, Yi-Ming
; Jing, Lian-hua; Pang, Da-wen: Study of properties of a-Si1-xGex:H prepared by SAP-CVD method, V1361,325-335(1991)
See Chen, Liang-Hui: V1361,60-73(1991)

Wang, Yigong See Guo, Maolin: V1554A,310-312(1991)

Wang, Yingting See Yu, Bing Kun: V1530,363-369(1991)

Wang, Yi P.
; Wang, Yu J.: Fractal simulation of aggregation in magnetic thin film, V1519,605-608(1991)

Wang, Ynjiun P.
; Pavlidis, Theo; Swartz, Jerome: Analysis of one-dimensional barcode, V1384,145-160(1991)
; Pavlidis, Theo; Swartz, Jerome: High-density two-dimensional barcode, V1384,169-175(1991)

Wang, Yong B.
See Yuan, Ren K.: V1519,831-834(1991)
See Yuan, Ren K.: V1519,396-399(1991)

Wang, Yongjiang
; Pan, Bailiang; Ding, Xiande; Qian, Yujun; Shi, Shuyi: Multiple-spectral structure of the 578.2 nm line for copper vapor laser, V1412,60-66(1991)
; Shi, Shuyi; Qian, Yujun; Pan, Bailiang; Ding, Xiande; Guo, Qiang; Chen, Lideng; Fan, Ruixing: Adsorption of bromine in CuBr laser, V1412,67-71(1991)

Wang, Yonglin
See Wu, Zhao P.: V1519,194-198(1991)
See Zhou, Shixun: V1519(1991)

Wang, Yong Q.
See Liao, Chang G.: V1519,152-155(1991)
See Xiang, Jin Z.: V1519,683-687(1991)

Wang, Youguan See Yang, Zhiguo: V1572,252-257(1991)

Wang, Yuan-Fang
: New method for sensor data fusion in machine vision, V1570,31-42(1991)
; Wang, Jih-Fang: Surface reconstruction method using deformable templates, V1383,265-276(1991)

Wang, Yue
; Tang, Zhi J.; Zhuang, Wei S.; He, Jing F.: Investigation of defects in HgCdTe epi-films grown from Te solutions, V1519,428-433(1991)

Wang, Yuh-Lin
; Temkin, Henryk; Harriott, Lloyd R.; Hamm, Robert A.: Focused ion-beam vacuum lithography of InP with an ultrathin native oxide resist, V1392,588-594(1991)
See Harriott, Lloyd R.: V1465,57-63(1991)

Wang, Yu J. See Wang, Yi P.: V1519,605-608(1991)

Wang, Yun-Shan See Wang, Ji-Zhong: V1554B,188-192(1991)

Wang, Yun Z.
; Pan, Yao L.: Defects in SIPOS film studied by ESR, V1519,860-863(1991)

Wang, Yuren
: Laser high-speed photography systems used to ammunition measures and tests, V1346,331-337(1991)

Wang, Yutian
; Jiang, G.; Yu, J.: Fiber microbend sensor and instrumentation for fluid-level measurement, V1572,230-234(1991)
; Shi, Jinshan; Li, Zhiquan: Two-colour ratio pyrometer with optical fiber, V1572,192-196(1991)
See Li, Zhiquan: V1572,185-188(1991)
See Shi, Jinshan: V1572,175-179(1991)

Wang, Yuzhu
; Li, Yongqing; Liu, Yashu; Lu, Baolong: Nd-glass microspherical cavity laser induced by cavity QED effects, V1501,40-48(1991)

Wang, Z. H. See Hua, Zhong Y.: V1519,2-7(1991)

Wang, Zhao See Zhao, Mingjun: V1558,529-534(1991)

Wang, Zhen-he See Wang, Rui-zhong: V1421,203-207(1991)

Wang, Zhengrong See Wang, Guozhi: V1358,73-81(1991)

Wang, Zhi-Jiang
See Shen, Ai D.: V1519,656-659(1991)
See Tang, Xue F.: V1519,96-98(1991)
See Wu, Zhouling: V1441,214-227(1991)
See Wu, Zhouling: V1441,200-213(1991)
See Zhang, Yudong: V1463,456-463(1991)
See Zhang, Yudong: V1463,688-694(1991)

Wang, Zhi C. See Gu, Shu L.: V1519,175-178(1991)

Wang, Zhiguang See Wang, Anbo: V1584,294-303(1991)

Wang, Zhongcai See Yang, Quanzu: V1513,264-269(1991)

Wang, Zhulun See Chu, Benjamin: V1403,258-267(1991)

Wani, M. A.
; Batchelor, Bruce G.: Two-dimensional boundary inspection using autoregressive model, V1384,83-89(1991)

Wannop, Neil M.
; Charlton, Andrew; Dickinson, Mark R.; King, Terence A.: Interaction of erbium laser radiation with corneal tissue, V1423,163-166(1991)

Wanuga, Stephen
; Ackerman, Edward I.; Kasemset, Dumrong; Hogue, David W.; Chinn, Stephen R.: Low-loss L-band microwave fiber optic link for control of a T/R module, V1374,97-106(1991)
See Kasemset, Dumrong: V1371,104-114(1991)

Warbrick, K. J. See Henshall, Gordon D.: V1418,286-291(1991)

Ward, Barry
; Emmony, David C.: Interactions of laser-induced cavitation bubbles with a rigid boundary, V1358,1035-1045(1991)
; Emmony, David C.: Laser generation of Stoneley waves at liquid-solid boundaries, V1358,1228-1236(1991)

Ward, Kenneth J. See Blair, Dianna S.: V1437,76-79(1991)

Ward, Matthew O.
; Rajasekaran, Suresh: Image segmentation using domain constraints, V1381,490-500(1991)

Ward, Michael B.
See Deason, Vance A.: V1332,868-876(1991)
See Deason, Vance A.: V1554B,390-398(1991)

Ward, Michael J. See Welstead, Stephen T.: V1565,482-491(1991)

Ward, Richard S. See Salmon, J. T.: V1542,459-467(1991)

Ward, Steven D. See Ostrout, Wayne H.: V1392,151-164(1991)

Ward, T. C. See DiFrancia, Celene: V1588,44-49(1991)

Warde, Cardinal
See Hillman, Robert L.: V1562,136-142(1991)
See Horsky, Thomas N.: V1540,527-532(1991)
See O'Mara, Daniel M.: V1527,110-117(1991)
See Rotman, Stanley R.: V1442,205-215(1991)

Warden, Victor N. See Banash, Mark A.: V1590,8-13(1991)

Wardosanidze, Zurab V.
: Spectrally nonselective holographic objective, V1574,218-226(1991)
: Zone plate of anisotropic profile, V1574,109-120(1991)
See Kakichashvili, Shermazan D.: V1238,134-137(1991)

Wareberg, P. G.
; Scholes, R.; Taylor, R.: Linear LED arrays for film annotation and for high-speed high-resolution printing on ordinary paper,V1538,112-123(1991)

Wargo, Michael J.
: Real-time quantitative imaging for semiconductor crystal growth, control, and characterization,V1557,271-282(1991)
See Carlson, D. J.: V1557,140-146(1991)

Waring, George O. See Ren, Qiushi: V1423,129-139(1991)

Warken, D. See Krehl, Peter: V1358,162-173(1991)

Warmack, R. J.
See Goudonnet, Jean-Pierre: V1400,116-123(1991)
See Lowndes, Douglas H.: V1394,150-160(1991)

Warman, John M.
; Jonker, Stephan A.; de Haas, Matthijs P.; Verhoeven, Jan W.; Paddon-Row, Michael N.: Photon-induced charge separation in molecular systems studied by time-resolved microwave conductivity: molecular optoelectric switches,V1559,159-170(1991)

Warner, Joseph D. See Leonard, Regis F.: V1394,114-125(1991)

Warnes, Richard H.
See Hemsing, Willard F.: V1346,133-140(1991)
See Mathews, Allen R.: V1346,122-132(1991)

Warren, David W. See Chambers, Robert J.: V1488,312-326(1991)

Warren, Roger W.
See Chan, Kwok-Chi D.: V1552,69-78(1991)
See Newnam, Brian E.: V1552,154-174(1991)

Wartenberg, Mark
See Jones, Philip J.: V1456,6-14(1991)
See Reamey, Robert H.: V1455,39-44(1991)

Wasa, Kiyotaka
See Mizuno, Koichi: V1477,197-204(1991)
See Setsune, Kentaro: V1394,79-88(1991)

Washington, Randolph T. See Brown, James C.: V1488,300-311(1991)

Washwell, Edward R. See Huang, Chao H.: V1564,427-438(1991)

Waskiewicz, Warren K.
See Raab, Eric L.: V1343,104-109(1991)
See White, Donald L.: V1343,204-213(1991)
See Windt, David L.: V1343,292-308(1991)
See Windt, David L.: V1343,274-282(1991)

Wassick, Thomas A.
: Laser processes for repair of thin-film wiring,V1598,141-148(1991)

Watabe, Akinori
; Yamada, Ichiro; Yamamoto, Manabu; Katoh, Kikuji: Three-beam overwritable magneto-optic disk drive,V1499,226-235(1991)

Watakabe, Yaichiro
See Miyazaki, Junji: V1464,327-335(1991)
See Wiley, James N.: V1464,346-355(1991)

Watanabe, F. See Yamaguchi, H.: V1499,29-38(1991)

Watanabe, Fumito See Takahashi, Fumiho: V1490,255-268(1991)

Watanabe, Hiroshi
; Sano, Masaharu; Mills, F.; Chang, Sheng-Huei; Masuda, Shoichi: Airborne and spaceborne thermal multispectral remote sensing,V1490,317-323(1991)
; Singhal, Sharad: Windowed motion compensation,V1605,582-589(1991)

Watanabe, Hisashi
; Todokoro, Yoshihiro; Hirai, Yoshihiko; Inoue, Morio: Transparent phase-shifting mask with multistage phase shifter and comb-shaped shifter,V1463,101-110(1991)

Watanabe, Hitoshi
; Koyama, Eiji; Nunomura, T.; Taii, T.; Miura, Michio; Gotoh, Akira; Nishida, T.; Horigome, Shinkichi; Ohta, Norio: Highly reliable 7-GB, 1.2- to 2.2-MB/s, 12-inch write-once optical disk,V1499,21-28(1991)

Watanabe, Isao See Shimamoto, Masayoshi: V1499,393-400(1991)

Watanabe, Kazuhiro See Kashiwabara, S.: V1397,803-806(1991)

Watanabe, Kenjirou See Aratani, Katsuhisa: V1499,209-215(1991)

Watanabe, Miki See Onishi, Randall M.: V1406,171-178(1991)

Watanabe, Mikio
; Saito, Osamu; Okamoto, Satoru; Ito, Kenji; Moronaga, Kenji; Hayashi, Kenkichi; Nishi, Seiki: Signal processing LSI system for digital still camera,V1452,27-36(1991)

Watanabe, Nobuko See Gemma, Takashi: V1332,77-84(1991)

Watanabe, S. See Hane, Kazuhiro: V1332,577-583(1991)

Watanabe, Tamishige See Sakurada, Masami: V1425,158-164(1991)

Watanabe, Tetsu See Fujita, Goro: V1499,426-432(1991)

Watanabe, Tsutomu See Suzaki, Shinzoh: V1374,126-131(1991)

Watanabe, Yutaka See Niibe, Masahito: V1343,2-13(1991)

Watari, Toshihiko See Murano, Hiroshi: V1390,78-90(1991)

Wataya, Hideo
; Tsuchiya, Toshiyuki: Optical subscriber line transmission system to support an ISDN primary-rate interface,V1363,72-84(1991)

Waters, Joe W.
: Microwave limb sounder experiments for UARS and EOS,V1491,104-109(1991)

Waters, Robert G.
See Baumann, John A.: V1418,328-337(1991)
See Coleman, James J.: V1418,318-327(1991)

Waters, Ruth A.
; Thomas, J. M.; Clement, R. M.; Ledger, N. R.: Comparison of carbon monoxide and carbon dioxide laser-tissue interaction,V1427,336-343(1991)

Watkins, Arthur D. See Deason, Vance A.: V1332,868-876(1991)

Watkins, E. T.
; Yu, K. K.; Yau, W.; Wu, Chan-Shin; Yuan, Steve: Ultralinear low-noise amplifier technology for space communications,V1475,62-72(1991)

Watkins, James F. See Detwiler, Paul W.: V1425,149-155(1991)

Watkins, Mark See Kramer, Charles J.: V1454,434-446(1991)

Watkins, Wendell R.
; Billingsley, Daniel R.; Palacios, Fernando R.; Crow, Samuel B.; Jordan, Jay B.: Characterization of the atmospheric modulation transfer function using the target contrast characterizer,V1486,17-24(1991)
; Clement, Dieter; eds.: *Characterization, Propagation, and Simulation of Sources and Backgrounds,*V1486(1991)
See Crow, Samuel B.: V1486,333-344(1991)

Watlington, John See Bove, V. M.: V1605,886-893(1991)

Watowich, Stanley J.
; Gross, Leon J.; Josephs, Robert: Intermolecular contacts within sickle hemoglobin fibers,V1396,316-323(1991)

Watson, A. J. See Morris, David R.: V1358,254-261(1991)

Watson, Carolyn K. See Rundle, Debra A.: V1446,379-387(1991)

Watson, Catherine E. See Fishman, Jack: V1491,348-359(1991)

Watson, Graham M.
: Range of modalities for prostate therapy,V1421,2-5(1991)
; Steiner, Rudolf W.; Pietrafitta, Joseph J.; eds.: *Lasers in Urology, Laparoscopy, and General Surgery,*V1421(1991)

Watson, James E.
See Kanofsky, Alvin S.: V1374,59-66(1991)
See Smith, Terrance L.: V1512,92-100(1991)

Watson, John
; Kilpatrick, J. M.: Optical aberrations in underwater holography and their compensation,V1461,245-253(1991)

Watson, Ken See Curran, Robert J.: V1492(1991)

Watson, Larry J. See Mechtenberg, Monica L.: V1496,124-155(1991)

Watt, B. See Puetz, Norbert: V1361,692-698(1991)

Watt, David W.
; Philbrick, Daniel A.: Images of turbulent, absorbing-emitting atmospheres and their application to windshear detection,V1467,357-368(1991)

Watteau, Jean P. See Andre, Michel L.: V1502,286-298(1991)

Watts, Bob See Johnson, R. B.: V1354,669-675(1991)

Watts, Jeffrey W. See Greene, Ben: V1492,126-139(1991)

Watts, Richard N.
See Gluskin, Efim S.: V1548,56-68(1991)
See Slaughter, Jon M.: V1343,73-82(1991)
See Windt, David L.: V1343,274-282(1991)

Wawer, Janusz
See Makuchowski, Jozef: V1391,348-350(1991)
See Makuchowski, Jozef: V1391,79-86(1991)

Wawrzynek, Paul A. See Andrews, John R.: V1461,110-123(1991)

Waxman, Allen M. See Seibert, Michael: V1383,10-19(1991)

Weaver, Harry T. See Carson, Richard F.: V1378,84-94(1991)

Weaver, William L. See Gustafson, Terry L.: V1403,545-554(1991)

Webb, Colin E. See Knowles, Martyn R.: V1410,195-201(1991)

Webb, Curtis M.
See Brown, James C.: V1488,300-311(1991)
See D'Agostino, John A.: V1488,110-121(1991)
See Hoover, Carl W.: V1488,280-288(1991)

Webb, David J.
; Benedict, Melvin K.; Bona, Gian-Luca; Buchmann, Peter L.; Daetwyler, K.; Dietrich, H. P.; Moser, A.; Sasso, G.; Vettiger, Peter; Voegeli, O.: Mirror fabrication for full-wafer laser technology,V1418,231-239(1991)
See Tang, Wade C.: V1418,338-342(1991)

Webb, Jennifer M.
; Amini, Zahra H.: Characteristics of gate oxide surface material after exposure to magnetron-enhanced reactive ion etching plasma,V1392,47-54(1991)

Webb, Paul P.
: Planar InGaAs APD (avalanche photodiode) for eyesafe laser rangefinding applications,V1419,17-23(1991)
; Dion, Bruno: High-speed 128-element avalanche photodiode array for optical computing applications,V1563,236-243(1991)

Webb, Roderick P. See Barnes, Nigel: V1389,477-483(1991)
Webb, Watt W. See Strickler, James H.: V1398,107-118(1991)
Webber, Stephen E. See Haruvy, Yair: V1590,59-70(1991)

Weber, Andrew M.
See Gambogi, William J.: V1555,256-267(1991)
See Zager, Stephen A.: V1461,58-67(1991)

Weber, Charles L.
See Ghosh, Monisha: V1565,188-195(1991)
See Sullivan, Daniel P.: V1476,234-245(1991)

Weber, Hans-Georg See Schunk, Nikolaus: V1362,391-397(1991)
Weber, Hans M. See Miller, Kurt: V1421,108-113(1991)

Weber, Heinz P.
See Frenz, Martin: V1427,9-15(1991)
See Romano, Valerio: V1427,16-26(1991)

Weber, Robert G. See Weiman, Carl F.: V1386,102-110(1991)
Weber, W. J. See Lesh, James R.: V1522,27-35(1991)

Webster, Chris R.
; May, R. D.: Aircraft laser infrared absorption spectrometer (ALIAS) for polar ozone studies,V1540,187-194(1991)

Wee, William G.
See Amladi, Nandan G.: V1472,165-176(1991)
See Hu, Yong-Lin: V1468,653-661(1991)
See Jiang, Dareng: V1468,16-25(1991)
See Krovvidy, Srinivas: V1468,216-226(1991)
See Niu, Aiqun: V1469,495-505(1991)
See Nolan, Adam R.: V1472,157-164(1991)
See Tang, Yonghong: V1469,113-120(1991)
See Wan, Liqun: V1469,432-440(1991)

Weed, Harrison See Huff, Howard R.: V1464,278-293(1991)

Weeks, Arthur R.
; Myler, Harley R.; Jolson, Alfred S.: Calibration issues in the measurement of ocular movement and position using computer image processing,V1567,77-87(1991)
See Myler, Harley R.: V1565,57-68(1991)
See Wenaas, Holly: V1569,410-421(1991)

Weeks, Richard A.
: Campus fiber optic enterprise networks,V1364,222-227(1991)

Wegerif, Daniel G. See Vargo, Rick C.: V1387,72-81(1991)

Wegmann, Ulrich
See Doerband, Bernd: V1332,664-672(1991)
See Freischlad, Klaus R.: V1332,18-24(1991)

Wegner, Gerhard See Neher, Dieter: V1560,335-343(1991)

Wegner, Paul J.
; Henesian, Mark A.: Precision high-power solid state laser diagnostics for target-irradiation studies and target-plane irradiation modeling,V1414,162-174(1991)
See Henesian, Mark A.: V1415,90-103(1991)

Wehrli, Felix W. See Rundle, Debra A.: V1446,379-387(1991)

Wei, Cailin
: Surface reflection coefficient correction technique for a microdisplacement OFS,V1572,42-46(1991)

Wei, David T.
; Kaufman, Harold R.: New ion-beam sources and their applications to thin film physics,V1519,47-55(1991)

Wei, G. See Goldner, Ronald B.: V1536,63-69(1991)

Wei, Guang P.
: Amorphous silicon thin film x-ray sensor,V1519,225-233(1991)

Wei, M. Z. See Wu, Zi L.: V1519,618-624(1991)

Wei, Mingzhi See Robinson, Lloyd B.: V1447,214-228(1991)

Wei, Ta-Sheng
; Yuce, Hakan H.; Varachi, John P.; Pellegrino, Anthony: Mechanical testing and reliability considerations of fusion splices,V1366,235-240(1991)
See Zemon, Stanley A.: V1373,21-32(1991)

Wei, Tai-Huei See Soileau, M. J.: V1441,10-15(1991)

Wei, Yan-nian
; Jiang, Pei-sheng; Zhang, Zeng-xiang; Wu, Ming-da; Shi, Gao-yi; Ye, Niao-ting: New development of friction speeding mechanism for high-speed rotating mirror device,V1358,457-460(1991)

Weichel, Hugo
ed.: *Selected Papers on Laser Design,* VMS29(1991)

Weichold, Mark H. See Gopalakrishnan, G. K.: V1476,270-275(1991)

Weidel, Edgar See Maile, Michael: V1563,188-196(1991)

Weideman, William E.
: Image quality, dollars, and very low contrast documents,V1454,382-390(1991)

Weidenheimer, Douglas M. See Huttlin, George A.: V1407,147-158(1991)

Weigand, R. See Martin, E.: V1397,835-838(1991)

Weigler, William See Sandborn, Peter A.: V1389,177-186(1991)

Weil, Gary J.
; Graf, Richard J.: Infrared-thermography-based pipeline leak detection systems,V1467,18-33(1991)

Weiland, Timothy L.
See Henesian, Mark A.: V1415,90-103(1991)
See Kyrazis, Demos T.: V1441,469-477(1991)
See Richards, James B.: V1346,384-389(1991)

Weill, Andre P.
; Amblard, Gilles R.; Lalanne, Frederic P.; Panabiere, Jean-Pierre: Wet-developed, high-aspect-ratio resist patterns by 20-keV e-beam lithography,V1465,264-270(1991)
See Paniez, Patrick J.: V1466,583-591(1991)
See Paniez, Patrick J.: V1466,336-344(1991)

Weiman, Carl F.
; Weber, Robert G.: Video-rate image remapper for the PC/AT bus,V1386,102-110(1991)
See King, Steven J.: V1388,190-198(1991)

Weimer, Wayne A.
; Cerio, Frank M.; Johnson, Curtis E.: Plasma parameters in microwave-plasma-assisted chemical vapor deposition of diamond,V1534,9-13(1991)

Wein, Deborah S. See Zucker, Oved S.: V1378,22-33(1991)

Weinberg, Wolfram S.
; Hayrapetian, Alek; Cho, Paul S.; Valentino, Daniel J.; Taira, Ricky K.; Huang, H. K.: X-window-based 2K display workstation,V1446,35-39(1991)
; Loloyan, Mansur; Chan, Kelby K.: On-line acquisition of CT and MRI studies from multiple scanners,V1446,430-435(1991)
See Hayrapetian, Alek: V1446,243-247(1991)
See Lou, Shyh-Liang: V1446,302-311(1991)
See Taira, Ricky K.: V1446,451-458(1991)

Weinberger, Alex
; Weinberger, Ervin: Long-term reliability and performance testing of fiber optic sensors for engineering applications,V1367,30-45(1991)

Weinberger, Doreen A. See Kamal, Avais: V1516,137-153(1991)

Weinberger, Ervin See Weinberger, Alex: V1367,30-45(1991)

Weiner, Maurice See Kim, A. H.: V1378,173-178(1991)

Weinhaus, Frederick M. See Walterman, Michael T.: V1567,204-214(1991)

Weinreb, Sander
: Monolithic GaAs integrated circuit millimeter wave imaging sensors,V1475,25-31(1991)

Weinshall, Daphna
: Direct computation of geometric features from motion disparities and shading,V1570,274-285(1991)

Weinstein, Ehud See Shalvi, Ofir: V1565,143-152(1991)

Weinstein, Samuel See Libutti, Steven K.: V1421,169-172(1991)

Weinswig, Shepard A.
; Hookman, Robert A.: Optical analysis of thermal-induced structural distortions,V1527,118-125(1991)

Weir, Kenneth
; Boyle, William J.; Palmer, Andrew W.; Grattan, Kenneth T.; Meggitt, Beverley T.: Novel optical processing scheme for interferometric vibration measurement using a low-coherence source with a fiber optic probe,V1584,220-225(1991)

Weir, Michael P. See Tomlinson, Harold W.: V1541,178-186(1991)

Weisenbach, Lori
; Zelinski, Brian J.; O'Kelly, John; Morreale, Jeanne; Roncone, Ronald L.; Burke, James J.: Influence of processing variables on the optical properties of SiO2-TiO2 planar waveguides,V1590,50-58(1991)
See Roncone, Ronald L.: V1590,14-25(1991)
See Zaugg, Thomas C.: V1590,26-35(1991)

Weisenberger, M. See Morgan, Alan R.: V1426,350-355(1991)

Weisfield, Richard L. See Kobayashi, Kenichi: V1448,157-163(1991)

Weisman, Andrew See Dougherty, Edward R.: V1567,88-99(1991)

Weiss, Aryeh M. See Rotman, Stanley R.: V1442,194-204(1991)

Weiss, Howard
; Cebula, Richard P.; Laamann, K.; Hudson, Robert D.: Evaluation of the NOAA-11 solar backscatter ultraviolet radiometer, Mod 2 (SBUV/2): inflight calibration,V1493,80-90(1991)

Weiss, M. See Oertel, Heinrich K.: V1465,244-253(1991)

Weiss, Markus R. See Siahmakoun, Azad: V1396,190-192(1991)

Weiss, Matania See Biderman, Shlomo: V1535,27-34(1991)

Weiss, Richard S. See Grupen, Roderic A.: V1383,189-201(1991)

Weiss, Robert See Costello, Kenneth A.: V1449,40-50(1991)

Weiss, Rudolf M.
; Sills, Robert M.: Application of an electron-beam scattering parameter extraction method for proximity correction in direct-write electron-beam lithography,V1465,192-200(1991)

Weiss, S. See Rochkind, Simeone: V1428,52-58(1991)

Weissbach, Severin
; Wyrowski, Frank: Application of error diffusion in diffractive optics,V1507,149-152(1991)

Weisskopf, Martin C.
; Elsner, Ronald F.; Novick, Robert; Kaaret, Philip E.; Silver, Eric H.: Predicted performance of the lithium scattering and graphite crystal polarimeter for the Spectrum-X-Gamma mission,V1343,457-468(1991)
See Holley, Jeff: V1343,500-511(1991)
See Kaaret, Philip E.: V1548,106-117(1991)

Weissman, Harold
: Beryllium galvanometer mirrors,V1485,13-19(1991)

Weisz, John R.
: Software calibration of scan system distortions,V1454,265-271(1991)

Weit, Scott K. See Kutal, Charles: V1466,362-367(1991)

Weitekamp, Daniel P. See Bowers, C. R.: V1435,36-50(1991)

Welch, Ashley J.
See Beyerbacht, Hugo P.: V1427,117-127(1991)
See LeCarpentier, Gerald L.: V1427,273-278(1991)
See Porindla, Sridhar N.: V1427,267-272(1991)

Welch, Charles W. See Spiller, Eberhard: V1343,134-144(1991)

Welch, David F.
See Mehuys, David G.: V1418,57-63(1991)
See Yaeli, Joseph: V1442,378-382(1991)
See Zucker, Erik P.: V1563,223-228(1991)

Welch, Eric B.
; Moorhead, Robert J.; Owens, John K.: Converting non-interlaced to interlaced images in YIQ and HSI color spaces,V1453,235-243(1991)

Welch, Sharon S.
; Cox, David E.: Characteristics of a dynamic holographic sensor for shape control of a large reflector,V1480,2-10(1991)

Welford, W. T. See MacAndrew, J. A.: V1500,172-176(1991)

Welkowsky, Murray S. See Efron, Uzi: V1455,237-247(1991)

Weller, Scott W.
: Artificial neural net learns the Sobel operators (and more),V1469,219-224(1991)
: Neural network optimization, components, and design selection,V1354,371-378(1991)

Wellfare, Michael R.
: Two-dimensional encoding of images using discrete reticles,V1478,33-40(1991)

Welling, Herbert
See Rahe, Manfred: V1441,113-126(1991)
See Schink, Harald: V1441,327-338(1991)

Welling, Larry W. See Morris, Stephen J.: V1428,148-158(1991)

Wells, Alan A.
; Castelli, C. M.; Holland, Andrew D.; McCarthy, Kieran J.; Spragg, J. E.; Whitford, C. H.: CCD focal-plane imaging detector for the JET-X instrument on spectrum R-G,V1549,357-373(1991)

Wells, Gregory M.
; Chen, Hector T.; Engelstad, Roxann L.; Palmer, Shane R.: Parametric studies and characterization measurements of x-ray lithography mask membranes,V1465,124-133(1991)

Wells, James A.
; Lombard, Calvin M.; Sloan, George B.; Moore, Wally W.; Martin, Claude E.: Lessons learned in recent beryllium-mirror fabrication,V1485,2-12(1991)

Wells, Lisa D. See Golobic, Robert A.: V1425,84-92(1991)

Wells, Mark E.
; Groff, Mary B.: Design and development of a transparent Bridgman furnace,V1557,71-77(1991)

Wells, Willard H.
: Indirect illumination to reduce veiling luminance in seawater,V1537,2-9(1991)

Wells, William M.
: Statistical approach to model matching,V1381,22-29(1991)

Welsch, Lawrence A.
: Multimedia courseware in an open-systems environment: a DoD strategy,VCR37,207-220(1991)

Welsh, Barry Y.
; Jelinsky, Patrick; Vedder, Peter W.; Vallerga, John V.; Finley, David S.; Malina, Roger F.: Results from the calibration of the Extreme Ultraviolet Explorer instruments,V1343,166-174(1991)

Welsh, Byron M.
: Imaging performance analysis of adaptive optical telescopes using laser guide stars,V1542,88-99(1991)
: Sensing refractive-turbulence profiles (Cn2) using wavefront phase measurements from multiple reference sources,V1487,91-102(1991)
See Johnston, Dustin C.: V1542,76-87(1991)
See Veronin, Christopher P.: V1469,281-291(1991)

Welsh, Kevin M.
See Hacker, Nigel P.: V1466,384-393(1991)
See Hacker, Nigel P.: V1559,139-150(1991)

Welstead, Stephen T.
; Ward, Michael J.; Keefer, Christopher W.: Neural network approach to multipath delay estimation,V1565,482-491(1991)

Welter, Rudy See Sessa, William B.: V1372,208-218(1991)

Wen, Baicheng
See Hecker, Friedrich W.: V1554A,151-162(1991)
See Pindera, Jerzy T.: V1554A,196-205(1991)

Wen, Chuanju See Goutsias, John I.: V1606,174-185(1991)

Wen, John T. See Murphy, Steve H.: V1387,14-25(1991)

Wen, Lin Ying
; Hua, Yun: Improvement of specular reflection pyrometer,V1367,300-302(1991)

Wen, Mei-Yuan
; Liu, Guang T.: New technique for multiplying the isoclinic fringes,V1332,673-675(1991)

Wen, Shengping See Li, Shaohui: V1572,543-547(1991)

Wen, Z. Q. See Barron, L. D.: V1403,66-75(1991)

Wen, Zhen-chu
See Chen, Ke-long: V1385,206-213(1991)
See Xu, Zhu: V1399,172-177(1991)

Wen, Zheng
; Tan, Yushan: Phase-stepping technique in holography,V1461,278-285(1991)

Wenaas, Holly
; Weeks, Arthur R.; Myler, Harley R.: Computer-generated correlated noise images for various statistical distributions,V1569,410-421(1991)

Wendt, H. R.
See de Vries, Mattanjah S.: V1437,129-137(1991)
See MacDonald, Scott A.: V1466,2-12(1991)

Weng, Hei M. See Li, Xiao Q.: V1519,14-17(1991)

Weng, Jian
; Zhang, M. Z.; Simons, K.; Chance, Britton: Measurement of biological tissue metabolism using phase modulation spectroscopic technology,V1431,161-170(1991)
See Sevick, Eva M.: V1431,264-275(1991)

Weng, Juyang
; Cohen, Paul: Fusion of stereo views: estimating structure and motion using a robust method,V1383,321-332(1991)

Weng, Zhicheng
; Chen, Zhiyong; Yang, Yu-Hong; Ren, Tao; Cong, Xiaojie; Yao, Yuchuan; He, Fengling; Li, Yuan-Yuan: Attempt to develop a zoom-lens-design expert system,V1527,349-356(1991)
; Chen, Zhiyong; Cong, Xiaojie: Design of an imaging spectrometer for observing ocean color,V1527,338-348(1991)

Wengler, Michael J.
; Pance, A.; Liu, B.; Dubash, N.; Pance, Gordana; Miller, Ronald E.: Submillimeter receiver components using superconducting tunnel junctions,V1477,209-220(1991)

Werblin, Frank S. See Teeters, Jeffrey L.: V1472,6-17(1991); 1473,102-113(1991)

Werby, Michael F.
; Gaunaurd, Guillermo C.: Critical frequencies for large-scale resonance signatures from elastic bodies,V1471,2-17(1991)
See Dean, Cleon E.: V1471,54-65(1991)
See George, Jacob: V1471,66-77(1991)

Werle, Peter W.
; Josek, K.; Slemr, Franz: Application of FM spectroscopy in atmospheric trace gas monitoring: a study of some factors influencing the instrument design,V1433,128-135(1991)
See Muecke, Robert J.: V1433,136-144(1991)

Werncke, W. See Lau, A.: V1403,212-220(1991)

Werner, Michael J. See Spehalski, Richard J.: V1540,2-14(1991)

Werner, Thomas R.
; Cox, James A.; Swanson, S.; Holz, Michael: Microlens array for staring infrared imager,V1544,46-57(1991)
See Cox, James A.: V1507,100-109(1991)
See Cox, James A.: V1555,80-88(1991)

Wernet, Mark P. See Pline, Alexander D.: V1557,222-234(1991)

Wertheimer, Alan See Vandenberg, Donald E.: V1542,534-542(1991)

Wesley, Michael
; Osterheld, Robert; Kyser, Jeff; Farr, Michele; Vandergriff, Linda J.: Surveillance test bed for SDIO,V1481,209-220(1991)

Wesly, Edward J.
: Teaching holography in an art school environment: the program at the School of the Art Institute of Chicago,V1396,71-77(1991)
See Jeong, Tung H.: V1396,60-70(1991)
See Jeong, Tung H.: V1238,298-305(1991)

Wessel, Frank J. See Rahman, Hafiz-ur: V1407,589-597(1991)

Wessel, K. See Wildermuth, Eberhard: V1397,367-371(1991)

Wessels, Bruce W.
; Zhang, Jiyue; DiMeo, Frank; Richeson, D. S.; Marks, Tobin J.; DeGroot, D. C.; Kannewurf, C. R.: Microstructure and superconducting properties of BiSrCaCuO thin films,V1394,232-237(1991)

Wessels, Jurina
See Schneckenburger, Herbert: V1403,646-652(1991)
See Schneckenburger, Herbert: V1525,91-98(1991)

Wessman, Susan C. See Oicles, Jeffrey A.: V1378,60-69(1991)

Wesson, Laurence N.
; Cabato, Nellie L.; Pine, Nicholson L.; Bird, Victor J.: Fiber optic pressure sensor system for gas turbine engine control,V1367,204-213(1991)
See Redner, Alex S.: V1332,775-782(1991)

West, Andrew A.
; Smith, Robin W.: Electron beam lithographic fabrication of computer-generated holograms,V1507,158-167(1991)

West, B. O. See Bell, J. M.: V1536,29-36(1991)

West, David L. See Freeman, J. L.: V1476,320-325(1991)

West, John L.
See McCargar, James W.: V1455,54-60(1991)
See Montgomery, G. P.: V1455,45-53(1991)

Westbrook, Edwin M. See Strauss, Michael G.: V1447,12-27(1991)

Westbrook, M. L. See Strauss, Michael G.: V1447,12-27(1991)

Westenberger, Gerhard
; Hoffmann, Hans J.; Jochs, Werner W.; Przybilla, Gudrun: Verdet constant and its dispersion in optical glasses,V1535,113-120(1991)

Westenskow, Glen A.
; Houck, Timothy L.; Ryne, Robert D.: Relativistic klystron research for future linear colliders,V1407,496-501(1991)

Westerman, Steven D.
; Drake, R. M.; Yool, Stephen R.; Brandley, M.; DeJulio, R.: Automated band selection for multispectral meteorological applications,V1492,263-271(1991)

Western, Arthur B.
See Boxler, Lawrence H.: V1396,85-92(1991)
See Brown, Glen: V1396,164-173(1991)

Westmore, David B. See El-Hakim, Sabry F.: V1526,56-67(1991)

Weston, Norman E. See Wallenberger, Frederick T.: V1484,116-124(1991)

Westphalen, R. See Geurts, Jean: V1361,744-750(1991)

Westrick, Anthony See Staller, Craig O.: V1540,219-230(1991)

Westwater, Edward R. See Shaw, Joseph A.: V1540,681-686(1991)

Westwood, D. I. See Morley, Stefan: V1361,213-222(1991)

Wetterskog, Kevin See Garrett, Steven L.: V1367,13-29(1991)

Wettling, J. C. See David, Jean: V1397,697-700(1991)

Weverka, Robert T.
; Wagner, Kelvin H.: Staring phased-array radar using photorefractive crystals,V1564,676-684(1991)
; Wagner, Kelvin H.: Wide-angular aperture acousto-optic Bragg cell,V1562,66-72(1991)
See Wu, Kuang-Yi: V1563,168-175(1991)

Wey, Y. G. See Crawford, Deborah L.: V1371,138-141(1991)

Weymouth, Terry E.
See Roth-Tabak, Yuval: V1388,453-463(1991)
See Tehrani, Saeid: V1445,434-445(1991)

Whalen, P. V. See Leahy, Michael B.: V1387,148-158(1991)

Whalen, William E. See Neev, Joseph: V1427,162-172(1991)

Wharton, Charles B. See Butler, Jennifer M.: V1407,57-66(1991)

Wheat, T. A. See Pouskouleli, G.: V1590,179-190(1991)

Wheatley, Alvis A. See Bonometti, Robert J.: V1495,166-176(1991)

Wheeler, Frederick W.
; Vaz, Richard F.; Cyganski, David: Automated registration of terrain range images using surface feature level sets,V1606,78-85(1991)

Wheeler, J. R. See MacAndrew, J. A.: V1500,172-176(1991)

Wheeler, James R.
; Cook, William D.; Smith, Ron: Small-capacity low-cost (Ni-H2) design concept for commercial, military, and higher volume aerospace applications,V1495,280-285(1991)

Wheeler, John P.
: Market for low-power CO2 lasers,V1517,44-60(1991)

Wherrett, Brian S.
See Craig, Robert G.: V1505,76-86(1991)
See Wang, Chang H.: V1505,130-140(1991)

Whetstone, James R. See Roberts, James R.: V1392,428-436(1991)

White, Christopher J.
; Ramee, Stephen R.; Mesa, Juan E.; Collins, Tyrone J.; Kotmel, Robert F.; Godfrey, Maureen A.: Laser angioplasty with lensed fibers and a holmium:YAG laser in iliac artery occlusions,V1425,130-133(1991)
See Ramee, Stephen R.: V1420,199-202(1991)

White, Donald L.
; Bjorkholm, John E.; Bokor, J.; Eichner, L.; Freeman, Richard R.; Gregus, J. A.; Jewell, Tanya E.; Mansfield, William M.; MacDowell, Alastair A.; Raab, Eric L.; Silfvast, William T.; Szeto, L. H.; Tennant, Donald M.; Waskiewicz, Warren K.; Windt, David L.; Wood, Obert R.: Soft x-ray projection lithography: experiments and practical printers,V1343,204-213(1991)

White, Grady S. See Cranmer, David C.: V1330,152-163(1991)

White, John R.
: Extractive sampling systems for continuous emissions monitors,V1434,104-112(1991)

White, John V. See Tan, Oon T.: V1422(1991)

White, Julian D.
; Gell, Michael A.; Fasol, Gerhard; Gibbings, C. J.; Tuppen, C. G.: Raman scattering characterization of direct gap Si/Ge superlattices,V1361,293-301(1991)

White, Patricia L. See Wilson, Scott R.: V1441,82-86(1991)

White, Paul R. See Carmichael, I. C.: V1541,167-177(1991)

White, R. S. See Rahman, Hafiz-ur: V1407,589-597(1991)

White, Richard L.
; Hanisch, Robert J.: HST image processing: how does it work and what are the problems?,V1567,308-316(1991)
See Richards, Evan: V1494,40-48(1991)

White, Rodney A.
; Kopchok, George E.: Argon laser vascular tissue fusion: current status and future perspectives,V1422,103-110(1991)
See Tabbara, Marwan R.: V1425,208-216(1991)
See Tan, Oon T.: V1422(1991)

White, Sean T. See Opsasnick, Michael N.: V1499,276-280(1991)

White, Steven P. See Turner-Smith, Alan R.: V1380,75-84(1991)

White, William
: Precise method for measuring the edge acuity of electronically (digitally) printed hard-copy images,V1398,24-28(1991)

White, William E.
See Fittinghoff, David N.: V1413,81-88(1991)
See Gold, David: V1413,41-52(1991)

Whitehead, Mark See Kilcoyne, M. K.: V1389,422-454(1991)

Whitehouse, Colin R. See Ashley, Timothy: V1361,238-244(1991)

Whitehurst, Colin See Jiang, Zhi X.: V1421,88-99(1991)

Whitford, C. H. See Wells, Alan A.: V1549,357-373(1991)

Whiting, Bruce R.
See Jost, R. G.: V1446,2-9(1991)
See Kocher, Thomas E.: V1446,459-464(1991)

Whiting, James S.
; Honig, David A.; Carterette, Edward; Eigler, Neal: Observer performance in dynamic displays: effect of frame rate on visual signal detection in noisy images,V1453,165-175(1991)

Whitlock, S. V. See Jorgensen, Steven M.: V1346,180-191(1991)

Whitman, Robert A. See Cox, Jerome R.: V1446,40-51(1991)

Whitman, Tony
; Araghi, Mehdi N.: Electronic f-theta correction for hologon deflector systems,V1454,426-433(1991)

Whitmeyer, Charlie See Mosier, Marty R.: V1495,177-192(1991)

Whitney, Erich C. See Messner, Richard A.: V1381,261-271(1991)

Whitney, Raymond L. See Adams, Frank W.: V1541,24-37(1991)

Whittemore, Gerald R.
See Hemsing, Willard F.: V1346,133-140(1991)
See Mathews, Allen R.: V1346,122-132(1991)

Whitten, David G. See Spooner, Susan P.: V1436,82-91(1991)

Whitten, Gary E.
: Parametric optical flow without correspondence for moving sensors,V1468,167-175(1991)

Whitten, W. B. See Shaw, Robert W.: V1534,170-174(1991)

Whitworth, Ernest E.
See Block, Kenneth L.: V1483,62-65(1991)
See Block, Kenneth L.: V1481,32-34(1991)

Whitworth, Martin B.
; Huntley, Jonathan M.; Field, John E.: High-speed photography of high-resolution moire patterns,V1358,677-682(1991)
; Huntley, Jonathan M.; Field, John E.: Measurement of the dynamic crack-tip displacement field using high-resolution moire photography,V1554B,282-288(1991)

Whomes, Terence L. See Nwagboso, Christopher O.: V1386,30-41(1991)

Wible, Sheryl F. See Carey, Raymond: V1464,500-507(1991)

Wiborg, P. H. See Rimmele, Thomas: V1542,186-193(1991)

Wichansky, Anna M.
: User benefits of visualization with 3-D stereoscopic displays,V1457,267-271(1991)

Wickens, Christopher D. See Merwin, David H.: V1456,211-218(1991)

Wickersheim, Kenneth A.
: Application of fiber optic thermometry to the monitoring of winding temperatures in medium- and large-power transformers,V1584,3-14(1991)

Wickham, Michael
; Munch, Jesper: CCD holographic phase and intensity measurement of laser wavefront,V1414,80-90(1991)

Wickholm, David R.
: Design of an athermalized three-fields-of-view infrared sensor,V1488,58-63(1991)

Wiczk, Wieslaw M. See Lakowicz, Joseph R.: V1435,142-160(1991)

Widener, A. L. See Rimmele, Thomas: V1542,186-193(1991)

Wider, Todd M.
; Libutti, Steven K.; Greenwald, Daniel P.; Oz, Mehmet C.; Yager, Jeffrey S.; Treat, Michael R.; Hugo, Norman E.: Skin closure with dye-enhanced laser welding and fibrinogen,V1422,56-61(1991)

Widjaja, Joewono
; Uozumi, Jun; Asakura, Toshimitsu: Method for evaluating displacement of objects using the Wigner distribution function,V1400,94-100(1991)

Widmann, Klaus See Neger, Theo: V1507,476-487(1991)

Widmer, H. M. See Duveneck, G.: V1510,138-145(1991)

Widomski, L. See Konwerska-Hrabowska, Joanna: V1554B,225-232(1991)

Wieczorek, Casey J.
See Hart, Patrick W.: V1366,334-342(1991)
See Kilmer, Joyce P.: V1366,85-91(1991)

Wiedeger, S. See Samoriski, Brian: V1412,12-18(1991)

Wieder, Eckart See Wise, Timothy D.: V1346,79-85(1991)

Wiedmann, Wolfgang
See Doerband, Bernd: V1332,664-672(1991)
See Freischlad, Klaus R.: V1332,8-17(1991)

Wiedwald, Douglas J.
; Bell, Perry M.; Kilkenny, Joseph D.; Bonner, R.; Montgomery, David S.: Recent advances in gated x-ray imaging at LLNL,V1346,449-455(1991)
See Kilkenny, Joseph D.: V1358,117-133(1991)

Wiegmann, Thomas B. See Morris, Stephen J.: V1428,148-158(1991)

Wieland, Joerg B.
; Melchior, Hans M.: Optical receivers in ECL for 1 GHz parallel links,V1389,659-664(1991)

Wielogorski, A. L. See de Salabert, Arturo: V1521,74-88(1991)

Wielsch, U. See Gruska, Bernd: V1361,758-764(1991)

Wielunski, L. S. See Bell, J. M.: V1536,29-36(1991)

Wieman, T. J. See Mang, Thomas S.: V1426,188-199(1991)

Wiemokly, Gary D. See Cross, Edward F.: V1540,756-763(1991)

Wiepking, Mark
; LeVan, M.; Mayo, Phyllis: Dry etching of high-aspect ratio contact holes,V1392,139-150(1991)

Wiersma, Douwe A. See Fidder, Henk: V1403,530-544(1991)

Wiese, Gary E.
ed.: *Selected Papers on Optical Tolerancing*,VMS36(1991)

Wiesmann, Theo J. See Bopp, Matthias: V1522,199-209(1991)

Wiesner, J. See Frank, Klaus: V1431,2-11(1991)

Wiesneth, Peter See Bickel, P.: V1503,167-175(1991)

Wietzke, Joachim See Amor, Hamed: V1567,578-588(1991)

Wigdor, Harvey A. See Dederich, Douglas N.: V1424(1991)

Wigginton, Stewart C.
; Davidson, Scott E.; Harting, William L.: High-density chip-to-chip interconnect system for GaAs semiconductor devices,V1390,560-567(1991)

Wihardjo, Erning
; Silva, Donald E.: Accurate method for measuring oblique astigmatism and oblique power of ophthalmic lenses,V1529,57-62(1991)

Wijbrans, Klaas C.
; Korsten, Maarten J.: Flat plate project,V1386,197-205(1991)

Wika, Kevin G.
See Lawson, Michael A.: V1445,265-275(1991)
See Ramos, P. A.: V1443,160-170(1991)

Wike, Charles K.
; Lindacher, Joseph M.: Optical design and development of a small barcode scanning module,V1398,119-126(1991)

Wilbarg, Robert See Conley, Willard E.: V1466,53-66(1991)

Wilbers, A. T. See Meeusen, G. J.: V1519,252-257(1991)

Wilbrink, Jacob See Fielden, John: V1445,145-154(1991)

Wilcox, Brian H. See Bon, Bruce: V1387,337-342(1991)

Wilcox, Christopher D. See Healy, Donald D.: V1529,84-93(1991)

Wilcox, Jarka Z. See Jansen, Michael: V1418,32-39(1991)

Wilcox, William R.
: Commercial crystal growth in space,V1557,31-41(1991)

Wilczynski, Janusz S.
See Deutsch, Alina: V1389,161-176(1991)
See Spiller, Eberhard: V1343,134-144(1991)

Wildermuth, Eberhard
; Walz, B.; Wessel, K.; Schock, Wolfram: Characteristics of a compact 12 kW transverse-flow CO2 laser with rf-excitation,V1397,367-371(1991)

Wildes, Richard P.
: Qualitative three-dimensional shape from stereo,V1382,453-463(1991)

Wiley, James N.
ed.: *10th Annual Symp on Microlithography,* V1496(1991)
; Fu, Tao-Yi; Tanaka, Takashi; Takeuchi, Susumu; Aoyama, Satoshi; Miyazaki, Junji; Watakabe, Yaichiro: Phase-shift mask pattern accuracy requirements and inspection technology, V1464, 346-355(1991)

Wilk, Shalom
; Goldmunz, Menachem; Shahaf, Nachum; Klein, Yitschak; Goldman, Shmuel; Oren, Ehud: Line-of-sight alignment of a multisensor system, V1442, 140-148(1991)

Wilke, August See Panjehpour, Masoud: V1427, 307-315(1991)

Wilke, Ingrid
; Moix, Dominique; Herrmann, W.; Kneubuhl, F. K.: Submicron thin-film metal-oxide-metal infrared detectors, V1442, 2-10(1991)

Wilkerson, Gary W.
ed.: *Passive Materials for Optical Elements,* V1535(1991)

Wilkes, D. M. See Cadzow, James A.: V1486, 352-363(1991)

Wilkes, David R.
; Dudek, Gregory; Jenkin, Michael R.; Milios, Evangelos E.: Ray-following model of sonar range sensing, V1388, 536-542(1991)

Wilkes, Donald R. See Harada, Yoshiro: V1330, 90-101(1991)

Wilkes, Scott C.
: SCORPIUS: a successful image understanding implementation, V1406, 110-121(1991)

Wilkins, Belinda
; Goldgof, Dmitry B.; Bowyer, Kevin W.: Toward computing the aspect graph of deformable generalized cylinders, V1468, 662-673(1991)

Wilkins, Charles L. See Nuwaysir, Lydia M.: V1437, 112-123(1991)

Wilkins, Nathan A. See Caulfield, H. J.: V1564, 496-503(1991)

Wilkinson, Chris D. See MacLeod, Roderick W.: V1361, 562-567(1991)

Wilkinson, Timothy S.
; Goodman, Joseph W.: Slope histogram detection of forged handwritten signatures, V1384, 293-304(1991)

Wilks, Allan R.
See Becker, Richard A.: V1459, 48-56(1991)
See Becker, Richard A.: V1459, 150-154(1991)

Wilks, S. C.
; Kruer, William L.; Langdon, A. B.; Amendt, Peter; Eder, David C.; Keane, Christopher J.: Theory and simulation of Raman scattering in intense short-pulse laser-plasma interactions, V1413, 131-137(1991)
See Amendt, Peter: V1413, 59-69(1991)

Will, Ralph W.
; Rhodes, Marvin D.: Automated assembly system for large space structures, V1387, 60-71(1991)

Willand, Craig S.
See Penner, Thomas L.: V1560, 377-386(1991)
See Penner, Thomas L.: V1436, 169-178(1991)

Willemen, Patrick See Baert, Luc: V1421, 18-29(1991)

Willems, David
; Bahl, I.; Griffin, Edward: Accurate design of multiport low-noise MMICs up to 20 GHz, V1475, 55-61(1991)

Willett, Peter K. See Mahmood Reza, Syed: V1451, 298-308(1991)

Willey, Ronald R.
; Durham, Mark E.: Ways that designers and fabricators can help each other, V1354, 501-505(1991)

Willi, Oswald See Giulietti, Antonio: V1502, 270-283(1991)

Williams, Bryn See Zarnowski, Jeffrey J.: V1447, 191-201(1991)

Williams, D. S. See Leung, Dominic S.: V1471, 314-325(1991)

Williams, Dean R. See Fuchs, Elizabeth A.: V1467, 136-149(1991)

Williams, Doug L.
; Davey, Steven T.; Kashyap, Raman; Armitage, J. R.; Ainslie, B. J.: Photosensitive germanosilicate preforms and fibers, V1513, 158-167(1991)
; Davey, Steven T.; Kashyap, Raman; Armitage, J. R.; Ainslie, B. J.: UV spectroscopy of optical fibers and preforms, V1516, 29-37(1991)
See Kashyap, Raman: V1516, 164-174(1991)

Williams, E. J.
; Adams, Stephen D.: Three-dimensional vision system for peanut pod maturity, V1379, 236-245(1991)

Williams, Elmer F.
; Evans, Robert H.; Brant, Karl; Stockum, Larry A.: Optimization of a gimbal-scanned infrared seeker, V1482, 104-111(1991)
See Evans, Robert H.: V1406, 201-202(1991)

Williams, F. I. See Etienne, Bernard: V1362, 256-267(1991)

Williams, Gareth T.
; Bahuguna, Ramendra D.; Arteaga, Humberto; Le Joie, Elaine N.: Study of microbial growth I: by diffraction, V1332, 802-804(1991)
See Bahuguna, Ramendra D.: V1332, 805-807(1991)

Williams, Gary L. See Smith, Walter S.: V1434, 14-25(1991)

Williams, Geoffrey See Purvis, Alan: V1455, 145-149(1991)

Williams, Ken See Van Metre, Richard: V1369, 9-18(1991)

Williams, Mathew R.
See Auteri, Joseph S.: V1421, 182-184(1991)
See Bass, Lawrence S.: V1421, 164-168(1991)
See Libutti, Steven K.: V1421, 169-172(1991)
See Oz, Mehmet C.: V1422, 147-150(1991)

Williams, Memorie
; Hansen, Evan; Reyes-Mena, Arturo; Allred, David D.: Transmittances of thin polymer films and their suitability as a supportive substrate for a soft x-ray solar filter, V1549, 147-154(1991)

Williams, Paul F.
; Peterkin, Frank E.; Ridolfi, Tim; Buresh, L. L.; Hankla, B. J.: Surface flashover of silicon, V1378, 217-225(1991)
See Sardesai, Harshad P.: V1378, 237-248(1991)

Williams, R. H. See Morley, Stefan: V1361, 213-222(1991)

Williams, Richard A. See Zediker, Mark S.: V1418, 309-315(1991)

Williams, Rodney C. See Chimiak, William J.: V1446, 93-99(1991)

Williams, Rodney D.
; Donohoo, Daniel: Image quality metrics for volumetric laser displays, V1457, 210-220(1991)

Williams, Steven P. See Busquets, Anthony M.: V1457, 91-102(1991)

Williams, William J.
; Brown, Mark L.; Hero, Alfred O.: Uncertainty, information, and time-frequency distributions, V1566, 144-156(1991)

Williamson, David M. See Gortych, Joseph E.: V1463, 368-381(1991)

Williamson, Don L. See Gibart, Pierre: V1362, 938-950(1991)

Williamson, John W. See Sahai, Viveik: V1557, 60-70(1991)

Williamson, Ray
: Novel interferometer setup for evaluating the sum of surface contributions to transmitted wavefront distortion, V1527, 188-193(1991)
: Whenever two beams interfere, one fringe equals one wave in the plane of interference, always, V1527, 252-257(1991)

Willing, Steven L.
; Worland, Phil; Harnagel, Gary L.; Endriz, John G.: 10-watt cw diode laser bar efficiently fiber-coupled to a 381 um diameter fiber-optic connector, V1418, 358-362(1991)

Willis, Thomas D.
See DeForest, Craig E.: V1343, 404-414(1991)
See Walker, Arthur B.: V1343, 319-333(1991)

Willsch, Reinhardt
See Ecke, Wolfgang: V1511, 57-66(1991)
See Rasch, Andreas: V1511, 149-154(1991)

Willson, C. G.
See Kutal, Charles: V1466, 362-367(1991)
See MacDonald, Scott A.: V1466, 2-12(1991)
See Neher, Dieter: V1560, 335-343(1991)

Willson, Gregory B.
: Comparison of a multilayered perceptron with standard classification techniques in the presense of noise, V1469, 351-358(1991)

Wilmink, J. B. See Anema, P. C.: V1446, 352-356(1991)

Wilner, Kalman See Agassi, Eyal: V1442, 126-132(1991)

Wilson, Brian C.
See Chen, Qun: V1426, 156-161(1991)
See Farrell, Thomas J.: V1426, 146-155(1991)
See Madsen, Steen J.: V1431, 42-51(1991)
See Muller, Paul J.: V1426, 254-265(1991)

Wilson, Charles L. See Omidvar, Omid M.: V1452, 532-543(1991)

Wilson, David C. See Manseur, Zohra Z.: V1568, 164-173(1991)

Wilson, David L. See Andress, Keith M.: V1445, 6-10(1991)

Wilson, Donald K.
: Partitioning: the payoff role in telecommunication system design, V1390, 537-547(1991)
See Lalk, Gail R.: V1389, 386-400(1991)

Wilson, H. L. See Schimert, Thomas R.: V1484, 19-30(1991)

Wilson, J. M. See Humphries, Stanley: V1407, 512-523(1991)

Wilson, James N. See Brown, Neil H.: V1379,54-68(1991)

Wilson, Jeannie S.
: Thermal analysis of the bottle forming process,V1467,219-228(1991)

Wilson, Joseph N.
: Introduction to Image Algebra Ada,V1568,101-112(1991)
See Deng, Keqiang: V1568,304-312(1991)
See Deng, Keqiang: V1570,227-233(1991)
See Schmalz, Mark S.: V1474,212-234(1991)

Wilson, Kathleen See Bullock, Michael E.: V1471,291-302(1991)

Wilson, Keith E.
; Schwartz, Jon A.; Lesh, James R.: GOPEX: a deep-space optical
communications demonstration with the Galileo spacecraft,V1417,22-
28(1991)

Wilson, Leo T. See Bubb, Daniel: V1483,18-28(1991)

Wilson, LeRoy E. See Ulrich, Peter B.: V1408(1991)

Wilson, Michael A.
: Metamorphosis of laser writer,V1496,156-170(1991)

Wilson, P. B. See Farkas, Zoltan D.: V1407,502-511(1991)

Wilson, R. A. See Craig, Robert G.: V1505,76-86(1991)

Wilson, Raymond N. See Noethe, L.: V1542,293-296(1991)

Wilson, Russell S. See Read, Harold E.: V1441,345-359(1991)

Wilson, Scott R.
; Reicher, David W.; Kranenberg, C. F.; McNeil, John R.; White, Patricia L.;
Martin, Peter M.; McCready, David E.: Ion beam milling of fused silica
for window fabrication,V1441,82-86(1991)
See Reicher, David W.: V1441,106-112(1991)

Wilson, Stan
; Dozier, Jeff: Earth Observing System,V1491,117-124(1991)

Wilson, Stephen S.
: Order-statistic filters on matrices of images,V1451,242-253(1991)
: Unsupervised training of structuring elements,V1568,188-199(1991)

Wilson, Tony
: Coherent detection in confocal microscopy,V1439,104-108(1991)

Wilson, W. G. See Bradley, D. J.: V1488,186-195(1991)

Wilson, William See Chan, Kwok-Chi D.: V1552,69-78(1991)

Wilson, William L.
See Hofmann, Thomas: V1412,84-90(1991)
See Katz, Howard E.: V1560,370-376(1991)
See Tittel, Frank K.: V1397,21-29(1991)

Wiltshire, Michael C. See Purvis, Alan: V1455,145-149(1991)

Wimmers, James T. See Fischer, Robert C.: V1494,414-418(1991)

Win, Moe Z.
; Chen, Chien-Chung; Scholtz, Robert A.: Optical phase-locked loop for free-
space laser communications with heterodyne detection,V1417,42-
52(1991)
See Chen, Chien-Chung: V1417,170-181(1991)

Winakur, Eric See Spears, Kenneth G.: V1429,2-8(1991)

Winarsky, Norman
; Alexander, Joanna R.: Project DaVinci,V1459,67-68(1991)

Winch, P. J. See Earnshaw, J. C.: V1403,316-325(1991)

Windeln, Wilbert
See Beeck, Manfred-Andreas: V1507,394-406(1991)
See Stojanoff, Christo G.: V1507,426-434(1991)

Windham, William R.
; Gaines, Charles S.; Leffler, Richard G.: Moisture influence on near-infrared
prediction of wheat hardness,V1379,39-44(1991)

Windt, David L.
; Hull, Robert; Waskiewicz, Warren K.; Kortright, Jeffrey B.: Interface
characterization of XUV multilayer reflectors using HRTEM and x-ray and
XUV reflectance,V1343,297-308(1991)
; Waskiewicz, Warren K.; Kubiak, Glenn D.; Barbee, Troy W.; Watts, Richard
N.: XUV characterization comparison of Mo/Si multilayer
coatings,V1343,274-282(1991)
See Taylor, Gary N.: V1343,258-273(1991)
See White, Donald L.: V1343,204-213(1991)

Winebrenner, Dale P. See Tsang, Leung: V1558,260-268(1991)

Wineland, D. J. See Bergquist, James C.: V1435,82-85(1991)

Winer, Arthur M.
; Biermann, Heinz W.: Measurements of nitrous acid, nitrate radicals,
formaldehyde, and nitrogen dioxide for the Southern California Air
Quality Study by differential optical absorption spectroscopy,V1433,44-
55(1991)

Winfield, R. J. See Fieret, Jim: V1503,53-61(1991)

Wing, Omar See Huang, Wei-Xu: V1389,199-204(1991)

Wingard, Christopher J. See Swanson, Curtis J.: V1467,372-383(1991)

Wingo, Dennis R.
; Bankston, Cheryl D.: Complementary experiments for tether dynamics
analysis,V1495,123-133(1991)

Winkenbach, H. See Ackermann, F.: V1490,94-101(1991)

Winkle, Mark R. See Graziano, Karen A.: V1466,75-88(1991)

Winkler, Carl E.
; Dailey, Carroll C.; Cumings, Nesbitt P.: Advanced X-ray Astrophysics
Facility science instruments,V1494,301-313(1991)

Winkler, Richard See Mount, George H.: V1491,188-193(1991)

Winograd, Nicholas
; Hrubowchak, D. M.; Ervin, M. H.; Wood, M. C.: Multiphoton resonance
ionization of molecules desorbed from surfaces by ion beams,V1435,2-
11(1991)

Winston, Roland
: Nonimaging optics: optical design at the thermodynamic limit,V1528,2-
6(1991)
; Holman, Robert L.; eds.: *Nonimaging Optics: Maximum Efficiency Light
Transfer,*V1528(1991)
See Jacobson, Benjamin A.: V1528,82-85(1991)
See Lacovara, Phil: V1528,135-141(1991)
See Ries, Harald: V1528,7-14(1991)

Winter, Edwin M. See Stocker, Alan D.: V1481,156-169(1991)

Winterberg, F.
: Continuous inertial confinement fusion for the generation of very intense
plasma jets,V1407,322-325(1991)

Winterhalter, L. See Bays, Roland: V1525,397-408(1991)

Wirth, Allan See Bruno, Theresa L.: V1358,109-116(1991)

Wischow, Perry B. See Trenchard, Michael E.: V1456,318-326(1991)

Wise, M. L. See George, Steven M.: V1437,157-165(1991)

Wise, Timothy D.
; Wieder, Eckart: Latest advances in CAD data interfacing: a standardization
project of ISO/TC 172/SC1 task group "optical database",V1346,79-
85(1991)

Wishnow, Kenneth I. See Johnson, Douglas E.: V1421,36-41(1991)

Wisoff, P. J.
See Hofmann, Thomas: V1412,84-90(1991)
See Tittel, Frank K.: V1397,21-29(1991)

Wisotsky, Steve See Hadfield, Michael J.: V1478,126-144(1991)

Witanachchi, S.
; Lee, S. Y.; Song, L. W.; Kao, Yi-Han; Shaw, David T.: Critical current
enhancement in Y1Ba2Cu3O7-y/Y1Ba2(Cu1-xNix)3O7-y
heterostructures,V1394,161-168(1991)

Witherow, William K.
: Measuring residual accelerations in the Spacelab environment,V1557,42-
52(1991)
See Frazier, Donald O.: V1557,86-97(1991)
See Trolinger, James D.: V1557,250-258(1991)
See Vikram, Chandra S.: V1557,197-201(1991)

Witkin, Andrew See Nederlof, Michel A.: V1428,233-241(1991)

Witman, Sandy L. See Booth, Bruce L.: V1377,57-63(1991)

Witmer, Steve B. See Kanofsky, Alvin S.: V1374,48-58(1991)

Witt, August F. See Carlson, D. J.: V1557,140-146(1991)

Witte, G. See Graue, Roland: V1494,377-385(1991)

Wittekoek, Stefan See van den Brink, Martin A.: V1463,709-724(1991)

Wittels, Norman
See Bian, Buming: V1453,333-340(1991)
See Harding, Kevin G.: V1385(1991)

Witteman, Wilhelmus J.
; Ekelmans, G. B.; Trentelman, M.; van Goor, Frederik A.: Discharge
technology for excimer lasers of high-average power,V1397,37-45(1991)
See Bastiaens, H. M.: V1397,77-80(1991)
See Botma, H.: V1397,573-576(1991)
See Ekelmans, G. B.: V1397,569-572(1991)
See Trentelman, M.: V1397,115-118(1991)
See van Goor, Frederik A.: V1412,91-102(1991)

Wittig, Manfred E.
: Effect of microaccelerations on an optical space communication
system,V1522,278-286(1991)
See Oppenhaeuser, Gotthard: V1522,2-13(1991)

Witting, Harald L.
: Test results on pulsed cesium amalgam flashlamps for solid state laser pumping,V1410,90-98(1991)

Wittman, David See Wizinowich, Peter L.: V1542,148-158(1991)

Wittmeyer, Stacey A.
; Kaziska, Andrew J.; Shchuka, Maria I.; Topp, Michael R.: Hole-burning and picosecond time-resolved spectroscopy of isolated molecular clusters,V1435,267-278(1991)

Wittwer, Timothy Y. See Gates, James L.: V1540,262-273(1991)

Wizinowich, Peter L.
; Lloyd-Hart, Michael; McLeod, Brian A.; Colucci, D'nardo; Dekany, Richard G.; Wittman, David; Angel, J. R.; McCarthy, Donald W.; Hulburd, William G.; Sandler, David G.: Neural network adaptive optics for the multiple-mirror telescope,V1542,148-158(1991)

Wlodarczyk, Marek T.
: Environmentally insensitive commercial pressure sensor,V1368,121-131(1991)
See Lieberman, Robert A.: V1368(1991)

Wlodawski, Mitchell
; Nowakowski, Jerzy: Holographic image reconstruction from interferograms of laser-illuminated complex targets,V1416,241-249(1991)
See Nowakowski, Jerzy: V1416,229-240(1991)

Wnek, Gary E.
See LaPeruta, Richard: V1497,357-366(1991)
See McDonald, John F.: V1390,286-301(1991)

Wodkiewicz, K. See Bochove, Erik J.: V1497,338-347(1991)

Woehrle, D. See Kohl, M.: V1525,26-34(1991)

Woerner, K.-H. See Fabian, Heinz: V1513,168-173(1991)

Woerner, M. C. See Woolsey, G. A.: V1584,243-253(1991)

Woggon, Ulrike
; Rueckmann, I.; Kornack, J.; Mueller, Matthias; Cesnulevicius, J.; Kolenda, Jonas; Petrauskas, Mendogas: Growth, surface passivation, and characterization of CdSe microcrystallites in glass with respect to their application in nonlinear optics,V1362,888-898(1991)
See Rueckmann, I.: V1513,78-85(1991)

Wohlers, Martin R.
: Bounds on the performance of optimal four-dimensional filters for detection of low-contrast IR point targets,V1481,129-139(1991)
See Iannarilli, Frank J.: V1486,314-324(1991)
See Iannarilli, Frank J.: V1483,66-76(1991)
See Iannarilli, Frank J.: V1488,226-236(1991)
See Iannarilli, Frank J.: V1481,187-197(1991)

Woisetschlaeger, Jakob
See Neger, Theo: V1507,476-487(1991)
See Vukicevic, Dalibor: VIS08,160-193(1991)

Wojcik, Gregory L.
; Mould, John; Monteverde, Robert J.; Prochazka, Jaroslav J.; Frank, John R.: Numerical simulation of thick-linewidth measurements by reflected light,V1464,187-203(1991)
; Vaughan, David K.; Mould, John; Leon, Francisco A.; Qian, Qi-de; Lutz, Michael A.: Laser alignment modeling using rigorous numerical simulations,V1463,292-303(1991)

Wojcik, Waldemar
; Smolarz, Andrzej: Design of an optical fiber power supplying link,V1504,292-297(1991)

Wojiak, Joanna See Kujawinska, Malgorzata: V1554B,503-513(1991)

Wojtatowicz, Tomasz W. See Drobnik, Antoni: V1391,361-369(1991)

Wokaun, Alexander J. See Frank, Klaus: V1431,2-11(1991)

Wolbarsht, Myron L.
See Schwarzmaier, Hans-Joachim: V1427,128-133(1991)
See Shi, Dexiu: V1419,40-49(1991)

Wolber, John W.
: Measuring the geometric accuracy of a very accurate, large, drum-type scanner,V1448,175-180(1991)

Wolczak, Bohdan K.
: About contradictions among different criteria for evaluation of image (interference) elements width changes,V1391,318-328(1991)
See Wolinski, Wieslaw: V1391(1991)

Wolf, Barbara
; Fabricius, Norbert; Foss, Wolfgang; Dorsel, Andreas N.: Ion-exchanged waveguides in glass: simulation and experiment,V1506,40-51(1991)

Wolf, Edward D. See Lang, Robert J.: V1563,2-7(1991)

Wolf, Stuart A. See Nisenoff, Martin: V1394,104-113(1991)

Wolfbeis, Otto S.
ed.: *Chemical and Medical Sensors,*V1510(1991)
: Feasibility of optically sensing two parameters simultaneously using one indicator,V1368,218-222(1991)
See He, Huarui: V1368,175-180(1991)
See He, Huarui: V1510,95-103(1991)
See He, Huarui: V1368,165-171(1991)
See Moreno-Bondi, Maria C.: V1368,157-164(1991)

Wolfe, C. R. See Campbell, John H.: V1441,444-456(1991)

Wolfe, Ronald
; Laor, Herzel: Fiber optics in CATV networks,V1363,125-132(1991)

Wolfe, William J.
; Magee, Michael: Fusion of multiple views of multiple reference points using a parallel distributed processing approach,V1383,20-25(1991)
; Mathis, Donald W.; Anderson, C.; Rothman, Jay; Gottler, Michael; Brady, G.; Walker, R.; Duane, G.; Alaghband, Gita: Family of K-winner networks,V1382,240-254(1991)
See Chun, Wendell H.: V1388(1991)

Wolfe, William L.
: Optical materials for the infrared,VCR38,55-68(1991)
: Status and needs of infrared optical property information for optical designers,V1354,696-741(1991)
See Lundgren, Mark A.: V1354,533-539(1991)

Wolffer, Nicole See Abdelghani-Idrissi, Ahmed M.: V1362,417-427(1991)

Wolfrum, Juergen M.
See Hitzler, Hermine: V1503,355-362(1991)
See Seidel, Claus: V1432,105-116(1991)

Wolfson, Ronald I. See Smalanskas, Joseph P.: V1489,2-8(1991)

Wolinski, Tomasz R.
; Bock, Wojtek J.: Fiber optic liquid crystal high-pressure sensor,V1511,281-288(1991)
See Bock, Wojtek J.: V1511,250-254(1991)
See Bock, Wojtek J.: V1584,157-161(1991)

Wolinski, Wieslaw
; Kazmirowski, Antoni; Kesik, Jerzy; Korobowicz, Witold; Spytkowski, Wojciech: Argon dye photocoagulator for microsurgery of the interior structure of the eye,V1391,334-340(1991)
; Wolczak, Bohdan K.; Gajda, Jerzy K.; Gajda, Danuta; Romaniuk, Ryszard S.; eds.: *Laser Technology III,*V1391(1991)
; Wolski, Radoslaw; Janulewicz, K. A.: New construction of sealed-off CO2 laser,V1397,391-393(1991)
See Iwanejko, Leszek: V1391,98-100(1991)
See Szczepanski, Pawel: V1391,72-78(1991)
See Szczepanski, Pawel: V1391,65-71(1991)

Wollman, Yoram See Lubart, Rachel: V1422,140-146(1991)

Wollmann, Daphne
; Diaz, Art F.: Importance of proton transfer in contact charging,V1458,192-200(1991)

Wolski, Radoslaw See Wolinski, Wieslaw: V1397,391-393(1991)

Wolter, Joachim H. See Hendriks, Peter: V1362,217-227(1991)

Womack, G. See Blouke, Morley M.: V1447,142-155(1991)

Wombell, Richard J. See DeSanto, John A.: V1558,202-212(1991)

Wong, Albert W.
; Stewart, Brent K.; Lou, Shyh-Liang; Chan, Kelby K.; Huang, H. K.: Multiple communication networks for a radiological PACS,V1446,73-80(1991)
See Stewart, Brent K.: V1446,141-153(1991)

Wong, Alfred K.
; Doi, Takeshi; Dunn, Diana D.; Neureuther, Andrew R.: Experimental and simulation studies of alignment marks,V1463,315-323(1991)
See Chiu, Anita S.: V1466,641-652(1991)

Wong, Andrew K. See Chan, Keith C.: V1469,359-372(1991)

Wong, Brian
; Kitai, Adrian H.; Jessop, Paul E.: Pockels' effect in polycrystalline ZnS planar waveguides,V1398,261-268(1991)

Wong, C. M. See Asundi, Anand K.: V1400,80-85(1991)

Wong, D. M. See Harris, Neville W.: V1416,86-99(1991)

Wong, Edison See Forman, Scott K.: V1424,2-6(1991)

Wong, George K. See Allan, D. S.: V1560,362-369(1991)

Wong, H.
; Cheng, Y. C.; Yang, Bing L.; Liu, Bai Y.: Depopulation kinetics of electron traps in thin oxynitride films,V1519,494-498(1991)
See Yang, Bing L.: V1519,241-246(1991)
See Yang, Bing L.: V1519,269-274(1991)

Wong, H. L. See Tam, Siu-Chung: V1400,38-48(1991)

Wong, K. See Krongelb, Sol: V1389,249-256(1991)

Wong, K. S. See Hebden, Jeremy C.: V1443,294-300(1991)

Wong, Ka-Kha
ed.: *Integrated Optical Circuits,*V1583(1991)
ed.: *Integrated Optics and Optoelectronics II,*V1374(1991)
; Killian, Kevin M.; Dimitrov-Kuhl, K. P.; Long, Margaret; Fleming, J. T.; van de Vaart, Herman: Proton-exchange X-cut lithium tantalate fiber optic gyro chips,V1374,278-286(1991)
See Emo, Stephen M.: V1374,266-276(1991)
See Padden, Richard J.: V1474,148-159(1991)

Wong, Kam W.
; Ke, Ying; Lew, Michael; Obaidat, Mohammed T.: Three-dimensional gauging with stereo computer vision,V1526,17-26(1991)

Wong, Kon M. See Chen, Wei G.: V1566,464-475(1991)

Wong, Kwok-keung See Goldner, Ronald B.: V1536,63-69(1991)

Wong, Kwok Y.
; Torok, Georgia R.; Chang, Win-Chyi; Meisenzahl, Eric J.: 1536 x 1024 CCD image sensor,V1447,283-287(1991)

Wong, Ping W.
; Koplowitz, Jack: High-resolution reconstruction of line drawings,V1398,39-47(1991)

Wong, Tony See Smith, Walter S.: V1434,14-25(1991)

Wong, Vincent V.
; Swanson, Gary J.: Binary optic interconnects: design, fabrication and limits on implementation,V1544,123-133(1991)

Wong, Wallace K.
; Wang, Dexter; Benoit, Robert T.; Barthol, Peter: Comparison of low-scatter-mirror PSD derived from multiple-wavelength BRDFs and WYKO profilometer data,V1530,86-103(1991)
See Wang, Dexter: V1479,57-70(1991)

Wong, Wing H. See Chen, Chin-Tu: V1445,222-225(1991)

Wong, Woon-Yin See Forbes, Fred F.: V1532,146-160(1991)

Wong, Y. W. See Chan, W. K.: V1399,82-89(1991)

Wood, Andrew C. See Abram, Richard A.: V1361,424-433(1991)

Wood, Andrew P.
: Using hybrid refractive-diffractive elements in infrared Petzval objectives,V1354,316-322(1991)

Wood, Bobby E. See Young, Raymond P.: V1530,335-342(1991)

Wood, C. C. See George, John S.: V1443,37-51(1991)

Wood, Christopher M.
; Shaw, Michael M.; Harvey, David M.; Hobson, Clifford A.; Lalor, Michael J.; Atkinson, John T.: Absolute range measurement system for real-time 3-D vision,V1332,301-313(1991)
See Harvey, David M.: V1400,86-93(1991)

Wood, Craig P.
; Trolinger, James D.: Application of real-time holographic interferometry in the nondestructive inspection of electronic parts and assemblies,V1332,122-131(1991)

Wood, David
See Barnes, Nigel: V1389,477-483(1991)
See McKee, Paul: V1461,17-23(1991)

Wood, Hugh C.
; Vaidyanathan, C. S.: Sensor-based identification of control parameters for intelligent gripping,V1387,245-254(1991)
See Brown, Neil H.: V1379,54-68(1991)

Wood, M. C. See Winograd, Nicholas: V1435,2-11(1991)

Wood, Obert R. See White, Donald L.: V1343,204-213(1991)

Wood, Richard F.
See Mitchell, David C.: V1427,181-188(1991)
See Smith, Roy E.: V1425,116-121(1991)

Wood, Robert L. See Conley, Willard E.: V1466,53-66(1991)

Wood, Roger M.
; Greenham, A. C.; Nichols, B. A.; Nourshargh, Noorallah; Lewis, Keith L.: Fabrication and characterization of microwave-plasma-assisted chemical vapor deposited dielectric coatings,V1441,316-326(1991)

Wood, Samuel C.
; Apte, Pushkar P.; King, Tsu-Jae; Moslehi, Mehrdad M.; Saraswat, Krishna C.: Pyrometer modeling for rapid thermal processing,V1393,337-348(1991)
; Saraswat, Krishna C.; Harrison, J. M.: Economic impact of single-wafer multiprocessors,V1393,36-48(1991)

Woodall, Milton A. See Keeter, Howard S.: V1419,84-93(1991)

Woodell, G. W. See Fripp, Archibald L.: V1557,236-244(1991)

Woodford, Paul See Casasent, David P.: V1563,112-119(1991)

Woodgate, Bruce E. See Delamere, W. A.: V1447,288-297(1991); 1479,21-30(1991)

Woodruff, K. M. See Olsen, Gregory H.: V1419,24-31(1991)

Woodruff, William H.
; Dyer, R. B.; Einarsdottir, Oloef; Peterson, Kristen A.; Stoutland, Page O.; Bagley, K. A.; Palmer, Graham; Schoonover, J. R.; Kliger, David S.; Goldbeck, Robert A.; Dawes, T. D.; Martin, Jean-Louis; Lambry, J.-C.; Atherton, Stephen J.; Hubig, Stefan M.: Ultrafast and not-so-fast dynamics of cytochrome oxidase: the ligand shuttle and its possible functional significance,V1432,205-210(1991)
See Berg, John M.: V1435,331-337(1991)
See Dyer, R. B.: V1432,197-204(1991)

Woods, Andrew J.
; Docherty, Tom; Koch, Rolf: Use of flicker-free television products for stereoscopic display applications,V1457,322-326(1991)

Woods, Bruce W.
; Thomas, Ian M.; Henesian, Mark A.; Dixit, Sham N.; Powell, Howard T.: Large-aperture (80-cm diameter) phase plates for beam smoothing on Nova,V1410,47-54(1991)
See De Yoreo, James J.: V1561,50-58(1991)

Woods, Charles L. See Kane, Jonathan S.: V1564,511-520(1991)

Woods, John W.
; Han, Soo-Chul: Hierarchical motion-compensated deinterlacing,V1605,805-810(1991)

Woods, R. J. See Feit, Zeev: V1512,164-169(1991)

Woodward, J. R. See Gole, James L.: V1397,125-135(1991)

Woody, Loren M. See Elerding, George T.: V1479,380-392(1991)

Woolf, D. See Morley, Stefan: V1361,213-222(1991)

Woollam, John A. See He, Ping: V1499,401-411(1991)

Woolsey, G. A.
; Lamb, D. W.; Woerner, M. C.: Optical fiber sensing of corona discharges,V1584,243-253(1991)

Worland, Phil See Willing, Steven L.: V1418,358-362(1991)

Worrell, Steven W.
; Robertson, James A.; Varner, Thomas L.; Garvin, Charles G.: Prototype neural network pattern recognition testbed,V1382,219-227(1991)

Worsham, A. See Mears, Carl A.: V1477,221-233(1991)

Worth, Andrew J. See Lehar, Steve M.: V1469,50-62(1991)

Wortman, Jim J.
See Grider, Douglas T.: V1393,229-239(1991)
See Ozturk, Mehmet C.: V1393,260-269(1991)

Woychik, Charls G.
; Guo, Yi F.: Thermal strain measurements of solder joints in electronic packaging using moire interferometry,V1554B,461-470(1991)

Wright, Dan See Normile, James: V1452,480-484(1991)

Wright, David L. See Guggenhiemer, Steven: V1414,12-20(1991)

Wright, Michael J. See Kline-Schoder, Robert J.: V1489,189-200(1991)

Wright, S. W. See Crossland, William A.: V1455,264-273(1991)

Wright, William H.
; Sonek, Gregory J.; Numajiri, Yasuyuki; Berns, Michael W.: Measurement of light scattering from cells using an inverted infrared optical trap,V1427,279-287(1991)

Writer, Dean See Maltabes, John G.: V1463,326-335(1991)

Wtodkiewiez, A. See Giannini, Jean-Pierre: V1366,215-222(1991)

Wu, A. Y. See Zou, Lian C.: V1519,707-711(1991)

Wu, Bei X.
See Wang, Ren: V1519,146-151(1991)
See Zhang, Yun H.: V1519,729-734(1991)

Wu, Chan-Shin
See Lan, Guey-Liu: V1475,184-192(1991)
See Watkins, E. T.: V1475,62-72(1991)

Wu, Chao-Chia C. See Sun, C. K.: V1476,294-300(1991)

Wu, Chao-Wen
; Lin, Hao-Hsiung: Current transport in charge injection devices,V1362,768-777(1991)

Wu, Chengjiu
See Beeson, Karl W.: V1374,176-185(1991)
See McFarland, Michael J.: V1583,344-354(1991)

Wu, Chengke
; Mohr, Roger: Image representation by integrating curvatures and Delaunay triangulations,V1570,362-370(1991)

Wu, Chiung S. See Efron, Uzi: V1455,237-247(1991)

Wu, Chwan-Hwa
; Roland, David A.: Dynamically reconfigurable multiprocessor system for high-order-bidirectional-associative-memory-based image recognition,V1471,210-221(1991)

Wu, Doris I.
; Chang, David C.: Accurate and efficient simulation of MMIC layouts,V1475,140-150(1991)

Wu, Fang D.
See Nie, Chao-Jiang: V1584,87-93(1991)
See Nie, Chao-Jiang: V1579,264-267(1991)

Wu, Fang F.
; Su, Kai L.: AR layer properties for high-power laser prepared by neutral-solution processing,V1519,347-349(1991)

Wu, Frederick Y. See Rodriguez, Arturo A.: V1332,25-35(1991)

Wu, Gengsheng
; Yang, Fan; Li, Wen; Liao, Yan-Biao: Mixed-type optical fiber current sensor,V1572,497-502(1991)
See Liao, Yan-Biao: V1584,400-404(1991)

Wu, Guang M.
; Qian, Zheng X.: Optical properties of granular Sn films with coating Al,V1519,315-320(1991)

Wu, Guobing See Jiang, Xuping: V1358,1237-1244(1991)

Wu, Hsiang-Lung
: Weighted least-squared error method for object localization,V1384,90-99(1991)

Wu, Hsien-Huang
; Schowengerdt, Robert A.: Shape discrimination using invariant Fourier representation and a neural network classifier,V1569,147-154(1991)

Wu, Jan X.
See Cui, Jing B.: V1519,419-422(1991)
See Zhang, Wei P.: V1519,23-25(1991)

Wu, Jeong W.
; Valley, John F.; Stiller, Marc A.; Ermer, Susan P.; Binkley, E. S.; Kenney, John T.; Lipscomb, George F.; Lytel, Richard S.: Poled polyimides as thermally stable electro-optic polymer,V1560,196-205(1991)

Wu, Ji-Zong
; Deng, Jia-cheng; Chen, Ben-zhi: Study on Hadamard transform imaging spectroscopy,V1399,122-129(1991)

Wu, Jingshown
See Lee, Yang-Hang: V1372,140-149(1991)
See Lee, Yang-Hang: V1579,155-166(1991)
See Wu, Jyh-Horng: V1579,195-209(1991)
See Yang, Shien-Chi: V1372,128-139(1991)

Wu, Jiunn-Shyong See Leung, Chung-yee: V1572,566-571(1991)

Wu, Jyh-Horng
; Wu, Jingshown: Performance of Reed-Solomon codes in mulichannel CPFSK coherent optical communications,V1579,195-209(1991)

Wu, Ke Q. See Zhou, Shi P.: V1519,793-799(1991)

Wu, Kuang-Yi
; Weverka, Robert T.; Wagner, Kelvin H.; Garvin, Charles G.; Roth, Richard S.: Novel acousto-optic photonic switch,V1563,168-175(1991)

Wu, Lin
; Zhao, Jinglun; Zhang, Peng; Li, Shiqing: Research state-of-the-art of mobile robots in China,V1388,598-601(1991)

Wu, Mei See Grogan, Timothy A.: V1453,16-30(1991)

Wu, Ming-da See Wei, Yan-nian: V1358,457-460(1991)

Wu, Ping
: Study of p-ZnTe/n-CdTe thin film heterojunction,V1519,477-480(1991)

Wu, Rui-Lan See Wang, Ji-Zhong: V1554B,188-192(1991)

Wu, Ruiqi
: Research for mechanical properties of rock and clay by laser speckle photography,V1554A,690-695(1991)

Wu, Ru J. See Zhou, Bing: V1519,454-456(1991)

Wu, Samuel C. See Brossia, Charles E.: V1368,115-120(1991)

Wu, Sheng-li
See Zhang, Yue-qing: V1418,444-447(1991)
See Zhang, Yue-qing: V1400,137-143(1991)

Wu, Shixiong See Bao, Liangbi: V1332,862-867(1991)

Wu, Shudong
; Yin, Shizhuo; Rajan, Sumati; Yu, Francis T.: Multiple-channel sensing with fiber specklegrams,V1584,415-424(1991)

Wu, Siu W.
; Gersho, Allen: Enhancement of transform coding by nonlinear interpolation,V1605,487-498(1991)
; Gersho, Allen: HDTV compression with vector quantization of transform coefficients,V1605,73-84(1991)
; Gersho, Allen: Image compression for digital video tape recording with high-speed playback capability,V1452,352-363(1991)

Wu, T. S.
; Su, Yan K.; Juang, F. S.; Li, N. Y.; Gan, K. J.: Effects of TMSb/TEGa ratios on epilayer properties of gallium antimonide grown by low-pressure MOCVD,V1361,23-33(1991)
; Su, Yan K.; Juang, F. S.; Li, N. Y.; Gan, K. J.: Ohmic and Schottky contacts to GaSb,V1519,263-268(1991)
See Hsu, C. T.: V1519,391-395(1991)

Wu, Tai T.
See Fikioris, George: V1407,295-305(1991)
See Shen, Hao-Ming: V1407,286-294(1991)

Wu, Way-Chen See Yeh, Long-Ching: V1527,361-367(1991)

Wu, Wei See Kemp, Kevin G.: V1464,260-266(1991)

Wu, Wen-Rong
; Kundu, Amlan: New type of modified trimmed mean filter,V1451,13-23(1991)

Wu, Wenming See Li, Guozhu: V1358,1120-1122(1991)

Wu, Wennie H. See Bergman, Larry A.: V1364,14-21(1991)

Wu, Xiang
; Yao, He S.; Feng, Wei G.: Scaling properties of optical reflectance from quasiperiodic superlattices,V1519,625-631(1991)

Wu, Xiangchen
; Schwarzmann, Peter: Nonlinear approach to the 3-D reconstruction of microscopic objects,V1450,278-285(1991)

Wu, Xiao-Ping
See Dobbins, B. N.: V1554A,772-780(1991)
See Han, Lei: V1358,793-803(1991)
See Xu, Boqin: V1554A,789-799(1991)

Wu, Xiaoqing
; Zhu, Zhaoda: Application of neural networks to range-Doppler imaging,V1569,484-490(1991)

Wu, Xin D.
; Foltyn, Stephen R.; Muenchausen, Ross E.; Dye, Robert C.; Cooke, D. W.; Rollett, A. D.; Garcia, A. R.; Nogar, Nicholas S.; Pique, A.; Edwards, R.: Growth of high-Tc superconducting thin films for microwave applications,V1477,8-14(1991)
See Muenchausen, Ross E.: V1394,221-229(1991)

Wu, Xindong
: KEShell: a "rule skeleton + rule body" -based knowledge engineering shell*,V1468,632-639(1991)

Wu, Y. See Litvinenko, Vladimir N.: V1552,2-6(1991)

Wu, Yuehua See Yang, Jie: V1358,672-676(1991)

Wu, Yuzheng See Giger, Maryellen L.: V1445,101-103(1991)

Wu, Z. Q. See Guo, S. P.: V1519,400-404(1991)

Wu, Zhao P.
; Chen, Ru G.; Wang, Yonglin: New deposition system for the preparation of doped a-Si:H,V1519,194-198(1991)

Wu, Zhen-Chun See Shi, Yi-Wei: V1572,308-312(1991)

Wu, Zhensen
; Cheng, Denghui: IR laser-light backscattering by an arbitrarily shaped dielectric object with rough surface,V1558,251-257(1991)

Wu, Zhenyu
; Leahy, Richard M.: Graph theoretic approach to segmentation of MR images,V1450,120-132(1991)

Wu, Zhongming See Guo, Maolin: V1554A,657-658(1991)

Wu, Zhouling
; Reichling, M.; Fan, Zheng X.; Wang, Zhi-Jiang: Applications of pulsed photothermal deflection technique in the study of laser-induced damage in optical coatings,V1441,214-227(1991)
; Reichling, M.; Fan, Zheng X.; Wang, Zhi-Jiang: Understanding of the abnormal wavelength effect of overcoats,V1441,200-213(1991)

Wu, Zi L.
; Wei, M. Z.; Chen, Y. X.; Ren, Cong X.; Zhang, J. H.: Single crystallinity and oxygen diffusion in high-quality YBa2Cu3O7-delta films,V1519,618-624(1991)

Wu, Zou L. See Fan, Zheng X.: V1519,359-364(1991)

Wullert, John R. See Paek, Eung Gi: V1621,340-350(1991)

Wung, C. J. See Zhang, Yue: V1560,264-271(1991)

Wunsch, Donald C. See Caudell, Thomas P.: V1469,612-621(1991)

Wurtele, Jonathan S.
See Chen, Chiping: V1407,183-191(1991)
See Sharp, William M.: V1407,535-545(1991)

Wurzbach, Richard N.
: Thermographic monitoring of lubricated couplings,V1467,41-46(1991)

Wutke, John R.
: Automated visual imaging interface for the plant floor,V1386,180-184(1991)

Wyant, James C. See Creath, Katherine: V1332,2-7(1991)

Wyatt, John L. See Hakkarainen, J. M.: V1473,173-184(1991)

Wyatt, Richard See Massicott, Jennifer F.: V1373,93-102(1991)

Wybo, David R. See Kenue, Surender K.: V1468,538-550(1991)

Wyntjes, Geert J.
; Hercher, Michael: Wafer alignment based on existing microstructures,V1464,539-545(1991)
See Hercher, Michael: V1332,602-612(1991)
See Hercher, Michael: V1454,230-234(1991)

Wyrowski, Frank
: Digital holography as a useful model in diffractive optics,V1507,128-135(1991)
; Bernhardt, Michael: Marriage between digital holography and optical pattern recognition,V1555,146-153(1991)
See Weissbach, Severin: V1507,149-152(1991)

Wysocki, Paul F.
; Fesler, Kenneth A.; Liu, K.; Digonnet, Michel J.; Kim, Byoung-Yoon: Spectrum thermal stability of Nd- and Er-doped fiber sources,V1373,234-245(1991)
; Kalman, Robert F.; Digonnet, Michel J.; Kim, Byoung-Yoon: 1.55-um broadband fiber sources pumped near 980 nm,V1373,66-77(1991)
See Kalman, Robert F.: V1373,209-222(1991)

Xi, Xiao-chun
See Liu, Rui-Fu: V1572,180-184(1991)
See Liu, Rui-Fu: V1572,399-402(1991)
See Liu, Rui-Fu: V1572,189-191(1991)

Xi, Xiaoxing
; Venkatesan, T.; Etemad, Shahab; Hemmick, D.; Li, Q.: Anomalous optical response of YBa2Cu3O7-x thin films during superconducting transitions,V1477,20-25(1991)

Xia, Jia-zhi See Van Laethem, Marc: V1403,732-742(1991)

Xia, P. See Blair, William D.: V1482,234-245(1991)

Xia, Sheng-jie
; Yang, Ye-min; Tang, Di-zhu: Laser optical fiber high-speed camera,V1358,43-45(1991)

Xia, Yu-Xing See Liu, Wen-Qing: V1530,240-243(1991)

Xia, Zhong F.
; Jiang, Jian: Corona charged polychlorotrifluorotylene film electrets, and its charge storage and transport,V1519,866-871(1991)

Xian, Wu
See Wang, He-Chen: V1386,273-276(1991)
See Wang, He-Chen: V1384,133-136(1991)
See Zhang, Yu: V1384,60-65(1991)

Xiang, Caixin
; Xiang, Yang: On-line rapid testing of the optical transfer function,V1527,427-436(1991)

Xiang, Feng
; Yip, Gar L.: Simple and accurate technique to determine substrate indices by the multilayer Brewster angle measurement,V1583,271-277(1991)
See Yip, Gar L.: V1583,14-18(1991)

Xiang, Jin Z.
; Zheng, Zhi H.; Liao, Chang G.; Xiong, Jing; Wang, Yong Q.; Zhang, Fang Q.: Ion implantation of diamond-like carbon films,V1519,683-687(1991)

Xiang, Tingyuan
; Zheng, Yingna; Huang, Nanmin; Fan, Xinrui: Study on optical fibre sensor for on-line correlation velocity measurement,V1572,372-376(1991)

Xiang, Xiao L.
; Hou, Li S.; Gan, Fuxi: Sol-gel derived BaTiO3 thin films,V1519,712-716(1991)

Xiang, Yang See Xiang, Caixin: V1527,427-436(1991)

Xiao, Guohua
; Song, Ruhua H.; Hu, Zhiping; Le, Shixiao: Optimum design of optical array used as pseudoconjugator,V1409,106-113(1991)

Xiao, Guoqing See Chou, Ching-Hua: V1464,145-154(1991)

Xiao, Ji-Quang See Hong, Xiao-Yin: V1466,546-557(1991)

Xiao, Jing
: Recognition of contacts between objects in the presence of uncertainties,V1470,134-145(1991)

Xiao, Tongsan D. See Gohberg, I.: V1566,14-22(1991)

Xiao, Wen
; Chen, Yaosheng; Gao, Wei; Xue, Mingqiu: Fiber optic sensor applied to measure high temperature under high-pressure condition,V1572,170-174(1991)
See Chen, Yaosheng: V1572,124-128(1991)
See Luo, Nan: V1572,2-4(1991)

Xiao, Xiaodong
; Goel, Kamal K.; Sturm, James C.; Schwartz, P. V.: Data transmission at 1.3 um using silicon spatial light modulator,V1476,301-304(1991)

Xie, Gong-Wie
; Hua, Lifan; Patel, Sushil; Scott, Peter D.; Shaw, David T.: Using dynamic holography for iron fibers with submicron diameter and high velocity,V1385,132-141(1991)
; Scott, Peter D.; Shaw, David T.; Zhang, Yi-Mo: New optical technique for particle sizing and velocimetry,V1500,310-321(1991)
See Hua, Lifan: V1385,142-151(1991)

Xie, Haiming
See Wang, An: V1572,365-369(1991)
See Wang, An: V1572,444-449(1991)
See Wang, An: V1572,440-443(1991)

Xie, Jianping See Ming, Hai: V1572,27-31(1991)

Xie, Kan
; Van Eycken, Luc; Oosterlinck, Andre J.: Hierarchical motion-compensated interframe DPCM algorithm for low-bit-rate coding,V1567,380-389(1991)

Xie, Qin X.
See Liu, Wei J.: V1519,415-418(1991)
See Liu, Wei J.: V1519,481-488(1991)

Xie, Qi Y. See Sun, Yue Z.: V1519,234-240(1991)

Xie, Shan See Zhang, Wei P.: V1519,23-25(1991)

Xie, Tonglin
: Fused 3 x 3 single-mode fiber-optic couplers with stable characteristics for fiber interferometric sensors,V1572,132-136(1991)

Xie, Tuqiang
See Cheng, Xianping: V1572,216-219(1991)
See Liu, Bo: V1572,211-215(1991)

Xie, X. B. See Zhuang, Qi: V1397,157-160(1991)

Xie, Xiaoliang See Simon, John D.: V1432,211-220(1991)

Xie, Xiou-Qioun See Hao, Tianyou: V1584,32-38(1991)

Xie, Xuanli
; Beni, Gerardo: Validity criterion for compact and separate fuzzy partitions and its justification,V1381,401-410(1991)

Xing, Xiaozheng See Shi, Dexiu: V1419,40-49(1991)

Xing, Zhongjin See Zheng, Tian S.: V1519,339-346(1991)

Xiong, F. K.
; Zhu, Long D.; Wang, C. W.: Effect of parameter variation on the performance of InGaAsP/InP multiple-quantum-well electroabsorption/electrorefraction modulators,V1519,665-669(1991)

Xiong, Guiguang See Li, Yuchuan: V1572,382-385(1991)

Xiong, Guilan See Huang, Zong T.: V1519,788-789(1991)

Xiong, Jing See Xiang, Jin Z.: V1519,683-687(1991)

Xiong, Xian M. See Ge, Fang X.: V1554B,785-789(1991)

Xiong, Xiaoxiong
; Moore, Larry J.; Fassett, John D.; O'Haver, Thomas C.: Biomedical applications of laser photoionization,V1435,188-196(1991)

Xiong, Z. See Efstratiadis, Serafim N.: V1605,16-25(1991)

Xiong, Zhengjun
; Cheng, Xuezhong; Liu, Xiande: Research on enhancing signal and SNR in laser/IR inspection of solder joints quality,V1467,410-415(1991)

Xu, Boqin
; Wu, Xiao-Ping: Fringe formation in speckle shearing interferometry,V1554A,789-799(1991)

Xu, Bu Y. See Zhang, Su Y.: V1519,508-513(1991)

Xu, Chenglin See Huang, Weiping W.: V1583,268-270(1991)

Xu, Chun F. See Wang, Ji S.: V1519,857-859(1991)

Xu, Guang Z. See Zhong, Di S.: V1519,350-358(1991)

Xu, Hui J. See Zhou, Jian P.: V1519,559-564(1991)

Xu, Jiafang See Sun, Bingrong: V1606,1022-1026(1991)

Xu, Jian See Yuan, Ren K.: V1519,396-399(1991)

Xu, Jian-hua See Chen, Su-Shing: V1569,446-450(1991)

Xu, Jiangtong See Bao, Liangbi: V1332,862-867(1991)

Xu, Jieping P. See Young, Eddie H.: V1476,178-189(1991)

Xu, Jing
: Practicable fiber optic displacement sensor with subnanometer resolution,V1367,214-220(1991)

Xu, Jisen
; Zhang, Yi; Zhang, Xiang: High-quality 2 x 2 and 4 x 4 wavelength-flatted couplers for sensor applications,V1572,137-139(1991)

Xu, Jun See Chen, Kun J.: V1519,632-639(1991)

Xu, Ken
; Zhang, Zhipeng: Research on the photoelectricity-based swing measuring system,V1572,235-239(1991)

Xu, Kewei
; Chen, Jin; Gao, Runsheng; He, Jia W.; Zhao, Cheng; Li, Shi Z.: X-ray evaluation on residual stresses in vapor-deposited hard coatings,V1519,765-770(1991)

Xu, L. See Wang, Xianhua: V1358,775-779(1991)

Xu, Liangying See Qu, Zhi-Min: V1238,406-411(1991)

Xu, M. C. See She, K.: V1572,581-587(1991)

Xu, Ming See Wang, Tianmin: V1519,890-900(1991)

Xu, Qing
; Inigo, Rafael M.; McVey, Eugene S.: Combined approach for large-scale pattern recognition with translational, rotational, and scaling invariances,V1471,378-389(1991)

Xu, Sen-lu
; Luo, Gei-peng; Xu, Wei-dong: Recent developing status of fiber optic sensors in China,V1367,59-69(1991)
; Sheng, Lie-yi; Zhu, Lie-wei: Novel system for measuring extinction ratio on polarization-maintaining fibers and their devices,V1367,303-308(1991)
See Luo, Gei-peng: V1572,337-341(1991)
See Sheng, Lie-yi: V1572,273-278(1991)

Xu, Shi
; Evans, Brian L.: Multilayer reflectors for the "water window",V1343,110-121(1991)

Xu, Shi Z. See Shu, Ju P.: V1519,565-569(1991)

Xu, Tianning See Casasent, David P.: V1555,23-33(1991)

Xu, Wei See Tu, Meirong: V1554A,593-601(1991)

Xu, Wei-dong See Xu, Sen-lu: V1367,59-69(1991)

Xu, Xiaobing See Hopkins, John B.: V1432,221-226(1991)

Xu, Xin See McAulay, Alastair D.: V1564,685-690(1991)

Xu, Xu R.
; Lei, Gang; Xu, Zheng; Shen, Meng Y.: New apppproach to colored thin film electroluminescence,V1519,525-528(1991)

Xu, Y. B. See Zhai, H. R.: V1519,575-579(1991)

Xu, Yan See Wang, Defang: V1572,514-516(1991)

Xu, Yayong See Chen, Wei: V1450,198-205(1991)

Xu, Yian-sun
; Jackson, Jonathan A.; Price, Edwin P.; Walsh, John E.: Planar-grating klystron experiment,V1407,648-652(1991)

Xu, Yongzhi
: Shape reconstruction from far-field patterns in a stratified ocean environment,V1471,78-86(1991)

Xu, Zheng See Xu, Xu R.: V1519,525-528(1991)

Xu, Zhizhan
See Chen, Shisheng: V1552,288-295(1991)
See Li, Shiying: V1410,215-220(1991)
See Lin, Li-Huang: V1346,490-501(1991)

Xu, Zhu
; Chen, Ke-long; Wen, Zhen-chu; Yang, Han-Guo; He, Xiao-yuan; Zhang, Bao-he: Reconstruction of three-dimensional displacement fields by carrier holography,V1399,172-177(1991)
See Chen, Ke-long: V1385,206-213(1991)

Xue, Kefu
; He, Ping; Fu, Huimin; Bismar, Hisham: Ultrasonic b-scan image compounding technique for prosthetic socket design,V1606,675-684(1991)

Xue, L. A.
; Noh, T. W.; Sievers, A. J.; Raj, Rishi: Optical properties of ZnS/diamond composites,V1534,183-196(1991)

Xue, Liang-yan See Oleinick, Nancy L.: V1427,90-100(1991)

Xue, Mingqiu
See Chen, Yaosheng: V1572,124-128(1991)
See Xiao, Wen: V1572,170-174(1991)

Xue, Qing-yu
; Guo, Song; Zhao, Xue-peng; Nieu, Jian-guo; Zhang, Yi-ping: O3, NO2, NO3, SO2, and aerosol measurements in Beijing,V1491,75-82(1991)

Xue, Qing L. See Palakal, Mathew J.: V1468,456-466(1991)

Xue, Song S.
; Fan, Zheng X.; Gan, Fuxi: (Sb2Se3)1-x Nix alloy thin films and its application in erasable phase change optical recording,V1519,570-574(1991)

Xue, Wei See Collings, Neil: V1505,12-19(1991)

Xydis, Thomas G. See Svetkoff, Donald J.: V1385,113-122(1991)

Yablonovitch, Eli
: Epitaxial liftoff technology,V1563,8-9(1991)

Yacoubian, Araz
; Savant, Gajendra D.; Aye, Tin M.: Holographic recordings on 2-hydroxyethyl methacrylate and applications of water-immersed holograms,V1559,403-409(1991)

Yadav, M. S.
; Dumka, D. C.; Ramola, Ramesh C.; Johri, Subodh; Kothari, Harshad S.; Singh, Babu R.: Design optimization of three-stage GaAs monolithic optical amplifier using SPICE,V1362,811-819(1991)

Yadgiri, G. See Umadevi, P.: V1484,125-135(1991); 1485,195-205(1991)

Yadlowski, Michael See Bundy, Scott: V1475,319-329(1991)

Yaeli, Joseph
; Streifer, William S.; Cross, P. S.; Rozhyki, Alicia; Scifres, Donald R.; Welch, David F.: Mechanically stable external cavity for laser diodes,V1442,378-382(1991)

Yager, Jeffrey S. See Wider, Todd M.: V1422,56-61(1991)

Yager, Paul See Abrams, Susan B.: V1420,13-21(1991)

Yager, Ronald R.
: Fuzzy logic controller structures,V1381,368-378(1991)

Yagi, Hirofumi See Kimata, Masafumi: V1540,238-249(1991)

Yagyu, Eiji See Nishimura, Tetsuya: V1436,31-37(1991)

Yajima, Nobuyuki See Kurokawa, Haruhisa: V1332,643-654(1991)

Yak, A. S.
; Low, Toh-Siew; Lim, Siak-Piang: High-speed swing arm three-beam optical head,V1401,74-81(1991)

Yakimovich, A. P. See Odinokov, S. B.: V1238,109-117(1991)

Yakovlev, Evgeni B.
: Liquid as a deformed crystal: the model of a liquid structure,V1440,36-49(1991)
See Veiko, Vadim P.: V1544,152-163(1991)

Yakovlev, V. A. See Bazakutsa, P. V.: V1440,370-376(1991)

Yakovlev, Yurii P.
; Baranov, Alexej N.; Imenkov, Albert N.; Mikhailova, Maya P.: Optoelectronic LED-photodiode pairs for moisture and gas sensors in the spectral range 1.8-4.8 um,V1510,170-177(1991)
See Charykov, N. A.: V1512,198-203(1991)
See Mikhailova, Maya P.: V1361,674-685(1991)
See Titkov, A. N.: V1361,669-673(1991)

Yakovleva, S. V. See Skopinov, S. A.: V1403,680-681(1991)

Yakovuk, O. A. See Yusupov, I. Y.: V1238,240-247(1991)

Yakuoh, T. See Izawa, Takao: V1441,339-344(1991)

Yakushenkov, Yuri G.
: Choice of means for the adaptation of infrared systems figures of merit,V1540,455-459(1991)

Yakushevich, L. V.
: Experimental basis of the nonlinear models of the internal DNA dynamics,V1403,507-508(1991)

Yakymyshyn, Christopher P. See Perry, Joseph W.: V1560,302-309(1991)

Yamaba, Kazuo
; Miyake, Yoichi: Color character recognition method based on a model of human visual processing,V1453,290-299(1991)

Yamada, Akio See Tanimoto, Masayuki: V1605,394-405(1991)

Yamada, Akira See Suda, Yasumasa: V1560,75-83(1991)

Yamada, Ichiro See Watabe, Akinori: V1499,226-235(1991)

Yamada, Jun See Takada, Yutaka: V1332,571-576(1991)

Yamada, Koichi See Shimamoto, Masayoshi: V1499,393-400(1991)

Yamada, Mitsuho
; Hiruma, Nobuyuki; Hoshino, Haruo: Objective evaluation of the feeling of
depth in 2-D or 3-D images using the convergence angle of the
eyes,V1453,51-57(1991)
See Kato, Takahito: V1453,43-50(1991)

Yamada, Mitsuo See Tomie, Toshihisa: V1552,254-263(1991)

Yamada, Noboru See Ohno, Eiji: V1499,171-179(1991)

Yamada, Noelle See Gauduel, Yann: V1403,417-426(1991)

Yamada, Shinichi See Ohhashi, Akinami: V1444,63-74(1991)

Yamada, Tomoharu See Arai, Yasuhiko: V1554B,266-274(1991)

Yamada, Yukinori See Sukeda, Hirofumi: V1499,419-425(1991)

Yamada, Yukio
; Hasegawa, Yasuo: Simulation of time-resolved optical-CT
imaging,V1431,73-82(1991)

Yamagami, Tamotsu See Fujita, Goro: V1499,426-432(1991)

Yamagishi, Yasuo
; Ishitsuka, Takeshi; Kuramitsu, Yoko; Yoneda, Yasuhiro: New developing
process for PVCz holograms,V1461,68-72(1991)

Yamaguchi, H.
; Tsukamoto, Y.; Watanabe, F.; Sato, A.; Saito, M.; Honda, H.; Murahata, M.;
Yanagisawa, M.; Tsuno, Toshio: Extremely durable CD-ROM with a
novel structure,V1499,29-38(1991)

Yamaguchi, Ichirou
; Kobayashi, Koichi: Material testing by the laser speckle strain
gauge,V1554A,240-249(1991)

Yamaguchi, Koujiro See Ohhashi, Akinami: V1444,63-74(1991)

Yamaguchi, Masahiro
See Honda, Toshio: V1461,156-166(1991)
See Ohishi, Satoru: V1443,280-285(1991)

Yamaguchi, Naohiro See Yoda, Osamu: V1503,463-466(1991)

Yamaguchi, Naohito See Kanazawa, Hirotaka: V1397,445-448(1991)

Yamaguchi, S. See Yokozawa, T.: V1397,513-517(1991)

Yamaguchi, Shigeru See Tittel, Frank K.: V1397,21-29(1991)

Yamaguchi, Takao
; Yodoshia, Keiichi; Minakuchi, Kimihide; Tabuchi, Norio; Bessho, Yasuyuki;
Inoue, Yasuaki; Komeda, Koji; Mori, Kazushi; Tajiri, Atsushi; Tominaga,
Koji: 100-mW four-beam individually addressable monolithic AlGaAs
laser diode arrays,V1418,363-371(1991)

Yamaguchi, Toshikazu See Oyaizu, Ikuro: V1606,990-1001(1991)

Yamaguchi, Yasushi
; Tsu, Hiroji; Sato, Isao: Japanese mission overview of JERS and ASTER
programs,V1490,324-334(1991)

Yamakoshi, H.
; Kato, M.; Kubo, Y.; Enokizono, H.; Uchino, K.; Muraoka, K.; Takahashi, A.;
Maeda, Mitsuo: Thomson scattering diagnostics of discharge plasmas in
an excimer laser,V1397,119-122(1991)

Yamamoto, H. See Maeda, Koichi: V1536,138-148(1991)

Yamamoto, Kimiaki See Tsuchida, Hirofumi: V1354,246-251(1991)

Yamamoto, M. See Kojima, Arata: V1429,162-171(1991)

Yamamoto, Manabu See Watabe, Akinori: V1499,226-235(1991)

Yamamoto, Masaki
See Sudoh, Masaaki: V1343,14-24(1991)
See Tsunetomo, Keiji: V1513,93-104(1991)

Yamamoto, Shuhei
See Haemmerli, Jean-Francois: V1564,275-284(1991)
See Kato, Naoki: V1455,190-205(1991)
See Mitsuoka, Yasuyuki: V1564,244-252(1991)

Yamamoto, Yoshitaka
: Development of new beamsplitter system for real-time high-speed
holographic interferometry,V1358,940-951(1991)

Yamamura, N. See Izawa, Takao: V1441,339-344(1991)

Yamanaka, Chiyoe See Yamanaka, Masanobu: V1501,30-39(1991)

Yamanaka, Junko See Kato, Naoki: V1455,190-205(1991)

Yamanaka, Koji See Itoh, Katsuyuki: V1466,485-496(1991)

Yamanaka, Masanobu
; Naito, Kenta; Nakatsuka, Masahiro; Yamanaka, Tatsuhiko; Nakai, Sadao S.;
Yamanaka, Chiyoe: Diode-pumped, high-power, solid-state
lasers,V1501,30-39(1991)

Yamanaka, Tatsuhiko See Yamanaka, Masanobu: V1501,30-39(1991)

Yamanaka, Yutaka
; Katayama, Ryuichi; Komatsu, Y.; Ishikawa, S.; Itoh, M.; Ono, Yuzo:
Compact magneto-optical disk head integrated with chip
elements,V1499,263-268(1991)

Yamane, Nobumoto
; Morikawa, Yoshitaka; Hamada, Hiroshi: Effects of M-transform for bit-error
resilement in the adaptive DCT coding,V1605,679-686(1991)
See Morikawa, Yoshitaka: V1605,445-455(1991)

Yamaoka, Tsuguo See Ishikawa, Toshiharu: V1461,73-78(1991)

Yamasaki, Toru See Koshi, Yutaka: V1605,362-373(1991)

Yamashita, Akio See Amano, Tomio: V1452,330-339(1991)

Yamashita, H. See Misaka, Akio: V1465,174-184(1991)

Yamashita, Kazumi
See Hama, Hiromitsu: V1606,878-890(1991)
See Nakajima, Shigeyoshi: V1605,709-719(1991)

Yamashita, T. See Endo, Masamori: V1397,267-270(1991)

Yamashita, Tomihiro See Iannone, Mark A.: V1559,172-183(1991)

Yamashita, Toshiharu T. See Yanagita, Hiroaki: V1513,386-395(1991)

Yamashita, Tsukasa
; Aoyama, Shigeru: Micrograting device using electron-beam
lithography,V1507,81-93(1991)
See Aoyama, Shigeru: V1545,241-250(1991)
See Hosokawa, Hayami: V1559,229-237(1991)

Yamauchi, Masamitsu See Matsuoka, Masaru: V1549,2-8(1991)

Yamauchi, Takashi
: Pelliclizing technology,V1496,302-314(1991)

Yamazaki, Satomi
; Ishida, Shinji; Matsumoto, Hiroshi; Aizaki, Naoaki; Muramoto, Naohiro;
Mine, Katsutoshi: Highly sensitive microresinoid siloxane resist for EB
and deep-UV lithography,V1466,538-545(1991)

Yaminsky, Igor V. See Shapiro, Alexander G.: V1362,834-843(1991)

Yamout, Jihad S. See Bachnak, Rafic A.: V1457,27-36(1991)

Yan, Da-Peng
; He, Anzhi; Yang, Zu Q.; Zhu, Yi Yun: New method of increasing the
sensitivity of Schlieren interferometer using two Wollaston prisms and its
application to flow field,V1554B,636-640(1991)
; Wang, Hai-Lin; Miao, Peng-Cheng; He, Anzhi: Eliminating system errors of
a large-aperture and high-sensitivity moire deflector by real-time
holography,V1527,442-447(1991)
See He, Anzhi: V1527,334-337(1991)
See He, Anzhi: V1527,423-426(1991)
See He, Anzhi: V1554B,429-434(1991)
See Miao, Peng-Cheng: V1554B,641-644(1991)
See Wang, Hai-Lin: V1545,268-273(1991)
See Wang, Hai-Lin: V1545,274-277(1991)

Yan, Hongtao
; Li, Hanjie; Li, Yonghong: Optical fiber sensor for ammonia monitoring in
the blood,V1572,396-398(1991)

Yan, Jin-Li See Yu, Tong: V1572,469-471(1991); 1584,135-137(1991)

Yan, Li I.
See Kung, Chun C.: V1378,250-258(1991)
See Ling, Junda D.: V1454,353-362(1991)

Yan, Muolin See Luo, Fei: V1367,221-224(1991)

Yan, Ran H. See Coldren, Larry A.: V1362,24-37(1991)

Yan, Xiao-Hong
; Leahy, Richard M.: MAP image reconstruction using intensity and line
processes for emission tomography data,V1452,158-169(1991)

Yan, Yi S.
: New coating technology and ion source,V1519,192-193(1991)

Yan, Yixun See Zhang, Su Y.: V1519,508-513(1991)

Yan, Zu Q.
; Ruan, Ke F.: Transition radiation in foil stack and x-ray laser,V1519,183-191(1991)

Yanagihara, Kenji See Taira, Kazuo: V1466,570-582(1991)

Yanagihara, Mihiro See Sudoh, Masaaki: V1343,14-24(1991)

Yanagimachi, Masatoshi See Toriumi, Minoru: V1466,458-468(1991)

Yanagisawa, M. See Yamaguchi, H.: V1499,29-38(1991)

Yanagisawa, Toru
    See Isobe, Hiroshi: V1358,471-478(1991)
    See Isobe, Hiroshi: V1358,201-208(1991)
    See Sato, Eiichi: V1358,146-153(1991)
    See Sato, Eiichi: V1358,193-200(1991)
    See Sato, Eiichi: V1358,479-487(1991)
    See Sato, Eiichi: V1358,462-470(1991)
    See Shikoda, Arimitsu: V1358,154-161(1991)

Yanagishita, Yuichiro
    ; Ishiwata, Naoyuki; Tabata, Yasuko; Nakagawa, Kenji; Shigematsu,
      Kazumasa: Phase-shifting photolithography applicable to real IC
      patterns,V1463,207-217(1991)

Yanagita, Hiroaki
    ; Toratani, Hisayoshi; Yamashita, Toshiharu T.; Masuda, Isao: Diode-pumped
      Er3+ glass laser at 2.7 um,V1513,386-395(1991)

Yang, Ai H. See Chen, Lian C.: V1519,450-453(1991)

Yang, Bang C.
    ; Gou, Li; Jia, Yu M.; Ran, Jun G.; Zheng, Chang Q.; Tang, Xia: Temperature
      dependence of resistance of diamond film synthesized by microwave
      plasma CVD,V1519,864-865(1991)
    ; Jia, Yu M.: Ta/Al alloy thin film medium-power attenuator,V1519,156-
      158(1991)
    ; Wang, Ju Y.; Jia, Yu M.; Huang, Yong L.: Preparation of PbTiO3 thin film by
      dc single-target magnetron sputtering,V1519,725-728(1991)

Yang, Bin See Jiang, Xiang-Liu: V1534,207-213(1991)

Yang, Bing L.
    ; Liu, Bai Y.; Cheng, Y. C.; Wong, H.: High-field electron trapping and
      detrapping characteristics in thin SiOxNy films,V1519,241-246(1991)
    ; Liu, Bai Y.; Chen, D. N.; Cheng, Y. C.; Wong, H.: Study on the high-field
      current transport mechanisms in thin SiOxNy films,V1519,269-274(1991)
    See Wong, H.: V1519,494-498(1991)

Yang, Bing T.
    : Enhanced thematic mapper cold focal plane: design and testing,V1488,399-
      409(1991)

Yang, Binzhou See Hou, Xun: V1358,868-873(1991)

Yang, Chen-Jen See Meth, Jeffrey S.: V1560,13-24(1991)

Yang, Dao M. See Lin, Hong Y.: V1519,210-213(1991)

Yang, Darang
    ; Song, Ruhua H.; Hu, Zhiping; Le, Shixiao: Multistability, instability, and
      chaos for intracavity magneto-optic modulation output,V1417,440-
      450(1991)

Yang, Daren
    ; Jiang, Huilin; Li, Gongde; Yang, Huamin: Optimization of original lens
      structure type from optical lens data base,V1527,456-461(1991)

Yang, Datong See Holm-Kennedy, James W.: V1527,322-331(1991)

Yang, David See Kender, John R.: V1388,476-489(1991)

Yang, F. See Shaw, David T.: V1394,214-220(1991)

Yang, Fan See Wu, Gengsheng: V1572,497-502(1991)

Yang, Fanling H. See Marsh, Harry H.: V1447,298-309(1991)

Yang, Gen Q.
    See Feng, Yi P.: V1519,440-443(1991)
    See Wang, Xi: V1519,740-743(1991)

Yang, Guanglu
    ; Siahmakoun, Azad: Tunable holographic interferometer using
      photorefractive crystal,V1396,552-556(1991)

Yang, Guang Y.
    See Wang, Ren: V1519,146-151(1991)
    See Zhang, Yun H.: V1519,729-734(1991)

Yang, Guo-Zhen See Dong, Bizhen: V1429,117-126(1991)

Yang, Guoguang
    ; Gao, Wenliang; Cheng, Shangyi: Automatic inspection technique for optical
      surface flaws,V1332,56-63(1991)
    See Gao, Wenliang: V1530,118-128(1991)

Yang, H. See Soumekh, Mehrdad: V1452,104-113(1991)

Yang, H. P. See Zhuang, Qi: V1397,157-160(1991)

Yang, Han-Guo See Xu, Zhu: V1399,172-177(1991)

Yang, Hongbo See Lu, Eh: V1533,155-162(1991)

Yang, Hou-Min
    ; Wang, Xiaolin; Zhang, Yinxian: Microdisplacement positioning system for a
      diffraction grating ruling engine,V1533,185-192(1991)

Yang, Huamin See Yang, Daren: V1527,456-461(1991)

Yang, Jane
    See Anderson, Eric R.: V1417,543-549(1991)
    See Jansen, Michael: V1418,32-39(1991)
    See Largent, Craig C.: V1418,40-45(1991)

Yang, Jiahua
    : Study of welding residual stress by means of laser holography
      interference,V1554B,155-160(1991)

Yang, Jian
    ; Reeves, Adam J.: Harmonic oscillator model of early visual image
      processing,V1606,520-530(1991)

Yang, Jian R. See He, Jin: V1519,499-507(1991)

Yang, Jie
    ; Wu, Yuehua; Zhou, Yusheng; Uyemura, T.: Research on macroeffects and
      micromechanism of martensite phase transition of shape memory alloys
      by high-speed photography,V1358,672-676(1991)

Yang, Ji M. See He, Jin: V1519,499-507(1991)

Yang, Jing See Chen, Pu S.: V1519,258-262(1991)

Yang, Jing-Yu See Liu, Ke: V1567,720-728(1991)

Yang, Joon Mook See Ahn, Jae W.: V1358,269-277(1991)

Yang, Mao-hua See Sun, Fang-kui: V1488,2-5(1991)

Yang, N. P. See Zhang, Z. J.: V1519,790-792(1991)

Yang, Ping-Fai
    ; Maragos, Petros: Learnability of min-max pattern classifiers,V1606,294-
      308(1991)

Yang, Q.
    ; Butler, Clive: Three-dimensional fibre-optic position sensor,V1572,558-
      563(1991)

Yang, Qing See Qiang, Xue-li: V1567,670-679(1991)

Yang, Quanzu
    ; Wang, Zhongcai; Wang, Shizhuo: High-temperature Raman spectra of
      nioboborate glass melts,V1513,264-269(1991)

Yang, Renshu
    ; Yang, Yongqi: Dynamic photoelastic study on mechanism of short-delay
      blasting,V1554A,341-348(1991)

Yang, Sheng-Jenn See Sun, Tai-Ping: V1361,1033-1037(1991)

Yang, Sheng S. See Liao, Chang G.: V1519,152-155(1991)

Yang, Shien-Chi
    ; Tsao, Hen-Wai; Wu, Jingshown: Theoretical analysis of optical phase
      diversity FSK receivers,V1372,128-139(1991)

Yang, Shuwen See Ma, Junxian: V1572,377-381(1991)

Yang, Tsong-Yo See Chao, Shiuh: V1401,35-43(1991)

Yang, Tungsheng See Garofalo, Joseph G.: V1463,151-166(1991)

Yang, Woodward
    : Analog CCD processors for image filtering,V1473,114-127(1991)

Yang, X. See Wang, H. F.: V1519,405-410(1991)

Yang, Xiangyang
    See Yu, Francis T.: V1507,210-221(1991)
    See Yu, Francis T.: V1558,450-458(1991)

Yang, Xiao-zhi See Wang, Rui-zhong: V1421,203-207(1991)

Yang, Xiaobo See Liu, Dingyu: V1443,191-196(1991)

Yang, Xiao Y. See Zhang, Si J.: V1519,744-751(1991)

Yang, Xueyu
    ; Spears, Kenneth G.: Fiber probe for ring pattern,V1367,382-386(1991)

Yang, Y. J.
    ; Dziura, T. G.; Wang, S. C.; Fernandez, R.; Du, G.; Wang, Shyh: Low-
      threshold room temperature continuous-wave operation of a GaAs single-
      quantum-well mushroom structure surface-emitting laser,V1418,414-
      421(1991)
    See Wang, S. C.: V1563,27-33(1991)

Yang, Y. Y. See Imen, Kamran: V1365,60-64(1991)

Yang, Ya L. See Zhao, Shu L.: V1519,275-280(1991)

Yang, Yang
    ; Li, Naiji; Wang, Huiwen: Time characteristics in HTS rare-earth-doped
      optical fiber at high temperature,V1572,205-210(1991)
    See Li, Naiji: V1572,47-51(1991)

Yang, Ye-min See Xia, Sheng-jie: V1358,43-45(1991)

Yang, Yibing
    ; Wang, Houshu: Adaptive filtering and enhancement method of speckle
      pattern,V1554A,781-788(1991)

Yang, Yimin See Wang, Tianmin: V1519,890-900(1991)

Yang, Yong
; Guo, Jian Q.; Lu, Dong: Energy bands of graphite and CsC8-GIC, and CO physisorption on graphite basal surface,V1519,444-448(1991)

Yang, Yongqi
See Yang, Renshu: V1554A,341-348(1991)
See Zhu, Zhenhai: V1554A,472-481(1991)

Yang, Yu-Hong See Weng, Zhicheng: V1527,349-356(1991)

Yang, Yu X.
; Tan, Yushan: Influence of fixing stress on the sensitivity of HNDT,V1554B,768-773(1991)

Yang, Zhiguo
; Zhong, Hengyong; Cheng, Jubing; Wang, Youguan: Experimental research on optical fibre microbending sensors in on-line measuring of deep-hole drilling bit wear,V1572,252-257(1991)

Yang, Zishao
; Rosenbruch, Klaus J.: Corrections of aberrations using HOEs in UV and visible imaging systems,V1354,323-327(1991)

Yang, Zu Q. See Yan, Da-Peng: V1554B,636-640(1991)

Yanh, Jian R. See Ma, Ke J.: V1519,489-493(1991)

Yaniv, Zvi See den Boer, Willem: V1455,248-248(1991)

Yankelevich, Diego
; Knoesen, Andre; Eldering, Charles A.; Kowel, Stephen T.: Reflection-mode polymeric interference modulators,V1560,406-415(1991)

Yano, Jun-ichi See Kamemaru, Shun-ichi: V1564,143-154(1991)

Yano, M.
See Izawa, Takao: V1441,339-344(1991)
See Tomie, Toshihisa: V1552,254-263(1991)

Yano, Mitsuharu
See Mochizuki, Takashi: V1605,434-444(1991)
See Ohta, Mutsumi: V1605,456-466(1991)

Yano, T. See Imai, Masaaki M.: V1371,13-20(1991)

Yansen, Don E. See Bennett-Lilley, Marylyn H.: V1464,127-136(1991)

Yao, Cheng-Shan See Shi, Yi-Wei: V1572,308-312(1991)

Yao, He S. See Wu, Xiang: V1519,625-631(1991)

Yao, HuiHai See Keil, Norbert: V1559,278-287(1991)

Yao, J. Y. See Andersson, Thorvald G.: V1361,434-442(1991)

Yao, Jialing
; McStay, Daniel; Rogers, Alan J.; Quinn, Peter J.: Delayed fluorescence of eosin-bound protein: a probe for measurement of slow-rotational mobility,V1572,428-433(1991)

Yao, Kung See Erlich, Simha: V1565,47-56(1991)

Yao, Minyu
; Zhang, Xinyu: Experimental research on high-bifringence fiber phase retarder,V1572,148-150(1991)

Yao, Qi See Li, Chang J.: V1519,779-787(1991)

Yao, Shi-Kay
ed.: *Optical Technology for Microwave Applications V*,V1476(1991)
: TeO2 slow surface acoustic wave Bragg cell,V1476,214-221(1991)
See Young, Eddie H.: V1476,178-189(1991)

Yao, W. See Shi, W.: V1519,138-141(1991)

Yao, Wu
; Lian, Tongshu: Study on image-stabilizing reflecting prisms in the case of a finite angular perturbation,V1527,448-455(1991)

Yao, Yong See Baird, Bill: V1469,12-23(1991)

Yao, Yuchuan See Weng, Zhicheng: V1527,349-356(1991)

Yaou, Ming-Haw
; Chang, Wen-Thong: Design of M-band filter banks based on wavelet transform,V1605,149-159(1991)

Yap, Daniel
; Narayanan, Authi A.; Rosenbaum, Steven E.; Chou, Chia-Shing; Hooper, W. W.; Quen, R. W.; Walden, Robert H.: GaAs/GaAlAs integrated optoelectronic transmitter for microwave applications,V1418,471-476(1991)

Yarbrough, Walter A.
: Diamond growth on the (110) surface,V1534,90-104(1991)

Yardley, James T.
See Beeson, Karl W.: V1559,258-266(1991)
See Beeson, Karl W.: V1374,176-185(1991)
See McFarland, Michael J.: V1583,344-354(1991)

Yarling, Charles B.
; Cook, Dawn M.: Uniformity characterization of rapid thermal processor thin films,V1393,411-420(1991)
; Hahn, Sookap; Hodul, David T.; Suga, Hisaaki; Smith, W. L.: Investigation of rapid thermal process-induced defects in ion-implanted Czochralski silicon,V1393,192-199(1991)

Yaroshetskii, Ilya D.
See Beregulin, Eugene V.: V1362,853-862(1991)
See Sreseli, Olga M.: V1440,326-331(1991)
See Sreseli, Olga M.: V1545,149-158(1991)

Yaroslavsky, Ilya V.
See Medvedev, Boris A.: V1422,73-84(1991)
See Medvedev, Boris A.: V1403,682-685(1991)

Yarova, A. G. See Drozhbin, Yu. A.: V1358,1029-1034(1991)

Yarzev, V. See Antyuhov, V.: V1415,48-59(1991)

Yaseen, Mohammed
; Walker, Stuart D.: Optoelectronic sampling receiver for time-division multiplexed signal processing applications,V1562,319-326(1991)

Yasuda, Yasuhiko
See Chen, Yan-Ping: V1605,822-831(1991)
See Katto, Jiro: V1605,95-106(1991)
See Kimoto, Tadahiko: V1605,253-262(1991)

Yasuda, Yukio
; Tan, Oon T.; Kurban, Amal K.; Tsukada, Sadao: Electron microscopic study on black pig skin irradiated with pulsed dye laser (504 nm),V1422,50-55(1991)
See Kurban, Amal K.: V1422,43-49(1991)

Yasuoka, Koichi
; Ishii, Akira; Tamagawa, Tohru; Ohshima, Iwao: Newly developed excitation circuit for kHz pulsed lasers,V1412,32-37(1991)

Yatagai, Toyohiko
; Geiser, Martial; Tian, Ronglong; Tian, Xingkang; Onda, Hajime: CAD system for CGHs and laser beam lithography,V1555,8-12(1991)
; Ino, Tomomi: Use of high-resolution TV camera in photomechanics,V1554B,646-649(1991)
; Tian, Ronglong: Spatial light modulators and their applications,V1564,691-696(1991)
See Shimada, Wataru: V1332,525-529(1991)

Yates, Gigi See Eberlein, Susan: V1388,578-586(1991)

Yatsiv, Shaul
; Gabay, Amnon; Sterman, Baruch; Sintov, Yoav: Performance of CO2 and CO diffusively cooled rf-excited strip-line lasers with different electrode materials and gas composition,V1397,319-329(1991)

Yatsu, Masahiko
; Deguchi, Masaharu; Maruyama, Takesuke: Zoom lens with aspherical lens for camcorder,V1354,663-668(1991)

Yau, W. See Watkins, E. T.: V1475,62-72(1991)

Yavas, O. See Boneberg, J.: V1598,84-90(1991)

Yazawa, Y. See Minemura, Tetsuroh: V1361,344-353(1991)

Yaztsev, Vladimir P.
See Antyuhov, V.: V1397,355-365(1991)
See Bondarenko, Alexander V.: V1376,117-127(1991)

Ye, Biqing
; Abela, George S.: Decrease in total fluorescence from human arteries with increasing beta-carotene content,V1425,45-54(1991)

Ye, Miaoyuan
; Hu, Shichuang: Compensation of the phase shift in the optical fibre current sensor,V1572,483-486(1991)
See Hu, Shichuang: V1572,492-496(1991)

Ye, Naiqun See Hu, Yu: V1409,230-239(1991)

Ye, Niao-ting See Wei, Yan-nian: V1358,457-460(1991)

Ye, Qizheng See Jiang, Desheng: V1572,313-317(1991)

Ye, Qui-Yi See Tsu, Raphael: V1361,232-235(1991)

Ye, Shenhua See Liu, Ying: V1554A,610-612(1991)

Ye, Yizheng See Yu, He: V1469,412-417(1991)

Yearwood, M. See Brandstetter, Robert W.: V1385,173-189(1991)

Yeatman, Eric M. See Caldwell, Martin E.: V1505,50-58(1991)

Yeatman, Lawrence S. See Eton, Darwin: V1425,182-187(1991)

Yee, Conway See Haselgrove, John C.: V1431,30-41(1991)

Yee, David K.
; Lee, Woobin; Kim, Dong-Lok; Haass, Clark D.; Rowberg, Alan H.; Kim, Yongmin: RadGSP: a medical image display and user interface for UWGSP3,V1444,292-305(1991)

Yee, Jick H. See Mayhall, David J.: V1378,101-114(1991)

Yee, K. See Ishizuka, Hiroshi: V1407,442-455(1991)

Yee, Mark L.
; Casasent, David P.: Multiple target-to-track association and track estimation system using a neural network,V1481,418-429(1991)
; Casasent, David P.: Optimization neural net for multiple-target data association: real-time optical lab results,V1469,308-319(1991)

Yeh, Bao-Fuh See Chang, Po-Rong: V1606,456-469(1991)

Yeh, C. L. See Gopalakrishnan, G. K.: V1476,270-275(1991)

Yeh, C. S. See Ayekavadi, Raj: V1418,74-85(1991)

Yeh, Chao-Pin
; Umeagukwu, Charles; Fulton, Robert E.; Teat, William: Sensitivity study of printed wiring board plated-through-holes copper barrel voids,V1389,187-198(1991)
See Fulton, Robert E.: V1389,144-155(1991)

Yeh, Chia L.
See Chu, Frank J.: V1452,38-46(1991)
See Normile, James: V1452,480-484(1991)

Yeh, Long-Ching
; Wu, Way-Chen; Tang, Ru-Shyah; Tsai, Yong-Song; Chiang, Eugene; Chiao, Pat; Chu, Chu-Lin; Wang, Weng-Lyang: High-performance 400-DPI A4-size contact image sensor module for scanner and G4 fax applications,V1527,361-367(1991)

Yeh, Pochi A.
; Gu, Claire; Hong, John H.: Photorefractive devices for optical information processing,V1562,32-43(1991)
See Shellan, Jeffrey B.: V1545,179-188(1991)

Yeh, Smile See Sun, Ching-Cherng: V1564,199-210(1991)

Yeh, Yunhae
; Lee, J. H.; Lee, Chung E.; Taylor, Henry F.: In-line Fabry-Perot interferometric temperature sensor with digital signal processing,V1584,72-78(1991)

Yekozeki, Shunsuke See Arai, Yasuhiko: V1554B,266-274(1991)

Yelamarty, Rao V.
; Yu, Francis T.; Moore, Russell L.; Cheung, Joseph Y.: Liquid-crystal-television-based optical-digital processor for measurement of shortening velocity in single rat heart cells,V1398,170-179(1991)

Yelisseyev, Alexander P.
; Sinyakova, Elena F.: Nonstoichiometry effect on mercury thiogallate luminescence,V1512,204-212(1991)

Yellen, Steven L. See Baumann, John A.: V1418,328-337(1991)

Yen, Huan-chun See Chow, P. D.: V1475,48-54(1991)

Yen, William M. See Jaffe, Steven M.: V1435,252-257(1991)

Yen, Yi X.
See Feng, Weiting: V1535,224-230(1991)
See Feng, Weiting: V1519,333-338(1991)

Yeskov-Soskovetz, Vladimir M. See Dovgalenko, George Y.: V1508,110-115(1991)

Yestrebsky, Joseph T. See Basehore, Paul: V1470,190-196(1991)

Yeung, Keith K.
; Zakarauskas, Pierre; McCray, Allan G.: Neural network approach for object orientation classification,V1569,133-146(1991)

Yi, Sang-Yoon See Son, Yung-Sung: V1374,23-29(1991)

Yi, Xin J.
See Sun, Han D.: V1477,174-177(1991)
See Wang, Lingjie: V1540,738-741(1991)
See Zhao, Xing R.: V1519,772-774(1991)
See Zhou, Fang Q.: V1477,178-181(1991)

Yilbas, Bekir S.
See Alci, Mustafa: V1411,100-106(1991)
See Danisman, Kenan: V1412,218-226(1991)

Yin, Chen Z. See Sun, Yue Z.: V1519,234-240(1991)

Yin, Dong
; Guan, Wen X.; Sun, Shu Z.; Zhang, Zhao A.: Research on influence of substrate to crystal growth by ion beams,V1519,164-166(1991)

Yin, Fang-Fang
; Giger, Maryellen L.; Doi, Kunio; Vyborny, Carl J.; Schmidt, Robert A.; Metz, Charles E.: Computer vision system for the detection and characterization of masses for use in mammographic screening programs,V1396,2-4(1991)
See Giger, Maryellen L.: V1445,101-103(1991)

Yin, Lin
; Astola, Jaakko; Neuvo, Yrjo A.: Adaptive multistage weighted order-statistic filters for image restoration,V1451,216-227(1991)
; Astola, Jaakko; Neuvo, Yrjo A.: Optimal generalized weighted-order-statistic filters,V1606,431-442(1991)

Yin, Shizhuo See Wu, Shudong: V1584,415-424(1991)

Yin, Yuan-Cheng See Wang, Ji-Zhong: V1554B,188-192(1991)

Yin, Zhiqiang
; Stjerna, B. A.; Granqvist, Claes G.: Antireflection coatings of sputter-deposited SnOxFy and SnNxFy,V1536,149-157(1991)

Yin, Zhi Xiang
; Zhang, Shikun; Li, Zong Yan: Experimental simulation analysis of nonlinear problem: investigation into the mechanical and optical behavior of silver chloride of photoplastic material,V1559,487-496(1991)

Yin, Zongming See Pan, Jingming: V1572,403-405(1991)

Ying, Xingren
; Zeng, Nan: Controller implemented by recording the fuzzy rules by backpropagation neural networks,V1469,846-851(1991)

Ying, Xuantong
; Feldman, Albert; Farabaugh, Edward N.: Effect of oxygen on optical properties of yttria thin films,V1519,321-326(1991)

Ying, Zaisheng See Jin, Guoliang: V1513,50-55(1991)

Yionoulis, Steve M. See Heyler, Gene A.: V1481,198-208(1991)

Yip, Gar L.
: Characterization, modeling, and design optimization of integrated optical waveguide devices in glass,V1513,26-36(1991)
: Simulation and design of integrated optical waveguide devices by the BPM,V1583,240-248(1991)
; Kishioka, Kiyoshi; Xiang, Feng; Chen, J. Y.: Characterization of planar optical waveguides by K+ ion exchange in glass at 1.152 and 1.523 um,V1583,14-18(1991)
See Nikolopoulos, John: V1583,71-82(1991)
See Nikolopoulos, John: V1374,30-36(1991)
See Noutsios, Peter C.: V1374,14-22(1991)
See Xiang, Feng: V1583,271-277(1991)

Yitzchaik, Shlomo
See Berkovic, Garry: V1560,238-242(1991)
See Berkovic, Garry: V1442,44-52(1991)

Yla-Jaaski, Juha
; Yu, Xiaohan: Automatic digitization of contour lines for digital map production,V1472,201-207(1991)
See Yu, Xiaohan: V1468,834-842(1991)

Yli-Harja, Olli P. See Vepsalainen, Ari M.: V1568,2-13(1991)

Yoda, Osamu
; Miyashita, Atsumi; Murakami, Kouichi; Aoki, Sadao; Yamaguchi, Naohiro: Time-resolved x-ray absorption spectroscopy apparatus using laser plasma as an x-ray source,V1503,463-466(1991)

Yoder, Paul R.
: Axial stresses with toroidal lens-to-mount interfaces,V1533,2-11(1991)
: Optical engineering of an excimer laser ophthalmic surgery system,V1442,162-171(1991)

Yodoshi, Keiichi See Yamaguchi, Takao: V1418,363-371(1991)

Yogev, Amnon See Bernstein, Hana: V1442,81-88(1991)

Yokoi, Seiichi See Katayama, Norihisa: V1403,147-149(1991)

Yokoi, Takane
; Suzuki, Kenji; Oba, Koichiro: Ultraviolet light imaging technology and applications,V1449,30-39(1991)

Yokokura, Takashi See Cartlidge, Andy G.: V1423,167-174(1991)

Yokomori, Kiyoshi See Kihara, Tami: V1332,783-791(1991)

Yokota, Akira See Furuta, Mitsuhiro: V1466,477-484(1991)

Yokota, Tatsuya See Sasano, Yasuhiro: V1490,233-242(1991)

Yokota, Yoshiharu
See Isobe, Hiroshi: V1358,201-208(1991)
See Sato, Eiichi: V1358,479-487(1991)

Yokota, Yoshiro See Nishimura, Kazuhito: V1534,199-206(1991)

Yokoyama, M. See Hsu, C. T.: V1519,391-395(1991)

Yokoyama, Naoki
; Uyemura, Tsuneyoshi: Application of one-dimensional high-speed video camera system to motion analysis,V1358,351-357(1991)

Yokoyama, Ryouhei See Sudoh, Masaaki: V1343,14-24(1991)

Yokozawa, T.
; Nakajima, H.; Yamaguchi, S.; Ebina, K.; Ohara, M.; Kanazawa, Hirotaka; Yuasa, M.; Komatsu, Katsuhiko; Hara, Hiroshi: Performance characteristics of premixing gas-dynamic laser utilizing liquid C6H6 and liquid N2O,V1397,513-517(1991)

Yonaki, J. See Liu, Louis: V1475,193-198(1991)

Yoneda, Masahiro See Ogata, Shiro: V1544,92-100(1991)

Yoneda, Yasuhiro See Yamagishi, Yasuo: V1461,68-72(1991)

Yonekura, Fukuo See Ogura, Toshihiro: V1443,153-157(1991)

Yonekura, Yoshiharu See Minato, Kotaro: V1446,195-198(1991)

Yonezawa, Masaji See Tamura, Akira: V1465,271-281(1991)

Yong, Hongru See Chang, Zenghu: V1358,614-618(1991)

Yong, Jiang See Wang, He-Chen: V1386,273-276(1991)

Yonte, T.
; Quiroga, J.; Alda, J.; Bernabeu, Eusebio: Ophthalmic lenses testing by moire deflectometry,V1554B,233-241(1991)

Yoo, C. H. See Chu, Shirley S.: V1361,1020-1025(1991)

Yoo, Chue-San
; Jans, Jan C.: Thickness measurement of combined a-Si and Ti films on c-Si using a monochromatic ellipsometer,V1464,393-403(1991)

Yoo, Jisang
; Bouman, Charles A.; Delp, Edward J.; Coyle, Edward J.: Intensity edge detection with stack filters,V1451,58-69(1991)

Yoo, Kisuck See Thursby, Michael H.: V1588,218-228(1991)

Yoo, Young-Don See Lee, Hyouk: V1393,404-410(1991)

Yool, Stephen R. See Westerman, Steven D.: V1492,263-271(1991)

Yoon, Chun S.
: Ground target classification using moving target indicator radar signatures,V1470,243-252(1991)

Yoon, Hyun N. See Haas, David R.: V1371,56-67(1991)

York, Jim F.
; Nelson, Gary W.; Varshneya, Deepak: Characterization of macrobend sensitivity of step index optical fibers used in intensity sensors,V1584,308-319(1991)

Yorka, Christian M. See Eustace, John G.: V1584,320-327(1991)

Yorkey, Thomas J.
; Brase, James M.; Trebes, James E.; Lane, Stephen M.; Gray, Joe W.: X-ray holography for sequencing DNA,V1345,255-259(1991)

Yorkston, J. See Antonuk, Larry E.: V1443,108-119(1991)

Yoshida, A. See Matsuoka, Masaru: V1549,2-8(1991)

Yoshida, Harunobu See Kitaoka, Masaki: V1519,109-114(1991)

Yoshida, Hiroshi See Yoshida, Kunio: V1441,9-9(1991)

Yoshida, Kunio
; Yoshida, Hiroshi; Noda, T.; Nakai, Sadao S.: Multiple-pulse damage of BK-7 glass,V1441,9-9(1991)

Yoshida, Minoru See Yoshitake, Yasuhiro: V1463,678-687(1991)

Yoshida, S.
; Shimizu, K.: High-power chemical oxygen-iodine lasers and applications,V1397,205-212(1991)
See Kasuya, Koichi: V1397,67-70(1991)

Yoshida, Shigeru See Nakano, Yasuhiko: V1605,874-878(1991)

Yoshida, Tadashi See Tanazawa, Takeshi: V1435,310-321(1991)

Yoshida, Tomio See Ohara, Shunji: V1499,187-194(1991)

Yoshida, Toshifumi See Tanazawa, Takeshi: V1435,310-321(1991)

Yoshida, Yoshio
; Miyake, Takahiro; Kurata, Yukio; Ishikawa, Toshio: Three-beam CD optical pickup using a holographic optical element,V1401,58-65(1991)

Yoshikawa, Eiji See Kakizaki, Sunao: V1505,199-205(1991)

Yoshikawa, Hiroshi See St.-Hilaire, Pierre: V1461,254-261(1991)

Yoshikawa, Nobuo See Ohmi, Toshihiko: V1361,606-612(1991)

Yoshimi, Billibon See Allen, Peter K.: V1383,176-188(1991)

Yoshimoto, Takashi See Otsuki, Taisuke: V1420,220-224(1991)

Yoshimura, Hitoshi
; Giger, Maryellen L.; Matsumoto, Tsuneo; Doi, Kunio; MacMahon, Heber; Montner, Steven M.: Investigation of new filtering schemes for computerized detection of lung nodules,V1445,47-51(1991)
See Giger, Maryellen L.: V1445,101-103(1991)

Yoshimura, Motomu See Nishimura, Tetsuya: V1436,31-37(1991)

Yoshimura, Shunji See Fukumoto, Atsushi: V1499,216-225(1991)

Yoshimura, Takeaki
; Fujiwara, Kazuo; Miyazaki, Eiichi: Statistical properties of intensity fluctuations produced by rough surfaces under the speckle pattern illumination,V1332,835-842(1991)

Yoshinaga, Kazuomi See Ito, Katsunori: V1499,382-385(1991)

Yoshioka, Kajutoshi See Lauck, Teresa L.: V1464,527-538(1991)

Yoshioka, Nobuyuki
See Miyazaki, Junji: V1464,327-335(1991)
See Op de Beeck, Maaike: V1463,180-196(1991)

Yoshitake, T. See Takeuchi, Ichiro: V1394,96-101(1991)

Yoshitake, Yasuhiro
; Oshida, Yoshitada; Tanimoto, Tetsuzou; Tanaka, Minoru; Yoshida, Minoru: Multispot scanning exposure system for excimer laser stepper,V1463,678-687(1991)

Yoshizawa, Masayuki See Kobayashi, Takayoshi T.: V1403,407-416(1991)

Yoshizawa, Toru
; Tomisawa, Teiyu: Moire topography with the aid of phase-shift method,V1554B,441-450(1991)

You, Lu See McDonald, John F.: V1390,286-301(1991)

Youmans, Douglas G.
: Lidar backscatter calculations for solid-sphere and layered-sphere aerosols,V1416,151-162(1991)

Youmans, Robert J. See Kim, A. H.: V1378,173-178(1991)

Young, Donald S.
; Osche, Gregory R.; Fisher, Kirk L.; Lok, Y. F.: Dry/wet wire reflectivities at 10.6 um,V1416,221-228(1991)

Young, Eddie H.
; Ho, Huey-Chin C.; Yao, Shi-Kay; Xu, Jieping P.: Generalized phased-array Bragg interaction in anisotropic crystals,V1476,178-189(1991)

Young, J. D. See Vari, Sandor G.: V1426,111-120(1991)

Young, Kenneth C. See Lalk, Gail R.: V1389,386-400(1991)

Young, Lloyd M.
See Chan, Kwok-Chi D.: V1552,69-78(1991)
See Thode, Lester E.: V1552,87-106(1991)

Young, M. See Brown, David A.: V1369,2-8(1991)

Young, M. R. See Singh, Surendra: V1376,143-152(1991)

Young, Mark J. See Landy, Michael S.: V1383,247-254(1991)

Young, Niels O.
: Angle encoding with the folded Brewster interferometer,V1454,235-244(1991)

Young, Raymond P.
; Wood, Bobby E.; Stewart, P. L.: Infrared BRDF measurements of space shuttle tiles,V1530,335-342(1991)

Young, Richard
; Kitai, Adrian H.: Humidity dependence of ceramic substrate electroluminescent devices,V1398,71-80(1991)

Young, Richard A.
: Oh say, can you see? The physiology of vision,V1453,92-123(1991)

Young, William R. See Aikens, David M.: V1479,435-444(1991)

Youngner, D. W. See Chan, Lap S.: V1392,232-239(1991)

Yount, Larry J. See Todd, John R.: V1369,72-78(1991)

Yousfi, M. See Saissac, M.: V1397,739-742(1991)

Yovanof, Gregory S. See Pronios, Nikolaos-John B.: V1446,108-128(1991)

Yu, Bing Kun
; Chen, Xao Min: New preparation method and properties of diamondlike carbon films,V1534,223-229(1991)
; Li, Yu; Wang, Yingting: SERS used to study the effect of Langmuir-Blodgett spacer layers on metal surface,V1530,363-369(1991)

Yu, Chang
; Sandhu, Gurtej S.; Mathews, V. K.; Doan, Trung T.: Applications of excimer lasers in microelectronics,V1598,186-197(1991)

Yu, Chongzhen See Zhou, Renkui: V1358,1245-1251(1991)

Yu, Daming
See Carender, Neil H.: V1555,182-193(1991)
See Casasent, David P.: V1555,23-33(1991)

Yu, Da W. See Yuan, Xiang L.: V1519,167-171(1991)

Yu, Dong X.
; Storti, George M.: Dual-eigenstate polarization preserving fiber optic sensor,V1584,236-242(1991)

Yu, Francis T.
: Liquid-crystal television optical neural network: architecture, design, and models,V1455,150-166(1991)
: Optical neural network: architecture, design, and models,V1558,390-405(1991)
; Lu, Taiwei; Yang, Xiangyang: Hetero-association for pattern translation,V1507,210-221(1991)
; Yang, Xiangyang; Gregory, Don A.: Polychromatic neural networks,V1558,450-458(1991)
See Wu, Shudong: V1584,415-424(1991)
See Yelamarty, Rao V.: V1398,170-179(1991)

Yu, Gu-Sheng See Nafie, Laurence A.: V1432,37-49(1991)

Yu, Gui Ying See He, Anzhi: V1554A,747-749(1991)

Yu, Guoping See Li, Yuchuan: V1572,382-385(1991)

Yu, He
; Zheng, XiangJun; Ye, Yizheng; Wang, LiHong: Acquiring rules of selecting cells by using neural network,V1469,412-417(1991)

Yu, Hongbin See Hou, Xun: V1358,868-873(1991)

Yu, J. See Wang, Yutian: V1572,230-234(1991)

Yu, Jia F. See Zhao, Shu L.: V1519,275-280(1991)

Yu, Jie
; Huang, Jianmin; Zhang, Kui; Zhang, Liang: Optical fibers in artificial joint,V1420,266-270(1991)
See Zhang, Renxiang: V1380,116-121(1991)

Yu, Jin-zhong See Zhuang, Weihua: V1361,980-986(1991)

Yu, Jing-ming See Nie, Chao-Jiang: V1584,87-93(1991)

Yu, Ju X.
; Tang, Jia T.: Study on the mechanism of ZnS antireflecting coating with high strength,V1519,308-314(1991)

Yu, K. K. See Watkins, E. T.: V1475,62-72(1991)

Yu, Kai-Bor
: Adaptive beamforming using recursive eigenstructure updating with subspace constraint,V1565,288-295(1991)

Yu, Luping
See Chen, Mai: V1409,202-213(1991)
See Shi, Youngqiang: V1559,118-126(1991)

Yu, Meiwen See Zhang, Jingfang: V1238,401-405(1991)

Yu, Ming H. See Fredricks, Ronald J.: V1367,127-139(1991)

Yu, Nai-Teng
; Nie, Shuming: Surface-enhanced hyper-Raman and near-IR FT-Raman studies of biomolecules,V1403,112-124(1991)

Yu, Paul K.
See Bradley, Eric M.: V1418,272-278(1991)
See Costa, Joannes M.: V1476,74-80(1991)
See Lam, Benson C.: V1371,36-45(1991)
See Pappert, Stephen A.: V1476,282-293(1991)
See Sun, C. K.: V1476,294-300(1991)

Yu, Peter K.
See Cheng, Andrew Y.: V1444,400-406(1991)
See Kwok, John C.: V1445,446-455(1991)

Yu, Phillip C.
See Goldner, Ronald B.: V1536,63-69(1991)
See Ma, Y. P.: V1536,93-103(1991)

Yu, Shan-qing See Zhang, Xiao P.: V1519,514-520(1991)

Yu, Soo-chang See Hopkins, John B.: V1432,221-226(1991)

Yu, Tian-Hu
; Mitra, Sanjit K.: New approach to image coding using 1-D subband filtering,V1452,420-429(1991)
; Mitra, Sanjit K.; Kaiser, James F.: Novel nonlinear filter for image enhancement,V1452,303-309(1991)
See Mitra, Sanjit K.: V1445,156-165(1991)

Yu, Tong
; Li, Qin; Chen, Rongsheng; Yan, Jin-Li: Magnet-sensitive optical fiber and its application in current sensor system,V1572,469-471(1991); 1584,135-137(1991)

Yu, Xiaohan
; Yla-Jaaski, Juha: Two-view vision system for 3-D texture recovery,V1468,834-842(1991)
See Yla-Jaaski, Juha: V1472,201-207(1991)

Yu, Xiaoli See Stocker, Alan D.: V1481,156-169(1991)

Yu, Xiaolin
; Chen, Chin-Tu; Bartlett, R.; Pelizzari, Charles A.; Ordonez, Caesar: Using correlated CT images in compensation for attenuation in PET (positron emission tomography) image reconstruction,V1396,56-58(1991)

Yu, Xinglong See Jin, Guofan: V1569,507-512(1991)

Yu, Yue H. See Zhu, Wen H.: V1519,423-427(1991)

Yu, Zenqi See Hattori, Shuzo: V1463,539-550(1991)

Yu, Zhen Z.
See Chen, Wei M.: V1519,521-524(1991)
See He, Jin: V1519,499-507(1991)
See Ma, Ke J.: V1519,489-493(1991)

Yuan, Haiji J.
; Li, Le; Palffy-Muhoray, Peter: Optical second-harmonic generation by polymer-dispersed liquid-crystal films,V1455,73-83(1991)

Yuan, Hong
See Yuan, Ren K.: V1519,831-834(1991)
See Yuan, Ren K.: V1519,396-399(1991)

Yuan, JinShan See Du, MingZe: V1361,699-705(1991)

Yuan, Lei See Deng, Hong: V1519,735-739(1991)

Yuan, Li-Ping
: Fast algorithm for size analysis of irregular pore areas,V1451,125-136(1991)

Yuan, Libo
; Shou, Reilan; Qiu, Anping; Lu, Zhiyi: Compensation mechanism of an optical fiber turning reflective sensor,V1572,258-263(1991)

Yuan, Q. N. See Zhuang, Qi: V1397,157-160(1991)

Yuan, Qiang See Li, Jie-gu: V1469,178-187(1991)

Yuan, Ren K.
; Gu, Zhi P.; Yuan, Hong; Yuan, Xue S.; Wang, Yong B.; Liu, Xiang N.; Shen, Xue C.: Electrochromic property and chemical sensitivity of conducting polymer PAn film,V1519,831-834(1991)
; Liu, Yu X.; Yuan, Hong; Wang, Yong B.; Zheng, Xiang Q.; Xu, Jian: Study of photoelectric transformation process at p-PAn/n-Si interface,V1519,396-399(1991)

Yuan, Ruixi
; Taylor, Henry F.: Amplified quantum fluctuation as a mechanism for generating ultrashort pulses in semiconductor lasers,V1497,313-319(1991)

Yuan, Shi X.
See Li, Jie: V1519,660-664(1991)
See Liu, Wei J.: V1519,415-418(1991)
See Liu, Wei J.: V1519,481-488(1991)

Yuan, Steve
See Lan, Guey-Liu: V1475,184-192(1991)
See Watkins, E. T.: V1475,62-72(1991)

Yuan, Weitao
; Zha, Kaide; Guo, Yili; Zhou, Bing-Kun: 34-Mb/s TDM photonic switching system,V1572,78-83(1991)

Yuan, X. See Qu, Dong-Ning: V1506,152-159(1991)

Yuan, Xiang L.
; Min, Szuk W.; Fang, Zhi Y.; Yu, Da W.; Qi, Lei: Growth of rf-sputtered selenium thin films,V1519,167-171(1991)

Yuan, Xiao L.
; Sayeh, Mohammad R.: Study of LiNbO3 in optical associative memory,V1396,178-187(1991)

Yuan, Xue S. See Yuan, Ren K.: V1519,831-834(1991)

Yuan, Xun-Hua See Ding, Zu-Quan: V1554A,898-906(1991)

Yuan, Y. F.
; Heavens, Oliver S.: Basis and applicaton of evanescent fluorescence measurement,V1519,434-439(1991)

Yuan, Yanrong
: Calculation of wave aberration in optical systems with holographic optical elements,V1354,43-52(1991)

Yuan, Youxin See Jin, Tianfeng: V1529,132-137(1991)

Yuasa, M. See Yokozawa, T.: V1397,513-517(1991)

Yuce, Hakan H.
; Kapron, Felix P.: Use of fatigue measurements for fiber lifetime prediction,V1366,144-156(1991)
; Key, P. L.; Chandan, Harish C.: Aging behavior of low-strength fused silica fibers,V1366,120-128(1991)
See Hart, Patrick W.: V1366,334-342(1991)
See Kilmer, Joyce P.: V1366,85-91(1991)
See Wei, Ta-Sheng: V1366,235-240(1991)

Yue, Alvin See Mankovich, Nicholas J.: V1444,2-8(1991)

Yuen, P. See Rosten, David P.: V1472,118-127(1991)

Yugo, Shigemi See Isshiki, Hideo: V1361,223-227(1991)

Yuille, Alan L.
; Peterson, Carsten; Honda, Ko: Deformable templates, robust statistics, and Hough transforms,V1570,166-174(1991)
; Vincent, Luc M.; Geiger, Davi: Statistical morphology,V1568,271-282(1991)
See Buelthoff, Heinrich H.: V1383,235-246(1991)
See Elfadel, Ibrahim M.: V1569,248-259(1991)
See Gennert, Michael A.: V1385,268-279(1991)

Yuk, Tung Ip
; Palais, Joseph C.: Analysis of wavelength division multiplexing technique for optical data storage,V1401,130-137(1991)

Yumoto, Yoshiji See Kajita, Toru: V1466,161-173(1991)

Yun, Dazhen See Wang, Lingli: V1554B,436-440(1991)

Yun, Huaishun See Wang, Guomei: V1590,229-236(1991)

Yun, Wen-Bing
; Chrzas, John J.; Viccaro, P. J.: Finite thickness effect of a zone plate on focusing hard x-rays,V1345,146-164(1991)
See Lai, Barry P.: V1550,46-49(1991)

Yung, S. See Juri, Hugo: V1422,128-135(1991)

Yur, Gung See Rahman, Hafiz-ur: V1407,589-597(1991)

Yurchenko, S. V. See Manenkov, Alexander A.: V1420,254-258(1991)

Yurevich, V. I. See Baloshin, Yu. A.: V1440,71-77(1991)

Yurkevich, Igor S. See Turko, Illarion V.: V1510,53-56(1991)

Yurkevith, B. M. See Shestakov, S. D.: V1440,423-435(1991)

Yuryshev, N. N.
: Pulsed chemical oxygen-iodine laser,V1397,221-230(1991)

Yusipov, N. Y. See Sadyigov, Z. Y.: V1621,158-168(1991)

Yusupov, I. Y.
; Mikhailov, M. D.; Herke, R. R.; Goray, L. I.; Mamedov, S. B.; Yakovuk, O. A.: Investigation of the arsenic sulphide films for relief-phase holograms,V1238,240-247(1991)

Yutani, Naoki See Kimata, Masafumi: V1540,238-249(1991)

Yuuki, Hayato
; Ito, Takeharu; Sugimoto, Tetsuo: Plastic star coupler,V1592,2-11(1991)

Yzuel, Maria J.
See Campos, Juan: V1574,141-147(1991)
See Campos, Juan: V1564,189-198(1991)
See Chalasinska-Macukow, Katarzyna: V1564,285-293(1991)
See Millan, Maria S.: V1507,183-197(1991)

Zaal, Gerard
: Market for industrial excimer lasers,V1517,100-116(1991)

Zaborov, A. N. See Ryabova, R. V.: V1238,166-170(1991)

Zabusky, Norman J. See Silver, Deborah E.: V1459,97-108(1991)

Zac, Yaacov See Keren, Eliezer: V1442,266-274(1991)

Zaccanti, Giovanni See Ferrari, Marco: V1431,276-283(1991)

Zaccone, Richard J. See Barlow, Jesse L.: V1566,286-294(1991)

Zachai, Reinhard See Menczigar, Ulrich: V1361,282-292(1991)

Zachorowski, Jan See Stankiewicz, Maria: V1391,174-180(1991)

Zachova, J.
; Shtepanek, J.; Zaitseva, N. P.: Application of high-rate crystal growth technique to single crystals of nucleic acid bases,V1402,216-222(1991)

Zack, Tim
See Anderson, Stephen J.: V1364,94-100(1991)
See Bulusu, Dutt V.: V1364,49-60(1991)

Zadeh, Lotfi A.
: Fuzzy logic: principles, applications, and perspectives,V1468,582-582(1991)

Zadkov, Victor N.
See Akhmanov, Sergei A.: V1402(1991)
See Grishanin, B. A.: V1402,44-52(1991)
See Vachev, V. D.: V1403,487-496(1991)

Zadorin, Anatoly See Bokov, Lev: V1505,186-198(1991)

Zaengel, Thomas T. See Hillen, Walter: V1443,120-131(1991)

Zafar, Sohail See Zhang, Ya-Qin: V1605,301-316(1991)

Zager, Stephen A.
; Weber, Andrew M.: Display holograms in Du Pont's OmniDex films,V1461,58-67(1991)

Zaghloul, A. I. See Sorbello, Robert M.: V1475,175-183(1991)

Zago, Tiziano See Gaboardi, Franco: V1421,73-77(1991)

Zagorskaya, Z. A. See Mikhailov, V. N.: V1238,144-152(1991)

Zagwodzki, Thomas W. See Johnson, C. B.: V1346,340-370(1991)

Zaharakis, Steven C. See Spencer, Paul E.: V1364,228-234(1991)

Zahn, Dietrich R.
See Hingerl, Kurt: V1361,383-393(1991)
See Morley, Stefan: V1361,213-222(1991)

Zahn, S. See Gilbert, Barry K.: V1390,235-248(1991)

Zahniser, Mark S. See Anderson, Stuart M.: V1433,167-178(1991)

Zaibel, Reuben See Levine, Alfred M.: V1376,47-53(1991)

Zaichenko, O. V.
; Komarov, Vyacheslav A.; Kuzmin, I. V.: Photothermoplastic media with organic and inorganic photosemiconductors for hologram recording,V1238,271-274(1991)

Zaidan, Jonathan T. See Kundu, Sourav K.: V1425,142-148(1991)

Zaidi, Saleem H.
; Naqvi, H. S.; Brueck, Steven R.: Submicrometer lithographic alignment and overlay strategies,V1343,245-255(1991)

Zaidi, Syed M. See Lachs, Gerard: V1474,248-259(1991)

Zaitsev, V. K. See Bouchenkov, Vyatcheslav A.: V1427,405-408(1991)

Zaitseva, N. P.
; Ganikhanov, Ferous S.; Katchalov, O. V.; Efimkov, V. F.; Pastukhov, S. A.; Sobolev, V. B.: Optical properties of KDP crystals grown at high growth rates,V1402,223-230(1991)
See Zachova, J.: V1402,216-222(1991)

Zajac, Marek
; Nowak, Jerzy: Effect of coma correction on the imaging quality of holographic lenses,V1507,73-80(1991)
; Nowak, Jerzy: Investigation of imaging quality of Fourier holographic lens,V1574,197-204(1991)
See Dubik, Boguslawa: V1574,227-234(1991)
See Nowak, Jerzy: V1574(1991)

Zajtsev, Viktor K. See Bouchenkov, Vyatcheslav A.: V1410,244-247(1991)

Zak, Michail
; Toomarian, Nikzad: Midcourse multitarget tracking using continuous representation,V1481,386-397(1991)

Zakarauskas, Pierre See Yeung, Keith K.: V1569,133-146(1991)

Zakaria, Marwan F.
; Ng, Terence K.: Labeled object identification for the mobile servicing system on the space station,V1386,121-127(1991)

Zakeri, Gholam-Ali
: Numerical wavefront propagation through inhomogeneous media,V1558,103-112(1991)

Zakery, A. See Ewen, Peter J.: V1512,101-111(1991)

Zakharov, N. S.
See Bugrov, N. V.: V1440,416-422(1991)
See Shainoga, I. S.: V1440,277-290(1991)

Zakharov, S. D. See Skopinov, S. A.: V1403,676-679(1991)

Zakhor, Avideh
See Han, Richard Y.: V1452,395-408(1991)
See Liu, Yong: V1463,382-399(1991)
See Rosenholtz, Ruth E.: V1452,116-126(1991)

Zaks, I. N. See Barachevsky, Valery A.: V1621,33-44(1991)

Zalessky, Viacheslav N.
; Bobrov, Vladimir; Michalkin, Igor; Trunov, Vitaliy: Effectiveness of porphyrin-like compounds in photodynamic damage of atherosclerotic plaque,V1426,162-169(1991)

Zaleta, David E. See Lin, Freddie S.: V1474,45-56(1991)

Zalinski, Charles M. See Kidd, Robert C.: V1454,414-424(1991)

Zamkovets, A. D. See Borisevich, Nikolai A.: V1500,222-231(1991)

Zammit, C. C.
; Sumner, Timothy J.; Lea, M. J.; Fozooni, P.; Hepburn, I. D.: Performance of milliKelvin Si bolometers as x-ray and exotic particle detectors,V1549,274-282(1991)

Zammit, Michael G. See Troyer, David E.: V1486,396-409(1991)

Zammit, Ugo See Pizzoferrato, R.: V1409,192-201(1991)

Zamorano, Lucia J.
; Dujovny, Manuel; Dong, Ada; Kadi, A. M.: Computer-assisted surgical planning and automation of laser delivery systems,V1428,59-75(1991)
; Dujovny, Manuel: ZD multipurpose neurosurgical image-guided localizing unit: experience in 103 consecutive cases of open stereotaxis,V1428,30-51(1991)
See Chavantes, Maria C.: V1428,99-127(1991)
See Chavantes, Maria C.: V1428,13-22(1991)

Zampetakis, Th. See Hourdakis, G.: V1503,249-255(1991)

Zandberg, E. Y.
; Knat'ko, M. V.; Paleev, V. I.; Sushchikh, M. M.: Photodissociation of single-adsorbed molecules of cesium halogenides,V1440,292-302(1991)

Zander, Mark E.
: Imagetool: image processing on the Sun workstation,V1567,9-14(1991)

Zanella, R. See Brandstetter, Robert W.: V1385,173-189(1991)

Zapka, Werner See Tam, Andrew C.: V1598,13-18(1991)

Zardini, Piero See Barbieri, Enrico: V1425,122-127(1991)

Zare, Richard N. See Sweedler, Jonathan V.: V1439,37-46(1991)

Zargham, Mehdi R.
; Sayeh, Mohammad R.: Neural network design for channel routing,V1396,202-208(1991)

Zarinetchi, F. See Smith, S. P.: V1367,103-106(1991)

Zarnowski, Jeffrey J.
; Williams, Bryn; Pace, M.; Joyner, M.; Carbone, Joseph; Borman, C.; Arnold, Frank S.; Wadsworth, Mark V.: Selectable one-to-four-port, very high speed 512 x 512 charge-injection device,V1447,191-201(1991)
See Carbone, Joseph: V1447,229-242(1991)

Zaroslov, D. Y. See Volkov, Gennady S.: V1503,146-153(1991)

Zarowin, Charles B. See Gratrix, Edward J.: V1544,238-243(1991)

Zarschizky, Helmut
; Karstensen, Holger; Gerndt, Christian; Klement, Ekkehard; Schneider, Hartmut W.: Holographic optical elements for free-space clock distribution,V1389,484-495(1991)

Zauberman, Hanan See Lewis, Aaron: V1423,98-102(1991)

Zaugg, Thomas C.
; Fabes, Brian D.; Weisenbach, Lori; Zelinski, Brian J.: Waveguide formation by laser irradiation of sol-gel coatings,V1590,26-35(1991)

Zavalishin, S. I. See Savostjanov, V. N.: V1554A,579-585(1991)

Zavecz, Terrence E. See Hershey, Robert R.: V1464,22-34(1991)

Zavidovique, Bertrand See Bonnin, Patrick: V1381,142-152(1991)

Zavislan, James M. See Brock, Phillip J.: V1463,87-100(1991)

Zawodny, J. M.
See Chu, William P.: V1491,243-250(1991)
See McCormick, M. P.: V1491,125-141(1991)

Zayakin, A. A. See Anisimov, N. R.: V1440,206-210(1991)

Zayas, Inna Y.
; Steele, James L.: Image analysis applications for grain science,V1379,151-161(1991)

Zec, Peter
: Aesthetic message of holography,V1238,355-364(1991)

Zediker, Mark S.
; Foresi, James S.; Haake, John M.; Heidel, Jeffrey R.; Williams, Richard A.; Driemeyer, D.; Blackwell, Richard J.; Thomas, G.; Priest, J. A.; Herrmann, Sandy: Design optimization of a 10-amplifier coherent array,V1418,309-315(1991)
See Haake, John M.: V1418,298-308(1991)
See Heidel, Jeffrey R.: V1418,240-247(1991)

Zeevi, Yehoshua Y.
See Assaleh, Khaled T.: V1606,532-540(1991)
See Gertner, Izidor: V1470,148-166(1991)
See Segall, Ilana: V1606,1048-1058(1991)
See Segman, Joseph: V1606,97-109(1991)
See Vitsnudel, Ilia: V1606,1086-1091(1991)

Zefferer, H. See Du, Keming: V1397,639-643(1991)

Zehnder, Alan T.
; Kallivayalil, Jacob A.: Temperature rise due to dynamic crack growth in Beta-C titanium,V1554A,48-59(1991)

Zeidler, James R. See Soni, Tarun: V1565,338-344(1991)

Zeigler, Bernard P. See Chi, Sung-Do: V1387,182-193(1991)

Zeijlemaker, H. See den Boggende, Antonius J.: V1345,189-197(1991)

Zeis, Joseph E. See Presuhn, Gary G.: V1479,249-258(1991)

Zeisberger, M. See Braeuer, Andreas: V1559,470-478(1991)

Zeise, Frederick F. See Stone, Richard V.: V1563,267-278(1991)

Zeitoun, David
; Tarabelli, D.; Forestier, Bernard M.; Truong, J. P.; Sentis, Marc L.: Effect of acoustic dampers on the excimer laser flow,V1397,585-588(1991)
See Tarabelli, D.: V1397,523-526(1991)

Zekak, A. See Ewen, Peter J.: V1512,101-111(1991)

Zelent, B. See Volkov, Alexander G.: V1436,68-79(1991)

Zelinski, Brian J.
See Roncone, Ronald L.: V1590,14-25(1991)
See Weisenbach, Lori: V1590,50-58(1991)
See Zaugg, Thomas C.: V1590,26-35(1991)

Zell, Andreas
; Mache, Neils; Sommer, Tilman; Korb, Thomas: Recent developments of the SNNS neural network simulator,V1469,708-718(1991)
See Mueller, Adrian: V1469,188-196(1991)
See Mueller, Adrian: V1468,875-881(1991)

Zelnio, Edmund G.
: Importance of sensor models to model-based vision applications,VIS07,112-121(1991)

Zeltzer, David L.
: Virtual environment technology,V1459,86-86(1991)

Zeltzer, G. L. See Nevorotin, Alexey J.: V1427,381-397(1991)

Zeman, Robert K.
See Horii, Steven C.: V1446,10-15(1991)
See Horii, Steven C.: V1446,475-480(1991)

Zemon, Stanley A.
; Lambert, Gary M.; Miniscalco, William J.; Davies, Richard W.; Hall, Bruce T.; Folweiler, Robert C.; Wei, Ta-Sheng; Andrews, Leonard J.; Singh, Mahendra P.: Excited state cross sections for Er-doped glasses,V1373,21-32(1991)

Zemtsov, S. S. See Golovin, A. F.: V1440,250-259(1991)

Zendzian, W. See Jankiewicz, Zdzislaw: V1391,101-104(1991)

Zeng, Bing
; Gabbouj, Moncef; Neuvo, Yrjo A.: Design of minimum MAE generalized stack filters for image processing,V1606,443-454(1991)
See Zhou, Hongbing: V1451,2-12(1991)

Zeng, Gengsheng L.
; Gullberg, Grant T.: Short-scan fan beam algorithm for noncircular detector orbits,V1445,332-340(1991)

Zeng, Guang L. See Huang, Zong T.: V1519,788-789(1991)

Zeng, Jing-gen See Wang, Ming: V1554B,242-246(1991)

Zeng, Ming See McAulay, Alastair D.: V1564,685-690(1991)

Zeng, Nan See Ying, Xingren: V1469,846-851(1991)

Zeng, Y. Y. See Zhang, Z. J.: V1519,790-792(1991)

Zenk, W. See Dietel, W.: V1403,653-658(1991)

Zenkevich, E. I. See Chirvony, V. S.: V1403,638-640(1991)

Zenteno, Luis A.
; Po, Hong: FM-cavity-dumped Nd-doped fiber laser,V1373,246-253(1991)

Zergioti, Y. See Hontzopoulos, Elias I.: V1397,761-768(1991)

Zernike, Frits
; Galburt, Daniel N.: Ultraprecise scanning technology,V1343,241-244(1991)

Zervaki, A. See Hontzopoulos, Elias I.: V1397,761-768(1991)

Zervakis, Michael E.
; Venetsanopoulos, Anastasios N.: Iterative algorithms with fast-convergence rates in nonlinear image restoration,V1452,90-103(1991)
See Kwon, Taek M.: V1569,317-328(1991)

Zervas, E.
; Proakis, John G.; Eyuboglu, Vedat: Effects of constellation shaping on blind equalization,V1565,178-187(1991)

Zerza, Gerald
; Knopp, F.; Kometer, R.; Sliwinski, G.; Schwentner, N.: UV-VIS solid state excimer laser: XeF in crystalline argon,V1410,202-208(1991)
; Kometer, R.; Sliwinski, G.; Schwentner, N.: High-gain tunable laser medium: XeF-doped Ar crystals,V1397,107-110(1991)

Zeto, Robert J. See Kim, A. H.: V1378,173-178(1991)

Zetzsche, Christoph
; Barth, Erhardt; Berkmann, J.: Spatio-temporal curvature measures for flow-field analysis,V1570,337-350(1991)
See Barth, Erhardt: V1570,86-95(1991)

Zeyfang, E.
; von Buelow, H.; Stoehr, M.: Gas-dynamically cooled CO laser with rf-excitation: optical performance,V1397,449-452(1991)
See von Buelow, H.: V1397,499-502(1991)

Zha, Kaide See Yuan, Weitao: V1572,78-83(1991)

Zhai, H. R.
; Xu, Y. B.; Lu, M.; Miao, Y. Z.; Hogue, K. L.; Naik, H. M.; Ahamd, M.; Dunifer, G. L.: Magneto-optical Kerr rotation of thin ferromagnetic films,V1519,575-579(1991)

Zhaiek, Sasson See Meidan, Moshe: V1540,729-737(1991)

Zhan, Xiaowei See Tammela, Simo: V1373,103-110(1991)

Zhan, Yun C. See Wang, Ren: V1519,146-151(1991)

Zhang, Bao-he See Xu, Zhu: V1399,172-177(1991)

Zhang, BaoLin See Du, MingZe: V1361,699-705(1991)

Zhang, Cai-Gen See Liu, Jian: V1540,744-755(1991)

Zhang, Chian-Fan See Birge, Robert R.: V1432,129-140(1991)

Zhang, Cunhao
    See Sang, Fengting: V1412,252-257(1991)
    See Zhuang, Qi: V1397,157-160(1991)

Zhang, Fang Q.
    See Jiang, Xiang-Liu: V1534,207-213(1991)
    See Xiang, Jin Z.: V1519,683-687(1991)

Zhang, Feng
    ; Lit, John W.: Optical coatings to reduce temperature sensitivity of polarization-maintaining fibers for smart structures and skins,V1588,100-109(1991)

Zhang, Feng P. See Zhang, Si J.: V1519,744-751(1991)

Zhang, Feng S. See Zhang, Su Y.: V1519,508-513(1991)

Zhang, Fu M. See Chen, You S.: V1519,56-62(1991)

Zhang, G. See Pessa, Markus: V1361,529-542(1991)

Zhang, H.
    ; Li, Ming-Jun; Najafi, S. I.; Schwelb, Otto: Fully planar proton-exchanged lithium niobate waveguides with grating,V1583,83-89(1991)

Zhang, Hong-zi See Zhang, Renxiang: V1380,116-121(1991)

Zhang, Honghai See Fan, Dapeng: V1572,11-14(1991)

Zhang, Honguo See Liu, Dingyu: V1443,191-196(1991)

Zhang, J.
    See Joyce, Bruce A.: V1361,13-22(1991)
    See Zhang, L. B.: V1572,240-242(1991)

Zhang, J. F. See Wang, Chin H.: V1559,39-48(1991)

Zhang, J. H.
    See Guan, Z. P.: V1519,26-32(1991)
    See Wu, Zi L.: V1519,618-624(1991)

Zhang, Jiajun See He, Anzhi: V1563,208-212(1991)

Zhang, Jianxing See Chen, Riqi: V1554A,407-417(1991)

Zhang, Jian Z. See Chen, Da Yong: V1435,161-167(1991)

Zhang, Jimin See Chen, William W.: V1535,199-208(1991)

Zhang, Jin-Long
    See Liu, Zhi-Shen: V1558,306-316(1991)
    See Liu, Zhi-Shen: V1558,379-383(1991)

Zhang, Jingfang
    ; Ma, Chunrong; Lang, Hengyuan: Color reflection holograms with photopolymer plates,V1238,306-310(1991)
    ; Yu, Meiwen; Tang, Shunqing; Zhu, Zhengfang: Chromaticity and color fidelity of images with multicolor rainbow holograms,V1238,401-405(1991)

Zhang, Jinghua
    ; Wang, Chunhua; Huang, Zhaoming: Alignment of principal axes between birefringent fiber by the spatial technique and its distribution-sensing effect,V1572,69-73(1991)
    See Huang, Zhaoming: V1572,140-143(1991)

Zhang, Jinru See Liu, Yanbing: V1572,61-64(1991)

Zhang, Ji Y. See Chen, Lian C.: V1519,450-453(1991)

Zhang, Jiyue See Wessels, Bruce W.: V1394,232-237(1991)

Zhang, Jun See Huang, Guang L.: V1519,179-182(1991)

Zhang, Kui
    See Yu, Jie: V1420,266-270(1991)
    See Zhang, Renxiang: V1380,116-121(1991)

Zhang, L. B.
    ; Shi, T. Y.; Zhang, J.: Application of an optical fibre sensor for weighing the truckload,V1572,240-242(1991)

Zhang, Li See He, Anzhi: V1563,208-212(1991)

Zhang, Liang
    See Yu, Jie: V1420,266-270(1991)
    See Zhang, Renxiang: V1380,116-121(1991)

Zhang, Lin
    ; Robinson, Michael G.; Johnson, Kristina M.: Fast optoelectronic neurocomputer for character recognition,V1469,225-229(1991)
    See Robinson, Michael G.: V1469,240-249(1991)

Zhang, M. Z. See Weng, Jian: V1431,161-170(1991)

Zhang, Mingde See Sun, Xiaohan: V1572,243-247(1991)

Zhang, Peng
    ; Martinez, Andrew B.; Barad, Herbert S.: Fractional Brownian motion and its fractal dimension estimation,V1569,398-409(1991)
    See Tian, Qi: V1606,260-268(1991)
    See Wu, Lin: V1388,598-601(1991)
    See Zhuang, Xinhua: V1569,434-445(1991)

Zhang, Peng-Gang
    ; Zhao, Huafeng; Liao, Yan-Biao: Theoretical analysis and design on optical fiber magneto-optic current sensing head,V1572,528-533(1991)
    See Zhao, Huafeng: V1572,503-507(1991)

Zhang, Qi See Luo, Bikai: V1554A,542-546(1991)

Zhang, Renji See He, Da-Ren: V1362,696-701(1991)

Zhang, Renxiang
    ; Yu, Jie; Lan, Zu-yun; Qu, Wen-ji; Zhang, Hong-zi; Zhang, Kui; Zhang, Liang: Biomechanical research of joints: IV. the biohinge of primates,V1380,116-121(1991)

Zhang, Rongyao See Zhuang, Qi: V1397,157-160(1991)

Zhang, Rui T.
    ; Ge, Ming; Luo, Wei G.: Growth mechanism of orientated PLZT thin films sputtered on glass substrate,V1519,757-760(1991)

Zhang, S. H. See Featherstone, John D.: V1424,145-149(1991)

Zhang, Shen See Shen, Weisheng: V1567,691-697(1991)

Zhang, Shengpei See Jiang, Desheng: V1572,313-317(1991)

Zhang, Shikun See Yin, Zhi Xiang: V1559,487-496(1991)

Zhang, Shu Y.
    ; Tan, Shun; Wang, Chang S.; Zhao, Te X.: Thermal stability and microstructure study of WSi0.6/GaAs by XRD and TEM,V1519,43-46(1991)

Zhang, Si J.
    ; Yang, Xiao Y.; Li, Xiao L.; Zhang, Feng P.: Preparation and magneto-optical properties of NdDyFeCoTi amorphous films,V1519,744-751(1991)

Zhang, Souyi See Luo, Zhishan: V1554B,339-342(1991)

Zhang, Su Y.
    ; Xu, Bu Y.; Zhang, Feng S.; Yan, Yixun: Preparation of Pb1-xGexTe crystal with high refractive index for IR coating,V1519,508-513(1991)

Zhang, T. J. See Allan, D. S.: V1560,362-369(1991)

Zhang, Wei
    ; Wang, Xiao-Ru: LCVD fabrication of polycrystalline Si pressure sensor,V1572,15-17(1991)
    ; Zhang, Zhipeng: Fiber optic magnetometer with stable and linear output,V1572,458-463(1991)
    See Hasegawa, Akira: V1621,374-379(1991)
    See Hasegawa, Akira: V1558,414-421(1991)

Zhang, Wei P.
    ; Chen, Jing; Fang, Rong C.; Hu, Ke L.: Optical emission spectroscopy in diamond-like carbon film deposition by glow discharge,V1519,680-682(1991)
    ; Cui, Jing B.; Xie, Shan; Song, Yi Z.; Wang, Chang S.; Zhou, Guien; Wu, Jan X.: Structure and optical properties of a-C:H/a-SiOx:H multilayer thin films,V1519,23-25(1991)
    See Cui, Jing B.: V1519,419-422(1991)

Zhang, Wei Q. See Luo, Tao: V1519,826-830(1991)

Zhang, Xhang H.
    ; Ma, Hong-Li; Lucas, Jacques: Properties and processing of the TeX glasses,V1513,209-214(1991)
    See Ma, Hong-Li: V1590,146-151(1991)

Zhang, Xi
    ; Wang, Baishi; Li, Yao W.: Whole-field stress fringe compensation using photoelastic carrier shifting and optical information processing,V1554A,444-448(1991)

Zhang, Xiang See Xu, Jisen: V1572,137-139(1991)

Zhang, Xiang D. See Huang, Xin F.: V1519,220-224(1991)

Zhang, Xiao See Gao, Wenliang: V1530,118-128(1991)

Zhang, Xiao-Chun
    ; Guo, Lu Rong; Guo, Yongkang: Measuring for thickness distribution of recording layer of PLH,V1461,93-96(1991)
    See Guo, Lu Rong: V1392,119-123(1991)
    See Guo, Lu Rong: V1463,534-538(1991)
    See Guo, Yongkang: V1461,97-100(1991)

Zhang, Xiao L. See Zhuang, Tian-ge: V1606,697-704(1991)

Zhang, Xiao P.
; Yu, Shan-qing; Ma, Min W.: ZnS/Me heat mirror systems,V1519,514-520(1991)

Zhang, Xiaoqiu
See Chang, Zenghu: V1358,614-618(1991)
See Hou, Xun: V1358,868-873(1991)

Zhang, Xingde
See Chen, Ruiyi: V1540,724-728(1991)
See Chen, Ruiyi: V1540,717-723(1991)

Zhang, Xinyu See Yao, Minyu: V1572,148-150(1991)

Zhang, Xitian
See Zhang, Yue-qing: V1418,444-447(1991)
See Zhang, Yue-qing: V1400,137-143(1991)

Zhang, Y. S.
See Kuo, Spencer P.: V1407,272-280(1991)
See Kuo, Spencer P.: V1407,260-271(1991)

Zhang, Y. Y. See Wang, Patrick S.: V1384,68-74(1991)

Zhang, Ya-Qin
; Loew, Murray H.; Pickholtz, Raymond L.: Combined-transform coding scheme for medical images,V1445,358-366(1991)
; Loew, Murray H.; Pickholtz, Raymond L.: Correlation model for a class of medical images,V1445,367-373(1991)
; Zafar, Sohail: Motion-compensated wavelet transform coding for color video compression,V1605,301-316(1991)

Zhang, Yan
; Li, Yao: Optically implementable algorithm for convolution/correlation of long data streams,V1474,188-198(1991)
; Li, Yao; Tolimieri, Richard; Kanterakis, Emmanuel G.; Katz, Al; Lu, X. J.; Caviris, Nicholas P.: Optoelectronic Gabor detector for transient signals,V1481,23-31(1991)
See Chang, Hsiao T.: V1388,442-446(1991)
See Li, Yao: V1474,167-173(1991)

Zhang, Yanxin
; Shen, Jinyuan: Optical/electronical hybrid three-layer neural network for pattern recognition,V1469,303-307(1991)

Zhang, Yaonan
: Matching in image/object dual spaces,V1526,195-202(1991)

Zhang, Yi See Xu, Jisen: V1572,137-139(1991)

Zhang, Yi-Mo See Xie, Gong-Wie: V1500,310-321(1991)

Zhang, Yi-ping See Xue, Qing-yu: V1491,75-82(1991)

Zhang, Yinxian See Yang, Hou-Min: V1533,185-192(1991)

Zhang, Yong-Tao See Jing, Xing-Liang: V1418,434-441(1991)

Zhang, Yongfeng See Chang, Zenghu: V1358,541-545(1991)

Zhang, You-Wen See Liu, Jian: V1540,744-755(1991)

Zhang, Yu
; Xian, Wu; Li, Li-Ping; Hall, Ernest L.; Tu, James Z.: New scheme for real-time visual inspection of bearing roller,V1384,60-65(1991)

Zhang, Yuan P.
: Application of electro-optic modulator in photomechanics,V1554B,669-678(1991)
: New method of thinning photoelastic interference fringes in image processing,V1554A,862-866(1991)

Zhang, Yudong
; Lu, Dunwu; Zou, Haixing; Wang, Zhi-Jiang: Excimer laser photolithography with a 1:1 broadband catadioptric optics,V1463,456-463(1991)
; Zou, Haixing; Wang, Zhi-Jiang: New family of 1:1 catadioptric broadband deep-UV high-Na lithography lenses,V1463,688-694(1991)

Zhang, Yue
; Cui, Y. P.; Wung, C. J.; Prasad, Paras N.; Burzynski, Ryszard: Sol-gel processed novel multicomponent inorganic oxide: organic polymer composites for nonlinear optics,V1560,264-271(1991)

Zhang, Yue-qing
; Wu, Sheng-li; Zhu, Lian; Zhang, Xitian; Piao, Yue-zhi; Li, Dian-en: Study on the mode and far-field pattern of diode laser-phased arrays,V1400,137-143(1991)
; Zhang, Xitian; Piao, Yue-zhi; Li, Dian-en; Wu, Sheng-li; Du, Shu-qin: High-power diffraction-limited phase-locked GaAs/GaAlAs laser diode array,V1418,444-447(1991)

Zhang, Yun H.
; Wu, Bei X.; Yang, Guang Y.; Wang, Ren: Study on different proportion W-Ti (C) binary alloy carbide thin film,V1519,729-734(1991)
See Wang, Ren: V1519,146-151(1991)

Zhang, Z. J.
; Luo, Wei A.; Zeng, Y. Y.; Yang, N. P.; Cai, Y. M.; Shen, X. L.; Chen, H. S.; Hua, Zhong Y.: Characterization and preparation of high-Tc YBa2Cu3O7-x thin films on Si with conducting indium oxide as a buffer layer,V1519,790-792(1991)

Zhang, Zaixuan
; Lin, Dan; Fang, Xiao; Jing, Shangzhong: Multimode fiber-optic temperature sensor system based on dual-wavelength difference absorption principle,V1572,201-204(1991)

Zhang, Zeng-xiang See Wei, Yan-nian: V1358,457-460(1991)

Zhang, Zhao A. See Yin, Dong: V1519,164-166(1991)

Zhang, Zhen See Lang, Zhengping: V1468,826-833(1991)

Zhang, Zhensheng See Torbey, Habib H.: V1605,650-666(1991)

Zhang, Zhiming
See Chen, Zhan: V1437,103-109(1991)
See Zheng, Tian S.: V1519,339-346(1991)

Zhang, Zhipeng
; Zhao, Zhi; Chong, Baoxin: Magneto-optical apparatus with comparator for the measurement of large direct current,V1572,464-468(1991)
See Li, Shaohui: V1572,543-547(1991)
See Xu, Ken: V1572,235-239(1991)
See Zhang, Wei: V1572,458-463(1991)

Zhang, Zhiyi
; Grattan, Kenneth T.; Palmer, Andrew W.: Novel signal processing scheme for ruby-fluorescence-based fiber-optic temperature sensor,V1511,264-274(1991)

Zhang, Zhongxian
; Gao, Hangjun; Nan, Zhilin: Spacially periodical stress method for measuring the beat length of a highly birefringent optical fiber,V1572,5-10(1991)

Zhang, Zuxun
: Modern optical method for strain separation in photoplasticity,V1554A,482-487(1991)

Zhao, Cheng See Xu, Kewei: V1519,765-770(1991)

Zhao, Dongming
: Characteristic pattern matching based on morphology,V1606,86-96(1991)

Zhao, Enyi
: Important position of optical fibre technology in shipboard equipment construction,V1572,65-68(1991)

Zhao, Feng
; Geng, Wanzhen; Jiang, Lingzhen; Hong, Jing: Modulation of absorption in DCG,V1555,297-299(1991)
; Geng, Wanzhen; Jiang, Lingzhen; Hong, Jing: New way of optical interconnection for VLSI,V1555,241-242(1991)
; Sun, Junyong; Geng, Wanzhen; Jiang, Lingzhen; Hong, Jing: New method of recording holographic optical elements applied to optical interconnection in VLSI,V1461,262-264(1991)

Zhao, Hanmin See Hur, Jung H.: V1378,95-100(1991)

Zhao, Huafeng
; Zhang, Peng-Gang; Liao, Yan-Biao: Optical fiber magneto-optic current sensor,V1572,503-507(1991)
See Liao, Yan-Biao: V1584,400-404(1991)
See Zhang, Peng-Gang: V1572,528-533(1991)

Zhao, Jing-Fu
; Li, Y. Q.; Chern, Chyi S.; Huang, W.; Norris, Peter E.; Gallois, B.; Kear, B. H.; Lu, P.; Kulesha, G. A.; Cosandey, F.: Growth and properties of YBCO thin films by metal-organic chemical vapor deposition and plasma-enhanced MOCVD,V1362,135-143(1991)
See Chern, Chyi S.: V1394,255-265(1991)

Zhao, Jinglun See Wu, Lin: V1388,598-601(1991)

Zhao, Ke. See Ling, Fuyun: V1565,296-306(1991)

Zhao, Li See Li, Yulin: V1385,200-205(1991)

Zhao, Mingjun
; Li, Yulin; Qin, Yuwen; Wang, Zhao: Photorefractive spatial light modulators and their applications to optical computing,V1558,529-534(1991)
See Li, Yulin: V1385,200-205(1991)

Zhao, Mingsheng
; Shen, Zhen-kang; Chen, Huihuang: New edge-enhancement method based on the deviation of the local image grey center,V1471,464-473(1991)
See Shen, Zhen-kang: V1482,325-336(1991)
See Shen, Zhen-kang: V1482,337-347(1991)

Zhao, Quingchun See He, Huijuan: V1409,18-23(1991)

Zhao, S.
See Pun, Edwin Y.: V1583,102-108(1991)
See Zhou, Lijia: V1469,404-411(1991)

Zhao, Shi-jie See Sun, Fang-kui: V1488,2-5(1991)

Zhao, Shu L.
; Li, Yuan J.; Yu, Jia F.; Yang, Ya L.; Liu, Shu Q.; Fan, Ya F.; Bao, Xiu Y.; Li, Zheng Q.: New process for improving reverse characteristics of platinum silicide Schottky barrier power diodes,V1519,275-280(1991)

Zhao, Te X. See Zhang, Shu Y.: V1519,43-46(1991)

Zhao, Tianji See Bovard, Bertrand G.: V1534,216-222(1991)

Zhao, Wei See Hou, Xun: V1358,868-873(1991)

Zhao, Xia See Lu, Yue-guang: V1332,287-291(1991)

Zhao, Xing R.
; Hao, Jian H.; Zhou, Fang Q.; Sun, Han D.; Wang, Lingjie; Yi, Xin J.: Ion-beam-sputtering deposition and etching of high-Tc YBCO superconducting thin films,V1519,772-774(1991)
See Sun, Han D.: V1477,174-177(1991)
See Wang, Lingjie: V1540,738-741(1991)
See Zhou, Fang Q.: V1477,178-181(1991)

Zhao, Xue-peng See Xue, Qing-yu: V1491,75-82(1991)

Zhao, Yanhan See Luo, Nan: V1572,2-4(1991)

Zhao, Yanzeng
; Hardesty, R. M.; Post, Madison J.: Use of a multibeam transmitter for significant improvement in signal-dynamic-range reduction and near-range coverage for incoherent lidar systems,V1492,85-90(1991)

Zhao, Yilin
; BeMent, Spencer L.: Heuristic search approach for mobile robot trap recovery,V1388,122-130(1991)

Zhao, Yin
See Guo, Maolin: V1554A,310-312(1991)
See Guo, Maolin: V1554A,657-658(1991)

Zhao, Yun F.
; Sun, Zhu Z.; Pang, Shi J.: Research on the temperature of thin film under ion beam bombarding,V1519,411-414(1991)

Zhao, Zhi See Zhang, Zhipeng: V1572,464-468(1991)

Zhao, Zongyao
: Possibility of keeping color picture in an image converter camera,V1358,1252-1256(1991)

Zhavoronkov, M. I. See Babaeva, N. A.: V1440,260-269(1991)

Zhavoronok, I. V.
; Nemchinov, V. V.; Litvin, S. A.; Skanavi, A. M.; Pavlov, V. V.; Evsenev, V. S.: Automatization of measurement and processing of experimental data in photoelasticity,V1554A,371-379(1991)
See Khesin, G. L.: V1554B,86-90(1991)

Zhdanov, Dmitriy D. See Gan, Michael A.: V1574,254-260(1991)

Zheleznykh, N. I. See Vorontsov, Michael A.: V1402,154-164(1991)

Zheltov, Georgi I.
; Glazkov, V. N.; Kirkovsky, A. N.; Podol'tsev, A. S.: Mathematical models of laser/tissue interactions for treatment and diagnosis in ophthalmology,V1403,752-753(1991)

Zhemerov, B. N. See Vishnevsky, G. I.: V1447,34-43(1991)

Zheng, Chang Q. See Yang, Bang C.: V1519,864-865(1991)

Zheng, Cheng-En
See Bollanti, Sarah: V1503,80-87(1991)
See Bollanti, Sarah: V1397,97-102(1991)

Zheng, Dayue See Chen, Ruiyi: V1540,724-728(1991)

Zheng, Fangqing See Liu, Dayou: V1468,37-49(1991)

Zheng, Gang
; Tian, Qian; Liang, Jinwen: Multifunction multichannel remote-reading optical fiber sensor system,V1572,299-303(1991)

Zheng, J. B. See Chen, Zhan: V1437,103-109(1991)

Zheng, J. P.
See Jiao, Kaili L.: V1361,776-783(1991)
See Kwok, Hoi-Sing: V1394,196-200(1991)

Zheng, Joe
; Valavanis, Kimon P.; Gauch, John M.: Decorrelation of color images using total color difference,V1606,1037-1047(1991)

Zheng, K. Q. See Shu, Q. Q.: V1519,675-679(1991)

Zheng, Kang See Shiau, T. H.: V1418,116-122(1991)

Zheng, S. See Cardenas-Garcia, Jaime F.: V1554B,210-224(1991)

Zheng, S. X.
; Hale, K. F.; Jones, Barry E.: Polarimetric monomode optical fibre sensor for monitoring tool wear,V1572,268-272(1991)
; McBride, R.; Hale, K. F.; Jones, Barry E.; Barton, James S.; Jones, Julian D.: Optical fibre interferometer for monitoring tool wear,V1572,359-364(1991)

Zheng, Tian S.
; Liu, Li Y.; Xing, Zhongjin; Wang, Wen C.; Shen, Yuanhua; Zhang, Zhiming: Study of C-H stretching vibration in hybrid Langmuir-Blodgett/alumina multilayers by infrared spectroscopy,V1519,339-346(1991)

Zheng, XiangJun See Yu, He: V1469,412-417(1991)

Zheng, Xiang Q. See Yuan, Ren K.: V1519,396-399(1991)

Zheng, Yingna See Xiang, Tingyuan: V1572,372-376(1991)

Zheng, Zhi H.
See Feng, Yi P.: V1519,440-443(1991)
See Liao, Chang G.: V1519,152-155(1991)
See Wang, Xi: V1519,740-743(1991)
See Xiang, Jin Z.: V1519,683-687(1991)

Zheng, Zhu H.
See Chen, Lian C.: V1519,450-453(1991)
See Guan, Z. P.: V1519,26-32(1991)

Zhengmin, Li
; Li, Min; Tang, Jinfa: Programmable holographic scanning,V1401,66-73(1991)

Zhilkina, V. M. See Silkis, E. G.: V1358,46-49(1991)

Zhilyaev, Yuri V.
; Rossin, Victor V.; Rossina, Tatiana V.; Travnikov, V. V.: Exciton-polariton photoluminescence in ultrapure GaAs,V1361,848-859(1991)

Zhivkova, Svetla See Tontchev, Dimitar A.: V1429,76-80(1991)

Zhivukcin, I. See Frolov, K.: V1397,461-468(1991)

Zhong, An See Li, Wen L.: V1554B,275-280(1991)

Zhong, Di S.
; Xu, Guang Z.; Liu, Wi: Fabrication of cold light mirror of film projector by direct electron-beam-evaporated TiO2 and SiO2 starting materials in neutral oxygen atmosphere,V1519,350-358(1991)

Zhong, Hengyong See Yang, Zhiguo: V1572,252-257(1991)

Zhong, Xian-Xin
; Li, Jianshu; Fu, Xin; Huang, Shang-Lian: Analysis on the characteristics of the optical fiber compensation network for the intensity modulation optical fiber sensors,V1572,84-87(1991)

Zhong, Y.
See Grider, Douglas T.: V1393,229-239(1991)
See Ozturk, Mehmet C.: V1393,260-269(1991)

Zhong, Zugen
See Deng, Xingzhong: V1572,220-223(1991)
See Liu, Bo: V1572,211-215(1991)

Zhorina, L. V. See Hianik, T.: V1402,85-88(1991)

Zhou, Bing
; Chen, Ju X.; Shi, Bao A.; Wu, Ru J.; Gong, Shuxing: Application of YBa2Cu3O7-x thin film in high-Tc semiconducting infrared detector,V1519,454-456(1991)

Zhou, Bing-Kun See Yuan, Weitao: V1572,78-83(1991)

Zhou, Changxin
: Design and fabrication of soft x-ray photolithography experimental beam line at Beijing National Synchrotron Radiation Laboratory,V1465,26-33(1991)
; Sun, Deming: Design and fabrication of x-ray/EUV optics for photoemission experimental beam line at Hefei National Synchrotron Radiation Lab.,V1345,281-287(1991)

Zhou, Fang Q.
; Sun, Han D.; Zhao, Xing R.; Wang, Lingjie; Yi, Xin J.: Infrared detectors from YBaCuO thin films,V1477,178-181(1991)
See Sun, Han D.: V1477,174-177(1991)
See Wang, Lingjie: V1540,738-741(1991)
See Zhao, Xing R.: V1519,772-774(1991)

Zhou, Feng-Shen See Hao, Tianyou: V1584,32-38(1991)

Zhou, Guien
See Cui, Jing B.: V1519,419-422(1991)
See Zhang, Wei P.: V1519,23-25(1991)

Zhou, Hongbing
; Zeng, Bing; Neuvo, Yrjo A.: Pyramid median filtering by block threshold decomposition,V1451,2-12(1991)

Zhou, Jian P.
; Shu, Ju P.; Xu, Hui J.: Organic-dye films for write-once optical storage,V1519,559-564(1991)
See Shu, Ju P.: V1519,565-569(1991)

Zhou, Jiu L.
See Ni, Xiao W.: V1554B,632-635(1991)
See Ni, Xiao W.: V1527,437-441(1991)
See Ni, Xiao W.: V1519,365-369(1991)

Zhou, Lijia
; Song, Hongjun; Zhao, S.: Constructing attribute classes by example learning: the research of attribute-based knowledge-style pattern recognition,V1469,404-411(1991)
See Song, Hongjun: V1469,581-584(1991)

Zhou, Menzhen See Lin, Wenzheng: V1358,29-36(1991)

Zhou, Mingyong See Nakajima, Shigeyoshi: V1605,709-719(1991)

Zhou, Qibo
: IR CCD staring imaging system,V1540,677-680(1991)

Zhou, Qihou L.
See Heflin, James R.: V1497,398-407(1991)
See Heflin, James R.: V1560,2-12(1991)

Zhou, Qing See Lutz, Marc: V1403,59-65(1991)

Zhou, Quan
; Tabacco, Mary B.; Rosenblum, Karl W.: Development of chemical sensors using plastic optical fiber,V1592,108-113(1991)

Zhou, Renkui
; Yu, Chongzhen; Ma, Jiankang; Zhu, Wenkai: Automatic film reading system for high-speed photography,V1358,1245-1251(1991)

Zhou, Shaoxiang
; Jiang, Jinyou; Wang, Qimin: Three-dimensional shape restoration using virtual grating phase detection from deformed grating,V1358,788-792(1991)

Zhou, Shi P.
; Wu, Ke Q.; Jabbar, A.; Bao, Jia S.; Lou, Wei G.; Ding, Ai L.; Wang, Shu H.: Microwave properties of YB2Cu3O7-x thin films characterized by an open resonator,V1519,793-799(1991)

Zhou, Shixun
; Wang, Yonglin; eds.: *Intl Conf on Thin Film Physics and Applications,*V1519(1991)

Zhou, Tian Ming See Du, MingZe: V1361,699-705(1991)

Zhou, Wen See Chen, Xiaoguang: V1572,294-298(1991)

Zhou, Wen-Sheng See Su, Xianyu: V1332,355-357(1991)

Zhou, Xingeng
; Gao, Jianxing: Digital speckle correlation search method and its application,V1554A,886-895(1991)

Zhou, Xiuli
See Chen, Ruiyi: V1540,724-728(1991)
See Chen, Ruiyi: V1540,717-723(1991)

Zhou, Yi-Tong
: Unsupervised target detection in a single IR image frame,V1567,502-510(1991)
; Crawshaw, Richard D.: Contrast, size, and orientation-invariant target detection in infrared imagery,V1471,404-411(1991)

Zhou, Yusheng See Yang, Jie: V1358,672-676(1991)

Zhou, Ziheng
; Venetsanopoulos, Anastasios N.: Hybrid bipixel structuring element decomposition and Euclidean morphological transforms,V1606,309-319(1991)

Zhou, Zu Y. See Wang, Xi: V1519,740-743(1991)

Zhu, Chang X. See Hou, Li S.: V1519,548-553(1991)

Zhu, Cui Y.
See Feng, Weiting: V1535,224-230(1991)
See Feng, Weiting: V1519,333-338(1991)

Zhu, Dayong See Hu, Yu: V1409,230-239(1991)

Zhu, Dongping
; Beex, A. A.; Conners, Richard W.: Stochastic field-based object recognition in computer vision,V1569,174-181(1991)
; Conners, Richard W.; Araman, Philip A.: CT image processing for hardwood log inspection,V1567,232-243(1991)
; Conners, Richard W.; Araman, Philip A.: Three-dimensional CT image segmentation by volume growing,V1606,685-696(1991)

Zhu, Guangkuan See Wang, Guozhi: V1554B,119-125(1991)

Zhu, Hui See Rao, Navalgund A.: V1443,81-95(1991)

Zhu, Huiping See Hopkins, John B.: V1432,221-226(1991)

Zhu, Jian T.
: Measurement of Poisson's ratio of nonmetallic materials by laser holographic interferometry,V1554B,148-154(1991)
: Optical nondestructive examination for honeycomb structure,V1554B,774-784(1991)

Zhu, Jing-Bing
See Liu, Wei J.: V1519,415-418(1991)
See Liu, Wei J.: V1519,481-488(1991)

Zhu, Lian See Zhang, Yue-qing: V1400,137-143(1991)

Zhu, Lie-wei
See Sheng, Lie-yi: V1572,273-278(1991)
See Xu, Sen-lu: V1367,303-308(1991)

Zhu, Long D. See Xiong, F. K.: V1519,665-669(1991)

Zhu, Minjun See Wang, Defang: V1572,514-516(1991)

Zhu, Q. S. See Zhuang, Qi: V1397,157-160(1991)

Zhu, Qin-Fan See Wang, Yao: V1605,667-678(1991)

Zhu, Qiuming
See Hung, Yau Y.: V1332,332-342(1991)
See Hung, Yau Y.: V1332,738-747(1991)
See Hung, Yau Y.: V1332,696-703(1991)
See Tang, Shou-Hong: V1332,731-737(1991)

Zhu, Shi-Yao
: Influence of atomic decay on micromaser operation,V1497,240-244(1991)
; Scully, Marlan O.; Su, Chang: Higher order effect of regular pumping in lasers and masers,V1497,255-262(1991)
; Scully, Marlan O.: Similarity and difference between degenerate parametric oscillators and two-photon correlated-spontaneous-emission lasers,V1497,277-282(1991)
See Scully, Marlan O.: V1497,264-276(1991)

Zhu, Wei
; Messier, Russell F.; Badzian, Andrzej R.: Morphological phenomena of CVD diamond (Part II),V1534,230-242(1991)

Zhu, Wei Y. See Huang, Xin F.: V1519,220-224(1991)

Zhu, Wen H.
; Lin, Cheng L.; Yu, Yue H.; Li, Aizhen; Zou, Shi C.; Hemment, Peter L.: Characterization of GaAs thin films grown by molecular beam epitaxy on Si-on-insulator,V1519,423-427(1991)

Zhu, Wenhua See Chang, Zenghu: V1358,541-545(1991)

Zhu, Wenkai See Zhou, Renkui: V1358,1245-1251(1991)

Zhu, Xioufang See Cao, Jianlin: V1345,225-232(1991)

Zhu, Yafei
See Kang, Songgao: V1527,376-379(1991)
See Kang, Songgao: V1527,409-412(1991)
See Kang, Songgao: V1527,400-405(1991)
See Kang, Songgao: V1527,406-408(1991)
See Lu, Kaichang: V1527,413-418(1991)

Zhu, Yang-ming See Zhuang, Tian-ge: V1606,697-704(1991)

Zhu, Yi Yun See Yan, Da-Peng: V1554B,636-640(1991)

Zhu, Youcai See Demeester, Piet M.: V1361,1132-1143(1991)

Zhu, ZeMin See Cui, DaFu: V1572,386-391(1991)

Zhu, Zhaoda See Wu, Xiaoqing: V1569,484-490(1991)

Zhu, Zhengfang See Zhang, Jingfang: V1238,401-405(1991)

Zhu, Zhenhai
; Qu, Guangjian; Yang, Yongqi; Shang, Jian: Analysis of diffracted stress fields around a noncharged borehole with dynamic photoelasticity and gauges,V1554A,472-481(1991)

Zhu, Zimin See Jin, Guofan: V1569,507-512(1991)

Zhuang, Qi
; Cui, T. J.; Xie, X. B.; Sang, Fengting; Yuan, Q. N.; Zhang, Rongyao; Yang, H. P.; Li, Li; Zhu, Q. S.; Zhang, Cunhao: Red emitter resulting from O2 (a1 delta g): a new lasing species?,V1397,157-160(1991)
See Sang, Fengting: V1412,252-257(1991)

Zhuang, Song Lin
; Chen, Huai'an: Calculation method of the Strehl Definition for decentral optical systems,V1354,252-253(1991)
; Jiang, Yingqui; Qiu, Yinggang; Gu, Lingjuan; Cai, Zhonghua; Chen, Wei: Fabrication and performance of CdSe/CdS/ZnS photoconductor for liquid-crystal light valve,V1558,28-33(1991)
; Qiu, Yinggang; Jiang, Yingqui; Tu, Yijun; Chen, Wei: Fast full-erasure laser-addressed smectic liquid-crystal light valve,V1558,149-153(1991)
; Qu, Zhijin: Nonlinear model of the optimal tolerance design for a lens system,V1354,177-179(1991)